# Fascinating Life Sciences

This interdisciplinary series brings together the most essential and captivating topics in the life sciences. They range from the plant sciences to zoology, from the microbiome to macrobiome, and from basic biology to biotechnology. The series not only highlights fascinating research; it also discusses major challenges associated with the life sciences and related disciplines and outlines future research directions. Individual volumes provide in-depth information, are richly illustrated with photographs, illustrations, and maps, and feature suggestions for further reading or glossaries where appropriate.

Interested researchers in all areas of the life sciences, as well as biology enthusiasts, will find the series' interdisciplinary focus and highly readable volumes especially appealing.

Gary Ritchison

# In a Class of Their Own

## A Detailed Examination of Avian Forms and Functions

Volume 1

Springer

Gary Ritchison
Richmond, KY, USA

ISSN 2509-6745 ISSN 2509-6753 (electronic)
Fascinating Life Sciences
ISBN 978-3-031-14851-4 ISBN 978-3-031-14852-1 (eBook)
https://doi.org/10.1007/978-3-031-14852-1



This Springer imprint is published by the registered company Springer Nature Switzerland AG
The registered company address is: Gewerbestrasse 11, 6330 Cham, Switzerland

Paper in this product is recyclable.

# Preface

This book represents my attempt to provide an up-to-date examination of what we know about the structure and function of birds and to do so in a way that is useful and understandable to a wide variety of readers. Readers will find that some chapters are more detailed than others, largely due to differences in the amount of relevant literature available for different topics and, to a much lesser degree perhaps, my interest in the different topics. Those who read multiple (or all) chapters in the book will find some overlap, with some topics covered in more than one chapter. In such cases, topics are covered in much more detail in particular chapters and to a lesser degree in others. I wanted each chapter to tell a particular story and, to do so, felt it is necessary to sometimes mention the same topic in different chapters (and, to emphasize important points, to sometimes mention the same topic more than once in individual chapters). In addition, with the e-version of the book, interested readers may only wish to read certain chapters. In such cases, they will learn something about topics that might be covered in more detail in chapters they do not read. In terms of organization, all chapters cover aspects of avian form and function, but the initial chapters (after a first chapter about the evolution of birds) focus more on structure (e.g., the skeleton, skeletal muscles, integument, and so on), whereas later chapters include more information about function. In many chapters, I also point out where I think additional study is needed. Each chapter includes multiple figures and photos that I hope complement and add to the information provided in the text. Finally, because scientific names are used more than common names in many parts of the world, and to aid those living in those countries, scientific names of species are often provided with common names more than once throughout each chapter.

Richmond, KY Gary Ritchison

# Acknowledgments

I sincerely thank the many reviewers for comments and suggestions that were incredibly helpful in many ways, including correcting errors (but, of course, any that remain are entirely my own and please, in case there might be another edition of this book sometime in the future, let me know of any such errors by email, i.e., ritchisong@gmail.com), providing useful references, suggesting ways to improve chapter organization, and improving the quality of the writing, among others. Thanks also to the many individuals who provided permission to use their figures and photos in the book. I also thank my editor, Lars Koerner, for providing assistance (and access to books published by Springer) whenever I asked and finding the many helpful reviewers. Writing this book has taken many years and so, most of all, I want to thank my family for their understanding and for listening to me when, during evening meals and on many other occasions, I felt the need to tell them something that I'd learned about birds during my research for the book. Although they often smiled, and sometimes even (briefly) laughed (oh no, not again!), they never once told me to please not continue (at least that's how I remember things). To be specific, thanks to my wonderful children, Brandon Tyler and Brianna Carol, for sharing my interest in birds and nature, for all the great times we had when you were kids, and for all the great times we've had since then. Some of my best memories include the many times our family spent in the woods and elsewhere hiking and enjoying nature. During the final stages of writing this book, Brandon's wife, Lisa, gave birth to their son, Carter James. Congratulations to the proud parents, and I look forward to introducing Carter to the wonderful world of birds (Brandon is already introducing him to the wonderful world of Star Trek)! Finally, a big, big thank you to my wife, Tameria. She never complained about the many times and many hours I was in the field studying birds, in the office writing about birds, or engaged in the many other things required for those in academics. After long days working on the book (and other assorted duties), I always looked forward to getting home to spend time with her. In addition to being a great wife, mother, and, more recently, grandmother, she also shares my love of birds and nature! Thanks Tam, you're the best! Finally, thanks to all of you, Tameria, Brandon, Brianna, Lisa, and Carter, for so many happy memories, and I look forward to making many more such memories in the future.

# Contents

# Contents for Volume 2

# Contents for Volume 3

# Origin and Evolution of Birds

# 1

## Contents




**Abstract**

*Archaeopteryx* lived about 155 million years ago and was a descendent of a long line of dinosaur and theropod ancestors. In this chapter, I review current ideas about the evolution of birds and discuss in detail how dinosaurs eventually gave rise to birds and why birds are considered to be dinosaurs. Over millions of years of dinosaur and theropod evolution, body sizes declined and limb lengths changed and theropods became more bird-like. Factors that likely contributed to such changes are




G. Ritchison, *In a Class of Their Own*, Fascinating Life Sciences,
https://doi.org/10.1007/978-3-031-14852-1_1

described in detail. How and why, during the evolution of birds, natural selection might have favored changes in digestive systems, including the loss of teeth, and reproductive systems is also explained. Information about the first birds, including *Archaeopteryx*, jeholornithids, confuciusornithids, sapeornithids, enantiornithids, and ornithuromorphs, is provided. Possible reasons why the ancestors of present-day birds survived the end-Cretaceous extinction event are also provided. Finally, I describe how birds quickly diversified after that extinction event and ultimately gave rise to the thousands of species of present-day birds.

## 1.1 Early Ideas About the Origin and Evolution of Birds

"If elegance of form, beauty of colouring, or sweetness of voice, were peculiarities which constituted the superiority of one class of beings over another, we should unquestionably assign to Birds the highest station in the scale of the animal creation. No shadow of fear mixes with those pleasurable sensations with which they are viewed; and those feelings, moreover, are heightened by the ethereal nature of the creatures themselves. In a moment they may spread their wings, launch into boundless air, and be seen no more... ." So wrote William Swainson in his book entitled '*On the Natural History and Classification of Birds*' and published in 1836. Swainson was a well-known ornithologist, author, and illustrator of his time and several species of birds were named in his honor, including Swainson's Warbler (*Limnothlypis swainsonii*), Swainson's Thrush (*Catharus ustulatus*), and Swainson's Hawk (*Buteo swainsoni*). Birds certainly are amazing in many ways, but when, where, and how did they evolve? In the 1830s, little was known about bird origins, but Swainson did point out in his book that 'Birds, in the vertebrate circle, occupy a station between Reptiles and Quadrupeds. Between Birds and the former there seems, in the living world, to be a wide hiatus; a gap, which nothing now known to exist in creation can fill up' (Fig. 1.1). He understood, however, that the key to filling this "wide hiatus" would be "studying the forms of extinct animals" and, as examples of that, noted the "union of the reptile and the bird" as exemplified by the discovery of three species of "birdlike reptiles" (pterosaurs) in limestone slates in Europe.

Just two years after the publication of *On the Origin of Species* (Darwin 1859), M. Hermann von Meyer (1861a) reported the discovery of the fossilized impression of a single feather in the limestone quarries at Solnhofen, Germany, that dated from the Jurassic period (Fig. 1.2). The scientific name given by von Meyer, a German paleontologist, to the species represented by the feather was *Archaeopteryx lithographica*. This name was derived from the ancient Greek αρχαίος (*archaīos*), meaning ancient, and πτέρυξ (*ptéryx*), meaning feather or wing. The first fossilized (partial) skeleton was found in 1861 by an unknown individual (although it may have been Johann Friedrich Ottman, the person who owned the quarry where the fossil was found) who subsequently traded it to Dr Karl Häberlein in return for medical services provided. The doctor was visited in the summer of 1861 by a fossil collector named Friedrich Ernst Witte who spent some time examining the fossil. Witt immediately understood the significance of the fossil and sent letters to von Meyer and Andreas Wagner, another paleontologist, to inform them of the discovery. Based on Witte's letter, von Meyer (1861b) published a letter to the editor of the journal *Neues Jahrbuch für Mineralogie, Geologie und Palaeontologie* in September 1861, reporting that, "*I heard from Mr. Obergerichtsrath Witte, that the almost complete skeleton of a feather-clad animal had been found in the lithographic stone. It is reported to show many differences with living birds.*"

Later in the summer of 1861, Mr Witt visited Andreas Wagner and provided him with a much more detailed description of the fossil than he had provided in his earlier letter. Shortly thereafter, as reported by Owen (1863), "At the Meeting of the Mathematico-Physical Class of the Royal

***Typical Characters.***

| | | |
|---|---|---|
| **VERTEBRATA.** | **Pre-eminently typical.** | **MAMMALIA.** |
| | **{ Body furnished with wings.** | **} AVES.** |
| | **Crawl upon their bellies.** | **REPTILIA.** |
| | **Changeable in form.** | **AMPHIBIA.** |
| | **Pre-eminently aquatic.** | **PISCES.** |

**Fig. 1.1** William Swainson placed birds (class Aves) between the classes Reptilia and Mammalia and realized that the "link" between reptiles and birds would have to be found in the fossil record. (Figure from Swainson 1836; CC0 Public Domain)

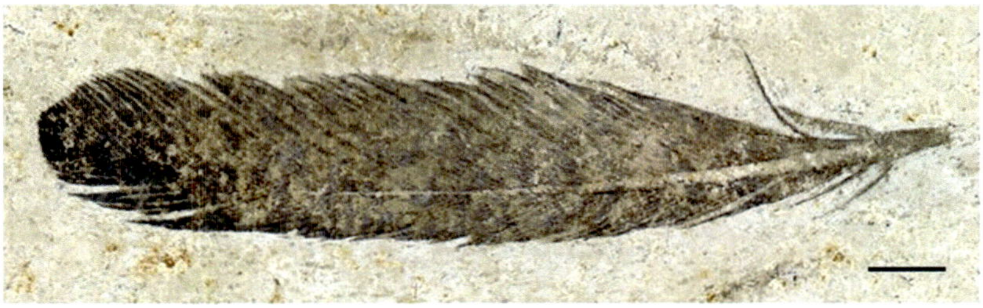

**Fig. 1.2** Photo of the fossil of a feather from *Archaeopteryx lithographica* reported by von Meyer in 1861. More recent analyses suggested that, rather than being an *Archaeopteryx* feather, this "... feather may belong to another basal avialan or even a non-avialan pennaraptoran ..." (Kaye et al. 2019b), but, upon further analysis, Carney et al. (2020) concluded that "... the most empirical and parsimonious conclusion is that the isolated feather represents a primary covert of *Archaeopteryx*." (Figure from Carney et al. 2012; © 2012 Springer Nature, used with permission)

Academy of Sciences of Munich, on the ninth of November, 1861, Professor Andreas Wagner communicated the discovery, in the lithographic slate of Solenhofen, of a considerable portion of the skeleton of an animal with impressions of feathers radiating fanwise from each anterior limb, and diverging obliquely in a single series from each side of a long tail." In April 1862, Wagner published a paper in which he provided Mr Witt's description of the fossil. Based on Witt's description, Wagner (1862) concluded that "a reptile ... with epidermic structures presenting a deceptive resemblance to birds' feathers, is far more comprehensible to me than a bird with the pelvis and vertebral column ... of a long-tailed Pterodactyle ... ." Wagner (1862) named the fossil *Griphosaurus* (from the Greek meaning an enigma). Wagner concluded his paper with "a few words to ward off Darwinian misinterpretations" of the new fossil. Knowing that Darwin and his adherents might view the fossil as an "intermediate creature" and as "justification of their strange views upon the transformations of animals," Wagner argued that Darwinians citing "... *Griphosaurus* as an intermediate creature undergoing a transformation from a reptile into a bird [must] show me, first of all, the intermediate steps by which the transition of some one living or extinct animal from one class to another was effected. If they cannot do this (as they certainly cannot), their views must be at once rejected as fantastic dreams ... ." On this point, Wagner failed because "Darwinists" did, and do, consider *Archaeopteryx* to be an important transitional fossil. As an early example of this, in a letter dated January 3, 1863, paleontologist Hugh Falconer wrote to Darwin about the fossil (Falconer 1863), "Had the Solnhofen quarries been commissioned—by august command—to turn out a strange being à la Darwin—it could not have executed the behest more handsomely—than in the *Archaeopteryx*."

In early 1862, the *Archaeopteryx* fossil was still in the possession of Dr Karl Häberlein. In

**Fig. 1.3** The London specimen of *Archaeopteryx lithographica*. (Figure from ICZN 2011; open-access article licensed under a Creative Commons Attribution 4.0 International License, https://creativecommons.org/licenses/by/4.0/)

February of that year, at the request of Richard Owen, Superintendent of the Natural History division of the British Museum of London, George Robert Waterhouse, a member of the geology department at the museum, sent a letter to Dr Häberlein asking if he would be willing to sell the fossil to the museum. After more than a year of negotiating, Dr Häberlein agreed to sell the *Archaeopteryx* fossil and the rest of his fossil collection to the museum for 8400 Bavarian guldens (which, currently, equals about £700 British pounds or $925 US dollars). For comparison, the doctor's annual salary at the time was only about 60 guldens.

After acquisition of the *Archaeopteryx* fossil (now referred to as the London specimen; Fig. 1.3), Henry Woodward, an assistant in the museum's geology department, published a paper in The Intellectual Observer in December 1862 entitled "*On a feathered fossil from the lithographic limestone of Solenhofen*." Woodward reported that Professor Richard Owen would soon be providing a formal description of the specimen and, further, that Owen believed that the specimen was a bird. Beyond that, however, Woodward (1862) provided readers of The Intellectual Observer a first look at *Archaeopteryx*, with the first published drawing of the fossil included with his paper (Fig. 1.4).

The first detailed written description of *Archaeopteryx* was provided by Owen (1863) in a paper entitled "On the *Archeopteryx* of von Meyer, with a description of the fossil remains of a long-tailed species, from the lithographic stone of Solenhofen." Owen (1863:46) provided a very detailed description of the fossil and concluded that "The best-determined parts of its preserved structure declare it unequivocally to be a Bird, with rare peculiarities indicative of a distinct order in that class." Owen, however, never embraced Darwin's concept of natural selection, as indicated by his critical review of *On the Origin of Species* (Owen 1860).

One of Darwin's early supporters was Ernst Haeckel, a well-known zoologist at the time. In a letter to Darwin written in 1864, Haeckel wrote "... what high esteem and profound respect I hold the discoverer of the 'Struggle for life' and of 'Natural selection.' Of all the books I have ever read, not a single one has come even close to making such an overpowering and lasting impression on me, as your theory of evolution of species. In your book I found all at once the harmonious solution of all the fundamental problems that I had continually tried to solve ever since I had come to know nature as she really is. Since then your theory—I can say so without exaggeration—has occupied my mind every day most pressingly, and whatever I investigate in the life of humans, animals or plants, your theory of descent always offers me a harmonious solution to all problems, however knotty" (Richards 2008:168).

In his book entitled *Generelle Morphologie der Organismen*, Haeckel (1866) was the first to suggest a possible evolutionary link between reptiles and birds, with a tree of life that suggested a common ancestor for reptiles and birds (Fig. 1.5). Shortly thereafter, Huxley (1867) proposed a new taxonomic category, Sauropsida, that

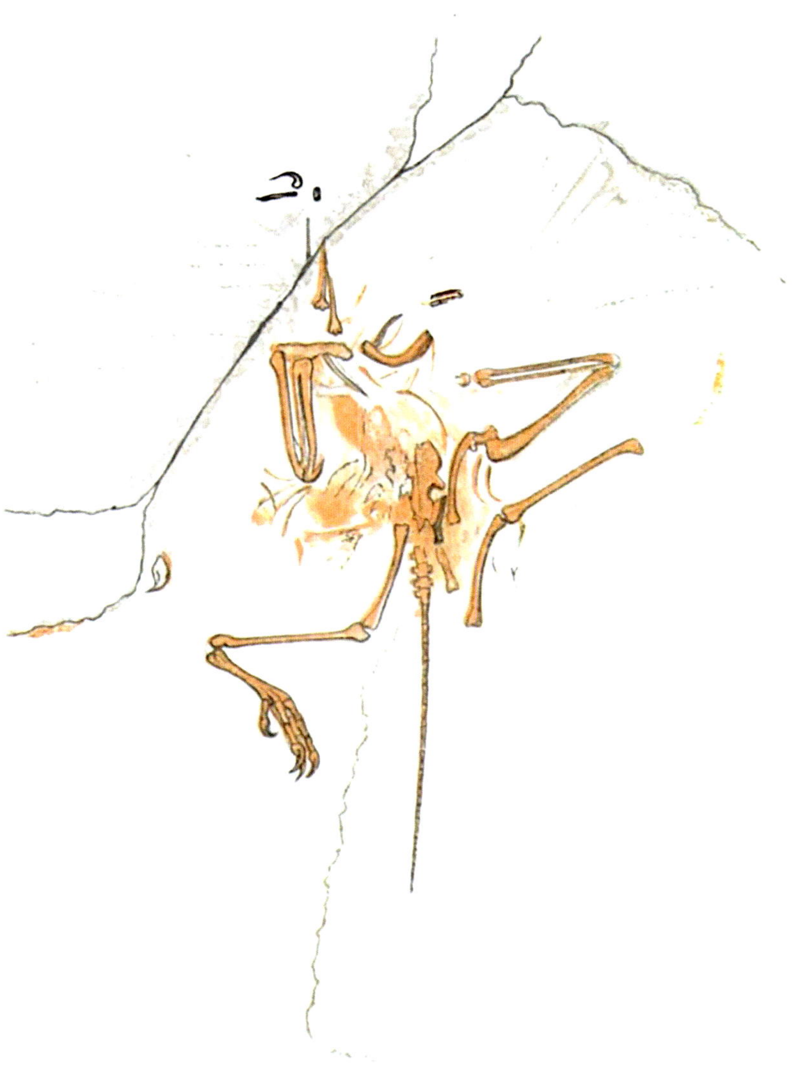

**Fig. 1.4** The first published drawing of the London specimen of *Archaeopteryx* that was included in a paper by Henry Woodward published in December 1862. (Figure from Woodward 1862; CC0 Public Domain)

included both reptiles and birds, noting a number of anatomical features shared by the two groups, but not by mammals. The next year, Huxley sent a letter to Haeckel and wrote "In scientific work the main thing just now about which I am engaged is a revision of the Dinosauria—with an eye to the *Descendenz Theorie*! The road from Reptiles to Birds is by way of Dinosauria to the Ratitae—the Bird 'Phylum' was Struthious, and wings grew out of rudimentary fore limbs. You see that among other things I have been reading Ernst Haeckel's *Morphologie*" (Switek 2010). Although clearly suggesting a belief that birds evolved from some dinosaur ancestor, Huxley was wrong in suggesting that that ancestor first gave rise to ratites, which then gave rise to other birds that could fly.

Huxley's general view that dinosaurs gave rise to birds received some support (e.g., Marsh 1877; Williston 1879; Baur 1886), but others disagreed. For example, in 1888, Max Fürbringer published a two-volume monograph on the anatomy and systematics of birds and suggested that birds did not evolve from dinosaurs, but, rather, from what

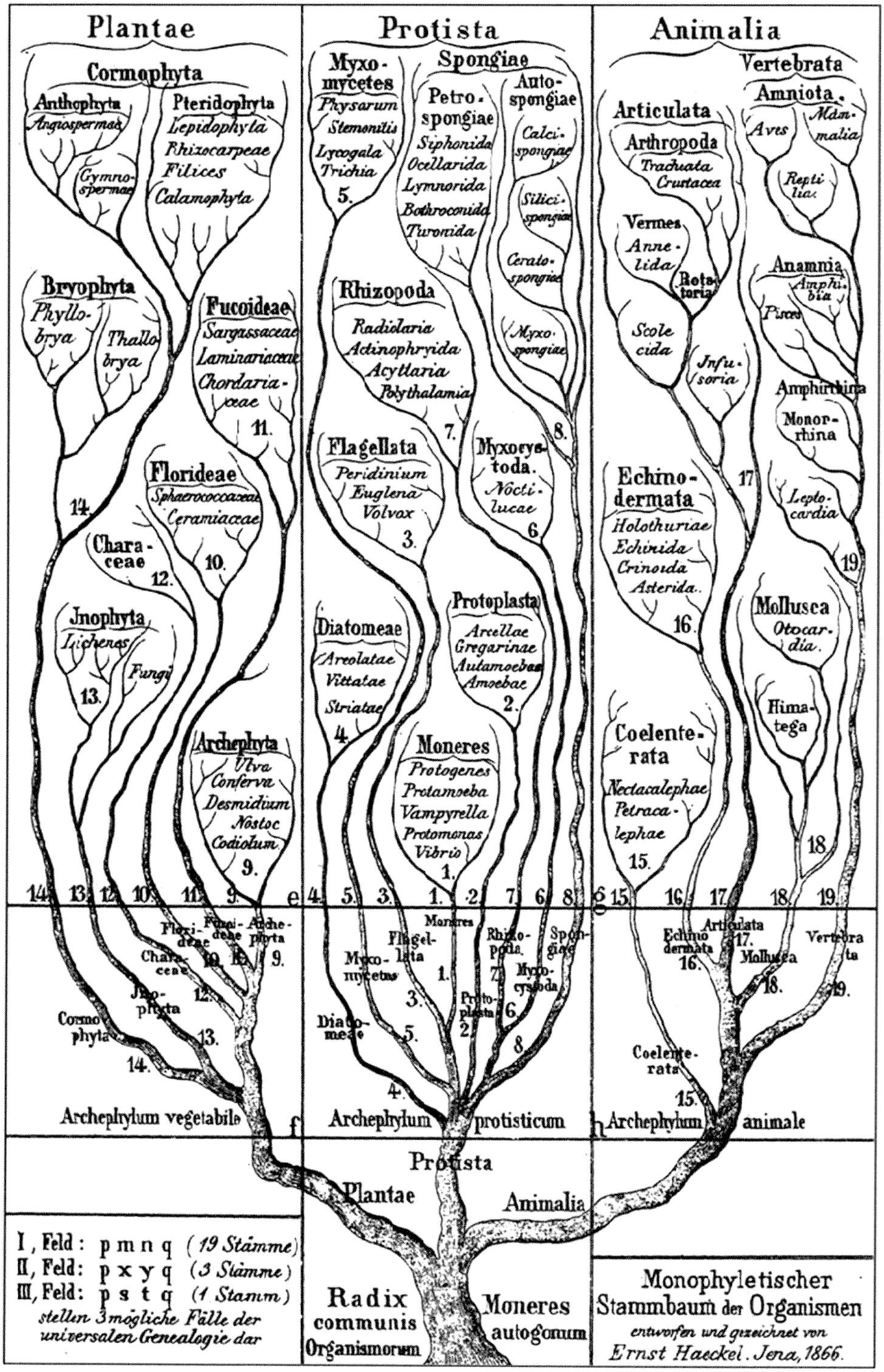


**Fig. 1.5** Haeckel's tree of life showing the Kingdoms Plantae (plants), Protista (microorganisms), and Animalia (animals). Note that near the top of the Animalian portion of the tree that reptiles and birds share a common ancestor. (Figure from Kutschera 2011 as adapted from Haeckel 1866; open-access article distributed under the terms of the Creative Commons Attribution License, https://creativecommons.org/licenses/by/4.0/)

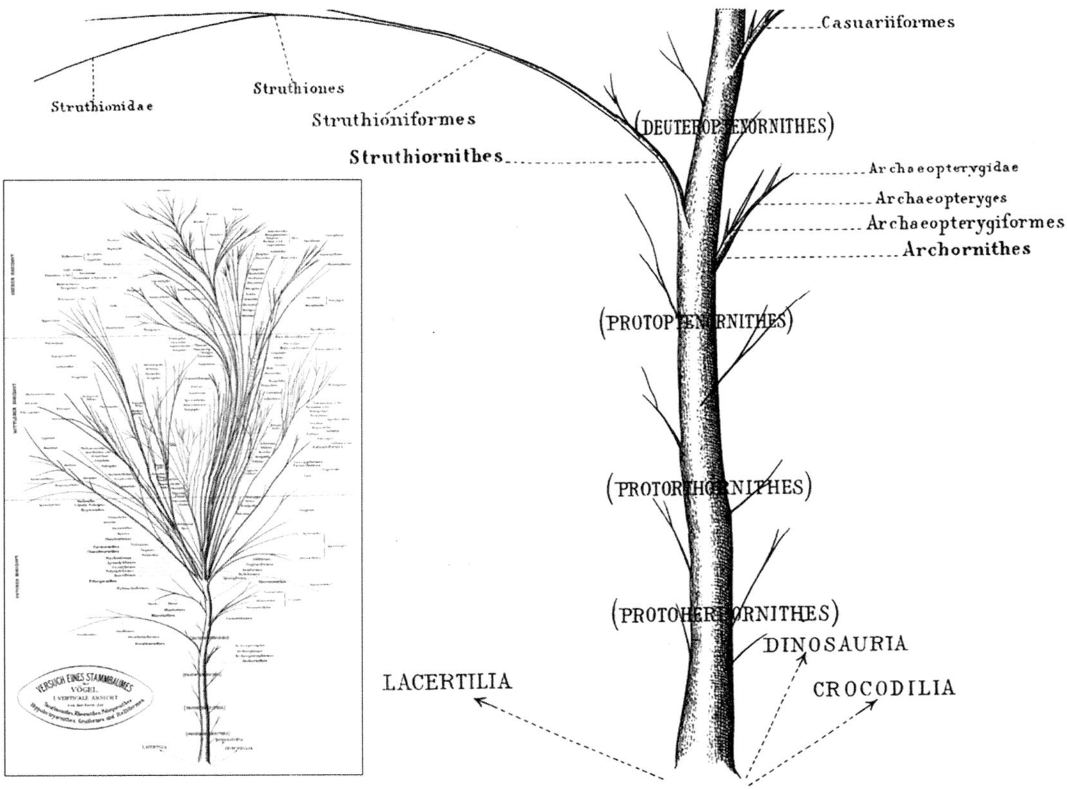


**Fig. 1.6** The base of Fürbringer's (1888) phylogeny of birds, with the entire phylogeny on the left. Fürbringer argued that birds did not evolve from dinosaurs, but from some common ancestor that also gave rise to the crocodilians, dinosaurians, and lactertilia (lizards). (Figure from Fürbringer 1888; CC0 Public Domain)

he called Protoherpornithes ("old reptilian birds"), small or moderate-sized reptiles (or early diverging Archosaurs) covered in a primitive type of down (Fig. 1.6). Any resemblance between birds and dinosaurs was simply due to convergent evolution. According to Fürbringer, next came the Protorthornithes, covered with feathers, bipedal, and flightless. As feathers increased in size and became stiffer, and with further modifications of the skeleton and muscles, birds, or the Protoptenornithes, were now able to fly. Fürbringer proposed that *Archaeopteryx* was a member of this group of birds. With additional time, these early diverging birds gave rise to Ostrich-like birds (Struthiornithes) and other ratites and, eventually, the modern taxa of birds, a group referred to by Fürbringer as the Deuteroptenornithes or the higher or better Birds of Flight. Subsequently, Fürbringer's hypothesis concerning the origin of birds would be referred to as the "Thecodont Hypothesis" (with thecodonts being archosaurs that originated about 250 million years ago and sometimes called the "ruling reptiles") or, more recently, the "Early-Archosaur Hypothesis" (James and Pourtless 2009).

Fürbringer only hypothesized what the non-dinosaur ancestor of birds might be like, but, in 1913, the doctor and paleontologist Robert Broom proposed that a likely candidate for the ancestors of birds was a pseudosuchian (literally, "false crocodile"). Broom knew that all bird-like dinosaurs known at the time had either lost their clavicles (collarbones) or had greatly reduced clavicles. Because present-day birds have clavicles that fuse to form the furcula, or wishbone, and are essential for flight, Broom reasoned that the ancestor of birds must also had to have a

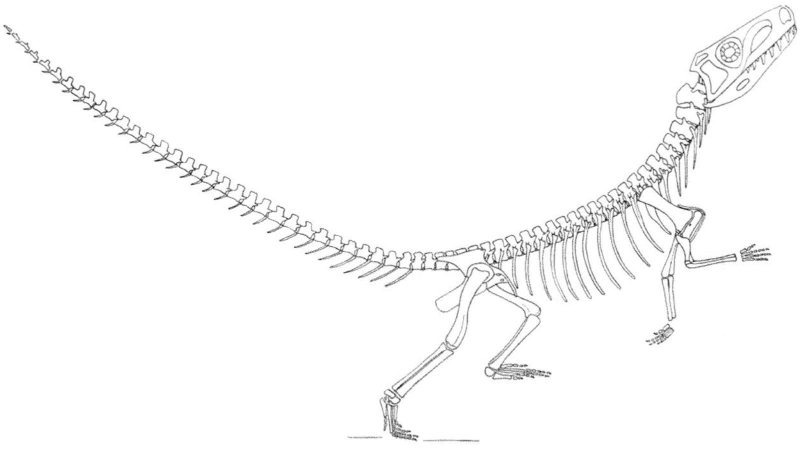

**Fig. 1.7** *Euparkeria capensis.* Ewer (1965) noted that "A study of the limbs and girdles indicates that *Euparkeria* was facultatively bipedal . . .." (Figure from Ewer 1965; used with permission)

clavicle and, acknowledging the characteristics shared by theropod dinosaurs and birds, concluded that the two groups must have had a common ancestor. In a paper published in 1913, Broom wrote, "There cannot, I think, be the slightest doubt that Pseudosuchia have close affinities with the Dinosaurs, or at least with the Theropoda... In fact there seems to me little doubt that the ancestral Dinosaur was a Pseudosuchian ... [and] there is still another group to which some Pseudosuchian has probably been ancestral, namely the Birds."

Broom suggested that one possible candidate for the common ancestor of birds and theropods was a small, 230-million-year-old pseudosuchian called *Euparkeria* that had recently been discovered in South Africa (Fig. 1.7). *Euparkeria*'s hind limbs were longer than its forelimbs, suggesting it may have been a facultative biped. Broom (1913) noted that "A Pseudosuchian which through a bipedal habit had developed a strengthened ankle-joint and a firm metatarsus, and had lost the fifth digit from the manus would meet all the requirements of the avian ancestor." Although *Euparkeria* was not particularly bird-like, Gregory (1916) suggested that "Their structure, however, was, on the whole, of so generalized a type that the diverse peculiarities of the birds, pterosaurs and other groups could readily be derived from this source."

By the 1920s, there was still no consensus concerning the ancestor of birds. The fossil record for the time prior to and after *Archaeopteryx* was still very limited, with only two fossil birds found from the age of dinosaurs—*Ichthyornis*, a gull-like bird, and *Hesperornis*, a large, flightless diving bird—and both from a period 60–80 million years after *Archaeopteryx*. Then, in 1926, Gerhard Heilmann published a book entitled *The Origin of Birds* (an earlier version, *Fuglenes Afstamning*, was published in Danish in 1916). Heilmann's book provided a detailed summary of avian anatomy, embryology, musculature, paleontology, and more. Importantly, however, he concluded, following Broom's earlier reasoning, that dinosaurs could not be the ancestors of birds because they lacked a clavicle. Again in agreement with Broom, Heilmann believed that pseudosuchians were the common ancestors of birds and dinosaurs. In his view, *Archaeopteryx* was not a missing link between reptiles and birds, but, rather, an early diverging bird descended from a much older ancestral bird he called "Proavis." In imagining what proavis would look like, Heilmann studied numerous skeletons, including those of *Archaeopteryx*, *Hesperornis*, present-day birds, and pseudosuchians like *Euparkeria*, and ultimately concluded that the skull, forearm, and the rest of the proavis skeleton would be "transitional" or intermediate in form between those of *Archaeopteryx* and those of pseudosuchians (Fig. 1.8). He also imagined that proavis would be able to climb trees and would have sufficiently developed feathers on its forelimbs to permit gliding flight. Heilmann's drawing of his hypothesized skeleton of Proavis,

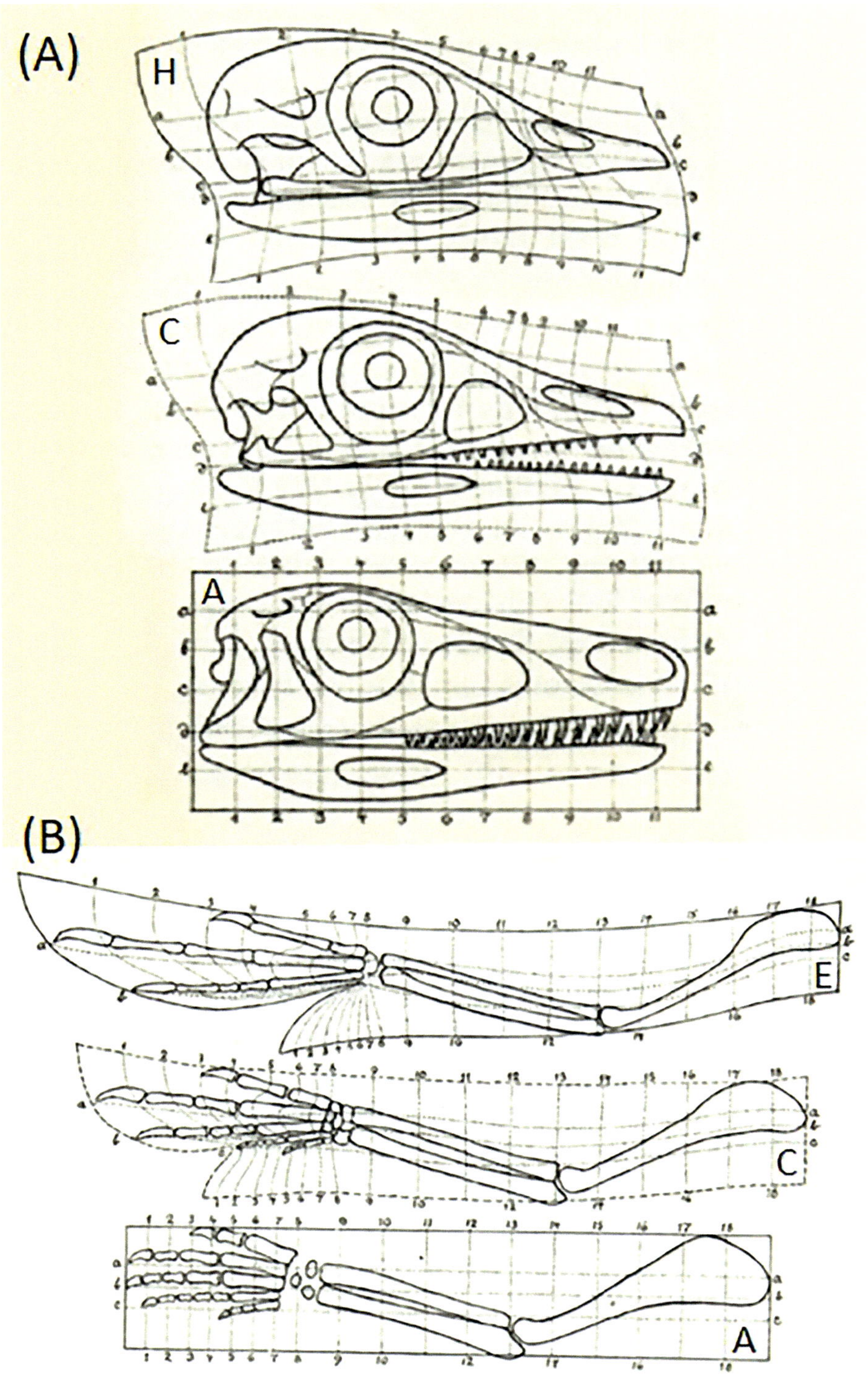


**Fig. 1.8** (**a**) The skull of proavis (labeled with a "C") as proposed by Heilmann was intermediate between those of bird ("H") and a pseuodsuchian such as *Euparkeria* ("A"). (**b**) Similarly, the forelimb of proavis ("C," middle drawing) was assumed to be intermediate between that of *Archaeopteryx* ("E," top) and a pseudosuchian ("A," bottom). (Figures from Heilmann 1916; CC0 Public Domain)

as well as a drawing of two proavises fighting, one climbing up a tree trunk, and two others in gliding flight (Fig. 1.9), was included in both the Danish and English versions of his book. In the caption of his drawing of the skeleton, Heilmann noted that "This skeleton has not yet been found in the ground."

For several decades after publication of *The Origin of Birds*, most ornithologists and paleontologists supported Heilmann's view that birds and dinosaurs evolved from a common pseudosuchian (or Early Archosaur) ancestor. However, based on examination of a sphenosuchian (Fig. 1.10) that revealed a number of characteristics that could be interpreted as being intermediate between crocodilians and birds, Alick Walker (1972) proposed that sphenosuchian crocodylomorphs and birds formed a monophyletic group. Based on several aspects of cranial anatomy, Walker (1977:320) stated that the common ancestor of birds and crocodylomorphs was at "a higher level of organization than that of the Thecodontia" and proposed a new class called Proquadrata that included Sphenosuchia, Crocodilia, and Aves as subclasses (Fig. 1.11).

In a series of papers published during the 1970s, John Ostrom (1973, 1975, 1976) resurrected the dinosaur hypothesis of bird origins that had been rejected by Broom and Heilmann. He first pointed out that the reason given for rejecting this hypothesis, i.e., theropod dinosaurs apparently did not have clavicles, was incorrect because clavicles had been reported in at least two theropods (Osborn 1924; Camp 1936). By the 1970s, five specimens of *Archaeopteryx* had been found and Ostrom noted numerous features of the skeletons of these specimens that were shared with theropods, and concluded toward the end of his 1975 paper that "Although *Archaeopteryx* has often been described as 'bird-like', it should be clear from the preceding sections that those specimens are more dinosaurian (theropodous) in their osteology than they are avian (or pseudosuchian). The only osteological feature of *Archaeopteryx* that is exclusively avian is the furcula." Since the 1970s, other examples of clavicles have been reported in dinosaurs (e.g., Bryant and Russell 1993; Chure and Madsen 1996; Makovicky and Currie 1998), so the objections of Broom and Heilmann are no longer relevant.

## 1.2 Current Views About the Origin and Evolution of Birds

Over the past few decades, three hypotheses—the Early Archosaur (or pseudosuchian) hypothesis, the crocodylomorph hypothesis, and the theropod hypothesis—have been proposed to explain the origin of birds. Some investigators have suggested that the early archosaur and crocodylomorph hypotheses are equally as plausible as the dinosaur (theropod) hypothesis (James and Pourtless 2009) and others have suggested that the early archosaur hypothesis is more plausible than the theropod hypothesis (e.g., Martin 2004; Feduccia et al. 2007; Czerkas and Feduccia 2014). However, the current, nearly universal, consensus is that birds have descended from theropod dinosaurs. For example, in the phylogenetic taxonomy system (i.e., using evolutionary relationships rather than morphological characters), dinosaurs can be defined as the least inclusive clade (or monophyletic group) that includes *Triceratops horridus* and *Passer domesticus* (house sparrow; Padian and May 1993; Sereno et al. 2005; Fig. 1.12) or as "*Triceratops horridus*, *Passer domesticus*, and all descendants of their most recent common ancestor" (Brusatte et al. 2010a, b).

## 1.3 Early Events in the Evolution of Birds

Life first developed on earth about 3.45 billion years ago (Allwood et al. 2007), the first chordates appear in the fossil record about 540 million years ago, the first vertebrates about 525 million years ago, the first terrestrial vertebrates (tetrapods) about 395 million years ago, and the first reptiles about 315 million years ago. By the end of the Permian (251 million years ago), the ancestors of mammals (cynodonts) and birds (archosauromorphs) were present along with a wide diversity of other tetrapods (almost

**Fig. 1.9** (**a**) Heilmann's drawing of the skeleton of proavis. (**b**) Two proavises fighting, another beginning to climb a tree, and two shown in gliding flight. (Figures from Ries 2010; Rights managed by Taylor & Francis, used with permission)

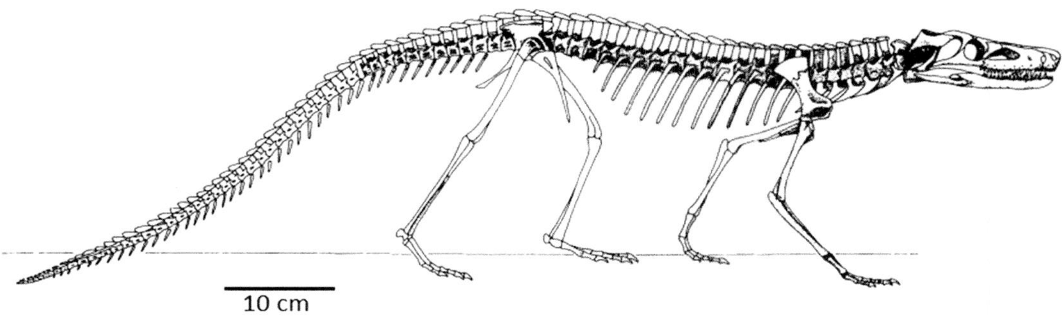


**Fig. 1.10** Example of a sphenosuchian (*Dibothrosuchus elaphros*). (Figure from Wu and Chatterjee 1993; Rights managed by Taylor & Francis, used with permission)

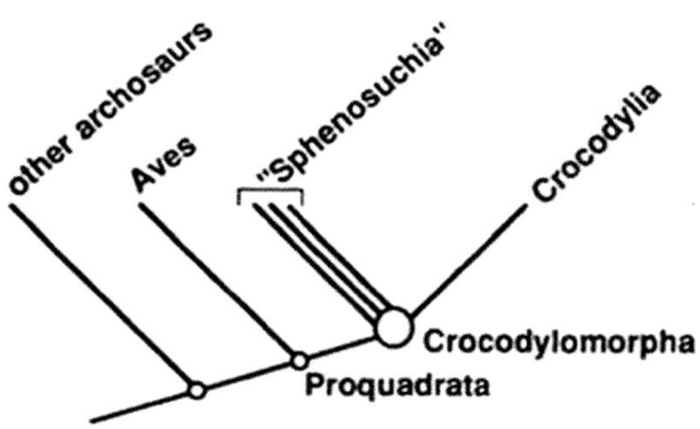


**Fig. 1.11** Phylogenetic relationships of birds as proposed by Walker (1972, 1977), with birds and crocodylomorphs sharing a common ancestor. "Sphenosuchia" is shown as paraphyletic as suggested by Benton and Clark (1988). (Figure from Witmer 1991; © 1991 Cornell University Press, used with permission via PLSclear)

40 different families are known). However, the greatest extinction event since the evolution of life occurred at the end of the Permian. Available evidence suggests that this end-Permian or Permo-Triassic extinction event was caused by large-scale volcanism in Siberia that caused major changes to the atmosphere and the collapse of terrestrial and marine ecosystems worldwide (Benton 2003). An estimated 80–96% of all marine species were lost as well as two-thirds of tetrapod families (Sahney and Benton 2008; Fig. 1.13). Among the tetrapod lineages that did survive were the therapsids (reptiles that ultimately gave rise to mammals) and archosauromorphs (reptiles that gave rise to crocodilians, pterosaurs, and dinosaurs, including birds) (Benton et al. 2004).

Another important event in the evolution of birds apparently occurred over a period of an estimated 15–20 million years extending from the late Permian into the early Triassic. Throughout most of the Permian, tetrapods, as indicated by fossilized trackways, had a sprawled posture, with limbs extending from the sides of the body like present-day alligators and crocodiles (Fig. 1.14). However, from the very end of the Permian into the early Triassic, that changed, with fossilized trackways revealing a substantial increase in erect postures, with limbs under the body. An erect limb posture has at least two important advantages, including reduced stress on joints and less of an impact on respiration (with a sprawled posture, the chest cavity can impact the ground and interfere with efficient respiration) (Carrier 1987). Among tetrapods that survived the end-Permian extinction event, both cynodonts and archosauromorphs had erect limb postures and were likely responsible for most of the fossilized trackways with the higher pace angulations (Kubo and Benton 2009; Fig. 1.15).

At the end of the Permian, fossil evidence indicates that therapsids and archosauromorphs coexisted, along with other terrestrial vertebrates (including other reptiles and amphibians), but therapsids were much more common than archosauromorphs. However, in yet another

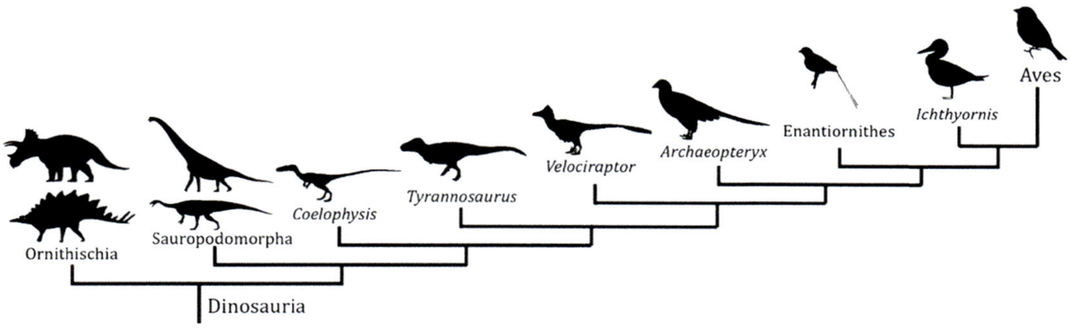


**Fig. 1.12** Phylogenetic taxonomy definition of dinosaurs—all descendants of the most recent common ancestor of *Passer domesticus* (House Sparrow) and *Triceratops horridus* (Sereno 2005). (Figure from Plotnick et al. 2015 [except silhouette of House Sparrow, CC0 Public Domain]; open-access article distributed under the terms of the Creative Commons Attribution 4.0 International License, http://creativecommons.org/licenses/by/4.0/)

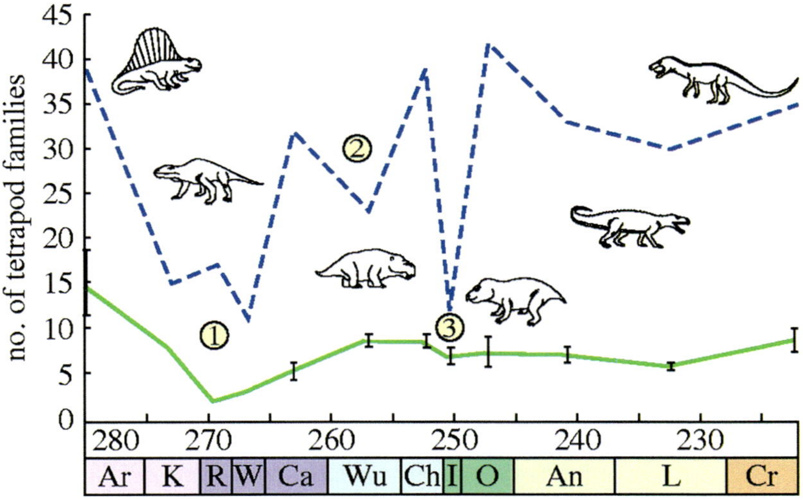


**Fig. 1.13** Global diversity (dashed line) and mean alpha diversity (solid line) of tetrapod families prior to and after the major extinction event that occurred 251 million years ago at the end of the Permian (circled number 3). Note also that there were other extinction events during the Permian, but diversity increased again after each of those events. Extinctions are labeled as 1, Olson's extinction; 2, end-Guadalupian extinction; and 3, end-Permian extinction. Geological stages are Ar, Artinskian; K, Kungurian; R, Roadian; W, Wordian; Ca, Capitanian; Wu, Wuchiapingian; Ch, Changhsingian; I, Induan; O, Olenekian; An, Anisian; L, Ladinian; Cr, Carnian. (Figure from Sahney and Benton 2008; © 2008 The Royal Society, used with permission)

important event in the evolution of birds, archosauromorphs became much more numerous relative to therapsids over a period of 15–20 million years after the end-Permian mass extinction. Through the rest of the Triassic and continuing into the Jurassic, archosauromorphs, including crocodilians, pterosaurs, and dinosaurs (including birds), eventually became the dominant terrestrial vertebrates while therapsids declined in diversity and number (Sookias et al. 2012; Fig. 1.16). One

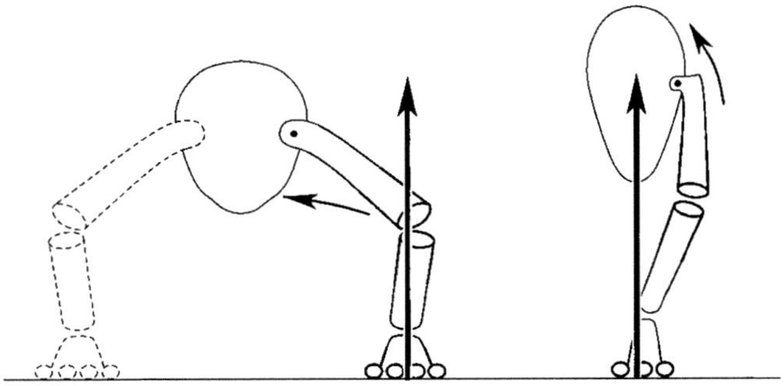


**Fig. 1.14** In contrast to extant crocodilians, archosauromorphs had upright postures. The gray arrows indicate the forces generated when the feet impact the ground. With a sprawled posture, the supporting muscles must exert constant force to hold the body off the ground. An upright posture places the feet closer to a body's center of gravity so there is less need for large limb muscles just to hold the body off the ground. All of the movement of the limb goes into rotation back and forward; moving forward rather than lifting the body off the ground. (Figure from Hutchinson and Gatesy 2000; © The Paleontological Society, used with permission)

important factor that contributed to this replacement of therapsids by archosauromorphs was the extinction of large-bodied therapsid predators and herbivores at the end of the Permian and in the middle Triassic, respectively (Huttenlocker 2014). These vacated niches were subsequently occupied by archosauromorphs. Additional factors likely contributing to the increasing dominance of archosauromorphs were physiological adaptions that first appeared in the early Triassic, including increasingly rapid growth rates and efficient unidirectional respiratory systems (Box 1.1 Unidirectional Lung Ventilation). Rapid growth rates and earlier sexual maturity likely meant that archosauromorphs, particularly the dinosaurs, had higher reproductive rates than therapsids (Sookias et al. 2012).

**Box 1.1 Unidirectional Lung Ventilation**

Based on comparisons with the rib and vertebral anatomy of present-day archosaurs (crocodilians and birds), all archosaurs likely had attached (to the rib cage), multichambered, unidirectionally ventilated lungs similar to those of present-day birds (Schachner et al. 2011). Farmer (2010) hypothesized that a selective factor important in the evolution of unidirectional ventilation was that it facilitated gas exchange during periods of apnea (e.g., when under water), with the beating heart causing gases to move through the lung. Regardless of what factor (s) favored its evolution, unidirectional lung ventilation would be particularly important for extracting oxygen from the air when oxygen levels are low.

Atmospheric levels of oxygen have fluctuated over time, with levels higher than current levels at the end of the Permian and declining substantially during the early Triassic. During this same time period, ecosystems were changing from therapsid-dominated to archosaur-dominated, and the efficient unidirectional respiratory system of archosaurs may have been a contributing factor. Like present-day mammals, therapsids had bidirectional, alveolar respiratory systems that are less efficient at extracting oxygen from the atmosphere (Farmer 2010). The respiratory system of archosaurs may have allowed the evolution of increased activity levels and, possibly,

(continued)

**Box 1.1** (continued)

higher metabolic rates that, in turn, may have contributed to their increasing dominance during the Triassic.

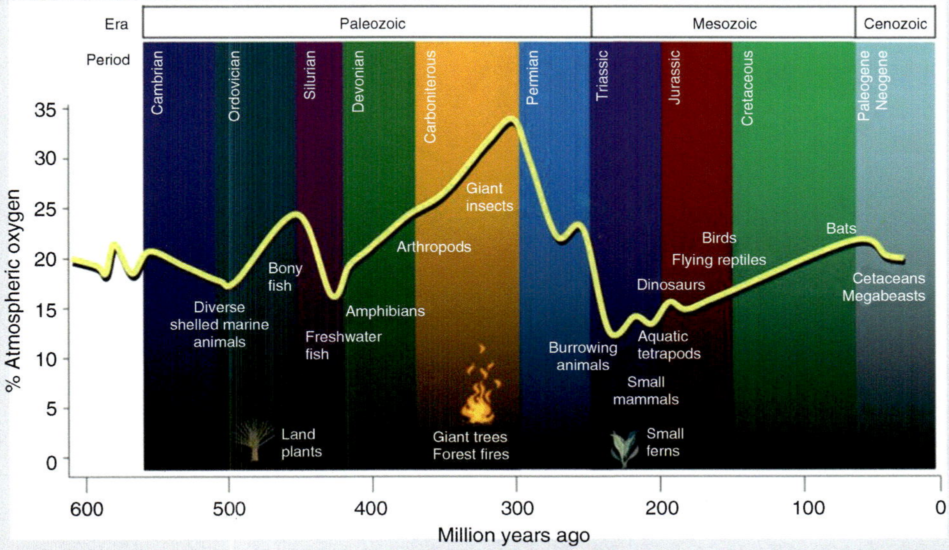


Variation in the percentage of Earth's atmosphere composed of oxygen over time. Note the low levels during the Triassic and continuing, although increasing, through the Jurassic at the time of the appearance of birds. The efficient archosaur respiratory system may have given them a competitive advantage over therapsids that had less efficient bidirectional, alveolar lungs (like present-day mammals). (Figure from Hsia et al. 2013; © 2013 American Physiological Society. All rights reserved, used with permission)

Like present-day birds, several archosaurs, including pterosaurs, sauropods, and theropods, also had pneumatic bones, suggesting the presence of air sacs (thin-walled outpocketings of the lungs). Basal archosaurs had lungs with unidirectional airflow, but not air sacs. However, skeletal pneumaticity, suggesting the presence of air sacs, first appeared in archosaurs (avemetatarsalia) from the late Triassic. Postulated functions of skeletal pneumatization include density reduction that would aid in conserving energy, weight reduction in large-bodied or flying archosaurs, and improved balance and agility (by minimizing changes in the center of mass during respiration) (Farmer 2006; Benson et al. 2012). Regardless of the selective pressures that contributed to their evolution, the presence of air sacs and the corresponding efficiency of lung ventilation may have allowed increased activity levels and (possibly) metabolic rates.

(continued)

**Box 1.1** (continued)

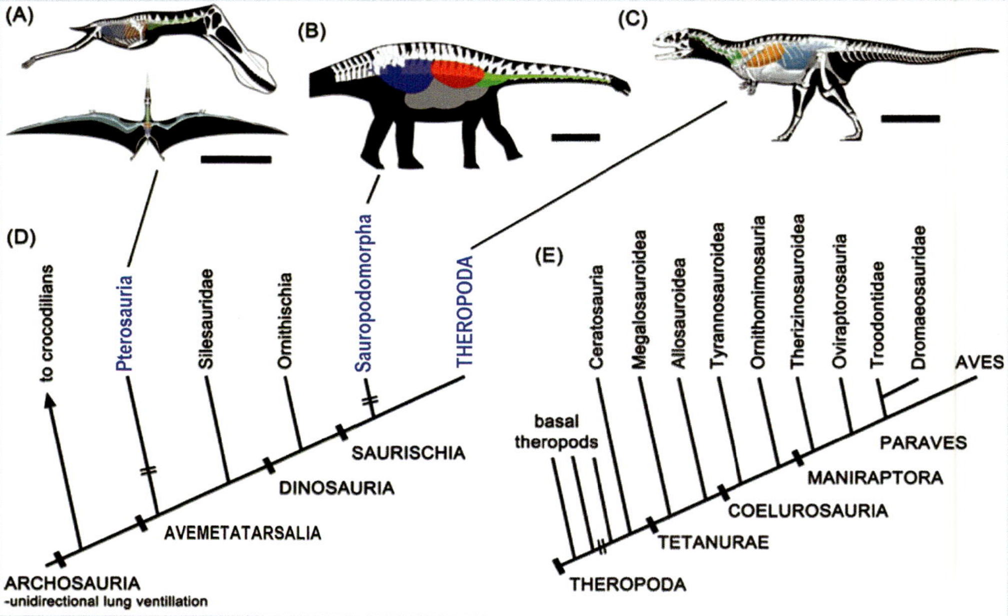


Available evidence indicates that all archosaurians, including theropods, had unidirectionally ventilated lungs. In addition, pneumatic (air-filled) bones evolved independently in several groups of archosaurs, including Pterosauria, Sauropodomorpha, and Theropoda (shown in blue font). Pneumatic bones and air sacs are not required for unidirectionally ventilated lungs (present-day crocodilians being an example; O'Connor and Claessens 2005) so other selective factors likely contributed to the evolution of skeletal pneumaticity, including weight reduction. (Figure modified from Benson et al. 2012; © 2011 The Authors. Biological Reviews © 2011 Cambridge Philosophical Society, used with permission)

During the early Triassic, archosauromorphs split into two major groups, the crurotarsi or "crocodile line" and the avemetatarsalia or "bird line" (Fig. 1.17). By the Late Triassic, both groups were increasingly diverse (in terms of number of taxa; Fig. 1.18) and abundant worldwide. In yet another important event in the evolution of birds, some archosaurs became bipedal during the Triassic. Kubo and Kubo (2012) estimated that bipedalism evolved at least twice and perhaps as many as six times among basal archosaurs (Fig. 1.19), including basal theropods that ultimately gave rise to birds. The selective advantage(s) that favored the evolution of archosaur bipedalism remain unclear, but possible advantages include freeing of the forelimbs for functions other than locomotion such has handling and manipulating food items, faster acceleration (Clemente et al. 2008) and, with longer hindlimbs, increasing speed (Kubo and Kubo 2012), and more efficient respiration because the lungs and ribcage are dissociated from locomotion (Ward and Berner 2011).

During the period from about 235 million years ago until the end of the Triassic, archosaurs filled a wide variety of ecological niches (Nesbitt 2007). However, at the end of the Triassic (201 million years ago), major volcanic eruptions

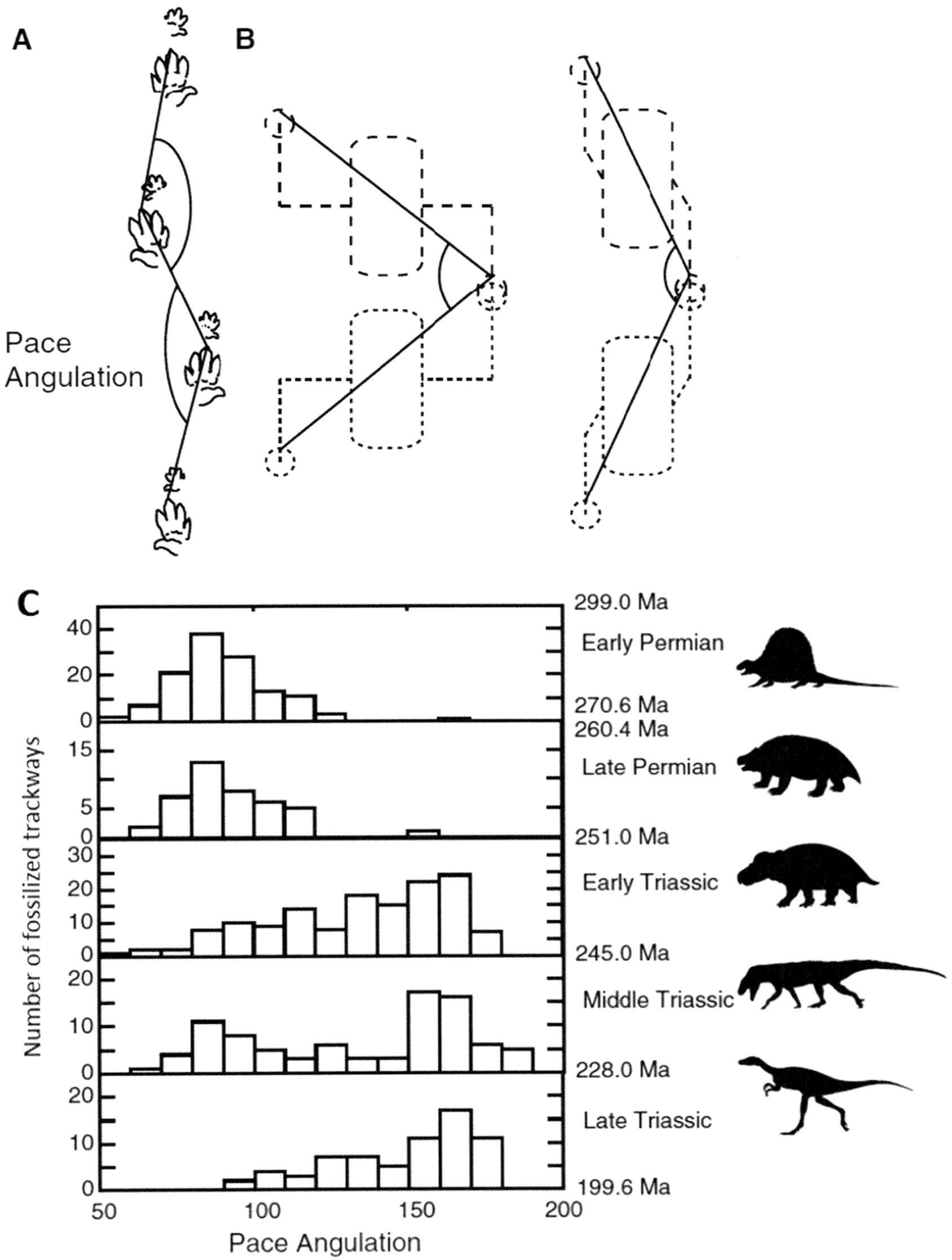


**Fig. 1.15** Pace angulation (the angle between successive steps) and limb posture. (**a**) Example of the measuring method using a fossilized archosaur trackway from the Early Triassic. (**b**) Left diagram illustrates the pace angle of a sprawling walker (with limbs extending from the sides) and the right diagram shows the pace angle of an erect walker (with limbs under the body). (**c**) Comparison of pace angles of fossilized trackways from different periods during the Permian and Triassic shows a shift from sprawling in the Permian to erect in the Triassic. (Figures modified from Kubo and Benton 2009; © The Palaeontological Association, used with permission)

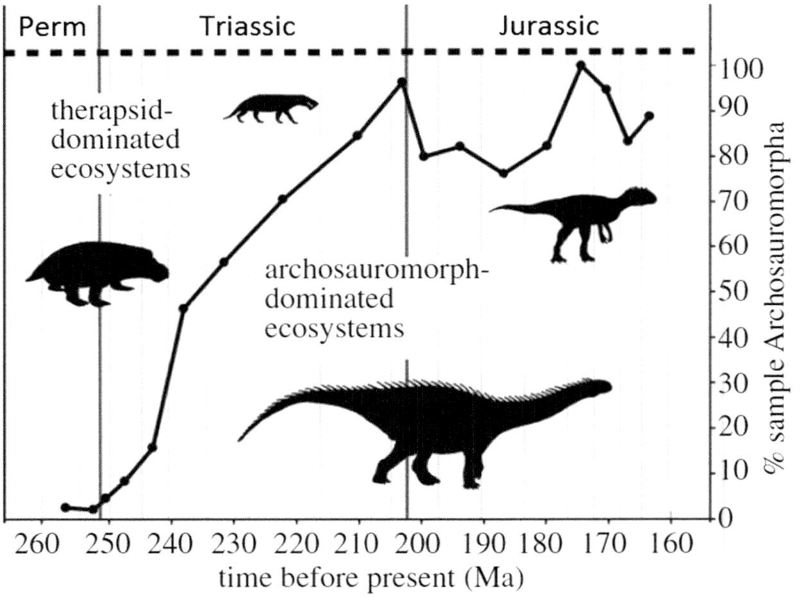


**Fig. 1.16** Proportion of archosauromorphs among archosauromorph and therapsid taxa from the end of the Permian through the Triassic and into the Jurassic. (Figure from Sookias et al. 2012; © 2012 The Royal Society, used with permission)

began and continued at irregular interval for the next 600,000 years (Schaller et al. 2012). These eruptions caused substantial increases in atmospheric carbon dioxide, a greenhouse gas, making oceans more acidic and causing a significant increase in global temperatures (Landwehrs et al. 2020). The resulting disruption of marine and terrestrial ecosystems caused a mass extinction called the end-Triassic extinction that impacted numerous plant and animal species and, in another important event in the evolution of birds, the only archosaurs that survived into the Jurassic were crocodylomorphs from the crocodile line and pterosaurs and dinosaurs from the bird line (Brusatte et al. 2010a, b).

## 1.4 The Age of Dinosaurs

The "Age of Dinosaurs" began at the beginning of the Jurassic and continued for 135 million years before coming to an end at the end of the Cretaceous. With their relatively rapid growth rates and high basal metabolic rates (Pontzer et al. 2009), dinosaurs radiated into a wide variety of forms. Dinosaurs are placed in two clades—Ornithischia and Saurischia (but see Box 1.2 A New Dinosaur Family Tree?)—and a key difference between these clades lies in the pelvic girdle and, specifically, the orientation of the pubic bone (Fig. 1.20). In ornithischians, the pubic bone projects backward and, in saurischians, it projects forward. Saurischia is derived from the Greek words "sauros" (lizard) and "ischion" (hip joint) and Ornithischia from Greek words "ornitheos" (bird) and ischion. So, literally, Saurischia means "lizard-hipped" and Ornithischia "bird-hipped." Although birds actually descended from saurischian (lizard-hipped) ancestors, ornithischians were so named because their hips (pelvic girdles) resemble those of present-day birds, and saurischians so-named because their hips resemble those of present-day lizards. All ornithischians were herbivores and the four major clades were Thyreophora, Ornithopoda, Ceratopsia, and Pachycephalosauria (Fig. 1.21).

**Box 1.2 A New Dinosaur Family Tree?**

In 1887, Harry Govier Seeley proposed that dinosaurs be divided into two major groups based on hip anatomy, with "lizard-hipped" saurischians (which included both theropods and sauropods) having pubic bones pointing forward like present-day lizards and "bird-hipped" Ornithischians having pubic bones pointing backward like present-day birds (Seeley 1887). This classification scheme is still used today. However, based on a study of 74 different taxa of dinosaurs scored on 457 different characteristics, Baron et al. (2017) found that theropods and ornithischians had 21 "unambiguous synapomorphies" (newly evolved characteristics shared by a particular group of species) and, therefore, hypothesized that they were more closely related and separate from sauropods. These authors further proposed that the clade (defined as a grouping that includes a common ancestor and all descendants of that ancestor) that includes the theropods and ornithischians be named Ornithoscelida, a name originally proposed by Thomas Huxley in 1870 for a grouping that included several taxa of theropods and ornithischians (Huxley 1870).

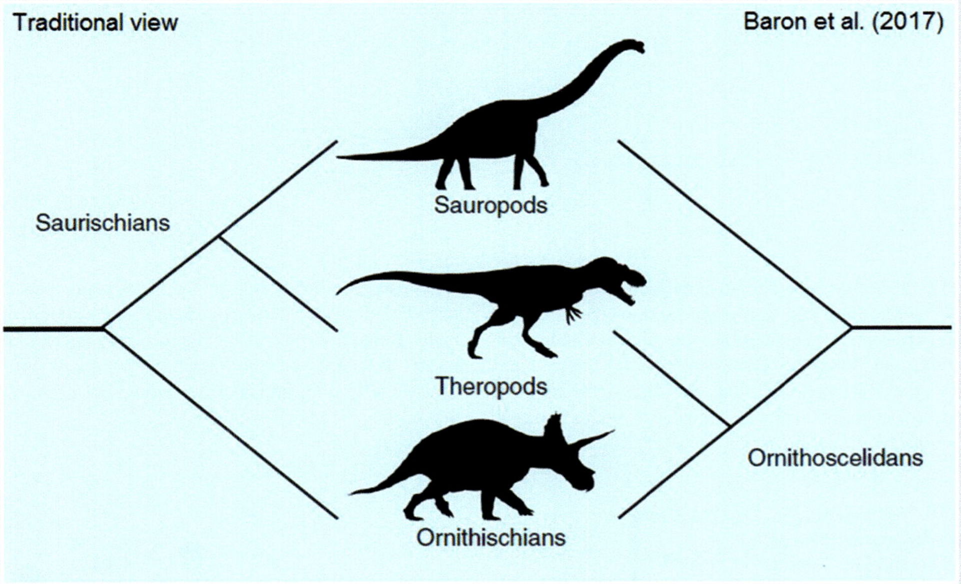


Difference between the "traditional" family tree of dinosaurs divided into Ornithischians and Saurischians (left) and that proposed by Baron et al. (2017) where Theropods and Ornithischians form a new group called the Ornithoscelidans that is separate from the Sauropods. (Figure from Brusatte 2017; © 2017 Elsevier Ltd., used with permission)

If correct (and see Langer et al. 2017 for reasons to be skeptical), Baron et al.'s (2017) hypothesis would "reshuffle" the dinosaur family tree (Brusatte 2017), but not our understanding of when and how birds evolved from theropod ancestors. In addition, the results of phylogenetic analyses depend on that data used. In the absence of dinosaur DNA, the data available to paleontologists studying dinosaur phylogeny consists of sets of morphological characteristics. Thus, whether or not Baron et al.'s (2017) hypothesis will be supported by other investigators who will analyze the same dataset or even larger, more comprehensive datasets remains to be determined. However, as noted by Brusatte (2017), "If something held to be canon for 130 years turns out to be wrong, then who knows what the next new fossil or new evolutionary analysis might tell us?"

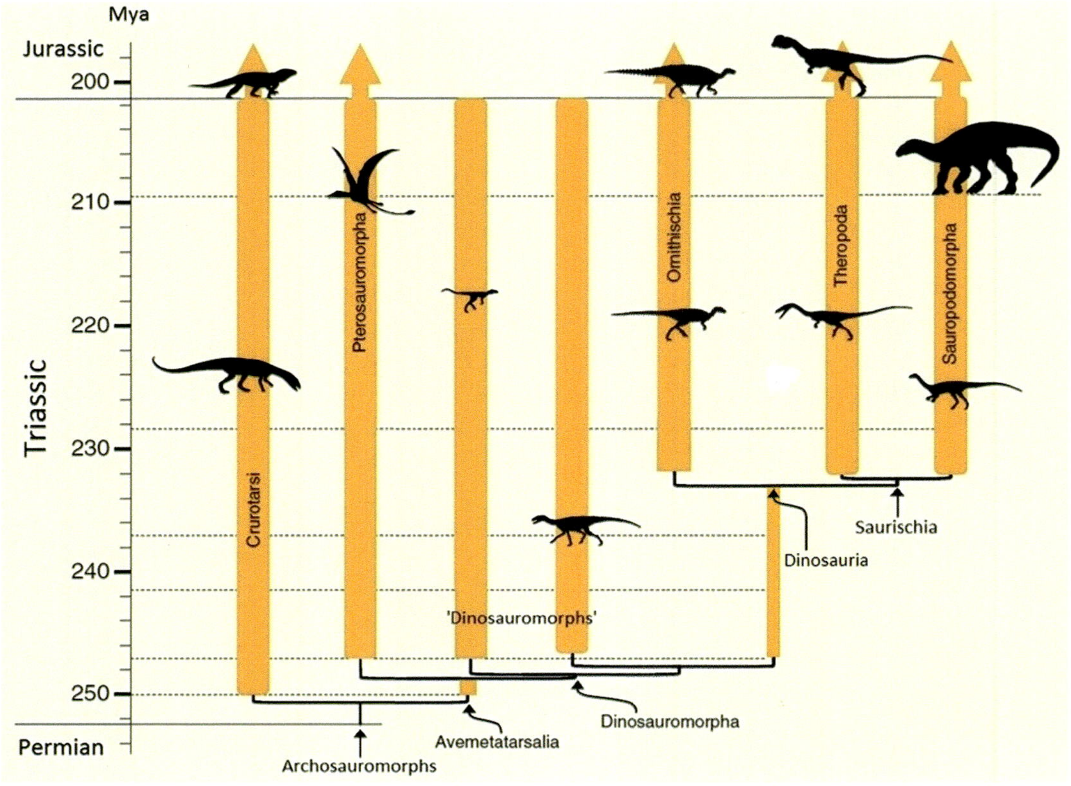


**Fig. 1.17** Major groups of archosaurs, including crurotarsi, or the crocodile line, and avemetatarsalia, or the bird line. Only the crurotarsi, pterosauromorphs, and dinosaurs survived into the Jurassic. The Herrerasauria were basal theropods that did not survive into the Jurassic. During the Triassic, Dinosauria diverged into two major groups, Ornithischia and Saurischia. The saurichians then diverged into Theropoda and Sauropodomorpha. (Figure modified from Benton et al. 2014; open-access article available under the terms of the Creative Commons Attribution License (CC BY), https://creativecommons.org/licenses/by/4.0/)

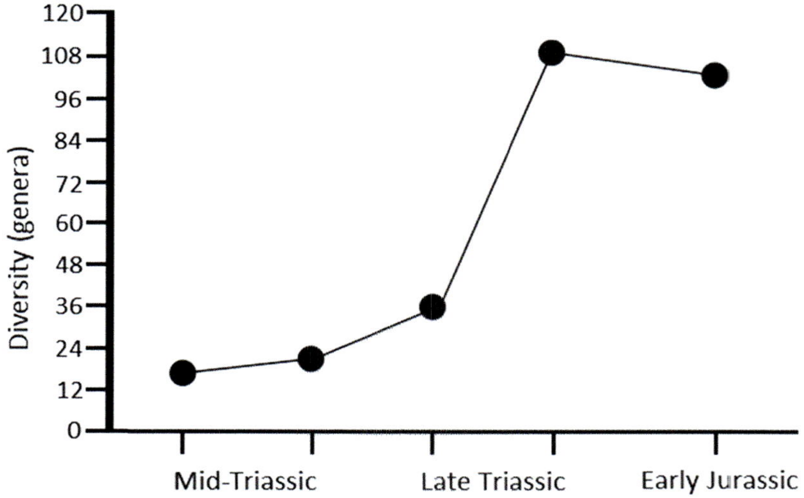


**Fig. 1.18** Archosaur diversity increased through the middle and late Triassic and into the early Jurassic. (Figure modified from Brusatte et al. 2010b; © Royal Society of Edinburgh 2011, used with permission)

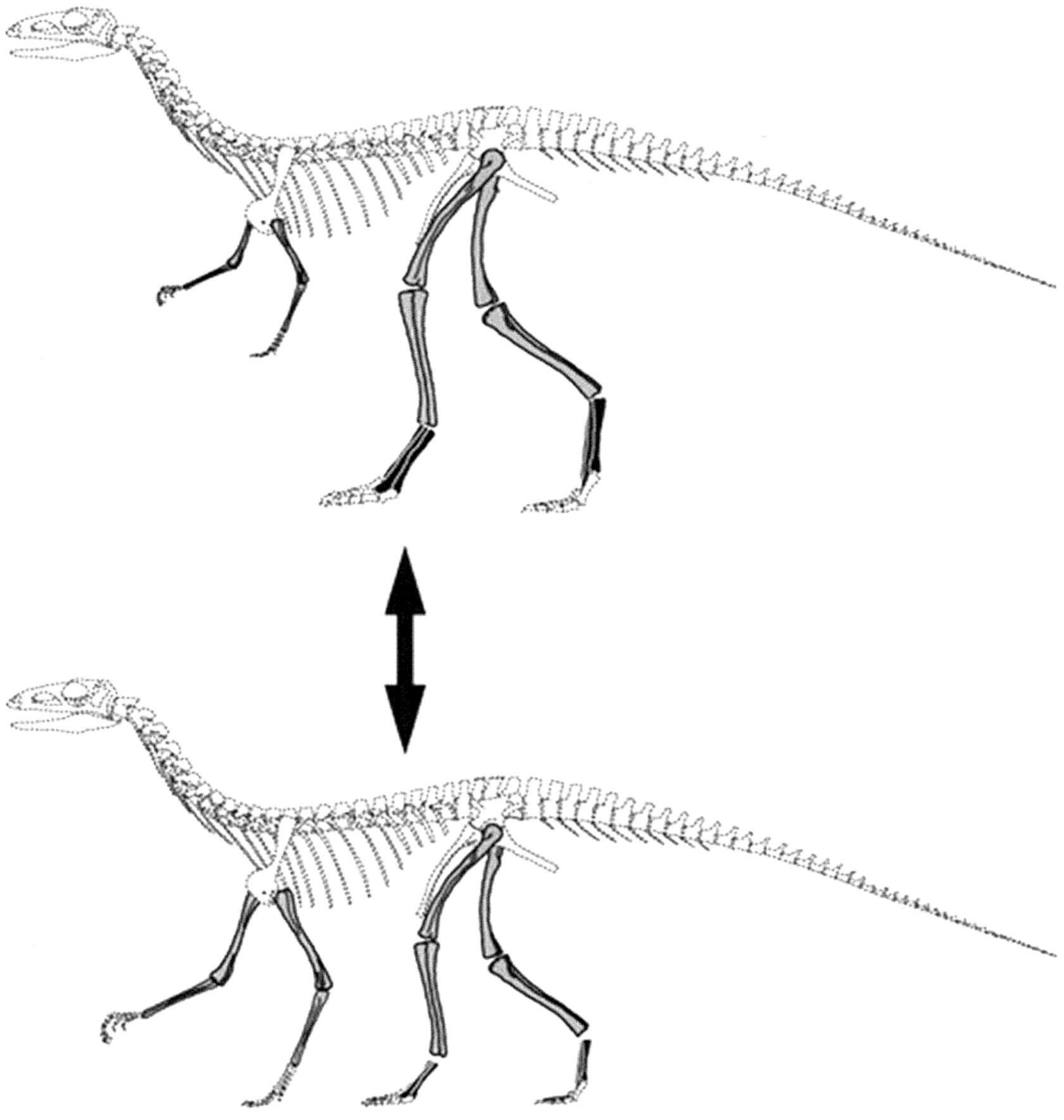

**Fig. 1.19** A general evolutionary trend among some Triassic archosaurs was the evolution of bipedalism with shorter forelimbs and longer metatarsals. Some bipedal archosaurs subsequently reversed this process, again becoming quadrupedal with the shortening of the metatarsals. (Figure from Kubo and Kubo 2012; © The Paleontological Society, used with permission)

The two main branches within the Saurischian clade, Theropoda and Sauropodomorpha, diverged over a period of a few million years during the middle Triassic (about 230 million years ago; Martinez and Alcober 2009). The sauropodomorphs included two groups, the prosauropods (i.e., early or basal sauropodomorphs) and sauropods, and all known species were long-necked herbivores (Fig. 1.22). The sauropods included the largest land animals ever known, including *Seismosaurus*, *Diplodocus*, *Apatosaurus*, and

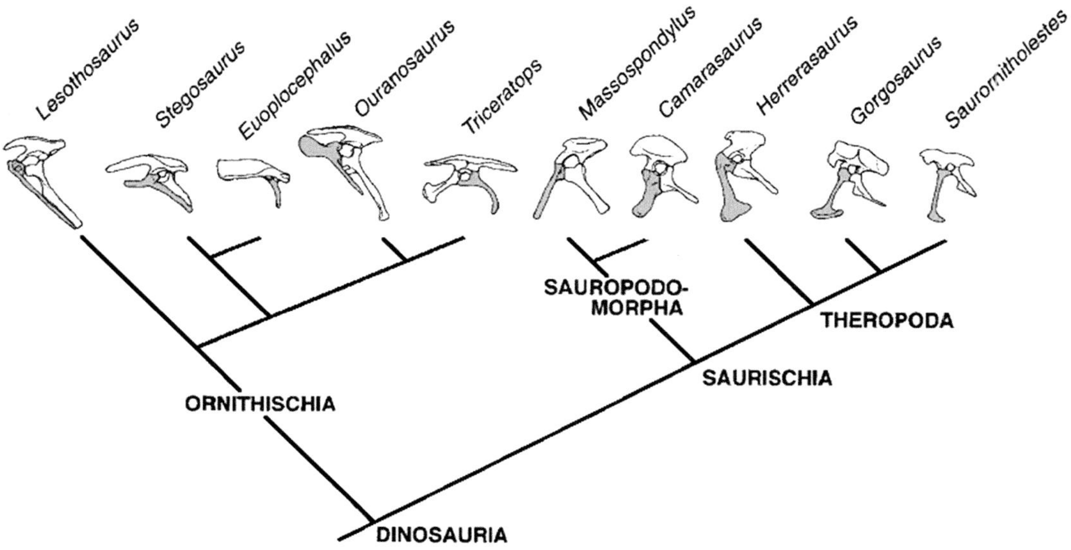


**Fig. 1.20** Pelvic girdles of representative ornithischians and saurischians. The pelvic girdle consists of three bones, the ilium, ischium, and pubis. The pubis, or pubic bone (shaded), projects backward in ornithischians and forward in saurischians. (Figure modified from Carrano 2000; © The Paleontological Society, used with permission)

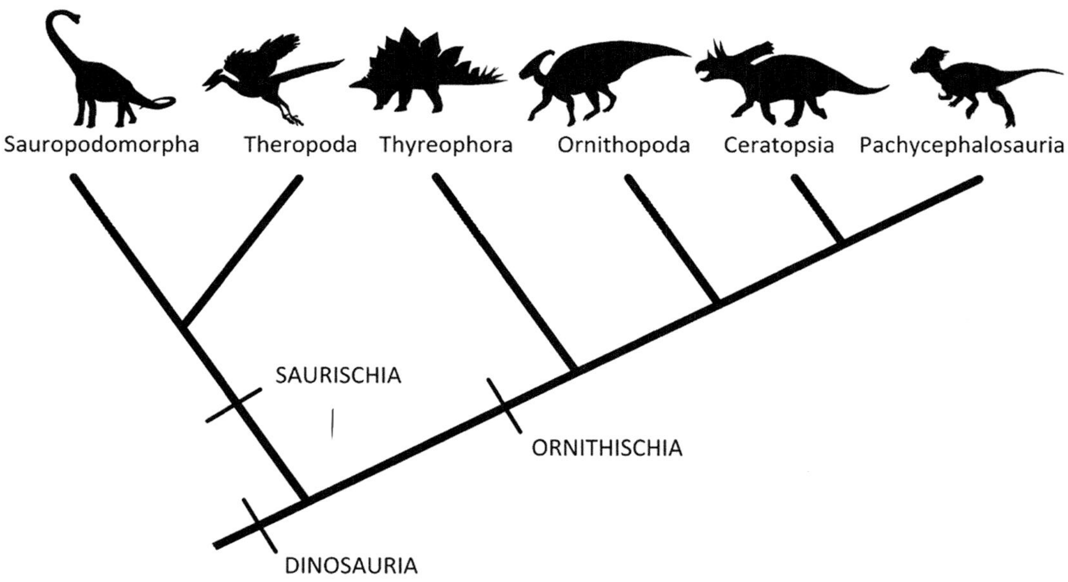


**Fig. 1.21** Major clades of the ornithischian dinosaurs included Thyreophora (e.g., ankylosaurs and stegosaurs), Ornithopoda (e.g., *Iguanodon* and hadrosaurs), Ceratopsia (e.g., *Triceratops*), and Pachycephalosauria (e.g., Pachycephalosaurus). (Figure modified from Maidment and Barrett 2014; © 2014 S.C.R. Maidment and P.M. Barrett, open-access article distributed under the terms of the Creative Commons Attribution License, https://creativecommons.org/licenses/by/4.0/)

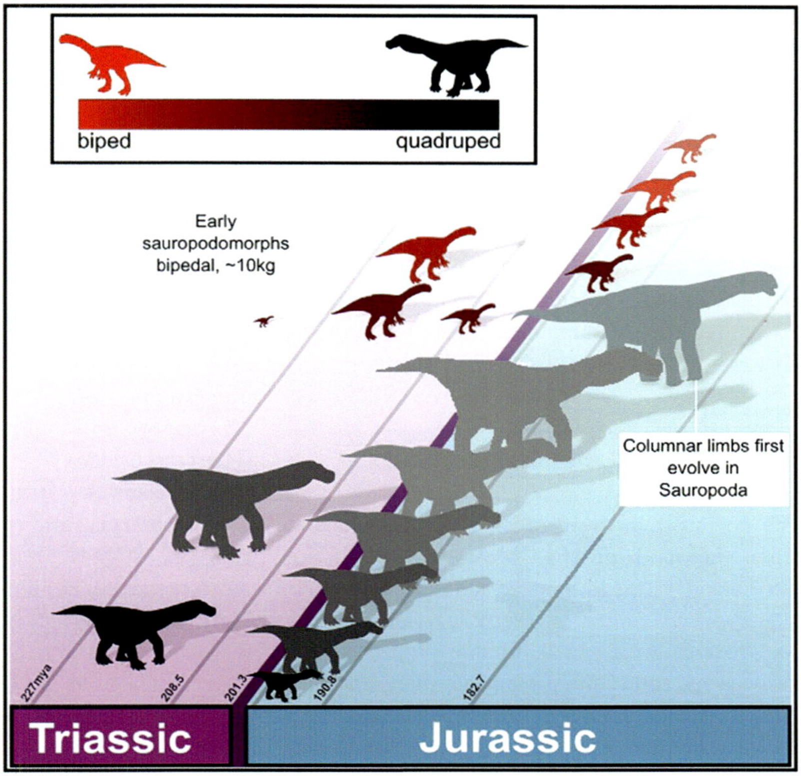


**Fig. 1.22** The sauropodomorphs included the prosauropods (early sauropodomorphs) and sauropods. (Figure modified from McPhee et al. 2018; © 2018 Elsevier Ltd., used with permission)

*Argentinosaurus*. Theropod dinosaurs, on the other hand, included some of the largest terrestrial predators in earth's history, including *Tyrannosaurus rex*. However, not all theropods were large predators; they ranged in size from a less than 200 grams to about 9000 kilograms (Lee et al. 2014; Persons et al. 2020) and exhibited a variety of diets, including omnivory and herbivory, as well as carnivory.

Theropods included two major lineages, Ceratosauria and Tetanurae, that diverged sometime during the late Triassic or early Jurassic (Carrano and Sampson 2008). The Ceratosauria were among the first theropod dinosaurs to appear in the fossil record and those fossils have revealed they already had some "bird-like" characteristics such as hollow bones and S-shaped necks, and some had highly reduced thumbs (or pollex), which have been an important step in theropod hand evolution (given that birds also have highly reduced thumbs; Xu et al. 2014). Ceratosaurians persisted into, but did not survive beyond, the Cretaceous period. In contrast, the Tetanurae persist even today because birds are part of this lineage of theropod dinosaurs! From the Jurassic and continuing through the Cretaceous, the Tetanurae exhibited much greater diversity than the ceratosaurians and consisted of three principal groups, the Megalosauroidea, Allosauroidea, and Coelurosauria (Carrano et al. 2012). Allosauroids and megalosauroids were primarily large carnivores and were widely distributed and abundant, particularly during the late Jurassic and early Cretaceous. Our focus, however, will be on Coelurosauria because birds are coelurosaurids.

Before proceeding, a quick review: the first tetrapods appeared in the fossil record about 395 million years ago and, by the end of the Permian (251 million years ago), the ancestors of birds (archosauromorphs) were present and had assumed an upright (rather than sprawling) posture. During the early Triassic, archosauromorphs split into two major groups, the crurotarsi or "crocodile line" and the avemetatarsalia or "bird line." During the Triassic, the avemetarsalia diverged into several groups, including the Dinosauria (Fig. 1.21), and Dinosauria subsequently diverged into the Ornithischians and Saurischians, with the Saurischians subsequently splitting into Sauropodomorpha and Theropoda. In addition, before the end of the Triassic, several dinosaurs became bipedal, including those in the theropod clade that would ultimately, during the Jurassic, give rise to the Coelurosauria and birds.

## 1.5 Theropods: Coelurosaurs

The coelurosaurs were much more diverse than other theropod groups. Most coelurosaurians were small to medium-sized animals, but some, like *Tyrannosaurus rex*, were very large. Many coelurosaurs were small, nimble predators with long arms, and sharp-clawed, three-fingered, grasping hands. However, Zanno and Makovicky (2011) examined known coelurosaur fossils and found that, although many were predators, many others were omnivores and herbivores (Fig. 1.23).

Basal coelurosaurs were bird-like in several ways, with unidirectional lung ventilation (Box 1.1 Unidirectional Lung Ventilation), increasingly knee-based locomotion, three-fingered hands, filamentous feathers, faster growth rates, and higher basal metabolic rates (Xu et al. 2014; Fig. 1.24). The possible advantages of unidirectional lung ventilation and a bipedal posture have already been discussed in this chapter and the evolution of feathers in Chap. 3, but what about these other characteristics?

## 1.6 Theropod Locomotion

Basal theropods, like present-day birds, were bipedal and terrestrial locomotion required flexing and extending of joints in their legs. Based on hindlimb anatomy, examination of fossil footprints/trackways, and biomechanical analysis, walking or running basal theropods exhibited hip/knee-based locomotion (Farlow et al. 2000). In other words, flexion/extension at both the hip and knee joints was important in flexing and extending the hindlimbs. In addition, when a basal theropod was standing, the femurs (thigh bones) were relatively vertical (Fig. 1.14). In contrast, terrestrial locomotion for present-day birds is knee-based, with relatively little flexion/extension at the hip joint than the knee joint and, in addition, when standing, a bird's femurs are relatively horizontal (Fig. 1.25). Allen et al. (2013) used three-dimensional digital reconstruction to estimate body shape from skeletal dimensions of 17 archosaurs, including 12 theropods ranging from the most basal (Herrerasauridae) to *Tyrannosaurus* (Coelurosauria), *Struthiomimus* (Maniraptora), *Archaeopteryx*, and present-day birds. Their analyses suggest a gradual development of more-crouched hindlimb postures (and increasingly knee-based locomotion) across much of theropod evolution, driven by a movement of the center of mass in a cranial direction. Important factors in this shift in the center of mass include a gradual reduction in tail length and mass plus a gradual increase in the size and mass of the forelimb. Analysis also revealed that this shift in the center of mass accelerated within Maniraptora, corresponding with the evolution of flight (and the increasing mass of flight muscles). With this shift in the center of mass from the hip to the knee, selection favored a shorter, stronger, and more horizontally positioned femur (Fig. 1.25). With a more horizontal femur required to keep the center of mass centered above the feet to maintain balance, even when walking or running, most of the flexion/extension occurs at the knee (i.e., knee-based locomotion) and ankle rather than at the hip.

Ancestral archosaurs were quadrupedal, but basal members of all dinosaurian lineages were bipedal (Sereno et al. 1993). For bipedal dinosaurs, including theropods, forelimbs no longer involved in locomotion could now serve nonlocomotor functions, and the possible functions of the forelimbs of non-avian theropods have been the focus of much speculation. What is known is that the forelimbs of non-avian theropods were relatively short, with a trend

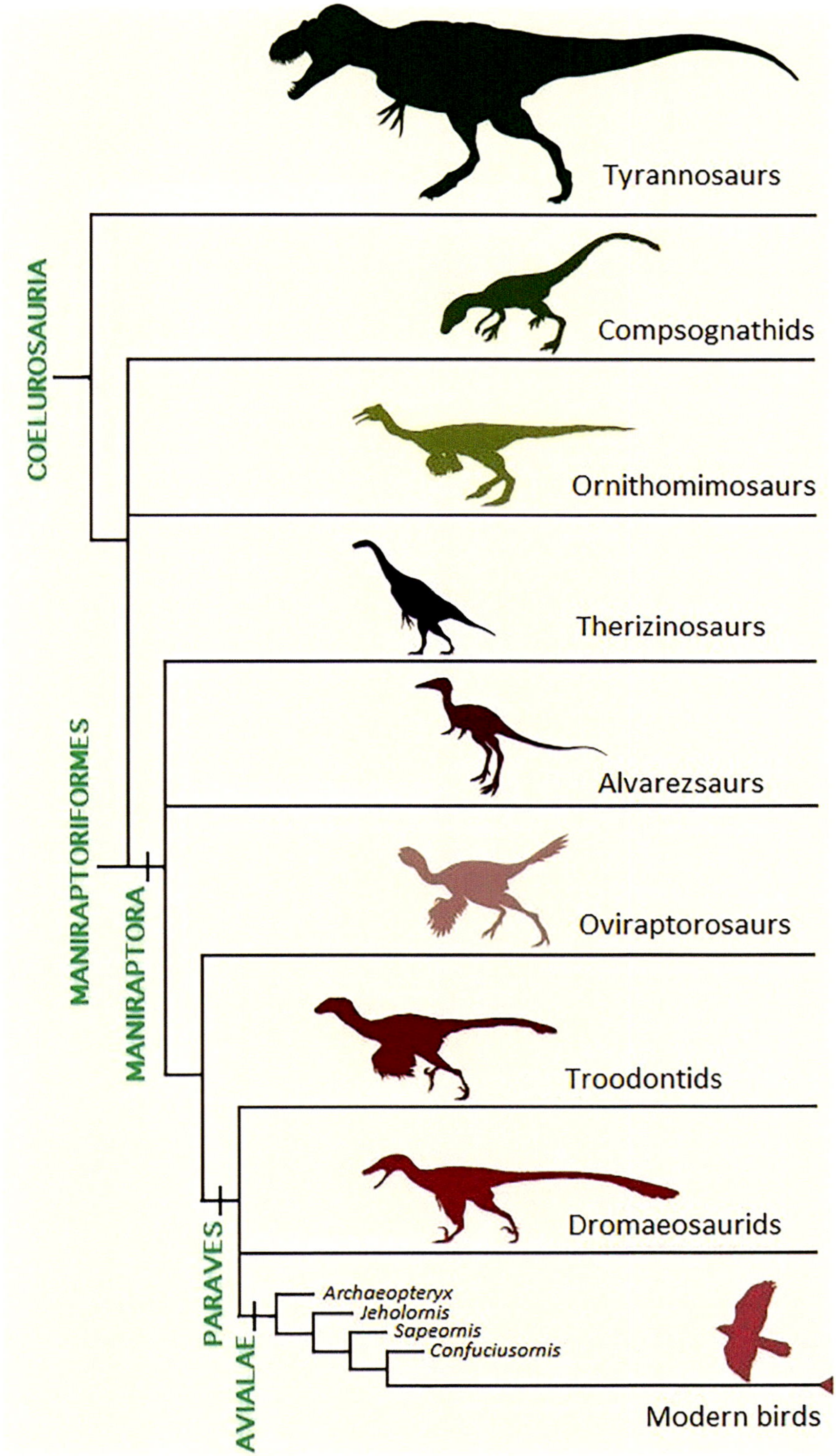


**Fig. 1.23** Interrelationships of the major taxa of the Coelurosauria. (Figure from Brusatte et al. 2014; © 2014 Elsevier Ltd., used with permission)

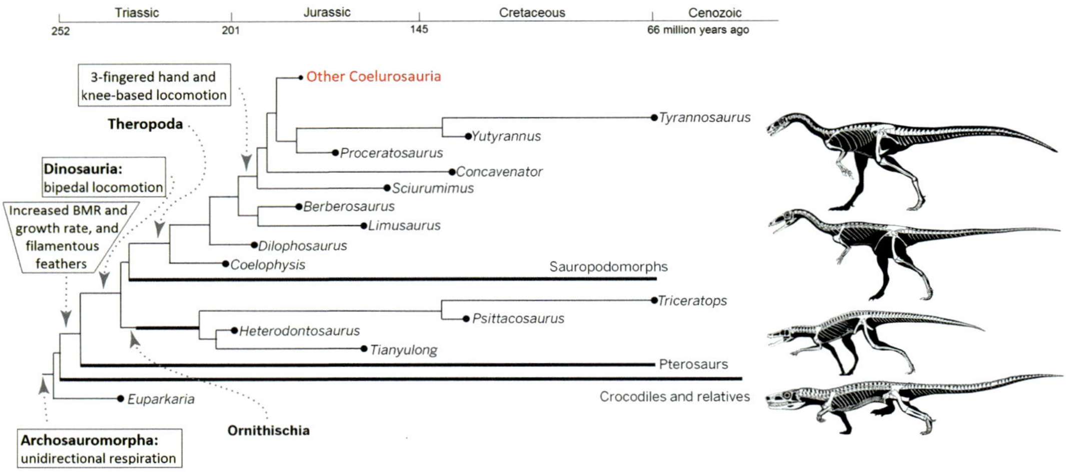


**Fig. 1.24** Selected species illustrating archosauromorphan phylogeny and the evolution of characteristics through the appearance of the Coelurosauria. Tyrannosaurs represent the basal coelurosaurs and they already had several bird-like characteristics, including being bipedal with increasingly knee-based locomotion, three-fingered hands, filamentous feathers, faster growth rates, and higher basal metabolic rates. (Figure modified from Xu et al. 2014; © 2014 AAAS, used with permission)

toward the loss and fusion of bones. For example, basal theropods, like *Herrerasaurus*, had five digits, but digits 4 or 5 were vestigial. Later theropods completely lost digits 4 and 5, and the trend for more fusion and loss of carpals and metacarpals continued (Box 1.3 Avian Forelimb Digits). Forelimb reduction may have been a result of disuse and becoming specialized for head-based predation (Lockley et al. 2009). However, some investigators have suggested that the forelimbs of predatory theropods were used to grasp and carry prey (e.g., Carpenter 2002; Senter 2006a, 2010; Fig. 1.26), and others have suggested that their forelimbs could have been important for digging and climbing (Xu et al. 2011). Other possible explanations for the reduced length and skeletal complexity of non-avian theropod forelimbs include a reduction of heat loss (i.e., smaller limb with less surface area) and better balance when running (especially with the smaller forelimbs held close to the body; Senter 2006a, b). Regardless of the selective factors involved, changes in the forelimbs of non-avian theropods were important in the subsequent evolution of birds and flight.

**Box 1.3 Avian Forelimb Digits**

Much debate has focused on the fingers, or forelimb digits, of birds, and for good reason. If birds evolved from theropods, then three-fingered birds and three-fingered theropods should have the same three fingers. In other words, basal archosaurs had five "fingers" and, during the evolution of theropods, two digits were lost; birds and three-fingered theropods should, therefore, have lost the same two digits. In the embryonic bird hand, the position of the developing fingers seems to indicate that they are digits 2, 3, and 4. However, paleontological evidence suggests that, during dinosaur-to-bird evolution, digits 4 and 5 became reduced in size and eventually lost (Xu and Mackem 2013). This apparent difference, with birds apparently having digits 2, 3, and 4, and

(continued)

**Box 1.3** (continued)

their theropod ancestors digits 1, 2, and 3, is sometimes used as evidence against the idea of dinosaur-to-bird evolution (e.g., Feduccia 2002).

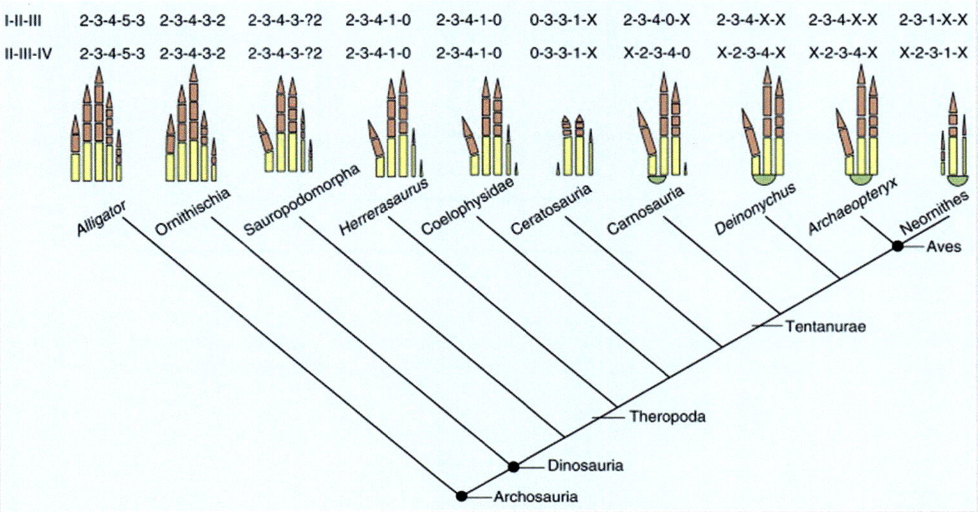


Hands (metacarpals and digits) of present-day birds (Neornithes) compared to those of their theropod ancestors. The basic tetrapod limb, and the limbs of basal archosaurs, has five digits. The trend among theropods, however, was for a reduction in both the number of forelimb digits and the number of bones (called phalanges) making up those digits. The numbers at the top of this figure represent the number of phalanges making up each digit and, given the debate concerning which digits are present in present-day birds, numbers are provided for both possibilities, i.e., digits 1, 2, and 3 (or I, II, and III) or digits 2, 3, and 4 (or II, III, and IV). (Figure from Xu and Mackem 2013; © 2013 Elsevier Ltd., used with permission)

Several hypotheses have been proposed to explain this apparent disagreement between embryological and paleontological views. The most widely supported hypothesis is referred to as the frameshift hypothesis and was first proposed by Wagner and Gauthier (1999). According to this hypothesis, digit 1 actually develops from the second position (where digit 2 should have arisen), and so on—a frame shift. In support of this hypothesis, Tamura et al. (2011) transplanted cells important in the growth of digit 4 from the feet to the hands (and vice versa) of domestic chickens. This experiment revealed that the last digit of the hand did not correspond to the last digit of the foot, supporting the hypothesis that the hand, unlike the foot, does not have a digit 4. Further, when they monitored digit development using cell-labeling techniques (enabling them to know where a certain cell ended up once it matured), they found that by day 3.5 of embryonic development, a shift occurs, causing cells in the region where digit 4 is developing to move forward and grow into digit 3. The same shift occurred for the digits that become 1 and 2. Other investigators have also reported evidence supporting the frameshift hypothesis (e.g., Salinas-Saavedra et al. 2014).

(continued)

**Box 1.3** (continued)

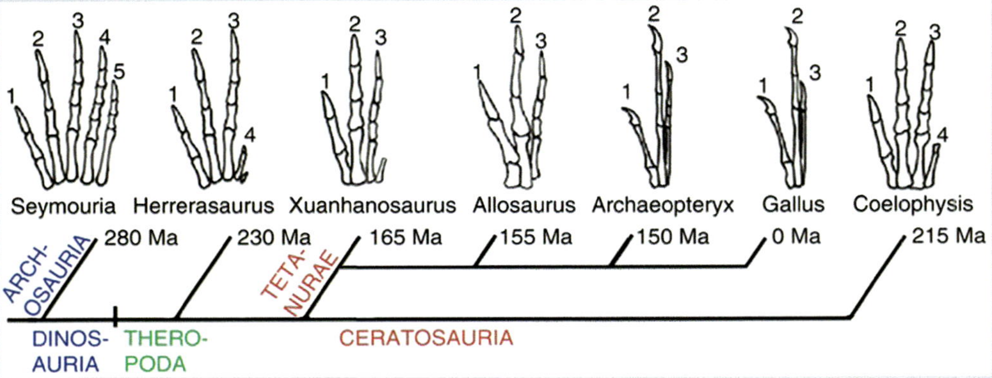


The fossil record indicates that birds evolved from basal theropods that gradually lost forelimb digits 4 and 5. (Figure modified from Towers et al. 2011; © 2011 Springer Nature, used with permission)

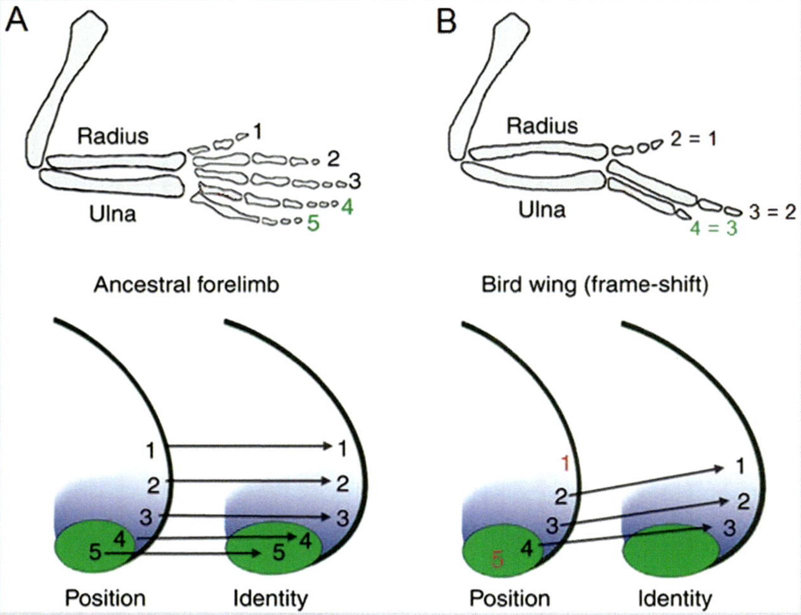


(**a**) Digits of the theropod ancestors of birds, with digits 4 and 5 gradually lost (as in Allosaurus, above). (**b**) The frameshift hypothesis proposes that what appear to be, based on their positions in developing embryos, digits 2, 3, and 4 are actually digits 1, 2, and 3. (Figure from Towers et al. 2011; © 2011 Springer Nature, used with permission)

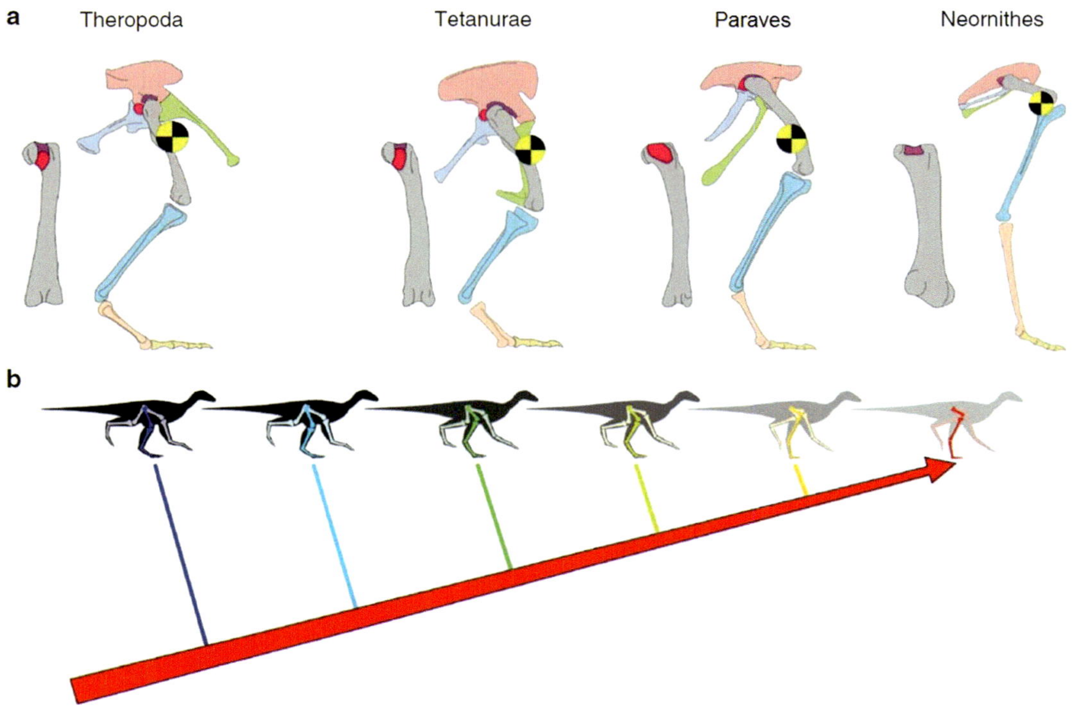


**Fig. 1.25** Evolution of theropod hindlimbs. (**a**) Right hindlimbs of representative theropods, including a basal theropod (Therapoda), a Tetanurae theropd, a paravian, and a present-day bird (Neornithes). The femurs (shown in gray) are shown to the left (with the colored areas at top indicating areas of articulation with the acetabulum of the pelvic girdle) and the black and yellow circles indicate the center of mass. Note the change in the orientation of the femur from nearly vertical in basal theropods to nearly horizontal in present-day birds. (**b**) Gradual transition from hip and knee-based locomotion to the knee-based locomotion of present-day birds. (Figure modified from Hutchinson and Allen 2009; © 2008 Springer-Verlag, used with permission)

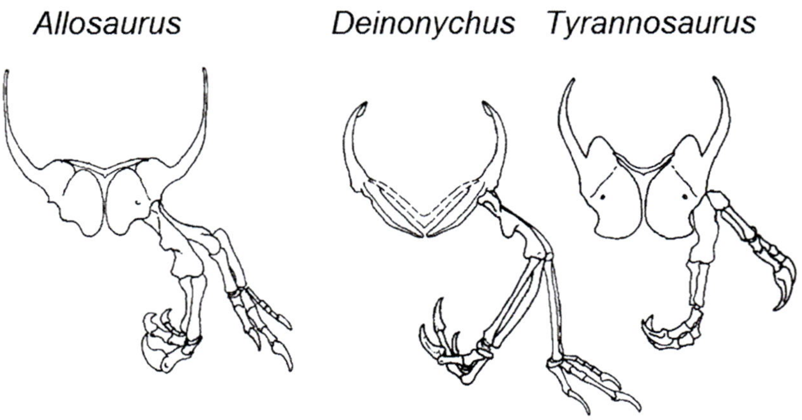


**Fig. 1.26** Forelimb motion of three theropods, with forelimbs shown both extended and flexed. With the aid of powerful muscles, flexing of the forelimb may have helped predatory dinosaurs to firmly grasp prey. (Figure modified from Carpenter 2002; © 2002 E. Schweizerbart'sche Verlagsbuchhandlung, used with permission)

## 1.7 Bones and Growth Rates

The term dinosaur originated with Richard Owen in 1842 and comes from the Greek words *deinos*, meaning "terrible" or "fearfully great," and *sauros*, meaning "lizard" (Owen 1842). Being reptiles, dinosaurs, like extant lizards and other reptiles, were originally assumed to be slow-growing ectotherms. For example, Hildebrand et al. (1930), commenting on locations where dinosaur fossils had been found at the time, noted that "Unless the dinosaurs were warm-blooded creatures—and it is supposed they were not—many of the regions where their fossil remains are now found must have had a warmer climate than they have now." However, Ostrom (1973), noting many similarities between theropods like *Deinonychus* and birds, suggested that "direct inheritance from a small coelurosaurian ancestor" best explained these similarities. This suggested relationship between dinosaurs and birds led to increased speculation concerning dinosaur growth and metabolic rates. The next year, de Ricqlès (1974) noted that many dinosaurs had fibrolamellar bone (Fig. 1.27) and, because present-day birds and mammals had the same type of bone and were endotherms, proposed that dinosaurs were also endothermic (Box 1.4 Evolution of Endothermy).

**Box 1.4 Evolution of Endothermy**

The best-supported hypothesis proposed to explain the evolution of endothermy is the aerobic capacity model (Bennett and Ruben 1979). This model proposes that natural selection favored the ability to sustain vigorous activity, which is limited by an organism's aerobic capacity (maximum rate of oxygen consumption) or, in other words, the maximum metabolic rate (MMR). This model assumes that there is a genetic correlation between resting metabolic rate (RMR) and MMR, so selection for increased MMR would necessarily result in an increased RMR or, in other words, endothermy. This assumption appears to be supported by the fact that physiological mechanisms important for MMR are similarly important for determining RMR (Hochachka and Burelle 2004; Clarke and Pörtner 2010).

The results of modeling and experimental studies suggest that efficient retention of metabolic heat, especially by smaller animals, would not be possible without some type of insulation (Clarke and Pörtner 2010). As such, the evolution of pennaceous feathers may be linked to an increase in MMR in dinosaurs closely related to birds and, indeed, fossils from China provide evidence for the presence of feathers in a wide variety of non-avian dinosaurs, including coelurosaurs (Norell and Xu 2005).

(continued)

**Box 1.4** (continued)

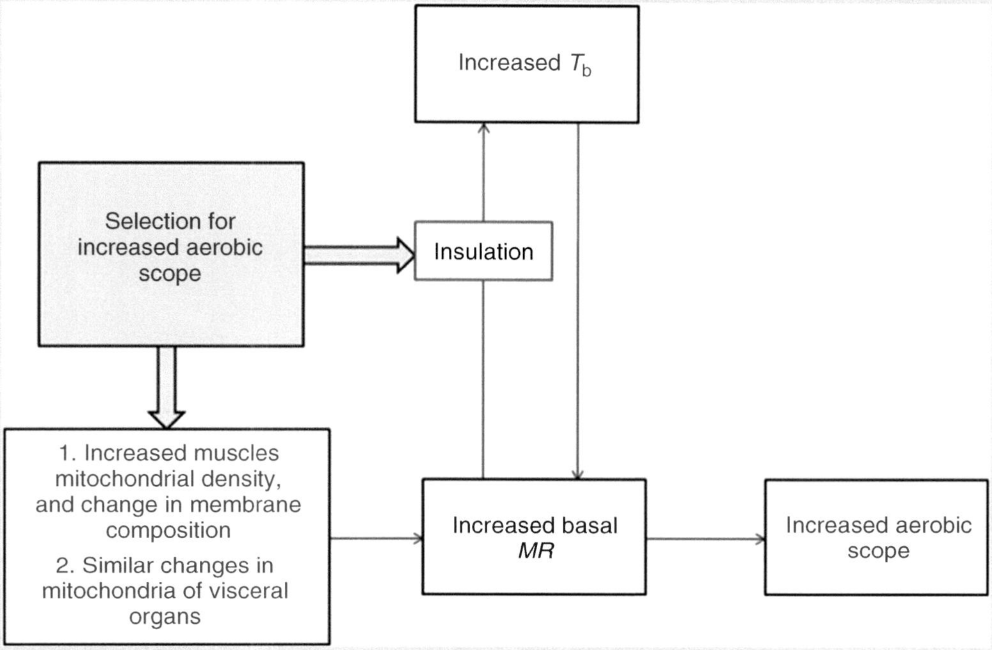


In support of the aerobic capacity model, studies have demonstrated that mitochondria can generate ATP faster when warm and skeletal muscles are more efficient at using ATP for contraction when warmer. As such, a higher resting body temperature ($T_b$) increases the capacity for sustained locomotor activity. Of course, a higher body temperature does come at the cost of the need for increased food consumption. (Figure from Clarke and Pörtner 2010; © 2010 The Authors. Biological Reviews © 2010 Cambridge Philosophical Society, used with permission)

Because growth rates and metabolic rates are correlated, one way to gain insight into dinosaur metabolic rates is to use growth lines in their bones to estimate growth rates (Fig. 1.28). Analysis of the bones and growth rates of several different species of dinosaurs has revealed that their growth rates were intermediate between those of extant reptiles and marsupials and extant birds and eutherian mammals (Fig. 1.29). This, in turn, suggests that the metabolic rates of dinosaurs, including theropods, may have also been intermediate, higher than those of today's reptiles and marsupials, but lower than those of birds and eutherian mammals. In further support of this "intermediate" status of dinosaurs, Eagle et al. (2015) compared the eggshells of two dinosaurs to those of 13 species of present-day birds using isotope analysis to examine shell chemistry (because the stable isotope composition of eggshells is influenced by temperatures in the oviduct) and the dinosaurs were able to elevate their body temperatures above environmental temperatures, but were not able to thermoregulate like modern birds (Fig. 1.30). Interestingly, the growth rates of *Archaeopteryx* were comparable to those of dinosaurs, suggesting that the first birds grew like and had metabolic rates like small dinosaurs, not like present-day birds (Erickson 2014). Importantly, however, the metabolic rates of dinosaurs appear

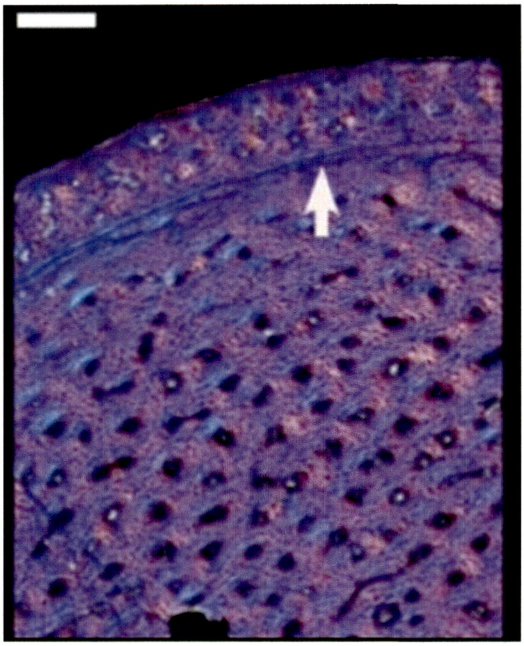

Fig. 1.27 Cross-section through fossilized bone (femur) of *Shuvuuia* (an Alvarezsaur; see Fig. 1.23) showing a growth line (arrow). This type of highly vascular bone is called fibrolamellar bone and is the type of bone found in present-day birds and mammals. (Figure from Erickson et al. 2009; © 2009 Erickson et al., open-access article distributed under the terms of the Creative Commons Attribution License, https://creativecommons.org/licenses/by/4.0/)

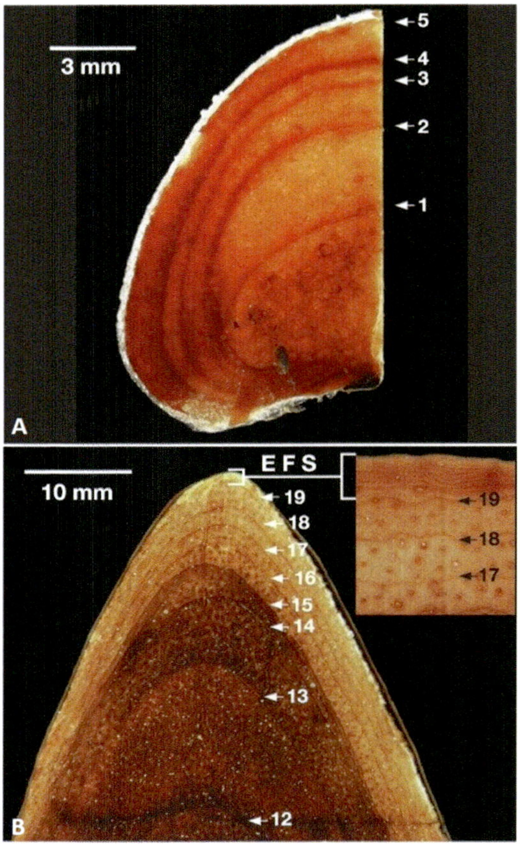


Fig. 1.28 (a) Thin-sectioned *Gorgosaurus* fibula showing five growth lines, indicating that it died early in its sixth year of life. (b) *Tyrannosaurus* rib showing the 12th to 19th growth lines. Inset, tightly spaced growth lines, indicating a slowing in growth rate. *Gorgosaurus* is a genus of bipedal, predatory dinosaurs in the family Tyrannosauridae. (Figure from Erickson et al. 2004; © 2004 Springer Nature, used with permission)

to have been sufficiently high to permit active lifestyles and to allow their avian descendants to take to the air.

## 1.8 Body Size

The oldest basal coelurosaurs pre-date the earliest known bird, *Archaeopteryx*, by about 25–30 million years and, during that time interval, the pace of evolution and morphological change increased (Brusatte et al. 2014). One such change was especially rapid (in an evolutionary sense), a reduction in body size (Fig. 1.20). Recall that theropods included two major lineages, Ceratosauria and Tetanurae, that diverged sometime during the late Triassic or early Jurassic. Early representatives of the Tetanurae (about 200 million years ago) were estimated to have weighed about 160 kg (Lee et al. 2014). However, by about 20–30 million years later, some coelurosaurs weighed as little as an estimated 27 kg, and this decrease in size continued over the next few million years with the maniraptorans (about 10 kg), paravians (about 3 kg), and the first birds (Aves, *Archaeopteryx*, about 0.8 kg) (Lee et al. 2014; Fig. 1.31). What factors may have contributed to this accelerated decline in size that began with the coelurosaurs? As noted earlier, coelurosaurs were already bird-like in many

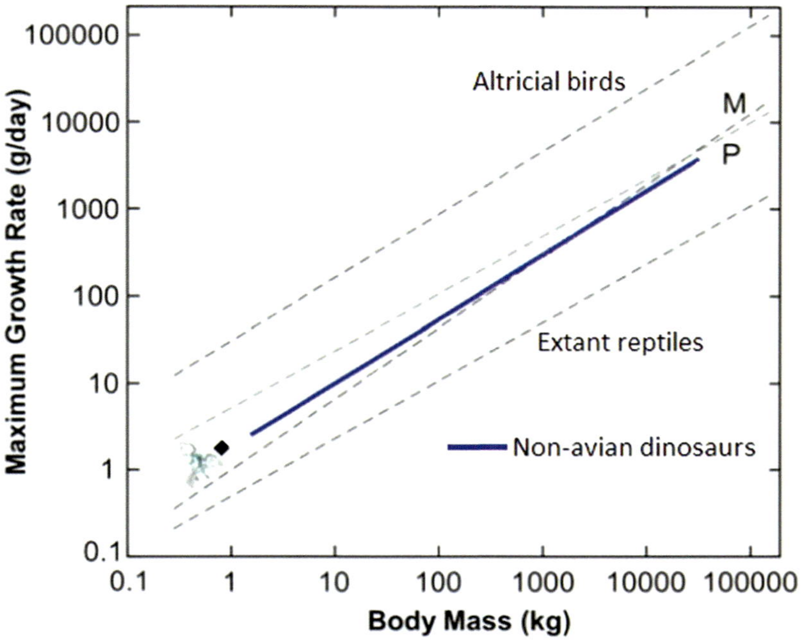


**Fig. 1.29** Growth rates of non-avian dinosaurs and *Archaeopteryx* (solid diamond) compared to those of altricial and precocial (P) birds, marsupials (M), and extant reptiles. (Figure modified from Erickson et al. 2009; © 2009 Erickson et al., open-access article distributed under the terms of the Creative Commons Attribution License, https://creativecommons.org/licenses/by/4.0/)

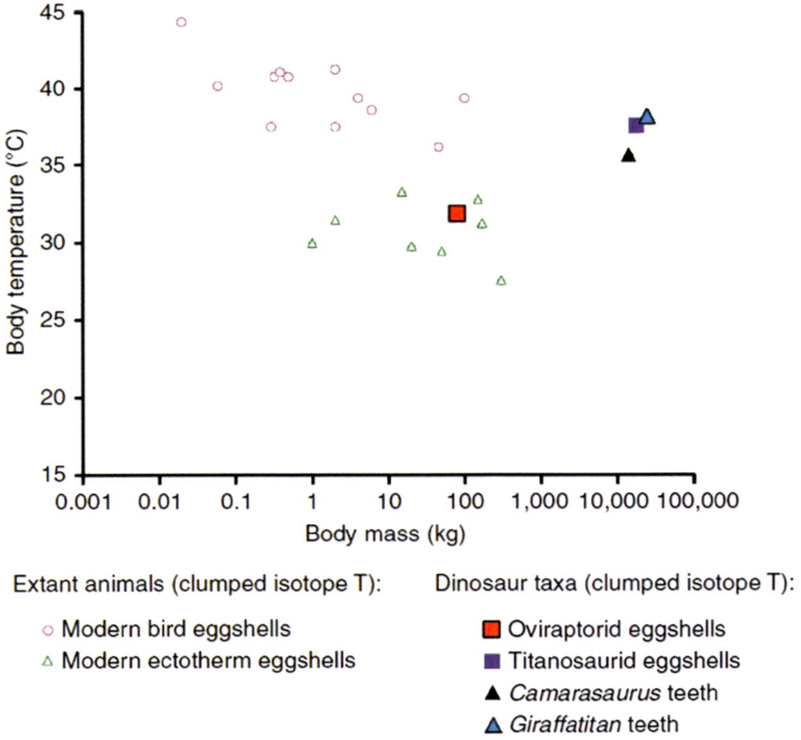


**Fig. 1.30** Body temperatures of 12 species of modern birds, 8 species of present-day ectotherms, and 4 species of dinosaurs based on isotopic analysis of egghells and/or teeth. (Figure modified from Eagle et al. 2015; © 2015 Springer Nature, used with permission)

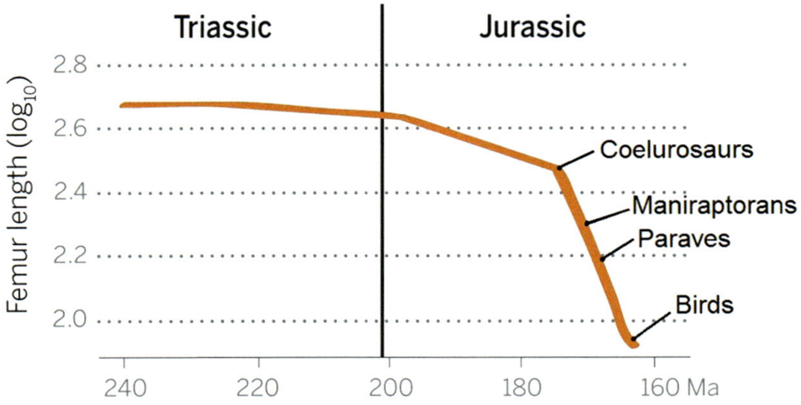


**Fig. 1.31** As shown by changes in femur lengths, the ancestors of birds decreased in size at a rapid evolutionary rate beginning about from 200 million years ago. The reduction in size, however, greatly accelerated beginning with the coelurosaurs. (Figure modified from Benton 2014 to match dates provided by Lee et al. 2014; © 2014 AAAS, used with permission)

ways, with bipedal and increasingly knee-based locomotion and three-fingered hands (Fig. 1.32), filamentous feathers, faster growth rates, unidirectional lung ventilation, and higher basal metabolic rates (Xu et al. 2014). All of these features

**Fig. 1.32** Examples of two therizinosaur theropods. (**a**) *Nanshiungosaurus brevispinus* and (**b**) *Falcarius utahensis*. Therizinosaurs were early maniraptorans and had several bird-like characteristics, including three-fingered hands and increasingly knee-based locomotion. Among the coelurosaurs, therizinosaurs were a bit unusual because they were herbivores. (Figure from Clark and Xu 2009; © 2009 Springer Nature, used with permission)

evolved and continued to evolve in unison. As pointed out by Kemp (2007:1668), an organism's traits are integrated so that "... each one is both dependent on and necessary for the functioning of all others in a well-adapted phenotype."

Trends for decreasing size, faster growth rates, and improving metabolic function among bird-like theropods continued for the next several million years, with multiple traits influenced by and likely contributing to these changes. For example, one contributing factor might have been the increasing complexity of feather structure, providing more effective insulation and facilitating selection for smaller body sizes and improved metabolic function. The reduction in size of bird-like theropods also shortened normal developmental patterns. In fact, the heads of birds can be viewed as dinosaur heads that retain juvenile features in adults (i.e., developmental truncation or paedomorphosis), with short snouts, smaller teeth, large, rotated brains, and large eyes (Bhullar et al. 2012; Lee et al. 2014; Figs. 1.33, 1.34, and 1.35). This reduction in size in combination with increased brain size (including the visual-associated areas) would likely have contributed to continuing improvement in agility and cursoriality and, ultimately, increasingly arboreal and aerial habits (Lee et al. 2014).

**Fig. 1.33** Skulls of three archosaurs. (**a**) Alligator 46-day embryo (left) and adult (right), (**b**) *Coelophysis* (early dinosaur) juvenile (left) and adult (right), and (**c**) young *Archaeopteryx* (left) and older adult (right). The skulls of young and adult alligators and the non-avian dinosaur *Coelophysis* are noticeably different. However, there is little difference between the skulls of a young and adult *Archaeopteryx*. The retention of characteristics of younger individuals or juveniles in adults is called paedomorphosis. (Figure modified from Bhullar et al. 2012; © 2012 Springer Nature, used with permission)

## 1.9 Limb Length

During the transition from non-avian theropods to birds, the relative length of the limbs also changed, with the forelimbs getting longer and the hindlimbs shorter (Dececchi and Larsson 2013). These changes accelerated during the late Jurassic among the first birds, e.g., *Archaeopteryx* (Figs. 1.36 and 1.37). One possible explanation for the increasing length of forelimbs is the relatively short length of the forelimb feathers of the earliest birds (Dececchi and Larsson 2013). The relatively large size of the first birds, in combination with relatively short forelimb feathers, may have favored the evolution of relatively longer forelimbs to increase wing surface area and generate additional lift. In contrast, the trend toward shorter hindlimbs may have been driven by the increasingly arboreal habits of the first birds in combination with the advantages of improved flight efficiency (Fig. 1.38). Shorter hindlimbs decrease torsional (twisting) forces when perching or moving along branches, e.g., when foraging (Cartmill 1985), and they are also lighter and can be tucked into the plumage and against the body and tail to reduce parasitic drag during flight (Tucker 1988).

Another important change during the evolution of birds was an increasing ability to abduct (or flex) the wrist joint (Figs. 1.39, 1.40, and 1.41). This mobility allows birds to partly fold the wing during the upstroke, reducing drag, and also allows birds to completely fold their wings when not flying, reducing the risk of damaging flight feathers and, with wings held out of the way and against the body, allowing more efficient terrestrial, aquatic, or arboreal locomotion (Sullivan et al. 2010). Just as importantly, this flexibility of the wrist is limited, allowing only hinge-like movement toward and away from the ulna as a wing closes and opens. When a bird is flying, this keeps the wing level and prevents the wrist from "rotating" (i.e., supinating or pronating).

The selective pressures contributing to the evolution of an abducting wrist remain unclear. However, one possibility is that the development of longer feathers on the manus (hand) favored the evolution of a wrist with an increasing ability to abduct to prevent damage to those feathers.

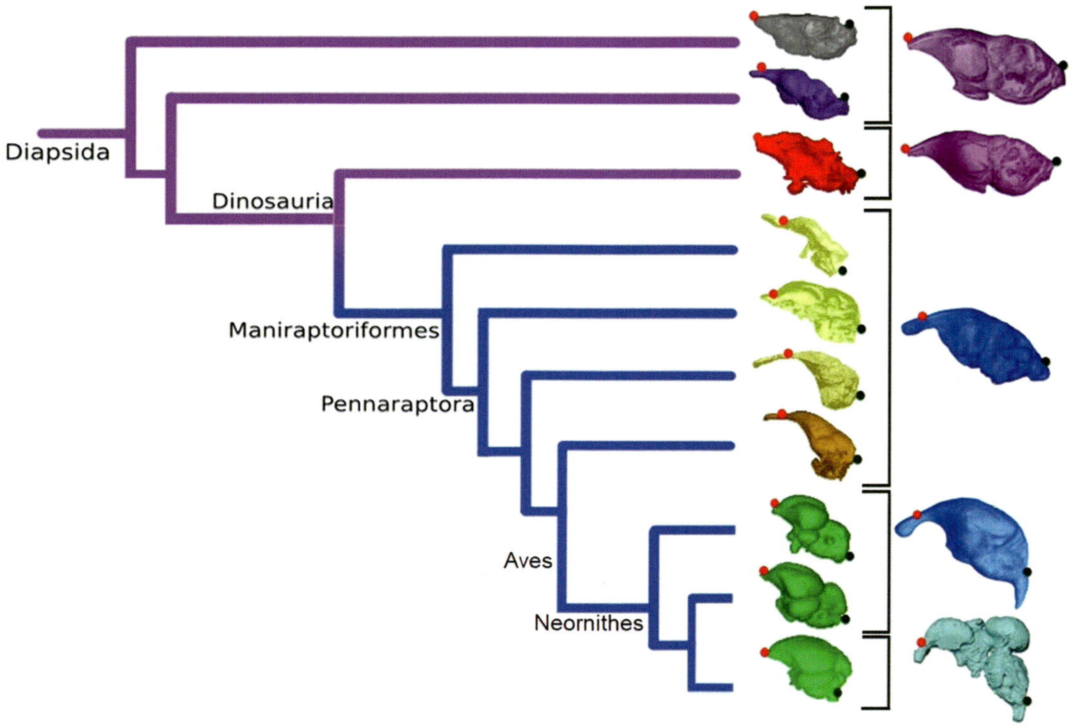


**Fig. 1.34** As birds evolved, the endocranium became less elongated and more bulbous and, as a result, the brain became less horizontally oriented, with the cerebrum above, rather than in front of, the rest of the brain. Dots indicate the positions of the anterior-most tip of the cerebrum and the opisthion (midpoint on the posterior margin of the foramen magnum). Brains (endocasts) on the right represent average orientations for each taxon. Brains are not to scale. (Figure modified from Beyrand et al. 2019; open-access article licensed under a Creative Commons Attribution 4.0 International License, https://creativecommons.org/licenses/by/4.0/)

However, it is also possible that other factors contributed to the evolution of abducting wrists, and longer feathers appeared later (Sullivan et al. 2010).

## 1.10 Digestive and Reproductive Systems

Fossils where ingested food items have been preserved can provide evidence concerning the structure of the digestive systems of non-avian theropods and some ancient birds. Based on evidence from available fossils, O'Connor (2019) inferred that two-part stomachs evolved outside of Aves, and the evolution of the crop and use of the esophagus to store food occurred much later. The most basal evidence of a crop is in *Sapeornis* (Zheng et al. 2011). Unfortunately, *Archaeopteryx* fossils discovered to date provide no information about their digestive system, but, given the available evidence, they likely had two-part stomachs (Fig. 1.42).

Fossils may also provide information about reproductive systems. For example, Bailleul et al. (2020) reported an enantiornithine from the early Cretaceous with fossilized imprints of what may be ovarian follicles (Fig. 1.43; but Mayr et al. 2020 believe that the putative ovarian follicles are actually plant propagules). Other investigators have also reported structures that appear to be ovarian follicles on the left side of the bodies of several specimens of enantiornithines and one confuciusorinthiform

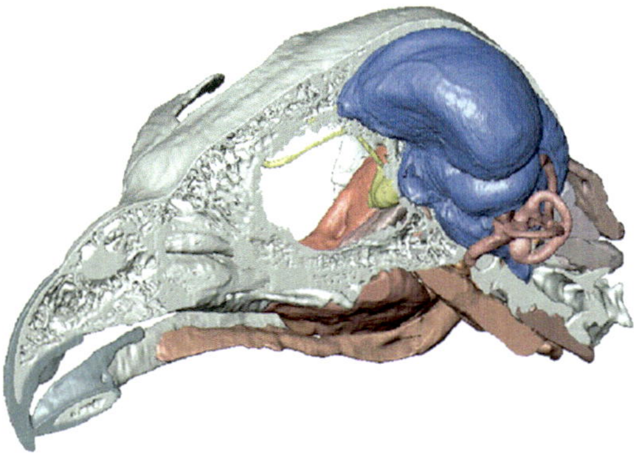

**Fig. 1.35** The brains of birds like those of a Common Buzzard (*Buteo buteo*), in contrast to the more "horizontally oriented" brains of their ancestors, are rotated posteroventrally in the skull, with the cerebrum located above (rather than in front of) the cerebellum, brainstem, and other parts of the brain. (Figure from Lautenschlager et al. 2014; open-access article available under the terms of the Creative Commons Attribution License, https://creativecommons.org/licenses/by/4.0/)

(O'Connor et al. 2013; Zheng et al. 2013, 2017). If these investigators are correct then these fossils indicate, as is also the case for present-day birds, the presence of only one functional ovary (the left one). In contrast, non-avian maniraptoran theropods were known to have two ovaries and two oviducts (Sato et al. 2005). The loss of an ovary may have been driven by the need for flying birds to minimize weight and, if so, may have occurred close to when the first birds appeared (O'Connor and Zhou 2015). However, as with the digestive system, *Archaeopteryx* fossils provide no information about their reproductive systems.

## 1.11 Avian and Non-avian Theropods

About 80 million years (from 235 to 155 million years ago) passed from the appearance of the first theropods until the appearance of the first birds. During that time, many of the characteristics associated with present-day birds first appeared in theropods (Fig. 1.44). Given this gradual (in an

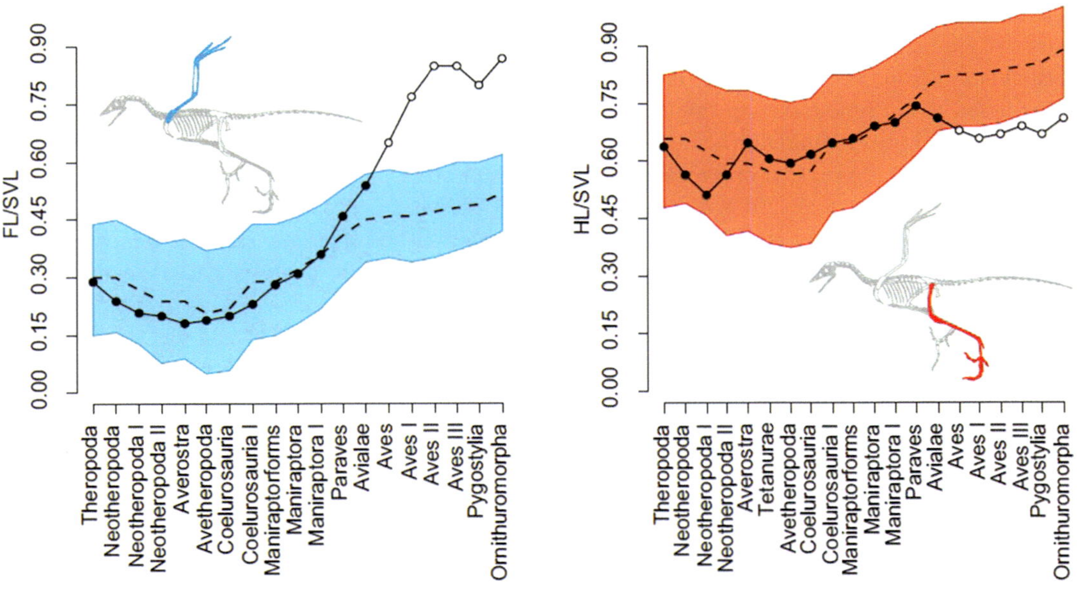


**Fig. 1.36** Evolution of forelimb (left) and hindlimb (right) lengths corrected for body size during the evolution of birds. Dashed lines indicate the estimated expected lengths based on data from non-avian theropods (shaded areas = 95% confidence intervals). FL/SVL = ratio of forelimb length to snout-vent length; HL/SVL = ratio of hindlimb length to snout-vent length. (Figure from Dececchi and Larsson 2013; © 2013 The Authors. Evolution © 2013 The Society for the Study of Evolution, used with permission)

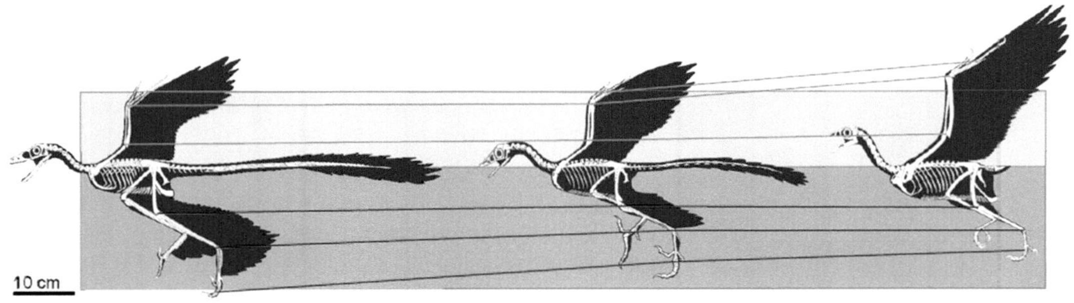


**Fig. 1.37** During the evolution of birds and powered flight, hindlimbs became relatively shorter and forelimbs relatively longer. (Figure from Dececchi and Larsson 2013; © 2013 The Authors. Evolution © 2013 The Society for the Study of Evolution, used with permission)

evolutionary sense) accumulation of bird-like traits among non-avian theropods, what trait or traits differentiate birds (Aves) from non-avian theropods?

Perhaps surprisingly, there is currently no clear answer to that question because the first birds were not significantly different from their closest non-avian theropod relatives. Characteristics once considered unique to birds such as feathers and a furcula were shared by non-avian theropods, and several investigators have noted that a time-traveling ornithologist or naturalist would likely find it difficult to differentiate basal birds (Aves), dromaeosaurids, and troodontids (Turner et al. 2012; Brusatte et al. 2014). Non-time-traveling investigators can also find it difficult, with investigators disagreeing about the status of *Archaeopteryx* as the most basal (or first) bird. For example, Godefroit et al. (2013) reported the discovery of a fossil (*Aurornis*, or "dawn bird") from the mid- to late Jurassic and that, based on a phylogenetic analysis involving 992 characteristics, supplants *Archaeopteryx* as the first bird (i.e., the most basal avialan) (Fig. 1.45). To illustrate the difficulty of determining relationships among the first birds and non-avian theropods, here is the description of the basis upon which Godefroit et al. (2013) considered *Aurornis* the most basal bird: "*Aurornis* is regarded as the basal-most avialan based on the presence of an acute anteroventral corner of premaxilla, on the absence of an external mandibular fenestra, a primordial sacral centra wider than other sacral centra; humerus elongate relative to tarsometatarsus; ischium proportionally broad, and the presence of a ventrodistal cleft on the ischium.... *Aurornis* is resolved as less derived than ... *Archaeopteryx* and more derived avialans because it retains a humeral shaft that is more slender than the femur, a humerus that is shorter than the femur, and the presence of shortened penultimate phalanges on pedal digits."

The year after Godefroit et al.'s (2013) paper was published, Brusatte et al. (2014) reported a new phylogenetic analysis based on more taxa that "demoted" *Aurornis*, classifying it as a troodontid (and, later, so did Pei et al. 2017), and again considered *Archaeopteryx* to be the first bird (Fig. 1.46). However, Brusatte et al. (2014) also noted that the first birds were not especially distinct relative to their non-avian theropod relatives (Fig. 1.45). Birds, therefore, are the result of a continuum of millions of years of theropod evolution, and there was no obvious difference between the first birds and their closest "non-bird" relatives. New fossils will continue to be found and *Archaeopterx* may at some point be

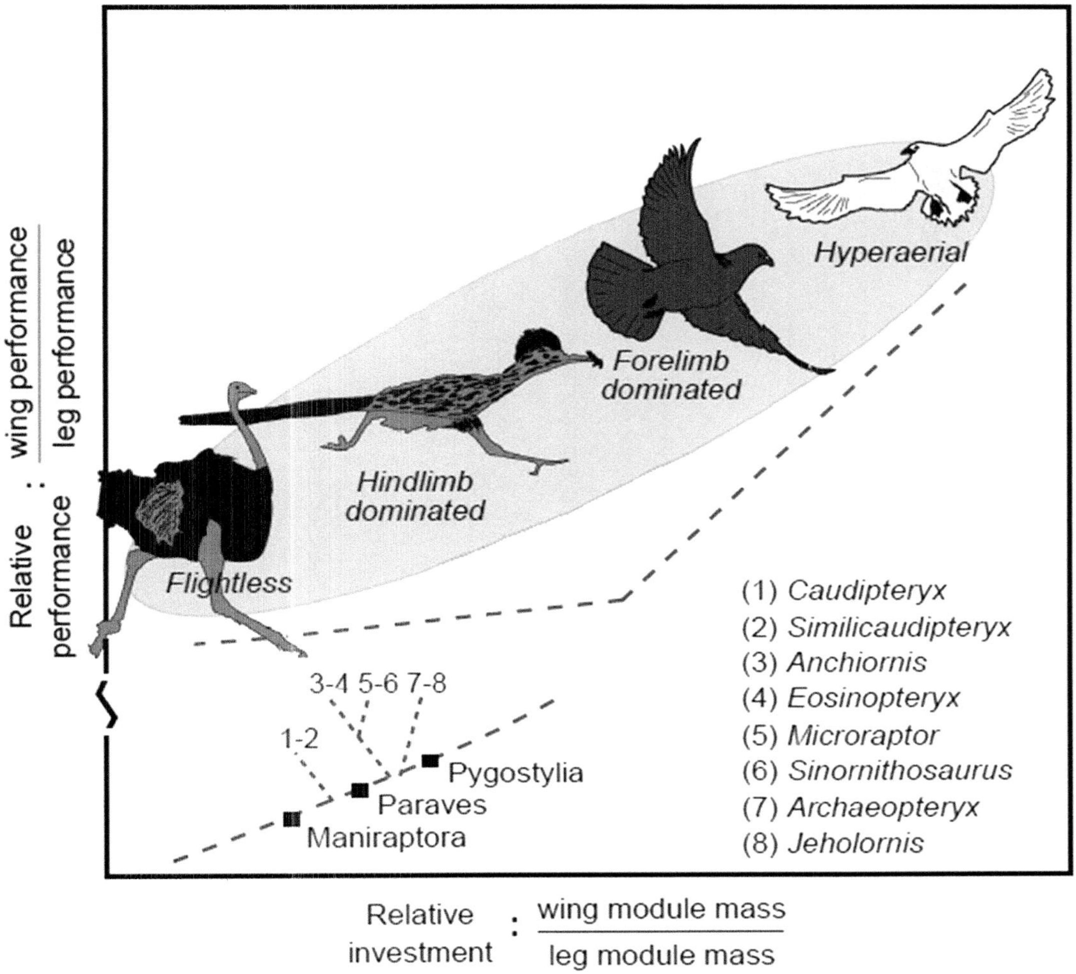


**Fig. 1.38** Expected tradeoffs between wings and legs among non-avian theropods and early birds (lower left) and present-day birds. Variation in forelimb versus hindlimb investment (and performance) may be influenced by tradeoffs between wings and legs; increased investment in one may contribute to decreased investment in the other. Among non-avian theropods and early birds, the apparent trend was for increased investment in the forelimbs (wings) and less investment in the hindlimbs (legs). Trend-line for non-avian theropods and early birds is shown below that of present-day birds to avoid overlap, not to imply that they had lower locomotor performance than some present-day birds. Also, relative investment and performance of theropods and early birds are inferred and so are only approximations. Pygostylia refers to avialians with a pygostyle and short tail, e.g., *Protopteryx*. (Figure modified from Heers and Dial 2015; © 2014 The Authors. Evolution © 2014 The Society for the Study of Evolution, used with permission)

"dethroned" from its position as the first bird, but what will likely not change is that the minor differences between the first birds and their close non-bird relatives will continue to make it difficult for those trying to understand their phylogenetic relationships

## 1.12 The First Birds

The first birds had many of the characteristics of present-day birds, but, not surprisingly, given that they existed about 150 million years ago, also differed from present-day birds in many ways.

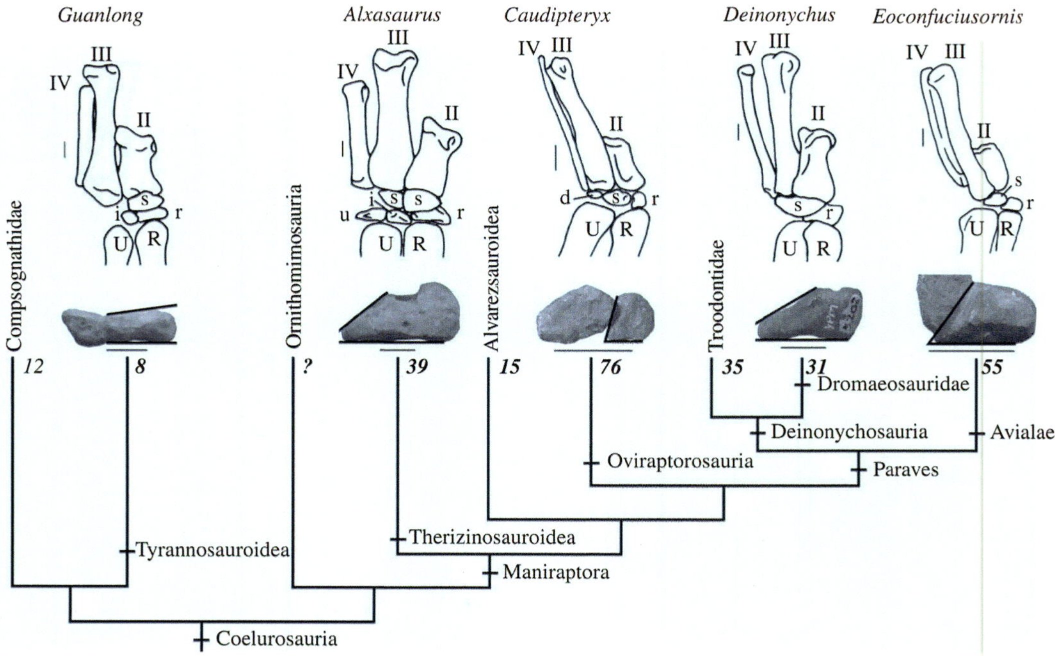


**Fig. 1.39** Evolution of wrist structure and the radial angle in theropods. Numbers indicate the radial angle (in degrees) between proximal and distal articular surfaces of the radial bone, with higher angles indicating greater ability to flex the wrist. II–IV, metacarpals II–IV; d, distal carpal; i, intermedium; R, radius; r, radial; s, semilunate carpal; U, ulna; u, ulnare. Scale bars to left of metacarpal IV = 0.25 cm for *Eoconfuciusornis*, 0.50 cm for *Caudipteryx*, and 1.00 cm for the other taxa. (Figure from Sullivan et al. 2010; © 2010 The Royal Society, used with permission)

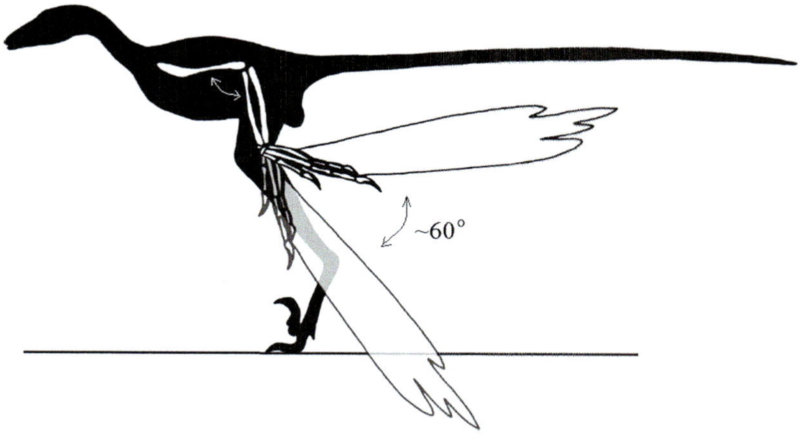


**Fig. 1.40** Wrist folding of the dromaeosaurid *Microraptor gui*. The wrist of *Microraptor*, with a range of movement of about 60°, was only about half as flexible as the wrists of present-day birds (see Fig. 1.41). Development of longer feathers on the hand may have favored the evolution of a more flexible (abducting) wrist or other factors may have favored a more flexible wrist, thus permitting the evolution of longer feathers. (Figure from Sullivan et al. 2010; © 2010 The Royal Society, used with permission)

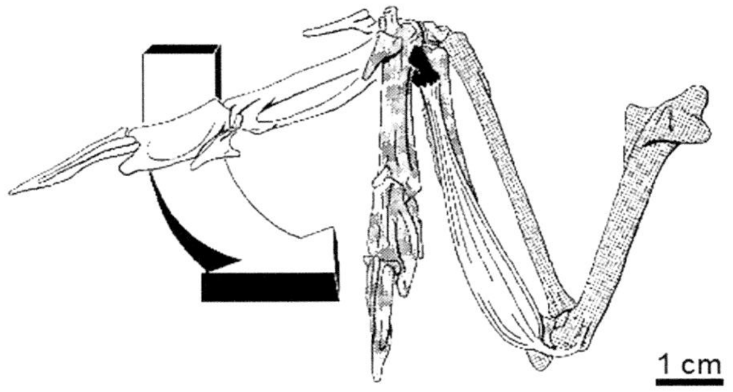


**Fig. 1.41** Right forelimb of a Rock Pigeon. Present-day birds can abduct (close) their wrist joints to a greater extent than non-avian theropods like *Microraptor gui* (see Fig. 1.40). (Figure from Vazquez 1994; © 1994 Springer-Verlag, used with permission)

For example, in contrast to present-day birds, *Archaeopteryx* had teeth, free fingers consisting of several bones (phalanges), no sternum (or, at best, a cartilaginous rather than bony sternum) and no keel, no synsacrum, a pubic bone not parallel with the ischium, and a long bony tail with many caudal vertebrae (Fig. 1.47). *Jeholornis* lived about 30 million years after *Archaeopteryx* (about 120 million years ago) and, like *Archaeopteryx*, had a long bony tail (but even longer and with more caudal vertebrae than the *Archaeopteryx* tail; Fig. 1.48). However, *Jeholornis* had also become more like present-day birds than *Archaeopteryx*, with a strut-like coracoid, a procoracoid process, narrow furcular, bony sternum, a shorter manus, and a shorter, more robust digit II that provided better support for the primary feathers (Zhou and Zhang 2003).

*Archaeopteryx* and *Jeholornis* were "long-tailed" birds, but those that followed in the fossil record, beginning with the sapeornithids and confuciusornithids, were "short-tailed" pygostylian birds (Fig. 1.49). To date, no fossil birds with "intermediate" tails have been found. Short-tailed birds have a few unfused caudal vertebrae that permit movement of the tail, and some (e.g., *Confuciusornis*) had a distal rod-like pygostyle consisting of several fused caudal vertebrae that help support the tail feathers (rectrices). The loss of the long bony tail was an important step in the evolution of birds, with fewer caudal vertebrae reducing the tail's weight and the attached rectrices important for improving maneuverability, reducing drag, serving as a break when landing, and, in some species, having an ornamental function. Available evidence suggests that this change in avian tail morphology occurred over a relatively short evolutionary timeframe (Rashid et al. 2018). In fact, Rashid et al. (2014) examined available fossil and genetic evidence and suggested that a single mutation could have both truncated bird tails and fused the last few into a pygostyle. That, of course, remains to be determined, but, if such a mutation (or such mutations) occurred, the advantages of a shorter bony tail with attached rectrices would have ensured the continued maintenance of this trait.

The short-tailed confuciusornithids are well represented in the fossil record of the Early Cretaceous (131–120 million years ago) in what is now China. Confuciusornithids were the first known birds without teeth (although other Cretaceous birds still had teeth; Box 1.5 Loss of Teeth). In addition, and in contrast to *Archaeopteryx*, confuciusornithids had bony sternums with, in some species, small keels (possibly cartilaginous; Fig. 1.50) and, as noted above, short tails with pygostyles (Fig. 1.51). The presence of keels in some confuciusornithids suggests the presence of larger flight muscles. However, both the orientation of shoulder joint in confuciusornithids (limiting their ability to elevate their humerus during an upstroke) and the structure of their flight

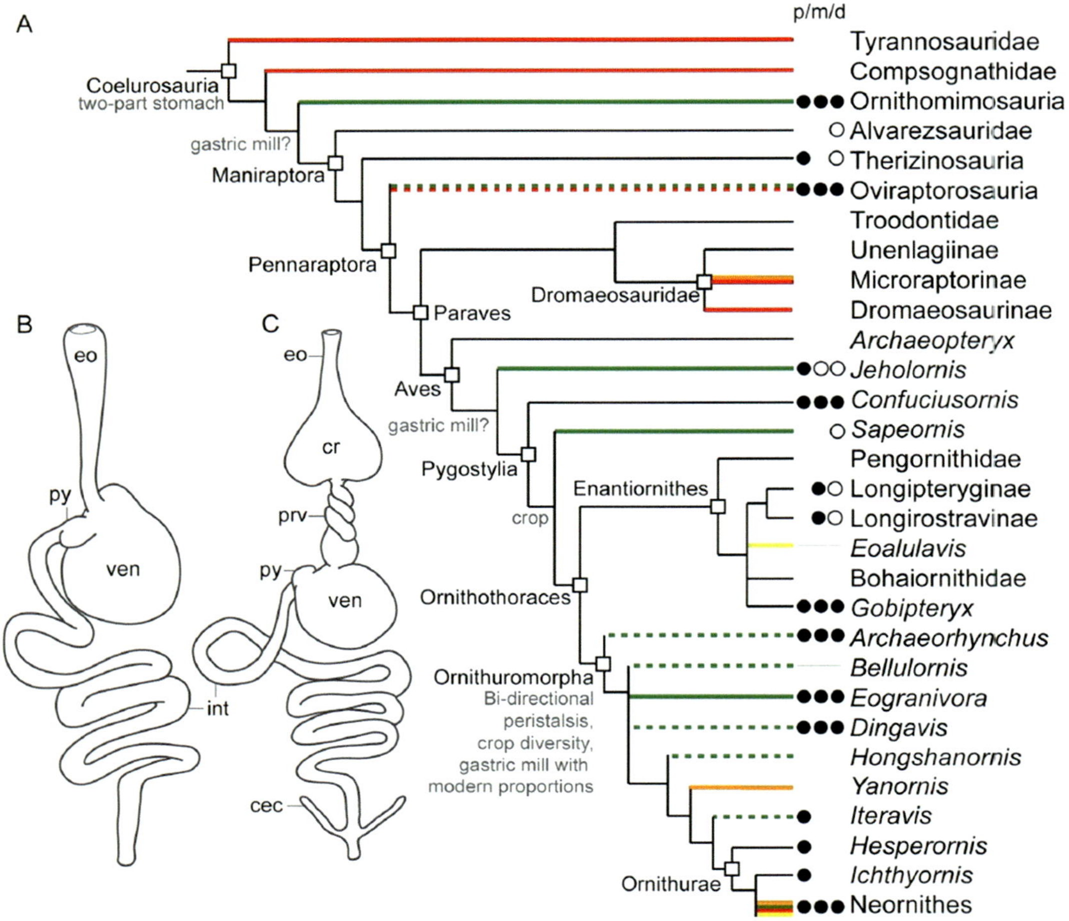


**Fig. 1.42** (**a**) Simplified cladogram of relationships within the Coelurosauria (Theropoda) highlighting lineages with some direct evidence of diet. Green lines indicate evidence of herbivory; dashed green lines indicate the presence of gizzard stones. Red lines indicate direct evidence of carnivory; dashed red lines indicate indirect associations. Orange indicates direct evidence of piscivory; yellow indicates evidence of a diet of invertebrates (just *Eoalulavis*). Filled circles indicate complete tooth loss, from left to right, in the premaxilla (p), maxilla (m), and dentary (d); open circles indicate tooth reduction. The inferred appearance of important traits of the modern avian alimentary canal is indicated in gray (e.g., two-part stomach, upper left). Simplified representations of the alimentary canal in extant archosaurs: (**b**) crocodilian; (**c**) neornithine. Anatomical abbreviations: cec, ceca; cr, crop; eo, esophagus; int, intestines; prv, proventriculus; py, pyloric chamber; ven, ventriculus. (Figure from O'Connor 2019; © 2018 Elsevier B.V., used with permission)

feathers, i.e., smaller trailing-edge barb angles than found in extinct enantiornithids and present-day birds (Ornithothoraces; Fig. 1.52) that would likely limit vane flexibility and create a less coherent wing surface (Fig. 1.53), suggest that they may have been gliders rather than active, flapping fliers (Senter 2006b; Feo et al. 2015). Based on examination of fossilized soft tissues (including the propatagium) as well as osteology and feathers, Falk et al. (2016) suggested that

*Confuciusornis* spp. were arboreal birds capable of short-term (nonmigratory) powered flight. More recently, Wang et al. (2022), based on a study of 11 fossils, concluded that "... diverse modes of flight adaptation existed among confuciusornithids ... ," with some apparently capable of long-distance flights (e.g., *E. zhengi*) and others resembling present-day facultative flap-gliding birds (e.g., *C. shifan*; Fig. 1.54).

**Box 1.5 Loss of Teeth**

Present-day birds have a distensible esophagus (and, in some species, a crop) and a two-part stomach, but no teeth. Although some have inferred that the selective factor most important in the loss of teeth and other features of the avian digestive system was the evolution of flight and a corresponding need to reduce weight, many of the birds that lived during the Cretaceous, including most enantiornithines and ornithuromorphs, had teeth and many of those birds were able to fly (Zhou and Li 2009). Indeed, Zhou et al. (2019) reported that tooth mass among Mesozoic birds represented less than 1.2% of total estimated body mass.

Brocklehurst and Field (2021) examined the pattern of tooth loss among Mesozoic birds and found no evidence of any selective trend. Rather, they found modularity among jaw regions and heterogeneous patterns of tooth loss, as well as the complete loss of teeth by crown birds by about 90 million years ago. The loss of teeth in the premaxilla of the upper jaw did not increase the likelihood of tooth loss in the maxilla of the upper jaw, and fossils reveal that, in the lower jay (dentary), teeth were sometimes entirely present or absent or present or absent in the anterior or posterior regions of the lower jaw. Brocklehurst and Field (2021) concluded that repeated "... transitions to toothlessness in Avialae reflect selection on a phylogenetically local scale combined with an underlying developmental propensity for tooth loss, instead of the outcome of long-term directional selection."

(continued)

**Box 1.5** (continued)

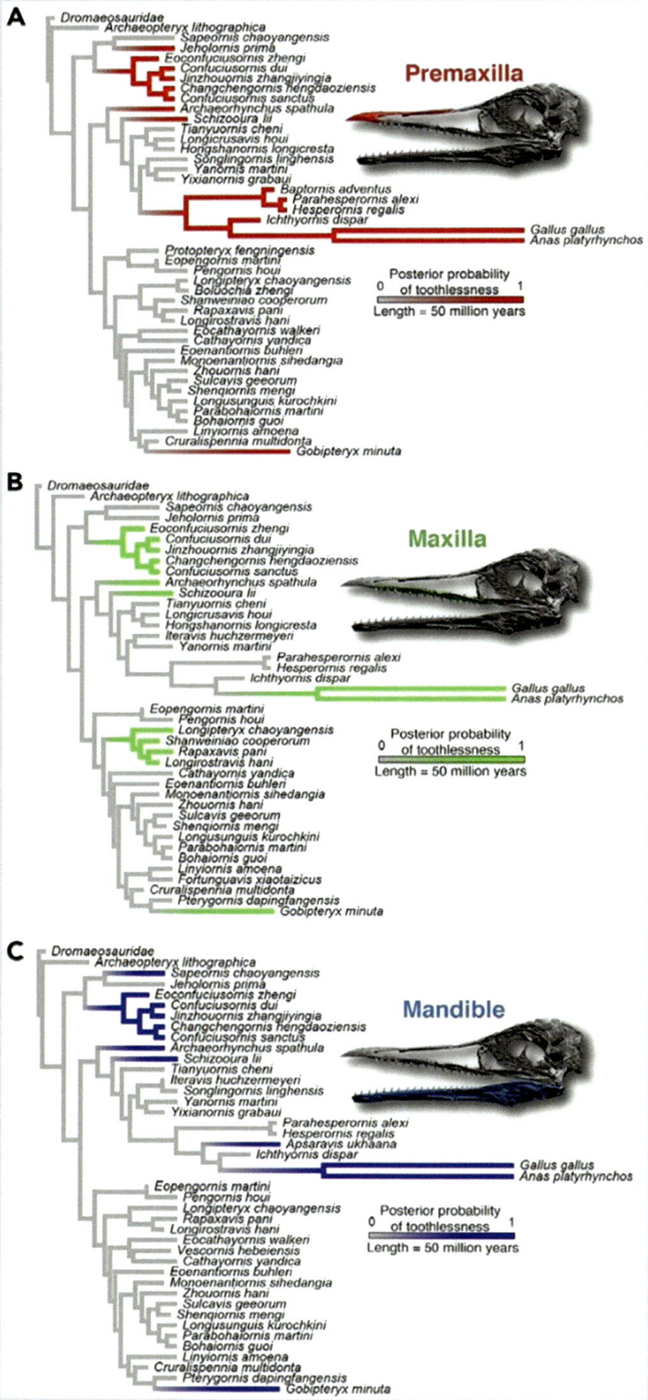


The extent and location of tooth loss varied among Mesozoic birds, with no apparent generalized selective trend. (Figure from Brocklehurst and Field 2021; © 2021 The Authors, used with permission)

(continued)

**Box 1.5** (continued)

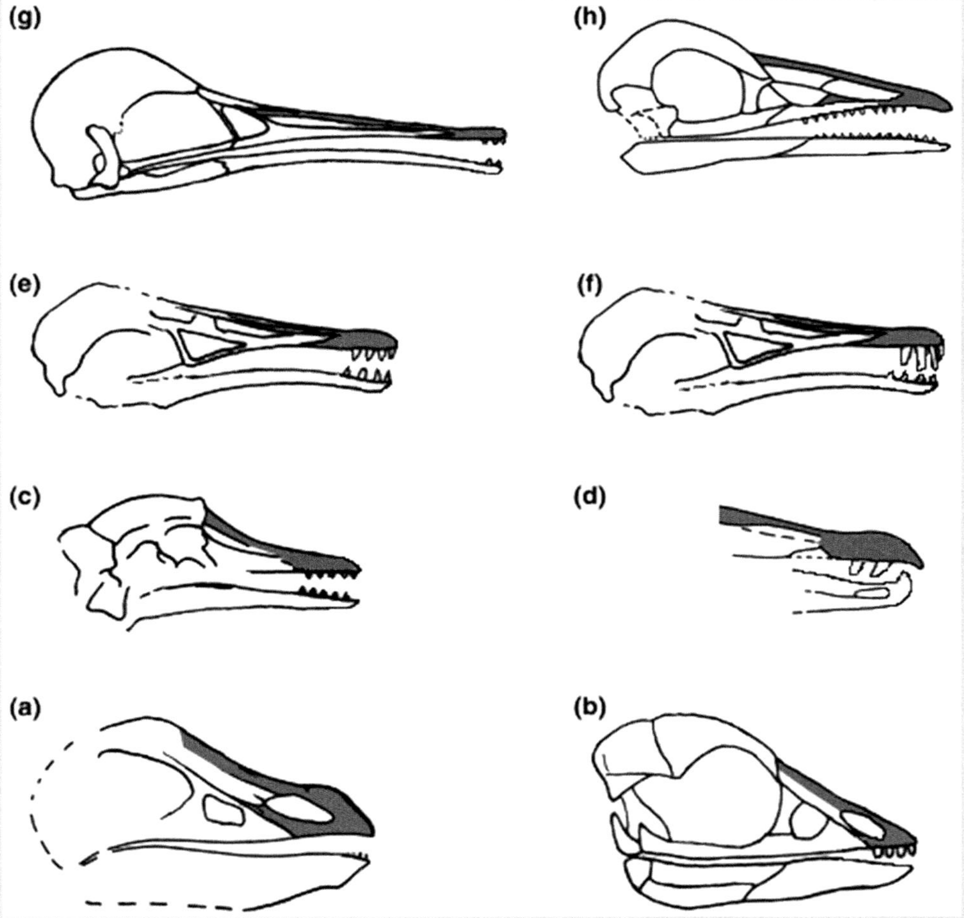


Examples of partial reduction of dentition in several species of Cretaceous birds. (**a**) *Jeholornis prima*, (**b**) *Sapeornis chaoyangensis*, (**c**) *Cuspirostrisornis houi*, (**d**) *Boluochia zhengi*, (**e**) *Longipteryx chaoyangensis*, (**f**) *Longipteryx* sp., (**g**) *Rapaxavis pani*, and (**h**) *Yanornis martini*. (Figure from Louchart and Viriot 2011; © 2011 Elsevier Ltd., used with permission)

A reduction in the number of teeth and eventually loss of all teeth may have been the result of a shift in diet from mainly carnivorous to more herbivorous or omnivorous (Zanno and Makovicky 2011; Li et al. 2020), perhaps, at least in part, due to competition of carnivorous non-avian theropods (Li et al. 2020). In support of this "dietary transition" hypothesis is the relationship between the presence of grinding gizzards and reduced dentition (Zheng et al. 2011). Muscular gizzards are always associated with partial to total loss of teeth and the presence of a partial or total rhamphotheca (Louchart and Viriot 2011). A muscular gizzard with ingested gastroliths (i.e., gastric mill) can efficiently process foods ingested by herbivorous and omnivorous birds (Louchart and Viriot 2011), rendering teeth redundant.

**Fig. 1.43** An enantiornithine fossil with purported ovarian follicles. (**a**) Purported follicles are inside the box and (**b**) close-up of the follicles. (Figure from Bailleul et al. 2020; open-access article licensed under a Creative Commons Attribution 4.0 International License, https://creativecommons.org/licenses/by/4.0/)

Although numerous specimens of confuciusornithiforms have been collected, none provide unequivocal evidence of their diet. Based on the possible presence of fish bones under the cervical vertebrae of one specimen, Dalsätt et al. (2006) suggested that some may have been piscivores; other investigators disagree, however, because, in contrast to fish remains associated with fossil enantiornithines (Wang et al. 2016), the fish remains did not appear to be integrated into a pellet. Other investigators have suggested that some may have been herbivores (Falk et al. 2016; Miller and Pittman 2021) or sally-striking predators (Hopson 2001; Elzanowski et al. 2018). They likely rested and probably nested in trees growing near freshwater lakes. When climbing trees and moving through tree canopies, they likely used their well-developed claws (Burnham et al. 2011; Pittman et al. 2022). The plumage of at least one species was rather dark (Fig. 1.55), and males of some species had a pair of elongated rectrices that were probably important in attracting mates (Zinoviev 2009; Chinsamy et al. 2013; Fig. 1.55). Evidence also suggests that confuciusornithiforms grew rapidly early in life, but growth later in the post-hatching period varied with body size, taxonomy, and the local environment and, for some, may not have reached skeletal maturity for at least three or four years (Chinsamy et al. 2020).

Coexisting with the confuciusornithids during the early Cretaceous were the sapeornithids. Sapeornithids were among the largest known Early Cretaceous birds, comparable in size to present-day turkeys. Their anatomy included some features shared with present-day birds, including a pygostyle, some fused limb bones, and a crop, as well some features shared with their dinosaur ancestors, including teeth and claws on their forelimb digits (Fig. 1.56). As with the confuciusornithids, the orientation of their shoulder joint and structure of their flight feathers (small trailing-edge barb angles) suggest that sapeornithids may have been gliders and soarers rather than active, flapping fliers (Senter 2006b; Feo et al. 2015; Serrano et al. 2020). In fact, based on a combination of computational modeling and morphofunctional analysis, Serrano and Chiappe (2017) suggested that one raven-sized specimen of sapeornithid (*Sapeornis*

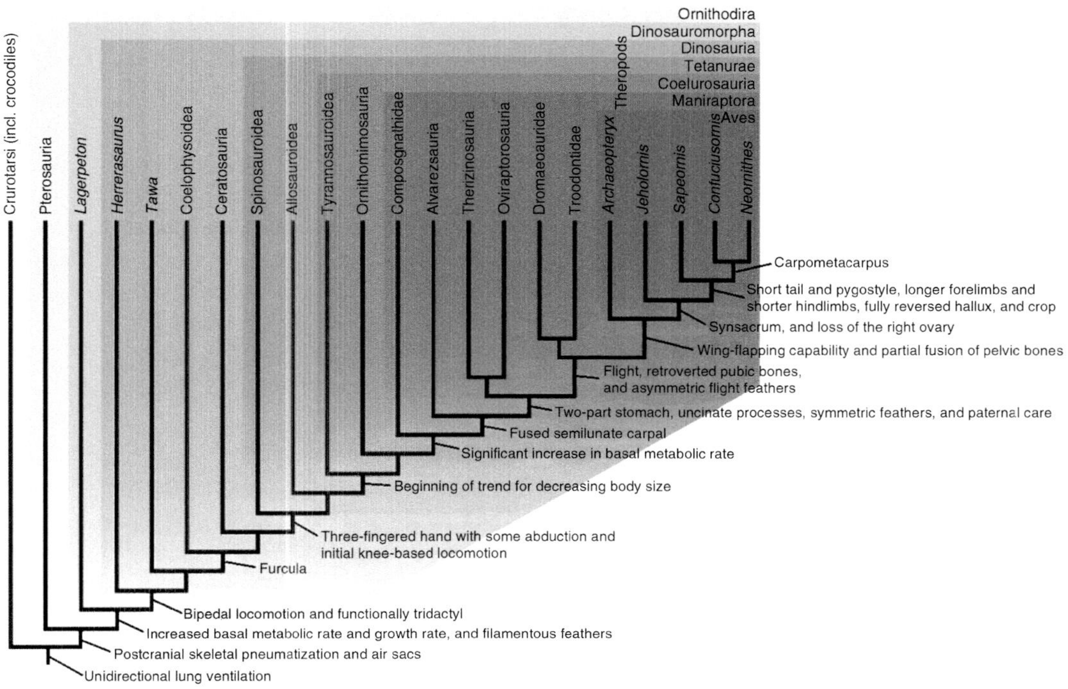


**Fig. 1.44** Appearance of key traits in the evolution of birds. Among birds (Aves), available fossils provide evidence for the presence of a crop and loss of the right ovary in *Jeholornis* (O'Connor and Zhou 2015), but such evidence is not available for *Archaeopteryx*. Information provided in this figure is from Butler et al. (2009), Makovicky and Zanno (2011), Heers and Dial (2011), Rashid et al. (2014), Han et al. (2014), Ksepka (2014a, b), Li et al. (2014a, b), O'Connor and Zhou (2015). (Figure modified from Makovicky and Zanno 2011; © 2011 John Wiley and Sons, used with permission)

*chaoyangensis*) would have exhibited aerial behavior similar to present-day thermal soarers like vultures and ravens (Fig. 1.57). Unlike confuciusornithids, however, sapeornithids had no sternum (Box 1.6 Evolution of the Avian Sternum). Sapeornithids were likely ground foragers and were known to eat seeds, but may have fed on other types of food as well (Zheng et al. 2011, Mitchell and Makovicky 2014; Fig. 1.58). More recently, Pittman et al. (2022) concluded that *Sapeornis* "... was an ecologically complex herbivorous thermal soarer that supplemented its diet with meat, perhaps analogous to the modern palm-nut vulture (*Gypohierax*)." In addition, sapeornithids had a fan-shaped tail (Fig. 1.59; O'Connor et al. 2016).

**Box 1.6 Evolution of the Avian Sternum**

The sternum, or breastbone, of present-day birds plays a critical role in both flight and respiration. Birds that fly have large keels, or carinas, extending from the sternum where flight muscles (pectoralis and supracoracoideus) attach and movements of the sternum and ribs move air into and out of the avian respiratory system. Perhaps surprisingly, then, *Archaeopteryx* likely had no sternum. In addition, among Paraves, troodontids had no sternum and, among Avialae (Aves), sapeornithids had no sternum. However, all other coelurosaurs, maniraptorans, and avialans had sternums, as do all present-day birds (although the sternums of some flightless birds have no keel).

The absence of a sternum in just three of the many lineages leading to present-day birds is puzzling, but the most likely explanation is that the presence of sternums is the basal condition and that they were lost, independently, either twice, once in the common ancestor of troodontids and *Archaeopteryx* and again in sapeornithids, or three different times in troodontids, *Archaeopteryx*, and sapeornithids (Lambertz and Perry 2015). Troodontids did not fly and *Archaeopteryx* and the sapeornithids were likely just gliders so did not have well-developed flight muscles, possibly minimizing the need for a sternum for attachment of flight muscles. However, the sternum of present-day birds also plays an important role in respiration so its absence requires an alternative mechanism for breathing.

Gastralia are bones found in the ventral body wall and can still be found in present-day crocodilians. They were also present in theropods, including non-avian theropods (e.g., troodontids), and in basal birds, including *Archaeopteryx* and the sapeornithids. Lambertz and Perry (2015) suggested that, in taxa without sternums, gastralia and their associated muscles could have served an important respiratory function (in addition to serving as the point of origin of the pectoralis muscle and providing support for the viscera). Similarly, O'Connor et al. (2015a, b) suggested that the large gastral baskets of *Sapeornis* may have compensated for the absence of a sternum by helping to support the flight muscles. Other investigators have suggested a role for gastralia in dinosaur respiration (Claessens 2004; Codd et al. 2008). Gastralia were lost in Neornithes and that, in combination with an improving ability to fly and larger flight muscles, favored the retention of a sternum and the development of a keel.

(continued)

**Box 1.6** (continued)

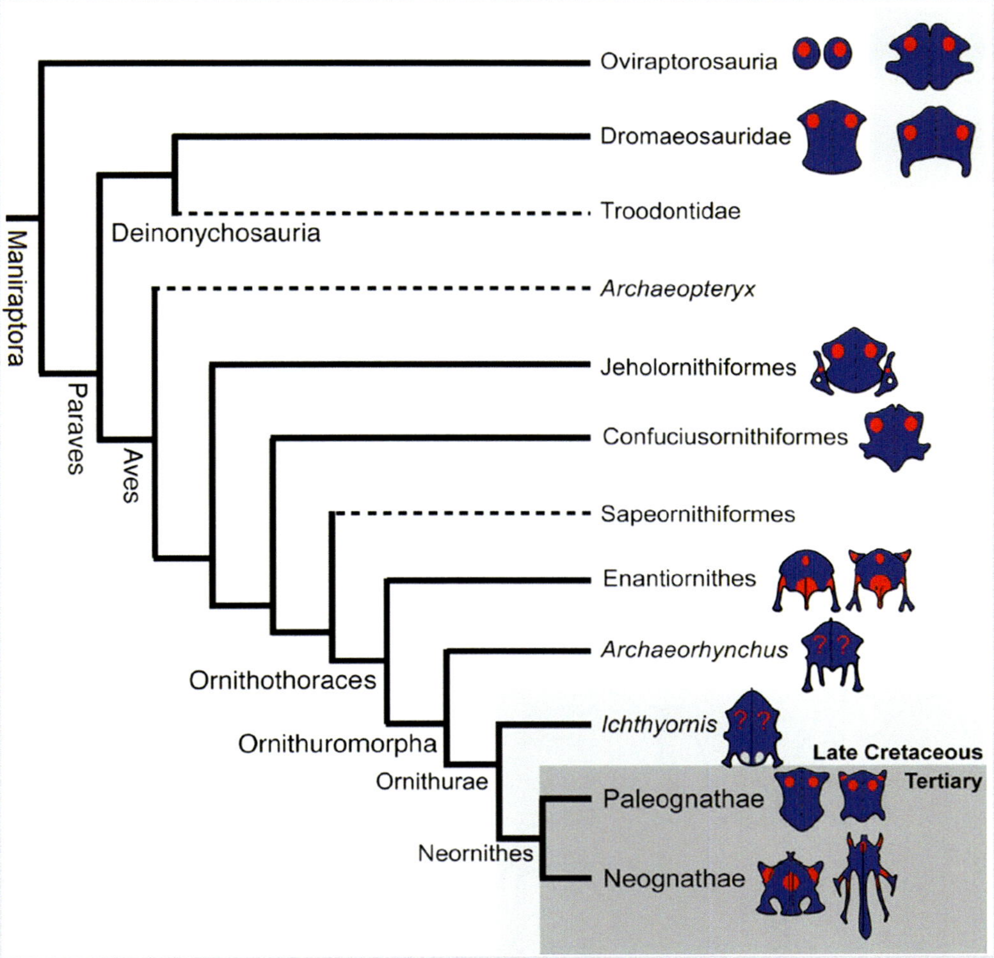


*Archaeopteryx* as well as the troodonts and sapeornithids had no sternum (as indicated by the dashed lines). However, other maniraptorans and all other taxa in the line leading to Neornithes had sternums. (Figure from O'Connor and Zhou 2015; © 2015 Springer Nature, used with permission)

(continued)

**Box 1.6** (continued)

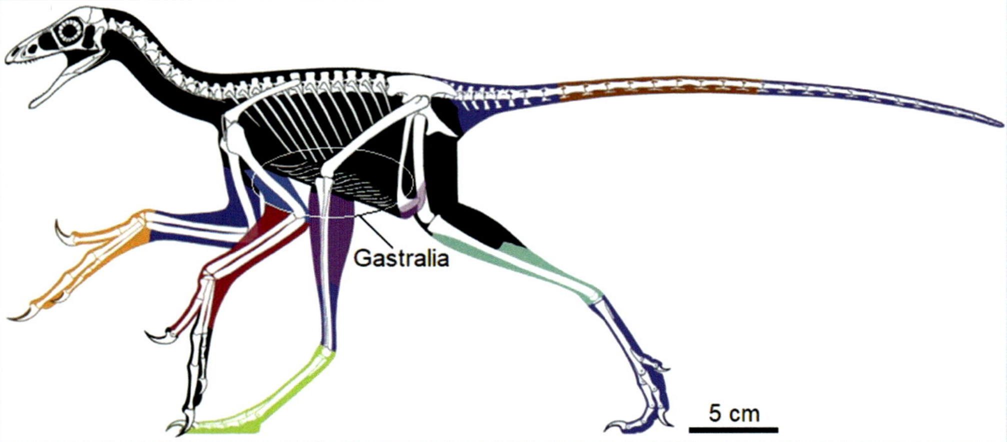


*Anchiornis*, like other troodontids, as well as sapeornithids and *Archaeopteryx*, did not have a sternum, but they did have gastralia. Gastralia and associated muscles may have played an important role in respiration in these taxa and may also have served as the point of origin for the pectoralis muscle. This reconstruction of *Anchiornis* was based on several different specimens, as indicated by the different colors. (Figure modified from Wang et al. 2017a, b; © The Authors 2017; licensed under a Creative Commons Attribution 4.0 International License, https://creativecommons.org/licenses/by/4.0/).

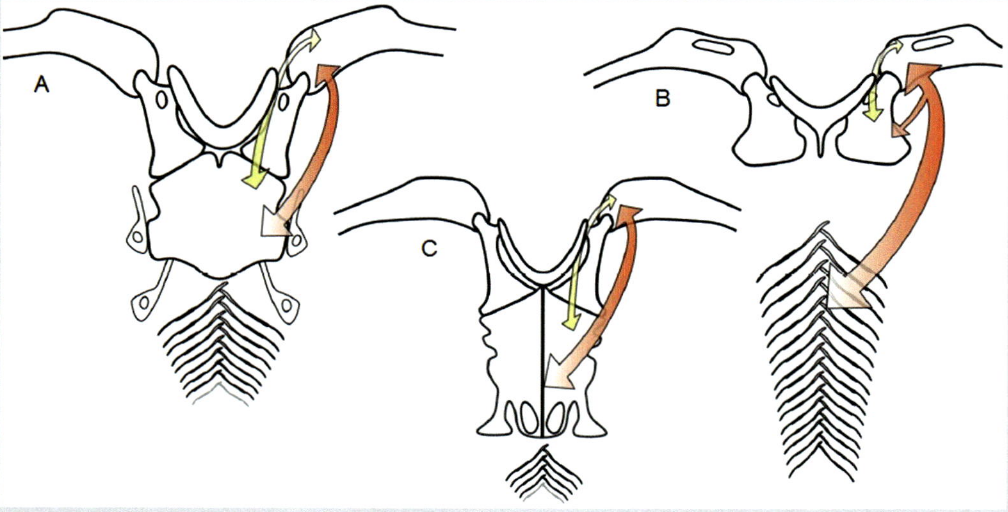


Possible points of origin of the major flight muscles of selected Pygostylia with and without a sternum; red arrow = pectoralis, yellow arrow = supracoracoideus. (**a**) *Jeholornis*, (**b**) *Sapeornis*, and (**c**) *Yanornis* (an early Ornithuromorph). (Figure from O'Connor et al. 2015a, b; © Vertebrata PalAsiatica, Published by the Chinese Academy of Sciences, used with permission)

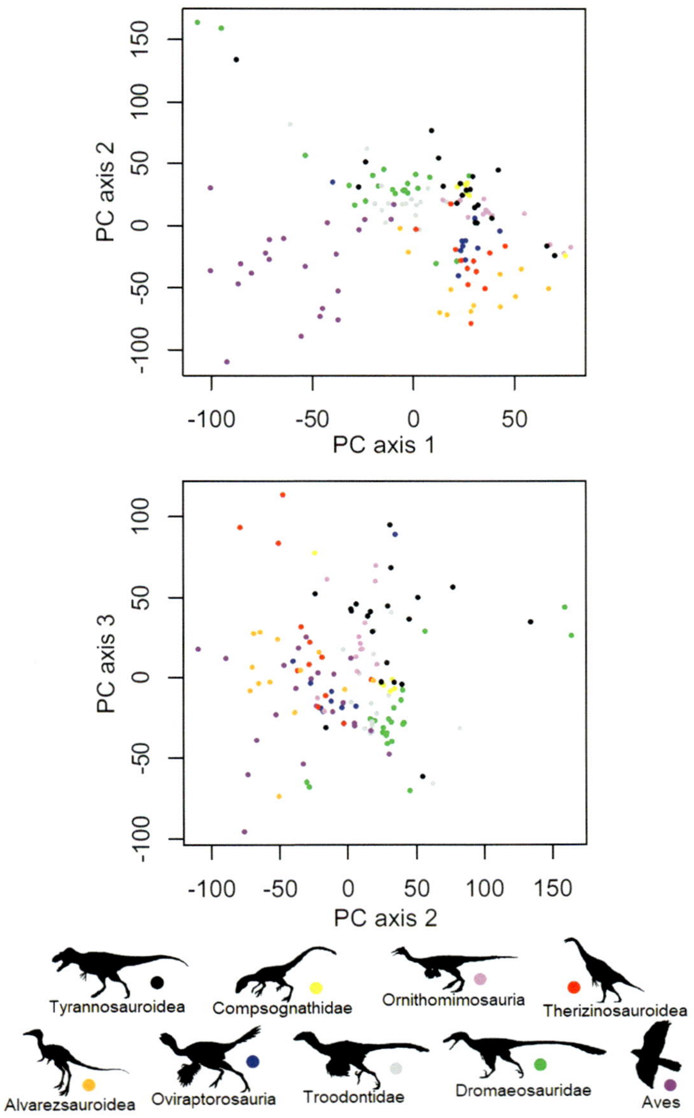


**Fig. 1.45** Discrete character morphospace of coelurosaurian theropods showing their anatomical variability. Birds (Aves) are relatively distinct from the other coelurosaurs on principal coordinate axis 1, but not on either axis 2 or 3. In other words, there is no clear evidence that birds group separately from the other coelurosaurs. In these figures, the first axis represents those character distances contributing most to the overall variability among theropods, and each additional axis represents distances of progressively less significance. Each theropod has a score on each axis, which together represents aspects of the overall form (based on discrete character scores) for each taxon. The axes define the multivariate morphospace, and the set of scores for each theropod on these axes therefore represent the taxon's position in the morphospace. (Figure from Brusatte et al. 2014; ©2014 Elsevier Ltd., used with permission)

Both confuciusornithids and saperornithids had pygostyles, a characteristic shared with two other groups, the enantiornithines and ornithuromorphs. Fossil records of confuciusornithids and saperornithids indicate that confuciusornithiforms lived from about 128 to 125 million years and sapeornithids about 120 million years ago (Fig. 1.60). In contrast, enantiornithines are present in the fossil record beginning about 131 million years ago (*Protopteryx*, He et al. 2006) and became extinct at the end of the Cretaceous (Fig. 1.60). Enantiornithines (Greek *einantios* "opposite" + *ornithes* "birds") have sometimes been called "opposite birds" because Walker (1981) thought that the articulation between the coracoid and scapula where the triosseal canal is located appeared to be opposite to that of present-day birds. This was later found to be an incorrect interpretation (Chiappe 1991).

Enantiornithines were widely distributed during the Cretaceous, with fossils found on every

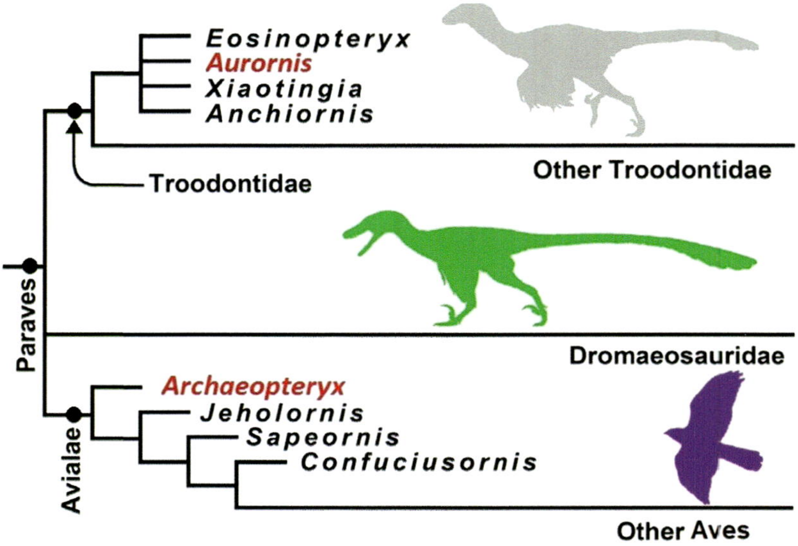


**Fig. 1.46** Phylogenetic relationships of Paraves, with *Aurornis* classified as a Troodont and *Archaeopteryx* as the first bird (Aves). (Figure modified from Brusatte et al. 2014; © 2014 Elsevier Ltd., used with permission)

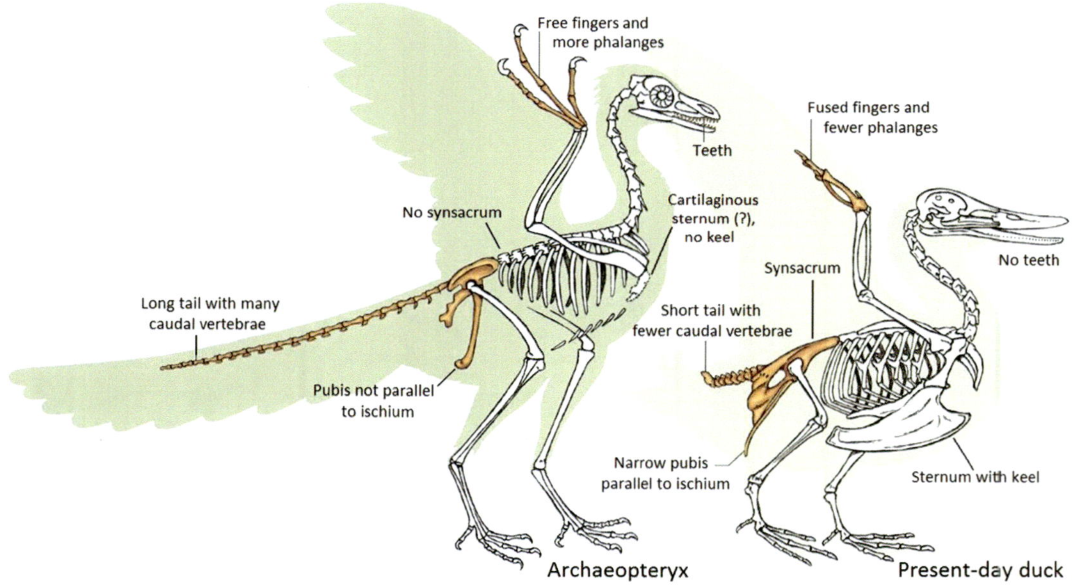


**Fig. 1.47** Comparison of some characteristics of *Archaeopteryx* compared to those of a present-day duck. Available evidence indicates that *Archaeopteryx* had no sternum, but it is possible they had a cartilaginous sternum that would not be apparent in fossils. (Figure modified from Dyke 2010; © 2010 Scientific American, Inc., used with permission)

continent except Antarctica (Fig. 1.61a). Based on these fossils, 32 genera (Fig. 1.61b) and more than 100 species have been described, and new species are being reported every year. Enantiornithines varied in size (Fig. 1.62) from a bit smaller than a House Wren (*Troglodytes aedon*) (i.e., 8-g

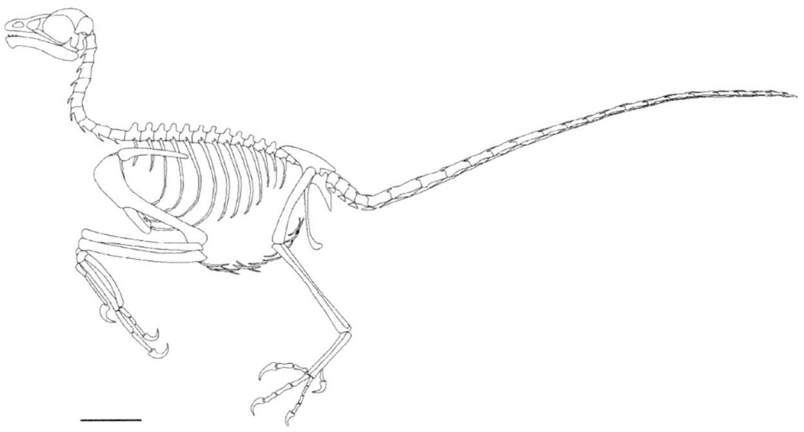

**Fig. 1.48** Skeletal reconstruction of *Jeholornis*, a seed-eating bird with a long bony tail, but, unlike *Archaeopteryx*, *Jeholornis* also had a bony sternum and a wing with a fused carpometacarpus. (Figure from Zhou 2004; © 2004 Springer-Verlag, used with permission)

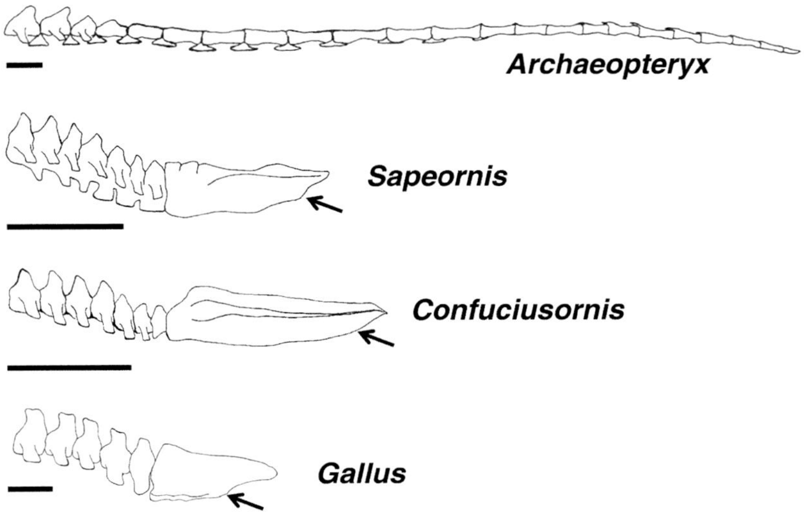


**Fig. 1.49** Tail skeletons of *Archaeopteryx, Sapeornis, Confuciusornis*, and *Gallus gallus* (Domestic Chicken), with pygostyles indicated by the arrows. (Figure from Rashid et al. 2014; open-access article distributed under the terms of the Creative Commons Attribution License, https://creativecommons.org/licenses/by/4.0/)

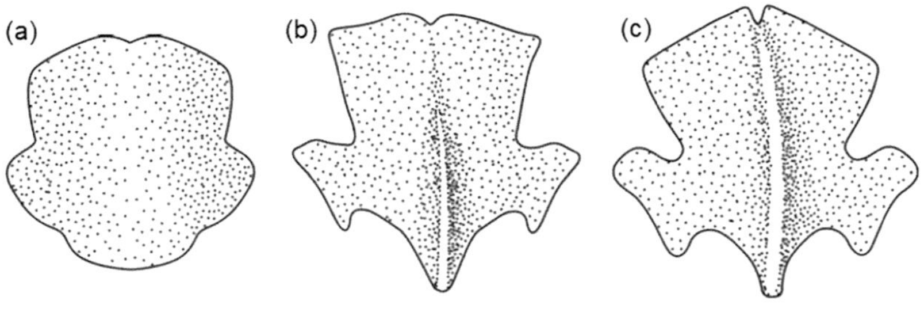


**Fig. 1.50** Sternums of (**a**) *Eoconfuciusornis*, (**b**) *Confuciusornis* sp., and (**c**) another *Confuciusornis* sp. Note that *Eoconfuciusornis*, an early diverging Confuciusornithid, had no keel and that the keels of the *Confuciusornis* sp. differed in extent of development. Larger keels suggest the presence of larger flight muscles. (Figure from Zhang et al. 2008; © 2008, Science in China Press and Springer-Verlag GmbH, used with permission)

*Dalingheornis liweii*) to as large as a Ross's Goose (*Anser rossii*) (i.e., 1500-g *Elsornis kini*) (Miller and Pittman 2021), and most had teeth (Chiappe and Walker 2002). The general consensus is that all enantiornithines were arboreal (Field et al. 2018), with feet specifically adapted for arboreality (e.g., textured plantar pads, long hallux, and long, pointed, recurved ungual sheaths; Clark and O'Connor 2021, Fig. 1.63), and their wing morphologies suggest that they exhibited a diversity of flight styles similar to those of present-day birds (Navalón et al. 2015; Fig. 1.64a; Box 1.7 Flying Ability of Mesozoic Birds). Based on analysis of the plumage of specimens preserved in amber, enantiornithines generally appear to exhibit cryptic color patterns (although they may have had color vision; Tanaka et al. 2017) that would have made them inconspicuous in forest habitats (Xing et al. 2020; Fig. 1.62). In addition, the flight feathers of the wing of most enantiornithines were comparable in dimension to those of present-day birds (Fig. 1.64; Sanz et al. 1996; Chiappe et al. 2014). Variability in proportions of forelimb segments of enantiornithines largely overlaps that of present-day birds, but hindlimb proportions were less variable for enantiornithines than present-day birds (Dyke and Nudds 2009; Fig. 1.64b).

**Box 1.7 Flying Ability of Mesozoic Birds**

How did the flying ability of Mesozoic birds compare to that of present-day birds? One approach to answering that question is to estimate the wing spans and body mass of Mesozoic birds and use those estimates to calculate parameters that provide information about flight modes: wing loading and aspect ratio. Wing loading is determined by dividing a bird's mass by the surface area of its two wings; aspect ratio is determined by dividing wing length by wing width. Present-day birds with certain aspect ratios and wing loading are known to use different flight modes, e.g., diving birds are heavy relative to wing surface area so have high wing-loading values and must beat their wings at high rates when flying.

Serrano et al. (2017) estimated the wing loading and aspect ratios of Mesozoic birds and found that they were comparable to those of some present-day birds, particularly the aspect ratios. Estimated wing-loading values for Confuciusornithidae, Jeholornithidae, and Archaeopterygidae were comparable to present-day birds that continuously flap their wings when flying. However, wing-loading values for sapeornithids were similar to those of present-day soaring birds, and those for enantiornithids overlapped those of present-day birds that soar or use flap-bounding flight (Liu et al. 2017; see Chap. 11 for more information about flight modes of present-day birds). No Mesozoic birds had aspect ratios and wing-loading values comparable to those of present-day dynamic soaring birds like albatrosses. However, most other flight modes of present-day birds were likely shared with their Mesozoic relatives. In addition, an enantiornithine bird, *Eoalulavis hoyasi*, is the oldest known bird with an alula, which would have enhanced its ability to fly and maintain lift during slow flight (e.g., taking off and landing).

(continued)

**Box 1.7** (continued)

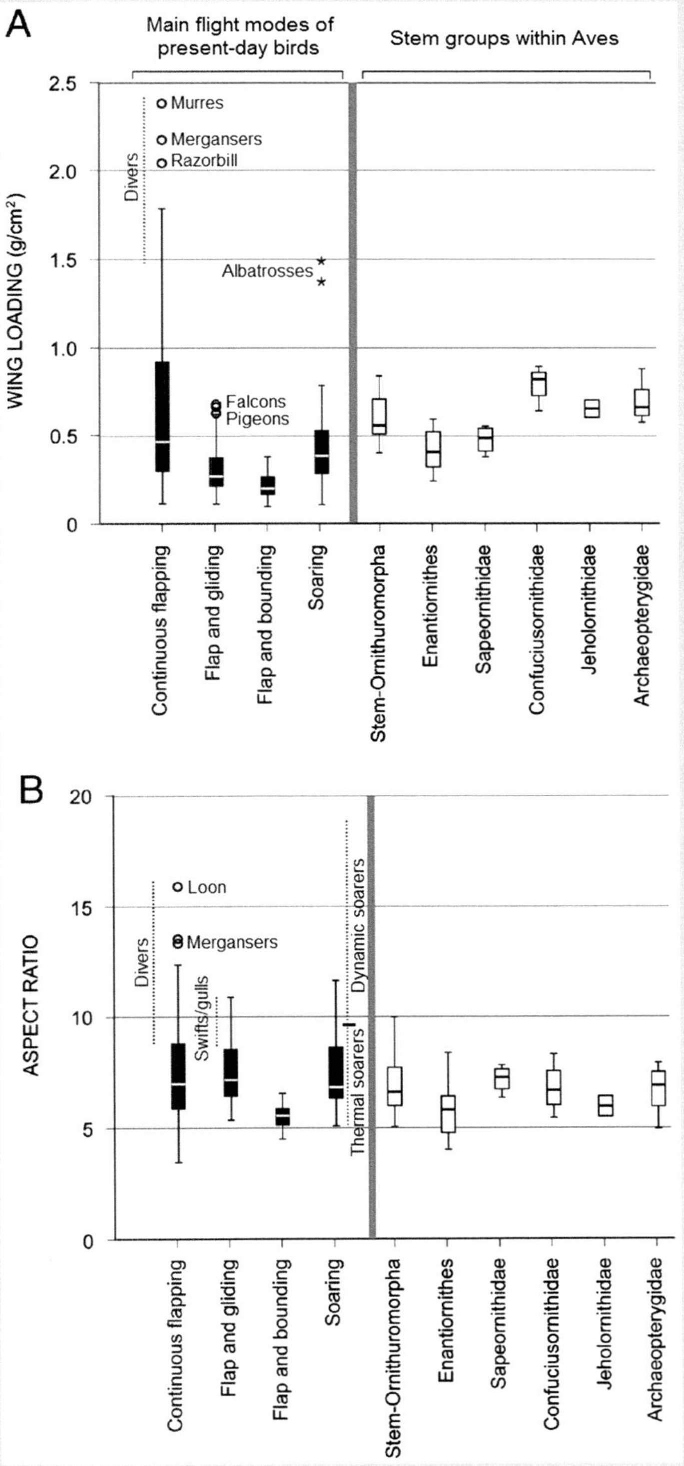


(continued)

**Box 1.7** (continued)

Comparison of the (**a**) wing loading and (**b**) aspect ratios of present-day birds and stem avian taxa from the Mesozoic. (Figure modified from Serrano et al. 2017; © 2016 The Paleontological Society, used with permission)

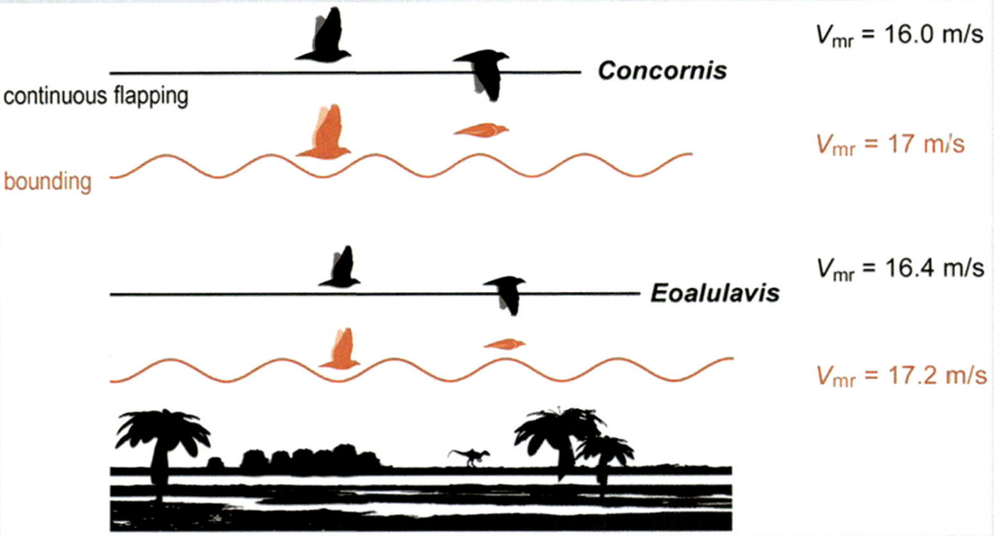


Hypothesized flight modes of two species of enantiornithine birds, *Concornis lacustris* and *Eoalulavis hoyasi*, flying at their cruising speeds (i.e., $V_{mr}$, or the speed with the minimum energetic cost of travel). (Figure modified from Serrano et al. 2018; © The Palaeontological Association, used with permission)

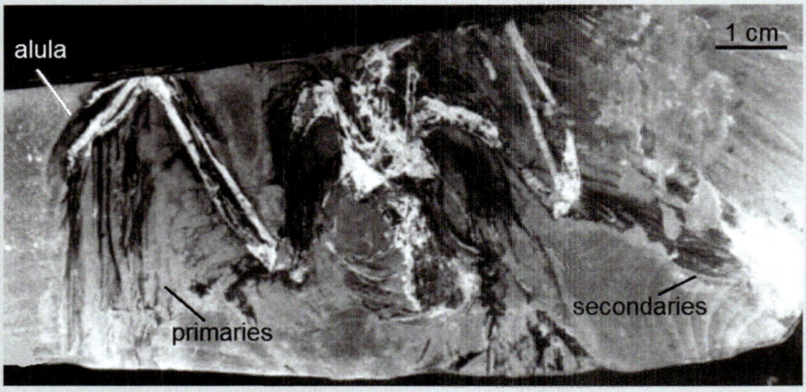


Fossil of *Eoalulavis hoyasi*, an enantiornithine from the early Cretaceous. This is the oldest fossil discovered to date showing evidence of an alula, feathers associated with the first digit that, when elevated, help maintain smooth airflow over the wing during slow flight. (Figure modified from Sanz et al. 1996; © 1996 Springer Nature, used with permission)

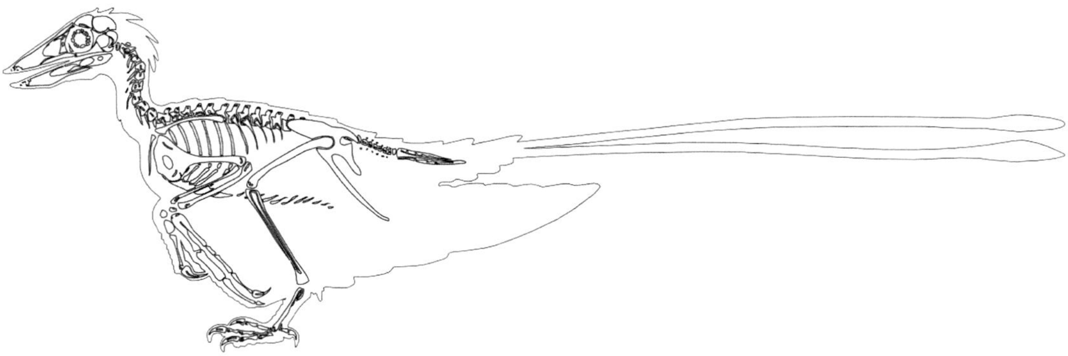

**Fig. 1.51** *Confuciusornis*. In contrast to *Archaeopteryx*, confuciusornithids had horny beaks with no teeth, short tails with pygostyles, a bony sternum, and a pubic bone that paralleled the ischium. (Figure from Guo et al. 2018; © 2018 by the Authors and Hans Publishers, Inc., licensed under the Creative Commons Attribution International License (CC BY), https://creativecommons.org/licenses/by/4.0/)

The bill morphology of enantiornithines exhibited much variation, suggesting a wide variety of food habits (Fig. 1.65). Most enantiornithines had teeth, with just one known exception (the Late Cretaceous *Gobiteryx minuta*; Elzanowski 1974). Some enantiornithines, like *Longirostravis*, were likely arboreal insectivores, gleaning and probing for insects in trees or shrubs (Morschhauser et al. 2009; O'Connor 2019), with one species apparently using elongated third toes to probe for larvae or other prey in holes and crevices in woody substrates (Xing et al. 2019). Other enantiornithines likely fed on arthropods and plant materials like present-day birds (Wang and Zhang 2011; Wang et al. 2016). Examination of the stomach contents of another enantiornithine (*Eoalulavis hoyasi*) revealed that it had eaten crustaceans (Sanz et al. 1996). More generally, Li et al. (2020) noted that the teeth of enantiornithines had relatively thick layers of enamel, suggesting that they had "durophagous dietary preferences (eating of tough materials)".

Little is known about the breeding biology of enantiornithines. However, Dyke et al. (2012) reported fossil evidence of a breeding colony containing thousands of eggshell fragments along with a few near-complete eggs and remains of embryos (Fig. 1.66), likely representing the drowned remains of a colony that had been swamped by rising water. Bailleul et al. (2019) examined the cuticle of an enantiornithine (*Avimaia schweitzerae*) eggshell and found nanospheres of calcium phosphate. Such nanospheres are also found in the cuticles of eggs of present-day birds that nest in areas that are wetter or warmer and, thus, more prone to flooding and where eggs and embryos may be more infection-prone (D'Alba et al. 2016). Because the presence of nanospheres in such areas can protect developing embryos by plugging eggshell pores and preventing water and bacteria from entering eggs (D'Alba et al. 2016), Bailleul et al. (2019) hypothesized that enantiornithines may have also nested in wet, warm areas prone to flooding. In addition, Fernández et al. (2013) reported another enantiornithine nesting colony, with evidence that adults produced scrape-like nests and positioned their eggs vertically in the substrate (Fig. 1.67); this latter feature suggests that enantiornithines did not turn their eggs like present-day birds. The extent to which adult enantiornithines cared for eggs is unknown.

Studies of enantiornithine embryos and juveniles have revealed well-developed feathers and relatively large brains, suggesting that they were precocial and possibly even super-precocial (Xing et al. 2017; Kaye et al. 2019a, b). As with present-day species of birds with super-precocial young (megapodes), some young

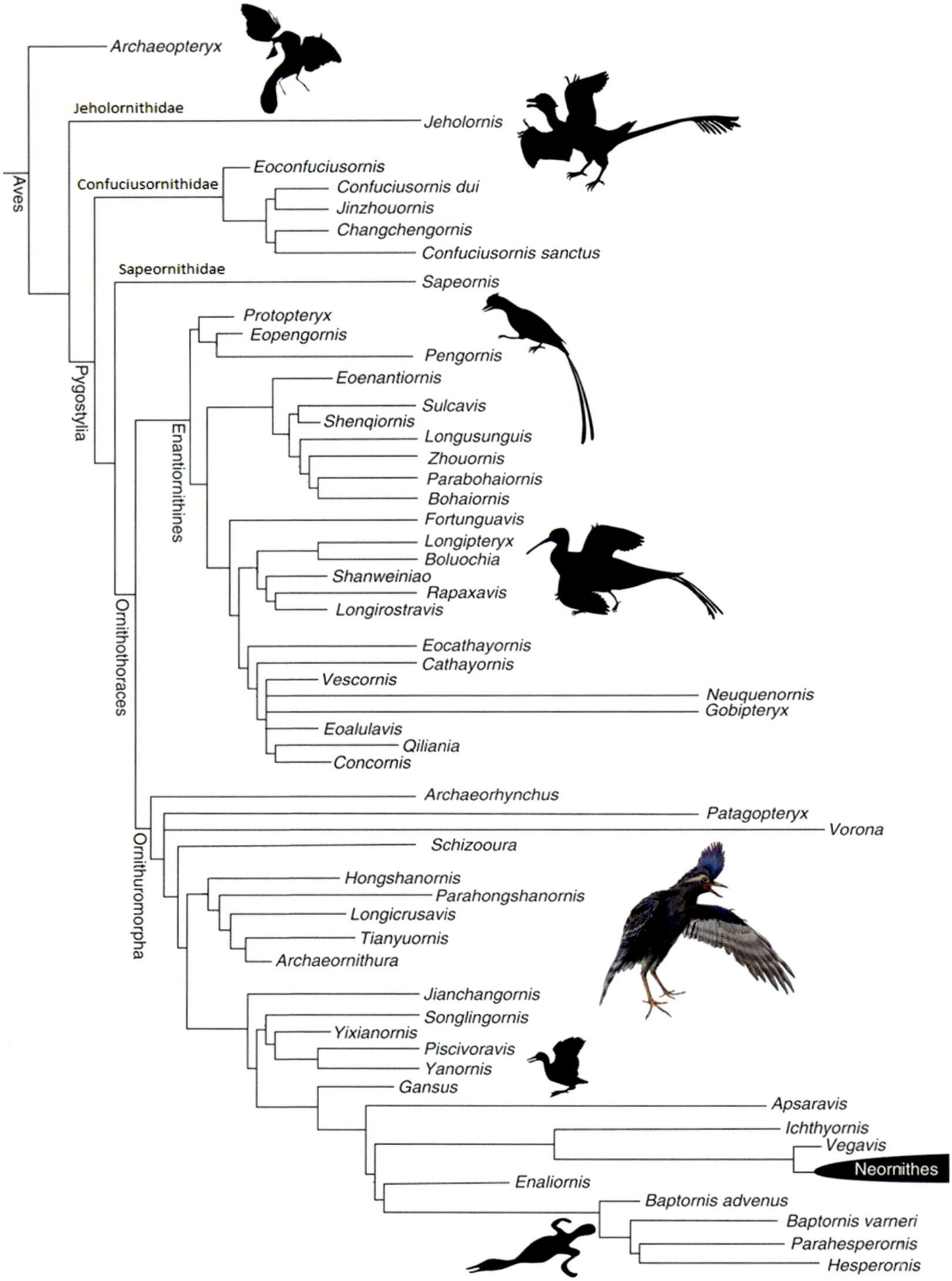


**Fig. 1.52** Major groups of Mesozoic birds included the Pygostylia (including the Confuciusornithiformes), enantiornithines, and Ornithuromorpha. The taxonomic status of saperornithids relative to confuciusornithiforms remains unclear. Most investigators consider sapeornithids basal to confuciusornithiforms, but others, as in this phylogeny, consider confuciusornithidiforms basal to sapeornithids. (Figure modified from Wang et al. 2015; open-access article under a Creative Commons Attribution 4.0 International License, https://creativecommons.org/licenses/by/4.0/)

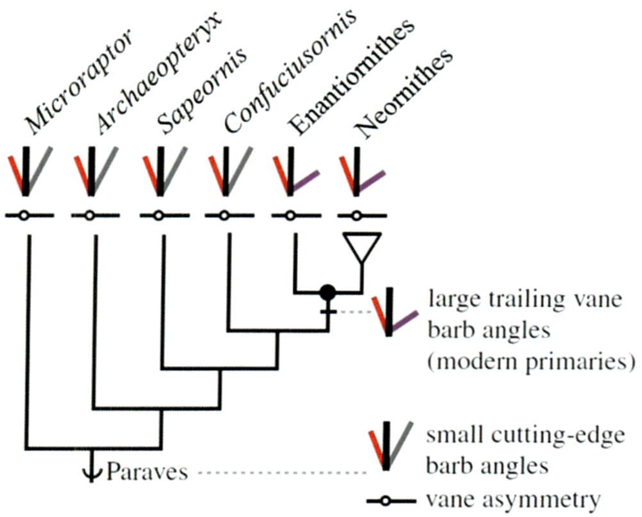


**Fig. 1.53** Examination of the flight feathers of representative Paraves reveals that all had (or have) asymmetrical vanes. In addition, measurements of barb angles (the angle between barbs and the feather shaft or rachis) of Mesozoic birds and present-day birds (crown birds) revealed that barb angles on the leading edge of flight feathers (red) of all measured birds were (are) small. However, barb angles on the trailing edge of flight feathers (purple) were initially small, but increased beginning with Enantiornithes (e.g., *Eopengornis*). Large trailing-edge barb angles help increase the flexibility of vanes and prevent separation of flight feathers during the flight stroke, maintaining a consistently smooth wing surfaces. The increasing forces acting on the flight feathers of increasingly active flyers (as opposed to gliders) during the upstroke and downstroke may have selected for this increase in trailing vane barb angle. (Figure modified from Feo et al. 2015; © The Royal Society, used with permission)

enantiornithines may have needed little or no post-hatching parental care (Mayr 2017a, b). Given the presence of embryos with feathers, some young enantiornithines may have been able to fly soon after hatching, but, others, with sternums still largely cartilaginous shortly after hatching and, therefore, potentially unable to withstand mechanical stress, may not have been able to fly until later in life (Knoll et al. 2018). Enantiornithines also appeared to have growth patterns different from that of present-day birds, with periods of rapid growth alternating with periods of much slower growth (Figs. 1.68 and 1.69). Analysis of patterns of bone growth along with the large number of fossils of juveniles that have been discovered suggests that at least some species of enantiornithines took much more time to reach adult size than present-day birds (Chinsamy et al. 1995, O'Connor et al. 2011, O'Connor et al. 2014, Atterholt et al. 2021).

## 1.13 Ornithuromorpha (Also Known as Euornithes)

The oldest known ornithuromorph fossil found to date was discovered in China in deposits estimated to be 131 million years old (Wang et al. 2015). Among the features that differentiated ornithuromorphs from the enantiornithines were a coracoid-based triosseal canal, U-shaped furcula (enantiornithines had a more V-shaped furcula), a sternum with a well-developed keel (with keels reduced in size in flightless species), a synsacrum that included more vertebrae, smaller, plough-shaped pygostyle, and, in crownward ornithuromorphs, faster growth rates (O'Connor et al. 2011, Zheng et al. 2012, Close and Rayfield 2012). Based on analyses of bone microstructure, Wang et al. (2020) demonstrated that an ornithuromorph bird (*Yanornis*) was fully grown within one or two years post-hatching. However, basal

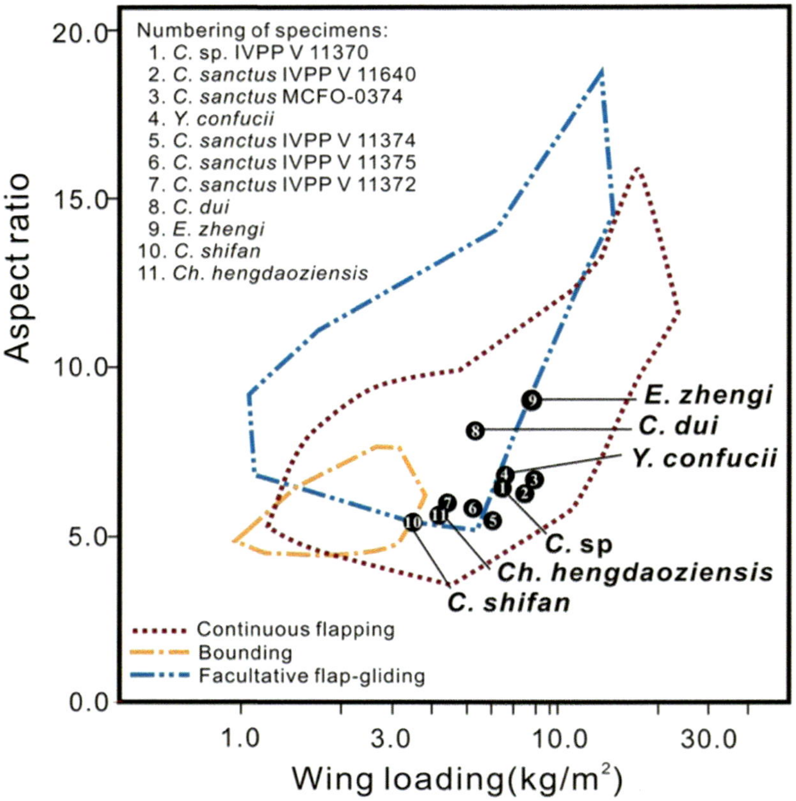


**Fig. 1.54** Locations of 11 confuciusornithid specimens in a morphospace based on wing loading (WL; ratio of body mass to the total area of the two wings) and aspect ratio (AR; ratio of wing length to wing width) of extant birds with different flight modes (as indicated by the areas encompassed by the colored lines). Note that all specimens fall within the continuous flapping morphospace, but some are also in the facultative flap-gliding morphospace. Note also that the 11 specimens vary in both their aspect ratios and wing loading. Those with high aspect ratios like *E. zhengi* were likely capable of fast, long-distance flights, whereas those with low aspect ratios and wing loading like *C. shifan* were better adapted for slower, but highly maneuverable, shorter-distance flights. The image of the morphospace was previously published by Serrano et al. (2018; © The Palaeontological Association, used with permission). Black circles indicate specimens that are holotypes of their respective species, unless only a specimen number is indicated. Specimens are numbered in descending order of estimated body mass. (Figure from Wang et al. 2022; © The Authors 2022, open-access article licensed under a Creative Commons Attribution 4.0 International License, https://creativecommons.org/licenses/by/4.0/)

ornithuromorphs (e.g., Hongshanornithidae; Fig. 1.70) also retained some primitive characteristics, including manual claws ("wing claws") and a pubic symphysis.

Cretaceous ornithuromorphs had a global distribution, occupied a wide diversity of habitats and niches, and had diverse diets. Many were likely excellent fliers (Fig. 1.71), others were more cursorial, e.g., the hindlimbs of *Hollanda luceria* resembled those of present-day Greater Roadrunners (*Geococcyx californianus*) and Wild Turkeys (*Meleagris gallopavo*) (Bell et al. 2010; Bell 2013), and still others were flightless (Fig. 1.72). Mayr (2017a) suggested that early ornithuromorphs nested on the ground and that newly hatched young likely required some parental care.

Several species occupied semi-aquatic or aquatic habitats as well as marine environments where, based on the presence of supraorbital

**Fig. 1.55** Top, Based on morphological and chemical evidence, the plumage of at least one species of *Confuciusornis* was likely dark in color with darker spots on some wing coverts. Bottom, (**a**) Fossil of *Confuciusornis sanctus*, and (**b**, **c**) close-ups of elongated rectrices that may have played a role in attracting mates. (Top figure from Li et al. 2018; open-access article distributed under the terms of the Creative Commons CC BY 4.0 license, https://creativecommons.org/licenses/by/4.0/; Bottom figure from O'Connor et al. 2012; © 2012 by the Authors; licensee MDPI, Basel, Switzerland, open-access article distributed under the terms and conditions of the Creative Commons Attribution license, https://creativecommons.org/licenses/by/4.0/)

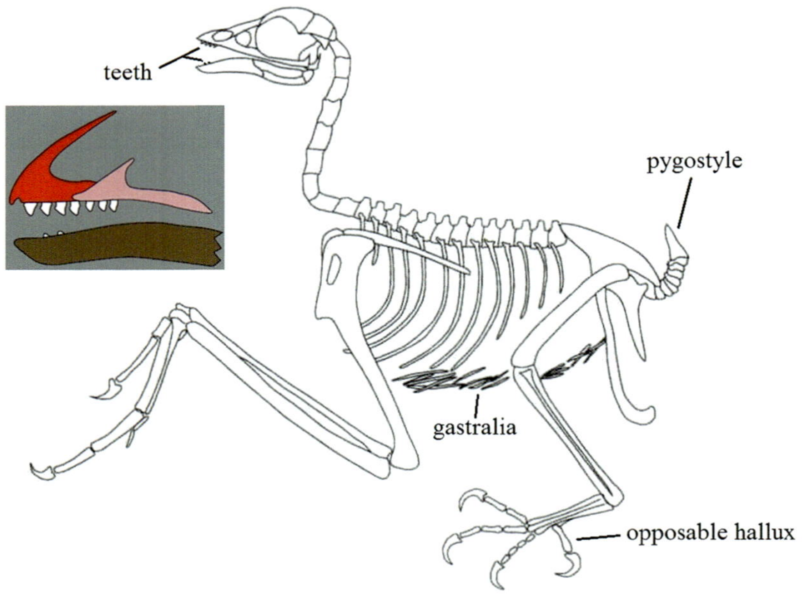


**Fig. 1.56** Reconstruction of a skeleton of *Sapeornis* and close-up of teeth. Sapeornithids had teeth (but only on the premaxilla and maxilla of the upper jaw, as well as two very small teeth on the lower jaw) and a pygostyle, but no sternum and no synsacrum. Note, however, the presence of gastralia. Sapeornithids also had very long wings (and digits), suggesting that they occupied open or semi-open habitats. (Figure of large skeleton from Zhou and Zhang 2003; © Canadian Science Publishing, used with permission; inset from Wang et al. 2017a, 2017b; © 2016 Elsevier Ltd. All rights reserved, used with permission)

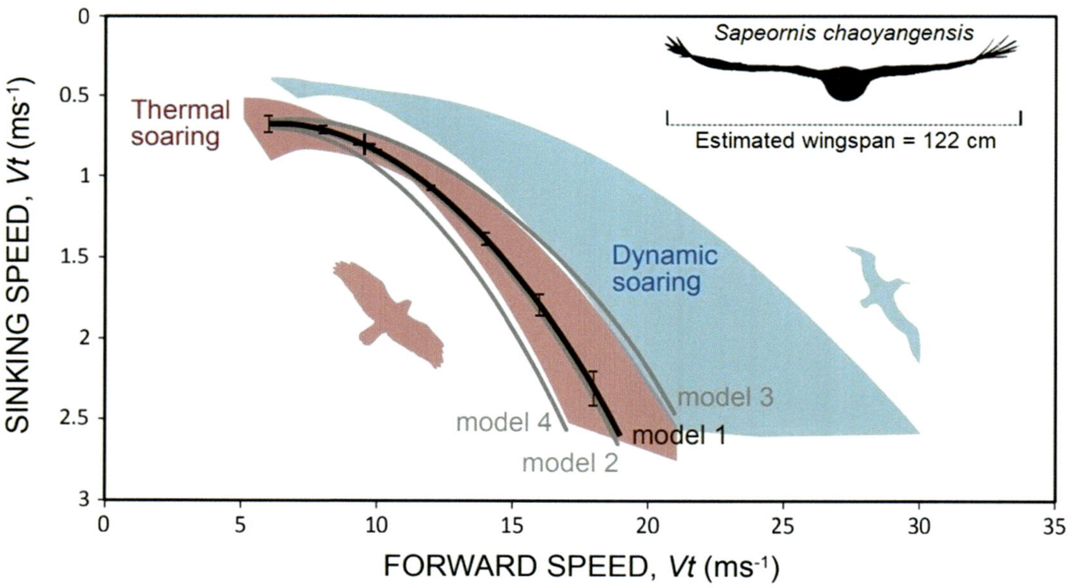


**Fig. 1.57** Based on a fossil where the body outline could be determined, and using four different models taking into account variables including the body mass, wingspan, body-drag coefficient, and flight speed, *Sapeornis chaoyangensis* was likely a thermal soarer. Its estimated gliding-forward velocity was 9.7 m/s and its sinking speed was estimated to be 0.82 m/s, values similar to those reported for present-day Red-tailed Hawks (*Buteo jamaicensis*). (Figure modified from Serrano et al. 2020; used with permission of the American Museum of Natural History)

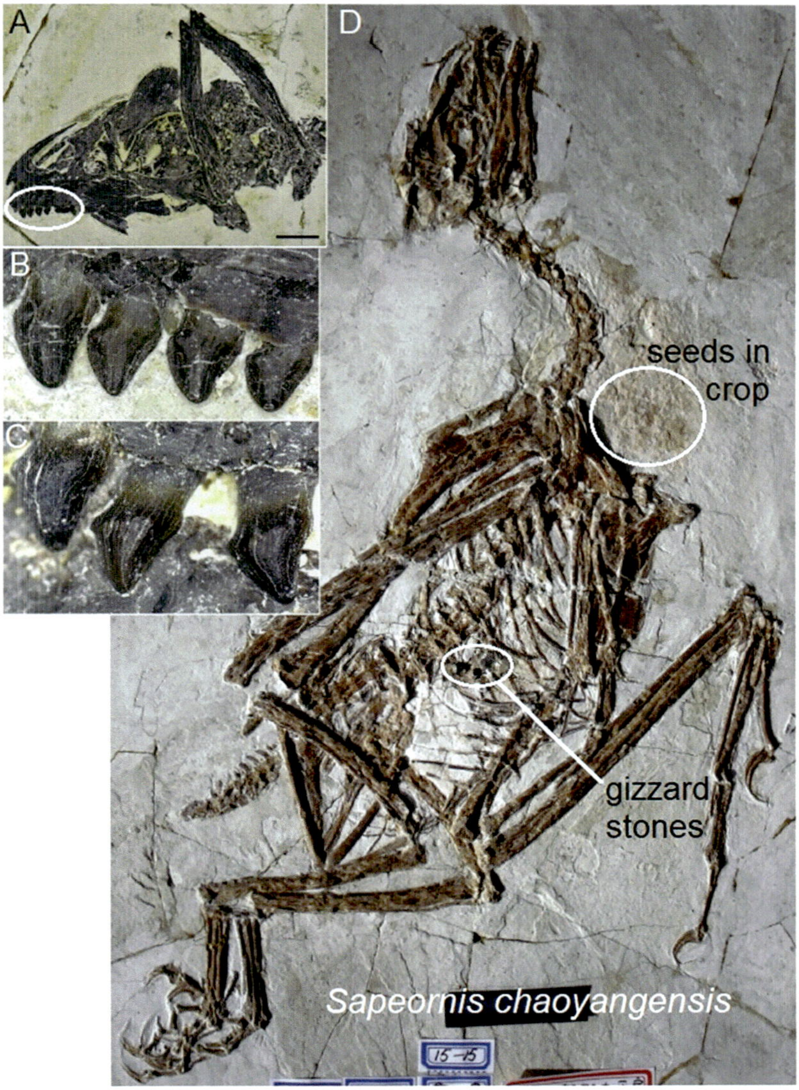


**Fig. 1.58** Fossil of a sapeornithid. (**a**) Close-up of the skull, with teeth circled. (**b**, **c**) Close-up views of the teeth. (**d**) Complete fossil showing preserved seeds in the crop and gizzard stones that would have aided in digestion. (Figure from O'Connor 2019; © 2018 Elsevier B.V., used with permission)

fossa, some were known to have salt glands (Wang et al. 2018a, b; Tanaka et al. 2020). Some ornithuromorphs were known to be piscivorous based on stomach remains (Zhou et al. 2014). Other species were more terrestrial and likely granivorous, and some may have been omnivorous or fed on insects or aquatic invertebrates (Zheng et al. 2018). Species in the family Hongshanornithidae had relatively long legs and were likely similar in habits to many present-day shorebirds (O'Connor et al. 2010; Fig. 1.73). Another ornithuromorph, *Gansus yumenensis*, was probably "... both volant and capable of diving to some degree suing either foot-propelled or, perhaps, both its wing and its feet for underwater locomotion" (Nudds et al. 2013). Species in the order Hesperornithiformes (about 25 reported to date) were birds specialized for diving and their fossils have been reported from sites dated from 113 to 66 million years old (Chiappe and Dyke 2006, Bell 2013). They are one of the earliest known groups of birds to have secondarily lost their ability to fly. The smallest hesperornithids were about the size of a

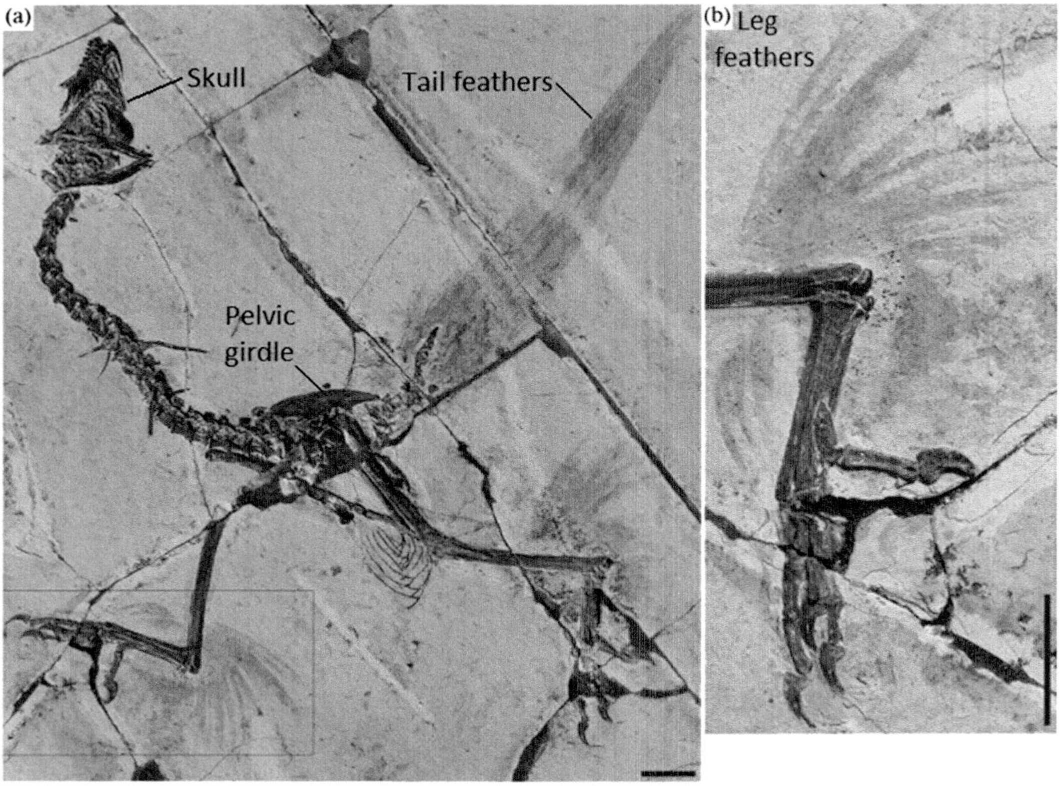


**Fig. 1.59** Sapeornithid fossil showing imprints of the fan-shaped tail. As was true for several non-avian theropods, basal birds, and some extant birds, sapeornithids also had pennaceous leg feathers. For some early gliders and flyers, such as *Microraptor*, pennaceous leg feathers may have provided additional lift. (Figure from O'Connor and Chang 2015; © 2015 Pleiades Publishing, Inc., used with permission)

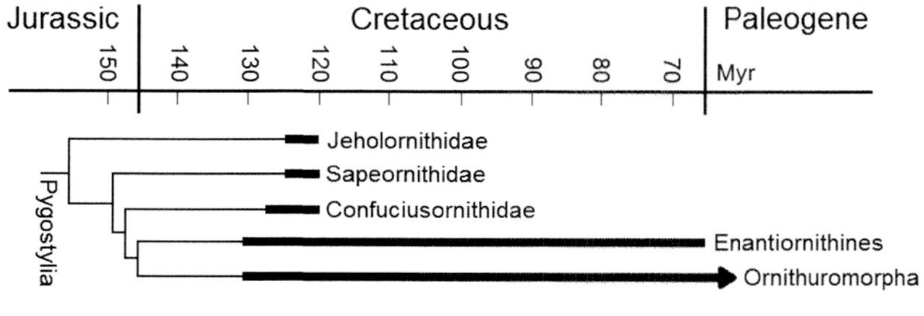


**Fig. 1.60** The phylogeny of the Pygostylia, birds with pygostyles. Enantiornithines were the dominant birds during the Cretaceous. The oldest ancestors of present-day birds, the ornithuromorphs, first appear in the fossil record about 131 million years ago (Yang et al. 2020). Note that, if these timelines are correct, over a period of about three million years just prior to 120 million years ago, all five groups of pygostylian birds may have coexisted. Jeholornithidae, Sapeornithidae, and Confuciusornithidae timelines are from Brusatte et al. (2015), basal Ornithuromorpha and enantiornithines from Yang et al. (2020). (Figure by G. Ritchison)

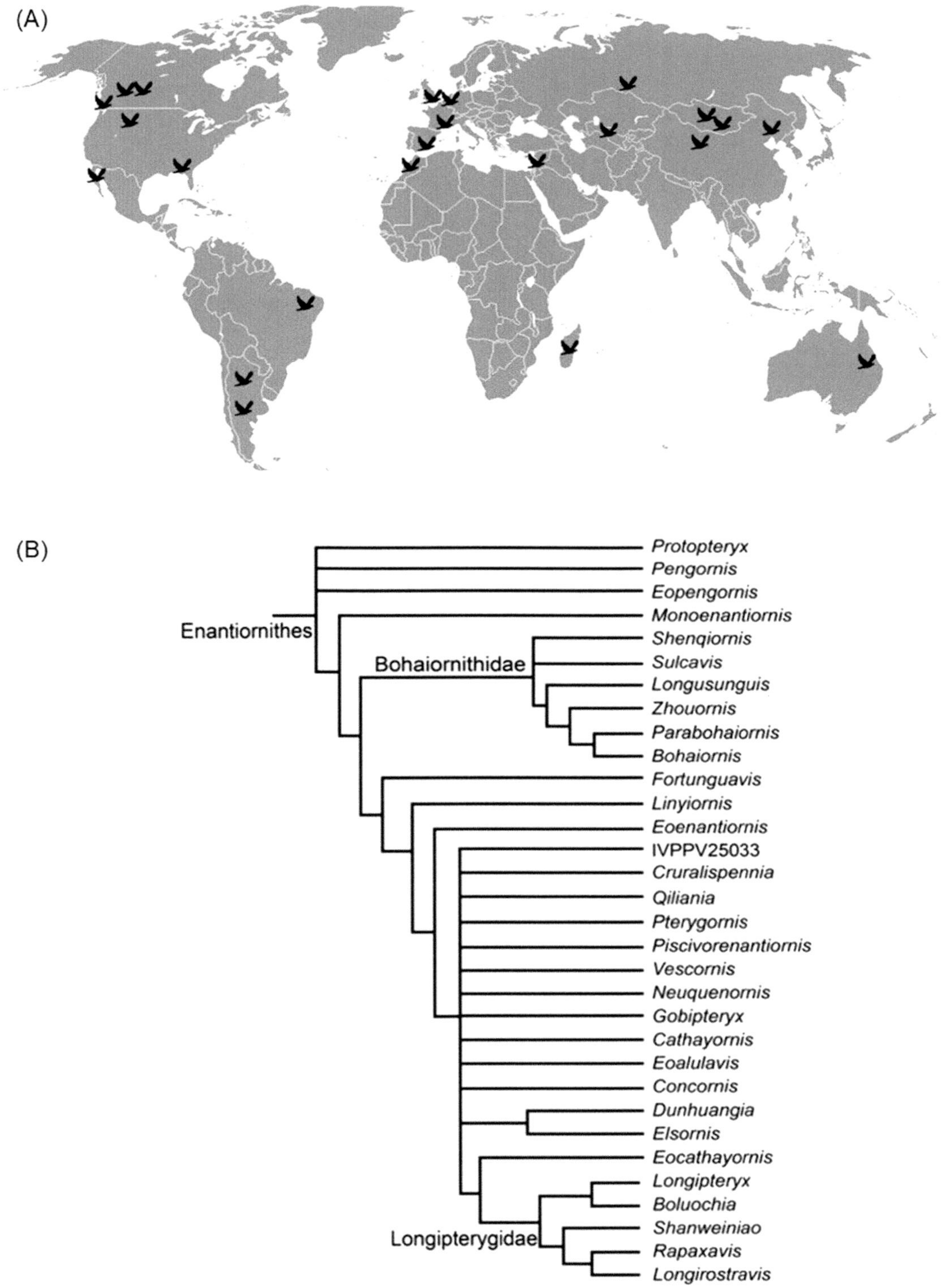


**Fig. 1.61** (**a**) Geographical distribution of known fossil enantiornithine birds. Of course, the location of the continents was quite different during the Cretaceous. (**b**) Enantiornithine families and genera. Enantiornithine phylogeny generally remains poorly resolved, but *Protopteryx* appears to be an early diverging taxa. Bohaiornithids exhibit less variation in size and morphology than longipterygids, suggesting that bohaiornithids

**Fig. 1.62** (**a**) Reconstruction of *Mirarce eatoni*, a turkey-sized enantiornithine from North America (Illustration by Brian Engh from Atterholt et al. 2018 [https://peerj.com/articles/5910/], open-access article distributed under the terms of the Creative Commons Attribution License, https://creativecommons.org/licenses/by/4.0/), (**b**) drawing of the enantiornithine bird *Shanweiniao cooperoruma* (Drawing by Nobu Tamura, CC BY 3.0, https://creativecommons.org/licenses/by/3.0/, Wikipedia), and (**c**) reconstructions of the plumage patterns on the wings of eight specimens of enantiornithines preserved in amber. (Figure from Xing et al. 2020; © 2020 Xing, O'Connor, Niu, Cockx, Mai and McKellar, open-access article distributed under the terms of the Creative Commons Attribution License [CC BY], https://creativecommons.org/licenses/by/4.0/

**Fig. 1.61** (continued) may have occupied a narrower range of ecotypes than longipterygids (Min et al. 2014). (Figure A modified from Dyke and Nudds 2009; © 2008 The Authors, Journal compilation © 2008 The Lethaia Foundation, used with permission; Figure B from Wang et al. 2018a, b, PNAS, used with permission)

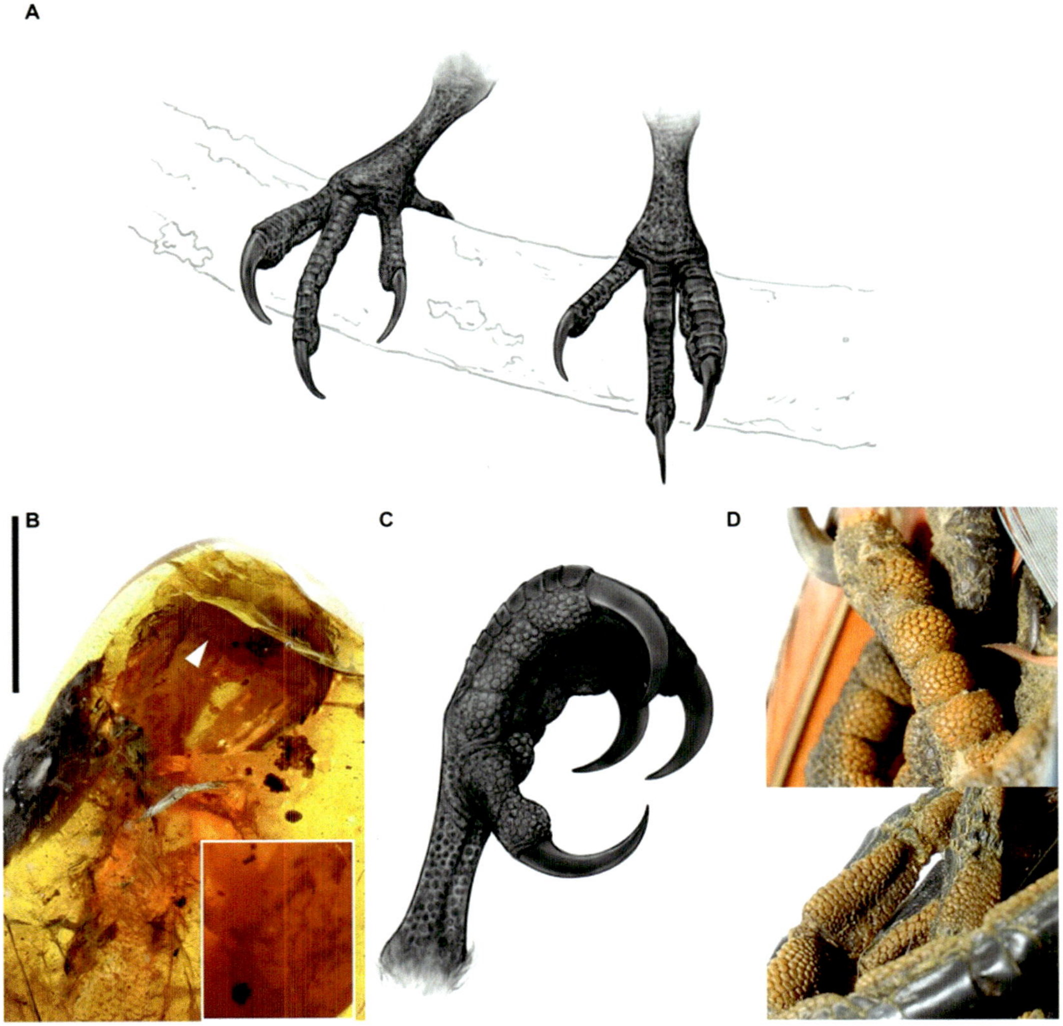


**Fig. 1.63** (**a**) Reconstruction of the feet of an enantiornithine bird (*Fortipesavis prehendens*) based on a foot preserved in approximately 100-million-year-old amber. The largest digit is laterally placed on each foot pes and the plantar pads have a slight pebbly texture. (**b**) Foot of *Fortipesavis prehendens* in amber. Inset shows larger image of plantar pad texture (indicated by the white arrow). Scale bar = 5 mm. (**c**) Reconstruction of a foot with the same posture as the amber specimen, showing the textured plantar pads similar to the those of some present-day birds, such as (**d**) a Red-and-Green Macaw (*Ara chloropterus*, top) and a Toco Toucan (*Rhamphastos toco*, bottom). (Figure from Clark and O'Connor 2021; © 2021 Clark and O'Connor, open-access article distributed under the terms of the Creative Commons Attribution License (CC BY), https://creativecommons.org/licenses/by/4.0/)

grebe, the largest as large as an average-sized human (O'Connor et al. 2011). From the Early to Late Cretaceous, hesperornithids increased in size (Fig. 1.74). The results of studies of present-day diving species suggest that increased body size improves diving duration via larger lungs so increasing size can be viewed as a diving specialization (Zavalga et al. 2007, Bell and Chiappe 2016).

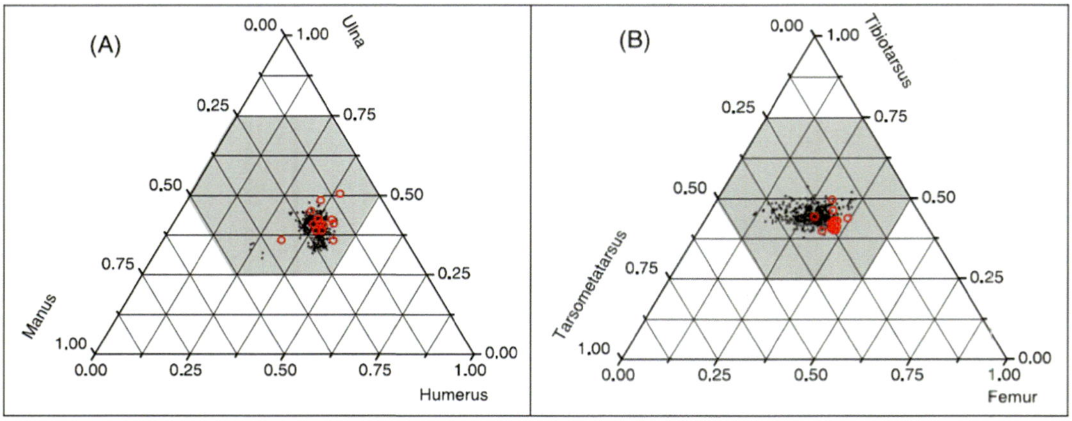


**Fig. 1.64** Relative proportions of the forelimb (**a**) and hindlimb (**b**) bones of enantiornithines (red) and present-day birds (black). These data suggest that variability in proportions of forelimb segments of enantiornithines largely overlaps that of present-day birds, but hindlimb proportions were less variable for enantiornithines than present-day birds. In (**a**), present-day birds on the lower left would be hummingbirds and swifts, with a long manus relative to the humerus and ulna; no enantiornithines discovered to date had forelimb morphology like that of hummingbirds and swifts (although one species indicated by the red circle on the lower left begins to approach that morphology). (Figure from Dyke and Nudds 2009; © 2008 The Authors, Journal compilation © 2008 The Lethaia Foundation, used with permission)

## 1.14 Neornithines

Key events in avian evolution include the appearance of the first "crown group birds" (Neornithines), i.e., the most recent common ancestor of all living birds (Gauthier 1986), and the subsequent diversification that ultimately, over many millions of years, resulted in the roughly 10,000 species of present-day birds. What remains to be determined is when these events occurred and more specifically when, during the Cretaceous, "modern" birds first appeared and whether extensive diversification began during the Cretaceous or after the Cretaceous/Paleogene (K/Pg) mass extinction that wiped out, among other groups, the non-avian dinosaurs, pterosaurs, and non-neornithine birds, including enantiornithes and hesperornithiforms (Box 1.8 The K/Pg Mass Extinction). Very few neornithine fossils date from the Cretaceous (Brocklehurst et al. 2012; Fig. 1.75), suggesting that diversification occurred early in the Paleogene and after the K/Pg mass extinction (e.g., Longrich et al. 2011; Feduccia 2014). In contrast, however, the results of some molecular studies (based on the rates at which mutations occur in nuclear or mitochondrial DNA, or "molecular clocks") suggest that diversification of neornithines began well before the end of the Cretaceous, with dates ranging from about 75 to more than 100 million years ago (e.g., Haddrath and Baker 2012; Jetz et al. 2012; Lee et al. 2014; Fig. 1.76). In this latter scenario, several major groups of birds such as ratites, galliforms, anseriforms, shorebirds, and even passerines originated during the Cretaceous and survived the K/Pg mass extinction (Lee et al. 2014).

**Box 1.8 The K/Pg Mass Extinction**

The end-Cretaceous mass extinction that occurred about 66.02 million years ago (Clyde et al. 2016) is associated with a clay layer containing abnormally high levels of iridium (Alvarez et al. 1980) along with impact ejecta, such as spherules and shocked minerals (Goderis et al. 2013). This layer resulted from the impact of a large bolide (meteor) that hit the Yucatan Peninsula of present-day Mexico, creating the large Chicxulub crater (~200 km diameter).

Artist's rendering of the asteroid impacting the shallow seas of the present-day Yucatan Peninsula. (Credit: Donald E. Davis, NASA/JPL-Caltech, Wikimedia Commons, CC0 Public Domain)

The impact would have triggered tsunamis that may have reached well inland, caused earthquakes and triggered volcanic eruptions (Adatte et al. 2014), and may have ignited large wildfires globally (Robertson et al. 2013). The impact also would have released huge amounts of sulfur into the atmosphere, causing acid rain and at least temporarily destroyed the ozone layer (Kring 2007). The particulates in the air would have caused rapid cooling of the Earth (i.e., "impact winter"; Vellekoop et al. 2014) and blocked much of the sunlight from reaching the surface, impacting the ability of plants to perform photosynthesis (Vajda and McLoughlin 2004). The impact of these events on plant and animal life is almost unprecedented in Earth history. Several major animal groups disappeared (e.g., archaic birds like the Enantiornithines, non-avian dinosaurs, and marine and flying reptiles), and several other major groups suffered considerable, but not complete, species-level extinction (e.g., land plants). The abrupt decline in photosynthesis due to the low light levels likely had a catastrophic effect on animals relying on primary

(continued)

**Box 1.8** (continued)

producers (e.g., the herbivorous dinosaurs) which, in turn, had a catastrophic effect on the carnivores that fed on the herbivores.

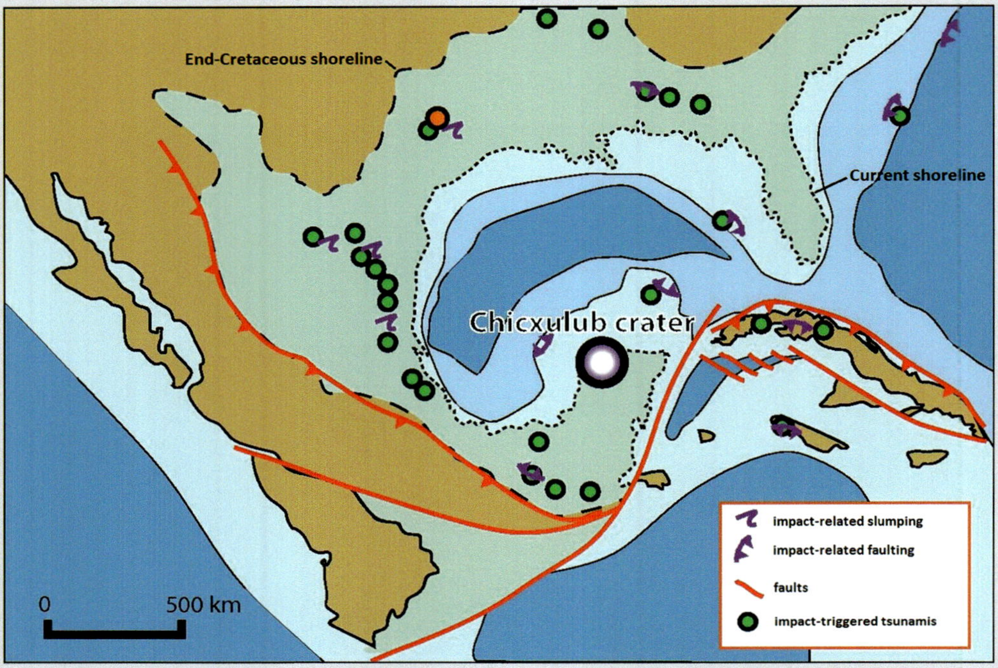


Map of the Gulf of Mexico at the end of the Cretaceous and the location of the Chicxulub crater. In addition to more local impacts caused by the tsunamis and impact-trigger earthquakes, the particulates and gases released into the atmosphere would have had global-wide impacts. (Figure from Vellekoop et al. 2014; used with permission of the United States National Academy of Sciences).

(continued)

**Box 1.8** (continued)

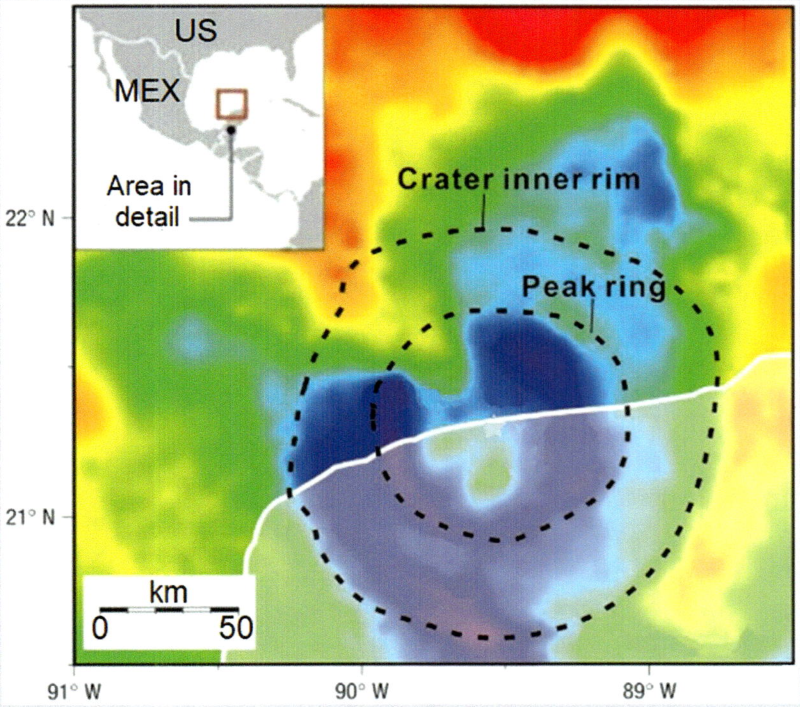


Location and size of the Chicxulub impact crater. White line indicates the shoreline of the present-day Yucatan Peninsula. (Figure modified from Zhao et al. 2020; © 2020 International Association for Gondwana Research. Published by Elsevier B.V., used with permission)

Clearly, examination of the fossil evidence and molecular clock data can lead to different conclusions concerning the evolution of crown group birds. However, some investigators have found general agreement between these two lines of evidence (Box 1.9 Avian Tree of Life). The results of studies based on genome-scale phylogenetic analyses (Jarvis et al. 2014; Prum et al. 2015) suggest that (1) crown group birds originated in the Cretaceous, (2) the Palaeognathae and Neognathae lineages split about 70–75 million years ago (Box 1.10 Palaeognathae Evolution), (3) the Galloanseres (waterfowl and gallinaceous birds) and Neoaves split about 66.7 million years ago (Field et al. 2020; Box 1.11 Galloanseres Phylogeny), and (4) rapid radiation within the Neoaves occurred over a period of about 10–15 million years after the Cretaceous–Paleogene mass extinction (Fig. 1.77; Box 1.12 Limits to Phylogenetic

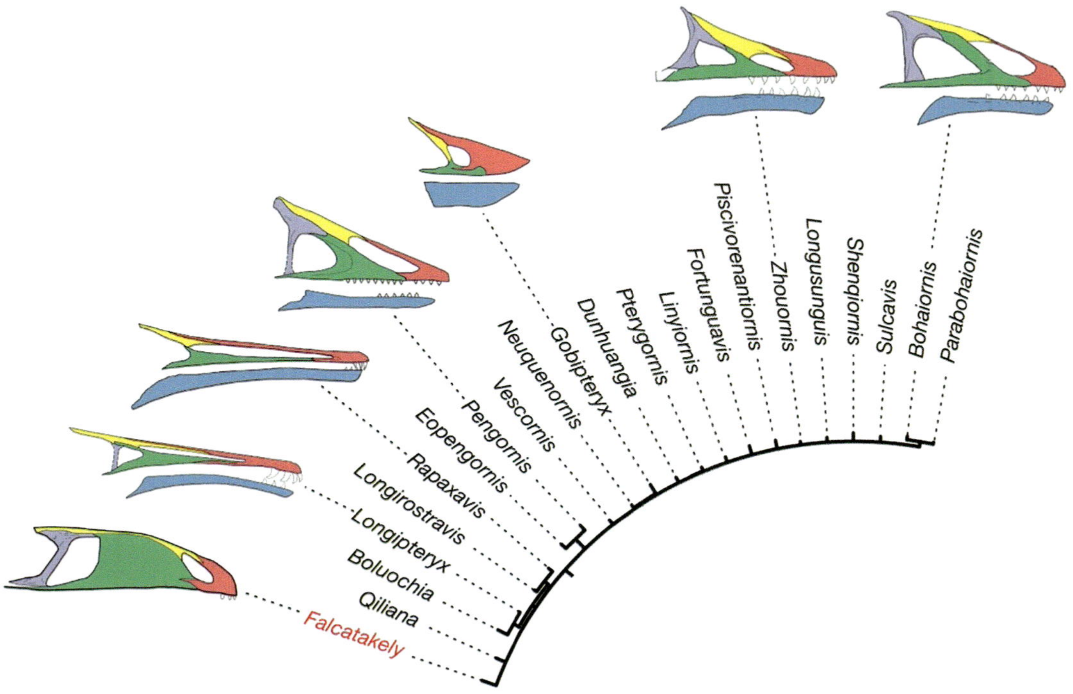


**Fig. 1.65** Examples of variation in bill morphology among enantiornithines. Illustrations are not to scale. Red, premaxilla; green, maxilla; yellow, nasal; lavender, lacrimal; blue, dentary. (Figure from O'Connor et al. 2020b; © 2020 The Authors, under exclusive license to Springer Nature Limited, used with permission)

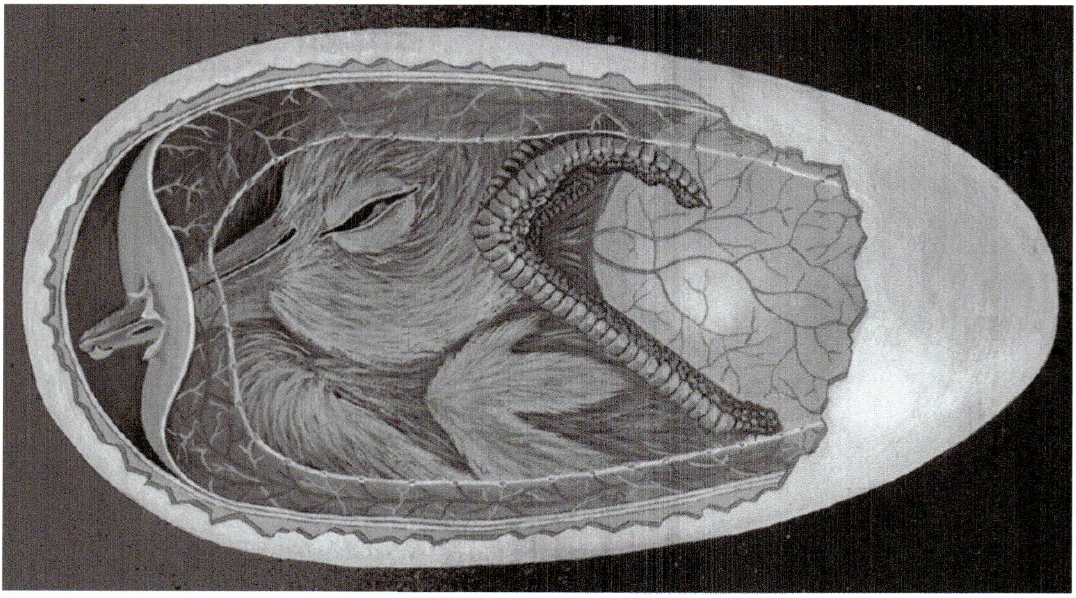

**Fig. 1.66** Drawing based on a fossil of an enantiornithine embryo in an egg. (Figure from Kurochkin et al. 2013, artist—Michael W. Nickell; © 2013 Pleiades Publishing, Ltd., used with permission)

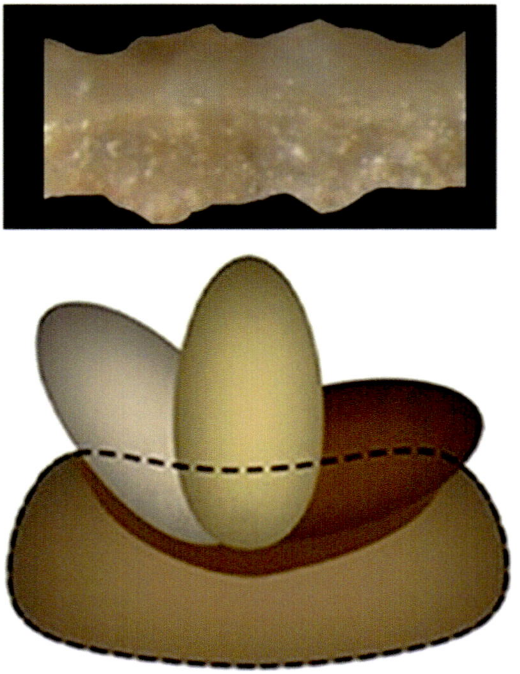

**Fig. 1.67** Available evidence suggests that eggs in the nests of enantiornithines were oriented vertically or nearly so and, based on analysis of pigments in fossil eggshells, were likely light brown in color. (Figure modified from Wiemann et al. 2018; © 2018 Springer Nature, used with permission)

Resolution). Similarly, Ksepka and Phillips (2015), using both mitochondrial and nuclear molecular clock analyses, suggested that palaeognaths and neognaths diverged before the end of the Cretaceous as did Galloanserae and Neoaves, but, given the uncertainties associated with assigning specific divergence dates, they were unable to determine if additional, more extensive diversification began near the very end of the Cretaceous or early in the Paleogene. However, based on two important factors, Ksepka and Phillips (2015) suggested that this diversification was more likely to have occurred during the early Paleogene. First, extensive diversification late in the Cretaceous would have required that several lineages were able to survive the K-Pg mass extinction that eliminated so many other taxa, including the previously successful Enantiornithes and Hesperornithes. In addition, the K-Pg mass extinction would have created vacant ecological niches, seemingly favoring diversification of birds during the early Paleogene to fill those niches. Probably not coincidentally, mammals experienced similar diversification during the early Paleogene (Fig. 1.78).

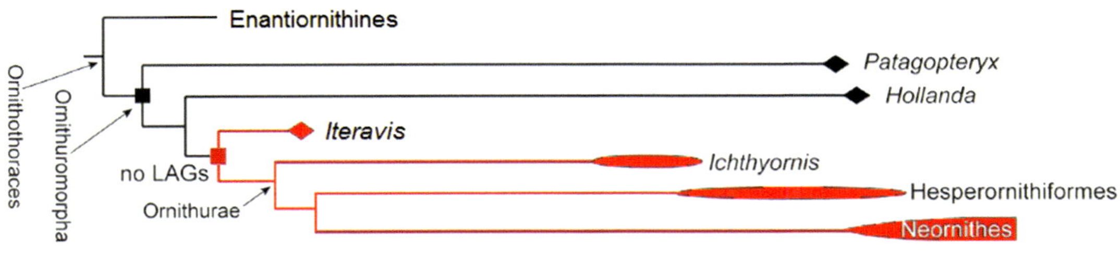


**Fig. 1.68** Examination of cross-sections of the bones of enantiornithines has revealed lines of arrested growth (LAGs), indicating that growth to adult size took more than a year. Bones of the more basal ornithuromorphs, e.g., the genera *Patagopteryx* and *Hollanda*, also have LAGs. However, bones of ornithuromorphs that existed after about 125 million years ago have no LAGs, indicating growth to adult size in less than a year. (Figure modified from O'Connor et al. 2015a, b, c; open-access article distributed under the terms of the Creative Commons Attribution License, https://creativecommons.org/licenses/by/4.0/)

**Box 1.9 Avian Tree of Life**

Although our understanding of the avian phylogenetic tree has certainly improved over the past several years, phylogenetic studies often produce conflicting results because investigators use different genes, sample different taxa, and use different types of data, i.e., exons (a segment of DNA containing information coding for a protein or peptide sequence), introns (a segment of DNA that does not code for proteins), noncoding ultraconserved elements (regions of DNA that are identical in different species and are abundant throughout the genomes of most organisms), conserved nonexonic elements (noncoding regions of the genome that evolve slower than other regions of the genome), and transposable element insertions (also known as transposons or "jumping genes," a DNA sequence that can change its position within a genome) (Braun et al. 2019). However, collectively, the results of these studies do suggest the existence of 10 major lineages of Neoaves birds (with the clade Neoaves consisting of all modern birds except Palaeognathae and Galloanseres), including seven that include multiple orders. Reddy et al. (2017) referred to these orders as the "magnificent seven" and the other three as "orphan orders," but, as noted by Braun et al. (2019), "interrelationships among these major lineages remain poorly resolved." Braun et al. (2019) further noted that "...whole-genome phylogenetics will only resolve the remaining relationships if we improve data quality, exploit information from other sources (i.e., rare genomic changes), and learn more about the functional and evolutionary landscape of avian genomes."

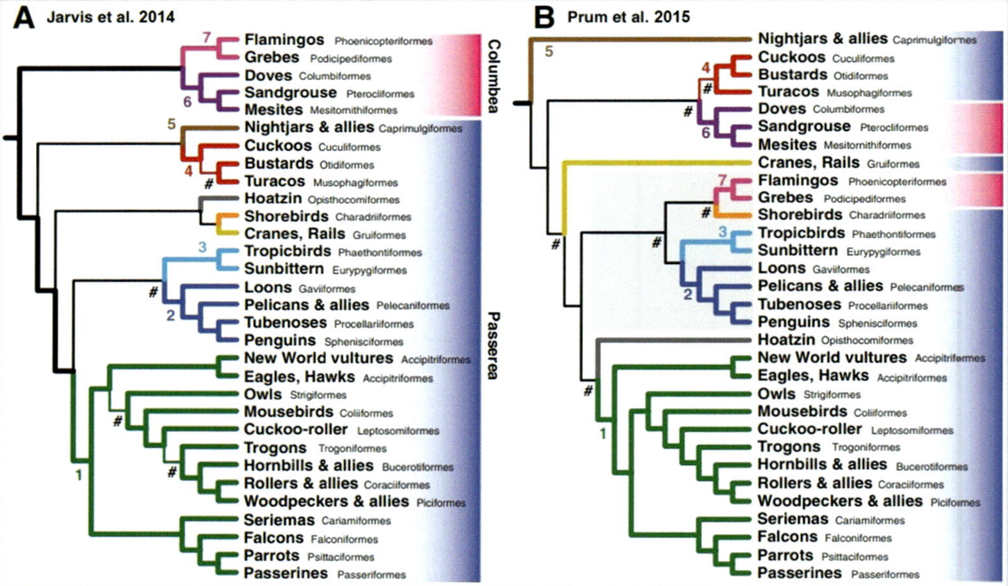


There are several differences between the phylogenetic trees of Jarvis et al. (2014) and Prum et al. (2015) at the base of Neoaves. (**a**) The Jarvis et al. (2014) tree, with branches having low support indicated by thin lines and those with even less support indicated with the symbol #. (**b**) The Prum et al. (2015) tree, with low support branches indicated the same way. An expanded waterbird clade (called Aequorlitornithes by Prum et al. 2015) is indicated by the gray shading. Numbers in both trees refer to the seven avian lineages that include multiple orders, i.e., the "magnificent seven." (Figure from Braun et al. 2019; © 2019 Springer Nature Switzerland AG, used with permission)

(continued)

**Box 1.9** (continued)

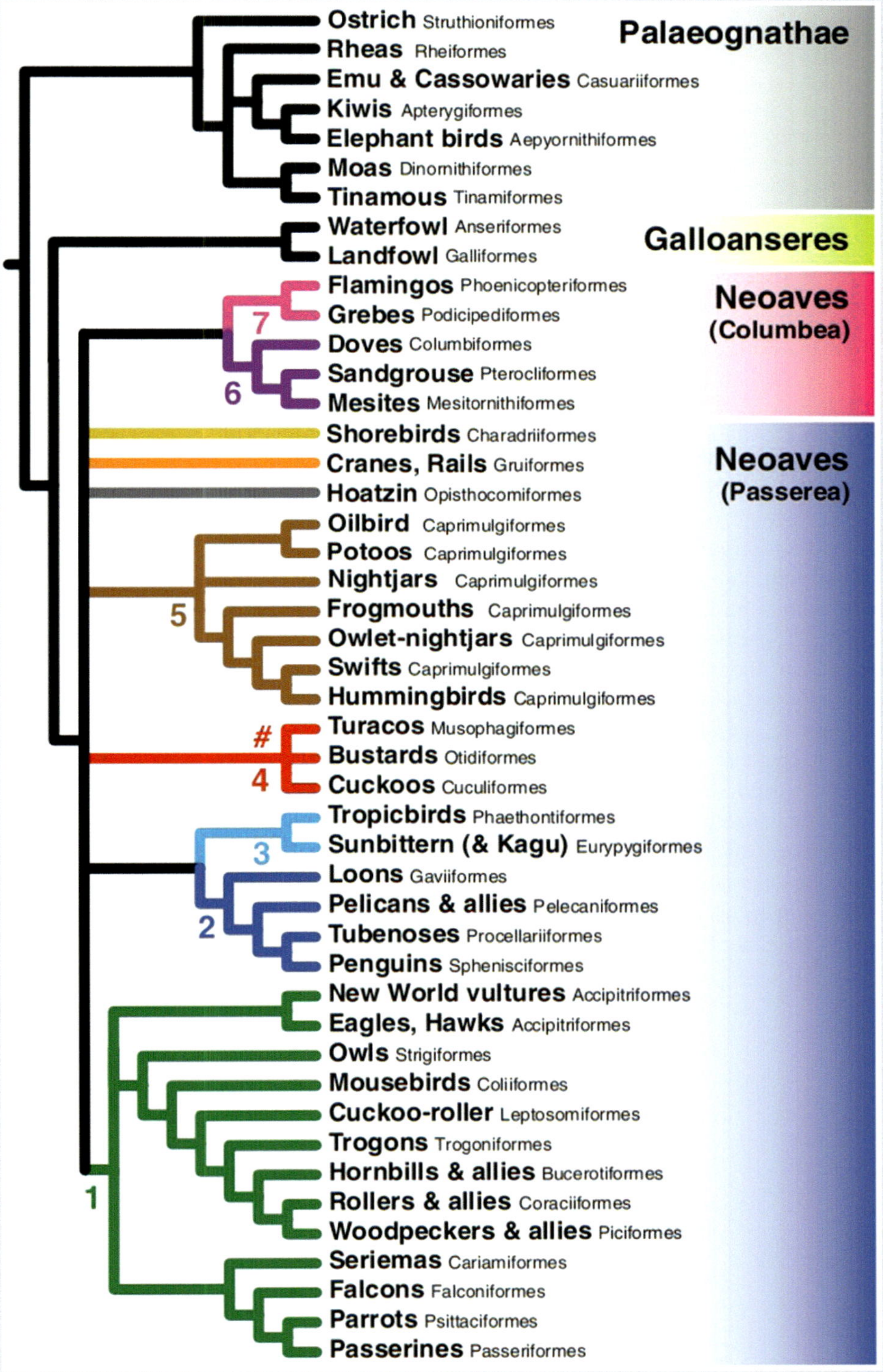


(continued)

**Box 1.9** (continued)

A phylogenomic tree of birds based on Jarvis et al. (2014) and Prum et al. (2015). The division of Neoaves into Columbea and Passerea is based on Jarvis et al. (2014) and is not included in the Prum et al. (2015) phylogeny. The numbered clades correspond to the "magnificent seven" of Reddy et al. (2017): (1) "core" landbirds (Telluraves), (2) "core" waterbirds (Aequornithes), (3) tropicbirds and sunbittern (Phaethontimorphae), (4) cuckoos, bustards, and turacos (Otidimorphae), (5) nightjars, swifts, hummingbirds, and allies (Caprimulgiformes), (6) doves, mesites, and sandgrouse (Columbimorphae), and (7) flamingos and grebes (Phoenicopterimorphae). Limited support for Otidimorphae is indicated by #. Reddy et al. (2017) called shorebirds, cranes, and the Hoatzin the "orphan orders"; shorebirds and cranes form a clade (Cursorimorphae) in Jarvis et al. (2014), but represent independent lineages in the Prum et al. (2015) phylogeny. (Figure from Braun et al. 2019; © 2019 Springer Nature Switzerland AG, used with permission).

**Box 1.10 Palaeognathae Evolution**

Different investigators have provided different hypotheses concerning the origin and evolution of the Palaeognathae (so-named because, in contrast to the Neognathae, they have a fused palate structure, i.e., pterygoid and palatine fused into a single element [Benito et al. 2022a, b]), and molecular and morphological analyses differ in terms of the interrelationships between palaeognathous birds (Mayr 2017b). According to one hypothesis proposed by Yonezawa et al. (2017), palaeognaths originated in the northern hemisphere during the late Cretaceous between about 105 and 115 Ma. The earliest divergence was between the ancestor of the present-day common ostrich (*Struthio camelus*) and the other palaeognaths (about 80 Ma). After this branching, the ancestor of palaeognaths other than the ostrich (collectively referred to as the Notopalaeognathae) migrated to the southern hemisphere (present-day South America) and may still have been capable of flight. Near the Cretaceous–Paleogene boundary, the Notopalaeognathae diverged rapidly (in an evolutionary sense), with the ancestor of present-day rheas likely diverging from the other notopalaeognaths, followed by the ancestor of present-day tinamous and now extinct moas, during the late Cretaceous. The ancestors of other present-day Palaeognaths may have migrated from South America via Antarctica (which had a temperate climate at the time; Pross et al. 2012) to present-day Australia and New Zealand.

(continued)

**Box 1.10** (continued)

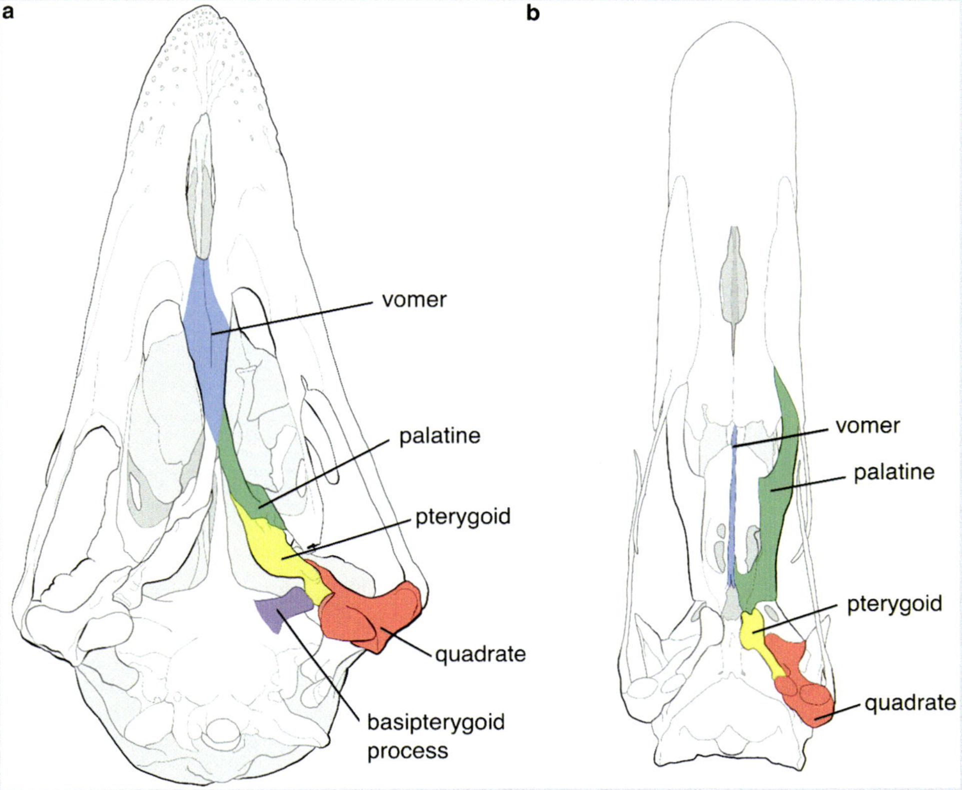


Palates of a palaeognathous and a neognathous bird. (**a**) Palate of the palaeognathous Emu (*Dromaius novaehollandiae*), with fused pterygoid and palatine bones and, therefore, an immobile upper jaw. (**b**) Palate of the neognathous Mute Swan (*Cygnus olor*), with the joint between the pterygoid and palatine bones allowing movement of the upper jaw. Birds with mobile palates are called neognaths, or "new jaws," whereas those with fused palate are palaeognaths, or "old jaws." A Late Cretaceous ornithurine (*Janavis finalidens*) was found to have a neognathous palate, suggesting that (despite the prefixes "palaeo" and "neo" that were based on the assumption that immobile palates were the ancestral condition) a mobile palate predates the origin of modern birds and that palaeognathous birds later evolved a fused palate (Benito et al. 2022a, b). (Figure from Widrig and Field 2022; 

Phylogenetic analysis indicates, surprisingly, that the now-extinct elephant birds of Madagascar are most closely related to New Zealand's kiwis. Given the current vast distance between Madagascar and New Zealand, this relationship is difficult to explain. However, Mitchell et al. (2014) suggested that the common ancestor of kiwis and elephant birds was likely able to fly and capable of long-distance dispersal, a view these authors suggest is supported by the discovery of "a small, possibly flighted kiwi relative from the Early Miocene of New Zealand" (Worthy et al. 2013).

(continued)

**Box 1.10** (continued)

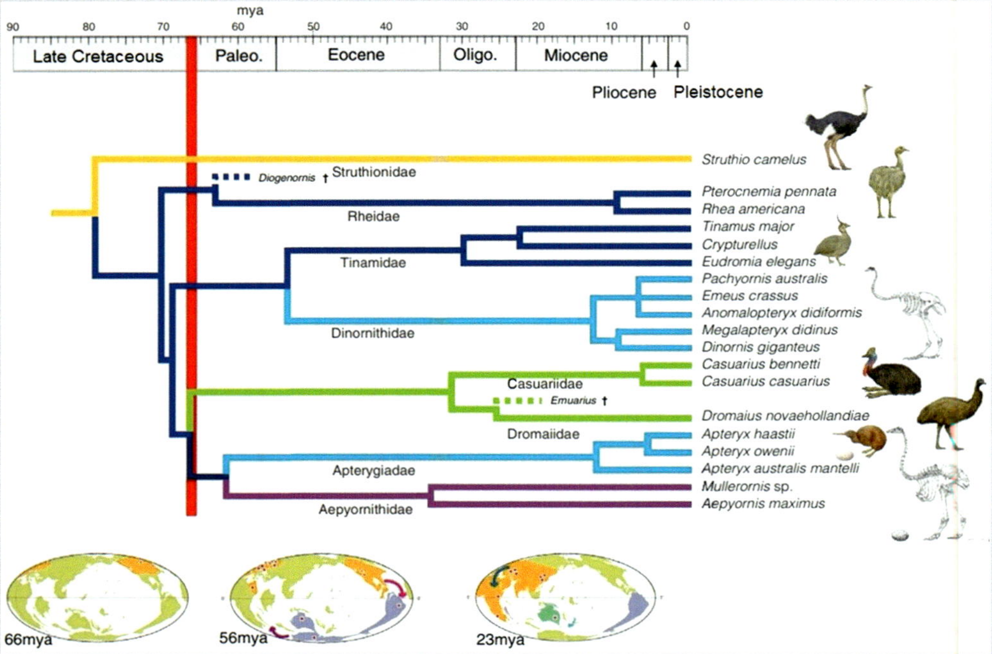


Possible Palaeognathae phylogeny. Phylogenetic positions of fossil species are indicated by dashed lines (their divergence times are arbitrary). The colors of the branches indicate geographic distribution. Hypothetical ancestral distribution and fossil records of the palaeognaths are indicated on the paleomaps. The thick red vertical line indicates the K–Pg boundary. (Figure modified from Yonezawa et al. 2017; © 2016 Elsevier Ltd., used with permission)

(continued)

**Box 1.10** (continued)

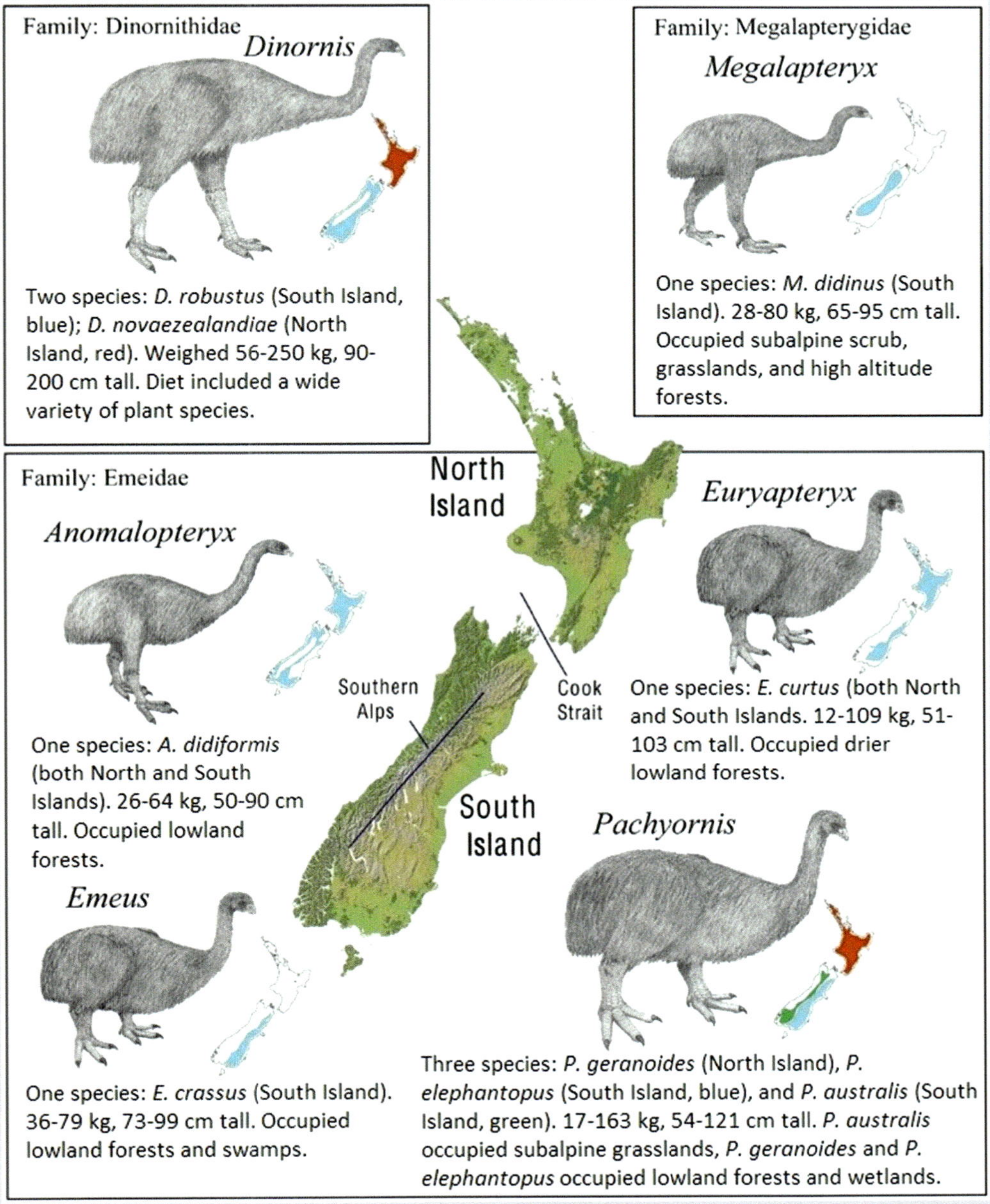


Sizes and estimated ranges of the estimated nine species of extinct moas (Dinornithiformes). Moas were flightless and diverse in size and in morphology. Species in the genus *Dinornis* were characterized by large body size, with males weighing as much as 250 kg and about twice the size of females. In contrast, coastal moas (*Euryapteryx curtus*) weighed as little as about 12 kg and females were about 20% larger than males. (Figure modified from Bunce et al. 2009; used with permission of the U.S. National Academy of Sciences)

**Box 1.11 Galloanseres Phylogeny**

The clade Galloanseres includes the present-day orders Galliformes and Anseriformes which are sister (i.e., the closest relative) to all other non-palaeognath birds (Neoaves). Among the last common ancestors of Galloanseres is the fossil *Asteriornis maastrichtensis* that dates from about 66.7 million years ago. Examination of the skull of *Asteriornis* revealed characteristics typical of Neornithes as well as both "galliform" features (e.g., weakly fused rostral elements and rostrally forked nasals) and "anseriform" features (e.g., a rostrally projecting postorbital process and a tall and strongly hooked retroarticular process) (Field et al. 2020).

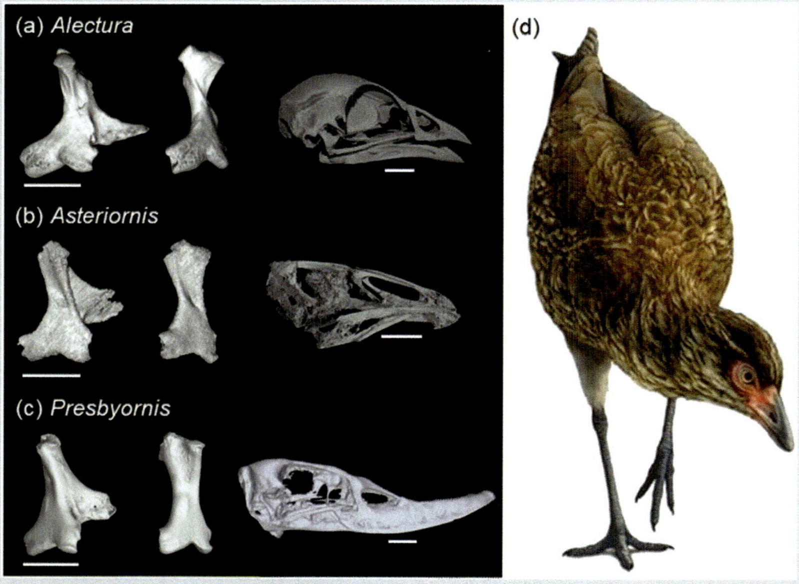


Comparison of the quadrates (left, medial view; middle, caudal view) and skulls of (**a**) present-day *Alectura* (Australian Brushturkey), (**b**) *Asteriornis*, and (**c**) *Presbyornis* (extinct anseriform; 61–62 million years ago). Note that the premaxillary bill of *Asteriornis* resembles that of extant Galliformes, whereas the quadrate is more similar to that of *Presbyornis*. (**d**) Possible appearance of *Asteriornis*. Scale bars, 5 mm (quadrates) and 1 cm (skulls) (Figures modified from Field et al. 2020; © 2020 The Authors, under exclusive license to Springer Nature Limited, used with permission).

There are also three extinct families of birds included in the clade Galloanseres: Dromornithidae (large, flightless birds also called mihirungs or thunder birds), Gastornithidae (also large, flightless bird with large bills), and Sylviornithidae (large, flightless bird found in New Caledonia that became extinct soon after the arrival of humans; Worthy et al. 2016).

(continued)

**Box 1.11** (continued)

*Dromornis stirtoni*, a flightless bird from the Late Miocene of Australia. (Drawing by Nobu Tamura, Wikipedia, CC BY 2.5, https://creativecommons.org/licenses/by/2.5/)

Worthy et al. (2017) examined the relationships of these extinct families to each other and to present-day Galloanseres and suggested that, in addition to Galliformes and Anseriformes, Galloanseres also included two additional clades, Gastornithiformes (which includes the families Dromornithidae and Gastornithidae) and the Cretaceous *Vegavis*. Although a member of Galloanseres, *Vegavis* was not included in the order Anseriformes, but, rather, Worthy et al. (2017) proposed a new order, Vegaviiformes, with *Vegavis* as a genus in that order.

(continued)

**Box 1.11** (continued)

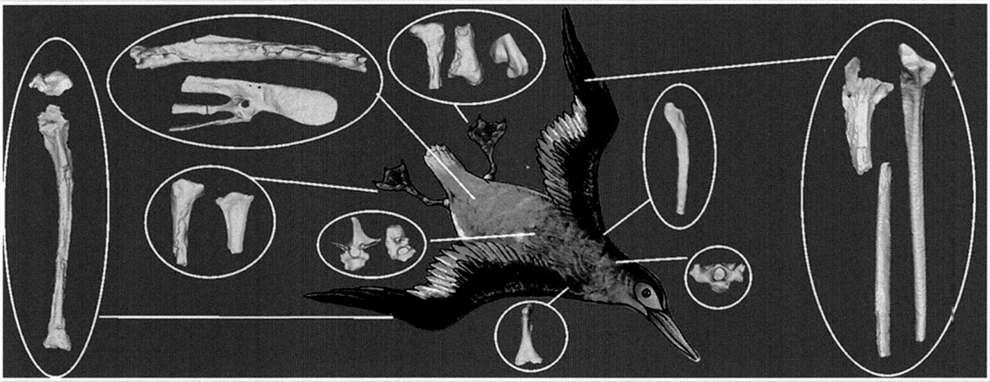

Bones identified from *Vegavis iaai* and an artist's reconstruction. The fossil of *Vegavis iaai* is the most complete of the few neornithine birds that existed at the end of the Cretaceous. Although clearly included in the clade Galloanseres, its relationship relative to Galliformes or Anseriformes within Galloanseres remains unresolved, largely because a skull of *Vegavis* has not been found. (Graphical abstract from Hospitaleche and Worthy 2021; © 2021 Elsevier Ltd. All rights reserved, used with permission)

Worthy et al. (2017) also suggested that *Brontornis* was not a Galloansere, but is more closely related to birds in the order Caramiformes and, specifically, the Phorusrhacids (also known as the terror birds in the family Phorusrhacidae). All of these clades apparently originated during the Late Cretaceous. Worthy et al. (2017) also suggested that Galloanseres (a) were or are largely herbivores or omnivores that fed primarily on plants, an exception being seaducks (Mergini) that are piscivorous, (2) were or are able to fly, but many large Galloanseres were flightless, and (3) varied substantially in size, with some in the extinct order Gastornithiformes being as large as the largest moas (about 2 m in height).

(continued)

**Box 1.11** (continued)

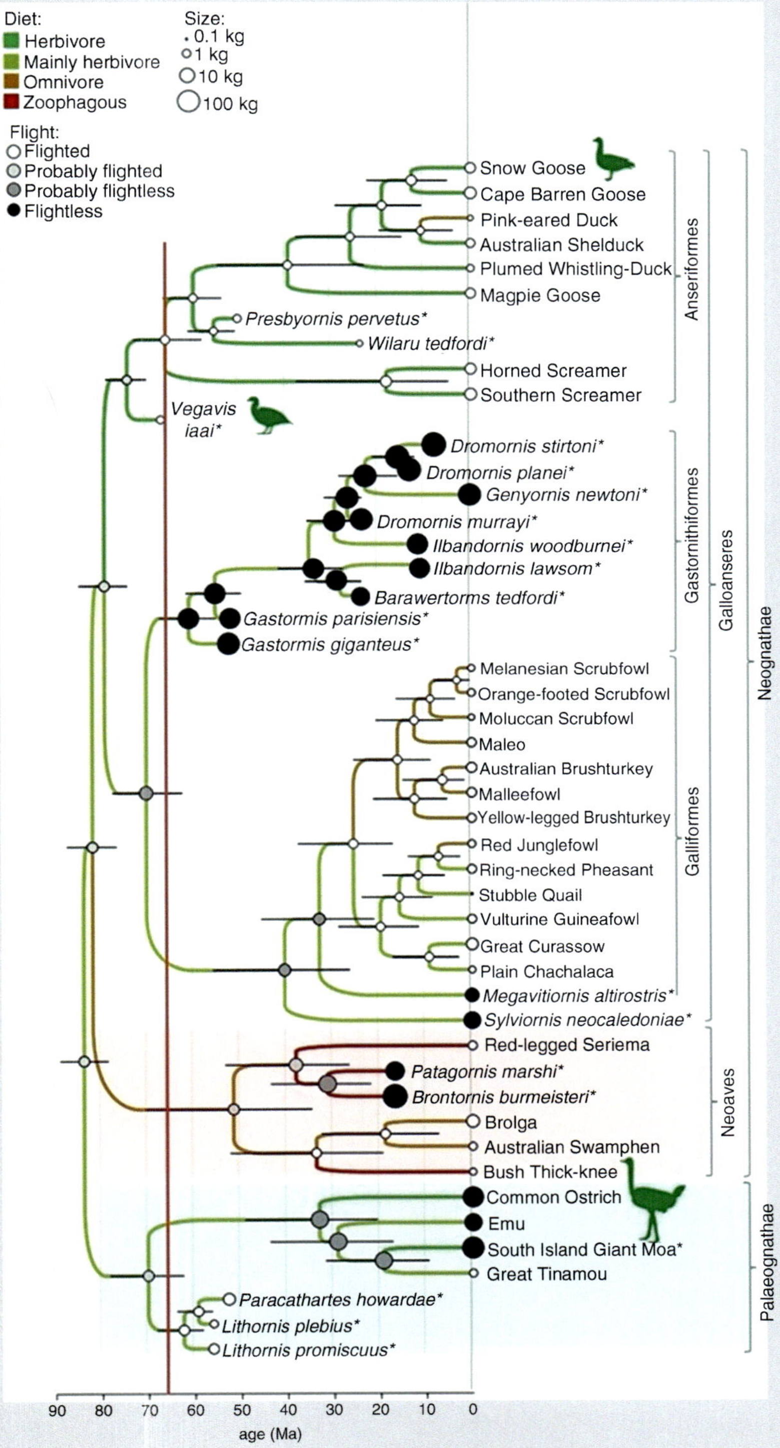


(continued)

**Box 1.11** (continued)

Possible phylogeny of Galloanseres, with information about diet, body size, and divergence ages of clades. Body mass is shown by circle size, flight ability by circle shading, and diet by branch color. Divergence dates are indicated by the scale at the bottom, confidence intervals shown as bars at nodes. The red vertical line indicates the K-Pg boundary. Species with an asterisk are extinct. (Figure from Worthy et al. 2017; ).

Worthy et al. (2017) also examined the relationships of three lithornithids, extinct chicken-sized palaeognaths that were able to fly. Their analyses suggest that lithornithids may be a sister taxon of either tinamous or all extant palaeognaths.

**Box 1.12 Limits to Phylogenetic Resolution**

When constructing phylogenies, certain a priori assumptions are made and one common assumption is that divergence of taxa involves a series of dichotomous branching events. However, ancestral species can give rise to several new species almost simultaneously, creating what is called a polytomous branching pattern.

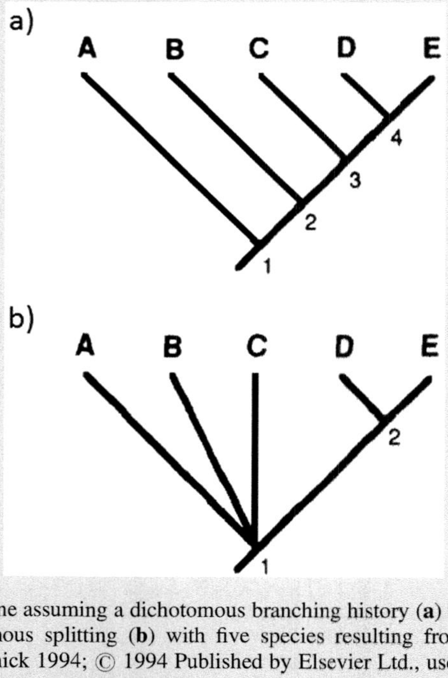


Alternative phylogenies with one assuming a dichotomous branching history (**a**) with four speciation events and the other assuming a polytomous splitting (**b**) with five species resulting from just two speciation events. (Figure from Hoelzer and Melnick 1994; © 1994 Published by Elsevier Ltd., used with permission)

Many investigators have constructed phylogenies of modern birds (Neornithes) based on molecular analyses (using nuclear and/or mitochondrial DNA), but often come to somewhat different conclusions concerning both the relationships among the deepest branches (orders) and when the branching occurred. Such differences can arise due to use of different or incomplete datasets, different methods of analysis (e.g. maximum parsimony vs. maximum

(continued)

**Box 1.12** (continued)

likelihood vs. Bayesian analysis), and, importantly, rapid speciation and polytomous lineages that can cause incomplete lineage sorting (Jarvis et al. 2014).

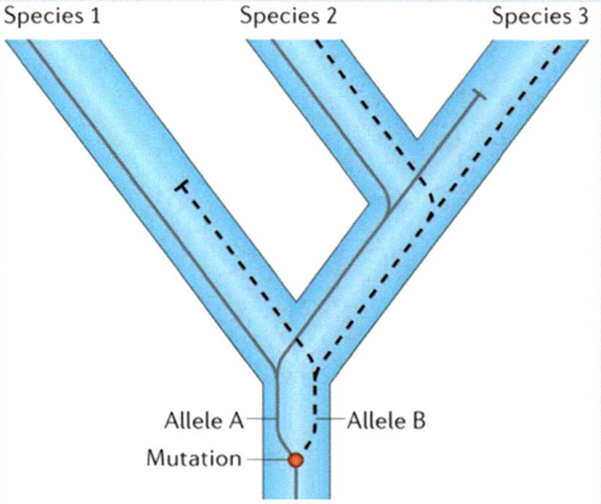


**Incomplete lineage sorting** can make it very difficult to generate accurate phylogenies, but what exactly is incomplete lineage sorting? As an example, assume three species have descended from a common ancestor, with Species 1 first splitting from the ancestor of Species 2 and 3 and, shortly thereafter, Species 2 and 3 diverge. Also assume that the ancestral species is polymorphic (with Alleles A and B) and, with the initial split into two descendent lineages, both alleles are retained in the two daughter lineages. Soon thereafter, one of these lineages again splits, with both resulting lineages still retaining both alleles. Over time, however, lineages can lose an allele due to genetic drift or selection. In the example above, Species 1 retains Allele A, but loses B, and Species 3 retains Allele B, but loses A. In a molecular phylogenetic analysis, Species 2 will seem to be more closely related to either Species 3 if it retains Allele B (which is actually the case). However, if Species 2 loses Allele B and retains Allele A, then it will seem to be more closely related to Species 1. If Species 2 and 3 had not diverged so soon after the initial divergence into two lineages, then it is likely that the loss of Allele A would have occurred prior to the split and both species would share Allele B, making it possible to generate an accurate (molecular) phylogeny. Thus, incomplete lineage sorting, with alleles retained in different lineages with different phylogenetic relationships due to rapid rates of divergence can make it difficult, if not impossible, to generate accurate phylogenies. In the example above, there is just a single pair of alleles. When multiple genomic segments are used to generate phylogenies, different branching patterns may emerge depending on the extent of and variation in incomplete lineage sorting in different segments. Available evidence suggests that the rapid diversification of birds in the early Paleogene resulted in relatively high levels of incomplete lineage sorting (Suh et al. 2015) that, in turn, limits the resolution of avian phylogenies (Figure from Rogers and Gibbs 2014; © 2014 Springer Nature, used with permission).

In lineages with polytomous branching, patterns of diversification are much more complex than in those with dichotomous branching. Suh et al. (2015) conducted a genome-level analysis of the retroposons (repetitive DNA fragments inserted into a host chromosome after reverse transcription from an RNA molecule) of 48 species of birds (the same species used by Jarvis et al. 2014) and their analysis revealed extremely rapid diversification of avian lineages shortly after the K-Pg mass extinction. The diversification was so rapid that Suh et al. (2015) found that it could be illustrated more accurately as a branching tree (below). After the mass extinction, numerous ecological niches suddenly became available and, as different populations of birds quickly specialized to fill these niches, they became reproductively isolated, creating the network-like pattern of diversification illustrated below. This initial "superradiation" produced

(continued)

**Box 1.12** (continued)

three core groups (landbirds, waterbirds, and all other taxa) that continued to diverge into the orders of present-day birds.

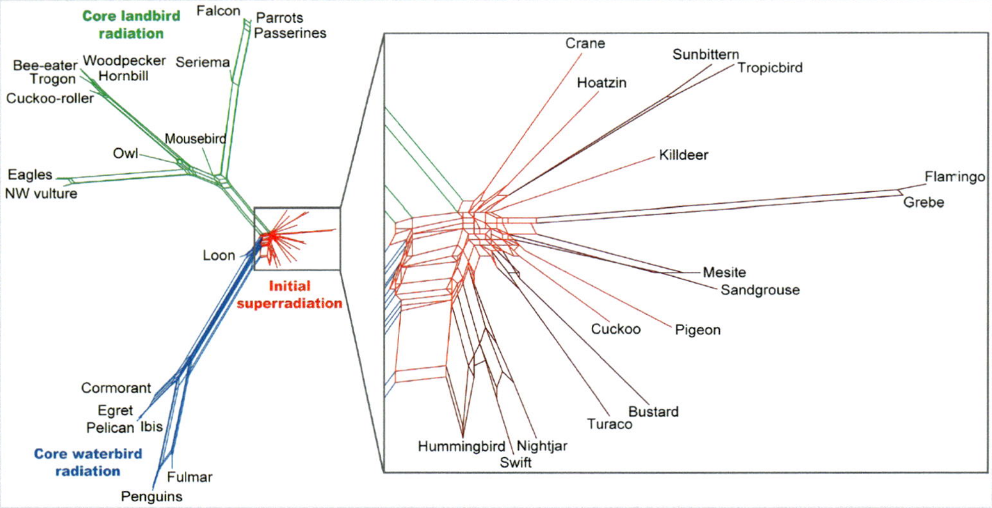


In the early Paleogene, after the K-Pg mass extinction, birds may have exhibited polytomous diversification as different populations specialized and dispersed, filling the many ecological niches that had become available. This "superradiation" gave rise to three core groups of birds, the core landbirds, core waterbirds, and all other lineages. Within a few million years, this radiation generated almost all orders of present-day birds. (Figure from Suh et al. 2015; © 2015 Suh et al., open-access article distributed under the terms of the Creative Commons Attribution License, https://creativecommons.org/licenses/by/4.0/)

New fossils will continue to improve our understanding of the early stages of Neornithine evolution. However, available fossil evidence, along with increasingly detailed phylogenetic analyses, suggests that a few basal lineages of present-day birds diverged before the end of the Cretaceous (i.e., Paleognathae, Galloanseres, and Neoaves), survived the K-Pg mass extinction (Box 1.13 Who Survived the End-Cretaceous Mass Extinction and Why?), and most present-day orders of birds first appeared in the early Paleogene, e.g., orders in the clade Strisores (Box 1.14 Strisores Phylogeny). The rapid (in an evolutionary sense) radiation and diversification of birds (and mammals as well) was likely a response, given the extinction of non-avian dinosaurs, lizards and snakes (Longrich et al. 2012), and archaic birds (i.e., enantiornithines and hesperornithines) among others, to the many vacant ecological niches that became available after the K-Pg mass extinction (Feduccia 2014; Ksepka and Phillips 2015). For example, analysis of forelimb bone proportions (humerus length/ ulna length, with lower ratios generally indicating increasing flight speed and maneuverability) suggests, in particular, that some enantiornithines occupied niche space now occupied by Neornithines (Nudds et al. 2004). Specifically, the lower ratios of some enantiornithines relative to basal Neornithines suggest that they occupied niches requiring fast, maneuverable flight and, with their extinction at the end of the Cretaceous, those niches were now available for Neornithines (Nudds et al. 2004). Interestingly, analysis of forelimb and hindlimb bone proportions suggest the absence of competitive exclusion between birds and pterosaurs during the Mesozoic (Jurassic and Cretaceous) and, further, that after the extinction of pterosaurs, birds did not evolve to occupy the niches formerly occupied by pterosaurs (McGowan and Dyke 2007).

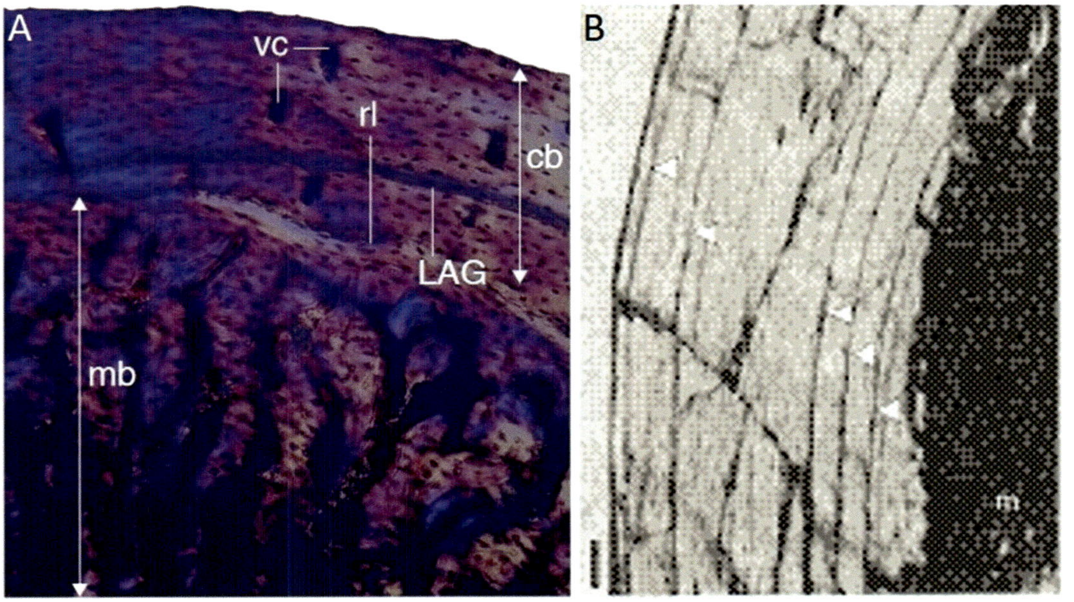


**Fig. 1.69** (A, B) Cross-sections of enantiornithine femora showing lines of arrested growth (LAGs). The lines indicate a period of much slower growth, likely indicating a change in season and, therefore, likely indicating the passing of one year. Examination of growth patterns of enantiornithines of different sizes and ages suggests that, unlike present-day birds that grow very quickly and reach adult size at a young age, enantiornithines exhibited periods of rapid growth followed by periods of much slower growth. However, too few fossils have been available to determine the timing and duration of these periods. (Figure A from O'Connor et al. 2018; open-access article distributed under the terms of the Creative Commons CC BY license; Figure B from Chinsamy and Elzanowski 2001; © 2001 Springer Nature, used with permission)

**Box 1.13 Who Survived the End-Cretaceous Mass Extinction and Why?**

With the exception of a few lineages of birds, dinosaurs did not survive the end-Cretaceous mass extinction event. Given their close phylogenetic relationship and presumably physiological and ecological similarity to birds that did survive (Neornithes), the extinction of small, bird-like theropods, including toothed birds, is particularly difficult to explain. Torres et al. (2021) suggested that one possible factor may have been the sensory systems of Neornithes, including their large brains and eyes. Another possible factor is differences in diet. Whereas most small theropods had teeth and were primarily carnivores, neornithines were toothless and had bills and likely had more diverse diets, including seeds.

As noted by Larson et al. (2016), the impact of the large bolide (meteor) at the end of the Cretaceous, along with increased volcanic activity, would have filled the atmosphere with particulates, blocking sunlight and, soon thereafter, the extinction of organisms, including herbivorous and carnivorous dinosaurs, dependent on photosynthesis. However, seeds, some of which would have likely remained viable for decades, would have still been available, providing an important source of food for noncarnivores like the neornithines. If so, this difference in the diet of neornithines and other dinosaurs may have been an important factor in

(continued)

**Box 1.13** (continued)

the survival of the ancestors of present-day birds through and after the end-Cretaceous mass extinction.

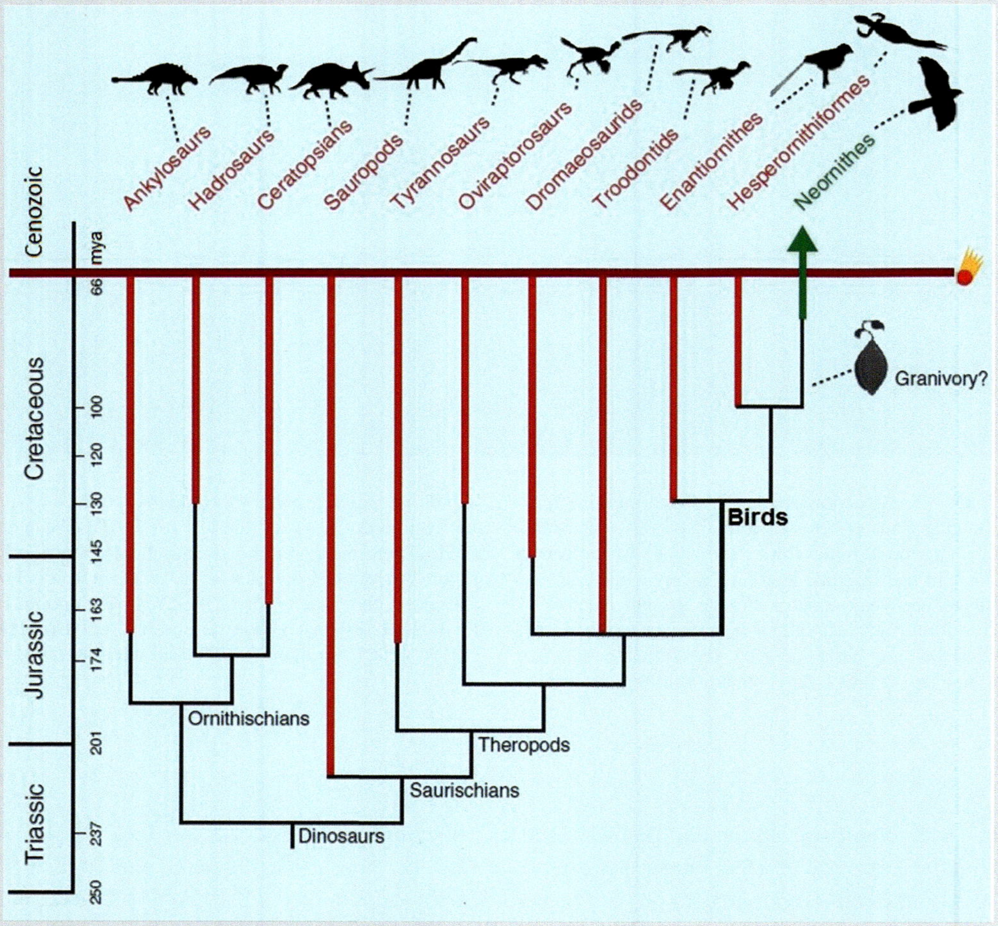


Among the many taxa of dinosaurs, only the neornithines survived the end-Cretaceous mass extinction and their ability to feed on seeds may have been an important contributing factor. (Figure modified from Brusatte 2016; © 2016 Elsevier Ltd., used with permission)

In addition, Field et al. (2018) hypothesized that the Chicxulub impact devastated forest communities, with the impact leveling trees within a radius of about 1500 km and global wildfires destroying most forests elsewhere. A result of this almost complete loss of forest habitat worldwide was the extinction of taxa of birds with arboreal lifestyles, i.e., enantiornithines, and a neornithine avifauna consisting of non-arboreal taxa, including Palaeognathae, Galloanserae, and terrestrial Neoaves. Early in the Paleocene, some of those ground-dwelling neoavians began filling the now newly available niches in the recovering forests, filling the niches vacated by Cretaceous enantiornithines and stem ornithurines.

(continued)

**Box 1.13** (continued)

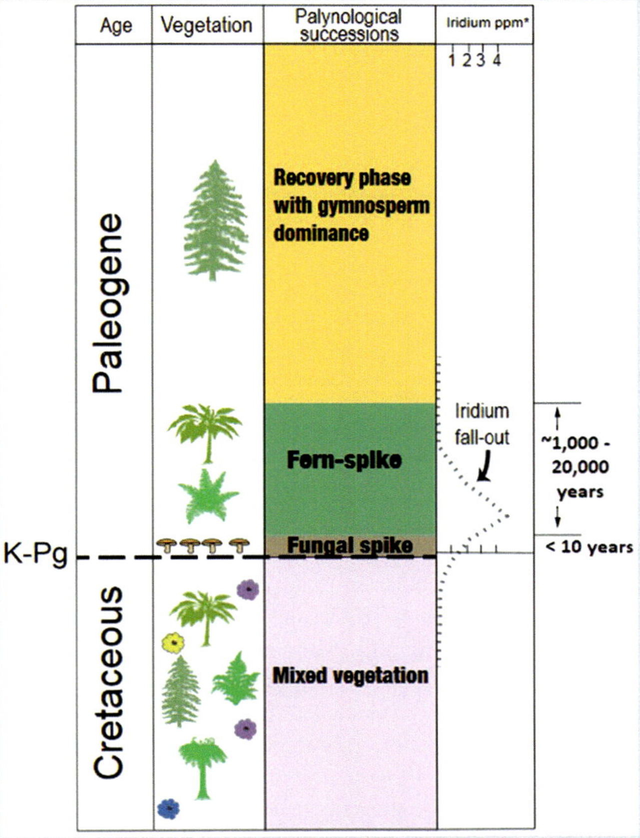


Hypothesized timeline for recovery of plant communities after the K–Pg extinction event. Decaying vegetation after the impact caused a spike in fungal abundance, followed by an increase in the abundance of ferns. Forest communities, initially dominated by gymnosperms, began to recover a few thousand years after the Chicxulub impact. (Supplemental figure from Field et al. 2018; © 2018 Elsevier Ltd., used with permission)

**Box 1.14 Strisores Phylogeny**

Strisores is a clade of birds that includes the order Apodiformes (swifts, treeswifts, and hummingbirds) as well as several orders of largely nocturnal birds, including the families Caprimulgidae (nightjars), Steatornithidae (Oilbird), Nyctibiidae (potoos), Podargidae (frogmouths), and Aegothelidae (owlet-nightjars) (Chen et al. 2019). Relationships among the taxa included in this clade remain unclear, with different possibilities suggested by different authors.

(continued)

**Box 1.14** (continued)

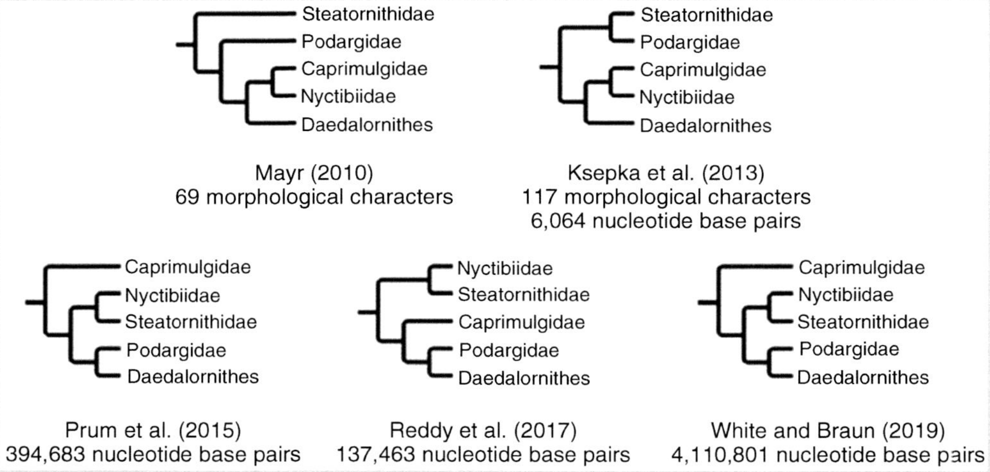


Alternative hypotheses for the phylogeny of taxa in the clade Strisores. (Figure from Chen et al. 2019; © 2019 The Authors. Licensee MDPI, Basel, Switzerland, open-access article distributed under the terms and conditions of the Creative Commons Attribution (CC BY) license, https://creativecommons.org/licenses/by/4.0/)

Chen et al. (2019) combined molecular and morphological data in an attempt to help clarify relationships of taxa in the Strisores clade. Their analysis suggests a rapid divergence in the early Paleocene and that caprimulgids (nightjars) are the sister group (i.e., the closest relatives) of other Strisores. Drawing other conclusions about relationships was found to be challenging because molecular and morphological analysis produced conflicting phylogenies. However, based on the molecular data and molecular plus morphological data combined, Caprimulgidae, Steatornithidae + Nyctibiidae, and Podargiae are more closely related to Daedalornithes (Aegotheliformes and Apodiformes) so they suggested a new name for that clade (Vanescaves). Chen et al. (2019) suggested that developing a phylogeny for Strisores is difficult, perhaps because of the rapid divergences in the early Paleocene during which the different groups "... had already acquired their distinctive morphologies, obscuring their ancestral character states and hampering congruence between morphology- and molecule-derived phylogenies despite their rich fossil record."

(continued)

**Box 1.14** (continued)

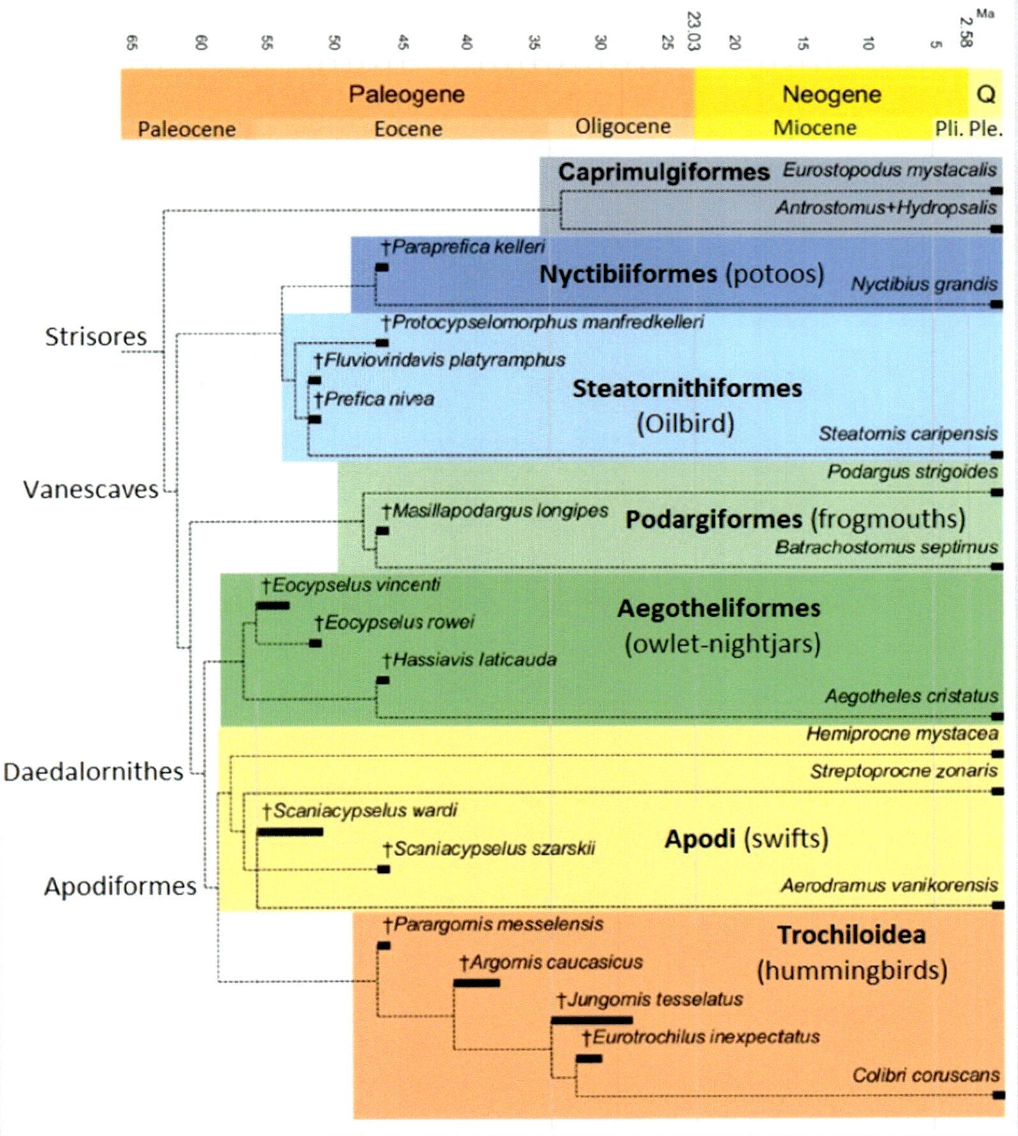


Proposed phylogeny of the clade Strisores. Dagger (†) indicates extinct taxa. Age ranges for fossil taxa are based on Mayr (2009, 2015) and Ksepka and Clarke (2015). Timing of divergence of *Eurostopodus* and other caprimulgids is based on Prum et al. (2015). The timings of other divergences are minimum estimates, although the results of several recent studies suggest a Paleocene origin for Strisores (Jarvis et al. 2014; Prum et al. 2015; Kimball et al. 2019). (Figure from Chen et al. 2019; © 2019 The Authors. Licensee MDPI, Basel, Switzerland, open-access article distributed under the terms and conditions of the Creative Commons Attribution (CC BY) license, https://creativecommons.org/licenses/by/4.0/)

(continued)

**Box 1.14** (continued)

An Australian Owlet-Nightjar (*Aegotheles cristatus*). (Photo by Ron Knight, Wikipedia, CC BY 2.0, https://creativecommons.org/licenses/by/2.0/)

The world at the end of the Cretaceous looked much different than today's, with continents more isolated and fragmented (Fig. 1.79). Climatic conditions throughout the Paleocene (65.5–55 Ma) are believed to have been relatively stable. In general, the Paleocene was thought to be a period of warmer temperatures and reduced latitudinal temperature gradients compared to the present (Fig. 1.80). Throughout much of the Paleogene, there was likely no marked seasonality except at the very highest latitudes, and forests dominated by oaks, laurels, and palms extended well to the north (Blondel and Mourer-Chauviré 1998). At the end of the Paleocene and beginning of the Eocene, global temperatures rose even higher during what is referred to as the Paleocene–Eocene Thermal Maximum (Fig. 1.81).

Several orders of waterbirds diverged during the early Paleogene, including the orders Gaviiformes (loons), Pelecaniformes, Sphenisciformes (penguins), and Procellariiformes (Figs. 1.82 and 1.83). As already noted, the availability of ecological niches that resulted from the extinction of archaic birds like the Hesperornithes (which were also waterbirds) likely contributed to this relatively rapid divergence. However, another factor may have been the Paleocene–Eocene Thermal Maximum. During the Paleocene–Eocene Thermal Maximum, global temperatures increased dramatically, causing increases in both sea-levels and sea-surface productivity (Crouch et al. 2001). This increase in productivity may have opened new niches that contributed to the diversification of waterbirds (Li et al. 2014a).

## 1.15 Diversification After the K-Pg Mass Extinction

Diversification of landbirds also began soon after the K-Pg mass extinction. With the sudden availability of open niches, a favorable climate,

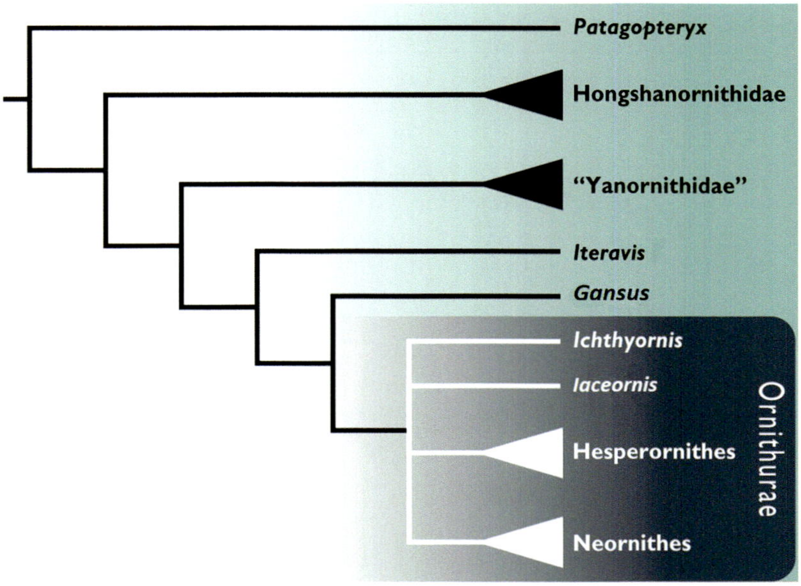


**Fig. 1.70** Ornithuromorpha phylogeny, with Ornithuromorpha defined as the "most inclusive clade containing extant birds but not the Mesozoic Enantiornithes" (Wang et al. 2015). Quotation marks for "Yanornithidae" are to indicate that "... this clade has not been consistently recovered in several recent phylogenetic analyses" and white branches indicate "... phylogenetic uncertainty within the major clade Ornithurae" (Benito et al. 2022b). (Figure modified from Benito et al. 2022a, b; open-access article distributed under the terms of the Creative Commons CC BY license, https://creativecommons.org/licenses/by/4.0/)

continents now separated with the breakup of Gondwanaland, but not completely isolated given their relative proximity (i.e., closer than some continents like South America and Africa are today; Fig. 1.84), conditions were likely ideal for the rapid divergence (i.e., within a few million years) of landbirds, e.g., the clade Columbaves (Fig. 1.85), the order Accipitriformes (Fig. 1.86), and the clade Coraciimorphae (Fig. 1.87). As birds diversified, movements among and within continents and islands likely led to the establishment of populations in different areas and further divergence into different species, genera, and families of birds (Fig. 1.88). For example, the common ancestor of present-day parrots likely lived in Australia, but trans-oceanic dispersal events during the Eocene and Oligocene gave rise to the ancestors of present-day parrots in Asia, IndoMalaya, Africa, Madagascar, South and Central America, Mexico, and the Caribbean (Schweizer et al. 2010; Ericson 2012). Similarly, trans-oceanic dispersal was important in the evolution and diversification of fruit doves (Cibois et al. 2014; Fig. 1.89), cuckoo-shrikes (Fuchs et al. 2007), the *Turdus* thrushes (Voelker et al. 2009), and other taxa of present-day birds.

During the past 65 million years, many new species of birds have evolved, but, of course, most are now extinct. Mayr (2009) reviewed the record on Paleogene birds (66–23 million years before present) and reported fossil evidence of 82 different families of birds, and all species in 48 of those families (58.6%) are extinct. Among the most iconic of the extinct Paleogene avifauna are the Presbyornithids, Pelagornithids (the "bony-toothed birds"), phorusrhacids, and teratorns (Figs. 1.90, 1.91, 1.92, 1.93). Presbyornithids were abundant in parts of what are now North and South America during the early Paleogene and were filter-feeders that

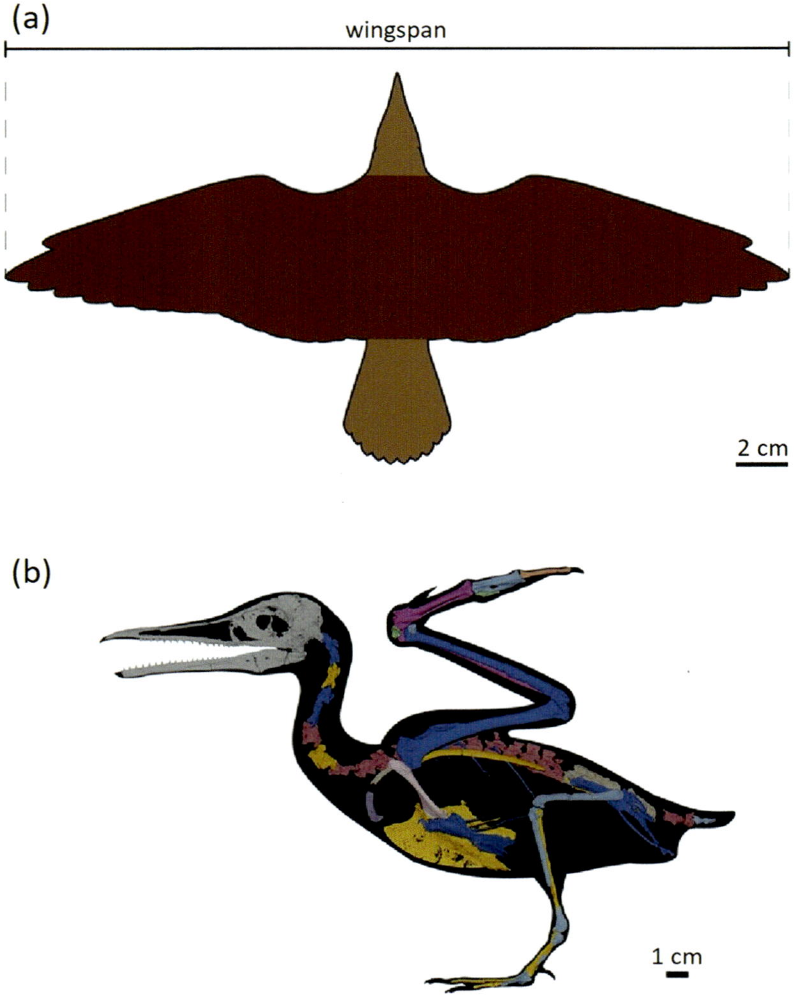


**Fig. 1.71** (a) Outline of the wings and feathered tail of a basal ornithuromorph, *Hongshanornis longicresta*. The similarity of the wings and tail to that of present-day birds suggests that *Hongshanornis longicresta* was capable of continuous flapping flight. (b) Reconstruction of the skeleton of *Ichthyornis dispar*. The temporal distribution of *Ichthyornis* ranges from 95 to 83.5 mya. As described by Porras-Múzquiz et al. (2014), Ichthyornis "... was a flying bird with powerful wings, a large keel in the sternum (breastbone), flexible scapulocoracoid joint with a triosseal canal, robust hindlimbs with well developed tibiotarsus and tarsometatarsus, and a highly kinetic skull; all of which highlight its evolutionary proximity to modern birds." Different colors in the drawing indicate bones recovered from different specimens. (Figure a from Chiappe et al. 2014; © 2014 Chiappe et al.; open-access article distributed under the terms of the Creative Commons Attribution License, https://creativecommons.org/licenses/by/4.0/. Figure b from Benito et al. 2022a, b; © 2022 Benito et al., open-access article distributed under terms of the Creative Commons Attribution License, https://creativecommons.org/licenses/by/4.0/)

probably foraged along the margins of shallow, saline lakes. With duck-like bills and long legs Presbyornithids are closely related to ducks (Anseriformes; Ericson 1997).

The Pelagornithids were unusual marine birds, with their beaks having spiny, bony projections along the edges (Fig. 1.91). This family of birds had a long history, first appearing in the

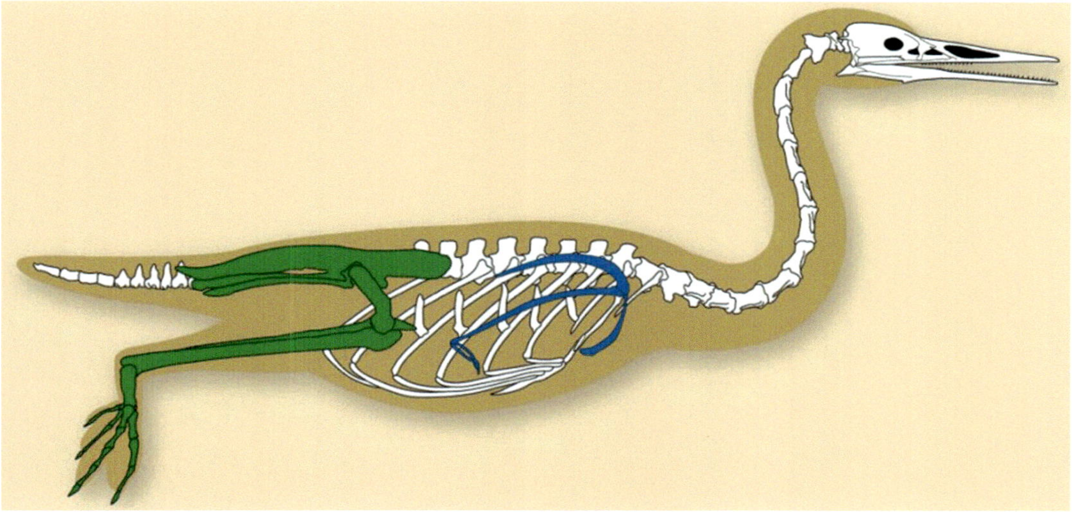

**Fig. 1.72** Typical bauplan of a hesperornithiform bird, including elongate skulls with sharp, toothed bills, small wings (they were flightless), long necks, and powerful hindlimbs for propelling them through the water. Fossils have been found over much of the northern hemisphere. Pectoral girdle is shown in blue; pelvic girdle and hindlimb in green. (Figure from Bell and Chiappe 2022; © 2022 by the authors. Licensee MDPI, Basel, Switzerland, open-access article distributed under the terms and conditions of the Creative Commons Attribution (CC BY) license, https:// creativecommons.org/licenses/ by/ 4.0/)

Paleocene (58 Ma) and not becoming extinct until about 2.6 million years ago. Pelagornithids varied in size, with the largest having wingspans estimated at about 6.4 meters (Fig. 1.91). With a worldwide distribution, Pelagornithids may have filled a niche comparable to that of present-day albatrosses, probably capturing prey-like fish or squid near the water's surface (Olson 1985; Zusi and Warheit 1992).

The phorusrhacids, or terror birds, were predators found primarily in South America and were a very successful group, with fossils of species in this family of birds present beginning in the late Paleocene (~58 million years ago) and continuing until the Pliocene (~5 million years ago). Phorusrhacids ranged in size from about 1.5 kg to about 160 kg (*Phorusrhacos longissimus*), but all had large skulls, raptor-like bills, long legs, and relatively small wings (Fig. 1.92). The largest terror birds were flightless, but the smaller ones may have still been able to fly. Terror birds likely chased and killed prey with their raptor-like bills (Tambussi 2011).

The teratorns (Teratornithidae) were large predators and/or scavengers found in North and South America during the period from the Late Oligocene (~28–23 Ma) to the Pleistocene (~10,000 years ago). To date, species belonging to five genera have been described and they range in size from *Teratornis merriami*, with a wingspan of about 3.5 m and weighing about 14 kg, to the world's large known flying bird, *Argentavis magnificens*, with a wingspan estimated at 7 m and weighing about 70 kg (Chatterjee et al. 2007; Fig. 1.93). *Argentavis* lived about six million years ago in what is now present-day Argentina and adjacent areas. Given their size, *Argentavis* was probably not capable of continuous flapping flight, but, rather, likely relied on lift generated by thermals to soar and search for prey or carrion. Chatterjee et al. (2007) suggested that they were excellent gliders with a cruising speed of 67 kph. *Argentavis* was likely unable to take off on level ground, but, rather, took off by running down slopes or launching from a perch.

Based on a whole-genome analysis, Prum et al. (2015) estimated that passerines

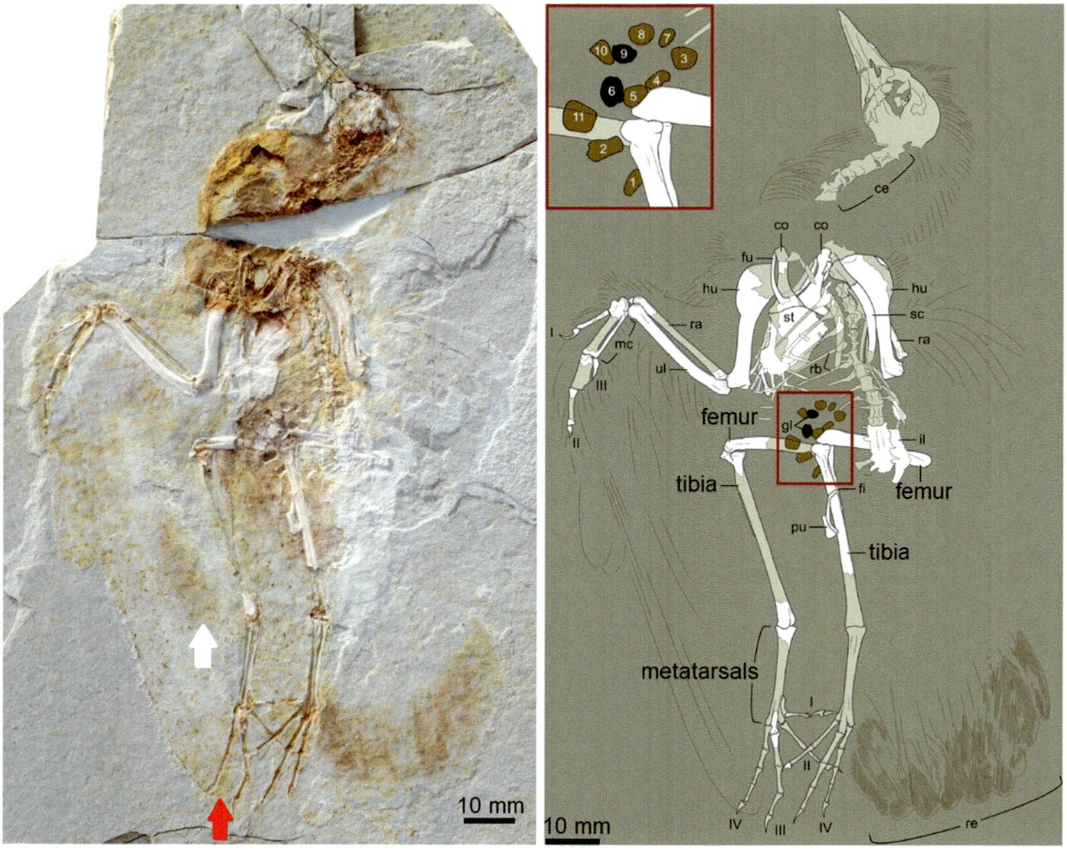


**Fig. 1.73** Fossil (left) and drawing (right) of *Hongshanornis longicresta*, a small ornithuromorph bird with unusually long legs. *Hongshanornis longicresta* was likely a wading bird and the presence of gastroliths (near the femurs and enlarged in inset of the right) suggests they consumed hard food items. The long, tapered wings (red arrow on the left indicates tip of primaries; white arrow indicates tip of secondaries) suggest that *H. longicresta* was able to fly continuously for significant distances. (Figures from Chiappe et al. 2014; © 2014 Chiappe et al., open-access article distributed under the terms of the Creative Commons Attribution License, https://creativecommons.org/licenses/by/4.0/)

(Passeriformes) diverged from their closest relatives, the parrots (Psittaciformes), about 56 million years ago. Basal parrots (Psittaciformes) and passerines (Passeriformes) first appeared in the southern hemisphere during the early Paleogene when the southern continents of South America, Antarctica, and Australia were much closer together than today. Basal parrots may have first appeared in Australasia, which is where the greatest diversity of parrots is found today. Parrots may have reached South America via Antarctica, and Africa and Madagascar via dispersal from Antarctica and Australia (Figs. 1.94 and 1.95).

## 1.16 Passeriformes

The order Passeriformes is by far the most speciose of extant orders, with three suborders, Passeri (Oscines), Tyranni (suboscines), and Acanthisitti (New Zealand wrens), and approximately 5900 species (Jarvis et al. 2014, Gill and Donsker 2015; Figs. 1.96 and 1.97). The oscines

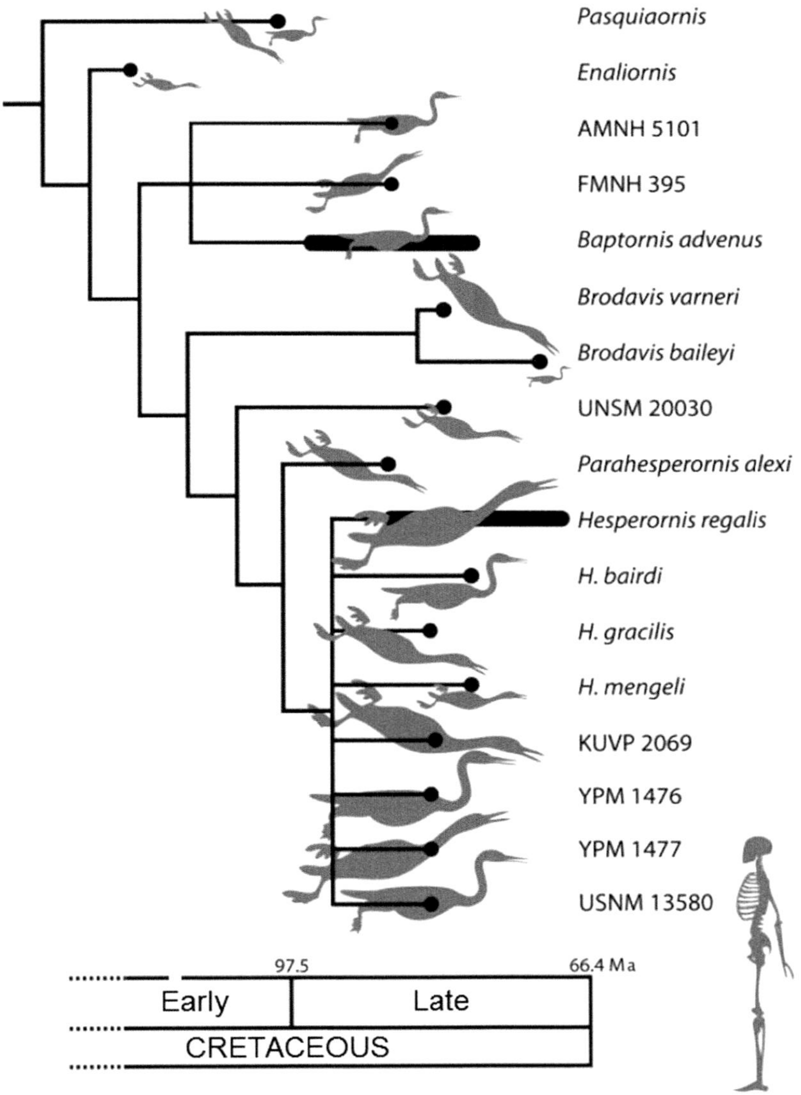


**Fig. 1.74** Phylogeny of the Hesperornithiformes illustrating the evolutionary trend for increasing body size. Human skeleton at lower right is to illustrate the size of different species of Hesperornithiformes. (Figure modified from Bell and Chiappe 2016; Rights managed by Taylor & Francis, used with permission)

likely originated in Australia (Jønsson et al. 2011; Ericson 2012), and the suboscines in Australia or Antarctica (Ericson 2012). The Old World suboscines (Eurylaimides) likely originated in Australia or, more broadly, Australasia, whereas the New World suboscines (Tyrannides) may have originated in Australia or Antarctica and dispersed to South America either from Australia via Antarctica or directly from Antarctica (Ericson 2012, Fjeldså 2013; Fig. 1.98). Available evidence suggests that Antarctic glaciation did not begin until ~33.6 million years ago (Houben et al. 2013; Fig. 1.99), so the ancestors of New World suboscines could have, during that period, either dispersed through or lived in Antarctica before then dispersing to South America. Prum et al. (2015) estimated that oscines and suboscines diverged about 45–50 million years ago.

The Old World Eurylaimides diversified and dispersed throughout Asia, Africa, Madagascar, and Australia, giving rise to several different

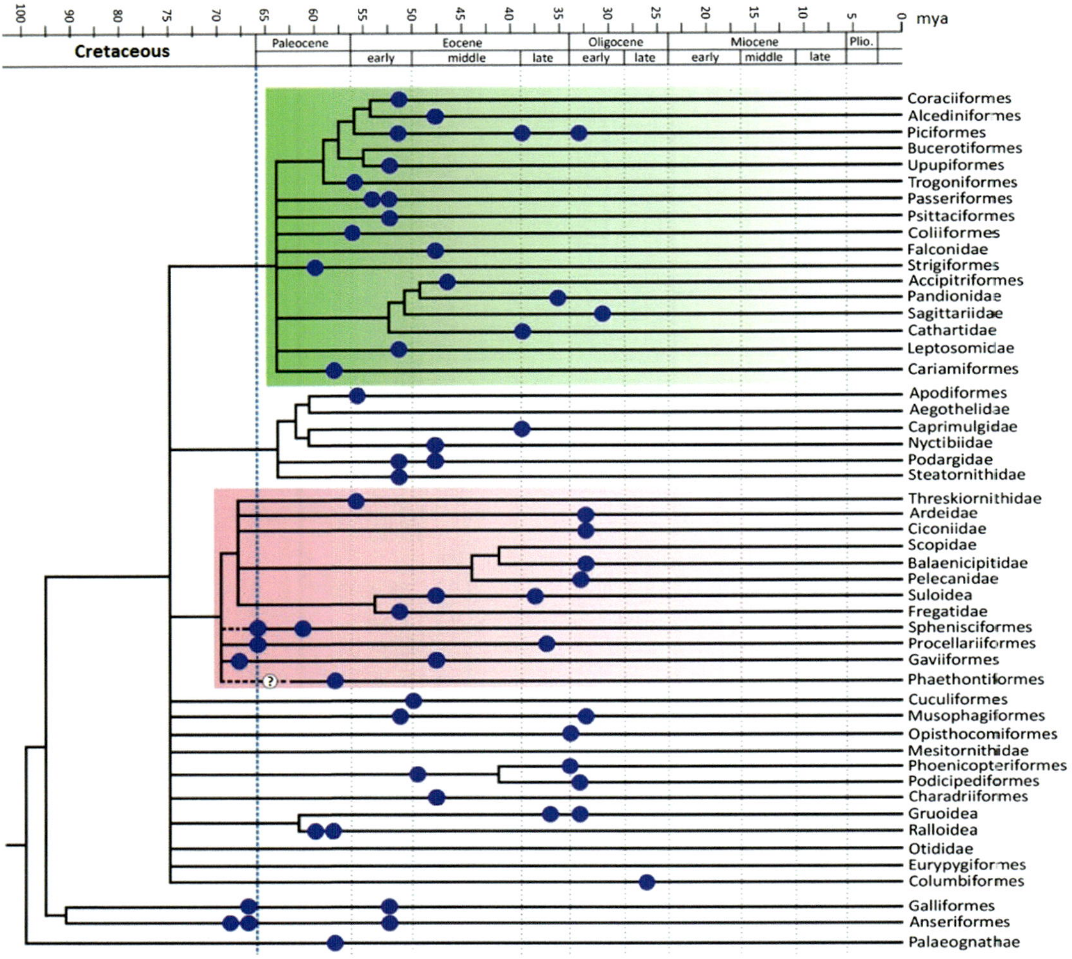


**Fig. 1.75** Neornithine phylogeny and evolution based on available fossil evidence. Blue circles indicate the earliest fossil evidence for each order or family of birds; question marks indicate fossils of uncertain identification. Most fossils date from deposits after the Cretaceous–Paleogene mass extinction (66.02 Ma), whereas most fossils from the Cretaceous have been found in deposits that date from the very late Cretaceous. (Figure modified from Mayr 2014; © The Palaeontological Association, used with permission)

genera and species (Fig. 1.100). Although referred to as the Old World suboscines, one species is actually found in the New World. This enigmatic species, the Broad-billed Sapayoa (*Sapayoa aenigma*), ranges from Panama, Colombia, and south into northern Peru. The ancestor of *Sapayoa aenigma* may have reached the Neotropics via Antarctica (Fjeldsa et al. 2003) or via a northern route across the North Atlantic or the Bering Sea and south through North and Central America (Moyle et al. 2006).

The New World suboscines (Tyrannides) likely reached South America via Antarctica and now, with about 1200 species, make up almost one-quarter of all passerine species in South America (Fig. 1.101). Surprisingly, only

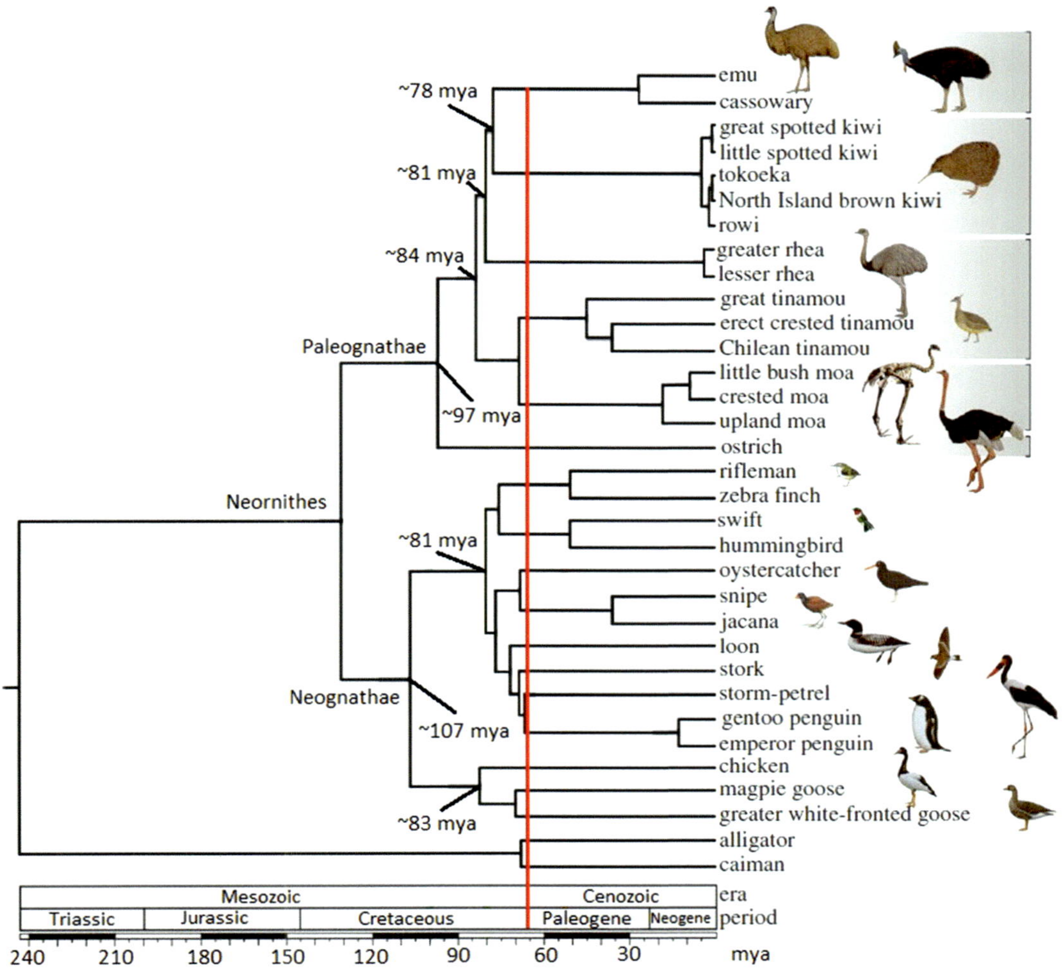


**Fig. 1.76** Timeline of Neornithine evolution based on molecular dating, with the origin of Neornithes estimated at about 131 million years ago. Based on this analysis, extensive diversification occurred before the end of the Cretaceous and most lineages survived the Cretaceous–Paleogene mass extinction that occurred 66 million years ago (red vertical line). (Figure modified from Haddrath and Baker 2012; © 2012 The Royal Society, used with permission)

one family (Tyrannidae) is present in both North America and South America. One factor likely contributing to the lower diversity in Central and North America is, prior to formation of the Isthmus of Panama (3–4 Ma), a water barrier that limited or prevented northward dispersal (Box 1.15 Avian Interchange Across the Panama Land Bridge). In addition, however, most New World suboscines are restricted to rainforest habitat and have relatively short, rounded wings, characteristics that limit their ability to disperse (Kennedy et al. 2014).

**Box 1.15 Avian Interchange Across the Panama Land Bridge**

North and South America were separated until the formation of the Panama Land Bridge. Although debate continues concerning when the land bridge was completed, available information about the timing of the interchange of birds and mammals between South and Central/North America suggests that the land bridge formed approximately 4–6 million years ago. The response of mammals to the formation of the land bridge can be inferred from their fossil record. However, bird bones are relatively thin and often pneumatic so do not fossilize very well. An alternative to the use of fossils is a molecular clock based on the estimated rates of change in mitochondrial DNA. For example, Smith and Klicka (2010) examined the results of 64 avian phylogenetic studies and applied a molecular clock to estimate the timing of trans-isthmus diversification events. Their analyses revealed that diversification events began increasing 5–6 million years ago, with some birds able to disperse across the water gap before the estimated time of completion of the land bridge. In addition, Smith and Klicka (2010) found that the numbers of birds dispersing across the land bridge in each direction were similar. However, most birds that moved north from South America remained in the Neotropics (Central American and southern Mexico). In contrast, many more birds that dispersed into South America from North and Central America became more broadly established. This asymmetry is consistent with the "time for speciation" effect (Stephens and Wiens 2003). Most lineages that originated in South America may not have had sufficient time to adapt to the temperate conditions of the Nearctic. In contrast, many "northern birds" have likely been able to invade the tropics because of their long exposure to tropical conditions in Central America well before the formation of the Panama Land Bridge (Smith and Klicka 2010).

(continued)

**Box 1.15** (continued)

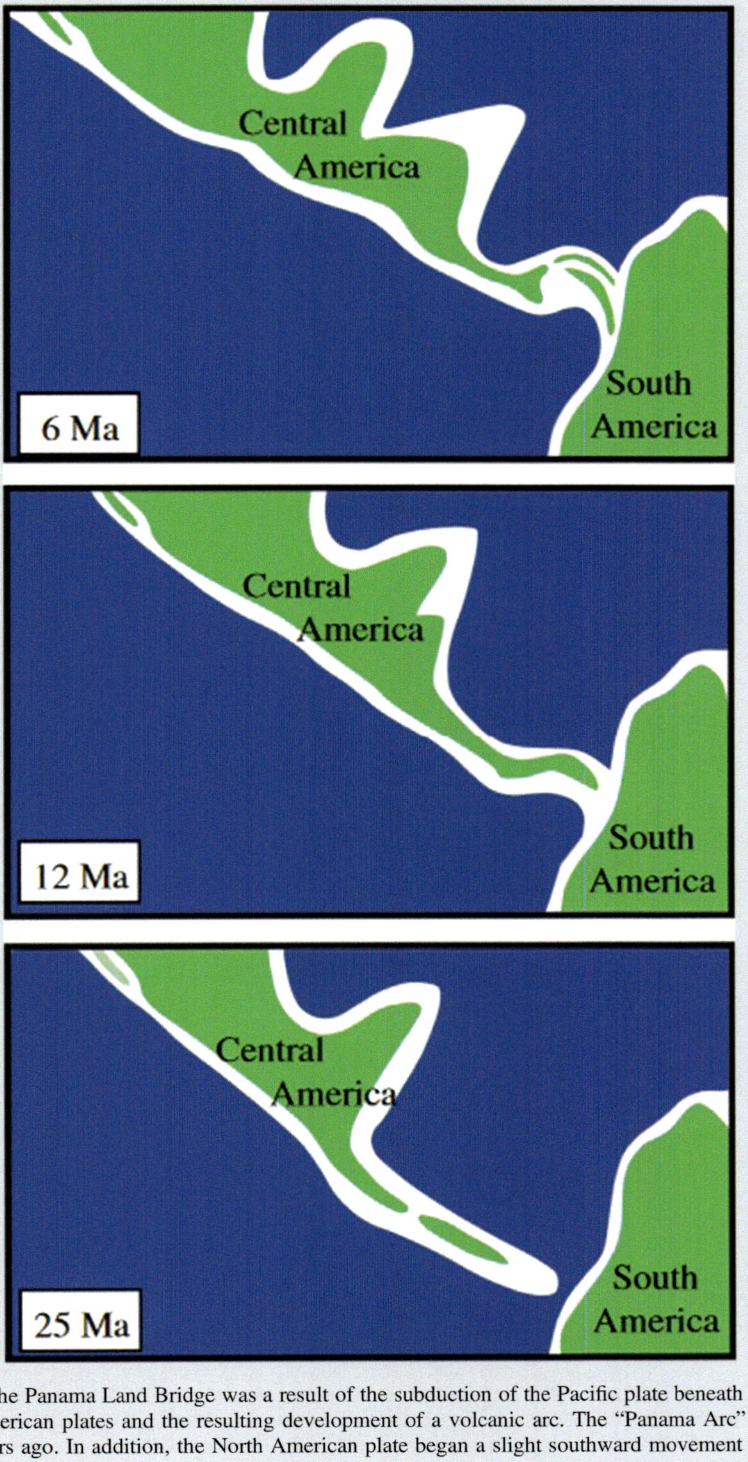


Formation of the Panama Land Bridge was a result of the subduction of the Pacific plate beneath the Caribbean and South American plates and the resulting development of a volcanic arc. The "Panama Arc" formed about 73 million years ago. In addition, the North American plate began a slight southward movement and the South

(continued)

**Box 1.15** (continued)

American plate a northward movement. The South American plate finally collided with the Panama Arc, causing uplift of the arc and, in conjunction with declining sea levels, formation of the Isthmus of Panama, i.e., the Panama Land Bridge. The timing of this formation continues to be debated, with estimates ranging from 30 to three million years ago. Data based on the interchange of birds between South and Central/North America suggest a more recent formation about 4–6 million years ago. (Figure from De Baets et al. 2016; )

Breeding distributions of several families of birds (inferred from molecular data) whose ancestors originated in North and South America. Many native North American species now breed in South America; many fewer native South American species now breed in North America. Of 16 families, identified as having South American origins, most invaded tropical Central America, but only two, Trochilidae and Tyrannidae, were able to move further north to colonize North America. (Table from Smith and Klicka 2010; © The Authors, used with permission)

| | Number of species | |
|---|---|---|
| North American origin | Neotropical | Nearctic |
| Trogonidae[a] | 25 | 0 |
| Momotidae | 9 | 0 |
| Mimidae | 34 | 8 |
| Vireonidae | 38 | 13 |
| Corvidae[a] | 18 | 17 |
| Polioptilidae | 11 | 4 |
| Troglodytidae | 66 | 9 |
| Thraupidae[b] | 402 | 0 |
| Parulidae | 65 | 49 |
| Icteridae | 76 | 20 |
| Cardinalidae | 33 | 15 |
| Emberizidae[a] | 81 | 35 |
| South American origin | | |
| Psittacidae[a] | 150 | 0 |
| Trochillidae | 331 | 18 |
| Ramphastidae | 35 | 0 |
| Semnornithidae | 2 | 0 |
| Capitonidae | 11 | 0 |
| Bucconidae | 32 | 0 |
| Galbulidae | 17 | 0 |
| Furnariidae | 235 | 0 |
| Thamnophilidae | 209 | 0 |
| Tityridae | 31 | 0 |
| Tyrannidae | 400 | 27 |
| Pipridae | 51 | 0 |
| Melanopareidae | 4 | 0 |
| Rhinocryptidae | 44 | 0 |
| Conopophagidae | 10 | 0 |
| Grallaridae | 49 | 0 |

[a]Old World members of family are excluded from this total
[b]Thraupidae is secondarily South American

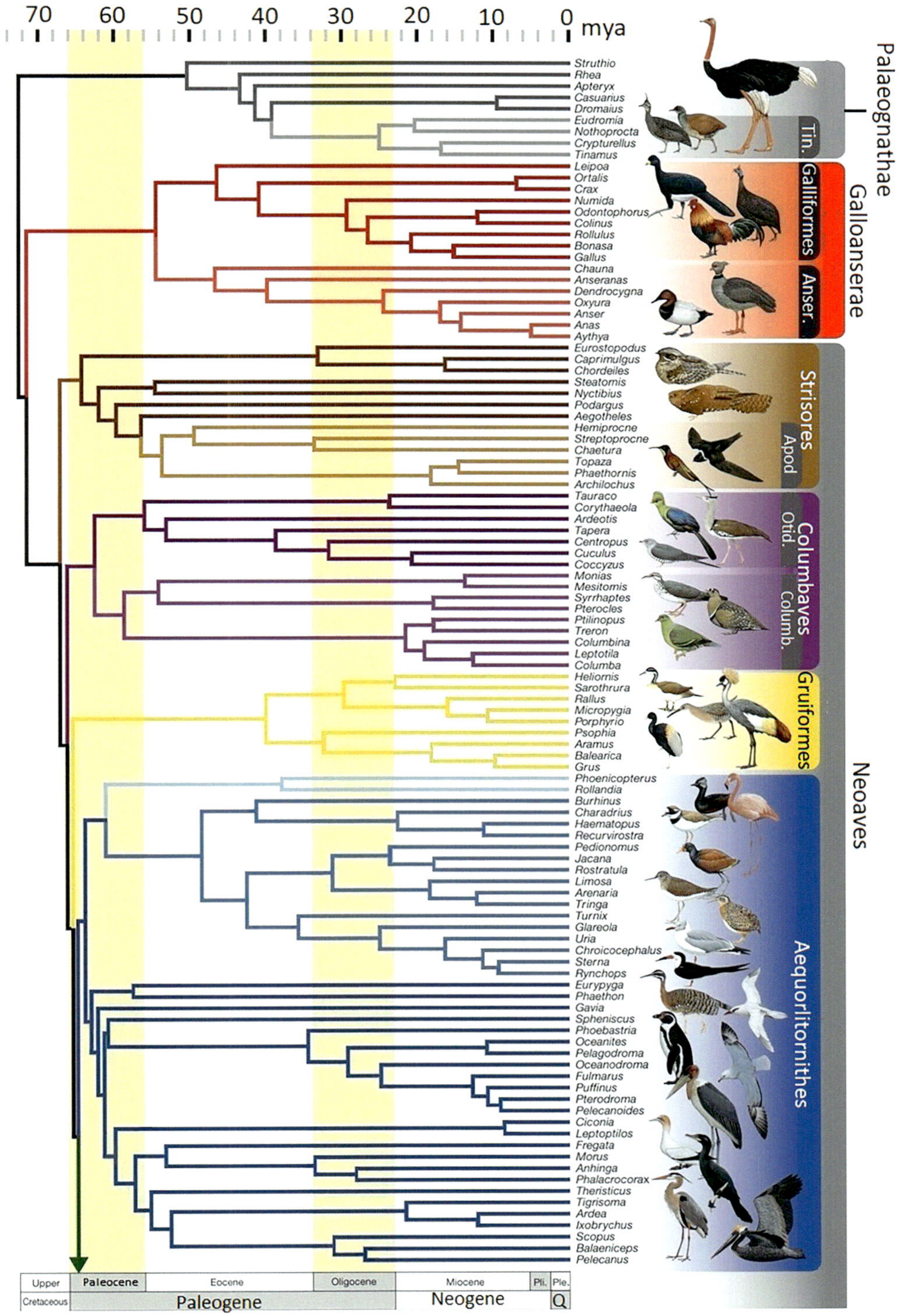


**Fig. 1.77** Proposed phylogeny of birds based on analysis of whole genomes of 198 species. The major Neoavian sister clades include Strisores (brown; nightjars, other caprimulgiforms, swifts, and hummingbirds), Columbaves (purple; a clade uniting cuckoos, bustards, and turacos with pigeons, mesites, and sandgrouse), Gruiformes (yellow; cranes and their relatives), Aequorlitornithes (blue; a waterbird clade that includes all diving, wading, and

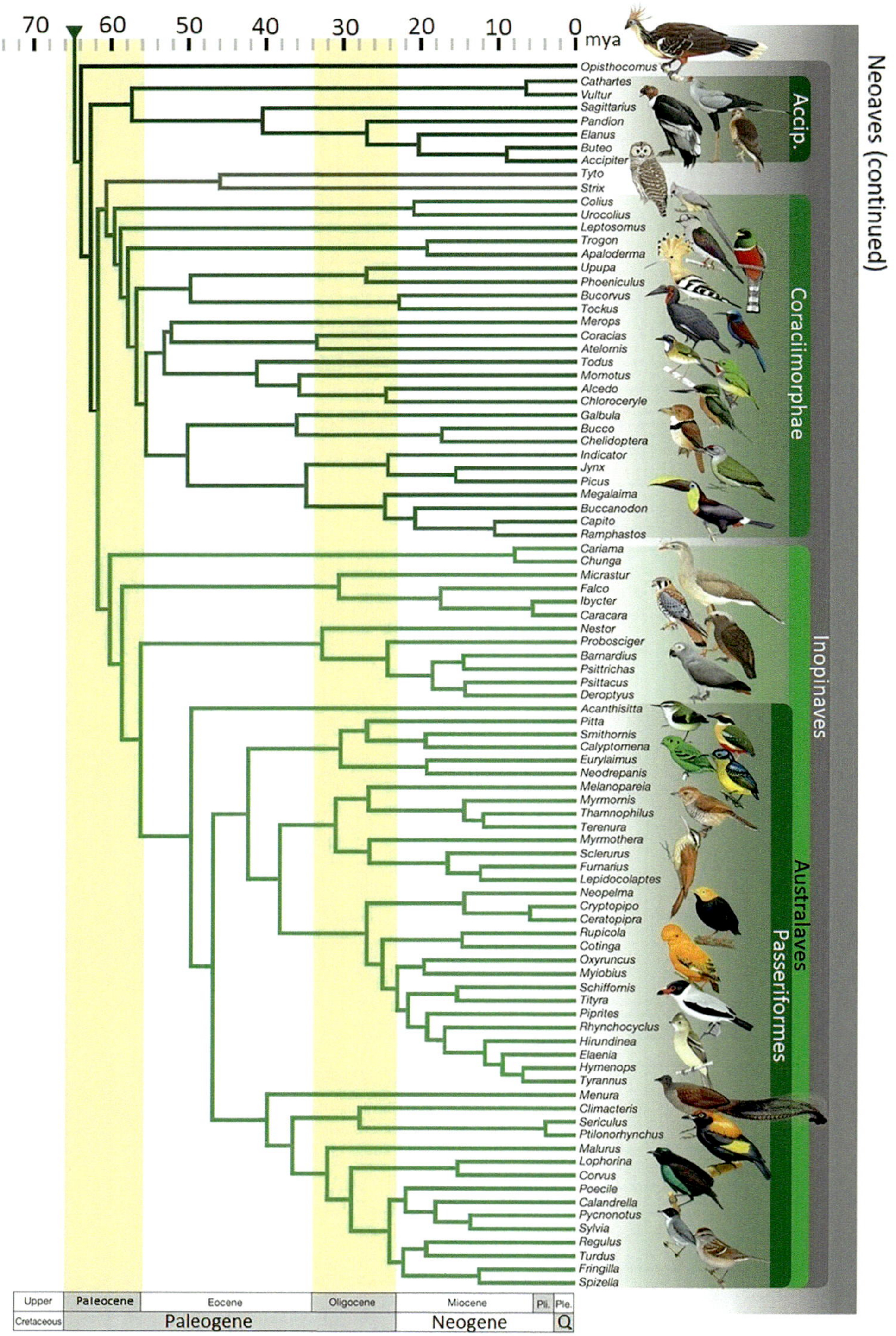


**Fig. 1.77** (continued) shorebirds), and Inopinaves (green; a landbird clade). Mya, million years ago; Tin, Tinamidae; Anser, Anseriformes; Apod, Apodiformes; Otid, Otidimorphae; Columb, Columbimorphae; Accip, Accipitriformes; Ple, Pleistocene; Pli, Pliocene; Q., Quaternary. (Figure from Prum et al. 2015; © 2015 Springer Nature, used with permission)

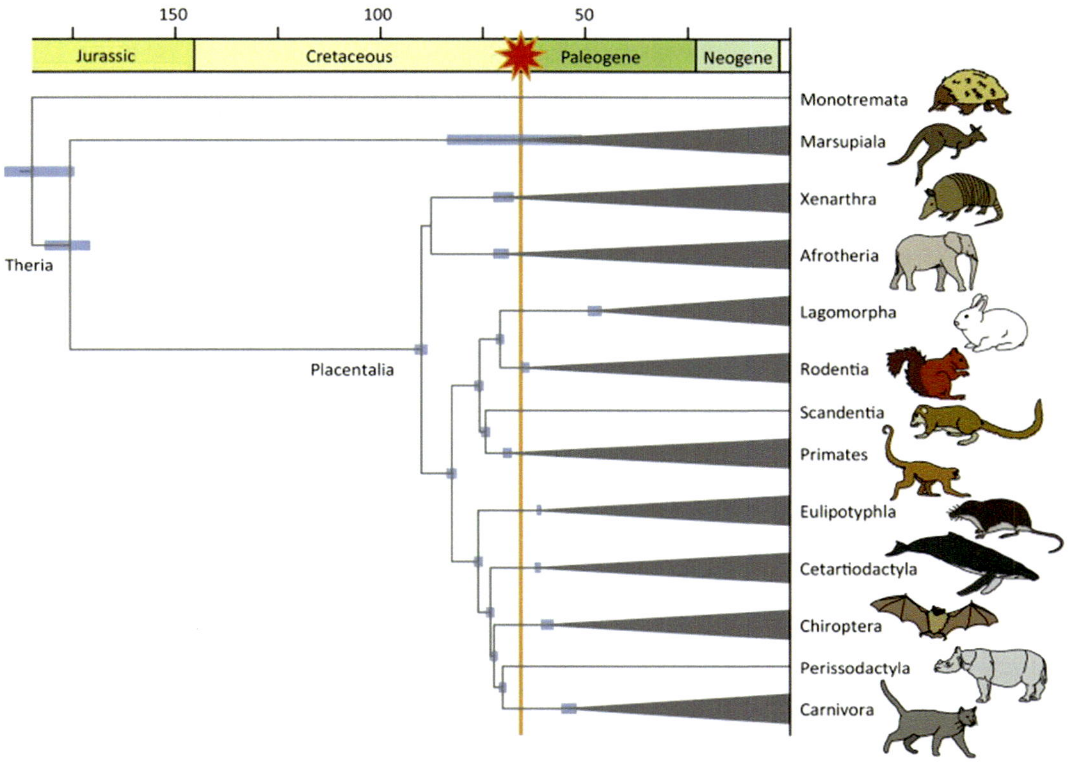


**Fig. 1.78** Timescale illustrating the diversification in mammals, based on a molecular clock analysis (dos Reis et al. 2012). Light-blue bars at nodes represent 95% credibility intervals of divergence-time estimates. The orange vertical line indicates the timing of the Cretaceous–Paleogene boundary. Most orders of placental mammals diversified in the Paleogene, but the basal divergences in Placentalia occurred during the Late Cretaceous. (Figure from Ho 2014; © 2014 Elsevier Ltd., used with permission)

Nearly half of all extant species of birds are oscine passerines or songbirds (about 4800 species). Songbirds originated in Australia about 25–30 million years ago and subsequently diversified in the Australo-Papuan region before lineages then dispersed into Asia, Europe, and North, Central, and South America (Barker et al. 2004; Jønsson et al. 2011). The ranges of present-day descendants of basal and transitional songbirds are still limited to Australia, New Guinea, and other areas in southeast Asia (Fig. 1.102). Songbirds are divided into two major groups (or parvorders), the more numerous Passerida (more than 3500 species) and the Corvida (about 740 species; Fig. 1.103). Both groups appear to have originated in the Australo-Papuan region, but they differ in their subsequent patterns of dispersal and diversification. Basal Passerida are thought to have initially dispersed into either Eurasia (Barker et al. 2004) or Africa (Jonsson and Fjeldsa 2006) and, in support of those hypotheses, most basal groups of Passerida are currently found in Africa and Asia (Fig. 1.104). However, with subsequent dispersal and diversification, all four superfamilies of Passerida now have worldwide distributions. Basal Corvida diversified in the Australo-Papuan region, with the presence of numerous islands including New Guinea (i.e., the transition zone between Australia and southeast Asia; Fig. 1.105), providing opportunities for dispersal and diversification. Additional Corvida lineages

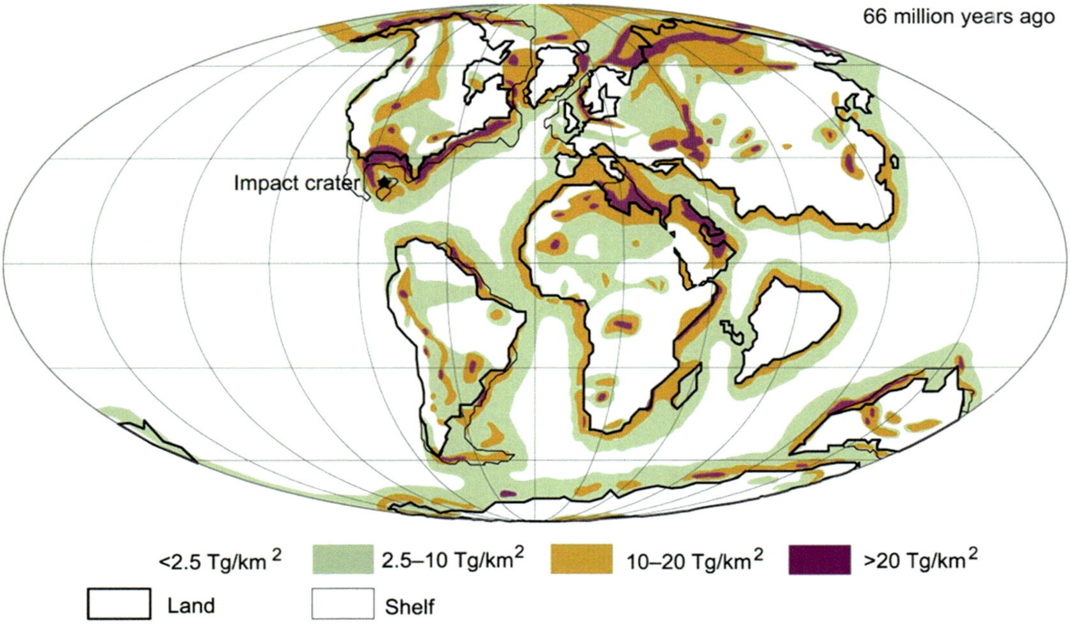


**Fig. 1.79** Location of the continents, islands, and the Chicxulub impact crater at the end of the Mesozoic (Cretaceous) and beginning of the Paleogene (Paleocene). Also shown is the amount of hydrocarbon-rich organic matter in sedimentary rocks at different locations (1 Tg = 1 trillion grams). Note that the impact location had large quantities of organic matter, likely resulting in the production of large amounts of stratospheric soot that contributed to global cooling and the mass extinction event. (Figure from Kaiho and Oshima 2017, open-access article licensed under a Creative Commons Attribution 4.0 International License, https://creativecommons.org/licenses/by/4.0/)

subsequently dispersed to and diversified in Asia, Africa, and the New World (Jønsson et al. 2011, Aggerbeck et al. 2014; Fig. 1.106).

## 1.17 Present-Day Birds

Birds have now existed for about 155 million years and, of course, given that species have variable, but limited, lifespans, most species that have existed during those millions of years are now extinct. Most present-day genera of birds originated during the Neogene period (1.8–23.8 Ma) and most extant species first appeared during the Plio-Pleistocene (0.08–5.3 Ma) (James 2005). For example, hummingbirds (Trochilidae) split from their sister group (swifts, Apodidae) about 50–55 million years ago (Prum et al. 2015) and the most recent common ancestor of all extant hummingbirds (Trochilidae) lived about 20–24 million years ago (McGuire et al. 2014; Fig. 1.107). Hummingbirds are thought to have first evolved in Eurasia, and subsequently dispersed into South America sometime between 20 and 40 million years ago. Present-day hummingbirds originated in South America, and the ranges of many taxa are still limited to South America. Some species, however, can also be found in Central and North America plus many Caribbean islands (McGuire et al. 2014). The results of molecular analyses suggest that most present-day species of hummingbirds (~340 species) originated in the past five million years (Fig. 1.107). Other taxa of birds exhibit a similar trend, with most species evolving within the past 5–10 million years. Examples include suboscine flycatchers

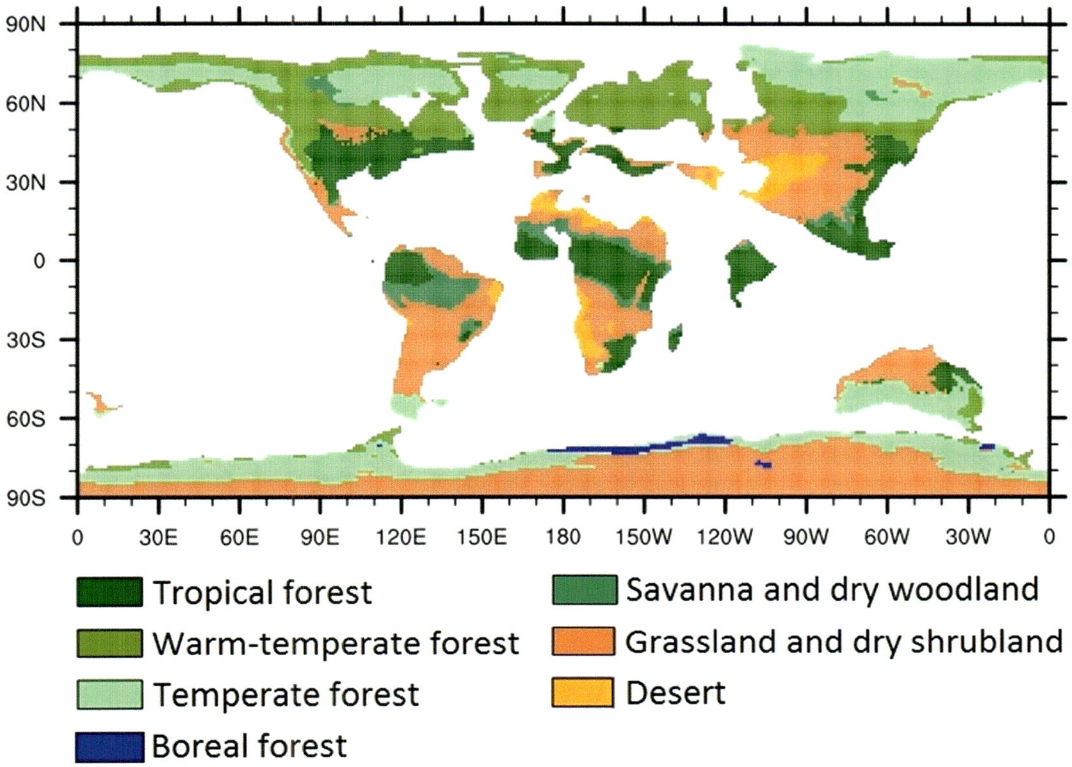


**Fig. 1.80** During the Eocene (56–50 Ma), warm temperate forests extended to much higher latitudes than today. (Figure from Herold et al. 2014; © Authors 2014, open-access article distributed under the Creative Commons Attribution 3.0 License, https://creativecommons.org/licenses/by/3.0/)

(Fig. 1.108) and songbirds in the parvorder Corvoidea (Fig. 1.109).

The evolution of birds continues to the present day. Species diversification rates have been found to vary, with mean species values ranging from 0.01 to 4.66 species per million years (Jetz et al. 2012). Species diversification rates are higher in the western hemisphere than the eastern hemisphere, but there is no difference between rates in the northern and southern hemispheres (Fig. 1.110). Diversification rates are particularly low in Australia, Southeast Asia, Africa, and Madagascar, and particularly high at higher latitude and in western North America, northeastern Asia, parts of South America and the Middle East, and several islands north and northeast of Australia (Fig. 1.110). A variety of factors can influence diversification rates (Fig. 1.111). For example, the low rates in Australia and Africa may in part be due to the early filling of ecological space by basal lineages that originated in those areas (Ericson 2012). In contrast, characteristics of areas with higher rates of diversification can include (1) high net primary productivity that increases the number of available niches (Evans et al. 2005), (2) isolation, e.g., islands can facilitate diversification by reducing gene flow and providing new niches (Grant and Grant 2011), (3) environmental harshness, e.g., populations may diverge faster because extreme weather events can cause local extinctions and range discontinuities (Botero et al. 2014), (4) presence of climate-induced refugia, e.g., glaciation can create refugia both directly (habitat refugia) and indirectly (due to changes in water levels during glacial and interglacial period) (Hosner et al.

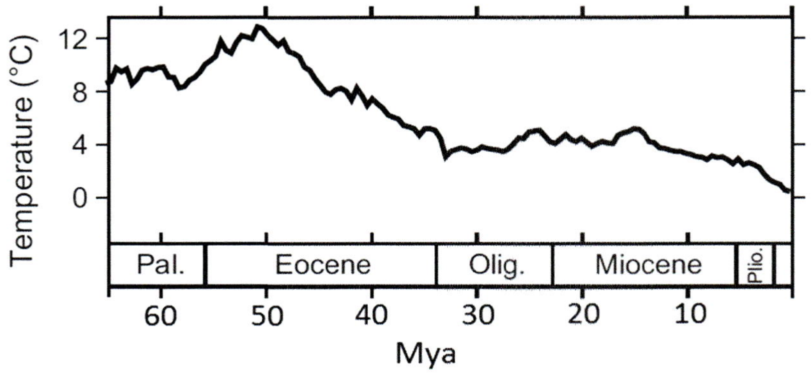


**Fig. 1.81** Estimated variation in global surface temperatures during the Cenozoic. Temperature is relative to global surface temperature prior to the Industrial Revolution when carbon dioxide levels began to increase. The increase in temperatures that began in the late Paleocene (Pal.) and continued into the early Eocene is referred to as Paleocene–Eocene Thermal Maximum. (Figure modified from Royer et al. 2012; © 2012 Blackwell Publishing Ltd., used with permission)

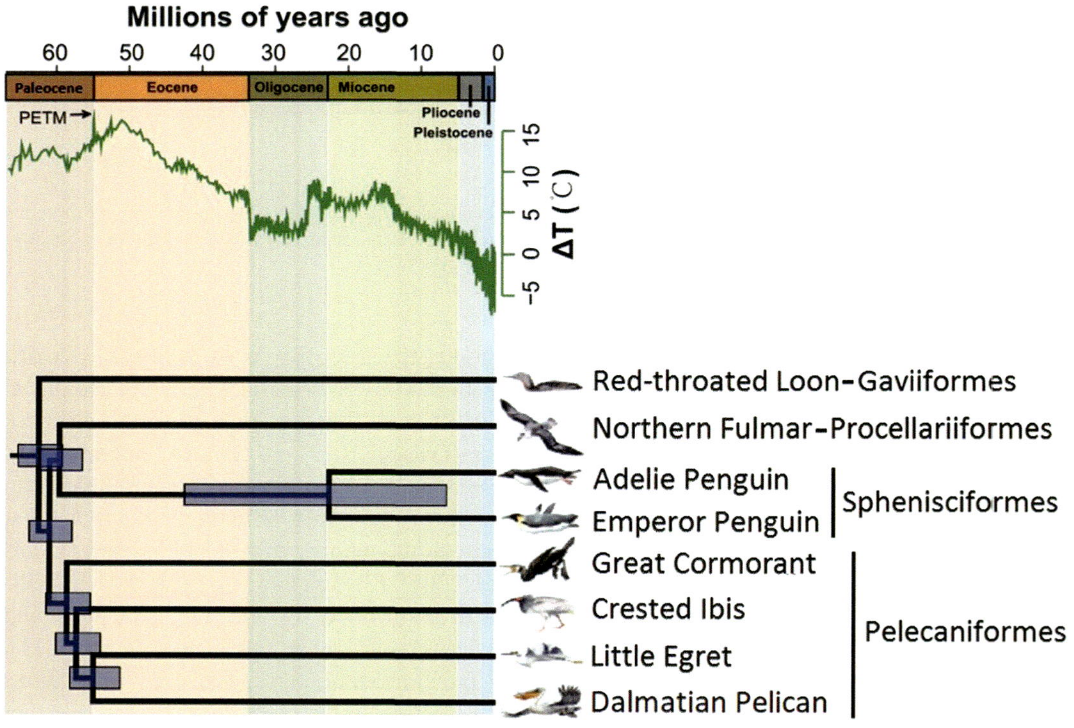


**Fig. 1.82** Phylogenetic relationships and timing of diversification of four orders of waterbirds. Horizontal bars at each node represent 95% credibility intervals of estimated divergence times. These four orders of waterbirds all diverged within an estimated 12–15 million years near the beginning of the Paleogene. At top are the timescale and changes in estimated global temperatures over the past 65 million years (relative to the global temperature prior to the industrial revolution). PETM, Paleocene–Eocene Thermal Maximum. Red-throated Loon, *Gavia stellata*; Northern Fulmar, *Fulmarus glacialis*; Adelie Penguin, *Pygoscelis adeliae*; Emperor Penguin, *Aptenodytes forsteri*; Great Cormorant, *Phalacrocorax carbo*; Crested Ibis, *Nipponia nippon*; Little Egret, *Egretta garzetta*; Dalmatian Pelican, *Pelecanus crispus*. (Figure modified from Li et al. 2014a; open-access article distributed under the terms of the Creative Commons CC BY license, https://creativecommons.org/licenses/by/4.0/)

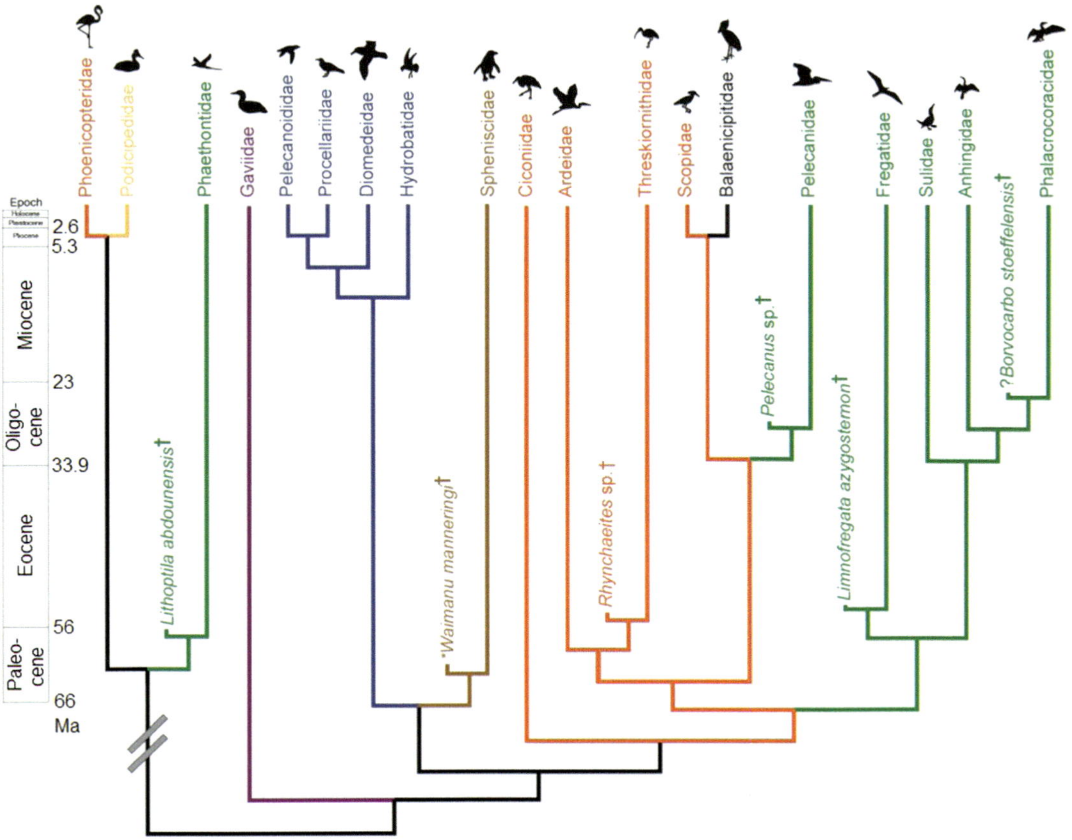


**Fig. 1.83** Possible phylogenetic tree of several taxa of waterbirds based on a phylogenomic analysis by Hackett et al. (2008) showing family-level relationships (with silhouettes of representatives of each family). Note the rapid divergence during the period prior to and after the end of the Cretaceous (66 Ma). Groups are color-coded by order: Pelecaniformes (green), Ciconiiformes (orange), Procellariiformes (blue), Sphenisciformes (brown), Podicipediformes (gold), Gaviiformes (purple). The double hash marks indicate that the clade containing Phaethon, Podiceps, and Phoenicopterus is actually recovered as distantly related to the waterbird clade (i.e., it is not its sister-taxon). (Figure from Smith and Ksepka 2015; open-access article licensed under a Creative Commons Attribution 4.0 International License, https://creativecommons.org/licenses/by/4.0/)

2014), and (5) geological events such as tectonic and volcanic events that create mountains or mountain ranges (Sweet and Johnson 2015; Fig. 1.111). Of course, low rates of diversification can sometimes be explained by the absence of factors that favor diversification.

Today, birds are the most species-rich of all terrestrial vertebrates (Fig. 1.112). More than 10,000 species of present-day birds occupy every continent (Box 1.16 Zoogeographical Realms) and roam every ocean (Box 1.17 Seabirds). They can be found flying above the highest mountains and diving deep into the oceans. Not surprisingly, given the evolutionary history of different taxa of birds plus variation in factors that influence the availability of niches and diversification rates, the planet-wide distribution of present-day species of birds varies (Fig. 1.113). However, with the exception of areas at the very highest latitudes, birds can be found almost everywhere. With a history extending back 150 million years to the

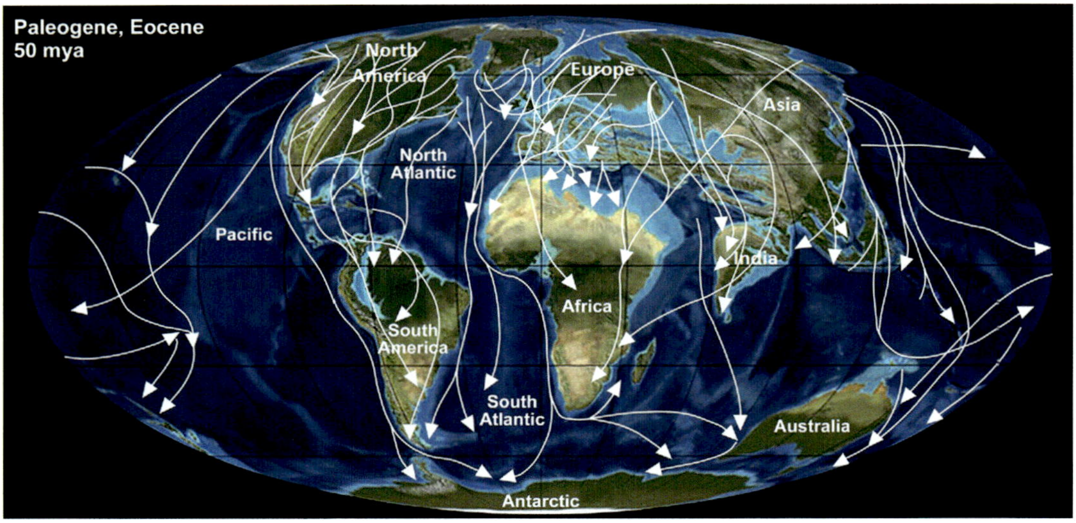


**Fig. 1.84** Relative location of the continents and larger islands during the Eocene (50 Ma) showing proximity of the continents. Lines and arrows indicate some migration routes of present-day birds to illustrate the likely ability of birds during the early Paleogene to move among continents. (Figure from Feduccia 2014; © 2014 Elsevier Ltd., used with permission)

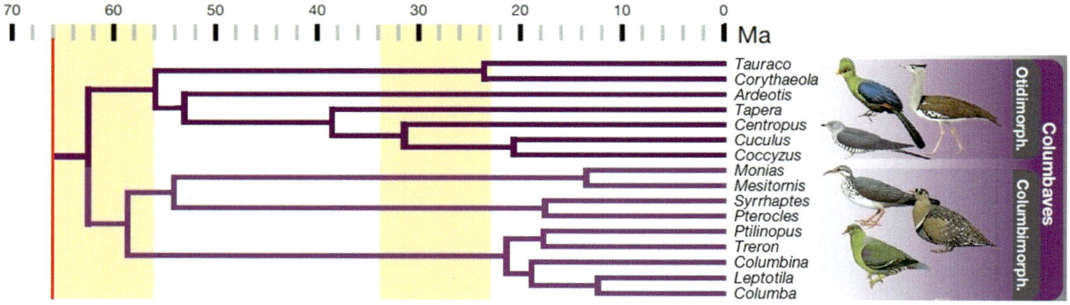


**Fig. 1.85** Proposed phylogeny of the clade Columbaves. This clade includes the clades Columbimorphae (pigeons, mesites, and sandgrouse) and Otidimorphae (bustards, cuckoos, and turacos). Note that the clades Columbimorphae and Otidimorphae likely diverged within a few million years after the end of the Cretaceous (indicated by the vertical red line at 66 Ma). (Figure from Prum et al. 2015; © 2015 Springer Nature, used with permission)

mid-Jurassic, birds have managed to survive mass extinctions, the movements of continents, ice ages, and more. They coexisted with dinosaurs and, indeed, the fossil record tells us that birds actually are dinosaurs! The reasons for the success of birds, including their impressively long evolutionary history and their current worldwide distribution and exceptional diversity, are many and varied and, in the chapters that follow, will be discussed in much more detail.

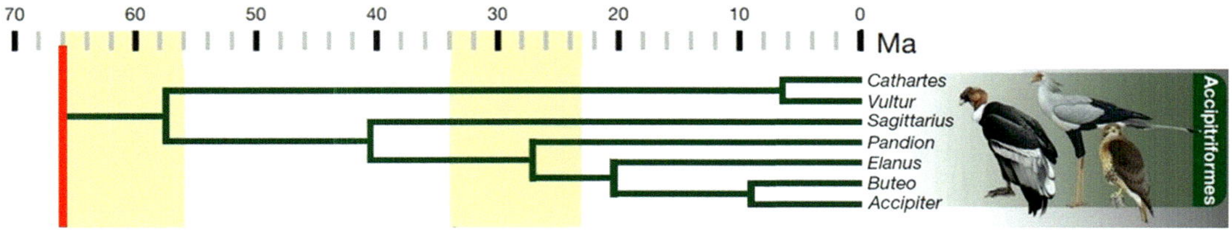


**Fig. 1.86** Proposed phylogeny of the order Accipitriformes. Age estimates for vultures (genera *Cathartes* and *Vultur*; order Cathartiformes) range from about 69 million years old (Johnson et al. 2016) to about 60 million years old (Jarvis et al. 2014; Prum et al. 2015). The split between Sagittariidae (genus *Sagittarius*; secretarybird) and the ancestor of present-day members of the families Pandionidae (Osprey, *Pandion haliaetus*) and Accipitridae (genera *Elanus*, *Buteo*, and *Accipiter*) occurred about 40 million years ago. Vertical red line at left indicates the end of the Cretaceous. (Figure from Prum et al. 2015; © 2015 Springer Nature, used with permission)

**Box 1.16 Zoogeographical Realms**

Over 10,000 species of birds occupy the eight zoogeographical realms: Nearctic, Neotropical, Palearctic, Afrotropical, Indomalayan, Australasian, Oceanian, and Antarctic. Each realm has their characteristic birds: avifaunas that represent a mixture of species of various ages and origins. The eight zoogeographical realms, delimited based on charactistic assemblages of plants and animals and separated to varying degrees by potential barriers to dispersal such as oceans and mountain ranges, also differ considerably in the number of breeding species present.

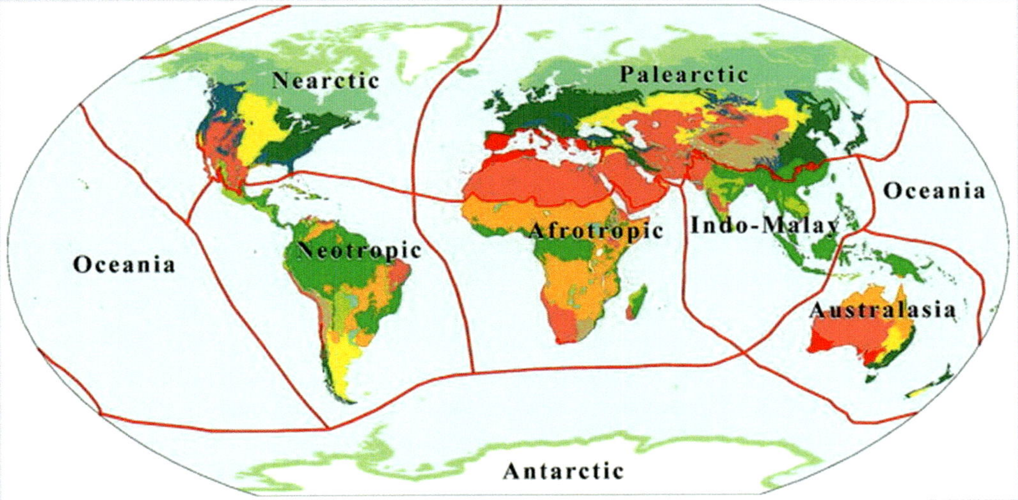


The eight zoogeographical realms. (Figure from Olson et al. 2001; © 2001 Oxford University Press, used with permission).

Variation among zoogeographical realms in (a) the number of different species, genera, families, and orders of birds, and (b) the number of endemic species, genera, families, and orders of birds. (c) Number of landbird species occurring in different numbers of zoogeographical realms. The numbers in parentheses are percentages of the total numbers of landbird species,

(continued)

**Box 1.16** (continued)

genera, families, and orders. (Tables from Newton and Dale 2001; © 2001 The Zoological Society of London, used with permission).

(a) Number of breeding landbirds

| | Palearctic | Indomalayan | Afrotropical | Australasian | Nearctic | Neotropical | Oceania | Overall |
|---|---|---|---|---|---|---|---|---|
| Species | 937 (10) | 1697 (18) | 1950 (21) | 1592 (17) | 732 (8) | 3370 (36) | 187 (2) | 9416 |
| Genera | 288 (14) | 431 (22) | 473 (24) | 457 (23) | 302 (15) | 893 (45) | 82 (4) | 2002 |
| Families | 58 (41) | 73 (52) | 75 (54) | 73 (52) | 52 (37) | 71 (51) | 23 (16) | 140 |
| Orders | 14 (61) | 17 (74) | 19 (83) | 16 (70) | 15 (65) | 18 (78) | 10 (43) | 23 |

(b) Number of endemic breeding landbirds

| | Palearctic | Indomalayan | Afrotropical | Australasian | Nearctic | Neotropical | Oceania | |
|---|---|---|---|---|---|---|---|---|
| Species | 442 (47) | 1184 (70) | 1807 (93) | 1415 (89) | 395 (54) | 3121 (93) | 163 (87) | |
| Genera | 26 (9) | 126 (29) | 293 (62) | 280 (61) | 58 (19) | 686 (77) | 31 (38) | |
| Families | 0 (0) | 3 (4) | 16 (21) | 18 (25) | 0 (0) | 20 (28) | 0 (0) | |
| Orders | 0 (0) | 0 (0) | 2 (11) | 0 (0) | 0 (0) | 2 (11) | 0 (0) | |

The largest realm, the Palearctic (about 46 million $km^2$), has relatively low species richness, probably due to its generally cold climate and the presence of large areas of homogeneous habitat (taiga). About 937 species of birds regularly breed in the Palearctic, and most are passerines (Finlayson 2011). In addition, most species are migratory. The accentor family (Prunellidae) is the only endemic family of birds in the Palearctic. The predominant groups among passerines in terms of numbers of species are the superfamilies Sylvioidea (old world warblers and allies; 127 species) and Passeroidea (sparrows and allies; 116 species) (Finlayson 2011).

The Nearctic realm includes over 4 billion hectares of boreal forest, the deciduous forests of the eastern United States and southeastern Canada, the grasslands of the midwestern United States and Canada, and the deserts of the southwestern United States and northern Mexico. This realm exhibits low species richness, with about 732 species of breeding birds. Predominant families of birds include Parulidae (New World warblers) and Emberizidae. Factors contributing to the low species richness are that (1) birds entering from Siberia must pass through areas of tundra and cold coniferous forest (with no access at all during periods of glaciation), (2) tropical-adapted species from the south are faced with an adverse climatic gradient, and (3) North American deciduous forests exhibit low species richness because they formed a continuous block (restricting opportunities for speciation) and have been compressed during glacial maxima (Mayr 1946).

The Neotropical realm has the most bird diversity of any realm with about 3370 breeding landbirds, 686 endemic genera, 20 endemic families, and 2 endemic orders (Newton and Dale 2001). The large number of endemics in the Neotropical realm suggests an ancient and distinctive evolutionary history (Mayr 1964). Some endemic families have undergone extensive evolutionary radiation and contain many species, including hummingbirds (Trochilidae; 341 species), ovenbirds (Furnariidae; 242 species), and antbirds, antshrikes, antwrens, and antvireos (Thamnophilidae; 208 species). Avian species richness in the Neotropics seems to be linked directly to habitat diversity that, in turn, is correlated with topographic heterogeneity (Rahbek and Graves 2001).

(continued)

**Box 1.16** (continued)

Ochre-rumped Antbird (*Drymophila ochropyga*), a species in the family Thamnophilidae that is endemic to the Neotropical realm. (Photo by C. H. L N. de Almeida, Wikipedia, CC BY 3.0, https://creativecommons.org/licenses/by/3.0/)

The Afrotropical realm is home to about 1950 species of birds. The Afrotropical realm has strong affinities with the Indomalayan realm at the generic level (about 30% of Afrotropical genera also occur in the Indomalayan realm), but not as much at the species level (about 2%). This suggests a long period of independent evolution in the two areas (Moreau 1952). The Afrotropical realm also serves as a wintering area for many Palearctic species. Over 70 families of land and fresh-water birds breed in the Afrotropics. Fifty-eight percent of Africa's resident birds are oscine songbirds (Fjeldså and Bowie 2008), and molecular analyses suggest that songbirds arrived in Africa in the Miocene (40 Ma). Madagascar is part of the Afrotropical realm, and the avifauna of Madagascar consists of 258 species (with 209 of breeding in Madagascar). Compared to land masses of similar size, Madagascar is rather species poor (Reddy et al. 2012), but nearly half of the species are endemic (109 species) and found nowhere else (McDowall 2008).

(continued)

**Box 1.16** (continued)

Examples of bird species endemic to Madagascar. (**a**) Velvet Asity, *Philepitta castanea*; (**b**) Madagascar Paradise Flycatcher, *Terpsiphone mutata*; (**c**) Pitta-like Ground Roller, *Atelornis pittoides*; (**d**) Madagascar Red Fody, *Foudia madagascariensis*; (**e**) Lesser Vasa Parrot, *Coracopsis nigra*; (**f**) Chabert's Vanga, *Leptopterus chabert*. (Photos by Ricardo Rocha and Figure from Rocha et al. 2015; © Ricardo Rocha, Tarmo Virtanen and Mar Cabeza, open-access article distributed under the Creative Commons Attribution 4.0 license, https://creativecommons.org/licenses/by/4.0/)

The Indomalayan realm is home to nearly 1700 species of breeding landbirds. The avifauna of the Indomalayan realm has closest affinities to the Palearctic, Australasian, and Afrotropical realms. The Himalayas represent a boundary between the Indomalayan and Palearctic regions, and water separates the Indomalayan and Australasian regions. Species richness is highest at the base of the Indochina Peninsula, particularly where Myanmar borders India and China. In this area, the Himalaya Mountains descend into tropical lowlands and topographic variation is high (Ding et al. 2006). Characteristic birds of the Indomalayan realm include many pheasants and pigeons, owls, woodpeckers, pittas, laughingthrushes, hoopoes, barbets, honeyguides, and sunbirds. The Indomalayan realm also includes numerous islands with many endemic species (Newton 2003).

(continued)

**Box 1.16** (continued)

Red-tailed Laughingthrush (*Trochalopteron milnei*). (Photo by Tom Friedel, Wikipedia, CC BY 3.0, https://creativecommons.org/licenses/by/3.0/deed.en)

The Australasian realm has a long history of isolation from other land masses, and Australia has been isolated from other landmasses for more than 55 million years. About 1600 species of birds are found in the Australasian realm. Many unique groups are found in the Australasian realm, including cassowaries and Emus (Casuariidae), kiwis (Apterygidae), megapodes (Megapodiidae), owlet-nightjars (Aegothelidae), scrubbirds (Atrichornithidae), and birds of paradise (Paradisaeidae) (BirdLife Australia 2018).

The continent of Antarctica encompasses about 14.3 million km$^2$, with most covered by glacial ice. During the Antarctic summer, about 13,700 km$^2$ of rock and soil is exposed, particularly in the Antarctic Peninsula (Lee et al. 2017). Much of the ice-free area is along the edges of the continent, but other ice-free areas include mountain tops, cliffs, and valleys (Lee et al. 2017). These ice-free areas support breeding populations of 20 species of birds, including five species of penguins (Harris et al. 2015).

The many oceanic islands (> 100,000) make up about 3.3% of the Earth's land area (Whittaker and Fernández-Palacios 2007). Oceanic islands vary in their geologic history, area, degree of isolation, elevation, and climatic conditions (Depraetere and Dahl 2007). As a result, species richness and composition of birds differ greatly among and between islands and mainland areas. Islands worldwide support fewer species of land and freshwater birds than equal areas of comparative mainland, but many of those species are endemics.

**Box 1.17 Seabirds**

Some birds spend much or most of their time at sea, with those in several species coming to land only to breed. Referred to as seabirds, these species are members of the orders Sphenisciformes (penguins), Pelecaniformes (pelicans), Suliformes (frigatebirds, boobies, and cormorants), Phaethontiformes (tropicbirds), Procellariiformes (shearwaters, petrels, prions, albatrosses, and storm petrels), and Charadriiformes (gulls, terns, puffins, guillemots, murres, and skuas). Many seabirds typically forage near shorelines, including cormorants, gulls, terns, and pelicans, whereas others range farther out toward the edges of continental shelves, including some storm petrels, shearwaters, boobies, penguins, alcids, larger gulls, and cormorants. Pelagic species that forage in the open ocean include some penguins, albatrosses, prions, petrels, and some boobies, shearwaters, and storm petrels (Karpouzi et al. 2007).

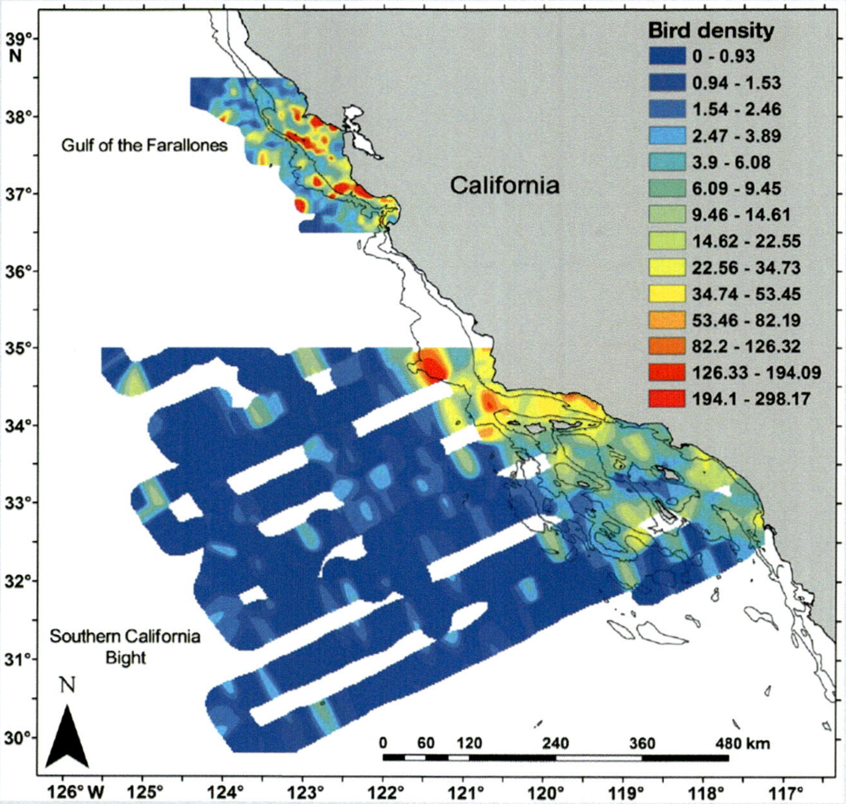


Based on at-sea surveys, seabird densities off the coast of California were found to be highest within about 150 km of the coast, the area where many species typically forage. (Figure from Suryan et al. 2012; © 2012 Inter-Research, used with permission)

Given that the oceans apparently present no barriers to movement and dispersal, seabirds might be expected to be rather homogeneous throughout the world's oceans. That, however, is not the case, with greater species richness and densities of seabirds in the southern hemisphere than the northern hemisphere. Areas with the highest seabird densities in the northern hemisphere include the area between Greenland and Europe and the waters around Alaska.

(continued)

**Box 1.17** (continued)

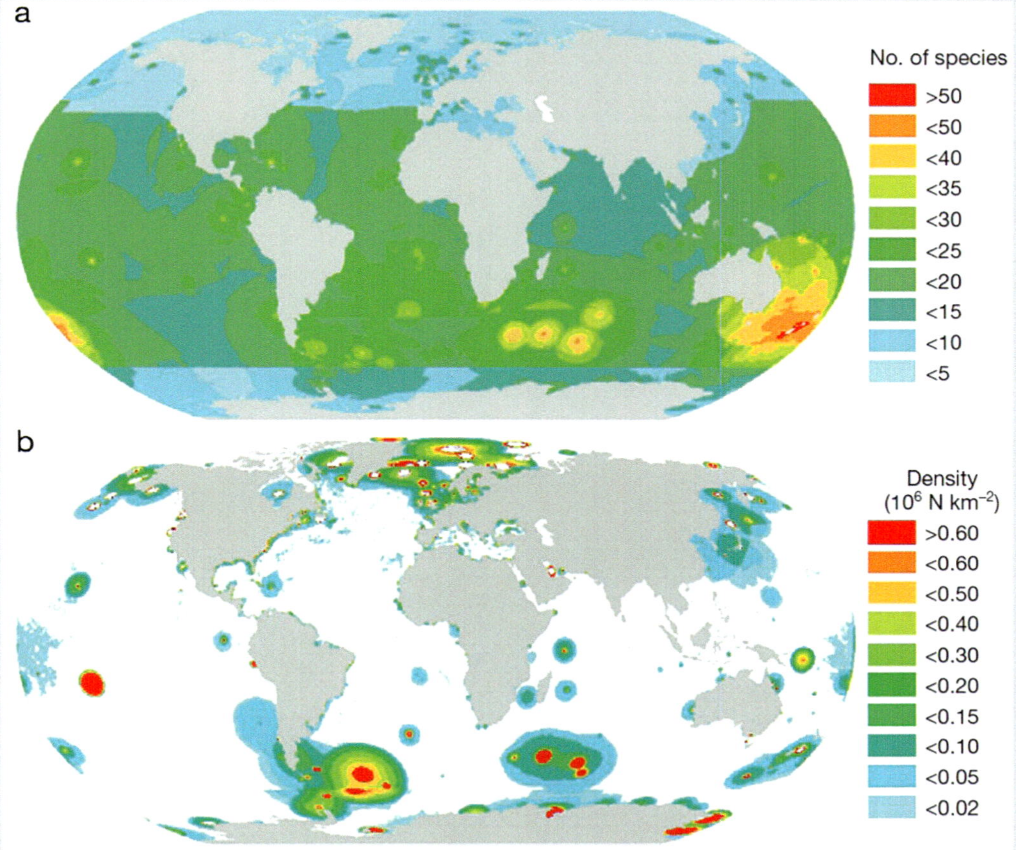


Estimated foraging distribution of seabirds indicated as (**a**) number of species per 0.5 × 0.5 degree areas, and (**b**) number of individuals per $km^2$. (Figure from Karpouzi et al. 2007; © 2007 Inter-Research, used with permission)

(continued)

**Box 1.17** (continued)

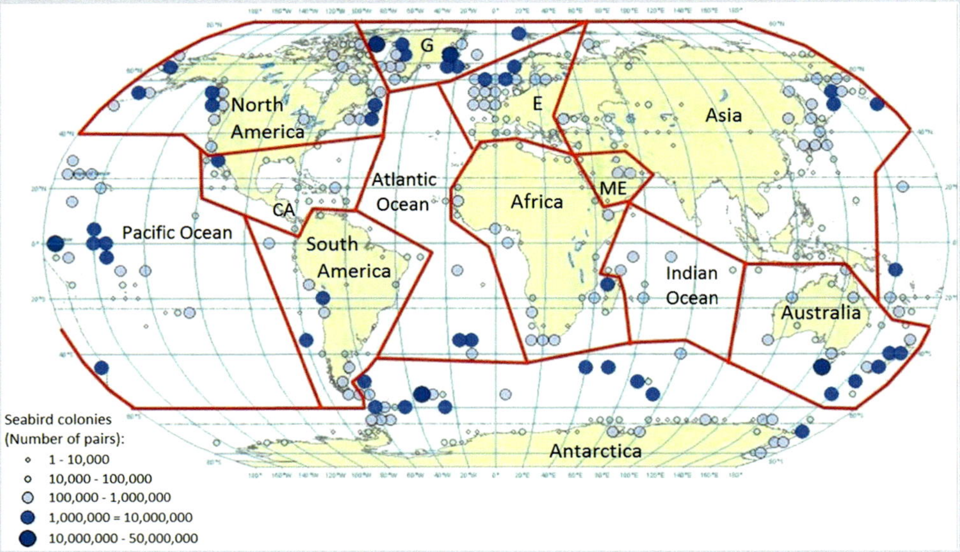


Global distribution of seabird colonies of different sizes. Red lines delineate regional boundaries. CA, Central America; G, Greenland; E, Europe; ME, Middle East. (Figure modified from Riddick et al. 2012; © 2012 Elsevier Ltd., used with permission)

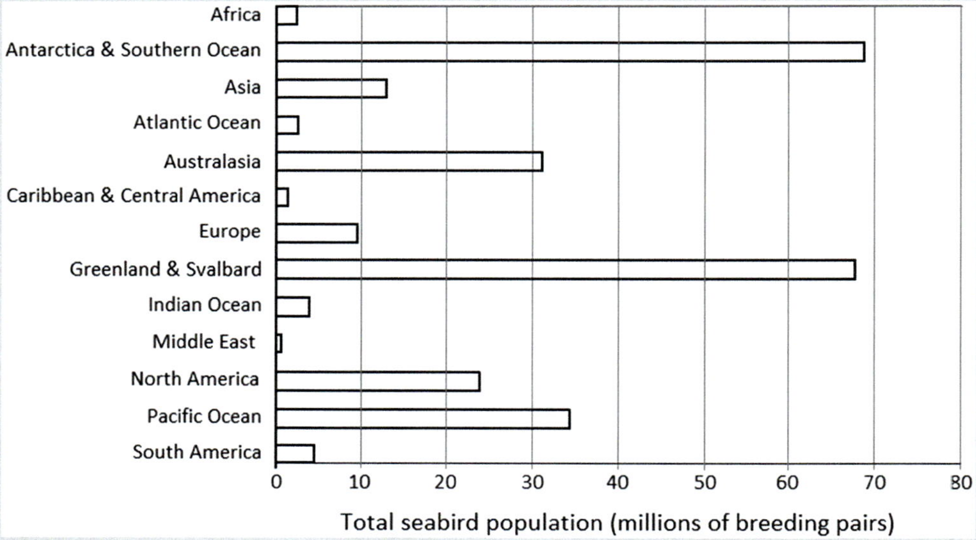


Estimated breeding population of seabirds in different regions of the world. (Figure modified from Riddick et al. 2012; © 2012 Elsevier Ltd., used with permission)

Several factors can influence the distribution of seabirds, including topographic features such as seamounts, sea surface temperature, and the locations of fronts (boundaries between water masses moving in different directions), eddies, and upwellings. What all of these things share in common is that they can influence productivity by bringing cooler water and nutrients to the

(continued)

**Box 1.17** (continued)

ocean's surface. Seamounts direct water and nutrients toward the surface, stimulating growth of phytoplankton that are at the base on the ocean's food chain. Fronts, eddies, and upwellings can do the same things. Variation in sea surface temperature can also enhance productivity, i.e., when surface waters are cooler, deeper water can rise to the surface, bringing nutrients to sunlit areas where phytoplankton can use them. In contrast, when surface water is warmer, the cooler, nutrient-rich water is trapped below. Areas with increased productivity are areas with more prey and often, as a result, more seabirds.

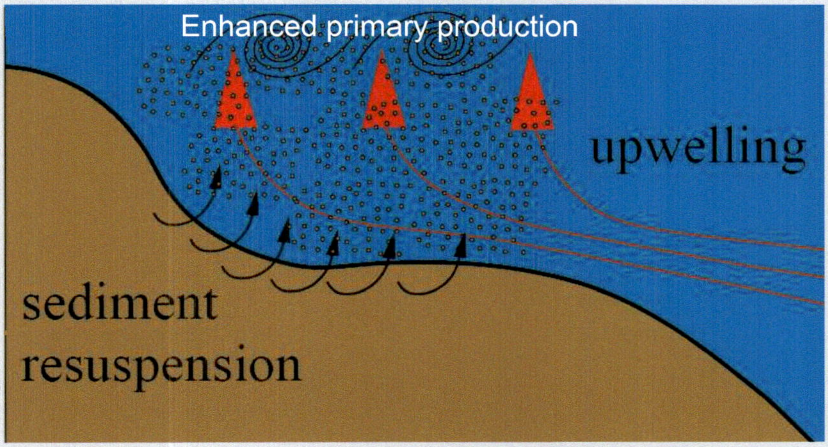


Seamounts can create upwellings that bring nutrients to the surface that in turn stimulate growth of the phyto- and zooplankton that are the base of the ocean's food chain. (Figure modified from Osadchiev et al. 2020; © 2020 by the Authors. Licensee MDPI, Basel, Switzerland, open-access article distributed under the terms and conditions of the Creative Commons Attribution (CC BY) license, https://creativecommons.org/licenses/by/4.0/)

Davies et al. (2010) examined several variables that potentially influenced the species richness of one group of seabirds, the procellariiforms (shearwaters, petrels, prions, albatrosses, and storm petrels). Most important were ocean expanse, or the distance of continuous ocean (uninterrupted by any major land mass) from any given location, and greater mean wind velocities. For procellariiform seabirds, particularly the larger species such as albatrosses (22 species) and larger shearwaters (about 25 species), these variables are important because wind in the open ocean creates swells that permit dynamic soaring, a mode of flying that allows them to travel great distances with minimal need to flap their wings. As a result, even if they must travel considerable distances to foraging locations, they can do so with minimal energy expenditure. Procellariiform species find conditions most favorable for dynamic soaring at higher latitudes in the southern hemisphere, an area with the highest annual mean wind velocities and large expanses of open ocean and, of course, the greatest species richness of these seabirds.

(continued)

**Box 1.17** (continued)

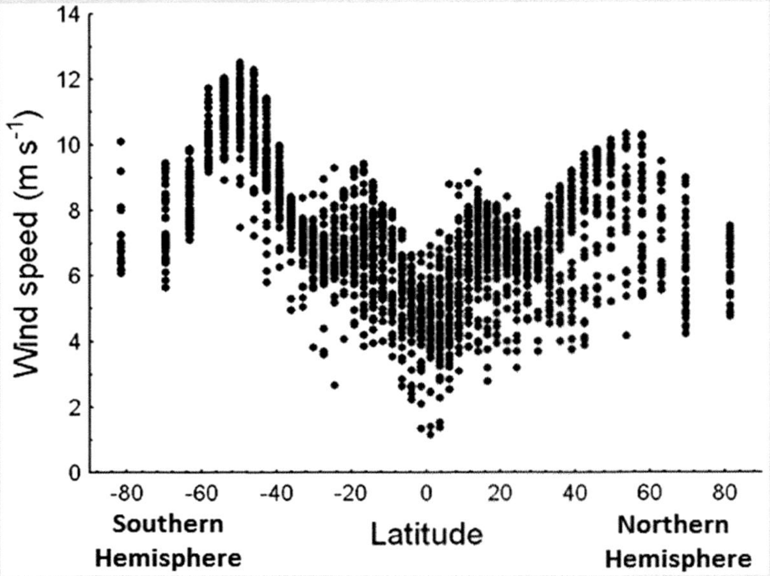


Mean annual wind speed (meters per second) over the ocean at different latitudes. Wind speeds are higher in the southern hemisphere, peaking at about 50 degrees south. (Figure from Davies et al. 2010; © 2009 Blackwell Publishing Ltd., used with permission)

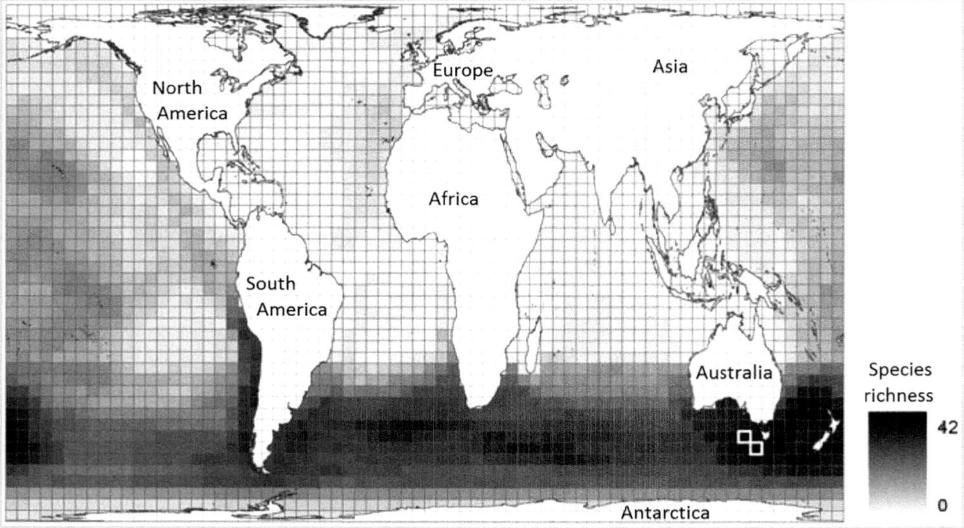


Global distribution of species in the order Procellariiformes (shearwaters, petrels, prions, albatrosses, and storm-petrels). The greatest species richness is clearly at higher latitudes in the southern hemisphere, with peak species richness in the vicinity of Australia and New Zealand. Areas with maximum species richness (42 species) are outlined in white (south of Australia). (Figure modified from Davies et al. 2010; Copyright 2009 Blackwell Publishing Ltd., used with permission)

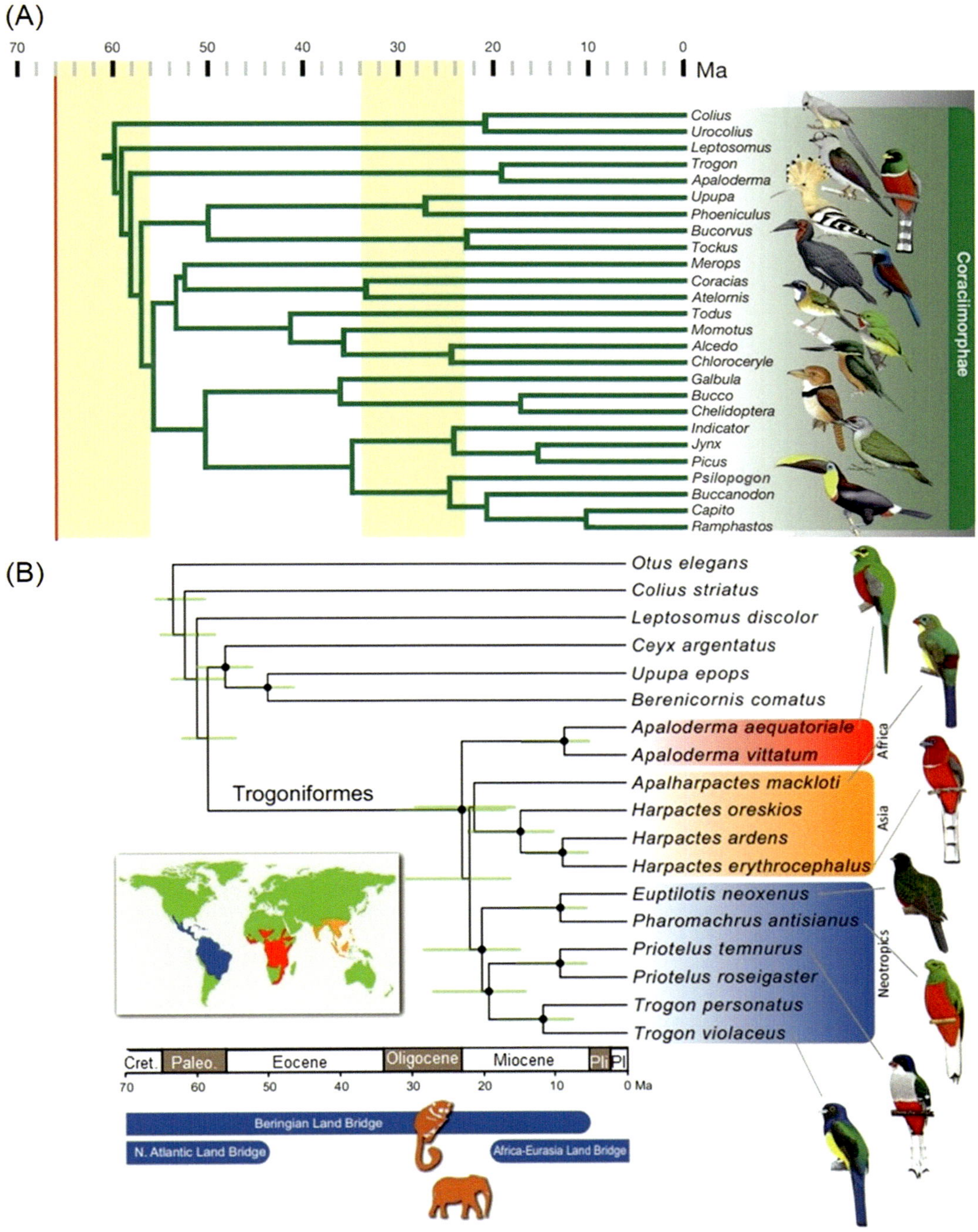


**Fig. 1.87** (**a**) Proposed phylogeny of the clade Coraciimorphae. Note the rapid diversification within a few million years after the end of the Cretaceous (as indicated by the vertical red line). Mousebirds (genera

**Fig. 1.87** (continued) Colius *and* Urocolius*; order Coliiformes) diverged from the other Coraciimorphae about 60 million years ago, as did the ancestor of the present-day cuckoo-roller (*Leptosomus discolor*; order Leptosomiformes) and present-day trogons (genera* Trogon *and* Apaloderma*; order Trogoniformes) within the next few million years. Other orders of present-day birds included in this phylogeny are Bucerotiformes (genera* Upupa *[hoopoes],* Phoeniculus, *wood hoopoes],* Bucorvus *[ground hornbills], and* Tockus *[hornbills]), Coraciiformes (genera* Merops *[bee-eaters],* Coracias *[rollers],* Atelornis *[ground rollers],* Todus *[todies],* Momotus *[motmots],* Alcedo *[kingfishers], and* Chloroceryle *[kingfishers]), Piciformes (genera* Galbula *[jacamars],* Bucco *[puffbirds],* Chelidoptera *[puffbird],* Indicator *[honeyguides],* Jynx *[wrynecks],* Picus *[woodpeckers],* Psilopogon *[Old World barbets],* Buccanodon *[spotted barbets],* Capito *[barbets], and* Ramphastos *[toucans]). (Figure from Prum et al. 2015; © 2015 Springer Nature, used with permission).* (***b***) *Proposed phylogenetic relationships of trogons. The current pantropical distribution of trogons likely resulted from dispersal across land bridges. Groups that diverged from present-day Trogoniformes within a few million years after the end of the Cretaceous include the present-day orders Bucerotiformes (*Upupa epops *[Eurasian Hoopoe] and* Berenicornis comatus *[White-crowned Hornbill] and, more generally, hornbills, ground hornbills, hoopoes, and wood hoopoes), Coraciiformes (*Ceyx argentatus *[Southern Silvery-Kingfisher] and, more generally, kingfishers, bee-eaters, rollers, motmots, and todies), Leptosomiformes (*Leptosomus discolor *[cuckoo-roller]), Coliiformes (*Colius striatus *[Speckled Mousebird] and other mousebirds), and Strigiformes (*Otus elegans *[Ryukyu Scops-Owl] and other strigiform owls). Bars across noted indicate 95% credible intervals for node age estimates. Inset map shows the present distribution of trogons. Abbreviations along the time scale: Cret., Cretaceous; Paleo., Paleocene; Pli., Pliocene; Pl, Pleistocene. (Figure from Oliveros et al. 2020; © 2019 British Ornithologists' Union, used with permission)*

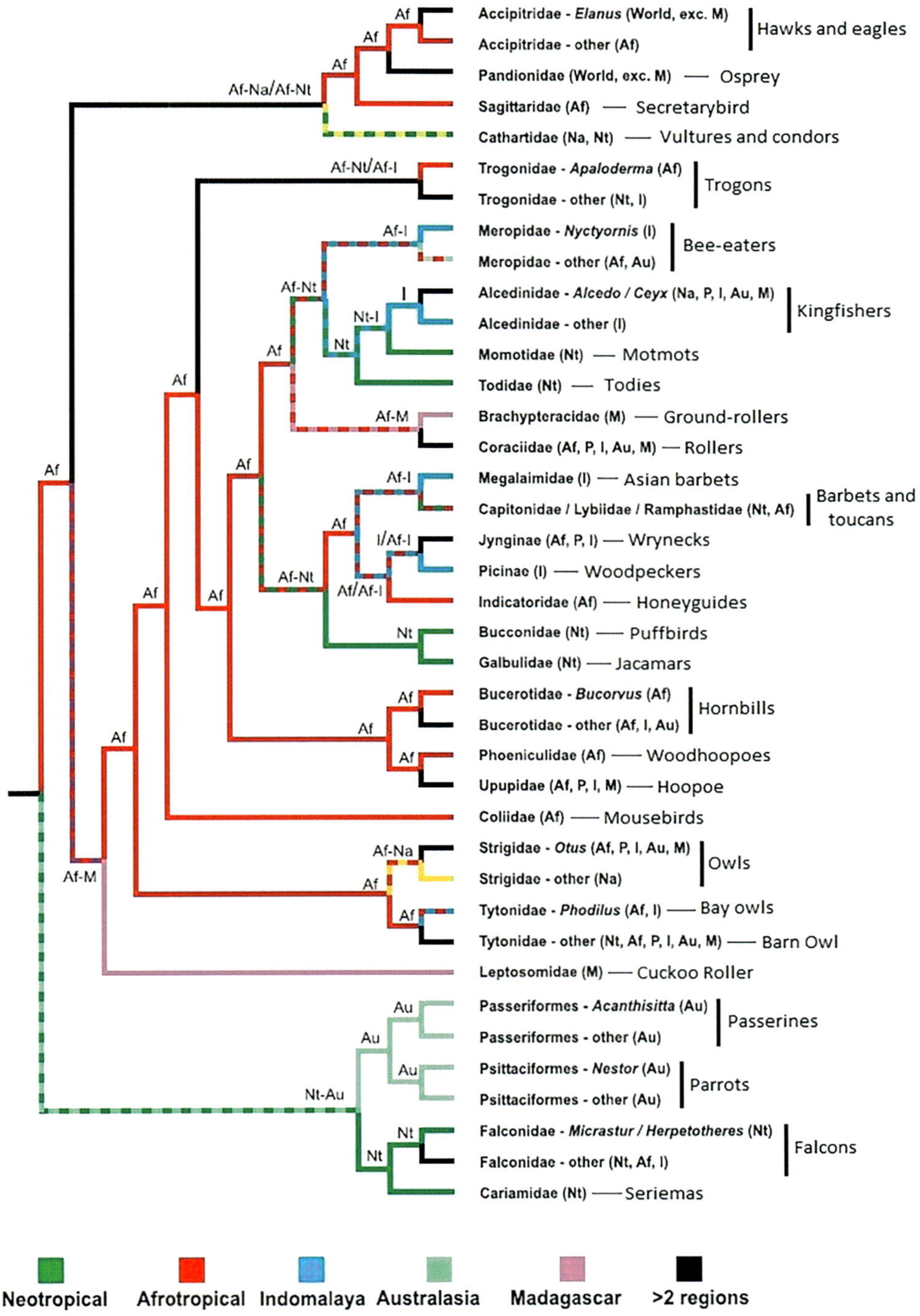


**Fig. 1.88** Possible scenario concerning where the evolution of different orders and families of present-day bird occurred during the early Paleogene. As orders and families diversified, movements of birds among and within continents and islands likely led to the establishment of populations in different areas that then could diverge into different families, genera, and species. For example, different genera of falcons (Falconidae) may have originated in the Neotropics, Africa, and Indomalaya. Abbreviations: Na, Nearctic; Nt, Neotropics; Af, Africa; P, Palearctic; I, Indomalaya; Au, Australia; M, Madagascar. (Figure from Ericson 2012; © 2011 Blackwell Publishing Ltd., used with permission)

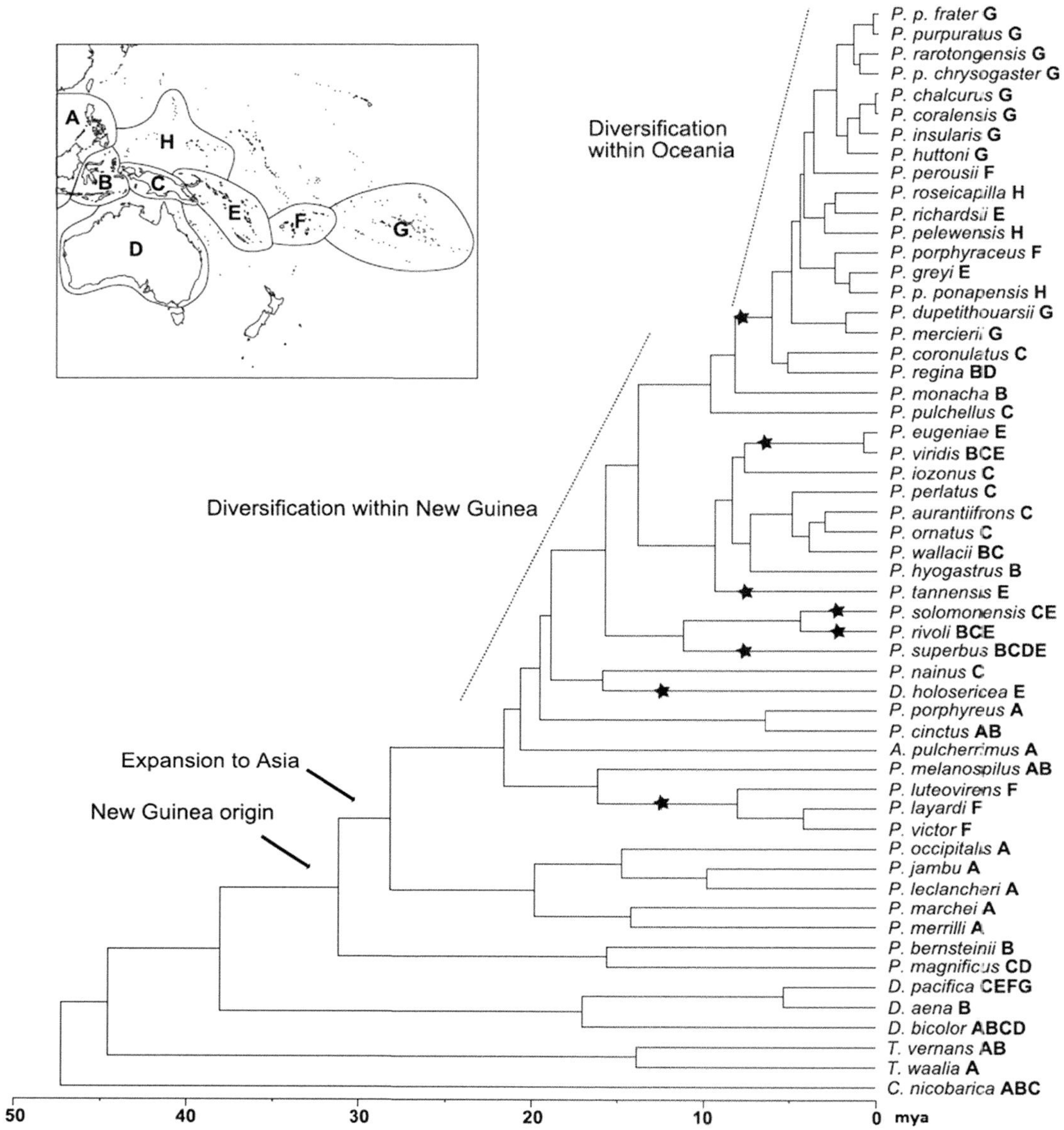


**Fig. 1.89** Phylogeny and biogeographic distribution of fruit doves. Eight areas of endemism include (**a**) Asia (including Philippine Islands), (**b**) Wallacea, (**c**) New Guinea, (**d**) Australia, (**e**) Melanesia, (**f**) Central Polynesia, (**g**) Eastern Polynesia, and (**h**) Micronesia. Stars indicate dispersals to Oceania, and letter after each species indicated the area(s) where they are currently found. (Figure from Cibois et al. 2014; © 2013 Elsevier Inc., used with permission)

**Fig. 1.90** Skeleton of *Presbyornis* showing the duck-like skull. (Figure from Zelenkov 2016; © 2016 Pleiades Publishing, Ltd., used with permission)

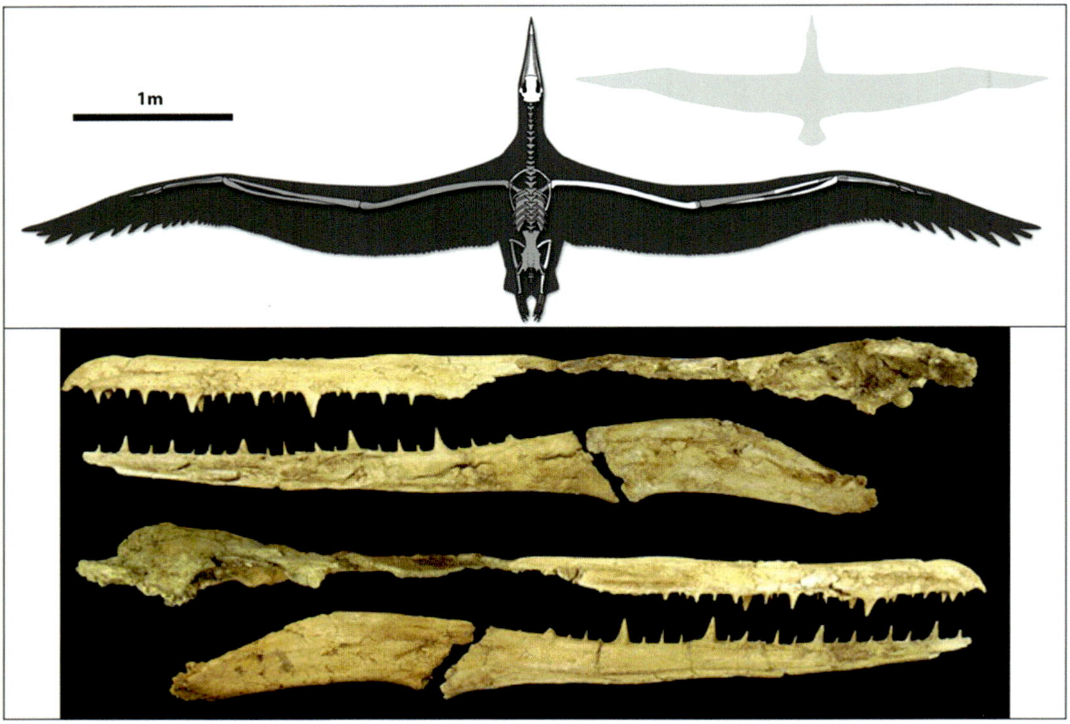


**Fig. 1.91** Top, comparison of the size of *Pelagornis sandersi* and a bird with one of the longest wingspans among present-day birds, a Royal Albatross (*Diomedea exulans*; upper right). Bottom, mandibles of a *P. sandersi* showing the pseudoteeth (spike-like protrusions of the jaw bones. (Figures from Ksepka 2014b; used with permission of the United States National Academy of Sciences)

**Fig. 1.92** Drawing of *Paraphysornis brasiliensis*, a phorusrhacid, from the Early Miocene of Brazil. Adults were about 2 m in height and likely weighed between 100 and 200 kg. (Drawing by Nobu Tamura, Wikipedia, CC BY 3.0, https://creativecommons.org/licenses/by/3.0/)

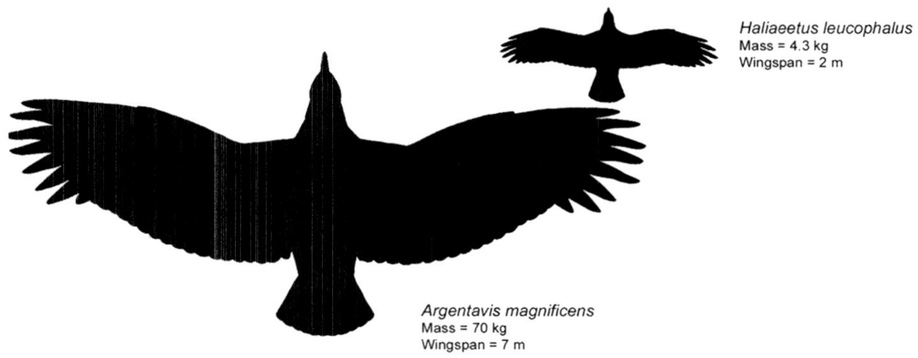


**Fig. 1.93** Comparison of the size of *Argentavis magnificens* and a Bald Eagle (*Haliaeetus leucocephalus*). Argentavis was about 16 times heavier than a Bald Eagle. (Figure from Chatterjee et al. 2007; Copyright 2007 National Academy of Sciences, USA., used with permission)

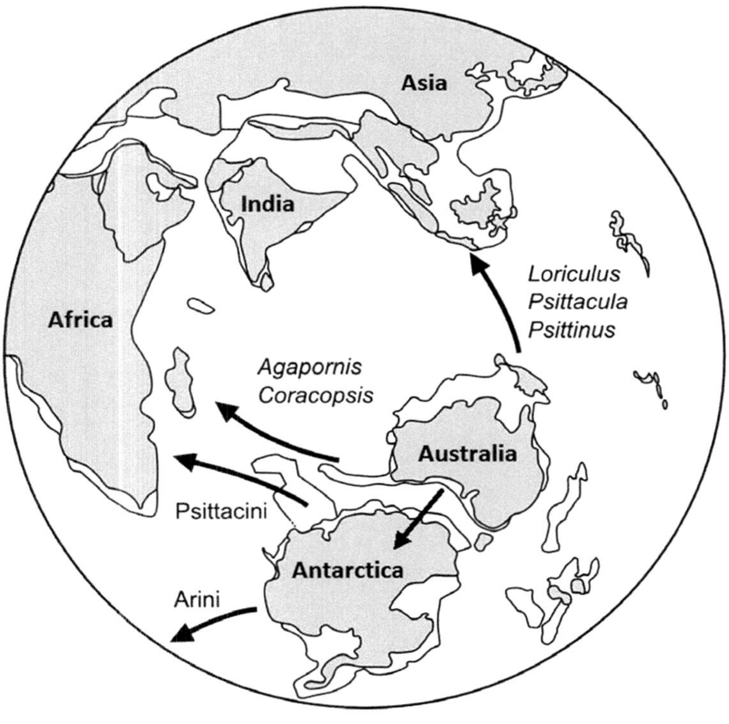


**Fig. 1.94** Parrots may have originated in Australia, with several trans-oceanic dispersal events possibly occurring and explaining their current distribution. Possible dispersal routes of parrots from Australia are indicated by the arrows. Gray areas are areas currently above sea level; black lines indicate additional areas that may have been above sea level during the early Paleogene. *Loriculus*, hanging parrots of tropical southern Asia; *Psittacula*, parakeets found primarily in southern Asia, but also in Africa and islands in the Indian Ocean; *Psittinus cyanurus*, blue-rumped parrot found in Thailand, Malaysia, Borneo, Sumatra, and nearby islands; *Agapornis*, lovebirds found in Africa and Madagascar; *Coracopsis*, vasa parrots endemic to Madagascar and other islands in the western Indian Ocean; Psittacini, Old World parrots (e.g., Gray Parrot, *Psittacus erithacus*) found in sub-Sahara Africa, Madagascar, and the Arabian Peninsula; Arini, New World parrots found throughout South and Central America, Mexico, and the Caribbean islands. The Psittacini and Arini dispersal events to the Africa and the Neotropics, respectively, were thought to be via Antarctica when it was ice-free and much warmer and wetter than today. (Figure modified from Schweizer et al. 2010; © 2009 Elsevier Inc., used with permission)

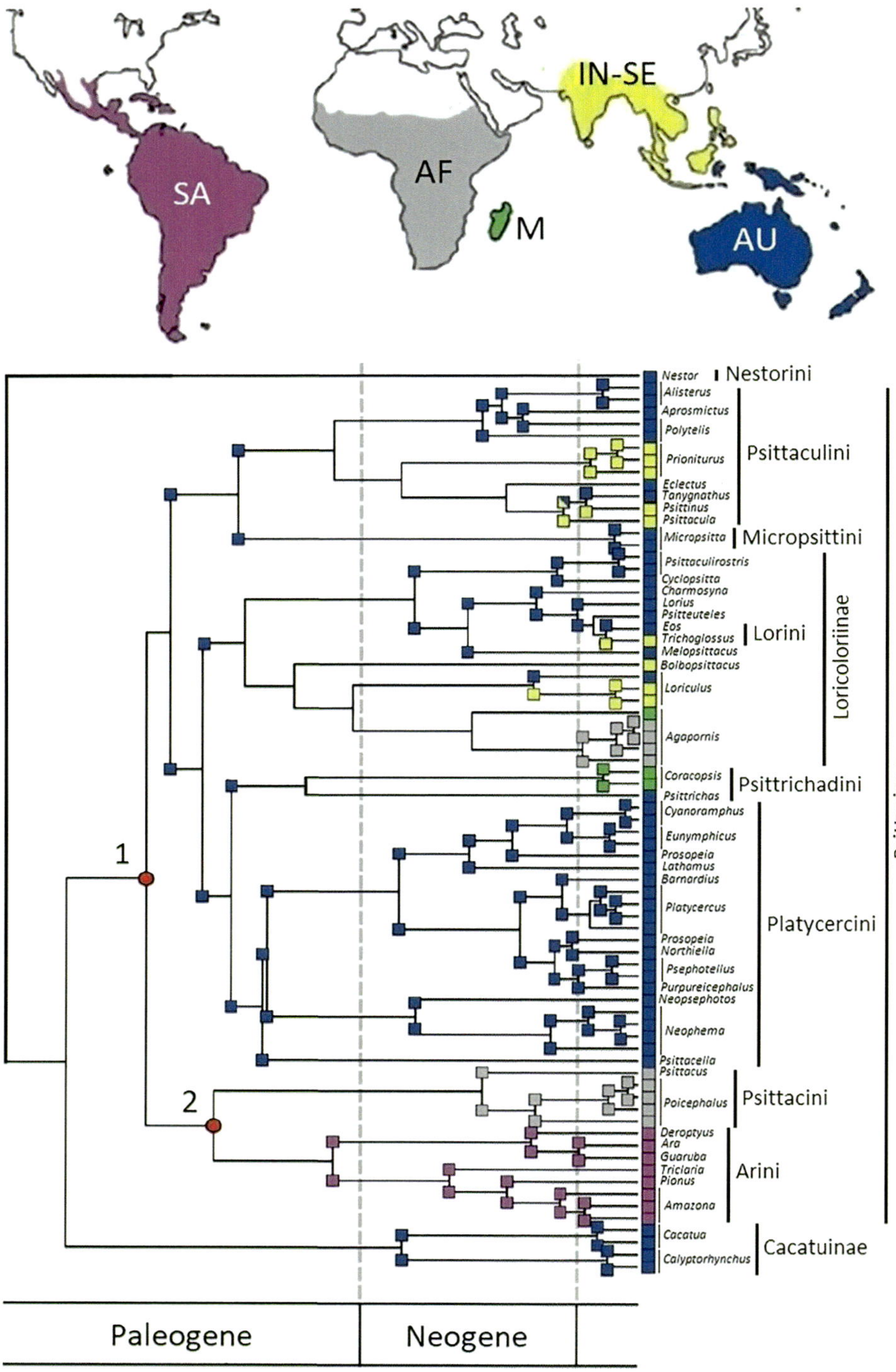


**Fig. 1.95** Phylogeny and biogeography of the Psittaciformes. Colors correspond to where species are presently found and where divergence events occurred. Node 1, divergence of South American (Arini) and African (Psittacini) parrots and Australasian parrots. Node 2, divergence of South American and African parrots. SA, South America; AF, Africa, M, Madagascar; IN-SE, India-Southeast Asia; AU, Australia. (Figure modified from Schweizer et al. 2011; © 2011 Blackwell Publishing Ltd., used with permission)

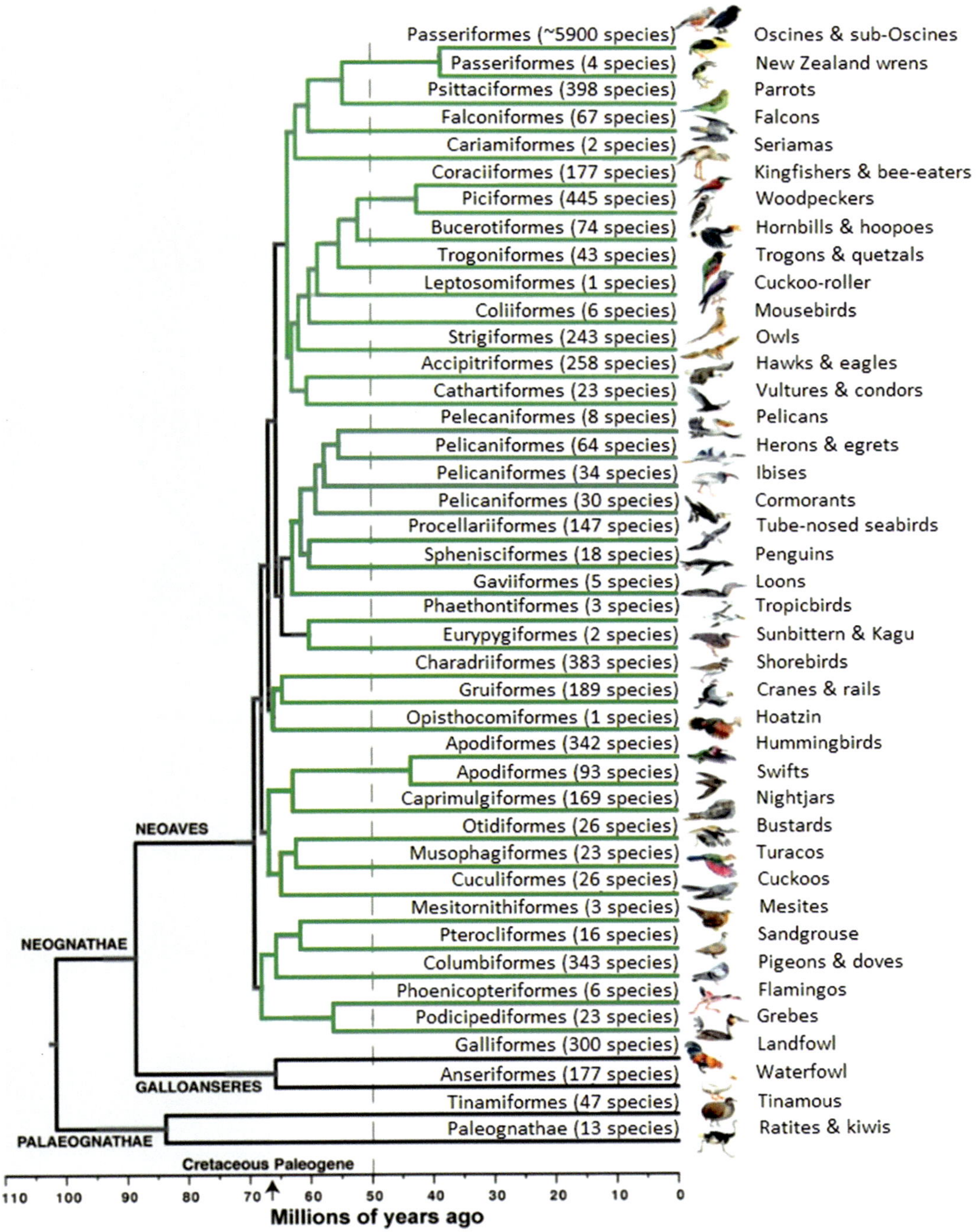


**Fig. 1.96** Phylogeny of birds with numbers of species in each order. The order Passeriformes includes more than half of all extant species of birds. (Figure modified from Jarvis et al. 2014; © 2014 The American Association for the Advancement of Science, used with permission)

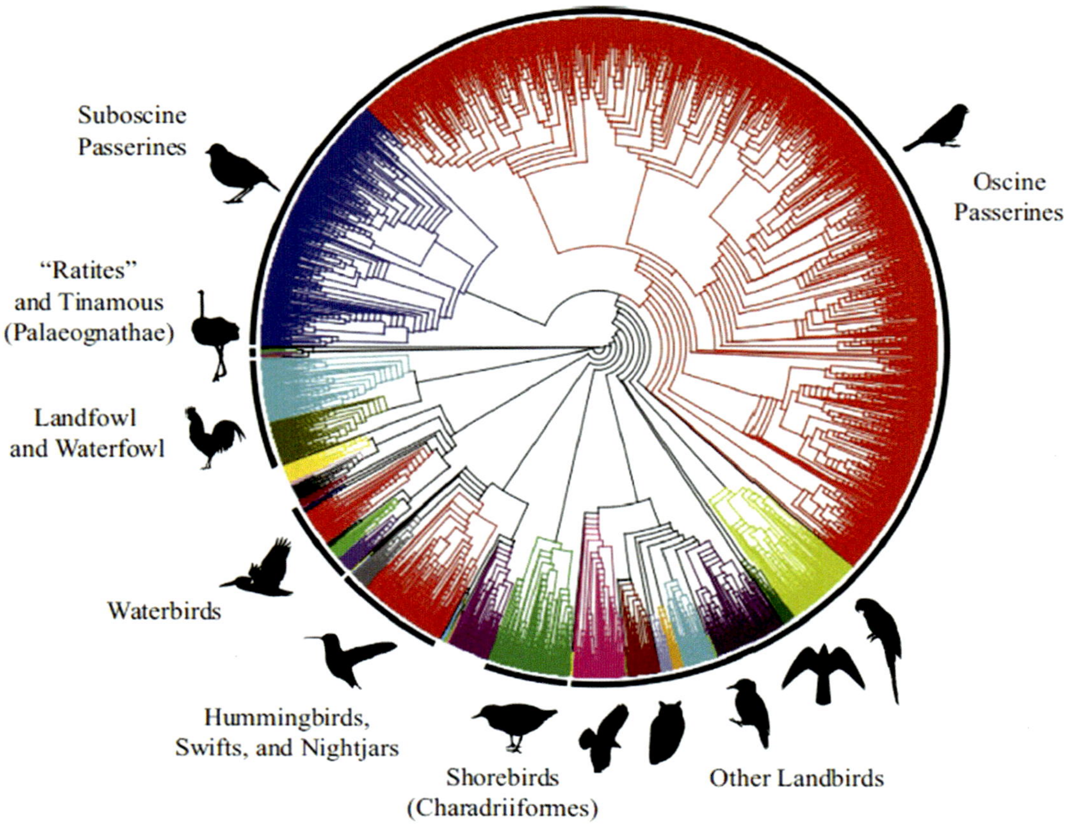


**Fig. 1.97** More than half of all extant species of birds are passerines. (Figure from Burleigh et al. 2015; © 2014 Elsevier Inc., used with permission)

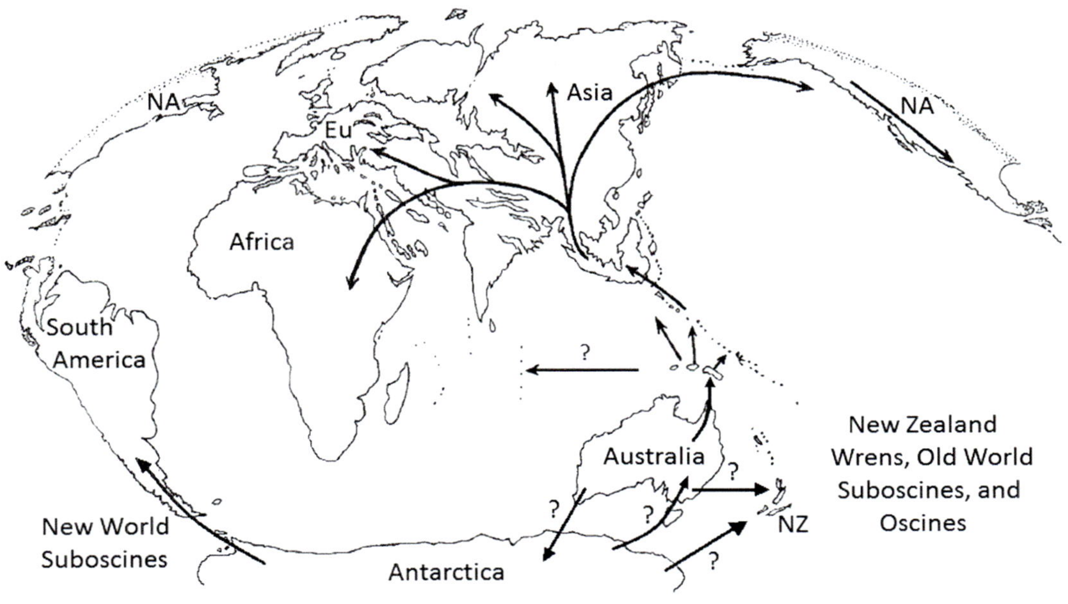


**Fig. 1.98** Distribution of continents and islands in the Early Oligocene (34 Ma) and likely or possible pathways of dispersal of songbirds from ancestral areas in Antarctica, Australia, and nearby islands. NA, North America; Eu, Europe. (Figure modified from Fjeldså 2013; open-access article distributed under the Creative Commons Attribution License 4.0, https://creativecommons.org/licenses/by/4.0/)

**Fig. 1.99** Topography of Antarctic 34 million years ago. Green areas are lowlands and light blue areas are shallow seas. Antarctica glaciation began at higher elevations about 33.6 million years ago during a period of global cooling, but areas at lower elevations remained forested until about 25 million years ago, so, prior to that, New World suboscines could have dispersed from Australia though Antarctica to South America or simply dispersed to South America from where they occurred in Antarctica. The black circle indicates 60 degrees south. (Figure modified from Houben et al. 2013; © 2013 American Association for the Advancement of Science, used with permission)

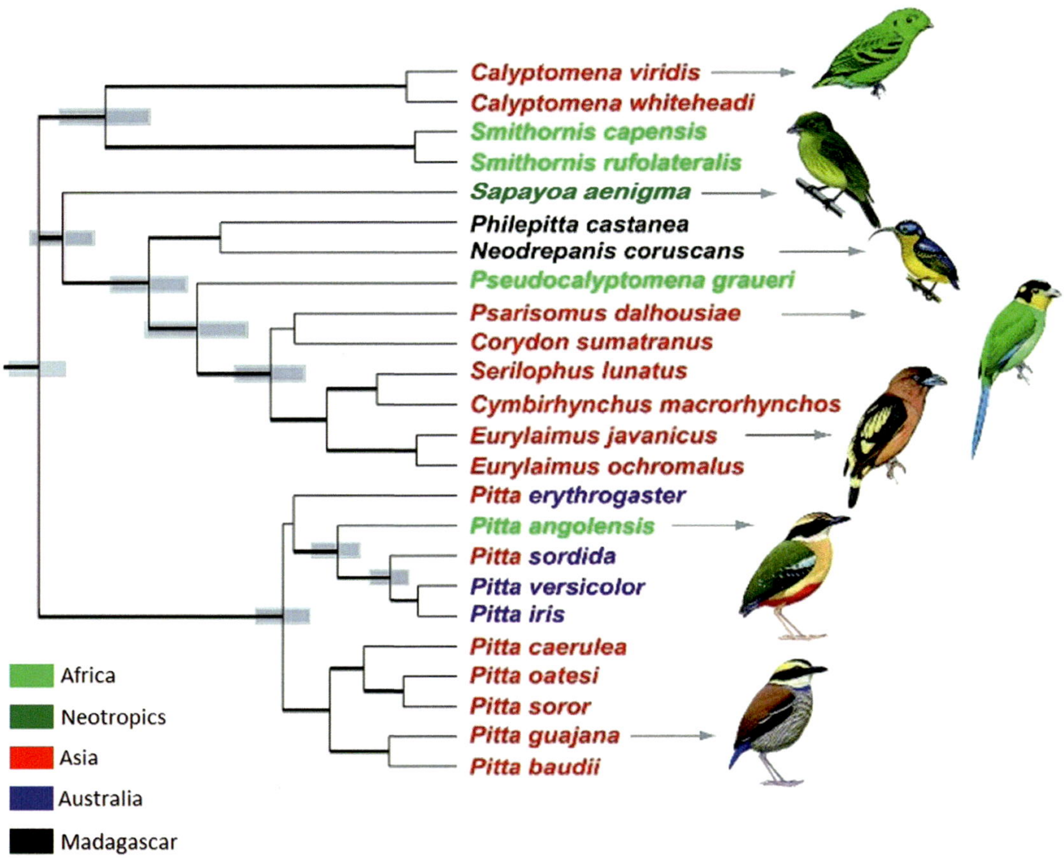


**Fig. 1.100** Phylogeny and biogeography of the Old World suboscines (Eurylaimides). Australasia is likely the ancestral area for the Eurylaimides, with dispersal throughout the Old World and giving rise to several different genera and species in Africa, Madagascar, Australia, New Zealand, and even the Neotropics (Broad-billed Sapayoa, *Sapayoa aenigma*). Horizontal gray bars at nodes indicate ±2 standard deviations of the age estimates. Species names with two colors indicate that the species is found in more than one region. (Figure modified from Moyle et al. 2006; used with permission of the American Museum of Natural History)

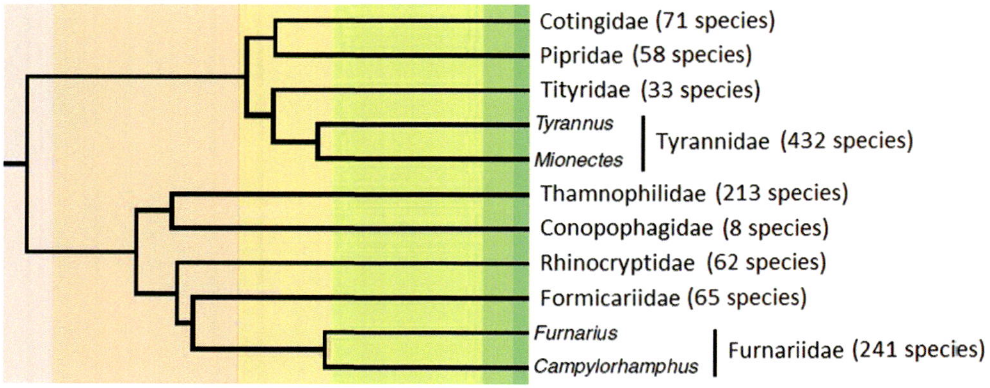


**Fig. 1.101** Phylogeny and species diversity of nine families of New World suboscines. (Figure modified from Ericson et al. 2014; open-access article distributed under the terms of the Creative Commons Attribution License, https://creativecommons.org/licenses/by/4.0/)

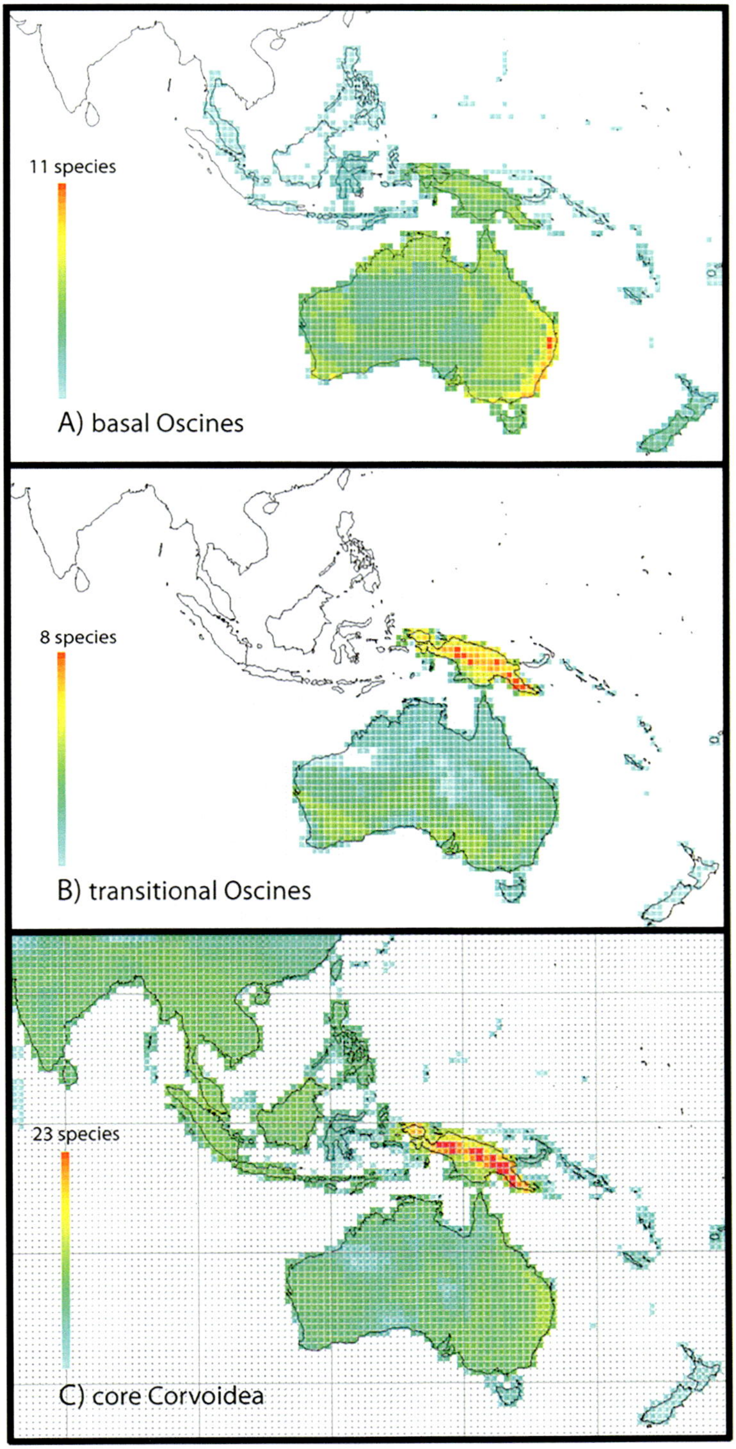


**Fig. 1.102** Diversity of three groups of oscines in Australia and southeast Asia. (**a**) Present distribution of 11 basal lineages (superfamilies Menuroidea and Meliphagoidea) that existed 25 Ma. (**b**) Present distribution of 12 transitional lineages (families Pomatostomidae, Orthonychidae, Callaeidae, Cnemophilidae, and Melanocharitidae; see figure above) that existed 25 Ma. (**c**) Present distribution of 34 lineages of core Corvoidea (also referred to as Corvides) that existed 25 Ma. (Figure from Jønsson et al. 2011; used with permission of the United States National Academy of Sciences)

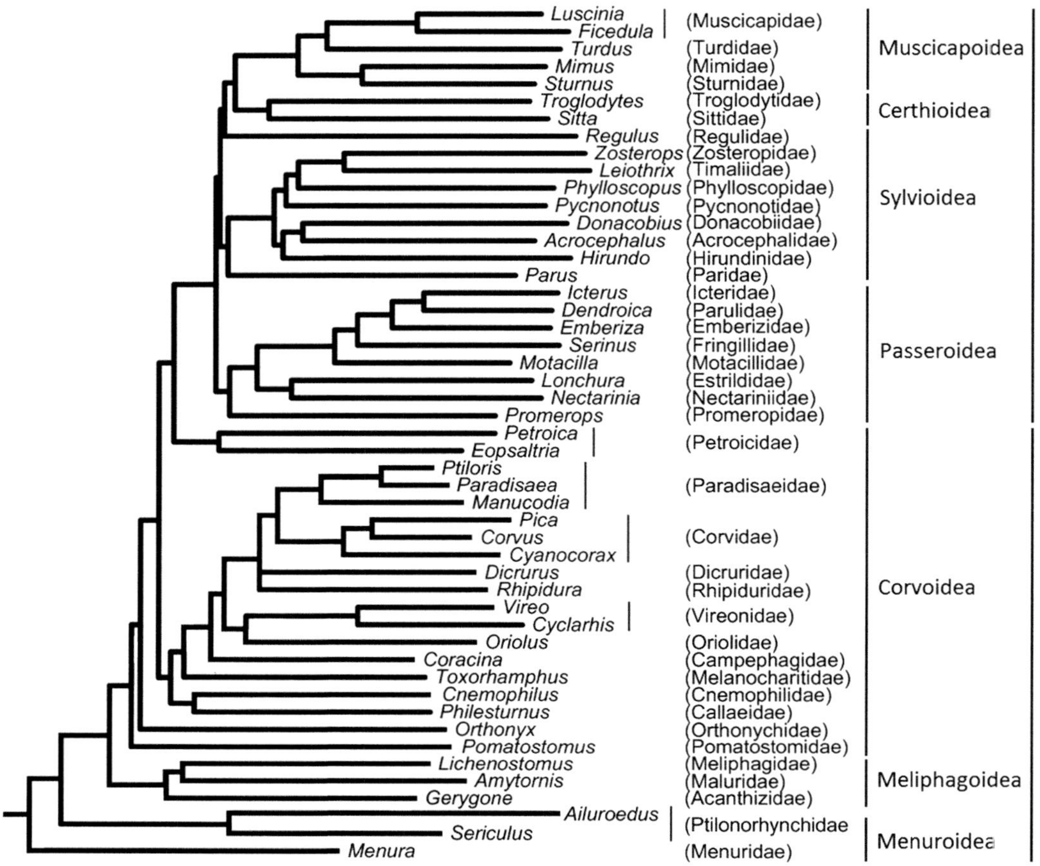


**Fig. 1.103** A simplified oscine phylogeny (not all families are listed) with the seven superfamilies and two parvorders (Corvida and Passerida). (Figure modified from Ödeen et al. 2011; open-access article distributed under the terms of the Creative Commons Attribution License, https://creativecommons.org/licenses/by/4.0/)

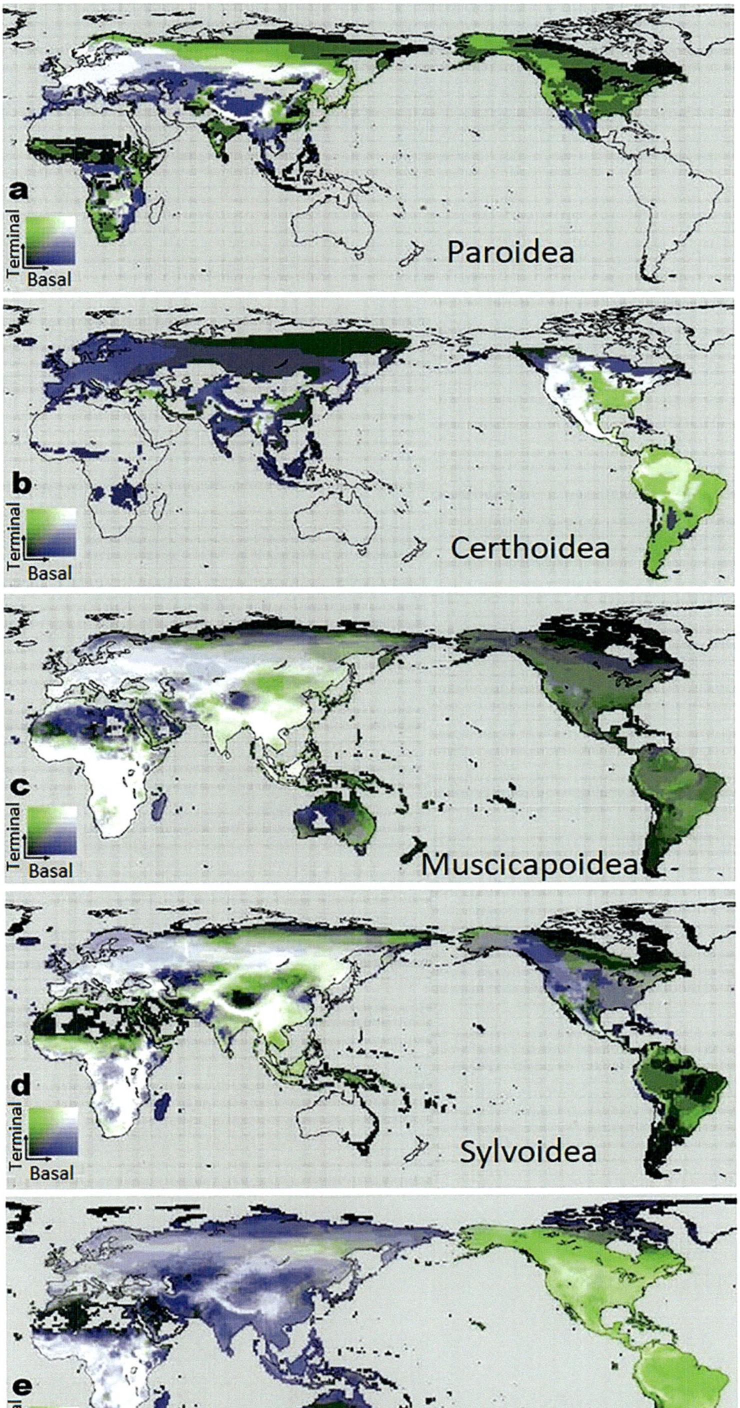


**Fig. 1.104** Worldwide distribution and diversity of five Passerida superfamilies. (**a**) Paroidae (72 species). Likely origin in Sino-Himalayan/Oriental region, but with early expansion west along Eurasian mountain ranges and further expansion to Africa and North America. (**b**) Certhoidea (124 species). Rates of speciation have been slower in Old World taxa (e.g., Certhidae and Sittidae) than in New World taxa (e.g., Troglodytidae). (**c**) Muscicapoidea (614 species). Areas with the greatest diversity include Africa, central Siberia, and mountainous areas in Asia. (**d**) Sylvoidea (1184 species). Basal taxa originated in Africa and areas of greatest diversity are in the Old World. (**e**) Passeroidea (1603 species). Basal taxa are found primarily in the Old World, and taxa that have diversified in the New World include the families Emberizidae and Parulidae. Species richness patterns of older taxa (basal) are indicated with violet color and more recent taxa (terminal) with green. Proportionally equal representation of the two groups is indicated by lighter colors (gray to white), and total species richness is indicated by relative brightness. (Figure from Fjeldså 2013; open-access article distributed under the Creative Commons Attribution License 4.0, https://creativecommons.org/licenses/by/4.0/)

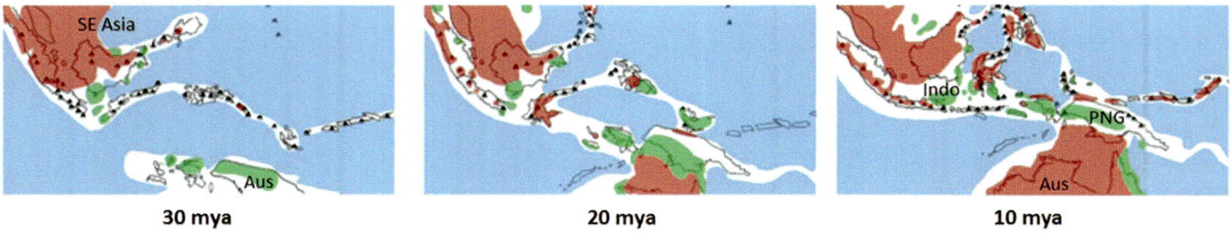


**Fig. 1.105** Transition zone between Australia and Southeast Asia over a 20-million-year period. The presence of numerous islands, including Papua New Guinea, likely provided opportunities for dispersal and colonization that led to diversification of the Corvida. SE Asia, Southeast Asia; Aust, Australia; PNG, Papua New Guinea; Indo, Indonesia. Red areas = land; green = low-elevation coastal land or islands; white = shallow sea; black triangles = areas of volcanism and emergence of islands. (Figure modified from Jønsson et al. 2011; used with permission of the United States National Academy of Sciences)

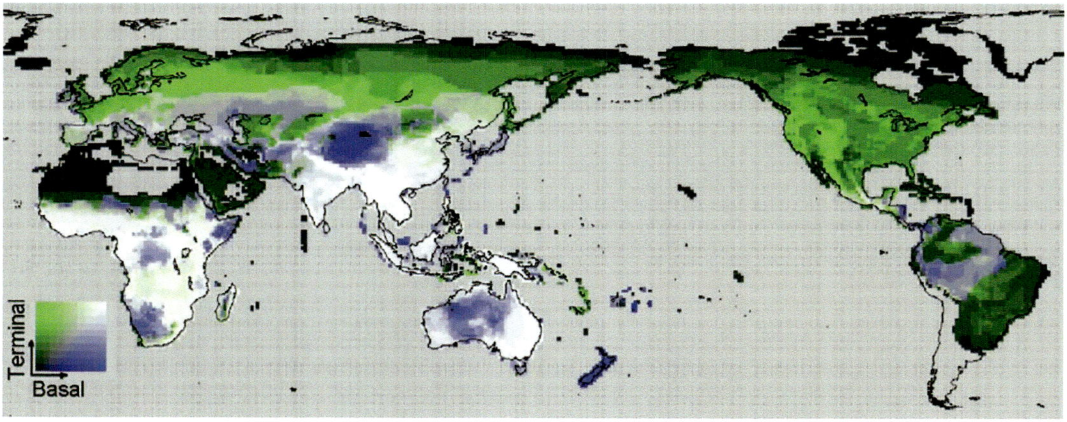


**Fig. 1.106** Worldwide distribution and diversity of Corvoidea (737 species). Basal taxa are found primarily in the Old World and more recent taxa primarily in the northern hemisphere. Crows (Corvidae) and shrikes (Laniidae) likely originated in Asia, but have expanded into North America and, from there, into Central and South America. Species richness patterns of older taxa (basal) are indicated with violet color and more recent taxa (terminal) with green. Proportionally equal representation of the two groups is indicated by lighter colors (gray to white), and total species richness is indicated by relative brightness. (Figure from Fjeldså 2013; © Avian Research, Published by BMC—Springer Nature, used with permission)

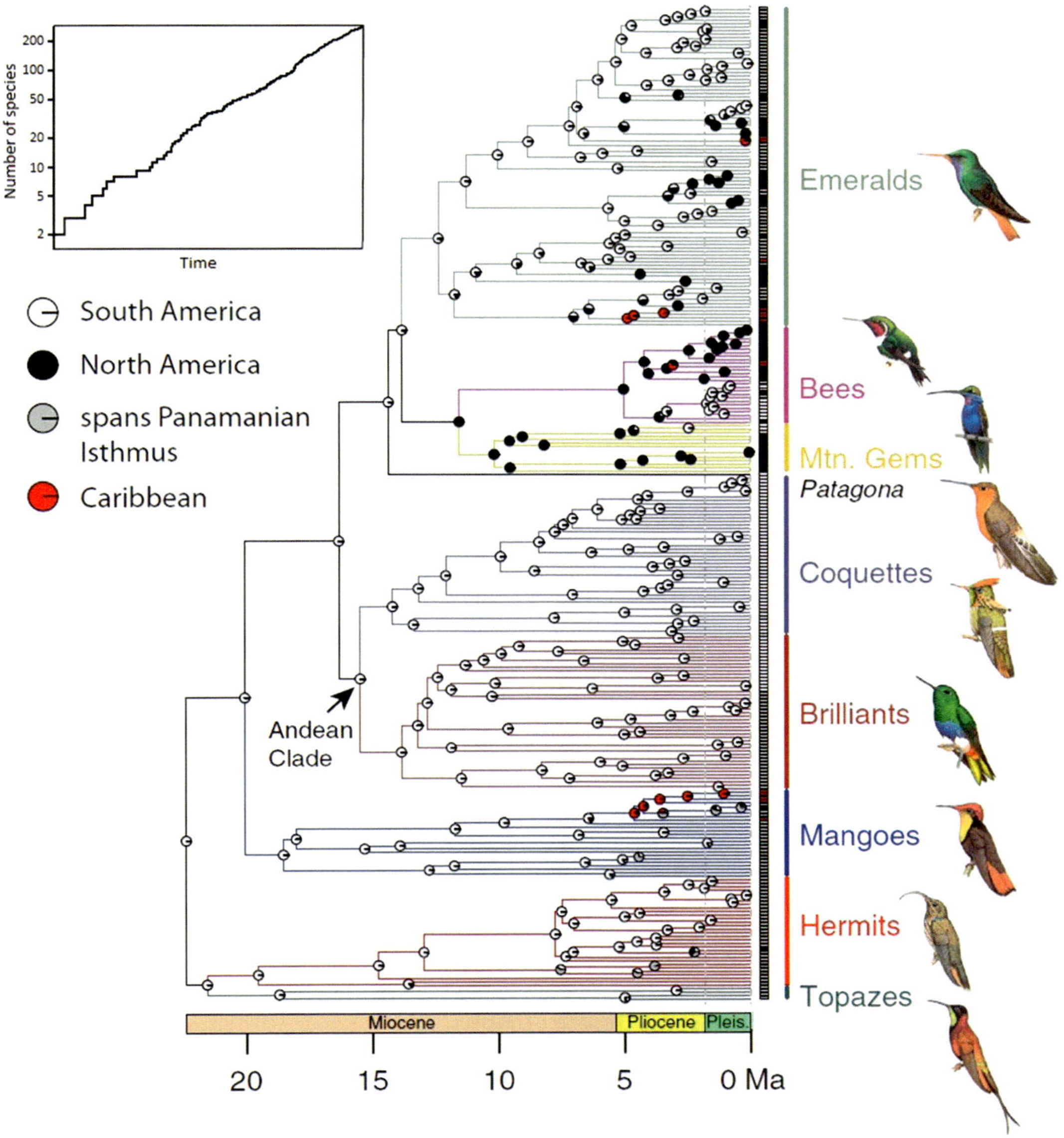


**Fig. 1.107** Phylogeny and biogeography of hummingbirds. Most species first appeared within the past five million years, with most evolving in South America and others in North America and on islands in the Caribbean. Top left: The number of species of hummingbirds has exhibited an increasingly upward trend for over 20 million years. (Figure from McGuire et al. 2014; © 2014 Elsevier Ltd., used with permission)

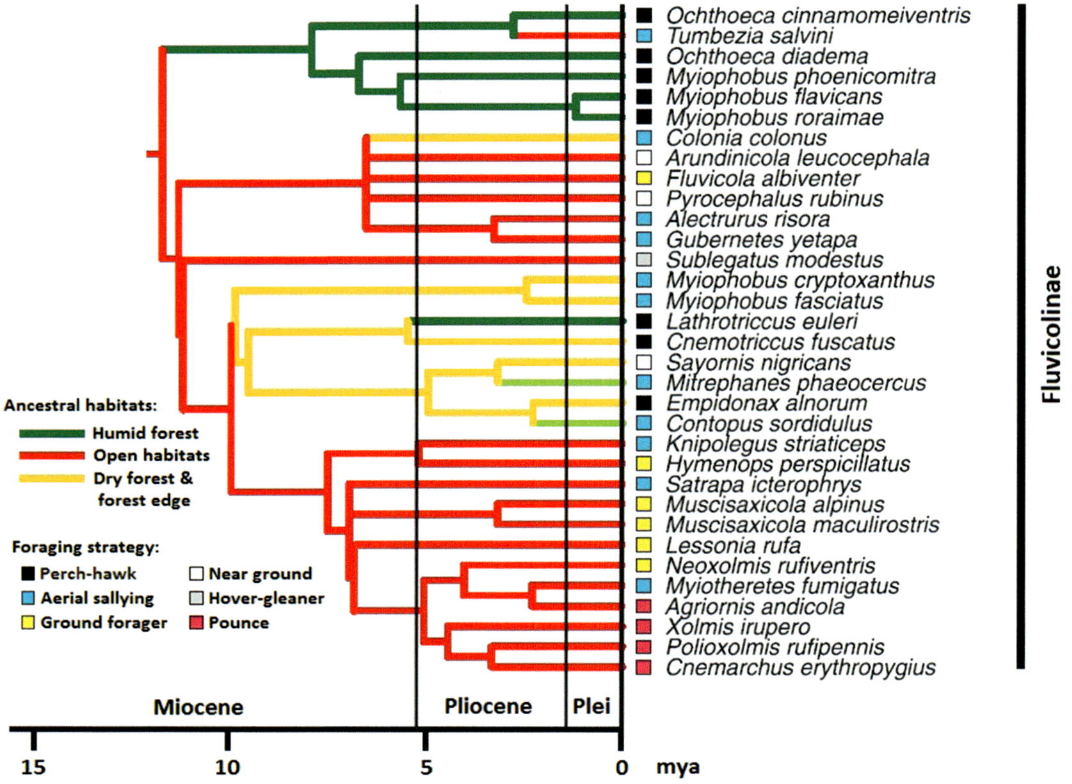


**Fig. 1.108** Estimated time when extant species in the subfamily Fluvicolinae (Family Tyrannidae: New World suboscines) first appeared, as well as their possible ancestral habitats and current foraging strategies. (Figure modified from Ohlson et al. 2008; © 2008 The Authors, used with permission)

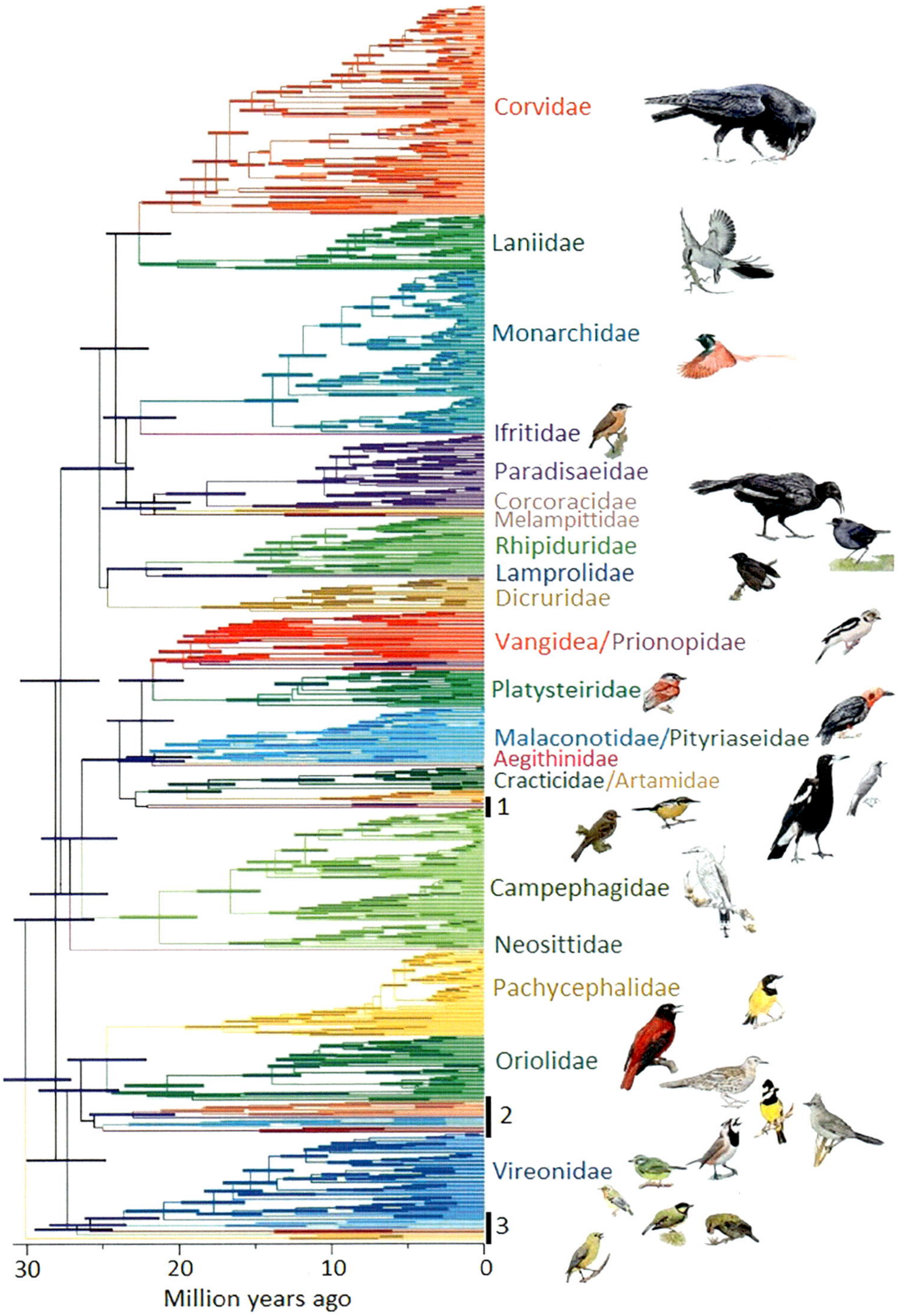


**Fig. 1.109** Phylogeny of the Corvoidea, with families indicated by different colors. Most extant species evolved within the last five million years. 1—Machaerirhynchidae (boatbills) and Rhagologidae (mottle whistler, *Rhagologus leucostigma*); 2—Cinclosomatidae (quail-thrushes), Falcunculidae (crested shriketits), Psophodidae (whipbirds, wedgebills, and jewel-babblers), and Oreoicidae (bellbirds); 3—Pteruthius (shrike-babblers), Paramythiidae (painted berrypeckers), Eulacestomatidae (ploughbills), and Mohouidae (*Mohoua* spp.). Horizontal bars indicate 95% highest probability. (Figure modified from Jønsson et al. 2016; © 2015 Elsevier Inc., used with permission)

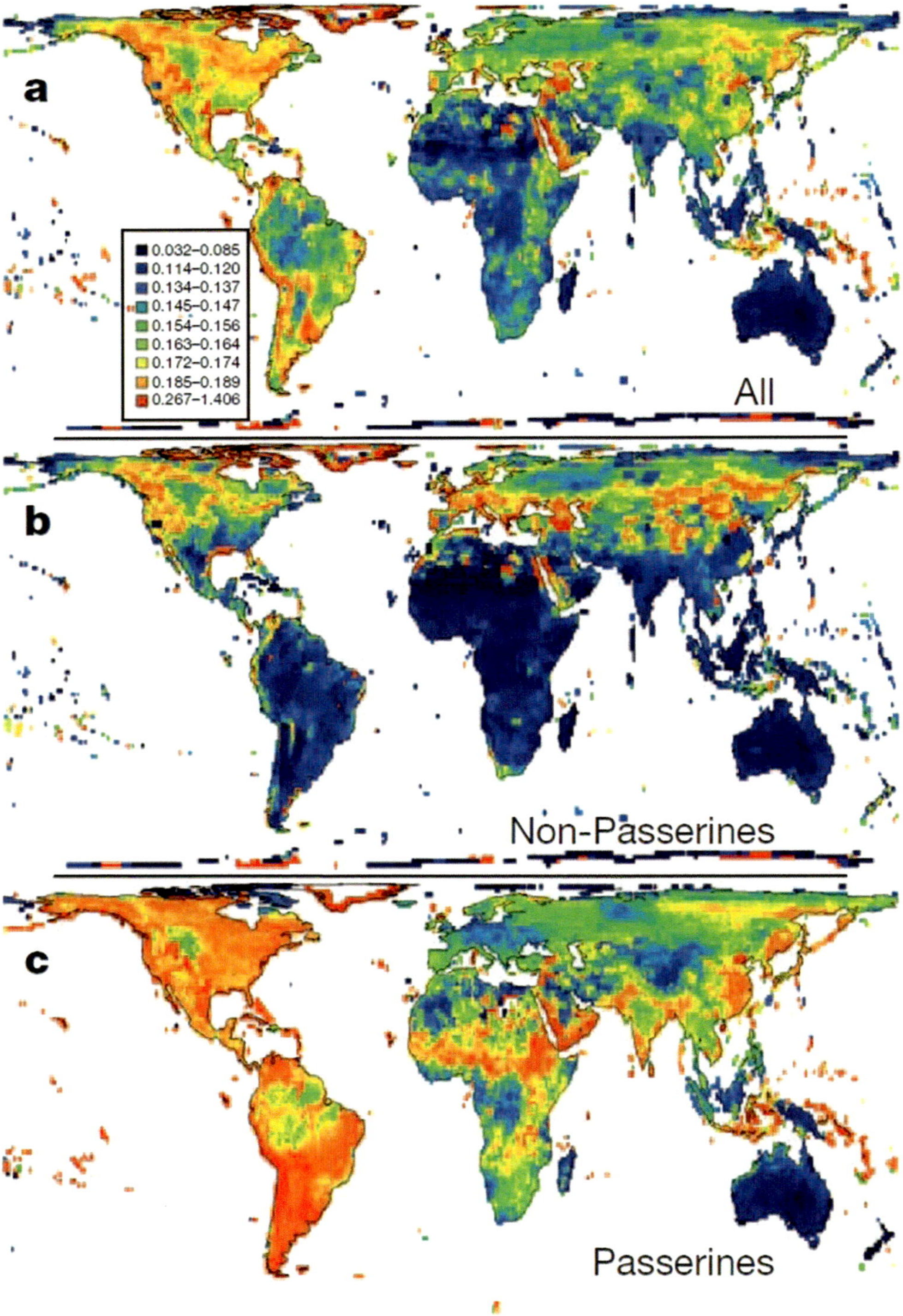


**Fig. 1.110** Geographic variation in species-level diversification rates. (**a**) All species of birds; (**b**) Non-passerines; (**c**) Passerines. Mean species assemblage diversification rates in the legend in (**a**) are based on the mean for all species in each grid cell; grid cell size is 110 × 110 km. (Figure from Jetz et al. 2012; © 2012 Springer Nature, used with permission)

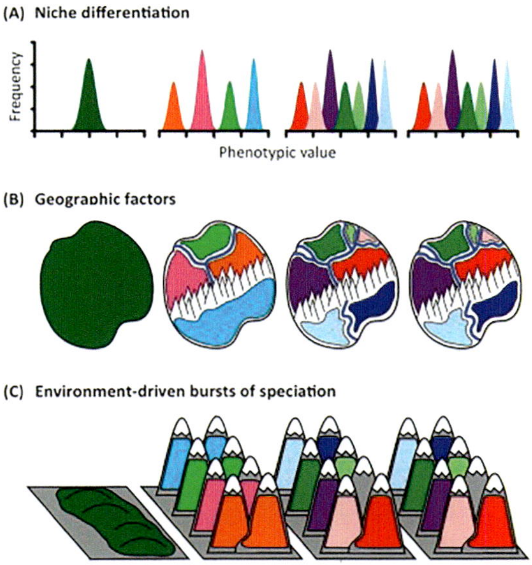


**Fig. 1.111** Factors that can contribute to species diversification. (**a**) Species may diverge (with different species indicated by different colors) when different populations begin to specialize in the use of different niches, e.g., different layers in a forest. (**b**) Geographical factors can contribute to species diversification, e.g., separation of populations by barriers like mountain ranges or large rivers. (**c**) Changes in the environment such as climate change or tectonic or volcanic events that create mountains can isolate populations and create new niches that lead to divergence and diversification. (Figure modified from Moen and Morlon 2014; © 2014 Elsevier Ltd., used with permission)

**Fig. 1.112** Species richness of various groups of terrestrial vertebrates. Bird species richness (Struthioniformes + Timaniformes + Galliformes + Anseriformes + Neoaves) exceeds that of amphibian, reptile, or mammal species richness. (Figure modified from Alfaro et al. 2009; used with permission of the United States National Academy of Science)

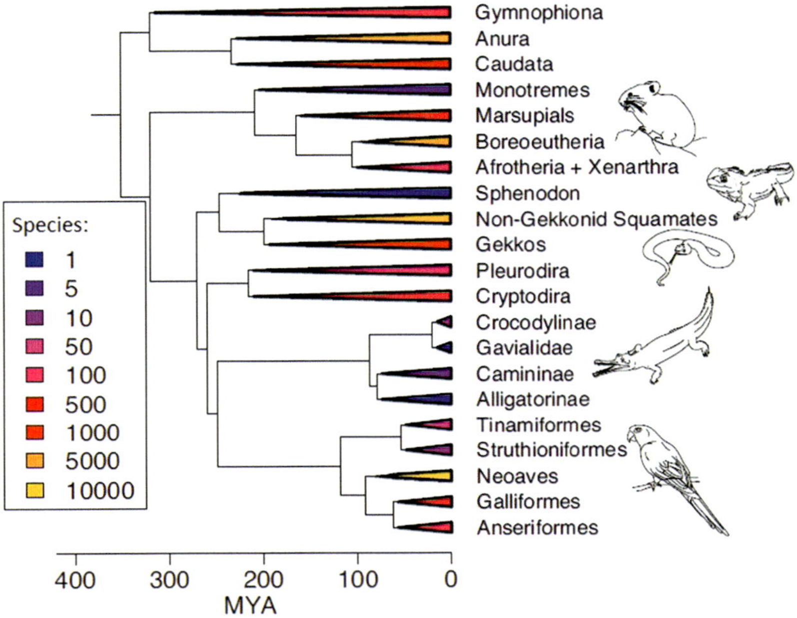

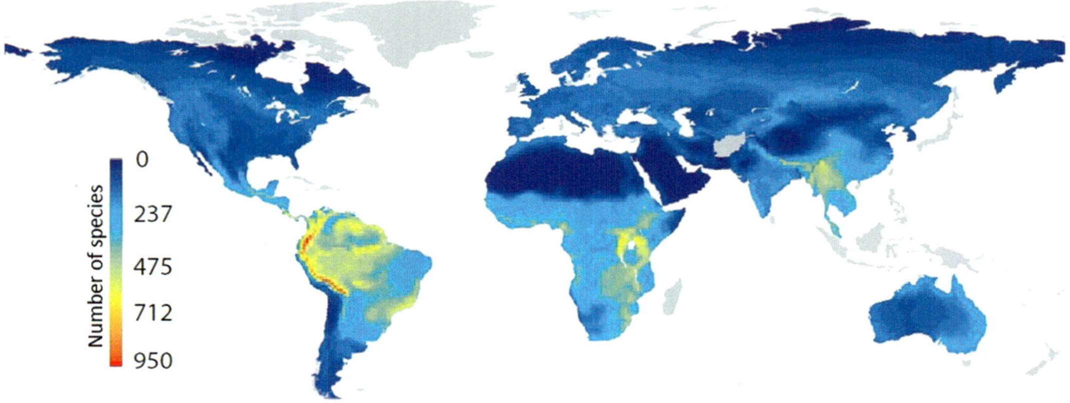


**Fig. 1.113** Worldwide species richness of birds. Areas of high species richness include South America, east-central Africa, and southeast Asia. Species richness in the Neotropics seems to be linked to habitat diversity that, in turn, is correlated with topographic heterogeneity. Rahbek and Graves (2001) examined bird diversity in South America and found that the areas with the highest were Andean Ecuador (845 species), southeastern Peru (782 species), and southern Bolivia (698 species). Species richness in Africa is greatest in areas with greater primary productivity and solar radiation, higher temperatures, and more rainfall. Specifically, more species of birds are found in the forested areas around Lake Victoria. Other areas of richness are the mountains of Kenya, Tanzania, and Malawi and the forest–savanna ecotone north of the Congo Basin into the Guinea forests (Brooks et al. 2001). In southeast Asia, species richness is highest at the base of the Indochina Peninsula, particularly where Myanmar borders India and China. In this area, the Himalaya Mountains descend into tropical lowlands and topographic variation is high (Ding et al. 2006). (Figure from Hawkins et al. 2012; © 2011 Blackwell Publishing Ltd., used with permission)

## References


Adatte T, Fantasia A, Samant B, Mohabey D, Font E, Keller G, Khozyem H, Gertsch B (2014) Deccan volcanism: a main trigger of environmental changes leading to the K/Pg mass extinction? Comunicações Geológicas 101:1435–1437

Aggerbeck M, Fjeldså J, Christidis L, Fabre P-H, Jønsson KA (2014) Resolving deep lineage divergences in core corvoid passerine birds supports a proto-Papuan island origin. Mol Phylogenet Evol 70:272–285

Alfaro ME, Santini F, Brock C, Alamillo H, Dornburg A, Rabosky DL, Carnevale G, Harmon LJ (2009) Nine exceptional radiations plus high turnover explain species diversity in jawed vertebrates. Proc Natl Acad Sci USA 106:13410–13414

Allen V, Bates KT, Li Z, Hutchinson JR (2013) Linking the evolution of body shape and locomotor biomechanics in bird-line archosaurs. Nature 497:104–107

Allwood AC, Walter MR, Burch IW, Kamber BS (2007) 3.43 billion-year-old stromatolite reef from the Pilbara Craton of Western Australia: ecosystem-scale insights to early life on Earth. Precambrian Res 158:198–227

Alvarez LW, Alvarez W, Asaro F, Michel HV (1980) Extraterrestrial cause for the Cretaceous-Tertiary extinction. Science 208:1095

Atterholt J, Hutchison JH, O'Connor JK (2018) The most complete enantiornithine from North America and a phylogenetic analysis of the Avisauridae. PeerJ 6: e5910

Atterholt J, Poust AW, Erickson GM, O'Connor JK (2021) Intraskeletal osteohistovariability reveals complex growth strategies in a Late Cretaceous enantiornithine. Front Earth Sci 9:640220

Bailleul AM, O'Connor J, Zhang S, Li Z, Wang Q, Lamanna MC, Zhu X, Zhou Z (2019) An Early Cretaceous enantiornithine (Aves) preserving an unlaid egg and probable medullary bone. Nat Commun 10:1275

Bailleul AM, O'Connor J, Li Z, Wu Q, Zhao T, Martinez Monleon MA, Wang M, Zheng X (2020) Confirmation of ovarian follicles in an enantiornithine (Aves) from the Jehol biota using soft tissue analyses. Commun Biol 3:399

Barker FK, Cibois A, Schikler P, Feinstein J, Cracraft J (2004) Phylogeny and diversification of the largest avian radiation. Proc Natl Acad Sci USA 101:11040–11045

Barker FK, Burns KJ, Klicka J, Lanyon SM, Lovette IJ (2015) New insights into New World biogeography: an integrated view from the phylogeny of blackbirds, cardinals, sparrows, tanagers, warblers, and allies. Auk 132:333–348

Baron MG, Norman DB, Barrett PM (2017) A new hypothesis of dinosaur relationships and early dinosaur evolution. Nature 543:501–506

Baur G (1886) Osteologische notizen uber reptilian. Zool Anzeiger 238:685

Bell AKA (2013) Evolution and ecology of Mesozoic birds: a case study of the derived Hesperornithiformes and the use of morphometric data in quantifying avian paleoecology. Ph.D. dissertation, University of Southern California, Los Angeles, CA

Bell A, Chiappe LM (2016) A species-level phylogeny of the Cretaceous Hesperornithiformes (Aves: Ornithuromorpha): implications for body size evolution amongst the earliest diving birds. J Syst Palaeontol 14:239–251

Bell A, Chiappe LM (2022) The Hesperornithiformes: a review of the diversity, distribution, and ecology of the earliest diving birds. Diversity 14:267

Bell AK, Chiappe LM, Erickson GM, Suzuki S, Watabe M, Barsbold R, Tsogtbaatar K (2010) Description and ecologic analysis of *Hollanda luceria*, a Late Cretaceous bird from the Gobi Desert (Mongolia). Cretac Res 31:16–26

Benito J, Chen A, Wilson LE, Bhullar BS, Burnham D, Field DJ (2022a) Forty new specimens of *Ichthyornis* provide unprecedented insight into the postcranial morphology of crownward stem group birds. PeerJ 10:e13919

Benito J, Kuo P-C, Widrig KE, Jagt JWM, Field DJ (2022b) Cretaceous ornithurine supports a neognathous crown bird ancestor. Nature 612:100–105

Bennett AF, Ruben JA (1979) Endothermy and activity in vertebrates. Science 206:649–654

Benson RBJ, Butler RJ, Carrano MT, O'Connor PM (2012) Air-filled postcranial bones in theropod dinosaurs: physiological implications and the 'reptile'-bird transition. Biol Rev 87:168–193

Benton MJ (2003) When life nearly died: the greatest mass extinction of all time. Thames & Hudson, London

Benton MJ (2014) How birds became birds. Science 345: 508–509

Benton MJ, Clark JM (1988) Archosaur phylogeny and the relationships of the Crocodylia. In: Benton MJ (ed) The phylogeny and classification of the tetrapods, Amphibians, reptiles, birds, vol 1. Clarendon Press, Oxford, pp 295–338

Benton MJ, Tverdokhlebov VP, Surkov MV (2004) Ecosystem remodelling among vertebrates at the Permian–Triassic boundary in Russia. Nature 432:97–100

Benton MJ, Forth J, Langer MC (2014) Models for the rise of the dinosaurs. Curr Biol 24:R87–R95

Beyrand V, Voeten DFAE, Bureš S, Fernandez V, Janáček J, Jirák D, Rauhut O, Tafforeau P (2019) Multiphase progenetic development shaped the brain of flying archosaurs. Sci Rep 9:1–15

Bhullar B-AS, Marugán-Lobón J, Racimo F, Bever GS, Rowe TB, Norell MA (2012) Birds have paedomorphic dinosaur skulls. Nature 487:223–226

Birdlife Australia [online] (2018) Working list of Australian birds. <http://www.birdlife.org.au/conservation/science/taxonomy>

Blondel J, Mourer-Chauviré C (1998) Evolution and history of the western Palaearctic avifauna. Trends Ecol Evol 13:488–492

Botero CA, Dor R, McCain CM, Safran RJ (2014) Environmental harshness is positively correlated with intraspecific divergence in mammals and birds. Mol Ecol 23:259–268

Braun EL, Cracraft J, Houde P (2019) Resolving the avian tree of life from top to bottom: The promise and potential boundaries of the phylogenomic era. In: Kraus RHS (ed) Avian genomics in ecology and evolution. Springer, Cham, pp 151–210

Brocklehurst N, Upchurch P, Mannion PD, O'Connor J (2012) The completeness of the fossil record of Mesozoic birds: implications for early avian evolution. PLoS ONE 7:e39056

Brocklehurst N, Field DJ (2021) Macroevolutionary dynamics of dentition in Mesozoic birds reveal no long-term selection towards tooth loss. iScience 24:102243

Brooks T, Balmford A, Burgess N, Fjeldsa J, Hansen LA, Moore J, Rahbek C, Williams P (2001) Toward a blueprint for conservation in Africa. BioScience 51:613–624

Broom R (1913) On the South-African pseudosuchian *Euparkeria* and allied genera. Proc Zool Soc Lond 83:619–633

Brusatte SL (2016) Evolution: how some birds survived when all other dinosaurs died. Curr Biol 26:R415–R417

Brusatte SL (2017) Evolution: uprooting the dinosaur family tree. Curr Biol 27:R390–R392

Brusatte SL, Nesbitt SJ, Irmis RB, Butler RJ, Benton MJ, Norell MA (2010a) The origin and early radiation of dinosaurs. Earth Sci Rev 101:68–100

Brusatte SL, Benton MJ, Lloyd GT, Ruta M, Wang SC (2010b) Macroevolutionary patterns in the evolutionary radiation of archosaurs (Tetrapoda: Diapsida). Earth Environ Sci Trans R Soc Edinb 101:367–382

Brusatte SL, Lloyd GT, Wang SC, Norell MA (2014) Gradual assembly of avian body plan culminated in rapid rates of evolution across the dinosaur-bird transition. Curr Biol 24:2386–2392

Brusatte SL, O'Connor JK, Jarvis ED (2015) The origin and diversification of birds. Curr Biol 25:R888–R898

Bryant HN, Russell AP (1993) The occurrence of clavicles within Dinosauria: implications for the homology of the avian furcula and the utility of negative evidence. J Vertebr Paleontol 13:171–184

Bunce M, Worthy TH, Phillips MJ, Holdaway RN, Willerslev E, Haile J, Shapiro B, Scofield RP, Drummond A, Kamp PJ, Cooper A (2009) The evolutionary history of the extinct ratite moa and New Zealand Neogene paleogeography. Proc Natl Acad Sci USA 106:20646–20651

Burleigh JG, Kimball RT, Braun EL (2015) Building the avian tree of life using a large-scale, sparse supermatrix. Mol Phylogenet Evol 84:53–63

Burnham DA, Feduccia A, Martin LD, Falk AR (2011) Tree climbing–a fundamental avian adaptation. J Syst Palaeontol 9:103–107

Butler RJ, Barrett PM, Gower DJ (2009) Postcranial skeletal pneumaticity and air-sacs in the earliest pterosaurs. Biol Lett 5:557–560

Camp CL (1936) A new type of small bipedal dinosaur from the Navajo Sandstone of Arizona. University of California Publications Department of Geological Sciences Bulletin 24:39–56

Carney RM, Vinther J, Shawkey MD, D'Alba L, Ackermann J (2012) New evidence on the colour and nature of the isolated *Archaeopteryx* feather. Nat Commun 3:637

Carney RM, Tischlinger H, Shawkey MD (2020) Evidence corroborates identity of isolated fossil feather as a wing covert of *Archaeopteryx*. Sci Rep 10:15593

Carpenter K (2002) Forelimb biomechanics of nonavian theropod dinosaurs in predation. Senechenbergiana lethaea 82:59–76

Carrano MT (2000) Homoplasy and the evolution of dinosaur locomotion. Paleobiology 26:489–512

Carrano MT, Sampson SD (2008) The phylogeny of Ceratosauria (Dinosauria: Theropoda). J Syst Palaeontol 6:183–236

Carrano MT, Benson RBJ, Sampson SD (2012) The phylogeny of Tetanurae (Dinosauria: Theropoda). J Syst Palaeontol 10:211–300

Cartmill M (1985) Climbing. In: Hildebrand M, Bramble D, Liem KF, Wake DB (eds) Functional vertebrate morphology. Harvard University Press, Cambridge, MA, pp 73–88

Chatterjee S, Templin RJ, Campbell KE Jr (2007) The aerodynamics of *Argentavis*, the world's largest flying bird from the Miocene of Argentina. Proc Natl Acad Sci USA 104:12398–12403

Chen A, White ND, Benson RBJ, Braun MJ, Field DJ (2019) Total-evidence framework reveals complex morphological evolution in nightbirds (Strisores). Diversity 11:143

Chiappe LM (1991) Cretaceous avian remains from Patagonia shed new light on the early radiation of birds. Alcheringa 15:333–338

Chiappe LM, Dyke GJ (2006) The early evolutionary history of birds. J Paleontol Soc Korea 22:133–151

Chiappe LM, Walker CA (2002) Skeletal morphology and systematics of the Cretaceous Enantiornithes (Ornithothoraces: Enantiornithes). In: Chiappe LM, Witmer LM (eds) Mesozoic birds: above the heads of dinosaurs. University of California Press, Berkeley, CA, pp 240–267

Chiappe LM, Zhao B, O'Connor JK, Chunling G, Wang X, Habib M, Marugan-Lobon J, Meng Q, Cheng X (2014) A new specimen of the Early Cretaceous bird *Hongshanornis longicresta*: insights into the aerodynamics and diet of a basal ornithuromorph. PeerJ 2:e234

Chinsamy A, Elzanowski A (2001) Bone histology: evolution of growth pattern in birds. Nature 412:402–403

Chinsamy A, Chiappe LM, Dodson P (1995) Mesozoic avian bone microstructure: physiological implications. Paleobiology 21:561–574

Chinsamy A, Chiappe LM, Marugán-Lobón J, Chunling G, Fenghiao Z (2013) Gender identification of the Mesozoic bird *Confuciusornis sanctus*. Nat Commun 4:1381

Chinsamy A, Marugán-Lobón J, Serrano FJ, Chiappe L (2020) Osteohistology and life history of the basal pygostylian, *Confuciusornis sanctus*. Anat Rec 303: 949–962

Chure DJ, Madsen JH (1996) On the presence of furculae in some non-maniraptoran theropods. J Vertebr Paleontol 16:573–577

Cibois A, Thibault J-C, Bonillo C, Filardi CE, Watling D, Pasquet E (2014) Phylogeny and biogeography of the fruit doves (Aves: Columbidae). Mol Phylogenet Evol 70:442–453

Claessens LPAM (2004) Dinosaur gastralia: origin, morphology, and function. J Vertebr Paleontol 24:89–106

Clark AD, O'Connor JK (2021) Exploring the ecomorphology of two Cretaceous enantiornithines with unique pedal morphology. Front Ecol Evol 9:281

Clark JM, Xu X (2009) Evolutionary transitions among dinosaurs: examples from the Jurassic of China. Evol Educ Outreach 2:236–247

Clarke A, Pörtner H-O (2010) Temperature, metabolic power and the evolution of endothermy. Biol Rev 85: 703–727

Clemente CJ, Withers PC, Thompson G, Lloyd D (2008) Why go bipedal? Locomotion and morphology in Australian agamid lizards. J Exp Biol 211:2058–2065

Close RA, Rayfield EJ (2012) Functional morphometric analysis of the furcular in Mesozoic birds. PLoS ONE 7:e36664

Clyde WC, Ramezani J, Johnson KR, Bowring SA, Jones MM (2016) Direct high-precision U–Pb geochronology of the end-Cretaceous extinction and calibration of Paleocene astronomical timescales. Earth Planet Sci Lett 452:272–280

Codd JR, Manning PL, Norell MA, Perry SF (2008) Avian-like breathing mechanics in maniraptoran dinosaurs. Proc R Soc B 275:157–161

Crouch EM, Heilmann-Clausen C, Brinkhuis H, Morgans HE, Rogers KM, Egger H, Schmitz B (2001) Global dinoflagellate event associated with the late Paleocene thermal maximum. Geology 29:315–318

Czerkas SA, Feduccia A (2014) Jurassic archosaur is a non-dinosaurian bird. J Ornithol 155:841–851

D'Alba L, Maia R, Hauber ME, Shawkey MD (2016) The evolution of eggshell cuticle in relation to nesting ecology. Proc R Soc B 283:20160687

Dalsätt J, Zhou Z, Zhang F, Ericson PGP (2006) Food remains in *Confuciusornis sanctus* suggest a fish diet. Naturwissenschaften 93:444–446

Darwin C (1859) On the origin of species. John Murray, London

Davies RG, Irlich UM, Sl Chown, Gaston KJ (2010) Ambient, productive and wind energy and ocean extent predict global species richness of procellariform seabirds. Global Ecol Biogeog 19:98–110

De Baets K, Antonelli A, Donoghue PCJ (2016) Tectonic blocks and molecular clocks. Philos Trans R Soc B 371:20160098

Dececchi TA, Larsson HCE (2013) Body and limb size dissociation at the origin of birds: uncoupling allometric constraints across a macroevolutionary transition. Evolution 67:2741–2752

Depraetere C, Dahl AL (2007) Island locations and classifcations. In: Baldacchino G (ed) A World of Islands. An Island Studies Reader. Agenda Academic Publishers, Luqa, Malta, pp 57–105

Ding T-S, Yuan H-W, Geng S, Koh C-N, Lee P-F (2006) Macro-scale bird species richness patterns of the East Asian mainland and islands: energy, area and isolation. J Biogeogr 33:683–693

Dyke G (2010) Winged victory: modern birds now found to have been contemporaries of dinosaurs. Sci Am 1923:70–75

Dyke GJ, Nudds RL (2009) The fossil record and limb disparity of enantiornithines, the dominant flying birds of the Cretaceous. Lethaia 42:248–254

Dyke G, Vremir M, Kaiser G, Naish D (2012) A drowned Mesozoic bird breeding colony from the Late Cretaceous of Transylvania. Naturwissenschaften 99:435–442

Eagle RA, Enriquez M, Grellet-Tinner G, Pérez-Huerta A, Hu D, Tütken T, Montanari S, Lloyd SJ, Ramirez P, Tripati AK, Kohn MJ, Cerling TE, Chiappe LM, Eiler JM (2015) Isotopic ordering in eggshells reflects body temperatures and suggests differing thermophysiology in two Cretaceous dinosaurs. Nat Commun 6:8296

Elzanowski A (1974) Preliminary note on the palaeognathous bird from the Upper Cretaceous of Mongolia. Palaeontologica Polonica 30:103–109

Elzanowski A, Peters DS, Mayr G (2018) Cranial morphology of the Early Cretaceous bird *Confuciusornis*. J Vertebr Paleontol 38:e1439832

Erickson GM (2014) On dinosaur growth. Annu Rev Earth Planet Sci 42:675–697

Erickson GM, Makovicky PJ, Currie PJ, Norell MA, Yerby SA, Brochu CA (2004) Gigantism and comparative life-history parameters of tyrannosaurid dinosaurs. Nature 430:772–775

Erickson GM, Rauhut OWM, Zhou Z, Turner AH, Inouye BD, Hu D, Norell MA (2009) Was dinosaurian physiology inherited by birds? Reconciling slow growth in *Archaeopteryx*. PLoS ONE 4:e7390

Ericson PGP (1997) Systematic relationships of the palaeogene family Presbyornithidae (Aves: Anseriformes). Zool J Linnean Soc 121:429–483

Ericson PGP (2012) Evolution of terrestrial birds in three continents: biogeography and parallel radiations. J Biogeogr 39:813–824

Ericson PGP, Klopfstein S, Irestedt M, Nguyen JMT, Nylander JAA (2014) Dating the diversification of the major lineages of Passeriformes (Aves). BMC Evol Biol 2014:8

Evans KL, Warren PH, Gaston KJ (2005) Species-energy relationships at the macroecological scale: a review of the mechanisms. Biol Rev Camb Philos Soc 80:1–25

Ewer RF (1965) The anatomy of the thecodont reptile *Euparkeria capensis* Broom. Philos Trans R Soc B 248:379–435

Falconer H [online] (1863) Letter to Charles Darwin from the Darwin Correspondence Project. <http://www.darwinproject.ac.uk/letter/?docId=letters/DCP-LETT-3899.xml;query=falconer;brand=default>

Falk AR, Kaye TG, Zhou ZH, Burnham DA (2016) Laser fluorescence illuminates the soft tissue and life habits of the Early Cretaceous bird *Confuciusornis*. PLoS ONE 11:e0167284

Farlow JO, Gatesy SM, Holtz TR Jr, Hutchinson JR, Robinson JM (2000) Theropod locomotion. Am Zool 40:640–663

Farmer CG (2006) On the origin of avian air sacs. Respir Physiol Neurobiol 154:89–106

Farmer CG (2010) The provenance of alveolar and parabronchial lungs: insights from paleoecology and the discovery of cardiogenic, unidirectional airflow in the American alligator (*Alligator mississippiensis*). Physiol Biochem Zool 83:561–575

Feduccia A (2002) Birds are dinosaurs: simple answer to a complex problem. Auk 119:1187–1201

Feduccia A (2014) Avian extinction at the end of the Cretaceous: assessing the magnitude and subsequent explosive radiation. Cretac Res 50:1–15

Feduccia A, Martin LD, Tarsitano S (2007) *Archaeopteryx* 2007: Quo vadis? Auk 124:373–380

Feo TJ, Field DJ, Prum RO (2015) Barb geometry of asymmetrical feathers reveals a transitional morphology in the evolution of avian flight. Proc R Soc B 282: 20142864

Fernández MS, García RA, Fiorelli L, Scolaro A, Salvador RB, Cotaro CN, Kaiser GW, Dyke GJ (2013) A large accumulation of avian eggs from the Late Cretaceous of Patagonia (Argentina) reveals a novel nesting strategy in Mesozoic birds. PLoS ONE 8:e61030

Field DJ, Bercovici A, Berv JS, Dunn R, Fastovsky DE, Lyson TR, Vajda V, Gauthier JA (2018) Early evolution of modern birds structured by global forest collapse at the end-Cretaceous mass extinction. Curr Biol 28:1825–1831

Field DJ, Benito J, Chen A, Jagt JW, Ksepka DT (2020) Late Cretaceous neornithine from Europe illuminates the origins of crown birds. Nature 579:397–401

Finlayson C (2011) Avian survivors: the history and biogeography of Palearctic birds. T & AD Poyser, London

Fjeldså J (2013) The global diversification of songbirds (Oscines) and the build-up of the Sino-Himalayan diversity hotspot. Chin Birds 4:132–143

Fjeldså J, Bowie RCK (2008) New perspectives on the origin and diversification of Africa's forest avifauna. Afr J Ecol 46:235–247

Fjeldsa J, Zuccon D, Irestedt M, Johansson U, Ericson PGP (2003) *Sapayoa aenigma*: a New World representative of the 'Old World suboscines.' Proc R Soc B (Suppl) 270: S238–S241

Fuchs J, Cruaud C, Couloux A, Pasquet E (2007) Complex biogeographic history of the cuckoo-shrikes and allies (Passeriformes: Campephagidae) revealed by mitochondrial and nuclear sequence data. Mol Phylogenet Evol 44:138–153

Furbringer M (1888) Untersuchungen zur morphologie und systematik der vogel, volume II. Verlag von TJ. Van Holkema, Amsterdam

Gauthier J (1986) Saurischian monophyly and the origin of birds. Memoirs of the California Academy of Science 8, 1–55

Gill F, Donsker D (eds) [online] (2015) IOC World Bird List (v 5.3) <www.worldbirdnames.org>

Godefroit P, Cau A, Dong-Yu H, Escuillié F, Wenhao W, Dyke G (2013) A Jurassic avialan dinosaur from China resolves the early phylogenetic history of birds. Nature 498:359–362. (Supplementary information)

Goderis S, Tagle R, Belza J, Smit J, Montanari A, Vanhaecke F, Erzinger J, Claeys P (2013) Reevaluation of siderophile element abundances and ratios across the Cretaceous-Paleogene (K-Pg) boundary: implications for the nature of the projectile. Geochim Cosmochim Acta 120:417–446

Grant PR, Grant BR (2011) How and why species multiply: the radiation of Darwin's finches. Princeton University Press, Princeton, NJ

Gregory WK (1916) Theories of the origin of birds. Ann NY Acad Sci 27:31–38

Guo Y, Wang J, Zhang F, Lei L, Wang X, Zhao Y (2018) Research status and prospect of the flying ability and its related issues of *Confuciusornis*. Adv Geosci 8: 1338–1344

Haddrath O, Baker AJ (2012) Multiple nuclear genes and retroposons support vicariance and dispersal of the palaeognaths, and an Early Cretaceous origin of modern birds. Proc R Soc B 279:4617–4625

Hackett SJ, Kimball RT, Reddy S, Bowie RCK, Braun EL, Braun MJ, Chojnowski JL, Cox WA, Han KL, Harshman J, Huddleston CJ (2008) A phylogenomic study of birds reveals their evolutionary history. Science 320:1763–1768

Haeckel EHPA (1866) Generelle morphologie der organismen. G. Reimer, Berlin

Han G, Chiappe LM, Ji S-A, Habib M, Turner AH, Chinsamy A, Liu X, Han L (2014) A new raptorial dinosaur with exceptionally long feathering provides insights into dromaeosaurid flight performance. Nat Commun 5:4382

Harris CM, Lorenz K, Fishpool LDC, Lascelles B, Cooper J, Coria NR, Croxall JP, Emmerson LM, Fijn RC, Fraser WL, Jouventin P, LaRue MA, Le Maho Y, Lynch HJ, Naveen R, Fraser-Patterson DL, Peter H-U, Poncet S, Phillips RA, Southwell CJ, van Franeker JA, Weimerskirch H, Wienecke B, Woehler EJ (2015) Important bird areas in Antarctica 2015. BirdLife International and Environmental Research & Assessment Ltd., Cambridge

Hawkins BA, McCain CM, Davies TJ, Buckley LB, Anacker B, Cornell HV, Damschen EI, Grytnes J-A, Harrison S, Holt RD, Kraft NJB, Stephens PR (2012) Different evolutionary histories underlie congruent species richness gradients of birds and mammals. J Biogeogr 39:825–841

He H-Y, Wang X-L, Jin F, Zhou Z-H, Wang F, Yang L-K, Ding X, Boven A, Zhu R-X (2006) The 40Ar/39Ar dating of the early Jehol Biota from Fengning, Hebei Province, northern China. Geochem Geophys Geosyst 7:1–8

Heers AM, Dial KP (2011) From extant to extinct: locomotor ontogeny and the evolution of avian flight. Trends Ecol Evol 27:296–305

Heers AM, Dial KP (2015) Wings versus legs in the avian *bauplan*: development and evolution of alternative locomotor strategies. Evolution 69:305–320

Heilmann G (1916) Fuglenes Afstamning. Dansk Ornitologisk Tidsskrift, Copenhagen

Herold N, Buzan J, Seton M, Goldner A, Green JAM, Müller RD, Markwick P, Huber M (2014) A suite of early Eocene (~ 55 Ma) climate model boundary conditions. Geosci Model Dev 7:2077–2090

Hildebrand SF, Gilmore CW, Cochran DM (1930) Cold-blooded vertebrates: Part I. Fishes, Part II–III, Amphibians and reptiles. Smithsonian Scientific Series, Inc., New York, NY

Ho SYW (2014) The change face of the molecular evolutionary clock. Trends Ecol Evol 29:496–503

Hochachka PW, Burelle Y (2004) Control of maximum metabolic rate in humans: dependence on performance phenotypes. Mol Cell Biochem 256:95–103

Hoelzer GA, Melnick DJ (1994) Patterns of speciation and limits to phylogenetic resolution. Trends Ecol Evol 9: 104–107

Hopson JA (2001) Ecomorphology of avian and nonavian theropod phalangeal proportions: implications for the arboreal versus terrestrial origin of bird flight. In: Gauthier J, Gall LF (eds) New perspectives on the origin and early evolution of birds. Peabody Museum of Natural History, New Haven, CT, pp 211–235

Hosner PA, Sánchez-González LA, Peterson AT, Moyle RG (2014) Climate-driven diversification and Pleistocene refugia in Philippine birds: evidence from phylogeographic structure and paleoenvironmental niche modeling. Evolution 68:2658–2674

Hospitaleche CA, Worthy TH (2021) New data on the *Vegavis iaai* holotype from the Maastrichtian of Antarctica. Cretac Res 124:104818

Houben AJP, Bijl PK, Pross J, Bohaty SM, Passchier S, Stickley CE, Röhl U, Sugisaki S, Tauxe L, van de Flierdt T, Olney M, Sangiorgi F, Sluijs A, Escutia C, Brinkhuis H, and the Expedition 318 Scientists (2013) Reorganization of Southern Ocean plankton ecosystem at the onset of Antarctic glaciation. Science 340:341–344

Hsia CC, Schmitz A, Lambertz M, Perry SF, Maina JN (2013) Evolution of air breathing: oxygen homeostasis and the transitions from water to land and sky. Compr Physiol 3:849–915

Hutchinson JR, Gatesy SM (2000) Adductors, abductors, and the evolution of archosaur locomotion. Paleobiology 26:734–751

Hutchinson JR, Allen V (2009) The evolutionary continuum of limb function from early theropods to birds. Naturwissenschaften 96:423–448

Huttenlocker AK (2014) Body size reductions in nonmammalian eutheriodont therapsids (Synapsida) during the end-Permian mass extinction. PLoS ONE 9:e87553

Huxley TH (1867) On the classification of birds and on the taxonomic value of the modifications of certain of the cranial bones observable in that class. Proc Zool Soc London 1867:415–472

Huxley TH (1870) On the classification of the Dinosauria with observations on the Dinosauria of the Trias. Q J Geol Soc 26:32–51

International Commission on Zoological Nomenclature (ICZN) (2011) *Archaeopteryx lithographica* von Meyer, 1861 (Aves): conservation of usage by designation of a neotype. Bull Zool Nomencl 68:230–233

James HF (2005) Paleogene fossils and the radiation of modern birds. Auk 122:1049–1054

James FC, Pourtless JA IV (2009) Cladistics and the origin of birds: a review and two new analyses. Auk 66:1–78

Jarvis ED, Mirarab S, Aberer AJ, Li B, Houde P, Li C, Ho SY, Faircloth BC, Nabholz B, Howard JT, Suh A (2014) Whole-genome analyses resolves early branches in the tree of life of modern birds. Science 346:1320–1331

Jetz W, Thomas GH, Joy JB, Hartmann K, Mooers AO (2012) The global diversity of birds in space and time. Nature 491:444–448

Jonsson KA, Fjeldsa (2006) Determining biogeographical patterns of dispersal and diversification in oscine passerine birds in Australia, southeast Asia and Africa. J Biogeog 33:1155–1165

Johnson JA, Brown JW, Fuchs J, Mindell DP (2016) Multi-locus phylogenetic inference among New World vultures (Aves: Cathartidae). Mol Phylogenet Evol 105:193–199

Jønsson KA, Fabre P-H, Ricklefs RE, Fjeldså J (2011) Major global radiation of corvoid birds originated in the proto-Papuan archipelago. Proc Natl Acad Sci USA 108:2328–2333

Jønsson KA, Fabre P-H, Kennedy JD, Holt BG, Borregaard MK, Rahbek C, Fjeldså J (2016) A supermatrix phylogeny of corvoid passerine birds (Aves: Corvides). Mol Phylogenet Evol 94:87–94

Kaiho K, Oshima N (2017) Site of asteroid impact changed the history of life on Earth: the low probability of mass extinction. Sci Rep 7:14855

Karpouzi VS, Watson R, Pauly D (2007) Modelling and mapping resource overlap between seabirds and fisheries on a global scale: a preliminary assessment. Mar Ecol Prog Ser 343:87–99

Kaye TG, Pittman M, Marugán-Lobón J, Martín-Abad H, Sanz JL, Buscalioni AD (2019a) Fully fledged enantiornithine hatchling revealed by Laser-Stimulated Fluorescence supports precocial nesting behavior. Sci Rep 9:5006

Kaye TG, Pittman M, Mayr G, Schwarz D, Xu X (2019b) Detection of lost calamus challenges identity of isolated *Archaeopteryx* feather. Sci Rep 9:1182

Kemp TS (2007) The concept of correlated progression as the basis of a model for the evolutionary origin of major new taxa. Proc R Soc B 274:1667–1673

Kennedy JD, Wang Z, Weir JT, Rahbek C, Fjeldsa J, Price TD (2014) Into and out of the tropics: the generation of the latitudinal gradient among New World passerine birds. J Biogeog 41:1746–1757

Kimball RT, Oliveros CH, Wang N, White ND, Barker EK, Field DJ, Ksepka DT, Chesser RT, Moyle RG, Braun MJ, Brumfield RT (2019) A phylogenomic supertree of birds. Diversity 11:109

Knoll F, Chiappe LM, Sanchez S, Garwood RJ, Edwards NP, Wogelius RA, Sellers WI, Manning PL, Ortega F, Serrano FJ, Marugán-Lobón J (2018) A diminutive perinate European Enantiornithes reveals an asynchronous ossification pattern in early birds. Nat Commun 9:937

Kring DA (2007) The Chicxulub impact event and its environmental consequences at the Cretaceous–Tertiary boundary. Palaeogeogr Palaeoclimatol Palaeoecol 255:4–21

Ksepka DT (2014a) Evolution: a rapid flight towards birds. Curr Biol 24:R1052–R1055

Ksepka DT (2014b) Flight performance of the largest volant bird. Proc Natl Acad Sci USA 111:10624–10629

Ksepka D, Clarke J (2015) Phylogenetically vetted and stratigraphically constrained fossil calibrations within Aves. Palaeontol Electron 18:1–25

Ksepka DT, Phillips MJ (2015) Avian diversification patterns across the K-Pg boundary: influence of calibrations, datasets, and model misspecification. Ann Mo Bot Gard 100:300–328

Kubo T, Benton MJ (2009) Tetrapod postural shift estimated from Permian and Triassic trackways. Paléo 52:1029–1037

Kubo T, Kubo MO (2012) Associated evolution of bipedality and cursoriality among Triassic archosaurs: a phylogenetically controlled evaluation. Paleobiology 38:474–485

Kurochkin EN, Chatterjee S, Mikhailov KE (2013) An embryonic enantiornithine bird and associated eggs from the Cretaceous of Mongolia. Paleontol J 47:1252–1269

Kutschera U (2011) From the scala naturae to the symbiogenetic and dynamic tree of life. Biol Direct 6:33

Lambertz M, Perry SF (2015) Remarks on the evolution of the avian sternum, dinosaur gastralia, and their

functional significance for the respiratory apparatus. Zool Anz 255:80–84
Landwehrs JP, Feulner G, Hofmann M, Petri S (2020) Climatic fluctuations modeled for carbon and sulfur emissions from end-Triassic volcanism. Earth Planet Sci Lett 537:116174
Langer MC, Ezcurra MD, Rauhut OWM, Benton MJ, Knoll F, McPhee BW, Novas FE, Pol D, Brusatte SL (2017) Untangling the dinosaur family tree. Nature 551:E1–E3
Larson DW, Brown CM, Evans DC (2016) Dental disparity and ecological stability in bird-like dinosaurs prior to the end-Cretaceous mass extinction. Curr Biol 26: 1325–1333
Lautenschlager S, Bright JA, Rayfield EJ (2014) Digital dissection – using contrast-enhanced computed tomography scanning to elucidate hard- and soft-tissue anatomy in the Common Buzzard *Buteo buteo*. J Anat 224: 412–431
Lee MS, Cau A, Naish D, Dyke GJ (2014) Sustained miniaturization and anatomical innovation in the dinosaurian ancestors of birds. Science 345:562–566
Lee JR, Raymond B, Bracegirdle TJ, Chadès I, Fuller RA, Shaw JD, Terauds A (2017) Climate change drives expansion of Antarctic ice-free habitat. Nature 547: 49–54
Li C, Zhang Y, Li J, Kong L, Hu H, Pan H, Xu L, Deng Y, Li Q, Jin L, Yu H, Chen Y, Liu B, Yang L, Liu S, Zhang Y, Lang Y, Xia J, He W, Shi Q, Subramanian S, Millar CD, Meader S, Rands CM, Fujita MK, Greenwold MJ, Castoe TA, Pollock DD, Gu W, Nam K, Ellegren H, Ho SYW, Burt DW, Ponting CP, Jarvis ED, Gilbert MTP, Yang H, Wang J, Lambert DM, Wang J, Zhang G (2014a) Two Antarctic penguin genomes reveal insights into their evolutionary history and molecular changes related to the Antarctic environment. GigaScience 3:27
Li Q, Clarke JA, Gao K-Q, Zhou C-F, Meng Q, Li D, D'Alba L, Shawkey MD (2014b) Melanosome evolution indicates a key physiological shift within feathered dinosaurs. Nature 507:350–353
Li Q, Clarke JA, Gao KQ, Peteya JA, Shawkey MD (2018) Elaborate plumage patterning in a Cretaceous bird. PeerJ 6:e5831
Li Z, Wang C-C, Wang M, Chiang C-C, Wang Y, Zheng X, Huang E-W, Hsiao K, Zhou Z (2020) Ultramicrostructural reductions in teeth: implications for dietary transition from non-avian dinosaurs to birds. BMC Evol Biol 20:46
Liu D, Chiappe LM, Serrano F, Habib M, Zhang Y, Meng Q (2017) Flight aerodynamics in enantiornithines: information from a new Chinese Early Cretaceous bird. PLoS ONE 12:e0184637
Lockley M, Kukihara R, Mitchell L (2009) Why *Tyrannosaurus rex* had puny arms: an integral morphodynamic solution to a simple puzzle in theropod paleobiology. In: Larson P, Carpenter K (eds) *Tyrannosaurus rex*: the tyrant king. Indiana University Press, Bloomington, IN, pp 131–164
Longrich NR, Tokaryk TT, Field D (2011) Mass extinction of birds at the Cretaceous-Paleogene (K-Pg) boundary. Proc Natl Acad Sci USA 108:15253–15257
Longrich NR, Bhullar B-A, Gauthier JA (2012) Mass extinction of lizards and snakes at the Cretaceous-Paleogene boundary. Proc Natl Acad Sci USA 109: 21396–21401
Louchart A, Viriot L (2011) From snout to beak: the loss of teeth in birds. Trends Ecol Evol 26:663–673
Maidment SCR, Barrett PM (2014) Osteological correlates for quadrupedality in ornithischian dinosaurs. Acta Palaeontol Pol 59:53–70
Makovicky PJ, Currie PJ (1998) The presence of a furcula in tyrannosaurid theropods, and its phylogenetic and functional implications. J Vertebr Paleontol 18:143–149
Makovicky PJ, Zanno LE (2011) Theropod diversity and the refinement of avian characteristics. In: Dyke G, Kaiser G (eds) Living dinosaurs: the evolutionary history of modern birds. Wiley, New York, NY, pp 9–29
Martin LD (2004) A basal archosaurian origin of birds. Acta Zool Sin 50:978–990
Martinez RN, Alcober OA (2009) A basal sauropodomorph (Dinosauria: Saurischia) from the Ischigualasto Formation (Triassic, Carnian) and the early evolution of Sauropodomorpha. PLoS ONE 4: e4397
Marsh OC (1877) Notice of new dinosaurian reptiles from the Jurassic formation. Amer J Sci 14:514–516
Mayr E (1946) History of the North American bird fauna. Wilson Bull 58:3–41
Mayr E (1964) Inference concerning the Tertiary American bird faunas. Proc Natl Acad Sci USA 51:280–288
Mayr G (2009) Paleogene fossil birds. Springer, Berlin
Mayr G (2014) The origins of crown group birds: molecules and fossils. Palaeontology 57:231–242
Mayr G (2015) Skeletal morphology of the middle Eocene swift *Scaniacypselus* and the evolutionary history of true swifts (Apodidae). J Ornithol 156:441–450
Mayr G (2017a) Evolution of avian breeding strategies and its relation to the habitat preferences of Mesozoic birds. Evol Ecol 31:131–141
Mayr G (2017b) Avian evolution: the fossil record of birds and its paleobiological significance. Wiley Blackwell, Chichester
Mayr G, Kaye TG, Pittman M, Saitta ET, Pott C (2020) Reanalysis of putative ovarian follicles suggests that Early Cretaceous birds were feeding not breeding. Sci Rep 10:19035
McDowall RM (2008) Process and pattern in the biogeography of New Zealand – a global microcosm? J Biogeogr 35:197–212
McGowan AJ, Dyke GJ (2007) A morphospace-based test for competitive exclusion among flying vertebrates: did birds, bats and pterosaurs get in each other's space? J Evol Biol 20:1230–1236
McGuire JA, Witt CC, Remsen JV Jr, Corl A, Rabosky DL, Altshuler DL, Dudley R (2014) Molecular

phylogenetics and the diversification of hummingbirds. Curr Biol 24:910–916
McPhee BW, Benson RB, Botha-Brink J, Bordy EM, Choiniere JN (2018) A giant dinosaur from the earliest Jurassic of South Africa and the transition to quadrupedality in early sauropodomorphs. Current Biol 28:3143–3151
Min W, Zhong-He Z, O'Connor JK, Zelenkov NV (2014) A new diverse enantiornithine family (Bohaiornithidae fam. nov.) from the Lower Cretaceous of China with information from two new species. Vert Pal 52:31
von Meyer H (1861a) Vögel-Federn und *Palpipes priscus* von Solnhofen. Neues Jahrbuch fur Mineralogie, Geognosie, Geologie und Paläontologie 561
von Meyer H (1861b) *Archaeopteryx lithographica* (Vögel-Feder) und pterodactylus von Solnhofen. Neues Jahrbuch fur Mineralogie, Geognosie, Geologie und Paläontologie 678–679
Miller CV, Pittman M (2021) The diet of early birds based on modern and fossil evidence and a new framework for its reconstruction. Biol Rev 96:2058–2112
Mitchell JS, Makovicky PJ (2014) Low ecological disparity in Early Cretaceous birds. Proc R Soc B 281: 201440608
Mitchell KJ, Llamas B, Soubrier J, Rawlence NJ, Worthy TH, Wood J, Lee MSY, Cooper A (2014) Ancient DNA reveals elephant birds and kiwi are sister taxa and clarifies ratite bird evolution. Science 344:898–900
Moen D, Morlon H (2014) Why does diversification slow down? Trends Ecol Evol 29:190–197
Moreau RE (1952) Africa since the Mesozoic: with particular reference to certain biological problems. Proc Zool Soc London 121:869–913
Morschhauser EM, Varricchio DJ, Chunling G, Jinyuan L, Xuri W, Xiadong C, Qingjin M (2009) Anatomy of the Early Cretaceous bird *Rapaxavis pani*, a new species from Liaoning Province, China. J Vertebr Paleontol 29:545–554
Moyle RG, Chesser RT, Prum RO, Schikler P, Cracraft J (2006) Phylogeny and evolutionary history of Old World suboscine birds (Aves: Eurylaimides). Am Mus Novit Num 3544:1–22
Navalón G, Marugán-Lobón J, Chiappe LM, Sanz JL, Buscalioni ÁD (2015) Soft-tissue and dermal arrangement in the wing of an Early Cretaceous bird: implications for the evolution of avian flight. Sci Rep 5:14864
Nesbitt SJ (2007) The anatomy of *Effigia okeeffene* (Archosauria, Suchia), theropod-like convergence, and the distribution of related taxa. Bull Am Mus Nat Hist 302:1–84
Newton I (2003) The speciation and biogeography of birds. Academic Press, London
Newton I, Dale L (2001) A comparative analysis of the avifaunas of different zoogeographical regions. J Zool (Lond) 254:207–218
Norell MA, Xu X (2005) Feathered dinosaurs. Annu Rev Earth Planet Sci 33:277–299
Nudds RL, Dyke GJ, Rayner JMV (2004) Forelimb proportions and the evolutionary radiation of Neornithes. Proc R Soc B 271:S324–S327
Nudds RL, Atterholt J, Wang X, You H-L, Dyke GJ (2013) Locomotory abilities and habitat of the Cretaceous bird *Gansus yumenensis* inferred from limb length proportions. J Evol Biol 26:150–154
O'Connor JK (2019) The trophic habits of early birds. Palaeogeogr Palaeoclimatol Palaeoecol 513:178–195
O'Connor JK, Chang H (2015) Hindlimb feathers in paravians: primarily "wings" or ornaments? Biol Bull 42:616–621
O'Connor PM, Claessens LPAM (2005) Basic avian pulmonary design and flow-through ventilation in non-avian theropod dinosaurs. Nature 436:253–256
O'Connor J, Zhou Z (2015) Early evolution of the biological bird: perspectives from new fossil discoveries in China. J Ornithol 156:333–342
O'Connor JK, Gao K-Q, Chiappe LM (2010) A new ornithuromorph (Aves: Ornithothoraces) bird from the Jehol group indicative of higher-level diversity. J Vertebr Paleontol 30:311–321
O'Connor J, Chiappe LM, Bell A (2011) Pre-modern birds: avian divergences in the Mesozoic. In: Dyke G, Kaiser G (eds) Living dinosaurs: the evolutionary history of modern birds. Wiley, New York, NY, pp 39–114
O'Connor JK, Chiappe LM, Chuong C-M, Bottjer DJ, You H (2012) Homology and potential cellular and molecular mechanisms for the development of unique feather morphologies in early birds. Geosciences 2: 157–177
O'Connor JK, Zheng X-T, Wang X-L, Wang Y, Zhou Z-H (2013) Ovarian follicles shed new light on dinosaur reproduction during the transition towards birds. Natl Sci Rev 1:15–17
O'Connor JK, Wang M, Zheng X-T, Wang X-L, Zhou Z-H (2014) The histology of two female Early Cretaceous birds. Vertebr PalAsiatica 52:112–128
O'Connor JK, Wang M, Zhou S, Zhou Z (2015a) Osteohistology of the Lower Cretaceous Yixian Formation ornithuromorph (Aves) *Iteravis huchzermeyeri*. Palaeontol Electron 18.2.35A:1–11
O'Connor JK, Xiao-Ting Z, Xiao-Li W, Xiao-Mei Z, Zhong-He Z (2015b) The gastral basket in basal birds and their close relatives: size and possible function. Vertebr PalAsiatica 53:133–152
O'Connor JK, Zheng X-T, Wang X-L, Zhang X-M, Zhou Z-H (2015c) The gastral basket in basal birds and their close relatives: size and possible function. Vertebr PalAsiatica 53:133–152
O'Connor JK, Wang X, Zheng X, Hu H, Zhang X, Zhou Z (2016) An enantiornithine with a fan-shaped tail, and the evolution of the rectricial complex in early birds. Curr Biol 26:114–119
O'Connor J, Erickson GM, Norell M, Bailleul AM, Hu H, Zhou Z (2018) Medullary bone in an Early Cretaceous enantiornithine bird and discussion regarding its identification in fossils. Nat Commun 9:5169

O'Connor JK, Falk A, Wang M (2020a) First report of immature feathers in juvenile enantiornithines from the Early Cretaceous Jehol avifauna. Vertebr PalAsiatica 58:24–44

O'Connor PM, Turner AH, Groenke JR, Felice RN, Rogers RR, Krause DW, Rahantarisoa LJ (2020b) Late Cretaceous bird from Madagascar reveals unique development of beaks. Nature 588:272–276

Ödeen A, Håstad O, Alström P (2011) Evolution of ultraviolet vision in the largest avian radiation – the passerines. BMC Evol Biol 11:313

Ohlson J, Fjeldså J, Ericson PGP (2008) Tyrant flycatchers coming out in the open: phylogeny and ecological radiation of Tyrannidae (Aves, Passeriformes). Zool Scr 37:315–335

Oliveros CH, Andersen MJ, Hosner PA, Mauck WM III, Sheldon FH, Cracraft J, Moyle RG (2020) Rapid Laurasian diversification of a Pantropical bird family during the Oligocene-Miocene transition. Ibis 162:137–152

Olson SL (1985) The fossil record of birds. In: Farner DS, King JR, Parkes KC (eds) Avian biology, vol 8. Academic Press, New York, NY, pp 79–238

Olson DM, Dinerstein E, Wikramanayake ED, Burgess ND, Powell GV, Underwood EC, D'amico JA, Itoua I, Strand HE, Morrison JC, Loucks CJ (2001) Terrestrial ecoregions of the world: a new map of life on earth. Bioscience 51:933–938

Osadchiev A, Silvestrova K, Myslenkov S (2020) Wind-driven coastal upwelling near large river deltas in the Laptev and East-Siberian seas. Remote Sens 12:844

Osborn HF (1924) Three new theropoda, *Ornitholestes, Struthiomimus, Tyrannosaurus*. Am Mus Novit 144:1–12

Ostrom JH (1973) The ancestry of birds. Nature 242:136

Ostrom JH (1975) The origin of birds. Annu Rev Earth Planet Sci 3:55–77

Ostrom JH (1976) *Archaeopteryx* and the origin of birds. Biol J Linn Soc 8:91–182

Owen R (1842) Report on British fossil reptiles, Part II. In: Report of the eleventh meeting of the British Association for the Advancement of Science. John Murray, London, pp 60–204

Owen R (1860) Darwin on the origin of species. Edinburgh Review 3:487–532

Owen R (1863) On the *Archeopteryx* of von Meyer, with a description of the fossil remains of a long-tailed species, from the lithographic stone of Solnhofen. Philos Trans R Soc Lond 153:33–47

Padian K, May CL (1993) The earliest dinosaurs. In: Lucas SG, Morales M (eds) The nonmarine Triassic. New Mexico Museum of Natural History and Science, Albuquerque, NM, pp 379–380

Pei R, Li Q, Meng Q, Norell MA, Gao K-Q (2017) New specimens of *Anchiornis huxleyi* (Theropoda: Paraves) from the Late Jurassic of northeastern China. Bull Am Mus Nat Hist 411:1–67

Persons WS IV, Currie P, Erickson GM (2020) An older and exceptionally large adult specimen of *Tyrannosaurus rex*. Anat Rec 303:656–672

Pittman M, Bell PR, Miller CV, Enriquez NJ, Wang X, Zheng X, Tsang LR, Tse YT, Landes M, Kaye TG (2022) Exceptional preservation and foot structure reveal ecological transitions and lifestyles of early theropod flyers. Nat Commun 13:7684

Plotnick RE, Theodor JM, Holtz TR Jr (2015) Jurassic Pork: what could a Jewish time traveler eat? Evol Educ Outreach 8:17

Pontzer H, Allen V, Hutchinson JR (2009) Biomechanics of running indicates endothermy in bipedal dinosaurs. PLoS ONE 4:e7783

Porras-Múzquiz HG, Chatterjee S, Lehman TM (2014) The carinate bird *Ichthyornis* from the Upper Cretaceous of Mexico. Cretac Res 51:148–152

Pross J, Contreras L, Bijl PK, Greenwood DR, Bohaty SM, Schouten S, Bendle JA, Röhl U, Tauxe L, Raine JI, Huck CE (2012) Persistent near-tropical warmth on the Antarctic continent during the early Eocene epoch. Nature 488:73–77

Prum RO, Berv JS, Dornburg A, Field DJ, Townsend JP, Lemmon EM, Lemmon AR (2015) A comprehensive phylogeny of birds (Aves) using targeted next-generation DNA sequencing. Nature 526:569–573

Rahbek C, Graves GR (2001) Multiscale assessment of patterns of avian species richness. Proc Natl Acad Sci USA 98:4534–4539

Rashid DJ, Chapman SC, Larsson HCE, Organ CL, Rebin A-G, Merzdorf CS, Bradley R, Horner JR (2014) From dinosaurs to birds: a tail of evolution. EvoDevo 5:25

Rashid DJ, Surya K, Chiappe LM, Carroll N, Garrett KL, Varghese B, Bailleul A, O'Connor JK, Chapman SC, Horner JR (2018) Avian tail ontogeny, pygostyle formation, and interpretation of juvenile Mesozoic specimens. Sci Rep 8:9014

Reddy S, Driskell A, Rabosky DL, Hackett SJ, Schulenberg TS (2012) Diversification and the adaptive radiation of the vangas of Madagascar. Proc R Soc B 279:2062–2071

Reddy S, Kimball RT, Pandey A, Hosner PA, Braun MJ, Hackett SJ, Han K-L, Harshman J, Huddleston CJ, Kingston S, Marks BD (2017) Why do phylogenomic data sets yield conflicting trees? Data type influences the avian tree of life more than taxon sampling. Syst Biol 66:857–879

dos Reis M, Inoue J, Hasegawa M, Asher RJ, Donoghue PCJ, Yang Z (2012) Phylogenomic datasets provide both precision and accuracy in estimating the timescale of placental mammal phylogeny. Proc R Soc B 279: 3491–3500

Richards RJ (2008) The tragic sense of life: Ernst Haeckel and the struggle over evolutionary thought. University of Chicago Press, Chicago, IL

de Ricqlès A (1974) Evolution of endothermy: histological evidence. Evol Theory 1:51–80

Riddick SN, Dragosits U, Blackall TD, Daunt F, Wanless S, Sutton MA (2012) The global distribution of ammonia emissions from seabird colonies. Atmos Environ 55:319–327

Ries CJ (2010) Angels, demons, birds and dinosaurs: creativity, meaning and truth in the life, art and science of Gerhard Heilmann (1859–1946). Interdiscip Sci Rev 35:69–91

Robertson DS, Lewis WM, Sheehan PM, Toon OB (2013) K-Pg extinction: reevaluation of the heat-fire hypothesis. J Geophys Res 118:329–336

Rocha R, Virtanen T, Cabeza M (2015) Bird assemblages in a Malagasy forest-agricultural frontier: effects of habitat structure and forest cover. Trop Conserv Sci 8:681–710

Rogers J, Gibbs RA (2014) Comparative primate genomics: emerging patterns of genome content and dynamics. Nat Rev Genet 15:347–359

Royer DL, Pagani M, Beerling DJ (2012) Geobiological constraints on Earth system sensitivity to $CO_2$ during the Cretaceous and Cenozoic. Geobiology 10:298–310

Sahney S, Benton MJ (2008) Recovery from the most profound mass extinction of all time. Proc R Soc B 275:759–765

Salinas-Saavedra M, Gonzalez-Cabrera C, Ossa-Fuentes L, Botelho JF, Ruiz-Flores M, Vargas AO (2014) New developmental evidence supports a homeotic frameshift of digit identity in the evolution of the bird wing. Front Zool 11:33

Sanz JL, Chiappe LM, Pérez-Moreno BP, Buscalioni AD, Moratalla JJ, Ortega F, Poyato-Ariza FJ (1996) An Early Cretaceous bird from Spain and its implications for the evolution of avian flight. Nature 382:442–445

Sato T, Chang Y-N, Wu X-C, Zelenitsky DA, Hsiao Y-F (2005) A pair of shelled eggs inside a female dinosaur. Science 308:375

Schachner ER, Farmer CG, McDonald AT, Dodson P (2011) Evolution of the dinosauriform respiratory apparatus: new evidence from the postcranial axial skeleton. Anat Rec 294:1532–1547

Schaller MF, Wright JD, Kent DV, Olsen PE (2012) Rapid emplacement of the Central Atlantic Magmatic Province as a net sink for $CO_2$. Earth Planet Sci Lett 323:27–39

Schweizer M, Seehausen O, Güntert M, Hertwig ST (2010) The evolutionary diversification of parrots supports a taxon pulse model with multiple trans-oceanic dispersal events and local radiations. Mol Phylogenet Evol 54:984–994

Schweizer M, Seehausen O, Hertwig ST (2011) Macroevolutionary patterns in the diversification of parrots: effects of climate change, geological events and key innovations. J Biogeogr 38:2126–2194

Seeley HG (1887) On the classification of the fossil animals commonly named Dinosauria. Proc R Soc Lond 43:165–171

Sereno PC, Forster CA, Rogers RR, Monetta AM (1993) Primitive dinosaur skeleton from Argentina and the early evolution of Dinosauria. Nature 361:64–66

Senter P (2006a) Forelimb function in *Ornitholestes hermanni* Osborn (Dinosauria, Theropoda). Palaeontology 49:1029–1034

Senter P (2006b) Scapular orientation in theropods and basal birds, and the origin of flapping flight. Acta Palaeontol Pol 51:305–313

Senter P (2010) Vestigial skeletal structures in dinosaurs. J Zool 280:60–71

Sereno PC (2005) The logical basis of phylogenetic taxonomy. Syst Biol 54:595–619

Sereno PC, McAllister S, Brusatte SL (2005) TaxonSearch: a relational database for suprageneric taxa and phylogenetic definitions. PhyloInformatics 8:1–21

Serrano FJ, Chiappe LM (2017) Aerodynamic modelling of a Cretaceous bird reveals thermal soaring capabilities during early avian evolution. J R Soc Interface 14:20170182

Serrano FJ, Palmqvist P, Chiappe LM, Sanz JL (2017) Inferring flight parameters of Mesozoic avians through multivariate analysis of forelimb elements in their living relatives. Paleobiology 43:144–169

Serrano FJ, Chiappe LM, Palmqvist P, Figueirido B, Marugán-Lobón J, Sanz JL (2018) Flight reconstruction of two European enantiornithines (Aves, Pygostylia) and the achievement of bounding flight in Early Cretaceous birds. Palaeontology 61:359–368

Serrano FJ, Pittman M, Kaye TG, Wang XL, Zheng XT, Chiappe LM (2020) Laser-stimulated fluorescence refines flight modelling of the Early Cretaceous bird *Sapeornis*. In: Pennaraptoran theropod dinosaurs: past progress and new frontiers (M. Pittman and X. Xu, eds.). Bull Am Mus Nat Hist 420: 333–344

Smith BT, Klicka J (2010) The profound influence of the Late Pliocene Panamanian uplift on the exchange, diversification, and distribution of New World birds. Ecography 33:333–342

Smith ND, Ksepka DT (2015) Five well-supported fossil calibrations within the 'Waterbird' assemblage (Tetrapoda, Aves). Palaeontol Electron 18:1–21

Sookias RB, Butler RJ, Benson RBJ (2012) Rise of dinosaurs reveals major body-size transitions are driven by passive processes of trait evolution. Proc R Soc B 279:2180–2187

Stephens PR, Wiens JJ (2003) Explaining species richness from continents to communities: the time-for-speciation effect in Emydid turtles. Am Nat 161:112–128

Suh A, Smeds L, Ellegren H (2015) The dynamics of incomplete lineage sorting across the ancient adaptive radiation of Neoavian birds. PLoS Biol 13:e1002224

Sullivan C, Hone DWE, Xu X, Zhang F (2010) The asymmetry of the carpal joint and the evolution of wing folding in maniraptoran theropod dinosaurs. Proc R Soc B 277:2027–2033

Suryan RM, Santora JA, Sydeman WJ (2012) New approach for using remotely sensed chlorophyll a to identify seabird hotspots. Mar Ecol Prog Ser 451:213–225

Swainson W (1836) On the natural history and classification of birds, volume 1. Longman, Rees, Orme, Brown, Green, and Longman, London
Sweet AD, Johnson KP (2015) Patterns of diversification in small New World ground doves are consistent with major geologic events. Auk 132:300–312
Switek B (2010) Thomas Henry Huxley and the reptile to bird transition. Geol Soc Lond, Spec Publ 343:251–263
Tambussi CP (2011) Palaeoenvironmental and faunal inferences based on the avian fossil record of Patagonia and Pampa: what works and what does not. Biol J Linn Soc 103:458–474
Tamura K, Nomura N, Seki R, Yonei-Tamura S, Yokoyama H (2011) Embryological evidence identifies wing digits in birds as digits 1, 2, and 3. Science 331:753–757
Tanaka G, Zhou B, Zhang Y, Siveter DJ, Parker AR (2017) Rods and cones in an enantiornithine bird eye from the Early Cretaceous Jehol Biota. Heliyon 3: e00479
Tanaka T, Kobayashi Y, Ikuno K, Ikeda T, Saegusa H (2020) A marine hesperornithiform (Avialae: Ornithuromorpha) from the Maastrichtian of Japan: implications for the paleoecological diversity of the earliest diving birds in the end of the Cretaceous. Cretac Res 113:104492
Torres CR, Norell MA, Clarke JA (2021) Bird neurocranial and body mass evolution across the end-Cretaceous mass extinction: the avian brain shape left other dinosaurs behind. Sci Adv 7:abg7099
Towers M, Signolet J, Sherman A, Sang H, Tickle C (2011) Insights into bird wing evolution and digit specification from polarizing region fate maps. Nat Commun 2:426
Tucker VA (1988) Gliding birds: descending flight of the White-backed Vulture, *Gyps africanus*. J Exp Biol 344:325–344
Turner AH, Makovicky PJ, Norell MA (2012) A review of dromaeosaurid systematics and paravian phylogeny. Bull Am Mus Nat Hist 371:1–206
Vajda V, McLoughlin S (2004) Fungal proliferation at the Cretaceous-Tertiary boundary. Science 303:1489
Vazquez RJ (1994) The automating skeletal and muscular mechanisms of the avian wing (Aves). Zoomorphology 114:59–71
Vellekoop J, Sluijs A, Schouten S, Weijers JWH, Sinninghe Damsté JS, Brinkhuis H (2014) Rapid short-term cooling following the Chicxulub impact at the Cretaceous-Paleogene boundary. Proc Natl Acad Sci USA 111:7537–7541
Voelker G, Rohwer S, Outlaw DC, Bowie RCK (2009) Repeated trans-Atlantic dispersal catalyzed a global songbird radiation. Glob Ecol Biogeogr 18:41–49
Wagner A (1862) On a new fossil reptile supposed to be furnished with feathers. Ann Mag Nat Hist 52:261–267
Wagner GP, Gauthier JA (1999) 1,2,3 = 2,3,4: a solution to the problem of the homology of the digits of the avian hand. Proc Natl Acad Sci USA 96:5111–5116
Walker AD (1972) New light on the origin of birds and crocodiles. Nature 237:257–263
Walker AD (1977) Evolution of the pelvis in birds and dinosaurs. In: Andrews SM, Miles RS, Walker AD (eds) Problems in vertebrate evolution. Academic Press, London, pp 319–357
Walker CA (1981) New subclass of birds from the Cretaceous of South America. Nature 292:51–53
Wang X, Zhang Z (2011) Enantiornithine birds in China. Acta Geol Sin 85:1211–1223
Wang M, Zheng X, O'Connor JK, Lloyd GT, Wang X, Wang Y, Zhang X, Zhou Z (2015) The oldest record of ornithuromorpha from the early Cretaceous of China. Nat Commun 6:6987
Wang M, Zhou Z, Sullivan C (2016) A fish-eating enantiornithine bird from the Early Cretaceous of China provides evidence of modern avian digestive features. Curr Biol 26:1170–1176
Wang X, Pittman M, Zheng X, Kaye TG, Falk AR, Hartman SA, Xu X (2017a) Basal paravian functional anatomy illuminated by high-detail body outline. Nat Commun 8:14576
Wang Y, Hu H, O'Connor JK, Wang M, Xu X, Zhou Z, Wang X, Zheng X (2017b) A previously undescribed specimen reveals new information on the dentition of *Sapeornis chaoyangensis*. Cretac Res 74:1–10
Wang M, Stidham TA, Zhou Z (2018a) A new clade of basal Early Cretaceous pygostylian birds and developmental plasticity of the avian shoulder girdle. Proc Natl Acad Sci USA 115:10708–10713
Wang X, Huang J, Hu Y, Liu X, Peteya J, Clarke JA (2018b) The earliest evidence for a supraorbital salt gland in dinosaurs in new Early Cretaceous ornithurines. Sci Rep 8:3969
Wang J, Hao X, Kundrát M, Liu Z, Uesugi K, Jurašeková Z, Guo B, Hoshino M, Li Y, Menfroy Z, Zhou B (2020) Bone tissue histology of the Early Cretaceous bird *Yanornis*: evidence for a diphyletic origin of modern avian growth strategies within Ornithuromorpha. Hist Biol 32:1422–1434
Wang R, Hu D, Zhang M, Wang S, Zhao Q, Sullivan C, Xu X (2022) A new confuciusornithid bird with a secondary epiphyseal ossification reveals phylogenetic changes in confuciusornithid flight mode. Commun Biol 5:1398
Ward P, Berner R (2011) Why were there dinosaurs? Why are there birds? In: Dyke G, Kaiser G (eds) Living dinosaurs: the evolutionary history of modern birds. Hoboken, NJ, Wiley-Blackwell, pp 30–38
Whittaker RJ, Fernandez-Palacios JM (2007) Island biogeography: ecology, evolution, and conservation. Oxford Univ Press, Oxford, UK
Widrig K, Field DJ (2022) The evolution and fossil record of palaeognathous birds (Neornithes: Palaeognathae). Diversity 14:105
Wiemann J, Yang TR, Norell MA (2018) Dinosaur egg colour had a single evolutionary origin. Nature 563: 555–558

Williston SW (1879) Are birds derived from dinosaurs? Kansas City Rev Sci 3:457–460

Witmer LM (1991) Perspectives on avian origins. In: Schultze H-P, Trueb L (eds) Origins of the higher groups of tetrapods. Cornell University Press, Ithaca, NY, pp 427–466

Woodward H (1862) On a feathered fossil from the lithographic limestone of Solenhofen. Intellect Observ 2: 313–319

Worthy TH, Worthy JJP, Tennyson AJD, Salisbury SW, Hand SJ, Scofield RP (2013) Miocene fossils show that kiwi (*Apteryx*, Apterygidae) are probably not phyletic dwarves. In: Gölich UB, Kroh A (eds) Proceedings of the 8th International meeting of the society of avian paleontology and evolution. Natural History Museum Vienna, Vienna, pp 63–80

Worthy TH, Mitri M, Handley WD, Lee MSY, Anderson A, Sand C (2016) Osteology supports a stem-galliform affinity for the giant extinct flightless bird *Sylviornis neocaledoniae* (Sylviornithidae, Galloanseres). PLoS ONE 11:e0150871

Worthy TH, Degrange FJ, Handley WD, Lee MSY (2017) The evolution of giant flightless birds and novel phylogenetic relationships for extinct fowl (Aves, Galloanseres). R Soc Open Sci 4:170975

Wu X-C, Chatterjee S (1993) *Dibothrosuchus elaphros*, a Crocodylomorph from the Lower Jurassic of China and the phylogeny of the Sphenosuchina. J Vertebr Paleontol 13:58–89

Xing L, O'Connor JK, McKellar RC, Chiappe LM, Tseng K, Li G, Bai M (2017) A mid-Cretaceous enantiornithine (Aves) hatchling preserved in Burmese amber with unusual plumage. Gondwana Res 49:264–277

Xing L, O'Connor JK, Chiappe LM, McKellar RC, Carroll N, Hu H, Bai M, Lei F (2019) A new enantiornithine bird with unusual pedal proportions found in amber. Curr Biol 29:2396–2401

Xing L, O'Connor JK, Niu K, Cockx P, Mai H, McKellar RC (2020) A new enantiornithine (Aves) preserved in mid-Cretaceous Burmese amber contributes to growing diversity of Cretaceous plumage patterns. Front Earth Sci 8:264

Xu X, Mackem S (2013) Tracing the evolution of avian wing digits. Curr Biol 23:R538–R544

Xu X, Tan Q, Sullivan C, Han F, Xiao D (2011) A short-armed troodontid dinosaur from the Upper Cretaceous of Inner Mongolia and its implications for troodontid evolution. PLoS ONE 6:e22916

Xu X, Zhou Z, Dudley R, Mackem S, Chuong C-M, Erickson GM, Varricchio DJ (2014) An integrative approach to understanding bird origins. Science 346: 6215

Yang S, He H, Jin F, Zhang F, Wu Y, Yu Z, Li Q, Wang M, O'Connor JK, Deng C, Zhu R (2020) The appearance and duration of the Jehol Biota: constraint from SIMS U-Pb zircon dating for the Huajiying Formation in northern China. Proc Natl Acad Sci USA 117:14299–14305

Yonezawa T, Segawa T, Mori H, Campos PF, Hongoh Y, Endo H, Akiyoshi A, Kohno N, Nishida S, Wu J, Jin H (2017) Phylogenomics and morphology of extinct paleognaths reveal the origin and evolution of the ratites. Curr Biol 27:68–77

Zanno LE, Makovicky PJ (2011) Herbivorous ecomorphology and specialization patterns in theropod dinosaur evolution. Proc Natl Acad Sci USA 108:232–237

Zavalga CB, Benvenuti S, Dall'Antonia L, Emslie SD (2007) Diving behavior of Blue-footed Boobies *Sula nebouxii* in northern Peru in relation to sex, body size, and prey type. Mar Ecol Prog Ser 336:291–303

Zelenkov NV (2016) Stable morphological types and mosaicism in the macroevolution of birds (Neornithes). Biol Bull Rev 6:208–218

Zhang F, Zhou Z, Benton MJ (2008) A primitive confuciusornithid bird from China and its implications for early avian flight. Sci China Ser D Earth Sci 51: 625–639

Zhao J, Xiao L, Gulick SPS, Morgan JV, Kring D, Fucugauchi JU, Schmieder M, de Graaff SJ, Wittmann A, Ross CH, Claeys P (2020) Geochemistry, geochronology and petrogenesis of Maya Block granitoids and dykes from the Chicxulub Impact Crater, Gulf of México: implications for the assembly of Pangea. Gondwana Res 82:128–150

Zheng X, Martin LD, Zhou Z, Burnam DA, Zhang F, Miao D (2011) Fossil evidence of avian crops from the Early Cretaceous of China. Proc Natl Acad Sci USA 108: 15904–15907

Zheng X, Wang X, O'Connor J, Zhou Z (2012) Insight into the early evolution of the avian sternum from juvenile enantiornithines. Nature Comm 3:1116

Zheng X-T, O'Connor JK, Huchzermeyer FW, Wang X-L, Wang Y, Wang M, Zhou Z-H (2013) Preservation of ovarian follicles reveals early evolution of avian reproductive behaviour. Nature 495:507–511

Zheng X-T, O'Connor JK, Wang X-L, Pan Y-H, Wang Y, Wang M, Zhou Z-H (2017) Exceptional preservation of soft tissue in a new specimen of Eoconfuciusornis and its biological implications. Natl Sci Rev 4:441–452

Zheng X, O'Connor JK, Wang X, Wang Y, Zhou Z (2018) Reinterpretation of a previously described Jehol bird clarifies early trophic evolution in the Ornithuromorpha. Proc R Soc B 285:20172494

Zhou Z (2004) The origin and early evolution of birds: discoveries, disputes, and perspectives from fossil evidence. Naturwissenschaften 91:455–471

Zhou Z, Li FZZ (2009) A new Lower Cretaceous bird from China and tooth reduction in early avian evolution. Proc R Soc B 277:219–227

Zhou Z, Zhang F (2003) *Jeholornis* compared to *Archaeopteryx*, with a new understanding of the earliest avian evolution. Naturwissenschaften 90:220–225

Zhou S, Zhou Z, O'Connor J (2014) A new piscivorous ornithuromorph from the Jehol Biota. Hist Biol 26: 608–618

Zhou Y-C, Sullivan C, Zhang FC (2019) Negligible effect of tooth reduction on body mass in Mesozoic birds. Vertebr PalAsiatica 57:38–50

Zinoviev AV (2009) An attempt to reconstruct the lifestyle of confuciusornithids (Aves, Confuciusornithiformes). Paleontol J 43:444–452

Zusi RL, Warheit KI (1992) On the evolution of intraramal mandibular joints in pseudodontorns (Aves: Odontopterygia). In: Papers in avian paleontology honoring Pierce Brodkorb (K. E. Campbell, ed.), pp. 351–360. Natural History Museum of Los Angeles County, Science Series 36, Los Angeles, CA

# Skeleton and Skeletal Muscles

# 2

## Contents




**Abstract**

The avian skeleton and skeletal muscles have been modified by natural selection over millions of years to meet the demands of flight. Some birds have lost the ability to fly and their skeletons and skeletal muscles now differ to varying degrees from those of birds that fly. The skeleton of birds consists of bones and this chapter begins with an examination of the structure of bird bones and a discussion of the evolution of the avian skeleton. The bones of the pectoral and pelvic girdle plus forelimbs and hindlimbs are described and factors that have contributed to interspecific variation in those components of the avian skeleton are explained. The axial skeleton of birds, including the skull, vertebral column, sternum, and ribs, is described and




G. Ritchison, *In a Class of Their Own*, Fascinating Life Sciences,
https://doi.org/10.1007/978-3-031-14852-1_2

interspecific variation in these structures is described and explained. The structure and function of avian skeletal muscles is also explained and their role in avian locomotion, including flight and feeding, is described and explained.

## 2.1 Bone Structure

Bird bones, like those of mammals, contain both living and nonliving components. Cells called osteoblasts help form bone; cells called osteoclasts break down bone and these cells are supplied by blood vessels and nerves. Bone also contains nonliving materials, primarily collagen fibers and hydroxyapatite crystals made up of calcium, phosphate, and hydroxide. Collagen fibers, long fibrous protein molecules, make bones strong and flexible, and that strength is greatly enhanced by the presence of the hydroxyapatite. More precisely, the collagen fibers give bones tensile strength (ability to endure stretching forces) and the hydroxyapatite crystals give bones compressional strength (ability to endure squeezing or compression).

Bird bones vary in shape, with some being long (bones of the wings and legs), others short (e.g., phalanges in a bird's toes) or flat (e.g., bones of the skull and pelvic girdle), and still others irregular (e.g., vertebrae). Bones typically consist of an outer layer of cortical, or compact, bone that surrounds trabecular bone (Fig. 2.1, Box 2.1 Bone Microstructure). Both types of bone consist of collagen fibers and hydroxyapatite crystals. However, cortical bone is much denser and trabecular bone consists of a series of ridges or rod-like structures, or struts, with air spaces and, in some bones, marrow in between (Fig. 2.2). Many bird bones are also pneumatic, with diverticula, or outpocketings, of air sacs extending into the spaces within the bones (Fig. 2.3). Pneumatic bones do not contain marrow. Rather, bone marrow, where red blood cells are formed, is found only in the radius, ulna, femur, tibiotarsus, scapula, furcula (clavicles), and caudal vertebrae (Fig. 2.4; Schepelmann 1990). The proportion of the avian skeleton that consists of pneumatic bones varies among species, generally increasing with increasing bird size and decreasing for diving birds. Large flying birds like vultures and pelicans have the most pneumatic bones whereas diving specialists like penguins and auks have no pneumatic bones. Skeletal pneumaticity may relax constraints on body size evolution, allowing increases in body volume without a similar increase in body mass (O'Connor 2009).

The skeletons of birds need to be light to help minimize the metabolic costs of flying, but their bones also need to be strong enough to withstand the forces associated with flight. Light, however, is a relative term, so in what sense, or compared to what, are bird skeletons light? Given that many bird bones are pneumatic and that, during the evolutionary transition from theropods to birds, several bones were lost (see the next section for additional details), maybe bird skeletons are lighter than those of similar-sized mammals. However, Dumont (2010) compared similar-sized songbirds and rodents and found that, relative to total body mass, the mass of bird skeletons is similar to that of rodents (Fig. 2.5). So, despite having pneumatic bones and having fewer bones than rodents (e.g., fewer digits and phalanges), the mass of bird skeletons is comparable to that of rodents.

Relative to their theropod ancestors, bird skeletons have become lighter. However, the demands of flight also require strong bones and skeletons. Dumont (2010) compared the densities of several bird and rodent bones (mass relative to volume, or grams per cubic centimeter) and found that the mean density of bird bones was greater than that of rodent bones. In other words, bird bones are stronger (better able to resist fracture) and stiffer (able to resist deformation) relative to their weight than rodent bones. This means that bird bones are light, but relative to the need for those bones to also be strong and stiff rather than relative to the bones of similar-sized rodents. Given the importance of a skeleton that is both light and strong, the high strength- and stiffness-to-weight ratios of bird bones represent an elegant evolutionary compromise.

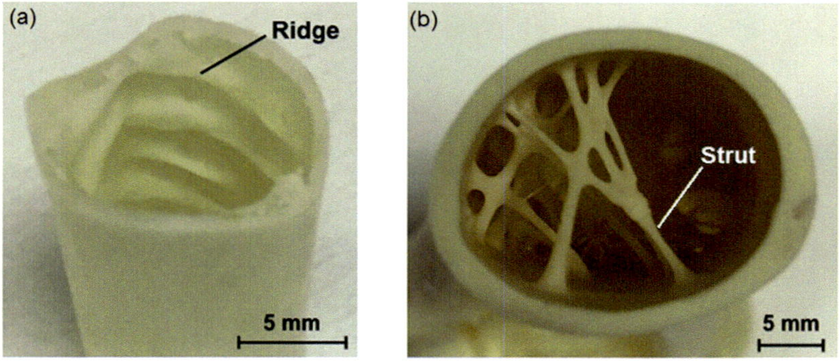


**Fig. 2.1** Reinforcing structures in avian bones. (**a**) Ridges in the ulna of a Turkey Vulture (*Cathartes aura*) and (**b**) struts (trabeculae) located toward the distal end of a humerus of a Cape Vulture (*Gyps coprotheres*). (Figure from Sullivan et al. 2017; © 2017 Elsevier Ltd., used with permission)

**Box 2.1 Bone Microstructure**

Bones are considered connective tissue and serve a variety of functions, including providing support for the body, protecting important organs like the brain and spinal cord, and serving as the site of attachment for tendons and muscles. Bones also represent a storage site for calcium and some bones are involved in the production of red blood cells. Bones consist of three types of cells, including osteoblasts, osteocytes, and osteoclasts. Osteoblasts are bone-forming cells, synthesizing the collagenous matrix that becomes mineralized by deposition of hydroxyapatite. Osteocytes are former osteoblasts that become trapped in the matrix they created and are located in small cavities called lacunae that are interconnected by canaliculi through which nutrients are supplied to osteocytes. Osteoclasts break down bone using proteolytic enzymes and acid, an important process in bone remodeling. Such remodeling is needed to adapt the skeleton to variation in mechanical loading and repair microfractures. In the process of remodeling, osteoclasts dissolve bone in one location and osteoblasts create bone elsewhere. When bones are increasingly stressed or exposed to greater forces, more bone is deposited and, when bones are exposed to less stress, more bone is dissolved.

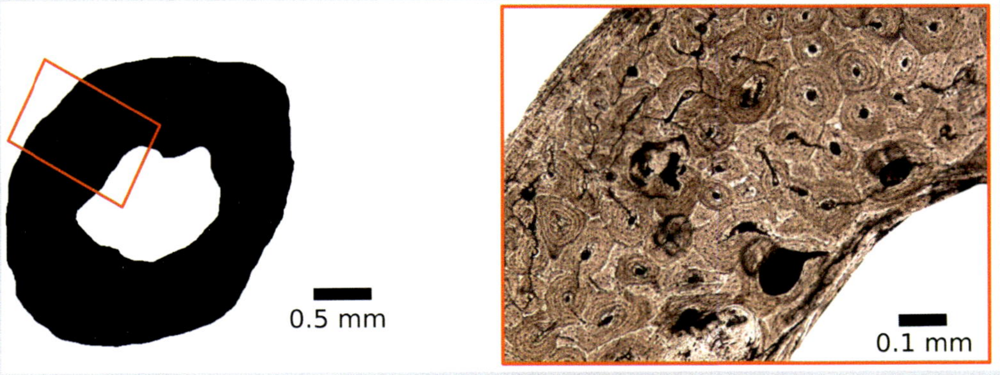


Left: Section of the furcula of a Lesser Black-backed Gull (*Larus fuscus*). The area in the red rectangle is enlarged in the image on the right. Right: Cortical bone. The numerous circular structures with dark centers are osteons. The dark centers are the Haversian canals. (Figure from Mitchell et al. 2017; © 2017 Elsevier GmbH, used with permission)

(continued)

**Box 2.1** (continued)

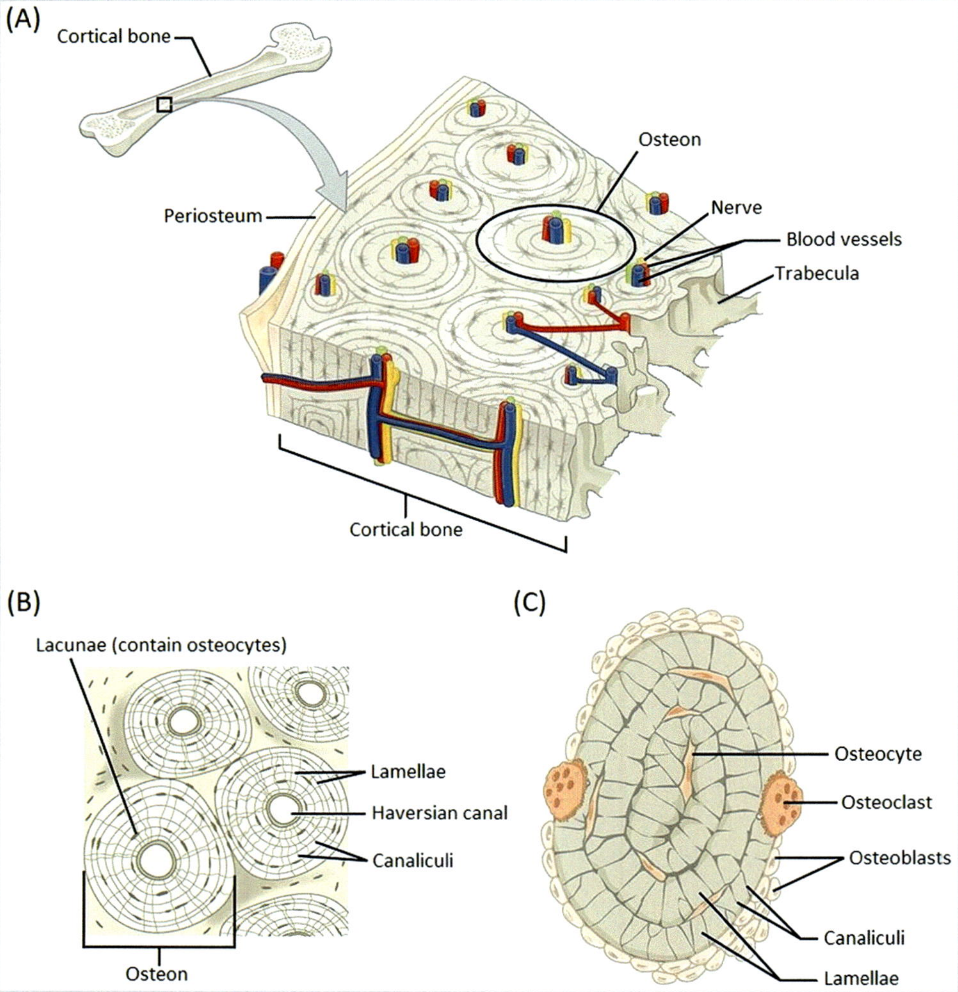


Microstructure of cortical (also called compact) bone. (A) The periosteum is the membrane that covers the outer surface of bones. Cortical, or compact, bone consists of osteons, each of which has a Haversian canal where nerves and blood vessels are located. (B) Osteons include lacunae where osteocytes are located; lamellae, which are thin sheets of bone formed by osteoblasts; and canaliculi through which nutrients are transported to osteocytes. (C) Microstructure of a trabecula. (Figure from CNX OpenStax, Wikipedia, CC BY 4.0, https://creativecommons.org/licenses/by/4.0/)

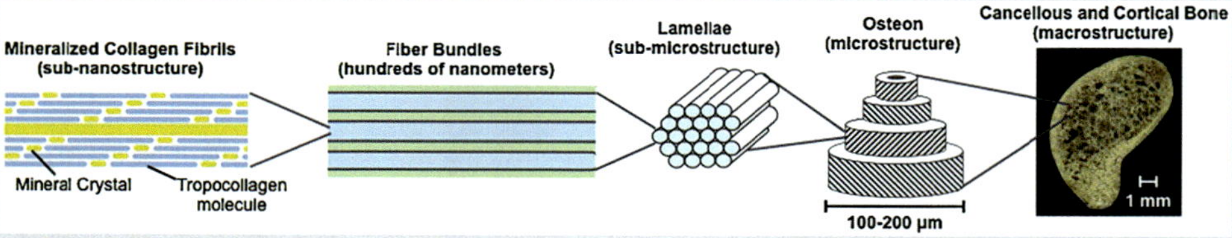


(continued)

**Box 2.1** (continued)

Hierarchical structure of bone. Tropocollagen molecules and mineral crystals organize to form fiber bundles. These bundles form lamellae that surround an osteon, and these form cancellous and cortical bone. (Figure from Sullivan et al. 2017; © 2017 Elsevier Ltd., used with permission)

## 2.2 Evolution of the Avian Skeleton

The ancestors of birds were bipedal, predatory theropods with teeth, small forelimbs, large hindlimbs, and long tails. With their large, muscular hindlimbs and long, muscular tails that served as dynamic stabilizers (Libby et al. 2012), theropods were fast and agile predators. Although present-day birds are also bipedal and capable, to varying degrees, of terrestrial locomotion, most also fly (and so did the ancestors of present-day flightless birds) and the evolutionary transition from terrestrial theropods to flying birds required numerous changes in skeletal morphology. One needed change was the conversion of the grasping forelimb of theropods to a forelimb capable of flapping flight. Other changes involved the loss or modification of bones and, in some cases, the fusion of bones. These modifications occurred gradually during the transition from theropods to birds and resulted in a skeleton well adapted for the demands of flight.

The theropod ancestors of birds were bipedal, and their hindlimbs and tails were both important in terrestrial locomotion (Fig. 2.6). Theropod tails

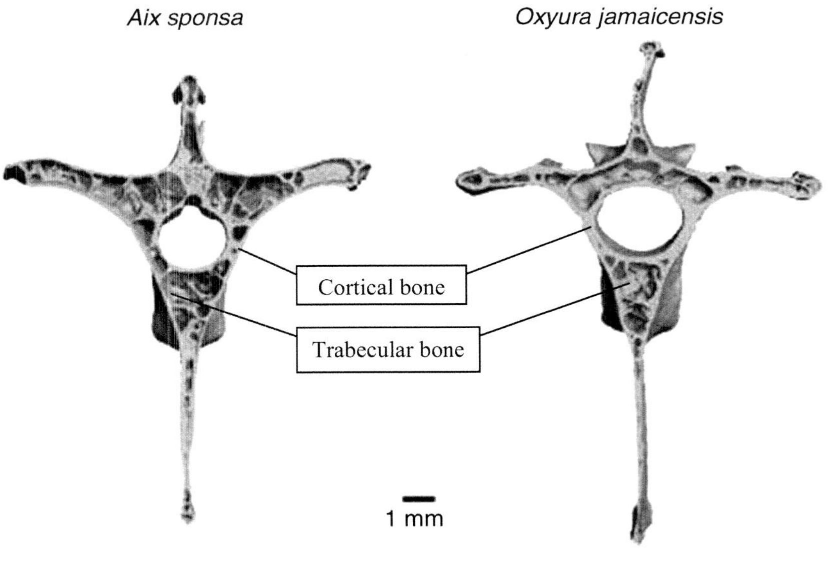


**Fig. 2.2** Vertebrae of a Wood Duck (*Aix sponsa*) and a Ruddy Duck (*Oxyura jamaicensis*) showing the compact cortical bone and trabecular bone. Note that the vertebra of the Ruddy Duck has more compact bone and less trabecular bone. Ruddy Ducks are diving ducks whereas Wood Ducks are dabbling ducks that forage at or near the water's surface. In general, the bones of diving ducks have more compact bone and less trabecular bone, making them heavier and, therefore, more efficient divers. (Figure from Fajardo et al. 2007; © Anatomical Society, used with permission)

**Fig. 2.3** Shaded bones are those that are commonly pneumatic in nondiving waterfowl; this can vary in other species of birds. Bones with marrow produce red blood cells in most birds, including the radius and ulna of the wing, femur and tibiotarsus of the leg, furcula and scapula of the pectoral girdle, and, in waterfowl, the pygostyle, are not pneumatic (compare with Fig. 2.4). (Figure from O'Connor 2004; © 2004 Wiley-Liss, Inc., used with permission)

served both to counterbalance the weight of the anterior portion of the body (keeping the center of gravity over the hindlimbs) and as dynamic stabilizers to help provide stability during fast turns. In contrast, birds only use their hindlimbs for terrestrial locomotion; their tails no longer play a role. In flight, birds use both their forelimbs (wings) and their tails, with tails important for steering, landing, and, to varying degrees, providing additional lift (Fig. 2.6). This evolutionary transition from the hindlimb/tail-based, terrestrial-only locomotion of theropods to both hindlimb-based terrestrial plus forelimb/tail-based aerial locomotion in birds required several changes to the skeleton, which will be discussed in the following sections.

## 2.3 Pectoral Girdle and Forelimb

The avian wing is braced by the pectoral girdle that consists of the clavicles (fused to form the furcula), coracoid, and scapula. The glenoid cavity, where the humerus of wing articulates with the pectoral girdle, is located where those three bones meet. During the transition from theropods to present-day birds, the triosseal canal developed. This canal, which is formed by the tip (acrocoracoid process) of the coracoid and the ends of the furcula and scapula that articulate with the coracoid, allows the tendon of the supracoracoideus muscle to insert on the humerus dorsally, pulling the wing upward (upstroke)

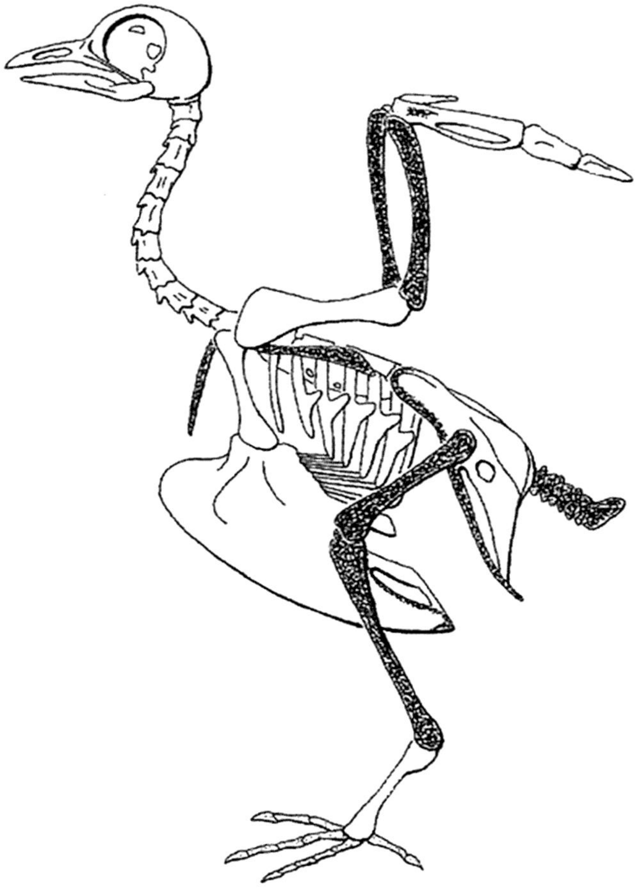

**Fig. 2.4** Skeleton of a Rock Pigeon (*Columba livia*) showing the bones (shaded) that contain red blood cell-producing marrow, including the radius and ulna of the wing, femur and tibiotarsus of the leg, and furcula and scapula of the pectoral girdle, pubis, and caudal vertebrae. The other bones are pneumatized (trabecular bone). (Figure from Schepelmann 1990; © 1990 Wiley-Liss, Inc., used with permission)

during avian flight. In addition, the glenoid cavity rotated so that it faces laterally rather than down, allowing greater elevation of the humerus, which is critical for avian flight. The coracoids also elongated and formed a rigid connection to the sternum (Fig. 2.7), acting as a strut between the wing and the sternum that resists forces generated by the pectoralis (or downstroke) muscle (Pennycuick 1967).

The clavicles of birds are fused to form the furcula (Fig. 2.8), but furculae appeared early in theropod history so are not a uniquely avian structure (Nesbitt et al. 2009). As the glenoid cavity rotated, avian furculae became narrower than those of theropods and, as a result, the scapula moved closer to and became parallel to the vertebral column (Senter 2006; Figs. 2.9 and 2.10). In flight, the avian furcula bends laterally during the downstroke, then recoils during the upstroke (Fig. 2.11). Jenkins et al. (1988) suggested that this alternate bending and recoiling of the furcula might serve a respiratory function, helping to move air into and out of the interclavicular air sac. Goslow et al. (1990) proposed that the energy stored by the furcula as it bends laterally during the downstroke might reduce the effort needed during the upstroke. However, Bailey and DeMont (1991) found little support for this hypothesis, and Baier et al. (2013) noted that there was little evidence to support either hypothesis.

The earliest theropods had forelimbs shorter than hindlimbs (e.g., *Sinosauropteryx*; Fig. 2.7), correspondingly short limb bones (humerus,

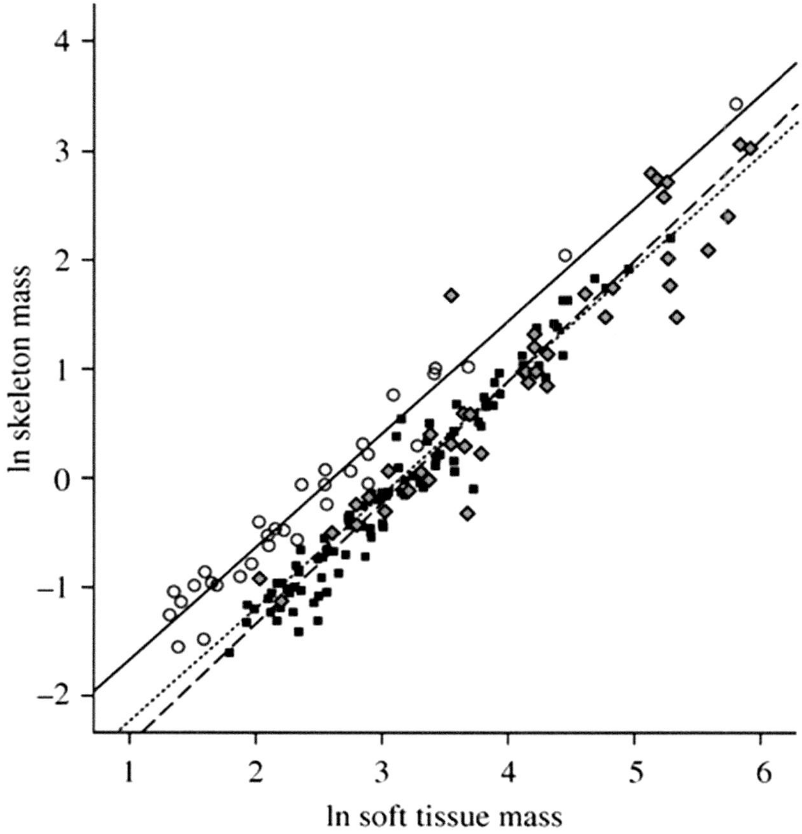


**Fig. 2.5** Regression of ln (natural log) skeletal mass against ln soft tissue mass (= total body mass minus skeletal mass) for 96 species of songbirds (black squares, dotted line), 39 species of rodents (gray diamonds, dashed line), and 34 species of bats (open circles, solid line) showing that the mass of bird skeletons is similar to that of rodents and less than that of bats. For mammal skeletons, mass values were increased by 15% because mammal bones are not pneumatized (Currey and Alexander 1985). (Figure from Dumont 2010; © 2010 The Royal Society, used with permission)

radius, and ulna), unfused carpal and metacarpal bones, and five digits consisting of numerous phalanges (Fig. 2.12). However, the forelimbs of theropods more closely related to birds, such as the deinonychosaurs, were very similar to those of *Archaeopteryx* (Middleton and Gatesy 2000; Dececchi and Larsson 2009), with longer forelimbs and fewer digits and phalanges than earlier theropods. This suggests either that the earliest bird wings were the result of exaptation with little to no skeletal changes needed for the origin of flight or that flight evolved sometime prior to *Archaeopteryx* (Dececchi and Larsson 2009; see Chaps. 1 and 11 for additional information about the evolution of flight).

If the theropod ancestors of *Archaeopteryx* did not fly (or fly very well), what selective factors might have favored the changes to their forelimb skeleton that ultimately made it possible for *Archaeopteryx* to fly? Among the possibilities are that the "wings" of some theropods may have aided them in climbing inclined surfaces such as trees (i.e., wing-assisted incline running; Dial et al. 2006) or over obstacles and may have allowed for controlled descending glides from elevated sites or perches (Dial et al. 2008). Another possibility is that predatory theropods such as *Deinonychus* may have used their feathered forelimbs for "stability flapping" and to mantle their prey (Fig. 2.13; Fowler et al. 2011). As prey struggled and attempted to escape following initial capture, predators like *Deinonychus* could have flapped their "wings" to help them maintain their balance and position on top of their prey. Present-day raptors like hawks and eagles are known to exhibit such behavior, vigorously flapping their wings as they try to subdue large prey. In addition, present-day raptors often encircle

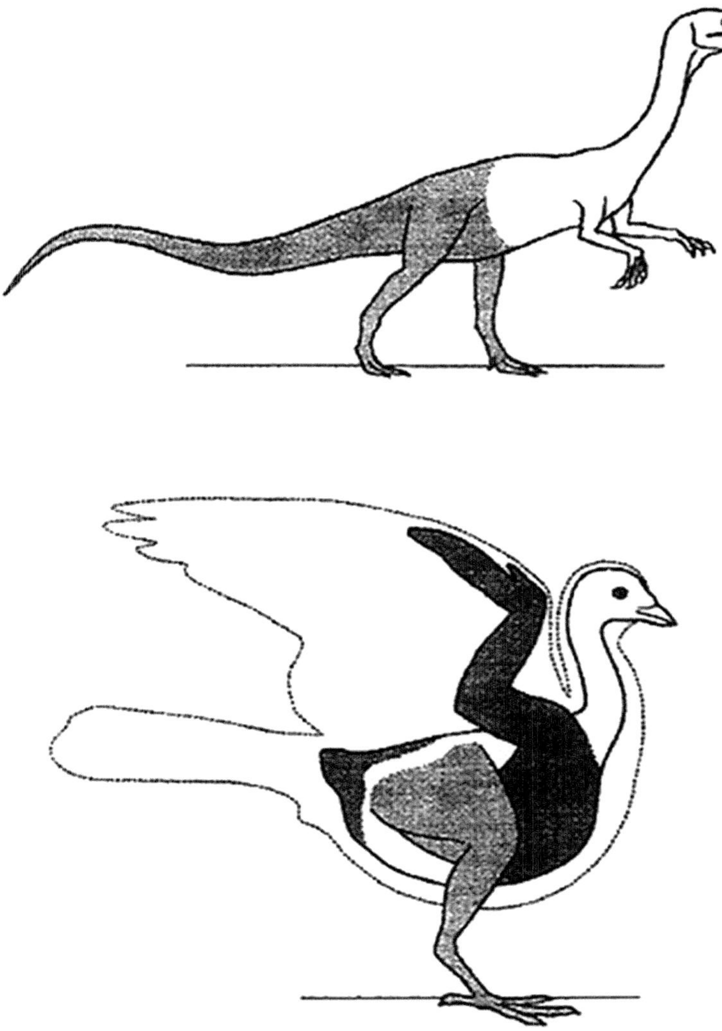

**Fig. 2.6** For efficient terrestrial locomotion, theropods depended on their hindlimbs and tails. In contrast, birds use their hindlimbs only for terrestrial locomotion and both their forelimbs (wings) and tails for aerial locomotion. (Figure from Gatesy and Dial 1996a; © 1996 The Society for the Study of Evolution, used with permission)

their prey with their wings, a behavior known as mantling. Such behavior may aid in preventing prey from escaping or, after prey have been killed, may help conceal a prey item from other predators that might attempt to steal it. Of course, the proto-wings of theropods may have served more than one of these functions, as well as others that remain to be determined.

During the evolutionary transition from theropods to *Archaeopteryx* to present-day birds, forelimbs became longer (relative to body size), several carpal and metacarpal bones were lost and others fused to form the carpometacarpus, two digits were lost (Box 2.2 Which Digits Were Lost), and the remaining digits consisted of fewer phalanges. In addition, the wrist was transformed from a universal joint to a swivel joint, with the swivel joint preventing extreme upward or downward movement of the manus during flight (Box 2.3 Evolution of the Avian Wrist). This restricted motion of the wrist is critical because excessive upward or downward movement of the manus (and the attached primary feathers) would alter the normally smooth airflow past the wing.

Several changes in the pectoral girdle and forelimb of birds, such as formation of the triosseal canal (also called the foramen

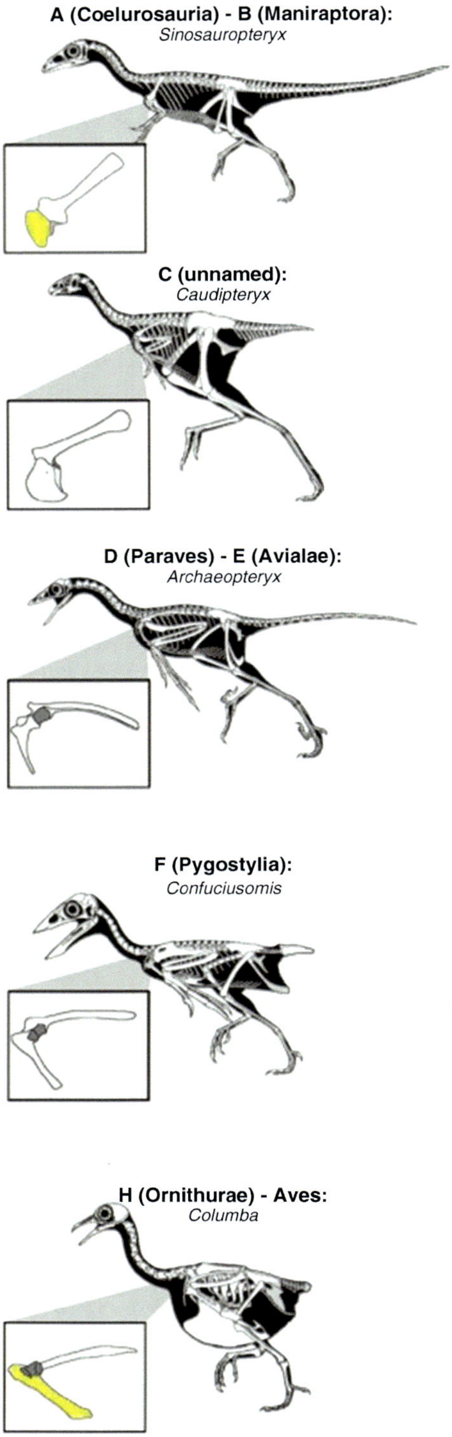


**Fig. 2.7** Changes in the pectoral girdle during the transition from theropods to present-day birds. Note the dramatic change in the coracoid (shaded yellow at top and bottom, but not in between), elongating to form a rigid connection to the sternum and acting as a strut between the wing and the sternum. (Figure from Heers and Dial 2012; © 2011 Elsevier Ltd., used with permission)

triosseum), loss of additional phalanges, and fusion of carpal and metacarpal bones to form the carpometacarpus, and reduced mobility of the wrist joint, occurred post-*Archaeopteryx* and, therefore, after the evolution of flight. This indicates that the changes were not essential for flight, but, rather, represent modifications or refinements that enhanced the flying ability of birds.

Although the basic forelimb skeleton of all present-day birds is similar, there is considerable variation in (1) the relative lengths of the brachium arm (humerus), antebrachium (radius and ulna), and hand or manus, and (2) bone strength and the relative cortical area of bones (ratio of cortical to trabecular bone). The relative lengths of the three components of the forelimb differ between ratites and other birds, with ratites having vestigial wings consisting of relatively long humeri, but much shorter radii/ulnas and manus. Even among flying birds, forelimb proportions vary considerably (Fig. 2.14). In general, birds that are more maneuverable in the air, such as swallows, have relatively shorter humeri than less maneuverable birds, such as loons, grebes, and albatrosses (Middleton and Gatesy 2000; Box 2.4 Avian Humeri). The most extreme examples of this are the highly maneuverable hummingbirds and swifts that have a very long manus relative to the rest of the wing (Fig. 2.15). The forelimb skeletons of hummingbirds are short relative to the length of their flight feathers (Figs. 2.14 and 2.15) and, as a result, about 50% of the wing mass of hummingbirds is less than 10% of the distance from the shoulder joint (Wells 1993). This shift in the concentration of mass reduces the moment of inertia (a measure of an object's resistance to changes in its rotation) and inertial work requirements for hummingbird wings (Tobalske 2010).

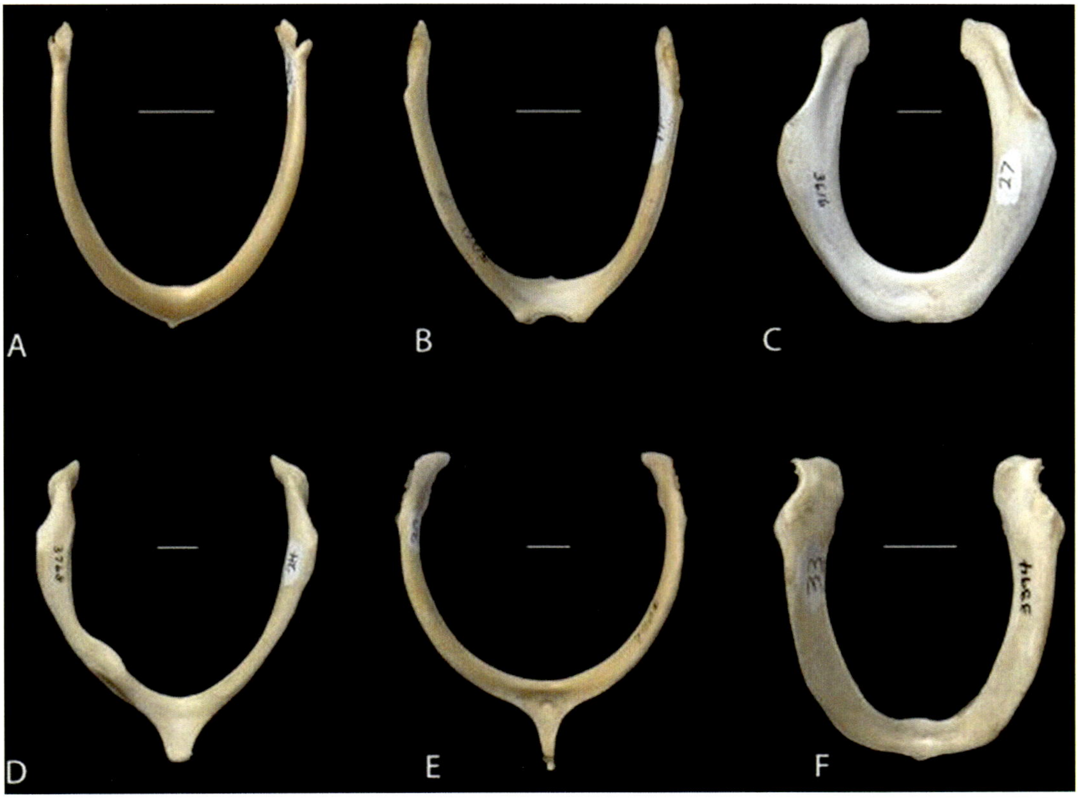


**Fig. 2.8** Furculae of six different species of birds. (**a**) Mallard, *Anas platyrhynchos*. (**b**) Red-tailed Tropicbird, *Phaethon rubricauda*. (**c**) Southern Screamer, *Chauna torquata*. (**d**) Wood Stork, *Mycteria americana*. (**e**) Secretarybird, *Sagittarius serpentarius*. (**f**) Peregrine Falcon, *Falco peregrinus*. (Figure from Nesbitt et al. 2009; © 2009 Wiley-Liss, Inc., used with permission)

The strength of forearm bones also varies among birds, as indicated by differences in the relative thickness of cortical bone (Fig. 2.16). The forearm bones of birds that use their wings to "fly" underwater, such as penguins, have much thicker cortical layers than the bones of other birds (Figs. 2.17 and 2.18). In addition, bones in the wings of penguins are flattened (Figs. 2.19 and 2.20), with narrower wings reducing resistance as they move through the water. Thick cortices are needed because water is much denser than air (about 850 times denser) and, as a result, the wing bones of wing-propelled divers must resist much greater forces than the wing bones of other birds. Foot-propelled diving birds, such as cormorants, also have relatively thick cortices compared to other birds. Because their wings are not used for propulsion underwater, the thicker cortices of the forearm bones of diving birds are likely an adaptation to increase density for more efficient diving.

## 2.4 Pelvic Girdle and Hindlimb

The pelvic girdles of birds consist of three bones, the ilium, ischium, and pubis, fused together into a single unit. The ilium, the largest of the three bones, and ischium, in turn, are fused with the synsacrum (Fig. 2.21). The pubis is a narrow, rod-like bone that extends caudally below and parallel to the ischium. The hip bones and

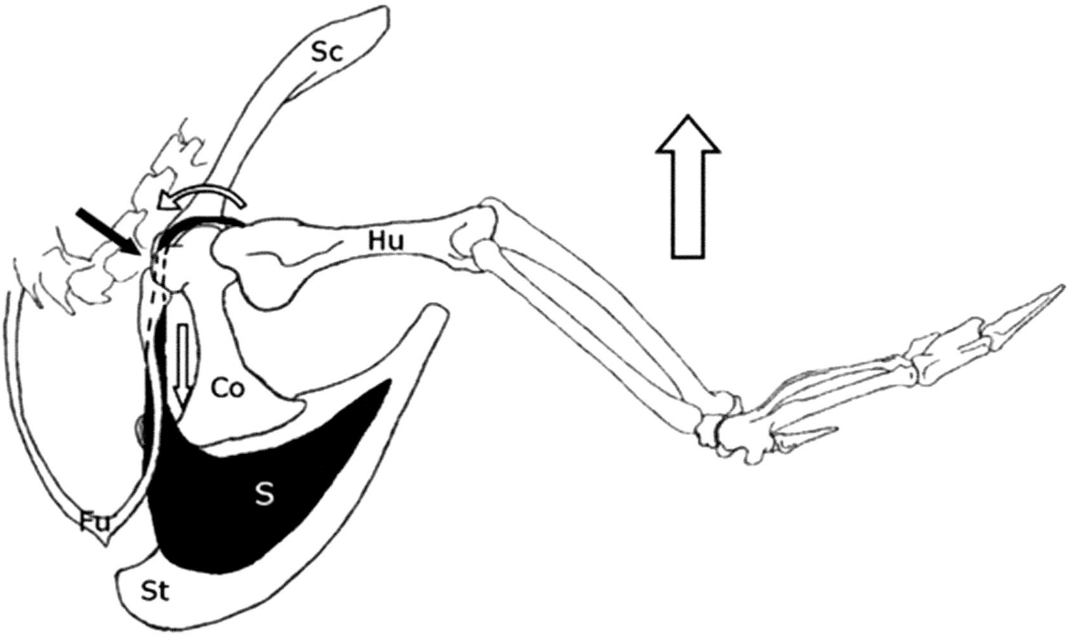


**Fig. 2.9** Avian pectoral girdle and forelimb, illustrating the action of the supracoracoideus muscle and associated tendon. When the supracoracoideus contracts, a tendon that passes through the triosseal canal and inserts on the humerus dorsally pulls the wing upward (upstroke; as indicated by the arrows). S indicates the supracoracoideus muscle; St, sternum; Fu, furcula; Co, coracoid; Sc, scapula; and Hu, humerus. (Figure from Beaufrère 2009; © 2011 Elsevier Ltd., used with permission)

synsacrum together make up the avian pelvic girdle or pelvis. A cup-shaped depression, the acetabulum, is located where the three hip bones meet and is where the head of the femur articulates with the pelvis. The acetabulum of birds is perforated (with an opening at the bottom called the acetabular foramen), a characteristic shared with many dinosaurs. The pelvic girdle supports the weight of birds when not in flight, serves as the site of origin of several skeletal muscles important in locomotion (muscles of the hindlimb), support, and movements of the tail, and encloses and protects organs like the kidneys.

The pelvis of present-day birds has a structure called the antitrochanter that is not found in other vertebrates, living or fossil. The antitrochanter is formed jointly by the ilium and ischium and is an elongated point of articulation with the neck and trochanter of the femur (Fig. 2.22). The function of the antitrochanter is to serve as a brace to prevent abduction of the leg (movement away from a bird's body) and reduce the stress placed on the head of the femur when a bird is standing, walking, hopping, or running (Hertel and Campbell 2007). Not surprisingly, the size of antitrochanters increases with bird body mass because heavier birds place more stress on the femur and need a better brace (Hertel and Campbell 2007).

The morphology of bird pelvic girdles does vary with lifestyle. Most birds have pelvic girdles that are roughly rectangular, about twice as long as they are wide, with a slight bend or iliac angle (the angle between the ilium and ischium; Fig. 2.21). Such pelvic girdles provide support for the hindlimbs and associated muscles among birds that are not primarily terrestrial and use their hindlimbs for walking or hopping and perching (i.e., most birds). However, the pelvic girdles of diving birds like cormorants, loons, grebes, and

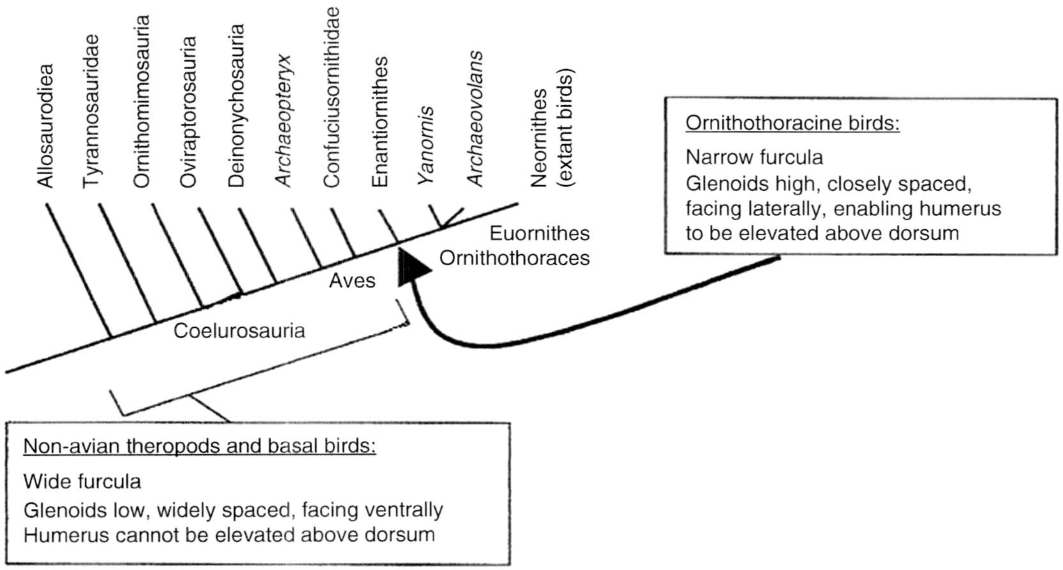


**Fig. 2.10** Evolution of the avian pectoral girdle. Compared to theropods and the earliest birds (like *Archaeopteryx*), the glenoid fossa (where the humerus articulates with the pectoral girdle) of present-day birds has rotated so that it faces laterally rather than ventrally. That rotation makes greater elevation of the humerus possible, a movement of obvious importance for flight. As the glenoid cavity rotated, the furcula of birds became narrower and the scapula moved closer to and became parallel to the vertebral column. (Figure from Senter 2006; open access following guidelines of the Budapest Open Access Initiative; Acta Palaeontologica Polonica articles are distributed under the terms of the Creative Commons Attribution License (CC BY), https://creativecommons.org/licenses/by/4.0/)

murres are long, flat, and narrow, giving their bodies a torpedo shape that likely reduces resistance and enhances their speed under water (Figs. 2.23, 2.24, and 2.25). In addition, the long pelvic girdles of foot-propelled divers provide more surface area for attachment of muscles that help hold the thigh relatively stationary and close to the body during dives (Hinić-Frlog and Motani 2010).

Birds specialized for terrestrial locomotion have wide, flat pelvic girdles that better support the hindlimbs and the large muscles of the upper leg (e.g., hindlimb muscles account for one-third of the body mass of a Common Ostrich, *Struthio camelus*; Smith et al. 2006; Fig. 2.23). Raptors also have relatively wide pelvic girdles, but with a noticeable bend (i.e., smaller iliac angle; Fig. 2.26). A wide pelvis provides more surface area for attachment of the powerful hindlimb muscles needed for capturing, killing, and carrying prey. An "angled" pelvis may provide better "leverage" for the muscles and further increase the force they can generate (Mattison 1998). The "angled" pelvis and resulting downward directed tail may also allow the tail to better serve as a brake when raptors attack their prey.

Among the ancestors of present-day birds, including *Archaeopteryx*, the pubic bone, or pubis, was either oriented in an anterior or cranial direction (i.e., propubic) or vertically (i.e., mesopubic) (Fig. 2.27). In contrast, the pubis of all present-day birds is oriented posteriorly (i.e., opisthopubic), a trait shared only with ornithischian dinosaurs, such as *Stegosaurus*, hadrosaurs (duck-billed dinosaurs), and iguanodons. Interestingly, during embryonic development of present-day birds, the pubis rotates from a vertical to the posterior direction found in adults (Goodrich 1986; Griffin et al. 2022).

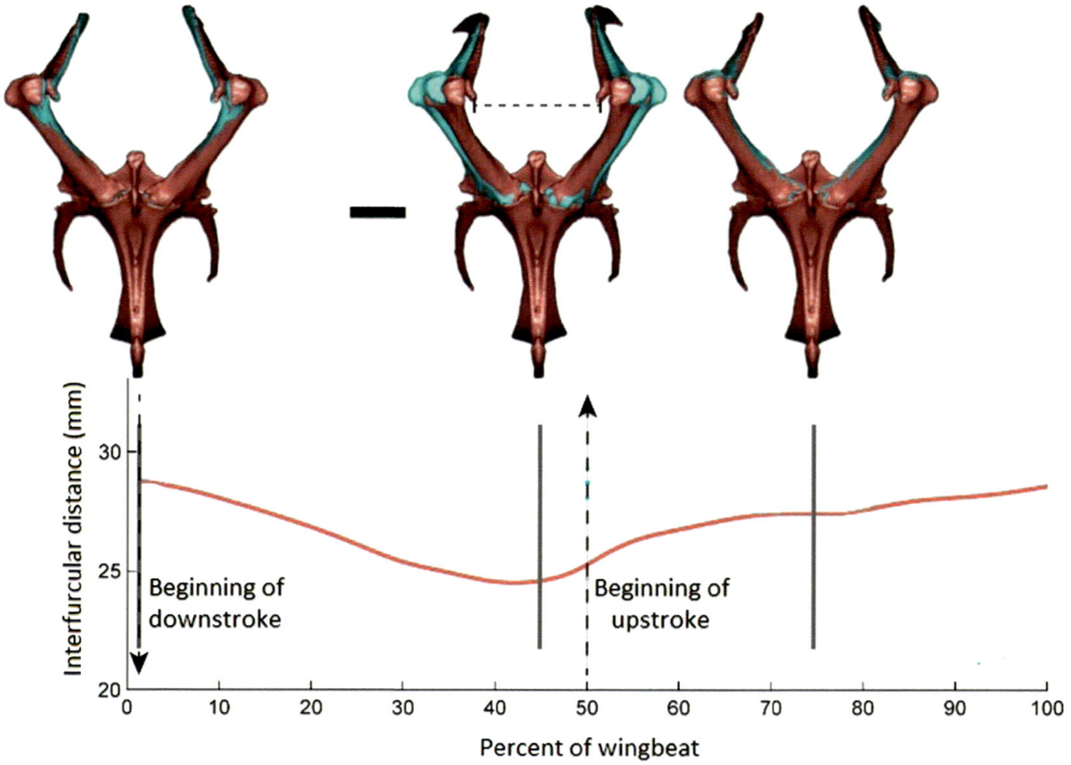


**Fig. 2.11** During flight, in this case the flight of a Chukar (*Alectoris chukar*), the interfurcular distance increases and decreases with each wingbeat. Scale bar = 10 mm. (Figure modified from Baier et al. 2013; © 2013 Baier et al., open-access article distributed under the terms of the Creative Commons Attribution License, https://creativecommons.org/licenses/by/4.0/)

Despite much conjecture, questions still remain concerning the possible reason(s) for this difference between archosaurs (nonavian dinosaurs and crocodilians) and ornithischians and birds in pubis orientation. One possible explanation for the difference in pubis orientation focuses on the absence of gastralia in birds and ornithischians, with gastralia being bones in the ventral body wall present in nonavian dinosaurs and crocodilians. These bones provide support for the abdomen and play an important role in crocodilian respiration (and may have also been important respiratory structures in nonavian dinosaurs) (Fig. 2.28). In crocodilians, contraction and relaxation of muscles originating on the anterior-directed pubis and attaching to the gastralia and sternum are known to generate movements of the abdominal wall important in changing the volume of the abdominal and thoracic cavities that, in turn, change the volume of air in the lungs and help move air into and out of the lungs (Claessens 2009). A similar process may have occurred in nonavian dinosaurs such as *Struthiomimus* (Fig. 2.28). If there are alternative mechanisms for changing air pressure in the abdominal and thoracic regions, then an anterior-directed pubis and gastralia would not be needed. In present-day birds, ventilation of the lungs occurs primarily via movements of the elongated sternum and ribs and the evolution of this alternative mechanism, which does not require gastralia and an adjacent (anterior-directed) pubis, may have led not only to the loss of gastralia, but to a reorientation of the pubis (Rasskin-Gutman and Buscalioni 2001).

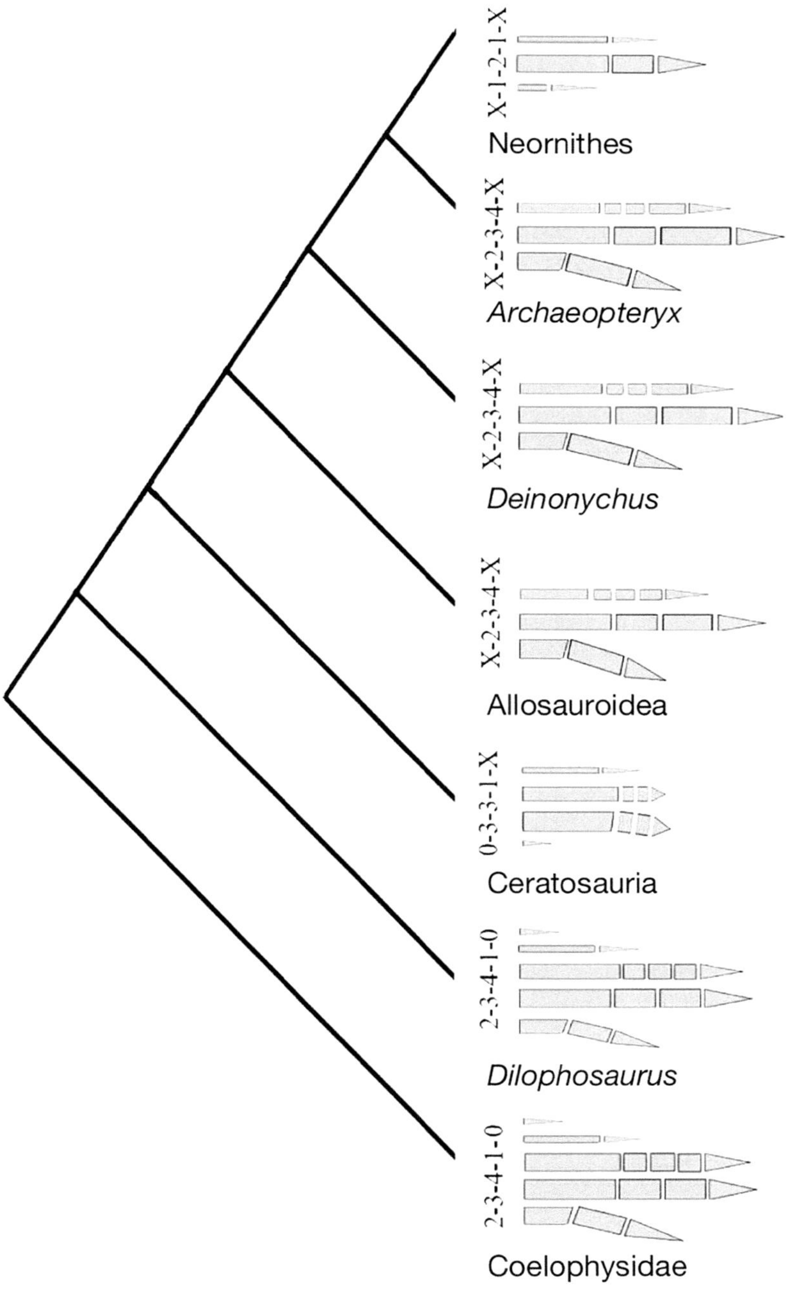


**Fig. 2.12** Phalanges in the digits of selected theropods, *Archaeopteryx*, and Neornithes showing a reduction in the number of digits and phalanges. The series of numbers to the left of digits refers to the number of phalanges per digit; X indicates that the digit has been lost. (Figure from Xu et al. 2009; © 2009 Springer Nature, used with permission)

Although based on limited fossil evidence, Paul and Leahy (1994) proposed that ornithischians possessed a muscular diaphragm similar to that of present-day mammals and, as in mammals, respiration driven by a muscular diaphragm would not require gastralia, an adjacent pubis, and associated musculature.

The hindlimb skeletons of theropod dinosaurs and many present-day birds are similar (Fig. 2.29), but with some differences. The fibula

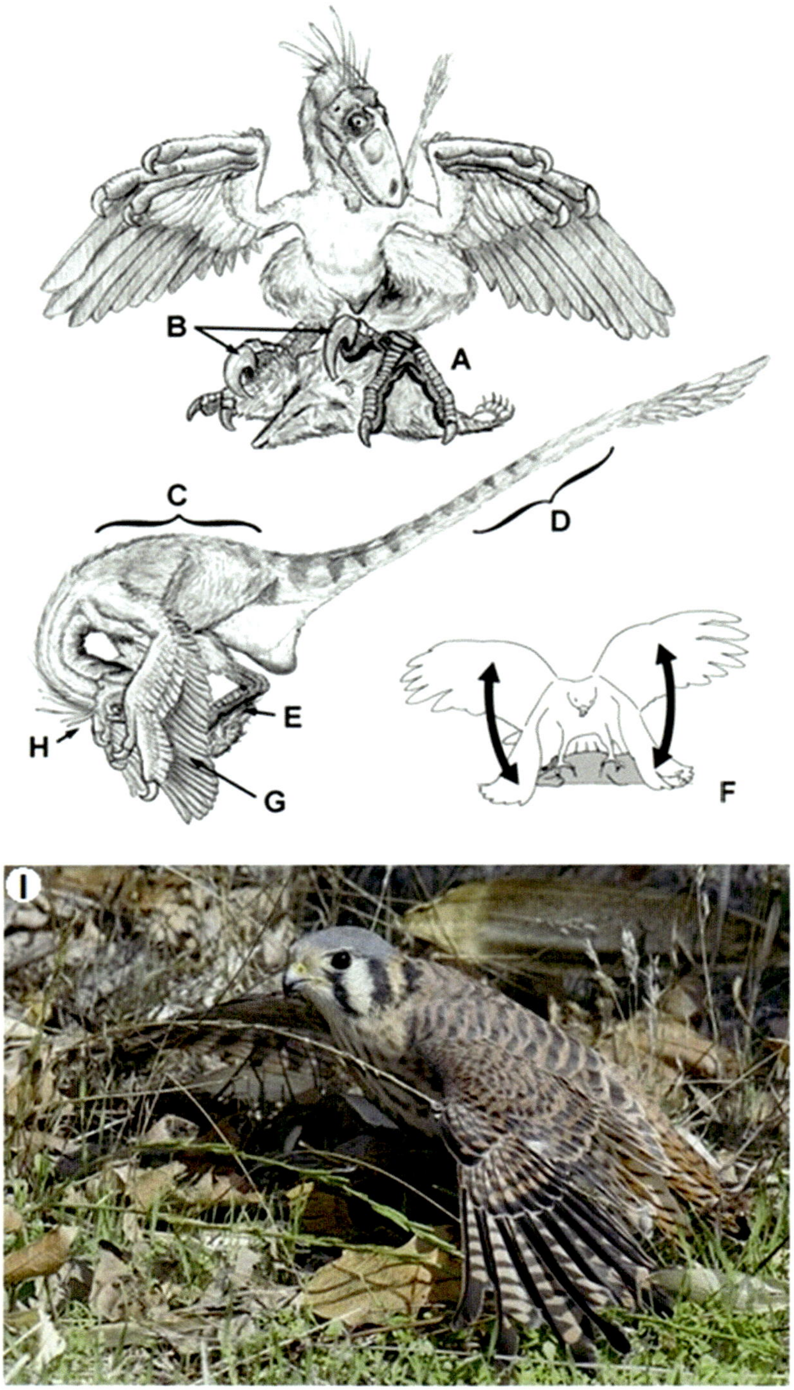


**Fig. 2.13** Possible advantage of the wing-like forelimbs of some theropod ancestors of birds, such as *Deinonychus*. (**a**) A grasping foot holds on to prey, (**b**) large claws used to maintain grip on prey, (**c**) a predator's body mass pins down victim, (**d**) the tail aids in maintaining balance, (**e**) the lower leg helps restrain prey, (**f**) "stability flapping" used to maintain position on top of prey, (**g**) the forelimbs encircle prey ("mantling"), reducing the likelihood of prey escaping, and (**h**) reaching down between feet and tearing off strips of flesh. (**i**) Present-day raptors, like this female American Kestrel (*Falco sparverius*), often mantle their prey. (Figure from Fowler et al. 2011; © 2011 Fowler et al., open-access article distributed under the terms of the Creative Commons Attribution License, https://creativecommons.org/licenses/by/4.0/. Photo of American Kestrel by Marlin Harms, Wikipedia, CC BY 2.0, https://creativecommons.org/licenses/by/2.0/)

**Box 2.2 Which Digits Were Lost?**

Bird wings have three highly modified digits, but their dinosaur ancestors had five digits. However, the more closely related theropod ancestors of birds had lost two of those digits and so, like most present-day birds, also had three digits. For nearly 150 years, investigators have attempted to answer an important question: are the three digits of theropods the same digits as those of present-day birds? This question has remained because paleontological evidence suggests that the three digits of theropods were digits I, II, and III, with IV and V being lost, whereas developmental evidence (i.e., the position of the early digit cartilages in bird embryos) seems to suggest that the digits of present-day birds are digits II, III, and IV, with I and V being lost (e.g., Wagner 2005).

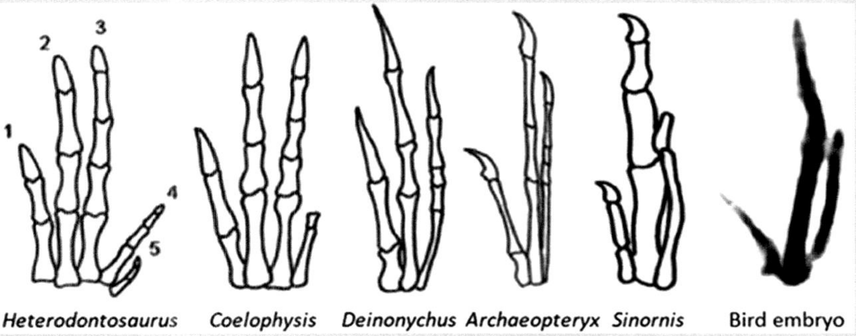


The dinosaur ancestors of birds, like *Heterodontosaurus* (an Ornithischian dinosaur that lived about 190 million years ago), had five digits. *Coelophysis*, a theropod, had four digits, and other theropods, like *Deinonychus*, had three, as did *Archaeopteryx* and *Sinornis*, an enantiornithine bird, and most present-day birds also have three digits. (Figure from Vargas and Fallon 2005; © 2004 Wiley-Liss, Inc., used with permission)

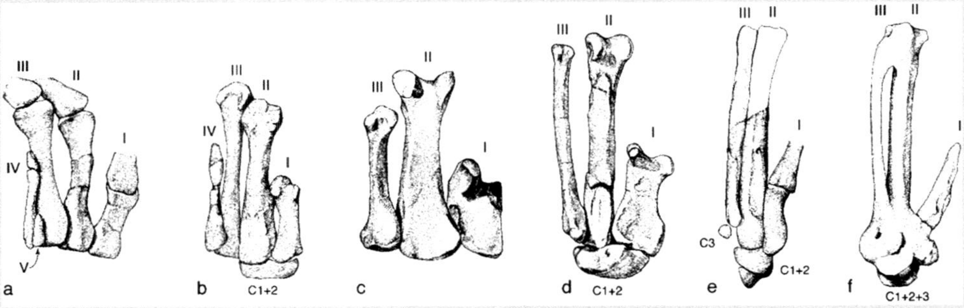


Ventral view of metacarpals and select wrist bones of theropods (**a**–**d**) and birds (**e** and **f**). These fossils provide evidence suggesting that digits IV and V were lost during the evolution of theropods and birds. (**a**) *Herrerasaurus ischigualastensis*, with vestigial digits IV and V, (**b**) *Coelophysis bauri*, with vestigial digit IV, (**c**) *Allosaurus fragilis*, (**d**) *Deinonychus antirrhopus*, (**e**) *Archaeopteryx lithographica*, and (**f**) spotted nothura (*Nothura maculosa*), a present-day tinamou. I–V, metacarpals; C1–3, distal carpals; r, radiale. (Figure from Wagner and Gauthier 1999; used with permission of the U.S. National Academy of Sciences)

(continued)

**Box 2.2** (continued)

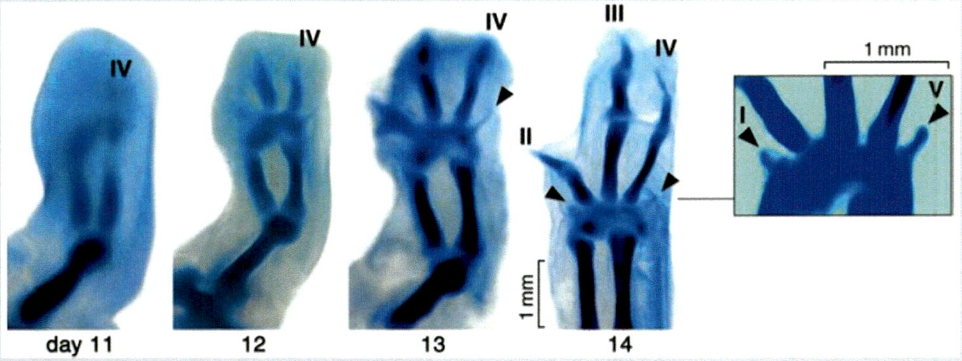


Development of the manus of a Common Ostrich (*Struthio camelus*) embryo, suggesting that the digits that eventually develop are numbers II, III, and IV. On the right is shown all that remains of metacarpals I and V at day 14 of development. (Figure from Feduccia and Nowicki 2002; © 2002 Springer Nature, used with permission)

One possible solution to the problem is to assume that birds did not evolve from theropod dinosaurs, but, of course, the evidence that they did is overwhelming. Another possible solution to the problem has been referred to as the "frame-shift" hypothesis, which posits that "... early morphogenesis and subsequent development of character identity are ... mechanistically independent" (Wagner and Gauthier 1999).

Frame-shift hypothesis. (a) The ancestral five-digit theropod hand had five metacarpals (C 1 – C V) and five digits (D I – D V), but only three were functional (D I – D III) (e.g., *Herrerasaurus*). (b) The next stage of evolution is the loss of digit V, but, again, only three digits were functional (D I – D III) (e.g., *Coelophysis*). At this stage, metacarpal V (C V) forms, but no digit forms. In present-day birds, C V still forms during development, but is eventually absorbed. (c) Transition from a four to a three-digit hand (e.g., *Allosaurus*) leads to the loss of condensation C I (as in present-day birds). The frame-shift hypothesis posits that the loss of C I shifts C II into the developmental trajectory of digit D I, C III into D II, and C IV into D III (Figure from Wagner and Gauthier 1999; © 1999 National Academy of Sciences, U.S.A., used with permission).

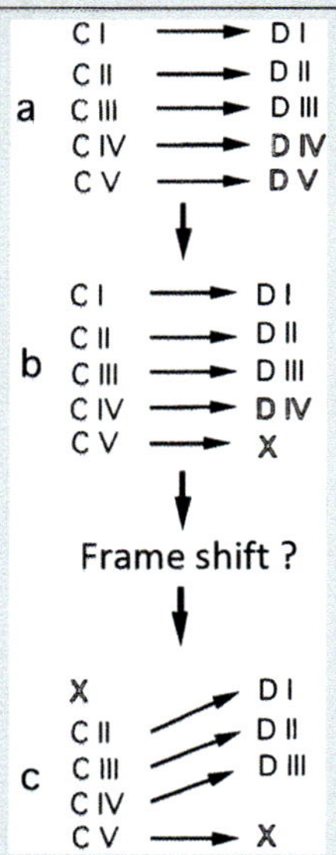


Frame-shift hypothesis. (**a**) The ancestral five-digit theropod hand had five metacarpals (C I–C V) and five digits (D I–D V), but only three were functional (D I–D III) (e.g., *Herrerasaurus*). (**b**) The next stage of evolution is the loss of digit V, but, again, only three digits were functional (D I–D III) (e.g., *Coelophysis*). At this stage, metacarpal V (C V) forms, but no digit forms. In present-day birds, C V still forms during development, but is eventually absorbed. (**c**) Transition from a four- to a three-digit hand (e.g., *Allosaurus*) leads to the loss of

(continued)

**Box 2.2** (continued)

condensation C I (as in present-day birds). The frame-shift hypothesis posits that the loss of C I shifts C II into the developmental trajectory of digit D I, C III into D II, and C IV into D III. (Figure from Wagner and Gauthier 1999; © 1999 National Academy of Sciences, U.S.A., used with permission)

A "limited" frame shift may also solve the problem. Stewart et al. (2019) identified a core set of transcription factors that are differentially expressed among digits, but they vary among species. For birds, these authors suggest that the three digits "reflect a combination of translocation and conserved digit identities." More specifically, the first digit of birds develops in position D2, but its gene expression profile matches that observed in D1 in other amniotes. However, the gene expression profiles of D3 and D4 are similar to those of D3 and D4 in other amniotes, suggesting no translocation. Thus, the "limited frame-shift" hypothesis of Stewart et al. (2019) posits that the digits in the wings of present-day birds correspond to digits D1, D3, and D4 of other amniotes. Other investigators have also suggested this possibility (Zhu et al. 2008). As noted by Stewart et al. (2019), their results "... indicate that diagnoses of digit identity from the paleontological record and hypotheses of digit identity based upon gene expression profiles have a more complex relationship than previously anticipated" and, as a result, additional data "... and a willingness to consider hypotheses that previously might have been regarded as heterodox, are required for the testing and refinement of integrative theories on the nature of limbs."

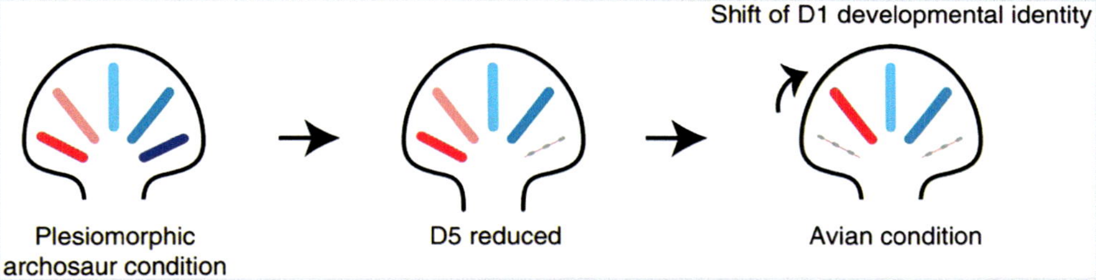


A limited frame-shift model for evolutionary origin of the avian wing, with the developmental identity of D1 translocated to position D2. (Figure from Stewart et al. 2019; open-access article licensed under a Creative Commons Attribution 4.0 International License, https://creativecommons.org/licenses/by/4.0/)

**Box 2.3 Evolution of the Avian Wrist**

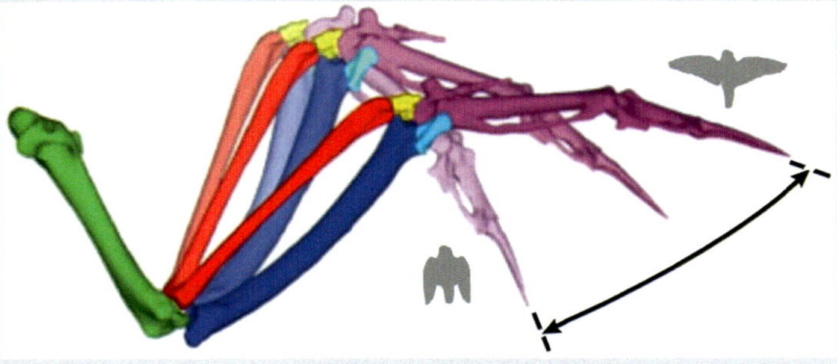

(continued)

**Box 2.3** (continued)

Birds can "morph" their wings via flexion and extension at the elbow and wrist joints. (Figure from Stowers et al. 2017; © 2017 The Authors. Published by the Royal Society, used with permission)

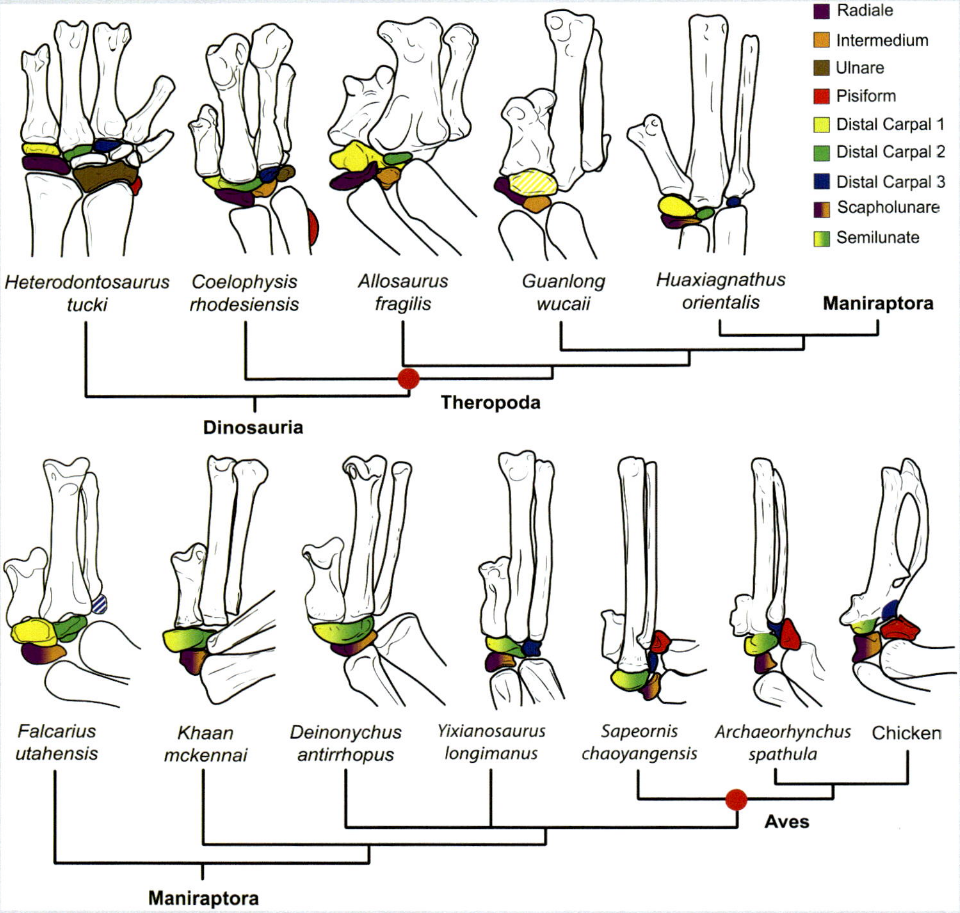


An important step in the evolution of birds and the ability of birds and some of their ancestors to fly was development of a light, flexible wrist that allows birds to fold their wings when not in use, but keeps wings rigid during flight, from the nine-bone wrist of quadripedal dinosaurs. Evolution of the avian wrist and hand involved the loss of many bones and fusion of others. In present-day birds such as the Domestic Chicken (*Gallus g. domesticus*), there are only four areas of ossification in the wrist—two distal ossifications fuse to each other and to metacarpal bones to form the carpometacarpus, and two proximal ossifications form two separate small bones. These small bones are commonly referred to as radiale and ulnare bones, but the actual derivation of these bones relative to bones in the wrists of avian ancestors remains unclear. Using new techniques to study embryonic development, Botelho et al. (2014) found that the avian radiale bone develops by fusion of radiale + intermedium cartilage and proposed that it be referred to as the scapholunare (purple–orange). One ossification of birds that fuses with the carpometacarpus (yellow–green) is homologous to the semilunate of dinosaurs, and the other (blue) is homologous to distal carpal 3. In early dinosaurs and most basal theropods, distal carpal 1 (yellow) and 2 (green) were separate bones. In maniraptoran dinosaurs such as *Deinonychus antirrhopus*, these two carpals fused and articulated with two metacarpal bones, which is consistent with evidence provided by Botelho et al. (2014) that this bone in present-day birds develops from a composite cartilage. Surprisingly, birds apparently "re-evolved" a large, ossified pisiform bone (red), perhaps by activation of a dormant gene. The pisiform and ulnare

(continued)

**Box 2.3** (continued)

were present only in early dinosaurs, suggesting that, in later dinosaurs, they were either absent or failed to ossify and so did not fossilize. In birds, evidence provided by Botelho et al. (2014) demonstrates that the ulnare was lost, but the pisiform is present. (Figure from Botelho et al. 2014; © 2014 Botelho et al., open-access article distributed under the terms of the Creative Commons Attribution License, https://creativecommons.org/licenses/by/4.0/)

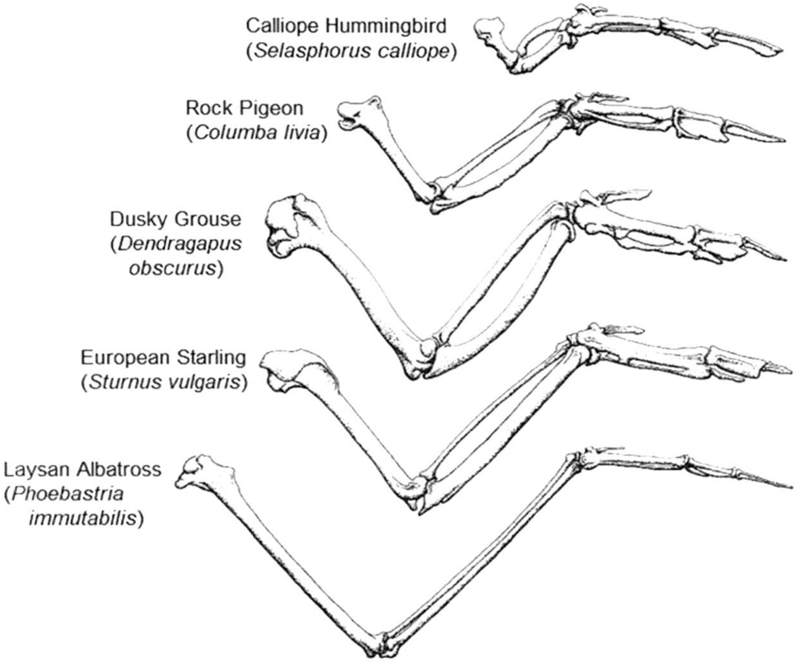


**Fig. 2.14** Wing skeletons of five species of birds showing variation in relative length of the forearm bones (scaled so that carpometacarpi are of equal length). (Figure from Dial 1992; © 1992 Oxford University Press, used with permission)

**Box 2.4 Avian Humeri**

The avian humerus plays a key role in locomotion for all birds except flightless, terrestrial birds like ratites. Serrano et al. (2020) examined morphological variation in the humeri of 153 species of birds in 71 families and 23 orders and found that humerus length increased faster with body mass than humerus width, i.e., humeri of flying birds tend to be longer and thinner with increasing body mass. For example, Southern Giant-Petrels (*Macronectes giganteus*, family Procellariidae) range in mass from about 3.8 kg (females) to 5 kg (males) and their humeri (as pictured below, C) are about 25 cm long and, at mid-shaft, only about 10 mm wide. Humeri of much smaller birds are generally much shorter and relatively thicker. For example, the humerus of White-throated Needletails (*Hirundapus caudacutus*, family Apodidae; mass = 110–120 grams) is only about 1.7 cm long, but 4.5 mm in width at mid-shaft (pictured below, E).

(continued)

**Box 2.4** (continued)

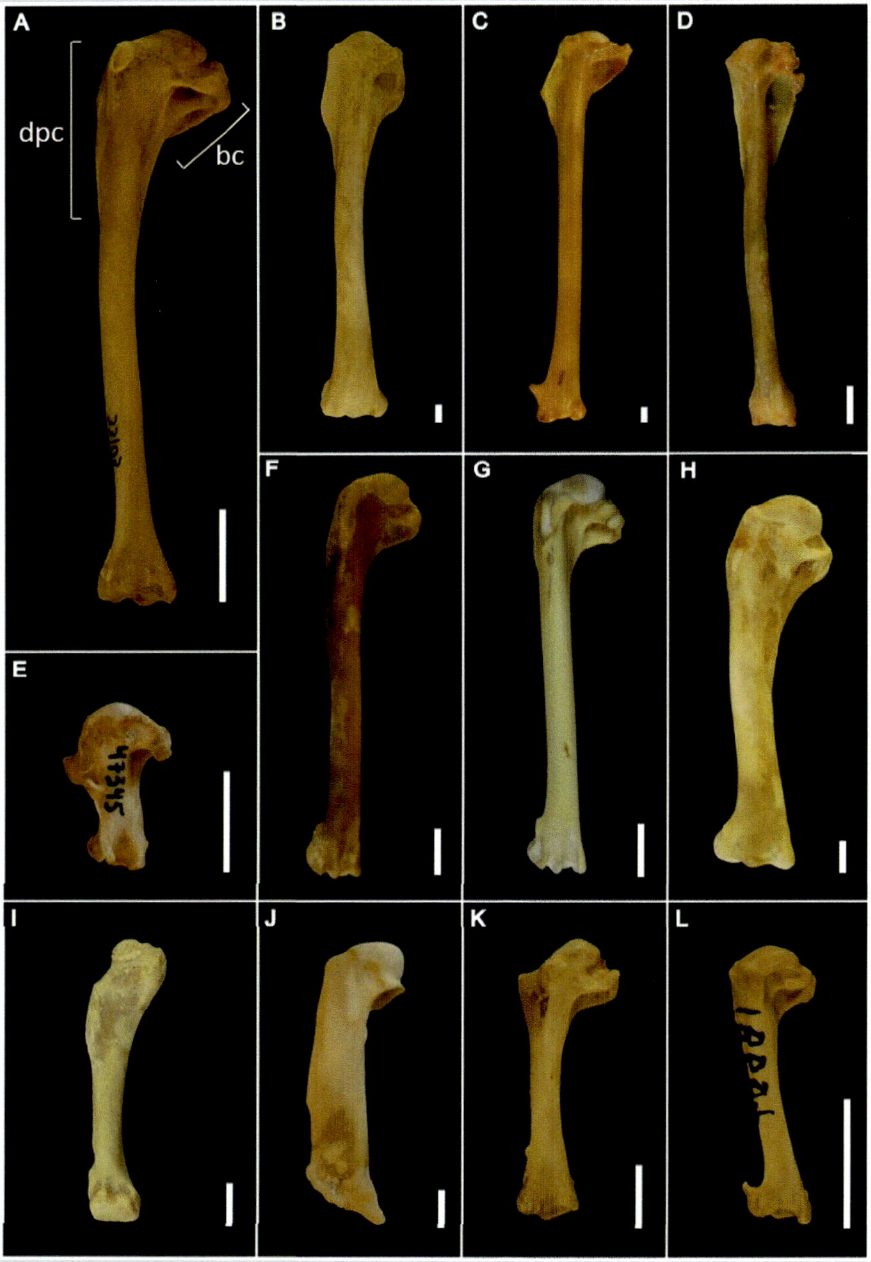


Humeri in caudal view for selected species of birds. (**a**) Ruddy Duck (*Oxyura jamaicensis*). Abbreviations: bc, bicipital crest; dpc, deltopectoral crest. (**b**) Bald Eagle (*Haliaeetus leucocephalus*). (**c**) Southern Giant-Petrel (*Macronectes giganteus*). (**d**) Flightless Cormorant (*Nannopterum harrisi*). (**e**) White-throated Needletail (*Hirundapus caudacutus*). (**f**) Common Murre (*Uria aalge*). (**g**) Horned Puffin (*Fratercula corniculata*). (**h**) Wild Turkey (*Meleagris gallopavo*). (**i**) Dwarf Cassowary (*Casuarius bennetti*). (**j**) Humboldt Penguin

(continued)

**Box 2.4** (continued)

(*Spheniscus humboldti*). (**k**) Hill Pigeon (*Columba rupestris*). (**l**) American Dipper (*Cinclus mexicanus*). Scale bars = 10 mm. (Figure from Serrano et al. 2020; © 2020 The Authors. Licensee MDPI, Basel, Switzerland, open-access article distributed under the terms and conditions of the Creative Commons Attribution (CC BY) license, https://creativecommons.org/licenses/by/4.0/)

Southern Giant-Petrel (Photo by Lieutenant Elizabeth Crapo, NOAA Corps, CC0 Public Domain)

Serrano et al. (2020) also found that, with a few exceptions, the humeri of most flying birds tended to overlap in morphospace (see figure below), suggesting the existence of a general morphology related to efficient flapping flight. Exceptions include swifts and hummingbirds (order Apodiformes), penguins (order Sphenisciformes), and some Alcidae (i.e., murres, guillemots, auklets, puffins, and murrelets). Species in the family Apodidae (e.g., swifts and swiftlets) are hyperaerials, spending most of their time (when not breeding) in the air and hummingbirds are the only birds capable of sustained hovering. These species have extremely robust humeri able to withstand the stresses of either prolonged flights or sustained hovering. Penguins and some alcids are wing-propelled divers and also have robust humeri able to resist the stress involved in flapping their wings in water, a medium that is much denser than air (Serrano et al. 2020).

(continued)

**Box 2.4** (continued)

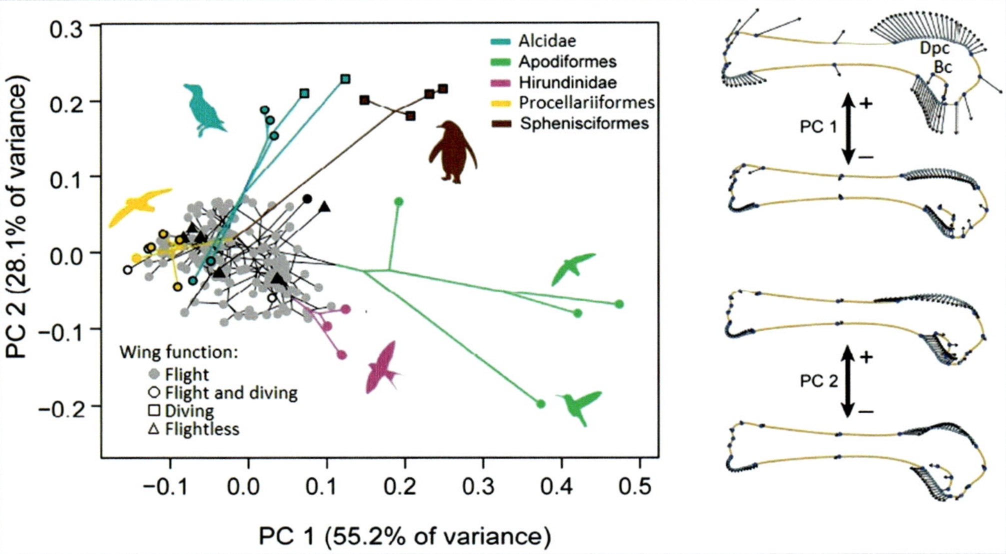


Phylomorphospace depicting the two first principal components of the PCA and vector graphs showing changes in shape of humeri associated with each of the axes. Shape change along PC1 is associated with general robustness of the bone; positive scores indicate humeri becoming shorter with broader ends. Positive scores in PC2 are associated with expanded deltopectoral crests (Dpc) and a bicipital crest (Bc) restricted to the proximal end of the humerus like in some penguins and some alcids; negative scores are associated with shorter deltopectoral crest and expanded bicipital crests like in hummingbirds (Trochilidae) and, to a lesser degree, swallows (Hirundinidae) and swifts (Apodidae). (Figure modified from Serrano et al. 2020; © 2020 The Authors. Licensee MDPI, Basel, Switzerland, open-access article distributed under the terms and conditions of the Creative Commons Attribution (CC BY) license, https://creativecommons.org/licenses/by/4.0/)

of birds is reduced in size, terminating in a sharp point that no longer articulates at the tarsal joint (Fig. 2.30). A reduction in the relative size of the fibula relative to the tibia began in quadripedal reptiles and this trend accelerated among bipedal theropods. With the evolution of flight and need to reduce weight, natural selection likely favored continued reduction in fibula size (Streicher and Müller 1992). In addition, whereas theropods and *Archaeopteryx* had tarsal bones that articulated with the tibia and fibula (though the tarsal bones were partially fused to each other and to the tibia and metatarsal bones), the proximal tarsals of present-day birds are completely fused with the tibia to form the tibiotarsus and the distal tarsals are completely fused to the metatarsals to form the tarsometatarsus. The tibiotarsus articulates with the tarsometatarsus at the intertarsal joint (Streicher and Müller 1992).

The theropod ancestors of birds were cursorial bipeds; extant birds are also bipedal and, with some exceptions such as swifts and hummingbirds, most are, to varying degrees, cursorial. A key difference between theropods and birds, however, is that theropod locomotion was "hip driven" whereas cursorial bird locomotion is "knee driven." In other words, when theropods walked and ran, the entire leg (including the femur) rotated (just as it does in humans). When birds walk, in contrast, the lower limb rotates, but

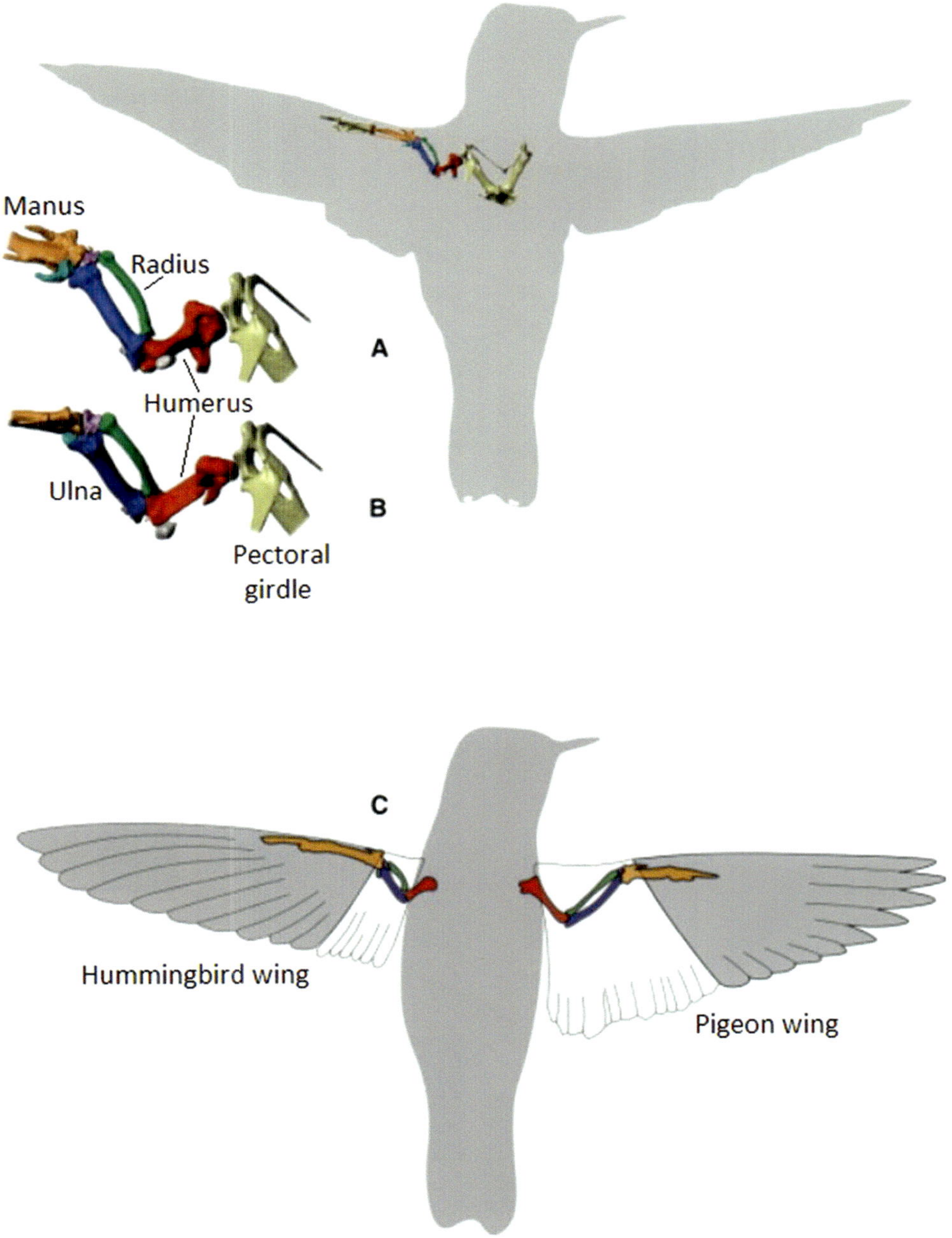


**Fig. 2.15** Forelimb skeleton of a hummingbird in flight. (**a**) Position at mid-downstroke. (**b**) Position at mid-upstroke. (**c**) Comparison of a hummingbird wing, with primaries shown in gray, and the wing of a pigeon wing scaled to hummingbird wing length to illustrate the proportionately smaller handwing. (Figure modified from Warrick et al. 2012; © 2012 Elsevier Ltd., used with permission)

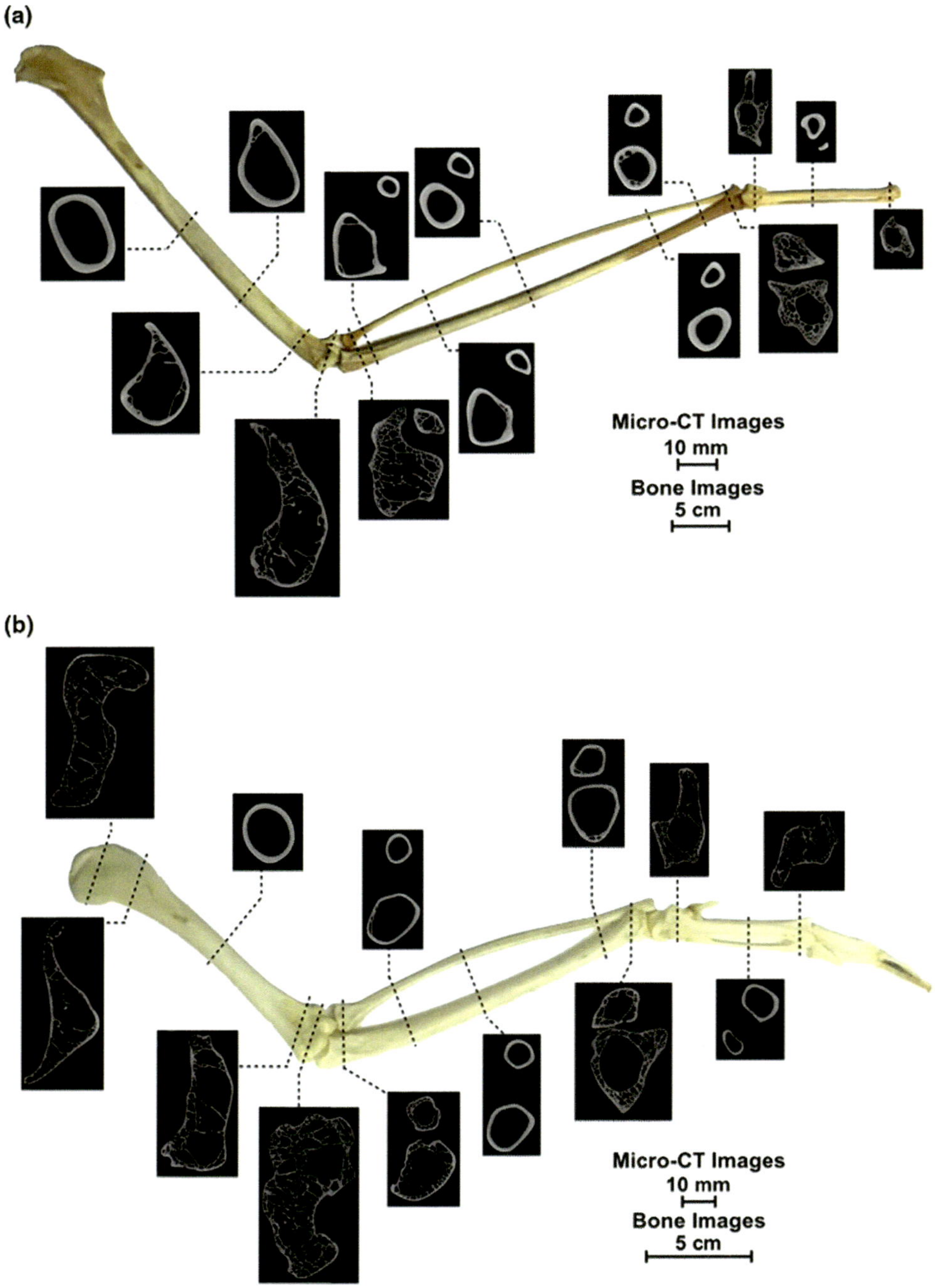


**Fig. 2.16** Micro-computerized tomography scans of the wing bones of (**a**) Wandering Albatross (*Diomedea exulans*), the species with the largest wingspan of any living bird, and (**b**) Turkey Vulture (*Cathartes aura*). Note that the bones have a hollow, circular cross-section near the middle and reinforcing structures toward the ends of the bones. (Figure from Sullivan et al. 2017; open-access article licensed under a Creative Commons License, https://creativecommons.org/licenses/by/4.0/)

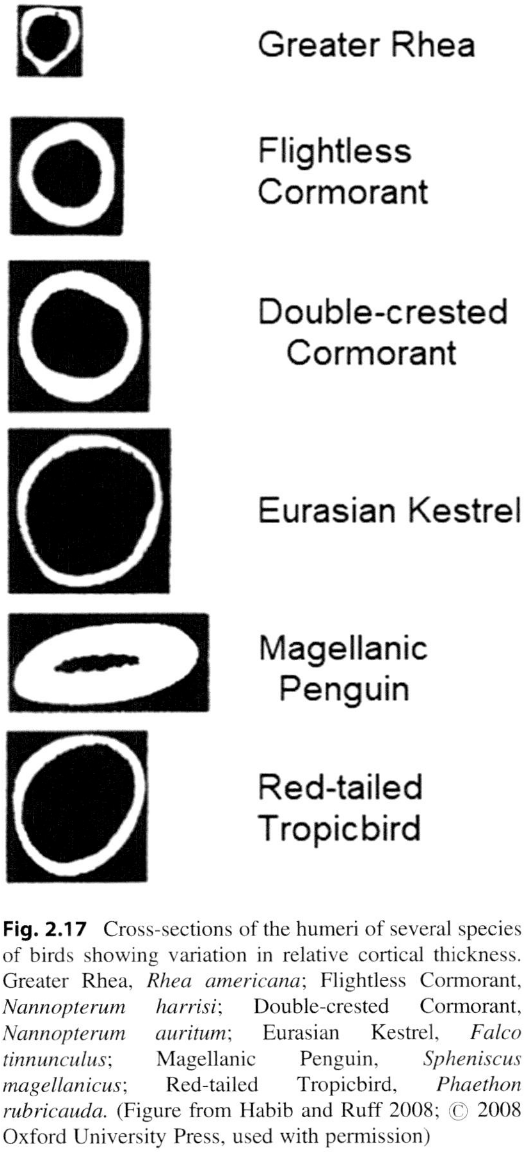


**Fig. 2.17** Cross-sections of the humeri of several species of birds showing variation in relative cortical thickness. Greater Rhea, *Rhea americana*; Flightless Cormorant, *Nannopterum harrisi*; Double-crested Cormorant, *Nannopterum auritum*; Eurasian Kestrel, *Falco tinnunculus*; Magellanic Penguin, *Spheniscus magellanicus*; Red-tailed Tropicbird, *Phaethon rubricauda*. (Figure from Habib and Ruff 2008; © 2008 Oxford University Press, used with permission)

the femur exhibits little movement (Figs. 2.31 and 2.32). When birds run, the femur exhibits more movement, but far less than was the case for theropods (Hutchinson and Allen 2008; Fig. 2.33). An important factor in this difference between birds and theropods is the location of the center of mass, with hip-driven locomotion apparently linked to a more caudal center of mass and knee-driven location with a more cranial center of mass (Hutchinson and Allen 2008).

Among birds, the angle between the femur and the pelvis (i.e., femoral splay angle) varies. Hertel and Campbell (2007) measured the femoral splay angle of 77 species of birds representing 26 families and found that the mean splay angle was about 40 degrees (the approximate splay angle of the Adelie Penguin [*Pygoscelis adeliae*] in Fig. 2.34), and ranged from about 27 degrees (falcons) to 77 degrees (grebes). Loons and grebes were found to be outliers, with femoral splay angles about 20–25 degrees greater than those of the species with the next highest splay angle (rhea, 52 degrees). Loons and grebes are foot-propelled divers with very narrow pelvic girdles (providing a streamlined body for more efficient movement through the water). The high femoral splay angle of these birds means that their feet are positioned farther posteriorly and farther apart, adaptations for efficient underwater propulsion (Hertel and Campbell 2007).

Among extant birds, the relative lengths of the femur, tibiotarsus, tarsometatarsus, and digits vary with leg length and leg function, or how a bird uses its legs (e.g., perching only, capturing prey, walking or running, climbing, swimming, or wading; Zeffer et al. 2003; Fig. 2.35). In long-legged birds, like flamingoes, the femur is relatively short and the tibiotarsus and tarsometatarsus are similar in length (Fig. 2.36), allowing the birds to maintain stability (keeping their center of mass over their feet) when they crouch. Aerial species such as swifts, hummingbirds, swallows, and caprimulgids that use their feet and legs primarily or only for perching have short legs (relative to their body size) and, in particular, relatively short tarsometatarsi (Zeffer et al. 2003). Short legs weigh less and, when pulled into the plumage, give these birds a streamlined shape. Also, for perching, a short tarsometatarsus helps birds maintain balance because it keeps the center of mass closer to the perch (Schulenberg 1983).

Raptors, including hawks, falcons, and owls, have relatively long legs for their size (Fig. 2.37), possibly because long legs extended at the time of attack increase reaching distance and make it possible to reach prey with the feet (and talons) first. This could potentially increase the element

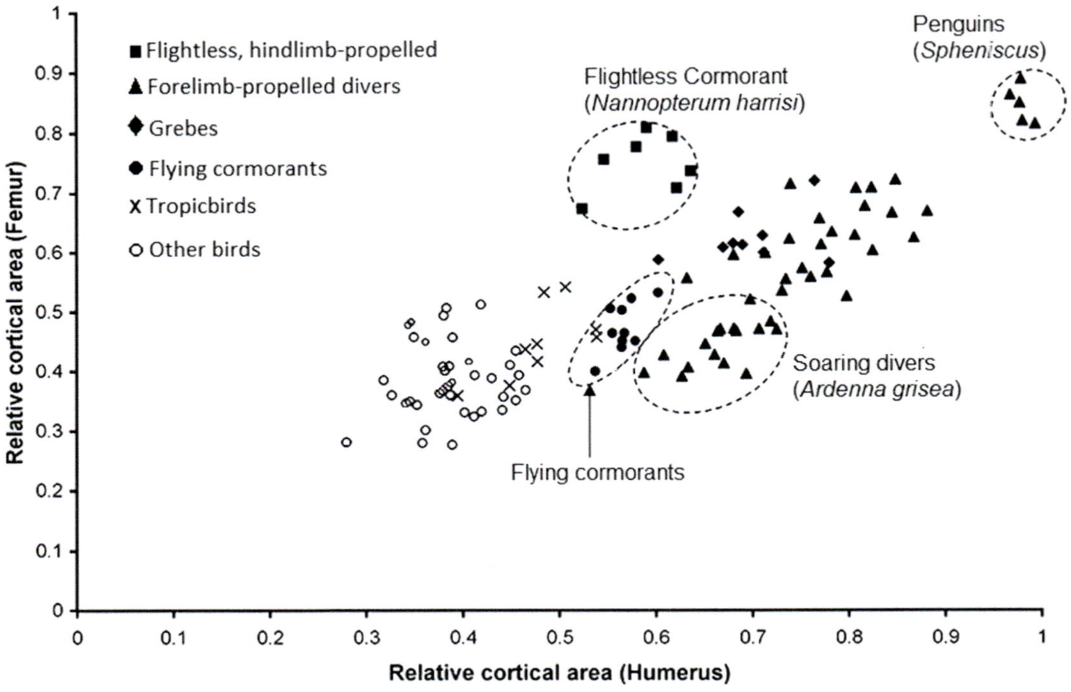


**Fig. 2.18** Percentage of bone cross-section composed of cortical (compact) bone in the humerus and femur of several taxa of birds. Higher percentages mean more bone relative to the hollow central section. Note that the humeri of forelimb-propelled divers like penguins and puffins have relatively more cortical bone than those of tropicbirds (oceanic soaring birds) and other birds that are excellent flying or soaring species, i.e., Barn Owls (*Tyto alba*), Golden Eagles (*Aquila chrysaetos*), Eurasian Kestrels (*Falco tinnunculus*), Sooty Shearwaters (*Ardenna grisea*), Wandering Albatrosses (*Diomedea exulans*), and Common Ravens (*Corvus corax*). Diving birds also benefit from being heavier, with leg bones, including the femur, having more bone relative to the hollow central section. (Figure modified from Habib and Ruff 2008; © 2008 Oxford University Press, used with permission)

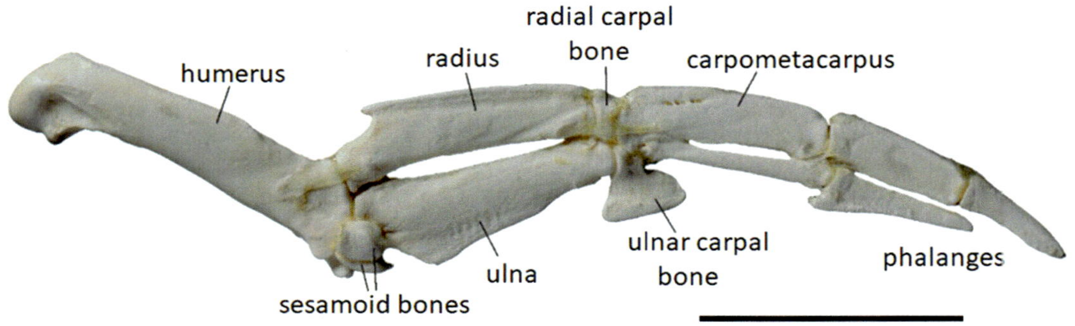


**Fig. 2.19** Wing bones of a Humboldt Penguin (*Spheniscus humboldti*). For wing-propelled swimming underwater, the "paddle" must be highly mobile at the shoulder (humerus-pectoral girdle articulation). However, the remaining joints must be relatively fixed to minimize the muscle contraction needed to maintain the proper position. (Figure from Mayr et al. 2020, © 2020 by the authors. Licensee MDPI, Basel, Switzerland, open-access article distributed under the terms and conditions of the Creative Commons Attribution (CC BY) license, http://creativecommons.org/licenses/by/4.0/)

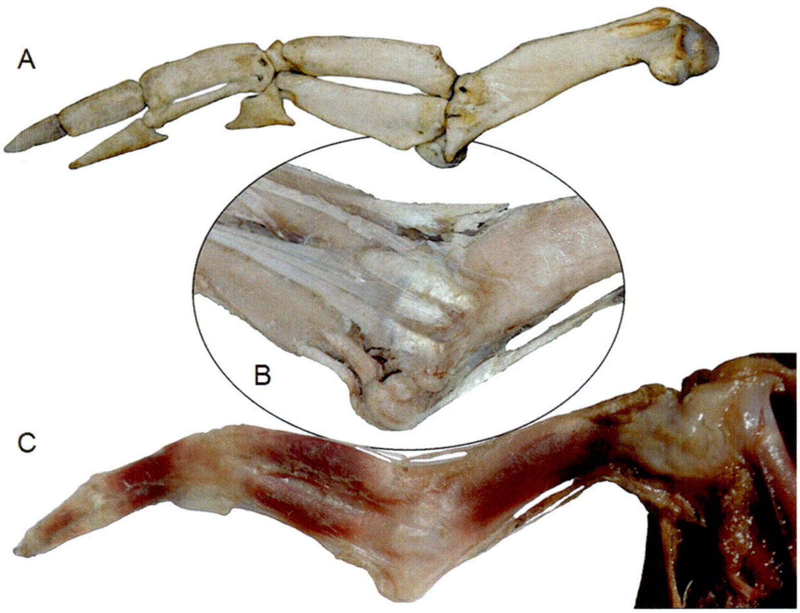


**Fig. 2.20** (**a**) Wing bones of an Adelie Penguin (*Pygoscelis adeliae*). (**b**) Close-up of "elbow" showing muscle and several large tendons. (**c**) Dissected wing of a Gentoo Penguin (*Pygoscelis papua*) chick. (Figure from Vargas et al. 2017; © 2017 Oxford University Press, used with permission)

of surprise and hunting success (Zeffer et al. 2003). The relative length of the legs of hawks and falcons exceeds that of owls, primarily due to a difference in the length of the tarsometatarsus. The owl tarsometatarsus is shorter and wider, whereas the hawk/falcon tarsometatarsus is longer and thinner (Fig. 2.38). This difference suggests that owls may depend on an especially forceful grip to subdue prey, with the more robust tarsometatarsus and associated muscles able to generate greater force and a stronger grip to quickly dispatch prey and minimize the chances

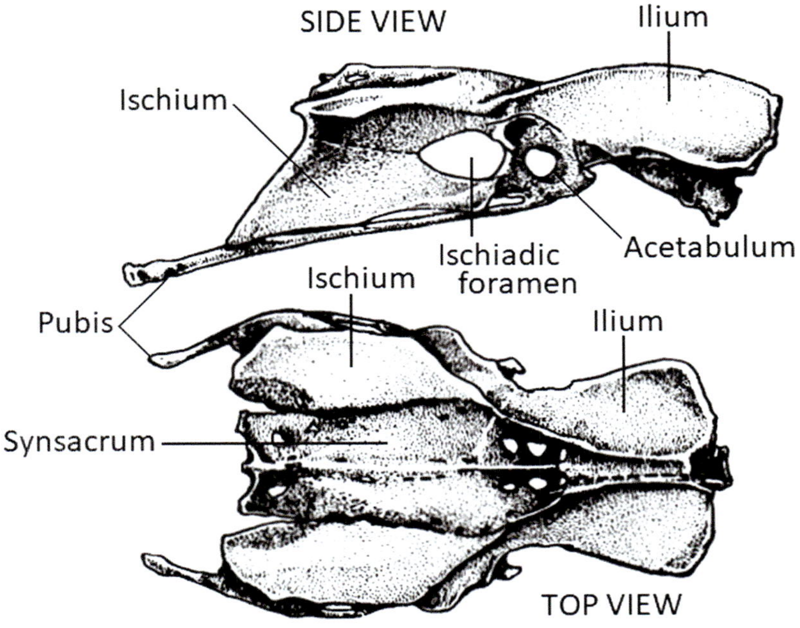


**Fig. 2.21** Bones and other structures of the pelvic girdle of a Wild Turkey (*Meleagris gallopavo*). The acetabulum is where the femur articulates with the pelvic girdle. The ischiadic foramen (also called the ischiatic foramen or ilioischiadic foramen) is where the ischiatic (or sciatic) nerve passes to muscles in the leg. (Figure modified from Hutchinson 2001; © 2008 Oxford University Press, used with permission)

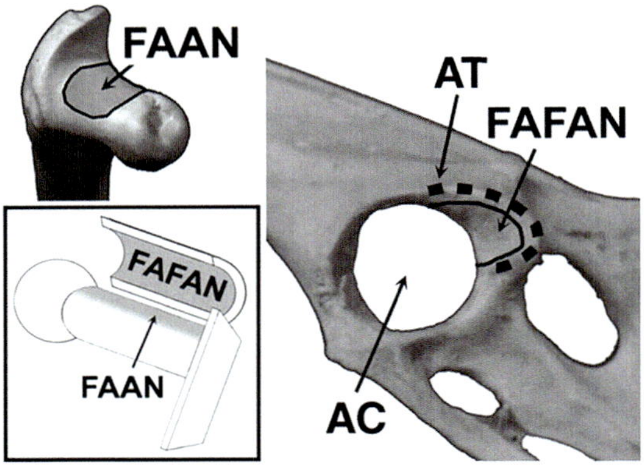


**Fig. 2.22** Diagram showing the proximal part of a left femur (left) with the antitrochanter (AT) of the pelvis (right). The concave facies articularis femoralis of the antitrochanter (FAFAN) articulates with the convex facies articularis antitrochanterica (FAAN) of the proximal end of the femur. AC = acetabulum. Inset shows the drum-in-trough-like nature of the antitrochanter–femur joint. (Figure from Hertel and Campbell 2007; © 2007 Oxford University Press, used with permission)

of prey escaping. In contrast, hawks and falcons may be able to use quick foot movements to capture and subdue prey that temporarily initially elude them when attacked (Ward et al. 2002). Some hawks, such as Eurasian Sparrowhawks (*Accipiter nisus*), occasionally chase prey on the ground and may also extend their legs into vegetation to capture prey (Newton 1979).

Birds that swim and dive typically have relatively short legs, with the tibiotarsus being the longest leg bone (Fig. 2.39). Shorter legs may minimize drag as the hindlimbs are used to propel swimming and diving birds through the water. In addition, a longer tibiotarsus may be advantageous because, in contrast to the other leg bones, it is directed backward and places the webbed feet further back on the body when a bird is in the water, a position that likely increases propulsive efficiency (Zeffer et al. 2003). Foot-propelled diving birds, like loons and grebes,

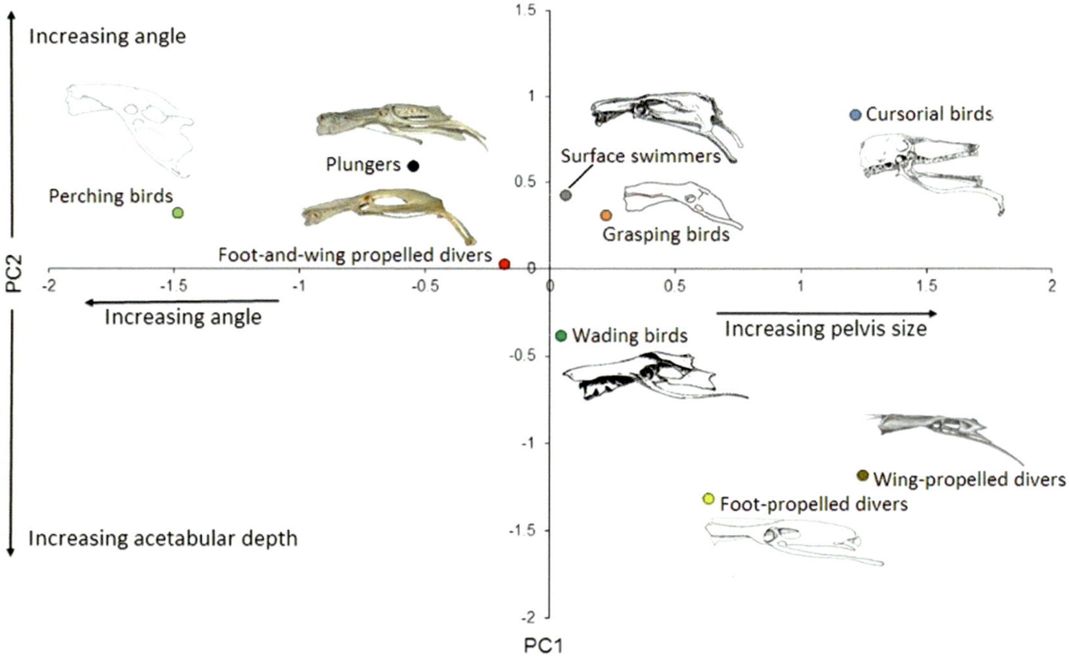


**Fig. 2.23** Average values of the first two principal components for the morphology of pelvic girdles for birds with different locomotory patterns. Note the difference in pelvic angles between foot- and wing-propelled divers and birds with other locomotory patterns. (Figure modified from Anten-Houston et al. 2017; © 2017 Anatomical Society, used with permission)

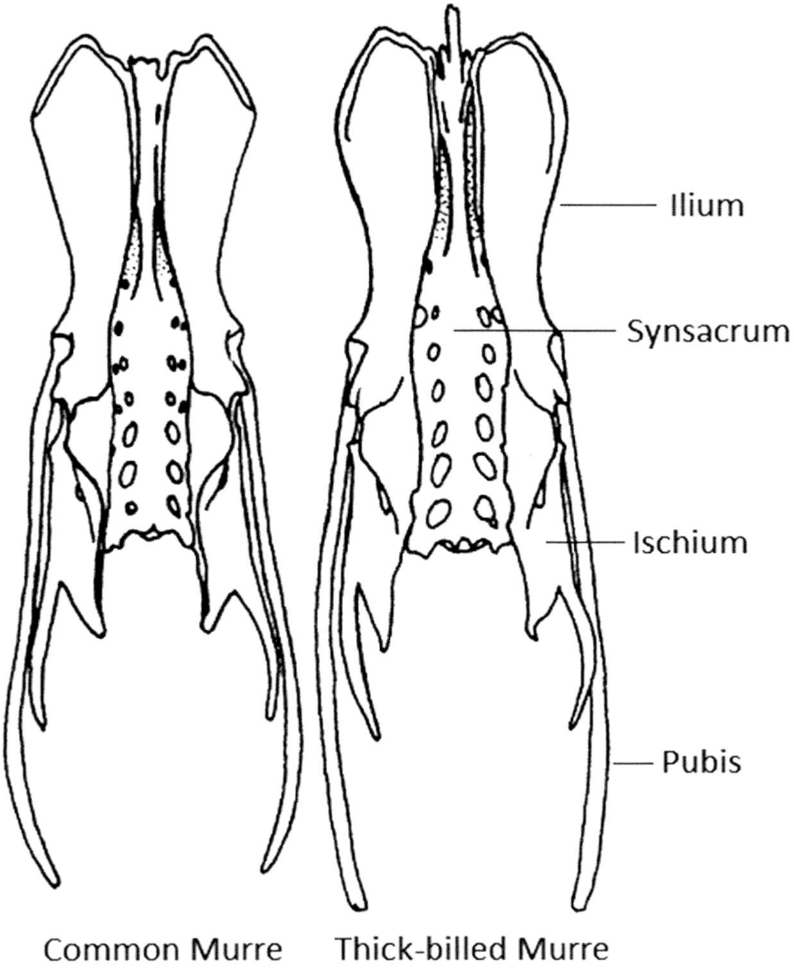


**Fig. 2.24** Long, relatively thin pelvic girdles of two species of diving birds, Common Murre (*Uria aalge*) and Thick-billed Murre (*Uria lomvia*). (Figure modified from Spring 1971; © 1899 CCC Republication, used with permission)

have an extension of the tibiotarsus called the cnemial crest that serves as an attachment site for muscles that help stabilize the knee, hold the tibiotarsus close to the body, and help increase the propulsive force generated by the tarsometatarsus (Fig. 2.40). Aquatic birds that also walk on land, e.g., ducks, geese, and swans, do have relatively longer legs than diving birds, e.g., grebes, cormorants, and loons, that do not (Zeffer et al. 2003), with longer legs permitting longer strides and more efficient locomotion on land.

Birds that climb trees vertically, such as woodpeckers, creepers, treecreepers, and woodcreepers, have relatively short hindlimbs (Carrascal et al. 1990; Fig. 2.41). Shorter legs reduce the distance between the center of gravity of birds and the vertical surface and reduce the effort needed to hold the body close to the trunk (Norberg 1979). In contrast and not surprisingly, wading birds such as herons, egrets, and storks have relatively long hindlimbs, allowing them to forage in deeper water.

Some birds, such as gallinaceous species, spend most of their time on the ground walking and running, and others, such as songbirds like crows, larks, and some thrushes, spend some of their time on the ground foraging. Although such species might be expected to have relatively longer hindlimbs (than less terrestrial species of similar size) to increase stride length and speed of

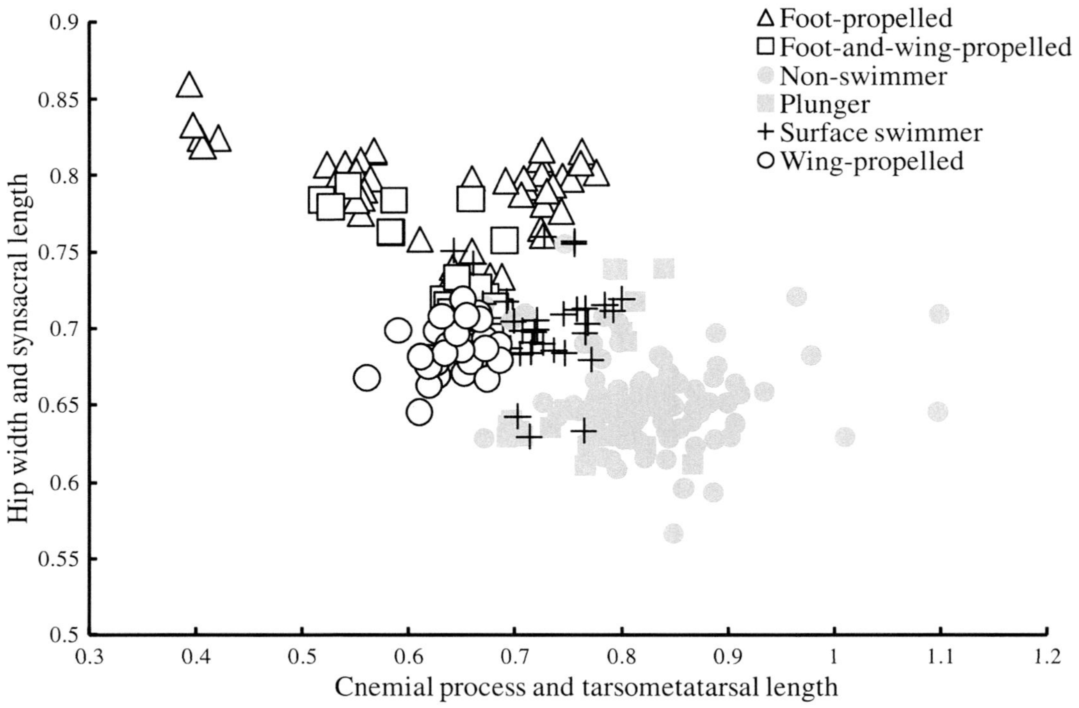


**Fig. 2.25** Relationship between hip and leg anatomy and type of avian locomotion. For the hip width and synsacral length axis, higher values indicate narrower hips and long synsacrums. For the cnemial process and tarsometatarsal length axis, lower values indicate longer cnemial processes and longer tarsometatarsi. Foot-propelled divers tend to have long, narrow hips and synsacrums and relatively long cnemial processes and tarsometatarsi. (Figure modified from Hinić-Frlog and Motani 2010; © 2009 The Authors. Journal Compilation © 2009 European Society for Evolutionary Biology, used with permission)

movement on the ground, analysis indicates no such specialization. Because these birds can also fly, there has likely been little selective pressure for increased running speed (e.g., to elude predators). In contrast, some flightless birds, such as Common Ostriches (*Struthio camelus*), Emus (*Dromaius novaehollandiae*), and rheas (*Rhea* spp.) have relatively long legs (particularly the tarsometatarsus, Gatesy and Middleton 1997; Fig. 2.42). Long legs increase stride length and allow increased running speed.

Most species of birds have four toes. However, one toe (the hind toe, or digit I) is reduced in size in several taxa of birds, including rails, flamingos, grebes, plovers, sandpipers, and their allies. Other species, e.g., Emu, cassowaries (Fig. 2.43), Sanderlings (*Calidris alba*), Three-toed Jacamars (*Jacamaralcyon tridactyla*), American Three-toed Woodpeckers (*Picoides dorsalis*), Spot-throated Flamebacks (*Dinopium everetti*), and Three-toed Parrotbills (*Cholornis paradoxus*), or other taxa (bustards and quails) have just three toes (tridactyl), having lost their hind toe, and Common Ostriches (*Struthio camelus*) have just two toes (didactyl; digits III and IV; Fig. 2.44). Toes are lost when they no longer serve any function. For example, for many woodpeckers, a backward projecting digit I (also called the hallux) can limit how close the tarsometatarsus can move toward the trunk of a tree so is often reduced in size or, in some species, absent (Zelenkov 2007; Fig. 2.45). For Common Ostriches, the only didactyl birds, this toe reduction creates a concentrated traction surface

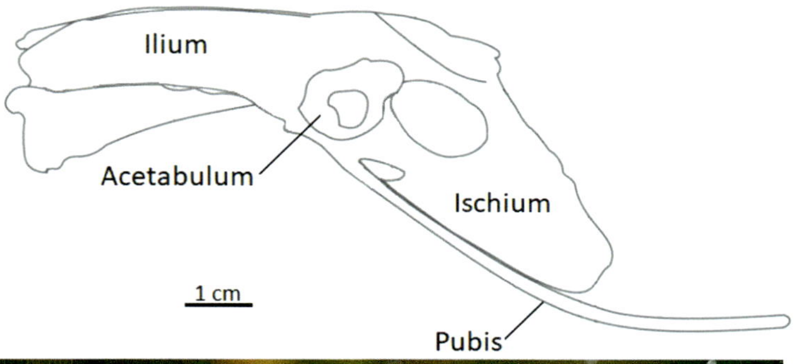

**Fig. 2.26** Top: Pelvic girdle of a raptor (Chimango Caracara, *Milvago chimango*). Raptors have wide pelvic girdles with a noticeable bend. A wide pelvis provides more surface area for attachment of hindlimb muscles and the bend in the pelvis may aid in moving the tail down to serve as a brake when raptors attack prey or land on a perch. Bottom: Harris's Hawk (*Parabuteo unicinctus*) moving its tail downward and spreading its rectrices to help lose speed as it approaches a perch. (Top figure modified from Mosto et al. 2013; © 2014 Wiley Periodicals, Inc., used with permission. Bottom photo by Peter K. Burian, Wikipedia, CC BY 4.0, https://creativecommons.org/licenses/by/4.0/)

comparable to the evolution of horses with reduction from five toes to one, and reduces distal limb mass via loss of phalanges and the corresponding musculature which, in turn, means an increase in stride rate and a corresponding increase in running speed.

Among most species of birds (88%), including all passerines and most (56%) nonpasserines (Raikow 1985), three toes point forward and one (digit I) backward, a condition referred to as anisodactyl (Fig. 2.46). Other toe orientations among nonpasserines can be categorized as zygodactyl, heterodactyl, syndactyl, and pamprodactyl (Fig. 2.46). The zygodactyl toe orientation is found in cuckoos (Cuculidae) and species in the orders Psittaciformes (parrots) and Piciformes (woodpeckers and allies). Only species in the order Trogoniformes (trogons) have a heterodactyl toe orientation, and only those in the order Coraciiformes (kingfishers and hornbills) have a syndactyl toe orientation. Although most species of birds are very limited in the extent to which they can change the relative positions of their toes, other species can facultatively change the orientation of their toes. For example, Ospreys (Pandionidae), owls (Strigiformes), and turacos (Musophagidae) can rotate digit IV so that digits II and III oppose digit I and the reversed digit IV to create a functionally zygodactyl foot (Fowler et al. 2009; Botelho et al. 2015; Fig. 2.47). Botelho et al. (2015) referred to these taxa as being semi-zygodactyl.

Species and taxa with toe orientations categorized, or categorized by some investigators, as pamprodactyl include swifts (Apodidae), mousebirds (Coliidae), and Oilbirds (*Steatornis*

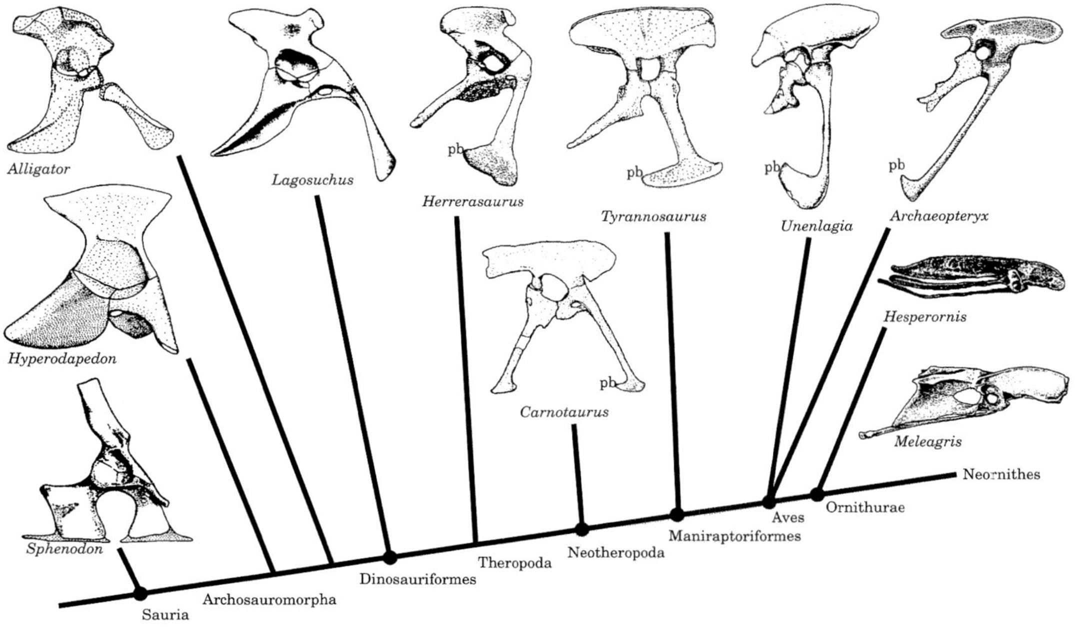


**Fig. 2.27** Evolution of the pelvic girdle from saurians to present-day birds (Wild Turkey, *Meleagris gallopavo*). Among present-day reptiles (like alligators) and the ancestors of present-day birds, including *Archaeopteryx*, the pubis (pb) is or was oriented either in an anterior direction or vertically. The pubis of present-day birds like Wild Turkeys is oriented posteriorly. (Figure from Hutchinson 2000; © 2008 Oxford University Press, used with permission)

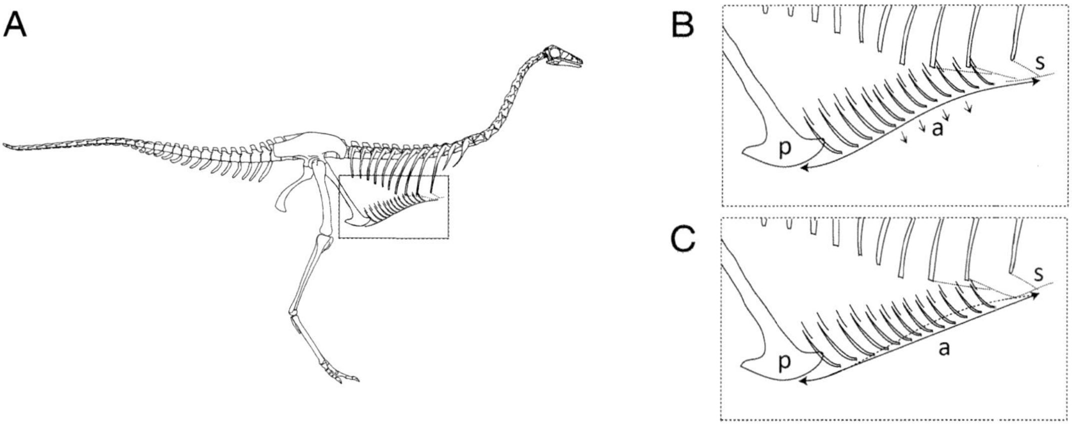


**Fig. 2.28** Possible mechanism for increasing and decreasing thoracic and lung volume during respiration in a nonavian dinosaur, *Struthiomimus*. Gastralia are bones in the ventral abdominal wall (shown in rectangular box, **a**). Muscles associated with the pubis (p) and sternum (s) may have caused movement (**b**, **c**) of the gastralia and abdominal wall (a). (Figure from Claessens 2004; Rights managed by Taylor & Francis, used with permission)

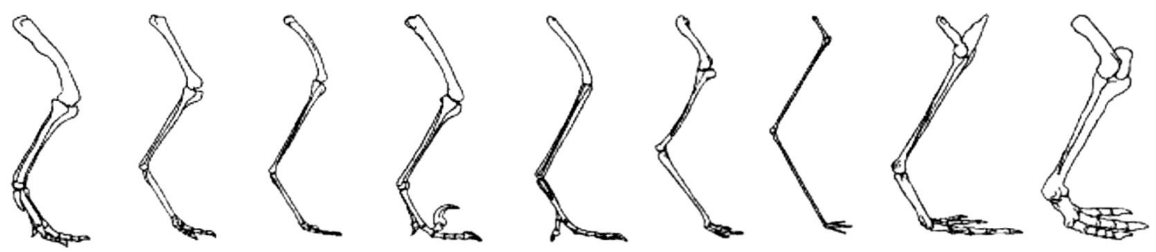


**Fig. 2.29** Hind limbs of representative theropods (*Allosaurus*, *Gallimimus*, *Sinornithoides*, and *Deinonychus*) and birds illustrating changes in relative lengths of bones and variation among living birds in bone size and thickness. (Figure from Gatesy and Middleton 1997; Rights managed by Taylor & Francis, used with permission)

*caripensis*). However, most swifts (Apodidae), including those in the subfamily Cypseloidinae and in two of three tribes in the subfamily Apodinae (Collocaliini and Chaeturini), are anisodactyl (Collins 1983; Fig. 2.48). Swifts in the third tribe, Apodini, can orient their toes in different ways depending on the substrate. Some species in this tribe roost and nest in trees and must be able to grasp branches, which requires opposable digits. In these species, as in species that are zygodactyl, digits I and IV oppose digits II and III. However, rather than two toes in front and two in back, these swifts have two toes on the medial side of the foot and two on the lateral side (Fig. 2.49), a method of grasping branches similar to chameleons and some primates (Holmgren 2004). Those species in the tribe Apodini that roost and nest on vertical or nearly vertical surfaces are able to rotate their toes so that all four face forward so, using the terminology of Botelho et al. (2015), are more accurately referred to as semi-pamprodactyl. Similarly, mousebirds can rotate digits I and IV so, at different times, they can be anisodactyl, zygodactyl, or pamprodactyl (Fig. 2.50). Oilbirds nest and roost in caves that have both vertical or nearly vertical surfaces and flatter ledges; when on a flat surface, all four digits point forward, with digit I (hallux) at a 45° angle relative to the other toes (i.e., "semi-pamprodactyl"). However, at steeper angles, digits II and III point forward, digit IV is held at a slight angle from those toes, and digit I rotates back to varying degree, sometimes backward (i.e., anisodactyl; Bock and Miller 1959).

The toe orientation of woodpeckers and their allies, including jacamars, puffbirds, toucans, barbets, and honeyguides, is generally considered zygodactyl. However, most species in one woodpecker subfamily, Picinae, rotate, to varying degrees, digits I and IV when climbing, creating a toe orientation that could be referred to as "ectropodactyl" ("turning-out-of-the-way" toe; Bock and Miller 1959; Fig. 2.51c, d). By rotating these digits, woodpeckers have a lateral grasp of a tree trunk that provides stability. This allows woodpeckers to move their body away from the trunk to make more forceful impacts with their bills (Zelenkov 2007). Some woodpeckers are able to rotate their first and fourth digits to the point where their toe orientation is basically pamprodactyl (Fig. 2.51).

Most birds have four toes and the typical phalangeal composition of those toes is indicated by the formula 2–3–4–5 (Fig. 2.52). The only exceptions are species in the families Caprimulgidae and Trochilidae, where digit IV has four phalanges rather than five (Livezey and Zusi 2006). The relative lengths of phalanges vary among birds. Among more terrestrial/cursorial species, distal phalanges tend to become shorter compared to proximal phalanges (Fig. 2.53). In species and taxa that are more arboreal, phalanges tend to be similar in length. Among species and taxa, where grasping is important, e.g., raptors like screech owls carrying prey in flight or swifts clinging to vertical surfaces, the most distal phalanges are longer

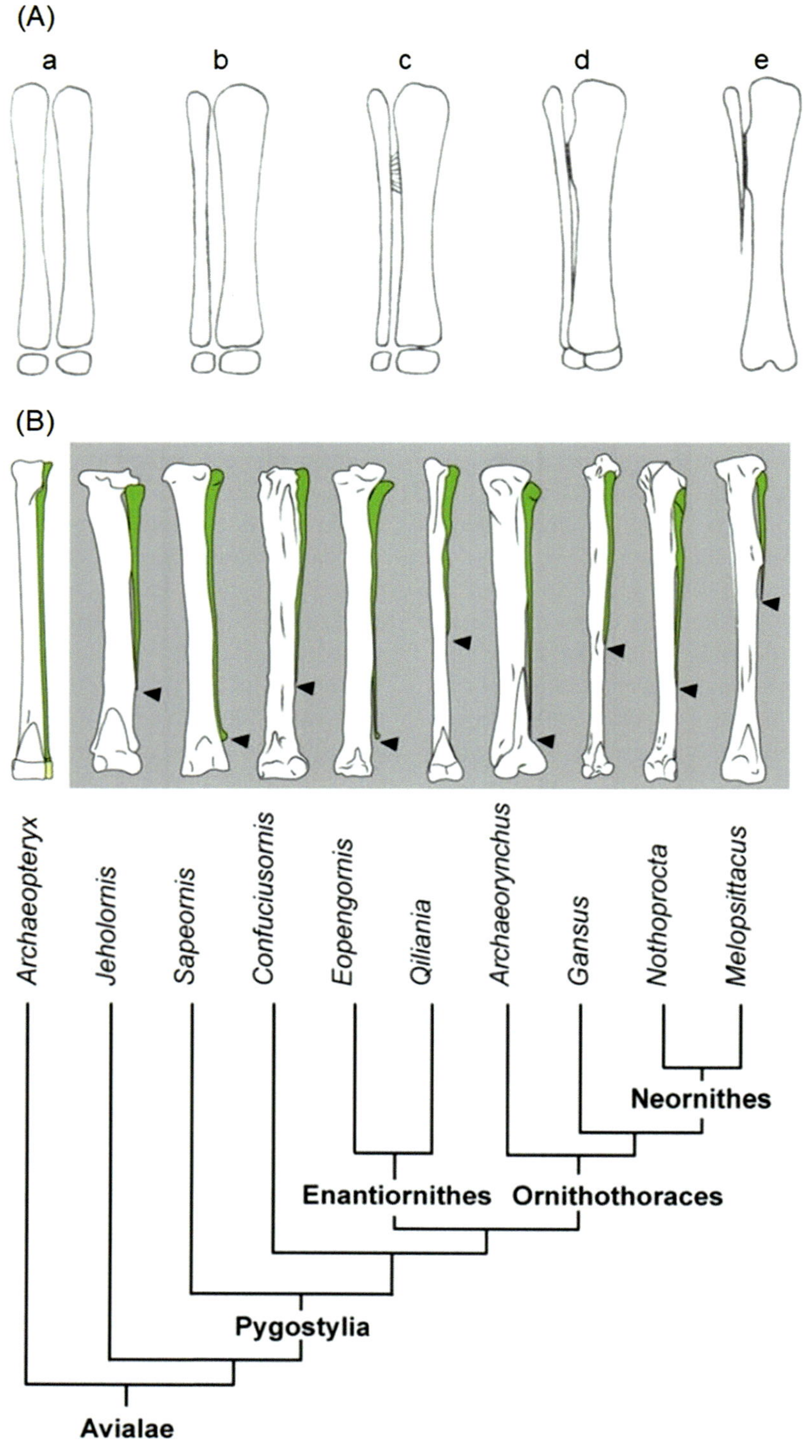


**Fig. 2.30** (**a**) Evolutionary transformation of the tibia, fibula, and proximal tarsal bones from reptiles to birds showing reduction in the fibula and fusion of the tarsal bones with the tibia to form the tibiotarsus. (a) Primitive

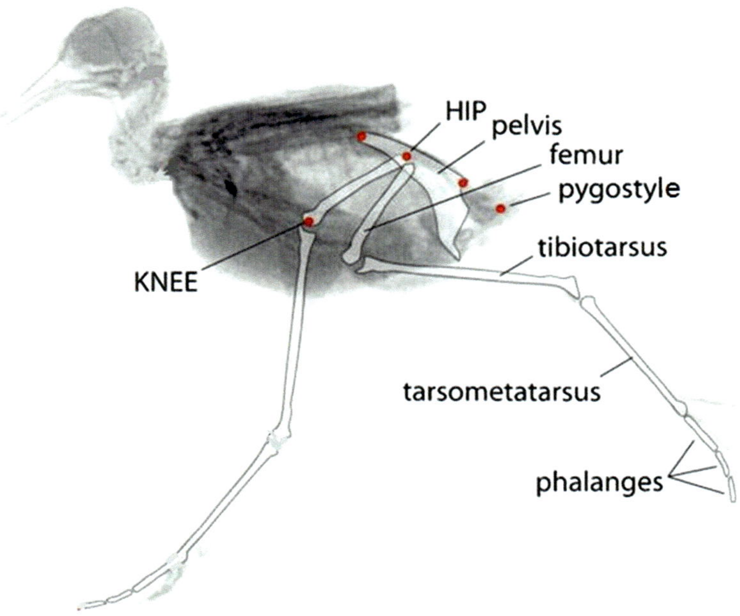


**Fig. 2.31** With short, light tails (due to reduction in the number of caudal vertebrae) and large flight muscles, natural selection has favored positioning of bird feet under a more cranial positioned center of mass. This is achieved by a subhorizontal orientation of the femur and, when walking and running, the knee acting as the main fulcrum near the bird's center of mass. (Figure from Nyakatura et al. 2012; © 2012 Wiley Periodicals, Inc., used with permission)

than the other phalanges (Figs. 2.53 and 2.54). Few investigators have proposed hypotheses to explain the possible biomechanical consequences of these differences. However, the terrestrial/cursorial ancestors of present-day birds also had distal phalanges shorter than proximal phalanges (Fig. 2.55). The reduced mass of shorter phalanges along with a corresponding reduction in muscle and other tissues at the distal end of hindlimbs may allow an increase in stride rate, and the increasingly shorter phalanges may increase flexibility of the toes that might be useful when walking and, especially, running (Fig. 2.56). For more arboreal species and taxa, phalanges of similar length may enable aid in encircling (or partially encircling) and firmly grasping the branches of trees and shrubs when perched. For birds like raptors, the shorter proximal phalanges may increase flexibility and allow them to efficiently encircle and grasp relatively small branches (Volkov 2004), whereas much longer distal phalanges lengthen the digits and increase the area encompassed by their four talons. This may be beneficial for both capturing prey and firmly grasping prey carried to perches or nests. In a study of humans that may be relevant to raptors, Trinkaus and Villemeur (1991) found that the distal phalanges in the fingers of Neanderthals

**Fig. 2.30** (continued) reptile with similar-sized tibia and fibula and free tarsal bones. (b and c) Dinosaurs with slimmer fibulas. (d) Theropod dinosaur with reduced fibula and proximal tarsal bones associated with the tibia. (e) Present-day bird with greatly reduced fibula and proximal tarsals completely fused with the tibia to form the tibiotarsus. (**b**) Fibulas of representative Avialae. The fibula of *Archaeopteryx* was as long as the tibia, but narrower. All other avialans had fibulas shorter than the tibia, including present-day birds. Fibulas shown in green. Arrowheads point to the tip of the fibulas. *Nothoprocta*, tinamous; *Melopsittacus*, Budgerigar. (Figure A from Müller and Streicher 1989; © 1989 Springer-Verlag, used with permission. Figure B from Botelho et al. 2016; open-access article distributed under the terms of the Creative Commons CC BY license, https://creativecommons.org/licenses/by/4.0/)

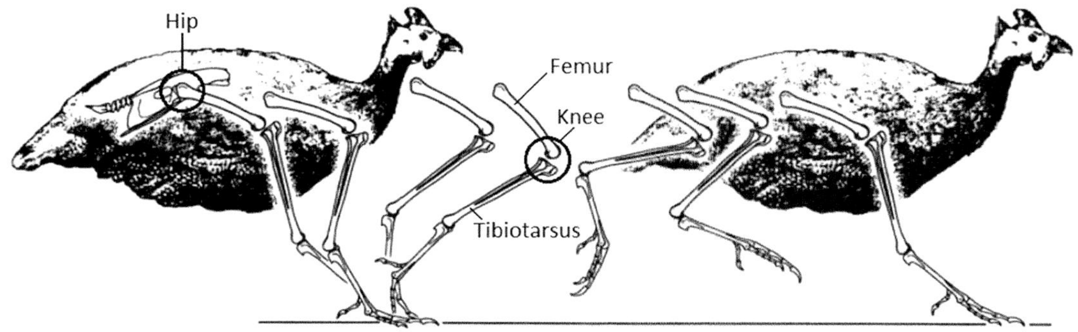


**Fig. 2.32** A running Helmeted Guineafowl (*Numida meleagris*). Note that most of the movement is at the knee joint (i.e., the angle between the femur and tibiotarsus changes dramatically) and there is much less movement at the hip (i.e., the angle between the femur and pelvic girdle changes much less). (Figure from Gatesy 1999; © 1999 Wiley-Liss, Inc., used with permission)

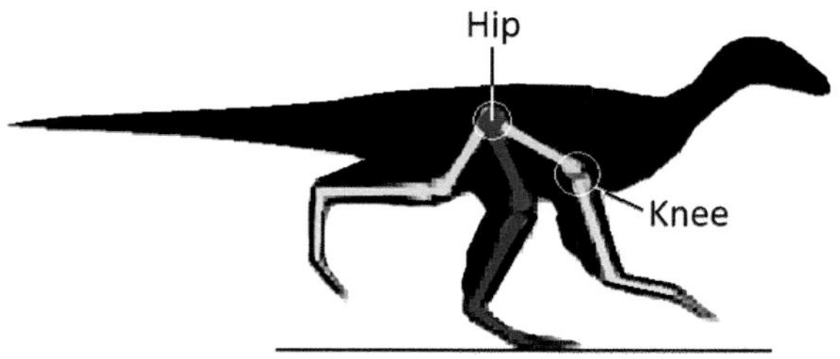


**Fig. 2.33** When a theropod walked or ran, most of the movement or rotation was at the hip joint (angle between the pelvic girdle and femur), with the femur moving back and forth. Much less movement occurred at the knee (angle between the femur and the tibia and fibula). (Figure modified from Hutchinson et al. 2008; © 2008 Springer Nature, used with permission)

were relatively longer compared to proximal phalanges than is the case in the fingers of modern Europeans and that this difference translated into a more powerful grip for Neanderthals than modern Europeans. Also contributing to the ability of raptors to grasp prey are large muscles whose tendons are attached to the distal phalanges (Fig. 2.57).

## 2.5 Axial Skeleton

The axial skeleton of birds includes the skull, rib cage, sternum, and vertebral column. In contrast to present-day birds, *Archaeopteryx* had teeth and the bones of the skull were not as tightly fused. In the skull of all present-day birds except tinamous, all bones are tightly fused (Zusi 1993).

### 2.5.1 Skull

The skull of birds consists of two major components, a beak and a braincase that protects the brain and special sense organs (eyes, middle and inner ears, and structures associated with olfaction), with all bones fused in adults. Like reptiles and, especially archosaurs, bird skulls have an elongated snout and a compressed orbital and temporal region (Marugán-Lobón and Buscalioni 2009; Fig. 2.58). An important step in the evolution of the avian skull appears to be

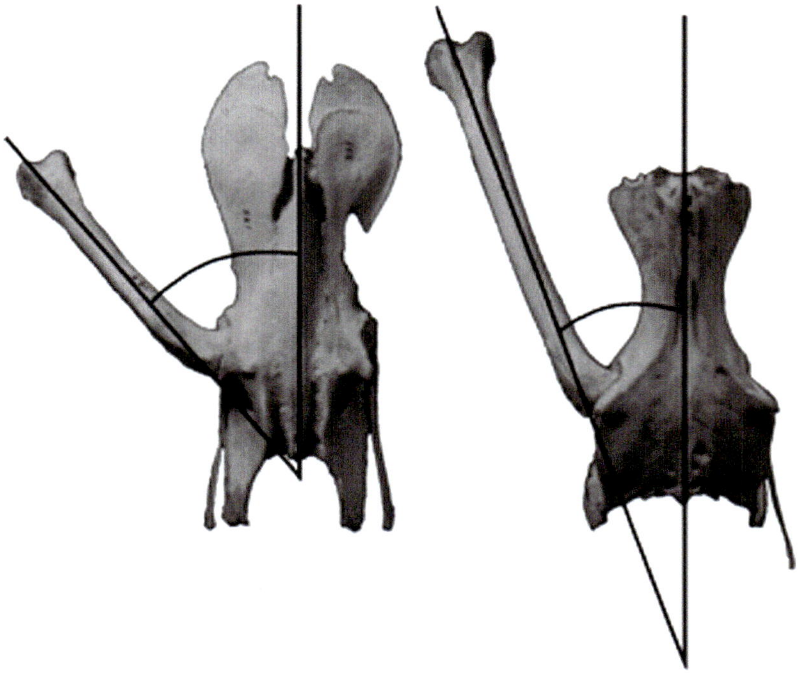

**Fig. 2.34** Dorsal view of the pelvic girdles of an Adelie Penguin (*Pygoscelis adeliae*, left) and a Great Blue Heron (*Ardea herodias*, right) illustrating differences in the femoral splay angle (i.e., the angle between the middle of the synsacrum and the femur). (Figure from Hertel and Campbell 2007; © 2007 Oxford University Press, used with permission)

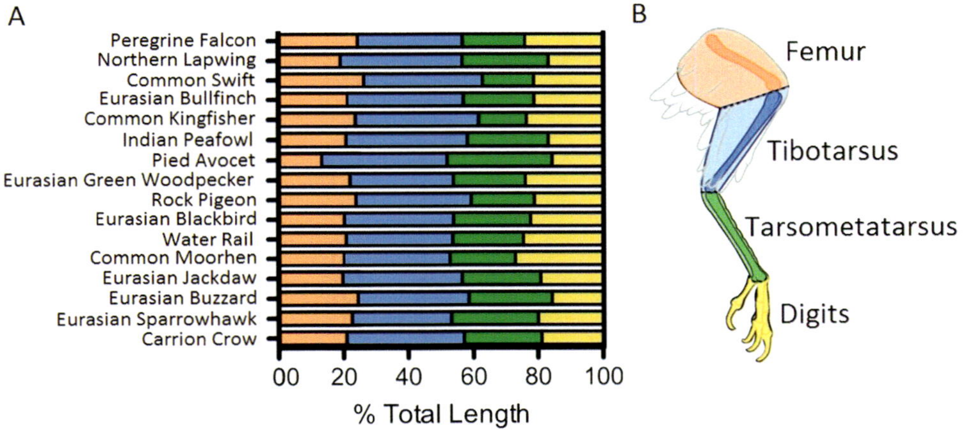


**Fig. 2.35** (**a**) Relative proportions of length of hindlimb segments for 15 species of birds. Digit length is only for digit III. Colors of boxes and bars correspond to segment in (**b**) this hindlimb. Peregrine Falcon, *Falco peregrinus*; Northern Lapwing, *Vanellus vanellus*; Common Swift, *Apus apus*; Eurasian Bullfinch, *Pyrrhula pyrrhula*; Common Kingfisher, *Alcedo atthis*; Indian Peafowl, *Pavo cristatus*; Pied Avocet, *Recurvirostra avosetta*; Eurasian Green Woodpecker, *Picus viridis*; Rock Pigeon, *Columba livia*; Eurasian Blackbird, *Turdus merula*; Water Rail, *Rallus aquaticus*; Common Moorhen, *Gallinula chloropus*; Eurasian Jackdaw, *Corvus monedula*; Eurasian Buzzard, *Buteo buteo*; Eurasian Sparrowhawk, *Accipiter nisus*; Carrion Crow, *Corvus corone*. (Figure modified from Kilbourne 2014; open-access article distributed under the terms of the Creative Commons Attribution License, http://creativecommons.org/licenses/by/4.0)

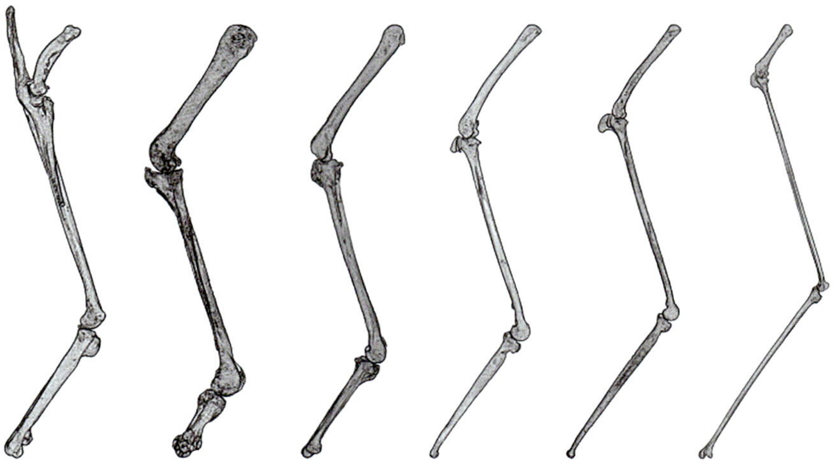

**Fig. 2.36** Leg bones (top to bottom: femur, tibiotarsus, and tarsometatarsus) of six different species of birds, showing variation in relative length. From left to right: Red-throated Loon (*Gavia stellata*), African Penguin (*Spheniscus demersus*), Common Wood-pigeon (*Columba palumbus*), Eurasian Magpie (*Pica pica*), European Robin (*Erithacus rubecula*), and Greater Flamingo (*Phoenicopterus ruber*). (Figure from Zeffer et al. 2003; © 2003 Oxford University Press, used with permission)

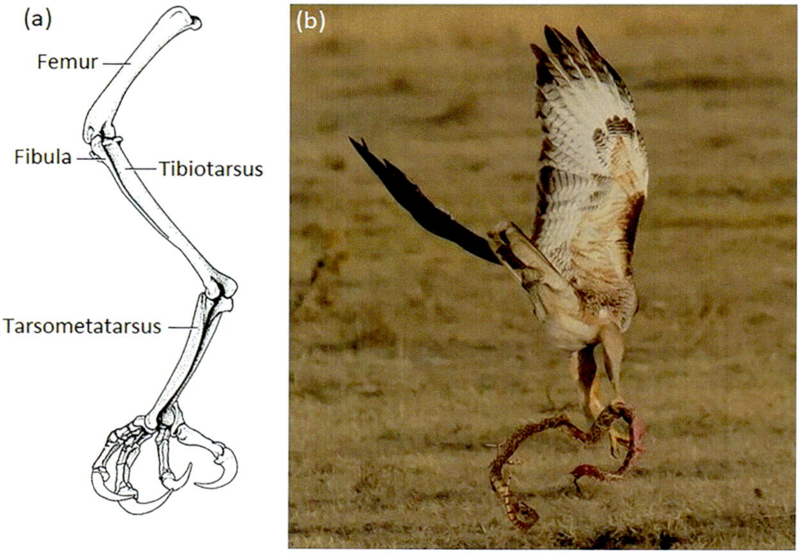


**Fig. 2.37** (**a**) Left hindlimb of a Red-tailed Hawk (*Buteo jamaicensis*). (**b**) Red-tailed Hawk attacking and capturing a snake with legs extended. (Figure a from Ward et al. 2002; © 2002 Oxford University Press, used with permission. Photo from U. S. Fish and Wildlife Service, Wikipedia, CC0 Public Domain)

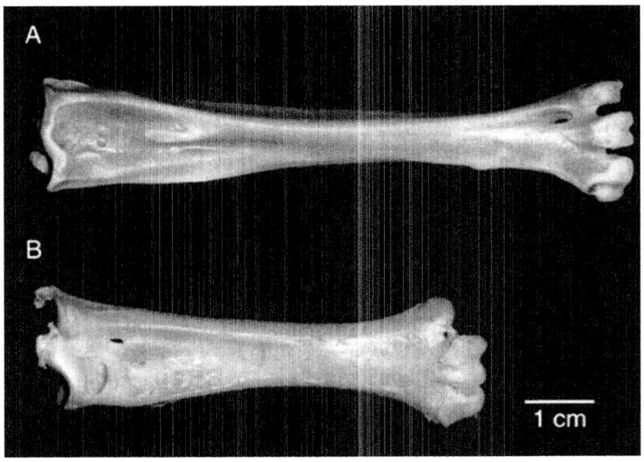


**Fig. 2.38** Tarsometatarsi of a similar-sized hawk and owl. (**a**) Red-tailed Hawk (*Buteo jamaicensis*). (**b**) Great Horned Owl (*Bubo virginianus*). The more robust tarsometatarsus and associated muscles of the owl can generate greater force and a stronger grip to quickly kill their prey. (Figure from Ward et al. 2002; © 2002 Oxford University Press, used with permission)

juvenilization, a process called progenetic padomorphosis that involves a shortening of the developmental process in terms of shape, size, and time (Bhullar et al. 2012). This process is evident in the shortened face, enlarged braincase, and elongate beak of birds, changes likely driven by enlargement of the avian eye and brain plus the posteroventral rotation of the brain (Bhullar et al. 2012; Figs. 2.59 and 2.60).

The avian skull consists of a neurocranium (or braincase), a splanchnocranium (upper and lower bills, plus the jugal, palate, and quadrates), and the hyoid apparatus, columella (middle ear bone), and a sclerotic ring. The splanchnocranium connects to the neurocranium via the quadrate and the craniofacial hinge. Bones that form the neurocranium include the occipital bones (posterior wall), the basioccipital, basisphenoid, and parasphenoid (ventral portion), the squamosals (lateral walls), plus the parietals and frontals (roof) (Zusi 1993; Fig. 2.61).

Changes in the skull during the evolution of birds included enlargement of the premaxilla and reduction in size of the maxilla, palatines, and pterygoids. The orbit increased in size, the jugal has been transformed into a horizontal rod, and the quadratojugal articulates with the quadrate. With the enlargement of the forebrain, the frontoparietal suture has moved backward, with the parietal bones now relatively small and the frontal bones much enlarged. The squamosal bones are now an important component of the side of an enlarged braincase, and the quadrate has developed a mobile joint with the squamosal, an important component of the avian kinetic skull. The lower mandible, still consisting primarily of the dentary, now articulates with the quadrate (Bhullar et al. 2016; Fig. 2.61, Box 2.5

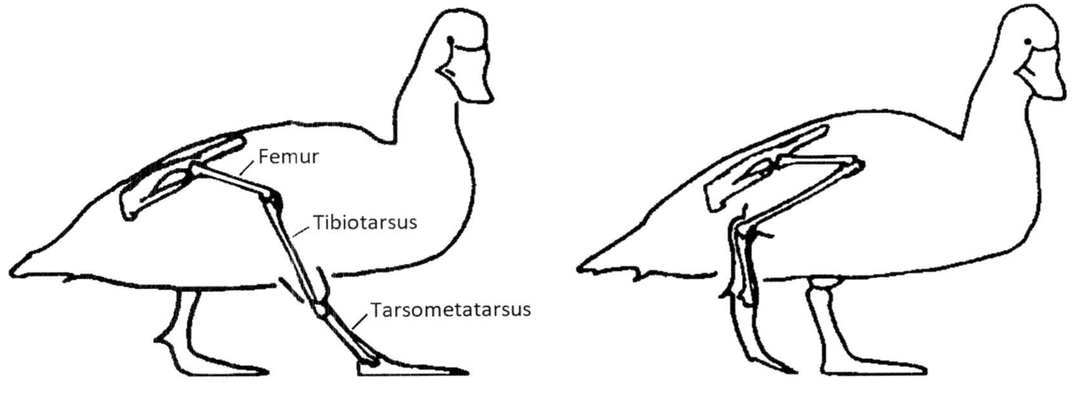


**Fig. 2.39** A walking Mallard (*Anas platyrhynchos*), showing the relative lengths of the femur, tibiotarsus, and tarsometatarsus. The longest leg bone is the tibiotarsus. (Figure modified from Abourachid 2000; © 1951 CCC Republication, used with permission)

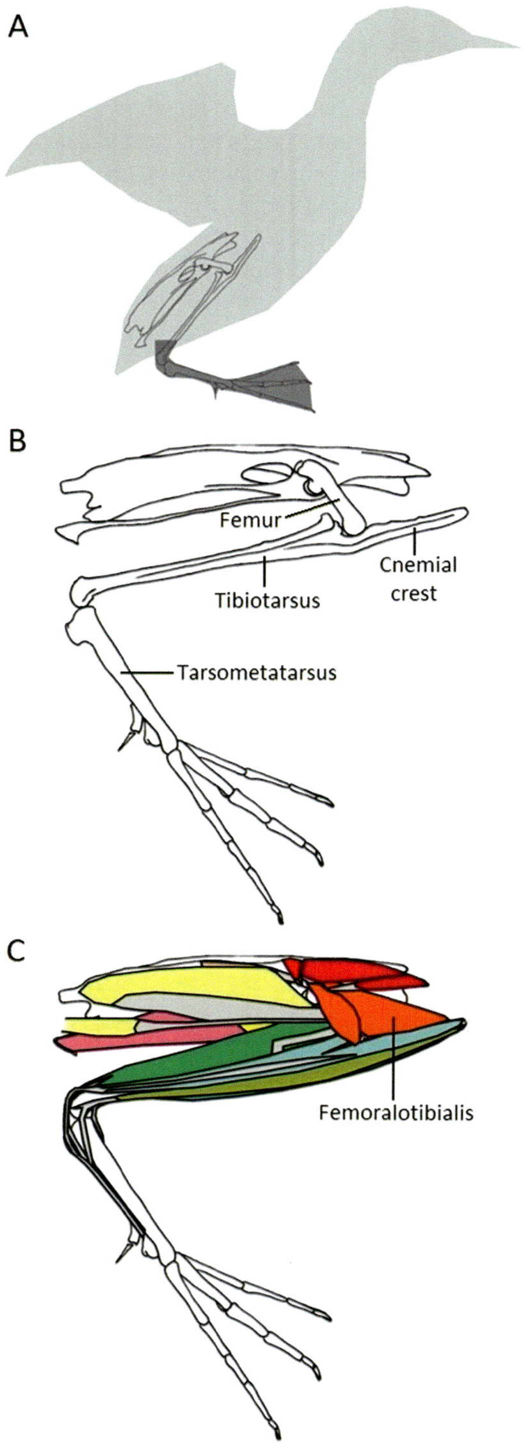


**Fig. 2.40** (**a**) A Red-throated Loon (*Gavia stellata*) showing the position of the pelvic girdle and hindlimb.

Mandibular Bowing of Pelicans). Although the skulls of present-day birds are similar with respect to these general changes, they still exhibit striking variation in shape and size (Fig. 2.62). This variation is possible because the avian skull is modular, with different modules evolving nearly independently of, and at different rates than, the others.

Based on the skull morphology of 352 species of birds representing 159 families and 320 genera, Felice and Goswami (2018) divided the avian skull into seven modules: rostrum (dorsal surface of the premaxilla, nasal, and jugal bar), naris (perimeter of the external naris), palate (ventral surface of premaxilla and palatine), cranial vault (frontal, parietal, and squamosal), occiput (supraoccipital, paraoccipital, and basioccipital), pterygoid-quadrate (ventral surface of pterygoid and articular surface of quadrate), and basisphenoid (Fig. 2.63). Their analysis revealed that, starting with a hypothetical ancestral Neornithine skull (Fig. 2.64), these modules have evolved at different rates over time and in different taxa of birds. All modules exhibited high rates of evolution during the early period of avian diversification during the late Cretaceous and at the Cretaceous–Paleogene (K-Pg) boundary (Fig. 2.65). This period corresponds with the estimated time of the origin of, and corresponding morphological changes in, several major clades of birds, including Strisores (e.g., nightjars, swifts, and hummingbirds), Aequorlitornithes (e.g., shorebirds, penguins, gulls, grebes, cormorants, herons, egrets, and pelicans), Gruiformes (e.g.,

**Fig. 2.40** (continued) (**b**) Bones of the hindlimb showing the cnemial crest of the tibiotarsus. (**c**) "Intermediate" layer of muscles of the hindlimb (superficial muscles have been removed). Note the femoralotibialis muscle that originates on the femur and inserts on the cnemial crest of the tibiotarsus. The femoralotibialis muscle attached to the long cnemial crest helps limit motion of the knee joint, stabilizing the tibiotarsus close to the body and increasing the propulsive forces generated by the tarsometatarsus. In foot-propelled divers like Red-throated Loons, the femur and knee joint remain almost stationary and the tarsometatarsi and feet provide most of the propulsive force. (Figures modified from Clifton et al. 2018; © 2017 The Anatomical Society, used with permission)

**Fig. 2.41** Birds the climb trees vertically like Brown Creepers (*Certhia americana*) have relatively short legs that reduce the effort needed to hold their bodies close to the trunk. (Photo by Alan Vernon, Wikipedia, CC BY 2.0, https://creativecommons.org/licenses/by/2.0/deed.en. Figure from Norberg 1986; © 1986 Wiley-Liss, Inc., used with permission)

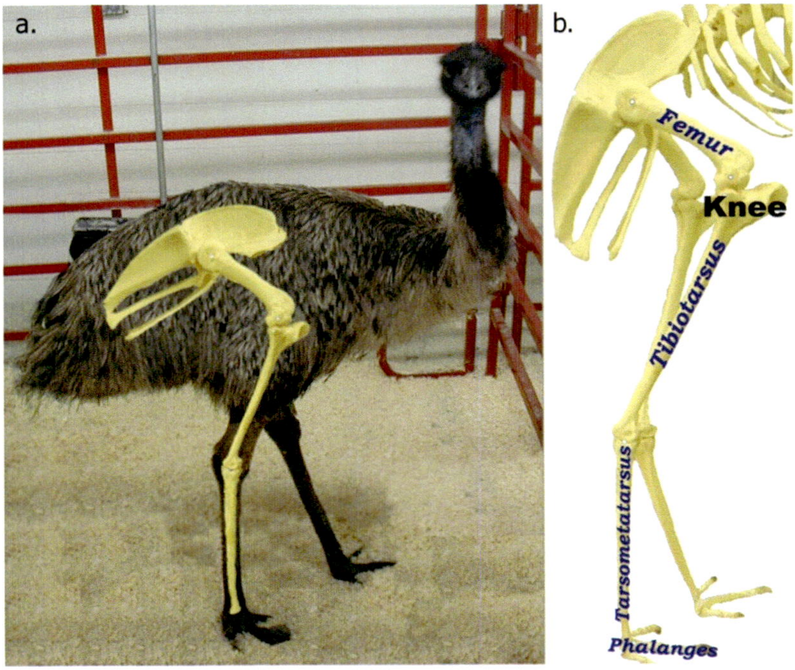


**Fig. 2.42** Pelvic girdle, leg bones, and phalanges of an Emu (*Dromaius novaehollandiae*). Note the relatively long tarsometatarsus and tibiotarsus. (Figure from Goetz et al. 2008; © 2007 Elsevier Ltd. All rights reserved, used with permission)

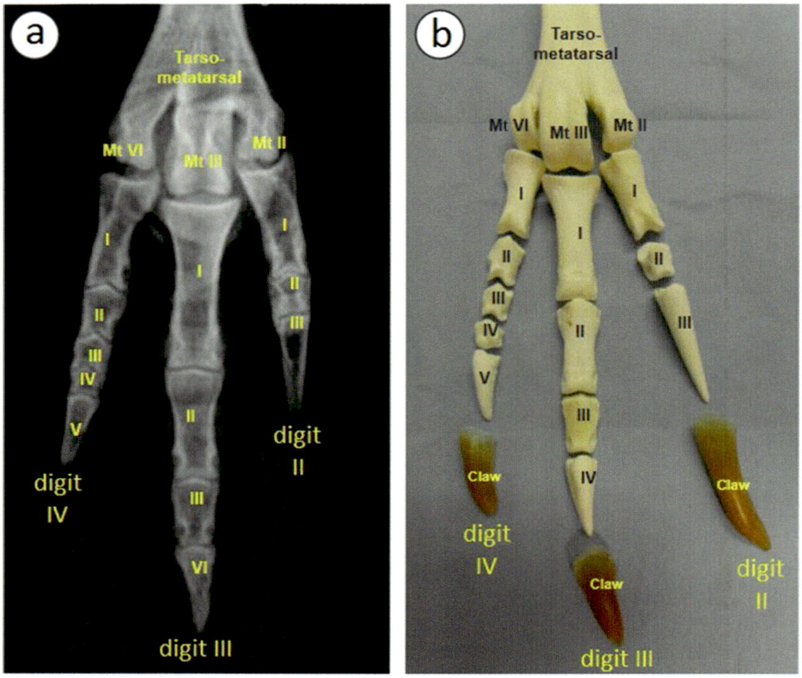


**Fig. 2.43** (**a**) X-ray image and (**b**) skeleton of the right foot of a cursorial Southern Cassowary (*Casuarius casuarius*) showing its three digits and the phalanges and claw of each digit. Note how the more distal phalanges are shorter than the proximal phalanges. (Figure modified from Saber and Hassanin 2014; © 2014 African Association of Veterinary Anatomists, used with permission)

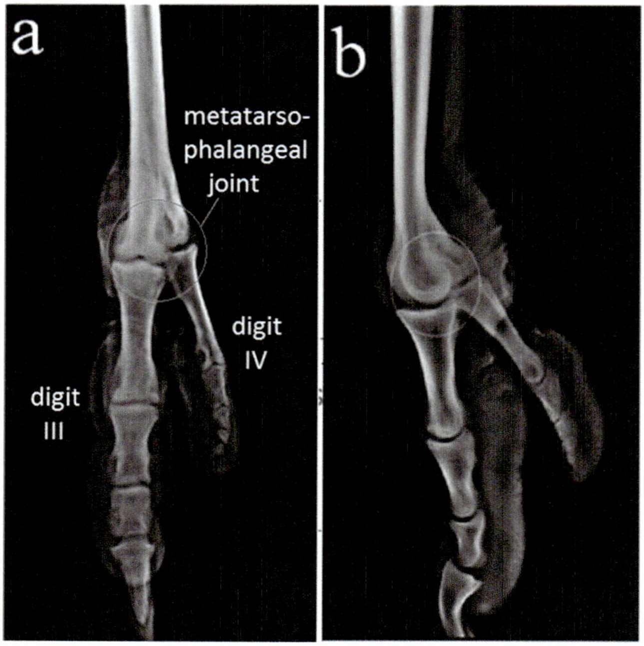


**Fig. 2.44** Right foot of a Common Ostrich (*Struthio camelus*). (**a**) Radiograph, dorsoplantar view and (**b**) lateromedial view. (Figure modified from Tehrani PR., Gilanpour H, and Veshkini A [2017]. Radiographic anatomy of the metatarsophalangeal joint and digits of the Ostrich. J. Avian Med. Surg.; 31(3): 198–205, used with permission)

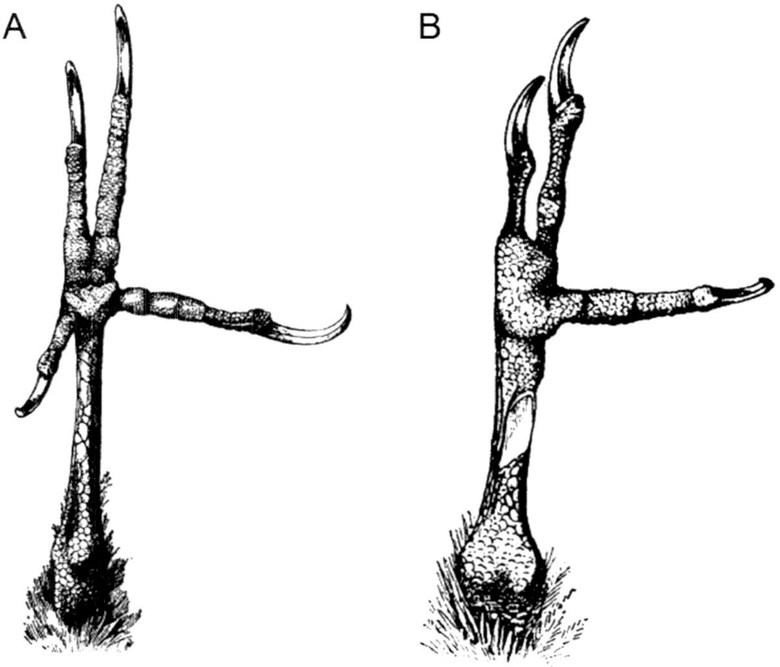


**Fig. 2.45** Toes of two species of woodpeckers. (**a**) Pileated Woodpecker (*Dryocopus pileatus*) and (**b**) Bamboo Woodpecker (*Gecinulus viridis*). On the vertical surface of a tree trunk, digits II and III, along with the tail, counterbalance the force of gravity, whereas digits II and III plus the laterally directed digit IV are most important for maintaining the woodpecker's position against the substrate. In some species of woodpeckers, such as Pileated Woodpeckers, digit I plays a limited role in holding them against substrates and so are reduced in size. In other species, such as Bamboo Woodpeckers, digit I became functionless and has been lost. (Figure from Bock and Miller 1959; used with permission of the American Museum of Natural History)

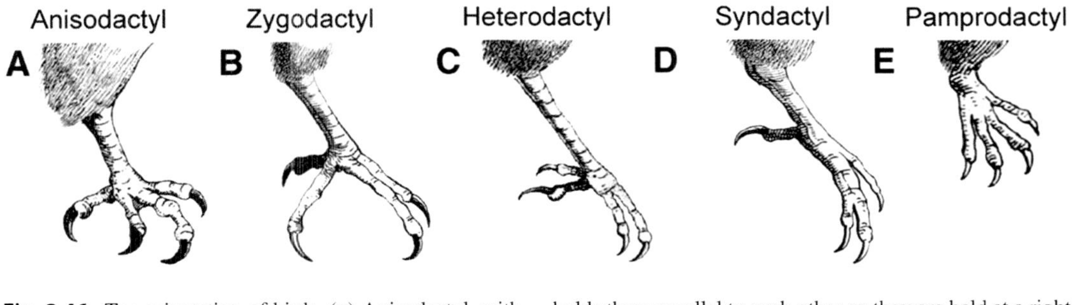


**Fig. 2.46** Toe orientation of birds. (**a**) Anisodactyl, with three toes in front and one (digit I, or hallux) in back, an adaptation for both perching and walking/running, (**b**) zygodactyl, with two toes in front (digits II and III) and two in back (digits I and IV), an adaptation for perching as is the (**c**) heterodactyl orientation, with two toes in front (digits III and IV) and two in back (digits I and II), (**d**) syndactyl, with the partial fusion of digits III and IV that holds them parallel to each other so they are held at a right angle to branches, which assures that the force of flexion is directed against the branches rather than being directed laterally (Bock and Miller 1959), and (**e**) pamprodactyl, with all four digits in front. (Figure from Botelho et al. 2015; © 2015 Springer Science Business Media New York, used with permission)

**Fig. 2.47** Feet of (**a**) a Red-tailed Hawk (*Buteo jamaicensis*) and (**b**) a Great Gray Owl (*Strix nebulosa*). Unlike hawks, owls can rotate digit IV so that digits I and IV are opposite digits II and III. (Figure modified from Fowler et al. 2009; © 2009 Fowler et al., open-access article distributed under the terms of the Creative Commons Attribution License, https://creativecommons.org/licenses/by/4.0/)

**Fig. 2.48** Anisodactyl toe orientation of a White-nest Swiftlet (*Aerodramus fuciphagus*). (Figure modified from Zuki et al. 2012; used with permission)

rails, coots, and cranes), Columbaves (e.g., bustards, cuckoos, pigeons, and doves), and Inopinaves (e.g., falcons, parrots, and songbirds) (Prum et al. 2015; Felice and Goswami 2018). Small peaks are apparent at about 60 million years ago, corresponding with the origin of several other orders of birds including Sphenisciformes, Coliiformes, Musophagiformes, Strigiformes, and Accipitriformes, and at about 45 million years ago, corresponding with (based on fossil evidence) the origin of the order Charadriiformes (Mayr 2011; Prum et al. 2015; Felice and Goswami 2018). In general, rapid evolutionary changes in skull morphology occur (1) at the origin of major clades such as orders, superfamilies, and families; (2) during the evolution of more diverse clades (e.g., Passeroidea); and (3) with the development of novel phenotypes (e.g., pelicans) (Felice and Goswami 2018).

Evolutionary change and diversification are perhaps most apparent in the rostrum and palate modules, two modules important in bill morphology. In some cases, rapid evolutionary change occurred at the origin of clades, e.g., at the origin of Anseriformes (waterfowl), Phoenicopteridae (flamingos; Fig. 2.66), and Trochilidae (hummingbirds; Fig. 2.67). In some cases, little change occurs after the evolution of highly specialized bill shapes (e.g., flamingos), suggesting that such specialization may place constraints on further morphological change (Cooney et al. 2017). High rates of evolutionary change in bill morphology are also apparent in clades with high speciation rates where new niches were being occupied, e.g., Psittaciformes, Furnariidae, and Passeroidea (Fig. 2.67). More recent examples of this include adaptive radiations of birds on islands, e.g., Malagasy vangas, Galapagos finches (Box 2.6 Darwin's Finches, Adaptive Radiation, and Evolution), and Hawaiian honeycreepers (Box 2.7 Hawaiian Honeycreepers; Cooney et al. 2017).

Rates of evolution of the cranial vault module of birds exhibit less variation than rates for the rostrum and palate modules (Box 2.9 Rapid Adaptive Evolution of Bill Length). However, high rates of evolution are apparent at the origin of the clade Strisores (which includes the present-day family Caprimulgidae and the order Apodiformes; Fig. 2.68), the family Trochilidae (Fig. 2.69), and the order Strigiformes (owls;

**Fig. 2.49** Photograph showing a young Common Swift (*Apus apus*) grasping a wooden rod (4-mm diameter). Digits II and III are on the left, with digits I (bottom) and IV on the right. (Figure from Holmgren 2004; © British Ornithologists' Union, used with permission)

Figs. 2.70, and 2.71). High rates of cranial vault evolution are also apparent in genera that have cranial ornaments, including species in the genera *Casuarius* (cassowaries), *Numida* (guineafowl), *Balearica* (two species of cranes), and *Bucorvus* (hornbills) (Felice and Goswami 2018; Box 2.8 Bony Cranial Protuberances of Birds).

Along with the limited variation in rates of evolution of the cranial vault of birds, rates of evolution of the other components of the braincase (occiput, basisphenoid, and pterygoid-quadrate modules) are also relatively low. Variation in, and evolutionary rates of change in, these modules are limited because their morphologies must correspond to that of the brain that they surround and protect. Investigators have found similar integration between the braincase and brain across amniotes, suggesting a limit to the variation imposed by the structural design of the braincase–brain complex (Fabbri et al. 2017; Felice and Goswami 2018). In contrast to the braincase, the rostrum and palate of birds (i.e., the bill) exhibit great diversity and are capable of relatively rapid evolutionary change (Box 2.9 Rapid Adaptive Evolution of Bill Length). Additional information about variation in avian bill morphology can be found in Chap. 5.

### 2.5.2 Teeth

A reduction in tooth number occurred multiple times in the class Aves, and the total loss of teeth occurred sometime during the period from 146 to 65.5 million years ago (Louchart and Viriot 2011). No Cenozoic bird fossils (65.5 million years ago to the present) or present-day birds have teeth. Early investigators suggested that the loss of teeth (and associated heavy supporting jaws) may have been due, at least in part, to selective pressure for weight reduction, with reduced weight of the head particularly advantageous because it is further from a bird's center of gravity (e.g., Sibley 1957).

The need to minimize weight, however, may not have been the only or even most important selective factor in the loss of teeth by birds. During the Mesozoic, most birds had teeth, yet the

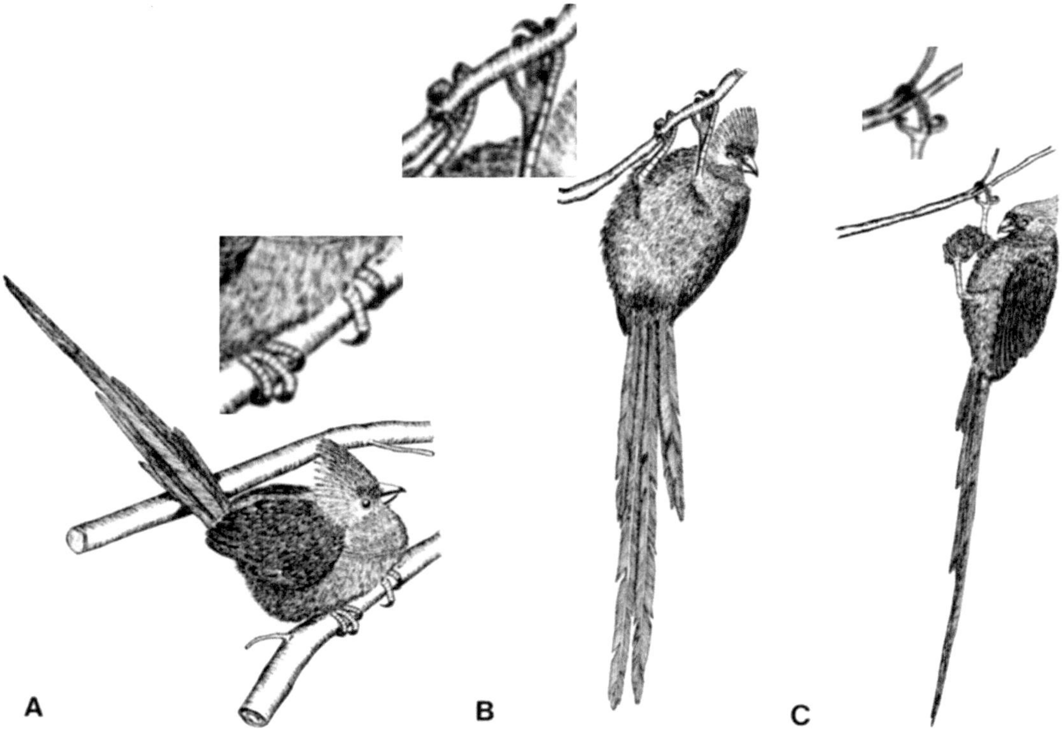


**Fig. 2.50** Mousebirds (Coliiformes, Coliidae) can rotate both their first and fourth digits to either a cranial or caudal position. As a result, they can be (**a**) anisodactyl or zygodactyl or (**b**) pamprodactyl. They position their toes in the way that is most effective in particular situations. For example, a mousebird holding a food item in its left foot (**c**) has a zygodactyl right foot. (Figure from Berman and Raikow 1982; © 1982 Oxford University Press, used with permission)

characteristics of some toothed birds (e.g., subclass Enantiornithes), including their skeletal features (such as the proportions of wing bones), small size, and the presence of an alula, suggest that their flying abilities may have been comparable to those of extant birds (Chiappe and Dyke 2002; Chiappe 2009; Liu et al. 2017). Zhou and Li (2010) also noted that some Mesozoic birds with teeth, such as *Yanornis* and *Yixianomis*, appeared to be strong flyers, reducing possible selection pressure for weight reduction. The ability, over millions of years, of birds with teeth to apparently fly as well as extant birds suggests that factors other than the need for weight reduction may have contributed to the loss of teeth by birds. In addition, several other groups of nonflying animals have also lost their teeth, including toads, turtles, echidnas, anteaters, and baleen whales, which indicate that selective factors other than a need to reduce weight may lead to the loss of teeth. An apparent prerequisite for tooth loss is the presence of a pre-adapted structure, such as an elongated sticky tongue (toads and anteaters), beak (turtles and echidnas), or baleen (baleen whales) that make teeth unnecessary (Davit-Béal et al. 2009).

For birds, pre-adapted structures that made teeth unnecessary were the rhamphotheca, the horny (keratinous) sheath or beak that covers the jaws of present-day birds and is important in food acquisition, and a muscular gizzard able to process foods efficiently preceded the complete loss of teeth in birds (Louchart and Viriot 2011). During the late Cretaceous, some birds had keratinized beaks covering a portion of the upper jaw (on the premaxilla bone) and teeth in

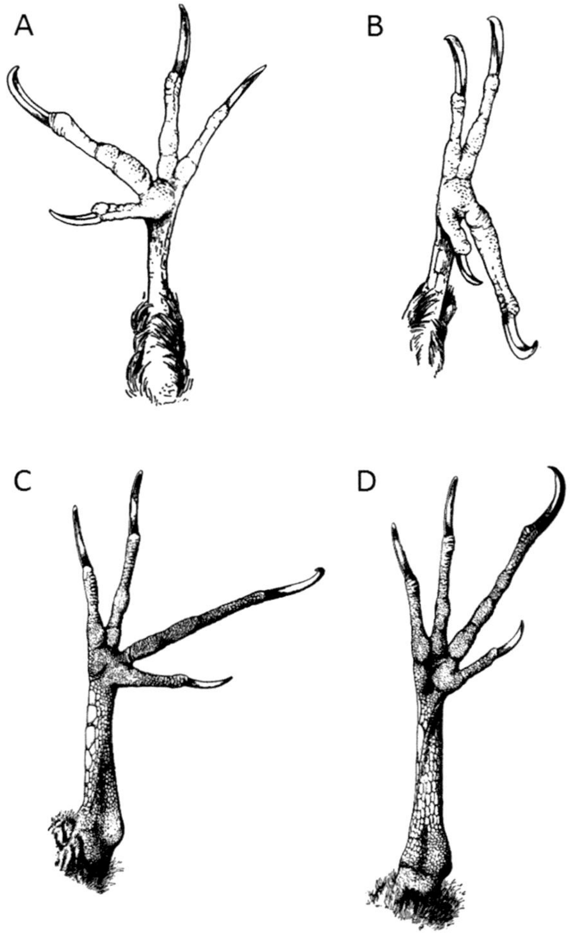


**Fig. 2.51** (**a**) Right foot of a Hairy Woodpecker (*Dryobates villosus*) showing position of the toes when climbing. (**b**) Left foot of a Hairy Woodpecker showing position of toes when perched on a branch. (**c**) Left foot of a Crimson-crested Woodpecker (*Campephilus melanoleucos*) and (**d**) a Red-necked Woodpecker (*Campephilus rubricollis*) showing position of toes when climbing. The first and fourth digits of Red-necked Woodpeckers rotate to face forward, a pamprodactyl toe orientation. (Figure from Bock and Miller 1959; used with permission of the American Museum of Natural History)

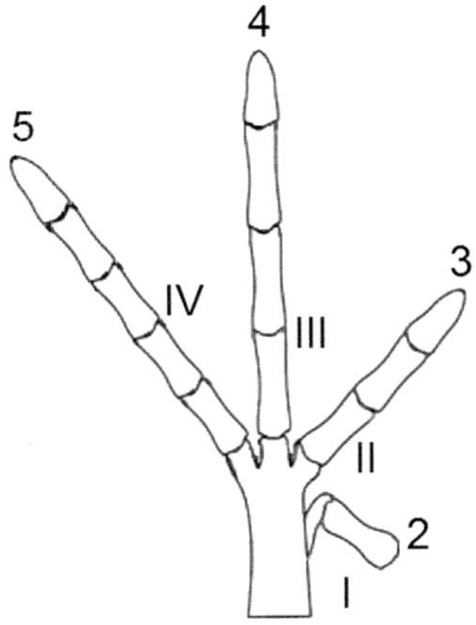


**Fig. 2.52** The typical number of phalanges in digits I through IV of birds. (Figure modified from Abourachid et al. 2017; © 2017 Anatomical Society, used with permission)

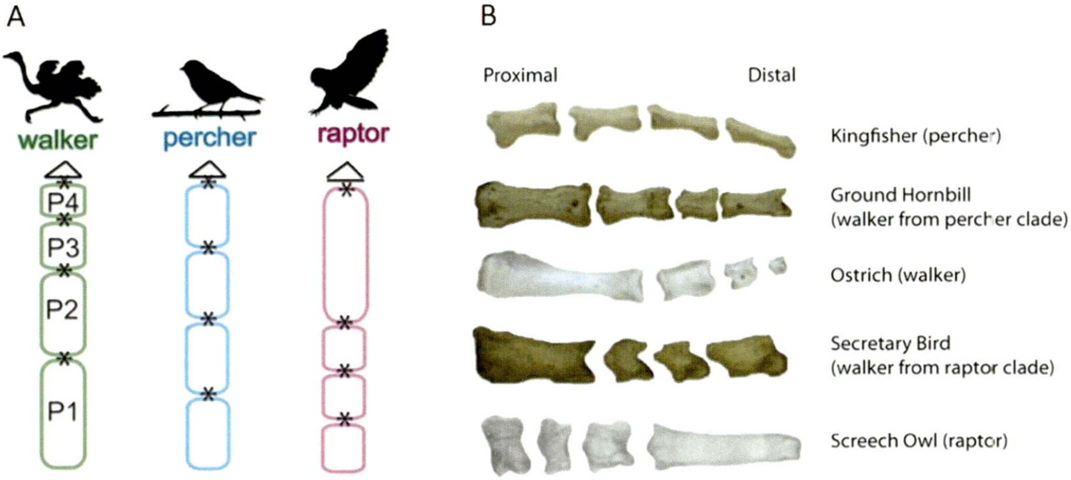


**Fig. 2.53** (**a**) The relative length of phalanges in the toes of birds varies with the functions of their feet. Birds that are primarily terrestrial have phalanges that get progressively shorter moving from the proximal to distal portions of toes. Phalanges in the toes of perching birds are similar in length whereas, among raptors, the most distal phalanx is much longer than the other phalanges. (**b**) Phalanges in the toes of five species of birds. The phalanges of the perching bird (kingfisher), walking bird (Common Ostrich, *Struthio camelus*), and raptor (screech owl) fit the pattern just described. However, the phalanges of some species of birds fall between these three main categories. For example, ground hornbills both perch and walk so their phalanges are "intermediate," i.e., becoming shorter at the distal ends of toes, but not as much shorter as with specialized for walking and running. Similarly, the phalanges of Secretarybirds (*Sagittarius serpentarius*) are also "intermediate"; they are raptors, but typically capture prey by walking and running on the ground rather than from perches. Kingfisher, family Alcedinidae; ground hornbills, *Bucorvus* spp.; screech owl, *Megascops* spp. (Figures A [modified] and B from Kavanagh et al. 2013; used with permission of the United States National Academy of Sciences)

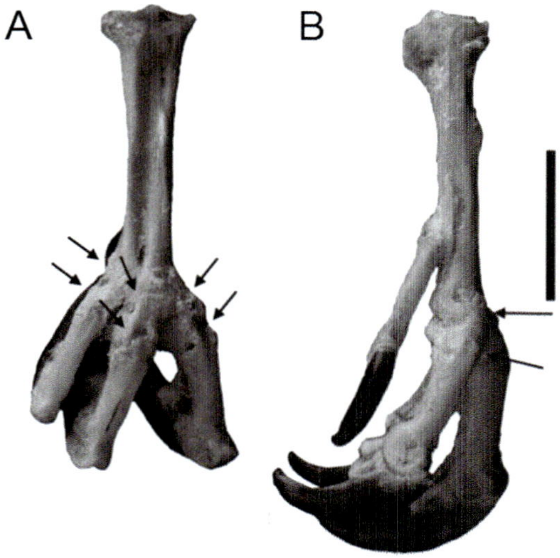


**Fig. 2.54** As with raptors, the distal phalanges of swifts in the subfamily Apodinae are also much longer than the proximal phalanges (indicated by the arrows). As a result, as illustrated by the (**a**) right and (**b**) left feet of a Common Swift (*Apus apus*), all four toes are nearly equal in length and this allows them to efficiently grasp branches and vertical surfaces. Scale bar = 5 mm. (Figure from Mayr 2015; © 2014 Springer Nature, used with permission)

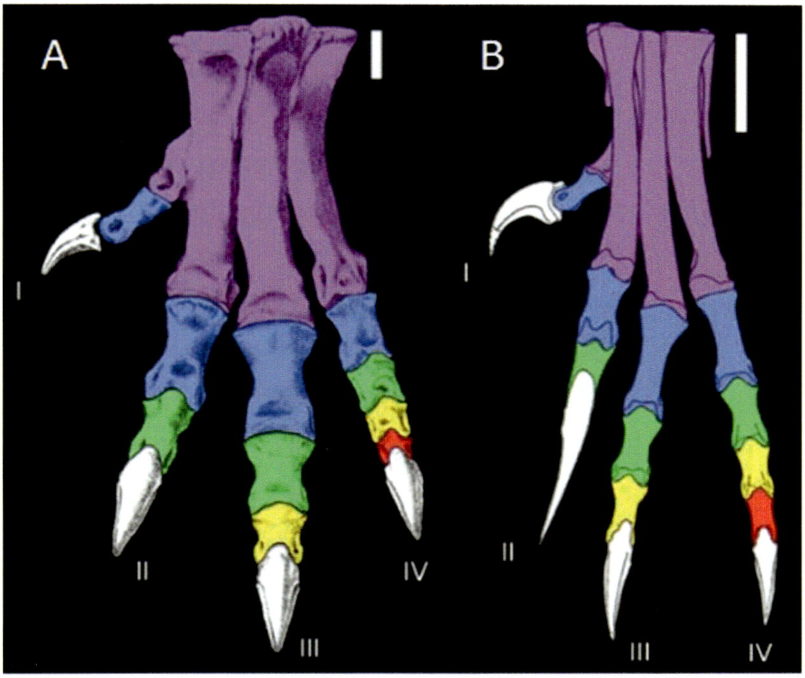


**Fig. 2.55** Feet of (**a**) *Allosaurus* and (**b**) *Deinonychus*. Phalanges in the toes of most theropods were comparable to those of present-day birds that are largely terrestrial/cursorial, with distal phalanges shorter than more proximal phalanges.
(Figure modified from Fowler et al. 2011; © 2011 Fowler et al., open-access article distributed under the terms of the Creative Commons Attribution License, https://creativecommons.org/licenses/by/4.0/)

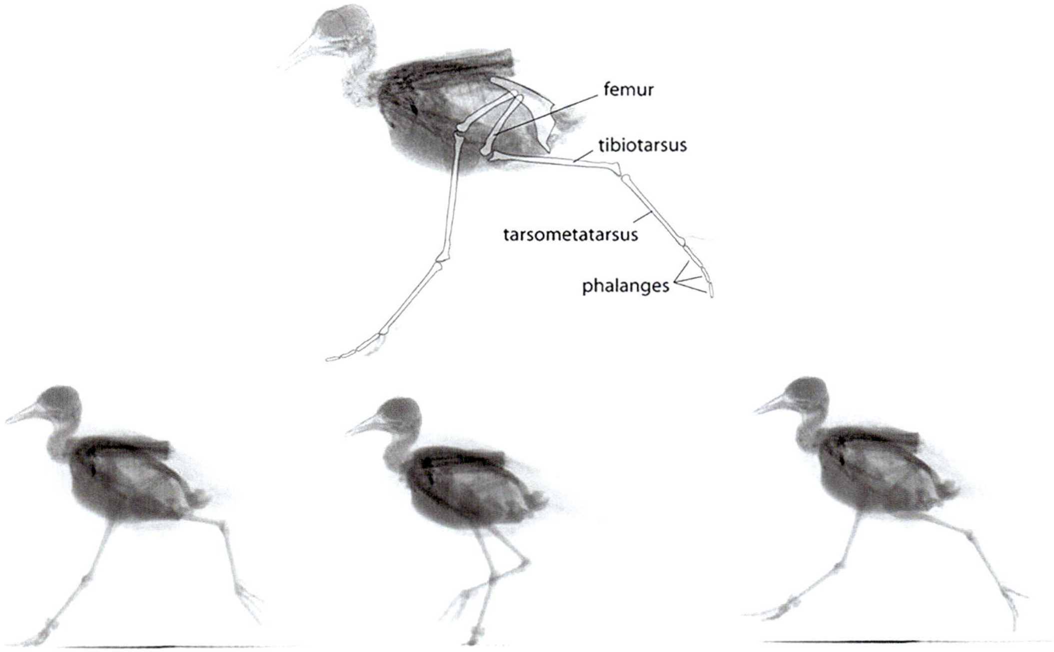


**Fig. 2.56** Top: Drawing superimposed on an x-ray of a running Northern Lapwing (*Vanellus vanellus*). Bottom: X-rays of a Northern Lapwing (from right to left) showing flexion of toes as it strides while running. (Figure made from two different figures from Nyakatura et al. 2012; © 2012 Wiley Periodicals, Inc., used with permission)

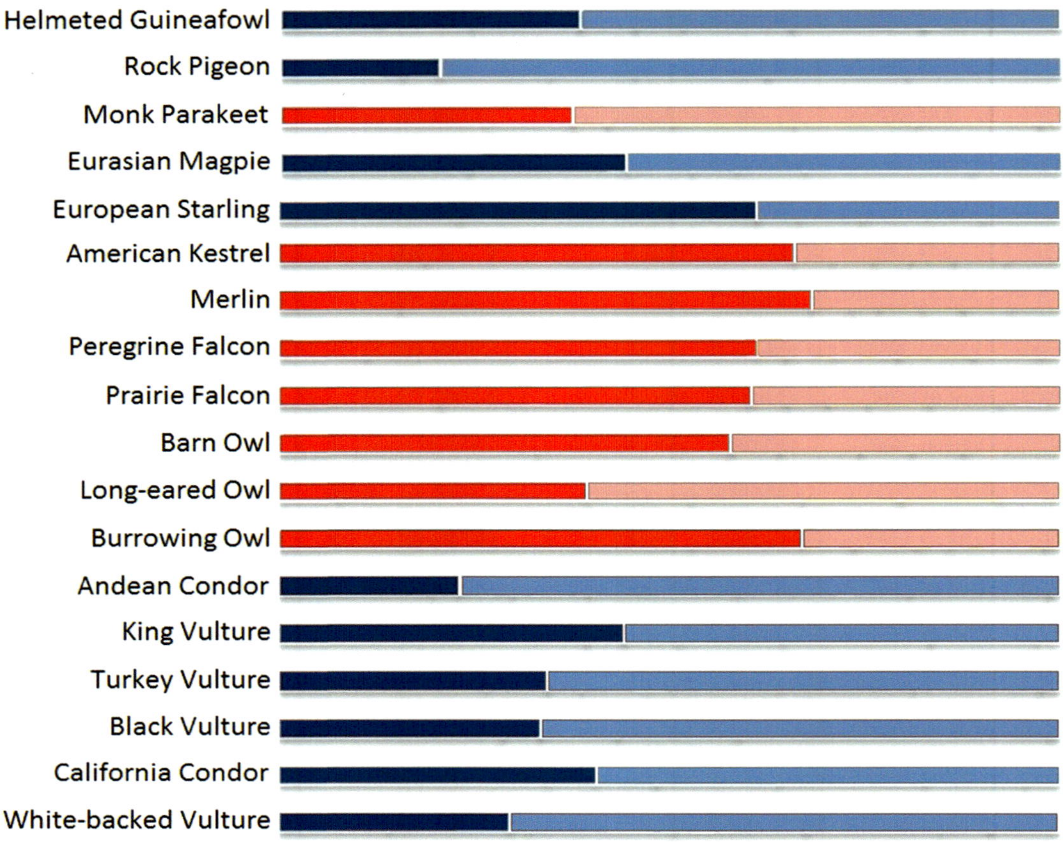


**Fig. 2.57** Relative cross-sectional area or mass (when area data were not available) of the distally and proximally inserted digital flexor muscles of several species of birds. "Grasping" species are represented by the red bars, and perching/walking species by the blue bars. Darker colors indicate the distally inserted flexor muscles, and the lighter colors the proximally inserted flexors. "Grasping" raptors have larger distally inserted muscles that aid in carrying prey in their feet; "perching/walking" birds have larger proximally inserted flexors. Helmeted Guineafowl, *Numida meleagris*; Rock Pigeon, *Columba livia*; Monk Parakeet, *Myiopsitta monachus*; Eurasian Magpie, *Pica pica*; European Starling, *Sturnus vulgaris*; American Kestrel, *Falco sparverius*; Merlin, *Falco columbarius*; Peregrine Falcon, *Falco peregrinus*; Prairie Falcon, *Falco mexicanus*; Barn Owl, *Tyto alba*; Long-eared o \Owl, *Asio otus*; Burrowing Owl, *Athene cunicularia*; Andean Condor, *Vultur gryphus*; King Vulture, *Sarcoramphus papa*; Turkey Vulture, *Cathartes aura*; Black Vulture, *Coragyps atratus*; California Condor, *Gymnogyps californianus*; White-backed Vulture, *Gyps africanus*. (Figure modified from Backus et al. 2015; 

the back of the upper jaw (maxilla bone) and on the lower jaw (e.g., *Hesperornis regalis*, Gingerich 1973; Fig. 2.72). In such birds and their ancestors, the rhamphotheca, perhaps modified in size and shape to efficiently acquire food just like those of present-day birds, might have become increasingly important in food acquisition relative to the teeth and, if so, tooth loss may have continued with no reduction in ability of birds to acquire food (Louchart and Viriot 2011).

Equally, if not more, important in the loss of teeth was the evolution of the gizzard. Animals with teeth may use them to process food. For example, many mammals chew their food to reduce the size of food particles, increasing the

**Fig. 2.58** Skulls of three archosaurs. (**a**) Alligator 46-day-old embryo (left) and an adult (right), (**b**) *Coelophysis* (primitive dinosaur) juvenile (left) and adult (right), and (**c**) *Archaeopteryx* juvenile (left) and adult (right). Note that the skull of the adult *Archaeopteryx* more closely resembles that of a juvenile than is the case for *Coelophysis* and the alligator, i.e., *Archaeopteryx* and present-day birds have paedomorphic skulls. (Figure from Bhullar et al. 2012; © 2012 Springer Nature, used with permission)

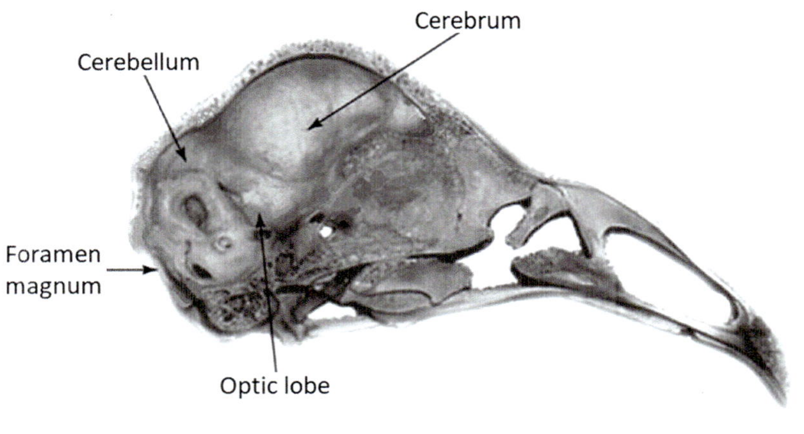


**Fig. 2.59** A bird skull with part of it removed to show the brain. Note the shortened braincase (neurocranium), relatively large braincase, long beak, and how the brain sits at an angle in the braincase (posteroventral rotation). (Figure from Marugán-Lobón 2010; © 2010 Springer-Verlag Berlin Heidelberg, used with permission)

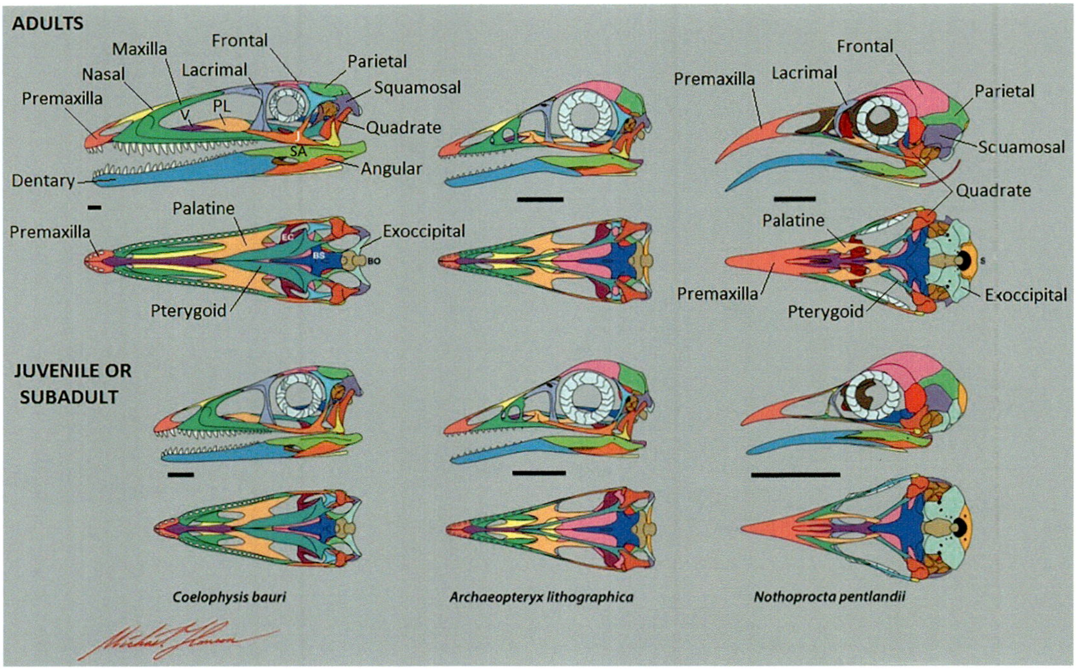


**Fig. 2.60** Comparison of the skulls of an early theropod (*Coelophysis bauri*), *Archaeopteryx lithographica*, and a present-day bird (Andean Tinamou, *Nothoprocta pentlandii*). Note the shortened braincase and enlarged eye sockets in the adult skulls. V, vomer; PL, palatine; J, jugal; SA, surangular. Scale bars = 1 cm. (Figure modified from Bhullar et al. 2016; © 2016 Oxford University Press, used with permission)

amount of surface area exposed and making enzymatic digestion more efficient. However, with the evolution of the avian muscular gizzard, this particle size reduction no longer required teeth (Louchart and Viriot 2011). Thus, although tooth loss may have enhanced the flying ability of birds via weight reduction and transfer of mass closer to the center of gravity (i.e., reducing mass of the head with a corresponding increase in mass of the gizzard), the evolution of the rhamphotheca and gizzard were likely more important selective factors in the loss of teeth by birds.

### 2.5.3 Cranial Kinesis

Cranial kinesis refers to the ability of birds to move all or part of their upper bill. This movement occurs in areas of the upper bill with thinner bone (Zusi 1984), and movement is caused by contraction of muscles located in the caudal parts of the orbits (eye sockets) with forces transmitted to the bill by bones in the palate (Bock 1964; Gussekloo et al. 2001). Birds exhibit three main types of cranial kinesis (Fig. 2.73): (1) prokinesis, where the upper bill is not flexible, but it flexes at the craniofacial hinge; (2) amphikinesis, where movement occurs at the craniofacial hinge as well as the distal part of the upper bill; and (3) rhynchokinesis, where movement only occurs at one or more points in the upper bill (Fig. 2.73). Many species of birds exhibit prokinesis, whereas only rails exhibit amphikinesis, and ratites, cranes, shorebirds, pigeons and doves, hummingbirds, and some passerines exhibit rhynchokinesis (Bühler 1981; Zusi 1984).

Parrots and cockatoos also have prokinetic skulls. However, the craniofacial hinge or joint of parrots and cockatoos evolved via secondary

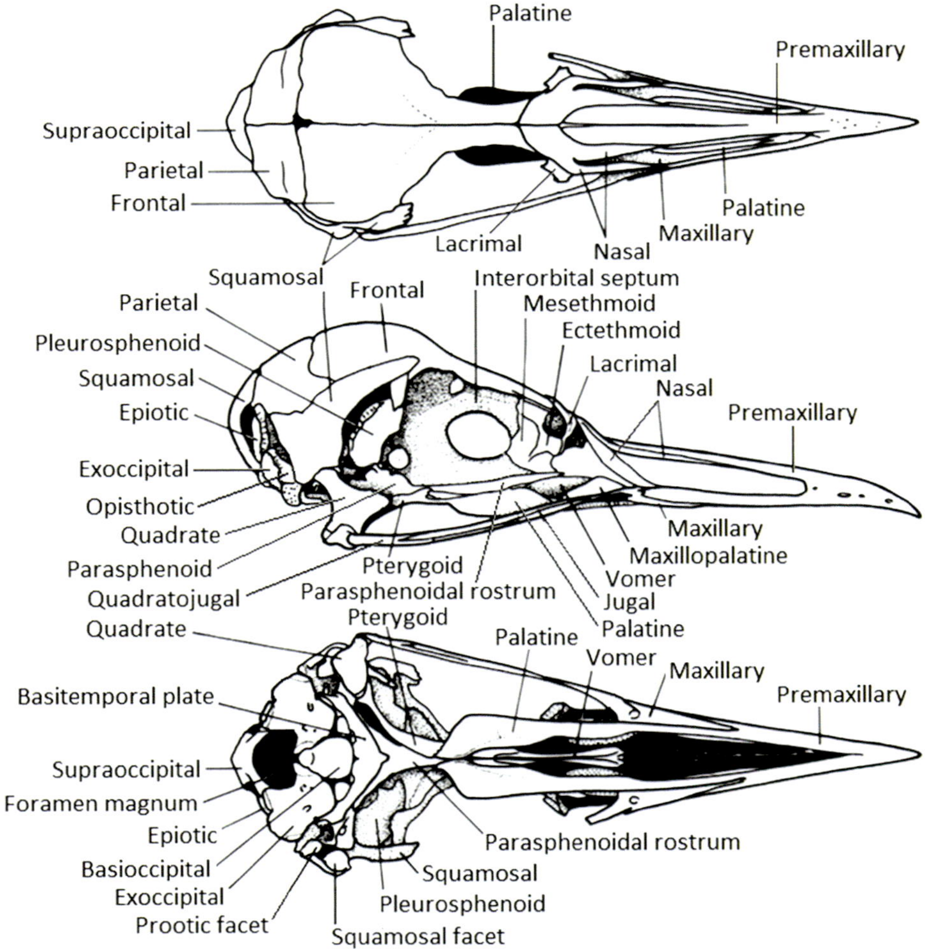


**Fig. 2.61** Top: Thick-billed Murres (*Uria lomvia*). Bottom: Dorsal, lateral, and ventral views of the skull of a young Thick-billed Murre. (Photo by Art Sowls, U. S. Fish and Wildlife Service, CC0 Public Domain; Bottom figure modified from Zusi 1993; © 1993 University of Chicago Press, used with permission)

**Box 2.5 Mandibular Bowing of Pelicans**

The gular pouches of pelicans are extremely distensible and, when a pelican is attempting to capture a fish, can hold several liters of water, e.g., up to 11 liters for Brown Pelicans (*Pelecanus occidentalis*, Schreiber et al. 1975). Holding that much water requires mandibular bowing. For example, the mandibles of Brown Pelicans are about 5 cm apart when their pouch is empty, but 15 cm or more apart when their pouch is filled with water (Schreiber et al. 1975). Meyers and Myers (2005) found two distinct bending areas in the mandibles of Brown Pelicans: the rostral regions near the distal end of the mandibles and the middle regions. The mandibles in the rostral regions are narrow and elliptical in shape and are areas of low mineralization (~20% mineral content), making these areas very flexible. The middle regions have an oval shape and are where the four bones that make up the mandibles meet, i.e., a syndesmosis joint where bones are joined by connective tissue (Meyers and Myers 2005). This joint allows these bones to move relative to each other, allowing the mandibles to bow outward when the pouch fills with water.

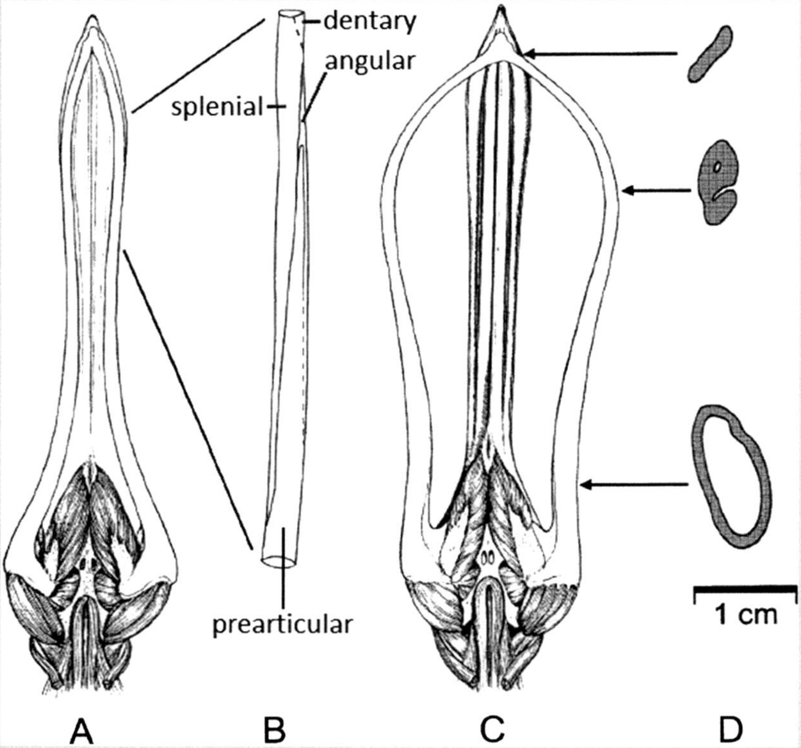


Mandibular bowing in the Brown Pelican (*Pelecanus occidentalis*). (**a**) Ventral view of pelican head and lower jaw in resting position (pouch removed). (**b**) Medial view of left mandible, showing the four lower jaw bones that make up the lateral bending zone. (**c**) Ventral view of pelican head and lower jaw in bowed position (pouch removed). (**d**) Cross-sections through three areas of the mandibles, including the top (rostral bending zone), middle (lateral bending zone), and bottom (a nonbending zone). (Figure from Meyers and Myers 2005; © 2005 Oxford University Press, used with permission)

(continued)

**Box 2.5** (continued)

Example of mandibular bowing by an Australian Pelican (*Pelecanus conspicillatus*). (Photo from pxhere.com, CC0 Public Domain)

**Fig. 2.62** Mosaic evolution produces an evolutionary mosaic, with impressive variation in cranial morphology. This variation in the morphology of avian heads and skulls is a product of mosaic evolution, with different regions of the skull evolving at different rates and by different modes. (Figure from Field 2018; Photos © D. J. Field; used with permission of the National Academy of Sciences)

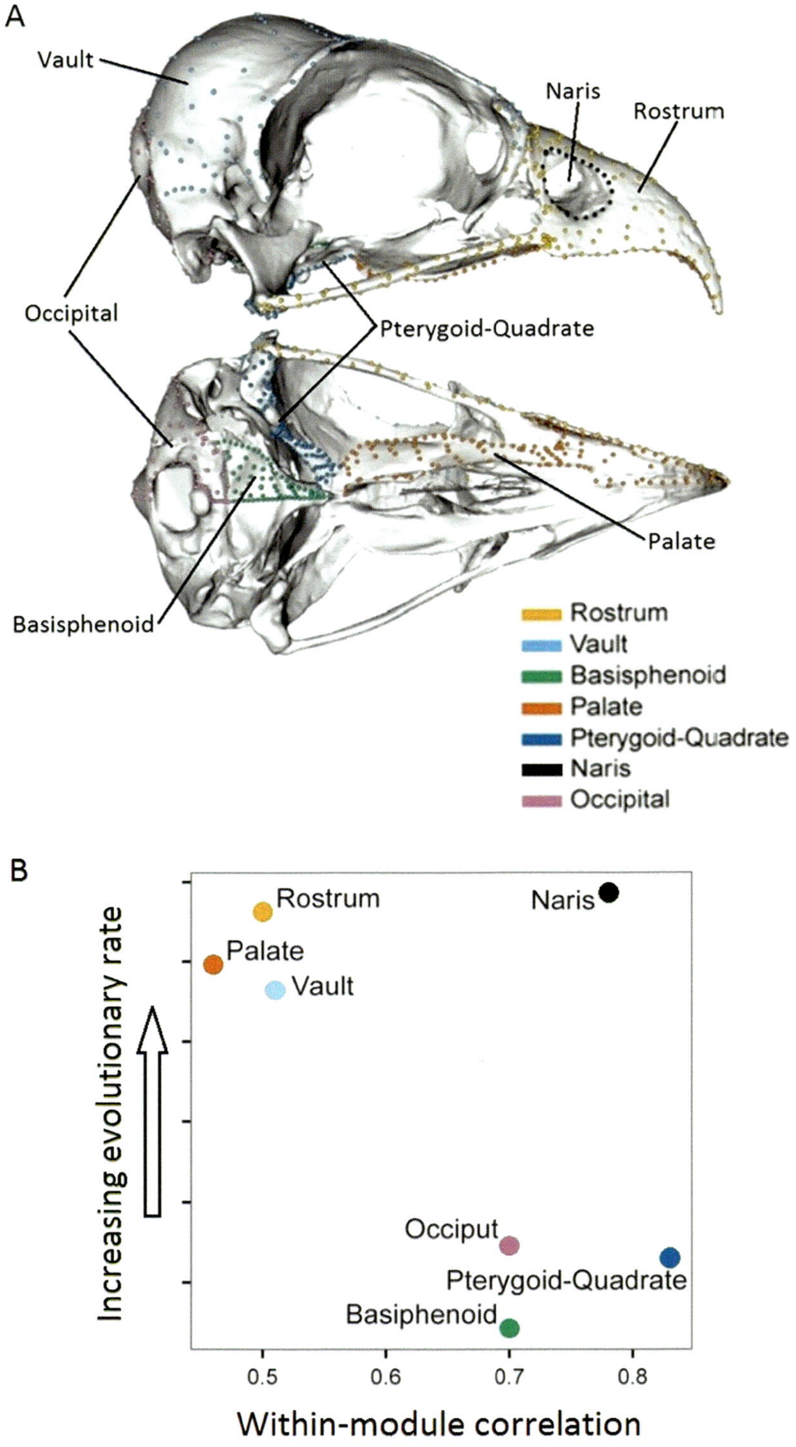


**Fig. 2.63** Modularity of the avian skull. (**a**) The seven modules of the skull and (**b**) variation in evolutionary rates of different modules and within-module correlations. (Figure made from two figures and modified from Felice and Goswami 2018; PNAS, used with permission of Ryan Felice)

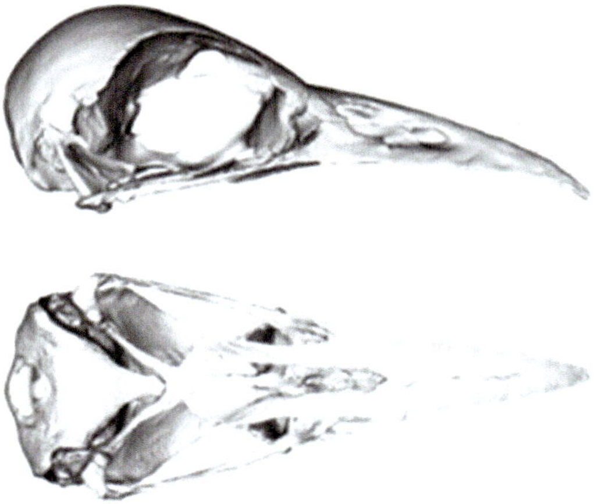

**Fig. 2.64** The hypothetical ancestral Neornithine skull. (Figure from Felice and Goswami 2018, PNAS, used with permission of Ryan Felice)

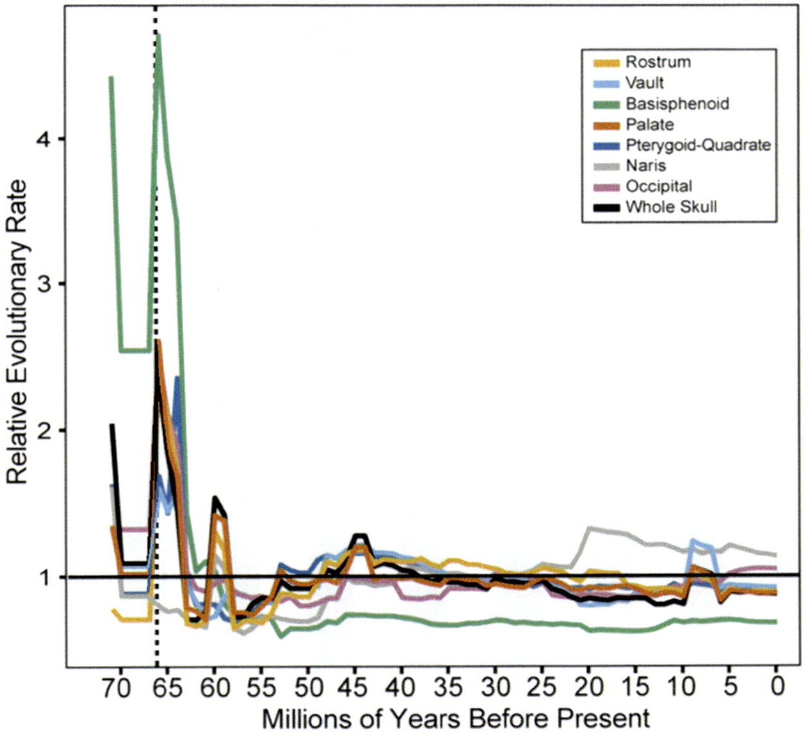


**Fig. 2.65** Relative evolutionary rates of the avian skull and the seven modules that make up the skull. Dashed vertical line indicates the Cretaceous–Paleogene (K-Pg) boundary. (Figure modified a bit from Felice and Goswami 2018; PNAS, used with permission of Ryan Felice)

**Fig. 2.66** Lesser, Andean, and James's flamingos have bills that are bulbous in cross-section and used to filter small food items like blue-green algae and diatoms. American, Greater, and Chilean flamingos have more compact bills better suited to filter larger food items like molluscs and crustaceans. (Figure from Torres et al. 2014; © 2014 Springer Nature, used with permission)

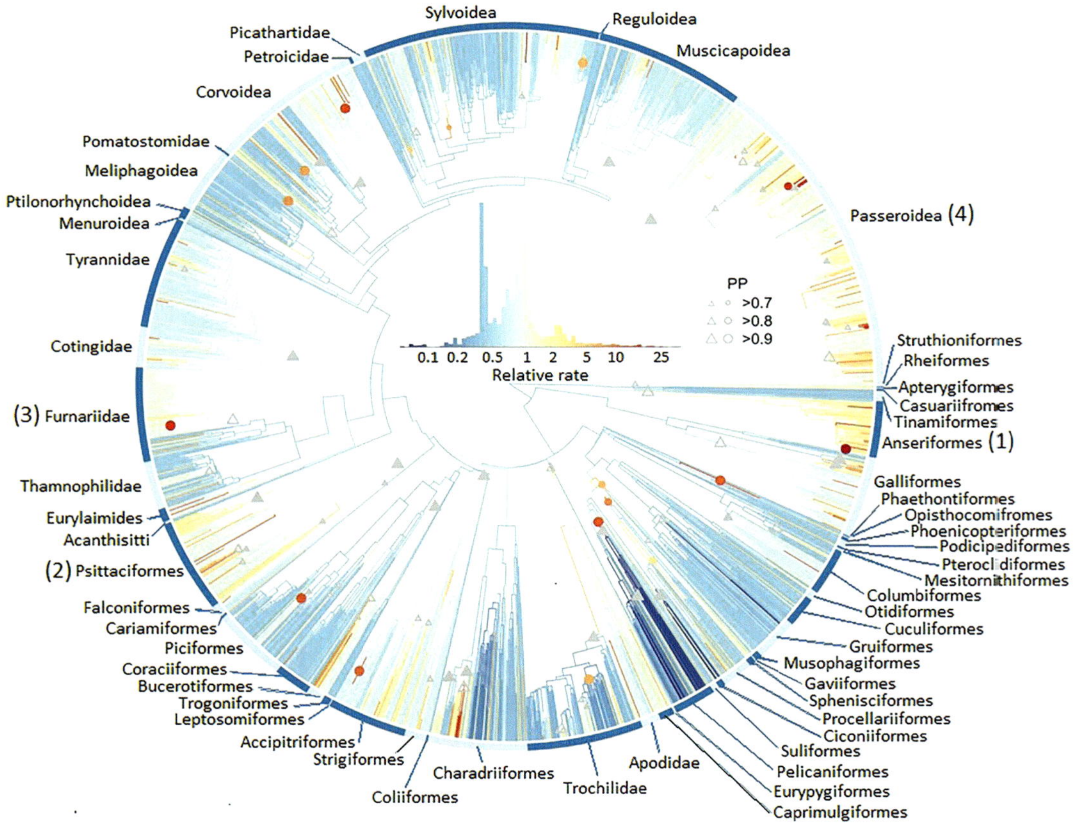


**Fig. 2.67** Avian phylogeny ($N$ = 2028 species) with estimates of the relative rates of bill shape evolution among different taxa. Note the particularly high rates of bill shape evolution and diversification in the orders Anseriformes (1), Psittaciformes (2), the family Furnariidae (3), and the superfamily Passeroidea (4). Gray triangles indicate stem branches

**Box 2.6 Darwin's Finches, Adaptive Radiation, and Evolution**

Charles Darwin spent 5 weeks collecting specimens in the Galapagos Islands during the voyage of the H. M. S. Beagle (1831–1836). Included among these specimens were nine of the 14 species now often referred to as Darwin's finches. These finches are often used as examples of adaptive radiation, with variation in their bill shapes illustrating how natural selection, in a complex process involving colonization, geographic isolation, speciation, and interisland dispersal, has resulted in divergent bill morphologies adapted for feeding on a variety of different food items. Some authors have also argued that Darwin's observations of these finches provided him with his first clue that species might be mutable and "inspired all his later theories by providing him with a decisive example of evolution in action" (see Sulloway 1982 and many references therein).

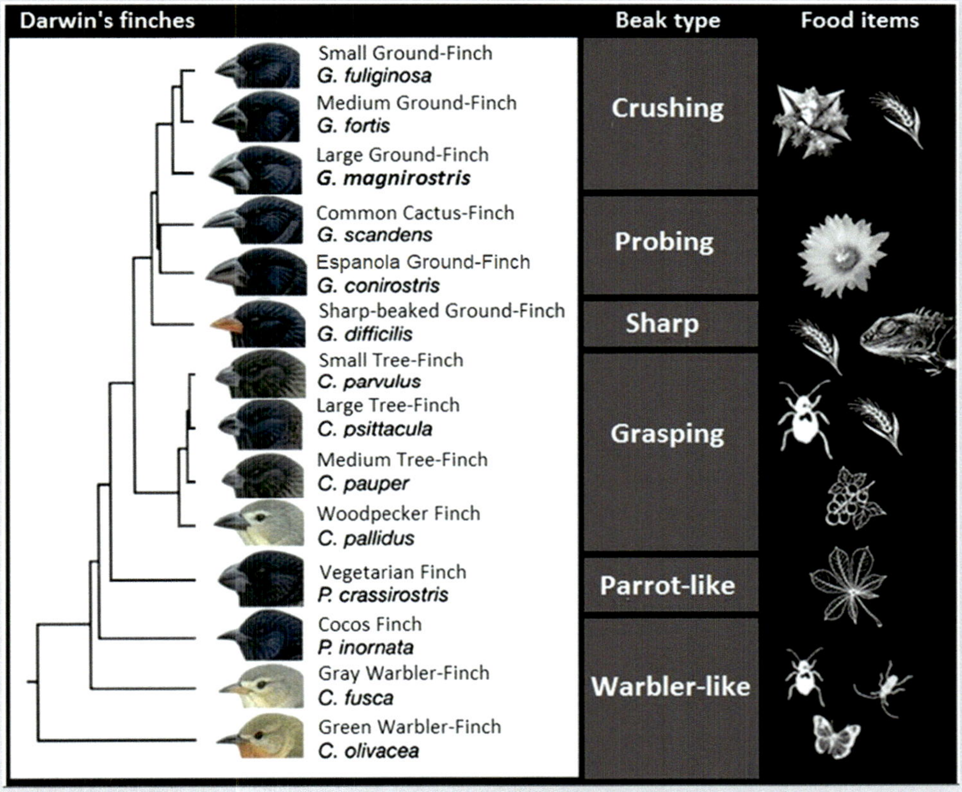


Darwin's finches. Note the diversity of bill shapes that allow different species to feed on different food items, including seeds, insects, berries, and leaves. (Figure modified from Rands et al. 2013; open-access article distributed under the terms of the Creative Commons Attribution License. Drawing of the finches from Tokita

(continued)

**Fig. 2.67** (continued) with support for whole clade shifts in evolutionary rate. Colored circles indicate rate shifts for individual internal branches (with the color indicating the rate estimate). The relative size of triangles and circles indicates the posterior probability of a rate shift. Triangles identify shifts on the focal node (filled) or shifts at the focal node or on one of its two daughter nodes (open). (Figure modified from Cooney et al. 2017; © 2017 Springer Nature, used with permission)

**Box 2.6** (continued)

et al. 2017; © 2016 The Authors. Published by the Royal Society under the terms of the Creative Commons Attribution License, https://creativecommons.org/licenses/by/4.0/)

Actually, Darwin did not even mention the finches in his diary during the voyage of the Beagle, making only a single brief mention in the first edition of his *Journal of Researches* (1839). In addition, Darwin found it difficult to separate the finches into different species and initially thought they belonged in different families or subfamilies of birds, referring to those with larger beaks as "Gross-beaks," those with smaller beaks as "Fringilla" (finches), to the cactus finch as "Icterus" (Icteridae, a family that includes meadowlarks, orioles, and blackbirds), and to one of the warbler-finches as a "wren" or "warbler" (Sulloway 1982). After returning to England, it was John Gould (1837a) who determined that the Galapagos finches were closely related, placing them all in a single genus and three subgenera. More importantly, however, Gould (1837b) identified three different species of Galapagos mockingbirds (Mimidae) on three different islands; all closely related to mockingbirds in South America. This, not Darwin's finches, first helped Darwin reach the conclusion that species must be mutable. It was only in retrospect that his growing understanding of evolution allowed Darwin, many years later, to understand the adaptive radiation of the finches (Sulloway 1982). Writing in 1859, Darwin noted that "the more diversified the descendants from any one species become in structure, constitution, and habits, by so much will they be better enabled to seize on many and widely diversified places in the polity of nature, and so be enabled to increase in number" (Darwin 1859:112). In writing this, however, Darwin was not specifically referring to finches and, in fact, the finches were not even mentioned in *On the Origin of Species* (Darwin 1859). How then did the Galapagos finches, not even mentioned in Darwin's famous book, become known as Darwin's finches and as a textbook example of adaptive radiation? The key was a book written by David Lack (1947) and entitled *Darwin's Finches*. Although Lowe (1936) was the first to refer to Darwin's finches, the success of Lack's book made Darwin's finches famous and led to their use in textbooks as "the" example of Darwin's theory of evolution.

**Box 2.7 Hawaiian Honeycreepers**

The Hawaiian honeycreepers (Fringillidae: Drepanidinae) represent the most impressive example of the adaptive radiation of birds. Among the approximately 50 species of honeycreepers (most of which are now extinct and known only from their bones) were seed eaters, fruit eaters, gleaning insectivores, aerial insectivores, nectarivores, and even a snail specialist (Poo-uli, *Melamprosops phaeosoma*), with bill morphologies adapted for each type of diet. The ancestors of the honeyeaters likely arrived in Hawaii, and specifically on the island of Kauai, from Asia about 6–7 million years ago, when the Hawaiian Islands looked very different than they do today. Based on a genetic analysis, Lerner et al. (2011) suggested that Hawaiian honeycreepers are a sister taxon to Eurasian rosefinches (*Carpodacus*) and probably came to Hawaii from Asia.

(continued)

**Box 2.7** (continued)

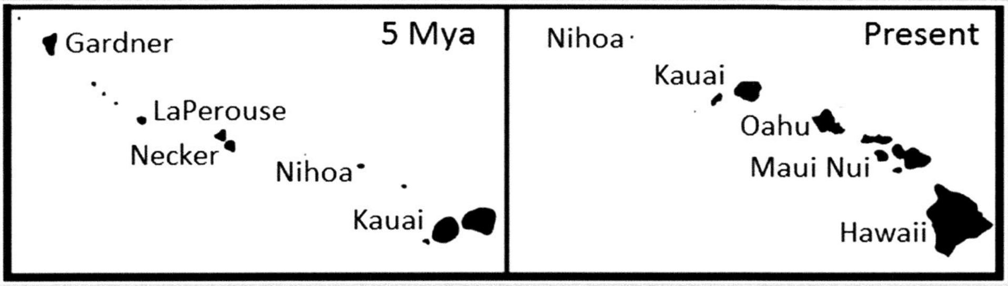


Relative size and locations of islands in the Hawaii archipelago five million years ago versus the present. (Figure modified from Price and Clague 2002; © 2002 The Royal Society, used with permission)

Diversification of the honeycreepers likely accelerated with the formation of the island of Oahu about four million years ago and the island of Maui Nui about 2.4 million years ago. These new, relatively nearby islands provided honeycreeper immigrants with new habitats and foraging niches (Price 2011). These habitats and niches, in combination with the initial absence of competitors, eventually led to the evolution of approximately 50 different species. Before humans colonized the Hawaiian Islands, as many as 24 species of honeycreepers could be found on just one Hawaiian island (James and Olson 1991).

(continued)

**Box 2.7** (continued)

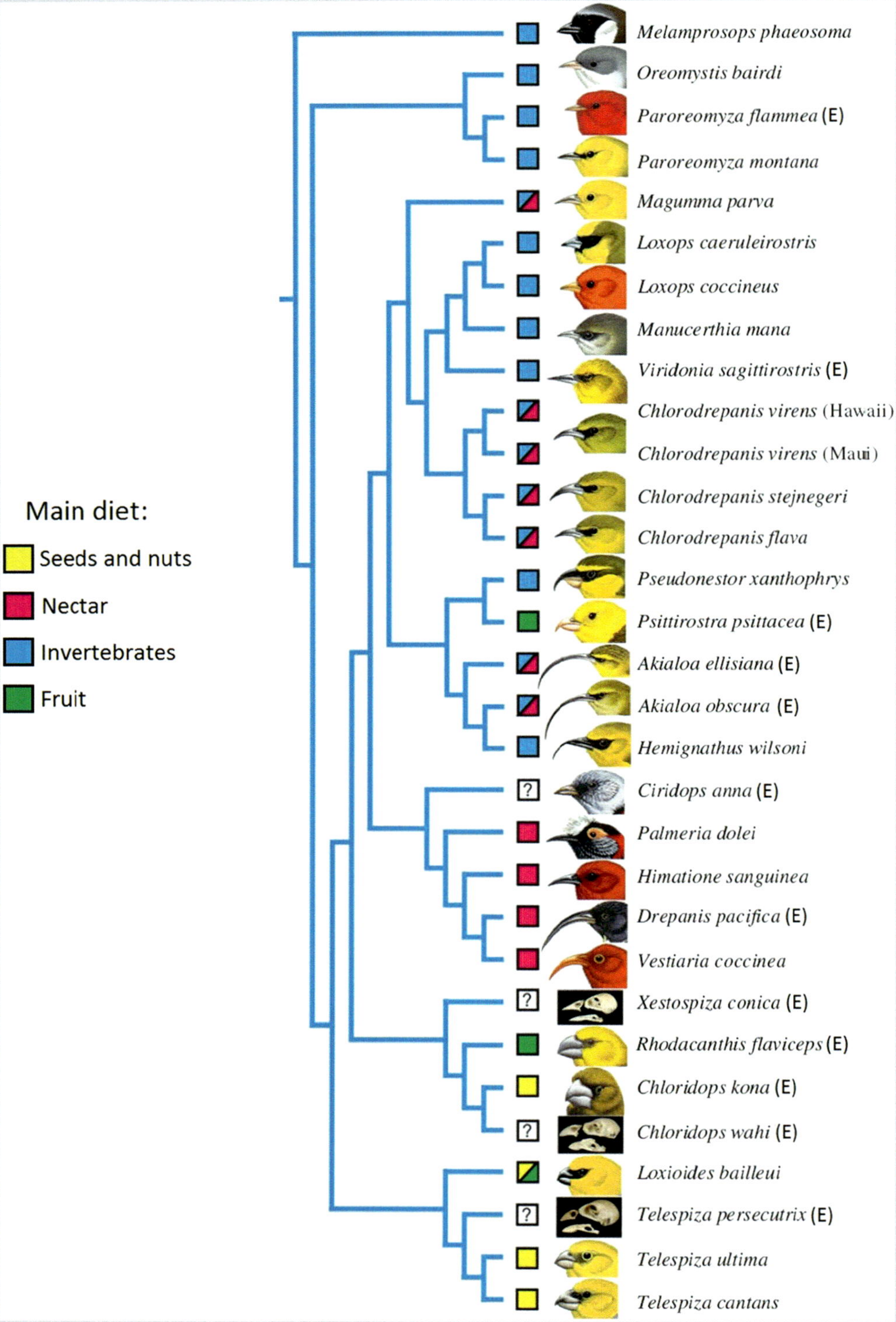


(continued)

**Box 2.7** (continued)

Phylogeny of Hawaiian honeycreepers (subfamily Drepanidinae), illustrating variation in bill morphology and diets. Diets of species that consume(d) two food types are indicated by two different colors. (E) indicates extinct species. (Figure modified from Tokita et al. 2017; © 2016 The Authors. Published by the Royal Society under the terms of the Creative Commons Attribution License, https://creativecommons.org/licenses/by/4.0/)

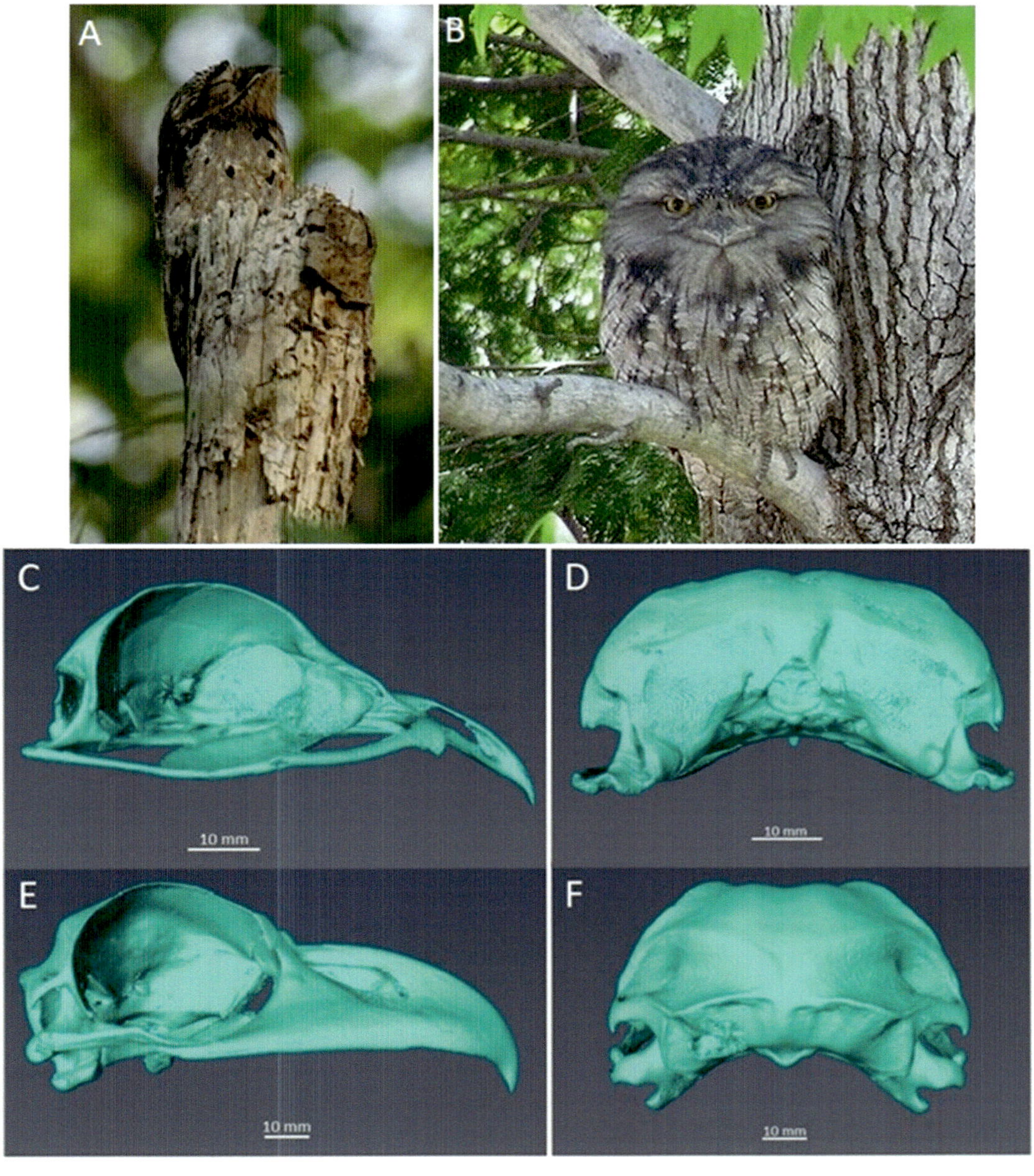


**Fig. 2.68** (**a**) Common Potoo (*Nyctibius griseus*), (**b**) Tawny Frogmouth (*Podargus strigoides*), skulls of a (**c**, **e**) Common Potoo and (**d**, **f**) Tawny Frogmouth. (**c**, **d**) Side views and (**e**, **f**) posterior views. Skulls of

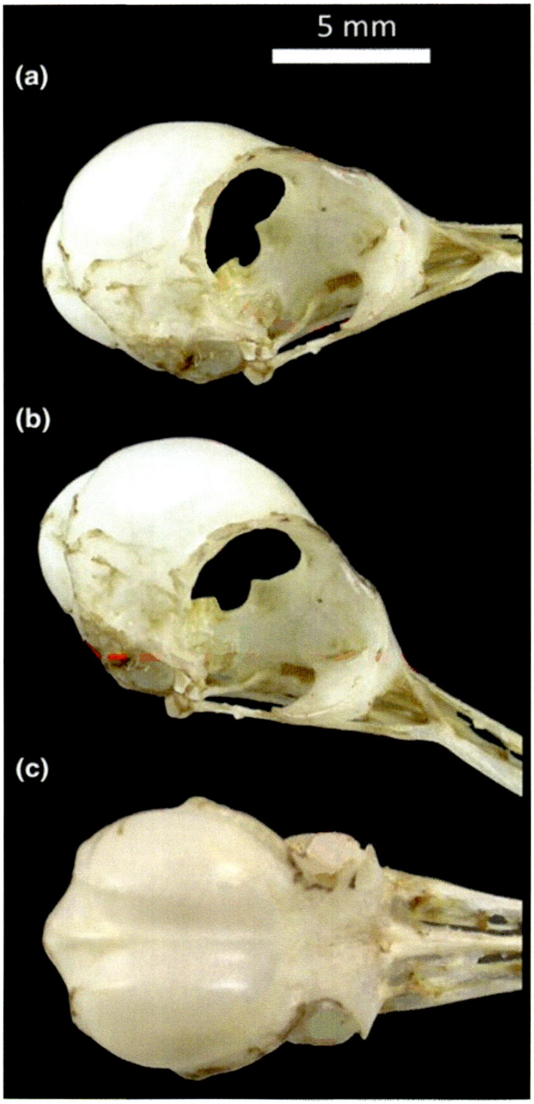


**Fig. 2.69** Skull of a Rufous-tailed Hummingbird (*Amazilia tzacatl*). (**a**, **b**) Lateral views and (**c**) dorsal view. Given their generally small size, selection pressures have modified hummingbird skulls to "accommodate larger brains and eyes without enlarging overall head size" (Ocampo Vargas et al. 2018). The braincase has become more spherical and the foramen magnum has a more ventral position. (Figure modified from Ocampo Vargas et al. 2018; open-access article under the terms of the Creative Commons Attribution License, https://creativecommons.org/licenses/by/4.0/)

**Fig. 2.68** (continued) species in the order Caprimulgiformes, nocturnal or crepuscular aerial insectivores, are relatively wide, with large nares and orbits and very large gapes. (Common Potoo photo by The Lilac Breasted Roller, CC BY 2.0, https://creativecommons.org/licenses/by/2.0/deed.en; Tawny Frogmouth photo, Wikipedia, CC0 Public Domain; skull figures from Chen et al. 2019; © 2019 The Authors, open-access article distributed under the terms and conditions of the Creative Commons Attribution (CC BY) license, https://creativecommons.org/licenses/by/4.0/)

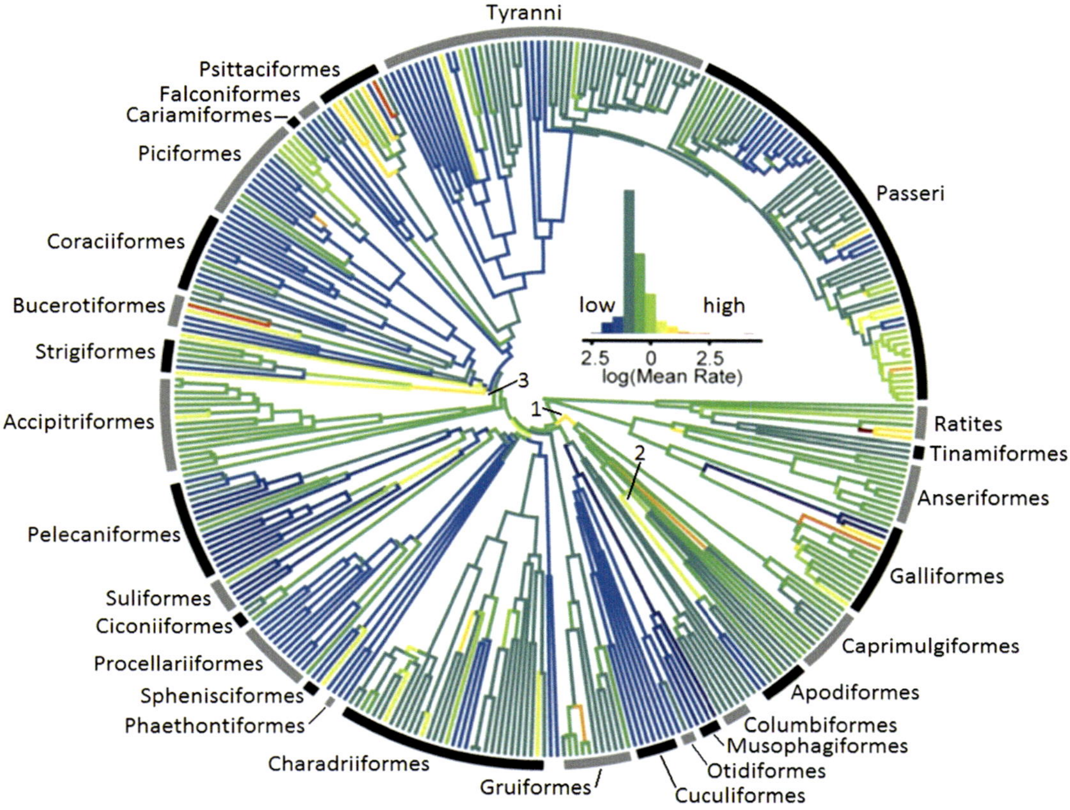


**Fig. 2.70** Rates of evolution of the cranial vault module of birds. Taxa with periods of rapid evolution include the clade Strisores (1) Caprimulgiformes and Apodiformes), hummingbirds (2), and the order Strigiformes (3). (Figure modified a bit from Felice and Goswami 2018; PNAS, used with permission of Ryan Felice)

transformation of dermal bones rather than via the nasal–frontal suture of other birds with prokinetic skulls (Tokita 2003). In addition, parrots and cockatoos exhibit uncoupled jaw movements, with the ability to move the upper and lower mandibles independently (Homberger 2017). In many birds, movement of the upper and lower mandibles is not independent because they are attached to each other by a postorbital ligament that attaches to the skull just behind the orbit and inserts on the mandible near where it articulates with the quadrate bone (Bock 1964). Other species of birds have this ligament, but the upper and lower mandibles are still capable of some independent, i.e., uncoupled, movement (e.g., Hoese and Westneat 1996). However, this ligament is not present in parrots and cockatoos, allowing completely independent movement of the upper and lower mandibles (Homberger 2017; Figs. 2.74 and 2.75).

Cranial kinesis may increase bill gap when birds feed on large prey (Zusi 1984) and, by allowing simultaneous movement of the upper and lower jaw, increase the speed of bill movement (Bout and Zweers 2001). In addition, rhynchokinesis likely reduces the force needed to open the bill, e.g., when a shorebird feeds on prey in sandy or muddy substrates (Bout and Zweers 2001; Fig. 2.76). However, shorebirds are known to use distal rhynchokinesis even

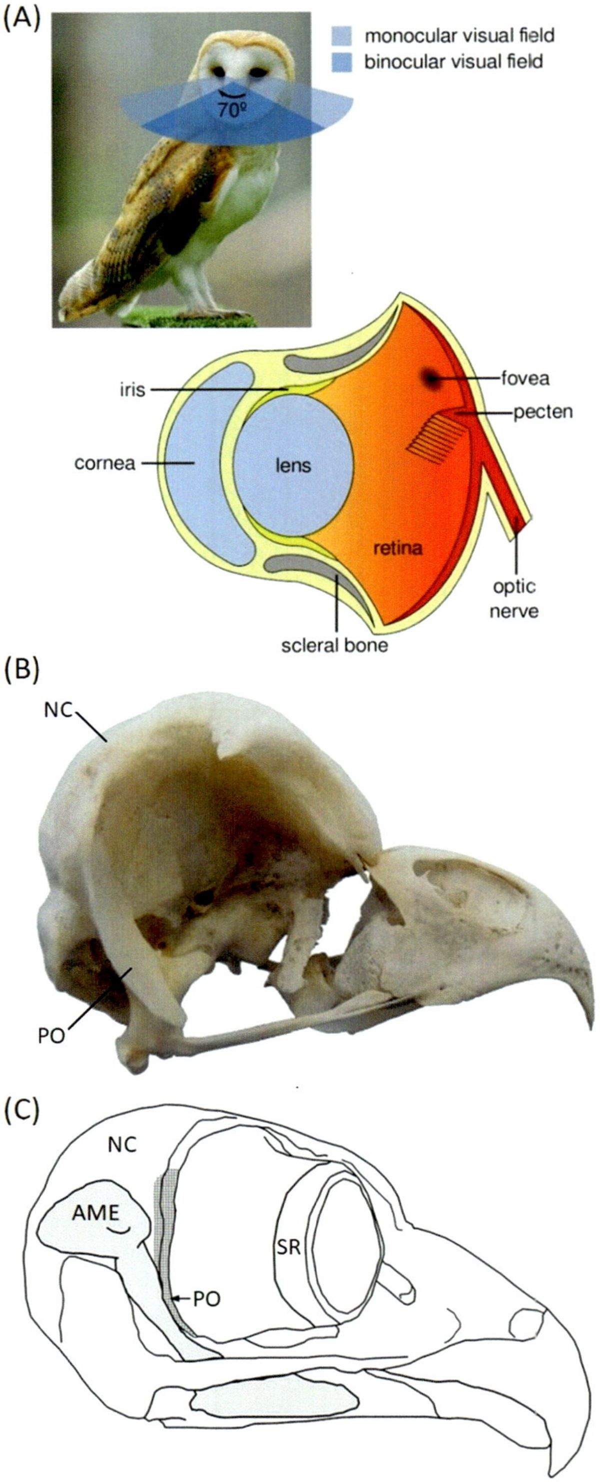


**Fig. 2.71** (**a**) Owls have very large, forward-facing eyes that provide them with large fields of binocular vision, an adaptation for increased visual acuity under low-light conditions. With the large orbits, the neurocranium has narrowed so the brain is tilted upward and the occipital condyle is ventrally located. (**b**) Skull of a Brown Wood Owl (*Strix leptogrammica*). Compared to other birds, the skulls of owls have very large and convergent orbits, plus expanded and plate-like postorbital processes (PO). (**c**) Drawing of a skull of a Great Horned Owl (*Bubo virginianus*) showing the postorbital process and the adductor mandibulae externus (AME) muscle that contracts to elevate the lower jaw. The postorbital process redirects the AME around the posterior-ventral margin of the large eyes of owls. PO, postorbital process; SR, scleral ring; NC, neurocranium; AME, adductor mandibulae externus muscle. (A, Figure from Borges et al. 2019; © 2019 Oxford University Press, used with permission; B, Figure modified from Choudhary et al. 2021; © 2020 ARCC Journals, used with permission; C, Figure from Menegaz and Kirk 2009; © 2009 Elsevier Ltd., used with permission)

when feeding on prey in water. Estrella and Masero (2007) suggested that, by using distal rhynchokinesis, shorebirds need not open the entire bill as wide, thus potentially reducing resistance when the bill enters the water and permitting a faster strike. In addition, distal rhynchokinesis may enhance use of capillary feeding (i.e., by slightly and repeatedly opening and closing the bill, drops of water with small prey are transported along the bill from the top to the mouth in a stepwise ratcheting fashion) by shorebirds (Fig. 2.77; Prakash et al. 2008; Bush et al. 2010).

All birds are thought to exhibit some form of cranial kinesis. Hummingbirds are also able to flex their lower bills. Yanega and Rubega (2004) found that hummingbirds catching small flying insects flex the distal end of their lower bills to increase their gap and make it easier to capture insects closer to the mouth, which in turn increases their foraging success rates (Fig. 2.78).

### 2.5.4 Sternum and Rib Cage

The sternum or breastbone of extant birds is a curved, dorsally concave plate of bone that forms the floor of the thoracic cavity (Box 2.10 Variation in Avian Sternums). The sternum of most extant birds has a bony plate called the carina or keel that projects ventrally along the midline. The keel is the point of origin of the two primary flight muscles, the pectoralis and supracoracoideus, and the size of the keel is proportional to the size of these muscles. Among flightless birds, the keel is reduced in size or lost and flight muscles are greatly reduced in size (Raikow 1985). For example, the sternum of the flightless ratites, such as common ostriches (*Struthio camelus*), is reduced in size and no keel is present.

Most of the earliest birds, including *Archaeopteryx*, had sterna with no keel (Fig. 2.79). However, birds with a keeled sternum are known from as early as the Early Cretaceous (~130 mya; e.g., Wang et al. 2015). In addition, many examples of birds both with and without keeled sterna have been reported from the Early Cretaceous, with those having keels exhibiting variation in the length and depth of the keel (Zhou 2002). This fossil evidence suggests that evolution of the keel was relatively rapid and keel morphology varied (and continues to vary in extant birds), with larger keels being favored by birds requiring greater muscle power to take off quickly or to fly at low speeds, and lower and longer keels for birds using their wings for faster sustained flight.

Extant birds have three to nine pairs of ribs and each rib has a dorsal part that articulates in a kinetic or movable joint with a thoracic vertebra (vertebral rib) and a ventral part that articulates

**Box 2.8 Bony Cranial Protuberances of Birds**

A few species of birds have bony outgrowths on their skulls and most can be placed into one of three categories: (1) protuberances of the upper bill like those of some curassows, (2) helmet-like casques on the dorsal surface of the skull like those of cassowaries, and (3) bulges at the base of the upper bill in the area of the frontal bone like those of some waterfowl (Mayr 2018). Most of these outgrowths consist of spongy bone and so are very light. Few investigators have examined that possible function(s) of these outgrowths. One possibility, however, is that they play a role in intra- or intersexual communication. For example, the size of the cranial bump of male magpie geese (*Anseranas semipalmata*) increases with age, and males with larger bumps were more likely to be found associating with nests (Whitehead 1998). Mayr (2018) suggested that the pneumatized frontal bumps of some waterfowl may serve some physiological, sensory, or acoustic function.

(continued)

**Box 2.8** (continued)

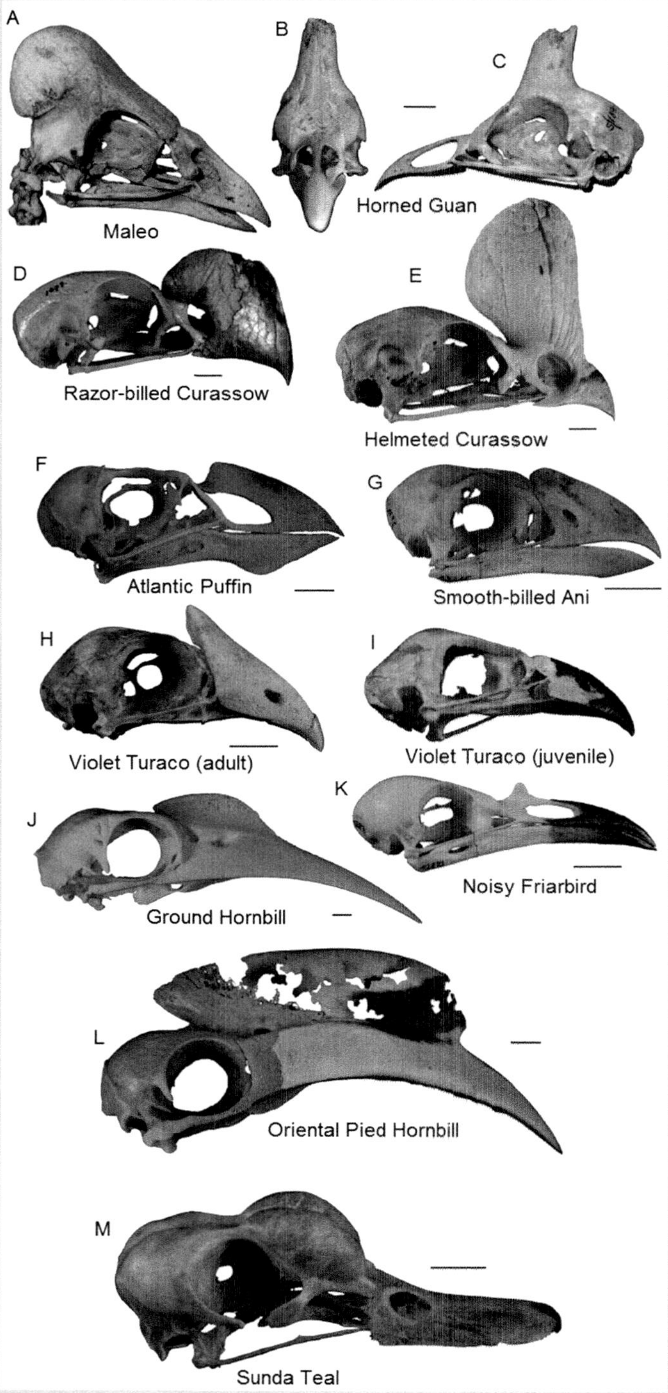


(continued)

**Box 2.8** (continued)

Examples of species of birds with bony cranial protuberances. (**a**) Maleo, *Macrocephalon maleo*, (**b**, **c**) Horned Guan, *Oreophasis derbianus*, (**d**) Razor-billed Curassow, *Mitu tuberosum*, (**e**) Helmeted Curassow, *Pauxi pauxi*, (**f**) Atlantic Puffin, *Fratercula arctica*, (**g**) Smooth-billed Ani, *Crotophaga ani*, (**h**, **i**) adult and juvenile Violet Turacos, *Musophaga violacea*, (**j**) Ground Hornbill, *Bucorvus* sp., (**k**) Noisy Friarbird, *Philemon corniculatus*, (**l**) Oriental Pied Hornbill, *Anthracoceros albirostris*, and (**m**) Sunda Teal, *Anas gibberifrons*. (Figure made from multiple figures from Mayr 2018; © 2018, Springer-Verlag GmbH Germany, part of Springer Nature, used with permission)

Several hypotheses have been proposed to explain the evolution and function of the casques of cassowaries (*Casuarius* spp.). One hypothesis is that they are the result of sexual selection, important in intersexual interaction, intrasexual interactions, or both (Green 2020). The casques could also serve as resonance chambers for their low-frequency vocalizations. In support of this hypothesis, cassowaries are known to vocalize during the mating season and, during interactions with the opposite sex, vocalizing birds lower their heads with their casques pointing toward their partner. This behavior suggests that casques might be used to direct vocalizations toward a partner, with casque size and structure possibly altering vocalizations in a way that provides information about the caller's fitness (Naish and Perron 2016). Yet another hypothesis is that casques serve as thermal windows, helping cassowaries lose heat at high ambient temperatures. In support of this hypothesis, Eastick et al. (2019) determined that cassowary casques have an extensive vascular network and serve as thermal radiators at high ambient temperatures.

Casques of three species of cassowaries. (**a**) Southern Cassowary (*Casuarius casuarius*), (**b**) Southern Cassowary in profile, (**c**) Northern Cassowary (*C. unappendiculatus*), and (**d**) Dwarf Cassowary (*C. bennetti*). (Figure from Naish and Perron 2016; Rights managed by Taylor & Francis, used with permission)

(continued)

**Box 2.8** (continued)

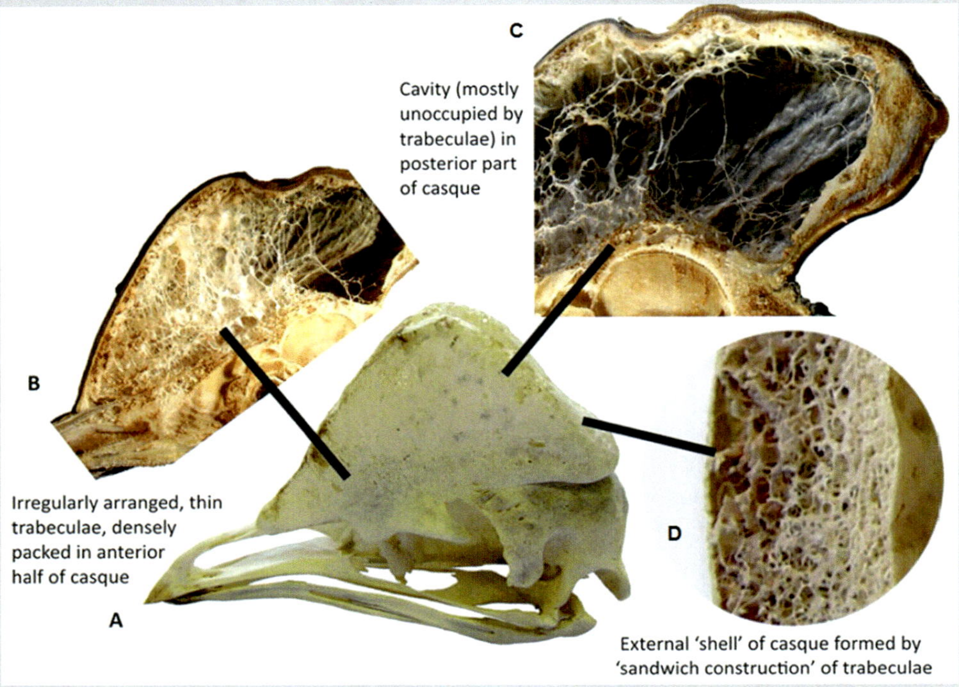


(**a**) Skull of a Dwarf Cassowary. (**b**) Anterior part of the interior of a casque and (**c**) posterior part of interior of sectioned casque of a Northern Cassowary. (**d**) Layer at the edge of a casque. (Figure from Naish and Perron 2016; Rights managed by Taylor & Francis, used with permission)

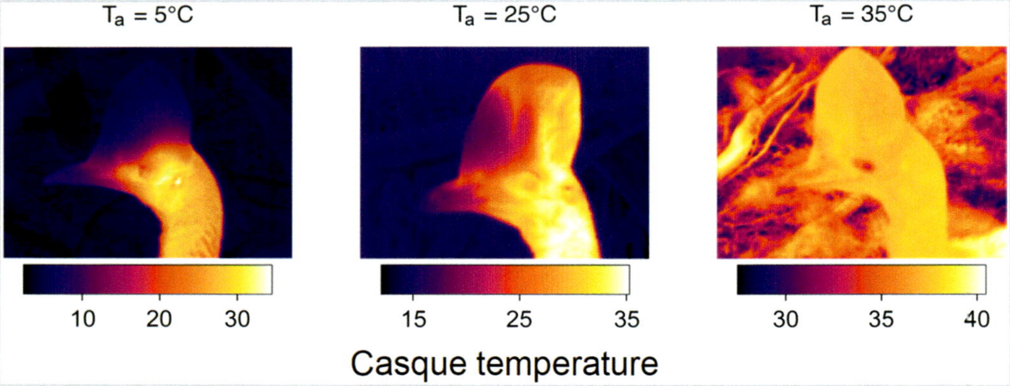


Thermal images showing changes in casque temperatures at different ambient temperatures. With increasing ambient temperature, casque temperature also increases, acting as a thermal radiator to increase heat loss and help cassowaries maintain their body temperature. (Figure from Eastick et al. 2019; open-access article licensed under a Creative Commons Attribution 4.0 International License, https://creativecommons.org/licenses/by/4.0/)

Among male Helmeted Hornbills (*Rhinoplax vigil*), casques appear to play a role in gaining access to resources, with their casques, consisting of solid keratin, used in aerial displays where males collide casque-to-casque. As described by Kinnaird et al. (2003:506), "In two instances,

(continued)

**Box 2.8** (continued)

males left their perches . . ., flew in opposite directions, circled, glided towards one another, and collided casque-to-casque in mid-air . . . When collisions occur, the resulting sound (a loud 'CLACK!') can be heard in the forest understorey at least 100 m away." Based on their observations, Kinnaird et al. (2003) suggested that this aerial jousting represented agonistic behavior, with males competing for food resources (especially fruiting figs) and nest sites.

Male Helmeted Hornbill. (Photo by CraigAnsibin, purchased from istockphoto)

(continued)

**Box 2.8** (continued)

Two Helmeted Hornbills engaging in aerial jousting, with the impact of the collision throwing one bird backwards. (Figure from Kinnaird et al. 2003; © British Ornithologists' Union, used with permission)

**Box 2.9 Rapid Adaptive Evolution of Bill Length**

The geographic range of Great Tits (*Parus major*) extends across Europe and through central Asia. Great Tits are omnivores, but, during the colder months, seeds are an important part of their diet. As such, when bird feeders with seeds are available, Great Tits are regular visitors, and this is particularly the case in the United Kingdom (UK) where bird feeding has been a popular activity since the nineteenth century. In fact, investigators have estimated that over half of all homeowners in urban areas feed birds in the UK (Orros and Fellowes 2015), and that twice as much is spent on bird seed in the UK than in mainland Europe (Jones and Reynolds 2008).

(continued)

**Box 2.9** (continued)

Geographic range of Great Tits is shown in red. (Map, CC0 Public Domain; https://en.wikipedia.org/wiki/Great_tit#/media/File:ParusMajorMap.svg; Great Tit photo by Andy Morffew, https://pxhere.com/en/photo/383699, https://creativecommons.org/licenses/by/2.0/)

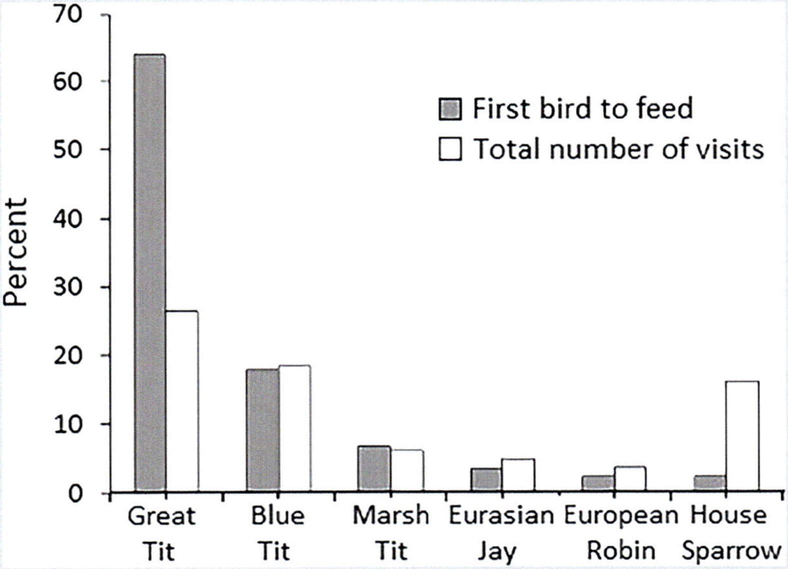


Differences among six species of songbirds in their tendency to first visit novel sources of food and in the subsequent use of those novel sources. More than 60% of first visits were by Great Tits and they all visited sources more frequently than the other species. Eurasian Blue Tit, *Cyanistes caeruleus*; Marsh Tit, *Poecile palustris*; Eurasian Jay, *Garrulus glandarius*; European Robin, *Erithacus rubecula*; House Sparrow, *Passer domesticus*. (Figure modified from Tryjanowski et al. 2015; open-access article under a Creative Commons Attribution 4.0 International License, https://creativecommons.org/licenses/by/4.0/)

(continued)

**Box 2.9** (continued)

In studying genomic variation between populations of Great Tits in the UK and, on mainland Europe, the Netherlands, Bosse et al. (2017) identified regions in the genomes involved in skeletal development and morphogenesis under divergent selection. More specifically, they identified a gene known to be associated with beak shape in Darwin's finches where there was evidence of selection in great tits in the UK, but not in the Netherlands. This suggested the possibility of differences in bill morphology between the two populations. Subsequent analysis revealed that the bills of great tits in the UK were significantly longer than those of great tits in the Netherlands and, further, that the average bill length of great tits in the UK had increased since 1982.

Interestingly, Bosse et al. (2017) also found that Great Tits with longer bills tended to visit bird feeders more frequently than those with shorter bills. Collectively, these data suggest that, in just a few decades, use of bird feeders has selected for an increase in the bill length of Great Tits in the UK, apparently because slightly longer bills improve their ability to access seeds in bird feeders which, in turn, for reasons yet to be determined, confers a fitness advantage.

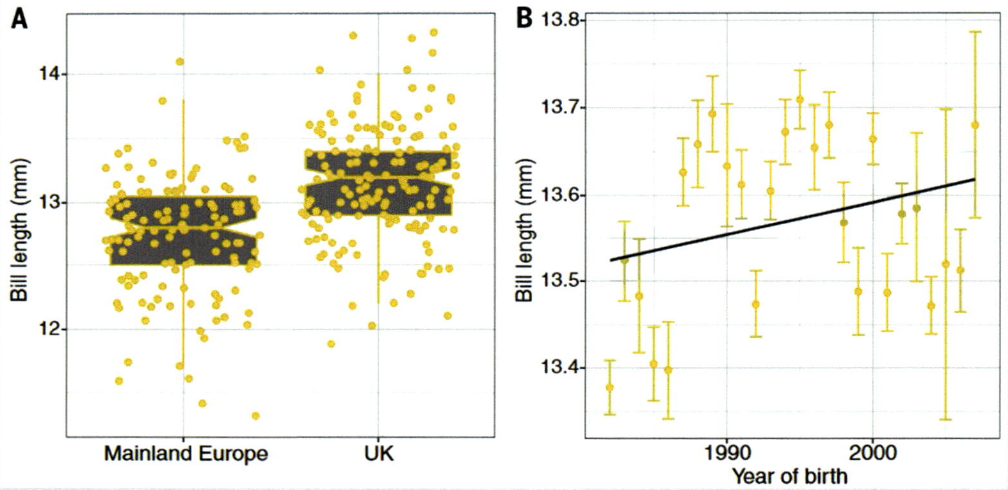


Spatiotemporal variation in bill length of Great Tits. (**a**) Bills of museum specimens of Great Tits from the UK were found to be significantly longer than those of specimens from mainland Europe. (**b**) During the period from 1982 to 2007, mean bill length of Great Tits has increased significantly in the UK. Note the different scales on the y-axes of the two figures. (Figure from Bosse et al. 2017; © 2017 The Authors, some rights reserved; exclusive licensee American Association for the Advancement of Science, used with permission)

with the sternum (sternal rib). Caudal to the ribs, there are usually a variable number of ribs that consist only of a short vertebral part (floating ribs). Most vertebral ribs have projections called uncinate processes that extend posteriorly (Fig. 2.80). The ribs of all extant birds except Emus (*Dromaius novaehollandiae*) and screamers (Anhimidae) (Bellairs and Jenkins 1960) have uncinate processes. Uncinate processes are not uniquely avian structures, having

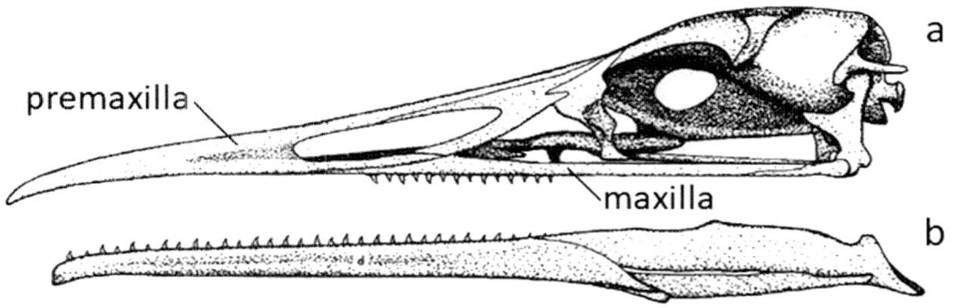


**Fig. 2.72** (**a**) Skull and (**b**) mandible of *Hesperornis regalis*. The premaxilla had a horny sheath (rhamphotheca) and teeth were confined to the maxilla and the mandible. (Figure from Gingerich 1973; © 1973 Nature Publishing Group, used with permission)

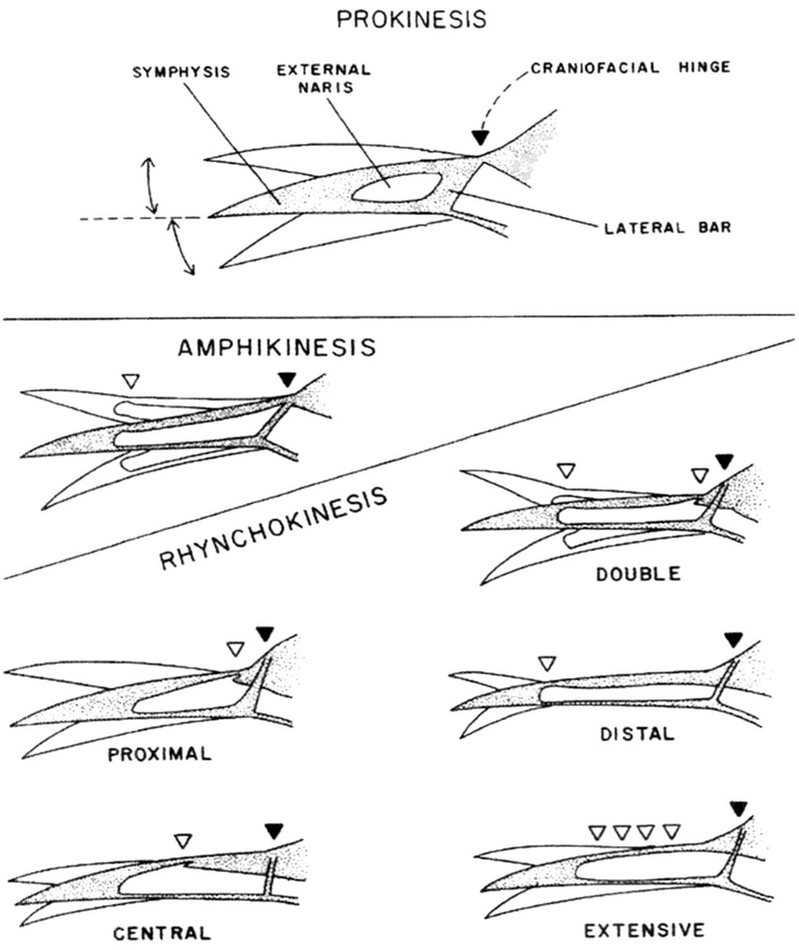


**Fig. 2.73** Outlines of upper bill and portion of a skull showing types of cranial kinesis in birds, including prokinesis, amphikinesis, and rhynchokinesis. The black triangles indicate the craniofacial hinge; open triangles indicate areas of movement in the upper bill. (Figure from Zusi 1984; CC0 Public Domain)

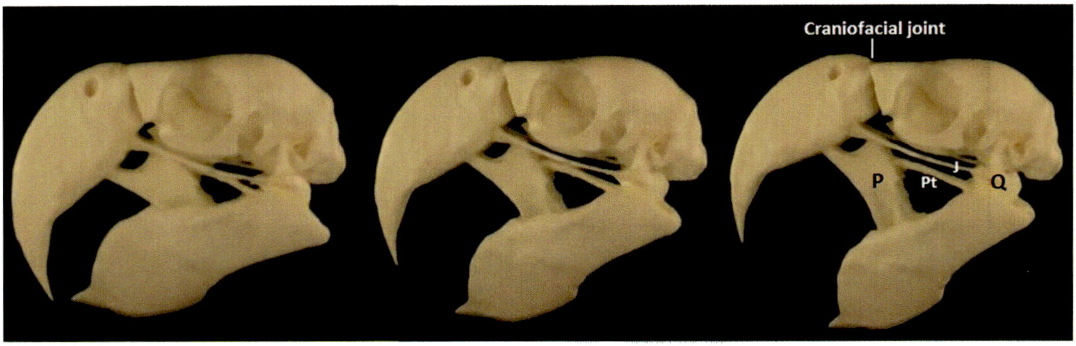


**Fig. 2.74** Series of images showing movement of the upper and lower mandibles of a macaw. P, palatine; Q, quadrate; Pt, pterygoid; J, jugal. (Figure made from screen captures from a YouTube video: https://www.youtube.com/watch?v=wLRSwiJ_Y9k; Images provided courtesy of Dr. Scott Echols and Avian Studios)

also been reported in several theropods, including *Velociraptor, Microraptor*, and *Deinonychus*, and represent yet another morphological character linking them to birds (Codd et al. 2007).

Uncinate processes may also provide mechanical support for and strengthen the rib cage and aid in respiration. Additional mechanical support might be important because of the pressures placed on the rib cage during flight or, for some birds, when diving underwater. However, not all birds that fly have ribs with uncinate processes (screamers) and other vertebrates that flew or fly (i.e., pterosaurs and bats) and dive (e.g., marine mammals) did or do not have ribs with uncinate processes, suggesting that flying does not require extra support of the rib cage. In addition, because uncinate processes overlap, but are not attached to, adjacent ribs and, in some species, are relatively fragile structures attached by fibroelastic tissue to vertebral ribs that allows their lateral deflection, it seems unlikely that their primary function is to provide increased support for or stability of the rib cage (Codd 2004).

Uncinate processes play an important role in respiration. In addition, the length of uncinate processes has been found to vary with locomotor mode, being shortest in walking species, intermediate in length in flying species, and longest in diving species (Tickle et al. 2007; Fig. 2.81). The length of uncinate processes is also influenced by body mass and metabolic rate; larger species of birds and those with higher metabolic rates tend to have longer uncinate processes (Tickle et al. 2009). Uncinate processes act as levers and contribute to the forward rotation of the dorsal ribs and the lowering of the sternum during inspiration. The length of these processes is important because longer uncinate processes serve as better levers. Flying birds may have longer uncinate processes than flightless (walking) birds because they have larger flight muscles and, during respiration, that muscle mass must be moved up and down. Diving birds may have the longest uncinate processes because, when they come to the surface after a dive, they may need to maximize gas exchange by increasing breathing frequency and their longer uncinate processes may be important because, during inspiration, the lowering of the sternum may be resisted by the pressure of water against the body (Tickle et al. 2007). Additional details concerning the role of the uncinate processes in respiration are provided in Chap. 7.

### 2.5.5 Vertebral Column

The vertebral column of birds consists of cervical vertebrae, thoracic vertebrae, a synsacrum, caudal vertebrae, and a pygostyle. The synsacrum

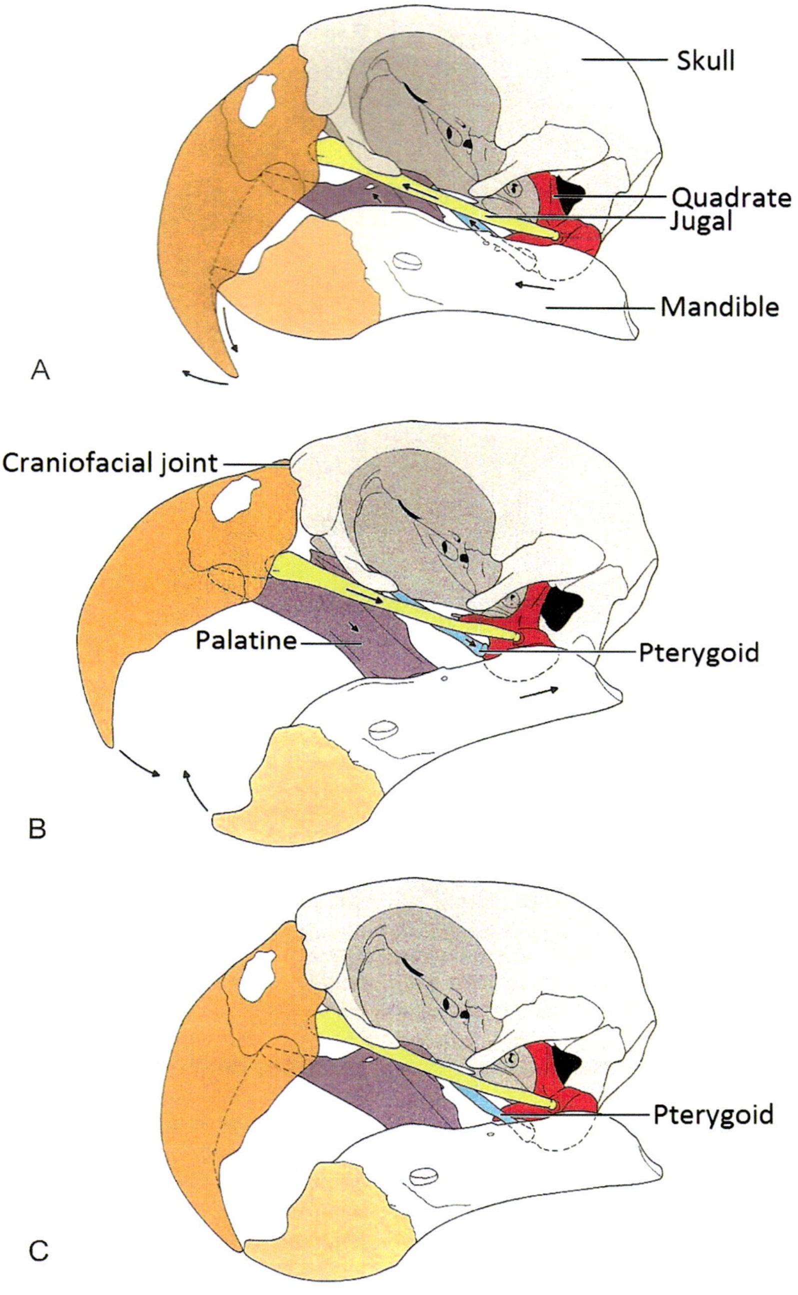


**Fig. 2.75** Examples of the uncoupled cranial kinesis of a Gray Parrot (*Psittacus erithacus*). (**a**) Closed-bill position with upper mandible lowered and lower mandible raised; arrows indicate movements needed to open the bill. (**b**) Open-bill position with upper mandible raised and lower mandible lowered; arrows indicate movements needed to close the bill. (**c**) Pincered-bill position for picking up items, with both mandibles lowered. (Figure modified from Homberger 2017; © 2017 Springer International Publishing AG, used with permission)

consists of fused thoracic, lumbar, sacral, and caudal vertebrae, and the pygostyle is formed by fusion of several caudal vertebrae. The unfused vertebrae of birds are heterocoelous vertebrae, with the points of articulation between vertebrae having a saddle-shaped joint at the base of the vertebral body and two sliding joints at the top (Fig. 2.82). These joints are uniquely avian and allow both large dorsoventral and more limited lateral rotation (van der Leeuw et al. 2001).

**Fig. 2.76** Example of distal rhynchokinesis by a Common Snipe (*Gallinago gallinago*). (Photos by Amar-Singh HSS, used with permission)

*Archaeopteryx* had amphicoelous vertebrae (both points of articulation with adjacent vertebrae were concave), but other birds from the Lower Cretaceous, including *Hesperornis*, had heterocoelous vertebrae like those of extant birds.

The cervical vertebrae along with associated ligaments and muscles form the avian neck. The neck moves the head and bill and, as such, plays a critical role in foraging, preening, nest building, displaying, and other behaviors that involve head movement. Most birds have 14 or 15 cervical vertebrae, but some have as few as 10 (some parrots) and others as many as 25 (swans). The number of cervical vertebrae that birds have is influenced by phylogenetic relatedness, but, even within individual taxonomic groups, the number of cervical vertebrae varies. For example, among Galloanseriformes, the number of cervical vertebrae ranges from 14 to 23 and, in the order Accipitriformes, the number varies from 12 to 15 (Böhmer et al. 2019). Despite this variation among birds in the number of cervical vertebrae,

**Fig. 2.77** Top: A young Wilson's Phalarope (*Phalaropus tricolor*) using the ratchet mechanism while feeding. Bottom: Note the prey item in the drop of water in the bird's beak. (Figure from Bush et al. 2010, Photo credit: Robert Lewis; © 2010 Published by Elsevier B.V., used with permission)

Marek et al. (2021) analyzed the cervical region of 48 species of birds differing in locomotory ecology, head mass, body mass, and neck length and found that differences in the degree of vertebral elongation were more important than the number of cervical vertebrae in explaining variation among species in neck length. Marek et al. (2021) also found that neck length of birds varied isometrically with both body mass and head mass (Box 2.11 Neck Length and Body Mass: Birds vs. Mammals).

Variation among birds in the number and length of cervical vertebrae and the length and flexibility of the neck is also influenced by a taxa's ecological niche and feeding technique. Based on examination of 103 species of birds representing 34 orders and 68 families, Böhmer et al. (2019) found a trend toward shorter necks in terrestrial taxa and longer necks in aquatic taxa (Fig. 2.83). For example, some ducks and geese have rather long necks that enhance their ability to reach food items under water, whereas terrestrial birds that feed by probing such as woodcocks and kiwis have relatively short necks.

Because all birds have a flexible S-shaped neck (Fig. 2.84), the number of cervical vertebrae present is not always apparent from a bird's

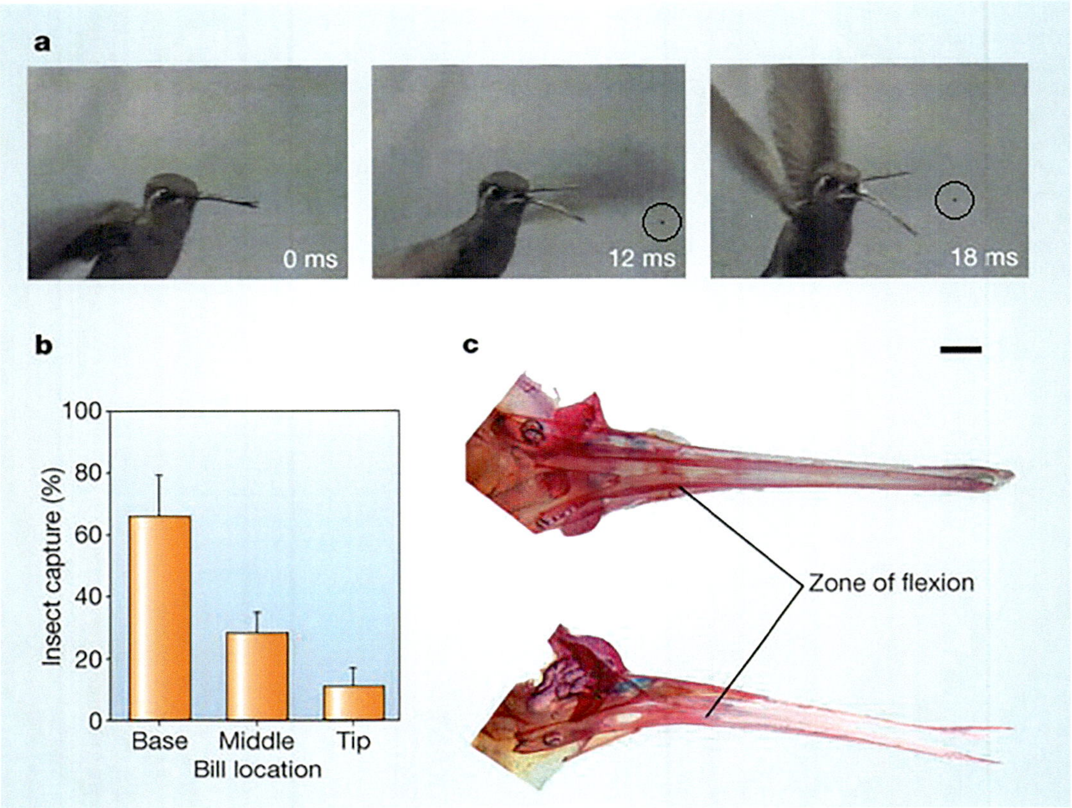


**Fig. 2.78** (**a**) A female Magnificent Hummingbird (*Eugenes fulgens*) attacking a fruitfly (*Drosophila*; dark spot within circle) and showing the jaw opening, followed by bowing of the mandible at the base to widen the gap, and flexion of the distal half of the mandible. (**b**) Success of insect capture (percentage caught) by a Ruby-throated Hummingbird in relation to where the prey first strikes the bill; prey caught at the tip of the bill are often lost during transport to the mouth. (**c**) Ventral (top) and right lateral (bottom) view of a stained ruby-throated hummingbird skull. The mandible is a single, fused unit and has to deform to bend as seen in **a**. Scale bar: 1 mm. (Figure from Yanega and Rubega 2004; © 2004 Springer Nature, used with permission)

appearance. "External" neck length is influenced as much or more by degree of flexion (how much the neck is "folded") than by the number of cervical vertebrae present. Although neck length also varies among mammals, most have the same number of cervical vertebrae (usually 7). Compared to other extant vertebrates, the avian neck is generally very flexible, capable of both dorsoventral and lateral (side-to-side) flexion (Figs. 2.85, 2.86, and 2.87).

Interestingly, Böhmer et al. (2019) found a correlation between neck length and leg length (Fig. 2.88). One possible explanation for this correlation is that birds with longer legs need longer necks so their heads can reach the ground when foraging and drinking. Bipedal animals like birds could potentially get their heads to the ground by rotating their bodies. However, as birds evolved from theropods, they lost their tails. For theropods, their tails served as a counterbalance to their upper bodies and placed their center of gravity over their legs, allowing them to move their heads to the ground by rotating their bodies (Fig. 2.89). However, without tails to

**Box 2.10 Variation in Avian Sternums**

The avian sternum, like those of mammals, helps, along with ribs, protect the organs in the thoracic cavity. In addition, however, the sternum of birds that fly has a median bony projection called that keel that serves to anchor the major flight muscles—the pectoralis and supracoracoideus. The sternums of birds vary in size and shape and, for birds that have one, the height and curvature of keels.

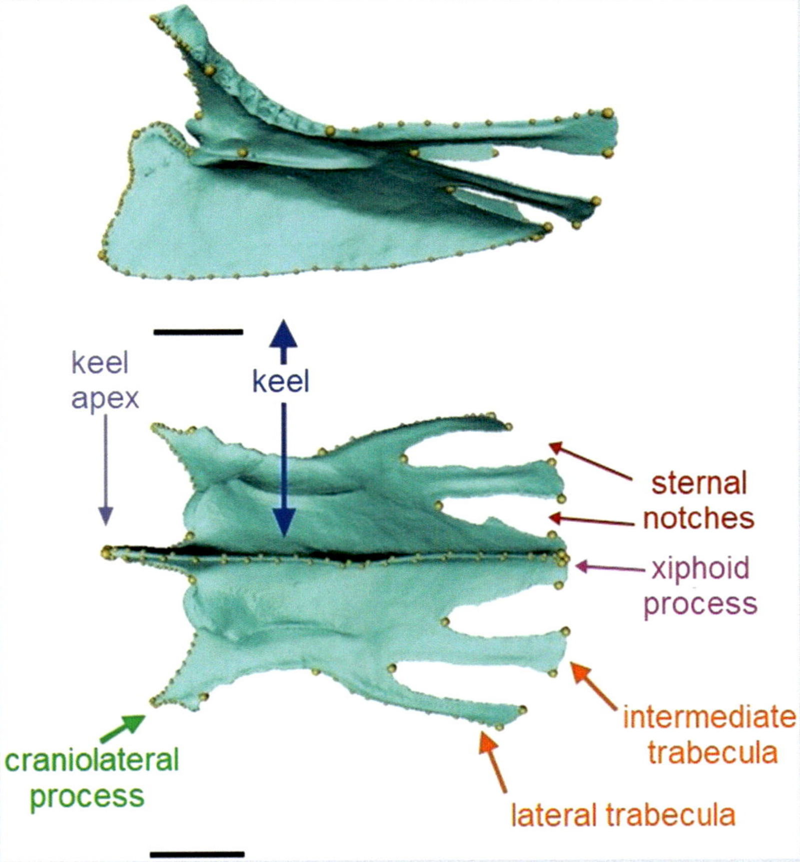


A 3D model of the sternum of a Yellow-throated Toucan (*Ramphastos ambiguus*). (Figure from Lowi-Merri et al. 2021; © The Authors 2021, open-access article licensed under a Creative Commons Attribution 4.0 International License, https://creativecommons.org/licenses/by/4.0/)

(continued)

**Box 2.10** (continued)

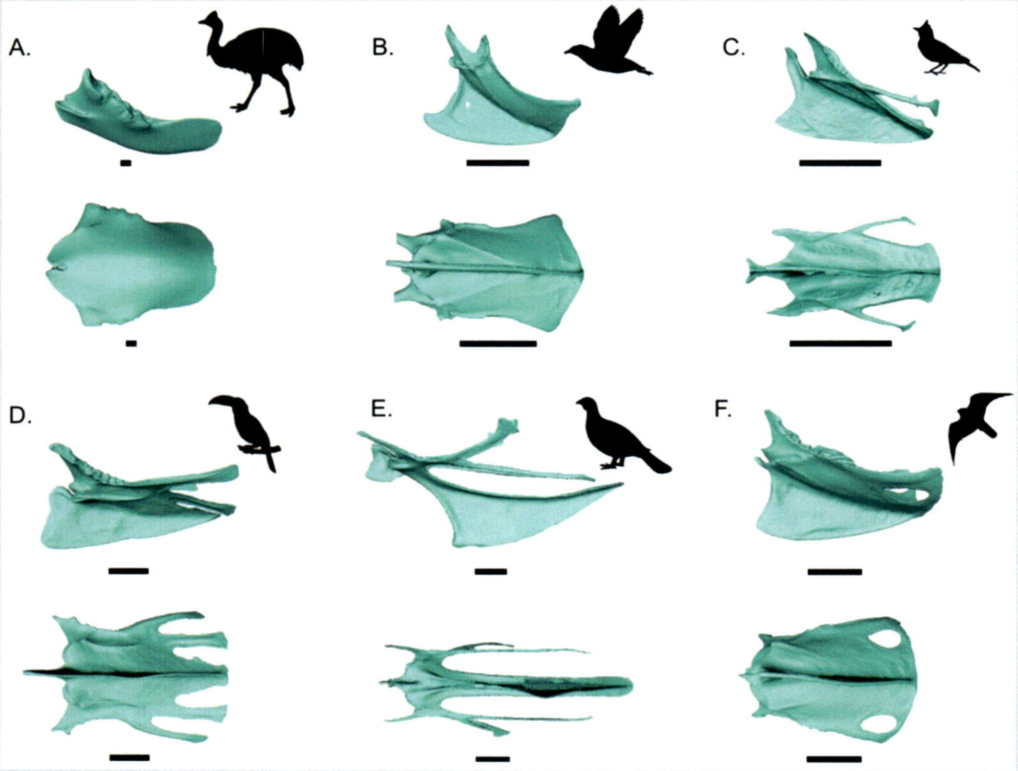


Examples of variation in avian sternums. Sternums are shown in both lateral (above) and ventral (below) views). (**a**) Southern Cassowary (*Casuarius casuarius*; note the absence of a keel in this flightless species), (**b**) Leach's Storm-Petrel (*Hydrobates leucorhoa*), (**c**) Red-capped Lark (*Calandrella cinerea*), (**d**) Yellow-throated Toucan (*Ramphastos ambiguus*), (**e**) Chukar (*Alectoris chukar*), and (**f**) American Kestrel (*Falco sparverius*). Scale bars = 10 mm. Silhouettes sourced from phylopic.org, credited to (in order of appearance): *Casuarius casuarius* uncredited; Procellariifomes by Juan Carlos Jerí; Aluadidae uncredited; Ramphastidae by FJDegrange; Phasianidae by Elisabeth Östman; *Falco* by Liftarn. (Figure from Lowi-Merri et al. 2021; © The Authors 2021, open-access article licensed under a Creative Commons Attribution 4.0 International License, https://creativecommons.org/licenses/by/4.0/)

Lowi-Merri et al. (2021) quantified variation in the sternums of 105 species of birds representing 62 families and found differences in sternum morphology among species with different methods of locomotion. Specifically, they found that "Deeper keels tend to be correlated with slower but stronger wingbeats, and robust caudal sternal borders provide attachment surfaces for muscles involved in ventilation, which may strengthen under higher-powered flapping, or flight requiring higher metabolic rates. Sternal size and shape are both most strongly influenced by body size and forelimb propulsion ability, with shape also being strongly influenced by sternal size. Certain shapes are notably associated with locomotory modes, specifically the short sternal body and trabeculae associated with soaring, the elongated sternal body and lateral trabeculae associated with burst flight, the expanded trabeculae associated with continuous flapping, and the narrower sternal body found in terrestrially and cursorially adapted

(continued)

**Box 2.10** (continued)

birds." More fundamentally, the authors found that prominent keels are essential for avian flight. The keels of flightless, wing-propelled diving birds like penguins are elongated and relatively narrow, with the increased length possibly providing more surface area for muscle attachment to compensate for the narrow keel that results in a more aerodynamic shape important in "flying" underwater (Lowi-Merri et al. 2021).

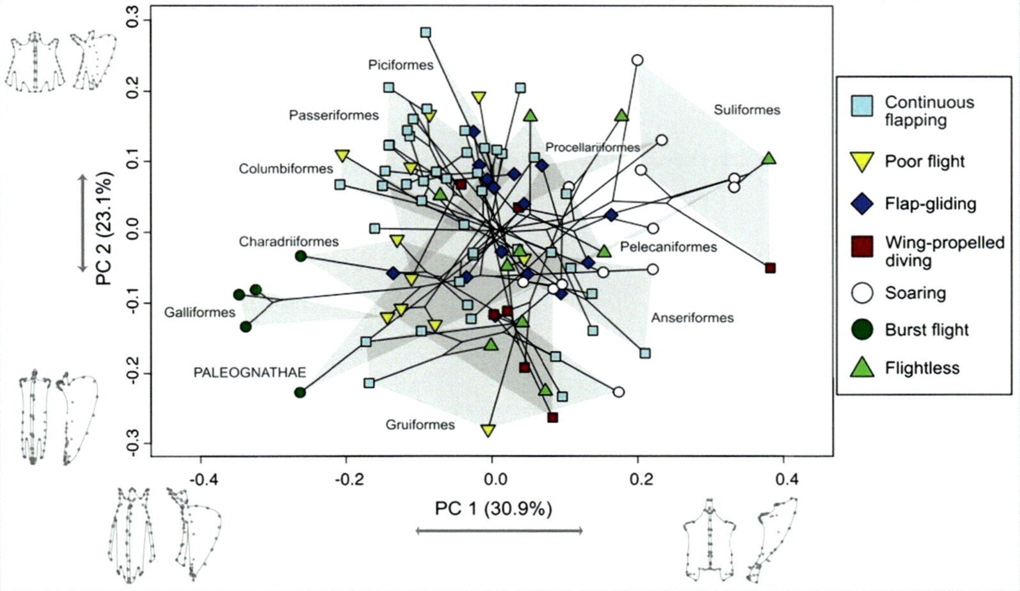


Morphospaces showing principal component axes PC1 and PC2 plotted against each other, with variation in sternum shape illustrated with the extremes of each principal component indicated along each axis in ventral view (left) and lateral view (right). PC1 describes variation in sternal elongation and PC2 also describes sternal elongation, but without the concave caudal edge apparent in PC1. Point colors and symbols indicate the primary locomotory mode of each species. Major taxonomic groups that separate out at the periphery of the morphospace are indicated by the gray polygons. Note that flightless birds and wing-propelled diving birds are placed at the extreme right, with narrow, elongated sternums and keels. (Figure from Lowi-Merri et al. 2021; © The Authors 2021, open-access article licensed under a Creative Commons Attribution 4.0 International License, https://creativecommons.org/licenses/by/4.0/)

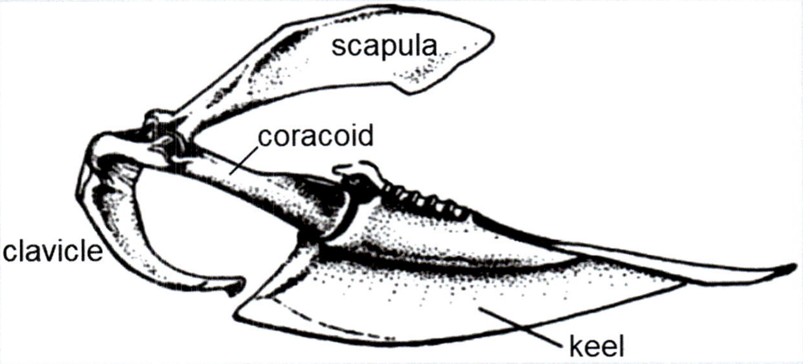


(continued)

**Box 2.10** (continued)

Pectoral girdle and sternum of a wing-propelled diving penguin. Note that the keel is rather narrow, but elongated, with a cranial projection that provides more surface area for muscle attachment. (Figure from Bannasch 1994; © 1993 Cambridge University Press, used with permission)

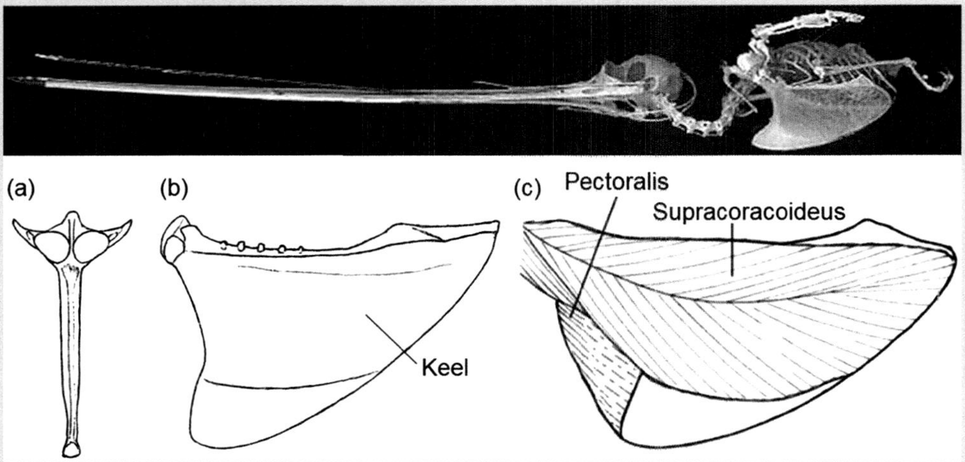


Top: Skeleton of a Sword-billed Hummingbird (*Ensifera ensifera*). Note, in addition to the impressively long bill, the sternum with a very wide keel for attachment of large flight muscles. Below: (**a**) Rostral and (**b**) lateral view of the sternum of a hummingbird, again showing the wide keel. (**c**) To sustain their hovering flight, hummingbirds have large flight muscles that originate on the keel, including the supracoracoideus and pectoralis (most of which is not shown because it would completely cover the underlying supracoracoideus muscle. (Image of the skeleton is from the Slater Museum of Natural History, used with permission; bottom figures are modified from Zusi 2013; © 2013 American Ornithological Society, used with permission)

serve as a counterbalance, birds have kept their center of gravity over their legs by adopting a "crouched" posture with their femurs more horizontal than those of their theropod ancestors. As such, a bird rotating their body forward so their head could reach the ground (rather than bending their neck) would move their center of mass forward and be unable to maintain their balance.

The first cervical vertebra, the atlas, articulates with the skull by a single occipital condyle, and with the second cervical vertebra, the axis (Fig. 2.90). The joint between the atlas and axis allows much more rotational movement than the joints between other cervical vertebrae (Fig. 2.91). However, rotational movement between cervical vertebrae, as well as between the skull and atlas, is sufficient to allow birds to reach around and preen their back feathers with their bills. In contrast to many other birds, joints between the cervical vertebrae of owls allow considerable rotational movement (Shufeldt 1900) and, as a result, owls can rotate their heads up to about 280 degrees (although rarely more than 180 degrees), compensating for their limited fields of binocular vision (Lynch 2007). The cervical vertebrae of birds are generally divided into three sections, with the articulations between the more cranial vertebrae allowing ventroflexion and high axial and lateral movement, the caudal cervical vertebrae allowing more lateral flexion, and those in the middle allowing more dorsoflexion and varying among species in the amount of axial

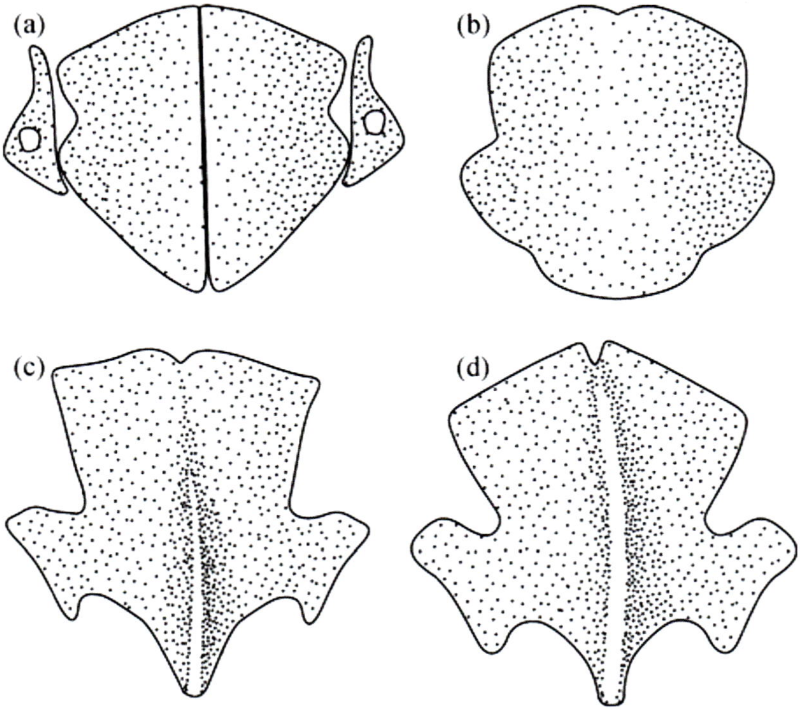


**Fig. 2.79** Variation in the morphology of the sternum and keel of some early birds. (**a**) Unfused sternal plates (*Jeholornis*, 120 mya), (**b**) fused sternum (*Eoconfuciusornis*; 131 mya), (**c**) fused sternum with small keel (*Confuciusornis*, 125–120 mya), (**d**) elongated and deeper keel (*Confuciusornis*, 125–120 mya). (Figure from Zhang et al. 2008; © 2008 Science in China Press and Springer-Verlag GmbH, used with permission)

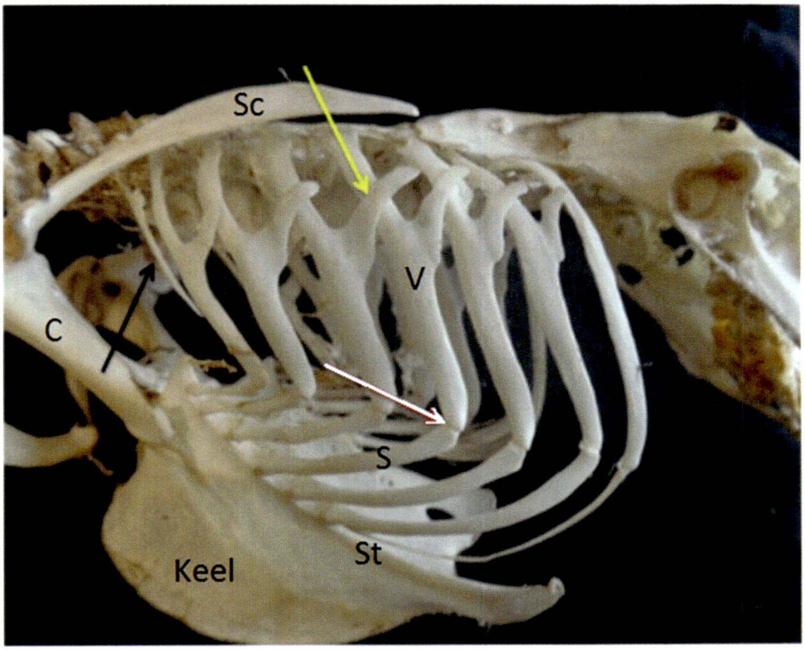


**Fig. 2.80** Ribs of a Black Kite (*Milvus migrans*) showing uncinate processes (yellow arrow), a floating rib (black arrow), and the acute angle between the vertebral and sternal ribs. V, vertebral rib; S, sternal rib; Sc, scapula; C, coracoid; St, sternum. (Figure from John et al. 2015; used with permission of Applied Biological Research)

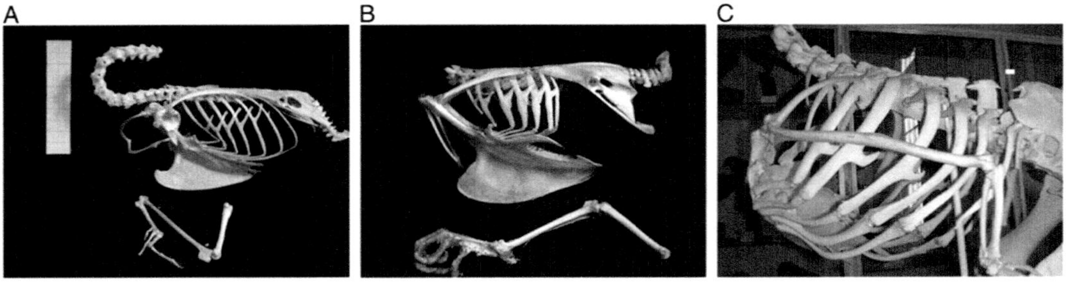


**Fig. 2.81** Rib cages of three species of birds showing differences in relative length of the uncinate processes in (**a**) a diving bird (Common Tern, *Sterna hirundo*) are long, (**b**) a nonspecialist bird (Rock Pigeon, *Columba livia*) are intermediate in length, and (**c**) a walking/running bird (Common Ostrich, *Struthio camelus*) are relatively short. Scale bar in A, 5 cm. (Figure from Codd 2010; © 2009 Elsevier Inc., used with permission)

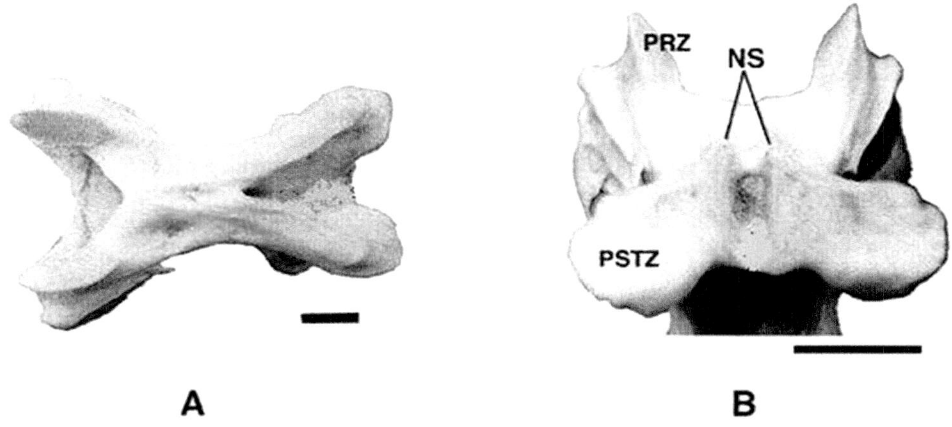


**Fig. 2.82** Cervical vertebra (14th) of a Greater Rhea (*Rhea americana*). (**a**) Top view. (**b**) Dorsal view. NS, neural spine; PRZ, prezygapophysis; PSTZ, postzygapophysis. Pre- and postzygapophyses are the points where the vertebra articulates with an adjacent vertebra. (Figure from Tsuihiji 2004; Rights managed by Taylor & Francis, used with permission)

rotation (Zweers et al. 1994; Kambic et al. 2017; Fig. 2.92).

Thoracic vertebrae are those that articulate with complete ribs that, in turn, articulate with the sternum. Most theropods, as well as *Archaeopteryx*, had 14 unfused thoracic vertebrae (also called dorsal vertebrae). Among extant birds, the number of thoracic vertebrae ranges from up to seven in Columbiformes, eight in Passeriformes, and 12 in Anseriformes (Samejima and Otsuka 1984), but only 4–6 are not part of the synsacrum. This shortening of the vertebral column in the trunk region (along with the reduction in the number of caudal vertebrae discussed below) helps shift the center of mass toward the wings. In contrast to the cervical vertebrae of birds, motion between the thoracic vertebrae is very limited because of long spinous processes or, in some cases, impossible because of fusion. Thoracic vertebrae that are fused to each other, but

**Box 2.11 Neck Length and Body Mass: Birds vs. Mammals**

Among birds, neck length (i.e., length of all cervical vertebrae) scales isometrically with body mass (Marek et al. 2021), but, among mammals, neck length scales with negative allometry against body mass (Arnold et al. 2017). For mammals, head mass is what primarily limits neck length because "head mass scales at a faster rate than the cross-sectional area of the neck which must resist the stress of the weight of the head" (Marek et al. 2021). Among birds, head mass has been reduced via pneumatization of the skull and because evolution of the gizzard has allowed a reduction in jaw musculature. In addition, the "S-shaped" curvature of the avian neck means that head mass is located closer to the body's center of mass, which lessens constraints on head mass compared to mammals and other vertebrates (Marek et al. 2021).

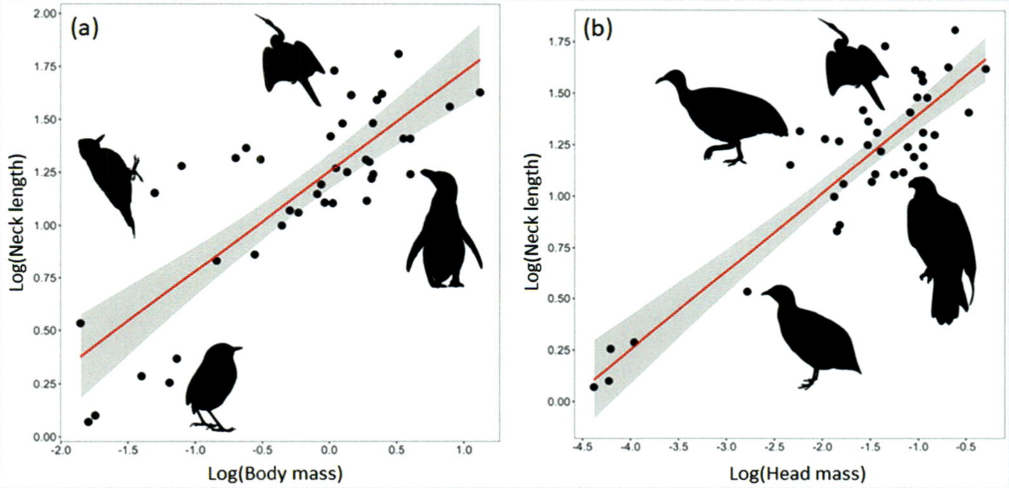


Neck length of birds scales with (**a**) body mass and (**b**) head mass according to isometry. Inset silhouettes (clockwise beginning at the top): (**a**) Anhinga (*Anhinga anhinga*), Humboldt Penguin (*Spheniscus humboldti*), Bush Wren (*Xenicus longipes*), and Yellow-bellied Sapsucker (*Sphyrapicus varius*), and (**b**) Anhinga, Little Eagle (*Haliaeetus morphnoides*), Tataupa Tinamou (*Crypturellus tataupa*), and Black Guineafowl (*Agelastes niger*). (Figure from Marek et al. 2021; © 2021 The Authors. Published by the Royal Society under the terms of the Creative Commons Attribution License, https://creativecommons.org/licenses/by/4.0/)

(continued)

**Box 2.11** (continued)

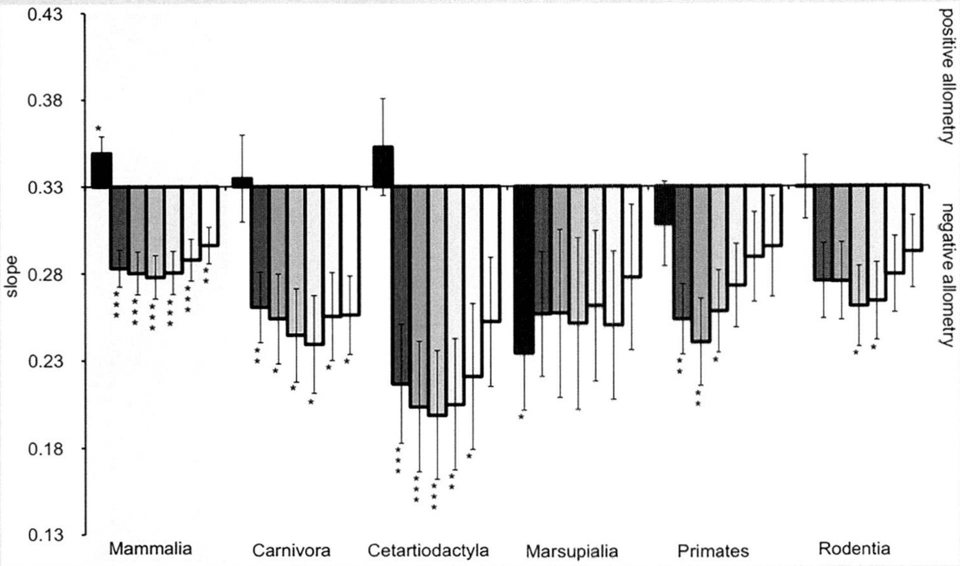


Comparison of the slopes for the regressions of log vertebral length against log body mass for cervical vertebra 1 (dark gray) to cervical vertebra 7 (white) in mammals in general and in the major mammalian clades. Error bars represent the standard error of the slopes. Asterisks indicate the significance of slope tests for allometry (i.e., significantly different from isometry, significance levels: $*P < 0.05$; $**P < 0.01$; $***P < 0.001$). (Figure from Arnold et al. 2017; © 2017 The Authors. Evolution © 2017 The Society for the Study of Evolution, used with permission)

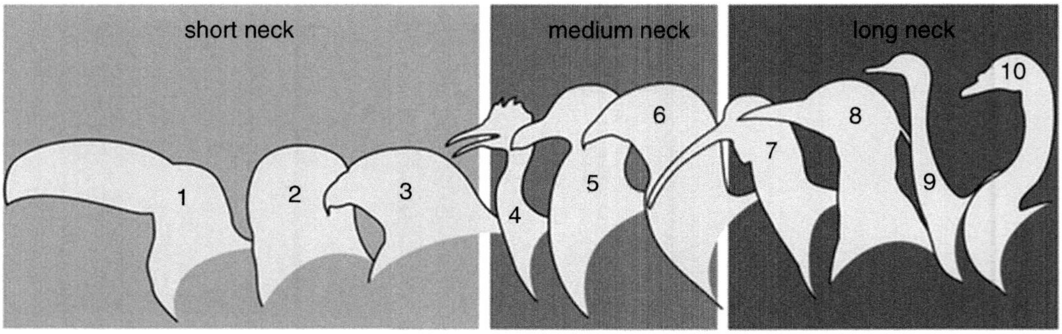


**Fig. 2.83** Variation in neck length among different species of birds. Much of the variation is due to differences in the extent of neck-folding rather than differences in number of cervical vertebrae. 1—toucan, 2—owl, 3—eagle, 4—tinamous, 5—gull, 6—seriema, 7—ibis, 8—egret, 9—rhea, and 10—swan. (Figure from Tambussi et al. 2012; © 2012 Tambussi et al., open-access article distributed under the terms of the Creative Commons Attribution License, https://creativecommons.org/licenses/by/4.0/)

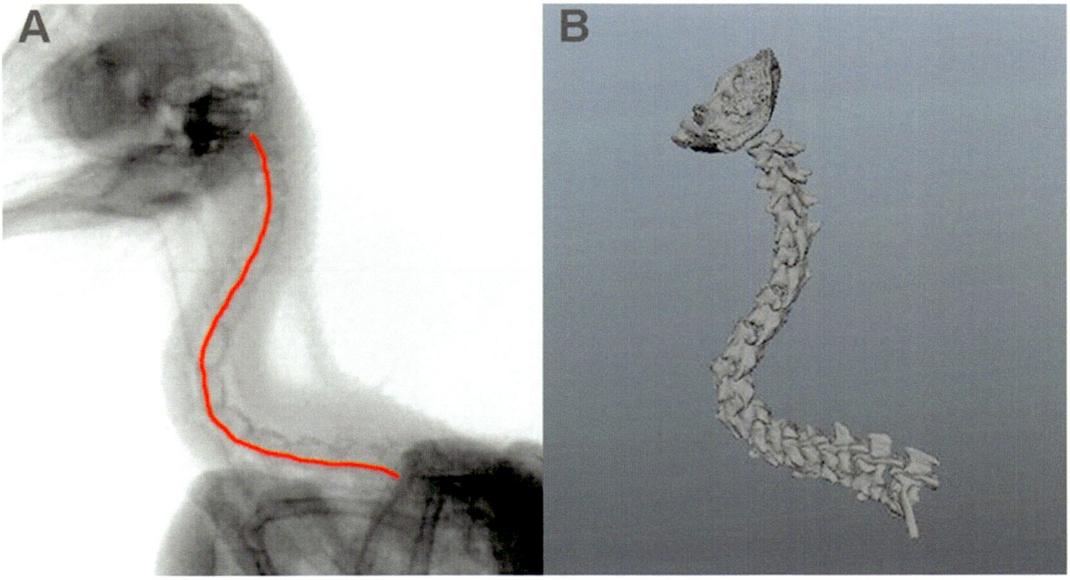


**Fig. 2.84** (**a**) X-ray showing the cervical vertebrae of a Barn Owl (*Tyto alba*) and how they approximate an S-shape. (**b**) More detailed view of the cervical vertebrae of a Barn Owl. The caudal portion is almost horizontal whereas the central section is almost vertical. (Figure modified from Krings et al. 2014; © 2014 Krings et al., open-access article distributed under the terms of the Creative Commons Attribution License, https://creativecommons.org/licenses/by/4.0/)

not to the synsacrum form a structure called the notarium (Fig. 2.93). Notaria occur in several groups of birds, including species in the orders Galliformes, Columbiformes, Gruiformes, Podicipediformes, Falconiformes, and Passeriformes (James 2009), and have evolved independently at least 17 times, including at least 12 times in oscine songbirds (James 2009). Muscles that attach to thoracic vertebrae play important roles in breathing, flight, hindlimb locomotion, and movements of the neck, so notaria provide a stable point of attachment (James 2009). Movement between thoracic vertebrae may be limited in species without notaria via expanded spinous processes that are in contact with each other or a series of overlapping tendons along the dorsal surface of the vertebra (James 2009).

More recently, Aires et al. (2022) analyzed the notaria of 270 specimens (representing 80% of avian orders), noting that the structure originated independently several times and predominantly in ground-dwelling taxa of birds. These authors noted that ground-dwelling birds, like those in the orders Galliformes and Tinamiformes, that take off from the ground require "specialized mechanics," i.e., using energy and acceleration from the hindlimbs to take flight. As such, Aires et al. (2022) hypothesized that the notaria of ground-dwelling birds, positioned near the center of mass, provide a rigid, stable structure that better allows the musculature of the legs to use "the ground as a catapult bringing energy to the wings."

In extant birds, 10–23 vertebrae, including some thoracic vertebrae, all lumbar and sacral vertebrae, and some caudal vertebrae, are fused to form the synsacrum (Fig. 2.94). The theropod ancestors of birds also had synsacrums. However, all known nonavian theropod dinosaurs had fewer than eight vertebrae in the synsacrum (Chiappe 1996), but all birds, with the exception of the primitive birds *Archaeopteryx* (five sacral vertebrae) and *Confuciusornis* (seven sacral

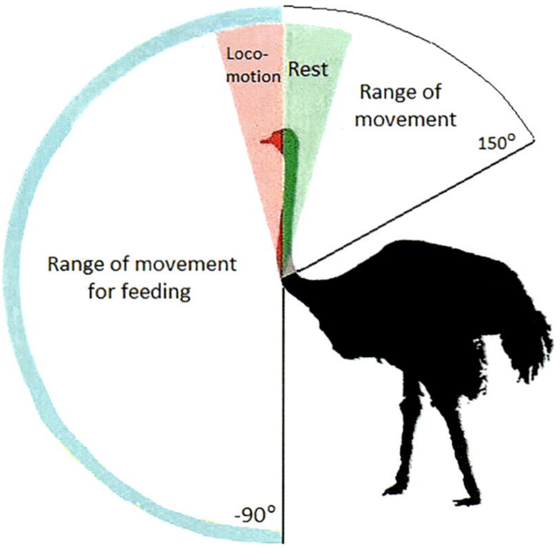


**Fig. 2.85** With 18 cervical vertebrae and the dorsoventral movement allowed by heterocoelous vertebrae, a Common Ostrich (*Struthio camelus*) neck is capable of movement across an arc spanning 240 degrees. (Figure from Dzemski and Christian 2007; © 2007 Wiley-Liss, Inc., used with permission)

vertebrae) (Chiappe et al. 1999), have more than eight. The fusion of vertebrae to form the synsacrum coincided with the loss of intrinsic vertebral muscles (Organ 2006) with a corresponding reduction in body mass. The synsacrum is fused with the pelvic girdle and, together, these structures provide a strong, stable base for attachment of thigh and tail muscles (Proctor and Lynch 1993). In addition, muscles attached to the posterior end pull the synsacrum toward the femur and this helps hold the anterior end of birds (where most of the body mass is located) up (Feduccia 1996).

### 2.5.6 Vertebral Column—Tail

The theropod ancestors of birds and the earliest birds had long bony tails consisting of numerous caudal vertebrae (Fig. 2.95). The tail of *Archaeopteryx* consisted of about 23 caudal vertebrae, with a pair of tail feathers, or rectrices, attached to each of the distal 15–20 vertebrae (Gatesy and Dial 1996a, b). The rectrices angled away from the vertebrae to form a leaf- or frond-shaped tail (Fig. 2.96). In contrast, extant birds have short bony tails consisting of just a few caudal vertebrae plus a pygostyle formed by the fusion of distal caudal vertebrae. Four processes contributed to the shortening of the avian tail, including a reduction in the number of caudal vertebrae, shortening of the centra (or bodies) of caudal vertebrae, fusion of proximal caudal vertebra with other vertebrae to form the synsacrum, and fusion of distal caudal vertebrae to form the pygostyle (Gatesy and Dial 1996a, b; Gao et al. 2008).

Advantages of this reduction in the length of the bony tail included some weight reduction (due to loss of some vertebrae and shortening of others) and a transfer of mass closer to the center of gravity. However, as with the loss of teeth, factors other than weight reduction may have been more important in favoring a reduction in the length of bird tails. One hypothesis is that the first birds, like *Archaeopteryx*, were less efficient flyers, with wings that were less able to generate lift and to control and stabilize flight. For those

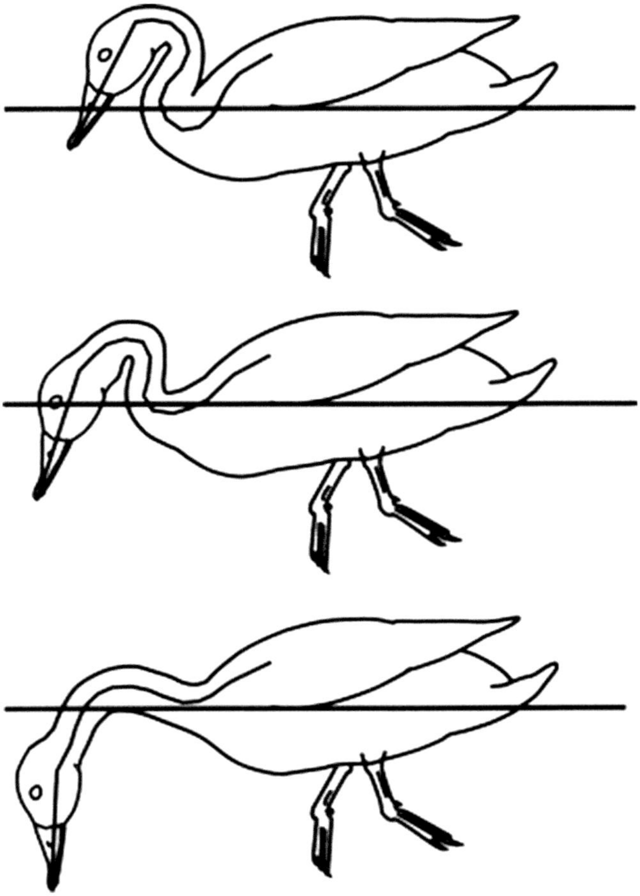

**Fig. 2.86** Neck movements of a Mallard (*Anas platyrhynchos*) when foraging underwater. (Figure from van der Leeuw et al. 2001; © 2015 Oxford University Press, used with permission)

birds, a long frond-shaped tail (1) added a caudal surface that provided at least some lift and may have enhanced stability in flight by shifting the center of lift further back (Bleiweiss 2009) and, (2) via its elevation or depression, aided in pitch control (Caple et al. 1983). As bird wings became better able to control and stabilize flight, the need for a long tail diminished and selection likely favored a reduction in tail length and transformation of the tail from frond- to fanned-shape. Critical in that change in shape was the formation of the pygostyle by fusion of the most distal caudal vertebrae. The pygostyle provides a firm foundation for the rectrices, and associated muscles allow control of tail shape and position (Box 2.12 Pygostyle Morphology).

Among extant birds, the relative importance of tails varies. For many species, such as hawks, swallows, and frigatebirds, tails provide lift and enhance maneuverability. However, other birds have relatively short tails and are either less maneuverable in flight (e.g., waterfowl) or rely more on wing control (e.g., swifts) (Gatesy and Dial 1996a, b). Among some species where tails became less critical aerodynamically, sexual selection favored alteration of tail morphology, e.g., the long tail-streamers of Black-billed (*Trochilus scitulus*) and Red-billed streamertails (*Trochilus polytmus*; Fig. 2.97). Although such alterations in tail morphology might have some effect on flight performance and maneuverability, analysis suggests that, at least for some species,

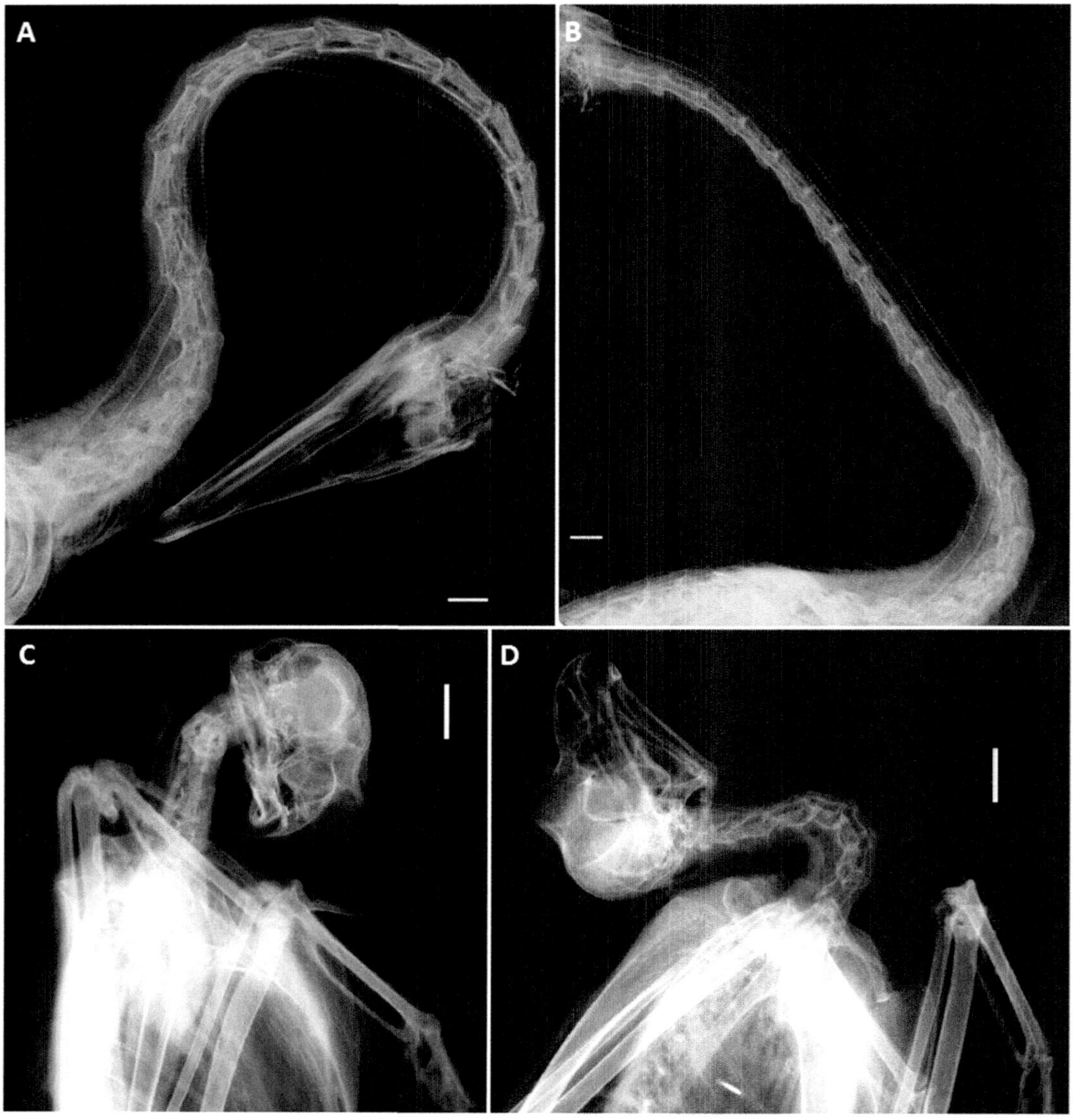


**Fig. 2.87** Cervical ventro- and dorsoflexion of a Trumpeter Swan (*Cygnus buccinator*) (**a**, **b**) and a Snowy Owl (*Bubo scandiacus*) (**c**, **d**) illustrating a difference in the extent to which such flexion is possible among difference taxa of birds with different numbers of cervical vertebrae. Scale bars = 2 cm. (Figures from Samman 2006; used with permission of Tanya Samman. Bird specimens were provided by the University of Calgary Department of Biological Sciences, Calgary, AB, Canada; radiography was performed at the Southern Alberta Institute of Technology non-destructive testing facility (SAIT NDT), Calgary, AB, Canada, and the radiographs were digitized by Acuren Inc., Calgary, AB, Canada)

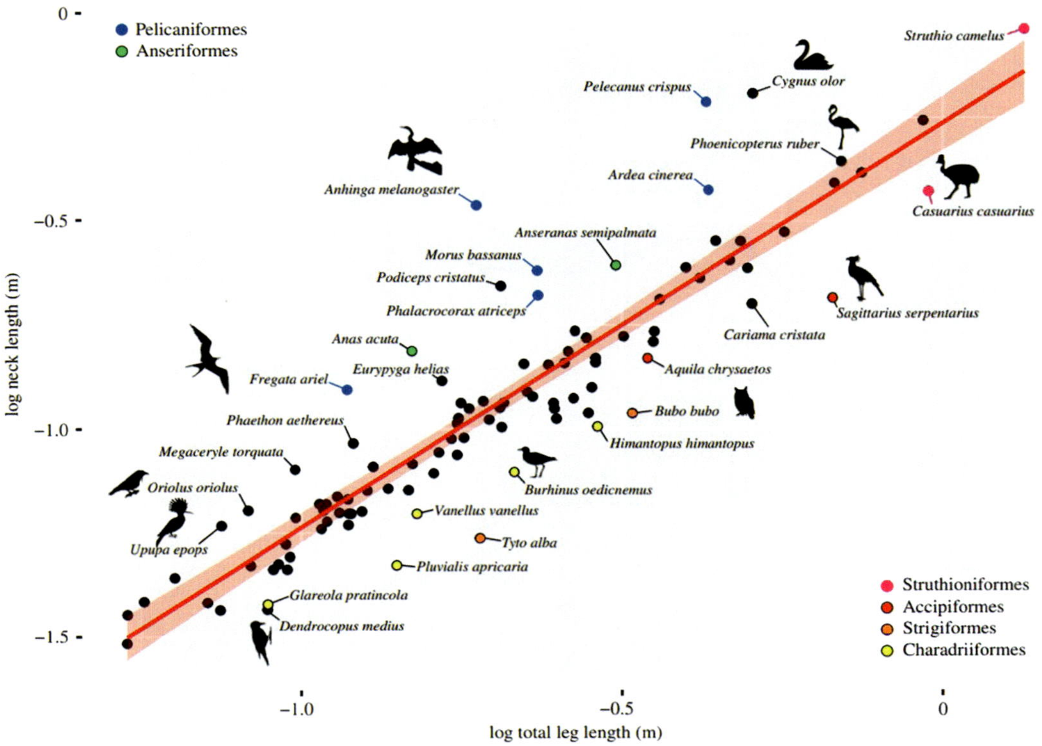


**Fig. 2.88** Relationship between neck length and total leg length for 103 species of birds. Species that have relatively longer necks lie above the linear regression line (red line with 95% confidence interval), whereas those with relatively shorter necks are below it. Note that many of the species with relatively longer necks are aquatic, whereas many with relatively shorter necks are terrestrial. From bottom left: *Dendrocopos medius*, Middle Spotted Woodpecker; *Glareola pratincola*, Collared Pratincole; *Pluvialis apricaria*, European Golden-Plover; *Tyto alba*, Barn Owl; *Upupa epops*, Eurasian Hoopoe; *Vanellus vanellus*, Northern Lapwing; *Oriolus oriolus*, Eurasian Golden Oriole; *Megaceryle torquata*, Ringed Kingfisher; *Burhinus oedicnemus*, Eurasian Thick-knee; *Phaethon aethereus*, Red-billed Tropicbird; *Himantopus himantopus*, Black-winged Stilt; *Bubo bubo*, Eurasian Eagle Owl; *Fregata ariel*, Lesser Frigatebird; *Eurypyga helias*, Sunbittern; *Anas acuta*, Northern Pintail; *Aquila chrysaetos*, Golden Eagle; *Cariama cristata*, Red-legged Seriema; *Phalacrocorax* atriceps, Imperial Cormorant; *Podiceps cristatus*, Great Crested Grebe; *Morus bassanus*, Northern Gannet; *Sagittarius serpentarius*, Secretarybird; *Anseranas semipalmata*, Magpie Goose; *Anhinga melanogaster*, Oriental Darter; *Ardea cinerea*, Gray Heron; *Casuarius casuarius*, Southern Cassowary; *Phoenocopterus ruber*, American Flamingo; *Pelecanus crispus*, Dalmatian Pelecan; *Cygnus olor*, Mute Swan; *Struthio camelus*, Common Ostrich. (Figure from Böhmer et al. 2019; © 2019 The Authors. Published by the Royal Society under the terms of the Creative Commons Attribution License, https://creativecommons.org/licenses/by/4.0/)

elongated tail feathers have minimal effect on flight performance (Clark 2010, 2011).

## 2.6 Avian Skeletal Muscles

The skeletal muscles of birds provide the force needed for movement, including locomotion. By way of contractile proteins, muscles respond to nervous impulses and convert chemical energy into mechanical force (Weeks 1989; Box 2.13 Skeletal Muscle Anatomy and Function). In addition to locomotion, or movement of birds through the air, water, and on land, skeletal muscles also move the avian head, bill, and neck—movements important in obtaining food and, for some

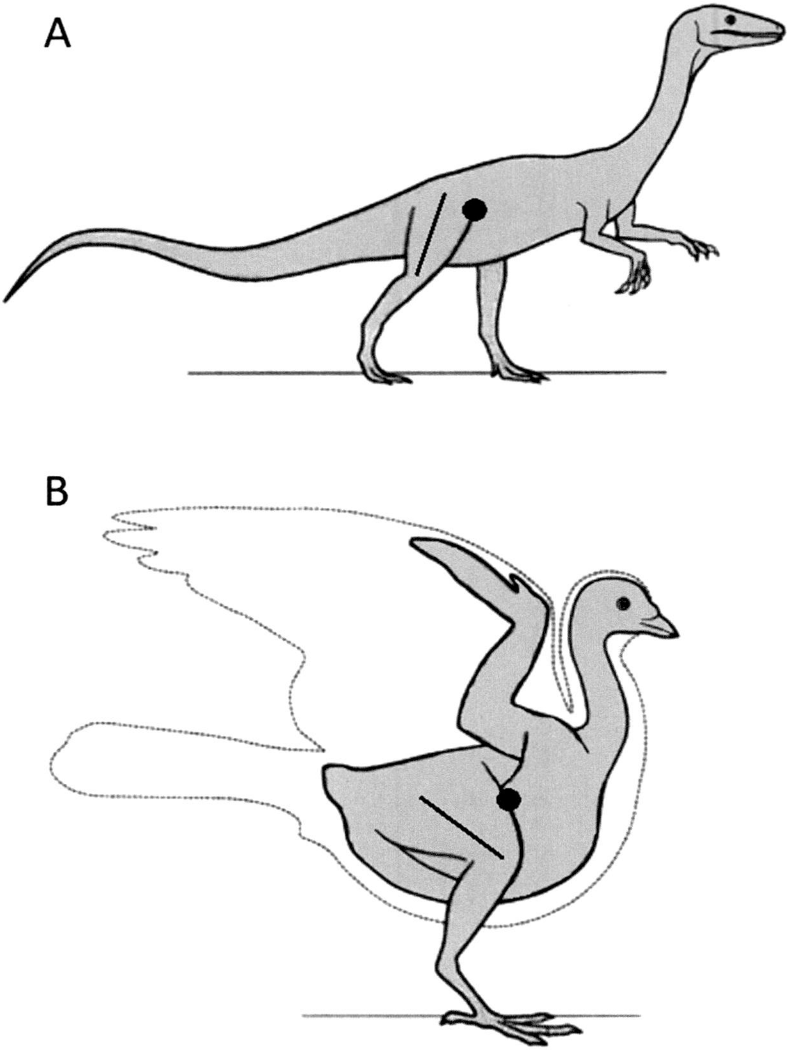


**Fig. 2.89** A theropod (**a**) and a bird (**b**) with the black circles indicating their approximate center of mass. The theropod could rotate its body forward to its head and could reach the ground, with its long muscular tail serving as a counterbalance to keep its center of mass over its legs. However, birds, with the loss of the muscular tail, must maintain a "crouched" posture with a less vertical femur (dark line) so, if they rotate their body forward, their center of gravity also moves forward and would no longer be over their legs. As a result, they would tend to fall forward. (Figure modified from Gatesy and Dial 1996a; © 1996 The Society for the Study of Evolution, used with permission)

species, mating and other displays (Box 2.14 Superfast Muscles of Some Manakins). Skeletal muscles are also important for vocalizing (e.g., syringeal muscles), perching (movement of the legs and toes), maintaining posture, and thermoregulation (e.g., shivering).

An important adaptation for flight is the concentration of body mass, much of which is skeletal muscle, between the wings and near the center of gravity. In most birds, this concentration is apparent, with the largest muscles, the primary flight muscles (pectoralis and supracoracoideus), located on the sternum. The muscle that moves the avian wing downward, the pectoralis, is located on the sternum, as it is in all vertebrates. However, in contrast to other vertebrates where the muscles that elevate the forearm are positioned on the back (e.g., latissimus dorsi and deltoideus), the primary upstroke muscle of birds, the supracoracoideus, is also located on the sternum, just below the pectoralis. In primitive birds, such as *Archaeopteryx*, other muscles, primarily the deltoideus muscle, powered the upstroke (Ruben 1991; Fig. 2.98), and the primary function of the supracoracoideus was probably to protract the humerus (i.e., move the wing forward, not upward). The conversion of the supracoracoideus from a wing protractor to a

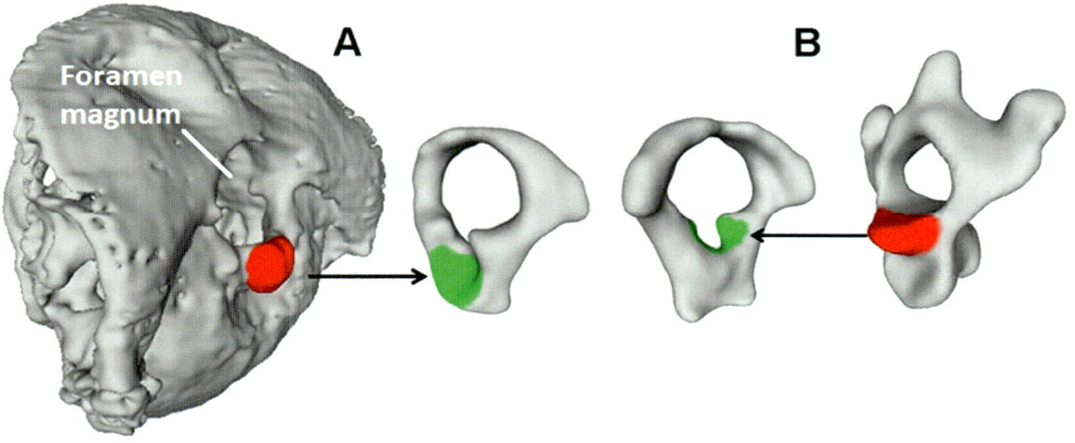


**Fig. 2.90** Base of skull and first two cervical vertebrate of a Barn Owl (*Tyto alba*). The atlas forms a ball-and-socket joint with the occiput of the skull and a pivot joint with the axis. (**a**) The ball-and-socket joint between occiput of the skull (left) and atlas (right). The ball is marked red, and the socket is marked green. The center of rotation in this joint is the center of the ball. (**b**) The pivot joint between the atlas (left) and axis (right). The pivot is marked red, and the ring-like socket is green. (Figure from Krings et al. 2014; © 2014 Krings et al., open-access article distributed under the terms of the Creative Commons Attribution License, https://creativecommons.org/licenses/by/4.0/)

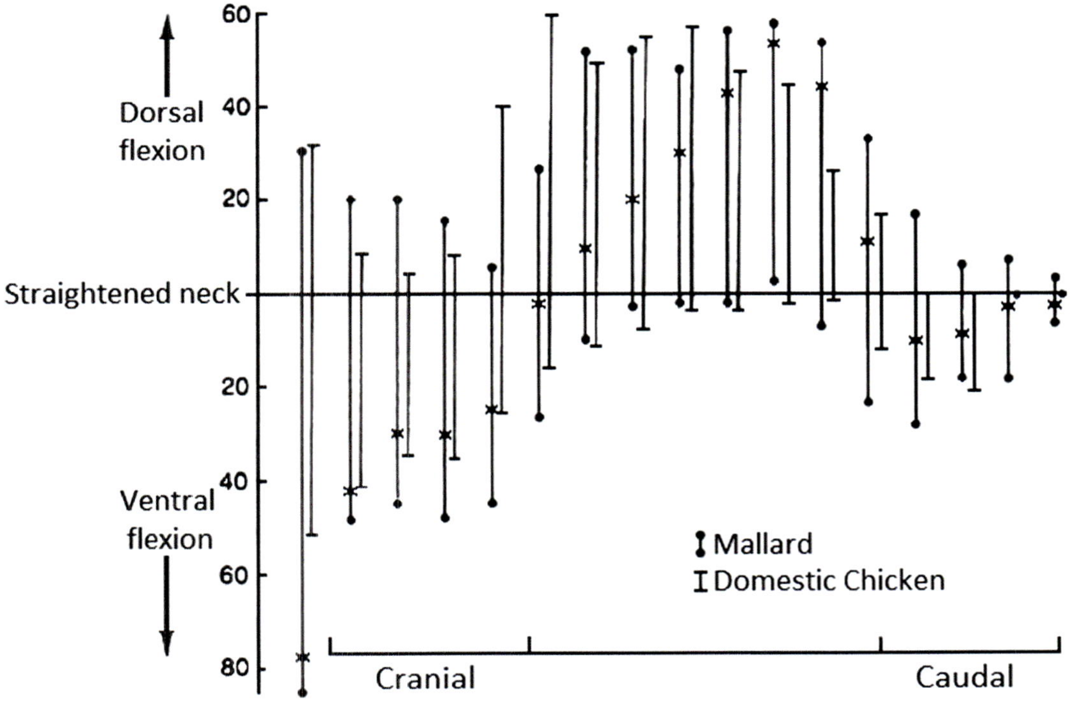


**Fig. 2.91** Dorso- and ventroflexion of the cervical vertebrae of a Mallard (*Anas platyrhynchos*) and domestic chicken. Note that the more cranial vertebrae allow more ventroflexion and the centrally located vertebrae allow more dorsoflexion. The position of the vertebrae of a Mallard when in a resting position is indicated by the crosses. (Figure modified from Zweers et al. 1994; © 1994 Springer-Verlag Berlin, used with permission)

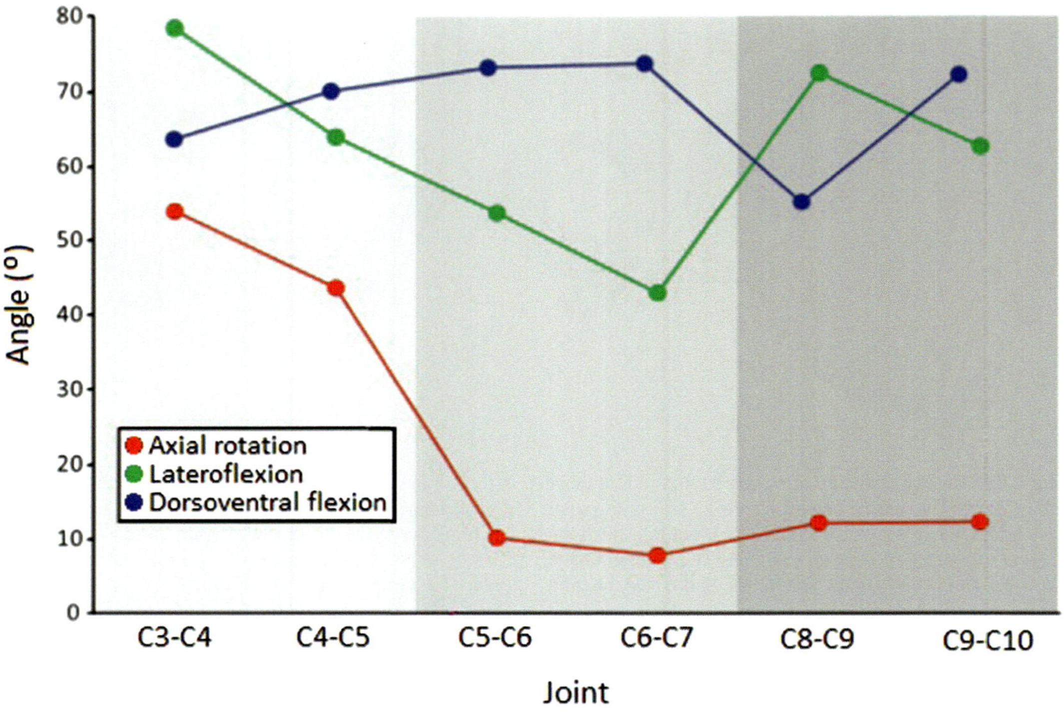


**Fig. 2.92** Degree of axial rotation, lateroflexion, and dorsoventral flexion for the cervical vertebrae of Wild Turkeys (*Meleagris gallopavo*). The different sections are indicated by different shades of gray (i.e., cranial, middle, and caudal). (Figure modified from Kambic et al. 2017; open-access article distributed under the terms of the Creative Commons Attribution 4.0 International License, https://creativecommons.org/licenses/by/4.0/)

wing elevator involved elongation of the coracoid and development of a process called the acrocoracoid (Figs. 2.99 and 2.100). These changes created a foramen (foramen triosseum or triosseal canal) above the glenoid cavity (where the head of the humerus articulates with the pectoral girdle) through which the supracoracoideus tendon passed and inserted on the dorsal surface of the humerus (Fig. 2.98), allowing the supracoracoideus to elevate the wing. This change in primary function of the supracoracoideus may have occurred within a few million years after *Archaeopteryx*. Although a definitive, articulate foramen triosseum has not been reported in any enantiornithine bird (Mesozoic birds that occurred after *Archaeopteryx*), elongated coracoids as well as connections between the scapula and coracoid and the position of the scapula in some specimens (e.g., *Eoalulavis*) strongly suggest that the foramen triosseum was present (Poore et al. 1997; Chiappe and Walker 2002).

### 2.6.1 Flexibility in Muscle Mass

For some species of birds, muscle mass, particularly the mass of the pectoralis muscle, can vary. For example, pectoralis muscle mass has been found to increase during the winter in several species of birds at higher latitudes (Swanson and Merkord 2013; Petit et al. 2014; Vézina et al. 2021). This increase is likely important in thermoregulation, with a larger pectoralis aiding in increased heat production via shivering (Swanson and Liknes 2006). Among other species, pectoralis muscle mass has been found to vary during migration, allowing birds to match the

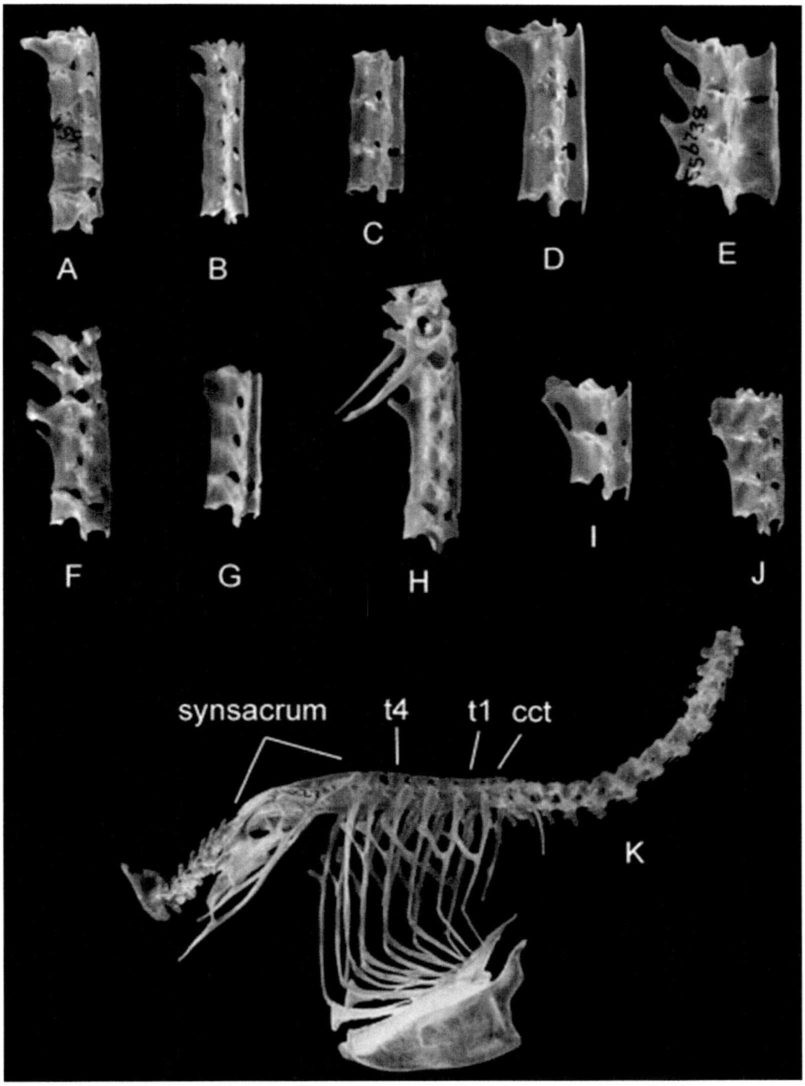


**Fig. 2.93** Examples of notaria in several species of songbirds, including (**a**) Ashy Woodswallow (*Artamus fuscus*), (**b**) Chabert's Vanga (*Leptopterus chabert*), (**c**) Phainopepla (*Phainopepla nitens*), (**d**) European Starling (*Sturnus vulgaris*), (**e**) California Thrasher (*Toxostoma redivivum*), (**f**) Cape Penduline Tit (*Anthoscopus minutus*), (**g**) Eurasian Penduline Tit (*Remiz pendulinus*), (**h**) Verdin (*Auriparus flaviceps*), (**i**) Red Crossbill (*Loxia curvirostra*), and (**j**) White-winged Crossbill (*Loxia leucoptera*). (**k**) Vertebral column of a Black-capped Chickadee (*Poecile atricapillus*), a species without a notarium. Labeled are the last cervical (cct) and the first through fourth thoracic vertebrae (t1–t4) and the synsacrum. (Figure from James 2009; © 2009 Oxford University Press, used with permission)

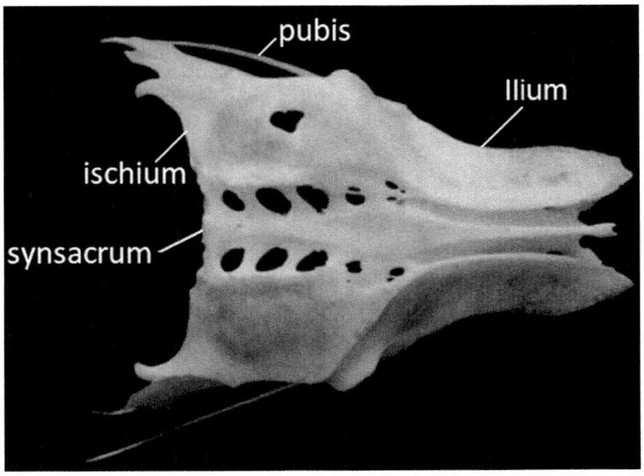


**Fig. 2.94** The avian pelvic girdle consists of three fused bones, ilium, ischium, and pubis, that, in turn, are fused with the synsacrum. (Figure modified from Stoessel et al. 2013; © 2013 Wiley Periodicals, Inc., used with permission)

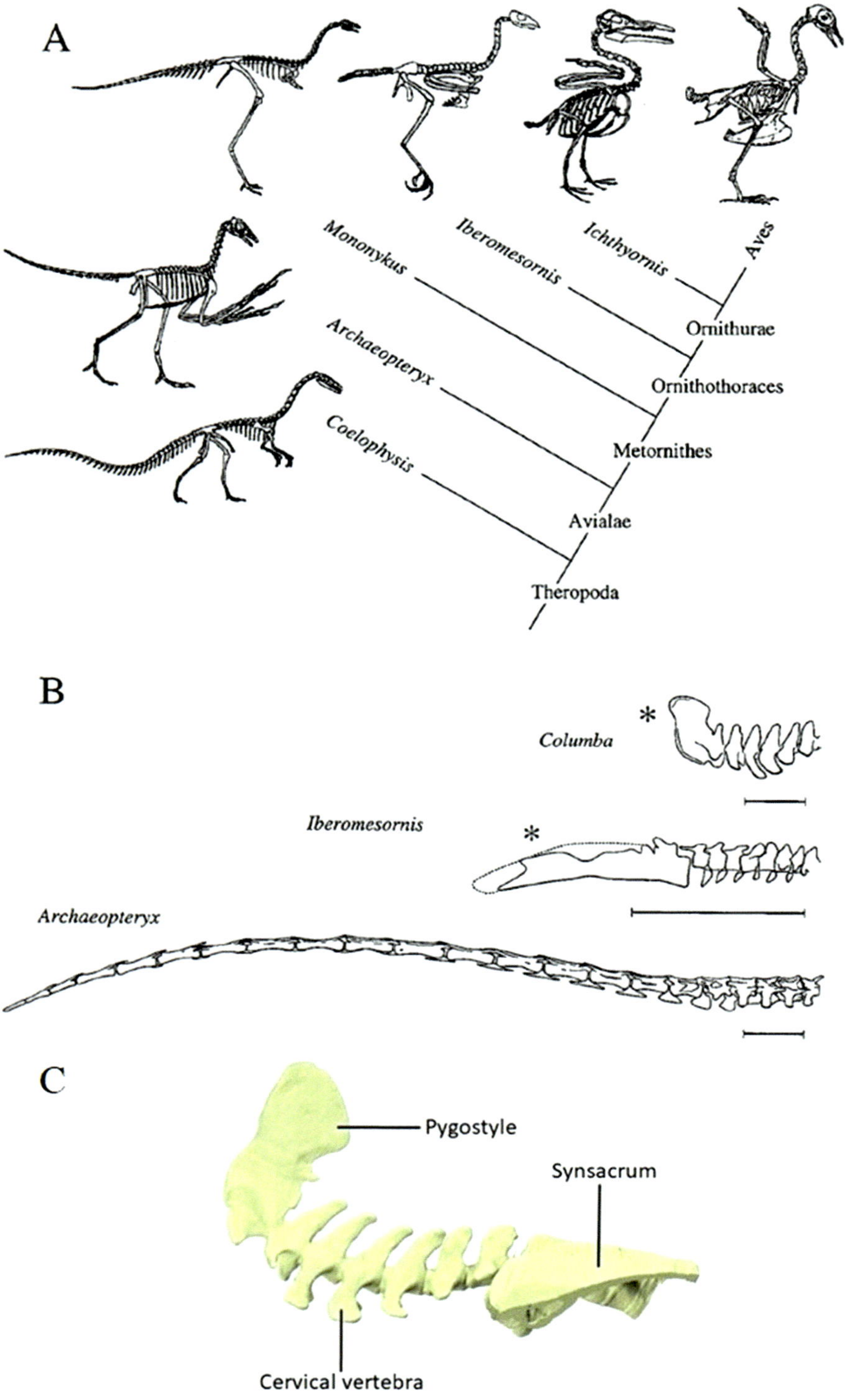


**Fig. 2.95** (**a**, **b**) Tail evolution among theropods and birds. *Iberomesornis* (an enantiornithine bird from the Cretaceous) and *Columba* (pigeons and doves) possess pygostyles. * Indicates a pygostyle. Scale bars = 2 cm. (**c**) Typical caudal skeleton of a bird. (Figures A and B from Gatesy and Dial 1996b; © 1996 The Society for the Study of Evolution, used with permission; Figure C from Felice 2014; © 2014 Wiley Periodicals, Inc., used with permission)

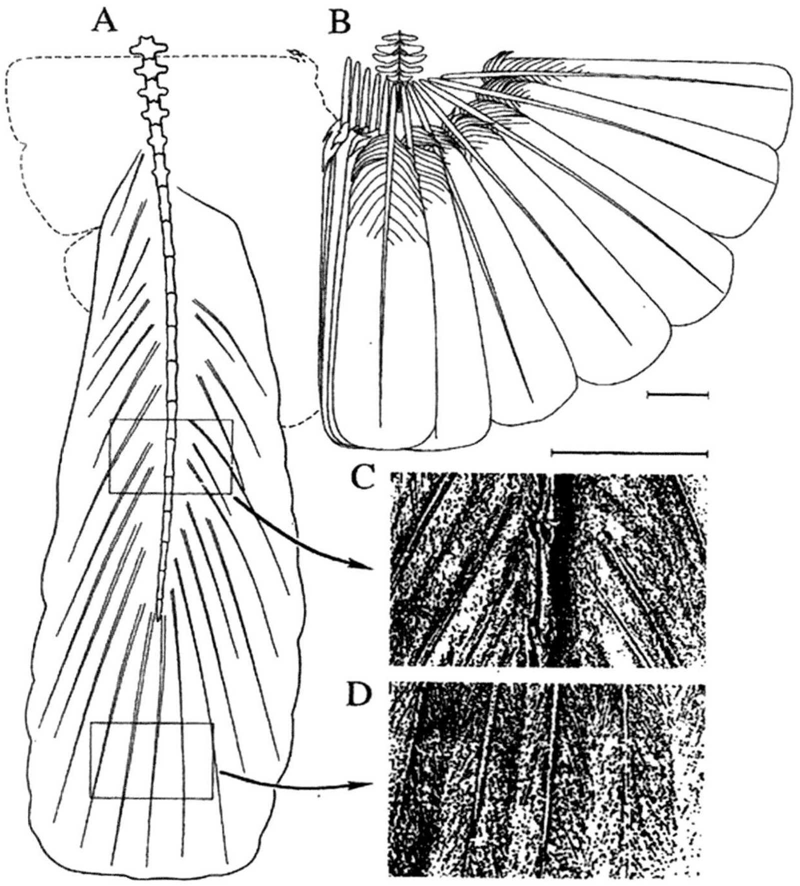


**Fig. 2.96** Tails of (**a**) *Archaeopteryx* and (**b**) a pigeon, with impressions of different areas of the tail (**c** and **d**) from an *Archaeopteryx* fossil. *Archaeopteryx* had a long leaf- or frond-shaped tail consisting of numerous caudal vertebrae with a pair of rectrices attached to the most distal 15–20 vertebrae. Present-day birds like pigeons have tails consisting of just a few caudal vertebrae and a pygostyle with attached rectrices that can be spread like a fan. Scale bars equal 2 cm. (Figure from Gatesy and Dial 1996a, b; © 1996 The Society for the Study of Evolution, used with permission)

power requirements of flight as body mass changes prior to and during migration (Marsh 1984; Lindström et al. 2000; Price et al. 2011). Larger pectoralis muscles of migrating birds can also serve as a source of energy, particularly during long flights when lipid stores have been depleted (Schwilch et al. 2002), and as a source of metabolic water (Gerson and Guglielmo 2011). Proteins in larger than usual pectoralis muscles can also serve as a source of nutrients for egg production. Bolton et al. (1993), e.g., found that clutch sizes of female Lesser Black-backed Gulls (*Larus fuscus*) were positively correlated with levels of flight muscle protein at the start of egg

**Box 2.12 Pygostyle Morphology**

The caudal skeleton of birds consists of several "free" caudal vertebrae (range = 5–9) plus a pygostyle formed by the fusion of the last several caudal vertebrae (range = 3–7) (Baumel and Witmer 1993; Felice 2014). Pygostyles serve as the point of attachment for the caudal muscles and rectrices for all birds, but their shapes vary among species.

(continued)

**Box 2.12** (continued)

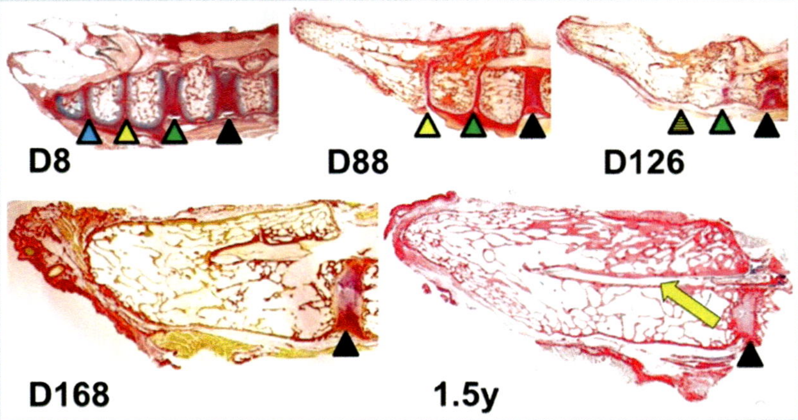


Formation of the pygostyle in a Domestic Chicken from day 8 (D8) posthatching to 1.5 years (1.5y) posthatching. The process takes about five months and occurs from a distal to proximal direction (left to right in the above images). At day 8, no fusion of the four caudal vertebrae has occurred and the intervertebral discs are indicated by the color-coded triangles. By day 168, all four ossified caudal vertebrae have fused and the intervertebral disc between the pygostyle and synsacrum is indicated by the black triangle. The yellow arrow at 1.5y indicates the spinal cord channel. (Figure from Rashid et al. 2018; open-access article licensed under a Creative Commons Attribution 4.0 International License, https://creativecommons.org/licenses/by/4.0/)

Among some species in the order Piciformes, pygostyles have an expanded ventral surface that increases the surface area for attachment of tail muscles and the rectrices. Woodpeckers used their stiff tail feathers as a prop when climbing trees as they forage, and large caudal muscles that originate on the expanded pygostyles provide the needed force to support their bodies.

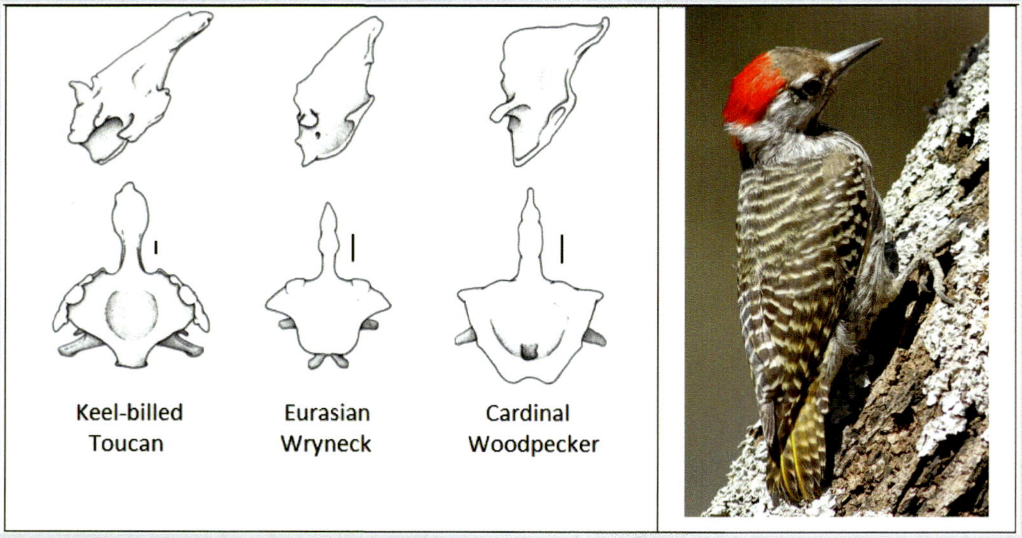


(continued)

**Box 2.12** (continued)

Left: Pygostyles of three species in the order Piciformes. Top: Side view. Bottom: View from above. Keel-billed Toucan, *Ramphastos sulfuratus*; Eurasian Wryneck, *Jynx torquilla*; Cardinal Woodpecker, *Dendropicos fuscescens*. Scale bars = 5 mm. Right: A Cardinal Woodpecker using its tail as a prop against the trunk of a tree. (Pygostyle figure modified from Manegold and Töpfer 2013; © 2012 Blackwell Verlag GmbH, used with permission; Photo of Cardinal Woodpecker by Derek Keats from Johannesburg, South Africa, Wikipedia, CC BY 2.0, https://creativecommons.org/licenses/by/2.0/)

Felice (2014) examined the pygostyles of 51 species of aquatic birds and shorebirds and found that the shapes varied with foraging style. Birds that forage underwater (i.e., divers), like cormorants, penguins, puffins, gannets, and tropicbirds, had long straight pygostyles (e.g., see the pygostyles of the Great Cormorant, Adelie Penguin, and Common Loon below). Birds that do not forage underwater (aerial and terrestrial) have craniocaudally restricted, dorsally oriented pygostyles (e.g., see the pygostyle of the great frigatebird below).

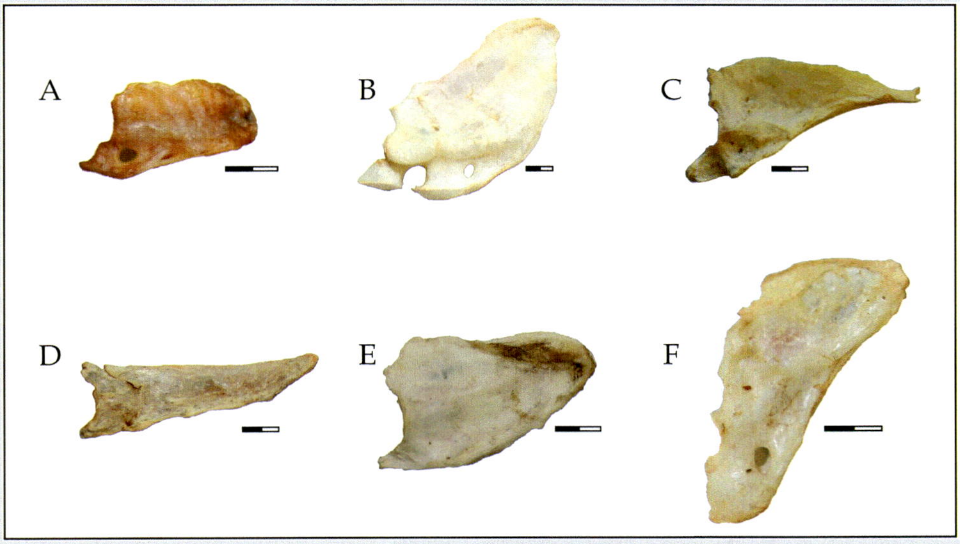


Variation in pygostylemorphology among several species of aquatic birds. (**a**) Northern Fulmar (*Fulmaris glacialis*), (**b**) American White Pelican (*Pelecanus erythrorhynchos*), (**c**) Great Cormorant (*Phalacrocorax carbo*), (**d**) Adelie Penguin (*Pygoscelis adeliae*), (**e**) Common Loon (*Gavia immer*), and (**f**) Great Frigatebird (*Fregata minor*). Scale bar = 5 mm. (Figure from Felice 2014; © 2014 Felice, O'Connor, open-access article distributed under the terms of the Creative Commons Attribution License, https://creativecommons.org/licenses/by/4.0/)

In the air, tails resist a dorsally directed force (lift). However, underwater, tails experience both dorsally and ventrally (i.e., upward force due to their buoyancy) directed forces. This may favor longer, straighter, more symmetrical pygostyles as an attachment site for muscles that elevate and depress the tail (Felice 2014).

(continued)

**Box 2.12** (continued)

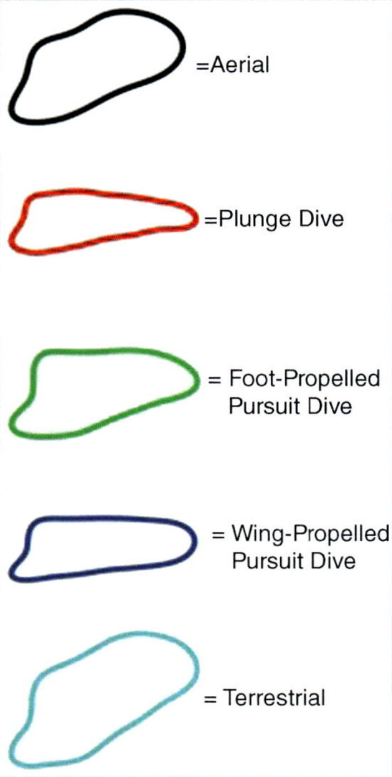


Pygostyle shape varies with foraging style among several species of aquatic birds and shorebird. Species were categorized into one of five foraging style groups: aerial, terrestrial, plunge dive, foot-propelled pursuit dive, and wing-propelled pursuit dive. The aerial foraging group included species that typically forage in the air by hawking, dipping, pattering, and kleptoparasitism. Aerial foragers had pygostyles with a rounded caudal margin that were narrower midway along their length, giving them an hourglass shape. Foot-propelled pursuit divers had pygostyles that were wider at the cranial end and thinned toward the caudal end. Wing-propelled pursuit divers had long pygostyles that did not taper. Plunge divers had pygostyles similar to those of wing-propelled pursuit divers, but thinner at the caudal end. Terrestrial foragers had pygostyles similar to those of aerial foragers, but without the constriction. (Felice 2014; © 2014 Felice, O'Connor, open-access article distributed under the terms of the Creative Commons Attribution License, https://creativecommons.org/licenses/by/4.0/)

Falcons and some hummingbirds have accessory bones alongside their pygostyles and provide additional surface area for attachment of caudal muscles and rectrices. Falcons and hummingbirds often fly at relatively high speeds and, at times, must brake quickly, e.g., when falcons approach potential prey or landing sites or when hummingbirds are involved in territorial disputes or quickly approach flowers seeking nectar. These accessory bones, called accessory pygostyle bones in falcons and oval bones in hummingbirds, provide additional surface area for attachment of muscles that aid in depressing the tail downward as they brake (Richardson 1972).

(continued)

**Box 2.12** (continued)

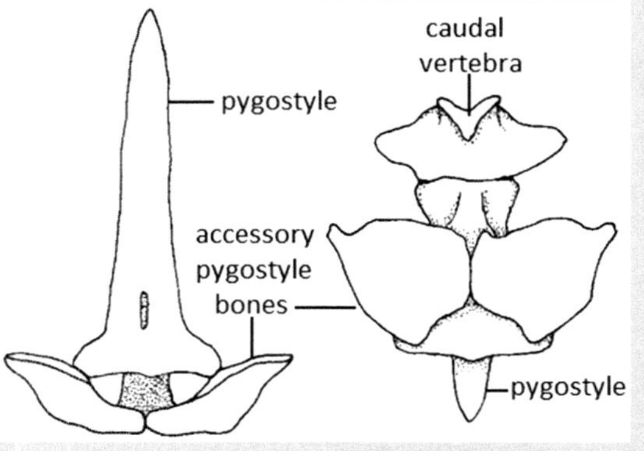


Posterior (left) and ventral (right) views of the pygostyle and accessory pygostyle bones of a falcon. The last free caudal vertebra is also shown on the right. (Figure from Richardson 1972; © 1972 Oxford University Press, used with permission)

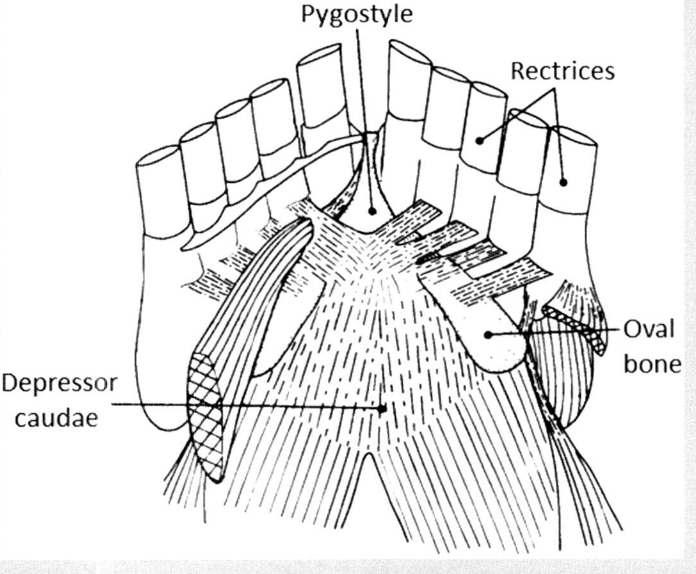


Ventral view of muscles in the tail of a Purple-throated Carib (*Eulampis jugularis*) hummingbird. Fibers of the depressor caudae muscle attach to the pygostyle as well as to flat oval bones that are embedded in a thin sheet of connective tissue. Several flat tendinous sections of connective tissue extend from the oval bones to the base of the rectrices. When the depressor caudae muscle contracts, the rectrices rotate downward, increasing drag and helping to reduce speed. (Figure from Zusi and Bentz 1984; CC0 Public Domain)

(continued)

**Box 2.12** (continued)

Hummingbird spreading its rectrices and contracting its depressor caudae muscle to depress its tail (rotate it downward) as it approaches a flower. This creates more drag and acts as a brake to help slow its forward momentum. (Photo from pxhere.com, CC0 Public Domain)

laying. Pectoralis mass may also increase when birds are molting flight feathers. A reduction in wing area during molt can result in increased wing loading and impaired flight ability. However, for some species of birds, an increase in pectoralis mass during molt helps maintain normal wing loading values and flight ability (Lind and Jakobsson 2001). Among waterfowl and grebes that molt all of their flight feathers simultaneously, pectoralis mass may decline at the onset of molt, reducing the metabolic costs of maintaining flight muscle during a flightless period, then rapidly increase as flight feather molt is completed and they are again able to fly (Piersma 1988; Ndlovu et al. 2017).

### 2.6.2 Fiber Types

The skeletal muscles of birds are composed of different types of muscle cells or fibers that vary in contraction speed, energy source, color, and how long they can contract (Table 2.1; Box 2.15 Myosin Isoforms). Slow oxidative fibers are also called red fibers, with the red color caused by a high concentration of myoglobin (a molecule similar to hemoglobin that allows muscle cells to store oxygen). These fibers have numerous mitochondria for generating ATP and, compared to the other fiber types, contract more slowly and generate comparatively less force. However, slow oxidative fibers can contract for long periods and are important for slower, repetitive movements, e.g., maintaining posture or gliding flight where a spread-wing posture must sometimes be maintained for extended periods. Sustained contraction is possible because oxidative fibers use lipids (fats) as their primary energy source and these lipids can be stored in the muscle fibers or transported to the fibers from fat deposits elsewhere.

Glycolytic fibers use glycogen (carbohydrate) as an energy source. Because little glycogen is stored in fast glycolytic fibers, there are fewer

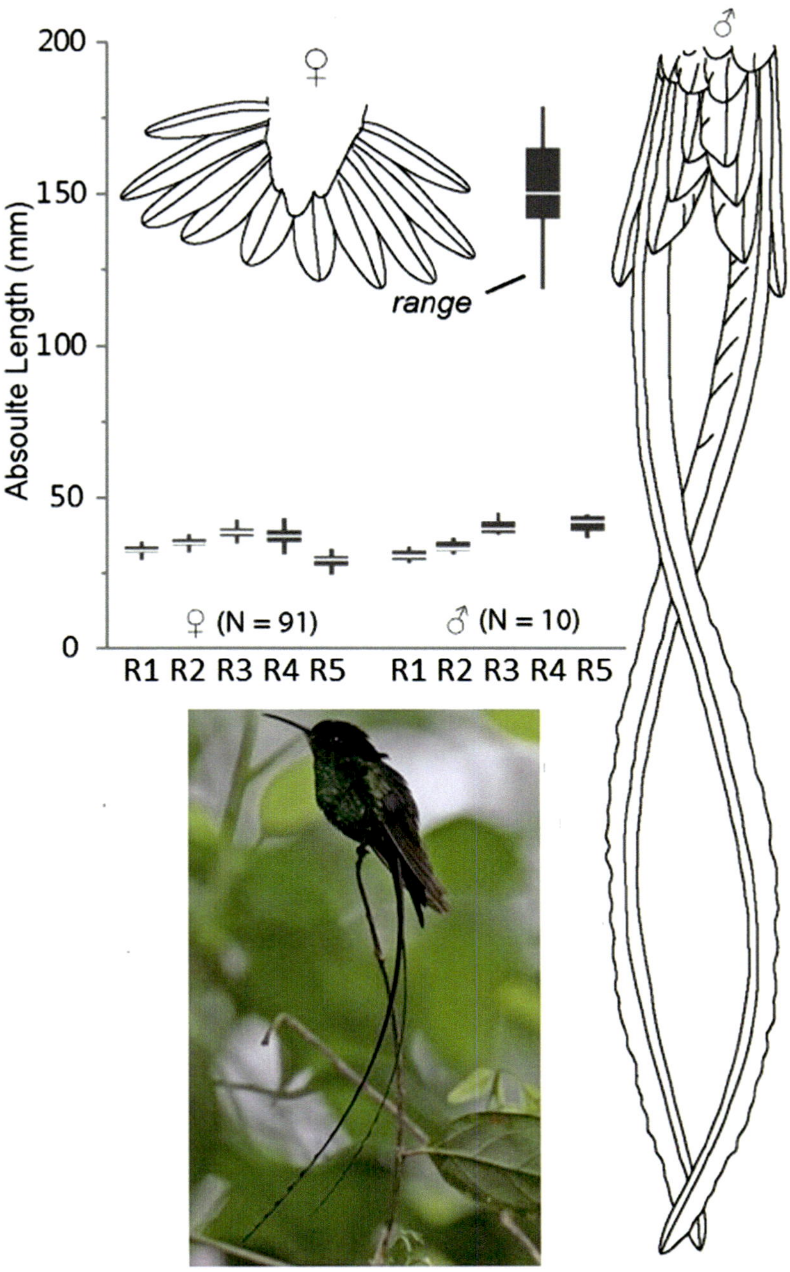


**Fig. 2.97** Male Red-billed Streamertails have greatly elongated tail-streamers that are the longest tail feathers of any hummingbird (Clark 2010). Top: Bar and whisker plot of tail feather lengths (rectrices 1–5) of male and female Red-billed Streamertails. Bar indicates 25th and 75th percentiles, and whiskers indicate ranges. Bottom: Photo of a male Black-billed Streamertail. (Figure from Clark and Rankin 2019, © 2019 The Authors. Evolution © 2019 The Society for the Study of Evolution, used with permission. Photo by Ron Knight, Wikipedia, CC By 2.0, https://creativecommons.org/licenses/by/2.0/deed.en)

mitochondria, and they have less myoglobin (making them appear white in color), fast glycolytic fibers are more susceptible to fatigue. However, glycolytic fibers contract faster and with greater force than oxidative fibers. Fast oxidative-glycolytic fibers also contract rapidly, with a force of contraction intermediate between slow oxidative and fast glycolytic fibers. With moderate numbers of mitochondria and levels of myoglobin and the ability to use both lipids and glycogen as energy sources, fast oxidative-glycolytic fibers are also long-endurance fibers.

**Box 2.13 Skeletal Muscle Anatomy and Function**

The structure and function of avian skeletal muscles are like that of mammals. Each muscle consists of several subunits called fascicles within which are numerous long, thin muscle cells (also called muscle fibers). Muscle cells, like all cells, have membranes called the sarcolemma. The interior of muscle cells contains numerous subunits called myofibrils that are made up of long, thin proteins called the thick and thin myofilaments. Muscle cells also contain large amounts of modified endoplasmic reticulum called the sarcoplasmic reticulum (SR) that forms a complex network around the myofibrils, storing and releasing calcium ions ($Ca^{2+}$) that are needed for muscle contraction. Transverse (T)-tubules invaginate the sarcolemma, surrounding the myofibrils and in close contact with sarcoplasmic reticulum. The function of the T-tubules is to conduct impulses traveling along the sarcolemma down into the interior of the muscle cell and to membranes of the sarcoplasmic reticulum. This causes the release of calcium ions that will then trigger muscle contraction.

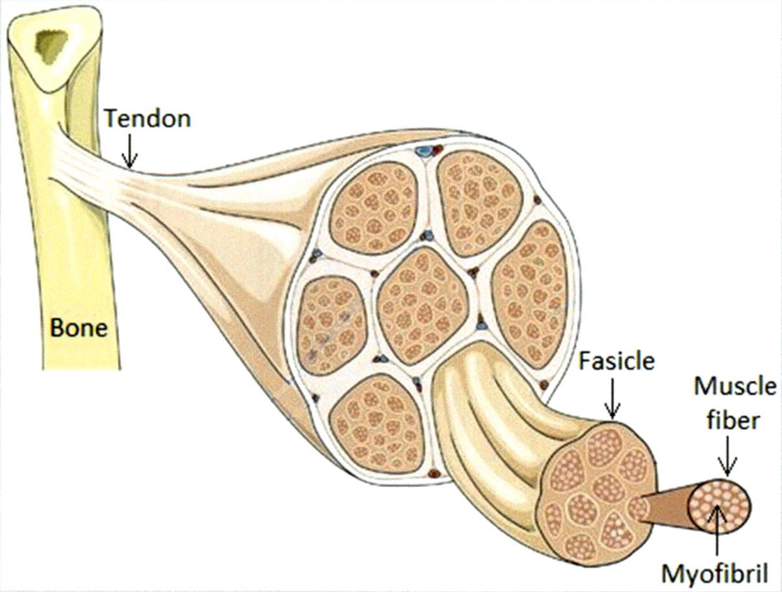


Avian skeletal muscles consist of subunits called fascicles that are surrounded by connective tissue. Fascicles are made up of numerous muscles fibers (or cells) which, in turn, consist of numerous even smaller subunits called myofibrils. (Figure modified from Sciorati et al. 2016; © 2016 Springer Nature, used with permission)

(continued)

**Box 2.13** (continued)

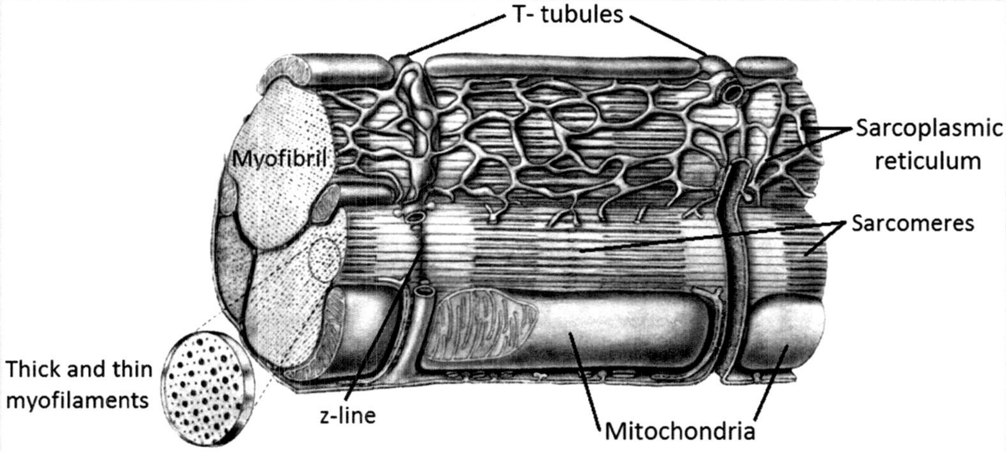


Drawing of the interior of an avian skeletal muscle cell. Myofibrils consist of subunits called sarcomeres that line up end-to-end along the length of myofibrils. Myofibrils are covered with sarcoplasmic reticulum that, during contraction, releases calcium ions when stimulated by impulses transmitted from the T-tubules. Contraction requires substantial amounts of energy (adenosine triphosphate, or ATP) so muscle cells have many mitochondria, the organelle that produces most of the ATP used by cells. (Figure modified from Sommer 1995; © 1995 Published by Elsevier Ltd., used with permission)

Myofibrils are contractile units that consist of very precisely arranged thick and thin myofilaments. Thick myofilaments are made from a protein called myosin, whereas thin myofilaments are made from three different proteins: actin, tropomyosin, and troponin. Thick and thin myofilaments form sarcomeres—the functional units of skeletal, so named because sarcomeres are the smallest structures in skeletal muscle that shorten when muscles contract. Sarcomeres in a myofibril are arranged end-to-end in myofibrils and adjacent sarcomeres share structures called z-lines, lines produced in two-dimensional views of myofibrils at the point where filaments from adjacent sarcomeres overlap slightly.

(continued)

**Box 2.13** (continued)

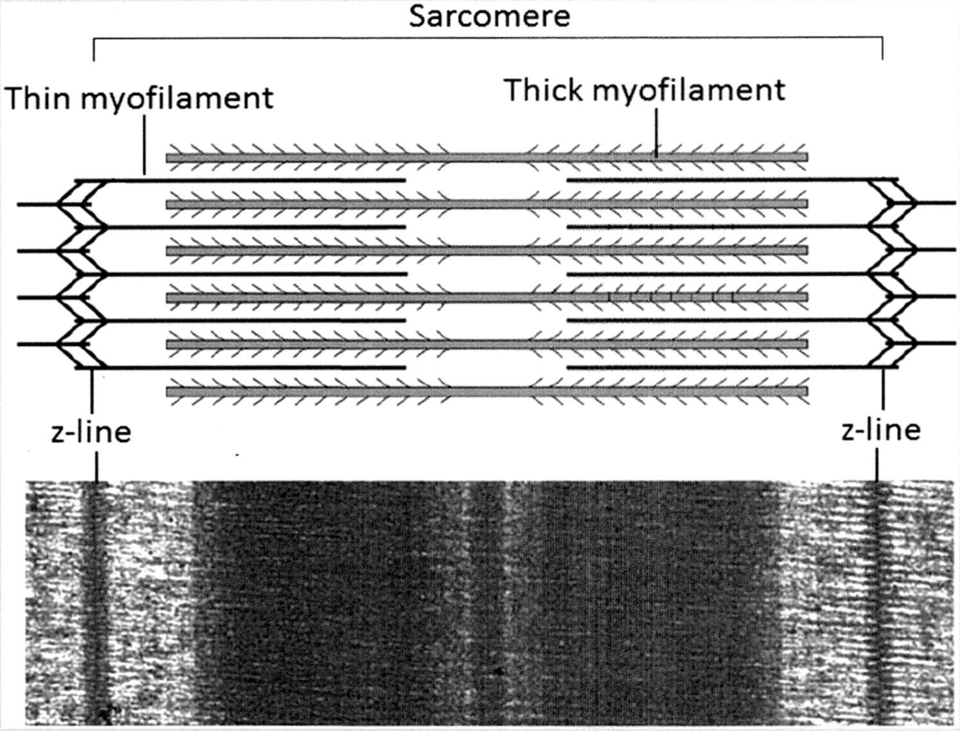


A drawing (top) and photomicrograph (bottom) and a sarcomere. Sarcomeres contain both thick and thin myofilaments, with the thin myofilaments extending from the z-lines and thick myofilaments centrally located in sarcomeres and between thin myofilaments. Thick myofilaments have projecting structures called cross-bridges at each end. These projections "bridge the gap" between thick and thin filaments and, although not shown in the diagram above, the cross-bridges are in contact with adjacent thin myofilaments. When skeletal muscles contract and shorten, sarcomeres get shorter as the thick and thin myofilaments slide past each other and the z-lines move closer together. (Figure modified from Luther 2009; © 2009 The Authors. Published by Springer Nature, used with permission)

(continued)

**Box 2.13** (continued)

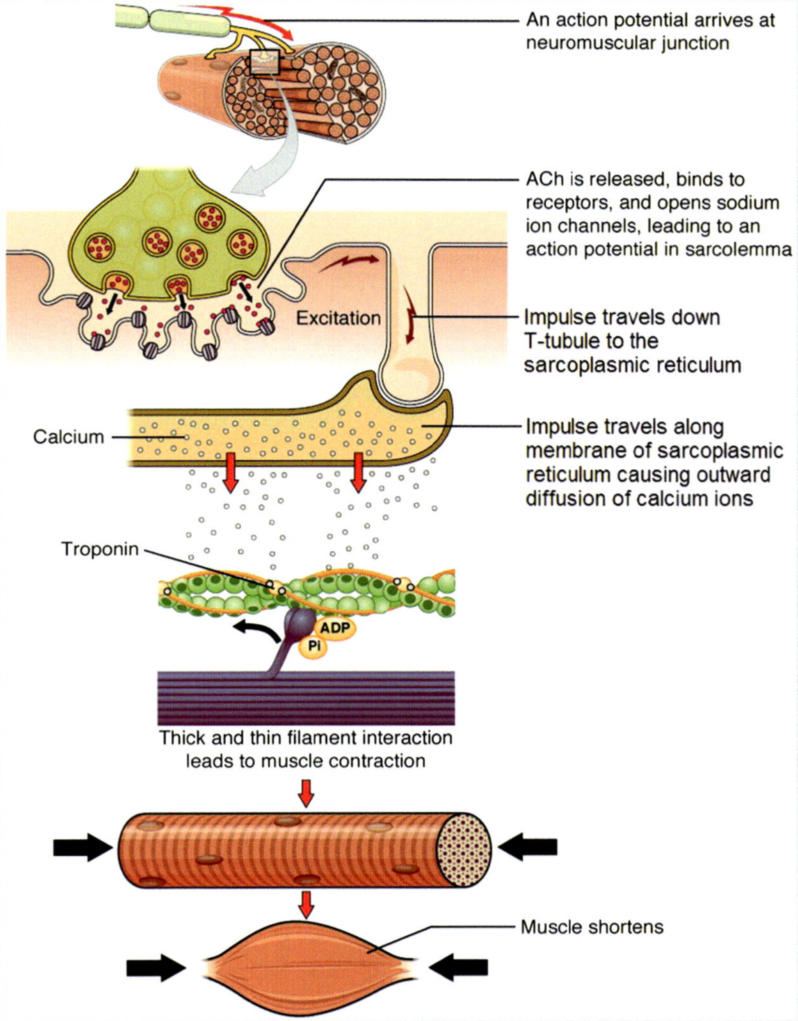


Skeletal muscle is voluntary muscle so contraction begins when nervous impulses that originate in motor areas of the brain travel along motor neurons and stimulate muscle cells. The stimulation begins at a neuromuscular junction where a neurotransmitter (acetylcholine, or Ach) is released, binds to receptors in the membrane of the muscle cell, and causes an action potential that then travels along the muscle cell membrane (called the sarcolemma) as an impulse. Impulses then travel along the sarcolemma, down the transverse tubules, and to the membrane of the sarcoplasmic reticulum. Impulses traveling along the membrane of the sarcoplasmic reticulum (SR) cause calcium channels to open and calcium ions diffuse out of the SR and into the myofibrils and sarcomeres. Calcium ions then bind to a protein on thin myofilaments called troponin. This reaction, as explained below, allows cross-bridges to contact actin. This contact causes the cross-bridge to "swivel" forward, pulling the actin, and the thin filament, with it. The bond between the cross-bridge and actin is broken at that point and the cross-bridge quickly binds to a molecule of ATP. The ATP then breaks down to ADP+P, releasing energy that swivels the cross-bridge back to its original position. The cross-bridge then binds to another actin and the process is repeated as long as the muscle is contracting. The collective "swiveling" of many cross-bridges causes the thick and thin filaments to "slide" past each other and sarcomeres (and myofibrils, muscle cells, fascicles, and the entire

(continued)

**Box 2.13** (continued)

muscle) get shorter. A muscle begins to relax when no longer being stimulated by a motor neuron. In response, the membrane of the SR transports calcium ions back into the SR, removing the ions from the troponin molecules (step 5), and, as explained below, causing the cross-bridge to lose contact with actin. (Figure from Wikipedia, licensed under the Creative Commons Attribution 4.0 International license, https://creativecommons.org/licenses/by/4.0/)

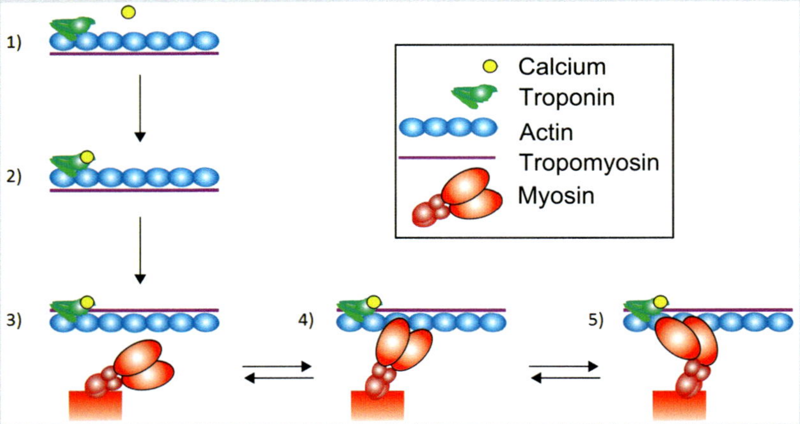


Calcium ions released from the sarcoplasmic reticulum (1) bind to troponin (2). This causes a change in shape of the troponin molecule that, because they are connected, moves another protein that is part of thin myofilaments called tropomyosin (3). Before moving, the tropomyosin molecules were located between a cross-bridge and actin molecules, preventing contact and so the muscle was relaxed. However, movement of the tropomyosin allows contact between the cross-bridge and actin (4). In response the cross-bridge swivels forward, pulling the actin (and the thin filament) forward (5). The bond then breaks and energy released by the breakdown of ATP causes the cross-bridge to swivel back (3). The cross-bridge then bonds to another actin molecule and the process repeats until the muscle is no longer being stimulated by a motor neuron. (Figure modified from Fenwick et al. 2017; © 2017 Fenwick et al., open-access article distributed under the terms of the Creative Commons Attribution License, https://creativecommons.org/licenses/by/4.0/)

**Box 2.14 Superfast Muscles of Some Manakins**

In the tropical forests of Central and South America, males in several species of manakins form leks and perform complex, fast-moving displays to impress females. During these displays, males open and elevate their wings, causing the wrists to collide, at incredibly fast rates (about 45 times per second). These rapid wing claps produce a loud "snap" sound and is an important part of their display. The rate at which the muscle that moves the wings upward (scapulohumeralis caudalis; SHC) must contract and relax is much faster than the rate at which the main flight muscles (pectoralis and supracoracoideus) contract and relax when the birds fly (about 25–30 times per second). The rate of contraction of the scapulohumeralis caudalis of these manakins is the fastest reported for a vertebrate limb muscle (Fuxjager et al. 2016).

(continued)

**Box 2.14** (continued)

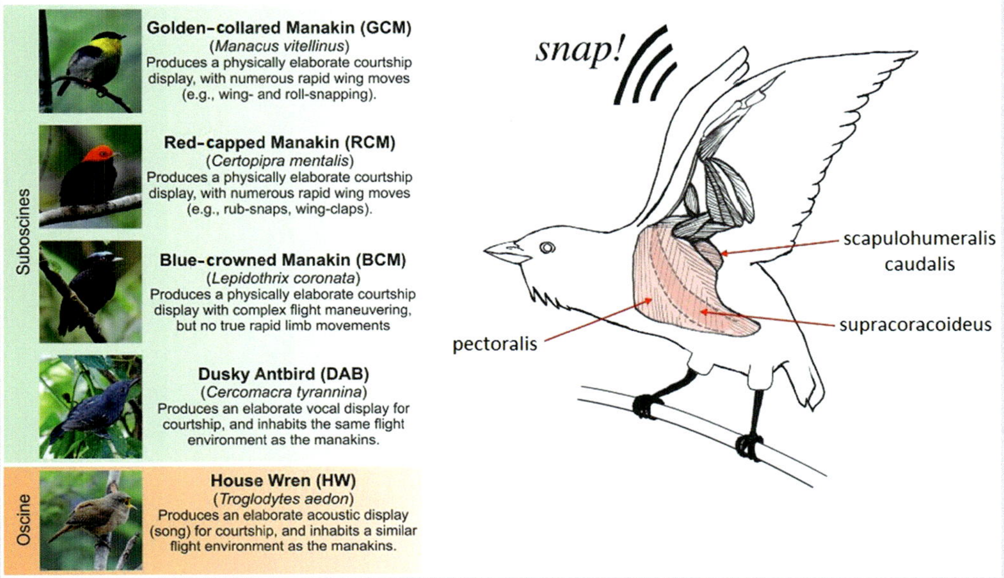


Golden-collared Manakins and Red-capped Manakins perform displays that involve superfast movements of their wings to create "snap" sounds. Blue-crowned Manakins, Dusky Antbirds, and House Wrens do not perform such displays. (Figure from Fuxjager et al. 2016; open-access article under the Creative Commons Attribution (CC-BY) license, https://creativecommons.org/licenses/by/4.0/)

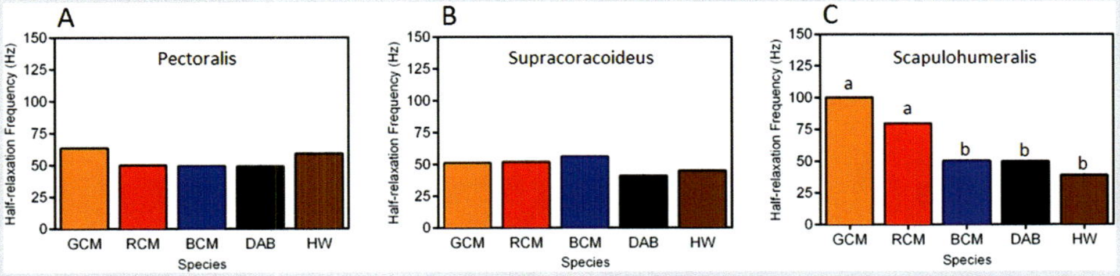


Speed of relaxation (and contraction) of the pectoralis, supracoracoideus, and scapulohumeralis caudalis muscles of five species of small species of songbirds. The half-relaxation frequency is a measure of how rapidly a muscle can reach a state of being half-relaxed in a period of one second (Hz). Once at least half-relaxed, a muscle is again able to contract so a muscle that is able to reach a half-relaxed state with greater frequency is also able to contract at a higher rate. For the two main flight muscles, the pectoralis and supracoracoideus, there was no difference among species in half-relaxation frequency. However, for the muscle that retracts the wing, the scapulohumeralis caudalis, the half-relaxations frequency was significantly greater for Golden-collared Manakins and Red-capped Manakins than for the other three species (GCM, Golden-collared Manakin; RCM, Red-capped Manakin; BCM, Blue-crowned Manakin; DAB, Dusky Antbird; HW, House Wren). (Figure from Fuxjager et al. 2016; open access under the Creative Commons Attribution (CC-BY) license, https://creativecommons.org/licenses/by/4.0/)

(continued)

**Box 2.14** (continued)

Series of screenshots from a high-speed video of a male Golden-collared Manakin showing the rapid wing movements. In the column on the right, note that successive wing snaps (at 64 and 82 ms, respectively, and indicated by orange bursts) were just 18 ms apart. (Figures from Pease et al. 2022; used with permission of the U. S. National Academy of Sciences)

A key ingredient that allows the scapulohumeralis caudalis (SHC) of these manakins to contract with such speed is testosterone. Working with captive Golden-collared Manakins,

(continued)

**Box 2.14** (continued)

Fuxjager et al. (2017) enhanced endogenous testosterone levels for three males (by implanting small tubes from which testosterone was gradually released) and reduced endogenous testosterone levels for three other males (by implanting small tubes from which a testosterone antagonist was gradually released). The SHC of these males was then stimulated at three different frequencies and, at each frequency, the contraction speed (as measured by the percent of relaxation relative to the resting length of the muscle) was significantly faster for males treated with testosterone.

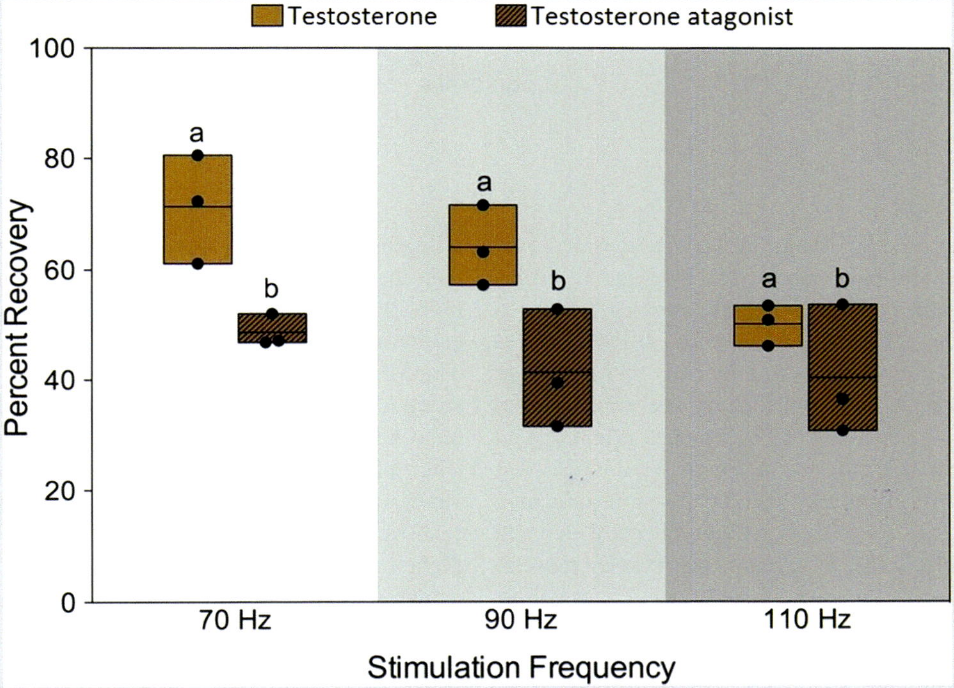


The scapulohumeralis caudalis muscle of male Golden-collared Manakins was stimulated at different frequencies and, at each frequency, the muscle of males treated with testosterone contracted at significantly higher rates than those of males treated with a testosterone antagonist. Different letters above boxes indicate that mean percent recovery values for the two treatment groups (middle lines) were significantly different. Boxes (and filled circles) indicate the range of values for the three males for each treatment. (Figure modified from Fuxjager et al. 2017; © 2017 Oxford University Press, used with permission)

The specific mechanism(s) by which testosterone enhances contraction rate of the scapulohumeralis caudalis muscle of Golden-collared Manakins remain(s) to be determined. However, one possibility is the testosterone increases cross-bridge detachment rates (the rate at which cross-bridges of thick myofilaments detach from actin molecules in the thin myofilaments), a process that sets the speed at which thick and thin filaments slide past each other during contraction. Another possibility is that testosterone improves the ability of the muscle cells to rapidly release (from the sarcoplasmic reticulum) and bind (to troponin molecules in the thin myofilaments) calcium ions, two critical steps in muscle contraction (Fuxjager et al. 2017, 2022).

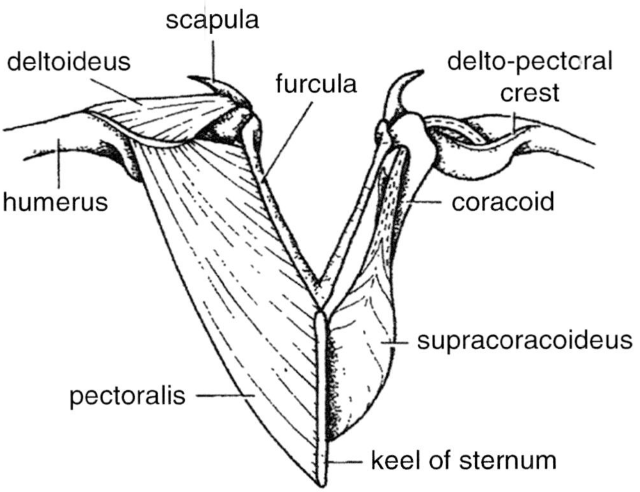


**Fig. 2.98** The major flight muscles of a bird. The supracoracoideus is the primary upstroke muscle, and the pectoralis is the primary downstroke muscle. In *Archaeopteryx*, the deltoideus was probably the primary upstroke muscle. (Figure from Ruben 1991; © 1991 The Society for the Study of Evolution, used with permission)

The types of fibers in skeletal muscles vary among different muscles and also among different species of birds. Many muscles contain a mixture of fiber types, with different types allowing the muscles to efficiently serve different functions. For example, leg muscles (i.e., gastrocnemius muscle) of Zebra Finches (*Taeniopygia guttata*) were found to consist of 27% fast oxidative-glycolytic fibers, 13% slow oxidative fibers, and 60% fast glycolytic fibers (Welch and Altshuler 2009). Many birds use their leg muscles to generate lift during takeoff, with as much as 80–90% of the initial takeoff force produced by those muscles (Earls 2000). Fast glycolytic fibers are well suited to provide such forces. However, leg muscles are also important for perching or standing, sometimes for extended periods, and fast oxidative-glycolytic and slow oxidative fibers, in contrast to fast glycolytic fibers, are capable of prolonged contraction. Leg muscles may also be important for thermoregulation, shivering to generate heat when body temperatures begin to drop. Because of their resistance to fatigue and high mitochondrial density, slow oxidative fibers are excellent heat producers (Block 1994; Jurgens 2002).

Fiber types in the flight muscles of birds (i.e., pectoralis, supracoracoideus, and other muscles of the wing) vary among species. The flight muscles of some birds contain a mixture of the different fiber types, with proportions varying with the frequency, type, and typical duration of flight. For example, the flight muscles of many gallinaceous birds, like pheasants, partridges, and chickens, consist either primarily or, in some species, entirely of fast glycolytic fibers, allowing rapid takeoff, but limiting the duration and distance of flight (Kiessling 1977; Pyörnilä et al. 1998). Birds that that regularly soar or glide, such as pelicans, vultures, and albatrosses, have flight muscles with more high-endurance slow oxidative fibers, allowing them to hold their wings outstretched for extended periods (Goldspink 1981). In fact, albatrosses have more slow muscle fibers than any other birds studied to date, with some extensor muscles (e.g., extensor metacarpi radialis dorsalis) consisting completely of slow fibers, and 38% of the fibers in the cranial region of the pectoralis muscle are slow fibers (Meyers and McFarland 2016).

Although flight muscles often consist of more than one type of fiber, the pectoralis muscles of birds in a wide variety of taxa consist entirely of fast oxidative-glycolytic fibers. Rosser and George (1986) examined the pectoralis muscles of 43 species and found only fast oxidative-glycolytic fibers in the muscles of 40 species representing 13 orders of birds (Anseriformes, Apodiformes, Caprimulgiformes, Charadriiformes, Ciconiiformes, Columbiformes,

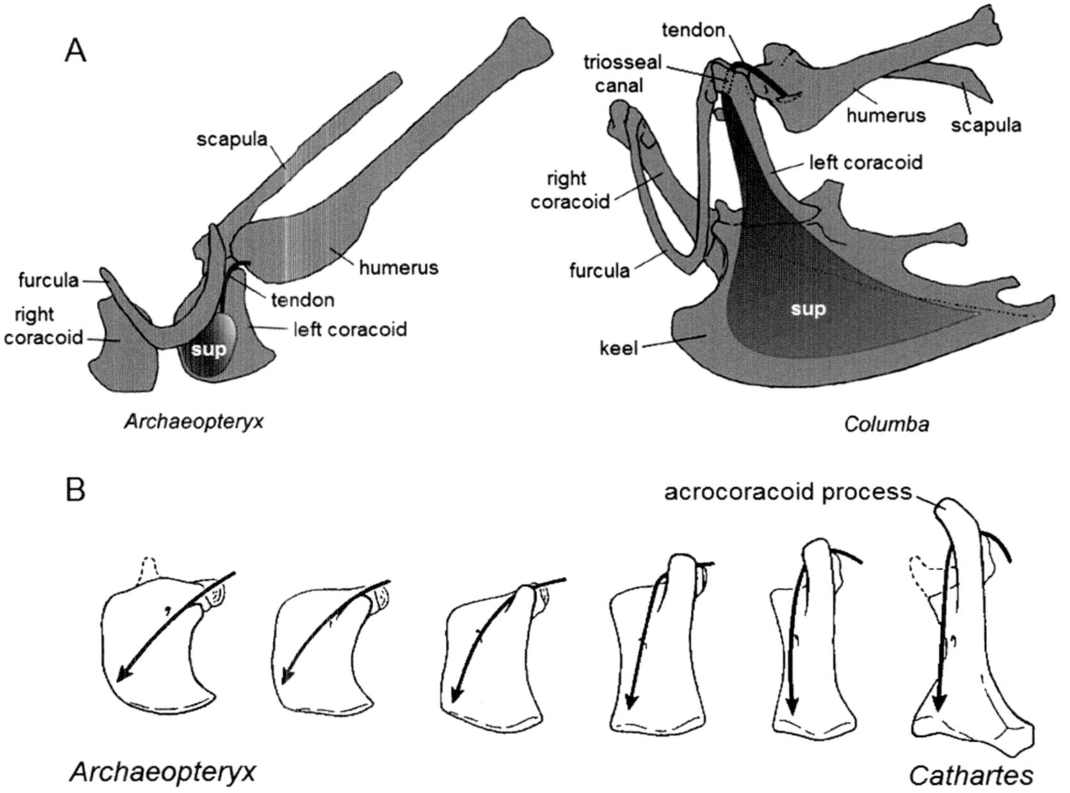


**Fig. 2.99** Possible transition in supracoracoideus orientation from *Archaeopteryx* to a present-day bird. (**a**) Supracoracoideus (sup) of *Archaeopteryx* and a present-day bird (*Columba*). In present-day birds, the coracoid is elongated and has a process called the acrocoracoid that, where the coracoid meets the scapula and clavicle, forms the triosseal canal (also called the foramen triosseum) through which the supracoracoideus tendon passes in a pulley-like arrangement and inserts on the dorsal side of the humerus (the pectoralis has been removed). (**b**) Arrow indicates the position of the tendon of the supracoracoideus and the orientation of the muscle fibers in the supracoracoideus and how the supracoracoideus pulls its tendon on the humerus of *Archaeopteryx* (left), hypothetical intermediates, and a present-day bird, a *Cathartes* vulture (right). Elongation of the coracoid and development of the acrocoracoid process converts the supracoracoideus from a humeral protractor (moving the humerus forward as in *Archaeopteryx*) to a humeral elevator (as in present-day birds). (A—modified from Mayr 2017; © 2017 Dt. Ornithologen-Gesellschaft e.V., used with permission; B - modified from Goslow et al. 1989; © 2015 Oxford University Press, used with permission)

Coraciiformes, Falconiformes, Galliformes, Passeriformes, Podicipediformes, Psittaciformes, and Strigiformes). Species where pectoralis muscles included multiple fiber types included Double-crested Cormorants (*Nannopterum auritum*), Red-tailed Hawks (*Buteo jamaicensis*), and Domestic Chickens. The absence of slow oxidative and fast glycolytic fibers in the pectoralis muscles of many birds suggests either that (1) those birds rarely perform behaviors that require either fast, forceful contractions (fast glycolytic fibers) or sustained (isometric) contractions (slow oxidative fibers), or (2) fast oxidative-glycolytic fibers permit such behaviors (Welch and Altshuler 2009). In support of point 2, Meyers and McFarland (2016) found that the flight muscles of Golden Eagles (*Aquila chrysaetos*) had few or, in some cases, no slow

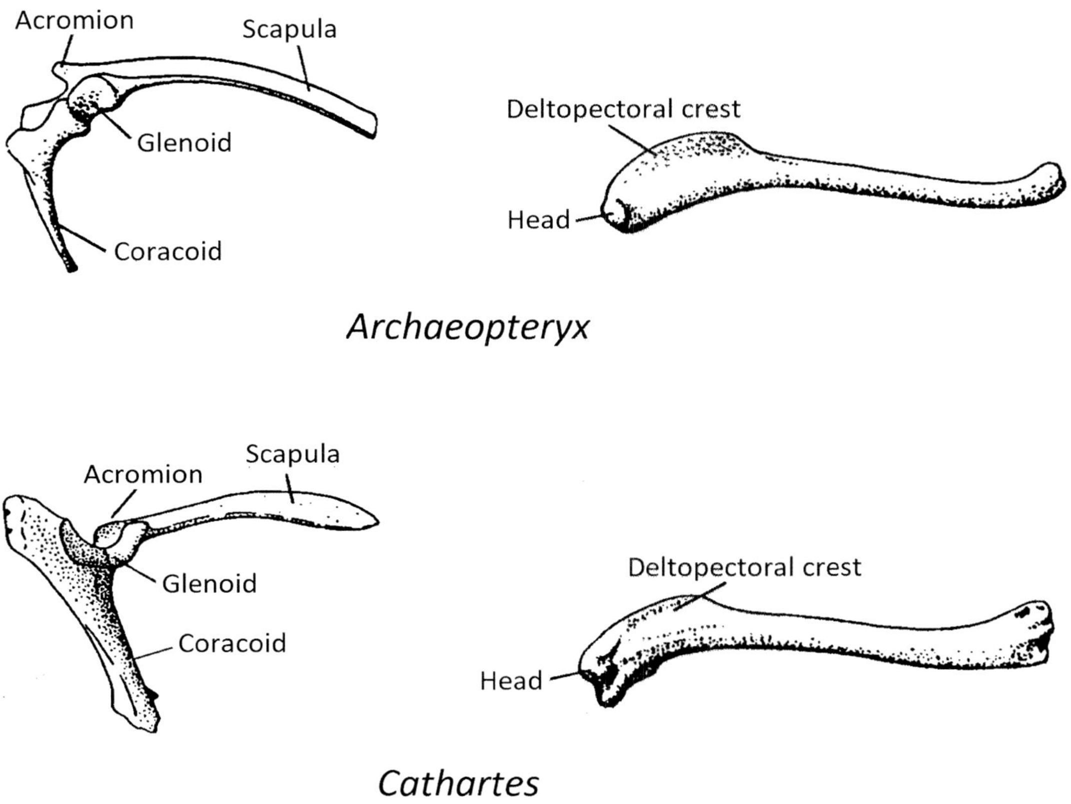


**Fig. 2.100** The coracoids, scapulas, and humeri of *Archaeopteryx* and a present-day bird (vulture). The coracoids of present-day birds are elongated and have a conspicuous process, the acrocoracoid, that, along with the scapula and clavicle (not shown), forms the foramen triosseum through which the tendon of the supracoracoideus passes and inserts on the humerus. *Archaeopteryx* had a shorter coracoid with no acrocoracoid process and, therefore, no foramen triosseum. However, *Archaeopteryx* had a well-developed deltopectoral crest where its primary upstroke muscle (deltoideus) likely inserted on the humerus. (Figure adapted from Ruben 1991; © 1991 The Society for the Study of Evolution, used with permission)

fibers. One possible explanation for this is that, because of their slow wingbeats, the fast fibers of eagles have a lower optimal contraction frequency that, although still faster than slow fibers, is more efficient at isometric contractions than the fast muscle fibers of smaller birds with faster wingbeat frequencies.

### 2.6.3 Locomotion

Birds can fly, walk, run, hop, climb, swim, and dive and all of these modes of locomotion require skeletal muscles. Most birds, such as songbirds, use multiple modes of locomotion. However, some specialized birds depend much more, if

**Table 2.1** Types of muscle fibers that make up the skeletal muscles of birds (Adapted from Hohtola and Visser 1998)

| Fiber type | Contraction speed | Energy source | Mitochondrial density | Color | Endurance |
|---|---|---|---|---|---|
| Slow oxidative | Slow | Lipids | High | Red | High |
| Fast oxidative-glycolytic | Fast | Lipids/ glycogen | Moderate | Intermediate | High |
| Fast glycolytic | Fast | Glycogen | Low | White | Low |

**Box 2.15 Myosin Isoforms**

Protein isoforms are similar to each other and have similar roles in cells. They can result from single genes encoding different isoforms (a process called alternative splicing) or they can be produced by two closely related genes (Gunning and Hardeman 2018). Myosin, a key protein in skeletal muscle, has isoforms and, in mammalian muscle, these forms are referred to as slow type 1 and fast types 2a, 2x, and 2b. These different isoforms of myosin differ in speed of contraction and rate of ATP use, affecting their power and resistance to fatigue (Robinson et al. 2021). Less is known about the myosin isoforms of birds, but Robinson et al. (2021) noted that terrestrial vertebrates (and birds) would likely have the slow type 1 isoform (also referred to as myosin heavy chain-1, or MHC-1) that has greater endurance, whereas birds that fly might have the MHC-2a isoform that provides more power. Importantly, Robinson et al. (2021) also noted that MHC isoforms are not oxidative or glycolytic; other components of muscle cells determine metabolic preferences. Given our limited knowledge of avian myosin isoforms, additional studies are needed to identify those isoforms and to determine the extent to which birds that differ in their flying abilities might also differ in their myosin isoforms.

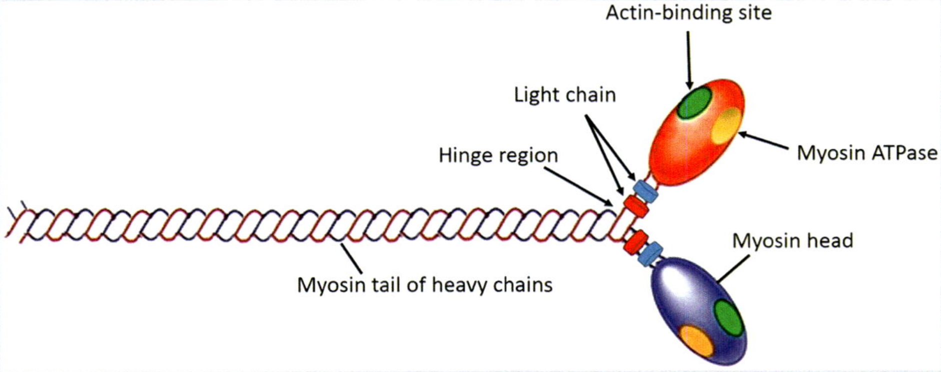


Myosin is a protein consisting of four light chains and two heavy chain subunits. The heavy chain subunits have been found to influence the contractile properties of muscle fibers, including shortening velocity and force generation, and these properties can vary among different isoforms of myosin. (Figure from Jena 2020; © 2020 Springer Nature Switzerland AG, used with permission)

Characteristics of different myosin isoforms (table from Robinson et al. 2021; © 2021 Oxford University Press, used with permission).

| Myosin type | Definition |
|---|---|
| Type 1 myosin | "Slow" myosin found most often in fatigue-resistant muscles and linked to aerobic sources of metabolism. Commonly found in muscles used for terrestrial locomotion and seems less suited for sustained flapping flight, but may perform well in locking wing joints of soaring birds. |
| Type 2a myosin | Faster contractile speed than type 1, but present in muscle cells that may support highly aerobic metabolism and has excellent fatigue resistance. An important component of flight muscles, and apparently used in hovering flight rather than type 1. |
| Type 2x, 2b myosin | Expression in bird flight muscles not well documented and additional study needed. |
| Other myosins embryonic and neonatal | Importance for most Neotropical birds still unknown. Prevalent in avian literature including generally as an artifact of using domestic chickens as an avian model system. Much of the chicken literature based on studies of immature or juvenile birds. These isoforms are found in other vertebrates during development and when muscles have been damaged, but relevance to adult birds and Neotropical populations has not been studied. |

not entirely, on certain types of locomotion. For example, flightless birds as well as rails and gallinaceous birds are entirely or primarily terrestrial. Other birds, like aerial foragers such as nighthawks and frigatebirds and hovering specialists (hummingbirds), are excellent flyers, but use their legs only to make limited movements on the ground or perches (e.g., nighthawks and frigatebirds) or just for perching (hummingbirds). The skeletal muscles most important in locomotion are those of the forelimbs, hindlimbs, and tail and, as the relative importance of each mode of locomotion varies for different species (Fig. 2.101), so does the associated musculature.

Muscles associated with the wings include the primary flight muscle, the pectoralis major, and the primary upstroke muscle, the supracoracoideus (Fig. 2.102). The pectoralis muscle generates the force for the downstroke, or power stroke, of the wing and is the largest single muscle in the avian body (Box 2.16 Divided Pectoralis of Soaring Birds). However, other muscles are also used when a bird is flying (Figs. 2.103 and 2.104). The relative size or mass of the pectoralis muscle varies among species, ranging from about 7 to 25% of total body mass (Table 2.2). For most birds, the pectoral muscles make up about 12–20% of total body mass. However, this percentage is generally lower for heavy-bodied diving birds, like grebes (Podicipedidae) and cormorants (Phalacrocoracidae), and birds that fly infrequently, like rails (Rallidae) (Table 2.2). Birds with pectoralis muscles

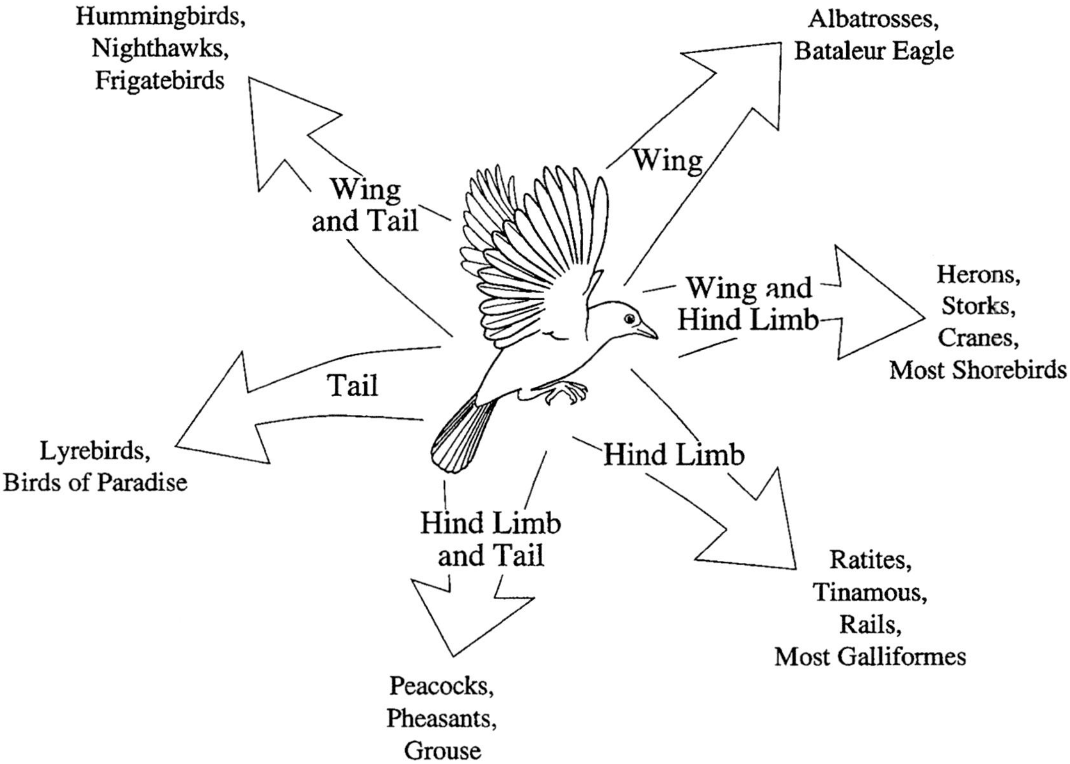


**Fig. 2.101** Birds vary in the extent to which the wings, legs, and tail play a role in locomotion. For most birds, all three are important. For others, such as hummingbirds, nighthawks, and frigatebirds, the hindlimbs are used little except for perching; still other birds, like ratites, rails, and gallinaceous birds are primarily or exclusively terrestrial and have well-developed hindlimbs. The musculature of the wings, tail, and hindlimbs varies with the relative importance of different types of locomotion. (Figure from Gatesy and Dial 1996a; © 1996 The Society for the Study of Evolution, used with permission)

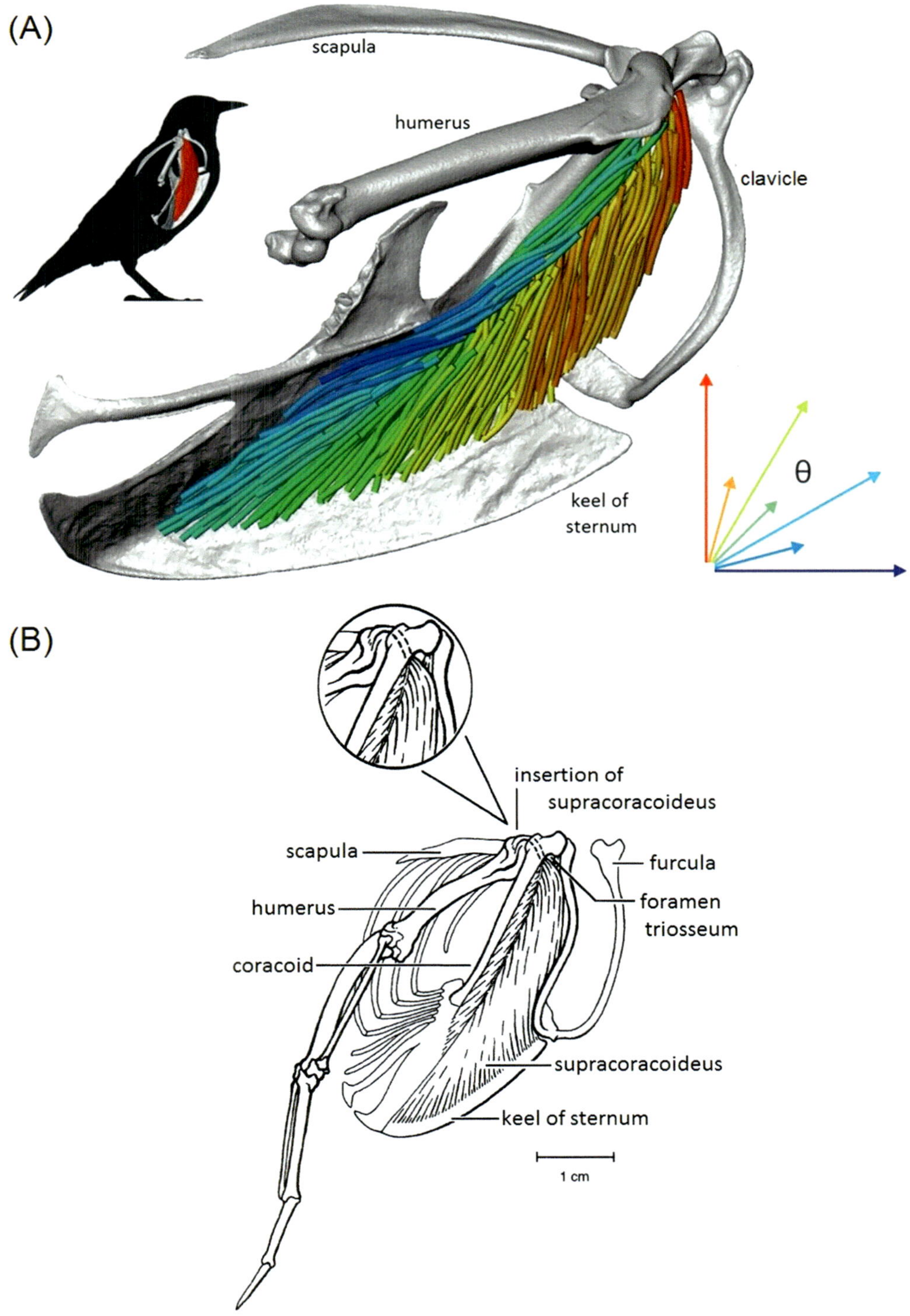


**Fig. 2.102** Supracoracoideus muscle of a European Starling (*Sturnus vulgaris*). (**a**) Orientation of fascicles within the muscle. All fascicles are oriented toward the tendon that extends from the supracoracoideus, passes through the

**Box 2.16 Divided Pectoralis of Soaring Birds**

Birds that regularly soar, such as albatrosses, petrels, frigatebirds, and vultures, must hold their wings outstretched for extended periods, an action that requires several different muscles. In addition, however, muscles are needed to serve as struts. Just as airplanes with wings attached high on the fuselage often have wing struts to help keep the wings level, birds that soar need "struts" to resist the air pressure from below the wings and hold them in the position needed to continue soaring.

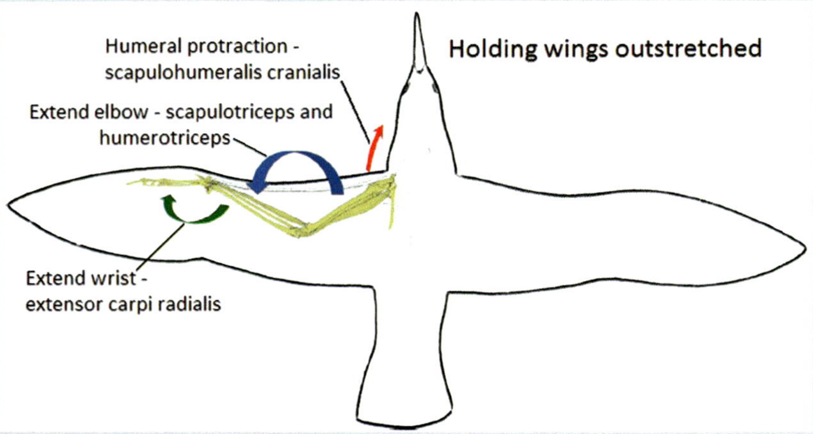


Skeletal muscles involved in holding wings outstretched. (Figure modified from Meyers 2019; © 2019 Deutsche Ornithologen-Gesellschaft e.V., used with permission)

A high-winged plane with struts that anchor the wings and keep them level in flight. (Photo from pxhere.com, CC0 Public Domain)

(continued)

**Fig. 2.102** (continued) foramen triosseum (also called the triosseal canal), and inserts on the humerus. (Figure modified from Sullivan et al. 2019; open-access article distributed under the terms of the Creative Commons CC BY License, https://creativecommons.org/licenses/by/4.0/). (**b**) The tendon of the supracoracoideus passes through the foramen triosseum and inserts on the humerus. Contraction of the supracoracoideus elevates the humerus during the upstroke phase of flight. (Figure from Poore et al. 1997; © 1997 Springer Nature, used with permission)

**Box 2.16** (continued)

Soaring birds have pectoralis muscles divided (or partially divided) into superficial and deep layers and referred to by Meyers (2019) as the pectoralis superficialis and pectoralis profundus. The pectoralis profundus serves as the strut, inserting on the humerus and contracting during soaring flight to maintain the position of the wing. A pectoralis profundus muscle has been found in several orders of soaring birds, including Cathartiformes (e.g., vultures), Pelecaniformes (e.g., American White Pelican [*Pelecanus erythrorhynchos*] and spoonbills), Suliformes (e.g., Anhinga [*Anhinga anhinga*], boobies, cormorants, and Magnificent Frigatebird [*Fregata magnificens*]), Ciconiiformes (storks), Procellariiformes (e.g., albatrosses, prions, petrels, and shearwaters), and Phaethontiformes (tropicbirds) (Meyers 2019).

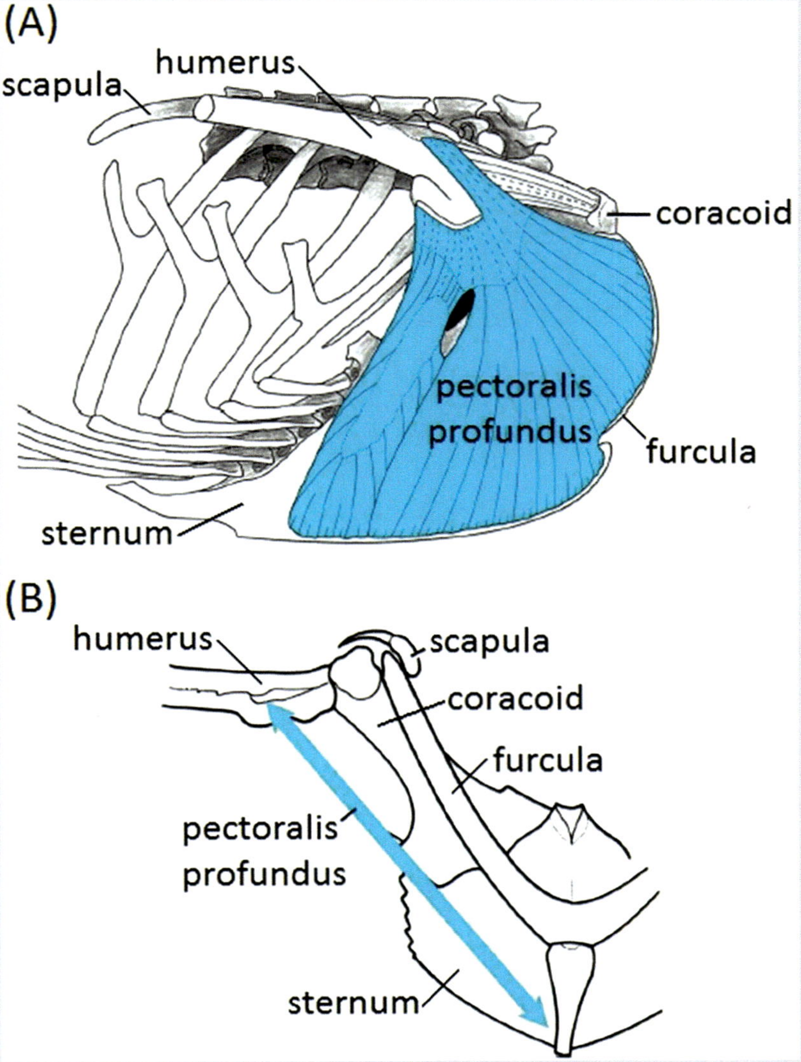


(**a**) Right side of the breast region of a Laysan Albatross (*Phoebastria immutabilis*) after removal of the pectoralis superficialis. The pectoralis profundus muscle inserts on the humerus and serves as a "strut" to maintain the position of the wing during soaring flight. (**b**) A simplified diagram showing how the pectoralis profundus serves

(continued)

**Box 2.16** (continued)

as a strut. (Figure modified from Meyers 2019; © 2019 Deutsche Ornithologen-Gesellschaft e.V., used with permission)

Laysan Albatross. (Photo by Kristi Lapenta/U. S. Fish and Wildlife Service, CC0 Public Domain)

sometimes making up more than 20% of total body mass include some tinamous (Tinamidae), pheasants (Phasianidae), hummingbirds (Trochilidae), and flycatchers (Tyrannidae) (Table 2.2). Although primarily terrestrial, tinamous and pheasants can take off quickly with powerful wingbeats, then glide away to avoid predators. Flycatchers and hummingbirds are among the most accomplished flyers, capable of extended and, when foraging, often acrobatic, flight.

Because the upstroke, or elevation of the wing, requires less force than the down- or power stroke, the supracoracoideus is a smaller muscle than the pectoralis. In most species, the supracoracoideus is much smaller than the pectoralis, often just 10–20% or even less of the mass of the pectoralis (Table 2.2). There are, however, some exceptions. Among penguins, the supracoracoideus can be 35–50% of the mass of the pectoralis (Hui 1988; Baldwin 1988; Bribiesca-Contreras et al. 2021) and, among hummingbirds, the supracoracoideus can be 40–45% of the mass of the pectoralis (Savile 1950; Fig. 2.105, Table 2.2). The supracoracoideus muscles of penguins and other wing-propelled diving birds are relatively large because, when wing-propelled diving (or aquaflying), thrust is generated on both the downstroke and the upstroke (Habib 2010; Figs. 2.106 and 2.107); generating thrust on the upstroke requires substantial force and, hence, a relatively large supracoracoideus. Similarly, hovering hummingbirds generate lift during both the downstroke and, by inverting or supinating their wings, the upstroke (Warrick et al. 2005), and a large supracoracoideus is needed to power the upstroke (Fig. 2.107).

Birds that are entirely or primarily terrestrial, including ratites, tinamous, rails, and most gallinaceous birds, have well-developed hindlimb

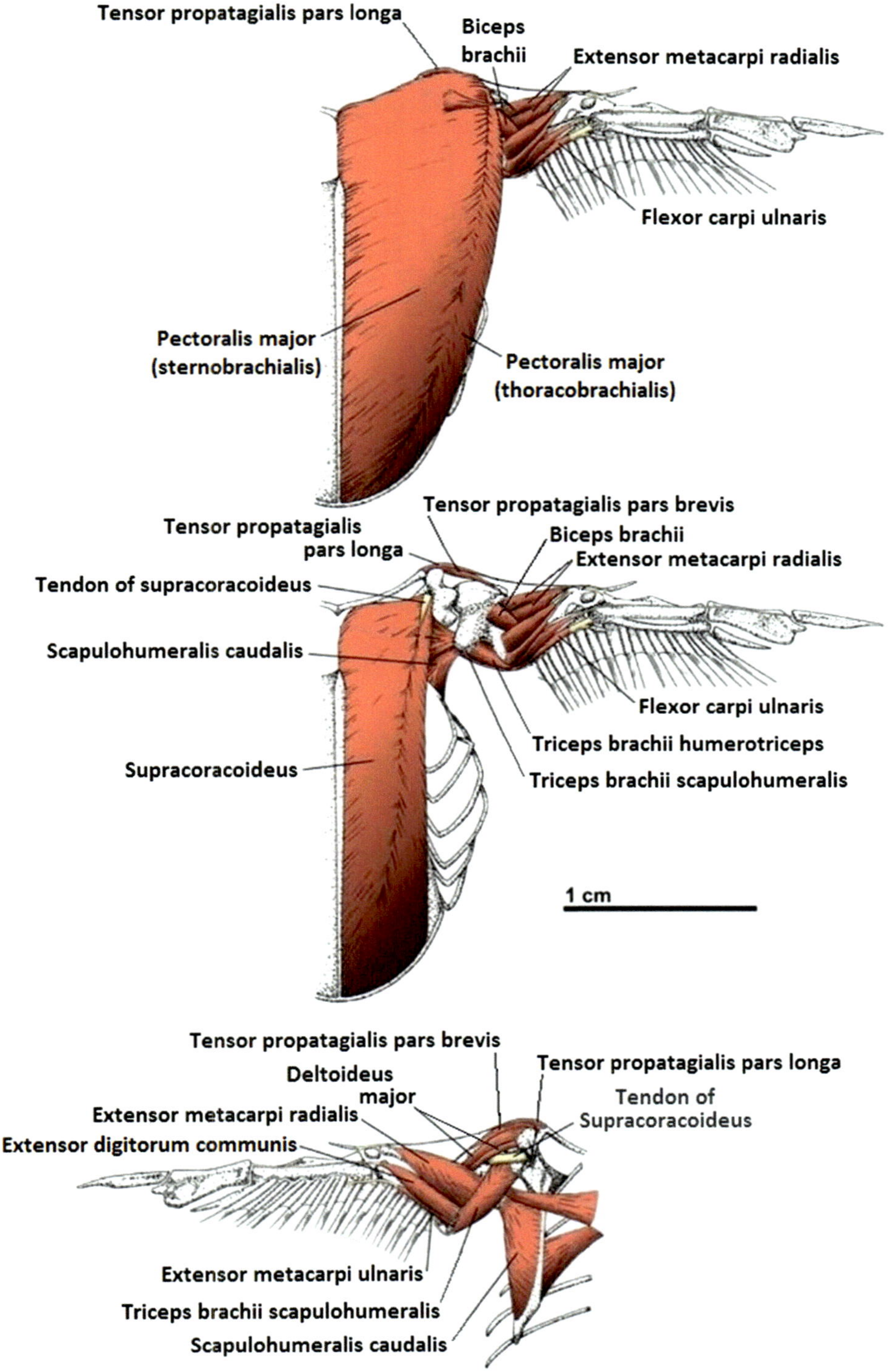


**Fig. 2.103** Superficial muscles (selected muscles only) of the wing and pectoral girdle of an Anna's Hummingbird (*Calypte anna*). The pectoralis and supracoracoideus are the primary depressors and elevators of the wings, respectively; other muscles of the forelimb are important in flexing and extending the bones and the shoulder, elbow, and wrist as a bird flies. (Figure modified from Welch Jr. and Altshuler 2009; © 2008 Elsevier Inc., used with permission)

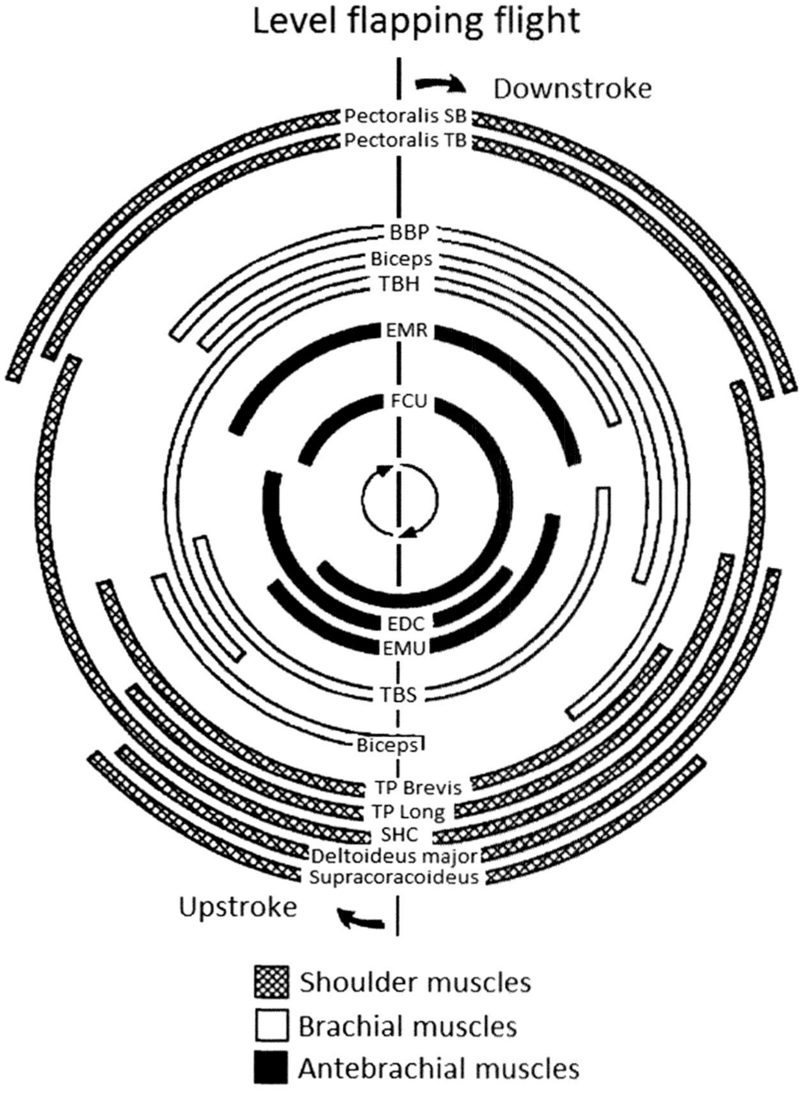


**Fig. 2.104** Activity of the wing muscles of a Rock Pigeon (*Columba livia*) during level flapping flight. The vertical line at the top indicates the point during the wing-beat cycle when the humerus (brachium) is fully elevated (i.e., the beginning of the downstroke), and the vertical line at the bottom indicates when the humerus is fully depressed (i.e., the beginning of the upstroke). Abbreviations: Pectoralis SB and Pectoralis TB, pectoralis major pars thoracicus sternobrachialis and thoracobrachialis, respectively; TP brevis, tensor propatagialis pars brevis; BBP, biceps brachii, pars propatagialis (not present in all species of birds); Biceps, biceps brachii; TBH, triceps brachii humerotriceps; EMR, extensor metacarpi radialis; FCU, flexor carpi ulnaris; EDC, extensor digitorum communis; EMU, extensor metacarpi ulnaris; TBS, triceps brachii scapulohumeralis; TP Brevis, tensor propatagialis pars brevis; TP Long, tensor propatagialis pars longa; SHC, scapulohumeralis caudalis. (Figure modified from Dial 1992; © 1992 Oxford University Press, used with permission)

musculature (Fig. 2.108). For example, the hindlimb muscles of two flightless birds, Common Ostriches (*Struthio camelus*) and Emus (*Dromaius novaehollandiae*), represent 29% and 25%, respectively, of their total body mass (Table 2.2). The hindlimb muscles of groups of birds that are primarily terrestrial, but still capable of flight, such as tinamous (Tinamidae),

**Table 2.2** Variation among various families of birds in the relative mass (percent of total body mass) of the primary flight muscles and leg muscles

| Family | Pectoralis | Supracoracoideus | Leg muscles |
|---|---|---|---|
| Struthionidae (ostrich) | – | – | 29% |
| Dromaiidae (emu) | – | – | 25% |
| Tinamidae | 22–25% | 7–8% | 13–18% |
| Podicipedidae | 10% | 1–2% | 15–19% |
| Phalacrocoracidae | 11–12% | – | 10–12% |
| Anatidae | 13–21% | 1.5–2.1% | 4.5–7% |
| Accipitridae | 13–18% | 0.4–0.8% | 7.5–17% |
| Phasianidae | 15–22% | 5–7% | 12–17% |
| Tytonidae | – | – | 14.1% |
| Rallidae | 7–10% | 1–2% | 12–24% |
| Laridae | 12–14% | 1–1.5% | 3–7% |
| Caprimulgidae | 15–20% | 1.5–2% | 3–6% |
| Trochilidae | 16–25% | 7–12% | 1–2% |
| Tyrannidae | 15–23% | 1.5–2.5% | 3–7% |
| Hirundinidae | 14–18% | 1–2% | 1.7–3% |
| Paridae | 16–18% | 1.5–2% | 7–7.5% |
| Parulidae | 14–18% | 1.5–2.5% | 6–8% |
| Fringillidae | 15–19% | 1.5–2% | 5.5–8.5% |

Note that hummingbirds (Trochilidae) have a relatively large supracoracoideus compared to most other taxa of birds (Data from Hartman 1961, Alexander et al. 1979 [Common Ostrich, *Struthio camelus*], Patak and Baldwin 1993 [Emu, *Dromaius novaehollandiae*], and Mosto 2017 [Tytonidae])

pheasants (Phasianidae), and rails (Rallidae) are also relatively massive, representing 12–24% of total body mass (Table 2.2). Diving birds that propel themselves with their hindlimbs, such as grebes (Podicipedidae) and cormorants (Phalacrocoracidae), also have well-developed hindlimb muscles (Table 2.2). The hindlimb muscles of raptors that kill prey with feet and

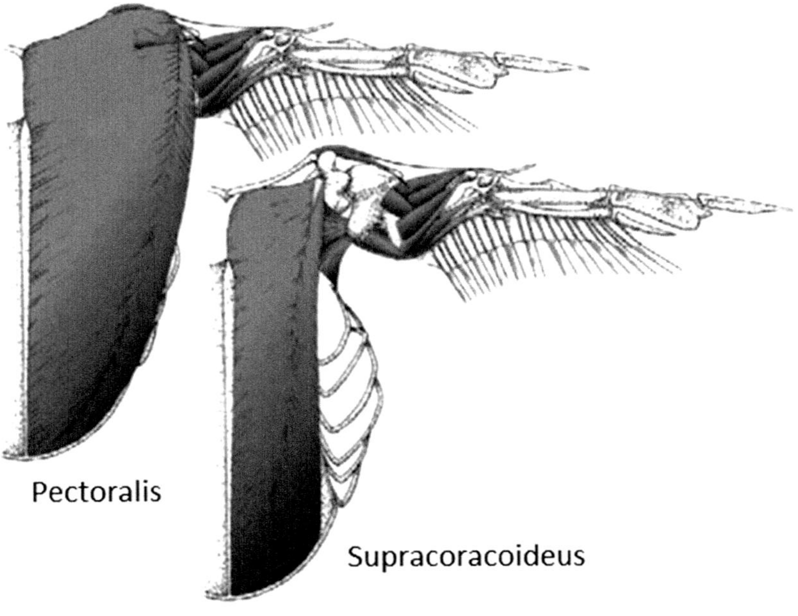


**Fig. 2.105** Main flight muscles of an Anna's Hummingbird (*Calypte anna*) (pectoralis) and a Rufous Hummingbird (*Selasphorus rufus*) (supracoracoideus). Note the relatively large size of the supracoracoideus compared to the pectoralis. (Figure modified from Warrick et al. 2012; © 2012 Elsevier Ltd., used with permission)

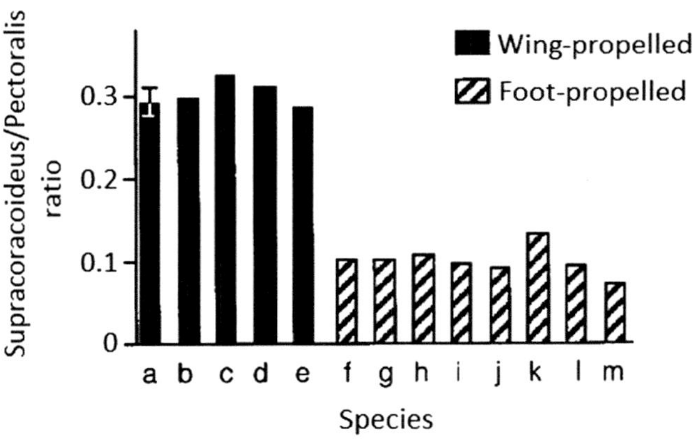


**Fig. 2.106** Ratio of supracoracoideus mass to pectoralis mass for several species of wing-propelled and foot-propelled diving birds. a. Atlantic Puffin, *Fratercula arctica*; b. Razorbill, *Alca torda*; c. Common Murre, *Uria aalge*; d. another Atlantic Puffin; e. Dovekie, *Alle alle*; f. Red-breasted Merganser, *Mergus serrator*; g. Common Merganser, *Mergus merganser*; h. Smew, *Mergellus albellus*; i. Great Crested Grebe, *Podiceps cristatus*; j. Red-necked Grebe, *Podiceps grisegena*; k. Little Grebe, *Tachybaptus ruficollis*; l. Red-throated Loon, *Gavia stellata*; m. Common Loon, *Gavia immer*. (Figure modified from Kovacs and Meyers 2000; © 2000 Wiley-Liss, Inc., used with permission)

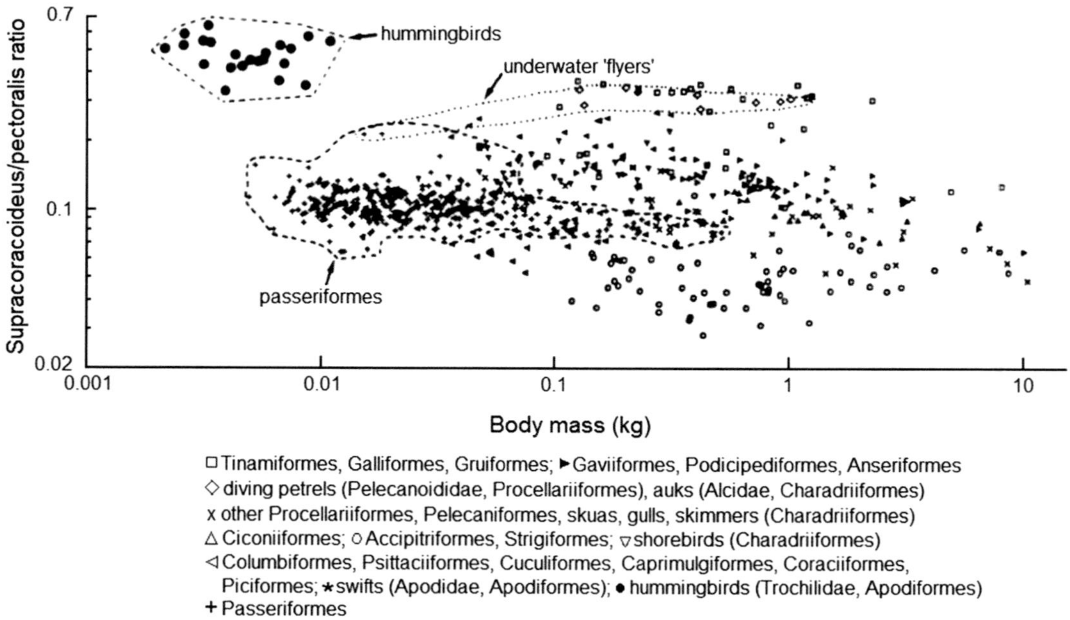


**Fig. 2.107** Ratio of the mass of the supracoracoideus muscle and the pectoralis muscle for birds that can fly. Hummingbirds have relatively large supracoracoideus muscles and very large ratios; underwater "flyers" also have relatively large ratios. (Figure modified from Rayner 1988; © 1988 Plenum Press, Published by Springer Nature, used with permission)

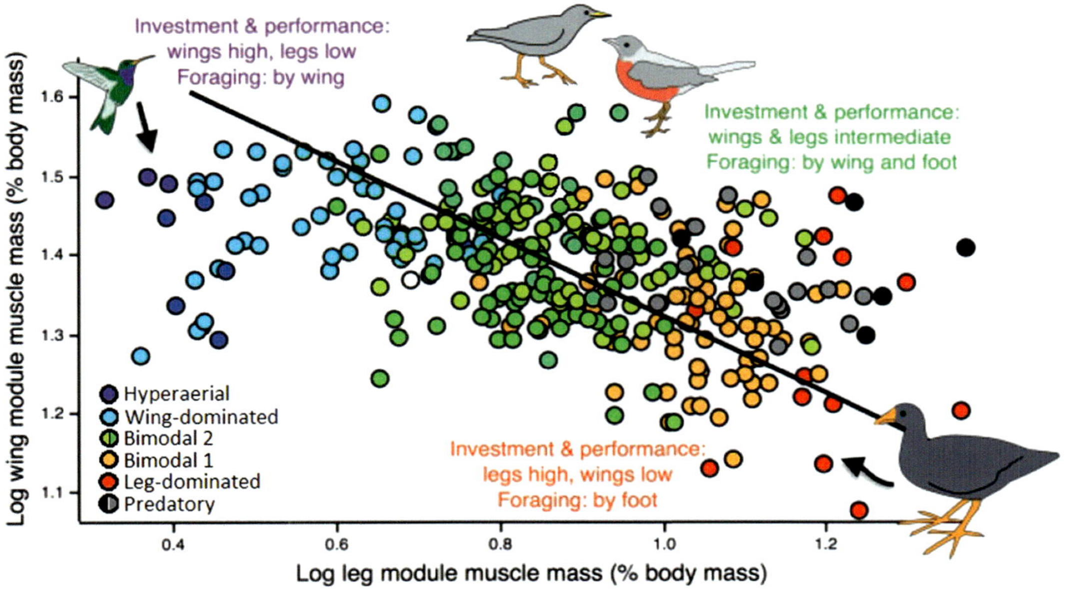


**Fig. 2.108** Across different taxa and species of birds, the relative importance of the wings and legs for locomotion is negatively correlated. For birds where flight is more important, flight (wing) muscle mass exceeds that of leg muscle mass (e.g., feeding on nectar and aerial insects). More terrestrial birds have greater leg muscle mass. Most birds are intermediate (or bimodal) and, at different times, need to fly and move about on the ground or in trees and shrubs (e.g., hopping and flying among tree branches to glean insects). (Figure modified from Heers 2016; © 2016 Oxford University Press, used with permission)

talons, such as eagles, hawks, and owls, are also well developed (Table 2.2, Fig. 2.109). The hindlimb muscles of birds tend to be located more proximally, with long tendons permitting movement of distal portions of the hindlimb (Fig. 2.110). This proximal concentration of muscle mass keeps most of the muscle nearer a bird's center of mass, which is, as noted previously, an important adaptation for flight.

The tail feathers of birds vary in number, relative length, shape, color, and function. For most birds, tails are an important part of their flight apparatus, helping birds maintain balance and stability, turn, and, when landing, brake. Other possible functions of tails include serving to brace birds that climb or roost on vertical or nearly vertical surfaces (e.g., woodpeckers and swifts) and, in several species where tail morphology has been influenced by sexual selection, playing an important role in female mate choice (Fig. 2.111). Regardless of function, however, the skeleton and muscles of bird tails exhibit little variation (Fig. 2.112). The two most medial rectrices are supported by the pygostyle and the others by rectricial bulbs, structures consisting primarily of connective tissue and fat (Box 2.17 Marvelous Tails (and Rectricial Bulbs) of Marvelous Spatuletails). Skeletal muscles originating on the pelvis and synsacrum attach to the caudal vertebrae and rectricial bulbs and control the movements of the tail (Fig. 2.112). The bulbi rectricium muscles are used for tail fanning, allowing birds to spread their tail feathers to varying degrees. The remaining caudal muscles are used to elevate (levator caudae), depress (depressor caudae), rotate, laterally deviate, or stabilize the variable-sized tail fan with respect to the body (Gatesy and Dial 1996a, b; Fig. 2.112), movements important for adjusting the amount of lift generated by the tail when a bird is flying as well as for turning and braking. For many species of birds, the levator caudae muscles that

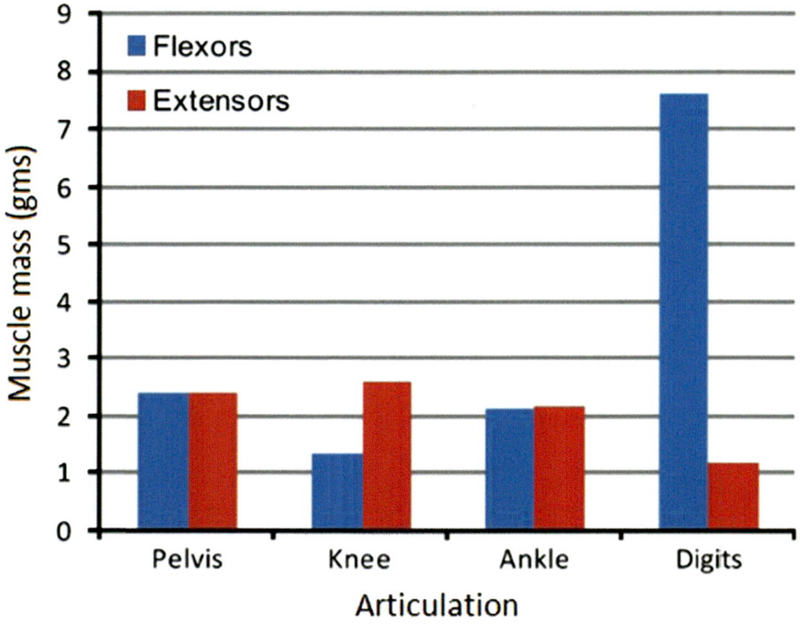


**Fig. 2.109** Muscle mass of the hindlimb of a Barn Owl (*Tyto alba*). Note the large mass of the muscles that flex the digits and are used to kill prey. (Figure from Mosto 2017; © 2016 Blackwell Verlag GmbH, used with permission)

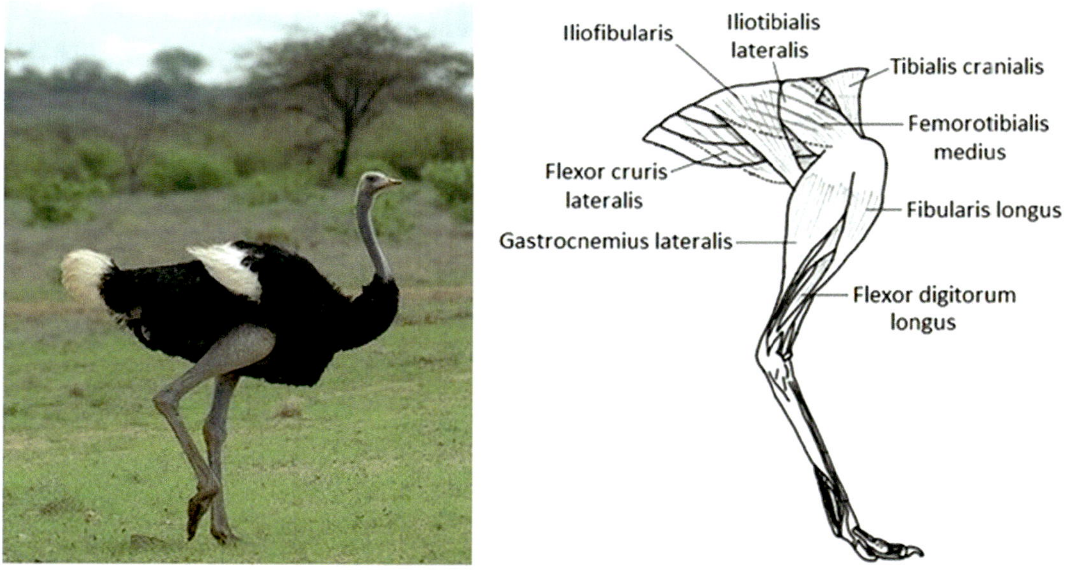


**Fig. 2.110** Left: Common Ostrich. Right: Hindlimb musculature of a Common Ostrich (*Struthio camelus*). Most of the muscle mass is on the proximal half of the hindlimb. (Figure modified from Smith et al. 2006; © Anatomical Society, used with permission. Photo by Donna Brown, Wikipedia, CC BY 2.0, https://creativecommons.org/licenses/by/2.0/)

**Fig. 2.111** Examples of species of birds where sexual selection has favored the evolution of tails with elongated feathers. (**a**) Booted Racket-tail (*Ocreatus underwoodii*), (**b**) Marsh Widowbird (*Euplectes hartlaubi*), (**c**) Violet-tailed Sylph (*Aglaiocercus coelestis*), (**d**) Pin-tailed Whydah (*Vidua macroura*), and (**e**) White-tailed Tropicbird (*Phaethon lepturus*). (Booted Racket-tail photo by Andy Morffew, pxhere.com, CC BY 2.0, ttps://creativecommons.org/licenses/by/2.0/; Marsh Widowbird photo by Davidk79, Wikipedia, CC BY 3.0, https://creativecommons.org/licenses/by/3.0/deed.en; Violet-tailed Sylph photo by Andy Morffew, pxhere.com, CC BY 2.0, https://creativecommons.org/licenses/by/2.0/; Pin-tailed Whyda photo by Derek Keats, Wikipedia CC BY 2.0, https://creativecommons.org/licenses/by/2.0/; White-tailed Tropicbird photo by USFWS/USFWS Pacific Islands, CC0 Public Domain)

elevate the tail and the depressor caudae muscles that depress that tail are the tail muscles with the greatest mass, indicating the importance of elevating and depressing the tail during flight and when landing (Mosto et al. 2020; Fig. 2.113).

### 2.6.4 Feeding

Some skeletal muscles play critical roles in obtaining and ingesting food. For raptors, hindlimb muscles that close their talons are critical for capturing and subduing prey (Fig. 2.114). In general, these muscles generate greater force in larger raptors that tend to take larger prey. However, even among similar-sized raptors, force production may vary depending on the types and sizes of their typical prey. Grip force also differs between owls and hawks, with owls producing significantly greater force during talon closure than hawks of similar size (Ward et al. 2002; Fig. 2.115). This difference between hawks and owls may be due to differences in when and how they hunt. Because owls typically hunt under low-light conditions and may be unable to relocate and capture prey that avoid an initial capture

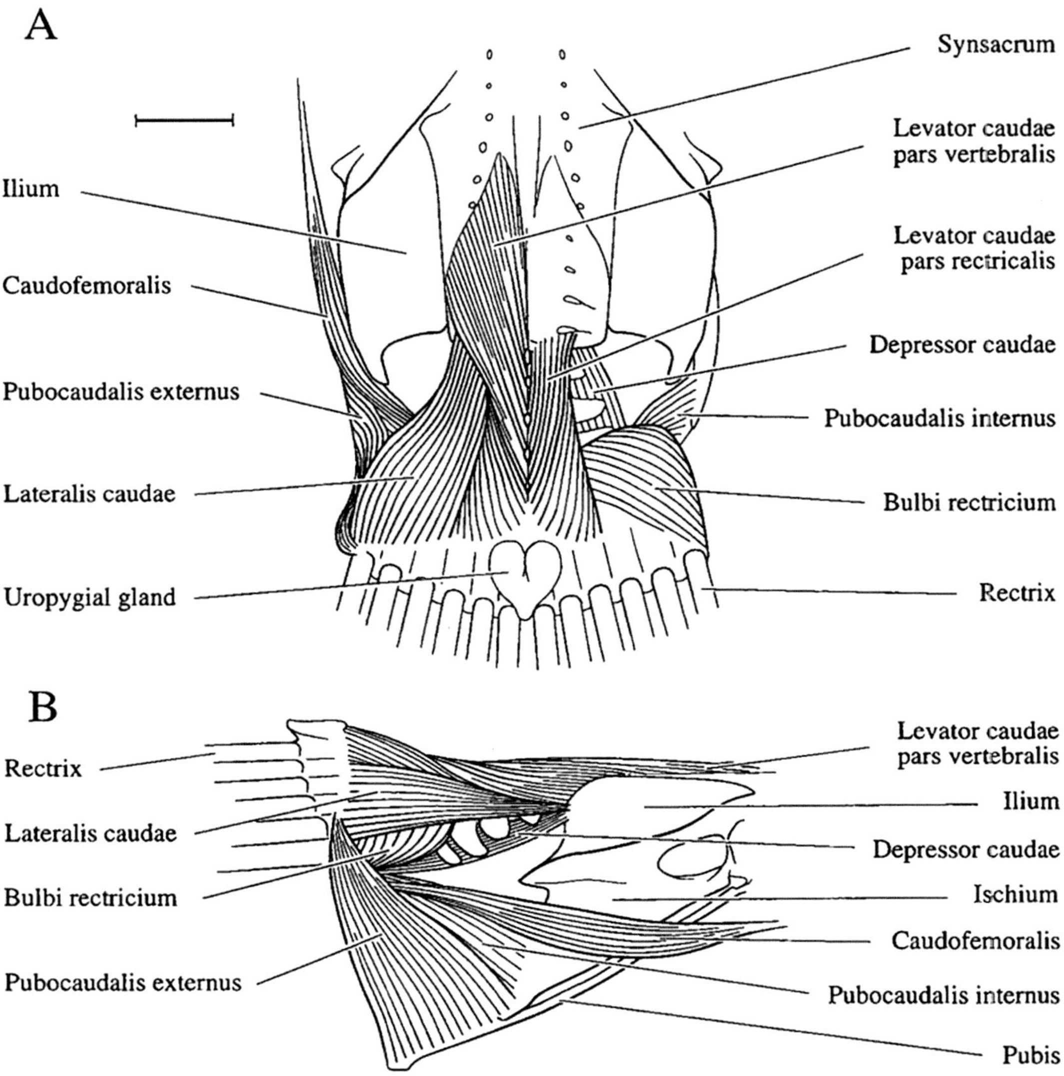


**Fig. 2.112** Tail muscles of a Rock Pigeon (*Columba livia*), showing dorsal (**a**) and right lateral (**b**) views. (**a**) Superficial muscles are shown on the left; deeper muscles on the right after the removal of the levator caudae pars vertebralis, lateralis caudae, pubocaudalis externus, and caudofemoralis. In pigeons, tail fanning is controlled by the bulbi rectricum muscles; the lateralis caudae muscles help maintain and move the fan. The caudofemoralis flexes the thigh (femur) and moves the tail laterally. The pubocaudalis muscles help depress or rotate the tail. Scale bar = 1 cm. (Figure from Gatesy and Dial 1996b; © 1996 The Society for the Study of Evolution, used with permission)

attempt, high force production at the time of impact and upon capture may reduce the likelihood that prey will be able to escape an initial attack (Usherwood et al. 2014; Madan et al. 2017). Hawks, in contrast, are diurnal hunters and can use visual cues to relocate and capture prey that avoid an initial capture attempt. In addition, hawks may attack prey more horizontally than owls, using rapid movement of their legs and talons to grab prey as they pass over it (Ward et al. 2002).

**Box 2.17 Marvelous Tails (and Rectricial Bulbs) of Marvelous Spatuletails**

Marvelous Spatuletails (*Loddigesia mirabilis*) are an endangered species of hummingbird found only in a small area in the Peruvian Andes. Males have very long tails (11–13 cm) and the curved outermost rectrices end with rackets. When displaying to conspecifics, adult males can rapidly move their outer rectrices and iridescent rackets in a wide variety of directions and configurations.

Male Marvelous Spatuletails can elevate the outermost rectrices when performing displays (Photo by Dubi Shapiro, used with permission of Dubi Shapiro)

Bird tails consist of caudal vertebrae, pygostyle, plus rectricial bulbs with feather follicles and the calami of rectrices and upper- and undertail coverts. The rectrices and rectricial bulbs are supported by the pygostyle, connective tissue, caudal vertebrae, and associated muscles, and movements of the rectrices are controlled by those muscles, including the levator caudae, depressor caudae, lateralis caudae, and bulbi rectricium (Zusi and Gill 2009).

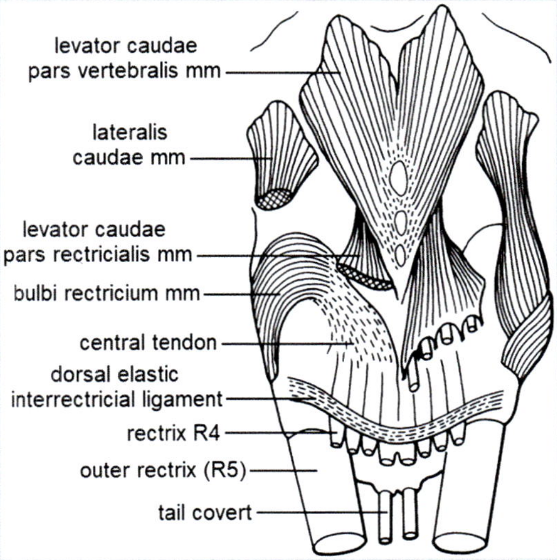


(continued)

**Box 2.17** (continued)

Dorsal view of the muscles of the tail of an adult male Marvelous Spatuletail (some muscles having been removed). The levator caudae muscles elevate the rectrices and the lateralis caudae muscles move them from side to side. (Figure modified from Zusi and Gill 2009; © 2009 Oxford University Press, used with permission)

Rectricial bulbs consist of rectrix calami, coverts, bulbi rectricium muscles, and connective tissue. In the tail of Marvelous Spatuletails, tendons from the levator caudae muscle attach to the dorsal side rectrix follicles and, ventrally, tendons from the depressor caudae muscles attach just distal to those of the bulbi rectricium muscles. In addition, the lateralis caudae muscles attach to the lateral side of rectrix follicles. Collectively, these muscles allow dorsal, ventral, lateral, and tilting movements of the rectrices as well as fanning of the rectrices (i.e., lateral spreading). The two rectricial bulbs can act either in unison or independently to allow more complex movements of the rectrices on each side of the tail (Zusi and Gill 2009).

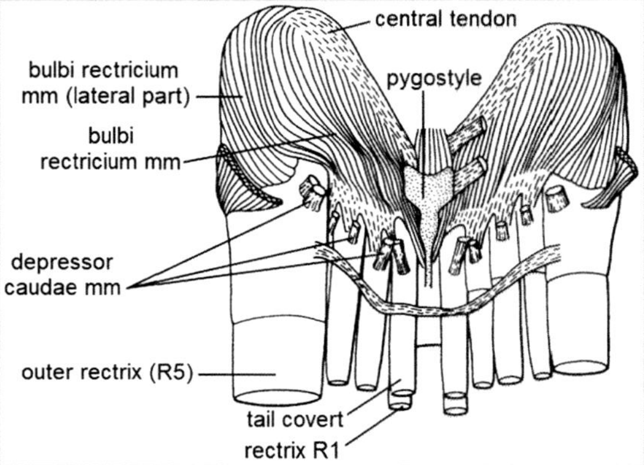


Recrticial bulbs of an immature male Marvelous Spatuletail with one adult outer rectrix (left side). The depressor caudae muscles depress (or lower) the rectrices and the bulbi rectricium muscles control tail fanning. (Figure modified from Zusi and Gill 2009; © 2009 Oxford University Press, used with permission)

Raptors, and many other birds, have a digital tendon-locking mechanism in their toes that helps them sustain grip force (Fig. 2.116). At the end of the toe flexor muscles (flexor digitorum longus) are tendons with thousands of small projections called tubercles (Fig. 2.116). The adjacent portion of the surrounding tendon sheath contains a series of folds, or plicae, that typically have a proximal slant (toward the base of the toe; Figs. 2.116 and 2.117). When the toes flex and the flexor tendons are pulled, the tubercles move proximally over the stationary plicae on the sheath. If there is resistance to flexion (e.g., a struggling prey attempting to escape the grip of a raptor or simply the toes of a perching bird as they close and grip a branch), the locking elements (tubercles and plicae) "catch" and the friction prevents slipping of the tendons. Such ratchet-like gripping helps birds maintain a tight grip or simply maintain the position of toes (e.g., when web-footed, foot-propelled birds flex their toes to enhance cupping of their feet as they swim) (Quinn and Baumel 1990; Einoder and Richardson 2006).

Related to the digital tendon-locking mechanism, several investigators have suggested that birds also have an automatic perching mechanism that holds them on their perches even when sleeping (Fig. 2.118). As proposed, when a bird settles

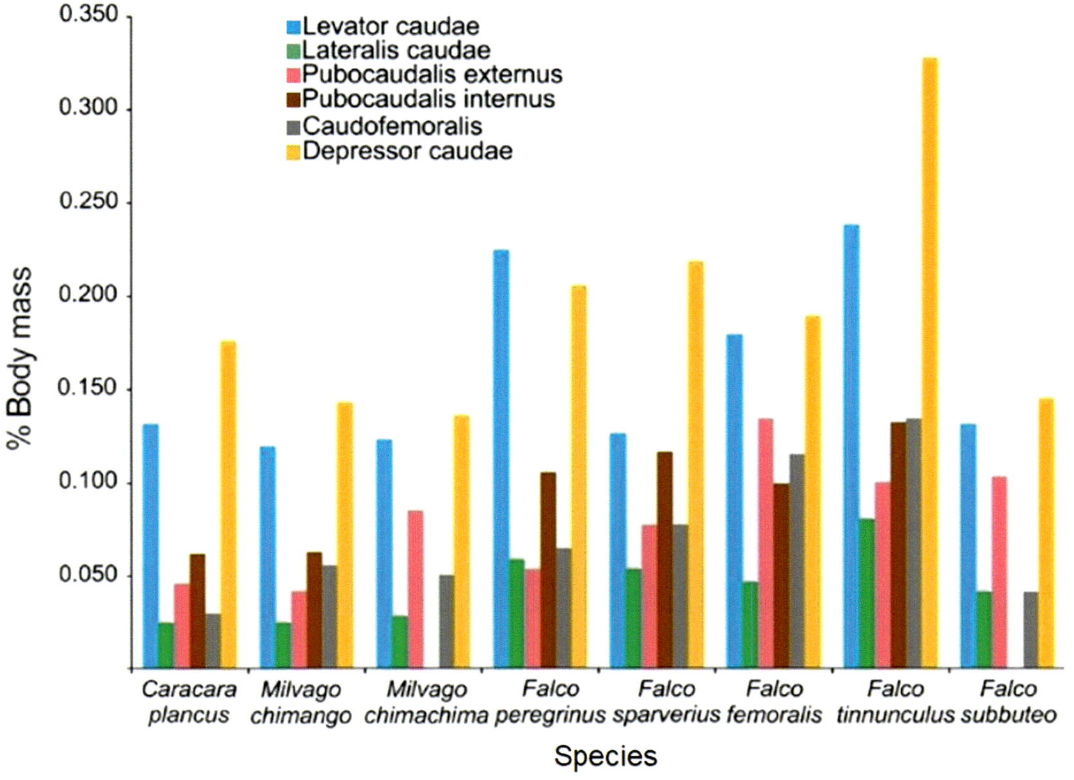


**Fig. 2.113** Percent of muscle mass of six tail muscles relative to total body mass for eight species of birds in the family Falconidae. Note the high values for the levator caudae and depressor caudae muscles. *Caracara plancus*, Crested Caracara; *Milvago chimango*, Chimango Caracara; *Milvago chimachima*, Yellow-headed Caracara; *Falco peregrinus*, Peregrine Falcon; *Falco sparverius*, American Kestrel; *Falco femoralis*, Aplomado Falcon; *Falco tinnunculus*, Eurasian Kestrel; *Falco subbuteo*, Eurasian Hobby. (Figure modified from Mosto et al. 2020; © 2019 The Royal Swedish Academy of Sciences, used with permission)

down on a perch, the tendons attached to muscles of the upper leg that extend behind the tarsometatarsus and attach to the toes are automatically pulled because the muscles and tendons are not sufficiently long or extensible. This increased tension then automatically flexes or closes the toes without any muscular contraction. The digital tendon-locking mechanism then locks the toes in place, allowing a bird to maintain a grip on a branch without continuous muscular effort even when sleeping.

Experimental evidence to date has not provided conclusive evidence for the existence of this automatic perching mechanism. Bock (1965) conducted experiments with Rock Pigeons (*Columba livia*) and American Crows (*Corvus brachyrhynchos*) and found that flexing the legs did not cause flexing of the toes because the leg muscles were sufficiently stretched to compensate for the extra distance around the heel joint (where the tibiotarsus articulates with the tarsometatarsus). Bock (1965) suggested that previous investigators may have conducted experiments with preserved birds or dead birds where the muscles were in rigor and would be less extensible. Similarly, Galton and Shepherd (2012) found that leg flexion did not cause toe flexion in living European Starlings (*Sturnus vulgaris*), but did so in sacrificed starlings after about 15–20 min and the onset of rigor mortis. More recently,

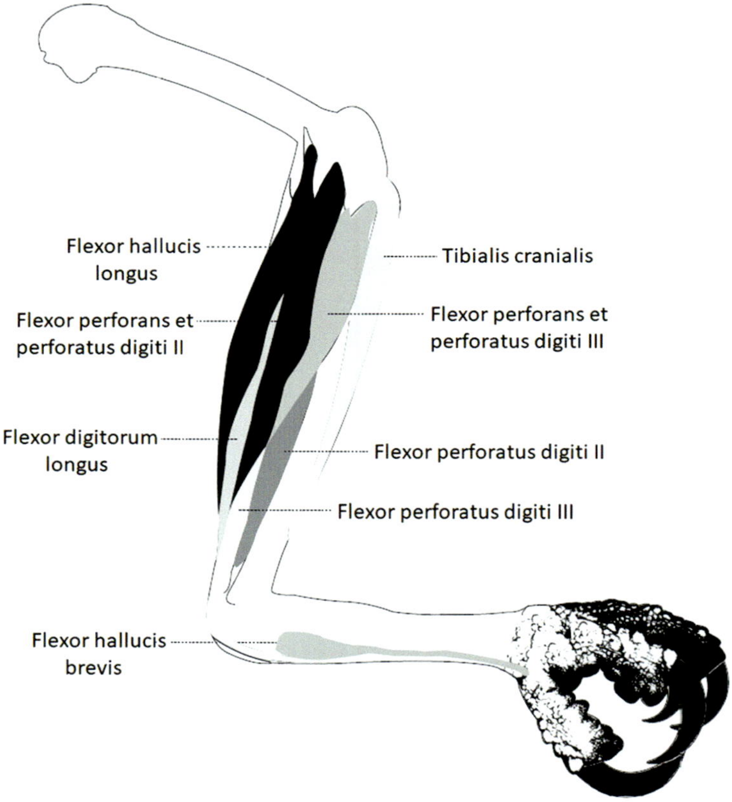


**Fig. 2.114** Digital flexor muscles of a White-tailed Eagle (*Haliaeetus albicilla*). Muscles with key roles in grasping and holding prey (primarily fish) include the flexor hallucis longus, flexor digitorum longus, tibialis cranialis, and flexor hallucis brevis. All have large cross-sectional areas of both muscle and tendons, with the flexor hallucis longus (that inserts on digit I, the hallux) being the largest. The flexor digitorum longus flexes the talons of digits II, III, and IV. The tibialis cranialis muscle flexes the ankle joint, which, in raptors, also indirectly helps flex the talons. Proximal muscles like the flexor hallucis longus have greater moment arms (i.e., a measure of the effectiveness of a muscle at contributing to a particular motion) than more distally located muscles. Long talons plus powerful muscles help White-tailed Eagles capture and hold their elusive fish prey. (Figure from Mosto et al. 2021; © 2021 The Zoological Society of London, used with permission. Photo of White-tailed Eagle by Jacob Spinks, Wikipedia, CC BY 2.0, https://creativecommons.org/licenses/by/2.0/)

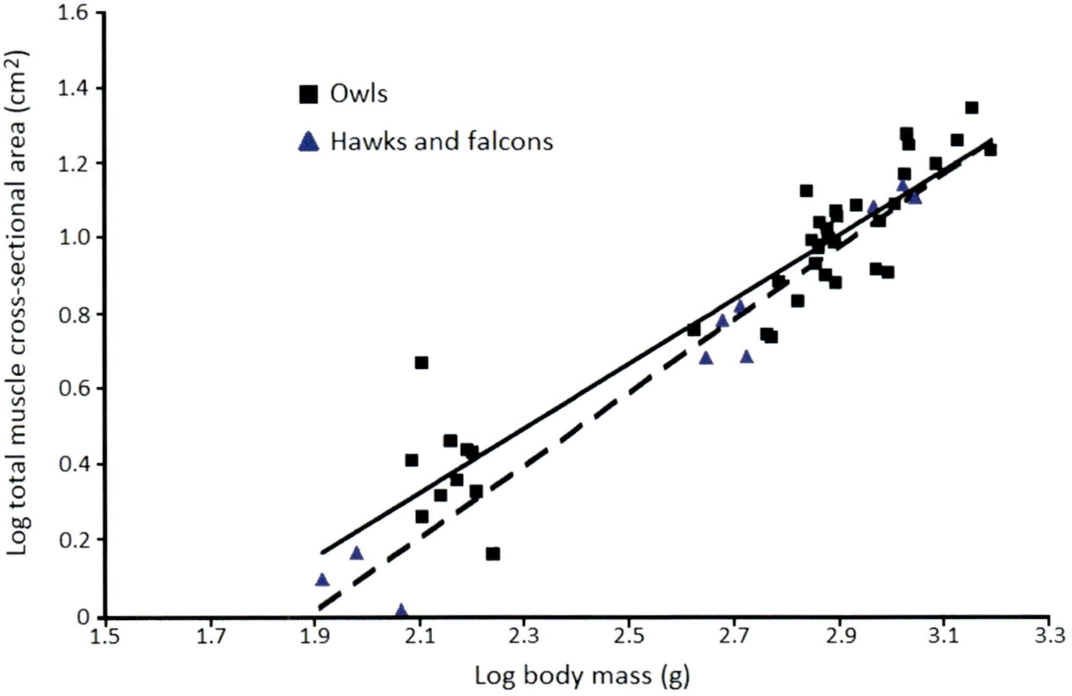


**Fig. 2.115** Log hindlimb muscle cross-sectional area relative to log body mass for 10 species of hawks and falcons and 39 species of owls. Note that hindlimb muscle cross-sectional area of owls generally exceeds that of hawks and falcons. (Figure modified from Ward et al. 2002; © 2002 Oxford University Press, used with permission)

Trbojević Vukičević et al. (2018) examined the hind limbs of Domestic Chickens and eight species of parrots (order Psittaciformes) and identified apparent automatic digital flexor mechanisms in all nine species. These authors suggested that, for these species, the digital-locking mechanism would likely reduce the need for muscle contraction when perching, even if not completely eliminating the need for contraction.

Even among nonraptors, differences in foraging strategies correspond to differences in the hindlimb muscles. For example, some birds, including some parids (tits, chickadees, and titmice) and mousebirds (Coliiformes), often hang below branches as they forage. In these species, the muscles that flex the hindlimbs to counteract the force of gravity are relatively larger than in other similar-sized birds that do not hang below branches as they forage (Berman and Raikow 1982; Moreno 1990; Moreno and Carrascal 1993). Similarly, birds such as woodcreepers, treecreepers, and woodpeckers that regularly climb vertical tree trunks when foraging must also flex their hindlimbs to maintain their position and have well-developed hindlimb flexor muscles (Moreno 1991; Raikow 1994).

The force generated by the muscles that close the avian bill also varies among species, with bite force positively correlated with body and head size and length of the moment arm (Anderson et al. 2008; Fig. 2.119, Table 2.3) and, for a given body and head size, with jaw muscle mass (van der Meij and Bout 2004; Fig. 2.120, Table 2.4). Species that must generate greater bite force to crack open large seeds, such as many parrots, have greater jaw muscle mass relative to their size than species that require less bite force (Fig. 2.121). In contrast, species that feed on small food items or soft-bodied prey such as

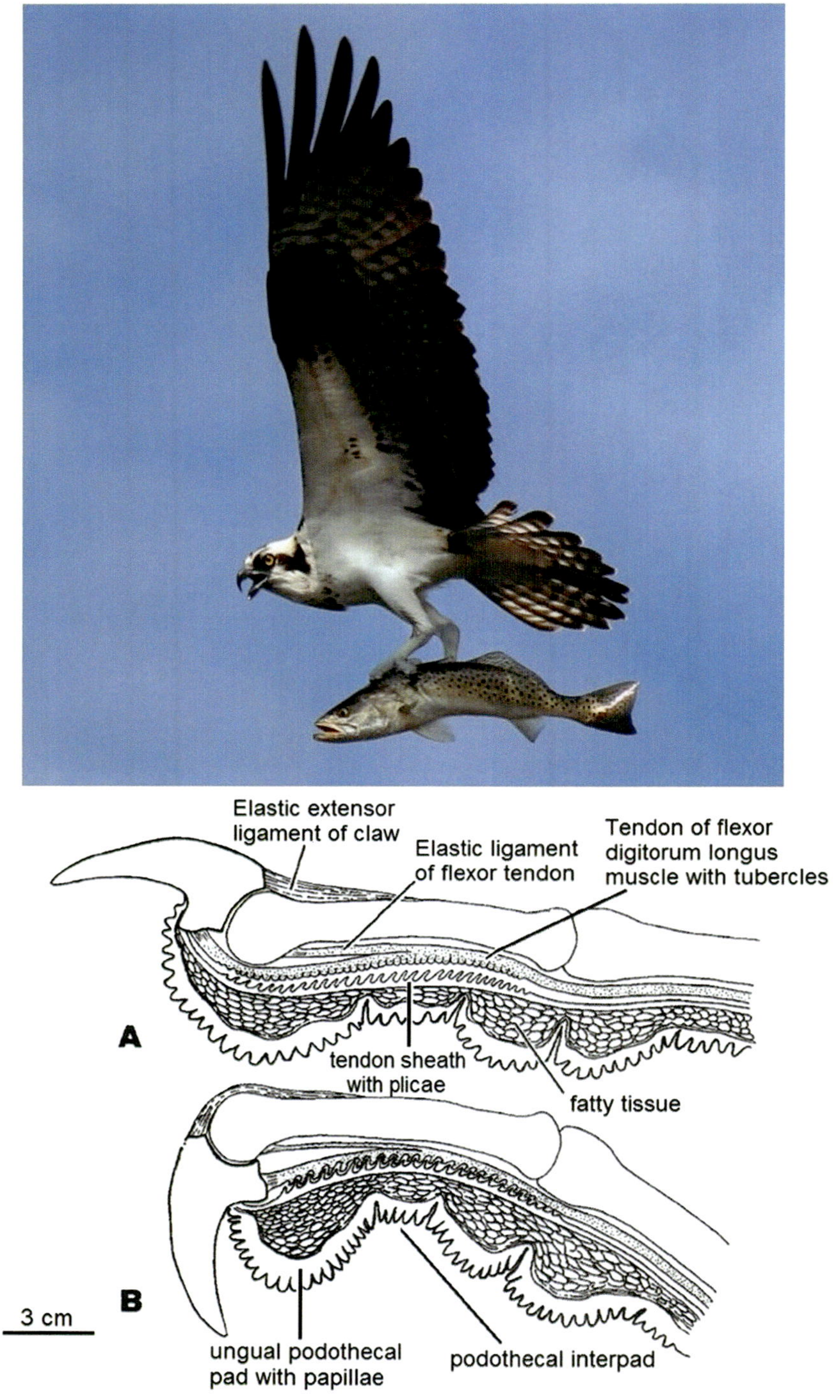


**Fig. 2.116** Top: Osprey (*Pandion haliaetus*) holding a captured fish in its talons. Bottom: (**a**) Claw extended. (**b**) Claw flexed. When flexed, the tubercles move over the plicae on the sheath. If there is resistance to flexion, the tubercles and plicae "catch" and the friction prevents slipping of the tendon. Friction also increases because, during flexion, the podothecal interpad and underlying tissue are forced up against the sheath. (Top photo by Mike Weimer, U. S. Fish and Wildlife Service, CC0 Public Domain. Bottom figure from Quinn and Baumel 1990; © 1990 Springer-Verlag, used with permission)

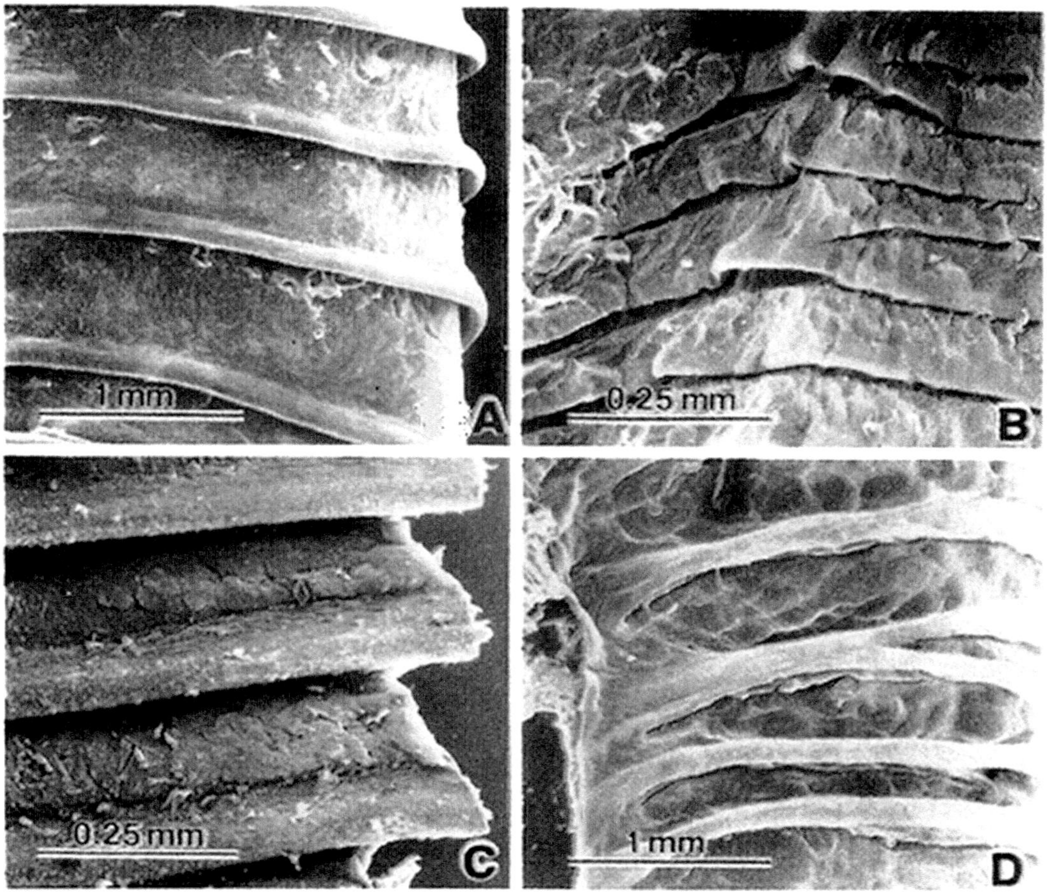


**Fig. 2.117** Scanning electron micrographs of tendon sheath plicae of four species of birds. (**a**) Great Blue Heron (*Ardea herodias*), (**b**) Hoatzin (*Opisthocomus hoazin*), (**c**) Red-shouldered Hawk (*Buteo lineatus*), and (**d**) Common Loon (*Gavia immer*). (Figure from Quinn and Baumel 1990; © 1990 Springer-Verlag, used with permission)

Mallards (*Anas platyrhynchos*, Anatidae) have reduced jaw muscle mass (Table 2.4).

The muscles of the avian tongue permit movement of the tongue, with the extent and importance of those movements varying among species (see Chap. 5 for additional details). However, for most birds, tongue mobility is limited to movements within or just short distances out of the oral cavity. For species that ingest smaller food items, back-and-forth movements of the tongue are important for moving food items through the oral cavity and toward the esophagus so they can be swallowed. Among some other birds, movement of the tongue short distances out of the oral cavity is also important when feeding. For example, some birds such as lorikeets sometimes feed on pollen or nectar by extending their tongue a short distance into flowers. Among species with limited tongue mobility, the muscles that protract and retract the tongue are relatively short, originating on the mandible or the bottom of the skull (Fig. 2.122).

Some birds, however, regularly extend their tongues considerable distances beyond the tip of the bill when foraging. In these birds, including woodpeckers, hummingbirds, and crossbills, the

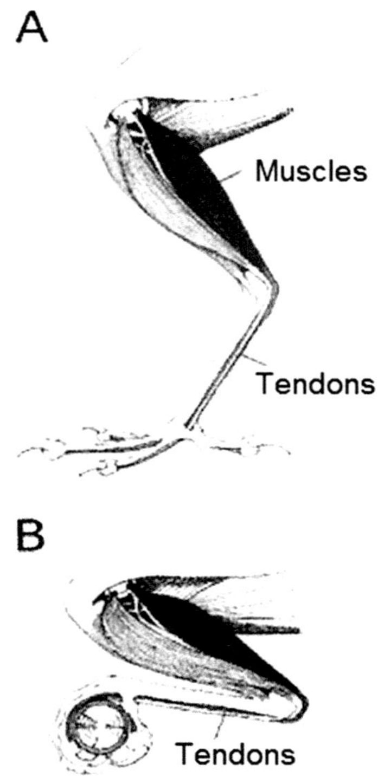


**Fig. 2.118** (**a**, **b**) The possible automatic perching mechanism of birds. The tendons that extend from the muscles of the upper leg (tibiotarsus) to the toes pass behind the tarsus (tarsometatarsus) so, when a bird lowers itself on a branch, it has been suggested that the tendons are automatically pulled, causing the toes to close around the branch. The tendons are then locked in placed by the digital tendon-locking mechanism. (Figure from Galton and Shepherd 2012; © 2012 Wiley Periodicals, Inc., used with permission)

protractor and retractor muscles of the tongue (and the hyoids) are considerably longer than in most birds, extending around the back of the skull and originating on top of the skull (Figs. 2.123 and 2.124). These muscles allow these birds to extend their tongues much further than most birds, allowing them to efficiently forage on insects or insect larvae (most woodpeckers), nectar (hummingbirds), and seeds within the cones of conifers (crossbills).

The neck or cervical musculature of birds is complex, consisting of numerous muscles, most of which cross more than one vertebra (Fig. 2.125). The musculature of the avian neck generally permits similar and complex movements of the head (controlled by muscles that originate on vertebrae and insert on the skull) and neck for most species (Fig. 2.126). These muscles can play important roles in such activities as food acquisition, preening, display, maintaining the position of the head during the flight (e.g., for birds that hover) and when walking (e.g., head-bobbing), and any other behavior that requires movement of the head or neck (e.g., drumming by woodpeckers; Box 2.18 Woodpecker Drumming Muscles). Although the anatomy and function of avian neck muscles has not been widely studied, some investigators have reported modifications in species where specialized movements are needed. For example, a neck muscle of swifts and hummingbirds exhibits a unique modification that may permit rapid and precise movements of the head in flight and when foraging (Fig. 2.127). Other examples

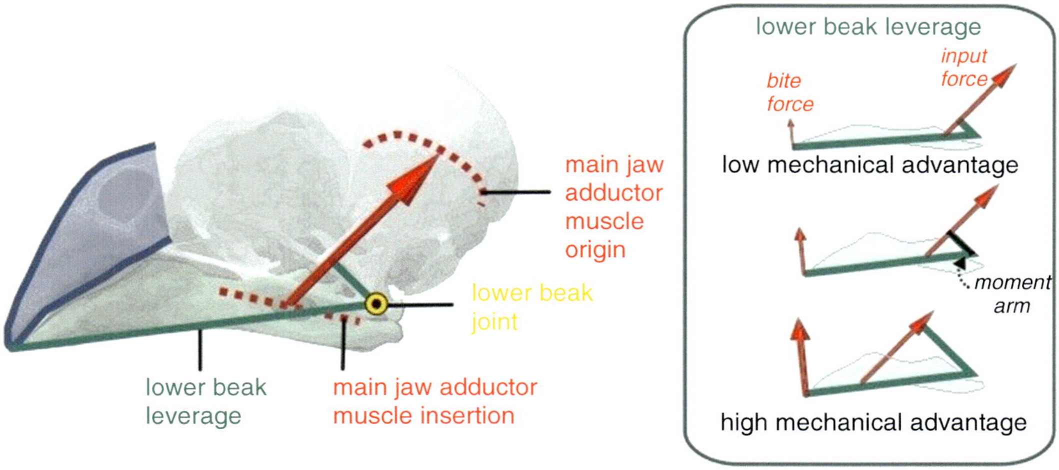


**Fig. 2.119** Higher bite forces are achieved as the moment arm increases in length and creates a higher mechanical advantage (moment arm = perpendicular distance from an axis to the line of action of a force). (Figure modified from Van Wassenbergh and Baeckens 2019; © 2019 The Author(s). Evolution © 2019 The Society for the Study of Evolution, used with permission)

**Table 2.3** Bite force (in Newtons) and body mass (grams) of several species of birds

| Common name | Scientific name | Order | Bite force | Body mass |
|---|---|---|---|---|
| American Kestrel | *Falco sparverius*[a] | Falconiformes | 3.50 | 78.8 |
| Prairie Falcon | *Falco mexicanus*[a] | Falconiformes | 16.50 | 487.7 |
| Merlin | *Falco columbarius*[a] | Falconiformes | 5.26 | 137.0 |
| Peregrine Falcon | *Falco peregrinus*[a] | Falconiformes | 16.90 | 683.6 |
| Sharp-shinned Hawk | *Accipiter striatus*[a] | Accipitriformes | 2.73 | 113.5 |
| Cooper's Hawk | *Accipiter cooperii*[a] | Accipitriformes | 3.90 | 342.7 |
| Java Sparrow | *Lonchura oryzivora*[b] | Passeriformes | 9.60 | 30.4 |
| Collared Grosbeak | *Mycerobas affinis*[b] | Passeriformes | 38.40 | 70.0 |
| Monk Parakeet | *Myiopsitta monachus* | Psittaciformes | 16.74 | 120.0 |

[a]From Sustaita (2008)

[b]From van der Meij and Bout (2004)

Note the impressive bite force of Collared Grosbeaks, a seed-eating bird. (Table modified from Carril et al. 2015; © 2015 Anatomical Society, Published by John Wiley and Sons, used with permission)

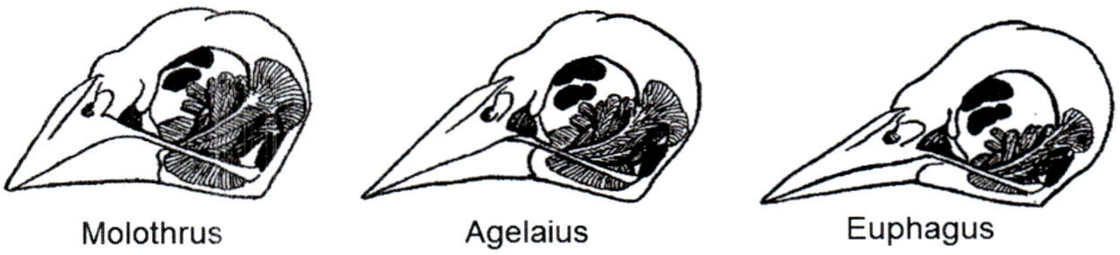


**Fig. 2.120** Variation in size of jaw muscles among different species of blackbirds that differ in food habits. Brown-headed Cowbirds (*Molothrus ater*) and Red-winged Blackbirds (*Agelaius phoeniceus*) are primarily seed eaters, but Red-winged Blackbirds feed on insects more than cowbirds. Rusty Blackbirds (*Euphagus carolinus*) feed on seeds and insects to similar degrees. Greater force is needed to crack open seeds and so blackbirds that are primarily seed eaters have larger jaw muscles than more omnivorous species. (Figure from Beecher 1951; © 1884 CCC Republication, used with permission)

**Table 2.4** Jaw muscle mass and body mass for species in different families that vary in body size (Data from van der Meij 2004; van der Meij and Bout 2004)

| Family | Species | Body mass (g) | Jaw muscle mass (g) | Percent total body mass of jaw muscles |
|---|---|---|---|---|
| Rheidae | *Rhea americana* | 12,500 | 19.8 | 1.6% |
| Anatidae | *Anas platyrhynchos* | 998 | 7.2 | 0.7% |
| Psittacidae | *Poicephalus senegalus* | 149 | 4.1 | 2.8% |
| Columbidae | *Columba livia* | 537 | 1.8 | 0.3% |
| Rallidae | *Fulica atra* | 450 | 1.5 | 0.3% |
| Charadriidae | *Calidris canutus* | 131 | 0.4 | 0.3% |
| Laridae | *Larus argentatus* | 416 | 4.4 | 1.1% |
| Paridae | *Parus major* | 15 | 0.12 | 0.8% |
| Passeridae | *Passer luteus* | 13 | 0.17 | 1.3% |
| Ploceidae | *Euplectus hordeacea* | 19 | 0.27 | 1.4% |
| Fringillidae | *Eophona migratoria* | 52 | 1.4 | 2.7% |

of birds with modifications of the neck musculature related to foraging techniques include grebes (for dislodging and crushing crayfish; Zusi and Storer 1969), skimmers (for skimming the water's surface with their lower jaw; Zusi 1962; Fig. 2.128), plunge divers (Chang et al. 2016;

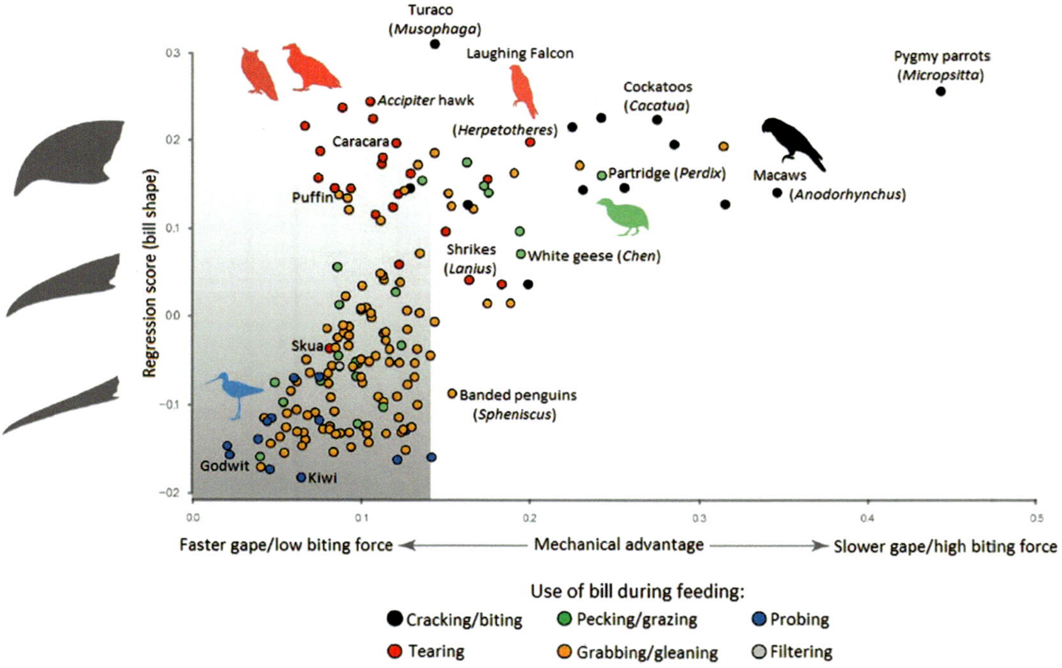


**Fig. 2.121** Relationship between beak shape and function. Most species, including grabbing/gleaning species, probing species, and even some tearing species, have fast gapes, but low biting force. Cracking/biting species have slower gapes, but high biting force. The gray shaded area indicates where 80% of species fall. (Figure modified from Navalón et al. 2019; © 2018 The Authors. Evolution © 2018 The Society for the Study of Evolution, used with permission)

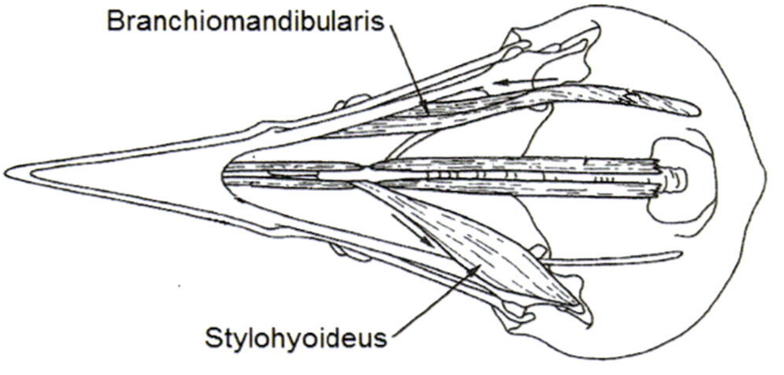


**Fig. 2.122** Top: Hawaii Amakihi (*Chlorodrepanis virens*). Bottom: Ventral view of the tongue muscles of a Hawaii Amakihi. As is typical of many birds, the primary protractor (branchiomandibularis) and retractor (stylohyoideus) muscles are relatively short and originate at the bottom of the skull or the mandible. (Figure modified from Richards and Bock 1973; © 1973 American Ornithological Society, used with permission. Photo by Bettina Arrigoni, Wikipedia, CC BY 2.0, https://creativecommons.org/licenses/by/2.0/)

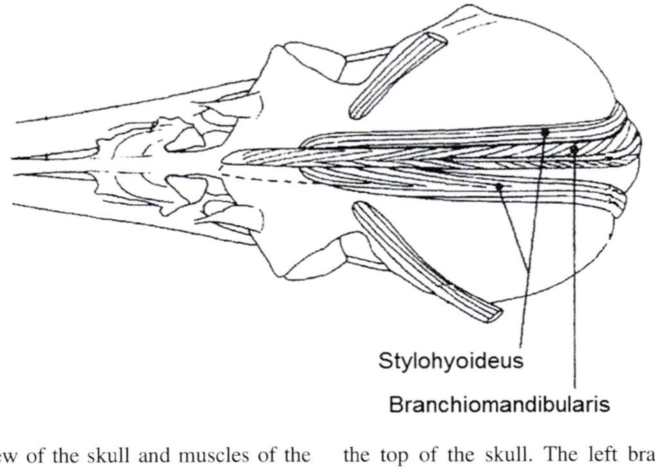


**Fig. 2.123** Dorsal view of the skull and muscles of the tongue and hyoid apparatus of a hummingbird, the Purple-throated Carib (*Eulampis jugularis*). As in woodpeckers, the muscles that extend (protract; branchiomandibularis) and retract (stylohyoideus) the tongue are very long, extending around the back of the skull and attaching to the top of the skull. The left branchiomandibularis has been removed. (Figure modified from Zusi and Bentz 1984; Reprinted from SCZ, Number 385 [1984] by Richard L. Zusi & Gregory Dean Bentz with the permission of the Smithsonian Institution)

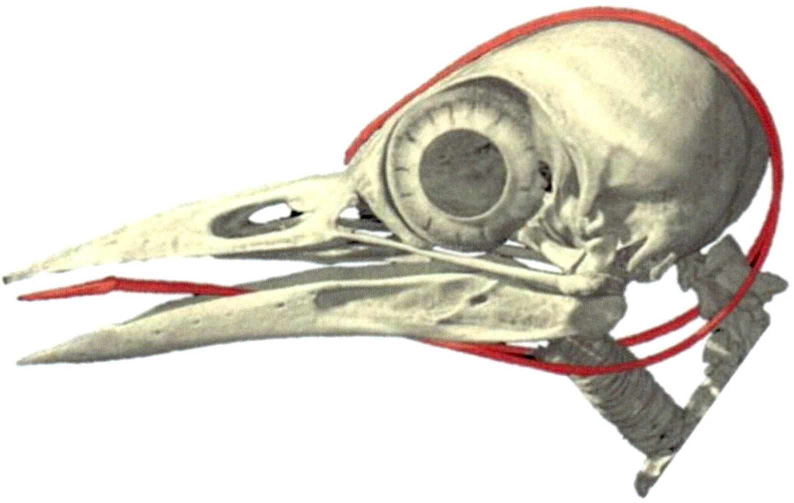

**Fig. 2.124** Skull and hyoid apparatus of a Red-bellied Woodpecker (*Melanerpes carolinus*) with the hyoid apparatus colored in red. Note that the hyoid in a living woodpecker would be supported by skeletal muscle, wraps around the skull, and inserts between the eyes at the base of frontal bone. (Figure from Jung et al. 2016; © 2016 Acta Materialia Inc. Published by Elsevier Ltd., used with permission)

Fig. 2.129), and herons and egrets (for the darting movements of their head and neck to capture prey; Kral 1965).

### 2.6.5 Extrinsic Eyeball Muscles

Birds have several extrinsic eye muscles that permit some rotation or movement of the eyes in the sockets (Figs. 2.130 and 2.131). In addition, extrinsic eye muscles (quadratus nictitantis and pyramidal nictitates) also control movements of the nictitating membrane that helps protect and moisten the eye. The extent of motion of the avian eye in the socket varies among species and ranges from movements of $\leq 1°$ to as much as 40° (Table 2.5). Rotation of eyes in the sockets serves two primary functions, gaze stabilization (i.e., keeping eyes focused on an object while the head moves) and scanning the environment. Variation among species in the extent of eye rotation is likely due to variation in the relative importance of these two functions. Factors that might contribute to this variation include food habits, foraging behavior, habitat, and variation in predation risk. The limited mobility of owl eyes is a result of selection favoring large tubular eyes (Fig. 2.132). Such eyes enhance nocturnal vision, but allow for very little movement within the sockets.

### 2.6.6 Vocalizing

Avian skeletal muscles also play a role in sound production by controlling the activity of the syrinx. The number of syringeal muscles varies among different groups of birds, as does the complexity of the sounds produced by the syrinx. Additional detail concerning the role of the syringeal muscles in sound production is provided in Chap. 12.

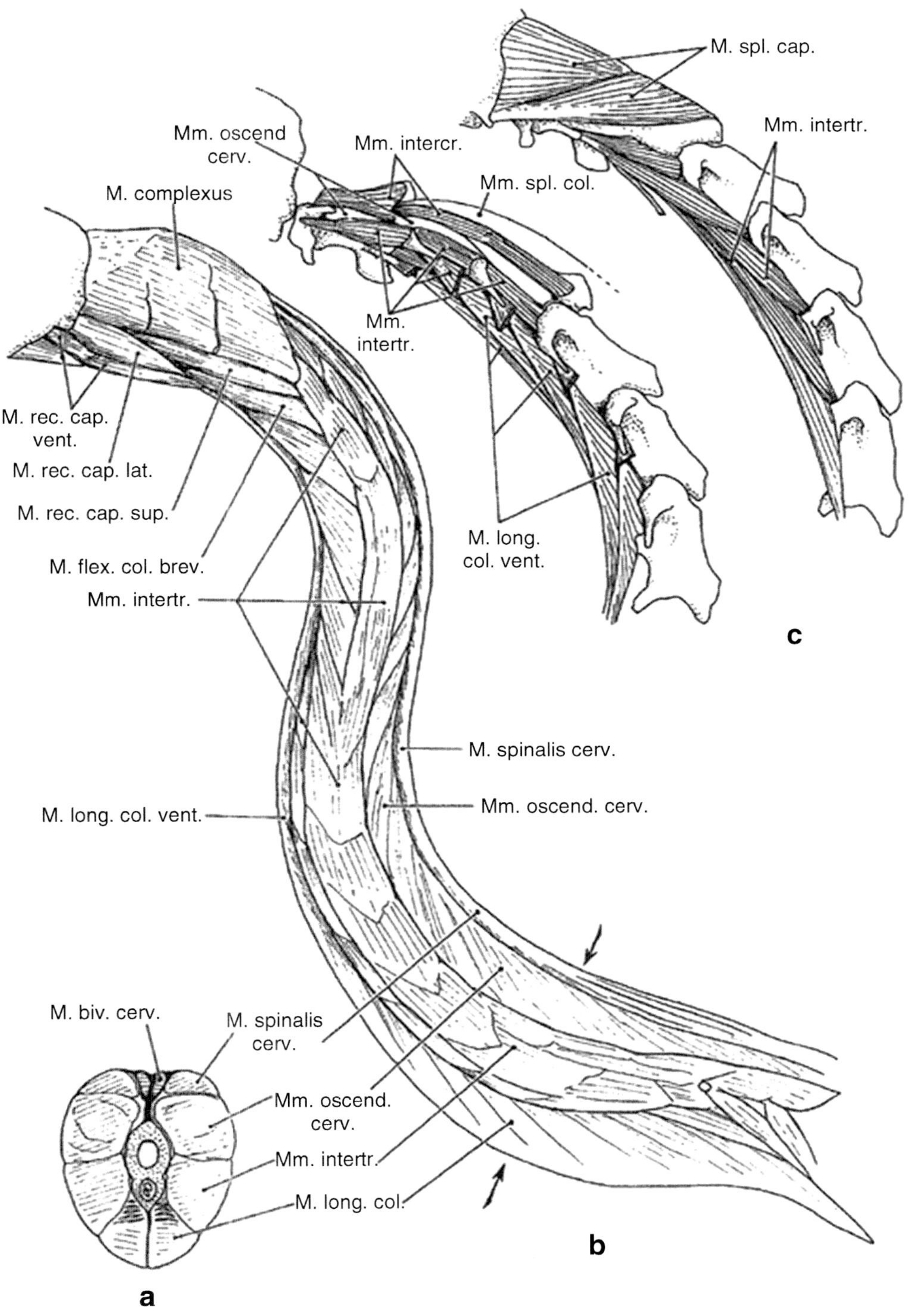


**Fig. 2.125** Neck or cervical muscles of a Pied-billed Grebe (*Podilymbus podiceps*). (**a**) Cross-section through vertebra number 15. (**b**) Superficial neck muscles. Arrows indicate location of cross-section shown in **a**. (**c**) Deep neck muscles. Abbreviations: *ascend* ascendentes, *biv* biventer, *brev* brevis, *cap* capitis, *cerv* cervicis, *col* colli, *flex* flexor, *intercr* intercristales, *intertr* intertransversarius, *lat* lateralis, *long* longus, *M* musculus, *Mm* musculi, *rec* rectus, *spl* splenii, *sup* superior, and *vent* ventralis. (Figure from Zusi and Storer 1969; used with permission)

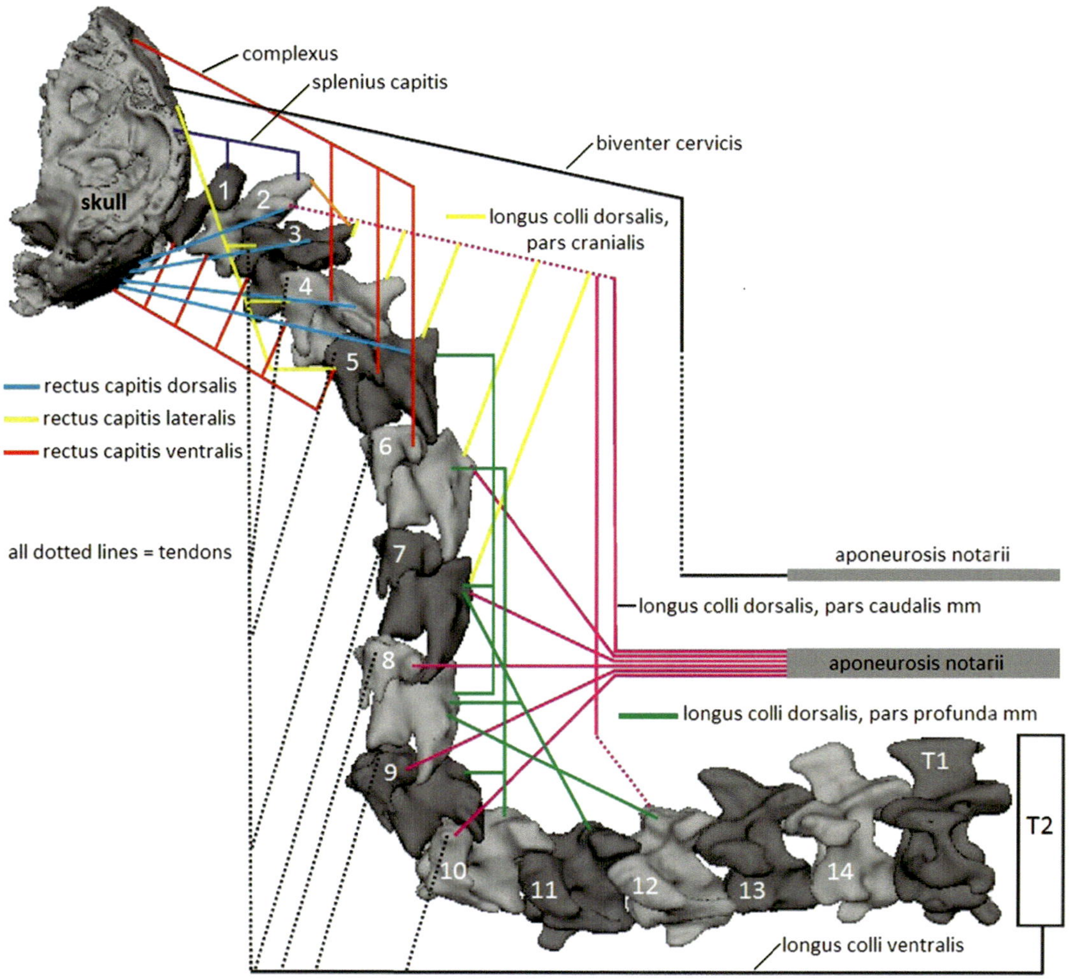


**Fig. 2.126** Representative muscles of the cervical region of the vertebral column of a Barn Owl, with likely functions of some muscles. The complexus muscle is important in the lateroflexion, dorsiflexion, and rolling of the head. The biventer cervicis muscle raises the neck and extends the head. The splenius capitis muscle moves the head upward and plays a role in rotating the head. The rectus capitis lateralis muscle appears to be important for lateroflexion (rotation) of the head. The rectus capitis ventralis muscle is important for the ventroflexion and lateroflexion of the head. The rectus capitis dorsalis ventroflexes the head. The longus colli dorsalis, pars cranialis muscle is likely important in protraction of the head. The longus colli ventralis muscle causes downward rotation of the neck and head. (Figure modified from Boumans et al. 2015; © 2015 Boumans et al., open-access article distributed under the terms of the Creative Commons Attribution License, https://creativecommons.org/licenses/by/4.0/)

### 2.6.7 Thermoregulation

In addition to the roles already described, skeletal muscles can also play an important role in thermoregulation via their ability to generate heat by shivering (shivering thermogenesis is discussed in more detail in Chap. 10). The skeletal muscles most important in generating heat in birds are the pectoralis muscles. The relatively large size of these muscles and proximity to the heart and other vital organs likely explain their important role in heat production (Olson 1994).

**Box 2.18 Woodpecker Drumming Muscles**

Woodpeckers defend territories and attract mates by drumming, rapidly striking their bills against resonant substrates. For example, drumming Downy (*Dryobates pubescens*) and Red-bellied (*Melanerpes carolinus*) woodpeckers hit substrates with their bills at rates of about 16 or 17 times per second. As noted by Schuppe and Fuxjager (2018), this means that the neck muscles of these woodpeckers, including the longus colli ventralis muscle, must move their heads back and forth about every 50–60 milliseconds. As such, within these muscles, vast numbers of calcium ions must diffuse out of (contraction) and be transported back into the sarcoplasmic reticulum (relaxation) every 50–60 milliseconds. Schuppe et al. (2018) thought that this rapid diffusion and transport of calcium ions might require more transport proteins in the membranes of the sarcoplasmic reticulum.

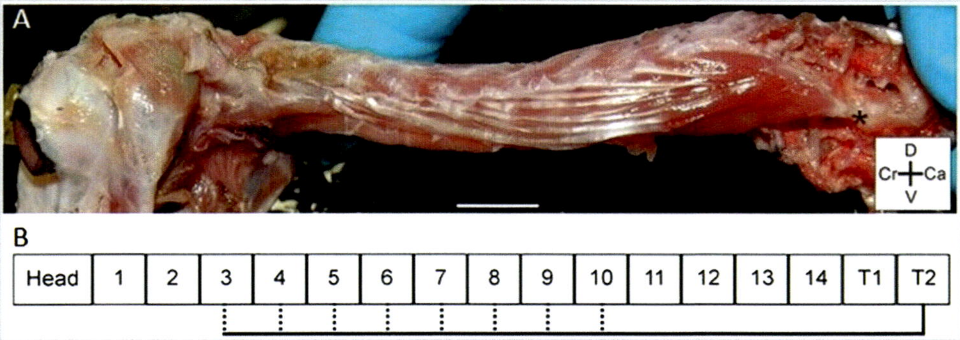


(**a**) Lateral view of the longus colli ventralis muscle that is important for woodpecker drumming. (**b**) Origin and insertion sites of the longus colli ventralis in the vertebral column; 1–14 are cervical vertebrae and T1 and T2 are thoracic vertebrae. (Figure from Boumans et al. 2015; © 2015 Boumans et al., open-access article distributed under the terms of the Creative Commons Attribution License, https://creativecommons.org/licenses/by/4.0/)

(continued)

**Box 2.18** (continued)

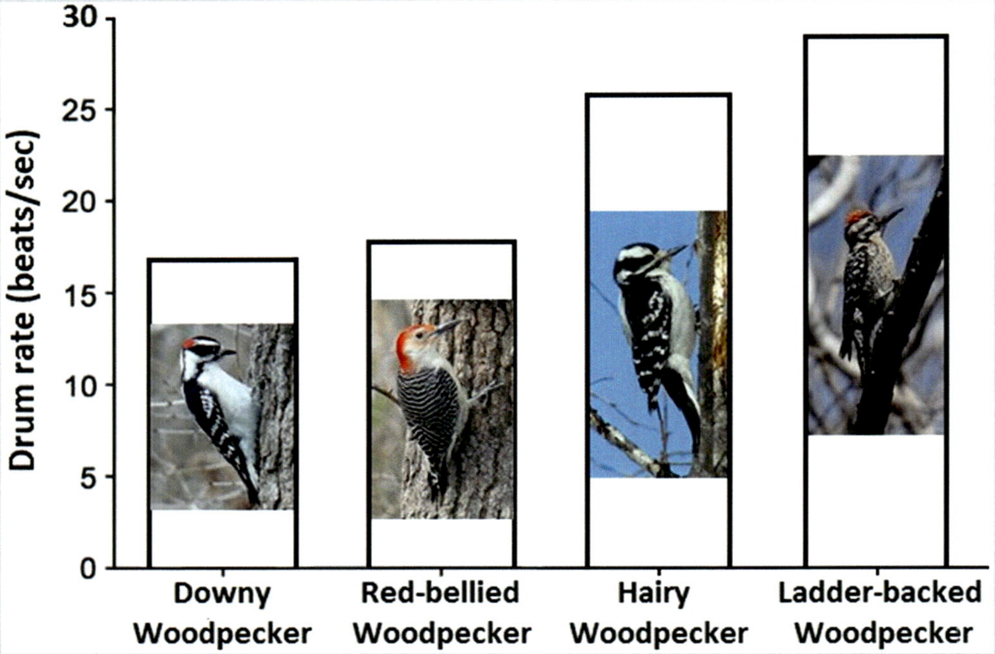


Drum rates of four species of woodpeckers: Downy Woodpecker (*Dryobates pubescens*), Red-bellied Woodpecker (*Melanerpes carolinus*), Hairy Woodpecker (*Dryobates villosus*), and Ladder-backed Woodpecker (*Dryobates scalaris*). (Drum rates from Stark et al. 1998; Downy Woodpecker and Red-bellied Woodpecker photos by Ken Thomas, Wikipedia, CC0 Public Domain; Hairy Woodpecker photo by Dewitp, Wikipedia, CC BY 3.0, https://creativecommons.org/licenses/by/3.0/; Ladder-backed Woodpecker photo by Andy Reago and Chrissy McClarren, Wikipedia, CC BY 2.0, https://creativecommons.org/licenses/by/2.0/deed.en)

To test this, Schuppe et al. (2018) compared the relative expression of genes needed to synthesize proteins important in the "calcium handling dynamics" of two muscles—one important in drumming and one that plays no role in drumming. They also compared gene expression of two species of woodpeckers that drum with a species of bird that does not drum (White-breasted Nuthatch, *Sitta carolinensis*). For the two species of woodpeckers, these authors found that the relative expression of a gene needed to produce a protein important in the uptake of calcium by the sarcoplasmic reticulum (as a muscle relaxes) was significantly higher in the muscle important in drumming (longus colli ventralis) than in the muscle not important for drumming (scapulohumeralis). In addition, the relative expression of that gene in the longus colli ventralis of the woodpeckers was significantly higher than in the same muscle of the nondrumming White-breasted Nuthatch. Thus, not surprisingly, selection favoring an ability to rapidly contract and relax muscles (i.e., drumming) has simultaneously favored the molecular mechanisms needed for such behavior.

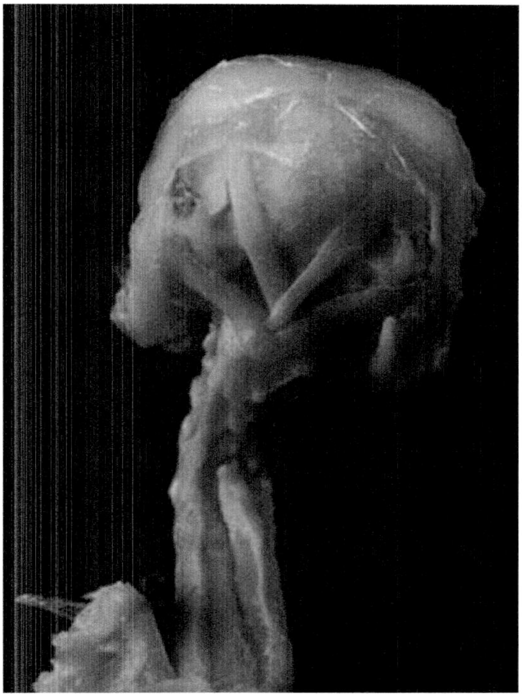

**Fig. 2.127** A muscle that originates on cervical vertebrae and inserts on the back of the head of swifts (like the Common Swift, *Apus apus*, shown here) and hummingbirds (splenius capitis muscle). Note how the different sections of the muscle criss-cross before inserting on the skull. This arrangement may permit rapid movements of the head during flight, potentially important when capturing insects (swifts and hummingbirds) or inserting a bill into a small flower to obtain nectar (hummingbirds). (Figure from Brause et al. 2009; © 2009 The Authors. Journal compilation © 2009 British Ornithologists' Union, used with permission)

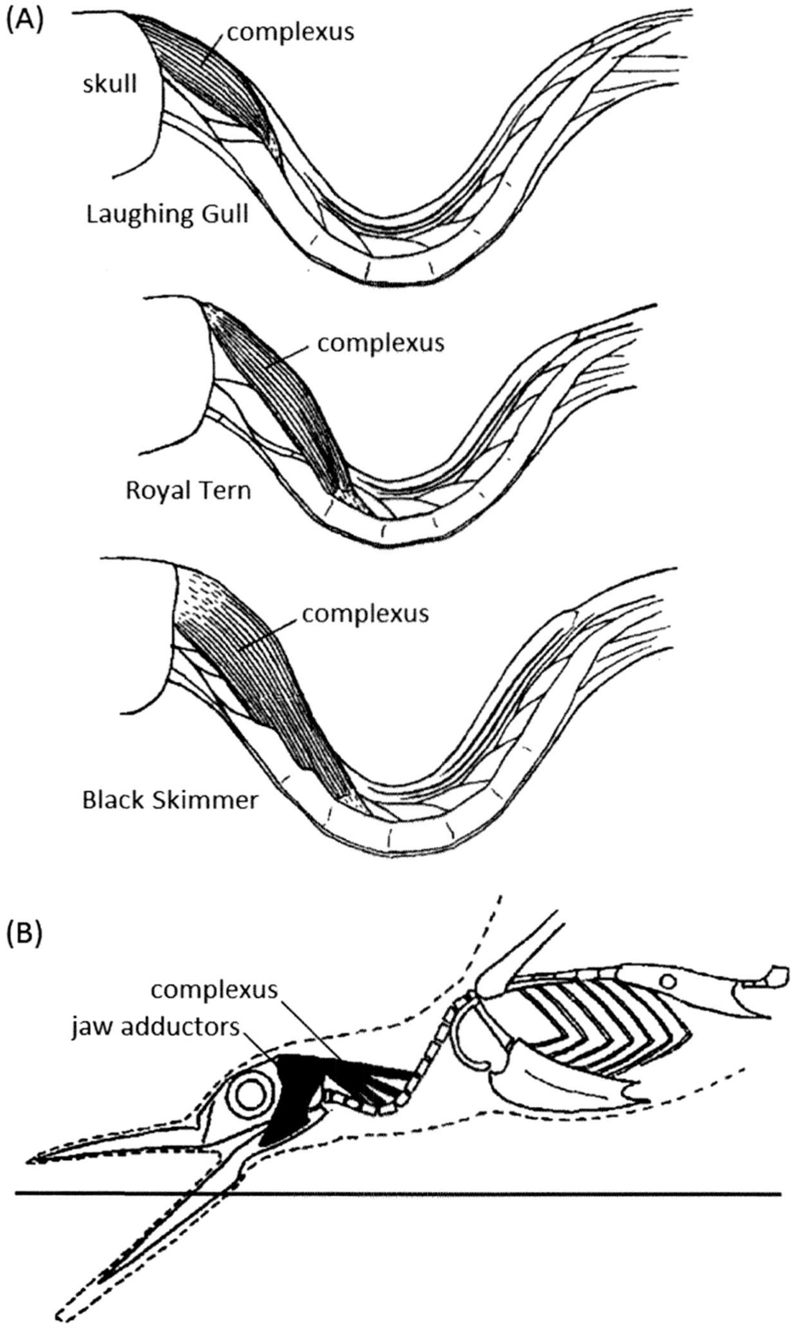


**Fig. 2.128** (**a**) Relative size of the complexus muscles of a Laughing Gull (*Leucophaeus atricilla*), Royal Tern (*Thalasseus maximus*), and Black Skimmer (*Rynchops niger*). Note the much larger complexus muscle of the Black Skimmer. (**b**) The large complexus muscle of Black Skimmers helps support and maintain the location of the head as they forage with their lower bills in the water. (Figures from Zusi 1962; used with permission of the Nuttall Ornithological Club)

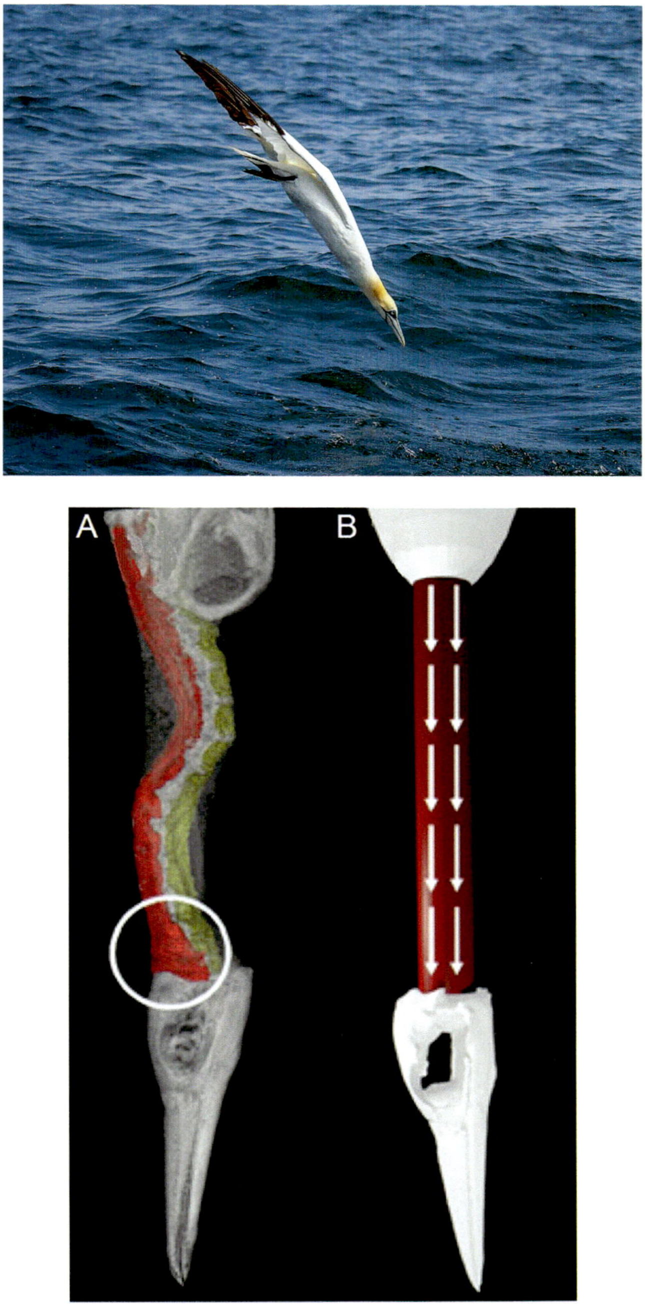


**Fig. 2.129** Top: A plunge-diving Northern Gannet (*Morus bassanus*) with wings retracted. Bottom: (**a**) CT scan of the head and neck of a Northern Gannet. A bundle of muscles is circled behind the skull. (**b**) Contracting muscles along the cervical vertebrae help to keep the neck straight and stable during a plunge-dive. (Figure from Chang et al. 2016; used with permission of the U. S. National Academy of Sciences. Photo by Equilibrium00, purchased from istockphoto.com)

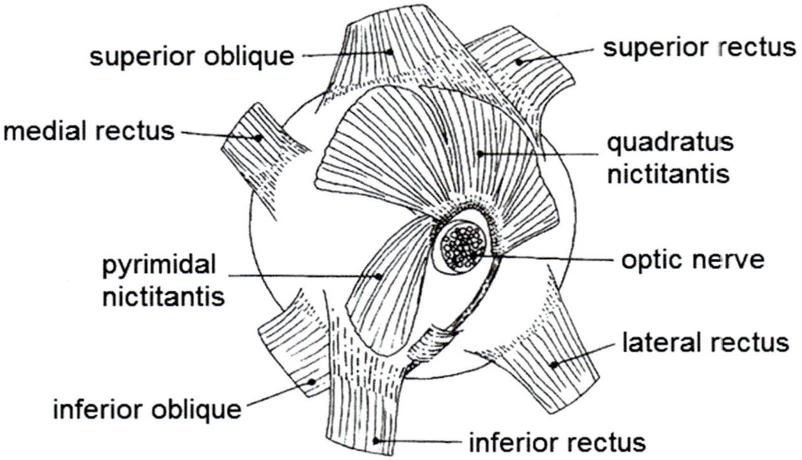


**Fig. 2.130** Extrinsic eyeball muscles of a hummingbird, the Purple-throated Carib (*Eulampis jugularis*). The oblique and rectus muscles move or rotate the eye in the socket; the nictitantis muscles move the nictitating membrane. (Figure modified from Zusi and Bentz 1984; Reprinted from SCZ, Number 385 (1984) by Richard L. Zusi & Gregory Dean Bentz, used with permission of the Smithsonian Institution)

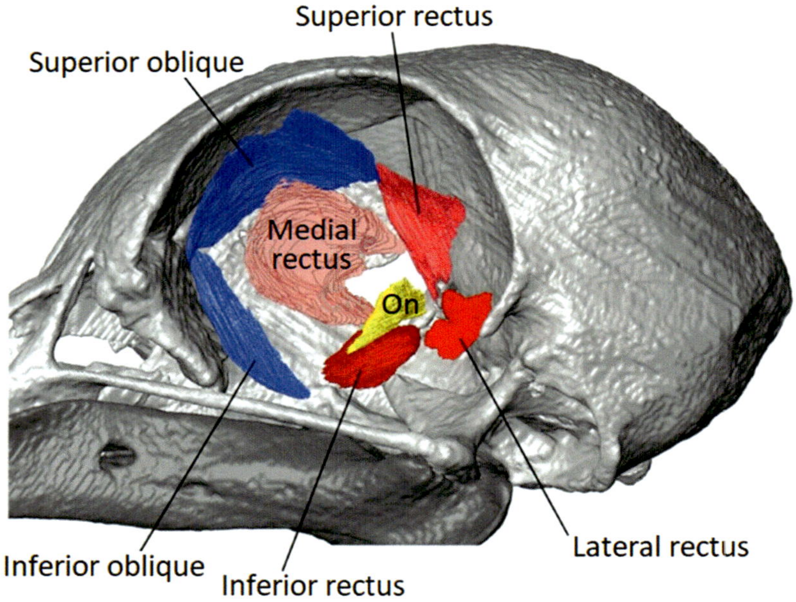


**Fig. 2.131** Extrinsic eyeball muscles of a Rock Pigeon (*Columba livia*). On, optic nerve. (Figure modified from Jones et al. 2019; open-access article distributed under the terms of the Creative Commons Attribution 4.0 International License, http://creativecommons.org/licenses/by/4.0/)

**Table 2.5** Extent of movement or rotation of eyes in the sockets of different species of birds

| Species/group | Maximum eye movement or rotation | Reference |
|---|---|---|
| Canada Goose (*Branta canadensis*) | 24.5° | Fernández-Juricic et al. (2011b) |
| Ibises and spoonbills | 14° | Martin and Portugal (2011) |
| Red-tailed Hawk (*Buteo jamaicensis*) | 5° | O'Rourk et al. (2010) |
| Cooper's Hawk (*Accipiter cooperii*) | 8.5° | O'Rourk et al. (2010) |
| American Kestrel (*Falco sparverius*) | 1° | O'Rourk et al. (2010) |
| Great Cormorant (*Phalacrocorax carbo*) | 15° | Martin et al. (2008) |
| Black-crowned Night-Heron (*Nycticorax nycticorax*) | 10–13.5° | Katzir and Martin (1998) |
| King Penguin (*Aptenodytes patagonicus*) | 10–15° | Martin (1999) |
| Gray-headed (*Thalassarche chrysostoma*) and Black-browed (*T. melanophris*) albatrosses | 20–25° | Martin (1998) |
| Southern Ground (*Bucorvus leadbeateri*) and Southern Yellow-billed (*Tockus leucomelas*) hornbills | 30–40° | Martin and Coetzee (2004) |
| Great Horned Owl (*Bubo virginianus*) | 1.5° | Steinbach and Money (1973) |
| Mourning Dove (*Zenaida macroura*) | 3° | Blackwell et al. (2009) |
| Senegal Parrot (*Poicephalus senegalus*) | 24° | Demery et al. (2011) |
| Black Phoebe (*Sayornis nigricans*) | 18° | Gall and Fernández-Juricic (2010) |
| American Crow (*Corvus brachyrhynchos*) | 17° | Fernández-Juricic et al. (2010) |
| California Scrub-Jay (*Aphelocoma californica*) | 8° | Fernández-Juricic et al. (2010) |
| White-crowned Sparrow (*Zonotrichia leucophrys*) | 12° | Fernández-Juricic et al. (2011a) |
| California Towhee (*Melozone crissalis*) | 11° | Fernández-Juricic et al. (2011a) |
| Brown-headed Cowbird (*Molothrus ater*) | 19° | Blackwell et al. (2009) |
| House Finch (*Haemorhous mexicanus*) | 11° | Fernández-Juricic et al. (2008) |
| House Sparrow (*Passer domesticus*) | 16° | Fernández-Juricic et al. (2008) |

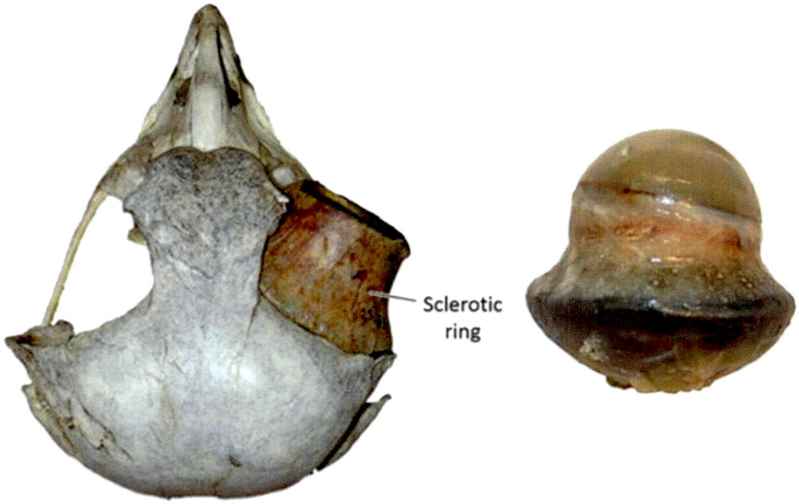


**Fig. 2.132** Left: Skull of a Great Horned Owl (*Bubo virginianus*), including one of the sclerotic rings that provides structural support for the tubular eyeballs (on the right; an eyeball of a Northern Saw-whet Owl, *Aegolius acadicus*) of owls that extend beyond the eye sockets. The large tubular eyeballs of owls limit the mobility of the eyes in the eye sockets so owls use head and neck movements to view their surroundings. (Figure on the left [skull] from Plochocki et al. 2018; © 2018 American College of Veterinary Ophthalmologists, used with permission. Figure on the right [tubular eye] from Lisney et al. 2012; © 2012 S. Karger AG, Basel, used with permission)

## References


Abourachid A (2000) Bipedal locomotion in birds: the importance of functional parameters in terrestrial adaptation in Anatidae. Can J Zool 78:1994–1998

Abourachid A, Fabre A-C, Cornette R, Höfling E (2017) Foot shape in arboreal birds: two morphological patterns for the same pincer-like tool. J Anat 231:1–11

Aires AS, Reichert LM, Müller RT, Andrade MB (2022) Review of morphology, development, and evolution of the notarium in birds. Anat Rec 305:2079–2098

Alexander RM, Maloiy GMO, Njau R, Jayes AS (1979) Mechanics of the running of the Ostrich (*Struthio camelus*). J Zool 187:169–178

Anderson RA, McBrayer LD, Herrel A (2008) Bite force in vertebrates: opportunities and caveats for use of a nonpareil whole-animal performance measure. Biol J Linn Soc 93:709–720

Anten-Houston MV, Ruta M, Deeming DC (2017) Effects of phylogeny and locomotor style on the allometry of body mass and pelvic dimensions in birds. J Anat 231: 342–358

Arnold P, Amson E, Fischer MS (2017) Differential scaling patterns of vertebrae and the evolution of neck length in mammals. Evolution 71:1587–1599

Backus SB, Sustaita D, Odhner LU, Dollar AM (2015) Mechanical analysis of avian feet: multiarticular muscles in grasping and perching. R Soc Open Sci 2: 140350

Baier DB, Gatesy SM, Dial KP (2013) Three-dimensional, high-resolution skeletal kinematics of the avian wing and shoulder during ascending flapping flight and uphill flap-running. PLoS ONE 8:e63982

Bailey JP, DeMont ME (1991) The function of the wishbone. Can J Zool 69:2751–2758

Baldwin J (1988) Predicting the swimming and diving behaviour of penguins from muscle biochemistry. Hydrobiologia 165:255–261

Bannasch R (1994) Functional anatomy of the 'flight' apparatus in penguins. In: Maddock L, Bone Q, Rayner JMV (eds) Mechanics and physiology of animal swimming. Cambridge University Press, Cambridge, pp 163–192

Baumel JJ, Witmer LM (1993) Osteologia. In: Baumel JJ, King AS, Breazile JE, Evans HE, Vanden Berge JC (eds) Handbook of avian anatomy: nomina anatomica avium, second ed. Publications of the Nuttall Ornithological Club, No. 23, Cambridge, MA, pp 45–132

Beaufrère H (2009) A review of biomechanic and aerodynamic considerations of the avian thoracic limb. J Avian Med Surg 23:173–185

Beecher WJ (1951) Adaptations for food-getting in the American blackbirds. Auk 68:411–440

Bellairs A, Jenkins CR (1960) The skeleton of birds. In: Marshall AJ (ed) Biology and comparative physiology of birds. Academic Press, New York, NY, pp 241–300

Berman SL, Raikow RJ (1982) The hindlimb musculature of the mousebirds (Coliiformes). Auk 99:41–57

Bhullar B-AS, Marugán-Lobón J, Racimo F, Bever GS, Rowe TB, Norell MA, Abzhanov A (2012) Birds have paedomorphic dinosaur skulls. Nature 487:223–226

Bhullar B-A, Hanson M, Fabbri M, Pritchard A, Bever GS, Hoffman E (2016) How to make a bird skull: major transitions in the evolution of the avian cranium, paedomorphosis, and the beak as a surrogate hand. Integr Comp Biol 56:389–403

Blackwell BF, Fernández-Juricic E, Seamans TW, Dolan T (2009) Avian visual system configuration and behavioural response to object approach. Anim Behav 77:673–684

Bleiweiss R (2009) The tail end of hummingbird evolution: parallel flight system development in living and ancient birds. Biol J Linn Soc 97:467–493

Block BA (1994) Thermiogenesis in muscle. Annu Rev Physiol 56:535–577

Bock WJ (1964) Kinetics of the avian skull. J Morphol 114:1–42

Bock WJ (1965) Experimental analysis of the avian passive perching mechanism. Am Zool 5:681

Bock WJ, Miller WD (1959) The scansorial foot of the woodpeckers, with comments on the evolution of perching and climbing feet in birds. American Museum Novitates No. 1931. American Museum of Natural History, New York, NY

Böhmer C, Plateau O, Cornette R, Abourachid A (2019) Correlated evolution of neck length and leg length in birds. R Soc Open Sci 6:181588

Bolton M, Monaghan P, Houston DC (1993) Proximate determination of clutch size in Lesser Black-backed Gulls: the roles of food supply and body condition. Can J Zool 71:273–279

Borges R, Fonseca J, Gomes C, Johnson WE, O'Brien SJ, Zhang G, Gilbert MTP, Jarvis ED, Antunes A (2019) Avian binocularity and adaptation to nocturnal environments: genomic insights from a highly derived visual phenotype. Genome Biol Evol 11:2244–2255

Bosse M, Spurgin LG, Laine VN, Cole EF, Firth JA, Gienapp P, Gosler AG, McMahon K, Poissant J, Verhagen I, Groenen MA (2017) Recent natural selection causes adaptive evolution of an avian polygenic trait. Science 358:365–368

Botelho JF, Ossa-Fuentes L, Soto-Acuña S, Smith-Paredes D, Nuñez-León D, Salinas-Saavedra M, Ruiz-Flores M, Vargas AO (2014) New developmental evidence clarifies the evolution of wrist bones in the dinosaur-bird transition. PLoS Biol 12:e1001957

Botelho JF, Smith-Paredes D, Vargas AO (2015) Altriciality and the evolution of toe orientation in birds. Evol Biol 42:502–510

Botelho JF, Smith-Paredes D, Soto-Acuña S, O'Connor J, Palma V, Vargas AO (2016) Molecular development of fibular reduction in birds and its evolution from dinosaurs. Evolution 70:543–554

Boumans MLLM, Krings M, Wagner H (2015) Muscular arrangement and muscle attachment sites in the cervical region of the American Barn Owl (*Tyto furcate pratincola*). PLoS ONE 10:e0134272

Bout RG, Zweers GA (2001) The role of cranial kinesis in birds. Comp Biochem Physiol 131A:197–205

Brause C, Gasse H, Mayr G (2009) New observation of the splenius capitus and rectus capitis ventralis muscles of the Common Swift *Apus apus* (Apodidae). Ibis 151: 633–639

Bribiesca-Contreras F, Parslew B, Sellers WI (2021) Functional morphology of the forelimb musculature reflects flight and foraging styles in aquatic birds. J Ornithol 162:779–793

Bühler P (1981) Functional anatomy of the avian jaw apparatus. In: King AS, McLelland J (eds) Form and function in birds, vol 2. Academic Press, London, pp 439–468

Bush JWM, Peaudecerf F, Prakash M, Quéré D (2010) On a tweezer for droplets. Adv Colloid Interf Sci 161:10–14

Caple G, Balda RP, Willis WR (1983) The physics of leaping animals and the evolution of preflight. Am Nat 121:455–476

Carrascal LM, Moreno E, Telleria JL (1990) Ecomorphological relationships in a group of insectivorous birds of temperate forests in winter. Holarct Ecol 13:105–111

Carril J, Degrange FJ, Tambussi CP (2015) Jaw myology and bite force of the Monk Parakeet (Aves, Psittaciformes). J Anat 227:34–44

Chang B, Croson M, Straker L, Gart S, Dove C, Gerwin J, Jung S (2016) How seabirds plunge-dive without injuries. Proc Natl Acad Sci USA 113:12006–12011

Chen A, White ND, Benson RB, Braun MJ, Field DJ (2019) Total-evidence framework reveals complex morphological evolution in nightbirds (Strisores). Diversity 11:143

Chiappe LM (1996) Late Cretaceous birds of southern South America: anatomy and systematics of Enantiornithes and *Patagopteryx de-ferrariisi*. Miinchner Geowiss Abh (A) 30:203–244

Chiappe LM (2009) Downsized dinosaurs: the evolutionary transition to modern birds. Evol Educ Outreach 2: 248–256

Chiappe LM, Dyke GJ (2002) The Mesozoic radiation of birds. Annu Rev Ecol Syst 33:91–124

Chiappe LM, Walker CA (2002) Skeletal morphology and systematics of the Cretaceous Euenantiornithes (Ornitholthoraces: Enantiornithes). In: Chiappe LM, Witmer LM (eds) Mesozoic birds: above the heads of dinosaurs. University of California Press, Berkeley, CA, pp 240–267

Chiappe LM, Ji S-A, Ji Q, Norell MA (1999) Anatomy and systematics of the Confuciusornithidae (Theropoda:

Aves) from the Late Mesozoic of northeastern China. Bull Am Mus Nat Hist 242:1–89

Choudhary OP, Choudhary P, Kalita PC, Arya RS, Rajkhowa T, Kalita A, Doley PJ, Keneisenuo K (2021) Morphometric and radiographic characteristics of the skull in Crested Serpent Eagle (*Spilornis cheela*) and Brown Wood Owl (*Strix leptogrammica*). Indian J Anim Res 55:426–432

Claessens LPAM (2004) Dinosaur gastralia: origin, morphology, and function. J Vertebr Paleontol 24:89–106

Claessens LPAM (2009) A cineradiographic study of lung ventilation in *Alligator mississippiensis*. J Exp Zool 311A:563–585

Clark CJ (2010) The evolution of tail shape in hummingbirds. Auk 127:44–56

Clark CJ (2011) Effects of tail length on an escape maneuver of the Red-billed Streamertail. J Ornithol 152:397–408

Clark CJ, Rankin D (2019) Subtle, pervasive genetic correlation between the sexes in the evolution of dimorphic hummingbird tail ornaments. Evolution 74:528–543

Clifton GT, Carr JA, Biewener AA (2018) Comparative hindlimb myology of foot-propelled swimming birds. J Anat 232:105–123

Codd JR (2004) The uncinate processes in birds and their implications for the breathing mechanics of maniraptorian dinosaurs. Ph. D. dissertation, University of Bonn, Bonn, Germany

Codd JR (2010) Uncinate processes in birds: morphology, physiology and function. Comp Biochem Physiol A 156:303–308

Codd JR, Manning PL, Norell MA, Perry SF (2007) Avian-like breathing mechanics in maniraptoran dinosaurs. Proc R Soc B 275:157–161

Collins CT (1983) A reinterpretation of pamprodactyly in swifts: a convergent grasping mechanism in vertebrates. Auk 100:735–737

Cooney CR, Bright JA, Capp EJR, Chira AM, Hughes EC, Moody CJA, Nouri LO, Varley ZK, Thomas GH (2017) Mega-evolutionary dynamics of the adaptive radiation of birds. Nature 542:344–347

Currey JD, Alexander RM (1985) The thickness of the walls of tubular bones. J Zool 206:453–468

Darwin CR (1859) On the origin of species by means of natural selection, or, the preservation of favoured races in the struggle for life. John Murray, London

Davit-Béal T, Tucker AS, Sire Y-Y (2009) Loss of teeth and enamel in tetrapods: fossil record, genetic data and morphological adaptations. J Anat 214:477–501

Dececchi TA, Larsson HCE (2009) Patristic evolutionary rates suggest a punctuated pattern in forelimb evolution before and after the origin of birds. Paleobiology 35:1–12

Demery ZP, Chappell J, Martin GR (2011) Vision, touch and object manipulation in Senegal Parrots *Poicephalus senegalus*. Proc R Soc B 278:3687–3693

Dial KP (1992) Avian forelimb muscles and nonsteady flight: can birds fly without using the muscles in their wings? Auk 109:874–885

Dial KP, Randall RJ, Dial TR (2006) What use is half a wing in the ecology and evolution of birds? BioScience 56:437–445

Dial KP, Jackson BE, Segre P (2008) A fundamental avian wing-stroke provides a new perspective on the evolution of flight. Nature 451:985–989

Dumont ER (2010) Bone density and the lightweight skeletons of birds. Proc R Soc B 277:2193–2198

Dzemski G, Christian A (2007) Flexibility along the neck of the Ostrich (*Struthio camelus*) and consequences for the reconstruction of dinosaurs with extreme neck length. J Morphol 268:701–714

Earls KD (2000) Kinematics and mechanics of ground take-off in the starling *Sturnis vulgaris* and the quail *Coturnix coturnix*. J Exp Biol 203:725–739

Eastick DL, Tattersall GJ, Watson SJ, Lesku JA, Robert KA (2019) Cassowary casques act as thermal windows. Sci Rep 9:1966

Einoder L, Richardson A (2006) An ecomorphological study of the raptorial digital tendon locking mechanism. Ibis 146:515–525

Estrella SM, Masero JA (2007) The use of distal rhynchokinesis by birds feeding in water. J Exp Biol 210:3757–3762

Fabbri M, Koch NM, Pritchard AC, Hanson M, Hoffman E, Bever GS, Balanoff AM, Morris ZS, Field DJ, Camacho J, Rowe TB (2017) The skull roof tracks the brain during the evolution and development of reptiles including birds. Nature Ecol Evol 1:1543–1550

Fajardo RJ, Hernandez E, O'Connor PM (2007) Postcranial skeletal pneumaticity: a case study in the use of quantitative microCT to assess vertebral structure in birds. J Anat 211:138–147

Feduccia A (1996) The origin and evolution of birds. Yale University Press, New Haven, CT

Feduccia A, Nowicki J (2002) The hand of birds revealed by early Ostrich embryos. Naturwissenschaften 89: 391–393

Felice RN (2014) Coevolution of caudal skeleton and tail feathers in birds. J Morphol 275:1431–1440

Felice RN, Goswami A (2018) Developmental origins of mosaic evolution in the avian cranium. Proc Natl Acad Sci USA 115:555–560

Fenwick AJ, Wood AM, Tanner BCW (2017) Effects of cross-bridge compliance on the force-velocity relationship and muscle power output. PLoS ONE 12: e0190335

Fernández-Juricic E, Gall MD, Dolan T, Tisdale V, Martin GR (2008) The visual fields of two ground-foraging birds, House Finches and House Sparrows, allow for simultaneous foraging and anti-predator vigilance. Ibis 150:779–787

Fernández-Juricic E, O'Rourke C, Pitlik T (2010) Visual coverage and scanning behavior in two corvid species:

American Crow and Western Scrub Jay. J Comp Physiol A 196:879–888
Fernández-Juricic E, Gall MD, Dolan T, O'Rourke C, Thomas S, Lynch JR (2011a) Visual systems and vigilance behaviour of two ground-foraging avian prey species: White-crowned Sparrows and California Towhees. Anim Behav 81:706–713
Fernández-Juricic E, Moore BA, Doppler M, Freeman J, Blackwell BF, Lima SL, DeVault TL (2011b) Testing the terrain hypothesis: Canada Geese see their world laterally and obliquely. Brain Behav Evol 77:147–158
Field DJ (2018) Endless skulls most beautiful. Proc Natl Acad Sci USA 115:448–450
Fowler DW, Freedman EA, Scannella JB (2009) Predatory functional morphology in raptors: interdigital variation in talon size is related to prey restraint and immobilisation technique. PLoS ONE 4:e7999
Fowler DW, Freedman EA, Scannella JB, Kambic RE (2011) The predatory ecology of *Deinonychus* and the origin of flapping in birds. PLoS ONE 6:e28964
Fuxjager MJ, Goller F, Dirkse A, Sanin GD, Garcia S (2016) Select forelimb muscles have evolved superfast contractile speed to support acrobatic social displays. eLife 5:e13544
Fuxjager MJ, Miles MC, Goller F, Petersen J, Yancey J (2017) Androgens support male acrobatic courtship behavior by enhancing muscle speed and easing the severity of its tradeoff with force. Endocrinology 158: 4038–4046
Fuxjager MJ, Fusani L, Schlinger BA (2022) Physiological innovation and the evolutionary elaboration of courtship behaviour. Anim Behav 184:185–195
Gall MD, Fernández-Juricic E (2010) Visual fields, eye movements, and scanning behavior of a sit-and-wait predator, the Black Phoebe (*Sayornis nigricans*). J Comp Physiol A 196:15–22
Galton PM, Shepherd JD (2012) Experimental analysis of perching in the European Starling (*Sturnus vulgaris*: Passeriformes; Passeres), and the automatic perching mechanism of birds. J Exp Zool A 317:205–215
Gao C, Chiappe LM, Meng Q, O'Connor JK, Wang X, Cheng X, Liu J (2008) A new basal lineage of Early Cretaceous birds from China and its implications on the evolution of the avian tail. Palaeontology 51:775–791
Gatesy SM (1999) Guineafowl hind limb function. I: Cineradiographic analysis and speed effects. J Morph 240:115–125
Gatesy SM, Dial KP (1996a) Locomotor modules and the evolution of avian flight. Evolution 50:331–340
Gatesy SM, Dial KP (1996b) From frond to fan: *Archaeopteryx* and the evolution of short-tailed birds. Evolution 50:2037–2048
Gatesy SM, Middleton KM (1997) Bipedalism, flight, and the evolution of theropod locomotor diversity. J Vertebr Paleontol 17:308–329
Gerson AR, Guglielmo CG (2011) Flight at low ambient humidity increases protein catabolism in migrating birds. Science 333:1434–1436
Gingerich PD (1973) Skull of Hesperornis and early evolution of birds. Nature 243:70–73
Goetz JE, Derrick TR, Pedersen DR, Robinson DA, Conzemius MG, Baer TE, Brown TD (2008) Hip joint contact force in the Emu (*Dromaius novaehollandiae*) during normal level walking. J Biomech 41:770–778
Goldspink G (1981) The use of muscles during flying, swimming, and running from the point of view of energy saving. Symp Zool Soc London 48:219–238
Goodrich ES (1986) Studies on the structure and development of vertebrates. University of Chicago Press, Chicago, IL
Goslow GE, Dial KP, Jenkins FA Jr (1989) The avian shoulder: an experimental approach. Am Zool 29: 287–301
Goslow GEJ, Dial KP, Jenkins FA (1990) Bird flight: insights and complications. BioScience 40:108–115
Gould J (1837a) Remarks on a group of ground finches from Mr. Darwin's collection, with characters of the new species. Proc Zool Soc London 5:4–7
Gould J (1837b) Three species of the genus *Orpheus*, from the Galapagos, in the collection of Mr. Darwin. Proc Zool Soc London 5:27
Green TL (2020) Ontogeny, disparity, and function of the enigmatic casques of cassowaries (*Casuarius*): a case study of cranial ornamentation in archosaurs. Ph.D. dissertation, Oklahoma State University, Stillwater, OK
Griffin CT, Botelho JF, Hanson M, Fabbri M, Smith-Paredes D, Carney RM, Norell MA, Egawa S, Gatesy SM, Rowe TB, Elsey RM (2022) The developing bird pelvis passes through ancestral dinosaurian conditions. Nature 608:346–352
Gunning PW, Hardeman EC (2018) Isoforms: fundamental differences. eLife 7:e34477
Gussekloo SWS, Vosselman MG, Bout RG (2001) Three-dimensional kinematics of skeletal elements in avian prokinetic and rhynchokinetic skulls determined by roentgen stereophotogrammetry. J Exp Biol 204: 1735–1744
Habib M (2010) The structural mechanics and evolution of aquaflying birds. Biol J Linn Soc 99:687–698
Habib MB, Ruff CB (2008) The effects of locomotion on the structural characteristics of avian limb bones. Zool J Linnean Soc 153:601–624
Hartman FA (1961) Locomotor mechanisms of birds. Smithsonian Miscell Collect 143:1–140
Heers AM (2016) New perspectives on the ontogeny and evolution of avian locomotion. Integ Comp Biol 56:428–441
Heers AM, Dial KP (2012) From extant to extinct: locomotor ontogeny and the evolution of avian flight. Trends Ecol Evol 27:296–305
Hertel F, Campbell KE Jr (2007) The antitrochanter of birds: form and function in balance. Auk 124:789–805
Hinić-Frlog S, Motani R (2010) Relationship between osteology and aquatic locomotion in birds: determining

modes of locomotion in extinct Ornithurae. J Evol Biol 23:372–385

Hoese WJ, Westneat MW (1996) Biomechanics of cranial kinesis in birds: testing linkage models in the White-throated Sparrow (*Zonotrichia albicollis*). J Morphol 227:305–320

Hohtola E, Visser GH (1998) Development of locomotion and endothermy in altricial and precocial birds. In: Starck JM, Ricklefs RE (eds) Avian growth and development: evolution within the altricial-precocial spectrum. Oxford University Press, New York, NY, pp 157–173

Holmgren J (2004) Roosting in tree foliage by Common Swifts *Apus apus*. Ibis 146:404–416

Homberger DG (2017) The avian lingual and laryngeal apparatus within the context of the head and jaw apparatus, with comparisons to the mammalian condition: functional morphological and biomechanics of evaporative cooling, feeding, drinking, and vocalization. In: Maina JN (ed) The biology of the avian respiratory system: evolution, development, structure and function. Springer, Cham, pp 27–98

Hui CA (1988) Penguin swimming. I. Hydrodynamics. Physiol Zool 61:333–343

Hutchinson JR (2000) The evolution of pelvic osteology and soft tissues on the line to extant birds (Neornithes). Zool J Linnean Soc 131:123–168

Hutchinson JR (2001) The evolution of pelvic osteology and soft tissues on the line to extant birds (Neornithes). Zool J Linnean Soc 131:123–168

Hutchinson JR, Allen V (2008) The evolutionary continuum of limb function from early theropods to birds. Naturwissenschaften 96:423–448

Hutchinson JR, Miller E, Fritsch G, Hildebrandt T (2008) The anatomical foundation for multidisciplinary studies of animal limb function: example from dinosaur and elephant limb imaging studies. In: Endo H (ed) Frey R. Springer, Cham, Switzerland, pp 23–38

James HF (2009) Repeated evolution of fused thoracic vertebrae in songbirds. Auk 126:862–872

James HF, Olson SL (1991) Descriptions of thirty-two new species of birds from the Hawaiian Islands. Part II. Passeriformes. Ornithological Monographs No. 46

Jena BP (2020) Myosin: cellular molecular motor. In: Cellular nanomachines. Springer Nature Switzerland, Cham, pp 79–89

Jenkins FA Jr, Dial KP, Goslow GE Jr (1988) A cineradiographic analysis of bird flight: the wishbone is a spring. Science 241:1495–1498

John MA, Mamde CS, Rohankar RU, Baba OK (2015) Comparative anatomical studies on rib cages of kite (*Millvus migrans*) and crow (*Corvus splendens*). Appl Biol Res 17:210–213

Jones DN, Reynolds SJ (2008) Feeding birds in our towns and cities: a global research opportunity. J Avian Biol 39:265–271

Jones ME, Button DJ, Barrett PM, Porro JB (2019) Digital dissection of the head of the Rock Dove (*Columba livia*) using contrast-enhanced computed tomography. Zool Lett 5:17

Jung J-Y, Naleway SE, Yaraghi NA, Herrera S, Sherman VR, Bushong EA, Ellisman MH, Kisailus D, McKittrick J (2016) Structural analysis of the tongue and hyoid apparatus in a woodpecker. Acta Biomater 37:1–13

Jurgens KD (2002) Etruscan shrew muscle: the consequences of being small. J Exp Biol 205:2161–2166

Kambic RE, Biewener AA, Pierce SE (2017) Experimental determination of three-dimensional cervical joint mobility in the avian neck. Front Zool 14:37

Katzir G, Martin GR (1998) Visual fields in the Black-crowned Night Heron *Nycticorax nycticorax*: nocturnality does not result in owl-like features. Ibis 140:156–162

Kavanagh KD, Shoval O, Winslow BB, Alon U, Leary BP, Kan A, Tabin CJ (2013) Developmental bias in the evolution of phalanges. Proc Natl Acad Sci USA 110: 18190–18195

Kiessling KH (1977) Muscle structure and function in the goose, quail, pheasant, guinea hen and chicken. Comp Biochem Physiol B 57:287–292

Kilbourne BM (2014) Scale effects and morphological diversification in hindlimb segment mass proportions in neognath birds. Front Zool 11:37

Kinnaird MF, Hadiprakarsa Y-Y, Thiensongrusamee P (2003) Aerial jousting by Helmeted Hornbills *Rhinoplax vigil*: observations from Indonesia and Thailand. Ibis 145:506–508

Kovacs CE, Meyers RA (2000) Anatomy and histochemistry of flight muscles in a wing-propelled diving bird, the Atlantic Puffin, *Fratercula arctica*. J Morphol 244: 109–125

Kral B (1965) Functional adaptations of Ciconiiformes to the darting stroke. Acta Societatis Zoologicae Bohemoslovenicae 29:377–391

Krings M, Nyakatura JA, Fischer MS, Wagner H (2014) The cervical spine of the American Barn Owl ( *Tyto furcata pratincola*): I. Anatomy of the vertebrae and regionalization in their S-shaped arrangement. PLoS ONE 9:e91653

Lack D (1947) Darwin's finches: an essay on the general biological theory of evolution. Cambridge University Press, Cambridge

van der Leeuw AHJ, Bout RG, Zweers GA (2001) Control of the cranio-cervical system during feeding in birds. Am Zool 41:1352–1363

Lerner HRL, Meyer M, James HF, Hofreiter M, Fleischer RC (2011) Multilocus resolution of phylogeny and timescale in the extant adaptive radiation of Hawaiian honeycreepers. Curr Biol 21:1838–1844

Libby T, Moore TY, Chang-Siu E, Li D, Cohen DJ, Jusufi A, Full RJ (2012) Tail-assisted pitch control in lizards, robots and dinosaurs. Nature 481:181–184

Lind J, Jakobsson S (2001) Body building and concurrent mass loss: flight adaptations in Tree Sparrows. Proc R Soc B 268:1915–1919

Lindström Å, Kvist A, Piersma T, Dekinga A, Dietz MW (2000) Avian pectoral muscle size rapidly tracks body mass changes during flight, fasting and fuelling. J Exp Biol 203:913–919

Lisney TJ, Iwaniuk AN, Bandet MV, Wylie DR (2012) Eye shape and retinal topography in owls (Aves: Strigiformes). Brain Behav Evol 79:218–236

Liu D, Chiappe LM, Serrano F, Habib M, Zhang Y, Meng Q (2017) Flight aerodynamics in Enantiornithines: information from a new Chinese Early Cretaceous bird. PLoS ONE 12:e0184637

Livezey B, Zusi R (2006) Higher-order phylogeny of modern birds (Theropoda, Aves: Neornithes) based on comparative anatomy. I. Methods and characters. Bull Carnegie Mus Nat Hist 37:1–544

Louchart A, Viriot L (2011) From snout to beak: the loss of teeth in birds. Trends Ecol Evol 26:663–673

Lowe PR (1936) The finches of the Galapagos in relation to Darwin's conception of species. Ibis 6:310–321

Lowi-Merri TM, Benson RB, Claramunt S, Evans DC (2021) The relationship between sternum variation and mode of locomotion in birds. BMC Biol 19:165

Luther PK (2009) The vertebrate muscle Z-disc: sarcomere anchor for structure and signaling. J Muscle Res Cell Motil 30:171–185

Lynch W (2007) Owls of the United States and Canada: a complete guide to their biology and behavior. Johns Hopkins University Press, Baltimore, MD

Madan MA, Rayfield EJ, Bright JA (2017) Scaling and functional morphology in strigiform hind limbs. Sci Rep 7:44920

Manegold A, Topfer T (2013) The systematic position Hemicircus and the stepwise evolution of adaptations for drilling, tapping and climbing up in true woodpeckers (Picinae, Picidae). J Zool Syst Evol Res 51:72–82

Marek RD, Falkingham PL, Benson RB, Gardiner JD, Maddox TW, Bates KT (2021) Evolutionary versatility of the avian neck. Proc R Soc B 288:20203150

Marsh RL (1984) Adaptations of the Gray Catbird *Dumetella carolinensis* to long-distance migration: flight muscle hypertrophy associated with elevated mass. Physiol Zool 57:105–117

Martin GR (1998) Eye structure and amphibious foraging in albatrosses. Proc R Soc B 265:665–671

Martin GR (1999) Eye structure and foraging in King Penguins *Aptenodytes patagonicus*. Ibis 141:444–450

Martin GR, Coetzee HC (2004) Visual fields in hornbills: precision-grasping and sunshades. Ibis 146:18–26

Martin GR, Portugal SJ (2011) Differences in foraging ecology determine variation in visual fields in ibises and spoonbills (Threskiornithidae). Ibis 153:662–671

Martin GR, White CR, Butler PJ (2008) Vision and the foraging technique of Great Cormorants *Phalacrocorax carbo*: pursuit or close-quarter foraging? Ibis 150:485–494

Marugán-Lobón J (2010) Combining shape data and traditional measurements with the 2B-PLS: testing the covariation between avian brain size and cranial shape variation as an example. In Elewa AMT (ed) Morphometrics for nonmorphometricians. Lecture Notes in Earth Sciences, vol. 124, Springer, Berlin, Heidelberg, pp 179–190

Marugán-Lobón J, Buscalioni ÁD (2009) New insight on the anatomy and architecture of the avian neurocranium. Anat Rec 292:364–370

Mattison RG (1998) Quantifying the avian pelvis: statistical correlations of lifestyle to pelvic structure among birds. Ph.D. dissertation, University of Massachusetts, Amherst, MA

Mayr G (2011) The phylogeny of charadriiform birds (shorebirds and allies)-reassessing the conflict between morphology and molecules. Zool J Linnean Soc 161: 916–934

Mayr G (2015) Skeletal morphology of the middle Eocene swift *Scaniacypselus* and the evolutionary history of true swifts (Apodidae). J Ornithol 156:441–450

Mayr G (2017) Pectoral girdle morphology of Mesozoic birds and the evolution of the avian supracoracoideus muscle. J Ornithol 158:859–867

Mayr G (2018) A survey of casques, frontal humps, and other extravagant bony cranial protuberances in birds. Zoomorphology 137:457–472

Mayr G, De Pietri VL, Love L, Mannering AA, Bevitt JJ, Scofield RP (2020) First complete wing of a stem group sphenisciform from the Paleocene of New Zealand sheds light on the evolution of the penguin flipper. Diversity 12:46

Menegaz RA, Kirk EC (2009) Septa and processes: convergent evolution of the orbit in haplorhine primates and strigiform birds. J Hum Evol 57:672–687

Meyers RA (2019) Comparative anatomy of the postural mechanisms of the forelimbs of birds and mammals. J Ornithol 160:869–882

Meyers RA, McFarland JC (2016) Anatomy and histochemistry of spread-wing posture in birds. 4. Eagles soar with fast, not slow muscle fibres. Acta Zool 97: 319–324

Meyers RA, Myers RP (2005) Mandibular bowing and mineralization in Brown Pelicans. Condor 107:445–449

Middleton KM, Gatesy SM (2000) Theropod forelimb design and evolution. Zool J Linnean Soc 128:149–187

Mitchell J, Legendre LJ, Lefèvre C, Cubo J (2017) Bone histological correlates of soaring and high-frequency flapping flight in the furculae of birds. Zoology 122: 90–99

Moreno E (1990) The musculi flexor perforatus digiti II and flexor digitorum longus in Paridae. Condor 92: 634–638

Moreno E (1991) Musculature of the pelvic appendages of the treecreepers (Passeriformes: Certhiidae): mycological adaptations for tail-supported climbing. Can J Zool 69:2456–2460

Moreno E, Carrascal LM (1993) Leg morphology and feeding postures in four *Parus* species: an experimental ecomorphological approach. Ecology 74: 2037–2044

Mosto MC (2017) The hindlimb myology of *Tyto alba* (Tytonidae, Strigiformes, Aves). Anat Histol Embryol 46:25–32

Mosto MC, Carril J, Picasso MBJ (2013) The hindlimb myology of *Milvago chimango* (Polyborinae, Falconidae). J Morphol 274:1191–1201

Mosto MC, Picasso MBJ, Montes MM, Krone O (2020) Tail myology and flight behaviour: differences between caracaras, falcons and forest falcons (Aves, Falconiformes). Acta Zool 101:292–301

Mosto MC, Cassini GH, Picasso MBJ, Krone O (2021) Grasping behavior in the White-tailed Sea Eagle (Accipitridae, Aves) explained by muscle architecture. J Zool 314:234–244

Müller GB, Streicher J (1989) Ontogeny of the syndesmosis tibiofibularis and the evolution of the bird hindlimb: a caenogenetic feature triggers phenotypic novelty. Anat Embryol 179:327–339

Naish D, Perron R (2016) Structure and function of the cassowary's casque and its implications for cassowary history, biology and evolution. Hist Biol 28:507–518

Navalón G, Bright JA, Marugán-Lobón J, Rayfield EJ (2019) The evolutionary relationship among beak shape, mechanical advantage, and feeding ecology in modern birds. Evolution 73:422–435

Ndlovu M, Cumming GS, Hockey PAR (2017) Body mass and pectoral muscle size changes in African waterfowl during moult. Afr J Wildl Res 47:24–31

Nesbitt SJ, Turner AH, Spaulding M, Conrad JL, Norell MA (2009) The theropod furcula. J Morphol 270:856–879

Newton I (1979) Population ecology of raptors. T & A Poyser, Berkhamstedt

Norberg UM (1979) Morphology of the wings, legs and tail of three coniferous forest tits, the goldcrest, and the treecreeper in relation to locomotor pattern and feeding station selection. Phil Trans Roy Soc B 287:131–165

Norberg RA (1986) Treecreeper climbing: mechanics, energetics, and structural adaptations. Ornis Scand 17:191–209

Nyakatura JA, Andrada E, Grimm N, Weise H, Fischer MS (2012) Kinematics and center of mass mechanics during terrestrial locomotion in Northern Lapwings (*Vanellus vanellus*, Charadriiformes). J Exp Zool A 317:580–594

O'Connor PM (2004) Pulmonary pneumaticity in the postcranial skeleton of extant Aves: a case study examining Anseriformes. J Morphol 261:141–161

O'Connor PM (2009) Evolution of archosaurian body plans: skeletal adaptations of an air-sac-based breathing apparatus in birds and other archosaurs. J Exp Zool A 311A:629–646

O'Rourk CT, Hall MI, Pitlik T, Fernández-Juricic E (2010) Hawk eyes I: diurnal raptors differ in visual fields and degree of eye movement. PLoS ONE 5: e12802

Ocampo Vargas D, Montero GB, Uy JA (2018) Morphological adaptations for relatively larger brains in hummingbird skulls. Ecol Evol 8:10482–10488

Olson JM (1994) The ontogeny of shivering thermogenesis in the Red-winged Blackbird (*Agelaius phoeniceus*). J Exp Biol 191:59–88

Organ CL (2006) Thoracic epaxial muscles in living Archosaurs and ornithopod dinosaurs. Anatom Record Part A 288A:782–793

Orros ME, Fellowes MDE (2015) Wild bird feeding in an urban area: intensity, economics and numbers of individuals supported. Acta Ornithologica 50:43–58

Patak A, Baldwin J (1993) Structural and metabolic characterization of the muscles used to power running in the Emu (*Dromaius novaehollandiae*), a giant flightless bird. J Exp Biol 175:233–249

Paul GS, Leahy GD (1994) Terramegathermy in the time of titans: restoring the metabolics of colossal dinosaurs. Paleontol Soc Spec Publ 7:177–198

Pease JB, Driver RJ, de la Cerda DA, Day LB, Lindsay WR, Schlinger BA, Schuppe ER, Balakrishnan CN, Fuxjager MJ (2022) Layered evolution of gene expression in "superfast" muscles for courtship. Proc Natl Acad Sci 119:e2119671119

Pennycuick CJ (1967) The strength of the pigeon's wing bones in relation to their function. J Exp Biol 46:219–233

Petit M, Lewden A, Vézina F (2014) How does flexibility in body composition relate to seasonal changes in metabolic performance in a small passerine wintering at northern latitude? Physiol Biochem Zool 87:539–549

Piersma T (1988) Breast muscle atrophy and constraints on foraging during the flightless period of wing moulting Great Crested Grebes. Ardea 76:96–106

Plochocki JH, Segev T, Grow W, Hall MI (2018) Extraocular muscle architecture in hawks and owls. Vet Ophthalmol 21:595–600

Poore SO, Sánchez-Haiman A, Goslow GE Jr (1997) Wing upstroke and the evolution of flapping flight. Nature 387:799–802

Prakash M, Quéré D, Bush JWM (2008) Surface tension transport of prey by feeding shorebirds: the capillary ratchet. Science 320:931–934

Price TD (2011) Adaptive radiations: there's something about finches. Curr Biol 21:R953–R955

Price JP, Clague DA (2002) How old is the Hawaiian biota? Geology and phylogeny suggest recent divergence. Proc R Soc B 269:2429–2435

Price ER, Bauchinger U, Zajac DM, Cerasale DJ, McFarlan JT, Gerson AR, McWilliams SR, Guglielmo CG (2011) Migration-and exercise-induced changes to flight muscle size in migratory birds and association with IGF1 and myostatin mRNA expression. J Exp Biol 214:2823–2831

Proctor NS, Lynch PJ (1993) Manual of ornithology: avian structure and function. Yale University Press, New Haven, CT

Prum RO, Berv JS, Dornburg A, Field DJ, Townsend JP, Lemmon EM, Lemmon AR (2015) A comprehensive phylogeny of birds (Aves) using targeted next-generation DNA sequencing. Nature 526:569–573

Pyörnilä AEI, Putaala AP, Hissa RK (1998) Fibre types in breast and leg muscles of hand-reared and wild Grey Partridge (*Perdix perdix*). Can J Zool 76:236–242

Quinn TH, Baumel JJ (1990) The digital tendon locking mechanism of the avian foot (Aves). Zoomorphology 109:281–293

Raikow RJ (1985) Locomotor system. In: King AS, McLelland J (eds) Form and function in birds, vol 3. Academic Press, New York, NY, pp 57–147

Raikow RJ (1994) Climbing adaptations in the hindlimb musculature of the woodcreepers (Dendrocolaptinae). Condor 96:1103–1106

Rands CM, Darling A, Fujita M, Kong L, Webster MT, Clabaut C, Emes RD, Heger A, Meader S, Hawkins MB, Eisen MB (2013) Insights into the evolution of Darwin's finches from comparative analysis of the *Geospiza magnirostris* genome sequence. BMC Genomics 14:95

Rashid DJ, Surya K, Chiappe LM, Carroll N, Garrett KL, Varghese B, Bailleul A, O'Connor JK, Chapman SC, Horner JR (2018) Avian tail ontogeny, pygostyle formation, and interpretation of juvenile Mesozoic specimens. Sci Rep 8:9014

Rasskin-Gutman D, Buscalioni AD (2001) Theoretical morphology of the Archosaur (Reptilia: Diapsida) pelvic girdle. Paleobiology 27:59–78

Rayner JMV (1988) Form and function in avian flight. In: Johnston RF (ed) Current ornithology, vol 5. Springer, Boston, MA, pp 1–66

Richards LP, Bock WJ (1973) Functional anatomy and adaptive evolution of the feeding apparatus in the Hawaiian honeycreeper genus *Loxops* (Drepanididae). Ornithological Monograph, no. 15. American Ornithologists' Union

Richardson F (1972) Accessory pygostyle bones of Falconidae. Condor 74:350–351

Robinson WD, Rourke B, Stratford JA (2021) Put some muscle behind it: understanding movement capacity of tropical birds. Ornithology 138:ukaa068

Rosser BWC, George JC (1986) The avian pectoralis: histochemical characterization and distribution of muscle fiber types. Can J Zool 64:1174–1185

Ruben J (1991) Reptilian physiology and the flight capacity of *Archaeopteryx*. Evolution 45:1–17

Saber AS, Hassanin A (2014) Some morphological studies on the wing and foot of the Southern Cassowary (*Casuarius casuarius*). J Vet Anat 7:17–32

Samejima M, Otsuka JI (1984) Observations on the number of thoracic vertebrae and the presence of notarium in birds. Jpn J Ornithol 33:29–38

Samman T (2006) Craniocervical functional morphology of several North American coelurosaurian dinosaurs. Ph.D. dissertation, University of Calgary, Calgary, AB, Canada

Savile DBO (1950) The flight mechanism of swifts and hummingbirds. Auk 67:499–504

Schepelmann K (1990) Erythropoietic bone marrow in the pigeon: development of its distribution and volume during growth and pneumatization of bones. J Morphol 203:21–34

Schreiber RW, Woolfenden GE, Curtsinger WE (1975) Prey capture by the Brown Pelican. Auk 92:649–654

Schulenberg TS (1983) Foraging behavior, eco-morphology, and systematic of some antshrikes (Formicariidae: *Thamnomanes*). Wilson Bull 95:505–521

Schuppe ER, Fuxjager MJ (2018) High-speed displays encoding motor skill trigger elevated territorial aggression in Downy Woodpeckers. Funct Ecol 32:450–460

Schuppe ER, Petersen JO, Fuxjager MJ (2018) Woodpecker drumming behavior is linked to the elevated expression of genes that encode calcium handling proteins in the neck musculature. J Exp Biol 221: jeb180190

Schwilch R, Grattarola A, Spina F, Jenni L (2002) Protein loss during long-distance migratory flight in passerine birds: adaptation and constraint. J Exp Biol 205:687–695

Sciorati C, Rigamonti E, Manfredi AA, Rovere-Querini P (2016) Cell death, clearance and immunity in the skeletal muscle. Cell Death Diff 23:927–937

Senter P (2006) Scapular orientation in theropods and basal birds, and the origin of flapping flight. Acta Palaeontol Pol 51:305–313

Serrano FJ, Costa-Pérez M, Navalón G, Martín-Serra A (2020) Morphological disparity of the humerus in modern birds. Diversity 12:173

Shufeldt RW (1900) On the osteology of the Striges (Strigidae and Bubonidae). Proc Am Philos Soc 39: 665–722

Sibley CG (1957) The loss of teeth in birds. Auk 74:103–104

Smith NC, Wilson AM, Jespers KJ, Payne RC (2006) Muscle architecture and functional anatomy of the pelvic limb of the Ostrich (*Struthio camelus*). J Anat 209:765–779

Sommer JR (1995) Comparative anatomy: in praise of a powerful approach to elucidate mechanisms translating cardiac excitation into purposeful contraction. J Mol Cell Cardiol 27:19–35

Spring L (1971) A comparison of functional and morphological adaptations in the Common Murre (*Uria aalge*) and Thick-billed Murre (*Uria lomvia*). Condor 73:1–27

Stark RD, Dodenhoff DJ, Johnson EV (1998) A quantitative analysis of woodpecker drumming. Condor 100: 350–356

Steinbach MJ, Money KE (1973) Eye movements of the owl. Vis Res 13:889–891

Stewart TA, Liang C, Cotney JL, Noonan JP, Sanger TJ, Wagner GP (2019) Evidence against tetrapod-wide digit identities and for a limited frame shift in bird wings. Nat Commun 10:3244

Stoessel A, Kilbourne BM, Fischer MS (2013) Morphological integration versus ecological plasticity in the avian pelvic limb skeleton. J Morphol 274:483–495
Stowers AK, Matloff LY, Lentink D (2017) How pigeons couple three-dimensional elbow and wrist motion to morph their wings. J R Soc Interface 14:20170224
Streicher J, Müller GB (1992) Natural and experimental reduction of the avian fibula: developmental thresholds and evolutionary constraint. J Morphol 214:269–285
Sullivan TN, Wang B, Espinosa HD, Meyers MA (2017) Extreme lightweight structures: avian feathers and bones. Mater Today 20:377–391
Sullivan SP, McGechie FR, Middleton KM, Holliday CM (2019) 3D muscle architecture of the pectoral muscles of European Starling (*Sturnus vulgaris*). Integr Organis Biol 1:oby010
Sulloway FJ (1982) Darwin and his finches: the evolution of a legend. J Hist Biol 15:1–53
Sustaita D (2008) Musculoskeletal underpinnings to differences in killing behavior between North American accipiters (Falconiformes: Accipitridae) and falcons (Falconidae). J Morphol 269:283–301
Swanson DL, Liknes ET (2006) A comparative analysis of thermogenic capacity and cold tolerance in small birds. J Exp Biol 209:466–474
Swanson DL, Merkord C (2013) Seasonal phenotypic flexibility of flight muscle size in small birds: a comparison of ultrasonography and tissue mass measurements. J Ornithol 154:119–127
Tambussi CP, de Mendoza R, Degrange FJ, Picasso MB (2012) Flexibility along the neck of the Neogene Terror Bird *Andalgalornis steulleti* (Aves Phorusrhacidae). PLoS ONE 7:e37701
Tehrani PR, Gilanpour H, Veshkini A (2017) Radiographic anatomy of the metatarsophalangeal joint and digits of the Ostrich (*Struthio camelus*). J Avian Med Surg 31:198–205
Tickle PG, Ennos AR, Lennox LE, Perry SF, Codd JR (2007) Functional significance of the uncinate processes in birds. J Exp Biol 210:3955–3961
Tickle P, Nudds R, Codd J (2009) Uncinate process length in birds scales with resting metabolic rate. PLoS ONE 4:e5667
Tobalske BW (2010) Hovering and intermittent flight in birds. Bioinspir Biomim 5:1–10
Tokita M (2003) The skull development of parrots with special reference to the emergence of a morphologically unique cranio-facial hinge. Zool Sci 20:749–758
Tokita M, Yano W, James HF, Abzhanov A (2017) Cranial shape evolution in adaptive radiations of birds: comparative morphometrics of Darwin's finches and Hawaiian honeycreepers. Philos Trans R Soc B 372: 20150481
Torres CR, Ogawa LM, Gillingham MA, Ferrari B, Van Tuinen M (2014) A multi-locus inference of the evolutionary diversification of extant flamingos (Phoenicopteridae). BMC Evol Biol 14:36
Trbojević Vukičević T, Galić S, Horvatek Tomić D, Kužir S (2018) The morphological characteristics of the passive flexor mechanism of birds with different digit layout. Veterinarski Arhiv 88:125–138
Trinkaus E, Villemeur I (1991) Mechanical advantage of the Neanderthal thumb in flexion: a test of an hypothesis. Amer J Phys Anthropol 84:249–260
Tryjanowski P, Skorka P, Sparks TH, Biadun W, Brauze T, Hetmanski T, Martyka R, Indykiewicz P, Myczko L, Kunysz P, Kawa P (2015) Urban and rural habitats differ in number and type of bird feeders and in bird species consuming supplementary food. Env Sci Pollution Res 22:15097–15103
Tsuihiji T (2004) The ligament system in the neck of *Rhea americana* and its implication for the bifurcated neural spines of sauropod dinosaurs. J Vertebr Paleontol 24: 165–172
Usherwood JR, Sparkes EL, Weller R (2014) Leap and strike kinetics of an acoustically 'hunting' Barn Owl (*Tyto alba*). J Exp Biol 217:3002–3005
van der Meij MAA (2004) A tough nut to crack: adaptations to seed cracking in finches. Ph.D. dissertation,. Leiden University, Leiden, The Netherlands
van der Meij MAA, Bout RG (2004) Scaling of jaw muscle size and maximal bite force in finches. J Exp Biol 207:2745–2753
Van Wassenbergh S, Baeckens S (2019) Digest: evolution of shape and leverage of bird beaks reflects feeding ecology, but not as strongly as expected. Evolution 73: 621–622
Vargas AO, Fallon JF (2005) Birds have dinosaur wings: the molecular evidence. J Exp Zool B 304:86–90
Vargas AO, Ruiz-Flores M, Soto-Acuña S, Haidr N, Acosta-Hospitaleche C, Ossa-Fuentes L, Muñoz-Walther V (2017) The origin and evolutionary consequences of skeletal traits shaped by embryonic muscular activity, from basal theropods to modern birds. Integr Comp Biol 57:1281–1292
Vézina F, O'Connor RS, Le Pogam A, De Jesus AD, Love OP, Gabriela Jimenez A (2021) Snow Buntings preparing for migration increase muscle fiber size and myonuclear domain in parallel with a major gain in fat mass. J Avian Biol 52:jav.02668
Volkov SV (2004) The hindlimb musculature of the true owls (Strigidae: Strigiformes): morphological peculiarities and general adaptations. Ornithologia 31: 154–174
Wagner GP (2005) The developmental evolution of avian digit homology: an update. Theory Biosci 124:165–183
Wagner GP, Gauthier JA (1999) 1, 2, 3= 2, 3, 4: a solution to the problem of the homology of the digits in the avian hand. Proc Natl Acad Sci USA 96:5111–5116
Wang M, Zheng X, O'Connor JK, Lloyd GT, Wang X, Wang Y, Zhang X, Zhou Z (2015) The oldest record of ornithuromorpha from the Early Cretaceous of China. Nat Commun 6:6987
Ward AB, Weigl PD, Conroy RM (2002) Functional morphology of raptor hindlimbs: implications for resource partitioning. Auk 119:1052–1063

Warrick DR, Tobalske BW, Powers DR (2005) Aerodynamics of the hovering hummingbird. Nature 435: 1094–1097

Warrick D, Hedrick T, Fernández MJ, Tobalske B, Biewener A (2012) Hummingbird flight. Curr Biol 22:R472–R477

Weeks OI (1989) Vertebrate skeletal muscle: power source for locomotion. BioScience 39:791–799

Welch KC Jr, Altshuler DL (2009) Fiber type homogeneity of the flight musculature in small birds. Comp Biochem Physiol B 152:324–331

Wells DJ (1993) Muscle performance in hovering hummingbirds. J Exp Biol 178:39–57

Whitehead PJ (1998) Boofheads with deep voices: sexual dimorphism in the Magpie Goose *Anseranas semipalmata*. Wild 49:72–91

Xu X, Clark JM, Mo J, Choiniere J, Forster CA, Erickson GM, Hone DW, Sullivan C, Eberth DA, Nesbitt S, Zhao Q (2009) A Jurassic ceratosaur from China helps clarify avian digital homologies. Nature 459:940–944

Yanega GM, Rubega MA (2004) Hummingbird jaw bends to aid insect capture. Nature 428:615

Zeffer A, Johansson LC, Marmebro A (2003) Functional correlation between habitat use and leg morphology in birds (Aves). Biol J Linn Soc 79:461–484

Zelenkov NV (2007) The structure and probable mechanism of evolutionary formation of the foot in Piciform birds (Aves: Piciformes). Paleontol J 41:290–297

Zhang F, Zhou Z, Benton MJ (2008) A primitive confuciusornithid bird from China and its implications for early avian flight. Sci China Ser D Earth Sci 51: 625–639

Zhou Z (2002) A new and primitive Enantiornithine bird from the Early Cretaceous of China. J Vertebr Paleontol 22:49–57

Zhou Z, Li FZZ (2010) A new Lower Cretaceous bird from China and tooth reduction in early avian evolution. Proc R Soc B 277:219–227

Zhu J, Nakamura E, Nguyen M-T, Bao X, Akiyama H, Mackem S (2008) Uncoupling Sonic hedgehog control of pattern and expansion of the developing limb bud. Develop Cell 14:624–632

Zuki ABZ, Abdul Ghani MM, Khadim KK, Intan-Shameha AR, Kamaruddin MI (2012) Anatomical structures of the limb of White-nest Swiftlet (*Aerodramus fuciphagus*) and White-headed Munia (*Lonchura maja*). Pertanika J Trop Agric Sci 35:613–622

Zusi RL (1962) Structural adaptations of the head and neck in the Black Skimmer *Rhynchops nigra* L. Publications of the Nuttall Ornithological Club 3:1–101

Zusi RL (1984) A functional and evolutionary analysis of rhynchokinesis in birds. Smithsonian Contrib Zool 395:1–40

Zusi RL (1993) Patterns of diversity in the avian skull. In: Hanken J, Hall BK (eds) The skull, volume 2: patterns of structural and systematic diversity. University of Chicago Press, Chicago, IL, pp 391–437

Zusi RL (2013) Introduction to the skeleton of hummingbirds (Aves: Apodiformes, Trochilidae) in functional and phylogenetic contexts. Ornithol Monogr 77:1–94

Zusi RL, Bentz GD (1984) Myology of the Purple-throated Carib (*Eulampis jugularis*) and other hummingbirds (Aves: Trochilidae). Smithsonian Contributions to Zoology, no. 385. Smithsonian Institution Press, Washington, DC

Zusi RL, Gill FB (2009) The marvelous tail of *Loddigesia mirabilis* (Trochilidae). Auk 126:590–603

Zusi RL, Storer RW (1969) Osteology and myology of the head and neck of the Pied-billed Grebe (*Podilymbus*). Miscellaneous Publications of the Museum of Zoology, University of Michigan 139:1–49

Zweers G, Bout R, Heidweiller J (1994) Motor organization of the avian head-neck system. In: Davies MNO, Green PR (eds) Perception and motor control in birds. Springer, Berlin, pp 201–221

# Integument

# 3

## Contents





G. Ritchison, *In a Class of Their Own*, Fascinating Life Sciences,
https://doi.org/10.1007/978-3-031-14852-1_3

**Abstract**

The skin of birds keeps out pathogens and other potentially harmful substances, retains vital fluids and gases, serves as a sensory organ, and produces and supports feathers. This chapter describes the structure of avian skin and explains the functions of unfeathered areas of skin found in some species of birds, like vultures. Interspecific variation in the structure of avian claws and rhamphotheca and the factors that contribute to such variation are discussed. Next, the structure and function of specialized structures like wattles and combs are explained, as are the structure and function of integument glands. Next, the evolution of feathers is discussed, and the structure and function of the different types of feathers are described. Also described in detail is skin and feather color, including the role of pigments and structure. Colors produced by thin- and multi-film interference and photonic structures are also explained. The chapter closes with a discussion of feather parasites and the defenses used by birds to combat those parasites.

## 3.1 Skin: Structure and Function

The functions of bird skin are the same as for other vertebrates—to keep out pathogens and other potentially harmful substances, retain vital fluids and gases, and serve as a sensory organ. The continual renewal of the skin acts to repel parasitic microorganisms. The skin of birds also produces and supports feathers. With feathers, the skin also plays an important role in thermoregulation. Although largely covered by feathers, the integument is unfeathered on the beak, feet, and, in some species, other areas that play important roles in thermoregulation (e.g., van Vuuren et al. 2020) or communication (e.g., Negro et al. 2006). Avian skin does not have sweat glands and sebaceous glands; these glands evolved only in mammals. Avian skin consists of two layers, the epidermis and dermis. The outer layer, the epidermis, is generally very thin and pliable. The dermis is thicker than the epidermis and contains blood vessels, fat deposits, nerves, free nerve endings, several neuroreceptors, and smooth muscles that move the feathers (Lucas and Stettenheim 1972).

The epidermis, the most superficial layer of the skin, is thinner in birds than in mammals of comparable size, flexible, and smooth, and this is due, at least in part, to selective pressures to minimize body weight for more efficient flight (Spearman 1966). The epidermis is thinnest in areas covered by feathers (both feather tracts and apteria; Fig. 3.1, Box 3.1 Evolution of Avian Skin) and thickest in exposed, featherless areas, including the covering of the beak (rhamphotheca) and feet (podotheca). The epidermis has two main layers—a superficial stratum corneum and a deeper strateus germinativium (Figs. 3.2 and 3.3). The stratum corneum consists of flattened, keratinized cells called corneocytes. These cells are called keratinized because they contain a protein called keratin (and, specifically, beta-keratin; Lillywhite 2006) that, along with extracellular lipids (fats) produced by epidermal cells, provides a tough, permeability barrier that prevents excessive evaporative water loss. The stratum corneum can be viewed as having a "brick-and-mortar" organization, with the keratin-enriched cells

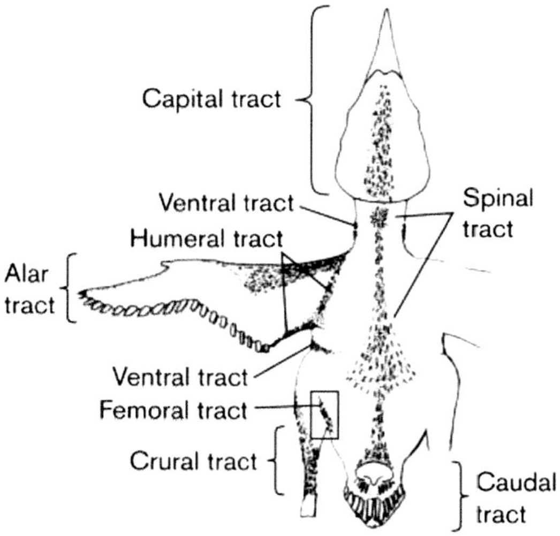


**Fig. 3.1** Feathered (feather tracts) and unfeathered (apteria) areas of the avian integument. (Figure from Chuong et al. 2000; © 2000 Elsevier Science Ltd., used with permission)

**Box 3.1 Evolution of Avian Skin**

Skin similar to that of present-day birds evolved as long ago as the late Jurassic or early Cretaceous. The stratum corneum of birds has an outer layer consisting of flattened corneocytes and, in present-day birds, corneocytes are shed individually or in patches up to 0.5 $mm^2$ that can be temporarily trapped in feathers. McNamara et al. (2018) discovered evidence of the skin-shedding process in basal birds (*Confuciusornis*) and nonavian maniraptoran dinosaurs, confirming that at least some nonavian dinosaurs also shed their skin in small patches.

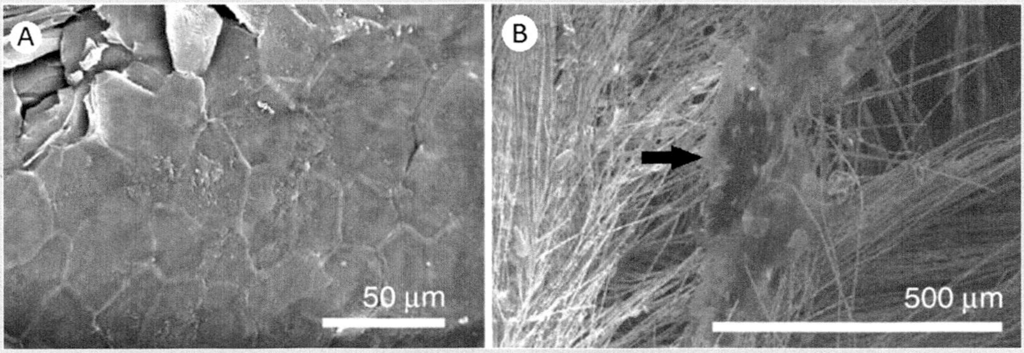


Scanning electron micrographs of a Zebra Finch shed skin (*Taeniopygia guttata*). (**a**) Corneocytes with polygonal shapes, and (**b**) a shed skin flake (indicated by the arrow) in a feather (Figure modified from McNamara et al. 2018; open-access article distributed under the terms of the Creative Commons CC BY license, https://creativecommons.org/licenses/by/4.0/).

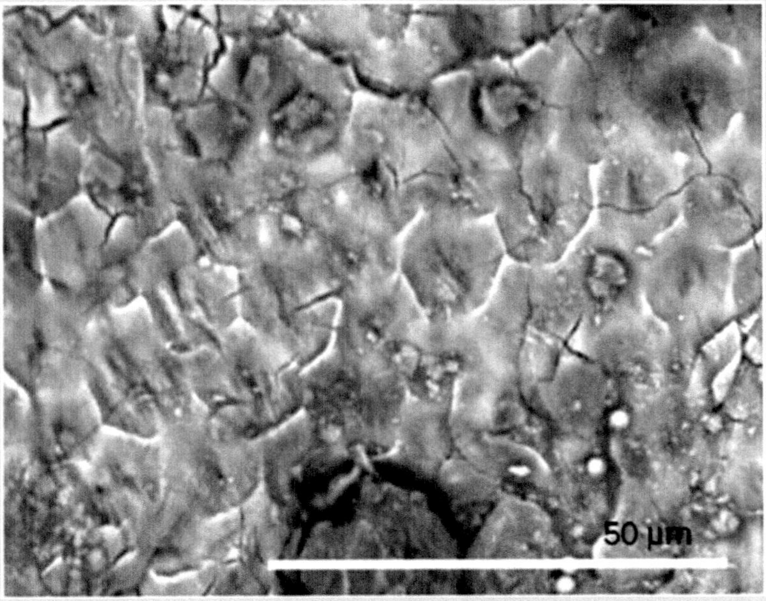


Fossilized skin of *Confuciusornis* showing polygonal structure of corneocytes similar to that of present-day birds (Figure from McNamara et al. 2018; open-access article distributed under the terms of the Creative Commons CC BY license, https://creativecommons.org/licenses/by/4.0/).

(continued)

**Box 3.1** (continued)

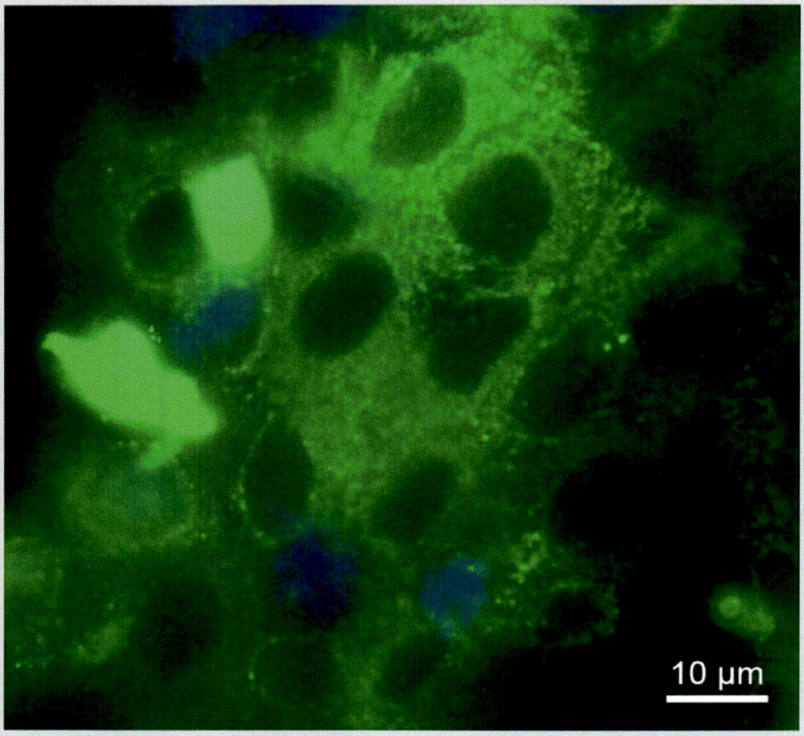


Corneocytes in the stratum corneum of a Java Sparrow (*Lonchura oryzivora*) after staining. The blue areas are the cell nuclei; the green color indicates the presence of the protein keratin, which, in turn, indicates the presence of keratin tonofibrils that are too small to see at this magnification (supplemental figure from McNamara et al. 2018; open-access article distributed under the terms of the Creative Commons CC BY license, https://creativecommons.org/licenses/by/4.0/).

In one important respect, however, the skin and corneocytes of present-day birds differ from that of feathered dinosaurs and early birds. Whereas the keratin tonofibrils in the corneocytes of present-day birds are relatively dispersed within intracellular lipids, tonofibrils in the corneocytes of feathered dinosaurs and early birds were densely packed, filling the interior of the cells (McNamara et al. 2018). The dispersed tonofibrils in corneocytes mean that present-day birds can more efficiently lose excess heat via evaporative cooling. In contrast, the density of tonofibrils in the corneocytes of feathered dinosaurs and early birds likely means that they had lower metabolic rates and less need for evaporative cooling (Menon et al. 1996). This, in turn, lends support to the hypothesis that dinosaurs like *Microraptor* (Dyke et al. 2013) and, possibly, *Confuciusornis* (Nudds and Dyke 2010) were not capable of powered flight, or at least powered flight for extended periods (Falk et al. 2016).

(continued)

**Box 3.1 (continued)**

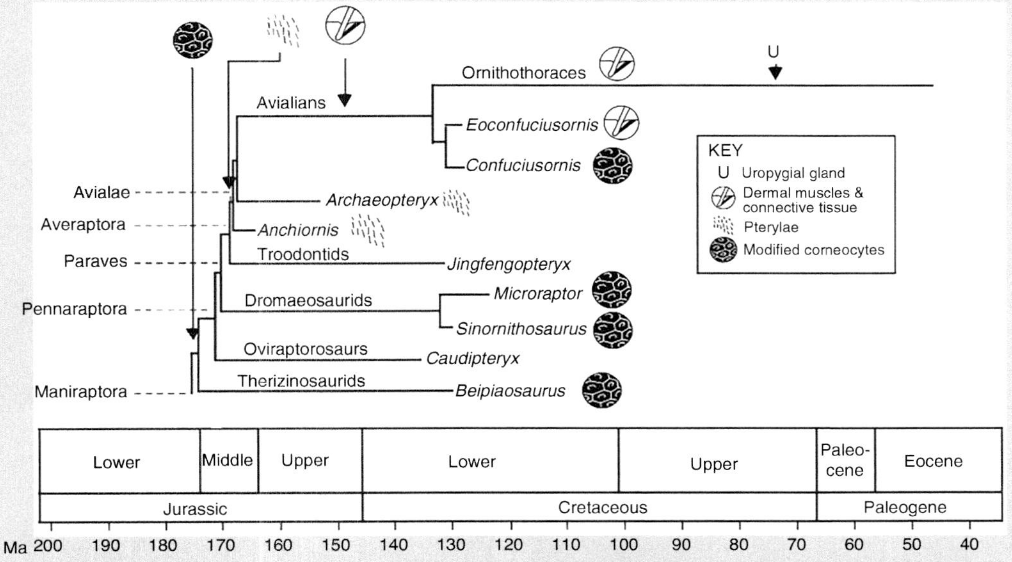


Skin similar to that of present-day birds may have appeared as early as the middle Jurassic. Other features of avian skin, including the presence of pterylae and apteria, dermal muscles that allow control of feather position, and the uropygial gland, evolved later (Figure modified from McNamara et al. 2018; open-access article distributed under the terms of the Creative Commons CC BY license, https://creativecommons.org/licenses/by/4.0/).

forming the "bricks" and the extracellular lipids the "mortar" (Elias and Menon 1991). However, compared to reptiles and even mammals, cells in the stratum corneum have less keratin and, as a result, this barrier is less stringent and can facilitate evaporative cooling while retaining the capacity for facultative waterproofing (Menon et al. 1996). The high body temperatures of birds, increased heat production during flight, insulation by plumage, and the lack of sweat glands, require a higher rate of evaporative cooling through a relatively "leaky" epidermal permeability barrier. Importantly, however, the relative permeability of the avian epidermis can be modified. For example, Menon et al. (1996) found that, within 16 h of water deprivation, adult Zebra Finches (*Taeniopygia guttata*) can reduce water loss via the epidermis by 50% through the rapid secretion of epidermal lipids. A similar ability to influence water loss by regulating the secretion of epidermal lipids has been reported in larks (Haugen et al. 2003), House Sparrows (*Passer domesticus*; Muñoz-Garcia and Williams 2005), and the tropical Dusky Antbird (*Cercomacroides tyrannina*; Muñoz-Garcia and Williams 2007).

Besides forming a dynamic barrier that regulates water loss through the skin, epidermal lipids may also have antimicrobial properties and offer protection against ultraviolet light (Menon 1984). In addition, epidermal lipids are used for cosmetic coloration in the Crested Ibis (*Nipponia nippon*; Wingfield et al. 2000). Before breeding, the skin of the neck and head starts secreting a black substance that the ibises apply to their white plumage (Fig. 3.4). The extent of the secretory skin area and how much of the plumage is covered by the "cosmetic" varies among individuals

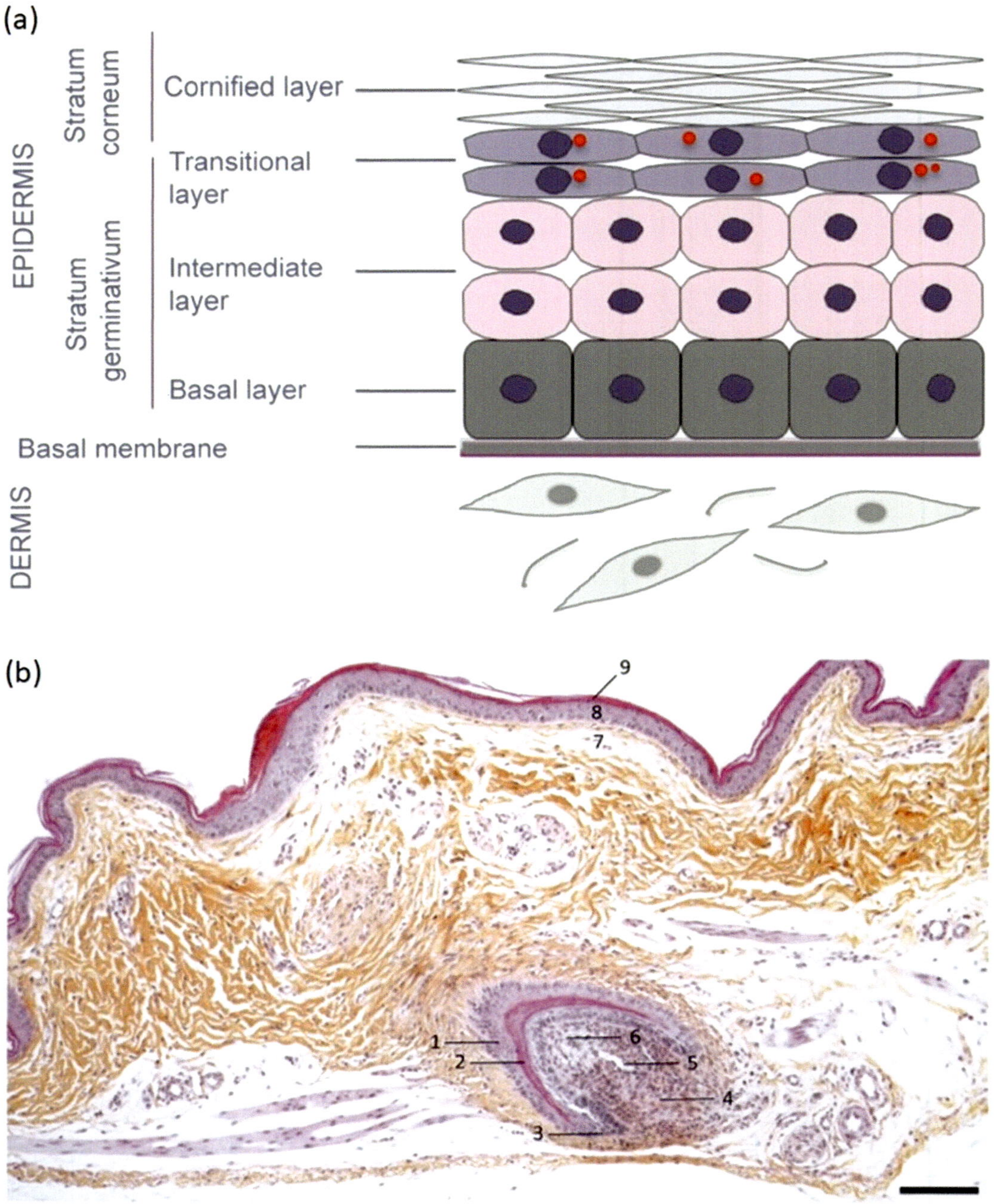


**Fig. 3.2** (**a**) Drawing of a section of Domestic Chicken (*Gallus g. domesticus*) skin with no feather follicle. The epidermis consists of four layers with keratinocytes at various stages of differentiation. The red circles represent lipid droplets. (**b**) Skin with a feather follicle from an adult domestic chicken (stained). Feather follicle = 1 to 6, and skin = 7 to 9: 1, feather follicle; 2, cornified cells; 3, epidermal collar; 4, dermal papilla; 5, axial blood vessel; 6, feather pulp; 7, dermis of the skin; 8, stratum germinativum of the skin; 9, stratum corneum of the skin. (figures from Couteaudier and Denesvre 2014; © 2014 Couteaudier and Denesvre; licensee BioMed Central Ltd., an open-access article distributed under the terms of the Creative Commons Attribution License, https://creativecommons.org/licenses/by/4.0/)

**Fig. 3.3** Epidermis and dermis of an adult Greater Rhea (*Rhea americana*). (**a**) Stained section showing the arrangement of epidermis and dermis of an adult, at: adipose tissue, bv: blood vessel, ep: epidermis, pn: peripheral nerve, sc: stratum corneum; sg, stratum germinativum. Scale bar, 400 µm. (**b**) Stained section showing the epidermal layers, sco: stratum corneum, si: stratum intermedium, and sb: stratum basale. Scale bar = 40 µm. (**c**) Stained section showing the stratum corneum (sc) and stratum germinativum (sg). Scale bar = 400 µm. (Figure from Picasso et al. 2016; © 2015 The Royal Swedish Academy of Sciences, used with permission)

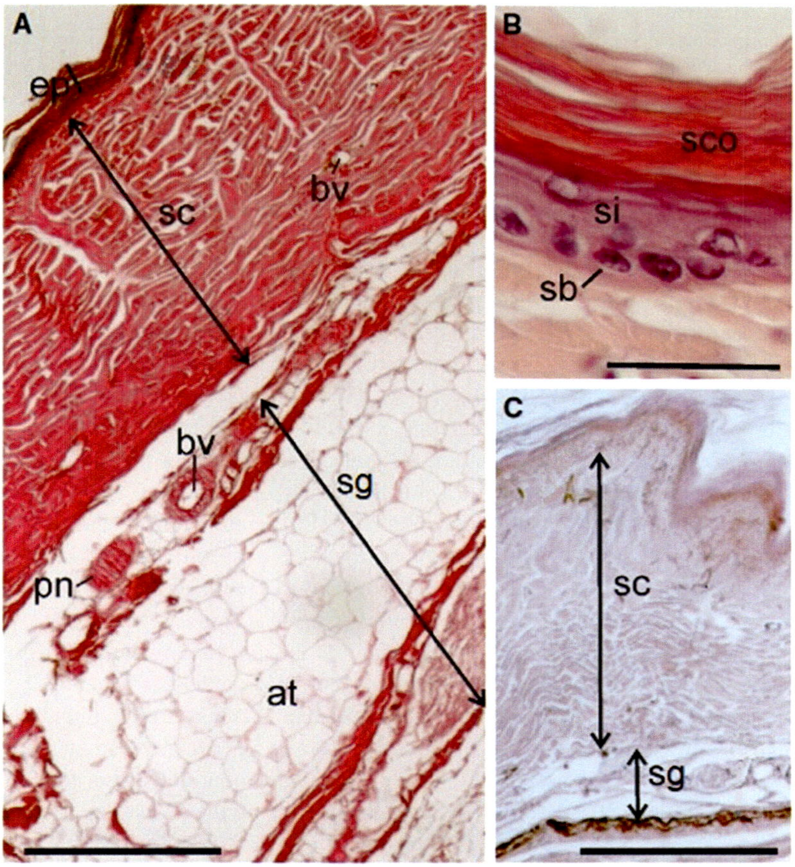

**Fig. 3.4** Plumages of the Crested Ibis (*Nipponia nippon*). (Left) Typical nonbreeding plumage. (Right) "Cosmetically colored" ibis. (Figure from Sun et al. 2020; open-access article distributed under the Creative Commons Attribution License, https://creativecommons.org/licenses/by/4.0/)

and this variation plays a role in mate choice. Young birds do not produce black secretion at all.

## 3.2 Unfeathered and Colored Skin

Among many species of birds, the integument exhibits specialized modifications. For example, the skin on the head is unfeathered to varying degrees and distinctively colored in guineafowl, vultures, colies (*Colius*), and many storks, ibises, spoonbills, and cranes. The skin around the eyes is unfeathered and distinctively colored in other birds, such as cariamas, falcons, sheathbills (*Chionis*), parrots, cuckoos, broadbills, bare-eyes (*Phlegopsis*), lyrebirds (*Menura*), and helmet-shrikes (*Prionops*) (Stettenheim 2000). More generally, patches of bare skin, other than the bill and legs, can be found in birds belonging to at least 19 different orders and 62 families (Negro et al. 2006). Colored unfeathered areas on the head and necks of birds may be important in (1) intra- and intersexual communication, e.g., advertising status or quality, (2) thermoregulation, (3) protection against ultraviolet irradiation (Nicolaï et al. 2020), and (4) preventing the soiling of feathers for species that sometimes extend their heads into carcasses when feeding (e.g., vultures and condors). Among birds with colorful skin (Fig. 3.5), the coloration is due either to pigments, structural mechanisms in the epidermis, or to blood (and, specifically, hemoglobin in the red blood cells) in the superficial capillary network (Lucas 1970; Prum and Torres 2003a; Justyn et al. 2023).

Some species with bare skin on the head and neck alter skin color by changing blood flow to the area. For example, skin in the unfeathered areas of a Crested Caracara's (*Caracara plancus*) head has a much denser supply of blood vessels than skin in feathered areas (Figs. 3.6 and 3.7). Although they sometimes feed on carrion, the bare skin of caracaras is most important for thermoregulation. Caracaras are relatively large birds with generally dark plumage, typically found in relatively hot areas. When their body temperature increases, blood flow to the areas of bare skin increases as vessels dilate. This increased blood flow causes the skin to become deeper red in color, but, most importantly, enhances the loss of heat across the bare skin. Bare skin on the head and neck likely serves a similar function for many species of birds because many such birds are relatively large, dark-plumaged birds that occur at low latitudes where heat dissipation may be of great importance (Negro et al. 2006).

For those species of birds with bare skin on the head and neck and where skin coloration is altered by changes in blood flow ("flushing"), thermoregulation was almost certainly the primary selective factor. However, in some species, flushing occurs in contexts unrelated to thermoregulation, such as during courtship or agonistic encounters. For example, the skin of turkeys become redder when courting females and when engaged in agonistic encounters with other males. This suggests that "flushing" can also have a signaling function for some species and in some contexts. The ability to generate a deeper red coloration or maintain redder coloration for longer periods may be correlated with individual quality if doing so is energetically costly or potentially damaging to the body (Negro et al. 2006).

The colored skin of many birds is due to pigments, molecules that differentially absorb and emit wavelengths of visible light. One such pigment is melanin, which, when present, results in dark or black skin (Fig. 3.8). Nicolaï et al. (2020) examined the skin of 2247 species of birds representing all families and >99% of bird genera and found 138 genera (6%) and 59 families (23%) had species with black skin on the head, but few species had black skin in the ventral portions of their bodies. In addition, analyses revealed that black skin was more prevalent at lower latitudes. Because melanin can protect skin against DNA-damaging ultraviolet (UV) irradiation, these results suggest that selection has favored black skin in species of birds found at lower latitudes with greater exposure to UV radiance. Furthermore, the general absence of black skin ventrally, but its presence on the head where exposure to UV is higher, further supports the UV protection hypothesis (Nicolaï et al. 2020; Justyn et al. 2023).

**Fig. 3.5** Structurally colored ornaments of a sample of birds. (**a**) Velvet Asity (*Philepitta castanea*) (Photo by Frank Vassen, Wikipedia, CC BY 2.0, https://creativecommons.org/licenses/by/2.0/deed.en), (**b**) Bali Mynah (*Leucopsar rothschildi*) (Photo by Brian, Wikipedia, CC BY 2.0, https://creativecommons.org/licenses/by/2.0/deed.en), (**c**) Capuchinbird (*Perissocephalus tricolor*). (Wikipedia, Creative Commons CC0 1.0 Universal Public Domain Dedication), (**d**) Toco Toucan (*Ramphastos toco*) (Photo by Ivan Sitnikov, pxhere.com, Creative Commons CC0), (**e**) Channel-billed Toucan (*Ramphastos vitellinus*) (Photo by Snowmanradio, Wikipedia, CC BY 2.0, https://creativecommons.org/licenses/by/2.0/deed.en), and (**f**) Madagascar Paradise-Flycatcher (*Terpsiphone mutata*) (Photo by Frank Wouters, Wikipedia, CC BY 2.0, https://creativecommons.org/licenses/by/2.0/deed.en)

Carotenoids are the pigments responsible for colorful skin (as well as feathers) in many birds and typically generate a red, orange, or yellow hue. However, birds cannot synthesize carotenoids and, therefore, must acquire them in their diet (Fig. 3.9). As a result, variation in carotenoid-based skin and feather coloration, can provide conspecifics with information about individual quality, perhaps via a role in mitochondrial function (Hill et al. 2019; Cantarero et al. 2020; Powers and Hill 2021; Fig. 3.10), and the ability of different individuals to acquire a limited resource (Negro et al. 2002). For example, Red-legged Partridges (*Alectoris rufa*) have bills and eye rings (bare skin, not feathers) that are reddish due to the presence of carotenoids and

**Fig. 3.6** A Crested Caracara (*Caracara plancus*) showing unfeathered and feathered areas of the head. (Photo by Keenan Adams, USFWS, Wikipedia, CC BY 2.0, https://creativecommons.org/licenses/by/2.0/deed.en)

individuals with redder bills and eye rings are in better physical condition (Pérez-Rodríguez and Viñuela 2008). Similarly, the yellow-orange skin on the legs, feet, and ceres (skin at the base of the upper bill) of Eurasian Kestrels (*Falco tinnunculus*) is due to carotenoids and studies have revealed that male kestrels with more brightly colored skin are better hunters and have better quality territories (Casagrande et al. 2006).

For many birds, skin coloration is the result of optical interactions with biological nanostructures or, in other words, the microscopic structure of the skin (Fig. 3.11). Such structural colors occur in the skin, bill (rhamphotheca), legs, and feet (podotheca) in about 129 avian genera in 50 families from 16 avian orders. Structurally colored skin is present in more than 250 bird species, or about 2.5% of all bird species (Prum and Torres 2003a). Examination of the color, anatomy, and nanostructure of structurally colored skin, rhamphotheca, and podotheca from several different species of birds indicates that color, including ultraviolet, dark blue, light blue, green, and yellow hues, is produced by coherent scattering (i.e., constructive interference) of light from arrays of parallel collagen fibers in the skin (dermis) (Figs. 3.12 and 3.13). Scattering, in this

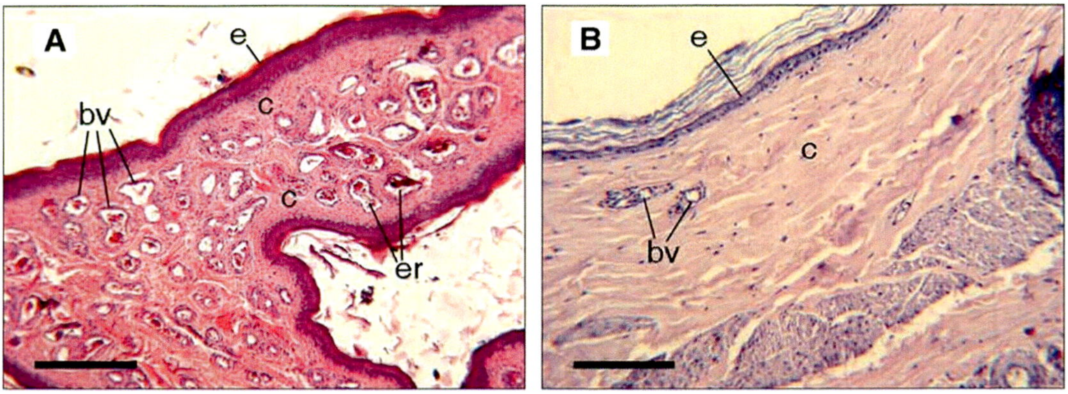


**Fig. 3.7** Micrographs of cross-sections of the skin of a Crested Caracara (*Caracara plancus*), a species with unfeathered areas on the head. (**a**) Unfeathered area (bare skin) on the face, and (**b**) feather-covered area on the head. Note the greater number of blood vessels in the unfeathered skin. Scale bars = 25 µm. e, epidermis; c, collagen; er, erythrocytes; bv, blood vessels. (Figure from Negro et al. 2006; Copyright © 2005 Elsevier Inc. All rights reserved, used with permission)

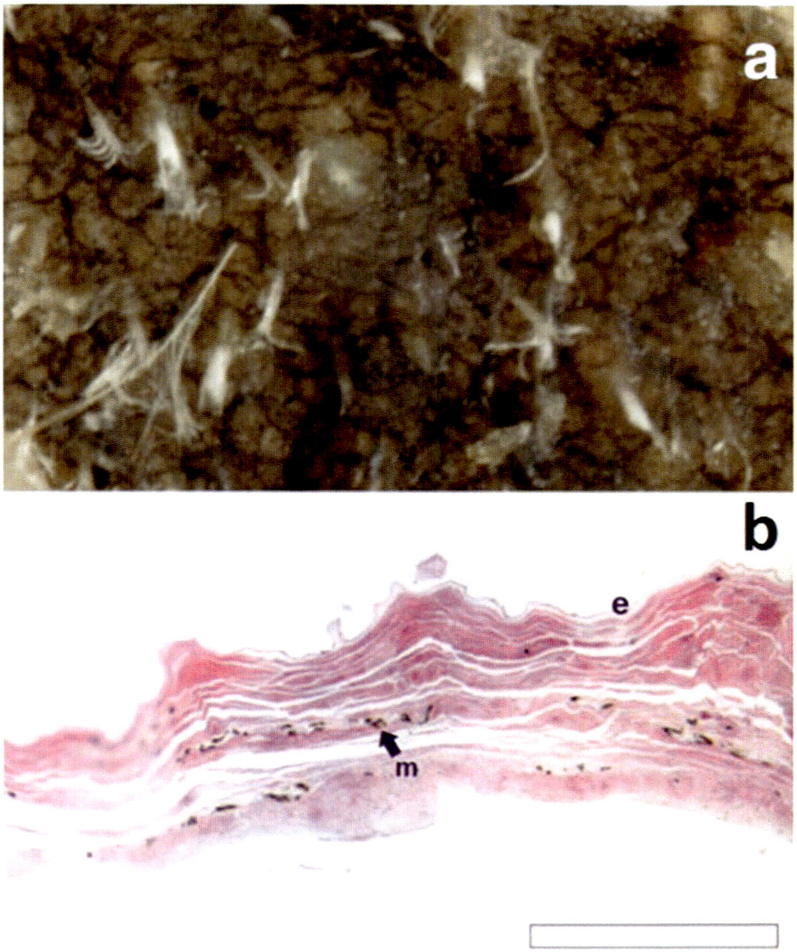


**Fig. 3.8** (**a**) Photo of the black skin of a Northern Gannet (*Morus bassanus*). (**b**) Magnified view of a section of Northern Gannet skin showing the presence of melanin (m), e, epidermis, Scale bar in b = 50 μm. (Figure from Nicolaï et al. 2020; open-access article licensed under a Creative Commons Attribution 4.0 International License, https://creativecommons.org/licenses/by/4.0/)

case, simply means that light deviates from a straight path.

Visible light is, of course, composed of many colors of light with distinct wavelengths. Red light has a long wavelength (~700 nm), whereas violet and blue light have a much shorter wavelength (~400 nm). When visible light encounters particles with the same or larger diameter than its component wavelengths, those specific light photons are reflected. For example, particles that are 400 nm or slightly larger will selectively reflect blue light photons while allowing other light photons to pass. In the structurally colored skin of birds, light is reflected by collagen fibers (long, string-like protein molecules) that are arranged in a much more highly organized manner than in normal skin. For patches of skin that are a particular color (e.g., blue or green), all collagen fibers are the same thickness (Fig. 3.12). As a result, each fiber scatters wavelengths of light that are in phase (Fig. 3.14) and, therefore, are reinforced, producing very bright colors. In addition to the reflection of light of certain wavelengths, skin structural coloration also requires a means to prevent the reflection or scattering of white light by deeper tissues below the color-producing nanostructures. In bird skin, this light-absorbing layer consists of a thick layer of melanin granules (melanosomes; Fig. 3.8).

In some groups of birds (e.g., Phasianidae, Eurylaimidae, Cotingidae, Paradisaeidae, and Cnemophilidae), sexual selection is likely

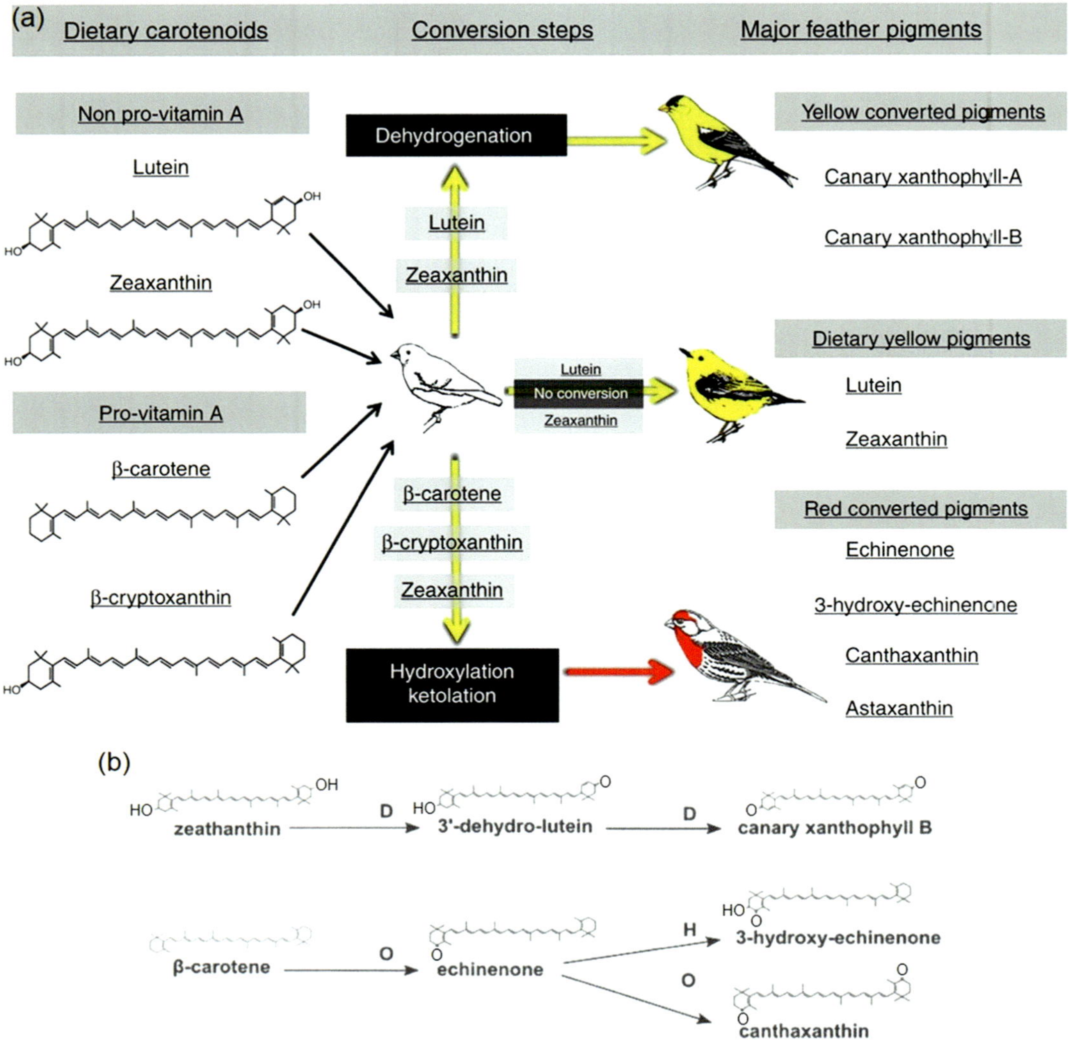


**Fig. 3.9** (**a**) Carotenoids responsible for skin and feather coloration. For feather coloration, songbirds either use unaltered dietary carotenoids or use carotenoid pigments metabolically altered from dietary carotenoids. Shown here are proposed metabolic pathways that convert carotenoids in bird diets into the red and yellow carotenoids found in feathers. For example, House Finches (*Haemorhous mexicanus*; bottom) use converted red carotenoids, American Goldfinches (*Spinus tristis*; top) used converted yellow carotenoids, and wood warblers (middle) used dietary yellow carotenoids. Some species of birds directly obtain red carotenoids, such as astaxanthin, from food items in their diet (not shown). (**b**) Chemical structure of two carotenoids and their conversion pathways. Zeathanthin is dehydrogenated (D) to form 3′-dehydro-lutein, which is then dehydrogenated (D) to form canary xanthophyll B. β-carotene is oxidized (O) to form echinenone, which can either be hydroxylated (H) to form 3-hydroxy-echinenone or oxidized (O) to form canthaxanthin. Other carotenoids undergo similar conversions. (Figure a from Weaver et al. 2018; open-access article licensed under a Creative Commons Attribution 4.0 International License, https://creativecommons.org/licenses/by/4.0/. Figure b from Koch et al. 2016; © 2016 Wilson Ornithological Society, used with permission)

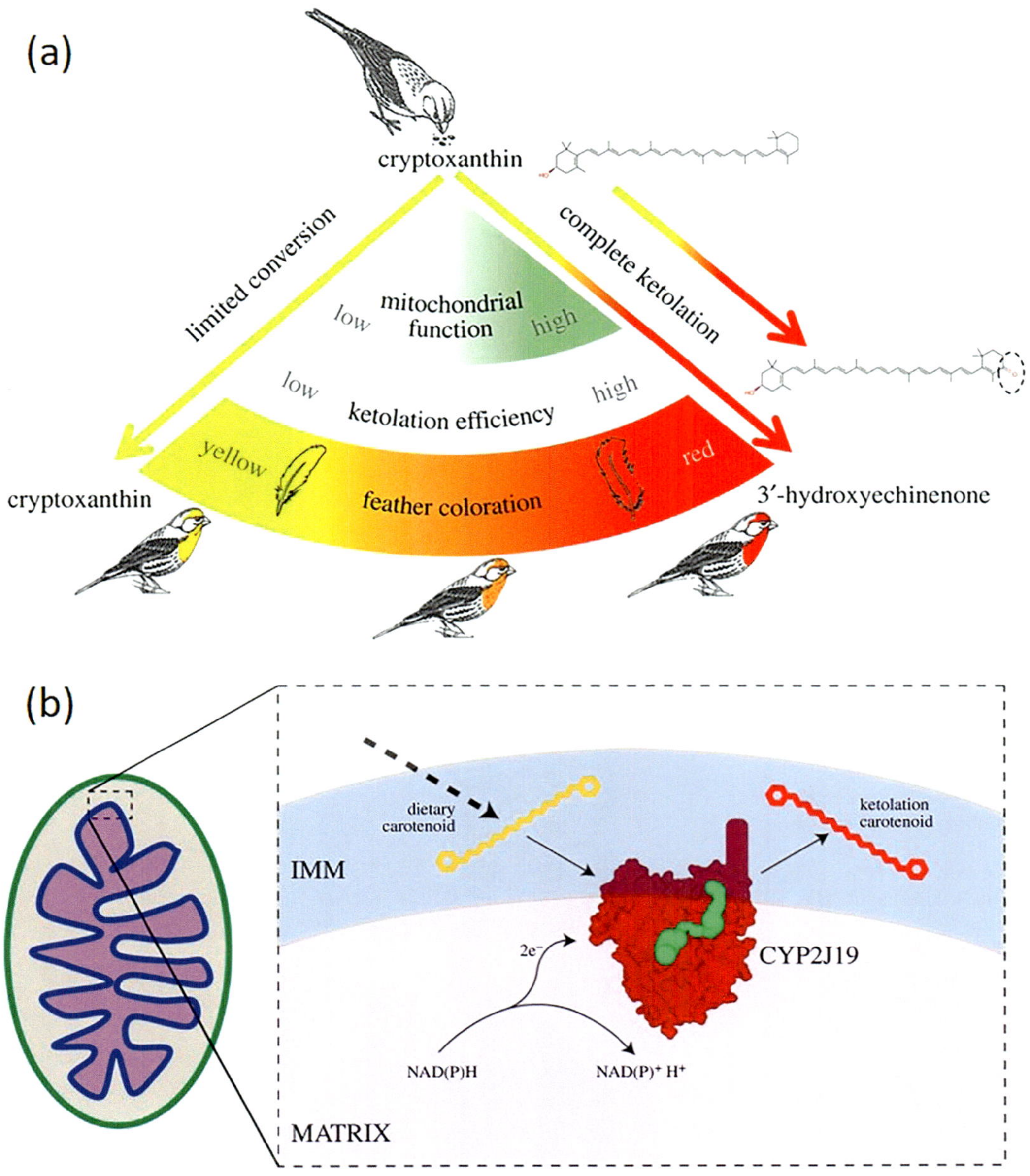


**Fig. 3.10** (**a**) Hypothesized links between red skin, feather coloration and mitochondrial function. To produce red feathers, birds like House Finches (*Haemorhous mexicanus*) ingest the yellow carotenoid cryptoxanthin and oxidize it to the red pigment 3-hydroxyechinenone. The efficiency of this process is hypothesized to be linked to mitochondrial bioenergetics. This hypothesis posits that birds with low mitochondrial function have limited ketolation capacity (i.e., ability to produce red carotenoids from yellow carotenoids) and produce yellow feathers. Birds with high mitochondrial function have improved ketolation capacity and produce red feathers. By linking feather coloration to mitochondrial function, this hypothesis, referred to as the Shared-Pathway Hypothesis, establishes a link between coloration and individual condition or quality. (**b**) A possible mechanism whereby the ketolation reaction caused by the enzyme CYP2J19 links red coloration to mitochondrial performance. Red carotenoids are present in the inner mitochondrial membrane, and CYP2J19 is hypothesized to convert yellow carotenoids into red carotenoids in a reaction requiring an electron donor (NAD(P)H) and oxygen. If CYP2J19 is anchored to the inner mitochondrial membrane (IMM), then its function as an enzyme would likely be affected by mitochondrial performance, particularly the mitochondrial redox state, that is, $NAD(P)^+$/NAD(P)H redox couples that help regulate ATP production by mitochondria. (Figures from Hill et al. 2019; © 2019 The Authors. Published by the Royal Society, used with permission)

**Fig. 3.11** Top, a male Velvet Asity (*Philepitta castanea*) showing the green and less visible blue skin above the eye. Fleshy structures of birds like this are called caruncles. Bottom, (**a**) Section of a caruncle viewed under a dissecting scope. Each small circular structure is called a papilla. (**b**) Image showing the collagenous material inside a papilla. (**c**) Light micrograph of a cross-section of stained papillae. Each papilla consists of sheath(s), a capsule filled with collagenous macrofibrils (cm), and a bottom layer of melanocytes (m) in the dermis of the skin. The stained collagen appears blue. The gap below the collagen macrofibrils on the right is an artifact. Uc, unordered collagen. Scale bar = 175 μm. (**d**) Light micrograph of the core of a single papilla showing irregularly packed collagen macrofibrils (cm). Scale bar = 75 μm. (Photo of Velvet Asity by Frank Vassen, Wikipedia, CC BY 2.0, https://creativecommons.org/licenses/by/2.0/deed.en. Bottom figure from Prum et al. 1994; © 1994 Wiley-Liss, Inc., used with permission)

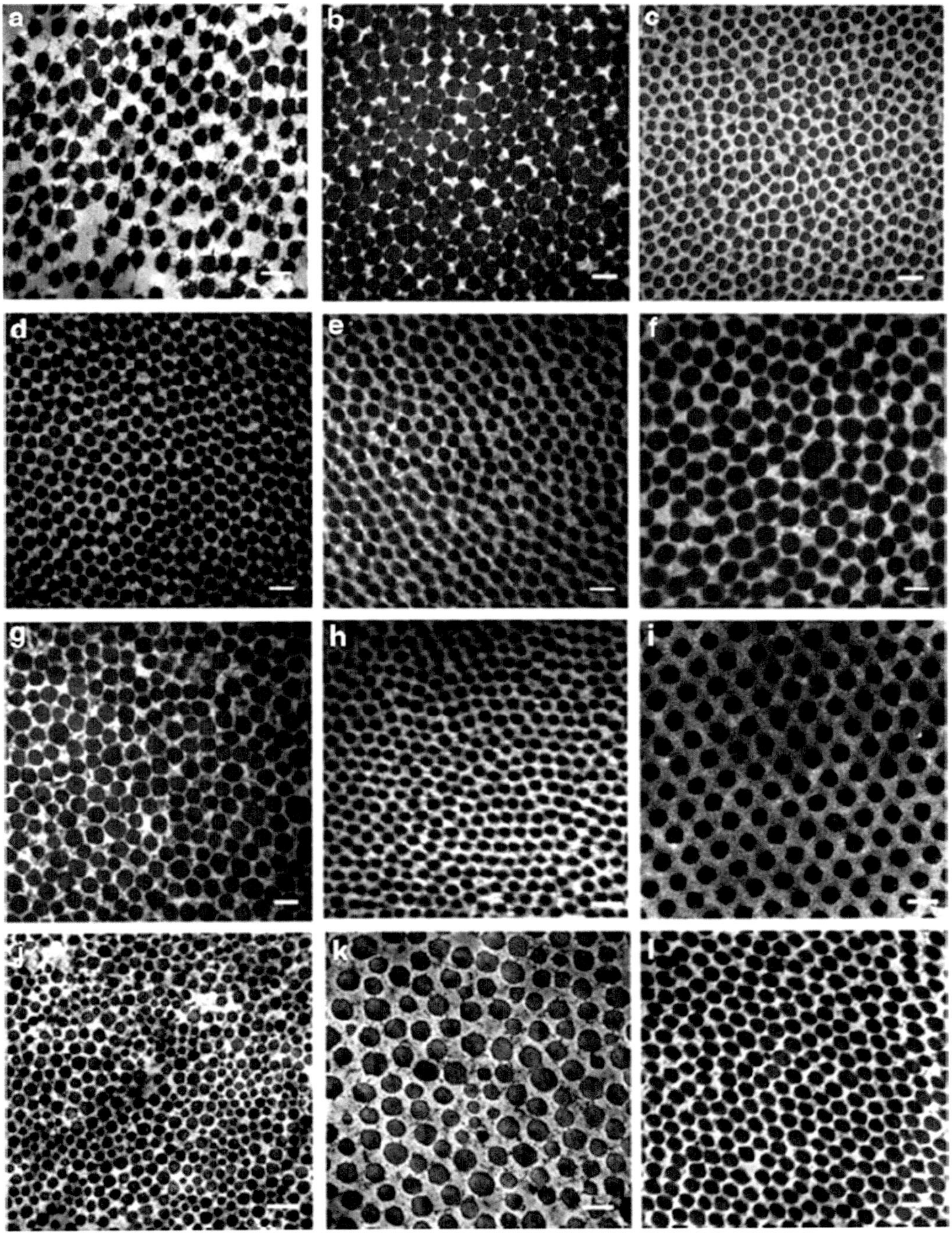


**Fig. 3.12** Transmission electron micrographs showing the very regular nanostructured arrays of dermal collagen fibers from: (**a**) Ruddy Duck (*Oxyura jamaicensis*), light blue; (**b**) Helmeted Guineafowl (*Numida meleagris*), dark blue; (**c**) Satyr Tragopan (*Tragopan satyra*), dark blue; (**d**) Cabot's Tragopan (*Tragopan caboti*), dark blue; (**e**) Cabot's Tragopan, light blue; (**f**) Cabot's Tragopan, orange; (**g**) Whistling Heron (*Syrigma sibilatrix*), blue; (**h**) Toco Toucan (*Ramphastos toco*), dark blue; (**i**) Velvet Asity (*Philepitta castanea*), light blue; (**j**) White-cheeked Antbird (*Gymnopithys leucaspis*), light blue; (**k**) Bare-throated Bellbird (*Procnias nudicollis*), green and (**l**) Madagascar Paradise-Flycatcher (*Terpsiphone mutata*), dark blue. All images were taken at 30,000×. All scale bars = 200 nm. (Figure from Prum and Torres 2013; © 2013 Birkhäuser Boston, used with permission)

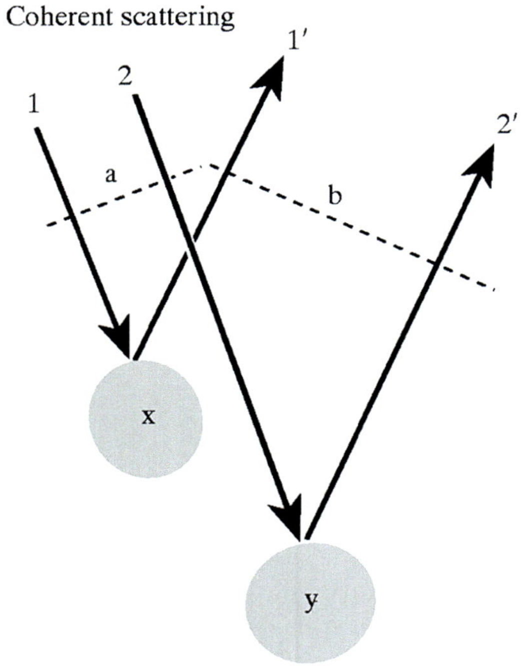


**Fig. 3.13** Coherent scattering is differential interference or reinforcement of wavelengths scattered by multiple light-scattering objects (x, y). The coherent scattering of specific wavelengths is determined by the phase relationships among the scattered waves. Scattered wavelengths that are out of phase will cancel each other out, but scattered wavelengths that are in phase will be constructively reinforced and coherently scattered. Phase relationships of wavelengths scattered by two different objects (x and y) are given by the differences in the path lengths of light scattered by the first object (x: 1–1′) and a second object (y: 2–2′) as measured from planes perpendicular to the incident (a) and reflected (b) waves in the mean refractive index of the media. (Figure from Prum and Torres 2003b; © 2003 Oxford University Press, used with permission)

responsible for the structurally colored skin of males because most species are polygynous. In other groups of birds, including Ardeidae, Cariamidae, Bucerotidae, Ramphastidae, Meliphagidae, and Monarchidae, both sexes have integumentary structural colors, suggesting that such coloration may be important in both inter- and intrasexual communication (Prum and Torres 2003a). Many birds with structurally colored skin are found in rainforests. For example, nearly all species with structurally colored skin in the orders Casuariiformes, Galliformes, Opisthocomiformes, Cuculiformes, Trogoniformes, Coraciiformes, Piciformes, and Passeriformes occur in tropical forests. The quality of ambient light in tropical forest habitats may favor the evolution of communication signals in the shorter wavelength portion of the visible spectrum (blue and green) and, if so, selection might favor structural colors because vertebrates, including birds, have no pigments that generate such colors (Prum and Torres 2003a).

## 3.3 Specialized Epidermal Structures

The skin of the throat or the neck of some non-passerine birds is bare, distensible, and forms a pouch. Pelicans use their pouches to capture fish and for thermoregulation (gular fluttering enhances heat loss by evaporation). Pouches also occur on the throat or sides of the neck of male frigatebirds, certain storks (*Leptotilos*), male grouse of several species, and others. These,

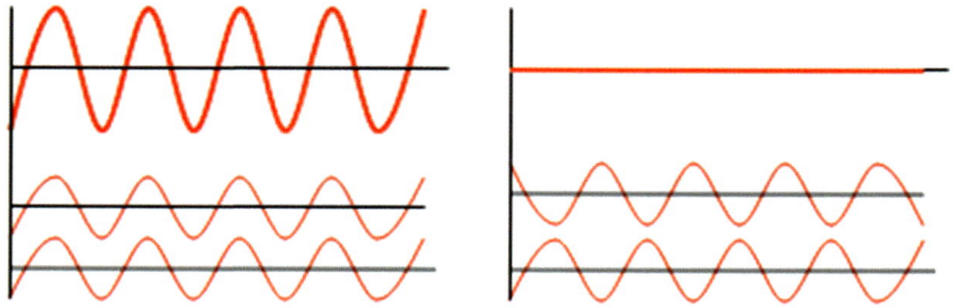

**Fig. 3.14** Left, In constructive interference (coherent scattering), two light waves (on bottom) are in phase so the combined waveform (top) is enhanced. Right, In destructive interference, two light waves (bottom) out of phase so they cancel out (top). (Figure from Wikipedia, CC BY 3.0, https://creativecommons.org/licenses/by/3.0/)

however, are inflated by swelling of the upper end of the esophagus as mouth action directs air into the glottis (Johnsgard 1983). The bare skin is normally constricted or even concealed, inflated, and displayed in courtship or social displays. Bright coloring in some species enhances its role as a visual signal. Additionally, in grouse and Greater Painted-Snipe (*Rostratula benghalensis*), the pouches augment their vocalizations by enlarging the sound-resonating chamber (Stettenheim 2000).

Birds exhibit various integumentary outgrowths on the head and upper neck that vary in size, shape, color, and location. Caruncles are simple, rounded protuberances found on, for example, turkeys, Magpie Geese (*Anseranas semipalmata*), Southern Ground-Hornbills (*Bucorvus cafer*), and certain curassows, megapodes, and cathartid vultures (Fig. 3.15). In some species, caruncles play a role in intersexual communication (e.g., Buchholz 1995). For example, adult males of all asities (family Philepittidae and endemic to Madagascar) have colored facial caruncles during the breeding season (Fig. 3.16) that play an important role in intersexual communication (Prum and Razafindratsita 1997). In other species, such as Ross's Geese (*Anser rossii*), facial caruncles may play a role in intrasexual aggressive interactions (McLandress 1983).

Wattles are the most common soft integumentary outgrowth in birds. These protuberances or flaps are generally located on the sides of the head or neck, especially the base of the bill and around the eyes, but, in some birds, they hang beside the mouth or under the throat. Birds with wattles include cassowaries, many cracids, megapodes, ptarmigan, pheasants, guineafowl, turkeys, rails, jacanas, lapwings, alcids, cotingas, starlings, honeyeaters, and wattlebirds (Stettenheim 2000; Fig. 3.17). Wattles may serve as signals of aggression by birds defending territories (e.g., South Island Saddlebacks, *Philesturnus carunculatus*; Lloyd-Jones and Briskie 2016) and, more generally, convey dominance status (Iverson and Karubian 2017). In some species, wattle size is also correlated with mating success (e.g., Parker and Ligon 2003).

Combs are thick, upright wattles on the top of the head (Fig. 3.18). Single, mid-dorsal combs occur in junglefowl (including Domestic Chickens), certain brush turkeys (Wattled Brushturkey, *Aepypodius arfakianus,* and Waigeo Brushturkey, *A. bruijnii*), Comb-crested Jacanas (*Irediparra gallinacea*), and male Andean Condors (*Vultur gryphus*). Among Andean Condors, changes in blood flow and cause the color of combs and bare skin on their heads to become more yellow, orange, or red in just a few seconds. The intensity of these colors appears to be important during interactions among condors at carcasses and in sexual contexts (Blanco et al. 2013). Adult grouse have paired combs above the eyes that are larger in males than females (Johnsgard 1983). Combs are red, orange, or yellow, with color caused by blood (combs are highly vascular) and carotenoid pigments in the epidermis (Stettenheim 2000). For many, if not most species of birds, caruncles, wattles, and combs are ornaments that likely evolved as signals between males (intrasexual selection) or to help females assess males (intersexual selection; Zuk 1991). For example, wattle size is correlated with the ability of male Ring-necked Pheasants (*Phasianus colchicus*) to obtain a territory during the breeding season (Papeschi et al. 2003), and comb size in lekking Black Grouse (*Tetrao tetrix*) was found to be positively correlated with copulatory success (Rintamäki et al. 2000). More generally, in the families Phasianidae and Rallidae, comb and wattle size is positively correlated with male dominance status (e.g., Zuk and Johnsen 2000; Dey et al. 2014). Mate choice by female Red Junglefowl (*Gallus gallus*) is influenced by the color and size of male combs (Zuk et al. 1995). More generally, the color and size of male combs and wattles of males in the family Phasianidae have often been found to be correlated with success in achieving copulations (e.g., Rintamäki et al. 2000; Parker and Ligon 2003). Although sexual selection has likely been a factor in the evolution of the various integumentary outgrowths of other species of birds, empirical studies are needed with many species of birds that possess such structures to either confirm, or not, that assumption.

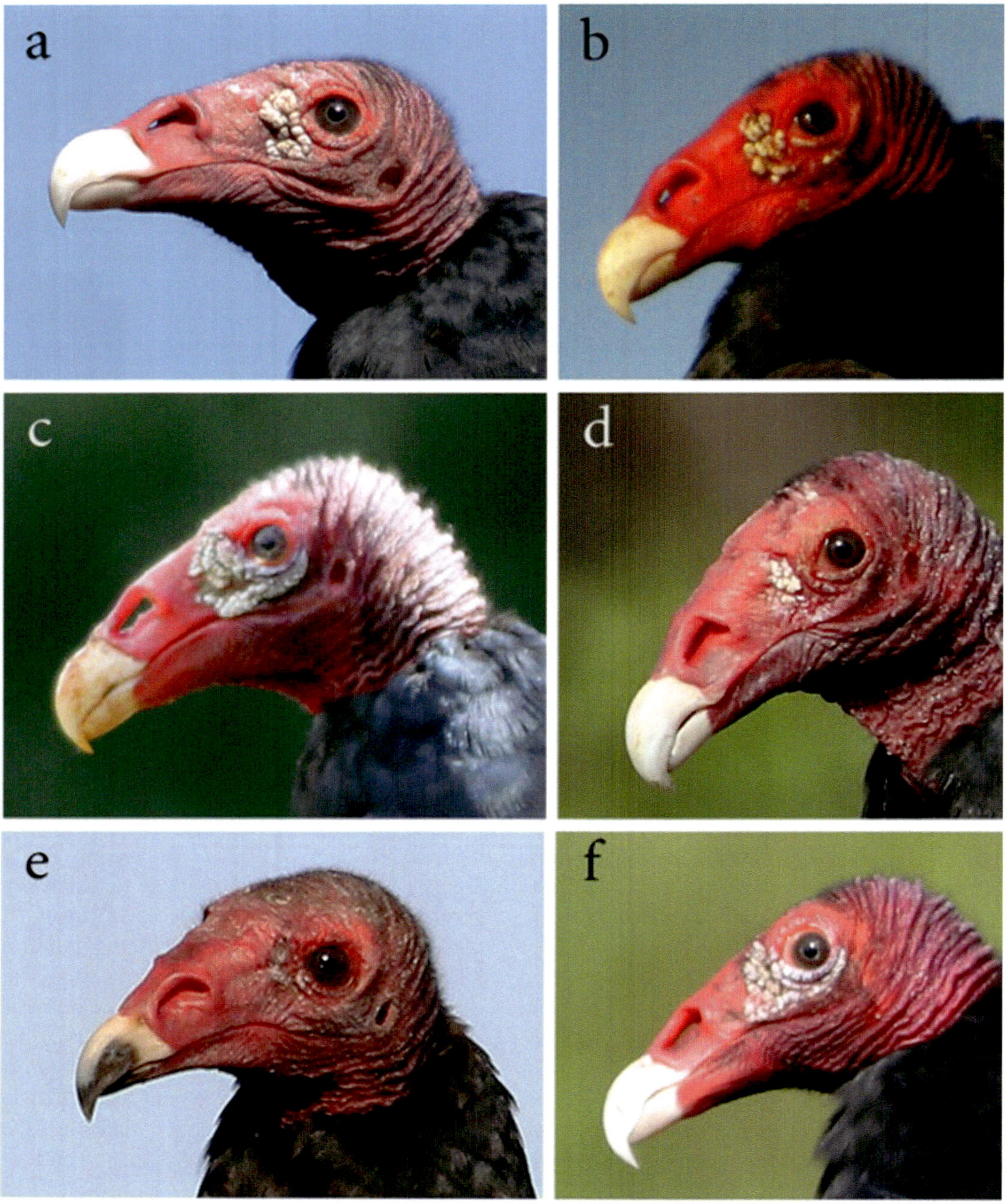


**Fig. 3.15** (**a**–**f**) Examples of preorbital facial caruncles of Turkey Vultures (*Cathartes aura*) in Jamaica that are thought to function in intra- and interspecific signaling. (Figure from Graves 2019; open-access article licensed under a Creative Commons Attribution License. https://creativecommons.org/licenses/by/4.0/)

**Fig. 3.16** Yellow-bellied Sunbird-Asity (*Neodrepanis hypoxanthus*). Note the colorful facial caruncles that develop during the breeding season and play an important role in intersexual communication. (Photo by Dubi Shapiro, Wikipedia, CC BY 3.0, https://creativecommons.org/licenses/by/3.0/)

## 3.4 Cutaneous Nervous System

For the integument to be able to react appropriately to the ever-changing and unpredictable air currents during flight, a feedback mechanism through an intricate nervous network with sensory receptors and motoric innervation is necessary. The feather muscles are extensively innervated (Lucas and Stettenheim 1972; Bennett 1974), and the skin, especially near the follicles of contour feathers, is especially rich and diverse in sensory neurons (Brown and Fedde 1993; Saxod 1996). There are more touch receptors in the skin of birds than in that of mammals (Dorward 1970), and the density of touch receptors is significantly higher in the feather-bearing skin of flying birds than in flightless birds, suggesting that the complex sensory innervation of avian skin is related to the demands of flight (Homberger and de Silva 2000).

Sensory neurons (mechanoreceptors) located on or near feather follicles likely provide birds with feedback from all body feathers (Altshuler et al. 2015). For flying birds, these receptors likely translate feather movements into information about airspeed and, for covert feathers on the wings, the separation of airflow from the wing's surface that occurs during a stall (Brown and Fedde 1993). Altshuler et al. (2015) speculated that gliding birds whose wings are maintained at a shallow angle of attack may be able to monitor the separation of airflow over the wing via deflection of wing coverts to help prevent stalling and, more generally, birds may depend on such sensory information when intentionally stalling as they land (additional information about cutaneous receptors is provided in Chap. 4).

## 3.5 Podotheca

The feet and, in most birds, tarsometatarsal areas are covered by plates of keratin, or scutes. Scales are flat, rounded, or conical raised thickenings of the highly keratinized epidermis, separated by inward folds of thinner, less keratinized epidermis. Bird scales vary in size, shape, amount of overlap, and degree of fusion on different parts of the foot and among species. Scales on the anterior and caudal surfaces of the tarsometatarsus and the dorsal surface of the toes (scutellate, or scutate, scales) tend to be larger, more rectangular, and

**Fig. 3.17** Examples of species of birds with wattles. (**a**) North Island Saddleback (*Philesturnus rufusater*), (**b**) North Island Kokako (*Callaeas wilsoni*), and (**c**) Masked Lapwing (*Vanellus miles*). (**a**, Wikipedia, CC0 Public Domain; **b**, Photo by Matt Bins, Wikipedia, CC BY 2.0, https://creativecommons.org/licenses/by/2.0/; **c**, Photo by Bernard Spragg, Wikipedia, CC0 Public Domain)

**Fig. 3.18** Examples of species with combs. (**a**) Comb-crested Jacana (*Irediparra gallinacea*), (**b**) Red Junglefowl (*Gallus gallus*), and (**c**) Andean Condor (*Vultur gryphus*). (**a**, Photo by Jean and Fred Hort, Wikipedia, CC BY 2.0, https://creativecommons.org/licenses/by/2.0/deed.en: **b**, Photo by Jason Thompson, Wikipedia, CC BY 2.0, https://creativecommons.org/licenses/by/2.0/deed.en; **c**, Photo by Michael Gäbler, Wikipedia, CC BY 3.0, https://creativecommons.org/licenses/by/3.0/)

more regularly arranged than those on the other surfaces (reticulate scales; Fig. 3.19). However, in some species, including Brown Pelicans (*Pelecanus occidentalis*), all of the scales on the tarsometatarsus are reticulate. Tarsal scales, not needing flexibility, are sometimes fused (booted condition), especially on the anterior surface (Stettenheim 2000). In areas of the leg and feet that exhibit greater movement, scales are usually smaller (Fig. 3.20).

The bottom of the foot has a more flexible layer of the epidermis, and the reticulate scales are covered with a thin layer of hard keratin that rests on a thick layer of more pliable keratin. The epidermis on the bottom of the foot also forms pads—thick layers of reticulate scales, many cells thick. The horny layer (stratum corneum) of the pads also contains a large amount of lipid, making the pads spongy and better able to withstand compression (Spearman and Hardy 1985; Fig. 3.21).

Specializations of the bird foot include heel pads, webs between the toes, and tarsal spurs. Heel pads are thickenings of the integument at the heel (joint of tibiotarsus and tarsometatarsus) and occur in nestlings of, primarily, cavity-nesting species, including some parrots, owls, coraciiformes, toucans, and woodpeckers (Stettenheim 1972; Figs. 3.22 and 3.23). The skeleton and muscles of the leg above the heel pad develop faster than those in the foot below the pad and, as a result, the heel pad functions as a nestling's "foot" during the early nestling period. For example, concerning the heel pads of young toucans, Josselyn Van Tyne (1929) stated that their function was ". . . simply to form a pair of substitute feet during the long period of helpless nest life." Consequently, heel pads are reduced or lost after young leave the nest.

Webs between toes vary in size and in the number of toes they connect. Pelecaniforms have webbing connecting all four toes (totipalmate), whereas loons, procellariforms, anseriforms, larids, and alcids have full webs between the three forward toes. Plovers and some sandpipers have partial webs between the proximal portions of the front three toes. Rather than webbing that connects the toes, each of the front three toes of grebes, coots, and rails have lobes (Stettenheim 2000; Fig. 3.24). Among grebes (Podicipedidae), the lobing is asymmetrical, with the lateral lobe on each toe smaller than

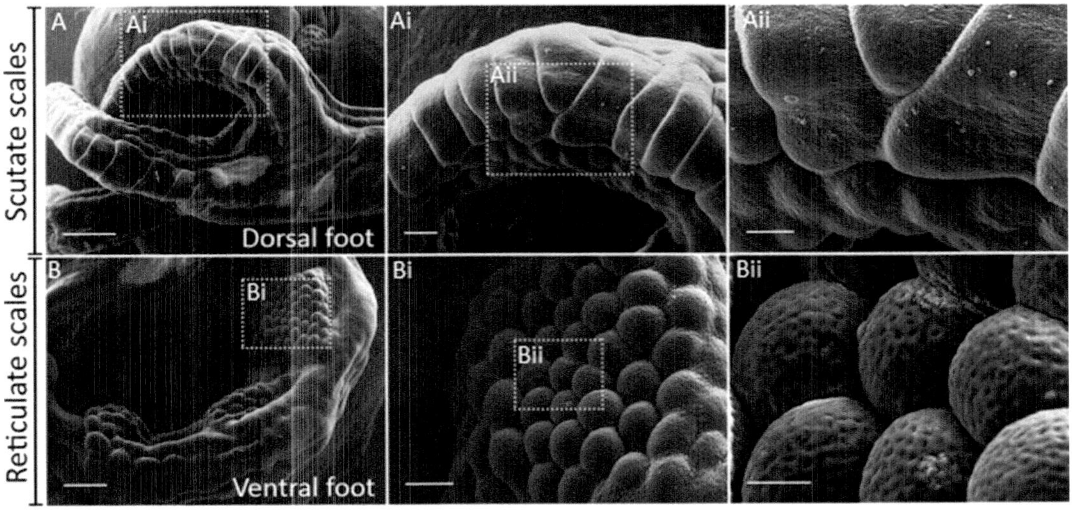


**Fig. 3.19** (**a**) Scutate (or scutellate) and (**b**) reticulate scales of birds. (Figure from Cooper et al. 2019; open-access article distributed under the terms of the Creative Commons Attribution 4.0 International License, https://creativecommons.org/licenses/by/4.0/)

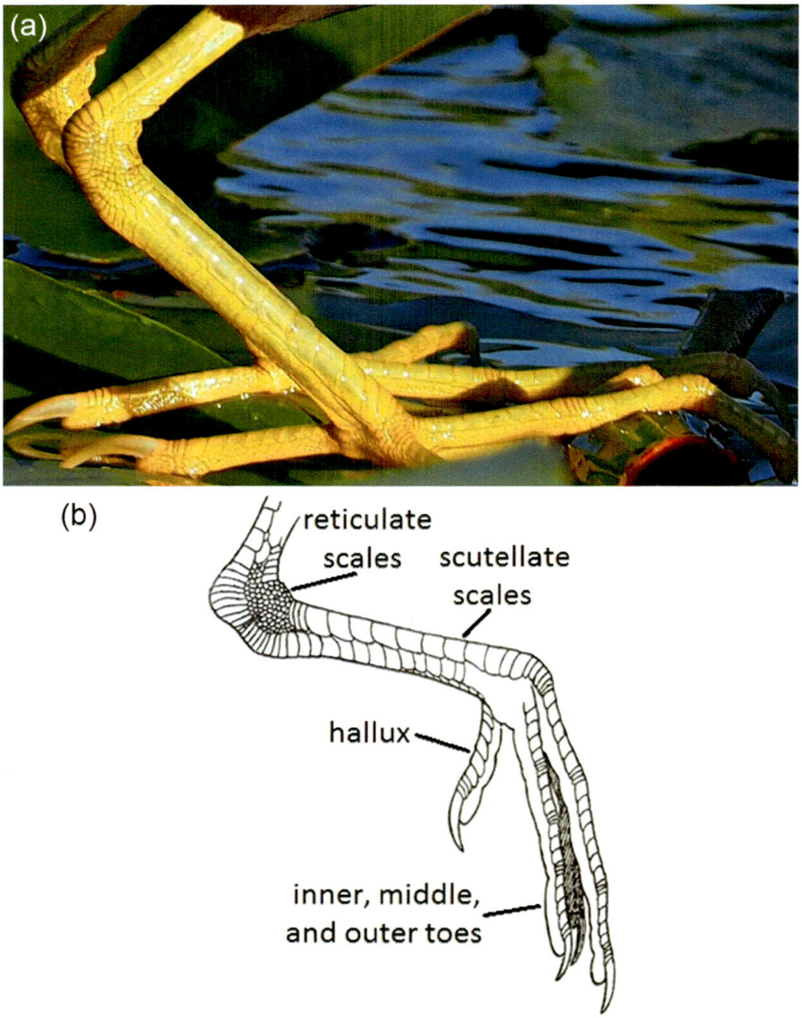


**Fig. 3.20** (**a**) Legs and feet of a Purple Gallinule (*Porphyrio martinica*) showing the different types of scales. (**b**) Right tarsus and foot of a Purple Gallinule. Note the smaller scales in areas with the greatest movement. (Photo a by Judy Gallagher, Wikipedia, CC BY 2.0, https://creativecommons.org/licenses/by/2.0/deed.en. Figure b modified from Wikipedia, CC0 Public Domain)

the medial lobe (Stolpe 1935). This asymmetry helps to increase speed and decrease the energy needed for foot-propelled underwater swimming (Johansson and Norberg 2001).

Each fall, grouse and ptarmigan develop fringes of scales along their toes (Johnsgard 1983) that, along with an increase in the number of feathers on the toes, increase the surface area of their feet and better allow them to walk on snow (Höhn 1977; Fig. 3.25). In fact, the word *Lagopus* (the generic name for ptarmigan) literally means "harefoot." The increase in the number of feathers on the feet also reduces the loss of heat from the feet (Stokkan 1992). The fringes and feathers on the toes are molted in the spring (Stettenheim 2000).

## 3.6 Spurs

Spurs are projections from either leg or wing bones that consist of a heavily cornified sheath over a bony core (Fig. 3.26). In cassowaries, pheasants, guineafowl, and turkeys, spurs are found on the back of the tarsometatarsus and are generally well-developed in males and either small or absent in females (Stettenheim 2000). In gallinaceous birds, males in both monogamous

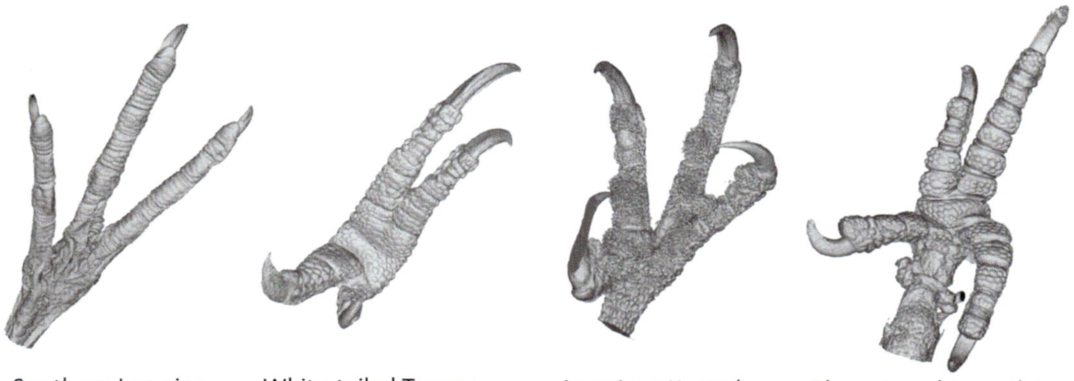


**Fig. 3.21** Reticulate scales on the plantar surface of the feet of four species of birds. Scale density varies among species, and scales tend to be flatter in aquatic and terrestrial birds and sharper, more pointed in raptors like American Kestrels and some woodpeckers. Scientific names: Southern Lapwing, *Vanellus chilensis*; White-tailed Trogon, *Trogon chionurus*; American Kestrel, *Falco sparverius*; Blue-winged Parrotlet, *Forpus xanthopterygius*. (Figure modified from Höfling and Abourachid 2021; © 2020 John Wiley and Sons, used with permission)

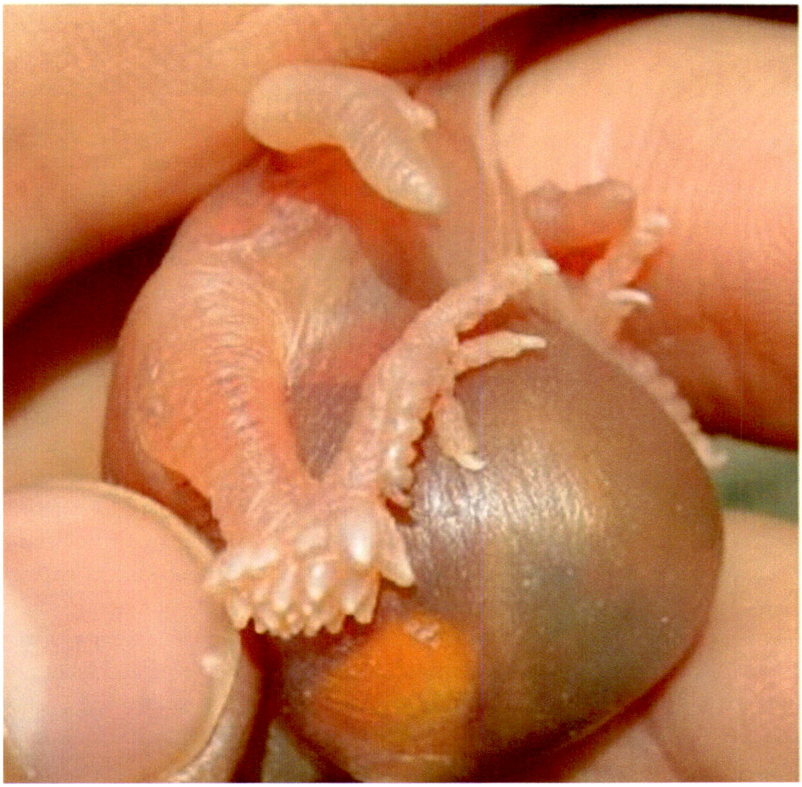

**Fig. 3.22** Heel pad of a nestling Crested Barbet (*Trachyphonus vaillantii*). (Figure from Engelbrecht 2010; used with permission of Derek Engelbrecht)

**Fig. 3.23** Nestling Red-breasted Toucan (*Ramphastos dicolorus*) supporting itself, in part, with its heel pads. (Figure from Perrella and Guida 2019; open-access article distributed under the terms of the Creative Commons Attribution License, https://creativecommons.org/licenses/by/4.0/)

and polygynous species have spurs, but spurs are longer and sharper in polygynous species where sexual selection is more intense (Davison 1985). Spurs are used as either weapons or ritualized ornaments in male–male (intrasexual) competition and appear to play no role in the female choice of mates (intersexual selection; Mateos and Carranza 1996). Males with longer spurs are generally more successful in interasexual competition for mates and have greater mating success.

Wing, or carpal, spurs are found in all species of screamers (Anhimidae) and in some species of geese (Anseridae), jacanas (Jacanidae), and plovers and lapwings (Charadriidae) (Figs. 3.27 and 3.28). They are usually well-developed in both sexes, but, except for jacanas, males usually have slightly longer spurs than females (Rand 1954; Prater et al. 1977). Seven of the eight species of jacanas exhibit sex-role reversal, with males incubating eggs and caring for young. In these species, females are larger than males and have longer wing spurs (Emlen and Wrege 2004). Wing spurs, like tarsal spurs, are used as either weapons or ritualized ornaments during aggressive encounters with conspecifics. For example, when defending breeding territories, Wattled Jacanas (*Jacana jacana*) respond to intruders by "lowering the head and holding the wings aloft with spurs and yellow underwings exposed" (Osborne and Bourne 1977).

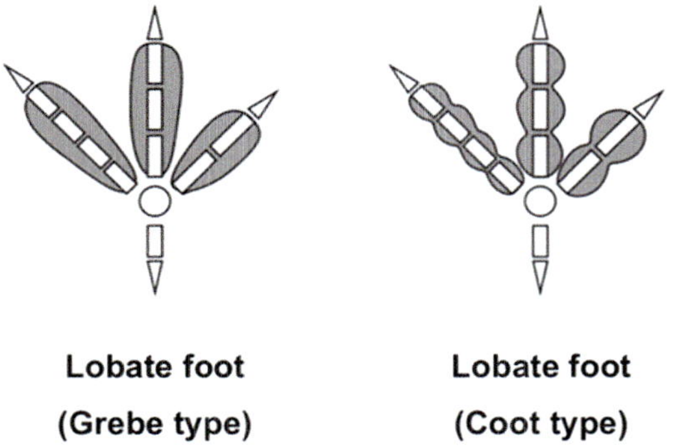


**Fig. 3.24** Lobate feet of grebes and coots. Note that the lobes of coots are constricted at the toe joints. The lobes of lobate feet have "hinges" at their base so that, during a power stroke (moving the feet backward to propel a bird forward), the lobes flare out, and the increased surface area produces greater thrust and forward propulsion. During the recovery stroke, the lobes fold down along the toes, reducing toe surface area and drag as a foot moves forward. (Figure from Tokita et al. 2020; © 2020 The Authors, open-access article licensed under a Creative Commons Attribution 4.0 International License, https://creativecommons.org/licenses/by/4.0/)

## 3.7 Claws

Claws are found at the distal end of all toes of all birds and cover the bones of terminal phalanges. Claws are formed by dorsal (unguis) and ventral (subunquis) keratin plates that cover the terminal phalanx (ungual bone) (Lucas and Stettenheim 1972). The two plates mesh in a longitudinal direction to form a claw's tip (Hahn et al. 2014). Claws represent specialized skin, with the epidermis consisting of the plates and underlying layers (stratum corneum and stratum germinativum) (Figs. 3.29 and 3.30). The keratin of the unguis is tougher and more compact than that of the subunguis. Cells in the stratum germinativum proliferate and undergo a progressive maturation called keratinization as they migrate to the surface. The keratin sheaths of bird claws consist of β-keratin, a form of keratin harder than the alpha keratin found in humans and other mammals (Box 3.2 Keratins). The keratin sheath of claws is constantly abraded, especially at the tip, but is continuously renewed as cells in the stratum germinativum divide and mature. For example, Bearhop et al. (2003) examined the growth of the claws of the middle front toes of five species of songbirds and found an average growth rate of 0.04 mm per day. The mean growth rates of the claws of Mallards (*Anas platyrhynchos*) and Tundra Swans (*Cygnus columbianus*) were 0.068 and

**Fig. 3.25** Foot of a grouse showing the fringes of scales that increase toe surface area and make it easier to walk on snow. (Photo by Mary Holland, used with permission)

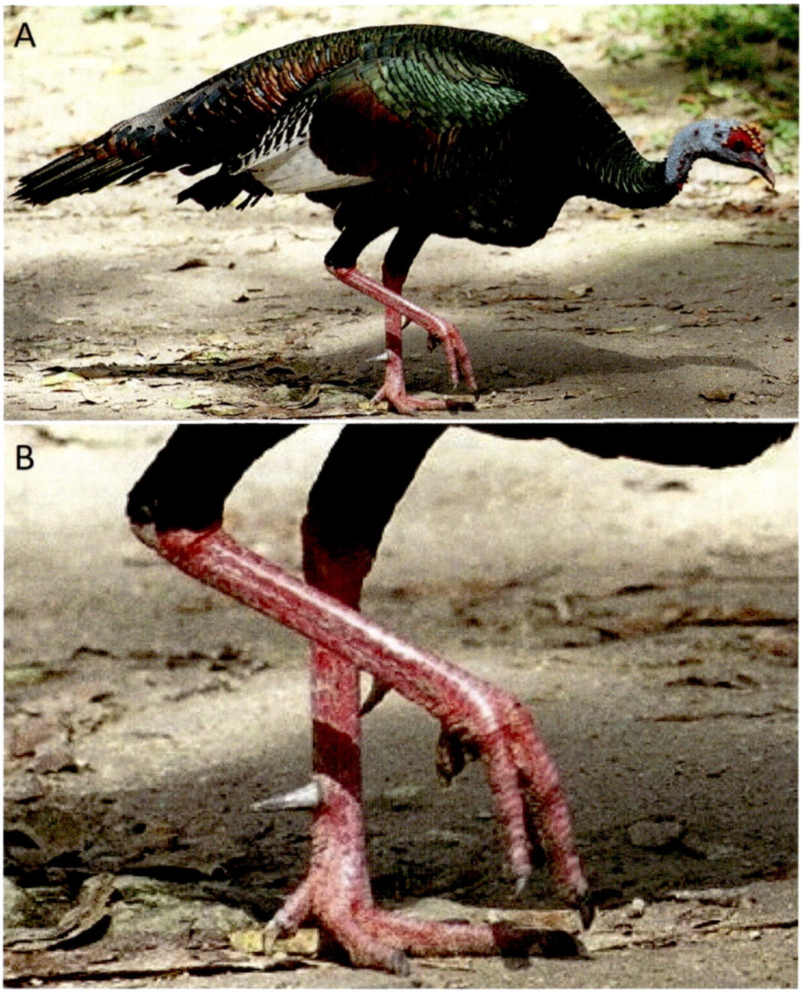


**Fig. 3.26** (**a**) Ocellated Turkey (*Meleagris ocellata*), and (**b**) close-up view of its spurs. (Photo by Somarinoa, Wikipedia, CC BY 2.0, https://creativecommons.org/licenses/by/2.0/deed.en)

0.076 mm per day, respectively (Hahn et al. 2014).

Claws are curved to varying degrees because the dorsal portion grows faster than the ventral portion (Fig. 3.31). In addition to variation in curvature, claws also vary in relative length and pointedness (Stettenheim 2000). Among diving and swimming birds with webbed or partially webbed feet, such as gannets, waterfowl, and gulls, claws tend to be small, less curved, and flatter. At the extreme, grebes have very flat claws that increase foot surface area and aid in underwater propulsion. A number of investigators have examined possible relationships between claw morphology and different ecological modes. Pike and Maitland (2004) found that claw angle increased with body mass for predatory and climbing birds (i.e., bigger birds have relatively more curved claws) and decreased with body mass for ground-dwelling birds (i.e., bigger birds have flatter claws). However, the mode of life (predatory, climbing, perching, or ground-dwelling) could not be predicted based on measurements of claw angle. Birn-Jeffery et al.

**Fig. 3.27** Southern Lapwing (*Vanellus chilensis*) with wing spurs indicated by the white circles. (Photo by Reyes, Wikipedia, CC BY 2.0, https://creativecommons.org/licenses/by/2.0/deed.en)

(2012) found that the claws of ground-dwelling birds were less curved, but, for other ecological modes, there was substantial overlap in claw morphology. To explain such overlap, Birn-Jeffery et al. (2012) noted that most birds are not limited to a single mode of life, for example, predatory birds also perch, and birds that perch and climb may also spend time foraging on the ground. These authors did add, however, that the claws of some taxa of birds that are well suited for climbing, for example, woodpeckers and treecreepers, have a contricted region near the tip that may aid in climbing (Fig. 3.32). However, these constrictions are not found on the claws of other species of birds that regularly climb, e.g., Pearled Treerunners (*Margarornis squamiger*) and Sri Lanka Hanging-Parrots (*Loriculus beryllinus*), suggesting that, although potentially advantageous, the constrictions are not essential for climbing. In yet another study, Hedrick et al. (2019) examined the claw morphology of 145 species of birds representing 21 orders and concluded that there were no significant associations between claw shape and ecological mode. These authors did, however, note some (albeit not significant) separation between predatory and ground-dwelling birds (Fig. 3.33), with predatory birds using highly curved claws to capture and kill prey and ground-dwelling birds having flatter claws that would likely be beneficial for terrestrial locomotion.

Several species of birds representing 17 families of birds in eight orders, including herons, frigatebirds, and pratincoles, have pectinate middle claws with comb-like edges that are used for grooming and preening feathers (Stettenheim 2000; Moyer and Clayton 2003; see the section on "Preening and other defenses against ectoparasites"). For example, Fiery-necked Nightjars (*Caprimulgus pectoralis*) use their pectinated claw to groom their rather long rictal bristles (Jackson 2007), and nocturnally foraging nightjars, more generally, may use their pectinated claw to clean spider webs from their plumage (Masterson 1979).

Wing claws can be found at the tip of the alular digit in several groups of birds, including loons, storks, owls, and some shorebirds, but they are

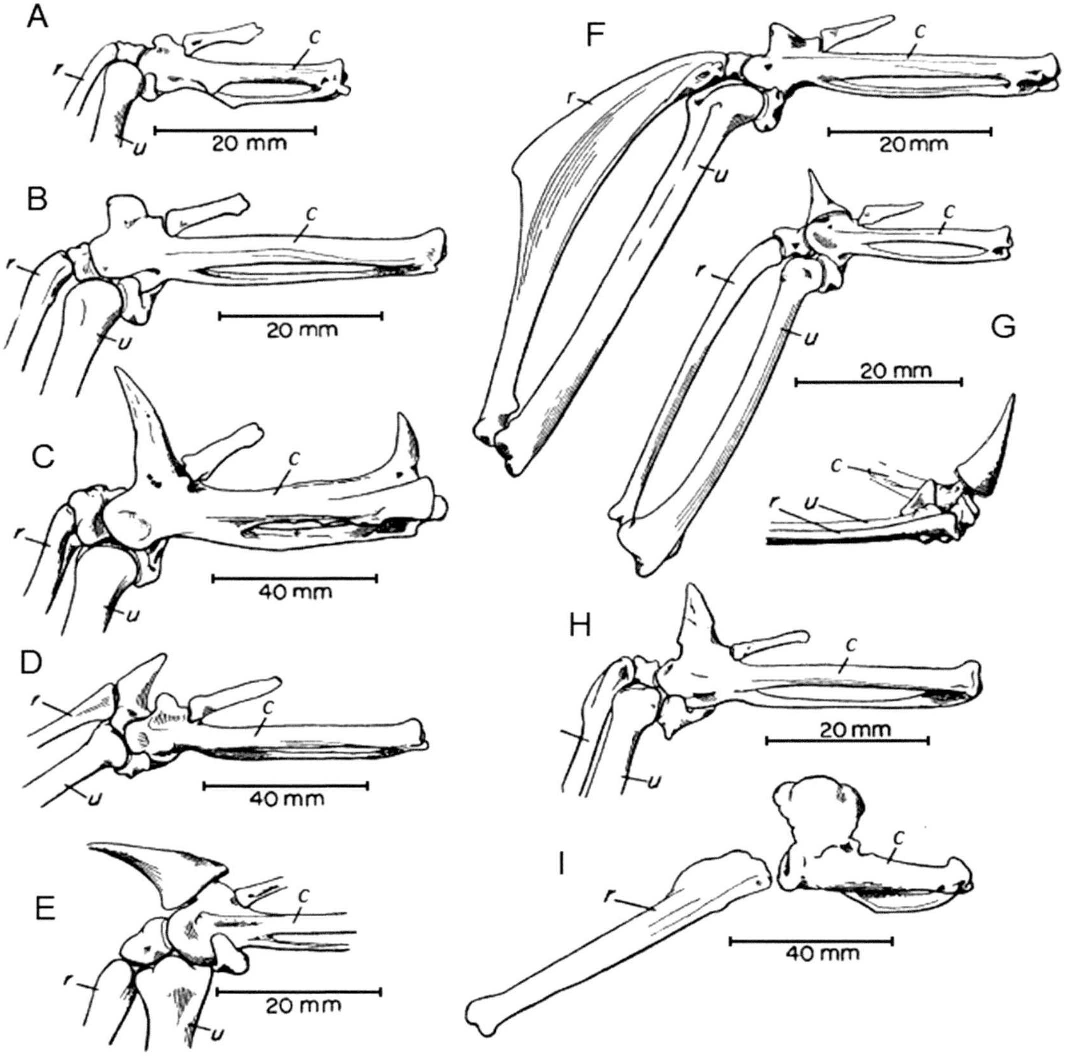


**Fig. 3.28** Bony structures in the wrist region of several species of birds (r = radius, u = ulna, and c = carpometacarpus). **a** and **b**, "normal" metacarpals of a Red-legged Partridge (*Alectorus rufa*) and a Eurasian Oystercatcher (*Haematopus ostralegus*); **c**, wing spur of a Northern Screamer (*Chauna chavaria*); **d**, wing spur of a Spur-winged Goose (*Plectropterus gambensis*); **e**, wing spur of a Torrent Duck (*Merganetta armata*), including the horny sheath; **f**, thickened radius of an African Jacana (*Actophilornis africana*); **g**, two views of the wing spur of a Northern Jacana (*Jacana spinosa*), including the horny sheath; **h**, wing spur of a Southern Lapwing (*Vanellus chilensis*); **i**, wing spur of the extinct Rodrigues Solitaire (*Pezophaps solitaria*). (Figure from Rand 1954; © 1954 Wilson Ornithological Society, used with permission)

very small and nonfunctional. However, young Hoatzins (*Opisthocomus hoazin*) have two well-developed claws on each wing and use them for climbing shrubs and trees (Fig. 3.34). These claws have an important function for Hoatzins, a Neotropical species, because they typically nest over water, and nestlings sometimes jump from nests when threatened by a predator. Once in the water, they swim to nearby shrubs and trees and climb upwards using their wing claws and feet.

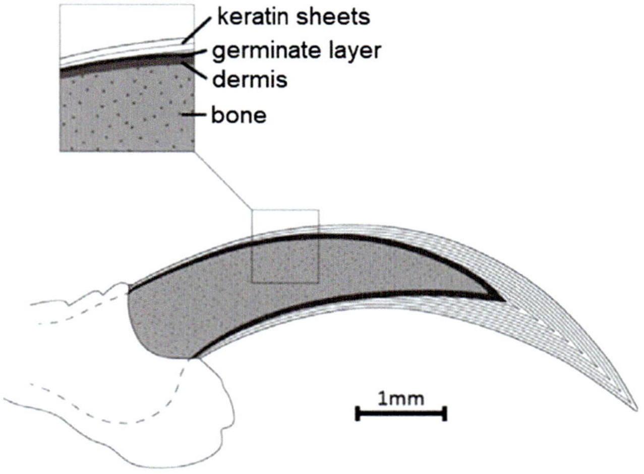


**Fig. 3.29** Longitudinal section of the back toe claw of a Coal Tit (*Periparus ater*). The claw consists of an ungual bone (terminal phalanx) and keratin. Keratin is part of the epidermis and is derived from underlying layers of the epidermis (germinate layer that consists of the stratum germinativum and stratum corneum). (Figure from Hahn et al. 2014; © 2014 The Authors, used with permission)

Young Hoatzins shed their wing claws when 70 to 100 days old (Thomas 1996).

## 3.8 Rhamphotheca

Bird bills consist of bones that form the cores of the upper and lower mandibles. However, the outer surface and part of the inner surface of these bones are covered with a modified integument called the rhamphotheca (Fig. 3.35). The epidermis of the rhamphotheca is relatively thick, hard, and consists of heavily cornified cells (Lucas and Stettenheim 1972). These cells produce beta-keratin like that found in avian scales and claws, and calcium deposited between the keratin proteins generally makes the rhamphotheca hard and strong (Homberger and Brush 1986; Bonser 1996). The epidermis is tightly bound to the bone by a thin dermal layer that contains numerous collagen fibers. The dermis also contains sensory receptors, including

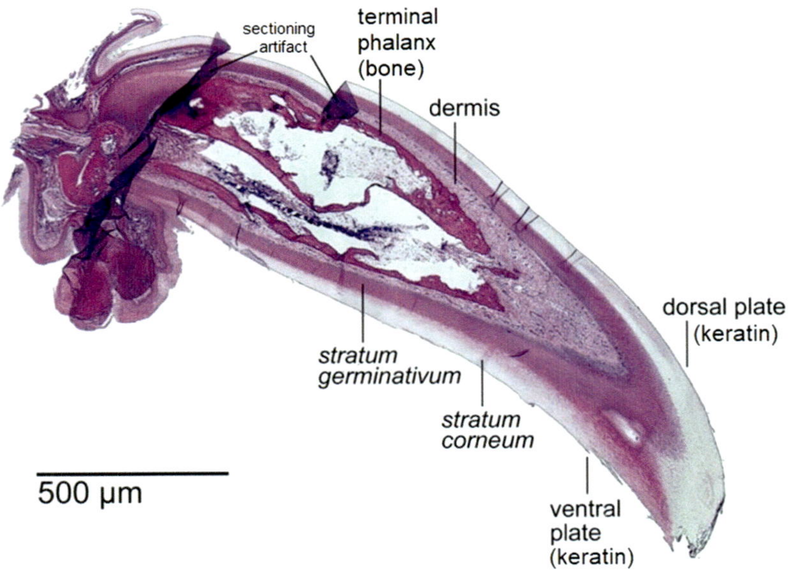


**Fig. 3.30** Stained mid-sagittal section of a claw of a Black-capped Chickadee (*Poecile atricapillus*) showing the bone (terminal phalanx), dermis, and the stratum germinativum and stratum corneum of the epidermis. (Figure from Van Hemert et al. 2012; © 2011 Wiley Periodicals, Inc., used with permission)

**Box 3.2 Keratins**

Avian integumentary structures, including feathers, scales, claws, and bills, consist mainly of materials made of α-keratins in all vertebrates and β-keratins found only in reptiles and birds. The α-keratins are coiled structures stabilized by hydrogen bonds that cause the coiling. The β-keratins form when a protein chain folds to form frou lateral beta-strands held together by hydrogen bonds. The sheet then twists or distorts to form a helical β-sheet, and two of these sheets combine to form a filament.

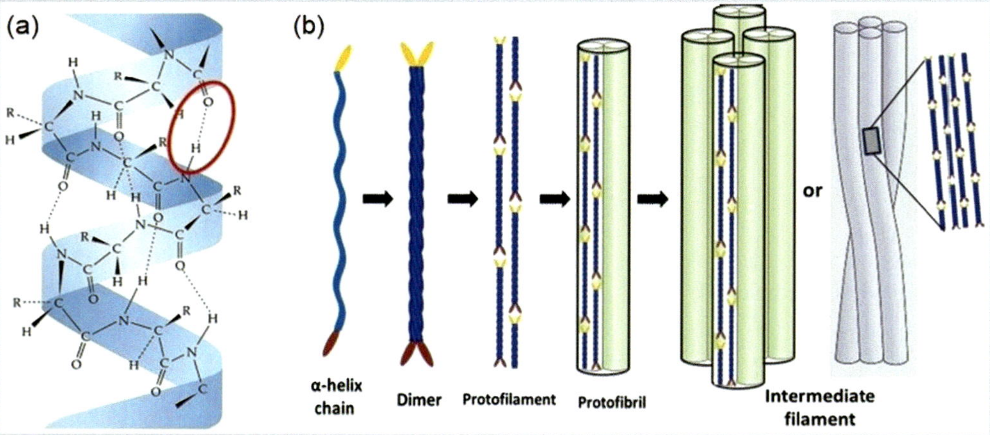


Filament structure of α-keratin. (**a**) α-helix chain (hydrogen bond shown in red ellipse). (**b**) Drawings (from left to right) of a single α-helix chain, twisted to form dimers that assemble to form protofilaments, protofibrils, and intermediate filaments (formed by four protofilaments). Intermediate filaments are the basic subunit of α-keratins) (figure modified from Wang et al. 2016a; © 2015 Elsevier Ltd. All rights reserved, used with permission).

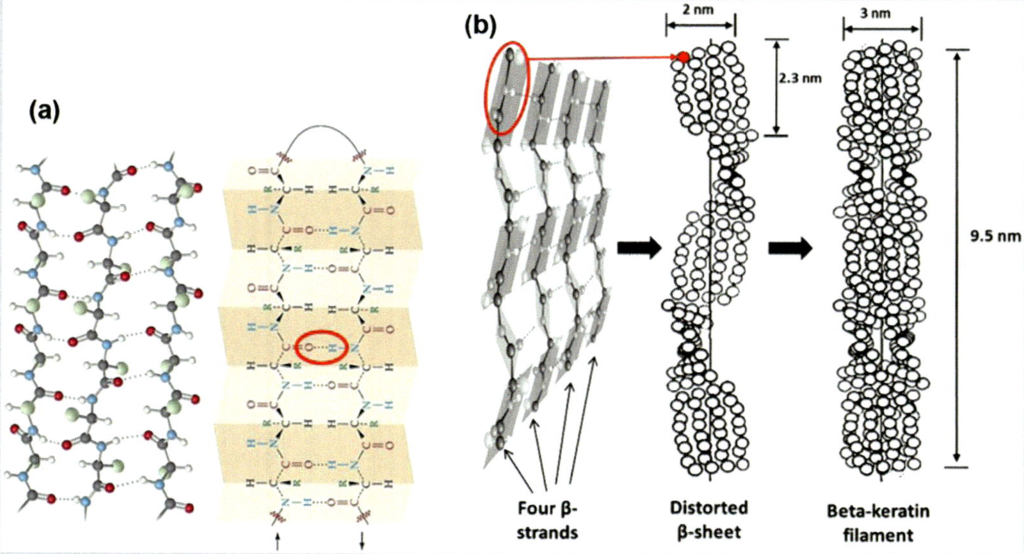


Structure of β-keratin filaments. (**a**) Model of the protein chain (left) and drawing of the pleated beta-sheet (right). Hydrogen bond is shown in the red elipse. (**b**) Drawing showing the formation of a β-keratin filament, with one chain folding to form four β-strands that twist to form

(continued)

**Box 3.2** (continued)

a distorted β-sheet. Two sheets combine to form a β-keratin filament (Figure from Wang et al. 2016a; © 2015 Elsevier Ltd. All rights reserved, used with permission).

Comparative studies have revealed a diversity of α- and β-genes in the avian genome (Greenwold et al. 2014; Ng et al. 2014). Different combinations of α- and β-keratins form the different integumentary structures of birds. Wu et al. (2015) examined the different expression of α- and β-keratin genes in different structures in the integument of Domestic Chickens and found that different combinations of α- and β-keratins contribute to differences among different types of feathers. These authors also found that the expression of five β-keratin genes on chromosomes 25 and 27 differed among, and even within, different structures. For example, different β-keratins can be found in feathers and scales, and different parts of scales and claws consist of different α- and β-keratins. Thus, morphological and structural differences of different avian skin structures results from expression of different α- and β-keratin genes (Wu et al. 2015).

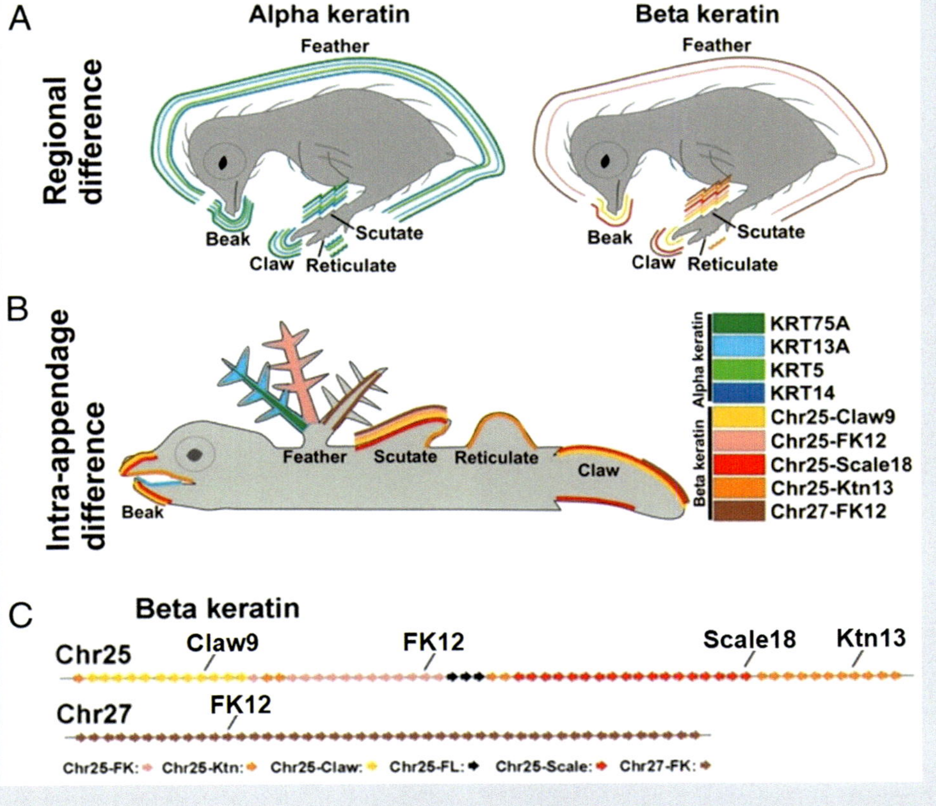


Expression patterns of α- and β-keratin genes in different skin structures. (**a**) Regional differences among different skin structures. Each line indicates the expression of different keratin genes in different structures. (**b**) Intra-structure differences of keratin expression. Different colors represent the indicated keratin genes. (**c**) β-keratin gene arrangements on Chromosome 25 (Chr25) and Chromosome 27 (Chr27). Claw, claw keratin; FK, feather keratin; FL, feather-like keratin; Ktn, keratinocyte keratin; Scale, scale keratin (Figure from Wu et al. 2015; used with permission of the U.S. National Academy of Sciences).

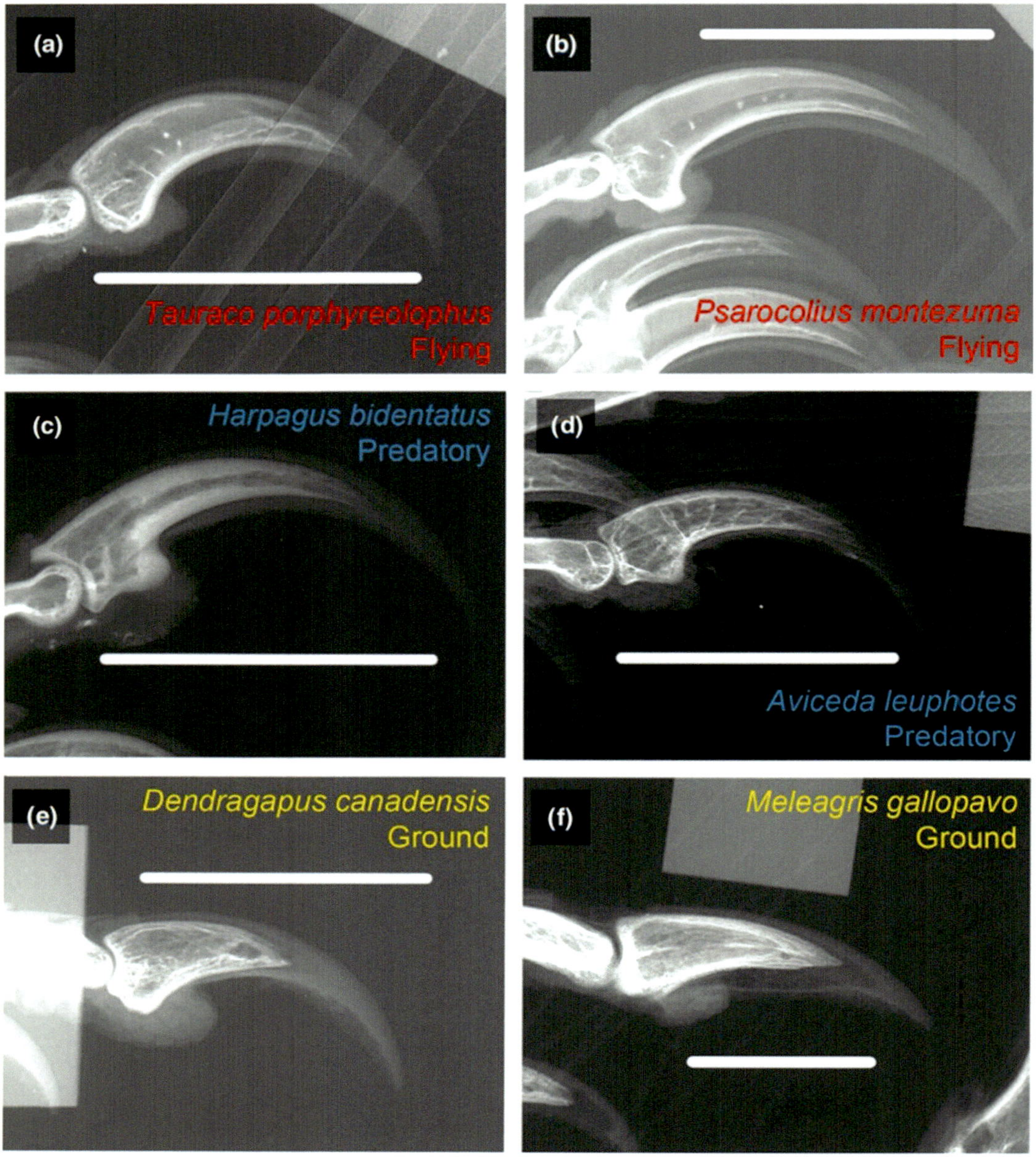


**Fig. 3.31** Third pedal claws showing bony cores and keratinous sheaths of representative species in three ecological groups. (**a**, **b**) Flying birds, (**c**, **d**) predatory birds, and (**e**, **f**) cursorial, or ground-based, birds. *Tauraco porphyreolophus*, Purple-crested Turaco; *Psarocolius montezuma*, Montezuma Oropendola; *Harpagus bidentatus*, Double-toothed Kite; *Aviceda leuphotes*, Black Baza; *Dendragapus canadensis*, Spruce Grouse; *Meleagris gallopavo*, Wild Turkey. (Figure from Hedrick et al. 2019; open-access article published under the terms of the Creative Commons Attribution License, https://creativecommons.org/licenses/by/4.0/)

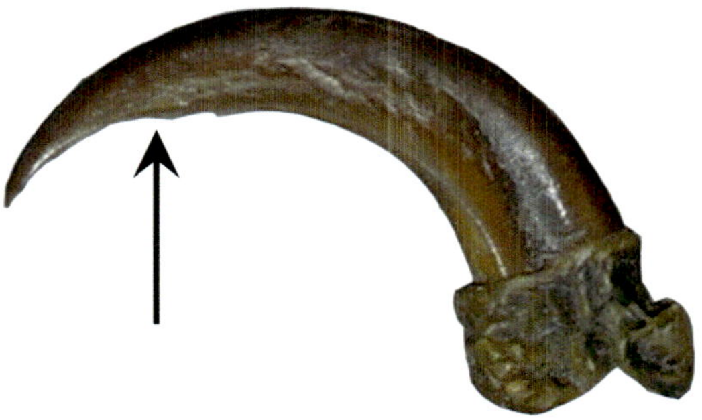

**Fig. 3.32** Pedal claw of a Crimson-crested Woodpecker (*Campephilus melanoleucus*) with the arrow indicating the point of constriction. (Figure from Birn-Jeffery et al. 2012; © 2012 Birn-Jeffery et al., open-access article distributed under the terms of the Creative Commons Attribution License, https://creativecommons.org/licenses/by/4.0/)

Herbst and Grandry corpuscles that are sensitive to touch and vibration (Figs. 3.35, 3.36, and 3.37). Grandry corpuscles occur only in waterfowl, whereas Herbst corpuscles occur in several species and are particularly abundant at the tip of the bill of many shorebirds and are used to locate prey in the sand and other soft substrates (Box 3.3 Birds "Feel" Their Prey Under the Sand). The dermis of some birds also contains Merkel corpuscles, mechanoreceptors that respond to pressure applied to the beak or skin (Fig. 3.38; Halata et al. 2003). The integument of the bill grows continually from the base and culmen, with growth directed rostrally so that there is a continuous movement of the horny beak from base to tip. The rhamphotheca, particularly at the tomia, is worn away by abrasion from food and other materials and by friction where the upper and lower bills meet.

The shape of the rhamphotheca varies with the shape of the underlying bones and, in several species of birds, is also modified by knobs, ridges, and other projections (Stettenheim 2000). For example, a highly fibrous, epidermal plate extends vertically from the upper bill of American White Pelicans (*Pelecanus erythrorhynchos*; Fig. 3.39). This plate varies in size among pelicans, ranging from 40 to 80 mm long and

**Fig. 3.33** Principal component analysis of total claw shape showing separation between predatory and ground birds with flying birds spreading across morphospace. Blue = predatory, red = flying, and yellow = cursorial or ground-based. Note that claw angle or curvature tends to increase in predatory birds and decrease for cursorial birds. (Figure modified from Hedrick et al. 2019; open-access article published under the terms of the Creative Commons Attribution License, https://creativecommons.org/licenses/by/4.0/)

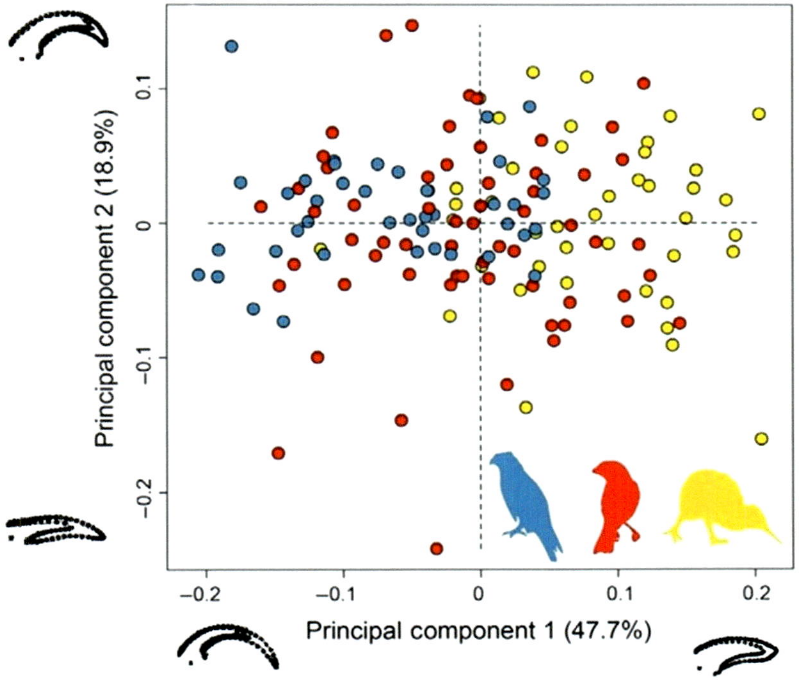

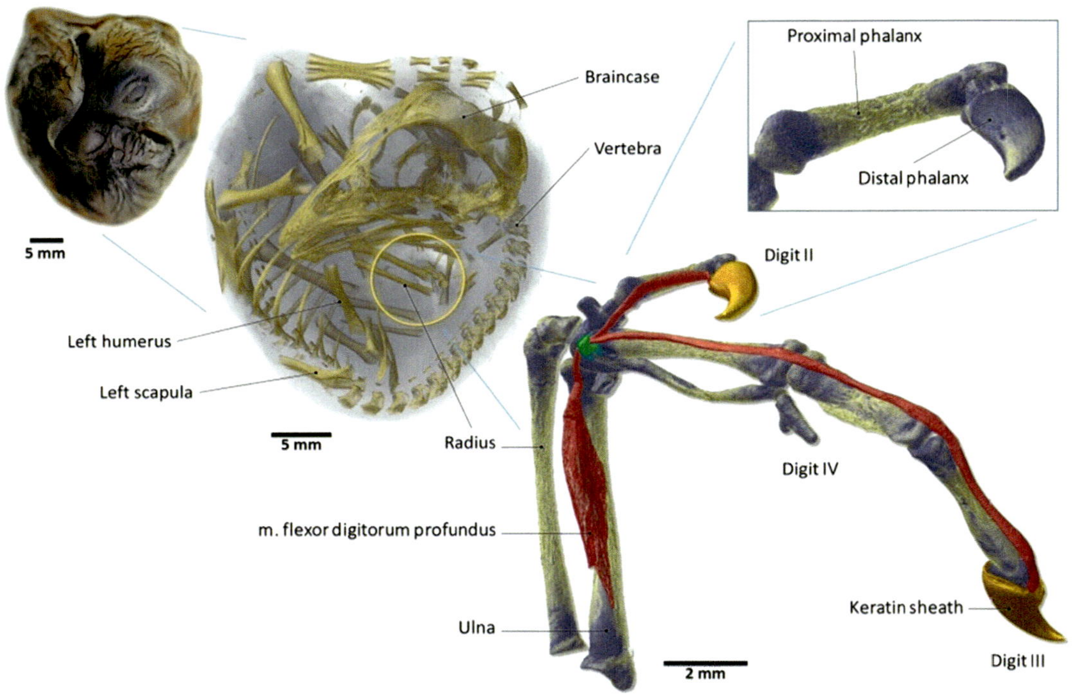


**Fig. 3.34** Musculoskeletal anatomy of a Hoatzin (*Opisthocomus hoazin*) before hatching. Left: Soon-to-hatch young in the egg. Middle: Mineralized parts of the skeleton, showing the position of the wing skeleton (yellow circle). Right: Detailed view of the wing showing the flexor digitorum profundus muscle and tendon attaching to the alula digit (II). Inset: Alula digit, with the keratin sheet to show the claw-like distal phalanx. Blue, cartilage; yellow, bone; red, muscle; green, connective tissue sling of the muscle-tendon; orange, keratin. (Figure from Abourachid et al. 2019; © 2019 American Association for the Advancement of Science, used with permission)

from just a few millimeters to more than 60 mm tall. The function of this plate is not clear, but it is probably involved in courtship or agonistic behavior during pairing and territory establishment; the plate is shed after the breeding season (Knopf and Evans 2020).

The edges of the rhamphotheca (tomia) are relatively sharp in most bird species. However, edges are finely serrated for straining small food particles in filter-feeding birds like flamingoes and some waterfowl. Species in a number of bird families, including Ardeidae, Cuculidae, Coraciidae, Picidae, and several others, have scopate tomia with very small (typically only 0.3–0.7 mm high and difficult to see without magnification) brush-like ridges that likely create friction and aid birds in capturing and holding food items (Fig. 3.40). Tiny serrations along the tomia of one (just upper bill) or both tomia of some hummingbirds likely aid in capturing and holding insect prey and, in some species of nectar-robbers, piercing the base of flowers to gain access to nectar (Ornelas 1994; Rico-Guevara et al. 2019; Fig. 3.41). The tomia of fish-eating mergansers have numerous pointed projections that aid in capturing and holding their primary prey (Fig. 3.42). Falcons and shrikes have single, sharp projections of the tomia on each side of their upper bill (Fig. 3.43). Several authors have suggested that falcons use these "tomial teeth" to break the neck of their prey, but there is little or no evidence to support this hypothesis. Rather, "tomial teeth" are likely used to help firmly grip and pull the flesh from

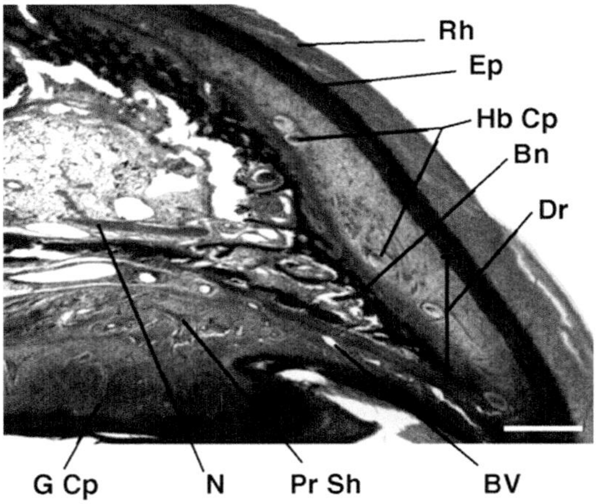


**Fig. 3.35** A sagittal section near midline of the upper beak of a 2-week-old Domestic Chicken (*Gallus g. domesticus*). The bill tip is to the right. The dorsal region shows the upper smooth portion of the bill covered by the rhamphotheca (Rh), whereas the ventral region shows the tomium or cutting edge of the upper bill. Beneath the rhamphotheca is the epidermis (Ep) which provides a constant supply of cellular material to form the outer, hardened covering of the beak. Internal to the epidermis is the dermis (Dr) layer, the most heterogeneous of all tissue layers. It extends from the epidermal to bone layers (Bn). Prominent structures found in this region include mechanoreceptors (Herbst [Hb Cp]and Grandry [G Cp] corpuscles), blood vessels (BV), perineural sheaths, and free nerve endings, or nociceptors (pain) Scale bar, 400 μm. N = nerve; Pr Sh = perineural sheaths. (Figure from Kuenzel 2007; © 2007 Oxford University Press, used with permission)

prey (Csermely et al. 1998) and, for large falcons, to help grip and break long bones of wings and legs of smaller prey before swallowing (White et al. 2020).

In some birds, including raptors (Falconiformes), owls (Strigiformes), parrots, cracids, and pigeons, the rhamphotheca at the base of the upper bill is called the cere (Stettenheim 1972). Some species of birds, including pigeons, also have an operculum, a small disc of cartilage or membranous keratin over the nostril that keeps foreign objects out of the nasal cavity (Fig. 3.44). The cere is a thickened, often brightly colored portion of the integument that straddles the base of the nasal region (Lucas 1979). Lucas and Stettenheim (1972) suggested that the cere may provide protection for the underlying elongated nasal fossa (cavities). However, because ceres are often brightly colored, individual variation in cere color may also convey information about individual quality. For example, the cere of male Montagu's Harriers (*Circus pygargus*) reflects light at wavelengths corresponding to yellow-orange (500–600 nm), but also reflects ultraviolet (UV; 300–400 nm) wavelengths (Fig. 3.45). Differences among males in the UV peak were found to be associated with differences in body mass and condition, suggesting that UV reflectance conveys information about individual quality (Mougeot and Arroyo 2006).

## 3.9 Integument Glands

The uropygial (or preen) gland is a bilobed structure located that the base of the tail between the skin and body muscles (Figs. 3.46, 3.47, and 3.48). Secretions are transported by ducts that open at the top of papillae that, in some species, have tufts of feathers that presumably aid in distributing the secretions to the bill or head during preening (Jacob and Ziswiler 1982). Preen glands are found in most birds, but are absent in some columbids (Columbiformes), parrots (Psittaciformes), and woodpeckers (Picidae), as well as Common Ostriches (*Struthio camelus*, Struthioniformes), rheas (Rheidae), cassowaries (Casuariidae), Emus (*Dromaius novaehollandiae*, Dromaiidae), mesites (Mesitornithidae), bustards (Otidae), and frogmouths (Podargidae) (Elder 1954; Jacob and Ziswiler 1982). Uropygial gland secretions are a mixture of waxes (monoester and diester), triglycerides, fatty acids, alcohols, and hydrocarbons, with the specific composition varying among species (Jacob and Ziswiler 1982; Praveenkumar et al. 2023). One important function of these secretions, sometimes called preen oil, is preserving feather structure by keeping the proteins in feathers (keratin) flexible (Jacob and Ziswiler 1982). For example, Moyer et al. (2003) removed the uropygial glands from some Rock Pigeons (*Columba livia*) and, after several weeks, found that the plumage of glandless birds was in significantly poorer condition, with more missing barbules, than the plumage of control birds with

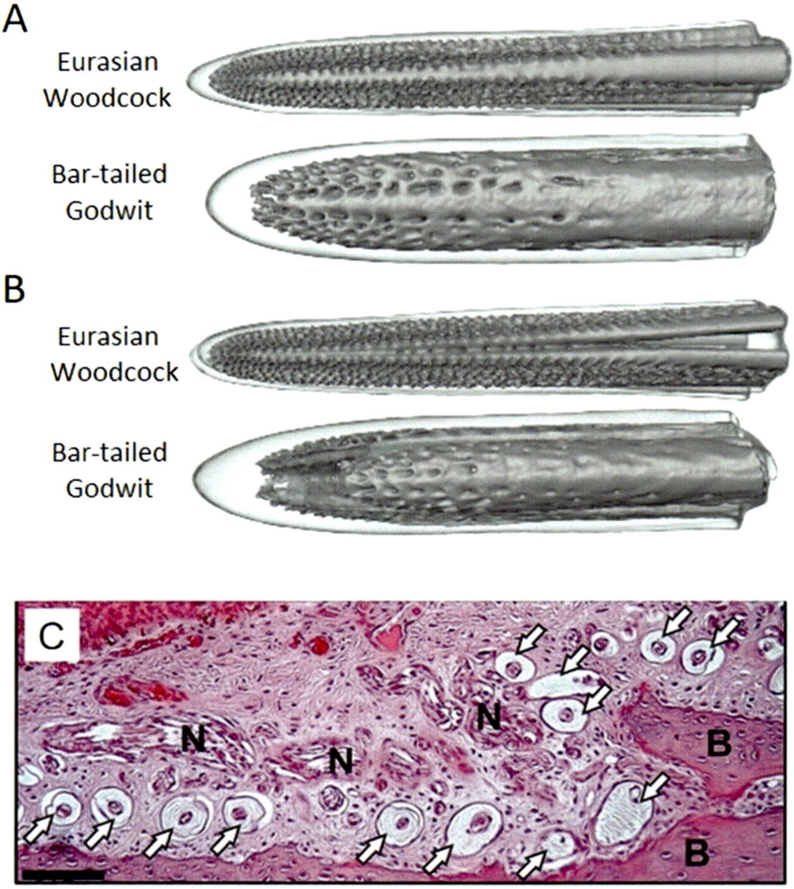


**Fig. 3.36** Sensory pits in the bill tips of a Eurasian Woodcock (*Scolopax rusticola*) and Bar-tailed Godwit (*Limosa lapponica*). (**a**) Upper bill, and (**b**) lower bill. Sensory pits, in the bone, create more surface area and, in the overlying dermis layer of the skin, Herbst corpuscles are found in high densities. (**c**) A sensory pit in the dentary bone (lower bill) of a Bar-tailed Godwit dentary (stained). N, nerves; B, bone. Herbst corpuscles are indicated by the white arrows. Scale bar = 100 μm. (Figure modified from Cunningham et al. 2013; © 2013 Cunningham et al., open-access article distributed under the terms of the Creative Commons Attribution License, https://creativecommons.org/licenses/by/4.0/)

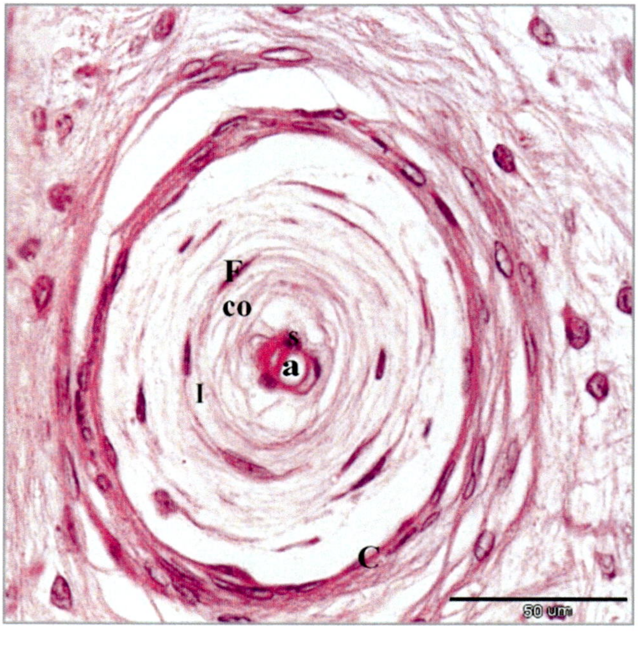


**Fig. 3.37** Herbst corpuscle. C, capsule; I, collagen fibers; F, concentric lamellar layers of fibroblast-like cells; co, collagen fibers, s, sensory cell; a, central axon. (Figure from Soliman and Madkour 2017; ©2017 Soliman SA., open-access article distributed under the terms of the Creative Commons Attribution License, https://creativecommons.org/licenses/by/4.0/)

**Box 3.3 Birds "Feel" Their Prey Under the Sand**

Red Knots (*Calidris canutus*) can locate their favorite food (shellfish) in wet sand by inserting their beak half a centimeter into the sand for a few seconds (Piersma et al. 1998). This ability was demonstrated in experiments where researchers hid small stones in the sand. Because stones do not send out any signals, the ability of Red Knots to detect them must be based on the sensitivity of their beaks to differences in currents in the water in wet sand between the individual grains, stones, or shells. Red Knots used in the experiments could not find stones placed in dry sand. At the end of their beak, knots have clusters of 10 to 20 Herbst corpuscles that are sensitive to differences in pressure. When a bird sticks its sensitive beak into the sand at low tide, it produces a pressure wave because of the inertia of the water in the interstices between sand particles. The pattern thus created indicates the presence of objects larger than the grains of sand. The rapid up-and-down movements of the bird's beak loosen the grains of sand, so they become more tightly packed together, displace the interstitial water, and cause the pressure around the object to increase. This ability means that knots cannot distinguish between stones and shellfish in the sand, so they rarely look for food in areas where sand contains stones, no matter how many shellfish could be found there.

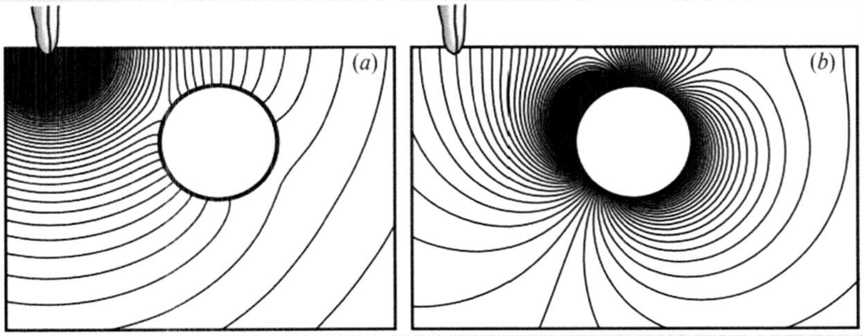


(**a**) Pressure waves of water are created when a bill is inserted into a moist substrate like wet sand or mud. (**b**) Pressure waves strike objects like a shellfish (shown as a round object) and reflect back toward the bill. Those waves are then detected by pressure receptors in the bill tip, providing a bird with information about the location of a possible prey item (Figure from Piersma et al. 1998; © 1998 The Royal Society, used with permission).

(continued)

**Box 3.3** (continued)

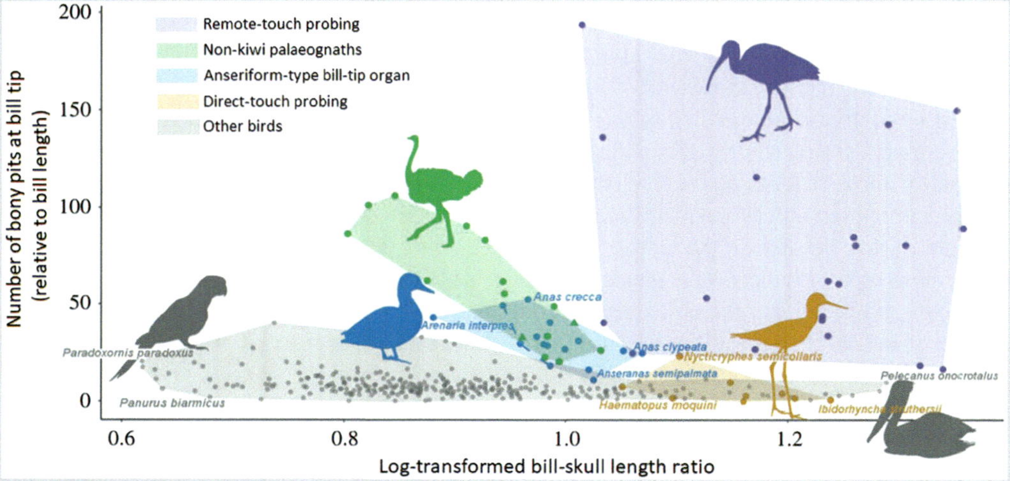


Several other species of birds also forage, like Red Knots, probing moist substrates with their relatively long bills and using pressure waves to locate their prey. These "remote-touch probing" birds tend to have higher concentrations of bony sensory pits with pressure receptors at their bill tips. Interestingly, some paleognaths also have relatively high concentrations of bony sensory pits at their bill tips, suggesting that the common ancestor of present-day paleognaths likely foraged by probing moist substrates (Figure modified from du Toit et al. 2020; © 2020 The Authors. Published by the Royal Society, used with permission).

glands. Similar results have been reported in studies of waterfowl. Possible additional functions include protecting feathers from feather-degrading fungi and bacteria (e.g., Alt et al. 2020), playing a role in olfactory communication (e.g., Grieves et al. 2019a), and acting as a cosmetic (e.g., Møller and Mateos-González 2019). Some authors have suggested that preen oil is important for waterproofing feathers (e.g., Møller and Laursen 2019). However, clean keratin is as water-repellant as preen gland secretions, and the components of those secretions do not appear to have unique water-repellant properties compared to other hydrophobic oils and waxes (Bakken et al. 2006). Thus, the primary role of preen oil in water repellency is likely to help preserve and maintain feather structure.

For some birds, preen oil can retard or inhibit the growth of bacteria and fungi. For example, a fatty acid in the preen oil of pelicans and allies (Pelicaniformes) inhibits the growth of fungi (Jacob et al. 1997), and preen oil was found to have an antibacterial effect on feather bacteria of Great Tits (*Parus major*) and European Pied Flycatchers (*Ficedula hypoleuca*) (Alt et al. 2020). Similarly, the preen gland secretions of Wild Turkeys (*Meleagris gallopavo*) were found to inhibit the growth of a wide variety of bacteria and fungi (Braun et al. 2018). Interestingly, Verea et al. (2017) found that the uropgyial secretions of Spectacled Thrushes (*Turdus nudigenis*) reduced the extent of feather degradation by bacteria, but by serving as a barrier between the bacteria and feathers rather than having an antibacterial effect.

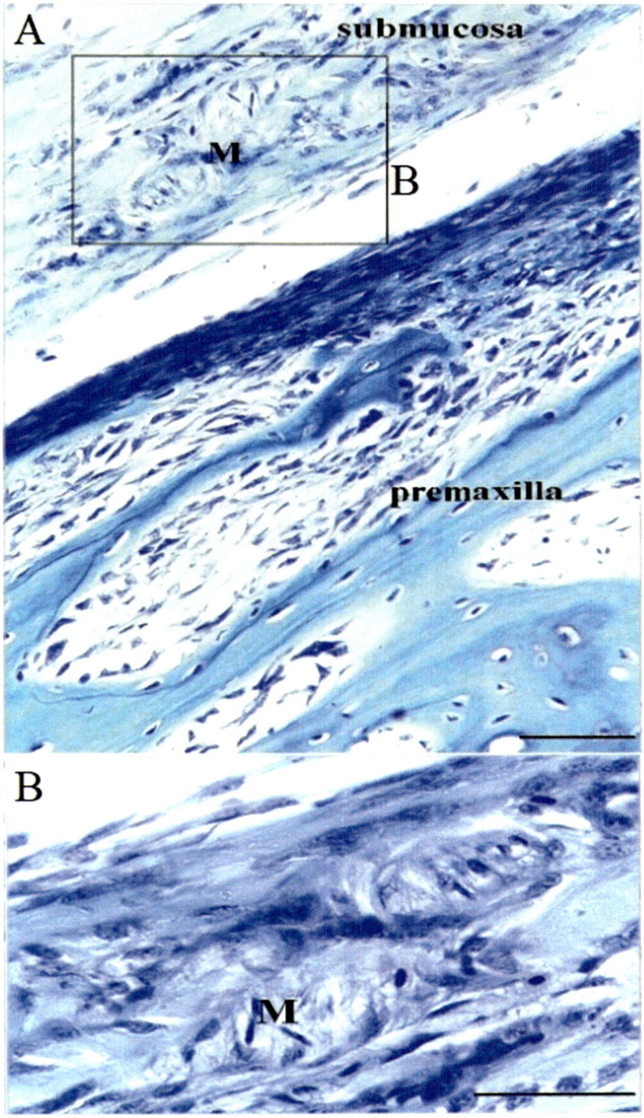


**Fig. 3.38** (**a**) Merkel corpuscles in the skin (submucosa) of a Muscovy Duck's (*Cairina moschata*) bill. (**b**) Merkel corpuscle at higher magnification. Scale bars = 50 μm. (Figure from Soliman and Madkour 2017; ©2017 Soliman SA., open-access article distributed under the terms of the Creative Commons Attribution License, https://creativecommons.org/licenses/by/4.0/)

The preen oil of two species of birds in the order Coraciiformes, Eurasian Hoopoes (*Upupa epops*) and Green Woodhoopoes (*Phoeniculus purpureus*), also serves an antimicrobial function. However, the substance in the preen oil of hoopoes and woodhoopoes that kills bacteria is actually produced by other bacteria! These birds have a symbiotic association with bacteria that live inside their uropygial glands, bacteria that break down the preen oil secretions into different compounds that are toxic to a wide range of pathogenic bacteria. Interestingly, these birds use their preen oil to protect not only their feathers, but their eggs as well. Soon after laying, female hoopoes smear their eggs with preen oil that has been found to limit the growth of bacteria on eggs and increase hatching success (Soler et al. 2008). The preen oil of Green Woodhoopoes also emits an odor that repels potential vertebrate predators of eggs (Law-Brown 2001).

For some birds, the odors of preen oils may also be important intraspecific signals (Box 3.4 Possible Sexual and Social Functions of Female Uropygial Gland Secretions). For example, odors

**Fig. 3.39** Note the epidermal plates extending vertically from the upper bill of these American White Pelicans (*Pelecanus erythrorhynchos*). (Photo from pxhere.com, CC0 Public Domain)

that are likely to arise from preen oil may be used by Leach's Storm-Petrels (*Hydrobates leucorhous*) to locate nest sites after returning to breeding areas at night (Grubb 1994). Adults in other species of petrels also appear to use odors to distinguish between their own nests and those of neighbors (Bonadonna et al. 2003, 2004). Mobile chicks of European Storm-Petrels (*Hydrobates pelagicus*) also use odors, again likely arising at least in part from preen oil, to identify their own burrows, with home burrows critical to chick survival because they provide protection and are

**Fig. 3.40** Scopate tomia with brush-like ridges that likely create friction and aid birds in capturing and holding food items. (Photo from pxhere.com, CC0 Public Domain)

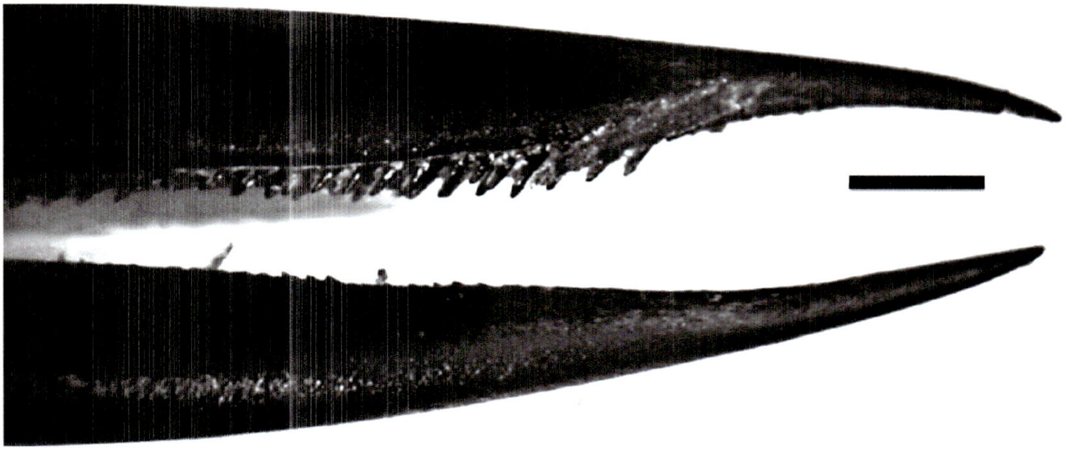

**Fig. 3.41** Serrate tomia on the bill of a male Sparkling Violetear (*Colibri coruscans*) hummingbird. Scale bar = 1 mm. (Figure from Rico-Guevara et al. 2019; open-access article distributed under the terms of the Creative Commons CC BY license, https://creativecommons.org/licenses/by/4.0/)

the only place where young are fed by adults (Mínguez 1997). For male Dark-eyed Juncos (*Junco hymalis*), uropygial gland secretions may function as signals of aggressive intent, with volatile compounds produced by the gland potentially providing conspecifics with information about an individual's level of aggression (Whittaker et al. 2017). Additional study is needed to determine the extent to which other birds might use odors and, specifically, those of preen gland secretions in intraspecific signaling.

**Fig. 3.42** Serrate tomia of a female Common Merganser (*Mergus merganser*). (Photo by Mark Medcalf, Wikipedia, CC BY 2.0, https://creativecommons.org/licenses/by/2.0/)

**Fig. 3.43** "Tomial tooth" of a Loggerhead Shrike (*Lanius ludovicianus*). (Figure from Sustaita and Rubega 2014; © 2014 Oxford University Press, used with permission)

There is some evidence that preen oil plays a role in interspecific communication. For example, uropygial secretions of hoopoes (Upupidae) are malodorous and appear to serve an antipredator function (Burger et al. 2004). Other reports of a possible antipredator function of preen oils are anecdotal and require additional study (e.g., Ligon and Ligon 1978). However, the results of experiments conducted with Carolina (*Poecile carolinensis*) and Black-capped (*P. atricapillus*) chickadees, species with adjacent geographical ranges and a narrow hybrid zone, suggest that differences in their preen gland odors may be important in mate choice and promote premating reproductive isolation (Van Huynh and Rice 2019).

In several species of shorebirds, including the Eurasian Oystercatcher (*Haematopus ostralegus*), six species of plovers (Charadriidae), and at least 19 species of sandpipers, the composition of preen wax changes prior to breeding, becoming less volatile and more difficult to smell. More generally, Grieves et al. (2022) found that changes in preen oil composition were more likely in the incubating than the nonincubating sex among ground-nesting species of birds. These changes may reduce predation risk during incubation, with mammalian predators that use olfactory cues to locate prey less likely to locate incubating birds and their eggs (Reneerkens et al. 2005).

Secretions of the uropygial gland can also act as a cosmetic either by making feathers glossy and causing them to appear brighter or by differentially absorbing or reflecting light of a certain wavelength and changing feather color (Delhey et al. 2007; Box 3.5 Cosmetic Coloration). Several species (8 of 54) of hornbills (Bucerotidae) produce a colored preen oil (ranging from yellow to red, depending on the species) that they apply to their plumage, bill, and casque (a hard projection on top of the head or bill) and change their color (Kemp 2001). The colors produced by the preen oil likely serve some sexual signaling function, but no data are available concerning specific functions (Montgomerie 2006). Some pelicans also have colored preen oil that they apply to their plumage during the breeding season. For example, Great White Pelicans (*Pelecanus onocrotalus*) produce an orange-red preen oil (Stegmann 1956; Delhey et al. 2007), and Brown Pelicans (*P. occidentalis*) have yellowish preen oil that is used to enhance the yellow plumage on their heads (Schreiber et al. 1989). The preen-oil-colored plumage of these pelicans likely plays a role in mate choice, but this has not been tested experimentally. Although preen oils can alter feather color, preen oils do not

**Fig. 3.44** Top, head of a Rock Pigeon (*Columba livia*). Bottom, close-up view of the cere, operculum, and external nasal opening of a Rock Pigeon. The operculum is a small disc of cartilage centered in the nostril that keeps foreign objects out of the nasal cavity. (Photo by Dori, Wikipedia, CC BY 3.0, https://creativecommons.org/licenses/by/3.0/)

modify plumage ultraviolet reflectance that plays an important role in sexual signaling in many species of birds (Delhey et al. 2008).

Other glands associated with the integument of birds are the ear and vent glands. The ear gland is located on the floor of the ear canal and produces a waxy substance that traps particles and keeps the ear canal open. Vent glands, also called anal or proctodeal glands, are present in most, but possibly not all groups of birds (Quay 1967 and are located either just outside of or deeper within the vent, depending on the species, and secrete mucoproteins. Vent glands are larger in males and increase in size during the breeding season. The function of these mucoprotein secretions remains unclear. However, males in the quail genus *Coturnix* have particularly well-developed vent glands that are sometimes called foam glands because the viscous mucoprotein they produce is "whipped" into foam by the action of the cloacal

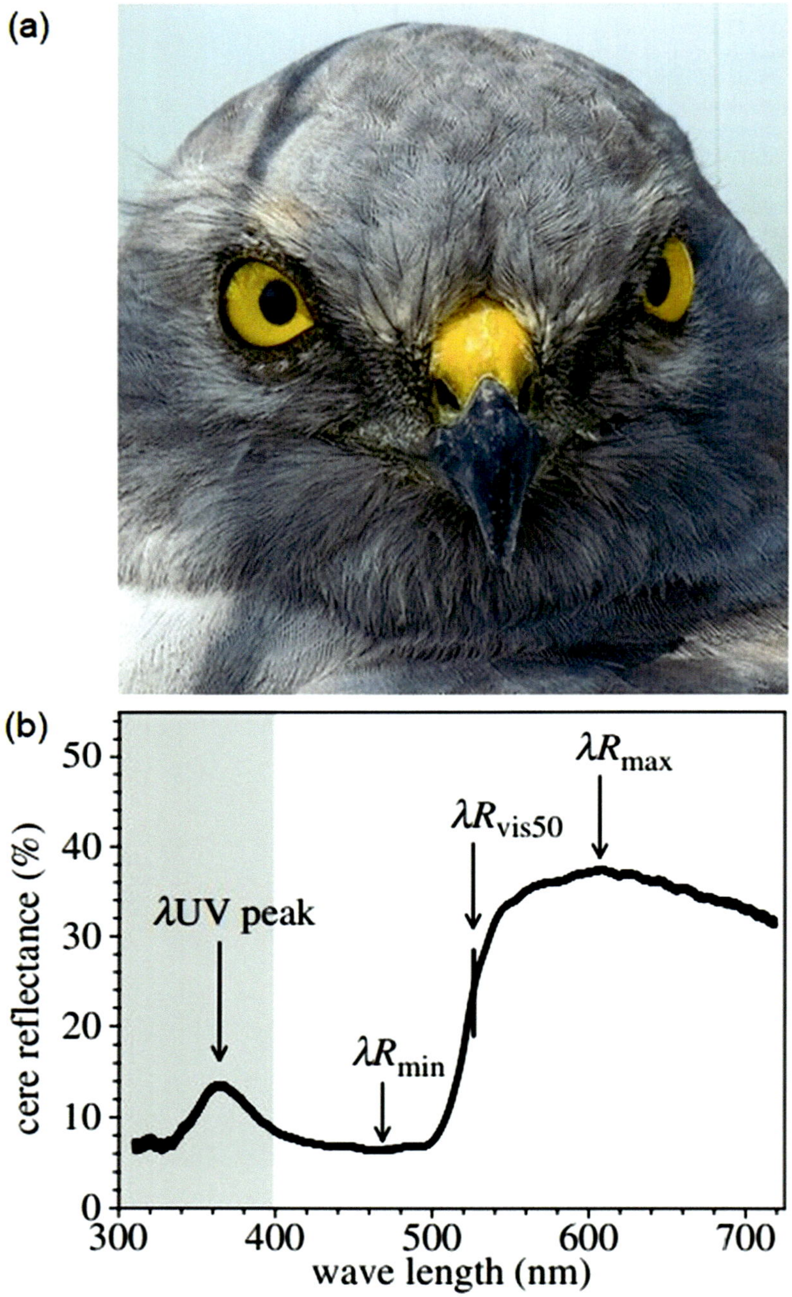


**Fig. 3.45** (**a**) Male Montagu's Harrier (*Circus pygargus*) showing the cere (skin area above the beak and between the eyes). (**b**) Reflectance pattern of a typical male cere showing peaks in the yellow-orange wavelengths (500–600 nm) and in the ultraviolet (300–400nm). (Photo a by Sameer Mudaye, used with permission. Figure **b** from Mougeot and Arroyo 2006; © 2006 The Royal Society, used with permission)

sphincter (Fig. 3.49; Seiwert and Adkins-Regan 1998). During copulation, male quail introduce both semen and foam into the female, and available evidence suggests that the foam enhances sperm survival and the ability of sperm to enter sperm storage tubules (Adkins-Regan 1999). The presence of foam from one male also reduces the likelihood of successful fertilization by other males (Finseth et al. 2013).

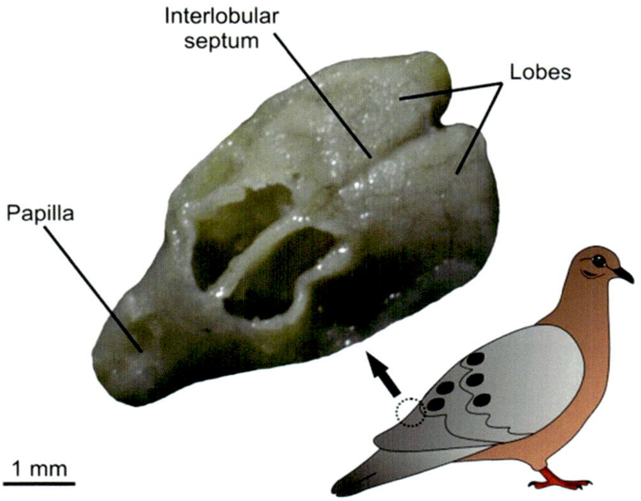


**Fig. 3.46** Location and external anatomy of the uropygial gland of an Eared Dove (*Zenaida auriculata*). The gland has a pear-like shape and consists of two lobes and a conical papilla with no feather tuft. (Figure from Chiale et al. 2019; © 2019 Springer Nature, used with permission)

## 3.10 Feather Evolution

Among the various integumentary structures of vertebrates, feathers are the most complex (Fig. 3.50). Feathers are unique in their complex branching and impressive variation in size, shape, color, and texture (Prum 1999; Prum and Williamson 2001; Fig. 3.50; Box 3.6 Feather

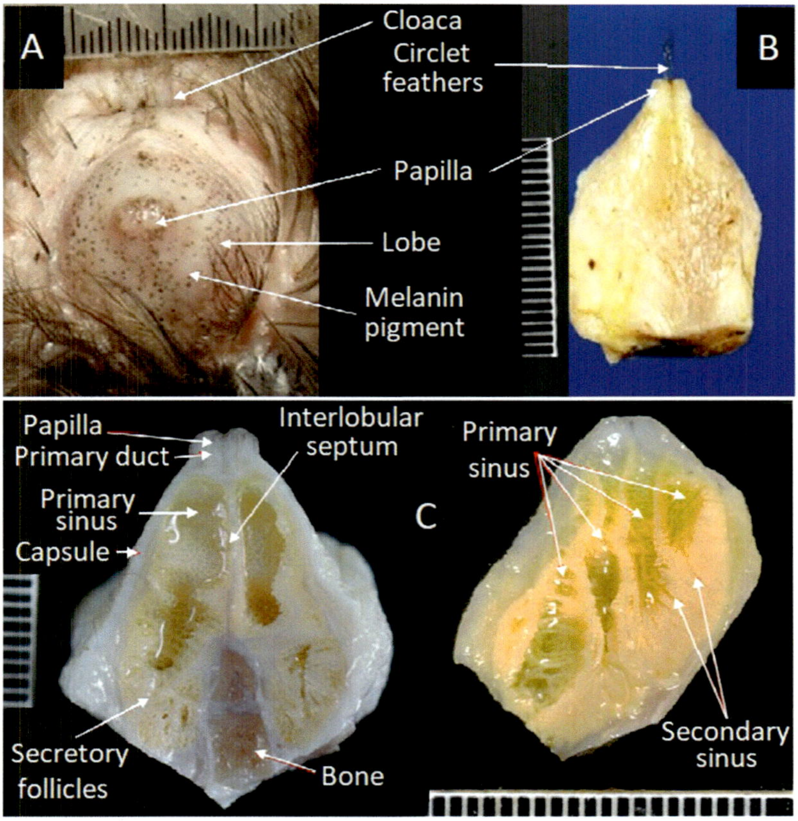


**Fig. 3.47** (**a**) Uropygial gland of a North Island Brown Kiwi (*Apteryx mantelli*) showing spots of melanin pigmentation on the surface. (**b**) Ventral view of the gland showing two lobes, a papilla, and circlet feathers. (**c**) Left, Transverse section through the uropygial gland. Right, Lateral sagittal section through one of the lobes showing four primary sinuses and some secondary sinuses. Scales = millimeters. (Figure from Reynolds et al. 2017; used with permission of the Editorial Board of the *European Journal of Anatomy*)

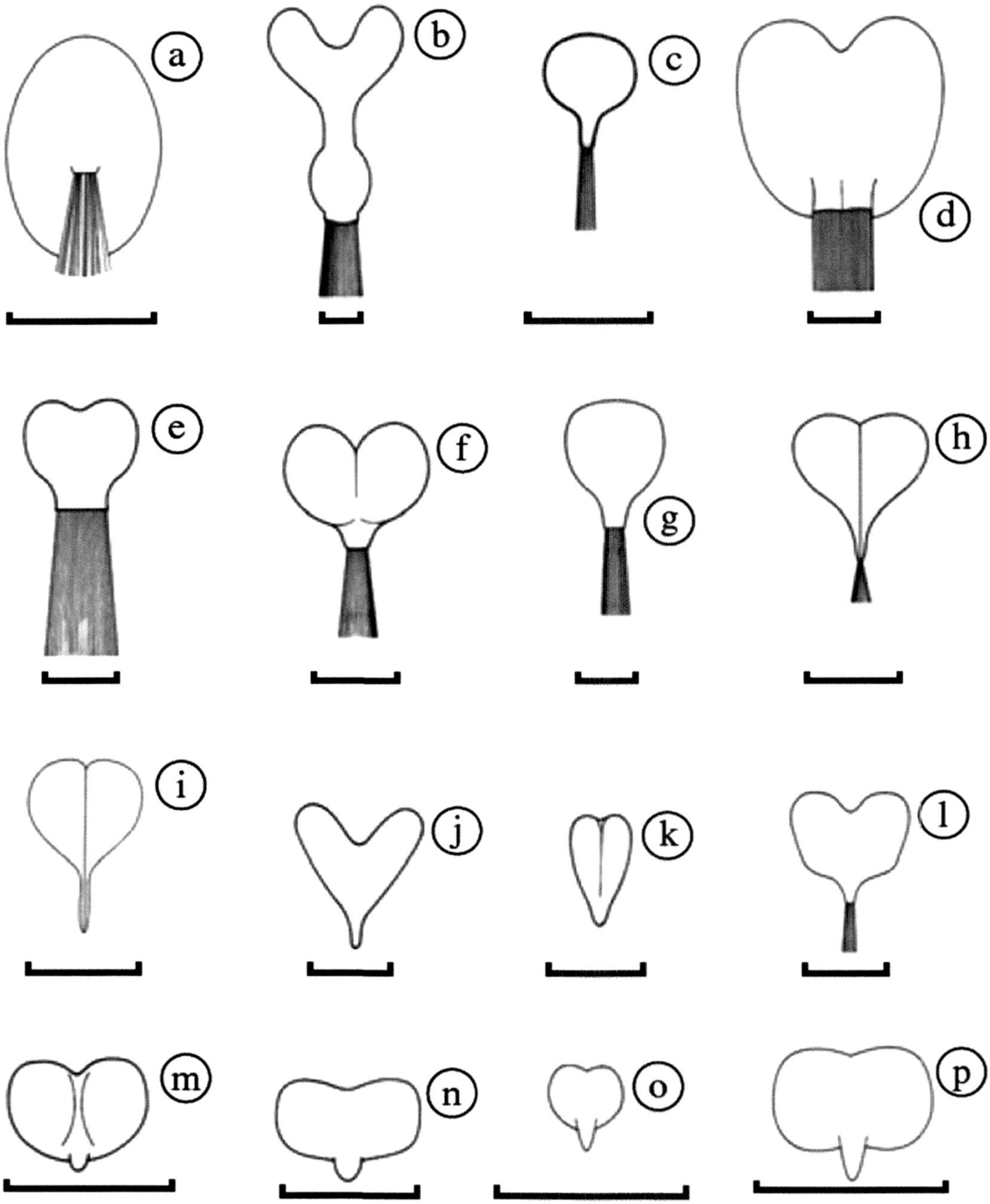


**Fig. 3.48** Morphological variation in the uropygial glands of different species of birds. Scale bars = 1 cm. (**a**) Spotted Nothura, *Nothura maculosa*, (**b**) Adelie Penguin, *Pygoscelis adeliae*, (**c**) Wilson's Storm-Petrel, *Oceanites oceanicus*, (**d**) Neotropic Cormorant, *Nannopterum brasilianum*, (**e**) South Polar Skua, *Stercorarius maccormicki*, (**f**) Brown-hooded Gull, *Chroicocephalus maculipennis*, (**g**) Scarlet Macaw, *Ara macao*, (**h**) Barn Owl, *Tyto alba*, (**i**) Burrowing Owl, *Athene cunicularia*, (**j**) Guira Cuckoo, *Guira guira*, (**k**) Rock Pigeon, *Columba livia*, (**l**) Campo Flicker, *Colaptes campestris*, (**m**) Rufous Hornero, *Furnarius rufus*, (**n**) Great Pampa-Finch, *Embernagra platensis*, (**o**) Hooded Siskin, *Spinus magellanica*, and (**p**) Grayish Baywing, *Agelaioides badius*. (Figure from Saliban and Montalti 2009; used with permission of the Instituto Internacional de Ecologia, publisher of the *Brazilian Journal of Biology*)

**Box 3.4 Possible Sexual and Social Functions of Female Uropygial Gland Secretions**

The results of a number of studies suggest that the uropygial gland secretions of birds may serve a variety of functions in social or sexual contexts (e.g., Amo et al. 2012a, b; Whittaker et al. 2017; Shutler 2019) and, further, that the secretions of female uropygial glands are more abundant and diverse than those of males, particularly during the breeding season (Whittaker and Hagelin 2021). In addition, in studies to date, the size of uropygial glands of males and females differed in over 60% (33/54) of the species, with females having larger glands in 23 of 30 species and males having larger glands in just seven species (Whittaker and Hagelin 2021). Given these differences between males and females, Whittaker and Hagelin (2021) proposed that the uropygial gland secretions (also called preen oils) of female birds could play important roles in intersexual attraction, intrasexual competition, and parental behavior. The results of some studies suggest that the preen oils of females may aid in attracting males (e.g., Song Sparrows, *Melospiza melodia*; Grieves et al. 2019b) and provide information about a female's breeding condition (e.g., Gray Catbirds, *Dumetella carolinensis*; Shaw et al. 2011; Dark-eyed Juncos, *Junco hyemalis*; Whittaker et al. 2011). In addition, Hirao et al. (2009) found that male Domestic Chickens were less like to mate with females that had their uropygial glands removed.

The results of another study suggest a role for preen oil in intrasexual competition and, specifically, as a signal of intrasexual aggression for female and male Dark-eyed Juncos (Whittaker et al. 2017). Finally, preen oil may also be important in parent-offspring, and even parent-egg, recognition in some species of birds. For example, nestling Zebra Finches (*Taeniopygia guttata*) from cross-fostered eggs were found to recognize their genetic mothers by their scent (Caspers et al. 2017), and Golüke et al. (2016) found that female Zebra Finches recognized the scent of their own eggs and preferred them to the eggs of other females.

Although the above-cited studies suggest potentially important roles of preen oil secretions and their odors, much remains to be learned about the importance of olfaction in the social and sexual interactions of birds. As noted by Whittaker and Hagelin (2021), "The field of bird chemical communication . . . lags behind studies on other taxa with regard to understanding social function."

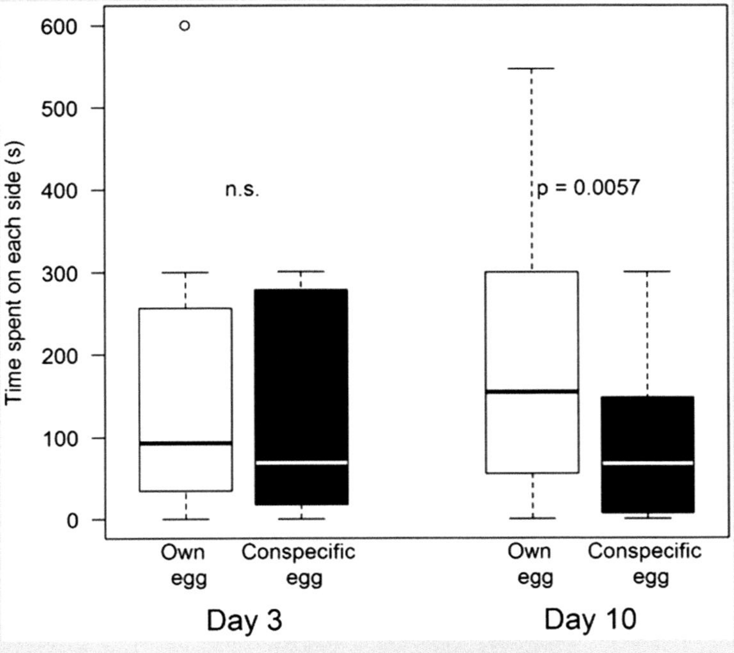


(continued)

**Box 3.4** (continued)

Female Zebra Finches (*Taeniopygia guttata*) spent more time with their own egg odors than those of a conspecific's eggs during the later incubation period (Day 10), but not earlier in the incubation period (Day 3). Therefore, females may not recognize the odors of their eggs early during incubation because embryos do not produce sufficient odor cues until later in development, and the amount of female preen gland secretions on eggs must increase during incubation to allow recognition, or some combination of these two explanations (Figure from Golüke et al. 2016; 

Protein Evolution). Therefore, feathers were long considered the defining anatomical feature of birds. However, many specimens of nonavian dinosaurs have been discovered in China, showing that feathers were not restricted to birds (Fig. 3.51).

In taxa more distantly related to birds, such as *Sinosauropteryx* (Fig. 3.52), multiple tufts projecting a few millimeters from the skin have been discovered that resemble hypothesized early stages in avian feather development. These filamentous "feathers" (or "protofeathers"; there is disagreement concerning whether or not these integumentary structures were true feathers, e.g., Unwin 1998; Lingham-Soliar et al. 2007) were about 20 (5–40) mm long and appeared to be rather homogenous over the body rather than originating in specific tracts. To some investigators, the filaments appeared to be like down feathers and were probably used for insulation. They were hollow, and appeared to have a short shaft with barbs, but no barbules. In 2009, a fossil of another feathered dinosaur, *Beipiaosaurus* (a coelurosaurian theropod), with even simpler feathers was reported (Xu et al. 2009). These feathers consisted of single broad (about 2 mm wide) filament, were 10 to 15 cm long, and were only present on the head, neck, and tail (Fig. 3.53). In taxa more closely related to birds, such as the oviraptorid *Caudipteryx* and dromaeosaurid *Sinornithosaurus*, elongate pinnate wing and tail feathers structurally identical to the feathers of present-day birds and comprised of a central rachis, branching barbs, and barbules, have been found. In addition, fossils of a dromaeosaurid (*Microraptor*) have revealed asymmetrically veined pennaceous feathers on both the forelimbs and hindlimbs (Clarke and Middleton 2006).

Because birds evolved from reptiles and the integument of present-day reptiles (and most extinct reptiles, including most dinosaurs) is characterized by scales, early hypotheses concerning the evolution of feathers began with the assumption that feathers developed from scales, with scales elongating, then growing fringed edges and, ultimately, producing hooked and grooved barbules (Regal 1975). The problem with that scenario is that scales are basically flat folds of the integument whereas feathers are tubular structures. A pennaceous feather becomes "flat" only after emerging from a cylindrical sheath (Prum and Brush 2002). In addition, the type and distribution of protein (keratin) in feathers and scales differ (Sawyer et al. 2000). The only feature shared by feathers and scales is that they both begin development as a morphologically distinct placode—an epidermal thickening above a condensation, or congregation, of dermal cells. Feathers, then, are not derived from scales but, rather, are evolutionary novelties with numerous unique features, including the feather follicle, tubular feather germ (an elevated area of epidermal cells), and a complex branching structure (Prum and Brush 2002).

Feathers are branched structures. The main branch of a typical feather is the rachis, and barbs, consisting of a barb ramus and projections called barbules, branch off the rachis (Fig. 3.54). Feathers grow from the base and the different

**Box 3.5 Cosmetic Coloration**

The color of a bird's plumage can be altered by the application of various substances. Such "cosmetic coloration" has been reported in at least 13 families of birds, including hornbills (Bucerotidae), ibises (Threshiornithidae), bustards (Otididae), vultures (Accipitredae), pelicans (Pelecanidae), cranes (Gruideae), and ptarmigans (Tetraonidae) (Delhey et al. 2007). Most species use substances they produce as the cosmetic, including secretions of uropygial glands or other skin secretions, and, in many cases, the cosmetic coloration serves as a sexual signal (Delhey et al. 2007). Other species use other substances, but primarily soil. For example, as the snow melts and the white plumage of male Rock Ptarmigans (*Lagopus muta*) becomes more conspicuous, males "bathe" in muddy soil, and their dirty plumage makes them less conspicuous to potential predators like Gyrfalcons (*Falco rusticolus*; Montgomerie et al. 2001). Bearded Vultures (*Gypaetus barbatus*) typically have orange- or reddish-colored feathers on their ventral surfaces even though the feathers are actually white. They change the color of their plumage by "bathing" in red soil containing iron oxide deposits, and actively rubbing their ventral surfaces against the soil (Negro et al. 1999). Their colored plumage appears to serve as a signal of dominance, with adults generally more intensely colored than juveniles and slightly larger females more intensely colored than males (Delhey et al. 2007). Other investigators have suggested that colored feathers may play a role in pair formation and the maintenance of pair bonds (Margalida et al. 2019).

Changing plumage coloration of Rock Ptarmigans (*Lagopus muta*). (**a**) Cryptic male in white winter plumage, (**b**) cryptic incubating female in brown summer plumage, (**c**) conspicuous male

(continued)

**Box 3.5** (continued)

in white winter plumage early in the breeding season when their white plumage plays a role in female mate choice, and (**d**) cryptic male with dirty plumage later in the breeding season (Figure from Montgomerie et al. 2001; © 2001 Oxford University Press, used with permission).

Bearded Vultures (*Gypaetus barbatus*) "bathe" in reddish soils containing iron oxide deposits to change the color of feathers on their ventral surface from white to orange or reddish (Figure from Tributsch 2016; open-access article distributed under the terms of the Creative Commons CC BY license, https://creativecommons.org/licenses/by/4.0/).

branches are generated by various mechanisms in the feather follicle. Feather growth depends on nutrients provided via the follicular cavity (dermal pulp), and the feather structure develops on the follicular (or follicle) collar (inner epidermal layer; Fig. 3.55). The production of the complex branched structure involves the interaction of several processes of cellular differentiation that occur on the follicle collar (Calcott 2009). Among present-day birds, variation in the shape and structure of the rachis, barbs, and barbules generates a variety of feather types, including flight (contour) feathers, semiplumes, bristles, down feathers, filoplumes, and powder-downs (Fig. 3.56; Box 3.7 Structure and Properties of the Primary Flight Feathers of Birds).

Based on fossil evidence, we know that nonavian theropods had feathers with a complex branching structure like those of present-day birds (pennaceous feathers). These fossils raise

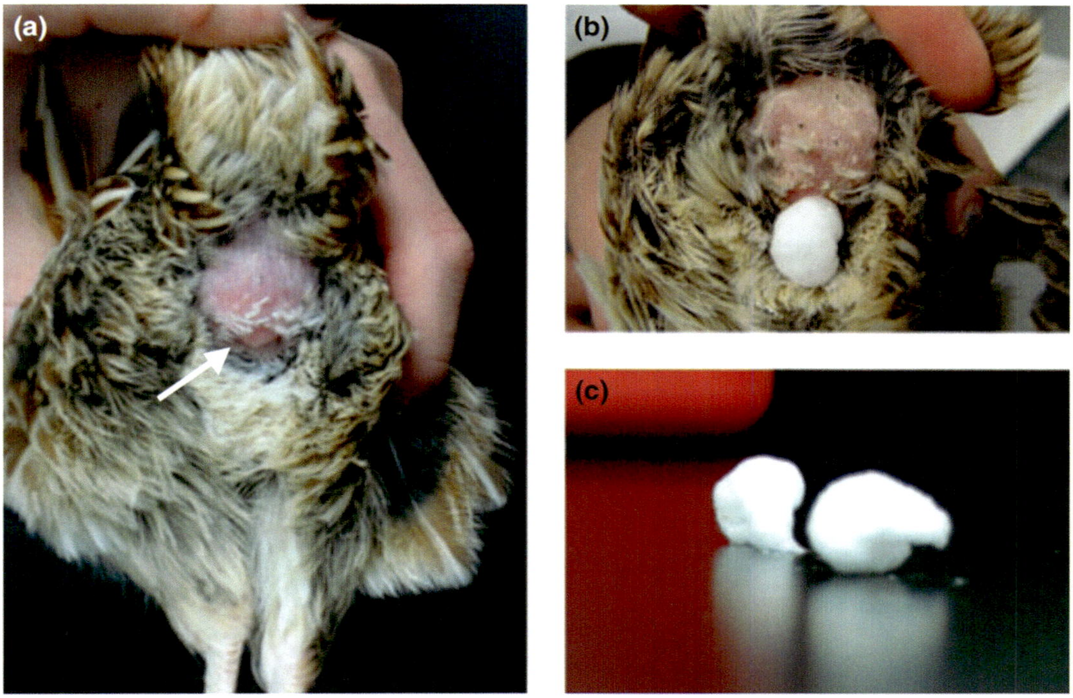

**Fig. 3.49** (**a**) The foam gland of a male Japanese Quail (*Coturnix japonica*). Arrow indicates a cloacal opening. Feathers around the foam gland were plucked. (**b**) Manual expression of foam from a foam gland. (**c**) Foam from two male Japanese Quail. (Figure from Finseth et al. 2013; © 2013 The Authors. Journal of Evolutionary Biology © 2013 European Society For Evolutionary Biology, used with permission)

two important questions. First, if not derived from scales, how did feathers evolve and, second, how did simple, single-filament feathers evolve to become much more complex pennaceous feathers? Of course, the related question is, given that nonavian theropods did not fly, what function or functions did these feathers serve?

Both fossil and developmental evidence suggests that feathers evolved through a series of transitional stages, each the result of a developmental evolutionary novelty or, in other words, a new mechanism of growth (Prum 1999; Prum and Brush 2002, 2003; Box 3.8 A New Mechanism of Growth: Genes and Proteins). The first feathers, like those of *Beipiaosaurus*, were unbranched, hollow cylinders that developed from the tubular elongation (the feather germ) of a placode. The advantage of a tubular feather germ is that the growth of a structure (in this case, a feather) can occur without an increase in the size of the skin itself (in contrast to, for example, scales; Prum 2005). An important step in the evolution of the first feathers was a change in the characteristics of the placode. Both scales and feathers begin development from placodes, but feather development, in contrast to scale development, requires the generation of suprabasal cell populations (dermal condensations) to form the follicle (Fig. 3.51). The development of placodes where dermal condensations occur, an evolutionary novelty, required changes in gene expression and timing. However, such changes are known to be an important mechanism in the origin of morphological innovations in many other organisms (True and Carroll 2002; Prum 2005).

Based on Prum's (1999) model of feather evolution, the next step after the origin of the feather follicle was the differentiation of the follicle collar into barb ridges to generate barbs (stage II; Fig. 3.57). The resulting feather would consist of

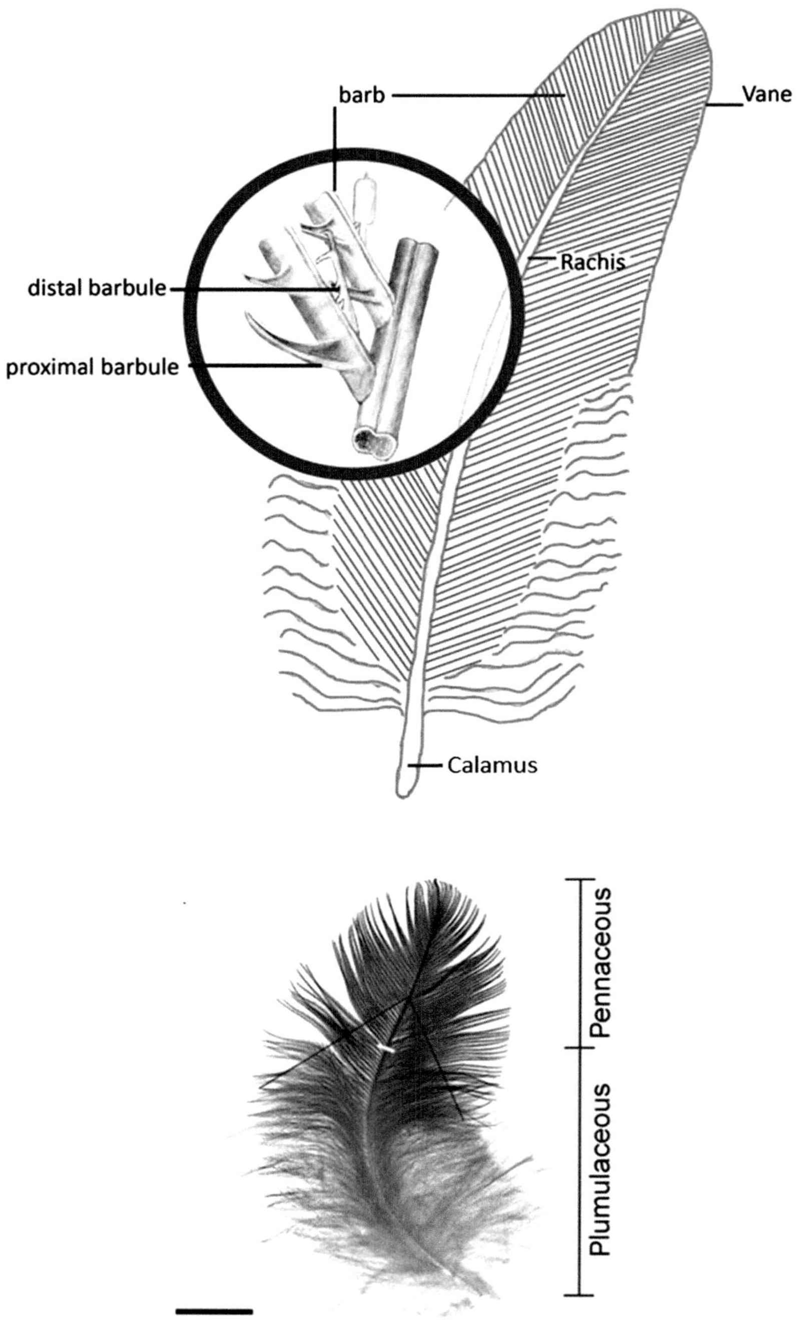


**Fig. 3.50** The two basic feather types are pennaceous and plumulaceous (or downy). Both types have a calamus. The pennaceous feather also has a rachis from which the barbs branch. Branching from the barbs (upper right) are barbules. Hooklets of the barbules on the distal side of barbs interlock with the barbules on the proximal side of adjacent barbs. The "interlocked" barbs on each side of the rachis form the feather vanes. Plumulaceous feathers have several noninterlocked barbs that form either from the calamus (e.g., down feather) or rachis (e.g., semiplume). (Top figure from Riedler et al. 2014; Rights managed by Taylor and Francis, used with permission; Bottom figure from Pap et al. 2017 © 2016 The Authors. Functional Ecology © 2016 British Ecological Society, used with permission)

**Box 3.6 Feather Protein Evolution**

The mature feathers of present-day birds consist primarily of β-keratins, proteins found only in birds and reptiles (Alibardi 2017). Another family of proteins, α-proteins, in contrast, are found in all vertebrates. Based on their chemical makeup, the β-keratins of birds are divided in subfamilies, including the claw and scale subfamilies and the feather subfamily. Compared to those in the other subfamilies, the feather β-keratin is more flexible, a characteristic of critical importance for powered flight (Pan et al. 2019). Phylogenetic analysis suggests that avian scale and feather β-keratins evolved from archosaurian claw β-keratins (Greenwold and Sawyer 2011, 2013). To further examine the evolution of feather β-keratins, Pan et al. (2019) compared feathers on the forelimb of *Anchiornis* to feathers of fossils of taxa that occur much later in the fossil record as well as to those of present-day birds.

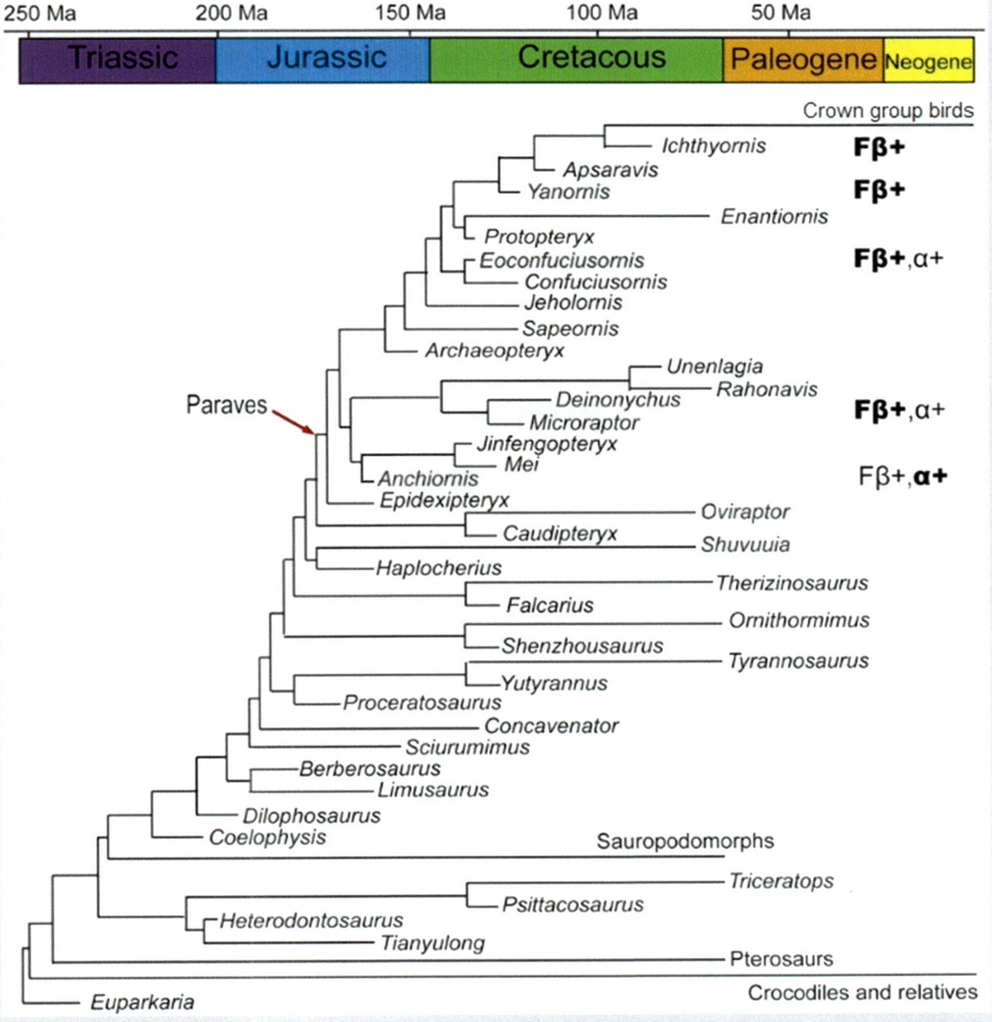


(continued)

**Box 3.6** (continued)

Evolution of the molecular composition and ultrastructure of feathers within a simplified Mesozoic avian and nonavian phylogeny. The feathers of *Anchiornis* were composed of both feather β-keratins and α-keratins but dominated by α-keratins. In contrast, feathers of later theropods, like *Microraptor*, were dominated by β-keratins, as are the feathers of present-day birds. Fβ+ = positive reaction to specific feather β-keratins, α+ = positive reaction to antipan cytokeratin antiserum (consistent with the presence of α-keratins), Fβ+ in bold = thin β-keratin filaments are dominant in ultrastructure, and α+ in bold = thick α-keratin filaments are dominant in ultrastructure (Figure modified from Pan et al. 2019; used with permission of the U.S. National Academy of Sciences).

Pan et al. (2019) found that *Anchiornis* feathers were composed primarily of α-keratins, but also contained some β-keratins. Because the mature feathers of present-day birds are composed of β-keratins, these results suggest that *Anchiornis* feathers represent an intermediate stage in feather evolution and, further, that their biomechanical properties may not have been suitable for powered flight. Feathers continued to evolve during the Cretaceous and, as revealed by analysis of the feathers of *Microraptor* and *Deinonychus*, feather β-keratins became the dominate protein in feathers during the Late Cretaceous. Finally, the results of Pan et al.'s (2019) study also provide support for the hypothesis proposed by Greenwold and Sawyer (2013) that feather β-keratins diverged from other β-keratins by about 143 million years ago.

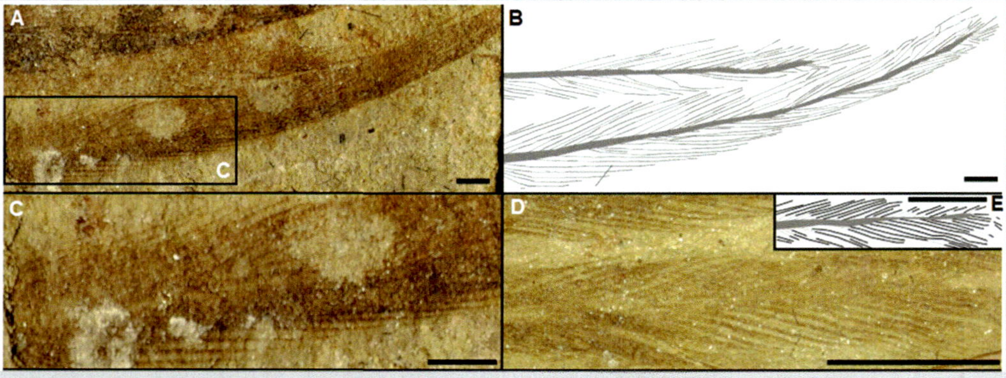


*Anchiornis* wing feathers. (**a**) Photograph and (**b**) corresponding drawing of an *Anchiornis* primary remex; the feather crossing the panel from the bottom-left and extending to the top-right has a curved rachis (dark gray) and unzipped barbs. (**c**) Close-up of the basal section of the primary feather in **a**. (**d**) Barbs of major covert and secondary remiges of *Anchiornis*. (**e**) Drawing of major covert shown in **d** (Figure from Saitta et al. 2018; © The Palaeontological Association, used with permission).

a tuft of barbs extending from the calamus (Fig. 3.57). Such a feather is hypothesized to have evolved before the origin of the rachis (stage IIIA) because the rachis is initially formed by the fusion of barb ridges. In addition, barbs are hypothesized to evolve before barbules because barbules develop within layers of pre-existing barb ridges (Prum 1999). Feathers comparable

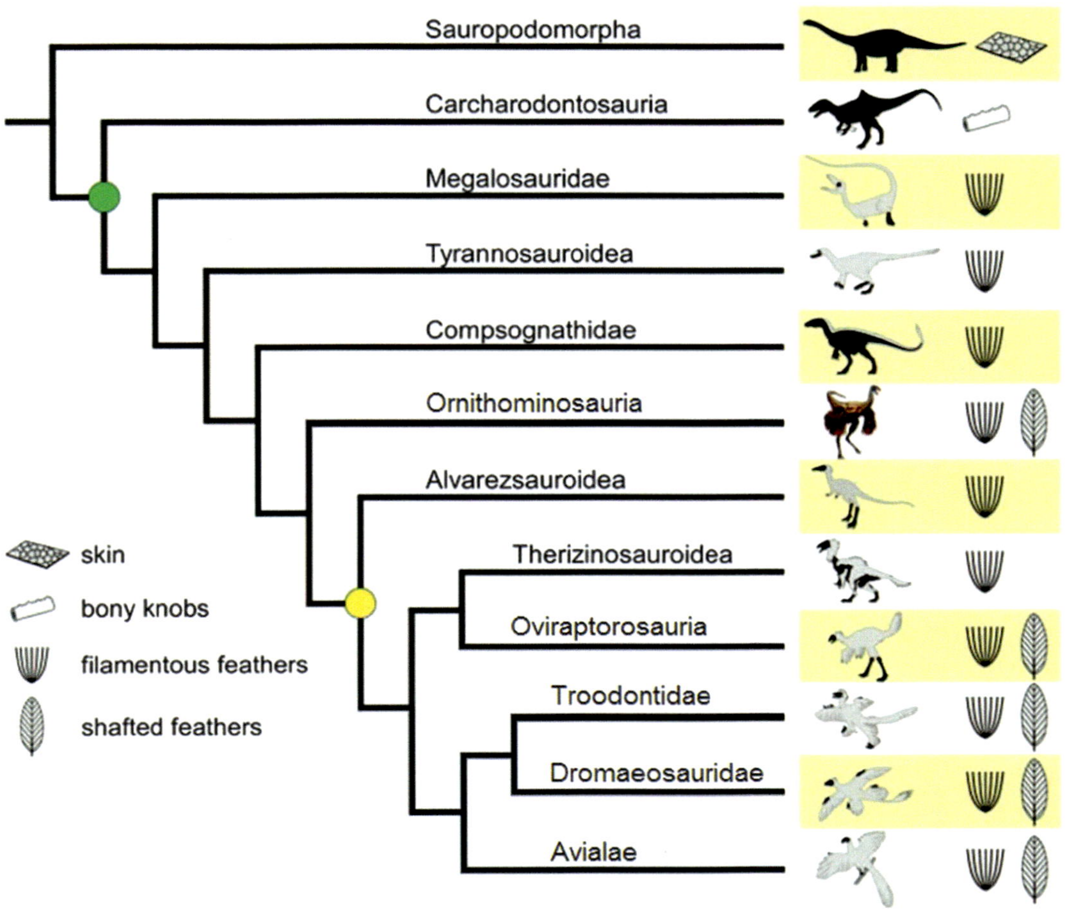


**Fig. 3.51** Phylogenetic distribution of major feather types among theropods. Filamentous feathers refer to all feathers that lack a rigid shaft (types 1, 2, and 3b of Prum and Brush 2002), whereas shafted feathers refer to all feathers that possess a rigid shaft (types 3a, 3a+b, 4, and 5 of Prum and Brush 2002). Reported occurrences of feathers are from Xu and Guo (2009) and Rauhut et al. (2012). Green node = Theropoda; Yellow node = Maniraptora. Avialae includes *Archaeopteryx* and present-day birds. (Figure modified from Zelenitsky et al. 2012; © 2012 American Association for the Advancement of Science, used with permission)

in structure to hypothesized stage II feathers have been reported from fossils of nonavian theropods, such as *Sinornithosaurus mellenii* (Figs. 3.58 and 3.59; Xu et al. 2001; Norell and Xu 2005).

The next step in feather evolution could have involved either the development of a rachis via the fusion of barbs or the development of barbules that branched from the tufts of barbs. Perrichot et al. (2008) discovered feathers from the early Cretaceous (and preserved in amber) that had shafts (rachis) consisting of incompletely fused, still distinguishable, partially superimposed barbs. This represents an intermediate stage between Prum's (1999) stages II and IIIa and suggests the possibility that rachis development may have preceded barbule development (Fig. 3.60).

With the development of the rachis, the next stage in feather evolution would likely have been the development of barbules (without hooklets) to generate a bipinnate, open pennaceous structure (Fig. 3.61, stage 11; Box 3.9 Feathers from the Mid-Cretaceous). Subsequent evolution of differentiated proximal and distal barbules

**Fig. 3.52** Restoration of *Sinosauropteryx* preying on *Dalinghosaurus*. (Painting by Robert Nicholls, Wikipedia, CC BY 4.0, https://creativecommons.org/licenses/by/4.0/)

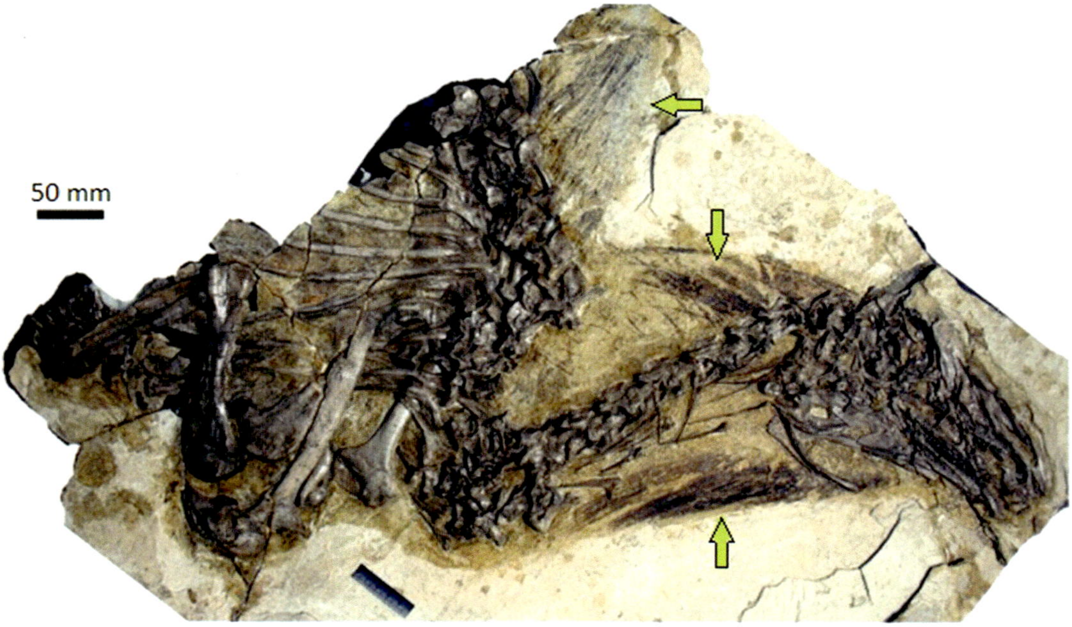


**Fig. 3.53** The elongated, single-filament feathers of *Beipiaosaurus*. The yellow arrows point to feathers on the head and neck (right), and tail (above). (Figure from Xu et al. 2009; used with permission of the U.S. National Academy of Sciences)

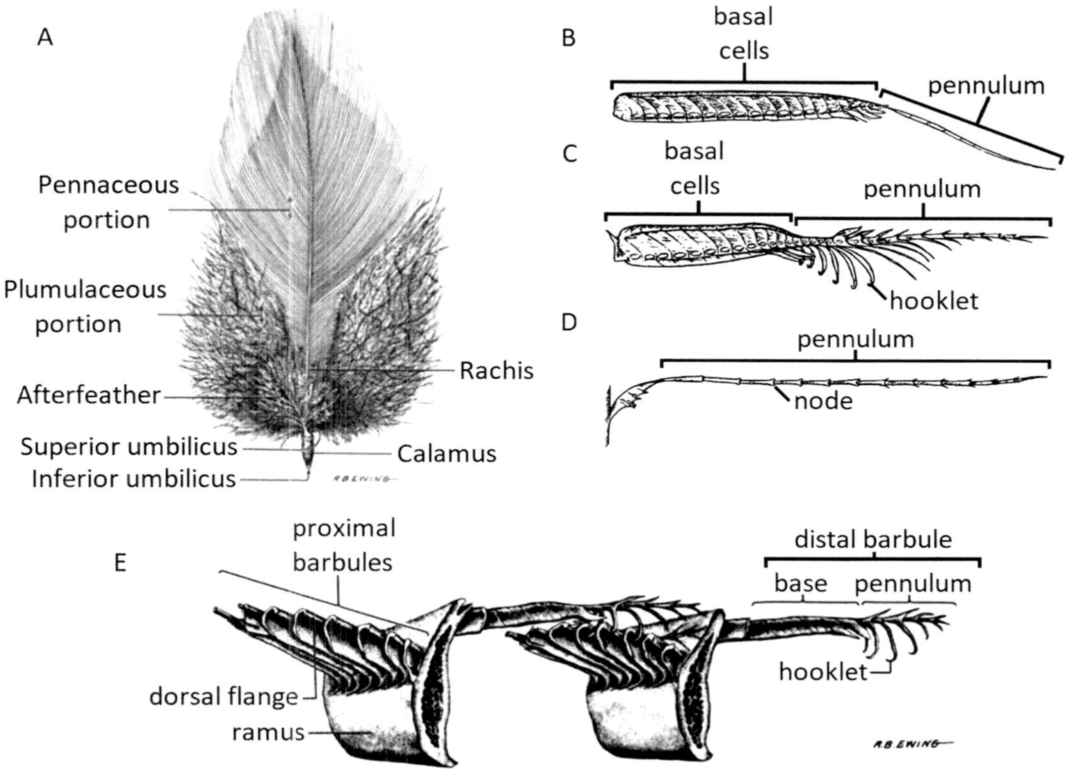


**Fig. 3.54** (**a**) Contour feather with pennaceous and plumulaceous portions and an afterfeather, (**b**) proximal barbule of a pennaceous feather, (**c**) distal barbule of a pennaceous feather, (**d**) plumulaceous barbule, and (**e**) detail of proximal barbules and distal barbules. (Figures from Lucas and Stettenheim 1972, CC0 Public Domain)

would then generate the first closed, pennaceous vane, with distal barbules growing hooklets to attach to the simpler, grooved proximal barbules of the adjacent barb (Fig. 3.62). Finally, lateral displacement of the new barb locus by differential new barb ridge addition to each side of the follicle led to the growth of a closed pennaceous feather with an asymmetrical vane resembling modern remiges (Fig. 3.61, stage 13; Fig. 3.62).

## 3.11 Evolution of Feather Function

Early functional hypotheses for the origin of feathers focused on their importance for flight (Steiner 1917; Heilmann 1926). However, the discovery of filamentous (and pennaceous) feathers on flightless nonavian theropods provides clear evidence that feathers evolved before the origin of flight and that the first feathers did not serve an aerodynamic function. The earliest tuft-like feathers could have served a variety of functions, including insulation, heat shielding (Regal 1975), communication (Mayr 1960), crypsis (Prum 1999), water repellency (Dyck 1985; Box 3.10 Water and Ice Repellency of Contour Feathers), and defense (Prum 1999).

The first cylindrical, filamentous feathers could have provided insulation if they were sufficiently numerous. Such feathers have been found on fossils of *Beipiaosaurus* (a coelurosaurian theropod; Xu et al. 2009). However, these primitive feathers consisted of single broad (about 2 mm wide) filaments and were only present on the head, neck, and tail. Given their morphology and distribution on the body, these feathers likely

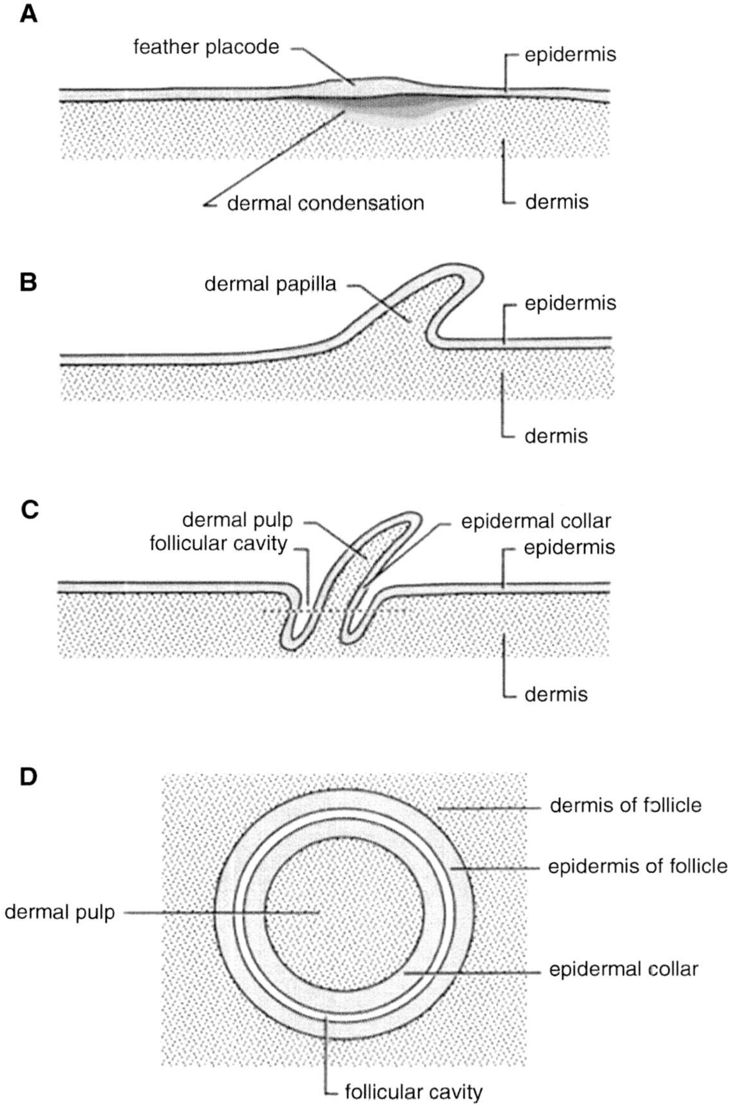


**Fig. 3.55** Diagram showing the development of a feather follicle. (**a**) Development of the epidermal feather placode and dermal condensation. (**b**) Development of a feather papilla via proliferation of dermal cells. (**c**) Feather follicle forms by the invagination of a cylinder of epidermal tissue around the base of the feather papilla. (**d**) Cross-section of the feather follicle through the horizontal plane where the dotted line is located in **c**. The follicle consists of a series of tissue layers, including the dermis of the follicle, the epidermis of the follicle (outer epidermal layer), follicle cavity or lumen (the space between epidermal layers), follicle collar (inner epidermal layer or ramogenic zone), and dermal pulp (tissue at the center of the follicle). The proliferation of feather keratinocytes occurs in the follicle collar of the inner epidermal layer. (Figure from Prum and Williamson 2001; © 2001 Wiley-Liss, Inc., used with permission)

did not serve a thermoregulatory function. Instead, their localized distribution and morphology (relatively long and probably rather stiff) suggest they served as display structures (Xu et al. 2009). However, other types of filamentous feathers in nonavian theropods more likely served a thermoregulatory function (Norell and Xu 2005). For example, the presence of dense filamentous feathers on *Sinosauropteryx* suggests that these theropods were endothermic and that heat retention was the primary function of the feathers (Chen et al. 1998).

The fossil of a pigeon-sized theropod, *Epidexipteryx hui*, found in sediments from the Middle to Late Jurassic (152–168 million years ago) of northern China revealed two pairs of elongate ribbon-like tail feathers that probably served as ornaments (although they could have also helped *E. hui* maintain balance when moving along tree branches). These long feathers had a central shaft (rachis) but, unlike the rectrices of present-day birds, the vanes were not branched into individual filaments. Rather, they consisted of a single ribbon-like sheet. Shorter feathers also

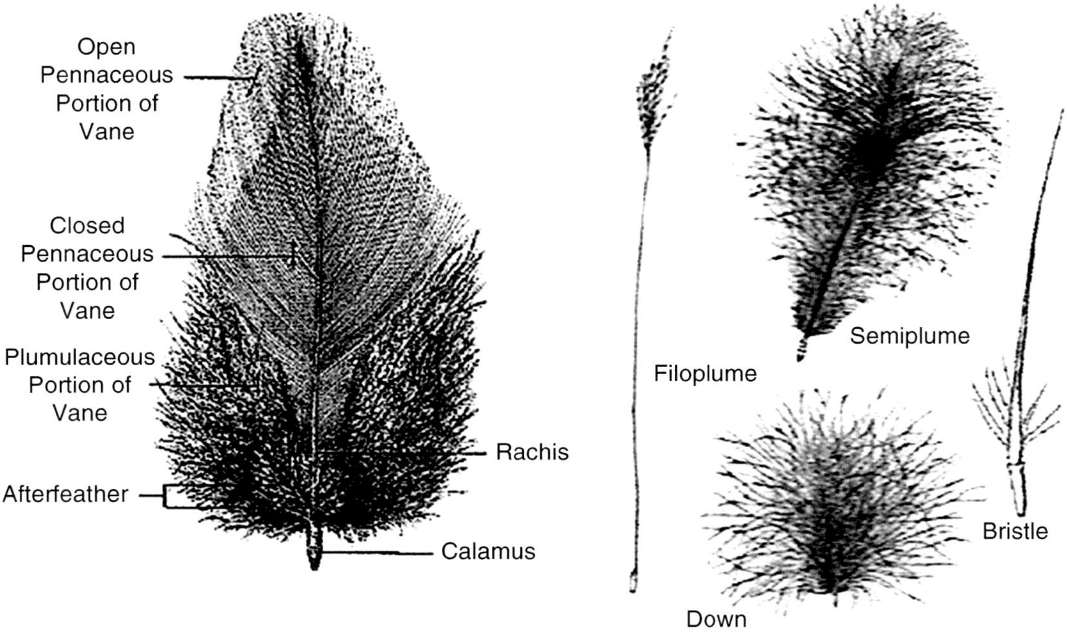


**Fig. 3.56** Various types of feathers of present-day birds, including the contour feather (left) plus filoplumes, semiplumes, down feathers, and bristles. (Figures from Lucas and Stettenheim 1972; CC0 Public Domain)

**Box 3.7 Structure and Properties of the Primary Flight Feathers of Birds**

The primary flight feathers of birds must be very light, but, at the same time, very strong to cope with the stresses of flight. They must also be somewhat flexible so that, during the downstroke, they provide the thrust needed to propel flying birds forward. The central feather shaft consists of a calamus at the base of the feather and a rachis—the calamus anchors flight feathers to the second digit and carpometacarpus of the forearm. The feather shaft consists of a solid keratinous (β-keratin) shell called the cortex, and inside the cortex is a foamy core called the medulla consisting of medullary cells.

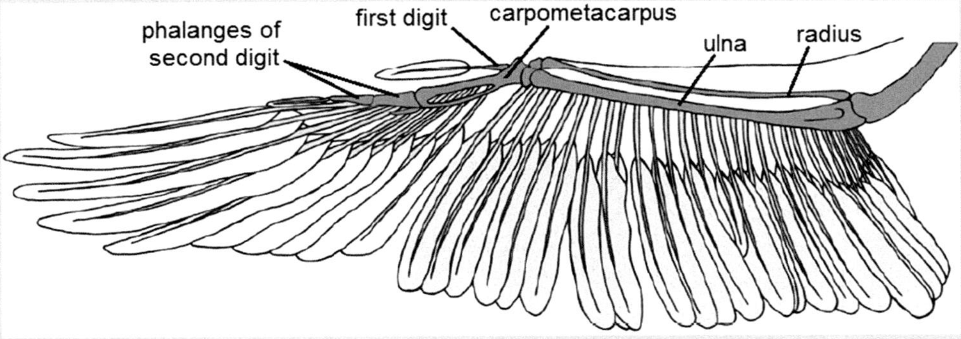


(continued)

**Box 3.7** (continued)

Wing anatomy of a Brown Pelican (*Pelecanus occidentalis*) showing attachment sites of the primary feathers (phalanges of second digit and the carpometacarpus) and the secondary feathers (ulna) (Figure modified from Simons 2009; used with permission of Erin Simons).

Top view of the wings of three species of birds showing the primary and secondary flight feathers. (**a**) American Tree Sparrow, *Spizelloides arborea*; (**b**) Double-crested Cormorant, *Nannopterum auritus*; (**c**) Laysan Albatross, *Phoebastria immutabilis* (figure from Wang and Clarke 2015; © 2015 The Authors. Published by the Royal Society, used with permission).

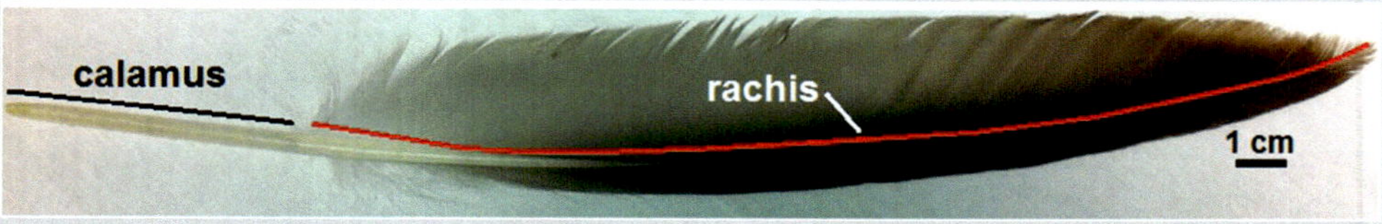


Primary flight feather of a California Gull (*Larus californicus*) (Figure modified from Wang and Meyers 2017; © 2016 Acta Materialia Inc. Published by Elsevier Ltd. All rights reserved, used with permission).

(continued)

**Box 3.7** (continued)

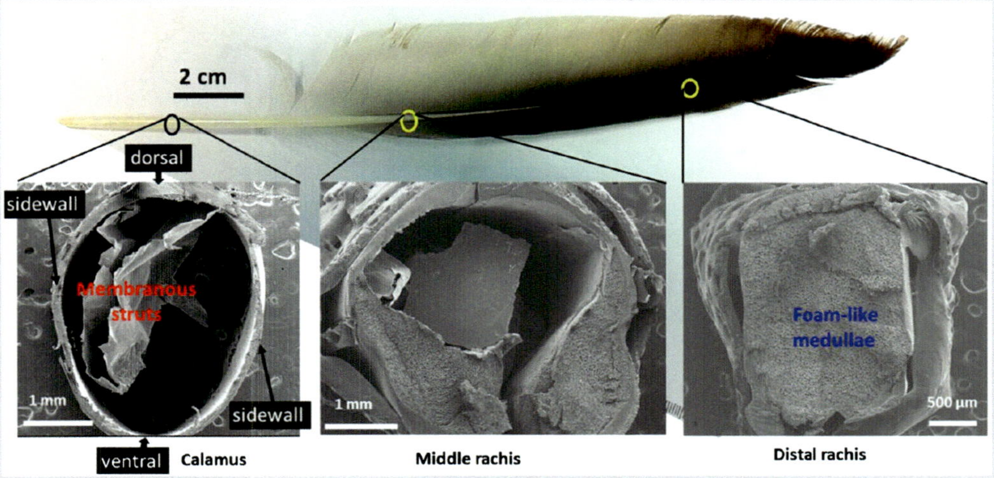


Structure of a primary feather of a California Gull, with micrographs of transverse cross-sections along the feather shaft (the dorsal, ventral, and sidewalls are indicated). The calamus is a hollow cylinder filled with struts; struts are reduced, with more foam-like medullae (also called medulloid pith) in the middle rachis, and all foam-like medullae in the distal rachis (Figure from Wang et al. 2016a; Copyright © 2015 Elsevier Ltd. All rights reserved, used with permission).

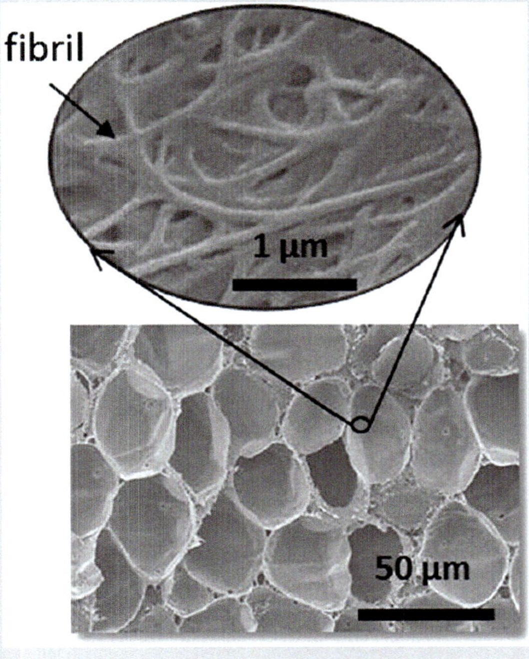


Foam-like medullae of the feather core showing its fibrous structure (Figure modified from Wang et al. 2016a; Copyright © 2015 Elsevier Ltd. All rights reserved, used with permission).

(continued)

**Box 3.7** (continued)

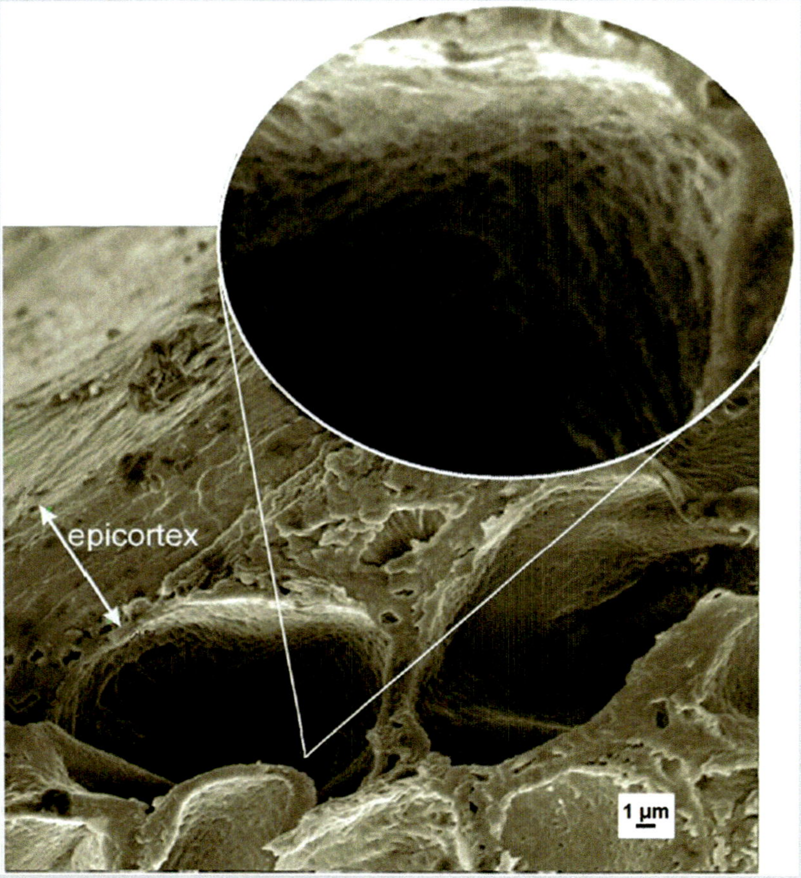


A medullary pith cell adjacent to the epicortex (lateral wall of the rachis), with the inset showing the trabeculae-like support of β-keratin fibers in the walls (Figure modified from Lingham-Soliar 2014; Copyright © 2013, Dt. Ornithologen-Gesellschaft e.V., used with permission).

The cortex of a feather's shaft is a layered structure with β-keratin filaments oriented in different directions in the different layers to provide greater structural support. In addition, the dorsal and ventral walls of the rachis are thicker than the lateral walls, creating a beam-like structure that is easier to twist. The medulla consists of a foam-like material made up of numerous closed cells (keratinocytes) filled with air. The walls of these cells are very thin, made of numerous β-keratin fibers, and the cells are interconnected to form a continuous inner "skeleton." The innermost fibers in the walls form trabeculae-like structures that provide structural support.

(continued)

**Box 3.7** (continued)

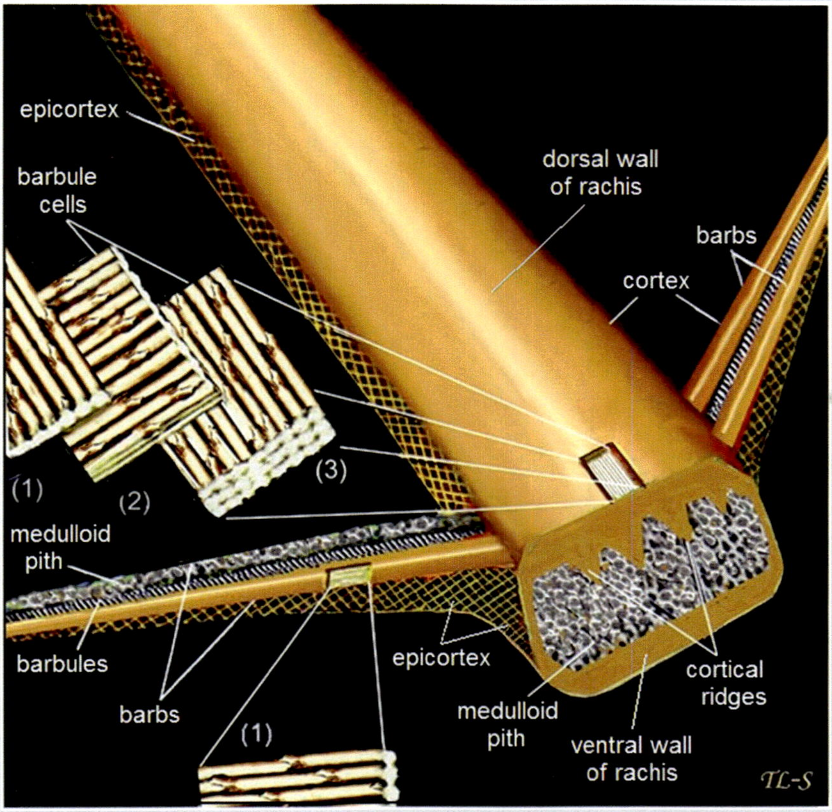


(continued)

**Box 3.7** (continued)

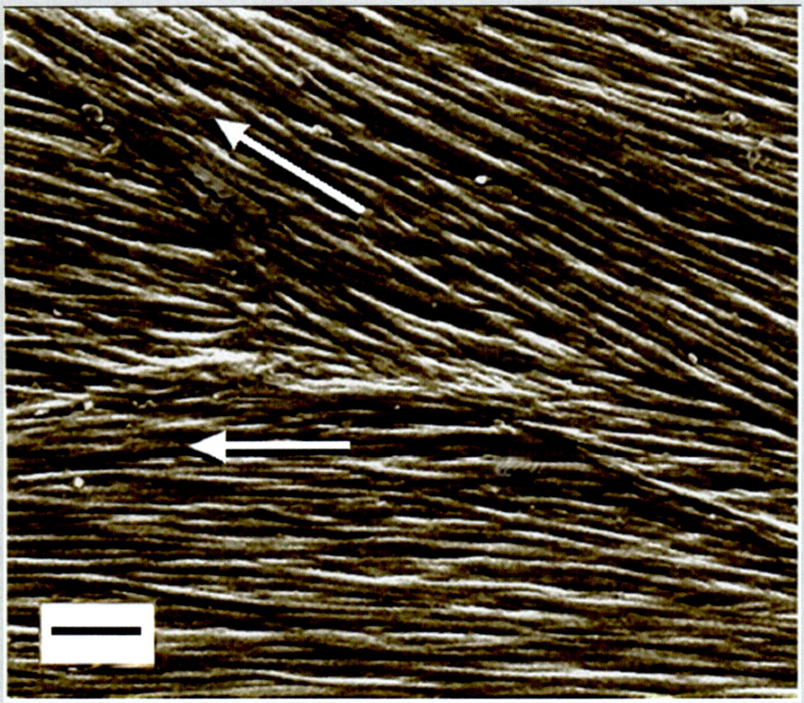

Top, The structure of the rachis and barbs of a feather showing the three layers of barbule cells or fibers (composed of β-keratin) in the rachidial cortex (the dorsal and ventral walls of the rachis) and a single layer in the barb cortex. The interior of the rachis and barbs consists of medullary pith cells. The lateral walls of the rachis and barbs, called the epicortex, consist of layers of crossed fibers (below), but no barbule cells. The overlapping layers of fibers make the rachis and barbs of feathers strong and capable of some torsion (twisting), yet resistant to buckling and still relatively light in weight. Arrows in the bottom figure follow the direction of fibers in different layers. Scale bar = 2 μm (Figures modified from Lingham-Soliar 2014; Copyright © 2013, Dt. Ornithologen-Gesellschaft e.V., used with permission).

(continued)

**Box 3.7** (continued)

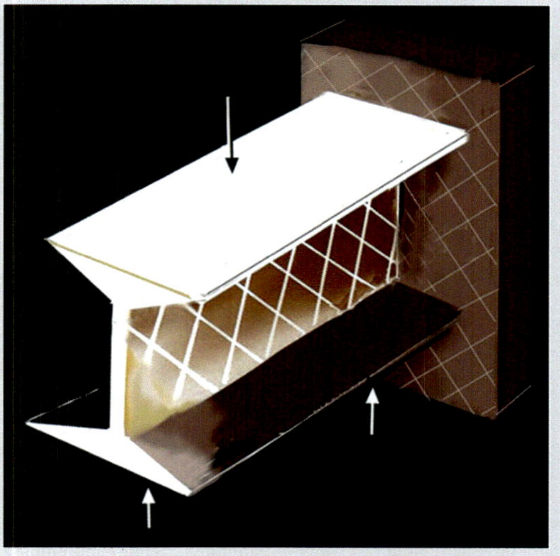

The rachis is a hollow shaft of beta-keratin filled with medullary foam. However, the dorsal and ventral parts of the rachis are typically thicker than the lateral portions. This produces an I-beam structure that is resistant to dorsal and ventral forces. As a result, when those forces increase, feathers respond by twisting (Figure modified from Lingham-Soliar 2014; © 2013, Dt. Ornithologen-Gesellschaft e.V., used with permission).

In addition to becoming smaller from the proximal to distal ends, the cross-sectional shape of the shaft of flight feathers varies along their length; the calamus is round or nearly so, and the rachis becomes progressively more rectangular or square toward its distal end. This change is beneficial because a square or rectangular shape provides greater rigidity and greater resistance to ovalization and buckling than a hollow round shape. Not surprisingly, the feather shafts of nonflight feathers, such as the wing feathers of flightless birds like Common Ostriches (*Struthio camelus*), do not change from round to square. In addition to this change in shape, the rachis of flight feathers becomes stiffer, or more difficult to deform or bend, toward the distal end. Relative stiffness is measured in Young's modulus units, with higher values indicating greater resistance to deformation.

(continued)

**Box 3.7** (continued)

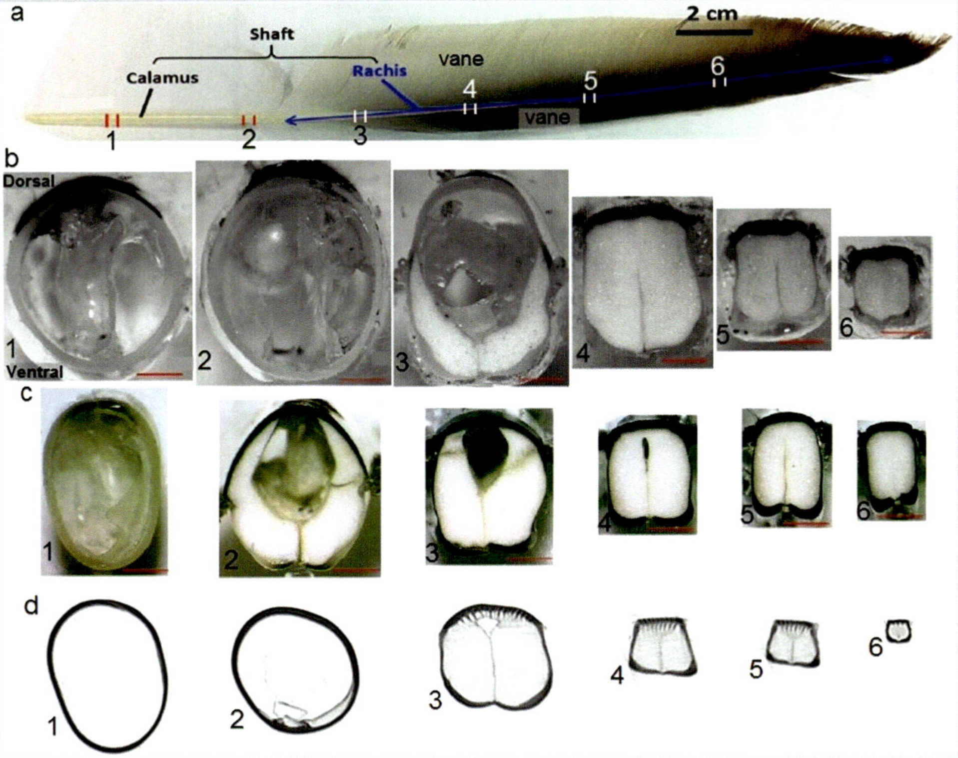


(**a**) A primary flight feather showing where sections were made (1–6) in a primary flight feather of (**b**) a California Gull (*Larus califormicus*), (**c**) an American Crow (*Corvus brachyrhynchos*), and (**d**) a California Condor (*Gymnogyps californianus*). Note how the shape of the shafts of primary flight feathers change from one end to the other, becoming less round and more rectangular or square toward the end of the rachis (Figure modified from Wang et al. 2016a; © 2015 Elsevier Ltd., used with permission).

(continued)

**Box 3.7** (continued)

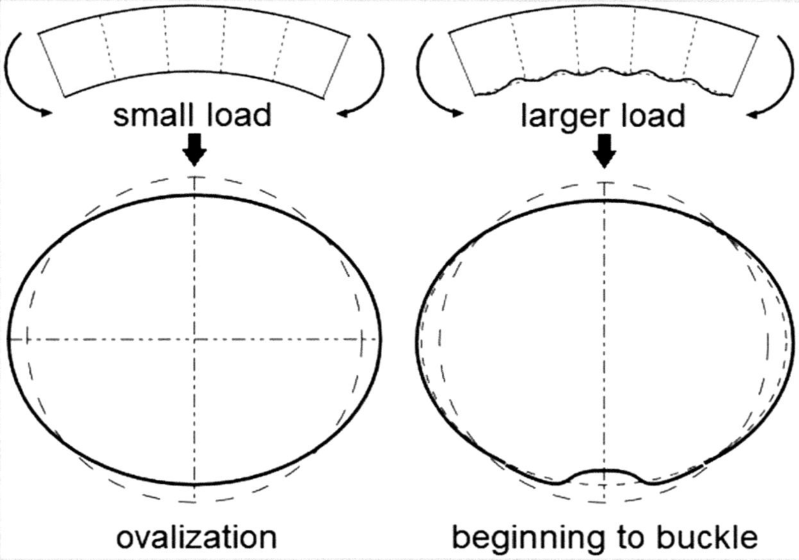


Round tubes, or feather shafts, are less resistant to ovalization and buckling than are square or rectangular tubes (Figure modified from Wadee et al. 2007; © 2006 Elsevier Ltd., used with permission).

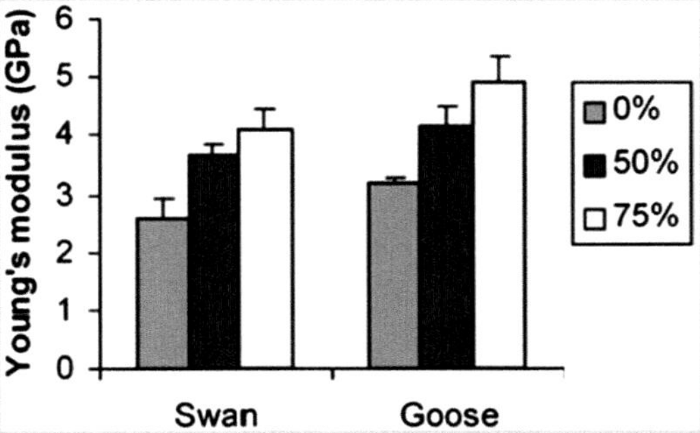


Variation in the mean (±SD) Young's (or elastic) modulus from specific locations from the calamus and along the rachis of primary flight feathers of a swan and a goose. Young's modulus measures stiffness or how easy it is to deform a structure. The increase in Young's modulus, or stiffness, toward feather tips, allows a reduction in mass (i.e., the rachis is thinner near feather tips) so birds can beat their wings faster and potentially save energy (Figure modified from Cameron et al. 2003; © 2003 Elsevier Inc., used with permission).

(continued)

**Box 3.7** (continued)

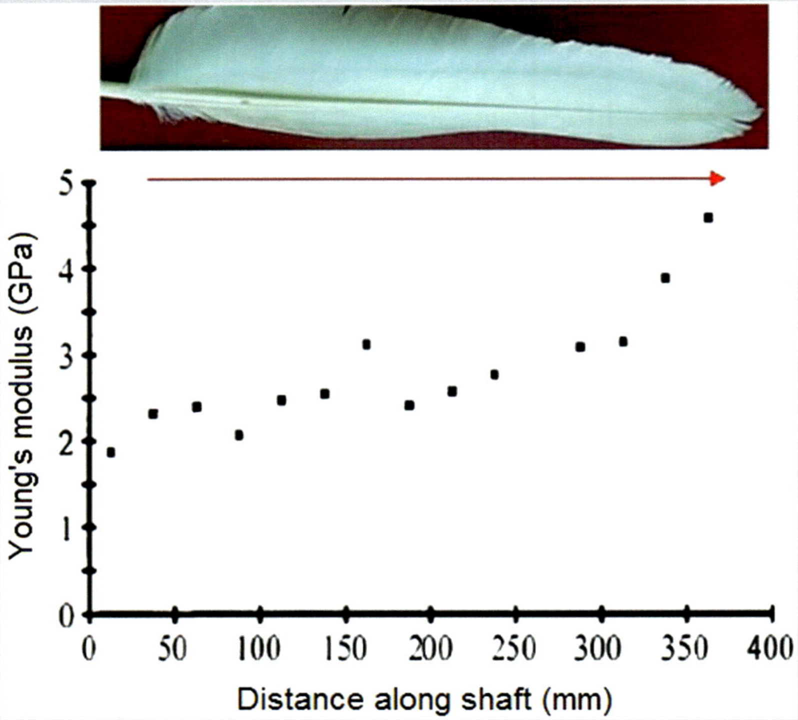


Variation in Young's modulus along the length of a primary feather of a Mute Swan (*Cygnus olor*) (Figure modified slightly from Wang et al. 2016a; © 2015 Elsevier Ltd., used with permission).

Therefore, the rachis of primary flight feathers is ideally suited for flight. The narrowing of the rachis toward the feather tip reduces weight and drag, but the increase in Young's modulus toward the feather's tip ensures the stiffness necessary to prevent buckling due to stresses resulting from flight. The square or rectangular shape of the rachis, in combination with the thinner lateral wall of the cortex, also results in lower torsional resistance (compared to a round shape), so the distal end of the feather can slightly twist during the power- or downstroke of the wings (Wang and Meyers 2017). Finally, the primary feathers of some birds have small grooves on their ventral surface, making the feathers easier to twist. This twisting is crucial because it allows the primaries to generate the backward thrust that propels a bird forward in flight.

(continued)

**Box 3.7** (continued)

A Kea (*Nestor notabilis*) in flight. Note how the outer primaries are bent and twisted during the downstroke. This change in shape helps provide the thrust needed to propel birds through the air (Photo by Christian Mehlführer, Wikipedia, CC BY 2.5, https://creativecommons.org/licenses/by/2.5/)

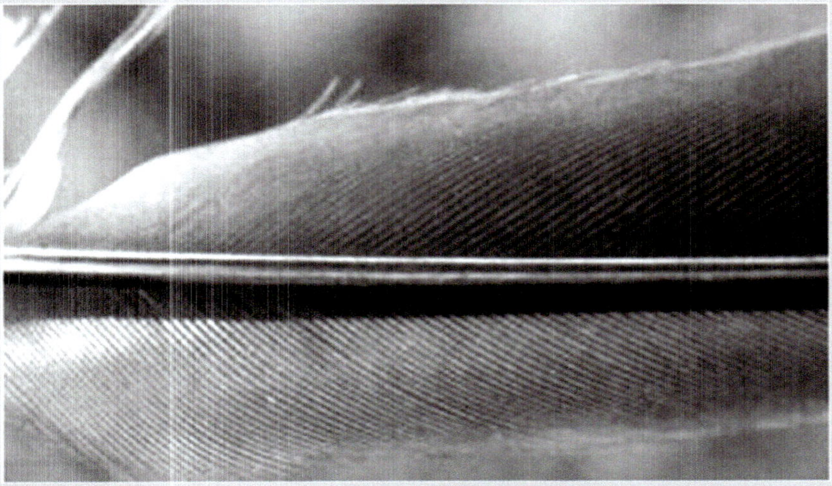

The ventral surface of the primary flight feathers of some species of birds has grooves that vary in width and depth. These grooves enhance the twisting ability of the feathers (Figure from Vogel 2007; © 2007 Indian Academy of Sciences, used with permission).

(continued)

**Box 3.7** (continued)

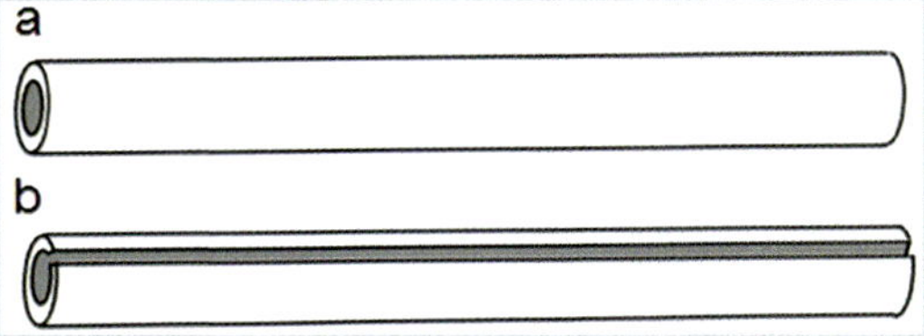


Two tube-like structures, one without a slit (**a**) and one with a slit (**b**). The tube with the slit would clearly be more easily twisted than the tube without the slit. Similarly, the groove along the ventral surface of primary flight feathers makes the distal end of the feathers easier to twist (Figure modified from Vogel 2007; © 2007 Indian Academy of Sciences, used with permission).

The rachis of flight feathers varies among species of birds with different flight styles and among flightless species, like terrestrial Common Ostriches (*Struthio camelus*) and wing-propelled diving penguins. In general, among species of birds that engage in frequent or sustained flight, natural selection has favored lighter, stronger rachises (Chang et al. 2019).

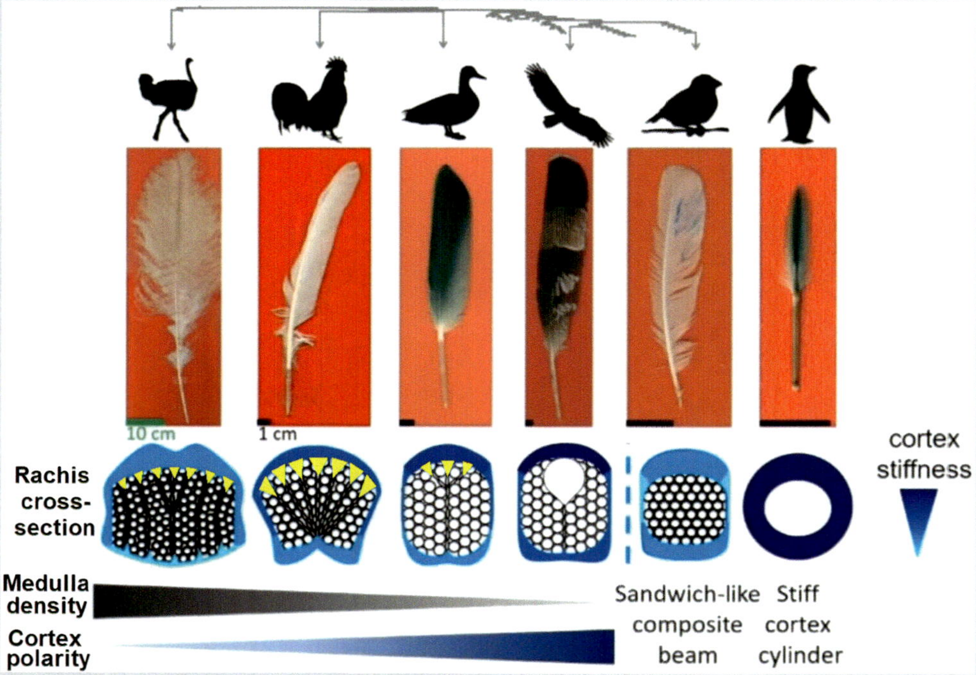


The rachis of flight feathers differs among birds with different flight styles. Top, left to right: Common Ostrich (flightless), Domestic Chicken (short distance flight, high wing loading), ducks (flapping flight), eagles (soaring flight), sparrows (high wing flapping frequency and bounding flight), and penguin (wing-propelled diver). For the Ostrich and chicken, vacuolated keratinocytes (open circles) form bands aligned with cortical ridges (colored yellow) within a dense medulla, increasing the ability of the rachis to support mechanical loads. Ducks and eagles

(continued)

**Box 3.7** (continued)

show bilateral symmetry of keratinocytes that have relatively large pores that increase the stiffness-to-weight ratio of the rachis. Eagles also have a dorsally located hollow zone in the medulla, which likely increases the ability of the rachis to twist. Sparrows and other small birds have relatively homogeneous medullas, with small keratinocytes within a relatively dense medulla. The medulla provides sufficient structural support, but the keratinocyte pores minimize rachis weight. A relatively stiff, but light, rachis is important for small birds that rapidly flap their wings when flying. The cortex of flight feathers also varies with flight style, with dorsal-ventral polarity higher in ducks and, especially, eagle flight feathers than in ostrich and chicken feathers. In general, birds that engage in sustained flight have thicker, stronger ventral and, especially, dorsal cortices and, therefore, stronger, stiffer rachises. However, the strongest, stiffest cortices and rachises are those of penguins. This structural support is needed because, for wing-propelled diving penguins, the density of water is much greater than that of air (Figure is a screenshot from the video abstract of Chang et al. 2019; © 2019 Elsevier BV, used with permission).

covered the body and could have served as insulation (Zhang et al. 2008a, b). *Epidexipteryx* is the oldest theropod known to have feathers that apparently served a display function (Fig. 3.63).

Pennaceous (or contour) feathers have been reported for several theropods, including the maniraptor *Protarchaeopteryx* (early Cretaceous; 120–122 million years ago), the oviraptorid *Caudipteryx*, and the dromaeosaurids *Sinornithosaurus* and *Microraptor gui* (Fig. 3.64). Both *Protarchaeopteryx* and *Caudipteryx* had pennaceous feathers (with barbules) on the forearms and tail (as well as semiplumes and down-like feathers on the rest of the body). However, the arms of these small theropods (about 0.4–0.7 kg) were relatively short, and all pennaceous feathers were symmetrical, indicating that these dinosaurs could not fly or glide effectively. Some investigators have suggested that these theropods, with relatively long legs and an elevated hallux, were ground-dwelling runners (Qiang et al. 1998). However, the forelimbs of *Protarchaeopteryx* and *Caudipteryx*, although short relative to their hindlimbs, were longer than those of other theropods. Some investigators have argued that such elongation (along with other characteristics, including recurved claws) suggests a more (but not exclusive) arboreal lifestyle. For example, Chatterjee and Templin (2004) argued that these theropods were largely arboreal and that their small "protowings" (in combination with the pennaceous feathers on the tail) enhanced arboreal maneuvering and permitted parachuting from branch to branch or from branch to ground. Feathers on the "protowings" and tail would increase drag when parachuting and, to some extent, slow the rate of descent, permitting a safer landing. Another possible function of the forearm feathers could have been to increase hindlimb traction in the same manner that some present-day birds, such as Chukars (*Alectoris chukar*), flap their wings to improve hindlimb traction when they climb inclined surfaces like the trunk of a tree (i.e., wing-assisted incline running; Dial 2003; Clarke and Middleton 2006).

Other small theropods from the early Cretaceous (124–128 million years ago), including *Sinornithosaurus* and *M. gui*, had both plumulaceous and pennaceous feathers. *Sinornithosaurus* weighed about 1.5 kg, were likely arboreal, and, in contrast to *Protarchaeopteryx* and *Caudipteryx*, their forelimbs were as long as their hindlimbs. Although unlikely, the longer wings and greater wing surface area, in combination with feathers

**Box 3.8 A New Mechanism of Growth: Genes and Proteins**

The epidermal appendages of reptiles and birds, including claws, scales, and feathers, comprise beta (β) keratin proteins. Phylogenetic analyses have revealed that the evolution of archosaurian epidermal appendages in the lineage leading to birds involved duplication and divergence of an ancestral β-keratin gene. This resulted in novel β-keratin genes and a novel epidermal appendage—feathers.

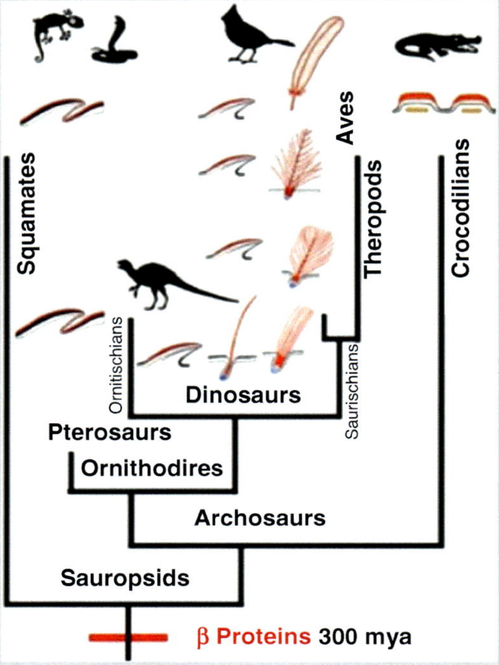


The epidermal appendages of reptiles and birds are made of different β-keratins. The initial ancestral β-keratins are thought to have first appeared in the archosaur lineage about 300 million years ago (Figure from Dhouailly et al. 2019; © 2017 John Wiley & Sons A/S. Published by John Wiley & Sons Ltd, used with permission).

However, the formation of claws, scales, and feathers requires more than just β-keratins. Also necessary are transcriptase factors, genes that turn other genes, like those that produce different β-keratins, off and on or, more specifically, bind to specific sites on DNA to activate transcription of genetic information from DNA to messenger RNA which is then, via the process of translation, used to produce a protein. Different transcriptase factors are needed to activate different genes that in turn produce the different β-keratins in claws, scales, and feathers. By comparing the genomes of four species of birds, two crocodilians, two turtles, a lizard, four mammals, a frog, and five fish, Lowe et al. (2015) found that, surprisingly, what we now call "feather-development genes" were present even before the origin of dinosaurs and feathers.

(continued)

**Box 3.8** (continued)

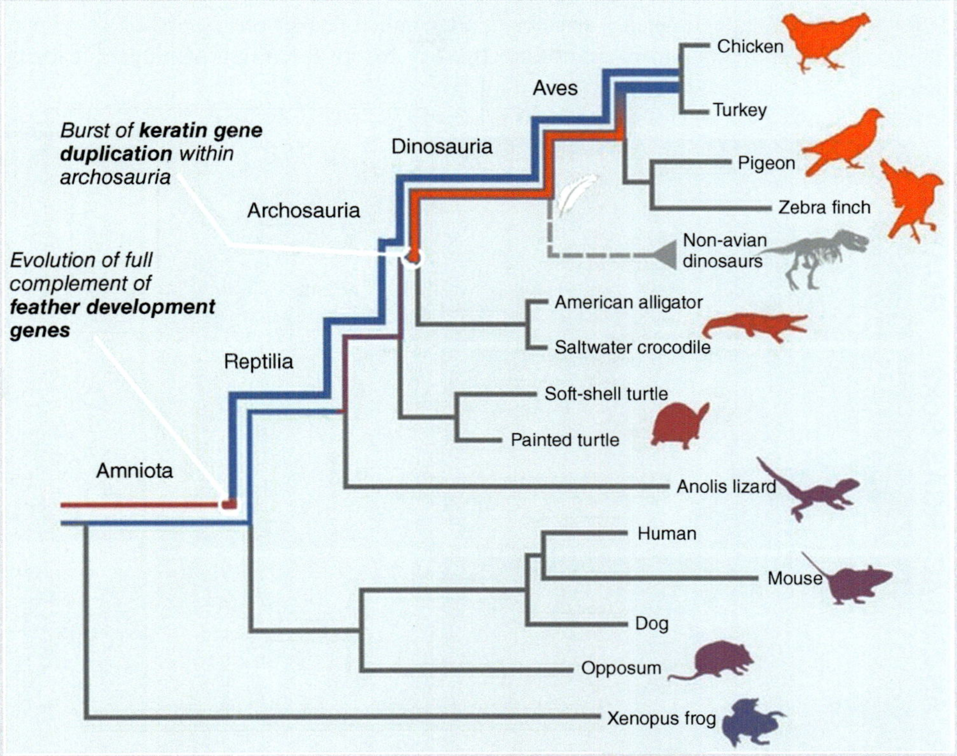


Important genomic events in the evolution of feathers, including the evolution of feather-development genes and diversification and divergence of β-keratin genes (Figure modified from Lowe et al. 2015; © 2014 Oxford University Press, used with permission).

In an interesting experiment, Wu et al. (2017) injected genes associated with feather development in Domestic Chickens into the amnionic cavity and legs of chicken embryos to examine their effects on the scales developing on the legs. In addition, those genes were also injected into alligator eggs to see if the alligator genes for scales could be overridden by switching on the chicken feather genes. In the chicken embryos, different feather-development genes led to the development of several intermediate types of shapes, ranging from scales to more complex forms of filamentous feathers. Some of the shapes resembled the filamentous appendages associated with feathered dinosaur fossils, whereas other shapes formed have similar characteristics to those found in the feathers of modern birds. Overall, the results of these studies suggest the possibility that the evolution of feathers involved selection favoring new β-keratins and the expression of much older genes able to guide the developmental process necessary to produce the novel structures called feathers. As noted by Lowe et al. (2015) concerning these ancient genes, "These results are also consistent with new data on integumentary innovation and diversity in Archosauria: filamentous or bristle structures either originated once early in the clade, or three or more times (Clarke 2013) in pterosaurs (Kellner et al. 2010), and ornithischian (Zheng et al. 2009; Godefroit et al. 2014) and theropod dinosaurs (Norell and Xu 2005). Thus, the genic and regulatory complement identified in the ancestral archosaur was either a flexible toolkit co-opted

(continued)

**Box 3.8** (continued)

in multiple origins of new structures, including feathers, or indicates an ancient origin in that clade for filamentous integumentary structures, often called feather precursors, on some part of the body or stage in development more than 100 My before the origin of pinnate feathers in dinosaurs."

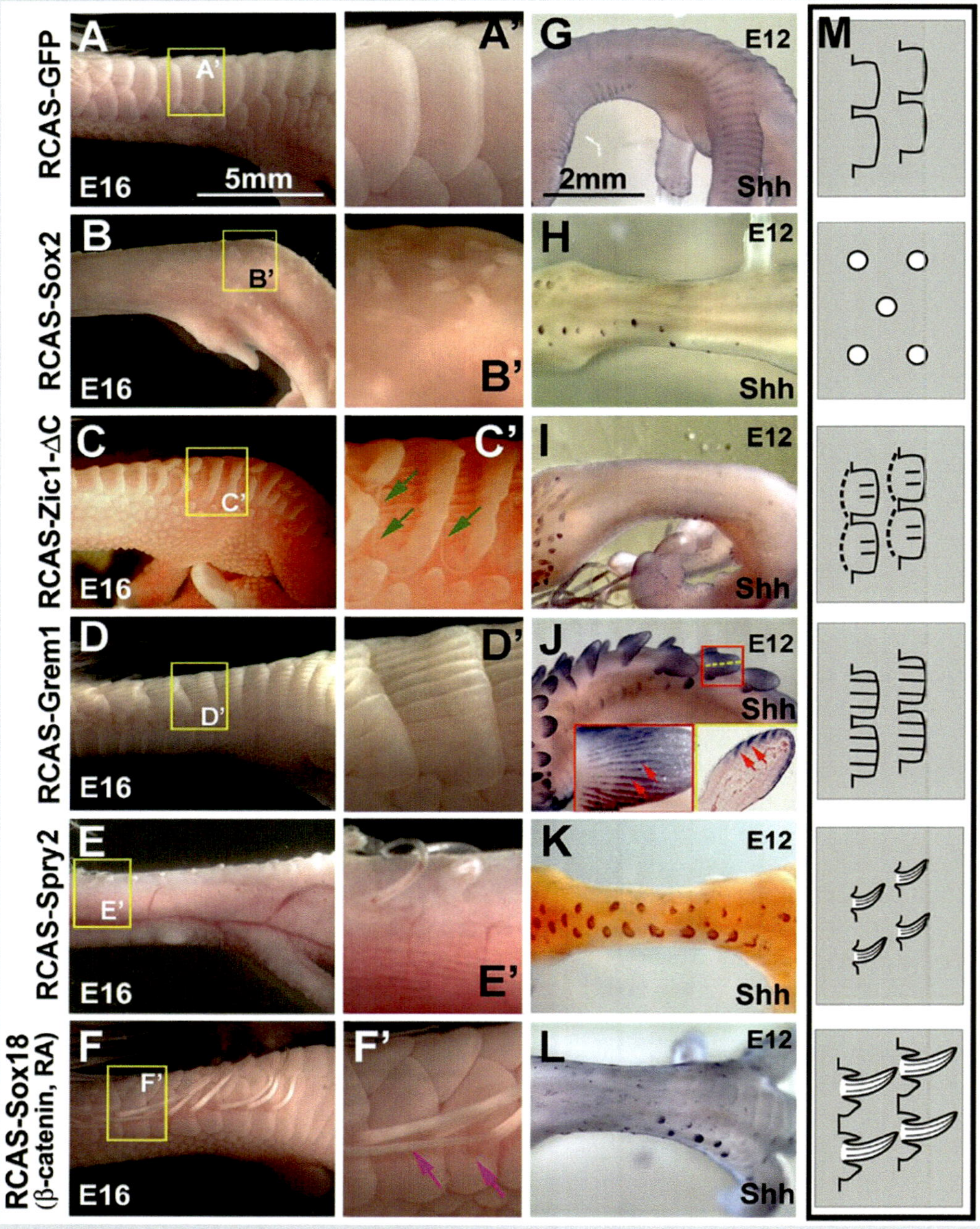


(continued)

**Box 3.8** (continued)

Different feather-development genes (indicated to the left) caused different changes to the scales on the legs of embryonic Domestic Chickens. At the top, RCAS-GFP was control and scales developed normally. In C', the green arrows indicate the invagination of scales. In F', the purple arrows point to feather filaments. Note the different feather-bud-like structures form in J, K, and L. In J, the inset with the red frame shows the enlarged bud, and the inset with the yellow frame is a longitudinal section at the point indicated by the yellow dotted line; these insets show the formation of barb ridges. Column M provides drawings of the different effects of different genes on the scales (Figure from Wu et al. 2017; © 2017 Oxford University Press, used with permission).

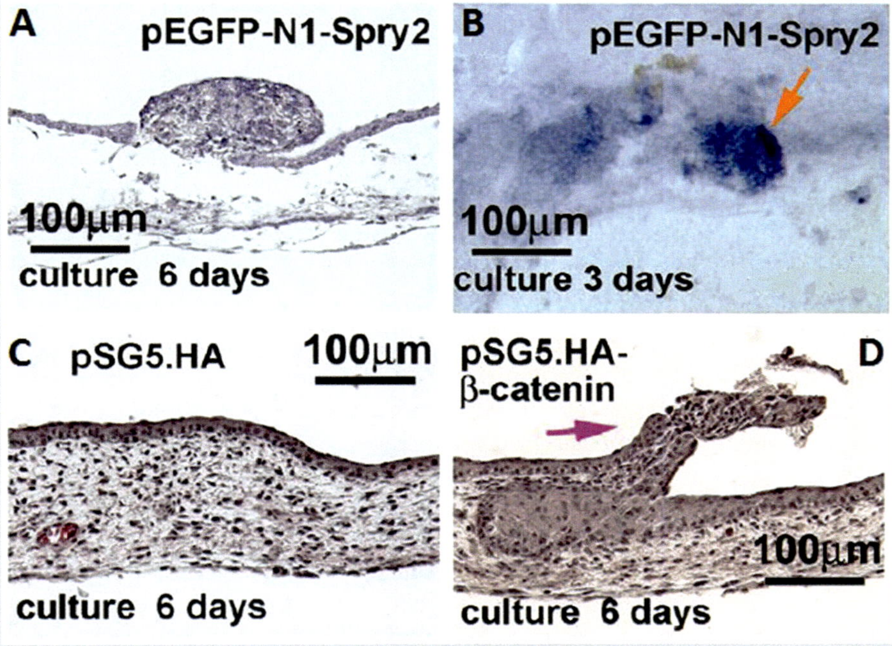


(**a**) One feather-development gene caused the formation of bud-like structures that resembled feather buds, but had no invagination or follicle formation. (**b**) Arrow points to the area where a feather-development gene is being expressed. (**c**) Control scale with normal development. (**d**) A feather-development gene caused the formation of elongated bud-like structures with some invagination (pink arrow) (Figure modified from Wu et al. 2017; © 2017 Oxford University Press, used with permission).

on the tail, may have allowed *Sinornithosaurus* to glide between perches and from elevated perches to the ground (Chatterjee and Templin 2004).

*M. gui* was covered by plumulaceous feathers about 25–30 mm long, and feathers on the top of the head were up to 40 mm long. Some feathers on the head were pennaceous and probably served a display function. Large, asymmetric pennaceous feathers were also attached to the distal tail, forelimb, and, surprisingly, the

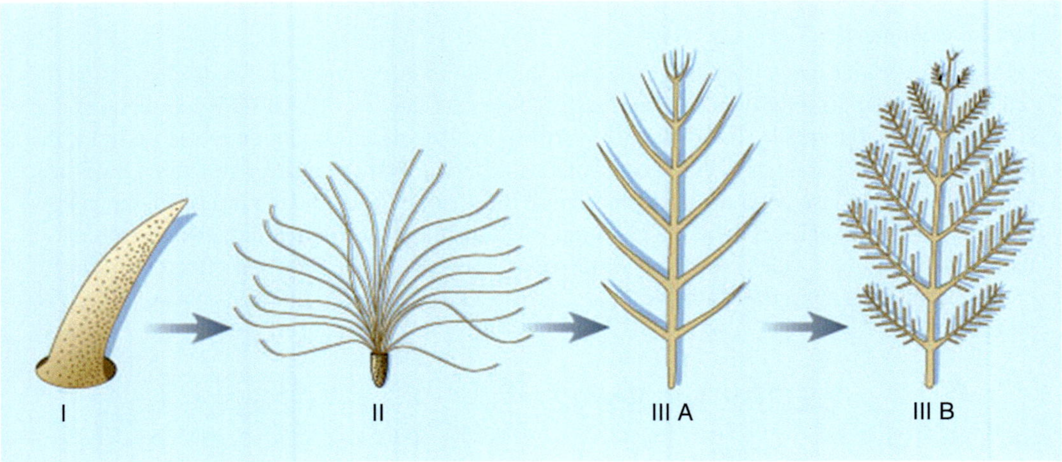


**Fig. 3.57** Hypothesized stages I–III of feather evolution. Stage I of this model assumes an unbranched, hollow filament developed from a cylindrical invagination of the epidermis around a papilla. In stage II, a tuft was formed by the fusion of several filaments at their bases. Stage III represents the formation of a central rachis and the development of serially fused barbs (III A)—to which, at a slightly later stage (III B), secondary barbs (barbules) were added. The two other stages, IV (bipinnate feathers with elaborate barbules and a closed vane) and V (the asymmetrical flight feathers of modern flying birds), are not shown. (Figure from Sues 2001; © 2001 Springer Nature, used with permission)

hindlimb. *Microraptor* had both primary and secondary flight feathers. This pattern was mirrored on the hind legs, with flight feathers attached to the upper foot bones and the upper and lower leg. When first described, Xu et al. (2003) proposed that *Microraptor* was arboreal and glided from tree to tree with four "wings"—two forelimb wings and two hindlimb wings. However, Xu et al. (2003) proposed that the legs extended out to the side (Fig. 3.65), and other investigators pointed out that such a leg position was unlikely because no known bird or theropod could extend their legs in such a manner without dislocating the hip joint (Padian 2003). Thereafter, Chatterjee and Templin (2007) proposed that the wings of *Microraptor gui* would have been split-level (like a biplane) and not spread as originally proposed, with the hindlimb flight feathers extending horizontally and able to generate lift along with the forelimb wings. *Microraptor* may have employed a phugoid (from the Greek, meaning take flight) style of gliding flight—launching itself from a perch, swooping downward in a deep U-shaped curve, and moving upward to land on a perch in another tree.

## 3.12 Feather Types and Functions

Birds have six types of feathers: contour feathers, semiplumes, filoplumes, down feathers, bristles, and powder-down feathers. All feathers have a basal calamus anchored in a follicle and barbs; the structure of different types of feathers varies. The surface plumage consists of contour feathers characterized by a long, relatively rigid rachis from which extend barbs that, along with the barbules that extend from barbs, form vanes (Box 3.11 Shape and Strength Recovery of Feathers). Most vanes are composed of stiff, pennaceous barbs that interlock and plumulaceous barbs that do not interlock and have a fluffy appearance. Pennaceous barbs are typically at the distal end of the rachis, and plumulaceous barbs are at the proximal end (Fig. 3.66). Unlike most contour feathers, remiges

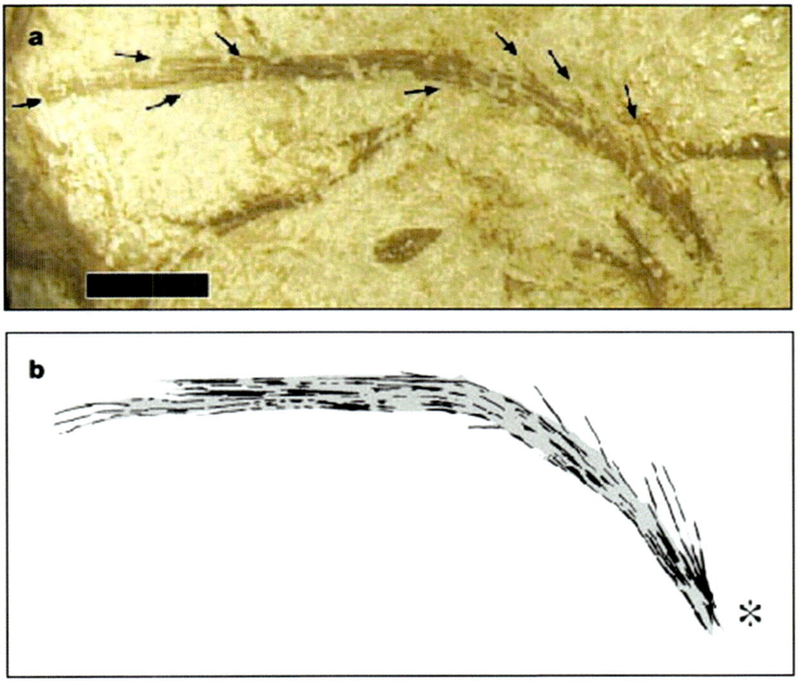


**Fig. 3.58** The filamentous integumental structure of *Sinornithosaurus millenii* has compound structures composed of multiple filaments. These structures exhibit two types of branching structures unique to avian feathers: the filaments are joined in a basal tuft, and the filaments are joined at their bases in series along a central filament. **a**, Arrows indicate the distal tips of some component filaments. **b**, Drawing of the structure showing the positions of the observed filaments (lines) and the inferred outline of the appendage (shading). Asterisk, the proximal end of the appendage. The curved position of the appendage reveals its compound structure. Each filament converges on the center of the appendage at its base. Scale bar, 5 mm. (Figure from Xu et al. 2001; © 2001 Springer Nature, used with permission)

**Fig. 3.59** Skeletal reconstruction of *Sinornithosaurus millenii*. (Figure from Wikipedia, CC BY 3.0, https://creativecommons.org/licenses/by/3.0/)

**Fig. 3.60** Three-dimensional virtual reconstruction of a fossil feather from the early Cretaceous (about 100 million years ago) preserved in amber. This feather could be from either a bird or a nonavian theropod. (**a–c**) long barbs form two vanes on each side of a relatively flattened shaft; (**d**) the shaft is flattened and composed of incompletely fused bases of the barbs, a stage in feather evolution that was hitherto unknown in fossil records and corresponding to an intermediate stage between the very distinct stages II and IIIa defined by Prum (1999). Scale bars, 100 μm. (Figure from Perrichot et al. 2008; © 2008 The Royal Society, used with permission)

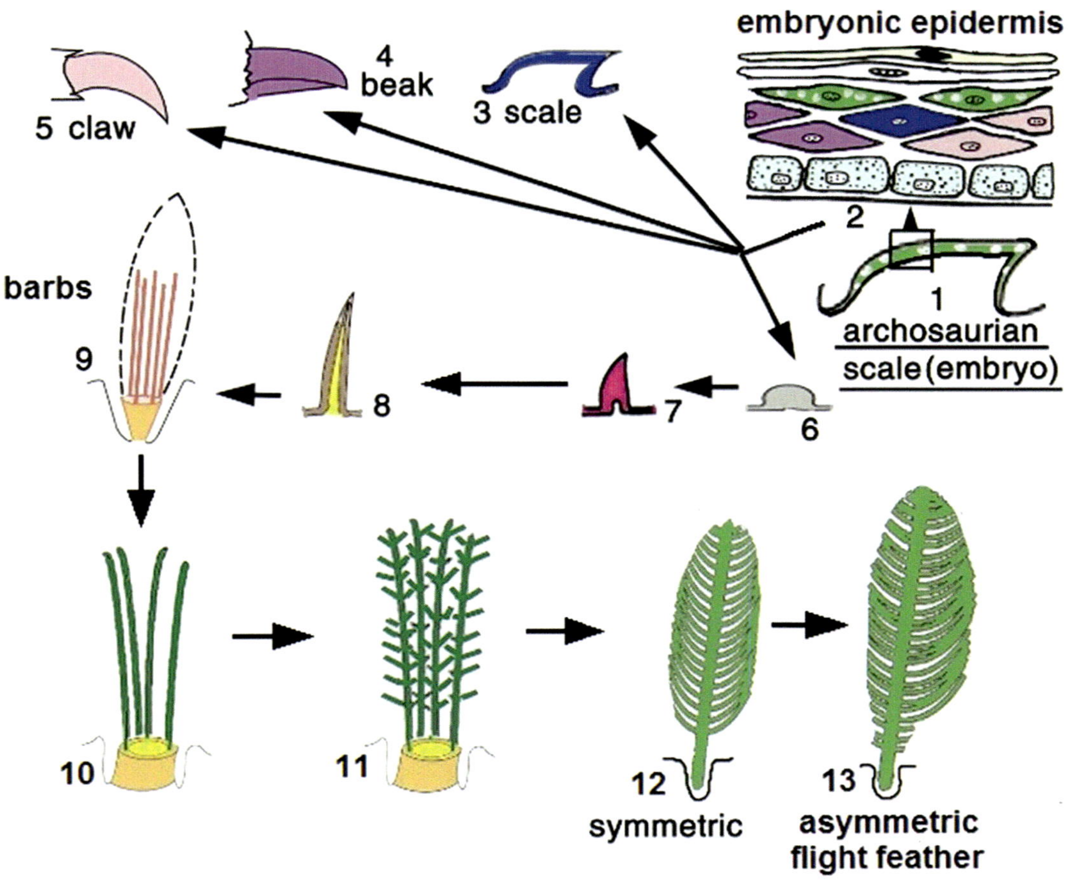


**Fig. 3.61** Evolution of feathers from the scaled integument of basal archosaurs. (1) Embryonic epidermis can (2) activate genes to generate (3) scale, (4) beak, or (5) claw beta-keratins. (6) From a tuberculate scale, (7) a cone-shaped appendage became elongated and, later, (8) developed a vascular mesenchymal core (yellow). Then, (9) one or more barb ridges (shown in red) were formed to generate (10) down feathers. (11) Further elaboration of barb ridges resulted in barbules. (12) Merging of barb ridges produced the feather shaft and, with the development of barbicels, different types of pennaceous feathers evolved. (13) As the ancestors of birds took to the air, so did the evolution of asymmetric flight feathers. (Figure modified from Alibardi 2017; © 2016 Springer-Verlag Wien, used with permission)

and rectrices typically have few, if any, plumulaceous barbs. Vanes that consist of interlocked pennaceous barbs form flat sheets. The distal barbules of one barb overlap the proximal barbules of the next barb, with the hooklets of the distal barbs grasping the dorsal flanges of the proximal barbules and "locking" the barbules together (Fig. 3.67).

Contour feathers serve several functions. The flight feathers of the wings (remiges) have a rigid rachis and well-developed vanes. These feathers, along with the wing coverts, transform bird wings into cambered airfoils that generate lift; the primary remiges (with obviously asymmetrical vanes) also serve to propel birds forward (or, for hummingbirds, backward) during flapping flight. The tail feathers, or rectrices, are also contour feathers that can help create lift and, when fanned out during landing, function as an air brake. The numerous contour feathers that

**Box 3.9 Feathers from the Mid-Cretaceous**

A piece of amber with a portion of the tail (3.7 cm long) of a nonavian theropod preserved inside was discovered in a market in Myitkyina, Myanmar, in 2015. The partial tail belonged to a recently hatched coelurosaur that got stuck in some tree resin 99 million years ago (mid-Cretaceous). The partial tail includes some caudal vertebrae, soft tissue features (presumably muscle, ligaments, and skin that have been reduced to a carbon film), plus several feathers. The feathers have a thin rachis with barbs extending asymmetrically from each side. Barbules branch off from the barbs, but, in addition, from the rachis as well (rachidal barbules). Based on our current understanding of feather evolution, the feathers of this young coelurosaur were at an intermediate stage between stages IIIa (rachis with naked barbs) and IIIb (barbs with barbules, but no rachis). In the direct ancestors of birds, barbules became restricted to the barbs and just the proximal portion of the rachis, perhaps to allow an increase in the number and density of barbs. Thus, these coelurosaur feathers suggest that nonavian theropods had a greater variety of feather morphologies than expected (Xing et al. 2016).

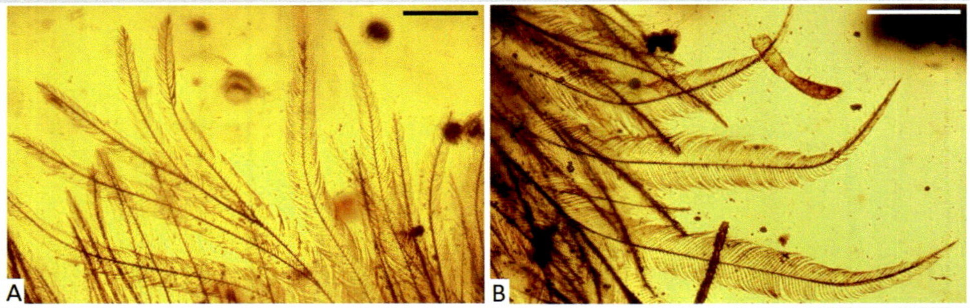


(**a**, **b**) Photomicrographs of the feathers of a nonavian theropod from the mid-Cretaceous. Feathers have slender rachis with alternating barbs and a series of barbules. Scale bar in **a**, 1 mm; Scale bar in **b**, 0.5 mm (Figure modified from Xing et al. 2016; © 2016 Elsevier Ltd., used with permission).

(continued)

**Box 3.9** (continued)

Drawing and photo of one of the largest feathers. (**a**) Barbules have been omitted on the upper side and on one side of a barb (black arrow) to show a relationship between barbs and barbules. Note that barbules branch directly from the rachis as well as from the barbs. (**b**) White arrows point to the base of two feathers (Figure from Xing et al. 2016; © 2016 Elsevier Ltd., used with permission).

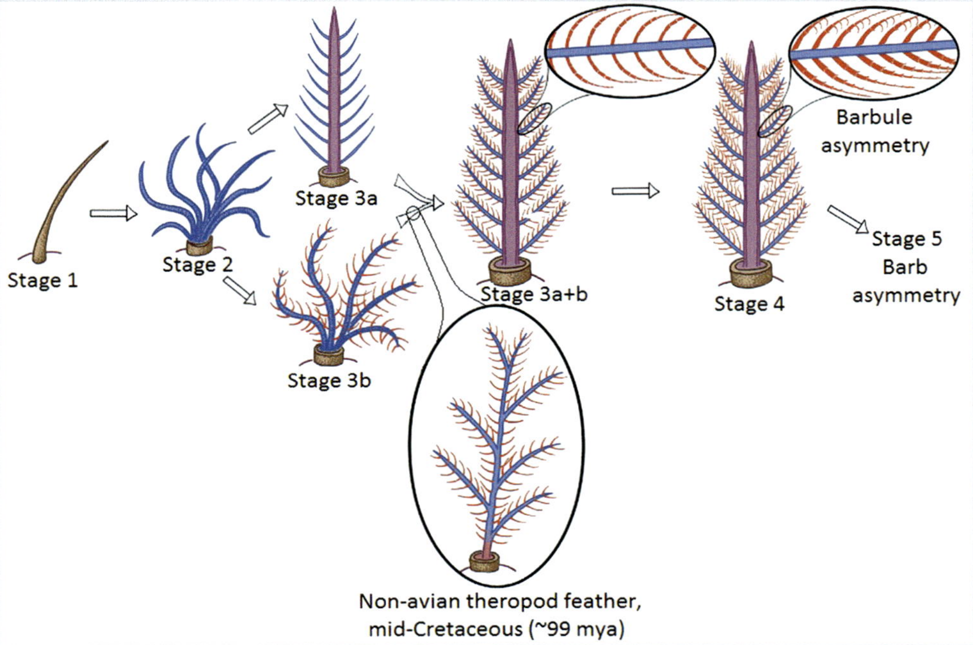


Model illustrating possible steps in the evolution of feathers. The feathers of the nonavian theropod represent an intermediate stage between Stage 3b and Stage 3a+b. Brown indicates the calamus, purple the rachis, blue the barb, and red the barbules (Figure modified from Xing et al. 2016; © 2016 Elsevier Ltd., used with permission).

cover most of the rest of a bird's body give flying birds a streamlined shape important in reducing friction or drag as they move through the air. Contour feathers also play an important role in thermoregulation (see Chap. 10) and, depending on their color, contour feathers can also have important display functions or help camouflage birds.

Down feathers are entirely plumulaceous, and the rachis is either short (shorter than the longest barb) or absent. The barbules of downy feathers are slender and flexible, with small outgrowths that help keep them from becoming entangled and matted (Stettenheim 2000; Figs. 3.68, 3.69, and 3.70). As a result, down feathers remain "fluffy" and are better able to trap air in the plumage for thermal insulation (Stettenheim 2000). The natal down of hatchling birds is generally simpler in structure than adult, or definitive, down, with barbs that have fewer barbules. Down feathers

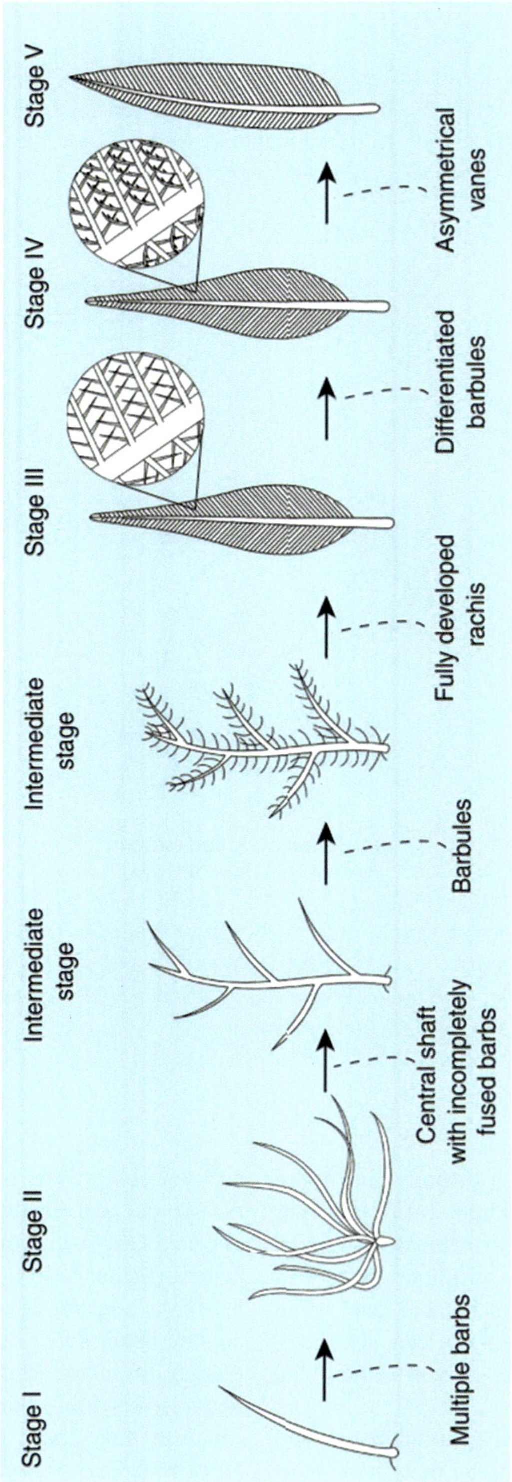


**Fig. 3.62** Summary of the five possible stages of feather evolution, with two intermediate stages that have been observed in feathers preserved in amber. (Figure from Ksepka 2020; © 2020 Elsevier Inc., used with permission)

**Box 3.10 Water and Ice Repellency of Contour Feathers**

The hydrophobicity, or water repellency, of a porous material, like feathers, is enhanced because drops of water sit partially on trapped air (Cassie and Baxter 1944). The contour feathers of birds are hydrophobic mainly because of the width and spacing of barbs and barbules that minimizes contact between water and hydrophilic keratin and maximizes contact between water and air. However, this hydrophobicity does not mean that contour feathers cannot be cleaned by contact with water because water droplets that bead up and roll off the surface carry surface contaminants with them (Zhang et al. 2008a, b).

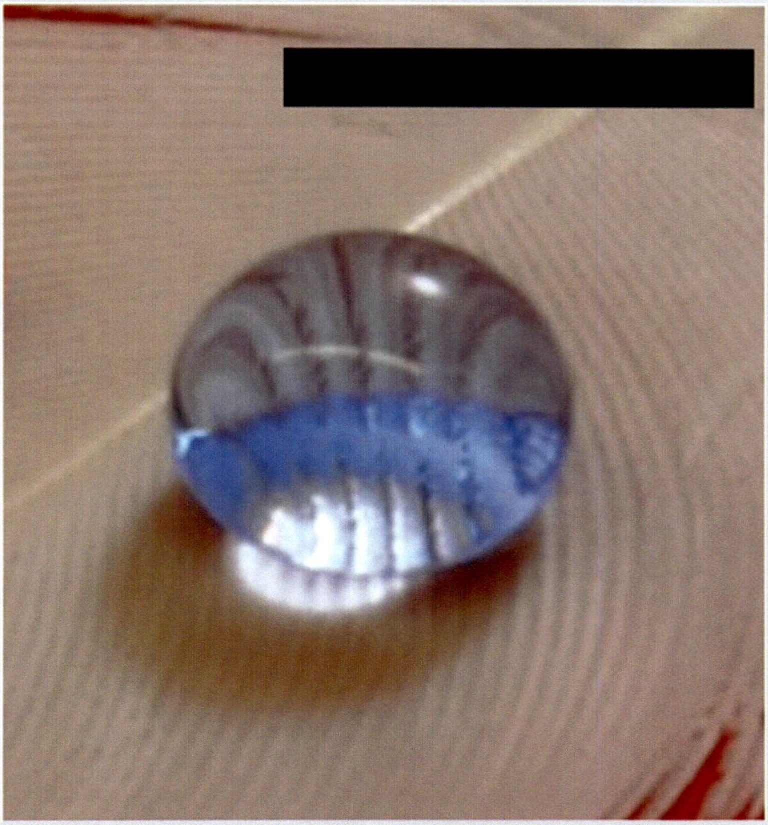


Drop of water on a contour feather of a Common Shelduck (*Tadorna tadorna*). Scale bar = 0.5 cm (Figure from Srinivasan et al. 2014; © 2014 The Authors. Published by the Royal Society, used with permission).

(continued)

**Box 3.10** (continued)

Drop of water on the edge of cut barbs from a contour feather of a Rock Pigeon (*Columba livia*) (Figure from Bormashenko et al. 2007; © 2007 Elsevier Inc., used with permission).

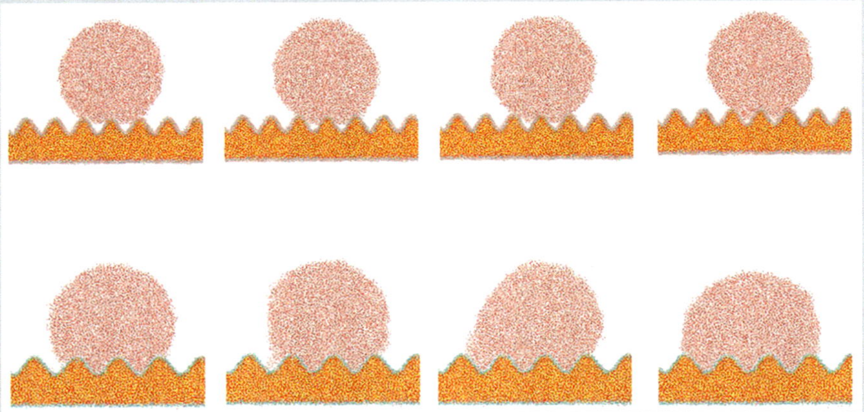

Contour feathers are hydrophobic due to the specific width and spacing of barbs and barbules. To illustrate, note how changing the width and spacing of "barbs" in this series of illustrations (from upper left to lower right) gradually changes the "feather" from hydrophobic (with air between the water drop and the space between "barbs") to wettable (with water in direct contact with barbs and barbules) (Figure from Katasho et al. 2015; open-access article distributed under the terms of the Creative Commons CC BY 4.0 license, https://creativecommons.org/licenses/by/4.0/).

Trapped air is critical for maintaining a feather's hydrophobicity. However, among diving birds, water pressure increases with increased depth, and that pressure may be sufficient to "collapse" the protective layer of air. If so, water can then penetrate into the feather. However, Srinivasan et al. (2014) found that, if this happens to some species of cormorants (Phalacrocoracidae), a thin layer of hydrophobic preen oil that coats a feather's surface prevents water from fully penetrating the barbs and barbules. As a result, when a bird returns to the surface

(continued)

**Box 3.10** (continued)

and leaves the water, the great reduction in pressure means that most of the water "trapped" in the plumage is quickly ejected (a process referred to by Srinivasan et al. [2014] as "spontaneous dewetting"). The "wing-spreading" behavior of cormorants may facilitate this "dewetting" process. The feathers of other diving birds may respond in a similar manner.

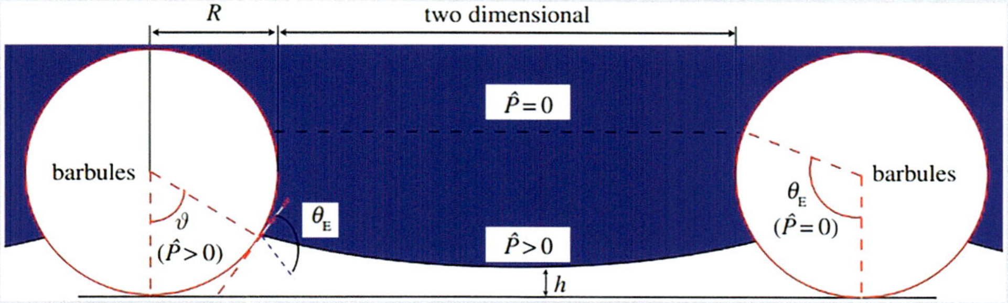


With increasing water depth and increasing water pressure, water is forced further down along the barbules and closer to the skin's surface (solid line at the bottom). The dashed line indicates the position of water relative to barbules near the water's surface; the solid line indicates the position of the water relative to barbules at a depth of 30 m where water pressure is much higher. With additional pressure, the protective air layer may collapse. *R*, radius of barbule; $P = 0$, pressure near the surface; $P > 0$, pressure at a depth of 30 m; $\theta_E$, the equilibrium contact angle of the liquid; $\vartheta$, the angular coordinate of the location of the contact line at a given pressure; *h*, altitude of the bottom of the curved meniscus about the skin (Figure from Srinivasan et al. 2014; © 2014 The Authors. Published by the Royal Society, used with permission).

(continued)

**Box 3.10** (continued)

The spread-wing posture of cormorants, like this Double-crested Cormorant (*Microcarbo melanoleucos*), may aid in the "dewetting" of their feathers (Photo by Alan Levine, pxhere.com, CC BY 2.0, https://creativecommons.org/licenses/by/2.0/).

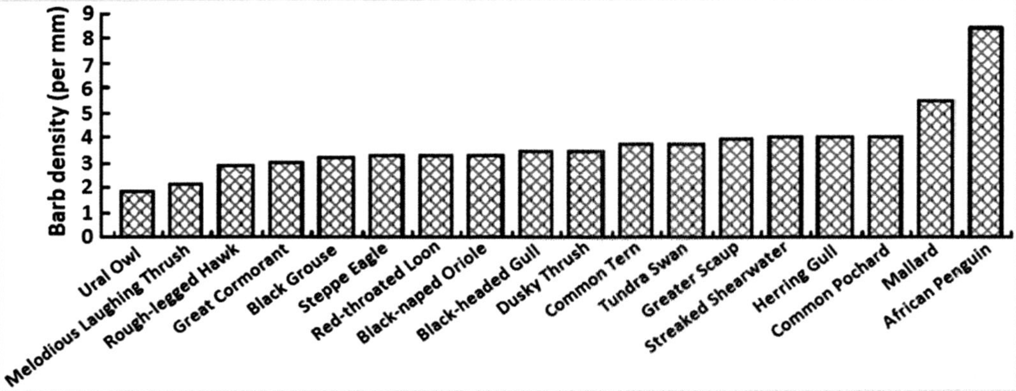


As already noted, the hydrophobicity of contour feathers depends on the specific width and spacing of barbs and barbules. In general, a greater density of barbs and barbules means greater hydrophobicity, and so, not surprisingly, aquatic birds typically have feathers with greater barb densities than terrestrial birds. Ural Owl, *Strix uralensis*; Melodious Laughing Thrush (also referred to as Chinese Hwamei), *Garrulax canorus*; Rough-legged Hawk, *Buteo lagopus*; Great Cormorant, *Phalacrocorax carbo*; Black Grouse, *Tetrao tetrix*; Steppe Eagle, *Aquila nipalensis*; Red-throated Loon, *Gavia stellata*; Black-naped Oriole, *Oriolus chinensis*; Black-headed Gull, *Larus ridibundus*; Dusky Thrush, *Turdus naumanni*; Common Tern, *Sterna hirundo*; Tundra Swan, *Cygnus columbianus*; Greater Scaup, *Aythya marila*; Streaked Shearwater, *Puffinus leucomelas*; Herring Gull, *Larus argentatus*; Common Pochard, *Aythya ferina*; Mallard, *Anas platyrhynchos*; African Penguin, *Spheniscus demersus* (Figure modified from Shu-hui et al. 2006; © 2006 Northeast Forestry University and Ecological Society, used with permission).

The contour feathers of most aquatic birds are, to varying degrees, hydrophobic. Despite that, those feathers can potentially be coated with ice if water droplets on feathers freeze. Wang et al. (2016b) found that the width and spacing of barbs and barbules in the contour feathers of Humboldt Penguins (*Spheniscus humboldti*), in combination with nanoscale irregularities and grooves along the surface of barbules, reduced the ability of ice to adhere. Similar results have been reported for Gentoo Penguins (*Pygoscelis papua*; Wood et al. 2023). In addition, trapped air, a critical feature of hydrophobic feathers, is warmed by the underlying skin, which also helps prevent ice from adhering to feathers.

(continued)

**Box 3.10** (continued)

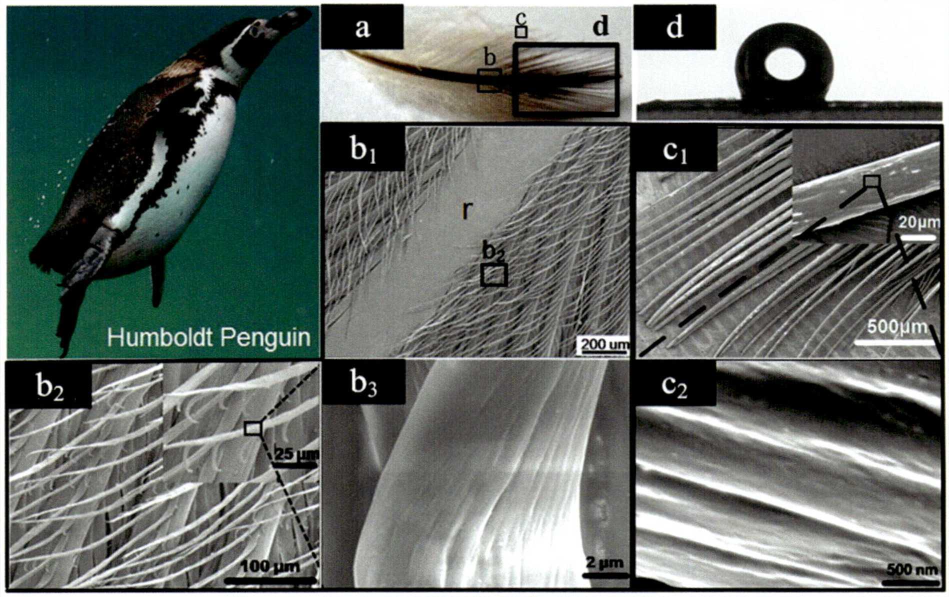


Micro- and nanostructures of the body feathers of a Humboldt Penguin. (a) A body (contour) feather, ($b_1$) electron micrograph of the rachis (r) and barbs, ($b_2$) barbules with the hamuli (hooklets), ($b_3$) wrinkles on the barbules and hamuli, ($c_1$) tips of barbules without hamuli, ($c_2$) oriented nanoscaled grooves on the barbules (~100 nm deep), and (d) water droplet on feather demonstrating hydrophobicity (Figure modified from Wang et al. 2016b; Reprinted (adapted) with permission from Wang et al. 2016b. Copyright 2016 American Chemical Society).

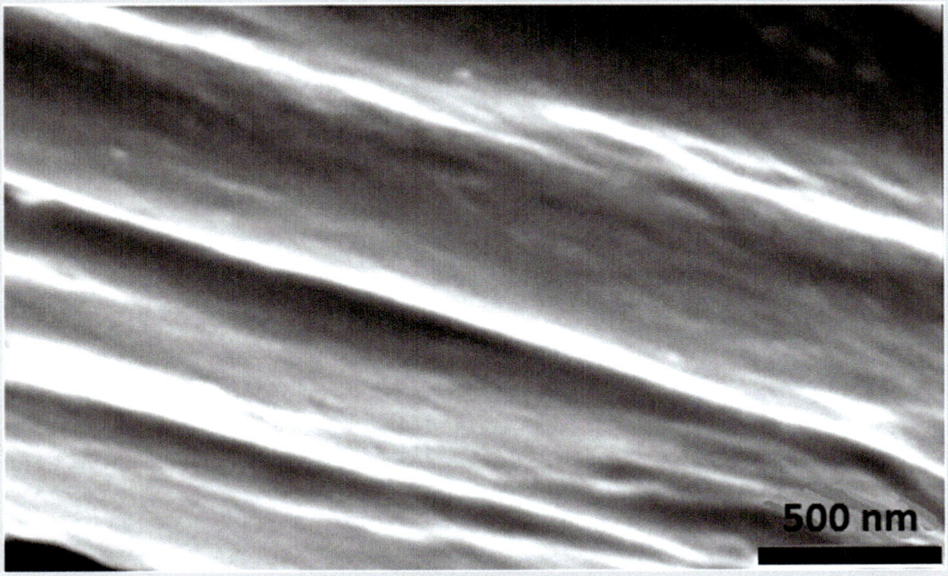


(continued)

**Box 3.10** (continued)

A more highly magnified view of the nanoscaled grooves on the barbules of Humboldt Penguin contour feathers (Figure from Wang et al. 2016b; Reprinted (adapted) with permission from Wang et al. 2016b. Copyright 2016 American Chemical Society).

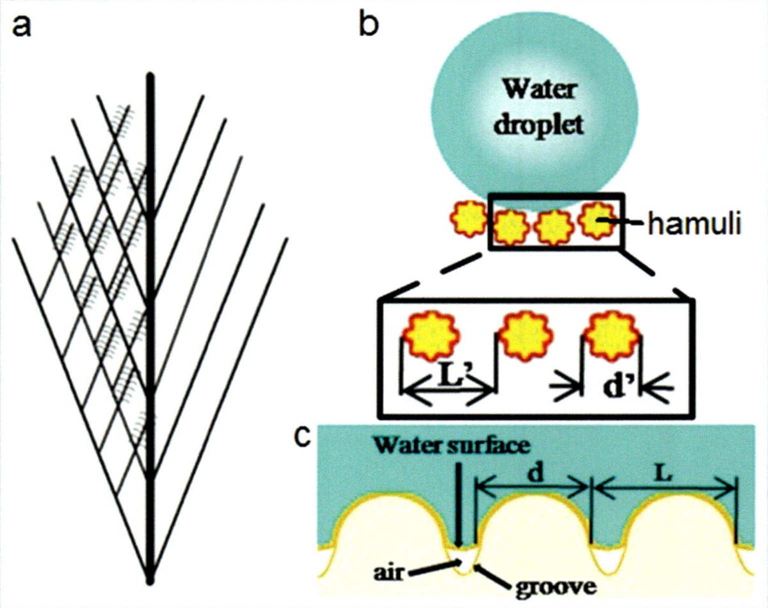


Model illustrating the micro- and nanostructures of the body feathers of a Humboldt Penguin. (**a**) Diagram showing a rachis, barbs, barbules, and hamuli (hooklets), (**b**) contact between a water microdroplet and hamuli, and (**c**) contact between water microdroplets and nanogrooves. L' = mean distance between hamuli; d' = mean diameter of the hamuli (Figure modified from Wang et al. 2016b; Reprinted (adapted) with permission from Wang et al. 2016b. Copyright 2016 American Chemical Society).

are not evenly distributed on the body, depending on a bird's need for insulation. For example, the apteria of finches that occur at higher latitudes is densely covered with down feathers, whereas those of finches at lower latitudes have no down feathers (West 1962). Two high-latitude species, Hoary (*Acanthis hornemanni*) and Common (*A. flammea*) redpolls, have down feathers in the apteryia, but, when temperatures increase, and there is less need for insulation, they pluck those feathers (Brooks 1968). During the breeding season, female ducks and geese pluck down feathers from their breasts and abdomen and add the feathers to their nests to provide insulation for eggs, and help conceal eggs when females are absent from the nest during incubation.

The first plumage of most young birds (with the exception of young megapodes that have contour feathers at hatching) is natal down (also called neoptile feathers, Foth 2011; Figs. 3.71, 3.72, 3.73, 3.74, and 3.75). Both the amount and color of natal down feathers vary among young birds. Although there is variation among species in the amount of down present, in general, the

**Fig. 3.63** *Epidexipteryx hui* fossil showing the ribbon-like tail feathers that likely served as ornaments and played a role in intra- and intersexual interactions. (**a**) The main slab with an arrow pointing to impressions of elongated tail feathers. (**b**, **c**) Close-up of a skull in main slab and counter-slab. (**d**) Close-up of a section of the tail feathers. (**b'**, **c'**) line drawings of teeth in upper and lower jaws, respectively. (Figure from Zhang et al. 2008a, b; © 2008 Springer Nature, used with permission)

young of altricial species hatch with little or no down, whereas the young of semialtricial, semiprecocial, and precocial species are covered with down at hatching (Starck and Ricklefs 1998). Downy plumage varies in color. For example, young rails have black down, young swans have white down, young Domestic Chickens have yellow down, young eiders have brown down, and the downy plumage of many young birds is mottled or striped (Kilner 2006). Down color in most species of birds is rather drab, likely functioning to camouflage young and reduce predation risk. Among species where downy plumage is largely light or dark, plumage color may have a thermoregulatory function (Stoutjesdijk 2002). The downy plumage of some ducks and geese is dark above and light below and this countershading may provide camouflage and protection from both aerial and aquatic predators (Bradbury and Vehrencamp 1998).

Semiplumes combine features of both pennaceous contour feathers and plumulaceous downy feathers, with long rachis and plumulaceous vanes (Fig. 3.76). Semiplumes are found under contour feathers, providing insulation and helping form the smooth, aerodynamic body shape of flying birds (Proctor and Lynch 1993). Unlike downy feathers, semiplumes are found only in and at the margins of tracts of contour feathers (Hudon 2005). Not surprisingly, given their importance in providing insulation, Osváth et al. (2018), based on feathers sampled from 152 species of birds, found a negative relationship between the combined mass of down and semiplume feathers and minimum winter temperature (Fig. 3.77).

**Fig. 3.64** Fossil of *Microraptor gui* showing areas where imprints of feathers can be seen (white arrows) and areas where the imprints are absent (black arrows). Scale bar = 5 cm. (Figure from Hone et al. 2010; © 2010 Hone et al., open-access article distributed under the terms of the Creative Commons Attribution License, https://creativecommons.org/licenses/by/4.0/)

**Fig. 3.65** A computer-generated reconstruction of *Microraptor gui* in gliding flight. The hindlimbs of *Microraptor* would likely not have been extended as much to the side as shown (especially the left hindlimb) in this reconstruction. (Illustration credit: MR1805, purchased from istockphoto.com)

Filoplumes are hair-like feathers, with a narrow rachis and a tuft of short barbs at the tip (Figs. 3.78 and 3.79). Filoplumes are widely distributed on a bird's body and are usually hidden by contour feathers (Figs. 3.80 and 3.81). For example, filoplumes are found under the contour feathers in all feather tracts of House Wrens (*Troglodytes aedon*), and there is at least one filoplume at the base of each of the remiges and rectrices (Boulton 1927). The follicles of filoplumes are typically tightly anchored to the follicles of contour feathers, and vibration-sensitive mechanoreceptors (primarily Herbst corpuscles) are located near the filoplume follicles (Fig. 3.82). These receptors respond to feather movement and provide birds with information about feather position. For example, mechanoreceptors in the filoplume follicles associated with the secondary remiges respond to vibrations of the secondaries caused by changes in the velocity of airflow over the wing and supply a bird with information about flight speed (Brown and Fedde 1993). Filoplumes associated with contour feathers on the head and nape that, in some species in the orders Procellariformes and Passeriformes protrude beyond the contour feathers (Fig. 3.83), likely provide birds with information about the position of contour feathers in areas not visible to birds and not easily accessible for preening with the bill. Wind can alter feather position on the head and nape that might, for example, increase heat loss. Based on sensory information from mechanoreceptors in the filoplume follicles, birds can quickly restore the integrity of the feather coat by, for example, turning into the wind (Clark and de Cruz 1989).

Bristles are hair-like feathers with stiff, tapered rachis and few or no short barbs near the base. Some birds also have "semibristles," similar to bristles, but have more barbs (Fig. 3.84). Most, but not all (e.g., Rock Pigeons, *Columba livia*), birds have bristles, and they are generally located around the base of the bill and nares, the lores, malar region (cheek and below the eyes), and rictal (corner of the mouth; Fig. 3.85) regions, and sometimes resemble eyelashes (Proctor and Lynch 1993; Cunningham et al. 2011). Bristles and semibristles located around the eyes and nostrils (narial bristles; Fig. 3.85) likely serve a protective function, deflecting objects to protect the eyes (Fig. 3.86) and preventing particles from entering the nostrils. Some investigators have suggested that bristles around the base of the bill (rictal bristles) might facilitate the capture of prey by aerial insectivores by creating and larger, effective gap and "funneling" insects into the mouth. However, examination of species in several bird families reveals no apparent relationship between the presence, dimensions, or number of rictal bristles and tendencies toward aerial foraging. In addition, analysis of prey-capture techniques of several species of flycatchers revealed that insects are captured using the bill, and typically at the bill tip, rather than being "funneled" into the oral cavity (Lederer 1972). A more likely function for rictal bristles is that they help protect the eyes of aerial insectivores.

**Box 3.11 Shape and Strength Recovery of Feathers**

Feather shafts (calamus and rachis) are composed of keratinized cortical cells (dead epidermal cells) that, in turn, consist of macrofibrils composed of β-keratin fibers embedded in amorphous matrix proteins. The amorphous matrix material can be considered a shape memory polymer, a substance that can recover its shape under the right conditions. For the amorphous matrix material in feathers, recovering its shape requires exposure to water.

(continued)

**Box 3.11** (continued)

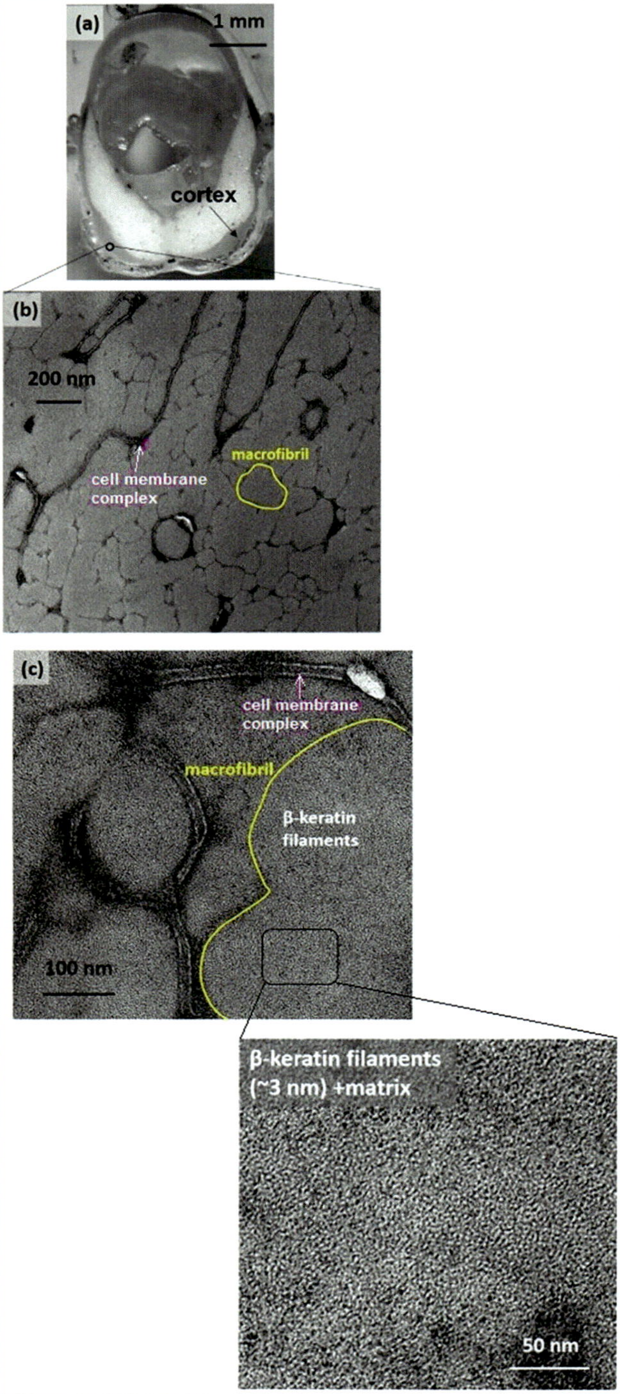


Structure of the wall (cortex) of the rachis of a feather. (**a–c**) Electron micrographs of sections showing the cortex, keratinized cortical cells separated by a cell membrane complex,

(continued)

**Box 3.11** (continued)

macrofibrils (outlined in yellow) composed of multiple β-keratin filaments, and β-keratin filaments embedded in amorphous matrix proteins (Figure modified from Wang and Meyers 2017; © 2016 Acta Materialia Inc. Published by Elsevier Ltd., used with permission).

Sullivan et al. (2018) examined the effect of hydration on the flight feathers of Cape Vultures (*Gyps coprotheres*) after being deformed. They found that, after exposure to water for 24 h, the rachis of these feathers recovered about 86% of their bending stiffness. In addition, the amorphous matrix material was found to swell when exposed to water, causing the rachis to return to its original shape and, after drying, regain much of its strength. For free-living birds, these results suggest that exposure to water, e.g., when bathing, may help deformed feathers regain much of their stiffness and strength.

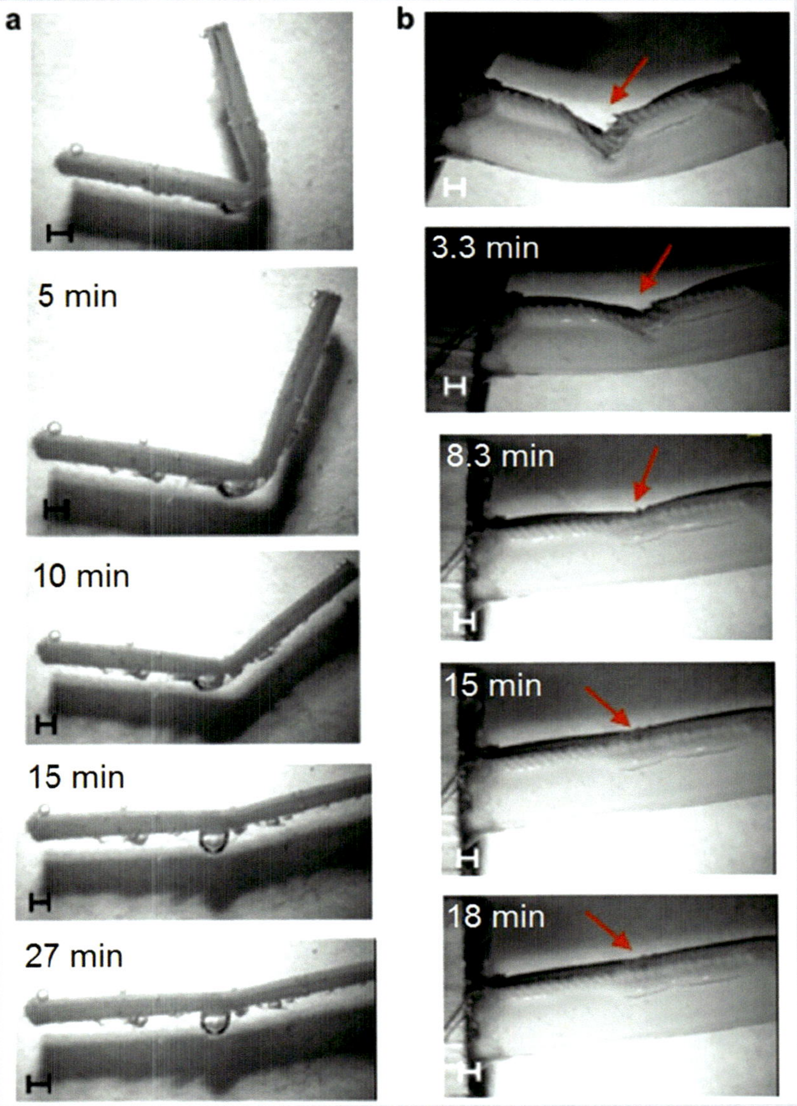


(continued)

**Box 3.11** (continued)

Recovery of the rachis of a feather in water. (**a**) A strip of the feather shaft cortex is severely deformed but recovers in water. (**b**) Section of a rachis (with the vane removed) is bent and recovers in water; red arrows indicate the location of deformation. Time is indicated in seconds on the top left of each image. Scale bars = 1 mm (Figure from Sullivan et al. 2018; © 2018 WILEY-VCH Verlag GmbH & Co. KGaA, Weinheim, used with permission).

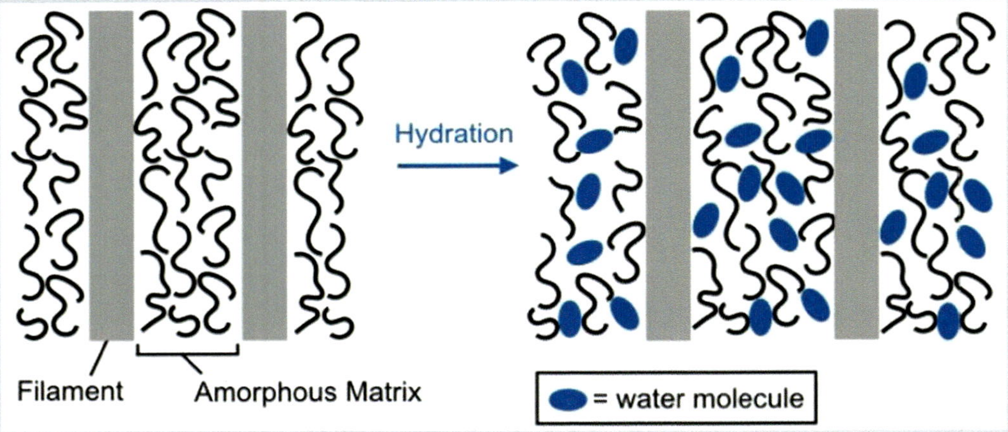


Drawings showing swelling of the amorphous matrix (consisting of protein) with the addition of water. The addition of water molecules causes the amorphous matrix to swell, and this, in turn, would cause the β-keratin filaments to straighten if they had been deformed (Figure is a supplemental figure [3] from Sullivan et al. 2018; © 2018 WILEY-VCH Verlag GmbH & Co. KGaA, Weinheim, used with permission).

Wind tunnel experiments with Willow Flycatchers (*Empidonax traillii*) revealed that rictal bristles prevented food items (missed prey or prey items that break apart) from striking the eyes; specimens with bristles removed had more particles hit the eyes than specimens with intact bristles (Conover and Miller 1980).

Some species of birds have longer bristles on the face and head that likely serve a tactile function. For example, rictal bristles as long as 41.38 mm have been reported in Silky-tailed Nightjars (*Antrostomus sericocaudatus*), and those of Oilbirds (*Steatornis caripensis*) averaged 39 mm in length (Delaunay et al. 2022). Some investigators have suggested that the bristles of some burrow- and hole-nesting species are important for tactile navigation, with bristles serving as mechanosensory structures and providing information about a bird's head position relative to the walls of burrows and cavities (Lucas and Stettenheim 1972). Although not bristles, Seneviratne and Jones (2008) showed experimentally that long contour feathers on the heads of Whiskered Auklets (*Aethia pygmaea*; Fig. 3.87) have a sensory function, helping auklets detect and avoid obstacles in a maze that simulated their dark, underground breeding crevices. Bristles likely have a similar function for some species of birds, for example, nocturnal species and burrow-, crevice-, and cavity-nesting species. In support of this hypothesis, Cunningham et al. (2011) found that three species of nocturnal and hole-nesting species (North Island Brown Kiwi [*Apteryx mantelli*], Morepork [*Ninox*

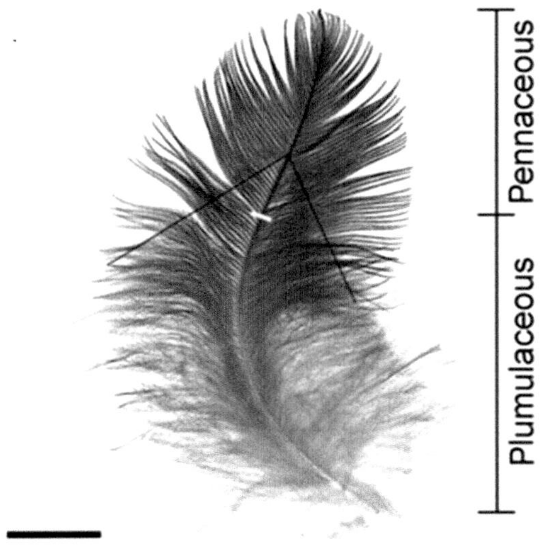


**Fig. 3.66** A contour feather showing the pennaceous and plumulaceous portions above and below the white strip on the rachis, respectively. Black lines represent the boundaries between the pennaceous and plumulaceous section of the vanes. (Figure from Pap et al. 2017; © 2016 The Authors. Functional Ecology © 2016 British Ecological Society, used with permission)

*navaeseelandiae*, an owl], and Stitchbird [*Notiomystis cincta*, cavity-nester]) had significantly more encapsulated Herbst corpuscles (mechanoreceptors) associated with their bristle follicles than did two diurnal open-nesting species (New Zealand Fantail [*Rhipidura fuliginosa*] and South Island Robin [*Petroica australis*]). However, the presence of Herbst corpuscles associated with the bristle follicles of diurnal open-nesting species suggests that their bristles may also have a sensory function. Cunningham et al. (2011) suggested that the rictal bristles of insectivorous birds might provide tactile information useful in manipulating captured prey, and this might be particularly important for nocturnal aerial insectivores like species in the family Caprimulgiformes (Fig. 3.88). These authors also suggested that facial bristles could provide flying birds with information about airflow around the head and, for aerial insectivores, information about the position of potential prey based on air movements caused by flying prey.

Powder-down feathers (pulviplumes) are found under contour feathers, typically on the breast, belly, and sides, in several species of birds in six different families, including pigeons (Columbidae), herons (Ardeidae), Marabou Storks (*Leptotilos crumeniferus*; Ciconiidae), parrots (Psittacidae), cockatoos (Cacatuidae), and bustards (Otididae) (Stettenheim 1972; Delhey et al. 2007). These feathers grow continuously and disintegrate into a talcum-like powder that consists of keratinized cells that are presumptive barbule cells stuck together by an adherent material (Figs. 3.89 and 3.90) that is lipoid in nature (Menon and Menon 2000). The powder is usually colorless, but several species of herons (Pelecaniformes: Ardeidae) have powder-down feathers that are yellow due to carotenoid pigmentation (Thomas and McGraw 2018; Fig. 3.89). When preening, birds sometimes use their bill or head to apply powder to the rest of their plumage. The powder appears to be important for feather maintenance and waterproofing, similar in function to secretions of uropygial glands. In fact, Johnston (1988) noted that the uropygial glands of birds with powder-down feathers might be absent or reduced in size.

## 3.13 Pterylae and Apteria

With the exception of their feet and tarsus (and, of course, the bill and eyes), the bodies of most birds are entirely covered by feathers. Exceptions include vultures, and a few other birds, such as Marabou Storks (*Leptoptilos crumenifer*) and African Sacred Ibis (*Threskiornis aethiopicus*), that lack feathers on their heads and, sometimes, necks. The number and size of feathers vary with the size of birds, with total numbers ranging from fewer than 1000 for small hummingbirds to more than 25,000 for swans (Stettenheim 2000). However, feathers (specifically their follicles), with the exception of down feathers, are usually confined

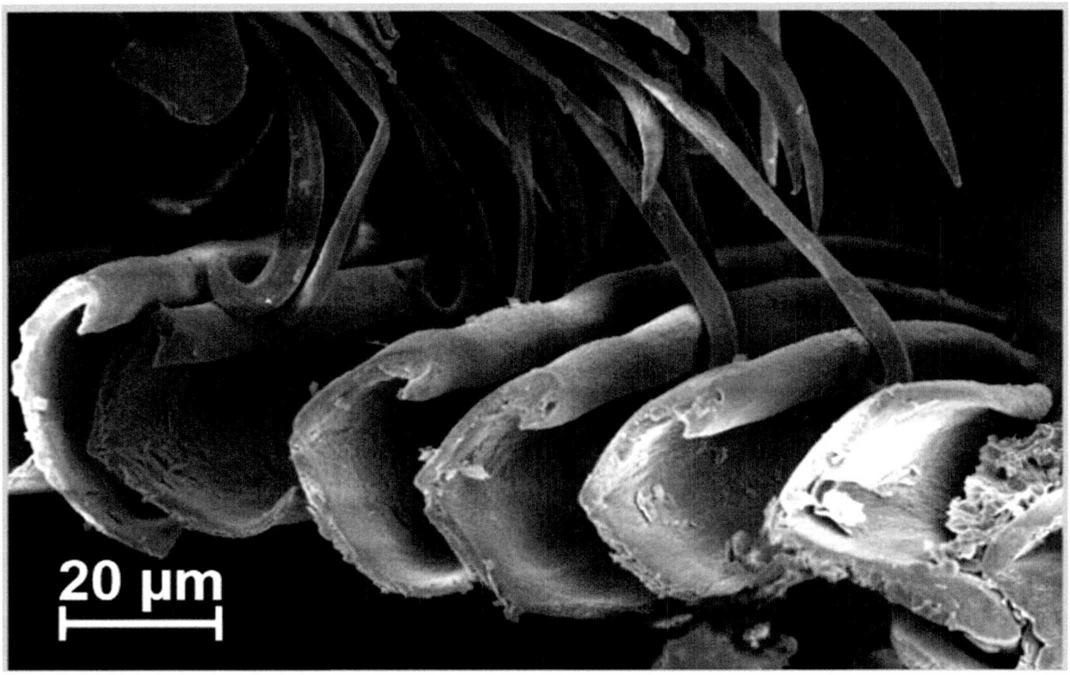


**Fig. 3.67** Proximal and distal barbules of a contour feather of a Razor-billed Curassow (*Mitu tuberosum*). Distal barbules have hooklets, and proximal barbules have grooves. (Figure from Sullivan et al. 2016; © 2016 Acta Materialia Inc. Published by Elsevier Ltd., used with permission)

to discrete areas called pterylae (or feather tracts; Box 3.12 Positioning Skin Follicles); areas lacking feathers (follicles) are called apteria (Figs. 3.91 and 3.92). Ratites, including Common Ostriches (*Struthio camelus*) and rheas, penguins (Spheniscidae), and toucans (Ramphastidae), lack apteria (Proctor and Lynch 1993). Pterylae cover about one-half of the skin of land birds, and more than half of waterbirds (Stettenheim 2000). Although the shape and distribution of pterylae vary among different families and orders of birds (Lucas and Stettenheim 1972), eight major pterylae or feather tracts can generally be identified. These include the: (1) alar tract, including primary and secondary flight feathers plus coverts, (2) caudal tract that includes the rectrices and coverts, (3) capital tract, including all feathers on the head, (4) spinal tract that extends along the mid-dorsal line and includes the cervical, interscapular, dorsal, and pelvic regions; Fig. 3.91), (5) humeral tract that extends from where the leading edge of the wing meets the body over the dorsal surface to the trailing edge of the wing, (6) femoral tract located on the dorsal surface of the thigh, (7) crural tract located on the lower leg, and (8) ventral tract that includes cervical, sternal, and abdominal regions (Fig. 3.92). The presence of pterylae and apteria, and differences among birds in their shape and distribution, may (1) be adaptations for reducing the total weight of the feathering, (2) better accommodate the movements of the body and the feathers, and (3) aid in thermoregulation via loss of body heat from apteria (Stettenheim 2000).

## 3.14 Feather Color: Pigments

Feathers, particularly contour feathers, exhibit impressive variation in color and pattern (Fig. 3.93). Colors and patterns are typically

**Fig. 3.68** Scanning electron micrograph of a down feather. 1, calamus, 2, barb, and 3, barbules. Magnified 101×. (Figure from Fuller 2015; image source: University of Leeds/Matthew Fuller, used with permission)

species-specific and may also vary within a species with sex, age, and season (Box 3.13 Male vs. Female Plumage Coloration). Feather color results from either chemical pigments or the physical interactions of light waves with biological structures. Pigments are molecules that

**Fig. 3.69** Barbs branching from the top of the calamus of a down feather. Magnified 150×; Scale bar = 100 μm. (Figure from Fuller 2015; image source: University of Leeds/Matthew Fuller, used with permission)

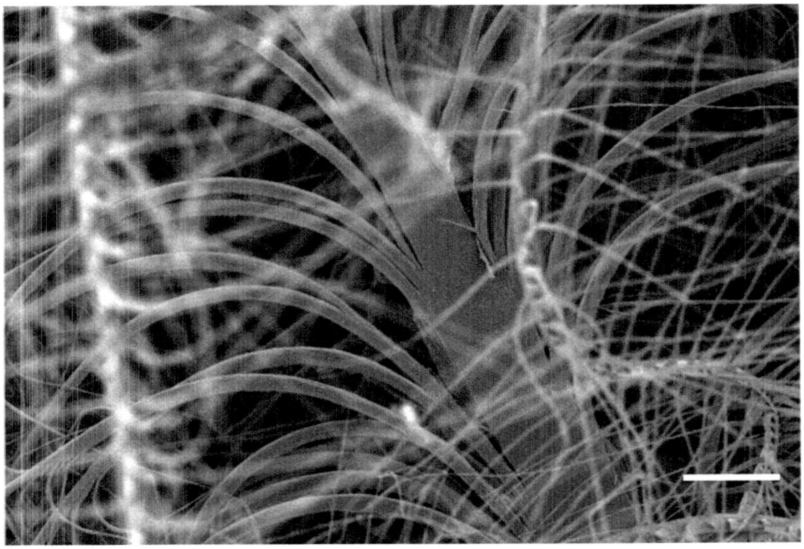

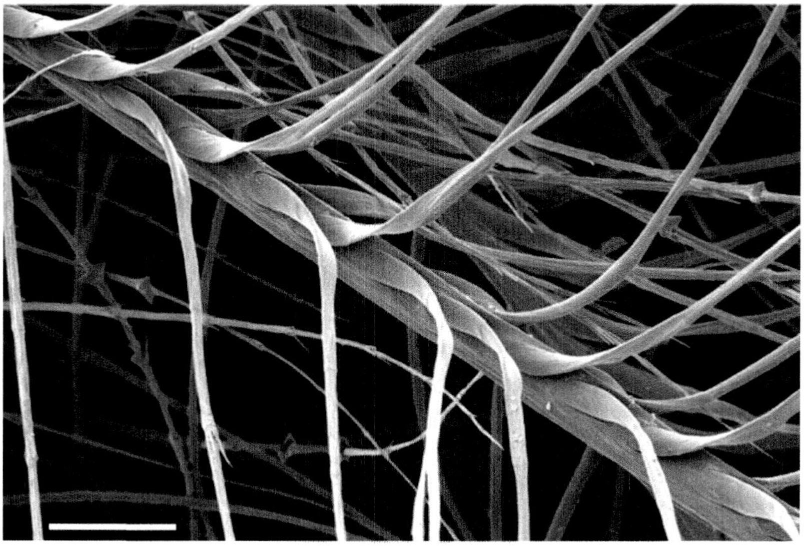

**Fig. 3.70** Barbules branching from a barb of a down feather. Magnified 400×; scale bar = 50 μm. (Figure from Fuller 2015; image source: University of Leeds/Matthew Fuller, used with permission)

differentially absorb and reflect certain wavelengths of visible light, whereas structural colors are produced when light interacts with nanometer-scale variation in feather structure (Prum 2006).

Chemical pigments contributing to feather color in birds include melanins, carotenoids, psittacofulvins, and porphyrins. These pigments differ in their chemical composition and in how they are produced and incorporated into feathers.

Melanins are the most common and widespread pigment in birds and give feathers black, brown, reddish-brown, or gray colors. There are two categories of melanin pigments—eumelanins and phaeomelanins—and most, if not all, melanin-containing feathers contain both pigments but in different proportions (McGraw 2006a). Variation in the color of melanin-containing feathers is largely determined by that proportion. Eumelanins are more abundant in feathers that are brown or black, whereas phaeomelanins dominate in feathers that are chestnut, rufous, or yellow. These melanins are not obtained from a bird's diet but are synthesized by cells in the epidermis of the skin. The cells, called melanocytes, synthesize melanin in organelles called melanosomes. Melanocytes at the base of developing feathers transfer melanin granules to the cells that eventually make up the feather (Hudon 2005).

Feathers and, more broadly, plumage colored by melanins can serve a number of functions. Melanins strengthen feathers and make them more resistant to wear by abrasion (Burtt 1979) and to damage by bacterial degradation (Goldstein et al. 2004; Gunderson et al. 2008). The importance of melanin in protecting feathers from bacteria may explain why birds in humid climates tend to have darker plumage (Gloger's rule; Gloger 1833). Humid environments may promote the growth of bacteria and, if so, increased melanin levels in feathers for additional protection may be the evolutionary response (Burtt and Ichida 2004). For some birds, plumage color and, specifically, the extent of dark melanin-containing plumage plays an important role in thermoregulation, with darker plumage absorbing and acquiring greater heat loads from solar radiation than lighter plumages (see Chap. 10 for additional information).

For many birds, melanin-containing plumage plays an important role in sexual or social advertisement. In cases of melanin-based signaling, an important question has been whether the signals are reliable or honest. To be honest, a signal must be costly so that only higher-quality individuals can display the most exaggerated forms of the signal (Zahavi 1975; Grafen 1990). The results of some studies seem to suggest that melanin-

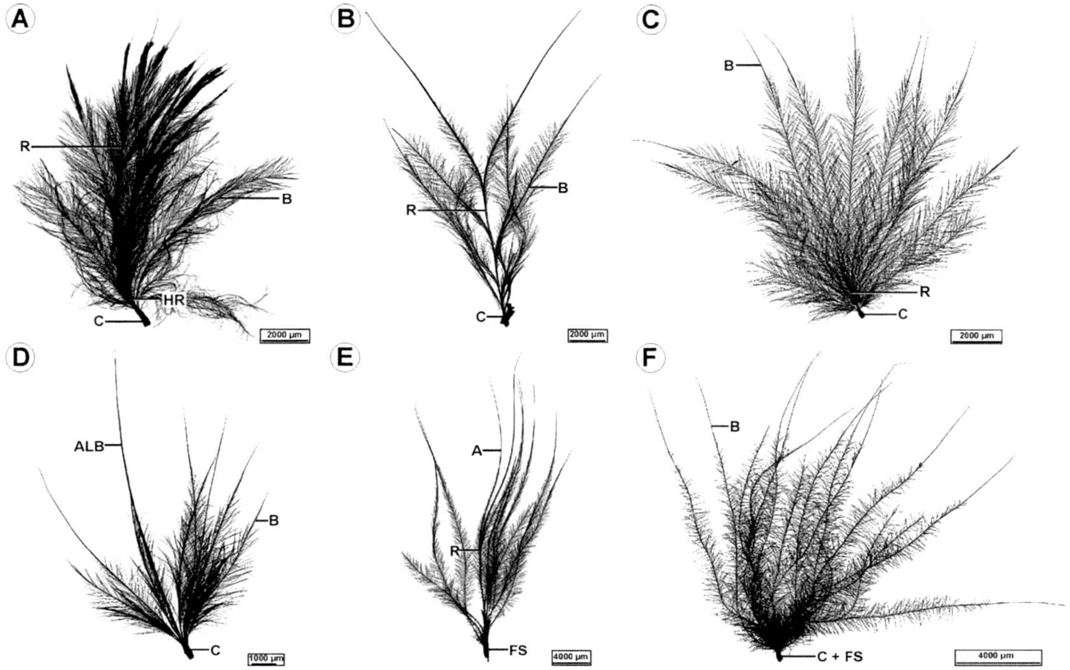


**Fig. 3.71** Examples of neoptile feathers of several species of birds. (**a**) Elegant Crested-Tinamou (*Eudromia elegans*), with a long rachis, (**b**) Southern Cassowary (*Casuarius casuarius*), with a long rachis, (**c**) Domestic Chicken (*Gallus g. domesticus*), with a short rachis, (**d**) Greater Rhea (*Rhea americana*), with one long barb, (**e**) Greater Rhea neoptile feather from the wing, and (**f**) Common Redshank (*Tringa tetanus*), with no rachis, ALB, anterior long barb; B, barbs; C, calamus, FS, feather sheath; HR, hyporachis; R, rachis. (Figure modified from Foth 2011; © 2011 Wiley-Liss, Inc., used with permission)

based plumage signals are condition-dependent (Griffith et al. 2006), whereas others suggest they are not (McGraw et al. 2002). However, melanin deposition appears to be influenced by sex hormones and by the mineral and amino acid content of a bird's diet (McGraw 2008). For example, male Zebra Finches (*Taeniopygia guttata*) have a patch of black feathers on their breast, and adding more calcium to their diet increases the number of black feathers and the size of the black patch (McGraw 2007). Male House Sparrows (*Passer domesticus*) also have a black patch on their throat and breast, and a reduction in the amounts of certain amino acids (phenylalanine and tyrosine) in their diet caused the patch to become lighter, probably by reducing the eumelanin concentration (Poston et al. 2005; McGraw 2008). Such results indicate that diet can influence the expression of melanin-based signals and, because individual quality can influence foraging behavior, that they are likely condition-dependent. In addition, melanin-based signals are influenced by circulating levels of testosterone in some species of birds, which also suggests that such signals can be reliable indicators of condition. This would be the case if increasing testosterone levels to enhance signal development also has physiological costs, such as impaired immune function (Folstad and Karter 1992; Buchanan et al. 2003). If so, only high-quality individuals in good physical condition would be able to generate high-quality signals.

Carotenoids have been found in the feathers of about 415 species of birds in the orders Columbiformes, Phoenicopteriformes, Charadriiformes, Pelecaniformes, Piciformes, Trogoniformes, and Passeriformes (Thomas

**Fig. 3.72** Natal down of newly hatched Eagle Owls (*Bubo bubo*). (Figure from Penteriani et al. 2005; © British Ornithologists' Union, used with permission)

et al. 2014), and further study will almost certainly add to that list. Carotenoids are highly unsaturated hydrocarbons that produce red, orange, and yellow feathers (colors that are also produced by psittacofulvin pigments in parrots; Fig. 3.94). Birds (and other animals) lack the enzymes needed to manufacture carotenoids and must, therefore, obtain them by feeding on plants

**Fig. 3.73** Natal down of a Chestnut-crowned Antpitta (*Grallaria ruficapilla*). (Figure from Greeney 2012; used with permission of Harold F. Greeney)

**Fig. 3.74** Natal plumage of a just-hatched Cinereous Mourner (*Laniocera hypopyrra*; family Tityridae). The bright orange down feathers with bright white tips, in combination with side-to-side head movements when disturbed, means that these young nestlings resemble a hairy, aposematic caterpillar. This appears to be an example of Batesian mimicry, resembling a noxious prey item to avoid predation and possibly enhancing survival for a tropical songbird with a relatively long nestling period for its size (20 days). (Figure from Londoño et al. 2015; used with permission of Duvan A. Garcia)

**Fig. 3.75** American Coot (*Fulica americana*) chicks have conspicuous orange, waxy-tipped feathers around their heads and necks. Lyon et al. (1994) found that adults feed more ornamented chicks preferentially over less ornamented chicks, resulting in higher growth rates and greater survival of more ornamented chicks. (Figure from Mock 2016; © 2016 Springer Nature, used with permission)

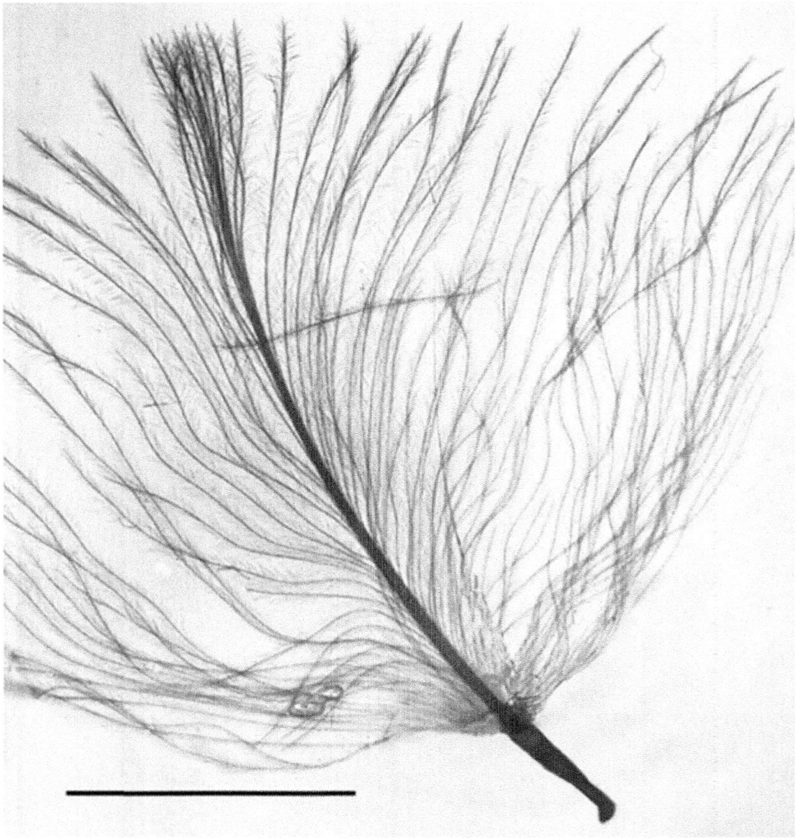

**Fig. 3.76** Semiplume from an Adelie Penguin (*Pygoscelis adeliae*). (Figure from Chiale and Montalti 2013; open-access article licensed under a Creative Commons Attribution 4.0 International License, https://creativecommons.org/licenses/by/4.0/)

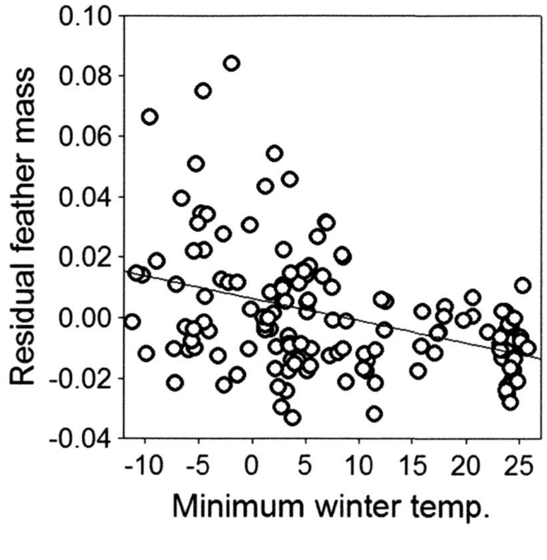


**Fig. 3.77** Relationship between the combined mass of downy and semiplume feathers and minimum winter temperature for 152 species of birds. (Figure from Osváth et al. 2018; © 2017 The Authors. Functional Ecology © 2017 British Ecological Society, used with permission)

or animals whose tissues contain carotenoids. Birds can, however, metabolize ingested carotenoids into different forms (with 26 different carotenoids identified in bird tissues; McGraw 2006b).

Because carotenoids serve functions other than coloring feathers (e.g., antioxidants and stimulating immune systems), they are present in all birds. However, only some birds deposit carotenoids in their feathers, and most of those birds deposit carotenoids in only some of their feathers and, often, only in parts of those feathers (e.g., House Finches, *Haemorhous mexicanus*; Fig. 3.95). During feather development, carotenoids dissolve in lipid droplets located in cells in the feather follicle. These carotenoids are transported by binding proteins and incorporated into and bind to the feather proteins (keratin) (Fig. 3.96). However, how these binding proteins are regulated to selectively color certain feathers

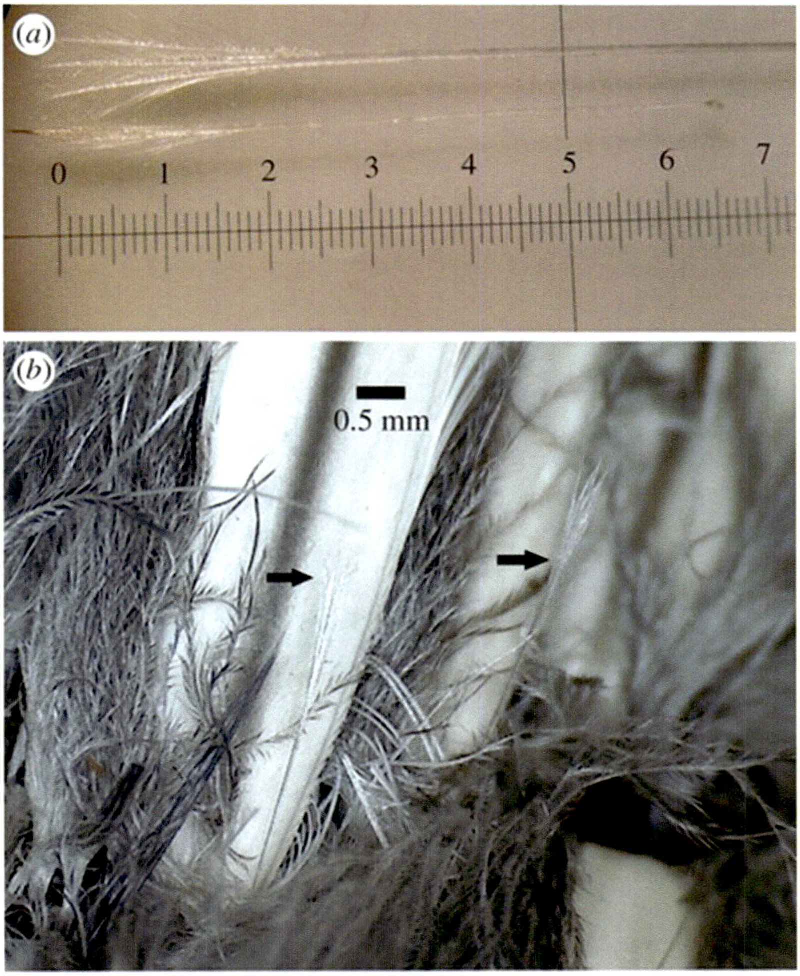


**Fig. 3.78** (**a**) Two filoplumes from an Emperor Penguin (*Aptenodytes forsteri*). (**b**) Two filoplumes (indicated by the black arrows) at the base of two contour feathers. (Figure from Williams et al. 2015; © 2015 The Authors. Published by the Royal Society, used with permission)

or certain parts of feathers is currently unknown (McGraw 2006b).

Carotenoid-based colored feathers can play important roles in both intra- and, especially, intersexual interactions. An important feature of carotenoid-based signals is that they, like melanin-based signals, are reliable and honest. Because carotenoids are acquired exclusively through the diet and, as noted above, serve a number of important functions (so that not all carotenoids acquired can be deposited in plumage), only high-quality individuals in good condition can display high-quality carotenoid-based plumage.

Although most plumage displays of birds that serve as intrasexual signals of status are melanin-based, some status signals are carotenoid-based. For example, Red-shouldered Widowbirds (*Euplectes axillaris*) are black with red (carotenoid-based) epaulets or shoulder-patches that serve as status signals. Male widowbirds display their epaulets during aggressive encounters with other males, and those with redder and larger epaulets are dominant over males with smaller, more yellow epaulets (Pryke and Andersson 2003). Male Red-collared Widowbirds (*E. ardens*), Red-winged Blackbirds (*Agelaius phoeniceus*), and Yellow-headed Blackbirds (*Xanthocephalus xanthocephalus*) also have carotenoid-based feather patches that serve as status signals (Blount and McGraw 2008).

**Fig. 3.79** Filoplumes associated with the primary feathers of (left to right) a Green-winged Teal (*Anas crecca*), Ring-billed Gull (*Larus delewarensis*), Turkey Vulture (*Cathartes aura*), Red-tailed Hawk (*Butio jamaicensis*), and Red-winged Blackbird (*Agelaius phoeniceus*). (Figure from Rohwer et al. 2021; © 2021 Oxford University Press, used with permission)

In several species of birds, carotenoid-based plumage ornaments are important in mate choice. Female mate choice decisions are known to be influenced by the carotenoid-based plumage of males in House Finches (*Haemorhous mexicanus*), American Goldfinches (*Spinus tristis*), Eurasian Siskins (*Spinus spinus*), Red-backed Fairywrens (*Malurus melanocephalus*), Village Weavers (*Ploceus cucullatus*), Yellowhammers (*Emberiza citronella*), and Zebra Finches (*Taeniopygia guttata*) (Blount and McGraw 2008), and additional study will certainly reveal other such species. In House Finches, for example, plumage redness and saturation (a measure of the concentration of carotenoids in feathers) are the primary criteria used by females when selecting mates (Hill 2006). In an experimental test, a higher proportion of male House Finches with red dye added to plumage paired with females than did males whose plumage was made drabber with hair lightener (Hill 1991).

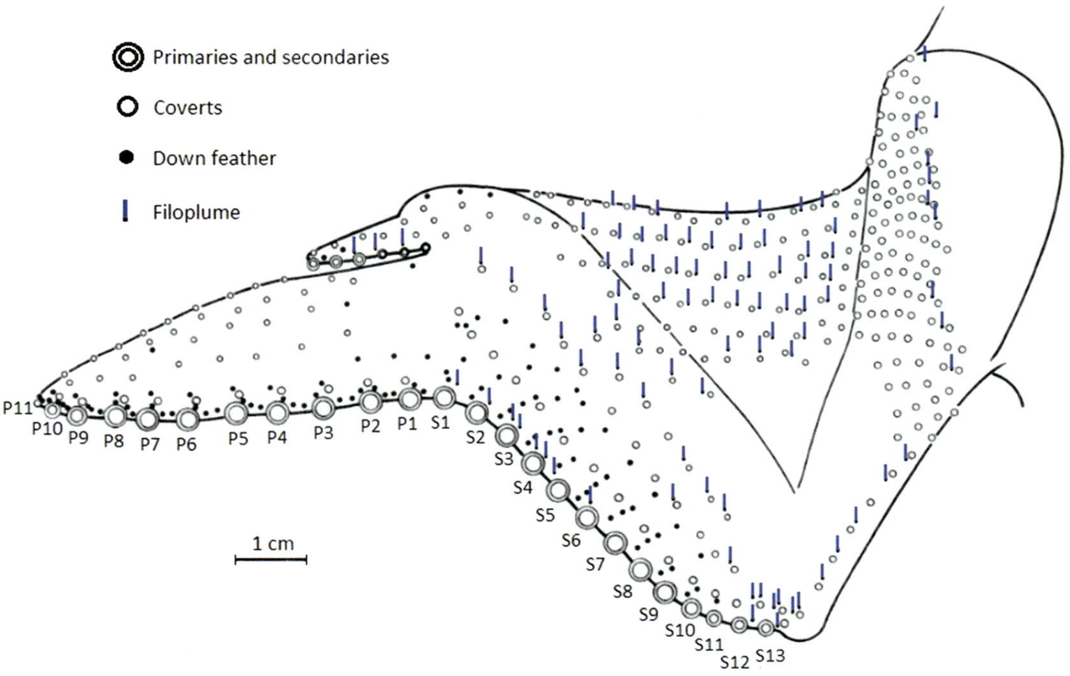


**Fig. 3.80** Feathers on the dorsal surface of the wing of a Superb Lyrebird (*Menura novaehollandiae*). Note the numerous filoplumes, most associated with coverts, including those making up the alula; the ventral surface has many fewer filoplumes. P1–P11 = primaries; S1–S13 = secondaries. (Figure modified from Morlion 1985; © 1985 Australian Museum, used with permission)

Although melanins and carotenoids are the most common pigments in bird feathers, a few other feather pigments have been identified. These include porphyrins, psittacofulvins, pterins, turacin, turacoverdin, and as yet unidentified pigments in the downy plumage of several species of birds (McGraw 2006c; McGraw et al. 2007). Porphyrins have been found in the red and brown feathers of birds in 13 orders, including owls (Strigiformes), goatsuckers (Caprimulgiformes), and bustards (Gruiformes), and these pigments (unlike carotenoids) are manufactured by the birds. Several species of birds with porphyrin-based plumage coloration are nocturnal (e.g., owls and goatsuckers) and, in some of these species, the downy plumage of their young is also colored by porphyrins. The possible adaptive value of porphyrins in feathers remains unknown. However, porphyrins tend to be unstable and fade when exposed to bright light (Lucas and Stettenheim 1972; Hudon 2005), perhaps explaining their occurrence in the feathers of several nocturnal species. In addition, porphyrins, unlike most dark pigments, do not absorb infrared radiation (Bakken et al. 1978). In many species of birds, porphyrins are deposited in eggshells (generating the spots and streaks associated with the eggs of many species), and one possible function of these pigments, because they do not absorb infrared radiation, is to prevent eggs from over-heating (Bakken et al. 1978). The porphyrin-colored plumage of some nocturnal species may similarly help both adults and young thermoregulate. That is, birds active at night are likely adapted to cooler temperatures than diurnal species, and feathers with porphyrin that do not absorb infrared radiation may help nocturnal birds stay cooler if they are exposed to sunlight during daylight hours (McGraw 2006c).

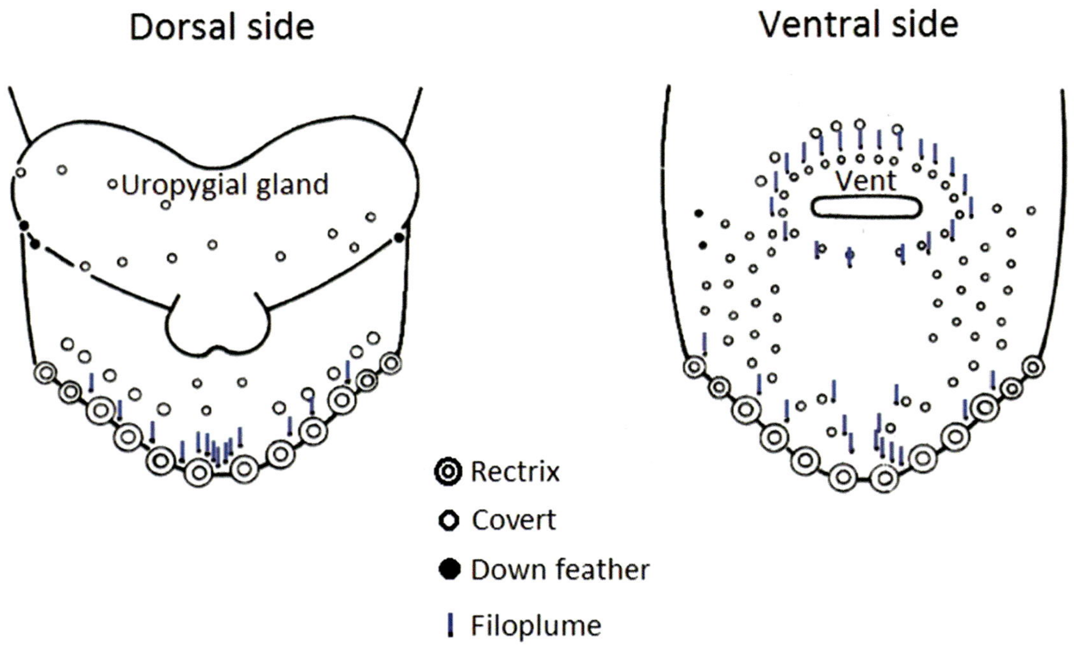


**Fig. 3.81** Feathers of the tail region of a Noisy Scrub-bird (*Atrichornis clamosus*). Note the numerous filoplumes associated with the rectrices and with the coverts surrounding the vent. Information about the position of the rectrices is critical during flight and when landing, and information about the position of the coverts surrounding the vent is important when a bird defecates (to minimize soiling of the feathers). (Figure modified from Morlion 1985; © 1985 Australian Museum, used with permission)

Psittacofulvins are red, orange, and yellow pigments found only in the feathers of parrots (Order Psittaciformes; Fig. 3.97). Unlike carotenoids, psittacofulvins are synthesized by parrots. However, the mechanisms and possible costs of producing psittacofulvin-based

**Fig. 3.82** Light micrograph of Common Ostrich (*Struthio camelus*) skin showing a Herbst corpuscle (h) located near a feather follicle sheath (fs), and the sheath of a filoplume (fp). f filoplume. The epidermis is visible on the right side of the micrograph. (Figure from Weir and Lunam 2011; © 2011, Springer-Verlag Berlin Heidelberg, used with permission)

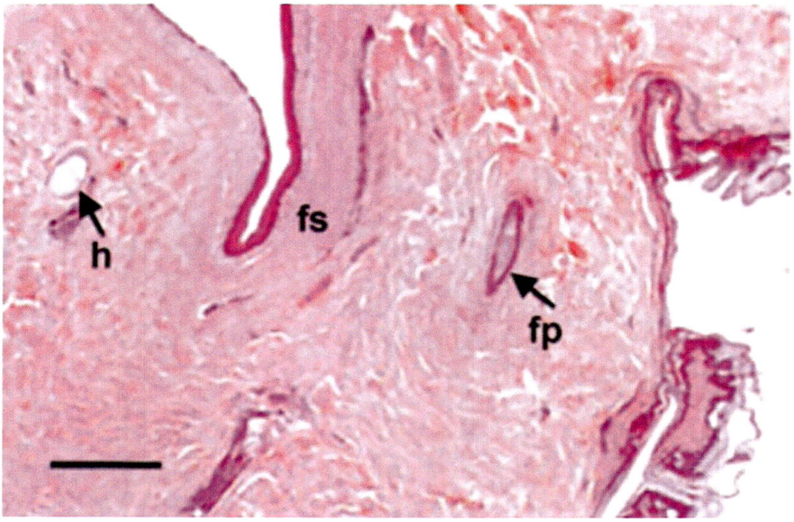

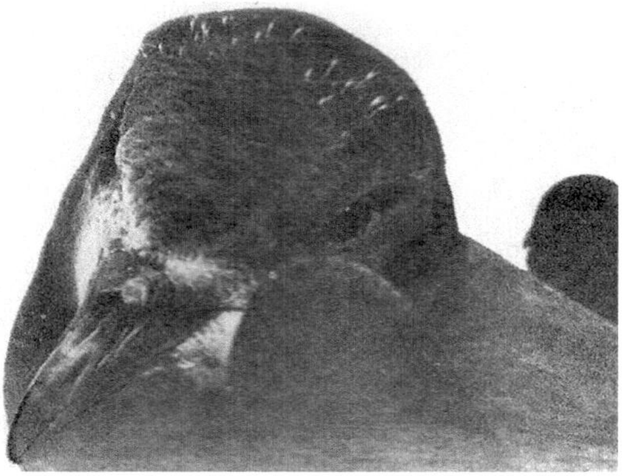

**Fig. 3.83** Filoplumes extending through the contour feathers on the head of a male Gray-faced Petrel (*Pterodroma gouldi*) likely provide information about the position of contour feathers in areas not visible to birds and not easily accessible for preening with the bill. (Figure from Imber 1971; Rights managed by Taylor & Francis, used with permission)

coloration remain to be identified. Morelli et al. (2003) found that psittacofulvins are potent antioxidants, suggesting that parrots might face a trade-off between using these pigments for ornamentation and immune functions. Based on a limited number of studies of one species (Burrowing Parrot, *Cyanoliseus patagonus*; Masello and Quillfeldt 2003; Masello et al. 2008), psittacofulvin-based plumage colors can provide information about individual quality.

Pterins are commonly red, orange, and yellow pigments in a wide variety of animals, ranging from insects to birds. In birds, pterins are known primarily as pigments that color the eyes (irises), and the yellow or yellow-orange feathers of several species of penguins (*Aptenodytes* spp., *Eudyptes* spp., and the Yellow-eyed Penguin, *Megadyptes antipodes*) were initially thought to be colored by pterins (McGraw et al. 2007). However, Thomas et al. (2013) found that the yellow pigment in these penguins' feathers is a unique chemical called spheniscin that is synthesized directly by penguins (Fig. 3.98). The yellow feathers of these penguins have been found to be important to both sexes for mate selection, providing information about individual quality (Massaro et al. 2003; McGraw et al. 2009).

Turacin and turacoverdin pigments have been found only in the feathers of turacos (Order Musophagiformes, Family Musophagidae), medium-sized, arboreal birds found in sub-Saharan Africa. Turacin is a red pigment limited to feathers on the wing or head of 18 species of turacos. Turacoverdin is a green pigment found in ventrally located feathers of 21 species of turacos (McGraw 2006c; Fig. 3.99). Both turacin and turacoverdin contain copper, and manufacturing these pigments requires that they obtain large quantities of copper in their diets (McGraw 2006c). This suggests that plumage colored by these pigments, if they serve as sexual or social advertisements, would be, like those that are melanin- or carotenoid-based, reliable and honest. However, this remains to be tested experimentally.

## 3.15 Feather Structural Colors

The color of feathers can result from selective absorption of light by pigments, as just described, but also by coherent or incoherent scattering of light (structural color; Fig. 3.100) (Shawkey et al. 2009a, b). Incoherent scattering of light "creates a highly diffuse reflectance spectrum and produces white coloration" (Shawkey et al. 2006a, b). Coherent scattering is the differential interference or reinforcement of wavelengths scattered by multiple light-scattering objects, and the scattering of specific wavelengths is determined by the phase relationships among the scattered waves (Fig. 3.101).

Colors can also be categorized as either iridescent or noniridescent, with iridescent colors changing in appearance depending on the angle of observation or illumination and noniridescent colors remaining the same regardless of angle of observation (Osorio and Ham 2002). Iridescent colors can be produced by both interference and diffraction, and noniridescent colors by coherent scattering caused by feather structure (Sun et al. 2013). Among birds, the nanostructural organization of keratin, melanin, and air in feather barbs and barbules can produce iridescence in several

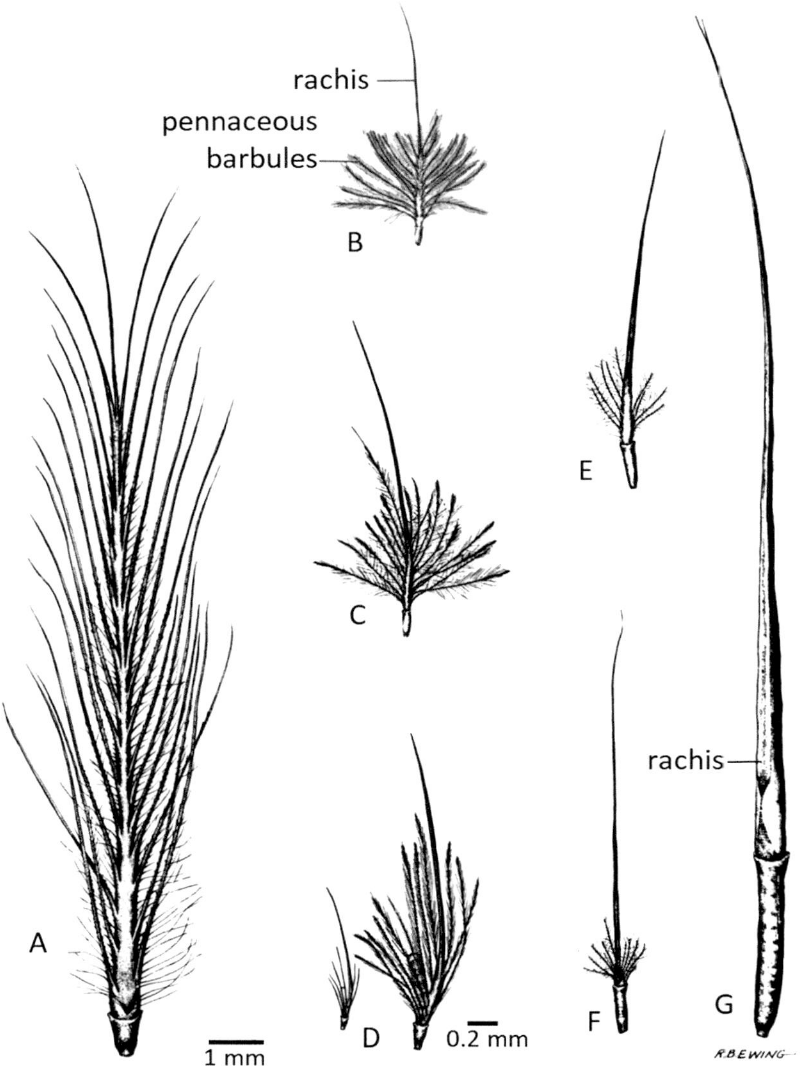


**Fig. 3.84** Examples of a semibristle and bristles of several species of birds. (**a**) Lorial semibristle of a Great Horned Owl (*Bubo virginianus*), (**b**) malar bristle of an American Robin (*Turdus migratorius*), (**c**) rictal bristle of a European Starling (*Sturnus vulgaris*), (**d**) upper eyelid bristle of a European Starling, (**e**) rictal bristle of an American Robin, (**f**) upper eyelid bristle of a Red-tailed Hawk (*Buteo jamaicensis*), and (**g**) rictal bristle of a Band-winged Nightjar (*Systellura longirostris*). 0.2-mm scale for **d**; 1-mm scale for all others. (Figure from Lucas and Stettenheim 1972; CC0 Public Domain)

ways, including thin-film and multi-film (or multilayer) interference and photonic crystals (Sun et al. 2013). For all structural colors, peak reflectance in visible wavelengths depends on three general characteristics, including the nano-scale organization, size of the scattering structures, and refractive index (RI) contrast of the structural components (Eliason and Shawkey 2012).

Many species of birds have blue plumage, a structural color resulting from a process called coherent light-scattering. Scattering simply means that light strikes objects (scatterers) that change its trajectory. Blue feathers have barbs with a "spongy layer" consisting of tiny granules of keratin interspersed with pockets of air, above which is located a layer of dark-colored melanin (Figs. 3.102 and 3.103). Within the spongy layer, the very small and similar sizes of these granules means that short-wavelength light (blue and violet) are reflected, whereas incoherently scattered white light is absorbed by

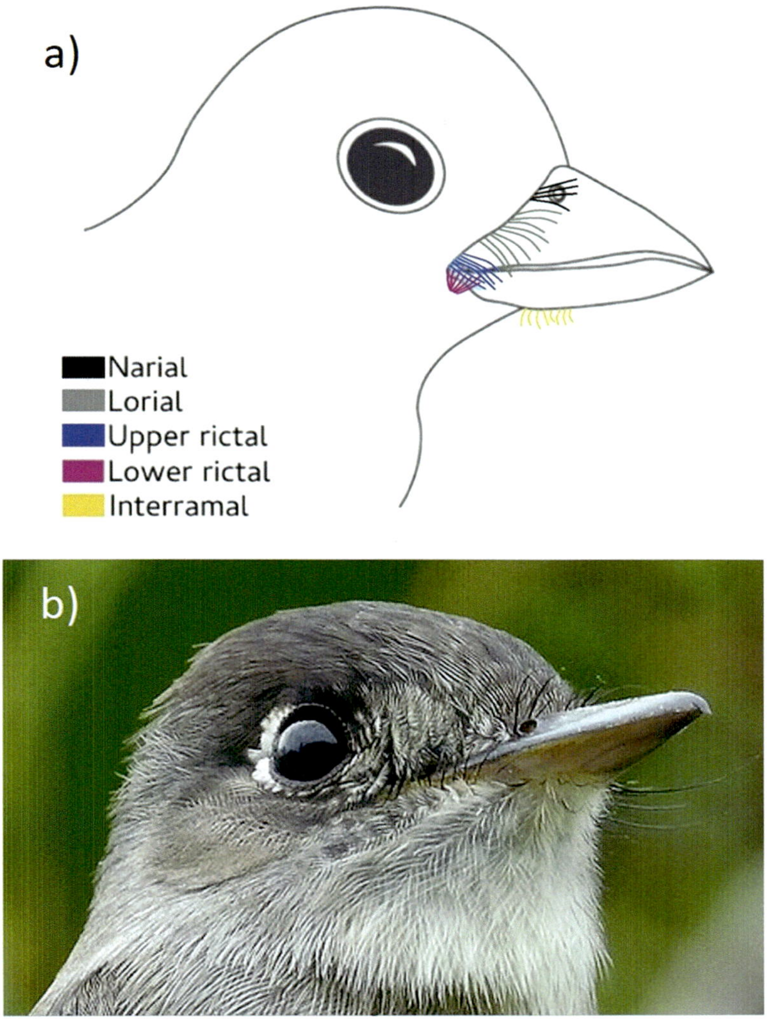


**Fig. 3.85** (**a**) Possible locations of bristles located at the base of bird bills. Delaunay et al. (2022) examined 1022 species of birds representing 418 genera, 91 families, and 29 orders and found that 365 of those species (35.7%) had lorial and upper rictal bristles (together grouped as "rictal bristles" by the authors). (**b**) Bristles of a Cuban Peewee (*Contopus caribaeus*). (Figure (**a**) from Delaunay et al. 2022; © The Authors 2022, open-access article licensed under a Creative Commons Attribution 4.0 International License, http://creativecommons.org/licenses/by/4.0. Photo (**b**) by Gail Hampshire, Wikipedia, CC BY 2.0, https://creativecommons.org/licenses/by/2.0/)

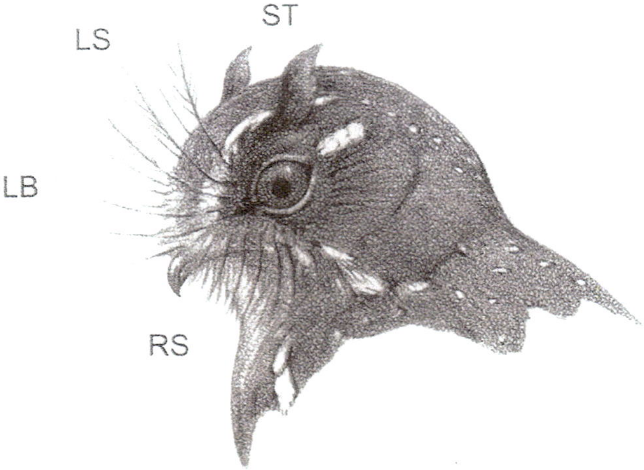


**Fig. 3.86** Profile of a Starry Owlet-nightjar (*Aegotheles tatei*) showing a pair of ear tufts (ST), plus three pairs of loral semibristles (LS), two pairs of loral bristles (LB), and rictal semibristles (RS). The numerous bristles and semibristles likely help protect the eyes. Drawing by James Coe. (Figure from Pratt 2000; © 2000 Oxford University Press, used with permission)

**Fig. 3.87** Whiskered Auklet (*Aethia pygmaea*) with long specialized contour feathers extending above, behind, and to the sides of the head. (U.S. Geological Survey, CC0 Public Domain)

melanin, increasing saturation of the blue wavelengths. Different shapes and sizes of the air pockets and keratin granules make different shades of blue and can potentially generate other colors as well (Fig. 3.104). White is also a structural color resulting from incoherent light-scattering. In white feathers, the light-scattering keratin granules vary in size and are randomly distributed so that all wavelengths of light are reflected or, in other words, there is an incoherent scattering of all wavelengths or white light.

## 3.16 Iridescent Structural Color: Thin-Film Interference

Feather color can also result from thin-film interference. When light strikes the surface of a thin transparent film, some light is reflected, and some transmitted through the film. The transmitted light is then reflected from a bottom of the film and reflects back out of the film. Thus, two waves emerge from the thin film, one reflected from the top surface and another from the bottom surface (Fig. 3.105). The two waves have different

**Fig. 3.88** Australian Owlet-nightjar (*Aegotheles cristatus*. Note the numerous bristles around the bill. (Photo by Patrick Kavanagh, Wikipedia, CC BY 2.0, https://creativecommons.org/licenses/by/2.0/)

path optical lengths that is determined by the thickness and refractive index (i.e., speed of light as it travels through a substance compared to its speed in a vacuum) of the film. The two waves will interfere either constructively or destructively and the interference pattern depends on the thickness of the film as well as the angle at which light strikes the thin film (Figs. 3.106 and 3.107). For example, if the film is exactly the right thickness for the two waves of emerging red light to undergo constructive interference, the film will appear red at that location, and the same for other colors at different locations of the film that differ in thickness.

## 3.17 Structural Color: Thin- and Multi-film Interference

Iridescence is defined as "a lustrous rainbowlike play of color caused by differential refraction of light waves . . . that tends to change as the angle of view changes" (Merriam-Webster.com), and has been reported in 14 different orders of birds (Prum 2006). Iridescence is produced by nanostructures in feather barbules that can create a wide variety of bright colors (Doucet et al. 2006; Stoddard and Prum 2011; Fig. 3.108). Melansomes can produce iridescent colors when arranged into nanostructures and noniridescent colors when randomly arranged (Fig. 3.109). More generally, iridescent colors are produced through coherent light-scattering by laminar or crystal-like arrays by layers of materials with different refractive indices, that is, keratin, melanin, and sometimes air, in feather barbules (Fig. 3.110). For example, iridescence in four species of manakins is produced by "periodic matrices of air and β-keratin in feather barbs" (Igic et al. 2016). Iridescence can result from either thin-film (Fig. 3.111) or multi-film (Fig. 3.112) interference. The arrangement of nanostructures like melanosomes can vary in

**Fig. 3.89** Contour feathers moved aside to reveal the yellow powder-down feathers of a Green Heron (*Butorides virescens*). (Photo provided by Powdermill Nature Reserve, used with permission)

different areas of barbules to create either iridescent or noniridescent colors (Fig. 3.113).

## 3.18 Structural Colors Produced by Photonic Structures

The feathers of different species of birds contain a diversity of photonic structures, including thin films and multilayers, as well as diffraction gratings and photonic crystals. Photonic crystals (so-named because light interacts in these structures in a way that is analogous to the way that electrons interact with a periodic crystal of ions) are microstructures that consist of ordered (or amorphous) "particles" that scatter incident light (Fig. 3.114). The scattered light waves then interfere with each other so that, depending on the angle at which incident light strikes the surface, one specific wavelength of light is emitted (Fig. 3.115).

Photonic crystals in bird feathers can be one-, two-, or three-dimensional (Vigneron et al. 2006; Stavenga et al. 2011; Saranathan et al. 2021; Fig. 3.116). One-dimensional photonic crystals consist of two or more very thin layers of materials and, as described above, produce color by multilayer interference. The feathers of many species of birds contain two-dimensional photonic crystals, with a lattice consisting of keratin or melanin rods or spheres and air (Fig. 3.117). For all structural colors, achieving peak reflectance in visible wavelengths depends on three factors: the nanoscale organization, size of scattering structures, and refractive index (RI) contrast of its structural components. Slight variation in the organization, size, or RI contrast in these nanostructures can lead to dramatic changes in color (Figs. 3.118 and 3.119).

## 3.19 Feather Color: Pigment Plus Structure

Pigment-based and structural colors can interact with each other to produce colors not possible by either mechanism alone. For example, most green feathers are thought to result from a combination of structural blue color and yellow pigments (Fox 1976). Captive breeding of Budgerigars (*Melopsittacus undulatus*) has produced a diversity of colors not found in wild populations, including several that result from an interaction between feather nanostructure and pigment (Fig. 3.120). D'Alba and Shawkey (2012) examined a variety of colored Budgerigar feathers and found that, regardless of color, barbs had quasi-ordered air-keratin spongy layers. However, the organization of air and keratin in these layers differed among different-colored feathers. In addition, barbs and barbules in feathers of all colors except gray and yellow had a layer of melanosomes below the spongy layer. When present, the melanosomes likely increase the intensity or saturation of feather color; when absent, gray or yellow colors are produced. Yellow pigment (psittacofulvin) was present in the barbs of all feathers of all colors except gray and purple. The contribution of this pigment, present in both barbs and barbules, to feather color, became apparent when the pigment was chemically removed. The effect was most apparent for yellow feathers; after the removal of the pigment, they appeared white. In addition, green feathers appeared blue after pigment removal. However, spectrometry revealed that these depigmented feathers still reflected light in green wavelengths, indicating that the nanostructure of green barbs is "tuned" to the color green (Prum et al. 1999), but selective absorption of blue wavelengths by the

**Box 3.12 Positioning Skin Follicles**

In the pterylae of birds, feather follicles occur in a spaced array, but what determines that spacing? The skin of avian embryos initially consists of a flat sheet of epithelial cells. However, this flat sheet is quickly transformed into a sheet with regularly spaced aggregates of cells (i.e., bumps), each of which becomes a follicle. Some investigators have suggested that differential gene expression in the epithelial cells controls this process (e.g., Noramly and Morgan 1998). Shyer et al. (2017), however, reported evidence that structural changes in the skin initiate the process of follicle formation. These investigators first noted that (A–C), in embryonic skin, cells in the dermis layer contract (and can do so because they contain the protein myosin, a protein also found in muscle cells), creating aggregrations of cells, or bumps, in the overlying epidermis. This compression and clumping of cells in the epidermis causes is sensed through the protein β-catenin, which responds to the force by moving into the nuclei of the cells. β-catenin, in turn, activates the follicle gene expression pathway. The distance between aggregrates (and follicles) is determined by the amount of contraction, with more contraction resulting in greater spacing, and the amount of resistance to contraction by the epidermis, with more resistance leading to shorter spacing (Shyer et al. 2017; Grill 2017).

(continued)

**Box 3.12** (continued)

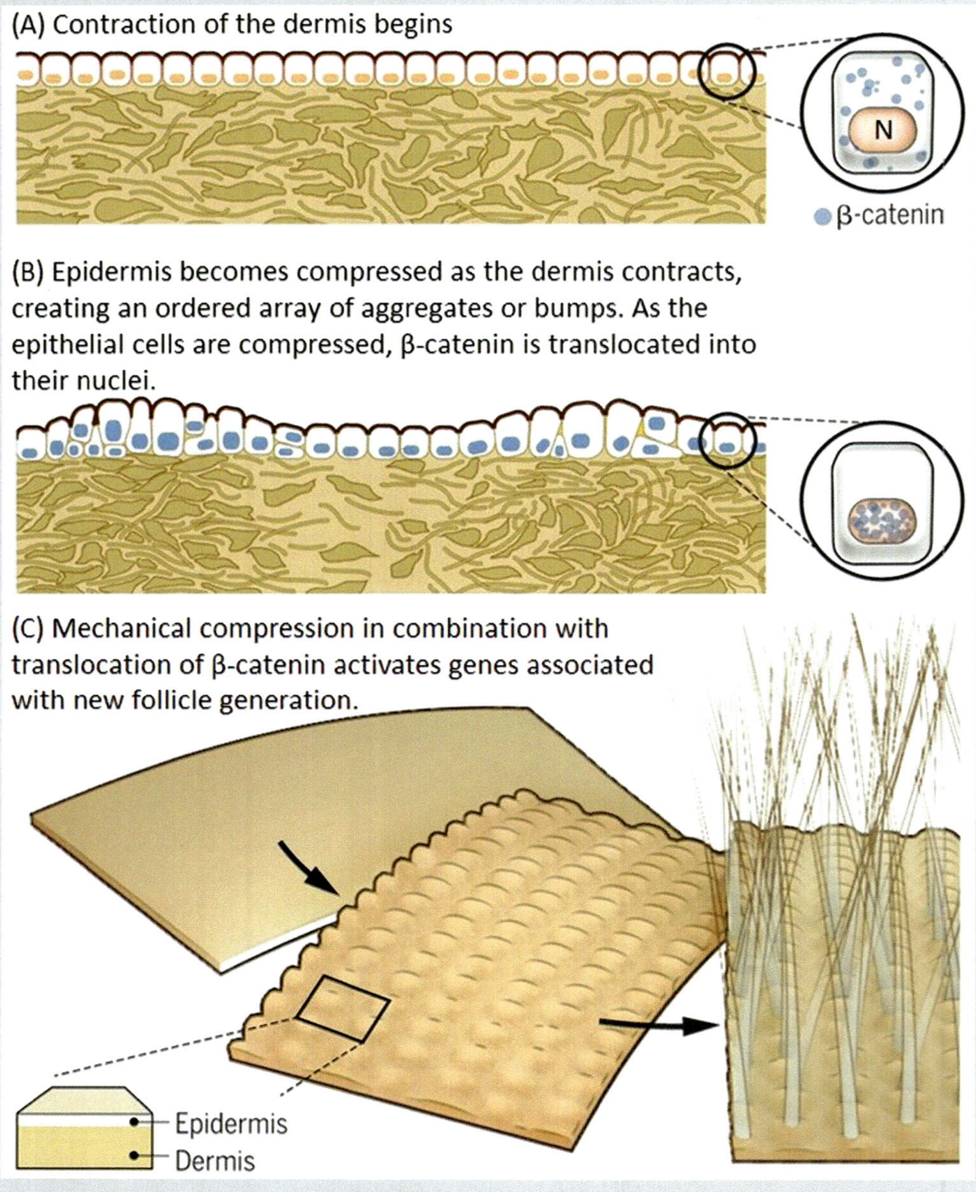


Follicle placement in the skin results from mechanical forces in the dermis, with regions with cell aggregates being surrounded by regions with fewer cells. The distance between aggregates is determined by the amount of contraction, with more contraction increasing the distance between aggregates (Figure from Grill 2017; © 2017 American Association for the Advancement of Science, used with permission).

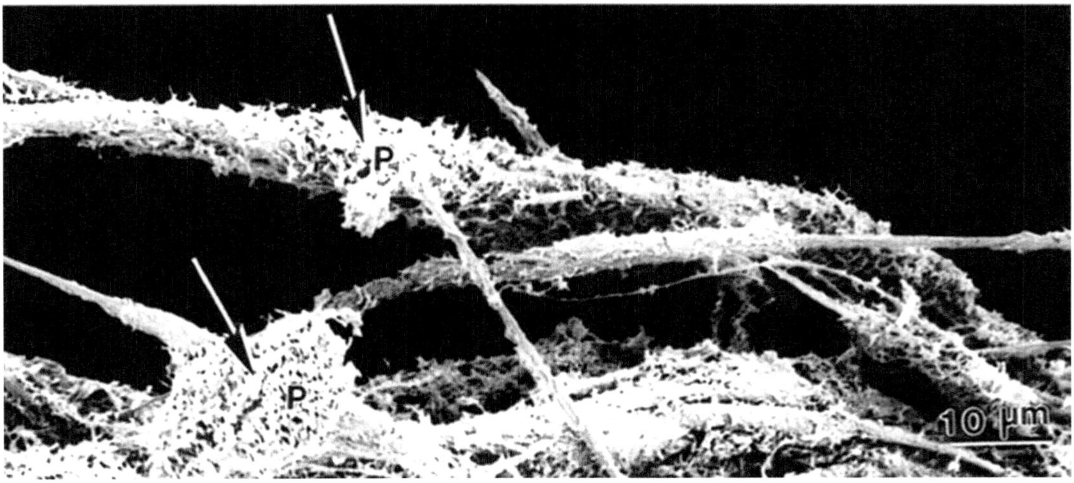


**Fig. 3.90** Scanning electron micrograph of a portion of powder-down feather from a Cattle Egret (*Bubulcus ibis*). Arrows point to the clump of a powder-lipoid adherent mixture (P). (Figure from Menon and Menon 2000; © 2015 Oxford University Press, used with permission)

pigments is required for the perception of green (D'Alba and Shawkey 2012). Removal of the pigment had little effect on the appearance of gray, dark blue, and purple feathers, suggesting that little yellow pigment was present. Finally, after pigment removal, turquoise feathers appeared bluer and light blue feathers a bit less blue and, as a result, were similar in color. Thus, differences in the color of Budgerigar feathers result from differences in feather nanostructure, pigment concentrations, and the presence or absence of melanosomes (D'Alba and Shawkey 2012). Other examples of an interaction between feather microstructure and pigments producing different colors include glossy red feathers of several species of birds, including Red-crested Cardinals (*Paroaria coronata)* and Scarlet Minivets (*Pericrocotus speciousus*, Iskandar et al. 2016; Fig. 3.121), the yellow plumage of American Goldfinches (*Spinus tristis*, Shawkey and Hill 2005), and the green feathers of many species of parrots and other species of birds (e.g., Surmacki et al. 2011; Tinbergen et al. 2013; Zhang et al. 2014).

## 3.20 Feather Parasites

Many bacteria can damage feathers by degrading keratin, the principle building block of feathers (i.e., keratinolytic bacteria) (Burtt and Ichida 1999; Gunderson 2008; Fig. 3.122). These feather-degrading bacteria have been found in the plumage of numerous species of birds, and the damage they cause can potentially affect the ability of birds to thermoregulate, signal, or even fly (Swaddle et al. 1996; Shawkey et al. 2007). Feather-degrading bacteria can also affect the color of feathers via the alteration of feather microstructures and the consumption or modification of feather pigments (Shawkey and Hill 2004). Different species of bird may have different bacterial communities in their plumage (Muza et al. 2000; Kent and Burtt 2016), and some of this variation may be due to interspecific differences in foraging behavior (Kent and Burtt 2016) and habitat use (Bisson et al. 2007). Kent and Burtt (2016) found that a greater proportion of ground-foraging and fly-catching species of birds had feather-degrading bacteria (*Bacillus*

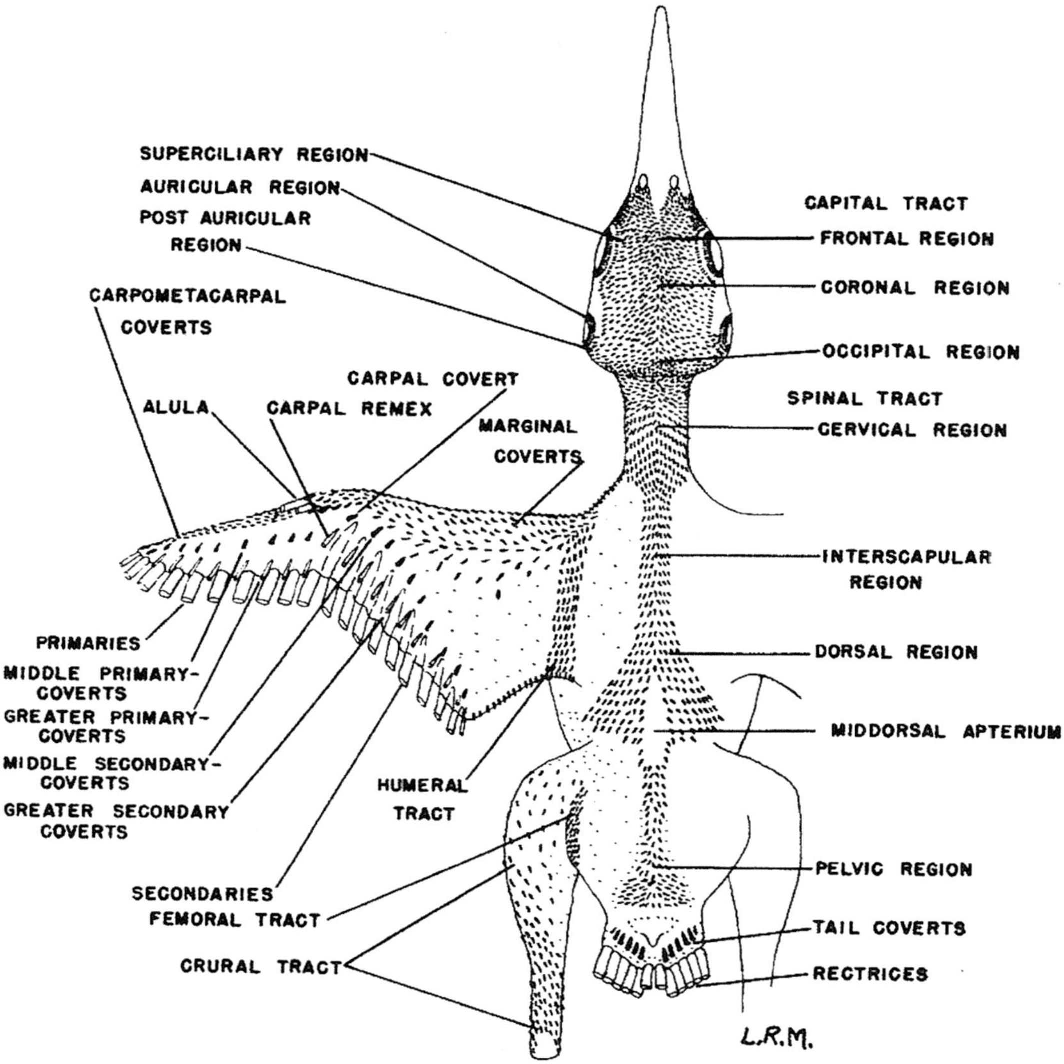


**Fig. 3.91** Pterylae and apteria of a Clark's Nutcracker (*Nucifraga columbiana*), dorsal view. (Figure from Mewaldt 1958; © 1958 Oxford University Press, used with permission)

spp.) than did marine-foraging, nectivorous, and tree-probing species (Fig. 3.123). Reasons for these differences among species in different habitats and in different foraging guilds remain to be determined.

Birds can potentially prevent or minimize the negative effects of feather-degrading bacteria in several ways, including such behaviors as sunning, bathing, and preening (e.g., Saranathan and Burtt 2007). The results of some in vitro experiments suggest that secretions of uropygial glands (preen oils) have inhibitory effects on feather-degrading bacteria (e.g., Shawkey et al. 2003; Ruiz-Rodríguez et al. 2016). However, an in vivo study of Mallards (*Anas platyrhynchos*) revealed no differences in plumage bacterial loads between birds with and without access to their preen glands (Giraudeau et al. 2013). In addition,

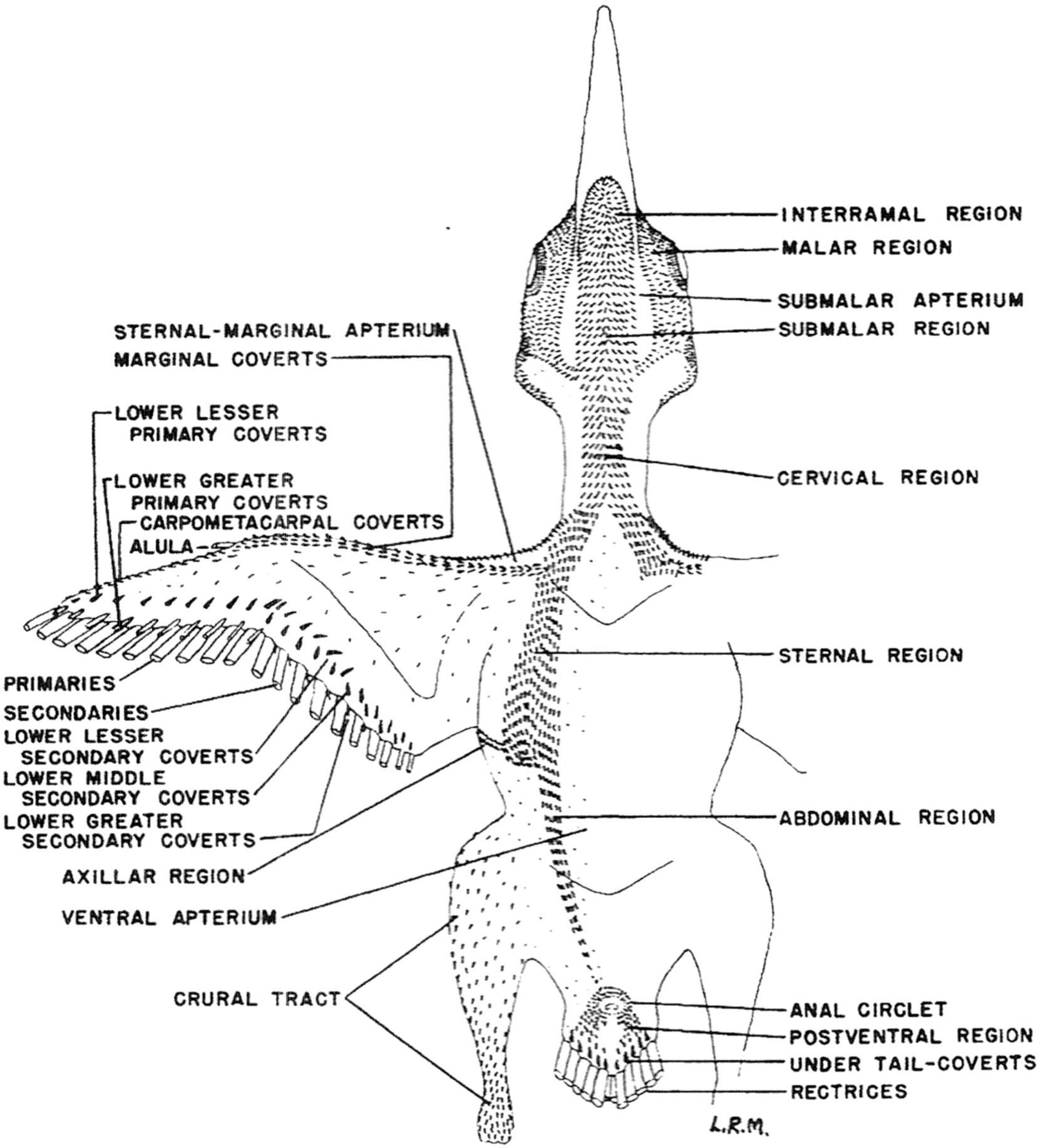


**Fig. 3.92** Pterylae and apteria of a Clark's Nutcracker (*Nucifraga columbiana*), ventral view. (Figure from Mewaldt 1958; © 1958, Oxford University Press, used with permission)

Czirják et al. (2013) found no difference between House Sparrows (*Passer domesticus*) with and without uropygial glands in the relative abundance of feather-degrading bacteria. Possible explanations for differences in the results of in vitro versus in vivo studies include: (1) investigators in the in vitro studies may have over-estimated the antibacterial effects of preen oil by using unnaturally large amounts, (2) preen oils may have bactericidal effects of some bacteria, but not others, (3) different species of birds have preen oils that differ in their bactericidal

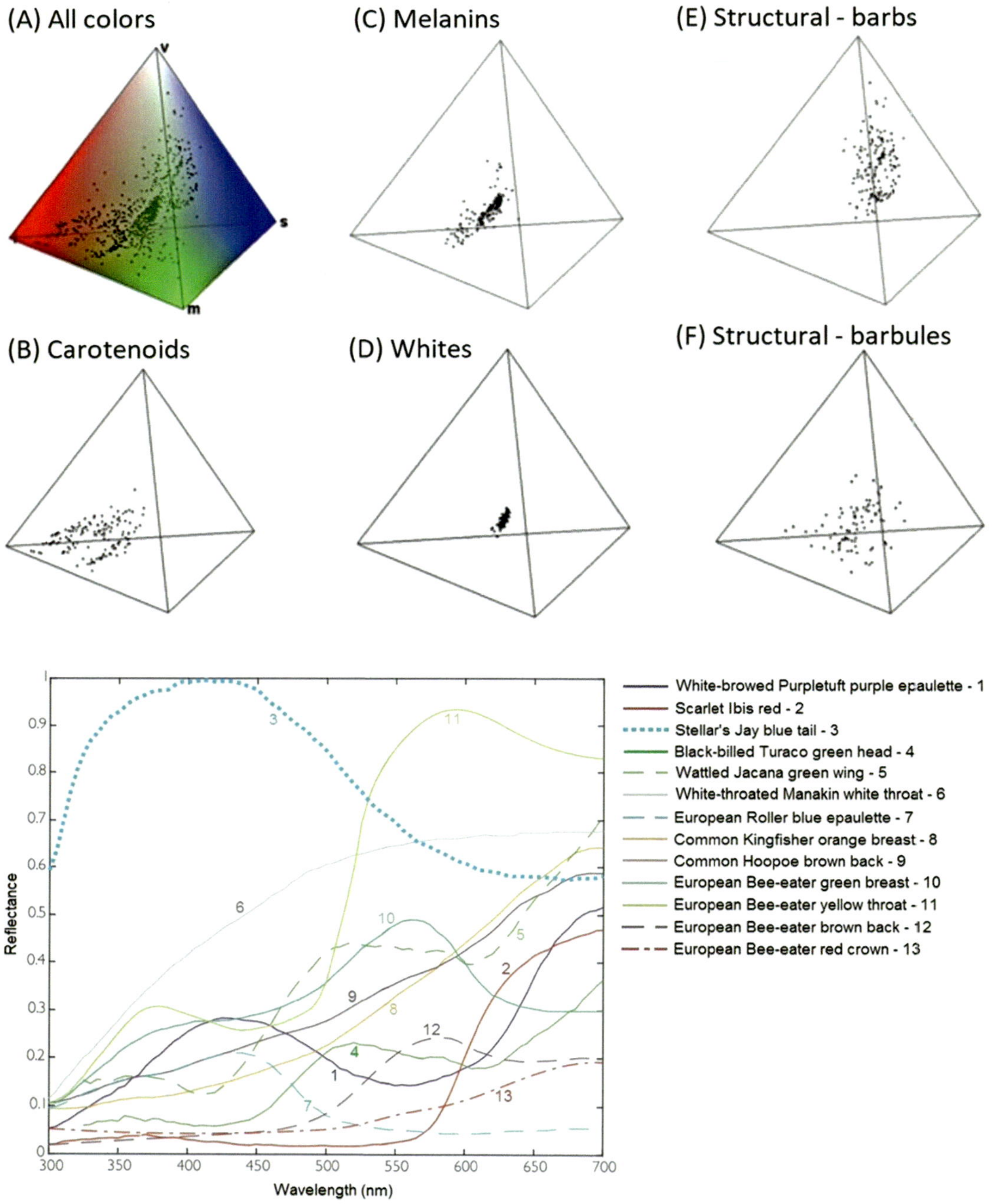


**Fig. 3.93** Top, (**a**) Complete range of all avian plumage colors in avian tetrahedral color space, (**b**) Carotenoid pigments, (**c**) Melanin pigments, (**d**) Nonpigmented whites, (**e**) Structural colors produced by barb rami, and (**f**) Structural colors produced by barbules. Location of color points is determined by the relative stimulation of ultraviolet or violet (uv/v), blue (s), green (m), and red (l) retinal cones. (Figure from Stoddard and Prum 2011; © 2011 Oxford University Press, used with permission). Bottom, Examples of the many colors of feathers among and within different species of birds. Visible light for humans extends from about 400 to 700 nm; many species of birds are, in addition to seeing the light at those wavelengths, also able to perceive ultraviolet light. White-browed Purpletuft, *Iodopleura isabellae*; Scarlet Ibis, *Eudocimus ruber*; Steller's Jay, *Cyanocitta stellari*; Black-billed Turaco, *Tauraco schuettii*; Wattled Jacana, *Jacana jacana*; White-throated Manakin, *Corapipo gutturalis*; European Roller, *Coracias garrulous*; Common Kingfisher, *Alcedo atthis*; Common (or Eurasian) Hoopoe, *Upupa epops*; European Bee-eater, *Merops apiaster*. (Figure modified from Tedore and Johnsen 2017; © 2016 Oxford University Press, used with permission)

**Box 3.13 Male vs. Female Plumage Coloration**

Many species of songbirds are sexually dichromatic, with males more colorful than females. However, the extent to which the plumage of males and females differs varies and, in some species, there is little or no difference between the sexes, and females are as colorful or nearly as colorful as males. Possible explanations for the colorful plumage of some female songbirds include: (1) genetic correlation, that is, female traits evolve as a genetically correlated response to selection on males (sexual or natural) and, because most genes are shared by the two sexes, females inherit the same genetic basis for ornamentation as males (Lande 1980); this correlation will persist unless there is strong selection pressure to suppress the trait in females, (2) selection may favor colorful female plumage for signaling purposes, and (3) some combination of explanations 1 and 2.

Monochromatic Altamira Orioles (*Icterus gularis*) and dichromatic Baltimore Orioles (*I. galbula*) (Figure from Hofmann et al. 2008; © 2008 Oxford University Press, used with permission).

Dale et al. (2015) scored the plumage of males and females in 5831 species of songbirds (order Passeriformes) based on relative "dullness" and "brightness" and found a positive correlation, with the plumage of females tending to be brighter or more colorful in species where male plumage was also brighter and more colorful. Such results provide support for the genetic correlation hypothesis. However, many species of songbirds are dichromatic so, to varying degrees among these species, selection must favor less colorful plumage for females. Factors that appear to favor less colorful plumage include small size, breeding at higher latitudes (either as residents or migrants), and male-biased sexual selection.

(continued)

**Box 3.13** (continued)

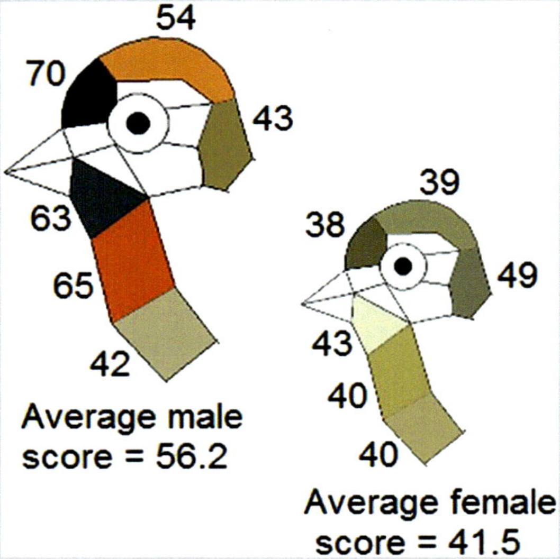


An example of how Dale et al. (2015) assigned color scores to different areas on the bodies of males and females of the same species. In general, dull green, olive, and brown colors were given lower plumage scores, whereas richer, more contrasting colors like black, purple, blue, red, and yellow were given higher scores (Figure from Dale et al. 2015; © 2015 Springer Nature, used with permission).

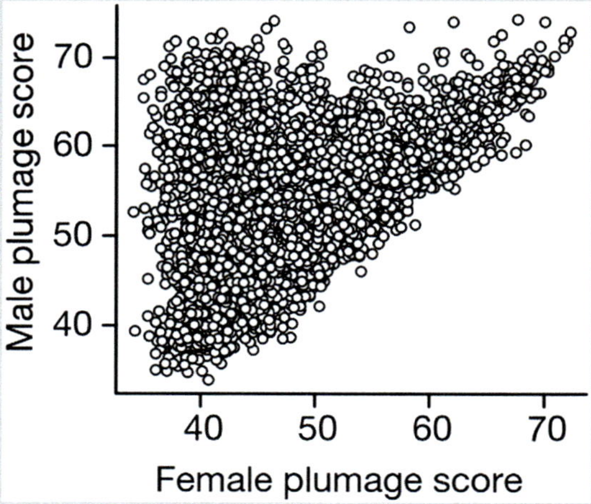


Among species of songbirds ($N = 5983$; each circle = one species), the plumage scores of males and females were found to be correlated. Note, however, that the plumage scores of males were higher than those of females for many species (Figure from Dale et al. 2015; © 2015 Springer Nature, used with permission).

Dale et al. (2015) found that larger species of birds are less sexually dichromatic than smaller species, which is consistent with the hypothesis that the risk of predation for less cryptic larger species is lower than that for less cryptic (more colorful) smaller species. In addition, species in tropical latitudes are generally less dichromatic than those breeding (either as residents or

(continued)

**Box 3.13** (continued)

migrants) at more temperate latitudes. A possible explanation for this latitudinal difference is that competition for resources is greater in the tropics than at higher latitudes. As a result, selection has favored the signaling of aggression via more colorful female plumage for tropical species. Finally, species with more male-biased sexual selection are more dichromatic with less colorful females than species where sexual selection acts to a similar degree on males and females. With male-biased sexual selection, the parental roles of males and females often differ, with females playing a greater role than males in incubation, brooding, and provisioning young. In such species, selection may favor more cryptic plumage for females to reduce the likelihood of being detected by predators when at and near nest sites (Dale et al. 2015). Selection for divergent parental roles may simultaneously select for differences in hormone levels that may then contribute to differences between the sexes in ornamentation (Kraaijeveld 2014).

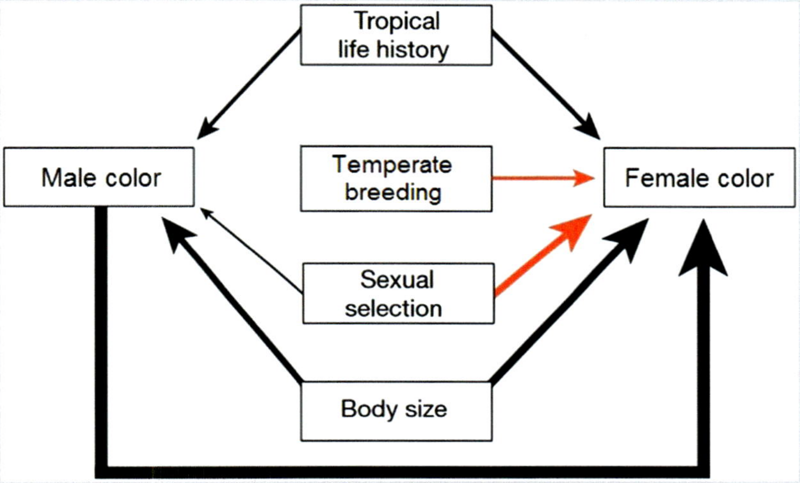


Arrows indicate the effects of several variables on the plumage ornamentation of male and female songbirds. The extent of effect is indicated by the thickness of lines, and the arrow color indicates whether the effect is positive (black) or negative (red) (Figure modified from Dale et al. 2015; © 2015 Springer Nature, used with permission).

(continued)

**Box 3.13** (continued)

Changes in gene expression can contribute to differences in the plumage of males and females. For example, genes that contribute to the development of the comb of Domestic Chickens are sensitive to testosterone (T), so males, with much higher circulating levels of testosterone than females, develop much larger combs than females (Photo from pxhere.com, CC0 Public Domain).

properties, and (4) preen oils in some species may not be bactericidal, but still provide some protection of feathers by forming a physical barrier between microbes and feathers (Reneerkens et al. 2008; Giraudeau et al. 2013).

About 2000 species of feather mites have been described (Proctor 2003; e.g., Fig. 3.124). Some investigators have suggested that feather mites are parasites and have a detrimental effect on their bird hosts (e.g., Figuerola et al. 2003). However, other investigators have concluded that feather mites are probably not parasites. Rather, the symbiotic relationship between feather mites and birds can be better described as commensalism or mutualism (e.g., Pap et al. 2005a, 2010; Doña et al. 2019) because the body condition of most species of birds does not seem to be affected by the presence of feather mites on their feathers (Galván et al. 2012). Feather mites may benefit by feeding on the wax secretions of the uropygial gland on feathers (Proctor 2003), with no apparent effect on birds (Galván et al. 2012). Doña et al. (2019) examined the gut contents of 1300 mites collected from 190 bird species representing 72 families and 19 orders and found that fungi and bacteria were their main

**Fig. 3.94** Examples of carotenoid-based colors in birds. **a**. Male Black-and-gold Cotinga (*Tijuca atra*) with a yellow wing patch. **b**. Male Golden-breasted Fruiteater (*Pipreola aureopectus*) with a yellow breast. **c**. Male Guianan Cock-of-the-Rock (*Rupicola rupicola*) with bright orange plumage. **d**. Male Purple-throated Fruitcrow (*Querula purpurata*) with burgundy throat patch. **e**. Male Pompadour Cotinga (*Xipholena punicea*) with burgundy body plumage. **f**. Male Lovely Cotinga (*Cotinga amabilis*) with purple throat and belly patches pigmented by a carotenoid called cotingin. (Figure from Prum et al. 2012; © 2012 Springer-Verlag, used with permission)

food sources. They found no evidence that the mites fed on either bird skin or blood and concluded that mites and birds have a commensalistic-mutualistic relationship.

The feather, or chewing, lice (Insecta: Phthiraptera) of birds spend their entire life cycles on the bodies of their hosts and some occupy different areas, or "microhabitats," on their hosts, including the head, wing, and body. Other lice are generalists and can be found in different areas on birds (Johnson et al. 2012; Fig. 3.125). More than 2700 species of feather lice in 140 genera have been described (Price et al. 2003). Avian chewing lice are divided into the suborders Ischnocera and Amblycerea, with Ischnocerans feeding on feathers and skin debris and Amblycerans on feathers as well as blood (Clayton et al. 1992).

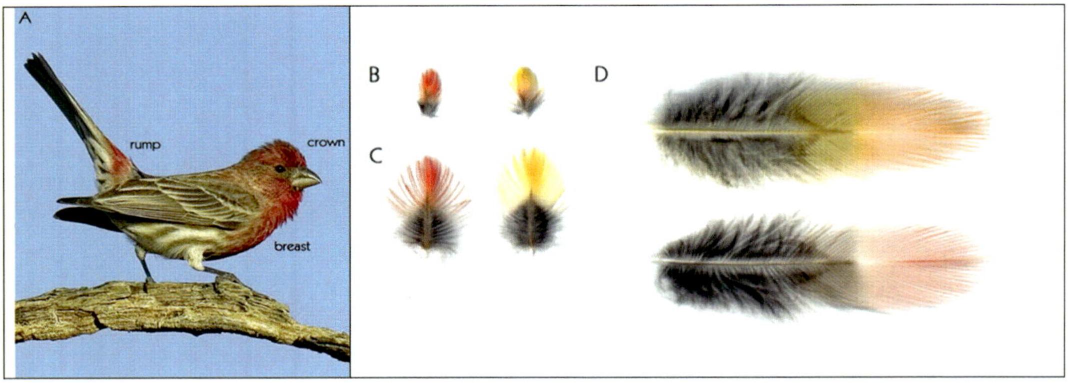


**Fig. 3.95** Carotenoid-colored feathers of male House Finches (*Haemorhous mexicanus*) (**a**) differ in size, shape, and pigmentation. Shown are feathers from the (**b**) crown, (**c**) breast, and (**d**) rump. Feather coloration is determined by the pigmentation of each individual barb ridge during growth with carotenoids delivered to the follicle by the centrally located pulp. (Figure modified from Badyaev and Landeen 2007; © 2007 Oxford University Press, used with permission)

By feeding on feathers, chewing lice can potentially reduce the quality of the plumage, creating small holes in feathers (Vas et al. 2008), reducing a bird's ability to thermoregulate (Booth et al. 1993), and increasing the chance of feathers breaking (Kose and Møller 1999). The effects of chewing lice on birds appear to vary among species and with the degree of infestation. Some investigators have reported that the damage caused by feather-chewing lice had little or no effect on the breeding success of Rock Pigeons (*Columba livia*; Clayton and Tompkins 1995), Common Swifts (*Apus apus*; Tompkins et al. 1996), and Tree Swallows (*Tachycineta bicolor*; Lombardo et al. 2015). However, with increasing numbers of lice and more feather damage, potential impacts on birds include delayed arrival dates for migrating birds (Møller et al. 2004), later initiation of breeding (Pap et al. 2005b), reduced likelihood of obtaining a mate (Moreno-Rueda and Hoi 2012), reduced body condition (Hoi et al. 2012), and even lower survival rates (Brown et al. 1995; Pap et al. 2005b).

Birds combat lice in a variety of ways, including dusting, sunning, scratching (with their feet; Fig. 3.126), allopreening, and preening. The importance of scratching and preening in controlling lice is apparent when birds are unable, either due to injury or in experiments, to engage in these behaviors. Birds unable to scratch due to leg or foot injuries often have large numbers of lice on their heads and neck, but not in areas where they can preen (Clayton 1991). Numbers of lice also typically increase when birds are prevented from preening in lab experiments (Clayton 1991). However, even scratching and preening may not remove all lice because feather lice have morphologies that make removal more difficult. Avian wing lice have long, slender bodies (Fig. 3.125) and try to evade removal by inserting themselves between the feather barbs of the primary wing feathers. Head lice remain on the heads of their hosts where birds cannot preen with their bills, but can do so with their feet. Head lice have short, rounded bodies with a rostral groove on their head that allows them to

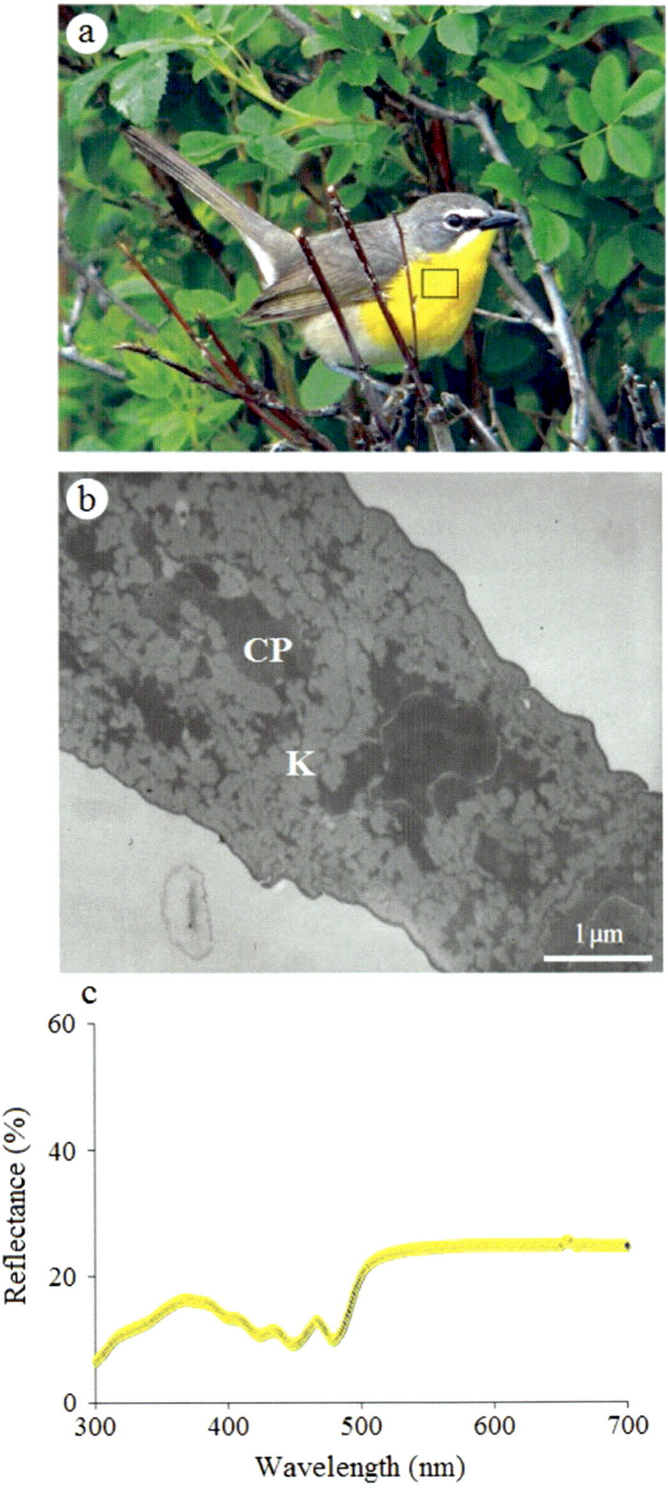


**Fig. 3.96** (**a**) A Yellow-breasted Chat (*Icteria virens*) (photo by Denny Granstrand), with the black box indicating the region where the transmission electron microscope (TEM) image and spectral data were obtained. (**b**) TEM image of a noniridescent barb from a yellow feather of a Yellow-breasted Chat, showing the diffuse distribution of carotenoid pigments (CP) in a keratin (K) matrix. (**c**) Spectral data for a yellow feather of a Yellow-breasted Chat; in the visible light spectrum, the color yellow corresponds to a wavelength range of about 570–590 nm. (Figure from Shawkey et al. 2009a, b; © 2009 The Royal Society, used with permission)

**Fig. 3.97** Bright red coloration of a Red-and-green Macaw (*Ara chloropterus*). Of about 350 species of parrots, 80% have red in their plumage. (Photo by Arjan Haverkamp, Wikipedia, CC BY 2.0, https://creativecommons.org/licenses/by/2.0/)

tightly grip feather barbs with their mandibles (Fig. 3.125). Body lice escape preening birds by moving to the downy portions of body feathers or by moving to different feathers (Clayton 1991). Lice that are generalists can be found over most of a bird's body, wings, and sometimes heads and likely escape preening by running through the feathers (Clay 1949). These lice have an intermediate body shape and rounded heads (Johnson et al. 2012).

## 3.21 Preening and Other Defenses against Ectoparasites

To keep feathers in good condition, birds must preen on a regular basis. An examination of the time budgets of 62 different species revealed that birds spent nearly 10% of the day preening and bathing (Cotgreave and Clayton 1994). Preening, and for some species of birds, allopreening (Fig. 3.127), is critical for straightening and oiling feathers and removing dirt, debris, and ectoparasites. Birds unable to preen (e.g., because of a bill deformity) have more ectoparasites and may, as a result, have reduced survival and mating success (Clayton 1990). For example, after investigators dirtied the plumage of males with sand and soil, female Budgerigars (*Melopsittacus*

**Fig. 3.98** The yellow and yellow-orange feathers of some penguins, including this King Penguin (*Aptenodytes patagonicus*), contain a unique pigment called spheniscin (so-named because penguins are in the order Sphenisciformes). (Photo from pxhere.com, CC0 Public Domain)

**Fig. 3.99** Red-crested Turaco (*Tauraco erythrolophus*) with red feathers (turacin pigment) and green feathers (turacoverdin pigment). (Photo from pxhere.com, CC0 Public Domain)

*undulatus*) spent more time near males that were able to preen than near males unable to preen (because a collar was placed around their neck and not removed until just before they were presented to females), suggesting that male plumage quality is influenced by their preening behavior and, indirectly then, preening can influence mate choice decisions (Zampiga et al. 2004).

Beaks play an important role in preening for most birds. Although beak morphology is adapted primarily for feeding, most birds also use their beaks for preening. For example, comparative studies suggest that some features of beak morphology may be adapted to deal with harmful ectoparasites. For example, species of birds with long maxillary overhangs have fewer feather lice than species with short overhangs (Clayton and Walther 2001). Similarly, within species, populations with long overhangs have fewer lice than populations with short overhangs (Moyer et al. 2002). Experimental trimming of the tiny (1–2 mm) maxillary overhang of Rock Pigeons (*Columba livia*; Fig. 3.128) had no effect on their feeding efficiency, but resulted in a dramatic increase in feather lice and the feather damage they caused. Electronic recordings of pigeon beak movements revealed that the overhang generates a shearing force against the tip of the lower mandible, and this force damages parasite exoskeletons (Clayton et al. 2005).

Additional work is still needed to determine the role of the overhang in other species of birds. For example, although the presence of an overhang does not impair feeding in Rock Pigeons, this may not be the case for other species. In addition, some birds, including woodpeckers, hummingbirds, herons, skimmers, and oystercatchers, have no overhang, presumably because it would hinder their feeding ability. Because these birds all have ectoparasites, they

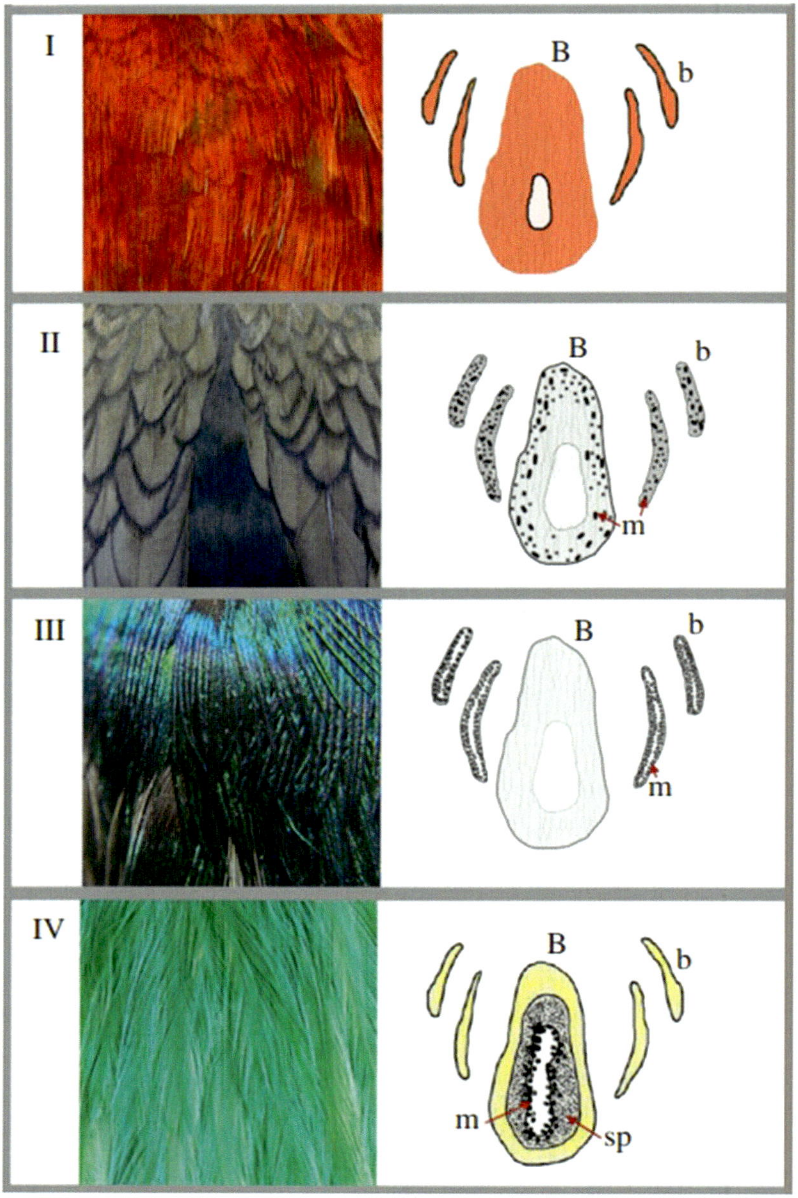


**Fig. 3.100** Different mechanisms for producing feather colors. (I) Colors produced by red or yellow pigments (e.g., carotenoids and pteridines), (II) dark browns and blacks produced by melanins, (III) predominantly structural coloration, and color produced with a combination of pigmentary and feather structures. Drawings on the right represent cross-sections of feather barbs and barbules. B = feather barb, b = barbule, m = melanosome, and sp = keratinous spongy layer. (Figure from Shawkey and D'Alba 2017; © 2017 The Authors. Published by the Royal Society, used with permission)

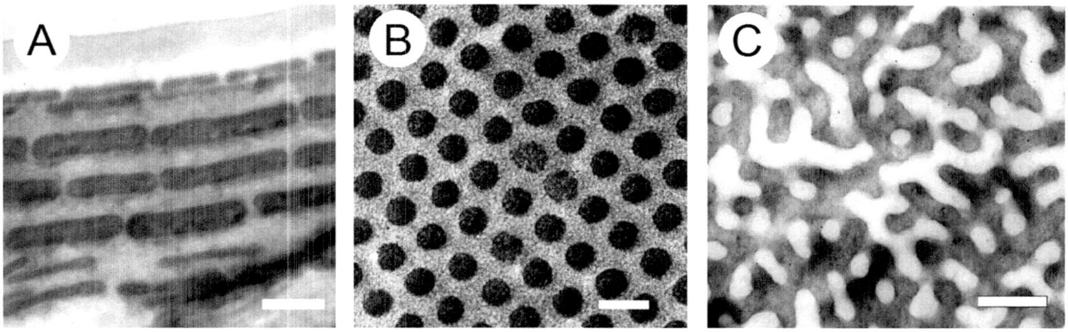


**Fig. 3.101** Three types of coherently scattering nanostructures. (**a**) Laminar array of plate-shaped melanosomes in the barbules of iridescent green feathers of a Superb Sunbird (*Cinnyris superbus*). (**b**) Crystal-like, hexagonal array of parallel collagen fibers from the green facial caruncle of a Velvet Asity (*Philepitta castanea*). (**c**) Quasi-ordered arrays of keratin and air vacuoles from the blue feather barbs of a Rose-faced Lovebird (*Agapornis roseicollis*). Scale bars = 200 nm. (Figure from Prum and Torres 2003a; © 2003 Oxford University Press, used with permission)

apparently have other mechanisms, such as scratching, sunning, dusting, and feather chemistry, for eliminating ectoparasites or minimizing their damage (Clayton and Moore 1997; Moyer and Clayton 2003). For example, the feathers and skin of several species of birds in the genus *Pitohui* contain a neurotoxin (homobatrachotoxin) that likely deters predators and also appears to deter ectoparasites (Dumbacher 1999).

Birds may also use their feet to groom feathers. In fact, birds with a deformed or missing leg often have large numbers of ectoparasites, especially around their head and neck (areas that cannot be preened with the bill; Clayton 1991). Some birds are actually more dependent on their feet than their bill for grooming and cleaning feathers. For example, hummingbirds and toucans have relatively long bills that are less useful for preening, particularly their upper bodies. Such long-billed species spend relatively more time preening with their feet than with their bill (Clayton and Cotgreave 1994).

Several species of birds have pectinate claws on their middle toenails that may make ectoparasite removal from feathers more efficient (Fig. 3.129). The serrations on these claws resemble teeth of combs and may help remove lice and other ectoparasites from feathers (Moyer and Clayton 2003; Bush et al. 2012).

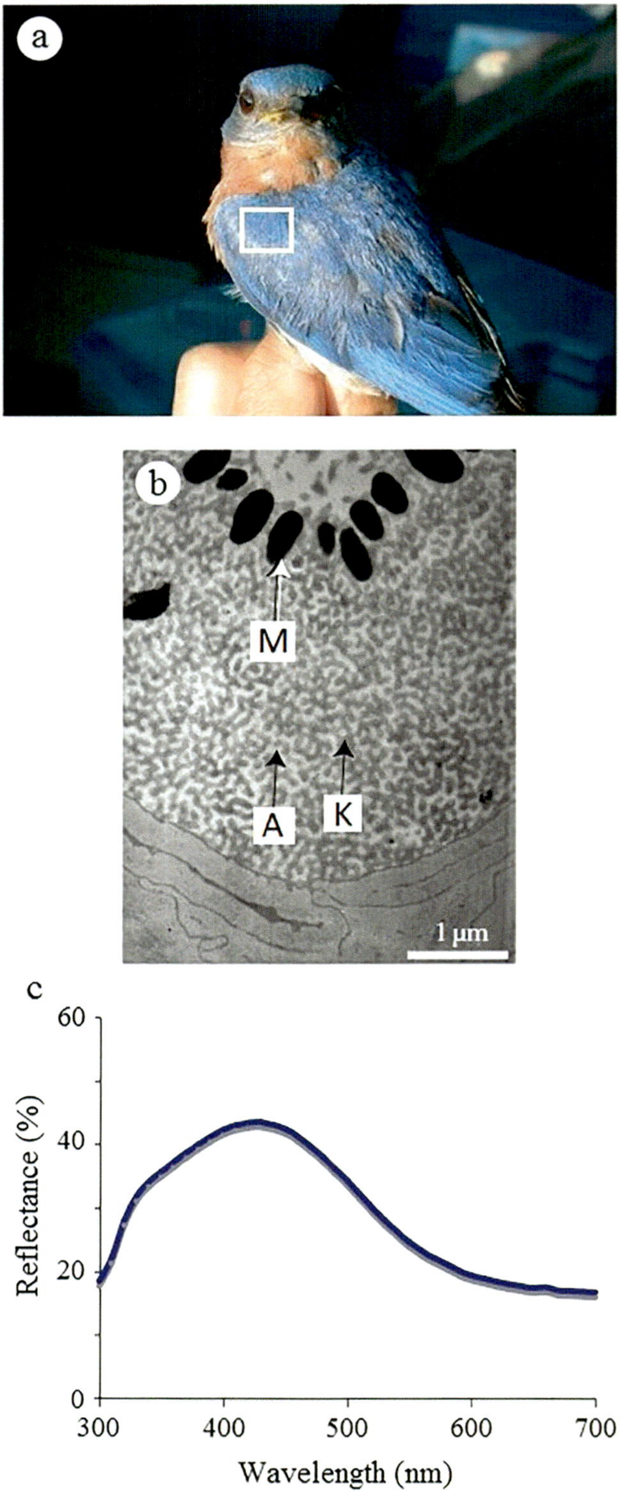


**Fig. 3.102** (**a**) A male Eastern Bluebird (*Sialia sialis*) (Photo by Mark Liu), with the box indicating the area where the transmission electron microscope (TEM) image and spectral data were obtained. (**b**) TEM image of a noniridescent barb of a blue feather of an Eastern Bluebird showing a solid keratin layer surrounding a spongy layer, or matrix of keratin (K) and air (A) above a layer of melanin granules (M) that absorb incoherently backscattered light. (**c**) Spectral data from a blue feather of an Eastern Bluebird; in the visible spectrum, the wavelengths for the blue color range from 450 to 495 nm. (Figure from Shawkey et al. 2009a, b; © 2009 The Royal Society, used with permission)

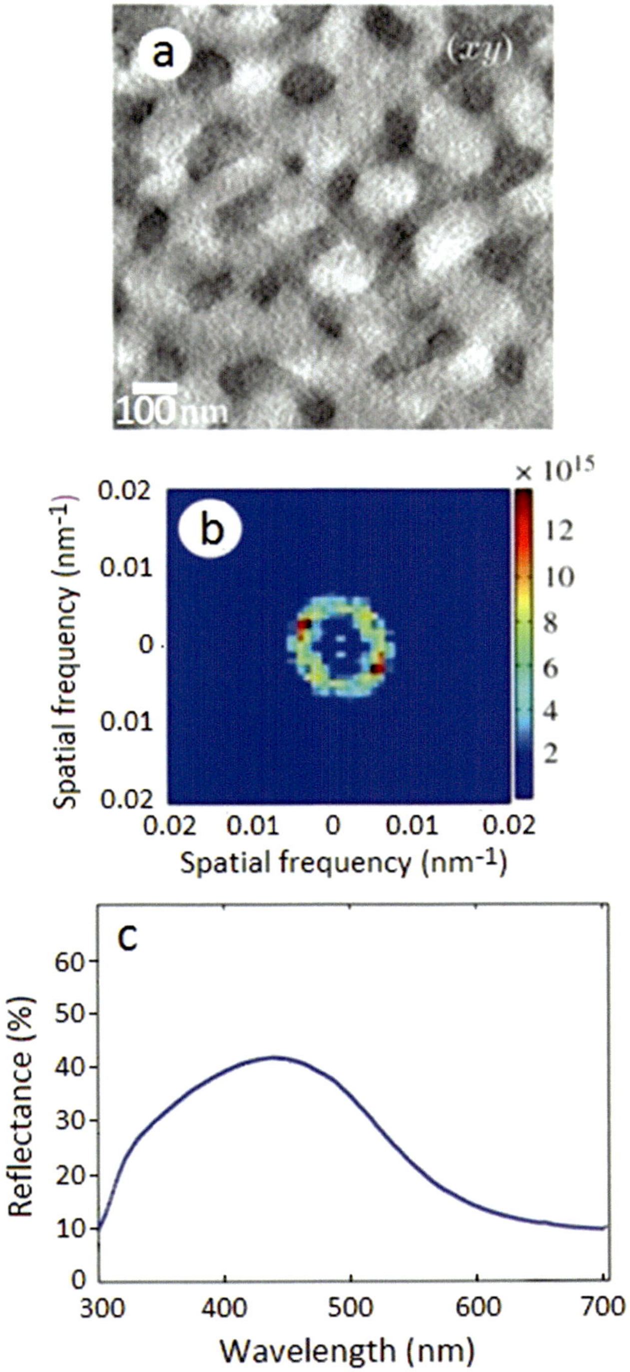


**Fig. 3.103** A three-dimensional analysis of the blue feather barbs of an Eastern Bluebird (*Sialia sialis*). (**a**) Spongy medullary tissue of a barb; dark areas are keratin and light areas are air. (**b**) A three-dimensional Fourier power spectrum, with the ring shape indicating an orderly arrangement of keratin and air over short spatial scales. Relative magnitude is indicated on the right. (**c**) Measured reflectance spectrum from spongy tissue with a peak in the blue wavelength (475 nm). (Figures from Shawkey et al. 2009a, b; © 2009 The Royal Society, used with permission)

**Fig. 3.104** Effect of varying the size and spacing of "scatterers" on the wavelengths of light absorbed or reflected. Variation in the diameter and spacing of holes in a three-layer sheet of silver and silica differentially absorb and reflect light to produce different colors. (Figure from Cheng et al. 2015; open-access article licensed under a Creative Commons Attribution 4.0 International License, https://creativecommons.org/licenses/by/4.0/)

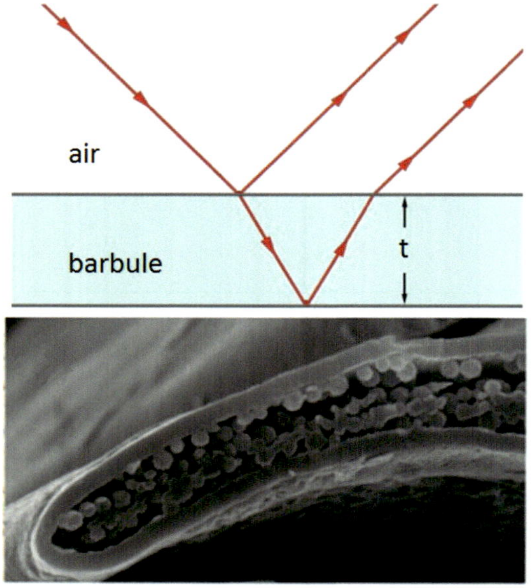


**Fig. 3.105** Light striking a thin film or structure like the barbule of a feather (below) will be partially reflected and partially refracted at the top surface. The refracted ray is partially reflected at the bottom surface and emerges as a second light ray. These rays will interfere either constructively or destructively depending on the thickness (t) of the film and the indices of refraction of the substances that make up the barbules. (Top figure modified from Matin et al. 2010; © 2010 AIP Publishing, used with permission; Bottom, photo of barbule modified from Kinoshita et al. 2008; © 2008 IOP Publishing Ltd., used with permission)

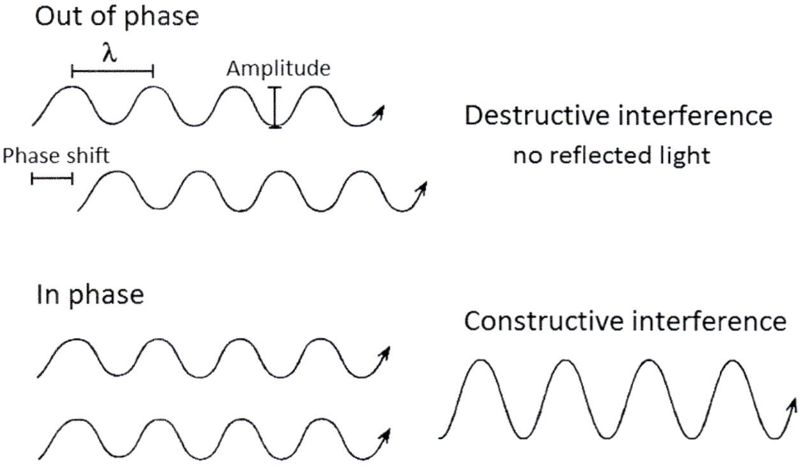


**Fig. 3.106** Destructive and constructive interference. Reflected light waves traveling different distances (path lengths) may either be out of phase or in phase with each other. Light waves that are completely out of phase cancel each other out, a condition called destructive interference. Light waves that are perfectly in phase sum together and generate a wave with twice the amplitude. $\lambda$ = wavelength. (Figure modified from Prum et al. 1994; © 1994 Wiley-Liss, Inc., used with permission)

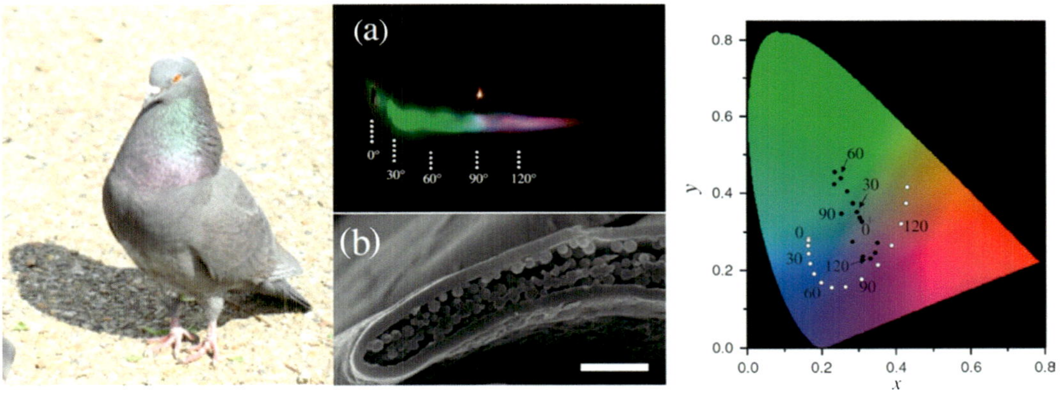


**Fig. 3.107** Rock Pigeon (*Columba livia*) and (**a**) the reflection pattern of a barbule, where 0° indicates the direction of incident white light. (**b**) Scanning electron micrograph image of the cross-sections of the barbule. The right figure indicates the chromaticity diagram showing the loci of the reflection in thin-layer interference at various angles $2\theta$ for the layer thickness of 650 nm (closed circle) and 400 nm (open circle) with the refractive index of 1.5, where $\theta$ is the angle of incidence. (Figure from Kinoshita et al. 2008; © 2008 IOP Publishing Ltd., used with permission).

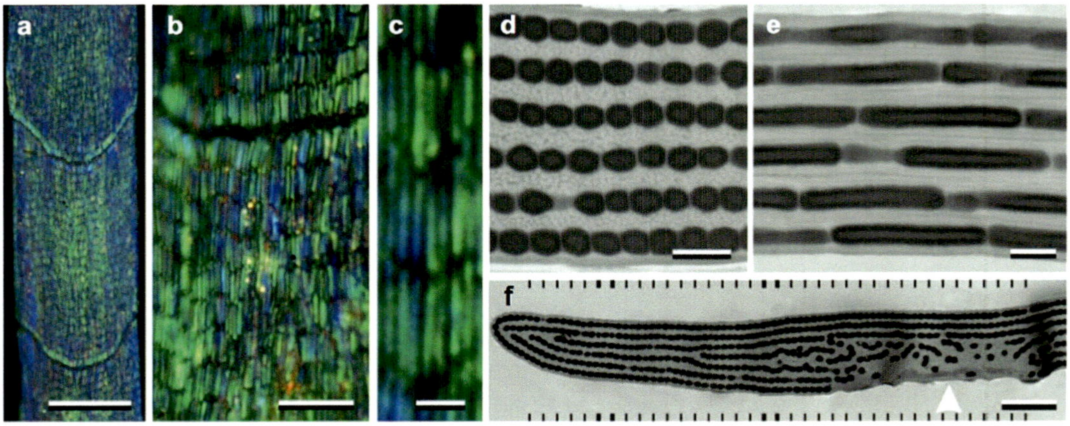


**Fig. 3.108** Distal barbules of the occipital feathers of a male Lawes's Parotia (*Parotia lawesii*). (**a**) A single barbule consisting of a row of cells. (**b**) At higher magnification, individual melanin rodlets can be discerned. (**c**) Close-up view showing the single melanin rodlets arranged in parallel bands. (**d**) Transmission electron micrograph of a cross-section, revealing ordered layers of rodlets (diameter = ~250 nm). (**e**) Longitudinal section of the rodlets. (**f**) A cross-section perpendicular to the barbule axis. The layered order of the rodlets is distorted in the area indicated by the arrowhead. Scale bars: (**a**) = 20 μm, (**b**) = 5 μm, (**c**) = 1 μm, (**d**) and (**e**) = 0.5 μm, and (**f**) = 2 μm. (Figure modified from Stavenga et al. 2015; © 2015 Springer Nature, used with permission)

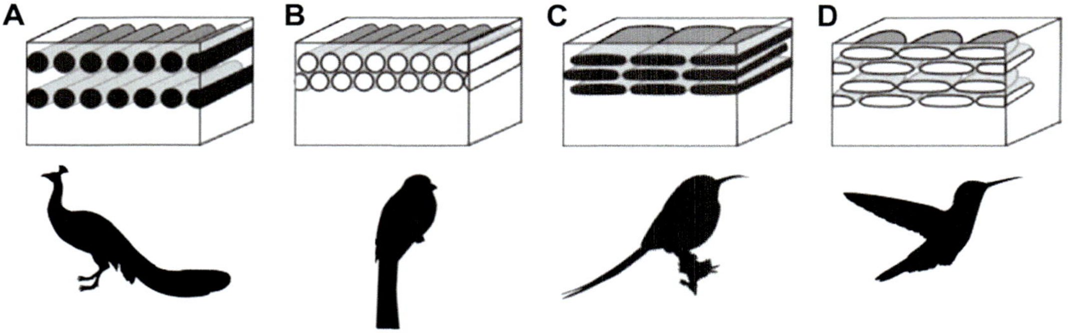


**Fig. 3.109** Melanosome morphologies found in the iridescent feathers of several taxa of birds. Silhouettes represent examples of bird families with each type of morphology. (**a**) Solid cylindrical melanosomes, (**b**) hollow cylindrical melanosomes, (**c**) flat and solid melanosomes, (**d**) flat and hollow melanosomes. Silhouettes: (**a**) Phasanidae, (**b**) Trogonidae, (**c**) Nectariniidae (sunbirds), and (**d**) Trochilidae (hummingbirds). The most common type of iridescent feathers has solid cylindrical melanosomes. Hollow melanosomes have more interfaces for scattering light, increasing the saturation and brightness of colors. (Figure from Nordén et al. 2019; open-access article distributed under the terms of the Creative Commons CC BY license, https://creativecommons.org/licenses/by/4.0/. Image credits for silhouettes: Natasha Sinegina and Katerina Ryabtseva, downloaded under a Creative Commons license, https://creativecommons.org/licenses/by/4.0/)

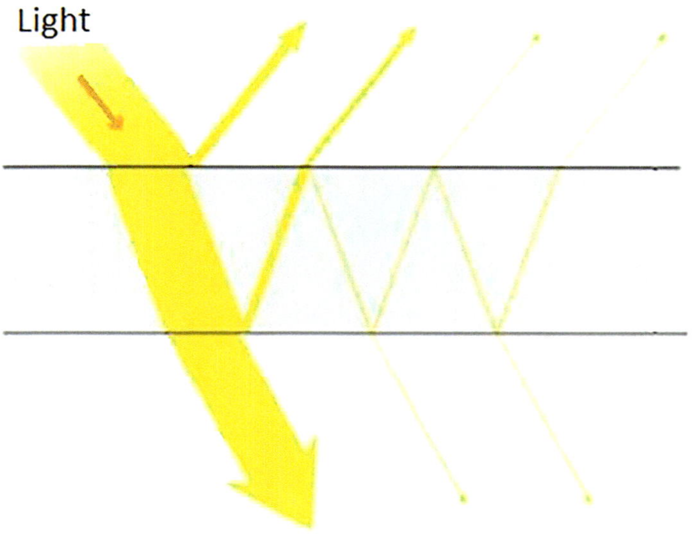


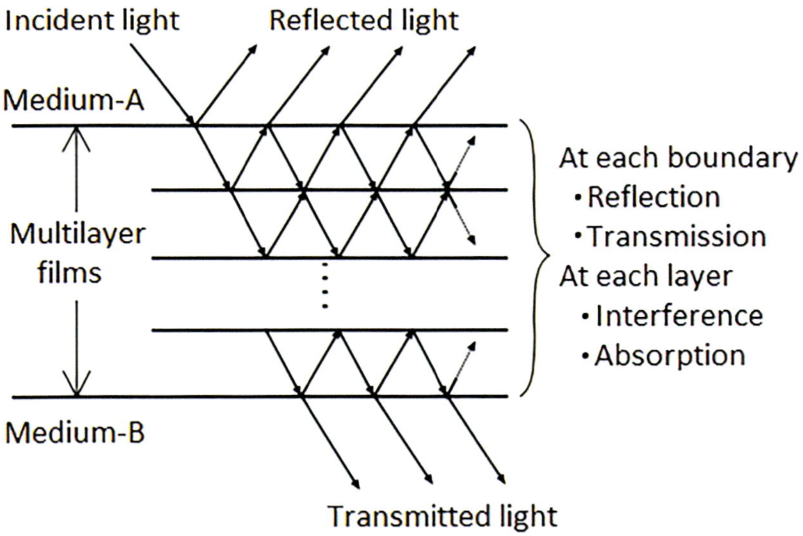


**Fig. 3.110** Top, Thin-film interference. Light waves reflected by the upper and lower boundaries of a thin-film interfere with each other. Bottom, Multi-film interference. Light waves reflected by multiple boundaries interfere with each other. (Top figure modified from Sun et al. 2013; © 2013 The Royal Society of Chemistry, used with permission; Bottom figure from Hirayama et al. 2001; © 2001 Elsevier Science Ltd., used with permission)

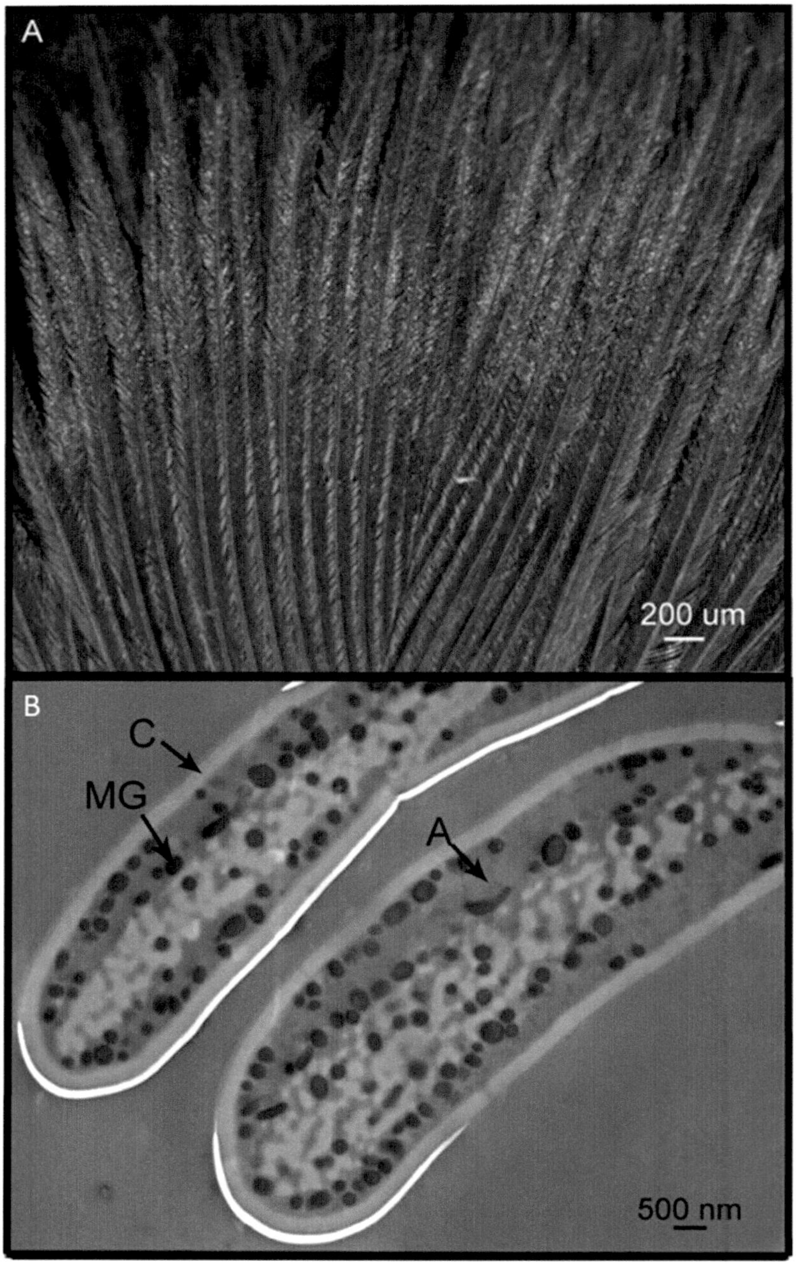


**Fig. 3.111** (**a**) Iridescent feathers of a Mourning Dove (*Zenaida macroura*). (**b**) Cross-section of an iridescent barbule. The iridescence is produced by thin-film interference from a single (~335 nm) layer of keratin around the edge of feather barbules, with a layer of air and melanosomes below the keratin layer. A, air; C, keratin cortex, and MG = melanosome granules. (Figure from Shawkey et al. 2011; © 2011 Published by Elsevier GmbH, used with permission)

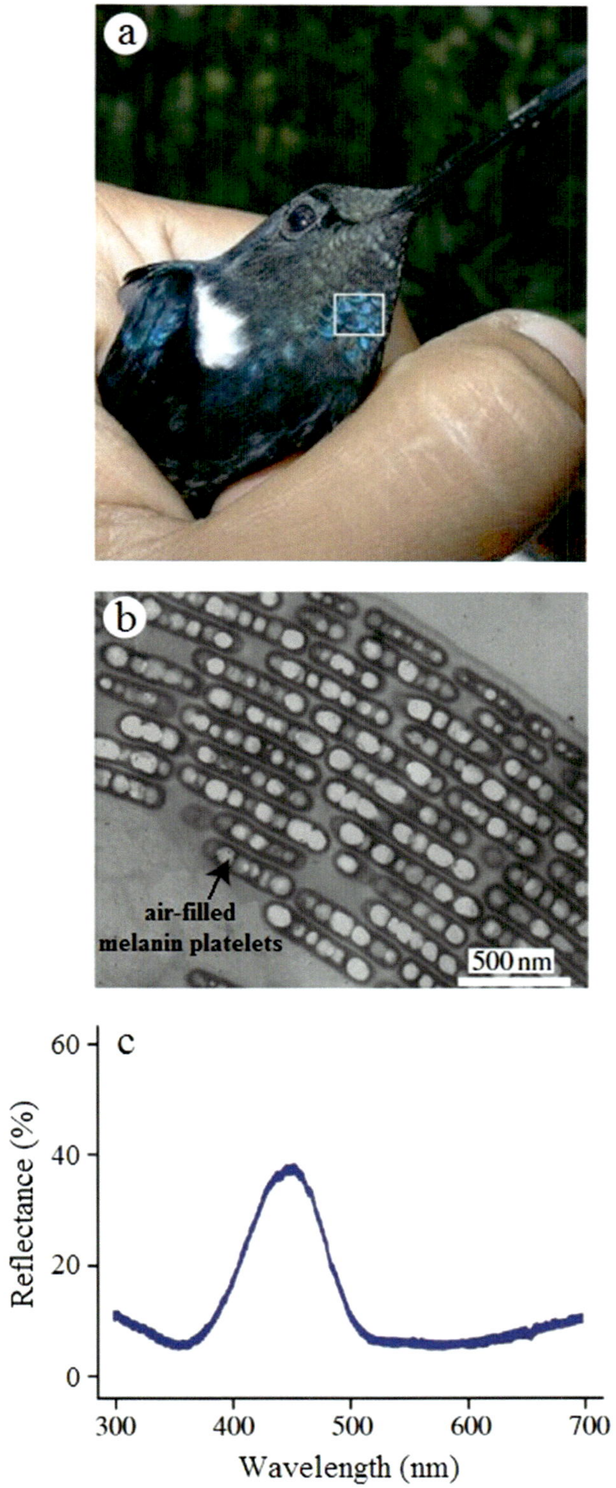


**Fig. 3.112** (**a**) A Black Inca (*Coeligena prunellei*) hummingbird, with the white box indicating the colored area where the spectral data were derived. (**b**) Transmission electron microscope image of an iridescent green barbule from a Rainbow Starfrontlet (*Coeligena iris*) hummingbird showing highly ordered layers of air-filled melanin platelets in a keratin matrix. (**c**) Spectral data for the Black Inca hummingbird. (Figure from Shawkey et al. 2009a, b; © 2009 The Royal Society, used with permission)

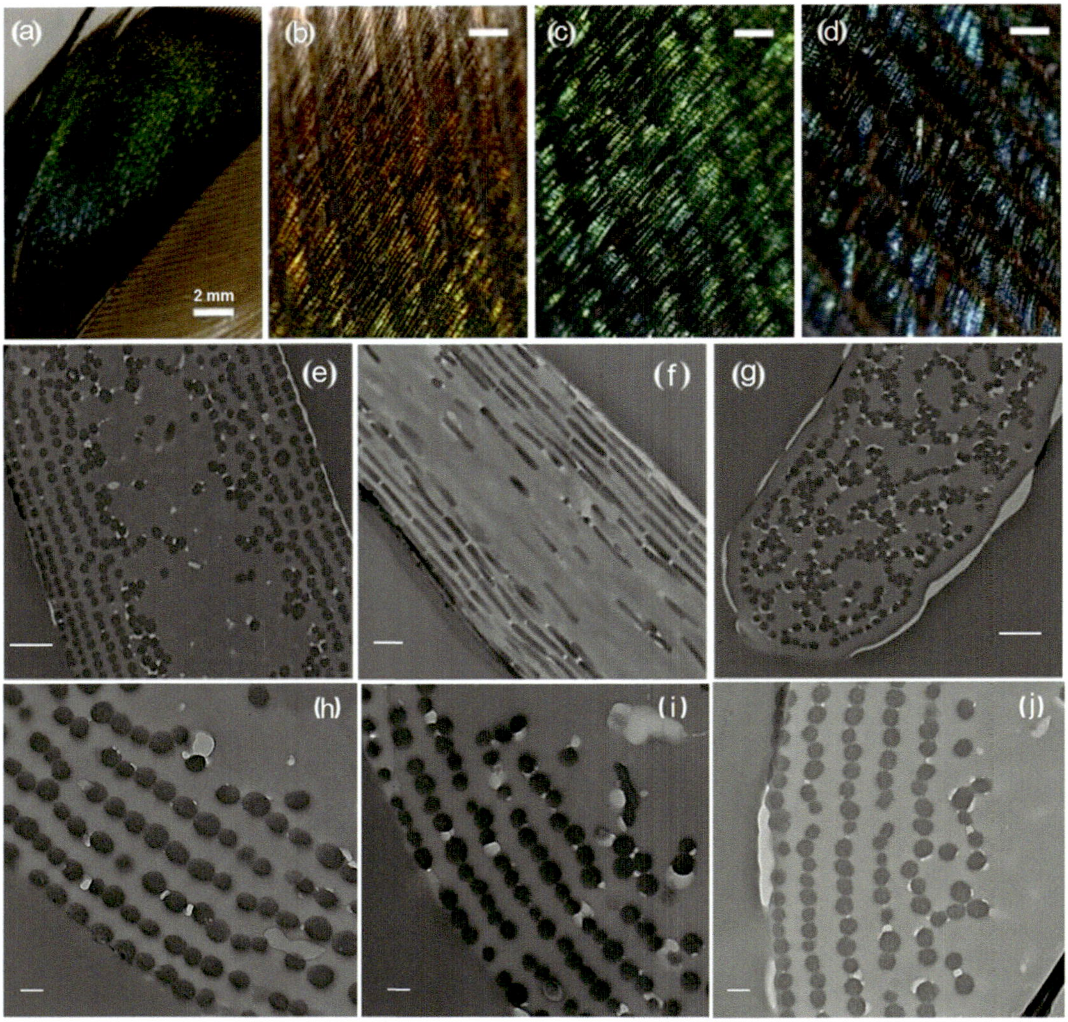


**Fig. 3.113** (**a**) Iridescent covert feather of a male Common Bronzewing (*Phaps chalcoptera*; Columbiformes). (**b–d**) Barbules in different regions of the covert feather show reddish, green, and blue/purple coloration. Scale bars = 200 µm. (**e**) Cross-section through a red barbule, (**f**) a green barbule, and (**g**) a noniridescent brown barbule. Scale bars = 500 nm. (**h**) Cross-section through a red barbule, (**i**) a green barbule, and (**j**) a blue barbule. Scale bar = 100 nm. (Figure modified from Xiao et al. 2014; Reprinted with permission of © The Optical Society)

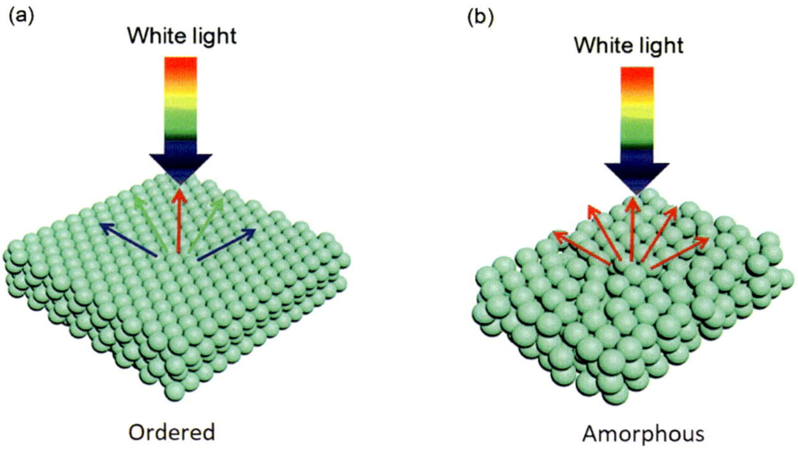


**Fig. 3.114** Optical properties of (**a**) an ordered photonic structure and (**b**) an amorphous (or quasi-ordered) photonic structure. The structural color of an ordered photonic structure changes with the viewing angle; the structural color of the amorphous array does not. (Figure modified from Liu et al. 2019; © 2019 Royal Society of Chemistry, used with permission)

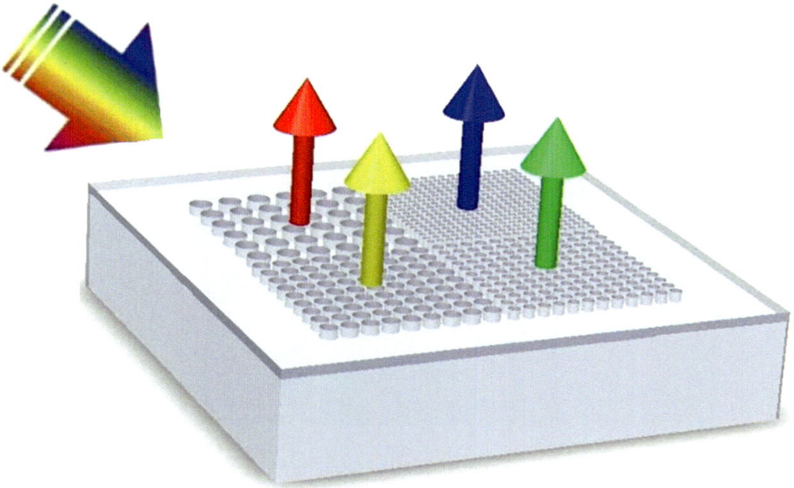

**Fig. 3.115** Illustration of spatially resolved, wavelength-selective outcoupling of wave-guided light using an array of photonic crystals. Lightwave guided into the edge of the substrate is outcoupled by photonic crystal patterns corresponding to different bands of wavelengths. (Figure modified from Zhou et al. 2016 [arrow]; © 2016 Elsevier Ltd., used with permission, and Pervez et al. 2010; Reprinted with permission of © The Optical Society)

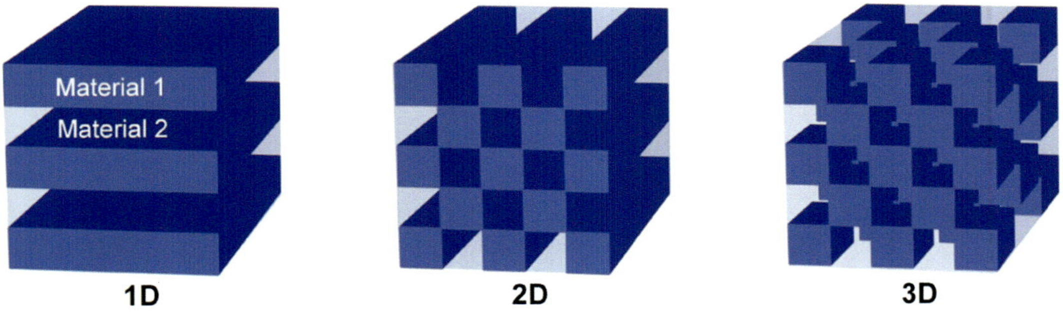


**Fig. 3.116** Schematic representation of one-, two-, and three-dimensional photonic crystals consisting of materials with different refractive indices. Among birds, material 1 is typically melanin, and material 2 is air. Three-dimensional photonic crystals are found in some beetles and butterflies, and are also found in at least one species of bird (Blue-winged Leafbird, *Chloropsis cochinchinensis*; Saranathan et al. 2021). (Figure from Chan et al. 2013; © 2013 WILEY-VCH Verlag GmbH & Co. KGaA, Weinheim, used with permission)

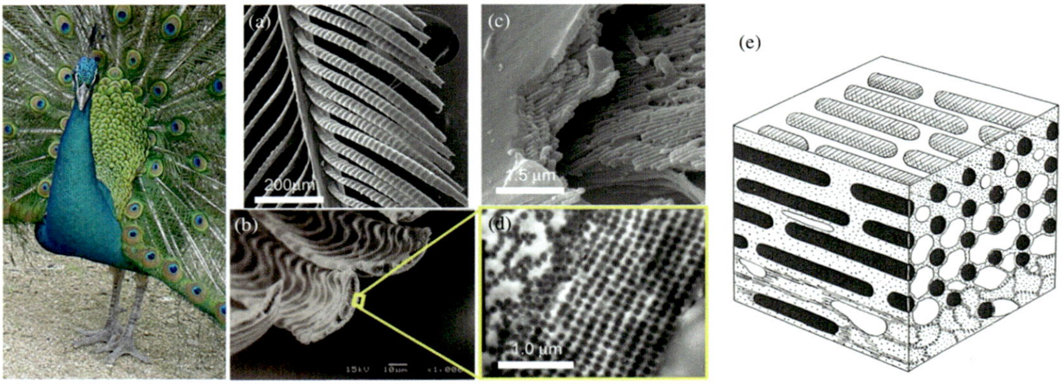


**Fig. 3.117** Indian Peafowl (*Pavo cristatus*). Scanning electron microscope images of (**a**) barbules, (**b**) the cross-section, and (**c**) interior of a barbule. (**d**) Transmission electron microscope image of the cross-section of a barbule. (**e**) Schematic illustration of a two-dimensional photonic crystal in a peacock barbule. (Figure from Kinoshita et al. 2008; © 2008 IOP Publishing, used with permission)

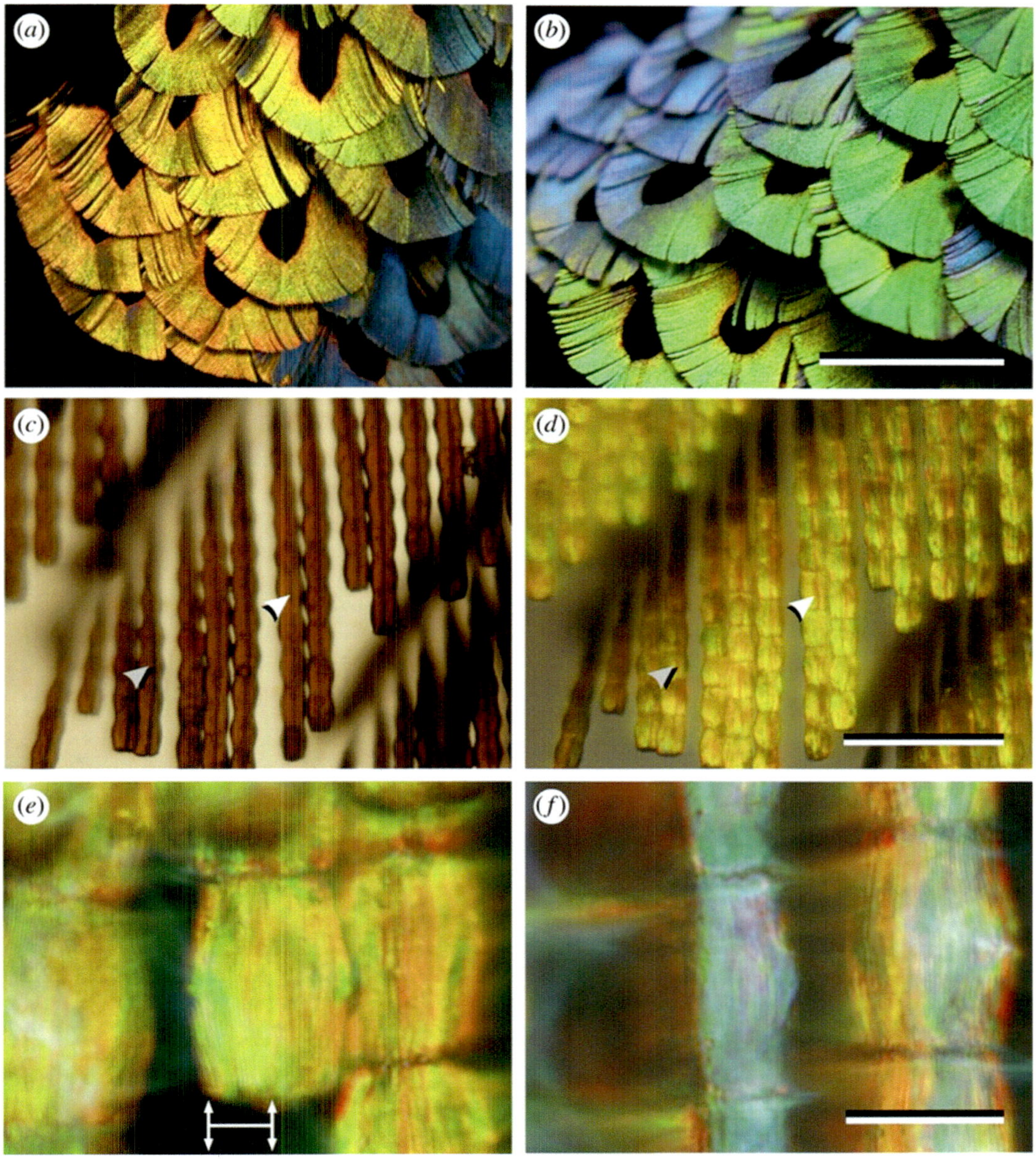


**Fig. 3.118** Colors of male Lawes' Parotia (*Parotia lawesii*) breast feathers. (**a**) With an approximately fronto-parallel view and illumination behind the observer, the reflection is mainly yellow-orange. (**b**) An oblique view yields blue and green reflections. (**c**) A transmitted-light microscope photograph of a few barbs shows brown, segmented barbules with a central dark line, indicated by arrows. (**d**) The same barbs photographed with epi-illumination show yellow barbules, also with a central dark line, indicated as in (**c**). (**e**) At high magnification, the cushion-shaped, yellow barbule segments show longitudinal stripes separated by 0.5–0.6 μm (approximately 15 lines are visible over a distance of 8 μm, as indicated by the horizontal line with vertical arrows). (**f**) Slight rotation of the barbules yields a blue reflection from one side of each segment. Scale bars (**a**, **b**) = 1 cm; (**c**, **d**) = 100 μm; (**e**, **f**) = 20 μm. (Figure from Stavenga et al. 2011; © 2010 The Royal Society, used with permission)

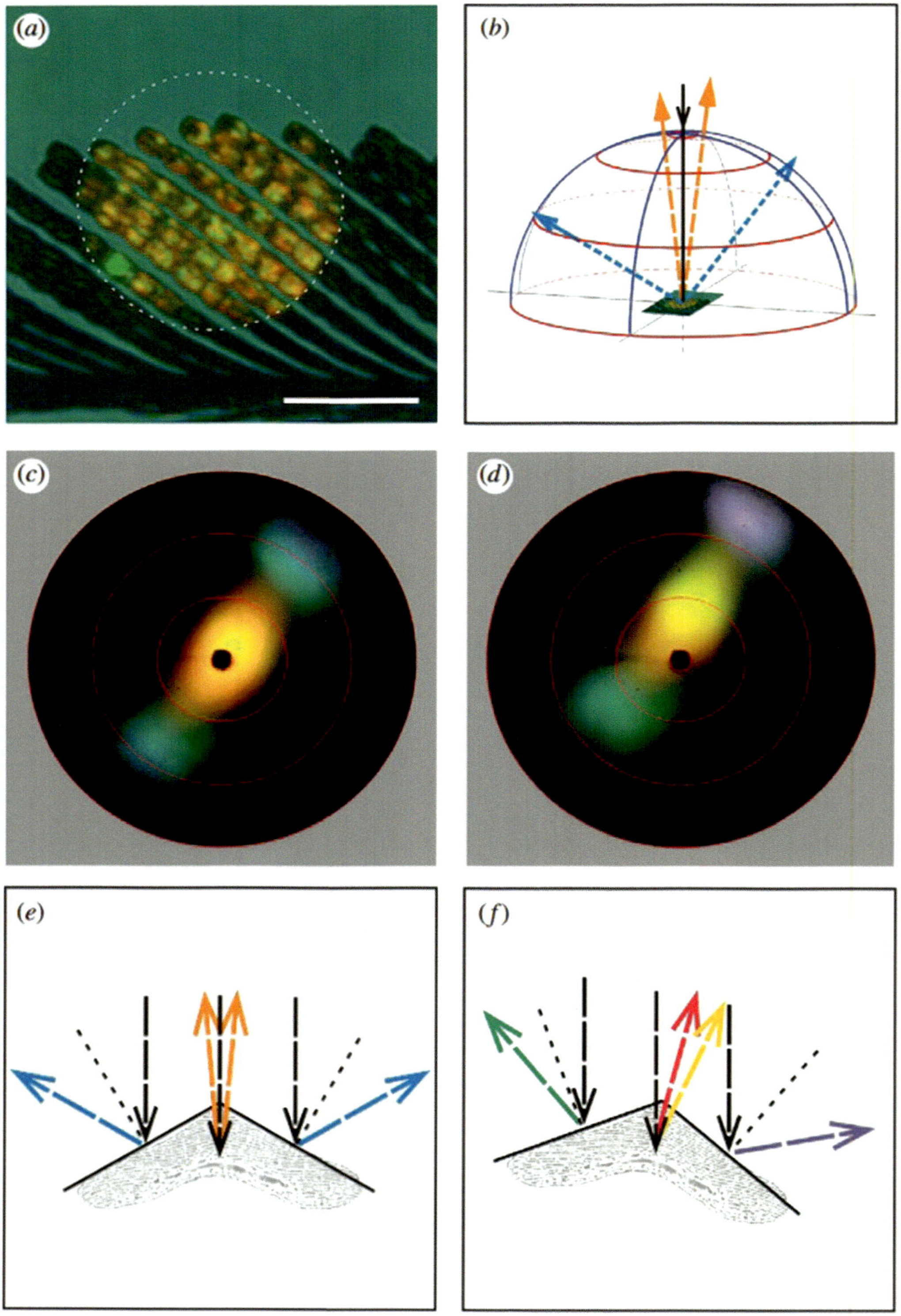


**Fig. 3.119** Image showing how a single barbule reflects light from a point source. The barbule acts as three separate colored mirrors, with each reflecting in a different direction. (**a**) A 200-μm-diameter spot is illuminated with a narrow beam of white light; scale bar, 100 μm. (**b**) Diagram of how normally incident light is reflected.

**Fig. 3.120** Captive breeding has resulted in Budgerigars (*Melopsittacus undulatus*) with a wide variety of colors not found in wild populations. (Photo for pxhere.com, CC0 Public Domain)

**Fig. 3.119** (continued) Yellow-orange light is reflected about normally, and bluish light is reflected in two opposite directions with angles of 60°; (**c**) Polar scatterograms, showing the angular distribution of the scattered light by the barbule with about normal illumination. The red circles indicate inclination angles of 5°, 30°, 60° and 90°. The central black spot (approximating the circle with polar angle 5°) is due to the axial hole in the ellipsoidal mirror of the scatterometer. (**d**) The angular distribution of the scattered light from the barbule rotated over a 10° angle. (**e, f**) Diagrams explaining the polar scattergrams in **c** and **d**. (Figure from Stavenga et al. 2011; © 2010 The Royal Society, used with permission)

**Fig. 3.121** Comparison of glossy and matte red feathers. (**a**) Red-crested Cardinal (*Paroaria coronata*), (**b**) Scarlet Minivet (*Pericrocotus speciosus*), (**c**) (feathers from the crown of a Red-crested Cardinal and (**d**) the breast of a Scarlet Minivet, (**e**) Electron microscopy image of a cross-section of a barb from a feather in the crown of a Red-crested Cardinal, and (**f**) of a barb and barbules from a feather from the breast of a Scarlet Minivet. (Figure from Iskandar et al. 2016; © 2016 Oxford University Press, used with permission)

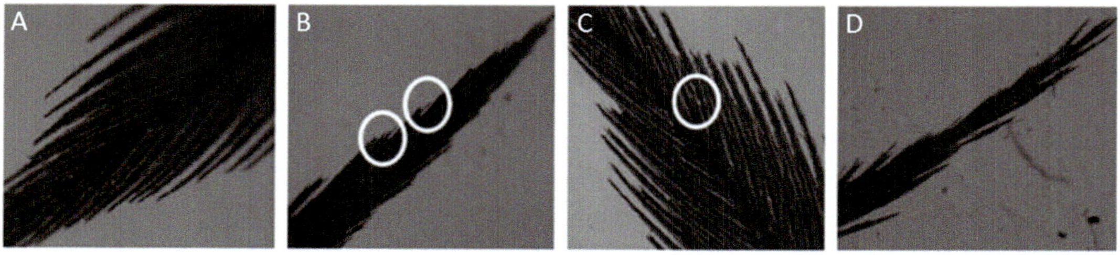


**Fig. 3.122** Examples of degradation of feathers of Spotless Starlings (*Sturnus unicolor*) by bacteria. (**a**) No degradation, (**b**) tips of barbs and slightly degraded, (**c**) holes in barbs, and (**d**) many barbs and barbules are degraded. (Figure modified from Ruiz-Rodríguez et al. 2015; © 2014 The Authors. Functional Ecology © 2014 British Ecological Society, used with permission)

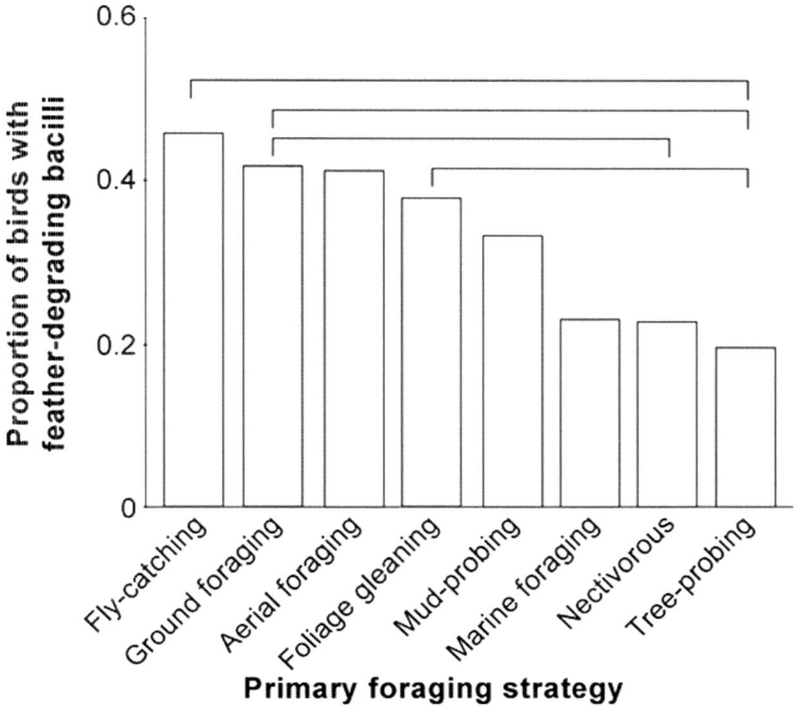


**Fig. 3.123** Proportion of birds in different primary foraging guilds with feather-degrading bacteria (*Bacillus* spp.). Brackets indicate foraging guilds that are significantly different from each other. (Figure from Kent and Burtt 2016; © 2016 Oxford University Press, used with permission)

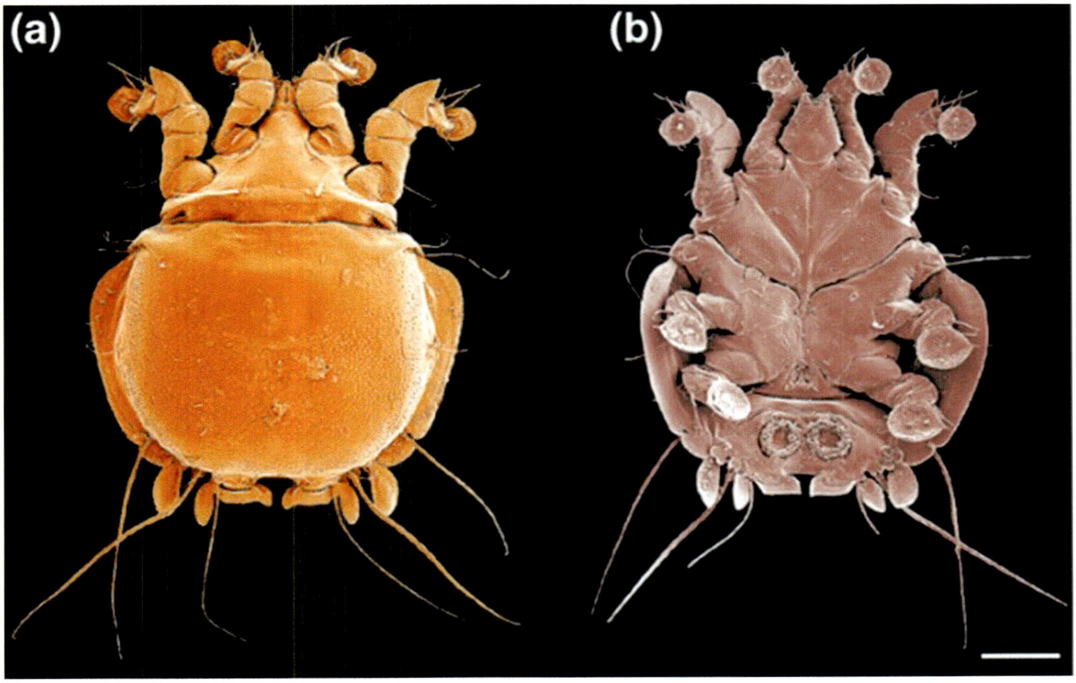


**Fig. 3.124** Dorsal (**a**) and ventral (**b**) views of a male feather mite (Astigmata; Freyanidae) showing the sucker-like ventral setae used by males to hold females during copulation. Scale bar = 100 μm. (Figure from Proctor and Owens 2000 and reproduced with permission from Dave Walter, University of Queensland; © 2000 Elsevier Science Ltd., used with permission)

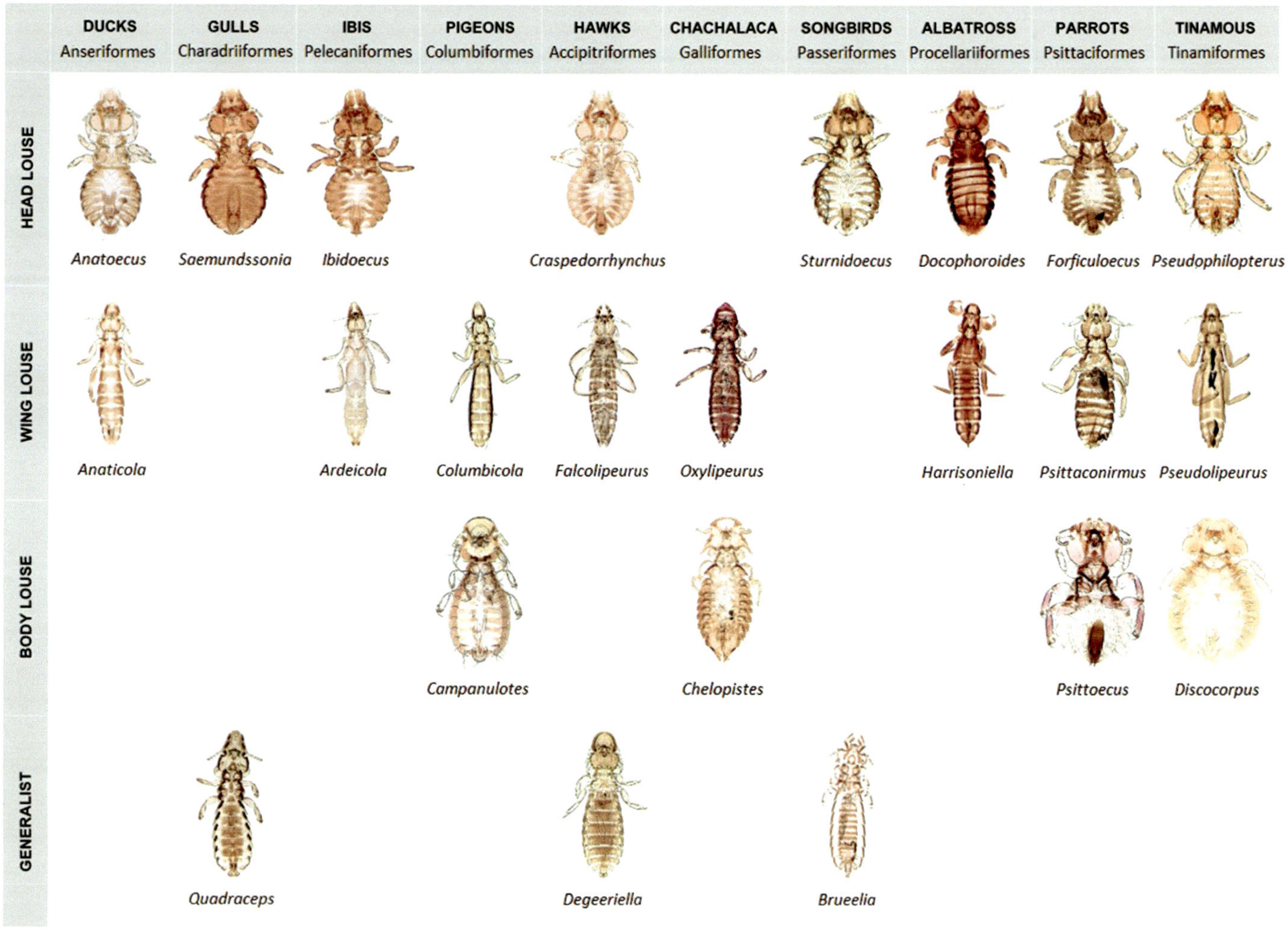


**Fig. 3.125** Examples of feather lice found in various taxa of birds and the "microhabitats" they occupy on birds. (Figure modified from Johnson et al. 2012; open-access article distributed under the terms of the Creative Commons CC BY 2.0 license, https://creativecommons.org/licenses/by/2.0/)

**Fig. 3.126** A Barn Owl (*Tyto alba*) scratching its head. (Figure from Clayton et al. 2010; open-access article licensed under the terms of the Creative Commons Attribution 4.0 International Public License (CC-BY 4.0), https://creativecommons.org/licenses/by/4.0/)

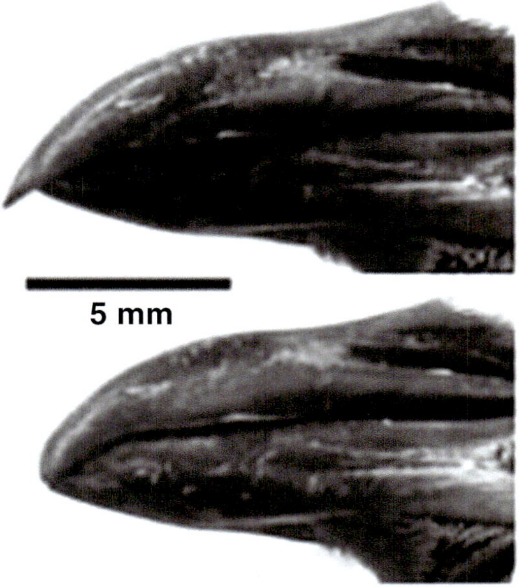


**Fig. 3.128** Rock Pigeon (*Columba livia*) beak before (above) and after (below) trimming of the overhang. (Figure from Waite et al. 2012; © 2012 Australian Society for Parasitology Inc., used with permission)

**Fig. 3.127** Allopreening Arrow-marked Babblers (*Turdoides jardineii*). Mutual preening, or allopreening, has been reported in over 50 families of birds and, in addition to helping to remove ectoparasites, this behavior may help reinforce pair bonds and social hierarchies (Clayton et al. 2010; Kenny et al. 2017; Fulmer and Hauber 2021). (Photo by Derek Keats, Wikipedia, CC BY 2.0, https://creativecommons.org/licenses/by/2.0/)

**Fig. 3.129** Pectinate claw of the middle toes of Barn Owls (*Tyto alba*). Owls that had claws with more "teeth" were less likely to be infested with lice, suggesting that larger pectinate claws provide more protection against lice (Bush et al. 2012). (Figure from Clayton et al. 2010; open-access article licensed under the terms of the Creative Commons Attribution 4.0 International Public License (CC-BY 4.0), https://creativecommons.org/licenses/by/4.0/)

## References


Abourachid A, Herrel A, Decamps T, Pages F, Fabre A-C, Van Hoorebeke L, Adriaens D, Amado MAG (2019) Hoatzin nestling locomotion: acquisition of quadrupedal limb coordination in birds. Sci Adv 5:eaat0787

Adkins-Regan E (1999) Foam produced by male *Coturnix* quail: what is its function? Auk 116:184–193

Alibardi L (2017) Review: cornification, morphogenesis and evolution of feathers. Protoplasma 254:1259–1281

Alt G, Mägi M, Lodjak J, Mänd R (2020) Experimental study of the effect of preen oil against feather bacteria in passerine birds. Oecologia 192:723–733

Altshuler DL, Bahlman JW, Dakin R, Gaede AH, Goller B, Lentink D, Segre PS, Skandalis DA (2015) The biophysics of bird flight: functional relationships integrate aerodynamics, morphology, kinematics, muscles, and sensors. Can J Zool 93:961–975

Amo L, Avilés JM, Parejo D, Peña A, Rodríguez J, Tomás G (2012a) Sex recognition by odour and variation in the uropygial gland secretion in starlings. J Anim Ecol 81:603–613

Amo L, López-Rull I, Pagán I, Macías Garcia C (2012b) Male quality and conspecific scent preferences in the House Finch, *Carpodacus mexicanus*. Anim Behav 84: 1483–1489

Badyaev AV, Landeen EA (2007) Developmental evolution of sexual ornamentation: model and a test of feather growth and pigmentation. Integr Comp Biol 47:221–233

Bakken GS, Vanderbilt VC, Buttemer WA, Dawson WR (1978) Avian eggs: thermoregulatory value of very high near-infrared reflectance. Science 200:321–323

Bakken GS, Banta MR, Higginbotham CM, Lynott AJ (2006) It's just ducky to be clean: the water repellency and water penetration resistance of swimming Mallard *Anas platyrhynchos* ducklings. J Avian Biol 37:561–571

Bearhop S, Furness RW, Hilton GM, Votier SC, Waldron S (2003) A forensic approach to understanding diet and habitat use from stable isotope analysis of (avian) claw material. Funct Ecol 17:270–275

Bennett T (1974) Peripheral and autonomic nervous systems. In: Farner DS, King JR, Parkes KC (eds) Avian biology, vol 4. Academic Press, New York, pp 1–77

Birn-Jeffery AV, Miller CE, Naish D, Rayfield EJ, Hone DWE (2012) Pedal claw curvature in birds, lizards and Mesozoic dinosaurs – complicated categories and compensating for mass-specific and phylogenetic control. PLoS ONE 7:e50555

Bisson IA, Marra PP, Burtt EH Jr, Sikaroodi M, Gillevet PM (2007) A molecular comparison of plumage and soil bacteria across biogeographic, ecological, and taxonomic scales. Microb Ecol 54:65–81

Blanco G, Hornero-Méndez D, Lambertucci SA, Bautista LM, Wiemeyer G, Sanchez-Zapata JA, Garrido-Fernández J, Hiraldo F, Donázar JA (2013) Need and seek for dietary micronutrients: endogenous regulation, external signalling and food sources of carotenoids in New World vultures. PLoS ONE 8: e65562

Blount JD, McGraw KJ (2008) Signal functions of carotenoid colouration. In: Britton G, Liaaen-Jensen S, Pfander H (eds) Carotenoids, volume 4: natural functions. Birkhäuser Verlag, Basel, pp 213–236

Bonadonna F, Cunningham GB, Jouventin P, Hesters F, Nevitt GA (2003) Evidence for nest-odour recognition in two species of diving petrel. J Exp Biol 206:3719–3722

Bonadonna F, Villafane M, Bajzak C, Jouventin P (2004) Recognition of burrow's olfactory signature in Blue Petrels, *Halobaena caerulea*: an efficient discrimination mechanism in the dark. Anim Behav 67: 893–898

Bonser BHC (1996) Comparative mechanics of bill, claw and feather keratin in the Common Starling *Sturnus vulgaris*. J Avian Biol 27:175–177

Booth DT, Clayton DH, Block BA (1993) Experimental demonstration of the energetic cost of parasitism in free-ranging hosts. Proc R Soc B 253:125–129

Bormashenko E, Bormashenko Y, Stein T, Whyman G, Bormashenko E (2007) Why do pigeon feathers repel water? Hydrophobicity of pennae, Cassie-Baxter wetting hypothesis and Cassie-Wenzel capillarity-induced wetting transition. J Colloid Interface Sci 311:212–216

Boulton R (1927) Ptilosis of the House Wren (*Troglodytes aedon aedon*). Auk 44:387–414

Bradbury JW, Vehrencamp S (1998) Principles of animal communication. Sinauer Associates, Sunderland

Braun MS, Sporer F, Zimmermann S, Wink M (2018) Birds, feather-degrading bacteria and preen glands: the antimicrobial activity of preen gland secretions from turkeys (*Meleagris gallopavo*) is amplified by keratinase. FEMS Microbiol Ecol 94:fiy117

Brooks WS (1968) Comparative adaptations of the Alaskan redpolls to the Arctic environment. Wilson Bull 80:253–280

Brown RE, Fedde MR (1993) Airflow sensors in the avian wing. J Exp Biol 179:13–30

Brown CR, Brown MB, Rannala B (1995) Ectoparasites reduce long-term survivorship of their avian hosts. Proc R Soc B 262:313–319

Buchanan KL, Evans MR, Goldsmith AR (2003) Testosterone, dominance signaling and immunosuppression in the House Sparrow, *Passer domesticus*. Behav Ecol Sociobiol 55:50–59

Buchholz R (1995) Female choice parasite load and male ornamentation in Wild Turkeys. Anim Behav 50:929–943

Burger BV, Reiter B, Borzyk O, Du Plessis MA (2004) Avian exocrine secretions. I. Chemical characterization of the volatile fraction of the uropygial secretion of the Green Woodhoopoe, *Phoeniculus purpureus*. J Chem Ecol 30:1603–1611

Burtt EH, Ichida JM (1999) Occurrence of feather-degrading bacilli in the plumage of birds. Auk 116: 364–372

Burtt EH Jr (1979) Tips on wings and other things. In: Burtt EH Jr (ed) The behavioral significance of color. Garland STPM Press, New York, pp 75–110

Burtt EH Jr, Ichida JM (2004) Gloger's Rule, feather-degrading bacteria, and color variation among Song Sparrows. Condor 106:681–686

Bush SE, Villa SM, Boves TJ, Brewer D, Belthoff JR (2012) Influence of bill and foot morphology on the ectoparasites of Barn Owls. J Parasitol 98:256–261

Calcott B (2009) Lineage explanations: explaining how biological mechanisms change. Br J Philos Sci 60: 51–78

Cameron GJ, Wess TJ, Bonser RHC (2003) Young's modulus varies with differential orientation of keratin in feathers. J Struct Biol 2:118–123

Cantarero A, Andrade P, Carneiro M, Moreno-Borrallo A, Alonso-Alvarez C (2020) Testing the carotenoid-based sexual signalling mechanism by altering CYP2J19 gene expression and colour in a bird species. Proc R Soc B 287:20201067

Casagrande S, Csermely D, Pini E, Bertacche V, Tagliavini J (2006) Skin carotenoid concentration correlates with male hunting skill and territory quality in the kestrel *Falco tinnunculus*. J Avian Biol 37:190–196

Caspers BA, Hagelin JC, Paul M, Bock S, Willeke S, Krause ET (2017) Zebra Finch chicks recognise parental scent, and retain chemosensory knowledge of their genetic mother, even after egg cross-fostering. Sci Rep 7:12859

Cassie ABD, Baxter S (1944) Wettability of porous surfaces. Trans Faraday Soc 40:546–551

Chan EP, Walish JJ, Urbas AM, Thomas EL (2013) Mechanochromic photonic gels. Adv Mater 25:3934–3947

Chang WL, Wu H, Chiu YK, Wang S, Jiang TX, Luo ZL, Lin YC, Li A, Hsu JT, Huang HL, Gu HJ (2019) The making of a flight feather: bio-architectural principles and adaptation. Cell 179:1409–1423

Chatterjee S, Templin RJ (2004) Feathered coelurosaurs from China: new light on the arboreal origin of avian flight. In: Currie PJ, Koppelhus EB, Shugar MA, Wright JL (eds) Feathered dragons. Indiana University Press, Bloomington, pp 251–281

Chatterjee S, Templin RJ (2007) Biplane wing planform and flight performance of the feathered dinosaur *Microraptor gui*. Proc Natl Acad Sci USA 104:1576–1580

Chen P-J, Dong Z-M, Zhen S-N (1998) An exceptionally well-preserved theropod dinosaur from the Yixian Formation of China. Nature 391:147–152

Cheng F, Gao J, Luk TS, Yang X (2015) Structural color printing based on plasmonic metasurfaces of perfect light absorption. Sci Rep 5:11045

Chiale MC, Montalti D (2013) The relationship between the feather tuft of the uropygial gland and terrestrial/aquatic birds. Revista Brasileira de Ornithologia 21: 162–167

Chiale MC, Carril J, Montalti D, Barbeito C (2019) The uropygial gland of the Eared Dove and its evolutionary history within the Columbiformes (Aves). J Ornithol 160:1171–1181

Chuong C-M, Chodankar R, Widelitz RB, Jiang T-X (2000) Evo-Devo of feathers and scales: building complex epithelial appendages. Curr Opin Genet Dev 10: 449–456

Clark GA, de Cruz JB (1989) Functional interpretation of protruding filoplumes in oscines. Condor 91: 962–965

Clarke J (2013) Feathers before flight. Science 340:690–692

Clarke J, Middleton K (2006) Bird evolution. Curr Biol 16:R350–R354

Clay T (1949) Some problems in the evolution of a group of ectoparasites. Evolution 3:279–299

Clayton DH (1990) Mate choice in experimentally parasitized Rock Doves: lousy males lose. Am Zool 30:251–262

Clayton DH (1991) Coevolution of avian grooming and ectoparasite avoidance. In: Loye JE, Zuk M (eds) Bird-parasite interactions: ecology, evolution, and

behaviour. Oxford University Press, Oxford, pp 258–289
Clayton DH, Cotgreave P (1994) Relationship of bill morphology to grooming behaviour in birds. Anim Behav 47:195–201
Clayton DH, Moore J (1997) Host–parasite evolution: general principles and avian models. Oxford University Press, Oxford
Clayton DH, Tompkins DM (1995) Comparative effects of mites and lice on the reproductive success of Rock Doves (*Columba livia*). Parasitology 110: 195–206
Clayton DH, Walther BA (2001) Influence of host ecology and morphology on the diversity of Neotropical bird lice. Oikos 94:455–467
Clayton DH, Gregory RD, Price RD (1992) Comparative ecology of Neotropical bird lice (Insecta: Phthiraptera). J Anim Ecol 61:781–795
Clayton DH, Moyer BR, Bush SE, Jones TG, Gardiner DW, Rhodes BB, Goller F (2005) Adaptive significance of avian beak morphology for ectoparasite control. Proc R Soc B 272:811–817
Clayton DH, Koop JAH, Harbison CW, Moyer BR, Bush SE (2010) How birds combat ectoparasites. Open Ornithol J 3:41–71
Conover MR, Miller DE (1980) Rictal bristle function in Willow Flycatcher. Condor 82:469–471
Cooper RL, Lloyd VJ, Di-Poï N, Fletcher AG, Barrett PM, Fraser GJ (2019) Conserved gene signalling and a derived patterning mechanism underlie the development of avian footpad scales. EvoDevo 10:19
Cotgreave P, Clayton DH (1994) Comparative analysis of the time spent grooming by birds in relation to parasite load. Behaviour 131:171–187
Couteaudier M, Denesvre C (2014) Marek's disease virus and skin interactions. Vet Res 45:36
Csermely D, Bertè L, Camoni R (1998) Prey killing by Eurasian Kestrels: the role of the foot and the significance of bill and talons. J Avian Biol 29:10–16
Cunningham SJ, Alley MR, Castro I (2011) Facial bristle feather histology and morphology in New Zealand birds: implications for function. J Morphol 272:118–128
Cunningham SJ, Corfield JR, Iwaniuk AN, Castro I, Alley MR, Birkhead TR, Parsons S (2013) The anatomy of the bill tip of kiwi and associated somatosensory regions of the brain: comparisons with shorebirds. PLoS ONE 8:e80036
Czirják GÁ, Pap PL, Vágási CI, Giraudeau M, Mureşan C, Mirleau P, Heeb P (2013) Preen gland removal increases plumage bacterial load but not that of feather-degrading bacteria. Naturwissenschaften 100: 145–151
D'Alba L, Shawkey MD (2012) Relative contributions of pigments and biophotonic nanostructures to natural color production: a case study in Budgerigar (*Melopsittacus undulatus*) feathers. J Exp Biol 215: 1272–1277
Dale J, Dey CJ, Delhey K, Kempenaers B, Valcu M (2015) The effects of life history and sexual selection on male and female plumage colouration. Nature 527:367–370
Davison GWH (1985) Avian spurs. J Zool 206:353–366
Delaunay MG, Brassey C, Larsen C, Lloyd H, Grant RA (2022) The evolutionary origin of avian facial bristles and the likely role of rictal bristles in feeding ecology. Sci Rep 12:21108
Delhey K, Peters A, Kempenaers B (2007) Cosmetic coloration in birds: occurrence, function, and evolution. Am Nat 169(Suppl):S145–S158
Delhey K, Peters A, Biedermann PHW, Kempenaers B (2008) Optical properties of the uropygial gland secretion: no evidence for UV cosmetics in birds. Naturwissenschaften 95:939–946
Dey CJ, Dale J, Quinn JS (2014) Manipulating the appearance of a badge of status causes changes in true badge expression. Proc R Soc B 281:20132680
Dhouailly D, Godefroit P, Martin T, Nonchev S, Caraguel F, Oftedal O (2019) Getting to the root of scales, feather and hair: as deep as odontodes? Exp Dermatol 28:503–508
Dial KP (2003) Wing-assisted incline running and the evolution of flight. Science 299:402–404
Doña J, Proctor H, Serrano D, Johnson KP, Oploo AOV, Huguet-Tapia JC, Ascunce MS, Jovani R (2019) Feather mites play a role in cleaning host feathers: new insights from DNA metabarcoding and microscopy. Mol Ecol 28:203–218
Dorward P (1970) Response patterns of cutaneous mechanoreceptors in the domestic duck. Comp Biochem Physiol 35:729–735
Doucet SM, Shawkey MD, Hill GE, Montgomerie R (2006) Iridescent plumage in Satin Bowerbirds: structure, mechanisms and nanostructural predictors of individual variation in colour. J Exp Biol 209:380–390
du Toit CJ, Chinsamy A, Cunningham SJ (2020) Cretaceous origins of the vibrotactile bill-tip organ in birds. Proc R Soc B 287:20202322
Dumbacher JP (1999) Evolution of toxicity in Pitohuis. I. Effects of homobatrachotoxin on chewing lice (order Phthiraptera). Auk 116:957–963
Dyck J (1985) The evolution of feathers. Zoologica Scripta 14:137–153
Dyke G, De Kat R, Palmer C, Van Der Kindere J, Naish D, Ganapathisubramani B (2013) Aerodynamic performance of the feathered dinosaur *Microraptor* and the evolution of feathered flight. Nat Commun 4:2489
Elder WH (1954) The oil gland of birds. Wilson Bull 66: 6–31
Elias PM, Menon GK (1991) Structural and biochemical correlates of the epidermal permeability barrier. Adv Lipid Res 24:1–26
Eliason CM, Shawkey MD (2012) A photonic heterostructure produces diverse iridescent colours in duck wing patches. J R Soc Interface 9:2279–2278
Emlen ST, Wrege PH (2004) Size dimorphism, intrasexual competition, and sexual selection in Wattled Jacana

(*Jacana jacana*), a sex-role reversed shorebird in Panama. Auk 121:391–403
Engelbrecht D (2010) Growth and development of Crested Barbet *Trachyphonus vaillantii* nestlings. Ornithol Obs 1:59–62
Falk AR, Kaye TG, Zhou Z, Burnham DA (2016) Laser fluorescence illuminates the soft tissue and life habits of the Early Cretaceous bird *Confuciusornis*. PLoS ONE 11:e0167284
Figuerola J, Domènech J, Senar JC (2003) Plumage colour is related to ectosymbiont load during moult in the Serin, *Serinus serinus*: an experimental study. Anim Behav 65:551–557
Finseth FR, Iacovelli SR, Harrison RG, Adkins-Regan EK (2013) A nonsemen copulatory fluid influences the outcome of sperm competition in Japanese Quail. J Evol Biol 26:1875–1889
Folstad I, Karter AJ (1992) Parasites, bright males, and the immunocompetence handicap. Am Nat 139:603–622
Foth C (2011) The morphology of neoptile feathers: ancestral state reconstruction and its phylogenetic implications. J Morphol 272:387–403
Fox DL (1976) Animal biochromes and structural colours: physical, chemical, distributional & physiological features of coloured bodies in the animal world, 2nd edn. University of California Press, London
Fuller ME (2015) The structure and properties of down feathers and their use in the outdoor industry. Ph.D. dissertation, University of Leeds, Leeds, UK
Fulmer AG, Hauber ME (2021) Autopreening behaviour may convey information about internal social state in Arabian Babbler (*Turdoides squamiceps*) allopreening dyads. Behaviour 158:427–446
Galván I, Aguilera E, Atiénzar F, Barba E, Blanco G, Cantó JL, Cortés V, Frías Ó, Kovács I, Meléndez L, Møller AP (2012) Feather mites (Acari: Astigmata) and body condition of their avian hosts: a large correlative study. J Avian Biol 43:273–279
Giraudeau M, Czirják GÁ, Duval C, Bretagnolle V, Gutierrez C, Guillon N, Heeb P (2013) Effect of preen oil on plumage bacteria: an experimental test with the Mallard. Behav Process 92:1–5
Godefroit P, Sinitsa SM, Dhouailly D, Bolotsky YL, Sizov AV, McNamara ME, Benton MJ, Spagna P (2014) A Jurassic ornithischian dinosaur from Siberia with both feathers and scales. Science 345:451–455
Goldstein G, Flory KR, Browne BA, Majid S, Ichida JM, Burtt EH Jr (2004) Bacterial degradation of black and white feathers. Auk 121:656–659
Golüke S, Dörrenberg S, Krause ET, Caspers BA (2016) Female Zebra Finches smell their eggs. PLoS ONE 11: e0155513
Grafen A (1990) Biological signals as handicaps. J Theor Biol 144:517–546
Graves GR (2019) Facial caruncles in Jamaican Turkey Vultures (*Cathartes aura*). J Caribb Ornithol 32:49–52
Greeney HF (2012) The natal plumages of antpittas (Grallariidae). Ornitología Colombia 12:65–68
Greenwold MJ, Sawyer RH (2011) Linking the molecular evolution of avian beta (β) keratins to the evolution of feathers. J Exp Zool B: Mol Dev Evol 316:609–616
Greenwold MJ, Sawyer RH (2013) Molecular evolution and expression of archosaurian β-keratins: diversification and expansion of archosaurian β-keratins and the origin of feather β-keratins. J Exp Zool B: Mol Dev Evol 320:393–405
Greenwold MJ, Bao W, Jarvis ED, Hu H, Li C, Gilbert MTP, Zhang G, Sawyer RH (2014) Dynamic evolution of the alpha (α) and beta (β) keratins has accompanied integument diversification and the adaptation of birds into novel lifestyles. BMC Evol Biol 14:249
Grieves LA, Bernards MA, MacDougall-Shackleton EA (2019a) Wax ester composition of songbird preen oil varies seasonally and differs between sexes, ages, and populations. J Chem Ecol 45:37–45
Grieves LA, Bernards MA, MacDougall-Shackleton EA (2019b) Behavioural responses of songbirds to preen oil odour cues of sex and species. Anim Behav 156: 57–65
Grieves LA, Gilles M, Cuthill IC, Székely T, MacDougall-Shackleton EA, Caspers BA (2022) Olfactory camouflage and communication in birds. Biol Rev 97:1193–1209
Griffith SC, Parker TH, Olson VA (2006) Melanin- versus carotenoid-based sexual signals: is the difference really so black and red? Anim Behav 71:749–763
Grill SW (2017) The mechanics of positioning skin follicles. Science 357:750–751
Grubb TC Jr (1994) Olfactory navigation to the nesting burrow in Leach's Petrel *Oceanodroma leucorrhoa*. Anim Behav 22:192–202
Gunderson AR (2008) Feather-degrading bacteria: a new frontier in avian and host-parasite research? Auk 125: 972–979
Gunderson AR, Frame AM, Swaddle JP, Forsyth MH (2008) Resistance of melanized feathers to bacterial degradation: is it really so black and white? J Avian Biol 39:539–545
Hahn S, Dimitrov D, Rehse S, Yohannes E, Jenni L (2014) Avian claw morphometry and growth determine the temporal pattern of archived stable isotopes. J Avian Biol 45:202–207
Halata Z, Grim M, Baumann KI (2003) The Merkel cell: morphology, developmental origin, function. Cas Lek Cesk 142:4–9
Haugen M, Williams JB, Wertz PW, Tieleman BI (2003) Lipids of the stratum corneum vary with cutaneous water loss among larks along a temperature-moisture gradient. Physiol Biochem Zool 76:907–917
Hedrick BP, Cordero SA, Zanno LE, Noto C, Dodson P (2019) Quantifying shape and ecology in avian pedal claws: the relationship between the bony core and keratinous sheath. Ecol Evol 9:11545–11556
Heilmann G (1926) The origin of birds. H. F. G. Whitherby, London
Hill GE (1991) Plumage coloration is a sexually selected indicator of male quality. Nature 350:337–339

Hill GE (2006) Female mate choice for ornamental coloration. In: Hill GE, McGraw KJ (eds) Bird coloration, volume II: function and evolution. Harvard University Press, Cambridge, pp 137–200

Hill GE, Hood WR, Ge Z, Grinter R, Greening C, Johnson JD, Park NR, Taylor HA, Andreasen VA, Powers MJ, Justyn NM (2019) Plumage redness signals mitochondrial function in the House Finch. Proc R Soc B 286: 20191354

Hirao A, Aoyama M, Sugita S (2009) The role of uropygial gland on sexual behavior in Domestic Chicken *Gallus gallus domesticus*. Behav Process 80: 115–120

Hirayama H, Kaneda K, Yamashita H, Monden Y (2001) An accurate illumination model for objects coated with multilayer films. Comput Graph 25:391–400

Höfling E, Abourachid A (2021) The skin of birds' feet: morphological adaptations of the plantar surface. J Morphol 282:88–97

Hofmann CM, Cronin TW, Omland KE (2008) Evolution of sexual dichromatism. 1. Convergent losses of elaborate female coloration in New World orioles (*Icterus* spp.). Auk 125:778–789

Höhn EO (1977) The 'snowshoe effect' of the feathering on ptarmigan feet. Condor 79:380–382

Hoi H, Krištofík J, Darolova A, Hoi C (2012) Experimental evidence for costs due to chewing lice in the European Bee-eater (*Merops apiaster*). Parasitology 139:53–59

Homberger DG, Brush AH (1986) Functional morphology and biochemical correlations of the keratinized structures of the African Gray Parrot, *Psittacus erithacus* (Aves). Zoomorphology 106:103–114

Homberger DG, de Silva KN (2000) Functional microanatomy of the feather-bearing integument: implications for the evolution of birds and avian flight. Am Zool 40:553–574

Hone DW, Tischlinger H, Xu X, Zhang F (2010) The extent of the preserved feathers on the four-winged dinosaur *Microraptor gui* under ultraviolet light. PLoS ONE 5:e9223

Hudon J (2005) Considerations in the conservation of feathers and hair, particularly their pigments. In: Brunn M, Burns JA (eds) Fur trade legacy. The preservation of organic materials. Canadian Association for Conservation of Cultural Property, Ottawa, pp 127–147

Igic B, D'Alba L, Shawkey MD (2016) Manakins can produce iridescent and bright feather colours without melanosomes. J Exp Biol 219:1851–1859

Imber MJ (1971) Filoplumes of petrels and shearwaters. N Z J Mar Freshwater Res 5:396–403

Iskandar J-P, Eliason CM, Astrop T, Igic B, Maia R, Shawkey MD (2016) Morphological basis of glossy red plumage colours. Biol J Linn Soc 119:477–487

Iverson ENK, Karubian J (2017) The role of bare parts in avian signaling. Auk 134:587–611

Jackson HD (2007) Measurements and functions of the pectinated claws and rictal bristles of Fiery-necked Nightjars *Caprimulgus pectoralis* and some congeners. Ostrich 78:641–643

Jacob J, Ziswiler V (1982) The uropygial gland. In: Farner DS, King JR, Parkes KC (eds) Avian biology, vol 6. Academic Press, New York, pp 199–324

Jacob J, Eigener U, Hoppe U (1997) The structure of preen gland waxes from pelecaniform birds containing 3,7-dimethyloctan-1-ol: an active ingredient against dermatophytes. Zeitschrift für Naturforschung C 52: 114–123

Johansson LC, Norberg UML (2001) Lift-based paddling in diving grebe. J Exp Biol 204:1687–1696

Johnsgard PA (1983) The grouse of the world. University of Nebraska Press, Lincoln

Johnson KP, Shreve SM, Smith VS (2012) Repeated adaptive divergence of microhabitat specialization in avian feather lice. BMC Biol 10:1

Johnston DW (1988) A morphological atlas of the avian uropygial gland. Bull Br Mus (Nat Hist) Zool 54:199–259

Justyn NM, Powers MJ, Hill GE, Alexander K, Naveda-Rodríguez A, Rush SA (2023) The mechanisms of color production in black skin versus red skin on the heads of New World vultures. Avian Res 14:100071

Katasho Y, Liang Y, Murata S, Fukunaka Y, Matsuoka T, Takahashi S (2015) Mechanisms for enhanced hydrophobicity by atomic-scale roughness. Sci Rep 5: 13790

Kellner AWA, Wang XL, Tischlinger H, Campos DD, Hone DWE, Meng X (2010) The soft tissue of *Jeholopterus* (Pterosauria, Anurognathidae, Batrachognathinae) and the structure of the pterosaur wing membrane. Proc R Soc B 277:321–329

Kemp AC (2001) Family Bucerotidae (Hornbills). In: del Hoyo J, Elliott A, Aargatal J (eds) Handbook of the birds of the world, Mousebirds to Hornbills, vol 6. Lynx Edicions, Barcelona, pp 436–623

Kenny E, Birkhead TR, Green JP (2017) Allopreening in birds is associated with parental cooperation over offspring care and stable pair bonds across years. Behav Ecol 28:1142–1148

Kent CM, Burtt EH Jr (2016) Feather-degrading bacilli in the plumage of wild birds: prevalence and relation to feather wear. Auk 133:583–592

Kilner RM (2006) Function and evolution of color in young birds. In: Hill GE, McGraw KJ (eds) Bird coloration: function and evolution, vol II. Harvard University Press, Cambridge, pp 201–232

Kinoshita S, Yoshioka S, Miyazaki J (2008) Physics of structural colors. Rep Prog Phys 71:076401

Knopf FL, Evans RM (2020) American White Pelican (*Pelecanus erythrorhynchos*), version 1.0. In: Poole AF (ed) Birds of the World. Cornell Lab of Ornithology, Ithaca

Koch RE, McGraw KJ, Hill GE (2016) Effects of diet on plumage coloration and carotenoid deposition in red and yellow Domestic Canaries (*Serinus canaria*). Wilson J Ornithol 128:328–333

Kose M, Møller AP (1999) Sexual selection, feather breakage and parasites: the importance of white spots in the tail of the Barn Swallow (*Hirundo rustica*). Behav Ecol Sociobiol 45:430–436

Kraaijeveld K (2014) Reversible trait loss: the genetic architecture of female ornaments. Annu Rev Ecol Syst 45:159–177

Ksepka DT (2020) Feathered dinosaurs. Curr Biol 30: R1347–R1353

Kuenzel WJ (2007) Neurobiological basis of sensory perception: welfare implications of beak trimming. Poult Sci 86:1273–1282

Lande R (1980) Sexual dimorphism, sexual selection, and adaptation in polygenic characters. Evolution 34:292–305

Law-Brown J (2001) Chemical defence in the Red-billed Woodhoopoe, *Phoeniculus purpureus*. M. S. thesis, University of Cape Town, Cape Town, South Africa

Lederer RJ (1972) The role of avian rictal bristles. Wilson Bull 84:193–197

Ligon JD, Ligon SH (1978) Communal breeding in Green Woodhoopoes as a case for reciprocity. Nature 276:496–498

Lillywhite HB (2006) Water relations of tetrapod integument. J Exp Biol 209:202–226

Lingham-Soliar T (2014) Feather structure, biomechanics and biomimetics: the incredible lightness of being. J Ornithol 155:232–336

Lingham-Soliar T, Feduccia A, Wang X (2007) A new Chinese specimen indicates that 'protofeathers' in the Early Cretaceous theropod dinosaur *Sinosauropteryx* are degraded collagen fibres. Proc R Soc B 274:1823–1829

Liu P, Bai L, Yang J, Gu H, Zhong Q, Xie Z, Gu Z (2019) Self-assembled colloidal arrays for structural color. Nanoscale Adv 1:1672–1685

Lloyd-Jones DJ, Briskie JV (2016) Mutual wattle ornaments in the South Island Saddleback (*Philesturnus carunculatus*) function as armaments. Ethology 122:61–71

Lombardo MP, Drake P, Olson A, Otieno S, Spadacene L, Thorpe PA (2015) Feather-chewing lice and Tree Swallow biology. Auk 132:551–561

Londoño GA, García DA, Sánchez Martínez MA (2015) Morphological and behavioral evidence of Batesian mimicry in nestlings of a lowland Amazonian bird. Am Nat 185:135–141

Lowe CB, Clarke JA, Baker AJ, Haussler D, Edwards SV (2015) Feather development genes and associated regulatory innovation predate the origin of Dinosauria. Mol Biol Evol 32:23–28

Lucas AM (1970) Avian functional anatomic problems. Fed Proc 29:1641–1648

Lucas AM (1979) Integumentum commune. In: Baumel JJ, King AS, Lucas AM, Breazile JF, Evans HE (eds) Nomina anatomica avium. Academic Press, New York

Lucas AM, Stettenheim PR (1972) Avian anatomy. Integument. Agriculture Handbook 362, U.S. Department of Agriculture, Washington, D.C.

Lyon BE, Eadie JM, Hamilton LD (1994) Parental choice selects for ornamental plumage in American Coot chicks. Nature 371:240–243

Margalida A, Braun MS, Negro JJ, Schulze-Hagen K, Wink M (2019) Cosmetic colouring by Bearded Vultures *Gypaetus barbatus*: still no evidence for an antibacterial function. PeerJ 7:e6783

Masello JF, Lubjuhn T, Quillfeldt P (2008) Is the structural and psittacofulvin-based coloration of wild Burrowing Parrots *Cyanoliseus patogonus* condition dependent? J Avian Biol 39:653–662

Masello JF, Quillfeldt P (2003) Body size, body condition and ornamental feathers of Burrowing Parrots: variation between years and sexes, assortative mating and influences on breeding success. Emu 103:149–161

Massaro M, Davis LS, Darby JT (2003) Carotenoid-derived ornaments reflect parental quality in male and female Yellow-eyed Penguins (*Megadyptes antipodes*). Behav Ecol Sociobiol 55:169–175

Masterson ANB (1979) Of birds, the war and the future. Honeyguide 100:7–11

Mateos C, Carranza J (1996) On the intersexual selection for spurs in the Ring-necked Pheasant. Behav Ecol 7: 362–369

Matin TR, Leong MK, Majlis BY, Gebeshuber IC, Abdullah M (2010) Correlating nanostructures with function: structural colors on the wings of a Malaysian bee. AIP Conf Proc 1284:5–14

Mayr E (1960) The emergence of evolutionary novelties. In: Tax S (ed) The evolution of life. University of Chicago Press, Chicago, pp 349–380

McGraw KJ (2006a) Mechanics of melanin-based coloration. In: Hill GE, McGraw KJ (eds) Bird coloration, volume I: mechanisms and measurements. Harvard University Press, Cambridge, pp 243–294

McGraw KJ (2006b) Mechanics of carotenoid-based coloration. In: Hill GE, McGraw KJ (eds) Bird coloration, volume I: mechanisms and measurements. Harvard University Press, Cambridge, pp 177–242

McGraw KJ (2006c) Mechanics of uncommon colors: pterins, porphyrins, and psittacofulvins. In: Hill GE, McGraw KJ (eds) Bird coloration, volume I: mechanisms and measurements. Harvard University Press, Cambridge, pp 354–398

McGraw KJ (2007) Dietary mineral content influences melanin-based ornamental coloration. Behav Ecol 18: 137–142

McGraw KJ (2008) An update on the honesty of melanin-based color signals in birds. Pigment Cell Melanoma Res 21:133–138

McGraw KJ, Mackillop EA, Dale J, Hauber ME (2002) Different colors reveal different information: how nutritional stress affects the expression of melanin- and structurally based ornamental plumage. J Exp Biol 205:3747–3755

McGraw KJ, Toomey MB, Nolan PM, Morehouse NI, Massaro M, Jouventin P (2007) A description of unique fluorescent yellow pigments in penguin feathers. Pigment Cell Res 20:301–304

McGraw KJ, Massaro M, Rivers TJ, Mattern T (2009) Annual, sexual, size- and condition-related variation in the colour and fluorescent pigment content of yellow crest-feathers in Snares Penguins (*Eudyptes robustus*). Emu 109:93–99

McLandress MR (1983) Winning with warts? A threat posture suggests a function for caruncles in Ross's Geese. Wildfowl 34:5–9

McNamara ME, Zhang F, Kearns SL, Orr PJ, Toulouse A, Foley T, Hone DWE, Rogers CS, Benton MJ, Johnson D, Xu X, Zhou Z (2018) Fossilized skin reveals coevolution with feathers and metabolism in feathered dinosaurs and early birds. Nat Commun 9: 2072

Menon GK (1984) Glandular functions of avian integument: an overview. J Yamashina Inst Ornithol 16:1–12

Menon GK, Menon J (2000) Avian epidermal lipids: functional considerations and relationship to feathering. Integr Comp Biol 40:540–552

Menon GK, Maderson PFA, Drewes RC, Baptista LF, Price LF, Elias PM (1996) Ultrastructural organization of avian stratum corneum lipids as the basis for facultative cutaneous waterproofing. J Morphol 227:1–13

Mewaldt LR (1958) Pterylography and natural and experimentally induced molt in Clark's Nutcracker. Condor 60:165–187

Mínguez E (1997) Olfactory nest recognition by British storm-petrel chicks. Anim Behav 53:701–707

Mock DW (2016) Animal behavior: some begging is actually bragging. Nature 532:180–181

Møller AP, Laursen K (2019) Function of the uropygial gland in eiders (*Somateria mollissima*). Avian Res 10: 24

Møller AP, Mateos-González F (2019) Plumage brightness and uropygial gland secretions in Barn Swallows. Curr Zool 65:177–182

Møller AP, de Lope F, Saino N (2004) Parasitism, immunity, and arrival date in a migratory bird, the Barn Swallow. Ecology 85:206–219

Montgomerie R (2006) Cosmetic and adventitious colors. In: Hill GE, McGraw KJ (eds) Bird coloration, volume I: mechanisms and measurements. Harvard University Press, Cambridge, pp 399–427

Montgomerie R, Lyon B, Holder K (2001) Dirty ptarmigan: behavioral modification of conspicuous male plumage. Behav Ecol 12:429–438

Morelli R, Loscalzo R, Stradi R, Bertelli A, Falchi M (2003) Evaluation of the antioxidant activity of new carotenoid-like compounds by electron paramagnetic resonance. Drugs Exp Clin Res 29:95–100

Moreno-Rueda G, Hoi H (2012) Female House Sparrows prefer big males with a large white wing bar and fewer feather holes caused by chewing lice. Behav Ecol 23: 271–277

Morlion ML (1985) Pterylosis of the wing and tail in the Noisy Scrub-bird, *Atrichornis clamosus*, and Superb Lyrebird, *Menura novaehollandiae* (Passeriformes: Atrichornithidae and Menuridae). Rec Aust Mus 37: 143–156

Mougeot F, Arroyo BE (2006) Ultraviolet reflectance by the cere of raptors. Biol Lett 2:173–176

Moyer BR, Clayton DH (2003) Avian defenses against ectoparasites. In: van Emden HF, Rothschild M (eds) Insect and bird interactions. Intercept Ltd., Andover, pp 241–257

Moyer BR, Peterson AT, Clayton DH (2002) Influence of bill shape on ectoparasite load in Western Scrub-Jays. Condor 104:675–678

Moyer BR, Rock AN, Clayton DH (2003) Experimental test of the importance of preen oil in Rock Doves (*Columba livia*). Auk 120:490–496

Munoz-Garcia A, Williams JB (2005) Cutaneous water loss and lipids of the stratum corneum in House Sparrows *Passer domesticus* from arid and mesic environments. J Exp Biol 208:3689–3700

Muñoz-Garcia A, Williams JB (2007) Cutaneous water loss and lipids of the stratum corneum in Dusky Antbirds, a lowland tropical bird. Condor 109:59–66

Muza MM, Burtt EH, Ichida JM (2000) Distribution of bacteria on feathers of some eastern North American birds. Wilson Bull 112:432–435

Negro JJ, Margalida A, Hiraldo F, Heredia R (1999) The function of the cosmetic coloration of Bearded Vultures: when art imitates life. Anim Behav 58:F14–F17

Negro JJ, Grande JM, Tella JA, Garrido J, Hornero D, Donázar JA, Sanchez-Zapata JA, Benítez JR, Barcell M (2002) An unusual source of essential carotenoids. Nature 416:807

Negro JJ, Sarasola JH, Fariñs F, Zorrilla I (2006) Function and occurrence of facial flushing in birds. Comp Biochem Physiol A 143:78–84

Ng CS, Wu P, Fan W-L, Yan J, Chen C-K, Lai Y-T, Wu S-M, Mao CT, Chen JJ, Lu MYJ, Ho MR (2014) Genomic organization, transcriptomic analysis, and functional characterization of avian α-and β-keratins in diverse feather forms. Genome Biol Evol 6:2258–2273

Nicolaï MP, Shawkey MD, Porchetta S, Claus R, D'Alba L (2020) Exposure to UV radiance predicts repeated evolution of concealed black skin in birds. Nat Commun 11:2414

Noramly S, Morgan BA (1998) BMPs mediate lateral inhibition at successive stages of feather tract development. Development 125:3775–3787

Nordén KK, Faber JW, Babarović F, Stubbs TL, Selly T, Schiffbauer JD, Štefanić PP, Mayr G, Smithwick FM, Vinther J (2019) Melanosome diversity and convergence in the evolution of iridescent avian feathers—implications for paleocolor reconstruction. Evolution 73:15–27

Norell MA, Xu X (2005) Feathered dinosaurs. Annu Rev Earth Planet Sci 33:277–299

Nudds RL, Dyke GJ (2010) Narrow primary feather rachises in *Confuciusornis* and *Archaeopteryx* suggest poor flight ability. Science 328:887–889

Ornelas JF (1994) Serrate tomia: an adaptation for nectar robbing in hummingbirds? Auk 111:703–710

Osborne DR, Bourne GR (1977) Breeding behavior and food habits of the Wattled Jacana. Condor 79:98–105

Osorio D, Ham AD (2002) Spectral reflectance and directional properties of structural coloration in bird plumage. J Exp Biol 205:2017–2027

Osváth G, Daubner T, Dyke G, Fuisz TI, Nord A, Pénzes J, Vargancsik D, Vágási CI, Vincze O, Pap PL (2018) How feathered are birds? Environment predicts both the mass and density of body feathers. Funct Ecol 32:701–712

Padian K (2003) Four-winged dinosaurs, bird precursors, or neither? BioScience 53:451–453

Pan Y, Zheng W, Sawyer RH, Pennington MW, Zheng X, Wang X, Wang M, Hu L, O'Connor J, Zhao T, Li Z (2019) The molecular evolution of feathers with direct evidence from fossils. Proc Natl Acad Sci USA 116:3018–3023

Pap PL, Tökölyi J, Szép T (2005a) Host-symbiont relationship and abundance of feather mites in relation to age and body condition of the Barn Swallow (*Hirundo rustica*): an experimental study. Can J Zool 83:1059–1066

Pap PL, Tökölyi J, Szep T (2005b) Frequency and consequences of feather holes in Barn Swallows *Hirundo rustica*. Ibis 147:169–175

Pap PL, Vágási CI, Osváth G, Mureşan C, Barta Z (2010) Seasonality in the uropygial gland size and feather mite abundance in House Sparrows *Passer domesticus*: natural covariation and an experiment. J Avian Biol 41: 653–661

Pap PL, Vincze O, Wekerle B, Daubner T, Vágási CI, Nudds RL, Dyke GJ, Osváth G (2017) A phylogenetic comparative analysis reveals correlations between body feather structure and habitat. Funct Ecol 31: 1241–1251

Papeschi A, Carroll JP, Dessì-Fulgheri F (2003) Wattle size is correlated with male territorial rank in juvenile Ring-necked Pheasants. Condor 105:362–366

Parker TH, Ligon JD (2003) Female mating preferences in Red Junglefowl: a meta-analysis. Ethol Ecol Evol 15: 63–72

Penteriani V, del Mar Delgado M, Maggio C, Aradis A, Sergio F (2005) Development of chicks and predispersal behaviour of young in the Eagle Owl *Bubo bubo*. Ibis 147:155–168

Pérez-Rodríguez L, Viñuela J (2008) Carotenoid-based bill and eye ring coloration as honest signals of condition: an experimental test in the Red-legged Partridge (*Alectoris rufa*). Naturwissenschaften 95:821–830

Perrella DF, Guida FJV (2019) Additional information on reproductive behavior of the Red-breasted Toucan, *Ramphastos dicolorus* (Aves: Piciformes: Ramphastidae). Biota Neotropica 19:e20180576

Perrichot V, Marion L, Néraudeau D, Vullo R, Tafforeau P (2008) The early evolution of feathers: fossil evidence from Cretaceous amber of France. Proc R Soc B 275: 1197–1202

Pervez NK, Cheng W, Jia Z, Cox MP, Edrees HM, Kymissis I (2010) Photonic crystal spectrometer. Opt Express 18:8277–8285

Picasso MBJ, Mario R, Barbeito CG (2016) The skin structure of Greater Rhea (Rheidae, Palaeognathae). Acta Zoologica 97:302–309

Piersma T, van Aelst R, Kurk K, Berkhoudt H, Maas LRM (1998) A new pressure sensory mechanism for prey detection in birds: the use of principles of seabed dynamics? Proc R Soc B 265:1377–1383

Pike AVL, Maitland DP (2004) Scaling of bird claws. J Zool 262:73–81

Poston JP, Hasselquist D, Stewart IRK, Westneat DF (2005) Dietary amino acids influence plumage traits and immune responses of male House Sparrows, *Passer domesticus*, but not as expected. Anim Behav 70:1171–1181

Powers MJ, Hill GE (2021) A review and assessment of the Shared-Pathway Hypothesis for the maintenance of signal honesty in red ketocarotenoid-based coloration. Integr Comp Biol 61:1811–1826

Prater AJ, Marchant JH, Vuorinen J (1977) Guide to the identification and aging of Holarctic waders. British Trust for Ornithology, Tring

Pratt TK (2000) Evidence for a previously unrecognized species of owlet-nightjar. Auk 117:1–11

Praveenkumar D, Vinothkumar A, Saravanan G, Selvakumar M, Vijayakumar AS, Kolanchinathan P, Kamalakkannan S, Achiraman S (2023) Symbiotic microbes play a role more important than preen gland in avian pheromone production—a review. Avian Biol Res 16:32–41

Price RD, Hellenthal RA, Palma RL, Johnson KP, Clayton DH (2003) The chewing lice: world checklist and biological overview. Illinois Natural History Survey Special Publication 24, Champaign

Proctor HC (2003) Feather mites (Acari: Astigmata): ecology, behavior, and evolution. Annu Rev Entomol 48: 185–209

Proctor NS, Lynch PJ (1993) Manual of ornithology: avian structure and function. Yale University Press, New Haven

Proctor H, Owens I (2000) Mites and birds: diversity, parasitism and coevolution. Trends Ecol Evol 15: 358–364

Prum RO (1999) Development and evolutionary origin of feathers. J Exp Zool 285:291–306

Prum RO (2005) Evolution of the morphological innovations of feathers. J Exp Zool 304B:570–579

Prum RO (2006) Anatomy, physics, and evolution of structural colors. In: Hill GE, McGraw KJ (eds) Bird coloration, vol. I: mechanisms and measurements. Harvard University Press, Cambridge, pp 295–353

Prum RO, Brush AH (2002) The evolutionary origin and diversification of feathers. Q Rev Biol 77:261–295

Prum RO, Brush AH (2003) Which came first, the feather or the bird? Sci Am 288:84–93

Prum RO, Razafindratsita VR (1997) Lek behavior and natural history of the Velvet Asity *Philepitta castanea* (Eurylaimidae). Wilson Bull 109:371–392

Prum RO, Torres R (2003a) Structural colouration of avian skin: convergent evolution of coherently scattering dermal collagen arrays. J Exp Biol 206:2409–2429

Prum RO, Torres RH (2003b) A Fourier tool for the analysis of coherent light scattering by bio-optical nanostructures. Integr Comp Biol 43:591–602

Prum RO, Torres RH (2013) Fourier blues: structural coloration of biological tissues. In: Andrews TD, Balan R, Benedetto JJ, Czaja W, Okoudjou KA (eds) Escursions in harmonic analysis, vol 2. Springer, New York, pp 401–422

Prum RO, Williamson S (2001) Theory of the growth and evolution of feather shape. J Exp Zool 291:30–57

Prum RO, Morrison RL, Ten Eyck GR (1994) Structural color production by constructive reflection from ordered collagen arrays in a bird (*Philepitta castanea*: Eurylaimidae). J Morphol 222:61–72

Prum RO, Torres R, Williamson S, Dyck J (1999) Two-dimensional Fourier analysis of the spongy medullary keratin of structurally coloured feather barbs. Proc R Soc B 266:13–22

Prum RO, LaFountain AM, Berro J, Stoddard MC, Frank HA (2012) Molecular diversity, metabolic transformation, and evolution of carotenoid feather pigments in cotingas (Aves: Cotingidae). J Comp Physiol B 182: 1095–1116

Pryke SR, Andersson S (2003) Carotenoid-based status signaling in Red-shouldered Widowbirds (*Euplectes axillaris*): epaulet size and redness affect captive and territorial competition. Behav Ecol Sociobiol 53:393–401

Qiang J, Currie PJ, Norell MA, Shu-An J (1998) Two feathered dinosaurs from northeastern China. Nature 393:753–761

Quay WB (1967) Comparative survey of the anal glands of birds. Auk 84:379–389

Rand AL (1954) On the spurs on bird's wings. Wilson Bull 66:127–134

Rauhut OW, Foth C, Tischlinger H, Norell MA (2012) Exceptionally preserved juvenile megalosauroid theropod dinosaur with filamentous integument from the Late Jurassic of Germany. Proc Natl Acad Sci USA 109:11746–11751

Regal PJ (1975) The evolutionary origin of feathers. Q Rev Biol 50:35–66

Reneerkens J, Piersma T, Damsté JSS (2005) Switch to diester preen waxes may reduce avian nest predation by mammalian predators using olfactory cues. J Exp Biol 208:4199–4202

Reneerkens J, Versteegh MA, Schneider AM, Piersma T, Burtt EH Jr (2008) Seasonally changing preen-wax composition: Red Knots' (*Calidris canutus*) flexible defense against feather-degrading bacteria. Auk 125: 285–290

Reynolds SM, Castro I, Alley MR, Cunningham SJ (2017) *Apteryx* spp. (Kiwi) possess an uropygial gland: anatomy and pathology. Eur J Anat 21:125–139

Rico-Guevara A, Rubega MA, Hurme KJ, Dudley R (2019) Shifting paradigms in the mechanics of nectar extraction and hummingbird bill morphology. Integr Org Biol 1:oby006

Riedler R, Pesme C, Druzik J, Gleeson M, Pearlstein E (2014) A review of color-producing mechanisms in feathers and their influence on preventive conservation strategies. J Amer Inst Cons 53:44–65

Rintamäki PT, Höglund J, Karvonen E, Alatalo RV, Björklund N, Lundberg A, Rätti O, Vouti J (2000) Combs and sexual selection in Black Grouse (*Tetrao tetrix*). Behav Ecol 11:465–471

Rohwer VG, Rohwer S, Kane L (2021) Filoplume morphology covaries with their companion primary suggesting that they are feather-specific sensors. Ornithology 138:ukab024

Ruiz-Rodríguez M, Tomas G, Martín-Gálvez D, Ruiz-Castellano C, Soler JJ (2015) Bacteria and the evolution of honest signals. The case of ornamental throat feathers in Spotless Starlings. Funct Ecol 29:701–709

Ruiz-Rodríguez M, Møller AP, Mousseau TA, Soler JJ (2016) Defenses against keratinolytic bacteria in birds living in radioactively contaminated areas. Sci Nat 103:71

Saitta ET, Gelernter R, Vinther J (2018) Additional information on the primitive contour and wing feathering of paravian dinosaurs. Palaeontology 61:273–288

Saliban A, Montalti D (2009) Physiological and biochemical aspects of the avian uropygial gland. Braz J Biol 69:437–446

Saranathan V, Burtt EH Jr (2007) Sunlight on feathers inhibits feather-degrading bacteria. Wilson J Ornithol 119:239–245

Saranathan V, Narayanan S, Sandy A, Dufresne ER, Prum RO (2021) Evolution of single gyroid photonic crystals in bird feathers. Proc Natl Acad Sci USA 118: e2101357118

Sawyer RH, Glenn T, French JO, Mays B, Shames RB, Barnes GL Jr, Rhodes W, Ishikawa Y (2000) The expression of beta-keratins in the epidermal appendages of reptiles and birds. Am Zool 40:530–539

Saxod R (1996) Ontogeny of the cutaneous sensory organs. Microsc Res Tech 34:313–333

Schreiber RW, Schreiber EA, Anderson DW, Bradley DW (1989) Plumages and molts of Brown Pelicans. Nat Hist Mus Los Angeles County Contrib Sci 402:1–43

Seiwert CM, Adkins-Regan E (1998) The foam production system of male Japanese Quail: characterization of structure and function. Brain Behav Evol 52:61–80

Seneviratne SS, Jones IL (2008) Mechanosensory function for facial ornamentation in the Whiskered Auklet, a crevice-dwelling seabird. Behav Ecol 19:784–790

Shaw CL, Rutter JE, Austin AL, Garvin MC, Whelan RJ (2011) Volatile and semivolatile compounds in Gray Catbird uropygial secretions vary with age and

between breeding and wintering grounds. J Chem Ecol 37:329–339
Shawkey MD, D'Alba L (2017) Interactions between colour-producing mechanisms and their effects on the integumentary colour palette. Philos Trans R Soc B 372:20160536
Shawkey MD, Hill GE (2004) Feathers at a fine scale. Auk 121:652–655
Shawkey MD, Hill GE (2005) Carotenoids need structural colours to shine. Biol Lett 1:121–124
Shawkey MD, Pillai SR, Hill GE (2003) Chemical warfare? Effects of uropygial oil on feather-degrading bacteria. J Avian Biol 34:345–349
Shawkey MD, Balenger SL, Hill GE, Johnson LS, Keyser AJ, Siefferman L (2006a) Mechanisms of evolutionary change in structural plumage coloration among bluebirds (*Sialia* spp.). J R Soc Interface 3:527–532
Shawkey MD, Hauber ME, Estep LK, Hill GE (2006b) Evolutionary transitions and mechanisms of matte and iridescent plumage coloration in grackles and allies (Icteridae). J R Soc Interface 3:777–786
Shawkey MD, Pillai SR, Hill GE, Siefferman LM, Roberts SR (2007) Bacteria as an agent for change in structural plumage color: correlational and experimental evidence. Am Nat 169(Suppl 1):S112–S121
Shawkey MD, Morehouse NI, Vukusic P (2009a) A protean palette: colour materials and mixing in birds and butterflies. J R Soc Interface 6(Suppl 2):S221–S231
Shawkey MD, Saranathan V, Palsdottir H, Crum J, Ellisman MH, Auer M, Prum RO (2009b) Electron tomography, three-dimensional Fourier analysis and colour prediction of a three-dimensional amorphous biophotonic nanostructure. J R Soc Interface 6:S213–S220
Shawkey MD, D'Alba L, Wozny J, Eliason C, Koop JAH, Jia L (2011) Structural color change following hydration and dehydration of iridescent Mourning Dove (*Zenaida macroura*) feathers. Zoology 114:59–68
Shu-hui Y, Yan-chun X, Da-wei Z (2006) Morphological basis for the waterproof characteristics of bird plumage. J For Res 17:163–166
Shutler D (2019) Some important overlooked aspects of odors in avian nesting ecology. J Avian Biol 50: e02003
Shyer AE, Rodrigues AR, Schroeder GG, Kassianidou E, Kumar S, Harland RM (2017) Emergent cellular self-organization and mechanosensation initiate follicle pattern in the avian skin. Science 357:811–815
Simons ELR (2009) The evolution of forelimb morphology and flight mode in extant birds. Ph.D. dissertation, Ohio University, Athens, OH
Soler JJ, Martín-Vivaldi M, Ruiz-Rodríguez M, Valdivia E, Martín-Platero AM, Martínez-Bueno M, Peralta-Sánchez JM, Méndez M (2008) Symbiotic association between Hoopoes and antibiotic-producing bacteria that live in their uropygial gland. Funct Ecol 22:864–871
Soliman SA, Madkour FA (2017) A comparative analysis of the organization of the sensory units in the beak of duck and quail. Histol Cytol Embryol 1:1–16
Spearman RIC (1966) The keratinization of epidermal scales, feathers, and hair. Biol Rev Camb Philos Soc 41:59–96
Spearman RIC, Hardy JA (1985) Integument. In: King AS, McLelland J (eds) Form and function in birds, vol 3. Academic Press, London, pp 1–56
Srinivasan S, Chhatre SS, Guardado JO, Park K-C, Parker AR, Rubner MF, McKinley GH, Cohen RE (2014) Quantification of feather structure, wettability and resistance to liquid penetration. J R Soc Interface 11: 20140287
Starck JM, Ricklefs RE (1998) Patterns of development: the altricial-precocial spectrum. In: Starck JM, Ricklefs RE (eds) Avian growth and development: evolution within the altricial-precocial spectrum. Oxford University Press, Oxford, pp 3–30
Stavenga DG, Leertouwer HL, Marshall NJ, Osorio D (2011) Dramatic colour changes in a bird of paradise caused by uniquely structured breast feather barbules. Proc R Soc B 278:2098–2104
Stavenga DG, Leertouwer HL, Osorio DC, Wilts BD (2015) High refractive index of melanin in shiny occipital feathers of a bird of paradise. Light: Sci Appl 4: e243
Stegmann B (1956) Über die herkunft de flüchtigen rosenroten federpigments. Journal für Ornithologie 97:204–205
Steiner H (1917) Das problem der diastataxie des vogelflügels. Jenaische Zeitschrift für Naturwissenschaft 55:222–496
Stettenheim P (1972) The integument of birds. In: Farner DS, King JR (eds) Avian biology, vol II. Academic Press, New York, pp 1–63
Stettenheim PR (2000) The integumentary morphology of modern birds – an overview. Am Zool 40:461–477
Stoddard MC, Prum RO (2011) How colorful are birds? Evolution of the avian plumage color gamut. Behav Ecol 22:1042–1052
Stokkan K-A (1992) Energetics and adaptations to cold in ptarmigan in winter. Ornis Scand 23:366–370
Stolpe M (1935) Colymbus, Hesperornis, Podiceps: ein vergleich ihrer hinteren extremität. J Ornithol 83: 115–128
Stoutjesdijk F (2002) The ugly duckling: a thermal viewpoint. J Therm Biol 27:413–422
Sues H-D (2001) Palaeontology: ruffling feathers. Nature 410:1036–1037
Sullivan TN, Pissarenko A, Herrera SA, Kisailus D, Lubarda VA, Meyers MA (2016) A lightweight, biological structure with tailored stiffness: the feather vane. Acta Biomaterialia 41:27–39
Sullivan TN, Zhang Y, Zavattieri PD, Meyers MA (2018) Hydration-induced shape and strength recovery of the feather. Adv Funct Mater 2018:1801250
Sun J, Bhushan B, Tong J (2013) Structural coloratoin in nature. RSC Adv 3:14862–14889

Sun L, Zhou T, Wan Q-H, Fang S-G (2020) Transcriptome comparison reveals key components of nuptial plumage coloration in Crested Ibis. Biomolecules 10:905

Surmacki A, Siefferman L, Yuan HW (2011) Effects of sunlight exposure on carotenoid-based and structural coloration of the Blue-tailed Bee-eater. Condor 113: 590–596

Sustaita D, Rubega MA (2014) The anatomy of a shrike bite: bill shape and bite performance in Loggerhead Shrikes. Biol J Linn Soc 112:485–498

Swaddle JP, Witter MS, Cuthill IC, Budden A, McCowen P (1996) Plumage condition affects flight performance in Common Starlings: implications for developmental homeostasis, abrasion and moult. J Avian Biol 27:103–111

Tedore C, Johnsen S (2017) Using RGB displays to portray color realistic imagery to animal eyes. Curr Zool 63:27–34

Thomas BT (1996) Family Opisthocomidae (Hoatzin). In: del Hoyo J, Elliott A, Sargatal J (eds) Handbook of the birds of the world, vol 3. Lynx Edicions, Barcelona, pp 23–32

Thomas DB, McGraw KJ (2018) Hidden carotenoids in the powder-down of herons. J Ornithol 159:785–792

Thomas DB, McGoverin CM, McGraw KJ, James HF, Madden O (2013) Vibrational spectroscopic analyses of unique yellow feather pigments (spheniscins) in penguins. J R Soc Interface 10:20121065

Thomas DB, McGraw KJ, Butler MW, Carrano MT, Madden O, James HF (2014) Ancient origins and multiple appearances of carotenoid-pigmented feathers in birds. Proc R Soc B 281:20140806

Tinbergen J, Wilts BD, Stavenga DG (2013) Spectral tuning of Amazon parrot feather coloration by psittacofulvin pigments and spongy structures. J Exp Biol 216:4358–4364

Tokita M, Matsushita H, Asakura Y (2020) Developmental mechanisms underlying webbed foot morphological diversity in waterbirds. Sci Rep 10:8028

Tompkins DM, Jones T, Clayton DH (1996) Effect of vertically transmitted ectoparasites on the reproductive success of swifts (*Apus apus*). Funct Ecol 10:733–740

Tributsch H (2016) Ochre bathing of the Bearded Vulture: a bio-mimetic model for early humans towards smell prevention and health. Animals 6:7

True JR, Carroll SB (2002) Gene co-option in physiological and morphological evolution. Annu Rev Cell Dev Biol 18:53–80

Unwin DM (1998) Feathers, filaments and theropod dinosaurs. Nature 391:119–120

Van Hemert C, Handel CM, Blake JE, Swor RM, O'Hara TM (2012) Microanatomy of passerine hard-cornified tissues: beak and claw structure of the Black-capped Chickadee (*Poecile atricapillus*). J Morphol 273:226–240

Van Huynh A, Rice AM (2019) Conspecific olfactory preferences and interspecific divergence in odor cues in a chickadee hybrid zone. Ecol Evol 9:9671–9683

Van Tyne J (1929) The life history of the toucan *Ramphastos brevicarinatus*. University of Michigan, Museum of Zoology, Miscellaneous Publications No. 19, Ann Arbor

van Vuuren AKJ, Kemp LV, McKechnie AE (2020) The beak and unfeathered skin as heat radiators in the Southern Ground-Hornbill. J Avian Biol 51:jav.02457

Vas Z, Csörgö T, Møller AP, Rózsa L (2008) The feather holes on the Barn Swallow *Hirundo rustica* and other small passerines are probably caused by *Brueelia* spp. lice. J Parasitol 94:1438–1440

Verea C, Vitelli-Flores J, Isturiz T, Rodríguez-Lemoine V, Bosque C (2017) The effect of uropygial gland secretions of Spectacled Thrushes (*Turdus nudigenis*) on feather degradation and bacterial growth in vitro. J Ornithol 158:1035–1043

Vigneron JP, Colomer J-F, Rassart M, Ingram AL, Lousse V (2006) Structural origin of the colored reflections from the Black-billed Magpie feathers. Phys Rev E 73: 021914

Vogel S (2007) Living in a physical world. XI. To twist or bend when stressed. J Biosci 32:643–655

Wadee MK, Wadee MA, Bassom AP (2007) Effects of orthotropy and variation of Poisson's ratio on the behaviour of tubes in pure flexure. J Mech Phys Solids 55:1086–1102

Waite JL, Henry AR, Clayton DH (2012) How effective is preening against mobile ectoparasites? An experimental test with pigeons and hippoboscid flies. Int J Parasitol 42:463–467

Wang X, Clarke JA (2015) The evolution of avian wing shape and previously unrecognized trends in covert feathering. Proc R Soc B 282:20151935

Wang B, Meyers MA (2017) Seagull feather shaft: correlation between structure and mechanical response. Acta Biomaterialia 48:270–288

Wang B, Yang W, McKittrick J, Meyers MA (2016a) Keratin: structure, mechanical properties, occurrence in biological organisms, and efforts at bioinspiration. Prog Mater Sci 76:229–318

Wang S, Yang Z, Gong G, Wang J, Wu J, Yang S, Jiang L (2016b) Icephobicity of penguins *Sphenicus humboldti* and an artificial replica of penguin feather with air-infused hierarchical rough structures. J Phys Chem C 120:15923–15929

Weaver RJ, Santos ESA, Tucker AM, Wilson AE, Hill GE (2018) Carotenoid metabolism strengthens the link between feather coloration and individual quality. Nat Commun 9:73

Weir KA, Lunam CA (2011) The structure and sensory innervation of the integument of ratites. In: Glatz P, Lunam C, Malecki I (eds) The welfare of farmed ratites. Springer, Berlin, Heidelberg, pp 131–145

West GC (1962) Responses and adaptations of wild birds to environmental temperature. Comp Physiol Temp Regul 3:291–324

White CM, Clum NJ, Cade TJ, Hunt WG (2020) Peregrine Falcon (*Falco peregrinus*), version 1.0. In: Billerman

SM (ed) Birds of the World. Cornell Lab of Ornithology, Ithaca

Whittaker DJ, Hagelin JC (2021) Female-based patterns and social function in avian chemical communication. J Chem Ecol 47:43–62

Whittaker DJ, Richmond KM, Miller AK, Kiley R, Bergeon Burns C, Atwell JW, Ketterson ED (2011) Intraspecific preen oil odor preferences in Dark-eyed Juncos (*Junco hyemalis*). Behav Ecol 22:1256–1263

Whittaker DJ, Rosvall KA, Slowinski SP, Soini HA, Novotny MV, Ketterson ED (2017) Songbird chemical signals reflect uropygial gland androgen sensitivity and predict aggression: implications for the role of the periphery in chemosignaling. J Comp Physiol A 204: 5–15

Williams CL, Hagelin JC, Kooyman GL (2015) Hidden keys to survival: the type, density, pattern and functional role of Emperor Penguin body feathers. Proc R Soc B 282:20152033

Wingfield JC, Ishii S, Kikuchi M, Wakabayashi S, Sakai H, Yamaguchi N, Wada M, Chikatsuji K (2000) Biology of a critically endangered species, the Toki (Japanese Crested Ibis) *Nipponia nippon*. Ibis 142:1–11

Wood MJ, Brock G, Kietzig A-M (2023) The penguin feather as inspiration for anti-icing surfaces. Cold Regions Sci Tech. https://doi.org/10.1016/j.coldregions.2023.103903

Wu P, Ng CS, Yan J, Lai Y-C, Chen C-K, Lai Y-T, Wu S-M, Chen JJ, Luo W, Widelitz RB, Li WH (2015) Topographical mapping of α-and β-keratins on developing chicken skin integuments: functional interaction and evolutionary perspectives. Proc Natl Acad Sci USA 112:E6770–E6779

Wu P, Yan J, Lai YC, Ng CS, Li A, Jiang X, Elsey R, Widelitz R, Bajpai R, Li WH, Chuong CM (2017) Multiple regulatory modules are required for scale-to-feather conversion. Mol Biol Evol 35:417–430

Xiao M, Dhinojwala A, Shawkey M (2014) Nanostructural basis of rainbow-like iridescence in Common Bronzewing *Phaps chalcoptera* feathers. Opt Express 22:14625–14636

Xing L, McKellar RC, Li G, Bai M, Persons WS IV, Miyashita T, Benton MJ, Zhang J, Wolfe AP, Yi O, Tseng K, Ran H, Currie PJ (2016) A feathered dinosaur tail with primitive plumage trapped in mid-Cretaceous amber. Curr Biol 26:3352–3360

Xu X, Guo Y (2009) The origin and early evolution of feathers: insights from recent paleontological and neontological data. Vertebrata PalAsiatica 47:311–329

Xu X, Zhou Z, Prum RO (2001) Branched integumentary structures in *Sinornithosaurus* and the origin of feathers. Nature 410:200–204

Xu X, Zhou Z, Wang X, Kuang X, Zhang F, Du X (2003) Four-winged dinosaurs from China. Nature 421: 335–340

Xu X, Zheng X, You H (2009) A new feather type in a nonavian theropod and the early evolution of feathers. Proc Natl Acad Sci USA 106:832–834

Zahavi A (1975) Mate selection – a selection for a handicap. J Theor Biol 53:205–214

Zampiga E, Hoi H, Pilastro A (2004) Preening, plumage reflectance and female choice in Budgerigars. Ethol Ecol Evol 16:339–349

Zelenitsky DK, Therrien F, Erickson GM, DeBuhr CL, Kobayashi Y, Eberth DA, Hadfield F (2012) Feathered non-avian dinosaurs from North America provide insight into wing origins. Science 338:510–514

Zhang X, Shi F, Niu J, Jiang Y, Wang Z (2008a) Superhydrophobic surfaces: from structural control to functional application. J Mater Chem 18:621–633

Zhang F, Zhou Z, Xu X, Wang X, Sullivan C (2008b) A bizarre Jurassic maniraptoran from China with elongate ribbon-like feathers. Nature 455:1105–1108

Zhang YF, Dong BQ, Shi L, Yin HW, Liu XH, Zi J (2014) Color production in blue and green feather barbs of the Rosy-faced Lovebird. Mater Today: Proc 1:130–137

Zheng X-T, You H-L, Xu X, Dong Z-M (2009) An Early Cretaceous heterodontosaurid dinosaur with filamentous integumentary structures. Nature 458:333–336

Zhou L, Li Y, Liu G, Fan Q, Shao J (2016) Study on the correlations between the structural colors of photonic crystals and the base colors of textile fabric substrates. Dyes Pigm 133:435–444

Zuk M (1991) Sexual ornaments as animal signals. Trends Ecol Evol 6:228–231

Zuk M, Johnsen TS (2000) Social environment and immunity in male Red Jungle Fowl. Behav Ecol 11:146–153

Zuk M, Johnsen TS, MacLarty T (1995) Endocrine-immune interactions, ornaments and mate choice in Red Jungle Fowl. Proc R Soc B 260:205–210

# Nervous System

# 4

## Contents




**Abstract**

The nervous system of birds is comparable in most respects to that of mammals, including relatively large brains. The central and peripheral nervous systems of birds are described in detail, with particular emphasis on the anatomy and function of avian brains. The impressive cognitive abilities and many species of birds are also described. The general and special senses of birds are discussed, including the special senses of olfaction, taste, vision, hearing, and static and dynamic equilibrium. The senses of smell and taste have been found to be important for many





G. Ritchison, *In a Class of Their Own*, Fascinating Life Sciences,
https://doi.org/10.1007/978-3-031-14852-1_4

species of birds and many examples are provided. Most birds have impressive vision and the structure and function of the eyes of both diurnal and nocturnal species of birds are discussed in detail. The structure and function of avian ears, and the relative importance of the sense of hearing in different species of birds, are also discussed. The ability of birds, especially some species of owls, to localize the source of sounds is explained. The use of echolocation by a few species of birds, including swiftlets and Oilbirds, is also described.

## 4.1 Cognitive Abilities

Birds, like mammals, have high metabolic rates and generally lead very active lives. The speed with which birds do things, and flying is just one obvious example, means that they must be able to efficiently and quickly obtain information about their external environment and, just as quickly, analyze and act upon that information. A Peregrine Falcon (*Falco peregrinus*) stooping at its prey at 300 km/h clearly must be capable of split-second decisions. Birds not only live fast, but many also lead complicated lives. For example, some birds, such as California Scrub-Jays (*Aphelocoma californica*), cache seeds at hundreds of different locations and their survival may depend on recalling those locations several weeks or months later. In laboratory experiments, several species of birds, especially corvids (family Corvidae) and parrots (order Psittaciformes), have demonstrated a variety of behaviors illustrating their cognitive abilities (Box 4.1 Examples of avian cognitive abilities).

Birds are capable of such behavior because they have nervous systems that quickly and efficiently obtain (via sensory receptors of various types) information about their environment, store and analyze information and, as needed, initiate responses (via muscles and glands). For many years, the nervous system of birds and, particularly, the avian brain have been viewed as being inferior to those of mammals and the ultimate insult was to be considered a "bird brain." Look it up in your dictionary; a "bird brain" is a person with confused ideas and incapable of serious thought. To be more specific, a person could be called "loony" (i.e., a foolish or crazy person). Such terms clearly give the impression that birds are a step (or more) below mammals in terms of their "mental abilities."

Several factors likely contributed to the widespread belief that birds were less "intelligent" than mammals. One obvious factor is that, for many humans, birds "look" less intelligent, with small eyes and heads, do "silly" things, e.g., a pigeon bobbing its head as it walks or a bird flying into a window, and many of their behaviors are widely considered to be hard wired or innate. In the scientific world, the problem goes back to the late nineteenth century, when German naturalist Ludwig Edinger (1896) conducted the first careful studies of avian neuroanatomy and labeled the various parts of the bird brain. Because of the existing consensus that evolution was a process that led to progressively more complex and better organisms and that mammals were a "step" above birds, Edinger assumed that the avian brain had to be less advanced than the mammalian (and, specifically, the human) brain. Basically, Edinger felt that the "higher" levels of the avian brain were derived from the basal ganglia (part of which is called the striatum), an area of the brain that is involved in species-specific instinctive behaviors, such as feeding and sexual and parental behavior. His names, therefore, included terms like "paleostriatum" and "archistriatum" to indicate the relatively primitive nature of the bird brain. In contrast, he believed that the highest area of the mammalian brain (cerebrum) was new (in an evolutionary sense) and called it the neoencephalon or pallium (derived from the Roman *pallium* or *palla* and referring to a cloak).

We now know that most of a bird's telencephalon is not derived from the striatum, but from the pallium. Indeed, the large area of forebrain that lies above the basal ganglia in birds is now recognized to be functionally and developmentally the same as the mammalian neocortex, derived in the same way from the pallial sector of the embryonic forebrain. However, rather than producing a layered cortex like that of mammals,

**Box 4.1 Examples of avian cognitive abilities**

As with mammals, the cognitive abilities of birds vary among species, with species in the family Corvidae (jays, jackdaws, crows, rooks, ravens, and magpies) and the order Psittaciformes (cockatoos and parrots) considered among the most "intelligent" birds. In laboratory experiments, some of these species have been found to succeed in a variety of tasks that require reasoning, prospection, imagination, planning, and memory (Clayton and Emery 2015).

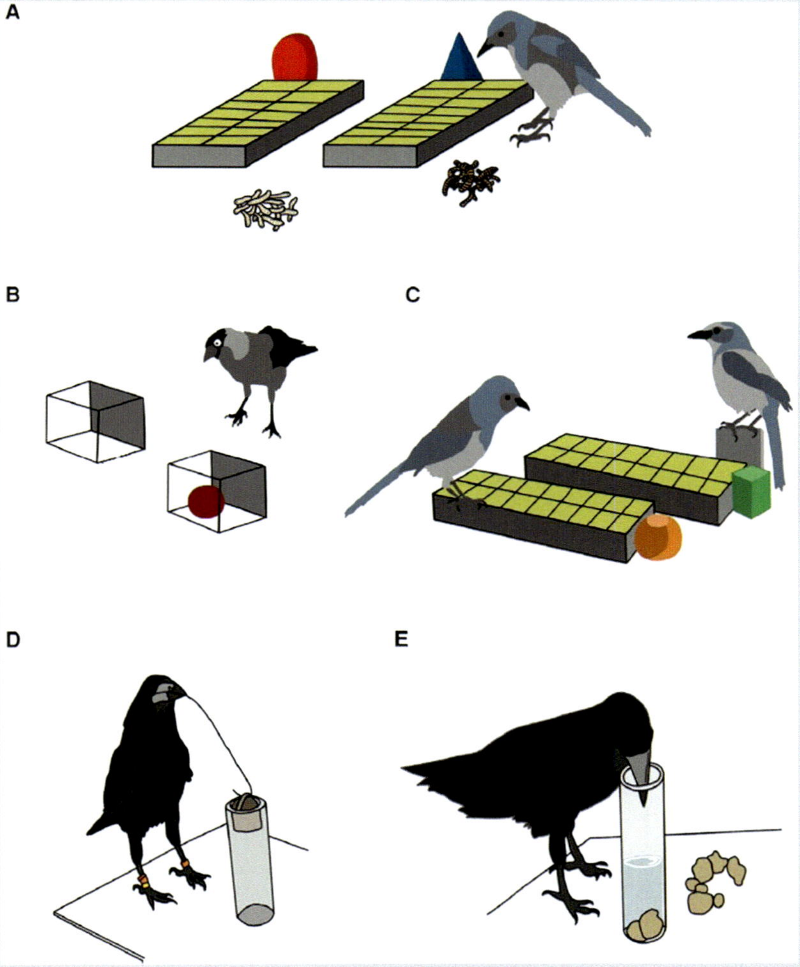


Some examples of the cognitive abilities of corvids. (**a**) California Scrub-Jays (*Aphelocoma californica*) that cached perishable food items chose not to retrieve those items if sufficient time had passed for the items to decay (Clayton and Dickinson 1998). (**b**) Eurasian Jackdaws (*Corvus monedula*) used the orientation of the eyes of a human to locate hidden food items (von Bayern and Emery 2009). (**c**) California Scrub-Jays that cached food items in view of other California Scrub-Jays subsequently re-hid the items in a different location when the other scrub-jays were no longer present, preventing the other scrub-jays from "stealing" their cached food when they returned (Clayton et al. 2007). (**d**) Rooks (*Corvus frugilegus*) modified the shape of a wire to create a hook that allowed them to obtain food from the bottom of a tube (Bird and Emery 2009a). (**e**) Rooks added stones to a tube-shaped container with water to raise the water level and gain access to a food item floating on top of the water (Bird and Emery 2009b) (Figure from Clayton and Emery 2015; © 2015 Elsevier Ltd., used with permission).

(continued)

**Box 4.1** (continued)

In another test of corvid cognitive abilities, Kabadayi et al. (2016) compared the performance of three species of corvids with that of coyotes and several species of primates in the "cylinder task." In this experiment, test subjects were first exposed to an opaque cylinder, open at both ends, within which was placed a food item. After test subjects learned to retrieve food from the opaque cylinder, each was then tested 10 times with a transparent cylinder. During these trials, the food item was placed in the transparent cylinder and investigators noted whether test subjects made contact with the cylinder (attempting to retrieve the food item directly), or if, having experience with the opaque cylinder, they correctly determined that they should use one of the openings at each end to retrieve the food item.

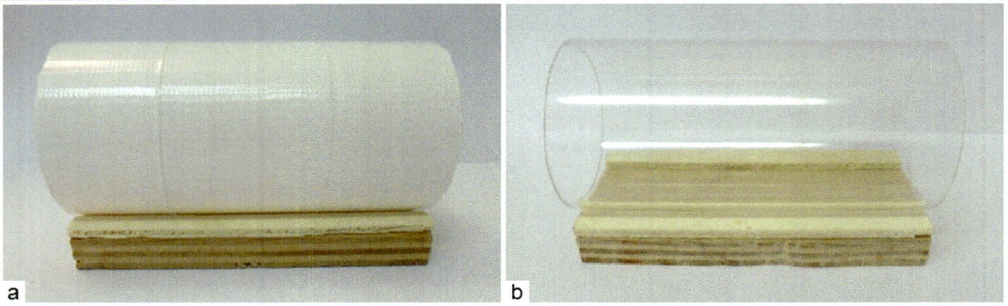


Cylinders used in the cylinder-task experiments. (**a**) Opaque cylinder with an opening on each end that allows test subjects to determine food item location within the cylinder. (**b**) Transparent cylinder that also has an opening at each end (Figure from Kabadayi et al. 2016; © 2016 The Authors. Published by the Royal Society under the terms of the Creative Commons Attribution License, https://creativecommons.org/licenses/by/4.0/)

Results of cylinder task experiments with several species of primates, coyotes, and three species of birds in the family Corvidae (Common Raven, *Corax*; Eurasian Jackdaw, *Corvus monedula*; and New Caledonian Crow, *Corvus moneduloides*). The cylinder task mean score indicates the mean percent times that test subjects made the correct decision (obtained the food item via the open ends of the cylinder rather trying to access it directly through the transparent cylinder) in 10 trials (Information in table from Kabadayi et al. 2016; © 2016 The Authors. Published by the Royal Society under the terms of the Creative Commons Attribution License, https://creativecommons.org/licenses/by/4.0/)

| Rank | Species | Cylinder task mean score | Endocranial volume (cm$^3$) |
|---|---|---|---|
| 1 | Chimpanzee | 100.0 | 368.4 |
| 1 | Common Raven | 100.0 | 14.5 |
| 3 | Orangutan | 99.1 | 377.4 |
| 4 | Eurasian Jackdaw | 97.0 | 5.2 |
| 5 | Capuchin monkey | 95.9 | 66.6 |
| 6 | Bonobo | 95.0 | 341.3 |
| 6 | Coyote | 95.0 | 85.2 |
| 8 | Gorilla | 94.4 | 490.4 |
| 9 | New Caledonian Crow | 92.0 | 7.3 |
| 10 | Rhesus macaque | 80.0 | 89.0 |

The three corvid species did as well or, in some cases, even better than the coyotes and primates with the cylinder task even though, as the table above indicates, the brain volumes of corvids are much smaller than those of the coyote and primates. These results suggest that the

(continued)

**Box 4.1** (continued)

cognitive abilities of corvids may be comparable to that of some species of primates (and coyotes) and, in addition, that absolute brain size may not be an accurate indicator of cognitive abilities. In addition to their cognitive abilities, crows may also have sensory consciousness (Nieder et al. 2020; Nieder 2021), i.e., "patterns of neuronal activity that represent mental content that drives behavior" (Herculano-Houzel 2020).

the avian brain is a nucleated structure with pockets of gray matter. The analogy would be to compare a club sandwich (mammalian) to a pepperoni pizza (avian) (Emery and Clayton 2005).

## 4.2 Avian Nervous System

The avian brain will be discussed in more detail later, but a bird's nervous system includes numerous other components. The brain and spinal cord make up the central nervous system, and all neurons, receptors, nerves, and ganglia outside of the brain and spinal cord make up the peripheral nervous system. As with other vertebrates, the avian spinal cord receives sensory information, and motor neurons originating in the cord innervate muscles and glands. The avian spinal cord has a unique appearance, with bulges in the cervical and lumbar regions and a relatively thin thoracic region (Fig. 4.1). Many sensory and motor neurons coming from and going to the wings create the cervical enlargement and those to and from the legs create the lumbar enlargement. The spinal cords of other tetrapods (amphibians, reptiles, and mammals) are thicker in the thoracic region than is the case for birds. A thicker cord means, of course, more neurons and other tetrapods have large numbers of neurons passing between the cervical and lumbar enlargements that help coordinate movements of the forelimbs and hindlimbs. Birds have fewer such neurons (interneurons) because less coordination between forelimbs and hindlimbs is needed. When landing and taking off, such coordination is necessary, but, in general, the wings and legs function independently.

Reflexes involving the spinal cord permit rapid responses, e.g., regulating muscle forces in the legs to maintain stable locomotion when walking or running across uneven terrain (Gordon et al. 2020) or responding to potentially harmful stimuli (e.g., extreme temperatures and high mechanical pressures) (Fig. 4.1). The perception of pain is called nociception and painful stimuli often trigger reflex responses that are obviously beneficial and help prevent or minimize injuries. Studies in the lab have revealed that increased levels of testosterone in male House Sparrows (*Passer domesticus*) reduce their responses to nociceptive stimuli, apparently by influencing sensory input in the spinal cord. This suggests that, when testosterone levels increase during the breeding season, male birds may become less sensitive to painful stimuli and, for example, may be more likely to engage in and continue intense aggressive encounters with conspecific males (Hau et al. 2004). Nociception, in this case, would still serve a protective function (signaling pain), but the pain threshold would be higher.

Testosterone can affect the spinal cord in other ways as well. Manakins are passerines common in the forests of Central and South America, and males of several species perform complex courtship displays involving short flights, jumping, and dancing movements. In some species, males also have special feather structures on their wings that produce loud snapping sounds when hit together or against other feathers (Fig. 4.2). Analysis has revealed that testosterone-accumulating cells are located in the cervical and lumbosacral enlargements of the spinal cord (Fig. 4.3). Because motor neurons in these areas control

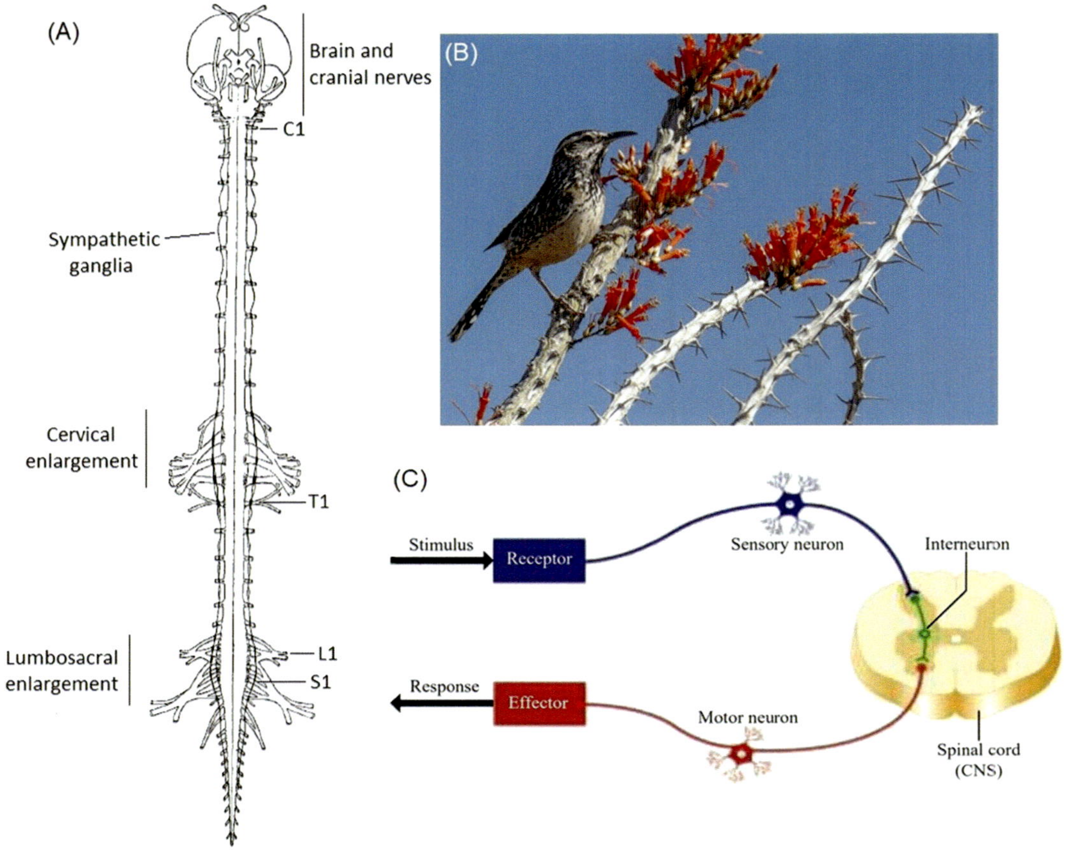


**Fig. 4.1** (**a**) Central nervous system of a Rock Pigeon (*Columba livia*). The spinal cord is the point of origin for numerous spinal nerves and a series of sympathetic ganglia are located parallel to the spinal cord. Sympathetic neurons are part of a bird's autonomic nervous system (and ganglia are structures within which are located the cell bodies of neurons). In the avian spinal cord, the enlargement in the cervical region is due to the many neurons needed to control the muscles of the wings. The lumbosacral enlargement is due to the many neurons that control the legs. The thoracic section is rather thin because movement of the wings and limbs need not be coordinated so fewer interconnecting, or interneurons, are needed. (**b**) Spinal reflexes allow very rapid responses to, for example, noxious, or painful, stimuli. If a Cactus Wren (*Campylorhynchus brunneicapillus*) accidentally steps, or begins to step, on a spine of an ocotillo (*Fouquiera splendens*) when attempting to perch, a spinal reflex will automatically cause contraction of skeletal muscles that will lift the wren's foot from the spine and prevent or minimize tissue damage. (**c**) Spinal reflexes are very fast because they occur without input from the brain. If the wren stepped, or began to step on a sharp spine, pain receptors (Stimulus > Receptor) would be stimulated, sending nervous impulses into the spinal cord along sensory neurons. Impulses are passed to neurons located in the spinal cord called interneurons, which, in turn, transmit impulses to motor neurons. Motor neurons will then stimulate skeletal muscles (Effector) and their contraction (response) will cause the wren's foot to move away from the spine. (Figure **a** modified from Huber 1936; © 1936 The Wistar Institute of Anatomy and Biology, used with permission. Photo **b** by Andrew Cattoir, National Park Service, CC0 Public Domain. Figure **c** from Li et al. 2014; © 2014 Jilin University, Published by Springer Nature, used with permission)

muscles of the wings and legs, these cells may have multiple behavioral functions and may innervate muscles controlling their elaborate displays (Schultz and Schlinger 1999). This suggests that sex steroids may control diverse behaviors in males of other species of birds, in part by acting directly on spinal neural circuits.

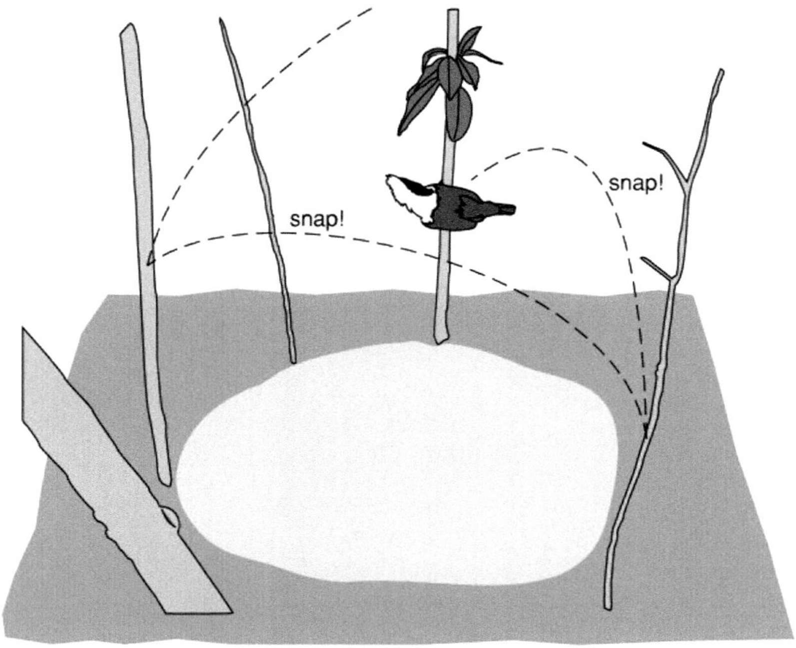


**Fig. 4.2** The snap-jump display of a male Golden-collared Manakin (*Manacus vitellinus*). Males fly down from a perch 2–5 m above their display court, land on one of the saplings, then jump toward another sapling while simultaneously generating a loud snap with a rapid upward movement of their wings that hit each other above the back. Males may repeat snap-jump displays up to 20 times in quick succession. (Figure from Fusani et al. 2007; © 2007 John Wiley and Sons, used with permission)

## 4.3 Avian Brains

The brains of birds are similar in many ways to the brains of other vertebrates, with the same five subdivisions (Fig. 4.4). The general functions of three of those subdivisions (myelencephalon, metencephalon, and mesencephalon) are similar among all classes of vertebrates. Collectively, these three areas are referred to as the brainstem. The myelencephalon, or medulla, and the pons (part of the metencephalon) are continuous with the spinal cord (Figs. 4.5 and 4.6) and contain the reticular formation and several centers involved in the regulation of critical systems, including the cardiovascular and respiratory centers. Swallowing is also controlled by neurons in the hindbrain. Cranial nerves V through to XII originate on the ventral surface of the medulla. The reticular formation helps to coordinate movements of the head and body and is also important in arousal (awakening from sleep) and attention or alertness (Butler and Hodos 1996).

The cerebellum (the other part of the metencephalon) of birds averages about 12% of total brain volume across species (Yopak et al. 2020) and serves to integrate sensory inputs from proprioceptors (located in skeletal muscles, tendons, and joints) and the vestibular system (receptors in the inner ear providing information about equilibrium and head position). Via motor neurons, the cerebellum is responsible for maintaining balance and coordinating the actions of skeletal muscles. Available evidence suggests that the cerebellum may also play a role in memory and learning (Strick et al. 2009) and, via cortico-cerebellar pathways, is "crucial for complex cognitive abilities in birds" (Gutiérrez-Ibáñez et al. 2018). More specifically, the cognitive abilities of some birds like parrots and corvids may be related to connections between the cerebellum and the telencephalon (and specifically the nidopallium caudolaterale that is thought to be analogous to the prefrontal cortex of mammals; Ito 2008) that allow them to "generate mental representations of objects thus allowing the mental manipulation of these objects and the prediction of possible outcomes" (Gutiérrez-Ibáñez et al. 2018).

The cerebellum has a unique appearance, with multiple lobes (or folia), and, among birds, there is variation in the relative sizes of various folia

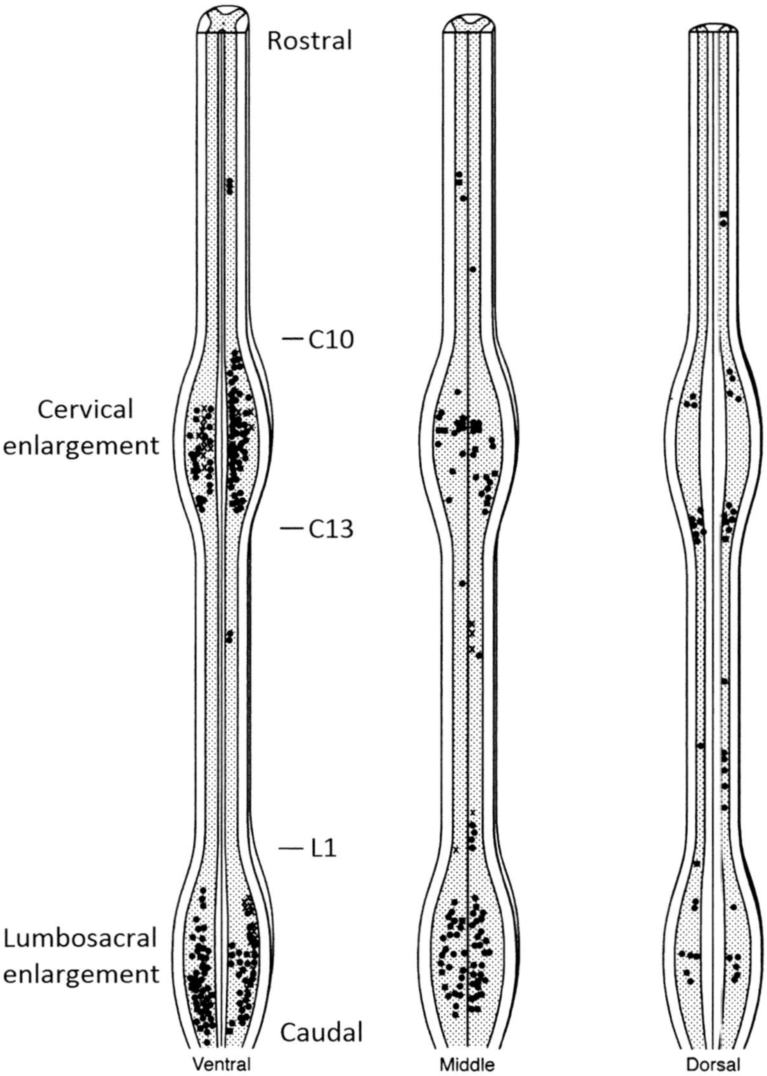


**Fig. 4.3** Distribution of sex-steroid accumulating cells in the spinal cord of a Golden-collared Manakin (*Manacus vitellinus*). Each dot represents three cells, and the cord is illustrated with three sections through the chord from front (ventral) middle, to back (dorsal). C10, level of 10th cervical vertebra; C13, level of 13th cervical vertebra; L1, level of 1st lumbar vertebra. (Figure from Schultz and Schlinger 1999; used with permission of the U. S. National Academy of Sciences)

and the relative depth of folds, or indentations, between the lobes (Fig. 4.7). Because different areas of the cerebellum, and different folia, are involved in coordinating muscular activity in different areas of the bird body (Fig. 4.8), differences in folia size and degree of folding can provide insight concerning those muscles and activities that are particularly important because larger folia and, especially, deeper folds means a greater surface area and, therefore, more nerve cells (Figs. 4.9 and 4.10). For example, crows, parrots, and woodpeckers have enlarged folia in the area of the cerebellum involved in coordinating movements of the head and beak. The enlargement of these folia corresponds well with the impressive abilities of these birds to use their bills and, for parrots and woodpeckers, tongues to manipulate objects (Sultan 2005). In addition, species of birds that use tools (Iwaniuk et al. 2009; Fig. 4.10) and build more complex nests (Fig. 4.11) or bowers (in the case of male bowerbirds, family Ptilonorhynchidae) exhibit a greater degree of cerebellar foliation, suggesting that the manipulative skills needed to construct

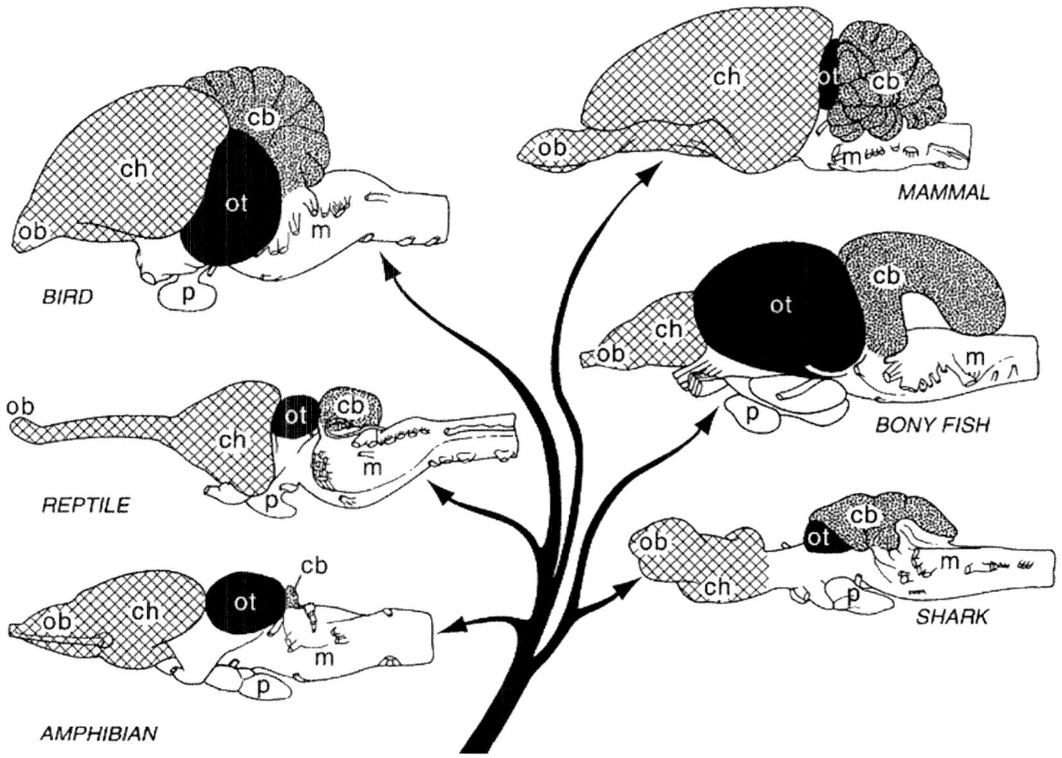


**Fig. 4.4** The brains of different vertebrates consist of the same five subdivisions, including the myelencephalon (medulla), metencephalon (pons and cerebellum), mesencephalon (midbrain), diencephalon (thalamus and hypothalamus), and telencephalon (cerebrum). m, medulla; cb, cerebellum; ot, optic tectum or lobe; p, pituitary gland; ch, cerebral hemispheres; ob, olfactory bulb. (Figure from Northcutt 2002; © 2002 Oxford University Press, used with permission)

more complex nests has favored selection for additional neurons and an increase in the processing ability of the cerebellum (Day et al. 2005; Hall et al. 2013, 2015). On the other hand, birds that have small hindlimb muscles, such as hummingbirds, swifts, potoos, and owlet-nightjars, have smaller folia in the area of the cerebellum involved in the coordination of hindlimb muscles (Larsell 1967; Iwaniuk et al. 2006). Surprisingly, birds that are excellent flyers, e.g., swifts and falcons, do not have unusually large cerebellums, suggesting that well-developed motor skills do not require an increase in cerebellum size and contradicting the common idea that the large cerebella of birds are related to flight (Sultan 2005; Walsh et al. 2013). In general, however, available evidence indicates that foliar size and structure varies among species, and that variation is often correlated with behavioral and morphological differences (Iwaniuk et al. 2006).

The midbrain (mesencephalon) includes the tectum and the tegmentum. The tegmentum contains a portion of the reticular formation and so is involved in motor control. The tectum contains the inferior colliculus and optic tectum. In the optic tectum, auditory information is combined with visual input to form a topographic map of space. This spatial map allows birds to locate objects using visual or auditory cues or both. The optic tectum of Barn Owls (*Tyto alba*) has been studied intensively (e.g., Zahar et al. 2012, 2018). The retina of the eye preprocesses the image to enhance motion and texture information and passes the signals to the optic tectum. Meanwhile,

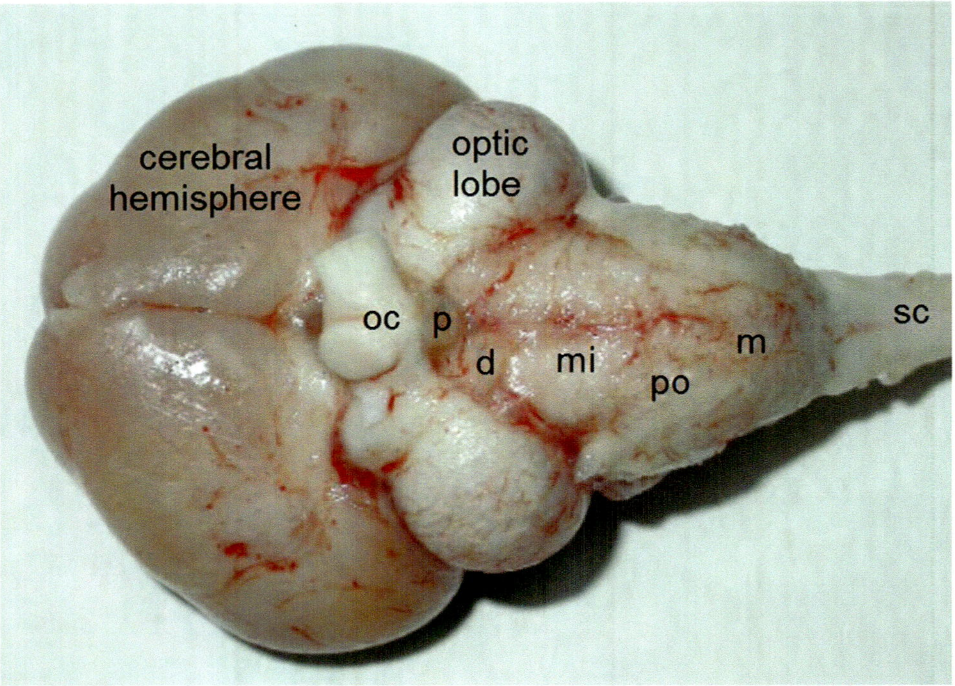


**Fig. 4.5** Ventral view of the brain of a Common Ostrich (*Struthio camelus*) showing the spinal cord (sc), medulla (m), pons (po), midbrain (mi), diencephalon (d), pineal gland (p), optic chiasma (oc, where the two optic nerves cross), optic lobe, and cerebral hemispheres. (Figure modified from Peng et al. 2010; © 2010 TÜBİTAK, Scientific and Technological Research Council of Turkey, used with permission)

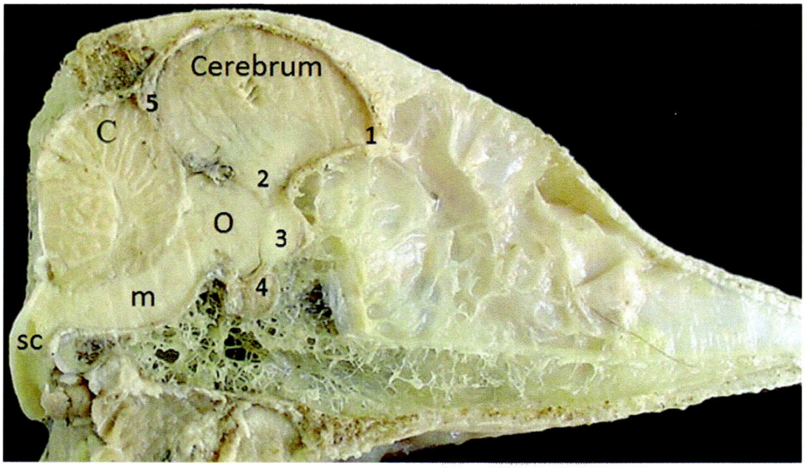


**Fig. 4.6** Section of the head and brain of a Common Ostrich (*Struthio camelus*). C, cerebellum; O, optic lobe; m, medulla; sc, spinal cord; 1, olfactory nerve; 2, thalamus; 3, optic chiasma; 4, pituitary gland; 5, pineal gland. (Figure modified from Karkoura et al. 2015; open-access article distributed under the terms of the Creative Commons Attribution License, https://creativecommons.org/licenses/by/4.0/)

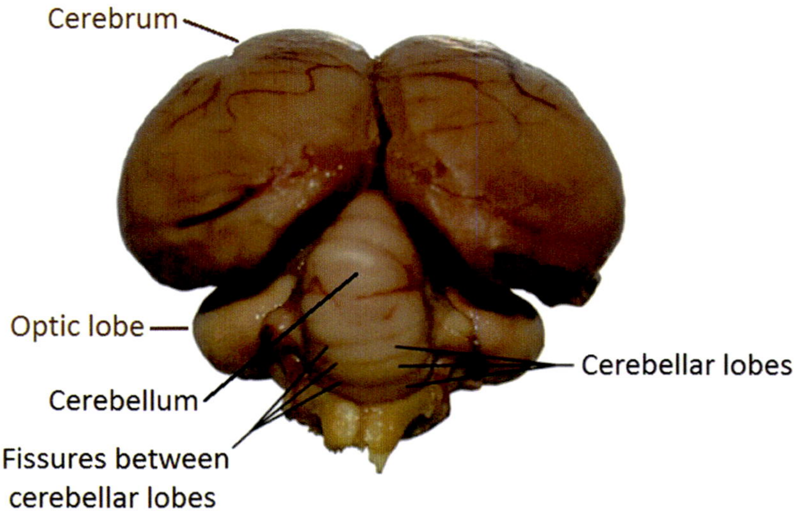


**Fig. 4.7** Brain of a Monk Parakeet (*Myiopsitta monachus*) as viewed from the back. The cerebellum consists of multiple lobes (also called folia) separated by fissures. (Figure modified from Carril et al. 2016; © 2016 John Wiley and Sons, used with permission)

more prominently than in other species, the owl's ears receive auditory signals that differ in arrival times (at the two ears) and frequency depending on the location of the sound source. These differences are mapped to target azimuth and elevation at the external nucleus of the inferior colliculus (ICx). The ICx consists of a spatial array of neurons that translate the auditory information into a low-resolution "image" of a sound's location. Impulses from the ICx then travel to the optic tectum where they may be processed in spatial registration with visual information (depending on light levels) (Fig. 4.12). This permits the joint stimulation of acoustic and visual neurons that correspond to the same location in space and enhances the owl's ability to distinguish a noisy, moving target against a background of other moving targets and uncorrelated noise.

The diencephalon of the forebrain includes the epithalamus, thalamus, and hypothalamus. The epithalamus of birds includes the habenular complex and the pineal gland. Although little studied in birds, the habenula is found in all vertebrates and is thought to be important in making decisions when engaged in behaviors such as foraging as well as in selecting and adapting strategies to cope with aversive stimuli (Stephenson-Jones et al. 2012). The avian pineal gland is located on top of the cerebrum (Fig. 4.13), has both photoreceptor cells and secretory cells, and interacts with the hypothalamus and retinas of the eyes to maintain circadian (daily) and seasonal rhythms. The pineal gland and retinas secrete melatonin only at night, with production and secretion inhibited by cells in the hypothalamus (suprachiasmatic nucleus) during the day. This daily variation or oscillation in melatonin secretion establishes the avian circadian "clock," or the daily rhythm of behavioral, physiological, and biochemical processes (Cassone 2014; Fig. 4.14).

The diencephalon also includes the thalamus and hypothalamus. The functions of the thalamus vary among vertebrates, but appear to be similar for birds and mammals (Csillag and Montagnese 2005). One function of the avian thalamus is to serve as a relay station, receiving sensory impulses from throughout a bird's body and "relaying"those impulses along different neurons to sensory areas in the cerebrum. More specifically, the thalamus can be subdivided into three areas based on the types of sensory information being relayed. These include the sensory thalamus that relays sensory information from skin and skeletal muscles, the motor thalamus relays sensory information from the cerebellum and basal ganglia (areas of the brain important in coordinating muscle activity), and the limbic thalamus that relays information from the limbic system (an area of the brain important in feeding

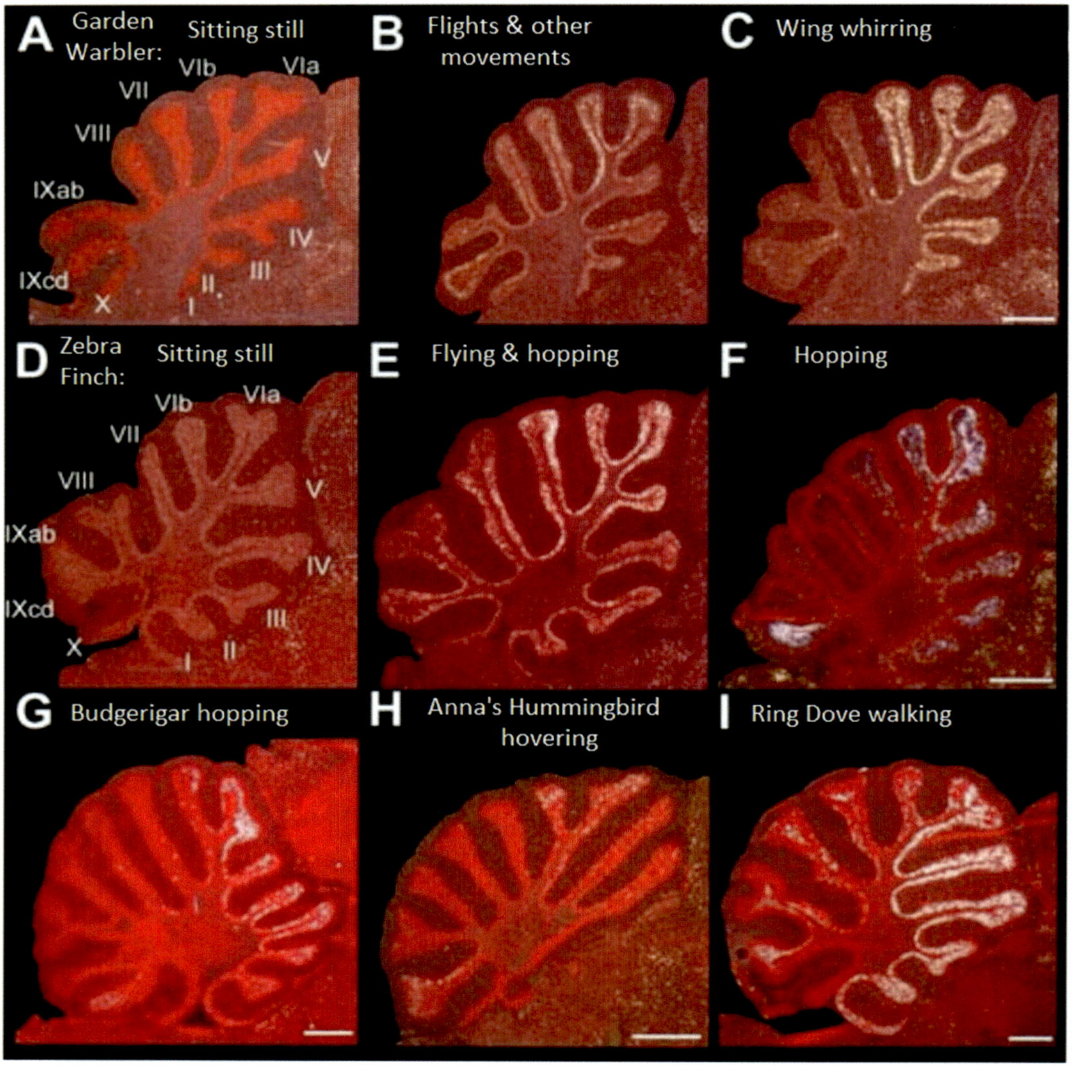


**Fig. 4.8** Sections of the cerebellums of several species of birds with lighter areas indicating stimulation of neurons in those areas of the cerebellums. (**a**)–(**c**) Neuron activity in cerebellums of Garden Warblers (*Sylvia borin*) when sitting still, flying and making other movements, and when wing whirring (rapid movements of wings associated with migratory restlessness). (**d**, **e**, and **f**) Neuron activity in cerebellums of Zebra Finches (*Taeniopygia guttata*) when sitting still, flying and hopping, and hopping. (**g**, **e**, and **f**) Neuron activity when a Budgerigar (*Melopsittacus undulatus*) is hopping, an Anna's Hummingbird (*Calypte anna*) is hovering, and an African Collared-Dove (*Streptopelia roseogrisea*: also referred to as the Ringed Turtle Dove) is walking. (Figure modified from Feenders et al. 2008; © 2008 Feenders et al., open-access article distributed under the terms of the Creative Commons Attribution License, https://creativecommons.org/licenses/by/4.0/)

behavior, memory, and basic emotions such as fear). The thalamus also appears to be important in activating the rhythmic muscle contractions needed for the flapping flight of birds (Gold et al. 2016; Fig. 4.15) and, among songbirds, the thalamus also plays a role in song production, contributing to the ability of songbirds to produce

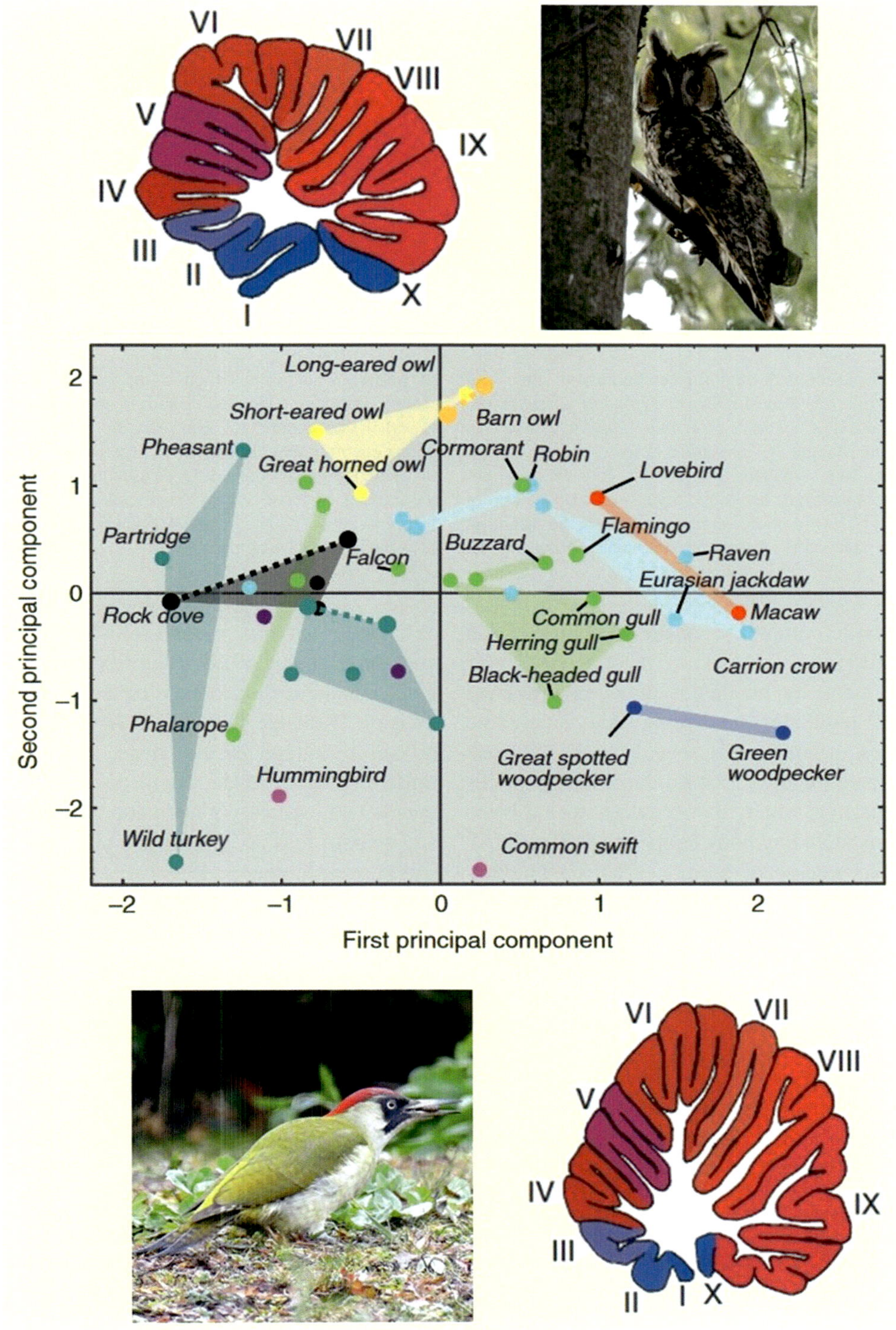


**Fig. 4.9** Woodpeckers (e.g., Eurasian Green Woodpecker [*Picus viridis*] pictured below), corvids (e.g., Carrion Crows [*Corvus corone*], ravens, and Eurasan Jackdaws [*Corvus monedula*]), and parrots (e.g., macaws and lovebirds) have longer, larger cerebellar lobes IV, VI, VII, VIII, and IX (shown in red at bottom) than many other

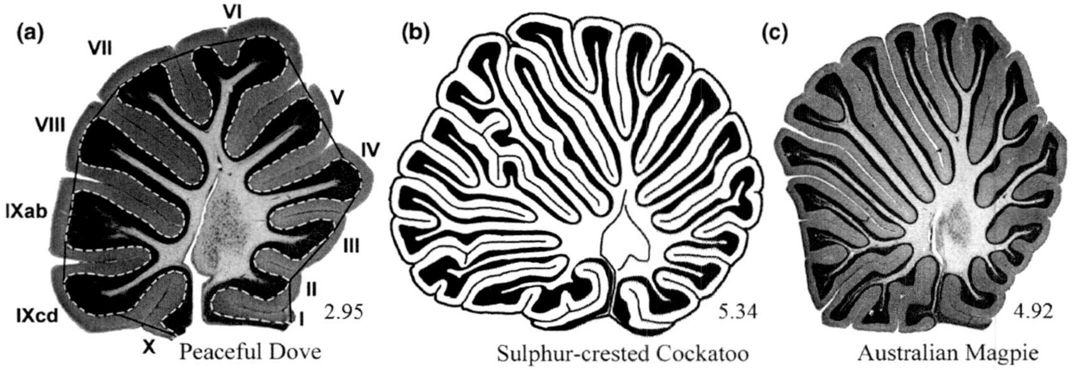


**Fig. 4.10** Sections through the cerebellums of three species of birds: (**a**) Peaceful Dove (*Geopelia placida*), (**b**) Sulfur-crested Cockatoo (*Cacauta galerita*), and (**c**) Australian Magpie (*Gymnorhina tibicen*). Some species of parrots, like cockatoos, and corvids, like magpies, are known to sometimes use tools; pigeons and doves do not use tools. The numbers represent a cerebellar foliation ratio, i.e., the ratio of two measurements shown in (**a**): the dotted white line, which is the outline nerve cells (called Purkinje cells and found in dark areas of the lobes) in the cerebellar lobes, and the solid black line, which encompasses all the lobes of the cortex. The higher the ratio, the greater of folding, or foliation, and the higher the number of nerve cells in the cerebellum. (Figure from Iwaniuk et al. 2009; © 2009 Canadian Psychological Association, used with permission)

songs with complex acoustic structure (Chen et al. 2014).

The avian hypothalamus plays an important role in regulating endocrine and autonomic functions. It consists of several nuclei (clusters of nerve cell bodies) that produce neuropeptides and neurotransmitters that regulate several basic functions including body temperature (Fig. 4.16), thirst (via release of hormones by the pituitary gland—see Chap. 9—Osmoregulation), hunger (Fig. 4.17), and, as the circannual "clock" of birds found at higher latitudes, sexual behavior, and reproduction (Fig. 4.18). The hypothalamus also serves as the primary control center for the autonomic nervous system (ANS). The ANS consists of motor neurons and has two divisions: the sympathetic division and the parasympathetic division (Kuenzel 2000; Fig. 4.19). These two divisions innervate the heart, intrinsic eye muscles that regulate pupil diameter, several glands (e.g., salivary glands and adrenal gland), and smooth muscle in the walls of blood vessels and in several other body systems (e.g., digestive system, respiratory system, reproductive system, and excretory system). The hypothalamus and ANS play critical roles in regulating cardiovascular function, respiration, digestion, excretion, and the reproductive organs. In many cases, both

**Fig. 4.9** (continued) birds. These areas of the cerebellum (First principal component) help coordinate visual and beak-related movements, and woodpeckers, corvids, and parrots are generally very adept when it comes to using their beaks and/or tongues to manipulate and explore external objects. Cerebellums of owls (e.g., Long-eared Owl [*Asio otus*] pictured above) have larger lobes I, II, and X (shown in blue at top), areas with important vestibular and tail somatosensory functions and likely beneficial given their specialization as noctural raptors. Short-eared Owl, *Asio flammeus*; Barn Owl, *Tyto alba*; Great Horned Owl, *Bubo virginianus*; Rock Dove (= Rock Pigeon), *Columba livia*; Common Gull, *Larus canus*; Herring Gull, *Larus argentatus*; Black-headed Gull, *Chroicocephalus ridibundus*; Carrion Crow, *Corvus corone*; Great Spotted Woodpecker, *Dendrocopos major*; Common Swift, *Apus*; Wild Turkey, *Meleagris gallopavo*. (Photo of Long-eared Owl from pxhere.com, CC0 Public Domain; Photo of Green Woodpecker by Remyymer, Wikipedia, CC BY 2.0, https://creativecommons.org/licenses/by/2.0/; Other figures from Sultan 2005; © 2005 Elsevier Ltd., used with permission)

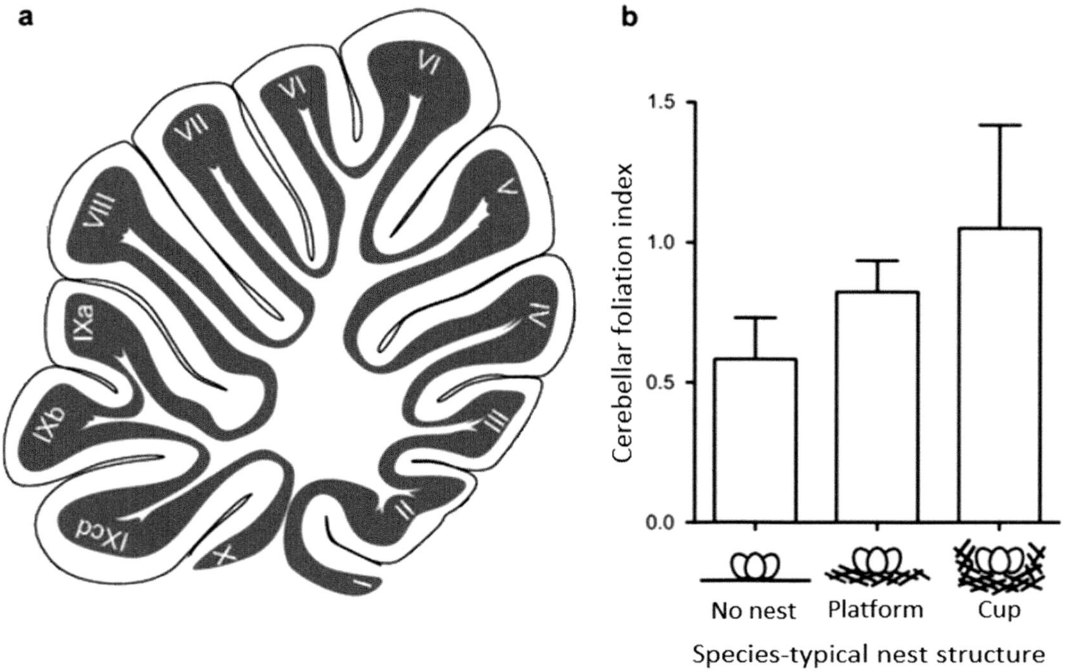


**Fig. 4.11** Relationship between cerebellar foliation (the degree of folding of the cerebellum relative to a hypothetical smooth-surfaced, or unfolded, cerebellum) and species-typical nest structure of birds. (**a**) A sagittal section of a bird cerebellum. The roman numerals simply refer to different lobes of the cerebellum. (**b**) Mean cerebellar foliation index (±SE) of bird species that build no nests, platform nests, and cup nests. (Figure from Hall et al. 2015; open-access article distributed under the terms of the Creative Commons CC BY, https://creativecommons.org/licenses/by/4.0/)

divisions of the ANS innervate the same glands, organs, and muscles and typically have opposite, or antagonistic, effects. For example, sympathetic stimulation will increase heart rate whereas parasympathetic stimulation reduces heart rate. The parasympathetic division is particularly important in regulating the digestive system (Figs. 4.17 and 4.19), and the sympathetic division in initiating and maintaining the physiological changes important when birds must quickly become very physically active in what is referred to as the "fight-or-flight" response. In response to stress, e.g., during a territorial interaction or being pursued by a predator, the sympathetic nervous system is activated via the release of norepinephrine by sympathetic neurons and epinephrine (also called adrenalin) from the adrenal medulla (Wingfield 2013). These substances trigger a body-wide response that includes an increased heart rate, increased blood pressure, and increased blood flow to skeletal muscles and the central nervous system, physiological changes that quickly prepare a bird for intense physical activity.

The telencephalon of vertebrates consists of three main areas: pallium, striatum, and pallidum (Fig. 4.20). The pallidum and striatum regions (together also called the basal ganglia) of birds and mammals (and other vertebrates) are similar in structure and function. The basal ganglia is important in motor learning, and the retention and recall of learned or natural motor skills or programs (Graybiel 2008). Among songbirds, an excellent example of this is the learning of songs, with young birds "practicing" (subsong) before being able to consistently produce the song or songs in their vocal repertoire. Other examples of behaviors that might require development of

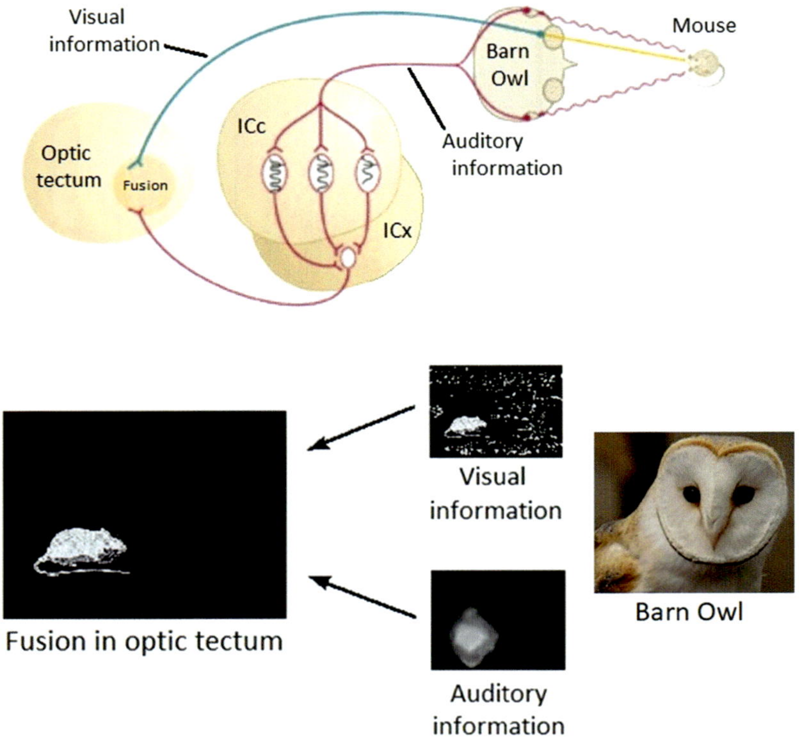


**Fig. 4.12** The auditory and visual space maps of Barn Owls (*Tyto alba*) are connected, allowing them to precisely locate prey. The sight of a mouse stimulates rod cells in the retina that send impulses to the optic tectum so the owl can orient toward the prey. The sound of the mouse stimulates the two ears with a slight interaural time difference that is detected by neurons in the central nucleus of the inferior colliculus (ICc). The auditory space map is then generated by neurons in the external nucleus of the inferior colliculus (ICx) and these neurons then activate the same region of the optic tectum as visual stimuli. The visual and auditory space maps fuse in the optic tectum and the Barn Owl can then pinpoint the location of the mouse and initiate an attack. (Top figure modified from Stryker 1999; AAAS, used with permission. Bottom image from Goodridge 1997; used with permission of Steven Goodridge. Photo of Barn Owl from pxhere.com, CC0 Public Domain)

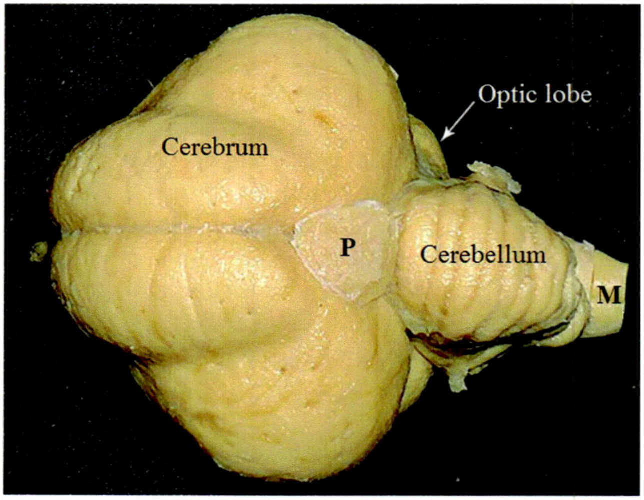


**Fig. 4.13** Dorsal view of the brain of a Common Ostrich (*Struthio camelus*) showing the location of the pineal gland (P), cerebrum, optic lobe, cerebellum, and medulla oblongata (M). (Figure modified from Karkoura et al. 2015; open-access article distributed under the terms of the Creative Commons Attribution License, https://creativecommons.org/licenses/by/4.0/)

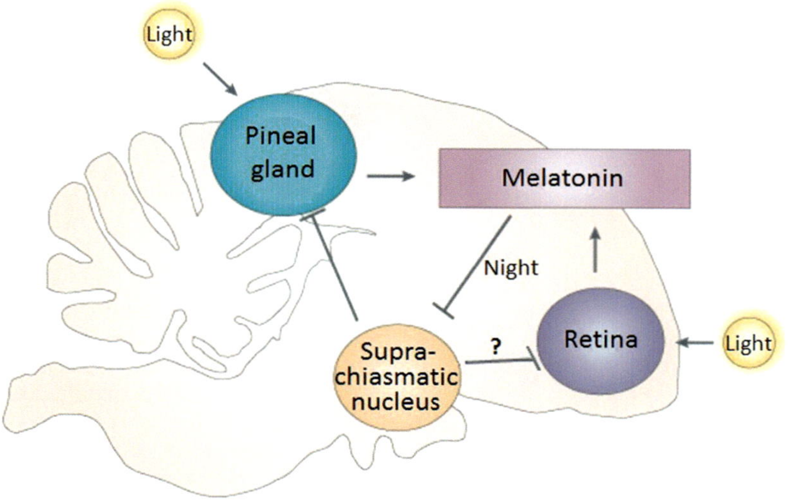


**Fig. 4.14** Avian photoreceptors are located in the retina of the eye, the suprachiasmatic nucleus (SCN) of the hypothalamus (deep brain receptors), and the pineal gland. At night, the pineal gland and retinas of the eyes secrete melatonin into the blood that inhibits activity of the SCN. During the day, the SCN inhibits melatonin production by the pineal gland (and possibly by the retinas as well). The pineal gland and retinas, via daily variation in release of melatonin, are important in maintenance of the circadian (daily) rhythms of birds. (Figure modified from Bell-Pedersen et al. 2005; © 2005 Springer Nature, used with permission)

motor "programs" would include mating displays, nest building, and foraging behavior. To varying degrees among different species, such behaviors require the coordinated actions of many muscles, and being able to quickly and efficiently perform these behaviors can clearly be important in terms of a bird's survival and fitness.

In contrast to the basal ganglia, the palliums of birds and mammals are organized very differently (Fig. 4.20). The mammalian pallium consists of a outer cortex with six layers of nerve cell bodies (gray matter) and an inner region called the white matter that consists of myelinated axons (myelin is a fatty material that occurs at regular intervals

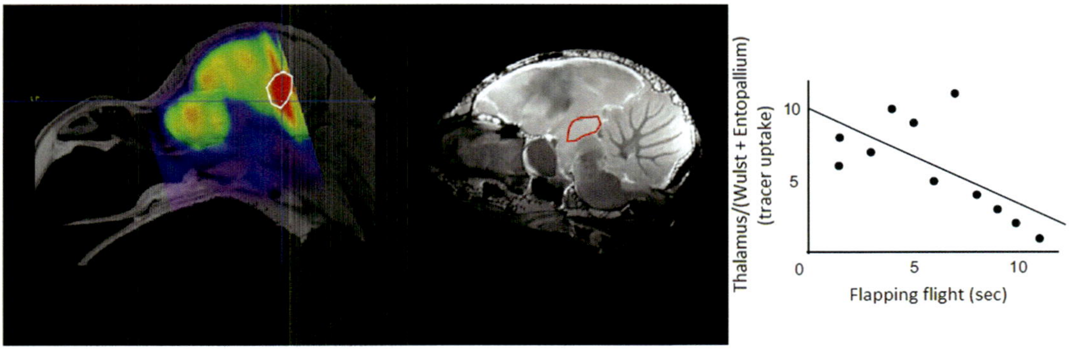


**Fig. 4.15** The thalamus (indicated by the red area in the brain of a European Starling (*Sturnus vulgaris*) on the left and the area outlined in red in the brain on the right) became more active when a starling first began flying in a wind tunnel (as indicated by rate of uptake of a tracer material), but then became less active with continued flapping flight. (Figure modified from Gold et al. 2016; © 2016 Elsevier Ltd., used with permission)

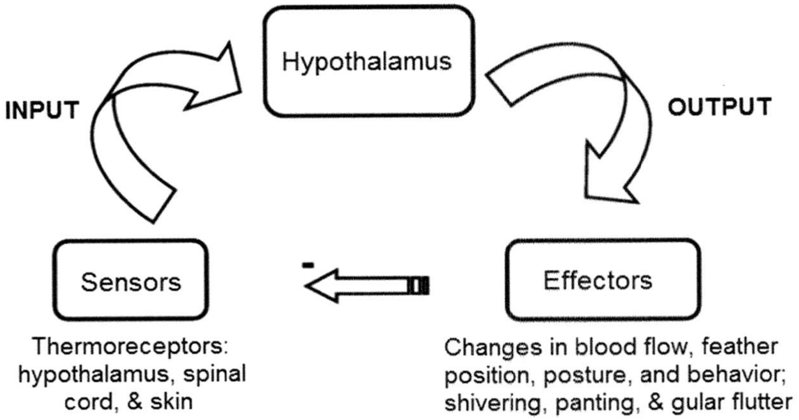


**Fig. 4.16** The hypothalamus helps regulate bird body temperatures. Receptors in the hypothalamus, spinal cord, and skin monitor body temperature (via blood temperature) and impulses from the hypothalamus can, when necessary, influence various "effectors" whose actions can increase or decrease heat loss and increase heat production (shivering) to help maintain normal body temperature. (Figure modified from Bicego et al. 2007; © 2006 Elsevier Inc., used with permission)

on the outside of axons). In contrast, the avian pallium includes four major subdivisions called the hyperpallium, mesopallium, nidopallium, and arcopallium, and within these areas are found clusters (also called nuclei) of nerve cell bodies (Fig. 4.21; Jarvis et al. 2005). Despite this organizational difference, the mammalian cortex and avian pallium have functional similarities, with specific areas serving sensory and motor functions (Fig. 4.22). Another similarity is that both the avian pallium and mammalian cortex have an area called the hippocampus that is known to be important in consolidation of short-term memory into long-term memory. The hippocampus is connected to many other parts of the brain, with input from visual, auditory, olfactory, and somatosensory areas (Box 4.2 The avian hippocampus).

The pallium of adult birds comprises about 75% of the telencephalic volume, which is comparable to mammals, and processes information in a manner similar to mammalian sensory and motor cortices (Jarvis et al. 2005). The basic function of the avian pallium is to link sensory inputs and motor outputs, serving as an interface between sensory and perceptual processing and analysis and the processes that modulate behavior. This is also the basic function of the mammalian cortex. Although the mammalian cortex is laminar and the avian pallium is nucleated, the organization and connectivity are similar. Both have areas that are functionally specialized (e.g., motor, sensory, and associative areas), but integrative processes and interactions among these areas make possible cognitive processes such as visual recognition, vocal communication, and complex social interactions. Such integration requires communication among neurons in different areas and recent work suggests that both avian and mammalian brains are "richly connected" and, to varying degrees among different species, provide the basis for complex cognition, which means reasoning, flexibility, problem solving, prospection, and declarative knowledge (ability to recall factual knowledge and information) (Shanahan et al. 2013; Clayton and Emery 2015; Fig. 4.23).

In mammals, different areas of the cortex serve different functions, with some areas receiving sensory input, other areas initiating motor impulses, and still other areas important in analyzing sensory inputs and storing information

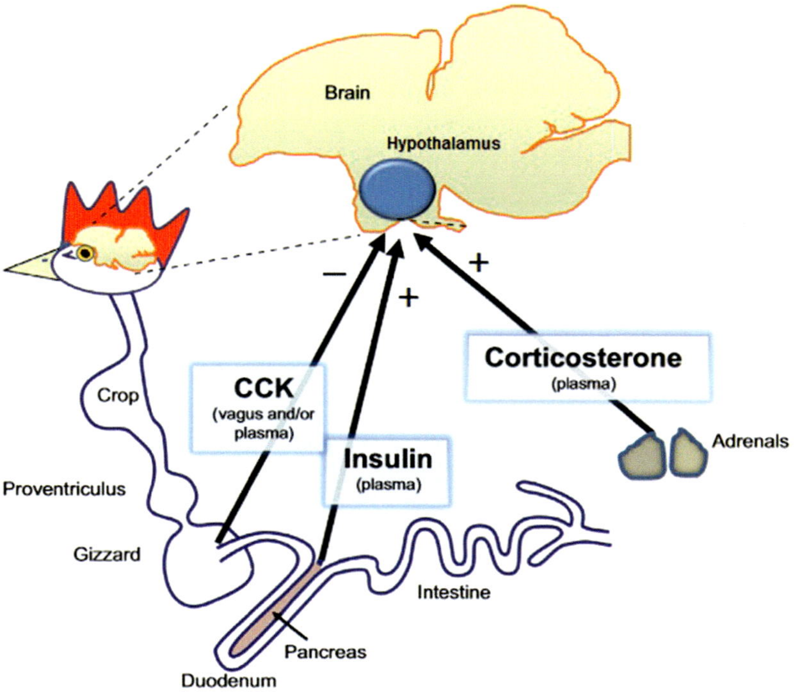


**Fig. 4.17** The hypothalamus helps regulate hunger levels via input from the gizzard, pancreas, and adrenal glands. The presence of food in the gizzard reduces hunger levels via direct nervous stimulation (vagus nerve) of the hypothalamus by receptors in the walls of the gizzard or detection by cells in the hypothalamus of CCK (cholecystokinin) released into the blood plasma by cells in the stomach. Detection by cells in the hypothalamus of increased levels of insulin (indicating reduced levels of glucose) in the blood and increased levels of corticosterone (indicating increasing stress) in the blood causes an increase in hunger levels. (Figure modified from Boswell and Dunn 2015; © 2015 Elsevier Inc., used with permission)

(association or associative areas; Fig. 4.22). These functional areas extend through all six cortical layers in the mammal cortex. In birds, different areas of the pallium also serve sensory, motor, and associative functions, with neurons in these areas forming columns of various depths that extend through different areas of the pallium and, in some cases, the striatum (Jarvis et al. 2013; Fig. 4.24). Nervous impulses with sensory information from the special senses (e.g., vision, hearing, and olfaction) as well as from the skin and skeletal muscles (somatosensory) are conducted to the sensory areas of the avian brain, and this information must be analyzed within the context of relevant memory to determine needed motor responses or generate plans for future responses. In primates, the area that serves as the center of such multimodal integration is the prefrontal cortex (PFC; Fig. 4.22), The PFC of the primate brain is assumed to be the focus of working memory, action planning, and intelligence, with intelligence defined as "mental or behavioral flexibility or the ability of an organism to solve problems occurring in its natural and social environment, culminating in the appearance of novel solutions that are not part of the animal's normal repertoire (Dicke and Roth 2016)."

The avian equivalent of the PFC of primates appears to be the caudolateral nidopallium (also called the nidopallium caudolaterale, or NCL; Fig. 4.22). Both the PFC of mammals and the NCL of birds are multimodal association areas that appear ideally positioned to integrate sensory

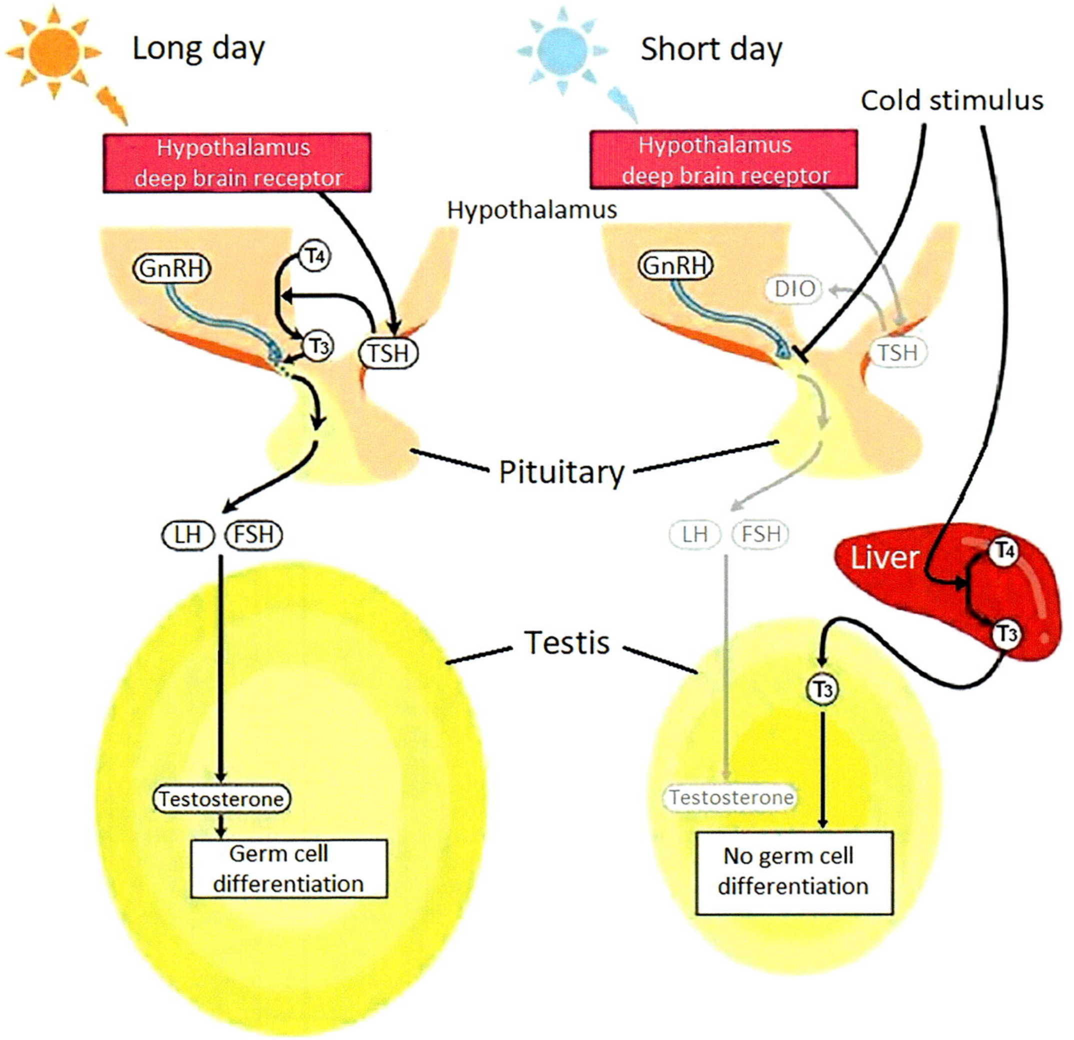


**Fig. 4.18** The hypothalamus of birds contains photoreceptors called deep brain receptors that are responsive to changes in day length. As days get longer, those receptors stimulate the pituitary which then releases thyroid-stimulating hormone (TSH). TSH converts precursor thyroxine (T4) into triiodothyronine (T3) which, in turn, triggers the release of gonadotropin-releasing hormone (GnRH) from the hypothalamus. GnRH triggers the production and release of luteinizing hormone (LH) and follicle-stimulating hormone (FSH) by the pituitary. These hormones then stimulate hormone production by the testes of males (and, not shown, by the ovary of females) and gamete production (sperm in males and ova in females). With shorter, colder days, the hypothalamus stops the production and release of GnRH and the liver produces triiodothyronine that inhibits hormone and gamete production by the testes of males (and ovary of females). (Figure modified from Ikegami and Yoshimura 2016; © 2015 Elsevier Inc., used with permission)

input and project to motor output (Veit and Nieder 2013). The NCL has reciprocal connections with other areas of the pallium, and receives and projects connections to and from motor and limbic areas (Box 4.3 Avian mesolimbic reward system and social behavior network). The NCL of birds, therefore, appears to be, as with the PFC of primates, the focus of action, planning and intelligence. However, as

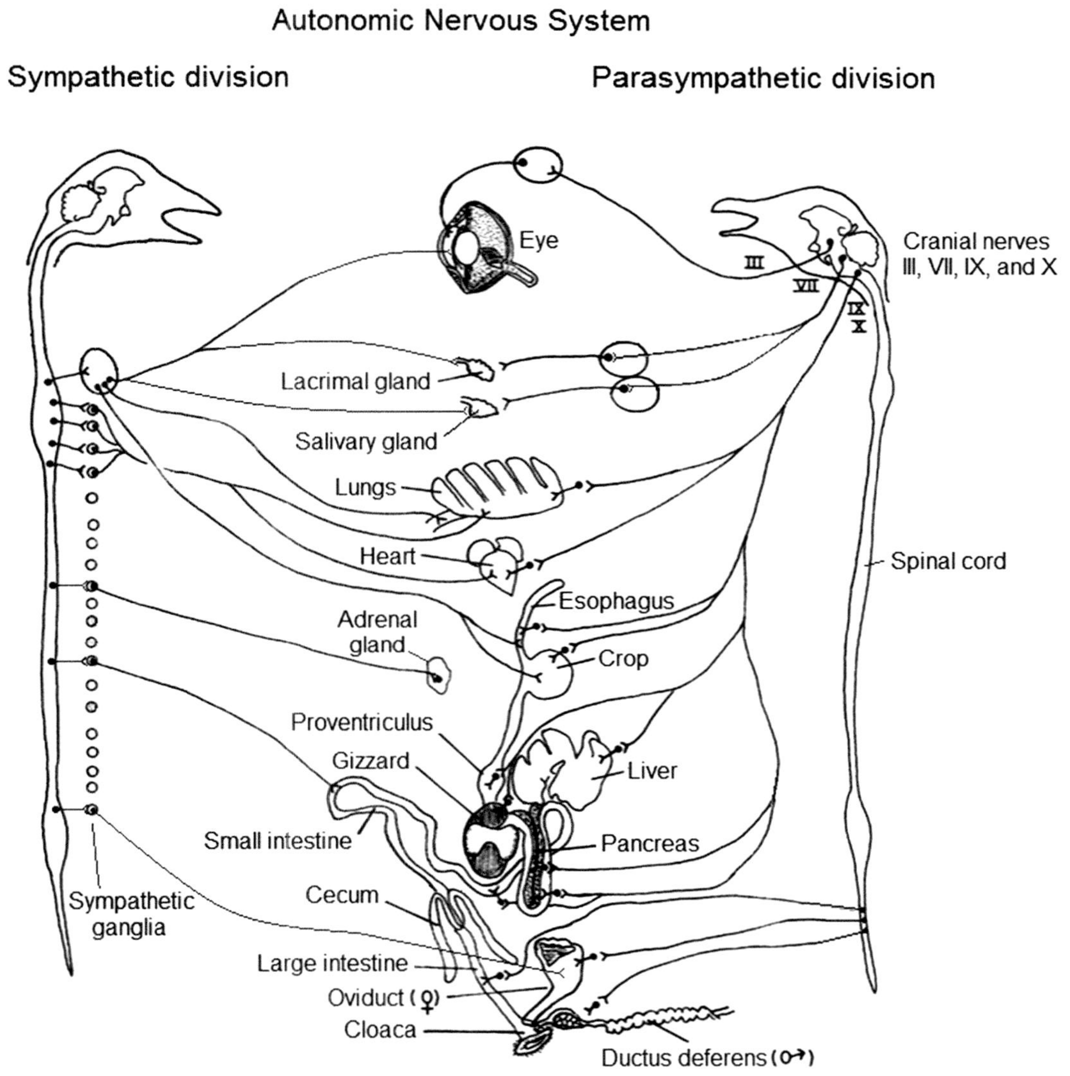


**Fig. 4.19** The avian autonomic nervous system has two divisions, sympathetic and parasympathetic, that innervate and regulate the activities of glands, smooth muscle, and cardiac muscle. (Figure modified from Kuenzel 2000; © 2000 Elsevier Ltd., used with permission)

with mammals, different species of birds differ in their apparent levels of intelligence.

For their size, birds and mammals have relatively larger brains than other vertebrates (Fig. 4.25; Box 4.4 Avian brain size evolution). Although there is no clear correlation between absolute or relative brain size and intelligence (i.e., the ability of an organism to solve problems in their natural and social environments), the size of those areas of the brain generally associated with intelligence in mammals (the cortex) and birds (the meso-nidopallium) are also relatively large. This suggests that natural selection has favored an increase in "intelligence" during the evolution of mammals and birds, which, in turn, suggests that, over time, the fitness of individuals with greater "intelligence" has generally exceeded that of less intelligent individuals.

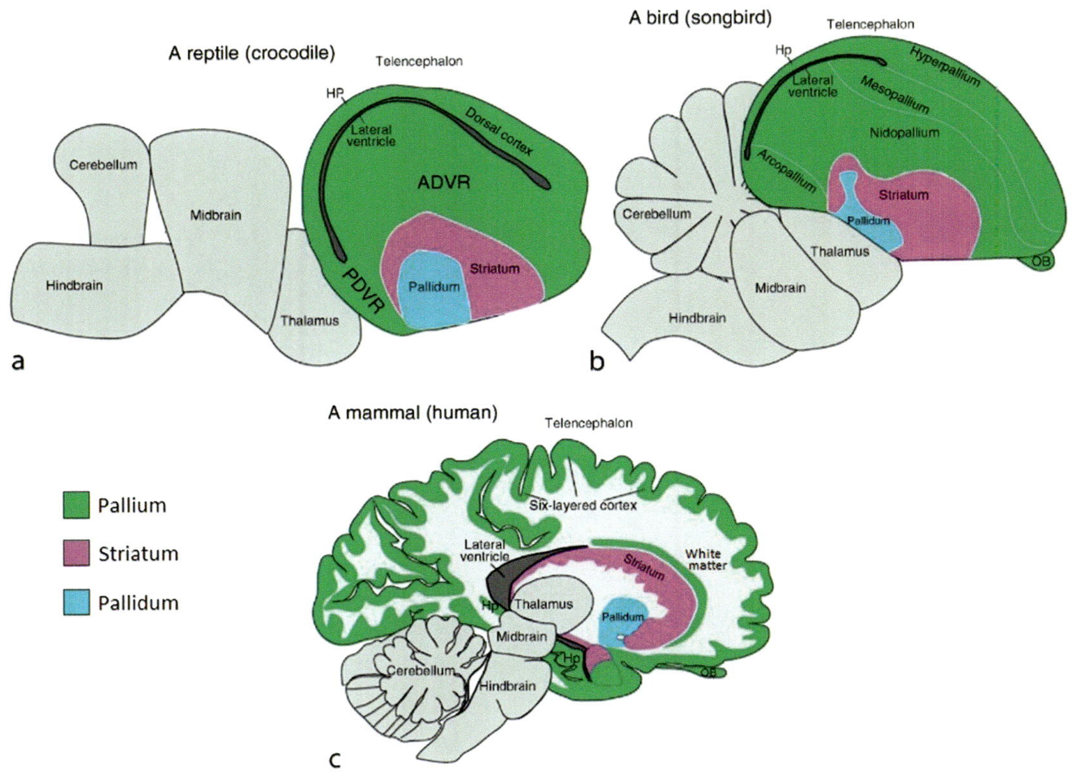


**Fig. 4.20** Organization of the telencephalon of (**a**) a reptile, (**b**) songbird, and (**c**) mammal (human). The mammalian pallium consists of an outer cortex with six layers of nerve cell bodies (gray matter) and an inner region called the white matter that consists of myelinated axons. The avian pallium includes four major subdivisions called the hyperpallium, mesopallium, nidopallium, and arcopallium, and within these areas are found "clusters" of nerve cell bodies. As with mammals, reptiles have layered cortex, but with three layers rather than six and with many fewer subdivisions than the cortex of mammals. (Figure modified from Jarvis 2009; © 2009 Springer-Verlag GmbH Berlin Heidelberg, used with permission)

Numerous characteristics can affect an individual's fitness, and one of those is their "intelligence," or cognitive ability (Fig. 4.26).

Cognitive abilities vary not only among individuals but also, and to a much greater degree, among species. Among birds, species in the family Corvidae and in the order Psittaciiformes generally considered among the most intelligent birds, as indicated both by their behavior, and the relative size of, and neuronal densities in, their meso-nidopalliums, areas of the avian brain most associated with intelligence (Fig. 4.27; Box 4.5 Neuronal densities in the avian brain and cognition). Although many questions still remain concerning the evolution of intelligence, several factors appear to have contributed to the differences among different avian taxa in the relative size of meso-nidopalliums and their associated cognitive abilities.

In general, altricial species of birds have larger brains (and palliums) than precocial species (Charvet and Streidter 2011; West 2014). In contrast to precocial young, altricial young and their brains are relatively immature at hatching and altricial young require extensive parental care. However, this means that their brains can continue to mature after hatching, and the food provided by parents provides the energy needed for such maturation (Isler and van Schaik 2009). This continued growth of brains after hatching,

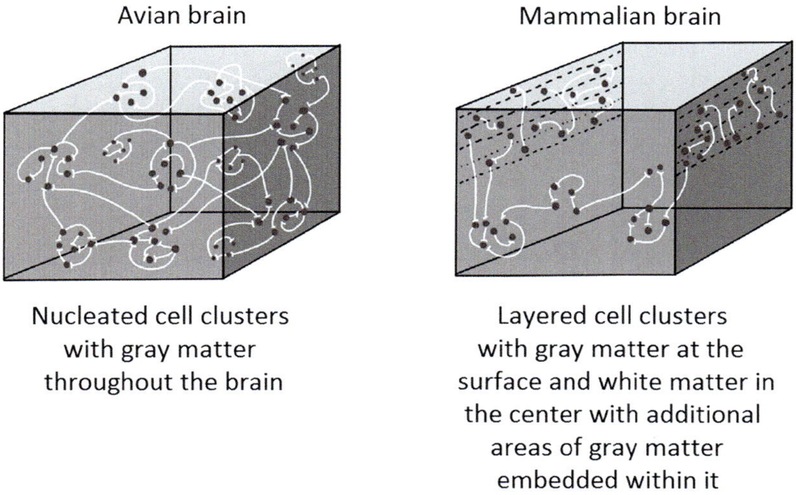


**Fig. 4.21** Organization of avian and mammalian pallial and subpallial areas. In mammal brains, the cortex consists of layers of tightly packed nerve cell bodies (called the gray matter), with connections between layers and with additional nerve cell bodies embedded within the white matter (myelinated axons) located under the cortex. In avian brains, the pallium is organized into nuclei (masses of nerve cell bodies), with no large areas of white matter. (Figure modified from Clayton and Emery 2015; © 2015 Elsevier Inc., used with permission)

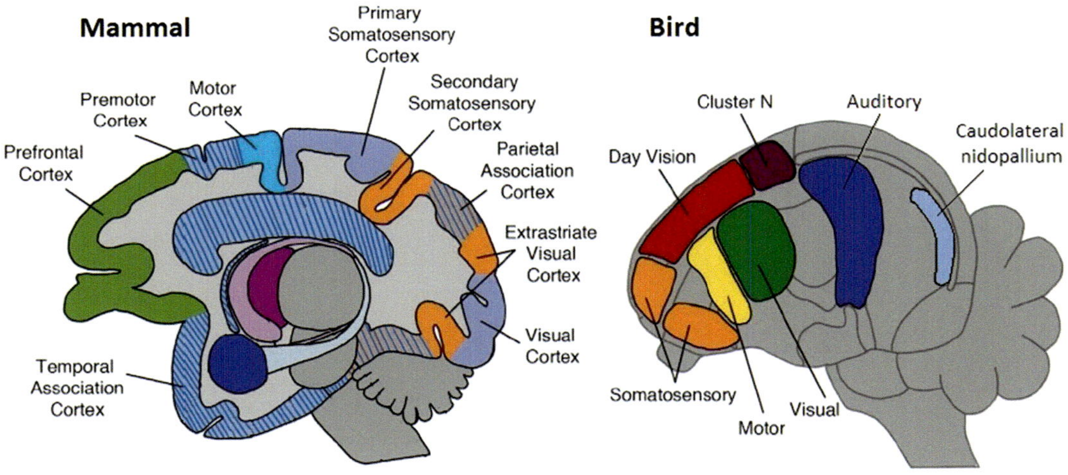


**Fig. 4.22** Representative bird (American Crow, *Corvus brachyrhynchos*) and mammal (rhesus macaque, *Macaca mulatta*) brains showing locations of areas of nerve cells in the mammalian cortex and avian pallium with particular functions. Motor areas stimulate and control skeletal muscles, somatosensory areas receive sensory information from the skin, skeletal muscles, tendons, and joints, visual areas (including cluster N in birds) receive nervous impulses from the eyes, and auditory areas receive nervous impulses from the inner ear. In mammals, association areas integrate sensory information and information stored as memories. (Figure from Clayton and Emery 2015; © 2015 Elsevier Inc., used with permission)

**Box 4.2 The avian hippocampus**

The hippocampus of birds is known to be important in memory formation and spatial navigation. Perhaps not surprisingly then, the volume of the hippocampus relative to the rest of the telencephalon is larger in species more dependent on memory and spatial navigation, such as food-storing birds, brood parasites, hummingbirds, and homing pigeons. Food-storing birds such as chickadees, nuthatches, and jays must remember (1) the different locations where food has been stored or cached, (2) which caches they have previously emptied, (3) which caches they discovered were empty due to loss to other animals, and (4) what type of food is stored at each cache site (Sherry 1984). Some food-storing birds store their food at one location, e.g., Acord Woodpeckers (*Melanerpes formicivorus*), but scatter-hoarders cache food items at many different locations. For example, Clark's Nutcrackers (*Nucifraga columbiana*) have been estimated to store up to 100,000 seeds each fall (Vander Wall 1990) so, with several seeds placed at each cache site, may need to then relocate as many as 20,000 different cache locations (Pravosudov and Smulders 2010). Even more remarkable, species of tits (*Parus* spp.) cache only one seed per location and may cache anywhere from 100,000 to 500,000 seeds per year (Brodin 1994).

Female brood parasites must remember the locations and status (e.g., nest building vs. start of egg laying) of nests that they seek to parasitize. Many hummingbirds use a foraging strategy called trap-lining that requires them to remember both the location of sources of nectar and the optimum time intervals between visits to those locations (to allow replenishment of the nectar). Hummingbirds, such as Green-backed Firecrowns (*Sephanoides sephanoides*), are also able to remember the location of the flowers, among others with the same appearance, that provide the most nectar (González-Gómez and Vásquez 2006). Homing pigeons that use landmarks and landscape features to "home" must remember the locations of those landmarks and landscape features relative to each other and to their goal location. For all of these species, the importance of memory has led, via natural selection, to an increase in hippocampal volume.

Mean volume of the hippocampal formation (HF; expressed as a percentage of telencephalon volume) of food-caching (N = 18 species) and non-food-caching songbirds (N = 53 species), woodpeckers (N = 4 species), and hummingbirds (N = 4 species). Ranges for food-caching songbirds = 2.05–6.86%, for non-food-caching songbirds = 1.04–5.47%, for woodpeckers = 3.50–5.12%, and for hummingbirds = 6.94–9.60% [Data from Ward et al. 2012; all photos are CC0 Public Domain from pxhere.com, including, from left to right, Red-winged Blackbird (*Agelaius phoeniceus*), Mountain Chickadee (*Poecile gambeli*), Downy Woodpecker (*Dryobates pubescens*), and Anna's Hummingbird (*Calypte anna*)]

(continued)

**Box 4.2** (continued)

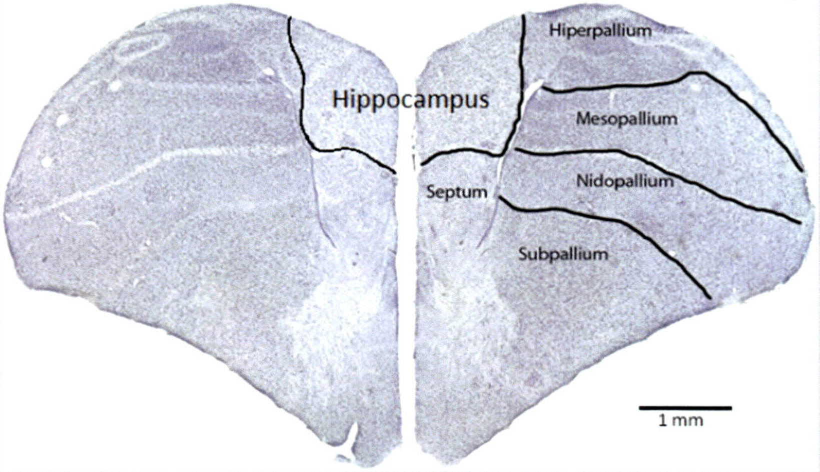


Coronal section through the telencephalon of a Green-backed Firecrown hummingbird showing the hippocampus. Note the large area occupied by the hippocampus relative to the overall size of the telencephalon (Figure modified from Gonzáles-Gómez et al. 2014; © 2014 González-Gómez et al., open-access article distributed under the terms of the Creative Commons Attribution License, https://creativecommons.org/licenses/by/4.0/)

For some species, there is seasonal and/or geographical variation in the importance of memory formation and spatial navigation. For example, food-storing birds at higher latitudes typically only cache food during late summer and the fall, food that will be essential to surviving the winter. During the spring and summer, food is readily available so caching is not required. The results of some studies suggest that hippocampal volume varies seasonally in these species, with volume increasing in the fall when food is being cached, then decreasing in the spring when the birds no longer need to relocate cached food (Sherry and MacDougall-Shackleton 2015; but see Pravosudov et al. 2015). More recently, Lange et al. (2022) reported seasonal changes in the number of neurons in the hippocampal formation of both caching (Willow Tit, *Poecile montanus*) and non-caching (Great Tit, *Parus major*) birds in Finland, with numbers increasing for both species in August and September. Based on these results, Lange et al. (2022) suggested that hippocampal plasticity may not be limited to food-caching species of birds, but a more general phenomenon. More specifically, seasonal changes in the hippocampus may be driven by different factors, e.g., the need to remember cache sites for food-caching species like Willow Tits and changes in spatial (i.e., larger home ranges) and social (i.e., formation of flocks) environments for non-food-caching species like Great Tits.

Other investigators have also found that variation in environmental conditions can influence the avian hippocampus. As climates become harsher, either at higher latitudes or higher altitudes, chickadees (*Poecile* spp.) need more caches to survive the winter and, therefore, a better memory, favoring selection for a larger hippocampus with more neurons (Roth and Pravosudov 2009; Freas et al. 2012).

(continued)

**Box 4.2** (continued)

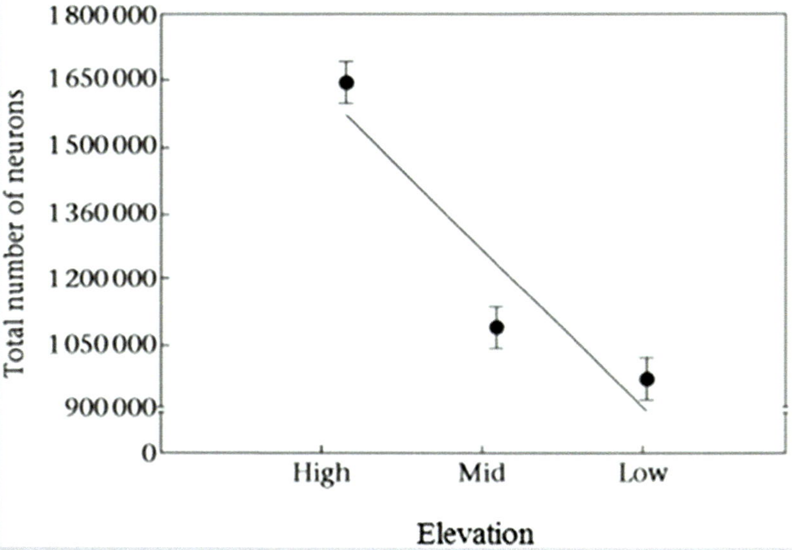


Total number of neurons in the hippocampus of Mountain Chickadees (*Poecile gambeli*) from three elevations collected during late September and early October. High = 2400 m, Mid = 1800 m, and Low = 1200 m (Figure from Freas et al. 2012; © 2012 The Association for the Study of Animal Behaviour. Published by Elsevier Ltd., used with permission. Photo of Mountain Chickadee by Andy Reago and Chrissy McClarren, Wikipedia, CC BY 2.0, https://creativecommons.org/licenses/by/2.0/)

and particularly continued growth of the telencephalon (pallium and subpallium), is likely one of the more important factors contributing to the evolution of large brains, and enhanced cognitive abilities, of altricial species of birds, particularly parrots and songbirds (Fig. 4.28).

Another factor influencing the brain size of birds is migratory status, with resident birds and short-distance migrants tending to have relatively larger brains than long-distance migrants (Sol et al. 2010; Fuchs et al. 2014; Fig. 4.29). Reasons for this difference are unclear, but possible explanations include (1) selection favors larger

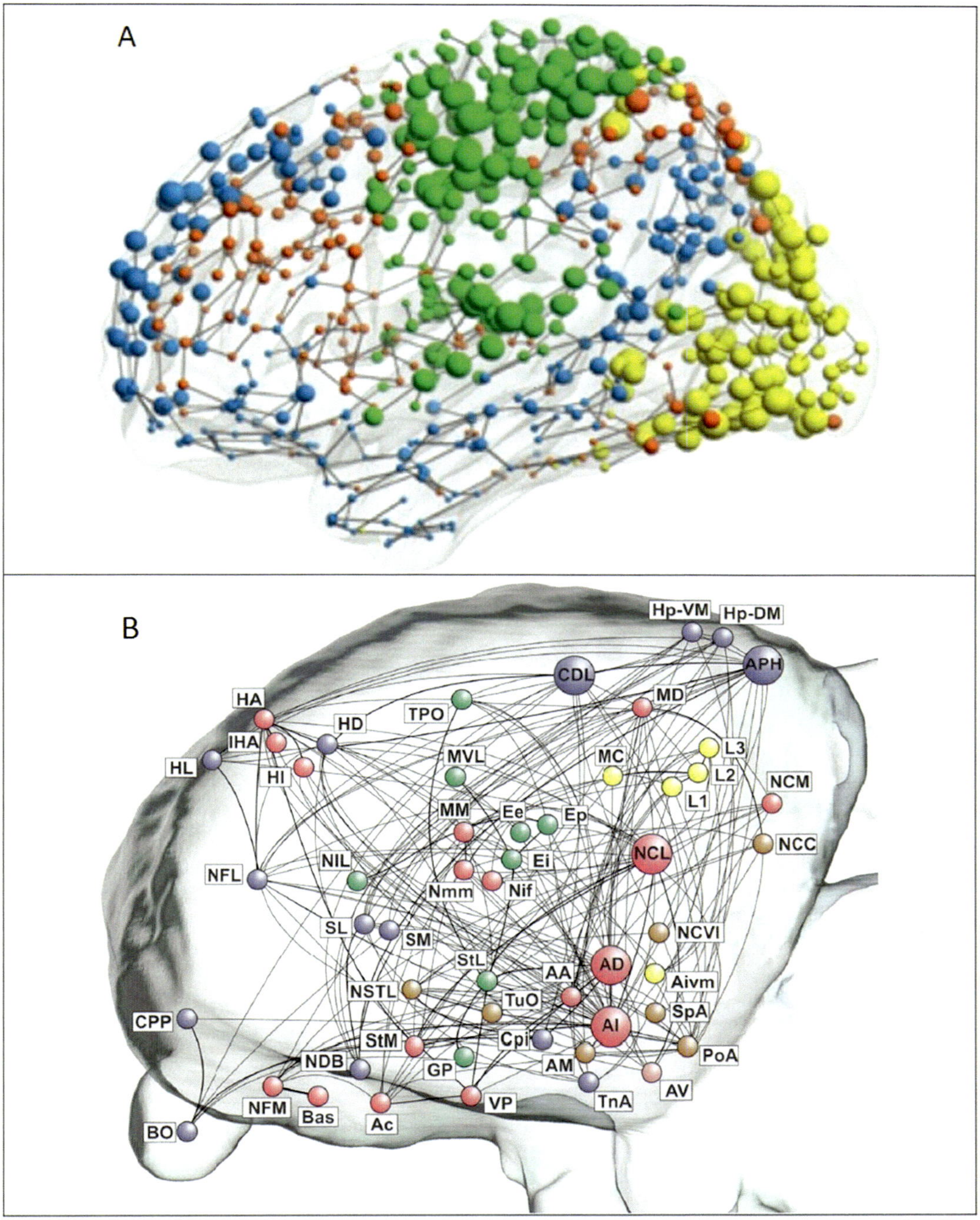


**Fig. 4.23** Complex networks of connections between different areas of the (**a**) mammalian and (**b**) avian telencephalons are essential for the flow of information and cognition. Connections in the laminar mammalian cortex are via neurons that conduct impulses between and among layers; connections in the avian nuclear pallium are via neurons that conduct impulses between and among different nuclei. (Top figure from Vértes et al. 2014; © 2014 The Authors. Published by the Royal Society, used with permission; Bottom figure from Shanahan et al. 2013; © 2013 Shanahan, Bingman, Shimizu, Wild and Güntürkün, open-access article distributed under the terms of the Creative Commons Attribution License, https://creativecommons.org/licenses/by/4.0/)

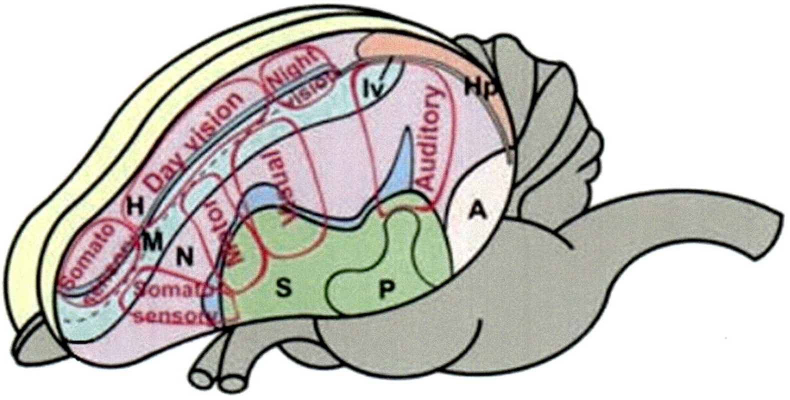


**Fig. 4.24** Gene expression patterns in subdivisions of the dorsal and ventral avian pallium. Areas are considered "functional columns" (Jarvis et al. 2013). Mammalian and avian "functional columns" both contain similar specialized groups of neurons, but their developmental origins are different. In the avian brain, the cells of the coactivated networks are derived from different sectors of the pallial neuroepithelium. H, hyperpallium; M, mesopallium; N, nidopallium; S, striatum; P, pallidum; Hp, hippocampus; lv, lateral ventricle. (Figure modified from Montiel and Molnár 2013; © 2013 Wiley Periodicals Inc., used with permission)

**Box 4.3 Avian mesolimbic reward system and social behavior network**

Birds must often make decisions concerning how to respond to environmental stimuli, some of which represent challenges (e.g., risk of predation or aggressive encounters with conspecifics) and other opportunities (e.g., foraging, habitat use, and reproduction) (O'Connell and Hofmann 2011). Sensory information must be processed and, within the context of their current physiological state, the social context (e.g., territory defense, mate choice, and parental care), and prior experiences, birds decide how to respond. Behavioral ecologists and ornithologists have studied these behavioral responses for many years and much is known about how behavioral decisions made by birds impact their survival and fitness. However, much less is known about the neural processes involved in avian decision-making.

Recent studies, mainly of mammals, but of some species of birds and other vertebrates as well, have revealed complex neural circuits that appear to underly the behavioral decision-making process. This "social decision-making circuit" largely lies within the telencephalon (basal, or limbic, forebrain) and midbrain and the areas that make up this circuit appear to be involved in many types of social behavior, including reproduction, aggression, and parental care, and may also play a role in other types of behavior such as foraging and habitat selection (O'Connell and Hofmann 2011).

Two neural circuits appear to be involved in avian decision-making: the social behavior network (yellow regions in the figure below) and the mesolimbic reward system (blue regions in the figure below). The social behavior network includes areas in the hypothalamus (preoptic area, anterior hypothalamus, and ventrolateral hypothalamus), forebrain (lateral septum, bed nucleus of the stria terminalis, and medial amygdala), and midbrain (central gray) that are sensitive to sex steroid hormones and are important in sexual, aggressive, and parental behaviors. The mesolimbic reward system includes multiple areas in the forebrain plus an area in the

(continued)

**Box 4.3** (continued)

midbrain (ventral tegmental area) and, together with the social behavior network, appear to be important in evaluating the relative importance of environmental stimuli and reinforcing, or "rewarding," appropriate responses. Behavioral contexts in which the mesolimbic reward system and social behavior networks play important roles include copulation, pair bonding, positive outcomes in aggressive contexts, parental care, singing, and positive social interactions (Goodson and Kabelik 2009; O'Connell and Hofmann 2011, 2012). These neural circuits, along with neuroendocrine modulators and the effect of external and internal stimuli on gene expression in the social decision-making areas of the brain, play critical roles in the expression of the complex social behavior of birds (Rubenstein and Hofmann 2015).

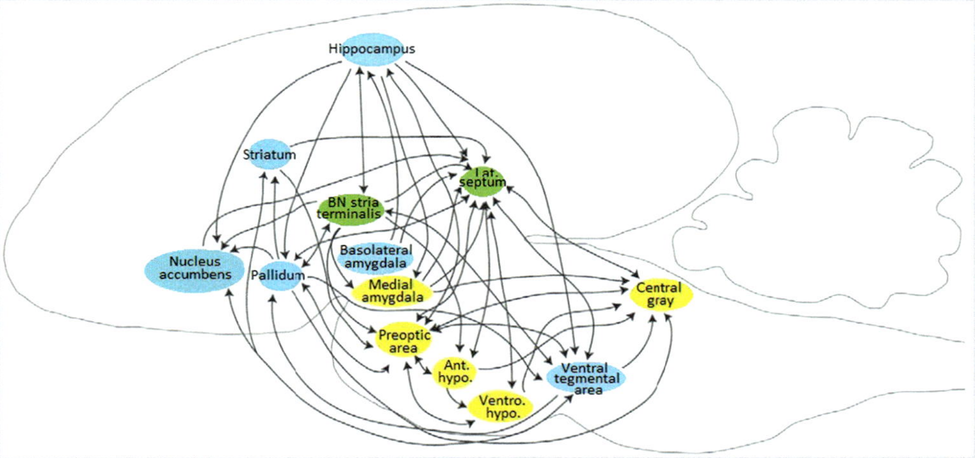


The social decision-making circuit of birds. Arrows indicate neural connections. Brain regions in the social behavior network are colored yellow, those in the mesolimbic reward system are blue, and brain regions shared by both networks are colored green. BN stria terminalis = Bed nucleus of the stria terminalis; Lat. Septum = Lateral septum; Central gray is also referred to as the Periaqueductal gray; Ant. hypo. = Anterior hypothalamus; Ventro. hypo. = Ventromedial hypothalamus (Figure modified from O'Connell and Hofmann 2011; © 2011 Wiley-Liss, Inc., used with permission)

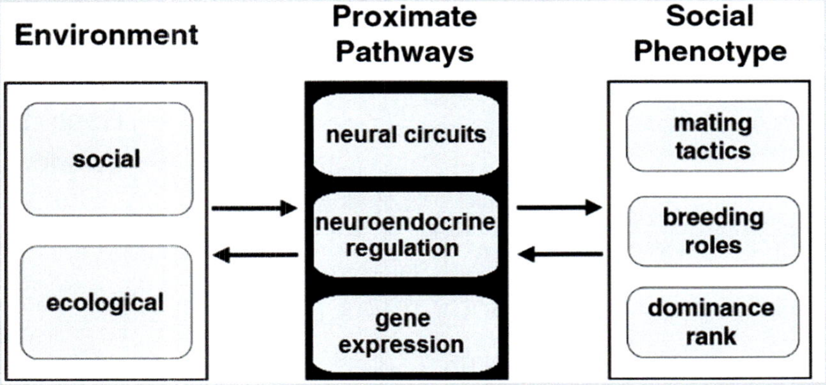


The behavior of birds results from the integration of information from their external and internal (physiological) environments via at least three proximate pathways. The neural circuits are the social behavior network that

(continued)

**Box 4.3** (continued)

regulates social behavior and the mesolimbic reward system that evaluates the importance of, and need to respond to, environmental stimuli. Social behaviors are also influenced by neuroendocrine modulators, including hormones such as testosterone and serotonin that regulate aggressive and defensive behaviors. The social and reproductive behavior of birds is also influenced by changes in gene expression across brain regions that result from changes in the activity of both neural circuits and neuroendocrine modulation. These three proximate pathways are not independent, but interact to influence avian behavior (Figure modified from Rubenstein and Hofmann 2015; © 2015 Elsevier Ltd., used with permission)

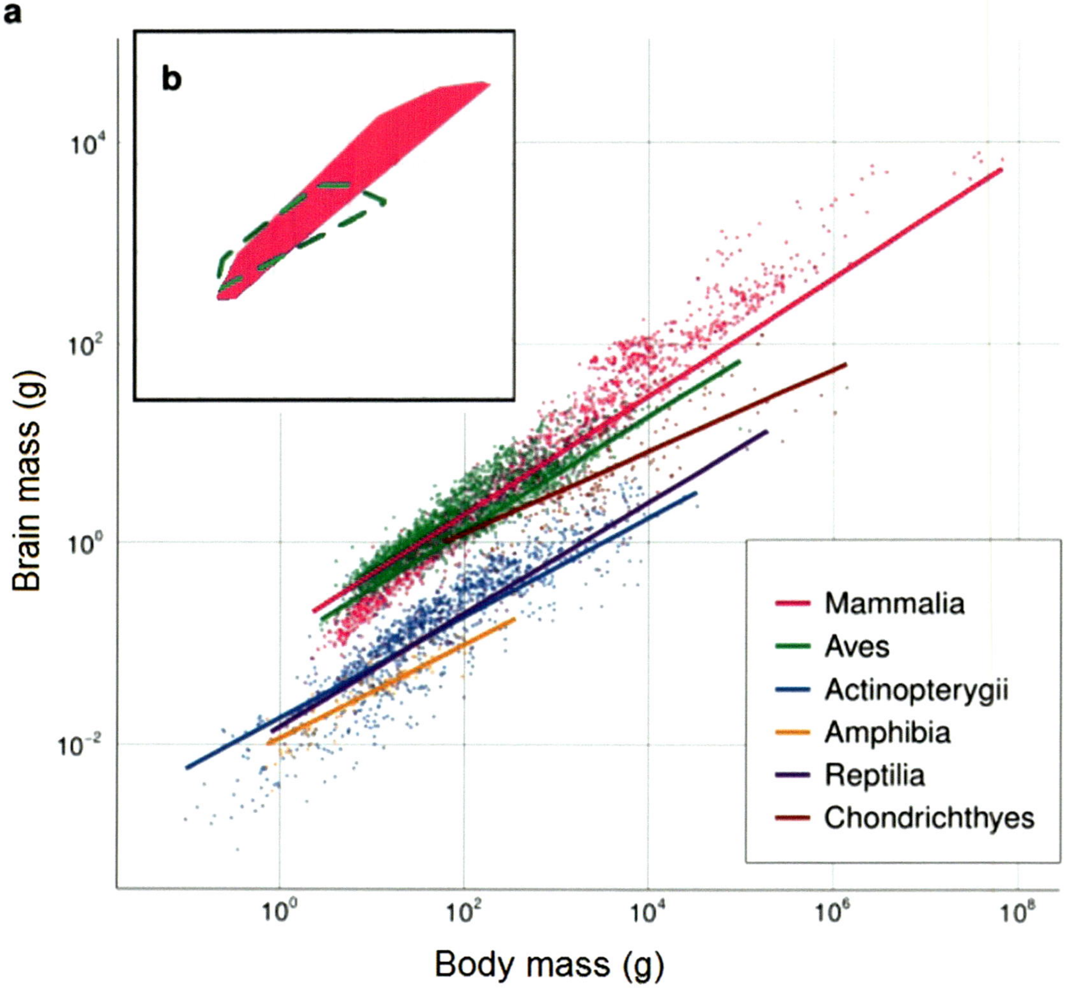


**Fig. 4.25** (**a**) Relationship between brain mass and body mass for six classes of vertebrates. Relative to their body mass, birds and mammals have larger brains than reptiles and other vertebrates. (**b**) Mininum convex polygons of the morphospaces occupied by birds (N = 1902 species) and mammals (N = 1409 species). (Figure from Tsuboi et al. 2018 with **b** modified to include only birds and mammals; © 2018 Springer Nature, used with permission)

brains in residents because better cognitive abilities help them survive seasonally changing and sometimes harsh environmental conditions (Vincze et al. 2015; Sayol et al. 2016; Fig. 4.30, Box 4.6 Brain size and latitude), (2) the energetic needs of a larger brain and energetic costs of long-distance migration require a trade-off that has resulted in smaller, less-energy-demanding brains (Vincze et al. 2015; Vincze 2016), and (3) the young of long-distance migrants have less time available for development and brain maturation after hatching because they must prepare for migration and this has led to a reduction in brain size compared to residents and short-distance migrants (Fuchs et al. 2014).

**Box 4.4 Avian brain size evolution**

The size of bird brains relative to their body size has generally increased (with some exceptions) over the past 150 million years. Initially, this increase appears to be the result of a decrease in body size, with body size decreasing to a greater degree than brain volume was decreasing and resulting in greater average brain volumes for a given body mass (Ksepka et al. 2020). Rates of brain-body size evolution were high for theropods and at the origins of early-diverging crown birds (Palaeognathae, Galloanserae, Phoeicopteriformes, and Columbimorphae). However, rates of brain-body size evolution for these crown birds plateaued, and rates for most other taxa slowed or plateaued at the beginning of the Paleogene radiation of Neoaves.

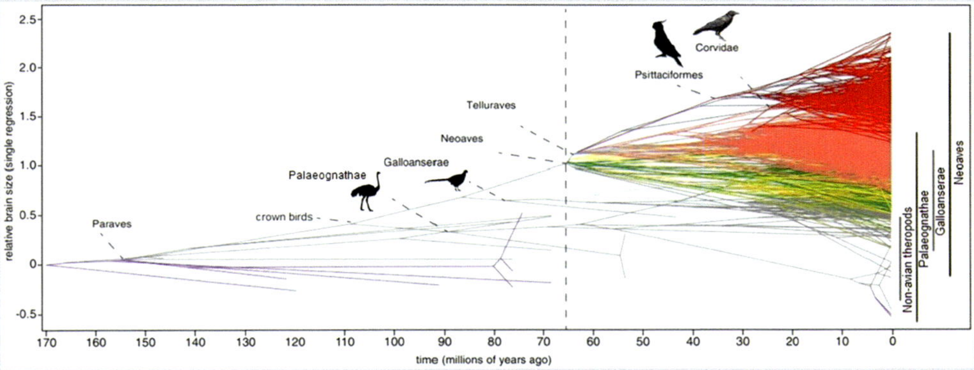


Evolution of variation in relative brain size in birds. Different colors indicate different relationships between body mass and brain volume among different taxa. Dashed vertical line indicates the end of the Cretaceous (and extinction of non-avian dinosaurs). Crown birds include the order Palaeognathae (ratites and tinamous), Galloanserae (gallinaceous birds and waterfowl), Phoenicopterimorphae (grebes and flamingoes), and Columbimorphae (pigeons and doves, sandgrouse, and mesites). Neoaves includes all present-day birds except the Palaeognathae and Galloanserae. Telluraves are also referred to as landbirds or core landbirds (Figure from Ksepka et al. 2020; © 2020 Elsevier Inc., used with permission)

One factor apparently contributing to this slowing at the origin of Neoaves was a shift to a carnivorous diet in several orders of birds, including Accipitriformes (e.g., eagles and hawks), Strigiformes (owls), Falconiformes (falcons), and Cariamiformes (seriemas). Species in these orders still have relatively large brains compared to many other taxa of birds, but the rate of increase in brain volume relative to body size has been lower than that of several other taxa. One possible explanation for this is that selection has favored an increase in body size, but not brain volume, especially for raptors that primarily take larger prey (Ksepka et al. 2020).

(continued)

**Box 4.4** (continued)

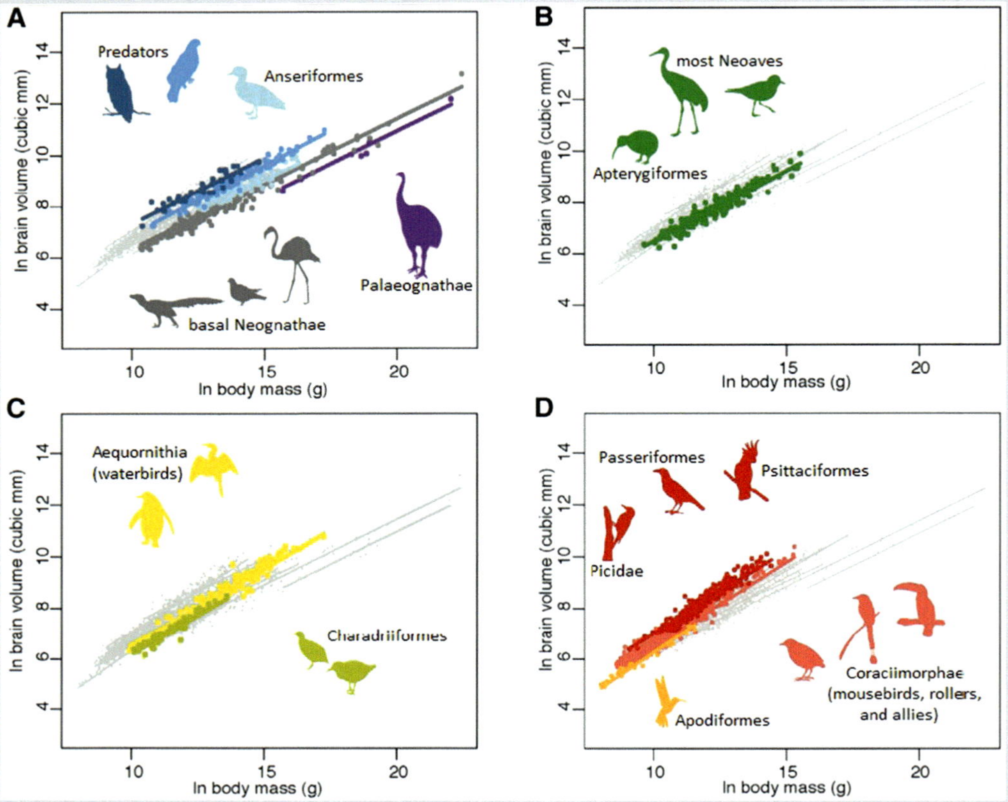


Relationship between brain volume and body mass for different taxa of birds, with greater slopes indicating greater brain volume relative to body mass. (**a**) Five low-slope taxa, including Paleognathae, basal Neognathae (e.g., Galliformes, grebes, flamingoes, and pigeons and doves), Anseriformes (waterfowl), and predatory taxa including the orders Accipitriformes (e.g., hawks), Strigiformes (owls), Cariamiformes (seriemas), and Falconiformes (falcons). (**b**) Intermediate-slope taxa, including most Neoavians (e.g., cuckoos, Gruiformes [cranes and allies], and most Charadriformes), and Apterygiformes (kiwis). (**c**) Higher-slope taxa, including some Charadriiformes and waterbirds (Aequornithia). (**d**) Highest-slope taxa, including Apodiformes (swifts and hummingbirds), Coraciimorphae, Picidae (woodpeckers), Psittaciformes (parrots), and Passeriformes (especially Corvidae) (Figure from Ksepka et al. 2020; © 2020 Elsevier Inc., used with permission; Silhouettes from http://phylopic.org)

As Neoaves diversified at the beginning of the Paleogene, so did relative brain size. For many Neoaves, brain volume relative to body size ratio has remained stable or even declined (e.g., taxa in B and C above). However, for other Neoaves, relative brain sizes continued to increase (e.g., taxa in D above). One possible contributing factor was the environmental disruption caused by the K-Pg impact. Ksepka et al. (2020) suggested that "the aftermath of the K-Pg mass extinction created conditions ripe for the preferential survival and subsequent diversification of larger brained birds." Among the birds where brain size relative to body mass has continued to increase

(continued)

**Box 4.4** (continued)

since the beginning of the Paleogene are hummingbirds, but this is likely due to their small size. The taxa exhibiting the highest rates of brain-body size evolution are Psittaciformes (e.g., parrots) and Corvidae (e.g., crows, ravens, and jays). Not only do they have relatively large brains, but parrots and corvids also have higher neuronal densities in their cerebrums than other birds. A number of factors have likely contributed to the evolution of parrot and corvid brains, all of which are discussed in the accompanying text.

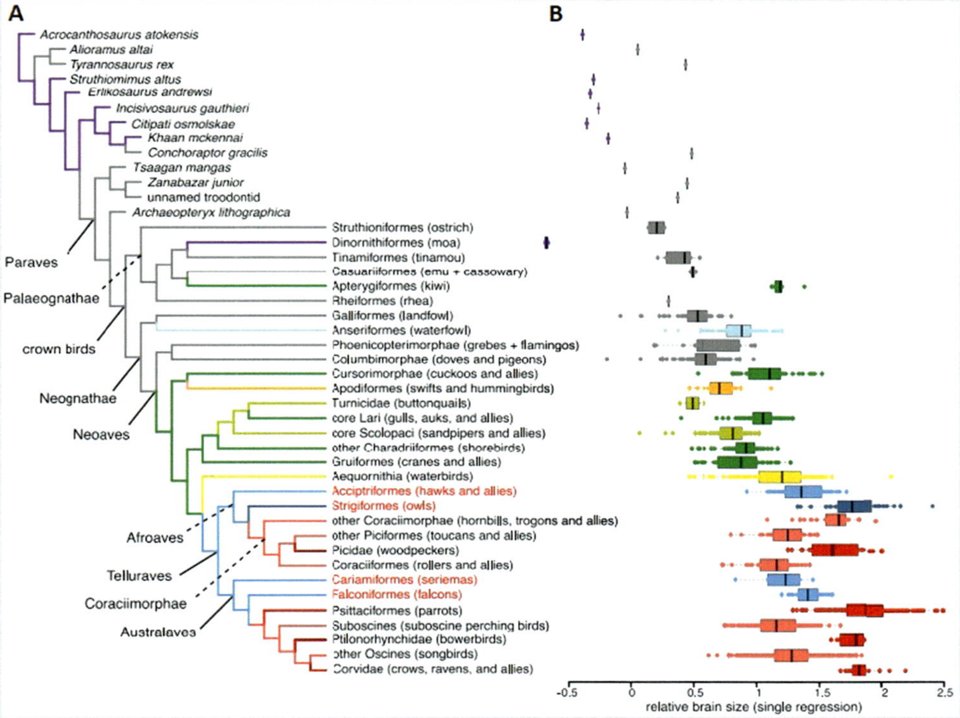


Avian brain-body size evolution. (**a**) Phylogeny of non-avian theropods and birds, with different colors indicating "shifts" in relative brain-body size. (**b**) Brain size relative to body size for the different taxa in (**a**). Predatory birds are indicated by red font (Figure from Ksepka et al. 2020; © 2020 Elsevier Inc., used with permission)

In addition to mode of development (altricial vs. precocial) and migratory status, other factors that appear to have influenced the brain size of birds include the duration of pair bonds and participation in, and the relative stability of, foraging associations (Shultz and Dunbar 2010; West 2014; Fig. 4.31). The apparent relationship between larger brains and long-term pair bonds may be due to the increased need for individuals to be able to successfully interact and negotiate investment decisions over long periods to maximize their reproductive success (Shultz and Dunbar 2010). Participation in persistent, stable foraging associations requires recognition of and repeated interactions with other group members, coordination of foraging activities, and recognition of interindividual relationships (e.g., dominance hierarchies,

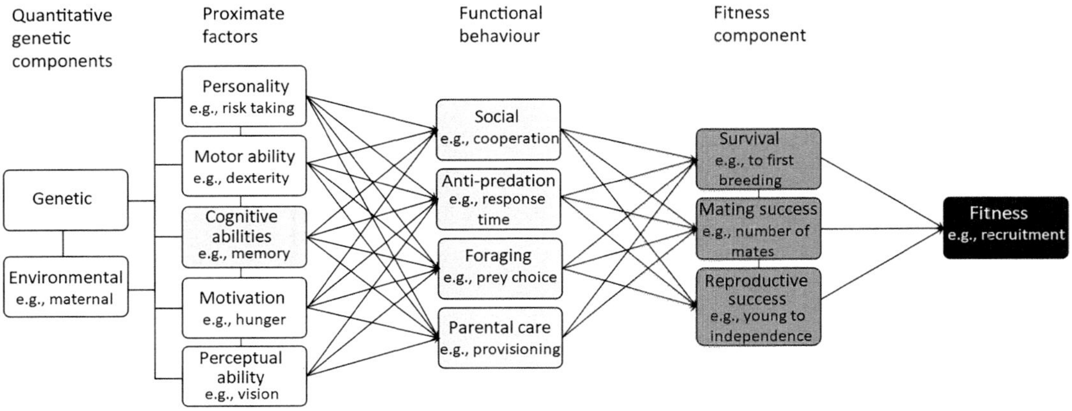


**Fig. 4.26** The cognitive abilities of birds are a product of the interaction between genes and the environment in which they develop. Those abilities interact with other characteristics to influence their behavior which, in turn, influences the components of fitness that will determine their overall fitness, i.e., the number of their young recruited into future generations. (Figure modified from Morand-Ferron et al. 2016; © 2015 Cambridge Philosophical Society, used with permission)

familial relationships, and pair bonds), and these complex interactions may favor the evolution of larger brains (Shultz and Dunbar 2010).

Another factor potentially influence brain size evolution in birds is diet diversity, with generalist diets apparently favoring selection for larger brains (van Horik et al. 2012; Ducatez et al. 2015; but see Overington et al. 2011; Fig. 4.32). Diets that include a variety of different food items distributed throughout complex environments, with some only available at certain times, may require use of a variety of different foraging techniques as well as cognitive capacities for spatial and temporal memory. Collectively, these factors may be important drivers in the cognitive abilities of omnivorous birds (van Horik et al. 2012). Improved cognitive abilities may, in turn, influence diet by way of innovation, such as the use of tools (Box 4.7 Tool use by New Caledonian Crows).

Brain size and cognitive abilities vary among different taxa of birds, with parrots (order Psittaciformes; Box 4.8 Parrots vs. primates) and corvids (family Corvidae) having relatively larger brains and considered more intelligent than species in other taxa (Fig. 4.33). As described above, variation in a number of ecological, social, and life-history factors has contributed to variation in the relative cognitive abilities of different taxa of birds. For some taxa such as parrots and corvids, a combination of multiple selective factors, such as being altricial species, having long-term pair bonds, membership in stable groups, and omnivorous diets, have likely contributed to the evolution of their relatively large brains and impressive cognitive abilities (Fig. 4.34).

## 4.4 Avian Sleep

Sleeping animals, including birds, are behaviorally quiescent and exhibit a reduced responsiveness to stimuli (Lima et al. 2005). In addition, sleeping animals often assume a species-specific posture that usually includes closing of one or both eyes (Flanigan 1972; Fig. 4.35). Because animals are typically immobile and either less aware or unaware of their surrounding environment, they are more vulnerable to predation when asleep. Because selection has clearly favored the evolution of sleep in birds and many other animals, its benefits must outweigh the potential cost of such increased

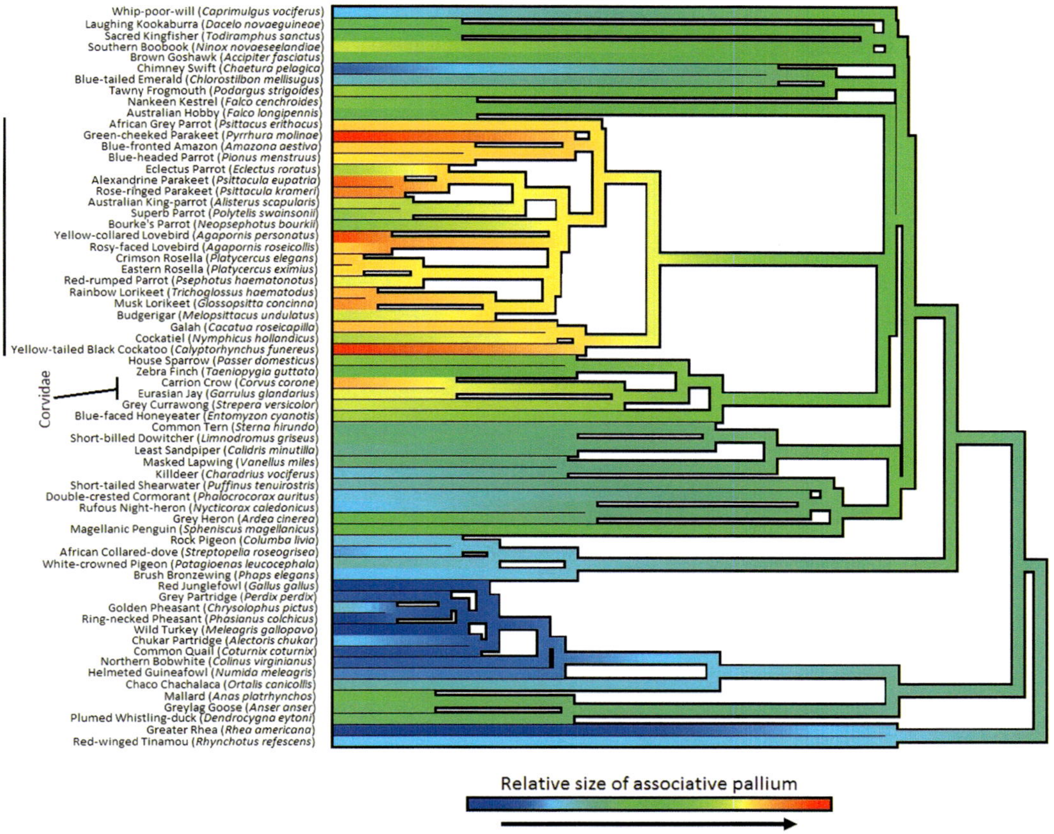


**Fig. 4.27** Phylogenetic variation in the size of the associative pallium (nidopallium plus mesopallium) relative to the rest of the pallium in several species of birds. Red-yellow coloration indicates relatively larger associative pallium. Note that species in the order Psittaciformes and the family Corvidae have the largest associative palliums relative to the rest of their palliums. (Figure modified from Sayol et al. (2016a); © 2016 S. Karger AG, Basal, used with permission)

vulnerability. Although those benefits remain to be determined, proposed functions for sleep include memory consolidation, maintaining high neurobehavioral performance (e.g., cognitive processing, attention, and reaction time) when awake, optimizing the temporal use of energy (i.e., energy conserved during sleep is available when awake), aiding in the maturation of the central nervous system (because young typically sleep for longer periods than adults), and selection for periods of inactivity when being awake is dangerous or non-productive (Aulsebrook et al. 2016). In addition, Mussoi et al. (2022) suggested that, for birds, "... quality sleep is likely essential when learning new vocalizations and that sleep disturbances will have especially strong effects on learned vocalizations." In support of that hypothesis, Johnsson et al. (2022) found that sleep deprivation impaired the motivation and cognitive performance of adult Australian Magpies (*Cracticus tibicen*) and also affected their singing behavior (e.g., sang fewer songs at different times of day than non-sleep-deprived magpies).

Until 2016, it was thought that only birds and mammals experienced periods of electrical activity in the brain called non-rapid eye movement sleep (NREM; also called slow-wave sleep) and rapid eye movement (REM) sleep, an apparent example of convergent evolution. However, Shein-Idelson et al. (2016) reported both NREM and REM sleep in a lizard (Australian dragon, *Pogona vitticeps*), suggesting that these electrophysiological features of sleep evolved in a common ancestor of present-day reptiles, birds, and mammals more than 300 million years ago. Sleeping reptiles, birds, and mammals differ in the frequency and duration of NREM and REM sleep. For example, compared to mammals where periods of REM sleep can last minutes or even tens of minutes, REM sleep of birds usually occurs as short bouts lasting less than 10 s, and birds spend much less time than mammals in

**Box 4.5 Neuronal densities in the avian brain and cognition**

The brains of corvids and psittacids are relatively small compared to those of mammals also considered to be intelligent (e.g., primates), ranging from just 8–15 g in corvids and up to 24.7 g in psittacids (Mlikovsky 2003; Iwaniuk et al. 2005). How can birds with such small brains have cognitive abilities comparable, or even superior, to those of mammals with much larger brains? The answer to that question is that relative brain size is not directly correlated with intelligence. Rather, the number of neurons and number of connections between neurons appear to be the key factors. Olkowicz et al. (2016) found that bird brains have more neurons than mammalian brains, including primate brains, of similar mass. Among species of songbirds (order Passeriformes), including some corvids, that ranged in body mass from 4.5 to 1070 g, they found that the total number of neurons in their brains ranged from 136 million to 2.17 billion. For parrots, body mass ranged from 23 to 1008 g and numbers of brain neurons from 227 million to 3.14 billion. The relatively small brains of birds have more neurons than the larger brains of comparably sized mammals because avian neurons are smaller, resulting in higher densities of neurons, particular in the avian pallium. The higher densities and, therefore, shorter interneural distances likely also result in faster information processing that would further enhance cognitive abilities (Olkowicz et al. 2016).

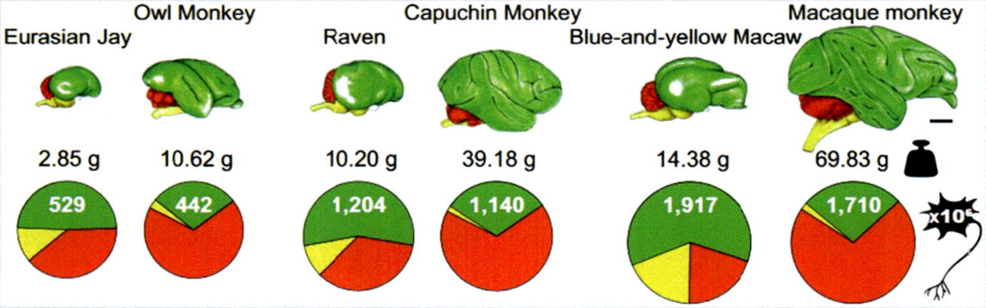


Brains of corvids (Eurasian Jay [*Garrulus glandarius*] and Common Raven [*Corvus corax*]), a parrot (Blue-and-yellow Macaw, *Ara ararauna*), and primates (owl monkey [*Aotus* spp.], capuchin monkey [family Cebidae], and macaque monkey [*Macaca* spp.]) drawn at the same scale. Numbers under brains indicate brain mass (in grams) and the total number of neurons in the pallium (green), cerebellum (red), and the rest of the brain (yellow). The palliums of these birds have more neurons than those of primates, even though the primate brains are much larger (Figure from Olkowicz et al. 2016; used with permission of the U. S. National Academy of Sciences)

(continued)

**Box 4.5** (continued)

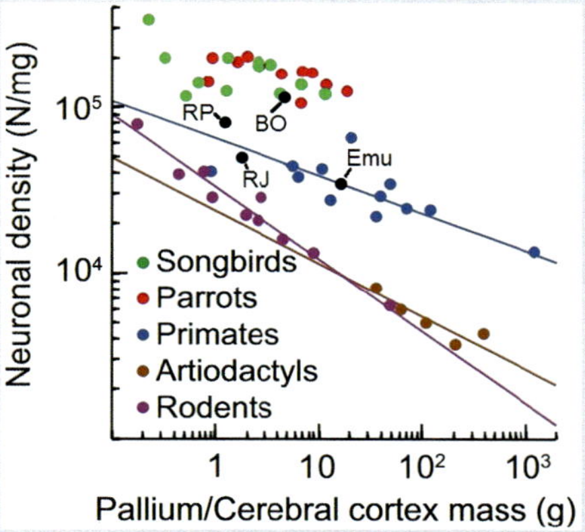


Neuronal densities in the palliums of songbirds and parrots are generally higher than in other groups of bird and also higher than neuronal densities in the cerebral cortex of mammals, including primates, artiodactyls (ungulates), and rodents. Note also that neuronal densities tend to decline with increasing pallium/cerebral cortex mass. RP, Rock Pigeon (*Columba livia*); BO, Barn Owl (*Tyto alba*); RJ, Red Junglefowl (*Gallus gallus*); Emu (*Dromaius novaehollandiae*) (Figure from Olkowicz et al. 2016; used with permission of the U. S. National Academy of Sciences)

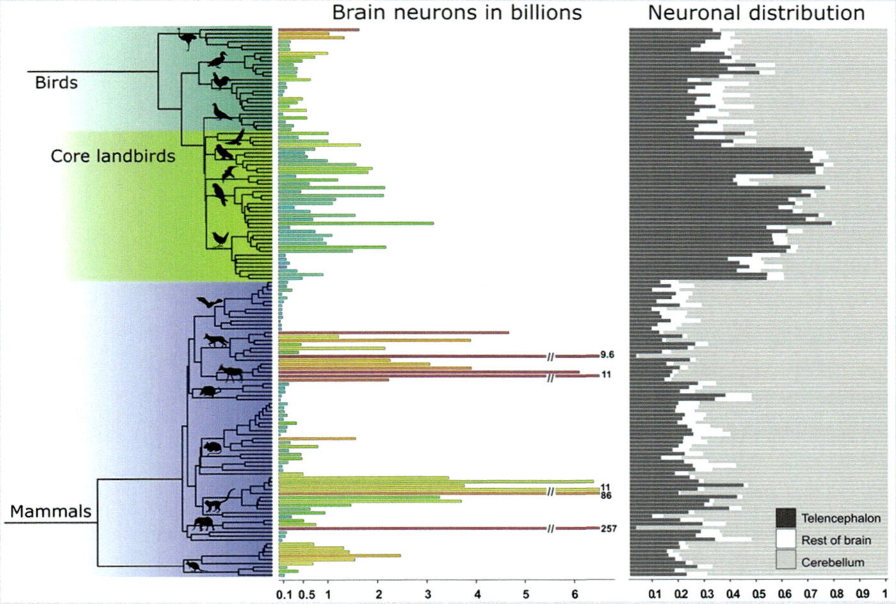


Number of neurons and neuronal distribution in the brains of birds and mammals (note that the bars for five species of mammals with the highest numbers of neurons have been truncated). The gray-scale bars indicate the proportion of brain neurons found in the telencephalon, rest of brain, and cerebellum. For mammals, most neurons are located in the cerebellum. For birds there are two patterns, with more neurons in the cerebellum in basally diverging birds and more in the telencephalon in core land birds (Telluraves; hawks, eagles, owls, falcons,

(continued)

**Box 4.5** (continued)

songbirds, and parrots). The increase in telencephalic neurons in core land birds and anthropoid primates "might provide the neural substrate for their remarkably similar cognitive feats" (Kverková et al. 2022) (Figure modified from Kverková et al. 2022; © 2022 the Authors. Published by PNAS, open-access article distributed under Creative Commons Attribution License 4.0 (CC BY), https://creativecommons.org/licenses/by/4.0/)

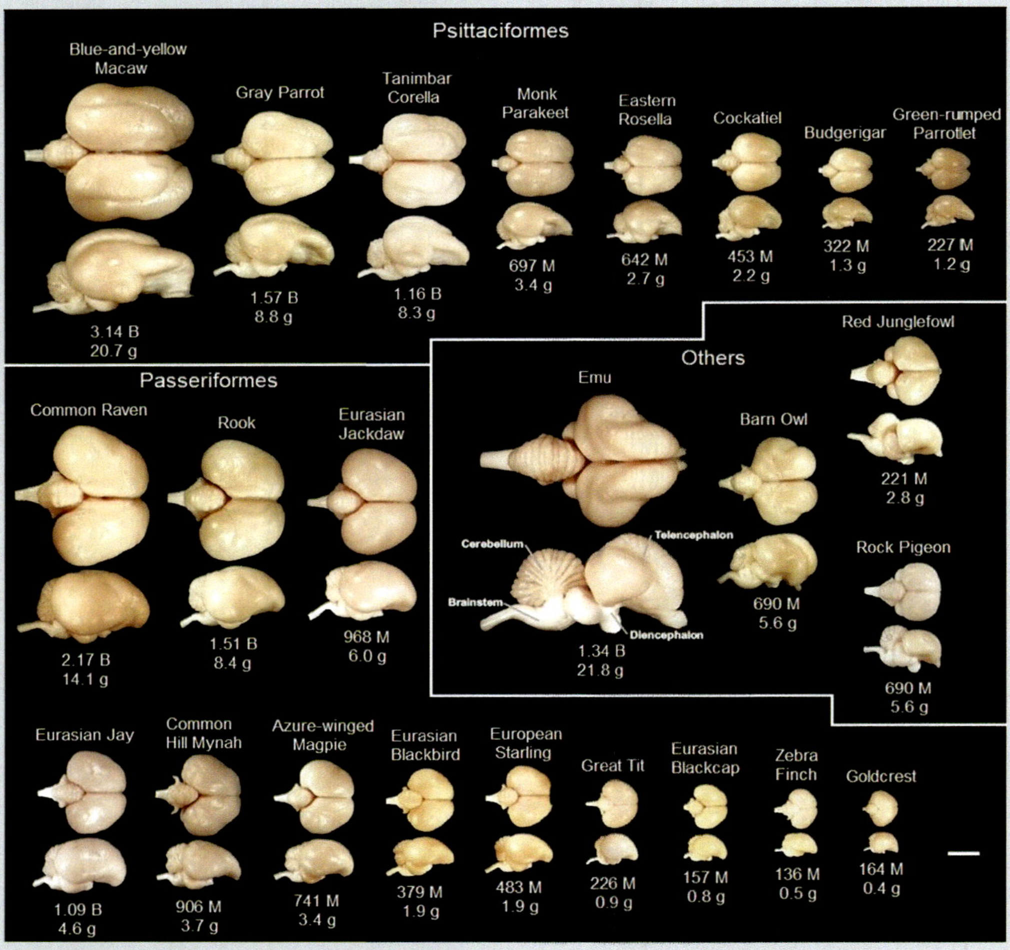


Brain size, morphology, and number of neurons for several species of birds. Numbers below the dorsal and lateral views of brains are the total number of neurons in the brains and brain mass. *B* billion, *M* million; Scale bar = 10 mm. Black-and-yellow Macaw, *Ara ararauna*; Gray Parrot, *Psittacus erithacus*; Tanimbar Corella, *Cacatua goffiniana*; Monk Parakeet, *Myiopsitta monachus*; Eastern Rosella, *Platycercus eximius*; Cockatiel, *Nymphicus hollandicus*; Budgerigar, *Melopsittacus undulatus*; Green-rumped Parrotlet, *Forpus passerinus*; Common Raven, *Corvus corax*; Rook, *Corvus frugilegus*; Eurasian Jackdaw, *Corvus monedula*; Emu, *Dromaius novaehollandiae*; Barn Owl, *Tyto alba*; Red Junglefowl, *Gallus gallus*; Rock Pigeon, *Columba livia*; Eurasian Jay, *Garrulus glandarius*; Common Hill Mynah, *Gracula religiosa*; Azure-winged Magpie, *Cyanopica cyanus*; Eurasian Blackbird, *Turdus merula*; European Starling, *Sturnus vulgaris*; Great Tit, *Parus major*; Eurasian Blackcap, *Sylvia atricapilla*; Zebra Finch, *Taeniopygia guttata*; Goldcrest, *Regulus regulus* (Figure modified from Olkowicz et al. 2016; used with permission of the U.S. National Academy of Sciences)

REM sleep (Lesku and Rattenborg 2014; Fig. 4.36). REM sleep in birds is characterized by eye movements under closed eyelids, twitching, wake-like brain activity, wobbling and head drops due to reduced muscle tone, variable cardiac and respiratory rhythms, and suspension of thermoregulatory responses (Blumberg et al. 2020). Despite much study, particularly of sleeping birds and mammals, the respective functions of these two stages of sleep, and possible reasons for the differences among taxa in the frequency and duration of these stages, remain unclear. However, REM sleep in birds and mammals is thought to be involved in the maturation of the central nervous system because young birds and mammals exhibit more REM sleep than older individuals (Scriba et al. 2013; Blumberg et al. 2020). Interestingly, a number of

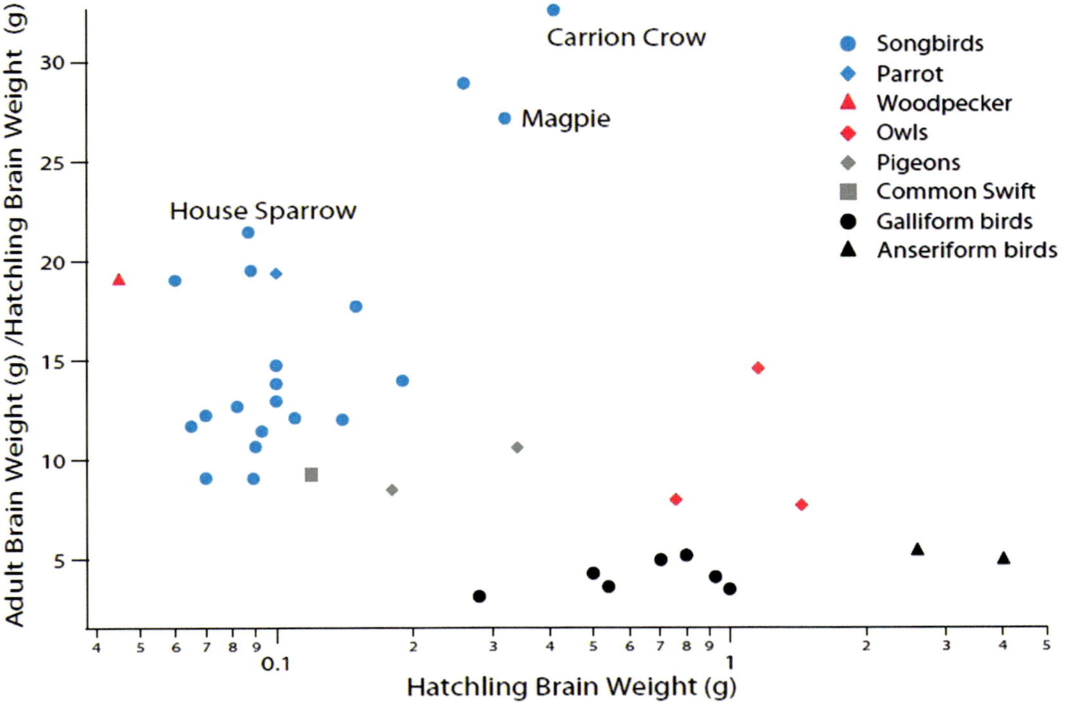


**Fig. 4.28** Compared to precocial species (e.g., Galliforms and Anseriforms), most brain growth in altricial species of birds occurs after hatching, and this is especially the case for songbirds and parrots. Among songbirds, post-hatching growth and maturation of brains are particularly extensive for species in the family Corvidae, including Carrion Crows (*Corvus corone*) and magpies. Most post-hatching growth of the brains of altricial species is due to expansion of the telencephalon. House Sparrow, *Passer domesticus*; Common Swift, *Apus apus*. (Figure from Charvet and Striedter 2011; © 2011 Charvet and Striedter, open-access article subject to an exclusive license agreement between the authors and Frontiers Media SA)

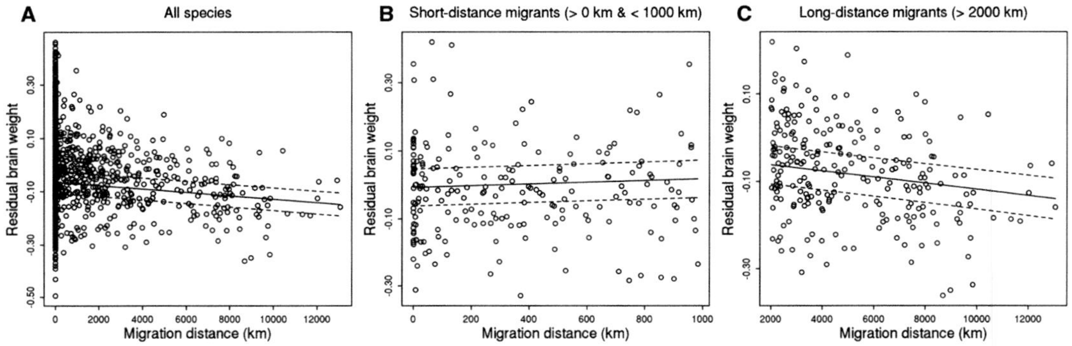


**Fig. 4.29** Relationship between migration distance and body-mass controlled residual brain weight for (**a**) 1466 species of birds, (**b**) 142 species of short-distance migrants, and (**c**) 146 species of long-distance migrants. Species that migrate long distances tend to have smaller brains relative to their body size. (Figure from Vincze 2016; © 2016 The Author. Evolution © 2016 The Society for the Study of Evolution, used with permission)

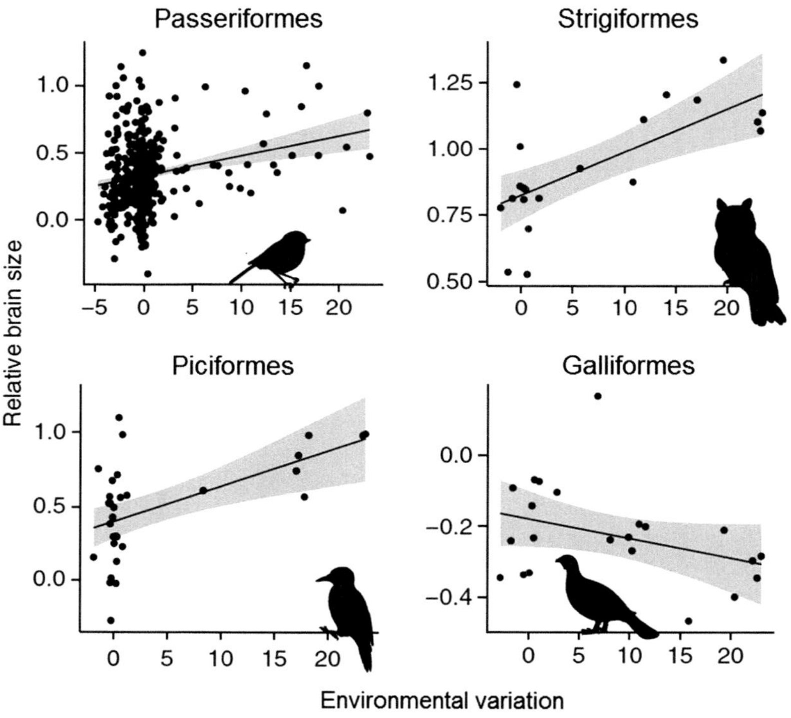


**Fig. 4.30** Relationships between environmental variation and relative brain size (brain size relative to body size) in four orders of birds. Environmental variation refers to the extent to which the environment varies seasonally and among years and the duration of snow cover. The relationship between environmental variation and brain size holds for most orders of birds, with Galliformes being an exception. A possible explanation for this is that gallinaceous species survive harsh conditions due to their specialized adaptions, such as reduced metabolism and ability to survive on low-quality foods like conifer needles, rather than their ability to explore and learn. (Figure modified from Sayol et al. 2016; open-access article distributed under the terms of the Creative Commons CC BY license, https://creativecommons.org/licenses/by/4.0/)

studies have revealed an interaction between REM sleep and thermoregulation, with the ability to thermoregulate impaired during REM sleep (Glotzbach and Heller 1976; Martelli et al. 2014). One possible explanation for this linkage is that REM sleep reduces the energy devoted to thermoregulation "to enhance energy appropriation for somatic and CNS-related processes" (Schmidt 2014; Komagata et al. 2019). Yet another possible function of REM sleep is that it plays a role in memory consolidation (Brawn et al. 2018; van der Meij et al. 2020).

Available evidence indicates that sleeping birds wake up frequently. For example, sleeping Eurasian Blue Tits (*Cyanistes caeruleus*) were found to wake up anywhere from 23 to 230 times during the night, with most individuals waking up an average of about 3–10 times per hour (Steinmeyer et al. 2010). Similar results have been reported for other species of birds (Jones et al. 2008; Wellmann and Downs 2009; Fig. 4.37). When awake, Blue Tits usually preened, scratched, stretched legs and wings, or moved inside their nest boxes (Steinmeyer et al. 2010). During periods of wakefulness and drowsiness, birds exhibit greater awareness of their surroundings, suggesting that their intermittent occurrence during periods of sleep may reduce the risk of predation.

Birds, like some aquatic mammals (cetaceans and pinnipeds), are capable of unihemispheric sleep, sometimes keeping one eye open (and one brain hemisphere awake) (Fig. 4.38). This allows birds to visually monitor their environment while sleeping, and this can be especially important when threatened by predators. For example, Rattenborg et al. (1999a) found that Mallards (*Anas platyrhynchos*) sleeping at the edge of a group spent more time with one eye open (the eye facing away from the group), and more time awake, than Mallards sleeping near the center of a group (Fig. 4.39). Similar observations have been reported for Black-tailed Godwits (*Limosa limosa*), with birds on the periphery of roosts spending more time with eyes open than those more centrally located (Dominguez 2003). Birds sleeping at the edge of a group are more susceptible to predation, so spending more time with one or both eyes open likely increases the likelihood of detecting a predator and reduces the risk of predation.

The time spent sleeping by birds can exhibit substantial variation throughout the year. During the breeding season, particularly at higher latitudes where periods of daylight are long, birds spend much less time sleeping than during the non-breeding season. For example, during the summer breeding season, White-crowned Sparrows (*Zonotrichia leucophrys*) spent an average of 14% of each day sleeping compared to 36% during the winter (Jones et al. 2010; Fig. 4.40). Pectoral Sandpipers (*Calidris melanotos*) are polygynous, Arctic-breeding shorebirds and males have been found to greatly reduce their time spent sleeping during the period when they are competing with other males for

**Box 4.6 Brain size and latitude**

In general, resident birds that occupy cold, seasonal, and unpredictable high-latitude habitats have relatively large brains that permit flexible and innovative behavioral responses that aid in survival (e.g., Sayol et al. 2016). However, not all resident birds with high-latitude distributions have relatively large brains, e.g., grouse (subfamily Tetraoninae, order Galliformes). More generally, Fristoe and Botero (2019) used a sample of 1280 resident species of birds and found that relative brain size is approximately normally distributed throughout the world, with the exception of resident land birds at high latitudes in the Northern Hemisphere. At high latitudes, relative brain size exhibits a bimodal distribution, with relative large and relative small brain sizes overrepresented.

(continued)

**Box 4.6** (continued)

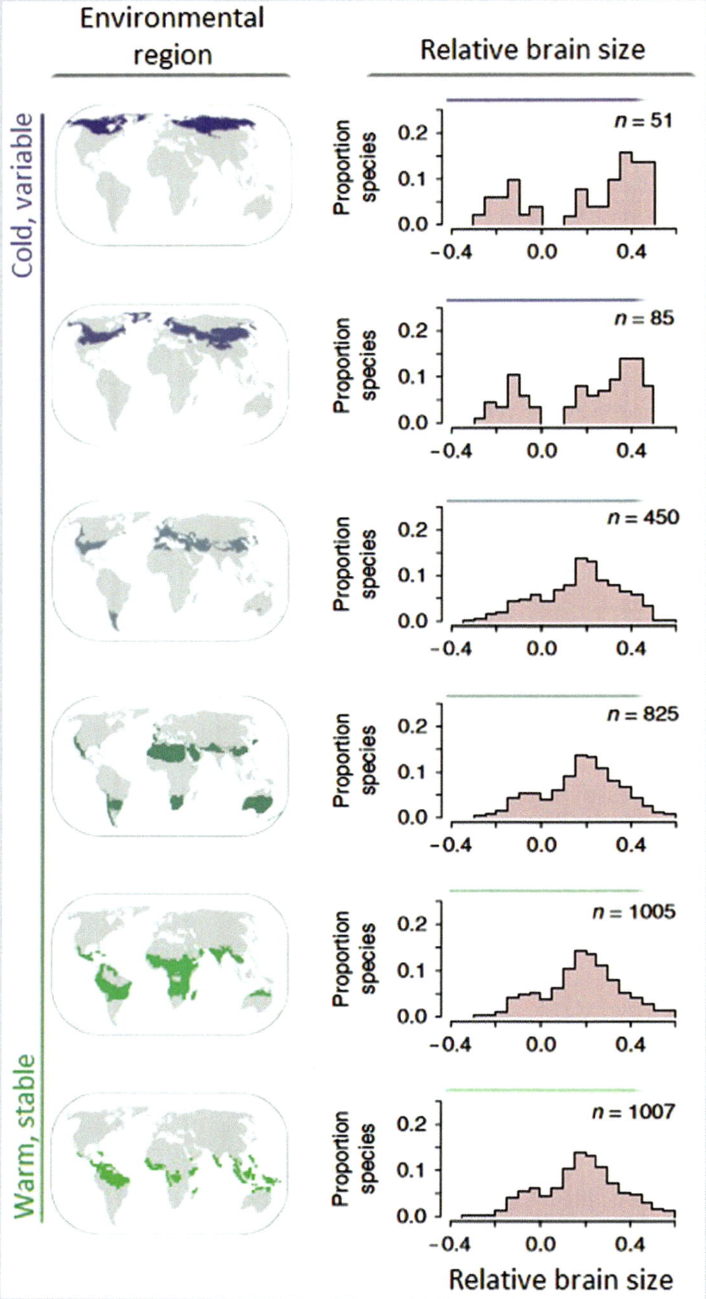


Global distribution of brain size for 1280 resident species of land birds at different latitudes with different environments. At lower latitudes, brain sizes are approximately normally distributed, but brain size exhibits a

(continued)

**Box 4.6** (continued)

bimodal distribution at higher latitudes (Figure from Fristoe and Botero 2019; open-access article licensed under a Creative Commons Attribution 4.0 International License, https://creativecommons.org/licenses/by/4.0/)

To explain this bimodal distribution of brain size at high latitudes, Fristoe and Botero (2019) suggested that different species have evolved different strategies for coping with generally harsh and unpredictable environmental conditions. As noted above, selection has favored the development of larger brains that permit flexible and innovative behavioral responses to harsh conditions in some species (cognitive buffer hypothesis). Because larger brains are metabolically costly, these species tend to have high-quality diets (sometimes aided by caching food and social foraging) and, during the breeding season, smaller clutches (Ricklefs 2004). However, for other species, selection has favored smaller brains that are less energetically costly and, in contrast to species with larger brains, an ability to withstand harsh environmental conditions by increasing in size, developing digestive systems able to hold and digest large amounts of readily available, but difficult to digest, low-quality food, and, during the breeding season, larger clutches.

copulations with fertile females (Lesku et al. 2012; Fig. 4.41). In an extreme case, one male was active more than 95% of the time over a period of 19 days (Lesku et al. 2012). Going without sleep was found to be beneficial because males that slept the least had the greatest reproductive success (Lesku et al. 2012).

Long-distance migrants that fly at night may also get very little sleep. In the lab, for example, migratory White-crowned Sparrows spent 67%

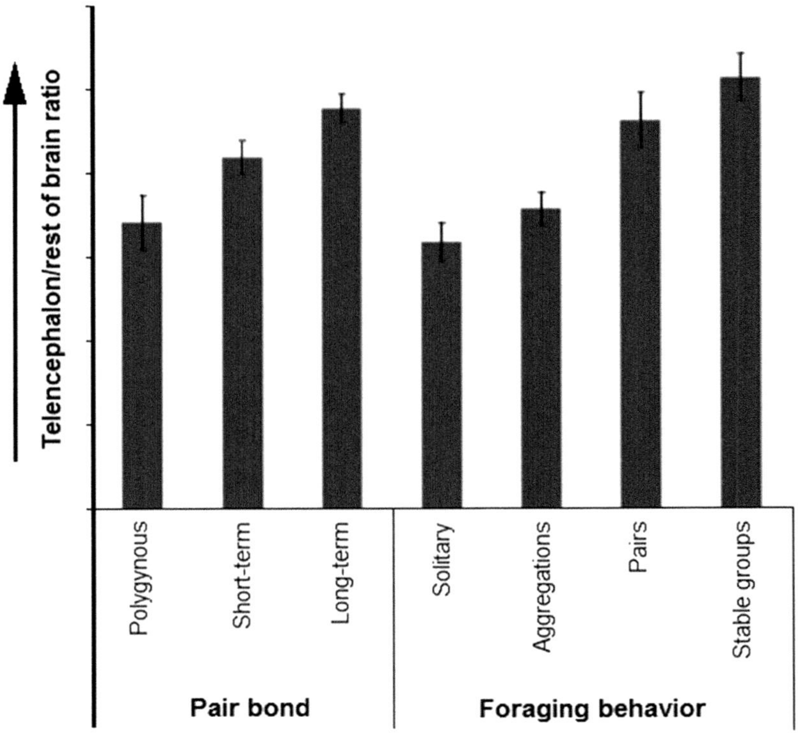


**Fig. 4.31** Relationship between the duration of pair bonds and stability of foraging associations on the telencephalon/rest of brain ratio. This ratio tends to increase with the increasing duration of pair bonds and increasing stability of foraging associations. (Figure modified from Shultz and Dunbar 2010; © 2010 Oxford University Press, used with permission)

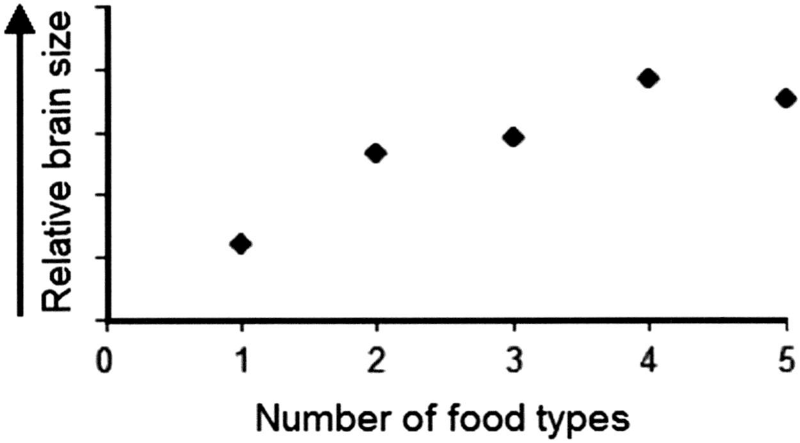


**Fig. 4.32** Relationship between relative brain size and the number of different food types in the diets of 780 species of birds. Food types included vertebrate prey, invertebrate prey, vertebrate carrion, fruits or seeds, and nectar or pollen. (Figure modified from Ducatez et al. 2015; © 2014 The Authors. Journal of Animal Ecology © 2014 British Ecological Society, used with permission)

less time sleeping than non-migratory individuals (Rattenborg et al. 2004; Fig. 4.42). Some species of birds make non-stop migratory flights that may last for several days. For example, Bar-tailed Godwits (*Limosa lapponica*) migrating from Alaska to New Zealand may fly continuously for as long as 7 or 8 days (Battley et al. 2012) and Great Snipes (*Gallinago media*) for as long as 4 days (Klaassen et al. 2011). More amazingly, Great Frigatebirds (*Fregata minor*) have been found to remain in flight—soaring, gliding, and, less than 20% of the time, flapping their wings—for periods as long as 2.1 months (Weimerskirch et al. 2016), Alpine Swifts (*Tachymarptis melba*) can remain in the air for more than 6 months as they migrate, forage, and roost (Liechti et al. 2013), and some Common Swifts (*Apus apus*) remain airborne for 10 months during the non-breeding season (Hedenström et al. 2016).

How are some birds able to apparently spend little or no time sleeping for extended periods of time? Birds like Great Frigatebirds and Alpine Swifts that spend long periods in flight have short periods of sleep when soaring (frigatebirds) and gliding (frigatebirds and swifts) (Liechti et al. 2013; Weimerskirch et al. 2016; Rattenborg et al. 2016; Fig. 4.43). Some night-migrating birds have been found to take short naps during the day. For example, migrating Swainson's Thrushes (*Catharus ustulatus*) sleep for short periods (less than 15 s on average) during the day (Fuchs et al. 2009) and a Hooded Warbler

**Box 4.7 Tool use by New Caledonian Crows**

A crucial stage in hominin evolution was the development of metatool use—the ability to use one tool on another. Although great apes can solve metatool tasks, monkeys have been less successful. Taylor et al. (2007) provided experimental evidence that New Caledonian Crows (*Corvus moneduloides*) can spontaneously solve a demanding metatool task where a short tool is used to extract a longer tool that can then be used to obtain meat. Six of seven crows initially attempted to extract the long tool with the short tool. Four successfully obtained meat on the first trial. The experiments revealed that the crows did not solve the metatool task by trial-and-error learning during the task or through a previously learned rule. The sophisticated physical cognition shown appears to have been based on analogical reasoning. The ability to reason analogically may explain the exceptional tool-manufacturing skills of New Caledonian Crows.

(continued)

**Box 4.7** (continued)

(**a**) A New Caledonian Crow using a stick tool. (**b**) The tool was used to extract a beetle larva from its burrow in this snag (Figure from Rutz and St Clair 2012; © 2012 Elsevier B.V., used with permission)

(continued)

**Box 4.7** (continued)

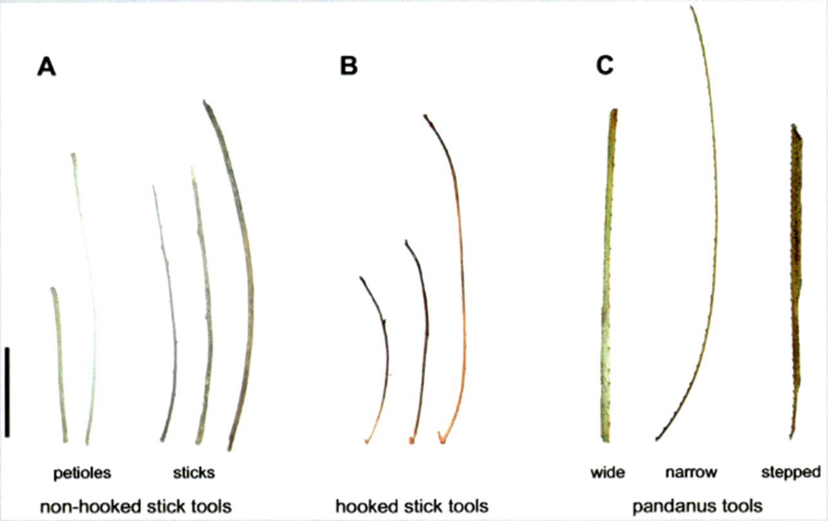


Tools used by New Caledonian Crows. (**a**) Non-hooked tools, e.g., leaf petioles and sticks, (**b**) Hooked tools made from twigs, and (**c**) Stick tools made from leaf edges of screw pines (*Pandanus* spp.) (Figure from Rutz and St Clair 2012; © 2012 Elsevier B. V., used with permission)

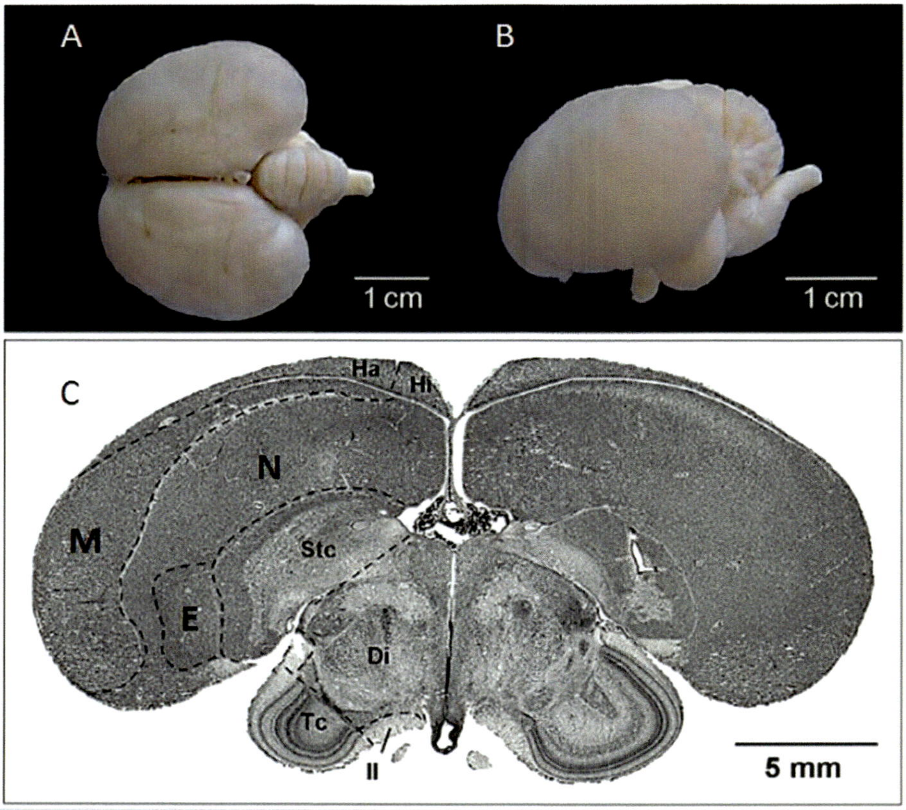


(continued)

**Box 4.7** (continued)

Brain of a New Caledonian Crow with noticeably large cerebral hemispheres and, specifically, large meso- and nidopalliums. (**a**) Dorsal view. (**b**) Lateral view. (**c**) Section through the brain showing some specific areas. Ha, hyperpallium apicale, Hi, hippocampus; M, mesopallium; N, nidopallium; E, entopallium; Stc, striatopallidal complex; Di, diencephalon; Tc, tectum opticum; II, tractus opticus (Figure modified from Mehlhorn et al. 2010; © 2010 S. Karger AG, Basel, used with permission)

(*Setophaga citrina*) that apparently had just arrived on the Gulf Coast of the United States after a flight across the Gulf of Mexico appeared to sleep for several minutes between foraging bouts (Németh 2009). Other species of migrating birds have also been observed sleeping after long-distance (≥500 km) flights (Schwilch et al. 2002). Birds may also be able to sleep unihemispherically for short periods while active (Rattenborg et al. 1999a). Finally, some species of birds may simply be able to go without sleep for long periods, or at least to sleep very little, with little or no effect on their neurobehavioral performance (Rattenborg and Martinez-Gonzalez 2015).

## 4.5 Sense Organs: General Receptors

Birds, like humans, have receptors that provide information about their external and internal (body) environments. These receptors are special neurons able to respond to various types of stimuli and are placed in two general categories: general (temperature, pain, touch, stretch, and pressure) and special (taste, olfaction, vision, magnetoreception, equilibrium, and hearing).

The epidermis of birds contains only free (naked) nerve endings that allow the perception of temperature (thermoreceptors) and pain

**Box 4.8 Parrots vs. primates**

Complex cognitive abilities do not arise from a single area of the brain, but, rather, result from particular neural pathways in the brain. For example, the complex cognitive abilities of primates are thought to be aided by their ability to generate mental representations of objects and, by mentally manipulating those objects, predict possible outcomes (Ito 2008). This is thought to occur via neural pathways between the cerebral cortex, areas in the pons (called pontine nuclei), the thalamus, and motor and associative areas of the prefrontal cortex, i.e., the cortico-ponto-cerebellar pathway. The prefrontal cortex has "executive control" (i.e., developing a plan to achieve a goal) of the process, with the cerebellum generating neural representations of the essential properties of objects that can be used to predict outcomes.

The avian "equivalent" of the prefrontal cortex is the nidopallium caudolaterale (NCL) and the area is thought to have "executive control" of a similar process in birds. However, compared to mammals, the cortico-ponto-cerebellar pathway is not as well developed in birds. Instead, the medial spiriform nuclei (SpM in the midbrain) are thought to act as the primary relay between the NCL and wulst and the cerebellum. Gutiérrez-Ibáñez et al. (2018) found that the SpM of parrots (Order Psittaciformes) is relatively larger than that of species in several other orders, including the order Passeriformes. The larger SpM, and the corresponding enhanced communication between the NCL and cerebellum, may help explain the complex cognitive abilities of parrots.

(continued)

**Box 4.8** (continued)

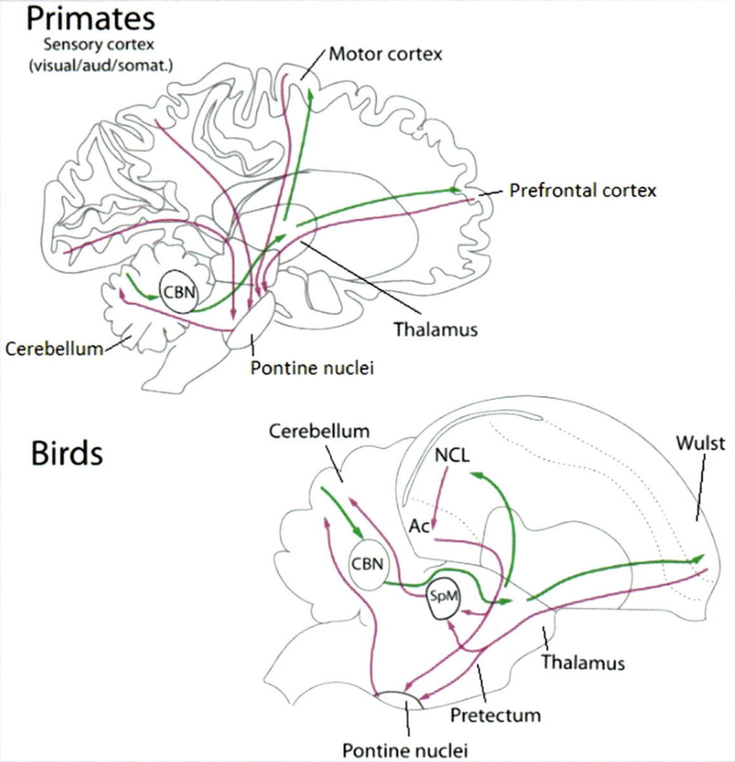


CBN, cerebellar nuclei; NCL, nidopallium caudolaterale; Ac, arcopallium; SpM, medial spiriform nuclei; green arrows—impulses to cortex in primates and the wulst and NCL in parrots; purple arrows—impulses originating in the cortex in primates and the wulst and NCL in parrots (Figure from Gutiérrez-Ibáñez et al. 2018; open-access article distributed under the terms of the Creative Commons CC BY license, https://creativecommons.org/licenses/by/4.0/)

(continued)

**Box 4.8** (continued)

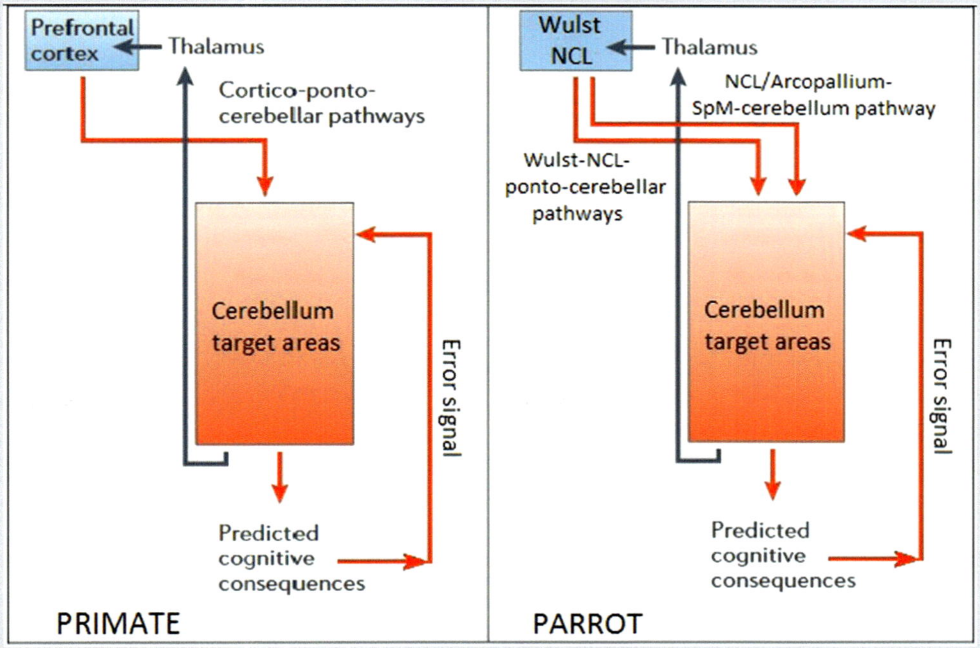


As noted above, complex cognitive abilities may involve generating mental representations of objects, then mentally manipulating those objects to predict possible outcomes. In primates, this is thought to occur via cortico-ponto-cerebellar pathways (i.e., between the cerebral cortex, pontine nuclei in the pons, and the cerebellum). In parrots and other birds, the cortico-ponto-cerebellar pathway is not as well developed. Instead, the medial spiriform nuclei (SpM) are thought to act as the primary relay between the NCL and wulst and the cerebellum. For both primates and birds, feedback loops may allow consideration of multiple predicted outcomes (Figure modified from Ramnani 2006; © Springer Nature, used with permission)

(continued)

**Box 4.8** (continued)

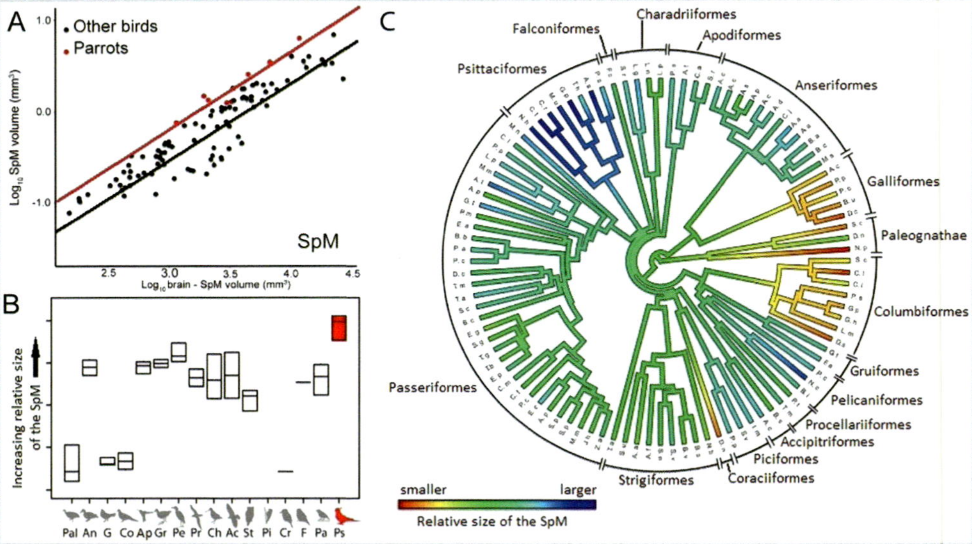


Relative size of the medial spiriform nucleus (SpM) of species of birds in several different orders. (**a**) The volume (log-transformed) of the SpM is plotted as a function of the brain volume minus the volume of SpM. The black line represents all species examined, and the red line represents parrots. (**b**) Box plots showing the relative size of the SpM for several different orders of birds. Parrots (Psittaciformes) are shown in red. Solid horizontal lines represent the median; the two boxes represent the limits of the second and third quartiles. (**c**) Relative SpM volumes plotted onto an avian phylogeny. Ac, Accipitriformes, An, Anseriformes; Ap, Apodiformes; Ch, Charadriiformes; Co, Columbiformes; Cr, Coraciiformes; F, Falconiformes; G, Falliformes; Gr, Gruiformes; Pa, Passeriformes; Pal, Paleognathae; Pe, Pelecaniformes; Pe, Piciformes; Pr, Procellariiformes; Ps, Psittaciformes; St, Strigiformes (Figure from Gutiérrez-Ibáñez et al. 2018; Published by Springer Nature, open-access article distributed under the terms of the Creative Commons CC BY license, https://creativecommons.org/licenses/by/4.0/)

(nocioceptors; Box 4.9 Pain in birds). Other receptors, called mechanoreceptors, are located in the dermis and include Herbst corpuscles, Grandry corpuscles, Merkel nerve endings, and Ruffini endings. These receptors respond to touch and pressure and are located primarily around feather follicles, in foot pads, and in the beak (especially the beaks of some waterfowl and shorebirds; Figs. 4.44 and 4.45). Herbst corpuscles are the most widely distributed receptors in bird skin and are particularly abundant in the bills of some birds, such as waterfowl (up to 140 per square millimeter; Ziolkowski et al. 2022) and shorebirds, and in the tongues of other birds, such as woodpeckers (Figs. 4.46 and 4.47). High densities of sensory pits that contain large numbers of Herbst corpuscles have been found at the tip of the bill of several sandpipers (*Calidris* spp.; Gerritsen and Meiboom 1986). These corpuscles detect changes in pressure and aid in the capture of infaunal prey (in the sand) (Box 4.10 Pressure sensory mechanism for prey

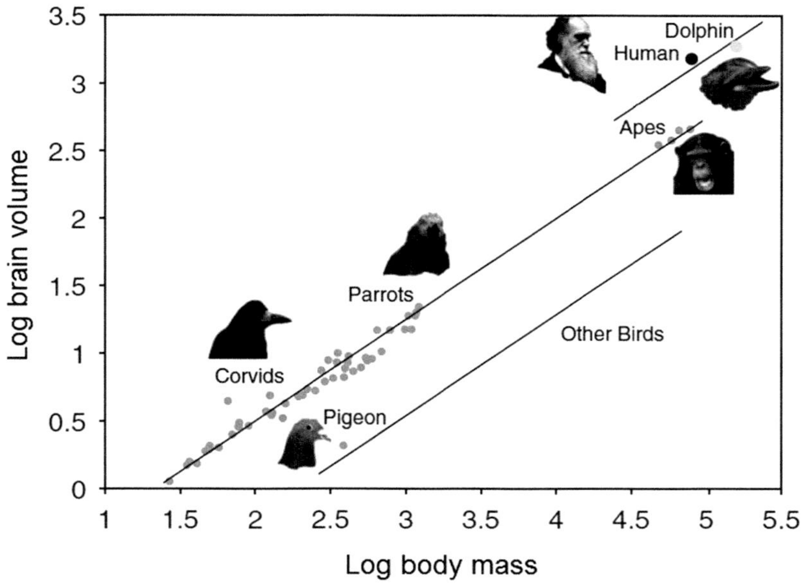


**Fig. 4.33** Relative brain size among birds and dolphins, apes, and humans. Among birds, parrots, and corvids have relatively larger brains than most other species. (Figure modified from van Horik and Emery 2011; © 2011 John Wiley & Sons, Ltd., used with permission)

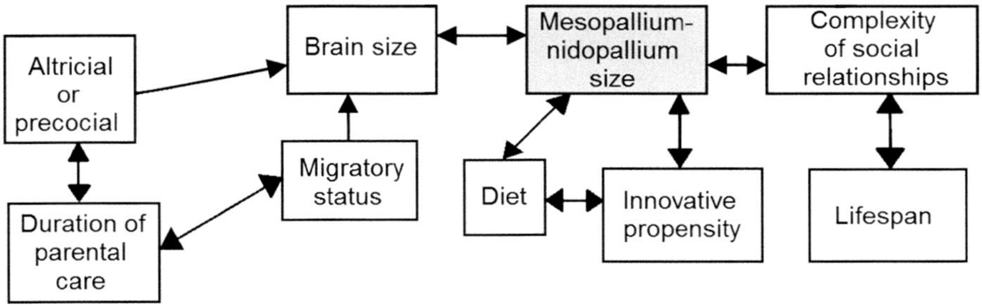


**Fig. 4.34** Multiple factors have likely interacted during the evolution of avian brains. Mode of development (altricial) appears to be a key driver in the evolution of larger brains, but other factors, including the complexity and duration of social relationships (that might be influenced by lifespan; longer-lived birds may develop more complex, long-lasting relationships), diet (omnivorous), and migratory status (residents or short-distance migrants exposed to greater environmental variation), also appear to have played an important role in the evolution of larger brains and, specifically, the areas most associated with cognitive ability—the mesopallium and nidopallium. With increasing cognitive abilities, innovations, such as use to tools, may enhance foraging efficiency and lead to improved diets. (Figure by G. Ritchison)

detection; Piersma et al. 1998). Similarly, Black-faced Spoonbills (*Platalea minor*) have high densities of sensory pits on the lateral and insides of their bills, aiding them in the detection of prey as they swing their bills back-and-forth through the water (Swennen and Yu 2004). Herbst corpuscles are also found in the in the legs of birds, and likely permit a bird to detect any vibration or movement of their perch. This may be helpful in predator detection (Dorward and McIntyre 1971; Shen and Xu 1994) or simply for maintaining balance on a moving perch. Herbst corpuscles in the wings of birds, particularly around the leading edge of the alula, may serve as airflow sensors that aid in determining air speed (Hörster 1990; Brown and Fedde 1993).

Additional touch/pressure receptors (Merkel cells) are found in the dermis (skin) of birds

**Fig. 4.35** Some typical sleeping postures of birds. (**a**) Barn Owl (*Tyto alba*; pxhere.com, CC0 Public Domain), (**b**) Domestic Chicken (*Gallus g. domesticus*; pxhere.com, CC0 Public Domain), (**c**) Greater Flamingo (*Phoenicopterus roseus*) with bill on its back (pxhere.com, CC0 Public Domain), (**d**) Gyrfalcon (*Falco rusticolus*). (Photo by Sheila Sund, pxhere.com, CC BY 2.0, https://creativecommons.org/licenses/by/2.0/), (**e**) Zebra Finches (*Taeniopygia guttata*; Photo by Ingrid Melichar, CC0 Public Domain, Pixabay), and (**f**) Mallard (*Anas platyrhynchos*) with bill under its scapulars (pxhere.com, CC0 Public Domain)

(Fig. 4.48). Only waterfowl have Grandry corpuscles and these are located in the dermis of the bill (Figs. 4.49 and 4.50). Merkel nerve endings are also found in the featherless skin of the legs of birds. The tarsometatarsal and digital regions of the skin have keratin scales and, below these scales, are Merkel nerve endings. Merkel cells are mechanoreceptors that respond to pressure applied to the beak or skin (Halata et al. 2003).

## 4.6 Olfaction

Olfaction, or the sense of smell, involves the detection of airborne chemical compounds and an increasing number of studies are revealing that olfaction plays an important role in the lives of birds. That role differs among species, but can include locating food, orientation, courtship, mate and offspring recognition, and locating nests or burrows. The olfactory system of birds includes the nares, or nostrils, and the nasal cavities (Fig. 4.51), and there is much variation among birds in the structure of each. Nares are, of course, usually open, but the nares of some birds in the order Pelicaniformes are covered by a secondary growth of the horny beak and breathe through a "secondary nares" at the angle of the mouth (Bang and Wenzel 1985). The two elongated nasal cavities are divided by a septum and each cavity is divided into three compartments. The first one (anterior concha) warms and humidifies inspired air, and the second (middle concha) is

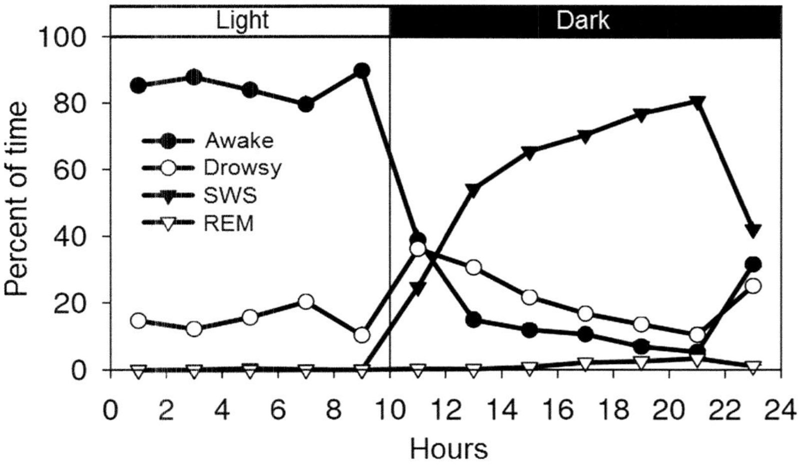


**Fig. 4.36** Relative amount of time in different states by White-crowned Sparrows (*Zonotrichia leucophrys*) during a 24-h period. Note the periods of drowsiness during the day as well as the periods of wakefulness during the night. White-crowned Sparrows, like other species of birds, spend much more time in slow-wave sleep (SWS) than in rapid-eye movement (REM) sleep. (Figure modified from Jones et al. 2008; © 2008 Springer Nature, used with permission)

lined with ciliated epithelial tissue that collects inhaled particles and prevents their passage into the lungs and air sacs. In several species of aquatic birds, including penguins, loons, grebes, waterfowl, and others, a nasal valve is present. This "valve" is a crescent-shaped fold of mucous membrane extending from the roof that is deflected when a bird is underwater and prevents water from passing further into the nasal cavities (Bang and Wenzel 1985).

The third compartment (caudal, or olfactory, concha) varies considerable among different birds in the amount of surface area and amount of olfactory epithelium (Fig. 4.52). The amount of surface area varies with the size of this compartment as well as the presence and arrangement of thin turbinate bones. More surface area and more olfactory epithelium means a better sense of smell. For example, Black Vultures (*Coragyps atratus*) have relatively simple caudal concha with limited surface area, whereas Turkey Vultures (*Cathartes aura*) have complex caudual concha with coiled turbinates and substantial surface area. These two scavengers also have different strategies for locating carrion, with Black Vultures more dependent, but not entirely, on vision, and Turkey Vultures primarily using their excellent sense of smell to locate carrion (Bang and Wenzel 1985). At short distances (≤30 m), Black Vultures can use their sense of smell to locate carrion, but Turkey Vultures can do so over much greater distances (Santos et al. 2023).

The olfactory epithelium is densely packed with chemoreceptors that transmit impulses via the olfactory nerve to the olfactory bulb (Figs. 4.53 and 4.54). For most species of birds, the number of receptors is not known. However, the number ranges from about 3 to 7.4 million in pigeons (Wenzel 1987), and from about 1 to 3.5 million in auklets (Hagelin 2007). Cells in the olfactory bulb (mitral cells) receive impulses from the chemoreceptors and the number of the cells varies among birds. For example, the olfactory bulb of fulmars has about 120,000 mitral cells, whereas pigeons have about 20,000 mitral cells. By comparison, the olfactory bulbs of rats and mice have about 60,000 and 20,000 mitral cells, respectively (Wenzel 1987). Impulses from the olfactory bulb then travel to the olfactory tubercle in the subpallial region of the brain where the perception of odors occurs.

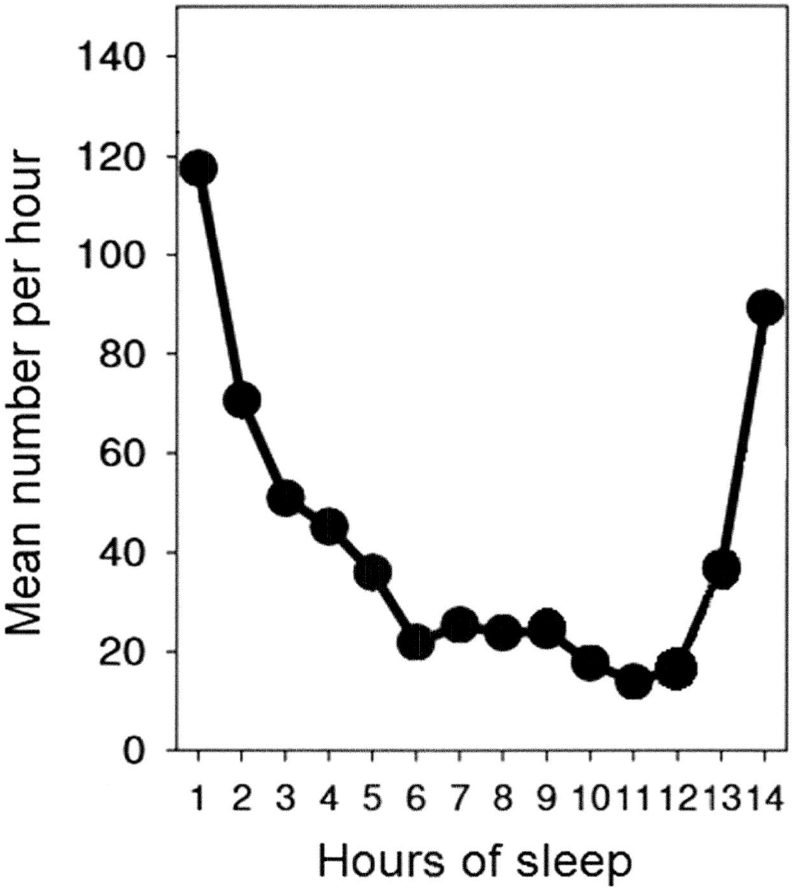


**Fig. 4.37** Mean number of times sleeping White-crowned Sparrows (*Zonotrichia leucophrys*) woke up while sleeping. Periods of wakefulness were brief (<16 s) and preceded and followed by at least one 4-s period of drowsiness or slow-wave sleep. (Figure modified from Jones et al. 2008; © 2008 Springer Nature, used with permission)

Among birds, the size of olfactory bulbs exhibits much variation (Corfield et al. 2015). Because birds vary so much in size, the size of the olfactory bulb relative to brain size (olfactory bulb ratio) is a better indicator of olfactory abilities. Bang and Cobb (1968) suggested that Snow Petrels (*Pagodroma nivea*) hade the largest olfactory bulb ratio (37%), and Black-capped Chickadees (*Poecile atricapillus*) had the smallest (3%). More recently, based on a comparison with 143 other species, Grigg et al. (2017) found that Turkey Vultures (*Cathartes aura*) had the largest olfactory bulbs relative to brain volume. In general, birds more dependent on olfaction have larger olfactory bulb ratios (Fig. 4.55). Among orders, those with the highest olfactory bulb ratios include Anseriformes, Apterygiformes (kiwis) and Procellariiformes, and among those with the lower ratios are Galliformes, Sphenisciformes, Strigiformes, Psittaciformes, Passeriformes, and Piciformes (Bang and Cobb 1968; Corfield et al. 2015). More generally, species with the highest olfactory bulb ratios tend to be ground-nesting, water-associated, colonial-breeding, and carnivorous or piscivorous birds (Avilés and Amo 2018; Fig. 4.56). One possible explanation for the tendency of aquatic birds to have larger olfactory bulbs than terrestrial species is that aquatic birds tend to have larger uropygial glands than terrestrial species, possibly due to the need to waterproof their plumage and/or protect it from microbial degradation resulting from humid conditions (Vincze et al. 2013). Because uropygial gland secretions are considered a main odor source in birds (Campagna et al. 2012; Caro et al. 2015), species living close to aquatic

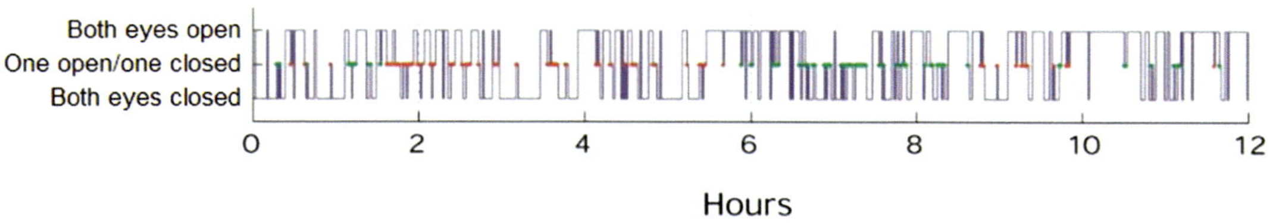


**Fig. 4.38** Changes in eye opening during a 12-h period of sleep by a Rock Pigeon (*Columba livia*). For nine pigeons, both eyes were open an average of 37% of the time, both were closed 31% of the time, and one eye was open and the other closed 32% of the time. (Figure from Rattenborg et al. 2001; © 2002 S. Karger AG, Basel, used with permission)

environments may be more likely to rely on olfaction than terrestrial species (Avilés and Amo 2018). Frugivorous and granivorous birds tend to have relatively small ratios (Bang and Wenzel 1985). Although larger ratios likely reflect greater olfactory acuity, even birds with much smaller ratios are able to detect odors. Larger olfactory bulbs and more olfactory epithelium may permit birds to detect a greater variety of odors (Adrian 1951) or increase sensitivity (permitting detection of odors at lower concentrations).

Given that all birds, to varying degrees, can perceive odors, how do they use that information? Numerous studies have revealed that birds may use olfactory information to locate food, nests and burrows (e.g., Bonadonna et al. 2003a, b), for orientation and navigation, avoid predators (e.g., Amo et al. 2008), and odors can also serve as signals that play an important role in reproduction. The importance of olfaction in finding food has been described in a number of species, most often kiwis, Turkey Vultures (*Cathartes aura*), and many seabirds, but a variety of other birds also exhibit some ability of locate food using their sense of smell, including Common Ravens (*Corvus corax*; Harriman and Berger 1986), Black-

**Fig. 4.39** Percent time spent both eyes closed and with one eye open and one closed by Mallards (*Anas platyrhynchos*) sleeping near the center of group and those sleep near the edge. During the remaining time, Mallards were awake with both eyes open. (Figure from Rattenborg et al. (1999b); © 1999 Elsevier Science B.V., used with permission)

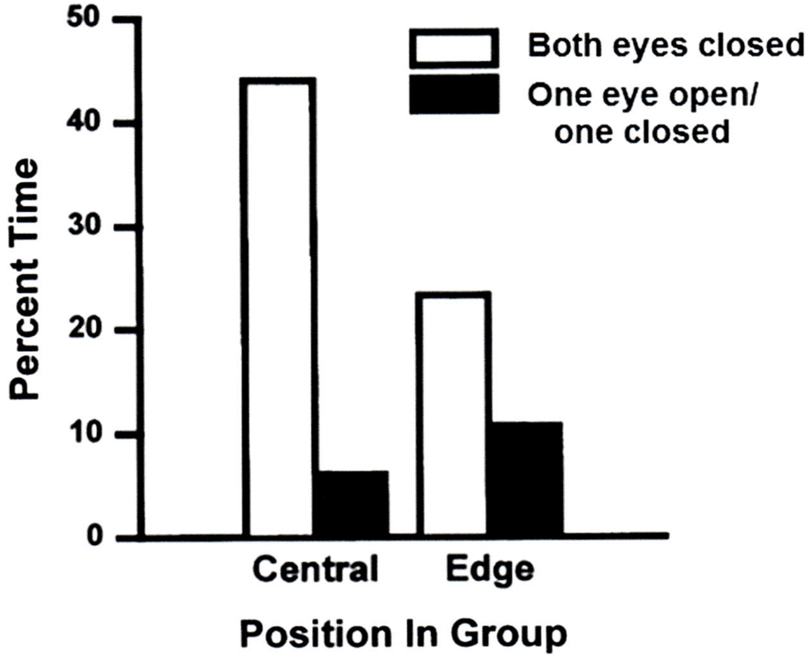

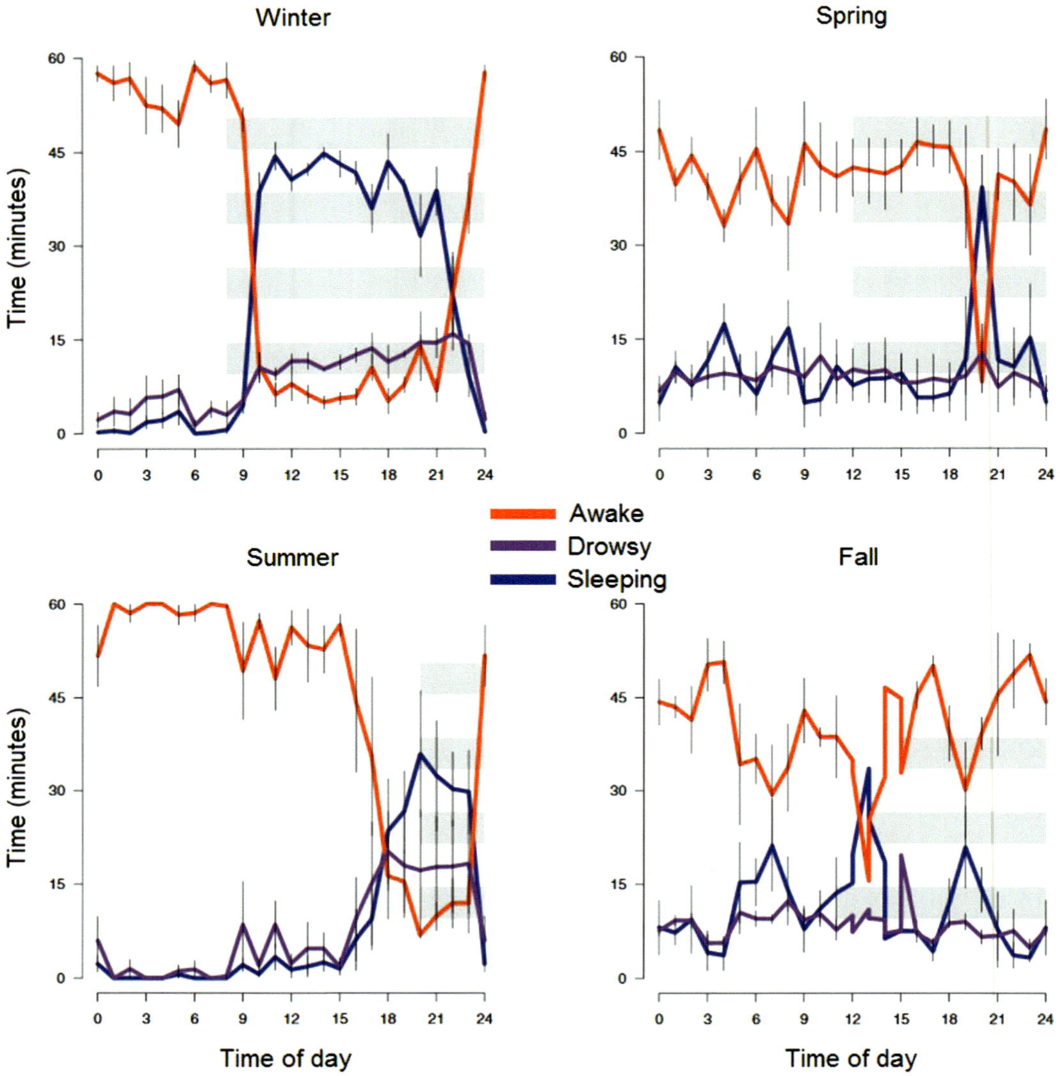


**Fig. 4.40** Seasonal variation in average time spent awake, drowsy, and sleeping by White-crowned Sparrows (*Zonotrichia leucophrys*). Values indicate mean number of minutes (± SE) in each state at 1-h intervals. Time 0 = lights on; shaded areas on the right = dark periods. In the fall, birds were not tested on the same day and so each shaded area represents the dark period for each of three birds. (Figure modified from Jones et al. 2010; Published under license to BioMed Central Ltd., open-access article distributed under the terms of the Creative Commons Attribution License, https://creativecommons.org/licenses/by/4.0/)

billed Magpies (*Pica hudsonia*; Buitron and Nuechterlein 1985; Molina-Morales et al. 2020), Kakapos (*Strigops habroptilus*; Hagelin 2004), Chattering Lories (*Lorius garrulous*; Roper 2003), Scarlet Macaws (*Ara macao*), Keel-billed Toucans (*Ramphastos sulfuratus*), and Yellow-throated Toucans (*Ramphastos ambiguus*) (Hernández et al. 2023), and, perhaps (because it is not clear whether they use olfaction, UV vision, or both), Willow Warblers (*Phylloscopus*

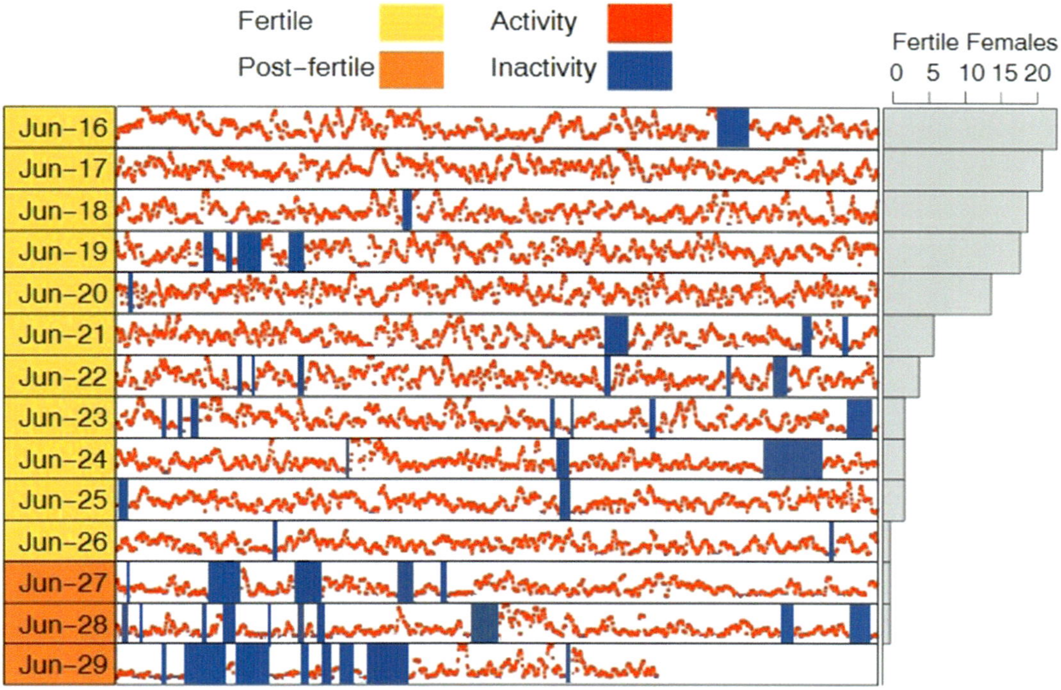


**Fig. 4.41** Activity patterns of male Pectoral Sandpipers (*Calidris melanotos*) during a portion of the breeding season. During the periods when most fertile females are present in the populations, males are rarely inactive (sleeping). The activity patterns of female Pectoral Sandpipers are similar to those of males. (Figure modified from Lesku et al. 2012; © 2012 American Association for the Advancement of Science, used with permission)

*trochilus*; Mäntylä et al. 2004). Some species of insectivorous birds have been found to use odors to locate prey. For example, Hitpold and Shriver (2018) found that significantly more imitation insects were attacked by birds when located close to dispensers releasing herbivore-induced plant volatiles (HIPV; chemicals released by plants being fed on by herbivores) than close to dispensers with non-HIPV chemicals. Similar results have been reported for Great Tits (*Parus major*) and Eurasian Blue Tits (*Cyanistes caeruleus*, Mäntylä et al. 2020). Graham et al. (2021) found that male Eurasian Blue Tits with higher plasma levels of testosterone were better at detecting HIPVs, suggesting that males may, as breeding seasons approach and testosterone levels increase, be able to identify areas (and potential breeding territories) with a greater abundance of prey. In addition, Wikelski et al. (2021) determined that White Storks (*Ciconia ciconia*) located mowed fields, where insect and rodent prey are most exposed, by the smell of volatile organic compounds released by freshly cut grass [(Z)-3-hexenal, (Z)-3-hexenol, hexenyl acetate]. These results suggest that birds can use plant volatiles as a cue to locate insect prey foraging on plants or locate areas that will provide good foraging opportunities. Other investigators have reported that some insectivorous species of birds can locate Lepidopteran prey using the pheromones emitted by females to attract males (Saavedra and Amo 2018; Amo and Saavedra 2021; Díaz-Siefer et al. 2021).

Some species of birds can detect the odors of potential competitors or predators and respond by exhibiting avoidance behavior. Kim et al. (2021) found that, in the absence of visual cues, Allen's (*Selasphorus sasin*), Anna's (*Calypte anna*),

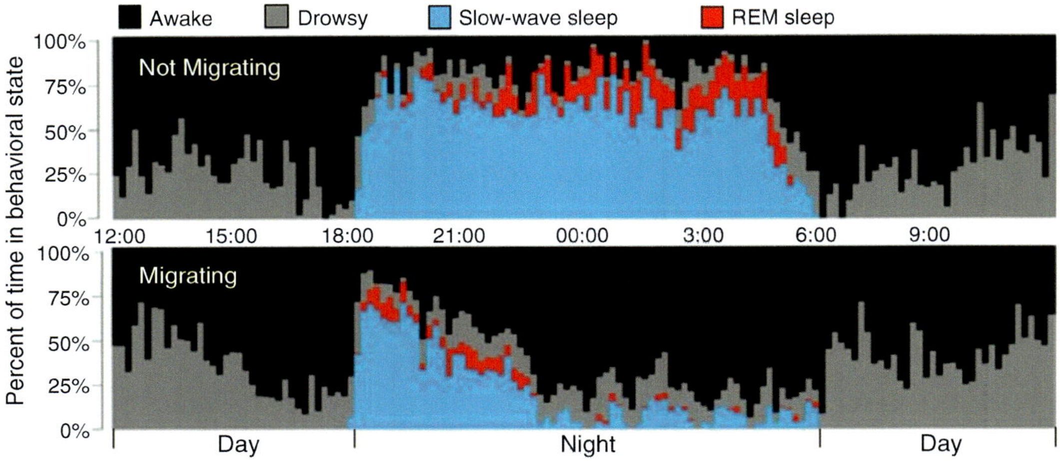


**Fig. 4.42** Proportion of time in each behavioral state (awake, drowsy, slow-wave sleep, and REM sleep) by non-migrating and migrating birds. Note that migrating birds spend much less time in slow-wave and REM sleep. (Figure modified from Rattenborg et al. 2004; © 2004 Rattenborg et al., open-access article distributed under the terms of the Creative Commons Attribution License, https://creativecommons.org/licenses/by/4.0/)

Costa's (*Calypte costae*), and Black-chinned (*Archilochus alexandri*) hummingbirds avoided foraging at feeders where the areas around feeding ports (but not the sugar water) had been treated with formic acid, a defensive chemical found in the venoms of many Formicine ants. This avoidance behavior suggests that these hummingbirds use olfactory cues to avoid ant competitors at nectar sources. To examine the possibility that olfactory cues might be used to avoid predation, Amo et al. (2008) placed absorbent papers with the scents of a potential predator (ferret, *Mustelo furo*), a non-predator (Japanese Quail, *Coturnix japonica*), and a control (water) in the nest boxes of Eurasian Blue Tits (*Cyanistes caeruleus*) and found that, compared with the non-predator and control, the birds either delayed their entry into nest boxes or refused to enter nest boxes more often when exposed to the scent of a predator (ferret). The Blue Tits also spent less time in nest boxes with the predator scent when provisioning nestlings. Similar avoidance behavior in response to the odors of potential predators has been described in several species of ducks (Eichholz et al. 2012), House Finches (*Carpodacus mexicanus*, Amo et al. 2015), Great Tits (*Parus major*, Amo et al. 2011), Red-legged Partridges (*Alectoris rufa*; Mahr and Hoi 2018), and Crested Auklets (*Aethia cristatella*, Hagelin et al. 2003), but no such responses were found for Eastern Bluebirds (*Sialia sialis*, Godard et al. 2007; Stanback et al. 2019), House Wrens (*Troglodytes aedon*, Johnson et al. 2011), or European Starlings (*Sturnus vulgaris*; Blackwell et al. 2020). These mixed results suggest that, although some species of birds can use odors to detect and avoid predators, others do not.

Seabirds in the order Procellariiformes (petrels, albatrosses, and shearwaters) are well-known for their generally excellent sense of smell. These birds routinely travel long distances over the world's oceans in search of ephemeral prey. When foraging, procellariiform seabirds are often observed flying crosswind, a behavior that maximizes the chance of detecting odor plumes; specifically, detecting a substance called dimethylsulphide (DMS; Fig. 4.57). DMS is produced by bacteria that metabolize decaying plankton and marine plants and by phytoplankton in response to zooplankton grazing. Where there is more plankton, there is more DMS and more potential prey for seabirds. Once a plume of DMS is detected, seabirds can then "track" it back to its

**Fig. 4.43** Drawing of a Great Frigatebird (*Fregata magnificens*) equipped with a head-mounted data logger to record electroencephalograms from both cerebral hemispheres. While soaring, frigatebirds sleep asymmetrically, with one eye closed and one hemisphere asleep while the other eye is open and the other hemisphere is awake. (Figure from Rattenborg 2017; © 2016 The Author, used with permission)

source. Because these "sources" can vary in area from a few square kilometers to hundreds of square kilometers, seabirds must use their sense of smell, plus visual cues (e.g., other foraging seabirds or marine mammals), to pinpoint a specific foraging area. At this smaller scale, seabirds may use odors associated with certain types of prey (e.g., fish) to locate the best foraging locations (Nevitt 2006).

Some petrels and albatrosses also appear to use olfactory navigation. This is possible because certain features, such as shelf breaks (at the edge of continental shelfs), upwelling zones, or seamounts (undersea mountain), and their associated life forms might create an "olfactory landscape" that, along with other features, like islands, could be learned by long-lived birds (like many seabirds) and used for navigational purposes (Bonadonna et al. 2003a, b). Similarly, homing pigeons are apparently able to use olfactory cues to find their way home from unfamiliar sites, as demonstrated by experiments where cutting the olfactory nerve greatly impaired their homing performance (Gagliardo et al. 2006) (see additional discussion of olfactory navigation in Chap. 14).

For several species of birds, olfaction plays a role in reproduction. The odors involved come from a variety of sources, including the uropygial gland, stomach oils, feces, and the plumage, and may be used for individual, sexual, and species recognition as well as for signals of quality or compatibility (Bonadonna et al. 2007). For example, several species of nocturnal, burrow-nesting petrels use odors in the secretions of their uropygial glands to identify their burrows

**Box 4.9 Pain in birds**

Three types of nociceptors, or pain receptors, have been identified in birds, including mechanothermal nociceptors, mechanical nociceptors, and thermal nociceptors (Gentle 1992; Paul-Murphy and Hawkins 2015). These pain receptors are found in the skin, muscle, joints, bones, and viscera of birds. A common definition of pain used for animals is that it is "an aversive sensory experience caused by actual or potential injury that elicits protective and vegetative reactions, results in learned behaviour, and may modify species specific behaviour" (Zimmerman 1986). Selection has likely favored a reduced tendency by birds to respond to pain because doing so might attract the attention of predators (Livingston 1994) or signal inferiority in interactions with conspecifics (Svärdson 1949). However, painful stimuli can generate a variety of physiological responses, such as an increased heart rate. Birds have also been found to respond behaviorally to noxious stimuli, e.g., moving a foot in response to high temperatures. Mazor-Thomas et al. (2014) examined the behaviors of Red-tailed Hawks (*Buteo jamaicensis*) suffering from orthopedic trauma and those that had healed from more minor injuries and found that those with orthopedic trauma exhibited less movement and head motion and produced fewer bill clacks.

Nociception, the capacity to respond to potentially damaging stimuli, allows animals to avoid injurious stimuli and protect themselves from further damage (Bateson 1991). In contrast to other senses that provide information, pain is an unpleasant sensory and emotional experience associated with actual or potential tissue damage (Council 2010). Pain is difficult to identify and quantify in birds and other animals because they cannot explain or describe their discomfort (Douglas et al. 2018). However, general areas in the avian cerebrum involved in the processing of persistent painful stimuli have been identified, suggesting that pain is a cognitive process for birds, just as it is for humans (Douglas et al. 2018).

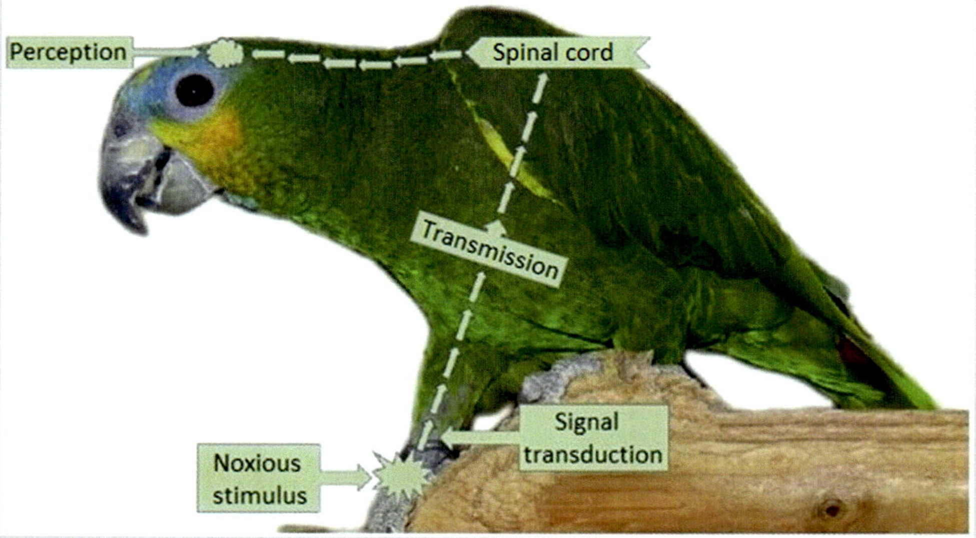


The ascending pain pathway of birds. A noxious stimulus causes signal transduction (conversion of the stimulus to a nervous impulse by the nociceptor, i.e., pain receptor). The impulse is transmitted along sensory neurons to the spinal cord and neurons in the spinal cord then conduct impulses to the brain. In the brain, nociceptive signals are processed for cognitive and emotional perception (Figure modified from Douglas et al. 2018; © 2017 Elsevier Inc., used with permission)

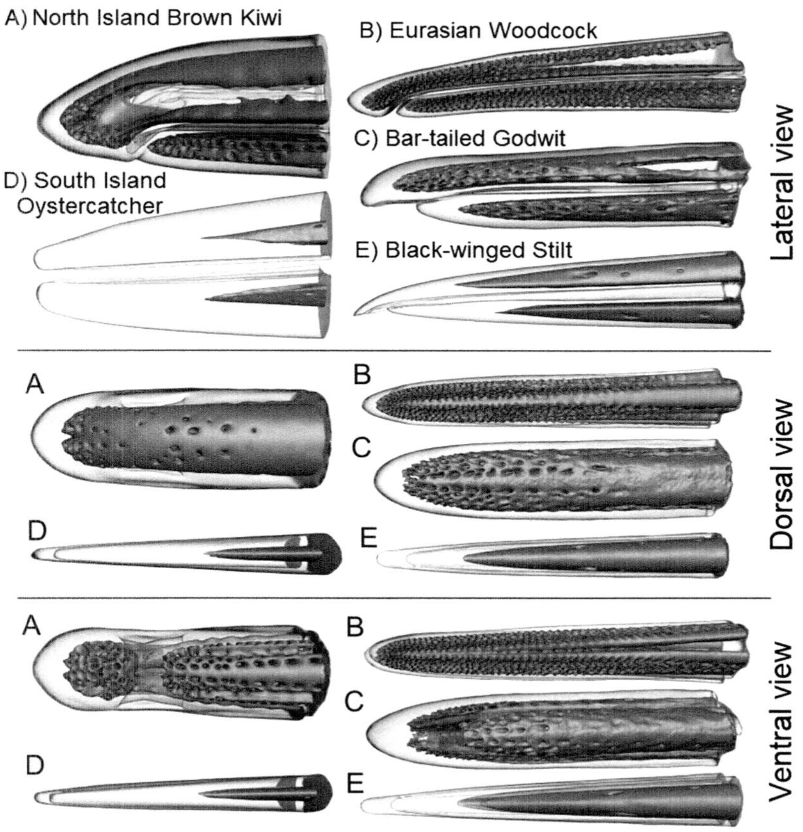


**Fig. 4.44** Sensory pits in the bills of a (**a**) North Island Brown Kiwi (*Apteryx mantelli*), (**b**) Eurasian Woodcock (*Scolopax rusticola*), and (**c**) Bar-tailed Godwit (*Limosa lapponica*), species that utilize tactile-based foraging. Bills of a (**d**) South Island Oystercatcher (*Haematopus finschi*) and (**e**) Black-winged Stilt (*Himantopus himantopus*), species that do not utilize tactile-based foraging and, therefore, have few sensory pits. Herbst corpuscles are found in high densities in these sensory pits. The dark gray structures represent bone and the transparent structure the keratin. (Figure modified from Cunningham et al. 2013; © 2013 Cunningham et al., open-access article distributed under the terms of the Creative Commons Attribution License, https://creativecommons.org/licenses/by/4.0/)

(Bonadonna and Bretagnolle 2002). Young European Storm-Petrels (*Hydrobates pelagicus*) also appear to use odor to identify their own burrows (Hagelin and Jones 2007).

Male Mallards (*Anas platyrhynchos*) whose olfactory nerves were cut exhibited less sexual and social behavior than control males with intact nerves (Balthazart and Schoffeniels 1979), suggesting that the ability to sense the odors of females is needed for normal courtship and mating behavior. In tests with individuals from a hybrid zone in the eastern United States, Van Huynh and Rice (2019) found that female Black-capped (*Poecile atricapillus*) and Carolina (*P. carolinensis*) chickadees exhibited significant preferences for conspecific over heterospecific uropygial gland odor cues in Y-maze experiments, suggesting that odors may serve as a mating cue for these species. Similarly, Antarctic Prions (*Pachyptila desolata*) tested in a Y-maze apparatus and exposed to the odors of their mate and an unknown conspecific were attracted to the odor of their mate. In addition, Crested Auklets (*Aethia cristatella*) produce a citrusy plumage odor that, particularly during the breeding season, may, along with visual displays, play a role in courtship and mate choice (Hagelin and Jones 2007). More generally, Whitaker and Hagelin (2021) noted that, of 30 species where investigators have compared the uropygial glands of males and females, females had larger uropygial glands in 23 of those species. Based on this female-bias, Whitaker and Hagelin (2021) proposed that volatile or semi-volatile compounds produced by female uropygial glands may advertise female quality or receptivity and have a stimulatory effect on males.

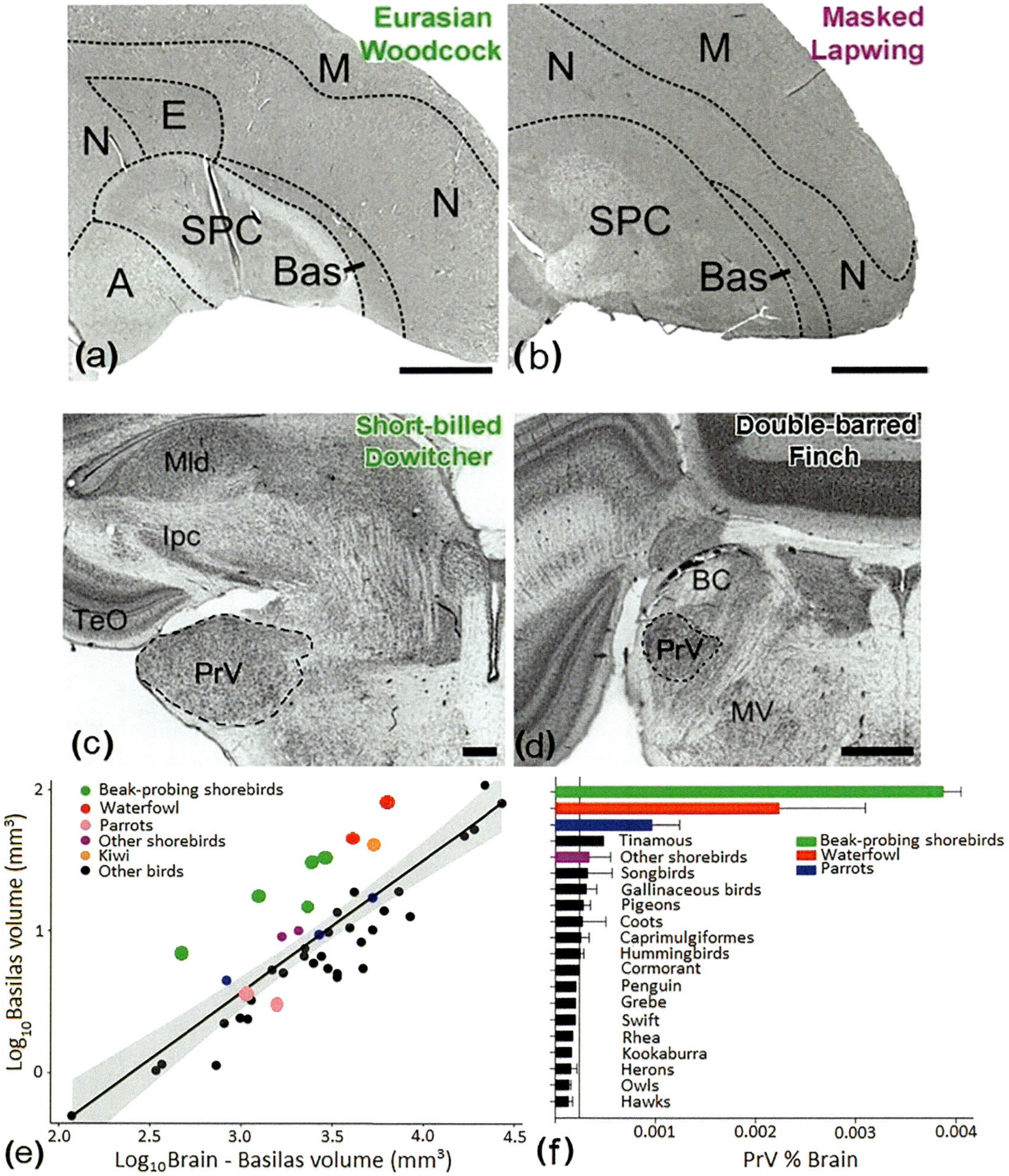


**Fig. 4.45** Tactile areas of the brains of birds that utilize tactile-based foraging behavior, including beak-probing shorebirds, some waterfowl, kiwis, and parrots, are relatively larger than those of birds that do not exhibit such foraging behavior. An area of the brain called the basorostralis (Bas) is critical for processing somatosensory information from the beak tip and is larger in species that use tactile-based foraging like (**a**) Eurasian Woodcocks (*Scolopax rusticola*) than in species like (**b**) Masked Lapwings (*Vanellus miles*) that do not use tactile-based foraging. Similarly, somatosensory information from the beak is transmitted to the principle sensory nucleus of the trigeminal nerve (PrV), and the PrV is larger in beak-probing birds like (**c**) Short-billed Dowitchers (*Limnodromus griseus*) than non-beak-probing birds like (**d**) Double-barred Finches (*Taeniopygia bichenovii*). (**e**) Volume of the nucleus basalis as a function of brain volume minus nucleus volume for beak-probing shorebirds (Family Scolopacidae), waterfowl, a parrot, other shorebirds (i.e., Charadriiformes other than

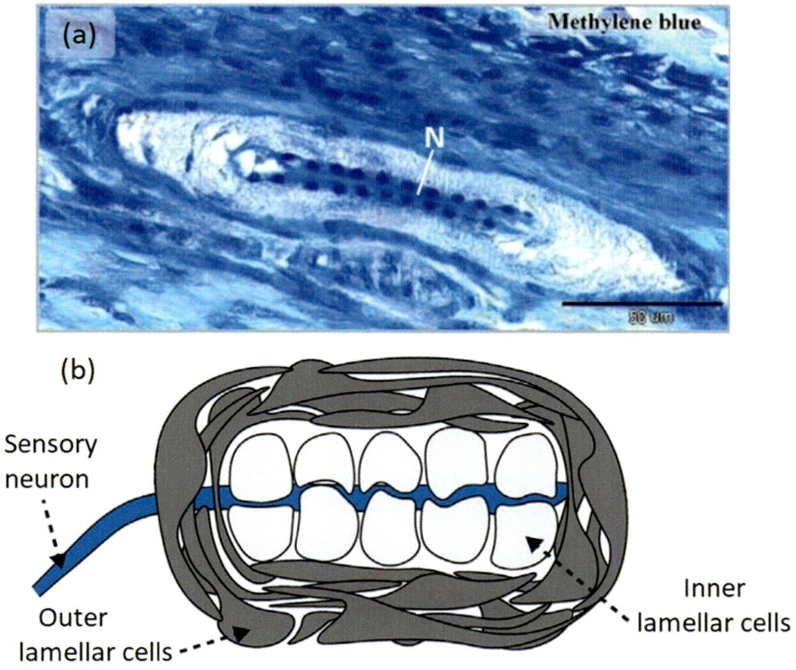


**Fig. 4.46** (**a**) Herbst corpuscle from a duck's bill (stained with methylene blue). N, neuron. Scale bar = 50 μm. (**b**) Drawing of a Herbst corpuscle. Herbst corpuscles consist of an outer capsule formed by outer core lamellar cells that enclose an inner core consisting of inner core lamellar cells. The inner core lamellar cells surround the sensory neuron. Lamellar cells are non-neuronal mechanosensors able to generate action potentials and may increase the sensitivity of Herbst corpuscles to pressure and vibration (Nikolaev et al. 2020). (Figure **a** from Soliman and Madkour 2017; © 2017 Soliman SA, open-access article distributed under the terms of the Creative Commons Attribution License, https://creativecommons.org/licenses/by/4.0/. Figure **b** from Ziolkowski et al. 2022; © 2022 Elsevier Ltd. All rights reserved, used with permission)

Prions given a choice between their own odor and that of an unknown conspecific avoided their self-odor (Bonadonna and Nevitt 2004). Such avoidance of self-odor suggests that, as in other vertebrates, prions, and perhaps other birds, might recognize kin by chemical profiles mediated by the major histocompatibility complex (MHC; highly polymorphic genes associated with immune function that likely influence other phenotypic traits, such as odor and condition; Zelano

**Fig. 4.45** (continued) Scolopacidae), a kiwi, and other birds. The solid line represents the least squares linear regression and the shaded region is the 95% confidence interval. (**f**) PrV volume expressed as a percentage of brain size for beak-probing shorebirds, waterfowl, parrots, other shorebirds, and other birds. Error bars represent standard deviations and the solid vertical line indicates the mean for all nonspecialists. Other abbreviations: A, arcopallium; BC, brachium conjunctivum; E, entopallium; lpc, nucleus isthmi parvocellularis; M, mesopallium; Mid, nucleus mesencephalicus lateralis pars dorsalis; MV, motor nucleus of the trigeminal nerve; N, nidopallium; SPC, striatopallidal complex; TeO, optic tectum. Scale bars = 2 mm in **a** and **b**; 600 μm in **c** and **d**. (Figure from Iwaniuk and Wylie 2020; © 2020 Wiley Periodicals, Inc., used with permission)

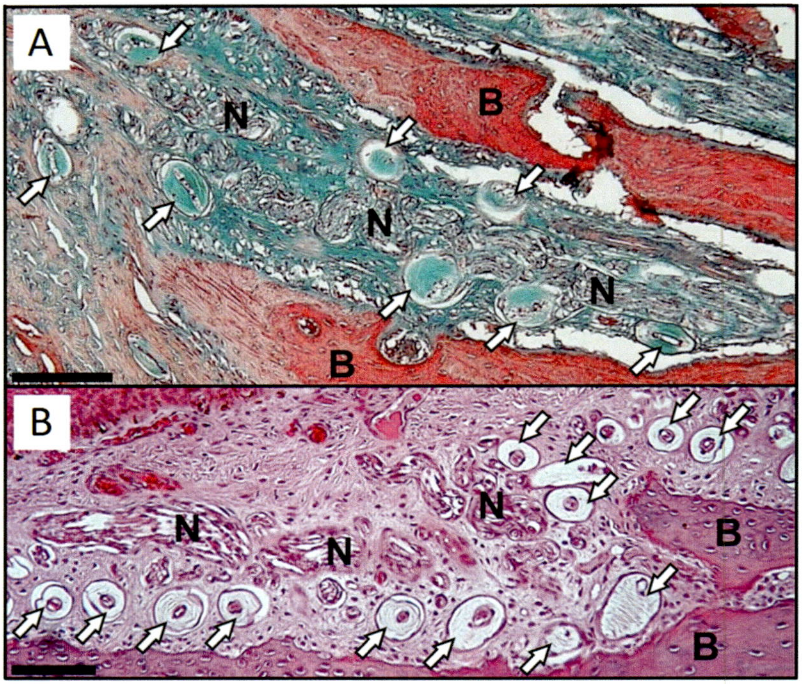


**Fig. 4.47** (**a**) Sagittal section through a sensory pit (stained) in the premaxilla bone of a North Island Brown Kiwi (*Apteryx mantelli*). (**b**) Sensory pit (stained) in the dentary bone of a Bar-tailed Godwit (*Limosa lapponica*). N, nerves; B, bone. White arrows are pointing to Herbst corpuscles. Scale bars = 100 μm. (Figure from Cunningham et al. 2013; © 2013 Cunningham et al., open-access article distributed under the terms of the Creative Commons Attribution License, https://creativecommons.org/licenses/by/4.0/)

and Edwards 2002). Such recognition remains to be tested, but recognition of kin either by odor or other cues could be important in inbreeding avoidance (Hagelin and Jones 2007). In addition, the relationship between MHC, odor, and phenotype suggests that odor could play a role in mate-choice decisions.

Kiwis (Apterygidae) are nocturnal, flightless, cursorial birds that forage mainly for soil and surface-dwelling invertebrates. They exhibit minimal reliance on vision as indicated by eye structure and brain structures and increased reliance on olfactory information. This lack of reliance on vision and increased reliance on olfactory information by kiwis is apparent based on the size of their olfactory bulbs (Fig. 4.58). This reliance on olfaction is similar to that of nocturnal mammals that exploit the forest floor. That kiwis and mammals evolved to exploit these habitats independently provides evidence for convergent evolution in their sensory capacities that are tuned to a common set of perceptual challenges found in forest floor habitats at night that cannot be met by the vertebrate visual system.

Kiwis are unique among birds in having the opening of their nostrils close to the tip of the maxilla (Fig. 4.59). In all other birds, the nostrils open externally close to the base of the bill, or internally in the roof of the mouth. Martin et al. (2007) provided evidence that kiwi bill tips are the focus of both olfactory and tactile information. Inspection of prepared skulls shows that clustered around the tips of both the maxilla and mandible, on both internal and external surfaces, is a high concentration of sensory pits. Such pits house clusters of mechanoreceptors (Herbst and Grandry corpuscles) protected by a soft rhamphotheca. These sensory pits function in foraging to detect objects touching or close to the bill tips. In kiwis, the sensory pits cover the entire surface of the tip of the maxilla and almost encircle the nostrils that open laterally about 3 mm behind the bill tip, suggesting that the bill tip is a focus for gaining both tactile and olfactory information for guiding the bill when foraging.

**Box 4.10 Pressure sensory mechanism for prey detection**

Red Knots (*Calidris canutus*) can locate their favorite food (shellfish) in wet sand by inserting their beak half a centimeter into the sand for a few seconds (Piersma et al. 1998). This ability was demonstrated in experiments where small stones were hid in sand. Because stones do not send out any signals, the ability of Red Knots to detect them must be based on the sensitivity of their beaks to differences in currents in the water in wet sand between the individual grains, stones, or shells. Red Knots used in the experiments were unable to find hidden stones in dry sand. At the end of their beak, knots have clusters of 10–20 Herbst corpuscles that are sensitive to differences in pressure. Sticking a beak into wet sand creates a pressure wave because of the inertia of the water in the interstices between sand particles. The pattern thus created betrays the presence of objects larger than the grains of sand. The rapid up-and-down movements of the bird's beak loosen the grains of sand, which then become packed together more tightly, displace the interstitial water, and cause the residual pressure around the object to increase. Other birds that probe wet substrates for prey, e.g., ibises, also have high concentrations of sensory receptors at the tip of their bills and likely locate prey in a similar fashion.

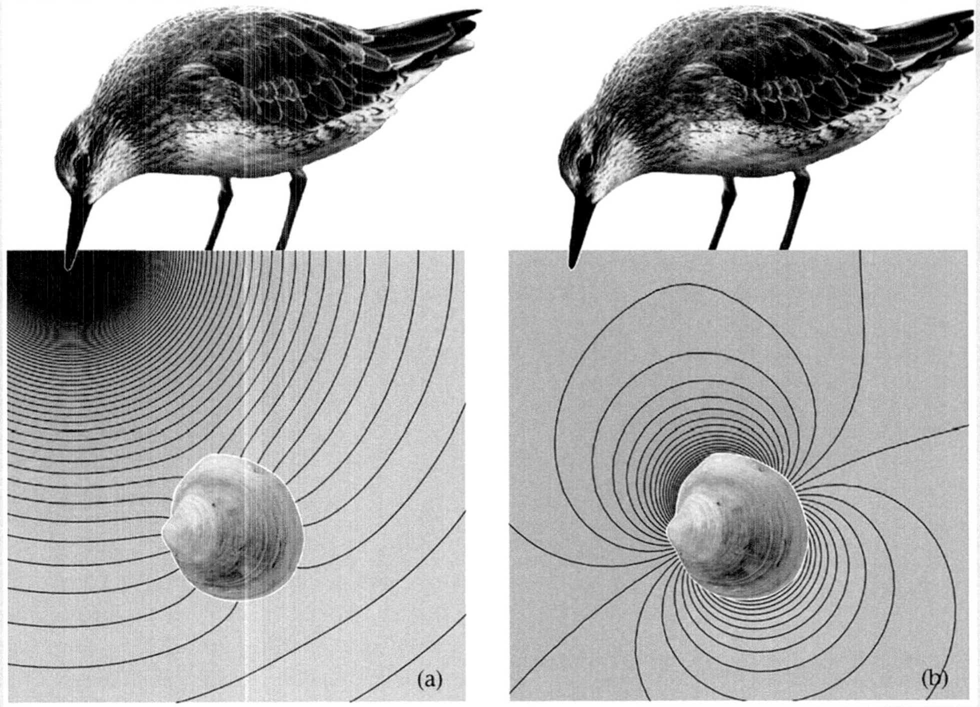


(**a**) Pressure waves produced by a shorebird's bill probing into sand. (**b**) Resulting pressure waves caused by potential prey item in the sand. Red Knots, with pressure receptors located in the bill tip, should be able to detect these pressure waves and determine the location of the possible prey item (Figure from de Fouw et al. 2016; © 2016 The Association for the Study of Animal Behaviour. Published by Elsevier Ltd., used with permission)

(continued)

**Box 4.10** (continued)

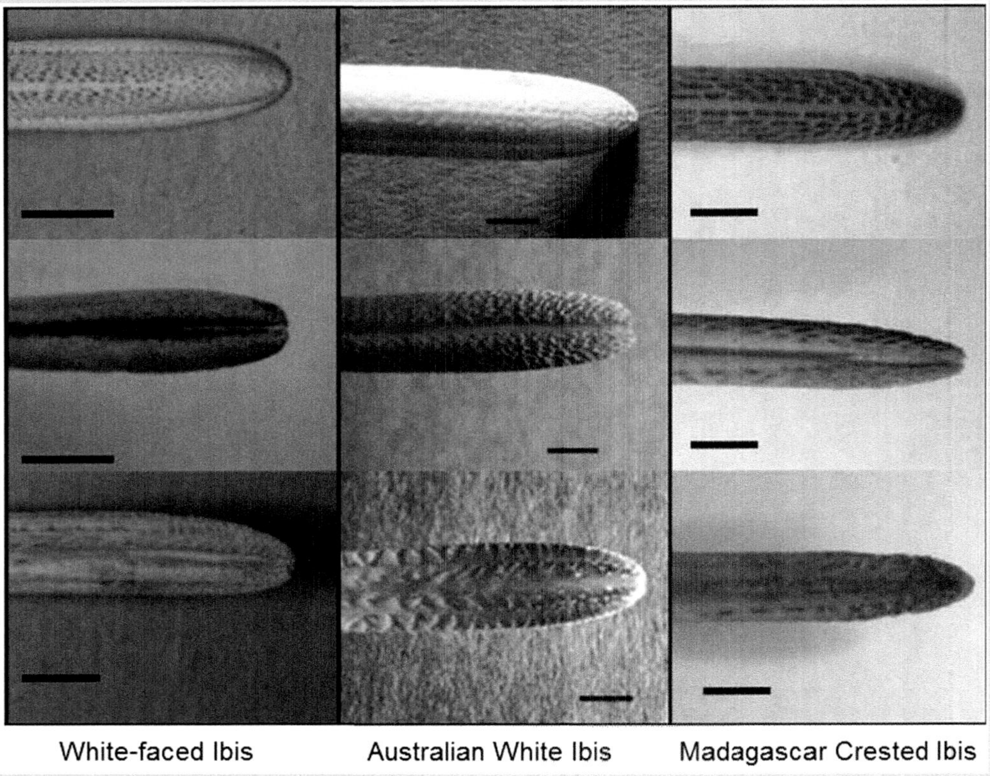


Sensory pits in the bills of three ibis species (top to bottom: dorsal tip of premaxilla, ventral tip of mandible, and ventral tip of premaxilla). White-faced Ibis, *Plegadis chihi*; Australian White Ibis, *Threskiornis moluccus*; Madagascar Crested Ibis, *Lophotibis cristata* (Images from Cunningham et al. 2010; © 2010 Oxford University Press, used with permission)

## 4.7 Taste

Taste is the ability to detect dissolved ions and molecules. The sense of taste can provide birds with important dietary information. Among vertebrates, there are five basic tastes: sweet, umami (a savory, meaty taste), bitter, salty, and sour. Sweet and umami tastes allow recognition of liquids and foods with nutritious carbohydrates, amino acids, and proteins, bitter and sour tastes may indicate potentially harmful foods, and the salty taste allows detecting sodium and other minerals needed to varying degrees by different species (Zhao et al. 2015). Surprisingly, birds have apparently lost a gene responsible for perception of one of these basic vertebrate tastes—sweet (Zhao et al. 2010; but see Box 4.11 Sugar tastes sweet to hummingbirds and songbirds). However, experimental studies have revealed that many species of birds, particularly frugivores and nectarivores, are able to detect differences in solutions with different concentrations of sugars. Given these results,

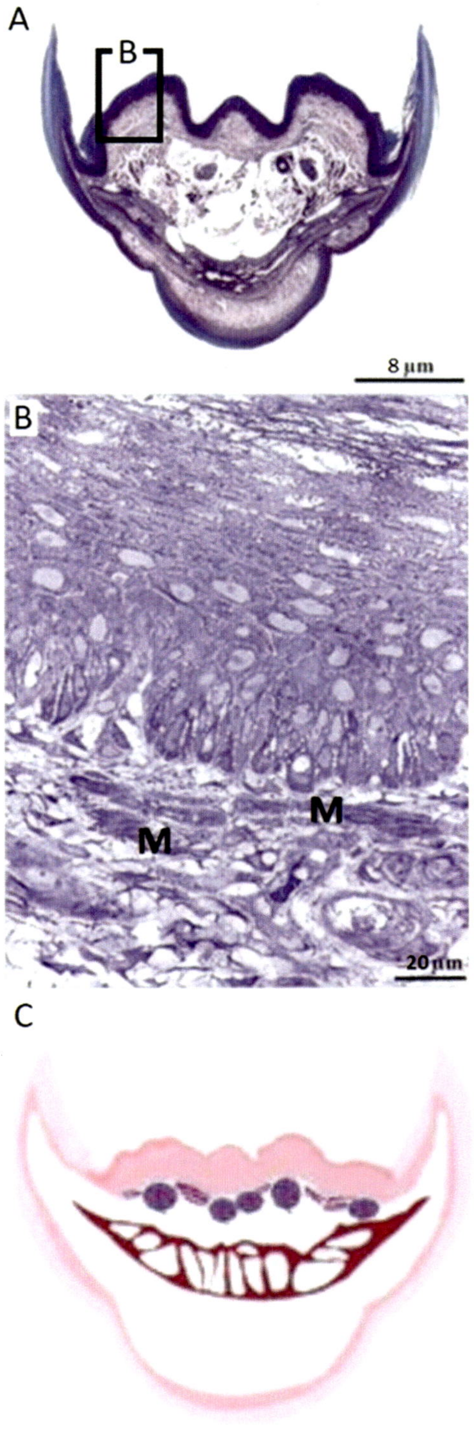


**Fig. 4.48** (**a**) Section through the anterior section of the lower bill of a Japanese Quail (*Coturnix japonica*). (**b**) Merkel cells (M) are located just below the mucosal epithelium. (**c**) A series of Merkel cells distributed just below the mucosal epithelium. (Figure from Soliman 2019; open-access article distributed under the terms of the Creative Commons Attribution License, https://creativecommons.org/licenses/by/4.0/)

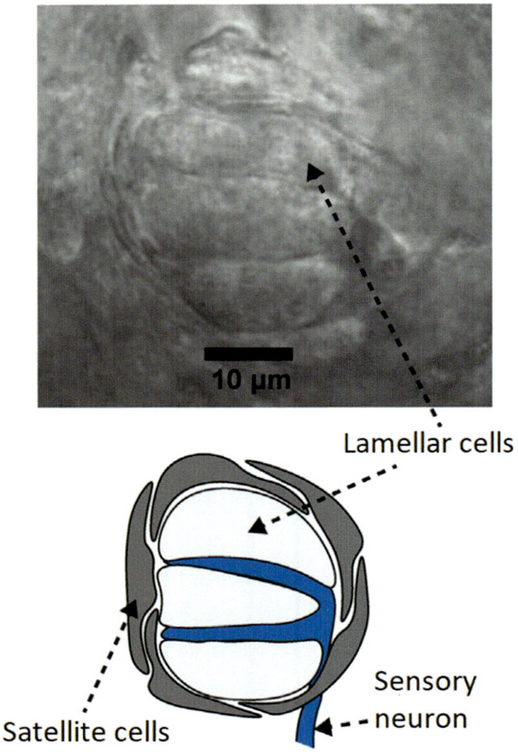


**Fig. 4.49** Image and diagram of a Grandry corpuscle. Grandry corpuscles consist of 2–12 lamellar cells located above and below the terminals of a sensory neuron. Satellite cells surround the entire structure. (Figure from Ziolkowski et al. 2022; © 2022 Elsevier Ltd. All rights reserved, used with permission)

Clark et al. (2015) suggested that other receptors may exist that permit birds to perceive the sweet taste (as is the case for hummingbirds and songbirds; Figs. 4.60 and 4.61). Among some taxa of birds, other tastes have been lost, e.g., penguins have also lost the sweet, umami, and bitter tastes (Zhao et al. 2015; Fig. 4.61). The loss of these tastes by penguins is likely a result of their carnivorous diet and habit of swallowing prey whole (Zhao et al. 2015).

Birds, like other vertebrates, detect taste with taste buds located primarily on the posterior tongue and pharynx as well as the upper palate and base of the tongue (Niknafs and Roura 2018; Fig. 4.62). The taste buds of birds are often associated with salivary ducts, which is important because saliva transports taste stimuli to the receptors (Mason and Clark 2000; Fig. 4.63). Compared to other vertebrates, birds have few taste buds, with most apparently having about 75–125. The greatest number of taste buds reported for a bird is about 350–400 in Mallards (*Anas platyrhynchos*) and 300–400 in parrots (Berkhoudt 1985; Kare and Mason 1986). Domestic Chickens (*Gallus g. domesticus*) have between 240 and 360 taste buds, depending on the breed (Kudo et al. 2008). On the low end, Eurasian Blue Tits (*Cyanistes caeruleus*) have only 20–25 taste buds and Eurasian Bullfinches (*Pyrrhula pyrrhula*) only 46 (Gentle 1975). By comparison, humans have about 9000 taste buds.

Avian taste birds are generally ovoid in shape and consist of several types of cells. Follicular, or sheath, cells (not found in the taste buds of mammals) separate the inner sensory portion of taste buds from the surrounding epithelial tissue. Dark cells (so named because they appear dark when stained) are scattered throughout the interior of taste buds and are generally considered to be supporting cells. Light cells (so named because they resist staining) are also found along the length of taste buds, are innervated by sensory neurons, and are the main sensory cells of taste buds (Northcutt 2004). Unlike the taste buds of mammals, some avian taste buds also have taste canals (also called tubules or channels) that may extend to the basal region (Figs. 4.64 and 4.65), perhaps enhancing contact between dissolved molecules and the taste receptors.

Despite the limited number of taste buds, most birds use their sense of taste, to varying degrees, to determine whether to accept or avoid food and liquids. For example, nectar-feeding birds, like hummingbirds, sunbirds, and honeyeaters, can distinguish among solutions with different sugar concentrations and different types of sugars. Rufous Hummingbirds (*Selasphorus rufus*) have been found to discriminate between solutions where sugar content varied by just 1% (Blem et al. 2000). Some fruit-eating birds can also discriminate between solutions that differ in sugar content. For example, four species of tanagers were found to prefer a 12% glucose-sucrose solution over 8% and 2.5% solutions (Schaefer et al. 2003). Two shorebirds,

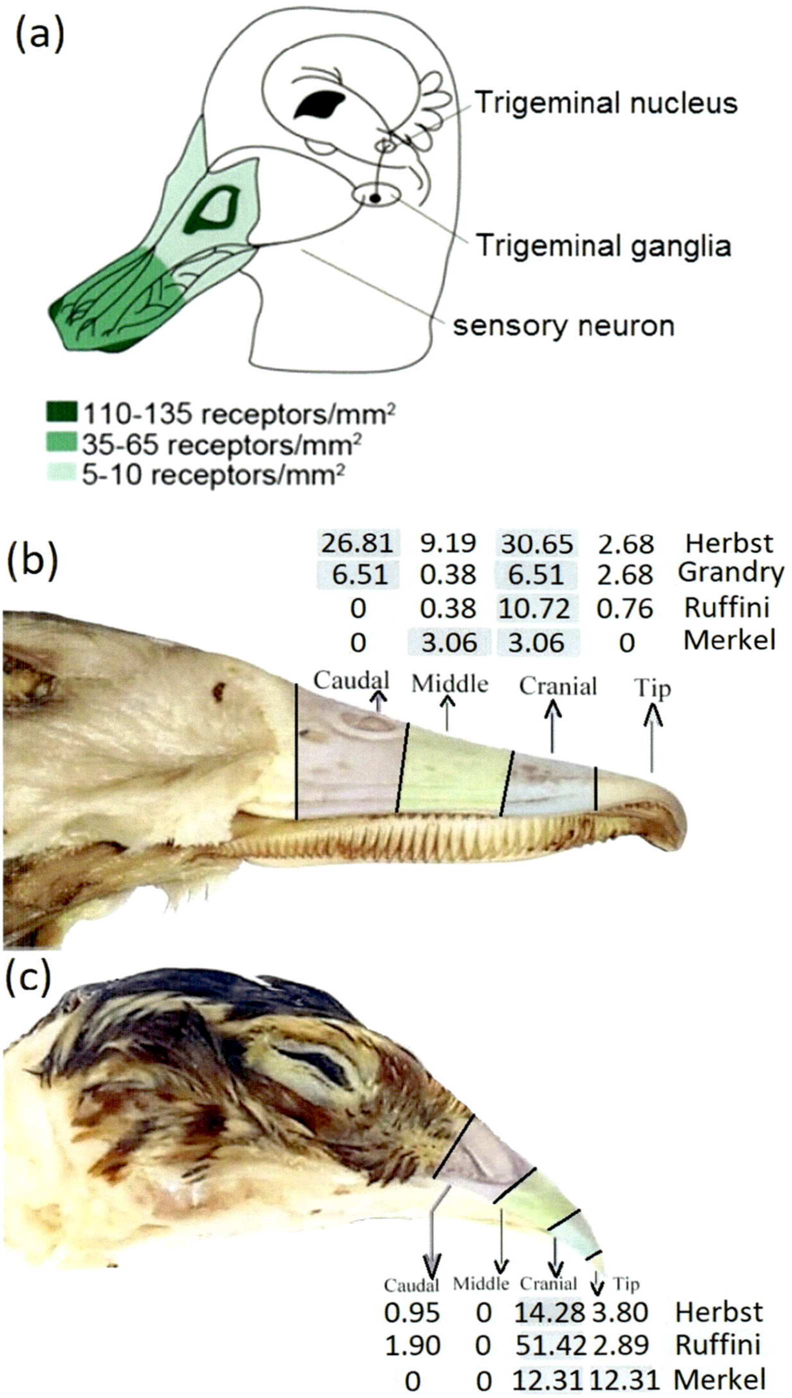


**Fig. 4.50** (**a**) Concentrations of touch receptors in the bill of a "tactile foraging" duck. The cell bodies of these neurons are located in the trigeminal ganglia and the sensory information is processed in the trigeminal nucleus in the brainstem. (**b**) Percentage of each type of receptor in the bill of a Muscovy Duck (*Cairina moschata*). (**c**) Percentage of each type of receptor in the bill of a Japanese Quail (*Coturnix japonica*). (Figure **a** modified from Schneider et al. 2014, used with permission of the U. S. National Academy of Sciences. Figures **b** and **c** from Soliman and Madkour 2017; ©2017 Soliman SA, open-access article distributed under the terms of the Creative Commons Attribution License, https://creativecommons.org/licenses/by/4.0/)

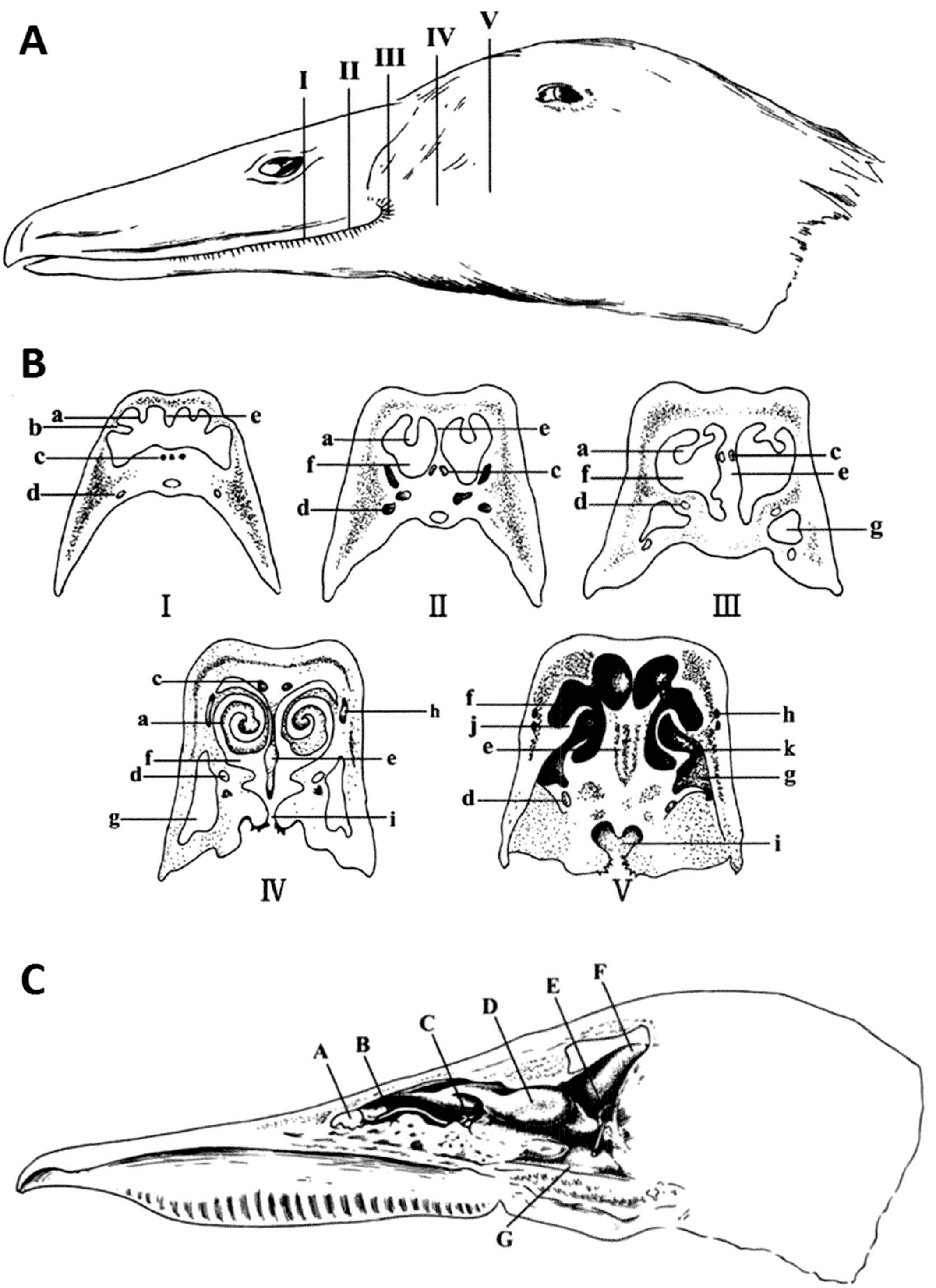


**Fig. 4.51** Nasal cavity of a Mallard (*Anas platyrhynchos*). (**A**) The positions of the five cross-sections with the same interval used for showing the general anatomical structure and morphological characteristics of the nasal cavity of ducks. (**B**) Transverse views of the five cross-sections (I–V): (a) concha nasalis

Sanderlings (*Calidris alba*) and Dunlins (*C. alpina*) can distinguish between sand where no worms had been present and sand where worms had been present (van Heezik et al. 1983). Laboratory experiments have also revealed that birds can taste and reject prey entirely on the basis of the prey's level of chemical defense (i.e., bitter taste; Skelhorn and Rowe 2006). For example, the sensitivity of Cockatiels (*Nymphicus hollandicus*) to the bitter compound quinine is similar to that of humans and higher than that of some mammals (Matson et al. 2004). In the field, many species of birds are selective in both attacking and consuming unpalatable, distasteful insect prey including several species of butterflies (e.g., Brower 1984; Chai 1996). Among herbivorous birds, Canada Geese (*Branta canadensis*) will avoid plants with high levels of secondary plant compounds (Buchsbaum et al. 1984). Other investigators have found that some species of birds are able to discriminate among foods differing in their carbohydrate, lipid, and amino acid composition (e.g., Murphy and King 1989; Schaefer et al. 2003).

Based on the presence or absence of relevant gene sequences, some investigators have sought to determine the role of taste in a number of species of birds. As an example, Wang and Zhao (2015) found a positive correlation between the presence of genes important in detecting the bitter taste and diet, with herbivorous and insectivorous birds having more such genes than carnivorous birds, perhaps because plants and insects are more likely to contain bitter toxins than noninsect prey.

## 4.8 Vision

Birds, with the possible exception of diurnal primates (e.g., humans), are the vertebrates that may rely most on vision to function in their environment. Flight is an obvious visually dependent behavior, but birds also exhibit an impressive range of visually guided behaviors other than flight, e.g., foraging, predator detection, and mate choice. One clue to the importance of vision for most birds is the relative size of their eyes. Whereas human eyes make up about 1% of the total mass of the head, European Starling (*Sturnus vulgaris*) eyes make up about 15% of the mass of their head. Larger eyes can have more photoreceptors, can collect more light, and provide larger and sharper images on the retina (Fig. 4.66).

In general, the size of bird eyes is correlated with body size; larger birds have larger eyes (Fig. 4.67). There are, however, some exceptions. For example, Ausprey et al. (2021) found that, after controlling for body size, eyes of birds in a cloud forest community that were found in darker microenvironments had significantly larger eyes that those found in microenvironments with more light (Fig. 4.68). In addition, nocturnal birds (Figs. 4.69 and 4.70) and raptors have larger eyes than expected based on their size. The eyes of raptors and owls are 1.4 times and 2.2 times larger, respectively, than those of other birds with the same mass (Snyder and Miller 1978; Martin 1983, 1985). For diurnal raptors, one possible disadvantage of larger eyes and corneas is that more peripheral light can enter the eye, potentially reducing the contrast in the target image ("disability glare"; Martin and Katziir 2000). The supraorbital ridges (bony protuberances above the eye sockets) of raptors may help reduce

**Fig. 4.51** (continued) media, (b) concha nasalis rostralis, (c) optic nerve of nervus trigeminus, (d) maxillary nerve of nervus trigeminus, (e) nasal septum, (f) nasal passage, (g) infraorbital sinus, (h) nasolacrimal duct, (i) choanal cleft, and (j) concha nasalis caudalis. (**C**) Anatomical structure of the nasal cavity from the duck's skull: (A) nostril, (B) concha nasalis rostralis, (C) opening of ductus nasolacrimalis, (D) concha nasalis media, (E) opening of nasolacrimal duct, (F) concha nasalis caudalis, and (G) choanal cleft. (Figures fom Kang et al. 2014; © 2014 Wiley Periodicals, Inc., used with permission)

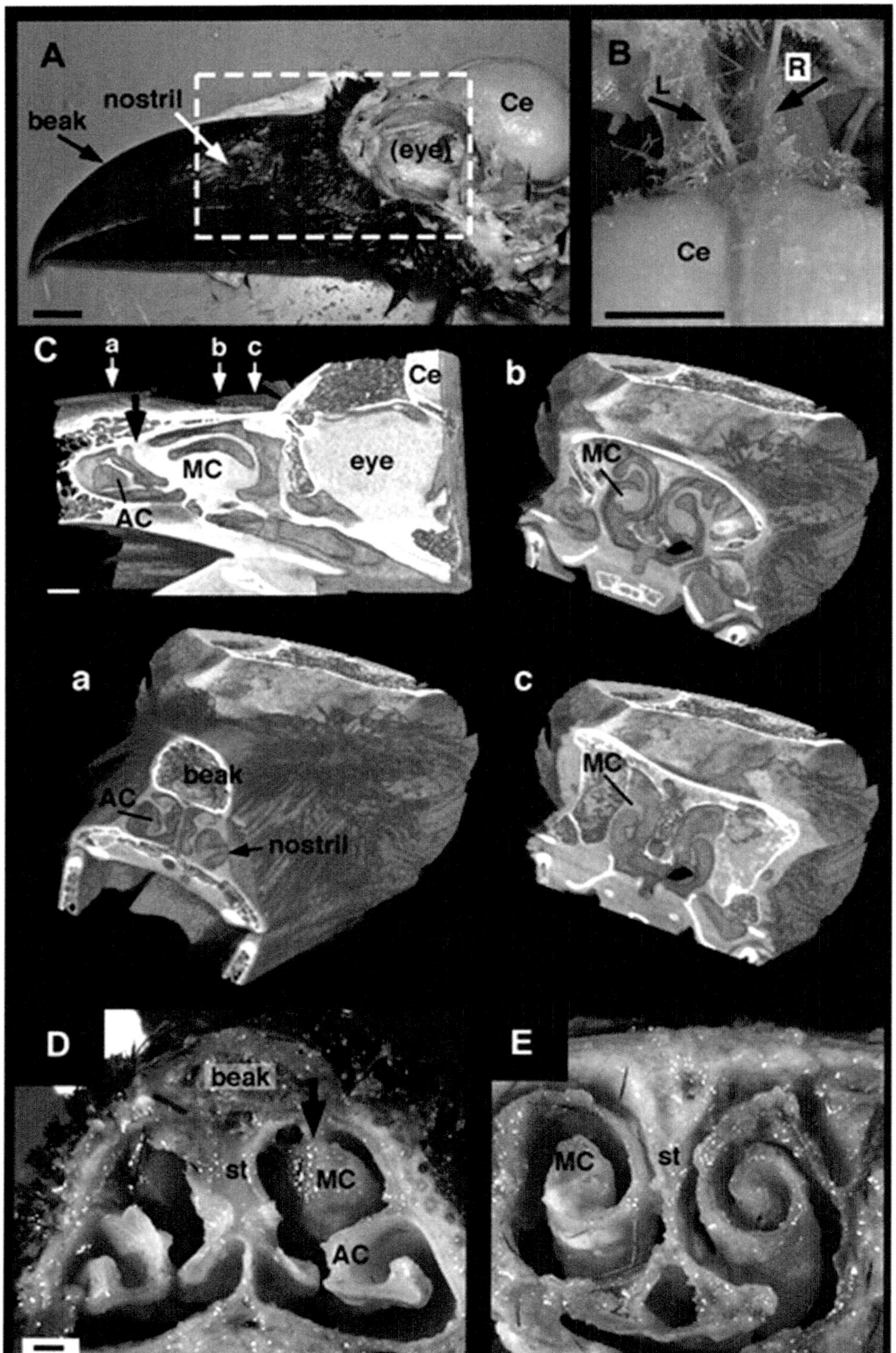


**Fig. 4.52** Nasal cavity of a Large-billed Crow (*Corvus macrorhynchos*). (**A**) Head of the crow (the cranial bone and eyeball have been removed). Scale bar = 10 mm. (**B**) The left and right olfactory nerve bundles independently

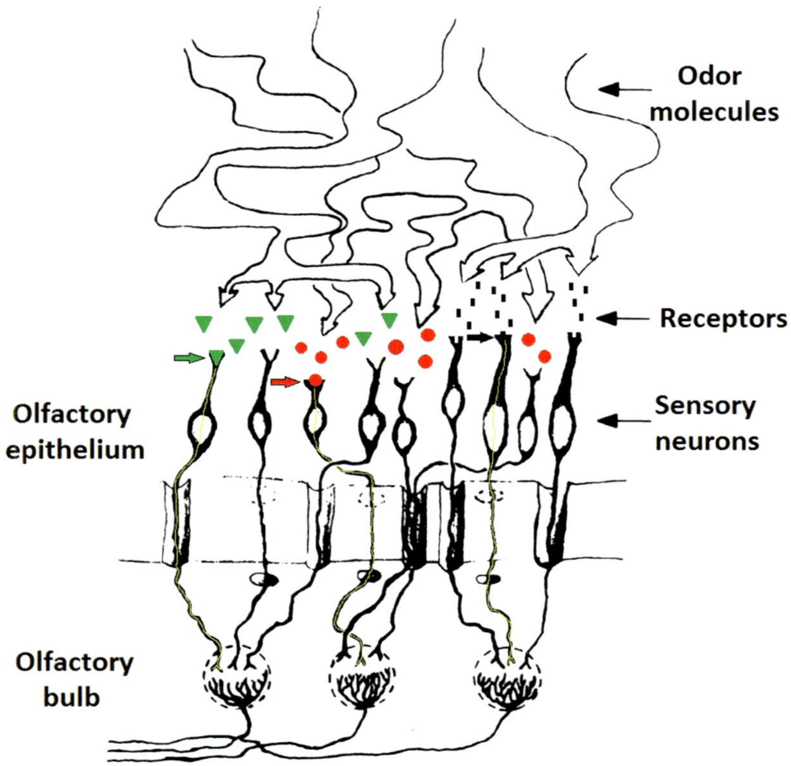


**Fig. 4.53** Olfactory sensory neurons in the olfactory epithelium detect odors via odor molecules filling receptor sites (indicated by arrows) and conduct impulses to the olfactory bulb. From the olfactory bulb, impulses travel to other areas of the brain where odor is perceived. (Figure modified from Nef 1998; © 1998 The American Physiological Society, used with permission)

this glare by shading the eyes from direct sunlight (Martin and Katziir 2000; Fig. 4.71).

The avian eye is similar to that of other vertebrates, with light passing through the cornea, aqueous humor, lens, and vitreous humor before striking the retina where the sensory cells (rods and cones) are located (Fig. 4.72). Behind the retina are two additional layers, the choroid and the sclera that provides structural support (Box 4.12 Scleral rings). The choroid aborbs light that has passed through the retina (except for species in the order Caprimulgiformes that have a reflective layer called the tapetum lucidum). Avian corneas are transparent and have no blood vessels so nutrients are provided by the aqueous humor (Jones et al. 2007). The iris regulates the amount of light entering the eye by adjusting the diameter of the pupil. This is mediated by iris muscles (sphincter and dilator pupillae) that are innervated by autonomic neurons. Changes in pupil area are related, to varying degrees among different species, to light levels. Lind et al. (2008) found that the pupils of two species of owls (Snowy [*Bubo*

**Fig. 4.52** (continued) project to the brain. L, left olfactory bundle; R, right olfactory bundle. Scale bar = 10 mm. (**C**) Sagittal and coronal three-dimensional structural computed tomography (CT) images of the nasal cavity. The small white arrows (a–c) in the sagittal image indicate the positions through which the planes of the transverse CT images pass. Scale bar = 10 mm. (**D** and **E**) Transverse sections of the nasal cavity through the rostral end (**D**) and middle portion (**E**) of the maxillary conchae (MC). The bold black arrows in (**C** and **D**) indicate the rostral end of the MC. Scale bar in (**D**) = 1 mm. Ce, cerebrum; st, olfactory septum. (Figure from Yokosuka et al. 2009; © 2009 Oxford University Press, used with permission)

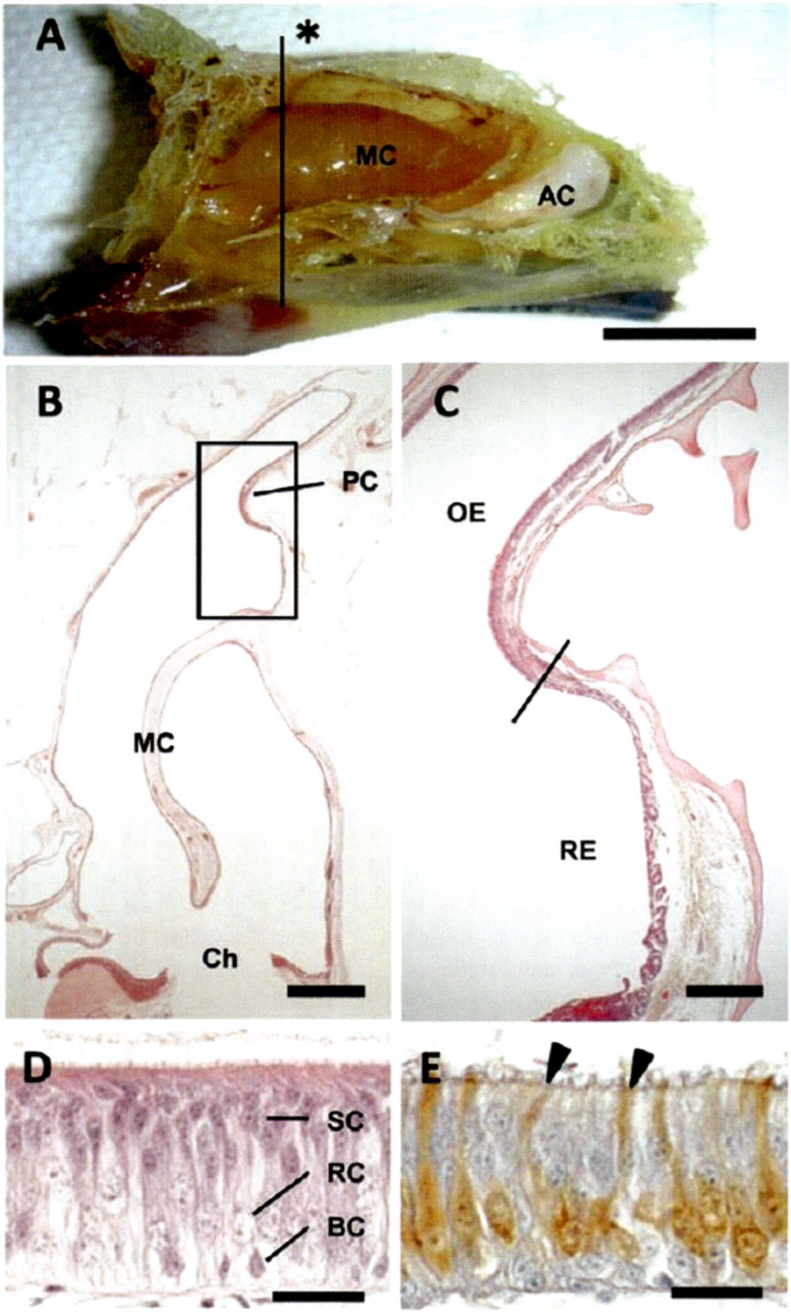


**Fig. 4.54** (**a**) Left half of the nasal cavity of a Large-billed Crow (*Corvus macrorhynchos*). The solid line indicated the by asterisk corresponds to b (frontal section). AC, anterior concha, MC, middle concha. Scale bar = 10 mm. (**b**) Frontal section in the caudal region of the nasal cavity. The boxed area is the area magnified in **c**. Ch, choana; MC, middle concha; PC, posterior concha. Scale bar = 1000 μm. (**c**) Higher magnification view around the posterior concha. The solid line represents the border between the olfactory epithelium (OE) and the respiratory epithelium (RE). Scale bar = 400 μm. (**d**) Olfactory epithelium. BC, nuclei of basal cells; RC, nuclei of receptor cells; SC, nuclei of supporting cells. Scale bar = 20 μm. (**e**) Olfactory epithelium. Arrows are pointing to the nerve endings (dendrites) of receptor cells. Scale bar = 50 μm. (Figure from Kondoh et al. 2011; Reproduced with permission of the Japanese Society of Veterinary Science from Kondoh et al. Ultrastructure and histochemical properties of the olfactory system in a Japanese Jungle Crow, *Corvus macrorhynchos*. J. Vet. Med. Sci. 73: 1007–1014, 2011)

*scandiacus*] and Ural [*Strix uralensis*] owls) constricted at lower light levels than those of two species of parrots (Turquoise-fronted [*Amazona aestiva*] and Gray [*Psittacus erithacus*] parrots), but the extent of maximum constriction was less than that of several species of mammals. Bird pupils constricted to about 30–50% of maximum area, whereas the pupils

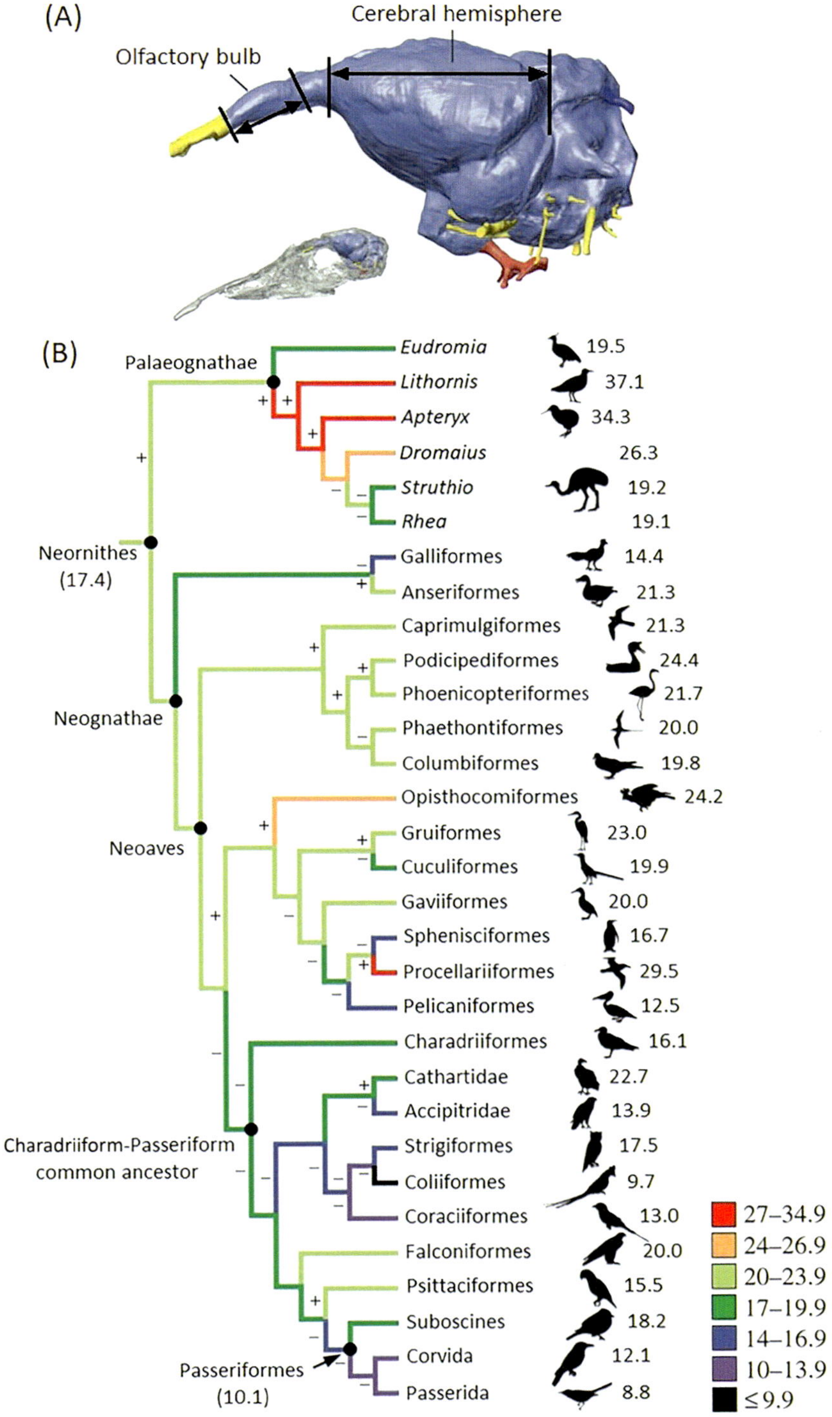


**Fig. 4.55** (**a**) The olfactory ratio is determined by the length of the olfactory bulb relative to that of the cerebral hemispheres. (**b**) Olfactory ratios of birds in genera in the infraclass Palaeognathae, orders of birds in the infraclass Neognathae, the suborder Suboscines, plus the parvorders Corvida and Passerida. Taxa with the highest olfactory ratios include the genus *Lithornis* (extinct) and *Apteryx* (kiwis), and the order Procellariiformes (albatrosses, petrels, shearwaters, and storm petrels). Plus and minus signs indicate whether olfactory ratios have increased (+) or decreased (−) compared to ancestral conditions. (Figure modified from Zelenitsky et al. 2011; © 2011 The Royal Society, used with permission)

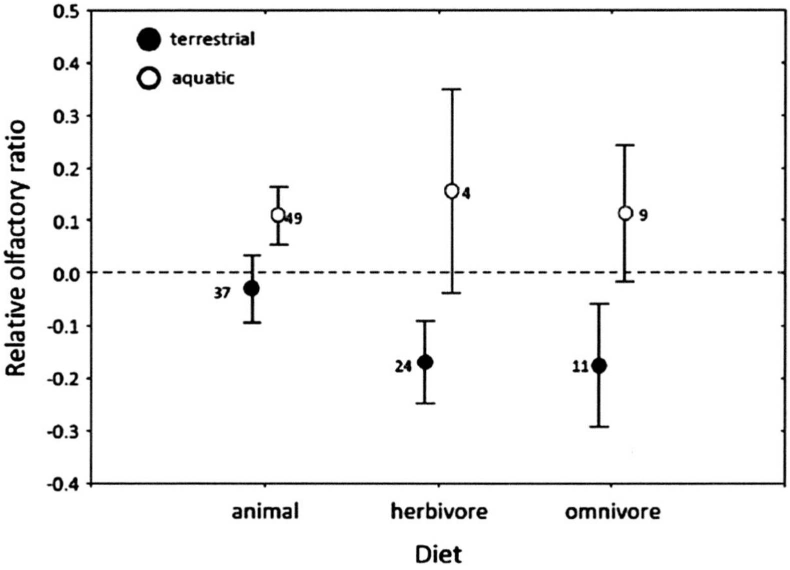


**Fig. 4.56** Relative size of olfactory bulbs varies with diet and between terrestrial and aquatic birds. Herbivorous and omnivorous birds in aquatic environments have larger olfactory bulbs than terrestrial birds, whereas aquatic and terrestrial species of birds with animal diets have similar-sized olfactory bulbs. Selection may have favored larger olfactory bulbs and an improved sense of smell in terrestrial birds with animal diets because mobile animal prey may be more difficult to locate and capture than non-mobile vegetation. Relative olfactory ratios were calculated as the ratio between the size of the olfactory bulb and that of the cerebral hemispheres. Sample sizes (number of species) are provided next to the mean values. (Figure from Avilés and Amo 2018; © 2017 Springer Science Business Media, LLC, used with permission)

of some mammals constricted to less than 5% of maximum area (Lind et al. 2008). In contrast, the pupils of some species of diving birds (e.g., Emperor [*Aptenodytes forsteri*] and King [*A. patagonicus*] penguins) can constrict to "pinholes" in bright daylight (Martin 1999). One possible explanation for the limited constriction of the pupils of some birds, even in relatively bright light, is that large pupils cause less diffraction of light than very small pupils and can enhance visual acuity (Land and Nilsson 2002). When bird pupils constrict, they do so rapidly, responding to a single flash of light in about 50 ms and reaching maximum constriction in about 100–150 ms (Kallähn 2014). By comparison, human pupils take longer to respond to a single flash of light (about 230 ms) and maximum constriction takes about 800 ms (Barbur et al. 2002). The pupillary reflex of birds is relatively fast because the avian iris contains striated muscle as well as some smooth muscle, and striated muscles contract faster than smooth muscle. The iris of mammals includes only smooth muscle (Douglas 2018). The dilation of bird pupils in response to lower light conditions is slower than pupillary constriction because the dilator muscle contains more smooth muscle (Douglas 2018). In general, however, the relatively fast pupillary reflexes of birds "may be an adaptation to the rapid changes in light levels that might be experienced during flight" (Douglas 2018).

The pupils of most birds are circular in shape; exceptions include some species with iris heterochromia (asymmetrical iris pigmentation; Fig. 4.73), skimmers (*Rhynchops* spp.) with pupils that can become vertical slits (Fig. 4.74), and King (*Aptenodytes patagonicus*) and Emperor (*A. forsteri*) penguins with pupils that can become very small and rhomboid-shaped

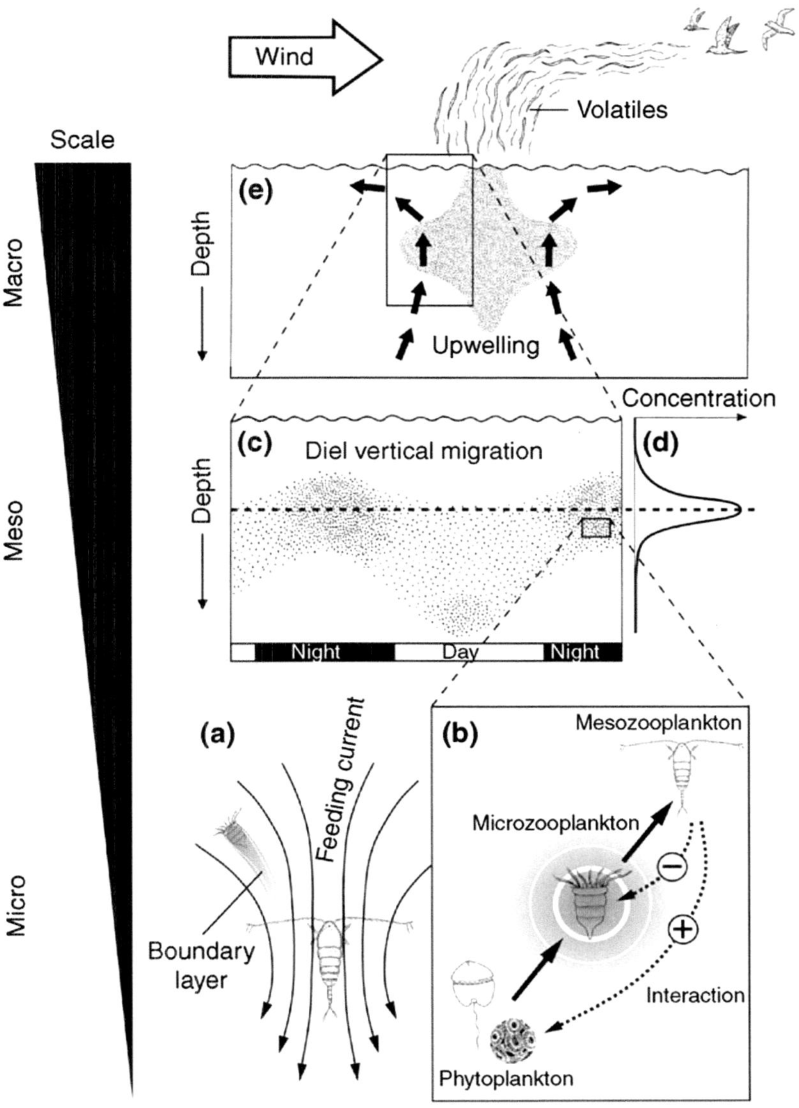


**Fig. 4.57** DMS and seabird foraging. The function of DMS (dimethyl sulfide) and other volatile infochemicals in aquatic trophic interactions. (**a**) A suspension-feeding copepod (mesozooplankton) produces a feeding current that deforms the boundary layer at the surface of microzooplankton. This layer is enriched with infochemicals such as DMS (produced by bacteria metabolizing dead plankton). (**b**) The interaction between phyto- and microzooplankton produces more DMS and is represented by the circular shaded area around the microzooplankton. Such infochemicals are exploited by higher-order mesozooplankton when searching for microzooplankton prey. Dotted lines indicate the mode of the trophic interaction with negative effects on the microzooplankton and indirect positive effects for the phytoplankton. (**c**, **d**) The accumulation of infochemicals from microzooplankton prey (**d**) provides chemical cues to migrating mesozooplankton [individual dots in (**c**)]. (**e**) Local upwelling generates biological activity (dotted area) that increases the sea–air flux of DMS and other volatile infochemicals. The resulting chemical gradient serves as a directional cue to seabirds. (Figure from Pohnert et al. 2007; © 2007 Elsevier Ltd., used with permission)

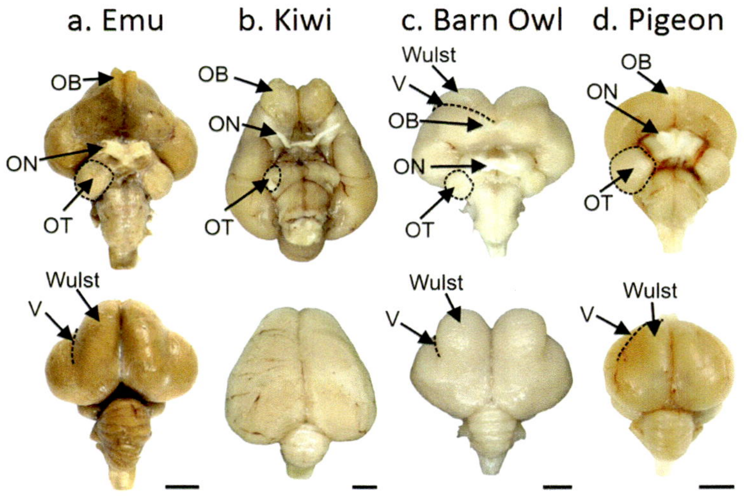


**Fig. 4.58** Olfactory bulbs of four species of birds. Ventral and dorsal views of the brains of (**a**) Emu (*Dromaius novaehollandiae*), (**b**) North Island Brown Kiwi (*Apteryx mantelli*), (**c**) Barn Owl (*Tyto alba*), and (**d**) Rock Pigeon (*Columba livia*). OB, olfactory bulb; OT, optic tectum; ON, optic nerve; V, vallecula. Note the large size of the olfactory bulb of the Kiwi compared to that of the three other species. Scale bars = Emu, 1 cm; Kiwi, Barn Owl and Rock Pigeon, 0.5 cm. (Figure from Martin et al. 2007; © 2007 Martin et al., open-access article distributed under the terms of the Creative Commons Attribution License, https://creativecommons.org/licenses/by/4.0/)

(González-Martín-Moro et al. 2014). When fully constricted, the pupil area of a King Penguin is just 0.42 mm$^2$, but, when fully dilated, the pupil area increases to 127 mm$^2$, a 300-fold increase. By comparison, pupils of Rock Pigeons and humans are limited to a 16-fold change in area (Martin 1999). Pupils that become very small, like those of King and Emperor penguins and other diving birds like cormorants (Fig. 4.75), allow a minimal amount of light into the eye when maximally constricted in daylight. A possible advantage of this is that the retina may then be pre-adapted for a rapid change in light conditions, e.g., penguins diving to forage at depths where there is minimal light (Martin 1999). Because the process of full adaptation to dark conditions by bird eyes can take up to 40 min (Blough 1956), such "pre-adaptation" may greatly improve the vision of King and Emperor penguins and other deep-diving birds as their pupils fully dilate during their deep dives.

Skimmers (Rynchopidae) are the only birds with pupils that can assume the shape of a vertical slit, but this characteristic is shared by a number of other vertebrates including some amphibians, reptiles, and mammals. Banks et al. (2015) categorized 214 species of terrestrial animals based on time of day when active (diurnal, nocturnal, or active both day and night), foraging mode (predator or prey and, for predators, whether active foragers that chase prey or ambush predators), and pupil shape. They found that circular pupils were generally associated with diurnal activity and active foraging, horizontal pupils with being prey, and vertically elongated pupils with nocturnal activity and ambush foraging. Skimmers forage by flying along the water's surface with their lower bill in the water and snapping their bill shut when contacting prey (or other objects that might be near the water's surface); in addition, they are often crepuscular or nocturnal. Skimmers must be able to reliably and accurately estimate distances at they forage, e.g., to avoid objects protruding from the water that, if hit, could, at minimum, damage their lower bill. Distance, or depth, perception can be determined by stereopsis (binocular vision) and defocus blur (because the degree to which objects blur depends

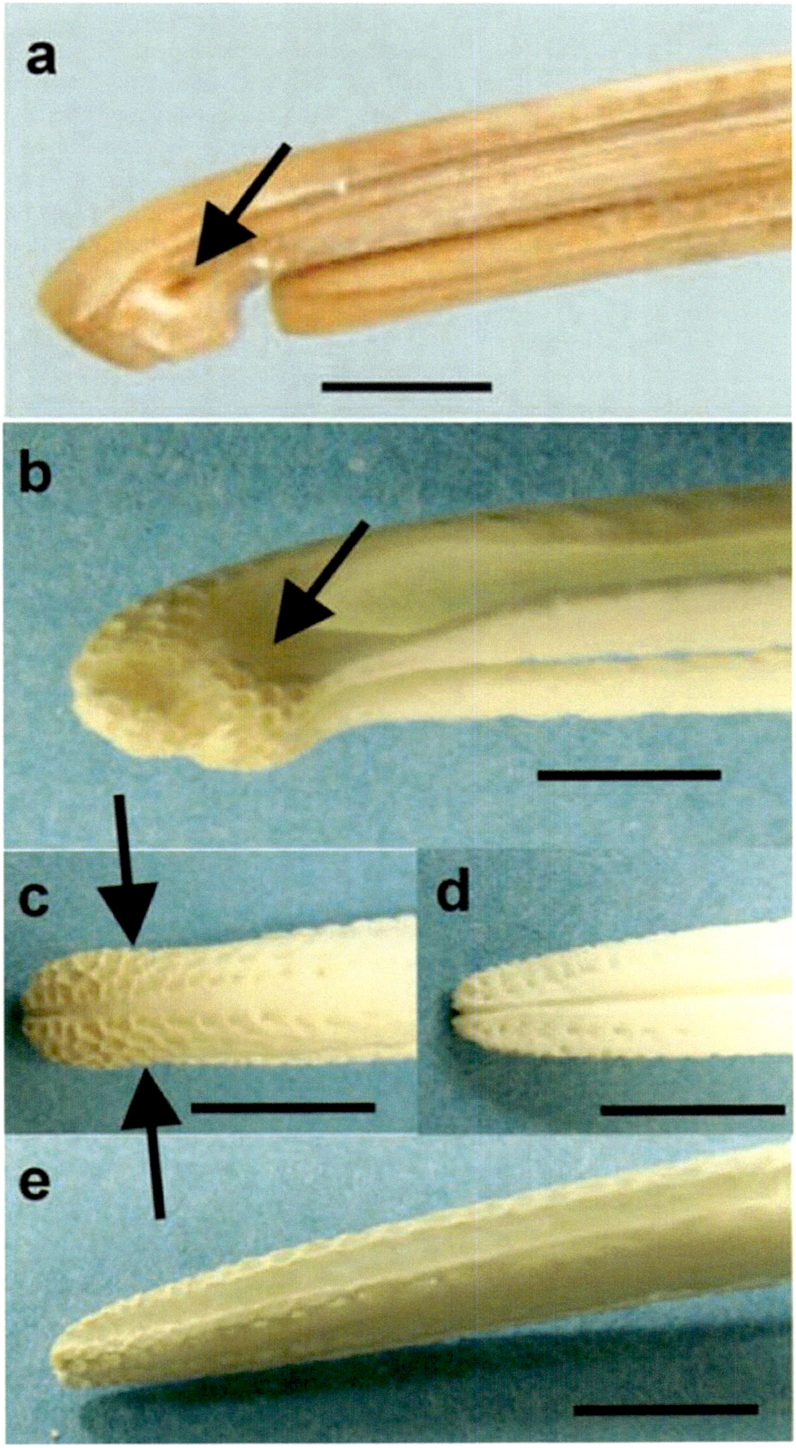


**Fig. 4.59** Nostrils and sensory pits at the bill tip of a North Island Brown Kiwi (*Apteryx mantelli*). (**a**) Lateral view of the bill tip of a museum skin specimen with the rhamphotheca intact and showing the position of the nostril (arrow). (**b**)–(**e**) Bones of the bill tip. (**b**) Latero-ventral view of the maxilla showing the complex blunt shape of the bill tip whose surface is covered with closely packed sensory pits; the approximate position of the nostril is indicated (arrow). (**c**) Dorsal view of the maxilla with the approximate positions of the nostrils indicated (arrows), (**d**) dorsal view of the mandible showing that sensory pits are found at the bill tips within the mouth. All scale bars 5 mm. (Figure from Martin et al. 2007; © 2007 Martin et al., open-access article distributed under the terms of the Creative Commons Attribution License, https://creativecommons.org/licenses/by/4.0/)

**Box 4.11 Sugar tastes sweet to hummingbirds and songbirds**

In all bird genomes examined thus far, the sweet taste receptor gene (*Tas1r2*) is not present (Zhao et al. 2010, 2015; Baldwin et al. 2014). Present-day reptiles, including crocodilians, have the sweet taste receptor gene, which means that the gene was likely lost in the early ancestors of birds, the dinosaurs. Experiments with hummingbirds, however, have revealed that they do have a sweet taste receptor and are able to discriminate among solutions that vary in sugar content. The molecular basis for this ability is a structural change in the hummingbird umami receptor. Among other vertebrates, the shape of the umami receptor is such that it allows them to taste various amino acids and proteins. However, the shape of the umami receptor of hummingbirds is different and allows them to taste sugars. The evolution of the unique umami receptors of hummingbirds was likely a key adaptation for nectar feeding that led to the extensive radiation of hummingbird species.

Hummingbirds, of course, also eat insects as a source of amino acids and proteins. Baldwin (2015) conducted taste trials with hummingbirds and found that they exhibited preferences for solutions with higher concentrations of some amino acids. Such results suggest that the sweet taste receptor may also serve as an umami taste receptor, with detection of either taste by the receptor serving as a pleasurable stimulus (Baldwin 2015).

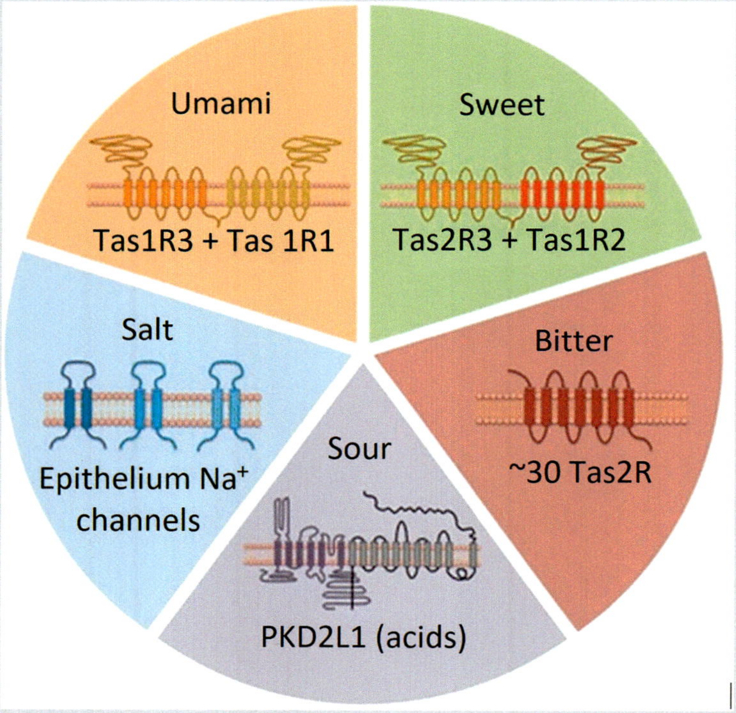


Receptors responsible for the five basic tastes. Note the relatively similarity of the umami and sweet receptors. The unique "sweet" taste receptor of hummingbirds required several mutations in the umami receptor that now allow the binding, and tasting, of sugars. (Figure from Martin and Dupré 2016; © 2016 Springer Nature, used with permission)

(continued)

**Box 4.11** (continued)

The shape of the umami receptor of songbirds has also been altered (although in a slightly different manner than for hummingbirds) to provide them with a sweet-sensing ability. These changes appear to have occurred early in the evolution of songbirds and, although some species may have later lost the the ability to perceive the sweet taste, Toda et al. (2021) noted that "songbirds appear to have broadly retained an ancestral sensory ability as they radiated out of Australia and colonized diverse habitats across the planet." This sweet-sensing ability allows even non-nectar specialists "... to opportunistically exploit novel or seasonally varying food sources, such as the nectar consumed by many insectivorous species during migration" (Toda et al. 2021).

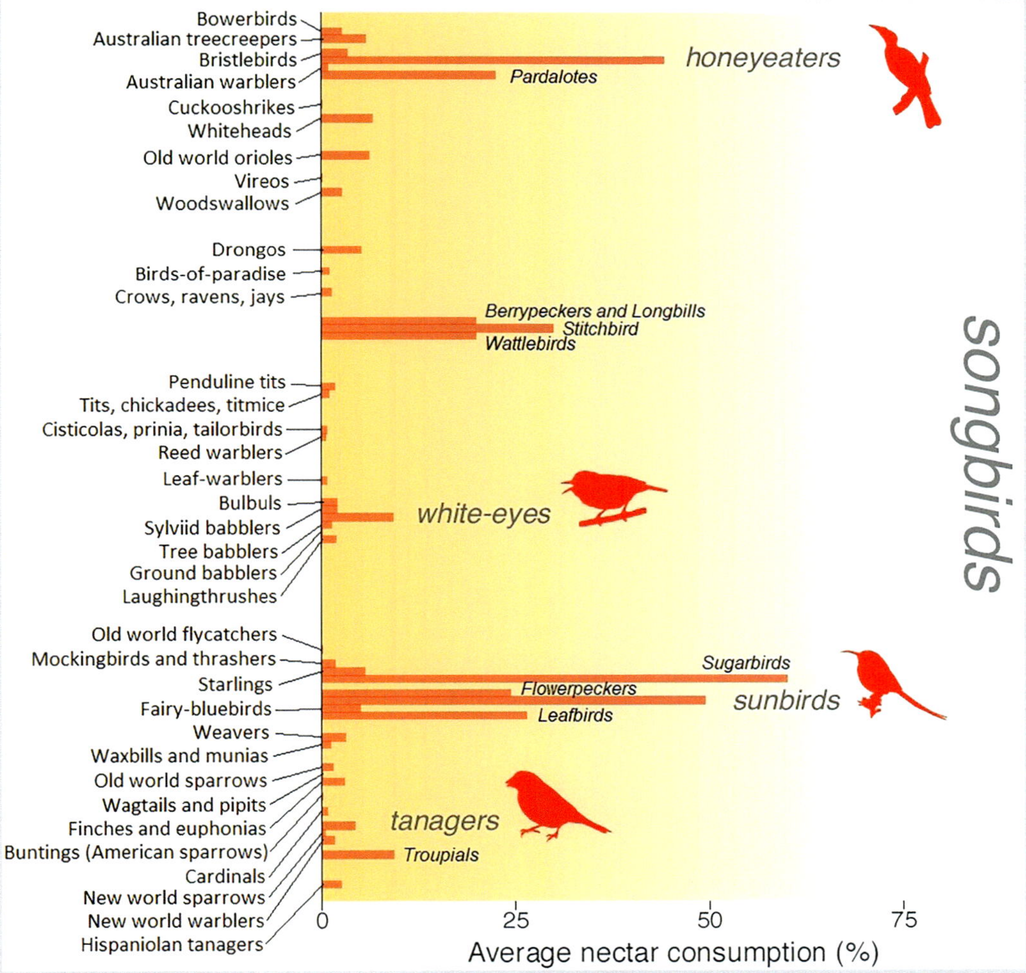


Of 122 families of songbirds, 49 (40%) have species that consume nectar. By comparison, 14 (11%) of 126 families of non-songbird families consume nectar. The average nectar consumption is the percentage of the total diet per family. Among families with species that consume the most nectar are the Honeyeaters, family Meliphagidae; Pardalotes, family Pardalotidae; Berrypeckers and Longbills, family Melanocharitidae; Stitchbird,

(continued)

**Box 4.11** (continued)

*Notiomystis cincta*, family Notiomystidae; Wattlebirds, family Callaeidae; White-eyes, family Zosteropidae; Sugarbirds, family Promeropidae; Flowerpeckers, family Dicaeidae; Sunbirds, family Nectariniidae; Leafbirds, family Chloropseidae; Tanagers, family Thraupidae; and Troupials, family Icteridae. Additional families with species that consume less nectar are listed on the left. (Figure modified from Toda et al. 2021; © 2021 The American Association for the Advancement of Science, used with permission)

**Fig. 4.60** Black-throated Blue Warbler (*Setophaga caerulescens*) eating granulated sugar at an outdoor restaurant in its wintering area in Kingston, Jamaica. Grains of sugar can be seen at the tip of its bill. Consumption of sugar suggests that these warblers have receptors that perceive the sweet taste. (Photo by Gary R. Graves in Graves 2014; used with permission of Gary R. Graves)

in part on the distance to an object relative to where an eye is focused), and Banks et al. (2015) suggested that vertically elongated pupils, like those of skimmers, facilitate the use of both stereopsis and blur for accurately estimating distances to objects (Fig. 4.76).

The color of avian eyes is the color of the iris, the structure that regulates pupil size. Avian irises contain chromatophores, specialized pigment cells that contain crystalline purines (especially guanine), pteridines, or other pigments, including melanins and carotenoids (Oliphant 1987; Oliphant and Hudon 1993; Hudon and Muir 1996; Prum 1999). Interestingly, chromatophores play important roles in the integumentary coloration of fish, amphibians, and reptiles, but, in birds, are only found in irises (Oliphant 1987; Oliphant et al. 1992; Prum 2006). The color of bird irises can also be influenced by the degree of vascularization (i.e., hemoglobin; McCulloch 1963). Most birds have dark irises (brown and black; Craig and Hulley 2004), but some species have brightly colored irises ranging from red to yellow to white (Fig. 4.77). Davidson et al. (2017) categorized the iris color of birds (1582 species) as either bright or dark and found that species that nest in the open, in contrast to cavity-nesting species, appear to be under strong selection to evolve dark eyes, perhaps because bright eyes might make birds more easily detected by predators. Some birds have multicolored eyes. For example, Greater Roadrunners (*Geococcyx californianus*) have a dark iris, but with a lighter ring around the pupil.

Few species of birds exhibit a sex difference in iris color. Among the species where males and females differ in iris color are Saddle-billed

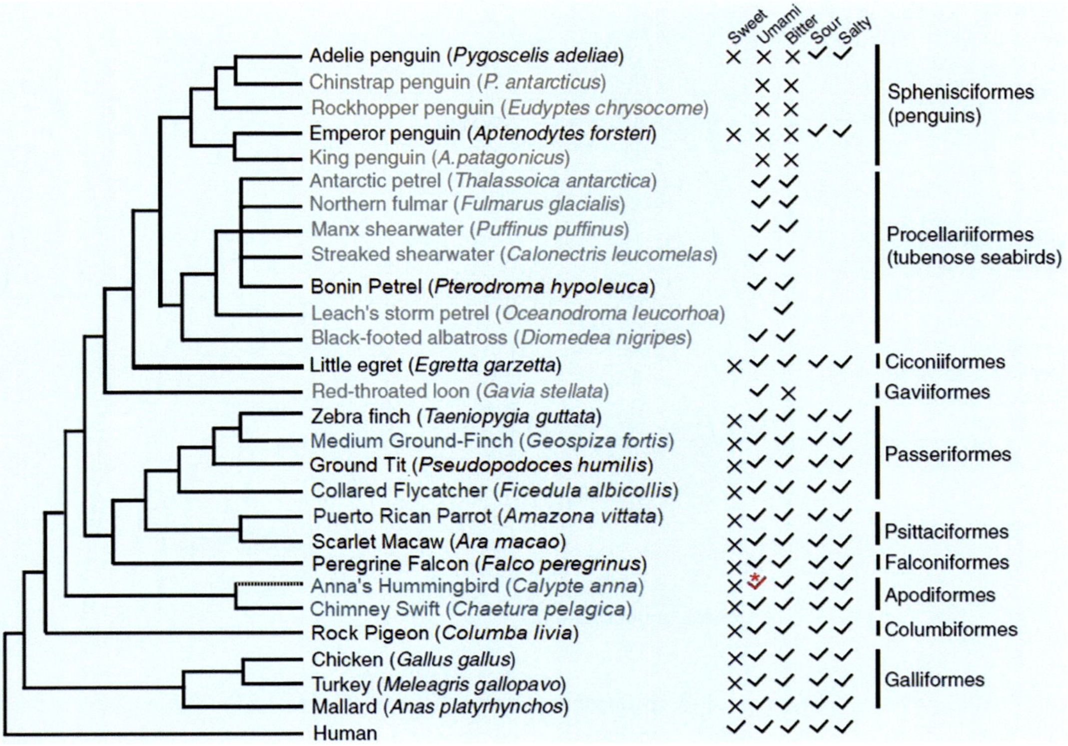


**Fig. 4.61** Species tree of 27 birds and humans showing presence (✓) and absence (x) of sweet, umami, bitter, sour, and salty tastes inferred from genes for taste receptors. Blank spaces indicate that there are no data available. Species with available genome sequences are shown in black, whereas those without available genome sequences are shown in gray. (*) In hummingbirds, including Anna's Hummingbirds (*Calypte anna*), the umami taste receptor has evolved to detect sweet. As a result, hummingbirds have a sweet taste, plus a weak umami taste. (Figure modified from Zhao et al. 2015; © Elsevier Ltd., used with permission)

Storks (*Ephippiorhynchus senegalensis*), Bushtits (*Psaltriparus minimus*; Fig. 4.78), some Icterids, including Brewer's Blackbirds (*Euphagus cyanocephalus*), and some species of weavers in the genus *Ploceus* (Craig and Hulley 2004). In addition, based on a survey of published papers, personal observations, and verbal descriptions by Worthy (2000) of 4918 species of birds, Negro et al. (2017) reported sexual dimorphism in iris color in 25 species in eight families, including 9 species of waterfowl and 4 species of raptors. The possible function(s) of sex differences in iris colors remain to be determined. Even rarer among birds is a seasonal change in iris color. Negro et al. (2017) reported such changes in five species of birds, including Red-throated Loons (*Gavia stellata*), Southern Giant Petrels (*Macronectes giganteus*), Brown Pelicans (*Pelecanus occidentalis*), Cattle Egrets (*Bubulcus ibis*), and Shoebills (*Balaeniceps rex*). For example, the irises of Brown Pelicans are brown during the non-breeding season, but change to light blue prior to breeding (Schreiber et al. 1989). In addition, the irises of several species of herons and egrets become redder during the breeding season (Hancock and Elliott 1978; Menkhorst et al. 2017).

Changes in iris color with age have been reported in many species of birds and, typically, young birds have dull-colored irises that transition to the more brightly colored irises of adults. Negro et al. (2017) reported that changes in iris

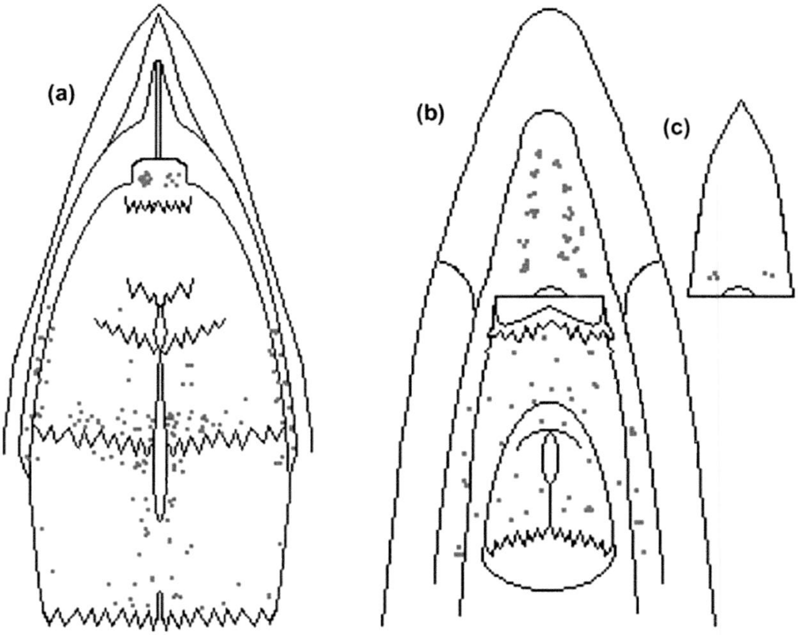


**Fig. 4.62** Distribution of taste buds in the oral cavity of a Domestic Chicken (*Gallus g. domesticus*). Each dot = one taste bud. (**a**) Palate, (**b**) tongue and floor of the oral cavity, and (**c**) tip of the tongue moved to the side to show taste buds on the floor of the oral cavity. (Figure from Kudo et al. 2008; © 2008 The Authors. Journal compilation © 2008 Japanese Society of Animal Science, used with permission)

color with age have been reported in all lineages of extant birds (Fig. 4.79), includings non-passerines (at least 24 families) and passerines (at least five families). Changes with age include from darker hues to lighter or brighter hues, brown to yellow, brown to red, and yellow to red (Figs. 4.80 and 4.81). These changes may be due to the presence of different pigments, with melanins prevalent in young birds with darker irises and other lighter or brighter pigments such as carotenoids increasingly available from ingested foods as birds age (Negro et al. 2017).

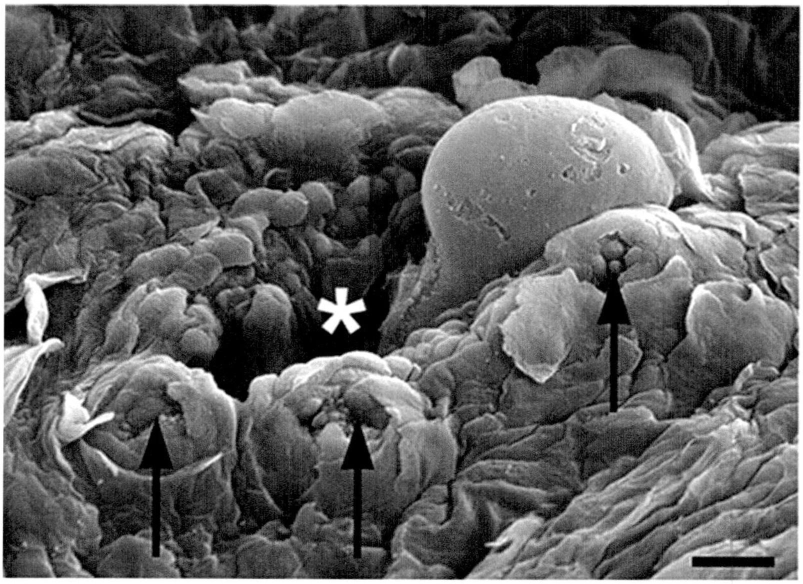

**Fig. 4.63** Electron photomicrograph of the palate of a Domestic Chicken (*Gallus g. domesticus*) showing a cluster of three taste buds (arrows) located adjacent to a salivary gland opening (asterisk). (Figure from Rajapaksha et al. 2016; open-access article licensed under a Creative Commons Attribution 4.0 International License, https://creativecommons.org/licenses/by/4.0/)

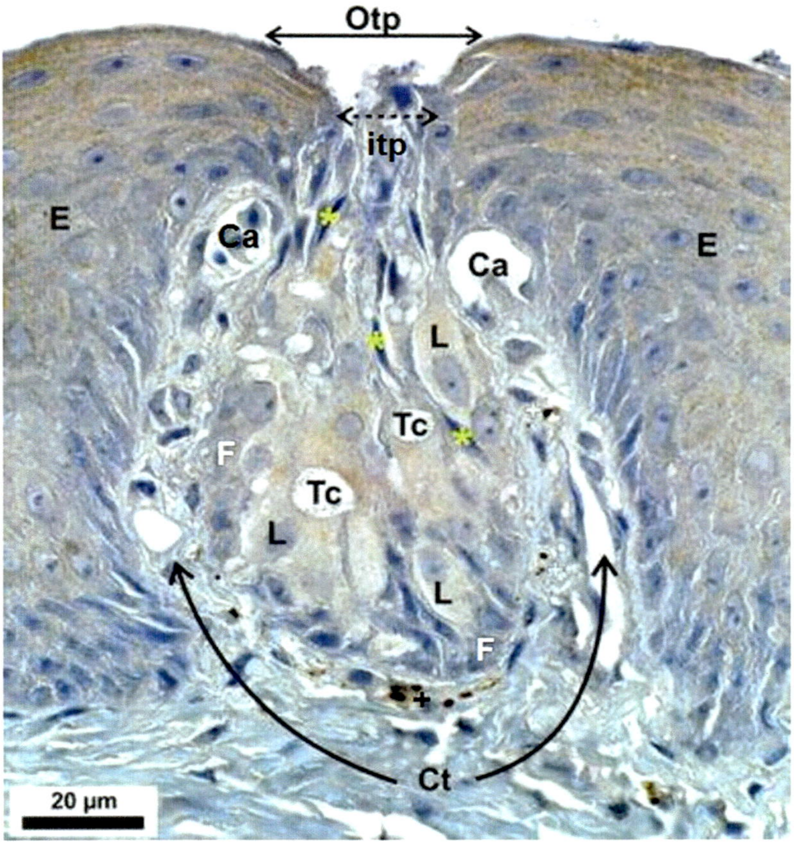


**Fig. 4.64** Taste bud from the oropharyngeal roof of an Emu (*Dromaius novaehollandiae*). Otp, outer taste pore; itp, inner taste pore; E, epithelial tissue; Ca, capillary; L, light cell; F, follicular cell; Tc, taste canal; Ct, connective tissue; yellow asterisk, dark cell; black +, neurofilament. (Figure from Crole and Soley 2015; © 2015 Springer-Verlag Berlin, used with permission)

Little is known about the possible role played by iris color in interactions with conspecifics. Rosenfield et al. (2003) studied three populations of Cooper's Hawks (*Accipiter cooperii*) and found no significant relationship between male eye color and brood size, suggesting that male eye color does not serve as either an indicator of male fitness or a sexual trait for assortative mating. However, Boal (2001) suggested that the iris color of adult Cooper's Hawks may be one of several cues used by conspecifics to assess potential mates. Among Yellow-eyed Penguins (*Megadyptes antipodes*), Massaro et al. (2003) found that the intensity of yellow eye coloration was correlated with breeding success, even within age classes, suggesting that eye color is correlated with parental quality (Fig. 4.82). Similar results have been reported for female Little Owls (*Athene noctua*), with the extent of iris yellowness positively associated with nest success (Passarotto et al. 2020).

Other examples of iris color playing a role in intraspecific communication include the threat displays of Purplish-backed Jays (*Cyanocorax beecheii*) where "all feathers of the body and head are strongly sleeked except those of the crest, which are strongly erected. The pupils are constricted, so that the bright yellow irides are boldly evident" (Hardy 1974: 254). Similarly, Davidson et al. (2014) found that the conspicuous eyes of Eurasian Jackdaws (*Corvus monedula*) appear to act as warning signals and deter intrusion by conspecifics into nest cavities (Fig. 4.83). In an apparent case of iris color playing a role in interspecific interactions, Trnka et al. (2012) found that Great Reed Warblers (*Acrocephalus*

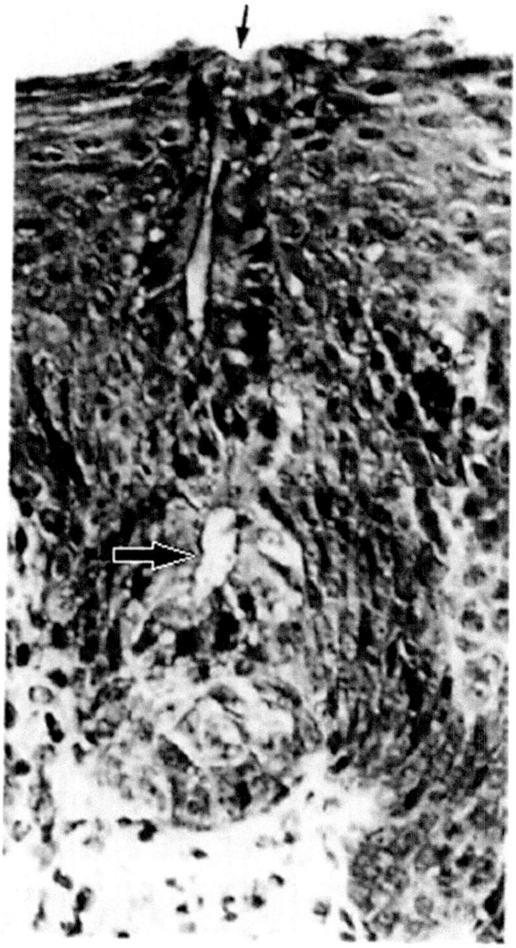

**Fig. 4.65** Taste bud of a Domestic Chicken showing the opening into the taste pore (small arrow at top) and a tubule extending from the opening toward the base of the pore (larger horizontal arrow). (Figure from Ganchrow and Ganchrow 1987; © 1987 Wiley-Liss, Inc., used with permission)

*arundinaceus*) recognized Common Cuckoos (*Cuculus canorus*), brood parasites that often parasitize the nests of Great Reed Warblers, primarily by focusing on the cuckoo's yellow eyes (Fig. 4.84).

Light entering the eye is refracted by the cornea and lens and strikes the retina where the sensory cells (the cells that convert light into nervous impulses) are located (Fig. 4.85). Clear vision (visual acuity) requires that images focus on the retina and the process of focusing on objects located at different distances from the eyes is called accommodation. Most birds accommodate by changing the curvature of both the cornea and lens (Glasser and Howland 1996; Ott 2006), whereas mammals can only change the curvature of the lens. The muscles involved are two divisions of the ciliary muscle: Crampton's muscle and Brücke's muscle (Fig. 4.86). Crampton's muscle originates in the sclera below the scleral ossicles and inserts on the cornea. Contraction of Crampton's muscle causes the cornea to bulge outward and increases its refractive power (Fig. 4.87). A second muscle, Brücke's muscle, also originates in the sclera, but inserts on the ciliary body. Contraction of Brücke's muscle pulls the ciliary body forward to reduce outward tension on the lens by the ciliary body and tenacular ligament. This causes the circumferential muscle fibers of the iris to exert pressure against the ciliary process that is fused with the annular pad of the lens. This squeezes the anterior surface of lens and causes it to bulge forward to increase its refractive power (Ott 2006; Fig. 4.86). Relaxation of these muscles flattens the cornea and returns the lens to its original shape.

Owls in the family Strigidae do not have a Brücke's muscle or lens accommodation so rely on Crampton's muscle and corneal accommodation. Schobert and Dubielzig (2014) suggested that the absence of Brücke's muscle might be due to the relatively large eyes and scleral ossicle of owls. In contrast, several species of aquatic birds, including penguins, cormorants, loons, and puffins, do not have Crampton's muscle and rely on Brücke's muscle and lens accommodation (Dubielzig et al. 2014). In air, the cornea is the principal refracting agent of light because the cornea is bordered on its inner surface by the aqueous humor, with a refractive index of 1.33, and on its outer surface by air, with a refractive index of 1.0. However, the refractive power of the cornea is minimal when a bird is underwater because aqueous humor and water have similar refractive indices. As a result, when a bird is underwater, accommodation involves only changes in the shape of the lens, which likely explains the loss of Crampton's muscle in many

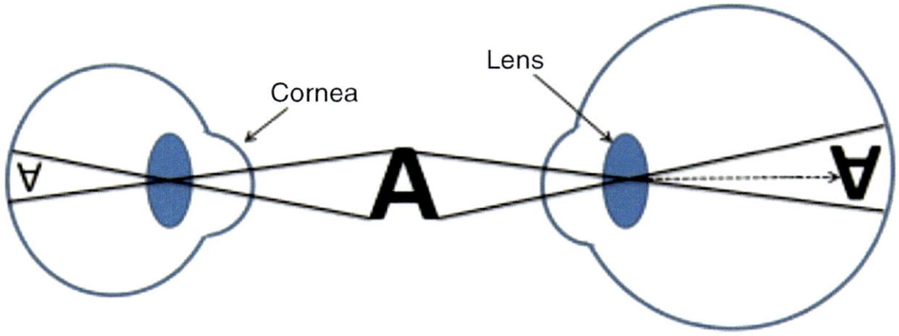


**Fig. 4.66** Relationship between length of eyes and the relative size of images on the retina, with images being larger on the retinas of longer eyes (eye on right). Note that images are actually inverted on the retina, but, after processing in the brain, they are not perceived that way. (Figure from Hall and Heesy 2010; © 2010 The Authors. Journal of Zoology © 2010 The Zoological Society of London; used with permission)

species of aquatic birds. What exactly is the refractive index? When light passes from one transparent medium into another, it changes speed and bends. How much this happens depends on the difference in the refractive indices of the two mediums (the greater the difference the greater the change in speed and bending of light) and the angle of the light ray.

Avian eyes have a characteristic shape, with front and back sections being spheres of unequal radii. Eye shapes vary among birds, ranging from globose or flattened to tubular (Fig. 4.88). Eye size and shape vary with the time of day when a bird is active. Nocturnal birds have larger corneas than birds active during the day, and corneas of crepuscular (active at dawn and dusk) and cathemeral (equally likely to be active at any time of day) birds are intermediate in size (Hall and Ross 2007). A larger cornea, and a correspondingly larger pupil, increases the amount of light entering the eye. Light entering the eye

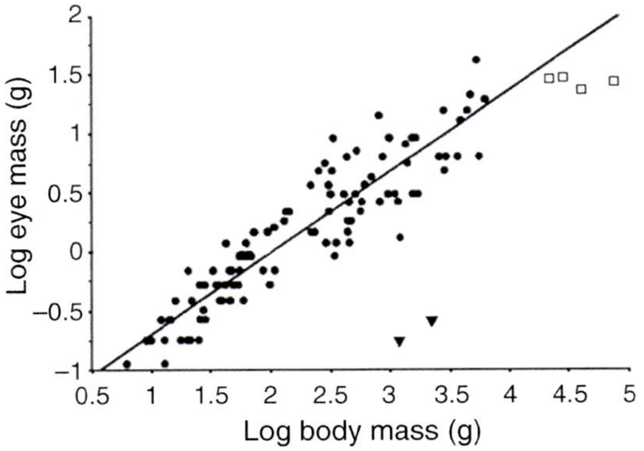


**Fig. 4.67** Relationship between body mass and eye mass for 104 species of flying birds. For comparison, four species of ratites (open squares) and two species of kiwis (*Apteryx* spp.; inverted triangles) are also included. Note that kiwis, which are nocturnal and more dependent on their sense of smell than vision, have exceptionally small eyes for their size. (Figure from Brooke et al. 1999; © 1999 The Royal Society, used with permission)

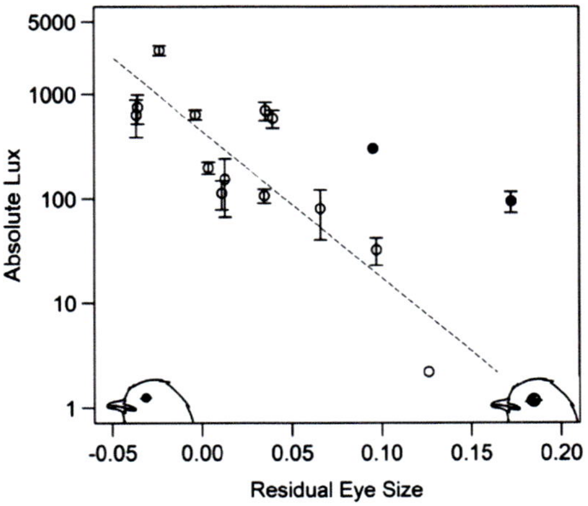


**Fig. 4.68** In a cloud forest bird community in the Peruvian Andes, species of birds typically found in darker microenvironments had larger eyes than those typically found in microenvironments with more light. Clear circles are myopic foragers (feeding on items close to them, e.g., gleaners and frugivores), and filled circles are hyperopic foragers (feeding on items farther away, e.g., flycatchers). (Figure from Ausprey et al. 2021; © 2020 by the Ecological Society of America, used with permission)

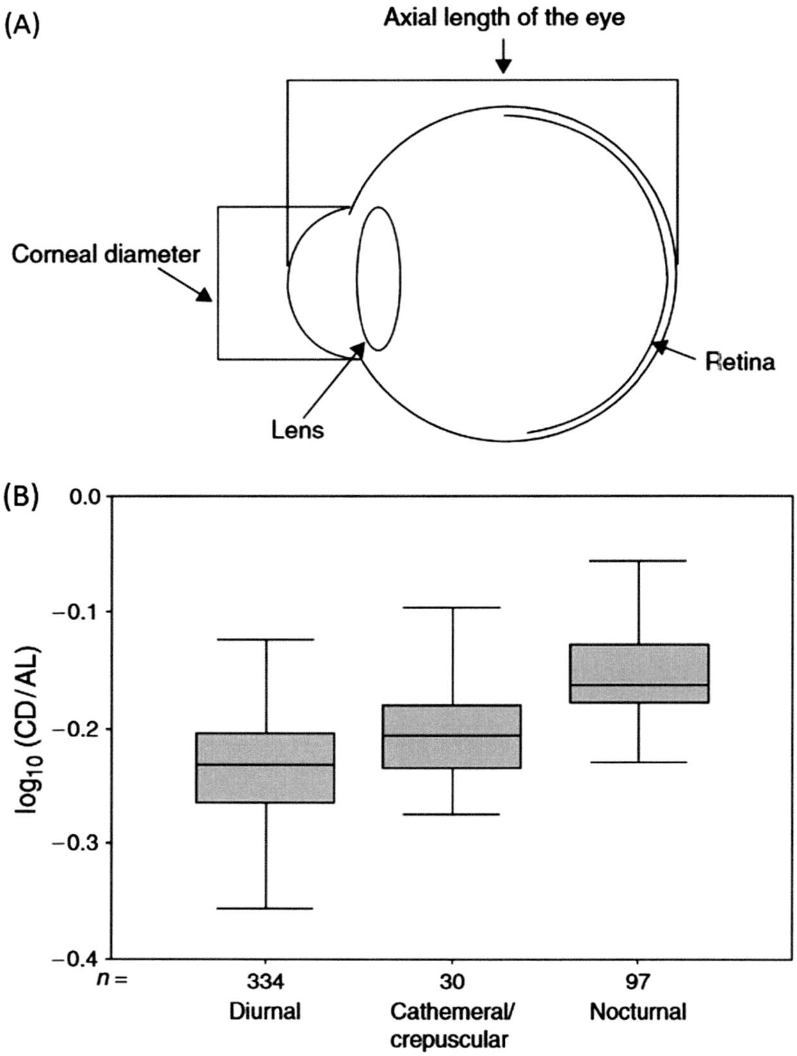


**Fig. 4.69** (**a**) Eye size as determined by corneal diameter and axial length. (**b**) Box-and-whiskers plot reveals that nocturnal birds have relatively larger eyes (corneal diameter [CD]/axial length [AL]) than diurnal birds and birds that are either crepuscular or cathemeral (active at various times during the day and night). Larger corneas allow more light into the eye, and longer axial lengths mean larger images on the retina and more photoreceptors stimulated. (Figures from Hall and Ross 2007; © 2006 The Authors. Journal compilation © 2006 The Zoological Society of London, used with permission)

strikes photoreceptors in the retina and the clarity of that image depends not only on the amount of light (number of photons), but on the length of the eye (distance from the lens to the retina) and the number and density of photoreceptors. Eye length (referred to as axial length) determines the size of an image on the retina and larger images improve visual acuity because each photoreceptor processes a smaller portion of the image (Hall and Ross 2007). Owls (Order Strigiformes) not only have large corneas, but also have relatively longer axial lengths than other birds of comparable size (Figs. 4.89, 4.90, and 4.91).

Birds also have three eyelids, with one upper and one lower eyelid plus a nictitating membrane (Fig. 4.92). The nictitating membrane is located between the other two eyelids and the cornea and has its own lubricating duct equivalent to the human tear duct. The nictitating membrane, or third eyelid, is unique to vertebrates, although not found in all groups. In many species, it represents the principal mechanism of ocular cleansing, and also serves to protect the cornea (Collin and Collin 2020). In birds, the external eyelids possess smooth muscle, and may close only during sleep. The nictitans, on the other

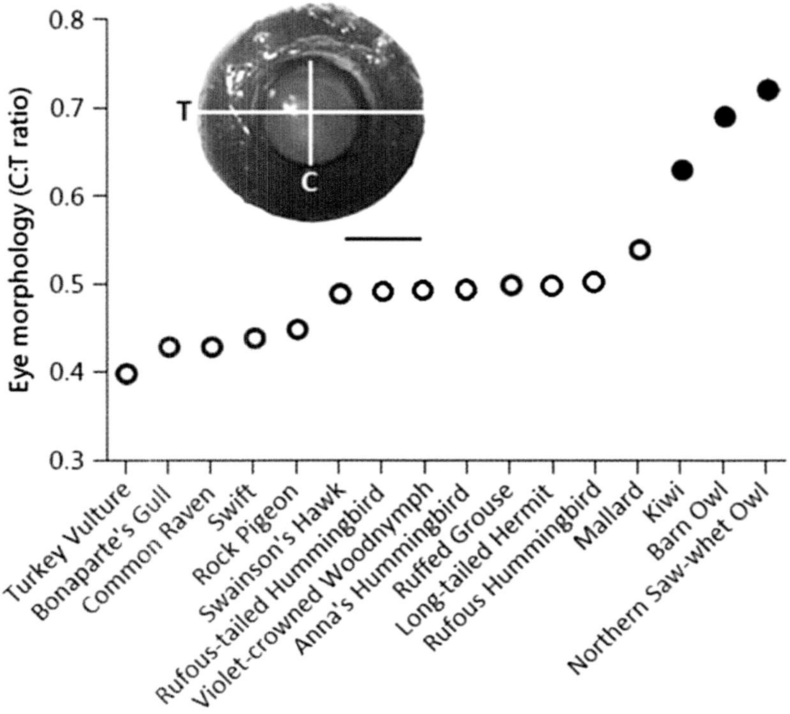


**Fig. 4.70** Eye morphology expressed as the ratio of the mean corneal diameter (C) to the mean transverse eye diameter (T), i.e., the C:T ratio several diurnal species of birds (open circles) and three nocturnal nocturnal species of bird (closed circles). Nocturnal species have larger corneas relative to the diameter of their eyes. Eye shown is from a Rufous-tailed Hummingbird (*Amazilia tzacatl*). Violet-crowned Woodnymph is also referred to as the Crowned Woodnymph (*Thalurania colombica*). Turkey Vulture, *Cathartes aura*; Bonaparte's Gull, *Chroicocephalus philadelphia*; Common Raven, *Corvus corax*; Rock Pigeon, *Columba livia*; Swainson's Hawk, *Buteo swainsoni*; Rufous-tailed Hummingbird, *Amazilia tzacatl*; Anna's Hummingbird, *Calypte anna*; Ruffed Grouse, *Bonasa umbellus*; Long-tailed Hermit, *Phaethornis longirostris*; Rufous Hummingbird, *Selasphorus rufus*; Mallard, *Anas platyrhynchos*; kiwi, *Apteryx* spp., Barn Owl, *Tyto alba*; Northern Saw-whet Owl, *Aegolius acadicus*. Scale bar = 2 mm. (Figure modified from Lisney et al. 2015; © 2015 S. Karger AG, Basel, used with permission)

hand, is operated by two striated muscles and is capable of extremely rapid sweeps across the ocular surface to clear the cornea of debris. Ocular surface lubrication originates from two secretory glands. The lacrimal gland is located in the inferior temporal quadrant associated with the more active lower eyelid, although in many species this gland is absent or rudimentary (Fig. 4.93). Additionally, most birds, especially cormorants and falcons, have a second secretory gland called a Harderian gland (first described by Johann Harder in 1694) located in the posterior and nasal aspect of the orbit associated with base of the nictitating membrane (Fig. 4.93). Secretions of the Harderian gland lubricate the nictitating membrane and the cornea (Beheiry et al. 2020). In falcons, this secretory gland produces a viscous solution to moisten the cornea during the stoops that are the falcon's trademark. Although the composition of these secretions is not known, a compound such as hyaluronic acid would moisten the surface without the rapid evaporation seen with a more dilute tear film. Such a coating would maintain a smooth surface, but might pose other difficulties by collecting debris (Schwab and Maggs 2004).

The nictitating membrane of many such birds has a cartilaginous-like connective tissue fold

**Fig. 4.71** Supraorbital ridges of raptors are likely helpful in shading the eyes and reducing glare. (Photo from pxhere.com, CC0 Public Domain)

along the leading edge of the membrane called the marginal plait. With each blink, this flange-like flap collects the tear film and any associated debris to drain through the enlarged puncta (openings that act like little valves to take tears out of the eye) into the nasolacrimal system. It is probably aided in this corneal cleansing role by a layer of "feather epithelium." This adaptation of a surface epithelium is thought to be unique to the epithelium lining the nictitans of many birds and reptiles. Long microvilli with club-like termini and numerous secondary projections from their long axis extend from the apical membrane of epithelial cells lining the bulbar surface of the third eyelid. Those of Peregrine Falcons (*Falco peregrinus*) form a "histological feather duster" that sweeps the cornea clean with each blink of the clear nictitans (Schwab and Maggs 2004).

Some authors have suggested that, in certain aquatic birds, a transparent nictitating membrane

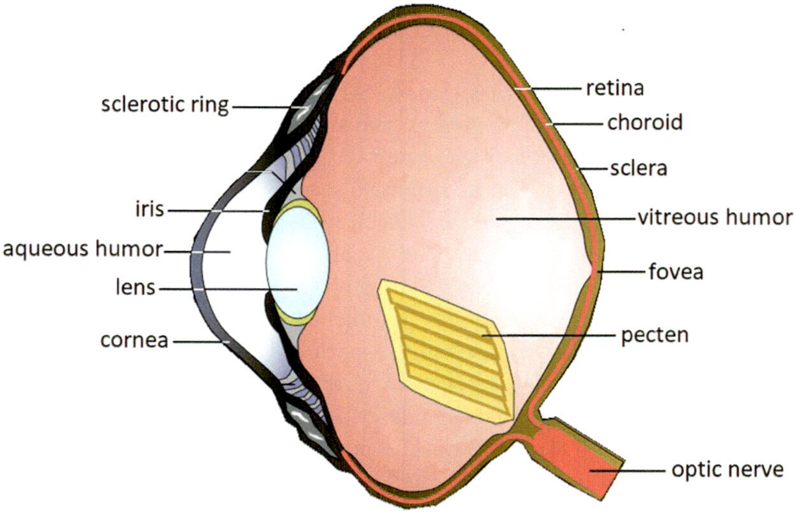


**Fig. 4.72** Avian eye. Light passes through the cornea, aqueous humor, lens, and vitreous humor before striking the retina. (Figure modified from UBC Wiki, https://wiki.ubc.ca/File:Avian_eye.png, CC0 Public Domain)

of high refractive index could compensate for the refractive loss of the cornea in water. However, measurements of refractive indices, refractive state, and curvature for several aquatic birds have revealed that the nictitating membrane does not have a refractive function; the refractive indices of the cornea and nictitating membrane are very similar, the nictitating membrane does not alter the refractive state of the eye, and its curvature, when it is in place in front of the globe, is virtually the same as that of the cornea (Sivak et al. 1978; Collin and Collin 2020). However, transparent nictitating membranes do provide

**Box 4.12 Scleral (or sclerotic) rings**

Scleral rings consist of a series of small bones or ossicles in the shape of a ring located in the scleral-corneal limbus that separates the cornea from the sclera (Franz-Odendaal 2020). The structure of sclerotic rings varies among birds in length, degree of curvature, number of ossicles, and thickness. The scleral rings of most birds consist of from 10 to 17 ossicles (Lima et al. 2009).

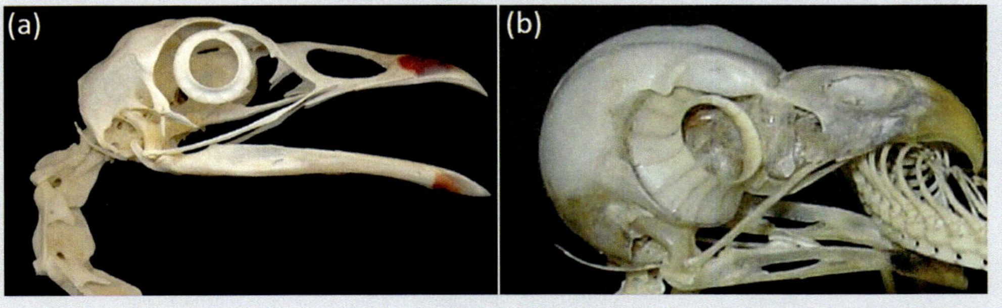


Examples of variation in avian scleral rings. (**a**) Wild Turkey (*Meleagris gallopavo*), and (**b**) Philippine Eagle-Owl (*Bubo philippensis*) (Figures from Franz-Odendaal 2020; © 2018 American Association for Anatomy, used with permission)

Scleral rings support and protect eyes at the corneoscleral junction and also serve as a site of attachment for ciliary muscles that play a role in acccommodation. Walls (1942) suggested that scleral ossicles help maintain the shape of the scleral-corneal limbus so that, as the shape of the cornea changes during accommodation, the ossicles prevent distortion of the retina. Walls (1942) also suggested that the scleral-corneal limbus would expand outward during accommodation if there were no scleral ossicles, with intraocular pressure increasing during the process of accommodation.

Few consistent differences in scleral ring morphology have been found. However, Hall (2008) did find a difference between the scleral rings of scotopic (active during low-light conditions) and photopic (active during the day) species of birds. The scleral rings of scotopic species have larger inner diameters than those of photopic species. Scotopic species such as most owls tend to have corneas that are larger in diameter than photopic species, with selection favoring larger diameters because they allow more light to enter the eye. The inner diameter of scleral rings is highly correlated with corneal diameter, explaining the difference between scotopic and photopic species of birds.

(continued)

**Box 4.12** (continued)

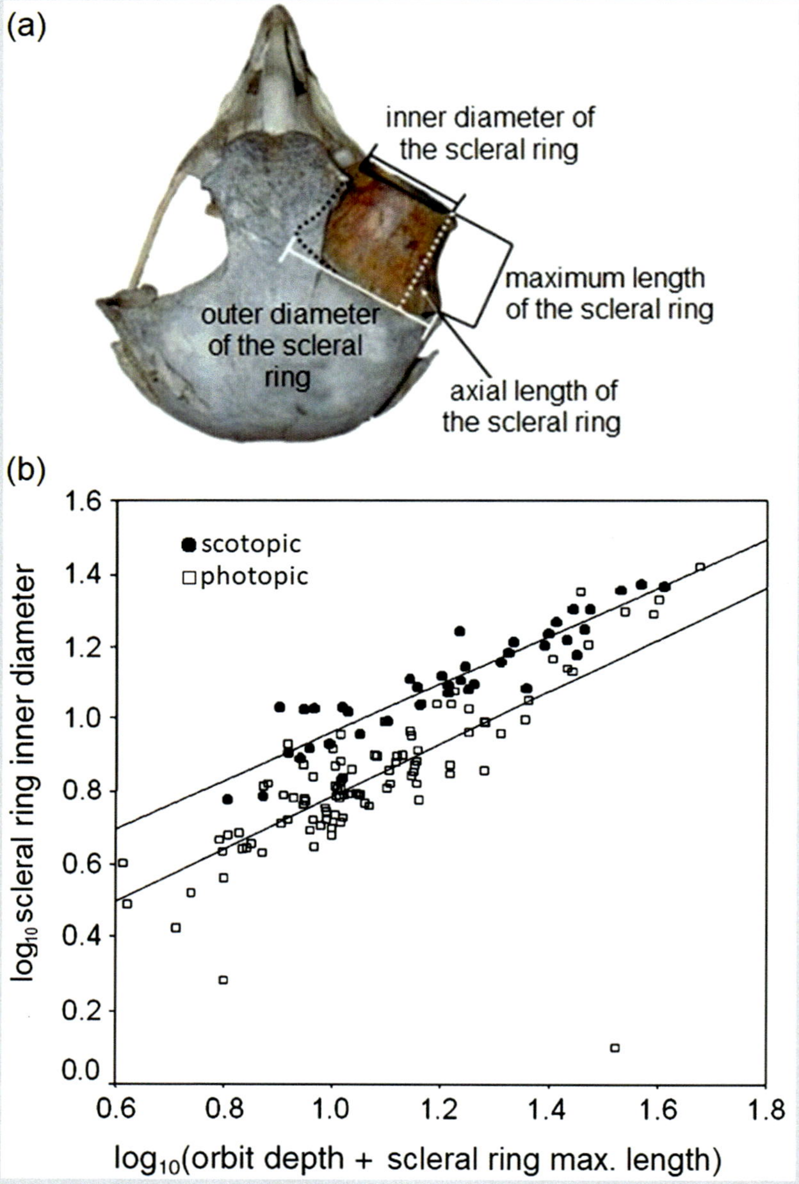


(**a**) Top view of the skull of a Snowy Owl (*Bubo scandiacus*), with a sclerotic ring in the right orbit. The sclerotic ring encloses the part of the eye that protrudes from the orbit. Measurement used for (**b**) are shown. (**b**) Scatter plot of $\log_{10}$ inner diameter of the sclerotic ring (bony correlate of corneal diameter) on the y-axis and $\log_{10}$ (orbit diameter + sclerotic ring maximum length) on the x-axis (a bony correlate of axial length of the eye). Analysis revealed a significant differencess between scotopic and photopic regression lines (Figures from Hall 2008; © 2008 The Author. Journal compilation © 2008 Anatomical Society of Great Britain and Ireland, used with permission)

(continued)

**Box 4.12** (continued)

Browne (2014) examined the scleral rings of several species of aquatic birds, including species that dive deep where light levels of low, and found no relationship between typical dive depths and the shape of scleral ring. However, examination of scleral rings revealed that deep-diving birds generally had more reduced ossicles, i.e., small ossicles located where adjacent larger, normal ossicles meet that may stabilize scleral rings in deep water with higher pressures. The thickened ossicles of deep-diving Emperor Penguins (*Aptenodytes forsteri*) may play a similar role. Diving birds, like Masked Boobies (*Sula dactylatra*), and pursuit divers like Thick-billed Murres (*Uria lomvia*), had "interwoven" ossicles, with the increased area of contact between adjacent ossicles potentially strengthening scleral rings for birds exposed to higher water pressure resulting from plunge dives or rapid swimming in pursuit to prey.

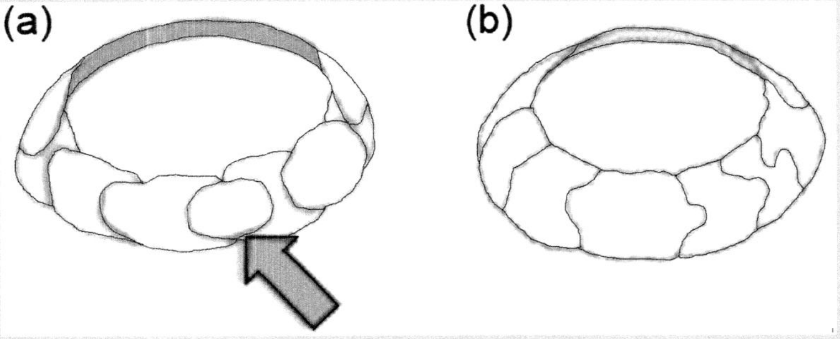


(**a**) Reduced ossicle (arrow) located where normal, adjacent ossicles meet. (**b**) Interwoven ossicles (Figure from Browne 2014; used with permission of K. Browne and the J. Clarke lab, University of Texas, Austin)

(continued)

**Box 4.12** (continued)

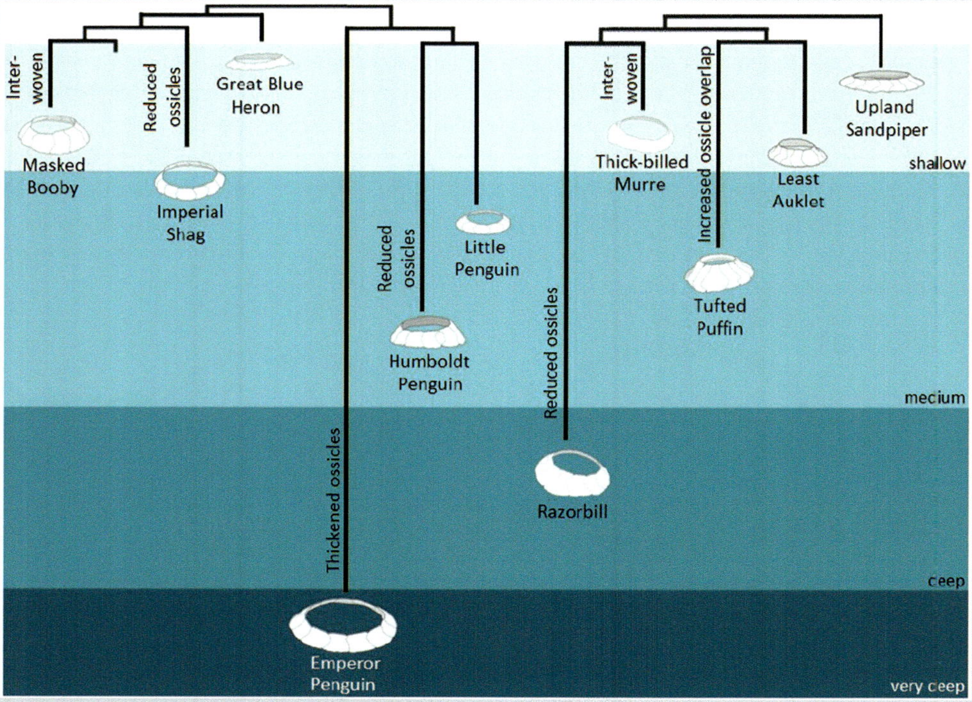


Variation in the characteristics of scleral rings of several species of aquatic birds. Scleral rings of birds that dive deeper or impact the water's surface have features that strengthen or reinforce the ring to better protect eyes from deformation. Imperial Shag is also referred to as the Imperial Cormorant (*Leucocarbo atriceps*). Masked Booby, *Sula dactylatra*; Great Blue Heron, *Ardea herodias*; Emperor Penguin, *Aptenodytes forsteri*; Humboldt Penguin, *Spheniscus humboldti*; Little Penguin, *Eudyptula* minor; Razorbill, *Alca torda*; Thick-billed Murre, *Uria lomvia*; Tufted Puffin, *Fratercula cirrhata*; Least Auklet, *Aethia pusilla*; Upland Sandpiper, *Bartramia longicauda* (Figure modified from Browne 2014; used with permission of K. Browne and the J. Clarke lab, University of Texas, Austin)

The scleral rings around the tubular eyes of owls must lengthen along with the developing eyes. In Barn Owls (*Tyto alba*), for example, eyes and scleral rings increase in length by a factor of 2.5 as they develop (Schaeffel and Wagner 1996) and most of this lengthening occurs after hatching (Schaeffel and Wagner 1996; Franz-Odendaal and Krings 2019). In contrast to the semi-altricial young of owls, the eyes and scleral rings of precocial species, like Domestic Chickens (*Gallus g. domesticus*), are fully developed at hatching.

(continued)

**Box 4.12** (continued)

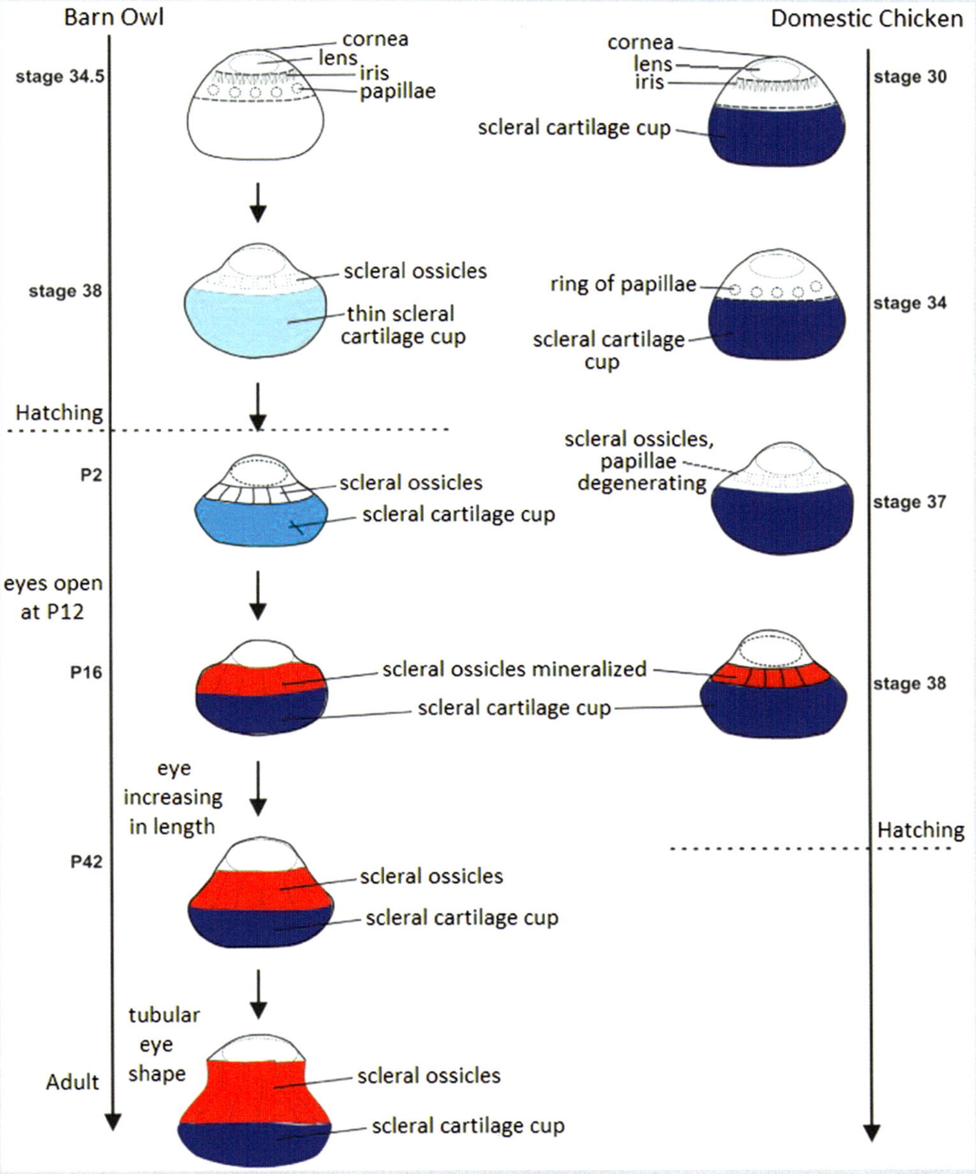


Timeline of development of the eye and scleral rings of a Barn Owl and Domestic Chicken. The eyes and scleral rings of Barn Owls are fully developed until several weeks after hatching. Eyes have been scaled to be similar in size. Dashed lines indicate the time of hatching (Figure modified from Franz-Odendaal and Krings 2019; © 2019 Wiley Periodicals, Inc., used with permission)

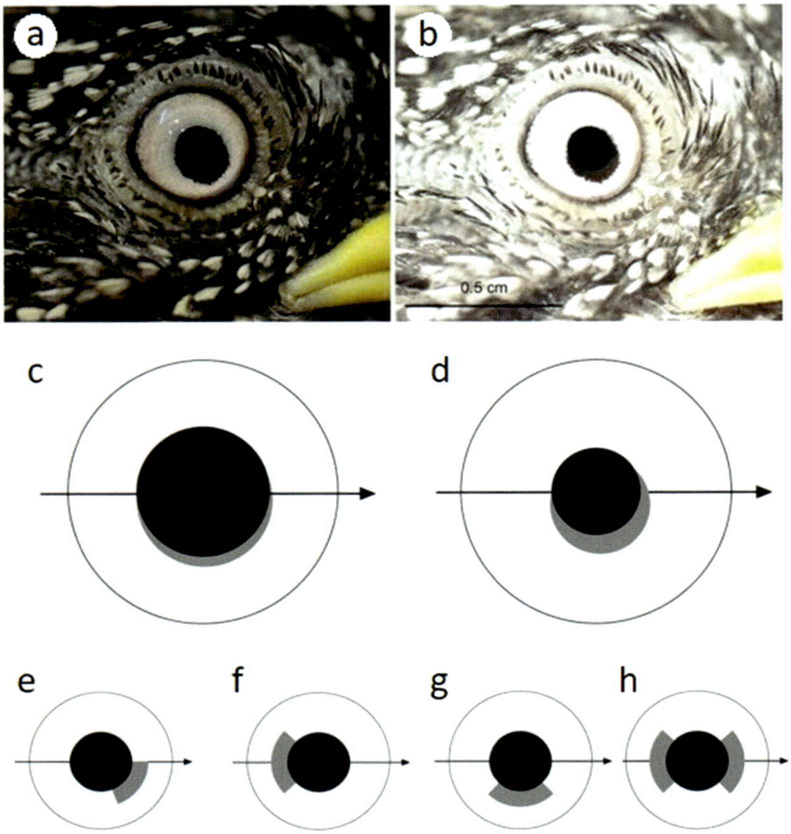


**Fig. 4.73** Asymmetrical iris pigmentation, i.e., iris heterochromia, of a Barred Buttonquail (*Turnix suscitator*) (**a**) before and (**b**) after the amount of light is increased, with the dark crescent oriented toward the bill. Drawings of the type and position of dark heterochromia in the eyes of a Plains-wanderer (*Pedionomus torquatus*) (**c**) during and (**d**) after pupil dilation, and of a (**e**) Madagascar Sacred Ibis (*Threskiornis bernieri*) as well as oystercatchers, woodpeckers, pigeons, and doves, (**f**) Rufous Potoo (*Nyctibius bracteatus*), (**g**) tiger-herons and bitterns, and (**h**) toucanets. These dark areas in pale irises may reduce the amount of light passing through the iris by absorbing the light and may also reduce glare. (Figure from Gutiérrez-Expósito 2019; © 2018, Dt. Ornithologen-Gesellschaft e.V., used with permission)

some aquatic birds with clear vision in both air and water (Collin and Collins 2020).

Typical bird eyes have a rounded cornea that, in air, plays an essential role in accommodation. Under water, however, the lens must compensate for the absence of corneal refraction (Katzir and Howland 2003). Among some diving birds, including diving ducks, cormorants, and likely other taxa of diving birds, the ciliary muscle pulls the lens forward so that it partially protrudes through the pupil, increasing its diameter and refractive ability (Levy and Sivak 1980; Cronin 2012; Figs. 4.94 and 4.95). Other birds, including penguins and albatrosses, have flatter corneas with refractive powers lower than those of species of birds with similarly sized eyes. These latter corneas lose little refractive power when submerged. However, the more spherical lenses of penguins and albatrosses change in shape and accommodate without needing to protrude through the pupil.

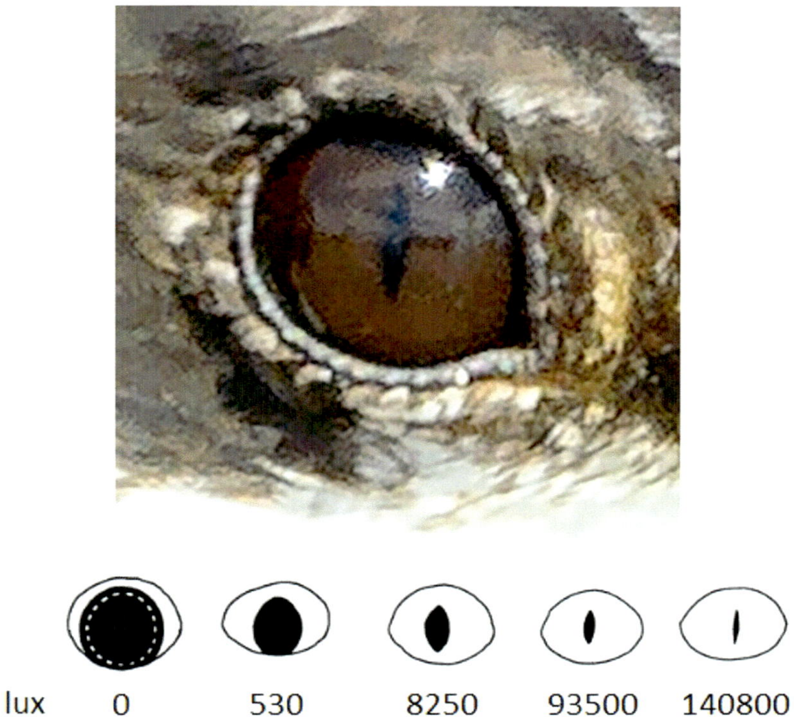


**Fig. 4.74** Top, The vertical slit pupil of a Black Skimmer (Top photo by David Sparks, used with permission of David Sparks). Bottom, Drawings of pupils of a Black Skimmer exposed to different light levels. (Figure from Zusi and Bridge 1981; © 1981 Association of Field Ornithologists, used with permission)

Pursuit-divers, such as penguins, petrels, loons, cormorants, many species of waterfowl, and other diving birds, rely on visual cues to locate and capture their prey and, therefore, have greater visual acuity when more light is available (Fig. 4.96). Given that many species dive to depths where light levels are very low (e.g., King Penguins, *Aptenodytes patagonicus*), others search for prey in turbid water, and, still others may dive throughout the night (Grémillet et al. 2005), many pursuit-divers may use a strategy similar to that of Great Cormorants (*Phalacrocorax carbo*). In full sunlight, the visual acuity of cormorants is similar to that of a human underwater without a face mask, i.e., about 3 cpd (or 20/200 using the Snellen notation, which is generally considered legally blind; Fig. 4.97). In other words, Great Cormorants have surprisingly poor visual acuity, even in full sunlight, and their acuity declines even further as they dive deeper and light levels decrease.

Birds in the orders Caprimulgiformes and Nyctibiiformes have a layer at the back of the eye called the tapetum lucidum (literally, "bright carpet") that acts as a mirror and reflects light back through the retina and, thereby, makes it more likely that light will strike sensory cells in the retina and greatly enhance their vision under low-light conditions. The presence of a tapetum lucidum produces the "eyeshine" of goatsuckers, potoos, and many other organisms (Fig. 4.98).

The position of eyes on the heads of birds exhibits much variation (Fig. 4.99). However, the eyes of most birds are on the side of their heads, allowing them to see objects on each side at the same time (monocular vision; Fig. 4.100). This provides birds with a wide field of view, e.g., Rock Pigeons can see 300 degrees without turning their head and American Woodcocks (*Scolopax minor*) can see 360 degrees! With monocular vision, birds may have a harder time judging distances and have poorer depth perception. Of course, these birds do have a limited field of binocular vision directly in front of them (and woodcocks also have a field of binocular vision behind them). Other birds, like owls, have their

**Fig. 4.75** Top, Eye and constricted pupil of a White-breasted Cormorant (some consider the White-breasted Cormorant to be a subspecies of the Great Cormorant [*Phalacrocorax carbo*] rather than a species; Hatch et al. [2020]). Bottom, Close-up of the eye and constricted pupil. (Photo by Peter Ryan; used with permission of Peter Ryan)

eyes located to the front, giving them a much wider field of binocular vision (Box 4.13 Stereopsis). These different visual fields have been categorized as: (a) Type 1. Large field of view with largely monocular vision (e.g., most birds), (b) Type 2. Very large field of view, with small binocular fields in front and back (e.g., woodcocks), and (c) Type 3. Larger frontal area with large binocular field (e.g., owls) (Fernández-Juricic et al. 2004).

Differences among species in the position of eyes result in differences in visual fields, i.e., "the space around the head . . . from which information can be extracted at any one instant" (Martin 2017). The eyes of most birds are on positioned somewhere on the sides of the head, but differences among species in the orientation of eyes mean that overlap in the horizontal visual fields of the two eyes ranges from as little as 5 to 10 degrees to as much as 60 degrees (Martin 2017). For most species of birds, however, horizontal visual fields of the two eyes overlap by about 20–30 degrees (Martin 2017). Visual fields of birds also vary vertically; the eyes of many

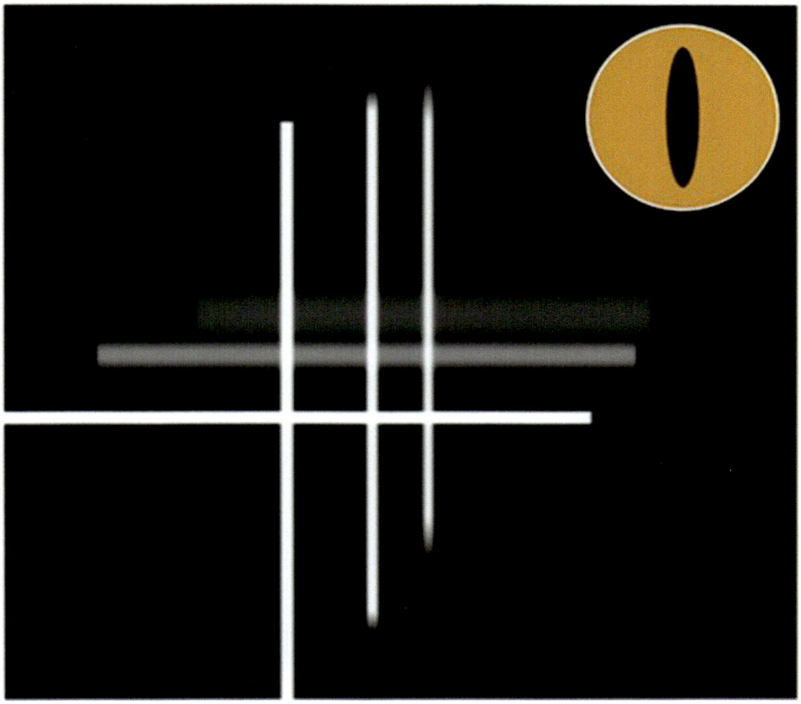

**Fig. 4.76** Three crosses located at different distances from hypothetical eyes focused on the closest cross. With vertical-slit pupils, there is less blurring of objects in the vertical than in the horizontal plane, and the extent of blurring of objects in the horizontal plane varies with distance. In combination, stereopsis and blurring allow eyes with vertical-slit pupils to accurately estimate distances. (Figure from Banks et al. 2015; Reprinted with permission of AAAS from Science Advances; © The Authors)

species of birds are located more toward to the top of the skull and thus provide large visual fields above and below their heads (Martin 2014; Fig. 4.101). The eyes of other species of birds are located lower on the skull and provide large visual fields below their heads (Fig. 4.102).

Importantly, birds can, to varying degrees, alter their visual fields by rotation of their eyes in their eye sockets (Table 4.1). The eye movements of some birds, including owls (Order Strigiformes) and other raptors (Orders Accipitriformes and Falconiformes), are limited so they depend much more on head movements than eye movements to change their visual fields. However, even among raptors, differences in the ability to move their eyes may be important. For example, raptors that search for fast-moving prey, and that need to shift focus between the central and temporal foveae when pursuing prey (Cunha et al. 2022), have greater eye mobility, e.g., up to 12° on the horizontal axis for Cooper's Hawks (*Accipiter cooperii*) and 10° for Red-tailed Hawks (*Buteo jamaicensis*), that may reduce retinal blur by allowing prey to be tracked without moving their heads (Tucker 2000b). Raptors less likely to engage in prey pursuit have more limited eye movements (~1° for American Kestrels, *Falco sparverius*, O'Rourke et al. 2010).

Most species of birds can use eye movements to alter their visual fields (Figs. 4.103 and 4.104) and, in some cases, rather substantially. For example, differences in the binocular visual fields of Carolina Chickadees (*Poecile carolinensis*), Tufted Titmice (*Baeolophus bicolor*), and White-breasted Nuthatches (*Sitta carolinensis*) when their eyes are converged compared to when their eyes are diverged range from 61° to 78° (Fig. 4.105) and Carolina Chickadees and Tufted Titmouse actually have blind areas in front of their heads when their eyes are fully diverged (Fig. 4.105). Moore et al. (2013) suggested that the relative wide binocular fields of these three species might improve foraging efficiency by increasing light sensitivity and contrast discrimination in the low-light forest habitats they occupy.

Most birds have relavely wide fields of monocular vision. In terms of visual attention, when

**Fig. 4.77** Examples of variation in the iris color of birds. (**a**) Barred Owl (*Strix varia*), (**b**) Eurasian Eagle-Owl (*Bubo bubo*), (**c**) Snowy Owl (*Bubo scandiacus*), (**d**) Little Grebe (*Tachybaptus ruficollis*), (**e**) Black-capped White-eye (*Zostertops atricapilla*), (**f**) Wood Duck (*Aix sponsa*), (**g**) Rainbow Lorikeet (*Pycnonotus atriceps*), (**h**) Female Satin Bowerbird (*Ptilonorhynchus violaceus*), and (**i**) Andean Cock-of-the-Rock (*Rupicola peruvianus*). (Barred Owl photo from U. S. Fish and Wildlife Service, Wikipedia, CC0 Public Domain; Eagle Owl photo from pxhere.com, CC0 Public Domain; Snowy Owl photo from pxhere.com, CC0 Public Domain; Little Grebe photo by J. M. Garg, Wikipedia, CC BY 3.0, https://creativecommons.org/licenses/by/3.0/; Black-capped White-eye photo from pxhere.com, CC0 Public Domain; Wood Duck photo from pxhere.com, CC0, Public Domain; Rainbow Lorikeet photo from pxhere.com, CC0 Public Domain; Satin Bowerbird photo by Will Brown, CC BY 2.0, https://creativecommons.org/licenses/by/2.0/; Andean Cock-of-the-Rock from Wikipedia CC BY 2.0, https://creativecommons.org/licenses/by/2.0/)

simultaneously scanning lateral visual fields (i.e., monocular vision), birds can apparently only process detailed images from one eye at a time (Voss and Bischof 2003). Voss and Bischof (2009) studied eye movements of Zebra Finches (*Taeniopygia guttata*) and found that, to view objects in front of them, they can simultaneously move both eyes toward the front for binocular vision. However, when viewing objects in the lateral (monocular) visual field, when one eye moved in a certain direction to view an object, the other eye simultaneously moved in the opposite direction. This suggests that the birds shift attention either to the right or left visual fields and that simultaneous visual processing of two images is unlikely. Voss and Bischof (2009) suggested that simultaneous movement of eyes in opposition directions allows birds to obtain information from their entire visual field, i.e., they may be focusing on a particular object with one eye, but are also able to detect potentially important objects such as an approaching predator with the other "unfocused" eye and, if necessary, quickly switch their focus of attention to that eye.

Visual fields vary among different species of birds and key factors in the evolution of these fields have likely been the need to acquire food and detect predators (Martin 2017). For most birds, acquiring food means being able to control

**Fig. 4.78** Male (top) and female (bottom) Bushtits differ in the color of their irises, with males having a dark iris and females a light-colored iris. (Top photo by Alex Critch, Wikipedia, CC BY 2.0, https://creativecommons.org/licenses/by/2.0/; Bottom photo by Becky Matsubara, Wikipedia, CC BY 2.0, https://creativecommons.org/licenses/by/2.0/)

bill position; for others, like raptors, it means controlling the position of their feet (talons). Although bill positioning is of critical importance for food acquisition in many species, this ability also allows birds to perform other tasks that require precise bill positioning such as feeding mates and young and building nests. The typically wide lateral visual fields of most birds allow them to scan for both predators and food items and, when close to those food items, converge the eyes to the front portion of the visual field to allow accurate positioning of the bill or feet to contact the target (Fig. 4.106). When not foraging, the eyes can diverge to provide larger visual fields and better detect predators or monitor the positions of conspecifics or heterospecifics, e.g., when in a flock (Fig. 4.106). Some species of birds, including some waterfowl and shorebirds, have rather narrow frontal visual fields and, therefore, more complete visual coverage around them

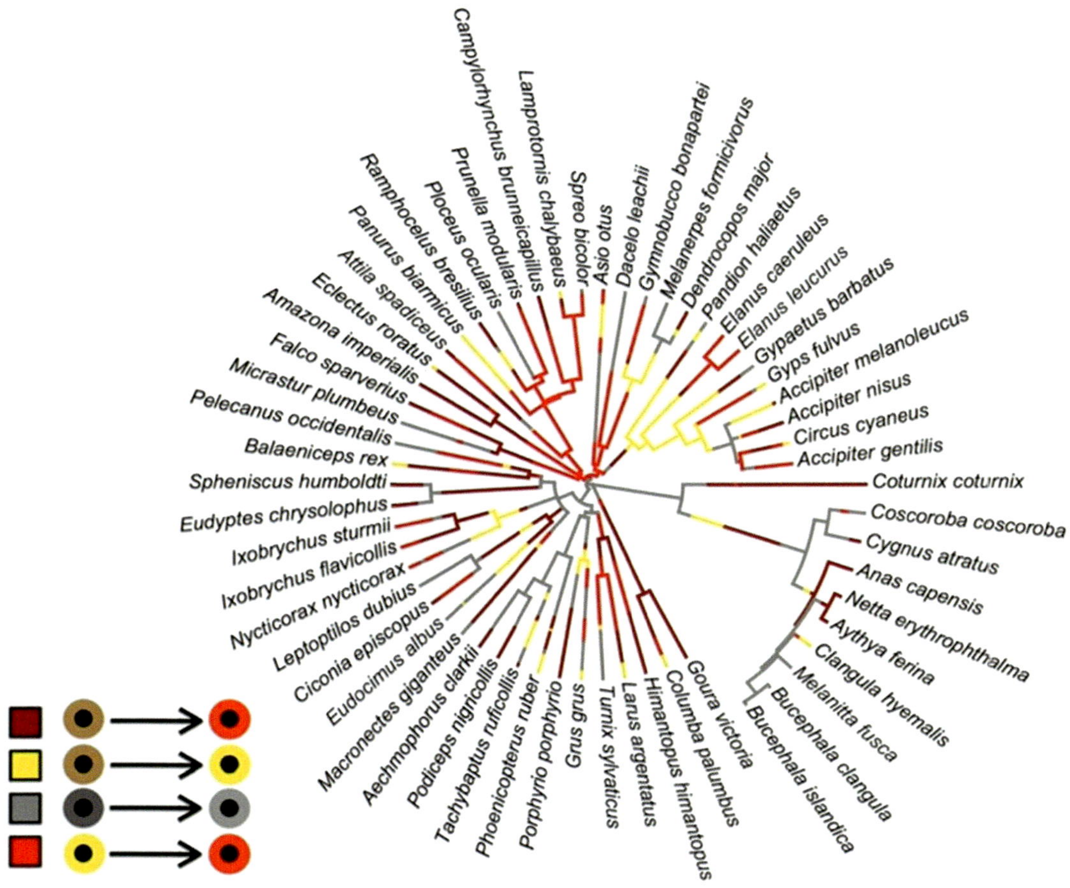


**Fig. 4.79** Phylogenetic relationships among 58 species of birds with four categories of age-related eye color changes, including brown to red, brown to yellow, yellow to red, and dark to light. (Figure from Negro et al. 2017; open-access article distributed under the terms of the Creative Commons Attribution 4.0 International License, https://creativecommons.org/licenses/by/4.0/)

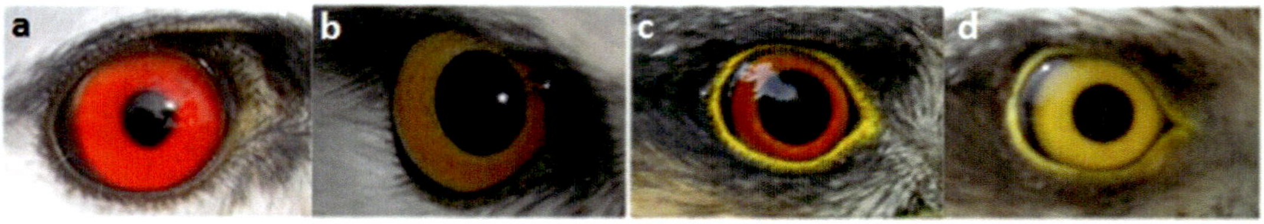


**Fig. 4.80** Examples of changes in iris color with age and sex. Eyes of (**a**) adult and (**b**) juvenile Black-winged Kite (*Elanus caeruleus*) and an (**c**) adult male and (**d**) adult female Eurasian Sparrowhawk (*Accipter nisus*). (Figure from Negro et al. 2017; open-access article distributed under the terms of the Creative Commons Attribution 4.0 International License, https://creativecommons.org/licenses/by/4.0/)

**Fig. 4.81** (**a–d**) Change in iris color with increasing age of Cooper's Hawks (*Accipiter cooperii*). The ages at which color changes varies geographically, between the sexes, and among individuals within sexes (Rosenfield et al. 2003). Because of this variation, iris color may not be a reliable means of ageing Cooper's Hawks. (Photos by Allison Miller and used with permission)

and smaller blind areas. These latter species tend to be those that are tactile foragers, detecting and acquiring food items using tactile receptors in their bills (Martin 2017).

Beyond food acquisition and predator detection, most birds fly and so, especially in complex habitats like forests, the frontal or binocular visual field is important. However, even species with very narrow binocular fields, like American

**Fig. 4.82** Top, Yellow-eyed Penguin (*Megadyptes antipodes*). Bottom, Relationship between eye saturation (a measure of the intensity of a color) and annual breeding success of 5-year-old male and female Yellow-eyed Penguins. (Top photo by B. Spragg, Wikipedia CC0; Bottom figure from Massaro et al. 2003; © 2003 Springer Nature, used with permission)

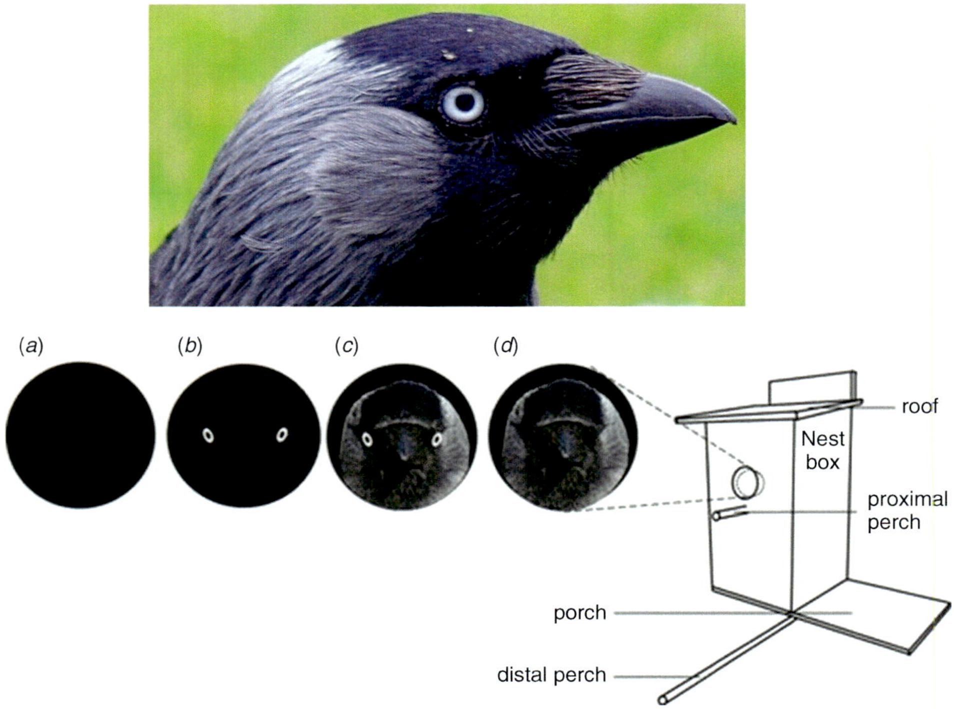


**Fig. 4.83** Top, A Eurasian Jackdaw (*Corvus monedula*) showing the light-colored iris. Bottom, The behavior of conspecifics responding to four possible stimuli placed at the entrance to a nest box), including (**a**) control (solid black circle), (**b**) bright eyes only, (**c**) image of jackdaw face with bright eyes, and (**d**) image of jackdaw face with dark eyes. The jackdaw face with bright eyes was most effective at deterring conspecifics from visiting next boxes, but both stimuli with bright eyes (**b** and **c**) reduced the amount of time conspecifics spend near nest boxes. (Top photo of a Eurasian Jackdaw by John Haslam, Wikipedia, CC BY 2.0, https://creativecommons.org/licenses/by/2.0/; Bottom figure from Davidson et al. 2014; © 2014 The Authors. Published by the Royal Society, used with permission)

Woodcocks (*Scolopax minor*), can fly through such habitats, suggesting that variation among species in visual fields is driven more by food acquisition and predator detection than the demands of flight (Martin 2017).

The avian retina consists of several different types of neurons arranged into three nuclear and two synaptic layers, with nuclear layers containing the nuclei of neurons (outer nuclear layer, inner nuclear layer, and ganglion cell layer) and synaptic layers (inner plexiform layer and outer plexiform layer) containing synapses (where impulses are transmitted between neurons) (Figs. 4.107 and 4.108). The avian retina, like those of other vertebrates, is inverted, with the photoreceptors (rods and cones) in the innermost layer of the retina so light must pass through the other layers before reaching the photoreceptors. Once stimulated, photoreceptor cells initiate nervous impulses that are then transmitted by the bipolar cells to the ganglion cells and then through the optic nerve to the brain. In addition to these neurons, the retina also contains Müller cells that span the entire thickness of the retina. Müller cells help maintain the retinal extracellular fluid, removing debris and waste products (like carbon dioxide) and providing energy for photoreceptor cells. Light striking the retina must pass through the ganglion cells, bipolar cells, and Müller cells before reaching the photoreceptor cells and, as a result, some "scattering" occurs as light strikes these cells. However,

**Fig. 4.84** Great Reed Warblers (*Acrocephalus arundinaceus*) attacked models of Common Cuckoos (*Cuculus canorus*) with yellow eyes significantly more vigorously than they attacked models with black eyes. Left, Common Cuckoo; Right, Great Reed Warbler. (Photo of Common Cuckoo by Andy Morffew, Wikipedia CC BY 2.0, https://creativecommons.org/licenses/by/2.0/; Photo of Great Reed Warbler by Agustin Povedano, Wikipedia CC BY 2.0, https://creativecommons.org/licenses/by/2.0/; Figure of cuckoo head images from Trnka et al. 2012; © 2012 Trnka et al., open-access article distributed under the terms of the Creative Commons Attribution License, https://creativecommons.org/licenses/by/4.0/)

despite this scattering, most birds, with their high concentrations of photoreceptor cells, have excellent visual acuity because Müller cells also serve as "light-guiding" structures, transmitting light through the retina to the photoreceptor cells (Figs. 4.109 and 4.110). Müller cells contain few organelles like mitochondria that would scatter light and have numerous long thin filaments oriented along the cell axis (Fig. 4.111). The ends of Müller cells cover the entire inner retinal surface and have a low refractive index so light passes efficiently from the vitreous humor into the Müller cells. These characteristics make Müller cells very effective light collectors and transmitters (Franze et al. 2007; Reichenbach et al. 2012; Zueva et al. 2014; Khmelinskii and Makarov 2019).

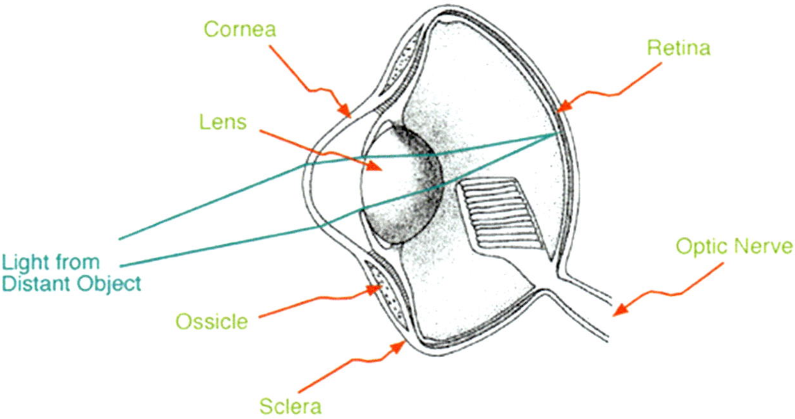


**Fig. 4.85** Typical avian eye. Light is refracted by the cornea and lens and focuses on the retina where sensory cells convert light into sensory impulses. (Figure modified from Meyer 1977; © 1977 Springer-Verlag Berlin Heidelberg, used with permission)

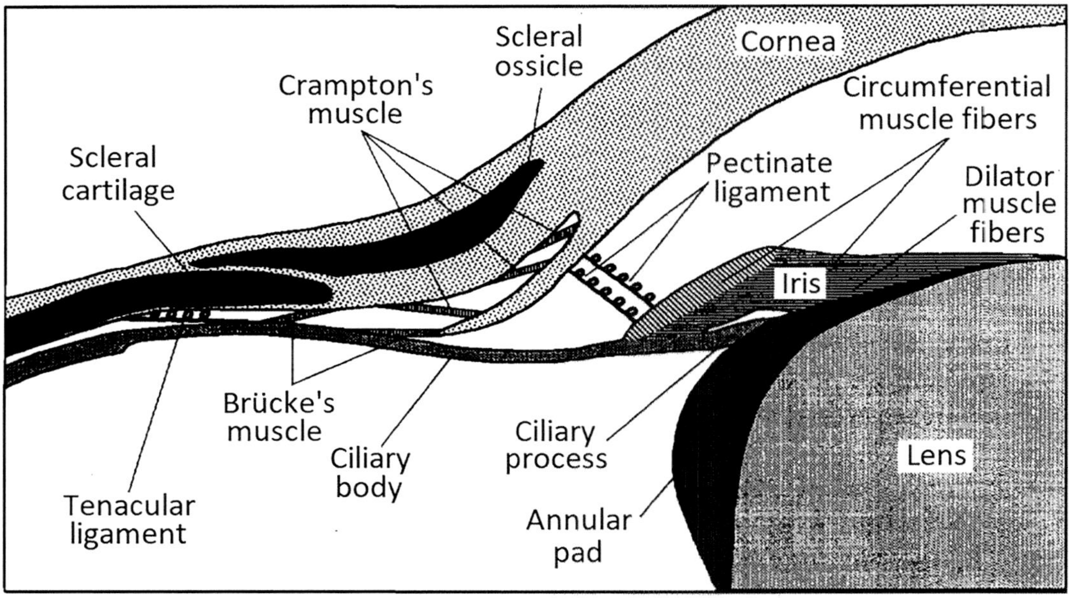


**Fig. 4.86** The ciliary region of the eye of a Domestic Chicken. The ciliary muscle has two divisions: Crampton's muscle and Brücke's muscle. The iris lies against the anterior surface of the lens and is attached to the lens by ciliary processes. The iris musculature consists primarily of circumferential muscle fibers. The ciliary body consists primarily of connective tissue and is located between the iris and the sclera. The ciliary body is attached to the sclera by Brücke's muscle and the tenacular ligament. The elastic pectinate ligament extends from base of the cornea to the iris. (Figure from Glasser et al. 1995; © 1995 Published by Elsevier Ltd., used with permission)

The retina of almost all birds has three types of photoreceptors that transform light into nervous impulses: rods for black and white vision, four single cones for color vision, plus double cones consisting of a larger principal and small accessory cone (Höglund et al. 2019; Fig. 4.112, Box 4.14 Phototransduction). The only exceptions are some species of penguins that have lost one type of cone (MWS cone) so only have three single cones (Bowmaker and Martin 1985). Double cones are the most abundant type of photoreceptor cell, usually making up 40–50% of all cones and ranging from 30% in the red field of the Rock Pigeon (*Columba livia*) retina (Galifret 1968) to 80% in the retina of the Magellanic Penguin (*Spheniscus magellanicus*; Suburo and Scolaro 1999). Double cones probably do not contribute to color vision, with proposed functions including helping to detect and perceive motion (Höglund et al. 2019), luminance detection (Lind et al. 2014a), polarization vision (Lind et al. 2014a, b; Kamermans and Hawryshyn 2011), and light-dependent magnetoreception (Günther et al. 2018; Chetverikova et al. 2022).

Rods are generally more abundant in the retinas of nocturnal species, are more sensitive to light than cones, and are important under scotopic (low light) conditions. Cones are more abundant in diurnal species of birds, are less sensitive to light, function best with brighter light (photopic) conditions, and are important for color vision. Although many species and taxa of birds are more nocturnal (e.g., Caprimulgiformes and Strigiformes) and others more diurnal (e.g., ratites and Galliformes), many others can be active at any time of day. For example, many birds considered diurnal must be active at night (e.g., night-migrating birds) or sometimes choose to be active at night (e.g., singing, foraging, or seeking-extra pair mates). Numerous genes have been found to influence avian vision, and those important in

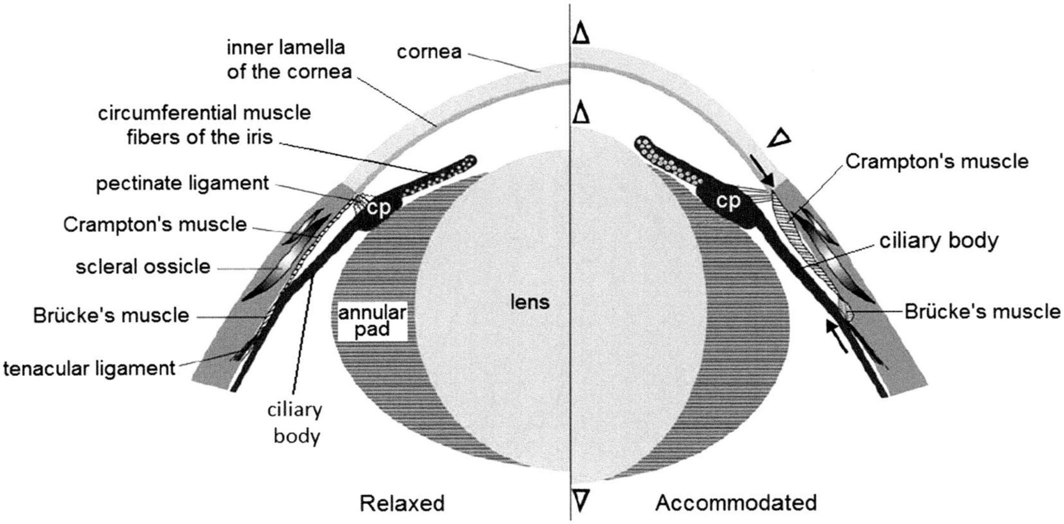


**Fig. 4.87** Crampton's muscle pulls down on the inner edge (lamella) of the cornea (as indicated by the black arrow) and causes the central portions of the cornea and lens to bulge outward. Contraction of Brücke's muscle pulls the ciliary body forward (as indicated by the black arrow), which reduces the tension applied to the lens by the ciliary body and tenacular ligament. This causes the circumferential muscle fibers of the iris to exert pressure against the ciliary process (cp) that is fused with the annular pad of the lens. This squeezes the anterior surface of lens and causes it to bulge forward to increase its refractive power. The open triangles show the resulting changes in the position of the cornea and the shape of the lens. Ciliary processes attached to the annular pad of the lens. (Figure modified from Ott 2006; © 2006 Springer Nature, used with permission)

enhancing rod dim-light vision have been found to be under positive selection in taxa of birds that are primarily or entirely nocturnal (Wu et al. 2016; Wu and Wang 2019). However, Wu (2019) found similar positive selection in several other orders of birds, and also provided evidence suggesting that the common ancestor of present-day birds was likely active under low-light conditions (Fig. 4.113). These results suggest that nocturnality among present-day birds was

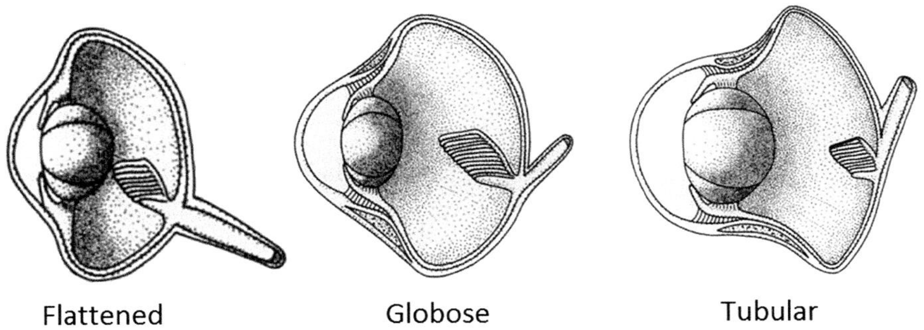


**Fig. 4.88** Basic shapes of avian eyes. Most birds have relatively flattened eyes that provide a large field of view. Many diurnal raptors have longer, globose eyes that provide a larger image on the retina. The tubular eyes of some nocturnal species, like owls, have a large cornea that allows more light into the eye. In addition, the length of tubular eyes means that images projected by the lens to the retina are relatively large. (Figure modified from Waldvogel 1990; © 1990 Sigma Xi, used with permission)

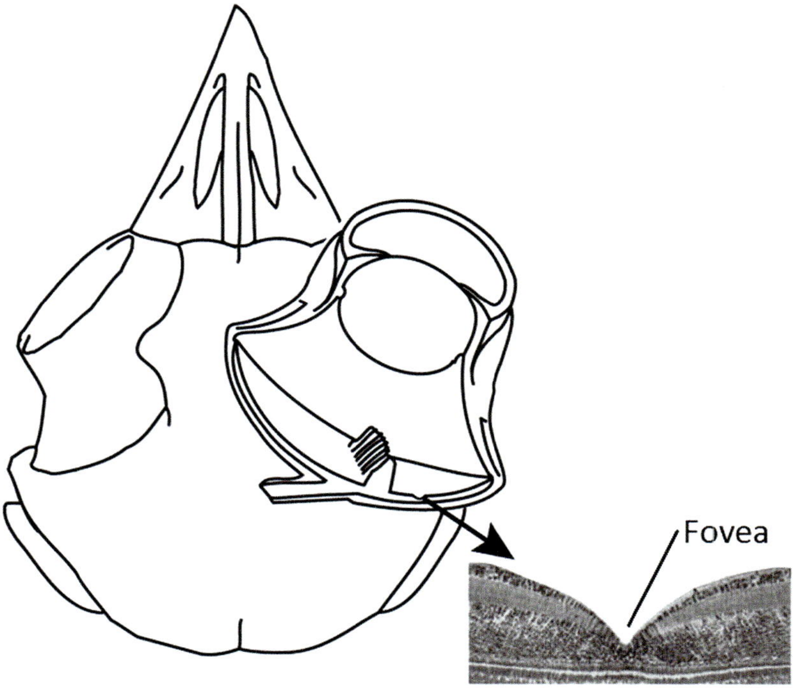


**Fig. 4.89** Nocturnal owls have long, tubular eyes, which translates into relatively large images projected onto the retina. The retina of owls also has a deep fovea with high concentrations of photoreceptors cells (rods) and increased visual acuity. (Figure from Moritz et al. 2014; open-access article distributed under the terms of the Creative Commons Attribution License (CC BY), https://creativecommons.org/licenses/by/4.0/)

likely a trait retained from common ancestors rather than being derived independently and that species of birds in several orders may have better low-light vision than previously thought (Wu 2019).

Avian visual opsins include rhodopsin (RH1) and four different conopsins (RH2, SW1, SW2, and LWS). LWS is long-wavelength sensitive (maximum sensitivity = 505–630 nm), RH2 is medium-wavelength sensitive (maximum sensitivity = 460–540 nm), SWS2 is short-wavelength sensitive type 2 (maximum sensitivity = 410–475 nm), and SWS1 is short-wavelength sensitive type 1 (Terakita 2005; Hart and Hunt

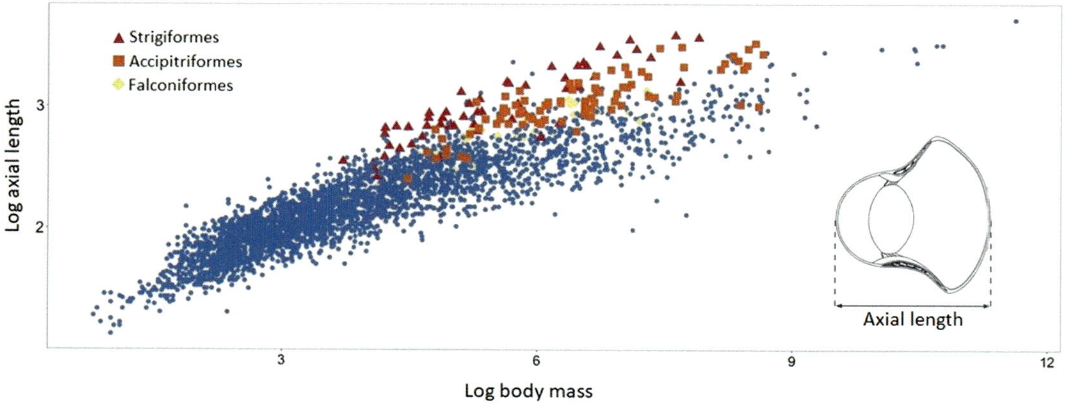


**Fig. 4.90** Axial length of the eyes relative to body mass for raptors compared to other birds. (Figure from Potier et al. 2020; © 2020 The Authors. Published by Elsevier Ltd., used with permission)

**Fig. 4.91** Dorsal and side views of excised eyeballs of four species of owls. (**a**) Northern Saw-whet Owl (*Aegolius acadicus*). (**b**) Short-eared Owl (*Asio flammeus*). (**c**) Great Gray Owl (*Strix nebulosa*). (**d**) Northern Hawk Owl (*Surnia ulula*). Scale bars = 10 mm. Photos (left to right): Northern Saw-whet Owl, Short-eared Owl, Great Gray Owl, and Northern Hawk Owl. (Figure from Lisney et al. (2012a, b); © 2012 S. Karger AG, Basel, used with permission. Photos: Northern Saw-whet Owl by Any Reago and Chrissy McClarren, Wikipedia, CC BY 2.0; Short-eared Owl by Bex Ross, Wikipedia, CC BY 2.0; Great Gray Owl by Eric Johnston/National Park Service, CC0 Public Domain; Northern Hawk Owl, Wikipedia, CC0 Public Domain)

2007). Among birds, there are two types of SWS1 opsins, including violet sensitive (VS; maximum sensitivity = 402–426 nm) and ultraviolet sensitive (UVS; maximum sensitivity = 355–380 nm), with the difference due to changes in just a few amino acids in the SWS1 opsin (Ödeen and Håstad 2013; Fig. 4.114). Humans, in contrast to birds, have only three different conopsins (Fig. 4.115). All cones except the accessory member of double cones have oil droplets pigmented with carotenoids that filter the spectrum of light on its way to the photosensitive outer segments and improve color discrimination (Olsson et al. 2016; Figs. 4.116 and 4.117). Rods do not have oil droplets. The oil droplets of cones have a relatively high refractive index and are spherical in shape (Wilby et al. 2015). Because the refractive index of the oil droplets is higher

**Fig. 4.92** Part of the nictitating membrane of a Eurasian Eagle-Owl (*Bubo bubo*) can be seen in its left eye. (Photo from pxhere.com, CC0 Public Domain)

than that of the surrounding cytoplasm, they act as microlenses. The pigmented oil droplets are largely transparent to long wavelengths of light, but absorb most short-wavelength light below a specific wavelength (λcut) (Toomey and Corbo 2017; Fig. 4.117). Avian spectral sensitivity is determined by the oil droplet spectral filtering and the photopigment sensitivity of each type of cone, with photopigment sensitivity determined by the type of opsin proteins in the cones.

Each cone type has a specific oil droplet: transparent (T, no pigment) in the UV/V cone, clear (C) in the blue cone, yellow (Y) in the green cone, and red (R) in the red cone. Double cones contain

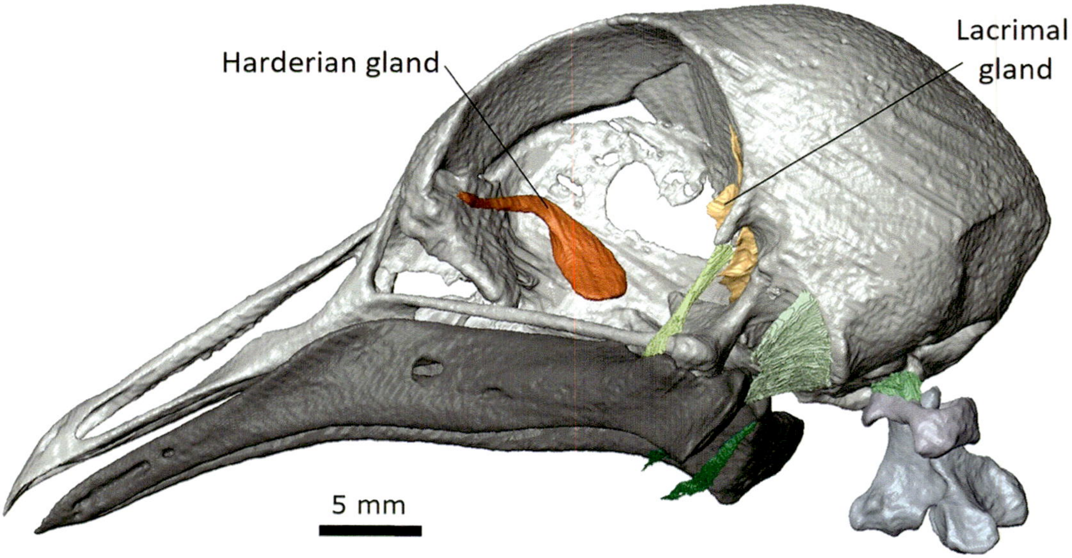


**Fig. 4.93** Harderian and lacrimal glands of a Rock Pigeon (*Columba livia*). (Figure modified from Jones et al. 2019; open-access article distributed under the terms of the Creative Commons Attribution 4.0 International License, https://creativecommons.org/licenses/by/4.0/)

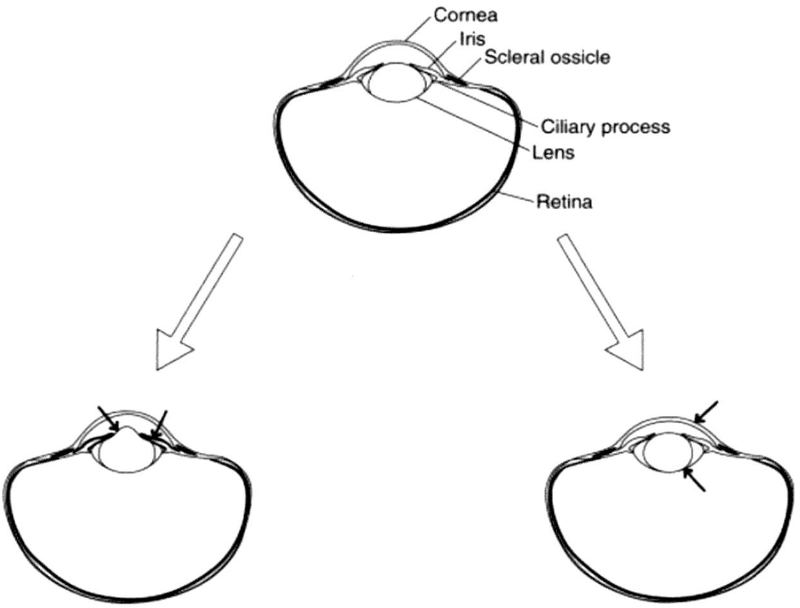


**Fig. 4.94** Mechanisms for accommodation under water. Left, Ciliary muscles contract and the lens protrudes through the pupil. Right, Flatter corneas like those of penguins and albatrosses, with a spherical lens with sufficient diameter to accommodate under water. (Figure from Cronin 2012; © 2012 Elsevier Ltd., used with permission)

the LWS opsin and pale (P) oil droplets. Oil droplets are categorized based on their cutoff wavelength (λcut) values, i.e., the wavelength below which no significant light is transmitted (Lipetz 1984; Fig. 4.118). Long-wavelength-sensitive (LWS) pigments are typically red (R-type). In double cones, the principal photoreceptor has an oil droplet that is colorless or pale (P-type). Cones categorized as middle-wavelength sensitive (or MWS) have yellow (Y-type) oil droplets, those categorized as short-wavelength sensitive type 2 (SWS2) have clear (C-type) oil droplets, and those categorized as short-wavelength sensitive type 1 (SWS1) have transparent (T-type) oil droplets with no carotenoid pigments (Porter et al. 2014). Although all birds have cones with oil droplets, the pigments they contain varies. For example, birds that are nocturnal or crepuscular have more pale or colorless oil droplets and, in addition, owls have no red cone droplets (Hart et al. 2006; Alix et al. 2017). In addition, King (*Aptenodytes patagonicus*) and Humboldt (*Spheniscus humboldti*) penguins have oil droplets that contain less pigment and, like owls, have no red cone droplets (Bowmaker and Martin 1985; Gondo and Ando 1995). Although not nocturnal, these penguins dive to depths where light is limited and droplets with less pigment may increase the sensitivity of the receptors (Toomey and Corbo 2017). The loss of red oil droplets is likely also beneficial for birds that are active in low light condition because red oil droplets absorb wavelengths of light at lower frequencies and reduce the amount of light available to the receptors.

Species of birds in most orders are violet sensitive (VS); orders of birds where species are ultraviolet sensitive (UVS) or where some species are violet sensitive and others are ultraviolet sensitive include Palaeognaths (Common Ostrich [*Struthio camelus*], Emu [*Dromaius novaehollandiae*], rheas, and kiwis), Pteroclidiformes (sandgrouse), Charadriiformes, Psittaciformes, and Passeriformes (Fig. 4.119). During the evolution of birds, there have been several shifts between VS and UVS vision, with shifts from VS to UVS vision occurring eight or nine times and shifts from UVS to VS occurring five or six times (Ödeen and Håstad 2013; Fig. 4.120). One factor that might limit shifts from VS to UVS vision is the ocular media, including the cornea, aqueous humor, lens, and vitreous humor because, to see ultraviolet light, that part of the spectrum must be able to pass

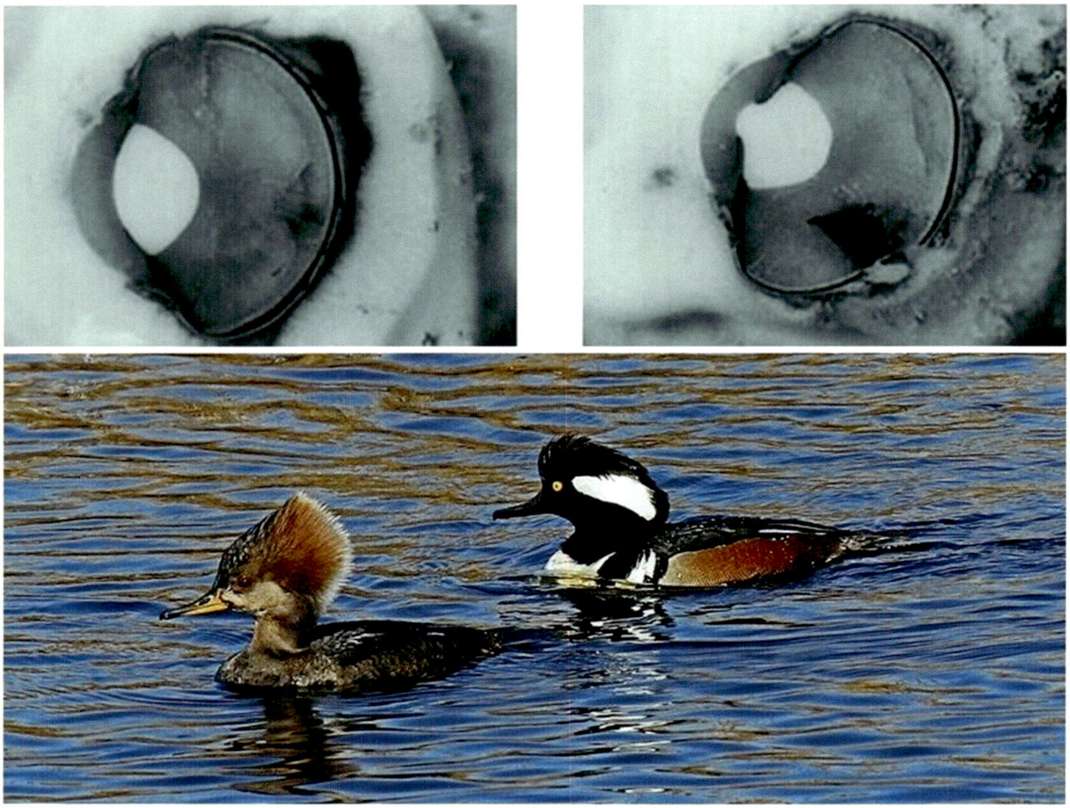

**Fig. 4.95** Top, Accommodation in the eye of a Hooded Merganser (*Mergus cucullatus*). (Left) Sagittal section of the eye of a Hooded Merganser without accommodation. (Right) Sagittal section of the eye of a Hooded Merganser stimulated with nicotine sulphate to approximate accommodation. When accommodating, muscle contraction (Brucke's muscle) forces the ductile lens though the rigid pupil creating a "bulge" (called the anterior lenticonus). Bottom, Female and male Hooded Mergansers. (Figure from Schwab 2002; © 2002 British Medical Journal, used with permission. Photo by Judy Gallagher, Wikipedia, CC BY 2.0, https://creativecommons.org/licenses/by/2.0/)

through the ocular media. Lind et al. (2014a, b) examined the ocular media transmittance of 38 species of birds from 11 orders and found that the ocular media of birds that are ultraviolet sensitive transmit shorter wavelengths of light than the ocular media of birds that are violet sensitive (Figs. 4.121). However, these investigators also noted that this difference was likely of limited importance in color discrimination except under low-light conditions like the cavity nests of songbirds and parrots. In dim light, absorbance by the ocular media would reduce the sensitivity of UVS and VS cones so birds with ocular media that transmit shorter wavelengths are likely better able to perceive ultraviolet light under those conditions than birds with ocular media that do not transmit those shorter wavelengths (Lind et al. 2014a, b).

The retinas of nocturnal birds like many species of owls are dominated by rods (e.g., Ault 1984; Alix et al. 2017), but cones may also be present. For example, the retina of Tawny Owls (*Strix aluco*) is dominated by rods, but they also have four types of cones (although not the T-type,

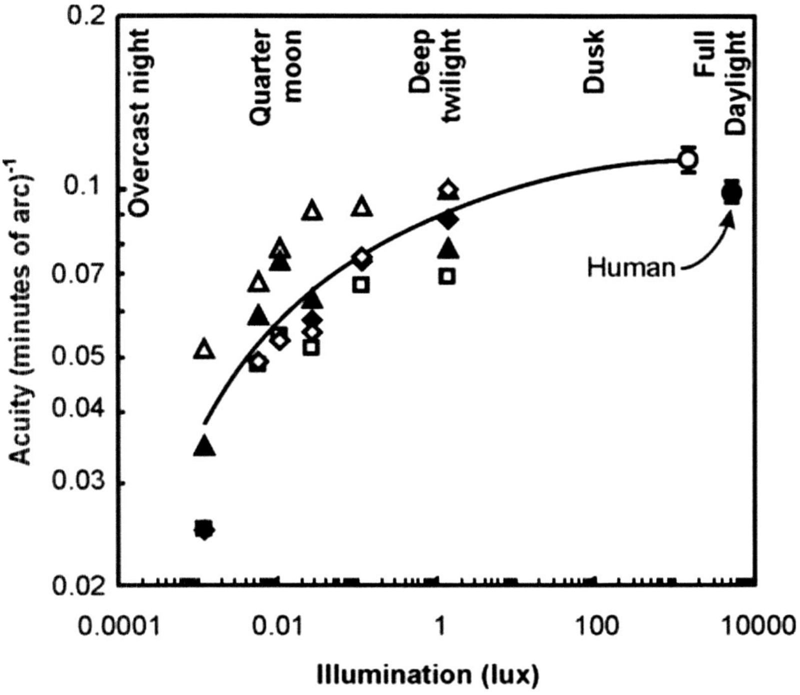


**Fig. 4.96** Effect of illumination (lux) on the visual acuity of five Great Cormorants (*Phalacrocorax carbo*) (symbols represent different birds. Visual acuity declines significantly with decreasing levels of available light. Acuity of 0.1 min of arc is equivalent to 3 cpd; a human with 20/20 vision has an acuity equivalent to 30 cpd. (Figure from White et al. 2007; © 2007 White et al., open-access article distributed under the terms of the Creative Commons Attribution License, https://creativecommons.org/licenses/by/4.0/)

or SWS1, important in ultraviolet vision), and so have at least some ability to discriminate colors (Bowmaker and Martin 1978). Similar results have been reported for Barred Owls (*Strix varia*) and Northern Spotted Owls (*Strix occidentalis caurina*) (Hanna et al. 2017). The retina of Eurasian Eagle-Owls (*Bubo bubo*) also has many cones, including some with T-type oil droplets (Alix et al. 2017). In contrast to owls, one group of nocturnal birds, the Caprimulgiformes, have retained the SWS1 opsin (violet sensitive; Ödeen and Håstad 2003). Surprisingly, although owls have lost the SWS1 opsin, the ocular media of owls transmit ultraviolet wavelengths (Fig. 4.121). Höglund et al. (2019) suggested that this represents an adaptation of increase the sensitivity of rods under low-light conditions (Fig. 4.122). Wu et al. (2016) found strong positive selection for long-wave-sensitive (LWS) and short-wave-sensitive (SWS2) genes in five species of owls and suggested that this might be due to an ancestral niche switch of owls from being diurnal to being crepuscular. Many owls are known to be active at dawn and dusk and the LWS and SWS2 visual pigments might help maximize visual sensitivity under low-light conditions.

A structure found only in the eyes of birds is the pecten oculi (Figs. 4.123 and 4.124). The pecten oculi is a highly vascular and pigmented structure that projects into the vitreous chamber in the lower portion of the retina, and its size and morphology vary among different taxa and species of birds (Fig. 4.125). With the exception of kiwi pectens that are smooth narrow cones (Sillman 1973), pectens have a base, bridge, and a variable number of folds or pleats (Fig. 4.123). Because about two-thirds of the total volume of a pecten consists of blood vessels (Kiama et al. 2001; Fig. 4.126) and the avian retina is avascular (contains to blood vessels; Figs. 4.127 and 4.128), the majority opinion concerning the function of pectens is that they supply the avian retina with oxygen and nutrients (e.g., Walls 1942; Kiama et al. 1998; Wolburg et al. 1999). The pleats of pecten increase the amount of surface area that enhances the diffusion of materials between the pecten, the vitreous humor, and

**Fig. 4.97** Simulations of what a Great Cormorant (*Phalacrocorax carbo*) sees underwater with different lighting conditions and different background contrast. (**a**) Great Cormorant, (**b**) fish as might be viewed by a Great Cormorant with different levels of background contrast at a light level of 10 lux, and (**c**) fish as might be viewed by a Great Cormorant at a distance of 1 m and with a background contrast of 60%, but with different light levels. With very low light levels, cormorants may not clearly see fish prey and, as a result, strike accuracy may be affected. (Figure modified from Martin 2012; ©2012 Springer Nature, used with permission)

retinal cells (Kiama et al. 2001). In addition, Pettigrew et al. (1990) suggested that eye movements (saccades) cause back-and-forth movements of the pecten that help optimize retinal perfusion. Beyond the apparent nutritive function of pectens, investigators have suggested many other possible functions, including regulation of intraocular pressure (Seaman and Himelfarb 1963), reduction of intraocular glare (Barlow and Ostwald 1972), and regulation of intraocular pH (Brach 1975), but little or no progress has been made in terms of testing these various hypotheses. However, van den Hout and Martin (2011) suggested that the pecten might be important in protecting the retina of some species of shorebirds from direct sunlight (Fig. 4.129).

In the avian retina, photoreceptors transduce light energy via stimulation of bipolar cells, horizontal cells, amacrine cells, and, finally, ganglion cells that conduct nervous impulses through the optic nerve to visual centers in the brain (Fig. 4.130). In the absence of light, bipolar

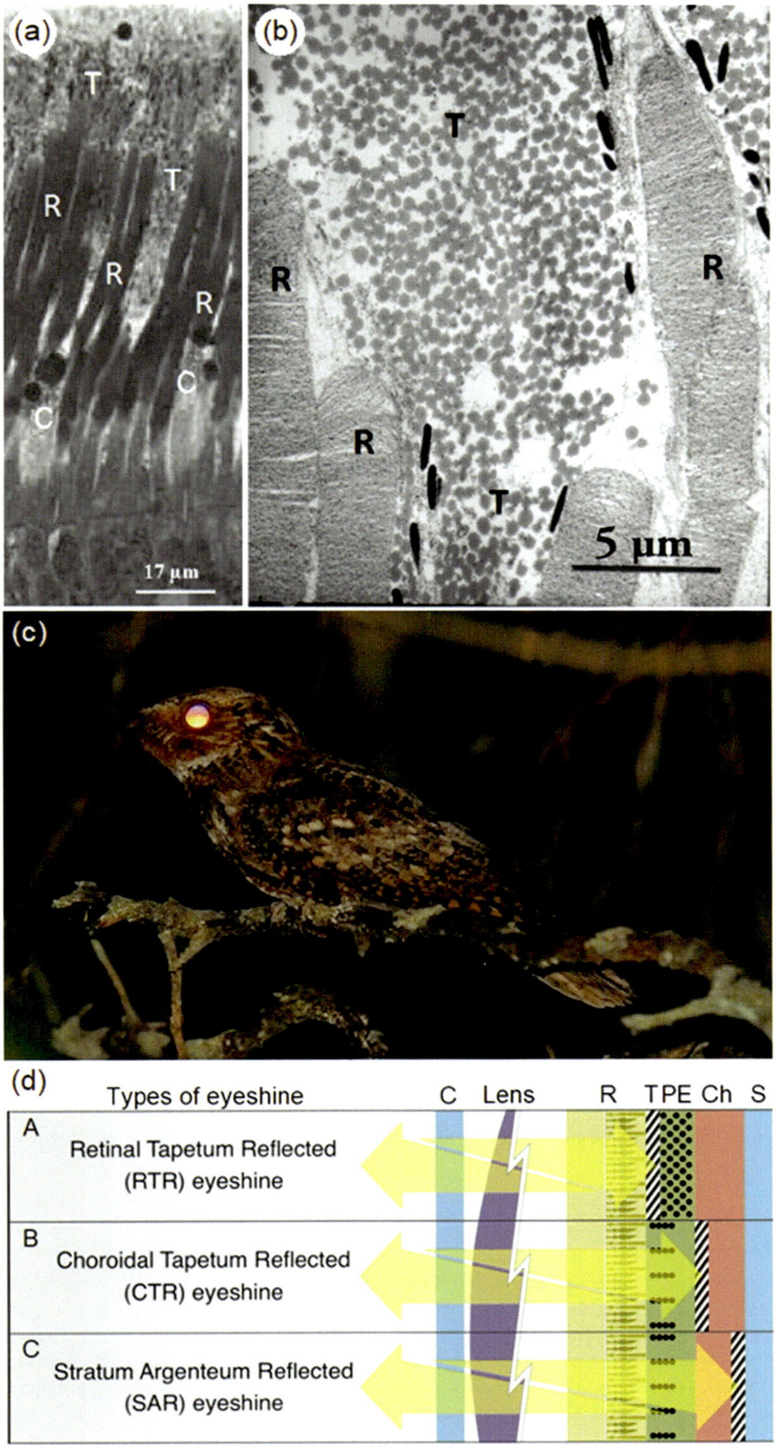


**Fig. 4.98** (**a**) Retina and tapetum of a Common Pauraque (*Nyctidromus albicollis*), with (**b**) a more highly magnified view. The tapetum of caprimulgids consists of numerous small spheres filled with pigmented lipid. The tiny spheres act as diffuse reflectors, reflecting light at a variety of angles back through the retina. T, tapetum lucidum;

neurons are inhibited by the rods and cones. However, exposure to light hyperpolarizes rods and cones, and bipolar cells, no longer inhibited, can stimulate ganglion cells. Horizontal cells can transmit signals from one rod or cone to other photoreceptors and to several bipolar cells. When stimulated, horizontal cells can inhibit adjacent photoreceptors and bipolar cells to enhance contrast in images by making regions receiving light appear lighter and dark surroundings appear darker. Amacrine cells regulate the transfer of signals from bipolar cells to ganglion cells and play an important role in motion detection (Kim et al. 2015).

The distribution of photoceptors and retinal ganglion cells in the retinas of birds varies among taxa and species. The number and distribution of retinal ganglion cells is important because they are the cells that send nervous impulses to the brain and, as such, the spatial distribution of retinal ganglion cells affects spatial resolving power, i.e., visual acuity. Also important is the ratio between the number of photoreceptor cells and number of retinal ganglion cells. If many photoreceptors connect with a single ganglion cell (high convergence ratio), sensitivity to light increases because light is more likely to be detected with more photoreceptors, but at the expense of visual resolution or acuity (Warrant 1999). In contrast, if fewer photoreceptors connect with a single ganglion cell (low convergence ratio), sensitivity is reduced, but visual resolution (i.e., being able to detect fine details of an object) is improved because impulses from more different photoreceptors are being transmitted to visual areas of the brain (Warrant 1999). These ratios vary among taxa and species and also within individual retinas. Within retinas, areas that provide the greatest visual acuity have lower convergence ratios than other areas (Fig. 4.131). Among species, convergence ratios in the area of retinas with the greatest visual acuity (foveas) have been found to vary from 1.2 to 5.9 (Table 4.2). By way of comparison, a typical convergence ratio in the fovea of humans is about 1:3, providing us with excellent visual acuity at the expense of sensitivity (Table 4.2).

In avian retinas, locations with higher concentrations of photoceptors or retinal ganglion cells are referred to as visual streaks, areas, or foveas (Fig. 4.132). Species in several orders of birds have visual streaks, linear sections in the retina with higher densities of photoreceptors and retinal ganglion cells (Table 4.3). Such streaks are characteristic of birds found in open habitats such as grasslands, tundra, deserts, and the open ocean with clear views of the horizon, providing a wide field-of-view of the horizon with high visual acuity without needing extensive eye or head movements (Collin 1999, 2008). Visual streaks, in combination with eyes that are laterally placed, provide large monocular visual fields that may be particularly important for detecting approaching predators (Lisney et al. 2013b). For waterfowl in flight, visual streaks may be important in orientation with respect to both the horizon and conspecifics during formation flight (Fernández-Juricic et al. 2011a, b). Some species of birds with visual streaks have either one (e.g., Common Ostrich [*Struthio camelus*] and species in the order Anseriformes) or two (e.g., flamingoes) areas within the streak with higher densities of

**Fig. 4.98** (continued) R, rods; C, cones. (**c**) The reflected light causes the eyes of caprimulgids to "glow" when viewed with a light at night, creating what is called "eyeshine" as shown with this Eastern Whip-poor-will (*Antrostomus vociferus*). (**d**) Different types of eyeshine include (A) retinal tapetum reflected, (B) choroidal tapetum reflected (e.g., many mammals), and (C) stratum argenteum reflected (e.g., some species of fish). Birds with tapetums have retinal tapetums. C, cornea; R, retina; T, tapetum lucidum; PE, retinal pigmented layer; Ch, choroid; S, sclera. (Figures **a** and **b** modified from Rojas et al. (2004a, b); © 2004 Neotropical Ornithological Society, used with permission; Photo of Whip-poor-will by Scott Somershoe, used with permission. Figure **d** modified from Fritsch et al. 2017; © 2017 The Authors, open-access article distributed under the terms of the Creative Commons Attribution 4.0 International License, https://creativecommons.org/licenses/by/4.0/)

**Fig. 4.99** Examples of variation in the position of eyes on the heads of birds. (Photos **a**, **b**, **c**, **e**, **f**, **g**, **i**, **j**, **k**, and **l** by Sandy Scott and used with permission; Photo **d** by Andy Morffew, CC0 Public Domain, pxhere.com, and photo **h**, CC0 Public Domain, pxhere.com)

**Fig. 4.100** Three species of birds with eye positions in the skull that result in very different visual fields. (**a**) Barn Owls (*Tyto alba*) and other owls have forward-facing eyes that provide relatively wide fields of binocular vision. (**b**) Most birds, like this European Robin (*Erithacus rubecula*), have less forward-facing eyes on the sides of the skull that provide a wider visual field, but with a narrower field of binocular vision and wider fields of monocular vision. (**c**) Few birds have eyes positioned on opposite sides of the skull, like this Eurasian Woodcock (*Scolopax rusticola*) facing away from the camera, that provide very narrow fields of binocular vision both in the front and back of the head, but with extremely wide fields of monocular vision. (Photo of Barn Owl from pxhere.com, CC0 Public Domain; Photo of European Robin from pxhere.com, CC0 Public Domain; Photo of Eurasian Woodcock by James Bamford O'Neill and used with permission; Bottom figure from Fernández-Juricic et al. 2004; © 2003 Elsevier Ltd., used with permission)

photoceptors and increased visual acuity (Table 4.3).

In constrast, the retinas of birds found in forest habitats or that are nocturnal are typically more radially symmetrical and have one or more areas with higher densities of photoreceptors referred to as areas or foveae. Without visual streaks, these birds rely more on eye and head movements to focus objects of interest on their areas or foveas (Collin 1999, 2008). Areas are simply locations on a retina with higher concentrations of receptors than other areas and, therefore, greater visual acuity, whereas a fovea is "a pitted invagination in the inner retina that overlies an area of densely packed photoreceptors specialized in high acuity vision" (Bringmann 2019; Fig. 4.133). Some birds have a single area or fovea, whereas others have two areas (e.g., Brown-eared Bulbuls, *Hypsipetes amaurotis*) and yet others have a fovea and an area (e.g., Acadian Flycatchers, *Empidonax virescens*) (Table 4.3). Birds with a single area centralis or central fovea can locate food items or other objects monocularly, then switch to frontal binocular vision to peck at those food items or objects (Bischof 1988; Güntürkün et al. 1993). Species of birds with foveas need high visual acuity for foraging and detecting predators and conspecifics, e.g., raptors

**Box 4.13 Stereopsis**

All birds have visual fields that include areas of binocular vision, although some species, like owls, have much wider binocular fields than other species. A common misconception is that binocular vision is advantageous because it allow depth perception, an ability referred to as stereopsis, i.e., the ability to perceive depth from views acquired simultaneously from different points in space (i.e., two eyes; Nityananda and Read 2017). Binocular vision is simply the ability to see one image with both eyes simultaneously, i.e., simultaneous perception. If the two retinal images are merged in the visual cortex of the brain, then a single image is perceived and, if the images are fused in the visual cortex, but with a slight horizontal disparity, then there is a perception of depth that is referred to as stereopsis. However, stereopsis is not synonymous with depth perception because other cues that enable depth perception are available, including light transport or shading (i.e., light creates shadows that provide cues about depth) and perspective (i.e., increase in retinal size of an approaching object).

Example of how shadows can provide depth perception (Photo from pxhere.com, CC0 Public Domain)

(continued)

**Box 4.13** (continued)

Example of how retinal size provides depth perception, with closer objects appearing larger (Photo from pxhere.com, CC0 Public Domain)

Early investigators (e.g., Walls 1942) believed that stereopsis was a natural by-product of binocular vision, but neuroscientists now know that is not the case. However, even without stereopsis, binocular vision does provide some advantages, including redundancy (i.e., frontal vision even if one eye is blinded), the ability to see around occluders (i.e., objects that partially or completely impede the light entering an eye) in cluttered habitats, and improved sensitivity under low-light conditions (i.e., receptors in two eyes are more likely to detect faint light than those in one eye) (Nityananda and Read 2017). Among birds, owls are capable of stereopsis, but evidence for stereopsis in other species is "mixed and hampered by a lack of sufficient electrophysiological, behavioral and anatomical data" (Iwaniuk and Wylie 2020). Studies in the 1970s revealed that the wulst of Barn Owls (*Tyto alba*) resembled the visual cortex of mammals, including the presence of binocular neurons that only responded to a stimulus with both eyes open (Pettigrew and Konishi 1976; Pettigrew 1979). Subsequent behavioral experiments provided further evidence that owls are capable of stereopsis (van der Willigen et al. 2010).

The wulst of owls is enlarged relative to those of many other species of birds, and Iwaniuk and Wylie (2006) suggested that the wulst mediates stereopsis. If so, other birds with enlarged wulsts may also be capable of stereopsis, including Tawny Frogmouths (*Podargus strigoides*) and Feline Owlet-nightjars (*Aegotheles insignis*) (Iwaniuk and Wylie 2020). However, confirmation of this will require additional behavioral and physiological studies. As far as other species of birds, Rock Pigeons (*Columba livia*; McFadden and Wild 1986; McFadden 1994) and American Kestrels (*Falco sparverius*; Fox et al. 1977) are reportedly capable of stereopsis, but, as noted by Graham (2017), the results of the kestrel study have not been replicated and, although Rock Pigeons exhibit convergent eye movements, neurons thought to be important in stereopsis in owls and mammals have not been found in pigeons.

(continued)

**Box 4.13** (continued)

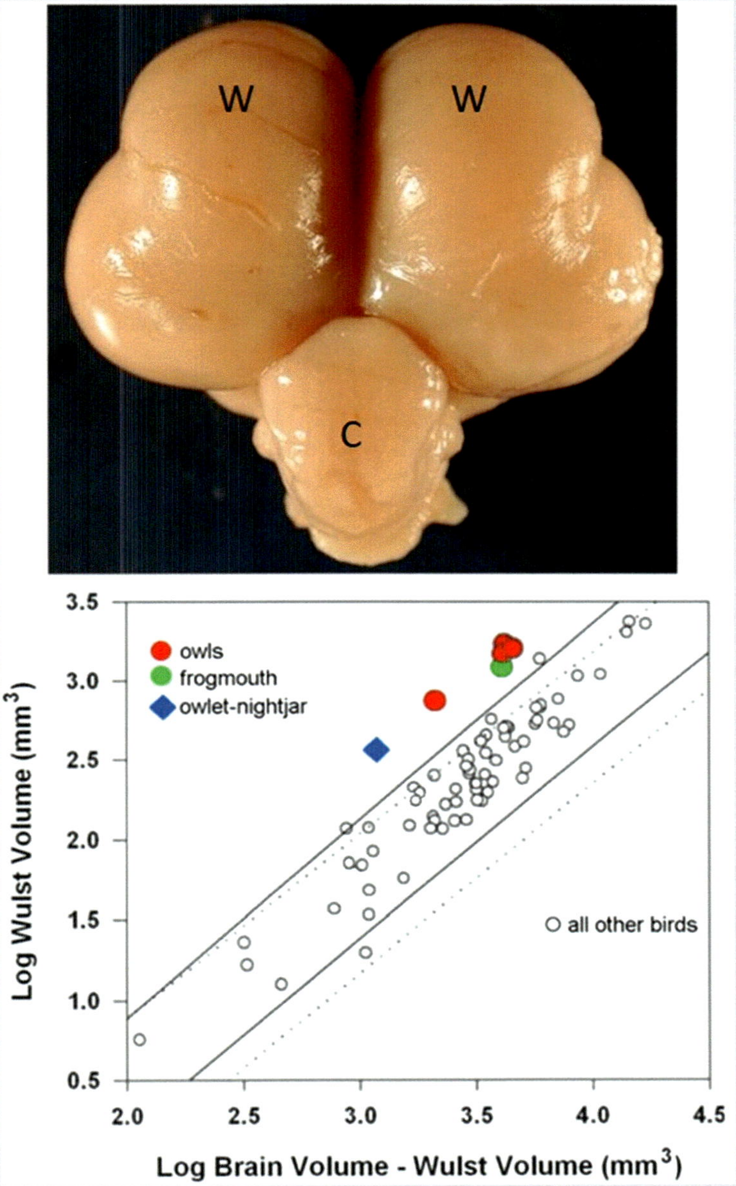


Top, Brain of a Barn Owl (*Tyto alba*) showing the enlarged wulsts (W) as well as the cerebellum (C) (wulst is a German word meaning "bulge"). Bottom, Relationship between wulst volume and brain volume—wulst volume. Note the relatively large volumes of the wulsts of owls as well as Tawny Frogmouths and Feline Owlet-nightjars compared to other birds (Figures from Iwaniuk and Wylie 2020; © 2020 Wiley Periodicals, Inc., used with permission)

(continued)

**Box 4.13** (continued)

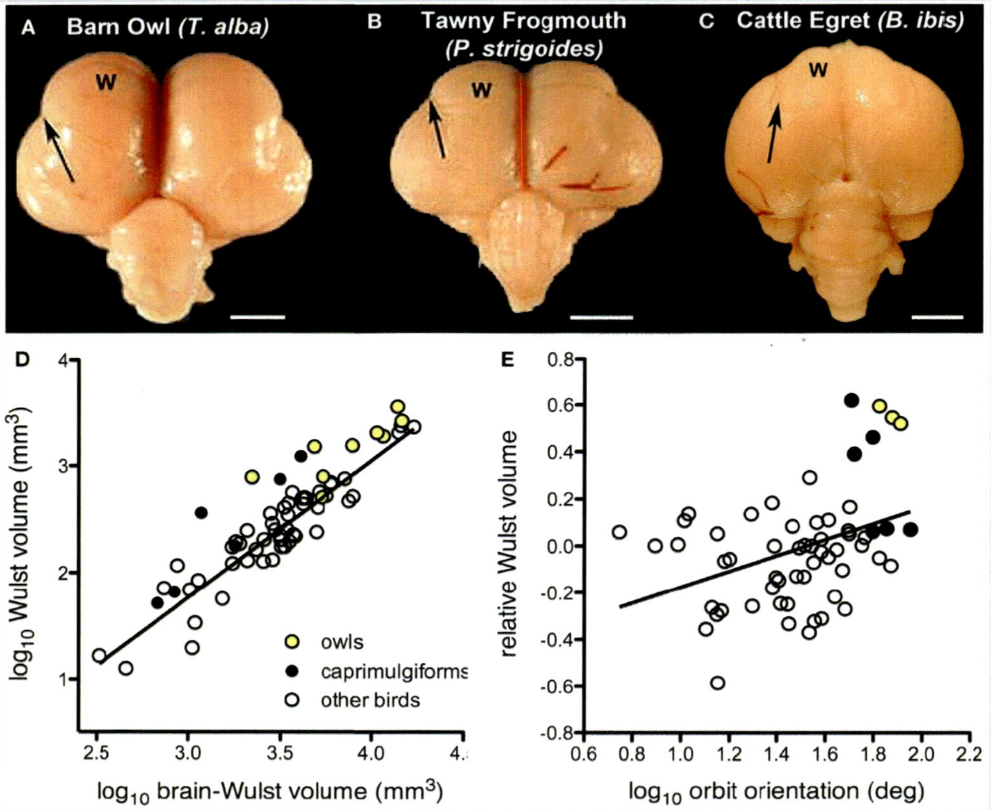


Variation in the size of the visual wulst (W) is related to binocular vision and stereopsis. (**a**–**c**) Dorsal views of the brains of a Barn Owl (*Tyto alba*), Tawny Frogmouth (*Podargus strigoides*), and Cattle Egret (*Bubulcus ibis*). The valecula, the lateral border of the wulst, is indicated by the arrow. Scale bars = 5 mm. (**d**) Relationship between Wulst volume and brain volume minus Wulst volume. (**e**) Relationship between Wulst volume and brain volume as a function of orbit orientation. Yellow circles = owls (Strigiformes), black circles = Caprimuligiformes, and open circles = other species. Caprimulgids with the largest wulst are the Oilbird (*Steatornis caripensis*), Feline Owlet-nightjar, and Tawny Frogmouth (Figure from Wylie et al. 2015; © 2015 Wylie, Gutiérrez-Ibáñez and Iwaniuk, open-access article distributed under the terms of the Creative Commons Attribution License (CC BY), https://creativecommons.org/licenses/by/2.0/)

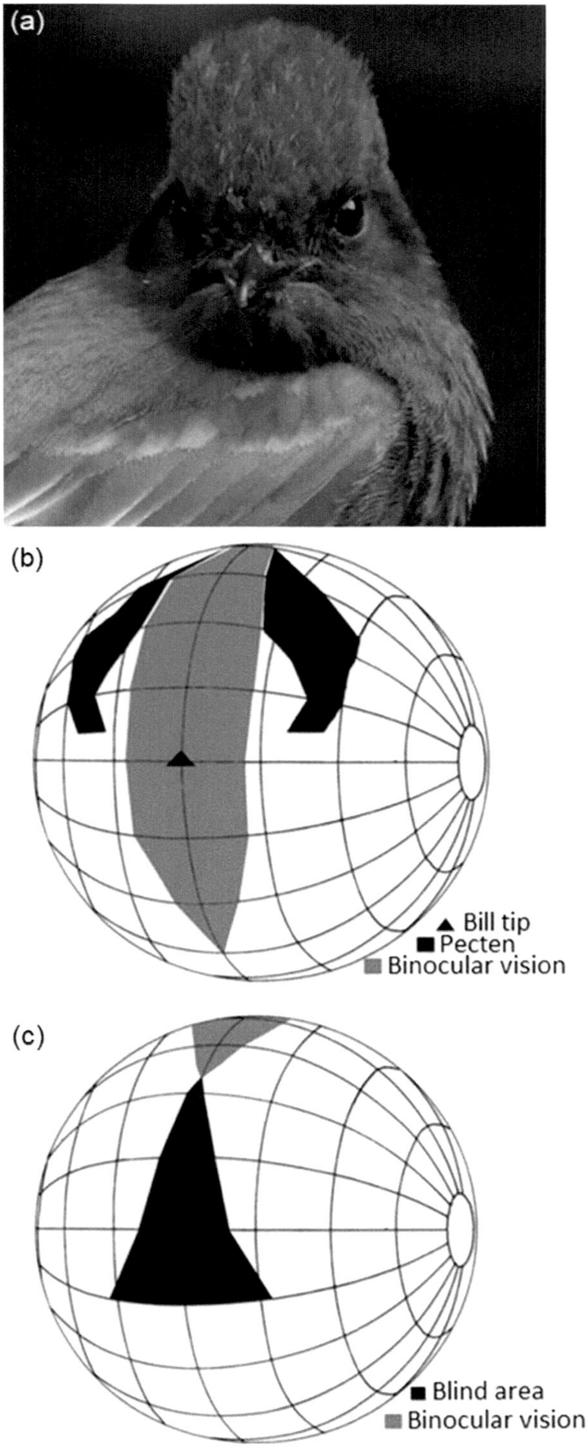


**Fig. 4.101** (**a**) Black Phoebe (*Sayornis nigricans*). (**b**) Three-dimensional representation of the visual field of a Black Phoebe when their eyes are at rest. The head would be at the center of the sphere, and the tip of the bill is indicated by the small black triangle. (**c**) The visual field behind the head. The grid is at 20° intervals. Unshaded areas indicated areas of lateral monocular vision that extend in both the horizontal and vertical planes. (Figures **b** and **c** from Gall and Fernández-Juricic 2010; © 2009 Springer Nature, used with permission. Photo of Black Phoebe by Mary K. Hanson, used with permission)

**Fig. 4.102** The position of the eyes of some birds that typically search for prey below their heads, like Black-crowned Night-Herons (*Nycticorax nycticorax*; top) and Cattle Egrets (*Bubulcus ibis*), provides them with a long, narrow frontal binocular field. (Photo of Black-crowned Night-Heron by Sandy Scott, used with permission; Cattle Egret figure from Martin 2007; © 2007 Springer Nature, used with permission)

**Table 4.1** Extent of eye movement in various species of birds

| Common name | Scientific name | Maximum eye movement (degrees) | Citation |
|---|---|---|---|
| Canada Goose | *Branta canadensis* | 24.6 | Fernández-Juricic et al. (2011b) |
| American Kestrel | *Falco sparverius* | ~1 | O'Rourke et al. (2010) |
| Red-tailed Hawk | *Buteo jamaicensis* | 5–10 | Tucker (2000b), O'Rourke et al. (2010) |
| Cooper's Hawk | *Accipiter cooperii* | 8–12 | Tucker (2000b), O'Rourke et al. (2010) |
| Great Horned Owl | *Bubo virginianus* | 1.5 | Steinbach et al. (1974) |
| Barn Owl | *Tyto alba* | 2 | Dulac and Knudsen (1990) |
| Little Eagle | *Hieraaetus morphnoides* | 2.8 | Wallman and Pettigrew (1985) |
| Great Cormorant | *Phalacrocorax carbo* | 15 | Martin et al. (2008) |
| Cattle Egret | *Bulbulcus ibis* | 18 | Martin and Katzir (1994) |
| Black-crowned Night-Heron | *Nycticorax* | 13.5 | Katzir and Martin (1998) |
| Southern Ground-Hornbill | *Bucorvus leadbeateri* | ~30–40 | Martin and Coetzee (2004) |
| Anna's Hummingbird | *Calypte anna* | 9 | Tyrrell et al. (2018) |
| Rufous Hummingbird | *Selasphorus rufus* | 12 | Tyrrell et al. (2018) |
| Senegal Parrot | *Poicephalus senegalus* | 24 | Demery et al. (2011) |
| Black Phoebe | *Sayornis nigricans* | 18 | Gall and Fernández-Juricic (2010) |
| Acadian Flycatcher | *Empidonax virescens* | 19 | Tyrrell and Fernández-Juricic (2017) |
| Tree Swallow | *Tachycineta bicolor* | 20 | Tyrrell and Fernández-Juricic (2017) |
| American Crow | *Corvus brachyrhynchos* | 16 | Fernández-Juricic et al. (2010) |
| California Scrub-Jay | *Aphelocoma californica* | 7 | Fernández-Juricic et al. (2010) |
| White-breasted Nuthatch | *Sitta carolinensis* | 74 | Moore et al. (2013) |
| Carolina Chickadee | *Poecile carolinensis* | 83 | Moore et al. (2013) |
| Tufted Titmouse | *Baeolophus bicolor* | 82 | Moore et al. (2013) |
| European Starling | *Sturnus vulgaris* | 18 | Martin (1986) |
| Chipping Sparrow | *Spizella passerina* | 31 | Moore et al. (2015) |
| Field Sparrow | *Spizella pusilla* | 36 | Moore et al. (2015) |
| White-crowned Sparrow | *Zonotrichia leucophrys* | 12 | Fernández-Juricic et al. (2011a, b) |
| California Towhee | *Melozone crissalis* | 11 | Fernández-Juricic et al. (2011a, b) |

and birds that forage on fast-flying insects (Bringmann 2019).

For birds with one fovea, it is centrally located in the retina, whereas birds with two foveas have both a central fovea and temporal fovea. Central foveas are deeper, have steeper walls, and provide greater visual acuity than shallower temporal foveas (Tucker 2000b; Potier et al. 2017; Fig. 4.134). Falcons have two foveas and appear to generally use their central foveas (monocular vision) when scanning for prey, but, once initiating an attack, switch to their temporal

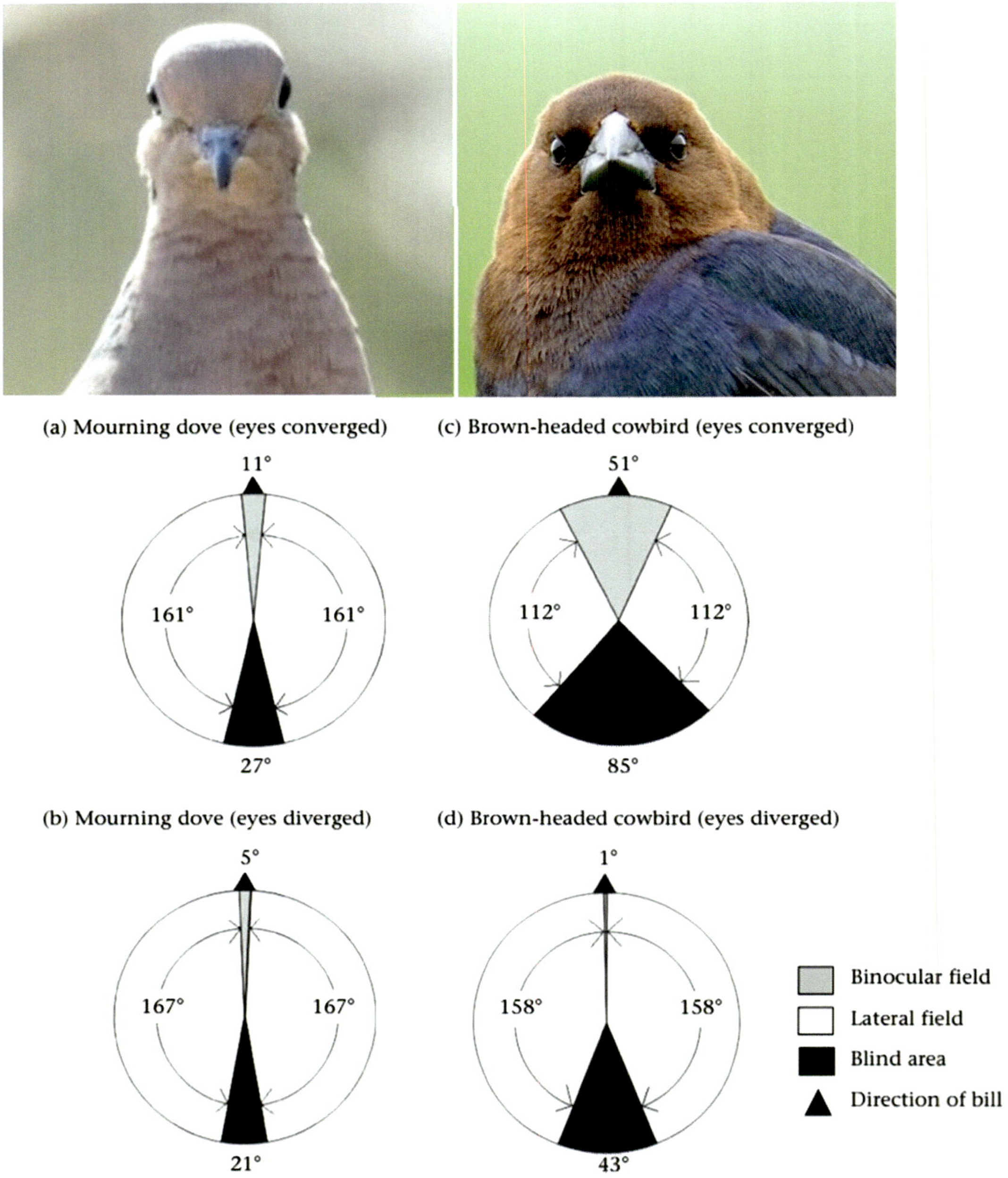


**Fig. 4.103** Top, Mourning Dove (*Zenaida macroura*) and male Brown-headed Cowbird (*Molothrus ater*). Bottom, Horizontal visual fields of a Mourning Dove (**a**, **b**) and Brown-headed Cowbird (**c**, **d**). For each species, the average visual fields when the eyes are fully converged (eyes rotated fully forward: **a**, **c**) and fully diverged (eye rotated fully backward: **b**, **d**) are provided. (Figure from Blackwell et al. 2009; Published by Elsevier Ltd., used with permission. Photo of Mourning Dove from pxhere.com, CC0 Public Domain; Photo of Brown-headed Cowbird from iStock by Getty Images, order number 2075757411)

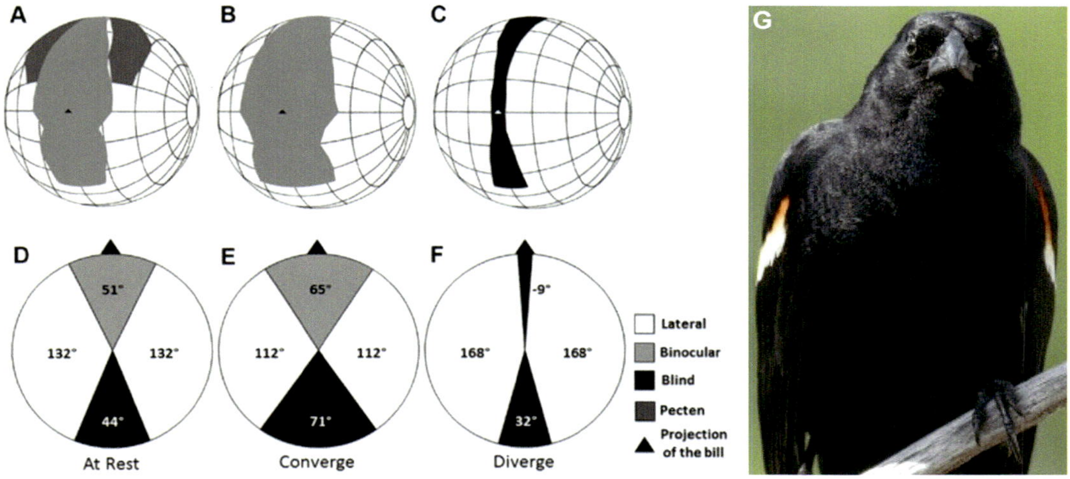


**Fig. 4.104** Visual fields of Red-winged Blackbirds (*Agelaius phoeniceus*). (**a**, **d**) Eyes at rest, (**b**, **e**) eyes converged, and (**c**, **f**) eyes diverged. (**a**–**c**) Spherical projection of the visual fields and pecten around the head. Lines correspond to latitude and longitude on a spherical coordinate system, which is centered on the head of the bird, in 20° increments. (**d**–**f**) Visual field configuration along the horizontal plane parallel to the ground (90–270°). (**g**) Male Red-winged Blackbird with eyes converged. (Figure from Fernández-Juricic et al. 2019; © 2019 Oxford University Press, used with permission. Photo by André Chivinski, pxhere.com, CC0 Public Domain)

foveas (binocular vision, but not necessarily stereoscopic) which reduces drag (i.e., as opposed to turning the head to focus prey on a central fovea) and allows a shorter interception path (Kane and Zamani 2014; Fig. 4.135). Another potential advantage of focusing images on the temporal foveas, especially with lower light levels, is an improved sensitivity due to binocular summation. As images are processed by visual areas in the brain, information from the two eyes focusing on the same image creates a single representation of that image. Campbell and Green (1965) reported that the mean sensitivity improvement was a factor of $\sqrt{2}$, which means an image viewed with one eye requires a contrast 1.4 times higher than the same image simultaneously viewed with both eyes to be equally detectable. Other investigators have reported that the binocular acuity of Little Owls (*Athene noctua*) and Rock Pigeons was 37% higher than their monocular acuity (Porciatti et al. 1989; Hodos et al. 2002).

In addition to providing greater visual acuity, foveas, especially deep foveas, may also magnify images (Fig. 4.136). Light striking a deep fovea is centrifugally refracted so produces larger images, with steeper-walled foveas providing greater magnification (Fig. 4.137). Müller cells in the walls of deep foveas refract the light to create the larger images (Zueva et al. 2014; Bringmann 2019). Among birds with two foveas, the deeper central fovea provides greater visual acuity, which may in part be due to the greater magnification of images by the deep central fovea (Bringmann 2019).

Visual acuity is a measure of the maximum resolving capacity of a visual system and can be measured in units of cycles per degree, i.e., the number of black-and-white parallel lines an animal can see in 1 degree of its vision. If you wanted to determine your visual acuity in terms of cycles per degree, you would start with two black lines where the white space between them is the same width as the black lines, e.g., 2 mm. That makes the length of one "cycle," i.e., one black line plus the white line, 4 mm. Then, walk away from the lines until you cannot see the space

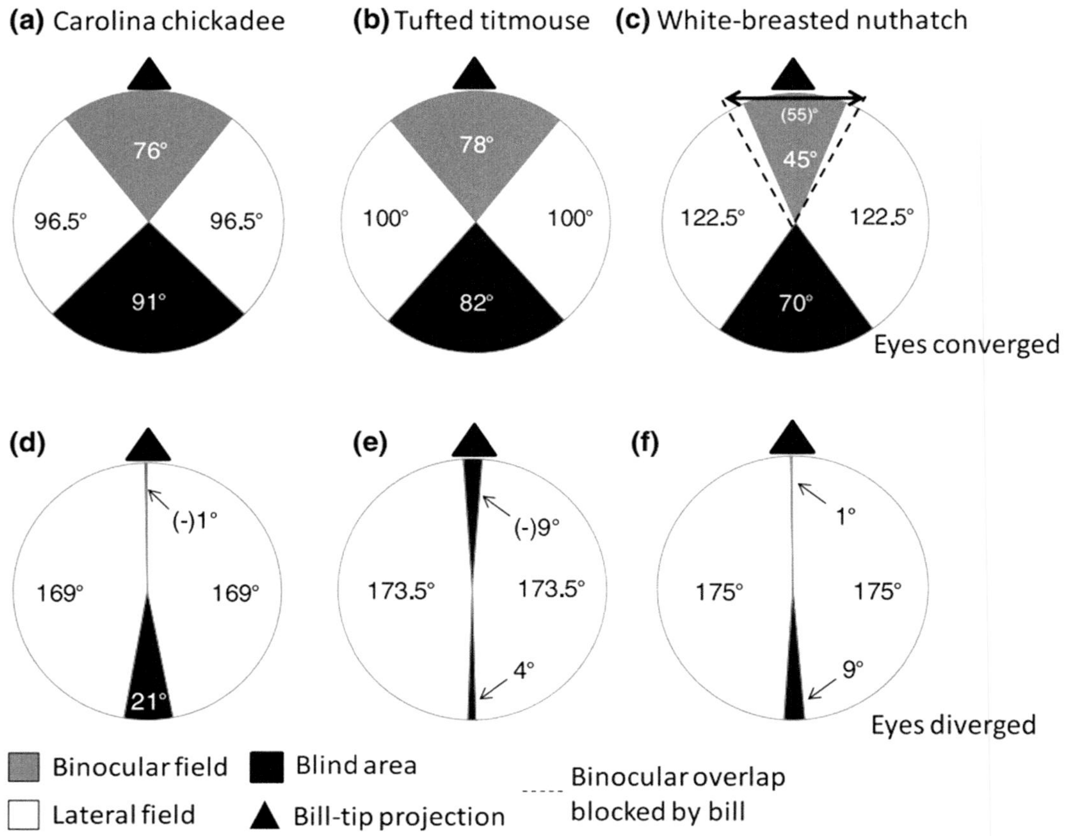


**Fig. 4.105** Visual fields in the horizontal plane (90–270) of (**a**, **b**) Carolina Chickadees (*Poecile carolinensis*), (**c**, **d**) Tufted Titmice (*Baeolophus bicolor*), and (**e**, **f**) White-breasted Nuthatches (*Sitta carolinensis*). (**a**–**c**) Visual fields when eyes were converged (rotated fully forward), maximizing the size of binocular and blind areas, and (**d**–**f**) when eyes were diverged (rotated fully backward), minimizing the size of the binocular and blind areas. (Figure from Moore et al. 2013; © 2013 Springer Nature, used with permission)

between them, e.g., 8 m. Next, divide the cycle width (4 mm) by the distance (6 m) and you get 0.0005. Take the inverse tangent of this number to convert it to degrees and you get 0.029 degrees per cycle. To get cycles per degree, divide one by that number and you get 34.5 cycles per degree. For humans, 20/20 vision corresponds to 30 cycles per degree; some people, however, have much better than 20/20 vision and the limits of human visual acuity ranges from about 60 to 75 cycles per degree (Curcio et al. 1990).

The visual acuity of most birds is below that of humans (Fig. 4.138). Caves et al. (2018) compiled data for 93 species of birds and found that acuity ranged from as low as 4 cycles per degree (cpd; Barn Owl, *Tyto alba*) to 138 cpd (Wedge-tailed Eagle, *Aquila audax*). However, the median acuity of these species was just 11 cpd, and most acuity estimates (84%) were below 30 cpd. In addition to Wedge-tailed Eagles, several other species of raptors and vultures have estimated visual acuities higher than humans, including Indian (*Gyps indicus*, 135 cpd), Griffon (*Gyps fulvus*, 104 cpd), and Egyptian (*Neophron percnopterus*, 135 cpd) vultures (Fischer 1969) and Condo Serpent-Eagles (*Dryotriorchis*

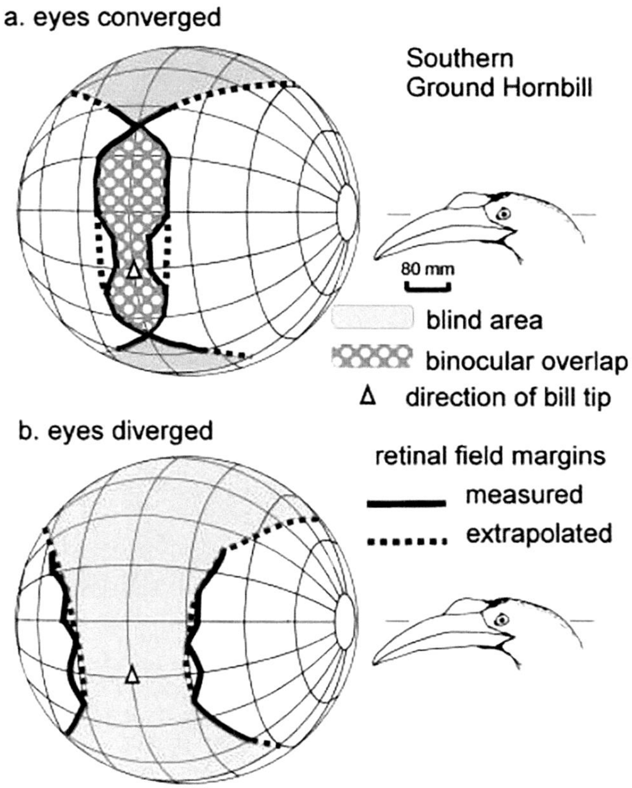


**Fig. 4.106** Visual fields of Southern Ground Hornbills (*Bucorvus leadbeateri*) when both eyes (**a**) are fully converged (eyes rotated fully forward), and (**b**) both eyes are fully diverged (eyes rotated fully backward). (Figure from Martin and Coetzee 2004; © British Ornithologists' Union, used with permission)

*spectabilis*, 120 cpd; Shlaer 1972). However, the estimated visual acuity of other species of raptors is lower than that of humans, including Black Kites (*Milvus migrans*, 25.9–32.9 cpd) and Harris's Hawks (*Parabuteo unicinctus*; 27.4–43.7 cpd) (Potier et al. 2016), Red-tailed Hawks (*Buteo jamaicensis*, 16.8 cpd; McIsaac 2001), and American Kestrels (*Falco sparverius*, 42–46 cpd; Hirsch 1982; Gaffney and Hodos 2003). One possible explanation for such differences in visual acuity among raptors is that searching for prey (or carrion) from high altitudes, such as vultures and some eagles, has favored selection for greater acuity than searching for prey over shorter distances by sit-and-wait predators, e.g., Harris's Hawks, Red-tailed Hawks, and American Kestrels (Potier et al. 2016).

How might visual acuity influence, for example, the ability of birds to detect prey? American Kestrels, with an estimated visual acuity of 42–46 cpd, are typically sit-and-wait predators, scanning for and attacking potential prey from distances as far away as 275 m (Balgooyen 1976). Gaffney and Hodos (2003) estimated that the minimum size of prey that American Kestrels could detect would range from about 2.3 mm at 12.2 m to 5.3 cm at 275 m. Birds with lower visual acuities must be much closer than that to clearly view potential prey (Fig. 4.139).

## 4.9 Avian Temporal Visual Acuity

For some birds, like diurnal birds of prey, their exceptional visual acuity is critically important in allowing them to detect and capture prey. However, for other birds, both visual and temporal acuity are important. Temporal acuity is determined by identifying the slowest rate at which a

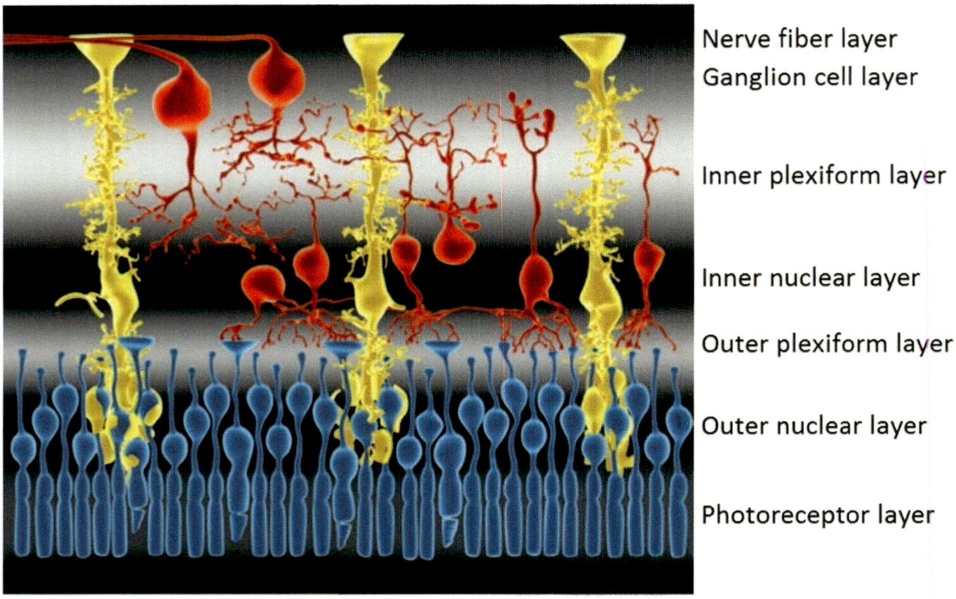


**Fig. 4.107** Drawing showing detailed view of the various layers and types of cells in the retina. Yellow, Müller cells; red, interneurons and ganglion cells; blue, photoreceptor cells (rods and cones). Müller cells maintain the retinal extracellular fluid, removing debris and waste products (like carbon dioxide) and providing energy for photoreceptor cells. (Figure modified from a supplemental figure from Lu et al. 2006; Copyright (2006) National Academy of Sciences USA, used with permission)

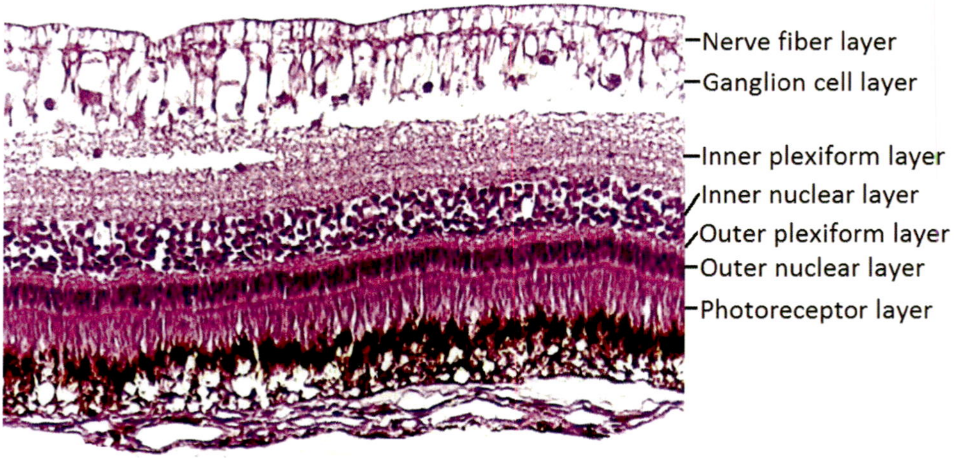


**Fig. 4.108** Cross-section of the retina of a Striped Owl (*Rhinoptynx clamator*) stained to better illustrate the various layers. Three layers contain cell nuclei (outer nuclear layer, inner nuclear layer, and ganglion cell layer), two layers, connecting the nuclear layers, contain nerve cell processes and synapses (outer plexiform layer and inner plexiform layer), the nerve fiber layer contains the axons of the ganglion cells that conduct impulses toward the optic nerve, and the photoreceptor layer contains the rods and cones. Magnified 40 times. (Figure modified from Jezler et al. 2010; © 2010 Formatex.org, used with permission)

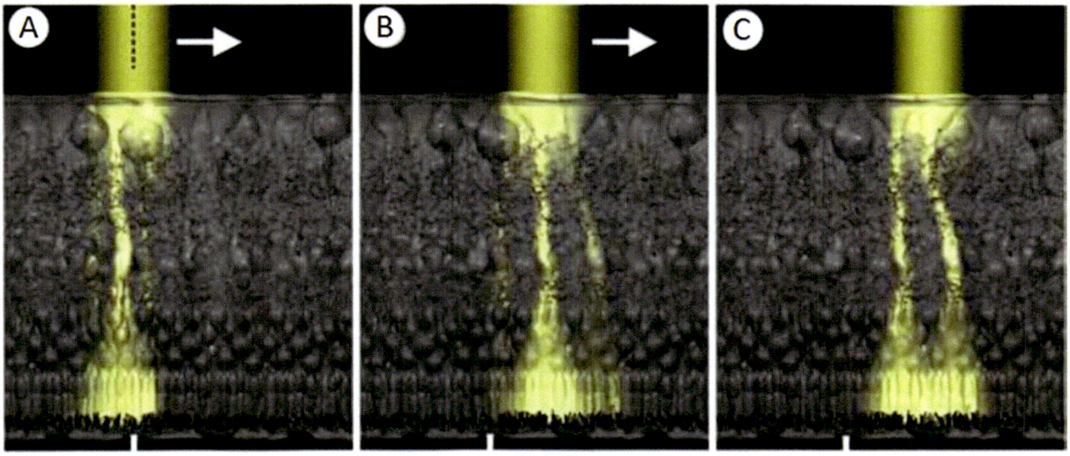


**Fig. 4.109** Each Müller cell distributes light to its own set of photoreceptors. In this example, movement of the light source relative to the retina's surface results in stimulation of a different array of photoreceptor cells. (Figure modified from Reichenbach et al. 2012; © 2012 The Author(s). Licensee IntechOpen, open-access article distributed under the terms of the Creative Commons Attribution 3.0 License, https://creativecommons.org/licenses/by/3.0/)

flickering light appears continuous, a measure referred to as the critical fusion frequency (CFF). Light flickering at a rate below a bird's CFF would be detected as flickering because the interval between "flashes" of light is longer than the time needed for the visual system to detect and perceive the light and be physiologically ready and able to detect the next "flash." This process is fast, but the perception of light does take time, with the light stimulating photoreceptors and nervous impulses then being transmitted to the visual areas of the brain. If light flickers at a rate equal to or faster than the CFF, the interval between "flashes" of light is shorter than the time needed for the visual system to detect and perceive the light and be physiologically ready to detect the next "flash" and the light appears to be continuous rather than flickering (Fig. 4.140).

Critical flicker fusion frequencies vary with the brightness of light, but, at best, the CFF for humans is about 50–60 Hz (light flashes per second; Jarvis et al. 2002). Although few species of birds have been tested in the lab, available evidence indicates that the CFFs of some species range between 100 and 150 Hz, depending on light levels (Boström et al. 2016; Fig. 4.141). Importantly, higher CFFs allow birds to more accurately detect, identify, and track fast-moving prey (Fig. 4.142). In addition, higher CFFs mean that fast-flying birds are better able to detect and track vegetation and other obstacles when flying in cluttered habitats.

## 4.10 Hearing

Birds, as a group, are among the most vocal of all animals, using songs, calls, and other sounds in a wide variety of contexts that serve a wide variety of functions. Given the importance of sounds to most birds, it would seem safe to assume that the ability to hear those sounds would be equally important and birds, in general, would have an excellent sense of hearing. They do. Hearing requires ears and, specifically, an inner ear where vibrations in the air can be translated into nervous impulses that are perceived as sounds by the avian brain (Fig. 4.143). The avian inner ear does that, but, as with other vertebrates, is also important for maintaining balance and equilibrium.

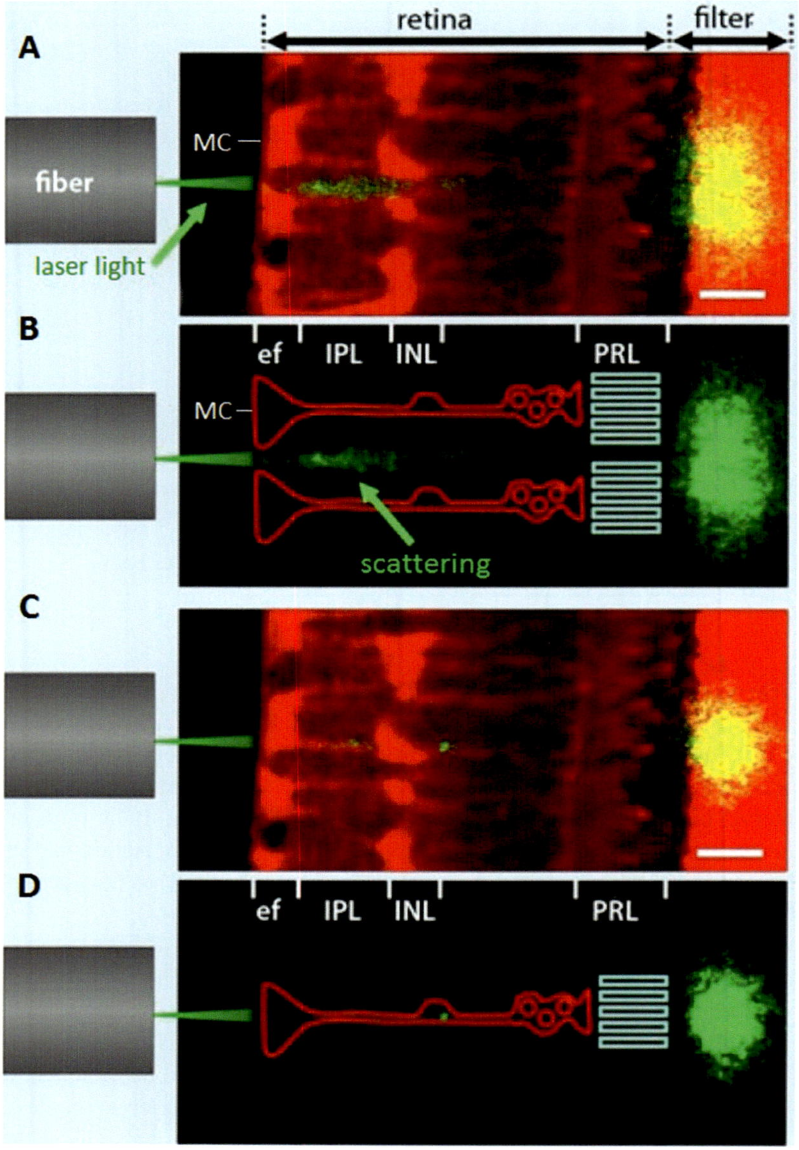


**Fig. 4.110** Light-guiding capability of Müller cells in the retina. (**a**) A laser beam (green) is used to illuminate the end of Müller cells (called endfeet, ef) in a section of retina attached to a filter membrane. The Müller cells and the membrane are stained with a red fluorescent dye, with filter membrane indicating where it is hit by the laser light. If the laser light passes between two Müller cells, some light gets scattered in the retina and the light spot on the filter membrane spreads out. (**b**) For a clearer view, outlines of the Müller cells are provided. MC, Müller cell; ef, Müller cell endfeet; IPL, inner plexiform layer; INL, inner nuclear layer; PRL, photoreceptor layer. (**c**) When a Müller cell is placed directly in front of the optical fiber, much less light is scattered as it passes through the retina so the image on the filter membrance is much sharper. (**d**) The outline of the Müller cell is provided for a clearer image. Scale bars = 20 μm. (Figure modified from Reichenbach et al. 2014; © 2014 Springer-Verlag Berlin Heidelberg, used with permission)

The avian ear, like that of mammals, has three sections (Fig. 4.144). The outer, or external, ear, unlike mammals, has no pinna and, as a result, is typically inconspicuous. The ear opening of most birds is covered by feathers (auriculars) that have a slender shaft (rachis) and barbs without barbules. These feathers help protect the tympanic membrane (eardrum; Fig. 4.145), but do not block sound. In addition, their structure may minimize turbulence (and noise) as air flows across the ear opening when a bird is flying (or when it is windy). The opening leads into a tube called the external acoustic meatus that ends at the tympanic membrane. For diving birds, water pressure several meters below the surface could potentially damage the tympanic

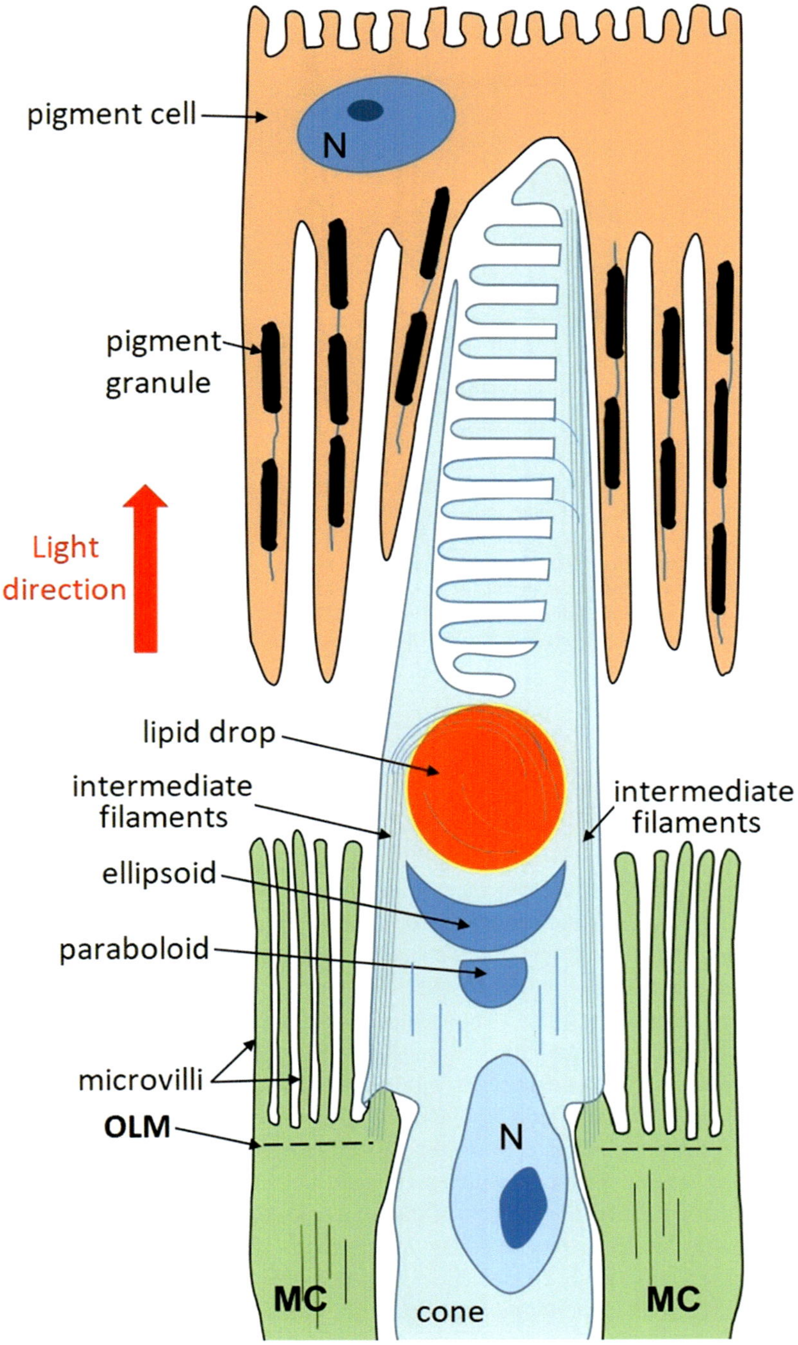


**Fig. 4.111** Longitudinal section of a portion of the foveolar retina of a European Pied Flycatcher (*Ficedula hypoleuca*). Light energy passes through the retina toward the pigment epithelium. The lower portion of the cone is surrounded by microvilli of Müller cells (MC). Some microvilli are connected to cones via intermediate filaments and some of these filaments may serve as conduits of light energy, servng as natural nanofibers that absorb light energy and transfer it to the photoreceptors. The paraboloid, containing glycogen, is located next to the ellipsoid that contains densely packed mitochondria. The top of the cone with the light-sensitive membranes (lamellae) is surrounded by pigment cells filled with pigment granules. N, nucleus. (Figure modified from Zueva et al. 2020; Published by Elsevier, used with permission)

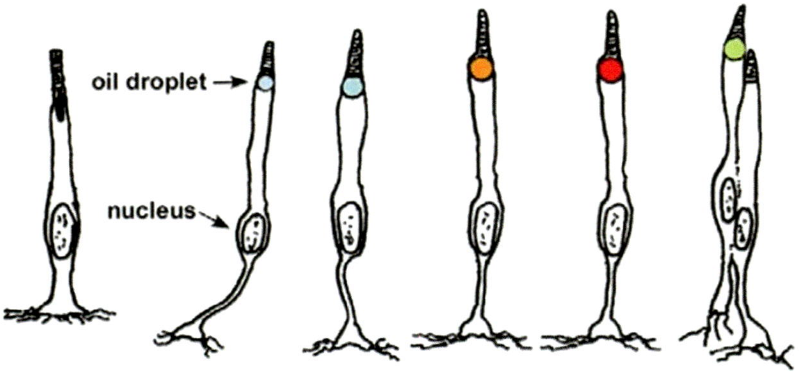


**Fig. 4.112** Types of photoreceptors in the retinas of Domestic Chickens, including rods, single cones, and double cones. The four types of single cones are UVS/VS cones (ultraviolet or violet sensitive; T-type), SWS cones (short wavelength sensitive; C-type), MWS cones (medium wavelength sensitive; Y-type), and LWS cones (long wavelength sensitive; R-type). The ellipsoid is where mitochondria are located, and the paraboloid (called hyperboloid in rods) is where glycogen is stored. (Figure from Kram et al. 2010; © 2010 Kram et al., open-access article distributed under the terms of the Creative Commons Attribution License, https://creativecommons.org/licenses/by/4.0/)

**Box 4.14 Phototransduction**

Perhaps surprisingly, the membranes of photoreceptors, both rods and cones, hyperpolarize (or are inhibited) and depolarize (or are stimulated) to varying degrees depending on the amount of light striking them and, more specifically, on the amount of light striking the visual pigments in rods and cones, i.e., rhodopsin (RH1) in rods and different conopsins (RH2, SW1, SW2, and LWS) in different cones. They are more inhibited in brighter light and more stimulated as light levels decline, so the rate of stimulation (or impulse conduction) varies with light levels. When stimulated (i.e., not exposed to light), rods and cones release a neurotransmitter called glutamate where they synapse with bipolar cells. Different bipolar cells have different receptors so respond differently to glutamate. Some bipolar cells are stimulated by glutamate and respond with what is called the OFF pathway, which permits the detection of dark images against lighter backgrounds (in color in the case of cones). Other bipolar cells are inhibited by glutamate, activating the ON pathway important in detecting light images (in color in the case of cones) against darker backgrounds (Kolb 2003). These responses by bipolar cells are further "processed" by other types of cells in the retina, including horizontal cells and amacrine cells, before transmission to visual centers of the brain via the optic nerve.

(continued)

**Box 4.14** (continued)

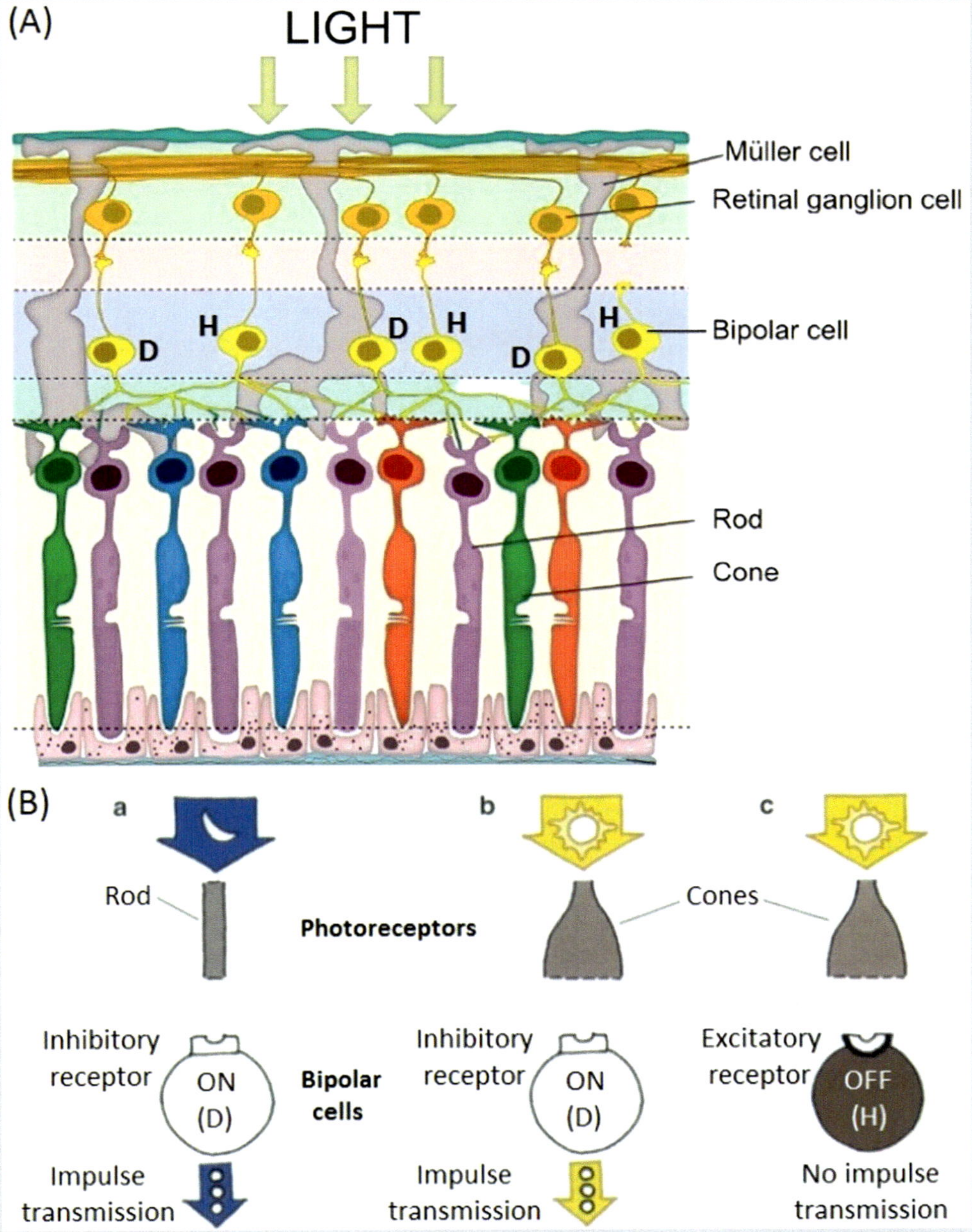


(**a**) Rod and cone cells are excited (depolarized) in the dark and inhibited (repolarized) by light (photons). Excitation (depolarization) triggers the release of the neurotransmitter glutamate at the synapse between the rods and cones and bipolar cells and either stimulates (depolarizes, D) or inhibits (hyperpolarizes, H) the bipolar cells. If stimulated, the impulses are then transmitted via other cells to the visual areas of the brain. (**b**) Photoreceptors synapse with two different types of bipolar cells, with some stimulated (depolarized; ON or D bipolar cells) and the other inhibited (hyperpolarized; OFF or H). In contrast to what might be expected, photoreceptors stop

(continued)

**Box 4.14** (continued)

releasing glutamate when stimulated by light, causing ON (D) bipolar cells to release glutamate and send the impulse to visual areas in the brain (Figure **a** modified from Hall and Chilcott 2018, open-access article distributed under the Creative Commons Attribution License, https://creativecommons.org/licenses/by/4.0/; Figure **b** modified from Kolb 2003; © 2003 Sigma Xi, used with permission)

membrane. As protection, the auricular feathers of some diving birds are less specialized and more like other feathers so that, during a dive, they are pressed against the ear opening and prevent water from entering. The ear openings of some diving birds are also relatively small and, therefore, more easily blocked by feathers (Larsen et al. 2020). Auks (Alcidae) have a flap of skin that is parallel to the initial section of the external acoustic meatus (that, in turn, is parallel to the skin for a short distance before turning and extending deeper). As a result, with increasing pressure during a dive, the flap of skin is forced against the opposite wall of the external acoustic meatus and prevents water from entering (Kuhne and Lewis 1985).

The tympanic membrane separates the external ear from the middle ear cavity. This membrane is very thin (improving its response to high-frequency sounds) and, for smaller birds, relatively large because sensitivity requires a sufficiently large membrane. Within the middle ear cavity is the single middle ear bone that inserts on the tympanic membrane, causing it to bulge outward. This bone has two segments, a cartilaginous extracolumella that inserts on the tympanic membrane and a bony columella that fits into the oval (vestibular) window of the cochlea (Figs. 4.146 and 4.147). Also located in the middle ear cavity is the paratympanic organ (Fig. 4.148). Most birds have paratympanic organs (owls are an exception), and some reptiles and mammals (bats) also have these organs. Paratympanic organs have mechanoreceptors (hair cells) that appear to provide birds with information about barometric (atmospheric) pressure (von Bartheld 1994; von Bartheld and Giannessi 2011; Fig. 4.149). Eastern pipistrelle (*Pipistrellus subflavus*) bats use their paratympanic (or Vitali) organ to adjust their metabolic rates and predict the behavior of their insect prey; leaving their cave hibernacula only when atmospheric pressure declines (with lower pressure meaning warmer, cloudy weather and more insects and higher pressure meaning colder weather, fewer or no insects, and the need to reduce metabolic rates; Paige 1995). In a similar manner, some birds appear to use changes in atmospheric pressure to track changing weather conditions and adjust their behavior accordingly (e.g., Shamoun-Baranes et al. 2006; Breuner et al. 2013). For example, adult Ferruginous Hawks (*Buteo regalis*) began delivering more prey to nestlings and exhibited an increase in nest maintenance activities during periods of falling barometric pressure that preceded storms (Laux et al. 2015), and White-crowned Sparrows (*Zonotrichia leucophrys*) foraged at higher rates in response to lower barometric pressures (Metcalfe et al. 2013). By providing birds with information about atmospheric pressure, paratympanic organs may also serve as altimeters (Giannessi et al. 2017), e.g., for migrating birds (Chmura et al. 2020).

The avian inner ear consists of three semicircular canals, saccule, utricle, lagena, and cochlea. The semicircular canals, saccule, and utricle are important for balance and equilibrium (as described in more detail later) and the cochlea for hearing. The cochlea of birds is a short, slightly curved, tube ranging in length from 2 to 11 mm (Manley 2000). The cochlea contains fluid called perilymph and, also inside the cochlea, are two membranes, the basilar membrane, and

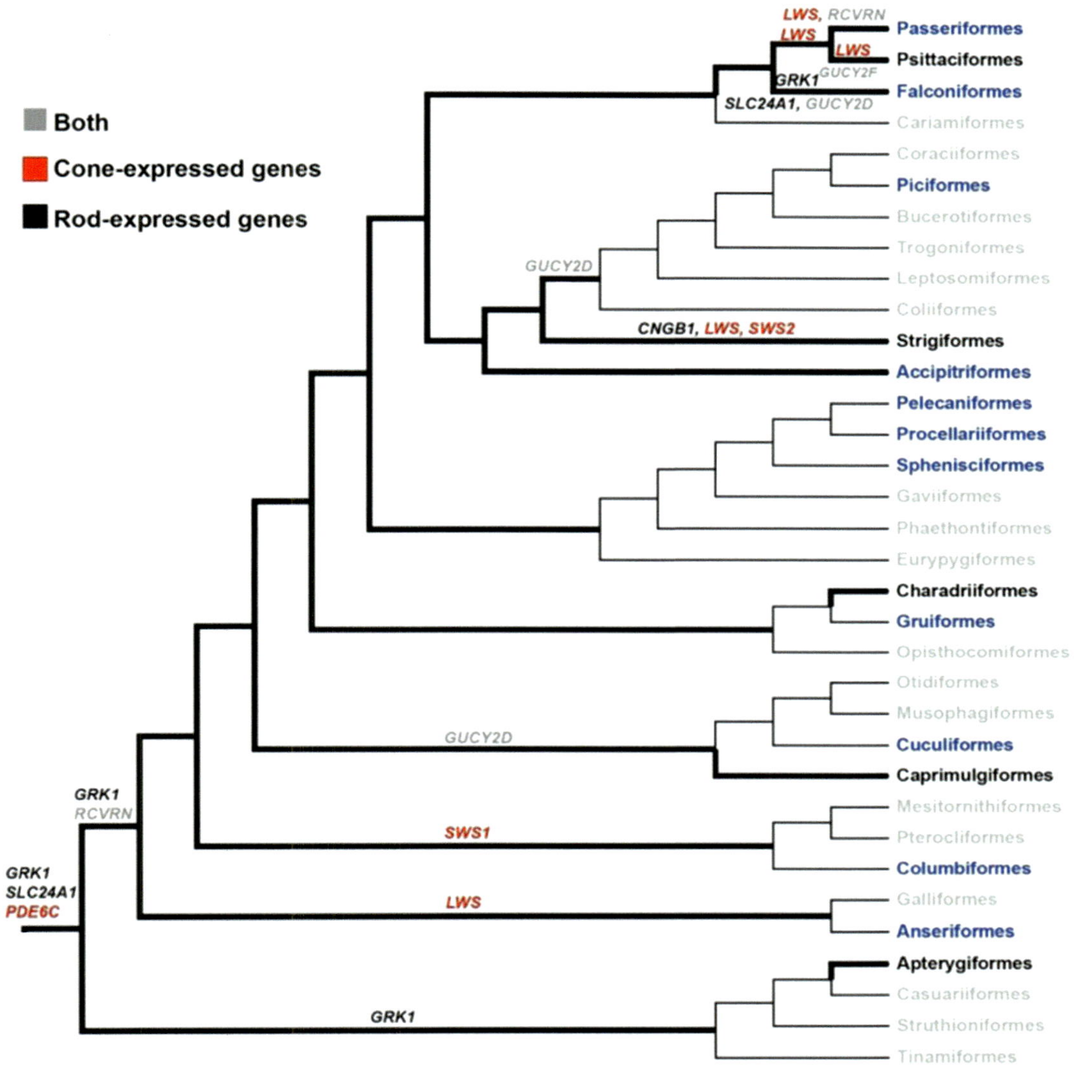


**Fig. 4.113** Positively selected genes identified in 34 orders of birds. Orders with species that are primarily diurnal, but occasionally nocturnal, are shown in blue, orders with species that are primarily or entirely nocturnal are shown in black, and orders with species that are diurnal and are not known to exhibit nocturnal activity are shown in gray. Bold lines indicate branches subjected to positive selection, and species genes subject to positive selection are listed along the branches. Genes shown in red are cone-expressed, those in black are rod-expressed, and those both rod- and cone expressed are gray. (Figure from Wu 2019; © The Author 2019, open-access distributed under the terms of the Creative Commons Attribution 4.0 International License, https://creativecommons.org/licenses/by/4.0/)

tectorial membrane (Fig. 4.150). On top of the basilar membrane are hair cells from which extend numerous hair-like structures called stereocilia, and the tectorial membrane is located immediately above the stereocilia. The avian cochlea contains between 3000 and 17,000 hair

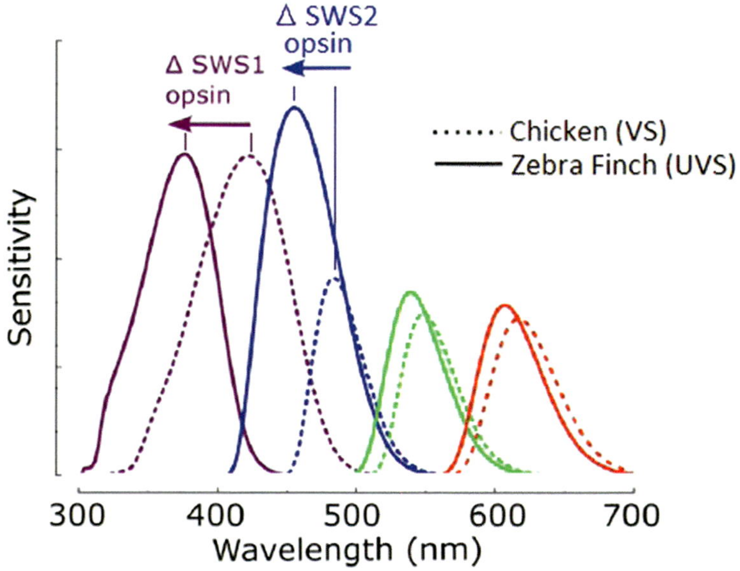


**Fig. 4.114** In UVS-sensitive species like Zebra Finches (*Taeniopygia guttata*), SWS1 cone sensitivity has been short-wavelength shifted compared to VS-sensitive species. Likely as a result, the sensitivity of SWS2 cones is also short wavelength. This shift helps maintain even spacing of cone sensitivities across the light spectrum, which is thought to enhance color discrimination. (Figure modified slightly from Toomey and Corbo 2017; © 2017 Toomey and Corbo, open-access article distributed under the terms of the Creative Commons Attribution License (CC BY), https://creativecommons.org/licenses/by/4.0/)

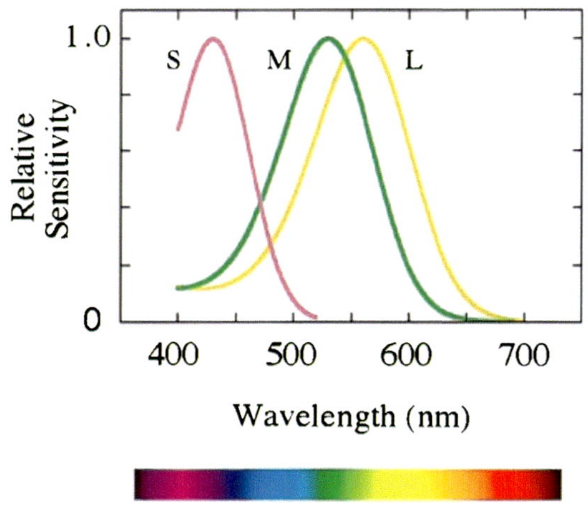


**Fig. 4.115** Sensitivities of the long-wave (L), middle-wave (M), and short-wave (S) photopigments in the retinas of humans. Peak absorption occurs at 420 nm for short-wavelength cones, 530 nm for medium-wavelength cones, and 560 nm for long-wavelength cones. (Figure from Mollon 1999; © 1999 National Academy of Sciences, USA, used with permission)

cells along the length of the cochlea (and basilar membrane), and the number of hair cells is positively correlated with body mass (except Barn Owls [*Tyto alba*] that have a longer cochlea and more hair cells than expected based on their mass) (Gleich and Langemann 2011).

The basilar and tectorial membranes extend the entire length of the cochlea, or from the basal (proximal) end to the apical (distal) end (Fig. 4.151). Stimulation of hair cells along the basilar membrane translates into perception of different frequency sounds, with those at the basal end associated with high-frequency sounds and progressively lower frequency sounds associated the hair cells toward the apical end. For most birds, the distance along the cochlea relative to frequency detected is less than 1 mm per octave (Manley 2000). Hair cells vary in morphology and the height of their stereocilia and, based on those differences, are often referred to as either tall hair cells or short hair cells.

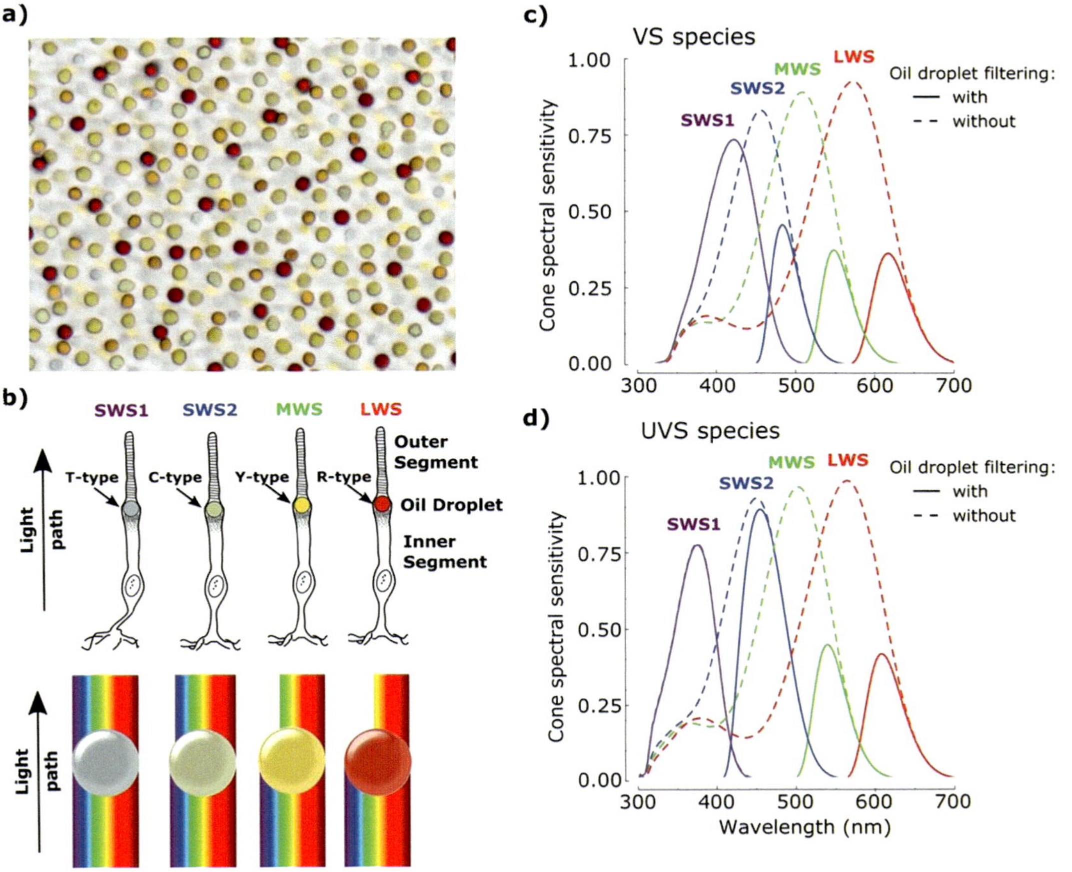


**Fig. 4.116** Avian color vision is mediated by four single cone photoreceptors that are tuned by cone oil droplet spectral filters. (**a**) Retina of a Domestic Chicken showing the cone oil droplets. (**b**) Four types of avian cones, with oil droplets just below the outer segments (where the opsin proteins are located), and showing how the different oil droplets filter the incoming light (bottom). (**c**) Spectral sensitivity of the cones of Domestic Chickens (a VS species) and (**d**) Zebra Finches (*Taeniopygia guttata*; UVS species) with (solid lines) or without (broken lines) the filtering of cone oil droplets. (Figure from Toomey et al. 2016; open-access article licensed under a Creative Commons Attribution License, https://creativecommons.org/licenses/by/4.0/)

However, recent work indicates that there is a continuous morphological gradient between the two types. Moving from the basal to the apical end of the cochlea, stereocilia become progressively shorter and the number of stereocilia per hair cell decreases (from about 300 to 50). Hair cells also differ in their innervation, with taller hair cells supplied by both efferent and afferent (sensory) neurons and shorter hair cells supplied only with efferent (motor) neurons (Fettiplace 2020). The role of the short hair cells in avian hearing is unclear. However, because they lack sensory innervation, their role must be limited to the basilar papilla. One possibility is that they somehow enhance vibration of the basilar membrane and improve the ability of birds to detect low volume sounds (Manley 2015).

Hearing involves the translation of vibrations in the air, or pressure waves, into nervous impulses that travel to the auditory areas of the brain (Fig. 4.152). Pressure waves pass into the external ear opening, through the external acoustic

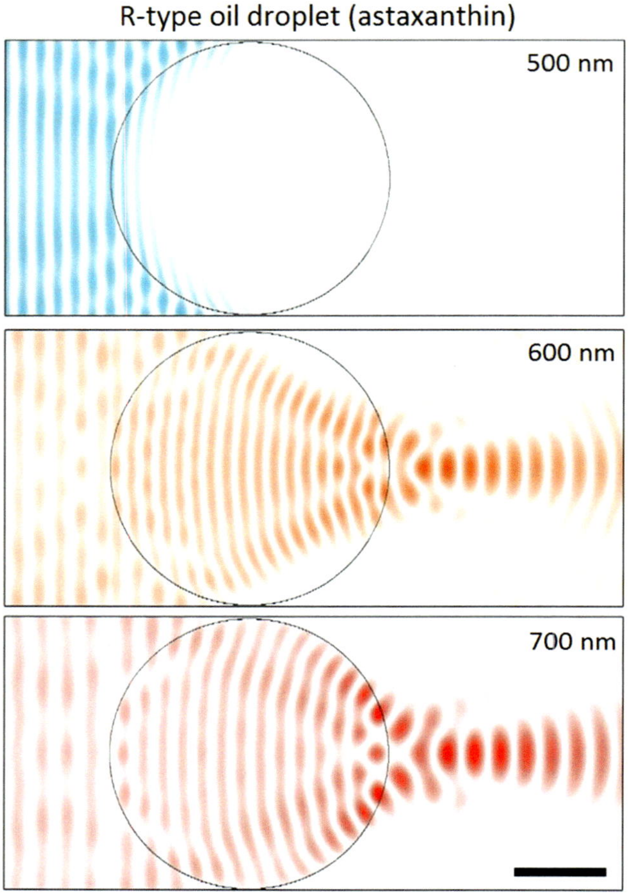


**Fig. 4.117** Light intensity of wavelengths 500, 600, and 700 nm after propagation through an R-type oil droplet containing the carotenoid pigment astaxanthin. The pigment in the oil droplet absorbs short-wavelength light (500 nm), but focuses light of longer wavelengths (600 and 700 nm). Scale bar = 1 μm. (Figure from Stavenga and Wilts 2014; © 2014 The Authors. Published by the Royal Society, used with permission)

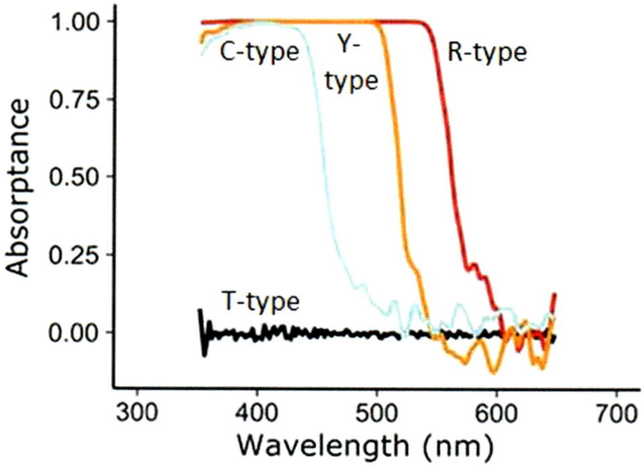


**Fig. 4.118** The cone droplets of three of the four single cones of the retina function of Domestic Chickens act as long-pass cutoff filters, absorbing nearly all light below the cut-off wavelength. (Figure from Toomey and Corbo 2017; © 2017 Toomey and Corbo, open-access article distributed under the terms of the Creative Commons Attribution License (CC BY), https://creativecommons.org/licenses/by/4.0/)

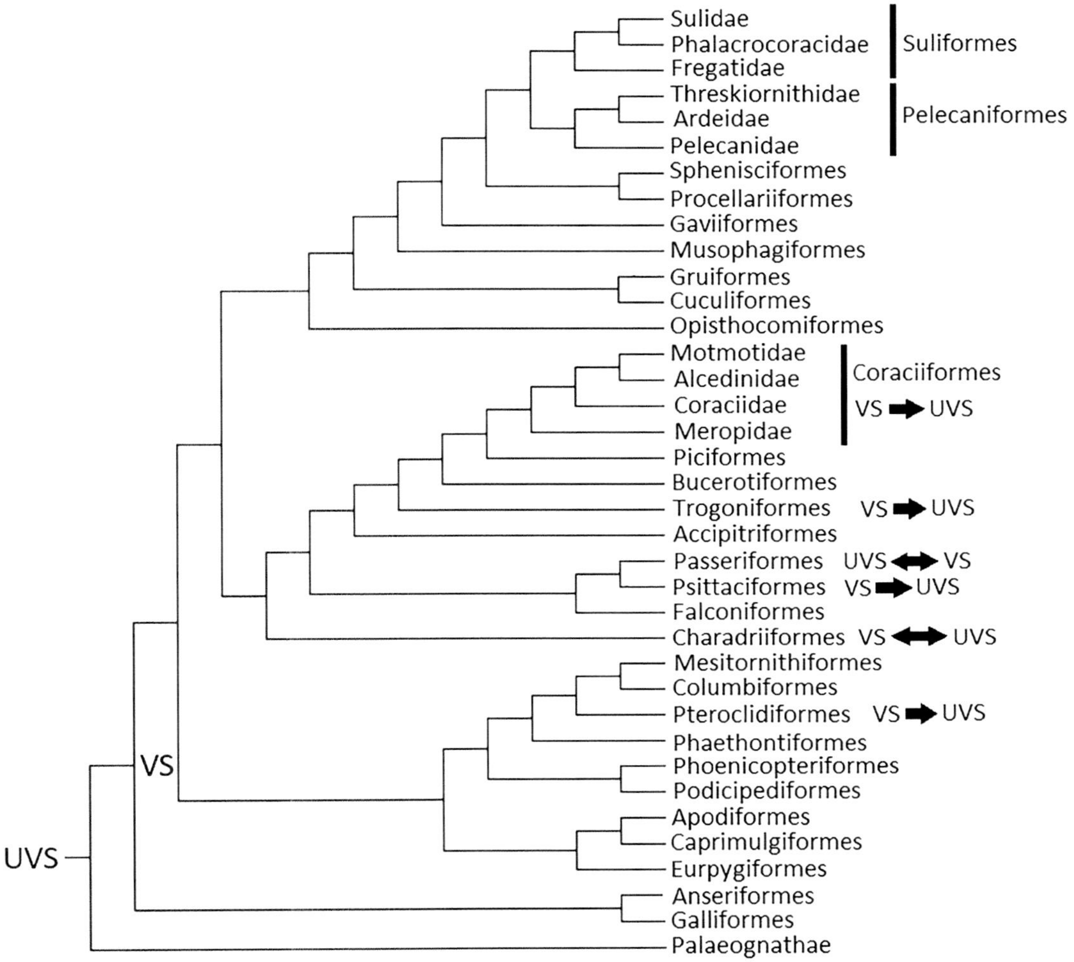


**Fig. 4.119** Evolution of UVS/VS SWS1 pigments in birds. The ancestral condition for birds was likely UVS vision (in fact, studies of agnathans and teleost fishes indicate that the ancestral vertebrate SWS1 pigment was ultraviolet sensitive with a peak around 360 nm; Hunt and Peichl 2014), with birds in most orders now having VS vision. VS↔UVS or UVS↔VS indicates some species have VS vision and some have UVS vision. (Figure modified from Ödeen and Håstad 2013; open-access article distributed under the terms of the Creative Commons Attribution License, https://creativecommons.org/licenses/by/4.0/)

meatus, and cause vibration of the tympanic membrane. That, in turn, causes vibration of the columella and the oval window (the membrane-covered opening into the cochlea). Sounds of differing frequencies cause the tympanic membrane to vibrate at different frequencies and create pressure waves in the fluid (perilymph) of the cochlea. These pressure waves travel through the perilymph and cause vibration of the basilar membrane that, in turn, causes the stereocilia of the hairs cells to move against the tectorial membrane. Displacement of the stereocilia generates nervous impulses that travel to the avian brain and are interpreted as sound. The tall hair cells respond to the vibration of the basilar membrane and their contact with the tectorial membrane by pushing back on it, exerting

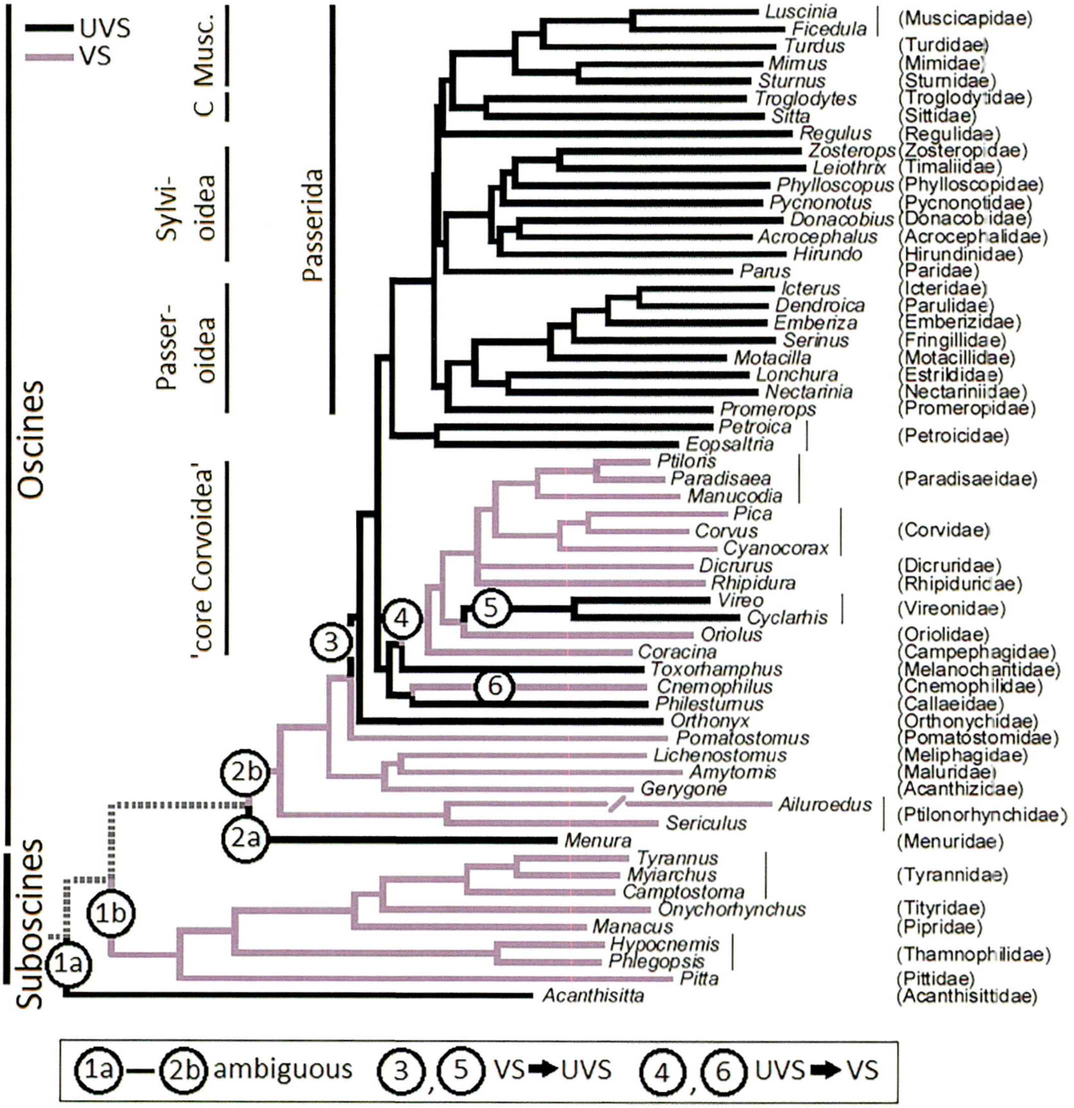


**Fig. 4.120** Phylogenetic recontruction of SWS1 opsin evolution among genera and families in the order Passeriformes. Based of this reconstruction, a minimum of six changes occurred, including two from VS to UVS, two from UVS to VS, and two with uncertain directions. C, Certhioidea; Musc., Muscicapoidea. (Figure modified from Ödeen et al. 2011; open-access article distributed under the terms of the Creative Commons Attribution License, https://creativecommons.org/licenses/by/4.0/)

force with just the right amplitude and phase to amplify the vibration, especially for faint sounds. Tall hair cells in particular areas of the cochlea only amplify sound of a particular frequency so that each area is tuned to a particular frequency and not to other frequencies.

## 4.11 Static and Dynamic Equilibrium

Along with the cochlea, the inner ear of birds contains three bony, endolymph-filled semicircular canals, plus three otolith organs, the saccule, utricle, and lagena (Figs. 4.153 and 4.154). The three semicircular canals provide information

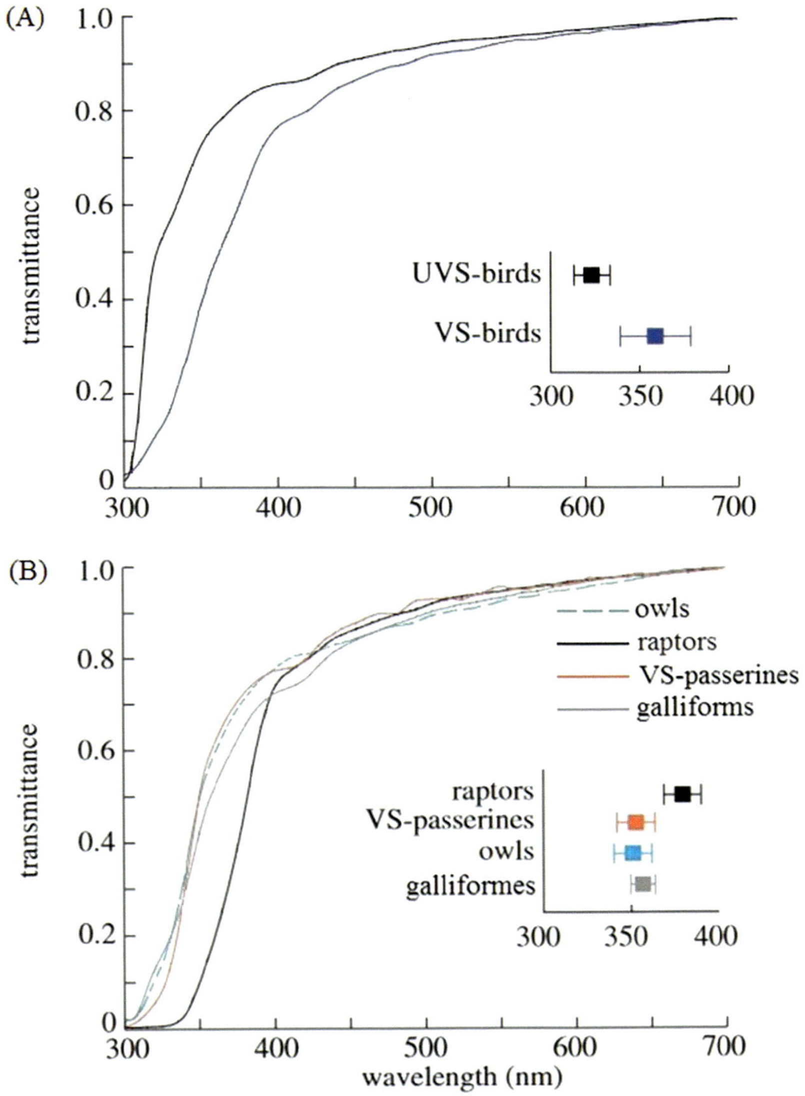


**Fig. 4.121** Variation in ocular media (cornea, aqueous humor, lens, and vitreous humor) transmittance (OMT) between bird groups. Average OMT for (**a**) birds with UVS and VS pigment, and (**b**) raptors, owls, galliformes, and VS-passerines. Insets show the average wavelength position of 50% transmittance ($\lambda_{T0.5}$) for each group. (Figure from Lind et al. (2014a, b); © 2013 The Authors. Published by the Royal Society under the terms of the Creative Commons Attribution License, https://creativecommons.org/licenses/by/4.0/)

about head movement and, more specifically, angular acceleration (roll, pitch, and yaw; Fig. 4.155). Each semicircular canal is filled with endolymphatic fluid, and each has an area slightly largely in diameter called the ampulla that contains a structure called the cupula. Hair cells at the base on the cupula have hair-like projections called stereocilia that are embedded in the cupula and, at the base of the stereocilia, are movement-sensitive neurons (Fig. 4.156). When a bird moves its head and the three semicircular canals (to varying degrees depending on the extent and direction of movement), the inertia of the endolymph in the semicircular canals resists movement and, as a result, the cupula moves (Fig. 4.157). This movement, in turn, triggers nervous impulses along the neurons at the base of the stereocilia, transmitting information to the brain that, combined with visual, proprioceptive, and other stimuli, provide birds with a multisensory impression of the head and body's

**Fig. 4.122** Expected spectral sensitivity of the rods and cones of Tawny Owls (*Strix aluco*). (Figure from Höglund et al. 2019; © 2019 Elsevier Ltd., used with permission)

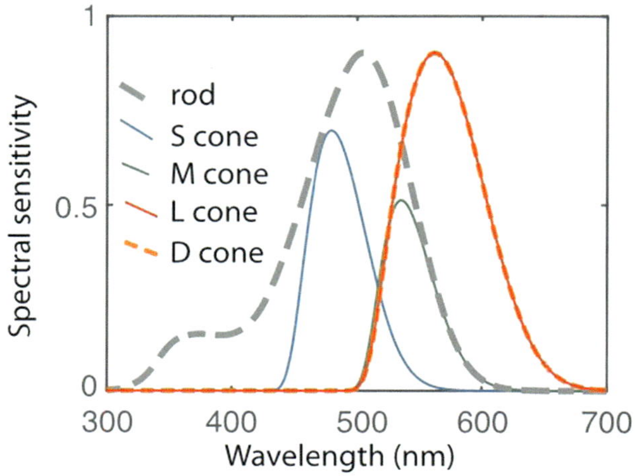


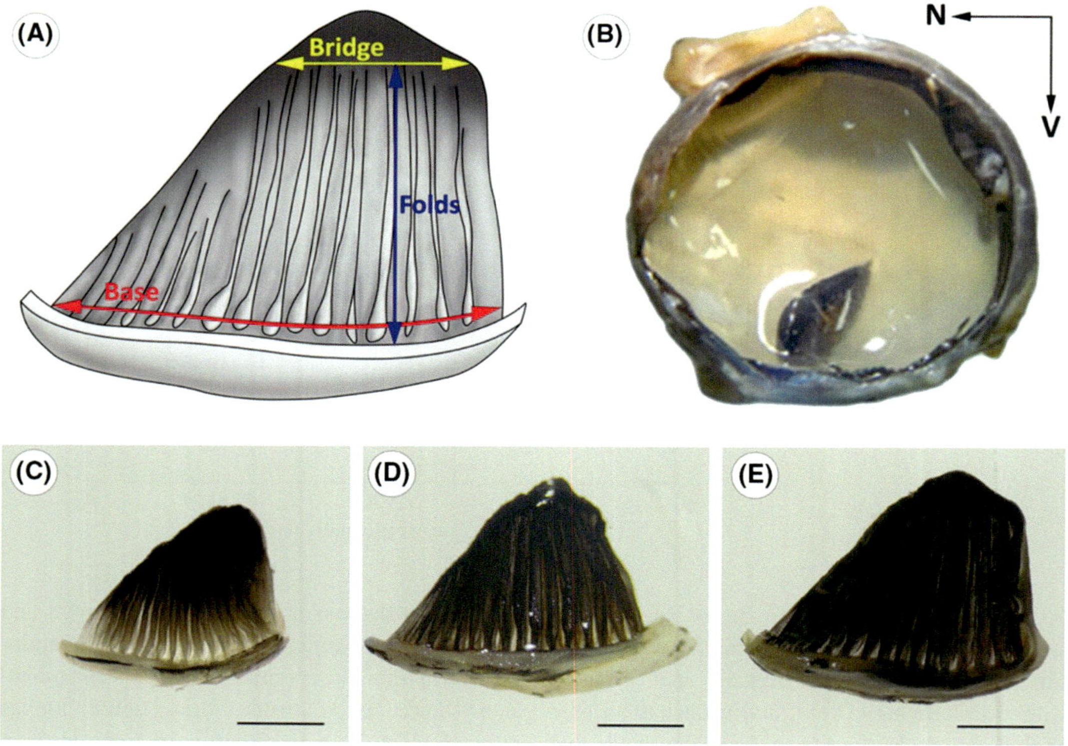


**Fig. 4.123** (**a**) Drawing of a pecten oculi of a Yellow-legged Gull (*Larus michahellis*) showing the base, folds, and bridge. (**b**) Pecten oculi in a nestling eye. Pecten oculi of a after removal from the eye of a (**c**) nestling, (**d**) juvenile, and (**e**) adult. Scale bars = 3 mm. (Figure from Segovia et al. 2020; © 2019 American College of Veterinary Ophthalmologists, used with permission)

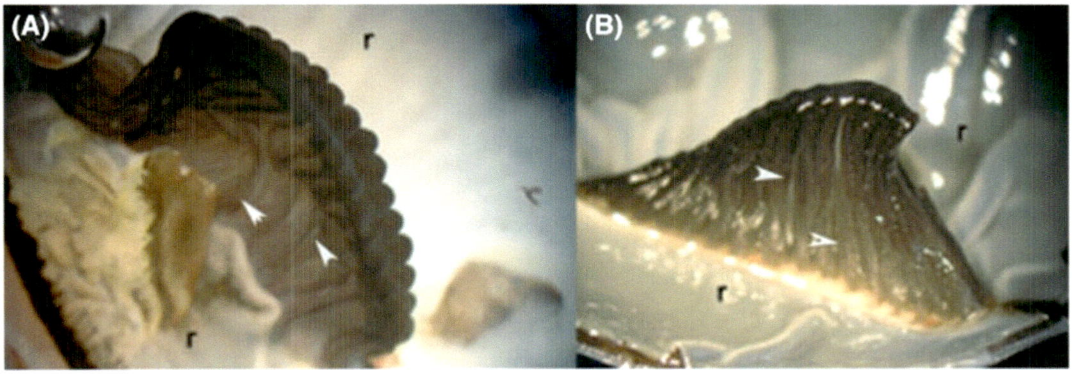


**Fig. 4.124** Pecten of a Common Buzzard (*Buteo buteo*). (**a**) Pecten as it appears through the vitreous humor of the eye and (**b**) after removal of the vitreous humor. r, retina; arrows are pointing to plications or folds in the pecten. (Figure from Gültiken et al. 2012; © 2011 American College of Veterinary Ophthalmologists, used with permission)

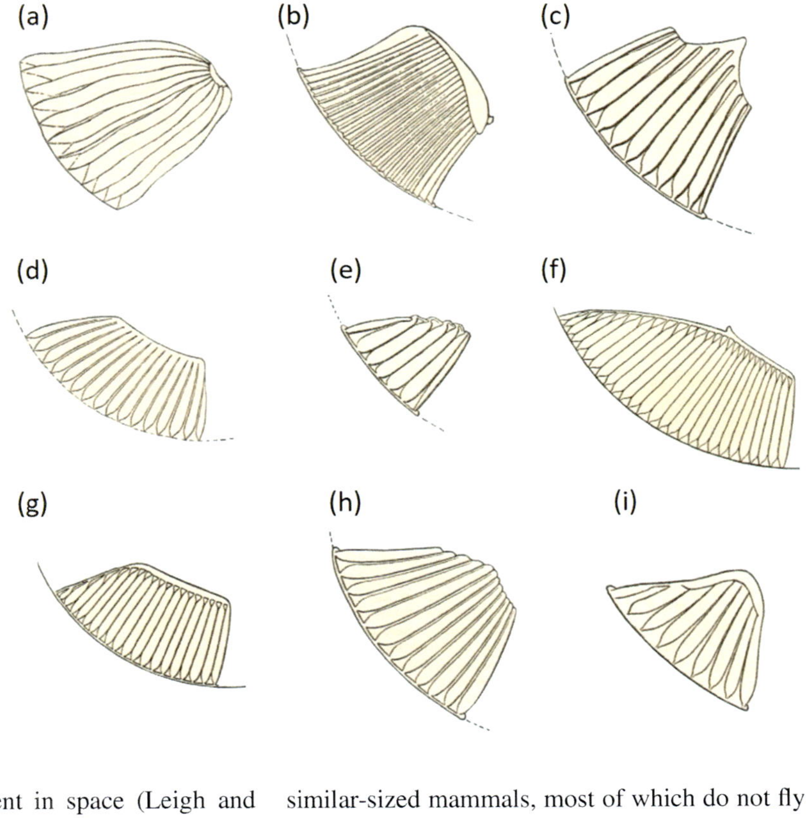


**Fig. 4.125** Examples of variation in the morphology of avian pectens. (**a**) Common Ostrich (*Struthio camelus*), (**b**) Red-winged Tinamou (*Rhynchotus rufescens*), (**c**) Great Bustard (*Otis tarda*), (**d**) Anna's Hummingbird (*Calypte anna*), (**e**) Tawny Owl (*Strix aluco*), (**f**) Steller's Jay (*Cyanocitta stelleri*), (**g**) Eastern Bluebird (*Sialia sialis*), (**h**) Yellow-billed Cuckoo (*Coccyzus americanus*), and (**i**) Greater Flamingo (*Phoenicopterus roseus*). (Figures from Wood 1917; CC0 Public Domain)

orientation and movement in space (Leigh and Zee 1999).

In general, birds, with most species able to fly, have semicircular canals with a larger radius than similar-sized mammals, most of which do not fly (Sipla 2007; Fig. 4.158). In addition, birds that fly tend to have larger-diameter semicircular canals than flightless birds (Fig. 4.159), and birds that

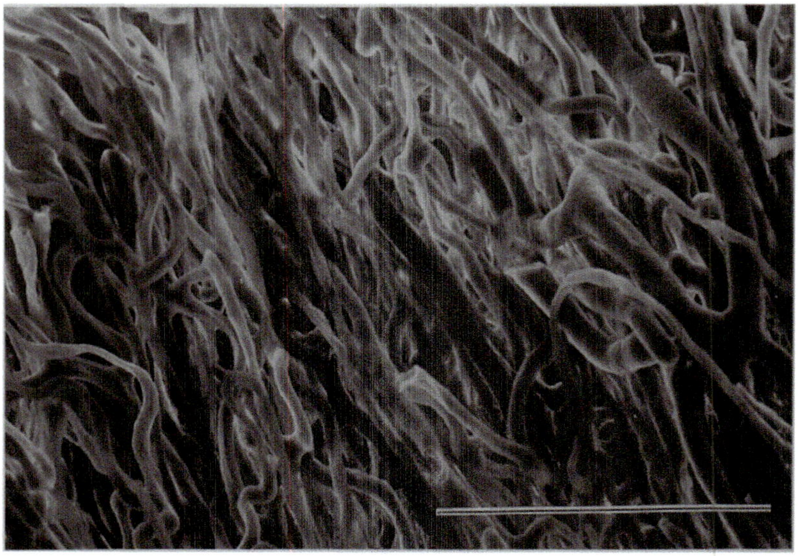

**Fig. 4.126** Complex capillary network inside a single fold of a pecten of a Domestic Chicken (magnified 200 times). Scale bar = 100 μm. (Figure from Matsunaga and Amemiya 1990; https://www.jstage.jst.go.jp/article/ofaj1936/67/4/67_263/_article/-char/en, used with permission)

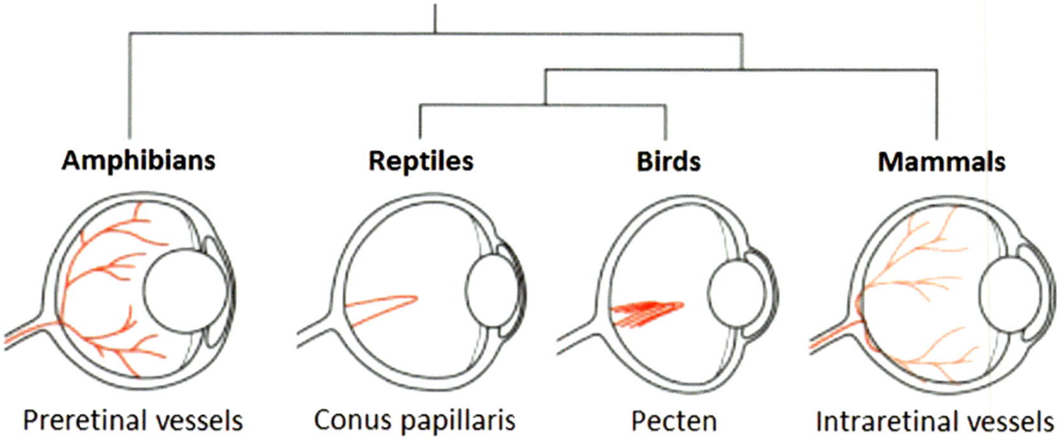


**Fig. 4.127** Adaptations for providing nutrients and oxygen to (and removing waste products and carbon dioxide from) vertebrate retinas. Among amphibians, the retinas of salamanders and newts are supplied by blood vessels in the choroid, whereas frogs and toads have blood vessels and capillaries on top of the retina (preretinal). Among reptiles, snakes have preretinal vessels, turtles, and crocodilians have blood vessels in the choroid, and lizards have a conus papillaris, a projection that extends into the vitreous humor with many small blood vessels via which nutrients and waste products are exchanged with the vitreous humor. Birds have pectens and mammals have blood vessels and capillaries embedded within the retina (intraretinal vessels). (Figure modified from Yu et al. 2009; © 2009 The Authors. Journal compilation © 2009 The Zoological Society of London, used with permission)

exhibit greater maneuverability and agility in flight have larger-circumference semicircular canals relative to their body size than those that are less maneuverable in flight (Sipla 2007; Fig. 4.160). Selection has likely favored enlarged semicircular canals in birds, and especially birds

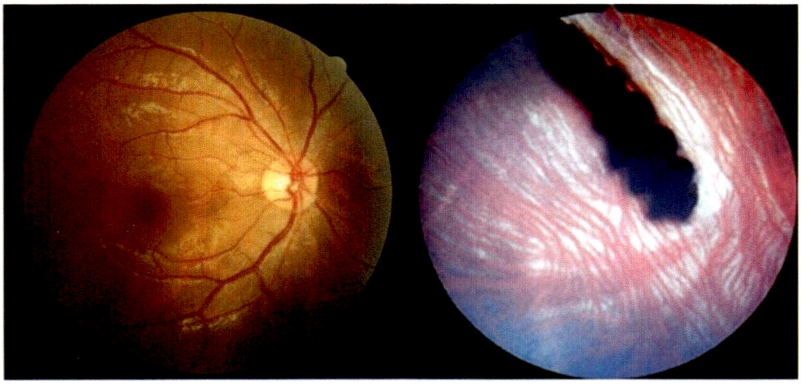

**Fig. 4.128** On the left, the fundus (back of the eye) of a human showing the retinal blood supply from central or cilioretinal arteries. On the right, the fundus of a screech-owl with no vessels in the retina. (Image of human fundus by ignis, Wikipedia, CC BY 3.0, https://creativecommons.org/licenses/by/3.0/; image of owl fundus from Bayón et al. 2008; © 2008 CAB International, used with permission)

that fly and birds that exhibit greater maneuverability in flight, because semicircular canals that are larger in circumference are more sensitive than those smaller in circumference. This is because longer "tubes" have more fluid and, as a result, greater variation in the amount of pressure exerted on the cupulas in the ampullae is possible. As a simple example, imagine that it takes the pressure equivalent to that exerted by the mass of 1 ml of fluid to move a cupula sufficiently for a bird to perceive a change in angular acceleration. If there is 10 ml of fluid in the semicircular canal, then a bird could only perceive changes in angular acceleration of at least 10% (i.e., 1 ml of 10 ml, 2 ml of 10 ml, and so on). With 20 ml of fluid, a bird could perceive changes of only 5% (i.e., 1 ml of 20 ml, 2 ml of 20 ml, and so on).

The vestibular otolith system functions to detect motion, particularly acceleration, and position of the head with respect to gravity. In birds, the otolith system consists of the utricular, the saccular, and the lagenar maculae. The saccule functions to detect tilts relative to gravity, with vertical linear accelerations of the head otolith structures of causing shearing motion relative to the underlying sensory epithelium. This shearing action provides an adequate mechanical stimulus for stereocilia of hair cells and in turn results in the activation of primary vestibular afferent neurons (Fernandez and Goldberg 1976). The otoliths are essentially masses of gel with embedded calcium carbonate otoliths supported on hair cells (Fig. 4.161). Tilting the head causes the cilia to bend under the sideways component of the gravitational force on the masses. This modulates the firing of the sensory nerves connected to the hair cells. Different hair cells respond to bending in different directions so that the total signal from all hair cells defines the direction of gravity with respect to the skull (Fig. 4.162).

Loss or damage of hair cells in the inner ear results in hearing impairment or impairment of vestibular function. These cells can be damaged by overexposure to intense sounds. In mammals, once these hair cells are lost, they are not replaced, so hearing loss is permanent. If a bird is exposed to intense sound that damages the cochlea, it can hear again in 3 weeks. When a hair cell is destroyed in a bird, one of the supporting cells receives a signal to divide and it makes two cells. One of those becomes a support cell and one of them a hair cell. Thus, it does two things: it replaces the hair cell and also replaces itself. Alternatively, mitotically derived daughter cells may convert directly into hair cells (Fig. 4.163).

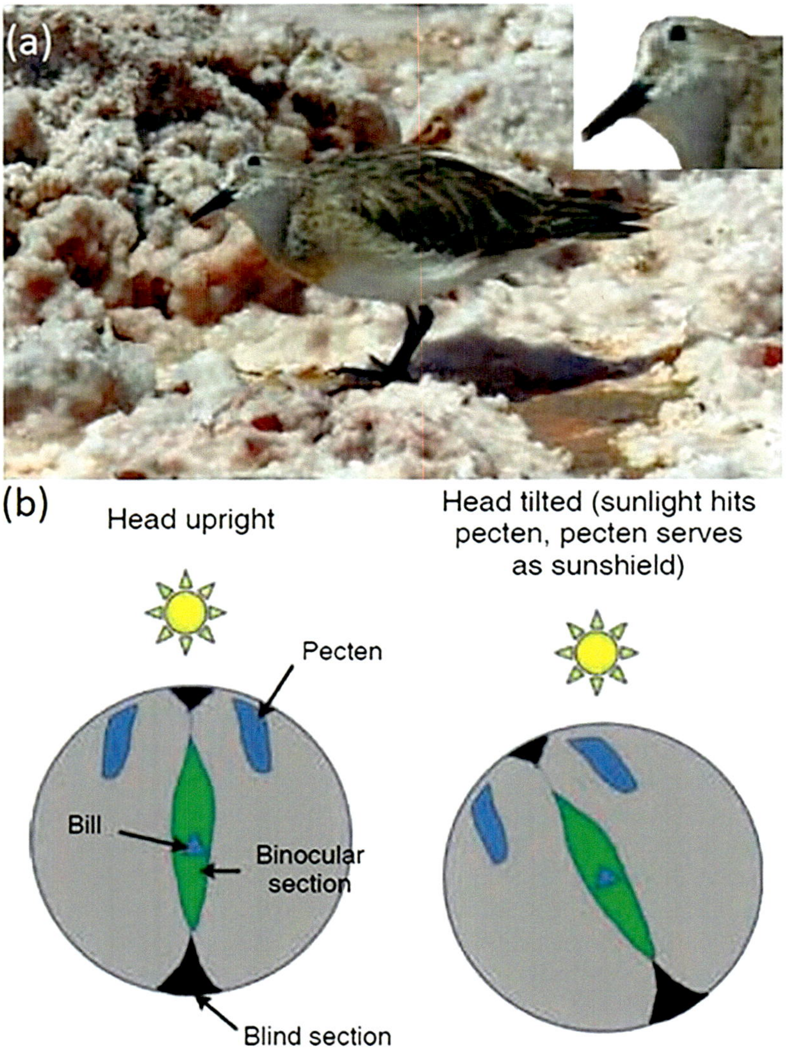


**Fig. 4.129** (**a**) Some species of shorebirds, like this Baird's Sandpiper (*Calidris bairdii*), have been observed to occasionally tilt their heads. Inset: Close-up of the sandpiper tilting its head. (**b**) One possible explanation for this is that they do so to focus light on areas of the retina with higher densities of photoreceptors to enhance visual acuity as the scan the sky to, for example, search for potential avian predators. In addition, however, head tilting may ensure that the image of the sun falls in the blind area formed by the pecten. For example, the image of the sun would fall on the retinas of both eyes if a sandpiper looked straight up with the sun directly overhead (left). By tilting its head like the Baird's Sandpiper in (**a**), the image of the sun now falls in the blind area formed by the pecten in the bird's left eye (**b**, right side) as its right eye faces toward the ground. This would provide the sandpiper with high visual acuity while simultaneously avoiding the sun's glare and protecting the retina. (van den Hout and Graham (2011); (**a**) Photo of Baird's Sandpiper by Piet van den Hout and used with permission; Figure **b** used with permission of Jesse Conklin, co-editor of Wader Study)

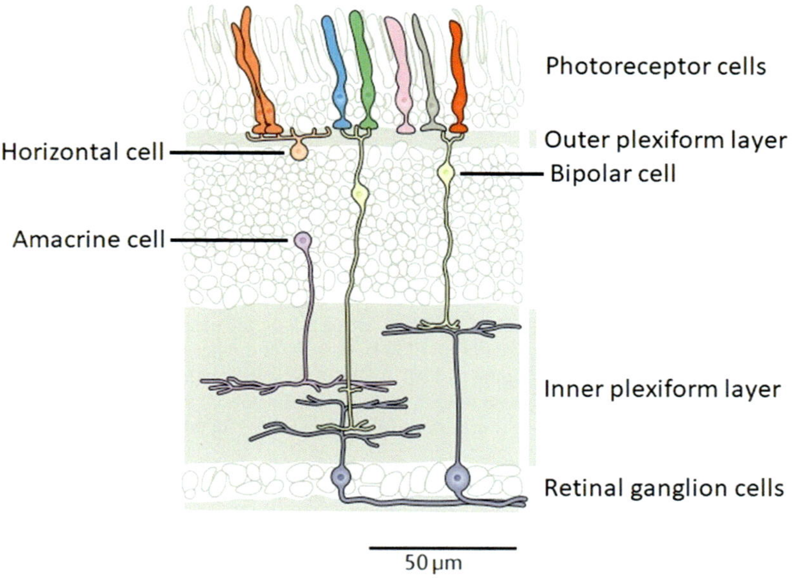


**Fig. 4.130** Simplified view of an avian retina showing the different types of cells. In the outer plexiform layer, photoreceptor cells synapse with bipolar cells and horizontal cells. In the inner plexiform layer, bipolar cells and amacrine cells synapse with each other and with retinal ganglion cells. The avian retina also contains Müller cells that are not shown in this drawing. (Figure from Baden et al. 2019; © 2019 Springer Nature, used with permission)

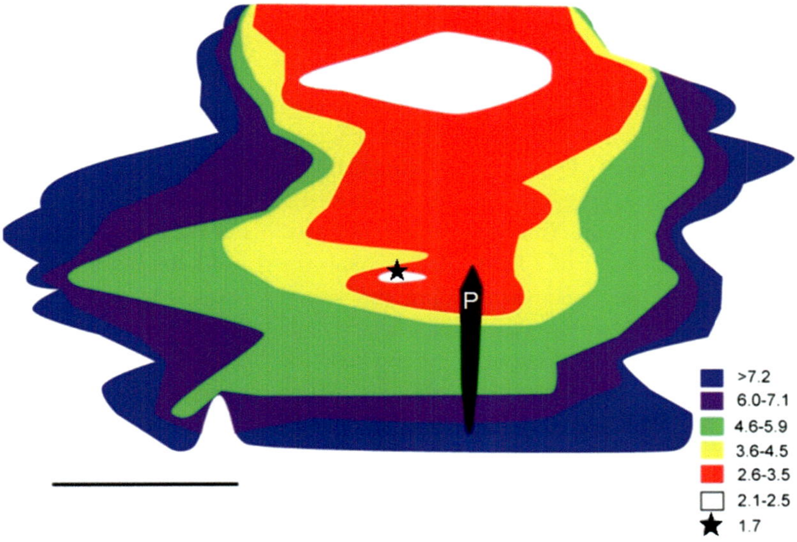


**Fig. 4.131** Ratio of cone photoreceptors to ganglion cells color-coded to show variation to show in convergence in different areas of the retina of a Rock Pigeon (*Columba livia*). The white and colored areas indicate the lowest convergence of photoreceptor to ganglion cells. The cooler colors, going toward blue, indicate a greater degree of convergence and higher ratios. P, pecten; Black star = fovea. (Figure from Querubin et al. 2009; © 2009 Wiley-Liss, Inc., used with permission)

## 4.12 Lumbosacral Organ

In addition to the inner ear, birds also appear to have a balance sensor located in the lumbosacral region of the vertebral column called the lumbosacral organ (LSO) that may be particularly important for terrestrial locomotion (Necker 2005, 2006; Fig. 4.164). The LSO consists of a specialized vertebral canal that has a series of transverse canals and a distinctive lumbosacral spinal cord (Stanchak et al. 2020; Kamska et al. 2020; Fig. 4.165). Also found in the lumbosacral region is a mass of glycogen-rich cells called the glycogen body located where the spinal cord splits dorsally. The possible function(s) of the glycogen body remain to be determined, but proposed hypotheses include secretion of glycogen into the cerebrospinal fluid (Azcoitia et al. 1985),

**Table 4.2** Photoreceptor (PR) to ganglion cell (GC) ratios for several species of birds and humans

| Species[a] | Ratio (PR:GC) central fovea | Ratio (PR:GC) temporal fovea/area dorsalis |
|---|---|---|
| Rock Pigeon | 2.6:1 | 3.5:1 |
| Red-tailed Hawk | 3.4:1 | 3.5:1 |
| Northern Goshawk | 5.1:1 | 6.6:1 |
| American Kestrel | 2.8:1 | 11.0:1 |
| Blue Jay | 3.6:1 | – |
| Least Tern | 5.9:1 | 2.4:1 |
| White-breasted Nuthatch | 1.2:1 | – |
| Southern Double-collared Sunbird | 2.0:1 | – |
| Human | 1:3 | – |

[a]Rock Pigeon data from Querubin et al. (2009), data for the other species of birds from Fite and Rosenfield-Wessels (1975), and human data from Curcio and Allen (1990). Scientific names: Rock Pigeon, *Columba livia*; Red-tailed Hawk, *Buteo jamaicensis*; Northern Goshawk, *Accipiter gentilis*; American Kestrel, *Falco sparverius*; Blue Jay, *Cyanocitta cristata*; Least Tern, *Sternula antillarum*; White-breasted Nuthatch, *Sitta carolinensis*; Southern Double-collared Sunbird, *Cinnyris chalybeus*

producing lipids that might be used to myelinate neurons (Benzo et al. 1975), or serving as a source of lactate or pyruvate that can be used as sources of energy (Dezza et al. 1970; Fink et al. 1975).

Investigators have long suggested that birds may need a balance-sensing organ for bipedal locomotion because their legs are located behind their center of gravity (e.g., Biederman-Thorson and Thorson 1973; Delius and Vollrath 1973), and Necker (2006) suggested that the lumbosacral organ was that organ. This hypothesis was based on the presence of fluid-filled, canal-like structures in the bone surrounding the spinal cord that appear similar to the semicircular canals in the inner ear. In addition to these canals, or lumbosacral transverse canals (LSTCs; Fig. 4.165), the lumbosacral organ also has a series of accessory lobes, protrusions of the spinal cord aligned with the transverse canals (Necker 1999; Fig. 4.166). Necker (2006) suggested, when a bird's body rotated, the inertia of the fluid in the lumbosacral canals stimulated neurons in the accessory lobes and the resulting impulses help birds maintain their balance. In support of this hypothesis, Stanchak et al. (2020) reported a trend toward larger LSTCs in birds that perch compared to birds that do not perch. In addition, Necker (2006) found that the ability of Rock Pigeons (*Columba livia*) with lesions of their lumbosacral to walk was much impaired when they were unable to see (Fig. 4.167). Finally, Matsushita et al. (2018) found that, when stimulated, neurons in the accessory lobes send sensory information to motor neurons in the spinal cord, suggesting that they may have a mechanosensory function.

## 4.13 Hearing Ranges of Birds

For most birds, hearing is best at frequencies of 1–4 kHz, but some birds can detect much higher frequencies (up to 10–12 kHz; Fig. 4.168). As a group, owls can detect lower volume sounds than other birds over the entire range of frequencies, an obvious advantage for nocturnal predators that often rely on sound to locate prey (Fig. 4.169). In general, mammals can hear lower and higher frequency sounds than birds (Fig. 4.170). The small size of most birds constrains sensitivity to lower frequency sounds. The difference in sensitivity of higher frequency sounds is likely related to the evolutionary accident that the three-ossicle middle ear of mammals transfers higher frequencies better than does the columellar middle-ear system of birds (Manley 1990). The ability of birds to discriminate differences in the frequency of sounds and to detect gaps between sounds is generally similar to humans (Dooling

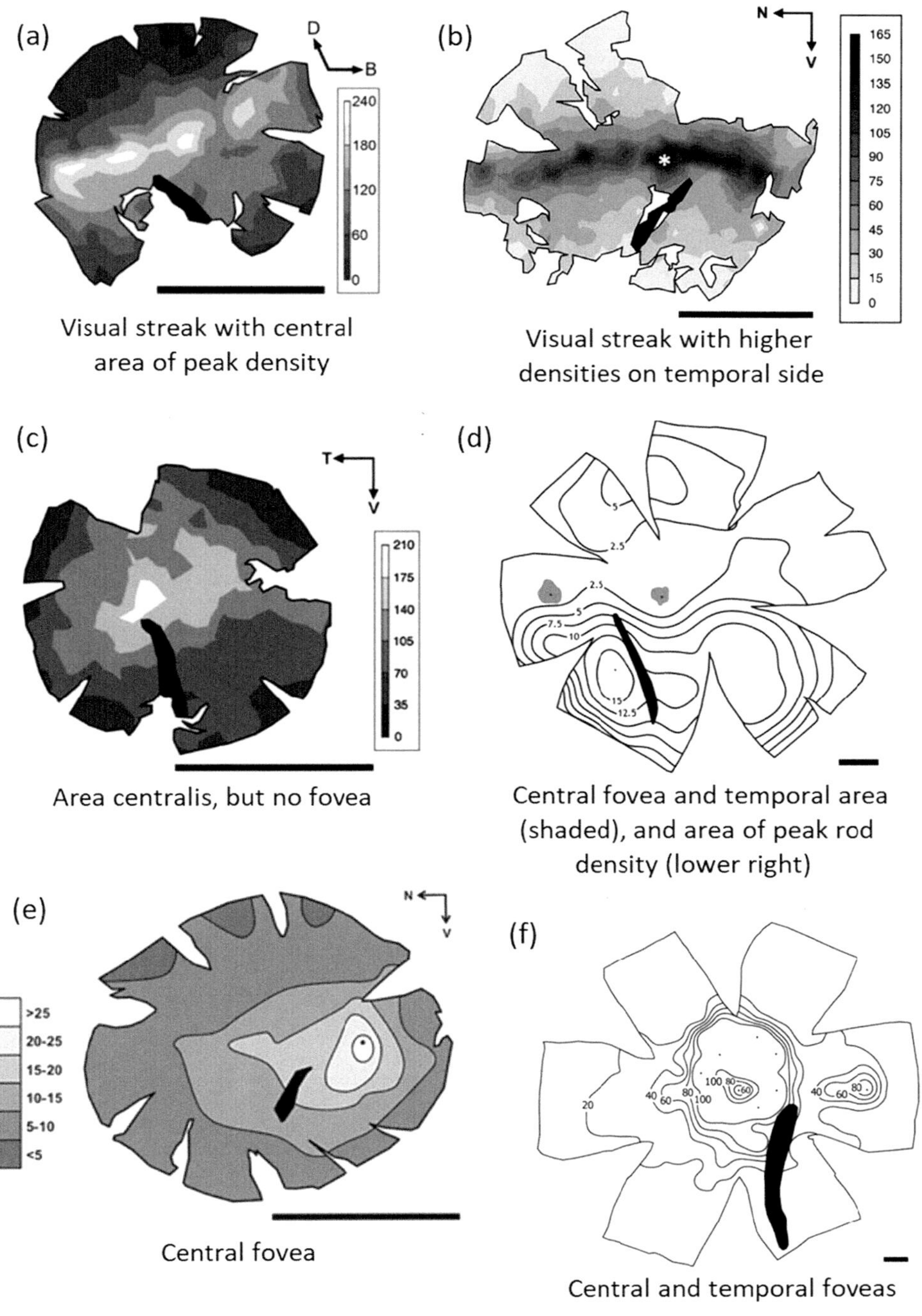


**Fig. 4.132** Examples of variation in the distribution of photoreptors or ganglion cells in the retinas of birds. (**a**) Greater Scaup (*Aythya marila*), (**b**) American Flamingo (*Phoenicopterus ruber*), (**c**) Gray Partridge (*Perdix perdix*), (**d**) Silvereye (*Zosterops lateralis*), (**e**) Northern Saw-whet Owl (*Aegolius acadicus*), and (**f**) Yellow-bellied Elaenia (*Elaenia flavogaster*) photoreceptors (numbers should be multiplied by 1000), scale bar = 1 mm. Dark streaks indicate the location of the pecten. (**a** from Lisney et al. (2013a, b); © 2013 Springer Nature, used with permission, **b** from Lisney et al. 2020; © Wiley Periodicals, Inc., used with permission; **c** from Lisney et al. 2012b; © 2012 Springer Nature, used with permission, **d** from Coimbra et al. 2015; © Wiley Periodicals, Inc., used with permission, and **e** from Lisney et al. (2012b); © 2012 S. Karger AG, Basel, used with permission)

**Table 4.3** Retinal topography and maximum densities of either retinal ganglion cells (RGC) or photoreceptors for several orders and species of birds

| Order | Species | Topography | Cells counted | Maximum density ($mm^2$) | Reference |
|---|---|---|---|---|---|
| Struthioniformes | Common Ostrich, *Struthio camelus* | Visual streak with peak density in central area | RGC | 9500 | Boire et al. (2001) |
| Tinamiformes | Chilean Tinamou, *Nothoprocta perdicaria* | Shallow single fovea | RGC and amacrine cells | ~58,000 | Krabichler et al. (2015) |
| Anseriformes | Canada Goose, *Branta canadensis* | Visual streak with central area of peak density | RGC and amacrine cells plus cones | 20,400 (RGC and amacrine cells) 22,000 cones | Lisney et al. (2013b) (RGC), Moore et al. (2012) (cones) |
| | Redhead, *Aythya americana* | " | RGC and amacrine cells | 24,898 | Lisney et al. (2013b) |
| Galliformes | Ruffed Grouse, *Bonasa umbellus* | Area centralis | RGC and amacrine cells | ~25,000 | Lisney et al. (2012b) |
| | Indian Peafowl, *Pavo cristatus* | Poorly defined visual streak with central area of increased cell density | RGC | 35,600 | Hart (2002) |
| Phoenicopteriformes | American Flamingo, *Phoenicopterus ruber* | Visual streak with higher densities in central and temporal regions | RGC and amacrine cells | 16,000 | Lisney et al. (2020) |
| | Chilean Flamingo, *Phoenicopterus chilensis* | " | RGC and amacrine cells | 14,000 | Lisney et al. (2020) |
| Caprimulgiformes | Band-winged Nightjar, *Systellura longirostris* | Oblique band with ventrotemporal area with highest cell density | RGC | 10,479 | Salazar et al. (2019) |
| | Oilbird. *Steatornis caripensis* | Uniform density across all areas of the retina; three-receptors deep | Rods | ~1,000,000 | Martin et al. (2004) |
| | Anna's Hummingbird, *Calypte anna* | Visual streak with two areas with higher density; central fovea and areas temporalis | RGC | ~48,000–53,000 | Lisney et al. (2015), Tyrrell et al. (2018) |
| Columbiformes | Rock Pigeon, *Columba livia* | Fovea and area dorsalis | Cones and RGC | 328,560 (cones) and 108,300 (RGC) | Querubin et al. (2009) |
| Sphenisciformes | King Penguin, *Aptenodytes patagonicus* | Prominent horizontal visual streak | RGC | 14,222 | Coimbra et al. (2012) |
| | Little Penguin, *Eudyptula minor* | " | RGC | 21,867 | Coimbra et al. (2012) |
| Procellariiformes | Leach's Storm-Petrel, *Hydrobates leucorhous* | Visual streak with central fovea | RGC and amacrine cells | 18,200–22,700 | Mitkus et al. (2016) |

(continued)

**Table 4.3** (continued)

| Order | Species | Topography | Cells counted | Maximum density (mm$^2$) | Reference |
|---|---|---|---|---|---|
| | Northern Fulmar, *Fulmarus glacialis* | " | RGC and amacrine cells | 19,200–25,200 | Mitkus et al. (2016) |
| Cathartiformes | Turkey Vulture, *Cathartes aura* | Single fovea | RGC and amacrine cells | ~21,000–24,000 | Lisney et al. (2013a) |
| Accipitriformes | Wedge-tailed Eagle, *Aquila audax* | Weak visual streak with central and temporal foveas | Cones | 450,000 | Reymond (1985) |
| | Black-chested Buzzard-Eagle, *Geranoaetus melanoleucus* | Central and temporal foveas | RGC | 62,000 | Inzunza et al. (1991) |
| Strigiformes | Barn Owl, *Tyto alba* | Visual streak with single fovea | RGC and amacrine cells | 16,780 | Lisney et al. (2012a, b) |
| | Snowy Owl, *Bubo scandiacus* | Visual streak with single fovea | RGC and amacrine cells | 34,157 | Lisney et al. (2012a, b) |
| | Barred Owl, *Strix varia* | " | RGC and amacrine cells | 33,500 | " |
| | Great Horned Owl, *Bubo virginianus* | " | RGC and amacrine cells | 26,150 | " |
| | Short-eared Owl, *Asio flammeus* | " | RGC and amacrine cells | 22,350 | " |
| | Great Gray Owl, *Strix nebulosa* | " | RGC and amacrine cells | 29,000 | " |
| Coraciiformes | Laughing Kookaburra, *Dacelo novaeguineae* | Central and temporal fovea | RGC | 140,000 | Moroney and Pettigrew (1987) |
| | Sacred Kingfisher, *Todiramphus sanctus* | Central and temporal fovea | RGC | 180,000 | Moroney and Pettigrew (1987) |
| Falconiformes | American Kestrel, *Falco sparverius* | Central and temporal foveas | RGC | 65,000 | Inzunza et al. (1991) |
| | Brown Falcon, *Falco berigora* | " | cones | 380,000 | Reymond (1987) |
| Psittaciformes | Bourke's Parrot, *Neophema bourkii* | Area centralis, but with two separate areas with higher densities | RGC | ~38,000 | Mitkus et al. (2014) |
| | Red-tailed Black Cockatoo, *Calyptorhynchus banksii* | Horizontal streak with a shallow central fovea | RGC | 52,500–55,000 | Coimbra et al. (2014) |

(continued)

**Table 4.3** (continued)

| Order | Species | Topography | Cells counted | Maximum density ($mm^2$) | Reference |
|---|---|---|---|---|---|
| | Long-billed Corella, *Cacatua tenuirostris* | " | RGC | 55,000–67,500 | Coimbra et al. (2014) |
| | Little Corella, *Cacatua sanguinea* | " | RGC | 65,000–73,500 | Coimbra et al. (2014) |
| Passeriformes | Yellow-bellied Elaenia, *Elaenia flavogaster* | Central and temporal foveas | RGC | >150,000 | Coimbra et al. (2009) |
| | Great Kiskadee, *Pitangus sulphuratus* | Inconspicuous horizontal streak with pronounced central fovea and temporal area | RGC | 55,000 | Coimbra et al. (2006) |
| | Rusty-margined Flycatcher, *Myiozetetes cayanensis* | " | RGC | 65,000 | Coimbra et al. (2006) |
| | Acadian Flycatcher, *Empidonax virescens* | Central fovea and an area temporalis | RGC | 64,000 | Tyrrell and Fernández-Juricic (2017) |
| | Tree Swallow, *Tachycineta bicolor* | Bifoveate | GRC | 58,000 | Tyrrell and Fernández-Juricic (2017) |
| | Yellow-rumped Thornbill, *Acanthiza chrysorrhoa* | Central fovea with a secondary peak in the temporal area | Cones | 440,000 | Coimbra et al. (2015) |
| | Red Wattlebird, *Anthochaera carunculata* | " | Cones | 360,000 | Coimbra et al. (2015) |
| | Silvereye, *Zostertops lateralis* | " | Cones | 320,000 | Coimbra et al. (2015) |
| | Large-billed Crow, *Corvus macrorhynchos* | Fovea centralis | | 92,109 (cones)—no. of photoreceptors 5X that of ganglion cells | Rahman et al. (2007a) |
| | Brown-eared Bulbul, *Hypsipetes amaurotis* | Central and temporal areas | Cones and RGC | 60,552 (cones) and 24,032 (RGC) | Rahman et al. (2008) |
| | Eastern Meadowlark, *Sturnella magna* | Central fovea | RGC | 33,000 | Tyrrell et al. (2013) |
| | Red-winged Blackbird, *Agelaius phoeniceus* | Central fovea | RGC and cones | 41,000 (RGC) and 66,000 cones | Fernández-Juricic et al. (2019) |

(continued)

**Table 4.3** (continued)

| Order | Species | Topography | Cells counted | Maximum density ($mm^2$) | Reference |
|---|---|---|---|---|---|
| | Chipping Sparrow, *Spizella passerina* | Single fovea | RGC | ~48,000 | Moore et al. (2015) |
| | American Tree Sparrow, *Spizelloides arborea* | Central fovea | RGC and cones | 26,192 and 50,731 | Rahman et al. (2007b) |
| | Dark-eyed Junco, *Junco hyemalis* | Single fovea | RGC | ~35,000 | Moore et al. (2015) |
| | American Goldfinch, *Spinus tristis* | Central fovea | RGC and cones | 38,500 (RGC) and 88,000 (cones) | Baumhardt et al. (2014) |
| | Zebra Finch, *Taeniopygia guttata* | Central fovea | RGC | 18,000 | Michael et al. (2015) |
| | House Sparrow, *Passer domesticus* | Central fovea | Cones | 82,000 | Ensminger and Fernández-Juricic (2014) |

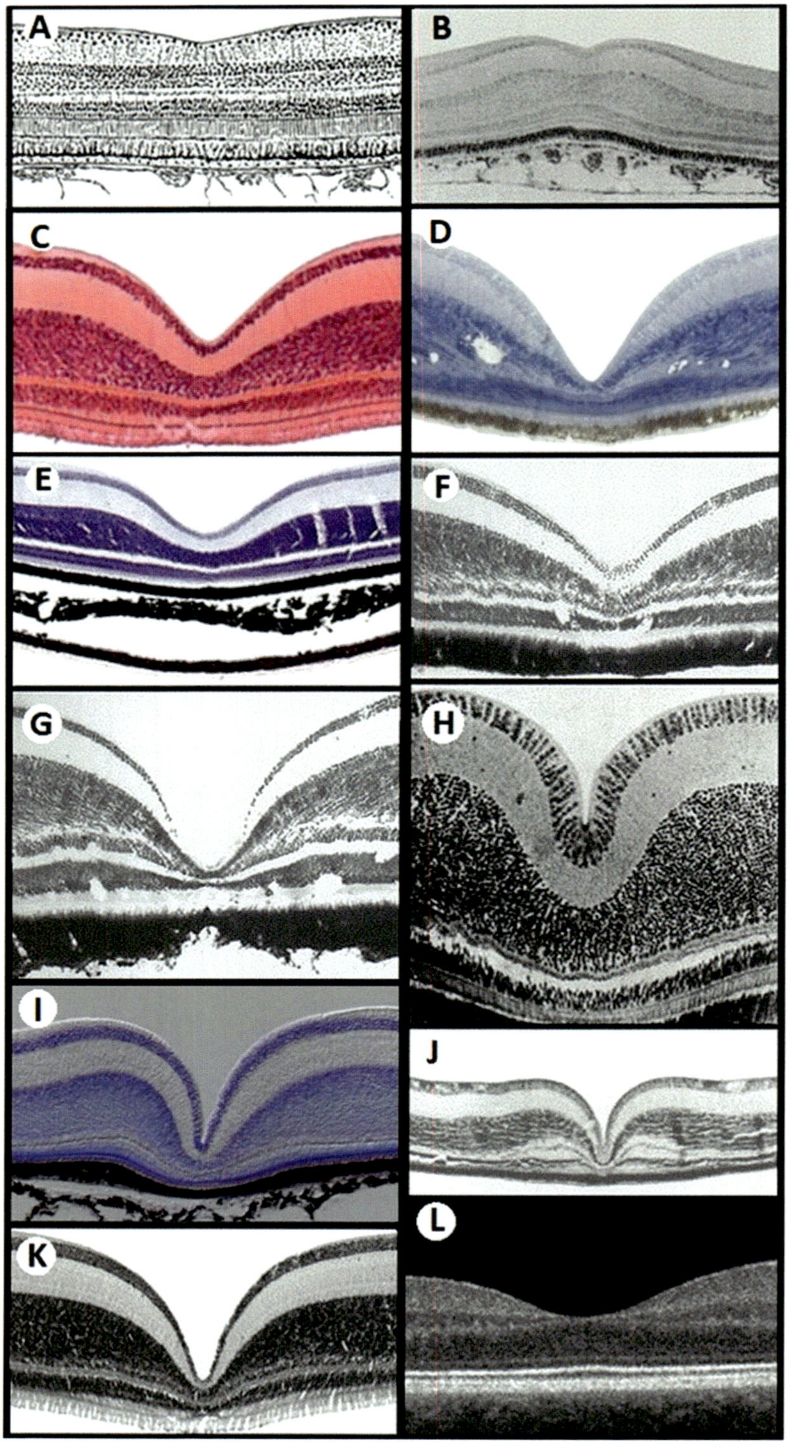


**Fig. 4.133** Examples of avian foveas. (**a**) Barn Owl (*Tyto alba*; Harmening and Wagner 2011; © 2011 Springer-Verlag, used with permission), (**b**) Rock Pigeon (*Columba livia*; Blough 1971; © 1971 Society for the Experimental Analysis of Behavior, used with permission), (**c**) White-crowned Sparrow (*Zonotrichia leucophrys*; Moore et al.

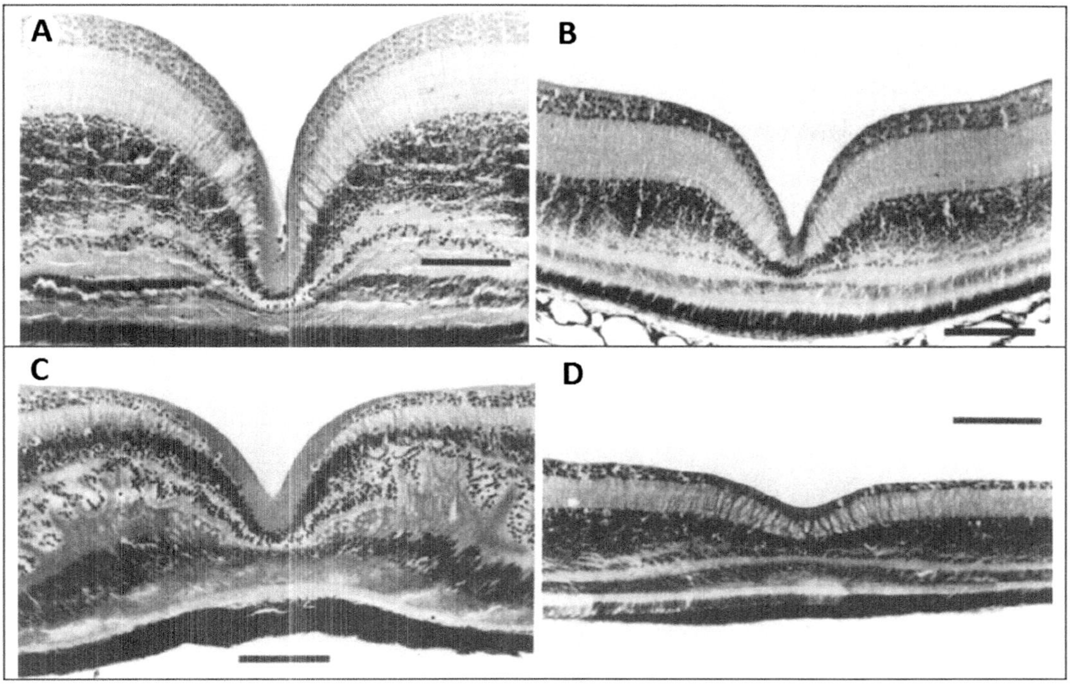


**Fig. 4.134** The two foveas in the eyes of a Sacred Kingfisher (*Todiramphus sanctus*) (**a** and **b**) and a Laughing Kookaburra (*Dacelo novaeguineae*) (**c** and **d**). (**a**, **c**) Deep foveas, (**b**, **d**) Shallow foveas. Scale bars = 200 μm in **a** and **b**, and 100 μm in **c** and **d**. (Figures from Moroney and Pettigrew 1987; © 1987 Springer-Verlag, used with permission)

**Fig. 4.133** (continued) 2016; © 2016 Moore et al., distributed under Creative Commons CC-BY 4.0, https://creativecommons.org/licenses/by/4.0/), (**d**) Eurasian Sparrowhawk (*Accipiter nisus*; Mitkus et al. 2017; © 2017 Wiley Periodicals, Inc., used with permission), (**e**) Anna's Hummingbird (*Calypte anna*; Lisney et al. 2015; © 2015 S. Karger AG, Basel, used with permission), (**f**) Rusty-margined Flycatcher (*Myiozetetes cayanensis*; Coimbra et al. 2006; © 2006 S. Karger AG, Basel, used with permission), (**g**) Great Kiskadee (*Pitangus sulphuratus*; Coimbra et al. 2006; © 2006 S. Karger AG, Basel, used with permission), (**h**) Shy Albatross (also referred to as the White-capped Albatross, *Thalassarche cauta*; O'Day 1940; CC0 Public Domain), (**i**) Zebra Finch (*Taeniopygia guttata*; Sugiyama et al. 2020; © The Authors 2020, open-access article licensed under a Creative Commons Attribution 4.0 International License, https://creativecommons.org/licenses/by/4.0/, (**j**) Sacred Kingfisher (*Halcyon sancta*; Moroney and Pettigrew 1987; © 1987 Springer-Verlag, used with permission), (**k**) Western Meadowlark (*Sturnella magna*; Tyrrell et al. 2013; © 2013 Springer Nature, used with permission), (**l**) Human (Image from Wies, Wikipedia, CC BY 3.0, https://creativecommons.org/licenses/by/3.0/)

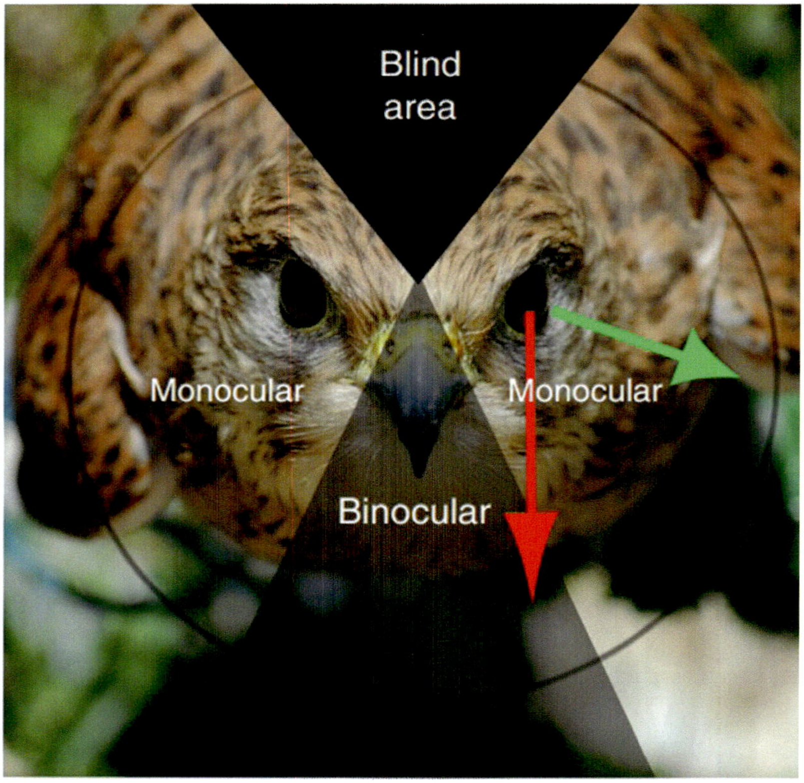


**Fig. 4.135** Visual fields of a Eurasian Kestrel (*Falco tinnunculus*). For monocular vision, light focuses on the central (deep) foveas; for binocular vision, light focuses on the temporal (shallow) foveas. (Figure from González-Martín-Moro et al. 2017; © 2017 Sociedad Española de Oftalmología. Published by Elsevier España, S.L.U., used with permission)

1980). Two sounds separated by a gap are recognized as separate if the gap is at least 2–13 ms in duration (Klump and Maier 1989; Okanoya and Dooling 1990; Klump and Geich 1991).

For most birds (with some predators, particularly owls, being an exception), the need to detect and discriminate among the acoustic signals of conspecifics has likely played a critical role in the evolution of their hearing abilities. Of course, there may be phylogenetic constraints, but, in general, correspondence between the characteristics of acoustic signals and the sensitivities of the auditory system would be expected (Wright et al. 2003). For example, the auditory sensitivities of Orange-fronted Parakeets (*Aratinga canicularis*) were compared with the average frequencies of their calls and both centered around 3 kHz (Wright et al. 2003).

Laboratory studies of the hearing ability of birds provide useful information, but, in the field, the hearing "environment" is more complicated because there are variable levels of noise (sounds not relevant to a bird). How do birds deal with such noise when attempting to communicate with conspecifics? Noise levels vary over time and birds can take advantage of breaks, or periods with less noise, to improve the signal-to-noise ratio (Popp 1989). Birds can also increase the amplitude (volume) of their signals.

## 4.14 Sound Localization

Birds must be able to localize the vocalizations (or other sounds) of their own and other species (Box 4.15 Intracranial cavities and directional hearing). To localize a sound source, a listening bird needs to compute three spatial coordinates:

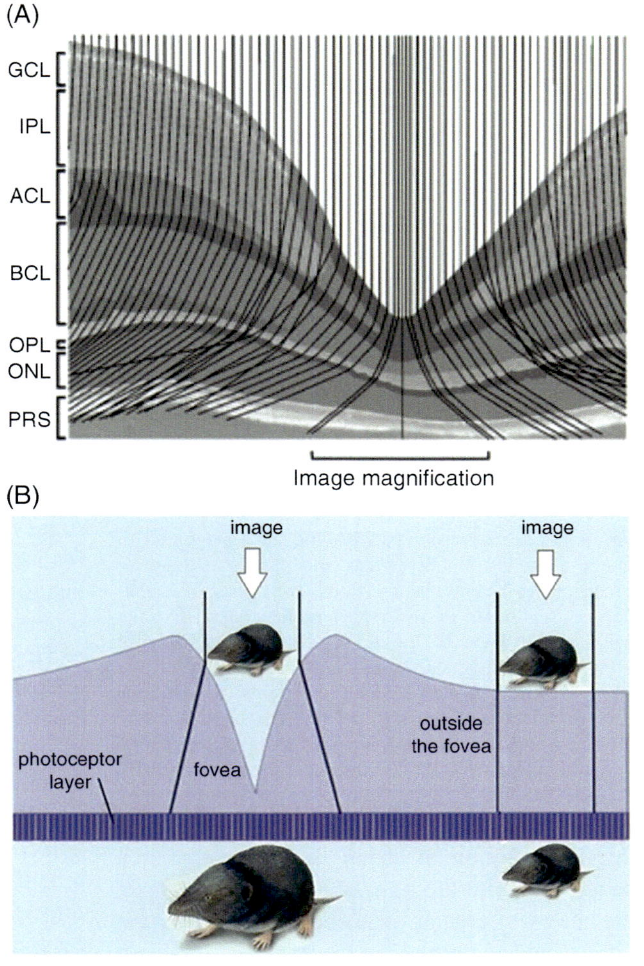


**Fig. 4.136** (**a**) The foveas of some species of birds like diurnal raptors are deep pits, providing more surface area and, therefore, space for more Müller cells. With the Müller cells acting as light cables, the foveal image is projected to photoreceptor cells both within and outside of the foveal region. (**b**) Because the foveal image is projected to photoceptors outside of the foveal region, the image is magnified on the photoceptor layer, enhancing visual acuity. Outside of the fovea, images are not magnified. (Figure **a** from Bringmann 2019; © 2019 Blackwell Verlag GmbH, used with permission; Figure **b** modified from Reichenbach et al. 2014; © 2014 Springer-Verlag Berlin Heidelberg, used with permission)

the horizontal angle, or azimuth, the vertical angle, or elevation, and the distance from listener to singer. The elevation coordinate seems important mainly at short distances and probably does not require a high degree of accuracy. The relative importance of these three coordinates varies with context. For example, for a territorial male monitoring the location of singing neighbors, elevation would generally be unimportant (except during close encounters), whereas azimuth and distance would be important for determining if a neighbor was trespassing and, if so, in which direction. Cues used to determine the distance of a sound source from a listener include sound intensity (volume) as well as the degree to which the sound signal degrades over distance. Birds, like humans, determine azimuth using two cues: interaural time differences and interaural sound level differences. The interaural time difference refers to the difference in time between when a sound reaches one ear relative to the other. The maximum interaural time difference (ITD) varies with the size of a bird's head, but is small. For a small bird, the maximum ITD is about 5–6 ms, and, for an owl, about 18 ms (Moiseff and Konishi 1981). Of course, the maximum ITD requires that a sound originates at an azimuth of either 90 or 270 degrees (with 0 degrees indicating the direction a bird's head is facing). For sounds originating from other locations along the horizontal plane, ITD's will be shorter. The interaural level difference arises

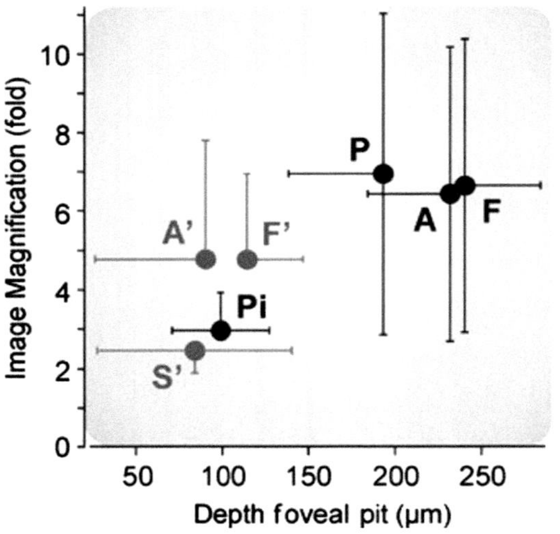


**Fig. 4.137** Relationship between the mean magnitude of image magnification by foveas and the mean depth of foveal pits in different taxa of birds. The degree of magnification was based on the spatial arrangement of Müller cells in the walls of the foveas. A, central fovea of accipitriform birds; A′, temporal fovea of accipitriform birds; F, central fovea of falconiform birds; F′, temporal fovea of falconiform birds; P, central fovea of passeriform birds; Pi, central fovea of Rock Pigeons; S′, temporal fovea of strigiform birds. (Figure from Bringmann 2019; © 2019 Blackwell Verlag GmbH, used with permission)

due to occlusion of sound by the head. For a sound source to the right of a bird, for example, the head will cast an "acoustic shadow" over the left ear, reducing the sound intensity.

Although all birds are able to localize the horizontal angle of sound cues to some degree, the accuracy with which they do so varies. Owls, especially Barn Owls (*Tyto alba*), are renowned for their ability to localize sounds (Fig. 4.171). In the lab, Barn Owls compute azimuth with an accuracy of as little as about 1–3° (the minimum resolvable angle, or MRA; Knudson et al. 1979; Bala et al. 2007). Northern Harriers (*Circus hudsonicus*) determine azimuth with an accuracy of as little as about 2° in behavioral tests (Rice 1982). Among other birds tested, azimuth resolution (MRA) was 23° for Great Tits, 27° for Budgerigars, 28° for canaries (*Serinus canarius*), and 57° for Zebra Finches (*Taeniopygia guttata*; Park and Dooling 1991). Surprising, Eastern Towhees (*Pipilo erythrophthalmus*) were found to compute the azimuth angle of a sound source with an accuracy of about 5° (Nelson and Stoddard 1998). Differences among species, and specifically the impressive ability of Barn Owls to localize sounds, are due, in part, to differences in the number of nerve cells in the midbrain that help create an "auditory map." Pigeons, for example, have about 1000 neurons, whereas Barn Owls have about 10,000 (Nishikawa 2002).

Because auditory neurons from the cochlea do not convey spatial information, the avian brain must somehow use the impulses coming from the two ears (cochleas) to determine the direction or azimuth of a sound source. This involves comparing the intensity (sound level) and time of arrival of the sound at the two ears. Available evidence suggests that time differences are more important for localizing lower frequency sounds, whereas intensity is more important for higher frequency sounds.

How are sounds translated into locations? The midbrain contains an auditory map of space such that stimulation of specific neurons tells a bird the relative location of a sound source. Stimulation of those neurons requires impulses that ultimately originate in the two cochleas and the "interpretation" of localization cues provided by interaural time differences and interaural level differences. Important components of the "localization" process are coincidence detectors. These detectors are neurons in the impulse pathway between the cochleas and the "auditory map" that are stimulated only when impulses arrive simultaneously from the two cochleas. Interaural time differences determine when cochlear neurons are stimulated and, in turn, which coincidence detectors will be stimulated. Impulses from these detectors, along with those from other neurons stimulated based on interaural level differences, then stimulate specific neurons in the auditory map to provide information about a sound's location (see section on the midbrain for more detail).

Among vertebrates, only some owls have asymmetrical ears that allow them to better determine the vertical position of a sound source (Fig. 4.172). This asymmetry occurs in several owl genera, including *Tyto*, *Strix*, *Bubo*, *Asio*, and *Aegolius*, and only involves the outer ear, from

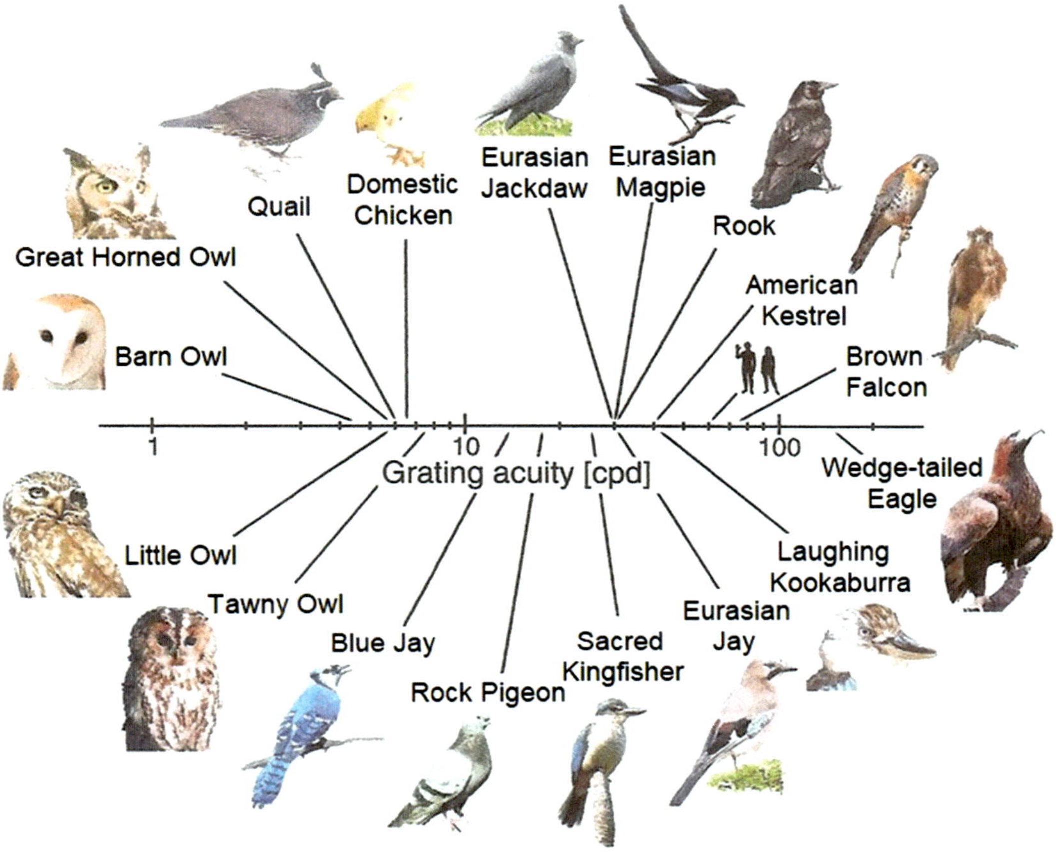


**Fig. 4.138** Grating acuity across bird species. Nocturnal species, like owls, are at the lower end of the acuity spectrum. In contrast, diurnal raptors have very high visual resolution. Typical value for humans is shown for comparison. Grating acuity is the smallest distance between single elements of a periodical pattern that is just resolved (and is reported as cycles per degree [cpd]). Although owl spatial acuity is below that of most other birds, they have excellent absolute sensitivity (i.e., the smallest amount of light that just elicits visual perception) as well as excellent depth perception. Scientific names: Barn Owl, *Tyto alba*; Great Horned Owl, *Bubo virginianus*; Quail, order Galliformes, Domestic Chicken, *Gallus g. domesticus*; Eurasian Jackdaw, *Corvus monedula*; Eurasian Magpie, *Pica*; Rook, *Corvus frugilegus*; American Kestrel, *Falco sparverius*; Brown Falcon, *Falco berigora*; Wedge-tailed Eagle, *Aquila audax*; Laughing Kookaburra, *Dacelo novaeguineae*; Eurasian Jay, *Garrulus glandarius*; Sacred Kingfisher, *Todiramphus sanctus*; Rock Pigeon, *Columba livia*; Blue Jay, *Cyanocitta cristata*; Tawny Owl, *Strix aluco*; Little Owl, *Athene noctua*. (Figure from Harmening and Wagner 2011; © 2011 Springer Nature, used with permission)

the tympanic membrane inward, structures exhibit bilateral symmetry. The most dramatic asymmetry is found in the genus *Aegolius*, including Northern Saw-whet Owls (*A. acadicus*) and Boreal Owls (*A. funereus*). In these owls, ear asymmetry is the result of different positions and orientations of the squamoso-occipital bones, giving the skull a very unique appearance (Fig. 4.172; Norberg 2002). In other owls, the asymmetry is less obvious. For example, asymmetry in Barn Owls is caused by soft anatomy structures only, with skin flaps in the front of the ears at different vertical levels. Ear asymmetry allows these owls to simultaneously localize a sound source in the horizontal (azimuth) and vertical planes, saving time and avoiding directional

**Fig. 4.139** The possible appearance of a butterfly (*Araschnia levana*) as viewed at different distances by three species of birds that would be potential predators. The estimated visual acuity of the three species (in cpd) is provided in parentheses after their common names. Reed Bunting, *Emberiza schoeniclus*; Eurasian Blackbird, *Turdus merula*; Eurasian Jay, *Garrulus glandarius*. (Figure modified from Caves et al. 2018; © 2018 Elsevier Ltd., used with permission. Photos of Reed Bunting by Wicken Fen, Wikipedia, CC BY 2.0, https://creativecommons.org/licenses/by/2.0/, Eurasian Blackbird by Tomáš Marek, Wikipedia, CC BY 3.0, https://creativecommons.org/licenses/by/3.0/, and Eurasian Jay by Pawel Kuźniar, Wikipedia, CC BY 3.0, https://creativecommons.org/licenses/by/3.0/)

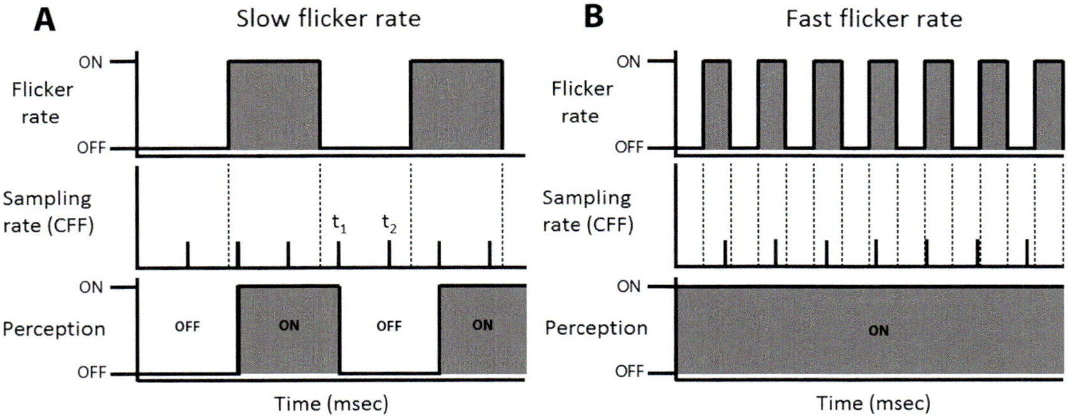


**Fig. 4.140** Difference in perception due to differences in flicker rates. (**a**) At flicker rates below the critical flicker fusion frequency (CFF), flickering is detected because there are sufficiently long intervals of time where the visual system perceives no light ($t_1$ to $t_2$). (**b**) At flicker rates at or above the CFF, no flickering is detected, i.e., images appear continuous, because light is perceived during each interval of time. (Figure modified slightly from Fellows 2013; used with permission of Tyee Fellows)

**Fig. 4.141** Mean flicker fusion frequencies for European Pied (*Ficedula hypoleuca*) and Collared (*F. albicollis*) flycatchers at five different light intensities. Peak flicker fusion frequency for both species of flycatchers varies with light intensity of 1500 candelas per square meter ($cdm^{-2}$). The candela is a measure of light intensity and a typical candle emits light with an intensity of one candela. (Figure modified from Boström et al. 2016; open-access article distributed under the terms of the Creative Commons Attribution License)

ambiguity in the location of moving prey (Norberg 2002).

Even with symmetrical ears, birds have some ability to determine the location of sounds in the vertical plane (i.e., elevation). This is possible because a bird's head absorbs, reflects, and refracts sound waves so that the volume of sounds reaching the two ears differs (Fig. 4.173). Sounds

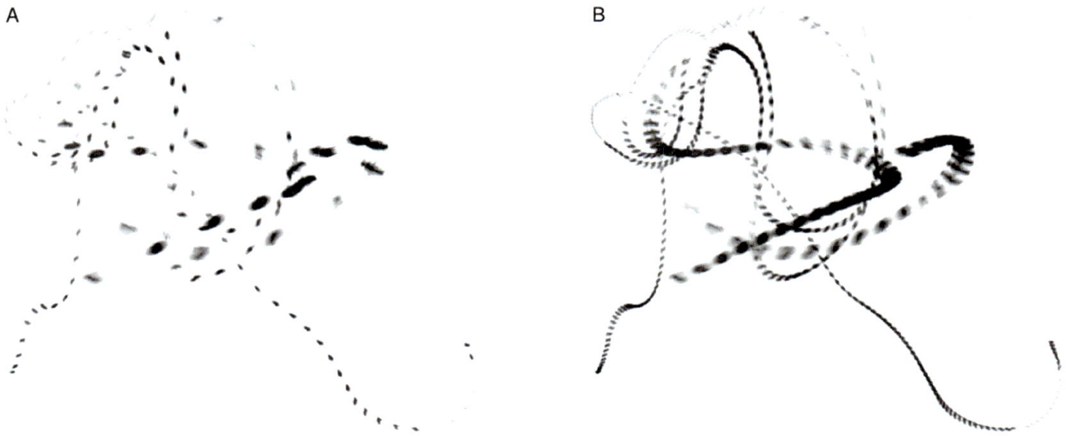


**Fig. 4.142** Flight paths of two blue bottle flies (*Calliphora vomitoria*) as viewed by (**a**) a human with a critical flicker fusion frequency (CFF) of 40 frames per second and by (**b**) a European Pied Flycatcher (*Ficedula hypoleuca*) with a CFF of 120 frames per second. The faster "refreshing" of visual input of European Pied Flycatchers provides them with a more detailed view of the flight paths of flies and other flying insects. (Figure from Boström et al. 2016; open-access article distributed under the terms of the Creative Commons Attribution License, https://creativecommons.org/licenses/by/4.0/)

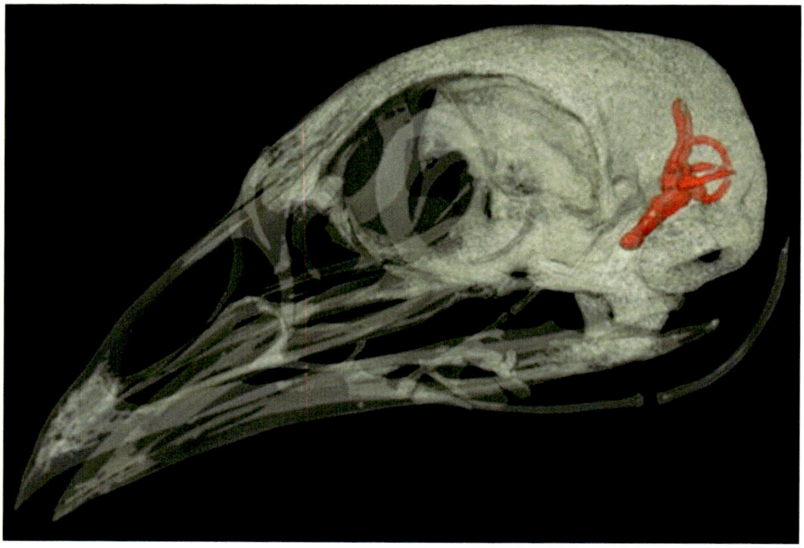

**Fig. 4.143** A semitransparent skull of a Common Quail (*Coturnix coturnix*) with the inner ear shown in red. (Figure from Bonsmann et al. 2016; © 2016 John Wiley and Sons, used with permission)

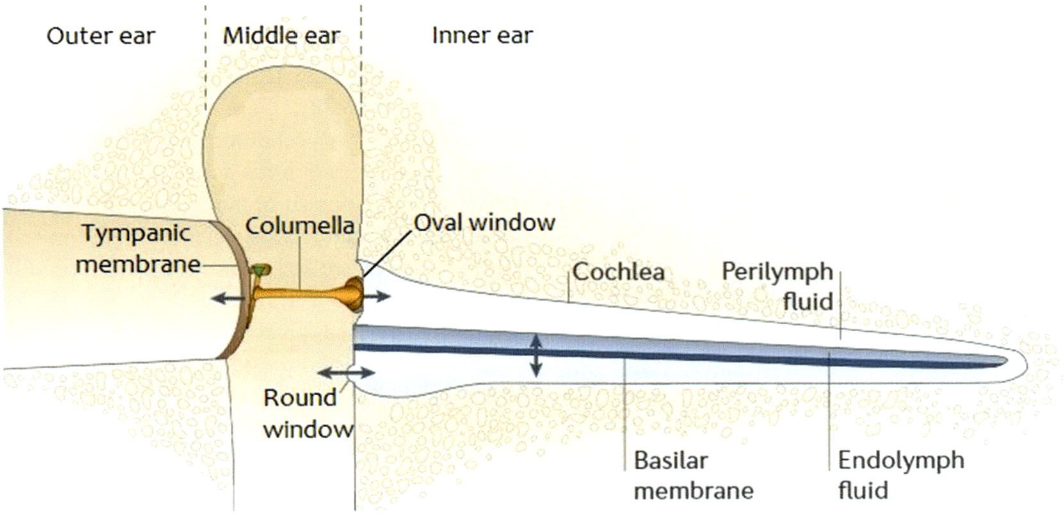


**Fig. 4.144** The three sections of the avian ear. Sound (pressure waves in the air) causes vibration of the tympanic membrane, which, in turn, causes vibration of the columella, which, in turn, causes vibration of the oval window. Vibration of the oval window creates pressure waves in the perilymph fluid that will stimulate hair cells on the basilar membrane. Stimulation of the hair cells stimulates neurons at the base of the hair cells, and these impulses are translated in the brain into the perception of sound. (Figure modified from Fettiplace and Hackney 2006; © 2006 Springer Nature, used with permission)

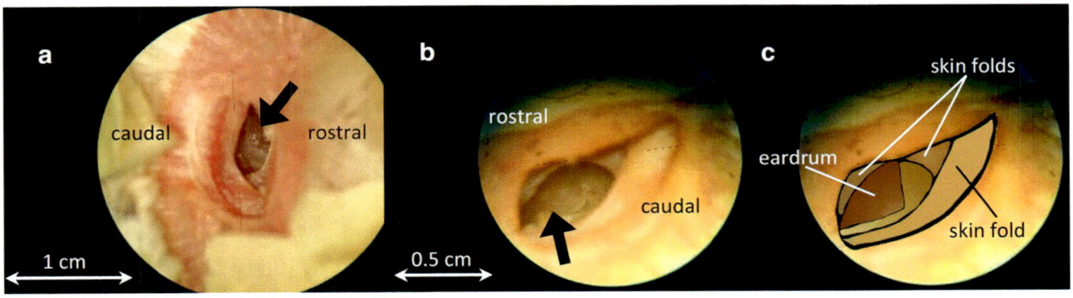


**Fig. 4.145** Tympanic membranes of a Barn Owl (*Tyto alba*). (**a**) Left and (**b**) right tympanic membranes indicated by the arrows. (**c**) Drawing of the right tympanic membrane. Note the different scales for (**a**) and (**b**). (Figure modified from Kettler et al. 2016; © 2016 Springer-Verlag Berlin, used with permission)

originating on the same side of an ear are similar in volume regardless of differences in elevation. However, the volume of sounds originating at different elevations varies for the ear on the other side of the head (Figs. 4.174 and 4.175).

## 4.15 Hearing Underwater

Although the in-air hearing of birds has been the focus of numerous studies, less is known about the ability of birds to hear underwater. Based on laboratory tests with a trained Great Cormorant (*Phalacrocorax carbo*), Hansen et al. (2017)

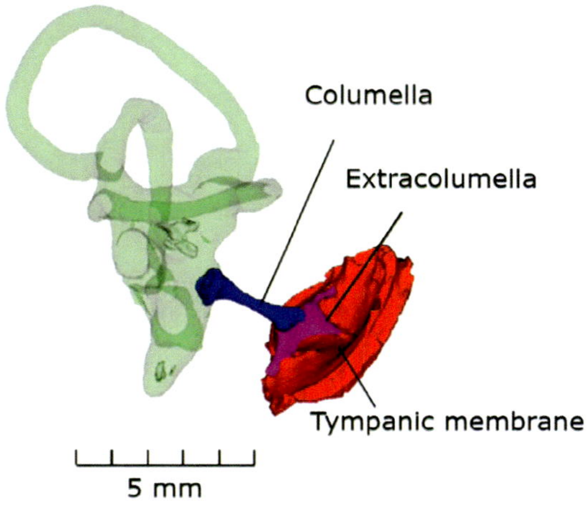


**Fig. 4.146** The tympanic membrane connects to the columella via the cartilaginous extracolumella, and vibrations of the columella are transmitted to fluid in the inner ear by the oval window. (Figure from Zeyl et al. 2020; © 2020 Cambridge Philosophical Society, used with permission)

found that its hearing underwater was comparable to that of seals and toothed whales at frequencies between 1 and 4 kHz, with the greatest sensitivity at 2 kHz (Fig. 4.176). Long-tailed Ducks (*Clangula hyemalis*) are also able to hear underwater (Crowell 2014), but with thresholds 20–25 decibels higher than that of the Great Cormorant. McGrew et al. (2022) examined the ability of Surf Scoters (*Melanitta perspicillata*), Long-tailed Ducks, and Common Eiders (*Somateria mollissima*) to hear underwater, and determined that all three species shared a common range of maximum auditory sensitivity from 1 to 3 kHz, with Long-tailed Ducks and Common Eiders at the high end of that range (2.96 kHz) and Surf Scoters at the low end (1 kHz). Gentoo Penguins (*Pygoscelis papua*) and Common Murres (*Uria aalge*) can also detect and react to underwater sounds (Sørensen et al. 2020; Hansen et al. 2020), and Gentoo Penguins may use sound stimuli for orientation and prey detection (Sørensen et al. 2020). The extent to which other aquatic birds can hear underwater remains to be determined, but being able to do so could clearly be useful for capturing sound-emitting prey, avoiding predators, and navigation (Hansen et al. 2017).

## 4.16 Echolocation

Swiftlets (Apodidae: Collocaliini) and Oilbirds (*Steatornis caripensis*) are the only birds that can navigate by echolocation. Swiftlets are small, insectivorous birds found throughout the Australasian region from the Indian Ocean to the South Pacific, and most species roost and nest in caves. Oilbirds are larger birds (30 cm) found in South American (from Trinidad and Guyana to Bolivia) that feed primarily on fruit and, like swiftlets, roost and nest in caves (Box 4.16 Oilbirds). Unlike the ultrasonic calls of bats, the echolocation clicks of swiftlets and oilbirds are within the human range of hearing (2–8 kHz; Fig. 4.177) and, as a result, are not suitable for detecting small objects. For example, the smallest objects detectable by echolocating swiflets range from about 6 to 10 mm. As a result, swiftlets and oilbirds use echolocation primarily for navigation and, specifically, to avoid obstacles when flying into and out of caves. Analysis has revealed that characteristics of the echo clicks of swiftlets are species-specific and so are likely also used to identify conspecifics. Although echo clicks must have certain properties (short in duration with sharp beginning and ending), some variation is clearly possible (Thomassen 2005).

Three genera of swiftlets are currently recognized. Species in the genus *Aerodramus* echolocate, those in the genus *Hydrochous* do not, and, in the genus *Collocalia*, one species (Pygmy Swiftlet, *C. troglodytes*) echolocates. Non-echolocating swiftlets use areas near cave entrances with enough light to navigate by sight, whereas echolocating swiftlets are found further from entrances where it is dark. As with the vocalizations of other birds, the echo clicks of swiftlets and oilbirds are generated in the syrinx and analysis has revealed no apparent morphological differences in the syrinxes of echolocating and non-echolocating swiftlets (Thomassen 2005). However, auditory nuclei in the brains of echolocating swiftlets are larger than those of non-echolating swiftlets, a possible adaptation for analyzing and interpreting the echoes produced by their "clicks" (Thomassen 2005).

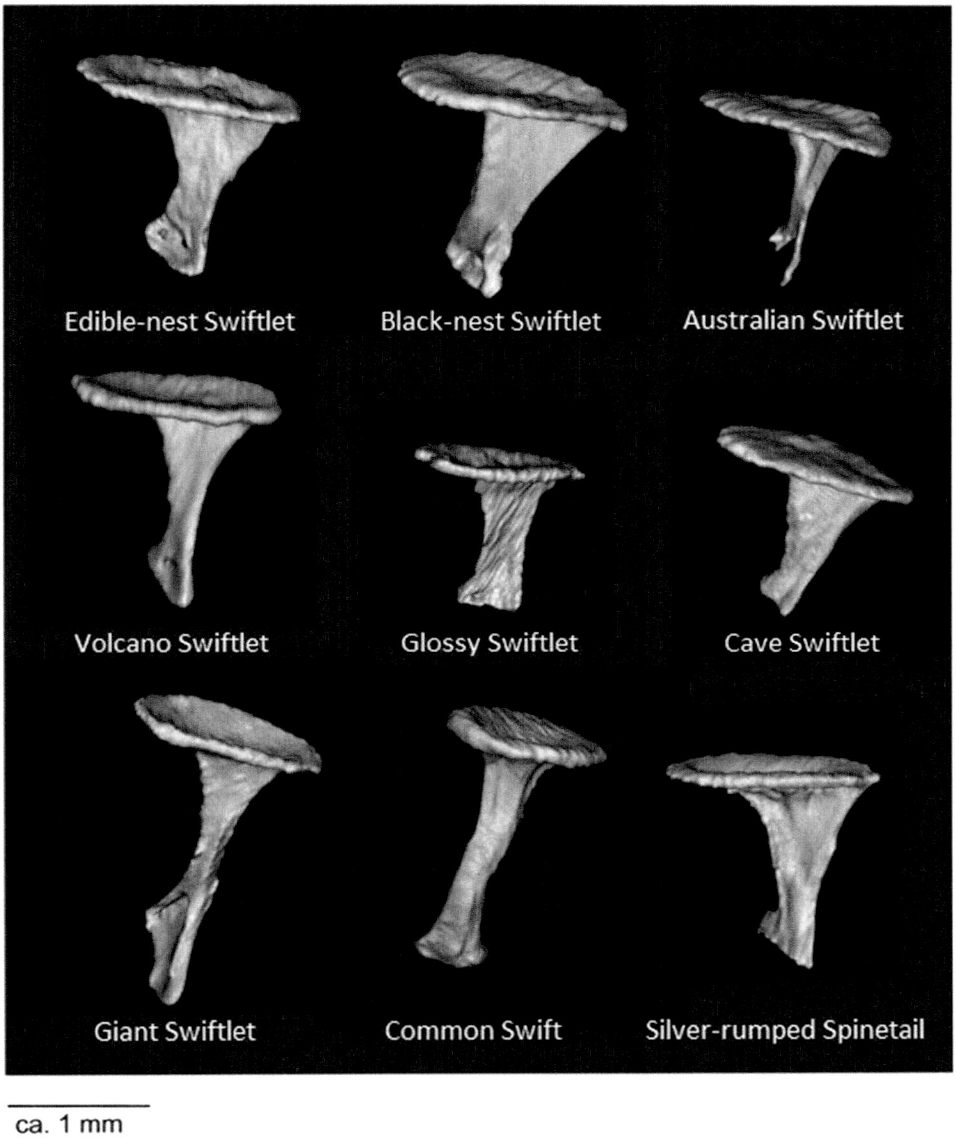


**Fig. 4.147** Columellae of several species of swifts and swiftlets. Note that each has two segments, a cartilaginous extracolumella (round structure at the top in these photos) that inserts on the tympanic membrane and a bony columella (longer thin structure below the extracolumella) that fits into the oval (vestibular) window of the cochlea. Giant Swiftlets are also referred to as Waterfall Swifts, and Silver-rumped Spinetails are also referred to as Silver-rumped Needletails. Edible-nest Swiftlet, *Aerodramus fuciphagus*; Black-nest Swiftlet, *A. maximus*; Australian Swiftlet, *A. terraereginae*; Volcano Swiftlet, *A. vulcanorum*; Glossy Swiftlet, *Collocalia esculenta*; Cave Swiftlet, *C. linchi*; Giant (or Waterfall) Swiftlet, *Hydrochous gigas*; Common Swift, *Apus apus*; Silver-rumped Spinetail (or Needletail), *Rhaphidura leucopygialis*. (Figure from Thomassen et al. 2007; © 2006 Elsevier B. V., used with permission)

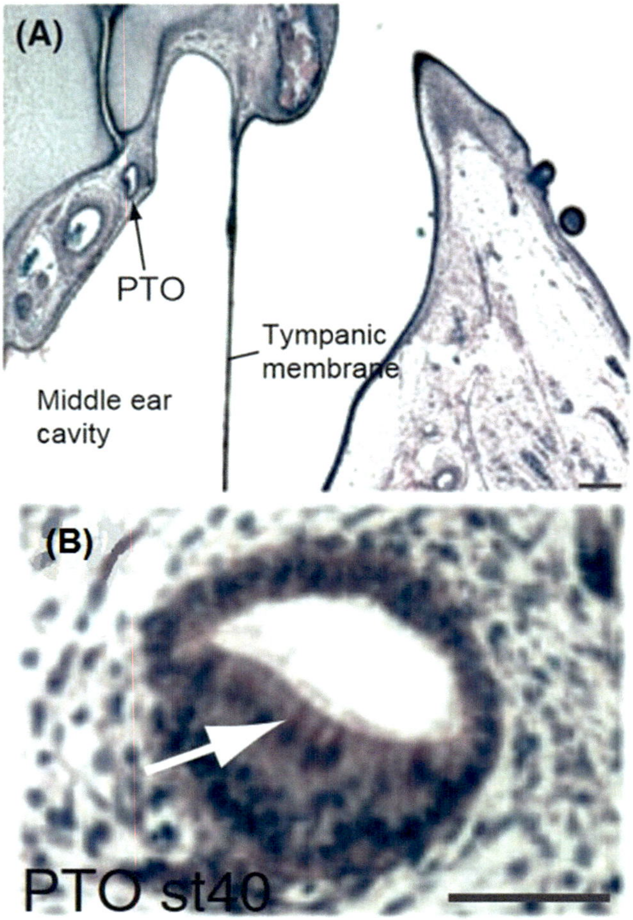


**Fig. 4.148** (**a**) Paratympanic organ (PTO) and tympanic membrane of a Domestic Chicken. (**b**) Cross-section through a paratympanic organ showing sensory hair cells. (Figure from O'Neill 2013; © 2012 The Author Development, Growth & Differentiation © 2012 Japanese Society of Developmental Biologists, used with permission)

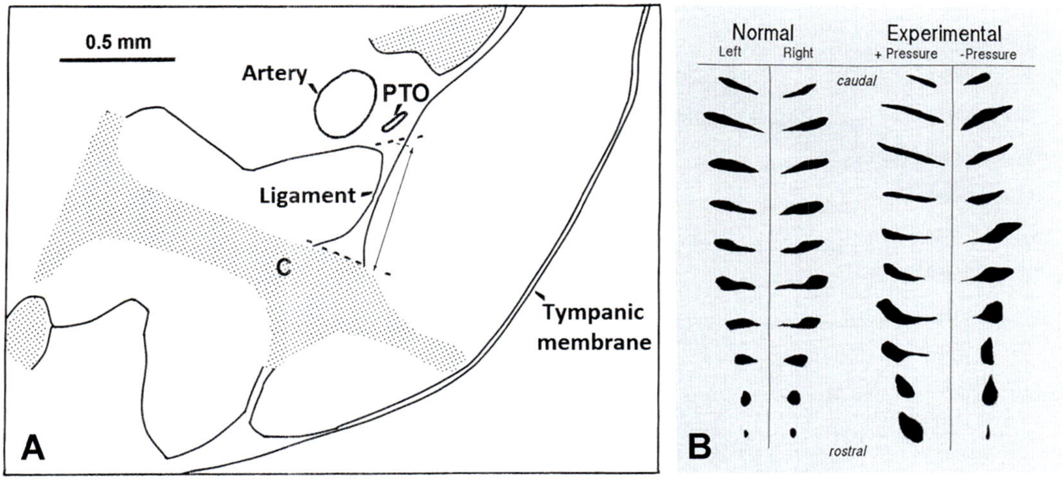


**Fig. 4.149** Possible mechanism by which the paratympanic organs might provide birds with information about barometric pressure. (**a**) Drawing showing a ligament that lies between the columella (C) and the paratympanic organ (PTO). (**b**) Differential pressure exerted on the tympanic membrane causes changes in the lumen (cavity) of the paratympanic organ. These changes may stimulate the hair cells and impulses to the brain could provide information about changes in barometric pressure. (Figure from von Bartheld and Giannessi 2011; © 2011 Wiley-Liss, Inc., A Wiley Company, used with permission)

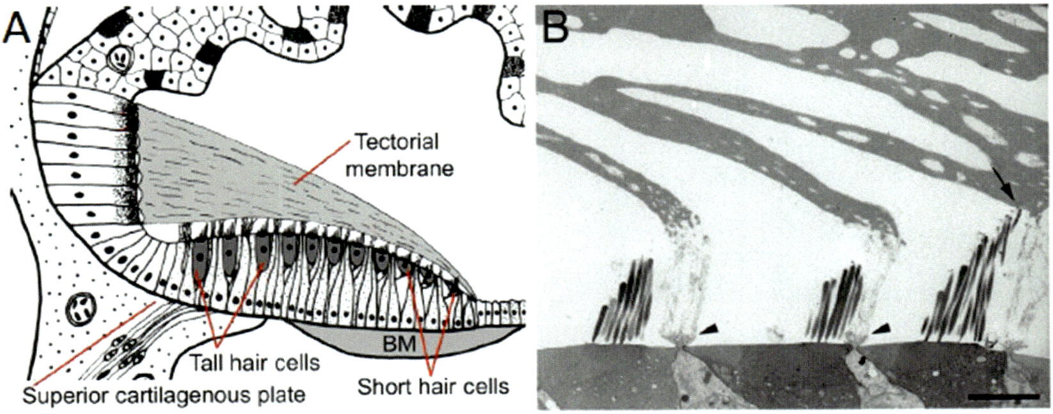


**Fig. 4.150** (**a**) Diagram showing the structure of the avian cochlea, with the tectorial and basilar (BM) membranes and hair cells. (**b**) Electron micrograph showing the structure of the tectorial membrane and attachment to hair cell bundles (arrow on right) and microvilli (arrowheads) on the supporting cells. (Figure from Goodyear and Richardson 2018; © 2018 Elsevier Inc., used with permission)

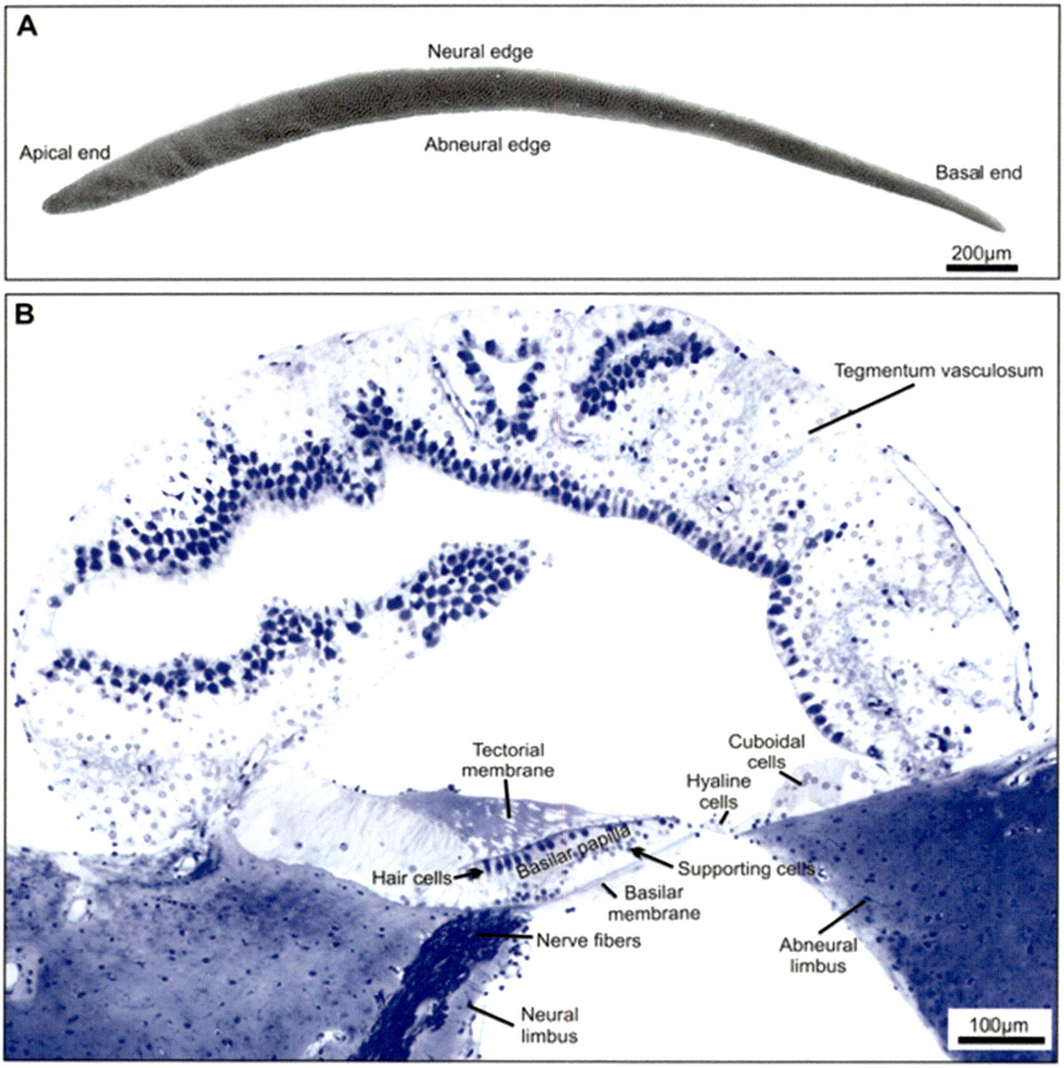


**Fig. 4.151** (**a**) Surface view of the basilar papilla of a North Island Brown Kiwi (*Apteryx mantelli*) obtained from scanning electron microscopy. (**b**) Cross section of a North Island Brown Kiwi cochlea. (Figure from Corfield et al. 2011; © 2011 Corfield et al., open-access article distributed under the terms of the Creative Commons Attribution License, https://creativecommons.org/licenses/by/2.0/)

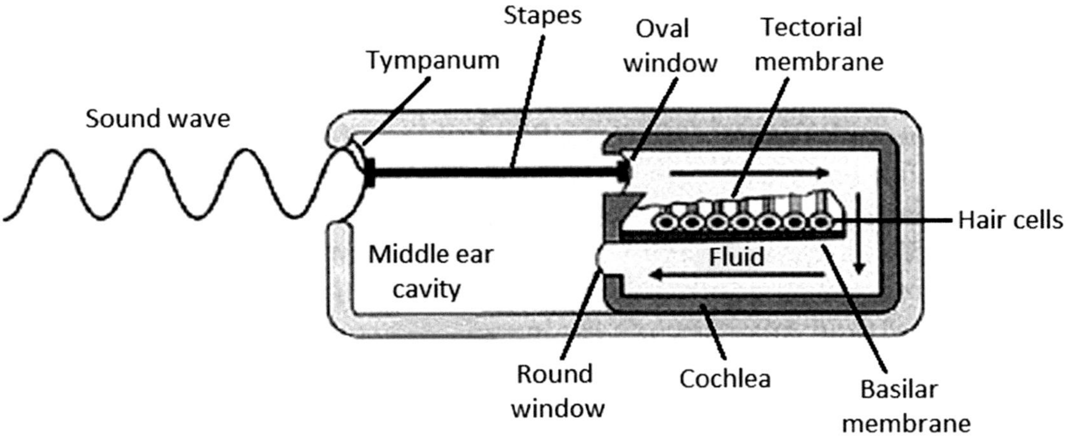


**Fig. 4.152** Process by which sound waves become nervous impulses in the avian ear. Sound waves cause vibration of the tympanum and those vibrations are transmitted to the oval window by the stapes. Vibration of the oval window generates pressure waves in fluid in the inner ear that cause vibration of the specific areas of the basilar membrane (depending on the frequency of the vibrations). Hair cells in those areas of the basilar membrane strike the tectorial membrane, generating nervous impulses that the auditory areas of the brain translate as sound. (CC0 Public Domain)

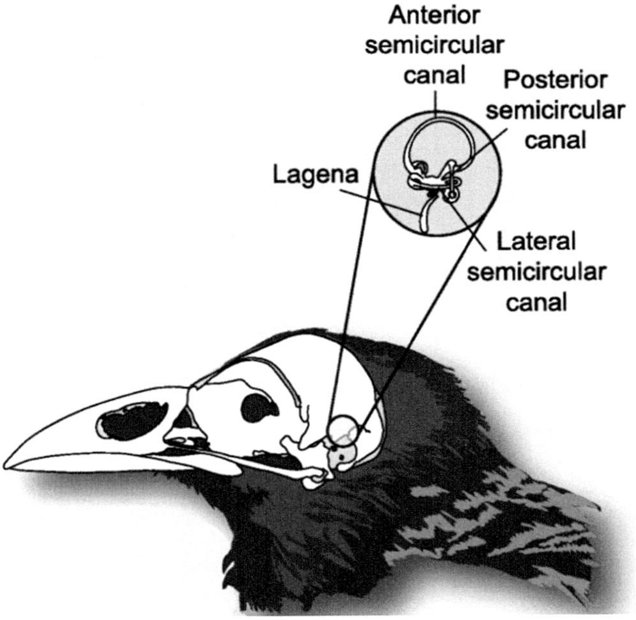


**Fig. 4.153** Inner ear of a bird showing the three semicircular canals and lagena. (Figure from Marugán-Lobón et al. 2013; open-access article distributed under the Creative Commons Attribution License, https://creativecommons.org/licenses/by/2.0/)

**Fig. 4.154** Inner ear of a Domestic Chicken. sc, semicircular canal. (Figure modified from Bok et al. 2011, used with permission of the U. S. National Academy of Sciences)

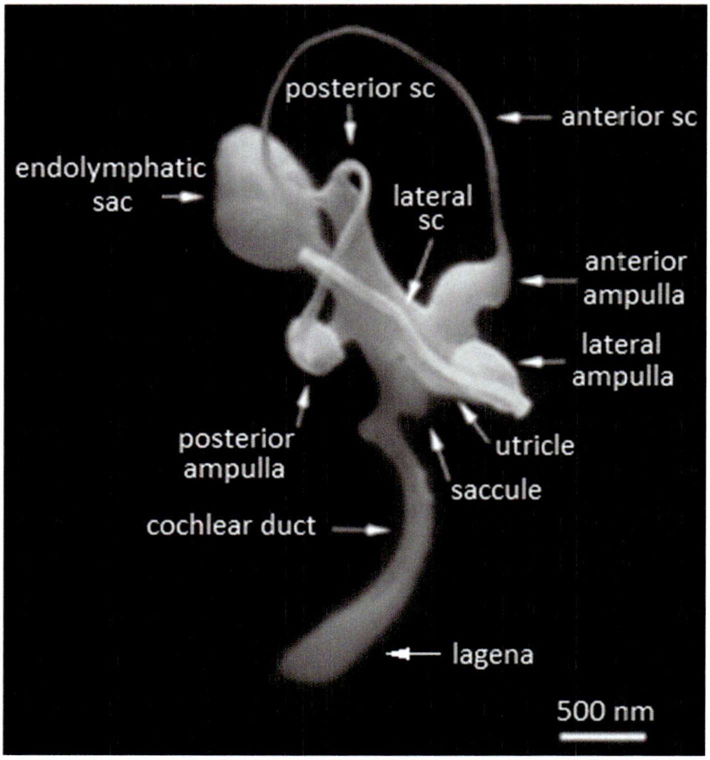


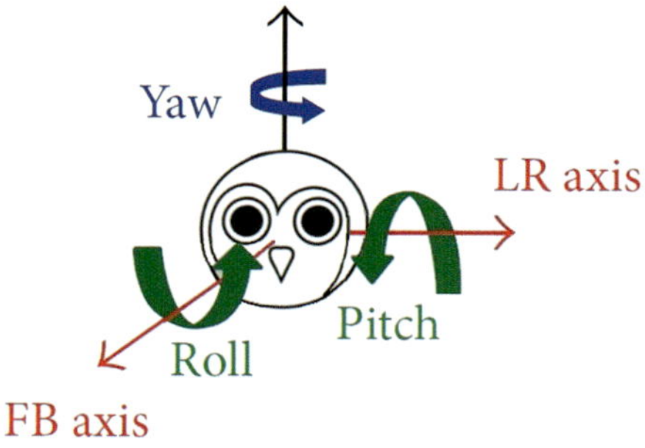


**Fig. 4.155** The three semicircular canals provide birds with information about angular acceleration in any direction along the left-right (LR) axis and/or front-back (FB) axis. (Figure modified from Lacquaniti et al. 2014; © 2014 Francesco Lacquaniti et al., open-access article distributed under the Creative Commons Attribution License, https://creativecommons.org/licenses/by/4.0/)

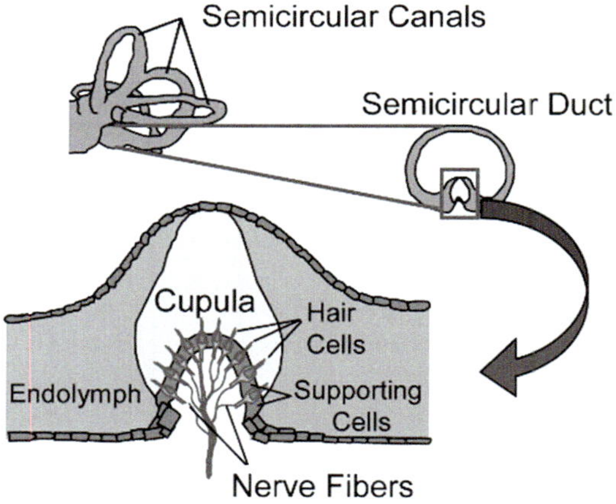


**Fig. 4.156** Location of a cupula in a semicircular canal. (Figure from Dernedde et al. 2014; © 2014 Dernedde et al., open-access article distributed under the terms of the Creative Commons Attribution License, https://creativecommons.org/licenses/by/2.0/)

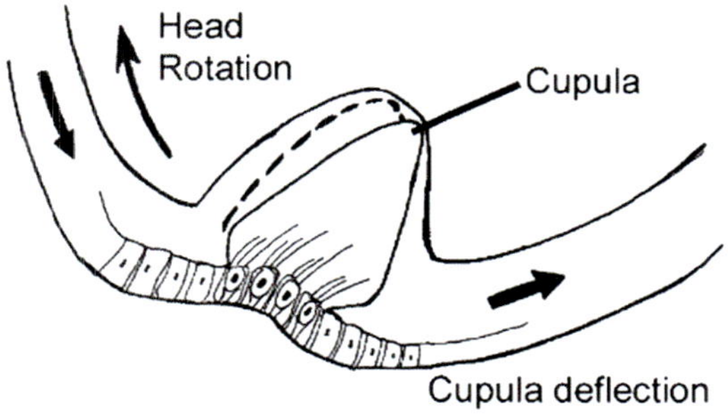


**Fig. 4.157** An ampulla of a semicircular canal contains a cupula (a gelatinous substance in which the stereocilia of hair cells are embedded). Movements of the head cause movements of the semicircular canals that cause the fluid in the canals to move the cupula, with the extent of movement in the three semicircular canals dependent on the direction and extent of head movement. Deflection of the cupula and movement of the embedded stereocilia generate nervous impulses along sensory neurons at the base of the stereocilia and these impulses are "interpreted" in the brain to provide information that helps birds maintain equilibrium. (Figure from Khan and Chang 2013; Reprinted from NeuroRehabilitation 32, Anatomy of the vestibular system: a review, Copyright 2013, with permission from IOS Press. The publication is available at IOS Press through https://doi.org/10.3233/NRE-130866)

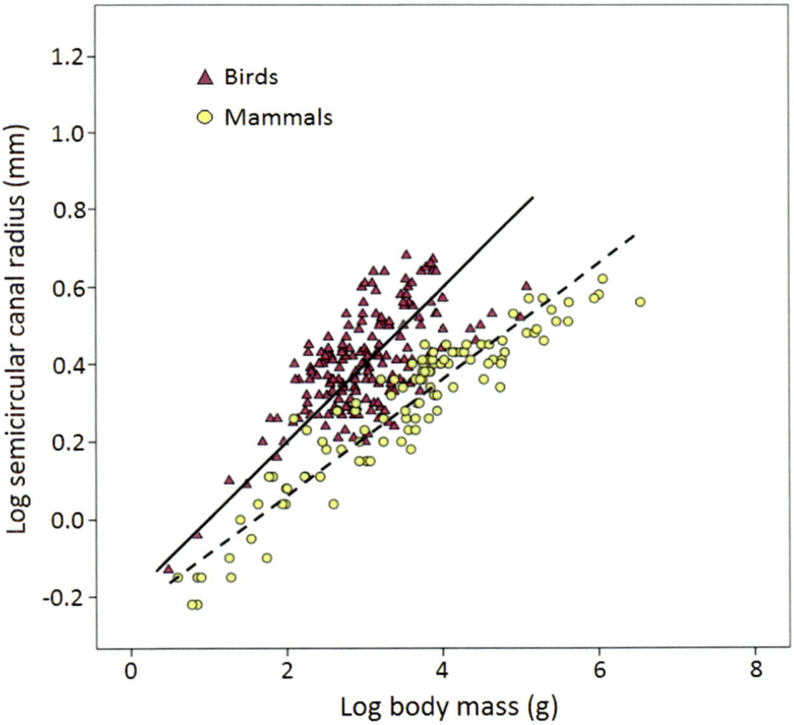


**Fig. 4.158** Relationship between body mass and the radius of semicircular canals for 178 species of birds and 106 species of mammals. Relative to body mass, the semicircular canals of birds are generally larger in radius than those of mammals. (Figure modified from Sipla 2007; used with permission of Justin S. Sipla)

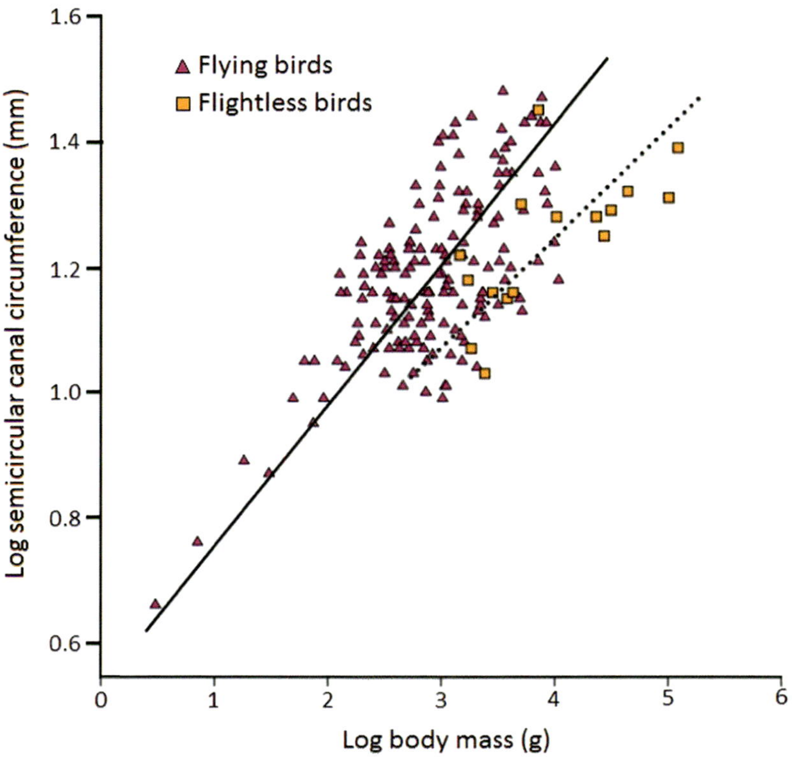


**Fig. 4.159** Circumference of semicircular canals relative to body mass for 162 species of birds that fly and 16 species of flightless birds. The circumference of the semicircular canals of flying birds is generally larger than that of flightless birds. (Figure modified from Sipla 2007; used with permission of Justin Sipla)

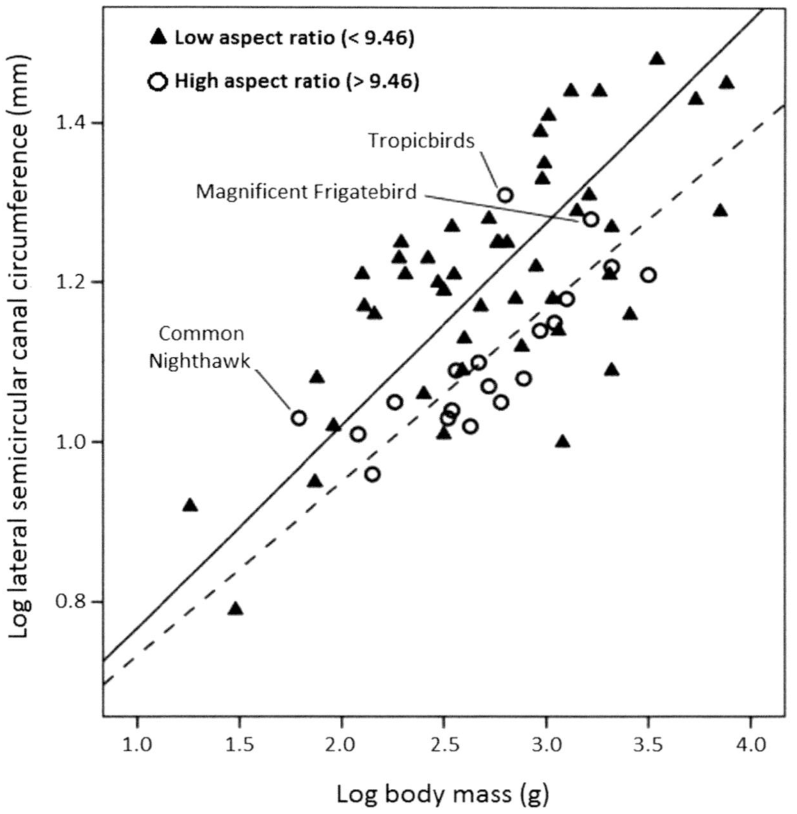


**Fig. 4.160** Relationship between body mass and the circumference of the lateral semicircular canal for birds with low aspect ratios and high aspect ratios. In most cases, the circumference of lateral semicircular canals is greater for more maneuverable birds, i.e., those with low aspect ratios, than for less maneuverable (those generally faster flying) birds of similar-size, but with higher aspect ratios. Some species of birds with relatively high aspect ratios, but with low wing loading, are also very maneuverable and have enlarged lateral semicircular canals, e.g., Common Nighthawks (*Chordeiles minor*), Magnificent Frigatebirds (*Fregata magnificens*), and tropicbirds (*Phaethon* spp.). (Figure modified from Sipla 2007; used with permission of Justin Sipla)

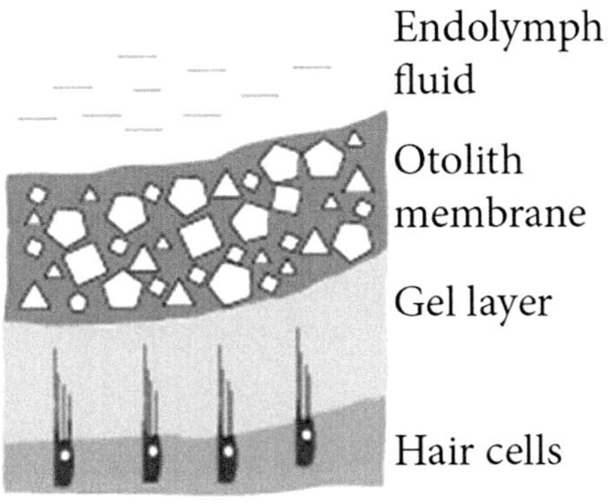


**Fig. 4.161** Generalized otolith organ showing the stereocilia of the hair cells embedded in the gel layer and the otolith membrane on top of the gel layer. (Figure from Lacquaniti et al. 2014; © 2014 Francesco Lacquaniti et al., open-access article distributed under the Creative Commons Attribution License, https://creativecommons.org/licenses/by/4.0/)

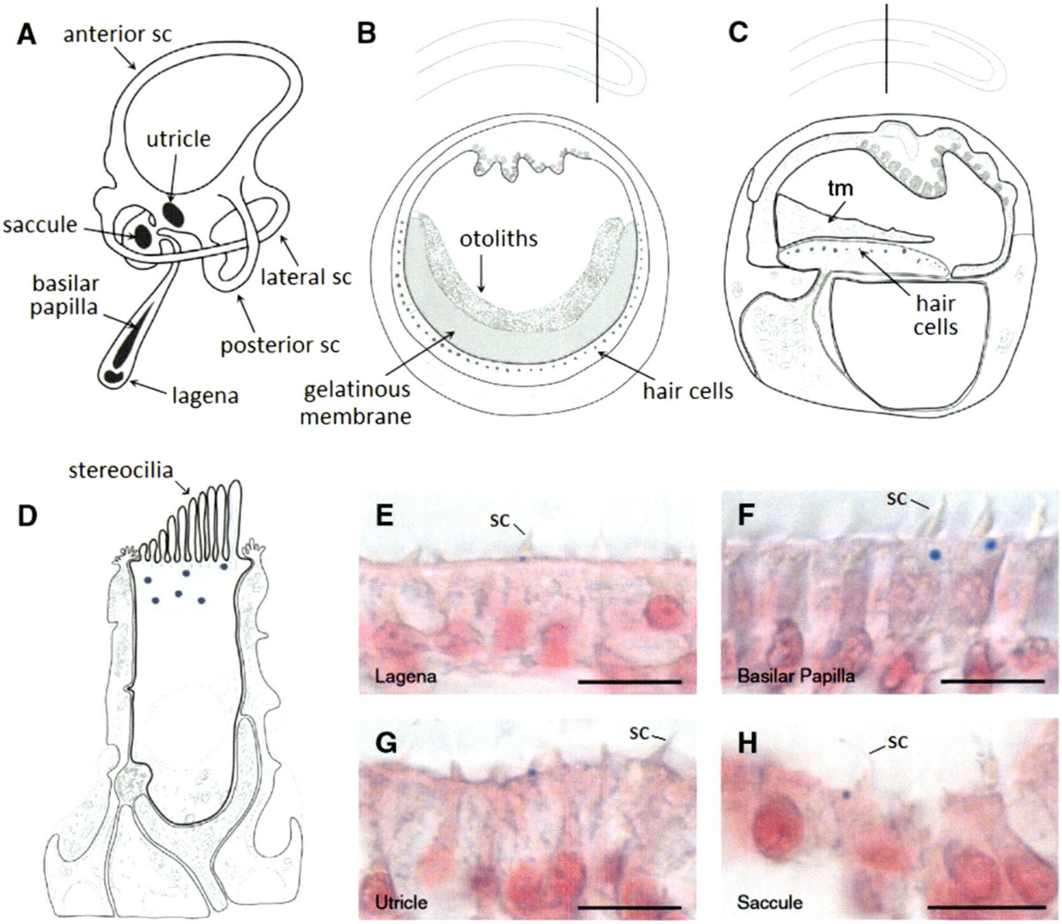


**Fig. 4.162** (**a**) Drawing of the inner ear of a Rock Pigeon (*Columba livia*). sc, semicircular canal. (**b**) Drawing of a cross-section through the lagena showing the calcium carbonate otoliths associated with a gelatinous membrane that lies on top of the sensory hair cells. (**c**) Drawing of a cross-section through the basilar papilla showing hair cells and the tectorial membrane (tm). (**d**) Drawing of a hair cell showing stereocilia that would be embedded in the gelatinous membrane. (**e**–**h**) Photomicrographs of the lagena, basilar papilla, utricle, and saccule of a Rock Pigeon showing stereocilia (sc) extending from the hair cells. In (**d**)–(**h**), the blue "dots" are iron-rich particles. Scale bars = 10 μm. (Figure modified from Lauwers et al. 2013; © 2013 Elsevier Ltd., used with permission)

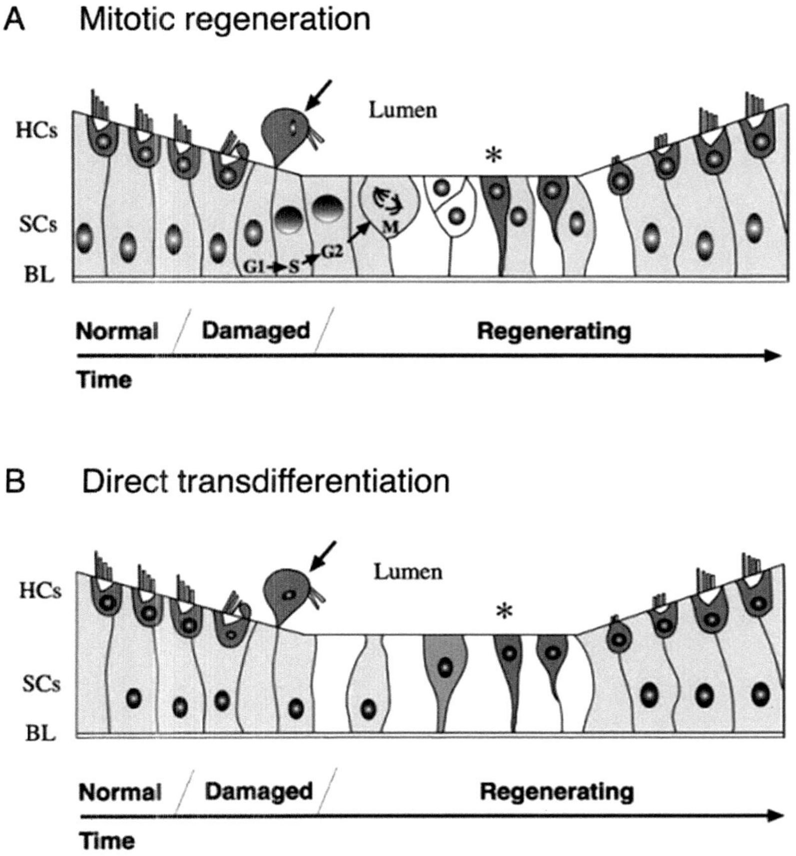


**Fig. 4.163** Mechanisms for hair cell regeneration in the avian auditory epithelium (basilar membrane). (**a**) After being damaged, hair cells are extruded from the epithelium and signal nearby supporting cells to divide and produce daughter cells that can become either hair cells or supporting cells. (**b**) Alternatively, supporting cells convert into hair cells. In both (**a**) and (**b**), events progress over time from left to right, with the normal epithelium shown on the left, the damaged and regenerating epithelium in the middle, and the epaired epithelium on the right. Mature and regenerating hair cells (HCs) are shown with darker cytoplasm than the supporting cells (SCs). Arrows point to damaged hair cells being extruded from the epithelium. Asterisks indicate regenerated hair cells. For mitotic regeneration in A, stages of the cell cycle are G1 (gap1), S (DNA synthesis), G2 (gap2), and M (mitosis). BL, basal lamina. [Figure from Oesterle and Stone 2008 (© 2008 Springer Science Business Media, LLC, used with permission) as modified from Bermingham-McDonogh and Rubel 2003 (© 2003 Elsevier Science Ltd., used with permission)]

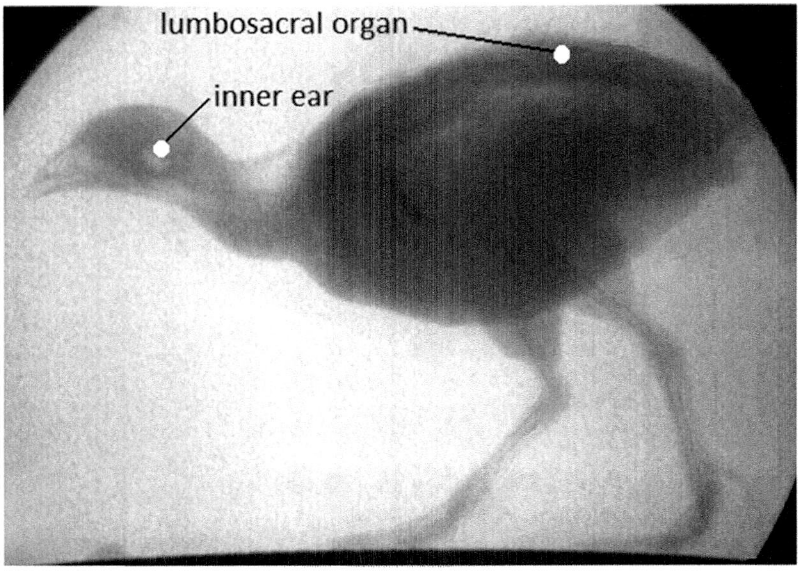


**Fig. 4.164** X-ray of walking Japanese Quail (*Coturnix japonica*) showing general locations of the inner ear and the lumbosacral organ. (Figure modified from Abourachid et al. 2011; © 2011 Elsevier GmbH, used with permission)

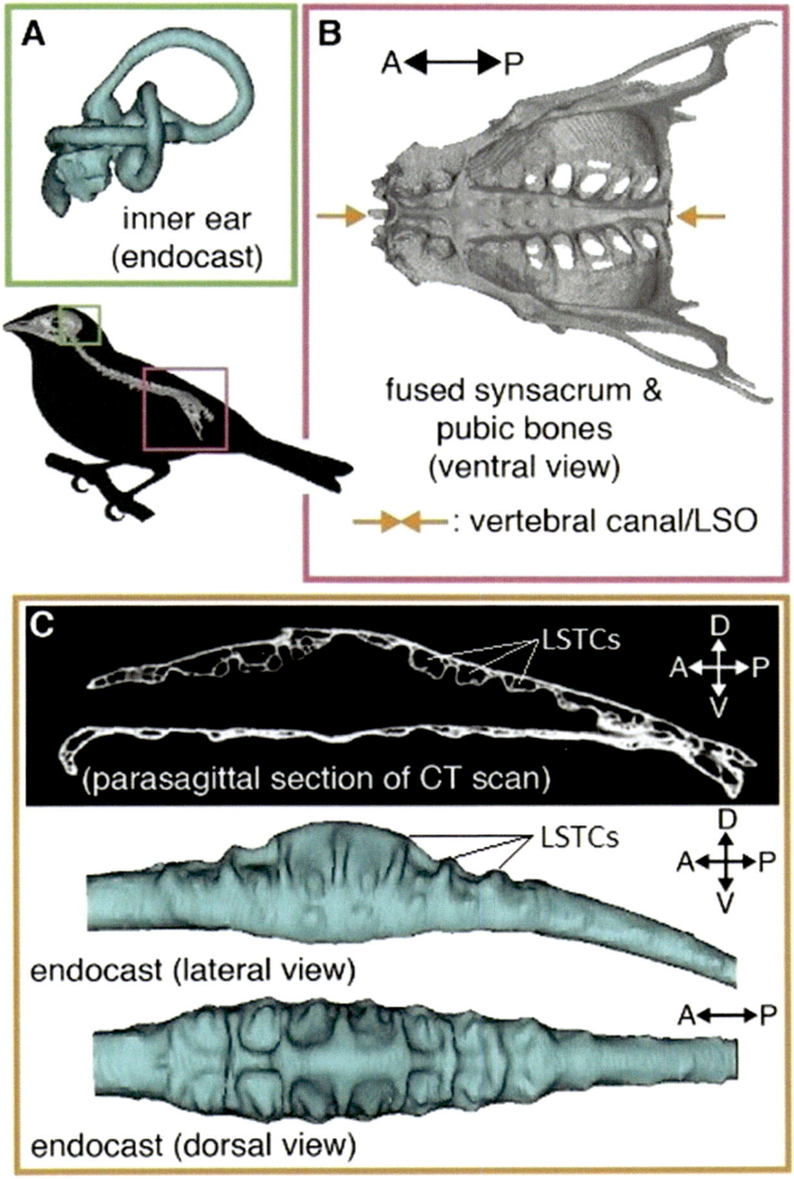


**Fig. 4.165** Birds may have two sets of balance organs. (**a**) One in the inner ear, and (**b**) a second within the synsacrum called the lumbosacral organ and (**c**) located within an expanded vertebral canal with a series of canal-like recesses (lumbosacral transverse canals, or LSTCs). (Figure from Stanchak et al. 2020; open-access article distributed under the terms of the Creative Commons CC BY license, https://creativecommons.org/licenses/by/4.0/)

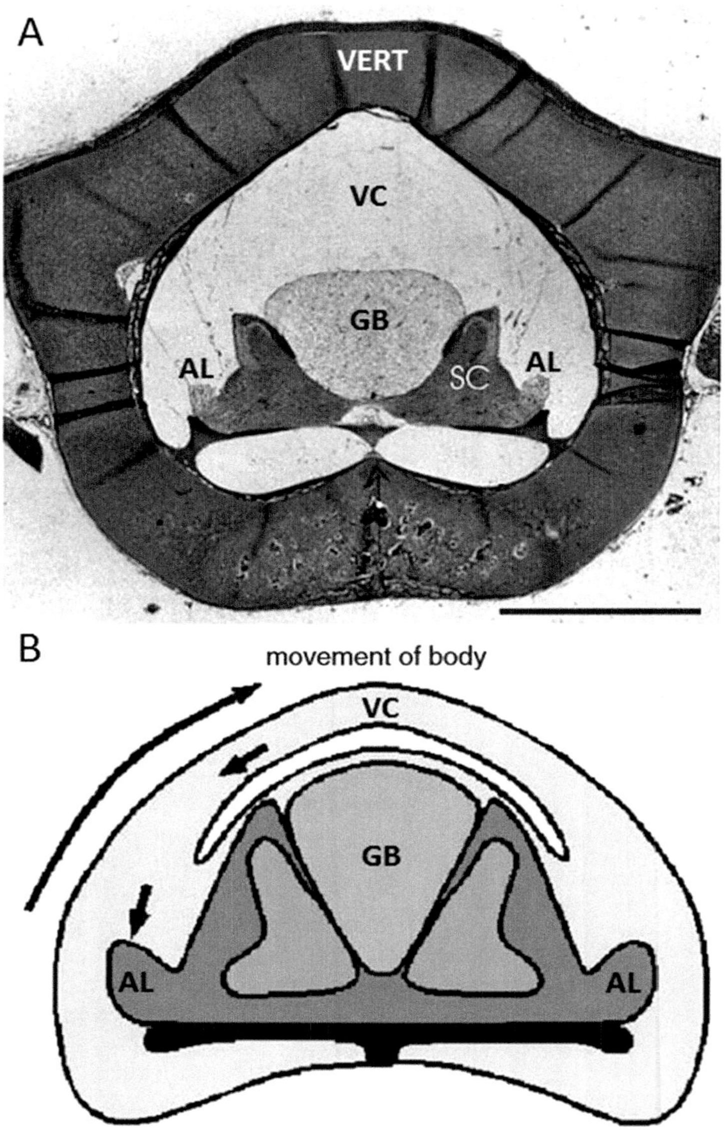


**Fig. 4.166** (**a**) Transverse section of the lumbosacral region of the vertebral column of a 1-week-old Rock Pigeon. (**b**) Possible function of the lumbosacral canals. During rotations of the body, inertia of the fluid in the lumbosacral canals and near the accessory lobes (AL) may mechanically distort the lobes, resulting in stimulation of finger-like processes of neurons in the lobe. AL, accessory lobe, GB, glycogen body, VC, vertebral canal, VERT, vertebra, SC, spinal cord. Scale bar = 2 mm in A. (Figure made from two figures and modified from Necker 2006; © 2006 Springer-Verlag, used with permission)

**Fig. 4.167** Effect of lesions of the lumbosacral canals on walking by a Rock Pigeon (*Columba livia*). When able to see, walking is just slightly impaired. However, when unable to see, the ability of pigeon to walk is very impaired. (Figure from Necker 2006; © 2006 Springer Nature, used with permission)

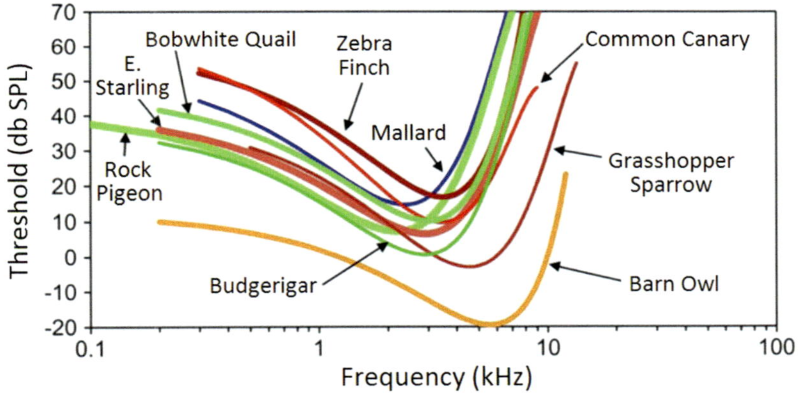


**Fig. 4.168** Audiograms of nine species of birds. All nine species are most sensitive to sound between about 2 and 7 kHz. Common Canary is also referred to as the Island Canary. Bobwhite Quail (Northern Bobwhite), *Colinus virginianus*; European Starling, *Sturnus vulgaris*; Rock Pigeon, *Columba livia*; Zebra Finch, *Taeniopygia guttata*; Budgerigar, *Melopsittacus undulatus*; Mallard, *Anas platyrhynchos*; Common Canary, *Serinus canaria*; Grasshopper Sparrow, *Ammodramus savannarum*; Barn Owl, *Tyto alba*. (Figure modified from Gleich and Langemann 2011; © Elsevier 2011, used with permission)

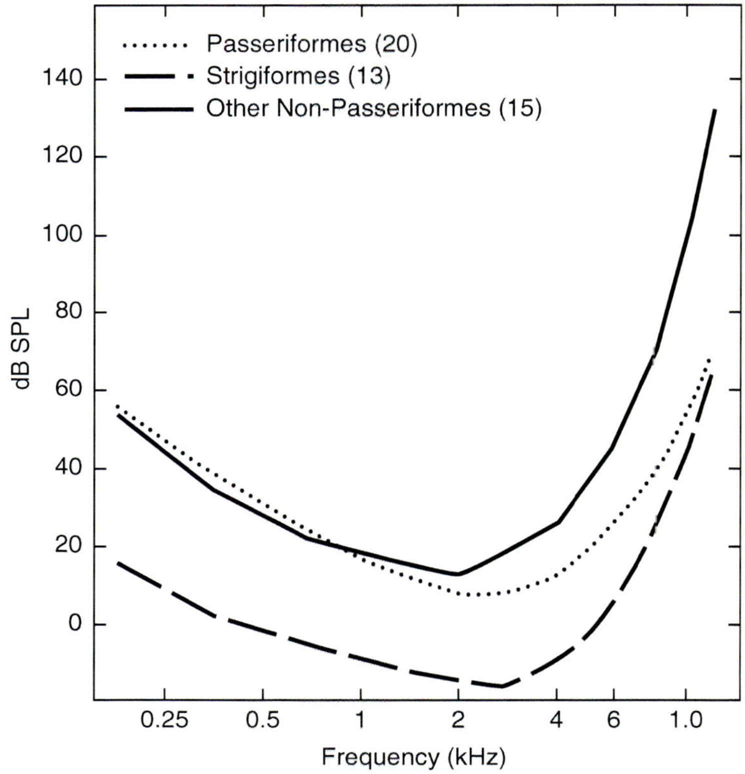


**Fig. 4.169** Median audiograms for 48 species of birds. In general, owls (Strigiformes) can detect softer sounds than other birds over the entire range of frequencies. dBSPL is a measurement of sound pressure level in decibels, where 0 dBSPL is the reference to the threshold of hearing for a typical person. (Figure from Dooling 2002; NREL, U. S. Department of Energy Laboratory, CC0 Public Domain)

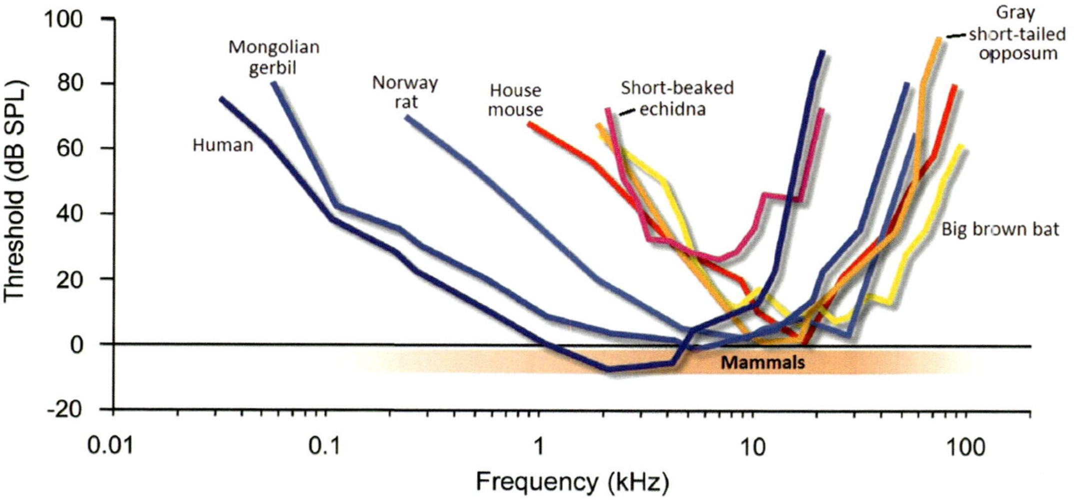


**Fig. 4.170** Audiograms of seven species of mammals and the overall range of hearing for mammals as a group (brown bar below the 0 line). Mongolian gerbil, *Meriones unguiculatus*; Norway rat, *Rattus norvegicus*; house mouse, *Mus musculus*; short-beaked echidna, *Tachyglossus aculeatus*; gray short-tailed opossum, *Monodelphis domestica*; big brown bat, *Eptesicus fuscus*. (Figure modified from Grothe and Pecka 2014; © 2014 Grothe and Pecka, open-access article distributed under the terms of the Creative Commons Attribution License (CC BY), https://creativecommons.org/licenses/by/4.0/)

**Box 4.15 Intracranial cavities and directional hearing**

Localizing sounds can be a challenge for birds and other small animals because, with their small heads, the interaural time difference (i.e., difference in time between when sounds reach each ear) is very short (<1 ms). However, birds and other small animals have internally coupled ears connected via an interaural canal so sound waves striking the external surface of the tympanic membrane of one ear pass through the canal and, with a slight (but important) delay because of the complex pathway through the skull and some attenuation, then strike the internal surface of the tympanic membrane of the other ear, enhancing a bird's ability to determine the source of incoming sounds.

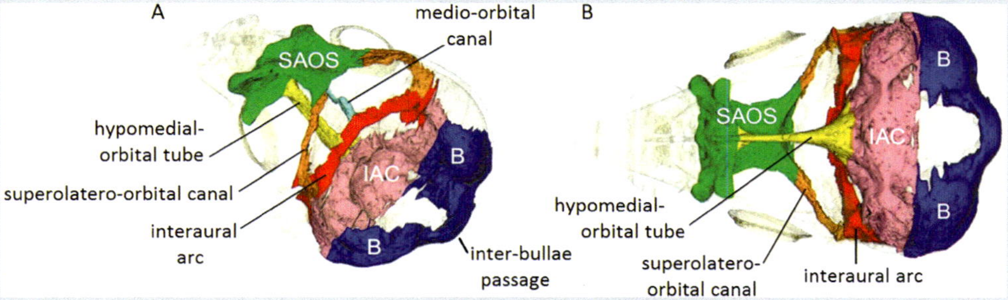


Visualization of air-filled canals in the head of a Zebra Finch (*Taeniopygia guttata*). (**a**) Reconstruction from micro-CT scanning. Posterior view from above and slightly to the left side. (**b**) Reconstruction from micro-CT scanning. View directly from below after removal of lower bill. SAOS, superoantero-orbital sinus; IAC, interaural canal; B, bullae (Figure modified from Larsen et al. 2016; © 2016 Springer-Verlag Berlin Heidelberg, used with permission)

Köppl (2019) examined the effect of the internally coupled ears of Domestic Chickens and found that, for sounds emitted on one side of the head, the interaural delay (time between pressure waves striking the outside of the ipsilateral tympanic membrane, traveling through the interaural canal, and striking the inside ot the contralateral tympanic membrane) was longer than the difference in time between when the pressure waves struck the outside of each tympanic membrane by an average of 264 μs. Studies of other species of birds have reported this average to range from 70 to 232 μs (Rosowski and Saunders 1980; Larsen et al. 2006; Kettler et al. 2016). Importantly, this additional time may allow birds to better localize sounds.

(continued)

**Box 4.15** (continued)

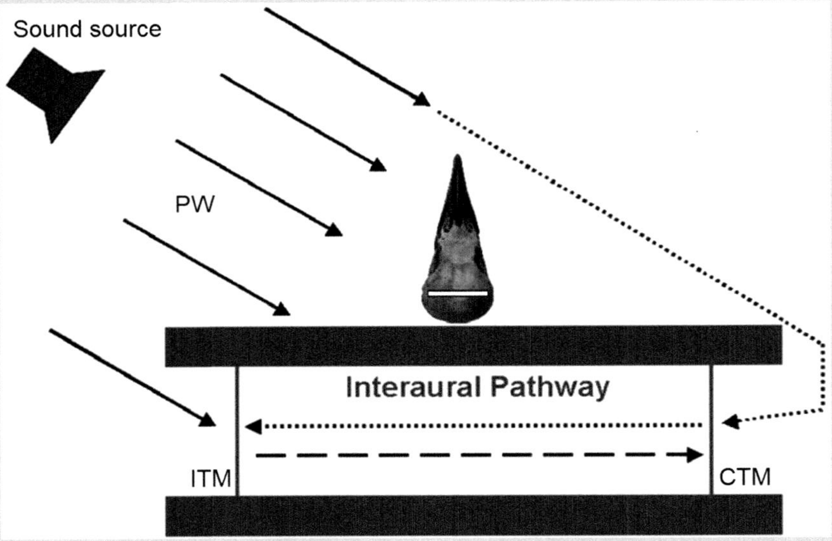


Sound transmission through the interaural canal turns bird ears into directional pressure difference receivers. Pressure waves strike the external surface of the ipsilateral tympanic membrane (ITM), its vibrations (dashed line) are transmitted through the interaural canal and, with a slight delay and some attenuation, strike the internal surface of the contralateral tympanic membrane (CTM). The same sequence of events occurs (dotted line) when the pressure waves strike the external surface of the contralateral tympanic membrane. The resulting movement of the two eardrums is determined by the net force working on their respective internal and external surfaces. Inset shows the approximate position of the interaural canal in the skull of a Hooded Crow (*Corvus cornix*). PW, pressure wave; ITM, ipsilateral tympanic membrane; CTM, contralateral tympanic membrane (Figure from Larsen 2004; open-access article distributed under the terms of the Creative Commons Attribution License (CC BY), https://creativecommons.org/licenses/by/4.0/)

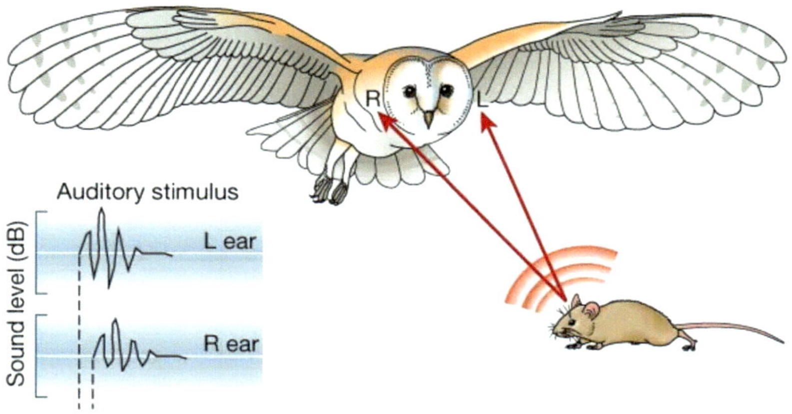


**Fig. 4.171** Barn Owls (*Tyto alba*) note the differences in intensity and timing of sounds between the two ears to locate prey. R, right; L, left. (Figure from Grothe 2018; © 2018 Elsevier Ltd. All rights reserved, used with permission)

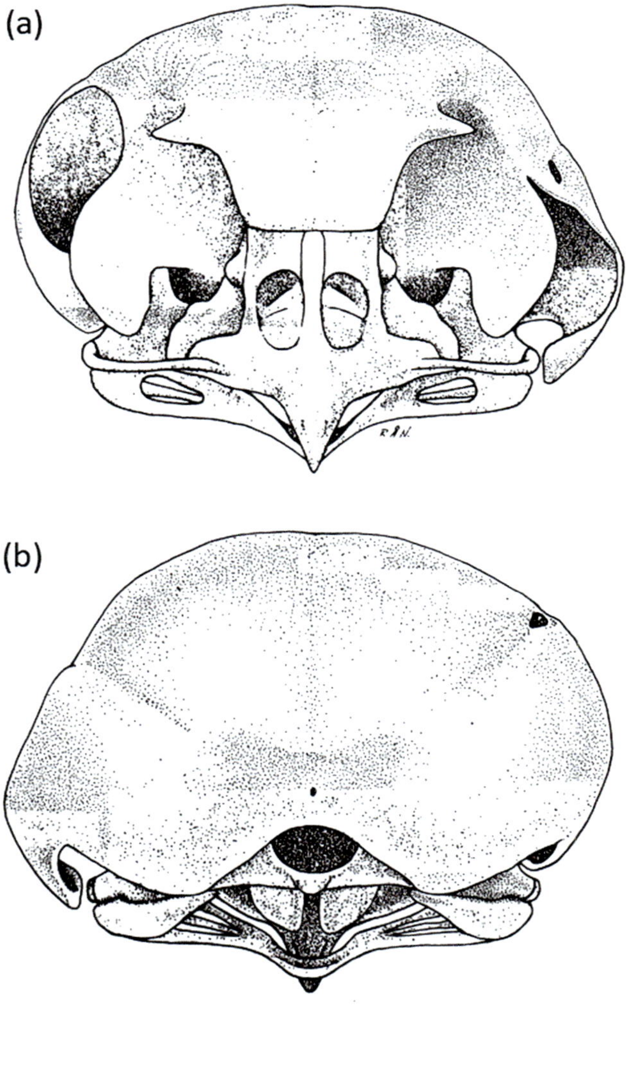


**Fig. 4.172** (**a**) Front view of the skull of a Boreal, or Tengmalm's, Owl (*Aegolius funereus*) illustrating the asymmetrical ear openings. (**b**) Back view. (Figure from Norberg 1978; © 2017 Royal Society, used with permission)

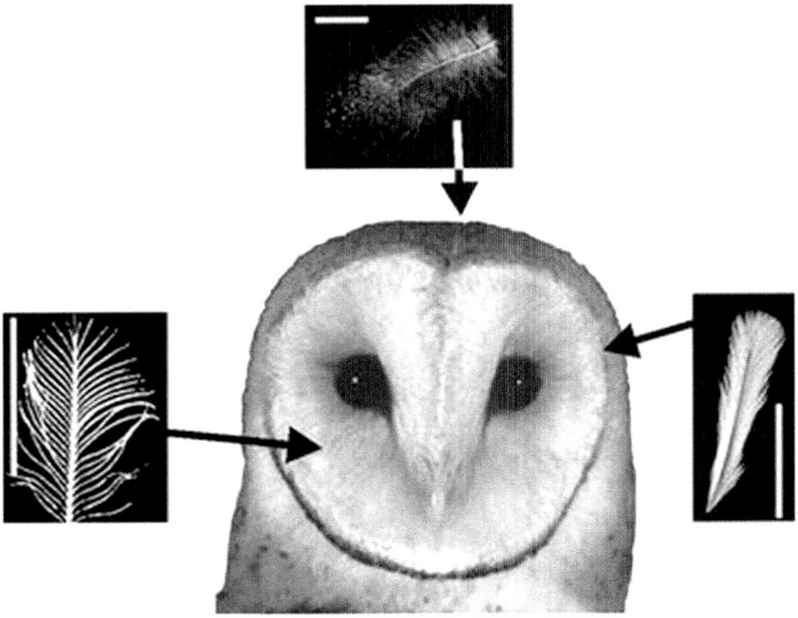

**Fig. 4.173** Feather types on the Barn Owl's (*Tyto alba*) head. Three types of feathers are shown (Koch and Wagner 2002). The contour feather (top) is the typical contour feather found everywhere on the head and body except in the ruff; it shows no specialization. Auricular feathers (left) fill the ruff. This feather type has a reduced ramification so that it becomes acoustically transparent, but is still effective in preventing the ruff from becoming dirty. Reflector feathers (right) form the border of the ruff. This feather type is very densely ramified and, thus, can influence the path of the sound. Scale bars = 1 cm. (Figure from von Campenhausen and Wagner 2006; © 2006 Springer-Verlag, used with permission)

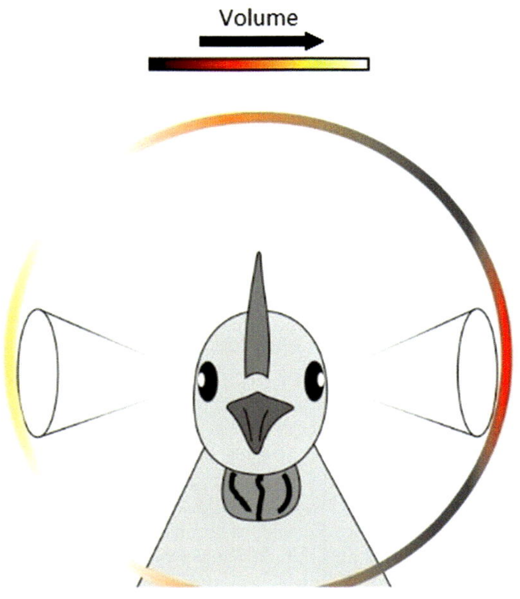


**Fig. 4.174** In this example, the volume of sounds originating on a bird's right side are perceived as similarly loud by the right ear regardless of elevation. However, the volume of that sound as perceived by the left ear varies with elevation, providing a bird with information about the location of the sound's source in the vertical plane. (Figure modified from Schnyder et al. 2014; © 2014 Schnyder et al., open-access article distributed under the terms of the Creative Commons Attribution License, https://creativecommons.org/licenses/by/4.0/)

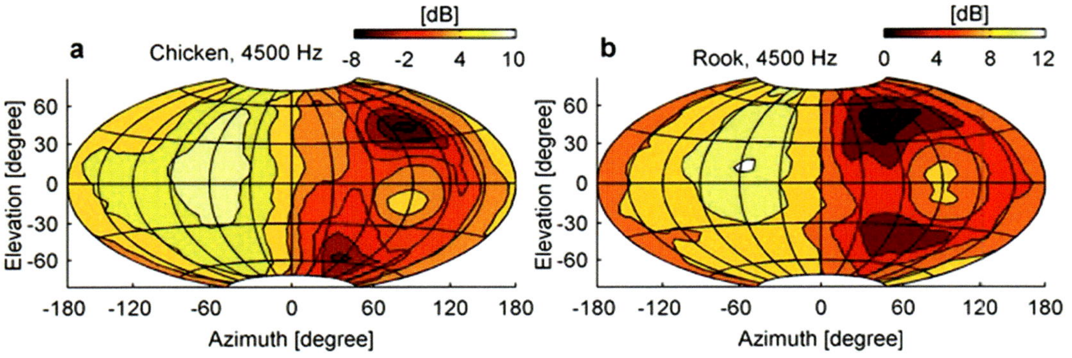


**Fig. 4.175** The heads of (**a**) Domestic Chickens (*Gallus g. domesticus*) and (**b**) Rooks (*Corvus frugilegus*) modify sound intensity dependent on elevation of a sound source. Volume as perceived by the right ear is shown for a sound (4500 Hz) originating at different elevations and different locations in the horizontal plane (azimuth). Sounds at coordinates of 0° azimuth and 0° elevation are in front of the bird's beak, and those at coordinates of −90° azimuth and 0° elevation are in front of the bird's right ear. The volume of sounds originating at different elevations on the same side as the right ear exhibits little variation, but the volume of sounds originating at different elevations on the opposite side of the head (+90° azimuth) exhibits considerable variation. At certain frequency ranges, spherical objects, like bird heads, produce an acoustic shadow, but also diffract sound in such a way that a "bright spot" is generated on the opposite side (Wiener 1947). This variation in volume provides birds with information about the elevation of a sound source. (Figure from Schnyder et al. 2014; © 2014 Schnyder et al., open-access article distributed under the terms of the Creative Commons Attribution License, https://creativecommons.org/licenses/by/4.0/)

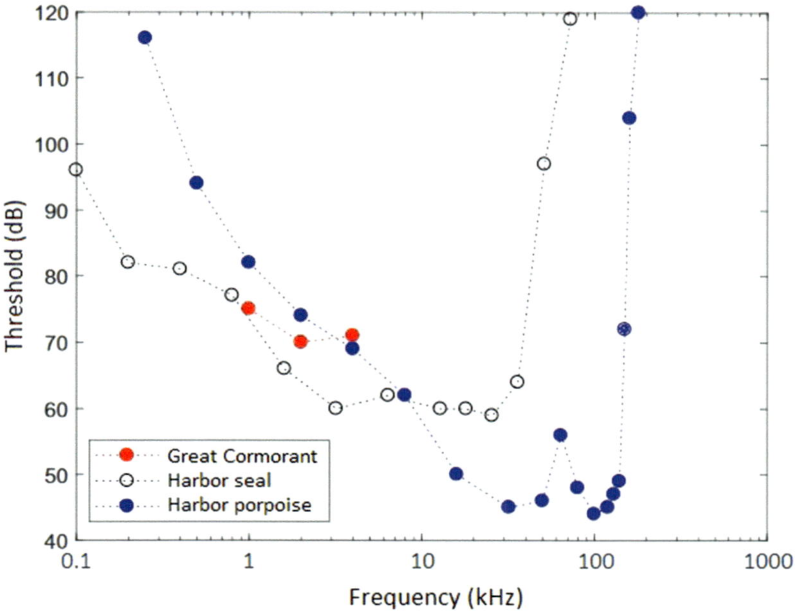


**Fig. 4.176** Underwater hearing thresholds of a Great Cormorant (*Phalacrocorax carbo*) compared to that of a harbor seal (*Phoca vitulina*) and harbor porpoise (*Phocoena phocoena*).
(Figure modified from Hansen et al. 2017; © 2017 Springer-Verlag Berlin Heidelberg, used with permission)

**Box 4.16 Oilbirds**

Oilbirds (*Steatornis caripensis*; Caprimulgiformes) are nocturnal birds found on the island of Trinidad as well as in northwestern South America. They roost and nest in caves and feed on fruit, mainly the nuts of oil palms. Oilbirds are about 30 cm long, with long broad wings, a hooked bill, and large dark eyes. As noted by Brinkløv and Warrant (2017), "Oilbirds were given their name by Alexander von Humboldt in 1817 because . . . they were harvested for cooking and lamp oil by natives and missionaries. Nestlings were particularly attractive for this purpose because these youngsters remain placidly in the nest, accumulating weight to a point that grossly exceeds the body mass of their parents." The Oilbird genus name (*Steatornis*) literally means "fat bird."

Oilbirds use echolocation to find their way in caves (and possibly to find fruit), with their "click-burst" calls ranging in frequency from 7 to 23 kHz (Brinkløv et al. 2017). Oilbirds use their sonar system when they approach or leave their dark roost and nest sites in caves, but switch it off when visual input is more reliable (Brinkløv et al. 2013). At night, oilbirds fly from their caves to feed, hovering while they pluck fruit from trees. In the past, nestlings were taken and rendered to produce an oil used for cooking and light, which is the reason for their common name. The eyes of Oilbirds have the highest light-gathering capacity of any terrestrial vertebrate, with a rod to cone ratio of 123:1 and a density of 1 million rods/mm$^2$ (Rojas et al. 2004a, b). Holland et al. (2009) fitted some Oilbirds with GPS loggers and found that they did not return to their caves every night, but made extended foraging trips over several days. Birds that did not return to caves roosted in trees during the day.

Oilbirds (*Steatornis caripensis*) roosting in a cave (Photo by The Lilac Breasted Roller, Wikipedia, CC BY 2.0, https://creativecommons.org/licenses/by/2.0/)

**Fig. 4.177** Examples of double and single clicks of two species of swiftlets. (**a**, **b**) White-nest Swiftlet, *Aerodramus fuciphagus*, and (**c**, **d**) Mossy-nest Swiftlet, *A. salanganus*. (Figure modified from Thomassen et al. 2004; © British Ornithologists' Union, used with permission)

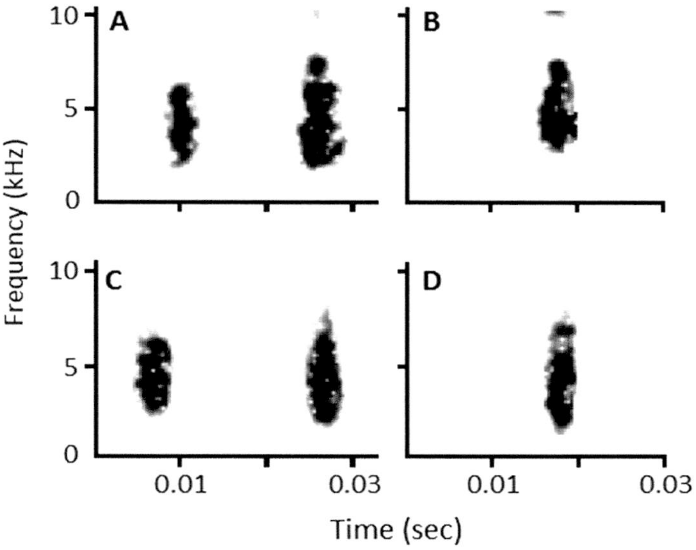


## References


Abourachid A, Hackert R, Herbin M, Libourel PA, Lambert F, Gioanni H, Provini P, Blazevic P, Hugel V (2011) Bird terrestrial locomotion as revealed by 3D kinematics. Zoology 114:360–368

Adrian ED (1951) Olfactory discrimination. L'Année Psychologie 50:107–113

Alix B, Segovia Y, García M (2017) The structure of the retina of the Eurasian Eagle-Owl and its relation to lifestyle. Avian Biol Res 10:36–44

Amo L, Galvan I, Tomas G, Sanz JJ (2008) Predator odour recognition and avoidance in a songbird. Funct Ecol 22:289–293

Amo L, López-Rull I, Pagán I, García CM (2015) Evidence that the House Finch (*Carpodacus mexicanus*) uses scent to avoid omnivore mammals. Rev Chil Hist Nat 88:5

Amo L, Saavedra I (2021) Attraction to smelly food in birds: insectivorous birds discriminate between the pheromones of their prey and those of non-prey insects. Biology 10:1010

Amo L, Visser ME, van Oers K (2011) Smelling out predators is innate in birds. Ardea 99:177–185

Aulsebrook AE, Jones TM, Rattenborg NC, Roth TC II, Lesku JA (2016) Sleep ecophysiology: integrating neuroscience and ecology. Trends Ecol Evol 31:590–599

Ault SJ (1984) Electroretinograms and retinal structure of the Eastern Screech Owl (*Otus asio*) and Great Horned Owl (*Bubo virginianus*). Raptor Res 18:62–66

Ausprey IJ, Newell FL, Robinson SK (2021) Adaptations to light predict the foraging niche and disassembly of avian communities in tropical countrysides. Ecology 102:e03213

Avilés JM, Amo L (2018) The evolution of olfactory capabilities in wild birds: a comparative study. Evol Biol 45:27–36

Azcoitia I, Fernandez-Soriano J, Fernandez-Ruiz B, Fernandez-Larios A (1985) Is the avian glycogen body a secretory organ? Int J Brain Res Neurobiol 26:651–657

Baden T, Euler T, Berens P (2019) Understanding the retinal basis of vision across species. Nat Rev Neurosci 21:5–20

Bala ADS, Spitzer MW, Takahashi TT (2007) Auditory spatial acuity approximates the resolved power of space-specific neurons. PLoS ONE 2:e675

Baldwin MW (2015) Evolution of sweet taste perception in hummingbirds. Ph.D. dissertation, Harvard University, Cambridge, MA

Baldwin MW, Toda Y, Nakagita T, O'Connell MJ, Klasing KC, Misaka T, Edwards SV, Liberles SD (2014) Evolution of sweet taste perception in hummingbirds by transformation of the ancestral umami receptor. Science 345:929–933

Balgooyen TG (1976) Behavior and ecology of the American Kestrel (*Falco sparverius* L.) in the Sierra Nevada of California. University of California Press, Berkeley, CA

Balthazart J, Schoffeniels E (1979) Pheromones are involved in the control of sexual behaviour of birds. Naturwissenschaften 66:55–56

Bang BG, Cobb S (1968) The size of the olfactory bulb in 108 species of birds. Auk 85:55–61

Bang BG, Wenzel BM (1985) Nasal cavity and olfactory system. In: King AS, McLelland J (eds) Form and function in birds, vol 3. Academic Press, New York, NY, pp 195–225

Banks MS, Sprague WW, Schmoll J, Parnell JAQ, Love GD (2015) Why do animal eyes have pupils of different shapes? Sci Adv 1:e1500391
Barbur JL, Prescott NB, Douglas RH, Jarvis JR, Wathes CM (2002) A comparative study of stimulus-specific pupil responses in the Domestic Fowl (*Gallus gallus domesticus*) and the human. Vis Res 42:249–255
Barlow HB, Ostwald TJ (1972) Pecten of the pigeon's eye as an inter-ocular eye shade. Nat New Biol 236:88–90
Bateson P (1991) Assessment of pain in animals. Anim Behav 42:827–839
Battley PF, Warnock N, Tibbitts TL, Gill RE Jr, Piersma T, Hassell CJ, Douglas DC, Mulcahy DM, Gartrell BD, Schuckard R, Melville DS, Riegen AC (2012) Contrasting extreme long-distance migration patterns in Bar-tailed Godwits *Limosa lapponica*. J Avian Biol 43:21–32
Baumhardt PE, Moore BA, Doppler M, Fernández-Juricic E (2014) Do American Goldfinches see their world like passive prey foragers? A study on visual fields, retinal topography, and sensitivity of photoreceptors. Brain Behav Evol 83:181–198
Bayón A, Almela RM, Talavera J (2008) Avian ophthalmology. Eur J Companion Anim Pract 17:253–266
Beheiry RR, Ali SA, Aref M, Emam H (2020) Harderian gland of flying and non-flying birds: morphological, histological, and histochemical studies. J Basic Appl Zool 81:35
Bell-Pedersen D, Cassone VM, Earnest DJ, Golden SS, Hardin PE, Thomas TL, Zoran MJ (2005) Circadian rhythms from multiple oscillators: lessons from diverse organisms. Nat Rev Genet 6:544–556
Benzo CA, De Gennaro LD, Stearns SB (1975) Glycogen metabolism in the developing chick glycogen body: functional significance of the direct oxidative pathway. J Exp Zool 193:161–166
Berkhoudt H (1985) Structure and function of avian taste receptors. In: McLelland K (ed) Form and function in birds, vol 3. Academic Press, London, pp 463–491
Bermingham-McDonogh O, Rubel EW (2003) Hair cell regeneration: winging our way towards a sound future. Curr Opin Neurobiol 13:119–126
Bicego KC, Barros RCH, Branco LGS (2007) Physiology of temperature regulation: comparative aspects. Comp Biochem Physiol A 147:616–639
Biederman-Thorson M, Thorson J (1973) Rotation-compensating reflexes independent of the labyrinth and the eye. J Comp Physiol 83:103–122
Bird CD, Emery NJ (2009a) Insightful problem solving and creative tool modification by captive nontool-using Rooks. Proc Natl Acad Sci USA 106:10370–10375
Bird CD, Emery NJ (2009b) Rooks use stones to raise the water level to reach a floating worm. Curr Biol 19: 1410–1414
Bischof H-J (1988) The visual field and visually guided behavior in the Zebra Finch (*Taeniopygia guttata*). J Comp Physiol A 163:329–337
Blackwell BF, Fernandez-Juricic E, Seamans TW, Dolan T (2009) Avian visual system configuration and behavioural response to object approach. Anim Behav 77:673–684
Blackwell BF, Seamans TW, Pfeiffer MB, Buckingham BN (2020) Predator scent and visual cue applied to nest boxes fail to dissuade European Starlings (*Sturnus vulgaris*) from nesting. Wilson J Ornithol 132:113–123
Blem CR, Blem LB, Felix J, van Gelder J (2000) Rufous Hummingbird sucrose preference: precision of selection varies with concentration. Condor 102:235–238
Blough DS (1956) Dark adaptation in the pigeon. J Comp Physiol Psychol 49:425–430
Blough PM (1971) The visual acuity of the pigeon for distant targets. J Exp Anal Behav 15:57–67
Blumberg MS, Lesku JA, Libourel P-A, Schmidt MH, Rattenborg NC (2020) What is REM sleep? Curr Biol 30:R38–R49
Boal CW (2001) Nonrandom mating and productivity of adult and subadult Cooper's Hawks. Condor 103:381–385
Boire D, Dufour JS, Théoret H, Ptito M (2001) Quantitative analysis of the retinal ganglion cell layer in the Ostrich, *Struthio camelus*. Brain Behav Evol 58:343–355
Bok J, Raft S, Kong K-A, Koo SK, Dräger UC, Wu DK (2011) Transient retinoic acid signaling confers anterior-posterior polarity to the inner ear. Proc Natl Acad Sci USA 108:161–166
Bonadonna F, Benhamou S, Jouventin P (2003a) Orientation in 'featureless' environments: the extreme case of pelagic birds. In: Berthold P, Gwinner E, Sonnenschein E (eds) Avian migration. Springer, Berlin, pp 367–377
Bonadonna F, Bretagnolle V (2002) Snelling home: a good solution for burrow-finding in nocturnal petrels? J Exp Biol 205:2519–2523
Bonadonna F, Hesters F, Jouventin P (2003b) Scent of a nest: discrimination of own-nest odours in Antarctic Prions, *Pachyptila desolata*. Behav Ecol Sociobiol 54: 174–178
Bonadonna F, Miguel E, Grosbois V, Jouventin P, Bessiere J-M (2007) Individual odor recognition in birds: an endogenous olfactory signature on petrels' feathers? J Chem Ecol 33:1819–1829
Bonadonna F, Nevitt GA (2004) Partner-specific odor recognition in an Antarctic seabird. Science 306:835
Bonsmann A, Stoffel MH, Burkhart M, Hatt J-M (2016) Anatomical atlas of the quail's ear (*Coturnix coturnix*). Anat Histol Embryol 45:399–404
Boswell T, Dunn IC (2015) Regulation of the avian central melanocortin system and the role of leptin. Gen Comp Endocrinol 221:278–283
Boström JE, Dimitrova M, Canton C, Håstad O, Qvarnström A, Ödeen A (2016) Ultra-rapid vision in birds. PLoS ONE 11:e0151099
Bowmaker JK, Martin GR (1978) Visual pigments and colour vision in a nocturnal bird, *Strix aluco* (Tawny Owl). Vis Res 18:1125–1130

Bowmaker JK, Martin GR (1985) Visual pigments and oil droplets in the penguin, *Spheniscus humboldti*. J Comp Physiol A 156:71–77

Brach V (1975) The effect of intraocular ablation of the pecten oculi of the chicken. Invest Ophthalmol Vis Sci 14:166–168

Brawn TP, Nusbaum HC, Margoliash D (2018) Sleep-dependent reconsolidation after memory destabilization in starlings. Nat Commun 9:3093

Breuner CW, Sprague RS, Patterson SH, Woods HA (2013) Environment, behavior and physiology: do birds use barometric pressure to predict storms? J Exp Biol 216:1982–1990

Bringmann A (2019) Structure and function of the bird fovea. Anat Histol Embryol 48:177–200

Brinkløv S, Elemans CPH, Ratcliffe JM (2017) Oilbirds produce echolocation signals beyond their best hearing range and adjust signal design to natural light conditions. R Soc Open Sci 4:170255

Brinkløv S, Fenton MB, Ratcliffe JM (2013) Echolocation in Oilbirds and swiftlets. Front Physiol 4:123

Brinkløv S, Warrant E (2017) Oilbirds. Curr Biol 27: R1145–R1147

Brodin A (1994) Separation of caches between individual Willow Tits hoarding under natural conditions. Anim Behav 47:1035–1037

Brooke ML, Hanley S, Laughlin SB (1999) The scaling of eye size with body mass in birds. Proc R Soc B 266: 405–412

Brower PP (1984) Chemical defense in butterflies. In: Vane-Wright RI, Ackery PR (eds) The biology of butterflies (Symposia of the Royal Entomological Society of London). Academic Press, New York, NY, pp 109–134

Brown RE, Fedde MR (1993) Airflow sensors in the avian wing. J Exp Biol 179:13–30

Browne KE (2014) Phylogenetic and ecological significance of variation in the scleral ring of aquatic foraging birds. M.S. thesis, University of Texas, Austin, TX

Buchsbaum R, Valiela I, Swain T (1984) The role of phenolic compounds and other plant constituents in feeding by Canada Geese in a coastal marsh. Oecologia 63:343–349

Buitron D, Nuechterlein GL (1985) Experiments on olfactory detection of food caches by Black-billed Magpies. Condor 87:92–95

Butler AB, Hodos W (1996) Comparative vertebrate neuroanatomy: evolution and adaptation. Wiley-Liss, New York, NY

Campagna S, Mardon J, Celerier A, Bonadonna F (2012) Potential semiochemical molecules from birds: a practical and comprehensive compilation of the last 20 years studies. Chem Senses 37:3–25

Campbell FW, Green DJ (1965) Optical and retinal factors affecting visual resolution. J Physiol 181:576–593

Caro SP, Balthazart J, Bonadonna F (2015) The perfume of reproduction in birds: chemosignaling in avian social life. Horm Behav 68:25–42

Carril J, Tambussi CP, Degrange FJ, Saldivar MJB, Picasso MBJ (2016) Comparative brain morphology of Neotropical parrots (Aves, Psittaciformes) inferred from virtual 3D endocasts. J Anat 229:239–251

Cassone VM (2014) Avian circadian organization: a chorus of clocks. Front Neuroendocrinol 35:76–88

Caves EM, Brandley NC, Johnsen S (2018) Visual acuity and the evolution of signals. Trends Ecol Evol 33:358–372

Chai P (1996) Butterfly visual characteristics and ontogeny of responses to butterflies by a specialized tropical bird. Biol J Linn Soc 59:37–67

Charvet CJ, Striedter GF (2011) Developmental modes and developmental mechanisms can channel brain evolution. Front Neuroanat 5:4

Chen JR, Stepanek L, Doupe AJ (2014) Differential contributions of basal ganglia and thalamus to song initiation, tempo, and structure. J Neurophysiol 111: 248–257

Chetverikova R, Dautaj G, Schwigon L, Dedek K, Mouritsen H (2022) Double cones in the avian retina form an oriented mosaic which might facilitate magnetoreception and/or polarized light sensing. J R Soc Interface 19:20210877

Chmura HE, Krause JS, Pérez JH, Ramenofsky M, Wingfield JC (2020) Autumn migratory departure is influenced by reproductive timing and weather in an Arctic passerine. J Ornithol 161:779–791

Clark L, Hagelin J, Werner S (2015) The chemical senses in birds. In: Whittow GC (ed) Sturkie's avian physiology, 6th edn. Academic Press, San Diego, CA, pp 89–111

Clayton NS, Dally JM, Emery NJ (2007) Social cognition by food-caching corvids. The Western Scrub-Jay as a natural psychologist. Philos Trans R Soc B 362:507–522

Clayton NS, Dickinson AD (1998) What, where and when: evidence for episodic like memory during cache recovery by Scrub Jays. Nature 395:272–278

Clayton NS, Emery NJ (2015) Avian models for human cognitive neuroscience: a proposal. Neuron 86:1330–1342

Coimbra JP, Collin SP, Hart NS (2014) Topographic specializations in the retinal ganglion cell layer correlate with lateralized visual behavior, ecology, and evolution in cockatoos. J Comp Neurol 522:3363–3385

Coimbra JP, Collin SP, Hart NS (2015) Variations in retinal photoreceptor topography and the organization of the rod-free zone reflect behavioral diversity in Australian passerines. J Comp Neurol 523:1073–1094

Coimbra JP, Marceliano MLV, da Silveira Andrade-da-Costa BL, Yamada ES (2006) The retina of Tyrant Flycatchers: topographic organization of neuronal density and size in the ganglion cell layer of the Great Kiskadee *Pitangus sulphuratus* and the Rusty-margined Flycatcher *Myiozetetes cayanensis* (Aves: Tyrannidae). Brain Behav Evol 68:15–25

Coimbra JP, Nolan PM, Collin SP, Hart NS (2012) Retinal ganglion cell topography and spatial resolving power in penguins. Brain Behav Evol 80:254–268

Coimbra JP, Trévia N, Marceliano MLV, da Silveira Andrade-da-Costa BL, Picanço-Diniz CW, Yamada ES (2009) Number and distribution of neurons in the retinal ganglion cell layer in relation to foraging behaviors of Tyrant flycatchers. J Comp Neurol 514: 66–83

Collin SP (1999) Behavioural ecology and retinal cell topography. In: Archer SN, Djamgoz MBS, Loew ER, Partridge JC, Vellarga S (eds) Adaptive mechanisms in the ecology of vision. Kluwer, Dordrecht, pp 509–535

Collin SP (2008) A web-based archive for topographic maps of retinal cell distribution in vertebrates. Clin Exp Optom 91:85–95

Collin SP, Collin HB (2020) The ultrastructure of the nictitating membrane of the Little Penguin (*Eudyptula minor*, Aves). Integr Organismal Biol 2:obaa048

Corfield J, Kubke MF, Parsons S, Wild JM, Köppl C (2011) Evidence for an auditory fovea in the New Zealand Kiwi (*Apteryx mantelli*). PLoS ONE 6: e23771

Corfield JR, Price K, Iwaniuk AN, Gutiérrez-Ibáñez C, Birkhead T, Wylie DR (2015) Diversity in olfactory bulb size in birds reflects allometry, ecology, and phylogeny. Front Neuroanat 9:102

Council NR (2010) Recognition and alleviation of pain in laboratory animals, vol 1. National Academies Press, Washington, DC

Craig AJFK, Hulley PE (2004) Iris colour in passerine birds: why be bright-eyed? S Afr J Sci 100:584–588

Crole MR, Soley JT (2015) Contrasting morphological evidence for the presence of taste buds in *Dromaius novaehollandiae* and *Struthio camelus* (Palaeognathae, Aves). Zoomorphology 134:499–507

Cronin TW (2012) Visual optics: accommodation in a splash. Curr Biol 22:R871–R873

Crowell SC (2014) In-air and underwater hearing in ducks. Ph.D. dissertation, Univ Maryland, College Park, MD

Csillag A, Montagnese CM (2005) Thalamotelencephalic organization in birds. Brain Res Bull 66:303–310

Cunha F, Gutiérrez-Ibáñez C, Brinkman B, Wylie DR, Iwaniuk AN (2022) The relative sizes of nuclei in the oculomotor complex vary by order and behaviour in birds. J Comp Physiol A 209:341–360

Cunningham SJ, Alley MR, Castro I, Potter MA, Cunningham M, Pyne MJ (2010) Bill morphology of ibises suggests a remote-tactile sensory system for prey detection. Auk 127:308–316

Cunningham SJ, Corfield JR, Iwaniuk AN, Castro I, Alley MR, Birkhead TR, Parsons S (2013) The anatomy of the bill tip of kiwi and associated somatosensory regions of the brain: comparisons with shorebirds. PLoS ONE 8:e80036

Curcio CA, Allen KA (1990) Topography of ganglion cells in human retina. J Comp Neurol 300:5–25

Curcio CA, Sloan KR, Kalina RE, Hendrickson AE (1990) Human photoreceptor topography. J Comp Neurobiol 292:497–523

Davidson GL, Clayton NS, Thornton A (2014) Salient eyes deter conspecific nest intruders in wild Jackdaws (*Corvus monedula*). Biol Lett 10:20131077

Davidson GL, Thornton A, Clayton NS (2017) Evolution of iris colour in relation to cavity nesting and parental care in passerine birds. Biol Lett 13:20160783

Day LB, Westcott DA, Olster DH (2005) Evolution of bower complexity and cerebellum size in bowerbirds. Brain Behav Evol 66:62–72

de Fouw J, van der Heide T, Oudman T, Maas LRM, Piersma T, van Gils JA (2016) Structurally complex sea grass obstructs the sixth sense of a specialized avian molluscivore. Anim Behav 115:55–67

Delius JD, Vollrath FW (1973) Rotation compensating reflexes independent of the labyrinth. J Comp Physiol 83:123–134

Demery ZP, Chappell J, Martin GR (2011) Vision, touch and object manipulation in Senegal Parrots *Poicephalus senegalus*. Proc R Soc B 278:3687–3693

Dernedde J, Weise C, Müller E-C, Hagiwara A, Bachmann S, Suzuki M, Reutter W, Tauber R, Scherer H (2014) Cupulin is a zona pellucida-like domain protein and major component of the cupula from the inner ear. PLoS ONE 9:e111917

Dezza MA, Rodríguez RR, Buschiazzo HD (1970) Pyruvic and lactic acids levels in glycogen body incubations. Life Sci 9:387–395

Díaz-Siefer P, Tapia-Gatica J, Martínez-Harms J, Bergmann J, Celis-Diez JL (2021) A larval aggregation pheromone as foraging cue for insectivorous birds. Biol Lett 17:20210360

Dicke U, Roth G (2016) Neuronal factors determining high intelligence. Philos Trans R Soc B 371:20150180

Dominguez J (2003) Sleeping and vigilance in Black-tailed Godwit. J Ethol 21:57–60

Dooling R (1980) Hearing in birds. In: Webster DB, Fay RR, Popper AN (eds) The evolutionary biology of hearing. Springer, New York, NY, pp 545–559

Dooling RJ (2002) Avian hearing and the avoidance of wind turbines. Technical Report NREL/TP-500-30844. National Renewable Energy Laboratory, Golden, CO

Dorward PK, McIntyre AK (1971) Responses of vibration-sensitive receptors in the interosseous region of the duck's hind limb. J Physiol 219:77–87

Douglas JM, Sanchez-Migallon Guzman D, Paul-Murphy JR (2018) Pain in birds: the anatomical and physiological basis. Vet Clin N Am Exotic Anim Pract 21:17–31

Douglas RH (2018) The pupillary light responses of animals; a review of their distribution, dynamics, mechanisms and functions. Prog Retin Eye Res 66: 17–48

Dubielzig RR, Musgrave K, Schobert CS (2014) Comparative anatomy of avian ciliary muscles among underwater sight hunters. Invest Ophthalmol Vis Sci 55: 4444

Ducatez S, Clavel J, Lefebvre L (2015) Ecological generalism and behavioural innovation in birds: technical intelligence or the simple incorporation of new foods? J Anim Ecol 84:79–89

Dulac S, Knudsen EI (1990) Neural maps of head movement vector and speed in the optic tectum of the Barn Owl. J Neurophysiol 1:131–146

Edinger L (1896) The anatomy of the central nervous system of man and of vertebrates in general, 5th edn. F. A. Davis Company, Philadelphia, PA

Eichholz MW, Dassow JA, Stafford JD, Weatherhead PJ (2012) Experimental evidence that nesting ducks use mammalian urine to assess predator abundance. Auk 129:638–644

Emery NJ, Clayton NS (2005) Evolution of the avian brain and intelligence. Curr Biol 15:R946–R950

Ensminger AL, Fernández-Juricic E (2014) Individual variation in cone photoreceptor density in House Sparrows: implications for between-individual differences in visual resolution and chromatic contrast. PLoS ONE 9:e111854

Feenders G, Liedvogel M, Rivas M, Zapka M, Horita H, Hara E, Wada K, Mouritsen H, Jarvis ED (2008) Molecular mapping of movement-associated areas in the avian brain: a motor theory for vocal learning origin. PLoS ONE 3:e1768

Fellows TK (2013) Visual resolution of Anna's Hummingbirds (*Calypte anna*) in space and time. M. S. thesis, University of British Columbia, Vancouver, BC, Canada

Fernandez C, Goldberg JM (1976) Physiology of peripheral neurons innervating otolith organs of the squirrel monkey. I. Response to static tilts and to long-duration centrifugal force. J Neurophysiol 39:970–984

Fernández-Juricic E, Baumhardt PE, Tyrrell LP, Elmore A, DeLiberto ST, Werner SJ (2019) Vision in an abundant North American bird: the Red-winged Blackbird. Auk 136:ukz039

Fernandez-Juricic E, Erichsen JT, Kacelnick A (2004) Visual perception and social foraging in birds. Trends Ecol Evol 19:25–31

Fernández-Juricic E, Gall MD, Dolan T, O'Rourke C, Thomas S, Lynch JR (2011a) Visual systems and vigilance behaviour of two ground-foraging avian prey species: White-crowned Sparrows and California Towhees. Anim Behav 81:705–713

Fernández-Juricic E, Moore BA, Doppler M, Freeman J, Blackwell BF, Lima SL, DeVault TL (2011b) Testing the terrain hypothesis: Canada Geese see their world laterally and obliquely. Brain Behav Evol 77:147–158

Fernández-Juricic E, O'Rourke C, Pitlik T (2010) Visual coverage and scanning behavior in two corvid species: American Crow and Western Scrub Jay. J Comp Physiol A 196:879–888

Fettiplace R (2020) Diverse mechanisms of sound frequency discrimination in the vertebrate cochlea. Trends Neurosci 43:88–102

Fettiplace R, Hackney CM (2006) The sensory and motor roles of auditory hair cells. Nat Rev Neurosci 7:19

Fink AS, Hefferan PM, Howell RR (1975) Enzymatic and biochemical characterization of the avian glycogen body. Comp Biochem Physiol 50B:525–530

Fischer VAB (1969) Laboruntersuchungen und freilandbeobachtungen zum sehvermögen und verhalten von altweltgeiern. Zoologische Jahrbücher/Abteilung für Systematik, Ökologie und Geographie der Tiere 96: 81–132

Fite KV, Rosenfield-Wessels S (1975) A comparative study of deep avian fovea. Brain Behav Evol 12:97–115

Flanigan WF (1972) Behavioral states and electroencephalograms of reptiles. In: Chase MH (ed) The sleeping brain: perspectives in the brain sciences. Brain Information Service/Brain Research Institute, University of California, Los Angeles, CA, pp 14–18

Fox R, Lehmkuhle SW, Bush RC (1977) Stereopsis in the falcon. Science 192:263–265

Franze K, Grosche J, Skatchkov SN, Schinkinger S, Foja C, Schild D, Uckermann O, Travis K, Reichenbach A, Guck J (2007) Müller cells are living optical fibers in the vertebrate retina. Proc Natl Acad Sci USA 104:8287–8292

Franz-Odendaal TA (2020) Skeletons of the eye: an evolutionary and developmental perspective. Anat Rec 303:100–109

Franz-Odendaal TA, Krings M (2019) A heterochronic shift in skeletal development in the Barn Owl (*Tyto furcata*): a description of the ocular skeleton and tubular eye shape formation. Dev Dyn 248:671–678

Freas CA, LaDage LD, Roth TC II, Pravosudov VV (2012) Elevation-related differences in memory and hippocampus in Mountain Chickadees, *Poecile gambeli*. Anim Behav 84:121–127

Fristoe TS, Botero CA (2019) Alternative ecological strategies lead to avian brain size bimodality in variable habitats. Nat Commun 10:3818

Fritsch R, Ullmann JFP, Bitton P-P, Collin SP, Michiels NK (2017) Optic-nerve-transmitted eyeshine, a new type of light emission from fish eyes. Front Zool 14:14

Fuchs R, Winkler H, Bingman VP, Ross JD, Bernroider G (2014) Brain geometry and its relation to migratory behavior in birds. J Adv Neurosci Res 1:1–9

Fuchs T, Maury D, Moore FR, Bingman VP (2009) Daytime micro-naps in a nocturnal migrant: an EEG analysis. Biol Lett 5:77–80

Fusani L, Giordano M, Day LB, Schlinger BA (2007) High-speed video analysis reveals individual variability in the courtship displays of male Golden-collared Manakins. Ethology 113:964–972

Gaffney MF, Hodos W (2003) The visual acuity and refractive state of the American Kestrel (*Falco sparverius*). Vis Res 43:2053–2059

Gagliardo A, Ioalè P, Savini M, Wild JM (2006) Having the nerve to home: trigeminal magnetoreceptor versus olfactory mediation of homing in pigeons. J Exp Biol 209:2888–2892

Galifret Y (1968) Les diverses aires fonctionnelles de la rétine du pigeon. Z Zellforsch Mikrosk Anat 86:535–545
Gall MD, Fernández-Juricic E (2010) Visual fields, eye movements, and scanning behavior of a sit-and-wait predator, the Black Phoebe (*Sayornis nigricans*). J Comp Physiol A 196:15
Ganchrow JR, Ganchrow D (1987) Taste bud development in chickens (*Gallus gallus domesticus*). Anat Rec 218:88–93
Gentle MJ (1975) Gustatory behaviour of the chicken and other birds. In: Wright P, Caryle PG, Vowles DM (eds) Neural and endocrine aspects of behaviour in birds. Elsevier, Amsterdam, pp 305–318
Gentle MJ (1992) Pain in birds. Anim Welf 1:235–247
Gerritsen AFC, Meiboom A (1986) The role of touch in prey density estimation by *Calidris alba*. Netherlands J Zool 36:530–562
Giannessi F, Ryskalin L, Ruffoli R (2017) Ultrastructural study of the neural microcircuits in the sensory epithelium of the paratympanic organ of the chicken. Arch Ital Biol 155:64–74
Glasser A, Howland HC (1996) A history of studies of visual accommodation in birds. Q Rev Biol 71:475–509
Glasser A, Murphy CJ, Troilo D, Howland HC (1995) The mechanism of lenticular accommodation in chicks. Vis Res 35:1525–1540
Gleich O, Langemann U (2011) Auditory capabilities of birds in relation to the structural diversity of the basilar papilla. Hear Res 273:80–88
Glotzbach SF, Heller HC (1976) Central nervous regulation of body temperature during sleep. Science 194: 537–539
Godard RD, Bowers BB, Wilson CM (2007) Eastern Bluebirds *Sialia sialis* do not avoid nest boxes with chemical cues from two common nest predators. J Avian Biol 38:128–131
Gold MEL, Schulz D, Budassi M, Gignac PM, Vaska P, Norell MA (2016) Flying starlings, PET and the evolution of volant dinosaurs. Curr Biol 26:R265–R267
Gondo M, Ando H (1995) Comparative histophysiological study of oil droplets in the avian retina. Jap J Ornithol 44:81–91
González-Gómez PL, Madrid-Lopez N, Salazar JE, Suárez R, Razeto-Barry P, Mpodozis J, Bozinovic F, Vásquez RA (2014) Cognitive ecology in hummingbirds: the role of sexual dimorphism and its anatomical correlates on memory. PLoS ONE 9: e90165
González-Gómez PL, Vásquez RA (2006) A field study of spatial memory in Green-backed Firecrown hummingbirds (*Sephanoides sephanoides*). Ethology 112:790–795
González-Martín-Moro J, Gómez-Sanz F, Sales-Sanz A, Huguet-Baudin E, Murube-del-Castillo J (2014) Pupil shape in the animal kingdom: from the pseudopupil to the vertical pupil. Arch Spanish Soc Ophthalmol 89: 484–494
González-Martín-Moro J, Hernández-Verdejo JL, Clement-Corral A (2017) The visual system of diurnal raptors: updated review. Archivos de la Sociedad Española de Oftalmología (English edition) 92:225–232
Goodridge SG (1997) Multimedia sensor fusion for intelligent camera control and human-computer interaction. Ph.D. dissertation, North Carolina State University, Raleigh, NC
Goodson JL, Kabelik D (2009) Dynamic limbic networks and social diversity in vertebrates: from neural context to neuromodulatory patterning. Front Neuroendocrinol 30:429–441
Goodyear RJ, Richardson GP (2018) Structure, function, and development of the tectorial membrane: an extracellular matrix essential for hearing. Curr Top Dev Biol 130:217–244
Gordon JC, Holt NC, Biewener A, Daley MA (2020) Tuning of feedforward control enables stable muscle force-length dynamics after loss of autogenic proprioceptive feedback. eLife 9:e53908
Graham JL, Charlier TD, Bonadonna F, Caro SP (2021) Olfactory detection of trace amounts of plant volatiles is correlated with testosterone in a passerine bird. Horm Behav 136:105045
Graves GR (2014) Questions raised by the consumption of cane sugar by a Black-throated Blue Warbler (*Setophaga caerulescens*). J Caribbean Ornithol 27: 27–30
Graybiel AM (2008) Habits, rituals, and the evaluative brain. Annu Rev Neurosci 31:359–387
Grémillet D, Kuntz G, Gilbert C, Woakes AJ, Butler PJ, Maho YI (2005) Cormorants dive through the polar night. Biol Lett 1:469–471
Grigg NP, Krilow JM, Gutierrez-Ibanez C, Wylie DR, Graves GR, Iwaniuk AN (2017) Anatomical evidence for scent guided foraging in the Turkey Vulture. Sci Rep 7:17408
Grothe B (2018) How the Barn Owl computes auditory space. Trends Neurosci 41:115–117
Grothe B, Pecka M (2014) The natural history of sound localization in mammals – a story of neuronal inhibition. Front Neural Circuit 8:116
Gültiken ME, Yildiz D, Onuk B, Karayiğit MÖ (2012) The morphology of the pecten oculi in the Common Buzzard (*Buteo buteo*). Vet Ophthalmol 15(Suppl 2):72–76
Günther A, Einwich A, Sjulstok E, Feederle R, Bolte P, Koch KW, Solov'yov IA, Mouritsen H (2018) Double-cone localization and seasonal expression pattern suggest a role in magnetoreception for European Robin cryptochrome 4. Curr Biol 28:211–223
Güntürkün O, Miceli D, Watanabe M (1993) Anatomy of the avian thalamofugal pathway. In: Zeigler HP, Bischof H-J (eds) Vision, brain, and behavior in birds. MIT Press, Cambridge, MA, pp 115–135
Gutiérrez-Expósito C (2019) Asymmetric iris heterochromia in birds: the dark crescent of buttonquails. J Ornithol 160:281–285

Gutiérrez-Ibáñez C, Iwaniuk AN, Wylie DR (2018) Parrots have evolved a primate-like telencephalic-midbrain-cerebellar circuit. Sci Rep 8:9960

Hagelin JC (2004) Observations on the olfactory ability of the Kakapo *Strigops habroptilus*, the critically endangered parrot of New Zealand. Ibis 146:161–164

Hagelin JC (2007) Odors and chemical signaling. In: Jamieson BGM (ed) Reproductive biology and phylogeny of birds. Science Publishers, Enfield, NH, pp 75–119

Hagelin JC, Jones IL (2007) Bird odors and other chemical substances: a defense mechanism or overlooked mode of intraspecific communication? Auk 124:741–761

Hagelin JC, Jones IL, Rasmussen LEL (2003) A tangerine-scented social odour in a monogamous seabird. Proc R Soc B 270:1323–1329

Halata Z, Grim M, Bauman KI (2003) Friedrich Sigmund Merkel and his Merkel cell, morphology, development, and physiology: review and new results. Anat Rec 271A:225–239

Hall CA, Chilcott RP (2018) Eyeing up the future of the pupillary light reflex in neurodiagnostics. Diagnostics 8:19

Hall MI (2008) The anatomical relationships between the avian eye, orbit and sclerotic ring: implications for inferring activity patterns in extinct birds. J Anat 212: 781–794

Hall MI, Heesy CP (2010) Eye size, flight speed and Leuckart's Law in birds. J Zool 283:291–297

Hall MI, Ross CF (2007) Eye shape and activity patterns in birds. J Zool 271:437–444

Hall ZJ, Meddle SL, Healy SD (2015) From neurons to nests: nest-building behaviour as a model in behavioural and comparative neuroscience. J Ornithol 156:133–143

Hall ZJ, Street SE, Healy SD (2013) The evolution of cerebellum structure correlates with nest complexity. Biol Lett 9:20130687

Hancock J, Elliott H (1978) Herons of the world. Harper & Row, New York, NY

Hanna ZR, Henderson JB, Wall JD, Emerling CA, Fuchs J, Runckel C, Mindell DP, Bowie RCK, DeRisi JL, Dumbacher JP (2017) Northern Spotted Owl (*Strix occidentalis caurina*) genome: divergence with the Barred Owl (*Strix varia*) and characterization of light-associated genes. Genome Biol Evol 9:2522–2545

Hansen KA, Hernandez A, Mooney TA, Rasmussen MH, Sørensen K, Wahlberg M (2020) The Common Murre (*Uria aalge*), an auk seabird, reacts to underwater sound. J Acoust Soc Am 147:4069–4074

Hansen KA, Maxwell A, Siebert U, Larsen ON, Wahlberg M (2017) Great Cormorants (*Phalacrocorax carbo*) can detect auditory cues while diving. Sci Nat 104:45

Hardy JW (1974) Behavior and its evolution in Neotropical jays (*Cissilopha*). Bird-Banding 45:253–268

Harmening WM, Wagner H (2011) From optics to attention: visual perception in Barn Owls. J Comp Physiol A 197:1031–1042

Harriman AE, Berger RH (1986) Olfactory acuity in the Common Raven. Physiol Behav 36:257–262

Hart NS (2002) Vision in the peafowl (Aves: *Pavo cristatus*). J Exp Biol 205:3925–3935

Hart NS, Hunt DM (2007) Avian visual pigments: characteristics, spectral tuning, and evolution. Am Nat 169(Suppl):S7–S26

Hart NS, Lisney TJ, Collin SP (2006) Cone photoreceptor oil droplet pigmentation is affected by ambient light intensity. J Exp Biol 209:4776–4787

Hatch JJ, Brown KM, Hogan GG, Morris RD, Orta J, Garcia EFJ, Jutglar F, Kirwan GM, Boesman PFD (2020) Great Cormorant (*Phalacrocorax carbo*), version 1.0. In: Billerman M (ed) Birds of the World. Cornell Lab of Ornithology, Ithaca, NY

Hau M, Dominguez OA, Evrard HC (2004) Testosterone reduces responsiveness to nociceptive stimuli in a wild bird. Horm Behav 46:165–170

Hedenström A, Norevik G, Warfvinge K, Andersson A, Bäckman J, Åkesson S (2016) Annual 10-month aerial life phase in the Common Swift *Apus apus*. Curr Biol 26:3066–3070

Herculano-Houzel S (2020) Birds do have a brain cortex—and think. Science 369:1567–1568

Hernández MC, Villada AM, Barja I (2023) Onto the sense of smell in macaws, amazons and toucans: can they use volatile cues of fruits to make foraging decisions? Integr Zool 18:762–771

Hiltpold I, Shriver WG (2018) Birds bug on indirect plant defenses to locate insect prey. J Chem Ecol 44:576–579

Hirsch J (1982) Falcon visual sensitivity to grating contrast. Nature 300:57–58

Hodos W, Ghim MM, Potocki A, Field JN, Storm T (2002) Contrast sensitivity in pigeons measured with behavioral and pattern ERG methods. Doc Ophthalmol 104:107–118

Höglund J, Mitkus M, Olsson P, Lind O, Drews A, Bloch NI, Kelber A, Strandh M (2019) Owls lack UV-sensitive cone opsin and red oil droplets, but see UV light at night: retinal transcriptomes and ocular media transmittance. Vis Res 158:109–119

Holland RA, Wikelski M, Kümmeth F, Bosque C (2009) The secret life of Oilbirds: new insights into the movement ecology of a unique avian frugivore. PLoS ONE 4:e8264

Hörster W (1990) Histological and electrophysiological investigations on the vibration-sensitive receptors (Herbst corpuscles) in the wing of the pigeon (*Columba livia*). J Comp Physiol A 166:663–673

Huber JF (1936) Nerve roots and nuclear groups in the spinal cord of the pigeon. J Comp Neurol 65:43–91

Hudon J, Muir AD (1996) Characterization of the reflective materials and organelles in the bright irides of North American blackbirds (Icterinae). Pigment Cell Res:96–104

Hunt DM, Peichl L (2014) S cones: evolution, retinal distribution, development, and spectral sensitivity. Vis Neurosci 31:115–138

Ikegami K, Yoshimura T (2016) Comparative analysis reveals the underlying mechanism of vertebrate seasonal reproduction. Gen Comp Endocrinol 227:64–68

Inzunza O, Bravo H, Smith RL, Angel M (1991) Topography and morphology of retinal ganglion cells in Falconiforms: a study on predatory and carrion-eating birds. Anat Rec 229:271–277

Isler K, van Schaik CP (2009) Why are there so few smart mammals (but so many smart birds)? Biol Lett 5:125–129

Ito M (2008) Control of mental activities by internal models in the cerebellum. Nat Rev Neurosci 9:304–313

Iwaniuk AN, Wylie DR (2006) The evolution of stereopsis and the Wulst in Caprimuliform birds: a comparative analysis. J Comp Phys A 192:1313–1326

Iwaniuk AN, Dean AM, Nelson JE (2005) Interspecific allometry of the brain regions of parrots (Psittaciformes): comparisons with other birds and primates. Brain Behav Evol 65:40–59

Iwaniuk AN, Hurd PL, Wylie DRW (2006) The comparative morphology of the cerebellum in Caprimulgiform birds: evolutionary and functional implications. Brain Behav Evol 67:53–68

Iwaniuk AN, Lefebvre L, Wylie DR (2009) The comparative approach and brain–behaviour relationships: a tool for understanding tool use. Can J Exp Psychol 63:150–159

Iwaniuk AN, Wylie DR (2020) Sensory systems in birds: what we have learned from studying sensory specialists. J Comp Neurol 528:2902–2918

Jarvis ED (2009) Evolution of the pallium in birds and reptiles. In: Binder MD, Hirokawa N, Windhorst U (eds) Encyclopedia of neuroscience. Springer, Berlin, pp 1390–1400

Jarvis ED, Güntürkün O, Bruce L, Csillag A, Karten H, Kuenzel W, Medina L, Paxinos G, Perkel DJ, Shimizu T, Striedter G, Wild M, Ball GF, Dugas-Ford J, Durand S, Hough G, Husband S, Kubikova L, Lee D, Mello CV, Powers A, Siang C, Smulders TV, Wada K, White SA, Yamamoto K, Yu J, Reiner A, Butler AB (2005) Avian brains and a new understanding of vertebrate brain evolution. Nat Rev Neurosci 6: 151–159

Jarvis ED, Yu J, Rivas MV, Horita H, Feenders G, Whitney O, Jarvis SC, Jarvis ER, Kubikova L, Puck AEP, Siang-Bakshi C, Martin S, McElroy M, Hara E, Howard J, Pfenning A, Mouritsen H, Chen C-C, Wada K (2013) Global view of the functional molecular organization of the avian cerebrum: mirror images and functional columns. J Comp Neurol 521:3614–3665

Jarvis JR, Taylor NR, Prescott NB, Meeks I, Wathes CM (2002) Measuring and modelling the photopic flicker sensitivity of the chicken (*Gallus g. domesticus*). Vis Res 42:99–106

Jezler PCOC, Braga MBP, Perlmann E, Squarzoni R, Borella MI, Barros PSM, Milanelo L, Antunes A (2010) Histological analysis of eyeballs of the Striped Owl *Rhinoptynx clamator*. In: Mendez-Vilas A, Diaz J (eds) Microscopy: science, technology, applications and education. Badajoz, Spain, Formatex Research Center, pp 1047–1054

Johnson LS, Murphy SM, Parrish GW (2011) Lack of predator-odor detection and avoidance by a songbird, the House Wren. J Field Ornithol 82:150–157

Johnsson RD, Connelly F, Mussoi JG, Vyssotski AL, Cain KE, Roth TC, Lesku JA (2022) Sleep loss impairs cognitive performance and alters song output in Australian Magpies. Sci Rep 12:6645

Jones ME, Button DJ, Barrett PM, Porro LB (2019) Digital dissection of the head of the Rock Dove (*Columba livia*) using contrast-enhanced computed tomography. Zoological Lett 5:17

Jones MP, Pierce KE, Ward D (2007) Avian vision: a review of form and function with special consideration to birds of prey. J Exotic Pet Med 16:69–87

Jones SG, Paletz EM, Obermeyer WH, Hannan CT, Benca RM (2010) Seasonal influences on sleep and executive function in the migratory White-crowned Sparrow (*Zonotrichia leucophrys gambelii*). BMC Neurosci 11:87

Jones SG, Vyazovskiy VV, Cirelli C, Tononi G, Benca RM (2008) Homeostatic regulation of sleep in the White-crowned Sparrow (*Zonotrichia leucophrys gambelii*). BMC Neurosci 9:47

Kabadayi C, Taylor LA, von Bayern AMP, Osvath M (2016) Ravens, New Caledonian Crows and Jackdaws parallel great apes in motor self-regulation despite smaller brains. R Soc Open Sci 3:160104

Kallähn SV (2014) Untersuchungen zur chromatischen pupillometrie am vogelauge. Ph.D. dissertation, Ludwig Maximilian University, Munich, Germany

Kamermans M, Hawryshyn C (2011) Teleost polarization vision: how it might work and what it might be good for. Philos Trans R Soc B 366:742–756

Kamska V, Daley M, Badri-Sprowitz A (2020) 3D anatomy of the quail lumbosacral spinal canal - implications for putative mechanosensory function. Int Organismal Biol 2:0baa037

Kane SA, Zamani M (2014) Falcons pursue prey using visual motion cues: new perspectives from animal-borne cameras. J Exp Biol 217:225–234

Kang H, Yan M, Yu Q, Yang Q (2014) Characterization of nasal cavity-associated lymphoid tissue in ducks. Anat Rec 297:916–924

Kare MR, Mason JR (1986) The chemical senses in birds. In: Sturkie PD (ed) Avian physiology, 4th edn. Springer, New York, NY, pp 59–73

Karkoura AA, Alsafy MAM, Elgendy SAA, Eldefrawy FA (2015) Morphological investigation of the brain of the African Ostrich (*Struthio camelus*). Int J Morphol 33:1468–1475

Katzir G, Howland HC (2003) Corneal power and underwater accommodation in Great Cormorants (*Phalacrocorax carbo sinensis*). J Exp Biol 206:833–841

Katzir G, Martin GR (1998) Visual fields in the Black-crowned Night Heron *Nycticorax nycticorax*: nocturnality does not result in owl-like features. Ibis 140:157–162

Kettler L, Christensen-Dalsgaard J, Larsen ON, Wagner H (2016) Low frequency eardrum directionality in the Barn Owl induced by sound transmission through the interaural canal. Biol Cybern 110:333–343

Khan S, Chang R (2013) Anatomy of the vestibular system: a review. NeuroRehabilitation 32:437–443

Khmelinskii I, Makarov V (2019) New unique optical and electric properties of intermediate filaments in Müller cells. Exp Eye Res 184:296–299

Kiama SG, Maina JN, Bhattacharjee J, Weyrauch KD (2001) Functional morphology of the pecten oculi in the nocturnal Spotted Eagle Owl (*Bubo bubo africanus*) and Domestic Fowl (*Gallus gallus* var. *domesticus*). J Zool 254:521–528

Kiama SG, Maina JN, Bhattacharjee J, Weyrauch KD, Gehr P (1998) A scanning electron microscope study of the luminal surface specializations in the blood vessels of the pecten oculi in a diurnal bird, the Black Kite (*Milvus migrans*). Ann Anat-Anat Anz 180:455–460

Kim AY, Rankin DT, Rankin EEW (2021) What is that smell? Hummingbirds avoid foraging on resources with defensive insect compounds. Behav Ecol Sociobiol 75:132

Kim T, Soto F, Kerschensteiner D (2015) An excitatory amacrine cell detects object motion and provides feature-selective input to ganglion cells in the mouse retina. eLife 4:e08025

Klaassen RHG, Alerstam T, Carlsson P, Fox JW, Lindström Å (2011) Great flights by Great Snipes: long and fast non-stop migration over benign habitats. Biol Lett 7:833–835

Klump GM, Gleich O (1991) Gap detection in the European Starling (*Sturnus vulgaris*). III. Processing in the peripheral auditory system. J Comp Physiol A Neuroethol Sensory Neural Behav Physiol 168:469–476

Klump GM, Maier EH (1989) Gap detection in the starling. I. Psychophysical thresholds. J Comp Physiol A 164:531–538

Knudson EI, Blasdel GG, Konishi M (1979) Sound localization by the Barn Owl measured with the search coil technique. J Comp Psychol 133:1–11

Koch UR, Wagner H (2002) Morphometry of auricular feathers of Barn Owls (*Tyto alba*). Eur J Morphol 40: 15–21

Kolb H (2003) How the retina works. Am Sci 91:28–35

Komagata N, Latifi B, Rusterholz T, Bassetti CLA, Adamantidis A, Schmidt MH (2019) Dynamic REM sleep modulation by ambient temperature and the critical role of the melanin-concentrating hormone system. Curr Biol 29:1976–1987

Kondoh D, Nashimoto M, Kanayama S, Nakamuta N, Taniguchi K (2011) Ultrastructural and histochemical properties of the olfactory system in the Japanese Jungle Crow, *Corvus macrorhynchos*. J Vet Med Sci 73:1007–1014

Köppl C (2019) Internally coupled middle ears enhance the range of interaural time differences heard by the chicken. J Exp Biol 222:jeb199232

Krabichler Q, Vega-Zuniga T, Morales C, Luksch H, Marín GJ (2015) The visual system of a Palaeognathous bird: visual field, retinal topography and retino-central connections in the Chilean Tinamou (*Nothoprocta perdicaria*). J Comp Neurol 523:226–250

Kram YA, Mantey S, Corbo JC (2010) Avian cone photoreceptors tile the retina as five independent, self-organizing mosaics. PLoS ONE 5:e8992

Ksepka DT, Balanoff AM, Smith NA, Bever GS, Bhullar BAS, Bourdon E, Braun EL, Burleigh JG, Clarke JA, Colbert MW, Corfield JR, Degrange FJ, De Pietri VL, Early CM, Field DJ, Gignac PM, Gold MEL, Kimball RT, Kawabe S, Lefebvre L, Marugán-Lobón J, Mongle CS, Morhardt A, Norell MA, Ridgely RC, Rothman RS, Scofield RP, Tambussi CP, Torres CR, van Tuinen M, Walsh SA, Watanabe A, Witmer LM, Wright AK, Zanno LE, Jarvis ED, Smaers JB (2020) Tempo and pattern of avian brain size evolution. Curr Biol 30:2026–2036

Kudo K, Nishimura S, Tabata S (2008) Distribution of taste buds in layer-type chickens: scanning electron microscopic observations. Anim Sci J 79:680–685

Kuenzel WJ (2000) The Autonomic Nervous System of avian species. In: Whittow GC (ed) Sturkie's avian physiology, 5th edn. Academic Press, New York, NY, pp 101–122

Kühne R, Lewis B (1985) External and middle ears. In: King AS, McLelland J (eds) Form and function in birds, vol 3. Academic Press, New York, NY, pp 227–271

Kverková K, Marhounová L, Polonyiová A, Kocourek M, Zhang Y, Olkowicz S, Straková B, Pavelková Z, Vodička R, Frynta D, Němec P (2022) The evolution of brain neuron numbers in amniotes. Proc Natl Acad Sci 119:e2121624119

Lacquaniti F, Bosco G, Garvano S, Indovina I, La Scleia B, Maffei V, Zago M (2014) Multisensory integration and internal models for sensing gravity effects in primates. Biomed Res Int 2014:615854

Land MF, Nilsson D-E (2002) Animal eyes. Oxford University Press, Oxford

Lange H, Walker L, Orell M, Smulders TV (2022) Seasonal changes in the hippocampal formation of hoarding and non-hoarding tits. Learn Behav 50:113–124

Larsell O (1967) The comparative anatomy and histology of the cerebellum from myxinoids through birds. University of Minnesota Press, Minneapolis, MN

Larsen ON (2004) Does the environment constrain avian sound localization? An Acad Bras Cienc 76:267–273

Larsen ON, Christensen-Dalsgaard J, Jensen KK (2016) Role of intracranial cavities in avian directional hearing. Biol Cybern 110:319–331

Larsen ON, Dooling RJ, Michelsen A (2006) The role of pressure difference reception in the directional hearing of Budgerigars (*Melopsittacus undulatus*). J Comp Physiol A 192:1063–1072

Larsen ON, Wahlberg M, Christensen-Dalsgaard J (2020) Amphibious hearing in a diving bird, the Great Cormorant (*Phalacrocorax carbo sinensis*). J Exp Biol 223:jeb217265

Lauwers M, Pichler P, Edelman NB, Resch GP, Ushakova L, Salzer MC, Heyers D, Saunders M, Shaw J, Keays DA (2013) An iron-rich organelle in the cuticular plate of avian hair cells. Curr Biol 23: 924–929

Laux CM, Nordell CJ, Fisher RJ, Ng JW, Wellicome TI, Bayne EM (2015) Ferruginous Hawks *Buteo regalis* alter parental behaviours in response to approaching storms. J Ornithol 157:355–362

Leigh RJ, Zee DS (1999) The neurology of eye movement, 3rd edn. Oxford University Press, Oxford

Lesku JA, Rattenborg NC (2014) Avian sleep. Curr Biol 24:R12–R14

Lesku JA, Rattenborg NC, Valcu M, Vyssotski AL, Kuhn S, Kuemmeth F, Heidrich W, Kempenaers B (2012) Adaptive sleep loss in polygynous Pectoral Sandpipers. Science 337:1654–1658

Levy B, Sivak JG (1980) Mechanisms of accommodation in the bird eye. J Comp Physiol A 137:267–272

Li G, Zhang H, Zhang J, Hildre HP (2014) An approach for adaptive limbless locomotion using a CPG-based reflex mechanism. J Bionic Eng 11:389–399

Liechti F, Witvliet W, Weber R, Bächler E (2013) First evidence of a 200-day non-stop flight in a bird. Nat Commun 4:2554

Lima FC, Vieira LG, Santos ALQ, De Simone SBS, Hirano LQL, Silva JMM, Romão MF (2009) Anatomy of the scleral ossicles in Brazilian birds. Braz J Morphol Sci 26:165–169

Lima SL, Rattenborg NC, Lesku JA, Amlaner CJ (2005) Sleeping under the risk of predation. Anim Behav 70:723–736

Lind O, Chavez J, Kelber A (2014a) The contribution of single and double cones to spectral sensitivity in Budgerigars during changing light conditions. J Comp Physiol A 200:197–207

Lind O, Mitkus M, Olsson P, Kelber A (2014b) Ultraviolet vision in birds: the importance of transparent eye media. Proc R Soc B 281:20132209

Lind OE, Kelber A, Kröger RHH (2008) Multifocal optical systems and pupil dynamics in birds. J Exp Biol 211:2752–2758

Lipetz LE (1984) A new method for determining peak absorbance of dense pigment samples and its application to the cone oil droplets of *Emydoidea blandingii*. Vis Res 24:597–604

Lisney TJ, Iwaniuk AN, Bandet MV, Wylie DR (2012a) Eye shape and retinal topography in owls (Aves: Strigiformes). Brain Behav Evol 79:218–236

Lisney TJ, Iwaniuk AN, Kolominsky J, Bandet MV, Corfield JR, Wylie DR (2012b) Interspecifc variation in eye shape and retinal topography in seven species of galliform bird (Aves: Galliformes: Phasianidae). J Comp Physiol A 198:717–731

Lisney TJ, Potier S, Isard PF, Mentek M, Mitkus M, Collin SP (2020) Retinal topography in two species of flamingo (Phoenicopteriformes: Phoenicopteridae). J Comp Neurol 528:2848–2863

Lisney TJ, Stecyk K, Kolominsky J, Graves GR, Wylie DR, Iwaniuk AN (2013a) Comparison of eye morphology and retinal topography in two species of New World vultures (Aves: Cathartidae). Anat Rec 296: 1954–1970

Lisney TJ, Stecyk K, Kolominsky J, Schmidt BK, Corfield JR, Iwaniuk AN, Wylie DR (2013b) Ecomorphology of eye shape and retinal topography in waterfowl (Aves: Anseriformes: Anatidae) with different foraging modes. J Comp Physiol A 199:385–402

Lisney TJ, Wylie DR, Kolominsky J, Iwaniuk AN (2015) Eye morphology and retinal topography in hummingbirds (Trochilidae: Aves). Brain Behav Evol 86:176–190

Livingston A (1994) Physiological basis for pain perception in animals. Vet Anaesth Analg 21:73–77

Lu Y-B, Franze K, Seifert G, Steinhäuser C, Kirchhoff F, Wolburg H, Guck J, Janmey P, Wei E-Q, Käs J, Reichenbach A (2006) Viscoelastic properties of individual glial cells and neurons in the CNS. Proc Natl Acad Sci USA 103:17759–17764

Mahr K, Hoi H (2018) Red-legged Partridges perceive the scent of predators and alarm scents of an avian heterospecific. Anim Behav 144:109–114

Manley GA (1990) Peripheral hearing mechanisms in reptiles and birds. Springer, Heidelberg

Manley GA (2000) Cochlear mechanisms from a phylogenetic viewpoint. Proc Natl Acad Sci USA 97:11736–11743

Manley GA (2015) Aural history. Scientist 29:36–42

Mäntylä E, Kipper S, Hilker M (2020) Insectivorous birds can see and smell systemically herbivore-induced pines. Ecol Evol 10:9358–9370

Mäntylä E, Klemola T, Haukioja E (2004) Attraction of Willow Warblers to sawfly-damaged mountain birches: novel function of inducible plant defences? Ecol Lett 7:915–918

Martelli D, Luppi M, Cerri M, Tupone D, Mastrotto M, Perez E, Zamboni G, Amici R (2014) The direct cooling of the preoptic-hypothalamic area elicits the release of thyroid stimulating hormone during wakefulness but not during REM sleep. PLoS ONE 9: e87793

Martin G, Rojas LM, Ramírez Y, McNeil R (2004) The eyes of Oilbirds (*Steatornis caripensis*): pushing at the limits of sensitivity. Naturwissenschaften 91:26–29

Martin GR (1983) Schematic eye models in vertebrates. Prog Sensory Physiol 4:43–81

Martin GR (1985) Eye. In: King AS, McClelland J (eds) Form and function in birds, vol 3. Academic Press, London, p 3110373

Martin GR (1986) The eye of a passeriform bird, the European Starling (*Sturnus vulgaris*): eye movement amplitude, visual fields and schematic optics. J Comp Physiol A 159:545–557

Martin GR (1999) Eye structure and foraging in King Penguins *Aptenodytes patagonicus*. Ibis 141:444–450

Martin GR (2007) Visual fields and their functions in birds. J Ornithol 148(Suppl 2):S547–S562

Martin GR (2012) Through birds' eyes: insights into avian sensory ecology. J Ornithol 153(Suppl 1):S23–S48

Martin GR (2014) The subtlety of simple eyes: the tuning of visual fields to perceptual challenges in birds. Trans Roy Soc B 369:20130040

Martin GR (2017) What drives bird vision? Bill control and predator detection overshadow flight. Front Neurosci 11:619

Martin GR, Coetzee HC (2004) Visual fields in hornbills: precision-grasping and sunshades. Ibis 146:18–26

Martin GR, Katzir G (2000) Sun shades and eye size in birds. Brain Behav Evol 56:340–344

Martin GR, Katzir G (1994) Visual fields and eye movements in herons (Ardeidae). Brain Behav Evol 44:74–85

Martin GR, White CR, Butler PJ (2008) Vision and the foraging technique of Great Cormorants *Phalacrocorax carbo*: pursuit or close-quarter foraging? Ibis 150:485–494

Martin GR, Wilson K-J, Wild JM, Parsons S, Kubke MF, Corfield J (2007) Kiwi forego vision in the guidance of their nocturnal activities. PLoS ONE 2:e198

Martin LTP, Dupré DJ (2016) Bitter taste receptors. In: Choi S (ed) Encyclopedia of signaling molecules. Springer, New York, NY

Marugán-Lobón J, Chiappe LM, Farke AA (2013) The variability of inner ear orientation in saurischian dinosaurs: testing the use of semicircular canals as a reference system for comparative anatomy. PeerJ 1: e124

Mason JR, Clark L (2000) The chemical senses of birds. In: Sturkie's avian physiology, 5th edn. Academic Press, San Diego, CA, pp 39–56

Massaro M, Davis LS, Darby JT (2003) Carotenoid-derived ornaments reflect parental quality in male and female Yellow-eyed Penguins (*Megadyptes antipodes*). Behav Ecol Sociobiol 55:169–175

Matson KD, Millam JR, Klasing KC (2004) Cockatiels (*Nymphicus hollandicus*) reject very low levels of plants secondary compounds. Appl Anim Behav Sci 85:141–156

Matsunaga N, Amemiya T (1990) Pecten of the chick eye demonstrated by vascular casts. Okajimas Folia Anatomica Japanica 87:263–270

Matsushita Y, Kitamura N, Higuchi M, Hosaka YZ, Shibuya I (2018) Neuron-like cells in the chick spinal accessory lobe express neuronal-type voltage-gated sodium channels. Biomed Res 39:189–196

Mazor-Thomas JE, Mann PE, Karas AZ, Tseng F (2014) Pain-suppressed behaviors in the Red-tailed Hawk (*Buteo jamaicensis*). Appl Anim Behav Sci 152:83–91

McCulloch D (1963) Colour change in the iris of the Black-bellied Starling *Lamprotornis corruscus*. Ostrich 34:177

McFadden SA (1994) Binocular depth perception. In: Davies MNO, Green PR (eds) Perception and motor control in birds. Springer, Berlin, pp 54–70

McFadden SA, Wild JM (1986) Binocular depth perception in the pigeon. J Exp Anal Behav 45:149–160

McGrew KA, Crowell SE, Fiely JL, Berlin AM, Olsen GH, James J, Hopkins H, Williams CK (2022) Underwater hearing in sea ducks with applications for reducing gillnet bycatch through acoustic deterrence. J Exp Biol 225:jeb243953

McIsaac HP (2001) Raptor acuity and wind turbine blade conspicuity. In: National avian-wind power planning meeting IV. Resolve, Inc., Washington, DC, pp 59–87

Mehlhorn J, Hunt GR, Gray RD, Rehkämper G, Güntürkün O (2010) Tool-making New Caledonia Crows have large associative brain areas. Brain Behav Evol 75:63–70

Menkhorst M, Rogers D, Clarke R, Davies J, Marsack P, Franklin K (2017) The Australian bird guide. CSIRO Publishing, Clayton, Victoria, Australia

Metcalfe J, Schmidt KL, Kerr WB, Guglielmo CG, MacDougall-Shackleton SA (2013) White-throated Sparrows adjust behavior in response to manipulations of barometric pressure and temperature. Anim Behav 86:1285–1290

Meyer DB (1977) The avian eye and its adaptations. In: Crescitelli F (ed) The visual system in vertebrates. Springer, Berlin, pp 549–611

Michael N, Löwel S, Bischof HJ (2015) Features of the retinotopic representation in the visual wulst of a laterally eyed bird, the Zebra Finch (*Taeniopygia guttata*). PLoS ONE 10:e0124917

Mitkus M, Chaib S, Lind O, Kelber A (2014) Retinal ganglion cell topography and spatial resolution of two parrot species: Budgerigar (*Melopsittacus undulatus*) and Bourke's Parrot (*Neopsephotus bourkii*). J Comp Physiol A 200:371–384

Mitkus M, Nevitt GA, Danielsen J, Kelber A (2016) Vision on the high seas: spatial resolution and optical sensitivity in two procellariiform seabirds with different foraging strategies. J Exp Biol 219:3329–3338

Mitkus M, Olsson P, Toomey MB, Corbo JC, Kelber A (2017) Specialized photoreceptor composition in the raptor fovea. J Comp Neurol 525:2152–2163

Mlikovsky J (2003) Brain size and foramen magnum area in crows and allies (Aves: Corvidae). Acta Societatis Zoologicae Bohemicae 67:203–211

Moiseff A, Konishi M (1981) Neuronal and behavioral sensitivity to binaural time differences in the owl. J Neurosci 1:40–48

Molina-Morales M, Castro J, Albaladejo G, Parejo D (2020) Precise cache detection by olfaction in a scatter-hoarder bird. Anim Behav 167:185–191

Mollon JD (1999) Color vision: opsins and options. Proc Natl Acad Sci USA 96:4743–4745

Montiel JF, Molnár Z (2013) The impact of gene expression analysis on evolving views of avian brain organization. J Comp Neurol 521:3604–3613
Moore BA, Doppler M, Young JE, Fernández-Juricic E (2013) Interspecific differences in the visual system and scanning behaviour of three forest passerines that form heterospecific flocks. J Comp Physiol A 199: 263–277
Moore BA, Baumhardt P, Doppler M, Randolet J, Blackwell BF, DeVault TL, Loew ER, Fernández-Juricic E (2012) Oblique color vision in an open-habitat bird: spectral sensitivity, photoreceptor distribution and behavioral implications. J Exp Biol 215: 3442–3452
Moore BA, Pita D, Tyrrell LP, Fernández-Juricic E (2015) Vision in avian emberizid foragers: maximizing both binocular vision and fronto-lateral visual acuity. J Exp Biol 218:1347–1358
Moore BA, Yoo I, Tyrrell LP, Benes B, Fernandez-Juricic E (2016) FOVEA: a new program to standardize the measurement of foveal pit morphology. PeerJ 4:e1785
Morand-Ferron J, Cole EF, Quinn JL (2016) Studying the evolutionary ecology of cognition in the wild: a review of practical and conceptual challenges. Biol Rev 91:367–389
Moritz GL, Melin AD, Tuh Yit Yu F, Bernard H, Ong PS, Dominy NJ (2014) Niche convergence suggests functionality of the nocturnal fovea. Front Integr Neurosci 8:61
Moroney MK, Pettigrew JD (1987) Some observations on the visual optics of kingfishers (Aves, Coraciformes, Alcedinidae). J Comp Physiol A 160:137–149
Murphy ME, King JR (1989) Sparrows discriminate between diets differing in valine or lysine concentrations. Physiol Behav 45:423–430
Mussoi JG, Stanley MC, Cain KE (2022) Importance of sleep for avian vocal communication. Biol Lett 18: 20220223
Necker R (1999) Specializations in the lumbosacral spinal cord of birds: morphological and behavioural evidence for a sense of equilibrium. Eur J Morphol 37:211–214
Necker R (2005) The structure and development of avian lumbosacral specializations of the vertebral canal and the spinal cord with special reference to a possible function as a sense organ of equilibrium. Anat Embryol 210:59–74
Necker R (2006) Specialization in the lumbosacral vertebral canal and spinal cord of birds: evidence of a function as a sense organ which is involved in the control of walking. J Comp Physiol A 192:439–448
Nef P (1998) How we smell: the molecular and cellular bases of olfaction. Physiology 13:1–5
Negro JJ, Blázquez MC, Galván I (2017) Intraspecific eye color variability in birds and mammals: a recent evolutionary event exclusive to humans and domestic animals. Front Zool 14:53
Nelson BS, Stoddard PK (1998) Accuracy of auditory distance and azimuth perception by a passerine bird in natural habitat. Anim Behav 56:467–477
Németh Z (2009) Observation of daytime sleep-like behavior in a migratory songbird during stopover. Wilson J Ornithol 121:644–646
Nevitt GA (2006) Olfactory foraging strategies of procellariiform seabirds. Acta Zool Sin 52 (Suppl):510–513
Nieder A (2021) Consciousness without cortex. Curr Opin Neurobiol 71:69–76
Nieder A, Wagener L, Rinnert P (2020) A neural correlate of sensory consciousness in a corvid bird. Science 369: 1626–1629
Niknafs S, Roura E (2018) Nutrient sensing, taste and feed intake in avian species. Nutr Res Rev 31:256–266
Nikolaev YA, Feketa VV, Anderson EO, Schneider ER, Gracheva EO, Bagriantsev SN (2020) Lamellar cells in Pacinian and Meissner corpuscles are touch sensors. Sci Adv 6:eabe6393
Nishikawa KC (2002) Evolutionary convergence in nervous systems: insights from comparative phylogenetic studies. Brain Behav Evol 59:240–249
Nityananda V, Read JCA (2017) Steropsis in animals: evolution, function and mechanisms. J Exp Biol 220: 2501–2512
Norberg R (2002) Independent evolution of outer ear asymmetry among five owl lineages: morphology, function and selection. In: Newton I, Kavanagh R, Olsen J, Taylor I (eds) Ecology and conservation of owls. CSIRO Publ, Clayton, pp 329–342
Norberg RÅ (1978) Skull asymmetry, ear structure and function, and auditory localization in Tengmalm's Owl, *Aegolius funereus* (Linné). Proc R Soc B 282: 325–410
Northcutt RG (2002) Understanding vertebrate brain evolution. Integr Comp Biol 42:743–756
Northcutt RG (2004) Taste buds: development and evolution. Brain Behav Evol 64:198–206
O'Connell LA, Hofmann HA (2011) The vertebrate mesolimbic reward system and social behavior network: a comparative analysis. J Comp Neurol 519: 3599–3639
O'Connell LA, Hofmann HA (2012) Evolution of a vertebrate social decision-making network. Science 336: 1154–1157
O'Day K (1940) The fundus and fovea centralis of the albatross (*Diomedea cauta* – Gould). Br J Ophthalmol 24:201–207
Ödeen A, Håstad O (2003) Complex distribution of avian color vision systems revealed by sequencing the SWS1 opsin from total DNA. Mol Biol Evol 20:855–861
Ödeen A, Håstad O (2013) The phylogenetic distribution of ultraviolet sensitivity in birds. BMC Evol Biol 13:36
Ödeen A, Håstad O, Alström P (2011) Evolution of ultraviolet vision in the largest avian radiation – the passerines. BMC Evol Biol 11:313
Oesterle EC, Stone JS (2008) Hair cell regeneration: mechanisms guiding cellular proliferation and differentiation. In: Salvi RJ, Popper AN, Fay RR (eds) Hair cell regeneration, repair, and protection. New York, NY, Springer Science, pp 141–198

Okanoya K, Dooling RJ (1990) Detection of gaps in noise by Budgerigars (*Melopsittacus undulatus*) and Zebra Finches (*Poephila guttata*). Hear Res 50:185–192

Oliphant LW (1987) Pteridines and purines as major pigments of the avian iris. Pigment Cell Res 1:129–131

Oliphant LW, Hudon J (1993) Pteridines as reflecting pigments and components of reflecting organelles in vertebrates. Pigment Cell Res 6:205–208

Oliphant LW, Hudon J, Bagnara JT (1992) Pigment cell refugia in homeotherms—the unique evolutionary position of the iris. Pigment Cell Res 5:367–371

Olkowicz S, Kocourek M, Lucan RK, Porte M, Fitch WT, Herculano-Houzel S, Nemec P (2016) Birds have primate-like numbers of neurons in the forebrain. Proc Natl Acad Sci USA 113:7255–7260

Olsson P, Wilby D, Kelber A (2016) Quantitative studies of animal colour constancy: using the chicken as model. Proc R Soc B 283:20160411

O'Neill P (2013) Magnetoreception and baroreception in birds. Dev Growth Differ 55:188–197

O'Rourke CT, Hall MI, Pitlik T, Fernández-Juricic E (2010) Hawk eyes I: diurnal raptors differ in visual fields and degree of eye movement. PLoS ONE 5: e12802

Ott M (2006) Visual accommodation in vertebrates: mechanisms, physiological response and stimuli. J Comp Physiol A 192:97–111

Overington SE, Griffin AS, Sol D, Lefebvre L (2011) Are innovative species ecological generalists? A test in North American birds. Behav Ecol 22:1286–1293

Paige KN (1995) Bats and barometric pressure: conserving limited energy and tracking insects from the roost. Funct Ecol 9:463–467

Park TJ, Dooling RJ (1991) Sound localization in small birds: absolute localization in azimuth. J Comp Psychol 105:125–133

Passarotto A, Cruz-Miralles Á, Avilés JM (2020) Iris yellowness relates to age and individual quality in two owl species. J Raptor Res 54:233–244

Paul-Murphy J, Hawkins MG (2015) Bird-specific considerations: recognizing pain behavior in pet birds. In: Gaynor JS, Muir WW III (eds) Handbook of veterinary pain management, 3rd edn. Elsevier Mosby, St. Louis, MO, pp 536–554

Peng K, Feng Y, Zhang G, Liu H, Song H (2010) Anatomical study of the brain of the African Ostrich. Turk J Vet Anim Sci 34:235–241

Pettigrew JD (1979) Binocular visual processing in the owl's telencephalon. Proc R Soc B 204:435–454

Pettigrew JD, Konishi M (1976) Neurons selective for orientation and binocular disparity in the visual wulst of the Barn Owl (*Tyto alba*). Science 193:675–678

Pettigrew JD, Wallman J, Wildsoet CF (1990) Saccadic oscillations facilitate ocular perfusion from the avian pecten. Nature 343:362–363

Piersma T, van Aelst R, Kurk K, Berkhoudt H, Maas LRM (1998) A new pressure sensory mechanism for prey detection in birds: the use of principles of seabed dynamics? Proc R Soc B 265:1377–1383

Pohnert G, Steinke M, Tollrian R (2007) Chemical cues, defence metabolites and the shaping of pelagic interspecific interactions. Trends Ecol Evol 22:198–204

Popp JW (1989) Temporal aspects of singing interactions among territorial Ovenbirds. Ethology 82:127–133

Porciatti V, Fontanesi G, Bagnoli P (1989) The electroretinogram of the Little Owl (*Athene noctua*). Vis Res 29: 1693–1698

Porter ML, Kingston ACN, McCready R, Cameron EG, Hofmann CM, Suarez L, Olsen GH, Cronin TW, Robinson PR (2014) Characterization of visual pigments, oil droplets, lens and cornea in the Whooping Crane *Grus americana*. J Exp Biol 217: 3883–3890

Potier S, Bonadonna F, Kelber A, Martin GR, Isard P-F, Dulaurent T, Duriez O (2016) Visual abilities in two raptors with different ecology. J Exp Biol 219:2639–2649

Potier S, Mitkus M, Bonadonna F, Duriez O, Isard PF, Dulaurent T, Mentek M, Kelber A (2017) Eye size, fovea, and foraging ecology in accipitriform raptors. Brain Behav Evol 90:232–242

Potier S, Mitkus M, Kelber A (2020) Visual adaptations of diurnal and nocturnal raptors. Semin Cell Dev Biol 106:116–126

Pravosudov VV, Smulders TV (2010) Integrating ecology, psychology and neurobiology within a food-hoarding paradigm. Phil Trans R Soc B 365:859–867

Pravosudov VV, Roth TC II, LaDage LD, Freas CA (2015) Environmental influences on spatial memory and the hippocampus in food-caching chickadees. Comp Cogn Behav 10:25–43

Prum RO (1999) The anatomy and physics of avian structural colours. In: Adams NJ, Slotow RH (eds) Proceedings of the 22nd International Ornithological Congress. BirdLife South Africa, Durban, pp 1–22

Prum RO (2006) Anatomy, physics, and evolution of avian structural colors. In: Hill GE, McGraw KJ (eds) Bird coloration, vol 1: Mechanisms and measurements. Harvard University Press, Cambridge, MA, pp 295–353

Querubin A, Lee HR, Provis JM, Bumsted O'Brien KM (2009) Photoreceptor and ganglion cell topographies correlate with information convergence and high acuity regions in the adult pigeon (*Columba livia*) retina. J Comp Neurol 517:711–722

Rahman ML, Aoyama M, Sugita S (2007a) Topography of retinal photoreceptor cells in the Jungle Crow (*Corvus macrorhynchos*) with emphasis on the distribution of oil droplets. Ornithol Sci 6:29–38

Rahman ML, Aoyama M, Sugita S (2007b) Regional specialization of the Tree Sparrow *Passer montanus* retina: ganglion cell density and oil droplet distribution. Ornithol Sci 6:95–105

Rahman ML, Aoyama M, Sugita S (2008) Ganglion cell density and oil droplet distribution in the retina of Brown-eared Bulbul (*Hysipetes amaurotis*). Anat Sci Int 83:239–246

Rajapaksha P, Wang Z, Venkatesan N, Tehrani KF, Payne J, Swetenburg RL, Kawabata F, Tabata S, Mortensen LJ, Stice SL, Beckstead R (2016) Labeling and analysis of chicken taste buds using molecular markers in oral epithelial sheets. Sci Rep 6:37247

Ramnani N (2006) The primate cortico-cerebellar system: anatomy and function. Nat Rev Neurosci 7:511–522

Rattenborg NC (2017) Sleeping on the wing. Interface Focus 7:20160082

Rattenborg NC, Amlaner CJ, Lima SL (2001) Unilateral eye closure and interhemispheric EEG asymmetry during sleep in the pigeon (*Columba livia*). Brain Behav Evol 58:323–332

Rattenborg NC, Lima SL, Amlaner CJ (1999a) Half-awake to the risk of predation. Nature 397:397

Rattenborg NC, Lima SL, Amlaner CJ (1999b) Facultative control of avian unihemispheric sleep under the risk of predation. Behav Brain Res 105:163–172

Rattenborg NC, Mandt BH, Obermeyer WH, Winsauer PJ, Huber R, Wikelski M, Benca RM (2004) Migratory sleeplessness in the White-crowned Sparrow (*Zonotrichia leucophrys gambelii*). PLoS Biol 2:e212

Rattenborg NC, Martinez-Gonzalez D (2015) Avian versus mammalian sleep: the fruits of comparing apples and oranges. Curr Sleep Med Rep 1:55–63

Rattenborg NC, Voirin B, Cruz SM, Tisdale R, Dell'Omo G, Lipp H-P, Wikelski M, Vyssotski AL (2016) Evidence that birds sleep in mid-flight. Nat Commun 7:12468

Reichenbach A, Agte S, Francke M, Franze K (2014) How light traverses the inverted vertebrate retina: no flaw of nature. e-Neuroforum 5:93–100

Reichenbach A, Franze K, Agte S, Junek S, Wurm A, Grosche J, Savvinov A, Guck J, Skatchkov SN (2012) Live cells as optical fibers in the vertebrate retina. In: Djamgoz MBA, Archer SN, Vallerga S (eds) Selected topics on optical fiber technology. Chapman & Hall, London, pp 395–416

Reymond L (1985) Spatial visual acuity of the eagle *Aquila audax*: a behavioural, optical and anatomical investigation. Vis Res 25:1477–1491

Reymond L (1987) Spatial visual acuity of the falcon, *Falco berigora*: a behavioural, optical and anatomical investigation. Vis Res 27:1859–1874

Rice WR (1982) Acoustical location of prey by the Marsh Hawk: adaptation to concealed prey. Auk 99: 403–413

Ricklefs RE (2004) The cognitive face of avian life histories: the 2003 Margaret Morse Nice Lecture. Wilson Bull 116:119–133

Rojas LM, Ramirez YM, Marin G, McNeil R (2004a) Capacidad visual en Caprimulgiformes. Ornitologia Neotropical 15(Suppl):251–260

Rojas LM, Ramirez Y, McNeil R, Mitchell M, Marin G (2004b) Retinal morphology and electrophysiology of two Caprimulgiformes birds: the cave-living and nocturnal Oilbird (*Steatornis caripensis*), and the crepuscularly and nocturnally foraging Common Pauraque (*Nyctidromus albicollis*). Brain Behav Evol 64:19–33

Roper TJ (2003) Olfactory discrimination in Yellow-backed Chattering Lories *Lorius garrulus flavopalliatus*: first demonstration of olfaction in Psittaciformes. Ibis 145:689–691

Rosenfield RN, Bielefeldt J, Rosenfield LJ, Stewart AC, Murphy RK, Grosshuesch DA, Bozek MA (2003) Comparative relationships among eye color, age, and sex in three North American populations of Cooper's Hawks. Wilson J Ornithol 115:225–231

Rosowski JJ, Saunders JC (1980) Sound transmission through the avian interaural pathways. J Comp Physiol 136:183–190

Roth TC, Pravosudov VV (2009) Hippocampal volumes and neuron numbers increase along a gradient of environmental harshness: a large-scale comparison. Proc R Soc B 276:401–405

Rubenstein DR, Hofmann HA (2015) Proximate pathways underlying social behavior. Curr Opin Behav Sci 6: 154–159

Rutz C, St Clair JJH (2012) The evolutionary origins and ecological context of tool use in New Caledonian Crows. Behav Process 89:153–165

Saavedra I, Amo L (2018) Insectivorous birds eavesdrop on the pheromones of their prey. PLoS ONE 13: e0190415

Salazar JE, Severin D, Vega-Zuniga T, Fernández-Aburto P, Deichler A, Sallaberry M, Mpodozis J (2019) Anatomical specializations related to foraging in the visual system of a nocturnal insectivorous bird, the Band-winged Nightjar (Aves: Caprimulgiformes). Brain Behav Evol 94:27–36

Santos LH, Santos CD, Da Silva ML (2023) The limits of olfactory perception in Black Vultures: a field experiment. Ethol Ecol Evol 35:340–347

Sayol F, Lefebvre L, Sol D (2016a) Relative brain size and its relation with the associate pallium in birds. Brain Behav Evol 87:69–77

Sayol F, Maspons J, Lapiedra O, Iwaniuk AN, Székely T, Sol D (2016) Environmental variation and the evolution of large brains in birds. Nat Commun 7:13971

Schaefer HM, Schmidt V, Bairlein F (2003) Discrimination abilities for nutrients: which difference matters for choosy birds and why? Anim Behav 65:531–541

Schaeffel F, Wagner H (1996) Emmetropization and optical development of the eye of the Barn Owl (*Tyto alba*). J Comp Physiol 178:491–498

Schmidt MH (2014) The energy allocation function of sleep: a unifying theory of sleep, torpor, and continuous wakefulness. Neurosci Biobehav Rev 47:122–153

Schneider ER, Mastrotto M, Laursen WJ, Schulz VP, Goodman JB, Funk OH, Gallagher PG, Gracheva EO, Bagriantsev SN (2014) Neuronal mechanism for acute mechanosensitivity in tactile-foraging waterfowl. Proc Natl Acad Sci USA 111:14941–14946

Schnyder HA, Vanderelst D, Bartenstein S, Firzlaff U, Luksch H (2014) The avian head induces cues for sound localization in elevation. PLoS ONE 9:e112178

Schreiber RW, Schreiber EA, Anderson DW, Bradley DW (1989) Plumages and molts of Brown Pelicans. Natural History Museum of Los Angeles County Contributions to Science 402, Los Angeles, CA

Schultz JD, Schlinger BA (1999) Widespread accumulation of [H]testosterone in the spinal cord of a wild bird with an elaborate courtship display. Proc Natl Acad Sci USA 96:10428–10432

Schwab IR (2002) I'll try to accommodate you. Br J Ophthalmol 86:715

Schwab IR, Maggs D (2004) The falcon's stoop. Br J Ophthalmol 88:4

Schwilch R, Piersma T, Holmgren NMA, Jenni L (2002) Do migratory birds need a nap after a long non-stop flight? Ardea 90:149–154

Scriba MF, Ducrest A-L, Henry I, Vyssotski AL, Rattenborg NC, Roulin A (2013) Linking melanism to brain development: expression of a melanism-related gene in Barn Owl feather follicles covaries with sleep ontogeny. Front Zool 10:42

Seaman AR, Himelfarb TM (1963) Correlated ultrafine structural changes of the avian pecten oculi and ciliary body of *Gallus domesticus*: preliminary observations on the physiology I. Effects of decreased intraocular pressure induced by intravenous injection of acetazolamide (diamox). Am J Ophthalmol 56:278–296

Segovia Y, Victory N, Navarro-Sempere A, Pinilla V, García M (2020) A comparative ultrastructural study of the pecten oculi in adult, juvenile, and nestling Yellow-legged Gulls, *Larus michahellis* (Naumann, 1840). Vet Ophthalmol 23:113–122

Shamoun-Baranes J, van Loon E, Alon D, Alpert P, Yom-Tov Y, Leshem Y (2006) Is there a connection between weather at departure sites, onset of migration and timing of soaring-bird autumn migration in Israel? Glob Ecol Biogeogr 15:541–552

Shanahan M, Bingman VP, Shimizu T, Wild M, Güntürkün O (2013) Large-scale network organization in the avian forebrain: a connectivity matrix and theoretical analysis. Front Comput Neurosci 7:89

Shein-Idelson M, Ondracek JM, Liaw H-P, Reiter S, Laurent G (2016) Slow waves, sharp waves, ripples, and REM in sleeping dragons. Science 352:590–595

Shen JX, Xu ZM (1994) Response characteristics of Herbst corpuscles in the interosseous region of the pigeon's hind limb. J Comp Physiol A 175:667–674

Sherry D (1984) Food storage by Black-capped Chickadees: memory for the location and contents of caches. Anim Behav 32:451–464

Sherry DF, MacDougall-Shackleton SA (2015) Seasonal change in the avian hippocampus. Front Neuroendocrinol 37:158–167

Shlaer R (1972) An eagle's eye: quality of the retinal image. Science 176:920–922

Shultz S, Dunbar RIM (2010) Social bonds in birds are associated with brain size and contingent on the correlated evolution of life-history and increased parental investment. Biol J Linn Soc 100:111–123

Sillman AJ (1973) Avian vision. In: Farner DS, King JR (eds) Avian biology, vol 3. Academic Press, New York, NY, pp 349–387

Sipla JS (2007) The semicircular canals of birds and non-avian theropod dinosaurs. Ph.D. dissertation, Stony Brook University, Stony Brook, NY

Sivak JG, Bobier WR, Levy B (1978) The refractive significance of the nictitating membrane of the bird eye. J Comp Physiol A 125:335–339

Skelhorn J, Rowe C (2006) Avian predators taste-reject aposematic prey on the basis of their chemical defence. Biol Lett 2:348–350

Snyder AW, Miller WH (1978) Telephoto lens system of falconiform eyes. Nature 275:127–129

Sol D, Garcia N, Iwaniuk A, Davis K, Meade A, Boyle WA, Székely T (2010) Evolutionary divergence in brain size between migratory and resident birds. PLoS ONE 5:e9617

Soliman SA (2019) Identification of Merkel receptors sheet in quail beak. Cytol Histol Int J 3:000107

Soliman SA, Madkour FA (2017) A comparative analysis of the organization of the sensory units in the beak of duck and quail. Histol Cytol Embryol 1:1–16

Sørensen K, Neumann C, Dähne M, Hansen KA, Wahlberg M (2020) Gentoo Penguins (*Pygoscelis papua*) react to underwater sounds. R Soc Open Sci 7:191988

Stanback MT, Dove CM, Fonda C, Parkes P, Ptaschinski J (2019) Eastern Bluebirds (*Sialia sialis*) do not avoid nest cavities containing predator odors. Wilson J Ornithol 131:680–686

Stanchak KE, French C, Perkel DJ, Brunton BW (2020) The balance hypothesis for the avian lumbosacral organ and an exploration of its morphological variation. Integr Organismal Biol 2:obaa024

Stavenga DG, Wilts BD (2014) Oil droplets of bird eyes: microlenses acting as spectral filters. Proc R Soc B 369:20130041

Steinbach MJ, Angus RG, Money KE (1974) Torsional eye movements of the owl. Vis Res 14:745–746

Steinmeyer C, Schielzeth H, Mueller JC, Kempenaers B (2010) Variation in sleep behavior in free-living Blue Tits, *Cyanistes caeruleus*: effects of sex, age and environment. Anim Behav 80:853–864

Stephenson-Jones M, Floros O, Robertson B, Grillner S (2012) Evolutionary conservation of the habenular nuclei and their circuitry controlling the dopamine and 5-hydroxytryptophan (5-HT) systems. Proc Natl Acad Sci USA 109:E164–E173

Strick PL, Dum RP, Fiez JA (2009) Cerebellum and nonmotor function. Annu Rev Neurosci 32:413–434

Stryker MP (1999) Sensory maps on the move. Science 284:925–926

Suburo AM, Scolaro JA (1999) Environmental adaptations in the retina of the Magellanic Penguin: photoreceptors and outer plexiform layer. Waterbirds 22:111–119

Sugiyama T, Yamamoto H, Kon T, Chaya T, Omori Y, Suzuki Y, Abe K, Watanabe D, Furukawa T (2020)

The potential role of Arhgef33 RhoGEF in foveal development in the Zebra Finch retina. Sci Rep 10: 21450
Sultan F (2005) Why some bird brains are larger than others. Curr Biol 15:R649–R650
Svärdson G (1949) Competition and habitat selection in birds. Oikos 1:157–174
Swennen C, Yu Y (2004) Notes on feeding structures of the Black-faced Spoonbill *Platalea minor*. Ornithol Sci 3:119–124
Taylor AH, Hunt GR, Holzhaider JC, Gray RD (2007) Spontaneous metatool use by New Caledonian Crows. Curr Biol 17:1504–1507
Terakita A (2005) The opsins. Genome Biol 1:1–9
Thomassen HA (2005) Swift as sound. Design and evolution of the echolocation system in swiftlets (Apodidae, Collocaliini). Ph.D. dissertation, Universiteit, Leiden, The Netherlands
Thomassen HA, Djasim UM, Povel GDE (2004) Echoclick design in swiftlets: single as well as double clicks. Ibis 146:173–174
Thomassen HA, Gea S, Maas S, Bout RG, Dirckx JJJ, Decraemer WF, Povel GDE (2007) Do swiftlets have an ear for echolocation? The functional morphology of swiftlets' middle ears. Hear Res 225:25–37
Toda Y, Ko MC, Liang Q, Miller ET, Rico-Guevara A, Nakagita T, Sakakibara A, Uemura K, Sackton T, Hayakawa T, Sin SYW (2021) Early origin of sweet perception in the songbird radiation. Science 373:226–231
Toomey MB, Corbo JC (2017) Evolution, development and function of vertebrate cone oil droplets. Front Neural Circuit 11:97
Toomey MB, Lind O, Frederiksen R, Curley RW Jr, Riedl KM, Wilby D, Schwartz SJ, Witt CC, Harrison EH, Roberts NW, Vorobyev M, McGraw KJ, Cornwall MC, Kelber A, Corbo JC (2016) Complementary shifts in photoreceptor spectral tuning unlock the full adaptive potential of ultraviolet vision in birds. eLife 5: e15675
Trnka A, Prokop P, Grim T (2012) Uncovering dangerous cheats: how do avian hosts recognize adult brood parasites? PLoS ONE 7:e37445
Tsuboi M, van der Bijl W, Kopperud BT, Erritzøe J, Voje KL, Kotrschal A, Yopak KE, Collin SP, Iwaniuk AN, Kolm N (2018) Breakdown of brain–body allometry and the encephalization of birds and mammals. Nat Ecol Evol 2:1492–1500
Tucker V (2000a) Gliding flight: drag and torque of a hawk and falcon with straight and turned heads, and a lower value for the parasite drag coefficient. J Exp Biol 203:3733–3744
Tucker V (2000b) The deep fovea, sideways vision and spiral flight paths in raptors. J Exp Biol 203:3745–3754
Tyrrell LP, Fernández-Juricic E (2017) The hawk-eyed songbird: retinal morphology, eye shape, and visual fields of an aerial insectivore. Am Nat 189:709–717
Tyrrell LP, Goller B, Moore BA, Altshuler DL, Fernández-Juricic E (2018) The orientation of visual space from the perspective of hummingbirds. Front Neurosci 12:16
Tyrrell LP, Moore BA, Loftis C, Fernández-Juricic E (2013) Looking above the prairie: localized and upward acute vision is a native grassland bird. Sci Rep 3:3231
van den Hout PJ, Martin GR (2011) Extreme head-tilting in shorebirds: predator detection and sun avoidance. Wader Study Group Bull 118:18–21
van der Meij J, Ungurean G, Rattenborg NC, Beckers GJL (2020) Evolution of sleep in relation to memory–a birds' brain view. Curr Opin Behav Sci 33:78–85
Vander Wall SB (1990) Food hoarding in animals. University of Chicago Press, Chicago, IL
van der Willigen RF, Harmening WM, Vossen S, Wagner H (2010) Disparity sensitivity in man and owl: psychophysical evidence for equivalent perception of shape-from-stereo. J Vis 10:10–11
van Heezik YM, Gerritsen AFC, Swennen C (1983) The influence of chemoreception on the foraging behaviour of two species of sandpiper, *Calidris alba* and *Calidris alpina*. Neth J Sea Res 17:47–56
van Horik J, Clayton N, Emery N (2012) Convergent evolution of cognition in corvids, apes and other animals. In: Vonk J, Shackelford TK (eds) The Oxford handbook of comparative evolutionary psychology. Oxford University Press, New York, NY, pp 80–101
van Horik J, Emery NJ (2011) Evolution of cognition. Wiley Interdiscip Rev Cogn Sci 2:621–633
Van Huynh A, Rice AM (2019) Conspecific olfactory preferences and interspecific divergence in odor cues in a chickadee hybrid zone. Ecol Evol 9:9671–9683
Veit L, Neider A (2013) Abstract rule neurons in the endbrain support intelligent behavior in corvid songbirds. Nat Comm 4:2878
Vértes PE, Alexander-Bloch A, Bullmore ET (2014) Generative models of rich clubs in Hebbian neuronal networks and large-scale human brain networks. Philos Trans R Soc B 369:20130531
Vincze O (2016) Light enough to travel or wise enough to stay? Brain size evolution and migratory behavior in birds. Evolution 70:2123–2133
Vincze O, Vágási CI, Kovács I, Galván I, Pap PL (2013) Sources of variation in uropygial gland size in European birds. Biol J Linn Soc 110:543–563
Vincze O, Vágási CI, Pap PL, Osváth G, Møller AP (2015) Brain regions associated with visual cues are important for bird migration. Biol Lett 11:20150678
von Bartheld CS (1994) Functional morphology of the paratympanic organ in the middle ear of birds. Brain Behav Evol 44:61–73
von Bartheld CS, Giannessi F (2011) The paratympanic organ: a barometer and altimeter in the middle ear of birds? J Exp Zool B 316B:402–408
von Bayern AMP, Emery NJ (2009) Jackdaws respond to human attentional states and communicative cues in different contexts. Curr Biol 19:602–606

von Campenhausen M, Wagner H (2006) Influence of the facial ruff on the sound-receiving characteristics of the Barn Owl's ears. J Comp Physiol A 192:1073–1082
Voss J, Bischof H-J (2003) Regulation of ipsilateral visual information within the tectofugal visual system in Zebra Finches. J Comp Physiol A 189:545–553
Voss J, Bischof H-J (2009) Eye movements of laterally eyed birds are not independent. J Exp Biol 212:1568–1575
Waldvogel JA (1990) The bird's eye view. Am Sci 78: 342–353
Wallman J, Pettigrew JD (1985) Conjugate and disjunctive saccades in two avian species with contrasting oculomotor strategies. J Neurosci 5:1418–1428
Walls GL (1942) The vertebrate eye and its adaptive radiation. Harper Publishing, New York, NY
Walsh SA, Iwaniuk AN, Knoll MA, Bourdon E, Barrett PM, Milner AC, Nudds RL, Abel RL, Sterpaio PD (2013) Avian cerebellar floccular fossa size is not a proxy for flying ability in birds. PLoS ONE 8:e67176
Ward BJ, Day LB, Wilkening SR, Wylie DR, Saucier DM, Iwaniuk AN (2012) Hummingbirds have greatly enlarged hippocampal formation. Biol Lett 8:657–659
Warrant EJ (1999) Seeing better at night: life style, eye design and the optimum strategy of spatial and temporal summation. Vis Res 39:1611–1630
Weimerskirch H, Bishop C, Jeanniard-du-Dot T, Prudor A, Sachs G (2016) Frigate birds track atmospheric conditions over months-long transoceanic flights. Science 353:74–78
Wellmann AE, Downs CT (2009) A behavioural study of sleep patterns in the Malachite Sunbird, Cape White-eye and Fan-tailed Widowbird. Anim Behav 77:61–66
Wenzel BM (1987) The olfactory and related systems in birds. Ann N Y Acad Sci 519:137–149
West RJD (2014) The evolution of large brain size in birds is related to social, not genetic, monogamy. Biol J Linn Soc 111:668-678/
White CR, Day N, Butler PJ, Martin GR (2007) Vision and foraging in cormorants: more like herons than hawks? PLoS ONE 7:e639
Whittaker DJ, Hagelin JC (2021) Female-based patterns and social function in avian chemical communication. J Chem Ecol 47:43–62
Wiener F (1947) Sound diffraction by rigid spheres and circular cylinders. J Acoust Soc Am 6:444–451
Wikelski M, Quetting M, Cheng Y, Fiedler W, Flack A, Gagliardo A, Salas R, Zannoni N, Williams J (2021) Smell of green leaf volatiles attracts White Storks to freshly cut meadows. Sci Rep 11:12912
Wilby D, Toomey MB, Olsson P, Frederiksen R, Cornwall MC, Oulton R, Kelber A, Corbo JC, Roberts NW (2015) Optics of cone photoreceptors in the chicken (*Gallus gallus domesticus*). J R Soc Interface 12: 20150591
Wingfield J (2013) The comparative biology of environmental stress: behavioural endocrinology and variation in ability to cope with novel, changing environments. Anim Behav 85:1127–1133
Wolburg H, Liebner S, Reichenbach A, Gerhardt H (1999) The pecten oculi of the chicken: a model system for vascular differentiation and barrier maturation. Int Rev Cytol 187:111–159
Wood CA (1917) The fundus oculi of birds especially as viewed by the opthalmoscope: a study in comparative antomy and physiology. The Lakeside Press, Chicago, IL
Worthy M (2000) Animal eye colors. Authors Choice Press, Lincoln, NE
Wright TF, Cortopassi KA, Bradbury JW, Dooling RJ (2003) Hearing and vocalizations in the Orange-fronted Conure (*Aratinga canicularis*). J Comp Psychol 117:87–95
Wu Y (2019) Widespread nocturnality of living birds stemming from their common ancestor. BMC Evol Biol 19:189
Wu Y, Hadly EA, Teng W, Hao Y, Liang W, Liu Y, Wang H (2016) Retinal transcriptome sequencing sheds light on the adaptation to nocturnal and diurnal lifestyles in raptors. Sci Rep 6:33578
Wu Y, Wang H (2019) Convergent evolution of bird-mammal shared characteristics for adapting to nocturnality. Proc R Soc B 286:20182185
Wylie DR, Gutiérrez-Ibáñez C, Iwaniuk AN (2015) Integrating brain, behavior, and phylogeny to understand the evolution of sensory systems in birds. Front Neurosci 9:281
Yokosuka M, Hagiwara A, Saito TR, Tsukahara N, Aoyama M, Wakabayashi Y, Sugita S, Ichikawa M (2009) Histological properties of the nasal cavity and olfactory bulb of the Japanese Jungle Crow *Corvus macrorhynchos*. Chem Senses 34:581–593
Yopak KE, Pakan JMP, Wylie D (2020) The cerebellum of nonmammalian vertebrates. In: Kaas JH (ed) Evolutionary neuroscience, 2nd edn. Academic Press, London
Yu CQ, Schwab IR, Dubielzig RR (2009) Feeding the vertebrate retina from the Cambrian to the Tertiary. J Zool 278:259–269
Zahar Y, Lev-Ari T, Wagner H, Gutfreund Y (2018) Behavioral evidence and neural correlates of perceptual grouping by motion in the Barn Owl. J Neurosci 38:6653–6664
Zahar Y, Wagner H, Gutfreund Y (2012) Responses of tectal neurons to contrasting stimuli: an electrophysiological study in the Barn Owl. PLoS ONE 7:e39559
Zelenitsky DK, Therrien F, Ridgely RC, McGee AR, Witmer LM (2011) Evolution of olfaction in non-avian theropod dinosaurs and birds. Proc R Soc B 278:3625–3634
Zelano B, Edwards SV (2002) An Mhc component to kin recognition and mate choice in birds: predictions, progress, and prospects. Am Nat 160(Suppl):S225–S237
Zeyl JN, den Ouden O, Köppl C, Assink J, Christensen-Dalsgaard J, Patrick SC, Clusella-Trullas S (2020) Infrasonic hearing in birds: a review of audiometry and hypothesized structure–function relationships. Biol Rev 95:1036–1054

Zhao H, Li J, Zhang J (2015) Molecular evidence for the loss of three basic tastes in penguins. Curr Biol 25: R141–R142
Zhao H, Zhou Y, Pinto CM, Charles-Dominique P, Galindo-González J, Zhang S, Zhang J (2010) Evolution of the sweet taste receptor gene Tas1r2 in bats. Mol Biol Evol 27:2642–2650
Zimmerman M (1986) Physiological mechanisms of pain and its treatment. Klinische Anäesthesiologie und Intensivtherapie 32:1–19
Ziolkowski LH, Gracheva EO, Bagriantsev SN (2022) Tactile sensation in birds: physiological insights from avian mechanoreceptors. Curr Opin Neurobiol 74: 102548
Zueva L, Golubeva T, Korneeva E, Resto O, Inyushin M, Khmelinskii I, Makarov V (2020) Electron microscopy study of the central retinal fovea in Pied Flycatcher: evidence of a mechanism of light energy transmission through the retina. Heliyon 6:e04146
Zueva L, Makarov V, Zayas-Santiago A, Golubeva T, Korneeva E, Savvinov A, Eaton M, Skatchkov S, Inyushin M (2014) Muller cell alignment in bird fovea: possible role in vision. J Neurosci Neureng 3:85–91
Zusi RL, Bridge D (1981) On the slit pupil of the Black Skimmer (*Rynchops niger*). J Field Ornithol 52:338–340

# Locating, Obtaining, Ingesting, and Digesting Food

# 5

## Contents







G. Ritchison, *In a Class of Their Own*, Fascinating Life Sciences,
https://doi.org/10.1007/978-3-031-14852-1_5

**Abstract**

Birds consume a wide variety of food items that must be digested and absorbed. In this chapter, I provide detailed information about avian diets and the different avian dietary guilds, including insectivores, frugivores, invertivores, granivores, carnivores, scavengers, nectarivores, herbivores, and omnivores. The anatomy and physiology of the avian digestive system are also discussed in detail, with information about the interspecific variation in the anatomy and functions of each component of the digestive system, including bills, the esophagus, two-part stomach, small and large intestine ceca, cloaca, and accessory organs, including the pancreas and liver. Information about the phenotypic plasticity of the avian digestive system and regulation of food intake is also provided.

## 5.1 Introduction

Because of their high metabolic rates, birds must locate, consume, and digest more food in proportion to their size than most animals. For example, a warbler might eat 80% of its body weight in a day. To meet their metabolic needs while remaining as light as possible (to be efficient flyers), the digestive system of birds has to be both light and efficient, processing food and liquids rapidly as possible (Fig. 5.1). The need to keep weight low also means that, except perhaps prior to migration, there is a limit to the amount of fat a bird can store. As such, birds may spend considerable time and effort locating and ingesting food.

Birds forage high (Peregrine Falcons [*Falco peregrinus*] may initiate stoops from altitudes greater than 1500 m) and low (to ocean depths greater than 500 m by Emperor Penguins [*Aptenodytes forsteri*]) and consume items that vary in size from the diatoms and blue-green algae consumed by Greater Flamingos (*Phoenicopterus roseus*) to a 6.5-kg red howler monkey (*Alouatta seniculus*) killed by a Harpy Eagle (*Harpia harpyja*) in Peru (Peres 1990). Each foraging strategy requires specific morphological adaptations and sensory abilities (discussed in detail in Chap. 4). For example, some carrion specialists, like Turkey Vultures

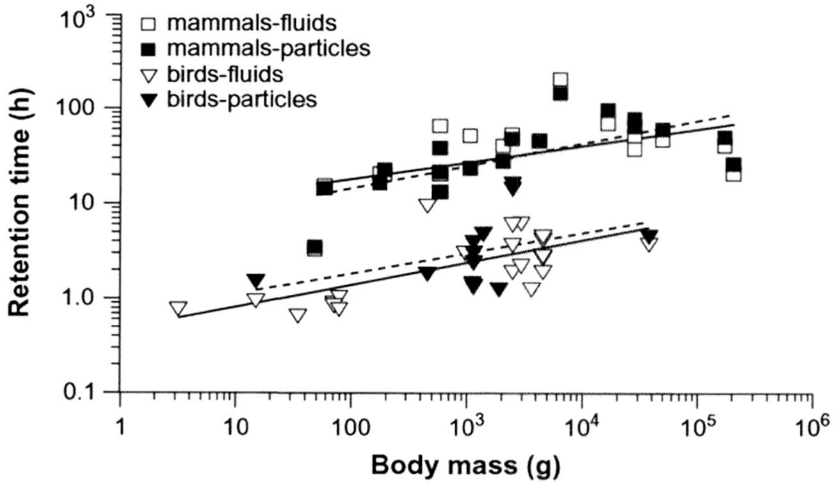


**Fig. 5.1** Retention times of fluids (solid lines) and particulate materials (dashed lines) markers in the gastrointestinal system of birds (triangles) are shorter than that of mammals (squares). Even with shorter retention times, the digestive efficiency (energy extracted relative to food intake) of birds is similar to that of mammals. (Figure from McWhorter et al. 2009; © 2009 John Wiley and Sons, used with permission)

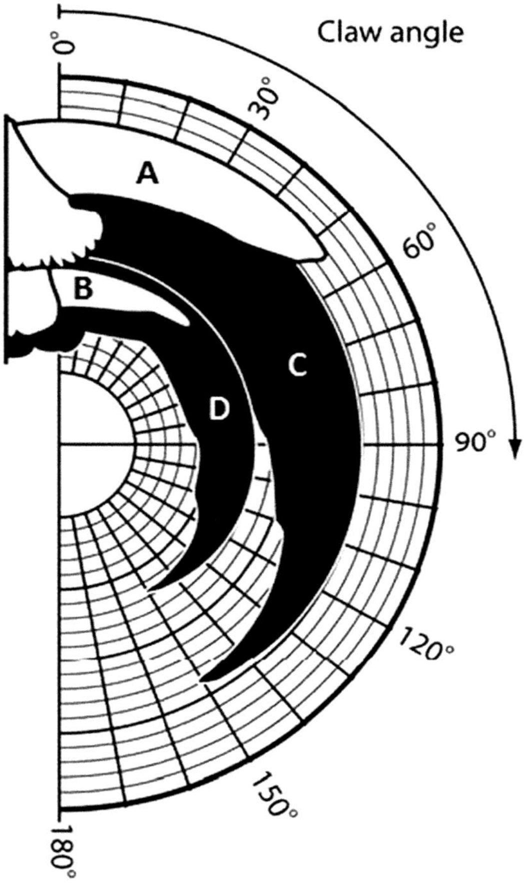


**Fig. 5.2** Examples of variation in claw or nail angles in birds. (**a**) Large ground-based bird (i.e., spends most of its time on the ground), (**b**) small ground-based bird, (**c**) large vertical-surface forager (i.e., forages on tree-trunks or other vertical surfaces), and (**d**) smaller vertical-surface forager. (Figure modified from Glen and Bennett 2007; © 2007 Elsevier Ltd., used with permission)

(*Cathartes aura*), have a well-developed sense of smell. Many foraging strategies require excellent vision and some require excellent hearing (some nocturnal hunters like Barn Owls [*Tyto alba*]) and tactile abilities (some shorebirds that feed by probing sand and mud with their bills). Morphological adaptations for efficient foraging range from the obvious (bill and foot morphology) to the less obvious (the asymmetrical ear openings of some owls). One characteristic that varies with foraging strategy and lifestyle is the curvature of claws (or talons). Climbing birds, such as woodpeckers, nuthatches, and creepers, have, on average, claws that exhibit the greatest degree of curvature (Pike and Maitland 2004). These birds, of course, rely on their curved claws to efficiently cling to vertical surfaces. Not surprisingly, predatory birds, including hawks and eagles, falcons, and owls, also have curved claws (although not, on average, as curved as those of climbing birds; Pike and Maitland 2004) and use them to capture, hold, and kill their prey. Perching birds that use their feet and claws to perch, but not to capture prey, have claws somewhat less curved than those of predatory birds, and ground-dwelling birds have claws with the least curvature (Pike and Maitland 2004). Longer, less-curved claws of ground dwellers, such as gallinaceous birds, larks, and pipits, distribute their weight over a wider area of contact with the ground, but are not so long that they hinder walking or running movements (Pike and Maitland 2004; Figs. 5.2 and 5.3).

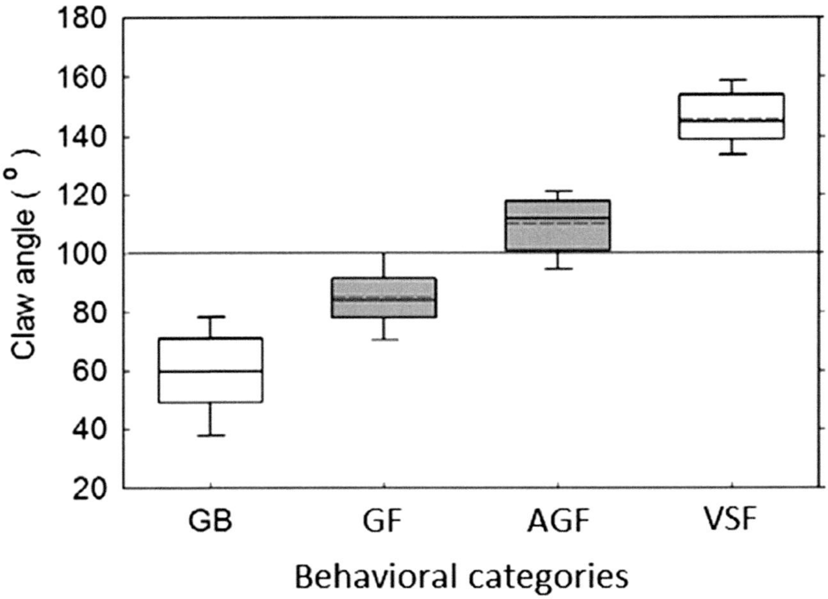


**Fig. 5.3** Relationship between foraging behavior and claw angle. GB, ground-based birds that spend most of their time on the ground like gallinaceous birds; GF, ground-foraging birds that forage primarily on the ground, but perch and roost in vegetation like some sparrows; AGF, above-ground foragers that forage and roost in vegetation like shrubs and trees like most warblers; VSF, vertical-surface foragers that forage on vertical surfaces like tree trunks like woodpeckers and nuthatches. (Figure modified from Glen and Bennett 2007; © 2007 Elsevier Ltd., used with permission)

## 5.2 Avian Diets

As a group, birds eat just about anything, including amphibians, ants, buds, crustaceans, fish, fruit, grass, insects, larvae, leaves, molluscs, nectar, pollen, other birds, reptiles, rodents, roots, sap, seeds, suet, snails, wax, and worms (Tables 5.1 and 5.2; Fig. 5.4) and use a variety of techniques to locate and obtain these various types of food (Table 5.3; Fig. 5.5; Box 5.1 Foraging behavior of waterfowl). The global distribution of birds with different food habits generally mirrors that for global bird species richness, with more species of each dietary guild in areas with more species of birds (Fig. 5.6). Latitudinal gradients in species richness are highest for terrestrial/arboreal insectivores, intermediate for frugivores, granivores, and carnivores, and lower for the other dietary guilds (herbivores, nectarivores, aerial insectivores, scavengers, and omnivores) (Fig. 5.6). Species richness of frugivores and insectivores is highest in areas with the highest net primary productivity and, therefore, more fruit and insects. Nectarivores were most abundant in areas with less seasonal variation in net primary productivity, which helps ensure a constant supply of flowers and nectar. Latitudinal gradients in species richness of granivores, carnivores, and omnivores are lower than for most other dietary guilds; their richness is influenced by both net primary productivity and habitat heterogeneity, factors that contribute to the diversity and abundance of seed-producing plants, vertebrate prey, and other food items. Scavenger richness is highest in east Africa, where the abundance of large ungulates ensures a constant supply of carrion. Herbivore species richness is relatively high in the Himalayas where there are many species of gallinaceous birds (grouse, quail, and pheasants) that feed on plants (Kissling et al. 2012).

**Table 5.1** Relative importance of dietary categories among avian orders ("P" indicates a primary item, "L" indicates less important item, and – indicates an item rarely or never eaten at the family level within orders). List based on Gill et al. (2021), Diets from Harris (2009) (Table modified from Wenny et al. 2011; © 2011 Oxford University Press, used with permission)

| Order | Families | Genera | Species | Terrestrial Invert | Terrestrial Vert | Aquatic Invert | Aquatic Vert | Carrion | Fruit | Seeds | Vegetation | Nectar |
|---|---|---|---|---|---|---|---|---|---|---|---|---|
| **Tinamiformes** (tinamous) | 1 | 9 | 46 | P | L | – | – | – | L | L | L | – |
| **Struthioniformes** (Common Ostrich, *Struthio camelus*) | 1 | 1 | 2 | L | – | – | – | – | L | – | P | – |
| **Rheiformes** (rheas) | 1 | 2 | 2 | L | L | – | – | – | – | – | P | – |
| **Casuariiformes** (cassowaries and Emu, *Dromaius novaehollandiae*) | 2 | 2 | 4 | P | L | – | – | L | P | L | L | – |
| **Apterygiformes** (kiwis) | 1 | 1 | 5 | P | – | – | – | – | – | – | – | – |
| **Galliformes** (grouse, ptarmigans, quail, partridges, pheasants, and curassows) | 5 | 83 | 300 | P | – | – | – | – | P | P | P | – |
| **Anseriformes** (geese, ducks, and mergansers) | 3 | 51 | 178 | P | – | P | P | – | – | P | P | – |
| **Gaviiformes** (loons) | 1 | 1 | 5 | – | – | L | P | – | – | – | – | – |
| **Sphenisciformes** (penguins) | 1 | 6 | 18 | – | – | P | P | – | – | – | – | – |
| **Procellariiformes** (albatrosses, petrels, shearwaters, and storm-petrels) | 4 | 26 | 134 | – | – | P | P | L | – | – | – | – |
| **Podicipediformes** (grebes) | 1 | 6 | 23 | – | – | P | P | – | – | – | – | – |
| **Phoenicopteriformes** (flamingoes) | 1 | 3 | 6 | – | – | P | – | – | – | – | – | – |
| **Phaethontiformes** (tropicbirds) | 1 | 1 | 3 | – | – | P | P | – | – | – | – | – |
| **Ciconiiformes** (storks) | 1 | 6 | 19 | – | – | P | P | P | – | – | – | – |
| **Pelecaniformes** (pelicans, ibises, spoonbills, herons, and egrets) | 5 | 35 | 111 | L | L | P | P | L | – | – | L | – |
| **Suliformes** (frigatebirds, boobies, gannets, and cormorants) | 4 | 8 | 55 | – | – | P | P | L | – | – | – | – |
| **Accipitriformes** (hawks, eagles, harriers, Osprey [*Pandion haliaetus*], condors, and vultures) | 4 | 72 | 266 | P | P | – | P | P | L | – | – | – |
| **Otidiformes** (bustards) | 1 | 11 | 26 | P | – | – | – | – | – | – | P | – |
| **Mesitornithiformes** (mesites) | 1 | 2 | 3 | P | – | – | – | – | L | L | – | – |
| **Eurypygiformes** (Kagu, *Rhynochetos jubatus*, and Sunbittern, *Eurypyga helias*) | 2 | 2 | 2 | P | L | L | P | – | – | – | – | – |
| **Gruiformes** (rails, coots, cranes, and trumpeters) | 6 | 42 | 189 | L | L | P | L | – | L | – | P | – |
| **Charadriiformes** (plovers, sandpipers, gulls, terns, skuas, and auks) | 19 | 94 | 379 | P | L | P | P | L | L | P | L | – |
| **Pteroclidoformes** (Sandgrouse) | 1 | 2 | 16 | L | – | – | – | – | – | P | – | – |

(continued)

**Table 5.1** (continued)

| Order | Families | Genera | Species | Terrestrial | | Aquatic | | Carrion | Fruit | Seeds | Vegetation | Nectar |
|---|---|---|---|---|---|---|---|---|---|---|---|---|
| | | | | Invert | Vert | Invert | Vert | | | | | |
| **Columbiformes** (doves and pigeons) | 1 | 42 | 344 | L | – | – | – | – | P | P | – | – |
| **Opisthocomiformes** (Hoatzin, *Opisthocomus hoazin*) | 1 | 1 | 1 | – | – | – | – | – | L | – | P | – |
| **Musophagiformes** (turacos) | 1 | 6 | 23 | L | – | – | – | – | P | – | L | – |
| **Cuculiformes** (cuckoos and roadrunners) | 1 | 32 | 149 | P | L | – | – | – | L | – | – | – |
| **Strigiformes** (owls) | 2 | 27 | 248 | P | P | – | L | – | – | – | – | – |
| **Caprimulgiformes** (nightjars, frogmouths, and potoos) | 4 | 21 | 117 | P | L | – | – | – | P | – | – | – |
| **Apodiformes** (treeswifts, swifts and hummingbirds) | 4 | 128 | 480 | P | – | – | – | – | – | – | – | P |
| **Coliiformes** (mousebirds) | 1 | 2 | 6 | – | – | – | – | – | P | – | P | – |
| **Trogoniformes** (trogons and quetzels) | 1 | 7 | 43 | P | – | – | – | – | P | – | – | – |
| **Leptosomiformes** (Cuckoo-Roller, *Leptosomus discolor*) | 1 | 1 | 1 | P | L | – | – | – | – | – | – | – |
| **Coraciiformes** (roller, ground rollers, kingfishers, bee-eaters, motmots, and todies) | 6 | 35 | 178 | P | L | P | P | – | L | – | – | – |
| **Bucerotiformes** (hornbills and hoopoes) | 4 | 17 | 74 | P | L | – | – | L | P | L | L | – |
| **Piciformes**[a] (woodpeckers and allies) | 9 | 67 | 445 | P | L | – | – | – | P | L | – | – |
| **Cariamiformes** (seriemas) | 1 | 2 | 2 | P | P | | | | | L | | |
| **Falconiformes** (falcons and caracaras) | 1 | 13 | 66 | P | P | | | L | | | | |
| **Psittaciformes** (parrots and cockatoos) | 4 | 92 | 398 | L | – | – | – | – | P | P | L | L |
| **Passeriformes** (songbirds) | 120 | 1278 | 6237 | P | L | L | – | L | P | P | L | P |

[a]Exceptionally, some honeyguides (family Indicatoridae) also feed on beeswax

**Table 5.2** Foraging guilds of North American birds. Many species are omnivores so are listed under more than one category (Source: DeGraaf et al. 1985, © 1985 Springer-Verlag New York Inc., used with permission)

| Category | Subcategory | Number of species | Examples |
|---|---|---|---|
| Carnivore (vertebrates) | Ground hawker | 46 | Hawks, kites, eagles, owls, shrikes, jaegers, skuas |
| | Air hawker | 11 | Accipiters, falcons, jaegers |
| | Ground chaser | 1 | Greater Roadrunner (*Geococcyx californianus*) |
| | Ground scavenger | 5 | Vultures, condors, caracaras, Parasitic Jaeger (*Stercorarius parasiticus*) |
| | Coastal beach scavenger | 13 | Gulls |
| | Coastal food pirate | 2 | Parasitic Jaeger, Great Skua |
| Crustaceovore (crustaceans) | Coastal beach/ rock gleaner | 7 | Plovers, turnstones, Surfbird (*Calidris virgate*), Purple Sandpiper (*Calidris maritima*), Rock Sandpiper (*Calidris ptilocnemis*) |
| | Coastal beach prober | 7 | Willet (*Tringa semipalmata*), Whimbrel (*Numenius phaeopus*), godwits, Sanderling (*Calidris alba*), Long-billed Curlew (*Numenius americanus*) |
| | Shallow water visual forager | ~10 | Phalaropes and other shorebirds with thin bills |
| | Marsh/mud prober | 8 | Ibises, King Rail (*Rallus elegans*) |
| | Mud strainer | 1 | Roseate Spoonbill (*Platalea ajaja*) |
| | Water ambusher | 9 | Herons, egrets, Wood Stork (*Mycteria americana*) |
| | Coastal bottom gleaner | 8 | Scoters, eiders, Harlequin Duck (*Histrionicus histrionicus*), Long-tailed Duck (*Clangula hyemalis*) |
| | Diver | 17 | Murrelets, auklets, Dovekie (*Alle alle*), Common Murre (*Uria aalge*), grebes |
| Frugivore (fruits) | Gleaners | 27 | Some pigeons, some woodpeckers, waxwings, a few flycatchers (during nonbreeding period) |
| Granivore (seeds, nuts) | Ground gleaner | 54 | Sparrows, doves, finches, quail, redpolls |
| | Shrub/thistle gleaner | 3 | Goldfinches |
| | Canopy gleaner | 13 | Woodpeckers, nuthatches, grosbeaks, crossbills, finches |
| Herbivore (grass, leaves, stems, roots) | Ground grazer/ forager | 23 | Geese, ptarmigans, quail |
| | Canopy forager | 5 | Grouse, Plain Chachalaca (*Ortalis vetula*) |
| | Water dabbler | 7 | Swans, Canada Goose (*Branta canadensis*), Gadwall (*Mareca strepera*), American Coot (*Fulica americana*) |
| Insectivore (insects) | Air sallier/ hawker | 49 | Flycatchers, hummingbirds, some warblers |
| | Air courser/ screener | 20 | Swallows, swifts, nighthawks, Chuck-will's-widow (*Antrostomus carolinensis*), Eastern Whip-poor-will (*Antrostomus vociferous*) |
| | Bark excavator/ gleaner/scaler | 30 | Woodpeckers, nuthatches, Brown Creeper (*Certhia americana*), some warblers |
| | Coastal beach/ rock gleaner | 15 | Plovers, some sandpipers, dowitchers |
| | Ground gleaner | 41 | Some warblers, pipits, wrens, some sparrows, meadowlarks, anis, some gulls, flickers |
| | Ground hawker | 10 | Small owls, kestrels, bluebirds |

(continued)

**Table 5.2** (continued)

| Category | Subcategory | Number of species | Examples |
|---|---|---|---|
| | Shrub/lower canopy gleaner | 53 | Many warblers, kinglets, titmice, chickadees, wrens |
| | Upper canopy gleaner | 26 | Many warblers, vireos, tanagers |
| Molluscovore (molluscs) | Coastal beach/ rock gleaner | 9 | Black-bellied Plover (*Pluvialis squatarola*), Semipalmated Plover (*Charadrius semipalatus*), Western Sandpiper (*Calidris mauri*) |
| | Coastal beach/ marsh, mud prober | 5 | Ibises, rails, several shorebirds |
| | Coastal bottom gleaner | 11 | Eiders, scoters, Harlequin Duck (*Histrionicus histrionicus*), Long-tailed Duck |
| | Pelagic diver | 2 | Razorbill (*Alca torda*), Pigeon Guillemot (*Cepphus columba*) |
| Nectarivore (nectar) | Floral hoverer/ feeder | 15 | Hummingbirds, sunbirds, honeyguides, flowerpeckers |
| Piscivore (fish) | Diver | 31 | Loons, grebes, cormorants, mergansers, murres |
| | Plunger | 23 | Boobies, Northern Gannet (*Morus bassanus*), Brown Pelican (*Pelecanus occidentalis*), Osprey (*Pandion haliaetus*), terns, kingfishers |
| | Surface gleaner | 4 | American White Pelican (*Pelecanus erythrorhynchos*), frigatebirds, Brown Noddy (*Anous stolidus*), Razorbill (*Alca torda*) |
| | Scavenger | 4 | Shearwaters |
| Vermivore (sandworms, earthworms) | Coastal beach gleaner/prober | 6 | Plovers, godwits, Short-billed Dowitcher (*Limnodromus griseus*), Dunlin (*Calidris alpina*), Western Sandpiper (*Calidris mauri*) |
| | Ground gleaner | 1 | American Robin (*Turdus migratorius*) |
| | Ground prober | 4 | Woodcocks, some shorebirds |

The diets of birds are also influenced by altitude. Analysis of data from 46 mountain regions revealed that the diets of birds varied little up to elevations of about 3500 m, with most species feeding primarily on invertebrates (Fig. 5.7a). At higher elevations, species of birds that feed carrion and seeds become more dominant, particularly in mountain ranges located at higher latitudes; at those locations, other food items become less available and species that feed on those items typically migrate to lower altitudes or latitudes (Fig. 5.7a). Overall, invertebrates and species of birds that feed on them become less abundant. However, in lower elevation mountains located at lower latitudes, e.g., the Sumatran Islands and Cameroon Mountains, the availability of invertebrates and abundance of invertebrate-eating species of birds do not decline with increasing altitude (Fig. 5.7b).

Some birds have highly specialized diets. For example, Snail Kites (*Rostrhamus sociabilis*) feed mainly on freshwater snails in the genus *Pomacea* and primarily (until about 2009) on one species (apple snail, *Pomacea paludosa*; Reichert et al. 2020). Beginning in about 2009, Snail Kites in Florida also began feeding on a non-native species of apple snail (*P. maculata*; Poli et al. 2022). At the other extreme are Ruffed Grouse (*Bonasa umbellus*) that may consume over 300 different species of plants and 100 different species of animals (Klasing 1998). Many birds, of course, fall somewhere between these extremes, and the diet of some birds changes during the year. For example, on their breeding

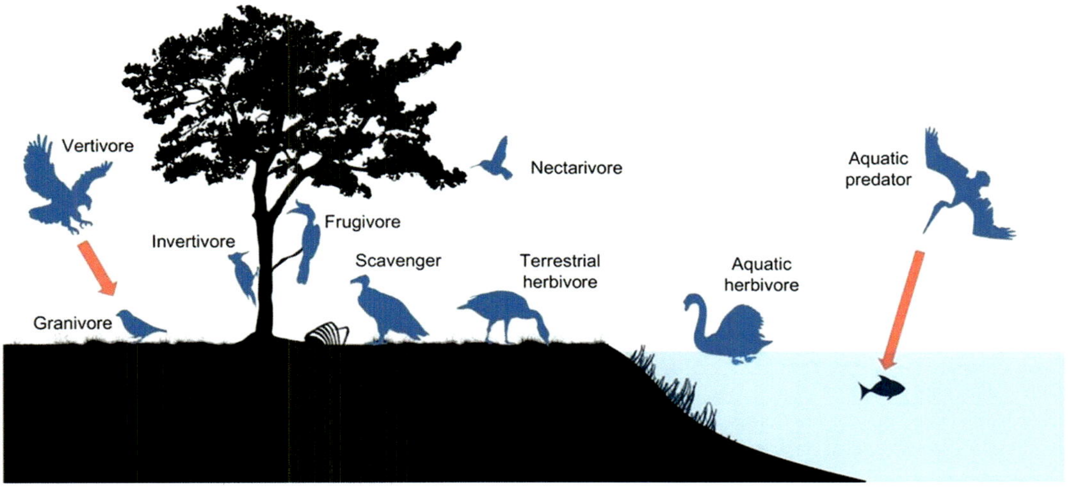


**Fig. 5.4** Major foraging guilds of birds. Vertivores prey primarily on vertebrates and invertivores on invertebrates, granivores are primarily seed eaters, frugivores feed primarily on fruit, scavengers feed on carrion, nectarivores primarily on nectar, herbivores on plants, and aquatic predators on a wide variety of aquatic prey. (Figure from Pigot et al. 2020; © 2020 Springer Nature, used with permission)

grounds, Hermit Thrushes (*Catharus guttatus*) feed primarily on insects and other small invertebrates; during migration and the winter, their diet is supplemented by a wide variety of fruits (Dellinger et al. 2020).

### 5.2.1 Insectivores

Insects are a common bird food (Table 5.2; Fig. 5.8; Box 5.3 Ecosystem services provided by avian insectivores), with many birds insectivorous year-round and others primarily insectivorous during warmer periods and supplementing their diet to varying degrees with fruit or seeds during colder periods. Insects are an excellent source of protein and, to a lesser degree, lipids (fats), and are also good sources of minerals and vitamins (Table 5.4). With the exception of the chitinous exoskeletion, insects are also rather easily digested (Klasing 1998). Chitin content varies considerably among insects (ranging from about 20% to 60% of body mass), birds deal with this largely (but not entirely; Jackson et al. 1992; Weiser et al. 1997) undigestible substance in a variety of ways. Birds may (1) select insects (and insect larvae) with less chitin, (2) remove body parts with lots of chitin (e.g., wings and legs) before ingesting insects, and (3) consume insects whole, but regurgitate the undigested chitin (Klasing 1998).

Among passerines, 46% of all families are primarily insectivorous, and nearly 80% feed at least to some degree on insects (Morse 1975). Although typically referred to as insectivores, many insectivorous birds eat arthropods other than insects (Class Insecta). Particularly favored as arthropod/insect prey by many insectivorous birds are those in the orders Coleoptera (beetles and their larvae), Lepidoptera (moths and butterflies and their larvae), and Orthoptera (grasshoppers, crickets, katydids, and locusts) and those in the order Araneae (spiders). These insects, along with spiders, are excellent sources of proteins and lipids (Razeng and Watson 2015).

Birds capture insects in a variety of different ways. Aerial insectivores fly rapidly at variable heights above ground vegetation or water and capture insects in their mouths (Fig. 5.9). Some do so during the day, including swifts and swallows, whereas others capture flying insects at night, including nightjars and poor-wills (Caprimulgidae; Fig. 5.10). Many other birds remain perched while searching for insects,

**Table 5.3** Some ways by which birds locate and obtain (or capture) food items[a]

| Method | Example |
|---|---|
| Ambush (or sit-and-wait)—sit or perch and wait for prey to appear | Many hawks and owls, shrikes |
| Chase—pursue prey on foot | Roadrunners |
| Cooperative hunting—several individuals work together to capture prey | Harris's Hawk (*Parabuteo unicinctus*), ravens, falcons[b] |
| Create sapwells—create shallow wells in trees that fill with sap | Sapsuckers (woodpeckers) |
| Create whirlpool—swim in circle, whirlpool lifts and concentrates prey | Phalaropes |
| Dabble—submerge head and neck in shallow water | Some ducks |
| Disturb and chase—run in shallow water, wing flick, strike disturbed fish | Reddish Egret (*Egretta rufescens*) |
| Dive—dive under water | Many ducks and grebes |
| Drop hard-shelled prey or food times | Crows, ravens, and raptors |
| Eat feces—eat the feces of other animals | Egyptian Vulture (*Neophron percnopterus*)[c] |
| Excavate—drill holes in substrate | Woodpeckers |
| Filter-feeding (or strainers)—straining small plants or animals from water or mud | Flamingoes and some ducks |
| Flight forage—search for flying insects while flying | Swallows and swifts |
| Flushing foragers—capture prey flushed by foraging army ants/large grazers | Antbirds, Cattle Egret (*Bubulcus ibis*), and cowbirds |
| Food "parasite"—steal food from other birds (kleptoparasitism) | Reported in 197 species (See Box 5.2 Kleptoparasites) |
| Fool flockmates—give "false" alarm calls to gain access to flushed insects | Bluish-slate Antshrike (*Thamnomanes schistogynus*) |
| Foot-paddling—rapidly stomping feet on ground bringing worms to surface | Some gulls[d] |
| Foot plunge—dive into water feet first, capture prey with talons | Osprey (*Pandion haliaetus*) |
| Glean—select food items from surface of vegetation substrate | Many warblers |
| Graze—feed on grass or sedges | Geese |
| Grub—dig up roots or tubers | Mute Swan (*Cygnus olor*) |
| Hawk (or sally)—fly from perch to capture flying insect | Flycatchers |
| Hover-glean—hover in air and take prey from substrate | Some warblers |
| Inter-class cooperation—using animals in other classes to help locate prey | Many seabirds search for prey along with marine mammals |
| Large mammal "parasite"—feed on ectoparasites/blood of large mammals | Oxpeckers |
| Leaf eating—eat leaves of trees and shrubs | Hoatzin (*Opisthocomus hoazin*) |
| Observe others—monitor behavior of other birds | Vultures |
| Olfaction—fly toward odors (e.g., DMS) indicating prey availability | Many seabirds and Turkey Vulture (*Cathartes aura*) |
| Pluck—removing seeds or fruits from plants | Many sparrows, doves, and waxwings |
| Plunge—plunge head first into water from above the surface | Terns, kingfishers, and gannets |
| Probe—insert bill into substrate (e.g., sand or mud) | Many shorebirds |
| Probe and smell—insert bill into mud, detect prey by smell | Kiwis |
| Scale—remove bark of trees to locate prey | Woodpeckers and nuthatches |
| Scoop—"scoop" fish with pouch | Some pelicans |
| Skim—fly along surface of water with lower bill in water | Skimmers |
| Sound localizing—focus attack on source sounds made by unseen prey | Great Gray Owl (*Strix nebulosi*) (e.g., rodents under snow) |
| Startle—flash black-and-white tails, startle insects and pursue | *Myioborus* warblers |

(continued)

**Table 5.3** (continued)

| Method | Example |
|---|---|
| Stoop—dive at prey at high speed from higher altitude | Peregrine Falcon (*Falco peregrinus*) |
| Strain—filter small food items from water | Flamingoes and some waterfowl |
| Surface tension transport—move drop of water with prey into mouth by repeatedly opening and closing bill | Phalaropes and other shorebirds with thin bills |
| Tool use—use "tools" like sticks to obtain prey | New Caledonian Crow (*Corvus moneduloides*), Woodpecker Finch (*Camarhynchus pallidus*) |
| Trap-line—visit known food sources on a regular schedule | Some hummingbirds |
| Turn stones—turn over stones looking for invertebrates | Ruddy Turnstone (*Arenaria interpres*) |
| Underwater pursuit—pursue prey propelled by feet or flippers | Penguins, loons, and cormorants |
| Upending—submerge most, but not all, of the body in water; tail pointing upward | Many waterfowl |
| Use bait—attract prey using "bait" (e.g., food particles) | Green Heron (*Butorides virescens*) |
| Water-surface gleaning—pick food items from water's surface | Frigatebirds and some petrels |

[a]List includes several methods from list in Klasing (1998)
[b]Bednarz (1988), Yosef and Yoset (2010)
[c]Negro et al. (2002)
[d]Tinbergen (1962), Mitra et al. (2009)

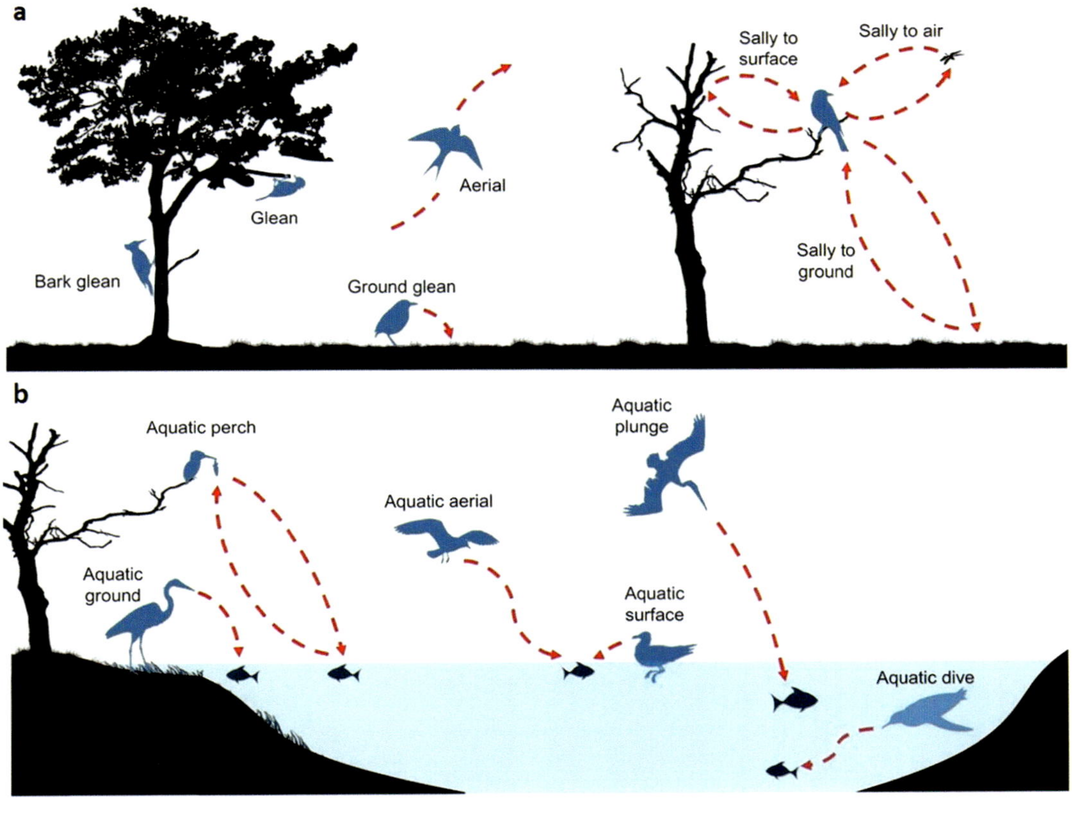


**Fig. 5.5** (**a**) Seven major foraging niches used by terrestrial invertivores and (**b**) six major foraging niches used by aquatic predators. (Figure from Pigot et al. 2020; © 2020 Spring Nature, used with permission)

**Box 5.1 Foraging behavior of waterfowl**

Waterfowl (Anseriformes) feed on a wide variety of items, including grass, aquatic vegetation, tubers, invertebrates, and fish (Johnsgard 1978, 2010). Differences in diets are associated with differences in feeding behavior and morphology, including differences in skull and bill morphology (Kurk 2008). Some larger species of geese forage by grubbing (i.e., digging for roots and tubers), whereas small and medium-sized geese feed by grazing. Waterfowl that forage in shallow water may feed by dabbling (i.e., submerging the head and neck in shallow water); in somewhat deeper water, most of the body may be submerged and tails point upwards (upending). Other waterfowl dive under water to forage on submerged vegetation or pursue prey such as fish.

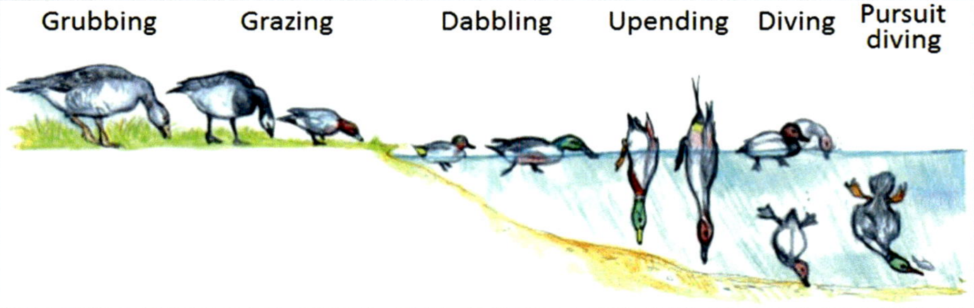


Feeding methods of the waterfowl: grubbing, e.g., Graylag Goose (*Anser anser*), grazing, e.g., Barnacle Goose (*Branta leucopsis*) and Eurasian Wigeon (*Mareca penelope*), dabbling, e.g., Eurasian Teal (*Anas crecca*) and Northern Shoveler (*Anas clypeata*), upending, e.g., Mallard (*Anas platyrhynchos*) and Northern Pintail (*Anas acuta*), diving, e.g., Common Pochard (*Aythya ferina*), and pursuit diving, e.g., Red-breasted Merganser (*Mergus serrator*) (Figure modified from Pecsics et al. 2017; used with permission of Tibor Pecsics)

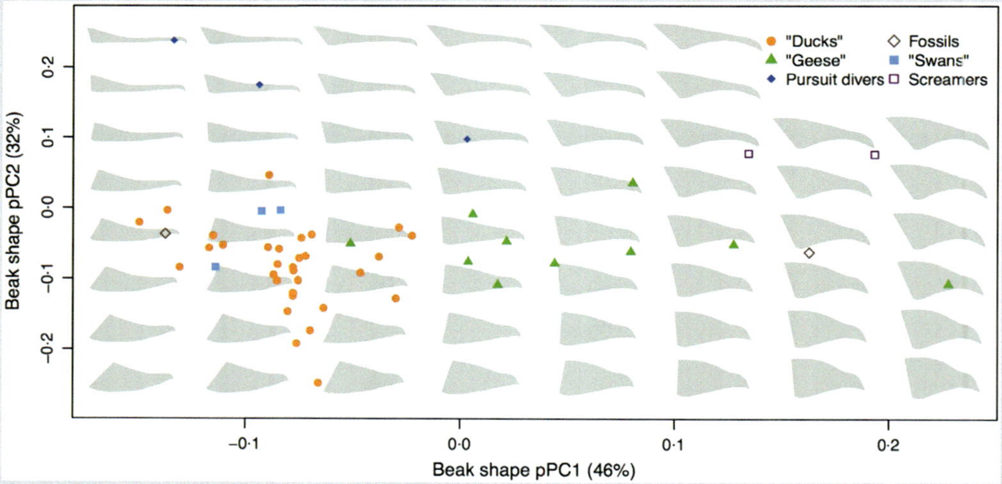


Low values along the pPC1 axis correspond to more duck-like beaks, with a ventrally arcing culmen and tomium and being relatively longer and shorter in height, whereas high pPC1 values correspond to more goose-like beaks with a dorsally arcing culmen and tomium and being relatively shorter and taller and wider at the base. Low values along the pPC2 axis correspond to the tall, broad beaks of browsing divers, such as steamer ducks, and high values correspond to the slender beaks of piscivorous pursuit divers like mergansers. The beak of an anseriform fossil (*Presbyornis*) suggests that they were filter-feeders, as are many present-day ducks. An increasing dependence on

(continued)

**Box 5.1** (continued)

a diet of grass and leaves led to the evolution of the goose-like beak because more goose-like beaks can generate greater bite force at the beak tip to crop plants (Figure modified from Olsen 2017; © 2017 The Author. Functional Ecology © 2017 British Ecological Society, used with permission)

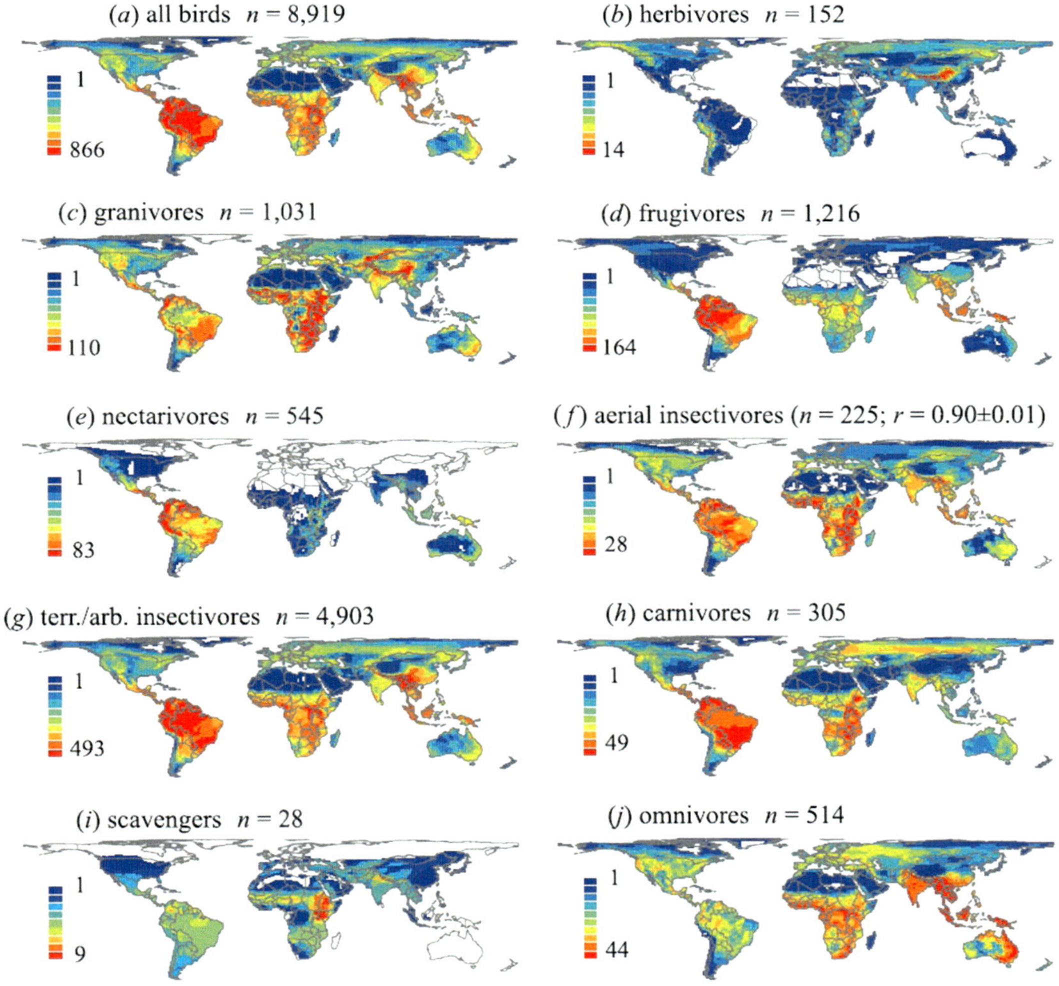


**Fig. 5.6** Geographic patterns of (**a**) overall bird species richness and (**b**)–(**j**) species richness of various dietary guilds. Note the impressive number of species that feed primarily on insects (terrestrial/arboreal insectivores plus aerial insectivores = 5126 species). (Figure modified from Kissling et al. 2012; © John Wiley and Sons, used with permission)

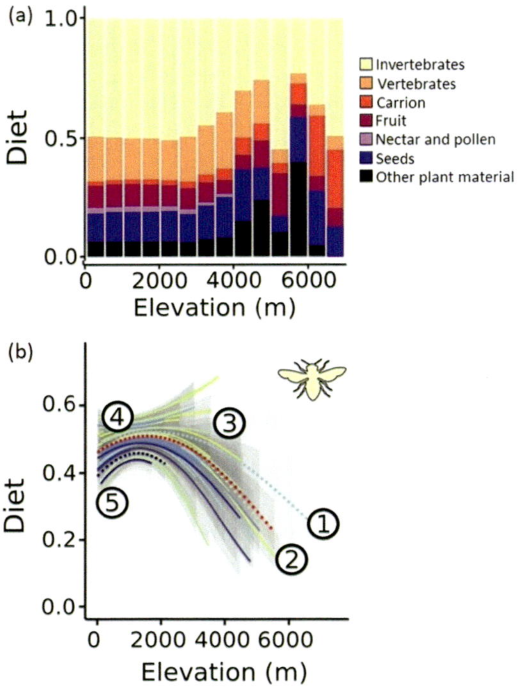


**Fig. 5.7** (**a**) Based on data from 46 mountain regions, the diets of birds exhibit little variation up to altitudes of about 3500 m, with invertebrates being the dominant food item. At higher elevations, invertebrates generally become less important in the diets of birds, with carrion and seeds becoming more important, particularly at high elevations in mountains located at higher latitudes. (**b**) The importance of invertebrates in the diets of birds declines dramatically with increasing elevation in higher elevation mountains such as the Hindukush-Himalaya and Andes mountains. Invertebrates remain an important food item for birds in lower elevation mountains, particular mountains located at lower latitudes such as the Sumatran Islands and Cameroon Mountains. Locations indicated by the numbers are the (1) Hindukush-Himalaya, (2) the Andes, (3) the Sumatran Islands, (4) the Cameroon Mountains, and (5) the Scandinavian Mountains. (Figure from Jarzyna et al. 2021; © 2020 John Wiley & Sons Ltd., used with permission)

make relatively short flights to capture them in the air (flycatching) or from a leaf or twig (hover-gleaning), then fly back to the same or another perch. Examples of birds that capture insects by flycatching or hover-gleaning include flycatchers (Tyrannidae), drongos (Dicruridae), bee-eaters (Meropidae), and jacamars (Galbulidae). Some birds do their flycatching at night, including frogmouths (Podargidae) and potoos (Nyctibiidae). By using different foraging methods and, for insectivorous birds in forest ecosystems, foraging at different heights, multiple species can coexist in the same habitat (Fig. 5.11).

Many insects are associated with the bark of trees and many birds either capture insects on the bark and in bark crevices (gleaners; Fig. 5.12) or actively search for insects by dislodging and removing bark and drilling into the wood. The best-known bark specialists are the woodpeckers (Picidae) and, among the over 200 species, some pick insects from the bark surface (gleaning), some remove bark and drill, and most use all of these techniques to capture insects. Woodpeckers use their strong feet and strongly curved claws to grasp the bark, chisel-like bills to drill in wood (Figs. 5.13 and 5.14; Box 5.4 How do woodpeckers avoid head impact injury?), and their bills and long, extensible tongues to capture insects. Other bark insectivores that glean and remove bark to varying degrees, but do not drill, include some warblers (Parulidae), nuthatches (Sittidae), creepers (Certhiidae), and woodcreepers (Dendrocolaptidae).

The largest category of insectivorous birds includes those that typically move through vegetation and capture insects on the surface of leaves, twigs, branches (gleaning insectivores). Approximately one-quarter of all bird species fall into this category, including the wood-warblers (Parulidae), vireos (Vireonidae), wrens (Troglodytidae), antbirds and allies (Thamnophilidae), old world flycatchers (Muscicapidae), old world warblers and allies (Sylviidae), and many others. Most gleaning insectivores are relatively small birds and have rather pointed, forceps-like bills for capturing insects. A final category of insectivores includes birds that typically walk or run along the ground to capture insects, including pipits and wagtails (Motacillidae), larks (Alaudidae), and many starlings (Sturnidae).

Although the bills of avian insectivores tend to be pointed and forceps-like with a small hook at the tip of the upper bill, bill morphology may, with some overlap, vary based on differences in foraging behavior. For example, among tyrannid

**Box 5.2 Kleptoparasites**

Kleptoparasitism is the stealing of food from another individual who already has the food item, often from an individual of another species (e.g., birds feeding on insects trapped in spider webs; Young 1971; Schronce and Davis 2022), but sometimes from conspecifics. Kleptoparasitism is particularly common among waterfowl (Anatidae), gulls and terns (Laridae), falcons (Falconidae), hawks and eagles (Accipitridae), storks (Ciconiidae), and frigatebirds (Fregatidae).

Great Frigatebirds (*Fregata minor*) chasing a Red-footed Booby (*Sula sula*) in an attempt to steal its prey. Chases by more than one frigatebird are more successful than chases by a single frigatebird (Gilardi 1994) (Photo by Duncan Wright, U. S. Fish and Wildlife Service, CC0 Public Domain)

Morand-Ferron et al. (2007) examined characteristics of birds and their food habits to determine which one might be important in the evolution of kleptoparasitism and found that families of birds where kleptoparasitic behavior was most common had good cognitive abilities (large brains relative to their body size), were carnivores (feeding on vertebrate prey), and typically occupied open habitats. Larger brains might allow improved perception and motor skills and may also be important in allowing birds to learn how best to steal food from other individuals. For example, young Black-headed Gulls (*Larus ridibundus*) are less adept at kleptoparasitism than adults (Hesp and Barnard 1989).

Vertebrate prey are typically relatively large (i.e., more energy) and can be difficult to locate and capture, features that can make kleptoparasitism a potentially important (and energetically profitable) behavior. Finally, visibility is better is open habitats, making it more likely that other individuals will be observed capturing prey (Morand-Ferron et al. 2007).

(continued)

**Box 5.2** (continued)

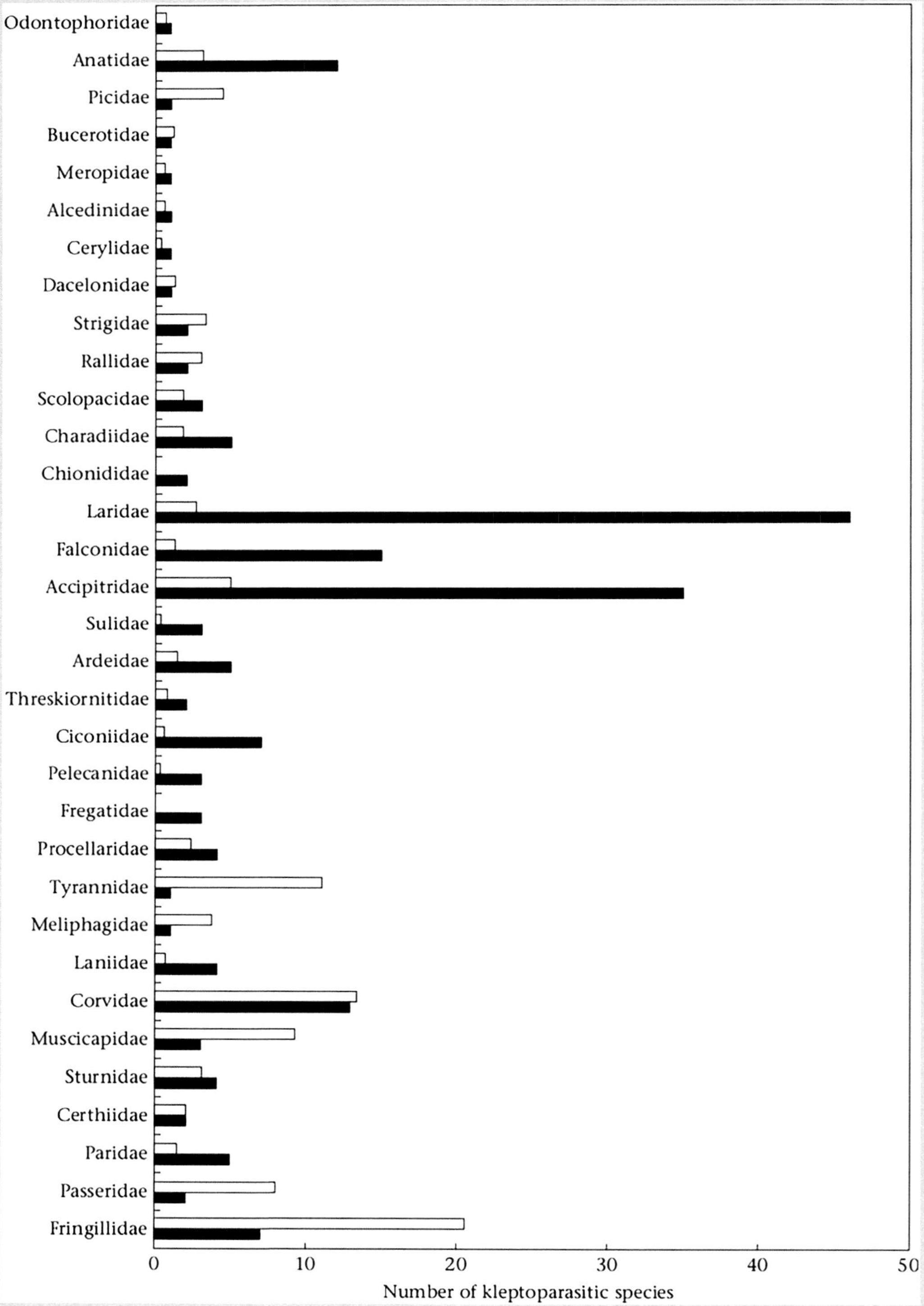


(continued)

**Box 5.2** (continued)

Kleptoparasitic species (N = 197) have been reported in 33 families of birds and is particularly common among species in the families Laridae, Falconidae, and Accipitridae. Expected frequencies are based on the total number of species per family (total number of species per family multiplied by 2.0%, the average proportion of kleptoparasitic species in the class Aves) (Figure from Morand-Ferron et al. 2007; © 2007 Elsevier Ltd., used with permission)

A Western Gull (*Larus occidentalis*) chasing an Elegant Tern (*Thalasseus elegans*) in an attempt to steal the tern's prey item (Photo by Kevin Cole, Wikipedia, CC BY 2.0 https://creativecommons.org/licenses/by/2.0/)

Among the more interesting examples of kleptoparasitism, Fork-tailed Drongos (*Dicrurus adsimilis*) act as sentinels in mixed-species foraging associations in sub-Saharan Africa, giving alarm calls when potential predators are detected (Ridley and Raihani 2007; Radford et al. 2011). However, drongos sometimes utter false alarm calls, sometimes mimicking the alarm calls of other species, which causes targeted individuals to seek cover and allowing the drongos to steal abandoned food from other species of birds and even from meerkats (*Suricata suricatta*). Such stealing has been found to provide, on average, 23% of the food mass intake of Fork-tailed Drongos (Flower and Gribble 2012; Flower et al. 2013).

(continued)

**Box 5.2** (continued)

Fork-tailed Drongo (Photo by Derek Keats, Wikipedia, CC BY 2.0 https://creativecommons.org/licenses/by/2.0/)

flycatchers, those that attack from below (upward-strikers) tend to have relatively wide bills that vary in length, but are often relatively long, perch gleaners and ground foragers tend to have narrower bills, and aerial hawkers and hover-gleaners have intermediate bill widths (Fig. 5.15). Upward-strikers make rapid sallies up toward the undersides of leaves and their wider bills may provide some room for error during split-second contact with the prey substrate when an insect is "scooped" from a leaf (Fitzpatrick 1985). Directional and temporal precision are less of an issue for perch-gleaners and ground foragers that use their narrower, pointed bills to pick insect prey from vegetation or the ground (Fitzpatrick 1985). Aerial-hawkers and hover-gleaners typically have relatively wide gapes, again leaving some room for error when striking flying insects or striking insects as they hover (Fig. 5.15).

Interestingly, tropical species of foliage-gleaners have longer and narrower bills than those at temperate latitudes (Fig. 5.16). Greenberg (1981) suggested that this difference is due to differences in their insect prey, particularly large soft-bodied prey that are often fed to nestlings during the breeding season. In tropical forests, Orthopterans are probably the large soft-bodied arthropods most abundant for foliage-gleaning birds during the breeding (rainy) season. In contrast, most temperate zone foliage-gleaning birds rely much more on caterpillars when feeding nestlings. The long, narrow bills of tropical foliage-gleaners can likely be closed more rapidly than shorter, broader bills and so may be better for efficiently capturing fast-moving Orthopterans. Caterpillars, on the other hand, are more easily captured, but can struggle with strong muscle contractions when seized in a bird's bill (Greenberg 1981). The typically broader, perhaps stronger bills of temperate-zone foliage-gleaning birds may allow them to hang on to struggling larvae. In addition, temperate-zone foliage-gleaners often carry several large prey items

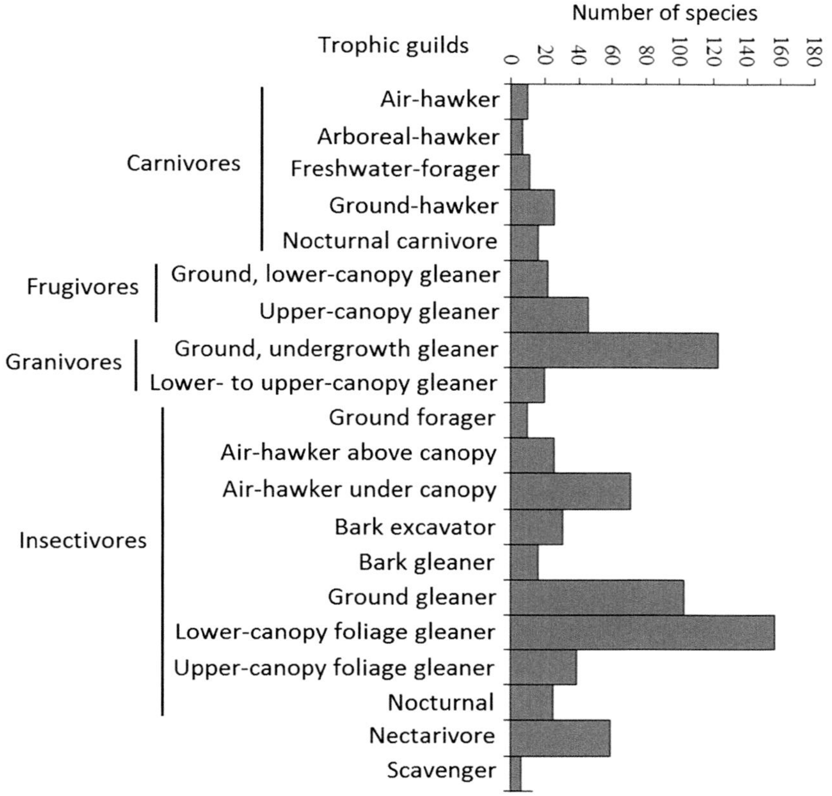


**Fig. 5.8** Common foraging guilds of 830 species birds in North America (excluding birds associated with marine and coastal environments and introduced species). Most species are primarily insectivores and forage for insects in a variety of ways. Many species are ground-foraging granivores and fewer frugivores, carnivores, nectarivores, or scavengers. (Figure modified from González-Salazar et al. 2014; used with permission of Constantino González-Salazar)

simultaneously when foraging to feed nestlings, likely because grubs are easily captured, even with one or more already in the bill; such multiple loading is less likely for tropical foliage-gleaners feeding on large, active Orthopterans (Greenberg 1981). Broader, stronger bills might also be better for carrying multiple prey items.

### 5.2.2 Invertivores

Many other birds feed on invertebrates other than insects. For example, as noted above, Snail Kites feed mainly on apple snails. Numerous other birds also feed on aquatic invertebrates. Shorebirds (Charadriiformes), including plovers (Charadriidae) and sandpipers (Scolopacidae), feed on a wide variety of invertebrates, including crustaceans, and different species have different feeding adaptations. Many shorebirds have long, thin bills for probing sand and mud, whereas others have shorter bills for capturing invertebrates on or just below the surface (Fig. 5.17; Box 5.6 Surface tension transport). Sanderlings (*Calidris alba*) feed primarily on or near the substrate surface and their bills are about 25 mm long (MacWhirter et al. 2020). Long-billed Dowitchers (*Limnodromus scolopaceus*), on the other hand, probe much deeper into substrates to locate invertebrate prey and have bills ranging in length from about 62 to 72 mm (Takekawa and Warnock 2020). Avocets

**Box 5.3 Ecosystem services provided by avian insectivores**

Ecosystem services are benefits obtained from the environment that increase human well-being. One such service involves the potential role of insectivorous birds in controlling arthropod populations and, by doing so, reducing damage to plants (including agricultural crops). For example, Orlowski et al. (2014) found that aerial insectivorous birds consumed common crop pests of nearby crops, likely reducing crop damage and improving crop yields. Kelly et al. (2013) estimated that a single species of aerial insectivore, Purple Martins (*Progne subis*), consume about 400 billion insects every year in North America. Most studies of bird–invertebrate interactions have been conducted at temperate latitudes, where seasonality increases the magnitude of insect population fluctuations. In the tropics, where reduced seasonality means fewer and less severe outbreaks, thousands of insectivorous bird species might be more significant year-around control agents.

Purple Martin (Photo used with permission of David Hemmings)

(continued)

Box 5.3 (continued)

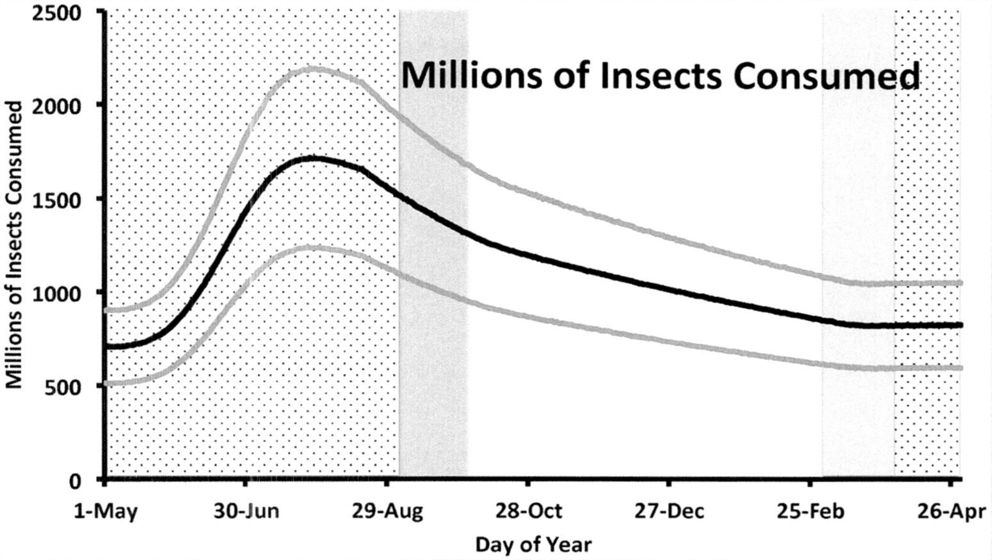


Estimated mean number (black line) ± standard deviation (gray lines) of insects consumed by Purple Martins. Consumption peaks when Purple Martins are feeding nestlings. Overall, Purple Martins are estimated to consume about 400 billion insects each year in North America. Stippling = breeding season; gray = migration; no shading = winter (Figure from Kelly et al. 2013; © 2013 Kelly et al., open-access article distributed under the terms of the Creative Commons Attribution License, https://creativecommons.org/licenses/by/4.0/)

(Recurvirostridae) have long, thin, recurved bills that are swept from side-to-side through the water to capture small invertebrates or use their bills like forceps to pick up small prey items (Fig. 5.18). Oystercatchers (Haematopodidae) have long, rather stout bills used to open oysters, clams, and other bivalves. American Oystercatchers (*Haematopus palliatus*) use a technique called "stabbing" where they insert their knife-like bill into the open valves, sever the adductor muscle holding the two valves together, then remove and consume the soft parts. They also use a "hammering" technique where they use their bill to hammer at the point where the adductor muscle lies inside the shell. Once they break through the shell, the adductor muscle is severed and the soft parts consumed (Working Group et al. 2020).

In addition to exhibiting variation in bill morphology, birds that feed on aquatic invertebrates also differ in leg morphology. Birds that feed primarily in mud and sand tend to have relatively short legs, whereas birds that typically feed in the water have legs that vary in length relative to the water depth where they typically forage.

### 5.2.3 Frugivores

Fruit is the second-most common food for birds, with about one-third of all birds in temperate and, especially, tropical areas consuming fruit as part of their diet (Figs. 5.19 and 5.20). Some birds, like Cedar Waxwings (*Bombycilla cedrorum*), are fruit-specialists and many other birds consume fruits opportunistically or as a regular part of an

**Table 5.4** Comparative composition of macronutrients of 11 arthropod taxa collected in southeastern Australia. Arthropod groups are arranged from most frequently consumed by ground-foraging insectivorous birds, based on a synthesis of published records by Razeng and Watson (2012). Values represent means of all individuals, with wet weight and dry weight relating to the mass of individual arthropods (mean of 50 randomly selected individuals; 100 for Araneae and Coleoptera). Moisture content is expressed on a wet matter (WM) basis, whereas crude protein (CP), crude fat (CF), and crude ash (CA) are expressed on a dry matter (DM) basis. Bold font indicates the three highest values for each factor (Table from Razeng and Watson 2015; © The Authors, used with permission)

| Prey group | Wet weight (mg) | Dry weight (mg) | Moisture (% WM) | Crude protein (% DM) | Crude fat (% DM) | Crude ash (% DM) |
|---|---|---|---|---|---|---|
| Coleoptera | 101.17 | **34.15** | **66.24** | **73.75** | **12.00** | **8.21** |
| Formicidae | 11.16 | 4.09 | 63.37 | 58.62 | 10.60 | **6.29** |
| Lepidoptera | 98.98 | 33.29 | **70.95** | 64.38 | **12.00** | 4.52 |
| Hemiptera | 15.36 | 7.67 | 50.09 | 63.75 | **16.50** | 3.04 |
| Orthoptera | **109.46** | **37.68** | 65.57 | **68.75** | 7.36 | 3.49 |
| Araneae | **195.49** | 29.45 | **69.88** | **76.25** | 9.55 | 4.19 |
| Diptera | 16.03 | 6.73 | 58.03 | 61.81 | 9.55 | 3.20 |
| Hymenoptera | 27.79 | 12.97 | 53.35 | 67.50 | – | – |
| Blattodea | **188.90** | **66.45** | 64.82 | 68.13 | 10.90 | **8.43** |
| Mantodea | 18.35 | 13.89 | 57.63 | – | – | – |
| Odonata | 35.05 | 13.91 | 60.32 | 68.13 | – | – |
| Mean | 83.54 | 23.66 | 60.48 | 66.93 | 11.06 | 5.17 |
| SD | 79.30 | 18.70 | 7.20 | 5.97 | 2.70 | 2.20 |

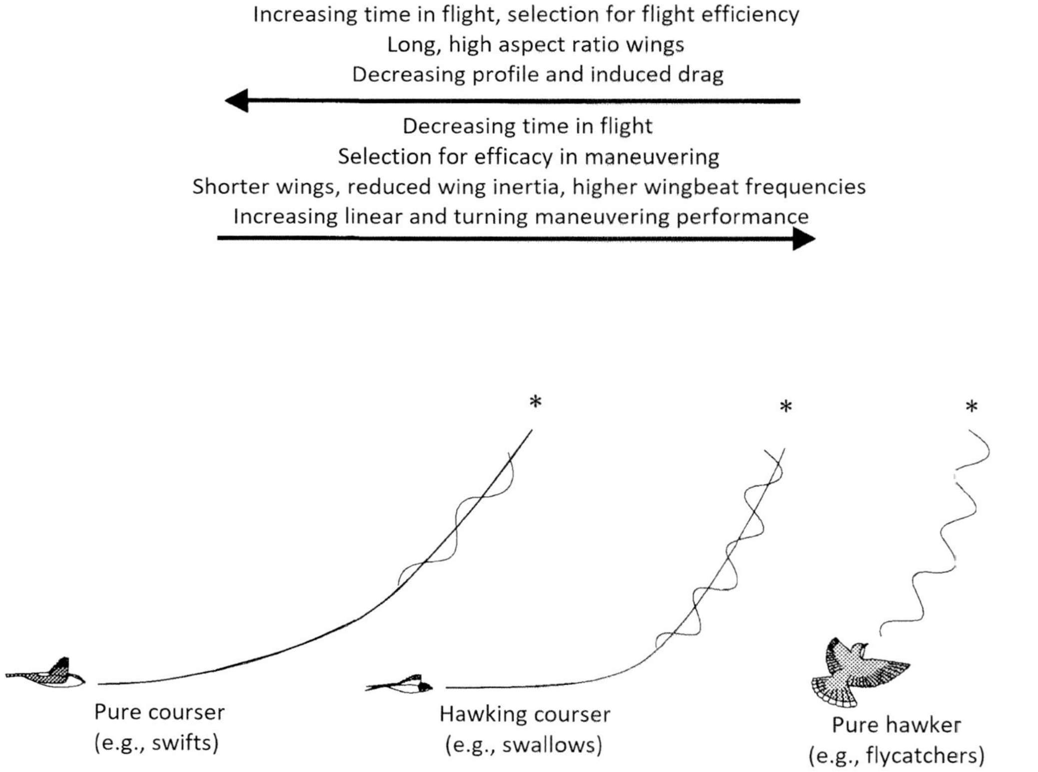


**Fig. 5.9** Swifts depend on flying efficiently and maintaining high speed. Hawking insectivores, like flycatchers, depend on perches located near prey, but must be able to accelerate rapidly and be very maneuverable. Swallows combine these two strategies, with fast, maneuverable flight and the ability to rapidly accelerate when necessary. (Figure from Warrick 1998; © 1951 CCC Republication, used with permission of Canadian Science Publishing)

**Fig. 5.10** Top, The gape of a Large-tailed Nightjar (*Caprimulgus macrurus*), a caprimulgid found in subtropical and tropical moist lowland forests in southeast Asia and Australia. The large gapes of caprimulgids likely enhance their success at capturing flying insects under low-light conditions. Bottom, Scatterplot of body mass (g, log-transformed), and gape area (square mm, log-transformed) of six orders of birds, representing 138 species of raptors and insectivorous birds. The line represents the average gape area among species in the data set. Individual points indicate species means. The relatively large gapes of birds in the order Caprimulgiformes represent an adaptation for more efficiently capturing insects in flight under low-light conditions. (Photograph of the nightjar is used with permission of Quang Nguyen Hao. Figure modified from Jones et al. 2012; © 1969, CCC Republication, used with permission)

omnivorous diet (Klasing 1998). Orders of birds with the most frugivorous species include Columbiformes (pigeons and doves), Psittaciformes (parrots), and Piciformes (woodpeckers and allies) (Kissling et al. 2009). A benefit of consuming fleshy fruits is that fruit can be an abundant, accessible source of energy and antioxidants (Witmer 1998; Schaefer et al. 2008), and some birds are able to discern differences in lipid and antioxidant contents of different fruits based on fruit color (Box 5.7 Fruit color and avian fruit selection). However, avian fruit preferences can also be influenced by sugar content (Levey 1987a), specific amino acids (Parrish and Martin 1977), fatty acid content (McWilliams et al. 2002), and pulp-to-seed ratio (Stanley and Lill 2002a).

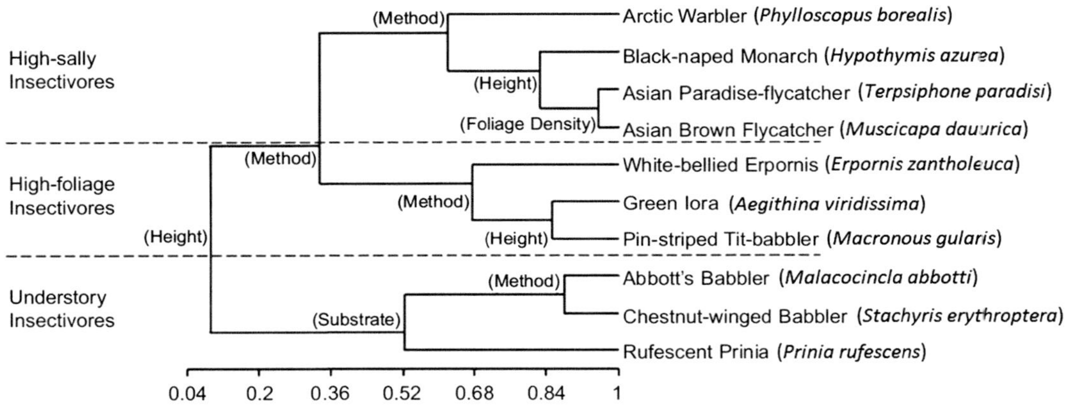


**Fig. 5.11** In a forest in Malaysia, 10 insectivores coexist with minimal competition with each other by foraging at different levels in the forest, using different foraging methods, and foraging in areas of the upper canopy that differ in foliage density. For different groups or pairs of species, foraging niches differ from those of other species to different degrees and based on differences in particular variables (i.e., height, method, or foliage density). The scale at the bottom indicates the relative degree to which the birds occupy different foraging niches, with lower values at branch points indicating less similarity in foraging niche, e.g., understory insectivores and high-sally insectivores occupy very different foraging niches based on their foraging heights. (Figure from Mansor and Sah 2012; © 2012 Museum and Institute of Zoology, Polish Academy of Sciences, used with permission)

The energy available in fleshy fruits is largely in the form of carbohydrates and lipids, and different types of fruits vary considerably in the relative proportions of these two substances. On that basis, fruits can be categorized as either carbohydrate-rich, lipid-poor (CRLP) or lipid-rich, carbohydrate-poor (LRCP; Levey and Martinez del Rio 2001). CRLP fruits are consumed by a wide variety of birds and can be digested quickly and efficiently. Because of their high lipid content, LRCP fruits take longer to digest and absorb. For example, American Robins (*Turdus migratorius*) retained high-lipid meals in their digestive tracts three times longer than low lipid meals (Zurovchak et al. 1999). Clearly, consuming foods that differ in digestion time (retention time) would be inefficient. So, some birds, particularly small-bodied frugivores and omnivores, consume primarily CRLP fruits, whereas others, insectivores with long retention times, such as thrushes, consume primarily LRCP fruits (Levey and Martinez de Rio 2001). In addition to retention time, digestibility is also a factor in fruit choice by birds. Some fruits, such as blueberries, contain high levels of sucrose and some birds, including thrushes and thrashers, lack the enzyme (sucrase) that digests this sugar. Without sucrase, ingestion of sucrose-rich can cause osmotic diarrhea and sucrose-intolerant birds quickly learn to avoid those fruits (Klasing 1998). Finally, fruit size can also influence use by birds. For example, bill size of flycatchers and tanagers in Panama was found to be correlated with the size of preferred fruits (Leck 1971). In Costa Rica, several species of fruit-eating birds attempted to feed on fruits of various sizes, but some birds were simply unable to consume larger fruits (Wheelwright 1985).

Beyond the potential for diarrhea, another possible cost of fruit consumption is the volume and mass of the indigestible portions of fruit (seeds and the fruit coat; Box 5.8 Birds as seed dispersers). Additional mass in the digestive system increases the energy needed for flight (Norberg 1990), and the space occupied by indigestible materials in the digestive system limits further consumption of food (Levey and Grajal 1991). These costs may be particularly important for smaller birds. As a result, birds must attempt to minimize the costs of fruit consumption while

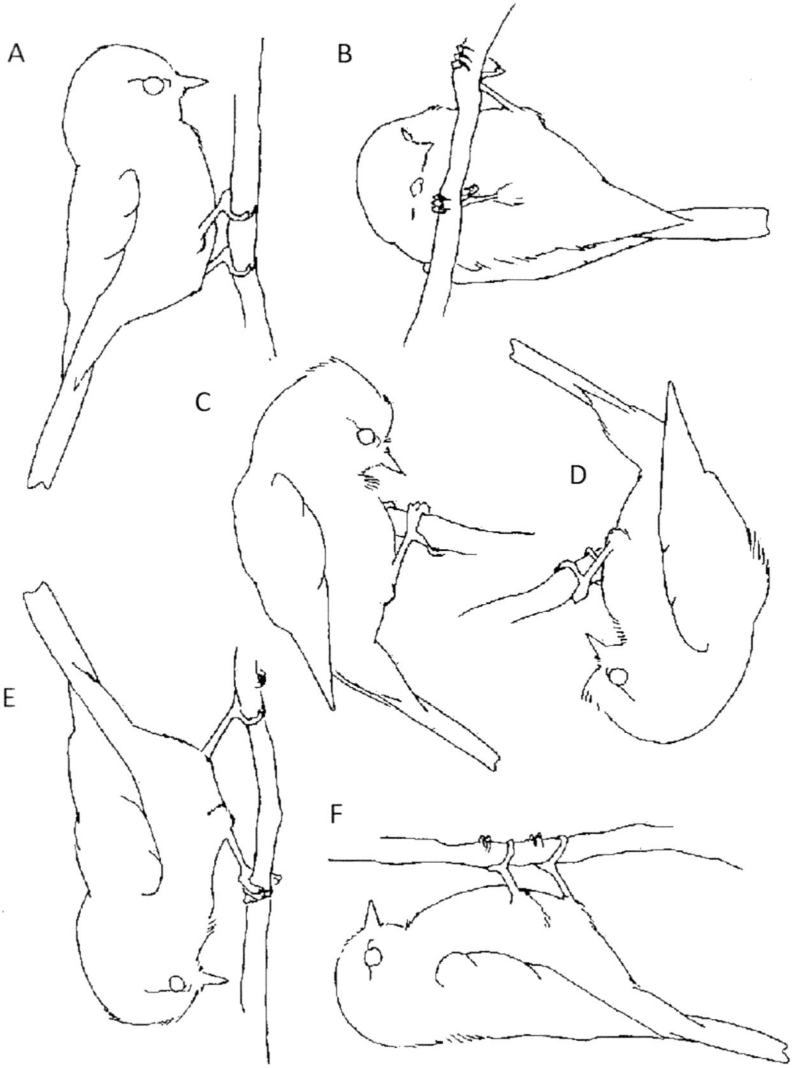


**Fig. 5.12** Small forest-dwelling insectivores use a wide variety of maneuvers when searching for insect prey, including a variety of hang maneuvers shown here. (**a**) Hang-up on vertical perch, (**b**) hang-sideways, (**c**) hang-up on horizontal perch, (**d**) hang-down on horizontal perch, (**e**) hang-down on vertical perch, and (**f**) hang-upside down. (Figure from Remsen and Robinson 1990; Drawing by Donna L. Dittman and used with her permission)

maximizing the rate of nutrient uptake and energy gain (Stanley and Lill 2002b).

Consuming fruit requires no specific morphological characteristics. Frugivores typically have relatively wide gapes and oral cavities, allowing them to consume large fruits relative to their body size (Fig. 5.21), but gape size of fruit-eating birds does influence the size of fruits consumed (Fig. 5.22). In addition, some fruit-eaters, such as toucans, have relatively long bills that allow them to better reach fruits from a perch (Box 5.9 Ballistic food transport). Also, some fruit-eaters, like toucans, and tanagers have strong beaks and jaw muscles that allow them to break-up or mash fruits so seeds can be removed (Klasing 1998).

### 5.2.4 Granivores

Seeds are yet another food item for some birds. Although about one-fifth of all bird families include species that consume seeds to varying degrees, only about 2% of bird families are

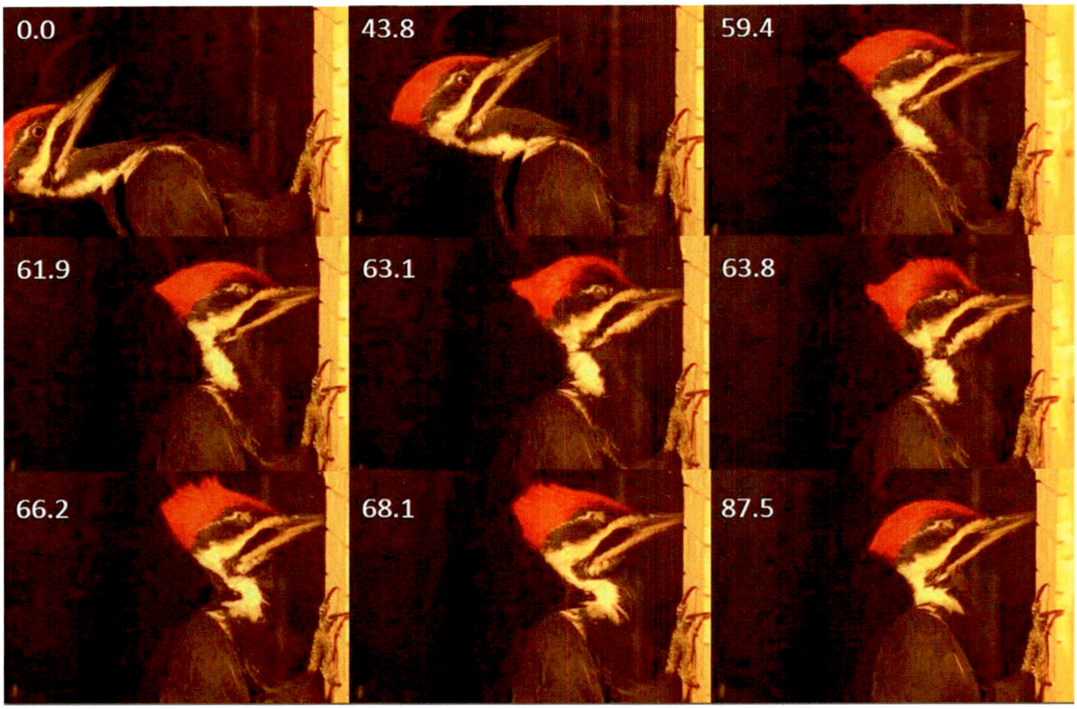


**Fig. 5.13** A Pileated Woodpecker (*Dryocopus pileatus*) striking a wood substrate with its bill (times are in milliseconds). Note how the force of the impact causes the woodpecker's crest to move forward (from 63.1 to 68.1 ms), after which the crest feathers return to their original position (87.5 ms). (Figure from Ross 2014; used with permission of Erica Ross)

primarily granivorous (Morse 1975). Seeds provide birds with a more "balanced" diet than fruits, containing lots of starch plus variable amounts of carbohydrates, protein, lipids, vitamins, and minerals (Fig. 5.23). On the negative side, however, seeds may also contain materials that interfere with digestion, such as alkaloids and tannins, as well as toxins (Klasing 1998). Many granivorous birds, however, tend to avoid seeds with high levels of these substances (Fig. 5.24). Consuming seeds requires a specialized beak (Fig. 5.25), a muscular gizzard, and the ability to coordinate the actions of the bill and tongue to remove a seed's husk (Fig. 5.26). The bills of seed-eating birds are generally deep and broad, and the bills of birds that eat seeds of different sizes and hardness vary in morphology. The jaw muscles of seed-eating birds are also well developed and able to generate the forces needed to crush seeds (Fig. 5.27). The upper bills of seed-eaters are reinforced with trabeculae and the density of those trabeculae varies depending on the size and hardness of seeds consumed and how birds manipulate and crush seeds (Fig. 5.25). The availability of seeds varies seasonally in temperate areas. In temperate areas, seed availability is high toward the end to the growing season, but low in early spring after seed-eating birds and other animals have been consuming seeds for several months and many remaining seeds germinate. Granivores may have to switch to other foods, such as insects, during periods of low seed availability.

A variety of factors can influence seed selection by granivorous birds, including seed morphology (e.g., size, shape, and characteristics, including hardness, of the hull), seed location, and seed content (e.g., carbohydrate, protein, and lipid content). Johansen et al. (2014) examined seed selection by 10 species of granivores visiting feeding stations and found that the

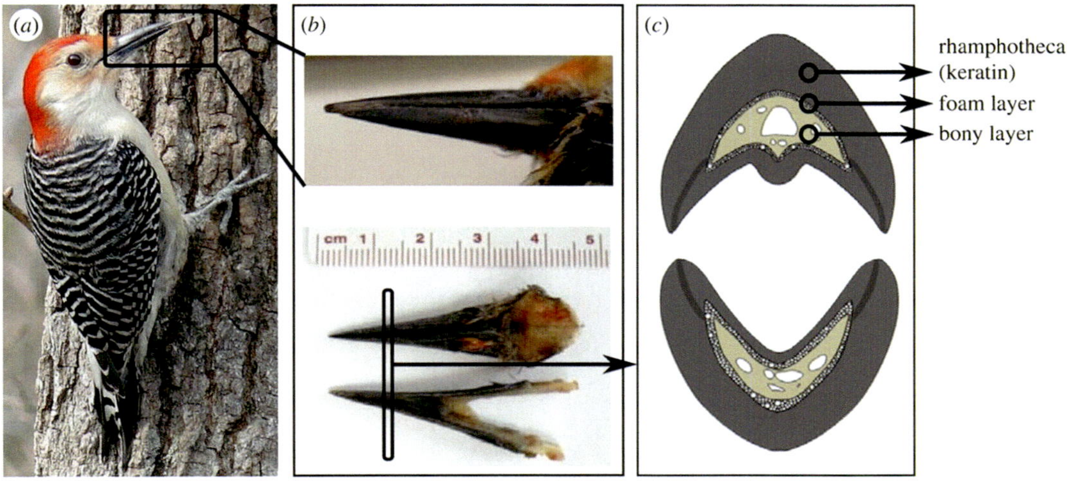


**Fig. 5.14** (**a**) A male Red-bellied Woodpecker (*Melanerpes carolinus*), (**b**) upper and lower beaks of the woodpecker, and (**c**) a schematic of the cross-sectional view of the woodpecker beaks comprising three layers: outer rhamphotheca, middle foam, and inner bony layers. In (**a**), note the toes of the woodpecker. The zygodactyl arrangement means that the toes are widely spaced and that, together with the curved nails, allow woodpeckers to firmly grip bark and other substrates. Also in (**a**), note the tongue protruding slightly from the bill. (Figure from Lee et al. 2014b; © 2014 The Authors. Published by the Royal Society, used with permission)

characteristics of seeds (the presence of an outer hull), nutritional content (more fat, fiber, and protein), and seed size relative to bill size were important to varying degrees for different species. Rios et al. (2012) found that seeds with a higher starch content were preferred by four of six species of granivorous birds (Fig. 5.28). More generally, Diaz (1990) studied seed selection by 10 species of granivores and found that seed selection was influenced by their body and bill sizes and that, given the limitations (in terms of seed size) posed by their body and bill size, smaller birds tended to select seeds with greater nutritional content (Fig. 5.29). Seed-eating birds may also refrain from eating seeds containing high levels of alkaloids and other plant secondary compounds as well as those infected with fungi (Madej and Clay 1991).

### 5.2.5 Carnivores

Birds that consume other vertebrates are carnivores, and avian carnivores include hawks, falcons, eagles, kites, owls, and shrikes. Some carnivores, including some eagles, owls, plus loons, mergansers, grebes, cormorants, herons and egrets, terns, pelicans, gulls, auks, gannets, some penguins, and kingfishers, prey primarily on fish and are called piscivores. Vertebrates, as a food source, provide a nutritionally complete diet and, as a bonus, also have a high water content (50–80%). As a result, avian carnivores may not require access to a source of drinking water (Klasing 1998). Small vertebrates may be consumed whole; larger prey are torn apart and consumed in pieces. For efficient digestion, food moves relatively slowly through a carnivore's digestive tract. For example, the mean retention time for a variety of raptors fed chicks (*Gallus g. domesticus*) was about 8 h. Averaged across species, the metabolizable energy coefficient (MEC; energy ingested minus energy excreted, divided by energy ingested) of vertebrate foods (including indigestible components like bone) is about 75% (Klasing 1998). Indigestible components, such as bone, fur, feathers, claws, and scales, are often collected in the gizzard and regurgitated as pellets (Box 5.10 Ingesting indigestible plastics).

**Box 5.4 How do woodpeckers avoid head impact injury?**

Woodpeckers are known to drum hard woody surfaces of trees at a rate of 18–22 times per second with a deceleration of 1200 g (humans can lose consciousness at g-forces as low as 4–6 g). Woodpeckers appear to have features that absorb mechanical shock and reduce the likelihood of brain damage, including a hyoid apparatus that rigidly supports the tongue and limited space for cerebrospinal fluid (CSF) between the skull and the brain. Some investigators have suggested that areas of spongy bone at the coup and contrecoup positions of the skull also help to absorb impacts (e.g., Wang et al. 2013), but Van Wassenbergh et al. (2022) concluded that, based on high-speed videography and kinematic analyses of three species of woodpeckers, the spongy bone "... probably serves an important role in 'resisting' impact forces without failing rather than "absorbing" impact energy by elastic deforming."

A woodpecker's hyoid extends posteriorly from the floor of the oral cavity, goes behind the neck, divides into two bands, goes around the back of the skull, and inserts at the front of the skull. This allows woodpeckers to extend their tongues well beyond the tip of the beak when foraging, but this arrangement also helps distribute mechanical vibrations when drumming. Finally, woodpeckers have little space for CSF between the skull and brain, and this limits movement of the brain in the braincase (Yoon and Park 2011, but see Ross 2014). In addition to these factors, the very short duration of a bill's impact plus the smooth brain of woodpeckers (i.e., no gyri or sulci so the force is more evenly distributed acros the surface of the brain) likely minimize the risk of concussion (Ross 2014).

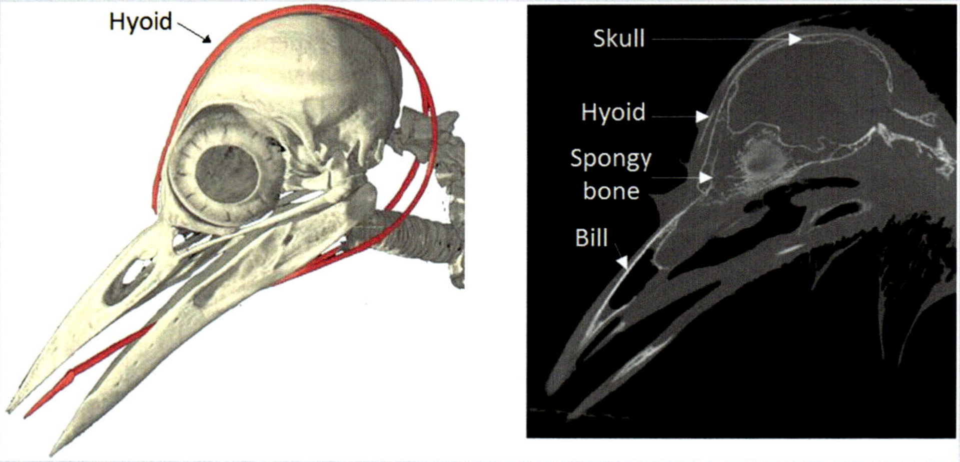


Skull and hyoid of a Golden-fronted Woodpecker. The CT scan on the right shows the location of the spongy bone directly behind the upper bill (Figures modified from Yoon and Park 2011; © IOP Publishing. Reproduced with permission. All rights reserved)

(continued)

**Box 5.4** (continued)

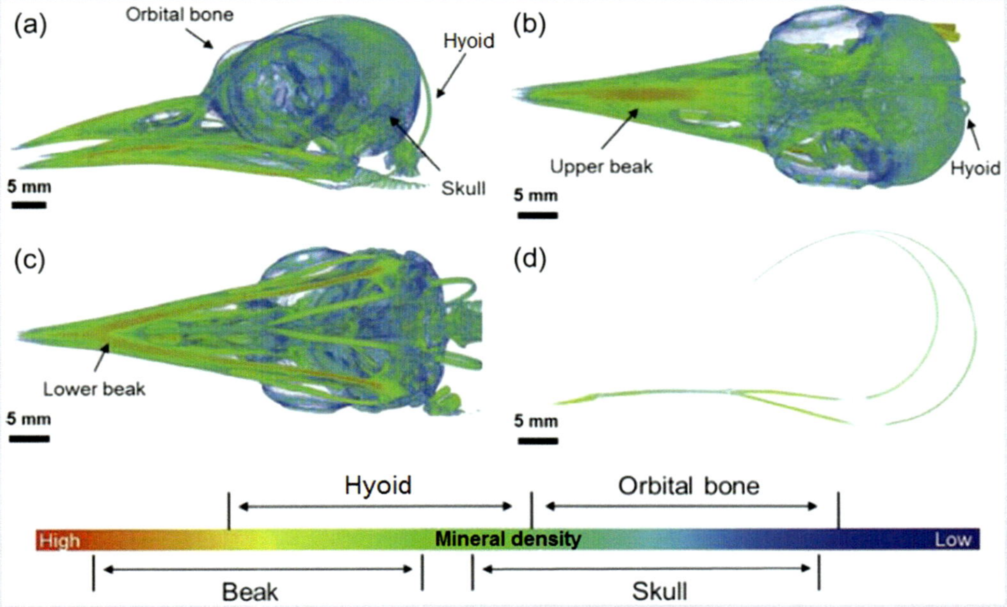


Skull and hyoid apparatus of an Acorn Woodpecker (*Melanerpes formicivorus*) showing that the upper and lower bills have the highest mineral densities, providing increased strength and stiffness (Figure from Jung et al. 2016; © 2016 Elsevier Ltd., used with permission)

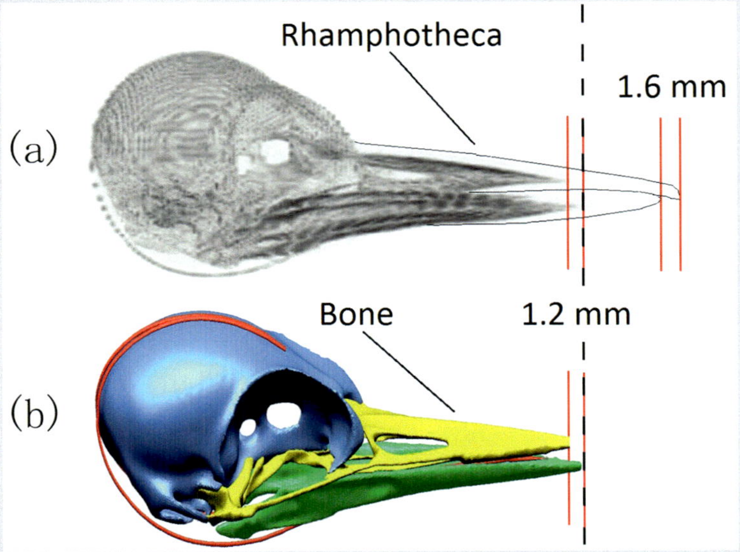


Skull of a Great Spotted Woodpecker (*Dendrocopos major*). The rhamphotheca, or outer tissue layer of the upper bill, is 1.6 mm longer than that of the lower bill, but the bone structure of the upper bill is about 1.2 mm shorter than the lower bill (Figure modified from Wang et al. 2011; © 2011 Wang et al., open-access article distributed under the terms of the Creative Commons Attribution License, creativecommons.org/licenses/by/4.0/)

(continued)

**Box 5.4** (continued)

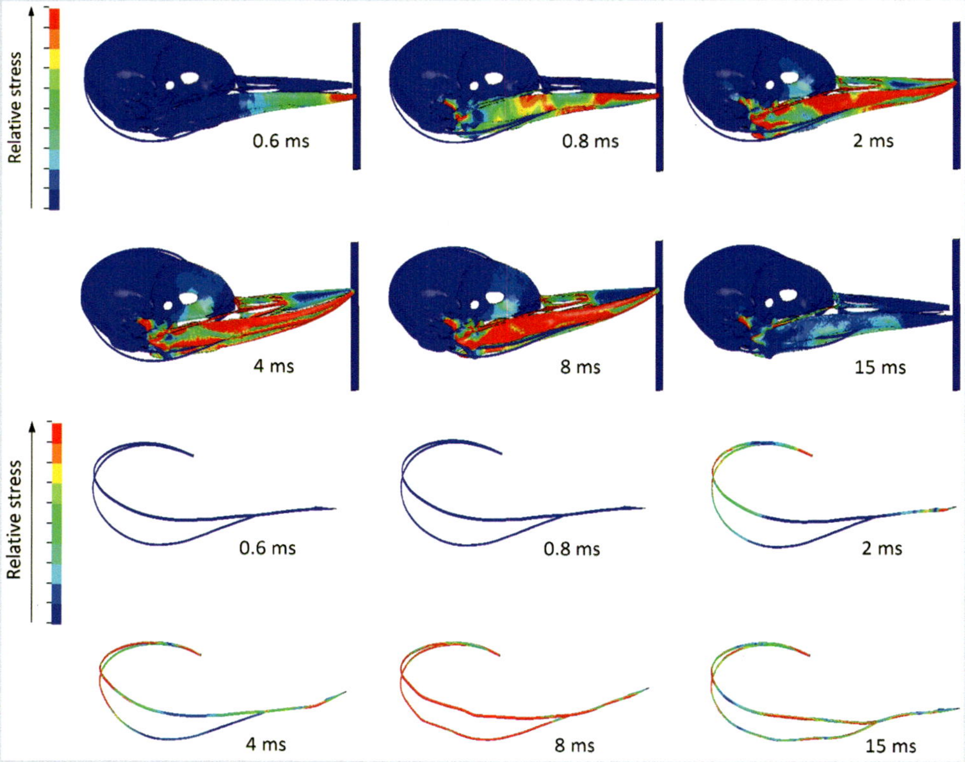


The unequal length of the upper and lower parts of woodpecker bills (the lower being longer) directs the force of impact downwards, away from the brain, when it hits a tree (red color indicates the greatest force or stress, blue the least). Time after impact proceeds from upper left (0.6 ms after impact) to lower right (15 ms after impact). The hyoid of woodpeckers loops over top of the skull and inserts at the front. The hyoid helps direct the force of impact below and around the skull (and brain). Because woodpeckers have little space for CSF between the skull and brain, brain movement is minimal during impact. This is critical because movement of the brain inside the skull during impact, more than the blow itself, is what causes concussions. If the brain is held in place, risk of injury is greatly reduced (Figures modified from Wang et al. 2011; © 2011 Wang et al., open-access article distributed under the terms of the Creative Commons Attribution License, creativecommons.org/licenses/by/4.0/)

Given the relatively small size of woodpeckers and their brains, Van Wassenbergh et al. (2022) provided data suggesting that shock absorption, to the extent that it might occur in woodpeckers, is not necessary and would even be maladaptive. These authors pointed out that any shock absorption would necessarily mean woodpecker bills would impact substrates with less force so they would need to pound even harder. As far as even being necessary, Van Wassenbergh et al. (2022) determined that, given the relatively small size and mass of woodpecker brains plus the minimal movement possible because there is little space between the skull and the brain, even the strongest pecks by the three species of woodpeckers in their study resulted in less than 60% of the pressure needed to cause a concussion for a human. However, these authors also noted that even subconcussion impacts may cause brain trauma. As such, features of woodpecker skulls like those described above may aid in preventing such trauma.

(continued)

**Box 5.4** (continued)

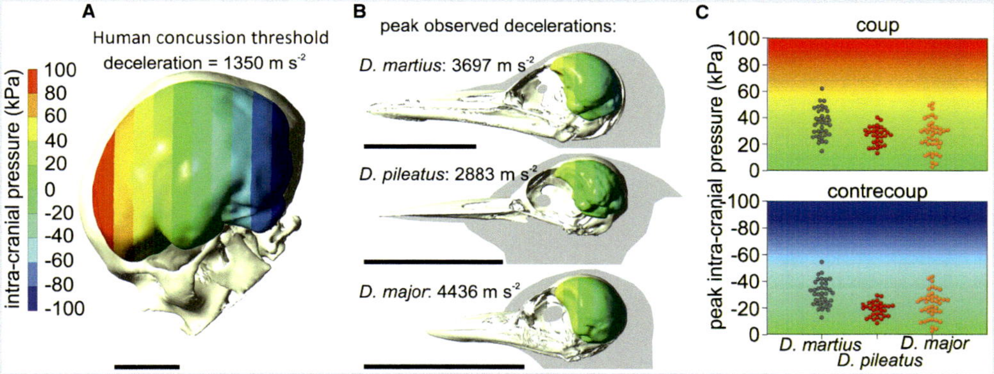


Comparison of intracranial pressures due to inertial loading (i.e., resistance of objects to a change in speed and direction) in humans and three species of woodpeckers. Computed brain cavity surface pressures for (**a**) an adult human at the concussion threshold deceleration and (**b**) the three woodpecker species at their measured peak deceleration. Intracranial pressure at the concussion threshold for a human with sagittal-place deceleration (1350 m/s) was 103 kilopascals (kPa; coup) and −101 kPa (contrecoup). (**c**) Even at the highest decelerations observed for the three species of woodpeckers, intracranial pressures were well below those that would result in concussions for humans (<60 and −60 kPa for coup and contrecoup, respectively). Each solid circled represents pressures for separate hits. *Dryocopus martius*, Black Woodpecker; *Dryocopus pileatus*, Pileated Woodpecker; *Dendrocopos major*, Great Spotted Woodpecker. Scale bars, 50 mm (Figure modified from Van Wassenbergh et al. 2022; © 2022 Elsevier Inc., used with permission)

Although woodpeckers have structures that may absorb mechanical shock and have small brains resistant to being shocked, Farah et al. (2018) detected tau protein accumulations in the brains of five species of woodpeckers. In humans, such accumulations can be associated with chronic brain trauma, e.g., as found in the brains of many former football players (Mez et al. 2017). The extent to which tau accumulations in the brains of woodpeckers might be pathological or occur for other, nonpathological, reasons is currently unknown. However, Tobiansky et al. (2021) suggested that, if woodpeckers do suffer some brain trauma due to repeated pecking and drumming, physiological systems exist that could help mitigate the effects of such trauma (described below).

(continued)

**Box 5.4** (continued)

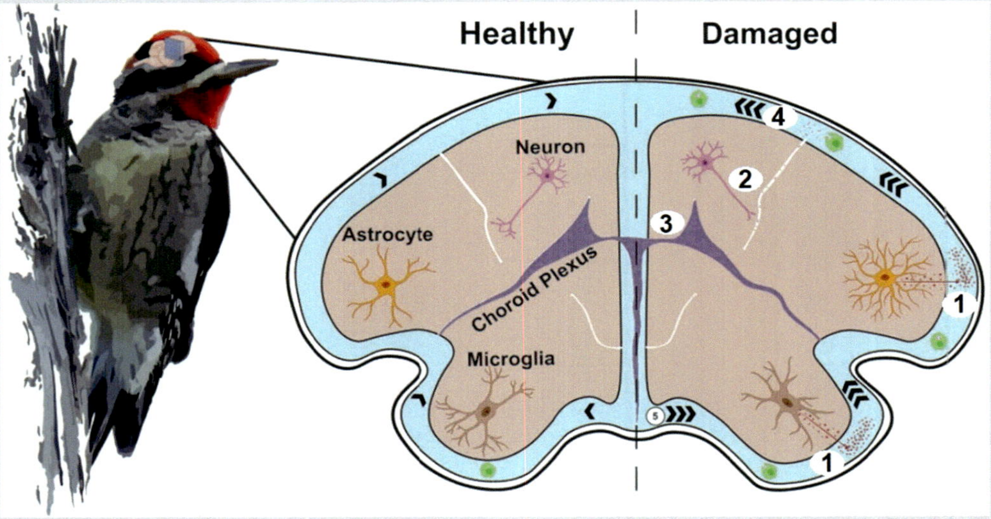


Possible physiological mechanisms for mitigating effects of any potential brain injury (i.e., damaged brain tissue on the right) in woodpeckers include: (1) release of proinflammatory chemicals called cytokines by microglia and astrocytes (types of white blood cells that are part of the avian immune system that aid in tissue repair, (2) removal of damaged cells from the brain via the perivascular spaces (fluid-filled spaces surrounding blood vessels; indicated by the white line), and (3 and 4) an increase in production of cerebrospinal fluid via an increase in number of cells in a network of capillaries called the choroid plexus, shown in purple) to better cushion the brain (Figure modified from Tobiansky et al. 2021; © 2021 Oxford University Press, used with permission)

Avian carnivores use a variety of methods to locate and attack prey (Box 5.11 Nonergodic hunting by raptors). Many are ambush, or sit-and-wait, predators, sitting on a perch and visually scanning for possible prey, e.g., many hawks, kingfishers, owls, and shrikes. When attacking terrestrial or aquatic prey, ambush predators fly from their perches and attack the prey on the ground (e.g., reptiles or small mammals) and or shallow water (e.g., small fish). When attacking other birds on the ground or on perches, however, ambush predators may be detected and must then pursue their prey in flight. For example, accipiters like Sharp-shinned (*Accipiter striatus*) and Cooper's (*A. cooperii*) hawks attacking smaller birds may not be detected and can capture their prey on the ground or pluck them from perches. If detected, however, and their targeted prey takes flight, the hawk's chances of a successful capture are lower. As a result, Cooper's Hawks are known to use "surprise attacks" where they use visual obstructions like vegetation or man-made structures to get close enough to prey so they are not detected (Roth and Lima 2003).

Some sit-and-wait predators, including some owls and harriers, rely on sound as well as vision to locate prey, and some can locate prey using only sound, e.g., Barn Owls (*Tyto alba*), Great Gray Owls (*Strix nebulosi*), Northern Saw-whet Owls (*Aegolius acadicus*), and others. Contributing to this ability is the asymmetrical ears of some owls, better allowing them to accurately pin-point the source of sounds (more detail about this in Chap. 4). However, the ability to capture prey by sound is enhanced by their relatively silent flight (compared to other birds), with potential prey less likely to hear their approach

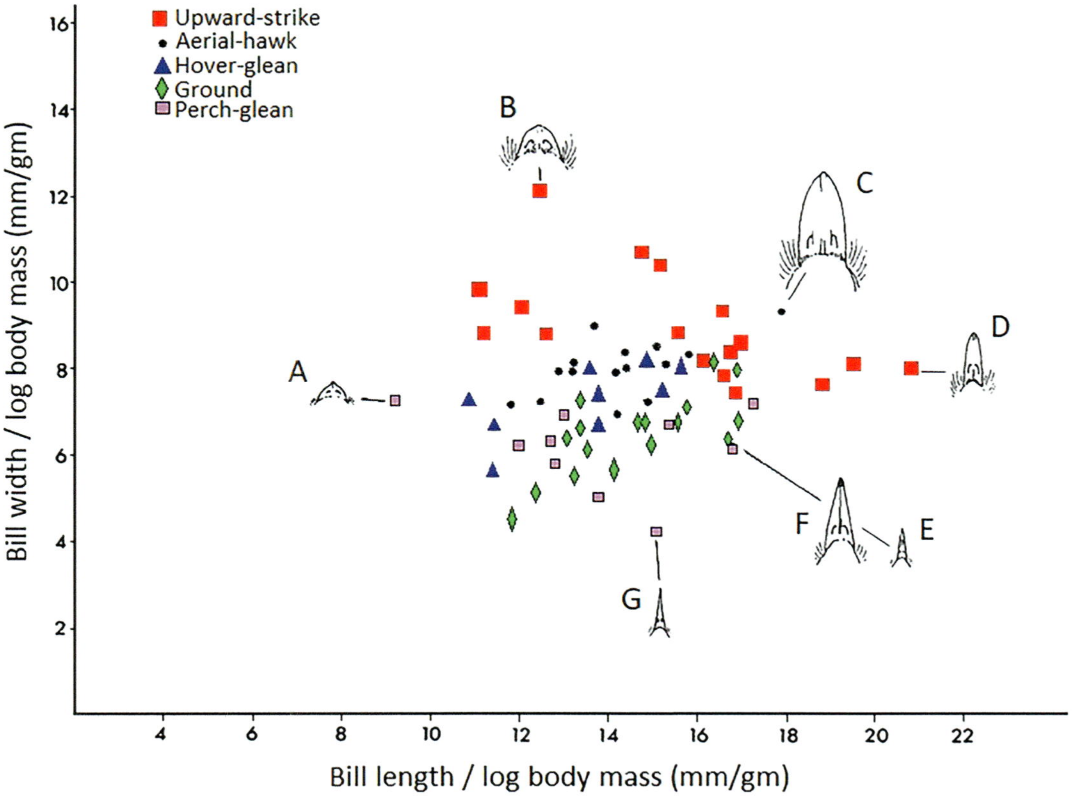


**Fig. 5.15** Relationship between relative bill width at base and relative bill length (from base) for 72 species of tyrant flycatchers, each assigned to one of five foraging categories (upper left). Extreme bill shapes drawn to scale = (A) Long-tailed Tyrant (*Colonia colonus*), (B) White-crested Spadebill (*Platyrinchus platyrhynchos*), (C) Boat-billed Flycatcher (*Megarynchus pitangua*), (D) Common Tody-Flycatcher (*Todirostrum cinereum*), (E) Tufted Tit-Tyrant (*Anairetes parulus*), (F) Black-billed Shrike-Tyrant (*Agriornis montanus*), and (G) Many-colored Rush Tyrant (*Tachuris rubrigastra*). (Figure modified from Fitzpatrick 1985; © American Ornithological Society, used with permission)

and the owls more likely to hear and accurately locate prey because they generate minimal sound as they fly. The wings of most owls (with the exception of fish-eating owls whose prey cannot hear them approach from above the water's surface) have four important adaptations that contribute to their silent flight: (1) serrations at the leading of the wing (generally the 10th primary and, in some cases, and 8th, 9th, and 10th primaries) (Figs. 5.30 and 5.31), (2) "fringes" at the trailing edge of the wings (Figs. 5.31b and 5.32), (3) a "velvety" upper surface of the wings (Figs. 5.33 and 5.34), and (4) the fringes on the primaries and secondaries merge smoothly into the grooves between barbs of adjacent feathers so there are no sharp edges to generate turbulence (Fig. 5.35; Winzen et al. 2008; Bachmann et al. 2012; Chen et al. 2012).

The serrations at the leading edge of owl wings do two things. First, they minimize turbulence by better maintaining smooth air flow over the wing (Fig. 5.36). The fringes on the primaries and secondaries that merge smoothly into the grooves between barbs of adjacent feathers also help minimize friction and help maintain smooth air flow. In addition, the serrations "separate" the air moving over wings that, at the trailing edge of the wing, will create fine longitudinal vortices or, in other words, a series of much smaller vortices rather than a single larger vortex (Fig. 5.37).

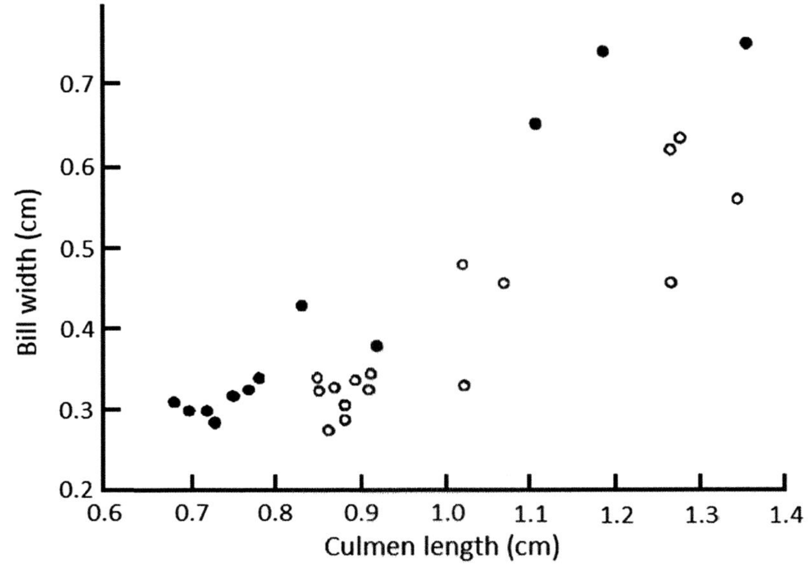


**Fig. 5.16** Relationship between bill width and culmen length for foliage-gleaning insectivores from tropical (open circles) and temperate (closed circles) latitudes. Tropical foliage-gleaners tend to have longer and narrower bills than temperate foliage-gleaners. (Figure modified from Greenberg 1981; © 1981 Springer-Verlag, used with permission)

Vortices created at the trailing edge of wings cause vibrations that create sound. However, the serrations at the leading edge of owl wings delay the formation of vortices (impacting less surface area) and create numerous small vortices. These vortices still can cause vibrations that create sound. However, the soft, velvety surface of the wing, together with the fringes at the trailing edge of the wing, act to "absorb" the impact of the vortices which, in turn, minimizes vibration. In combination, all of these features of owl wings, including the serrations at the leading edge, the absence of sharp edges between adjacent feathers, the soft texture of the upper wing (and the rest of the body), and the fringes at the trailing edge, greatly reduce the noise generated by owl wings as they move through the air.

Some avian carnivores actively pursue prey (other birds or bats) in flight, e.g., many falcons (Fig. 5.38). Kane and Zamani (2014) mounted video-cameras on the backs or heads of Peregrine Falcons (*Falco peregrinus*) and Gyrfalcon (*Falco rusticolus*)/Saker Falcon (*Falco cherrug*) hybrids to monitor how they attacked aerial prey. On many occasions, falcons hunt from high altitude and use high-velocity stoops (more than 300 km/h) when attacking prey. However, falcons sometimes pursue prey located at a similar altitude and

**Box 5.5 Flush-pursuit foragers**

Flush–pursuit foragers use exaggerated and animated foraging movements to flush potential insect prey that are then pursued and captured in flight. The *Myioborus* redstarts comprise 12 species of flush–pursuit warblers found in montane forests of the American tropics and subtropics. All members of the genus have contrasting black-and-white tail feathers that are exposed by spreading the tail during foraging. Mumme (2002) examined their plumage patterns and tail-spreading behavior to see how they affected flush–pursuit foraging performance of the Slate-throated Redstart (*Myioborus miniatus*) in Costa Rica.

(continued)

**Box 5.5** (continued)

Slate-throated Redstart (Photo by Ron Knight, Wikipedia, CC BY 2.0, https://creativecommons.org/licenses/by/2.0/)

Although flycatching was the most common foraging tactic used by Slate-throated Redstarts, flush–pursuit prey attacks occurred more frequently following hops in the spread-tail foraging posture than hops in more typical warbler-like posture, suggesting that tail-spreading behavior assists in startling and flushing potential insect prey. The hypothesis that the white tail feathers enhance flush–pursuit foraging was tested by means of a plumage-dyeing experiment. After locating nests, Mumme (2002) captured the male and female and assigned one member of each pair to the experimental treatment group; its mate served as a control. For experimental birds, a permanent marker was used to blacken the white tips of the three outer rectrices. For sham-darkened controls, the naturally black tips of the three inner rectrices were also "blackened." Experimental birds with darkened tail feathers were less successful in flush–pursuit foraging, with a lower rate of prey attack, and fed their nestlings at a significantly lower rate than their sham-darkened mates. For experimental birds, only 7.6% of hops in the spread-tail posture were followed by an attack on a prey item, compared to 20.9% of hops for controls. These results indicate that white tail feathers are critically important for startling potential prey.

(continued)

**Box 5.5** (continued)

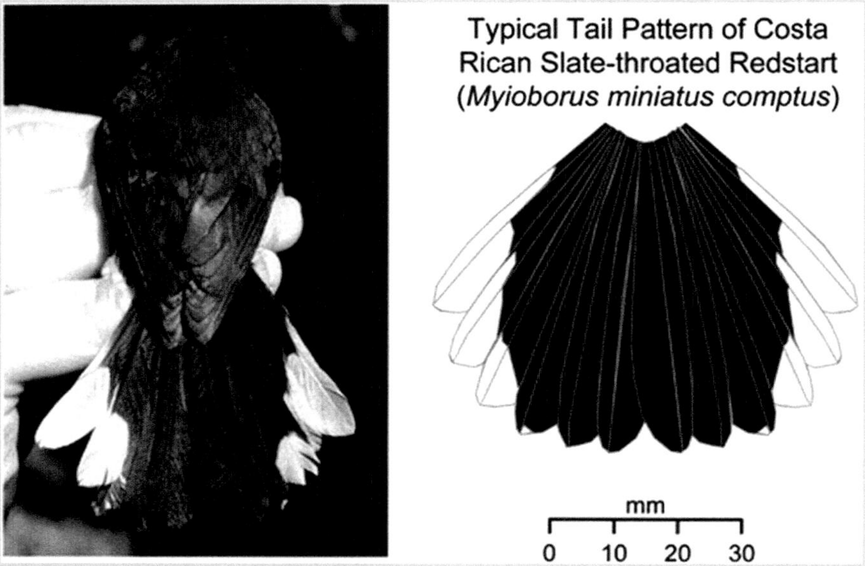


Typical tail pattern of a Slate-throated Redstart (Figure from Mumme 2002; © 2002 Oxford University Press, used with permission)

can potentially use a variety of other attack strategies, but analysis of videos from head-mounted cameras and data from GPS loggers suggests that the attack trajectories of Peregrine Falcons cannot be described by any single rule (Fig. 5.39). However, their trajectories appear to be best described by the proportional navigation guidance law (Brighton et al. 2017), where two moving objects will meet when the direct line-of-sight (compass bearing) between a pursuer and a target does not change as the distance between them decreases (Fig. 5.40). With this strategy, Peregrine Falcons keep their prey at a fixed visual angle and match the prey's velocity to achieve the shortest interception time (Kane and Zamani 2014; Figs. 5.40c and 5.41). By maintaining the same position or angle relative to prey being pursued, falcons may also be less likely to be detected (i.e., motion camouflage) because they would appear stationary (albeit appearing to increase in size) from the viewpoint of their prey (Brighton et al. 2017).

Some piscivores that forage in the open ocean locate foraging sites using their sense of smell, detecting odors like dimethyl sulfide (DMS) that suggest the presence of prey from considerable distances (additional information provided in Chap. 4). Additional strategies for locating fish (and other prey like squid and other invertebrates) in the open ocean are to use physical cues, such as smooth patches on the surface that indicate upwellings where small prey may accumulate (Box 5.12 Use of surface features as foraging cues by seabirds) or to monitor the behavior of other predators, including conspecifics and other species of birds, but also mammalian predators like dolphins and humans. Tremblay et al. (2014) attached both miniaturized video cameras and GPS tracking recorders to Cape Gannets (*Morus capensis*), and video evidence revealed that the

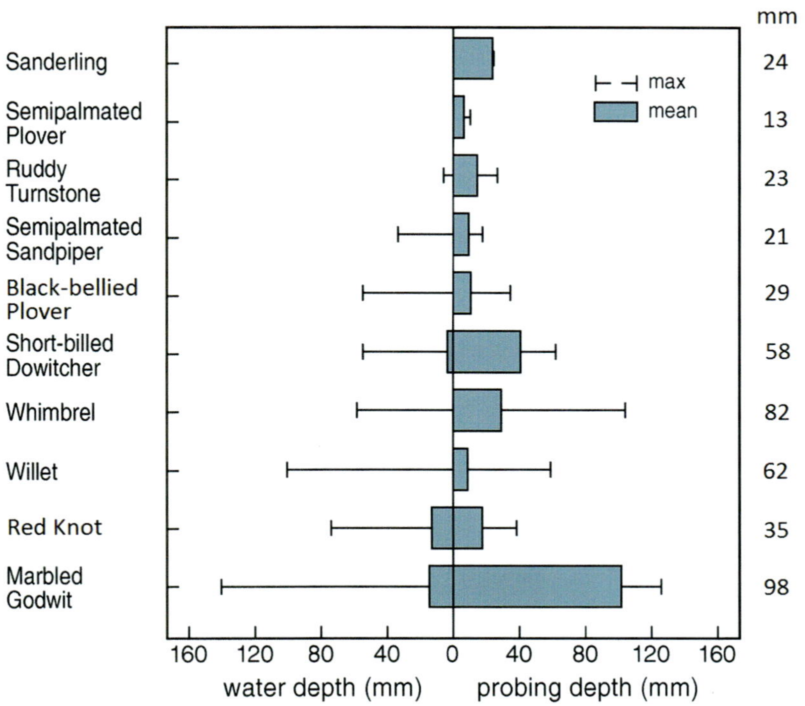


**Fig. 5.17** Observed water and foraging (or probing) depths of several species of shorebirds on the northeast coast of Brazil. Shorebirds with longer bills (and legs) tend to forage deeper into the substrate and in deeper water. Approximate bill lengths of the different species are shown on the right (in mm). Sanderling, *Calidris alba*; Semipalmated Plover, *Charadrius semipalmatus*; Ruddy Turnstone, *Arenaria interpres*; Semipalmated Sandpiper, *Calidris pusilla*; Black-bellied Plover, Pluvialis squatarola; Short-billed Dowitcher, *Limnodromus griseus*; Whimbrel, *Numenius phaeopus*; Willet, *Tringa semipalmata*; Red Knot, *Calidris canutus*; Marbled Godwit, *Limosa fedoa*. (Figure modified from Kober and Bairlein 2009; © Netherlands Ornithologists' Union, used with permission)

gannets used information obtained by observing dolphins, fishing boats, and other gannets to help locate prey (fish; Fig. 5.42).

Avian carnivores have morphological adaptations that enable them to capture vertebrate prey. Bills may be hooked for killing, holding, and tearing, pointed for capturing or spearing fish (loons, gannets, herons, and many others), long and broad with a gular sac for scooping (pelicans), serrated for capturing and holding fish (mergansers), or have "riblets" to reduce drag (skimmers; Box 5.13 Riblets reduce drag for skimmers). Raptors (hawks, falcons, eagles, kites, and owls) also have sharp, decurved talons for capturing, holding, and killing prey (Box 5.14 Kicking Secretarybirds). Piscivores often capture fish by diving or swimming. For example, kingfishers, terns, and gannets plunge into the water and capture fish in their bills (Figs. 5.43 and 5.44, Box 5.15 Fish eating and refraction). For gannets, studies have revealed that the angle of entrance into the water, dive depth, and their behavior once underwater (e.g., swimming in pursuit of prey) can be influenced by prey depth, prey type, and the presence of other predators such as dolphins (Garthe et al. 2000; Capuska et al. 2011). For example, if dolphins have "herded" fish into "prey balls" near the water's surface, gannets are more likely to use V-shaped dives directly into the "prey balls." For deeper prey, gannets are more likely to use U-shaped dives because they must swim deeper in pursuit

**Box 5.6 Surface tension transport**

By partially opening and closing their bills in a rapid, repetitive manner, some species of shorebirds move drops of water that contain prey items from the tip of their bill to their mouth in a stepwise ratcheting fashion. When the bill slightly closes, the drop's leading edge moves toward the mouth, while the trailing edge stays put. When the bill opens, the leading edge remains in place and the trailing edge moves toward the mouth. Surface tension transport (sometimes also called the capillary ratchet) requires that birds open and close their bill, varying the opening angle within a narrow range so that the small drop of water (~2 mm) does not break apart. Such transport also requires long and very narrow bills with a suitable contact angle between the bill and the drop of water (see figure below); water cannot form appropriately sized droplets when bills are wider (i.e., too much water in a wider bill and the surface tension is not sufficient to form a small drop). Red-necked Phalarpoes (*Phalaropus lobatus*) typically transport a drop of water (and prey) from the bill tip to the mouth by partially opening and closing their bill two or three times (Rubega and Obst 1993).

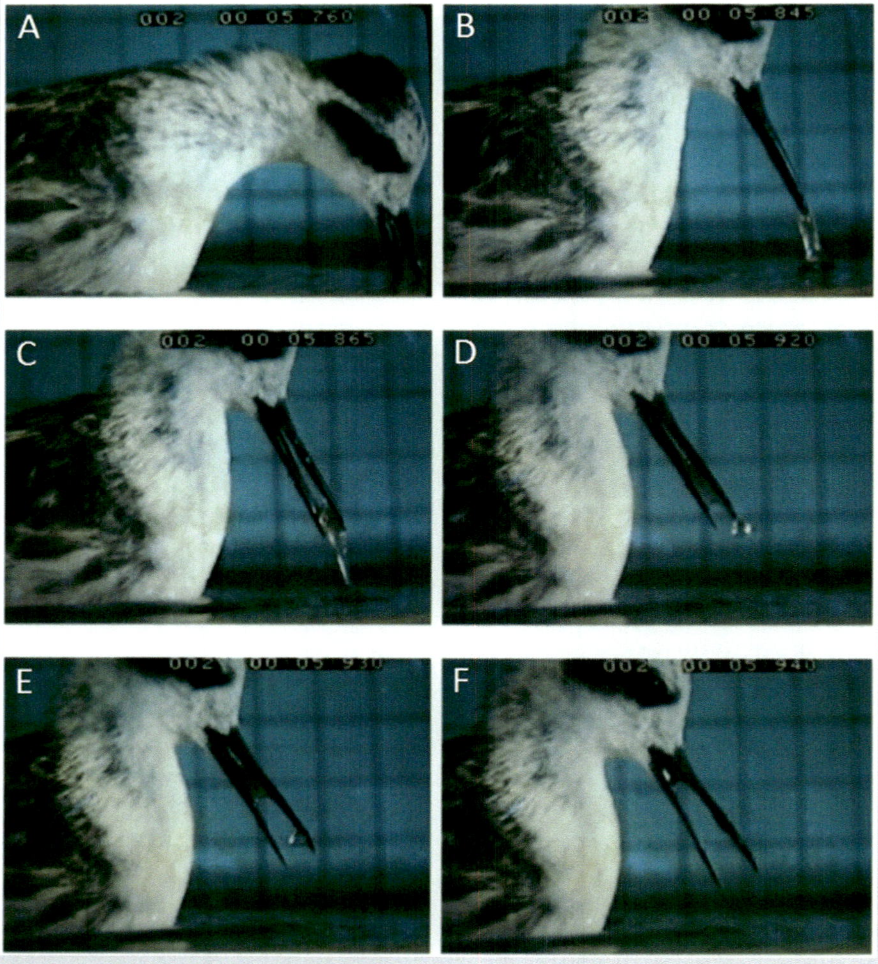


(continued)

**Box 5.6** (continued)

**Surface-tension prey transport by a Red-necked Phalarope**. This feeding mechanism uses the surface tension of water to rapidly transport prey from bill tip to mouth without use of suction or tongue movements by birds. After a bird captures prey with its bill tip (**a** and **b**), the water and prey move up the bill by rapid mandibular movements (**c** through **f**). This motion moves the drop up the bill and into the bird's mouth (Figure from Rubega and Obst 1993; © 1993 Oxford University Press, used with permission)

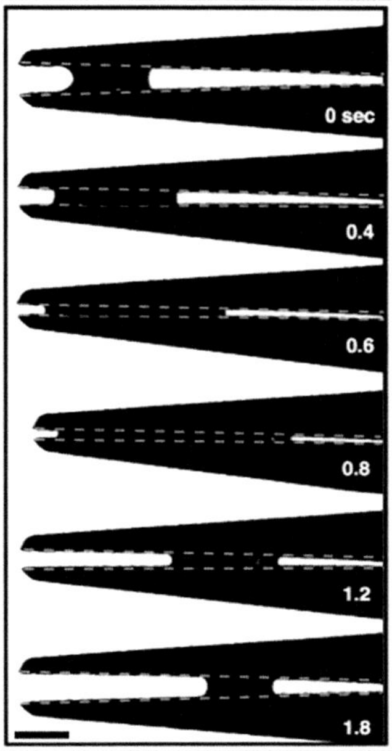


Movement of a drop of water in an artificial bill. The drop moves as the "bill" partially closes and opens. Time is shown on the right (in seconds). Scale bar (lower left) = 2 mm (Figure from Bush et al. 2010; © 2010 Elsevier Ltd., used with permission)

(continued)

**Box 5.6** (continued)

(**a**) A foraging juvenile Wilson's Phalarope with a prey item suspended in the droplet in its bill (inset, lower left) (Photo courtesy of Robert Lewis). (**b**) Shorebirds use a variety of foraging strategies and so exhibit much variation in beak size and shape. Shown here is a plot of bill length and width at the base of 18 species of shorebirds that have straight bills. Scale bars represent standard deviations. Red-necked Phalarope, *Phalaropus lobatus*; Wilson's Phalarope, *Phalaropus tricolor*; Common Snipe, *Gallinago gallinago*; Black-winged Stilt, *Himantopus himantopus*; Eurasian Oystercatcher, *Haematopus ostralegus*; Kentish Plover, *Charadrius alexandrinus*;

(continued)

**Box 5.6 (continued)**

Common Ringed Plover, *Charadrius hiaticula*; Dunlin, *Calidris alpina*; Little Stint, *Calidris minuta*; Sanderling, *Calidris alba*; Western Sandpiper, *Calidris mauri*; Spotted Redshank, *Tringa erythropus*; Black-tailed Godwit, *Limosa limosa*; Bar-tailed Godwit, *Limosa lapponica*; Gray (Black-belled) Plover, *Pluvialis squatarola*; Northern Lapwing, *Vanellus vanellus*; Ruff, *Calidris pugnax*; Ruddy Turnstone, *Arenaria interpres* (Figure modified from Prakash et al. 2008; © 2008 American Association for the Advancement of Science, used with permission)

of their prey than the maximum depth of their dive (about 4 m) (Capuska et al. 2011; Fig. 5.45). Other piscivores, including loons, mergansers, cormorants, and auks, pursue prey underwater, using their wings or webbed feet for propulsion (see Chap. 11 for more details).

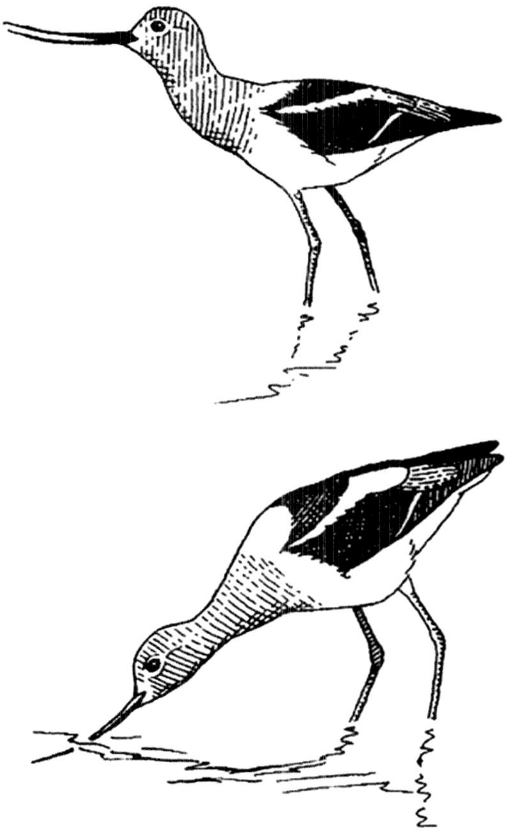

**Fig. 5.18** Foraging American Avocet (*Recurvirostra americana*). Avocets sweep their bills from side-to-side through the water to capture small invertebrates or use their bills like forceps to pick up small prey items. (Figure from Hamilton 1975; © American Ornithological Society, used with permission)

Although almost all raptors have sharp, decurved talons, analysis has revealed some differences in morphology (Fowler et al. 2009; Fig. 5.46). For example, accipiters like Sharp-shinned Hawks (*Accipiter striatus*) and Cooper's Hawks (*A. cooperii*) and falcons like Peregrine Falcons (*Falco peregrinus*) have more elongated toes than most other raptors, possibly because they prey primarily on birds (sometimes birds in flight) and a larger "grasp" may improve their chances of capturing and holding onto such elusive prey. Fish-eating raptors like Osprey (*Pandion haliaetus*), Bald Eagles (*Haliaeetus leucocephalus*), and, more generally, raptors that feed on larger prey items, have talons that are more curved and more resistant to stress than those of other raptors, likely an adaptation for better grasping slippery fish and larger, stronger prey items (Tsang et al. 2019; Fig. 5.47). Large owls that take larger prey like Great Horned Owls (*Bubo virginianus*) have long talons with less curvature than other raptors. The long talons (and zygodactyl feet) allow them to better grasp larger prey and less talon curvature is needed because owls, including Great Horned Owls, can grasp prey with greater force than most other raptors.

### 5.2.6 Scavengers

Although many species of birds are facultative scavengers, feeding on carrion opportunistically

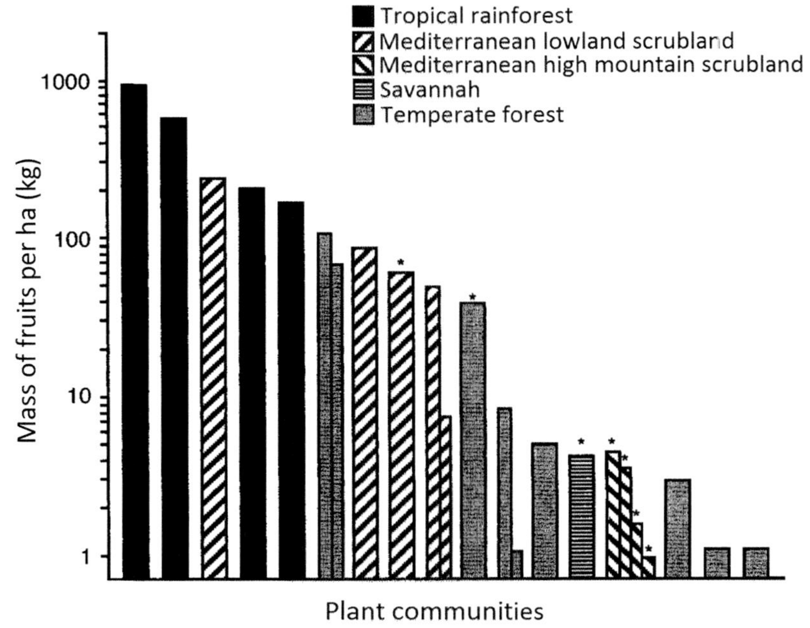


**Fig. 5.19** Production of fleshy fruits per hectare in different plant communities. The greatest number of frugivores is found in tropical rainforests where fruit is typically most abundant. Bars with asterisks indicate biomass expressed as wet mass; all others are expressed as dry mass. Multiple bars indicate data from more than one location. (Figure from Jordano 2000; © 2000 CAB International, used with permission)

(Fig. 5.48), some species specialize on carrion, notably vultures and condors. The importance of carrion as a source of energy has been little appreciated and inadequately studied (Shivik 1999; DeVault et al. 2003), and some authors have suggested that carrion may represent as much as 40–64% of the energy flow through ecosystems (DeVault et al. 2003). Because of the competition for carrion among scavengers, including vertebrates, invertebrates, and microorganisms, one important adaptation of avian scavengers is the ability to rapidly locate carrion. The ability of vultures and condors to search large areas at low energetic cost (by depending on soaring and gliding flight) is critical to their success (Shivik 2006; Fig. 5.49). As an example, vultures in east Africa may fly 240 km or more between nest sites and feeding areas (Pennycuick 1972). Andean Condors with flight recorders attached were found to flap their wings for only 1% of their flight time, specifically during takeoff and when close to the ground, and one bird flew 172 km over a period of more than 5 h without flapping its wings (Williams et al. 2020). In addition, some vultures, including Turkey Vultures (*Cathartes aura*), combine mobility with an excellent sense of smell. Interestingly, few avian carrion-specialists have a good sense of smell and this characteristic does influence their distribution. In open areas, the ability to search large areas, combined with great visual acuity (and, perhaps, a good sense of hearing; Jackson et al. 2020), permits carrion-specialists to efficiently locate carrion. However, in forested areas, excellent vision is not sufficient. As a result, vultures are only found in forested areas in the Nearctic and Neotropics because no old-world vultures have a good sense of smell. In the Neotropics, two of the most common vultures, Turkey Vulture and Greater Yellow-headed Vulture (*Cathartes melambrotus*), are known to have an excellent sense of smell (Houston 1986).

Once carrion is located, different species of vultures exhibit different feeding behaviors (Fig. 5.50). Those with the widest bills and strongest beaks ("rippers") twist and rip flesh from carcasses, feeding on skin, tendons, and other tough tissues, including bones (Bearded Vultures, *Gypaetus barbatus*; Margalida et al. 2020; Box 5.16 Prey dropping). Species with narrower skulls and intermediate beak strength ("gulpers")

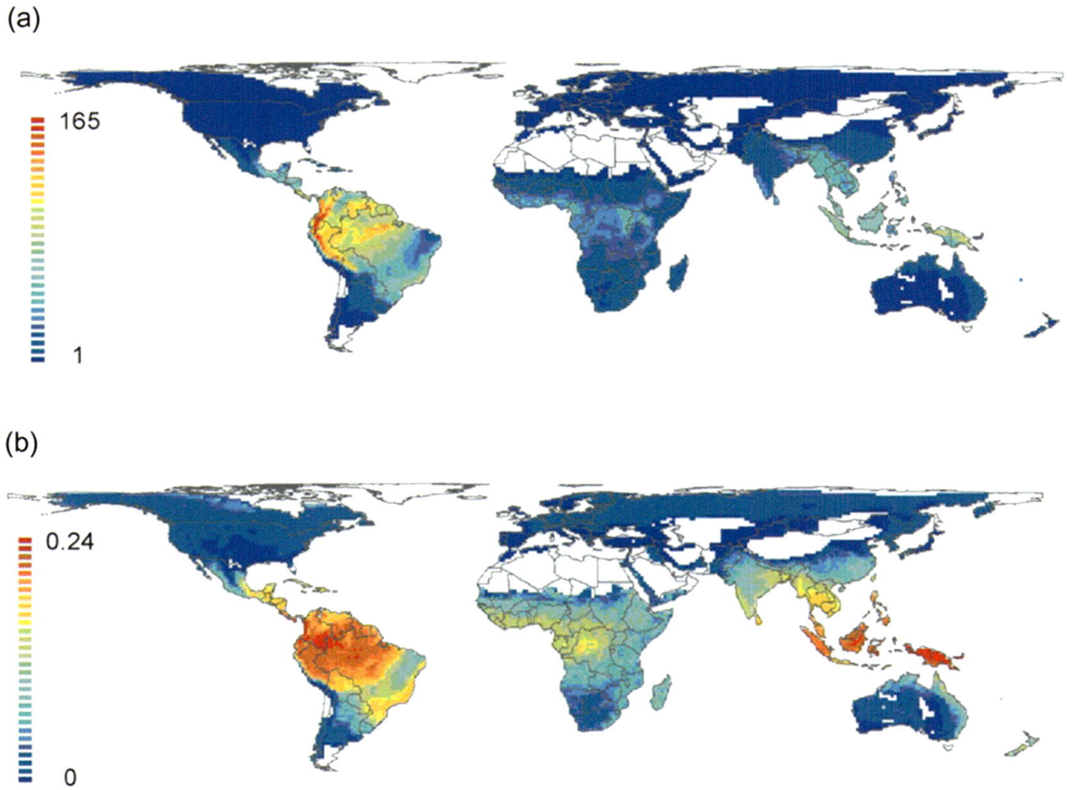


**Fig. 5.20** Global geographical patterns of avian frugivory. (**a**) Species richness of avian frugivores and (**b**) proportion of birds that are frugivorous. Species richness of avian frugivores is highest in the Neotropics, with most species found on the eastern slopes of the tropical Andes, the Guiana-Venezuela highlands, and along the Amazon River basin in Brazil. High numbers of frugivorous birds are also found in Indonesia and New Guinea. (Figure from Kissling et al. 2009; © 2009 John Wiley and Sons, used with permission)

prefer the softer viscera and often plunge their heads and necks deep into carcasses as they feed (Linde-Medina et al. 2021). Finally, vultures with the narrow, weak beaks ("scrappers") feed more on scraps on and around carcasses (Hertel 1994). "Gulpers" and "scrappers" that arrive first at carcasses may wait for a "ripper" to arrive and tear open the carcass of large vertebrates.

Another adaptive feature of carrion-specialists is the unfeathered head. Featherless heads and necks allow them to feed on carrion and reach deeply into large carcasses without getting fluids, pieces of carrion, and bacteria on their feathers. The digestive tract of carrion-specialists is similar to that of other avian carnivores, with a distensible esophagus that allows them to consume large quantities of carrion. For example, when fully distended, the crop of an African White-backed Vulture (*Gyps africanus*) can hold about 1200 g of carrion, or about 20% of its body weight (Houston and Cooper 1975).

### 5.2.7 Nectarivores

Several groups of birds are dependent to varying degrees on nectar from flowers, including hummingbirds (Trochilidae), sunbirds (Nectariniidae), sugarbirds (Promeropidae), honeyeaters (Meliphagidae), lorikeet parrots (Loriinae), honeycreepers (Thraupidae), flowerpiercers (Thraupidae, genus *Diglossa*),

**Box 5.7 Fruit color and avian fruit selection**

Several factors can influence choice of fruits by birds, including nutritional content (e.g., sugar, lipids, or certain amino acids), antioxidant content, and pulp-to-seed ratio. Because nutritional content and antioxidant content of fruits are often positively correlated with color, some, but not all (e.g., Fadzly et al. 2013), frugivorous birds may select fruits based on color (Valido et al. 2011). For example, Duan et al. (2014) found that four species of frugivorous birds in southwest China preferred red and black fruits over yellow, green. or blue fruits (see graph below), and other investigators have also reported avian preferences for red and/or black fruits (e.g., Larrinaga 2011; Barcelos et al. 2012). Preferences for fruits of certain color (e.g., red or strongly reflecting in the ultraviolet) might be a result of increased contrast with backgrounds, making them more likely to be spotted by birds (Camargo et al. 2013). However, most bird-dispersed fruits exhibit relatively little contrast with their backgrounds, suggesting that selection for increased detectability has not been a critical factor in the evolution of fruit color among fruits dispersed mainly by birds (Schaefer 2011).

For some species of birds, preferences for fruits of particular colors can provide nutritional and antioxidant benefits. For example, Eurasian Blackcaps (*Sylvia atricapillus*) prefer dark and ultraviolet-reflecting colored fruit, and these fruits have higher concentrations of an important plant antioxidant (anthocyanins; Schaefer et al. 2008). Schaefer et al. (2014) found that dark-colored fruits were preferred by Eurasian Blackcaps and Garden Warblers (*Sylvia borin*), likely because such fruits had a higher lipid content. In addition, Cazetta et al. (2012) examined the characteristics of 105 vertebrate-dispersed fruits from the Atlantic Forest of Brazil and found that a high sugar content of fruits was associated with darker colors (black or purple), colors produced by anthocyanins. Clearly, the choice of fruits by frugivorous birds can be influenced by a variety of factors, but, in some cases, fruit color provides birds with important information about the contents of those fruits.

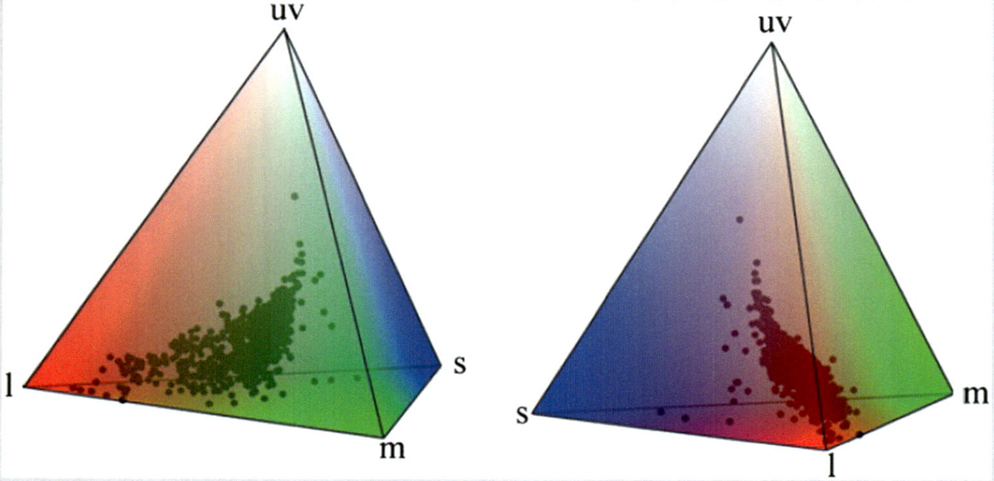


Fruit color gamut in the avian sensory color space, viewed from two different angles. Each point represents the fruit color of one plant species. Colors of 948 mostly bird-dispersed fruits occupied only 16.8% of the avian color space, i.e., the range of colors visible to birds with four cone types in their retinas, including ultraviolet-sensitive (uv), short-wave-length-sensitive (s), middle-wavelength-sensitive (m), and long-wavelength-sensitive (l). Fruit

(continued)

**Box 5.7** (continued)

colors may be limited by biochemical constraints or selection against certain colors (Figure from Stournaras et al. 2013; © 2013 John Wiley and Sons, used with permission)

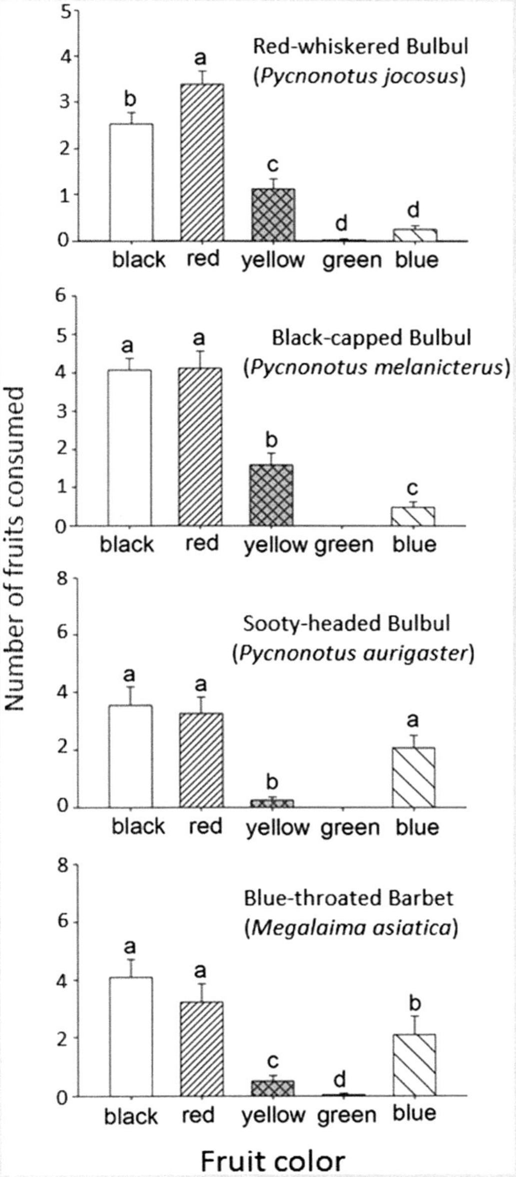


Mean number of fruits of various colors selected by four species of frugivorous birds in southwest China. Experiments were conducted with wild-caught birds and, during trials in the lab, birds were provided with 40 naturally ripe fruits (8 of each color) and the number fruits of each color selected by birds was noted. All four species selected red and black fruits more than yellow, green, or blue fruits (Figure modified from Duan et al. 2014; © 2014 Springer Nature, used with permission)

**Box 5.8 Birds as seed dispersers**

Approximately 33% of bird species disperse seeds by consuming fruits and by caching nuts and conifer seeds (Vander Wall 2001; Sekercioglu 2006). Perhaps as many as 30–50% of all plant species have vertebrate-dispersed fruit (80,000–140,000 species), and thousands of plant species benefit from the dispersal of their seeds by birds (Wenny et al. 2011). Although seeds can pass through a bird's digestive system in times ranging from less than an hour to several hours or more, birds can still disperse seeds over significant distances and, rarely, over thousands of kilometers (Popp et al. 2011) because they can fly at relatively high speeds.

California Scrub-Jay (*Aphelocoma californica*) with an acorn (Photo by Becky Matsubara, Wikipedia, CC BY 2.0, https://creativecommons.org/licenses/by/2.0/)

Seed dispersal is beneficial for both birds and plants. Fruit-eating birds obtain some nutritious fruit and, by dispersing seeds, help ensure the survival of plants that provide those fruits. For plants, seed dispersal may allow colonization of new areas and seedlings may face less competition from siblings when growing some distance from parent plants. Of course, not all birds that eat fruits, nuts, and seeds are seed-dispersers because some species typically destroy the seeds that they ingest. Such birds are referred to as seed-predators or fruit thieves (Janzen 1971; Howe 1980). The effectiveness of seed dispersal by birds can be influenced by the effect that their digestive systems have on seed viability (e.g., scarification) as well as the quality of sites where seeds are deposited.

(continued)

**Box 5.8** (continued)

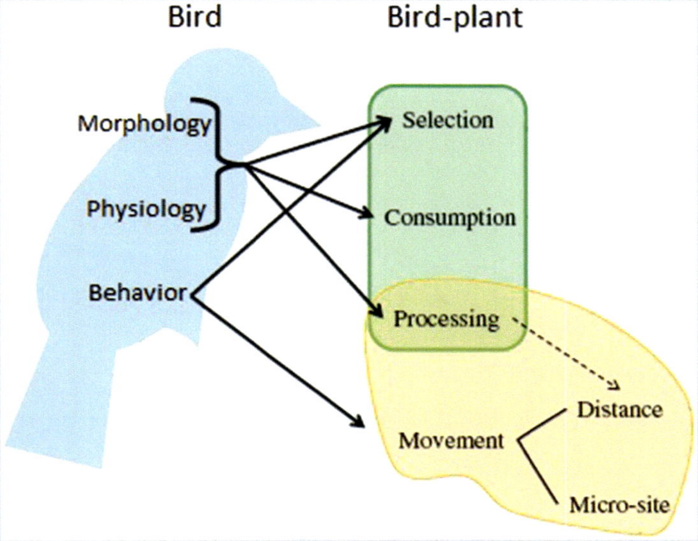


Characteristics of birds influence the fruits or seeds selected and consumed. How birds "process" fruits or seeds, i.e., how they are handled by bird bills, the effects of their digestive systems on seeds, and how long it takes for seeds to pass through the digestive system and be excreted, influences seed dispersal distances, as do the movement patterns of the bird that ingested the seed (Figure modified from Côrtes and Uriarte 2013; © 2013 John Wiley and Sons, used with permission)

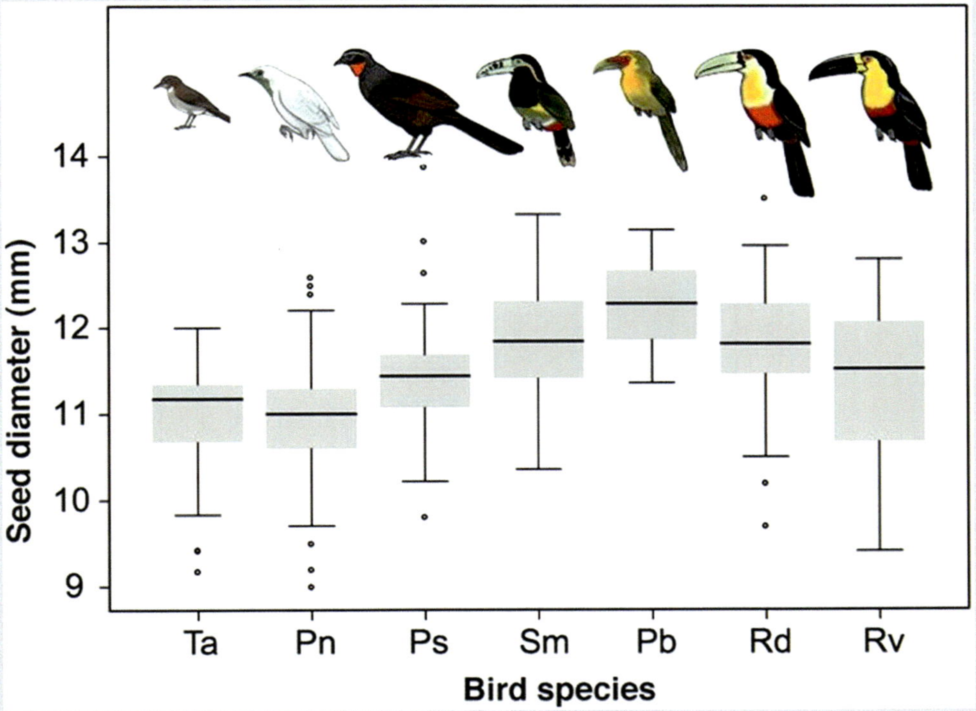


(continued)

**Box 5.8** (continued)

Birds vary in the preferred size of fruits and seeds and such differences influence the likelihood that fruits or seeds will be consumed and dispersed. For fruits and seeds ingested, seed retention times and distances moved by birds will influence seed dispersal distances. Values include means (horizontal black line), standard error (gray box), 95% confidence intervals (vertical lines), and outlier values (circles). From left to right: White-necked Thrush (Ta, *Turdus albicollis*), Bare-throated Bellbird (Pn, *Procnias nudicollis*), Rusty-margined Guan (Ps, *Penelope superciliaris*), Spot-billed Toucanet and Saffron Toucanet (Sm, *Selenidera maculirostris*; Pb, *Pteroglossus bailloni*), and Red-breasted and Channel toucans (Rd, *Ramphastos dicolorus*; Rv, *R. vitellinus*) (Figure from Galetti et al. 2013; © 2013 American Association for the Advancement of Science, used with permission)

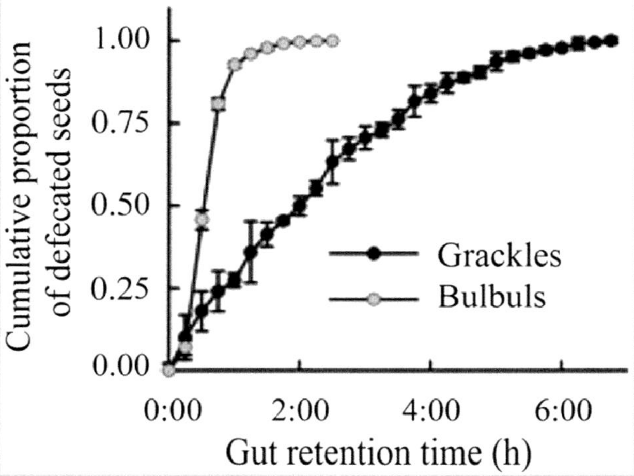


Cumulative proportion of defecated seeds as a function of gut retention time for Yellow-vented Bulbuls (*Pycnonotus xanthopygos*) and Tristram's Grackles (*Onychognathus tristramii*) (Figure from Spiegel and Nathan 2007; © 2007 John Wiley and Sons, used with permission)

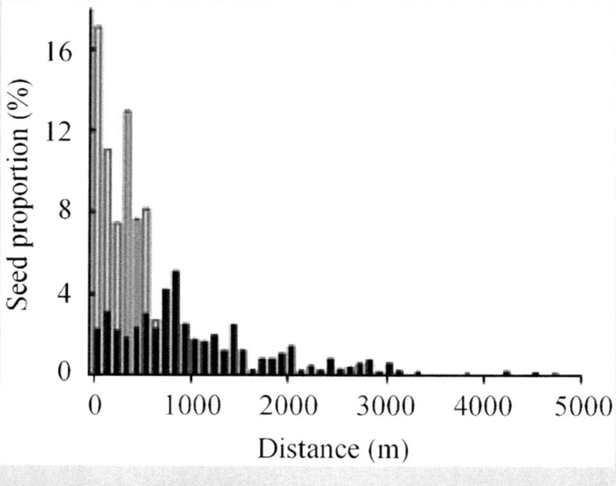


Because of the difference between species in gut retention times and the extent of their respective movements, Yellow-vented Bulbuls (gray bars) dispersed seeds closer to fruiting plants and Tristram's Grackles (black bars) dispersed seeds over greater distances. Differences among different species of birds in seed retention times and movement patterns and distances contribute to variation in distances that seeds are dispersed from source plants (Figure from Spiegel and Nathan 2007; © 2007 John Wiley and Sons, used with permission)

(continued)

**Box 5.8** (continued)

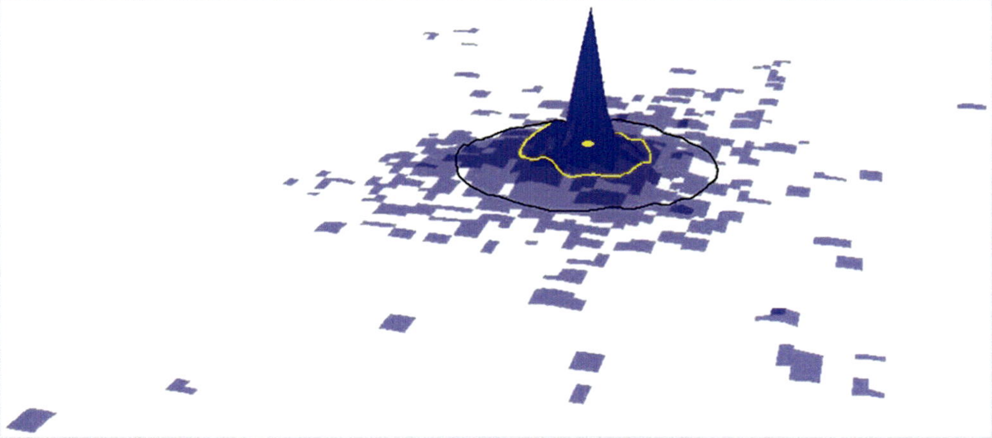

Dispersal of seeds of a fruiting tree (*Virola nobilis*; tree location = yellow dot) by toucans (*Ramphastos* spp.) in Panama. Fruits are swallowed whole, processed in the crop, and the seeds are then regurgitated. Data are based on movements by toucans (monitored by GPS tracking and accelerometers) over a 2-day period. Most seeds were regurgitated within 100 m of the tree (yellow line), but some were regurgitated well over 200 m (black line) from the tree (Figure from Kays et al. 2011; © 2011 Elsevier Masson SAS. All rights reserved, used with permission)

and flowerpeckers (Dicaeidae). These birds, called nectarivores, must supplement their diets, generally with insects and pollen, because nectar lacks essential amino acids (Box 5.17 Arthropods and pollen in the diet of nectarivorous birds). Nectar-dependent birds are generally smaller than birds with more varied or non-nectarivorous diets (Pyke 1980). Hummingbirds weigh from 2 to 20 g (Cotton 1996), sunbirds from 5 to 22 g (Cheke et al. 2001), and honeyeaters, less dependent on nectar and more dependent on insects, weigh from 8 to 250 g (Pyke 1980). Many birds also feed on nectar opportunistically, including orioles, barbets, louries, mousebirds, starlings, and warblers (Wolf and Gill 1986). The nectar of most flowers adapted for hummingbird pollination is rich in sucrose (Pyke and Waser 1981; Feinsinger 1987). Other sugars dominate in plants fed on by other nectarivores. For example, the nectar of some sunbird-pollinated plants is dominated by hexose. Interestingly, in laboratory studies, sunbirds and hummingbirds prefer more concentrated sugar solutions than found in natural nectars (Nicolson and Fleming 2003). Possible reasons for these lower-than-preferred sugar concentrations of the nectars of bird-pollinated plants are that the low concentration nectars discourage visits by bees because, for smaller bees, the costs of carrying water around are more prohibitive (Thomson 2003), and dilute nectars may encourage more visits by birds and, therefore, more efficient pollination (Martinez del Rio et al. 2001).

Nectar-feeding birds pollinate and collect nectar from plants in about 65 different families (Fig. 5.51), and flowers visited (and pollinated) primarily by avian nectarivores (called ornithophilous flowers) differ from those visited more often by insects and other pollinators such as bats (Fig. 5.52; Box 5.18 Birds as pollinators). For example, ornithophilous flowers are often red (but can also be orange, yellow, green, or blue), have nectar with high concentrations of sugar (particularly sucrose; Fig. 5.53) and lack characters associated with

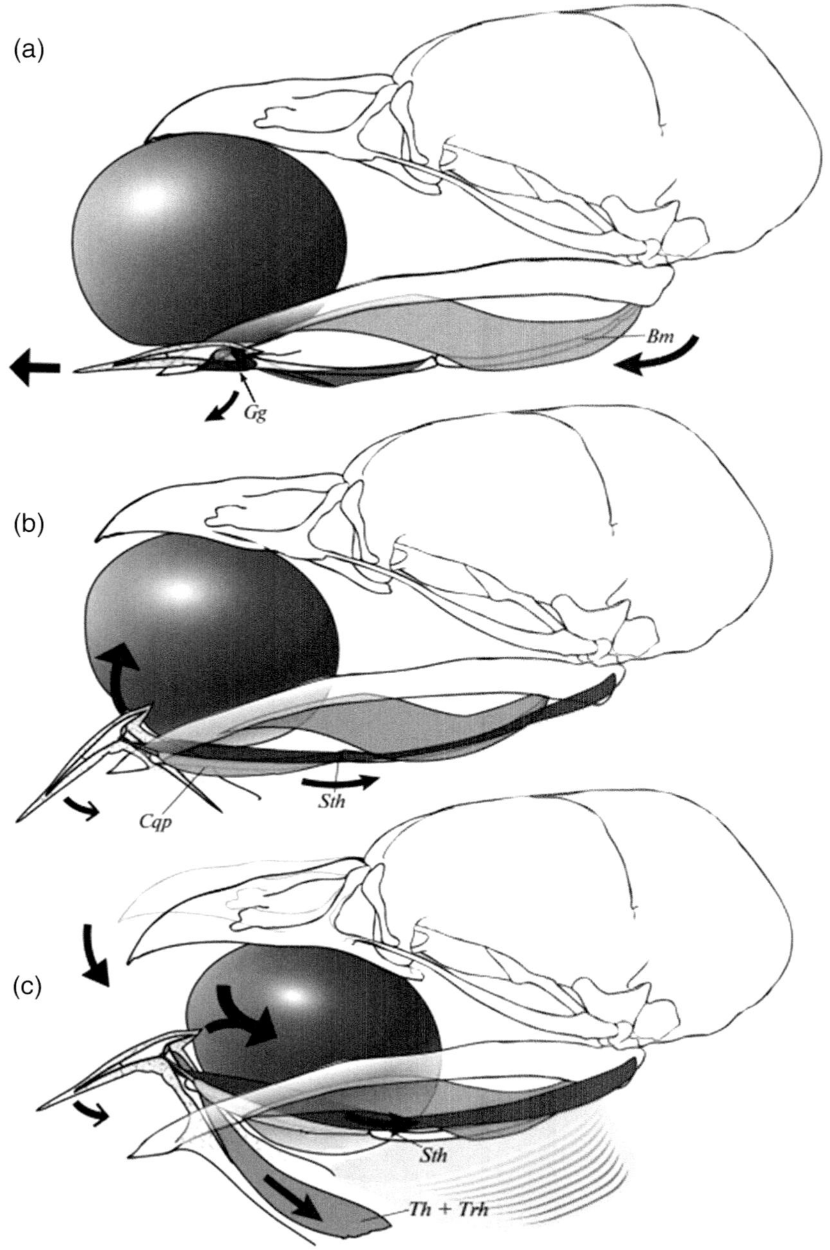


**Fig. 5.21** Drawings showing the process by which broadbills (Calyptomeninae) consume large fruits. Narrow arrows show the effect of the muscles on the hyoid apparatus during contraction, and wide arrows indicate movements of the tongue and maxilla. (**a**) The genioglossus muscle (Gg) moves the tongue to the bottom of the oral cavity and then, along with the branchiomandibular muscle (Bm), moves the tongue forward in front of the fruit. (**b**) The posterior ceratoglossus muscle (Cgp) lifts the front part of the hyoid apparatus (entoglossus) and the stylohyoid muscle (Sth) then pulls the tongue and the fruit back. (**c**) Contraction of the stylohyoid (Sth), tracheohyoid (Trh), and thyreohyoid (Th) muscles, along with pressure from the upper bill and palate push the fruit into swallowing position. (Figure from Zubkova 2019; © 2020 Pleiades Publishing, Inc., used with permission)

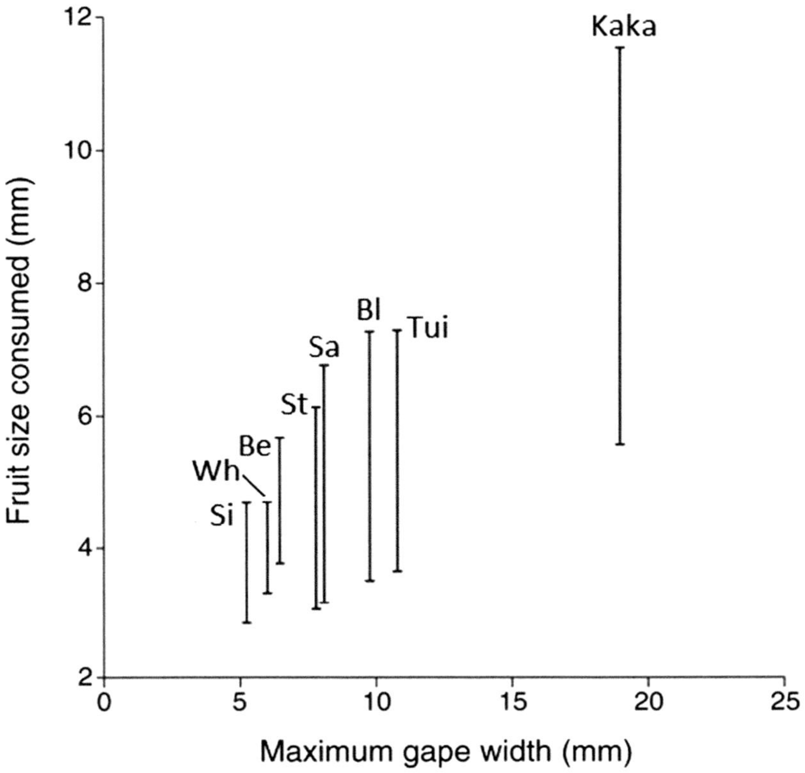


**Fig. 5.22** Relationship between maximum gape width and size of fruits consumed by eight species of frugivores in a New Zealand forest. Si, Silvereye (*Zostertops lateralis*); Wh, Whitehead (*Mohua albicilla*); Be, New Zealand Bellbird (*Anthornis melanura*); St, Stitchbird (*Notiomystis cincta*); Sa, South Island Saddleback (*Philesturnus carunculatus*); Bl, Eurasian Blackbird (*Turdus merula*); Tui (*Prosthemadera novaeseelandiae*); Kaka, New Zealand Kaka (*Nestor meridionalis*). (Figure modified from Burns 2013; © 2013 by the Ecological Society of America, used with permission)

flowers pollinated by insects or bats, such as scent. Other characteristic of bird-pollinated flowers include long floral tubes, either a downward orientation (for hummingbird-pollinated flowers; Fig. 5.54) or the presence of perches (for nonhovering birds like sunbirds; Fig. 5.55) (Cronk and Ojeda 2008), and a mechanism for transferring pollen on top of a bird's bill, head, or less often on a bird's back, feet, or tongue (Wester and Claßen-Bockhoff 2007; Cronk and Ojeda 2008; Figs. 5.56, 5.57, and 5.58).

Most nectarivores are found in the tropics where flowers and nectar are available year-round (Fig. 5.59). Temperate-zone nectarivores, like some hummingbirds, are migratory. Most nectarivores have relatively long narrow bills and tongues for probing flowers and extracting nectar. Bill shape, however, can be influenced by flower morphology (Fig. 5.60). Bill length does, of course, influence the ability of a nectarivore to potentially obtain nectar from a flower. However, flower width is also important. For example, both short- and long-billed hummingbirds can obtain nectar from long, wide flowers, but short-billed hummingbirds may not be able to obtain nectar from long, narrow flowers because the nectar might be beyond the reach of their bills (Temeles et al. 2002). Of course, some birds, called nectar robbers, "cheat" and remove nectar from flowers through a hole pierced or bitten in the corolla (Box 5.19 Nectar robbers). Many nectar robbers, including flowerpiercers (genus *Diglossa*) and some short-billed hummingbirds, have bill serrations (serrate tomia) and hooked bills to facilitate access to the nectar (Ornelas 1994; Fig. 5.61). The tongues of nectarivores are typically tubular and brushy-tipped for efficient ingestion of nectar. Some nectarivores, notably the hummingbirds, feed by hovering, but most others feed by simply hanging on to flowers and probing with their bills.

**Box 5.9 Ballistic food transport**

Toucans (Order Piciformes: Family Ramphastidae) and hornbills (Order Bucerotiformes: Family Bucerotidae) often eat whole fruits and play important roles in dispersing seeds. When consuming whole fruits, toucans and hornbills pick, or pick up, fruits with the tips of their bills, their heads are then rotated upward and backward and the angle between upper and lower bills increases to release the fruit that then drops to the back of the oral cavity where it can be swallowed (Baussart et al. 2009; Baussart and Bels 2011).

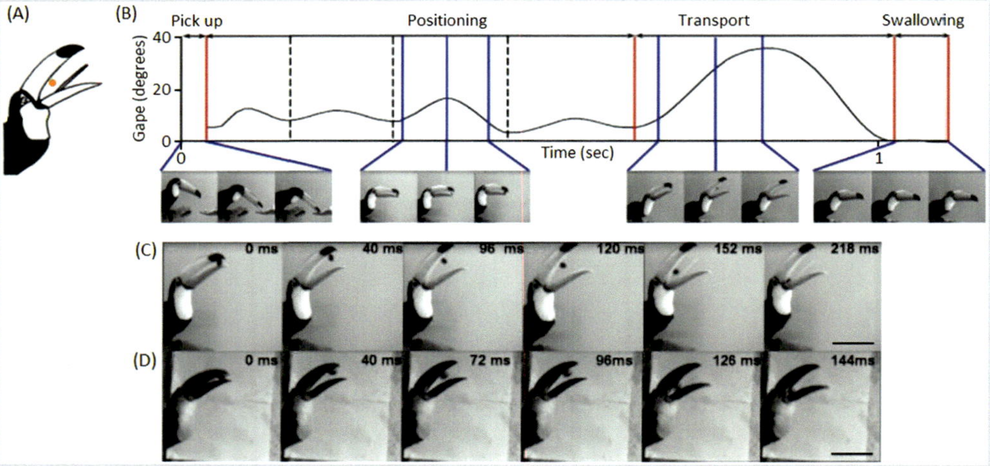


Feeding behavior of toucans. (**a**) Toucan "transporting" food item from bill tip to the back of the oral cavity. (**b**) Steps involved in ballistic food transport. Food items like fruits are picked up at the tip of the bill and properly positioned. The food item is then "transported" as the head rotates up and back while the bill opens, releasing the food item that then falls to the back of the oral cavity (pharynx) so it can be swallowed. (**c**) Ballistic food transport by a Toco Toucan (*Ramphastos toco*). (**d**) Ballistic food transport by a Channel-billed Toucan (*Ramphastos vitellinus*) (Figure from Baussart et al. 2009; © 2009 Wiley-Liss, Inc., A Wiley Company, used with permission)

Although primarily fruit-eaters, toucans, and hornbills also use ballistic food transport when they feed on arthropods, reptiles, eggs of reptiles and birds, and young birds. This method of feeding has presumably allowed the relatively large and long bills of these birds to also serve other important functions such as allowing them to reach further for fruits at the end of branches or into cavities or nests for other types of food (e.g., eggs or nestlings), intraspecific communication, and thermoregulation (Tatersall et al. 2009).

(continued)

**Box 5.9 (continued)**

Toco Toucans (*Ramphastos toco*) reaching to feed on (**a**) the buds of red silk-cotton tree (*Bombax ceiba*) and (**b**) the flowers of a ceiba tree (*Ceiba pubiflora*) (Figure from Silva 2019; © Neotropical Ornithological Society, used with permission)

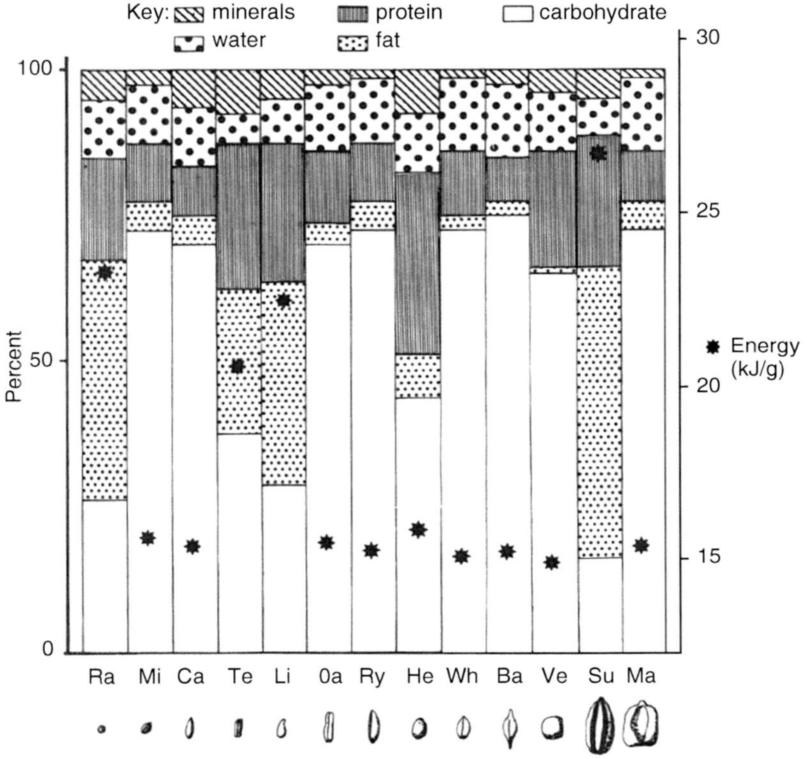


**Fig. 5.23** Nutrient analysis of several types of seeds. Ra, Rape (*Brassica rapa*); Mi, Millet (*Panicum miliaceum*); Ca, Canary grass (*Phalaris canariensis*); Te, Teasel (*Dipsacus sylvestris*); Li, Linseed (*Linum usitatissimum*); Oa, Oats (*Avena sativa*); Ry, Rye (*Secale cereale*); He, Hemp (*Cannabis sativa*); Wh, Wheat (*Triticum aestivum*); Ba, Barley (*Hordeum vulgare*); Ve, Common vetch (*Vicia sativa*); Su, Sunflower (*Helianthus annuus*); Ma, Maize (*Zea mays*). (Figure from Diaz 1990; © British Ornithologists' Union, Published by John Wiley and Sons, used with permission)

### 5.2.8 Herbivores

Fewer than 1% of birds are strictly herbivores (i.e., grazing on grasses, sedges and forbs, or browsing on shrubs, tree leaves, or needles; Gordon et al. 2019), including Common Ostriches (*Struthio camelus*), rheas (*Rhea* spp.), Hoatzins (*Opisthocomus hoazin*), and Kakapos (*Strigops habroptilus*), many species in the family Anatidae (ducks, geese and swans), screamers, grouse, and plantcutters (Morton 1978; Klasing 1998; Table 5.5). However, many other species graze or browse as part of a more varied, or omnivorous, diet. About 20 ducks, including whistling ducks, some diving ducks including pochards, and several dabbling ducks including wigeons, Gadwalls (*Anas strepera*), and Maned Ducks (*Chenonetta jubata*) are primarily herbivorous for at least part of the year, but have a more omnivorous diet during spring and summer to meet protein requirements for breeding (Sedinger 1997). Although often abundant, the high fiber content of plants makes digestion difficult and, this, in turn, means that plant material must remain in the digestive system for extended periods. For flying birds, particularly smaller

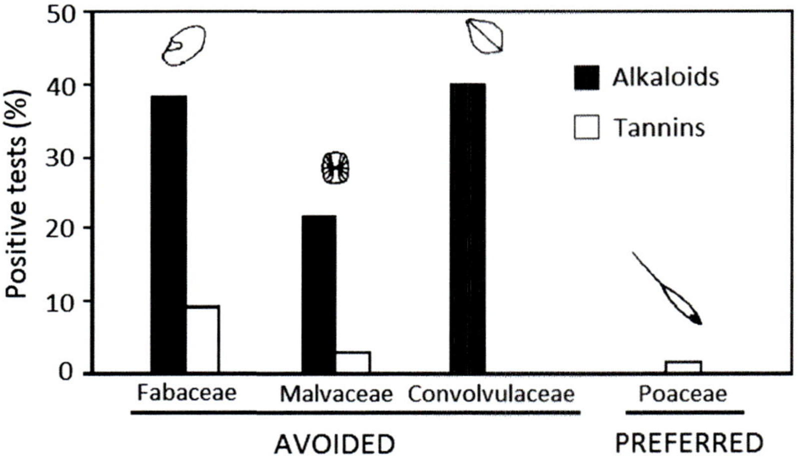


**Fig. 5.24** Percentage of species in three families of plants that have potentially toxic secondary compounds in their seeds and are typically avoided by finch species, plus percentage of species in the family Poaceae with those compounds that are preferred by finch species. Finches clearly prefer seeds of plants with lower levels of these secondary compounds. (Figure modified from Díaz 1996; © 1996 Published by Elsevier Inc., used with permission)

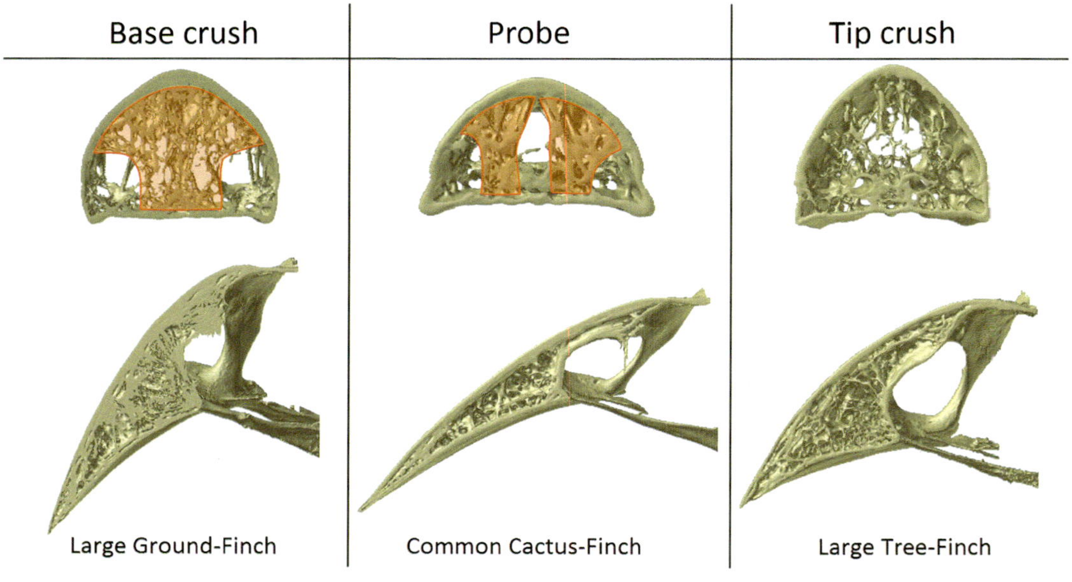


**Fig. 5.25** Bills and trabeculae of three Darwin's finches that vary in diet and how they crush seeds. Top, cross-sections at the base of the bills; Bottom, side views of the interior of the upper bills. Large Ground-Finches (*Geospiza magnirostris*) and Large Tree-Finches (*Camarhynchus psittacula*) are primarily seed-eaters, whereas Common Cactus-Finches (*Geospiza scandens*) feed primarily on the pollen and nectar of *Opuntia* cacti as well as *Opuntia* fruits and seeds. As a result, selection has favored elongation of the bill, making it more useful for probing. The bills of seed-eating birds are typically broad and deep and reinforced with trabeculae. Birds that crush seeds toward the back or base of the bill, like Large Ground-Finches have a greater density of trabeculae at the base of the bill, whereas those that crush seeds closer to the tip of the bill, like Large Tree-Finches, have a greater density of trabeculae near the tip. (Figure modified from Genbrugge et al. 2012; © 2012 John Wiley and Sons, used with permission)

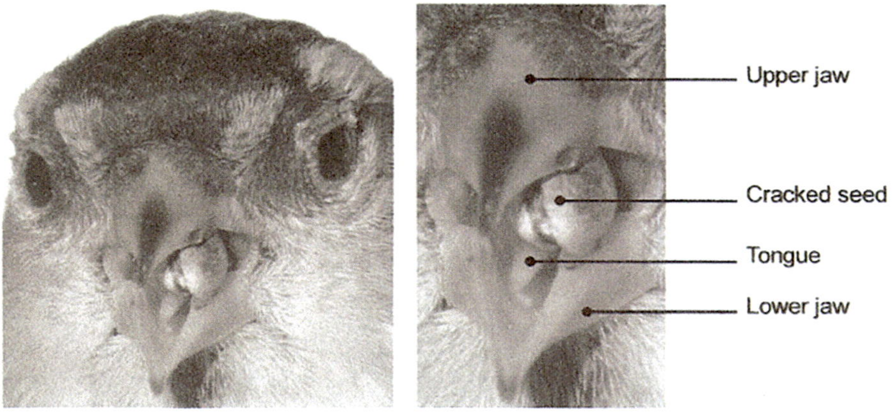


**Fig. 5.26** European Serin (*Serinus serinus*) with a seed positioned in its bill. Note how the tongue is used to hold the seed in position. Seed "processing" typically involves four phases, including positioning, biting, dehusking, and swallowing. During seed positioning, the lower bill of Domestic Canaries (*Serinus canaria domestica*) was found to move extremely fast, with an average frequency of 18 Hz (range = 10–25 Hz). During the positioning (mean = 64%) and swallowing (mean = 59%) phases, more movements of the canaries' lower bill were dorso-ventral than during the dehusking (mean = 49%) and biting (mean = 44%) phases. During the latter two phases, most movements of the lower bill were medio-lateral. Rapid movements in different directions by the upper and lower bills during the biting (seed cracking) and dehusking phases allow the removal of husks without breaking seeds into, and possibly losing, smaller pieces of seeds (Mielke and Van Wassenbergh 2022). (Figure from van der Meij 2005; used with permission of Marian van der Meij)

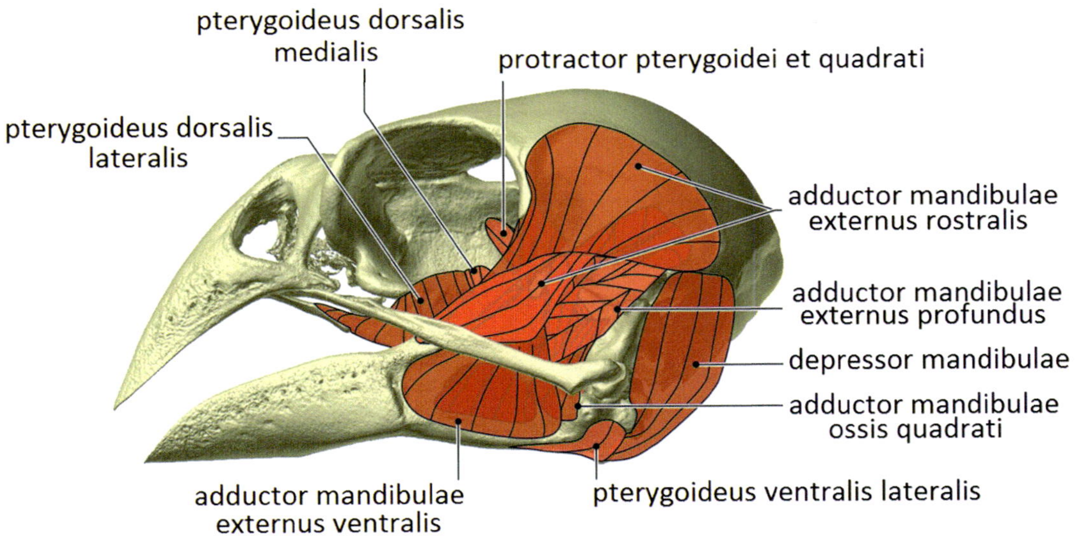


**Fig. 5.27** Jaw muscles of a Medium Ground Finch (*Geospiza fortis*). The adductor muscles are the primary jaw-closing muscles, with the pterygoideus dorsalis muscles also playing a role; the pterygoideus muscles also move the lower jaw back-and-forth. The depressor mandibulae opens the mouth. The hardest seeds eaten by Medium Ground Finches have an average hardness of about 45 N (or requiring a force of about 12 kg/cm$^2$ to crack open), with a maximum hardness of some of the seeds that are cracked by these birds reaching up to 100 N (or requiring about 27 kg/cm$^2$ to open) (De Leon et al. 2011). (Figure modified from Genbrugge et al. 2011; © 2011 John Wiley and Sons, used with permission)

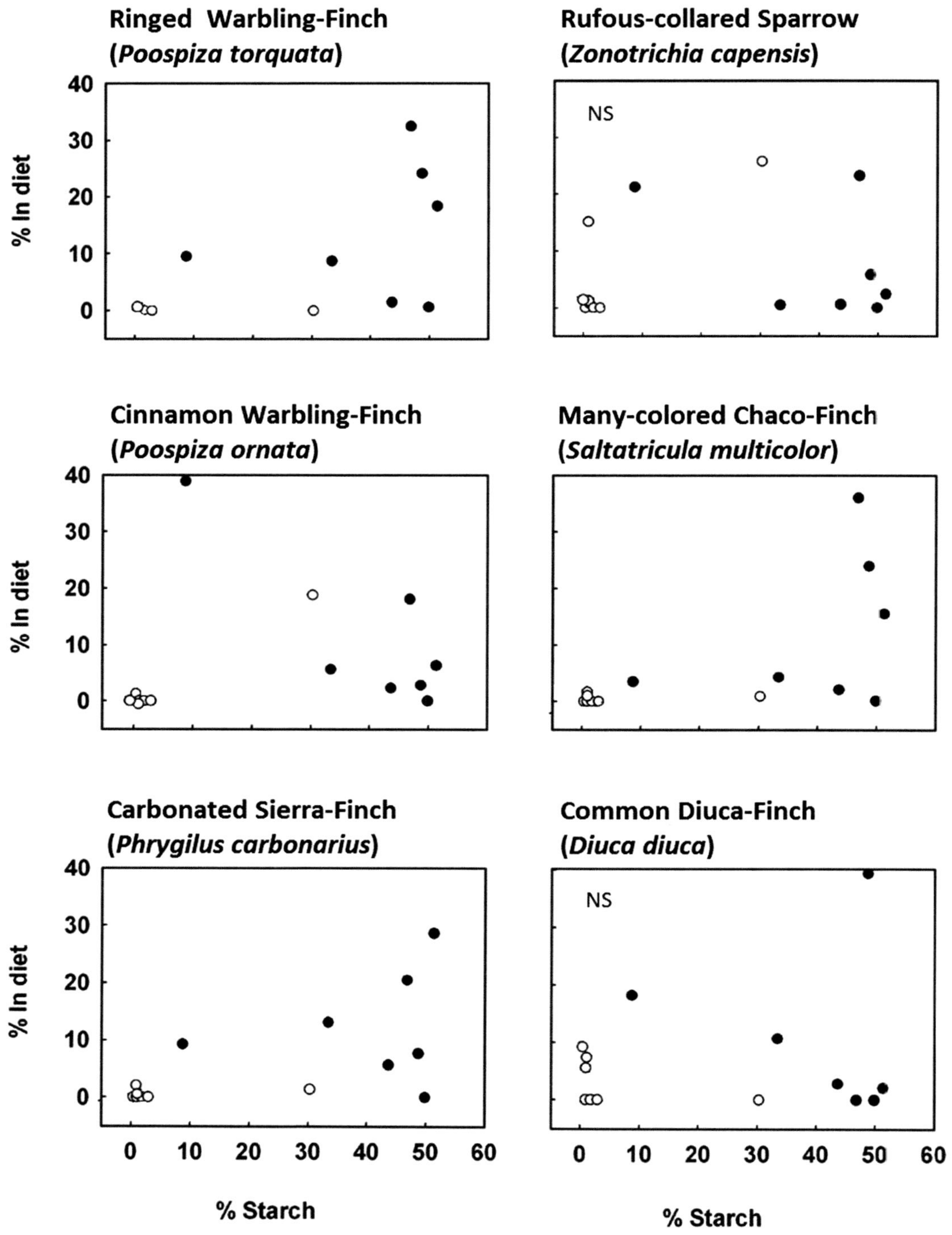


**Fig. 5.28** Relationships between the mean percentage of seeds in the diet of Ringed Warbling-Finches, Cinnamon Warbling-Finches, Carbonated Sierra-Finches, Rufous-collared Sparrows, Many-colored Chaco-Finches, and Common Diuca-Finches and the percent starch concentration in seeds of several species of plants. Four species had a significant preference for seeds with higher starch content; two did not (as indicated by NS in the upper-left corner of the graphs). Black circles represent grass species, and white circles forb species. (Figure modified from Ríos et al. 2012; © 2012 Oxford University Press, used with permission)

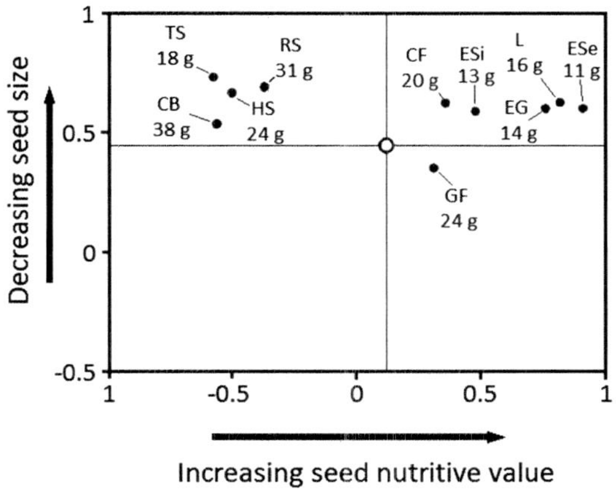


**Fig. 5.29** In experiments with captive birds of various sizes (typical mass of each species is provided) provided with a variety of seeds that varied in size and nutritive value, seed size selection was influenced by bill size (depth and width) (Diaz 1990). In addition, smaller seed-eating birds tended to select seeds with greater nutritive and energetic value. TS, Tree Sparrow (*Passer montanus*); CB, Corn Bunting (*Miliaria calandra*); HS, House Sparrow (*Passer domesticus*); RS, Rock Sparrow (*Petronia petronia*); CF, Chaffinch (*Fringilla coelebs*); ESi, Eurasian Siskin (*Carduelis spinus*); GF, Greenfinch (*Carduelis chloris*); EG, European Goldfinch (*Carduelis carduelis*); L, Linnet (*Carduelis canabina*); and ESe, European Serin (*Serinus serinus*). Open circle in graph = value for both variables indicating no selection at all. (Figure modified from Diaz 1990; © 1990 John Wiley and Sons, used with permission)

birds, this additional weight reduces flight efficiency. As a result, herbivorous birds tend to be less dependent on flight. For example, Common Ostriches are flightless, geese, swans, and diving ducks are aquatic, and grouse typically walk or run rather than fly. Small herbivores dependent on flight must be selective and consume vegetation high in protein and low in fiber. For example, White-tipped Plantcutters (*Phytotoma rutila*) depend entirely on leaves (91% of their diet), fruits (7%), and flowers (2%), but the leaf material is high in protein content and low in fiber content (Bucher et al. 2003).

Morphological adaptations for birds with plant-dominated diets are largely internal and involve specializations of the digestive tract discussed below. Externally, herbivorous birds have bills modified to permit the harvesting of plant material. For example, plantcutters have serrations on the edges of their bills (serrate tomia) that allow them to cut leaves (Klasing 1998). The anterior portion of the bill of Canada Geese (*Branta canadensis*) is specialized for grasping and holding grass, whereas the posterior portion is serrated and cuts the grass (Fig. 5.62). The bills of diving ducks (Anatidae) are flat and shovel-shaped for the efficient removal of rhizomes and tubers from mud (Klasing 1998).

### 5.2.9 Omnivores

Although most birds do specialize to varying degrees on specific types of food, many are also omnivorous and consume a wide variety of plant and animal foods. Burin et al. (2016) reviewed the literature on avian diets and, based on information collected for 9876 species of birds, identified 1158 species (11.7%) as omnivores. These authors also determined that the diets of all omnivores include some insects, most also consume some fruits and seeds, but very few feed on carrion. Although many species are omnivorous, Burin et al. (2016) assessed speciation and extinction rates within different avian dietary guilds and concluded that omnivory was associated with higher extinction rates and lower speciation rates than other dietary guilds. To explain this, these authors suggested that "generalist species might be at a disadvantage when competing with specialists. Such a 'jack of all trades is a master of none' mechanism (species that can utilize several resources while performing poorly at utilizing specific resources) could leave a signature at the macroevolutionary scale." A guild with high extinction rates and low speciation rates might be expected to become increasingly rare, but Burin et al. (2016) hypothesized that many species of extant birds are omnivous because species in other dietary guilds often transition to omnivory.

Because omnivores, such as many jays and crows (Corvidae), eat a variety of things, bill morphology tends to be intermediate between those of more specialized birds. For example, Steller's Jays (*Cyanocitta stelleri*) consume a wide variety of animal and plant food, including

**Box 5.10 Ingesting indigestible plastics**

Humans produce prodigious quantities of plastics, with an estimated 326 metric tons produced worldwide in 2018 and 334 metric tons in 2019 (PlasticsEurope 2020). A substantial portion of this plastic ends up in the oceans via river and atmospheric transport, beach littering, and fishing and shipping activities (Kershaw and Rochman 2015). Once in the ocean, larger pieces of plastic debris are broken down into microplastics and nanoplastics that become incorporated into the marine food web. In addition to plastics, other man-made debris can be found in oceans worldwide, including, but not limited to, nylon line, hooks, rope, foam, rubber, and fibers from textiles. Plastics and other debris are sometimes ingested by seabirds, either directly because they mistake plastics/debris for food (Derraik 2002) or indirectly because micro- and nano-plastics are present in the tissues of their prey (Amaral-Zettle et al. 2020). In a review of the literature, Kühn and van Franeker (2020) noted that 180 of 409 known species of seabirds have been found with plastics in their digestive systems. In addition, of 43,525 individual seabirds sampled, 12,064 (27.7%) had ingested plastics, with albatrosses, fulmarine petrels, and shearwaters having the highest percentage of individuals with ingested plastic.

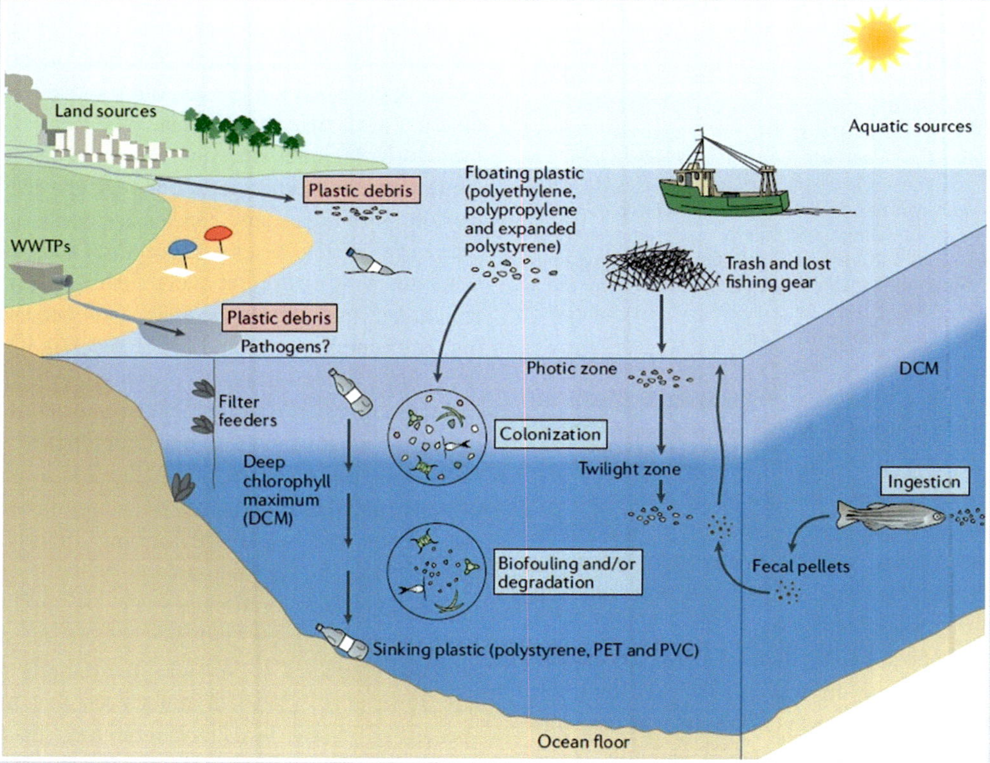


Plastic debris enters the ocean through both aquatic and land-based sources. Some plastic items will float on the surface for a while. As they become weighted down by biofouling, they begin to sink into the water and, ultimately, to the ocean bottom. Mechanical, photochemical, and biological forces break down plastic debris into microplastics and nanoplastics that subsequently become incorporated into the marine food web. Microorganisms can attach to plastic in the water, including disease-causing pathogens. An estimated 99% of plastic in the oceans is not on the surface, but either in the water column or the ocean bottom. WWTPs, waste-water treatment plants;

(continued)

**Box 5.10** (continued)

PET, polyethylene terephthalate; PVC, polyvinyl chloride (Figure from Amaral-Zettle et al. 2020; © 2020 Springer Nature Limited, used with permission)

Ingestion of plastics and other debris by seabirds can have both lethal and sublethal effects, but the effects of such ingestion, or lack thereof, vary with the type of debris ingested. For example, Roman et al. (2020a) examined the cause of death of 215 seabirds whose carcasses were recovered along beaches in Australia and New Zealand and found that debris ingestion caused the death of 10 of them (4.7%). However, many birds where the cause of death could not be determined had ingested plastic items, suggesting that ingested plastic may have sublethal effects in the short-term, but potentially lethal effects over longer periods of time. These authors also found that species of seabirds that feed primarily on cephalopods and crustaceans were more likely to ingest debris than those that feed on fish, likely because debris is more likely to resemble cephalopods and crustaceans than fish. In addition, immature seabirds were more likely to ingest debris than adults, suggesting that less-experienced birds are more likely to mistake debris for food than more-experienced adults (Roman et al. 2020a).

The amount of debris ingested also matters. Roman et al. (2019) collected 1733 carcasses of seabirds representing 51 species and found that those that died due to debris ingestion had more debris items in their digestive systems than those that died of other causes. Their analysis also revealed that obstruction of the gastrointestinal tract was the leading cause of death, and ingestion of balloons was more often lethal than ingestion of other types of debris.

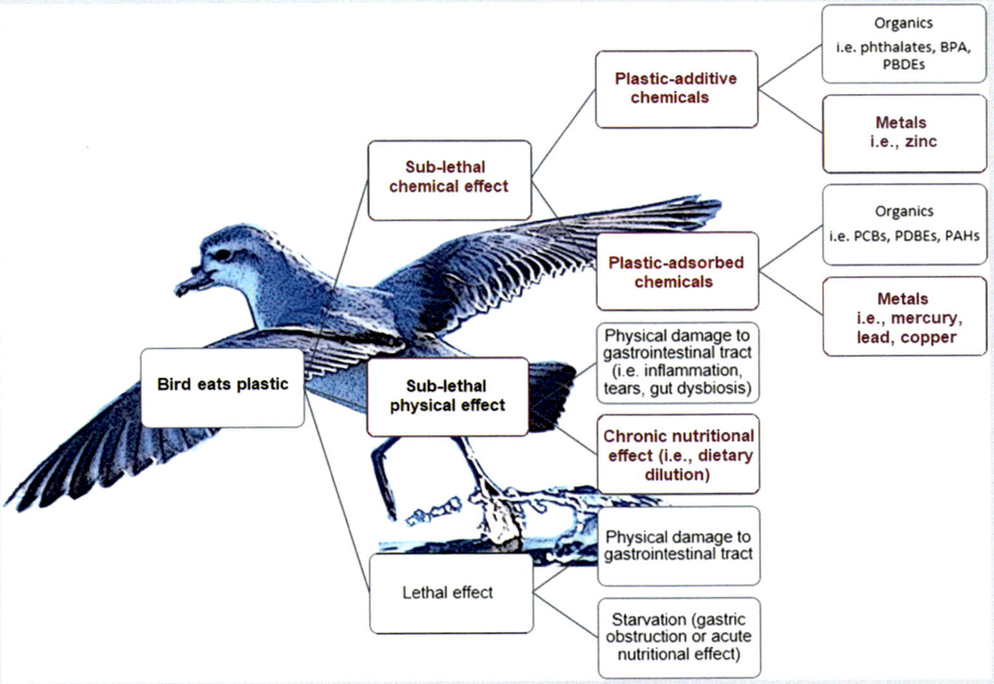


Ingestion of plastic by birds can have lethal and sublethal health effects, including physical effects, linked to physical damage and nutrition, and chemical effects due to plastic-additive and plastic-adsorbed chemicals (Figure modified from Roman et al. 2020b; open-access article licensed under a Creative Commons Attribution 4.0 International License, https://creativecommons.org/licenses/by/4.0/)

(continued)

**Box 5.10** (continued)

Mechanical, photochemical, and biological forces break down plastic debris in the ocean into microplastics and nanoplastics. Plastic may contain additives, such as phthalates, bisphenol A, nonylphenol, and brominated flame retardants, that can have numerous harmful effects on living organisms (e.g., Murata and Kang 2018). The small size of micro- and nanoplastics increases their surface area so they can adsorb marine pollutants, such as persistant organic pollutants, metals, and microrganisms (including potential pathogens), increasing their potential toxicity (Engler 2012). Micro- and nanoplastics enter marine food webs when ingested by zooplankton and filter-feeding organisms. For seabirds, toxins associated with plastic enter their bloodstreams and, depending on the toxin and their levels in the blood, can have negative impacts ranging from neurological, behavioral, to reproductive (Lavers et al. 2019).

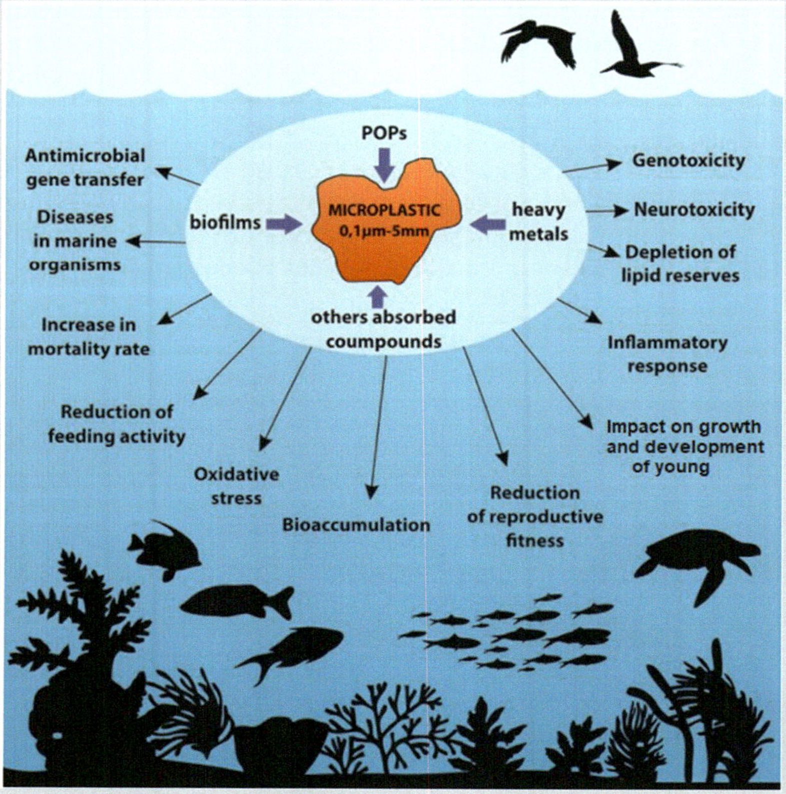


Potential ecotoxological effects of microplastics and associated pollutants on seabirds and other aquatic organisms. POPs, persistant organic pollutants (Figure from de Araujo et al. 2021; © 2021 Elsevier B.V., used with permission)

Quantifying the impacts of plastic/debris ingestion on seabird populations is difficult because many spend most of their time at sea, especially prebreeding juveniles, and ingestion of micro- and nanoplastics may have difficult-to-detect sublethal effects (Lavers et al. 2019). Importantly, some species of seabirds do not appear to ingest plastic or ingest plastic in small amounts (Kühn and van Franeker 2020). However, among those that do, e.g., albatrosses, some petrels, and

(continued)

**Box 5.10** (continued)

shearwaters, Lavers et al. (2019) noted that the negative impacts of plastic "are likely present at the molecular or cellular levels . . . and may not result in . . . death, but in poor health." Although uncertainty remains concerning the impacts of plastics and other debris on seabirds and their populations, as noted by Rochman et al. (2016), ". . . there appears to be enough evidence for policy makers to recognize the hazards and take a precautionary and/or anti-catastrophe approach . . ., by beginning to mitigate the problem now before there is any irreversible harm from such pervasive materials."

**Box 5.11 Nonergodic hunting by raptors**

If raptors searching for prey were ergodic, then an investigator observing the behavior of one raptor 50 times would obtain the same results by observing the behavior of 50 raptors a single time. In the same way, coin flipping (heads or tails) would be ergodic if the average result of one person flipping a coin 50 times would be the same as 50 people each flipping a coin once. Although coin flipping is ergodic, Vilk et al. (2022) determined that the behavior of three species of raptors searching for prey was nonergodic. Using an automated tracking system, these investigators tracked 44 Barn Owls (*Tyto alba*) over 4710 nights (with more than $5 \times 10^7$ locations), 16 Black-winged Kites (*Elanus caeruleus*) for 1619 days (and more than $10^7$ locations), and 10 Eurasian Kestrels (*Falco tinnunculus*) for 508 days (and more than $9 \times 10^6$ locations). Analyses revealed two types of behavior, with flights between search locations being ergodic (and predictable) and behavior at search locations (i.e., area-restricted searching) being nonergodic (and unpredictable). Whereas flights between search locations were relatively long and directional, behavior at search locations involved nonergodic waiting times, i.e., rather short and very long periods between successive movements (i.e., fat-tailed distribution) that varied among different individuals. In other words, ". . . there are many ways to hunt within a local patch but only a limited number of ways to commute between distant patches" (Vilk et al. 2022). Based on these results, Vilk et al. (2022) postulated that highly variable, individual-specific hunting tactics of raptors might prove more successful than less variable hunting tactics. Similarly, Bartumeus et al. (2016) noted the potential advantage of using different foraging behaviors for avian predators. In summary, Vilk et al. (2022) stated that ". . . as it is hard to catch highly apprehensive moving prey, adopting irreproducible (and thus unpredictable) movement tactics may prove beneficial for a predator, rather than committing to a single tactic or behavior."

(continued)

**Box 5.11** (continued)

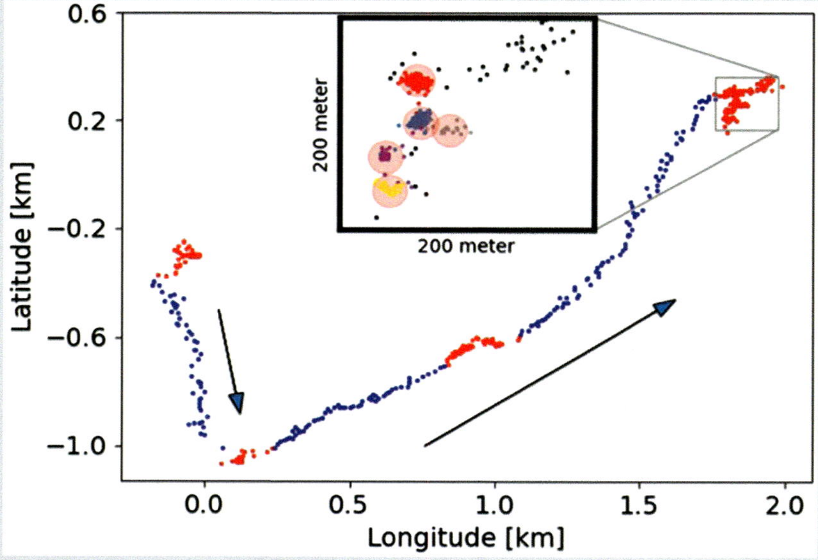


Foraging tracks of a female Barn Owl based on 2280 locations over a period of 75 min. During this time, the female actively searched for prey at four different locations (red dots) for periods of 4, 2, 4, and 56 min, respectively. Rapid movements between search locations are indicated by the blue dots and the arrows indicate the direction of movements. The inset shows more detailed movements in the area where the female searched for prey for 56 min, with five clusters (indicated by circles) where the female searched for 14, 36, 2, 3, and 1 min(s), respectively (Figure from Vilk et al. 2022; Published by the American Physical Society under the terms of the Creative Commons Attribution 4.0 International license, https://creativecommons.org/licenses/by/4.0/)

arthropods, nuts, seeds, berries, fruits, small vertebrates, and eggs and young of smaller birds (Walker et al. 2020), and have bills that function well as a forceps to pick up arthropods, but are thick and strong enough to use as a hammer to break open seeds.

## 5.3 Group or Flock Foraging

Many birds, particularly insectivores, frugivores, and granivores, forage in groups, particularly during the nonbreeding season. Some groups include only members of the same species; others have individuals from several species (mixed-species flocks). Two main nonexclusive hypotheses have been proposed to explain why birds often forage in flocks: (1) improved foraging efficiency (Fig. 5.63) and (2) reduced risk of predation (Morse 1977; Powell 1985). Being a flock member could improve foraging efficiency (number of food items obtained per unit time) because (1) a flock of birds might be better at locating food items (i.e., more eyes) than a single individual, (2) for birds that feed on mobile prey (e.g., insectivorous and piscivorous birds), the movements of other flock members might flush prey items that might have otherwise gone undetected (Kotagama and Goodale 2004), and (3) individuals foraging alone are more likely to visit sites where other birds might have recently foraged; individuals moving together in flocks may be less likely to do so (Cody 1971). Flock members may also have a reduced risk of

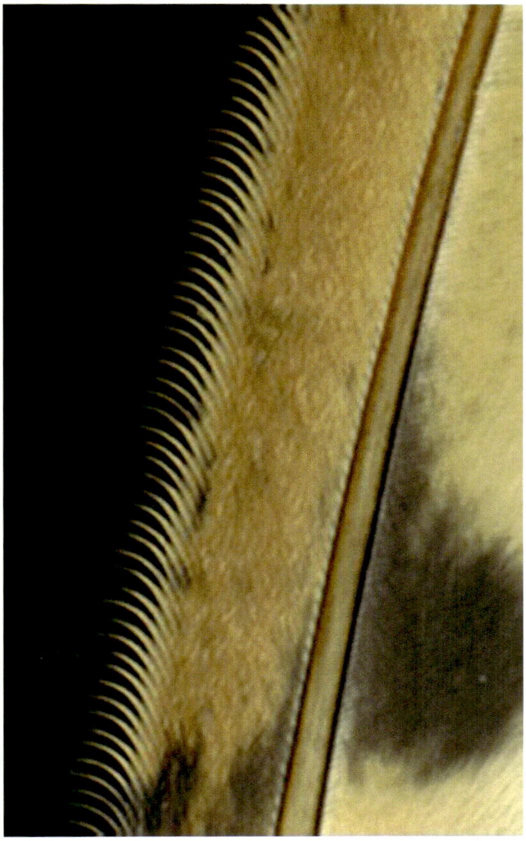

**Fig. 5.30** Leading-edge serrations on a primary feather of a Barn Owl (*Tyto alba*). The serrations are formed by single barbs (with very short barbules) that bend upwards. Serrations are about 1.8–2.7 mm long and about 600 μm apart. (Figure from Klän et al. 2010; Reprinted by permission of the American Institute of Aeronautics and Astronautics, Inc.)

predation (1) because a group of birds is more likely to detect an approaching predator (more eyes) than a single individual (Pulliam 1973), (2) due to a "dilution effect," i.e., an individual in a group of several individuals is less likely to be targeted by a predator (Foster and Treherne 1981), and (3) because of a "confusion effect," i.e., with many birds simultaneously attempting to avoid predation, a predator may be momentarily confused and thus less able to single out and attack an individual bird (Zoratto et al. 2009). In addition, because "more eyes" are available to scan for predators, each individual bird in a flock can spend less of their time scanning for predators; collectively, flock members will still be spending more time scanning for predators even though each individual spends less time doing so. Of course, that means that birds in a flock also have more time to forage, further increasing their foraging efficiency. Although foraging efficiency of flock members may be greater than that of nonflocking individuals, efficiency may decline as the number of birds in a flock increases. For example, Stolen et al. (2012) examined the foraging behavior of three species of largely piscivorous wading birds and found that, for two species (Great Egrets, *Ardea alba*, and Tricolored Herons, *Egretta tricolor*), the number of prey captures per minute was higher for birds in medium-sized groups than for individual birds or birds in small flocks (Fig. 5.64). However, foraging success was lower for birds in large flocks than in medium-sized flocks (Fig. 5.64), possibly due to interference competition (more individuals competing with each other for prey and thus reducing the likelihood of success) (Stolen et al. 2012).

## 5.4 Interclass Cooperative Feeding

Birds are sometimes able to increase their foraging efficiency by associating with other organisms, particularly mammals (both aquatic and terrestrial), but other organisms as well (e.g., army ants, e.g., Martinez et al. 2018). For example, as noted previously, Cape Gannets (*Morus capensis*) and other seabirds sometimes gain assistance in locating prey by associating with humans (fishing trawlers), dolphins, and whales. Another example, one that benefits both birds and humans, involves Greater Honeyguides (*Indicator indicator*). Honeyguides are so-named because, using vocal signals and visual clues, Greater Honeyguides "guide" human honey-hunters to the nests of wild bees (primarily those of *Apis mellifera scutellata*; Wood et al. 2014; Spottiswoode et al. 2016). Humans benefit by gaining access to honey, and honeyguides benefit by eating the beeswax left behind after humans have harvested the honey (Diamond and Place 1988; Lloyd-Jones et al. 2022; Spottiswoode

**Fig. 5.31** Magnified views of the primary feathers of an Eagle Owl (*Bubo bubo*) and, for comparison, a Common Buzzard (*Buteo buteo*). (**a**, **b**) Leading and trailing edges of an Eagle Owl primary. Note the serrated leading edge and fringes at the trailing edge. The barbs separate at the distal end because the barbules lack hooklets, creating the fringe. (**c**, **d**) Leading and trailing edges of a Common Buzzard primary. (Figure from Chen et al. 2012; © 2012 Jilin University, Published by Springer Nature, used with permission)

et al. 2022). Other examples of birds associating with mammals to increase foraging efficiency include (1) Great Egrets (*Ardea alba*) associating with strand-feeding dolphins (i.e., dolphins swim rapidly toward a shoreline, herding fish into shallower water where they can be captured more easily) and capturing fish at higher rates (Fox and Young 2012), (2) many species of birds capturing insects flushed by terrestrial and arboreal mammals (Dean and MacDonald 1981; Ruggiero and Eves 1998; Barnett and Shaw 2014; Ubaid and Beco 2022), (3) several species of seabirds feeding in association with tuna and other predatory fish (Hebshi et al. 2008), whales (Anderson and Loworn 2008), dolphins, fur seals, and tuna (Table 5.6), as well as marine turtles (i.e., seabirds feeding on small fish gathering under and around large turtles; Pitman 1993), (4) Bald Eagles (*Haliaeetus leucocephalus*) taking, or attempting to take, prey from bottlenose dolphins (*Tursiops truncates*), harbor porpoises (*Phocoena phocoena*), killer whales (*Orcinus orca*), and harbor seals (*Phoca vitulina richardii*) (Elliser et al. 2022), and, of course, (5) many species of birds obtain a wide variety of different food items from humans in both aquatic (e.g., Michel et al. 2022) and terrestrial (e.g., Dinsmore 1973) environments. Finally, some birds even use

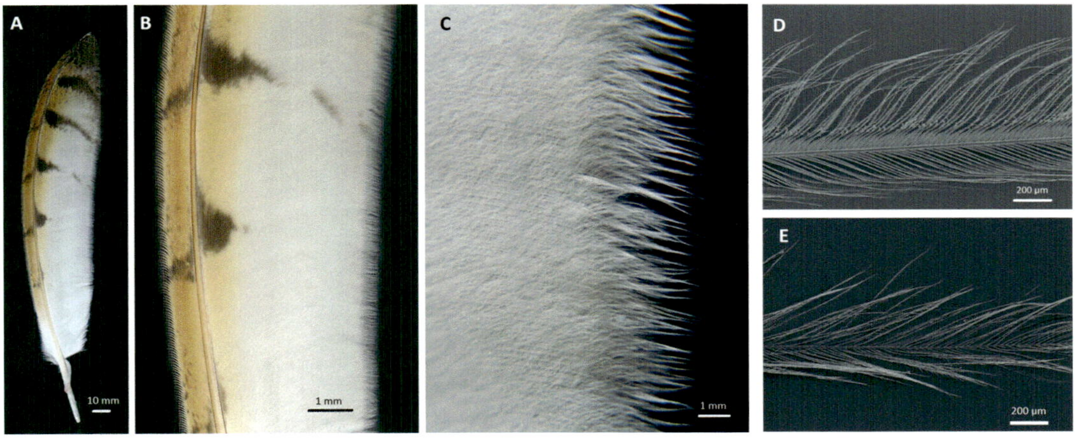


**Fig. 5.32** Inner (trailing) vane fringes of a 10th primary of a Barn Owl (*Tyto alba*) at different magnifications. (**a**, **b**) Dorsal surface. (**c**) Ventral surface. Note also the serrated leading edge of the outer vane (**b**) and the velvety dorsal surface of the feather. (**d**) Single barbs from the inner (trailing) vane and (**e**) Example of a fringe. (Figure from Bachmann et al. 2012; © 2012 The Authors. Journal of Anatomy © 2012 Anatomical Society, used with permission)

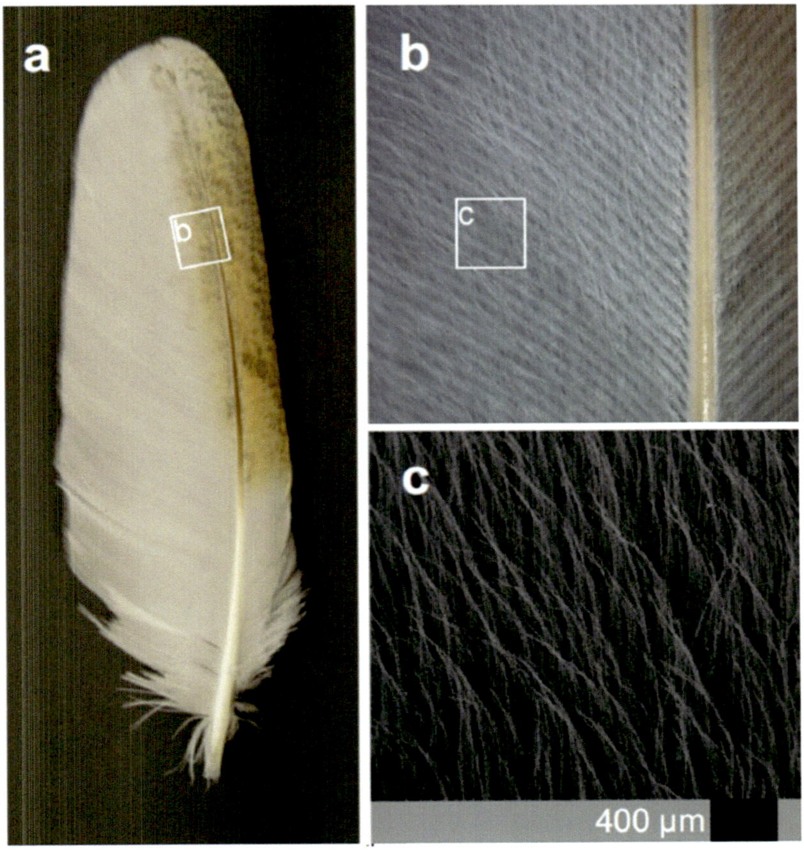


**Fig. 5.33** Velvety surface texture of a primary feather of a Barn Owl (*Tyto alba*) at different magnifications. Elongated hooklets, called pennula, create the velvety texture. (Figure from Winzen et al. 2008; 17th International Symposium on Applications of Laser Techniques to Fluid Mechanics, used with permission)

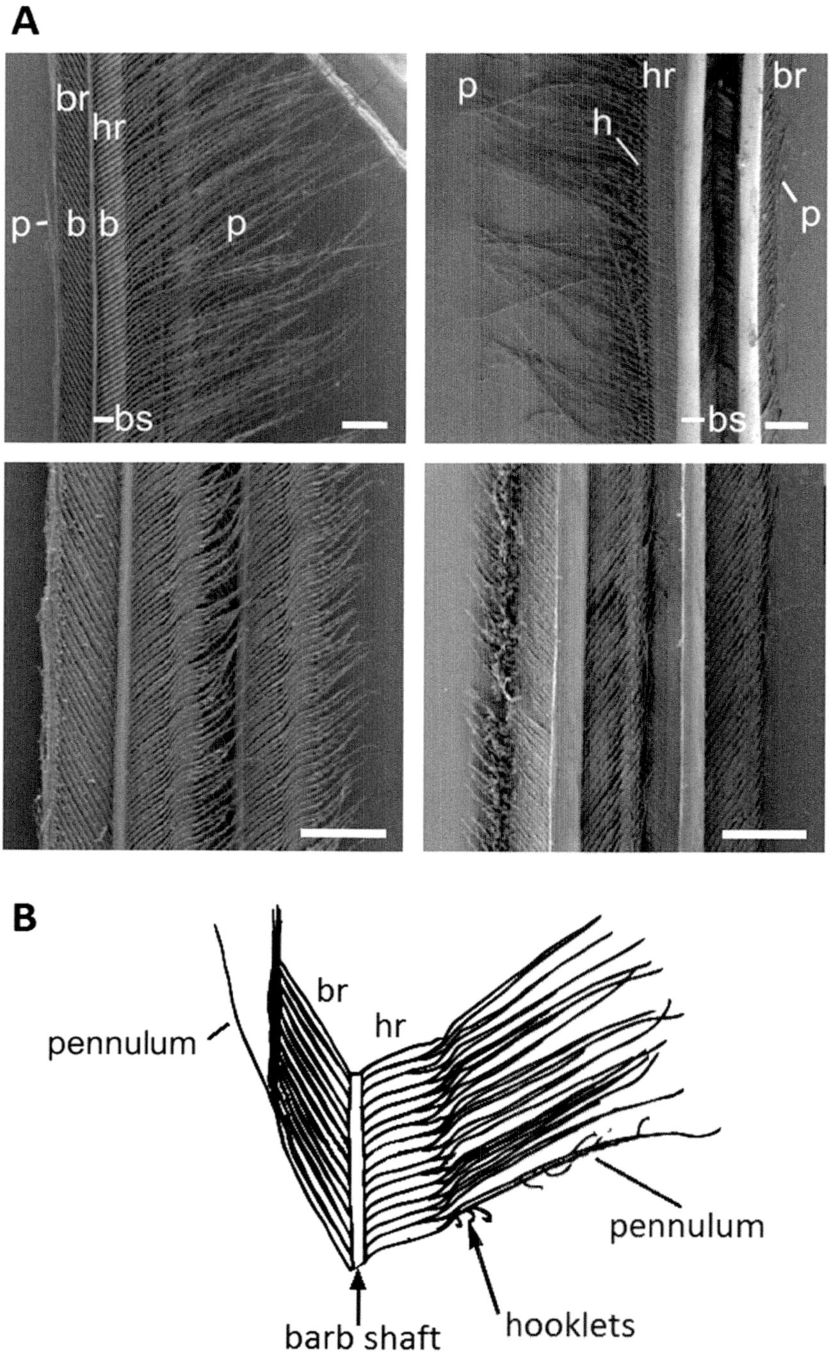


**Fig. 5.34** (**a**) Scanning electron micrograph two connected barbs from a primary feather of a Barn Owl (above) and a Rock Pigeon (below) with a dorsal view (left) and ventral view (right). Note the much longer pennula on the Barn Owl (*Tyto alba*) feather, creating the velvety texture of Barn Owl wing feathers (bs, barb shaft; hr, hook radiate (distal barbule); br, bow radiate (proximal barbule); p, pennulum; h, hooklet); Scale bars = 200 μm. (**b**) Drawing of a Barn Owl barb showing same structures. (Figure modified from Bachmann et al. 2007; open-access article distributed under the terms of the Creative Commons Attribution License, https://creativecommons.org/licenses/by/4.0/)

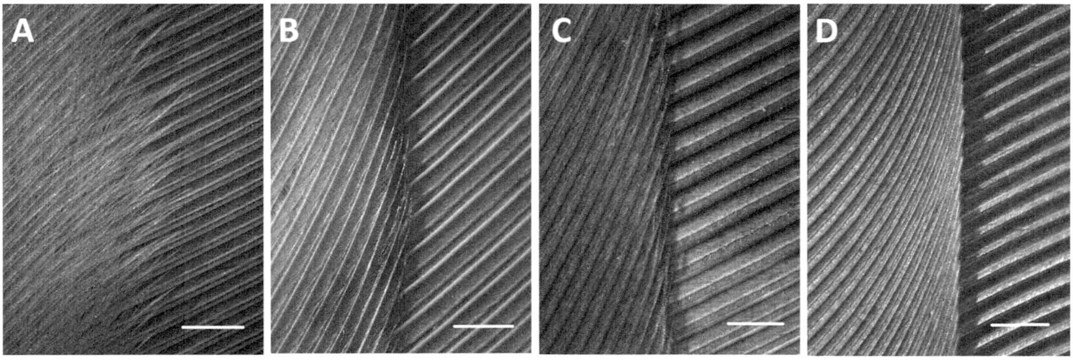


**Fig. 5.35** Bottom view of inner vanes of neighboring secondaries (s7 and s8). (**a**) Barn Owl (*Tyto alba*), (**b**) Common Buzzard (*Buteo buteo*), (**c**) Harris' Hawk (*Parabuteo unicinctus*), and (**d**) Rock Pigeon (*Columba livia*). Scale bars = 1 mm. (Figure from Bachmann et al. 2012; © 2012 The Authors. Journal of Anatomy © 2012 Anatomical Society, used with permission)

mammals as perches when foraging, e.g., birds using the backs of capybaras (*Hydrochoerus hydrochaeris*) as perches while hunting (Tomazzoni et al. 2005).

## 5.5 Avian Digestive System

The main components of the avian digestive system are the alimentary canal plus several accessory structures. The "canal" includes the oral cavity, pharynx, esophagus (which includes a crop in some birds), a two-part stomach (proventriculus and gizzard), small intestine, and large intestine (Figs. 5.65 and 5.66; Box 5.20 Evolution of the avian digestive system). The large intestine then empties into the cloaca. Important accessory structures of the digestive system include the beak, salivary glands, liver, and pancreas. The relative size and length and these organs vary with diet (Fig. 5.67). In general, birds with high protein diets (insectivores and carnivores) have a larger proventriculus (glandular stomach that secretes enzymes and acid important in the digestion of protein), less muscular ventriculus (because animal tissue does not require "grinding"), and shorter intestine (because some digestion of protein occurs in the proventriculus and, therefore, a long intestine is not needed to complete digestion and absorb the nutrients) than birds that feed primarily on seeds.

### 5.5.1 Bird Bills

One characteristic of all present-day birds is that they have no teeth, with the complete loss of teeth occurring about 116 million years ago (Meredith et al. 2014; Figs. 5.68 and 5.69). A bird's bill consists of a bony framework (upper and lower jaws) covered by a layer of keratin called the rhamphotheca. This cover is hard in most birds, but flamingoes and waterfowl (Anseriformes) have a thin, leathery rhamphotheca with a hard tip (called the nail) on the upper and lower bills. Sandpiper (Charadriiformes) bills are relatively soft, particularly at the very tip where there are sensitive touch and pressure receptors. The edges of the bill, the tomia, are typically relatively sharp and smooth. However, the tomia of several species are scopate, with a series of small, brush-like ridges extending along the cutting edges of the upper bill. These "brushy" tomia may create additional friction and make it easier for birds to hold food items in their bills (Gosner 1993). In addition, the tomia of flamingoes and some waterfowl

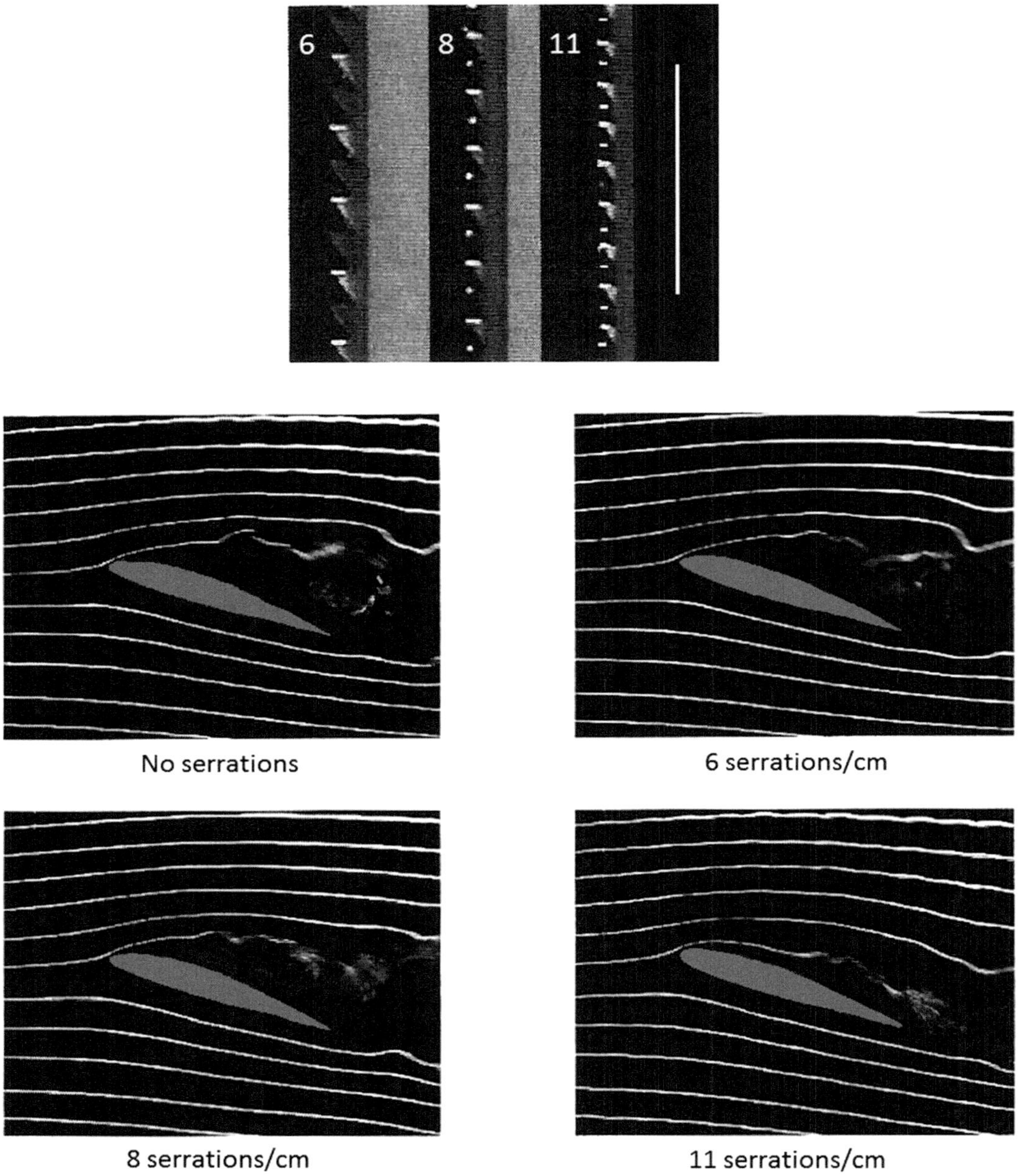


**Fig. 5.36** Air flow over simulated owl wings (Ito 2009). Top, Jigsaw blades with different concentrations of serrations were attached to the leading edge of airfoils (bottom) to mimic the serrations as the leading edge of owl wings. The control was an airfoil with no serrations. As the above images indicate, with increased number of serrations per centimeter (and more closely mimicking an actual owl wing), the smoke lines remained closer to the top of the airfoil. This helps maintain lift (very important for owls flying at relatively low speeds) and, importantly, reduces turbulence in the form of vortices near the trailing edge of the wing. Air speed across airfoils = 1.5 m/s and angle of attack = 19 degrees. (Figure modified from Ito 2009; © Japan Society of Mechanical Engineers, used with permission)

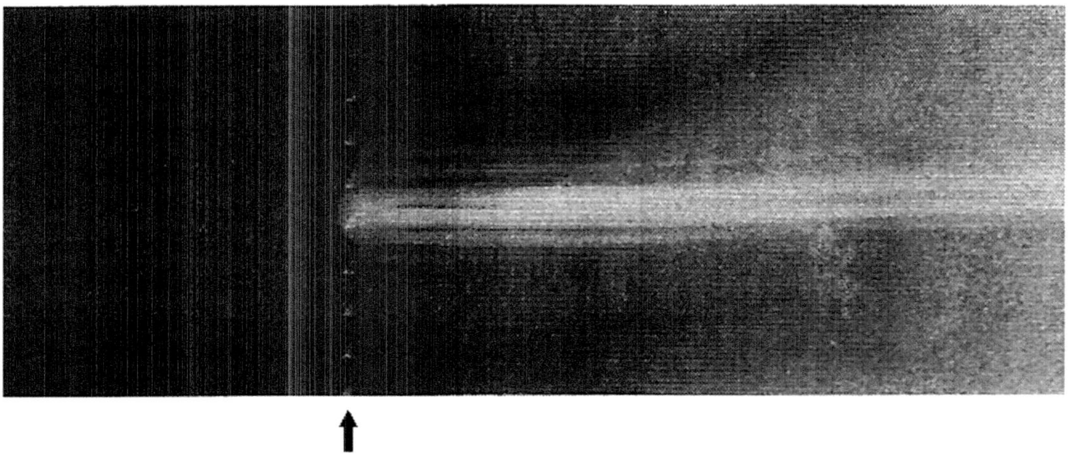


**Fig. 5.37** Smoke lines (coming from the left) divided into several smaller smoke lines by serrations (11/cm) at the leading edge of a simulated owl wing. (Figure modified from Ito 2009; © Japan Society of Mechanical Engineers, used with permission)

are finely serrated for filtering food particles from water (Fig. 5.70). In addition, some birds have projections from the tomia for holding or killing prey. For example, mergansers have numerous small projections that allow them to capture and hold fish, and falcons and shrikes have a single projection on each side of the upper bills (called tomial teeth) that may be used to puncture the skull and brain to damage the central nervous system, making it less likely that prey will escape, and to pull prey items apart (Csermely et al. 1998; Fig. 5.71). In addition, Peregrine Falcons (*Falco peregrinus*) appear to use the tomial tooth to break the long bones in the wings and legs of their avian prey (White et al. 2020).

The bill plays a critical role in food acquisition. In some birds, the rhamphotheca serves a sensory function. For example, sandpipers (Charadriiformes) and some waterfowl (Anseriformes) have bills with sensitive touch or tactile receptors that help them locate food items. A bird's bill is also used, of course, to capture, grasp, and prepare food items to be swallowed. Bill morphology varies widely among birds (Figs. 5.72, 5.73, and 5.74), with natural selection shaping bills, in part, based on food habits (Box 5.21 Evolution of bird bills). However, other factors, including thermoregulatory needs and ectoparasite control, also appear to have influenced bill morphology in some species of birds (Figs. 5.75 and 5.76).

### 5.5.2 Oral Cavity

The buccal, or oral, cavity of birds varies, as do birds, in size and shape and the inner surface is covered with mucosal epithelium. This surface includes specialized structures in some birds. For example, the roof of the mouth has papillae or spines that help fish-eating birds hold their prey, and there are longitudinal or transverse ridges along the roof of the mouth that help some seed-eating birds hold and break open seeds (Stettenheim 2000).

Associated with the buccal cavity are salivary glands that secrete mucous and, in some species, a carbohydrate-digesting enzyme called amylase. The mucous lubricates and moistens ingested food and may also help birds capture prey. For example, mucous coating the tongue of woodpeckers creates a "sticky" surface useful in

**Fig. 5.38** (a–j) A Peregine Falcon initiating an attack (unsuccessful) on a Mallard. (Figure made from screenshots from Movie S4 from Brighton et al. 2017; © 2017 The Authors. Published by PNAS, used with permission of the U. S. National Academy of Sciences)

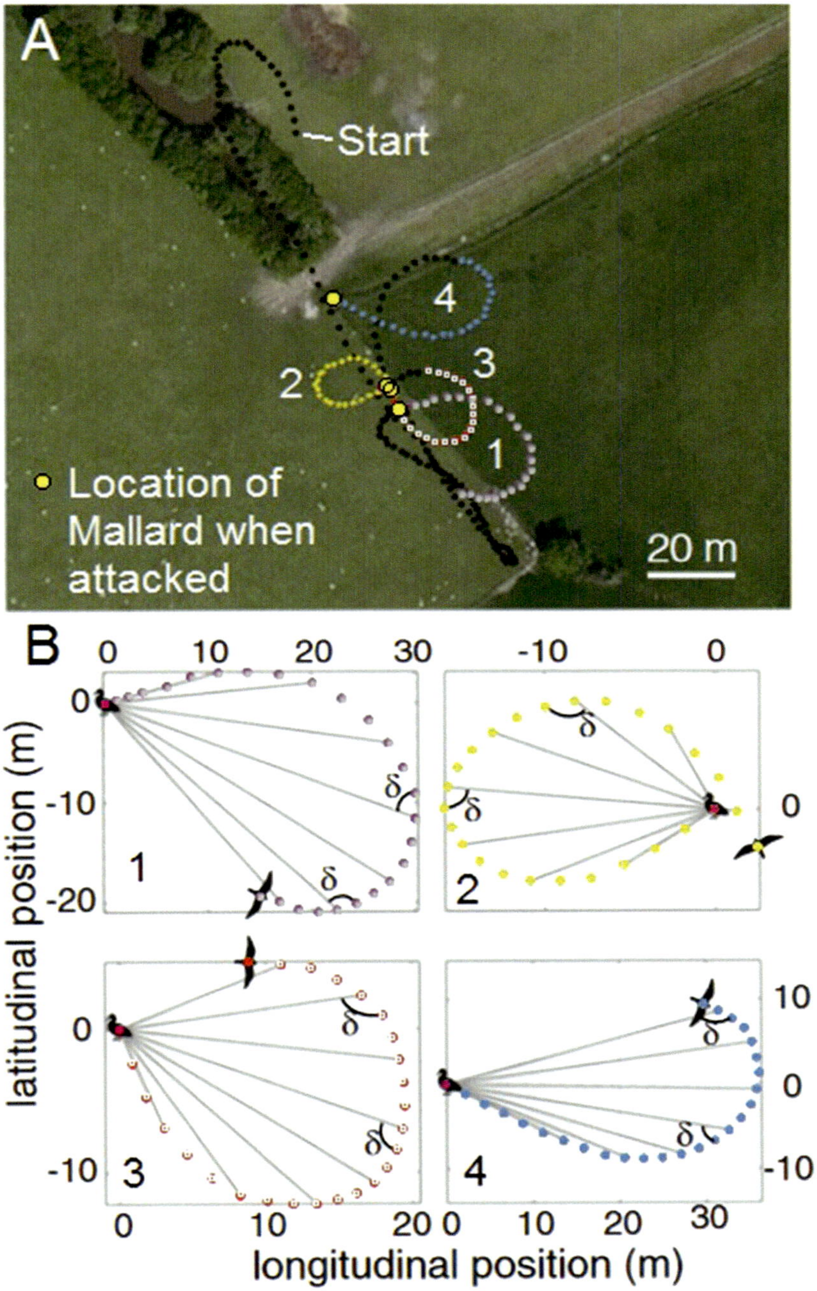


**Fig. 5.39** (**a**) Attack on Mallard (*Anas platyrhynchos*) by a Peregrine Falcon (*Falco peregrinus*) involving an aerial chase followed by four low-altitude swoops (colored points). (**b**) Flight paths of the Peregrine Falcon during the four low-altitude swoops. Neither the deviation angle (δ) nor the line-of-sight angle were held constant during the attacks on the Mallard so the behavior of the Peregrine Falcon did not follow any simple geometric rule. (Figure from Brighton et al. 2017; © 2017 the Author (s). Published by PNAS, used with permission of the U. S. National Academy of Sciences)

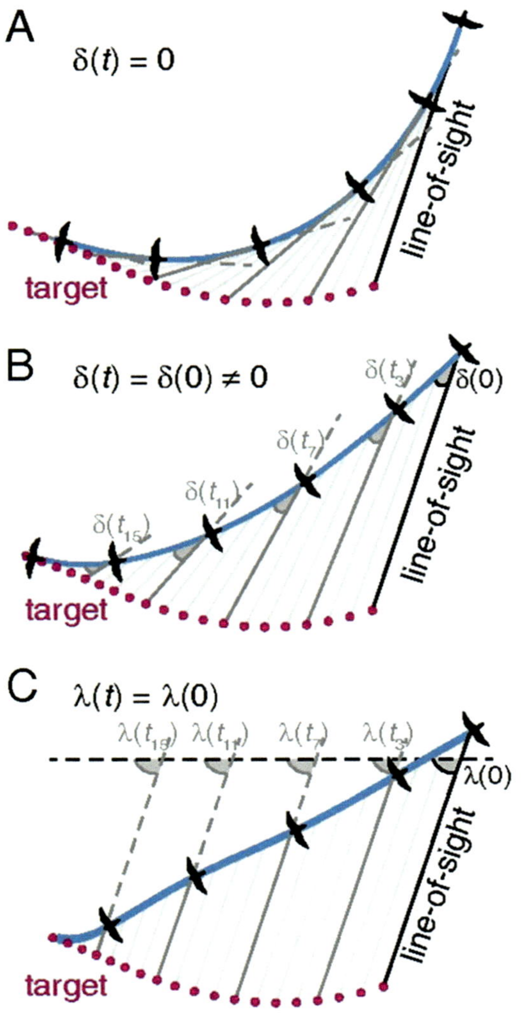


**Fig. 5.40** Three geometric rules potentially used to describe aerial attacks. (**a**) Direct, or Pure, pursuit, where the pursuer flies directly at its target at all times (i.e., deviation angle, or $\delta = 0$), (**b**) deviated pursuit, maintaining a constant bearing or angle relative to line-of-sight that is needed to intercept flying prey, and (**c**) parallel pursuit, keeping prey at a fixed visual angle and matching prey's velocity (called "motion camouflage" because, by maintaining the same position relative to the prey, the aerial predator is less likely to be detected). (Figure from Brighton et al. 2017; © 2017 the Author (s). Published by PNAS, used with permission of the U. S. National Academy of Sciences)

capturing insect prey. Mucous also coats the soft tissues of the oral cavity and both prevents dessication and provides protection by inhibiting bacterial activity (Samar et al. 1995). The saliva of some species, such as House Sparrows (*Passer domesticus*), contains amylase, but little or no amylase is found in the saliva of other species, such as Domestic Chickens and Wild Turkeys. Amylase, when present, may aid in carbohydrate digestion when food is in the esophagus and crop. Most carbohydrate digestion, however, occurs in the small intestine so the role of salivary amylase in that process is likely limited.

Among birds, the size of salivary glands and the amount of saliva produced vary with food habits. At the extremes, granivorous birds tend to have rather large salivary glands, and fish-eating species tend to have rather small salivary glands. Lubrication and moistening of food are important for birds eating seeds, whereas fish are naturally lubricated. The salivary glands of woodpeckers (Picidae) are also relatively large because of the importance of the sticky mucus secretion in capturing insect prey. Although unrelated to digestion, the salivary glands of swifts (Apodidae) increase dramatically in size during the breeding season because they use saliva as a "cement" to build their nests. Chimney Swift (*Chaetura pelagica*) salivary glands, for example, are more that 20 times larger during the breeding season than during the rest of the year (Johnston 1958; Fig. 5.77).

Also located in the buccal cavity of birds are taste buds. Avian taste buds are located in the epithelium of the tongue (primarily the base), palate, and the floor of the oral cavity. Compared to other vertebrates, birds have relatively few taste buds, with most, if not all birds, having fewer than 500 (McLelland 1979). For example, Domestic Chickens have about 300 taste buds (Ganchrow and Ganchrow 1985), Eurasian Blue Tits (*Cyanistes caeruleus*) only about 24, Rock Pigeons (*Columba livia*) about 37 (Kassarov 1999), Eurasian Bullfinches (*Pyrrhula pyrrhula*) about 42, European Starlings (*Sturnus vulgaris*) about 200, and Mallards (*Anas playrhynchos)* about 375 (Clark et al. 2015). In contrast, humans have about 7000 taste buds (Miller and Reedy 1990).

Taste buds are stimulated when molecules in solution enter a taste pore (the opening into the

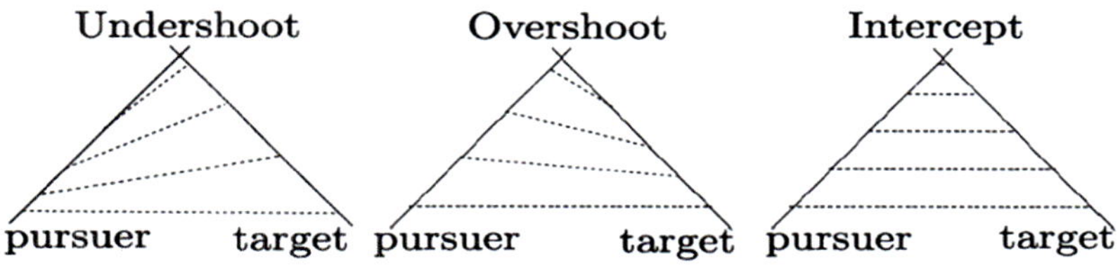


**Fig. 5.41** Three scenarios where a pursuer attempts to intercept a target. Assuming the pursuer and target have constant velocities, they will only intercept if the pursuer maintains a constant line-of-sight with the target (time passes from bottom to top). (Figure from Girard and Kabamba 2015; Copyright © 2015 Society for Industrial and Applied Mathematics. Reprinted with permission)

**Box 5.12 Use of surface features as foraging cues for seabirds**

What factors influence the foraging paths of seabirds as they fly over the ocean's surface? Lieber et al. (2021) used drones to track three species of terns, including Common Terns (*Sterna hirundo*), Sandwich Terns (*Thalasseus sandvicensis*), and Arctic Terns (*Sterna paradisaea*), as they foraged in a tidal channel and found that the terns were more likely to actively forage (i.e., exhibit foraging behaviors such as surface feeding and plunge diving) in areas where there were swirling vortices in the water. In contrast, terns engaged in transit behavior (i.e., flying faster while opportunistically searching for potential prey) when there were conspicuous upwellings ahead in their flight paths.

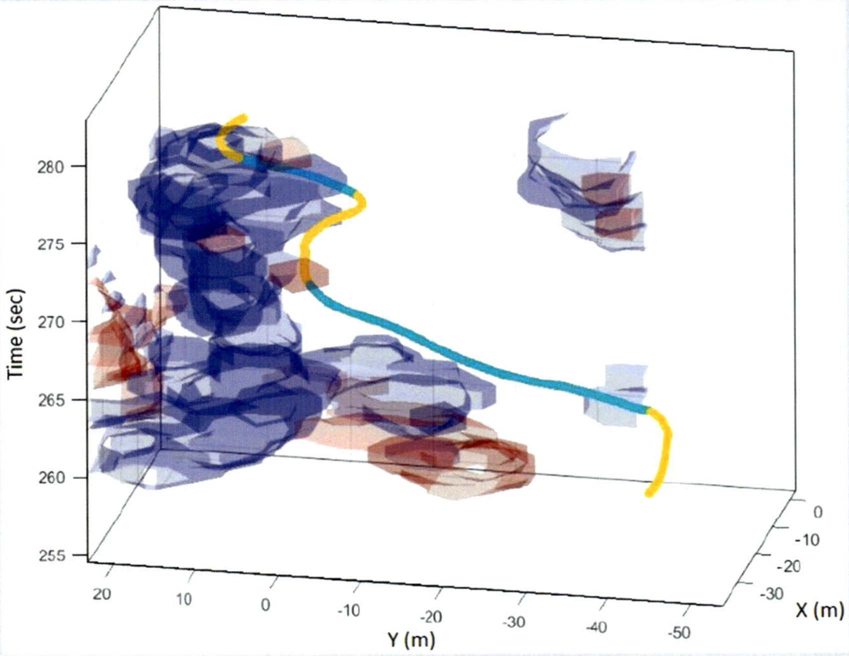


Movement of a tern (from lower right to upper left) over time in a three-dimensional (xy location over time) representation showing regions of upwellings (red) and vorticity (blue). Orange portions of the tern's flight path are areas of active foraging, whereas blue portions are areas of transit foraging (Supplemental figure from Lieber

(continued)

**Box 5.12** (continued)

et al. 2021; open-access article published by the Royal Society under the terms of the Creative Commons Attribution license, https://creativecommons.org/licenses/by/4.0/)

Common Tern (Photo by Alexis Lours, Wikipedia, CC BY 4.0, https://creativecommons.org/licenses/by/4.0/)

Terns likely engaged in active foraging in areas with strong vorticity because bouyant objects, possibly including potential prey, tend to accumulate at the centers of those areas. When upwellings were detected in their flight paths, identified by the formation of smooth patches of water on the surface, terns transitioned to transit behavior. Terns may increase their flight speeds when upwellings are detected because they may represent favorable foraging locations, with prey items potentially accumulating within or at the edges of these continuously evolving flow structures (Lieber et al. 2021).

taste bud) and contact receptor molecules (proteins) in the cell membranes of the taste receptor cells (see Chap. 4 for more information about taste buds). Taste is an important sense that can allow birds to discriminate between nutrient-rich and toxic food items. The basic taste categories include bitter (typically eliciting aversion), umami (the taste of amino acids), sour (acidity), salty, and sweet. Most birds have few taste buds so those that ingest food items whole (e.g., seeds) and swallow them quickly are unlikely to experience little, if any, "taste." Birds can, however, taste dissolved substances. For example, hummingbirds and other nectar

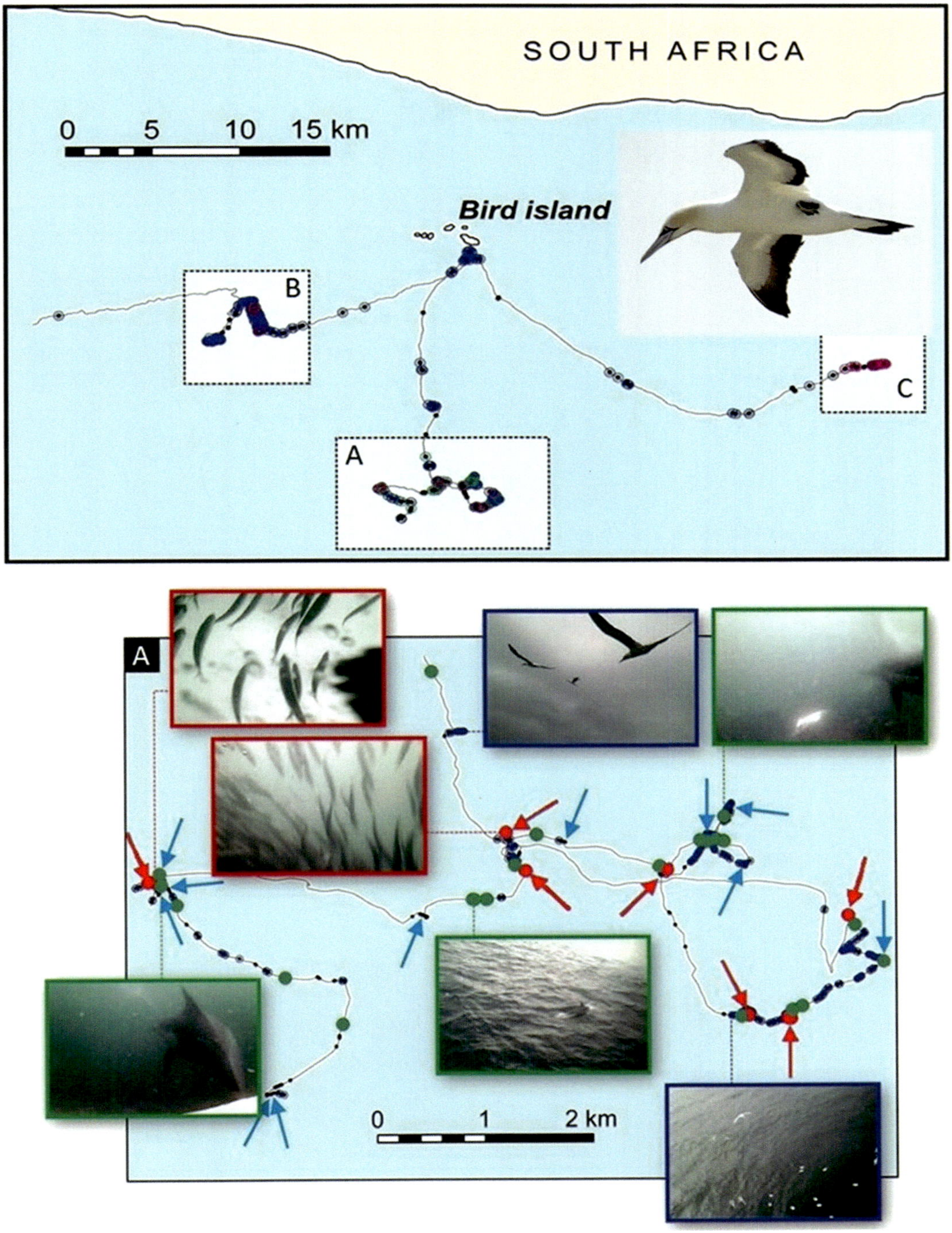


**Fig. 5.42** Portions of GPS tracks of Cape Gannets (*Morus capensis*) and video-camera images recorded concurrently. Dots represent events seen in the images. Detailed sections of the tracks of Cape Gannets are shown in (**a**), (**b**), and (**c**). Dotted lines from dots to images indicate locations (dots) where images were observed in the video. Arrows indicate dives; red arrows indicate dives where fish could be seen in the images. Red dots indicate where fish could be seen in the images, blue dots indicate where other gannets could be seen, green dots where dolphins could be seen, and magenta dots where a boat could be seen. In (**a**), a Cape Gannet spotted and followed

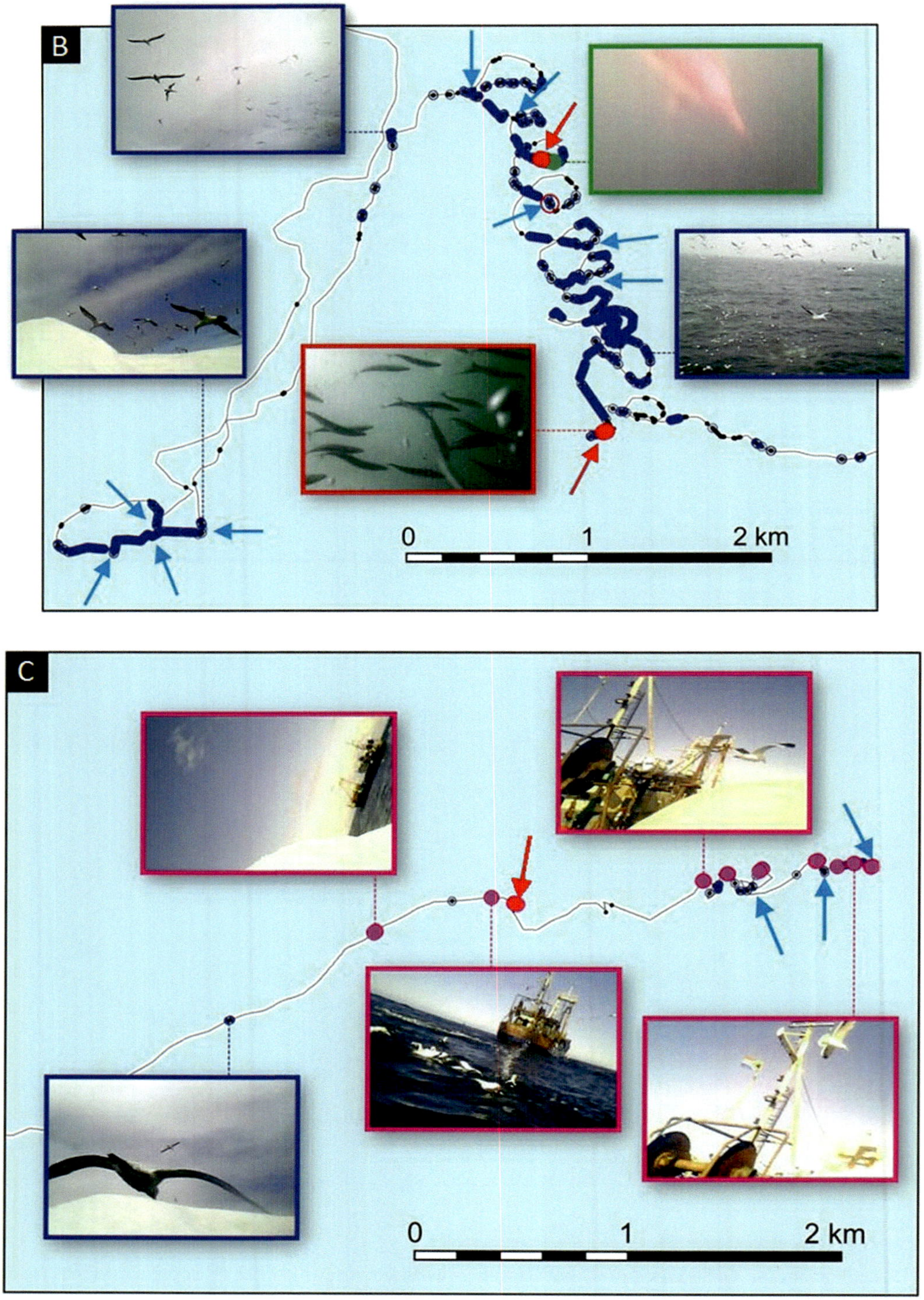


**Fig. 5.42** (continued) foraging dolphins, located fish, and made several dives to capture fish. In (**b**), a Cape Gannet observed other foraging gannets (>150) and flew toward them, located fish, and made several dives to capture fish. In (**c**), a Cape Gannet spotted a fishing boat in the distance, flew toward it, and again located fish and made several dives. (Figure modified from Tremblay et al. 2014; )

**Box 5.13 Riblets reduce drag for skimmers**

The three species of skimmers, Black Skimmers (*Rynchops niger*), African Skimmers (*R. flavirostris*), and Indian Skimmers (*R. albicollis*), have a unique method of foraging. They fly near the surface of the water with just their lower bills immersed, snapping their bills shut when the lower bill strikes a prey item.

Foraging Black Skimmer (Photo by Andy Morffew, pxhere.com, CC BY 2.0, https://creativecommons.org/licenses/by/2.0/)

As the lower bills of skimmers move through the water, friction between the bills and the water molecules creates drag and, the faster a skimmer flies, the greater the drag. Overcoming this drag requires energy so natural selection might be expected to favor any adaptation that would help reduce the drag. In this case, the adaptation is a series "riblets," slightly elevated ridges separated by grooves in the lower bill.

(continued)

**Box 5.13** (continued)

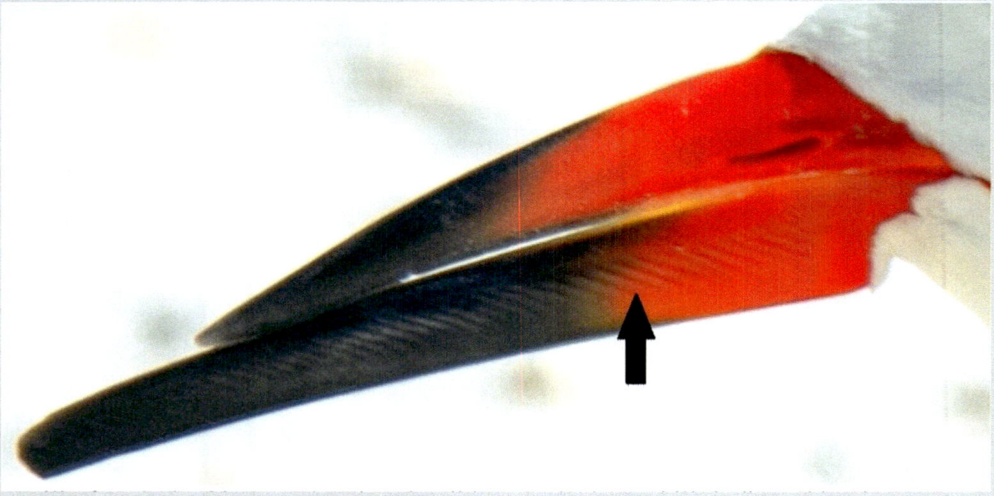

Bill of a Black Skimmer showing the riblets on the lower bill (indicated by the arrow) (Image modified from a National Park Service photo, CC0 Public Domain)

Martin and Bhushan (2016) examined how these riblets affect the flow of water along the lower bills of Black and African skimmers. They first made replicas of the lower bills of these skimmers and then measured the amount of drag (as measured by a variable called pressure drop) as water flowed past the bills held at two different angles, 0° and 45° (with 0° meaning that the riblets were parallel to the flowing water, and 45° indicating that the riblets were at a 45° angle relative to the water flowing past the replicas of the lower bills). At both angles, bills with riblets generated less drag than a bill with no riblets, with the difference increasingly pronounced at higher velocities. The riblets reduce friction by "lifting" the water vortices created by friction between water molecules and a bill's surface above the surface of the bill. Drag is reduced more when the riblets are at a 45° angle relative to the water because, at that angle, the vortices are "lifted" higher above the surface than when the angle is 0°.

(continued)

**Box 5.13** (continued)

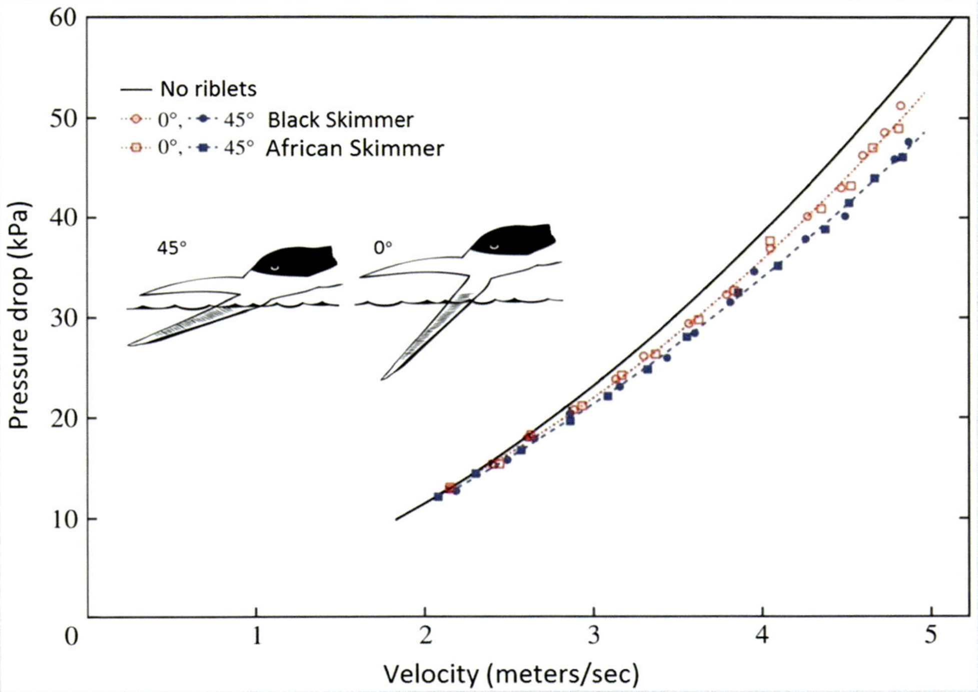


Pressure drop is a measure of drag, or the resistance of movement through water by an object (Figure from Bhushan 2018; © 2018 Springer Nature Switzerland AG, used with permission)

(continued)

**Box 5.13** (continued)

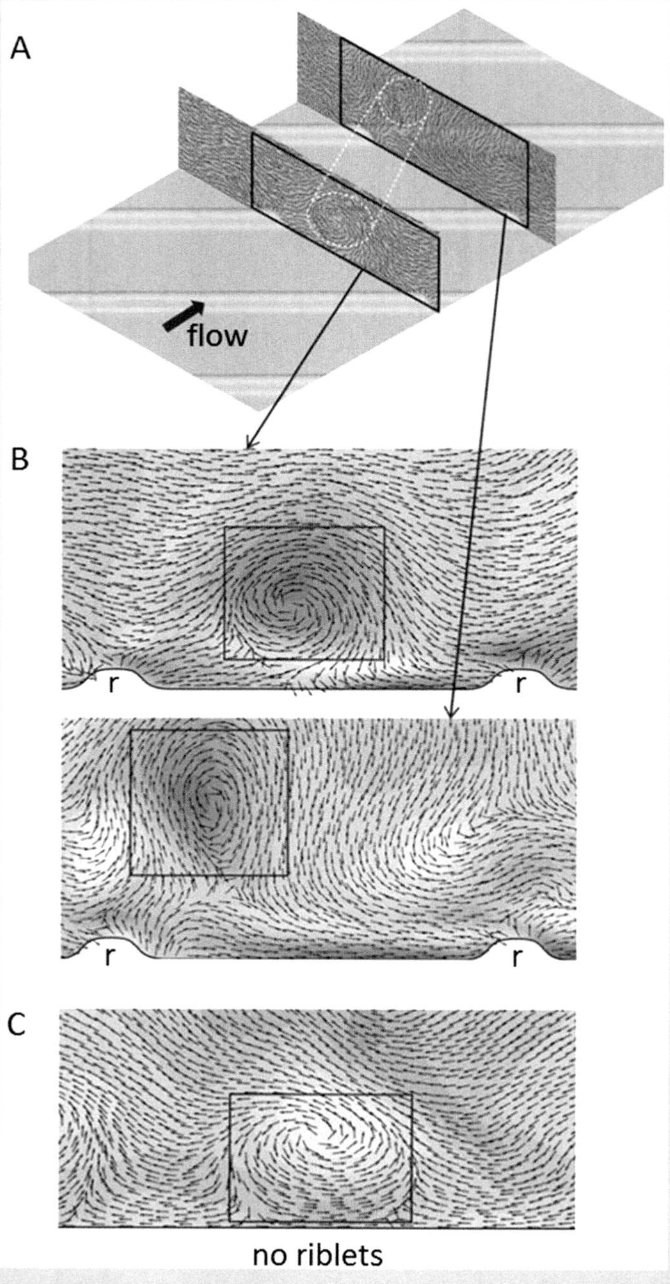


(**a** and **b**) Effect of the flow of water across a surface with riblets and (**c**) without riblets. As water moves across the surface, friction between the water and the surface generates vortices of swirling water (as indicated by the small arrows). With no riblets, the vortices interact with the surface, creating more friction. With riblets (r), note that the vortices are elevated about the surface, resulting in less friction (Figure from Bhushan 2018; © 2018 Springer Nature Switzerland AG, used with permission)

**Box 5.14 Kicking Secretarybirds**

Secretarybirds (*Sagittarius serpentarius*) are largely terrestrial raptors found in sub-Sahara Africa that feed primarily on snakes, lizards, and small mammals. Secretarybirds are usually found in grasslands and open savannah and, unlike other raptors, typically search for prey by walking through their grassland habitats. As a result, over evolutionary time and because of their terrestrial habits, selection has favored longer hindlimbs and talons that are a bit less curved and pointed. Another unique feature of Secretarybirds is that they typically kill prey by kicking and stamping their heads. Portugal et al. (2016) studied the kicking strikes of a captive Secretarybird (mass = 3.96 kg; height = 69.2 cm) that had been trained to "attack" a rubber snake (for public demonstrations). Using a portable force plate and high-speed video, Portugal and his colleagues found that the peak force generated by the bird's kick averaged 5.1 times the bird's body mass (or 195 N). The kicks were also fast, with a mean duration of contact with the snake of just 15 ms. These extremely fast, forceful kicks are clearly important given the potential negative consequences of failing to kill the venomous snakes frequently preyed on by Secretarybirds.

Secretarybird striking the head of a rubber snake (Figure from Portugal et al. 2016; ©2016 Elsevier Ltd, used with permission)

(continued)

**Box 5.14** (continued)

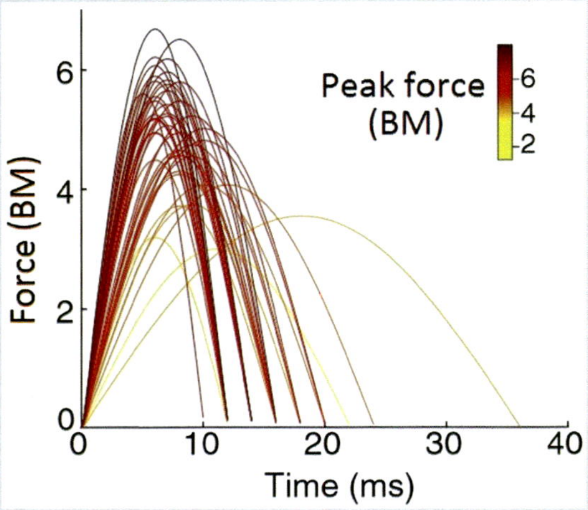


Forces generated when the Secretarybird kicked the snake's head. Peak force of 45 kicks averaged 5.1 times the body mass (BM) of the Secretarybird (Figure from Portugal et al. 2016; © 2016 Elsevier Ltd, used with permission)

feeders, including several species of songbirds, exhibit strong preferences for solutions with higher sugar concentrations (Stromberg and Johnsen 1990; Toda et al. 2021; Fig. 5.78). Fruit-eating Cedar Waxwings (*Bombycilla cedrorum*) exhibited a preference for simple sugars with higher absorption efficiencies (about 90%) over sucrose with a lower absorption efficiency (about 60%) (Martinez del Rio et al. 1989; Rowland et al. 2015), and Schaefer et al. (2003) determined that tanagers preferred a 12% glucose-sucrose mix over 8% and 2.5% mixes. Tanagers were also able to discriminate between food items differing by only 1% in carbohydrate concentration, 2% in lipid content, and 4% in cysteine (an amino acid) content (Schaefer et al. 2003).

Some birds are also able to detect differences among different salt solutions (Mason and Clark 2000). For example, Royal Penguins (*Eudyptes schlegeli*) prefer fresh water to salt water after extended periods at sea (Warham 1971). Rejection thresholds of 58 species of birds to various salt solutions ranged from 0.35% NaCl in a parrot to 37.5% in Pine Siskins (*Carduelis pinus*; Rensch and Neunzig 1925). Finally, Sanderlings (*Calidris alba*) and Dunlins (*C. alpina*) probing with their bills can distinguish between sand where no worms had been present and sand where worms had been present (van Heezik et al. 1983), possibly detecting amino acids from the mucus secretions of the worms (Mason and Clark 2000).

Reduced sensitivity to the taste of certain substances may be beneficial. For example, many birds are tolerant of acidic and alkaline solutions, and this may permit exploitation of otherwise unpalatable food resources like unripe fruit (Mason and Clark 2000). Similarly, some birds are tolerant of bitter solutions (Mason and Clark 2000), perhaps permitting them to eat potentially bitter prey like some insects. Other birds, however, have been found to reject bitter-tasting food items, and respond by beak wiping and head-shaking (Marples and Roper 1997), vomiting (Alcock 1970), or releasing captured prey (Wiklund and Järvi 1982; Chai 1996). Wang and Zhao (2015) examined 48 genomes representing all except three avian orders and

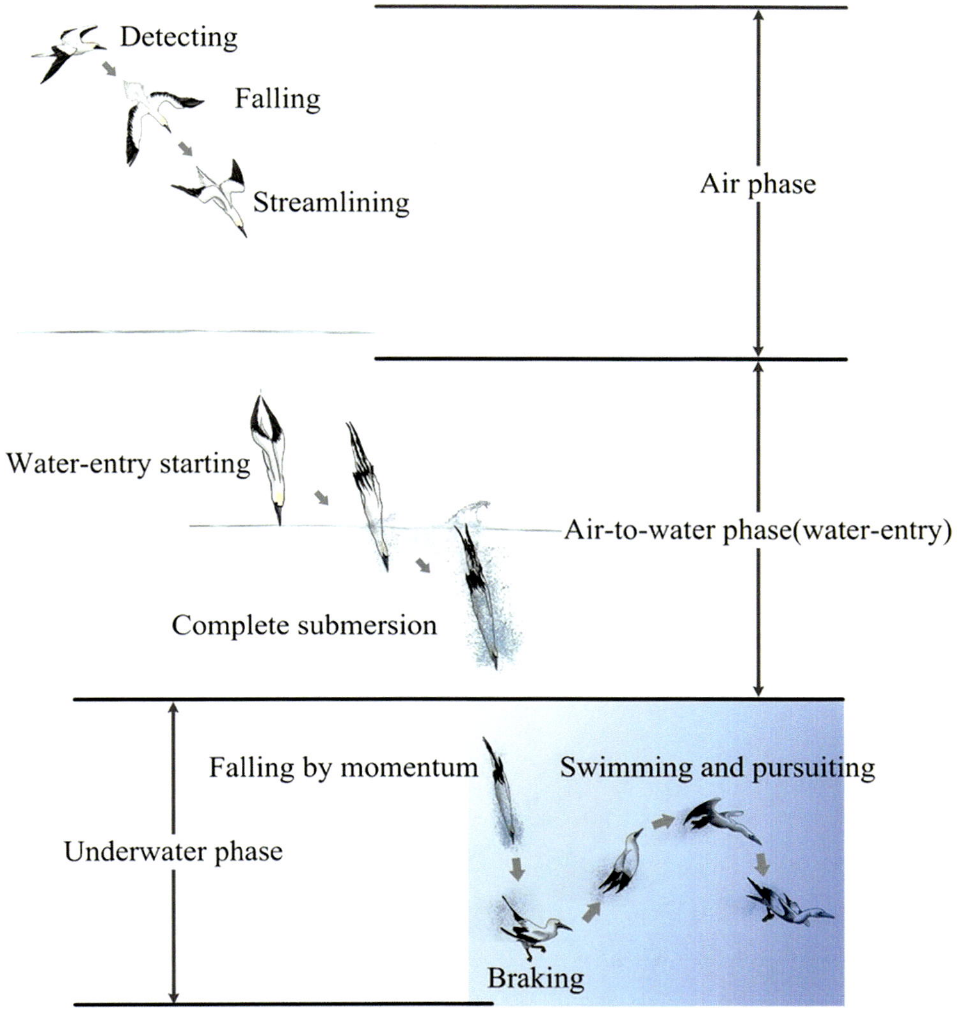


**Fig. 5.43** The plunge dive of a gannet. Gannets have one of the most spectacular prey-capture behaviors of all marine predators, plummeting from up to 30 m into the water, where they seize fish with their razor-sharp beaks. Ropert-Coudert et al. (2004) monitored the plunge diving of Cape Gannets (*Morus capensis*) using a rapid-sampling acceleration and depth recorder. Because gannets are so streamlined, no or a very low deceleration was noted when gannets entered the water. Gannets used their momentum to travel underwater to depths ranging from 0.3 to 9.7 m. Gannets sometimes used either their wings or feet for underwater propulsion during some dives. After chasing prey, birds developed an upward momentum before gliding passively back to the surface, making use of their buoyancy to complete the dive using as little energy as possible. (Figure from Wang et al. 2013; © 2013 IOP Publishing, used with permission)

determined that the average number of bitter taste-receptor genes (*Tas2rs*) was four and ranged from zero for Red-throated Loons (*Gavia stellata*) and Adelie (*Pygoscelis adeliae*) and Emperor (*Aptenodytes forsteri*) penguins (suggesting a loss of bitter taste perception) to 12 for Bar-tailed Trogons (*Apaloderma vittatum*). These authors also found a positive correlation

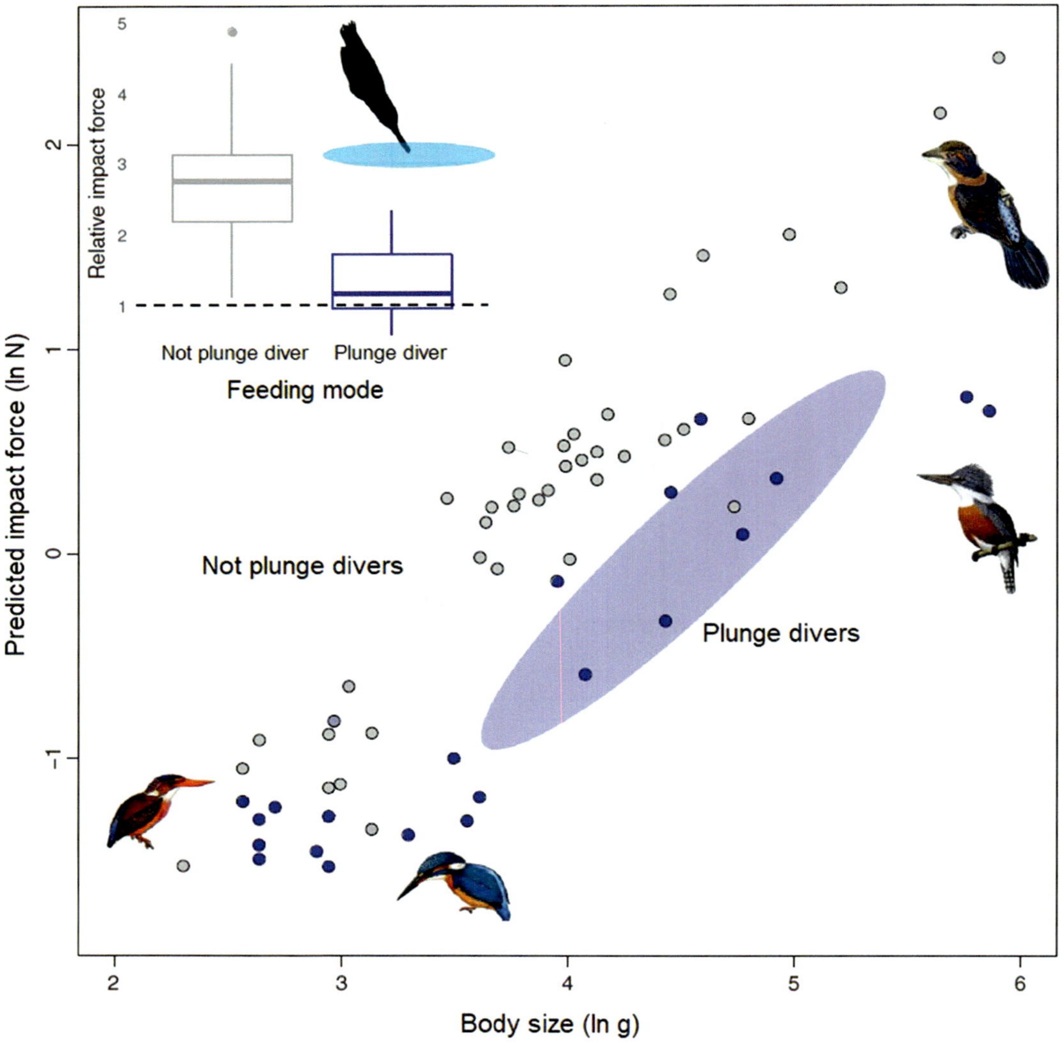


**Fig. 5.44** Kingfishers that capture prey by plunge diving (blue circle) into water have lower predicted impact forces for their body masses than kingfishers that do not plunge dive (gray circles). Predicted impact force versus body mass in plunge-diving (blue) and nonplunge-diving kingfisher species (gray). Inset (upper left) shows relative beak impact force calculated as the ratio between predicted impact force (in Newtons) and the force due to gravity (9.8 $m/s^2$ times body mass in kilograms). Horizontal dashed line shows 1:1 ratio between forces, indicating where a bird would theoretically experience no deceleration due to drag forces exerted by the water on the beak. Plunge-divers have narrow beaks and long, flat palates though to be important for swallowing prey that are less flexible like fish. The lower impact forces of plunge-divers minimize the risk of injury and help them maintain speed as they enter the water, likely improving their chances of capturing targeted prey. (Figure modified from Eliason et al. 2020; © 2020 The Authors. Evolution © 2020 The Society for the Study of Evolution, used with permission)

between the number of bitter taste-receptor genes and the abundance of potential toxins in bird diets, with herbivorous and insectivorous birds that feed on plants and insects, respectively, that are more likely to contain defensive toxins, generally having more *Tas2r* genes than noninsect

**Box 5.15 Fish eating and refraction**

Plunge-divers, e.g., gannets, boobies, terns, kingfishers, Osprey (*Pandion haliaetus*), and Brown Pelicans (*Pelecanus occidentalis*), and birds that prey on fish by striking from above the water's surface, e.g., herons and egrets, must contend with surface glare, wind, waves, the evasive tactics of prey, variable water transparency, and refraction. The degree of light refraction depends on the angle between a bird and its potential prey, with an increasing refraction at increasingly shallower angles. As such, birds might generally be expected to attack fish at steeper angles (i.e., closer to perpendicular). However, Little Egrets (*Egretta garzetta*) were found to attach fish at angles ranging from 30 to 90 degrees and capture success rates were higher at more shallow angles (Lotem et al. 1991). In addition, most plunge-dives by Great Cormorants (*Phalacrocorax carbo*) were at angles of about 35–40 degrees (Ropert-Coudert et al. 2005). Such results suggest that light refraction has little effect on the foraging behavior of these fish-eating birds. Some evidence suggests that this ability to adjust strikes due to refraction is, at least to some degree, innate. Katzir et al. (1999) examined the foraging success of four species of hand-reared (i.e., no previous experience at catching fish) herons and egrets, including Little Egrets, Squacco Herons (*Ardeola ralloides*), Black-crowned Night-Herons (*Nycticorax nycticorax*), and Cattle Egrets (*Bubulcus ibis*). The first three of those species normally forage in aquatic habitats, whereas Cattle Egrets are primarily terrestrial. Success rates of Little Egrets, Squacco Herons, and Black-crowned Night-Herons were much greater than those of Cattle Egrets, suggesting that prior experience was not necessary for the more aquatic species, and that Cattle Egrets, the terrestrial species, are losing their ability to cope with light refraction.

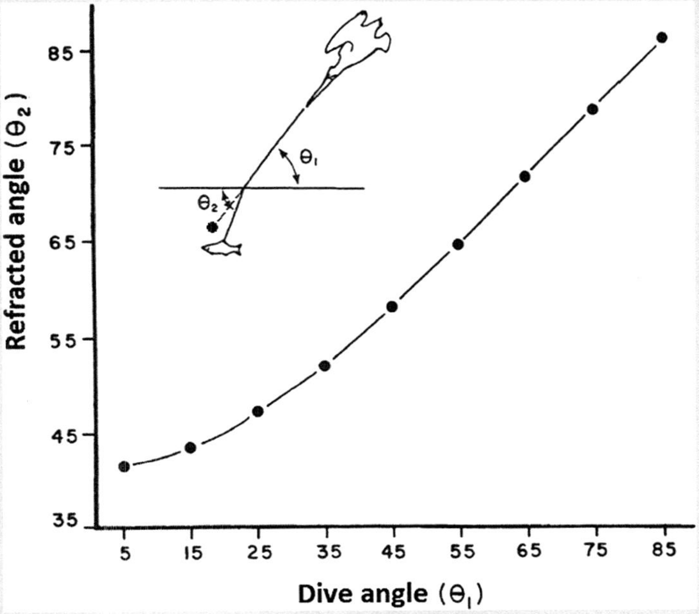


Refraction angles (in degrees, relative to the water surface) of prey as viewed by a plunge-diver like a Brown Pelican. The inset shows the actual position of a fish (solid circle) and the apparent position of the fish due to refraction (From Carl 1987; © 1987 Oxford University Press, used with permission)

(continued)

**Box 5.15** (continued)

A Brown Pelican initiating a dive in pursuit of a fish (Photo by Ingrid Taylor, Wikipedia, CC BY 2.0, https://creativecommons.org/licenses/by/2.0/)

eating carnivores (like Red-throated Loons and Adelie and Emperor penguins).

Birds exhibit much taxon-specific variation in the sense of taste, with the relative importance of perceiving different tastes varying with food habits. Although birds have fewer taste buds than mammals, for birds that ingest foods or liquids where molecules in solution can stimulate

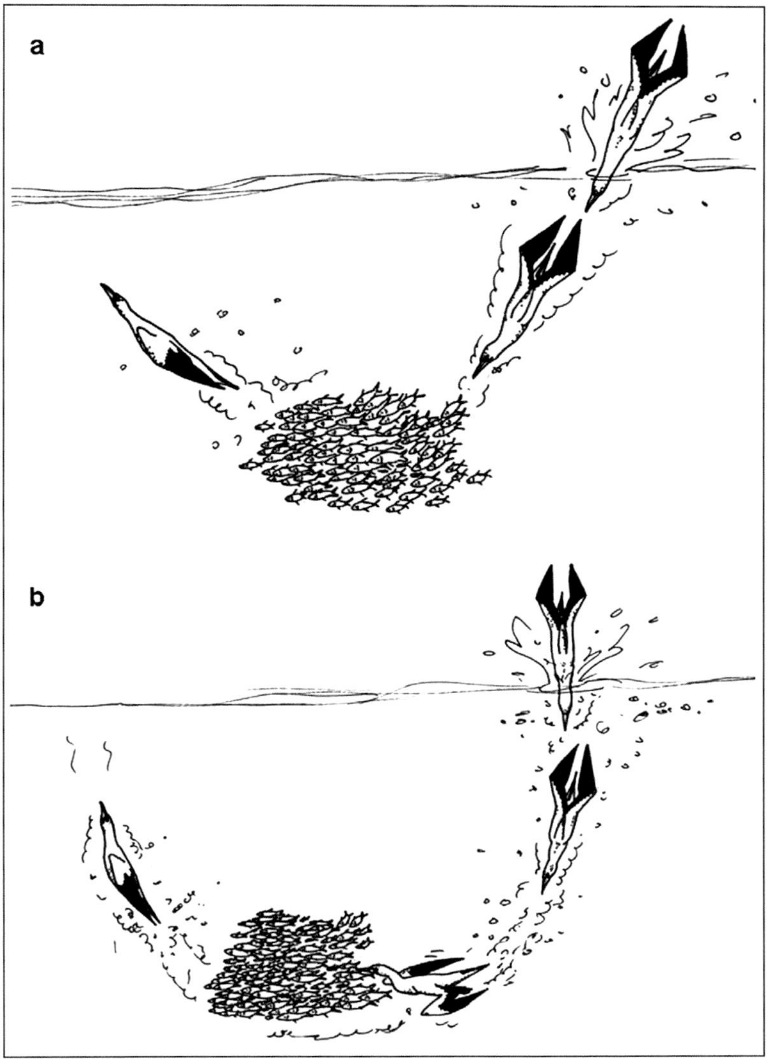


**Fig. 5.45** Australian Gannets (*Morus serrator*) and other plunge divers may use different strategies depending on the depth of their targeted prey, with (**a**) V-shaped dives used to capture prey (e.g., fish or squid) located closer to the surface and (**b**) U-shaped dives to capture prey further below the surface. (Figure from Capuska et al. 2011; Drawing by Laura van Zonneveld. © Inter-Research 2011, used with permission)

taste buds, the sense of taste is important in selection of food, nectar, water for drinking, and, for many species, avoiding toxins.

The tongues of most birds are attached to the floor of the oral cavity and exhibit substantial variation in size, shape, structure, and function. Tongues are generally not very muscular, but are reinforced by the hyoid apparatus. Parrots (Psittacidae) are a notable exception, with muscular tongues useful for manipulating seeds (Fig. 5.79). Avian tongues may be used to collect and manipulate food and may also play a role in swallowing. Based on the relative importance of these functions, tongue morphology varies widely among species (Fig. 5.80).

Tongues important for collecting food are long, narrow, and protrusible, and function as probes, hooks, spears, brushes, or fluid traps (McLelland 1979; Rico-Guevara and Rubega 2011). For example, among fish-eating birds, such as penguins and mergansers, tongues have numerous sharp, caudally directed projections (papillae) that help hold their slippery prey in place, and help ensure that prey only move further back into the oral cavity and pharynx prior to swallowing (Figs. 5.81 and 5.82). Woodpecker

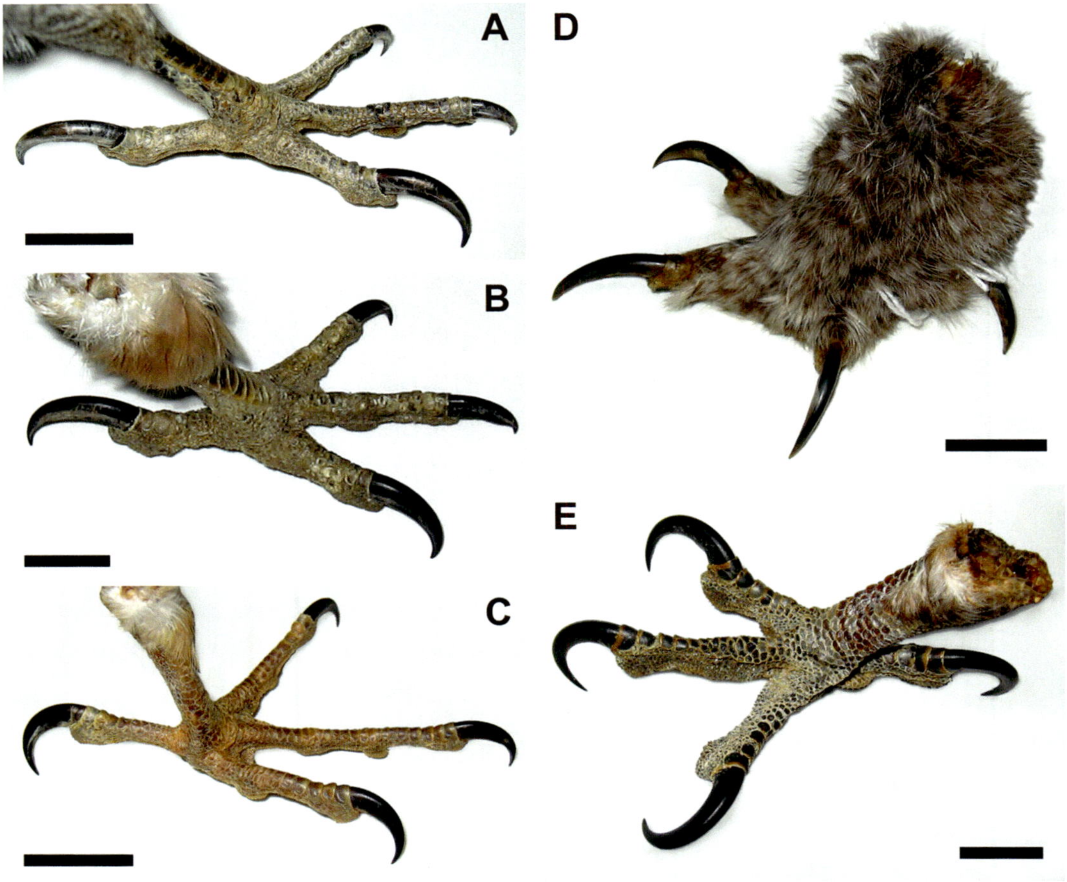


**Fig. 5.46** Feet and talons of five representative raptors. (**a**) Accipitridae: Northern Goshawk, *Accipiter gentilis*; (**b**) Accipitridae: Red-tailed Hawk, *Buteo jamaicensis*; (**c**) Falconidae: Peregrine Falcon, *Falco peregrinus*; (**d**) Strigiformes: Great Gray Owl, *Strix nebulosa*; (**e**) Pandionidae: Osprey, *Pandion haliaetus*. (Figure modified from Fowler et al. 2009; © 2009 Fowler et al., open-access article distributed under the terms of the Creative Commons Attribution License, https://creativecommons.org/licenses/by/4.0/)

tongues are long, extensible and most are "barbed" at the tip to facilitate the capture and extraction of prey (like insect larvae) from bark crevices (Fig. 5.83). Observations of a Guadeloupe Woodpecker (*Melanerpes herminieri*) revealed that insect larvae in holes were detected with the tip of the tongue (where numerous touch receptors called Herbst's corpuscles are located). Once detected, the woodpecker tried to hook the grub using the barbs at the tip of the tongue. Successfully capturing grubs typically required multiple tries, with the tongue extended and withdrawn several times before a grub was finally "hooked." In one case, the woodpecker made 29 attempts to "hook" a grub in 44.5 s, with the bird withdrawing its tongue completely back inside its mouth at the end of each attempt. In the 29 attempts, the woodpecker moved its tongue forward 601 times at an average rate of 13.5 times per second (range = 5–55 times per second). The velocity of the tongue's movements prevented the potential prey from

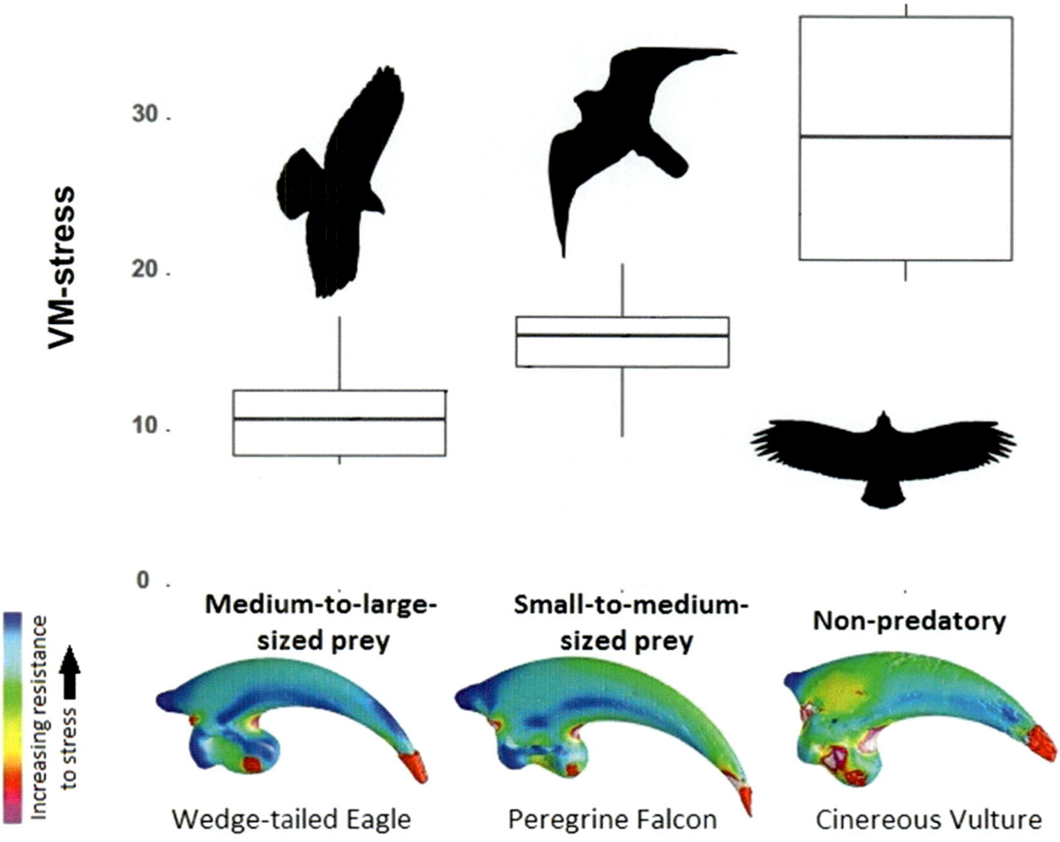


**Fig. 5.47** Resistance to stress of the hallux of raptors that feed on different-size prey and nonpredatory birds. VM-stress refers to Von Mises stress and is a measure of structural integrity or strength, with lower values (and the bluer colors below) indicating greater structural integrity or strength. Bottom and top of the boxes are the first and third quartiles, the horizontal black lines are the means, and whiskers indicate the minimum and maximum values. (Figure modified from Tsang et al. 2019; open-access article licensed under a Creative Commons Attribution 4.0 International License, https://creativecommons.org/licenses/by/4.0/)

escaping. When finally hooked, grubs were pulled out just far enough so they could be grabbed by the bill. These observations indicate that Guadeloupe Woodpeckers do not spear grubs with their tongue, but hook them with the tongue's barbed and saliva-coated tip. Other woodpecker species with similar tongues may use them the same way (Villard and Cuisin 2004).

The tongues of sap and nectar feeders are also highly protrusible (Box 5.22 Hummingbird hyoids), with tips specialized for collecting sap, nectar, pollen, and small insects. For example, the tongues of sapsuckers (Picidae) have fine bristles at the tip used as a "brush" to collect sap (Walters et al. 2020). Similarly, nectar-feeding Rainbow Lorikeets (*Trichoglossus haematodus*) have brush-tipped tongues (i.e., with numerous fine papillae) for feeding on nectar, pollen, and fruit (Fig. 5.84). Hummingbird tongues can collect nectar by acting as fluid traps (Fig. 5.85 and 5.86). When a hummingbird's tongue is inserted into nectar, the semicircular sections near the end

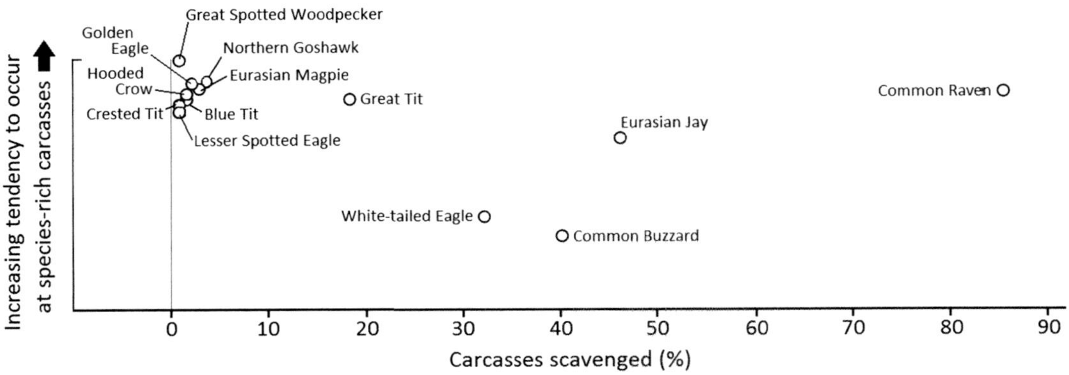


**Fig. 5.48** Species of birds observed to feed on carcasses (N = 137) of different species of ungulates in Poland. Facultative scavengers (upper left) were observed at a small percentage of carcasses and typically only at the carcasses that attracted the most species of birds. The main scavengers, species of birds observed at a greater percentage of carcasses, were more often observed at carcasses visited by fewer species of birds. Golden Eagle, *Aquila chrysaetos*; Great Spotted Woodpecker, *Dendrocopos major*; Hooded Crow, *Corvus corone*; Northern Goshawk, *Accipiter gentilis*; Crested Tit, *Lophophanes cristatus*; Eurasian Magpie, *Pica*; Eurasian Blue Tit, *Cyanistes caeruleus*; Lesser Spotted Eagle, *Aquila pomarina*; Great Tit, *Parus major*; White-tailed Eagle, *Haliaeetus albicilla*; Eurasian Jay, *Garrulus glandarius*; Common Buzzard, *Buteo*; Common Raven, *Corvus corax*. (Figure modified from Selva and Fortuna 2007; © 2007 The Royal Society, used with permission)

of the tongue close around the nectar as a result of surface tension (attraction of water molecules to each other increases when at the water's surface; this attraction "pulls" the semicircular portions of the tongue into circles). Nectar can then move into the tongue by capillary action. This would be important when, for example, nectar reservoirs are shallow. When nectar reservoirs are deeper, hummingbirds can extend their tongues further into the nectar and use them as fluid traps. Some less specialized nectar-feeding birds have tongues with brushy tips that collect nectar (Figs. 5.87 and 5.88).

Surprisingly, the tongue of Western Sandpipers (*Calidris mauri*) bears a striking resemblance to that nectar-feeding birds, with a dense fringe of long spines or bristles along the lateral margins and, particularly, the tip (Fig. 5.89). Brush-tipped tongues are generally associated with collecting liquid, like nectar or sap, or pollen. A possible explanation is that the bristles on the tongue of Western Sandpipers permit foraging on the biofilm present on the surface of the beaches and mudflats where sandpipers often forage. Biofilms are composed of communities of microorganisms that closely adhere to surfaces, particularly moist surfaces. Western Sandpipers, and perhaps other sandpipers, could "dab" the surface of foraging substrates and the spines on the tip of the tongue could either brush up or adsorb the mucilaginous biofilm (Elner et al. 2005). Western Sandpipers, like many other shorebirds, typically forage by pecking and probing to capture invertebrates. If Western Sandpipers, and perhaps other small shorebirds, do forage on biofilm, the nutrients obtained could represent a valuable dietary supplement (Elner et al. 2005).

Tongues adapted for manipulating food are typically not protrusible, with shapes and lengths that match that of the lower mandible. Parrots, finches, and crossbills have muscular tongues used to manipulate and extract seeds from husks or cones (Klasing 1998). For example, crossbills use their tongue to secure seeds in a lateral groove of the horny upper palate, then remove the seed coat and swallow the kernel (Benkman 1993). Similarly, finches use their tongue to hold seeds in place while cracking them open with their bills. More generally, tongues that are short, nonprotrusible, and have caudally directly papillae may be important in swallowing. Birds that

**Fig. 5.49** Top, An adult male Andean Condor (*Vultur gryphus*). Bottom, A portion of a flight path shown of an Andean Condor, with the orange arrow (bottom right) indicating the direction of travel. Condors and vultures are relatively large birds and make use of thermals to gain altitude (yellow-orange colors in the flight path), then glide (and lose altitude) as they search for carcasses and, if none are found, another thermal. The zig-zag flight paths indicate that the condor was circling to remain within the thermal. $V_z$ is the vertical velocity, with positive values indicating increasing altitude and negative numbers indicating declining altitude. (Photo by Michael Gäbler, Wikipedia, CC BY 2.0, https://creativecommons.org/licenses/by/2.0/. Figure from Shepard et al. 2011; © 2011 Shepard et al., open-access article distributed under the terms of the Creative Commons Attribution License, https://creativecommons.org/licenses/by/4.0/)

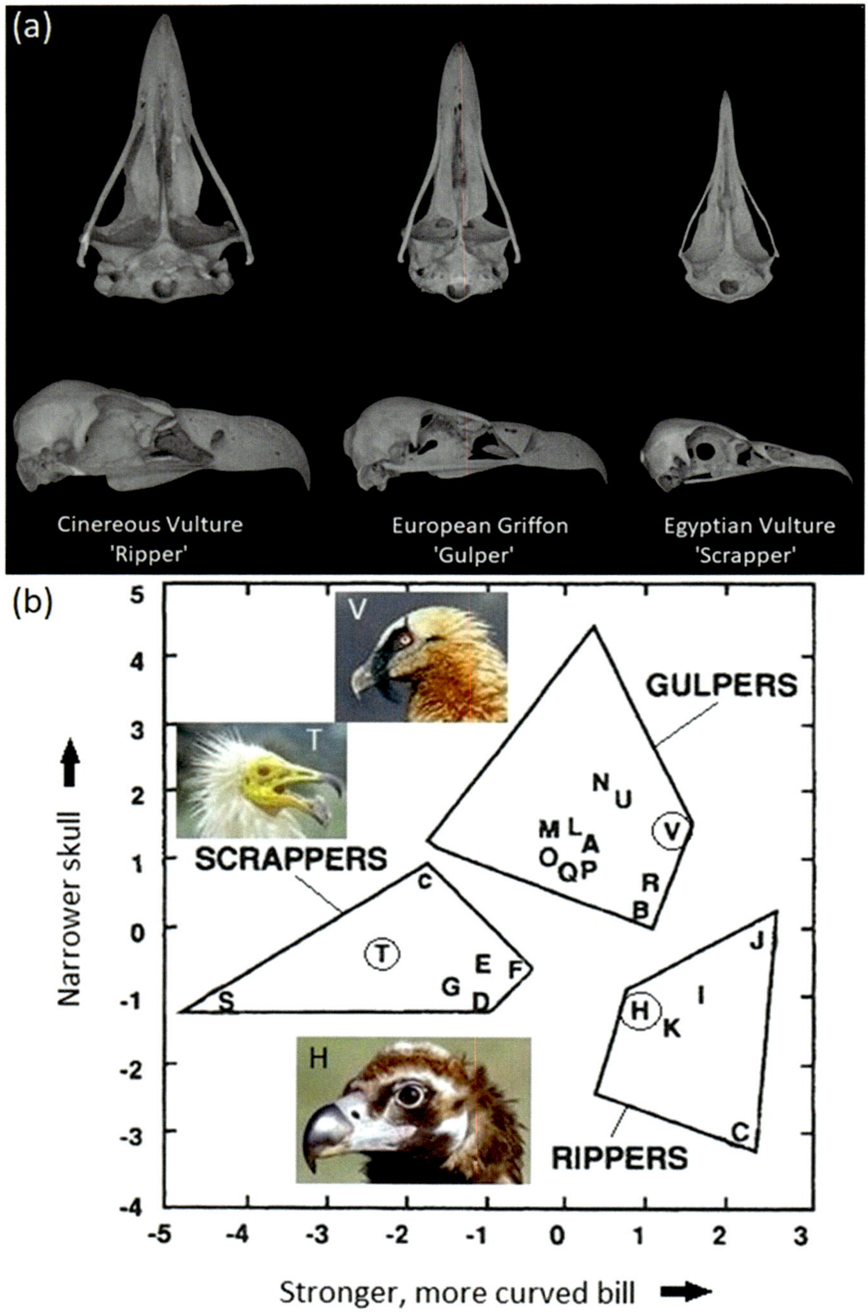


**Fig. 5.50** (**a**) Skulls of a "ripper," "gulper," and "scrapper." Rippers, gulpers, and scrappers have the strongest, intermediate, and weakest bills, respectively. Cinereous Vulture, *Aegypius monachus*; European (Eurasian) Griffon, *Gyps fulvus*; Egyptian Vulture, *Neophron percnopterus*. (**b**) Different species of vultures differ in bill morphology and strength, with "rippers" having the widest, strongest beaks, "gulpers" having narrower, intermediate-strength beaks, and "scrappers" having narrow, weak beaks. Vultures with these different bill morphologies exhibit different feeding behaviors at carcasses. European Vulture, *Gyps fulvus*. V—Bearded Vulture (*Gypaetus barbatus*, photo from Margalida et al. 2008; © 2007 Elsevier Inc., used with permission), T—Egyptian Vulture, photo from Negro et al. 2002; © 2002 Springer Nature, used with permission), H—Cinereous Vulture, photo by Koshy, Wikipedia, CC BY 2.0, https://creativecommons.org/licenses/by/2.0/). (Figure **a** modified from Linde-Median et al. 2021; © 2021 The Authors. Published by Elsevier GmbH, used with permission. Figure **b** modified from Hertel 1994; © 1994 John Wiley and Sons, used with permission)

**Box 5.16 Prey dropping**

Several species of birds are known to access food by dropping hard-shelled items onto hard substrates to break them open (Cristol and Switzer 1999). Birds using this technique must find suitable food items, identify suitable locations for dropping, and determine the dropping height that will be sufficient to break open the item (Zach 1979; Richardson and Verbeek 1986). For example, Zach (1978) found that whelks dropped by Northwestern Crows (*Corvus caurinus*) broke at heights ranging from 7 to 10 m, large bones dropped by Bearded Vultures (*Gypaetus barbatus*) broke at heights of 50 m or higher (Ferguson-Lees and Christie 2001), and Carrion (*Corvus corone*) and Hooded (*C. cornix*) crows dropped mussels from an average of 4.8 m, a height determined to ensure breaking on the first drop (Davenport et al. 2014). Multiple drops are typically required before items break. Bearded Vultures dropped bones an average of 4.5 times (range = 1–35) before they broke, which took an average of 5.3 min (range = 1–17 min) (Margalida and Bertran 2001). American Crows (*Corvus brachyrhynchos*) dropped English walnuts (*Juglans regia*) on asphalt an average of 13.9 times before the walnuts broke (Cristol and Switzer 1999), whereas Northwestern Crows dropped whelks (*Thais lamellosa*) an average of 4.4 times before they broke (Zach 1979).

Few investigators have examined the energetics of prey dropping, but Zach (1979) determined that Northwestern Crows gained an average of 1.49 kcal per whelk dropped, and Davenport et al. (2014) calculated that Hooded Crows lifting a medium-sized mussel (55-mm shell length) to a height of 5 m cost only 0.3% of the energy assimilated from that mussel. Taking into account the cost of flight, Hooded Crows flying to and from mussel beds to drop sites located 200 m away costs 5.8% of the energy contained in the mussel.

Bearded Vulture (Photo by Rolf Dietrich Brecher, Wikipedia, CC BY 2.0, https://creativecommons.org/licenses/by/2.0/)

(continued)

**Box 5.16** (continued)

Species of birds known to drop hard-shelled prey or seeds to access food inside (Table from Cristol and Switzer 1999; © 1999 Oxford University Press, used with permission)

| Common name | Scientific name | Prey types |
|---|---|---|
| American Crow | *Corvus brachyrhynchos* | Nuts, molluscs |
| Hooded Crow | *C. cornix* | Molluscs, nuts, crustaceans |
| Carrion Crow | *C. corone* | Molluscs, nuts |
| Rook | *C. frugilegus* | Molluscs |
| Eurasian Jackdaw | *C. monedula* | Nuts |
| New Caledonian Crow | *C. moneduloides* | Molluscs, nuts |
| Common Raven | *C. corax* | Molluscs, bones |
| White-necked Raven | *C. albicollis* | Turtles |
| Bearded Vulture | *Gypaetus barbatus* | Turtles, bones |
| Egyptian Vulture | *Neophron percnopterus* | Turtles |
| Osprey | *Pandion haliaetus* | Molluscs |
| Bald Eagle | *Haliaetus leucocephalus* | Turtles |
| Golden Eagle | *Aquila chrysaetos* | Turtles |
| Great Skua | *Stercorarius skua* | Eggs |
| Herring Gull | *Larus argentatus* | Molluscs, fish, crustaceans |
| Ring-billed Gull | *L. delawarensis* | Molluscs |
| Common Gull | *L. canus* | Molluscs |
| Kelp Gull | *L. dominicanus* | Molluscs |
| Glaucous-winged Gull | *L. glaucescens* | Molluscs |
| Great Black-baced Gull | *L. marinus* | Mammals, eggs |
| Western Gull | *L. occidentalis* | Molluscs |
| Pacific Gull | *L. pacificus* | Molluscs, echinoderms |
| Mediterranean Gull | *Ichthyaetus melanocephalus* | Molluscs |

[a]Additional species observed to drop prey since Cristol and Switzer (1999): Black-headed Gull (*Chroicocephalus ridibundus*; Vogrin 2017) and Sooty Gull (*Larus hemprichii*; Fraticelli 2014)

(continued)

**Box 5.16** (continued)

(**a**, **b**) Broken shells of snails (*Placostylus fibratus*) dropped by New Caledonian Crows. (**c**) Points where snails dropped by crows hit the rock. (**d**) New Caledonian Crow about to consume a snail after dropping it and breaking the shell (Figure from Tanaka et al. 2013; © 2013 Springer Nature, used with permission)

have tongues with these characteristics include pelicans, cormorants, storks, warblers, vireos, crows, and wrens (McLelland 1979; Klasing 1998).

The tongues of some waterfowl (Anseriformes) are used to both collect and manipulate food. In many ducks, geese, and swans the tongue is thick and muscular with double rows of overlapping projections or bristles along the lateral margins (Figs. 5.90 and 5.91). In filter-feeding ducks, those bristles serve to filter food particles from water. The tongue is first depressed against the lower bill to allow water in, then the tongue is pressed against the upper palate to eject water laterally through the bristles to trap food particles (McLelland 1979). The tongue is then retracted or pulled back, the bill closed, and food swallowed.

### 5.5.3 Esophagus

The esophagus is a muscular tube that connects the oral cavity and the stomach. The muscular walls of the esophagus produce wave-like contractions (peristalsis) that help propel food from the oral cavity to the stomach. The esophagus of birds is large in diameter compared to other

**Box 5.17 Arthropods and pollen in the diet of nectarivorous birds**

Nectar lacks important amino acids so nectivorous birds must obtain them from other sources and their most important sources are arthropods and pollen. Pollen contains highly digestible protein and numerous amino acids (Van Tets and Hulbert 1999) and, despite the presence of variable amounts of chitin, arthropods are also an excellent source of protein (generally 50–80% of dry mass) and lipids (5–35%) as well as vitamins and minerals (but little carbohydrate; Robel et al. 1995). The relative importance of pollen and arthropods in the diets of nectarivorous birds varies, with hummingbirds primarily consuming arthropods (Brice et al. 1989) and some lorikeets and sunbirds getting much of the protein they need from pollen (Wooller et al. 1988; van Tets and Nicolson 2000). The protein requirements of nectarivorous birds appear to be quite low (about 25% that of other birds, possibly because a nectar diet means they require fewer digestive enzymes and so less protein is lost in their feces and/or their large intestines are better at reabsorbing amino acids; Tsahar et al. 2006) so they likely need to spend relatively little time foraging on pollen and for arthropods. For example, Stiles (1995) observed two species of hummingbirds in Costa Rica during both the breeding and nonbreeding seasons and found they generally spent about 1–2% of the day foraging for arthropods. Breeding females (incubating or brooding), however, spent about 3–6% of their time foraging for arthropods, likely because of an increased need for protein in their diets and to feed nestlings.

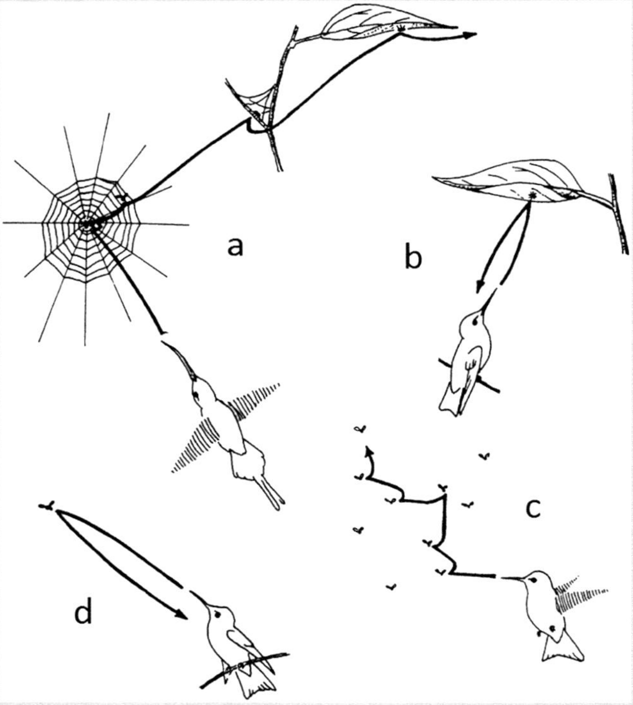


Techniques used by hummingbirds foraging for arthropods in Costa Rica. (a) hover-gleaning, (b) sally-gleaning, (c) hover-hawking, and (d) sally-hawking. Arthropods preyed on by 11 species of hummingbirds included small spiders, mites, mosquitoes, small wasps and ants, springtails (Collembola), small beetles, and others (Figure from Stiles 1995; © 1995 Oxford University Press, used with permission)

(continued)

**Box 5.17** (continued)

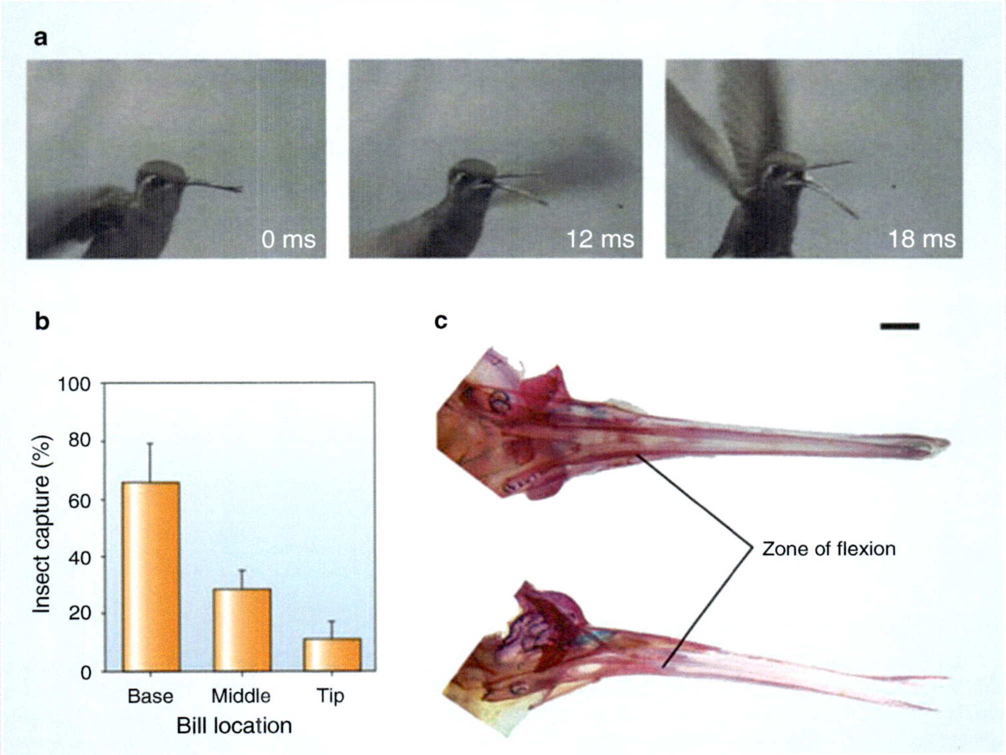


(**a**) High-speed video sequence of a female Magnificent Hummingbird (*Eugenes fulgens*) attacking a fruitfly (*Drosophila*; dark spot to the right). Note the jaw opening, followed by bowing of the mandible at the base to widen the gape and flexion of the distal half of the mandible. That downward bending places stress on the flexion point in the bone making up the lower mandible (shown in **c**) and, because its elastic nature, this causes the lower bill to quickly (in less than a hundredth of a second) flex upward, a phenomenon known as snap buckling. This extra-fast closing of the bill likely increases the likelihood of successfully catching insects. (**b**) Success of insect capture (percentage caught) by a Ruby-throated Hummingbird (*Archilochus colubris*) relative to where the insect first strikes the bill. More flies are captured at the flared base of the beak. (**c**) Ventral (top) and right lateral (bottom) view of a stained Ruby-throated Hummingbird skull. The intensity of the stain shows that the mandible is densely calcified up to the inflection point of dorsoventral bending. Scale bar = 1 mm (Figure from Yanega and Rubega 2004; © 2004 Springer Nature, used with permission)

vertebrates, and there is a positive correlation between esophageal diameter and the size of food items swallowed, the extent to which the esophagus is used to temporarily store food, or both. The esophagus is widest in carnivores and piscivores, such as cormorants, herons, and some raptors, that swallow large prey items, and in some granivores and frugivores, like jays and waxwings, that use the esophagus for temporary food storage. Longitudinal folds on the inner surface of the esophagus (plicae esophageales) make it distensible, with larger folds and greater

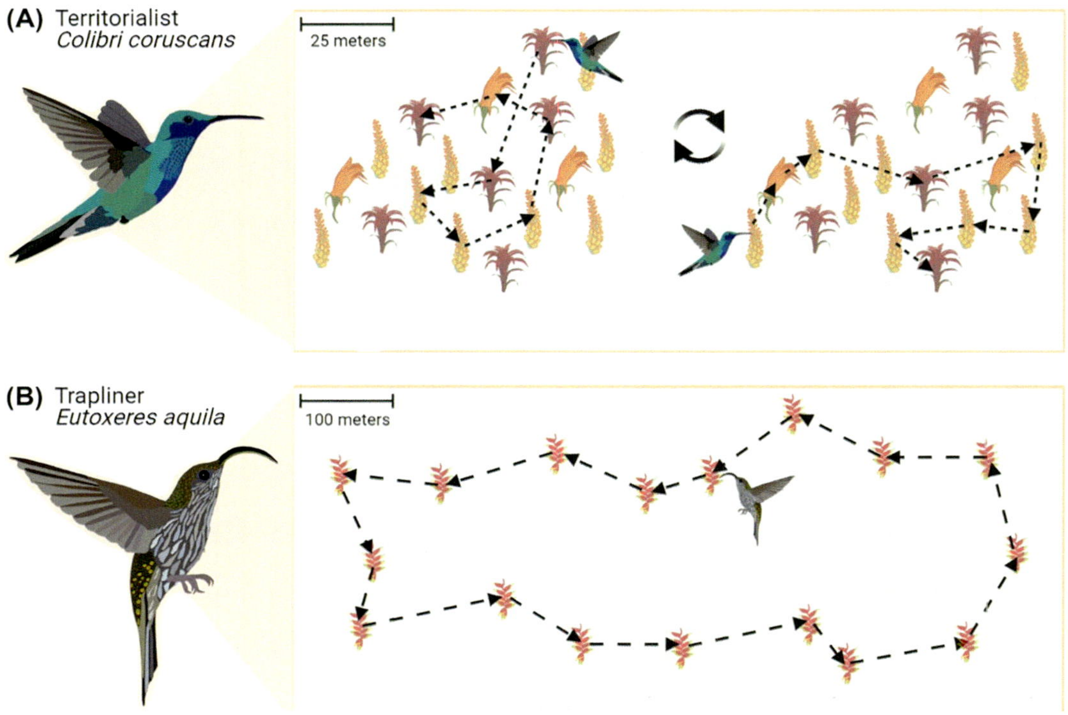


**Fig. 5.51** The nectar-foraging strategies of hummingbirds have often been placed in two categories: territorialists and trapliners. (**a**) Territorialist hummingbirds, like Sparkling Violetears (*Colibri coruscans*), feed on and defend smaller patches of flowers over extended periods by chasing and attacking any trespassing hummingbirds. Territorialists use a variety of different routes when visiting flowers. (**b**) Trapliners, like White-tipped Sicklebills (*Eutoxeres aquila*), regularly visit a series of flowers in a particular sequence across a larger landscape and do not attempt to defend those flowers from potential competitors. Traplining hummingbirds attempt to visit flowers on a schedule that allows flowers to replenish their nectar, but before competitors can do so. They also try to take the shortest routes between flowers, like bees do, to conserve energy (Ohashi et al. 2007). Although placing hummingbirds in just two broad categories is an oversimplification, it has been useful for identifying tendencies among different species (Stiles 1978; Altshuler et al. 2004). However, studies to date suggest that hummingbirds are not limited to just one strategy or the other. Rather, they regularly adjust their foraging strategies based on extrinsic factors such as availability of flowers and the presence (or absence) of competitors, as well as intrinsic factors like age, size, sex, energetic condition, breeding status, and season (e.g., before, during, or after migration) (Ewald and Carpenter 1978; Powers et al. 2003; Justino et al. 2012; Rodríguez-Flores and Arriaga 2016; Bribiesca et al. 2019; Márquez-Luna et al. 2019). Scale bars included to show examples of spatial scales of visited flowers. (Figure modified from Sargent et al. 2021; © 2021 Oxford University Press, used with permission)

distensibility in birds that swallow large items or store food in the esophagus (Fig. 5.92).

Some birds have a permanent distention or out-pocketing of the esophagus called the crop that functions as a temporary storage area (Fig. 5.93). Birds can quickly consume and store food in the crop, then fly to another, safer, location to continue the digestive process. Temporary storage of food in the crop is also useful during the breeding season when adults are feeding young. Retroperistaltic contractions of the esophagus allow adults to regurgitate food stored in the crop or esophagus and feed young of some birds (e.g., pigeons, penguins, and parrots; Paster 1992). Among the birds with the best-developed crops are pigeons and doves (Columbiformes), grouse and pheasants (Galliformes), and parrots (Psittaciformes) plus some raptors

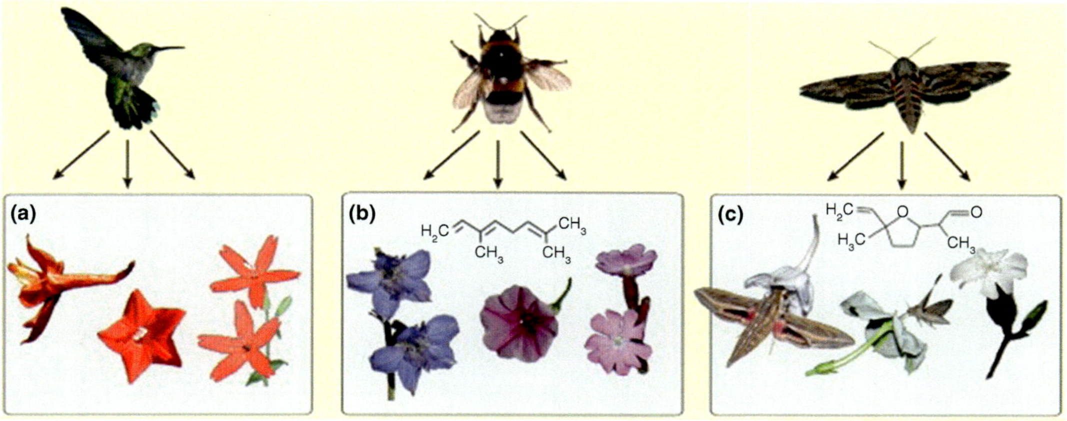


**Fig. 5.52** Flowers that are visited and pollinated by hummingbirds, bees, and moths tend to have different characteristics. Hummingbird-pollinated flowers are often red, bee-pollinated flowers often emit an odor that is attractive to bees (molecules of β-ocimene), and moth-pollinated flowers often emit an odor attractive to moths (molecules of lilac aldehyde). (Figure modified from Schiestl and Johnson 2013; © 2013 Elsevier Ltd, used with permission)

(Falconiformes) (McLelland 1979). In some species of birds, crops appear to play an important role in digestion. For example, the crops of seed-eating Green-rumped Parrotlets (*Forpus passerinus*) contain large concentrations of bacteria. High densities of amylase-producing bacteria are present in the crops of Green-rumped Parrotlets, suggesting that at least some starch digestion is occurring. These parrotlets retain food in their crops for periods long enough to facilitate starch digestion and enhance the efficiency of the microbial activity, and this appears to be an adaptation for processing difficult-to-digest, but energy-rich, seeds (Pacheco et al. 2004). In addition, the large numbers of bacteria in the crop might provide a source of protein, potentially important for birds like Green-rumped Parrotlets with protein-poor diets (Pacheco 2000).

Microbial fermentation (the enzymatic decomposition and utililization of foodstuffs, particularly carbohydrates, by microbes) also occurs in the crops of Hoatzins (*Opisthocomus hoazin*; Figs. 5.94 and 5.95), a South American folivorous bird. Hoatzins feed primarily on young leaves, twigs and shoots of trees and have well-developed crops where microbial fermentation produces high levels of volatile fatty acids (Grajal et al. 1989). The sternum and pectoral girdle of Hoatzins are modified to accommodate the filled crop, and there is a callosity on the skin of the breast where the heavy crop can be rested on a branch (Mackie 2002). Some fermentation may also occur in the crops of other species. For example, the Kakapo (*Strigops habroptila*), an herbivore endemic to New Zealand, eats plant roots, rhizomes, bulbs, buds, leaves, flowers, pollen cones, and berries from a variety of species (Powlesland et al. 1997). Although Kakapos have well-developed crops, no one to date has determined if microbial fermentation occurs (Morton 1978). Although additional study is needed, microbial activity in the crop of seed-eating and possibly leaf-eating birds could be a more general phenomenon than currently acknowledged and an important adaptation to a vegetable diet (Pacheco et al. 2004).

During the breeding season, the crop of pigeons and doves (Columbidae) also produces "crop milk" for feeding nestlings. Incubation and brooding behavior stimulate production of the hormone prolactin by the pituitary gland and prolactin stimulates proliferation and shedding of epithelial cells lining the crop that become the "milk" (Fig. 5.96). Crop milk, primarily protein

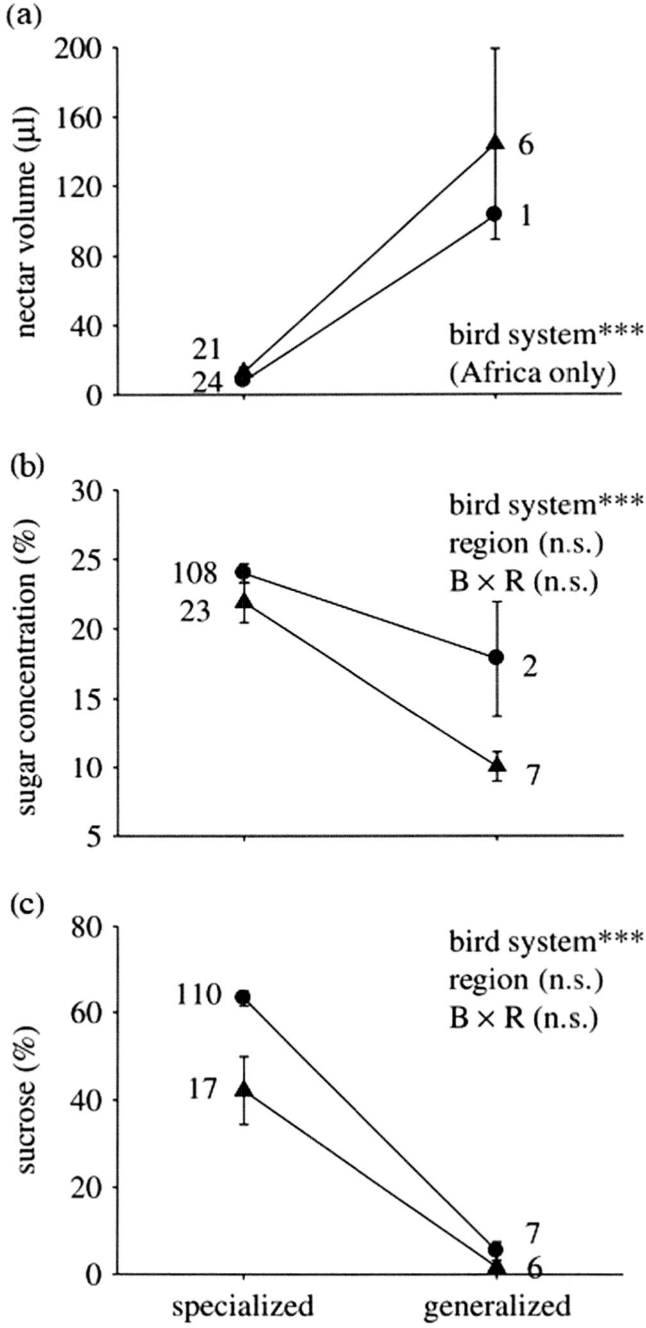


**Fig. 5.53** Nectar properties (volume, sugar concentration, and sugar composition) of flowers pollinated by nectar specialists (hummingbirds in the Americas and sunbirds in Africa) versus those pollinated by nectar generalists (birds that only feed on nectar occasionally). Flowers pollinated by nectar specialists have lower volumes of nectar (but still more than flowers pollinated by insects like bees, butterflies, and moths), with higher concentrations of sugar and sucrose in the nectar. The lower volumes of nectar means that nectarivorous birds must visit flowers more often, a plant "strategy" to help ensure pollination. Symbols are means ± SE. Sample sizes are number of genera. Significance values: $**P < 0.01$, $***P < 0.001$, n.s., not significant. (Figure from Johnson and Nicolson 2008; © 2007 The Royal Society, used with permission)

(50–60%), fat (32–45%), carbohydrate (1–3%), and minerals (calcium, potassium, sodium, and phosphorus) with a cheese-like texture, is produced by both males and females and is fed to nestlings beginning soon after hatching (Klasing 1998). Crop milk also contains antibodies and variable amounts of carotenoids (Goudswaard et al. 1979; Eraud et al. 2008). Crop milk is the

**Fig. 5.54** Typical hummingbird-pollinated flowers are red in color and the long floral tubes hang downward so only a hovering hummingbird can forage on the nectar (and collect the pollen). Left, Scarlet gilia (*Ipomopsis aggregate*); Right, Scarlet monkeyflower (*Mimulus cardinalis*). In laboratory tests, hummingbirds generally show no preferences for nectar sources of a particular color. However, the visual systems of bees may make red flowers more difficult to detect so natural selection may favor red coloration for flowers of plants that are hummingbird-pollinated because it reduces possible competition from bees. (Figure from Rodríguez-Gironés and Santamaria 2004; © 2004 Miguel A. Rodríguez-Gironés and Luis Santamaría, open-access article distributed under the terms of the Creative Commons Attribution, License https://creativecommons.org/licenses/by/4.0/)

primary food of young pigeons and doves early in the nestling period, but, as they age, adults begin to regurgitate a mixture of crop milk and seeds. By fledging, young pigeons and doves are fed primarily regurgitated seeds. For example, adult Mourning Doves (*Zenaida macroura*) feed their young primarily crop milk for the first 3–4 days after hatching, then include increasing numbers of seeds as production of crop milk begins to decline 5 or 6 days after hatching until, at the time of fledging (about 15 days after hatching), the diet of the young is essentially the same as that of the parents (Lewis 1993).

**Fig. 5.55** (**a**) The specialized bird perch (arrow) and ground-based flowers of a rat's tail babiana (*Babiana ringens*). (**b**) A male Malachite Sunbird (*Nectarinia famosa*) feeding on nectar while using the special bird perch. (Figure from Anderson et al. 2005; © 2005 Springer Nature, used with permission)

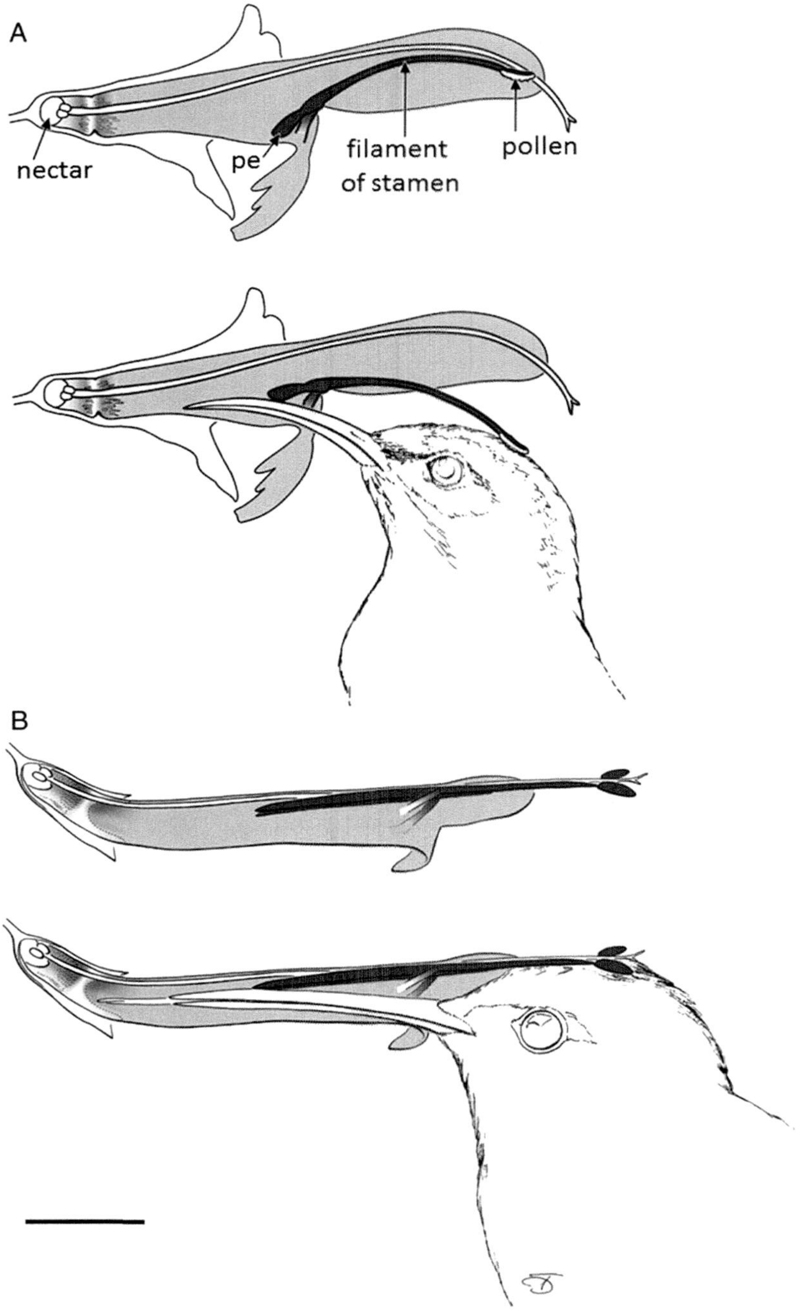


**Fig. 5.56** Pollen transfer with and without a staminal lever mechanism. (**a**) To access nectar at the base of a lance-leaved sage (*Salvia lanceolata*) flower (in a structure called the nectary), the bill of a Southern Double-collared Sunbird (*Nectarinia chalybea*) pushes back a posterior extension (pe) of the stamen, causing the pollen to move down onto the bird's head. (**b**) A prawn chorus (*Salvia haenkei*) flower has no lever mechanism. Pollen is deposited on the head of a Red-tailed Comet (*Sappho sparganurus*) as it feeds on nectar at the base of the flower. Scale bar = 1 cm. (Figure modified from Wester and Claßen-Bockhoff 2007; © 2007 Oxford University Press, used with permission)

Other species that feed young secretions of the esophagus include Emperor Penguins (*Aptenodytes forsteri*) and Greater Flamingos (*Phoenicopterus ruber*). Male Emperor Penguins feed young a secretion consisting of epithelial cells shed from the lining the esophagus mixed with mucus (Prevost and Vilter 1963). Greater Flamingoes feed their young a secretion produced by mucus glands in the epithelial lining of the esophagus. In addition to providing nutrients, this

**Fig. 5.57** (**a**) Transfer of pollen to the feet of a male Malachite Sunbird (*Nectarinia famosa*) perched on an orchid (*Disa chrysostachya*). Pollen is visible on the bird's feet (arrow). Scale bar = 10 mm. (**b**) Close-up of the feet of a captured Malachite Sunbird with pollen attached to the toes and the base of one of the claws (arrows). Scale bar = 5 mm. (Figure modified from Johnson and Brown 2004; © 2004 Springer Nature, used with permission)

regurgitated secretion also includes red pigments that will be incorporated into the plumage of the young flamingos (Wackernagel 1964).

### 5.5.4 Two-Part Stomach

The avian stomach is divided into a proventriculus and ventriculus (Figs. 5.97 and 5.98). The proventriculus is also called the glandular stomach because the inner wall contains cells that secrete mucus, and glands that secrete pepsinogen and hydrochloric acid (Fig. 5.99; Box 5.23 Stomach oils of Procellariiform seabirds). The mucus coats the cells that line the proventriculus and protects them from the enzyme pepsin and the acid. Pepsin is a protein-digesting enzyme that is produced when pepsinogen is exposed to hydrochloric acid. The proventriculus is relatively large in piscivores, frugivores, and, more generally, birds that feed on relatively large food items (Fig. 5.100). In these species, the proventriculus is a distensible sac-like structure (McLelland 1979). The proventriculus of carnivores, piscivores, and carrion-eaters also secretes large quantities of hydrochloric acid and pepsin that aid in the digestion of their high-protein diet (Fig. 5.101, Box 5.24 Preservation of stomach contents of King Penguins). The low pH also permits some digestion of bone. This is important because the soft tissues ingested by carnivores and piscivores tend to have limited amounts of calcium. Without the partial digestion of bone, many carnivorous birds would likely suffer from a calcium deficiency (Klasing 1998).

The walls of the ventriculus (also called the gizzard) are much thicker and more muscular than those of the proventriculus and serve as the avian equivalent of teeth (Fig. 5.102). Birds such as granivores and insectivores that feed on relatively hard items have larger gizzards with thicker, more muscular walls. Of course, birds that feed on

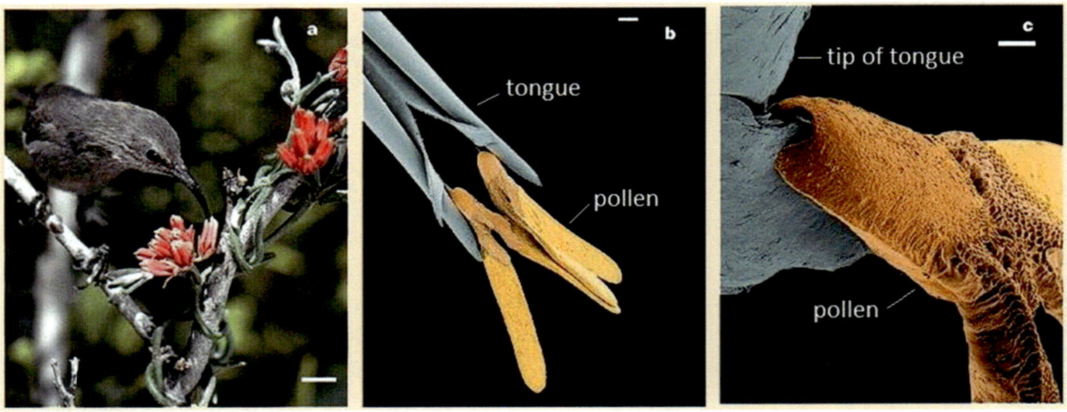


**Fig. 5.58** (**a**) A female Lesser Double-collared Sunbird (*Nectarinia chalybea*) feeding on nectar from flowers of *Microloma sagittatum*. Scale bar = 1 cm. (**b**) The frayed tip of a sunbird's tongue with two attached pollen grains. Scale bar = 100 μm. (**c**) Detail showing a groove in the pollen attaching it to the edge of the tongue. Scale bar = 20 μm. When a bird visits the next flower with the pollen attached to its tongue, tiny bristles orient the pollen so it slides into a small wedge at the base of the flower and breaks free as the tongue is withdrawn. (Figure modified from Pauw 1998; © 1998 Springer Nature, used with permission)

items requiring little mechanical digestion, such as carnivores, piscivores, and nectarivores, have smaller, less muscular gizzards (Fig. 5.103). Food habits also influence the ability of gizzards to sustain activity. Smooth muscle cells in the gizzards of herbivores, granivores, and some insectivores have higher concentrations of myoglobin than do muscle cells in the gizzards of carnivores (Enoki and Morimoto 2000). Because myogloblin stores oxygen, gizzards with more myoglobin are capable of stronger, more frequent contractions.

The main function of the gizzard is to break food into smaller particles with more surface area that enhances the action of digestive enzymes, a process called trituration. Strong contractions of

**Fig. 5.59** Most species of nectarivorous birds are found in the tropics. Hummingbirds (Trochildae) are found throughout the Neotropics (about 328 species), lorikeets (Psittacidae) are found in Africa (13 species) and Australasia (40 species), sunbirds (Nectariniidae) in Africa (88 species) and Australasia (2 species), flowerpeckers (Nectariniidae) in Australasia (44 species), honeyeaters (Meliphagidae) in Australasia (176 species), and white-eyes (Zosteropidae) in Africa (19 species) and Australasia (76 species). Specialists consume primarily nectar; generalists consume nectar plus fruit and/or insects. (Figure modified from Fleming and Muchhala 2008; © 2008 John Wiley and Sons, used with permission)

**Fig. 5.60** Variation in the bill morphology of hummingbirds that feed on different flowers. (**a**) Band-tailed Barbthroat (*Threnetes ruckeri*), (**b**) Long-billed Hermit (*Phaethornis longirostris*), (**c**) White-tipped Sicklebill (*Eutoxeres aquila*), (**d**) Scintillant Hummingbird (*Selasphorus scintilla*), (**e**) Violet Sabrewing (*Campylopterus hemileucurus*), and (**f**) Green-fronted Lancebill (*Doryfera ludovicae*). (Figure from Maglianesi et al. 2014; © 2014 Ecological Society of America, used with permission)

the ventricular muscles aid in this process, but a tough, abrasive keratin-like layer of koilin, also known as the cuticle or cutica gastrica, is also important (Fig. 5.104). Koilin consists of a carbohydrate-protein complex secreted by glands in the mucosa of the ventriculus. After secretion, koilin hardens in response to exposure to hydrochloric acid produced by the proventriculus. This produces a tough, wear-resistant material. The koilin layer is continually worn down by grinding action and replaced by glandular secretion. Insectivorous, herbivorous, and granivorous birds have gizzards with a thick, abrasive koilin layer. In contrast, the gizzards of carnivores, piscivores, and nectivores have a thinner, relatively soft layer of koilin (McLelland 1979; Akester 1986).

For birds where the grinding action of the gizzard is critical for efficient digestion, the lumen (or interior) typically has an elongated, biconvex shape creating two opposing walls on which koilin forms disc-like plates. These plates are hard so that the intraluminal pressure is efficiently transferred to the gizzard contents (rather than deforming the gizzard walls). In addition, as muscles contract, the two plates approach each other, creating compression, and simultaneously slide in opposite directions, generating additional friction (Moore 1998). Experiments with an artificial gizzard has revealed that sliding, or translational, movements break down plant materials such as grass more efficiently than just compression, particularly when grit is present (Moore et al. 1998).

Grit, or, for larger birds, small stones or pebbles, can, as just noted, further aid in the grinding process so many birds ingest grit (sand or small stones) on a regular basis (Fig. 5.105). Ingestion of grit has been reported in 54 different families of birds (Downs et al. 2019), but grit use does vary with diet (Fig. 5.106). Birds that eat seeds or other plant material generally have relatively large amounts of grit in their gizzards. Insectivores typically use less grit than granivores, and frugivores typically use little grit. The amount of grit ingested by granivores and insectivores likely depends on the hardness of the items consumed. For example, soft-bodied

**Box 5.18 Birds as pollinators**

Most plants are pollinated by insects, but more than 900 species of birds, primarily the nectarivores, pollinate approximately 500 of the 13,500 vascular plant genera (Nabhan and Buchmann 1997). The main families of pollinating birds are hummingbirds (Trochilidae), sunbirds (Nectariniidae), honeyeaters (Meliphagidae), and lorikeets (Psittaculidae), but species that pollinate flowering plants are also found in the families Philepittidae (asities), Promeropidae (sugarbirds), Fringillidae (Hawaiian honeycreepers), Icteridae (orioles and caciques in North and South America), Thraupidae (tanagers), Tyrannidae (flycatchers), Parulidae (warblers in Central and South America), and Zosteropidae (white-eyes in Africa (Kelly et al. 2004; Rocca and Sazima 2010). Most bird-pollinated plants are found at tropical latitudes, and bird pollination is particularly important in the Austral and Oceanic regions, where more than 10% of the flora can be bird pollinated (Kelly et al. 2004). Most birds that pollinate plants feed on a variety of different plant species, and other pollinators such as bees also visit bird-pollinated plants. However, nonbird pollinators like bees are generally much less effective as transferring the pollen of plants that are usually pollinated by birds (Krauss et al. 2017).

Examples of birds that pollinate flowering plants. (**a**) New Holland Honeyeater (*Phylidonyris novaehollandiae*) in Australia, (**b**) Tawny-bellied Hermit (*Phaethornis syrmatophorous*) in Ecuador, (**c**) Copper-rumped Hummingbird (*Amazilia tobaci*) in Mexico, (**d**) Eastern Spinebill (*Acanthorhynchus tenuirostris*) in Australia, (**e**) Dark-capped Bulbul (*Pycnonotus tricolor*) in South Africa, (**f**) female Amethyst Sunbird (*Chalcomitra amethystine*) in South Africa, and a (**g**) Rainbow Lorikeet (*Trichoglossus moluccanus*) in Australia (Figure from Krauss et al. 2017; © 2017 Elsevier Ltd, used with permission)

(continued)

**Box 5.18** (continued)

Because birds are so mobile and are active regardless of weather conditions (unlike insects), they can be particularly important pollinators in environments with harsh conditions such as at high altitudes (Cronk and Ojeda 2008). Plants in locations with few insects, such as isolated islands, can also be particularly dependent on bird pollinators.

Birds are much larger than most insects and require more energy, so bird-pollinated flowers typically produce larger quantities of nectar than insect-pollinated flowers (Cronk and Ojeda 2008). Even so, nectarivores like hummingbirds must typically spend much of the day foraging to meet their energetic needs. Because the bills (rhamphotheca) of birds are evolutionarily labile structures, some birds and flowers have evolved morphological coadaptations, with bill morphology co-evolving with flower morphology (Stiles 1978). However, flowers of most plant species are visited by several species of hummingbirds and all hummingbird species feed from more than a single plant species. As such, hummingbird–plant interactions are best described as diffuse co-evolution between functional groups at the population level (Feinsinger 1983; Fenster et al. 2004).

Green-headed Tanager (*Tangara seledon*) (Figure from Rocca and Sazima 2010; © 2010 Universidade Federal do Rio de Janeiro, used with permission)

(continued)

**Box 5.18** (continued)

(continued)

**Box 5.18** (continued)

The bill morphology of male and female Purple-throated Carib hummingbirds (*Eulampis jugularis*) differs, with females (**a**) having longer, more curved bills than males (**b**). This hummingbird is the only pollinator for *Heliconia caribaea* and *H. bihai* on the Caribbean islands of St. Lucia and Dominica, and these plants exhibit different morphs on the two islands. However, in all cases, the bill morphology of female Purple-throated Caribs allows them to forage efficiently on the nectar of *H. bihai* (with longer, curved flowers; **c**, **e**, **f**, and **h**) and the bill morphology of males allows them to forage efficiently on the nectar of *H. caribaeae* (with shorter, less curved flowers; **d**, **g**, **i**, and **j**). The bill dimorphism of male and female Purple-throated Caribs is among the most extreme of any hummingbird species and may have evolved because of the lack of many competing species on islands compared to mainland hummingbird communities (Feinsinger 1976) (Figure from Temeles and Kress 2003; © 2003 American Association for the Advancement of Science, used with permission)

larvae may require little grit, whereas hard-bodied insects, like beetles, may require large amounts of grit. Grit-sized insect fragments (e.g., exoskeletons) and small, hard seeds are sometimes retained in gizzards and serve as "grit substitutes" (Gionfriddo and Best 1996). Although the most common function of grit is to facilitate digestion, ingested sand or small stones may also provide birds with supplementary minerals, especially calcium, and serve other functions as well (Box 5.25 Avian gastroliths). Dietary calcium from grit or other sources, such as snail shells or even calcium-rich soil, can be especially critical for females during the breeding season because additional calcium may be needed to form eggshells.

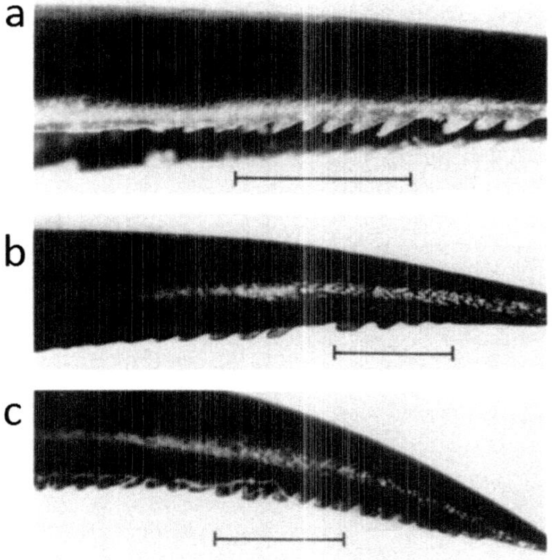


**Fig. 5.61** Upper bills of three species of hummingbird with serrate tomia that can be used to make holes near the base of long flower corolla tubes and extract nectar. (**a**) Sparkling Violetear (*Colibri coruscans*), (**b**) Mexican Woodnymph (*Thalurania ridgway*), and (**c**) Green-breasted Mango (*Anthracothorax prevostii*). Scale bars = 1 mm. (Figure modified from Ornelas 1994; © 1994 Oxford University Press, used with permission)

The retention of grit particles can be influenced by the rate at which grit is ingested (Tagami 1974; Trost 1981). If readily available, birds may ingest and eliminate considerable amounts of grit every day (Alonso 1985; Best and Stafford 2002), but, if not available, birds can retain grit particles in their gizzards for extended periods (Walter and Aitken 1961). Grit retention time has also been found to vary with grit size, but not in a consistent manner. For example, Gionfriddo and Best (1995) found that House Sparrows (*Passer domesticus*) retained smaller grit particles longer than larger particles when provided with either small or large grit particles. However, Smith (1960) found that grit retention by Domestic Chickens increased with increasing grit size, whereas Tagami (1974) reported that Domestic Chickens retained more medium-sized (1.2–2.4 mm) than large (2.4–3.4 mm) particles, and few small (0.6–1.2 mm) particles. Such differences may be due to differences in the types of food ingested as well as differences in the shape and surface texture of grit particles. Regardless of size, grit retention time may also be shorter for birds that

**Box 5.19 Nectar robbers**

Several species of birds that forage on nectar do so via robbery, i.e., extracting nectar at the base of flowers where the nectar-secreting tissue is located rather than via the flower entrance. Igic et al. (2020) divided nectar robbers into primary nectar robbers that create the hole in the flower tissue to obtain nectar and secondary nectar robbers that remove nectar via holes created by the primary nectar robbers. Avian nectar robbers include the primary nectar-robbing flowerpiercers (Passeriformes: *Diglossa* spp.), several species of primary and secondary nectar-robbing hummingbirds, and several secondary nectar-robbing songbirds, including orioles and bananaquits (Lasso and Naranjo 2003). In southern Africa, several species of birds have been reported as secondary nectar robbers, primarily sunbirds (e.g., *Anthothreptes* spp., *Nectarinia* spp., and *Phylloscopus* sp.) (Swynnerton 1916). Many insects, especially bees, wasps, and ants, are also nectar robbers (Irwin et al. 2010), and even some mammals are nectar robbers (e.g., mice, Inouye 1983).

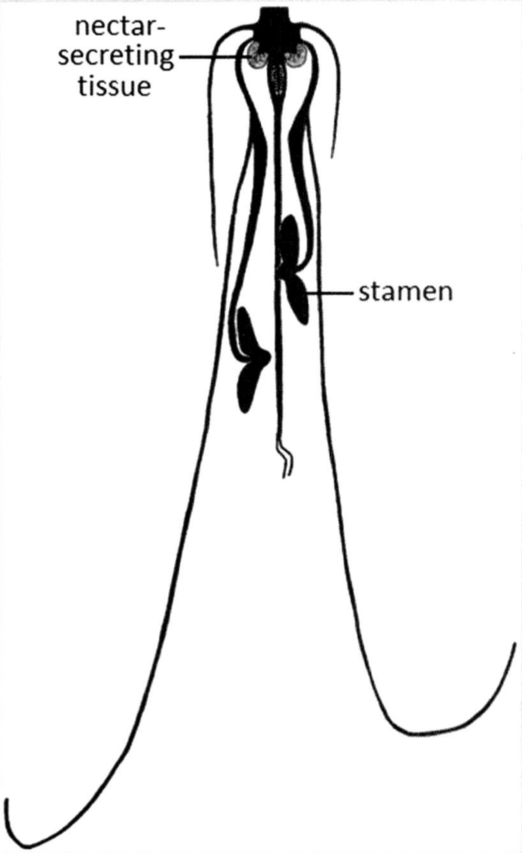


Cross-section through a tubular flower showing the location of the nectar-secreting tissue, which would be out of reach if attempting to access it from the flower entrance for birds with short bills (Figure from Galetto 1995; © 1995 Springer-Verlag, used with permission)

(continued)

**Box 5.19** (continued)

A Black Flowerpiercer (*Diglossa humeralis*) "robbing" nectar from a *Passiflora mixta* flower by creating a small hole at the base of the flower. The sharply hooked upper bill is used to hold the flower in place so the flower can be pierced by the lower bill (Figure from Rojas-Nossa et al. 2016; © 2015 The Authors. Published by John Wiley and Sons, used with permission)

Flowerpiercers (genera *Diglossa* and *Diglossopis*) are specialized nectar robbers of tubular hummingbird-pollinated flowers. There are 18 species of flowerpiercers distributed from Mexico to the Andes of northwestern Argentina (Isler and Isler 1987). They use the long hook at the tip of their bill to hold flowers firmly, stab the base of the corolla with their lower bill, and then extract nectar by inserting their tongues through the slit.

consume hard, more difficult to digest food items because such items cause faster disintegration and elimination of grit particles (Norris et al. 1975).

In some birds, indigestible materials are collected in the gizzard and regurgitated. The gizzards of carnivores, piscivores, and insectivores trap indigestible pieces of bones, feathers, fur, and chitin and form them into pellets that are then regurgitated (Fig. 5.107). Fruit-eating birds like manakins, waxwings, and thrushes that "gulp" fruits, or ingest whole fruits, must separate the pulp from the seeds in their digestive system and then eliminate those seeds. If relatively large (>4 mm), seeds are stripped of pulp in the stomach and then regurgitated. Smaller seeds (<2 mm) pass into the intestine and are defecated (Levey 1987b). Because regurgitation eliminates seeds faster than defecation, more "digestive space" becomes available more quickly when large seeds are regurgitated. Regurgitation requires reverse peristalsis, with contraction of the walls of the gizzard moving materials back into the proventriculus. However, the gizzards of all birds, not just those that regurgitate undigestible materials, exhibit distinct reverse peristaltic movements that are critical to optimal digestive function (Duke 1994). The gastric reflux allows material in the gizzard to reenter the proventriculus for additional treatment with acid and pepsin.

The posterior end of the gizzard of most birds tapers into a pyloric section that joins the small intestine. The pyloric part of the stomach is best developed in birds that eat fish (e.g., grebes, anhingas, herons, cormorants, penguins and pelicans) and aquatic invertebrates (e.g., waterfowl and rails). The mucosal epithelium of the pyloric region may contain projections (papillae; anhingas) or may trap feathers (grebes; Box 5.26 Feather eating by loons and grebes) that may help prevent undigested materials such as bones from passing into the intestine. By retaining these materials, additional time is available for digestion (McLelland 1979; Klasing 1998).

### 5.5.5 Small Intestine

From the gizzard, semi-digested material (called chyme) moves into the intestine. This movement is stimulated by the volume of chyme in the gizzard, with negative feedback generated by the presence of digestive by-products such as amino acids and fats in the first portion of the small intestine (duodenum). This feedback system ensures that the movement of food from the gizzard to the intestine proceeds at a rate that

**Table 5.5** Herbivorous species of birds, number of times herbivory evolved in each order, flight capabilities, and examples of species in each order. Examples include species where leaves, stems, buds, and bulbs make up more than 80% of stomach contents or foraging observations for at least a season (Table modified from Olsen 2015; © 2015 The Author. Ecology and Evolution published by John Wiley & Sons Ltd., open-access article under the terms of the Creative Commons Attribution License, https://creativecommons.org/licenses/by/4.0/)

| Order | Number of origins | Total species | Examples | Reference |
|---|---|---|---|---|
| Ratites | 1–2 | | | |
| Struthioniformes | – | 1 | Ostrich (*Struthio camelus*) | Milton et al. (1994) |
| Rheiformes | – | 2 | Greater Rhea (*Rhea americana*) | |
| Casuariiformes | – | 1 | Emu (*Dromaius novaehollandiae*) | Davies (1978) |
| Tinamiformes | – | <47 | Andean Tinamou (*Nothoprocta pentlandii*) | Mosa (1993) |
| Dinornithiformes[a] (moas) | – | 9 | – | Wood et al. (2008) |
| Aepyornithiformes[a] (elephant birds) | – | 2–7 | Elephant bird (*Aepyornis*) | Clarke et al. (2006) |
| Galliformes | 1+ | <19 | Rock Ptarmigan (*Lagopus muta*) | Sedinger (1997) |
| Anseriformes | 1+ | 20–30 | Screamers, geese, swans, moa-nalos | Morton (1978), James and Burney (1997) |
| Opisthocomoiformes | 1 | 1 | Hoatzin (*Opisthocomus hoazin*) | Grajal et al. (1989) |
| Gruiformes | 1 | 1 | Takahe (*Porphyria mantelli*) | Mills and Mark (1977) |
| Psittaciformes | 1 | 1 | Kakapo (*Strigops habroptila*) | Trewick (1996) |
| Passeriformes | 2+ | 2 | White-tipped Plantcutter (*Phytotoma rutila*) | Bucher et al. (2003) |
| | | | Thick-billed Saltator (*Saltator maxillosus*) | Munson and Robinson (1992) |
| Totals | 9+ | <120 | | |

[a]Extinct

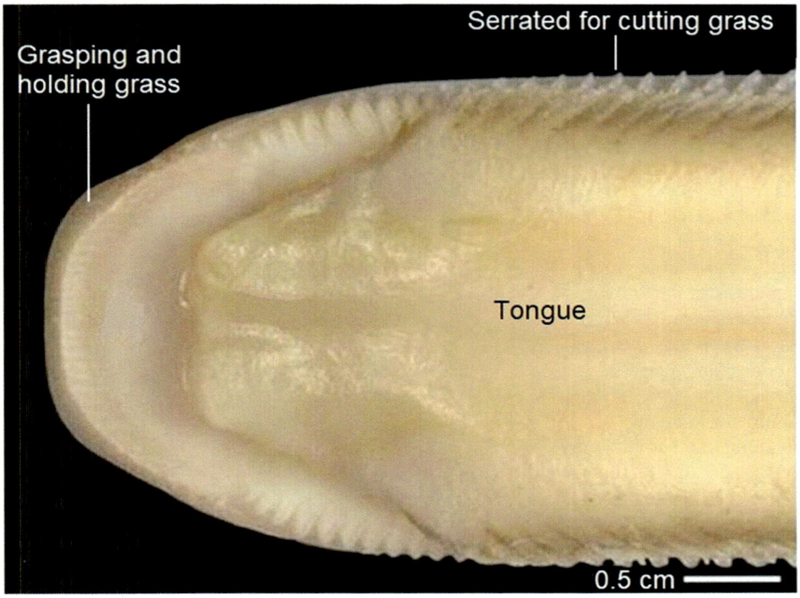


**Fig. 5.62** Lower bill and tongue of a Domestic Goose (*Anser anser domesticus*). Note that the anterior part of the bill is specialized for grasping and holding grass and the rest of the bill for cutting the grass. (Figure modified from Skieresz-Szewcyzk and Jackowiak 2016; open-access article distributed under the terms of the Creative Commons Attribution 4.0 International License, https://creativecommons.org/licenses/by/2.0/)

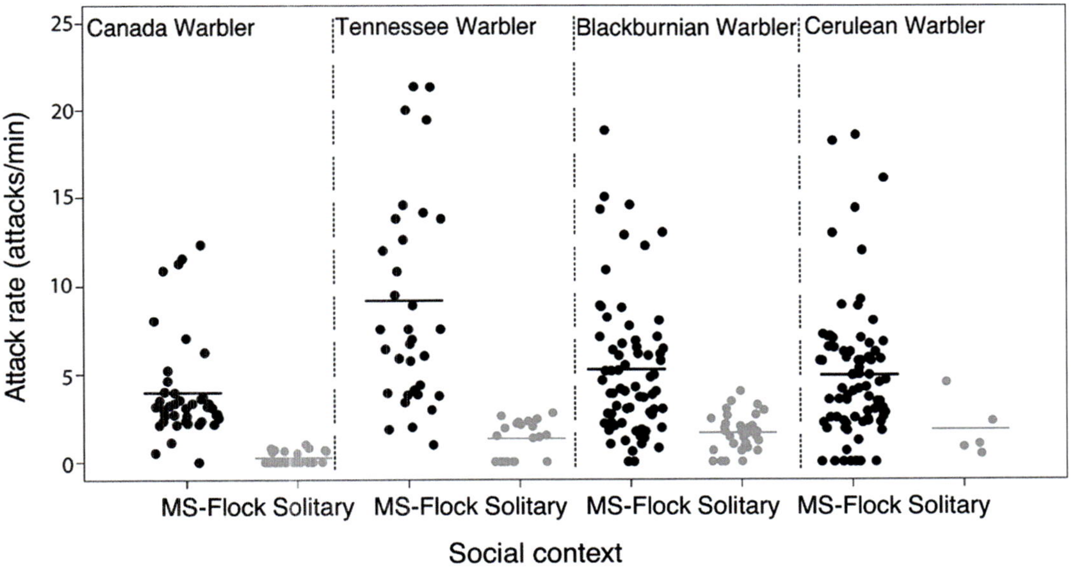


**Fig. 5.63** During the nonbreeding period, warblers in mixed-species flocks (MS-Flock) in shade-grown coffee plantations in the Colombian Andes had significantly higher attack rates than when foraging alone (Solitary). Horizontal lines indicate the mean, and circles indicate individual birds. Canada Warbler, *Cardellina canadensis*; Tennessee Warbler, *Leiothlypis peregrina*; Blackburnian Warbler, *Setophaga fusca*; Cerulean Warbler, *Setophaga cerulea*. (Figure from Munoz Z and Colorado Z 2021; © 2021 Association of Field Ornithologists, used with permission)

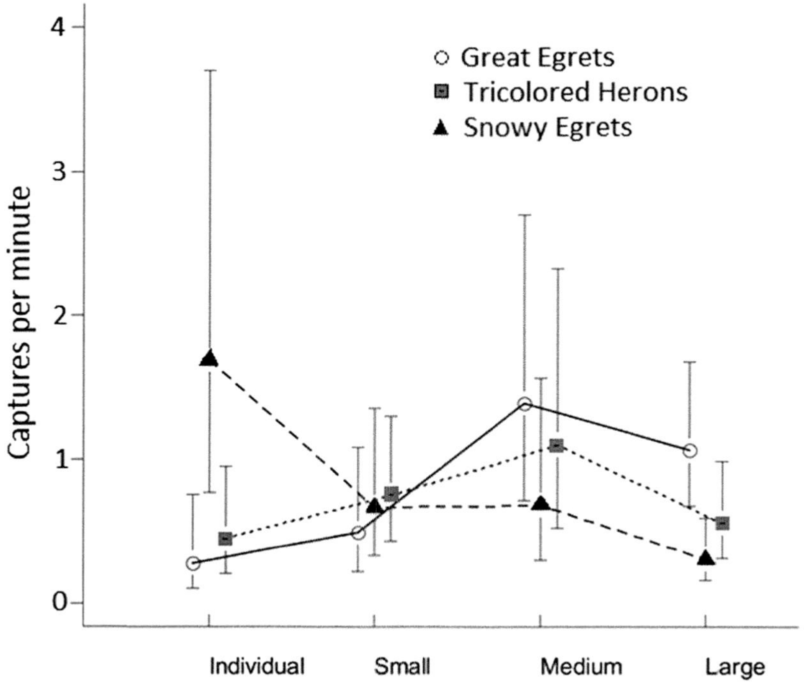


**Fig. 5.64** Rate of prey capture by herons foraging alone and in small (2–10 birds), medium (11–50 birds), and large (>50 birds) groups. Captures rates for Great Egrets (*Ardea alba*) and Tricolored Herons (*Egretta tricolor*) increased in medium and large groups, capture rates for Snowy Egrets (*Egretta thula*) were lower when they were in groups than when foraging alone. Error bars are 95% confidence intervals including the variation due to the random effect of site. Group-size classes are solitary, small (2–10 individuals), medium (11–50), and large (>50). (Figure modified from Stolen et al. 2012; © 2012 Oxford University Press, used with permission)

enhances digestion and absorption. The intestine is the primary organ of digestion and absorption, and is divided into a small intestine and a much shorter large intestine. Among vertebrates, the small intestine is usually smaller in diameter than the large intestine. However, there is rarely much difference in the diameter of avian small and large intestines. The small intestine is divided into three sections, including the duodenum, jejunum, and ileum, that do not differ either morphologically or histologically, whereas the large intestine usually consists of a rectum plus paired caeca. Birds exhibit much variation in the length of the small intestine, ranging from three times body length in about 17-cm long Common Swifts (*Apus apus*) to more than 20 times body length in 1.7–2.5 m tall Common Ostriches (*Struthio camelus*). On average, bird small intestines are shorter and have less surface area than those of nonflying mammals (bat small intestines are also shorter and with less surface area than those of nonflying mammals) (Caviedes-Vidal et al. 2007; Fig. 5.108). Duque-Correa et al. (2021, 2022) compared the total intestinal lengths of 519 species of mammals and 390 species of birds and found that the total intestinal length of birds (i.e., small and large intestine plus colon/rectum and ceca/cecum length) averages about 85% that of mammals. A reduction in intestinal volume (plus the corresponding mass of digesta in the intestine) is advantageous because it reduces the energetic costs of flight. However, flying is energetically costly and thus requires more nutrient assimilation so shorter intestines with less absorptive surface area would seemingly be problematic for birds. The avian solution to this is to rapidly absorb a high proportion (but not all) of water-soluble nutrients like glucose and amino acids via a paracellular pathway (diffusion through sieve-like openings between adjacent epithelial cells) rather than via transporter proteins (i.e., active transport or facilitated diffusion) (Fig. 5.109). This passive absorption may be advantageous because it requires little energy (unlike active transport) and allows matching of absorption rate with the rate of hydrolysis or breakdown (Pappenheimer 1993). However, a possible disadvantage of a high intestinal permeability is a greater likelihood of absorbing toxins from plant and animal material in the intestinal lumen

**Table 5.6** Examples of seabirds obtaining food by associating with whales, dolphins, fur seals, and tuna and other fish (Table modified from Veit and Harrison 2017; © 2017 Veit and Harrison, open-access article distributed under the terms of the Creative Commons Attribution License (CC BY), https://creativecommons.org/licenses/by/4.0/)

| Species association[a] | Local enhancement/ facilitation[b] | Reference |
|---|---|---|
| *Sula* spp., Wedge-tailed Shearwater (*Ardenna pacifica*), spotted dolphin (*Stenella attenuata*), spinner dolphin (*S. longirostris*), yellow-finned tuna (*Thunnus albacares*) | 1, 2 | Au and Pitman (1986) |
| Parkinson's Petrel (*Procellaria parkinsoni*), melon-headed whale (*Peponocephala electra*), false killer whale (*Pseudorca crassidens*) | 1, 2, 3 | Pitman and Balance (1992) |
| Wedge-tailed Shearwaters and Brown Noddies (*Anous stolidus*), skipjack tuna (*Katsuwonus pelarnis*) | 1, 2 | Hebshi et al. (2008) |
| Cory's Shearwaters (*Calonectris borealis*), Great Shearwater (*Ardenna gravis*), Atlantic spotted dolphin (*Stenella frontalis*) | 1, 2, 3 | Martin (1986) |
| Cory's Shearwaters, dolphins (*Delphinus* and *Stenella* spp.) and tuna (*Thunnus* spp.) | 1, 2 | Clus and Grosvalet (2001) |
| Black-browed Albatross (*Thalassarche melanophris*), Antarctic fur seals (*Arctocephalus gazella*), Macaroni Penguins (*Eudyptes chrysolophus*), *Pachyptila* spp. | 1, 2, 3 | Harrison et al. (1991) |
| Northern Gannet (*Sula bassana*), Atlantic white-sided dolphins (*Lagenorhynchus acutus*), harbor porpoise (*Phocoena phocoena*), minke whale (*Balaenoptera acutorostrata*) | 1, 2 | Guse (2013) |
| Black-legged Kittiwake (*Rissa tridactyla*), Northern Fulmar (*Fulmarus glacialis*), Red Phalarope (*Phalaropus fulicara*), Thick-billed Murre (*Uria lomvia*), California gray whale (*Eschrichtius robustus*) | 3 | Obst and Hunt (1990), Grebmeir and Harrison (1992) |
| Leach's Storm-Petrel (*Oceanodroma leucorhoa*), Manx Shearwater (*Puffinus puffinus*), pilot whale (*Globicephala melas*), bottlenose dolphins (*Tursiops truncatus*) | 1, 2, 3 | Skov et al. (1995) |
| Black-legged Kittiwakes, Northern Gannets, minke whales (*Balaenoptera acutostrata*), white-beaked dolphins (*Lagenorhynchus albirostris*) | 1, 2, 3 | Camphuysen and Webb (1999) |
| Wilson's Storm-Petrels (*Oceanites oceanicus*), rough-toothed dolphins (*Steno bredanensis*) | 3 | Olmos et al. (2013) |
| European Shag (*Gulosus aristotelis*), European sea bass (*Dicentrarchus labrax*) | 1 | Gatti et al. (2021) |

[a]Species frequently observed together, at core of interspecific association; not a full list of attendants
[b]Apparent basis of association: 1 = interspecific association aids participant(s) in location of prey patch, 2 = prey made available at the surface by diving species, and 3 = waste or fragmented prey made available for scavenging

(Diamond 1991). If so, exposure to toxins might be limited by avoiding unfamiliar food items and ingesting substances that inhibit absorption of toxins. For example, some birds intentionally consume soil (geophagy) and, for some species such as some parrots, soils may contain materials that adsorb and prevent absorption of some dietary toxins (Diamond et al. 1999; Gilardi et al. 1999; Box 5.27 Avian geophagy).

The cells that line the small intestine also have enzymes that help break down ingested foods into smaller molecules that can be absorbed (Table 5.7; Box 5.28 A missing enzyme). Carbohydrates are broken down into simple sugars (monosaccharides) like glucose and fructose, proteins into amino acids, and lipids into fatty acids, monoglycerides, disaccharides, and glycerol. Most absorption of monosaccharides, like glucose (Box 5.29 Avian blood glucose

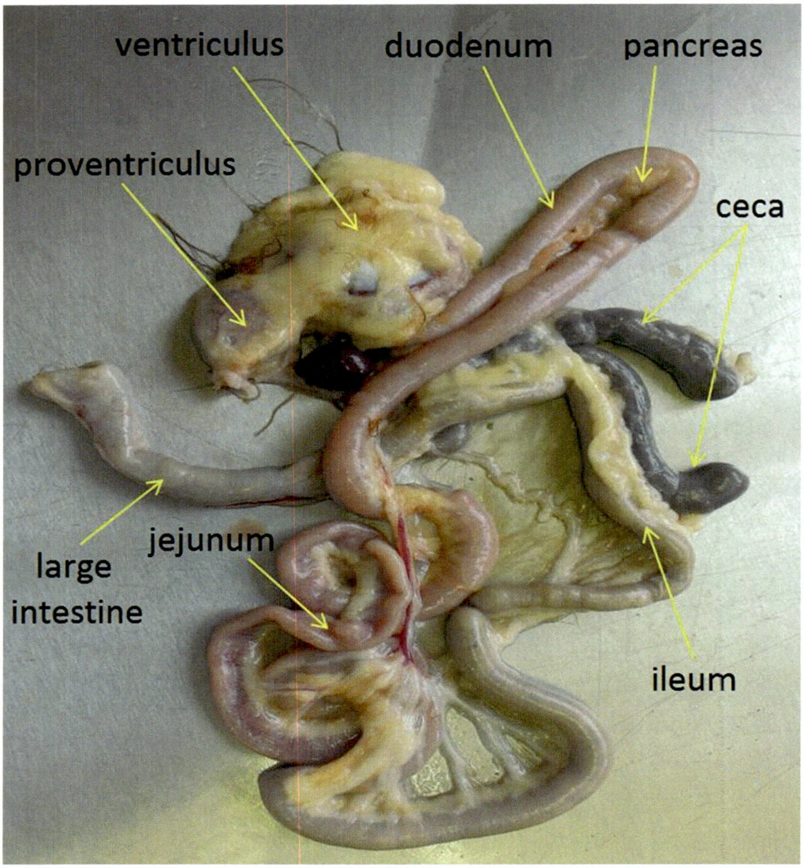


**Fig. 5.65** Digestive system of a Domestic Chicken (*Gallus g. domesticus*). The esophagus is not shown, and the dark spherical object below the ventriculus is the spleen. (Figure modified from Röhe 2013; used with permission of Ilen Röhe)

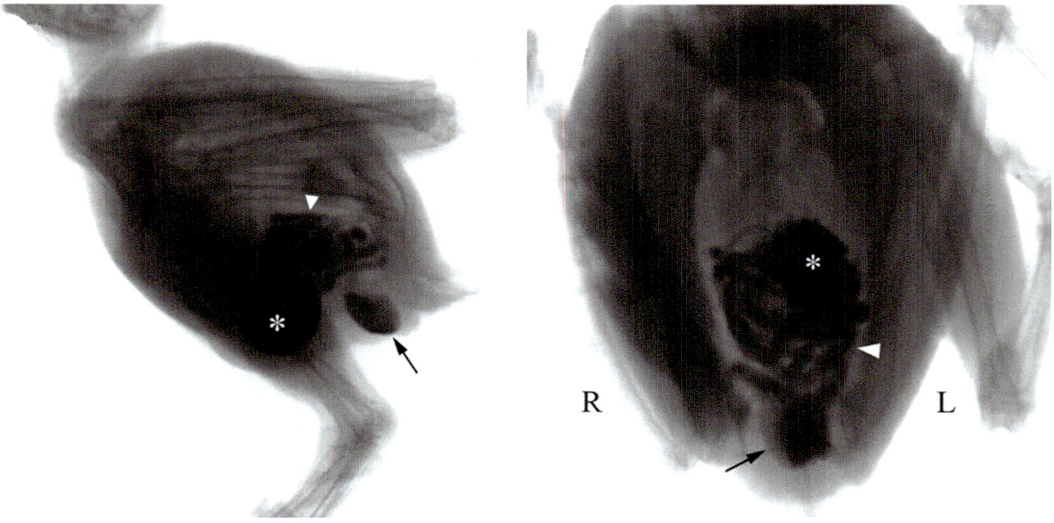


**Fig. 5.66** Relative position of the gizzard (asterisk), small intestine (white arrowhead), and cloaca (black arrow) of a Barred Owl (*Strix varia*). R right, L left. (Figure from Doss et al. 2017; © 2017 Association of Avian Veterinarians, used with permission)

**Box 5.20 Evolution of the avian digestive system**

Evidence of diet and digestive function is generally rare in the fossil record, but well-preserved fossils discovered in China have provided either direct evidence (stomach contents) or indirect evidence (e.g., gastroliths found in the abdomen and the presence or absence of teeth) about the diets and digestive systems of nonavian theropods and the first birds. For example, analysis of these fossils has revealed the two-part stomach (proventriculus and ventriculus) was already present in the early Jurassic (Coelurosauria) and that the "gastric mill" (muscular gizzard and ingested stones to breakdown food items) appeared as early as the middle Jurassic (Ornithomimosauria dinosaurs; shown below). Crops, outpocketings of the esophagus, appeared in the early Cretaceous, with fossils of *Sapeornis chaoyangensis* providing clear evidence of seeds stored in crops (shown below). The Ornithuromorpha, the lineage that includes present-day birds, include several taxa with evidence of crops, gastric mills (paired with complete tooth loss in some species), and even bidirectional peristalsis. As example, in one specimen of *Yanornis* (shown below), what appear to be pellets composed of macerated fish bones are located in the ventriculus and the crop. The pellet in the crop was likely transported up the esophagus from the ventriculus via reverse peristalsis in the process of being regurgitated.

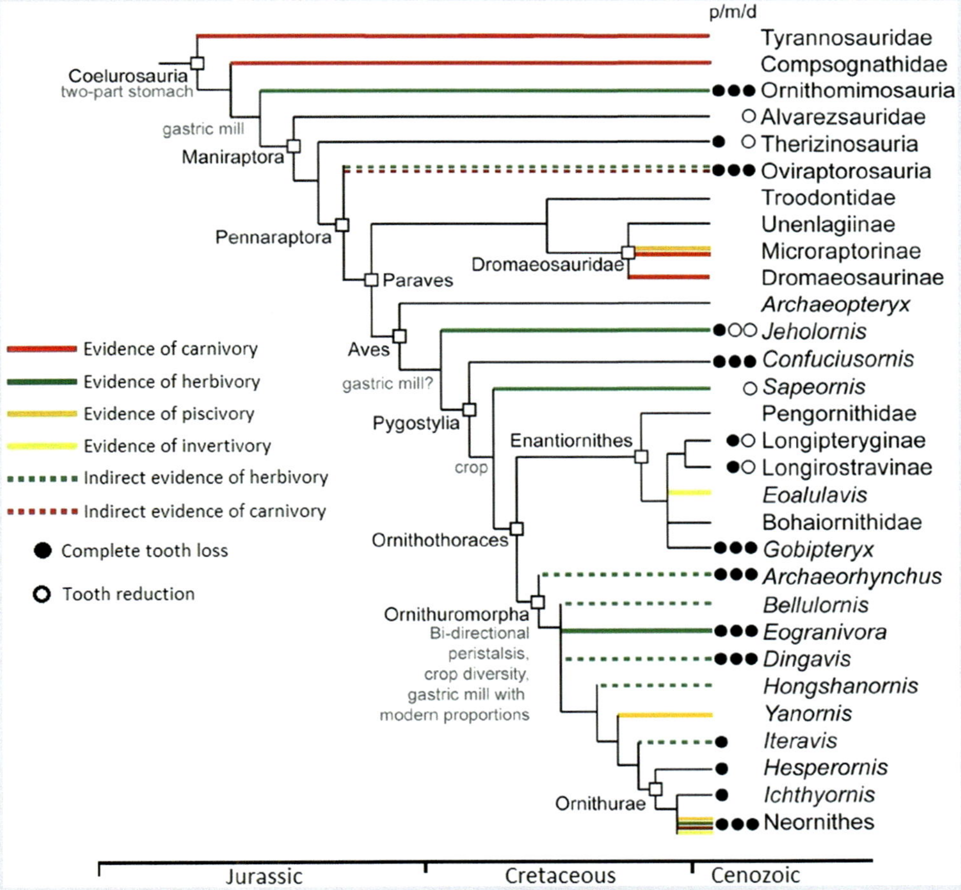


Avian lineages where evidence of their diets is available. p, premaxilla; m, maxilla; d, dentary (Figure modified from O'Connor 2019; ©2019 Elsevier Ltd., used with permission)

(continued)

**Box 5.20** (continued)

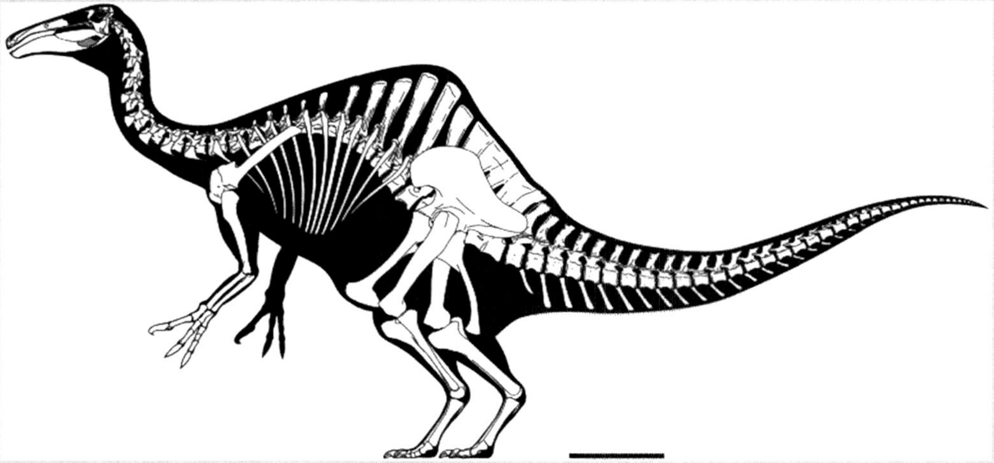

Reconstruction of *Deinocheirus mirificus* (an Ornithomimosaurian). More than 1400 gastroliths (range = 8–87 mm) were collected from inside the ribs as one fossil of this species was being excavated. *Deinocheirus* was likely an omnivore and, because they had no teeth, gastroliths were probably used to grind food in the gizzard (ventriculus). Scale bar = 1 m (Figure from Lee et al. 2014c: © 2014 Springer Nature, used with permission)

(continued)

**Box 5.20** (continued)

Photographs of two specimens of *Sapeornis chaoyangensis* with crops. (**a**) A partially articulated skeleton of the first specimen with seeds in the crop. (**b**) Close-up view of the crop. (**c**) The second specimen is a nearly complete articulated skeleton that again shows evidence of a crop and gastroliths. (**d**) Close-up view of the crop (Figure modified from Zheng et al. 2011; used with permission of the U. S. National Academy of Sciences)

(continued)

**Box 5.20** (continued)

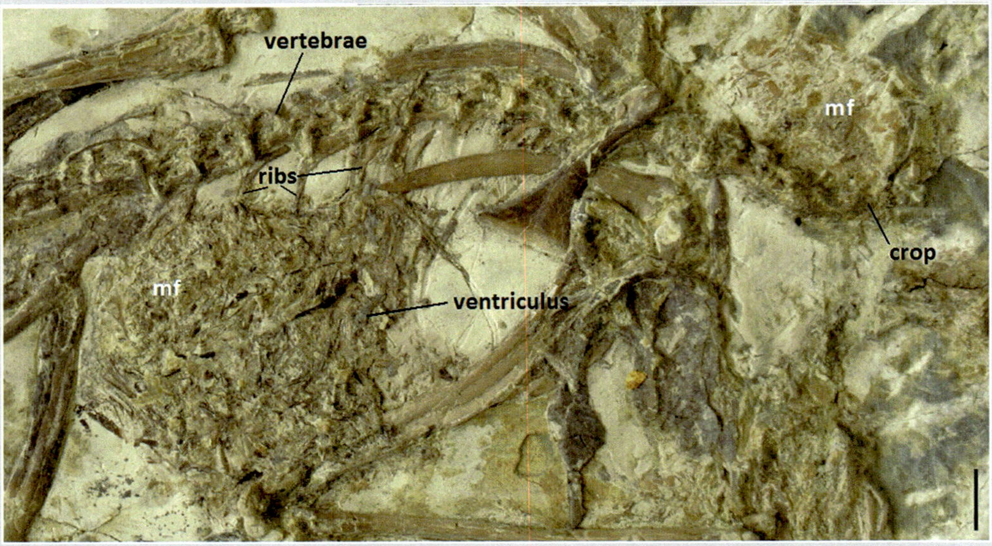


*Yanornis* fossil showing macerated fish remains in the ventriculus and the crop. The remains in the crop are much more compact, suggesting this represents a pellet in the process of being regurgitated. mf, macerated fish remains; Scale bar = 1 cm (Figure modified from O'Connor 2019; © 2018 Elsevier B.V., used with permission)

levels), by birds occurs in the duodenum and involves simple diffusion from the intestine, between intestinal cells (paracellular transport; Fig. 5.108), and into intestinal capillaries (Fig. 5.110). Some glucose and amino acids, plus other materials including fructose (another sugar), lipids, water, and several ions (like sodium and potassium) are absorbed by several different membrane transporters, special proteins that move materials across cell membranes by facilitated diffusion and active transport (Fig. 5.110).

Some variation in the length of the small intestines of birds is due to diet, although analyses have revealed much overlap in intestinal length among species of birds with different diets (Duque-Correa et al. 2022). Typically, intestines are generally longer in granivores, herbivores, and piscivores and shorter in frugivores, carnivores, insectivores, and nectivores. Birds that eat foods that are more difficult to digest, such as seeds and other plant material (with lots of cellulose), tend to have longer intestines than birds that eat more easily digested foods, such as fruit, animal material (high in protein), and, of course, nectar (Duque-Correa et al. 2022). However, the manner in which food is obtained may also play a role. For example, among raptors, species like falcons and accipiters that depend on rapid acceleration and maneuverability to capture prey have shorter intestines than those like buteos and eagles that depend more on soaring, searching flight (Hilton et al. 1999). The advantage is that a shorter intestine reduces weight (of both the intestine and its contents) (Hilton et al. 1999). A shorter intestine may also reduce digestive efficiency, but the benefits of reduced weight outweigh any such loss in efficiency (Barton and Houston 1994).

Most absorption of nutrients occurs in the small intestine and more surface area means

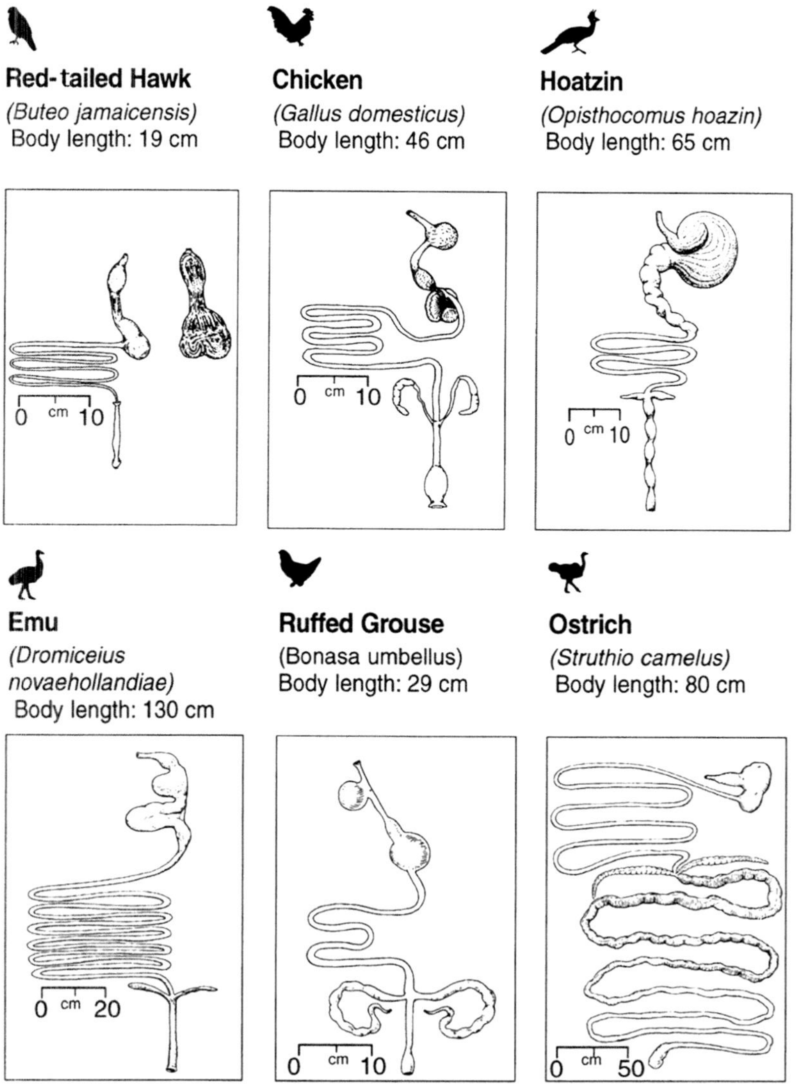


**Fig. 5.67** Gastrointestinal tracts of a carnivorous Red-tailed Hawk, an omnivorous Domestic Chicken, and four herbivorous birds. Note the larger size of crop in the omnivore and herbivores, and particularly in Hoatzins. Ceca are small in hawks and relatively large in grouse. Although ceca are relatively small in Hoatzins, Emus, and Common Ostriches, an expanded foregut (Hoatzins), a much longer midgut (Emus), or a much longer colon (Ostriches) compensates for this. (Figure from Stevens and Hume 1995; © Cambridge University Press, used with permission)

more efficient absorption. Intestine length obviously influences surface area, but folds of the inner lining of the intestine along with villi and microvilli can increase the amount of surface area by several orders of magnitude (Figs. 5.111 and 5.112). For example, the amount of surface area in the intestine of Rock Pigeons (*Columba livia*) is increased 8.1 times by villi (Karasov and Hume 1997). The surface of villi is covered with epithelial tissue that includes two types of cells: absorptive cells (enterocytes) and goblet cells. The absorptive cells are more numerous and are specialized for the absorption of nutrients across their plasma membranes and the subsequent export of these same nutrients across the plasma membrane at their base. After export, the nutrient molecules diffuse or are transported into capillaries and lymph vessels. The surface area of each absorptive cell is greatly increased by a dense array of microvilli, called the brush border. The goblet cells secrete mucus, a viscous substance composed of water and glycoprotein that coats the intestinal walls and serves to both protect the cells that line the intestine from autodigestion by

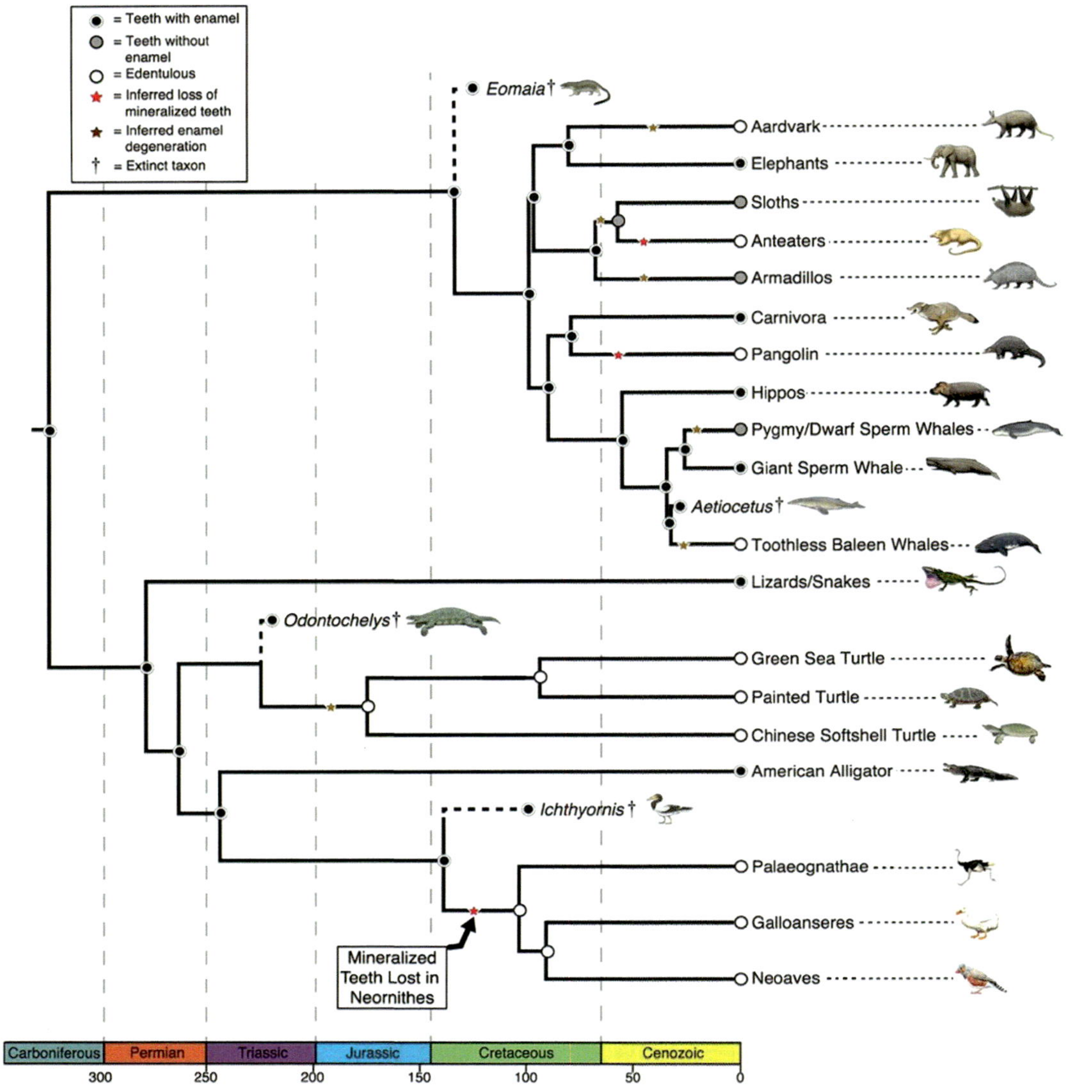


**Fig. 5.68** Time tree of toothless and enamel-less amniotes. Turtles lost teeth well before birds did; toothless mammals likely lost their teeth sometime after birds. Modern birds share a common ancestry with toothed, theropod dinosaurs, but, instead of teeth, modern birds have a rhamphotheca (horny beak) for food acquisition and, as part of the digestive tract, a muscular gizzard to aid in macerating and digesting food. Hypothesized loss of mineralized teeth on the stem Neornithine branch is assumed to have occurred about 116 million year ago, based on mutations of the genes that code for tooth formation in modern birds, including both dentin and enamel. (Figure modified from Meredith et al. 2014; 2012 American Association for the Advancement of Science, used with permission)

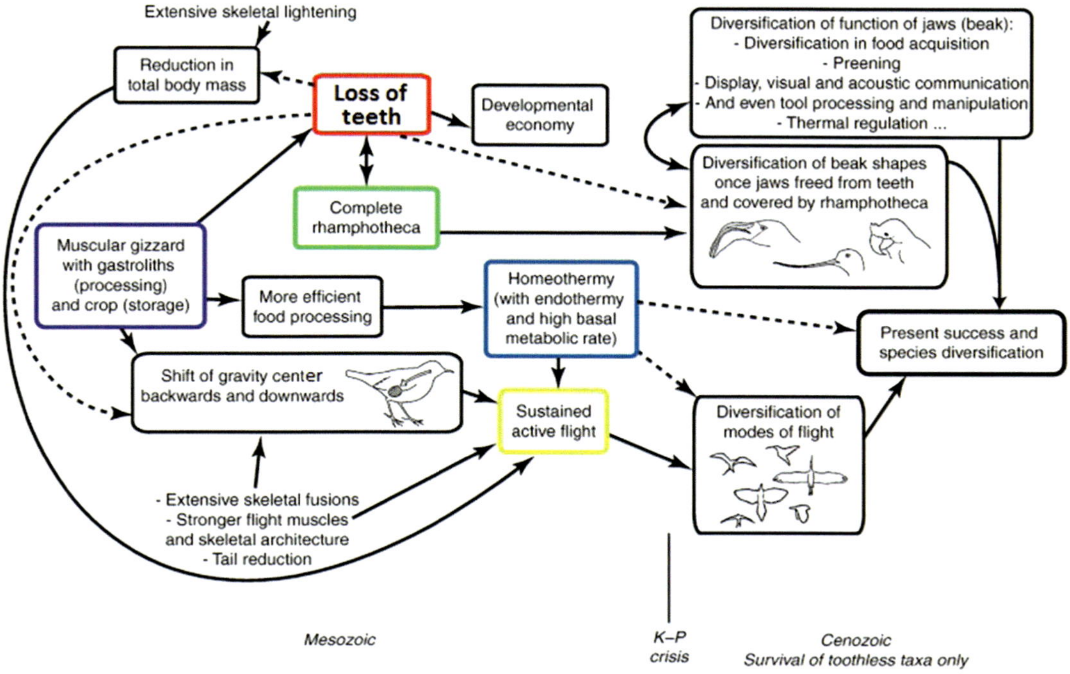


**Fig. 5.69** Proposed evolutionary interactions related to the loss of teeth by birds. Several major morphological, physiological, and behavioral innovations led to (arrows) the evolution of other innovations in a complex way; some facilitated the loss of teeth by birds, and others led to avian evolutionary success following. Dashed arrows represent less obvious influences. The horizontal distribution of events reflects the approximate time of occurrence, when known. (Figure from Louchart and Viriot 2011; © 2011 Elsevier Ltd., used with permission)

acid, pepsin, and other digestive enzymes, and also enhances the passage of materials through the intestine (lubrication). Mucus is also a barrier to bacterial and fungal invasion.

## 5.5.6 Liver and Pancreas

In the small intestine, secretions from two accessory organs, the liver (Box 5.30 The avian liver) and pancreas (Figs. 5.113 and 5.114), play important roles in digestion. Ducts transport bile from the liver or gall bladder and pancreatic juice from the pancreas to the first section of the small intestine (duodenum). Liver cells produce bile that, in most birds, is stored in the gall bladder before being released into the small intestine. In species with no gall bladder, including most pigeons and many parrots, bile drains directly from the liver to the duodenum. Bile emulsifies fats or, in other words, breaks fats into very small particles. This increases surface area (many small particles have more surface area than fewer large particles) and permits more efficient digestion of fats by enzymes (lipase). Bile also aids in the digestion of proteins by enhancing the action of proteases (protein-digesting enzymes) produced by the pancreas (Gass et al. 2007).

Pancreatic juice contains water and bicarbonate ions plus a variety of enzymes that aid in the digestion of carbohydrates (amylase), proteins (proteases), and fats (lipase). Because of the bicarbonate ions, pancreatic juice typically has a pH near 7 (Hulan et al. 1972; Duke 1986) and helps neutralize the acidic chyme moving into the intestine from the gizzard. This neutralization is important because it creates conditions favorable for the action of pancreatic enzymes. A bird's diet may influence levels of digestive enzymes in pancreatic juice. For example, levels of various

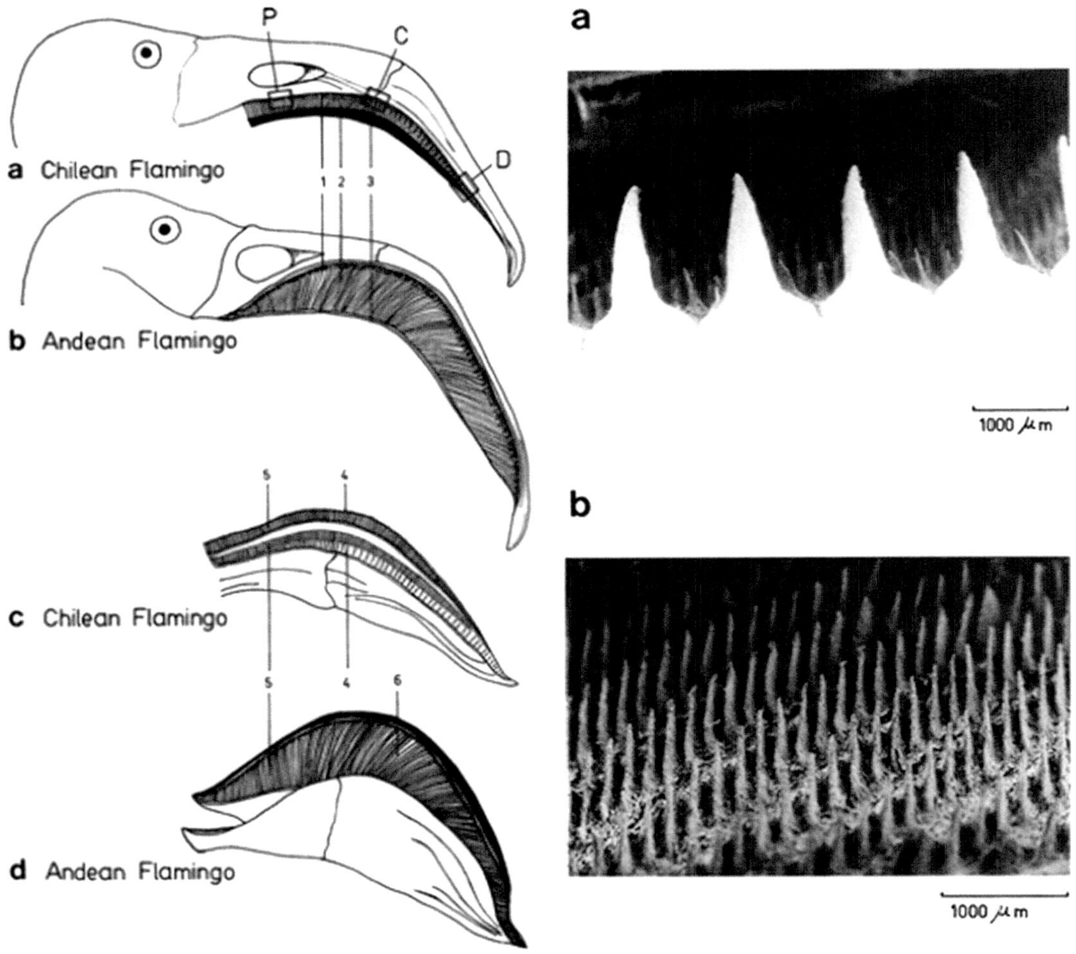


**Fig. 5.70** Left, Lateral view of the jaws of two flamingo species. Upper jaw of (**a**) a Chilean Flamingo (*Phoenicopterus chilensis*) and (**b**) an Andean Flamingo (*Phoenicoparrus andinus*). Lower jaw of (**c**) a Chilean Flamingo and (**d**) an Andean Flamingo. In Andean Flamingos, the edge of the upper jaw is almost completely covered by lamellae. The lower jaw of Andean Flamingos has a large lamellated area in the inner sides of the bill. (1) marginal lamellae, (2) keel covered by inner submarginal rows, (3) free edge of the keel, (4) serrated ridge-like outer submarginal rows, (5) hillock-like inner submarginals rows, and (6) expanded inflexed borders of the lower jaw. Letters indicate regions of the bill: (P) proximal, (C) curvature, (D) distal. Right, Scanning electron micrographs of the maxillary lamellae of a Chilean Flamingo. (**a**) Distal marginal hook-like lamellae from the outer side of the upper jaw. The lamellae are shown in feeding (inverted) position. Between two hook-like marginal lamellae lie 2–3 hillock-like outer submarginal lamellae. (**b**) Distal series of inner submarginal rows of the upper jaw. They typically run at a sharp angle relative to the free edge of the bill. (Figure from Mascitti and Kravetz 2002; © 2002 Oxford University Press, used with permission)

digestive enzymes (amylase, proteases, and lipase) in pancreatic juice varied in proportion to dietary content in chickens (Hulan and Bird 1972) and Pine Warblers (*Setophaga pinus*; Levey et al. 1999). Although such modulation would likely increase digestive efficiency, different species vary in the extent to which enzyme activity is modulated. For example, changes in diet had no significant effect on the content of the pancreatic juice of Yellow-rumped Warblers (*Setophaga*

**Fig. 5.71** Bill and tomial tooth (arrow) of a Loggerhead Shrike (*Lanius ludovicianus*). Scale bar = 5 mm. (Figure from Sustaita and Rubega 2014; © 2014 Oxford University Press, used with permission)

**Fig. 5.72** Variation in the morphology of passerine bills. (**a**) Food generalists (omnivores) include (from left to right): Rook (*Corvus frugilegus*), Great Tit (*Parus major*), and European Starling (*Sturnus vulgaris*). (**b**) Seed eaters include (clockwise starting at bottom left): Eurasian Tree Sparrow (*Passer montanus*), Common Redpoll (*Carduelis flammea*), Greater Short-toed Lark (*Calandrella brachydactyla*), Hawfinch (*Coccothraustes coccothraustes*), and Long-tailed Rosefinch (*Uragus sibiricus*). (**c**) Insectivores include (clockwise from bottom left): Dusky Warbler (*Phylloscopus fuscatus*), Siberian Rubythroat (*Luscinia calliope*), Eurasian Reed Warbler (*Acrocephalus scirpaceus*), and White Wagtail (*Motacilla alba*). (**d**) Specialists include (clockwise from bottom left): Reed Parrotbill (*Paradoxornis heudei*) that forages in dense reed beds, Eurasian Treecreeper (*Certhia familiaris*) that probes cavities and small holes in tree bark for tiny insects, Parrot Crossbill (*Loxia pytyopsittacus*) that opens pine cones to eat the seeds, and Great Gray Shrike (*Lanius excubitor*), a predatory species that preys on medium- to large-sized insects, rodents, and small birds. (Figure from Backström 2014; © 2014 John Wiley & Sons, Ltd., used with permission)

**Fig. 5.73** Many birds are omnivores, but often focus on particular types of food. Songbirds that are mainly seed-eaters generally have stouter, stronger bills (e.g., finches), and those that eat primarily insects generally have thinner, more forceps-like bills (e.g, Multicolored Tanager). Birds with generalist diets that eat a variety of different foods generally have bills that represent an evolutionary compromise, allowing them to efficiently feed on arthropods, fruit, and seeds (e.g., Common Cactus-Finch). Cinnamon-bellied Flowerpiercers have hooked bills that allow them to "pierce" the base of flower and gain access to the nectar. (Figure modified from Fritz et al. 2014; © 2014 Springer Nature, used with permission)

*coronata*; Ciminari et al. 2001). Such differences among species may be due to differences in food habits. Birds with more variable diets, such as consuming more insects during warmer periods of the year and fruit during colder periods, may better modulate pancreatic enzymes than those with less variable diets. Of course, differences among species in enzyme regulation

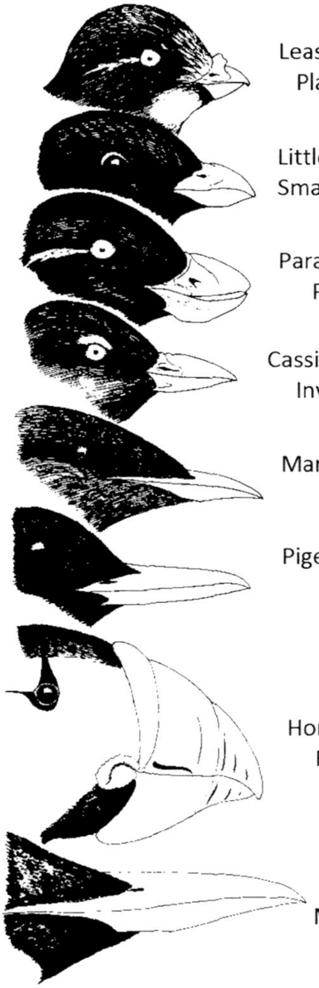


**Fig. 5.74** Variation in bill morphology and food habits of several species of alcids, pelagic species that only come ashore to breed. (Figure modified from Johnsgard 1987; used with permission of Paul Johnsgard)

may also be phylogenetically determined (Levey et al. 1999).

## 5.5.7 Absorption

Amino acids are absorbed into the blood by active transport, primarily in the small intestine, but also in the ceca. The absorption of lipids occurs primarily in the jejunum and first requires the breakdown (hydrolysis) of triglycerides into monoglycerides and free fatty acids by pancreatic lipase (after fats have been emulsified by bile). These monoglycerides and fatty acids then combine to form small droplets called micelles. Micelles that bump into the cells lining the intestine (enterocytes) then diffuse across the plasma membranes by simple diffusion. Some fatty acids also enter enterocytes by active transport. Once inside the enterocytes, fatty acids, and monoglycerides are converted back into triglyerides (reesterified) and packaged with cholesterol, lipoproteins, and other lipids into droplets called portomicrons. These small droplets (ranging from 47 to 128 nm in diameter in Domestic Chickens and >100 nm in diameter in Monk Parakeets, *Myiopsitta monachus*) are then secreted (by exocytosis) into intestinal capillaries for transport to the liver (Fraser et al. 1986; Hermier 1997; Ueda et al. 2019; Beaufrère et al. 2020). The droplets are called portomicrons because they are transported into capillaries that are part of the hepatic portal system that transports blood from the digestive system to the liver (Fig. 5.115). In mammals, by contrast, lipid droplets (called chylomicrons because their presence gives lymph fluid a "milky" [chylo-] appearance) are secreted from enterocytes into small lymph vessels. The process of fat absorption differs because the lymphatic system of birds is less developed than that of mammals (Hermier 1997).

### 5.5.8 Ceca

Two out-pocketings, or ceca, are located near the junction of the ileum and the large intestine of most birds (Fig. 5.116). Ceca are histologically similar to the small and large intestines and, based on function, are categorized as intestinal, lymphoid, or glandular ceca. In some groups of birds, including parrots (Psittaciformes), swifts and hummingbirds (Apodiformes), kingfishers and motmots (Coraciiformes), colies (Coliiformes), and most woodpeckers (Piciformes), ceca are absent. When present, ceca development is, at least in part, related to diet, with well-developed ceca in omnivores and some granivores and the longest ceca in herbivores (DeGolier et al. 1999). Hunt et al. (2019) examined possible relationships between cecal size and phylogeny, diet, and body mass and found that mean cecal length was longer in herbivores, but also that larger birds had longer ceca and cecal length was also influenced by phylogeny, e.g., species in the clade Gallcanserae have longer ceca than species in other clades.

In general, ceca length increases with the amount of cellulose in the diet. However, some insectivores and carnivores also have well-

**Box 5.21 Evolution of bird bills**

Can you always tell what a bird eats by looking at its bill? Not necessarily! Bird bills can serve a variety of functions other than feeding and foraging, including thermoregulation (Tattersall et al. 2009), preening, modifying vocalizations (Podos 2001), constructing nests (Sheard et al. 2023), sexually dimorphic weapons (Babbitt and Frederick 2007; Rico-Guevara and Araya-Salas 2015), and tool use (Laumer et al. 2017). Multiple selective pressures have, therefore, influenced the evolution of bird bills. To evaluate the possible importance of these different selective pressures in bill evolution, Navalón et al. (2019) examined the shape and size of beaks of 175 species of birds representing 94 families and 38 orders, and categorized their diet preferences as well as how they used their bills when feeding. They also quantified the mechanical advantage (MA) of the jaws of each species, with a high MA indicating the efficient transfer of force and slower jaw movement and a low MA indicating less efficient transfer of force and faster jaw movement. Analysis revealed that diet accounted for less than 12% of variation in beak shape, and relatively few taxa had high MAs (e.g., parrots). Most birds have relatively fast, weak bites (i.e., low MAs), even most raptors. The bills of most raptors likely have a low mechanical advantage because most raptors kill prey with their talons. Species that typically kill prey with their bills, such as falcons and shrikes, have relatively higher MAs. Interestingly, geese also have relatively high MAs, suggesting that grazing also requires a strong bite.

Overall, Navalón et al. (2019) found that birds with similar bill shapes sometimes have different diets, e.g., raptors and parrots, and penguins and insectivores. Also, many birds have relative thin, straight bills that can close fast with minimal force and are often "used as a versatile, tweezer-like clamp." Thus, evolution has generally favored bird bills that can serve multiple functions, not just feeding and foraging.

(continued)

**Box 5.21** (continued)

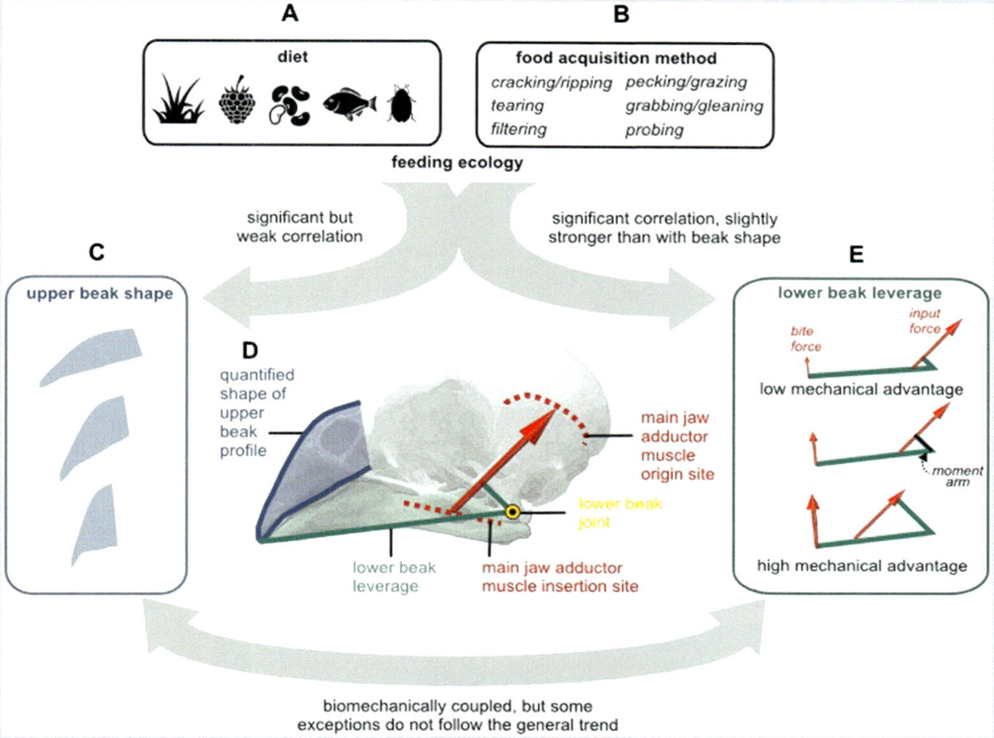


Overview of the analysis and results from Navalón et al. (2019). Birds were categorized based on (**a**) diet and (**b**) how they acquired food. Relationships between these categories and two features of cranial morphology were examined: (**c**, **d**) shape of the upper bill, and (**d**, **e**) leverage of the primary jaw closing muscles and resulting bite forces. Bite force increases when the moment arm of a muscle is relatively high (**e**) (Figure from Van Wassenbergh and Baeckens 2019; © 2019 The Author(s). Evolution © 2019 The Society for the Study of Evolution, used with permission)

(continued)

**Box 5.21** (continued)

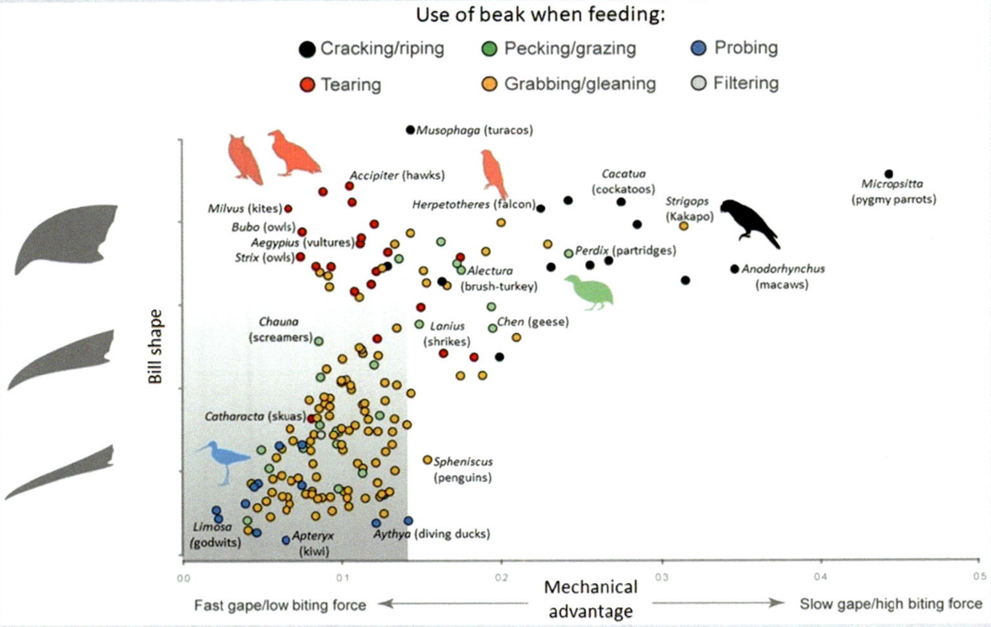


Birds with different diets and modes of feeding sometimes have bills with similar shapes. For example, birds that tear prey (e.g., raptors) and crack and bite (e.g., parrots) have similarly deep and curved bills. Birds that probe (e.g., many shorebirds) have relatively thin, straight bills. Bird that peck and graze (e.g., woodpeckers and geese) as well as many of those that grab and glean (e.g., many insectivorous birds) have relatively straight, flat bills. The shaded area on the left, indicating the lowest 20% of mechanical advantage, contains 80% of the species (Figure modified from Navalón et al. 2019; © 2018 The Authors. Evolution © 2018 The Society for the Study of Evolution, used with permission)

developed ceca. Examples include species in the orders Cuculiformes, Caprimulgiformes, and Trogoniformes that are primarily insectivores, and carnivorous owls (Strigiformes). One possible explanation for the well-developed ceca of some insectivorous birds is that the exoskeletons of insects contain chitin, a substance, like cellulose, that is difficult to digest. Although not yet demonstrated, the ceca of some insectivorous birds might contain chitinolytic microbes (DeGolier et al. 1999). Alternatively, the ceca of nonherbivorous birds may be important in water balance (see Chap. 9).

The most developed ceca are found in granivores and species with diets containing high levels of plant fiber, including tinamous, geese, gallinaceous birds (grouse, ptarmigans, and quail), and ratites other than the Emu. In these birds, the ceca are the primary site of microbial fermentation and contain numerous bacteria (Box 5.31 Avian gut microbiota). Although microbial activity in ceca is primarily fermentative, evidence to date suggests that there is little cellulose fermentation. Given the location of ceca and how materials move into and out of ceca, little cellulose probably enters the ceca (Vispo and Karasov 1997). In addition, many avian

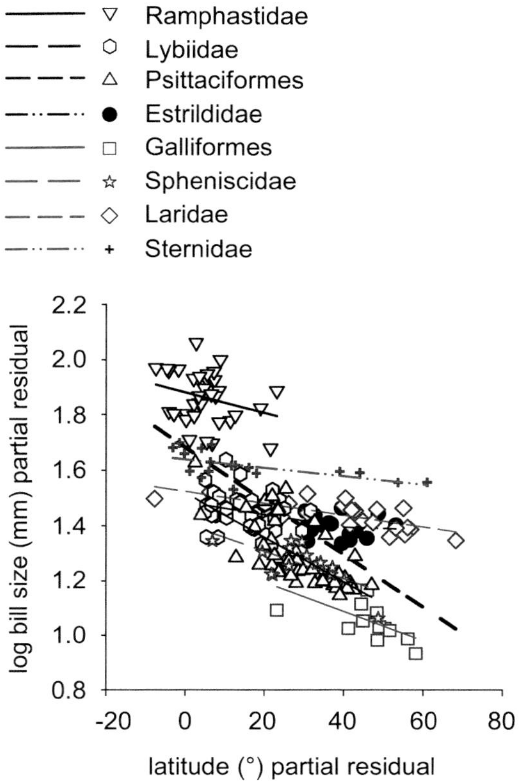


**Fig. 5.75** Among several families of birds, bill size tends to decrease at higher latitudes, suggesting that the thermoregulatory costs of larger bills become too high in colder environments, that larger bills may enable birds in warmer environments to better lose heat, or both. Regardless, the relationship between bill size and latitude suggests that factors other than food habits can influence beak morphology (Figure from Symonds and Tattersall 2010; © 1867 CCC Republication, used with permission)

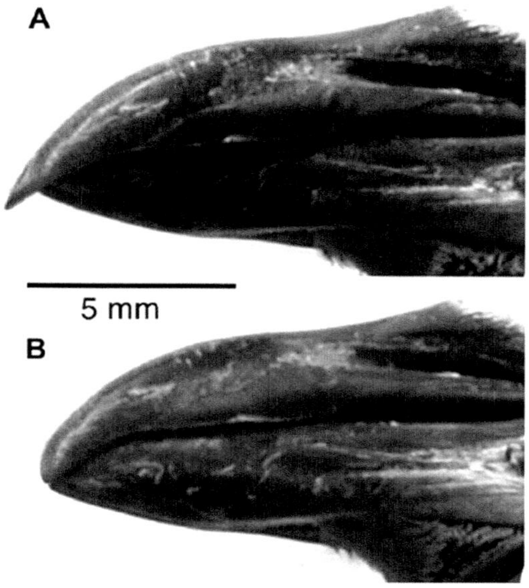


**Fig. 5.76** The bills of Rock Pigeons (*Columba livia*) are slightly hooked at the tip (**a**). Pigeons that had the tip trimmed (**b**) were able to eat seeds just as effectively as control pigeons. However, pigeons with trimmed bills were less efficient at removing ectoparasites from their plumage. Such results suggest that ectoparasite control has influenced the evolution of bill morphology in pigeons (Clayton et al. 2005). (Figure from Waite et al. 2012; © 2012 Australian Society for Parasitology Inc. Published by Elsevier Ltd., used with permission)

herbivores consume large quantities of plant material, but extract only the readily digestible components and the bulk of cell wall constituents are rapidly expelled without significant microbial fermentation (Mackie 2002).

The fermentative activity of microbes produces fatty acids that are absorbed by birds and used as a source of energy. However, the importance of this source of energy is limited for most birds. For example, even for grouse and ptarmigan with large ceca, microbial fermentation likely contributes less than 20% of their energetic needs (Mackie 2002). For birds, the reduction in body weight needed for flight limits the size and weight of "fermentation chambers." Not surprisingly then, the birds that rely most on fermentative digestion for energy are the flightless ratites. Emus (*Dromaius novaehollandiae*), for example, may obtain up to 50% of their energy needs from microbial fermentation (Herd and Dawson 1984). In Emus, however, fermentation occurs primarily in the small intestine rather than ceca (Stevens and Hume 1998). Young, growing Common Ostriches (*Struthio camelus*), may obtain up to 76% of needed energy from microbial fermentation that occurs in their large ceca and large intestine (Swart et al. 1993).

### 5.5.9 Large Intestine

For most birds, the large intestine is a relatively short, straight tube between the last section of the small intestine (ileum) and the cloaca.

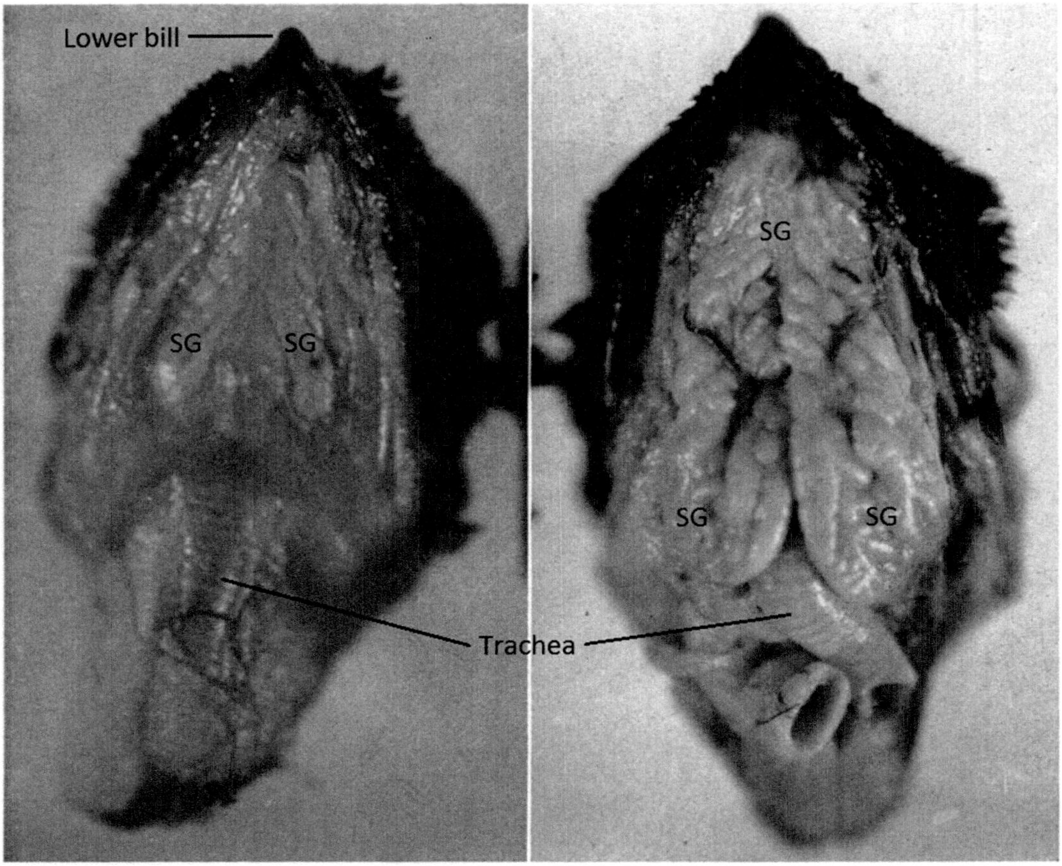


**Fig. 5.77** Salivary glands (SG) of Chimney Swifts (*Chaetura pelagica*). Left, prior to the breeding season (about 7 mm long and 2 mm wide and weighing about 3 mg). Right, during nest building (about 60 mg). (Figure modified from Johnston 1958; © 1899, CCC Republication, used with permission)

Exceptionally, the large intestine of Common Ostriches is longer than the small intestine and about 11–13 m long. Although typically short in length, the large intestine of birds is important in regulating water balance and regulating levels of sodium and chloride. Sodium is actively reabsorbed via intestinal microvilli (Fig. 5.117), chloride follows passively down the electrical gradient, and water follows osmotically. For Ostriches, as just noted, the large intestine is an important site for microbial fermentation. To minimize weight and remain as light as possible, the large intestine contains a minimum of nondigestible waste because wastes are voided as quickly as possible, and waste materials are sometimes voided in other, faster ways, such as owls regurgitating nondigestible materials as pellets and fruit eaters regurgitating fruit seeds or pits.

**Fig. 5.78** Eastern Spinebill (*Acanthorhynchus tenuirostris*) feeding on nectar. Mutations in the umami taste receptor genes, likely sometime during the Oligocene (23–34 million years ago), mean that oscine songbirds, just like hummingbirds, perceive sugars as sweet (Toda et al. 2021). (Photo by Gail Hampshire, Wikipedia, CC BY 2.0, https://creativecommons.org/licenses/by/2.0/)

### 5.5.10 Cloaca

The cloaca receives waste from the large intestine and materials from the urinary and reproductive systems. It is divided into three sections: the coprodaeum that receives waste from the large intestine, urodaeum that receives urine from the kidneys (via the ureters) and sperm and eggs from the gonads, and proctodaeum that stores (temporarily) and ejects material and is closed posteriorly by the muscular anus (Fig. 5.118). Nondigested materials moving from the large intestine into the cloaca are excreted along with materials from the kidneys. The kidneys of birds extract nitrogenous wastes from the bloodstream, but instead of excreting it as urea dissolved in urine like mammals, they excrete it in the form of uric acid. Uric acid has a very low solubility in water, so it emerges as a white paste. This material, as well as darker materials representing the output of the intestines, emerges from the cloaca.

## 5.6 Excretion

Although little studied, some investigators have considered the force with which excreta are expelled from the anus. Brooding Chinstrap (*Pygoscelis antarctica*) and Adélie (*P. adeliae*) penguins do not leave their stony nest to defecate, but move to the edge of it, stand up, turn their back nest-outward, bend forward, lift their tail, and shoot. The expelled material hits the ground an average of 40 cm away from the bird. Meyer-Rochow and Gal (2004) determined that the pressures involved could be approximated if they knew the (1) distance the feces traveled, (2) density and viscosity of the material, and (3) shape, aperture, and height of the anus above ground. They determined these parameters and calculated that penguins generated pressures of around 10 kPa (77 mm Hg) to expel watery material and 60 kPa (450 mm Hg) to expel material of higher viscosity similar to that of olive oil (forces well above those known for humans). How penguins choose the direction of defecation and how wind direction factors into that decision remain unknown.

## 5.7 Retention Time and Digestibility

The rate of passage through the digestive system, or retention time, varies with diet, and even different components ingested at the same time may

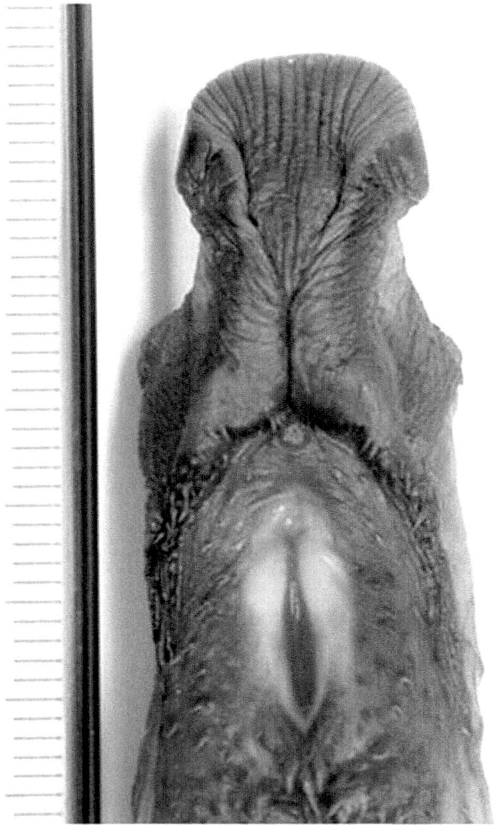

**Fig. 5.79** The muscular tongue of a Scarlet Macaw (*Ara macao*) used to manipulate and position seeds within the oral cavity to permit efficient cracking of seeds by their bills (Scale = 1 mm intervals). (Figure from Emura et al. 2012; used with permission of the Editorial Board of Okajimas Folia Anatomica Japonica; https://www.jstage.jst.go.jp/article/ofaj/89/3/89_57/_article/-char/en)

move at different rates. Retention time also varies with size; larger birds have longer digestive tracts and correspondingly longer retention times. In general, longer retention times enhance nutrient absorption, but, simultaneously, limit the amount of food that can be consumed (Afik and Karasov 1995). Fiber and fat tend to move more slowly through the digestive system, whereas certain fruits and, of course, nectar pass more quickly (Table 5.8).There are, however, interspecies differences in digestive "strategies." For example, some herbivores, such as geese, maximize food intake and the relatively rapid movement of digesta at the expense of digestive efficiency. By contrast, Common Ostriches (*Struthio camelus*) maximize digestive efficiency so retention time is relatively long. Small, fruit-eating passerines may pass seeds in less than 10 min, whereas seeds may be retained in the digestive system of a ratite for several weeks (Klasing 1998).

## 5.8 Phenotypic Plasticity

Organs and systems are often thought of as being static and unchanging. However, many species of birds experience different climates, occupy different habitats, consume different foods, and face variable energy demands during their annual cycle. Birds with flexible digestive systems, capable of changes in morphology and function, are better able to respond to, and survive, such variability. This ability of organs and systems to adapt to changing conditions is referred to as phenotypic plasticity, and such plasticity can be beneficial both in the short term (energy) and the long term (fitness). Energetically, tissues and organs that are either underutilized (e.g., more digestive capacity than needed) or underdeveloped (e.g., less digestive capacity than needed) are costly. If underutilized, more energy than necessary is used to maintain unneeded cells and tissues and, if underdeveloped, needed energy is not acquired because the capacity to effectively and efficiently digest food is limited. In either case, a bird's energy budget suffers and this, over the long-term, can have important fitness consequences.

In the digestive system of birds, phenotypic plasticity can involve morphological changes that may be macroscopic, such as changes in gizzard size and intestine length, or microscopic, such as changes in the number or ratios of different types of epithelial cells lining the small intestine. Functional changes can involve changes in the ability of the digestive system to digest particular foods via changes in enzyme production.

The benefits of phenotypic plasticity are clear, but how is such plasticity possible? Changes in

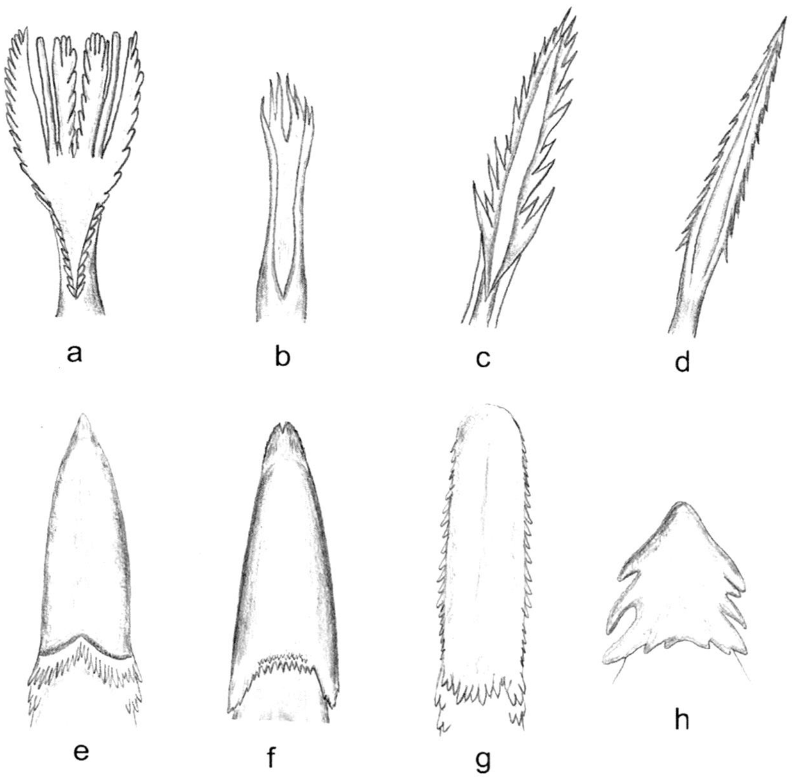


**Fig. 5.80** Diagrams of some tongue shapes. (**a**) Brush-like or fringe-shaped tongue (nectarivorous birds and flowerpeckers), (**b**) tube-like tongue (nectarivorous birds), (**c**) tube and brush-like tongue (sunbirds), (**d**) lancet-shaped tongue (woodpeckers), (**e**) dagger-like or triangular tongue (many birds), (**f**) bifid apex in tongue (magpies), (**g**) oval-shaped tongue (many waterfowl), and (**h**) star-shaped tongue (Emu, *Dromaius novaehollandiae*). (Figure from Erdoğan and Iwasaki 2014; © 2014 Elsevier Ltd., used with permission)

gizzard size are due to changes in the smooth muscle. With increasing amounts of nondigestible fiber or difficult-to-crush items (e.g., mussels) in the diet, smooth muscle cells in the walls of the gizzard increase in size (hypertrophy). With less fiber and softer foods, the smooth muscle cells hypotrophy (Piersma and Drent 2003). The small intestine of birds can change in length and absorptive capacity. Longer intestines provide additional surface area (and more time) for digestion and absorption, but, in addition, the mucosal epithelium lining the intestine can proliferate to generate additional surface area. This latter process is possible because the

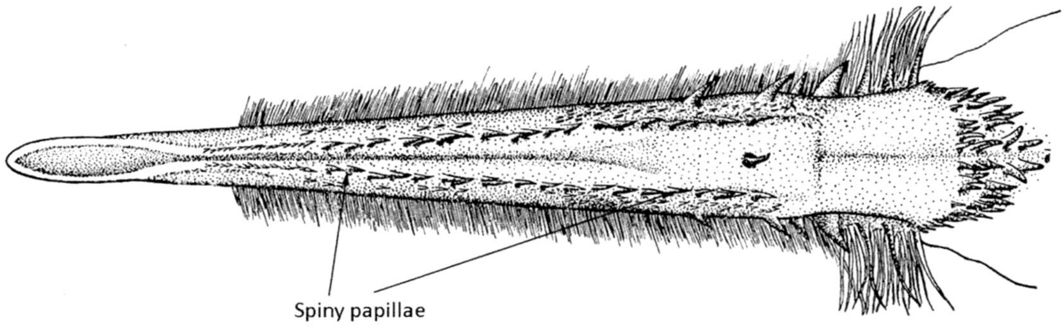


**Fig. 5.81** Tongue of a Red-breasted Merganser (*Mergus serrator*) with caudally directed papillae. (Figure from Gardner 1925; CC0 Public Domain)

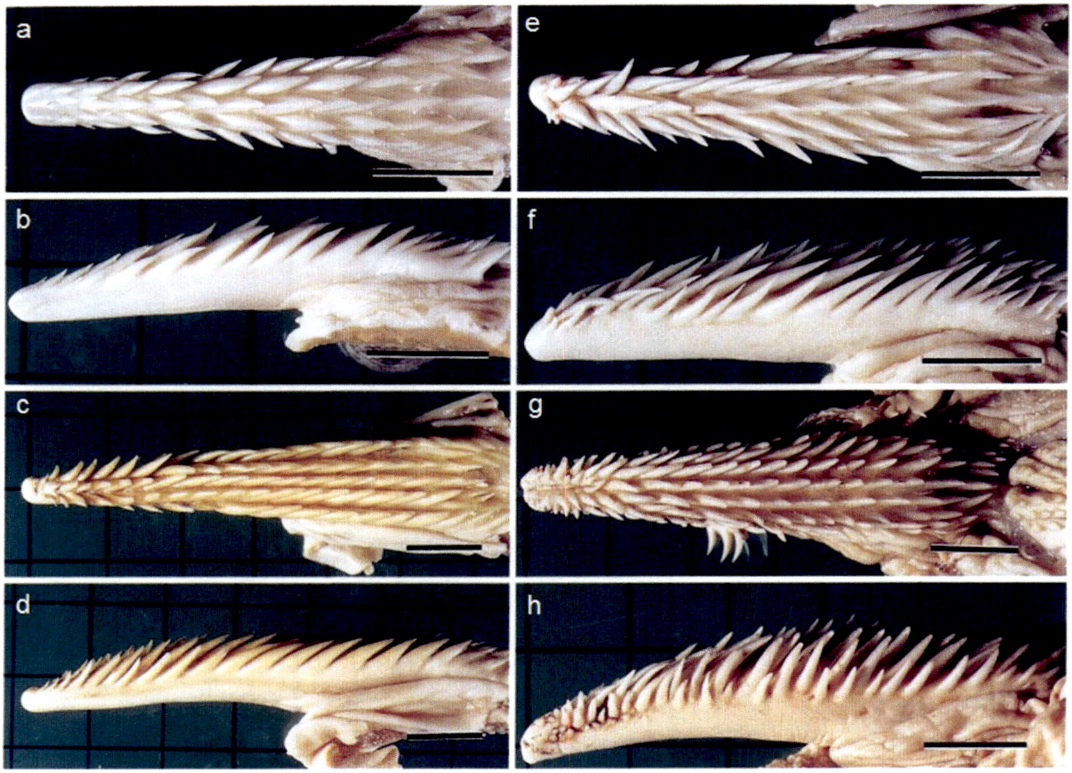


**Fig. 5.82** Tongues of penguins. Top (**a**, **c**, **e**, and **g**) and side (**b**, **d**, **f**, and **h**) views of the tongues of (**a** and **b**) an African Penguin (*Spheniscus demersus*), (**c** and **d**) a Gentoo Penguin (*Pygoscelis papua*), (**e** and **f**) a Humboldt Penguin (*Spheniscus humboldti*), and (**g** and **h**) a Macaroni Penguin (*Eudyptes chrysolophus*). The backward-projecting papillae allow food items to slide into, but not out of, the mouth. (Figure from Kobayashi et al. 1998; © Archives of Histology and Cytology, used with permission)

cells making up the mucosal epithelium undergo constant turnover. New cells are produced at the base of villi (the intestinal crypts), migrate up the villi to the tip, and are then shed into the lumen of the intestine. Depending on the age and size of birds, cellular turnover times in the intestine typically range from 4 days to 2 weeks (Starck 1999). With changes in diet, however, the rate of either cell proliferation or shedding can change, resulting in more or less surface area for digestion and absorption.

Concerning functional plasticity, the ability of avian digestive systems to modify enzyme production or activity with changes in diet has received less attention from investigators. Domestic chickens and turkeys are able to alter the activity of intestinal enzymes in response to changes in diet (Karasov 1996), but less in known about that ability in nondomesticated birds. In one study, however, Pine Warblers (*Setophaga pinus*) fed insect-based, fruit-based, and seed-based diets exhibited differences in enzyme activity. Enzymes that digest carbohydrates exhibiting more activity in warblers fed carbohydrates (fruit), protein-digesting enzymes exhibited greater activity in warblers fed insects (protein), and lipid-digesting enzymes exhibited greater activity in warblers fed diets with greater lipid content (insects and seeds) (Levey et al. 1999). Another study revealed increased activity of protein-digesting enzymes in Yellow-rumped Warblers (*Setophaga coronata*) fed insect-based diets. However, when fed diets high in carbohydrates, there was no change in the activity

**Fig. 5.83** Top, Tip of the tongue of a Red-bellied Woodpecker (*Melanerpes carolinus*) showing the rear-facing barbs (Photo by Frank L. Martin, used with permission). Bottom, detailed view of the horny tip (left), with barbs, of the Guadeloupe Woodpecker (*M. herminieri*) tongue *in vivo* position (From Villard and Cuisin 2004; © 2004 Oxford University Press, used with permission)

of carbohydrate-digesting enzymes (Afik et al. 1995). The results of these and a few additional studies (e.g., Martinez del Rio et al. 1995; Sabat et al. 1998) suggest that some birds can modify enzyme activity in response to changing diet, but that ability varies with species. One possible explanation for such variation is that modification of enzyme activity is more likely in species with diets that change on a regular basis. For example, diets of resident species in north-temperate areas may be primarily protein (insects) during the breeding season when it is warmer, and more

**Box 5.22 Hummingbird hyoids**

Nectar-feeding birds like hummingbirds must protrude their tongues to obtain nectar, and many hummingbirds can extend their tongue almost fully beyond the bill tip (e.g., Temeles et al. 2002). This extension requires a specialized hyoid apparatus, with long epibranchials that extend around the back of the skull and insert on either the frontal or nasal bones. Attached to the epibranchials are the branchiomandibularis muscles that originate on the lower jaw. Contraction of the branchiomandibularis muscles draws the base of the epibranchials forward and causes protraction of the tongue. Retraction of the tongue involves contraction of the stylohyoideus muscles that originate dorsally on the braincase and curves around the skull to insert on the basihyal.

(continued)

**Box 5.22** (continued)

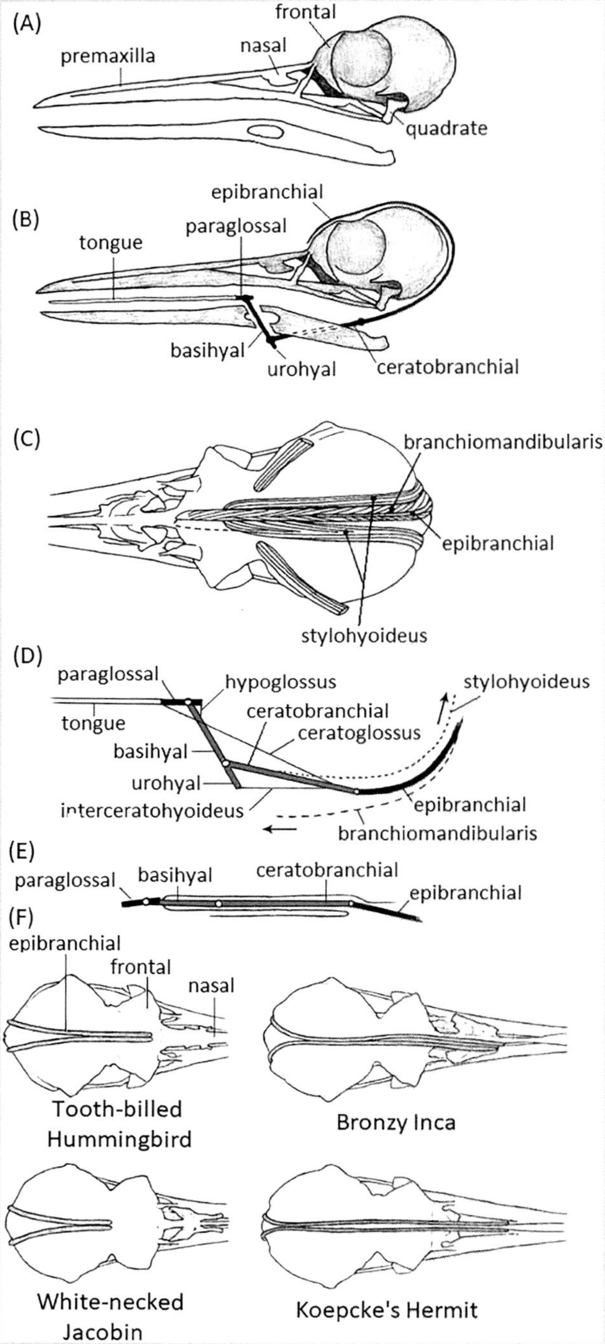


The hyoid apparatus of hummingbirds. (**a**) Hummingbird skull. (**b**) Hyoid components shown in black; mandibule removed from quadrate and separated at the rostral mandibular fenestra to better show the hyoid. (**c**) Primary

(continued)

**Box 5.22** (continued)

muscles involved in protracting (branchiomandibularis) and retracting (stylohyoideus) the hummingbird tongue. (**d**, **e**) Drawings showing the components of the hyoid apparatus and the muscles used to protract (extend) and retract the tongue. Arrows indicate the respective directions of movement due to contraction of the stylohyoideus and ceratoglossus (retraction) and of the branchiomandibularis, interceratohyoideus, and hypoglossus (protraction). (**f**) Dorsal view of the skulls of four species of hummingbirds showing point of attachment of the epibranchial to either the frontal bone or nasal bone of the skull. Tooth-billed Hummingbird, *Androdon aequatorialis*; Bronzy Inca, *Coeligena*; White-necked Jacobin, *Florisuga mellivora*; Koepcke's Hermit, *Phaethornis koepckeae* (Figure made from four different figures; **a**, **b**, **d**, **e**, and **f** modified from Zusi 2013; © 2013, used with permission of the American Ornithological Society, and **c** modified from Zusi and Bentz 1984, used with permission of Smithsonian Scholarly Press)

varied (fruit and some insects) during the cooler nonbreeding season. The ability to modulate enzyme activity would clearly be advantageous in such species. However, selection may be less likely to favor this ability in species with diets that typically exhibit little change, e.g., insectivorous birds in areas where insects are readily available throughout the year.

Migratory species may not only have to consume different foods during their annual cycle (and during migration) but also, particularly just prior to and during migration, consume foods in very different quantities. Prior to migration, birds

**Fig. 5.84** (**a**) Rainbow Lorikeet (*Trichoglossus moluccanus*) feeding on nectar. (**b**) Lingual apex of the tongue of a Rainbow Lorikeet (Scale = 10 mm). (**c**) Magnified view of the lingual apex. The numerous hair-like papillae allow lorikeets to lap up nectar and pollen. (Figures **b** and **c** from Emura et al. 2011; Published in Okajimas Folia Anatomica Japonica, https://www.jstage.jst.go.jp/article/ofaj/88/1/88_1_17/_article, used with permission. Photo from pxhere.com, CC0 Public Domain)

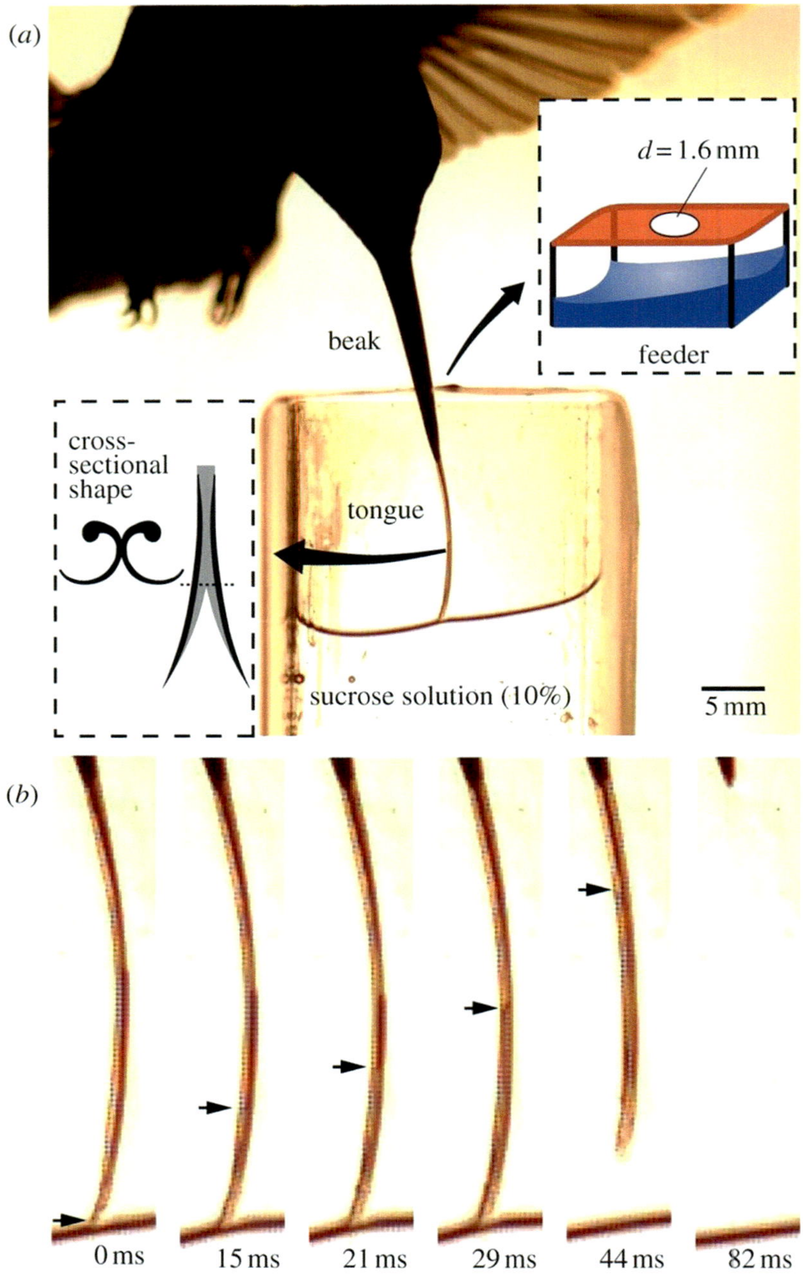


**Fig. 5.85** A Ruby-throated Hummingbird (*Archilochus colubris*) drinking from a transparent feeder. (**a**) A photograph of the drinking bird. Insets: schematic of the hummingbird's tongue (left) and feeder (right). The feeder is made of glass plates, and wrapped with red paper in order to attract the bird. (**b**) High-speed images of the hummingbird drinking from a feeder. Because the tongue is transparent, the nectar rising through the tongue can be seen (arrows). As the tongue tip touches the surface, the liquid moves along the tongue at speeds of approximately 20 cm/s. (Figure from Kim et al. 2011; © 2012 The Royal Society, used with permission)

may exhibit hyperphagy (higher feed rates needed to increase fat stores), but during migration they may fast for variable lengths of time and may have reduced food intake due to time limitations or limited food availability at stopover sites. For such species, phenotypic plasticity may be important, if not essential, for survival.

Eurasian Blackcaps (*Sylvia atricapilla*) breed in the Palearctic and winter in sub-Sahara Africa. Their migration, therefore, takes them across the

**Fig. 5.86** (Top) The protrusible tongues of hummingbirds, including Fawn-breasted Brilliants (*Heliodoxa rubinoides*) can be capillary tubes and fluid traps. (Bottom) The tongue is inserted well into some nectar and the two semicircular "forks" near the end of the tongue are pulled closed by surface tension. High-speed video observations showed that this tongue transformation occurs in as little as 1/20th of a second. The tongue can then be retracted, bringing the "trapped" nectar into the oral cavity. This tongue–fluid interaction occurs in both living and dead birds, demonstrating that this mechanism is a function of the tongue structure itself so requires no energy expenditure. (Photo of a Fawn-breasted Brilliant from Rico-Guevara 2017; open-access article distributed under the terms of the Creative Commons Attribution License, https://creativecommons.org/licenses/by/4.0/; Figure of tongue as nectar trap from Rico-Guevara and Rubega 2011, used with permission of the U. S. National Academy of Sciences)

Sahara Desert and, prior to and during migration, feeding rates and food availability vary considerably and, in response, their digestive systems exhibit impressive plasticity. In a laboratory study with Blackcaps captured during migration, the intestinal mass of fasted birds decreased by 45% in just 2 days, then recovered to the original mass after being fed for 2 days (Karasov et al.

**Fig. 5.87** (**a**) Male Olive-backed Sunbird (*Cinnyris jugularis*). (**b**) Scanning electron microscopy photograph of the tip of the tongue of a White–breasted Sunbird. Sunbird tongues, like those of hummingbirds, have a bifurcated tip, the edges coiling inward to form an open groove. When they lick nectar from flowers, the fluid is loaded into the tongue by capillary action. (Photo from pxhere.com, CC0 Public Domain. Figure **b** from Downs 2004; Rights managed by Taylor & Francis, used with permission)

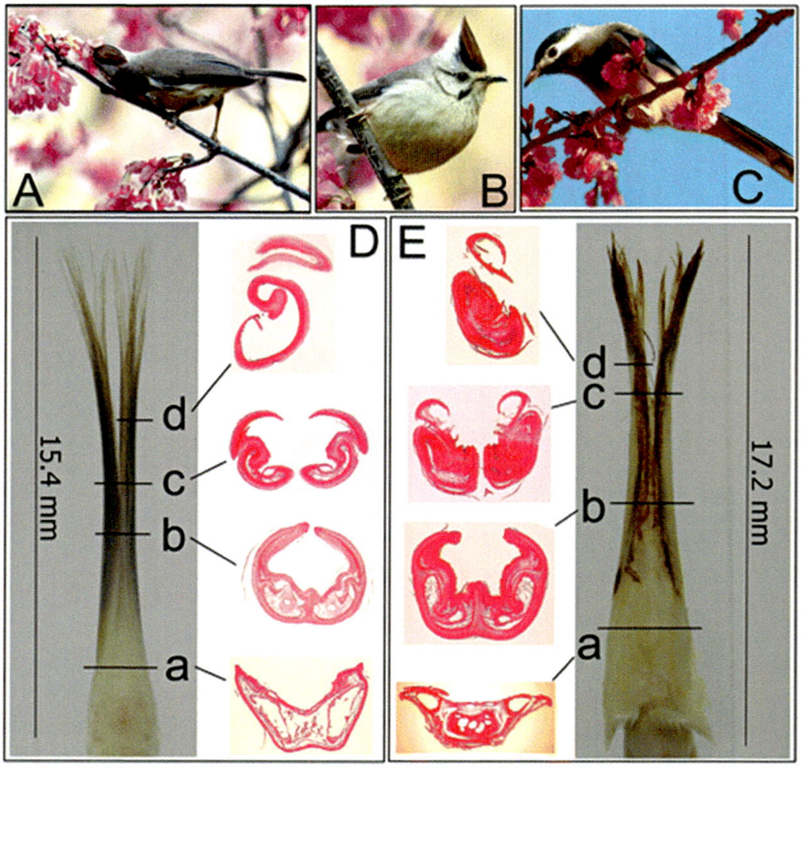


**Fig. 5.88** (**a**) A Taiwan Yuhina (*Yuhina brunneiceps*) feeding on the nectar of flowering cherries. (**b**) A Taiwan Yuhina and (**c**) a White-eared Sibia (*Heterophasia auricularis*). Gross morphology and appearance in transverse sections of the tongue of (**d**) a Taiwan Yuhina and (**e**) a White-eared Sibia. The brush-tipped tongues of these songbirds differ from those of other nectar-feeding birds, the bristles collect nectar and the tongue can then be retracted into the oral cavity. (Figure from Chang et al. 2013; © 2013 Wilson Ornithological Society, used with permission)

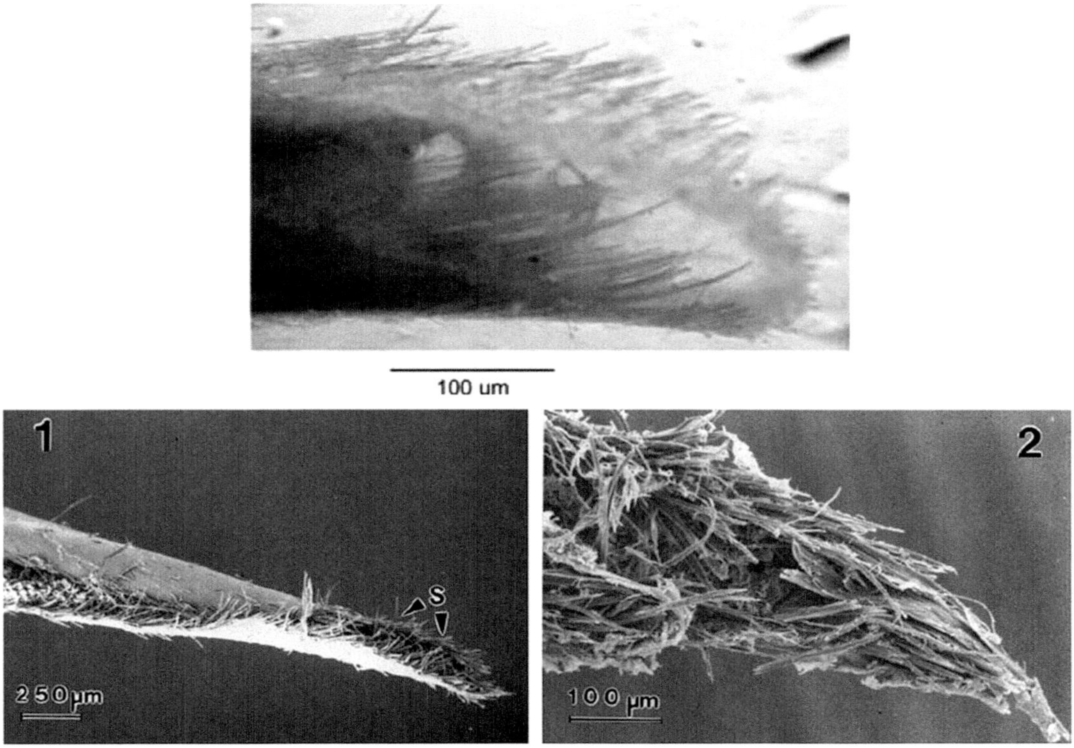


**Fig. 5.89** Tongue of a Western Sandpiper (*Calidris mauri*). Top, magnified view of the distal section of the tongue showing the velvety texture of the edges. Bottom, Scanning electron micrographs of excised tongue. (1) Distal end of tongue showing dense border of spines (S). (2) Detail of tongue tip, showing dense covering of fibrous spines and mucus. Western Sandpipers apparently use their specialized tongues to feed on the biofilm on the surface of moist substrates. Biofilm consists of a packed mat (10 μm to about 2 mm) of microbes, organic detritus, and sediment particles along with a variety of noncarbohydrate components (including amino acids, sugars, and proteins) secreted by benthic diatoms and bacteria. (Figure from Elner et al. 2005; © 2005 Springer Nature, used with permission)

2004). This change in intestinal mass was due to rapid alteration of the intestinal epithelium. When fasted, the length of villi decreased as cells at the tips were rapidly sloughed off and, after feeding, villi again increased in length as rapid cell division replaced cells that had been lost (Fig. 5.119). The loss of intestinal mass during fasting reduces body weight and makes flying more energetically efficient. Increasing intestinal mass when feeding resumes means more efficient digestion and absorption. For birds that typically weigh just 16–19 g, phenotypic flexibility that permits a quick reduction in mass or improved digestive efficiency, as needed, is clearly beneficial.

Red Knots (*Calidris canutus*), another migratory species, have large gizzards that typically make up about 10% of their body mass and, during much of the year, knots use their massive gizzards to crush hard-bodied prey like mollusks. However, gizzard mass decreases on their tundra breeding grounds where Red Knots eat soft-bodied arthropods and spiders. The gizzards of Red Knots temporarily held in captivity and alternately fed hard prey (mussels) and soft food pellets exhibited significant changes in size (Fig. 5.120), with gizzards rapidly adjusting to switches from a soft to a hard food type (Vézina et al. 2010).

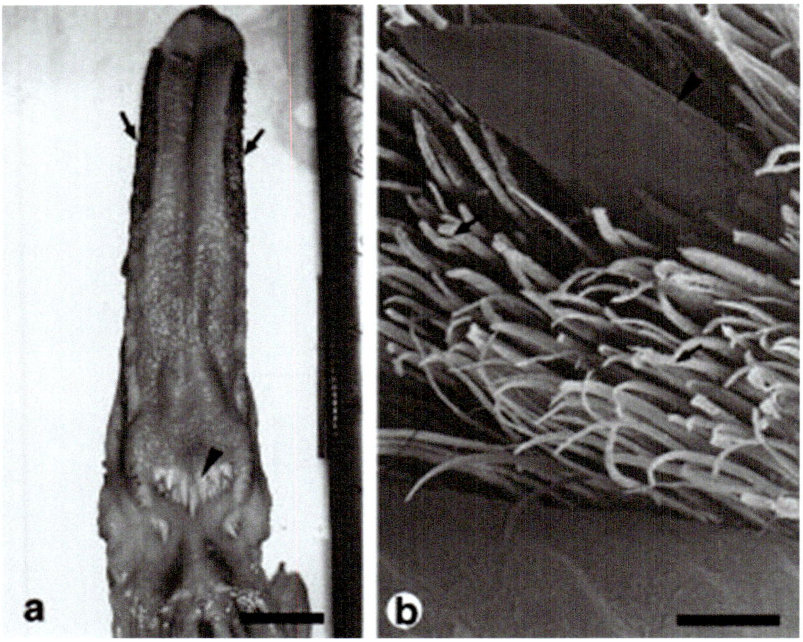


**Fig. 5.90** Surface structure and histology of the dorsal epithelium of the tongue of Middendorff's Bean Goose (*Anser fabalis middendorffii*). (**a**) Macroscopic dorsal view of the tongue. Arrows show lingual hairs on the lateral sides. (**b**) Scanning electron micrograph of the lateral side of the tongue. Lingual papillae (arrows) are compactly distributed on the tongue, and large conical papillae (arrowhead) are scattered among them. Scale bars = 10 μm (**a**) and 500 μm (**b**). (Figure from Iwasaki 2002; © 2002 John Wiley and Sons, used with permission)

## 5.9 Food Intake Regulation

Birds, like other animals, regulate their food intake. Such regulation occurs over varying time scales. For example, birds regulate the amount of food intake during a particular feeding bout. An owl that captures a large prey item may consume just a portion of that item and cache the rest. Regulation also occurs daily. Birds wintering in cold areas must regulate food intake so that they

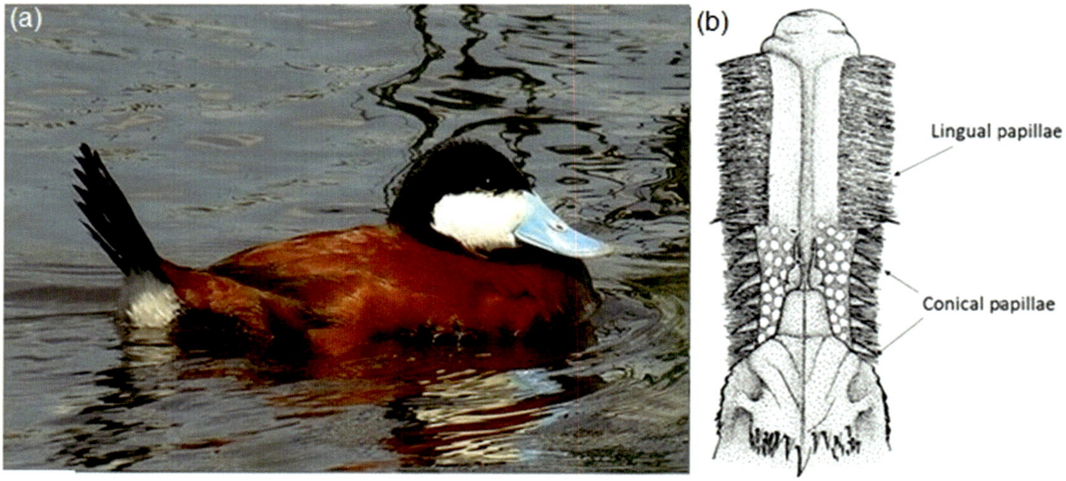


**Fig. 5.91** (**a**) Male Ruddy Duck (*Oxyura jamaicensis*). (**b**) Tongue of a Ruddy Duck showing papillae used for filter feeding. (Photo from U. S. Fish and Wildlife Service, CC0 Public Domain. Figure **b** from Gardner 1925; CC0 Public Domain)

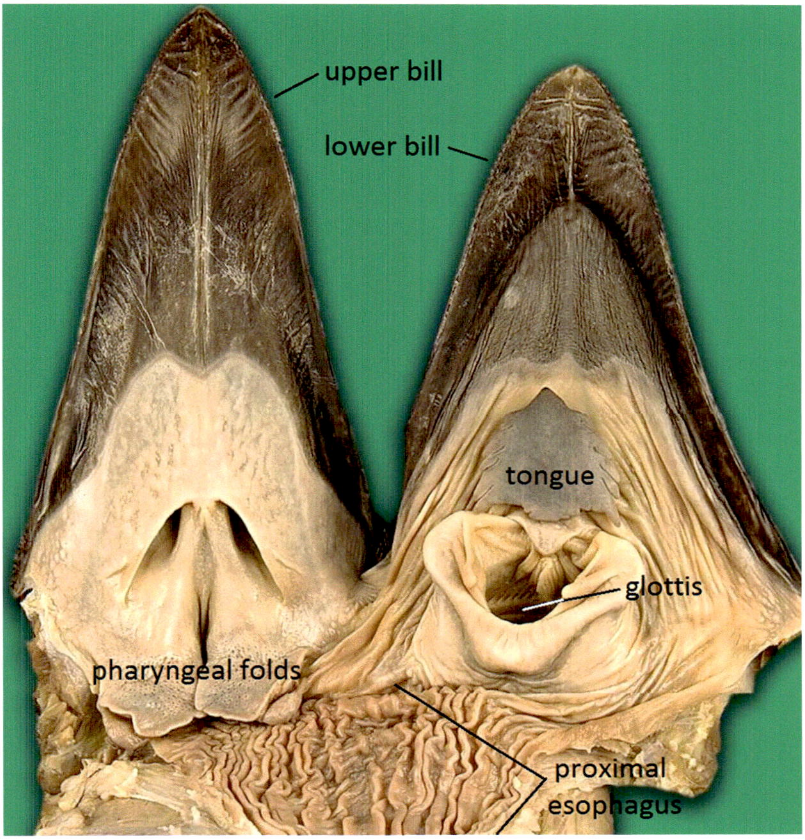


**Fig. 5.92** Oral cavity, pharynx, and proximal esophagus of an Emu (*Dromaius novaehollandiae*). Note the folds (plicae esophageales) on the wall of the esophagus that make the esophagus so distensible. (Figure modified from Crole 2011; used with permission of Martina Crole)

are not too heavy (because increased weight has a negative impact on flight performance and ability to elude predators), but have sufficient reserves to survive a long, cold night. Finally, regulation also occurs during the annual cycle. For example, migrating birds typically increase their body mass (and fat stores) prior to migration.

The short-term control of feeding involves the transmission of meal-related signals (such as the presence of food or specific nutrients) from the digestive tract and, to a lesser degree, the pancreas and liver to satiety centers in the brainstem (Jensen 2001). The presence of food in the digestive tract also stimulates the release of several peptides (proteins such as cholecystokinin, or CCK) that control gut motility and secretion. Nervous and chemical signals to the brainstem are relayed to the hypothalamus to activate neural pathways that signal satiety and regulate short-term food intake (Richards 2003).

Long-term regulation of feeding behavior and energy balance involves neural and neuroendocrine pathways activated by the hypothalamus in response to various signaling molecules. The ability of birds to regulate their body mass and other physiological processes at different levels at different times is referred to as poikilostasis (Kuenzel et al. 1999). The term poikilostasis literally means "various states of equilibrium" and indicates that equilibrium, or optimum, conditions (e.g., body mass) may vary during an annual cycle or a lifetime. Establishment of a bird's optimum feeding behavior and mass appears to be determined by changes in the levels of activity of the two divisions of the autonomic nervous system (parasympathetic and sympathetic) regulated by the hypothalamus. This control system, called the visceral forebrain system (or VFS), is thought to regulate food intake based on metabolic demands, environmental conditions,

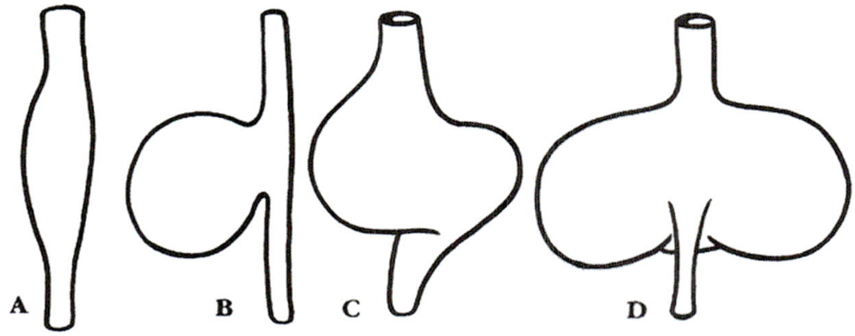


**Fig. 5.93** Bird crops vary in morphology. (**a**) Fusiform or spindle-shaped crop (e.g., many seed-eating birds and cormorants), (**b**) Unilobe-shaped crop (galliforms), (**c**) Unilobe-shaped crop (parrots), and (**d**) Bilobe-shaped crop (pigeons and doves). (Figure from King and McLelland 1984; © 1984 Baillière Tindall, used with permission)

and a bird's physiological state (Kuenzel et al. 1999).

The body mass of many birds that breed at higher latitudes and winter at lower latitude, temperate areas fluctuates throughout the year. For example, male White-throated Sparrows (*Zonotrichia albicollis*) that breed primarily in Canada and the northeastern United States and winter in the mid-western and eastern United States have a higher mean body mass during

**Fig. 5.94** Hoatzin (*Opisthocomus hoazin*), a folivorous bird of South America. (Photo by The Next Gen Scientist, Wikipedia, CC BY 2.0, https://creativecommons.org/licenses/by/2.0/)

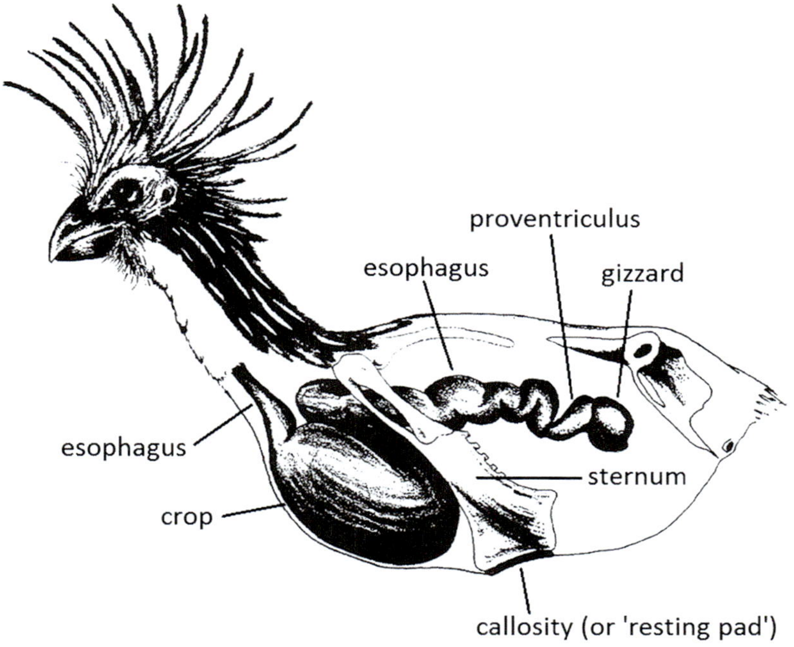


**Fig. 5.95** A portion of the digestive system of a Hoatzin (*Opisthocomus hoazin*) showing the esophagus, crop, proventriculus, and gizzard. The crop supports a wide diversity of bacteria, protozoans, and fungi that digest cellulose and provide Hoatzins with energy. These microbes also detoxify plant secondary compounds such as tannins. Hoatzins browse on young shoots and leaves of up to 52 different species of plants (Grajal et al. 1989). (Figure modified from Grajal 1995; © 1995 Oxford University Press, used with permission)

mid-winter (January) and prior to spring (April) and fall (October) migration than during late winter (March) and the breeding season (July; Kuenzel and Helms 1974; Kuenzel et al. 1999; Fig. 5.121). Such changes in optimum body mass are adaptive, with periods of greater body mass corresponding to periods when metabolic needs may vary unpredictably due to changing weather conditions (mid-winter) and when energy (fat) is needed to fuel migration. The loss of mass during other periods likely improves flight capabilities when metabolic needs are more predictable (late winter) and when caring for and provisioning young requires efficient flight and delivery (rather than consumption) of food. Of course, annual fluctuation in food intake and optimum body mass varies among species and with environmental conditions. Greenish Warblers (*Phylloscopus trochiloides*) wintering in a tropical climate (southern India) where metabolic needs are more

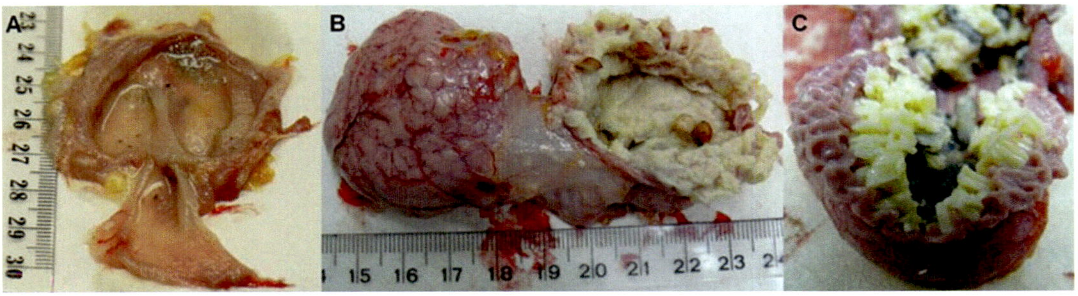


**Fig. 5.96** The non"lactating" pigeon crop (**a**) has a completely different appearance from that of the "lactating" crop. (**b**). The lactating crop is more than twice the size of the nonlactating' crop, with a thickened wall and two very obvious lateral lobes. When the "lactating" crop is opened, (**c**) the pigeon "milk" is seen as a bed of close-packed discrete rice-shaped pellets that are closely associated with the mucosal surface of the tissue. (Figure from Gillespie et al. 2011; © 2011 Springer Nature, used with permission)

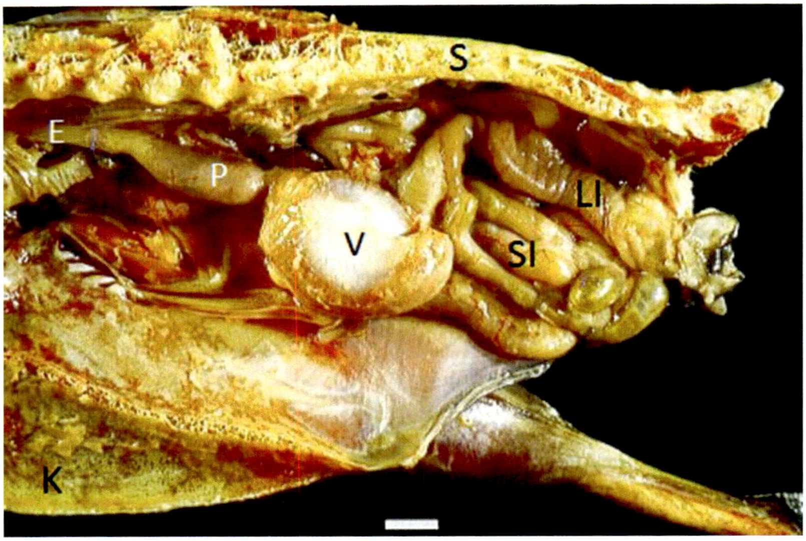


**Fig. 5.97** Right lateral view of the thoracic and abdominal cavities of a Houbara Bustard (*Chlamydotis undulata*). The liver has been removed. E, esophagus; P, proventriculus; V, ventriculus; SI, small intestine; LI, large intestine; K, keel of sternum; S, synsacrum. Scale bar = 12 mm. (Figure from Bailey et al. 1997; © 2007 John Wiley and Sons, used with permission)

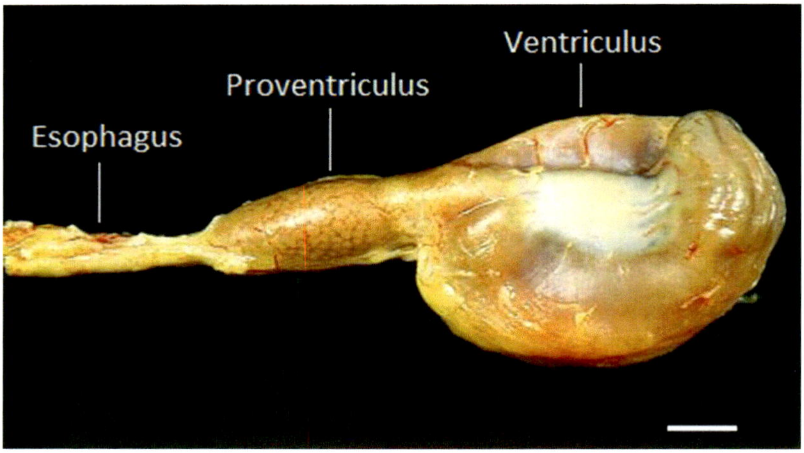


**Fig. 5.98** Stomach of a Houbara Bustard (*Chlamydotis undulata*) showing the esophagus leading into the glandular stomach, or proventriculus, and the muscular stomach, or ventriculus (also called the gizzard). Scale bar = 1 cm. (Figure modified from Bailey et al. 1997; © 2007 John Wiley and Sons, used with permission)

predictable exhibited no significant variation in mean body mass from October through March. Prior to migration in April, however, these warblers did exhibit a significant increase in mean body mass (Katti and Price 1999).

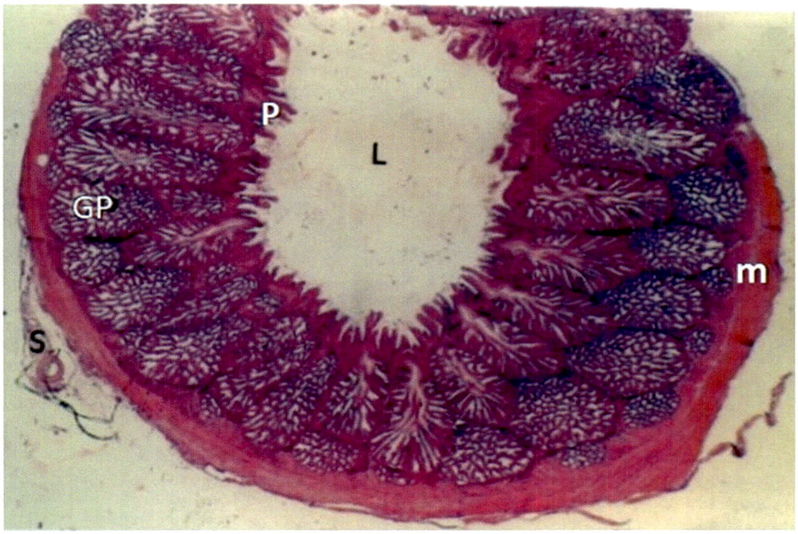


**Fig. 5.99** Photomicrograph (50×) of a cross section through a proventriculus showing folds of mucous membrane (P); deep proventricular glands (GP); capsule (connective tissue) around the glands (arrowhead); muscle layer (m); serosa (connective tissue) with blood vessels (S), and the lumen (L). (Figure from Catroxo et al. 1997; used with permission)

**Box 5.23 Stomach oils of Procellariiform seabirds**

Birds in the order Procellariiformes (petrels, shearwaters, and albatrosses) are unique because they store lipids, also called stomach oils, in their proventriculus. Stomach oil is derived from recently ingested prey (Wang et al. 2007) and is formed because the proventriculus is relatively large and exhibits little muscular activity. As a result, food remains in the proventriculus longer than in other birds and this allows separation of lipids and aqueous components. Aqueous components (water plus solutes) empty from the proventriculus faster than lipids, resulting in the accumulation of stomach oil. Connan et al. (2005) examined the stomach oils of Short-tailed Shearwaters (*Puffinus tenuirostris*) and identified 14 different oils, primarily wax esters and triacylglycerols with lesser amounts of cholesterol, diacylglyerols, free fatty acids, and polar lipids. Galluzzo et al. (2022) identified 20 different fatty acids in the stomach oil of Scopoli's Shearwaters (*Calonectris diomedea*). Similarly, the stomach oils of White-chinned Petrels (*Procellaria aequinoctialis*) consisted primarily of triacylglycerols, wax esters, and diacylglycerol-ethers (Connan et al. 2007).

(continued)

**Box 5.23** (continued)

White-chinned Petrel (Photo by Elizabeth Crapo, NOAA Corps, Wikipedia, CC0 Public Domain)

Stomach oils have been found in both breeding and nonbreeding birds and may serve multiple functions. Petrels and fulmars use stomach oils as a defense mechanism, ejecting or propelling stomach oils at competitors and predators. Mallory et al. (2020) noted that at least 20 different species of birds are believed or known to have been killed after their feathers were contaminated with fulmar stomach oil.

Storage of stomach oils can also be energetically advantageous because birds can, when needed, directly metabolize fatty acids in the oil rather than first storing them in adipose tissue and later using them as a source of energy. The process of storing fatty acids in adipose tissue and then later using them as a source of energy is estimated to require 25–30% of the assimilated energy, a loss of energy that might be critical if food supplies are limited (Ricklefs 1974; Spady et al. 1976; Roby et al. 1989). Finally, for breeding adults of some species, stomach oils may be an important source of nutrition for their young. For example, Obst and Nagy (1993) found that, by mass, 42% of the food provided to young Wilson's Storm-Petrels was stomach oil. The stomach oil of adults has also been found to be an important part of the diet of young in other species of Procellariiform seabirds (Roby et al. 1997; Oka 2011). Among some species, older chicks are also able to accumulate stomach oil, possibly serving as an important source of energy if adults are delayed in returning to provision their young (Oka 2011). For other species, adult stomach oil appears to be of limited importance as a source of food for young (Alonso et al. 2012). A possible explanation for differences among species in the importance of stomach oil as a source of food for young is the duration of foraging trips at sea. Warham (1977) suggested that stomach oil was an adaptation that allows longer foraging trips to exploit patchily distributed prey. Species that do so are able to accumulate stomach oil during those trips and then supplement captured prey with that oil when returning to nest sites and feeding young. Species

(continued)

**Box 5.23** (continued)

that forage on more predictable prey closer to breeding areas accumulate less stomach oil, but are able to provision young with prey at higher rates.

Finally, stomach oil is also used a defense mechanism against predators such as falcons, eagles, gulls, ravens, and foxes. The oil can be ejected to distances up to 3 m and the feathers and fur of potential predators covered with this oil provide less insulation that, during inclement weather, can result in death (Wang 2005).

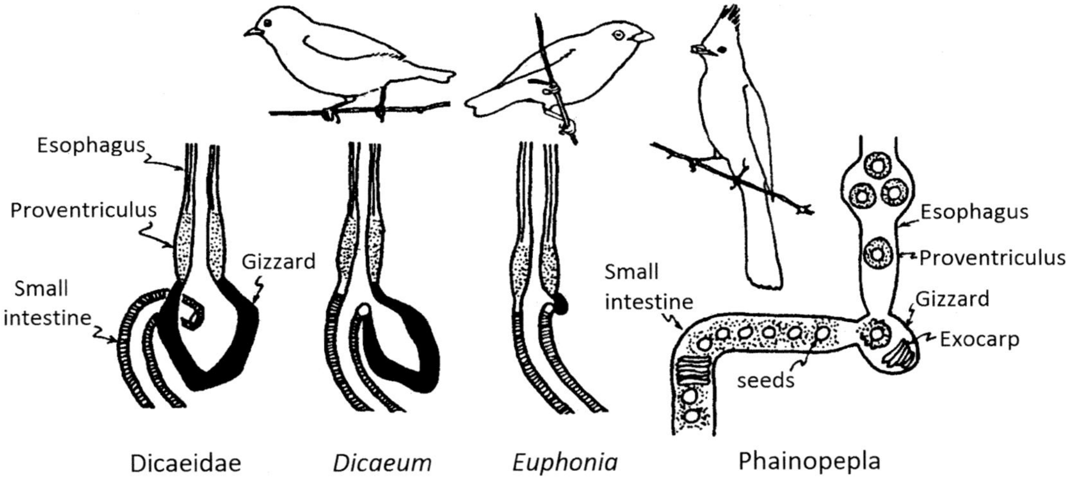


**Fig. 5.100** Digestive tracts of some frugivorous birds. The gizzards (ventriculus) of frugivores tend to be smaller and have thinner walls (less muscular). For some frugivores that consume whole fruits, like Phainopeplas (*Phainopepla nitens*), the exocarp (sometimes hard, outer skin of a fruit) may be removed in the gizzard and the carbohydrate-rich fruit is digested (stippled areas around the seeds) in the small intestine. Undigestible seeds pass through the digestive tract and are excreted in the feces. (Figure modified from Jordano 2013; © CAB International 2014, used with permission)

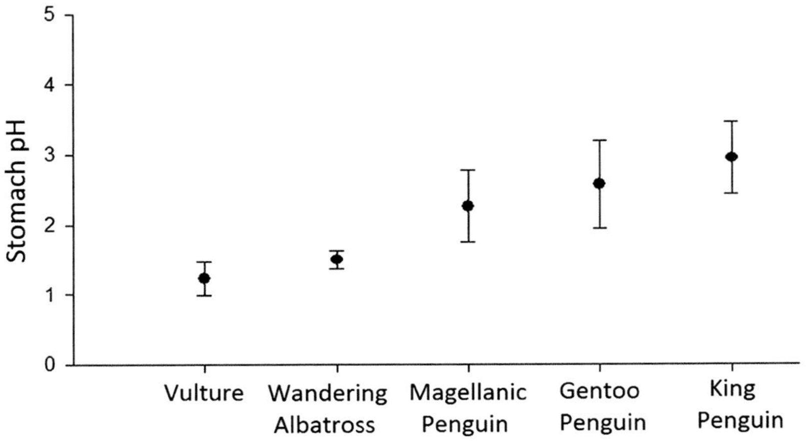


**Fig. 5.101** Stomach pH (mean ± SD) of African White-backed Vultures (*Gyps africanus*), Wandering Albatrosses (*Diomedea exulans*), Magellanic Penguins (*Sphenicus magellanicus*), Gentoo Penguins (*Pygyoscelis papua*), and King Penguins (*Aptenodytes patagonicus*). Carnivores, piscivores (e.g., penguins), and scavengers/carrion-eaters have a very low stomach pH to aid in the rapid digestion of protein. This rapid digestion is especially important (hence the very low stomach pH) for birds that must fly after ingesting large quantities of food, e.g., vultures and albatrosses. (Figure modified from Grémillet et al. 2012; © 2012 Grémillet et al., open-access article distributed under the terms of the Creative Commons Attribution License, https://creativecommons.org/licenses/by/4.0/)

**Box 5.24 Preservation of stomach contents of incubating King Penguins**

Male King Penguins (*Aptenodytes patagonicus*) store undigested food in their stomach for up to 3 weeks during their incubation fast. This ensures hatchling survival if their mate's return is delayed. Using small electronic recorders, Thouzeau et al. (2004) studied the change in gastric pH and motility during the first week of food storage. The pH could be maintained at values as high as 6 throughout the incubation fast, a pH too high for gastric proteinase (the enzyme that breaks down protein) activity. Gastric motility was also reduced, which also aids in the conservation of stomach content. These changes in pH and gastric motility in incubating King Penguins help inhibit digestive gastric processes. Mechanisms underlying these changes are probably complex, including a combination of neuronal, humoral, and/or hormonal factors.

(continued)

**Box 5.24** (continued)

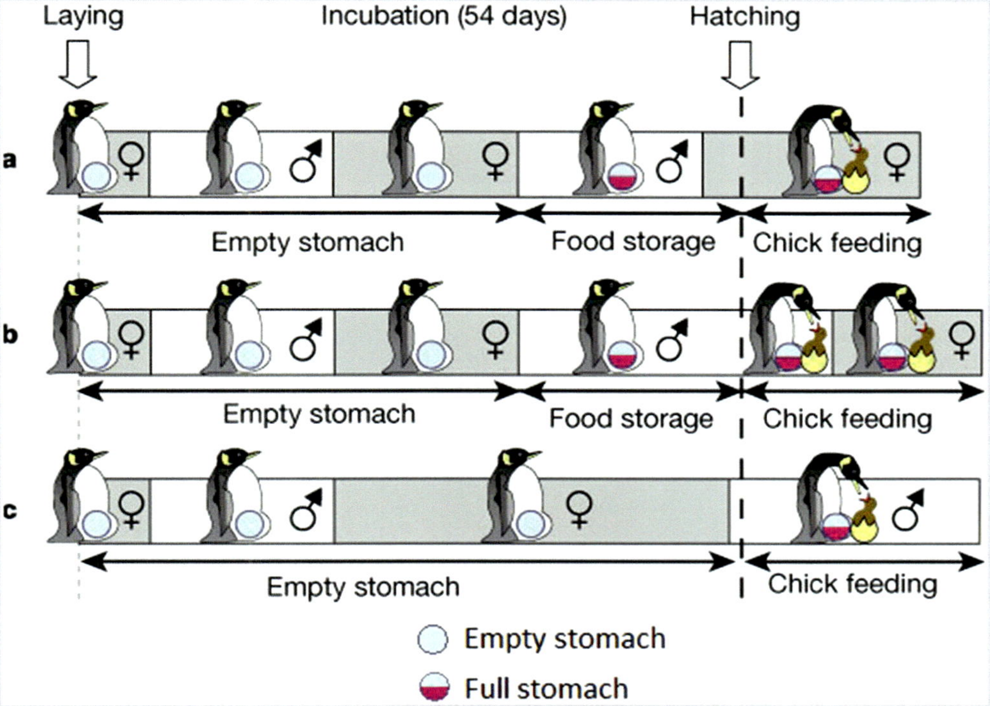


Stomach contents of King Penguins relative to the date of egg hatching and possible outcome of incubation due to the variable duration of a mate's time at sea. (**a**) Females usually relieve males a few days before hatching and his stored food is not needed. (**b**) The chick hatches before the female returns and the male can feed it with the food stored in his stomach. (**c**) Arriving just before or soon after hatching, the male brings more food in his stomach (Figure modified from Gauthier-Clerc et al. 2000; © 2000 Springer Nature, used with permission)

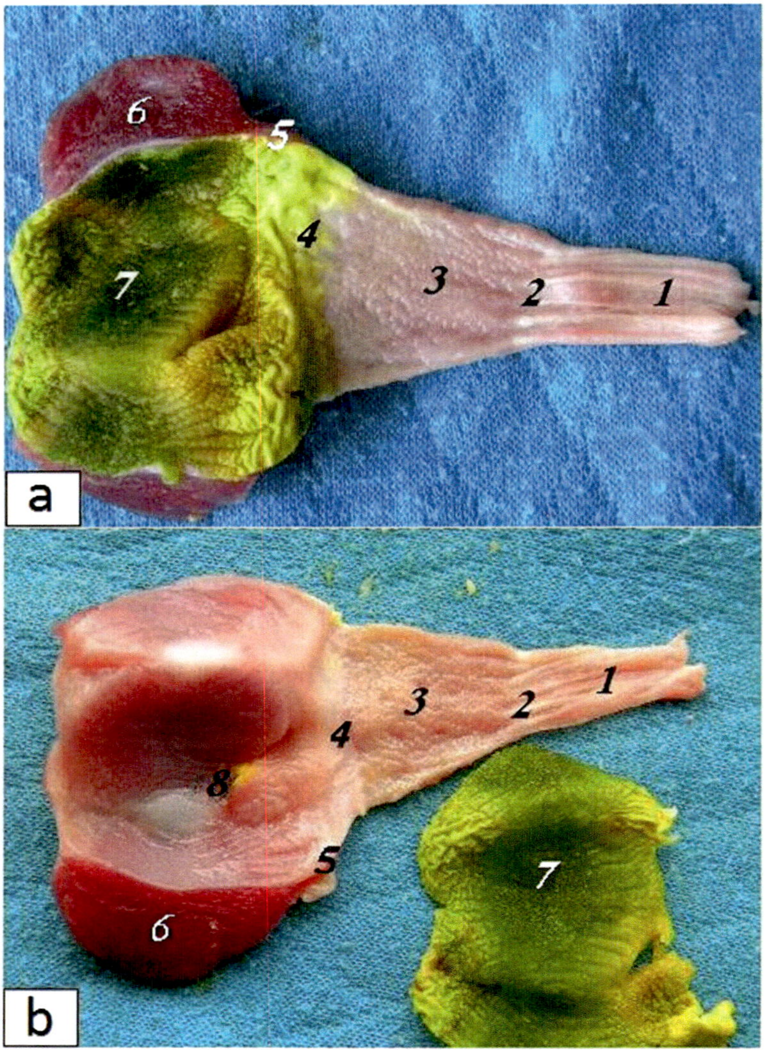


**Fig. 5.102** Esophagus, proventriculus, and ventriculus of a Rock Pigeon (*Columba livia*). (**a**) Stomach with cuticle (7), (**b**) Stomach with cuticle removed. (1) esophagus, (2) junction between esophagus and proventriculus, (3) proventriculus, (4) junction between proventriculus and ventriculus, (5) and (6), muscle in the wall of the ventriculus, (7) koilin (also called the cuticle), and (8) opening into the duodenum of the small intestine. (From Hassan and Moussa 2012; © 2012 African Association of Veterinary Anatomists, used with permission)

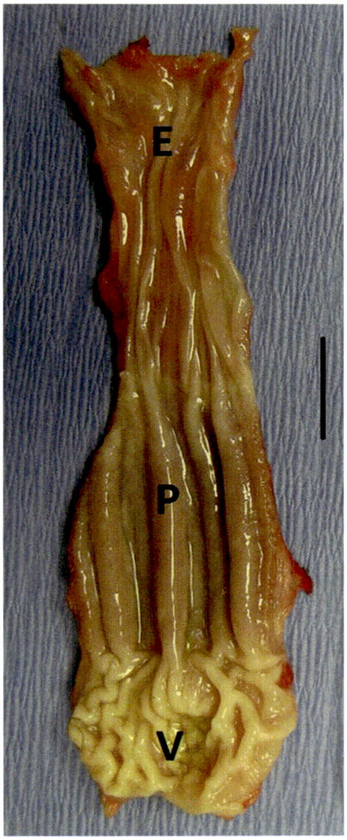


**Fig. 5.103** Esophagus (E), proventriculus (P), and ventriculus (V) of a Cooper's Hawk (*Accipiter cooperii*). Raptors have a less muscular ventriculus and a better developed proventriculus than seed-eating birds because animal prey contains more protein (that is partially digested by enzymes and acid produced by the proventriculus) and needs less mechanical grinding than seeds. Also, because raptors often ingest prey whole (e.g., owls) or in large pieces, their proventriculus is lined by mucosal ridges (plicae proventricularis) that allow substantial expansion. Scale bar = 1 cm. (Image from Murray 2014; © 2014 Elsevier Ltd., used with permission)

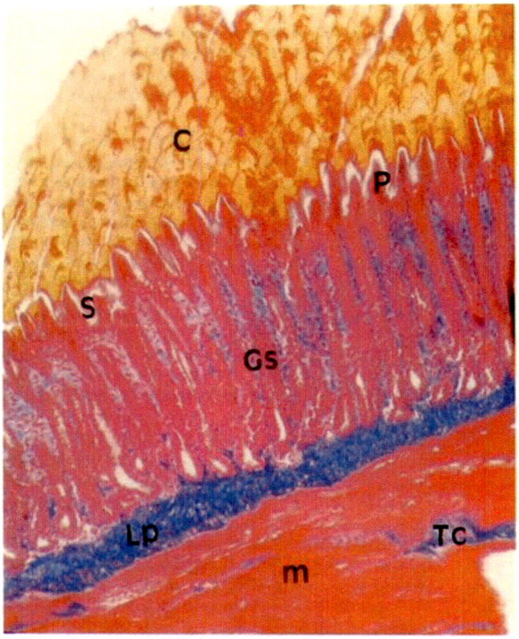


**Fig. 5.104** Photomicrograph (210×) of longitudinal section of the gizzard showing folds of mucous membrane lined by simple prismatic epithelium (P); simple tubular glands (Gs) in the lamina propria constituted by connective tissue (Lp); secretion of glands (S) that are continuous with the cuticle (or koilin) (C), part of muscle layer (m), interpersed with bundles of connective tissue (Tc). (Figure from Catroxo et al. 1997; used with permission)

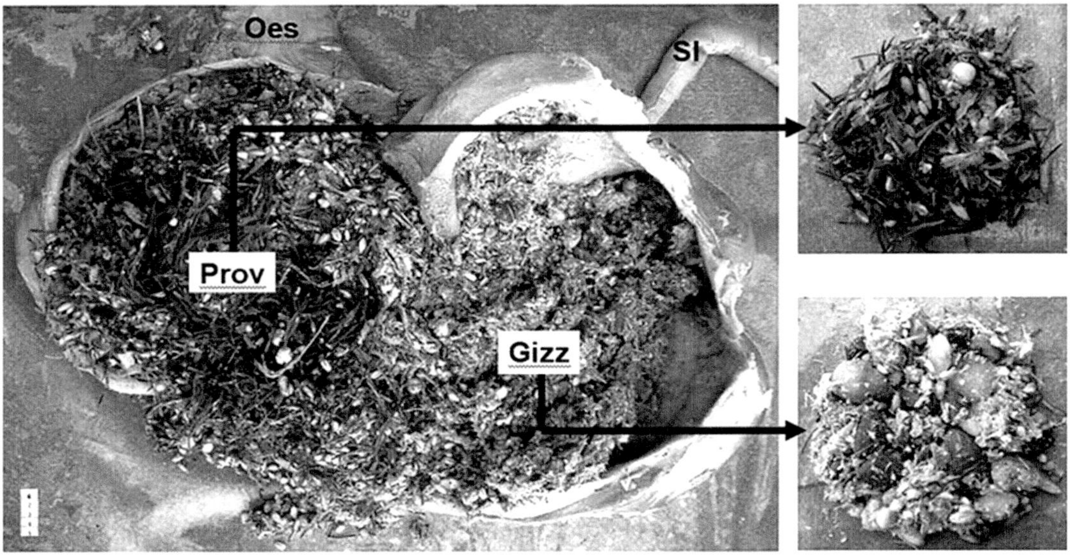


**Fig. 5.105** Common Ostrich (*Struthio camelus*) stomach. Note how particle size of material in the gizzard (ventriculus) is smaller than in the proventriculus due to the grinding action of the muscular walls plus small pebbles (gastroliths). The capacity to reduce particle size is related to the metabolic demands of a species. Therefore, particle size reduction is often considered the key digestive difference between ecto- and endotherms that allows endotherms to rely on shorter digesta retention times without losing digestive efficiency, and hence facilitate the high level of food intake necessary to meet their increased metabolic requirements. Oes, esophagus, Prov, proventriculus; Gizz, gizzard; SI small intestine. (Figure from Fritz et al. 2011; © 2011 Cambridge University Press, used with permission)

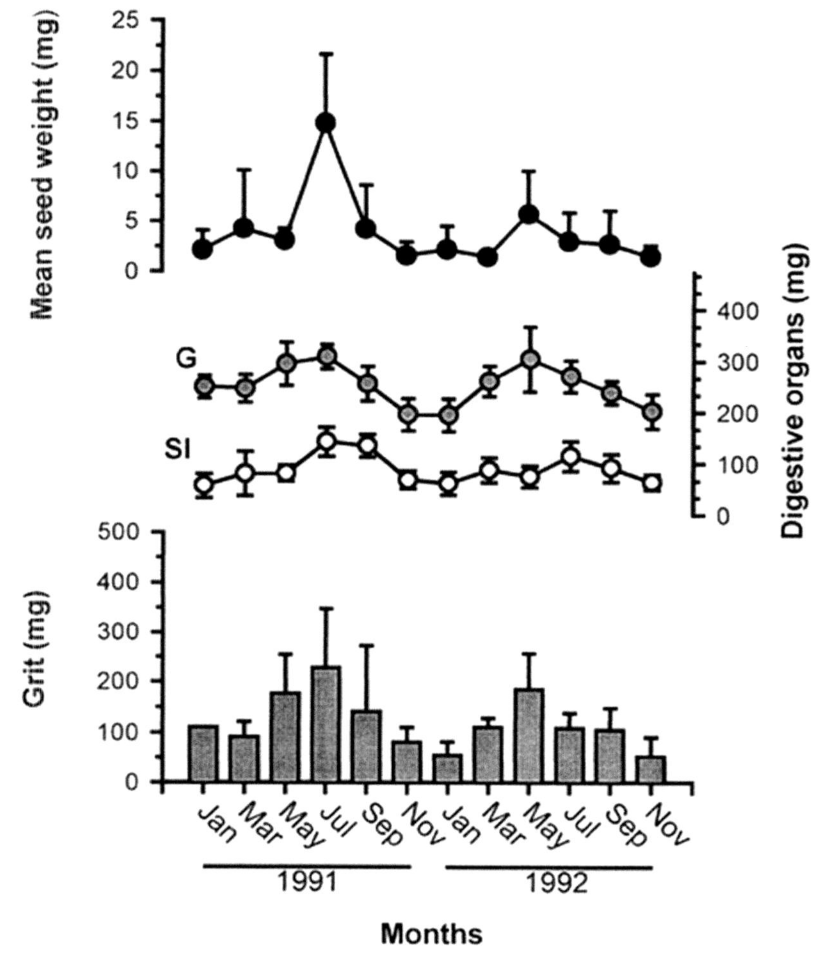


**Fig. 5.106** Seasonal variation in grit consumption, digestive organ mass (gizzard [G] and small intestine [SI]), and mean weight of seeds consumed by Rufous-collared Sparrows (*Zonotrichia capensis*). Variation in grit and food consumption corresponded with morphological changes in both the gizzard and small intestine. Grit consumption declined when sparrows were fed more soft food items (mealworms), and increased when sparrows were fed larger, harder seeds. Changes in the gizzard and small intestine corresponded with the variation in the digestibility of food items, with increases in mass when sparrows were fed less digestible large seeds and decreases when sparrows were fed more easily digested mealworms. Values are shown as mean ± SD. (Figure from López-Calleja et al. 2000; © 2000 Oxford University Press, used with permission)

**Box 5.25 Avian gastroliths**

Gastroliths are stones (or other hard objects) of various sizes (from sand grains to much larger objects in larger birds) that are retained in the digestive system, primarily in the ventriculus of birds. Of course, some gastroliths can be ingested by birds accidentily if mixed with their food. However, gastroliths are known to aid in digestion, particularly for seed-eating birds and grazers, by helping to macerate food. In Common Ostriches, gastroliths mix with food in the ventriculus and prevent agglutinated balls of grass from blocking the pyloric sphincter (where materials leave the ventriculus and enter the small intestine). In fact, Ostriches without access to stones may die of constipation (Wings 2007). Abrasion of gastroliths, along with the acid in the stomach, may also release minerals that can then be absorbed in the intestine, e.g., calcium, phosphorus, magnesium, iron, and others (Lee et al. 2004; Maya-García et al. 2021). Some investigators have suggested that gastroliths can also aid in stomach "cleaning." For example,

(continued)

**Box 5.25** (continued)

Fox (1976, 1995) suggested that gastroliths ingested by raptors, cormorants, and loons may help remove mucus and excess koilin (cuticle). Others have suggested that, for diving birds, gastroliths may serve a hydrostatic function, making them less buoyant (Beaune et al. 2009). There is, however, little empirical evidence to support the "buoyancy" hypothesis. Finally, gastroliths could serve an anti-parasite function. Robinson et al. (2008), for example, found that Double-crested Cormorants (*Phalacrocorax auritus*) with more gastroliths also had fewer nematodes. In sum, there are several hypotheses concerning possible functions of gastroliths in birds, but most of them need additional testing.

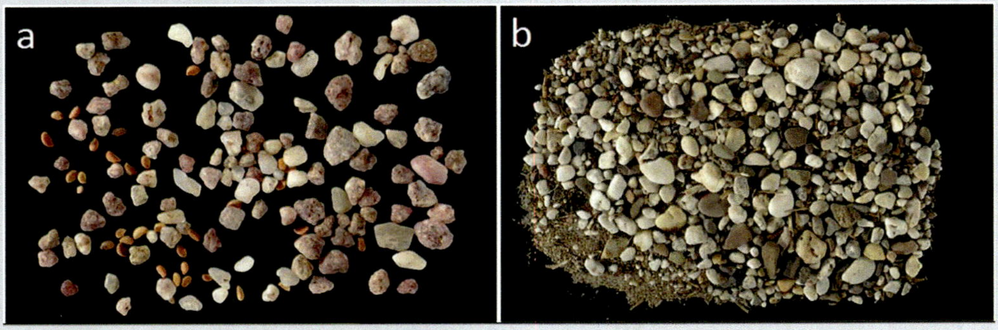


Gastroliths found in the ventriculus of a ptarmigan (**a**, *Lagopus* spp.) and an Ostrich (**b**, *Struthio camelus*) (Figure from Wings 2007; open-access article distributed under the terms of the Creative Commons Attribution License (CC BY), https://creativecommons.org/licenses/by/4.0/)

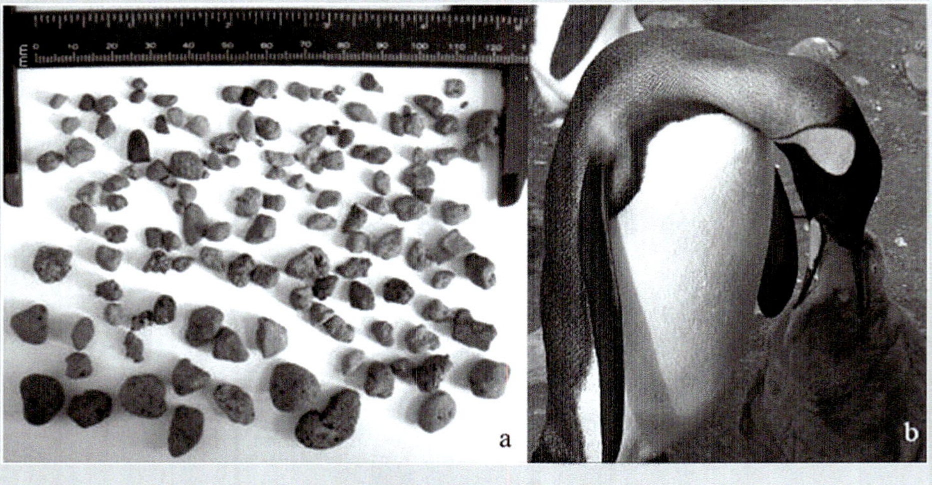


(**a**) Gastroliths from the ventriculus of a King Penguin (*Aptenodytes patagonicus*) chick. (**b**) King Penguin chick being fed by an adult. Chicks may obtain some gastroliths via transfer of food from their parents (Figure from Beaune et al. 2009; © 2009 Springer Nature, used with permission)

**Fig. 5.107** Pellets ejected by an owl and a shrike and a pellet being ejected by a bee-eater. (**a**) Pellet from a Spotted Wood Owl (*Strix seloputo*) with undigested fur and bones (scale bar = 5 mm). (**b**) Pellet from a Tiger Shrike (*Lanius tigrinus*) with undigestible remains of beetles. (**c**) A Blue-tailed Bee-eater (*Merops philippinus*) ejecting a pellet. (Images from Wang et al. 2009, photos by Wee Yeow-Chin and Chan Yoke Meng, used with permission)

**Box 5.26 Feather eating by loons and grebes**

Loons and grebes are known to consume feathers and to feed feathers to their young. Grebes eat their own feathers (generally breast and belly, but primarily flank feathers). Ingested feathers collect in the ventriculus (gizzard) and do not pass into the intestine (Meng et al. 2013). Hypotheses proposed to explain feather-eating include: (1) feathers in the ventriculus protect the stomach wall from sharp fish bones, (2) ingested feathers form a plug at the entrance to the small intestine (pylorus) and prevent sharp materials from entering the intestine, (3) ingested feathers help aggregate undigested materials (e.g., small, sharp bones) into pellets so they can be ejected (Piersma and van Eerden 1989), and (4) aid in digestion by helping retain small invertebrate prey in the gizzard until they can be fully liquefied (Jehl 2017). Available evidence suggests that hypothesis 1 can be rejected, with fish remains often found to contact the stomach wall (Hanzák 1952). However, the absence of bones or bone fragments in the intestine (Piersma and van Eerden 1989) supports hypotheses 2 and 3 (Piersma and van Eerden 1989; Meng et al. 2013). In support of hypothesis 4, Jehl (2017) noted that grebes forage on abundant, but hard-to-digest, small invertebrates (e.g., brine shrimp and alkali flies in the case of Eared Grebes,

(continued)

**Box 5.26** (continued)

*Podiceps nigricollis*), and the pyloric plugs that form from ingested feathers increase the time those small prey spend in the gizzard and greatly improve assimilation efficiency (e..g., >87% for Eared Grebes; Caudell and Conover 2006). Therefore, feather-eating by grebes and, although less studied, possibly loons, aids in digestion and in the ejection of undigested materials in the form of pellets. In addition, although additional evidence is needed, Meng et al. (2013) found feather-degrading bacteria in the stomachs of three species of grebes, including Great Crested Grebes (*Podiceps cristatus*), Black-necked Grebes (*Podiceps nigricollis*), and Little Grebes (*Tachybaptus ruficollis*), and suggested that these bacteria enzymatically digest some of the keratin that make up the feathers and, by doing so, produce essential amino acids (like cysteine and cystine) that can be absorbed in the intestine. As such, feather eating might also provide grebes with nutritional benefits.

Adult Eared Grebe with young that is (probably) about to swallow a feather (Photo taken at Sidewood, Saskatchewan, in July 2012 by William Burt and included in Burt's book "Water Babies" published in 2015, used with permission)

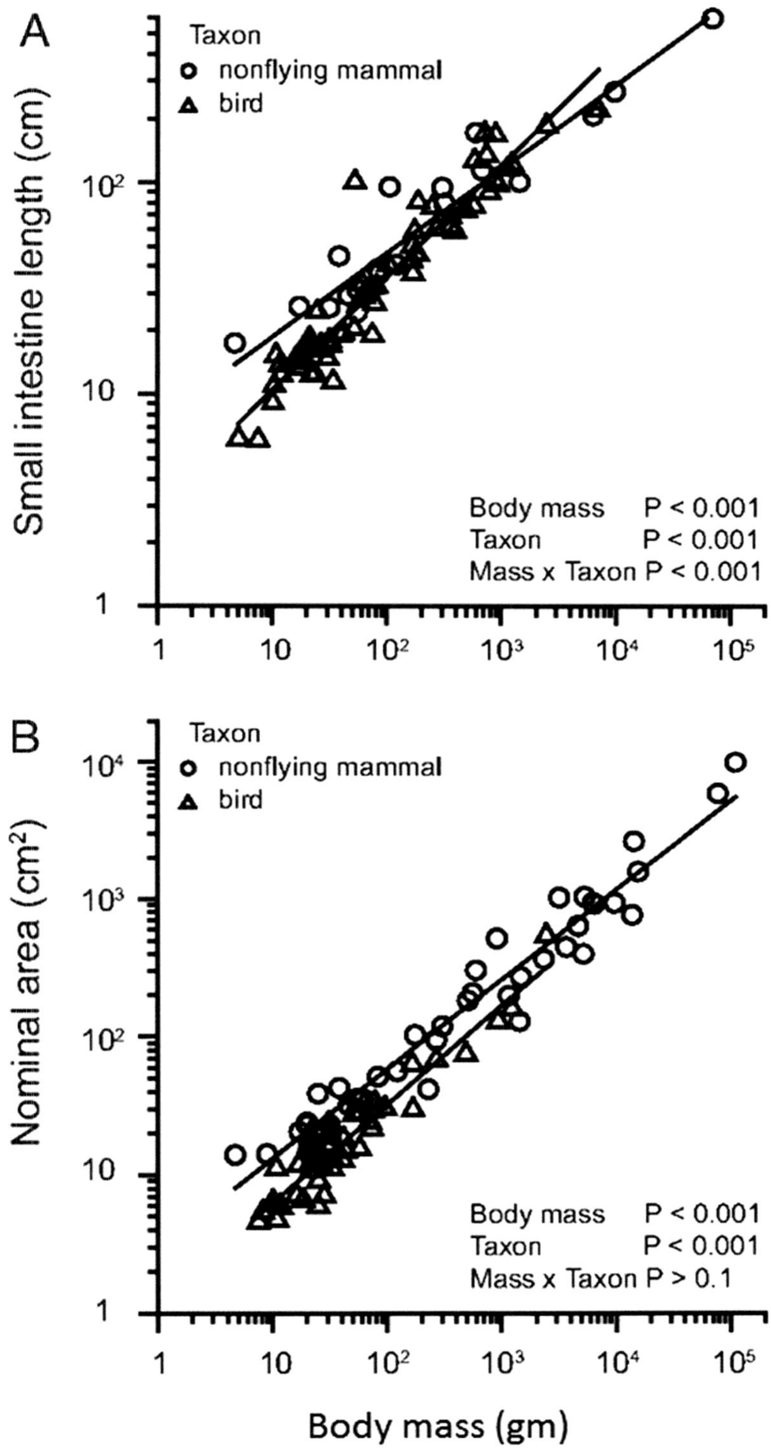


**Fig. 5.108** Small intestine length and nominal (smooth bore tube) surface area in omnivorous birds and nonflying mammals. (**a**) For length, the slopes of the relationships between body mass and small intestine length differ significantly (analysis based on 61 species of birds and 21 species of nonflying mammals). Birds that weigh less than 195 g have significantly shorter small intestines than nonflying mammals. (**b**) For nominal surface area, there was no difference in slope between birds and nonflying mammals (based on 46 species of bird and 41 species of mammals). However, small intestine nominal surface area for birds is about 36% lower than that for nonflying mammals. (Figure from Caviedes-Vidal et al. 2007; © 2007 National Academy of Sciences U.S.A., used with permission)

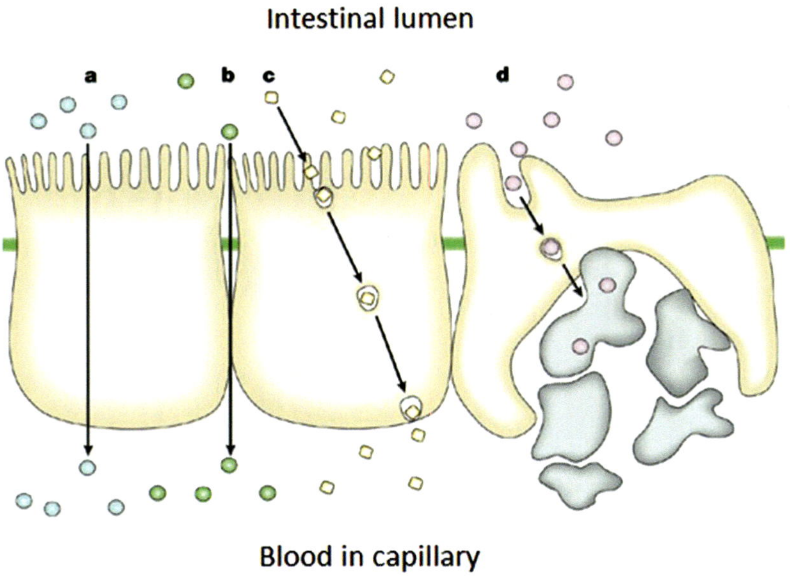


**Fig. 5.109** Possible pathways available for absorption of materials in the small intestine. (**a**) Transcellular pathway (through the epithelial cells, i.e., facilitated diffusion or active transport). (**b**) Paracellular pathway (in between adjacent cells). (**c**) Endocytosis (into the epithelial cell) and exocytosis (out of the cell and into the blood). (**d**) Absorption into the lymphatic circulation (lipids; this mechanism of lipid absorption from the small intestine into lymph vessels is usually assumed to operate in all birds, but, to date, has only been described in Domestic Chickens). Birds, more than nonflying mammals, rely on paracellular transport, which is a faster means of transport than the other pathways. (Figure from Goldberg and Gomez-Orellana 2003; © 2003 Springer Nature, used with permission)

**Box 5.27 Avian geophagy**

Geophagy (intentional consumption of soil) has been reported in 23 orders of birds and 54 families (36 non-passerine and 18 passerine) (Downs et al. 2019). Based on a review of the literature, Downs et al. (2019) determined that geophagy had been reported in about 260 species of birds, but these authors suspected that the incidence of geophagy among birds was likely higher. Excluding the ingestion of grit and small stones that aids in the mechanical digestion of foods in the gizzard, hypotheses concerning the ingestion of soil (e.g., clay) include: (1) buffering of pH in the digestive system (Gilardi et al. 1999), (2) protection of the digestive system from chemicals such as polyphenolics (plant secondary compounds found in the seeds, leaves, and unripe fruits of some plants) by stimulating mucous secretion (Gee and Johnson 1988; Ortiz et al. 1994), (3) protection from plant secondary compounds via "high adsorptive capacity and and binding to positively charged toxins and bitter polyphenolics" (Downs et al. 2019), and (4) acquisition of essential minerals such as sodium and calcium (Sanders and Koch 2018). The results of studies to date provide support for hypothesis 3, with ingestion of clay-dominated soil providing birds with sodium (e.g., Lee et al. 2014a; Sanders and Koch 2018; Brightsmith et al. 2018). The importance of sodium for many species of birds is further supported by their use

(continued)

**Box 5.27** (continued)

of both natural and artificial salt licks (e.g., Blake et al. 2011; Hon et al. 2020; Razali et al. 2022). There is less support for hypotheses 2 and 3. Howver, Gilardi et al. (1999) noted that parrots with diets that sometimes include polyphenolics have been observed consuming clay, which is consistent with hypothesis 2. In addition, some investigators have suggested that geophagy may protect birds from toxic plant secondary compounds (Gilardi et al. 1999; Symes et al. 2006). For example, in a laboratory study, Gilardi et al. (1999) determined that clay reduced the bioavailabililty of an ingested secondary plant compound (quinidine) in captive Turquoise-fronted Amazons (*Amazona aestiva*) by about 60%. Finally, little evidence is currently available to support hypothesis 1 (Gilardi et al. 1999; Downs et al. 2019).

Determining with some certainty the function(s) of geophagy is difficult because the composition of soil varies with location and actually observing avian geophagy is typically, with a few exceptions (e.g., parrots that regularly visit particular sites to ingest soil), difficult. As noted by Downs et al. (2019), "Additional work on birds' feeding ecology, perhaps aided by improved movement monitoring techniques, is needed to disentangle the contexts in which each of the functions of geophagy are important."

(continued)

**Box 5.27** (continued)

Blue-and-yellow Macaws (*Ara ararauna*), Scarlet Macaws (*Ara macao*), Mealy Parrots (*Amazona farinosa*), and Chestnut-fronted Macaws (*Ara severus*) at a clay lick in Peru (Photo by Brian Ralphs, Wikipedia, CC BY 2.0, https://creativecommons.org/licenses/by/2.0/deed.en)

**Table 5.7** Enzymes important in the digestive process for birds. The relative importance of different enzymes varies among species with different food habits. For example, pancreatic amylase and cholesterol 7-α-hydroxylase are more abundant among seed-eating and carnivorous birds, and pancreatic lipase is more abundant in non-carnivores (Table modified from Chen and Zhao 2019; )

| Enzyme | Full name | Site of secretion | Digestive function |
|---|---|---|---|
| *Carbohydrases* | | | |
| Hepatic AMY | Hepatic amylase | Liver | Facilitates hydrolysis of starch |
| Pancreatic AMY | Pancreatic amylase | Pancreas | Facilitates hydrolysis of starch |
| AGL | Glycogen debranching enzyme | Muscle, liver, heart | Facilitates breakdown of glycogen |
| C6PC | Glucose-6-phosphotase, catalytic subunit | Liver, kidney, intestine | Facilitates hydrolysis of glucose-6-phosphate in gluconeogenesis |
| GAA | α-1,4-glucosidase | Intestine | Participates in glycogen hydrolysis |
| GCK | Glucokinase | Gastrointestinal tract, liver, pancreas | Facilitates phosphorylation of glucose to glucose-6-phosphate |
| *Lipases/esterases* | | | |
| CYP7A1 | Cholesterol 7-α-hydroxylase | Liver | Participates in bile acid synthesis |
| PNLIP | Pancreatic lipase | Pancreas | Hydrolyzes ester linkages of triglycerides |
| *Proteases* | | | |
| CTRC | Chymotrypsin C | Pancreas | Hydrolyzes peptide bonds involving phenylalanine, tyrosine, and ryptophan[a] |
| PGC | Progastricsin | Stomach | Hydrolyzes peptide bonds involving phenylalanine, tyrosine, and leucine[a] |
| *Lysozymes* | | | |
| LYZ | c-Type lysozyme | Gastrointestinal tract, eggs, blood | Cleaves the β-1,4-glycosidic bond between NAM and NAG in peptidoglycan[b] |
| LYG | g-Type lysozyme | Intestine, tongue, eggs | Cleaves the β-1,4-glycosidic bond between NAM and NAG in peptidoglycan[b] |
| *Chitinases* | | | |
| CHIA | Acidic mammalian chitinase | Gastrointestinal tract, tongue, kidney | Participates in degrading chitin |

[a]Phenylalanine, tyrosine, tryptophan, and leucine are amino acids
[b]Peptidoglycan is a substance found in the membrane of most bacteria

**Box 5.28 A missing enzyme**

Trehalose is a disaccharide, or sugar, consisting of two glucose molecules that varies in abundance in the tissues of fungi and most taxa of animals except vertebrates. Although unable to synthesize trehalose, most vertebrates can digest it using the enzyme trehalase found in intestinal cells. Consisting of two molecules of glucose, trehalose can be an excellent source of energy.

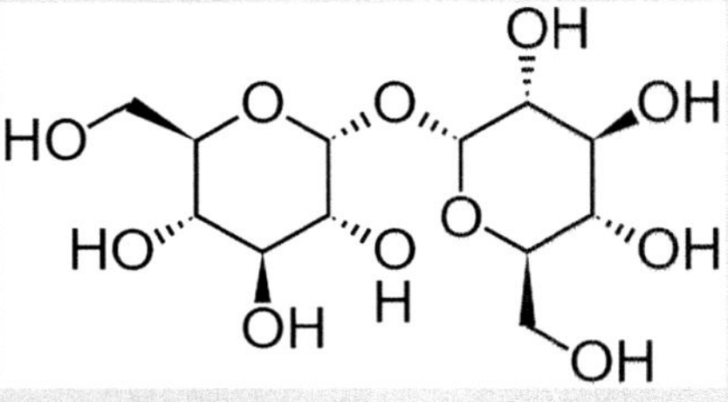


Trehalose molecule consisting of two molecules of glucose (Figure from Wikipedia, CC0 Public Domain)

However, the intestinal cells of birds do not contain trehalase, and the gene that codes for trehalase is not present in their genomes. Brun et al. (2022) determined that birds lost the trehalase gene due to "an inversion followed by a reinsertion of a large gene block" that "appears to have taken place after the split between crocodiles and birds and dinosaurs." Because so many species of birds are insectivorous and insects contain small amounts of trehalose, the absence of trehalase, as noted by Brun et al. (2022), results in "a diminished return in the total energy that birds assimilate from insects." In addition, fungi contain relatively high levels of trehalose (up to 20% of dry mass) and the fact that so few birds rely on fungi as a source of food (about 0.5% of species) may be due to the inability of birds to digest trehalose (Brun et al. 2022).

**Box 5.29 Avian blood glucose levels**

Birds maintain higher concentrations of glucose in the blood, often 1.5–2 times higher, than mammals (and other vertebrates) of comparable size (Sweazea 2022). Glucose in the blood of birds is either dissolved in the plasma or taken up by erythrocytes via glucose transport proteins (Diamond and Carruthers 1993). Why are avian blood glucose levels so high? In mammals, the hormone insulin (produced by cells in the pancreas) is released when blood levels of glucose increase (e.g., after ingesting and digesting food) and reduces blood glucose levels by stimulating the uptake of glucose by the cells and tissues of the body. In contrast, birds are "insulin resistant," i.e., insulin does not stimulate the uptake of glucose by the cells and tissues of birds as it does in mammals (e.g., Dupont et al. 2009, 2012). Possible advantages of insulin resistance and maintaining high levels of blood glucose include having a source of energy during periods of reduced food availability and reducing uptake of glucose by skeletal muscles to ensure the availability of glucose for the brain and other vital organs (Brand-Miller and Colagiuri 1999). Hickey et al. (2012) suggested that another potential benefit of hyperglycemia and insulin resistance relates to fuel storage and weight reduction. Insulin stimulates glycogen synthesis, but glycogen is hygroscopic (absorbs water) so adds mass where it is stored (especially the liver and skeletal muscles in mammals). Being insulin resistant, birds store little glycogen in their livers and skeletal muscles, making them correspondingly lighter than similar-sized mammals. For example, the liver of a mouse accounts for about 6% of its total body mass, whereas bird livers represent only about 2% of total body mass (Else et al. 2004; Hickey et al. 2012).

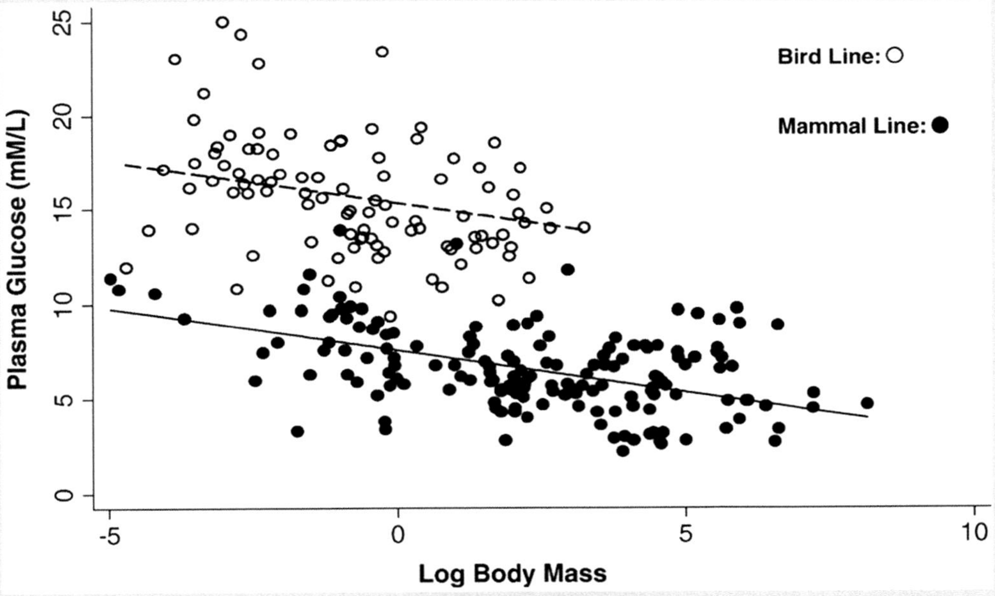


Birds have higher blood glucose levels than similar-sized mammals (Figure from Braun and Sweazea 2008; © 2008 Elsevier Ltd., used with permission)

High blood glucose levels in mammals can cause the formation of toxic molecules that can cause various health defects (e.g., cardiovascular disease and diabetes) (Sweazea 2022). Hyperglycemia can also increase production of reactive oxygen species (ROS) known to damage

(continued)

**Box 5.29** (continued)

DNA, lipids, and various protein molecules, and this damage is known to contribute to the aging process in mammals and other non-avian vertebrates (Sweazea 2022). However, birds are "largely resistant to oxidative stress" (Sweazea 2022) and, therefore, avoid age-related damage to their tissues (Holmes and Austad 1995a; Holmes et al. 2001) and, therefore, tend to live much longer than similar-sized mammals. As an example, a maximum house mouse (*Mus musculus*; 20 g) lifespan is about 4 years (Holmes et al. 2001), whereas Broad-billed Hummingbirds (*Selasphorus platycercus*; 5 g) can live up to 14 years (Holmes and Austad 1995b). Among birds, albatrosses are among the longest lived, with a banded female Laysan Albatross (*Phoebastria immutabilis*) known to be at least 70 years old and a Black-footed Albatross (*Phoebastria nigripes*) at least 62 years old (USGS Bird Banding Laboratory; https://www.usgs.gov/labs/bird-banding-laboratory/news/usgs-bird-banding-laboratory-curates-more-80-years-pacific). Important contributing factors to this "resistance" to oxidative stress, and the resulting long lifespans, include the maintenance of high concentrations of antioxidants in the blood of birds (e.g., Cohen et al. 2007; Smith et al. 2011; Ingram et al. 2017), and the release of fewer ROS by avian mitochondria than mitochondria of mammals and other vertebrates (Barja et al. 1994; Herrero and Barja 1998; Hickey et al. 2012).

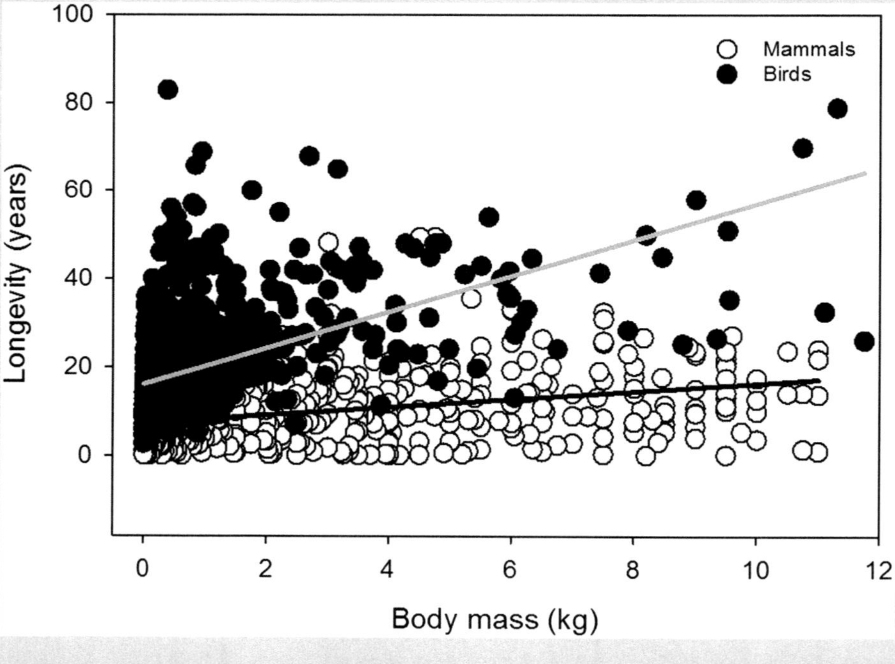


Relationship between reported lifespans of birds and mammals and body mass. Lifespans tend to increase with increasing body mass for both taxa, but birds generally live longer than similar-sized mammals (Figure from Jimenez 2018; © 2018 Oxford University Press, used with permission)

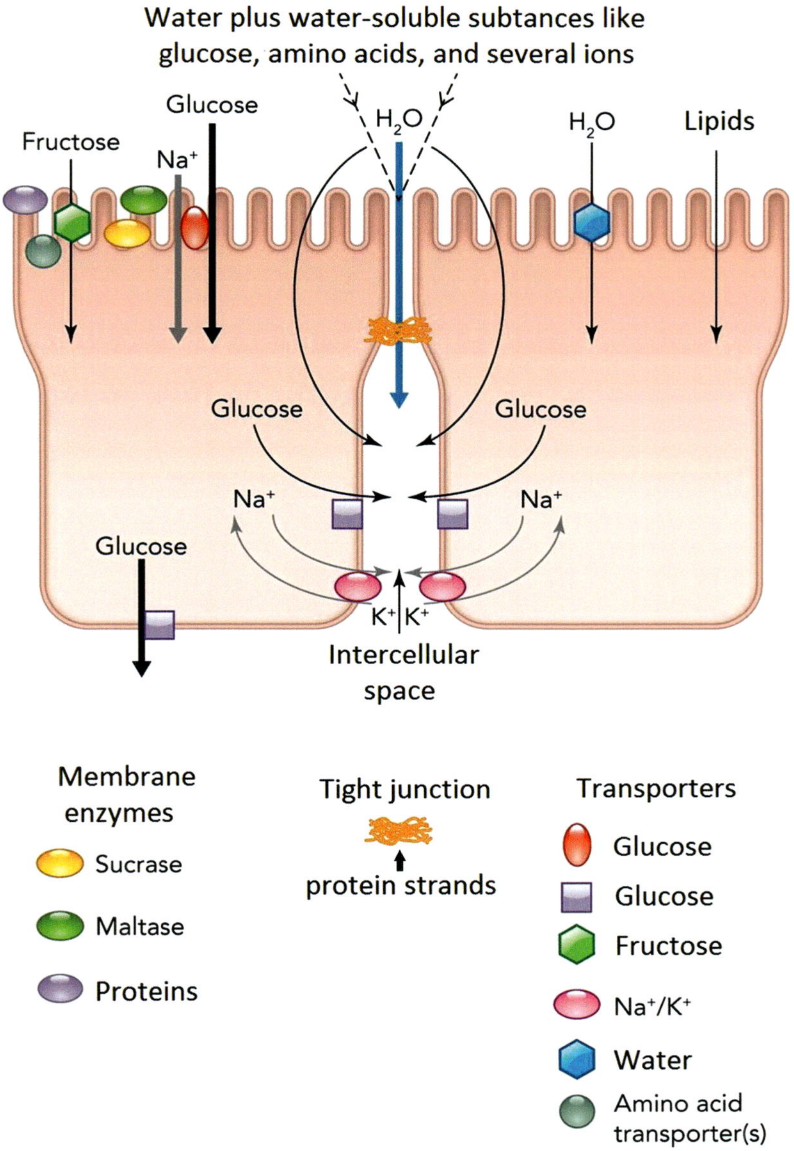


**Fig. 5.110** Enzymes present in the membranes of intestinal cells aid in the digestion of some carbohydrates (like sucrose and maltose) and proteins. Small, water-soluble substances including several ions like sodium and potassium can be transported across membranes by transporters in the membrane as well as through the tight junction (protein strands that limit movement of materials based on the size and charge of materials) by simple diffusion. Lipids like fatty acids can diffuse into and across the cell phospholipid bilayer membranes. Water can move into and out of cells via special proteins called aquaporins. (Figure modified from Price et al. 2015; © 2015 The American Physiological Society, used with permission)

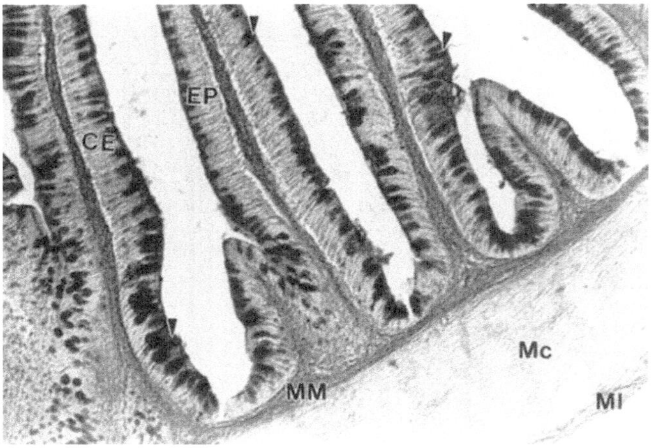


**Fig. 5.111** Cross section of the intestine (ileum) of a Spotted Tinamou (*Nothura maculosa*). Villi are lined with columnar epithelium (EP), including goblet cells (arrows) that secrete mucus. The muscle layer includes longitudinal fibers (MI) on the perimeter, circular fibers (Mc), and additional longitudinal fibers at the base of the villi (muscularis muscosae; MM). (Figure from Chikilian and de Speroni 1996; © 1996 Wiley-Liss, Inc., used with permission)

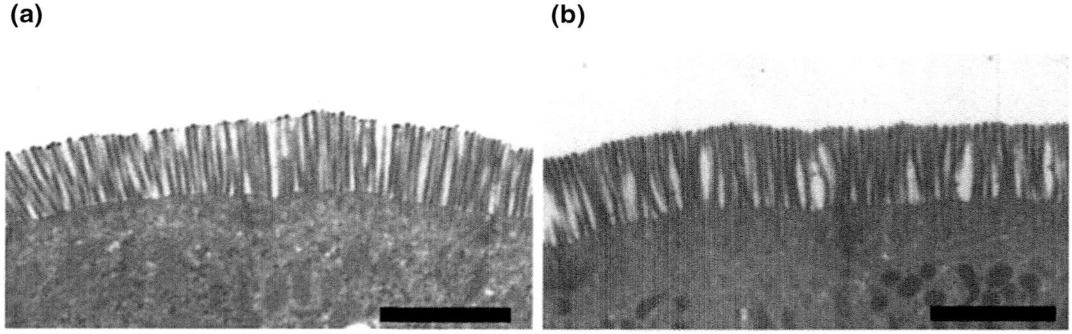


**Fig. 5.112** Intestinal microvilli ("brush border") of a (**a**) House Sparrow (*Passer domesticus*) and (**b**) Savannah Sparrow (*Passerculus sandwichensis*). Scale bars = 0.5 µm. (Figure from Casotti 2001; © 2001 Wiley-Liss, Inc., used with permission)

**Box 5.30 The avian liver**

The avian liver consists of two lobes and each lobe contains thousands of lobules, the functional unit of the liver. Lobules are typically hexagonal in shape, consist of several types of cells plus sinusoids, and have a centrally located central vein. Hepatocytes are the most common type of cell; other types include Kupffer cells that are found in the sinusoids, and epithelial cells that line the walls of the sinusoids. Kupffer cells are white blood cells that, by phagocytosis, remove pathogens and cellular debris from the liver.

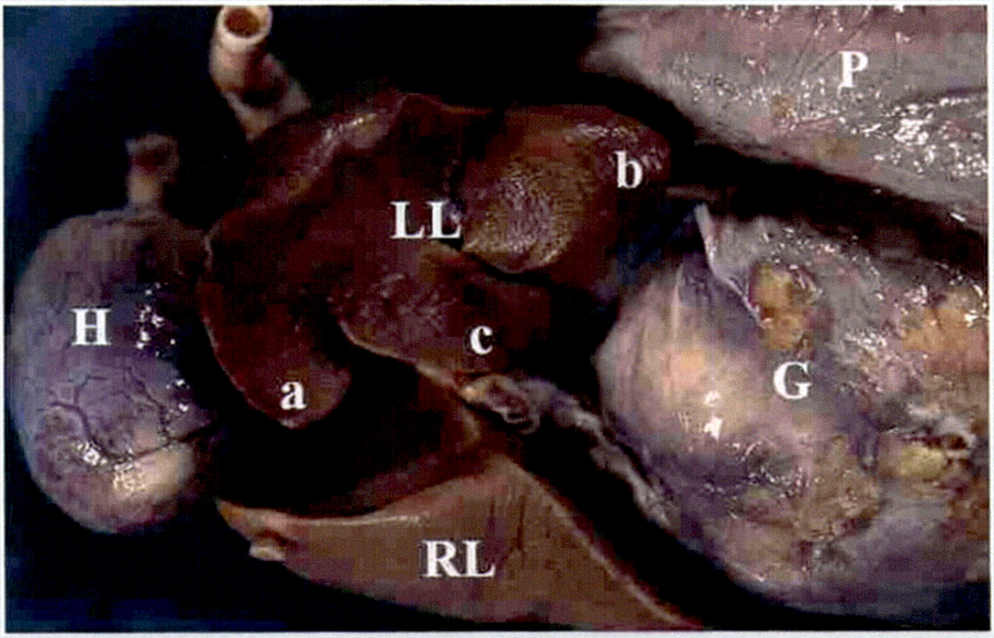


Common Ostrich liver. LL, left lobe (subdivided into a dorsal lobe [a], ventral lobe [b]. and intermediate lobe [c]); RL, right lobe. Also shown is the heart (H), proventriculus (P), and gizzard (G) (Figure from Stornelli et al. 2006; open-access article licensed under the terms of the Creative Commons Attribution 4.0 International Public License (CC-BY-4.0), https://creativecommons.org/licenses/by/4.0/)

(continued)

**Box 5.30** (continued)

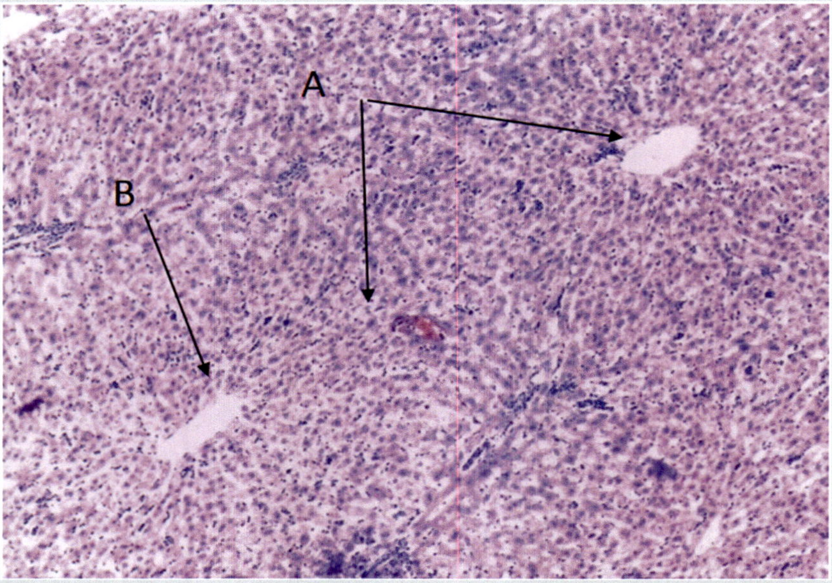


Liver lobules (A) of a Eurasian Coot (*Fulica atra*) with a central vein (B). Magnified 100× (Figure from Selman 2013; Published in the Basrah Journal of Veterinary Research, open-access journal)

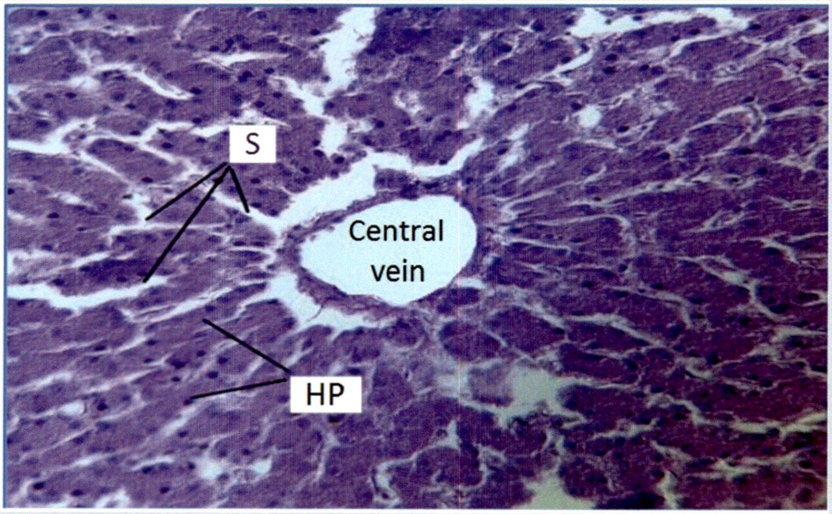


Portion of a liver lobule of a Mallard (*Anas platyrhynchos*) showing the central vein, sinusoids (S), and hepatic plates (HP) made up of hepatocytes. Magnified 40× (Figure modified from Khaleel et al. 2017; © 2017 Khaleel et al., open-access article distributed under the Creative Commons Attribution License, https://creativecommons.org/licenses/by/4.0/)

(continued)

**Box 5.30** (continued)

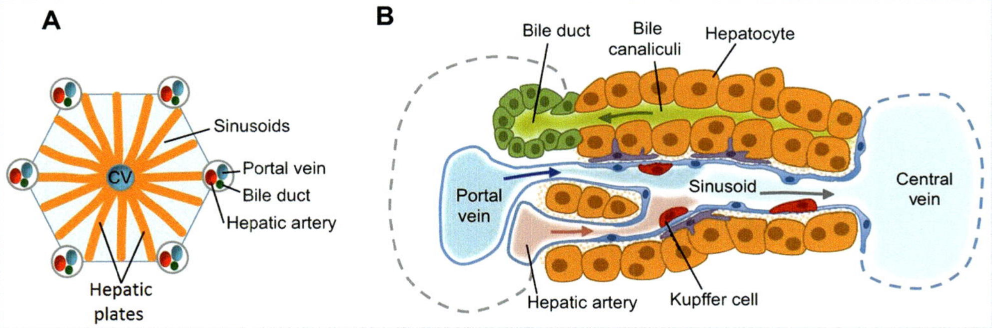


(**a**) The liver consists of numerous lobules, the functional units of the liver. Each lobule has a central vein (CV) around which hepatocytes radiate and form "plates." Hepatic plates are separated by vessels called sinusoids that carry blood from portal veins to the central veins. (**b**) Hepatocytes line the sinusoids and secrete bile salts into bile canaliculi that lead to bile ducts. Kupffer cells, specialized white blood cells (macrophages) found in the liver, are also present in sinusoids (Figure modified from Gordillo et al. 2015; © 2015. Published by The Company of Biologists Ltd., open-access article distributed under the terms of the Creative Commons Attribution License, https://creativecommons.org/licenses/by/4.0/)

Hepatocytes are arranged in rows called hepatic plates that radiate outward from the central vein of lobules. At the outer edge of each lobule are several "triads," so named because they contain three structures: a portal vein, hepatic artery, and bile duct. The liver receives oxygenated blood from the hepatic arteries and deoxygenated blood from the portal veins that contains nutrients and other materials absorbed by the digestive system. Blood passes from the portal veins and hepatic arteries through sinusoids, where hepatocytes absorb some of the nutrients as well as any potentially toxic materials that might have been absorbed by the digestive system, to the central veins and, finally, to the hepatic vein.

The liver has many functions, including removing waste products from the blood, detoxification of potentially harmful substances, storing vitamins and glycogen (made by absorbing excess glucose from the blood), synthesizing plasma proteins, storing and synthesizing lipids, and producing bile. Bile is produced by hepatocytes, secreted into bile canaliculi, and transported to and stored in the gall bladder. Some species, such as parrots, some pigeons (species in the genera *Ptilinopus*, *Lopholaimus*, *Hemiphaga*, *Ducula*, and *Gymnophaps* have gall bladders; Hagey et al. 1994), and Ostriches lack a gall bladder, and bile is released directly into the small intestine.

(continued)

**Box 5.30** (continued)

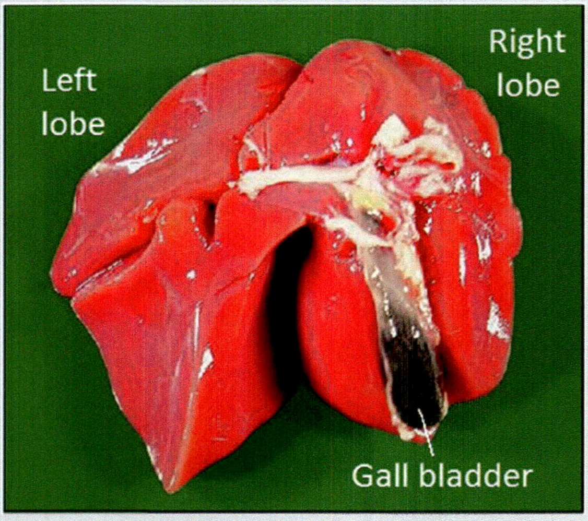


Liver and gall bladder of a Wild Turkey (*Meleagris gallopavo*) (Figure modified from Hünigen et al. 2016; © 2016 Oxford University Press, used with permission)

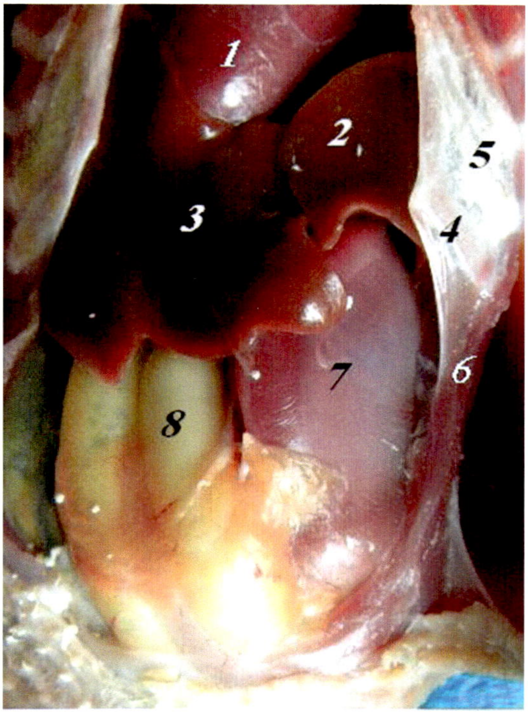


**Fig. 5.113** View into the thoracic and abdominal cavities of a Domestic Pigeon (*Columbo livia*) showing the heart (1), left lobe of liver (2), right lobe of liver (3), last two ribs (4,5), abdominal muscles (6), ventriculus (7), and duodenum (8). (Figure from Hassan and Moussa 2012; © 2012 African Association of Veterinary Anatomists, used with permission)

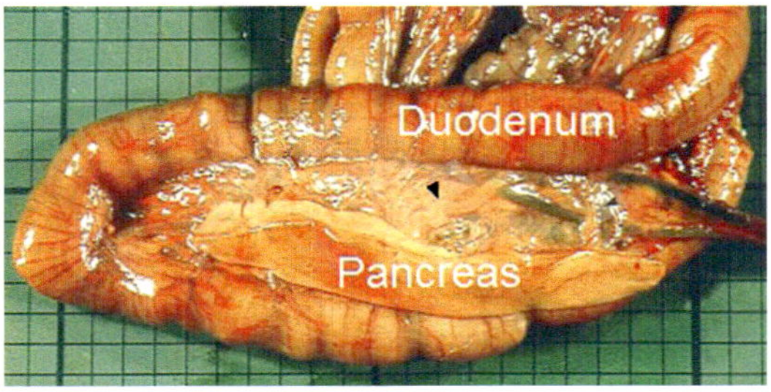


**Fig. 5.114** The pancreas is located between a looping section of the duodenum. A pancreatic duct (arrow) transports pancreatic secretions to the duodenum. (Figure modified from Bailey et al. 1997; © 1997 John Wiley and Sons, used with permission)

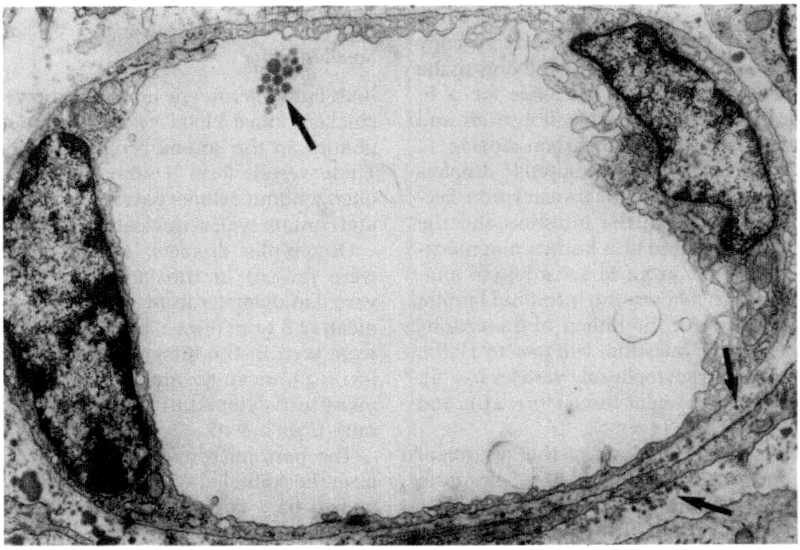

**Fig. 5.115** Electron micrograph of a portal capillary of a Domestic Chicken lined by endothelial cells. Portomicrons can be seen within the capillary lumen and in loose connective tissue surrounding the capillary (arrows). Portomicrons appear to enter the capillaries through intracytoplasmic vesicles in the endothelial cells rather than through gaps between those cells. (Figure from Fraser et al. 1986; © 1999 John Wiley & Sons, Inc., used with permission)

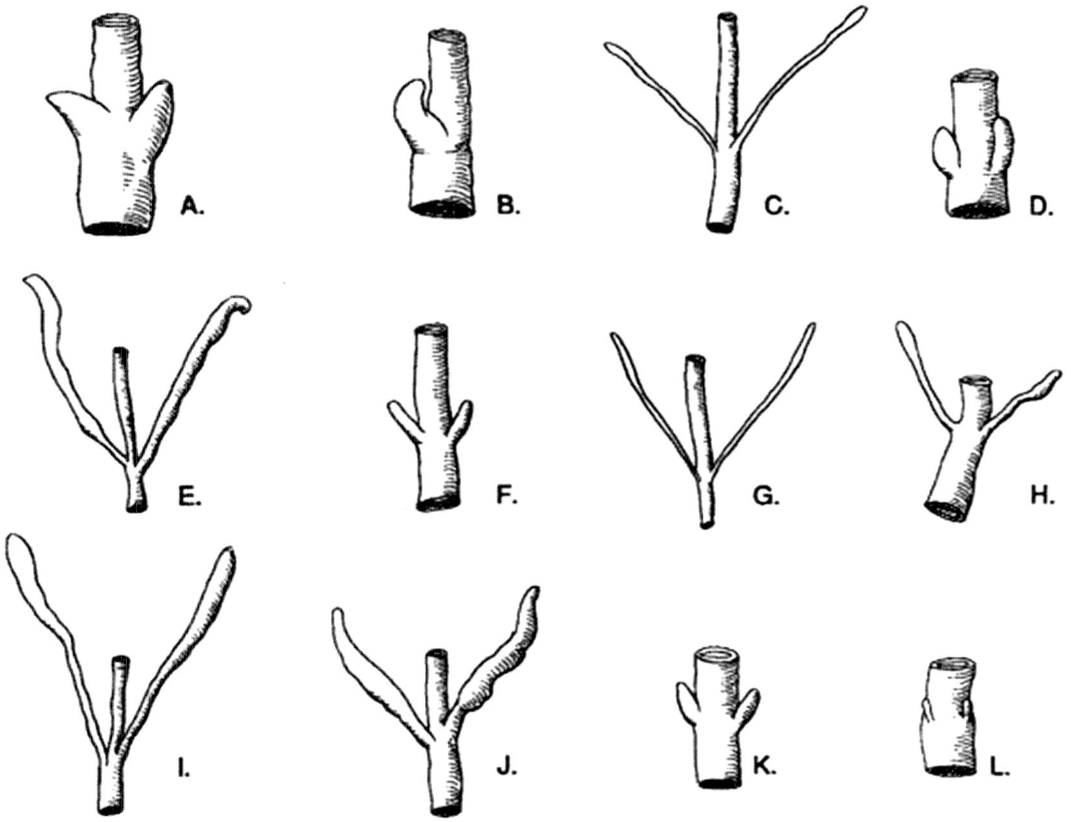


**Fig. 5.116** Variation in the morphology of avian ceca. (**a**) Little Cormorant, *Microcarbo niger*, (**b**) Cattle Egret, *Bubulcus ibis*, (**c**) Cotton Pygmy-Goose, *Nettapus coromandelianus*, (**d**) Crested Serpent-Eagle, *Spilornis cheela*, (**e**) Common Quail, *Coturnix coturnix*, (**f**) Eurasian Collared-Dove, *Streptopelia decaocto*, (**g**) Red-wattled Lapwing, *Vanellus indicus*, (**h**) Asian Koel, *Eudynamys scolopacea*, (**i**) Spotted Owlet, *Athene brama*, (**j**) Indian Roller, *Coracias benghalensis*, (**k**) Oriental Skylark, *Alauda gulgula*, and (**l**) Gray Wagtail, *Motacilla cinerea*. (Figure from Clench 1999; © 1999 John Wiley and Sons, used with permission)

**Box 5.31 Avian gut microbiota**

As with other vertebrates, the digestive tract of birds is home to a large community of microbes, with densities as high as $10^{11}$ colony-forming units/g in the ceca and large intestine (Barnes 1972). These microbes play essential roles in the digestive systems of birds. This microbiota of birds is dominated by four main phyla, including Firmicutes, Proteobacteria, Bacteroidetes, and Actinobacteria, with the predominance of each varying both intra- and interspecifically (Hird et al. 2015). The gut microbiota of birds can also vary seasonally (e.g., Greater Sage-Grouse, *Centrocercus urophasianus*; Drovetski et al. 2019) and with location (Hird et al. 2014).

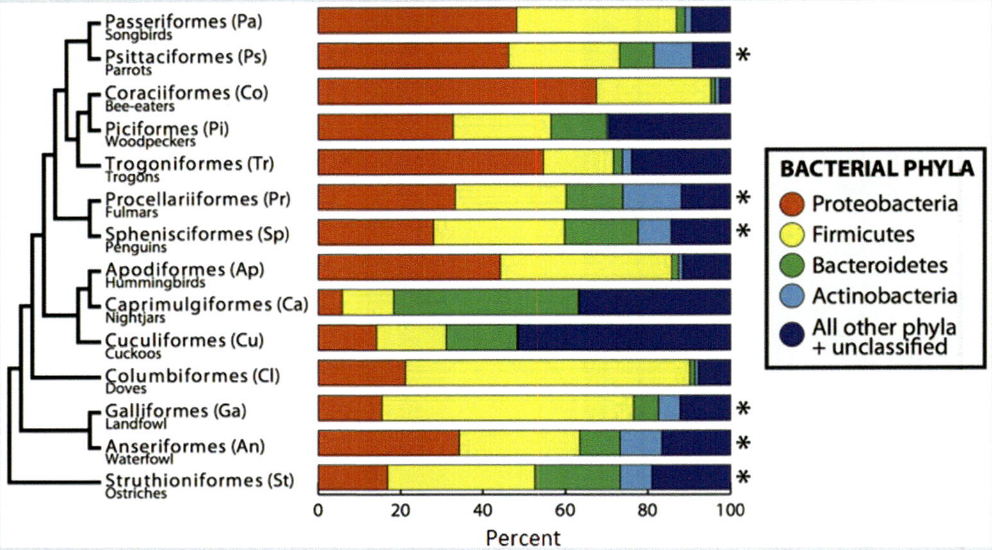


Phyla of bacteria in the digestive systems of 59 species of Neotropical birds in 14 different orders (Figure from Hird et al. 2015; © 2015 Hird, Sánchez, Carstens and Brumfield, open-access article distributed under the terms of the Creative Commons Attribution License (CC BY), https://creativecommons.org/licenses/by/4.0/. * indicates data obtained from Waite and Taylor 2014)

Studies of gut microbiota have largely focused on humans and other mammals. For example, Grond et al. (2018) conducted a literature review and found over 15,000 studies focusing on the gut microbiota of humans, and most studies of the gut microbiota of birds focused on domestic birds (i.e., chickens; ~1200 studies); only 32 studies focused on wild birds. Studies of the gut microbiota of chickens have revealed complex microbial communities, primarily bacteria, but also including fungi, protozoans, archaea, and viruses (Shang et al. 2018). More than 900 species of bacteria have been identified in the gastrointestinal tract of Domestic Chickens and these bacteria aid in digesting food, the breakdown of toxins, stimulating the immune system, and defending against pathogens (Borda-Molina et al. 2018). By attaching to the walls of the digestive tract, gut bacteria reduce the likelihood of colonization by pathogenic bacteria (Yegani and Korver 2008). Via their metabolism, bacteria produce fatty acids, amino acids, and vitamins, providing birds with important nutrients. Microbes in the ceca of chickens are known to play an important role in digesting cellulose and other polysaccharides (Borda-Molina et al. 2018).

Microbes are present along the entire length of avian digestive tracts, with different communities of microbes in different areas. In addition, gut microbiotas vary among species

(continued)

**Box 5.31** (continued)

as well as with age, geographic location, diet, and season. Although much remains to be learned about the gut microbiotas of wild birds, these microbiotas can clearly have a profound influence on their health.

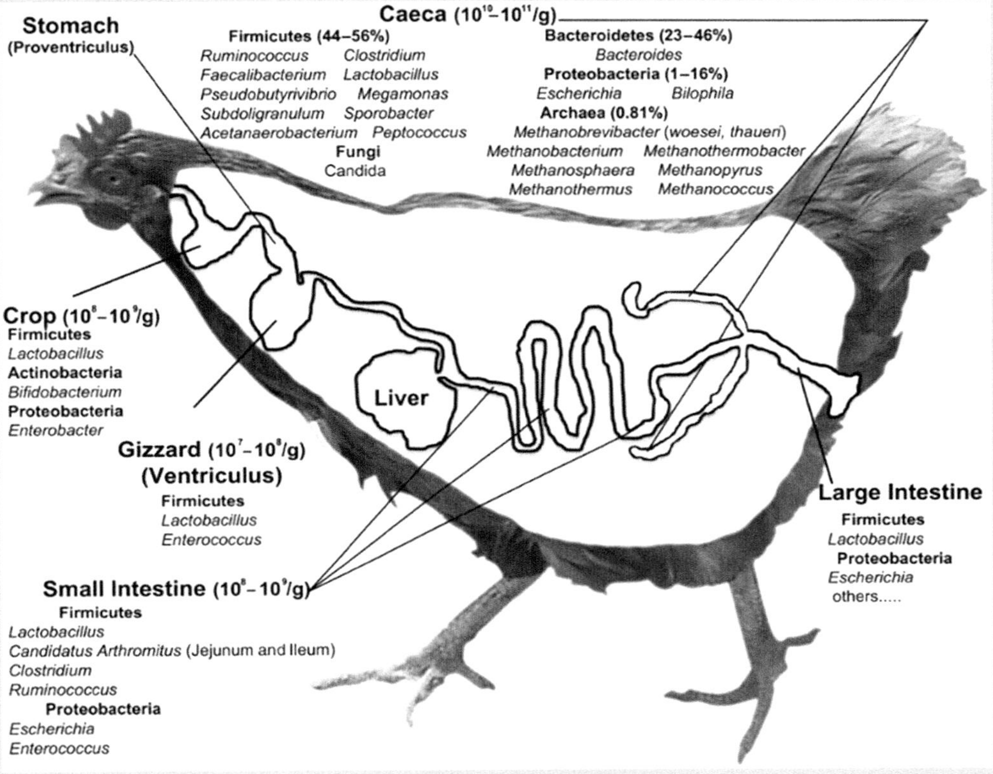


Some of the microbes found in different sections of the digestive tract of Domestic Chickens. In chickens, microbes are particularly abundant in the ceca, some of which aid in the digestion of cellulose (Figure from Yeoman et al. 2012; © 2012 Cambridge University Press, used with permission)

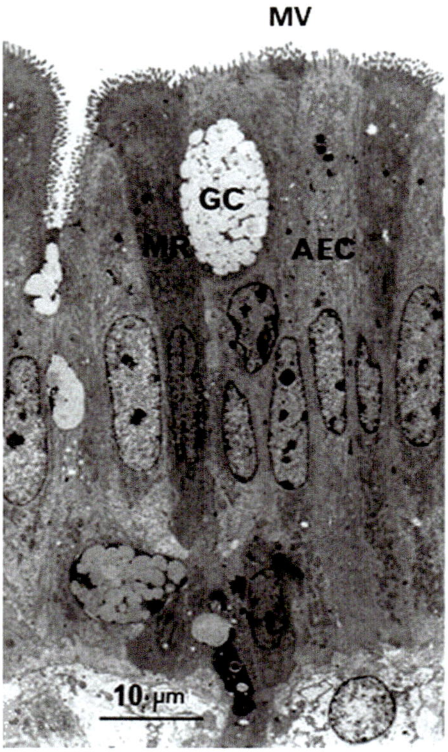


Fig. 5.117 Transmission electron micrograph image of a portion of the large intestinal wall of a Domestic Chicken. MV, microvilli; GC, goblet cell; MR, mitochondria-rich cell; AEC, absorptive epithelial cell. (Figure from Laverty et al. 2006; © 2006 Elsevier Ltd., used with permission)

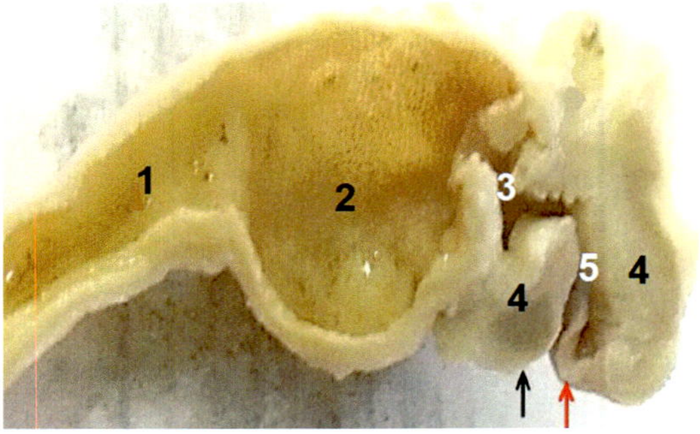


Fig. 5.118 Internal view of the cloaca of a Domestic Turkey (*Meleagris gallopavo domesticus*). (1) Last section of the large intestine, (2) coprodeum, (3) urodeum, (4) cloacal sphincter muscles, (5) proctodeum, and the dorsal labium (red arrow) and ventral labium (black arrow) of the cloaca. (Figure from Mohammed 2017; open-access article distributed under the terms of the Creative Commons Attribution License, https://creativecommons.org/licenses/by/4.0/)

**Table 5.8** Typical time food is retained in digestive systems after ingestion for birds with different diets[a]

| Diet category | Retention time (min) |
|---|---|
| Carnivore | 360–600 |
| Piscivore | 360–780 |
| Insectivore | 30–90 |
| Nectarivore | 30–50 |
| Granivore | 40–100 |
| Frugivore | 15–60 |
| Herbivore | |
| Flying | 50–300 |
| Nonflying | 300–1440 |

[a]Source: Klasing (1998)

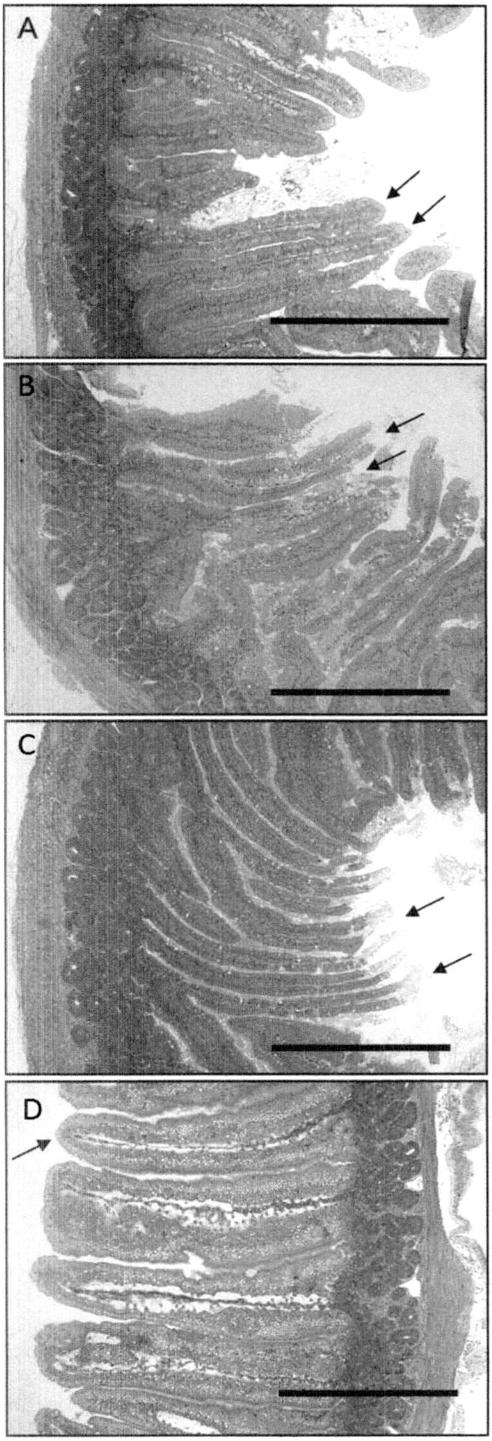


**Fig. 5.119** Changes in intestinal histology of Eurasian Blackcaps (*Sylvia atricapilla*) when fed versus when food restricted. (**a**) Well-fed bird. The arrows indicate areas of cell growth at the tip of the villi. (**b**) Fasted bird. Disintegration of the mucosal epithelium is apparent at the arrows. (**c**) Food-restricted bird. Villi resemble fasting

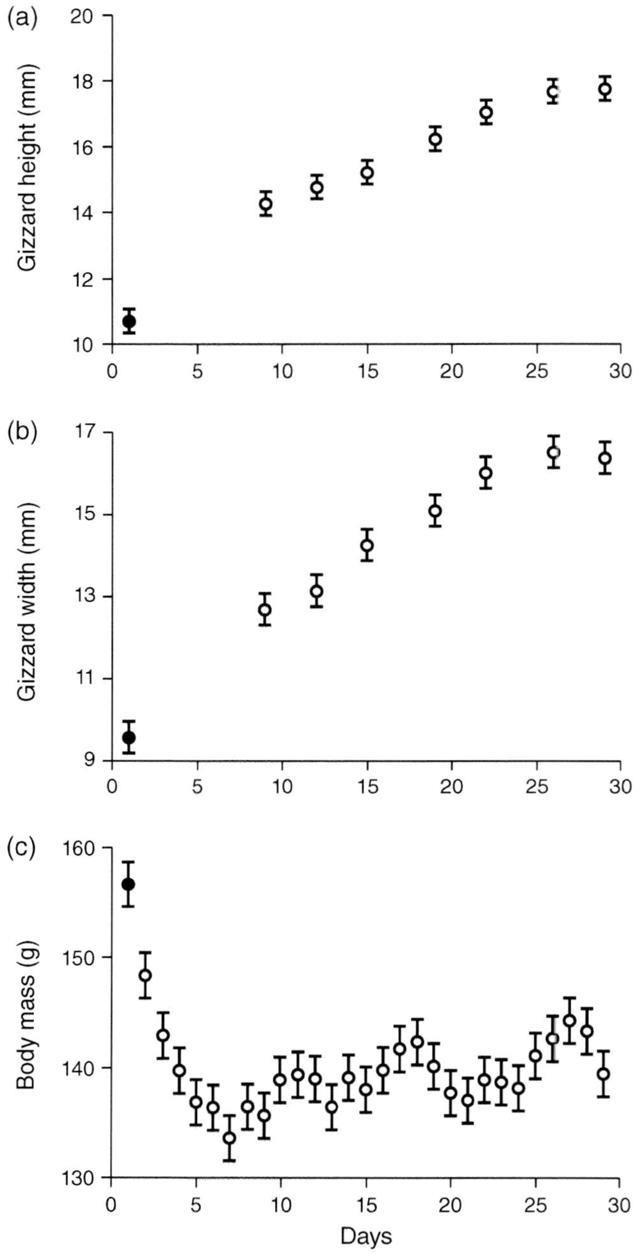


**Fig. 5.120** Variation in gizzard size and body mass of Red Knots (*Calidris canutus*) over a 29-day period following a change in diet from soft food (trout) to hard food (blue mussels). Filled circles on the far left of each graph represent values measured on the last day Red Knots were fed trout. Values are means ± SE. (Figure from Vézina et al. 2010; © 2010 The Authors, Published by John Wiley and Sons, used with permission)

**Fig. 5.119** (continued) birds; arrows indicate areas of villi disintegration. (**d**) Re-fed bird after fasting. Normal appearance of villi reestablished; extrusion zones reformed at the tip of villi (arrow). All micrographs are of the jejunum of the small intestine. Scale bars are 0.5 mm. (Figure modified from Karasov et al. 2004; © 2004 University of Chicago Press, used with permission)

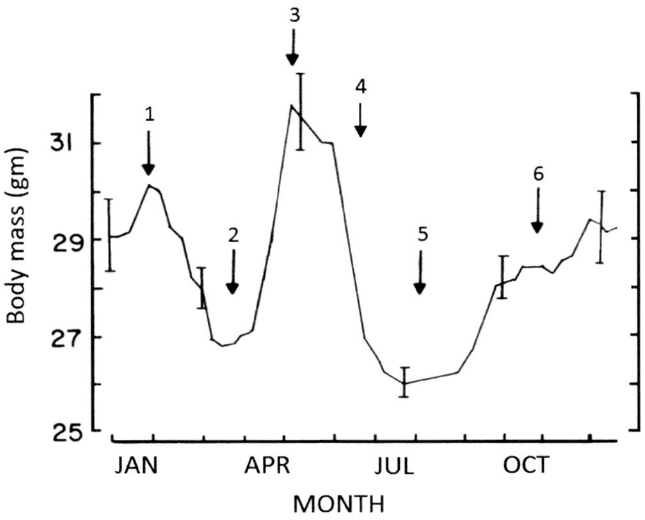


**Fig. 5.121** Annual variation in mean body mass (±SD) of male White-throated Sparrows (*Zonotrichia albicollis*; N = 8) held in an outdoor aviary in Pennsylvania. 1—Peak winter mass, 2—mass reduction during prenuptial molt, 3—mass prior to spring migration, 4—mass reduction during breeding, 5—further mass reduction during postnuptial molt, and 6—mass prior to fall migration. (Figure from Kuenzel et al. 1999; © 1999 John Wiley and Sons, used with permission)

## References


Afik D, Karasov WH (1995) The trade-offs between digestion rate and efficiency in warblers and their ecological implications. Ecology 76:2247–2257

Afik D, Vidal EC, Martinez del Rio C, Karasov WH (1995) Dietary modulation of intestinal hydrolytic enzymes in Yellow-rumped Warblers. Am J Physiol 269:R413–R420

Akester AR (1986) Structure of the glandular layer and koilin membrane in the gizzard of Domestic Fowl (*Gallus gallus domesticus*). J Anat 147:1–25

Alcock J (1970) Punishment levels and the response of Black-capped Chickadees (*Parus atricapillus*) to three kinds of artificial seeds. Anim Behav 18:592–599

Alonso H, Granadeiro JP, Paiva VH, Dias AS, Ramos JA, Catry P (2012) Parent-offspring dietary segregation of Cory's Shearwaters breeding in contrasting environments. Mar Biol 159:1197–1207

Alonso JC (1985) Grit in the gizzard of Spanish Sparrows (*Passer hispaniolensis*). Vogelwarte 33:135–143

Altshuler DL, Stiles FG, Dudley R (2004) Of hummingbirds and helicopters: hovering costs, competitive ability, and foraging strategies. Am Nat 163: 16–25

Amaral-Zettler LA, Zettler ER, Mincer TJ (2020) Ecology of the plastisphere. Nat Rev Microbiol 18:139–151

Anderson B, Cole WW, Barrett SCH (2005) Botany: specialized bird perch aids cross-pollination. Nature 435:41–42

Anderson EM, Lovvorn JR (2008) Gray whales may increase feeding opportunities for avian benthivores. Mar Ecol Prog Ser 360:291–296

Au DWK, Pitman RL (1986) Seabird associations with dolphins and tuna in the Eastern Tropical Pacific. Condor 88:304–317

Babbitt GA, Frederick PC (2007) Selection for sexual bill dimorphism in ibises: an evaluation of hypotheses. Waterbirds 30:199–206

Bachmann T, Klän S, Baumgartner W, Klaas M, Schröder W, Wagner H (2007) Morphometric characterisation of wing feathers of the Barn Owl *Tyto alba* and the pigeon *Columba livia*. Front Zool 4:23

Bachmann T, Wagner H, Tropea C (2012) Inner vane fringes on Barn Owl feathers reconsidered: morphometric data and functional aspects. J Anat 221:1–8

Backström N (2014) Adaptive evolution in passerine birds. In: eLS. Wiley, Chichester, UK

Bailey TA, Mensah-Brown EP, Samour JH, Naldo J, Lawrence P, Garner A (1997) Comparative morphology of the alimentary tract and its glandular derivatives of captive bustards. J Anat 191:387–398

Barcelos AO, Perônico C, Eutrôpio FJ (2012) Color and odor of artificial fruit used to signal potential dispersers in the Atlantic Forest in Brazil. Rev Biol Trop 60:925–931

Barnett AA, Shaw P (2014) More food or fewer predators? The benefits to birds of associating with a Neotropical primate varies with their foraging strategy. J Zool 294:223–233

Barja G, Cadenas S, Rojas C, Perez-Campo R, Lopez-Torres M (1994) Low mitochondrial free radical production per unit $O_2$ consumption can explain the simultaneous presence of high longevity and high aerobic metabolic rate in birds. Free Radic Res 21:317–327

Barnes EM (1972) The avian intestinal flora with particular reference to the possible ecological significance of the cecal anaerobic bacteria. Am J Clin Nutr 25:1475–1479

Barton NWH, Houston DC (1994) Morphological adaptation of the digestive tract in relation to feeding ecology of raptors. J Zool 232:133–150

Bartumeus F, Campos D, Ryu WS, Lloret-Cabot R, Méndez V, Catalan J (2016) Foraging success under uncertainty: search tradeoffs and optimal space use. Ecol Lett 19:1299

Baussart S, Bels V (2011) Tropical hornbills (*Aceros cassidix*, *Aceros undulatus*, and *Buceros hydrocorax*) use ballistic transport to feed with their large beaks. J Exp Zool A 315:72–83

Baussart S, Korsoun L, Libourel PA, Bels V (2009) Ballistic food transport in toucans. J Exp Zool A 311:465–474

Beaufrère H, Gardhouse S, Ammersbach M (2020) Lipoprotein characterization in Quaker Parrots (*Myiopsitta monachus*) using gel-permeation high-performance liquid chromatography. Vet Clin Pathol 49:417–427

Beaune D, Le Bohec C, Lucas F, Gauthier-Clerc M, Le Maho Y (2009) Stomach stones in King Penguin chicks. Polar Biol 32:593–597

Bednarz JC (1988) Cooperative hunting Harris' Hawks (*Parabuteo unicinctus*). Science 239:1525–1527

Benkman CW (1993) Adaptation of single resources and the evolution of crossbill (*Loxia*) diversity. Ecol Monogr 63:305–325

Best LB, Stafford TR (2002) Influence of daily grit consumption rate and diet on gizzard grit counts. J Wildl Manag 66:381–391

Bhushan B (2018) Skimmer bird beak (*Rynchops*) surface for fluid drag reduction in turbulent flow. In: Biomimetics. Springer, Cham, pp 563–576

Blake JG, Mosquera D, Guerra J, Loiselle BA, Romo D, Swing K (2011) Mineral licks as diversity hotspots in lowland forest of eastern Ecuador. Diversity 3:217–234

Borda-Molina D, Seifert J, Camarinha-Silva A (2018) Current perspectives of the chicken gastrointestinal tract and its microbiome. Comput Struct Biotechnol J 16:131–139

Brand-Miller JC, Colagiuri S (1999) Evolutionary aspects of diet and insulin resistance. Evol Aspects Nutr Health 84:74–105

Braun EJ, Sweazea KL (2008) Glucose regulation in birds. Comp Biochem Physiol B 151:1–9

Bribiesca R, Herrera-Alsina L, Ruiz-Sanchez E, Sánchez-González LA, Schondube JE (2019) Body mass as a supertrait linked to abundance and behavioral dominance in hummingbirds: a phylogenetic approach. Ecol Evol 9:1623–1637

Brice AT, Dahl KH, Grau CR (1989) Pollen digestibility in hummingbirds and psittacines. Condor 91:681–688

Brighton CH, Thomas AL, Taylor GK (2017) Terminal attack trajectories of Peregrine Falcons are described by the proportional navigation guidance law of missiles. Proc Natl Acad Sci USA 114:13495–13500

Brightsmith DJ, Hobson EA, Martinez G (2018) Food availability and breeding season as predictors of geophagy in Amazonian parrots. Ibis 160:112–129

Brun A, Gutiérrez-Guerrero Y, Magallanes ME, Caviedes-Vidal E, Karasov WH, Martínez del Rio C (2022) Opportunities lost? Evolutionary causes and ecological consequences of the absence of trehalose digestion in birds. Physiol Biochem Zool 95:340–349

Bucher EH, Tamburini D, Abril A, Torres F (2003) Folivory in the White-tipped Plantcutter *Phytotoma rutila*: seasonal variations in diet composition and quality. J Avian Biol 34:211–216

Burin G, Kissling WD, Guimarães PR, Şekercioğlu ÇH, Quental TB (2016) Omnivory in birds is a macroevolutionary sink. Nat Commun 7:11250

Burns KC (2013) What causes size coupling in fruit–frugivore interaction webs? Ecology 94:295–300

Bush JWM, Peaudecerf F, Prakash M, Quéré D (2010) On a tweezer for droplets. Adv Colloid Interf Sci 161:10–14

Capuska GEM, Vaughn RL, Würsig B, Katzir G, Raubenheimer D (2011) Dive strategies and foraging effort in the Australian Gannet *Morus serrator* revealed by underwater videography. Mar Ecol Prog Ser 442:255–261

Carl RA (1987) Age-class variation in foraging techniques by Brown Pelicans. Condor 89:525–533

Camargo MGG, Cazetta E, Schaefer HM, Morellato LPC (2013) Fruit color and contrast in seasonal habitats – a case study from a Cerrado savanna. Oikos 22:1335–1342

Camphuysen CJ, Webb A (1999) Multi-species feeding associations in North Sea seabirds: jointly exploiting a patchy environment. Ardea 87:177–198

Casotti G (2001) Luminal morphology of the avian lower intestine: evidence supporting the importance of retrograde peristalsis for water conservation. Anat Rec 263:289–296

Catroxo MHB, Lima MAI, Cappellaro CEMPDM (1997) Histological aspects of the stomach (proventriculus and gizzard) of the Red-capped Cardinal (*Faroaria gularris gularis*). Rev Chil Anat 15:19–27

Caudell JM, Conover MR (2006) Energy content and digestibility of brine shrimp (*Artemia franciscana*) and other prey items of Eared Grebes (*Podiceps nigricollis*) on the Great Salt Lake, Utah. Biol Conserv 130:251–254

Caviedes-Vidal E, McWhorter TJ, Lavin SR, Chedlack JG, Tracy CR, Karasov WH (2007) The digestive adaptation of flying vertebrates: high intestinal paracellular absorption compensates for smaller guts. Proc Natl Acad Sci USA 104:19132–19137

Cazetta E, Galetti M, Rezende EL, Schaefer HM (2012) On the reliability of visual communication in vertebrate-dispersed fruits. J Ecol 100:277–286

Chai P (1996) Butterfly visual characteristics and ontogeny of responses to butterflies by a specialized tropical bird. Biol J Linn Soc 59:37–67

Chang Y-M, Lin H-Y, Hatch KA, Yao C-T, Shiu H-J (2013) Brush-tipped tongue structure of the Taiwan Yuhina (*Yuhina brunneiceps*) and White-eared Sibia (*Heterophasia auricularis*). Wilson J Ornithol 125:204–208

Cheke RA, Mann CF, Allen R (2001) Sunbirds: a guide to the sunbirds, flowerpeckers, spiderhunters and sugarbirds of the world. Christopher Helm, London

Chen K, Liu Q, Liao G, Yang Y, Ren L, Yang H, Chen X (2012) The sound suppression characteristics of wing feather of owl (*Bubo bubo*). J Bionic Eng 9:192–199

Chen Y-H, Zhao H (2019) Evolution of digestive enzymes and dietary diversification in birds. PeerJ 7:e6840

Chikilian M, de Speroni NB (1996) Comparative study of the digestive system of three species of tinamou. I. *Crypturellus tataupa*, *Nothoprocta cinerascens*, and *Nothura maculosa* (Aves: Tinamidae). J Morphol 228:77–88

Ciminari ME, Afik D, Karasov WH, Caviedes-Vidal E (2001) Is diet-shifting facilitated by modulation of pancreatic enzymes? Test of an adaptational hypothesis in Yellow-rumped Warblers. Auk 118:1101–1107

Clark L, Hagelin J, Werner S (2015) The chemical senses of birds. In: Scanes CG (ed) Sturkie's avian physiology. Academic Press, Waltham, MA, pp 89–111

Clarke SJ, Miller GH, Fogel ML, Chivas AR, Murray-Wallace CV (2006) The amino acid and stable isotope biogeochemistry of elephant bird (*Aepyornis*) eggshells from southern Madagascar. Quat Sci Rev 25:2343–2356

Clayton DH, Moyer BR, Bush SE, Jones TG, Gardiner DW, Rhodes BB, Goller F (2005) Adaptive significance of avian beak morphology for ectoparasite control. Proc R Soc B 272:811–817

Clench MH (1999) The avian cecum: update and motility review. J Exp Zool 283:441–447

Clua É, Grosvalet F (2001) Mixed-species feeding aggregation of dolphins, large tunas and seabirds in the Azores. Aquat Living Resour 14:11–18

Cody ML (1971) Finch flocks in the Mohave desert. Theor Popul Biol 2:141–158

Cohen A, Klasing K, Ricklefs R (2007) Measuring circulating antioxidants in wild birds. Comp Biochem Physiol B 147:110–121

Connan M, Cherel Y, Mabille G, Mayzaud P (2007) Trophic relationships of White-chinned Petrels from Crozet Islands: combined stomach oil and conventional dietary analyses. Mar Biol 152:95–107

Connan M, Mayzaud P, Boutoute M, Weimerskirch H, Cherel Y (2005) Lipid composition of stomach oil in a procellariiform seabird *Puffinus tenuirostris*: implications for food web studies. Mar Ecol Prog Ser 290:277–290

Côrtes MC, Uriarte M (2013) Integrating frugivory and animal movement: a review of the evidence and implications for scaling seed dispersal. Biol Rev 88: 255–272

Cotton PA (1996) Body size and the ecology of hummingbirds. Symp Zool Soc Lond 69:239–258

Cristol DA, Switzer PV (1999) Avian prey-dropping behavior. II. American Crows and walnuts. Behav Ecol 10:220–226

Crole MR (2011) A gross anatomical and histological study of the oropharynx and proximal esophagus of the Emu (*Dromaius novaehollandiae*). Ph.D. dissertation, University of Pretoria, Pretoria, South Africa

Cronk Q, Ojeda I (2008) Bird-pollinated flowers in an evolutionary and molecular context. J Exp Bot 59: 715–727

Csermely D, Berti L, Camoni R (1998) Prey killing by Eurasian Kestrels: the role of the foot and the significance of bill and talons. J Avian Biol 29:10–16

Davenport J, O'Callaghan MJA, Davenport JL, Kelly TC (2014) Mussel dropping by Carrion and Hooded crows: biomechanical and energetic considerations. J Field Ornithol 85:196–205

Davies S (1978) The food of Emus. Aust J Ecol 3:411–422

Dean WRJ, MacDonald IAW (1981) A review of African birds feeding in association with mammals. Ostrich 52: 135–155

de Araujo FV, Castro RO, da Silva ML, Silva MM (2021) Ecotoxological effects of microplastics and associated pollutants. In: Kibenge FSB, Baldisserotto B, Chong RS (eds) Aquaculture toxicology. Elsevier, London, pp 189–227

DeGolier TF, Mahoney SA, Duke GE (1999) Relationships of avian cecal lengths to food habits, taxonomic position, and intestinal lengths. Condor 101:622–634

De Graaf RM, Tilghman NG, Anderson SH (1985) Foraging guilds of North American birds. Environ Manag 9: 493–536

De Leon LF, Raeymaekers JAM, Bermingham E, Podos J, Herrel A, Hendry AP (2011) Exploring possible human influences on the evolution of Darwin's finches. Evolution 65:2258–2272

Dellinger R, Wood PB, Jones PW, Donovan TM (2020) Hermit Thrush (*Catharus guttatus*), version 1.0. In: Poole AF (ed) Birds of the World. Cornell Lab of Ornithology, Ithaca, NY

Derraik JGB (2002) The pollution of the marine environment by plastic debris: a review. Mar Pollut Bull 44: 842–852

DeVault TL, Rhodes OE, Shivik JA (2003) Scavenging by vertebrates: behavioral, ecological, and evolutionary perspectives on an important energy transfer pathway in terrestrial ecosystems. Oikos 102:225–234

Diamond AW, Place AR (1988) Wax digestion by Black-throated Honey-guides *Indicator*. Ibis 130:558–561

Diamond DL, Carruthers A (1993) Metabolic control of sugar transport by derepression of cell surface glucose transporters. J Biol Chem 268:6437–6444

Diamond J (1991) Evolutionary design of intestinal nutrient absorption: enough but not too much. News Physiol Sci 6:92–96

Diamond J, Bishop KD, Gilardi JD (1999) Geophagy in New Guinea birds. Ibis 141:181–193

Díaz M (1990) Interspecific patterns of seed selection among granivorous passerines: effects of seed size, seed nutritive value and bird morphology. Ibis 132: 467–476

Díaz M (1996) Food choice by seed-eating birds in relation to seed chemistry. Comp Biochem Physiol 113A:239–246

Dinsmore JJ (1973) Foraging success of Cattle Egrets, *Bubulcus ibis*. Am Midl Nat 89:242–246

Doss GA, Williams JM, Mans C (2017) Determination of gastrointestinal transit times in Barred Owls (*Strix varia*) by contrast fluoroscopy. J Avian Med Surg 31: 123–127

Downs CT (2004) Some preliminary results of studies on the bill and tongue morphology of Gurney's Sunbirds and some southern African sunbirds. Ostrich 75:169–175

Downs CT, Bredin IP, Wragg PD (2019) More than eating dirt: a review of avian geophagy. Afr Zool 54: 1–19

Drovetski SV, O'Mahoney MJV, Matterson KO, Schmidt BK, Graves GR (2019) Distinct microbiotas of anatomical gut regions display idiosyncratic seasonal variation in an avian folivore. Anim Microbiome 1:2

Duan Q, Goodale E, Quan R-C (2014) Bird fruit preferences match the frequency of fruit colours in tropical Asia. Sci Rep 4:5627

Duke GE (1986) Alimentary canal: secretion and digestion, special digestive functions, and absorption. In: Sturkie PD (ed) Avian physiology. Springer, New York, NY, pp 289–302

Duke GE (1994) Anatomy and physiology of the digestive system in fowl. In: Proceedings of the 21st Annual Carolina Poultry Nutrition Conference. Charlotte, NC, pp 46–49

Dupont J, Métayer-Coustard S, Ji B, Ramé C, Gespach C, Voy B, Simon J (2012) Characterization of major elements of insulin signaling cascade in chicken adipose tissue: apparent insulin refractoriness. Gen Comp Endocrinol 176:86–93

Dupont J, Tesseraud S, Simon J (2009) Insulin signaling in chicken liver and muscle. Gen Comp Endocrinol 163: 52–57

Duque-Correa MJ, Clauss M, Hoppe MI, Buyse K, Codron D, Meloro C, Edwards MS (2022) Diet, habitat and flight characteristics correlate with intestine length in birds. Proc R Soc B 289:20220675

Duque-Correa MJ, Codron D, Meloro C, McGrosky A, Schiffmann C, Edwards MS, Clauss M (2021) Mammalian intestinal allometry, phylogeny, trophic level and climate. Proc R Soc B 288:20202888

Eliason CM, Straker L, Jung S, Hackett SJ (2020) Morphological innovation and biomechanical diversity in plunge-diving birds. Evolution 74:1514–1524

Elliser CR, Edison C, MacIver K, Rust LB (2022) Kleptoparasitic interactions by Bald Eagles (*Haliaeetus leucocephalus*) during marine mammal foraging events. Behaviour 159:1387–1399

Elner RW, Beninger PG, Jackson DL, Potter TM (2005) Evidence of a new feeding mode in Western Sandpiper and Dunlin based on bill and tongue morphology and ultrastructure. Mar Biol 146:1223–1234

Else PL, Brand MD, Turner N, Hulbert AJ (2004) Respiration rate of hepatocytes varies with body mass in birds. J Exp Biol 207:2305–2311

Emura S, Okumura T, Chen H (2011) Scanning electron microscopic study of the tongue in the Rainbow Lorikeet (*Trichoglossus haematodus*). Okajimas Folia Anat Jpn 88:17–21

Emura S, Okumura T, Chen H (2012) Scanning electron microscopic study on the tongue in the Scarlet Macaw (*Ara macao*). Okajimas Folia Anat Jpn 89:57–60

Engler RE (2012) The complex interaction between marine debris and toxic chemicals in the ocean. Environ Sci Technol 46:12302–12315

Enoki Y, Morimoto T (2000) Gizzard myoglobin contents and feeding habits in avian species. Comp Biochem Physiol A 125:33–43

Eraud C, Dorie A, Jacquet A, Faivre B (2008) The crop milk: a potential new route for carotenoid-mediated parental effects. J Avian Biol 39:247–251

Erdoğan S, Iwasaki S (2014) Function-related morphological characteristics and specialized structures of the avian tongue. Ann Anat 196:75–87

Ewald PW, Carpenter FL (1978) Territorial responses to energy manipulations in the Anna Hummingbird. Oecologia 31:277–292

Fadzly N, Burns KC, Zuharah WF (2013) Evaluating frugivore-fruit interactions using avian eye modeling. Trop Life Sci Res 24:31–50

Farah G, Siwek D, Cummings P (2018) Tau accumulations in the brains of woodpeckers. PLoS ONE 13:e0191526

Feinsinger P (1976) Organization of a tropical guild of nectarivorous birds. Ecol Monogr 46:257–291

Feinsinger P (1983) Coevolution and pollination. In: Futuyma DJ, Slatkin M (eds) Coevolution. Sunderland, MA, Sinauer, pp 282–310

Feinsinger P (1987) Approaches to nectarivore-plant interactions in the New World. Rev Chil Hist Nat 60: 285–319

Fenster CB, Armbruster WC, Wilson P, Dudash MR, Thomson JD (2004) Pollination syndromes and floral specialization. Annu Rev Ecol Evol Syst 35:375–403

Ferguson-Lees J, Christie DA (2001) Raptors of the world. Houghton Mifflin, New York, NY

Fitzpatrick JW (1985) Form, foraging behavior, and adaptive radiation in the Tyrannidae. Ornithol Monogr 36: 447–470

Fleming TH, Muchhala N (2008) Nectar-feeding bird and bat niches in two worlds: pantropical comparisons of vertebrate pollination systems. J Biogeogr 35:764–780

Flower TP, Child M, Ridley AR (2013) The ecological economics of kleptoparasitism: pay-offs from self-foraging versus kleptoparasitism. J Anim Ecol 82: 245–255

Flower TP, Gribble M (2012) Kleptoparasitism by attacks versus false alarms in the Fork-tailed Drongo. Anim Behav 83:403–410

Foster WA, Treherne JE (1981) Evidence for the dilution effect in the selfish herd from dish predation on a marine insect. Nature 293:466–467

Fowler DW, Freedman EA, Scannella JB (2009) Predatory functional morphology in raptors: interdigital variation in talon size is related to prey restraint and immobilization technique. PLoS ONE 4:e7999

Fox AG, Young RF (2012) Foraging interactions between wading birds and strand-feeding bottlenose dolphins (*Tursips truncates*) in a coastal salt marsh. Can J Zool 90:744–752

Fox N (1976) Rangle. Raptor Res 10:61–64

Fox N (1995) Understanding the bird of prey. Hancock House, Surrey, BC

Fraser R, Heslop VR, Murray FE, Day WA (1986) Ultrastructural studies of the portal transport of fat in chickens. Br J Exp Pathol 67:783–791

Fraticelli F (2014) Prey-dropping behavior in Sooty Gull, *Larus hemprichii*. Avocetta 38:89

Fritz JA, Brancale J, Tokita M, Burns KJ, Hawkins MB, Abzhanov A, Brenner MP (2014) Shared developmental programme strongly constrains beak shape diversity in songbirds. Nat Commun 5:3700

Fritz J, Hummel J, Kienzle E, Wings O, Streich WJ, Clauss M (2011) Gizzard vs. teeth, it's a tie: food-processing efficiency in herbivorous birds and mammals and implications for dinosaur feeding strategies. Paleobiology 37:577–586

Galetti M, Guevara R, Côrtes MC, Fadini R, Von Matter S, Leite AB, Labecca F, Ribeiro T, Carvalho CS, Collevatti RG, Pires MM, Guimarães PR Jr, Brancalion PH, Ribeiro MC, Jordano P (2013) Functional extinction of birds drives rapid evolutionary changes in seed size. Science 340:1086–1090

Galetto L (1995) Nectary structure and nectar characteristics in some Bignoniaceae. Plant Syst Evol 196:99–121

Galluzzo FG, Cumbo V, Cammilleri G, Calabrese V, Pulvirenti A, Cicero N, Pantano L, Mascetti A, Lo Cascio G, Bacchi E, Macaluso A (2022) Fatty acids composition of stomach oil of Scopoli's Shearwater (*Calonectris diomedea*) from Linosa's Colony. Animals 12:1069

Ganchrow D, Ganchrow JR (1985) Number and distribution of taste buds in the oral cavity of hatchling chicks. Physiol Behav 34:889–894

Gardner LL (1925) The adaptive modifications and the taxonomic value of the tongue in birds. In: Proceedings of the United States National Museum 67. Washington, DC

Garthe S, Benvenuti S, Montevecchi WA (2000) Pursuit plunging by Northern Gannets (*Sula bassana*) feeding on capelin (*Mallotus villosus*). Proc R Soc B 267: 1717–1722

Gass J, Vora H, Hofmann AF, Gray GM, Khosla C (2007) Enhancement of dietary protein digestion by conjugated bile acids. Gastroenterology 133:16–23

Gatti RC, Ugarkovic P, Tiralongo F (2021) New evidence of a fish–bird interspecific feeding association between the European seabass and the European Shag in the Mediterranean Sea. Aquat Ecol 55:1113–1119

Gauthier-Clerc M, Le Maho Y, Clerquin Y, Drault S, Handrich Y (2000) Penguin fathers preserve food for their chicks. Nature 408:928–929

Gee JM, Johnson IT (1988) Interactions between haemolytic saponins, bile salts, and small intestinal mucosa in the rat. J Nutr 118:1391–1397

Genbrugge A, Herrel A, Boone M, Van Hoorebeke L, Podos J, Dirckx J, Aerts P, Dominque A (2011) The head of the finch: the anatomy of the feeding system in two species of finches (*Geospiza fortis* and *Padda oryzivora*). J Anat 219:676–695

Genbrugge A, Adriaens D, De Kegel B, Brabant L, Van Hoorebeke L, Podos J, Dirckx J, Aerts P, Herrel A (2012) Structural tissue organization in the beak of Java and Darwin's finches. J Anat 221:383–393

Gilardi JD (1994) Great Frigatebird kleptoparasitism: sex-specific host choice and age-related proficiency. Condor 96:987–993

Gilardi JD, Duffey SS, Munn CA, Tell LA (1999) Biochemical functions of geophagy in parrots: detoxification of dietary toxins and cytoprotective effects. J Chem Ecol 25:897–922

Gill F, Donskar D, Rasmussen P (eds) (2021) IOC World Bird List, version 11.2. [online]. www.worldbirdnames.org/new/

Gillespie MJ, Haring VR, McColl KA, Monaghan P, Donald JA, Nicholas KR, Moore RJ, Crowley TM (2011) Histological and global gene expression analysis of the 'lactating' pigeon crop. BMC Genomics 12: 452

Gionfriddo JP, Best LB (1995) Grit use by House Sparrows: effects of diet and grit size. Condor 97:57–67

Gionfriddo JP, Best LB (1996) Grit-use patterns in North American birds: the influence of diet, body size, and gender. Wilson Bull 108:685–696

Girard AR, Kabamba PT (2015) Proportional navigation: optimal homing and optimal evasion. SIAM Rev 57: 611–624

Glen CL, Bennett MB (2007) Foraging modes of Mesozoic birds and non-avian theropods. Curr Biol 17: R911–R912

Goldberg M, Gomez-Orellana I (2003) Challenges for the oral delivery of macromolecules. Nat Rev Drug Discov 2:289–295

González-Salazar C, Martínez-Meyer E, López-Santiago G (2014) A hierarchical classification of trophic guilds for North American birds and mammals. Revista Mexicana de Biodiversidad 85:931–941

Gordillo M, Evans T, Gouon-Evans V (2015) Orchestrating liver development. 142:2094–2108

Gordon IJ, Prins HHT, Mallon J, Puk LD, Miranda EB, Starling-Manne C, van der Wal R, Moore B, Foley W, Lush L, Maestri R (2019) The ecology of browsing and grazing in other vertebrate taxa. In: Gordon IJ, Prins HHT (eds) The ecology of browsing and grazing II. Springer, Cham, pp 339–404

Gosner KL (1993) Scopate tomia: an adaptation for handling hard-shelled prey? Wilson Bull 105:316–324

Goudswaard J, van der Donk J, van der Gaag I, Noordzij A (1979) Peculiar IgA transfer in the pigeon from mother to squab. Dev Comp Immunol 3:307–319
Grajal A (1995) Structure and function of the digestive tract of the Hoatzin (*Opisthocomus hoazin*): a folivorous bird with foregut fermentation. Auk 112: 20–28
Grajal A, Strahl SD, Parra R, DomÍnguez MG, Neher A (1989) Foregut fermentation in the Hoatzin, a Neotropical avian folivore. Science 245:1131–1134
Grebmeir JM, Harrison NM (1992) Seabird feeding on benthic amphipods facilitated by gray whale activity in the northern Bering Sea. Mar Ecol Prog Ser 80:125–133
Greenberg R (1981) Dissimilar bill shapes in New World tropical versus temperate forest foliage-gleaning birds. Oecologia 49:143–147
Grémillet D, Prudor A, le Maho Y, Weimerskirch H (2012) Vultures of the seas: hyperacidic stomachs in Wandering Albatrosses as an adaptation to dispersed food resources, including fishery waste. PLoS ONE 7: e37834
Grond K, Sandercock BK, Jumpponen A, Zeglin LH (2018) The avian gut microbiota: community, physiology and function in wild birds. J Avian Biol 49:e01788
Guse N (2013) Habitat selection and foraging ecology of seabirds in the Gulf of St. Lawrence. Ph.D. dissertation, Christian-Albrechts-Universität zu Kiel, Kiel, Germany
Hagey LR, Schteingart CD, Ton-Nu HT, Hofmann AF (1994) Biliary bile acids of fruit pigeons and doves (Columbiformes): presence of 1-beta-hydroxychenodeoxycholic acid and conjugation with glycine as well as taurine. J Lipid Res 35:2041–2048
Hamilton RB (1975) Comparative behavior of the American Avocet and the Black-necked Stilt (Recurvirostridae). Ornithol Monogr 17:1–98
Hanzák J (1952) The Great Crested Grebe, *Podiceps c. cristatus* (L.), its ecology and economic significance. Sbornik Narodniho Musea v Praze 8:1–37
Harris T (ed) (2009) Complete birds of the world. National Geographic Society, Washington, DC
Harrison NH, Whitehouse MJ, Heinemann D, Prince PA, Hunt GL Jr, Veit RR (1991) Observations of multispecies seabird flocks around South Georgia. Auk 108: 801–810
Hassan SA, Moussa EA (2012) Gross and microscopic studies on the stomach of Domestic Duck (*Anas platyrhynchos*) and Domestic Pigeon (*Columba livia domestica*). J Vet Anat 5:105–127
Hebshi AJ, Duffy DC, Hyrenbach KD (2008) Associations between seabirds and subsurface predators around Oahu, Hawaii. Aquat Biol 4:89–98
Herd RM, Dawson TJ (1984) Fiber digestion in the Emu, *Dromaius novaehollandiae*, a large bird with a simple gut and high rates of passage. Physiol Zool 57:70–84
Hermier D (1997) Lipoprotein metabolism and fattening in poultry. J Nutr 127:805S–808S
Herrero A, Barja G (1998) $H_2O_2$ production of heart mitochondria and aging rate are slower in canaries and parakeets than in mice: sites of free radical generation and mechanisms involved. Mech Ageing Dev 103: 133–146
Hertel F (1994) Diversity in body size and feeding morphology within past and present vulture assemblages. Ecology 75:1074–1084
Hesp LS, Barnard CJ (1989) Gulls and plovers: age-related differences in kleptoparasitism among Black-headed Gulls (*Larus ridibundus*). Behav Ecol Sociobiol 24: 297–304
Hickey AJR, Jűllig M, Aitken J, Loomes K, Hauber ME, Phillips ARJ (2012) Birds and longevity: does flight driven aerobicity provide an oxidative sink? Ageing Res Rev 11:242–253
Hilton GM, Houston DC, Barton NWH, Furness RW, Ruxton GD (1999) Ecological constraints on digestive physiology in carnivorous and piscivorous birds. J Exp Zool 283:365–376
Hird SM, Carstens BC, Cardiff S, Brumfield RT (2014) Sampling locality is more detectable than taxonomy or ecology in the gut microbiota of the brood-parasitic Brown-headed Cowbird (*Molothrus ater*). PeerJ 2: e321
Hird SM, Sánchez C, Carstens BC, Brumfield RT (2015) Comparative gut microbiota of 59 Neotropical bird species. Front Microbiol 6:1403
Holmes DJ, Austad SN (1995a) The evolution of avian senescence patterns: implications for understanding primary aging processes. Am Zool 35:307–317
Holmes DJ, Austad SN (1995b) Birds as animal models for the comparative biology of aging: a prospectus. J Gerontol Ser A Biol Med Sci 50:B59–B66
Holmes DJ, Flückiger R, Austad SN (2001) Comparative biology of aging in birds: an update. Exp Gerontol 36: 869–883
Hon J, Shibata S, Samejima H (2020) Species composition and use of natural salt licks by wildlife inside a production forest environment in central Sarawak. In: Ishikawa N, Soda R (eds) Anthropogenic tropical forests. Springer, Singapore, pp 171–180
Houston DC (1986) Scavenging efficiency of Turkey Vultures in tropical forest. Condor 88:318–323
Houston DC, Cooper JE (1975) The digestive tract of the Whiteback Griffon Vulture and its role in disease transmission among wild ungulates. J Wildl Dis 11: 306–313
Howe HF (1980) Monkey dispersal and waste of a neotropical fruit. Ecology 61:944–959
Hulan HW, Bird FH (1972) Effect of fat level in isonitrogenous diets on the composition of avian pancreatic juice. J Nutr 102:459–468
Hulan HW, Moreau G, Bird FH (1972) A method for cannulating the main pancreatic duct of chickens: the continuous collection of avian pancreatic juice. Poult Sci 51:531–536
Hünigen H, Mainzer K, Hirschberg RM, Custodis P, Gemeinhardt O, Al Masri S, Richardson KC, Hafez

HM, Plendl J (2016) Structure and age-dependent development of the turkey liver: a comparative study of a highly selected meat-type and a wild-type turkey line. Poult Sci 95:901–911

Hunt A, Al-Nakkash L, Lee AH, Smith HF (2019) Phylogeny and herbivory are related to avian cecal size. Sci Rep 9:4243

Igic B, Nguyen I, Fenberg PB (2020) Nectar robbing in the trainbearers (*Lesbia*, Trochilidae). PeerJ 8:e9561

Ingram T, Zuck J, Borges CR, Redig P, Sweazea KL (2017) Variations in native protein glycation and plasma antioxidants in several birds of prey. Comp Biochem Physiol B 210:18–28

Inouye DW (1983) The ecology of nectar robbing. In: Beattey B, Elias T (eds) The biology of nectaries. Columbia University Press, New York, NY, pp 153–173

Irwin RE, Bronstein JL, Manson JS, Richardson L (2010) Nectar robbing: ecological and evolutionary perspectives. Annu Rev Ecol Evol Syst 41:271–292

Isler ML, Isler PR (1987) The tanagers: natural history, distribution, and identification. Smithsonian Institution Press, Washington, DC

Ito S (2009) Aerodynamic influence of leading-edge serrations on an airfoil in a low Reynolds number – a study of an owl wing with leading edge serrations. J Biochem Sci Eng 4:117–123

Iwasaki S (2002) Evolution of the structure and function of the vertebrate tongue. J Anat 201:1–14

Jackson CR, Maddox T, Mbise FP, Stokke BG, Belant JL, Bevanger K, Durant SM, Fyumagwa R, Ranke PS, Røskaft E, May R (2020) A dead giveaway: foraging vultures and other avian scavengers respond to auditory cues. Ecol Evol 10:6769–6774

Jackson S, Place AR, Seiderer LJ (1992) Chitin digestion and assimilation by seabirds. Auk 109:758–770

James HF, Burney DA (1997) The diet and ecology of Hawaii's extinct flightless waterfowl: evidence from coprolites. Biol J Linn Soc 62:279–297

Janzen DH (1971) Seed predation by animals. Annu Rev Ecol Syst 2:465–492

Jarzyna MA, Quintero I, Jetz W (2021) Global functional and phylogenetic structure of avian assemblages across elevation and latitude. Ecol Lett 24:196–207

Jehl JR Jr (2017) Feather-eating in grebes: a 500-year conundrum. Wilson J Ornithol 129:446–458

Jensen J (2001) Regulatory peptides and control of food intake in non-mammalian vertebrates. Comp Biochem Physiol A 128:471–479

Jimenez AG (2018) "The same thing that makes you live can kill you in the end": exploring the effects of growth rates and longevity on cellular metabolic rates and oxidative stress in mammals and birds. Integr Comp Biol 58:544–558

Johansen SM, Horn DJ, Wilcoxen TE (2014) Factors influencing seed species selection by wild birds at feeders. Wilson J Ornithol 126:374–381

Johnsgard P (1987) Diving birds of North America. University of Nebraska Press, Lincoln, NE

Johnsgard PA (1978) Ducks, geese, and swans of the world. Univ Nebraska Press, Lincoln, NE

Johnsgard PA (2010) The world's waterfowl in the 21st century: a 2010 supplement to Ducks, Geese, and Swans of the world. Papers in the Biol Sci. Univ of Nebraska, Linclon, NE

Johnson SD, Brown M (2004) Transfer of pollinaria on birds' feet: a new pollination system in orchids. Plant Syst Evol 244:181–188

Johnson SD, Nicolson SW (2008) Evolutionary associations between nectar properties and specificity in bird pollination systems. Biol Lett 4:49–52

Johnston DW (1958) Sex and age characters and salivary glands of the Chimney Swift. Condor 60:73–84

Jones LR, Black HL, White CM (2012) Evidence for convergent evolution in gape morphology of the Bat Hawk (*Macheiramphus alcinus*) with swifts, swallows, and goatsuckers. Biotropica 44:386–393

Jordano P (2000) Fruits and frugivory. In: Fenner M (ed) Seeds: the ecology of regeneration in plant communities, 2nd edn. CABI Publishing, Wallingford, pp 125–166

Jordano P (2013) Fruits and frugivores. In: Gallagher RS (ed) Seeds: the ecology of regeneration in plant communities, 3rd edn. CAB International, Wallingford, pp 18–61

Jung J-Y, Naleway SE, Yaraghi NA, Herrera S, Sherman VR, Bushong EA, Ellisman MH, Kisailus D, McKittrick J (2016) Structural analysis of the tongue and hyoid apparatus in a woodpecker. Acta Biomater 37:1–13

Justino DG, Maruyama PK, Oliveira PE (2012) Floral resource availability and hummingbird territorial behaviour on a Neotropical savanna shrub. J Ornithol 153:189–197

Kane SA, Zamani M (2014) Falcons pursue prey using visual motion cues: new perspectives from animal-borne cameras. J Exp Biol 217:225–234

Karasov WH (1996) Digestive plasticity in avian energetics and feeding ecology. In: Carey C (ed) Avian energetics and nutritional ecology. Chapman and Hall, New York, NY, pp 61–84

Karasov WH, Hume ID (1997) Vertebrate gastrointestinal system. In: Dantzler W (ed) Handbook of comparative physiology. American Physiological Society, Bethesda, MD, pp 409–480

Karasov WH, Pinshow B, Starck JM, Afik D (2004) Anatomical and histological changes in the alimentary tract of migrating Blackcaps (*Sylvia atricapilla*): a comparison among fed, fasted, food-restricted, and refed birds. Physiol Biochem Zool 77:149–160

Kassarov L (1999) Are birds able to taste and reject butterflies based on 'beak mark tasting'? A different point of view. Behaviour 136:965–981

Katti M, Price T (1999) Annual variation in fat storage by a migrant warbler overwintering in the Indian tropics. J Anim Ecol 68:815–823

Katzir G, Strod T, Schechtman E, Hareli S, Arad Z (1999) Cattle Egrets are less able to cope with light refraction than are other herons. Anim Behav 57:687–694

Kays R, Jansen PA, Knecht EMH, Vohwinkel R, Wikelski M (2011) The effect of feeding time on dispersal of *Virola* seeds by toucans determined by GPS tracking and accelerometers. Acta Oecol 37:625–631

Kelly D, Ladley JJ, Robertson AW (2004) Is dispersal easier than pollination? Two tests in New Zealand Loranthaceae. N Z J Bot 42:89–103

Kelly JF, Bridge ES, Frick WF, Chilson PB (2013) Ecological energetics of an abundant aerial insectivore, the Purple Martin. PLoS ONE 8:e76616

Kershaw PJ, Rochman CM (2015) Sources, fate and effects of microplastics in the marine environment: Part 2 of a global assessment. Reports and Studies-IMO/FAO/Unesco-IOC/WMO/IAEA/UN/UNEP Joint Group of Experts on the Scientific Aspects of Marine Environmental Protection (GESAMP) Eng No 93

Khaleel IM, Al-Khazraji KI, Al-Aameli MH (2017) A comparative evaluation of morphological and histological features of gull (*Larus canus*) and Mallard duck (*Anas platyrhynchos*) lungs. Adv Anim Vet Sci 5:307–311

Kim W, Peaudecerf F, Baldwin MW, Bush JWM (2011) The hummingbird's tongue: a self-assembling capillary syphon. Proc R Soc B 279:4990–4996

King AS, McLelland J (1984) Birds, their structure and function. Bailliere Tindall, East Sussex

Kissling WD, Böhning-Gaese K, Jetz W (2009) The global distribution of frugivory in birds. Glob Ecol Biogeogr 18:150–162

Kissling WD, Sekercioglu CH, Jetz W (2012) Bird dietary guild richness across latitudes, environments and biogeographic regions. Glob Ecol Biogeogr 21:328–340

Klän S, Klaas M, Schröder W (2010) The influence of leading-edge serrations on the flow field of an artificial owl wing. Proceedings of the 28th AIAA Applied Aerodynamics Conference, American Institute of Aeronautics and Astronautics, Reston, VA

Klasing KC (1998) Comparative avian nutrition. CAB International, New York, NY

Kobayashi K, Kumakura M, Yoshimura K, Inatomi M, Asami T (1998) Fine structure of the tongue and lingual papillae of the penguin. Arch Histol Cytol 61:37–46

Kober K, Bairlein F (2009) Habitat choice and niche characteristics under poor food conditions. A study on migratory Nearctic shorebird in the intertidal flats of Brazil. Ardea 97:31–42

Kotogama SW, Goodale E (2004) The composition and spatial organization of mixed-species flocks in a Sri Lankan rainforest. Forktail 20:63–70

Krauss SL, Phillips RD, Karron JD, Johnson SD, Roberts DG, Hopper SD (2017) Novel consequences of bird pollination for plant mating. Trends Plant Sci 22:395–410

Kuenzel WJ, Beck MM, Teruyama R (1999) Neural sites and pathways regulating food intake in birds: a comparative analysis to mammalian systems. J Exp Zool 283:348–364

Kuenzel WJ, Helms CW (1974) An annual cycle study of tan-striped and white-striped White-throated Sparrows. Auk 91:44–53

Kühn S, van Franeker JA (2020) Quantitative overview of marine debris ingested by marine megafauna. Mar Pollut Bull 151:110858

Kurk CD (2008) The bill of evolution. Ph.D. dissertation, Leiden University, Leiden, The Netherlands

Larrinaga AR (2011) Inter-specific and intra-specific variability in fruit color preferences in two species of *Turdus*. Integr Zool 6:244–258

Lasso E, Naranjo ME (2003) Effect of pollinators and nectar robbers on nectar production and pollen deposition in *Hamelia patens* (Rubiaceae). Biotropica 35:57–66

Laumer I, Bugnyar T, Reber S, Auersperg A (2017) Can hook-bending be let off the hook? Bending/unbending of pliant tools by cockatoos. Proc R Soc B 284: 20171026

Lavers JL, Hutton I, Bond AL (2019) Clinical pathology of plastic ingestion in marine birds and relationships with blood chemistry. Environ Sci Technol 53:9224–9231

Laverty G, Elbrønd VS, Árnason SS, Skadhauge E (2006) Endocrine regulation of ion transport in the avian lower intestine. Gen Comp Endocrinol 147:70–77

Leck CF (1971) Overlap in the diet of some Neotropical birds. Living Bird 10:89–106

Lee ATK, Brightsmith DJ, Vargas MP, Leon KQ, Mejia AJ, Marsden SJ (2014a) Diet and geophagy across a western Amazonian parrot assemblage. Biotropica 46: 322–330

Lee DE, Hamman MG, Black JM (2004) Grit-site selection of Black Brant: particle size or calcium content? Wilson Bull 116:304–313

Lee N, Horstemeyer MF, Rhee H, Nabors B, Liao J, Williams LN (2014b) Hierarchical multiscale structure-property relationships of the Red-bellied Woodpecker (*Melanerpes carolinus*) beak. J Roy Soc Inter 11:20140274

Lee Y-N, Barsbold R, Currie PJ, Kobayashi Y, Lee H-J, Godefroit P, Escuillié F, Chinzorig T (2014c) Resolving the long-standing enigmas of a giant ornithomimosaur *Deinocheirus mirificus*. Nature 515:257–260

Levey DJ (1987a) Sugar-tasting ability and fruit selection in tropical fruit-eating birds. Auk 104:173–179

Levey DJ (1987b) Seed size and fruit-handling techniques of avian frugivores. Am Nat 129:471–485

Levey DJ, Grajal A (1991) Evolutionary implications of fruit-processing limitations in Cedar Waxwings. Am Nat 138:171–189

Levey DJ, Martinez del Rio C (2001) It takes guts (and more) to eat fruit: lessons from avian nutritional ecology. Auk 118:819–831

Levey DJ, Place AR, Rey PJ, del Rio CM (1999) An experimental test of dietary enzyme modulation in Pine Warblers *Dendroica pinus*. Physiol Biochem Zool 72:576–587

Lewis JC (1993) Foods and feeding ecology. In: Baskett TS, Sayre MW, Tomlinson RE, Mirarchi RE (eds) Ecology and management of the Mourning Dove. Stackpole Books, Harrisburg, PA, pp 181–204

Lieber L, Langrock R, Nimmo Smith W (2021) A bird's eye view on turbulence: seabird foraging associations with evolving surface flow features. Proc R Soc B 288:20210592

Linde-Medina M, Guerra C, Alcover JA (2021) A revision of vulture feeding classification. Zoology 148:125946

Lloyd-Jones DJ, St. Clair JJ, Cram DL, Yassene O, van der Wal JE, Spottiswoode CN (2022) When wax wanes: competitors for beeswax stabilize rather than jeopardize the honeyguide–human mutualism. Proc R Soc B 289:20221443

López-Calleja MV, Soto-Gamboa M, Rezende EL (2000) The role of gastrolites on feeding behavior and digestive efficiency in the Rufous-collared Sparrow. Condor 102:465–469

Lotem A, Schechtman E, Katzir G (1991) Capture of submerged prey by Little Egrets, *Egretta garzetta*: strike depth, strike angle and the problem of light refraction. Anim Behav 42:341–346

Louchart A, Viriot L (2011) From snout to beak: the loss of teeth in birds. Trends Ecol Evol 26:663–673

Mackie RI (2002) Mutualistic fermentative digestion in the gastrointestinal tract: diversity and evolution. Integr Comp Biol 42:319–326

MacWhirter B, Austin-Smith P Jr, Kroodsma D (2020) Sanderling (*Calidris alba*), version 1.0. In: Billerman SM (ed) Birds of the World. Cornell Lab of Ornithology, Ithaca, NY

Madej CW, Clay K (1991) Avian seed preference and weight loss experiments: the effect of fungal endophyte-infected tall fescue seeds. Oecologia 88: 296–302

Maglianesi MA, Blüthgen N, Böhning-Gaese K, Schleuning M (2014) Morphological traits determine specialization and resource use in plant-hummingbird networks in the Neotropics. Ecology 95:3325–3334

Mallory ML, Hatch SA, Nettleship DN (2020) Northern Fulmar (*Fulmarus glacialis*), version 1.0. In: Billerman SM (ed) Birds of the World. Cornell Lab of Ornithology, Ithaca, NY

Mansor MS, Sah SAM (2012) Foraging patterns reveal niche separation in tropical insectivorous birds. Acta Ornithologica 47:27–36

Margalida A, Bertran J (2001) Function and temporal variation in use of ossuaries by Bearded Vultures (*Gypaetus barbatus*) during the nestling period. Auk 118:785–789

Margalida A, Negro JJ, Galván I (2008) Melanin-based color variation in the Bearded Vulture suggests a thermoregulatory function. Comp Biochem Physiol A 149: 87–91

Margalida A, Schulze-Hagen K, Wetterauer B, Domhan C, Oliva-Vidal P, Wink M (2020) What do minerals in the feces of Bearded Vultures reveal about their dietary habits? Sci Total Environ 728:138836

Marples NM, Roper TJ (1997) Response of domestic chicks to methyl anthranilate odour. Anim Behav 53: 1263–1270

Márquez-Luna U, Lara C, Corcuera P, Valverde PL (2019) Factors affecting the dominance hierarchy dynamics in a hummingbird assemblage. Current Zoology 65:261–268

Martin AR (1986) Feeding association between dolphins and shearwaters around the Azores Islands. Can J Zool 64:1372–1374

Martin S, Bhushan B (2016) Discovery of riblets in a bird beak (*Rynchops*) for low fluid drag. Phil Trans R Soc A 374:20160134

Martínez AE, Pollock HS, Kelley JP, Tarwater CE (2018) Social information cascades influence the formation of mixed-species foraging aggregations of ant-following birds in the Neotropics. Anim Behav 135:25–35

Martinez del Rio C, Brugger KE, Rios JL, Vergara ME, Witmer M (1995) An experimental and comparative study of dietary modulation of intestinal enzymes in European Starlings (*Sturnus vulgaris*). Physiol Zool 68:490–511

Martinez del Rio C, Karasov WH, Levey DJ (1989) Physiological basis and ecological consequences of sugar preferences in Cedar Waxwings. Auk 106:64–71

Martinez del Rio C, Schondube JE, McWhorter TJ, Hererra LG (2001) Intake responses in nectar feeding birds: digestive and metabolic causes, osmoregulatory consequences, and coevolutionary effects. Am Zool 41:902–915

Mascitti V, Kravetz FO (2002) Bill morphology of South American flamingos. Condor 104:73–83

Mason R, Clark J (2000) The chemical senses of birds. In: Whittow GC (ed) Sturkie's avian physiology. Academic Press, San Diego, CA, pp 39–56

Maya-García O, Ortega-Flores M, Schondube JE (2021) True grit: ingestion of small stone particles by hummingbirds in West Mexico. Avian Research 12:62

McLelland J (1979) Digestive system. In: King AS, McLelland J (eds) Form and function in birds, vol 1. Academic Press, New York, NY, pp 69–181

McWhorter TJ, Caviedes-Vidal E, Karasov WH (2009) The integration of digestion and osmoregulation in the avian gut. Biol Rev 84:533–565

McWilliams SR, Kearney SB, Karasov WH (2002) Diet preference of warblers for specific fatty acids in relation to nutritional requirements and digestive capabilities. J Avian Biol 33:167–174

Meng J, Xing L, Zhao X (2013) Isolation of a feather-degrading *Bacillus subtilis* strain from the alimentary tract of grebes. Afr J Microbiol Res 7:2550–2559

Meredith RW, Zhang G, Gilbert MTP, Jarvis ED, Springer MS (2014) Evidence for a single loss of mineralized teeth in the common avian ancestor. Science 346: 1254390

Meyer-Rochow VB, Gal J (2004) Pressures produced when penguins pooh—calculations on avian defaecation. Polar Biol 27:56–58

Mez J, Daneshvar DH, Kiernan PT, Abdolmohammadi B, Alvarez VE, Huber BR, Alosco ML, Solomon TM, Nowinski CJ, McHale L et al (2017) Clinicopathological evaluation of chronic traumatic encephalopathy in players of American football. J Am Med Assoc 318: 360–370

Michel L, Cianchetti-Benedetti M, Catoni C, Dell'Omo G (2022) How shearwaters prey. New insights in foraging behaviour and marine foraging associations using bird-borne video cameras. Mar Biol 169:1–11

Mielke M, Van Wassenbergh S (2022) Three-dimensional movement of the beak during seed processing in Domestic Canaries. J Exp Biol 225:jeb244360

Miller IJ Jr, Reedy FE Jr (1990) Variations in human taste bud density and taste intensity perception. Physiol Behav 47:1213–1219

Mills JA, Mark AF (1977) Food preferences of Takahe in Fiordland National Park, New Zealand, and the effect of competition from introduced red deer. J Anim Ecol 46:939–958

Milton SJ, Dean WRJ, Siegfried WR (1994) Food selection by Ostrich in southern Africa. J Wildl Manag 58: 234–248

Mitra O, Callaham MA, Smith ML, Yack JE (2009) Grunting for worms: seismic vibrations cause *Diplocardia* earthworms to emerge from soil. Biol Lett 5:16–19

Mohammed LE (2017) Morphological and histochemical features of the cloaca of turkey hen *Meleagris gallopavo*. Iraqi J Vet Med 41:28–33

Moore SJ (1998) The gizzard morphology of an avian herbivore: the Domestic Goose, *Anser*. Austral J Zool 46:345–357

Moore SJ, Lill CA, Sanson GD (1998) Functional morphology of the gizzard of the domestic goose: design of an artificial gizzard. J Zool 246:111–117

Morand-Ferron J, Sol D, Lefebvre L (2007) Food stealing in birds: brain or brawn? Anim Behav 74:1725–1734

Morse DH (1975) Ecological aspects of adaptive radiation in birds. Biol Rev 50:167–214

Morse DH (1977) Feeding behavior and predator avoidance in heterospecific groups. Bioscience 27:332–339

Morton ES (1978) Avian arboreal folivores: why not? In: Montgomery GG (ed) The ecology of arboreal folivores. Smithsonian Institution Press, Washington, DC, pp 123–130

Mosa SG (1993) Fall and winter diet and habitat preferences of the Andean Tinamou (*Nothura pentlandii*) in the northwest Argentina. Stud Neotropical Fauna Environ 28:123–128

Mumme RL (2002) Scare tactics in a Neotropical warbler: white tail feathers enhance flush-pursuit foraging performance in the Slate-throated Redstart (*Myioborus miniatus*). Auk 119:1024–1035

Munoz ZJ, Colorado JZG (2021) Importance of tropical mixed-species flocks for migratory birds in shade-grown coffee: implications of foraging together. J Field Ornithol 92:212–230

Munson ES, Robinson WD (1992) Extensive folivory by Thick-billed Saltators (*Saltator maxillosus*) in southern Brazil. Auk 109:917–919

Murata M, Kang J-H (2018) Bisphenol A (BPA) and cell signaling pathways. Biotechnol Adv 36:311–327

Murray M (2014) Raptor gastroenterology. Vet Clin: Exot Anim Pract 17:211–234

Nabhan GP, Buchmann SL (1997) Services provided by pollinators. In: Daily G (ed) Nature's services. Island Press, Washington, DC, pp 133–150

Navalón G, Bright JA, Marugán-Lobón J, Rayfield EJ (2019) The evolutionary relationship among beak shape, mechanical advantage, and feeding ecology in modern birds. Evolution 73:422–435

Negro JJ, Grande JM, Tella JL, Garrido J, Hornero D, Donazar JA, Sanchez-Zapata JA, Benitez JR, Barcell M (2002) An unusual source of essential carotenoids. Nature 416:807–808

Nicolson SW, Fleming PA (2003) Nectar as food for birds: the physiological consequences of drinking dilute sugar solutions. Plant Syst Evol 238:139–153

Norberg UM (1990) Vertebrate flight. Springer, Berlin

Norris E, Norris C, Steen JB (1975) Regulation and grinding ability of grit in the gizzard of Norwegian Willow Ptarmigan (*Lagopus lagopus*). Poult Sci 54:1839–1843

Obst BS, Hunt GL Jr (1990) Marine birds feed at gray whale mud plumes in the Bering Sea. Auk 107:678–688

Obst BS, Nagy KA (1993) Stomach oil and the energy budget of Wilson's Storm-Petrel nestlings. Condor 95: 792–805

O'Connor JK (2019) The trophic habits of early birds. Palaeogeogr Palaeoclimatol Palaeoecol 513:178–195

Ohashi K, Thomson JD, D'souza D (2007) Trapline foraging by bumble bees: IV. Optimization of route geometry in the absence of competition. Behav Ecol 18:1–11

Oka N (2011) Differential assimilation efficiency of prey and growth of Short-tailed Shearwater chicks. Ornithol Sci 10:39–50

Olmos F, Rotenberg E, Muscat E (2013) A feeding association between Wilson's Storm-Petrels *Oceanites oceanicus* (Kuhl, 1820) and rough-toothed dolphins *Steno bredanensis* (G. Cuvier in Lesson, 1828). Biota Neotropica 13:303–307

Olsen AM (2015) Exceptional avian herbivores: multiple transitions toward herbivory in the bird order Anseriformes and its correlation with body mass. Ecol Evol 5:5016–5032

Olsen AM (2017) Feeding ecology is the primary driver of beak shape diversification in waterfowl. Funct Ecol 31: 1985–1995

Orlowski G, Karg J, Karg G (2014) Functional invertebrate prey groups reflect dietary responses to phenology and farming activity and pest control services in three sympatric species of aerially foraging insectivorous birds. PLoS ONE 9:e114906

Ornelas JF (1994) Serrate tomia: an adaptation for nectar robbing in hummingbirds? Auk 111:703–710

Ortiz LT, Alzueta C, Treviÿo J, Castaÿo M (1994) Effects of faba bean tannins on the growth and histological structure of the intestinal tract and liver of chicks and rats. Br Poult Sci 35:743–754

Pacheco MA (2000) Aspectos energéticos y nutricionales del crecimiento en el periquito (*Forpus passerinus*). Ph.D. dissertation, Universidad Simón Bolívar, Caracas, Venezuela

Pacheco MA, García-Amado MA, Bosque C, Domínguez-Bello MG (2004) Bacteria in the crop of the seed-eating Green-rumped Parrotlet. Condor 106:139–143

Pappenheimer JR (1993) On the coupling of membrane digestion with intestinal absorption of sugars and amino acids. Am J Physiol 265:G409–G417

Parrish JW Jr, Martin EW (1977) The effect of dietary lysine on the energy and nitrogen balance of the Dark-eyed Junco. Condor 79:24–30

Paster MB (1992) A brief overview: the avian crop. J Assoc Avian Vet 6:229–230

Pauw A (1998) Pollen transfer on birds' tongues. Nature 394:731–732

Pecsics T, Laczi M, Nagy G, Csörgő T (2017) The cranial morphometrics of the wildfowl (Anatidae). Ornis Hungarica 25:44–57

Pennycuick CJ (1972) Soaring behaviour and performance of some east African birds, observed from a motor-glider. Ibis 114:178–218

Peres CA (1990) A Harpy Eagle successfully captures an adult male red howler monkey. Wilson Bull 102:560–561

Piersma T, Drent J (2003) Phenotypic flexibility and the evolution of organismal design. Trends Ecol Evol 18: 228–233

Piersma T, van Eerden MR (1989) Feather eating in Great Crested Grebes *Podiceps cristatus*: a unique solution to the problems of debris and gastric parasites in fish-eating birds. Ibis 131:477–486

Pigot AL, Sheard C, Miller ET, Bregman TP, Freeman BG, Roll U, Seddon N, Trisos CH, Weeks BC, Tobias JA (2020) Macroevolutionary convergence connects morphological form to ecological function in birds. Nat Ecol Evol 4:230–239

Pike AVL, Maitland DP (2004) Scaling of bird claws. J Zool 262:73–81

Pitman RL (1993) Seabird associations with marine turtles in the eastern Pacific Ocean. Colonial Waterbirds 16: 194–201

Pitman RL, Ballance LT (1992) Parkinson's petrel distribution and foraging ecology in the Eastern Pacific: aspects of an exclusive feeding relationship with dolphins. Condor 94:825–835

PlasticsEurope (2020) Plastic – the facts 2020. [online]. https://www.plasticseurope.org/en/resources/publications/4312-plastics-facts-2020

Podos J (2001) Correlated evolution of morphology and vocal signal structure in Darwin's finches. Nature 409: 185

Poli C, Robertson EP, Martin J, Powell AN, Fletcher RJ Jr (2022) An invasive prey provides long-lasting silver spoon effects for an endangered predator. Proc R Soc B 289:20220820

Popp M, Mirré V, Brochmann C (2011) A single mid-Pleistocene long-distance dispersal by a bird can explain the extreme bipolar disjunction in crowberries (*Empetrum*). Proc Natl Acad Sci USA 108:6520–6525

Portugal SJ, Murn CP, Sparkes EL, Daley MA (2016) The fast and forceful kicking strike of the Secretary Bird. Curr Biol 26:R58–R59

Powell GVN (1985) Sociobiology and adaptive significance of interspecific foraging flocks in the Neotropics. In: Buckley PA, Foster MS, Morton ES, Ridgely RS, Buckley FG (eds) Neotropical ornithology. Ornithological Monographs No. 36, American Ornithologists' Union, Washington, DC, pp 713–732

Powers DR, Brown AR, Van Hook JA (2003) Influence of normal daytime fat deposition on laboratory measurements of torpor use in territorial versus nonterritorial hummingbirds. Physiol Biochem Zool 76:389–397

Powlesland RG, Lloyd BD, Best HA, Merton DV (1997) Breeding biology of the Kakapo *Strigops habroptilus* on Stewart Island, New Zealand. Ibis 134:361–373

Prakash M, Quéré D, Bush JWM (2008) Surface tension transport of prey by feeding shorebirds: the capillary ratchet. Science 320:931–934

Prevost J, Vilter V (1963) Histologie de la sécrétion œsophagienne du Manchot Empereur. Proc Int Ornithol Congr 2:1085–1094

Price ER, Brun A, Caviedes-Vidal E, Karasov WH (2015) Digestive adaptations of aerial lifestyles. Physiology 30:69–78

Pulliam HR (1973) On the advantages of flocking. J Theor Biol 38:419–422

Pyke GH (1980) The foraging behaviour of Australian honeyeaters: a review and some comparisons with hummingbirds. Aust J Ecol 5:343–369

Pyke GH, Waser NM (1981) The production of dilute nectars by hummingbird and honeyeater flowers. Biotropica 13:260–270

Radford AN, Bell MBV, Hollén LI, Ridley AR (2011) Singing for your supper: sentinal calling by kleptoparasitites can mitigate the cost to victims. Evolution 65:900–906

Razali NB, Mansor MS, Ismail NA, Patah PA, Husin SM, Hussein MSR, Nor SM (2022) The use of salt licks by birds in Peninsular Malaysia. Global Ecol Conserv 38: e02210

Razeng E, Watson DM (2012) What do declining woodland birds eat? A synthesis of dietary records. Emu 112:149–156

Razeng E, Watson DM (2015) Nutritional composition of the preferred prey of insectivorous birds: popularity reflects quality. J Avian Biol 46:89–96

Reichert BE, Cattau CE, Fletcher RJ Jr, Sykes PW Jr, Rodgers JA Jr, Bennetts RE (2020) Snail Kite (*Rostrhamus sociabilis*), version 1.0. In: Poole AF

(ed) Birds of the World. Cornell Lab of Ornithology, Ithaca, NY
Remsen JV Jr, Robinson SK (1990) A classification scheme for foraging behavior of birds in terrestrial habitats. Stud Avian Biol 13:144–160
Rensch B, Neunzig R (1925) Experimentelle untersuchungen uber den geschmackssin der vogal II. J Ornithol 73:633–646
Richards MP (2003) Genetic regulation of feed intake and energy balance in poultry. Poult Sci 82:907–916
Richardson H, Verbeek NAM (1986) Diet selection and optimization by Northwestern Crows feeding on Japanese littleneck clams. Ecology 67:1219–1226
Ricklefs RE (1974) Energetics of reproduction in birds. In: Paynter RA Jr (ed) Avian energetics. Nuttall Ornithological Club Publication No. 15, Cambridge, MA, pp 152–297
Rico-Guevara A (2017) Relating form to function in the hummingbird feeding apparatus. PeerJ 5:e3449
Rico-Guevara A, Araya-Salas M (2015) Bills as daggers? A test for sexually dimorphic weapons in a lekking hummingbird. Behav Ecol 26:21–29
Rico-Guevara A, Rubega MA (2011) The hummingbird tongue is a fluid trap, not a capillary tube. Proc Natl Acad Sci 108:9356–9360
Ridley AR, Raihani NJ (2007) Facultative response to a kleptoparasite by the cooperatively breeding Pied Babbler. Behav Ecol 18:324–330
Ríos JM, Mangione A, Marone L (2012) Effects of nutritional and anti-nutritional properties of seeds on the feeding ecology of seed-eating birds of the Monte Desert, Argentina. Condor 114:44–55
Robel RJ, Press BM, Henning BL, Johnson KW (1995) Nutrient and energetic characteristics of sweepnet-collected invertebrates. J Field Ornithol 66:44–53
Robinson SA, Forbes MR, Hebert CE (2008) Is the ingestion of small stones by Double-crested Cormorants a self-medication behavior? Auk 110:782–785
Roby DD, Brink KL, Place AR (1989) Relative passage rates of lipid and aqueous digesta in the formation of stomach oils. Auk 106:303–313
Roby DD, Taylor JR, Place AR (1997) Significance of stomach oil for reproduction in seabirds: an interspecies cross-fostering experiment. Auk 114:725–736
Rocca MA, Sazima M (2010) Beyond hummingbird-flowers: the other side of ornithophily in the Neotropics. Oecologia Australis 14:67–99
Rochman CM, Browne MA, Underwood AJ, Van Franeker JA, Thompson RC, Amaral-Zettler LA (2016) The ecological impacts of marine debris: unraveling the demonstrated evidence from what is perceived. Ecology 97:302–312
Rodríguez-Flores CI, Arriaga MCA (2016) The dynamics of hummingbird dominance and foraging strategies during the winter season in a highland community in western Mexico. J Zool 299:262–274
Rodríguez-Gironés MA, Santamaria L (2004) Why are so many bird flowers red? PLoS Biol 2:e350
Röhe I (2013) Effects of feed structure on animal performance, gastrointestinal morphology, gut-associated lymphoid tissue and jejunal glucose transport in laying hens. Ph.D. dissertation, Free University of Berlin, Berlin
Rojas-Nossa SV, Sánchez JM, Navarro L (2016) Nectar robbing: a common phenomenon mainly determined by accessibility constraints, nectar volume and density of energy rewards. Oikos 125:1044–1055
Roman L, Hardesty BD, Hindell MA, Wilcox C (2019) A quantitative analysis linking seabird mortality and marine debris ingestion. Sci Rep 9:3202
Roman L, Hardesty BD, Hindell MA, Wilcox C (2020a) Disentangling the influence of taxa, behaviour and debris ingestion on seabird mortality. Environ Res Lett 15:12
Roman L, Kastury F, Petit S, Aleman R, Wilcox C, Hardesty BD, Hindell MA (2020b) Plastic, nutrition and pollution; relationships between ingested plastic and metal concentrations in the livers of two *Pachyptila* seabirds. Sci Rep 10:18023
Ropert-Coudert Y, Grémillet D, Kato A (2005) Diving angle of Great Cormorants. Polar Biosci 18:54–59
Ropert-Coudert Y, Grémillet D, Ryan P, Kato A, Naito Y, Le Maho Y (2004) Between air and water: the plunge dive of the Cape Gannet *Morus capensis*. Ibis 146: 281–290
Ross E (2014) Woodpeckers and the biomechanics of concussion. M.S. thesis, University of British Columbia, Vancouver, BC
Roth TC II, Lima SL (2003) Hunting behavior and diet of Cooper's Hawks: an urban view of the small-bird-in-winter paradigm. Condor 105:474–483
Rowland HM, Parker MR, Jiang P, Reed DR, Beauchamp GK, Doty RL (2015) Comparative taste biology with special focus on birds and reptiles. In: Doty RL (ed) Handbook of olfaction and gustation. Wiley, New York, NY, pp 957–982
Rubega MA, Obst BS (1993) Surface-tension feeding in phalaropes: discovery of a novel feeding mechanism. Auk 110:169–178
Ruggiero RG, Eves HE (1998) Bird-mammal associations in forest openings of northern Congo (Brazzaville). Afr J Ecol 36:183–193
Sabat P, Novoa F, Bozinovic F, Martinez del Rio G (1998) Dietary flexibility and intestinal plasticity in birds: a field and laboratory study. Physiol Zool 71:226–236
Samar ME, Avila RE, de Fabro SP, Centurion C (1995) Structural and cytochemical study of salivary glands in the Magellanic Penguin and the Kelp Gull. Mar Ornithol 23:153–157
Sanders TA, Koch RC (2018) Band-tailed Pigeon use of supplemental mineral. J Wildl Manag 82:538–552
Sargent AJ, Groom DJE, Rico-Guevara A (2021) Locomotion and energetics of divergent foraging strategies in hummingbirds: a review. Integr Comp Biol 61:736–748
Schaefer HM (2011) Why fruits go to the dark side. Acta Oecol 37:604–610

Schaefer HM, McGraw K, Catoni C (2008) Birds use fruit colour as honest signal of dietary antioxidant rewards. Funct Ecol 22:303–310

Schaefer HM, Schmidt V, Bairlein F (2003) Discrimination abilities for nutrients: which difference matters for choosy birds and why? Anim Behav 65:531–541

Schaefer HM, Valido A, Jordano P (2014) Birds see the true colours of fruits to live off the fat of the land. Proc R Soc B 281:20132516

Schiestl FP, Johnson SD (2013) Pollinator-mediated evolution of floral signals. Trends Ecol Evol 28:307–315

Schronce A, Davis AK (2022) Novel observation: Northern Cardinal (*Cardinalis cardinalis*) perches on an invasive jorō spider (*Trichonephila clavata*) web and steals food. Insects 13:1049

Sedinger JS (1997) Adaptations to and consequences of an herbivorous diet in grouse and waterfowl. Condor 99: 314–326

Sekercioglu CH (2006) Increasing awareness of avian ecological function. Trends Ecol Evol 21:464–471

Selman HA (2013) Morphological and histological study for liver in local coot birds *Fulica atra*. Basrah J Vet Res 12:152–158

Selva N, Fortuna MA (2007) The nested structure of a scavenger community. Proc R Soc B 274:1101–1108

Shang Y, Kumar S, Oakley B, Kim WK (2018) Chicken gut microbiota: importance and detection technology. Front Vet Sci 5:254

Sheard C, Street SE, Evans C, Lala KN, Healy SD, Sugasawa S (2023) Beak shape and nest material use in birds. Phil Trans R Soc B 378:20220147

Shepard ELC, Lambertucci SA, Vallmitjana D, Wilson RP (2011) Energy beyond food: foraging theory informs time spent in thermals by a large soaring bird. PLoS ONE 6:e27375

Shivik JA (1999) Carrion, context, and lure development: the relative importance of sensory modalities to foraging brown treesnakes (*Boiga irregularis*). Ph.D. dissertation, Colorado State University, Fort Collins, CO

Shivik JA (2006) Are vultures birds, and do snakes have venom, because of macro- and microscavenger conflict? Bioscience 56:819–823

Silva PA (2019) Flower eating by the Toco Toucan (*Ramphastos toco*) in an anthropogenic landscape during the dry season. Ornitología Neotropical 30:51–55

Skieresz-Szewczyk K, Jackowiak H (2016) Morphofunctional study of the tongue in the domestic duck (*Anas platyrhynchos f. domestica*, Anatidae): LM and SEM study. Zoomorphology 135:255–268

Skov H, Durinck J, Danielsen F, Bloch D (1995) Co-occurrence of cetaceans and seabirds in the Northeast Atlantic. J Biogeogr 22:71–88

Smith CL, Toomey M, Walker BR, Braun EJ, Wolf BO, McGraw K, Sweazea KL (2011) Naturally high plasma glucose levels in Mourning Doves (*Zenaida macroura*) do not lead to high levels of reactive oxygen species in the vasculature. Zoology 114:171–176

Smith RE (1960) The influence of size and surface condition of grit upon the digestibility of feed by the domestic fowl. Can J Anim Sci 40:51–56

Spady DW, Payne PR, Picou D, Waterlow JC (1976) Energy balance during recovery from malnutrition. Am J Clin Nutr 29:1073–1078

Spiegel O, Nathan R (2007) Incorporating dispersal distance into the disperser effectiveness framework: frugivorous birds provide complementary dispersal to plants in a patchy environment. Ecol Lett 10:718–728

Spottiswoode CN, Begg KS, Begg CM (2016) Reciprocal signaling in honeyguide-human mutualism. Science 353:387–389

Spottiswoode CN, Lund J, Lloyd-Jones DJ (2022) Honeyguides. Curr Biol 32:R1072–R1074

Stanley MC, Lill A (2002a) Importance of seed ingestion to an avian frugivore: an experimental approach to fruit choice based on seed load. Auk 119:175–184

Stanley MC, Lill A (2002b) Does seed packaging influence fruit consumption and seed passage in an avian frugivore? Condor 104:136–145

Starck JM (1999) Phenotypic flexibility of the avian gizzard: rapid, reversible and repeated changes of organ size in response to changes in dietary fibre content. J Exp Biol 202:3172–3179

Stettenheim PR (2000) The integumentary morphology of modern birds; an overview. Am Zool 40:461–477

Stevens CE, Hume ID (1995) Comparative physiology of the vertebrate digestive system, 2nd edn. Cambridge Univ Press, Cambridge

Stevens CE, Hume ID (1998) Contributions of microbes in vertebrate gastrointestinal tract to production and conservation of nutrients. Physiol Rev 78:393–427

Stiles FG (1978) Ecological and evolutionary implications of bird pollination. Am Zool 18:715–727

Stiles FG (1995) Behavioral, ecological and morphological correlates of foraging for arthropods by the hummingbirds of a tropical wet forest. Condor 97: 853–878

Stolen ED, Collazo JA, Percival HF (2012) Group-foraging effects on capture rate in wading birds. Condor 114:744–754

Stornelli MR, Ricciardi MP, Giannessi E, Coli A (2006) Morphological and histological study of the Ostrich (*Struthio camelus* L.) liver and biliary system. Ital J Anat Embryol 111:1–7

Stournaras KE, Lo E, Böhning-Gaese K, Cazetta E, Dehling DM, Schleuning M, Stoddard MC, Donoghue MJ, Prum RO, Schaefer HM (2013) How colorful are fruits? Limited color diversity in fleshy fruits on local and global scales. New Phytol 198:617–629

Stromberg MR, Johnsen PB (1990) Hummingbird sweetness preferences: taste or viscosity? Condor 92:606–612

Sustaita D, Rubega MA (2014) The anatomy of a shrike bite: bill shape and bite performance in Loggerhead Shrikes. Biol J Linn Soc 112:485–498

Sweazea KL (2022) Revisiting glucose regulation in birds – a negative model of diabetes complications. Comp Biochem Physiol B 262:110778

Swynnerton CFM (1916) Short cuts by birds to nectaries. Bot J Linn Soc 43:381–417

Symes CT, Hughes JC, Mack AL, Marsden S (2006) Geophagy in birds of Crater Mountain Wildlife Management Area, Papua New Guinea. J Zool 268:87–96

Symons MRE, Tattersall GJ (2010) Geographical variation in bill size across bird species provides evidence for Allen's Rule. Am Nat 176:188–197

Swart D, Mackie RI, Hayes JP (1993) Fermentative digestion in the Ostrich (*Struthio camelus* var. *domesticus*), a large avian species that utilizes cellulose. S Afr J Anim Sci 23:127–135

Tagami S (1974) On the variation of retention and form of grit in gizzard of growing chicks. Scientific Reports of the Faculty of Agriculture, Ibaraki University, Ami, Japan

Tanaka KD, Okahisa Y, Sato NJ, Theuerkauf J, Ueda K (2013) Gourmand New Caledonian Crows munch rare escargots by dropping. J Ethol 31:341–344

Takekawa JY, Warnock N (2020) Long-billed Dowitcher (*Limnodromus scolopaceus*), version 1.0. In: Poole AF, Gill FB (eds) Birds of the World. Cornell Lab of Ornithology, Ithaca, NY

Tattersall GJ, Andrade DV, Abe AS (2009) Heat exchange from the toucan bill reveals a controllable vascular thermal radiator. Science 325:468–470

Temeles EJ, Kress WJ (2003) Adaptation in a plant-hummingbird association. Science 300:630–633

Temeles EJ, Linhart YB, Masonjones M, Masonjones HD (2002) The role of flower width in hummingbird bill length-flower length relationships. Biotropica 34:68–80

Tobiansky DJ, Long KM, Hamden JE, Brawn JD, Fuxjager MJ (2021) Cost-reducing traits for agonistic head collisions: a case for neurophysiology. Integr Comp Biol 61:1394–1405

Toda Y, Ko MC, Liang Q, Miller ET, Rico-Guevara A, Nakagita T, Sakakibara A, Uemura K, Sackton T, Hayakawa T, Sin SYW (2021) Early origin of sweet perception in the songbird radiation. Science 373:226–231

Tomazzoni AC, Pedó E, Hartz SM (2005) Feeding associations between capybaras *Hydrochoerus hydrochaeris* (Linnaeus) (Mammalia, Hydrochaeridae) and birds in the Lami Biological Reserve, Porto Alegre, Rio Grande do Sul, Brazil. Revista Brasileira de Zoologia 22:712–716

Thomson J (2003) When is it mutualism? Am Nat 162:S1–S9

Thouzeau C, Peters G, Le Bohec C, Le Maho Y (2004) Adjustments of gastric pH, motility and temperature during long-term preservation of stomach contents in free-ranging incubating King Penguins. J Exp Biol 207:2715–2724

Tinbergen N (1962) Foot-paddling in gulls. Br Birds 55: 117–120

Tremblay Y, Thiebault A, Mullers R, Pistorius P (2014) Bird-borne video-cameras show that seabird movement patterns relate to previously unrevealed proximate environment, not prey. PLoS ONE 9:e88424

Trewick S (1996) The diet of Kakapo (*Strigops habroptilus*), Takahe (*Porphyrio mantelli*) and Pukeko (*P. porphyrio melanotus*) studied by faecal analysis. Notornis 43:79–112

Trost RE (1981) Dynamics of grit selection and retention in captive Mallards. J Wildl Manag 45:64–73

Tsahar E, Arad Z, Izhaki I, Martinez del Rio C (2006) Do nectar- and fruit-eating birds have lower nitrogen requirements than omnivores? An allometric test. Auk 123:1004–1012

Tsang LR, Wilson LAB, Ledogar J, Wroe S, Attard M, Sansalone G (2019) Raptor talon shape and biomechanical performance are controlled by relative prey size but not by allometry. Sci Rep 9:7076

Ubaid FK, Beco R (2022) Interspecific foraging association between the Cryptic Forest-falcon, *Micrastur mintoni* and primates in an Amazon rainforest in Brazil. Acta Amazon 52:49–52

Ueda K, Akashi F, Kawasaki M, Sugawara T, Manabe Y, Matsui T (2019) Effects of feeding on plasma concentrations of vitamin A in captive African Penguins (*Spheniscus demersus*). J Vet Med Sci 81: 1580–1585

Valido AH, Schaefer M, Jordano P (2011) Colour, design and reward: phenotypic integration of fleshy fruit displays. J Evol Biol 24:751–760

van der Meij MAA (2005) A tough nut to crack. Adaptations to seed cracking in finches. Ph.D. dissertation, Leiden University, Leiden, The Netherlands

Vander Wall SB (2001) The evolutionary ecology of nut dispersal. Bot Rev 67:74–117

van Heezik YM, Gerritsen AFC, Swennen C (1983) The influence of chemoreception on the foraging behavior of two species of sandpiper, *Calidris alba* and *Calidris alpina*. Neth J Sea Res 17:47–56

van Tets IG, Hulbert AJ (1999) A comparison of the nitrogen requirements of the eastern pygmy possum, *Cercartetus nanus*, on a pollen and on a mealworm diet. Physiol Biochem Zool 72:127–137

Van Tets IG, Nicolson SW (2000) Pollen and the nitrogen requirements of the Lesser Double-collared Sunbird. Auk 117:826–830

Van Wassenbergh S, Baeckens S (2019) Digest: Evolution of shape and leverage of bird beaks reflects feeding ecology, but not as strongly as expected. Evolution 73: 621–622

Van Wassenbergh S, Ortlieb EJ, Mielke M, Böhmer C, Shadwick RE, Abourachid A (2022) Woodpeckers minimize cranial absorption of shocks. Curr Biol 32:3189–3194

Veit RR, Harrison NM (2017) Positive interactions among foraging seabirds, marine mammals and fishes and implications for their conservation. Front Ecol Evol 5:121

Vézina F, Dekinga A, Piersma T (2010) Phenotypic compromise in the face of conflicting ecological demands: an example in Red Knots *Calidris canutus*. J Avian Biol 41:88–93

Vilk O, Orchan Y, Charter M, Ganot N, Toledo S, Nathan R, Assaf M (2022) Ergodicity breaking in area-restricted search of avian predators. Phys Rev X 12:031005

Villard P, Cuisin J (2004) How do woodpeckers extract grubs with their tongues? A study of the Guadeloupe Woodpecker in the French West Indies. Auk 121:509–514

Vispo C, Karasov WH (1997) The interaction of avian gut microbes and their host: an elusive symbiosis. In: Mackie RI, White BA (eds) Gastrointestinal microbiology. Chapman and Hall, New York, NY, pp 116–155

Vogrin M (2017) Prey dropping behaviour in Black-headed Gull *Chroicocephalus ridibundus*. Biharean Biologist 11:129

Wackernagel H (1964) Was futtern die flamingos ihren Jungen? Internationale Zeitschrift fuer Vitaminforschung 34:141–143

Waite DW, Taylor MW (2014) Characterizing the avian gut microbiota: membership, driving influences, and potential function. Front Microbiol 5:223

Waite JL, Henry AR, Clayton DH (2012) How effective is preening against mobile ectoparasites? An experimental test with pigeons and hippoboscid flies. Int J Parasitol 42:453–467

Walker LE, Pyle P, Patten MA, Greene E, Davison W, Muehter VR (2020) Steller's Jay (*Cyanocitta stelleri*), version 1.0. In: Rodewald PG (ed) Birds of the World. Cornell Lab of Ornithology, Ithaca, NY

Walter ED, Aiken JR (1961) The value of soluble and insoluble grit in all-mash and mash-grain rations for caged layers. Poult Sci 40:904–909

Walters EL, Miller EH, Lowther PE (2020) Red-breasted Sapsucker (*Sphyrapicus ruber*), version 1.0. In: Poole AF (ed) Birds of the World. Cornell Lab of Ornithology, Ithaca, NY

Wang K, Zhao H (2015) Birds generally carry a small repertoire of bitter taste receptor genes. Genome Biol Evol 7:2705–2715

Wang L, Cheung JT-M, Pu F, Li D, Zhang M, Fan Y (2011) Why do woodpeckers resist head impact injury: a biomechanical investigation. PLoS ONE 6:e26490

Wang LK, Chan M, Chan YM, Tan GC, Wee YC (2009) Pellet-casting by non-raptorial birds of Singapore. Nat Singapore 2:97–106

Wang SW (2005) Foraging patterns of Northern Fulmers in Alaska inferred from fatty acid signature analysis. M.S. thesis, University of Alaska, Fairbanks, AK

Wang SW, Iverson SJ, Springer AM, Hatch SA (2007) Fatty acid signatures of stomach oil and adipose tissue of Northern Fulmars (*Fulmarus glacialis*) in Alaska: implications for diet analysis of Procellariiform birds. J Comp Physiol B 177:893–903

Wang TM, Yang XB, Liang JH, Yao CC, Zhao WD (2013) CFD based investigation on the impact acceleration when a gannet impacts with water during plunge diving. Bioinspir Biomim 8:036006

Warham J (1971) Aspects of the breeding behaviour of the Royal Penguin *Eudyptes chrysolophus schlegeli*. Notornis 18:91–115

Warham J (1977) The incidence, functions and ecological significance of petrel stomach oils. Proc New Zeal Ecol Soc 24:84–93

Warrick DR (1998) The turning- and linear-maneuvering performance of birds: the costs of efficiency for coursing insectivores. Can J Zool 76:1063–1079

Weiser JI, Porth A, Mertens D, Karasov WH (1997) Digestion of chitin by Northern Bobwhites and American Robins. Condor 99:554–556

Wenny DG, Devault TL, Johnson MD, Kelly D, Sekercioglu CH, Tomback DF, Whelan CJ (2011) The need to quantify ecosystem services provided by birds. Auk 128:1–14

Wester P, Claßen-Bockhoff R (2007) Floral diversity and pollen transfer mechanisms in bird-pollinated *Salvia* species. Ann Bot 100:401–421

Wheelwright NT (1985) Fruit size, gape width, and the diets of fruit-eating birds. Ecology 66:808–818

White CM, Clum NJ, Cade TJ, Hunt WG (2020) Peregrine Falcon (*Falco peregrinus*), version 1.0. In: Billerman SM (ed) Birds of the World. Cornell Lab of Ornithology, Ithaca, NY

Wiklund C, Järvi T (1982) Survival of distasteful insects after being attacked by naive birds: a reappraisal of the theory of aposematic coloration evolving through individual selection. Evolution 36:998–1002

Williams HJ, Shepard ELC, Holton MD, Alarcón PAE, Wilson RP, Lambertucci SA (2020) Physical limits of flight performance in the heaviest soaring bird. Proc Natl Acad Sci 117:17884–17890

Wings O (2007) A review of gastrolith function with implications for fossil vertebrates and a revised classification. Acta Palaeontol Pol 52:1–16

Winzen A, Klaas M, Schröder W (2008) PIV measurements comparing natural and model owl wings. Exp Fluids 46:975–989

Witmer MC (1998) Do seeds hinder digestive processing of fruit pulp? Implications for plant–frugivore mutualisms. Auk 115:319–326

Wolf LL, Gill FB (1986) Physiological and ecological adaptations of high montane sunbirds and hummingbirds. In: Vuilleumier F, Monasteio M (eds) High altitude tropical biogeography. Oxford University Press and American Museum of Natural History, Oxford

Wood BM, Pontzer H, Raichlen DA, Marlowe FW (2014) Mutualism and manipulation in Hadza–honeyguide interactions. Evol Hum Behav 35:540–546

Wood JR, Rawlence NJ, Rogers GM, Austin JJ, Worthy TH, Cooper A (2008) Coprolite deposits reveal the diet and ecology of the extinct New Zealand megaherbivore moa (Aves, Dinornithiformes). Quat Sci Rev 27:2593–2602

Wooller RD, Richardson KC, Pagendham CM (1988) The digestion of pollen by some Australian birds. Austral J Zool 36:357–362

Working Group AO, Nol E, Humphrey RC (2020) American Oystercatcher (*Haematopus palliatus*), version 1.0. In: Poole AF (ed) Birds of the World. Cornell Lab of Ornithology, Ithaca, NY

Yanega GM, Rubega MA (2004) Hummingbird jaw bends to aid insect capture. Nature 428:615

Yegani M, Korver DR (2008) Factors affecting intestinal health in poultry. Poult Sci 87:2052–2063

Yeoman CJ, Chia N, Jeraldo P, Sipos M, Goldenfeld ND, White BA (2012) The microbiome of the chicken gastrointestinal tract. Anim Health Res Rev 13:89–99

Yoon S-H, Park S (2011) A mechanical analysis of woodpecker drumming and its application to shock-absorbing systems. Bioinspir Biomim 6:1–12

Yosef R, Yosef N (2010) Cooperative hunting in Brown-necked Raven (*Corvus rufficollis*) on Egyptian mastigure (*Uromastyx aegyptius*). J Ethol 28:385–388

Young AM (1971) Foraging for insects by a tropical hummingbird. Condor 73:36–45

Zach R (1978) Selection and dropping of whelks by Northwestern Crows. Behaviour 67:134–147

Zach R (1979) Shell dropping: decision-making and optimal foraging in Northwestern Crows. Behaviour 68: 106–117

Zheng X, Martin LD, Zhou Z, Burnham DA, Zhang F, Miao D (2011) Fossil evidence of avian crops from the Early Cretaceous of China. Proc Natl Acad Sci USA 108:15904–15907

Zoratto F, Santucci D, Alleva E (2009) Theories commonly adopted to explain the antipredatory benefits of the group life: the case of starling (*Sturnus vulgaris*). Rendiconti Lincei 20:163–176

Zubkova EN (2019) Functional morphology of the hyoid apparatus in Old World suboscines (Eurylaimides): 2. Functional analysis. Biol Bull 46:916–928

Zurovchak JG, Stiles EW, Place AR (1999) Effect of dietary lipid level on lipid passage rate and lipid absorption efficiency in American Robins (*Turdus migratorius*). J Exp Zool 283:408–417

Zusi RL (2013) Introduction to the skeleton of hummingbirds (Aves: Apodiformes, Trochilidae) in functional and phylogenetic contexts. Ornithol Monogr 77:1–94

Zusi RL, Bentz GD (1984) Myology of the Purple-throated Carib (*Eulampis jugularis*) and other hummingbirds (Aves: Trochilidae). Smithsonian Institution Press, Washington, DC

GPSR Compliance

*The European Union's (EU) General Product Safety Regulation (GPSR) is a set of rules that requires consumer products to be safe and our obligations to ensure this.*

*If you have any concerns about our products, you can contact us on ProductSafety@springernature.com*

In case Publisher is established outside the EU, the EU authorized representative is:

Springer Nature Customer Service Center GmbH
Europaplatz 3
69115 Heidelberg, Germany

**Batch number: 10756351**

Printed by Printforce, the Netherlands

Fascinating Life Sciences

This interdisciplinary series brings together the most essential and captivating topics in the life sciences. They range from the plant sciences to zoology, from the microbiome to macrobiome, and from basic biology to biotechnology. The series not only highlights fascinating research; it also discusses major challenges associated with the life sciences and related disciplines and outlines future research directions. Individual volumes provide in-depth information, are richly illustrated with photographs, illustrations, and maps, and feature suggestions for further reading or glossaries where appropriate. Interested researchers in all areas of the life sciences, as well as biology enthusiasts, will find the series' interdisciplinary focus and highly readable volumes especially appealing.

Gary Ritchison

# In a Class of Their Own

## A Detailed Examination of Avian Forms and Functions

Volume 2

Springer

Gary Ritchison
Richmond, KY, USA

ISSN 2509-6745 ISSN 2509-6753 (electronic)
Fascinating Life Sciences
ISBN 978-3-031-14851-4 ISBN 978-3-031-14852-1 (eBook)
https://doi.org/10.1007/978-3-031-14852-1



This Springer imprint is published by the registered company Springer Nature Switzerland AG
The registered company address is: Gewerbestrasse 11, 6330 Cham, Switzerland

Paper in this product is recyclable.

# Preface

This book represents my attempt to provide an up-to-date examination of what we know about the structure and function of birds and to do so in a way that is useful and understandable to a wide variety of readers. Readers will find that some chapters are more detailed than others, largely due to differences in the amount of relevant literature available for different topics and, to a much lesser degree perhaps, my interest in the different topics. Those who read multiple (or all) chapters in the book will find some overlap, with some topics covered in more than one chapter. In such cases, topics are covered in much more detail in particular chapters and to a lesser degree in others. I wanted each chapter to tell a particular story and, to do so, felt it is necessary to sometimes mention the same topic in different chapters (and, to emphasize important points, to sometimes mention the same topic more than once in individual chapters). In addition, with the e-version of the book, interested readers may only wish to read certain chapters. In such cases, they will learn something about topics that might be covered in more detail in chapters they do not read. In terms of organization, all chapters cover aspects of avian form and function, but the initial chapters (after a first chapter about the evolution of birds) focus more on structure (e.g., the skeleton, skeletal muscles, integument, and so on), whereas later chapters include more information about function. In many chapters, I also point out where I think additional study is needed. Each chapter includes multiple figures and photos that I hope complement and add to the information provided in the text. Finally, because scientific names are used more than common names in many parts of the world, and to aid those living in those countries, scientific names of species are often provided with common names more than once throughout each chapter.

Richmond, KY — Gary Ritchison

# Acknowledgments

I sincerely thank the many reviewers for comments and suggestions that were incredibly helpful in many ways, including correcting errors (but, of course, any that remain are entirely my own and please, in case there might be another edition of this book sometime in the future, let me know of any such errors by email, i.e., ritchisong@gmail.com), providing useful references, suggesting ways to improve chapter organization, and improving the quality of the writing, among others. Thanks also to the many individuals who provided permission to use their figures and photos in the book. I also thank my editor, Lars Koerner, for providing assistance (and access to books published by Springer) whenever I asked and finding the many helpful reviewers. Writing this book has taken many years and so, most of all, I want to thank my family for their understanding and for listening to me when, during evening meals and on many other occasions, I felt the need to tell them something that I'd learned about birds during my research for the book. Although they often smiled, and sometimes even (briefly) laughed (oh no, not again!), they never once told me to please not continue (at least that's how I remember things). To be specific, thanks to my wonderful children, Brandon Tyler and Brianna Carol, for sharing my interest in birds and nature, for all the great times we had when you were kids, and for all the great times we've had since then. Some of my best memories include the many times our family spent in the woods and elsewhere hiking and enjoying nature. During the final stages of writing this book, Brandon's wife, Lisa, gave birth to their son, Carter James. Congratulations to the proud parents, and I look forward to introducing Carter to the wonderful world of birds (Brandon is already introducing him to the wonderful world of Star Trek)! Finally, a big, big thank you to my wife, Tameria. She never complained about the many times and many hours I was in the field studying birds, in the office writing about birds, or engaged in the many other things required for those in academics. After long days working on the book (and other assorted duties), I always looked forward to getting home to spend time with her. In addition to being a great wife, mother, and, more recently, grandmother, she also shares my love of birds and nature! Thanks Tam, you're the best! Finally, thanks to all of you, Tameria, Brandon, Brianna, Lisa, and Carter, for so many happy memories, and I look forward to making many more such memories in the future.

# Contents

# Contents for Volume 1

# Contents for Volume 3

# Cardiovascular and Immune Systems 6

## Contents



**Abstract**

Birds have high metabolic rates and so require an efficient cardiovascular system that provides their cells and tissues with sufficient oxygen. Birds are also exposed to a wide variety of pathogens and so require an effective immune system. In this chapter, the structure and function of the avian cardiovascular system are explained, including how the blood transports oxygen and carbon dioxide. Adaptations that allow some birds to fly at high altitudes where there is less oxygen and



G. Ritchison, *In a Class of Their Own*, Fascinating Life Sciences,
https://doi.org/10.1007/978-3-031-14852-1_6

to dive under water when breathing is not possible are discussed. The specific functions of the formed elements in bird blood, including red blood cells, thrombocytes, and white blood cells, are also described. The various components of the avian immune system are described and the various ways that the immune system acts against pathogens and their toxins are explained in detail. The structure and function of the avian heart and blood vessels are explained, as is the process by which the flow of blood to different tissues and organs varies as needed to ensure that blood flow matches changes in metabolic activity.

## 6.1 Introduction

Endotherms like birds and mammals have higher metabolic rates than ectotherms, and birds, as a group, have higher metabolic rates than mammals. A high metabolic rate translates into rapid use of oxygen and nutrients, and sustaining high levels of metabolism requires a very efficient system for transporting materials throughout the body—getting oxygen from the lungs and nutrients, like glucose, from the digestive system to the body cells. Of course, metabolic rates vary among and within species and individual birds. Birds at rest have lower metabolic rates than birds in flight or engaged in other activities, so their cardiovascular systems must be flexible and able to quickly respond to substantial changes in levels of activity.

The avian cardiovascular system consists of a heart plus vessels that transport blood. In the blood are nutrients, oxygen and carbon dioxide, waste products, and a complex variety of other things such as hormones, proteins, ions of various types, and heat. Active birds, and their active skeletal muscles, can generate impressive amounts of heat and, by shifting blood to and from different parts of the body, the cardiovascular system plays a critical role in thermoregulation (see Chap. 13).

## 6.2 Avian Blood

Avian blood consists of plasma and a variety of formed elements. Plasma is largely water (~85%) plus lots of protein (~9–11%); other constituents of blood include glucose (blood glucose levels in birds are greater than in mammals; about 200–400 mg/dl), amino acids, waste products, hormones, antibodies, and electrolytes. The formed elements include red blood cells (or erythrocytes), white blood cells (or leucocytes), and thrombocytes. Bird red blood cells, unlike those of mammals, are, to varying degrees, elliptical in shape and nucleated (Fig. 6.1). In most species, red blood cells are ovoid in shape (typically about 7 × 14 microns)

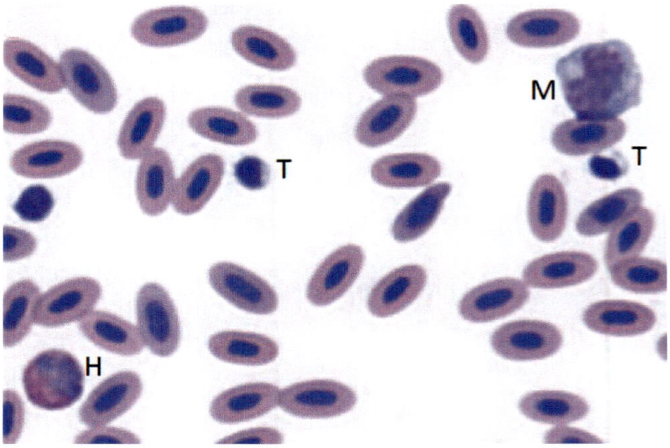


**Fig. 6.1** Avian red blood cells, plus white bloods cells in the lower left (H, heterophil) and upper right (M, monocyte) and two thrombocytes (T). The blood cells have been stained and so they appear light purple (and the nuclei of the red blood cells appear darker purple). (Figure modified from Beaufrére et al. 2013; © Journal of Avian Medicine and Surgery, used with permission)

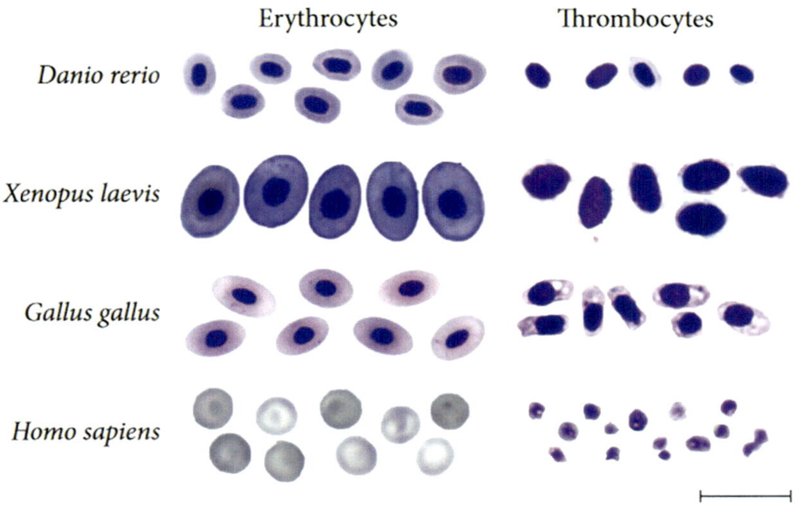


**Fig. 6.2** Red blood cells of different classes of vertebrates. The red blood cells of fishes, amphibians, reptiles, and birds have nuclei, whereas those of mammals do not. All red blood cells contain hemoglobin. Reasons for the loss of the nucleus in mammalian red blood cells are unclear. Enucleated red blood cells may have allowed smaller cells with more hemoglobin molecules (give the absence of nuclei) and increased surface area-to-volume ratios that facilitated efficient gas exchange efficiency, possibly during a period with reduced atmospheric oxygen levels. However, avian red blood cells contain similar levels of hemoglobin and birds, in general, have higher metabolic rates than mammals of comparable size and so questions remain concerning factors that might have contributed to the evolution of enucleated red blood cells in mammals. *Danio rerio*, zebrafish; *Xenopus laevis*, African clawed frog; *Gallus gallus*, Domestic Chicken; *Homo sapiens*, human. (Figure from Svoboda and Bartunek 2015; © 2015 Ondrej Svoboda and Petr Bartunek, open-access article distributed under the Creative Commons Attribution License, https://creativecommons.org/licenses/by/4.0/)

and a bit larger than those of mammals (typically 5.5–7.5 microns in diameter; Fig. 6.2). In area (length along the long axis × length along the short axis), based on data from over 600 species, erythrocyte size ranged from 38.59 $\mu m^2$ for Golden-crowned Kinglets (*Regulus satrapa*) to 143.70 $\mu m^2$ for Common Ostriches (*Struthio camelus*; Soulsbury et al. 2022; Box 6.1 Variation in the Size and Shape of Avian Erythrocytes). Typical concentrations are 2.5 to 4 million/cubic mm. Another way to express red blood cell concentration is the hematocrit, or the percentage of total blood volume that is red blood cells (Fig. 6.3).

## 6.3 Hematocrits

Hematocrit values of birds range from about 30% to 60%, with most species having values between 40 and 50% (Scanes 2015; Scoville and Doherty 2017). Minias (2020) determined that the average hematocrit for 279 species of birds was 48.3% and, in general, hematocrits averaged significantly higher in passerines (50.6%) than nonpasserines (46.0%). Hummingbirds have among the highest hematocrit values reported, with most species having values greater than 50% and some exceeding 60% (Johansen et al. 1987). A hematocrit value of 66.6% was reported for a Black Jacobin (*Florisuga fusca*), a South American hummingbird (Johansen et al. 1987). The hematocrit is a measure of the oxygen-carrying capacity of blood, a characteristic that can impact individual quality and fitness (Williams 2012). For example, Lownie et al. (2022) found that failed breeders of both sexes of Northern Giant Petrels (*Macronectes halli*) and Southern Giant Petrels (*M. giganteus*) had significantly lower hematocrit levels than successful breeders. However, as hematocrits increase so

**Box 6.1 Variation in the Size and Shape of Avian Erythrocytes**

The size and shape of erythrocytes influence their ability to deliver oxygen to the cells and tissues of the body. Soulsbury et al. (2022) examined possible relationships between the size and shape of avian erythrocytes and variables associated with differences in energy demands with data from 631 species of birds. Analysis revealed that larger birds had larger erythrocytes, likely at least in part because cell sizes in different tissues are intercorrelated and larger birds have larger cells (Kozłowski et al. 2010), and diving birds had larger erythrocytes than non-diving birds.

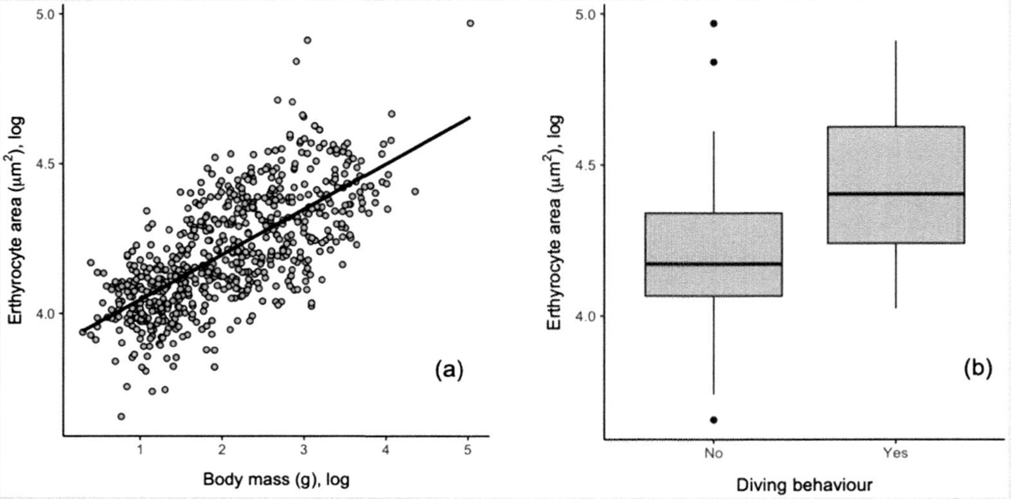


Among birds, (**a**) erythrocyte size (area) increases with increasing body mass and (**b**) diving birds have significantly larger erythrocytes than non-diving birds (Figure from Soulsbury et al. 2022; © 2021 Oxford University Press, used with permission)

Diving birds rely on oxygen stored in muscles (myoglobin) and in the blood (hemoglobin) and larger erythrocytes contain more hemoglobin molecules and, therefore, can store more oxygen. Other investigators have also reported that blood levels of hemoglobin are higher in diving than non-diving birds (Minias 2020). Soulsbury et al. (2022) also found variation among species in the shape of erythrocytes and, specifically, that erythrocytes were more elongated in species of birds that migrate longer distances and species found at higher altitudes and latitudes. More elliptical erythrocytes have a greater surface area than round one and more surface area facilitates the diffusion of oxygen into and out of erythrocytes. Migrating long distances is metabolically demanding and selection has likely favored elliptical erythrocytes in long-distance migrants to help meet the oxygen demands associated with long-distance flights.

Birds living at higher altitudes breathe air with less oxygen than those living at lower altitudes. The increased surface area of the more elliptical erythrocytes of birds at high altitudes is therefore beneficial, improving the ability of high-altitude birds to extract oxygen from the air in the lungs and deliver that oxygen to the cells and tissues of the body. Finally, birds living at high latitudes face the increased metabolic demands of thermoregulating during periods of cold weather and, as with long-distance migrants and birds at high altitudes, their more elliptical erythrocytes can deliver more oxygen to their cells and tissues and better enable them to meet those metabolic demands (Soulsbury et al. 2022).

(continued)

**Box 6.1** (continued)

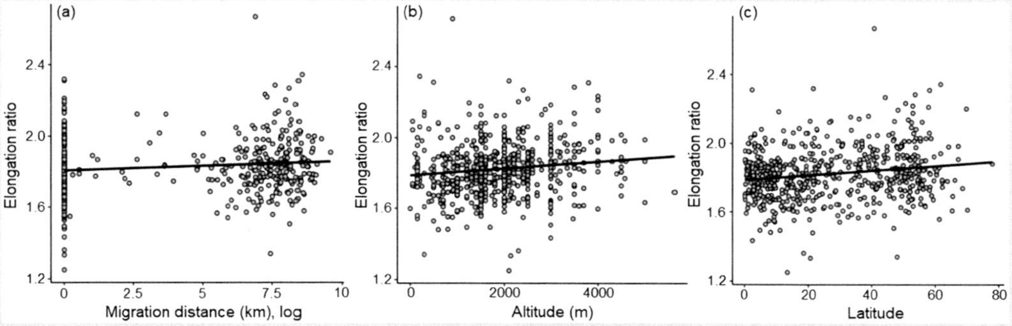


Avian erythrocytes tend to be more elongated among species with (**a**) increasing total migration distance (log km) and for species at (**b**) higher altitudes (m), and (**c**) higher latitudes (degrees) (Figure from Soulsbury et al. 2022; © 2021 Oxford University Press, used with permission)

does blood viscosity, with a corresponding increase in vascular resistance that could have a negative impact on cardiac output (Brown et al. 2021). As a result, the highest hematocrit is not necessarily the optimal hematocrit (Jensen et al. 2013b; Bowers et al. 2014; Fig. 6.4).

Numerous factors can influence hematocrits. Minias (2020) found a significant influence of phylogeny on hematocrits, as well as a negative correlation between body mass and hematocrits. Larger birds have lower metabolic rates and correspondingly lower hematocrits (i.e., blood oxygen-carrying capacity). Swanson (1990) found that hematocrits of Dark-eyed Juncos (*Junco hyemalis*) in Oregon averaged 11% higher during the winter than during the summer and suggested that, with the lower winter temperatures, more oxygen is needed for thermogenesis (e.g., shivering). Similar results, with higher hematocrits during winter than summer, have been reported for other temperate-latitude species, including American Goldfinches (*Spinus tristis*; Carey and Morton 1976), White-crowned Sparrows (*Zonotrichia leucophrys*; deGraw et al. 1979), and Black-capped Chickadees (*Poecile atricapillus*; Petit et al. 2017). More generally, Swanson and Vézina (2015) noted that hematocrits (as well as heart mass) increase in winter for most birds wintering in cold climates and that "... increasing the capacity for oxygen transport to shivering muscles is ... an important component of the winter phenotype in small birds."

Hematocrits also vary with migratory status, with species that migrate generally having higher hematocrits than species that are either nonmigratory or partial migrants (Yap et al. 2019; Fig. 6.5). In addition, hematocrits of migratory species typically increase prior to and during migration (e.g., Bairlein and Totzke 1992; Piersma et al. 1996). For example, the migration of Bar-tailed Godwits (*Limosa lapponica*; Fig. 6.6) covers 4000–5000 km as they travel between breeding grounds in Arctic Siberia and wintering areas in western Africa. At stopover sites in northern Europe during spring migration, hematocrits of Bar-tailed Godwits increase and peak just before departing on the next stage of their northward migration (Piersma et al. 1996; Landys-Ciannelli et al. 2002). Similar results have been reported for other species that exhibit long-distance migrations (Banerjee and Banerjee 1977). Along with an increased hematocrit, hemoglobin levels in the blood of migrating Bar-tailed Godwits also increase. These changes in the blood are clearly a

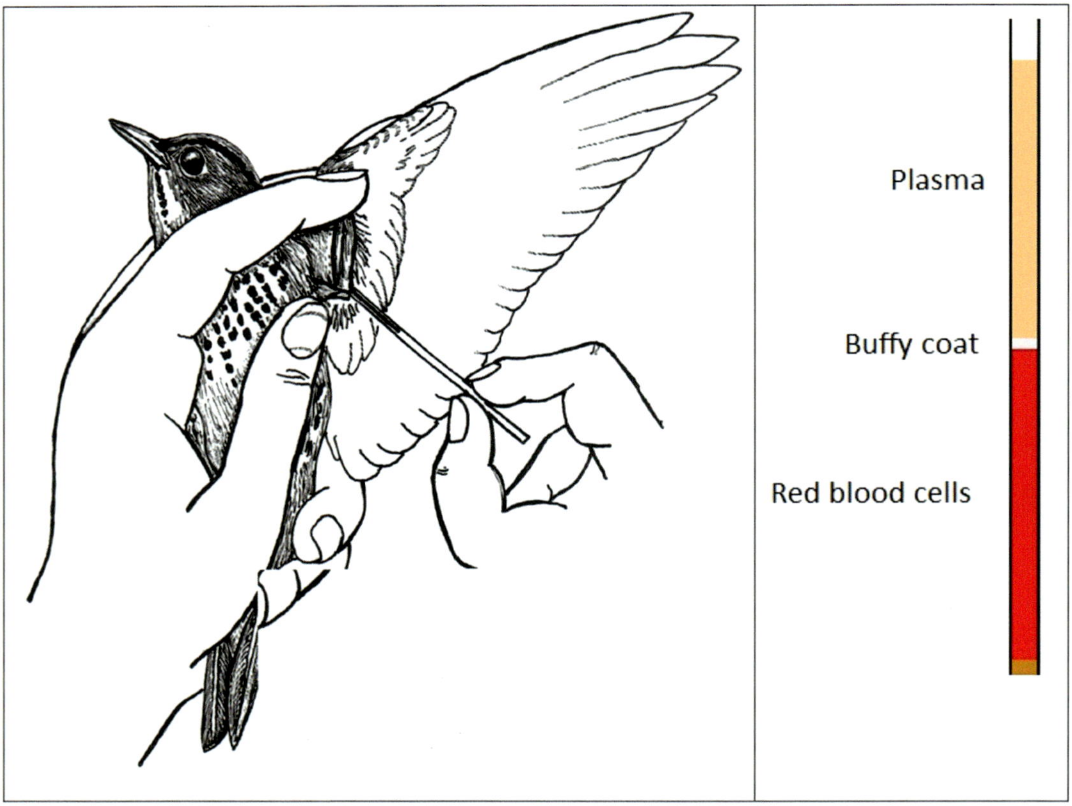


**Fig. 6.3** Left, to determine a bird's hematocrit, a small volume of blood is collected in a capillary tube. The tube is then spun in a centrifuge. Right, after centrifugation, the heavier blood elements, red blood cells, separate from the lighter plasma. A thin layer in between called the buffy coat consists of white blood cells and thrombocytes. The dark material at bottom is clay that is used to plug the bottom of the capillary tube before centrifugation. (Figure on left from Owen 2011; ©2011 The Author. Association of Field Ornithologists © 2011 Association of Field Ornithologists, used with permission. Figure on right by G. Ritchison)

response to the increased energy and oxygen demands of long-distance flights (Piersma et al. 1996).

However, at stopover sites during migration, hematocrit values may sometimes be lower. For example, Owen and Moore (2006) found that three species of migrating thrushes captured on the northern coast of the Gulf of Mexico had relatively low hematocrits (compared to values during the breeding season) and suggested it was likely because of the stress of their trans-Gulf flights and short-term food deprivation.

Based on data from studies of 160 species of birds, Yap et al. (2019) found that birds at higher altitudes did not have higher hematocrits than those at lower altitudes (Fig. 6.5). However, these authors also found that birds at higher altitudes did have higher concentrations of hemoglobin than those at lower altitudes, which would increase the oxygen-carrying capacity of the blood. In contrast to Yap et al. (2019), other investigators have reported that birds at higher elevations have higher hematocrits than those at lower elevations. Borras et al. (2010) found that the hematocrits of Citril Finches (*Carduelis citrinella*), an altitudinal migrant, at higher elevations averaged 3% higher than those at lower elevations. In addition, the hematocrits of

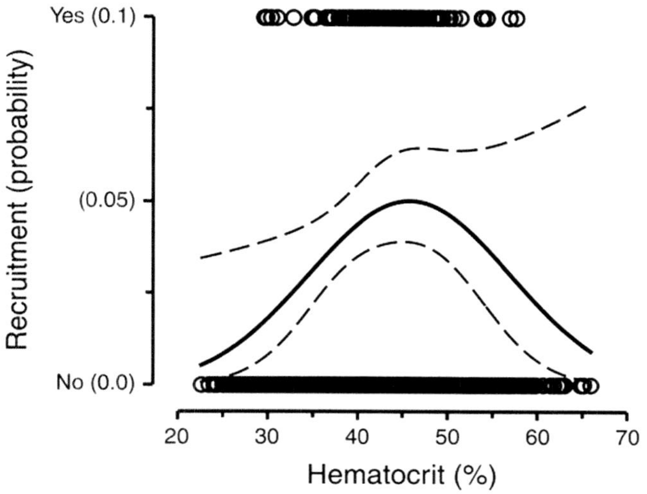


**Fig. 6.4** Relationship between hematocrits of nestling House Wrens (*Troglodytes aedon*) and probability of recruitment into the breeding population. Open symbols above and below curves represent nestlings that did and did not recruit to the breeding population, respectively, and the solid curve represents the logistic regression line (probability of recruitment) with 95% confidence limits (dashed lines). Nestlings with intermediate hematocrit values were more likely to recruit into the breeding population. High hematocrits may reflect dehydration or elevated metabolic activity and result in increased blood viscosity and higher peripheral vascular resistance and so both above- and below-optimal hematocrits can result in reduced oxygen transport which, in turn, can influence the likelihood of survival. (Figure from Bowers et al. 2014; © 2014 by the Ecological Society of America, used with permission)

captured Citril Finches that were subsequently moved to different elevations decreased by an average of 6% when moved to lower elevations and increased by the same amount when moved to higher elevations. An explanation for these apparently contradictory results was provided by Barve et al. (2016). They examined the hematocrits and hemoglobin levels of 15 species of Himalayan birds, with six being residents at different altitudes and nine being altitudinal migrants. For the altitudinal migrants, hematocrits and hemoglobin levels varied with altitude, increasing when at higher elevations and decreasing at lower elevations. In other words, hemoglobin levels of the migrant species were positively correlated with hematocrit values; to achieve either higher or lower hemoglobin levels, the migrant species either increased or decreased production of red blood cells (Box 6.2 Erythropoiesis). For resident species at different elevations, hemoglobin levels were not correlated with hematocrit values; rather, the red blood cells of residents at higher altitudes had higher concentrations of hemoglobin than those of residents at lower elevations. Thus, residents and migrants have different strategies for dealing with changes in availability of oxygen at different altitudes. Residents at higher elevations have more hemoglobin per red blood cell, increasing the oxygen-carrying capacity of their blood without increasing their hematocrits and increasing blood viscosity, whereas the oxygen-carrying capacity of the blood of migrants that spend time at different elevations during the year is adjusted as needed by varying the rate of red blood cell production (Barve et al. 2016). For migrants, in contrast to residents at high altitudes, these increases in blood viscosity are less likely to have negative effects because they are temporary and limited to relatively short breeding seasons at higher altitudes.

The hematocrits of female birds in many species have been found to decline during egg production (e.g., Williams et al. 2004; Sheridan et al.

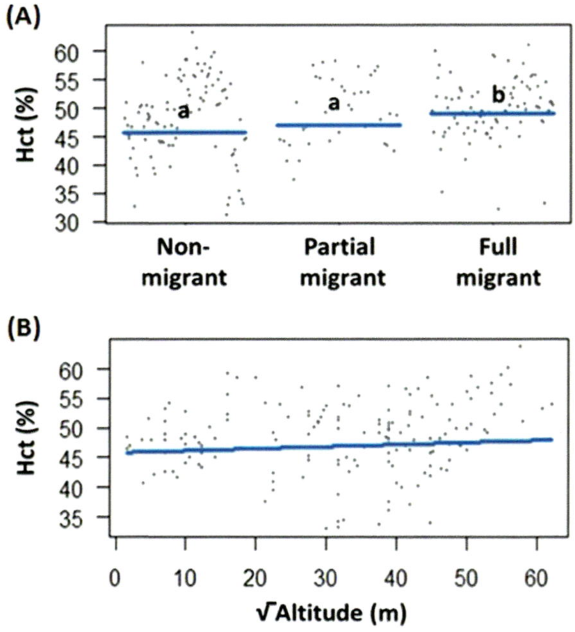


**Fig. 6.5** Relationships between (**a**) hematocrits (Hct) and migratory status (different letters above lines indicate significant differences), and (**b**) hematocrits and altitude (numbers represent square roots of altitude, e.g., 10 = 100 m, 30 = 900 m, and 60 = 3600 m). Migratory species have, on average, higher hematocrits than nonmigrants and partial migrants (where some individuals in a population migrate, but others do not). However, hematocrits do not vary with altitude. (Figure from Yap et al. 2019; open-access article licensed under a Creative Commons Attribution 4.0 International License, https://creativecommons.org/licenses/by/4.0/)

2004; Gayathri and Hegde 2006; Fig. 6.7), with reductions ranging from −1.5% in female Great Tits (*Parus major*; Hõrak et al. 1998) to −10% in female Red-billed Queleas (*Quelea quelea*, Jones 1983) relative to pre-breeding levels. The extent of this decline appears to be proportional to egg-laying effort; increasing clutch sizes cause increasing declines in hematocrits (Kalmbach et al. 2004). This decline is likely the result of increased production of estrogen, a hormone known to suppress erythropoiesis, prior to and during egg laying (Wagner et al. 2008). This reproductive anemia may represent an important "cost" of egg laying and its effects could potentially influence female behavior during incubation and chick rearing (e.g., Monaghan et al. 1998; Nager et al. 2001).

Hematocrits also change with changing metabolic demands. Even after just a few hours of flight, hematocrits can change. In homing pigeons (*Columba livia*), for example, hematocrits of birds that had flown for up to 22 hours averaged about 51% and those of control (nonflying) birds averaged about 54% (Bordel and Haase 1993). A lower hematocrit means less viscous blood that is easier for the heart to pump and more easily transported through the vessels. With the increased cardiac output associated with flight, less viscous blood is beneficial, but how can the hematocrit change in just a few hours? One likely possibility is that more fluids move into the blood and so plasma volume increases and the blood is "diluted" (Bordel and Haase 1993).

Many investigators have examined the possible relationship between hematocrits and the nutritional or body condition of birds (Table 6.1). Although some investigators have reported results suggesting a relationship between hematocrits and condition, others have not. Fair et al. (2007) concluded that data were too sparse

**Fig. 6.6** Bar-tailed Godwits (*Limosa lapponica*). (Figure from Conklin 2011, used with permission of Jesse Conklin)

**Box 6.2 Erythropoiesis**

The red blood cells of birds have limited lifespans because, even though they have a nucleus, they do not divide. Estimated lifespans of avian red blood cells range from about 30 to 48 days (Rodman et al. 1957; Glomski and Pica 2011). As old red blood cells wear out and continue to be lost, new ones must continuously be produced. In addition, rates of red blood cell production must sometimes increase, e.g., when altitudinal migrants move to higher elevations where there is less oxygen. The process of red blood cell production is called erythropoiesis and a hormone called erythropoietin plays a critical role in this process. Avian erythropoietin is a glycoprotein synthesized in the kidneys and stimulates erythropoiesis (red blood cell production) in the bone marrow.

In many birds, much of the skeleton becomes pneumatized during development, displacing the hemopoietic bone marrow. During the first months of life, bone marrow is distributed throughout the entire skeleton, except the skull. However, as birds age, bone marrow is displaced due to increasing pneumaticity. So, in adult birds, hemopoietic bone marrow is limited to the ulna, radius, femur, tibiotarsus, scapula, furcula, and the caudal vertebrae (Schepelmann 2005). Within these bones, red blood cells are produced at high, but variable, rates.

Normally, erythropoietin levels in the blood are such that the rate of red blood cell production matches the rate at which old red blood cells are being removed (e.g., by certain types of white blood cells, particularly in the spleen; Smith and Hunt 2004; Kozlu et al. 2011). However, if kidney cells that produce erythropoietin begin to get less oxygen from the blood (a condition

(continued)

**Box 6.2** (continued)

called hypoxia that could result, for example, from loss of blood or a bird moving to a higher altitude where there is less oxygen), they increase production of erythropoietin which, in turn, causes an increase in the rate of red blood cell production in the bone marrow (which is what would cause higher hematocrits in birds that migrate from lower to higher elevations). More red blood cells can then transport more oxygen which, with this feedback system, would cause the cells in the kidneys to reduce production of erythropoietin.

The process of erythropoiesis begins in the bone marrow with undifferentiated, or stem, cells that divide and produce rubriblasts (or proerythroblasts). These cells begin a maturation process, involving several different stages, that eventually results in mature erythrocytes (red blood cells). During this process, the nucleus gets smaller, the concentration of hemoglobin in the cell increases, and the shape of cells changes from round to ellipsoid (Campbell 2015).

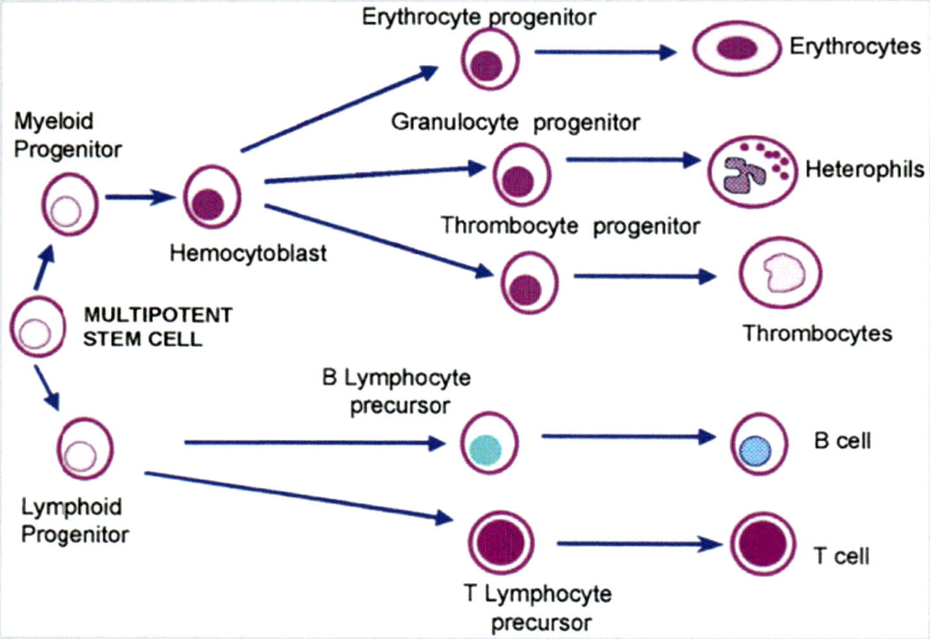


All formed elements develop from stem cells (rubriblasts) that ultimately become erythrocytes (red blood cells), white blood cells (heterophils, and B- and T-cells), and thrombocytes (Figure modified from Adair 2000; © 2000 Published by Elsevier Ltd., used with permission)

(continued)

**Box 6.2** (continued)

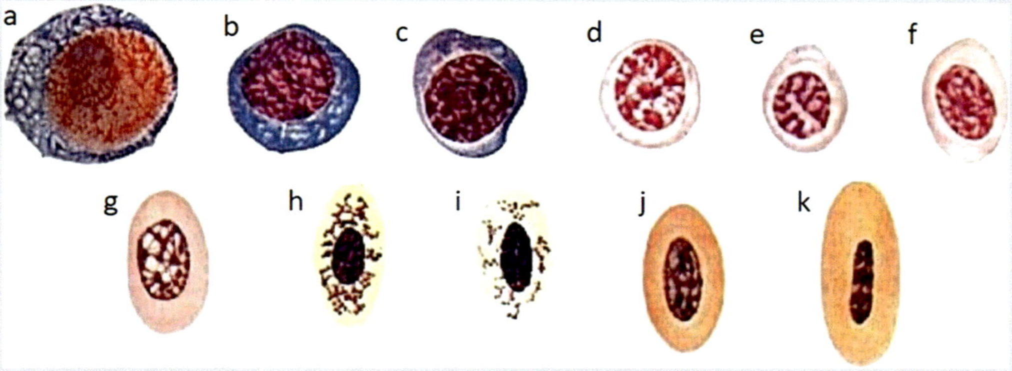


Erythropoiesis begins with undifferentiated cells called rubriblasts (**a**). Maturation proceeds through (**b**) prorubricytes, (**c**–**g**) several stages of developing polychromatophilic erythrocytes, (**h**, **i**) reticulocytes, and (**j**, **k**) mature erythrocytes. Orange color of mature erythrocytes is due to the presence of hemoglobin. Cells have been stained using Wright's-Leishman stain (Figure from Campbell 1994; used with permission of Terry Campbell)

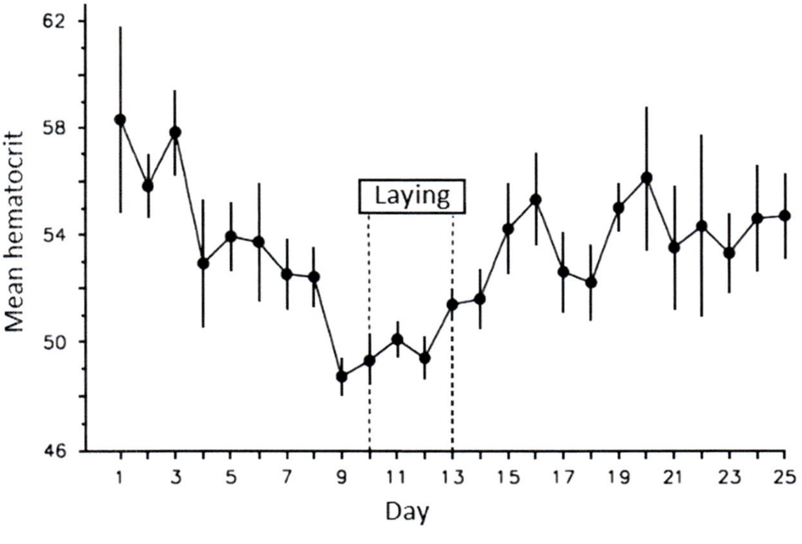


**Fig. 6.7** Mean (± SE) hematocrits of female White-crowned Sparrows (*Zonotrichia leucophrys*) during the period prior to, during, and after laying a four-egg clutch. First eggs are laid on day 10 and last eggs on day 13. (Figure modified from Morton 1994; © 1994 Oxford University Press, used with permission)

to conclude that hematocrits represent reliable indicators of nutritional condition. Some authors have concluded that hematocrits might best be used in conjunction with other hematological variables to ascertain condition. Because, as described above, multiple factors can affect hematocrits, sometimes over relatively short periods of time, specific values at specific times may not always provide accurate information about a bird's condition. In addition, red blood cells vary in size among species, so hematocrits may not accurately indicate red blood cell concentrations (Minias 2015). As suggested by Potti (2007), perhaps the ability to regulate hematocrits within certain limits is a better measure of individual quality.

**Table 6.1** Results of some representative studies examining the possible relationship between hematocrit and the nutritional or body condition of birds

| Common name | Scientific name | Age | Relationship | Citation |
|---|---|---|---|---|
| **Negative results** | | | | |
| Barn Swallow | *Hirundo rustica* | Nestling | Weakly related; use as estimate of condition not recommended | Cuervo et al. (2007) |
| Greater Flamingo | *Phoenicopterus roseus* | Nestling | No evidence that hematocrit was related to either nutritional or body condition | Amat et al. (2009) |
| Magellanic Penguin | *Spheniscus magellanicus* | Adult | No trend suggesting hematocrit would be an adequate index of health and condition | Moreno et al. (2002) |
| American Kestrel | *Falco sparverius* | Adult | Questionable, may lead to erroneous conclusions | Dawson and Bortolotti (1997) |
| Welcome Swallow | *Hirundo neoxena* | Nestling | Some caution is appropriate; advisable to measure a suite of erythrocyte variables | Lill et al. (2013a) |
| Welcome Swallow | *Hirundo neoxena* | Adult | Utility cannot be assumed; needs to be validated for each population or species | Lill et al. (2013b) |
| European Starling | *Sturnus vulgaris* | Adult and nestling | Results caution against using hematocrit as a measure of condition | Smith and Barber (2012) |
| Bonelli's Eagle | *Aquila fasciata* | Nestling | No effect of diet on hematocrit | Resano-Mayor et al. (2016) |
| Gray Catbird | *Dumetella carolinensis* | Adult | Hematocrit not correlated with body condition, but correlated with fat score; hematocrit may be indicative of aerobic capacity, but not body condition | Hatch et al. (2010) |
| Spotless Starling | *Sturnus unicolor* | Nestling | Hematocrit might not be a reliable estimate of phenotypic condition under certain nutritional circumstances | Cuervo et al. (2011) |
| Great Tit | *Parus major* | Adult | Hematocrit had no effect on adult survival | Kilgas et al. (2006) |
| House Wren | *Troglodytes aedon* | Nestling to adult | Nestlings with intermediate hematocrit values were most likely to be recruited to the population | Sakaluk et al. (2014) |
| Tree Swallow | *Tachycineta bicolor* | Nestling | Hematocrit was not related to nestling body condition | Morrison et al. (2009) |
| Noisy Miner | *Manorina melanocephala* | Adult | No significant relationship between any measure of body condition index (including mass) and hematocrit | Powell et al. (2013) |
| **Positive results** | | | | |
| European Pied Flycatcher | *Ficedula hypoleuca* | Fledgling | Fledglings in good condition as scored by mass and development of breast muscles had high hematocrits | Potti et al. (1999) |
| Eurasian Kestrel | *Falco tinnunculus* | Nestling | Nestlings had lower hematocrits in a year with reduced availability of prey | Laaksonen et al. (2004) |
| American Dipper | *Cinclus mexicanus* | Fledgling | Body mass and hematocrit were significantly correlated with rate of caloric intake | Donnelly and Sullivan (1998) |
| Great Tit | *Parus major* | Adult | Hematocrit is a sensitive indicator of the physiological condition of breeding females | Norte et al. (2010) |

## 6.4 Hemoglobin

Avian red blood cells contain hemoglobin (Fig. 6.8; Box 6.3 Avian Red Blood Cells), the molecule responsible for transporting oxygen throughout the body. Each red blood cell contains about 280 million molecules of hemoglobin and each molecule can bind to four molecules of oxygen. Not surprisingly then, hemoglobin increases the oxygen-carrying capacity of the blood of birds and other vertebrates by about 20 times compared to that due to oxygen dissolved in the water of the plasma. In contrast to the mixed results for hematocrits, hemoglobin levels in the blood appear to be, with few exceptions, "a relatively robust indicator of physiological condition in birds . . ." (Minias 2015). In several species of birds, investigators have found positive correlations between hemoglobin concentrations and either body mass or fat load (Table 6.2), and other investigators have reported positive effects of diet quality on hemoglobin concentrations (Table 6.2). However, food supplementation had no effect on hemoglobin concentrations of captive Zebra Finches (*Taeniopygia guttata*) and free-living Great Tits (*Parus major*; Table 6.2). However, for the Great Tits, the availability of natural food sources may explain why food supplementation had no effect on hemoglobin concentrations (Bańbura et al. 2011). Finally, individuals with higher hemoglobin concentrations may return earlier to breeding areas (Crossin et al. 2013), initiate breeding earlier in the breeding season (Minias 2014), and exhibit enhanced flight ability as fledglings (Cornell et al. 2017).

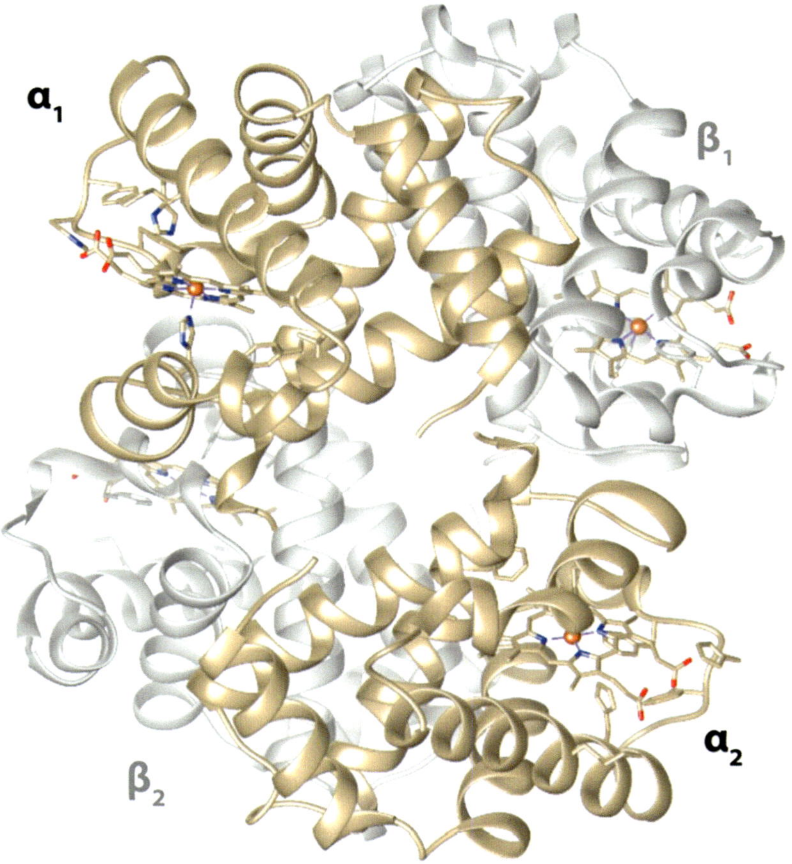


**Fig. 6.8** Schematic representation of the structure of Hemoglobin A in the deoxy state (not combined with any oxygen molecules). The globin chains are shown as solid, colored ribbons and the heme molecules as ball-stick structures. Oxygen molecules bind to the heme portions of the hemoglobin molecule. (Figure from Natarajan et al. 2018: © 2018 Natarajan et al., open-access article distributed under the terms of the Creative Commons Attribution License, https://creativecommons.org/licenses/by/4.0/)

**Box 6.3 Avian Red Blood Cells**

The red blood cells of birds and mammals are smaller than those of ectothermic vertebrates like reptiles and amphibians and, because smaller cells have more surface area relative to their volume, this may enhance the diffusion of oxygen into and out of red blood cells. However, in contrast to mammalian red blood cells, avian red blood cells are generally slightly larger, differ in shape, and have a nucleus. Interestingly, among and within vertebrate classes, there is a correlation between red blood cell size and genome size—organisms with larger genomes (more DNA) tend to have larger red blood cells (Gregory 2001). Avian genomes are smaller than those of other terrestrial vertebrates. Why this is so is unclear. Apparently, birds have either eliminated "genomic baggage" or never generated any. For example, there are polyploid fish and amphibians and even a polyploid mammal, the red viscacha rat (*Tympanoctomys barrerae*; Gallardo et al. 1999), but no polyploid birds. Thus, bird red blood cells have small nuclei compared to those of other vertebrates. In fact, some amphibians have cells with nuclei larger than an entire bird red blood cell (Gregory 2001). This difference in nucleus size may help explain why bird red blood cells have nuclei and those of mammals do not. The "ejection" of nuclei from mammalian red blood cells has reduced their size (and relative surface area) and provided room for more hemoglobin molecules. The red blood cells of birds, with their smaller genomes and nuclei, contain as many hemoglobin molecules as those of mammalian red blood cells and are still not very large.

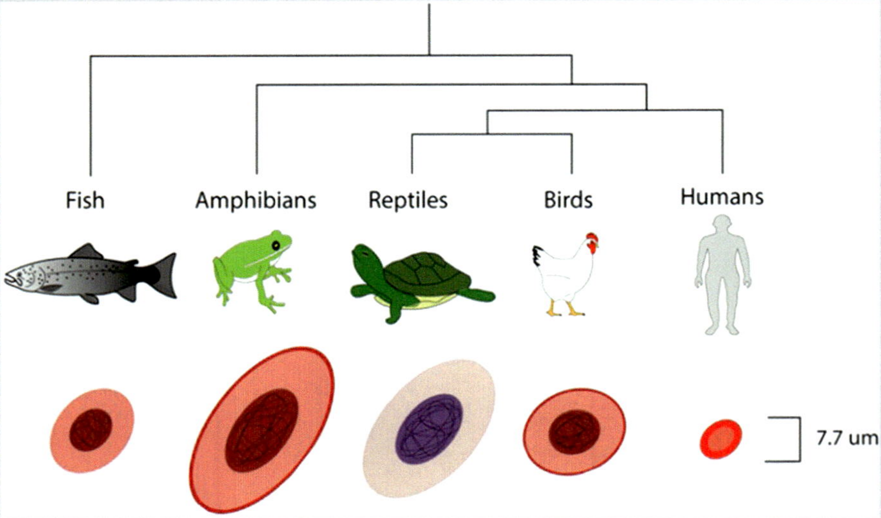


Comparison of the red blood cells of different vertebrates (Figure from Anderson et al. 2018; Copyright 2018. The American Association of Immunologists, Inc., used with permission)

Mean dimensions (μm) of red blood cells in several taxa of vertebrates. Means represent maximum and minimum diameters for oval-shaped red blood cells or, for the circular red blood cells of mammals, the diameter (Data from Snyder and Sheafor 1999).

| | |
|---|---|
| Agnatha | 21.4 × 14.2 |
| Cartilaginous fish | 20.1 × 14.9 |
| Bony fish | 12.0 × 8.5 |

(continued)

**Box 6.3** (continued)

| | |
|---|---|
| Salamanders | 48.1 × 30.8 |
| Anurans | 23.0 × 13.7 |
| Lizards | 18.4 × 12.1 |
| Turtles | 18.3 × 11.4 |
| Crocodylia | 20.1 × 10.6 |
| Birds | 13.5 × 7.5 |
| Mammals | 6.3 |

Red blood cells from a Red-tailed Black-Cockatoo (*Calyptorhynchus banksii*) (Figure from Clark et al. 2009; © 2009 John Wiley and Sons, used with permission)

Birds have two types of hemoglobin: Hemoglobin A (or HbA) and Hemoglobin D (or HbD). HbA is sometimes referred to as the "adult hemoglobin" because it is the predominant hemoglobin of adult birds (often about 90% of all hemoglobin molecules in the blood). HbD predominates in bird embryos (and so is called "embryonic hemoglobin"), but is also present in small amounts in the blood of many adult birds (Lutfullah et al. 2005). The advantage of having two types of hemoglobin is related to differences among different hemoglobins in their affinity for oxygen or, in other words, the ease with which oxygen can bind to and be released from a hemoglobin molecule. In this respect, the two avian hemoglobins are a bit different—HgD has a higher affinity for oxygen than does HbA. There are two ways of looking at hemoglobin and oxygen affinity. High affinity is potentially good because it means that, in the lungs, oxygen will firmly and quickly bind to hemoglobin. However, in the tissue capillaries, a lower affinity may be better because it means that hemoglobin is more likely to give up its oxygen.

The importance of oxygen affinity depends in part on how much oxygen is present. As oxygen pressure increases (e.g., in the lungs), hemoglobin in the red blood cells picks up oxygen and, as oxygen pressure decreases, as in tissue capillaries, hemoglobin molecules unload oxygen. This is

**Table 6.2** Relationships between hemoglobin concentrations and different measures of body condition and nutritional states of birds. Studies conducted in captivity are marked with an asterisk (Table modified from Minias 2015; © The Author 2015. Published by Oxford University Press and the Society for Experimental Biology, open-access article distributed under the terms of the Creative Commons Attribution License, https://creativecommons.org/licenses/by/4.0/)

| Common name | Scientific name | Order[a] | Age[b] | Parameter[c] | Hemoglobin concentration[d] | References |
|---|---|---|---|---|---|---|
| Welcome Swallow | *Hirundo neoxena* | PASS | Ad. | Body mass | + | Lill et al. (2013b) |
| Bar-tailed Godwit | *Limosa lapponica* | CHAR | Ad. | Body mass | + | Piersma et al. (1996) |
| Great Tit | *Parus major* | PASS | Ad. | Body mass | + | Dufva (1996) |
| Gray-headed Albatross | *Thalassarche chrysostoma* | PROC | Ad. | Body mass | + | Crossin et al. (2013) |
| Pale-breasted thrush | *Turdus leucomelas* | PASS | Ad. | Body mass | + | Lobato et al. (2011) |
| Welcome Swallow | *H. neoxena* | PASS | Pull. | SC body mass | + | Lill et al. (2013a) |
| Gould's Petrel | *Pterodroma leucoptera* | PROC | Pull. | SC body mass | + | O'Dwyer et al. (2007) |
| Bar-tailed Godwit | *L. lapponica* | CHAR | Ad. | SC body mass | + | Landys-Ciannelli et al. (2002) |
| Crimson Finch | *Neochmia phaeton* | PASS | Ad. | SC body mass/fat load | NR | Milenkaya et al. (2013) |
| Six wader species[e] | – | CHAR | Juv. | SC body mass/fat load | + | Minias et al. (2013) |
| Common Snipe | *Gallinago gallinago* | CHAR | Juv./Ad. | Plasma conc. of metabolites | NR TP | Minias et al. (2014) |
| Scopoli's Shearwater | *Calonectris dimedea* | PROC | Ad. | Plasma conc. of metabolites | NR TG, CH TP, UA, UR | Navarro et al. (2007) |
| Crimson Finch | *N. phaeton* | PASS | Ad. | Plasma conc. of metabolites | NR TP | Milenkaya et al. (2013) |
| Whiskered Tern | *Chlidonias hybrida* | CHAR | Ad. | Plasma conc. of metabolites | -GL | Minias et al. (2014) |
| Parrot Finch | *Erythrura trichroa* | PASS | Pull. | Diet quality | + | Pryke and Rollins (2012) |
| Gouldian Finch | *Erythrura gouldiae* | PASS | Pull. | Diet quality | + | Pryke et al. (2011) |
| Gouldian Finch | *E. gouldiae* | PASS | Ad. | Diet quality | + | Pryke et al. (2012) |
| Zebra Finch | *Taeniopygia guttata* | PASS | Ad. | Diet quality | NR | Wagner et al (2008) |
| Great Tit | *P. major* | PASS | Pull. | Food supplementation | NR | Bańbura et al. (2011) |

[a]CHAR, Charadriiformes; PASS, Passeriformes; PROC, Procellariiformes
[b]Ad., adult; Pull., pullus (= nestling); Juv., juvenile
[c]SC, Size corrected
[d]+, positive relationship; –, negative relationship, NR, no relationship; CH, cholesterol; GL, glucose; TG, triglycerides; TP, total protein; UA, uric acid; UR, urea
[e]Common Sandpiper (*Actitis hypoleucos*), Dunlin (*Calidris alpina*), Common Snipe (*Gallinago gallinago*), Ruff (*Philomachus pugnax*), Spotted Redshank (*Tringa erythropus*), and Green Sandpiper (*Tringa ochropus*)

apparent in the oxygen-hemoglobin dissociation curve, i.e., the relationship between oxygen levels (expressed in terms of pressure, mm Hg) and hemoglobin saturation (the percentage of hemoglobin molecules combined with and transporting oxygen in the blood) (Fig. 6.9). At a partial

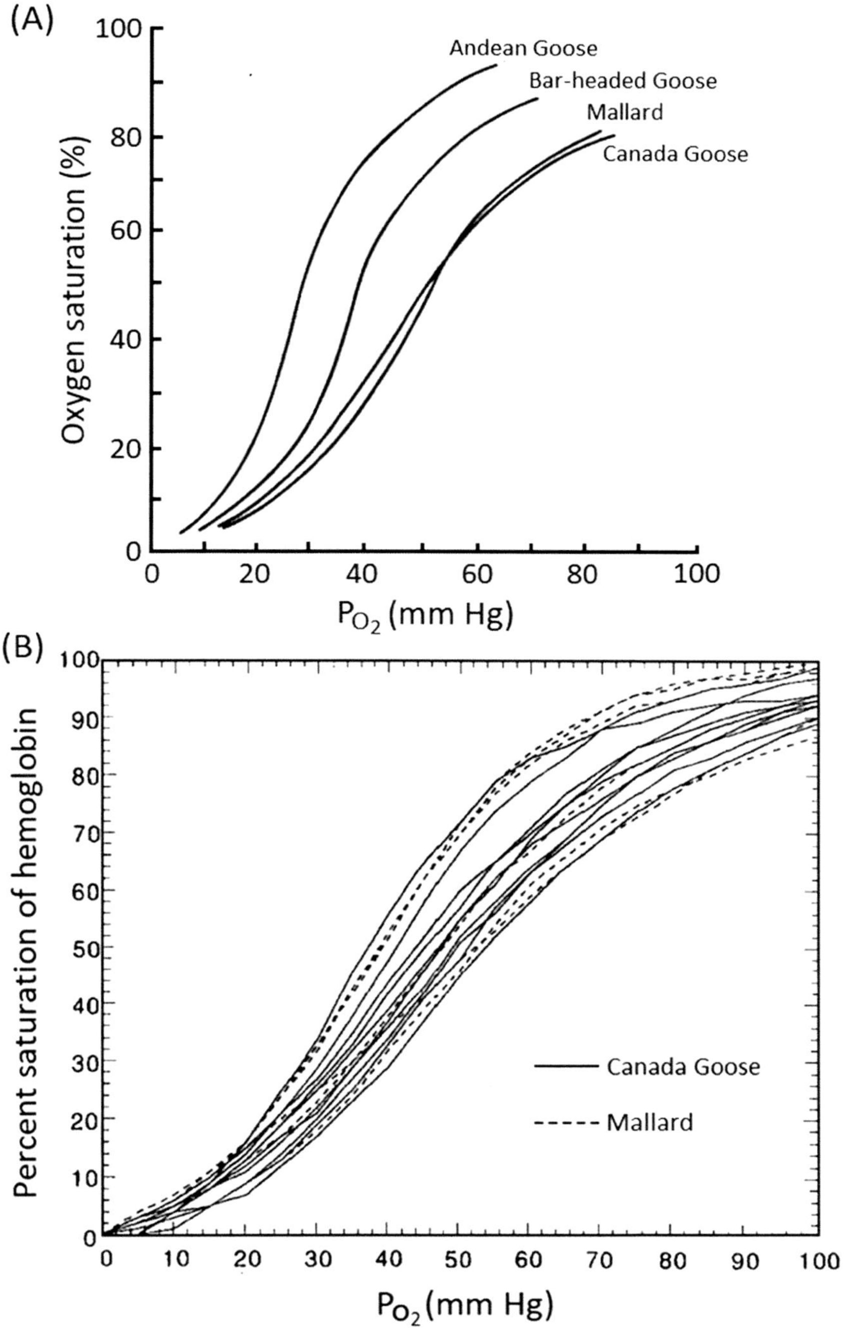


**Fig. 6.9** (**a**) Oxygen-hemoglobin dissociation curves for four species of waterfowl. Andean Goose, *Oressochen melanopterus*; Bar-headed Goose, *Anser indicus*; Mallard, *Anas platyrhynchos*; Canada Goose, *Branta canadensis*. (**b**) Oxygen-hemoglobin dissociation curves of Canada Geese ($N = 10$) and Mallards ($N = 10$) showing individual variation. Note the general s-shape of the curve, with oxygen dissociating from hemoglobin more rapidly at partial pressure of oxygen between about 70 and 30 mm Hg. (A modified from Black and Tenney 1980; © 1980 Published by Elsevier B.V., used with permission, B modified from Stephenson et al. 1996; © 1996 Springer Nature, used with permission)

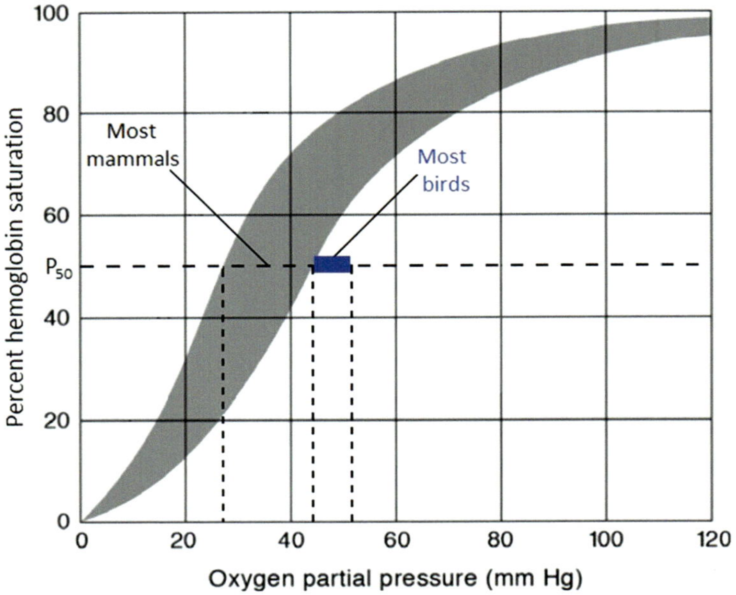


**Fig. 6.10** The oxygen-hemoglobin dissociation curves (and $P_{50}$ values) of most mammals are located to the left of those of most birds. (Figure modified from Storz 2007; © 2007 Oxford University Press, used with permission)

pressure of 100 mm Hg (in the lungs), nearly 100% of hemoglobin molecules are loaded with oxygen. When blood enters the tissue capillaries, oxygen levels are lower because nearby cells are constantly using oxygen (to make ATP). So, hemoglobin molecules unload the oxygen that then diffuses into those nearby cells. The relationship between oxygen levels (expressed in terms of pressure, mm Hg) and hemoglobin saturation (the percentage of hemoglobin molecules combined with and transporting oxygen in the blood) is expressed in the oxygen-hemoglobin dissociation curve (Fig. 6.9).

The shape of this curve varies among organisms and this curve tends to be further to the right for birds than for other organisms (Fig. 6.10). In other words, the $P_{50}$ values (partial pressure of oxygen at which hemoglobin is 50% saturated; Box 6.4 Partial Pressure) of most birds (44–52 mm Hg) are higher than those of most mammals (25–43 mm Hg) (e.g., Hirsowitz et al. 1977; Lenfant et al. 1970; Lutz 1980; Storz 2007). This means that most bird hemoglobin has a relatively low affinity for oxygen. For birds, to completely saturate hemoglobin (100%), the oxygen pressure must exceed 100 mm Hg (or 100 Torr). For other organisms, saturation occurs at lower oxygen pressures because their hemoglobin has a higher affinity for oxygen. At first glance, this seems to indicate that bird hemoglobins are less "efficient" than hemoglobins of other organisms, but that is not the case. For birds, hemoglobin with a low affinity for oxygen works fine because they have such an efficient respiratory system. As a result, oxygen pressure in the parabronchi of the avian lung is sufficiently high to "saturate" even "low-affinity" hemoglobin and, because of its low affinity, avian hemoglobin readily "unloads" its oxygen in the tissue capillaries.

**Box 6.4 Partial Pressure**

The partial pressure of a gas represents the pressure of a particular gas in a mixture of gases. The air that we breathe and that birds breathe is a mixture of gases, including oxygen, nitrogen, argon, and carbon dioxide. If the total air pressure in a container is 100 mm Hg, but that air includes molecules of these four gases, then each gas contributes to the total pressure based on the amount of each gas present. If 20% of the molecules are oxygen, then the oxygen

(continued)

**Box 6.4** (continued)

is responsible for 20% of the total pressure, which makes its partial pressure 20 mm Hg. If 79% of the molecules are nitrogen, then the partial pressure of nitrogen is 79 mm Hg, and so on. Because the diffusion of each gas from the blood capillaries to surrounding cells is not influenced by the presence of other gases in a mixture of gases, then what is important in terms of diffusion of a particular gas is its partial pressure.

The oxygen-hemoglobin dissociation curves of some birds are left-shifted, i.e., $P_{50}$ values are lower, meaning that their hemoglobin has a greater affinity for oxygen than the hemoglobin of most other birds (Fig. 6.11; Box 6.5 Bar-Headed Geese). Among those species with left-shifted oxygen-hemoglobin dissociation curves are penguins and many species of birds found at higher altitudes. Left-shifted hemoglobin is advantageous for these species because it means that more oxygen is available for any given partial pressure of oxygen. This is important for birds that experience low partial pressures of oxygen like diving birds and birds found at higher altitudes where partial pressures of oxygen are lower. For example, Emperor Penguins

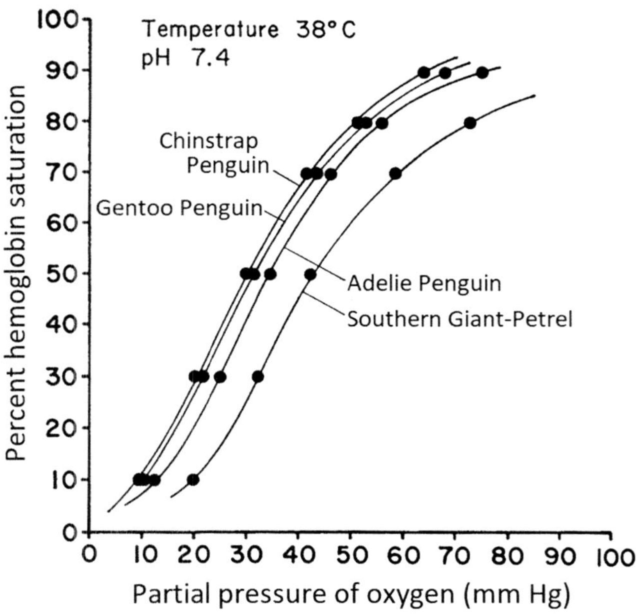


**Fig. 6.11** The oxygen-hemoglobin dissociation curves of diving birds and birds that live and fly at high altitudes are "left-shifted" compared to other birds. This means that as the partial pressure of oxygen in the blood declines, either because a diving bird is not breathing or a bird at high altitude is breathing air with less oxygen, oxygen availability remains relatively high. For example, imagine that the partial pressure of oxygen in the blood declines to 20 mm Hg either because a Southern Giant Petrel (*Macronectes giganteus*) has flown a long distance in search of prey or a Chinstrap Penguin (*Pygoscelis antarcticus*) is diving in search of prey. Extrapolating up from a partial pressure of 20 mm Hg to their respective oxygen-hemoglobin dissociation curves reveals that the Southern Giant Petrel has used 90% of its available oxygen, whereas the Chinstrap Penguin has only used 70%. Thus, a left-shifted oxygen-hemoglobin dissociation curve means that oxygen remains available even as the partial pressure of oxygen in the blood declines to very low levels. Gentoo Penguin, *Pygoscelis papua*; Adelie Penguin, *Pygoscelis adeliae*. (Figure modified from Milsom et al. 1973; © 1973 Oxford University Press, used with permission)

**Box 6.5 Bar-Headed Geese**

Bar-headed Geese (*Anser indicus*) migrate each spring and fall between breeding areas in Mongolia and wintering areas in India. That route requires that they cross the Himalayan mountains, reaching altitudes of 5000 to 6000 m where oxygen levels are about half that at sea level. Occasionally, they fly even higher, reaching altitudes over 7000 m (Takekawa et al. 2009; Butler 2016).

Bar-headed Geese. (Photo by Hari K. Patibanda, Wikipedia, CC BY 2.0, https://creativecommons.org/licenses/by/2.0/)

(continued)

**Box 6.5** (continued)

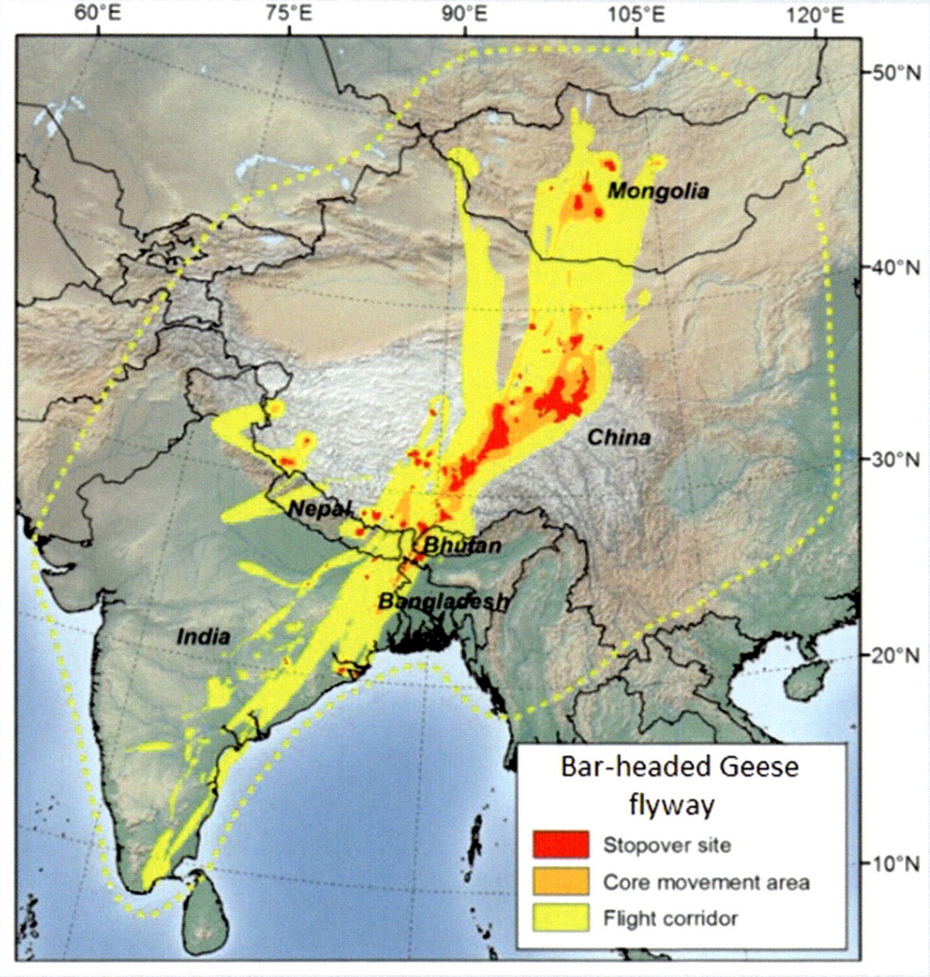


Biannual migration routes of Bar-headed Geese between India and China or Mongolia. (Figure modified from Palm et al. 2015; open-access article distributed under the terms of the Creative Commons Attribution License, https://creativecommons.org/licenses/by/4.0/)

For these geese and other birds that live at high altitudes, hemoglobin with a higher than usual affinity for oxygen would be advantageous because such molecules would be better at "loading" oxygen under the relatively hypoxic conditions found at high altitudes. In fact, Natarajan et al. (2018) examined the hemoglobin of Bar-headed Geese and closely related Graylag Geese (*Anser anser*; a low-altitude species) and found three amino-acid substitutions in Bar-headed Geese hemoglobin that increased its oxygen affinity relative to that of Graylag Geese hemoglobin. Other species of high-altitude waterfowl also have a slightly different form of hemoglobin with a greater affinity for oxygen (Weber et al. 1993, Lutfullah et al. 2005). High-affinity hemoglobin enhances oxygen loading in the lungs, but potentially makes "unloading" in the tissues more

(continued)

**Box 6.5** (continued)

difficult. Fortunately, this potential disadvantage can be overcome via the Bohr effect as well as by increasing the number of capillaries. More capillaries in active tissues such as muscle increase the surface area for exchange and shorten diffusion distance (Weber et al. 1993). In Slate-colored Coots (*Fulica ardesiaca*), for example, a comparison of skeletal muscle from individuals at sea level and those at an altitude of 4200 m revealed that the muscle of high-altitude birds had more capillaries per square millimeter (Leon-Velarde et al. 1993).

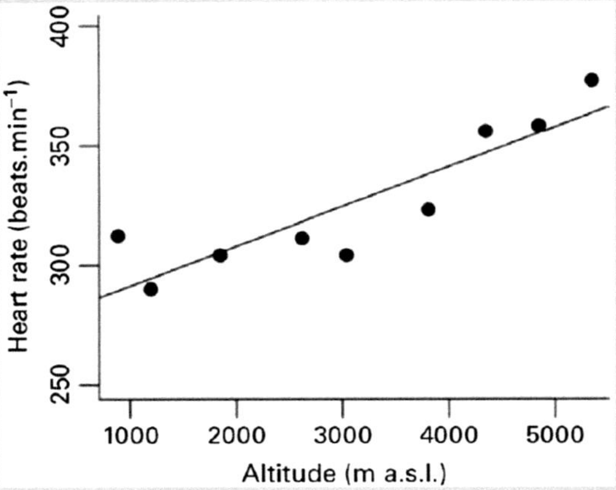


Relationship between flight altitude and heart rate for Bar-headed Geese flying across the Tibetan plateau. Note that heart rates increased with increasing altitude (Figure from Hawkes et al. 2017; © 2017 Cambridge University Press, used with permission via PLSclear)

(*Aptenodytes forsteri*) have a $P_{50}$ of about 28 mm Hg, which means that, compared to birds with higher values of $P_{50}$, more oxygen is available for vital tissues as the partial pressure of oxygen in the blood declines during dives.

Left-shifted hemoglobin resulted from mutations that slightly altered the amino-acid composition of the globin, or protein portion, of the hemoglobin molecule that increased its affinity for oxygen (Weber et al. 1993; Lutfullah et al. 2005; Signore et al. 2021). High-affinity hemoglobin enhances oxygen loading in the lungs, but potentially makes "unloading" in the tissues more difficult. This potential disadvantage can be overcome in part by increasing the number of capillaries. More capillaries in active tissues such as muscle increase the surface area for exchange and shorten diffusion distance (Weber et al. 1993).

Also important in the efficient "unloading" of oxygen by hemoglobin is the Bohr effect, i.e., hemoglobin's lower affinity for oxygen with increases in the partial pressure of carbon dioxide in the blood and/or a decrease in blood pH and so-named because it was first reported in the early 1900s by a Danish physiologist named Christian Bohr. When flying at high altitudes or diving underwater, skeletal muscles like the pectoralis and cardiac muscle in the heart are using oxygen and generating carbon dioxide at high rates and so, in these tissues, oxygen levels decrease and carbon dioxide levels increase. The increase in carbon dioxide levels also causes a decline in pH because carbon dioxide reacts with water to produce carbonic acid. Carbonic acid then dissociates or breaks down and releases bicarbonate plus a hydrogen ion. The release of these hydrogen ions creates a more acidic environment

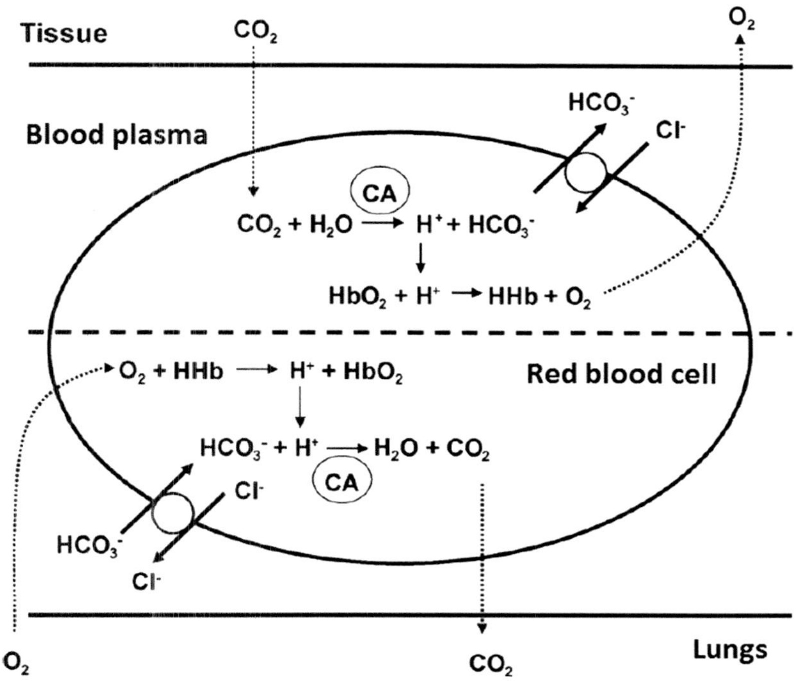


**Fig. 6.12** Transport of carbon dioxide (and oxygen) by the blood. Carbon dioxide ($CO_2$) diffuses into red blood cells and the enzyme carbonic anhydrase (CA) catalyzes its reaction with water ($H_2O$) to form carbonic acid ($H_2CO_3$) that quickly dissociates into a bicarbonate ion and a proton ($H^+$). The proton is buffered by hemoglobin (Hb) (the affinity of hemoglobin for oxygen drops after hemoglobin binds with a proton, so oxygen is released and diffuses from the red blood cell into the surrounding tissues) and the bicarbonate ion is transported out of the red blood cell into the plasma. In the lungs, these reactions are reversed and $CO_2$ diffuses from red blood cells into the lungs to be exhaled. (Figure modified from Esbaugh and Tufts 2006; ©2006 Elsevier B.V., used with permission)

(Fig. 6.12). Exposed to such conditions, including lower oxygen levels, more carbon dioxide, and a lower pH, the affinity of hemoglobin for oxygen decreases. This, of course, is beneficial because it means that hemoglobin passing through the capillaries in flight muscles and the heart will unload more oxygen to active cells that need more oxygen. Birds with hemoglobins that have higher affinities for oxygen also tend to have enhanced Bohr effects, so not only do they have enhanced extraction of oxygen in the lungs, they are also able to more efficiently "unload" that oxygen in the systemic capillaries (e.g., Signore et al. 2021). For example, diving Emperor Penguins (*Aptenodytes forsteri*) use almost all available oxygen in the blood during their dives, with end-of-dive values of the venous partial pressure of oxygen as low as 1 to 6 torr (Ponganis et al. 2007).

## 6.5 $CO_2$ Transport in the Blood

The blood also transports carbon dioxide (a waste product of ATP production) from the tissues of the body to the lungs where it can be exhaled (Fig. 6.11). Some carbon dioxide dissolves in, and is transported in, the water of the plasma (about 5%). The rest is transported as bicarbonate ion, with carbon dioxide reacting with water (a reaction catalyzed by carbonic anhydrase in red blood cells) to form carbonic acid. Carbonic acid then quickly dissociates to form bicarbonate ion plus a hydrogen ion:

$$CO_2 + H_2O \leftrightarrow H_2CO_3 \leftrightarrow HCO_3^- + H^+$$

This reaction is reversed in the lungs, with carbon dioxide diffusing from the blood into the lungs and eliminated by exhalation.

## 6.6 Flight at High Altitudes

How are birds able to fly at altitudes where oxygen levels are so much lower? In part, the answer to that question lies in the efficiency of the avian respiratory system (See Chap. 7), with oxygen levels in the lungs higher in birds than in mammals. As a result, at high altitudes where oxygen levels are low, bird blood and hemoglobin can more efficiently pick up oxygen in the lungs. In addition, as temperature decreases, the affinity between hemoglobin and oxygen increases (Meir and Milsom 2013). At high altitudes, temperatures can be very low. For example, a flock of Whooper Swans (*Cygnus cygnus*) was once observed flying at an altitude of 8200 m where the estimated ambient temperature was −48 °C (Elkins 1983). Breathing cold air is known to decrease arterial temperatures in pigeons (Maginniss et al. 1997). In addition, the body temperature of hypoxic birds exposed to cold temperatures tends to decrease (Bernstein 1989; Barnas and Rautenberg 1990). If breathing cold air also lowers the temperature of lung capillaries, then, with increasing altitude and decreasing temperature, the affinity of avian hemoglobin for oxygen would increase and the blood would pick up even more oxygen (Meir et al. 2019).

Other features of the avian circulatory system also help birds handle the hypoxic conditions at high altitude. Birds have higher densities of capillaries than mammals in skeletal muscles (Fig. 6.13), the heart (cardiac muscle), and the brain. This enhances the diffusion of oxygen from the blood to the cells in these tissues (Faraci 1991). The skeletal muscles and cardiac muscle of the heart of birds also contain substantial amounts of myoglobin.

Myoglobin is an oxygen-storage protein found in muscle and is so-named because of its functional and structural similarity to hemoglobin (Fig. 6.14). Myoglobin binds to and stores oxygen in skeletal and cardiac muscle, serving as a source of oxygen for cells (and mitochondria) during periods of increased metabolic activity and/or apnea (Ordway and Garry 2004). Myoglobin is particularly abundant in the muscles of diving birds, but is also found in the muscles of terrestrial, non-diving birds. Concentrations of myoglobin in muscle cells vary widely among birds. For example, gallinaceous birds may have concentrations of less than 1 mg per gram of muscle tissue (Kranen et al. 1999), whereas long-duration divers like Emperor Penguins (*Aptenodytes forsteri*) have concentrations of about 64 mg per gram of muscle tissue (Kooyman and Ponganis 1998).

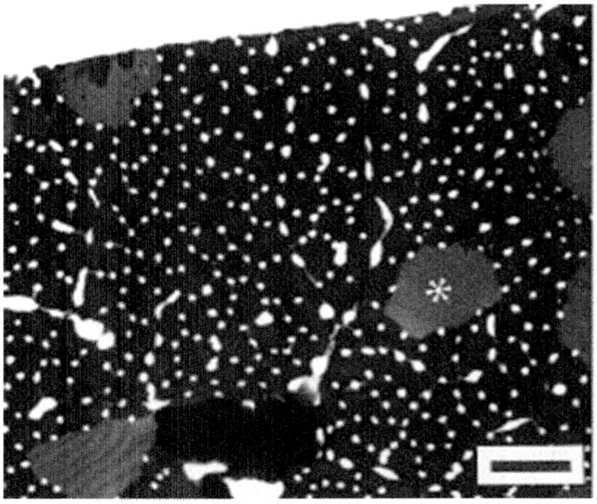

**Fig. 6.13** Light micrograph of a portion of a pectoralis muscle of a Rock Pigeon (*Columba livia*). Aerobic (dark, narrow) and glycolytic muscle fibers (wide; asterisk) can be clearly identified. Capillaries are empty. Scale bar, 50 μm. (Figure from Mathieu-Costello and Agey 1997; © 1997 Elsevier Science B.V. All rights reserved, used with permission)

Among different species of diving birds (and mammals), the concentration of myoglobin in muscles is positively correlated with the duration of their typical dives (Kooyman and Ponganis 1998). The structure of myoglobin molecules also varies among species and, as a result, so does the oxygen-binding affinity of these slightly different myoglobins. As with hemoglobin, the oxygen-binding affinity of myoglobin is

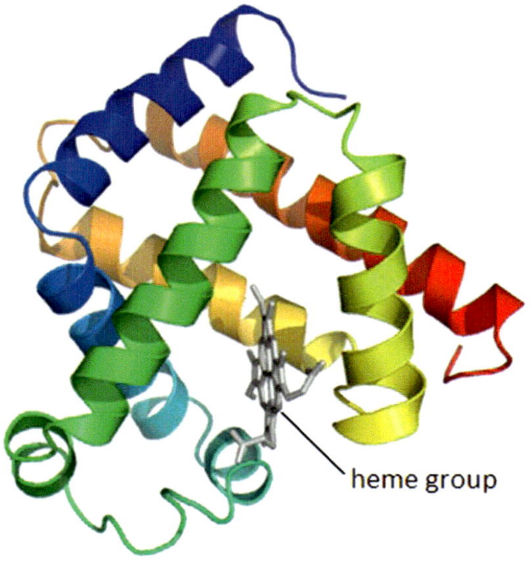


**Fig. 6.14** Three-dimensional model of a molecule of myoglobin. Myoglobin consists of a single polypeptide chain of 154 amino acids (blue). This chain wraps around a central pocket containing a heme group (red) that has a very high affinity for oxygen. (Figure from Wu et al. 2016; © 2015, Dt. Ornithologen-Gesellschaft e.V., used with permission)

expressed as $P_{50}$, or the partial pressure (mm Hg) of oxygen at which 50% of the molecules are still bound to oxygen (with partial pressure representing the pressure of a particular gas in a mixture of gases, e.g., the pressure of oxygen in air that also includes nitrogen and carbon dioxide). A lower $P_{50}$ indicates greater oxygen-binding affinity. Studies of multiple species of birds and mammals have revealed surprisingly little variation in the oxygen affinity of myoglobin. However, the myoglobins of several species of long-duration divers, including some seals (pinnipeds) and several species of penguins, do have significantly lower values of $P_{50}$ (Fig. 6.15). This increased oxygen affinity appears to be advantageous, enhancing the diffusion of oxygen from the blood into muscle cells (because the oxygen affinity of the myoglobin is greater than that of hemoglobin) even as blood oxygen levels decline and ensuring the availability of oxygen for mitochondria (and continued production of ATP) in muscle cells even under conditions of extreme hypoxia (Burmester and Hankeln 2014).

The nervous system of birds (and other vertebrates) also contains a respiratory protein called neuroglobin (Burmester and Hankein 2009). This protein, like hemoglobin and myoglobin, can bind to oxygen and may provide a way to store oxygen in nervous tissue in the same way that myoglobin stores oxygen in muscles (Burmester and Hankeln 2004). In addition, neuroglobin may serve as a "neuroprotector" during periods of ischemia or hypoxia in the nervous

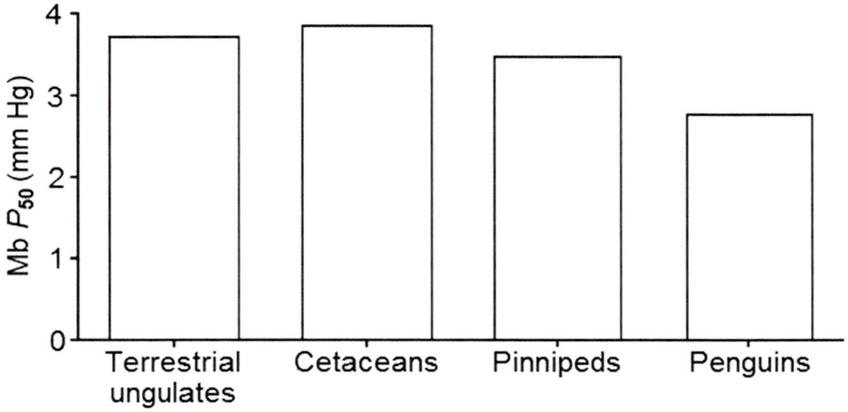


**Fig. 6.15** Mean oxygen affinity of myoglobin (Mb) for five species of terrestrial ungulates, eight species of cetaceans, six species of pinnipeds, and five species of penguins. Several species of penguins that make long-duration dives have myoglobins with significantly lower values of Mb $P_{50}$. (Figure based on data from Wright 2014)

system (Ramírez-Plascencia et al. 2019). Birds regularly exposed to hypoxic conditions, e.g., those found at higher altitudes or make extended dives, might have higher concentrations of neuroglobin and be able to store greater amounts of oxygen in nervous tissue, including the brain (e.g., Knower Stockard et al. 2005; Ponganis et al. 2007).

Another reason why birds can handle hypoxia so well is their unique cerebral circulation. For mammals, hypoxia causes hyperventilation (an increase in respiratory rate) and hyperventilation flushes more than the usual amounts of carbon dioxide from the lungs and the blood (hypocapnia). Hypocapnia, in turn, causes vasoconstriction of blood vessels to the brain and, in humans, for example, this can cause neurological impairment (altitude or mountain sickness; Harvey et al. 1988). Hypocapnia does not cause blood vessels supplying the brain of birds to vasoconstrict, so oxygen delivery is maintained during periods of hypoxia (Faraci et al. 1984). This ability to maintain oxygen delivery to the brain contributes to the exceptional tolerance of some birds to very high altitudes (Faraci 1991). Another aspect of cerebral circulation in birds is also exceptional. Behind the eyes of birds is a unique network of arteries and veins called the rete opthalmicum (Fig. 6.16). A rete is a dense network of blood vessels and, in Turkey Vultures (*Cathartes aura*), for example, the rete opthalmicum consists of about 100 arteries and 50 veins (Arad et al. 1989). Arteries traveling to the brain are immediately adjacent to veins carrying cool blood away from the buccopharyngeal (oral cavity) surfaces, beak, and eyes (Kilgore et al. 1973). This arrangement allows the rete to serve as a countercurrent heat exchanger and helps keep a bird's brain from overheating on hot days; heat from the arteries is transferred to the veins and, as a result, the brain receives cooler blood. It is likely, however, that this rete performs a second function. During hypoxic conditions, veins coming from the highly vascular eye and upper respiratory passageways may actually have more oxygen than the arteries. If so, oxygen could diffuse from the veins into the arteries and help provide the brain with additional oxygen (Bernstein 1989).

## 6.7 Embryonic Respiration

Birds at high altitudes must deal with hypoxic conditions, but so must bird embryos in eggs. Oxygen diffuses across the shell and the attached membranes into the capillary bed of the chorioallantois that serves as the "lung" of bird embryos. However, as the embryo grows, oxygen consumption continues to increase while the conductance of the shell to oxygen and carbon dioxide diffusion remains constant. As a result, the developing embryo is faced with increasing hypoxia and hypercapnia as the time of hatching approaches. In contrast to the red blood cells of adults, embryonic red blood cells are released into the circulation as erythroblasts (immature cells) and act as gas transporters while they complete the final maturation process. As this maturation occurs, enzymes in the red blood cell break down molecules of RNA that are no longer needed. As this process begins a few days before hatching, levels of several substances produced by this degradation of RNA (which, remember, is composed of four bases: adenine, guanine, cytosine, and uracil) increase, including ATP (adenosine triphosphate) and UTP (uridine triphosphate). ATP and UTP interfere with the ability of hemoglobin to bind with oxygen, but their levels decline as enzymes degrade them. At the same time, as the date of hatching approaches, red blood cells begin to produce a substance called 2,3-biphosphoglycerate (or BPG), a substance that binds to hemoglobin and causes it to more readily give up its oxygen in the tissue capillaries (Fig. 6.17). In addition, red blood cells increase production of carbonic anhydrase II (CAII), a substance that improves the ability of blood to transport $CO_2$ (Dragon and Baumann 2003).

As a result of the decrease in ATP and UTP levels, the affinity of hemoglobin for oxygen increases dramatically. In Domestic Chickens (*Gallus g. domesticus*), for example, as the day of hatching approaches, the $P_{50}$ declines from

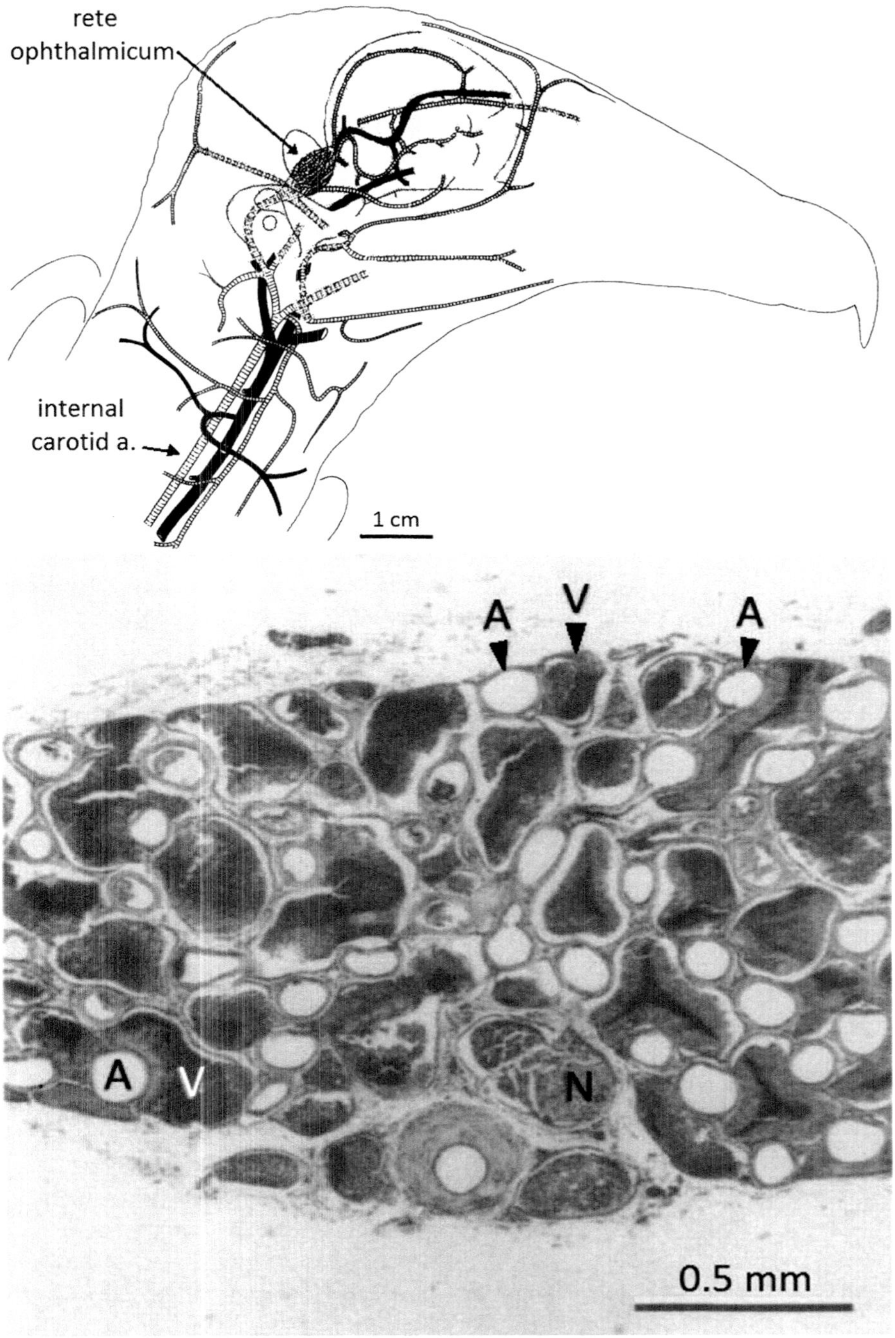


**Fig. 6.16** Top) The major arteries of the head in a Turkey Vulture (*Cathartes aura*) including the rete ophthalmicum. Bottom) A cross-sectional view of part of the rete ophthalmicum showing the close arrangement of arteries (A) and veins (V). N = nerve. (Figure modified from Arad et al. 1989; © 1989 Oxford University Press, used with permission)

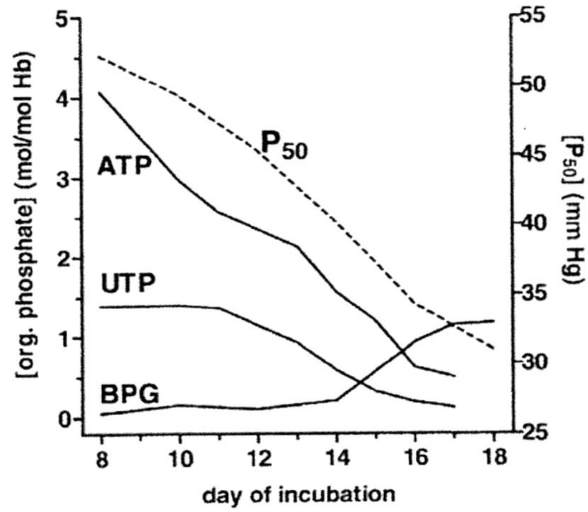


**Fig. 6.17** Levels of uracil triphosphate (UTP), adenosine triphosphate (ATP), 2,3-biphosphoglycerate (BPG), and partial pressure of oxygen at which 50% of hemoglobin molecules are bound to oxygen in Domestic Chicken (*Gallus g. domesticus*) embryos. Hatching will occur on about day 20 or 21. (Figure from Dragon and Baumann 2003; © 2003 The American Physiological Society, used with permission)

about 50 mm Hg on day 8 to about 31 mm Hg on day 18 (Fig. 6.17). So even with the hypoxic conditions in the egg, hemoglobin efficiently transports the available oxygen to the cells and tissues of the growing embryo.

## 6.8 Diving Birds

Flying or living at high altitude and living and growing in an egg require adaptations for dealing with hypoxia, but many birds forage underwater where, of course, no oxygen is available. Oxygen is, however, stored in a bird's body, primarily in three locations: the blood (hemoglobin), the muscles (myoglobin), and the respiratory system (e.g., air sacs and lungs). The extent to which oxygen is stored in each of these locations varies among species. For birds that typically forage in more shallow water and for shorter periods of time, more oxygen is stored in the respiratory system than in hemoglobin or myoglobin. In diving Tufted Ducks (*Aythya fuligula*), for example, 50–60% of the body's oxygen is in the respiratory system, 35–45% is in the blood (hemoglobin), and just 3–4% is in the muscles (myoglobin) (Butler and Jones 1997; Kooyman and Ponganis 1998).

Although cells can make energy (ATP) in the absence of oxygen (anaerobic metabolism), the process is relatively inefficient (compared to aerobic metabolism) and generates more waste products (like lactic acid). Some diving birds are sustained for variable lengths of time by anerobic metabolism, but most dives by most birds are likely aerobic (Butler and Jones 1997). The duration of dives, then, is determined by how efficiently birds can use their stored oxygen. Birds that depend on short duration, shallow dives, like many waterfowl, use changes in blood flow to minimize oxygen consumption. Tufted Ducks typically dive no deeper than about 5 meters and for no longer than 30–45 s (Butler 2004). Because these ducks are so buoyant, they must expend substantial effort to stay submerged. Oxygen consumption during a dive is about three times that during resting conditions and, as a result, dive duration is limited. To extend dive duration, blood flow in a diving Tufted Duck changes dramatically, with more blood going to leg muscles (about 57% of all blood!), the heart, and the Central Nervous System and less blood going to the rest of the body (Butler 2001, 2004). Also, movement of the legs (e.g., ducks) or wings (e.g., puffins and penguins) of diving birds causes pressure differences between the posterior and anterior air sacs that may move air through the lungs where gas exchange can occur (Boggs et al. 1998, 2001).

Although many species of birds forage by diving, most make relatively shallow, short-duration dives. In general, diving ability (depth and duration) increases with body mass (Halsey et al. 2006; Fig. 6.18), largely because larger birds have greater oxygen-storing capacity and smaller birds have higher metabolic rates and tend to use oxygen at faster rates (Lasiewski and Calder 1971; Butler and Jones 1982). More precisely, oxygen stores generally increase isometrically with body mass ($M^{1.00}$), whereas the rate of oxygen consumption is proportional to basal metabolic rate that scales at about $M^{0.75}$ (Halsey et al. 2006).

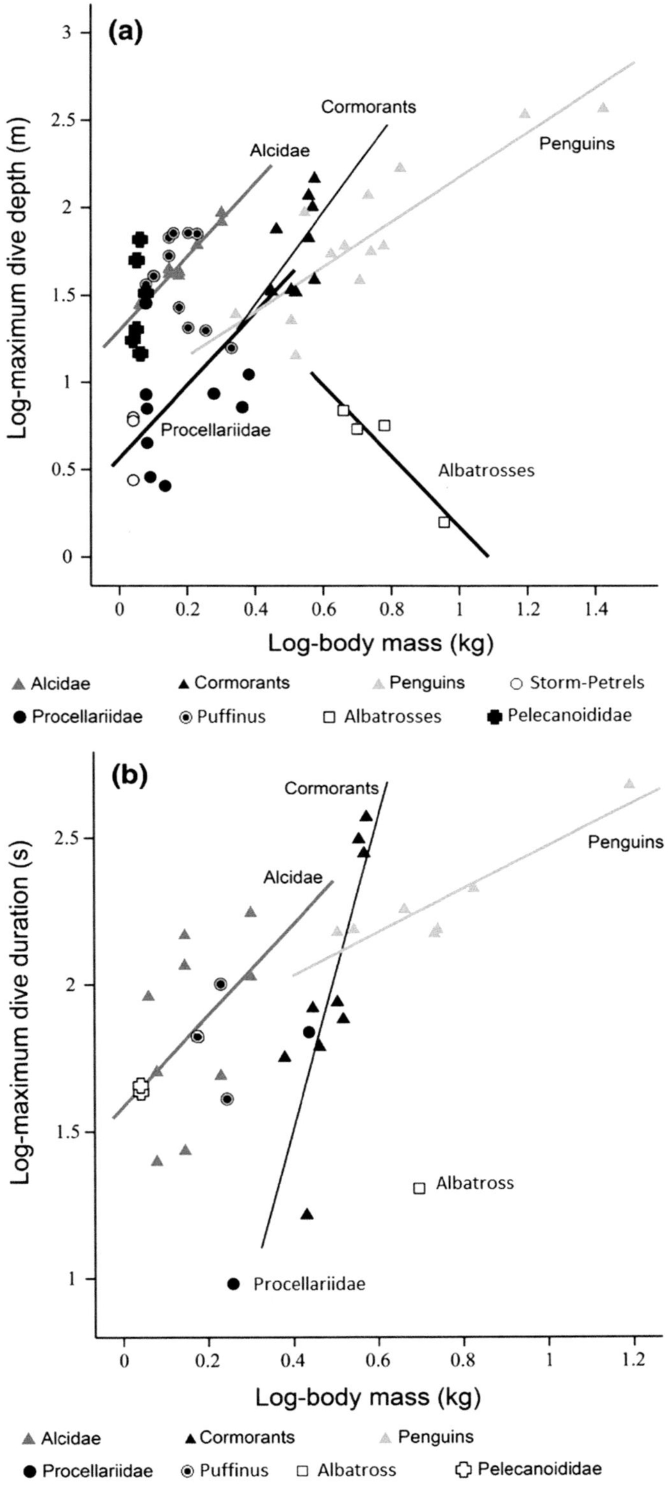


**Fig. 6.18** Relationships between body mass and dive depth (**a**) and duration (**b**) for several taxa of seabirds. In general, dive depth and duration increase with increasing body mass and an increasing ability to store oxygen. (Figure from Navarro et al. 2014; © 2014 Springer-Verlag Berlin Heidelberg, used with permission)

Some diving birds can reach impressive depths—Emperor Penguins (*Aptenodytes forsteri*) may dive as deep as 564 m—and stay submerged for several minutes (Kooyman and Ponganis 1998; Fig. 6.19). Among the longest dives ever recorded for a bird was an incredible 22-min dive by an Emperor Penguin (Kooyman and Ponganis 1998). Birds that dive so deep and for so long store more oxygen in the muscles (myoglobin) and less in the respiratory system. In King Penguins (*Aptenodytes patagonicus*), for example, only about 17% of the oxygen is in the respiratory system, with 29% in the blood and 54% in the muscles (Kooyman and Ponganis 1998). Less air and oxygen in the respiratory system of birds that make deep dives is advantageous because less nitrogen gas means less risk of decompression sickness ("the bends") (Kooyman and Ponganis 1998).

Although the best avian divers store less oxygen in the respiratory system than in the muscles, they appear to make good use of it. When Emperor Penguins dive, the amount of oxygen in the air sacs (e.g., $P_{O2} = 4$ mm Hg; Stockard et al. 2005) and blood decreases and can reach surprisingly low levels (6 mm Hg, Ponganis et al. 2010; Fig. 6.20). These levels indicate a remarkable hypoxic tolerance by Emperor Penguins. How Emperor Penguins avoid blacking out under such conditions is unclear. Oxygenation of blood going to the brain by way of the rete ophthalmicum is not possible during a dive, of course, because fresh air is not available to pass over the nasal and oral mucosa and oxygenate the venous blood in the rete. Other possible explanations for this remarkable tolerance of extreme hypoxia include the high concentration of capillaries in the brains of birds (Faraci 1986) and, as suggested previously, elevated concentrations of neuroglobin in the brain (Stockard et al. 2005).

Investigators have been curious about the physiology of diving birds for many decades. Early attempts to examine the physiological responses of diving birds involved captive birds being forced underwater (Scholander 1940). These studies revealed that heart rates dropped dramatically during "dives." Bradycardia, or reduced heart rate, also occurs when aquatic mammals dive and this almost immediate drop in heart rate after submersion is considered part of the "dive reflex" or "dive response" and helps diving animals conserve energy and oxygen. More recent studies have revealed that, although the heart rates of aquatic mammals do decline during dives, the heart rates of most diving birds apparently do not. For example, King Penguins (*Aptenodytes patagonicus*) equipped with data loggers that recorded diving depth and heart rates were monitored during several thousand dives and were found to have heart rates ranging from 50–150 beats per minute when ashore and 50–260 beats per minute when at sea. The mean heart rate of penguins was about 87.2 beats per minute when ashore and 111 beats per minute at sea (Froget et al. 2004).

Heart rates during dives vary among species. In Macaroni Penguins (*Eudyptes chrysolophus*), for example, heart rates are higher than when resting both before and after dives, but fall to a level close to or a bit lower than when resting during dives of longer duration (Green et al. 2003). Among diving birds that can also fly (e.g., diving ducks and cormorants), heart rates typically increase during dives, but, among most penguins, heart rates increase at the beginning of dives and progressively decline as the duration of dives increases (Ponganis 2015). Only Emperor Penguins exhibit true bradycardia, with very low heart rates during long-duration dives (Meir et al. 2008; Wright et al. 2014; Fig. 6.21). During the longest dives observed, Meir et al. (2008) reported a heart rate as low as 3 beats/min. The decline in heart rate with dives of longer duration may help conserve oxygen levels in the blood, while also suggesting an increasing dependence on myoglobin-bound oxygen (Ponganis 2015).

An increase in heart rate during the dives of many aquatic birds suggests an increase in metabolic rate and a corresponding increase in oxygen consumption. Based on rates of oxygen consumption calculated using heart rates, many diving birds exceed their aerobic diving limit (ADL) on a regular basis. For example, King Penguins

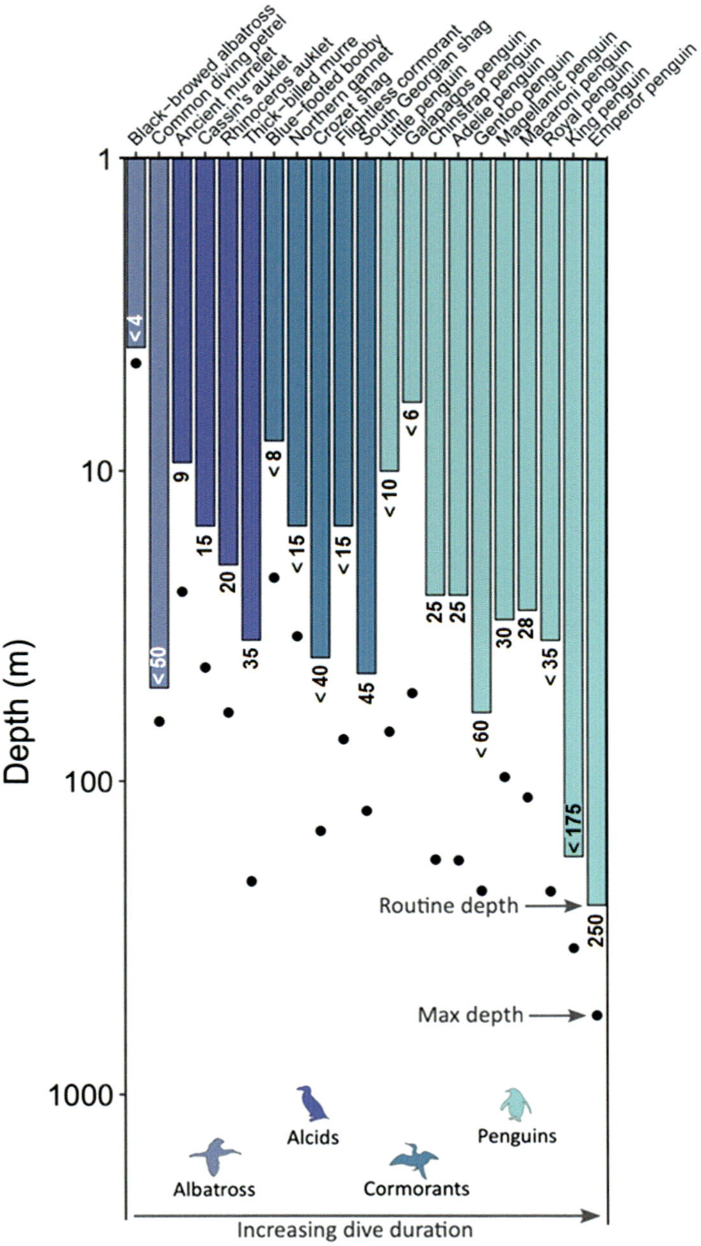


**Fig. 6.19** Routine and maximum dive depths for several species of birds. Routine dive depths are provided below (or at the end of) the bars (in meters). Species are grouped by orders as indicated by the different colors and ordered based on increased durations of routine dives. Albatross, order Procellariiformes; Alcids, order Charadriiformes; Cormorants, order Suliformes; Penguins, order Sphenisciformes. Scientific names: Black-browed Albatross, *Thalassarche melanophrys*; Common Diving Petrel, *Pelecanoides garnotii*; Ancient Murrelet, *Synthliboramphus antiquus*; Cassin's Auklet, *Ptychoramphus aleuticus*; Rhinoceros Auklet, *Cerorhinca monocerata*; Thick-billed Murre, *Uria lomvia*; Blue-footed Booby, *Sula nebouxii*; Northern Gannet, *Morus bassanus*; Crozet Shag, *Leucarbo melanogenis*; Flightless Cormorant, *Phalacrocorax harrisi*; South Georgian Shag, *Phalacrocorax georgianus*; Little Penguin, *Eudyptula minor*; Galápagos Penguin, *Spheniscus mediculus*; Chinstrap Penguin, *Pygoscelis antarcticus*; Adélie Penguin, *Pygoscelis adeliae*; Gentoo Penguin, *Pygoscelis*

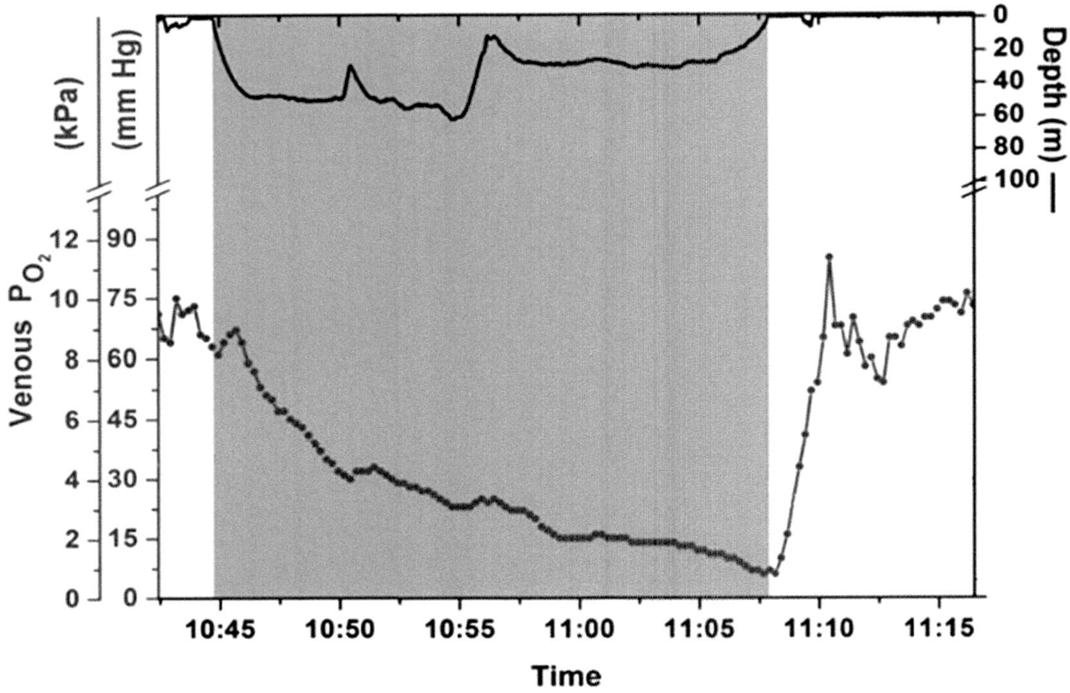


**Fig. 6.20** Dive depth and partial pressure of oxygen in the inferior vena cava of an Emperor Penguin (*Aptenodytes forsteri*) during a 23.1 min dive. Gray background = penguin diving. (Figure from Ponganis et al. 2010; © Inter-Research 2010, used with permission)

(*Aptenodytes patagonicus*) appear to exceed their ADL when dives exceed about 3.5 min in duration, and more than a third of their dives exceed 3.5 min (Handrich et al. 1997; Froget et al. 2004). Birds that exceed their ADL have used all available oxygen and begin to produce inorganic acids, like lactic acid, as by-products of anaerobic metabolism. Lactic acid dissociates into lactate and $H^+$ and an increase in $H^+$ (i.e., reduced pH or acidosis) can contribute to muscle fatigue. In addition, anaerobic metabolism in skeletal muscle generates inorganic phosphates that can also reduce muscle efficiency. Muscles depending on anaerobic metabolism, in other words, become increasingly inefficient and less able to sustain contraction.

One possible solution to this problem would be to accumulate lactic acid and other waste products during long dives and then recover (or pay off the "oxygen debt") by remaining at the surface as long as needed. King Penguins, however, typically do not remain at the surface long enough to permit recovery (Froget et al. 2004), and this is also true for other penguins (Handrich et al. 1997). The other possible "solution" is that perhaps no solution is actually

**Fig. 6.19** (continued) papua; *Magellanic Penguin,* Spheniscus magellanicus; *Macaroni Penguin,* Eudyptes chrysolophus; *Royal Penguin,* Eudyptes schlegeli; *King Penguin,* Aptenodytes patagonicus; *Emperor Penguin,* Aptenodytes forsteri. *(Figure modified from Favilla and Costa 2020; © 2020 Favilla and Costa, open-access article distributed under the terms of the Creative Commons Attribution License (CC BY), https://creativecommons.org/licenses/by/4.0/)*

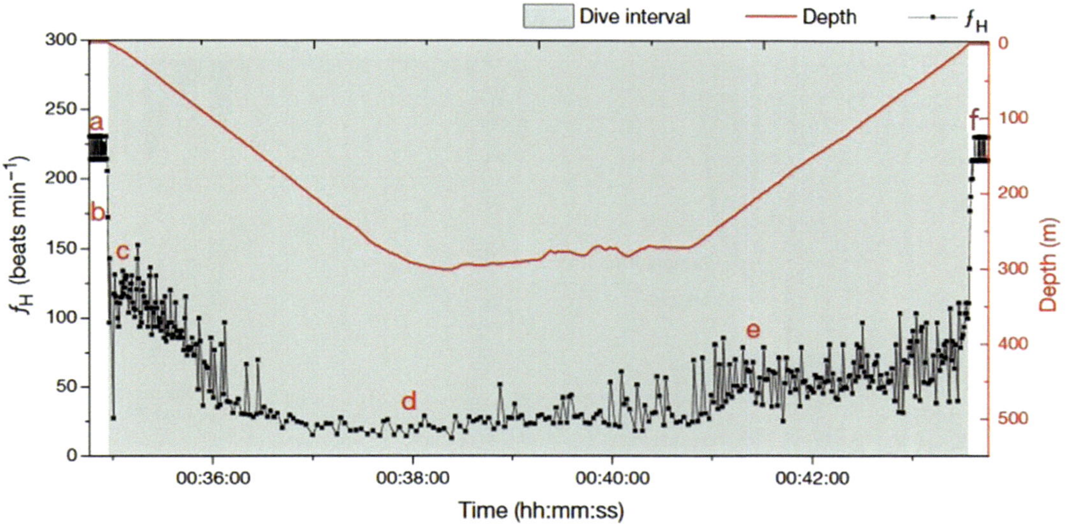


**Fig. 6.21** Heart rate ($f_H$) and dive depth of an Emperor Penguin (*Aptenodytes forsteri*). During a typical dive, heart rate is rapid prior to diving (**a**), declines immediately after entering the water (**b**), gradually declines as the penguin descends (**c**), extreme bradycardia at the maximum depth (**d**), slow increase during ascent (**e**), and rapid increase after leaving the water (**f**). (Figure from Wright et al. 2014; © Inter-Research 2014, used with permission)

needed. Deep-diving penguins, and perhaps other deep-diving birds, may lower their metabolic rates by lowering their body temperatures. The abdominal temperatures of King Penguins may fall as much as 10 or more degrees C during sustained deep dives (Fig. 6.22). The slower metabolism of cooler tissues might help explain how deep-diving birds can stay submerged so long (Handrich et al. 1997). For Gentoo Penguins (*Pygoscelis papua*), a drop in body temperature of only 2.4 degrees C would reduce oxygen consumption when diving to the point where all dives would be within their aerobic diving limit (Bevan et al. 2002).

## 6.9 Thrombocytes

Avian thrombocytes are, like avian red blood cells, nucleated and are the equivalent of the nonnucleated platelets in mammalian blood (Fig. 6.23). Thrombocytes are the second most numerous blood cell in bird blood (after red blood cells), with 20,000–30,000 per cubic mm in Domestic Chicken blood (Campbell 1988), an average of about 50,000 per cubic mm in Common Ostrich (*Struthio camelus*) blood (Bonadiman et al. 2009), and about 18,000 per cubic mm in the blood of Black-crowned Night-Herons (*Nycticorax nycticorax*; Yogamaya and Prafulla 2016). A primary function of thrombocytes is to aggregate at the site of a vessel injury or rupture. This "clumping" can create a "plug" that stops blood loss from damaged blood vessels, particularly very small vessels like capillaries. Thrombocytes are activated when tissues, like with walls of blood vessels, are damaged. This damage or rupture of a blood vessel wall exposes thrombocytes to collagen (a connective-tissue protein located in the walls of all blood vessels except capillaries and outside the walls of all blood vessels) and thrombin (a clotting protein formed in response to tissue damage). This exposure causes thrombocytes to change in shape (becoming much more irregular in shape with the extension of pseudopods) and release materials (such as a clotting protein called fibrinogen) that cause thrombocytes to adhere and aggregate (O'Toole et al. 1994; Fig. 6.24). In addition to forming "plugs" that can help stop

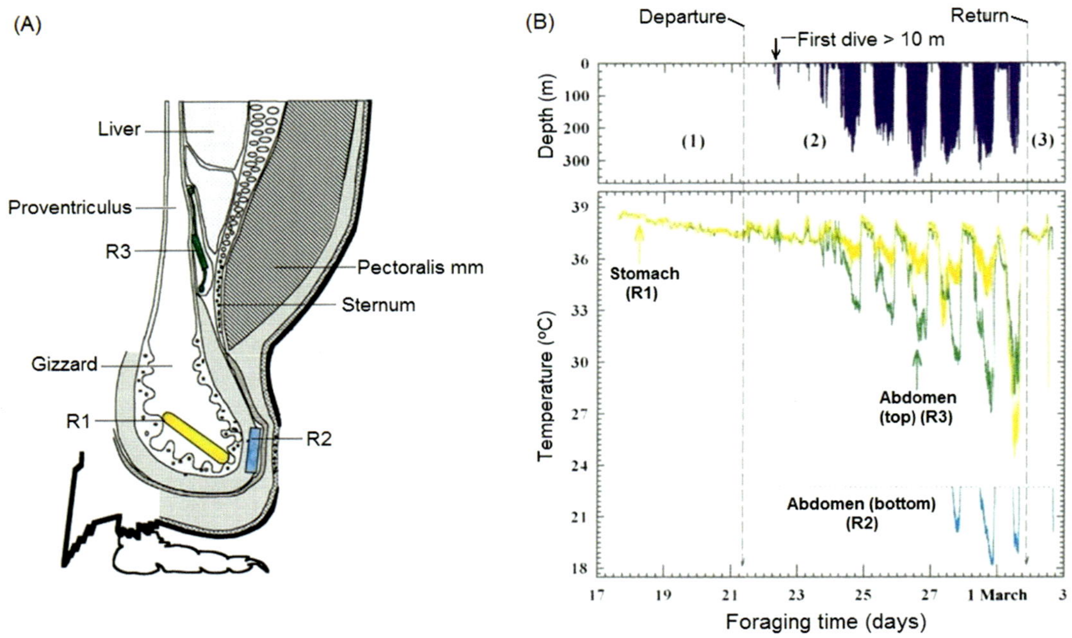


**Fig. 6.22** (**a**) Abdominal region of a King Penguin (*Aptenodytes patagonicus*) showing the location of recorders in the stomach (R1), abdominal fat of the lower abdomen (R2), and under the sternum and against the wall of the proventriculus (R3). (**b**) Data obtained from the three recorders during a foraging trip by a King Penguin. (Figures from Handrich et al. 1997; © 1997 Springer Nature, used with permission)

blood loss through small breaks in small vessels like capillaries, thrombocytes also play a critical role in blood clotting or coagulation. Thrombocytes adhere to the damaged area, aggregate, and create a surface, along with other cells, where blood coagulation, involving the interaction of several plasma proteins (factors plus the clotting proteins prothrombin and fibrinogen) and calcium ions, occurs. This process creates a complex mesh of protein (fibrin) and cells called a blood clot (Spronk et al. 2003; Fig. 6.24b).

Avian thrombocytes, unlike mammalian platelets, are also phagocytic and may play an important role in protecting birds from various types of bacteria and viruses (Cotter 2022). For example, thrombocytes become actively phagocytic when exposed to *Salmonella* and the Newcastle Disease Virus (Wigley et al. 1999; Fig. 6.25). Thrombocytes are, however, capable of phagocytosing a wide range of organisms and particles (such as cell debris), suggesting that they act as "scavenging" phagocytes in the blood (Wigley et al. 1999).

Thrombocytes also have protein receptors in their membranes that allow them to detect the presence of molecules (antigens) produced by pathogens like bacteria and viruses. Once detected, thrombocytes can produce and release a variety of substances that can play important roles in an immune response, including cytokines, chemokines, and other bioreactive mediators (Ferdous and Scott 2015; Bean and Lowenthal 2022; Lu et al. 2023; Fig. 6.26). These substances help mediate the immune response, e.g., attracting phagocytic heterophils, macrophages, and natural killer cells that can ingest and destroy pathogens (Lu et al. 2023).

## 6.10 White Blood Cells

Birds have several types of white blood cells, including lymphocytes, heterophils, monocytes,

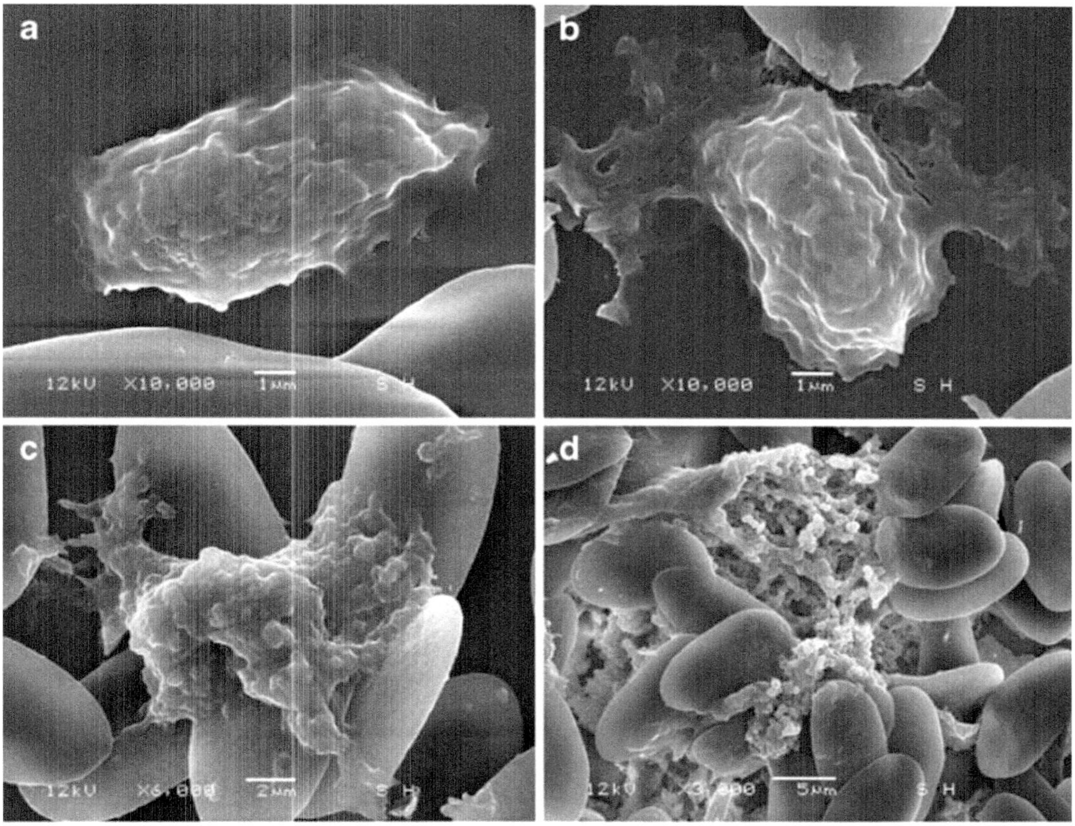


**Fig. 6.23** Thrombocytes of a White-bellied Sea Eagle (*Haliaeetus leucogaster*). (**a**) Normal thrombocyte; (**b**) "Spread" thrombocyte, with irregular shape assumed during blood clot formation; (**c**, **d**) Thrombocytes clumping with avian red blood cells to help form clots that would stop blood loss from a ruptured blood vessel. (Figure from Narkkong et al. 2010; © 2009, Springer-Verlag London Limited, used with permission)

eosinophils, and basophils (Fig. 6.27). Lymphocytes and heterophils are the most abundant, comprising up to 95% of all white blood cells (Davis et al. 2008). Lymphocytes are either T-lymphocytes (so-named because they develop in the thymus) or B-lymphocytes (so-named because they develop the bursa of Fabricius). As described in more detail below, B-lymphocytes produce antibodies and T-lymphocytes attack infected or abnormal cells. Heterophils are the avian equivalent of the mammalian neutrophil (Genovese et al. 2013). Heterophils are phagocytic and use their enzyme-containing granules to lyse, or digest, ingested materials. Heterophils are motile (or capable of moving independent of the blood by a process called ameboid movement) and can leave blood vessels to engulf foreign materials like bacteria. Monocytes are also phagocytic and motile (Qureshi et al. 2000). An important difference between heterophils and monocytes is that heterophils are found in the blood, but many monocytes remain in particular tissues or organs, such as the liver, lungs, and spleen, rather than circulating through the blood. In these organs, monocytes are referred to as macrophages and, by phagocytosis, they help eliminate foreign particles or pathogens. For example, macrophages help remove particles that enter the avian lungs during inspiration (Klika et al. 1996). In addition, as discussed in

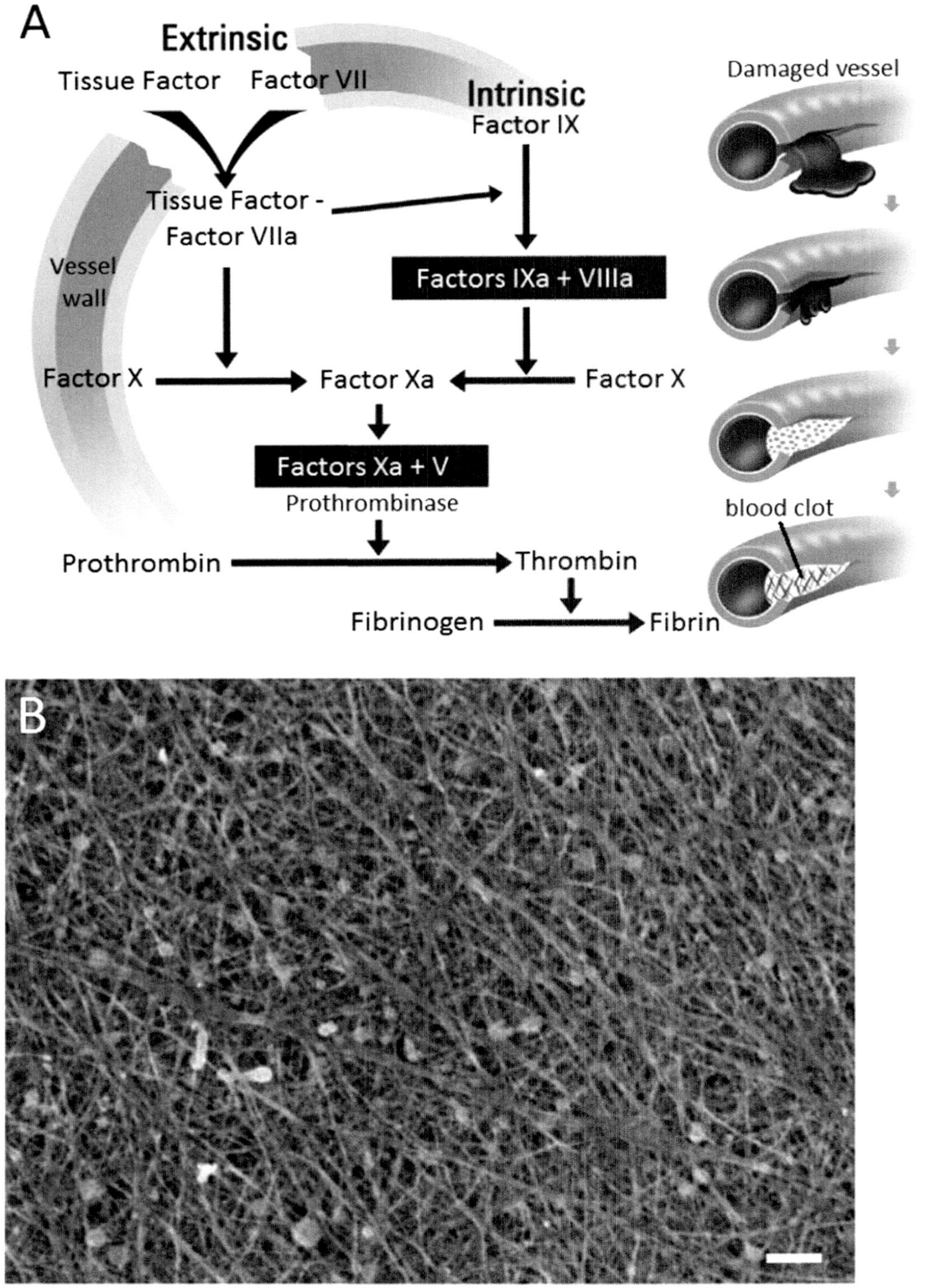


**Fig. 6.24** (**a**) Two interrelated series of reactions that usually occur simultaneously are referred to as the extrinsic and intrinsic pathways and the end result of both is production of the main component of a blood clot—fibrin threads. These reactions involve several types of proteins produced by the liver and released into the blood plasma,

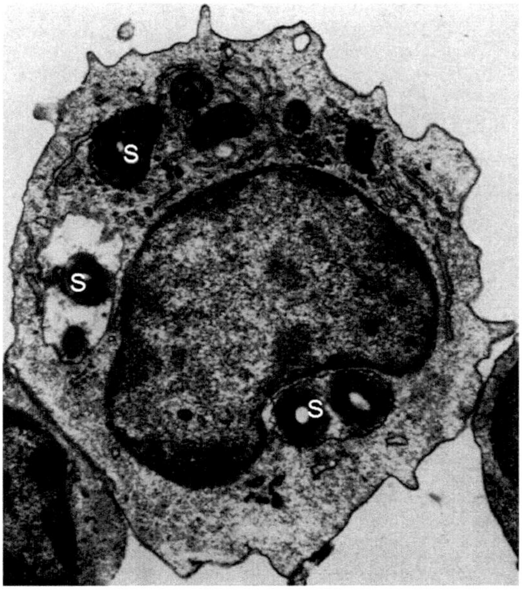


**Fig. 6.25** Thrombocyte of a Domestic Chicken that has phagocytosed three bacteria (indicated by S). Once phagocytosed, lysosome enzymes are used to "digest" the bacteria. The large object near the center of the thrombocyte is the nucleus. 21,000×. (Figure from Carlson et al. 1968; © 1968 American Association of Avian Pathologists, used with permission)

more detail later, monocytes play a critical role in initiating and maintaining an immune response.

Eosinophils make up about 2 to 3% of all white blood cells in healthy birds and appear to play some role in inflammation, particularly in the respiratory system and skin. Eosinophils also increase in numbers in response to a variety of infectious diseases such as West Nile virus, suggesting they may somehow enhance the immune response (Pendl and Tizard 2016). In non-avian vertebrates, eosinophils are thought to be important in parasite resistance (Rothenberg and Hogan 2006), and Domestic Chickens with lice and mite infestations had increased concentrations of eosinophils and heterophils (Al-Saffar and Al-Mawla 2008). Japanese Quail (*Coturnix japonica*) infected with blood parasites were also found to have higher blood concentrations of eosinophils than uninfected birds (Jubril et al. 2021). Although the mechanisms by which eosinophils might provide protection against parasites in birds are unknown, mammalian eosinophils do so by producing and releasing chemicals called interleukins that activate macrophages and neutrophils that then kill parasites via release of nitric oxide (Huang and Appleton 2016).

Avian basophils are motile, unlike those of mammals, and involved in inflammation reactions (Hodges 1979; Campbell and Ellis 2007). The abundance of basophils in the blood of birds appears to exhibit considerable variation, but generally greater than reported in mammals. Some investigators have reported that 0–5% of all white blood cells were basophils (most finches; Campbell and Ellis 2007). Others have reported percentages above 10% in some nonpasserines (Maxwell and Robertson 1995; Coke et al. 2004), and still others have reported mean percentages as high as 42% (adult Common Rosefinches, *Carpodacus erythrinus*) and 56% (nestling Common Rosefinches) (Vinkler et al.

**Fig. 6.24** (continued) including several different factors (indicated by different Roman numerals) plus prothrombin and fibrinogen. The extrinsic pathway (also called the tissue factor pathway) is activated when damaged cells in the wall of a blood vessel release a protein called tissue factor that in turn combines with Factor VII to form an enzyme that converts Factor X into Factor Xa. This enzyme (tissue factor combined with Factor VIIa) also activates the intrinsic pathway by reacting with Factor IX to produce an enzyme (Factor IXa combined with Factor VIIIa) that also converts Factor X into Factor Xa. Factor Xa then reacts with Factor V to produce the enzyme prothrombinase that converts the plasma protein prothrombin into an enzyme called thrombin. Thrombin then reacts with the plasma protein fibrinogen to produce a thread-like protein called fibrin. These fibrin threads form a mesh- or net-like structure that traps red blood cells and thrombocytes and, collectively, form a blood clot. (**b**) Blood clot of an African Penguin (*Spheniscus demersus*) showing numerous fibrin threads with "trapped" red blood cells and thrombocytes. Scale bar = 1 μm. (Figure **a** modified from Buzala et al. 2017; © 2016 Elsevier B. V., used with permission; Figure **b** from Pretorius et al. 2009; used with permission)

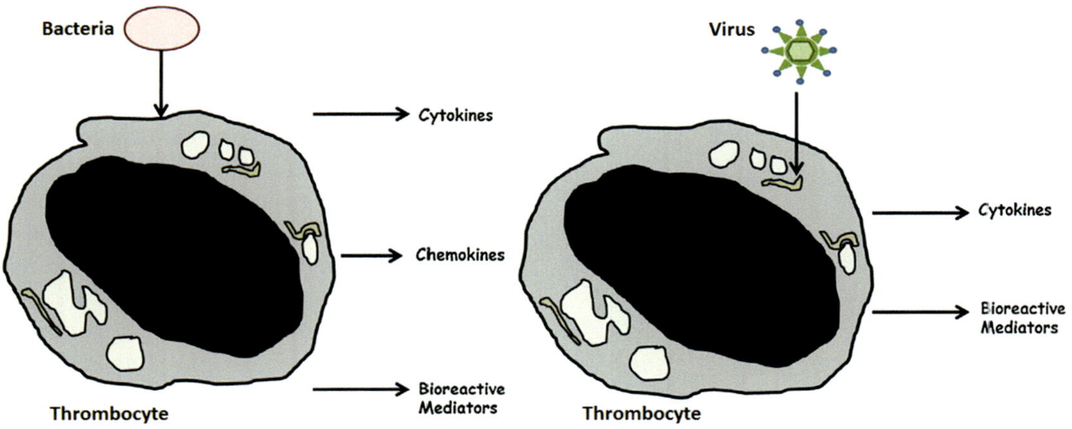


**Fig. 6.26** Avian thrombocytes detect pathogens via protein (antigen) receptors in their membranes and respond to producing and releasing a variety of substances that play important roles in eliminating those pathogens. (Figure modified from Ferdous and Scott 2015; © 2014 European Federation of Immunological Societies, used with permission)

2010). Vinkler et al. (2010) found that basophil counts were higher in birds with blood parasites (Haemoproteus) and suggested that basophils are an important defense against parasites in the blood and elsewhere. These authors further suggested that high basophil counts may represent an important anti-parasite defense for long-distance migrants.

## 6.11 Avian Immune System

White blood cells are an important component of the avian immune, or body defense, system. The function of the immune system is to provide protection from foreign substances, such as bacteria, viruses, and the toxic molecules they sometimes produce. Immune systems also provide protection from "abnormal" cells, such as cancerous cells. The avian immune system, like the immune systems of other animals, has two components: innate and acquired. The innate immune system, so-named because it provides protection from the moment of hatching (and even before hatching), has several components that provide protection against an array of potentially harmful organisms and toxins. The skin and mucous membranes act as barriers that prevent pathogens from entering the body. The digestive system is also part of the innate immune system because the lysozyme in saliva and mucus as well as the acid in the stomach can destroy bacteria and other microorganisms that might be present in food or water. For example, vultures feed on the decaying carcasses of animals, some that die from infectious diseases, and are not harmed. Several factors, including the presence of microorganisms that produce substances that inhibit the growth of other microorganisms, appear to help protect vultures from potential pathogens (de Carvalho et al. 2003; Zepeda Mendoza et al. 2018; Fig. 6.28).

Any time a bird is injured or suffers any sort of tissue damage, the area becomes inflamed and this response, called inflammation, is an important part of the innate immune system (Fig. 6.29). Imagine a crow chasing or mobbing a flying hawk and, wisely, stays above the hawk to avoid its talons. However, the crow gets too close and the hawk quickly rolls over with extended talons. Before the crow can respond, a single talon pierces its skin. The crow quickly backs off, but has suffered a puncture wound. Superficially, the damage may seem minor, but a hawk's talons are covered with bacteria and some of those bacteria have now been "injected" into the crow. This is

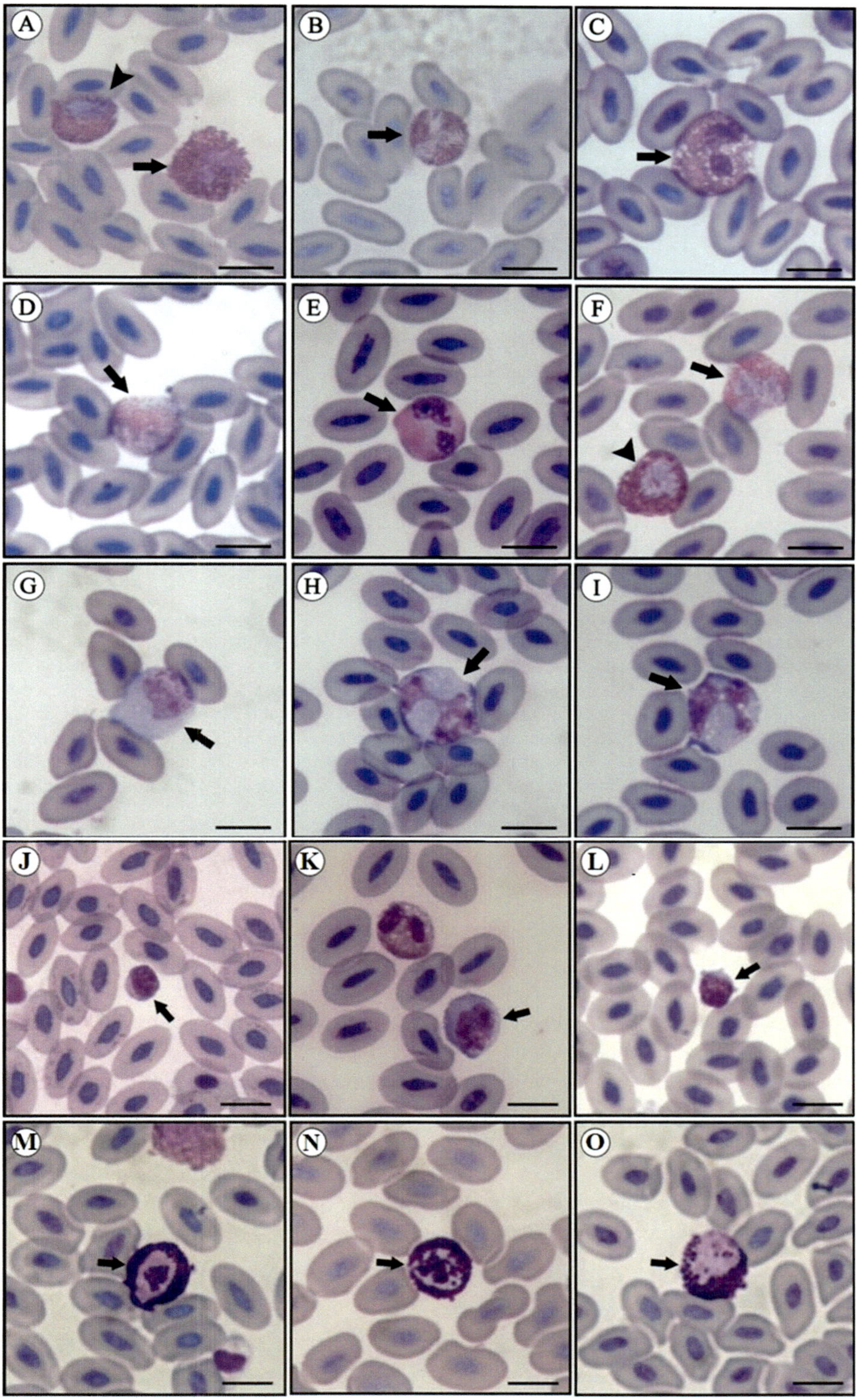


**Fig. 6.27** White blood cells of a Greater Rhea (*Rhea americana*). Stained blood smears showing (**a**–**c**) heterophils, (**d**–**f**) eosinophils, (**g**–**i**) monocytes, (**j**–**l**) lymphocytes, and (**m**–**o**) basophils. Scale bars = 10 μm. (Figure from Gallo et al.

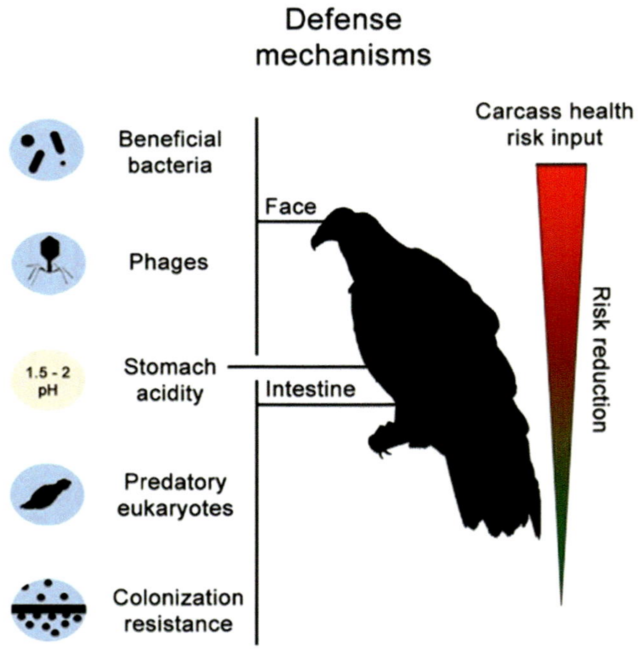


**Fig. 6.28** Vultures are exposed to a variety of pathogens, such as those that cause anthrax, pneumonia, brucellosis, and food poisoning. Factors that help protect vultures from these pathogens include (Top to bottom on left) beneficial bacteria present on facial skin that prevent colonization by potentially harmful bacteria (via biosynthesis of antibiotics), phages (viruses that infect bacteria, also called bacteriophages) on facial skin and in the digestive system can infect and eliminate potentially pathogenic bacteria, acid in the stomach kills some (but not all) potential pathogens, microbial eukaryotes in the intestine may feed on bacteria (additional study is need to confirm this), and assemblages of beneficial microbes (biofilms) lining the walls of the intestine may inhibit colonization by potential pathogens. (Figure and information from Zepeda Mendoza et al. 2018; © 2018 Springer Nature, used with permission)

where inflammation becomes crucial (Ashley et al. 2012; Carvalho et al. 2018; Fig. 6.29). The damaged cells, plus cells called mast cells, release chemicals that increase blood flow to the area, increase the permeability of capillaries in the area, attract white blood cells (heterophils and macrophages), and serve several other important functions (Box 6.6 Avian Mast Cells). Clotting proteins, heterophils, and macrophages move through the "leaky" capillaries into the surrounding tissues. While the clotting proteins create a barrier that makes it difficult for introduced bacteria to get into the blood, the heterophils and macrophages track down bacteria and, via phagocytosis, kill them. Inflammation, then, involves swelling (because of the "leaky" capillaries) and perhaps some pain (because the swelling causes stretching of the skin and other tissues), but reduces the chances that bacteria will spread to the rest of the body. In addition, the increased blood flow ensures delivery of the substances needed to repair and replace damaged cells and tissues.

**Fig. 6.27** (continued) 2015; open-access article distributed under the terms of the Creative Commons Attribution License, https://creativecommons.org/licenses/by/4.0/)

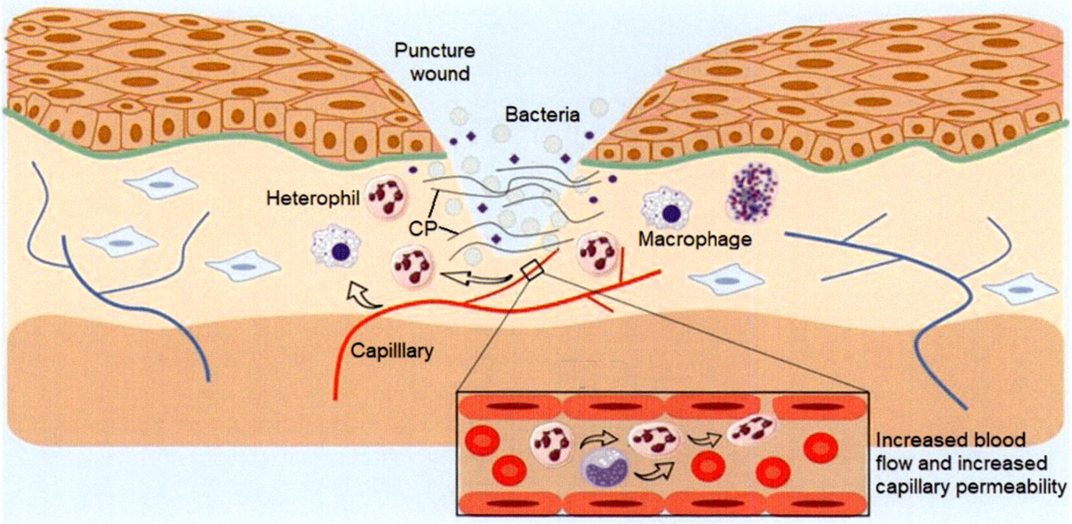


**Fig. 6.29** Example of inflammation. A puncture wound in the skin provides an entry route for bacteria. Damaged cells release chemicals called cytokines that cause vasodilation of blood vessels to increase blood flow, increase capillary permeability, and also attract heterophils and macrophages (phagocytic white blood cells). Increased blood flow brings more heterophils and macrophages to the area and they move out of capillaries to seek out and eliminate the bacteria by phagocytosis. The movement of clotting proteins (CP) out of capillaries and into surrounding tissues also creates a barrier that impedes the movement of bacteria from the site of the wound. (Figure modified from Carvalho et al. 2018; © 2018 Carvalho, Diniz, Suarez, Figueiredo, Zagmignan, Grisotto, Fernandes and da Silva, open-access article distributed under the terms of the Creative Commons Attribution License (CC BY), https://creativecommons.org/licenses/by/4.0/)

Another defense mechanism involves small proteins called host-defense peptides. For birds, contact with microbes typically first involves the skin or the epithelial lining of the digestive system, respiratory system, urinary system, or reproductive system. These epithelial cells produce host-defense peptides, or HDPs, as a defense mechanism. HDPs are also produced by some white bloods cells (heterophils). HDPs target any microbe with a negatively charged membrane, which includes many types of bacteria, fungi, and intracellular parasites (Scott and Hancock 2000). HDPs target microbial membranes, binding to the membranes in large numbers and creating channels or pores in the membranes that kill the microbes (Linde et al. 2008; Fig. 6.30). In addition to directly killing microbes, host-defense peptides attached to the membranes of microbes attract white blood cells, including heterophils, in a process called chemotaxis (Cuperus et al. 2013). The white blood cells then ingest (phagocytosis) the microbes and kill them with their lysosomal enzymes.

Avian red blood cells can also respond to, and help protect birds from, pathogens in several different ways (Chico et al. 2018; Fig. 6.31). Red blood cells can phagocytose bacteria and viruses and present their proteins (antigens) on their membranes to activate B- and T-cells. They also have receptor proteins in their membranes that bind to viral proteins and proteins in the membranes of bacteria and this, in turn, results in the production and release of substances that can prevent viruses from infecting other cells (interferon) and kill bacteria (cytokines and antimicrobial proteins). Other receptors in the membranes of avian red blood cells can bind to bacteria and viruses that have complement proteins on their membranes (proteins in the blood plasma activated by antibody-antigen reactions) and remove them from the circulation.

**Box 6.6 Avian Mast Cells**

Most avian mast cells are found in the skin and in the mucosal surfaces of the digestive system, reproductive system, lungs, and nervous system (Klika et al. 1996; Baccari et al. 2011). Because they are primarily located in areas most likely to be exposed to pathogens, mast cells are among the first to respond to the presence of pathogens. These cells have numerous receptors in their membranes that are able to recognize (bind with) pathogens or toxins. When activated by pathogens, mast cells release a variety of substances that, collectively referred to as cytokines, have a wide variety of antimicrobial effects (Rehman et al. 2017). Mast cells release these cytokines in two phases. First, some are released in just a few minutes after activation from granules (vesicles) located in the cytoplasm, a process called degranulation. Other cytokines are then released in a second phase within a few hours of activations. Collectively, these cytokines help activate and stimulate the various components of the avian immune system (Graham et al. 2015). In addition, however, mast cells play an important role in regulating the small intestinal, including muscle contraction, regulation of blood flow, peristalsis, and secretion of mucus by epithelial cells (Bischoff 2009).

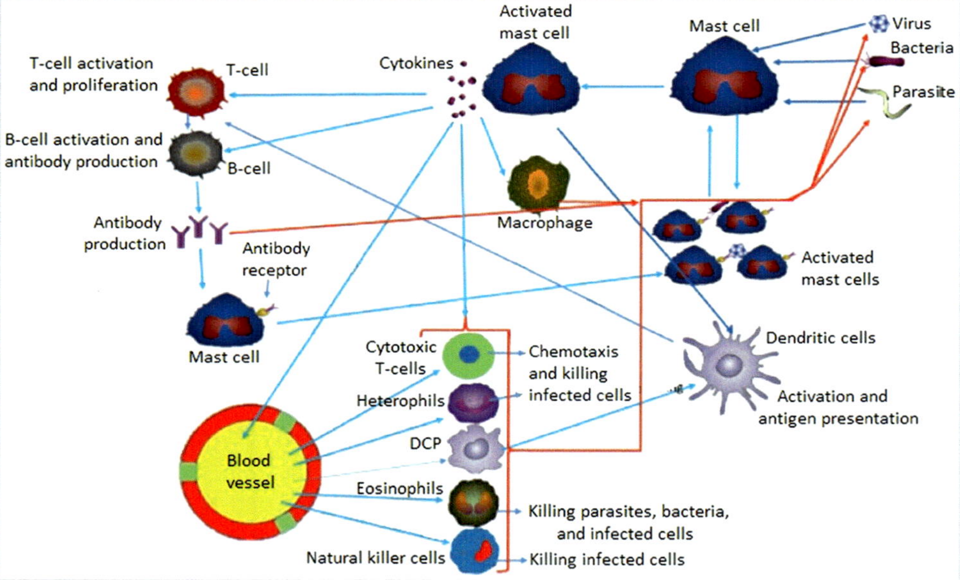


Mast cells activated by pathogens or toxins produced by pathogens release chemicals called cytokines that can activate and/or attract cells of the immune system, including B- and T-cells, macrophages, cytotoxic T-cells, heterophils, eosinophils, dendritic cell precursors (DCP), and natural killer cells, that act in a variety of ways to eliminate pathogens and cells infected by pathogens like viruses. B-cells produce antibodies that attach to membranes of mast cells, causing activation and proliferation that enhance the immune response. (Figure modified and information from Rehman et al. 2017; Reproduced with the kind permission of the World's Poultry Science Journal)

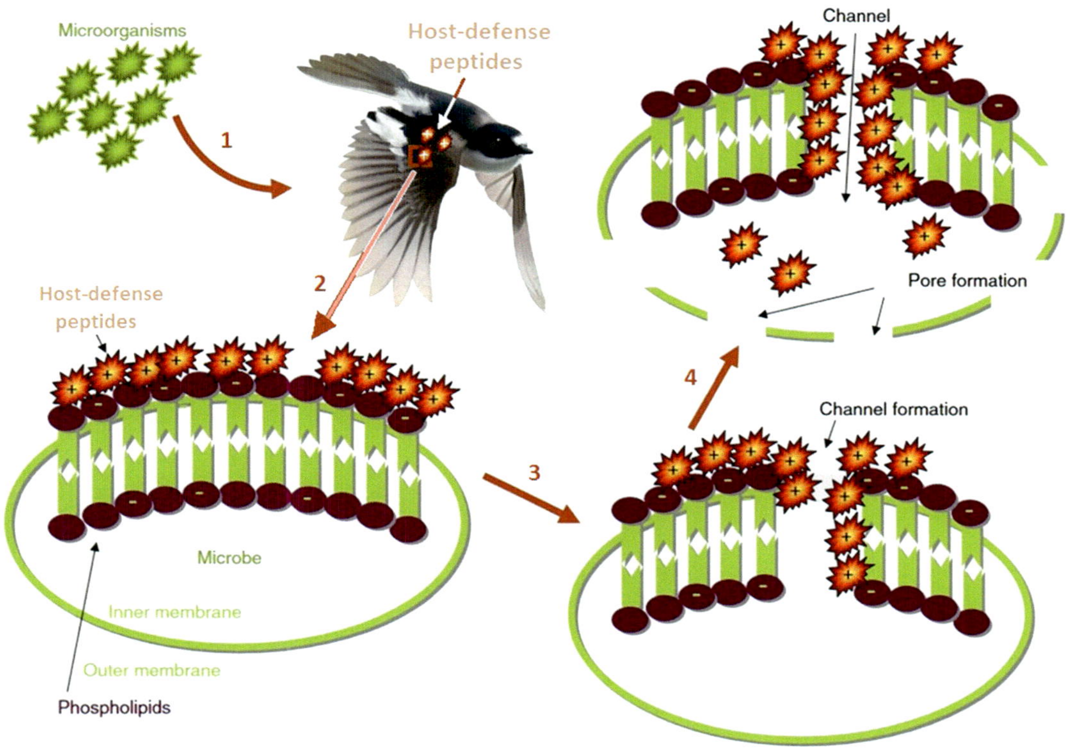


**Fig. 6.30** Host-defense peptides of birds. (1) A bird host is exposed to microorganisms like bacteria. (2) Positively charged host-defense proteins are attracted to the negatively charged microbial membrane. (3) Host-defense peptides form a carpet-like structure on the microbial membrane that leads to formation of channels or pores in the membrane. (4) The pores destabilize the membrane and kill the microbe. (Figure modified from Linde et al. 2008; © 2008 American College of Veterinary Internal Medicine, used with permission; Swallow image from Boström et al. 2016 © 2016 Boström et al., open-access article distributed under the terms of the Creative Commons Attribution License, https://creativecommons.org/licenses/by/4.0/)

Another component of the avian innate immune system is the complement system (Yatime et al. 2016). Complement is a complex network of about 25 circulating proteins, most of which are produced by the liver, and 10 or more cell membrane receptors that together represent one of the first lines of defense against foreign and altered (e.g., virus-infected) host cells. This "system" is so-named because, once activated, their actions "complement" the antibacterial actions of antibodies. All vertebrates, and even some invertebrates, have complement systems, but the components of those systems and how they interact vary among different taxa.

The avian complement system can be activated through three different pathways: classical, lectin, and alternative (Fig. 6.32). Activated complement protein can "defend" the avian body in a variety of ways. For example, one becomes a chemical called a "chemotaxin" that attracts heterophils (phagocytes) that, in turn, can ingest and destroy pathogens. Others join together to form a tube-like structure in the cell membranes of bacteria (membrane-attack complex) and this "tube" allows so much water to flow into bacteria that they "explode" or lyse (Fig. 6.32).

Interferons, another important component of avian innate immune systems, represent a key defense against viral infections (Goossens et al.

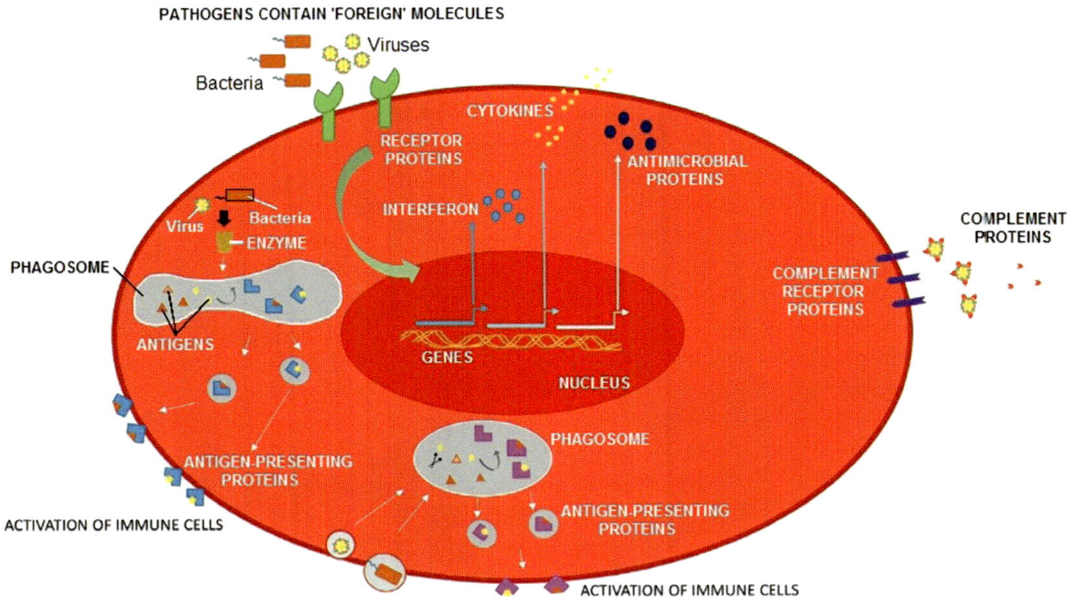


**Fig. 6.31** Avian red blood cells appear to play a role in immune responses. Receptor proteins in the membranes of red blood cells bind to pathogens (upper left) and trigger the production of interferon, cytokines, and antimicrobial proteins. In addition, pathogens can be taken in by phagocytosis, digested by enzymes, and their proteins (antigens) can be presented in the cell membrane by special antigen-presenting proteins. B- and T-cells that contact these antigens are then activated. Complement receptor proteins in the cell membranes can also bind to viruses and bacteria that are coated with complement proteins and remove them from the circulation. (Figure modified from Chico et al. 2018; © 2018 V. Chico, I. Nombela, S. Puente-Marín, and M. D. M. Ortega-Villaizan. Licensee IntechOpen, open-access article under the terms of the Creative Commons Attribution 3.0 License, https://creativecommons.org/licenses/by/3.0/)

2013; Santhakumar et al. 2017). A virus must invade a host cell to replicate; attaching to a cell membrane, injecting its genetic material (DNA or RNA), then, via translation (of viral RNA) or transcription and translation (of viral DNA), making copies of itself that can then invade other host cells. Infected host cells often die because viruses prevent them from performing their normal functions. Interferons are signaling proteins, produced by infected cells, released, and then binding to receptors in the membranes of other cells (Fig. 6.33). Production of interferons by a cell requires detection of viral genetic material by special receptor proteins called pattern recognition receptors (PRRs; so-named because they bind with, or recognize, the "patterns" or sequences of nucleic acids in viral genetic material), which then activate mitochondrial antiviral signaling proteins (MAVs) and toll-like receptors (TLRs; located in cell membranes). By way of interactions of MAVs and TLRs with a number of other proteins in the cytoplasm, genes that produce interferons (via transcription and translation) are activated. Interferons are then released from infected cells and bind to receptor proteins in the membranes of other cells. This "signals" cells to produce molecules that prevent infection by viruses either by preventing them from entering the cell or, if they do enter, by preventing them from replicating (by interfering with the ability of viruses to replication via transcription or translation) (Fig. 6.33). This "interference" is why the proteins are called interferons.

Natural killer cells are also part of the avian innate immune system (Vervelde et al. 2013; Paul and Lal 2017; Fig. 6.34). These cells are cytotoxic lymphocytes that identify and destroy cells that have been infected by a virus by detecting

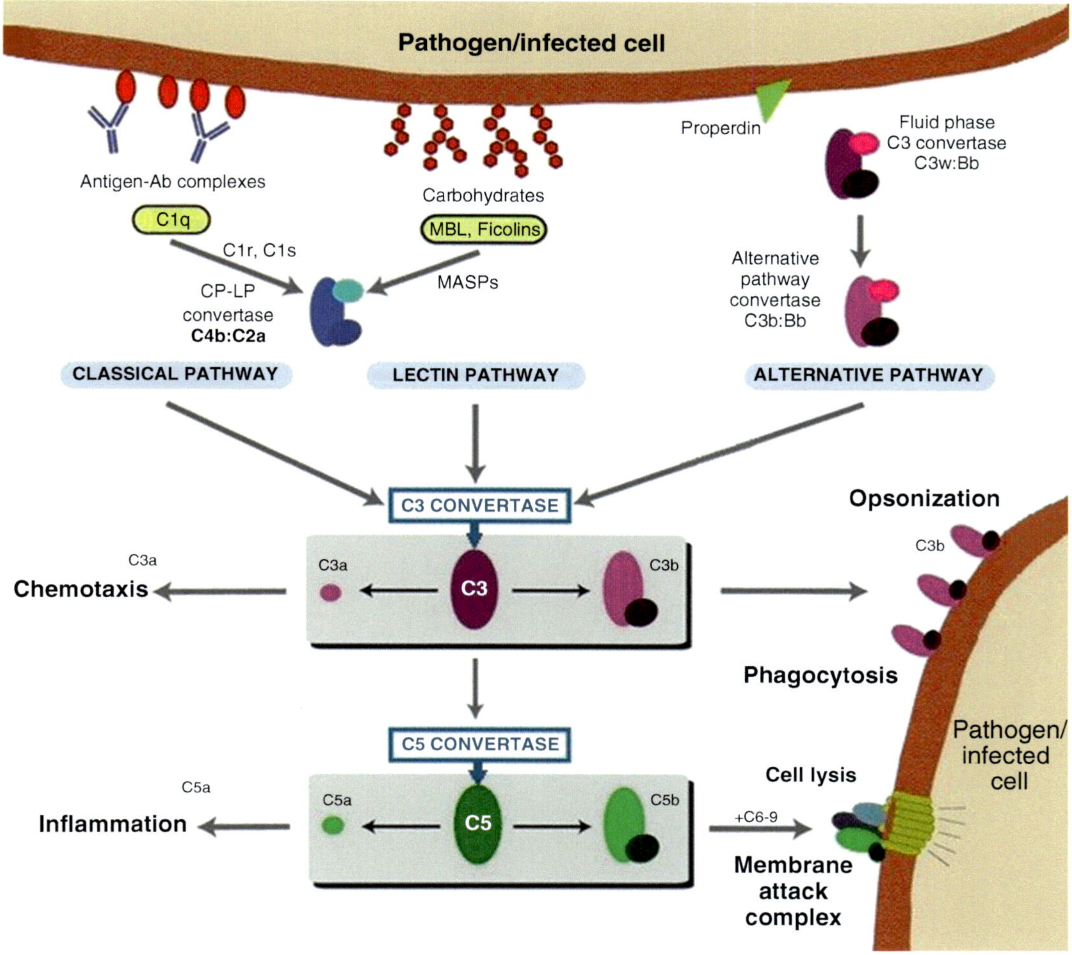


**Fig. 6.32** Overview of the complement cascade. Complement proteins defend against pathogens in several ways, including serving as chemotaxins (chemicals that attract phagocytes like heterophils and macrophages), stimulating phagocytes to ingest pathogens or infected cells (a process called opsonization), causing pathogens or infected cells to rupture by creating holes in cell membranes (i.e., cell lysis caused by complement protein that for the membrane-attack complex), and initiating inflammation (e.g., increased blood flow brings more phagocytes to the infected area). (Figure modified from Yatime et al. 2016; © 2016, Controlled Release Society, Published by Springer Nature, used with permission)

unusual or foreign proteins in the membranes of infected cells (Iannello et al. 2006). When viruses enter a cell, their outer structure remains at the membrane and their nucleic acid is basically "injected" into the cell. Natural killer cells are able to detect infected cells via proteins in the infected-cell membrane (Fig. 6.35). Once identified, the killer cell releases chemicals called cytokines that will kill infected cells. Natural killer cells can also identify and destroy "abnormal" cells, such as precancerous or cancerous cells.

Monocytes are yet another important part of the avian innate immune system (Fig. 6.36). These white blood cells can be found in the blood as well as in almost all organs, including the lungs, liver, and spleen, and digestive and respiratory systems, where they are called

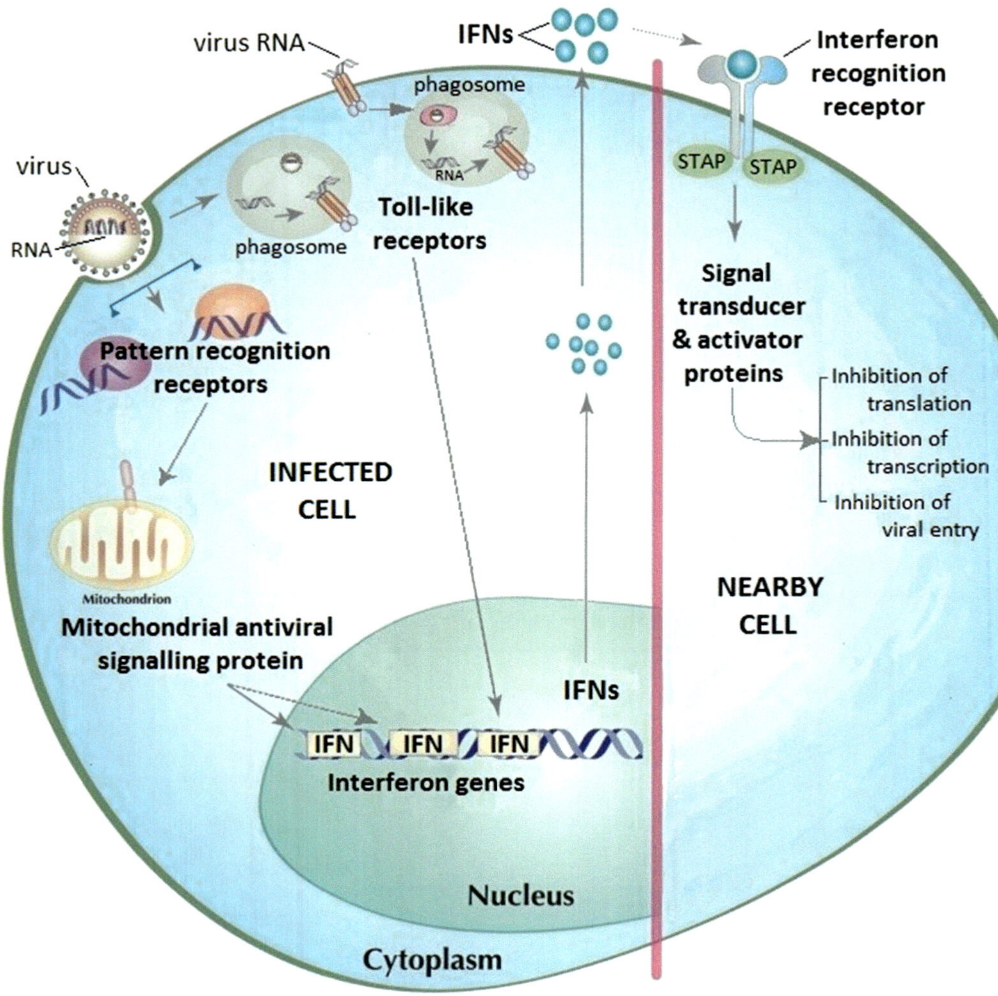


**Fig. 6.33** When viral genetic material (RNA or DNA) is detected by special receptor proteins, bird cells produce interferons (IFNs) that bind to receptor proteins in the membranes of other cells. This binding activates proteins that then prevent viral infections by preventing entry of viruses into cells or by preventing viruses that do enter from replicating. (Figure modified from Santhakumar et al. 2017; © 2017 Santhakumar, Rubbenstroth, Martinez-Sobrido and Munir, open-access article distributed under the terms of the Creative Commons Attribution License (CC BY), https://creativecommons.org/licenses/by/4.0/)

macrophages. The functions of macrophages in different organs vary, so, as a result, they have different names in different organs. In the respiratory system, macrophages (called free avian respiratory macrophages, or FARM) are present in the mucosa of the larger airways and the linings of parabronchi and eliminate invading pathogens by phagocytosis (de Geus et al. 2012; Mutua et al. 2016). In the liver, macrophages (called Kupffer cells) phagocytose pathogens, cell debris, and damaged red and white blood cells (Zaefarian et al. 2019; Fig. 6.37). In the digestive system, macrophages in the small intestine provide protection against bacteria and viruses and may also play a role in regulating neurons important in the process of peristalsis (Gabanyi et al. 2016; Dora et al. 2018; Fig. 6.38). The primary role of macrophages in the central nervous system

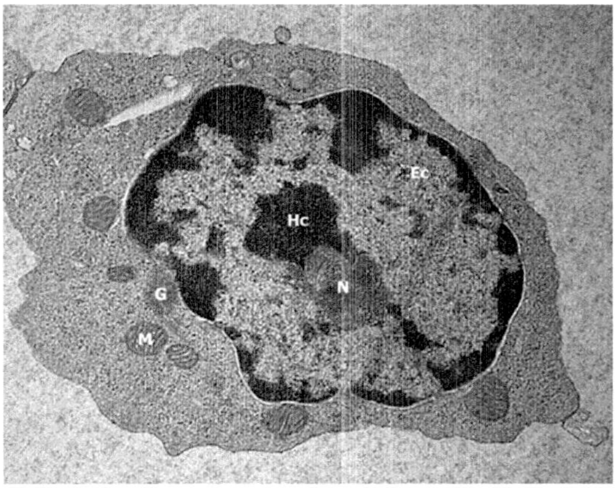


**Fig. 6.34** Natural killer cell of a Domestic Chicken. N, nucleus; Hc, heterochromatin; Ec, euchromatin, M, mitochondrion; G, granule. Heterochromatic is highly condensed DNA, and euchromatin is a lightly packed form of chromatin (DNA, RNA, and protein). (Figure from Abdolmaleki et al. 2018; © 2018 Abdolmaleki, Yeap, Tan, Satharasinghe, Bello, Jahromi, Bejo, Omar and Ideris, open-access article distributed under the terms of the Creative Commons Attribution License, https://creativecommons.org/licenses/by/4.0/)

(called microglia; Fig. 6.39) is to remove cell debris and pathogens, but microglia are also important for normal brain development and synaptic plasticity, i.e., forming new and eliminating previous connections between neurons (called synapses) that is critical in the processes of learning and memory and for normal brain activity (Larson 2018; Carvalho-Paulo et al. 2021). In bones, immature blood cells can become either mature macrophages or osteoclasts (Fig. 6.40).

Macrophages are considered part of the innate immune system because they are phagocytic. However, macrophages also play a variety of other roles in the avian immune system including removal of dead cells, pathogen recognition, phagocytosis and destruction of pathogens, and antigen presentation (Wynn et al. 2013). As antigen-presenting cells, macrophages "ingest" pathogens (by phagocytosis) and process associated their antigens. What "process" means in this case is that macrophages place antigens, or foreign proteins, in their cell membranes and these membrane-bound antigens can then be recognized by T-lymphocytes and B-lymphocytes. Such recognition activates the T- and B-lymphocytes, and the various actions of these activated cells are described later in this chapter.

Macrophages, along with other cells in the avian immune system such as T-lymphocytes, also synthesize and secrete a number of chemicals called cytokines (Klasing 1998; Lu et al. 2023). Cytokines are low molecular weight proteins that play essential roles in immune responses and act by binding to receptors in cell membranes. Among the cytokines produced by macrophages are molecules called interleukins. Interleukins 1β, 6, and 8 are pro-inflammatory cytokines, increasing blood flow and capillary permeability and attracting heterophils that clean up cell debris and eliminate bacteria by phagocytosis (Wigley and Kaiser 2003). Interleukin 2 stimulates the activation and proliferation of T-lymphocytes (Lu et al. 2023). Interleukins 15 and 16 are thought to play a role in activating natural killer cells as well as T- and B-lymphocytes (Al-Khalaifah and Al-Nasser 2018). Interleukin 12 stimulates proliferation of natural killer cells and T-cells (Trinchieri 2003). Interleukin-10 is an immunosuppressive cytokine, downregulating the inflammatory response and promoting wound repair (Burmeister and Marriott 2018). Another cytokine produced by macrophages, called transforming growth factor beta (TGF-β), plays a role in the initiation and resolution of inflammatory responses by regulating chemotaxis and activating natural killer cells and other cells in the avian immune system (Li et al. 2006). TGF-β, along with cytokines called epidermal growth factor and vascular endothelial growth factor, also helps facilitate the process of wound healing (Li et al. 2006; Barrientos et al. 2014). Macrophages also produce nitric oxide that is used to kill bacteria (Fig. 6.41).

B-lymphocytes (or B-cells), the cells that produce antibodies, are initially produced in the embryonic liver, yolk sac, and bone marrow, then move through the blood to the bursa of Fabricius (Ratcliffe and Härtle 2014). The bursa of Fabricius is a blind sac that extends from the dorsal side of the cloaca, the common portal of

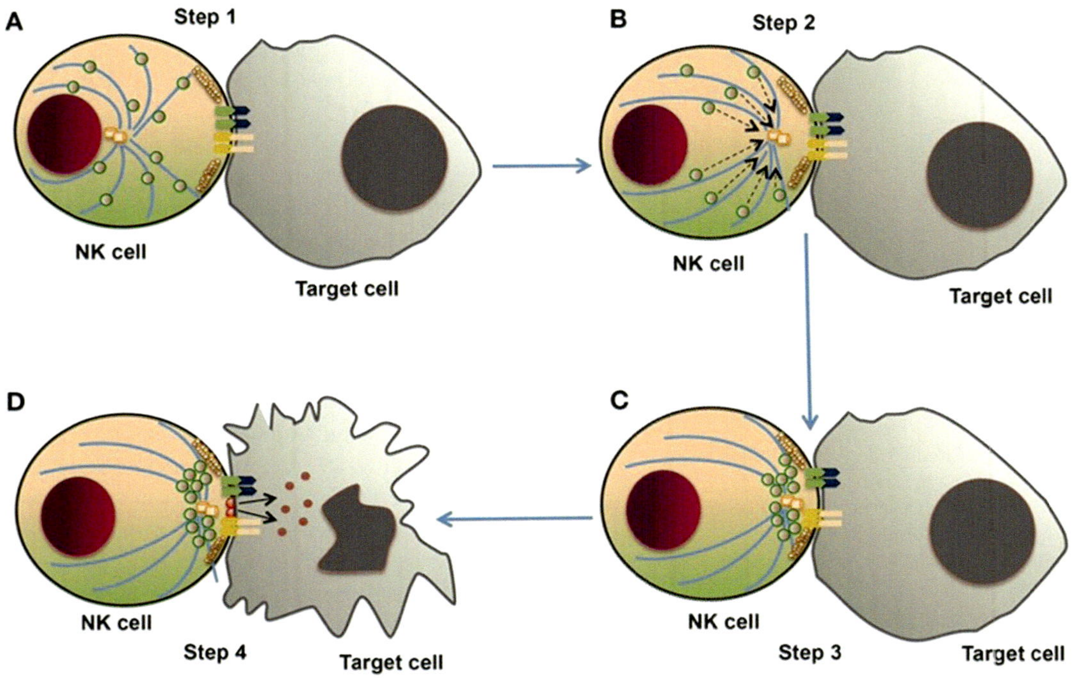


**Fig. 6.35** (**a**) Natural killer cells detect and bind to infected cells via membrane proteins. (**b**, **c**) Vesicles (granules) containing chemicals including perforin (that will rupture the target cell membrane) and enzymes (that will digest target cell components and cause its death) move toward the target cell guided by microtubules. (**d**) Chemicals are released, rupturing the cell membrane of, and killing, the target cell. (Figure from Paul and Lal 2017; 2017 Paul and Lal, open-access article distributed under the terms of the Creative Commons Attribution License, https://creativecommons.org/licenses/by/4.0/)

the reproductive, urinary, and digestive systems (Fig. 6.42). Within the walls of the bursas of young birds are bursal follicles (Fig. 6.42). Within these follicles, B-lymphocytes mature then migrate to other body tissues, including the blood, spleen, cecal tonsils, bone marrow, Harderian gland, and thymus. The bursa typically atrophies at about the time of sexual maturation. As a result, avian biologists can sometimes use the presence or absence of the bursa to determine a bird's age (at least prereproductive or juvenile versus adult) (e.g., Josephson 1980; Broughton 1994).

B-lymphocytes produce proteins called antibodies or immunoglobulins. Mammals, including humans, have five types or classes of immunoglobulins. Birds, however, have three classes or types of immunoglobulins (Ig): IgM, IgY, and IgA (Härtle et al. 2014; Figs. 6.43 and 6.44). The most abundant antibody in the blood of birds is IgY and is equivalent to the IgG antibody of mammals (Zhang et al. 2017; Fig. 6.43). These antibodies (IgY) are produced by B-lymphocytes (or B-cells) after exposure to a disease organism (Fig. 6.45). More precisely, B-cells are exposed to antigens or proteins associated with disease-causing organisms. B-cells respond to exposure to "foreign" proteins by transforming into cells called plasma cells that are specialized for the production of specific antibodies. Some of these transformed cells will, after several days, begin to produce and release large numbers of antibodies. Other transformed cells will not produce and release antibodies unless, at some time in the

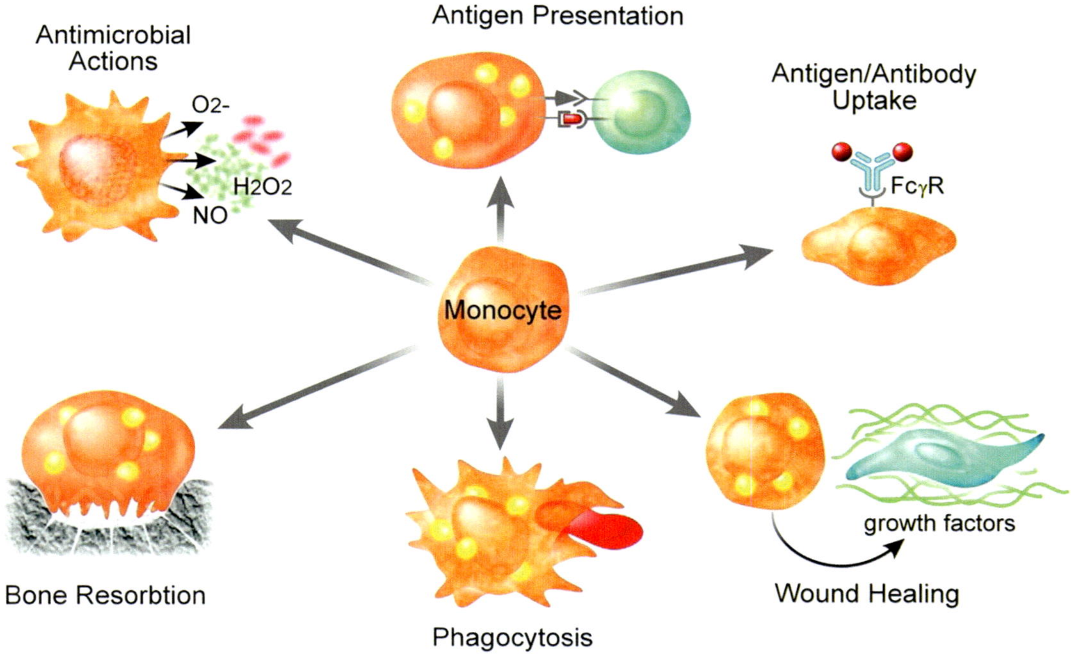


**Fig. 6.36** Monocytes are found in numerous tissues and organs where they are referred to as macrophages. Macrophages have a wide variety of functions, including (clockwise from top left) defense against intracellular pathogens by initiating inflammatory responses, activating other cells (e.g., T-lymphocytes) in the avian immune system via antigen presentation, downregulating immune responses by removing (by phagocytosis) immune (antigen/antibody) complexes, promoting wound healing by release of growth factors, and remodeling bones via resorption. (Figure from Chawla 2010; © 2010 Wolters Kluwer Health, used with permission)

future, they are again exposed to the same antigen. These cells are called "memory" cells and are responsible for immunity. This is the same process that occurs in humans. To create an "immune" person (e.g., immunity to polio), pathogens can be killed or altered in some way to weaken them then introduced into the body (sometimes, somewhat painfully, by injection); the immune system does the rest. Once memory cells have been produced, a bird (or human) is immune to that pathogen because antibodies are released so quickly and in such high numbers (Fig. 6.46).

Several days or more after exposure to a pathogen (antigen), the number of antibodies in the blood begins to increase. The first antibodies to appear are the IgM type. These antibodies may be found in the membranes of B-lymphocytes and aid those lymphocytes in recognizing antigens. IgM appears as soon as 3 days following exposure to a disease organism, with peak presence after 4 to 8 days and then declining rapidly (Tizard 2002). IgY is generally detected after 5 days following exposure, peaks at 3 to 3 1/2 weeks, and then slowly decreases. IgA is found primarily in the mucus secretions of the eyes, intestine, and respiratory tract and these antibodies provide protection from infection for these structures and tissues (Tizard 2002).

T-lymphocytes begin as the same stem cells as the B-cells, but are programmed in the thymus rather than the bursa of Fabricius. T-lymphocytes include a more heterogeneous population than the B-cells. Some T-cells act by producing lymphokines (also called cytokines; over 90 different ones have been identified); others directly destroy disease organisms (cytotoxic T-cells). Some T-cells (helper T-cells) act to enhance the

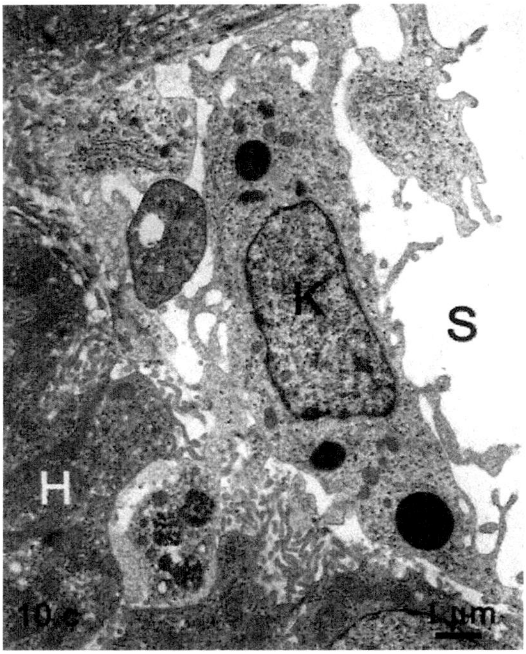


**Fig. 6.37** An electron micrograph of a chicken liver cell showing a Kupffer cell (K). S, sinusoid or cavity; H, hepatocyte or liver cell. (Figure from Ghoddusi and Kelly 2004; © 2004 Wiley-Liss, Inc., used with permission)

response of B-cells, macrophages, or other T-cells; others inhibit the activity of these cells (regulatory T-cells) (Fig. 6.47).

Antibodies do not kill viruses or bacteria directly, but their actions provide protection against pathogens in several different ways (Härtle et al. 2022; Fig. 6.48). Antibodies (especially IgY) can provide protection by attaching to disease organisms (like bacteria) or the toxic substances they produce and blocking their receptors. This prevents the pathogen from attaching to target cells or, in the case of toxins, effectively neutralizes the molecule. By attaching to pathogens, antibodies can also facilitate their destruction by phagocytes. Antibodies attached to proteins in the membrane of a bacterium, for example, become receptors for phagocytes like macrophages and heterophils that will, in turn, engulf (phagocytosis) and kill the bacterium. In the jargon of physiology, antibodies that enhance the process of phagocytosis like this are called opsonins. Antibodies also help sustain an immune response by, as described previously, activating the complement system.

## 6.12 Significance of the Avian Immune System

Individual birds, like all organisms, attempt to maximize their fitness or, in other words, produce as many offspring (that survive to reproduce) as possible during their lifetime. Successful reproduction requires that a bird stays alive and, in addition, is in good enough physical condition to obtain a mate and expend the energy needed to reproduce. Because parasites and pathogens are ubiquitous, remaining in good physical condition requires an efficient immune system. What is efficient may depend on a bird's lifestyle and life history. For example, some birds, such as scavengers, may be exposed to more pathogens and parasites than non-scavengers and their health and survival may require a "better" immune system. Scavengers do, in fact, have larger spleens (an organ where lymphocytes are produced) relative to body size than non-scavengers and, in addition, have higher concentrations of white blood cells in their blood (Blount et al. 2003). Other birds may require less robust immune systems because they occupy habitats, such as isolated islands that are relatively disease-free (Piersma 1997). For example, Darwin's finches on smaller islands in the Galapagos archipelago exhibit weaker immune responses to disease and foreign pathogens than those on larger islands (Lindström et al. 2004). More generally, Beadell et al. (2007) suggested that characteristics of the immune systems of island birds will vary due to differences in the pathogenic landscapes of different islands as well as to stochastic variation in mutations and genetic drift.

Because of associated differences in extent of contact with other individuals, colonial birds may be at greater risk of exposure to pathogens than non-colonial birds and colonial birds in larger colonies may be at greater risk of exposure than those in smaller colonies. Such differences in

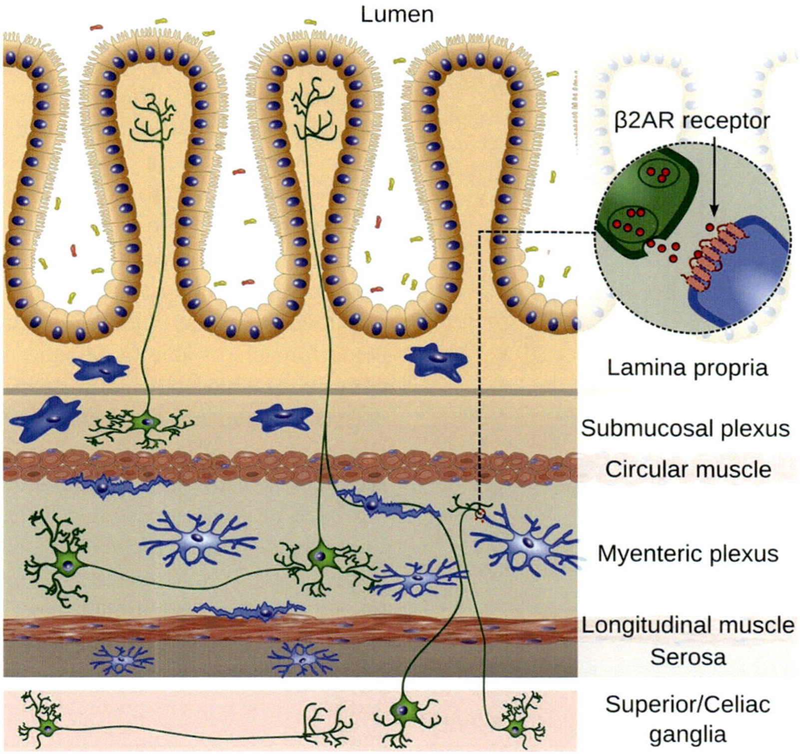


**Fig. 6.38** Macrophages in the small intestine (blue cells in the lamina propia and submucosal plexus) protect against bacterial infections. Infections in the intestine may cause activation of macrophages in myenteric plexus via sympathetic stimulation (inset: green sympathetic neuron releasing neurotransmitter molecules that activate macrophages via β2AR receptors in the macrophage membrane), and macrophages in the myenteric plexus may also help regulate neurons (green cells) important in the process of peristalsis. Lamina propia = thin layer of connective tissue; submucosal plexus = contains neurons that innervate circular muscle cells; myenteric plexus = contains neurons important in the regulation of digestion, e.g., secretion, absorption, and blood flow; serosa = epithelial tissue; Superior/celiac ganglia = sympathetic neurons that help regulate the actions of the small intestine. (Figure from Gabanyi et al. 2016; © 2016 Elsevier Inc., used with permission)

"risk" might translate into differences in immune systems. Cliff Swallows (*Petrochelidon pyrrhonota*) are colonial nesters and colonies differ considerably in size. During a period of cold, wet weather in Nebraska in May 1996, over 1700 Cliff Swallows died and were subsequently examined (Brown and Brown 2002). Examination of these birds revealed that swallows from larger colonies had more ectoparasites than those from smaller colonies and had significantly larger spleens. More generally, Møller and Erritzoe (1996) found that colonial species had larger spleens than species in the same genus that nested solitarily. These results indicate that immune systems do respond to differences in exposure to parasites and pathogens.

For all birds, regardless of lifestyle, an effective immune system is essential. Birds with weak immune systems will be less healthy and more likely to fall prey to predators. A study involving 18 species of songbirds revealed that those killed by domestic cats (*Felis catus*) had smaller spleens (an organ that plays an important role in the immune system) than birds that were not killed (Møller and Erritzoe 2000). Given the critical importance of the immune system, what factors can influence its quality? One factor is age. Immune function tends to deteriorate with age in

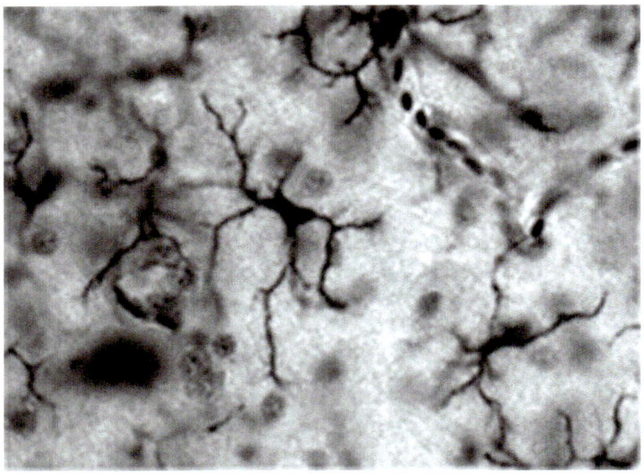

**Fig. 6.39** Microglia in the hippocampus of a Spotted Sandpiper (*Actitis macularius*). Microglia are macrophages, but, in the hippocampus, have also been found to be associated with important physiological functions in learning and memory by promoting learning-related synapse formation. (Figure from Diniz et al. 2016; open-access article distributed under the terms of the Creative Commons Attribution License, https://creativecommons.org/licenses/by/4.0/)

birds just as it does in humans. The immune response of older Collared Flycatchers (*Ficedula albicollis*) exposed to foreign proteins (antigens) is not as pronounced as that of younger individuals (Cichon et al. 2003). As a result, as birds age, they may become more susceptible to pathogens and parasites and this may affect their physical condition and ability to successfully reproduce.

Clearly, immune systems exhibit species- and population-specific differences in quality and effectiveness. The same is true among individual birds. Maintaining a good immune system requires energy, but so do other activities, such as reproduction. If food resources are limited, birds may have to decide how to allocate available energy. In this scenario, only birds able to obtain more energy (perhaps because they are better foragers or able to defend a territory with more food) would be able to successfully reproduce and maintain an effective immune system (Westneat and Birkhead 1998). The trade-off between reproductive effort and immune function has been nicely demonstrated in European Pied Flycatchers (*Ficedula hypoleuca*; Kilpimaa et al. 2004). Male Pied Flycatchers (Fig. 6.49) were forced to expend additional energy when their mates were removed during nest-building (and released elsewhere). These "widowed" males had higher hematocrit levels and larger forehead patches (a possible signal to potential mates) than non-widowed males, suggesting an increase in mating effort (e.g., increased testosterone secretion that stimulates red blood cell production). In addition, these widowed males exhibited reduced immune responses. These results indicate that

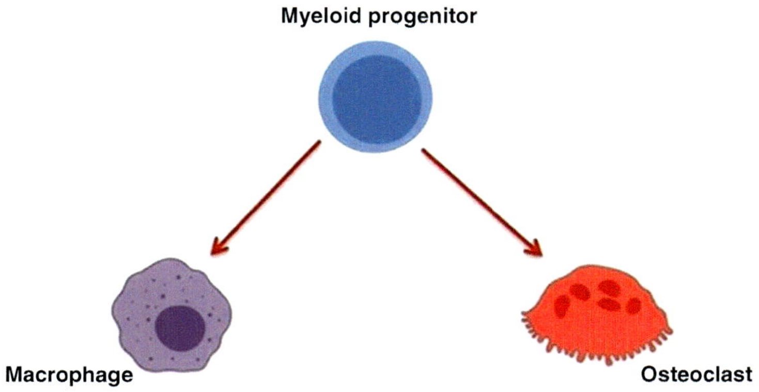


**Fig. 6.40** In bones, some immature blood cells become macrophages and some become osteoclasts. Osteoclasts are cells that break down bone tissue and play an important role in maintaining, repairing, and reshaping or remodeling bones in response to factors such as growth or variation in mechanical stress. (Figure modified from Yang and Wan 2019; © 2019 Springer-Verlag GmbH Germany, part of Springer Nature, used with permission)

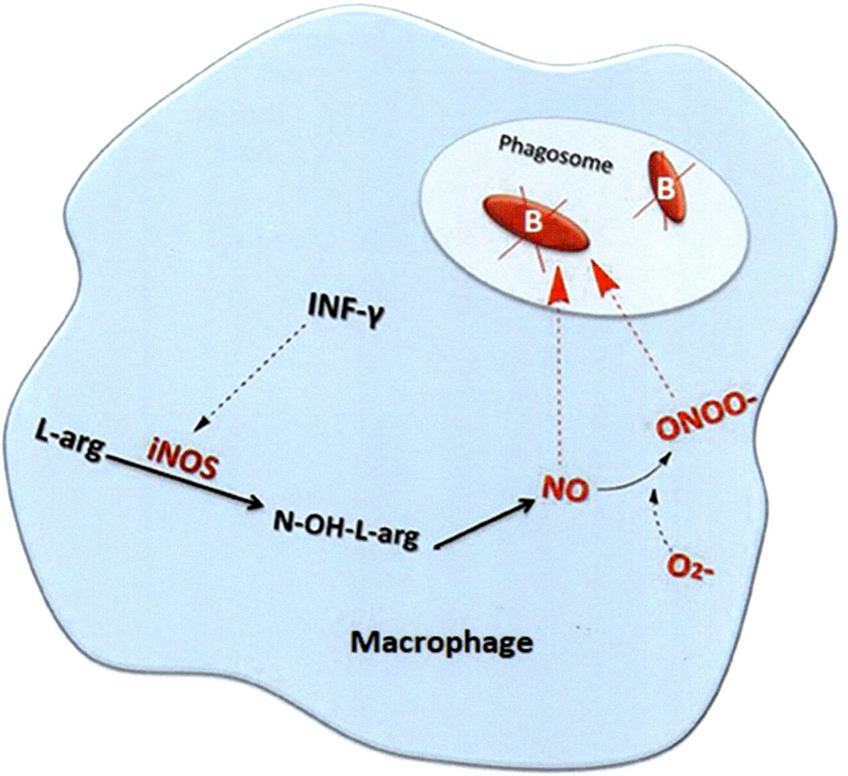


**Fig. 6.41** Macrophages produce nitric oxide (NO) from an amino acid (L-arginine, or L-arg). Interferon-y (INF-y) increases NO production by stimulating production of an enzyme called inducible nitric oxide synthase (INOS). NO kills bacteria (B) in phagosomes (membrane-bound vesicles with bacteria that have been "ingested" by macrophages by phagocytosis) directly or in combination with oxygen to form peroxynitrate (ONOO-). (Figure modified from Jamaati et al. 2017; © 2017 Jamaati, Mortaz, Pajouhi, Folkerts, Movassaghi, Moloudizargari, Adcock and Garssen, open-access article distributed under the terms of the Creative Commons Attribution License, https://creativecommons.org/licenses/by/4.0/)

male Pied Flycatchers can differentially allocate resources, but doing so requires trade-offs. More effort devoted to reproduction, e.g., maintaining a larger forehead patch, comes at the cost of a less effective immune system.

Female birds face the need to make trade-offs as well. Female Pied Flycatchers with larger broods must expend more energy and effort in provisioning nestlings. One result of this increased effort is a less effective immune system (Moreno et al. 1999). Clearly, a number of factors can influence clutch size, but females producing larger clutches and provisioning a greater number of nestlings must expend additional effort and assume the potential risk of a suppressed immune system. The possible costs associated with mounting an immune response when caring for young were demonstrated in a study where some female House Sparrows were injected with antigens to activate an immune response. These females, in contrast to control females, were more likely to desert their nests and those that did not desert fed nestlings at lower rates (Bonneaud et al. 2003). "Sick" females became less active and this reduced activity might conserve energy and permit tissue repair during infection (Hart 1988). However, "sick" females also had less energy to devote to caring for young and had to decide

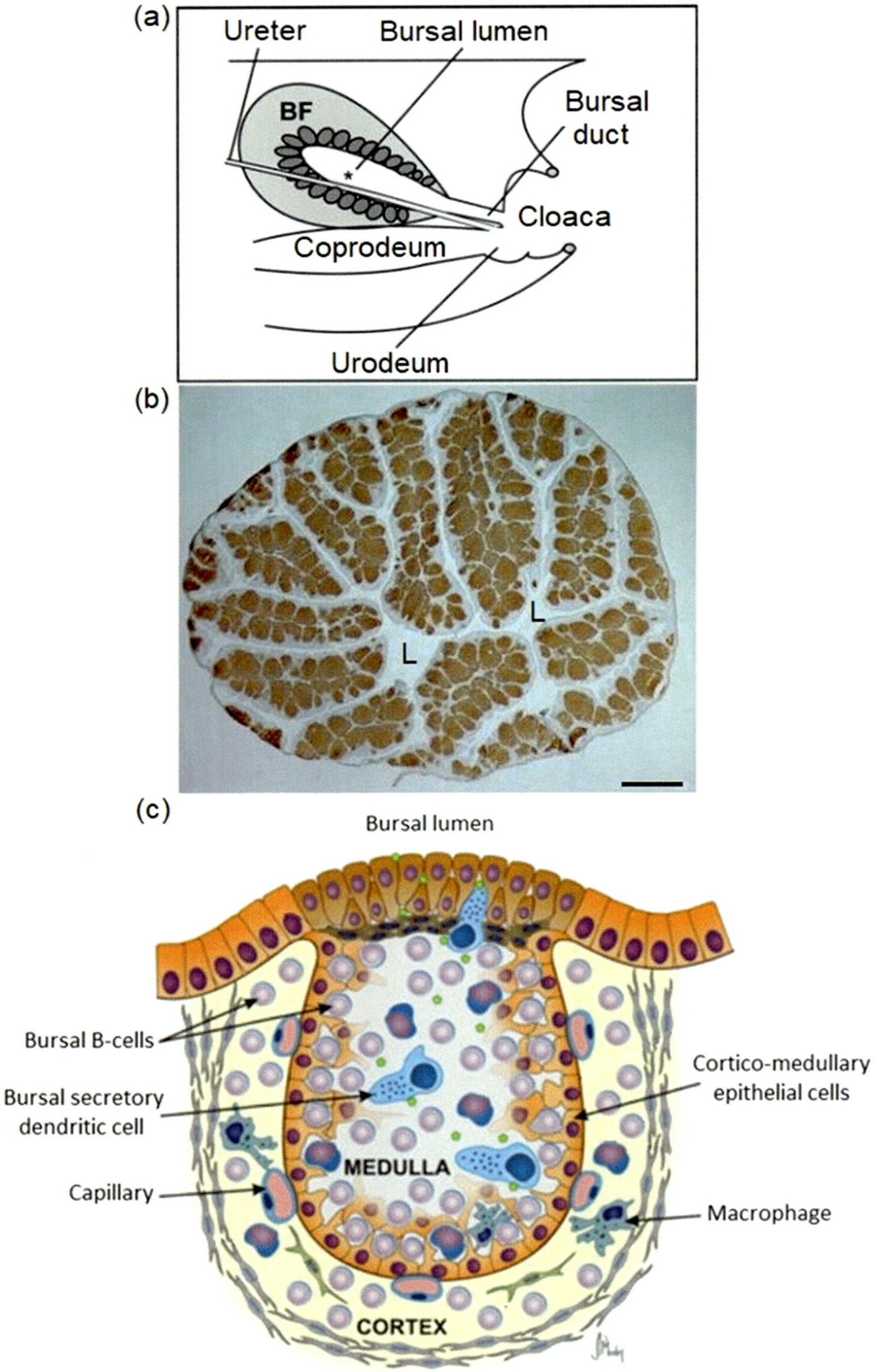


**Fig. 6.42** (**a**) Drawing of an avian cloaca (with coprodeum and urodeum) and the bursa of Fabricius (BF). (**b**) Histological section through a mature bursa. L, lumen. (**c**) Mature bursal follicle where B-lymphocytes (or just B-cells) mature and develop the ability to produce antibodies. Bursal secretory dendritic cells and cortico-

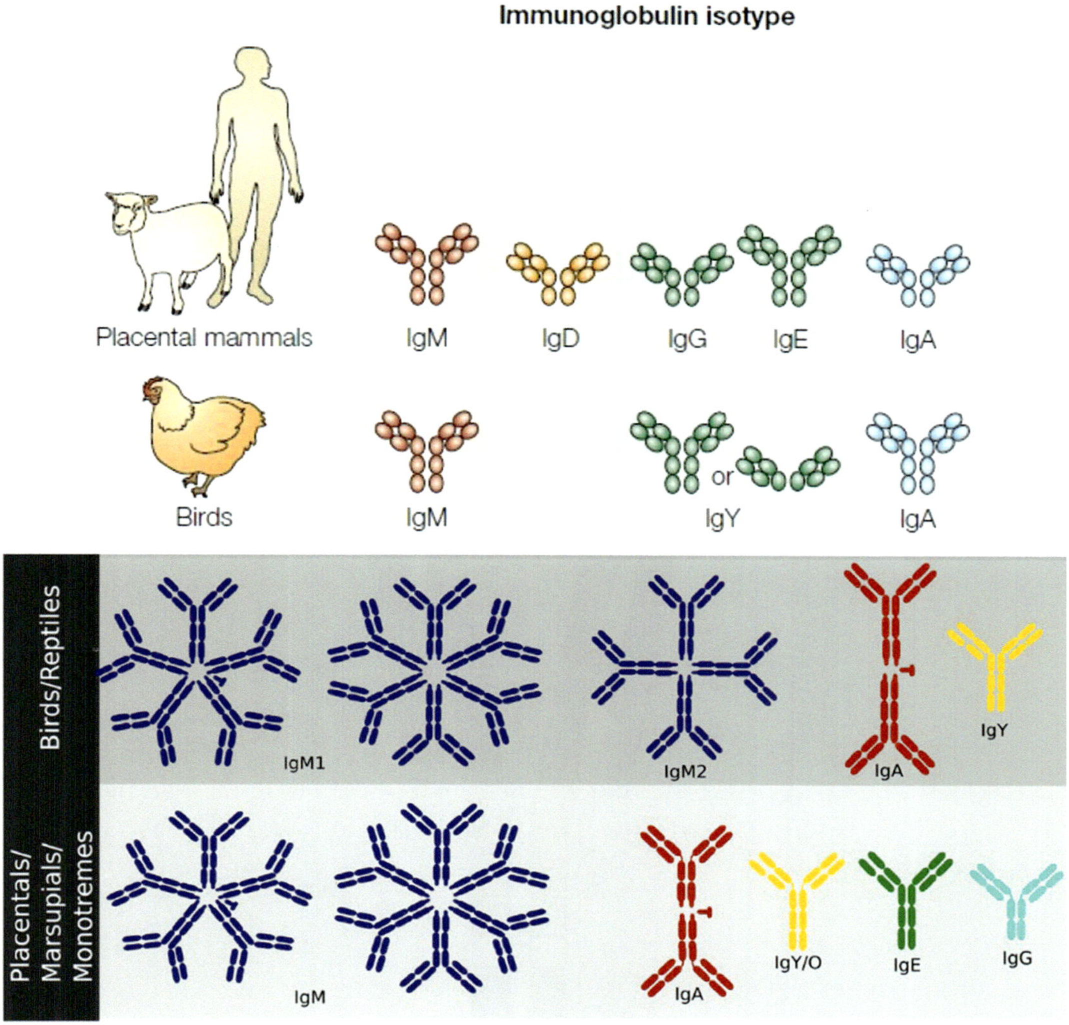


**Fig. 6.43** Top, Immunoglobulins of mammals and birds. (Top figure from Flajnik 2002; © 2002 Springer Nature, used with permission; Bottom figure from Zhang et al. 2017; © 2017 Cambridge Philosophical Society, used with permission)

whether to desert or not. "Sick" females with larger broods were less likely to desert and more likely to continue feeding nestlings. One explanation for this result is that females that produced large broods were higher-quality females and were more likely to stand the

**Fig. 6.42** (continued) medullary epithelial cells are thought to produce substances important in the maturation of B-cells. Macrophages are phagocytic cells. Scale bar in **b** = 500 μm. (Figures **a** and **b** from Ekino et al. 2012; © 2012 Springer-Verlag, used with permission. Figure c modified from Madej et al. 2013; © 2013 Blackwell Verlag GmbH, used with permission)

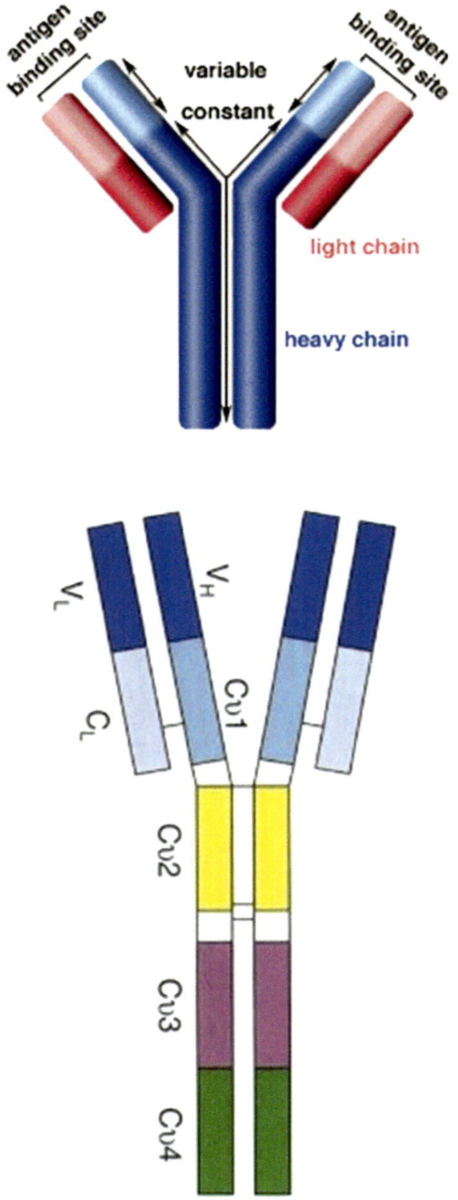


**Fig. 6.44** The most abundant antibody of birds (IgY) has a Y-shape and is composed of four "chains" or protein molecules. Heavy chains are so-named because they consist of more amino acids and have a higher molecular weight than light chains. The antibody also has constant and variable regions. The constant region does not vary and is similar in all IgY antibody molecules. However, the variable portion does change and each antibody produced in response to a different antigen, or foreign protein, has a variable region with a specific shape that allows it to bind to or combine with that antigen molecule. (Figure from Warr et al. 1995; © 1995 Published by Elsevier Ltd., used with permission)

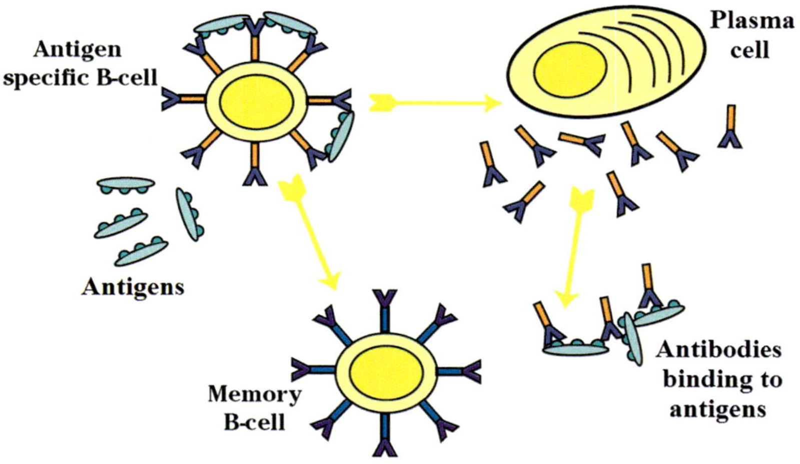


**Fig. 6.45** Avian B-cells (B-lymphocytes) exposed to and binding with an antigen (foreign protein) are transformed into plasma cells and memory cells. Plasma cells begin producing antibodies that bind to antigens, neutralizing them or targeting them for elimination by phagocytes. Memory B-cells will not produce antibodies unless exposed to the same antigen sometime in the future. (Figure from Grogan et al. 2007; used with permission of Gisela Erf)

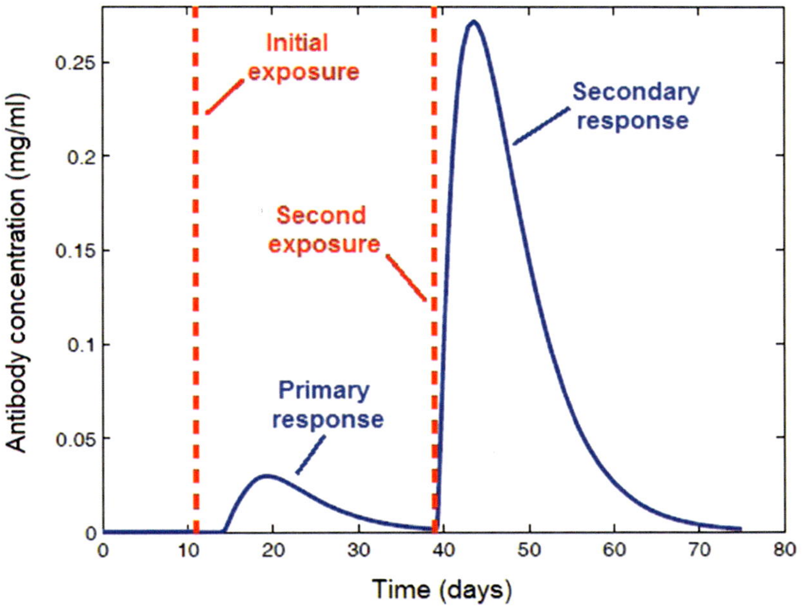


**Fig. 6.46** The difference between the response of a bird's immune system to the initial exposure to an antigen (primary humoral response) and a later, second exposure (secondary humoral response). Notice that antibody production does not peak until about 8 days after the initial exposure, whereas antibody levels in the blood increase almost immediately after a second exposure. Also note that antibody levels in the blood after second exposure are several times greater than levels after the first exposure. (Figure modified a from Dussan 2011; used with permission of F. A. O. Dussan)

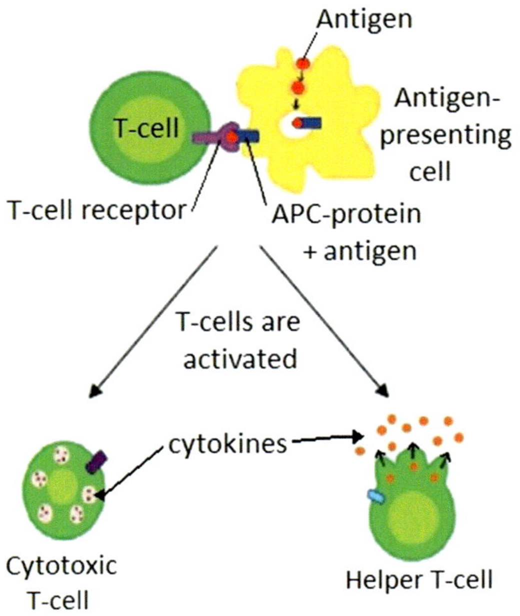


**Fig. 6.47** Activation of T-cells requires antigen-presenting cells (APC) like macrophages. Macrophages take in antigens (by phagocytosis), then attach them to one of their own proteins (APC-protein). This combination is then "presented" on the membrane of the APC and fits into a receptor protein in the membrane of a T-cell. This activates Cytotoxic and Helper T-cells. Cytotoxic T-cells recognize infected cells (e.g., infected by a virus) using membrane receptors and kill them by releasing cytokines (stored in vesicles in the cytoplasm). Helper T-cells use cytokines and stimulate B-cells (that produce antibodies) as well as other T-cells. (Figure modified from Warrington et al. 2011; open-access article distributed under the terms of the Creative Commons Attribution License. https://creativecommons.org/licenses/by/4.0/)

physiological costs of an immune response (Bonneaud et al. 2003). This study clearly reveals that mounting an immune response has costs and may require birds to make trade-offs. "Trade-offs," however, may be less costly in terms of fitness (lifetime reproductive success) when birds are in better condition.

For birds, the situation is even more complicated because the immune system can also be affected by stress and hormones. Birds face a variety of potential stressors including inclement weather, food deprivation, agonistic social interactions, migration, reproduction, and human disturbance (Fig. 6.50). Long-term exposure to stress can lead to immunosuppression (Apanius 1998). For example, Rufous-collared Sparrows (*Zonotrichia capensis*) can be found in both natural and urban environments in Chile, and sparrows in urban areas are lighter in weight and have fewer circulating lymphocytes than those in more natural habitats. This immunosuppression appears to be due to the greater stress associated with living in urban areas, including habitat fragmentation and aggressive interactions with House Sparrows. Rural birds held in captivity for two weeks exhibited the same immunosuppression as birds living in urban areas (Ruiz et al. 2002).

For male birds, the hormone testosterone may also cause immunosuppression. Of course, testosterone is also responsible for bringing males into breeding condition and for expression of traits important in female mate choice such as singing. The "trade-off" required by male birds during the breeding season has been expressed as the immunocompetence handicap hypothesis (Foo et al. 2017). According to this hypothesis, testosterone acts as a "double-edged sword" by promoting the expression of secondary sexual traits, but at the

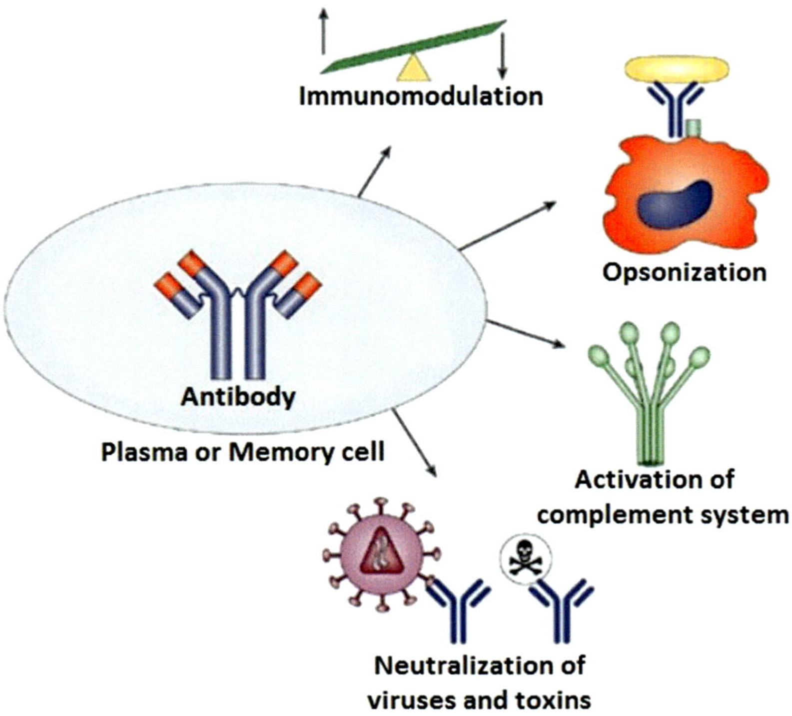


**Fig. 6.48** Antibodies do not have the capability to kill viruses or bacteria directly. Antibodies (especially IgY) perform their function by attaching to pathogens like viruses or toxic molecules produced by bacteria and neutralizing them, i.e., attaching to them to prevent them from entering or damaging host cells. The attached antibodies can also facilitate the destruction of pathogens by phagocytes, a process called opsonization. The binding of antibodies to pathogens also activates the complement system. Finally, antibodies also play an important role in immunomodulation, upregulating and downregulating the immune response as needed. (Figure modified from Casadevall et al. 2004; © 2004 Springer Nature, used with permission)

same time suppressing immune function. This physiological trade-off ensures honesty in signaling by allowing only those individuals with parasite resistance genes to effectively produce a high-quality signal. Thus, obligate immunosuppression by testosterone would prevent cheating by males with low viability or fitness because expression of the signal would lead to increased susceptibility to parasites and ultimately to poor health and decreased survivorship (Folstad and Karter 1992).

For female birds, estrogen is the primary sex hormone and studies of the possible effects of estrogen on immune function have provided mixed results, with some studies suggesting an immune-enhancing effect and others suggesting an immune-suppressing effect. Foo et al. (2017) examined the results of several studies and concluded that estrogen enhances humoral-mediated immunity (antibody production), but suppresses cell-mediated immunity (T-cells). One possible explanation for this difference is that humoral-mediated responses are less costly energetically than cell-mediated responses (Janeway et al. 1999), perhaps allowing females to maintain an effective immune system while diverting more energy to reproduction (Foo et al. 2017).

So, what does this all mean? Many factors contribute to evolutionary success or failure or, in other words, an individual's fitness, but staying alive to reproduce is essential and that makes an efficient immune system essential. The quality of immune systems is, in part, determined by genes.

**Fig. 6.49** Top, Male European Pied Flycatcher (*Ficedula hypoleuca*) showing the forehead patch that conveys information about male quality and positively influences male reproductive success. Bottom, Examples of variation in the size of forehead patches among males. (Top photo by Andy Morffew, pxhere.com, CC BY 2.0, https://creativecommons.org/licenses/by/2.0/. Bottom photos from Sirkiä and Qvarnström 2021; © 2021 The Authors. Ecology and Evolution published by John Wiley & Sons Ltd., open-access article distributed under the terms of the Creative Commons CC BY license, https://creativecommons.org/licenses/by/4.0/)

Although the relative roles of genetic and nongenetic factors in determining immunocompetence in birds remain unclear, a recent study provides some insight into the importance of the genetic component. Investigators performed a cross-foster experiment involving the transfer of young Zebra Finches (*Taeniopygia guttata*) among nests to try and determine how much of the variation in body condition and immunocompetence was due to genetic, maternal, and environmental effects. The investigators found a significant relationship between body condition and immune response and found that genetic variation in immune response explained 56% of the variation in body condition. In other words, the quality of a Zebra Finch's immune system is very much influenced by genetic factors, and differences in the quality of immune systems explain much of the variation in the physical condition of finches (Gleeson et al. 2005). Because secondary sexual ornaments are typically condition-dependent in Zebra Finches (Birkhead et al. 1998), this means that females that select males based on well-developed ornaments are likely to gain genes for their offspring that confer higher levels of immunocompetence. These results, therefore, are consistent with the hypothesis that the genes that determine parasite-resistance may be a major target of sexual selection in this species (Gleeson et al. 2005).

Of course, some variation in immunocompetence and condition is not explained by genetic factors. Regardless of lifestyle or life history, birds must deal with a variety of factors including season (e.g., breeding or not), hormone levels (e.g., testosterone levels in males), and those that can cause stress. All of these interacting factors require birds to balance the need to remain immunocompetent with the need to meet other energetic demands. Success, or not, at doing so will be a critical factor in determining a bird's lifetime fitness.

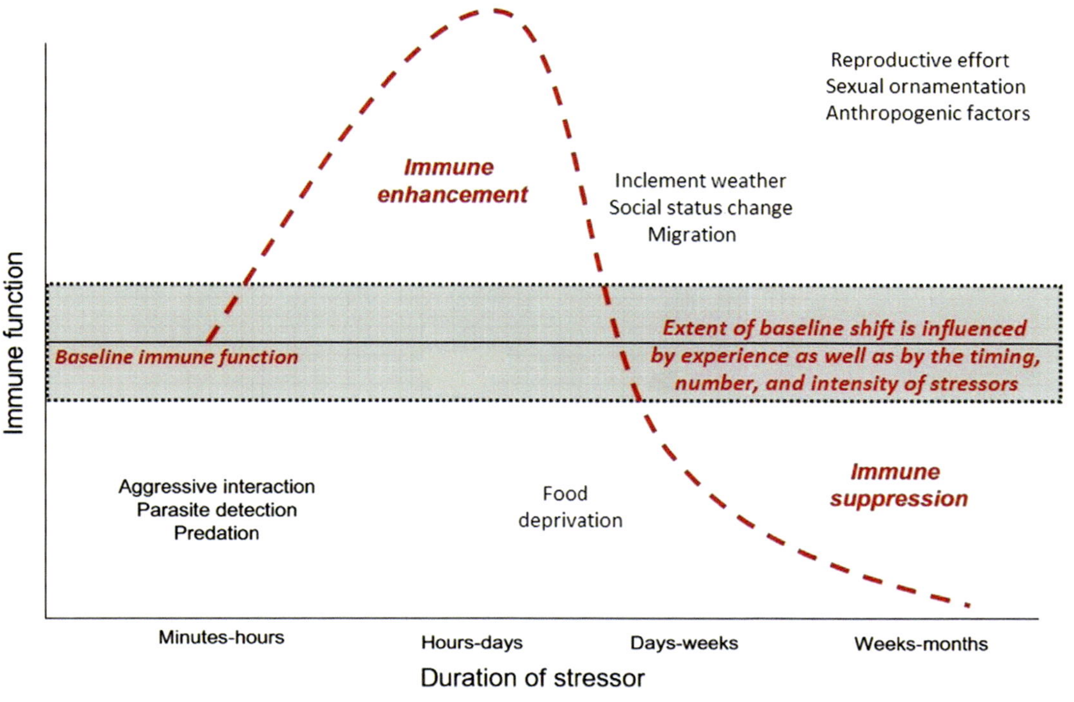


**Fig. 6.50** Stressors that birds can experience and that, over different time scales, can cause immunosuppression. The dashed line shows changes in immune function as a consequence of various stressors. Some stressors, such as brief aggressive interactions, detection of ectoparasites, or encounters with predators, can enhance immune function. Stressors that act over longer period can, in contrast, cause immunosuppression. Longer-duration stressors, such as reproduction and anthropogenic factors such as exposure to pollutants or urban environments, can potentially cause greater suppression than shorter-duration stressors such as inclement weather, migration, and food deprivation. Shaded area indicates a potential long-term increase or decrease in sensitivity to stressors due to previous experience and the timing, number, and intensity of stressors. (Figure modified from Martin 2009; © 2009 Elsevier Inc., used with permission)

## 6.13 Blood Parasites

A wide variety of parasites are commonly found in bird blood. A survey of nearly 15,000 songbirds representing 74 species in Europe revealed that 28.6% were infected with one or more blood parasites (Scheuerlein and Ricklefs 2004). Review of a database with information about sampling locations and avian host species resulted in the identification of 2585 avian blood parasites (Clark 2018). Among the more common blood parasites are five genera of haematoprotozoans: *Plasmodium, Haemoproteus, Leucocytozoon, Parahaemoproteus,* and *Trypanosoma* (Fig. 6.51). Species of *Haemoproteus* are the most common blood parasites in wild birds and are likely the least pathogenic (Atkinson and Van Riper 1991). The genus *Plasmodium* is the cause of malaria in birds. Other vector-borne blood parasites of birds include sporozoan parasites in the genera *Hepatozoon, Babesia,* and *Atoxoplasma* (Valkiūnas 2005). These parasites are transmitted from host to host by bloodsucking flies (Order Diptera), such as blackflies and mosquitoes, and have a cosmopolitan distribution (Atkinson and Van Riper 1991; Santiago-Alarcon et al. 2012). All species in these genera are potentially pathogenic in their natural hosts, but much remains to be learned about the short- and long-term effects of these parasites on wild birds. A

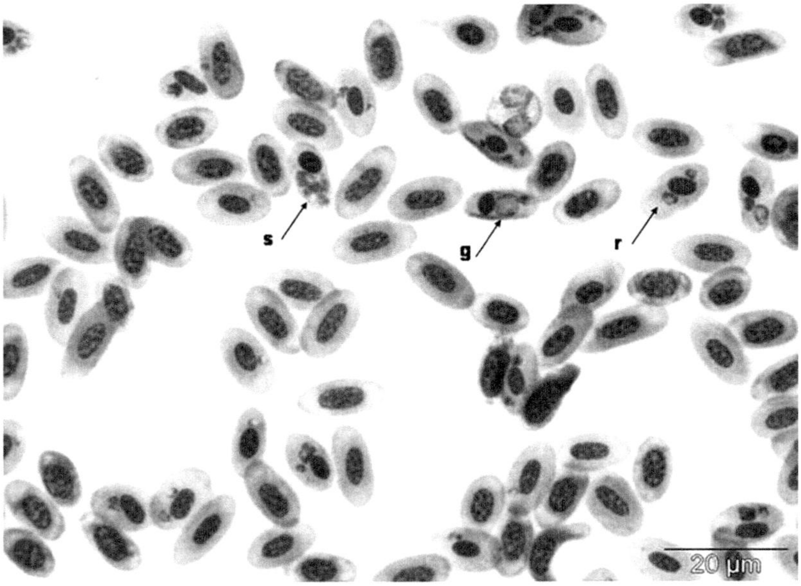


**Fig. 6.51** Blood smear from an Eastern Screech-Owl (*Megascops asio*) infected with Plasmodium subpraecox showing erythrocytes containing P. subpraecox in the vegetative or growing form (schizonts; s), gametocytes (capable of producing gametes; g), and ring forms (r). Scale bar = 20 µm. (Figure from Tavernier et al. 2005; © 2005 American Association of Avian Pathologists, used with permission)

complex variety of factors can influence the severity and outcome of infection, including age and nutritional condition (van Oers et al. 2010). Detecting possible effects of blood parasites on the fitness of infected hosts is difficult because such effects may vary during the course of an infection.

Infection by haemosporidian parasites follows a typical pattern, with an acute (symptomatic) phase with high levels of parasites in the blood (parasitemia) followed by a chronic phase with low (or zero) parasitemia (Valkiūnas 2005). Mortality can result during the acute phase of infection, either directly because of the negative effect of the parasite (Valkiūnas 2005) or indirectly, e.g., increased susceptibility to predation (Asghar et al. 2011). Infected birds may also experience reduced reproductive success (Marzal et al. 2005), delayed migration (DeGroote and Rodewald 2010; Emmenegger et al. 2021), accelerated senescence (Asghar et al. 2015), and reduced survival probabilities (Townsend et al. 2018). However, the results of several studies have also revealed that infected birds with low parasitemia may not be adversely affected, e.g., in reproductive success (Kilpatrick et al. 2006), aerobic performance (Hahn et al. 2018), body mass and likelihood of recapture in subsequent years (Piersma and van der Velde 2012), and nestling growth rates (Lee-Cruz et al. 2016). In general, infection by haemosporidian parasites can negatively impact birds in a variety of ways that, in turn, can reduce lifetime fitness. However, the risk of infection can be influenced by a number of factors (Gupta et al. 2020), e.g., location (zoogeographical realm; Fig. 6.52), body condition, degree of sociality, altitude (due to altitudinal stratification of vector abundance), foraging strata (due to vertical stratification of vector abundance), levels of sexual dimorphism in a species (greater sexual dimorphism is associated with higher parasite infection; Hamilton and Zuk 1982, Møller 1990), and phylogeny (susceptibility may vary depending on host-parasite evolutionary histories and closely related hosts may exhibit similar infection susceptibility or resistance, Medeiros et al. 2013; Fig. 6.53), and the costs of infection depend on the level of parasitemia that, in turn, are likely affected by the quality of host immune defenses (Schoenle et al. 2017).

Other factors can also influence the likelihood of infection by blood parasites. For example, migratory species were found to have a higher

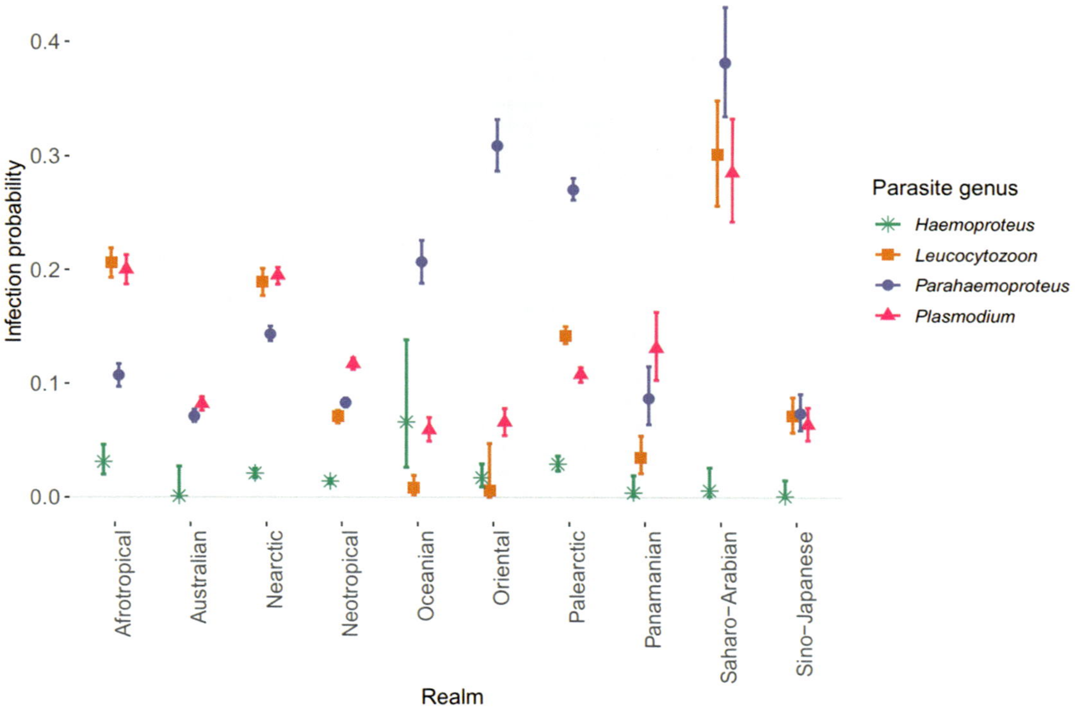


**Fig. 6.52** Realm-specific estimates of average infection probabilities of birds for four genera of haemosporidian blood parasites. For the three most common haemosporidian genera (*Leucocytozoon*, *Parahaemoproteus*, and *Plasmodium*), infection probabilities were highest in the Saharo-Arabian realm. Probability of *Leucocytozoon* infection was lowest in the Oceanian, Oriental, and Panamanian realms, whereas probability of *Parahaemoproteus* infection was lowest in the Australian, Neotropical, and Sino-Japanese realms. *Plasmodium* infection probability was lowest in the Australian, Oceanian, Oriental, and Sino-Japanese realms. In general, the probability of haemosporidian infection appears to be influenced by factors such as climate, avian host richness, and the locations of migratory flyways, as well as variation in region-scale habitats and climates (Fecchio et al. 2021). Error bars represent 95% credible intervals. (Figure from Fecchio et al. 2021; © 2021 John Wiley & Sons Ltd., used with permission)

prevalence of haemosporidian blood parasites than nonmigratory species, likely because migration may expose birds to a greater variety of blood parasites (de Angeli Dutra et al. 2021). Among passerines, larger species exhibit a higher prevalence of infection by blood parasites than smaller species. Larger birds might be more susceptible to infection than smaller birds because they produce more carbon dioxide, a signal used by some flies to locate hosts (Sutcliffe 1986). In addition, larger birds have more surface area for flies to attack (Atkinson and van Riper 1991). Yet another factor that influences the prevalence of infection by blood parasites is the duration of the incubation period. Species with longer incubation periods have lower rates of infection by several blood parasites, including those in the genera *Plasmodium*, *Haemoproteus*, and *Leucocytozoon* (Ricklefs 1992). This relationship between incubation period and prevalence of infection is likely due to the effect of incubation period on immunocompetence.

Antibodies represent an important defense against blood parasites because they enhance recognition of infected blood cells by phagocytes and also activate the complement system. To do these things, however, a specific antibody must be produced; one designed to attach to an antigen, a

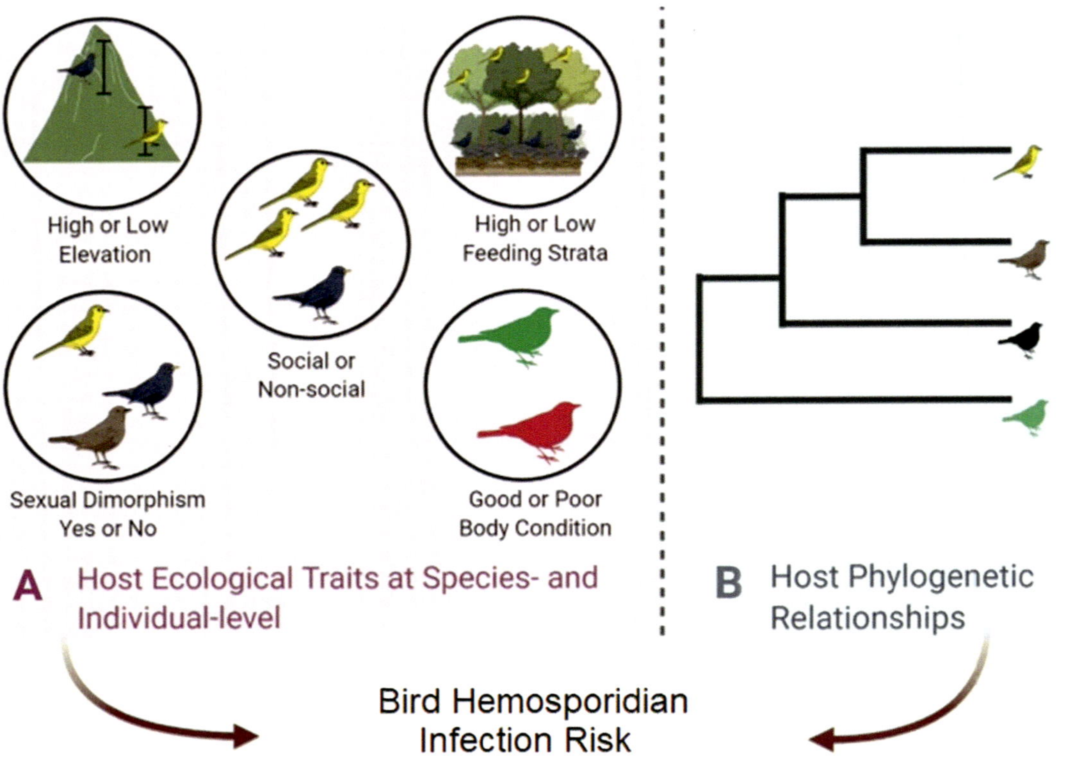


**Fig. 6.53** Numerous factors can influence the risk of Hemosporidian infection for birds. The abundance of parasites and vectors can vary with altitude (due to variation in temperature) and height above ground (feeding strata). Social species may be at greater risk of infection than nonsocial species, with the presence of more birds more likely to attract vectors. Birds in poor condition may be more susceptible to infection, and males in species with greater sexual dimorphism may also be more susceptive to infection (more investment in costly phenotypic traits may impact the quality of immune systems). Susceptibility to infection may also exhibit a phylogenetic signal because closely related hosts likely have similar susceptibility or resistance to infection. (Figure from Gupta et al. 2020; © 2020 Gupta et al., open-access article distributed under a Creative Commons Attribution 4.0 International License, https://creativecommons.org/licenses/by/4.0/)

foreign protein produced by a blood parasite that has a specific shape. Antibodies are produced by B-cells, but each B-cell produces only one type of antibody. An effective immune system has a greater variety of B-cells capable of producing a greater variety of antibodies. That is critical because if B-cells are incapable of producing the specific antibody needed to act against a particular antigen, then pathogens or parasites with that antigen are more likely to infect a bird.

For birds, different types of B-cells are produced only before hatching. Stem cells (or cells that will eventually become B-cells) in the bursa of Fabricius proliferate, or divide, several times before differentiating into specific types of B-cells. During this proliferation process, the genes (sections of DNA that code for proteins) responsible for producing antibodies (and, specifically, the variable regions that will bind to antigens) undergo a process called gene rearrangement. During this process, short-sections of nearby DNA are rearranged in a variety of ways to produce different antibody genes, and, therefore, different antibodies, capable of responding to different antigens and producing different antibodies (Box 6.7 Antibody

**Box 6.7 Antibody Diversity**

The antibodies produced by B-cells help protect birds from an incredible diversity of different antigens. In humans, the estimated potential number of different antibodies that can be produced by B-cells is $2 \times 10^{12}$ (Lefranc and Lefranc 2020). Antibodies, of course, are proteins and proteins are encoded by genes. Neither humans nor birds have enough genes to produce so many different proteins and antibodies. How then are so many different antibodies produced? During the process of B-cell maturation, segments of genes that produce the antibody proteins (and, specifically, the variable portions of the antibodies that actually bind to antigens) are rearranged in a process somewhat comparable to shuffling a deck of cards. In other words, each time a deck of cards is shuffled, the arrangement or order of the cards in the deck is different. Similarly, during the production of B-cells, gene segments in the DNA are rearranged in different ways to produce different genes that, in turn, produce different proteins and different antibodies. Also contributing to the impressive diversity of antibodies is somatic hypermutation, a process that allows B-cells to mutate the genes used to produce antibodies. This results in the production of antibodies better able to bind to antigens produced by bacteria and other pathogens

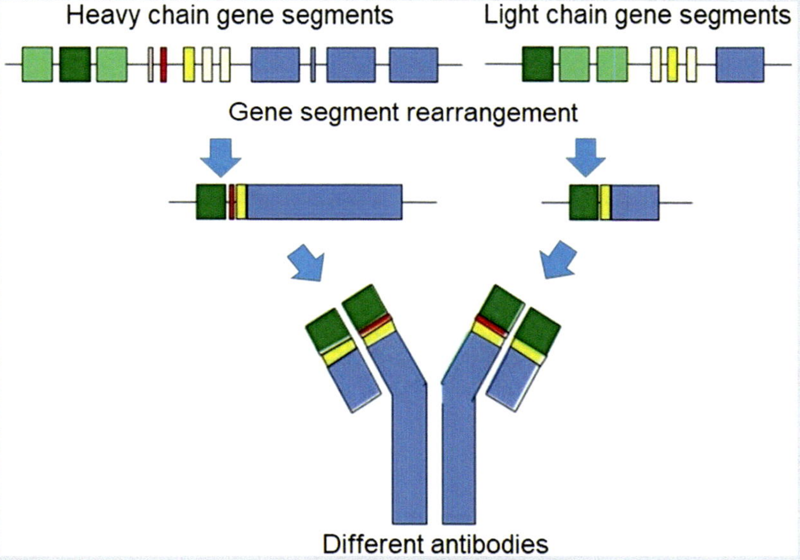


During B-cell development, a tremendous number of different antibodies are produced. An important part of this process involves a "shuffling" of gene segments to produce different genes that, in turn, will produce the different proteins that make up the binding sites of the antibodies. The different proteins have different three-dimensional shapes that will allow them to bind to a wide variety of different antigens (Figure modified from Lefranc and Lefranc 2020; © 2020 by the authors, open-access article distributed under the terms and conditions of the Creative Commons Attribution (CC BY) license, https://creativecommons.org/licenses/by/4.0/)

(continued)

**Box 6.7** (continued)

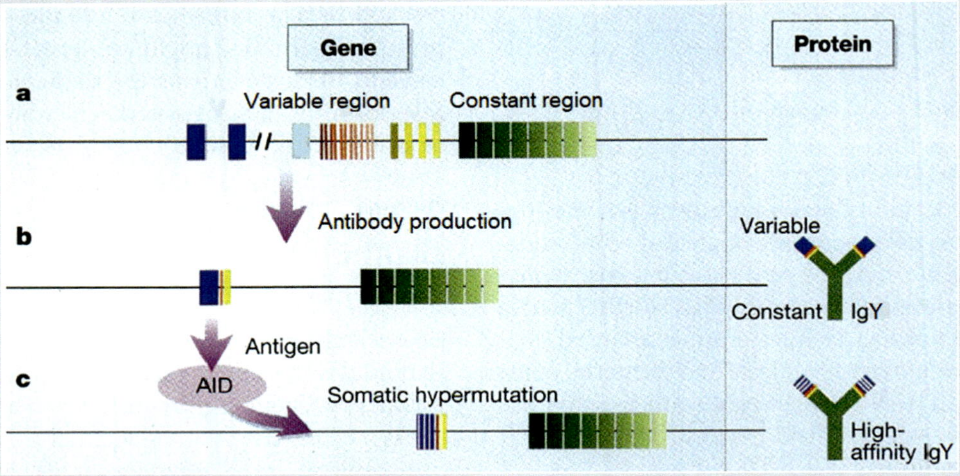


(**a**) In immature B-cells, gene segment rearrangement results in the production of an antibody with a variable region having a three-dimensional shape that allows it to bind to a particular antigen. (**b**) The binding of the antigen to the antibody in the membrane of a B-cell triggers a process where an enzyme called activation-induced deaminase (AID) causes mutations (thin-white lines) in the variable regions. These mutations alter the shape of the variable region so that the antibody now has a higher affinity (i.e., a better "fit") for its antigen (Figure modified from Ramiro and Nussenzweig 2004; © 2004 Nature Publishing Group, used with permission)

Diversity). The longer this proliferation phase, the greater the number of different types of B-cells produced. Because a longer incubation period means more time for B-cell proliferation, birds with longer incubation periods may, after hatching, have a greater diversity of B-cells and, therefore, more effective immune systems (Ricklefs 1992). As a result, these species may be better able to resist infection by blood parasites than those with shorter incubation periods and a lower diversity of B-cells.

In a much-cited paper, Hamilton and Zuk (1982) proposed a relationship between plumage brightness and parasite load. They found that species with the most brightly colored males had higher parasite loads and suggested that this was because males in these species had experienced strong selection pressure for ways to advertise their quality to females. Bright plumage is costly because it takes time and energy to obtain the needed pigments (e.g., carotenoids needed for red, orange, and yellow plumage) or to generate the necessary high-quality feathers (e.g., for colors due to feather structure such as blue and various iridescent colors). Males vary in quality and, as a result, in their ability to expend the time and energy needed for bright plumage. Lesser-quality males devoting time and energy to maximizing plumage quality may devote less energy to immune defense. Such males are more likely to be infected by parasites, which further impairs their ability to generate bright plumage and obtain a mate. This need, driven by sexual selection, to maximize plumage quality increases the proportion of males in the population susceptible (because of energetic trade-offs) to infection by parasites; hence, the positive relationship between prevalence of parasite infection and plumage brightness.

Avian malaria is caused by parasites in the genera *Plasmodium*, is transmitted by mosquitoes, and can have severe consequences for infected individuals. Avian *Plasmodium* species have lower host specificity than those of mammals. For example, *Plasmodium relictum* infects over 300 species of birds in 11 orders, and other *Plasmodium* species can also infect a broad range of avian hosts (Valkiūnas 2005; Bensch et al. 2009; Hellgren et al. 2015). In addition, avian malaria can be a more severe disease than human malaria, e.g., obstructive secondary exoerythrocytic stages of *Plasmodium* in brain capillaries that can lead to ischemia and rapid death in birds (Ilgūnas et al. 2016; Fig. 6.54). However, the severity of the disease caused by *Plasmodium* varies among different species of birds, ranging from no apparent symptoms to high mortality (Palinauskas et al. 2008; Atkinson et al. 2009; Dimitrov et al. 2015). For example, 43% of the individuals in a population of Great Reed Warblers (*Acrocephalus arundinaceus*) in Sweden were found to be infected with *Plasmodium* and investigators detected no detectable fitness cost associated with being infected (Westerdahl et al. 2005). However, some infected individuals may lay smaller clutches (Oppliger et al. 1996) or have a reduced probability of survival (Sol et al. 2003). Birds exposed to avian malaria for the first time, like birds on isolated islands, may be highly sensitive. For example, avian malaria (*Plasmodium relictum*) was first detected in Hawaii in the 1940s (and the mosquitoes that serve as the vector were introduced to Hawaii by whaling ships in the early 1800s) and has been a primary factor in the decline of many, and extinction of some, Hawaiian honeycreepers (Atkinson et al. 2001; Atkinson and LaPointe 2009; Fig. 6.55).

## 6.14 Blood Plasma

The liquid part of the blood is called plasma and, although primarily water, there are many other substances dissolved in that water. One is glucose. With their high metabolic rates, birds use energy at high rates and, therefore, typically maintain high concentrations of glucose in the blood. Typical levels of glucose in the blood plasma of birds range from about 200–400 mg/100 ml (Harr 2002) and, not surprisingly, are even higher in hummingbirds (up to 800 mg/100 ml; Diamond et al. 1986). By comparison, blood glucose levels in humans are typically about 70–100 mg/100 ml.

Flying birds use lots of energy and so it seems logical to assume that glucose levels in the blood would decrease when a bird is flying. However, comparison of blood glucose levels of homing pigeons (*Columba livia*) that had flown from between two and 22 hours and control pigeons that had not flown revealed no significant difference (Bordel and Haase 1993). Similar results have been reported for several species of songbirds (Jenni-Eiermann and Jenni 1991). Birds use blood glucose as an energy source for their flight muscles only during take-off and during short flights. During sustained flight, fats are the primary source of energy for the flight muscles (Jenni-Eiermann and Jenni 1998). After flying for several hours, plasma concentrations of free fatty acids and glycerol increase (Bordel and Haase 1993; Jenni-Eiermann et al. 2002). Fatty acids and glycerol are produced when triglycerides (a type of fat composed of molecules of fatty acids and glycerol) are metabolized or broken down in adipose or fat tissues and released into the blood (Jenni-Eiermann et al. 2002). In Red Knots (*Calidris canutus*), this transition from glucose to fats as the primary energy source occurs within one hour of taking flight (Jenni-Eiermann et al. 2002).

During sustained flight, levels of uric acid in the blood plasma also increase (Bordel and Haase 1993; Jenni-Eiermann et al. 2002). Uric acid is a waste product produced by the breakdown or catabolism of proteins, so this increase indicates that proteins are also used as a source of energy during prolonged flight. However, the extent to which protein is used during flight varies among species, with diet being one of the determining factors (Gannes 2001). During migration, fruit-eating omnivorous birds depend more on fats for energy and so have less uric acid in their plasma. Insectivorous species rely more on protein and have higher plasma levels of uric acid (Gannes 2001). Of course, a bird's condition also matters.

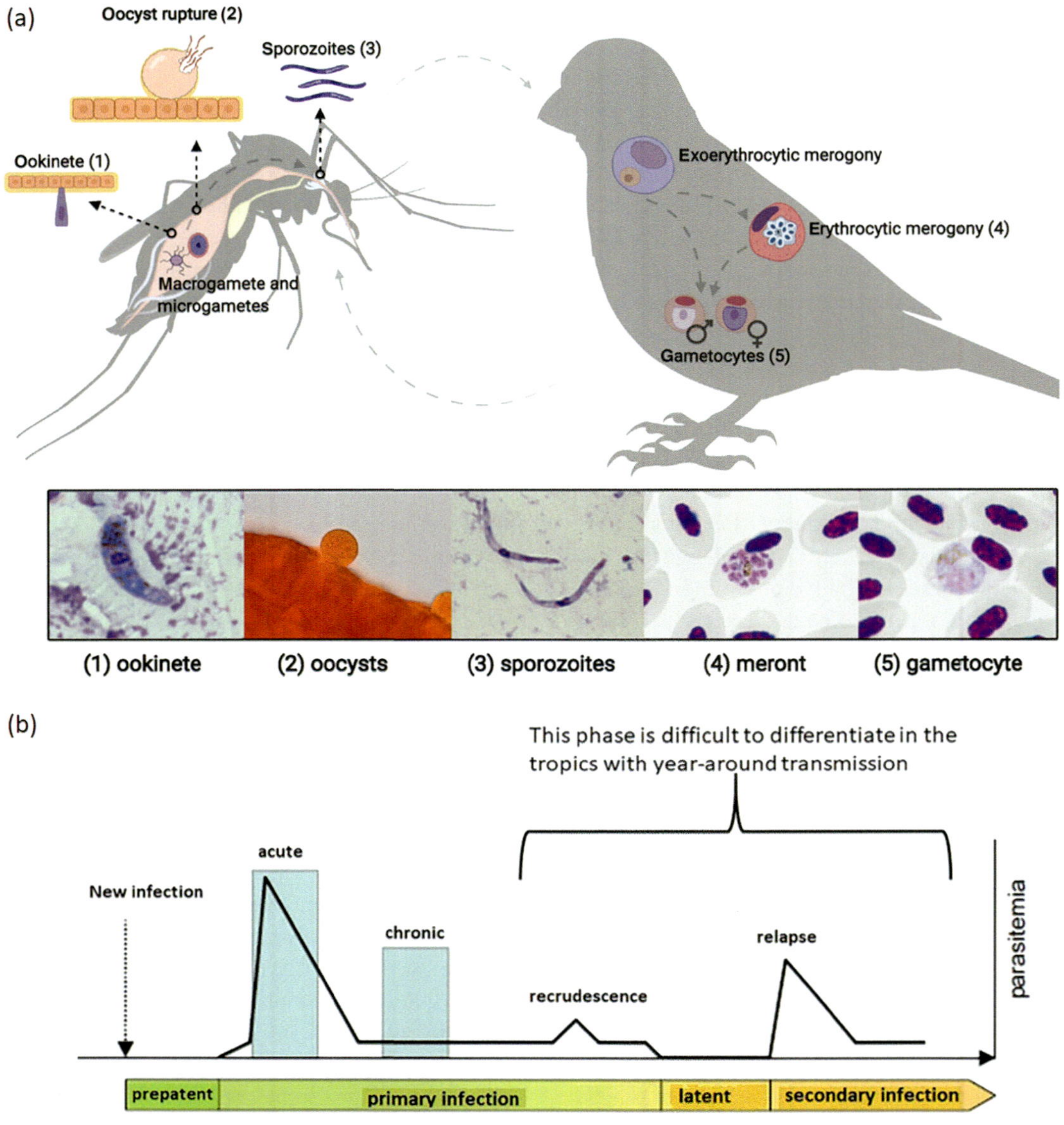


**Fig. 6.54** (**a**) *Plasmodium relictum* life cycle. *P. relictum* female and male gametes merge to form a zygote that quickly transforms into a motile ookinete that establishes an oocyst infection in the midgut epithelium. Several days later, oocysts release thousands of sporozoites that migrate to and invade the salivary glands and, when the mosquito bites an avian host, are injected into the host via the mosquito's saliva. During exoerythrocytic merogony, injected sporozoites invade macrophages, monocytes, and endothelial cells near the injection site and develop into exoerythrocytic meronts. Meronts multiply asexually to produce numerous small merozoites. When infected cells rupture and die, merozoites are released and spread through the blood stream and invade red blood cells (erythrocytes). Merozoites then further develop into either erythrocytic meronts that multiply and cause erythrocytes to rupture or into gametocytes. Merozoites that develop into gametocytes remain in red blood cells until a mosquito vector bites the host. Gametocytes then repeat the life cycle in the mosquito (Grilo et al. 2016; Valkiūnas and Iezhova 2018). (**b**) Infected birds go through an acute phase that is followed by a later chronic phase. The acute phase peaks about 6 to 12 days after parasites first appear in the blood. Chronic infections can persist for a bird's lifetime, with relapses possible during periods of physiological stress such as reproduction (Ishtiaq 2021). (Figure a from Martínez-de la Puente et al. 2021; © 2020 Elsevier Ltd., used with permission. Figure b from Ishtiaq 2021; © 2021 Indian Institute of Science, used with permission)

**Fig. 6.55** Densities (number of pairs/hectare) of four species of Hawaiian forest birds along an elevational gradient (1.0–1.9 km above sea level). Densities of all four species are higher at higher elevations because populations of mosquitos that transmit avian malaria are limited by the cooler temperature at higher elevations. 'Akiapölä'au, *Hemignathus wilsoni*; Hawai'i 'Akepa, *Loxops coccineus*; Hawai'i Creeper, *Loxops mana*; 'I'iwi, *Drepanis coccinea*. (Figure from Guillaumet and Paxton 2019; © 2019 by the Ecological Society of America, used with permission. Bird photos by Jack Jeffrey, U. S. National Park Service, CC0 Public Domain)

Those with less stored fat for migration will use more protein as a source of energy and will have higher levels of uric acid in their plasma than individuals in better condition and better able to store more fat (Bairlein and Totzke 1992; Gannes 2001).

Among birds that fly long distances during migration, some of the protein used as a source

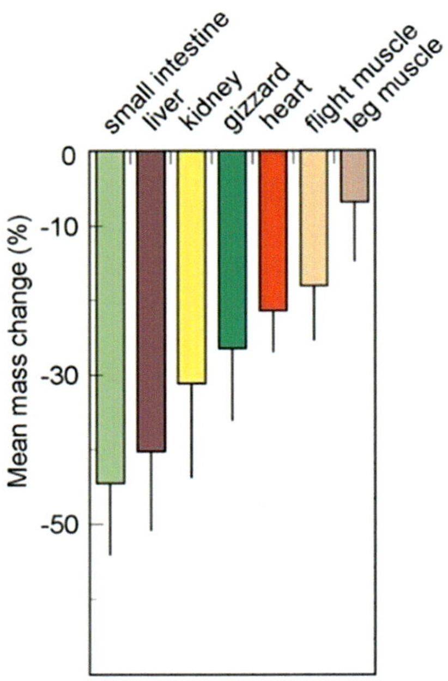


**Fig. 6.56** Mean dry mass (±95% CI) of different organs of five species of birds after long-distance migratory flights. The five species included Great Knots (*Calidris tenuirostris*), Eurasian Blackcaps (*Sylvia atricapilla*), Garden Warblers (*Sylvia borin*), European Pied Flycatchers (*Ficedula hypoleuca*), and Willow Warblers (*Phyloscopus trochilus*). (Figure modified from Bauchinger and McWilliams 2010; © 2010 John Wiley and Sons, used with permission)

of energy actually comes from the flight muscles (Fig. 6.56). For example, as migrating Red Knots use protein from their flight (pectoralis) muscles, muscle mass obviously decreases (Lindstrom et al. 2000). However, during a long-distance migration, a Red Knot's body mass also declines as fats are continuously used to sustain flight. So, the farther a Red Knot flies (without refueling), the lighter it gets, and a lighter bird requires less powerful flight muscles (Jenni-Eiermann et al. 2002). The mass of the flight muscles, in other words, continuously adapts to the mass of the flying bird (Pennycuick 1998).

Proteins in the plasma include albumin and the globulins (Scanes 2015). Typical concentrations in birds (3–6 mg/100 ml) are approximately half that of mammals. All plasma proteins, except immunoglobulins (or antibodies), are manufactured in the liver. Albumin is the most abundant plasma protein and is important in maintaining osmotic balance, serving as a source of energy during flight as just described, and transporting several important substances through the blood, including vitamins, minerals, hormones, and fatty acids. Osmotic balance simply means that the solute concentration of the blood matches that of body cells. This is critical because without this balance, water would, by osmosis, either tend to leave (with too little albumin in the plasma) or enter (with too much albumin in the plasma) the capillaries.

Although proteins and glucose are among the more abundant substances in the blood plasma of birds, many other substances are present as well (Scanes 2015). As noted above, plasma carries waste products like urea and uric acid. The plasma of birds also contains a variety of electrolytes or ions, including sodium, potassium, calcium, and magnesium. When too abundant, these ions can be waste products and may need to be excreted by the kidneys. However, these ions are also critically important for all body cells and their presence in the plasma means that they are always available to those cells.

## 6.15 Avian Heart

Birds, like mammals, and in contrast to other vertebrates, have a 4-chambered heart (two atria and two ventricles), with complete separation of oxygenated and oxygen-poor blood (Figs. 6.57 and 6.58). The right ventricle pumps blood to the lungs, while the left ventricle pumps blood to the rest of the body. Because the left ventricle must generate greater pressure to pump blood throughout the body (in contrast to the right ventricle that pumps blood to the lungs), the walls of the left ventricle are much thicker and more muscular (Figs. 6.59 and 6.60). The avian heart, like the mammalian heart, also has four valves that maintain a unidirectional flow of blood. The two atrioventricular valves, or AV valves (Figs. 6.61 and 6.62), close when the ventricles contract, preventing blood from flowing back up into the atria, and the pulmonary valve at the entrance to the pulmonary trunk (or artery) and the aortic

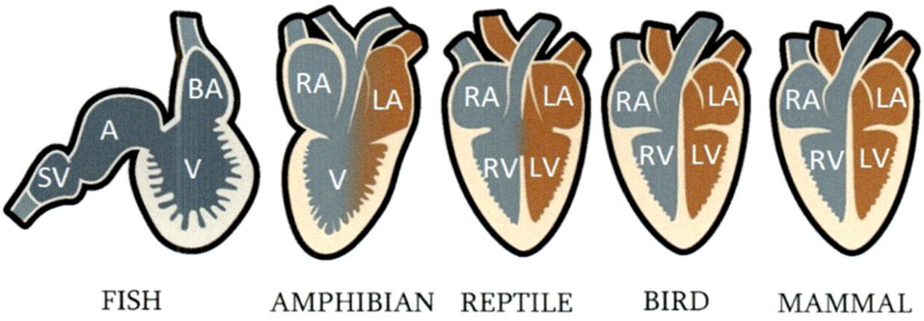


**Fig. 6.57** Hearts of vertebrates. Birds and mammals have four-chambered hearts with complete separation of deoxygenated (right atrium and ventricle) and oxygenated (left atrium and ventricle) blood. Fish hearts, with a single atrium and ventricle, pump blood to the gills. Amphibian heart has a single ventricle that has trabeculae that largely keep oxygenated and deoxygenated blood separate. Most reptiles have only a partial interventricular septum (the wall between ventricles), but have trabeculae in the ventricles that create "pockets" that help prevent mixing of oxygenated and deoxygenated blood. Crocodilians have complete interventricular septums, but also have a structure (Foramen of Panizza) that directs most blood away from the lungs (and, therefore, more blood and oxygen to the rest of the body) when they are under water. SV, sinus venosus; A, atrium; V, ventricle; BA, bulbus arteriosus; RA, right atrium; LA, left atrium; RV, right ventricle; LV, left ventricle. (Figure from Stephenson et al. 2017; © 2017 Anatomical Society, used with permission)

valve at the entrance to the aorta close when the ventricles relax and prevent blood from flowing back down into the ventricles.

Avian cardiac muscle cells (also called cardiomyocytes) are long (> 100 μm) and thin, with much smaller diameters than mammalian cardiomyocytes. Avian muscle cells are generally between 2–7 μm in diameter and those of mammals between 10–15 μm in diameter (Dzialowski and Crossley 2015). The long, thin shape of avian cardiomyocytes means that they have a large surface area-to-volume ratio, which is likely why the cardiac muscle cells of birds, in contrast to mammals, have no transverse (or t) tubules (Sommer and Johnson 1969). Because of the small size of their cardiac muscle cells, bird hearts have more muscle cells per unit volume than mammal hearts.

Avian cardiac muscle cells consist of subunits called myofibrils that, in turn, consist of repeating segments called sarcomeres (Fig. 6.63). Each sarcomere consists of thick and thin myofilaments that slide past each other during contraction. Contraction requires the release of calcium ions from the sarcoplasmic reticulum and, in the cardiac muscle cells of mammals (Fig. 6.64), this occurs when impulses travel along the sarcolemma (cell membrane of muscle cells), down structures called transverse tubules, or t-tubules, and finally to the membranes of the sarcoplasmic reticulum (SR). The SR stores calcium ions ($Ca^{++}$) and, as the impulse travels along the SR, calcium channels in the membrane open. Calcium ions then diffuse out among the myofilaments that make up the myofibrils and muscle cells contract (more detail about muscle contraction is provided in Chap. 6).

The cardiac muscle cells of birds do not have t-tubules (Fig. 6.65), so impulses from the sarcolemma are transmitted to the sarcoplasmic reticulum by way of peripheral couplings (Shiels 2022). Peripheral couplings consist of a series of membrane receptors (or, as referred to in the literature, L-type $Ca^{2+}$ channels [LTCCs]) closely associated with receptors in the membranes of sarcoplasmic reticulum (called ryanodine receptors, or RyRs) (Figs. 6.66 and 6.67). Several RyRs cluster together (from 14 up to 100) in the SR membrane to form Calcium Release Units, or CRUs (Shiels and Galli 2014). When an avian cardiac muscle cell is stimulated, the impulse travels along the sarcolemma. This opens the LTCCs, allowing entry of extracellular calcium ions that activate (open) the calcium channels in the membrane of the SR (RyRs). The calcium ions stored in the SR then diffuse among the

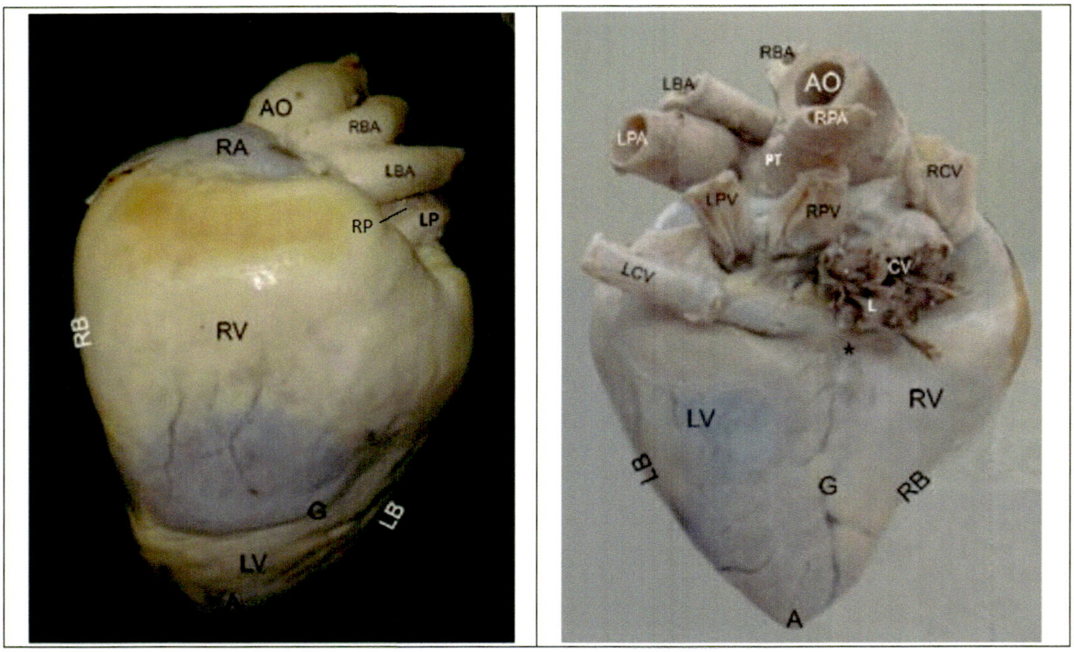


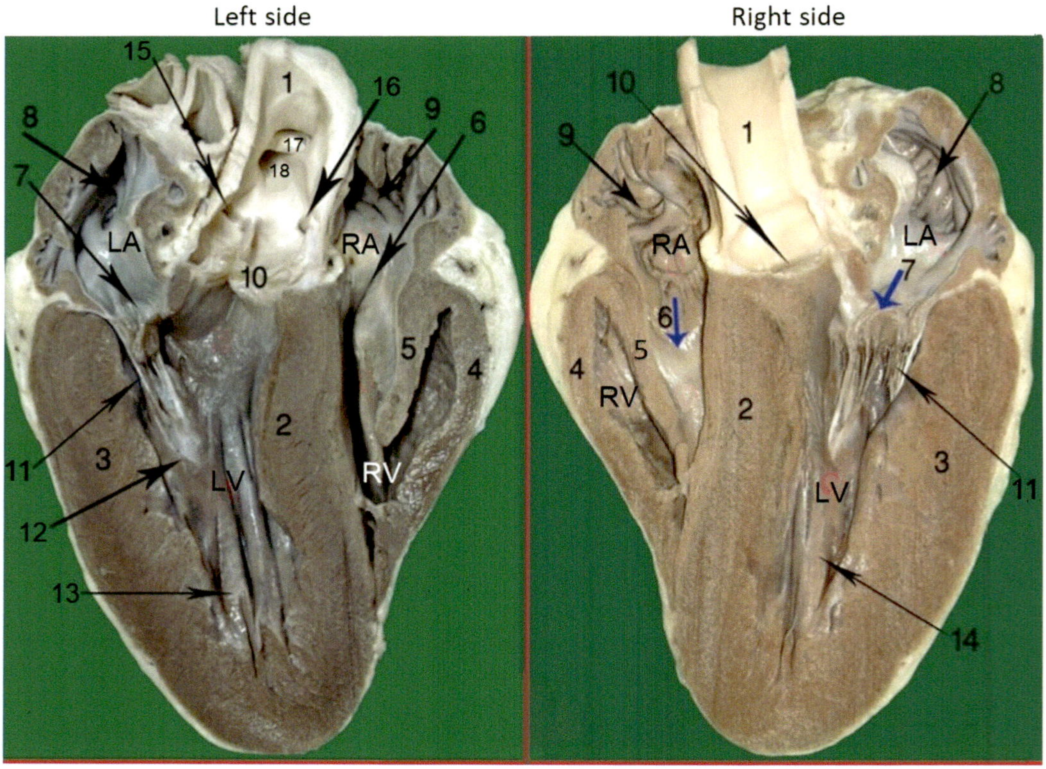


**Fig. 6.58** Top, Heart of a Common Ostrich (*Struthio camelus*). Left, cranioventral surface: RV) right ventricle, LV) left ventricle, RA) right atrium, A) apex of heart, RB) right border, LB) left border, G) longitudinal groove, AO) aorta, RBA) right brachiocephalic artery, LBA) left brachiocephalic artery, RP) right pulmonary artery, and

myofilaments that make up the myofibrils and contraction begins. Because bird hearts contract and relax at high rates (e.g., over 1000 beats per min for some hummingbirds), contraction must be quickly followed by relaxation. This requires the rapid active transport of calcium ions back into the SR. The hearts of smaller birds generally contract/relax at faster rates than those of larger birds and, as an adaptation to allow these faster rates, the cardiac muscle cells of smaller birds have more RyRs as well as CRUs that are closer together than those of larger birds, thus allowing faster contractions (Perni et al. 2012; Shiels 2022).

Birds tend to have larger hearts, and higher systemic blood pressures, than mammals (relative to body size and mass). The relatively large hearts of birds may be necessary to meet the high metabolic demands of flight. Among birds, smaller birds have relatively larger hearts (again relative to body mass) than larger birds. For example, small birds like swallows, finches, and swifts have relatively larger hearts than larger birds like hawks, cormorants, and storks (Fig. 6.68). Heart mass relative to body mass is also influenced by flight mode. Nespolo et al. (2018) categorized avian flight modes as short flight (e.g., most galliformes), flapping flight (most birds), gliding and soaring flight (e.g., many diurnal raptors and dynamic soarers like albatrosses), and hovering flight (hummingbirds) and found that hovering birds have substantially larger hearts relative to body mass than other birds, short-flight species have the smallest hearts, and species that use flapping flight and soaring and gliding flight have intermediate heart mass to body mass values. The energetic cost of hovering requires high cardiac output which, in turn, has clearly imposed selective pressure for increasing heart size among hummingbirds (Nespolo et al. 2018; Figs. 6.69 and 6.70).

Some groups of birds have particularly small hearts relative to their body mass. For example, tinamous (Family Tinamidae; 47 species) are secretive partridge-like birds largely restricted to tropical lowlands in South America and have very small hearts relative to their body mass (Figs. 6.69 and 6.70). Tinamous have relatively small wings and generally only fly when disturbed or when flying into a tree to roost. Tinamous typically attempt to escape from predators by walking and running and, when they do have to fly, they jump and just briefly flap their wings. Because of their low cardiac output (i.e., 2.5–3 times lower than that of Domestic Chickens; Joyce 2018), tinamous can only jump and flap their wings two or three consecutive times before they are unable to fly again (Altimiras et al. 2017). Because they fly infrequently, their metabolic demands are relatively low compared to birds that fly more often, and their small hearts can pump sufficient blood to meet those demands (Bishop 1997; Altimiras et al. 2017). Other birds that fly infrequently

**Fig. 6.58** (continued) LP) left pulmonary artery. Right, caudodorsal surface: LB) left border, RB) right border, A) apex of heart, G) longitudinal groove, *) coronary groove, LV) left ventricle, RV) right ventricle, AO) aorta, CV) caudal vena cava, RCV) right cranial vena cava, LCV) left cranial vena cava, LPV) left pulmonary vein, RPV) right pulmonary vein, RBA) right brachiocephalic artery, LBA) left brachiocephalic artery, A) apex of heart, PT) pulmonary trunk, RPA) right pulmonary artery and LPA) left pulmonary artery (Figures from Tadjalli et al. 2009; © Iranian Journal of Veterinary Research, used with permission). Bottom, Sagittal section of an Ostrich heart. RA, right atrium; LA, left atrium; RV, right ventricle; LV, left ventricle; 1, aorta, 2, interventricular septum; 3, wall of left ventricle; 4, wall of right ventricle; 5, right atrioventricular valve; 6, route that blood takes from right atrium into the right ventricle; 7, route that blood takes from the left atrium into the left ventricle; 8, ridges in the wall of the left atrium (sometimes called "pectinate" muscles); 9, ridges in the wall of the right atrium; 10, aortic valve; 11, chordae tendinae; 12, papillary muscle; 13, muscular ridges at the bottom of the left ventricle (sometimes called the trabeculae carneae); 14, muscular ridges at the bottom of the left ventricle; 15, opening into left coronary artery; 16, opening into right coronary artery; 17, opening into the right brachiocephalic artery; 18, opening into the left brachiocephalic artery. (Figure modified from Alsafy et al. 2009; used with permission of M. A. M. Alsafy)

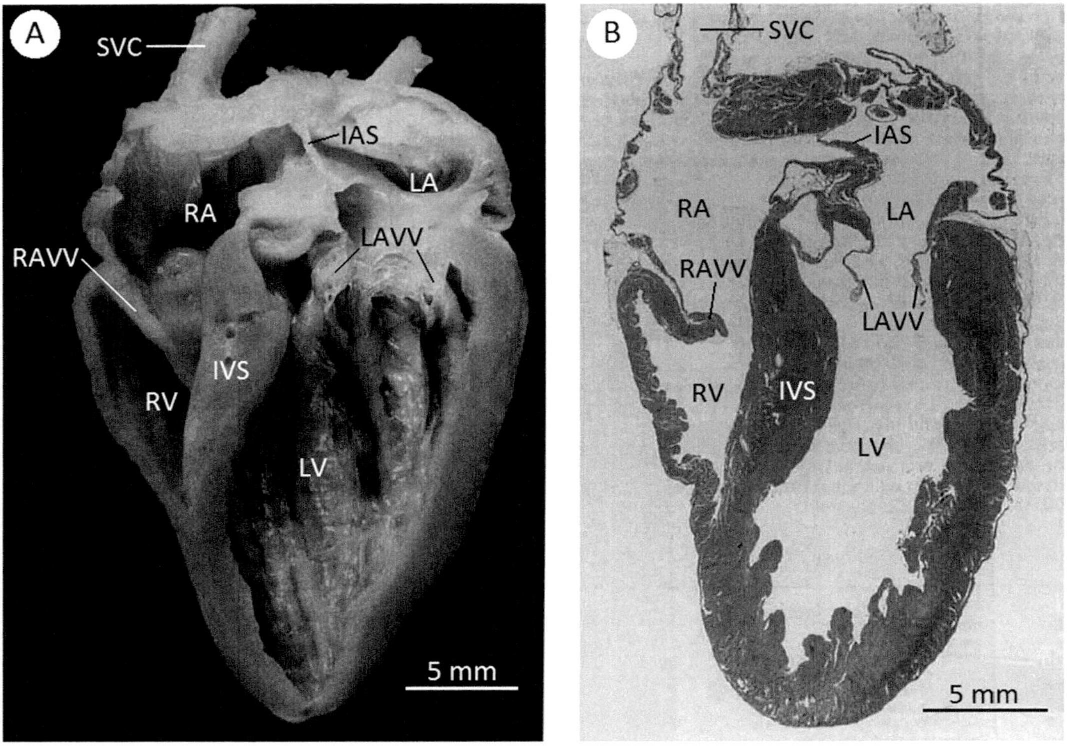


**Fig. 6.59** The four-chambered heart of a Domestic Chicken (*Gallus g. domesticus*). Note that the walls of the left ventricle are much thicker than the walls of the right ventricle. RA, right atrium; LA, left atrium; RV, right ventricle; LV, left ventricle; RAVV, right atrioventricular valve; LAVV, left atrioventricular valve, SVC, superior vena cava; IAS, interatrial septum; IVS, interventricular septum. (Figure modified from Lu et al. 1993; © 1993 Wiley-Liss, Inc., used with permission)

also have relatively small hearts, including rails, guineafowl, guans, and pheasants. Perhaps surprisingly, hawks and owls also have relatively small hearts (Fig. 6.70). Because most hawks and owls are sit-and-wait predators, they often spend considerable time on perches waiting for prey and, when they do fly, it is often just a short flight to another perch. Of course, many hawks are migratory and, at times, must fly long distances. Even then, however, many tend to soar rather than use flapping flight. The energetic demands of soaring are less than for flapping flight and, as a result, a relatively small heart can pump sufficient blood to meet metabolic needs. Raptors that depend more on flapping flight to capture prey, such as most species of falcons (Falconidae), have greater metabolic needs and have relatively larger hearts (Figs. 6.69 and 6.70).

Analysis has also revealed that, in several groups of birds, the relative size of male hearts is greater than female hearts. This sexual dimorphism in relative heart size has been reported for hummingbirds, tanagers, wrens, ovenbirds, tyrant flycatchers, and a few other groups (Hartman 1961). In these birds, males also have relatively larger flight muscles (mass of the pectoralis muscles relative to body mass), suggesting that these males might need additional cardiac output to support the additional muscle mass needed for courtship (Bishop 1997) or, perhaps, other activities such as territory defense.

Heart mass of birds can also vary seasonally or with changes in metabolic demands. Molting

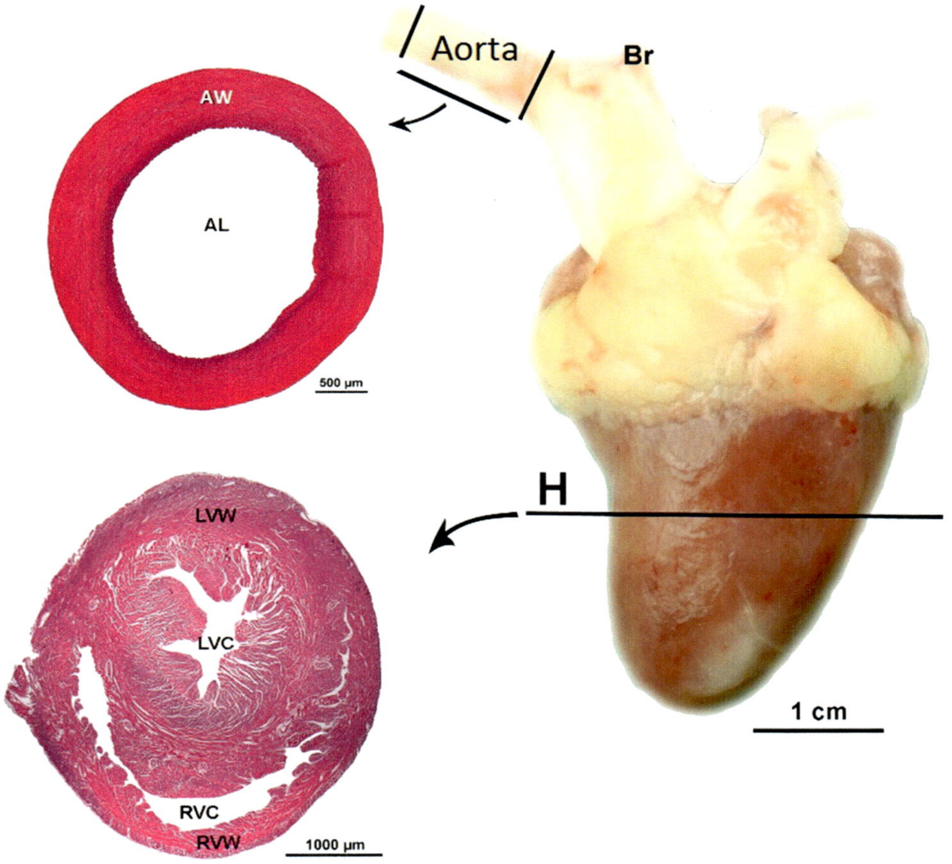


**Fig. 6.60** Right, Heart of a Domestic Chicken. Left top, cross section through the aorta. Left bottom, cross section through the ventricles showing the relative thickness of the walls of the left and right ventricles. Br, brachiocephalic artery; AW, aortic wall; AL, aortic lumen; LVW, left ventricular wall; LVC, left ventricular lumen; RVC, right ventricular cavity; RVW, right ventricular wall. (Figure from Harash et al. 2019; )

geese lose all their flight feathers simultaneously and, as a result, are incapable of flight during the time needed to replace those feathers. For Graylag Geese (*Anser anser*), heart mass during molt decreased by 37% before again increasing when they regained the ability to fly (Fox and Kahlert 2005). Heart size is also known to increase during migration in some species (Piersma et al. 1996, 1999; Guglielmo and Williams 2003) and during the winter in some species of resident birds at higher latitudes (Liknes and Swanson 2011). This increase in heart size may result from the need to increase cardiac output to meet the increased metabolic demands either of long-distance migratory flights or to better supply oxygen to shivering muscles during periods of cold weather (Liknes and Swanson 2011). Some of the increase in size might also be due to an increase in the concentration of red blood cells. Migrating Bar-tailed Godwits (*Limosa lapponica*) have larger hearts, but also have more red blood cells and, as a result, more viscous blood (Piersma et al. 1996). Blood that is more viscous is more difficult to pump and

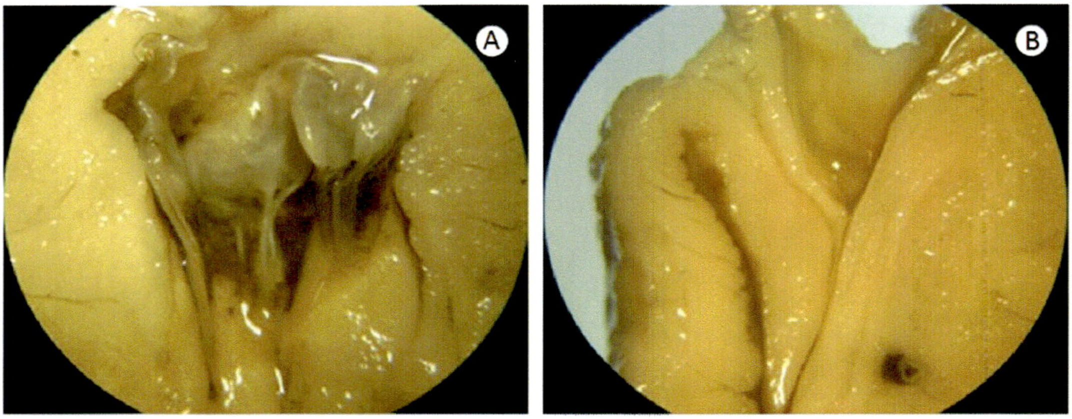


**Fig. 6.61** The left (**a**) and right (**b**) atrioventricular valves of Southern Crested Caracara (*Caracara plancus*). (Figure from Miguel et al. 2014; Morphological and morphometric analysis from the heart of *Caracara plancus* (Caracara). Journal of Morphological Science. Volume 31, Issue 4. Pp 225–227. Thieme. 2014, used with permission)

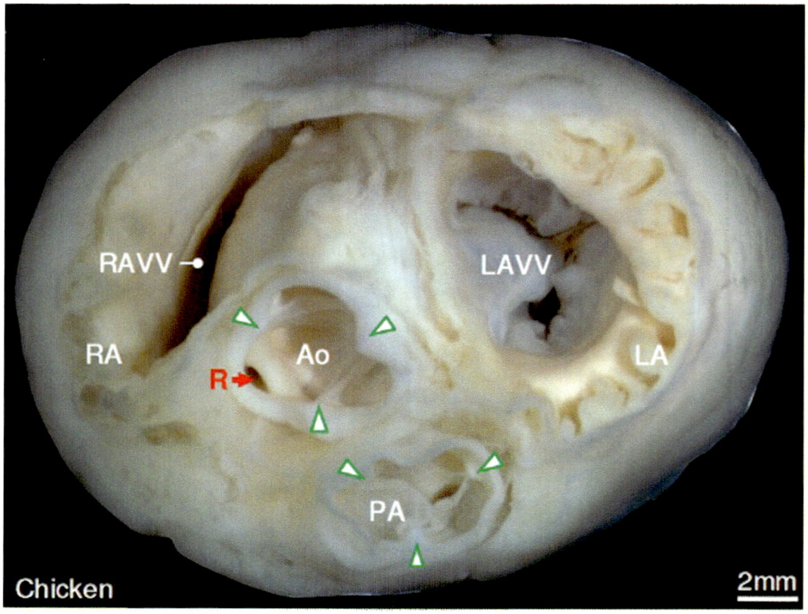


**Fig. 6.62** Cross section through the heart of a Domestic Chicken showing the right (RAVV) and left (LAVV) atrioventricular valves, the aorta (Ao) and pulmonary artery (PA), the aortic and pulmonary valves (indicated by the green arrows), and the right coronary artery (R). Note that the left AV valve has three cusps, whereas the right AV valve is a single section of myocardium. (Figure from Kroneman et al. 2019; open-access article available under the terms of the Creative Commons Attribution License (CC BY), https://creativecommons.org/licenses/by/4.0/)

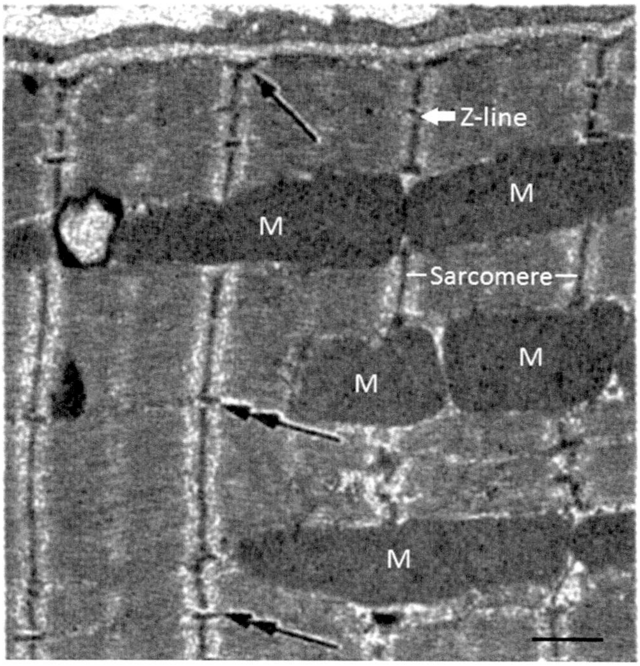


**Fig. 6.63** Avian hearts, like those of mammals, consist of cardiac muscle cells that, in turn, consist of numerous subunits called myofibrils. Myofibrils are made up of numerous subunits called sarcomeres, each of which is separated from adjacent sarcomeres by z-lines. Sarcomeres are made up of long, thin proteins called thick and thin myofilaments. During contraction, these filaments can slide past each other, causing sarcomeres and, therefore, muscle cells to shorten. Peripheral couplings (arrow at top) transmit impulses from the sarcolemma to the sarcoplasmic reticulum. Extended junctional sarcoplasmic reticulum (double arrows), although located micrometers from the sarcolemma, also release $Ca^{2+}$, with the initial $Ca^{2+}$ that enters the muscle cells causing the further release of $Ca^{2+}$ in a process called $Ca^{2+}$-induced $Ca^{2+}$ release. Muscle contraction requires substantial amounts of ATP, so myofibrils contain numerous mitochondria (indicated by M). Scale bar = 1 μm. (Figure modified from Sommer 1995; © 1995 Published by Elsevier Ltd., used with permission)

the additional work required likely causes hypertrophy of the heart muscle.

Heart size is important because a larger heart has more cardiac muscle and can generate more pressure and pump more blood (Box 6.8 Heart of an Acrobatic Bird). The amount of blood pumped by a heart per unit time is called cardiac output and is expressed in terms of the amount or volume of blood pumped per minute. This volume is determined by a bird's heart rate (beats per minute) and stroke volume (blood pumped with each beat or contraction). Even under resting conditions, the high metabolic rates of birds require a relatively high cardiac output. In general, avian hearts also tend to pump more blood per unit time than mammalian hearts (and much more than reptilian hearts) and, as a result, deliver more oxygen to the tissues (Smith et al. 2000; Hillman and Hedrick 2015; Lansford and Rugonyi 2020). Of course, a bird's metabolic needs change, both in the short-term and over longer periods of time. In the short-term, birds must be able to substantially and quickly alter cardiac output. A perched bird that takes flight must be able to increase cardiac output ten-fold or more in a matter of seconds. This is accomplished, primarily, by increasing heart rate in most species (Grubb et al. 1983, Fedde et al. 1989; Bishop and Butler 1995; Peters et al. 2005). For example, a Rock Pigeon (*Columba livia*) at rest has

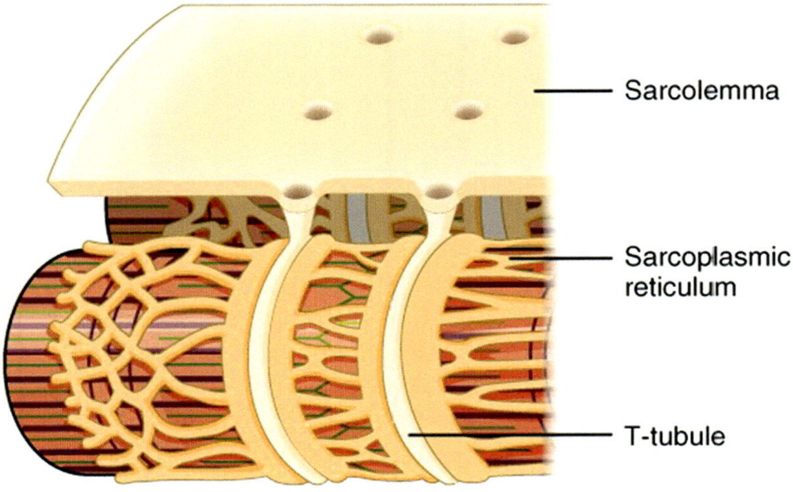


**Fig. 6.64** Cardiac muscle cell of a mammal. Impulses travel from the sarcolemma (cell membrane) down the T-tubules and to the sarcoplasmic reticulum. When the membrane of the sarcoplasmic reticulum is stimulated, calcium ions are released that will then bind to myofilaments to initiate muscle contraction. (Figure by OpenStax, Wikipedia, licensed under the Creative Commons Attribution 4.0 International license, https://creativecommons.org/licenses/by/4.0/)

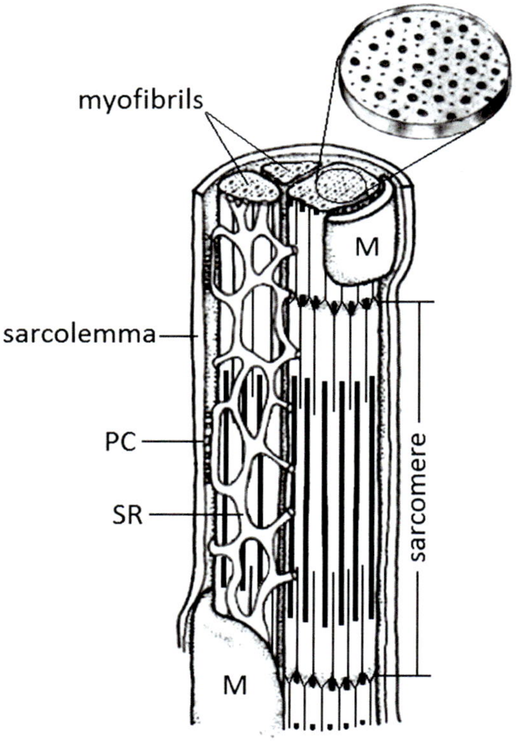


**Fig. 6.65** Avian cardiac muscle cell. Note the absence of T-tubules and the presence of peripheral coupling sites. SR, sarcoplasmic reticulum; PC, peripheral coupling site; M, mitochondria. (Figure modified from Smith et al. 2000; © 2000 Elsevier, Inc., used with permission)

a heart rate of about 175 beats per minute and a stroke volume of about 1.1 milliliters per beat, or a cardiac output of about 193 milliliters per minute. In flight, the cardiac output of that pigeon is about 1065 milliliters per minute (about 5.5 times the resting cardiac output). Although the stroke volume of a flying pigeon does increase to about 1.6 milliliters per beat (about 1.5 times the resting stroke volume), heart rate during flight increases to about 670 beats per minute (or 3.8 times the resting heart rate).

In other species of birds, increases in cardiac output result from changes in both heart rate and stroke volume. For example, Helmeted Guineafowl (*Numida meleagris*) walking and running at increasing speeds on a treadmill increased cardiac output by increasing both heart rate and stroke volume (Ellerby et al. 2005; Fig. 6.71). Among waterfowl, some species, like Bar-headed Geese (*Anser indicus*), increase cardiac output largely by increasing heart rate, but other species, like Andean Geese (*Oressochen melanopterus*), increase cardiac output primarily by increasing stroke volume (Laguë et al. 2017).

One benefit of increasing cardiac output by increasing heart rate is that heart rate can be altered quickly via sympathetic stimulation of the sinoatrial (SA) node. However, a faster heart rate also means that the cardiac muscle cells will need more oxygen. Alternatively, increasing cardiac output primarily by increasing stroke volume

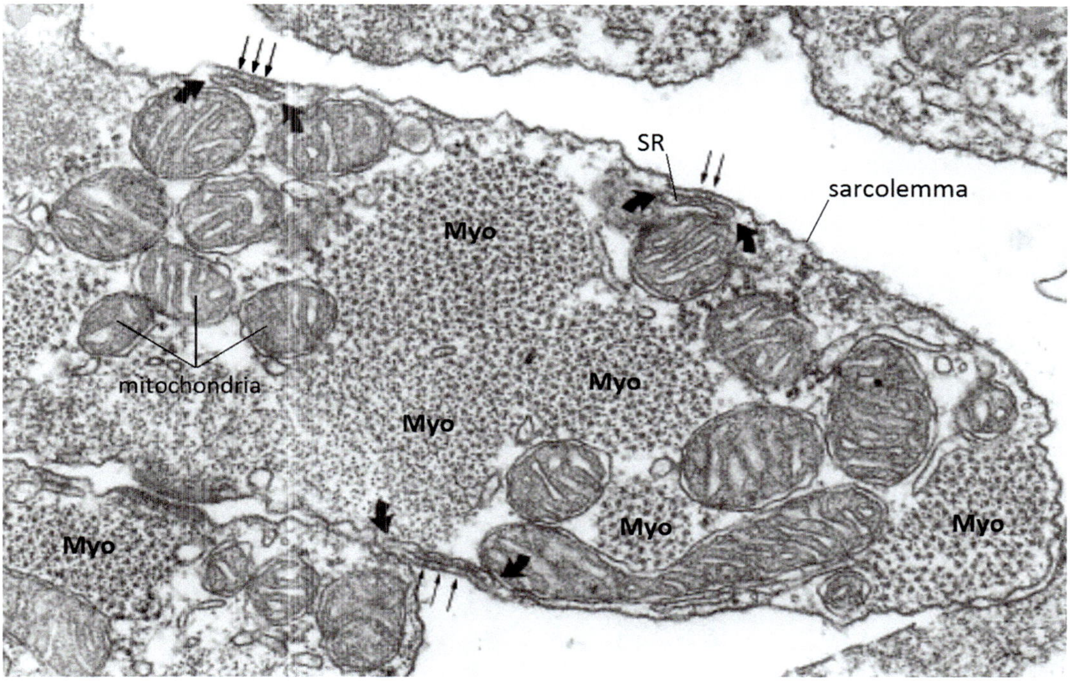


**Fig. 6.66** Cross section of an avian cardiac muscle cell showing (between curved arrows) the association between the sarcolemma (cell membrane of a muscle cell) and sections of sarcoplasmic reticulum (SR). Structures called peripheral coupling sites (small arrows) connect the sarcoplasmic reticulum and the sarcolemma, allowing impulses to travel from the sarcolemma to the sarcoplasmic reticulum. Impulses travelling along the sarcoplasmic reticulum open calcium channels allowing calcium ions to diffuse among the myofilaments in the myofibrils (Myo) and initiate muscle contraction. Large circular structures are mitochondria. (Figure modified from Protasi et al. 1996; © 1996 Academic Press, used with permission)

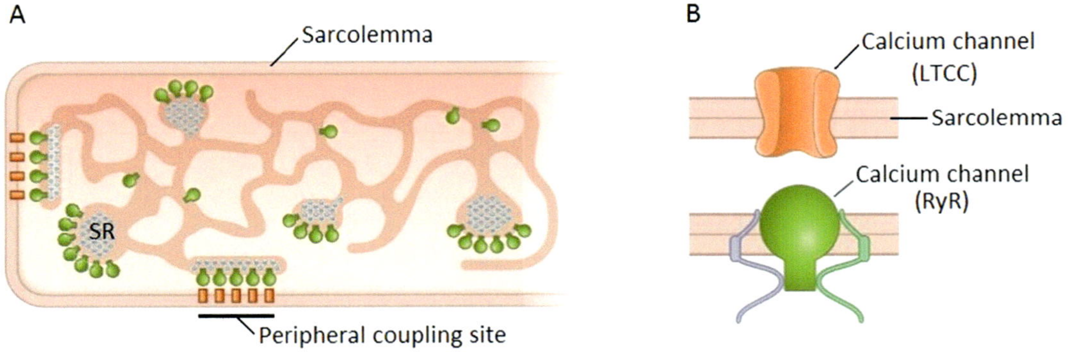


**Fig. 6.67** (**a**) Simplified drawing of an avian cardiac muscle cell showing the network of sarcoplasmic reticulum (SR) and peripheral coupling sites. Peripheral coupling sites consist of a series of calcium channels (called LTCCs) in the sarcolemma and adjacent calcium channels (called RyRs) in the membrane of the sarcoplasmic reticulum. Together, these two channels are referred to as couplons. (**b**) Close-up of a couplon. (Figure modified from Shiels and Galli 2014; © 2014 The American Physiological Society, used with permission)

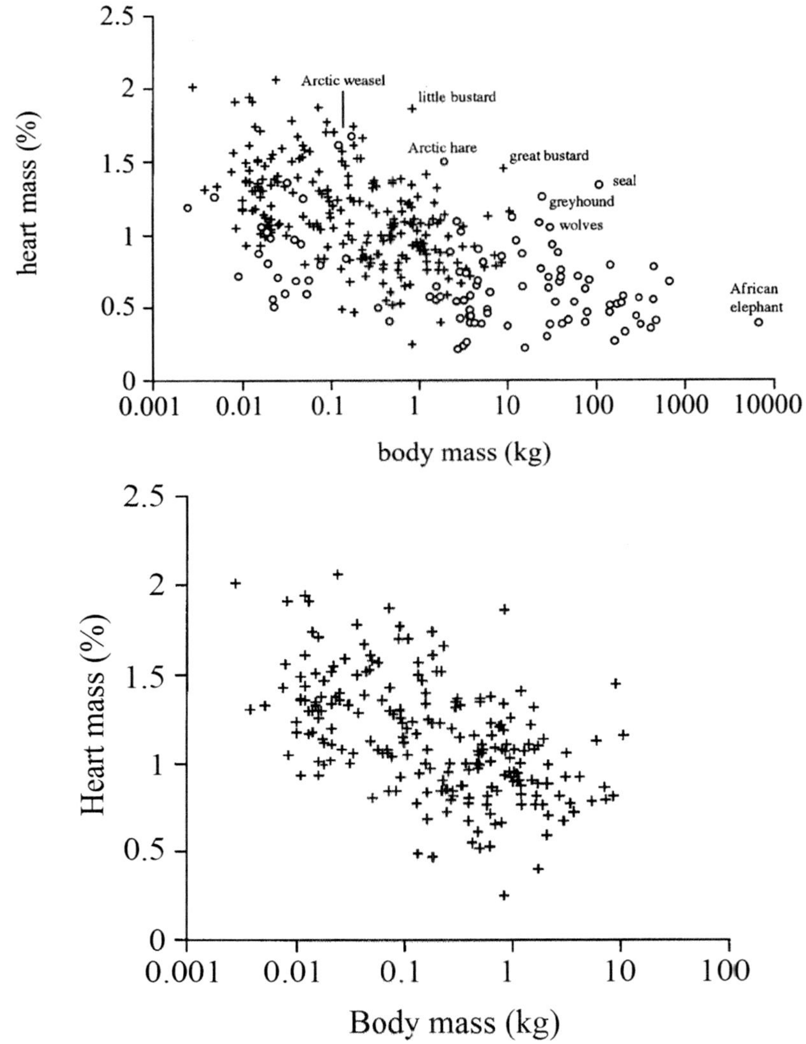


**Fig. 6.68** Top, Heart mass relative to body mass among several taxa of birds (+) and mammals (o). Bottom, Heart mass relative to body mass among several taxa of birds. Heart mass represents a greater percentage of body mass for smaller birds. Little Bustard, *Tetrax tetrax*; Great Bustard, *Otis tarda*; Arctic (least) weasel, *Mustela nivalis*; Arctic hare, *Lepus arcticus*; African elephant, *Loxodonta* sp. (Figure from Bishop 1997; © 1997 The Royal Society, used with permission)

translates into a slower heart rate and, therefore, less demand for oxygen by cardiac muscle cells. One possible disadvantage of increasing cardiac output by increasing stroke volume is that the ability to increase stroke volume can be limited by ventricular morphology, e.g., ventricular lumen volume, level of trabecularity, and cardiac muscle contractility and compliance (Laguë 2017). Given the relative advantages and disadvantages of the two ways to increase cardiac output (i.e., increased heart rate and/or increased stroke volume), natural selection has clearly favored different mechanisms in different taxa of birds.

In general, bird hearts pump more blood per "beat" than hearts of similar-sized mammals and, among birds, stroke volume is directly proportional to body size. For nine species of birds ranging from in mass from 35 grams (Budgerigar, *Melopsittacus undulatus*) to 37.5 kg (Emu, *Dromaius novaehollandiae*), the relationship between stroke volume and body size (mass) was expressed by the equation: stroke volume (ml) $= 1.72 \times$ mass (kg)$^{0.97}$ (Grubb 1983). Larger hearts, not surprisingly, pump more blood with

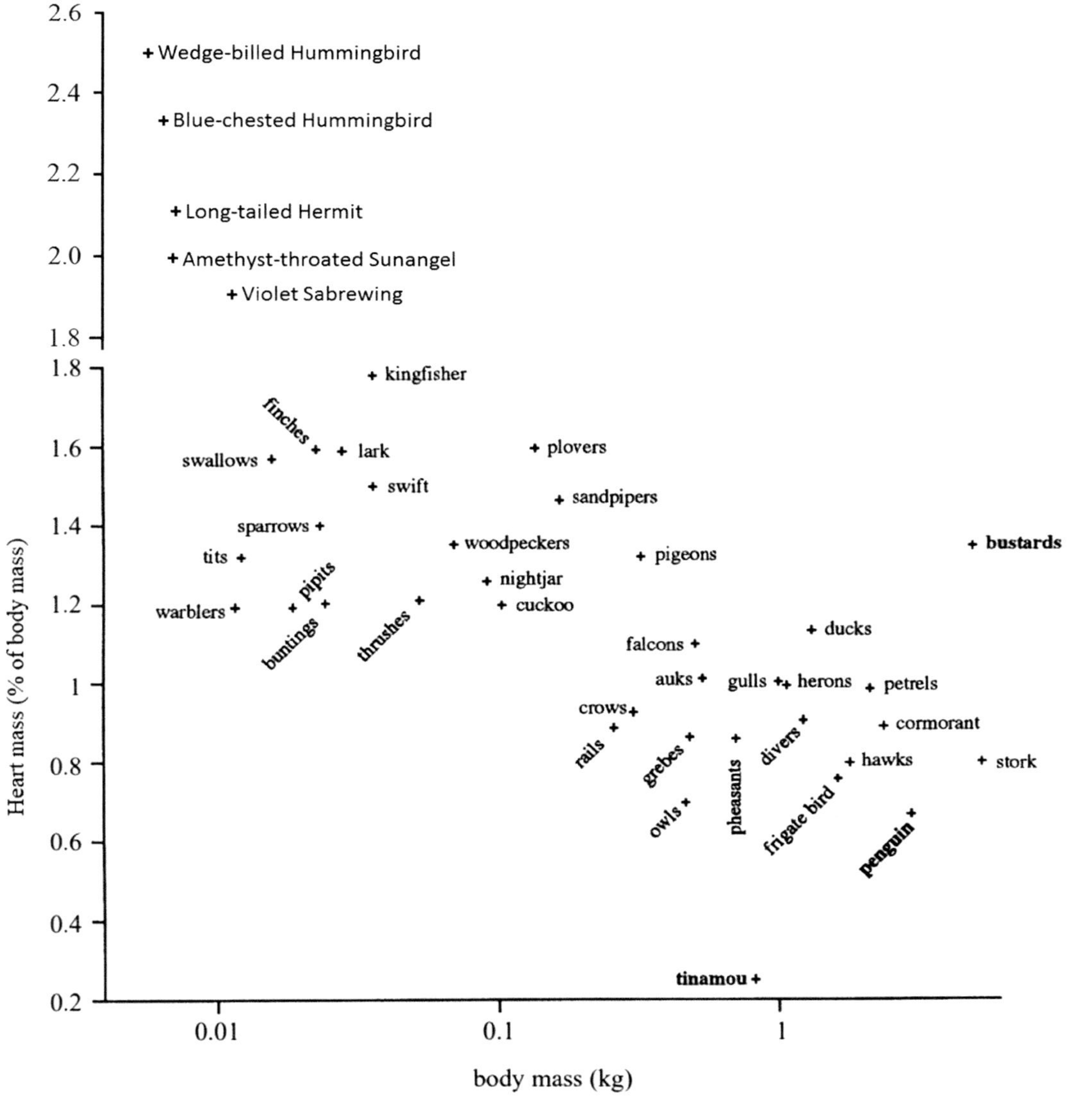


**Fig. 6.69** Bottom, Mean values for heart muscle mass as a percentage of body mass (vertical axis) plotted against body mass (horizontal axis; in kilograms) for select species in several families of birds, including (Top) five species of hummingbirds (Trochilidae). Note again that smaller birds have, for their mass, relatively larger hearts. Wedge-billed Hummingbird, *Schistes geoffroyi*; Blue-chested Hummingbird, *Amazilia amabilis*; Long-tailed Hermit, *Phaethornis superciliosus*; Amethyst-throated Sunangel, *Heliangelus amethysticollis*; Violet Sabrewing, *Campylopterus hemileucurus*. (Figure modified from Bishop 1997; © 1997 The Royal Society, used with permission. Additional hummingbird data from Altshuler and Dudley 2002)

each contraction. However, each bird's heart can pump more blood (increase stroke volume) when necessary. Direct stimulation of the heart muscle by the bird's nervous system (specifically the sympathetic division of the Autonomic Nervous System) causes the cardiac muscle cells in the ventricles to contract with more force and this ejects more blood. In addition, the nervous

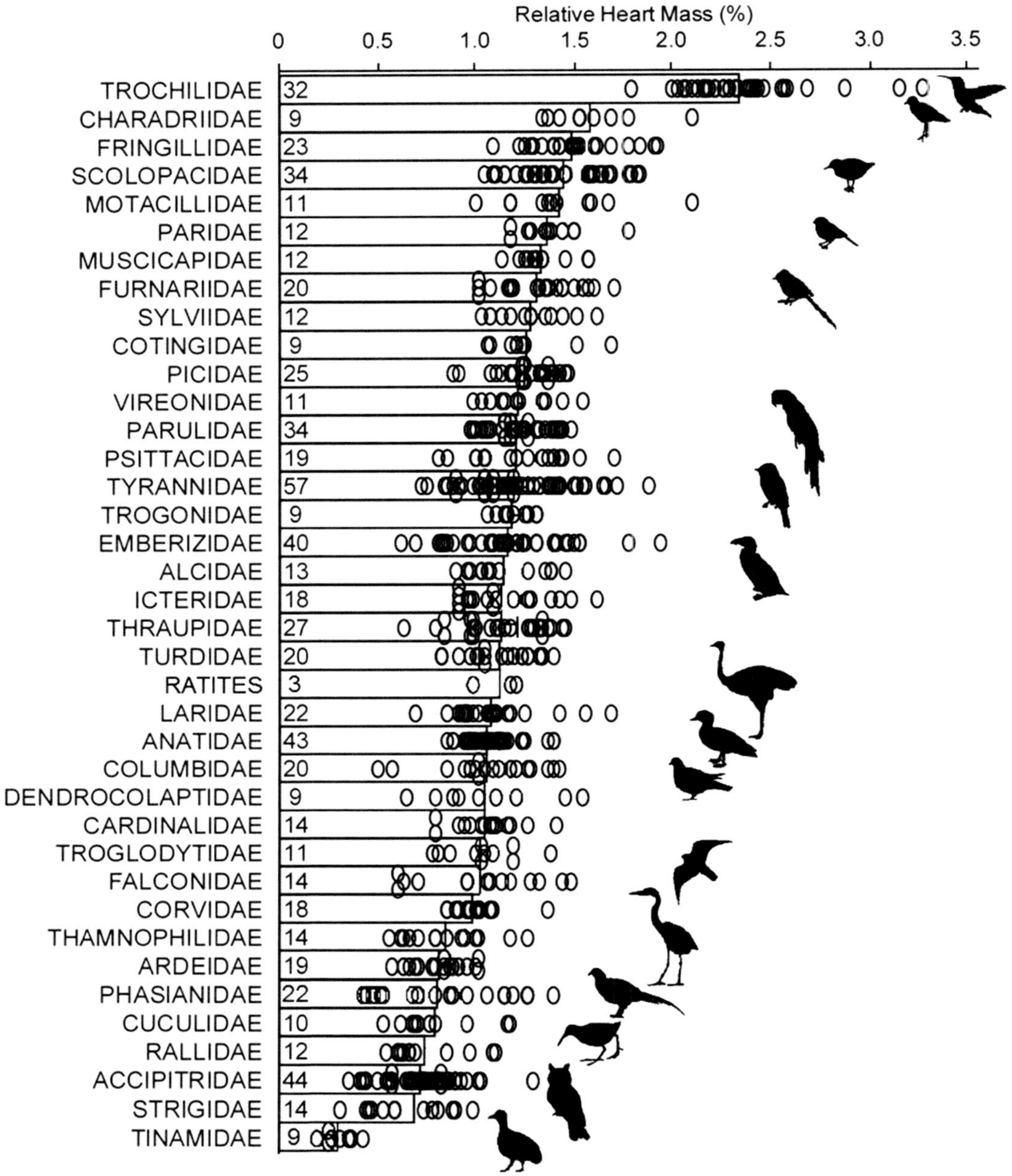


**Fig. 6.70** Percent heart mass relative to body mass for several different families of birds. Note the relatively high percent heart mass hummingbirds (Trochilidae) and the relatively low percent heart mass of tinamous (Tinamidae). (Figure from supplementary material from Altimiras et al. 2017; open-access article under a Creative Commons Attribution 4.0 International License, https://creativecommons.org/licenses/by/4.0/)

system (again the sympathetic division) causes veins delivering blood to the heart to vasoconstrict. This sends more blood back to the heart and that, along with the increased force of contraction, generates the increase in stroke volume.

Changes in heart rate have the greatest impact on cardiac output. In general, heart rates of birds at rest are similar to those of mammals of the same size (Fig. 6.72). However, when active, heart rates of birds tend to increase more than those of mammals (Fig. 6.73). This is especially so for small birds (<1 kg). When engaged in intense activity, like flying, the heart rates of small birds typically exceed those of small mammals (Fig. 6.73).

Heart rate varies with size, with larger birds having slower heart rates (Fig. 6.74). Among larger birds, Common Ostriches (*Struthio camelus*) have resting heart rates of about 52 beats per minute (Cornick and Jensen 1992), and Emus about 46 beats per minute (Grubb et al. 1983). Typically, heart rates at rest for small birds are about 400–600 beats per minute. For example, American Robins (*Turdus migratorius*, 80 gms) have a resting heart rate of about 520 beats per minute (Woodbury and Hamilton 1937) and House Sparrows (*Passer domesticus*) about 442 beats per minute (Odum 1945). Not surprisingly, the smallest birds—hummingbirds—have the highest heart rates. Resting heart rates of some hummingbirds are about 600–700 beats per minute.

The resting heart rates of birds are controlled by a special group of cells near the top of the right atrium called the sinoatrial node (Šolc 2007; Fig. 6.75). These cells, unlike most other cardiac muscle cells, are autorhythmic. They have an intrinsic, built-in rhythm (or rate of contraction) and "self-stimulate" or contract at a certain rate with no external stimulation. Each time these automatic cells "self-stimulate," electrical impulses spread along special conducting cells called Purkinje fibers and from cell to cell throughout the heart and cause contraction (Prosheva and Kaseva 2016). Because these

**Box 6.8 Heart of an Acrobatic Bird**

While performing their complex courtship displays, the heart rates of Golden-collared Manakins (*Manacus vitellinus*) can reach 1300 beats per minute, one of the highest heart rates ever recorded for any bird (or mammal). These manakins have a lek breeding system where males establish display areas on the forest floor. Their displays include wing- and roll-snaps. During wing-snap displays, males make short flights during which they lift their wings over their heads and strike them together to create a snapping sound. Males perform an average of nine wing-snaps per display. During roll-snap displays, perched males perform a rapid series of about 9–23 wing-snaps. Males may perform as many as 140 wing-snap displays and 315 roll-snap displays per day during a seven-month long breeding season (Barske 2013). Such intense activity suggests that Golden-collared Manakins might have specialized cardiovascular adaptations.

Barske et al. (2019) compared the hearts of Golden-winged Manakins to those of a less acrobatic species—Zebra Finches (*Taeniopygia guttata*)—and found that the hearts of the manakins had much thicker left ventricular walls and interventricular septa than those of Zebra Finches. Thicker walls may generate more pressure and help increase stroke volume (the amount of blood pumped with each ventricular contraction). In addition, the hearts of the manakins exhibited greater expression of androgen-receptor genes than those of Zebra Finches. High testosterone levels during the breeding season acting on the androgen receptors may increase the rate at which calcium ions move into and out of cardiac muscle cells during the contraction-relaxation cycle, permitting the extremely high heart rates of male Golden-collared Manakins during their courtship displays.

(continued)

**Box 6.8** (continued)

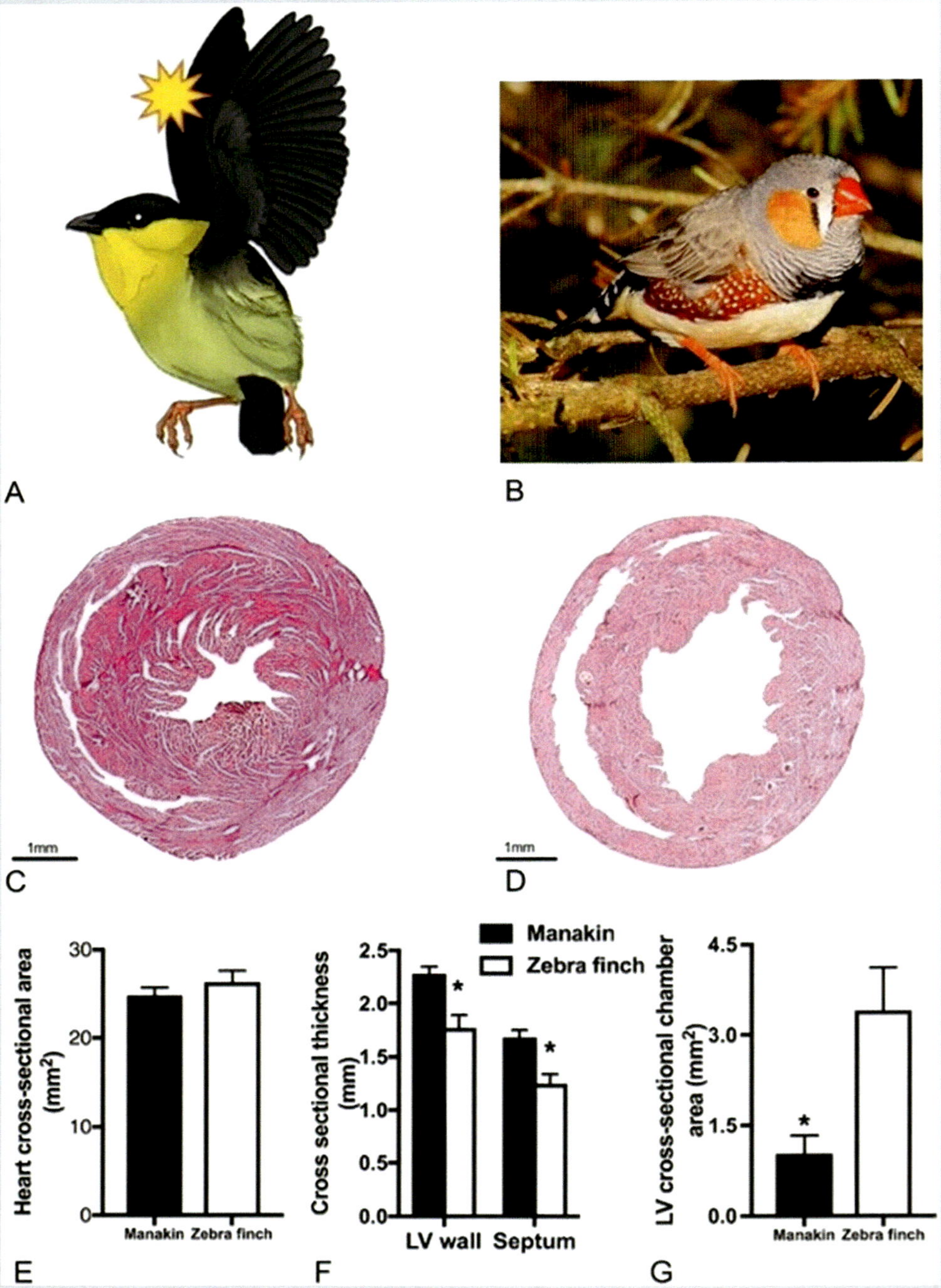


(**a**) Male Golden-collared Manakin wing-snap display, (**b**) Zebra Finch, (**c**) Cross section through the heart of a Golden-collared Manakin, (**d**) Cross section through the heart of a Zebra Finch, (**e**) the cross-sectional area of the hearts of the two species did not differ, (**f**) the wall of the left ventricle and the interventricular septum were

(continued)

**Box 6.8** (continued)

significantly thicker in the hearts of Golden-collared Manakins than those of Zebra Finches, and (**g**) the area of the left ventricles of Golden-collared Manakins was smaller than that of Zebra Finches (Figure **a** from Fuxjager and Schuppe 2018, © 2018 Elsevier Ltd., used with permission; **b** from pxhere.com, CC0 Public Domain, and **c–g** from Barske et al. 2019, © 2018 Elsevier Inc., used with permission)

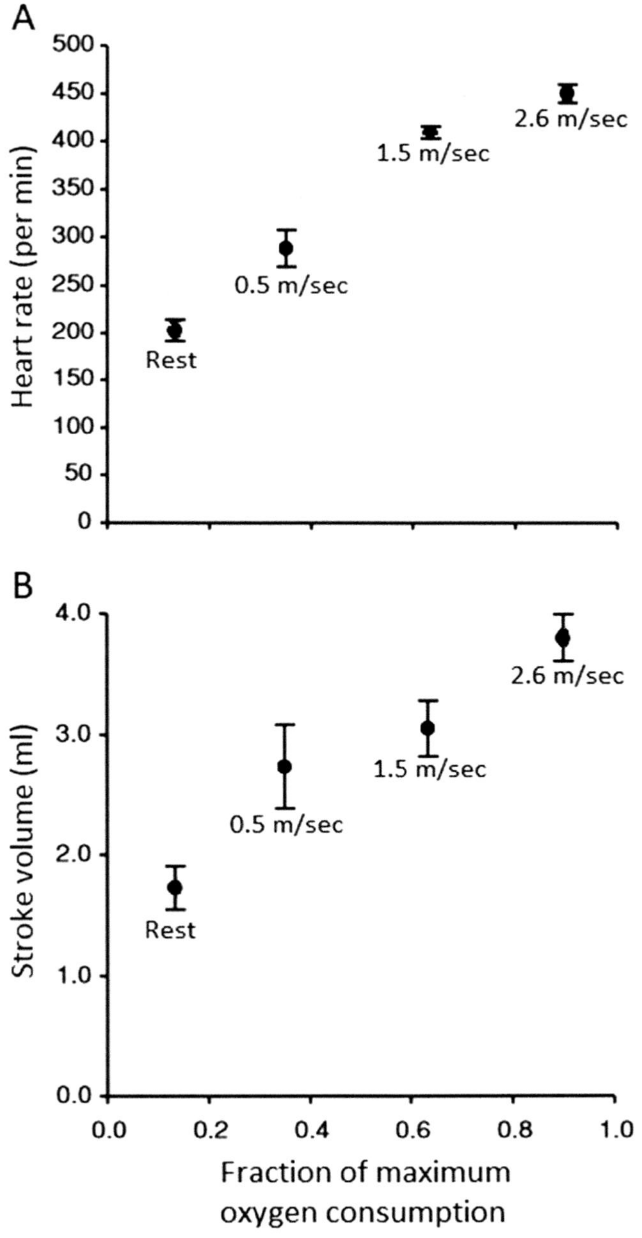


**Fig. 6.71** Changes in heart rate and stroke volume as Helmeted Guineafowl (*Numida meleagris*) walk and run faster on a treadmill. Cardiac output increased from 350 ml/min at rest to 1700 ml/min at the fastest running speed (2.6 m/s). Treadmill speed is indicated below the data points. (Figure modified from Ellerby et al. 2005; © 2005 The Journal of Physiology © 2005 The Physiological Society, used with permission)

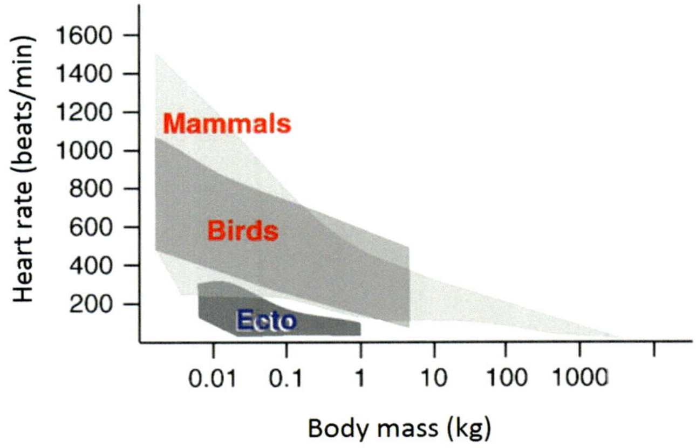


**Fig. 6.72** Heart rates of birds and mammals relative to body mass are similar and, of course, as endotherms, higher than those of ectotherms (Ecto). (Figure from (Jensen et al. 2013a, b; © 2012 Elsevier B.V., used with permission)

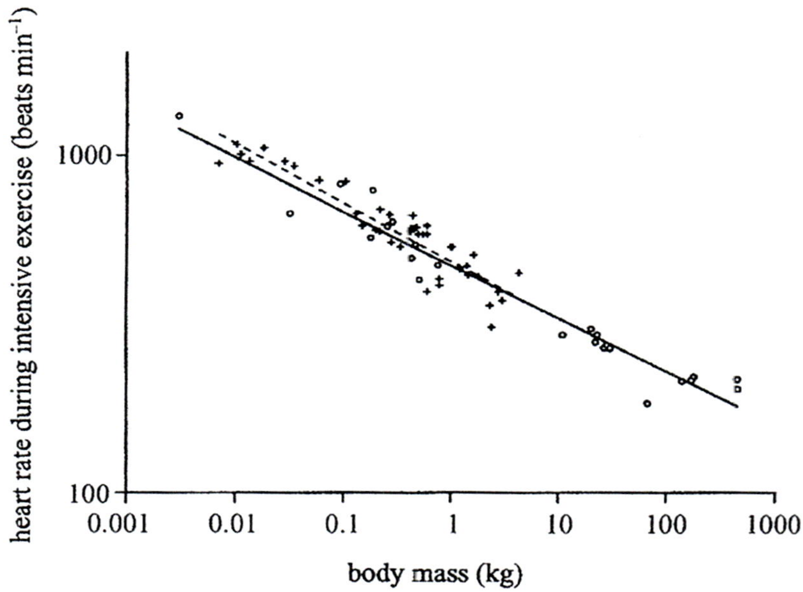


**Fig. 6.73** Maximum heart rate during intensive exercise for birds and mammals plotted against body mass. Birds indicated with + and dashed regression line; mammals with o and solid regression line. (Figure from Bishop 1997; © 1997 The Royal Society, used with permission)

sinoatrial node cells control the rate at which the heart contracts, they are called the "pacemaker" of the heart. To alter cardiac output, the normal or resting rhythm of the "pacemaker" must change. As with stroke volume, a bird's nervous system (again the sympathetic division of the Autonomic Nervous System; Fig. 6.76) can regulate the heart's pacemaker. Increased stimulation of the pacemaker cells increases heart rate. Depending on the level of activity, heart rate and cardiac output are adjusted.

This adjustment occurs automatically in birds, just as it does in humans and other mammals. Physical activity, hypoxia (reduced oxygen levels in the blood), and hypercapnia (increased carbon dioxide levels in the blood) are all known to influence the avian cardiovascular system, including heart rate. However, the mechanisms by which these factors alter heart rate remain unclear. Chemoreceptors (special neurons that detect chemicals like carbon dioxide and oxygen) are located in the carotid arteries of birds (the

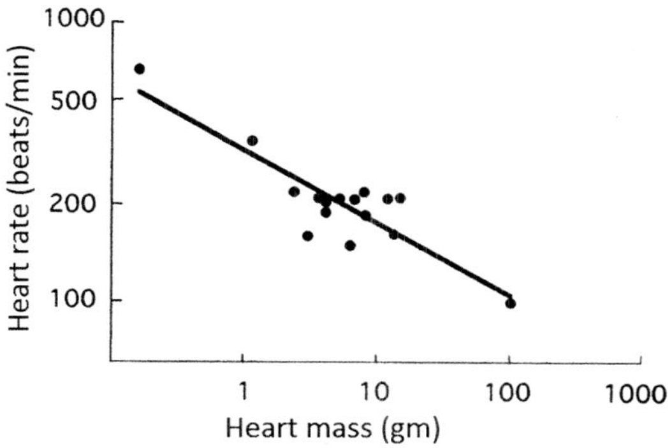


**Fig. 6.74** Relationship between heart weight and heart rate at rest given on a bilogarithmic scale for 14 species of birds. (Machida and Aohagi 2001; © 2001 American Association of Zoo Veterinarians, used with permission)

arteries that transport blood to the head) and may also be located in the aorta (Hempleman et al. 1992). These receptors are stimulated by reduced levels of oxygen in the blood, but are stimulated even more by increased levels of carbon dioxide. An increasingly active bird will generate more carbon dioxide that will then stimulate the chemoreceptors. When stimulated, these receptors then send impulses to the avian brain and this stimulation, in turn, causes increased sympathetic stimulation of the pacemaker and an increased heart rate.

Each time the pacemaker initiates an impulse, it spreads through the entire heart. This electrical activity can be detected by electrodes placed on a bird's skin and the resulting tracing is called an electrocardiogram, or ECG (Machida and Aohagi 2001; Figs. 6.77 and 6.78). Deflections in the tracing indicate electrical activity taking place in the heart and, during each heartbeat (or cardiac cycle), there are four primary events caused by waves of depolarization and repolarization traveling through different areas of the heart, the P wave, R wave, S wave, and T wave. The P wave is caused by depolarization of the atria, the R and S waves by ventricular depolarization, and the T wave by ventricular repolarization (Fig. 6.79; Box 6.9 Human vs. Bird Electrocardiograms). As the atria depolarize, they contract, pumping blood into the ventricles. As the impulse travels through the atria, it reaches another group of automatic cells at the junction of the atria and the ventricles called the atrioventricular, or AV, node. From this node, the impulse travels along specialized muscle fibers that quickly transmit the impulse to the many cardiac muscle cells in the ventricles. Depolarization of the ventricles generates the QRS (or just RS) complex, which is quickly followed by ventricular contraction and ejection of blood from the ventricles into the blood vessels.

## 6.16 Blood Vessels

Blood vessels transport blood and permit the exchange of gases, nutrients, and waste products between the blood and body cells. Within the avian body are two "circulations": the pulmonary circulation, taking blood to and from the lungs, and the systemic circulation, taking blood to and from the rest of the blood. Each of these circuits, so-named because blood follows a circuitous path from the heart to the cells and tissues and then back to the heart, consists of different types of blood vessels that serve different functions. The main categories of vessels are arteries, capillaries, and veins (Fig. 6.80). Arteries carry blood from the heart to the smallest of the vessels, the capillaries, where the exchange of materials between blood and cells takes place. Veins then conduct the blood from the capillaries back to the heart.

The systemic, or body, circulation begins as the aorta transports blood from the left ventricle of the heart. The first arteries to branch off the aorta are the coronary arteries that supply blood to the heart. Relative to its mass, the heart receives a rather high percentage of the total cardiac output (about 8%) even when a bird is at rest (Smith et al. 2000). Because cardiac muscle never rests, demand for oxygen and nutrients is high. When a bird becomes active, the heart receives a greater percentage of cardiac output.

The next arteries to branch off the aorta are the brachiocephalic arteries that, in turn, split to form the common carotid and subclavian arteries. The carotids deliver blood to the head (Figs. 6.81 and 6.82), and the subclavians (Fig. 6.82) lead into the

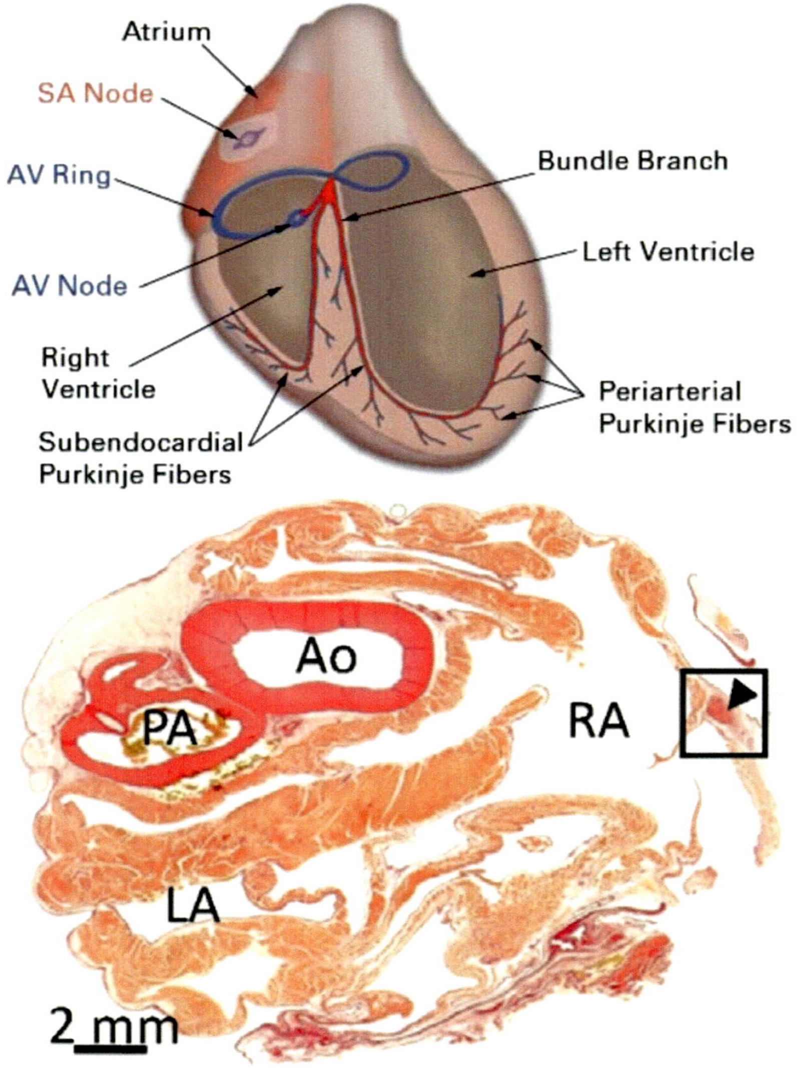


**Fig. 6.75** Top, The pacemaker (SA node) and conduction system of the avian heart. The AV node is sometimes called a "delay generator" because impulses traveling from the atria through this node to the ventricles travel rather slowly. This "delay," or slowing of the impulse, is beneficial because it ensures that the atria complete their contraction (and pump their blood into the ventricles) before the ventricles begin contracting. SA, sinoatrial; AV, atrioventricular. Bottom, SA node (Box, arrow head) in the heart of a Mallard. Ao, aorta; PA, pulmonary artery; RA, right atrium; LA, left atrium (Top figure from Harris et al. 2002, used with permission; Bottom figure from Kroneman et al. 2019, © 2019 The Authors. Journal of Morphology, open-access article under the terms of the Creative Commons Attribution License, https://creativecommons.org/licenses/by/4.0/)

axillary and then the brachial arteries that deliver blood to the wings (Fig. 6.83).

Because the brain requires a continuous supply of nutrients and oxygen, blood flow through the carotids must not be interrupted. Two factors help insure this flow. First, the carotid arteries are protected from compression by their location at the base of the neck (cervical) vertebrae and near to axis of rotation (Smith et al. 2000). Second, all birds, with the exception of those in the suborder Tyranni (that includes the tyrant flycatchers, antbirds, manakins, and others), have an intercarotid anastomosis (Fig. 6.84). This connection between the two common carotid arteries ensures continuous blood flow throughout the head and brain even if one carotid is momentarily compressed (Smith et al. 2000).

The aorta passes behind the heart and turns down into the abdominal cavity where it is called the abdominal, or descending, aorta (Fig. 6.85). In the abdominal cavity of many birds, the celiac, superior (or cranial) mesenteric, renal, external iliac, inferior (or caudal) mesenteric, internal iliac, and caudal arteries, in that order, branch

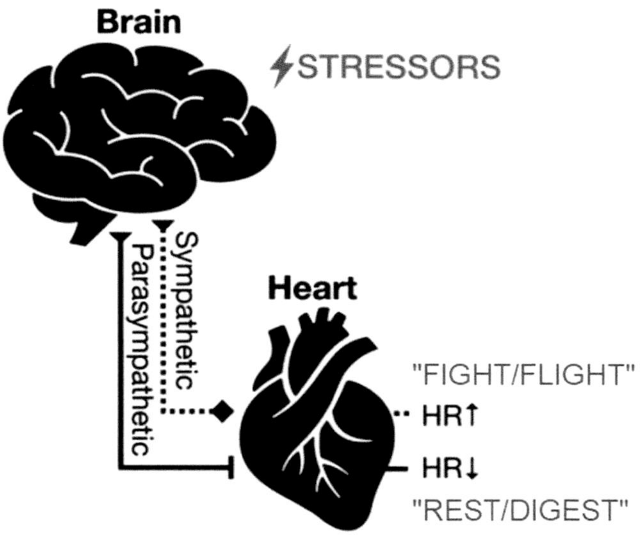


**Fig. 6.76** The avian autonomic nervous system innervates the heart's sinoatrial node (pacemaker) and can increase heart rate with increased sympathetic stimulation and decrease heart rates with increased parasympathetic stimulation. In response to stress, either physical (e.g., taking flight) and from external stimuli such as the approach of a potential predator, sympathetic stimulation (and reduced parasympathetic stimulation; Müller et al. 2017) increases heart rate and cardiac output. After a stressful event (e.g., landing on a perch after a flight or the potential predator has moved away and no longer represents a threat), parasympathetic stimulation lowers the heart rates to resting levels. The parasympathetic division is also important in regulating the digestive system and so is sometimes referred to as the "rest-and-digest" division of the autonomic nervous system. (Figure modified from Gaidica and Dantzer 2020; © 2020 Oxford University Press, used with permission)

from the descending aorta and supply various tissues and organs (Fig. 6.85). However, the branching patterns of blood vessels do exhibit some variation among different taxa of birds (Midtgård 1982; Figs. 6.86 and 6.87).

The pulmonary, or lung, circulation begins as the pulmonary artery leaves the right ventricle of the heart. Because the pulmonary circulation supplies only the lungs, pulmonary cardiac output is, of course, equal to systemic cardiac output. In other words, with each heartbeat, the lungs receive as much blood as the rest of the body. In the lungs, the pulmonary arteries branch into numerous smaller arteries that ultimately lead to interparabronchial arteries that are located on the surface of parabronchi, tiny tubes in the lungs through which air passes (Figs. 6.88 and 6.89). Branching from the interparabronchial arteries are the intraparabronchial arterioles. Blood flowing through the intraparabronchial arterioles then enters capillaries that lie adjacent to extremely tiny outpocketings of the parabronchi called atria or air capillaries. This is where gases are exchanged, with oxygen diffusing from the air capillaries into the blood capillaries and carbon dioxide from the blood capillaries into the air capillaries (Fig. 6.90). Additional details concerning gas exchange in the lungs of birds can be found in Chap. 7.

Veins return blood from the tissues and organs to the heart. Blood returning to the left atrium of the heart from the lungs travels through the pulmonary veins. The jugular veins that carry blood from the head back toward the heart merge with the subclavian veins that carry blood from the wings (via brachial and axillary veins) and the pectoralis muscles (pectoral veins) (Fig. 6.82). At the point of merger, the jugular and subclavian veins become the precavas (or right and left superior vena cavas) that empty into the right atrium. As with the common carotids, there is a bypass, or connecting vessel, between the two jugulars

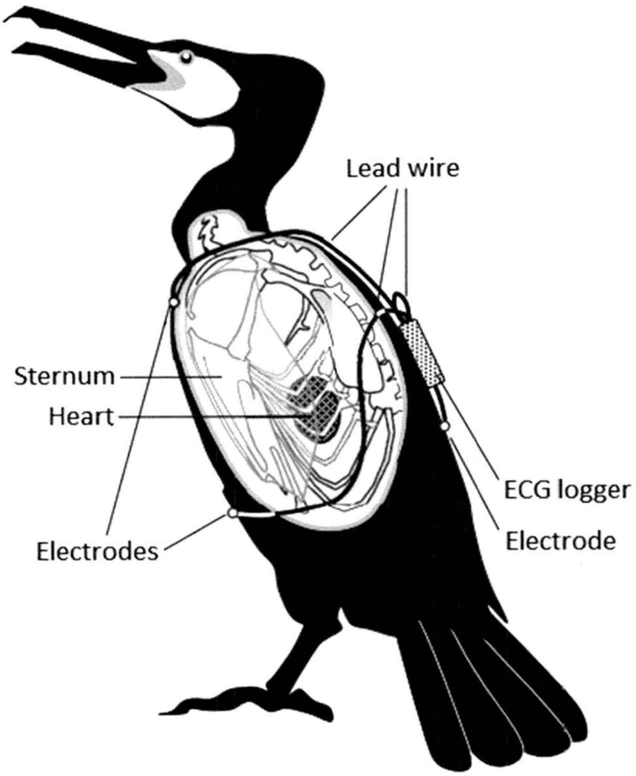


**Fig. 6.77** Placement of electrodes and electrocardiogram logger for monitoring the electrocardiogram of a Great Cormorant (*Phaloacrocorax carbo*). (Figure from Yamamoto et al. 2009; © 2009 Springer-Verlag, used with permission)

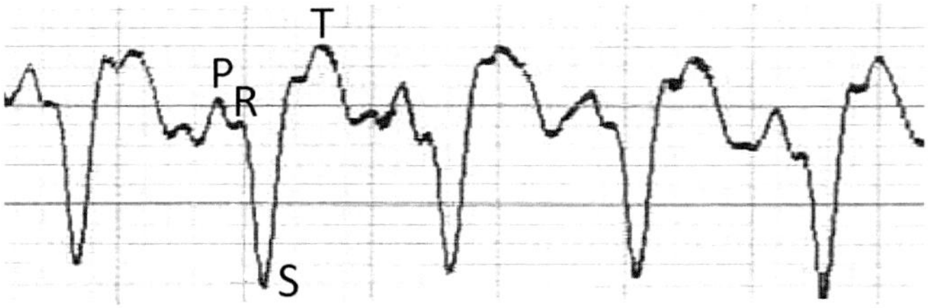


**Fig. 6.78** Electrocardiogram of a Laughing Dove (*Streptopelia senegalensis*). (Figure modified from Hassanpour et al. 2014; © Journal of Zoo and Wildlife Medicine, used with permission)

called the interjugular (or just jugular) anastomosis that provides an alternate pathway for blood if one of the jugular veins is temporarily compressed by movements of the neck.

All blood returning to the heart from the posterior portion of a bird eventually flows into the large postcava vein (or posterior vena cava; Fig. 6.91) that, like the superior vena cavas, empties into the right atrium. From the legs, blood travels through femoral veins to external iliac veins to common iliac veins and finally into the postcava. From the digestive system, blood flow is a bit more complicated. Rather than flowing directly to the posterior vena cava, blood from the digestive system, including the small intestine where the absorption of nutrients occurs, flows into the hepatic portal vein that takes the blood to the liver (Maher 2019). A portal system is one where blood passes through two sets of capillaries before returning to the heart. In the hepatic portal system of all vertebrates, including birds, blood flows through capillaries in the liver after already

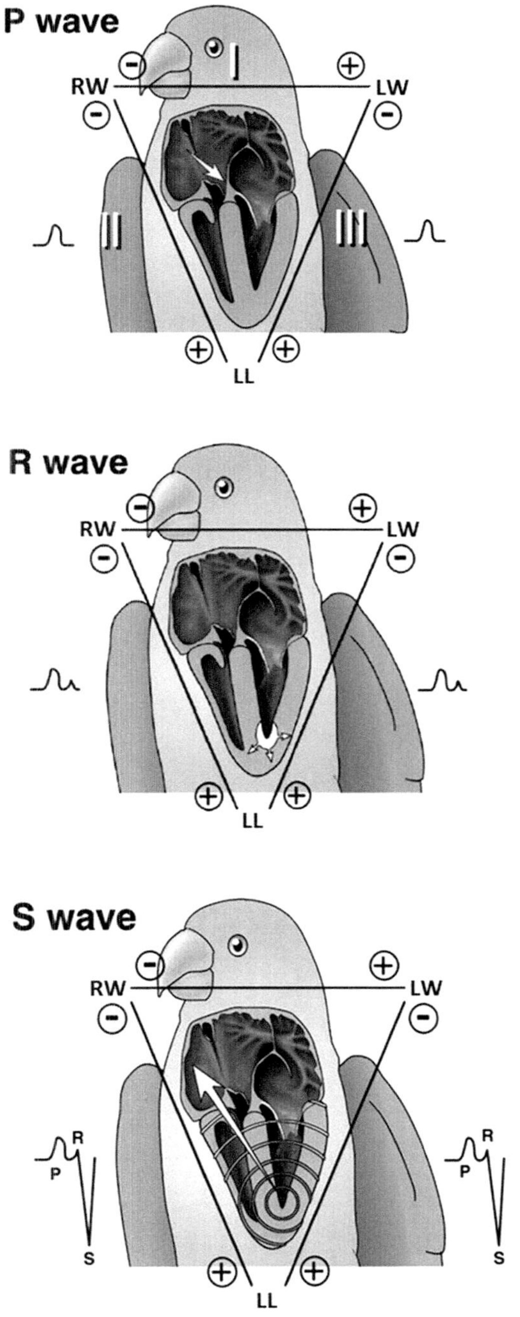


**Fig. 6.79** To understand the various components of an electrocardiogram, know that if a wave of depolarization travels towards an electrode attached to a positive (+) input terminal of an electrocardiograph and away from an electrode attached to a negative (−) terminal, there will be a positive (upward) deflection. If a wave of depolarization moves away from a positive electrode toward a negative electrode, there will be a negative (downward) deflection. The P wave occurs as the wave of depolarization travels from the SA node (the pacemaker) and continues through both atria (white arrow), the R wave occurs as the apex (bottom) of the left ventricle begins to depolarize, and the S wave is generated as the wave of depolarization travels upward through the ventricles. RW, right wing; LW, left wing; LL, left leg. (Figures from: Oglesbee BL, Hamlin RL, Klingaman H et al. Electrocardiographic reference values for macaws (*Ara* species) and cockatoos (*Cacatua* species). J. Avian Med Surg 2001; 15(1):17–22, used with permission)

**Box 6.9 Human vs. Bird Electrocardiograms**

The typical bird electrocardiogram differs from that of humans and, more generally, mammals. The human electrocardiogram begins with a P wave, corresponding to atrial depolarization, and ends with a T wave, corresponding to ventricular repolarization, and that is also the case with an avian electrocardiogram. What is different about the two electrocardiograms is what happens between those two events—Ventricular depolarization.

Electrocardiograms are obtained by placing sensitive electrodes on the surface of the body, typically one on each arm (or wing) and one on the left leg. Each pair of electrodes measures the electrical potential difference between them, and a frequently used comparison is between the electrodes on the right arm or wing and the left leg (referred to a "lead II").

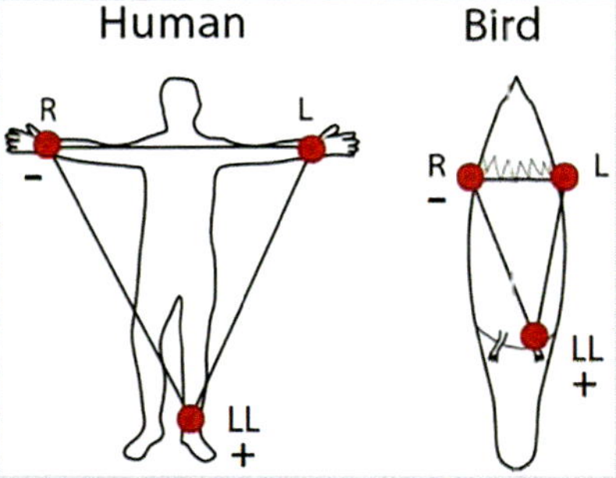


Electrodes are placed on the right and left arms of humans and right and left wings of birds and on the left legs of both human and birds. The electrocardiograph makes one electrode positive and one or more electrodes negative; the electrode on the right wing or arm is a negative electrode and the electrode on the left leg is a positive electrode. If a wave of depolarization travels toward the positive electrode, the electrocardiograph will generate a positive (upward) deflection in the electrocardiogram; if a wave of depolarization moves toward the negative electrode, the deflection will be negative (downward) (Figure from Boukens et al. 2019; © 2018 Elsevier Ltd., used with permission)

In humans and other mammals, depolarization of the ventricles begins at the top of the interventricular septum (1 in the figure below), the wall between the two ventricles. That wave then proceeds toward the apex or bottom of the septum and the ventricles. As a result, the "average" direction of the wave of depolarization through the ventricles is from the top to the bottom of the ventricles. This wave is moving toward the positive electrode (on the left leg), so it creates a positive (upward) deflection in the electrocardiogram.

(continued)

**Box 6.9** (continued)

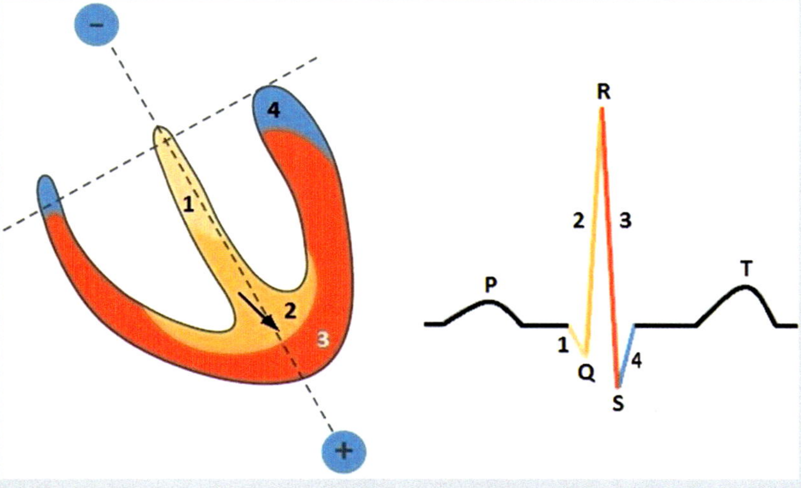


Depolarization of the ventricles in humans and mammals begins at the top of the interventricular septum and proceeds towards the bottom, or apex, of the ventricles. The numbers at left correspond to the sequence of electrical events and their corresponding deflections in the electrocardiogram on the right. The arrow indicates the "average" direction of the wave of depolarization (Figure modified from Peguero et al. 2017; © 2017 Elsevier Inc., used with permission)

In birds, the general pattern is for the wave of depolarization in the ventricles to spread from the apex (bottom) toward the upper right section of the right ventricle. This means that the "average" direction of the wave is from the lower left ventricle to the upper right ventricle or, in other words, away from the positive electrode and toward the negative electrode. This results in a downward deflection in the electrocardiogram.

(continued)

**Box 6.9** (continued)

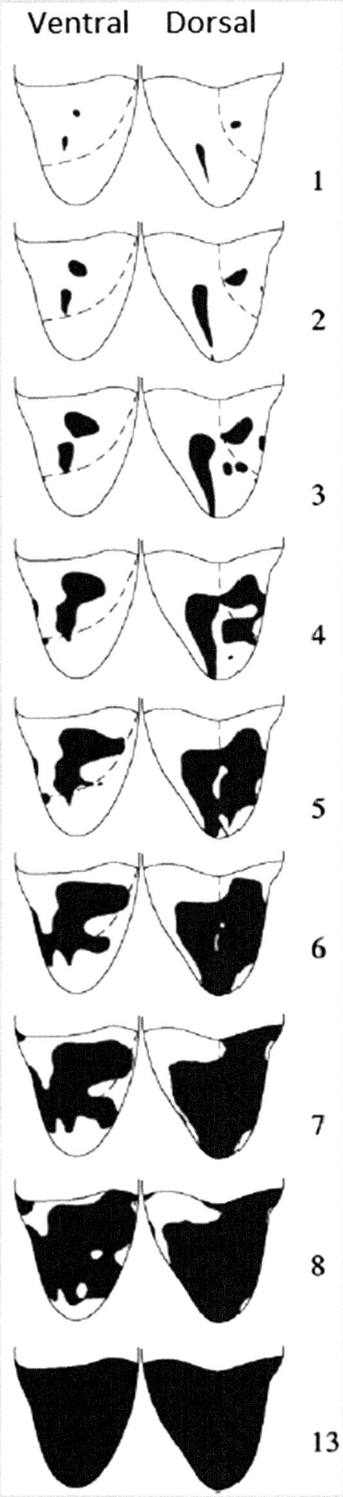


(continued)

**Box 6.9** (continued)

Sequence in which different areas in the ventricles of a Domestic Chicken depolarize (as indicated by the shaded areas). Note that the general pattern is for the wave of depolarization to spread from the apex (bottom) of the heart toward the upper right section of the right ventricle. Numbers of the right represent time in milliseconds from the beginning of ventricular depolarization (Figure modified from Kharin 2004; © 2003 Elsevier Inc., used with permission)

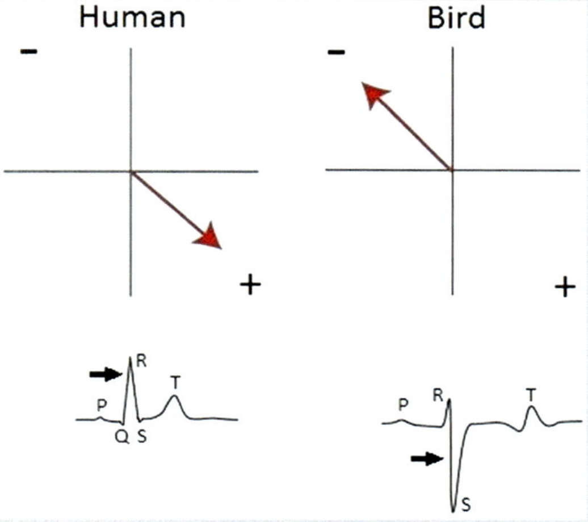


The "average" direction of the wave of depolarization through the ventricles of a human versus a bird. A wave moving toward the positive electrode creates a positive (upward) deflection in the electrocardiogram (human); a wave moving toward the negative electrode creates a negative (downward) deflection (bird) (Figure modified from Boukens et al. 2019; © 2018 Elsevier Ltd., used with permission)

flowing through capillaries in the digestive system. This portal system is important because blood from the digestive system contains nutrients, like glucose, amino acids, and, potentially, toxins and pathogens. As blood passes through capillaries in the liver, excess glucose is stored in liver cells as glycogen, excess amino acids are stored in liver cells, toxins can be detoxified, and pathogens removed by macrophages (Zaefarian et al. 2019).

Birds also have a renal portal system (Fig. 6.91). Venous blood from the tail and lower intestine, as well as some from the legs, enters the kidneys through the renal portal system. This venous blood mixes with arterial blood before eventually flowing into renal veins. The amount of blood flowing through the renal portal veins and into the kidneys varies, in part because of valves (controlled by the Autonomic Nervous System) located within the external iliac veins. Other vertebrates, with the exception of mammals, also have renal portal systems and questions still remain concerning the functional significance of this portal system. One likely function is that the renal portal system ensures adequate flow of blood to the kidneys during periods of dehydration. With dehydration, small arteries in the kidney tend to vasoconstrict (to minimize further loss of water). Although apparently adaptive (in terms of conserving water), such vasoconstriction, if prolonged, could cause the death of important kidney cells. Increased flow to the kidneys via the renal portal system during periods of dehydration helps

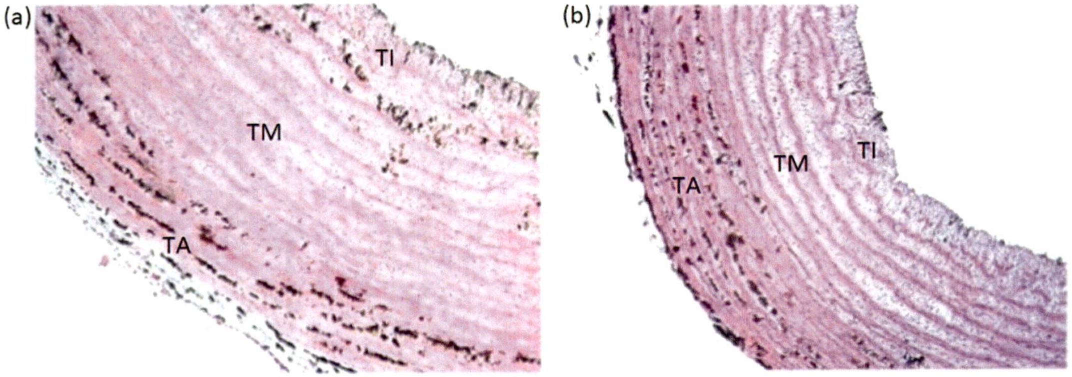


**Fig. 6.80** Cross sections of the walls of (**a**) an artery and (**b**) a vein of a Domestic Chicken (*Gallus g. domesticus*). The walls of arteries and veins consist of three layers: tunica intima (or interna, TI), tunica media (TM), and tunica adventitia (TA). The outermost layer, tunica adventitia, is connective tissue consisting primarily of longitudinally arranged collagen and elastic fibers. The tunica media consist of elastic connective tissue and concentrically arranged smooth muscle that alters the diameter of the vessels via stimulation by the Autonomic Nervous System. The tunica intima is the innermost endothelial layer of blood vessels. Note that the wall of the artery is thicker than that of the vein, primarily due to a difference in the thickness of the tunica media. With each contraction of the ventricles of the heart, pressure in the arteries increases and the elastic tissue and smooth muscles in the tunical media help the arteries withstand that pressure. In addition, as ventricles relax and blood pressure in the arteries drops, their elastic walls collapse in the remaining blood to present blood pressure from dropping too low (i.e., to maintain blood flow, blood pressure in the arteries must remain higher than that in the capillaries and veins), so arteries are sometimes referred to as "pressure reservoirs". (Figure from Nganvongpanit et al. 2020; )

maintain adequate flow of blood to these vital kidney cells (Dantzler 1989; Holz 1999).

## 6.17 Blood Pressure and Blood Flow

When ejected into the vessels by the ventricles, blood exerts a force against the walls of blood vessels called blood pressure. Blood pressure varies throughout the vascular system, being highest in the arteries and lower in the capillaries and veins. This difference in pressure is essential because blood flows along a pressure gradient, moving from areas of high pressure (arteries) to areas of lower pressure (veins). Although essential, blood pressure that is too high can damage vessels or cause rupture of small vessels, and blood pressure that is too low will reduce blood flow. The structure of arteries helps keep blood pressure within acceptable levels. When blood is ejected into the arteries by the ventricles of the heart, the elastic walls expand and keep the pressure from getting too high. The pressure does, of course, increase and reach its maximum during ventricular ejection and this maximum pressure is called the systolic pressure. After ejection and as the ventricles relax, with less blood in the arteries, blood pressure declines, but not substantially because the elastic walls of the arteries "collapse in" on the remaining blood (Fig. 6.92). In addition, because birds have rapid heart rates, the interval between successive ventricular contractions is relatively short, and that means that pressure does not drop too much before the next contraction. Although the drop in pressure is not large, the arterial pressure does drop to its lowest level when the ventricles relax and this minimum pressure is called the diastolic pressure. Importantly, even though the blood pressure drops, it remains higher than in the capillaries

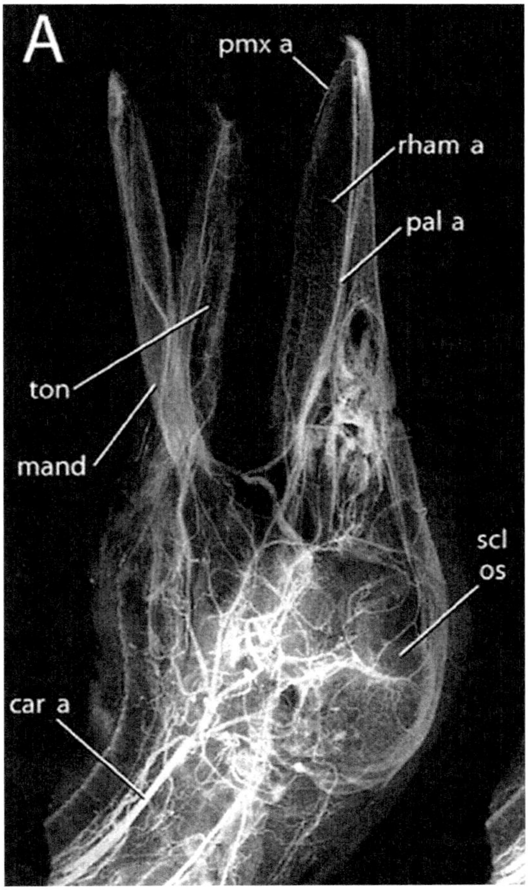


**Fig. 6.81** Head of a domestic duck showing blood vessels in the head region. Abbreviations: car a, common carotid artery; mand, mandible or lower bill; pal a, palatine artery; pmax a, premaxillary artery; rham a, rhamphothecal artery; sci os, scleral ossicles; ton, tongue. (Figure from Sedlmayr and Witmer 2002; © 2002 Wiley-Liss, Inc., used with permission)

and veins, so a "pressure gradient" is maintained and blood continues to flow.

Blood pressure of birds is typically higher than that of similar-sized mammals (Fig. 6.93). For most birds, systolic blood pressure at rest ranges from about 150 to 190 mm Hg (Seymour and Blaylock 2000). For mammals, systolic pressures typically range from about 80 to 150 mm Hg, but there are exceptions. For example, the blood pressure of giraffes (*Giraffa camelopardalis*) is generally about 235/161 (systolic/diastolic) mm Hg (Seymour and Blaylock 2000). Substantial pressure is clearly needed to get blood up the giraffe's long neck to the head! Among birds, blood pressure is not correlated with body mass. For example, typical blood pressure at rest for a Common Ostrich (*Struthio camelus*) is about 191/126 mm Hg and that of a House Sparrow (*Passer domesticus*) is about 180/140 mm Hg (Seymour and Blaylock 2000). Given the high blood pressure of giraffes, it seems that the long-necked Ostriches should also have higher blood pressure than smaller, shorter-necked birds. They likely do not because, being flightless, Ostriches have lower metabolic rates than smaller birds that fly (Seymour and Blaylock 2000).

Blood pressure is controlled automatically in birds (just as it is in mammals). In the walls of the aorta, the large artery that carries blood out of the left ventricle and pulmonary arteries are special nerve cells called baroreceptors (Taha et al. 1983, Smith 1994). The ends of these cells are stimulated to varying degrees depending on the blood pressure. Higher pressure causes the walls of the aorta and pulmonary arteries to stretch more and this causes greater stimulation of the baroreceptors. If pressure gets too high, the stimulated baroreceptors conduct more impulses to the brain stem which, in turn, initiates impulses that reduce cardiac output and, hence, reduce blood pressure. If blood pressure gets too low, less stimulation of the baroreceptors will result in increased cardiac output (Fig. 6.94).

Blood pressure drives blood through the arteries and those arteries get smaller and more numerous further from the heart. These smaller arteries are called arterioles and these less elastic and more muscular vessels "distribute" blood to the body cells and tissues. Distributing blood in a bird's body is a dynamic, every-changing process because the metabolic needs of tissues and organs change, sometimes dramatically. For example, the hindlimb muscles require much more oxygen and more nutrients when a duck or goose or other aquatic bird is swimming. Given that there is only so much blood in a bird's body, what happens when a duck begins swimming and much more of that blood must now go to the hindlimb muscles?

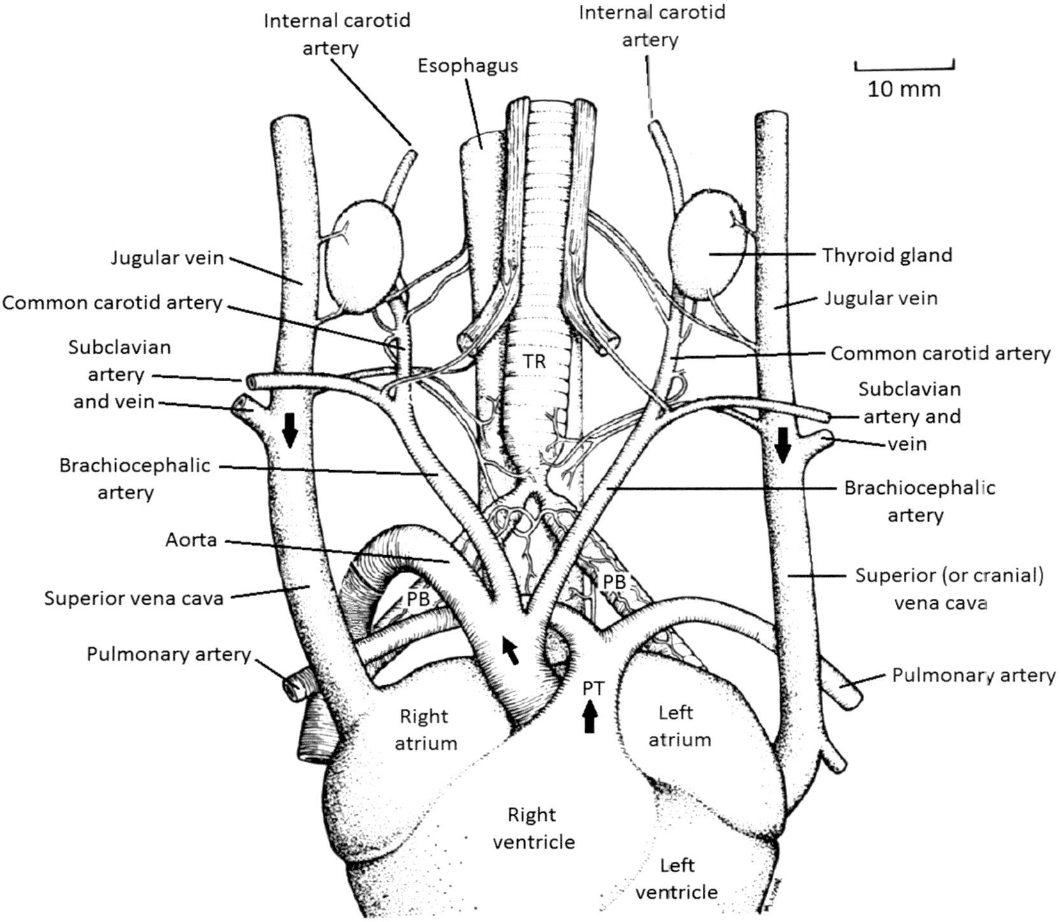


**Fig. 6.82** Blood vessels of the thoracic region of a Domestic Chicken. The right ventricle pumps blood into the pulmonary trunk (PT) which leads to the pulmonary arteries that take blood to the lungs. The left ventricle pumps blood into the aorta which branches into the two branchiocephalic arteries that then branch into the subclavian artery that takes blood to the wings and the internal carotid that takes blood to the head. The jugular veins drain blood from the head and the subclavian veins drain blood from the wings; those two veins merge and become the superior vena cavas that empty into the right atrium. Pulmonary veins (not shown) transport blood from the lungs to the left atrium. TR, trachea; PB, primary bronchi. Arrows indicate the direction of blood flow. (Figure modified from Abdel-Magied and King 1978; © 1978 John Wiley and Sons, used with permission via Copyright Clearance Center)

When swimming, cardiac output increases mainly through an increase in heart rate (Table 6.3). For Tufted Ducks (*Aythya fuligula*), for example, cardiac output at rest is about 276 ml/min and that increases to about 466 ml/min when swimming continuously (Butler et al. 1988). That helps meet the increased demands of the hindlimb muscles but, in addition, blood must be redistributed. When swimming, blood flow to vital organs, like the brain, liver, and respiratory muscles, changes little in Tufted Ducks. However, blood flow to the hindlimb muscles triples and more blood also goes, not surprisingly, to the heart (Fig. 6.95). In addition, blood flow to less

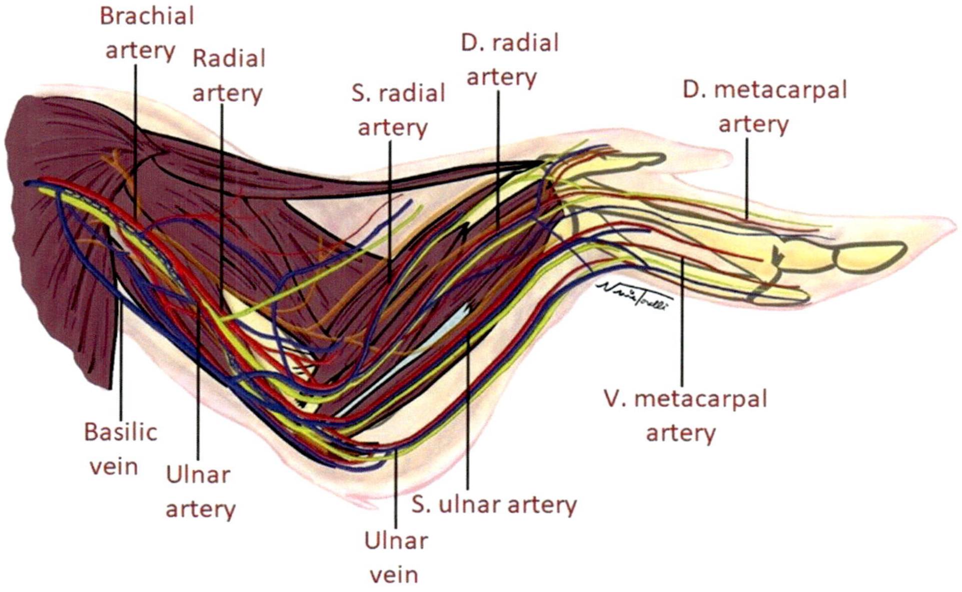


**Fig. 6.83** Blood vessels in the wing of a Rock Pigeon (*Columba livia*). The brachial artery branches into the radial and ulnar arteries, and the radial artery branches into the superficial (S) radial and deep (D) radial arteries. The deep (D) metacarpal artery branches off the deep radial artery. The superficial (S) ulnar artery branches from the ulnar artery, and the ventral (V) metacarpal artery then branches from the superficial ulnar artery. Nerves are shown as orange and yellow. (Figure modified from Navia et al. 2021; © 2020, Rights Managed by Georg Thieme Verlag KG Stuttgart • New York, used with permission)

critical areas (at least when a duck is swimming), such as the crop, stomach, intestine, and, especially, the flight muscles, decreases.

Similar changes occur when a bird takes flight. Cardiac output goes up, more blood is directed to the flight muscles (supracoracoideus and pectoralis muscles) and heart, and less blood flows to the digestive system and hindlimb muscles. Arterioles are responsible for such changes in blood flow and they alter flow simply by either vasoconstricting (smooth muscle in the walls of the vessel contracts and the vessel gets smaller in diameter) or vasodilating (the muscle relaxes and the vessels increase in diameter). Often, these changes in vessel diameter occur as a result of changes in the activity levels of the surrounding cells and tissues (Murrant and Sarelius 2015). For example, when a duck starts swimming, the muscle cells in its leg muscles start using more oxygen, using more energy, and releasing more waste products. So, in the muscle, oxygen levels decline and carbon dioxide (a waste product of energy [ATP] production) levels go up. More carbon dioxide also means that pH will decline (more acidic) because carbon dioxide reacts with water and produces carbonic acid. Carbonic acid then dissociates or breaks down and releases bicarbonate plus a hydrogen ion. More hydrogen ions mean a more acidic environment. These changing conditions in an active tissue or organ—declining oxygen levels, increasing carbon dioxide levels, and increasing acidity—cause smooth muscle in the walls of the arterioles to relax and this, of course, causes vasodilation and increased blood flow. This is an incredibly efficient system because, as the level of activity of a tissue or organ changes, there is a corresponding change in blood flow. A slight increase in activity causes a slight increase in blood flow. A more substantial increase in activity generates greater changes in levels of oxygen and so on and, therefore, a more

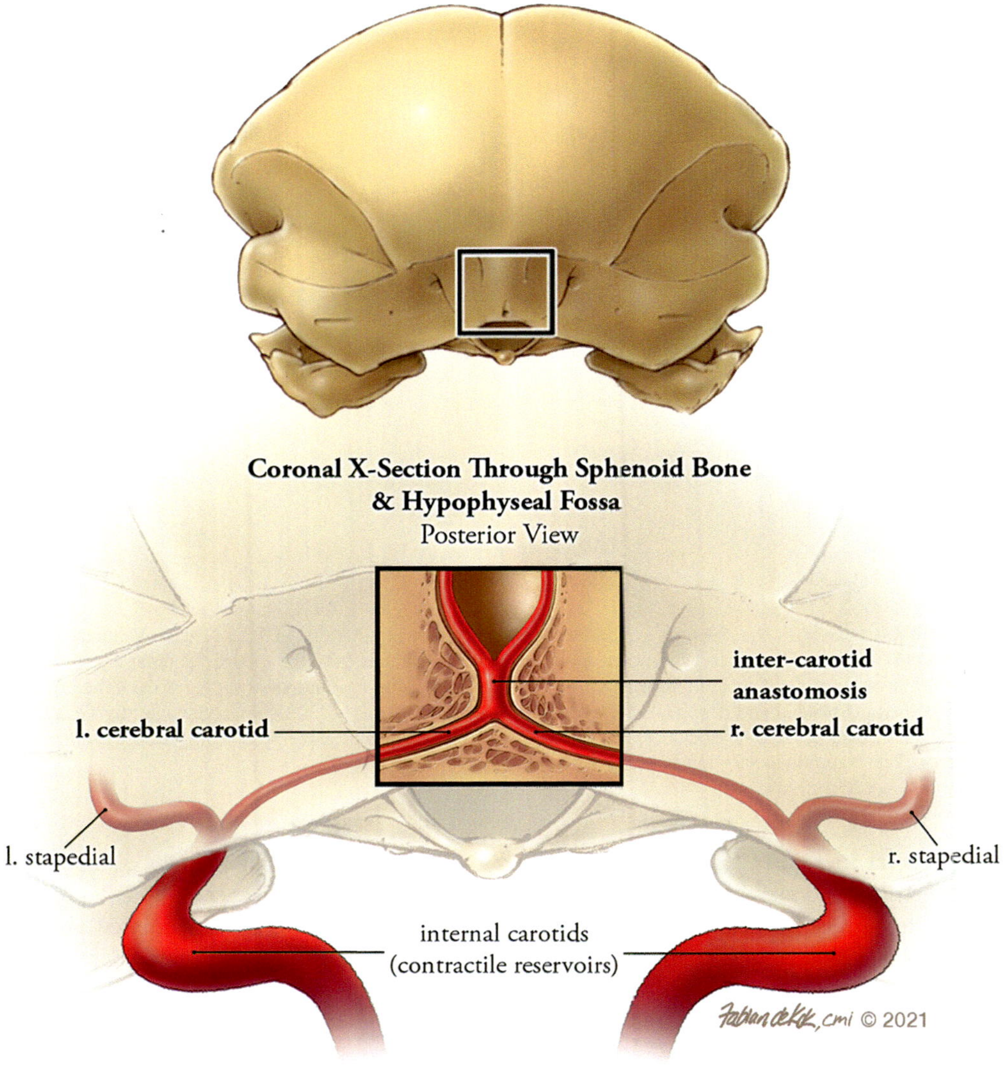


**Fig. 6.84** Intercarotid anastomosis of an owl. When birds turn their heads (and owls can rotate their heads 270 degrees), one of the carotid arteries may be constricted to varying degrees, potentially reducing blood flow to the brain. With an intercarotid anastomosis, however, blood from one carotid can pass to the other, ensuring continuous blood flow to the brain. Owls, in addition, have very distensible internal carotids that can act as temporary blood reservoirs if blood flow is temporarily restricted when owls rotate their heads. (Illustration by Fabian de Kok-Mercado, MA, CMI, FAMI © 2021, used with permission. Originally included as a figure in a poster created by de Kok-Mercado et al. 2012)

substantial increase in blood flow. In other words, blood flow matches metabolic needs.

Blood flow can also be altered by a bird's Autonomic Nervous System—the division of the nervous system that helps regulate the "automatic" activities that are not consciously controlled, e.g., digestion and blood flow (Romero and Gormally 2019). The sympathetic division of the Autonomic Nervous System is sometimes called the "flight-or-fight" division

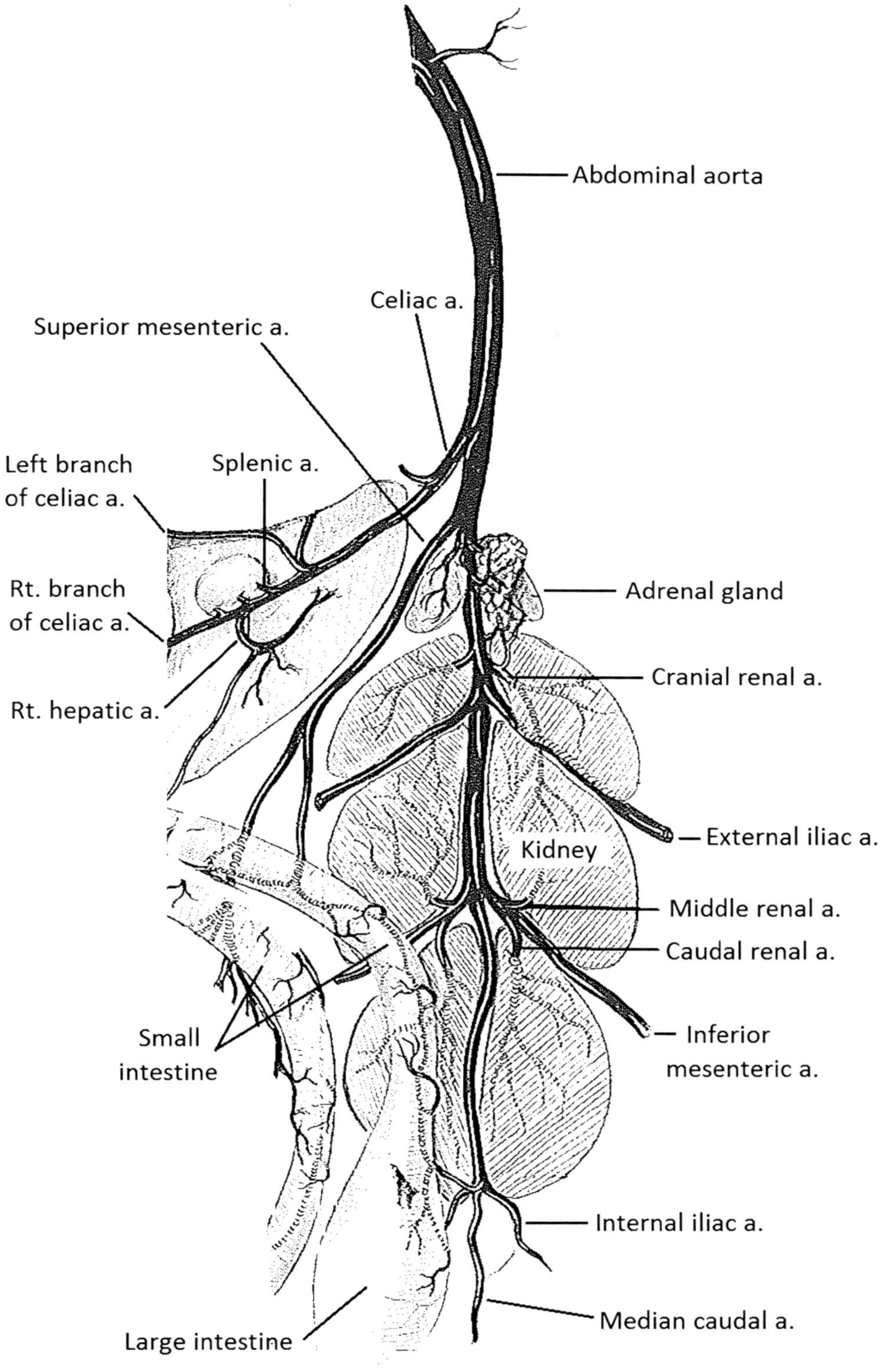


**Fig. 6.85** In the abdomen of parrots (*Amazona* spp.), the celiac artery is the first major branch of the abdominal (or descending aorta) and delivers blood to organs and tissues in the upper abdominal area, including the liver, spleen, proventriculus, ventriculus (gizzard), small intestine, and pancreas. The next branch is the superior, or

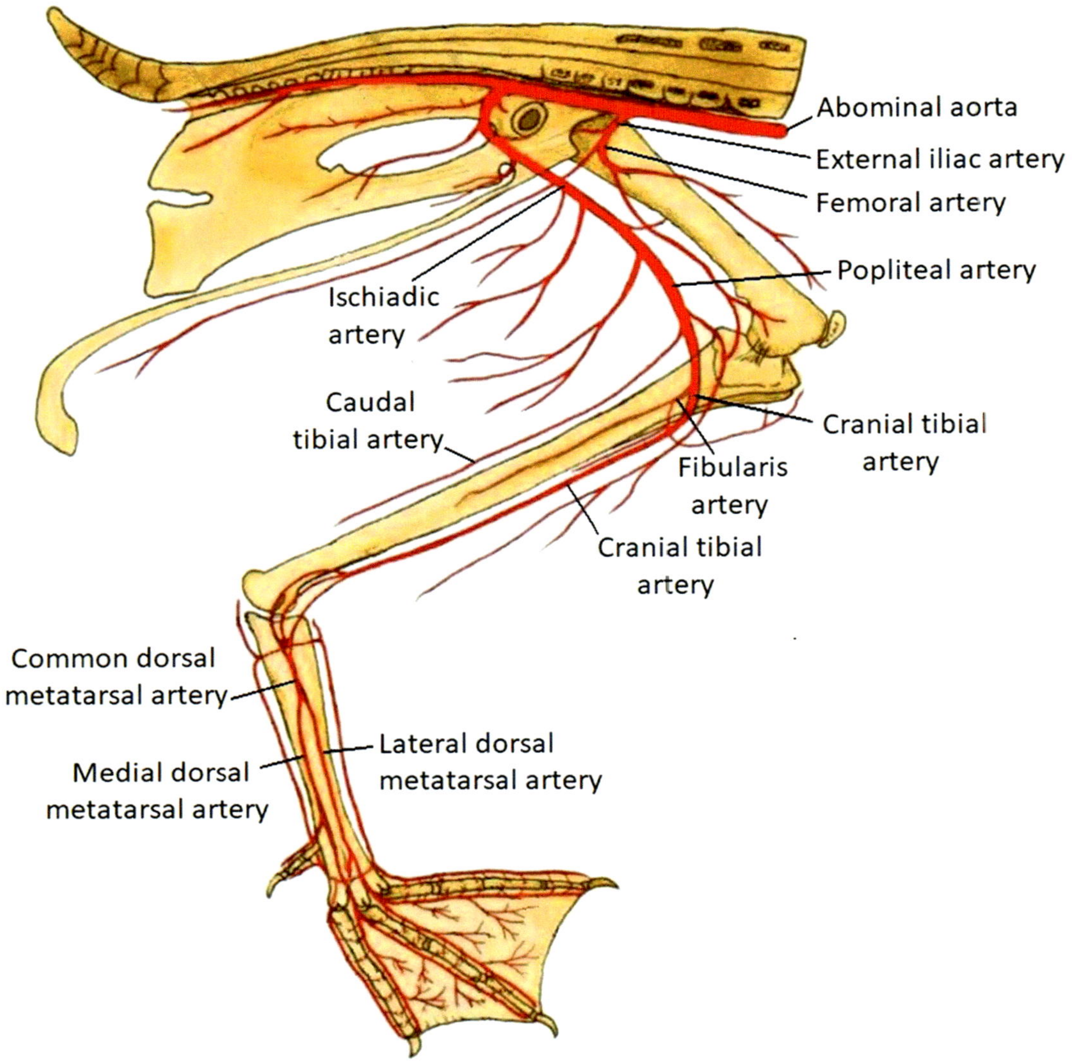


**Fig. 6.86** Major arteries of the hindlimb of an Egyptian Goose (*Alopochen aegyptiaca*). The two main arteries that branch from the abdominal aorta and supply the hindlimb are the external iliac and ischiadic arteries. (Figure modified from Ramsis et al. 2021; © 2021 El-Gammal et al., open-access article distributed under the Creative Commons Attribution License, https://creativecommons.org/licenses/by/4.0/)

**Fig. 6.85** (continued) cranial, mesenteric, artery that delivers blood to most of the small intestine and the pancreas. Renal arteries deliver blood to the kidneys, the external iliac arteries deliver blood to the legs and the median caudal artery takes blood to the tail, and the inferior, or caudal, mesenteric delivers blood to many organs and tissues in the lower abdominal area, including the large intestine and cloaca. (Figure modified from McKibben and Harrison 1986; used with permission via PLSclear Ref. No. 48781. Published by Elsevier B. V.)

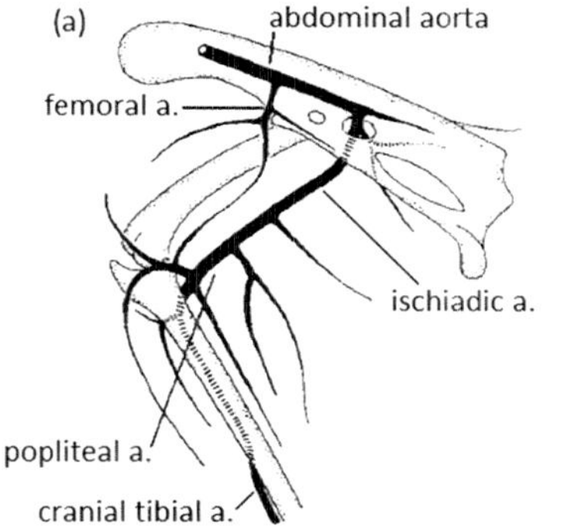


**Fig. 6.87** Arteries of the hindlimbs of (**a**) species in the orders Anseriformes and Galliformes and the genera *Puffinus*, *Uria*, *Columba*, and *Melopsittacus*, (b) species in the order Charadriiformes (except *Uria*) and Passeriformes (except *Procnias*, *Chiroxiphia*, and *Pipra*), and species in the genera *Picus*, *Falco*, and *Fulica*, and (c) species in the genus *Spheniscus*. Note the difference in the extent to which blood is supplied to the hindlimbs via the femoral versus the ischiadic arteries. (Figure modified from Midtgård 1982; © 2009 John Wiley and Sons, used with permission)

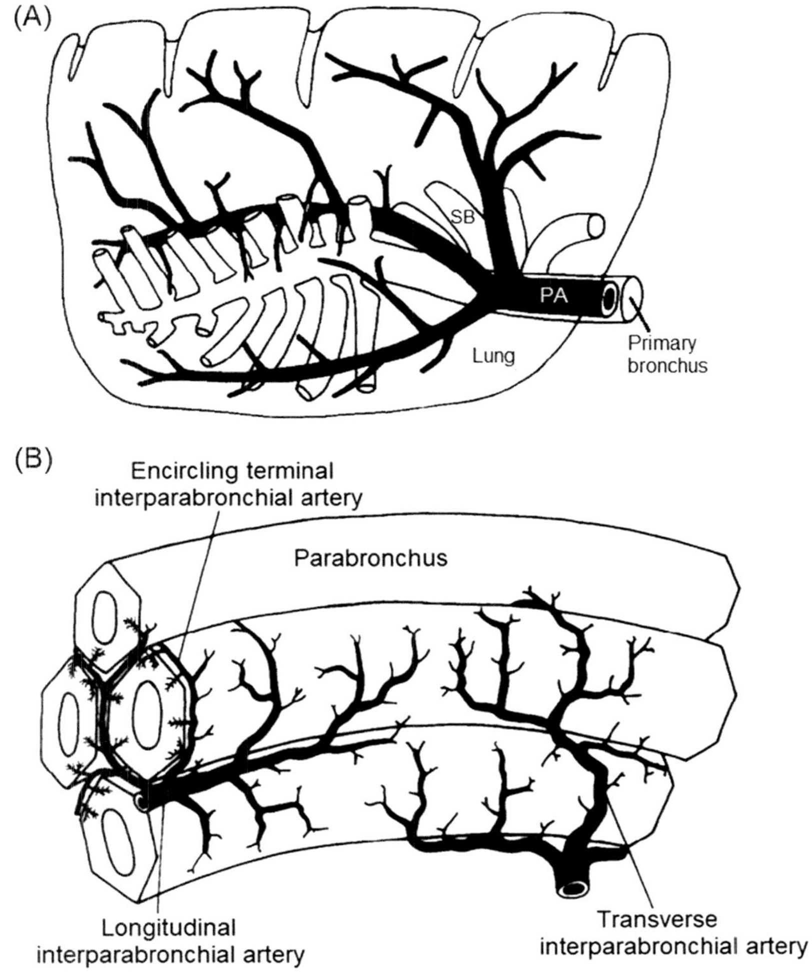


**Fig. 6.88** (**a**) Pulmonary arteries (PA) enter the lungs adjacent to the primary bronchi and divide into numerous smaller arteries. SB, secondary bronchi. (**b**) Interparabronchial arteries lie along the surface of parabronchi. (Figures modified from Abdalla and King 1975; © 1975 Published by Elsevier B.V., used with permission)

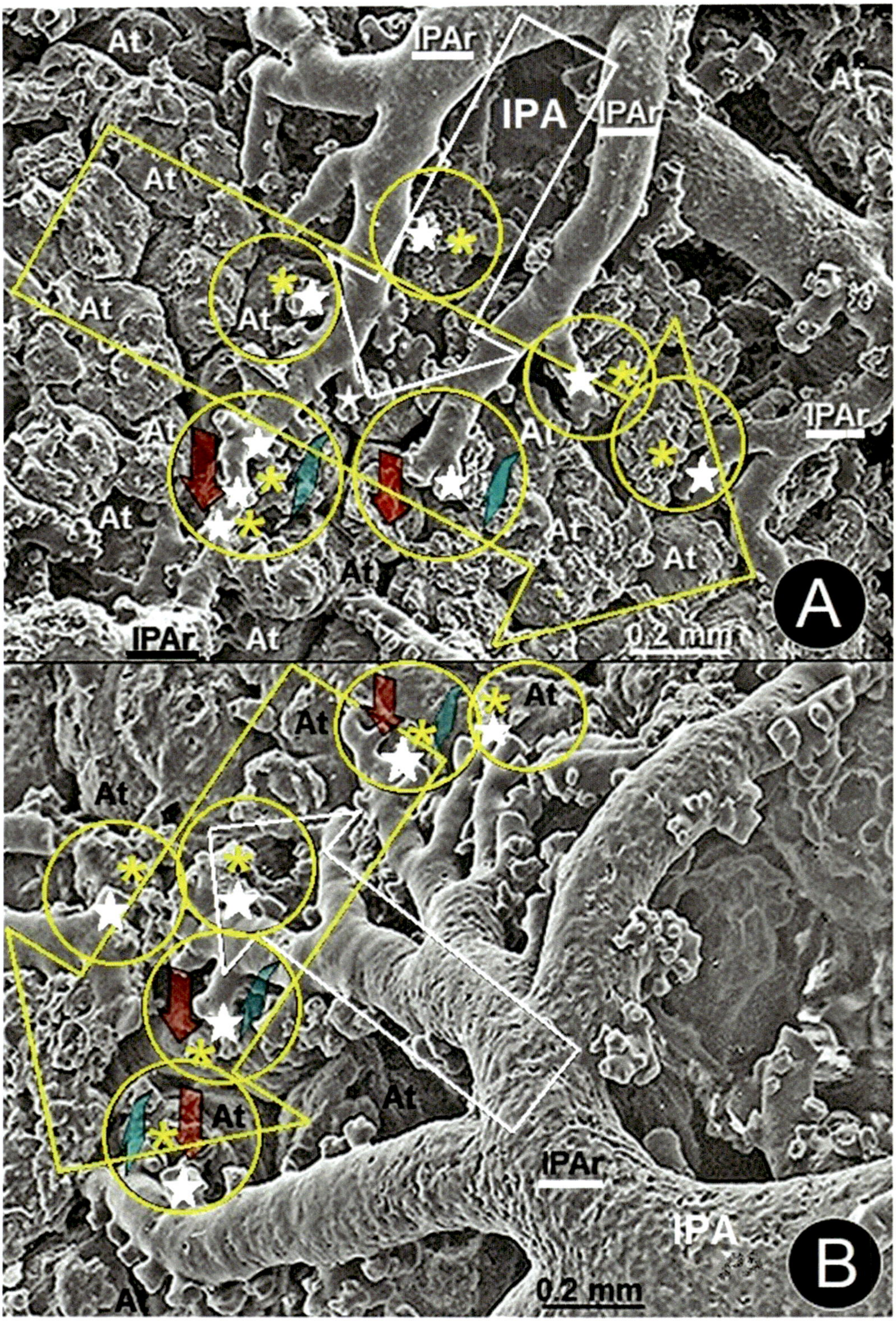


**Fig. 6.89** (**a**, **b**) Images generated by injecting latex into the airways and the blood vessels in the lung of a Domestic Chicken showing the relationships between the airways and blood vessels. The direction of air flow (large yellow

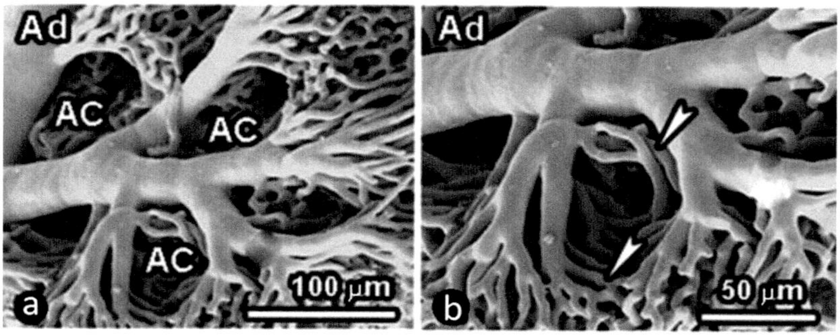


**Fig. 6.90** Scanning electron microscope micrographs of intravascular casts of blood capillaries in the lung of an adult (Ad) Domestic Chicken. (**a**, **b**) Vascular casts showing the blood capillaries forming smooth tubular structures (white arrowheads in **b**) that are perpendicular to the long axes of the air capillaries (AC), creating a crosscurrent flow of air and blood that enhances diffusion of oxygen and carbon dioxide. (Figure from Makanya 2017; © 2017 Springer International Publishing AG, used with permission)

because, when activated, this division sends impulses throughout the body that prepare a bird for intense physical activity, such as fighting (e.g., territorial defense) or fleeing (e.g., from a potential predator). If, for example, a foraging bird hears an alarm call from another flock member, the sympathetic division is automatically and immediately activated. Impulses to arterioles cause dramatic changes in blood flow, with blood directed to tissue and organs important in responding to this threat and away from those not so important. So, more blood goes to the flight muscles and heart, and less blood goes to the digestive system. Digestion can be delayed when a bird's life is at risk.

Arterioles control the flow of blood into the smallest of the blood vessels, the capillaries. This is where the exchange of materials between the blood and the cells of the body takes place (Fig. 6.96). Several characteristics of capillaries make this process of exchange very efficient. Importantly, capillaries have very thin walls, generally just one endothelial cell thick. In addition, capillaries are very small in diameter. As a result, the distance between the blood and the surrounding cells is minimized. Capillaries are also incredibly abundant, so no cell in the body is very far from its source of oxygen and nutrients, and capillary densities in birds are often higher than in other vertebrates. As an extreme example, the ratio of capillary surface area to the surface area of muscle cells in the pectoralis or flight muscle of a hummingbird is about twice that in a leg muscle of a small mammal (Matthieu-Costello et al. 1992). Even in the brain, the density of capillaries is higher in birds than it is in at least some mammals (West et al. 1981).

The exchange of gases, oxygen and carbon dioxide, occurs by simple (passive) diffusion across the walls of the capillaries (Fig. 6.96). Other substances, however, cannot easily move through the cells that make up the capillary walls. Some of these, like water, various ions, glucose,

**Fig. 6.89** (continued) arrows) in the parabronchi is perpendicular to the flow (large white arrows) of deoxygenated blood in the arteries. IPA, interparabronchial artery; underlined IPAr, intraparabronchial arterioles; circles indicate where gas exchange occurs. Asterisks air capillaries, stars blood capillaries. The diffusion of air in the air capillaries is shown by cyan-colored arrows and the flow of blood in the blood capillaries is shown by red-colored arrows. At, atria (branches of parabronchi, also called lung capillaries). (Figure from Maina 2015; © 2015 Springer Nature, used with permission)

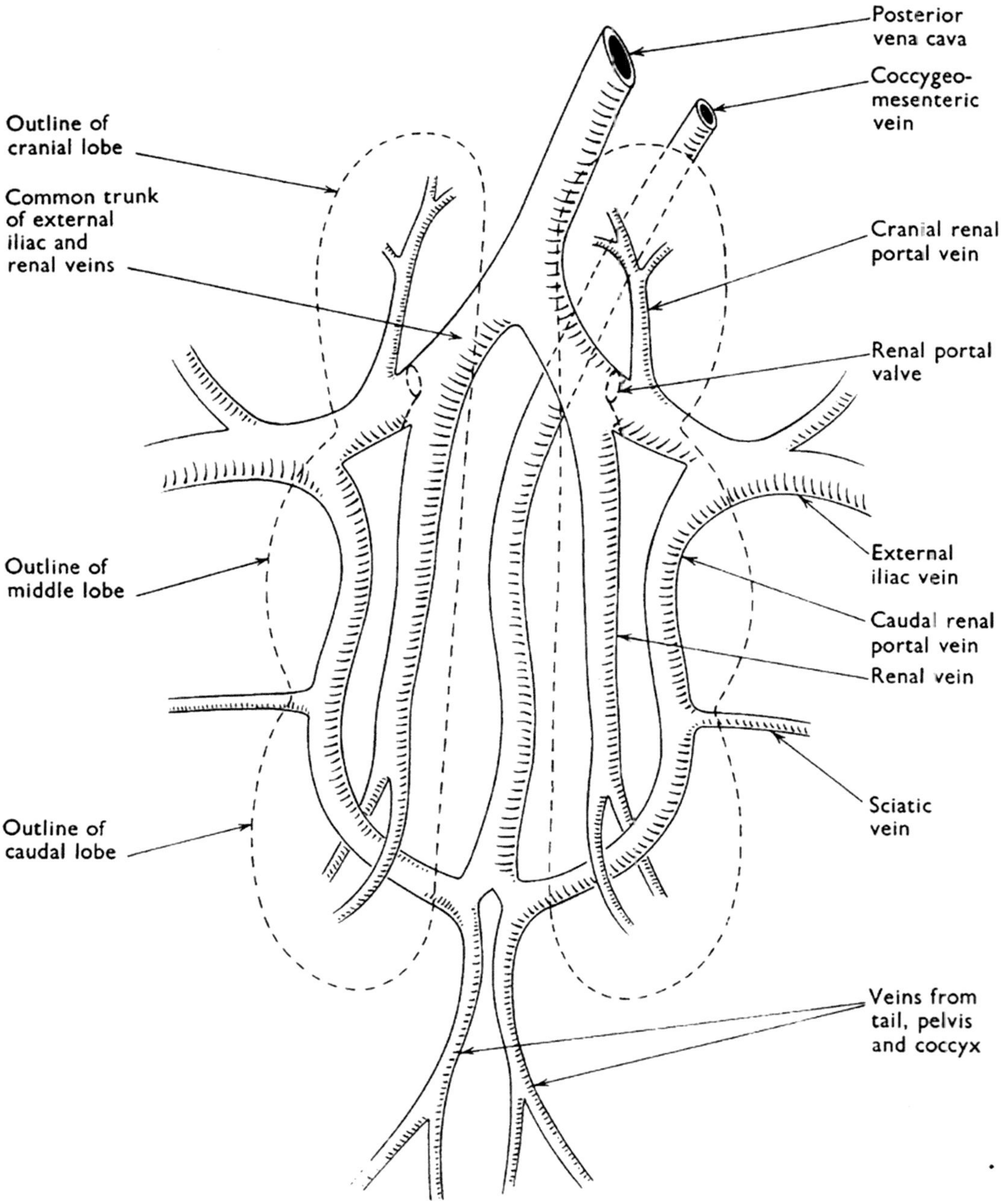


**Fig. 6.91** Veins in the abdominal region of birds. The posterior vena cava (or postcava) drains the posterior portion of the body. The coccygeomesenteric vein drains the posterior digestive system. The external iliac veins drain the legs. The sciatic veins drain the thigh region. The renal and caudal and cranial renal portal veins drain the kidneys. The outlines are the lobes of the kidneys. (Figure from Akester 1964; © 1964 John Wiley and Sons via Copyright Clearance Center, used with permission)

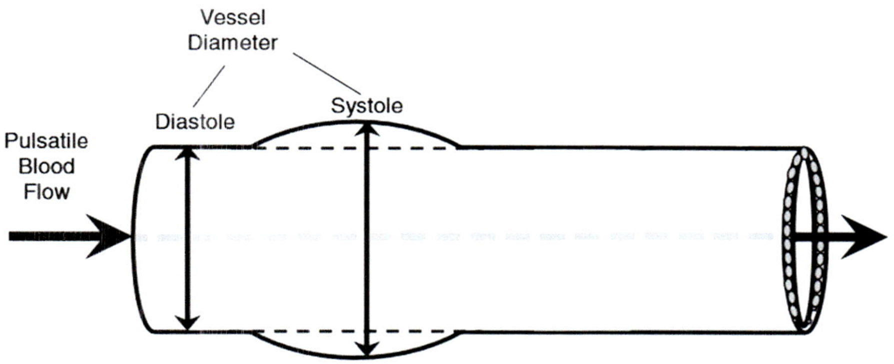


**Fig. 6.92** When the ventricles contract (systole), blood is ejected into the arteries and the walls expand because of the presence of elastic tissues (smooth muscle and connective tissue) in the walls. This expansion limits the increase in blood pressure. When the ventricles relax (diastole), no blood is being ejected, so the amount of blood in the arteries declines. As this occurs, the walls of the arteries "collapse" inward, limiting the drop in blood pressure. (Figure modified from Burggren 2013; © 2013, Biomedical Engineering Society, used with permission)

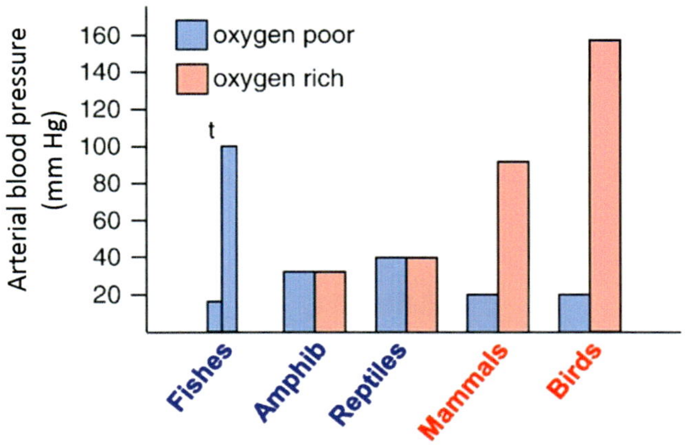


**Fig. 6.93** Typical arterial blood pressure of five vertebrate taxa. Oxygen poor refers to the pulmonary or gill circulation; oxygen rich to the systemic (body) circulation. Systemic blood pressure is generally higher in endotherms than ectotherms and is highest in birds. Some "active" fish like tuna (t) have much higher systemic blood pressures than most other fish. (Figure modified from Jensen et al. 2013a, b; © 2013 Elsevier B. V., used with permission)

and amino acids, can leave the blood through small pores located between the cells that make up the capillary walls, a process called paracellular transport, or can be transported through the cells by transporters, i.e., special proteins that move materials across cell membranes by facilitated diffusion or active transport (Fig. 6.96). Larger molecules (called macromolecules), such as proteins, are likely transported across capillary walls by moving through larger pores (paracellular transport) and by a process called transcytosis. In this process, vesicles form (by endocytosis) and materials move through the cells (in either direction depending on whether materials are moving into or out of the capillaries) to the other side where, by exocytosis, the contents of the vesicles are released (Michel and Curry 1999).

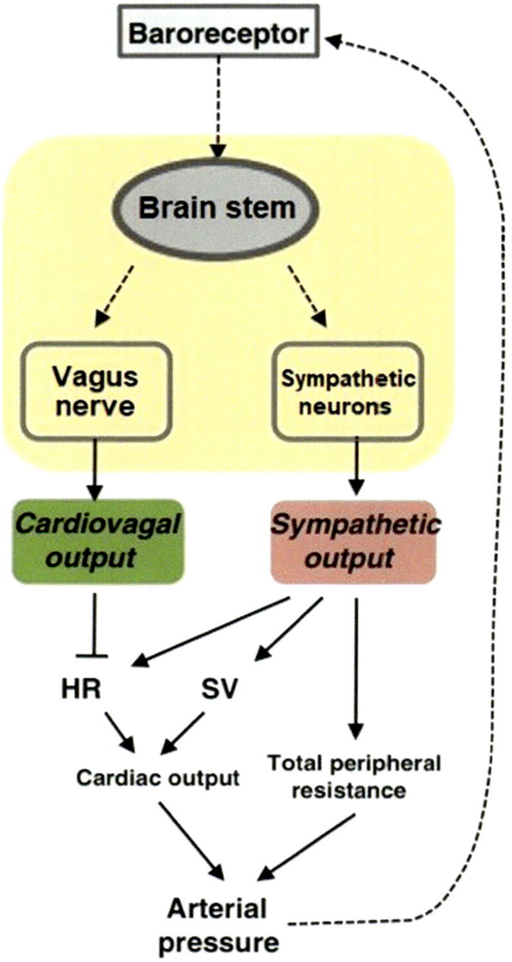


**Fig. 6.94** Arterial pressure that is too high stimulates baroreceptors that send more impulses to the brain. This results in increased parasympathetic stimulation (via parasympathetic neurons in the vagus nerve) of the pacemaker of the heart which reduces heart rate, and reduced sympathetic stimulation of the heart also contributes to a lower heart rate (HR) as well as a reduction in stroke volume (SV). As a result, cardiac output decreases and this, in addition to the reduced sympathetic stimulation that causes arterial vasodilation, reduces blood pressure. Arterial pressure that is too low reverses these actions, with reduced parasympathetic stimulation and increased sympathetic stimulation increasing cardiac output and increasing blood pressure. (Figure modified from Hong et al. 2013; © 2013 Springer-Verlag Italia, used with permission)

Blood passes from the capillaries to vessels called venules (small veins) and then to larger vessels called veins that carry blood back to the heart. The walls of veins are thinner than those of arteries and so are more distensible. As a result, under normal resting conditions, as much as 60–80% of total blood volume is contained in the veins (Smith et al. 2000). This blood can be thought of as a "reservoir" that can be used to increase cardiac output during periods of increased activity. When a bird takes flight, for example, nervous (sympathetic) impulses to the veins cause them to vasoconstrict. This reduces the volume of blood in the veins and sends more blood to the heart. This, as described previously, contributes to an increase in cardiac output.

**Table 6.3** Increases in heart rate with increased "excitement" (stress of being handled by someone) or activity in several species of birds. Values are means ± standard deviations (Table from Machida and Aohagi 2001; © 2001 American Association of Zoo Veterinarians, used with permission)

| Species | Heart rates (beats/min) | | Rate of increase (%) |
|---|---|---|---|
| | Rest | Excitement | |
| Whooper Swan | 85 ± 13 | 335 ± 19 | 401 ± 42 |
| Mallard | 155 ± 14 | 442 ± 14 | 286 ± 21 |
| Green-winged Teal | 152 ± 7 | 455 ± 9 | 300 ± 12 |
| Indian Spot-billed Duck | 140 ± 15 | 453 ± 23 | 325 ± 38 |
| Streaked Shearwater | 213 ± 22 | 443 ± 12 | 210 ± 21 |
| Great Crested Grebe | 116 ± 8 | 376 ± 14 | 324 ± 30 |
| Black-tailed Gull | 187 ± 15 | 479 ± 20 | 257 ± 27 |
| Black-legged Kittiwake | 184 ± 18 | 467 ± 27 | 256 ± 36 |
| Black-crowned Night-Heron | 211 ± 22 | 449 ± 19 | 210 ± 14 |
| Gray Heron | 211 ± 15 | 438 ± 22 | 208 ± 8 |
| Little Egret | 205 ± 20 | 433 ± 21 | 212 ± 16 |
| Intermediate Egret | 210 ± 18 | 439 ± 19 | 210 ± 10 |
| Great Egret | 213 ± 11 | 440 ± 24 | 207 ± 3 |
| Common Snipe | 349 ± 12 | ND[a] | – |
| Black Kite | 183 ± 13 | 413 ± 10 | 228 ± 14 |
| Ural Owl | 215 ± 19 | 395 ± 15 | 184 ± 10 |
| Short-eared Owl | 224 ± 19 | 445 ± 31 | 199 ± 11 |
| Carrion Crow | 224 ± 14 | 480 ± 17 | 215 ± 12 |
| Western Crowned Warbler | 647 ± 57 | ND | – |

Scientific names: Whooper Swan, *Cygnus cygnus*; Mallard, *Anas platyrhynchos*; Green-winged Teal, *Anas crecca*; Indian Spot-billed Duck, *Anas poecilorhyncha*; Streaked Shearwater, *Calonectris leucomelas*; Great Crested Grebe, *Podiceps cristatus*; Black-tailed Gull, *Larus crassirostris*; Black-legged Kittiwake, *Rissa tridactyla*; Black-crowned Night Heron, *Nycticorax nycticorax*; Gray Heron, *Ardea cinerea*; Little Egret, *Egretta garzetta*; Intermediate Egret, *Ardea intermedia*; Great Egret, *Ardea alba*; Common Snipe, *Gallinago gallinago*; Black Kite, *Milvus migrans*; Ural Owl, *Strix uralensis*; Short-eared Owl, *Asio flammeus*; Carrion Crow, *Corvus corone*; Western Crowned Warbler, *Phylloscopus occipitalis*

[a]ND = Not Determined

**Fig. 6.95** Blood flow (in ml/min) to different areas in Tufted Ducks before and while swimming at a speed of about 0.7 meters/s. Note the increase in cardiac output and blood flow to the heart and leg muscles (mm), and the decrease in blood flow to the gastrointestinal (GI) tract and flight muscles (mm). CO, cardiac output. (Data from Butler et al. 1988; Tufted Duck photo by Alexis Lours, Wikipedia, CC BY 4.0, https://creativecommons.org/licenses/by/4.0/)

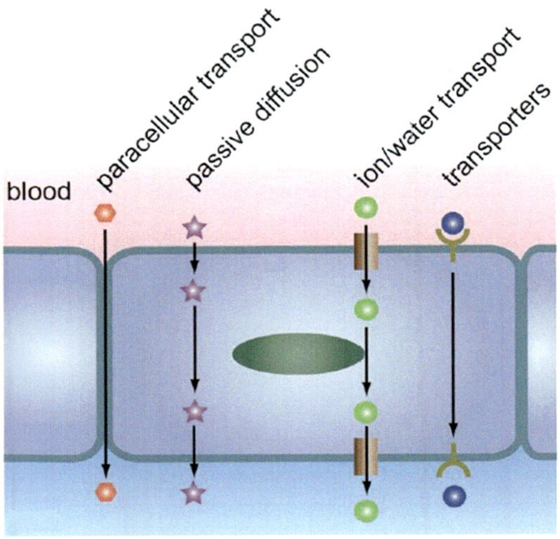


**Fig. 6.96** Pathways for exchange of materials across capillary walls. (Figure modified from Wong et al. 2013; © 2013 Wong, Ye, Levy, Rothstein, Bergles and Searson, open-access article distributed under the terms of the Creative Commons Attribution License (CC BY), https://creativecommons.org/licenses/by/4.0/)

## References


Abdalla MA, King AS (1975) The functional anatomy of the pulmonary circulation of the domestic fowl. Respir Physiol 23:267–290

Abdel-Magied EM, King AS (1978) The topographical anatomy and blood supply of the carotid body region of the Domestic Fowl. J Anat 126:535–546

Abdolmaleki M, Yeap SK, Tan SW, Satharasinghe DA, Bello MB, Jahromi MZ, Bejo MH, Omar AR, Ideris A (2018) Effects of Newcastle disease virus infection on chicken intestinal intraepithelial natural killer cells. Front Immunol 9:1386

Adair BM (2000) Immunopathogenesis of chicken anemia virus infection. Dev Comp Immunol 24:247–255

Akester AR (1964) Radiographic studies of the renal portal system in the Domestic Fowl (*Gallus domesticus*). J Anat 98:365–376

Al-Khalaifah H, Al-Nasser A (2018) Cytokines as effective elements of the avian immune system. J Microbiol Genet 2018:2574–7371

Al-Saffar TM, Al-Mawla ED (2008) Some hematological changes in chickens infected with ectoparasites in Mosul. Iraqi J Vet Sci 22:95–100

Alsafy MAM, El-Gendy SA, Enany S, Amine M (2009) Anatomical studies on the atrioventricular valves of the Ostrich heart (*Struthio camelus*). J Vet Anat 2:67–83

Altimiras J, Lindgren I, Giraldo-Deck LM, Matthei A, Garitano-Zavala Á (2017) Aerobic performance in tinamous is limited by their small heart. A novel hypothesis in the evolution of avian flight. Sci Rep 7:15964

Altshuler DL, Dudley R (2002) The ecological and evolutionary interface of hummingbird flight physiology. J Exp Biol 205:2325–2336

Amat JA, Rendón MA, Ramírez JM, Hortas F, Arroyo GM, Garrido A, Rendón-Martos M, Pérez-Hurtado A (2009) Hematocrit is related to age but not to nutritional condition in Greater Flamingo chicks. Eur J Wildl Res 55:179–182

Anderson HL, Brodsky IE, Mangalmurti NS (2018) The evolving erythrocyte: red blood cells as modulators of innate immunity. J Immunol 201:1343–1351

Apanius V (1998) Stress and immune defense. Adv Study Behav 27:133–153

Arad Z, Midtgard U, Bernstein MH (1989) Thermoregulation in Turkey Vultures: vascular anatomy, arteriovenous heat exchange, and behavior. Condor 91:505–514

Asghar M, Hasselquist D, Bensch S (2011) Are chronic avian haemosporidian infections costly in wild birds? J Avian Biol 42:530–537

Asghar M, Hasselquist D, Hansson B, Zehtindjiev P, Westerdahl H, Bensch S (2015) Hidden costs of infection: chronic malaria accelerates telomere degradation and senescence in wild birds. Science 347:436–438

Ashley NT, Weil ZM, Nelson RJ (2012) Inflammation: mechanisms, costs, and natural variation. Annu Rev Ecol Evol Syst 43:385–406

Atkinson CT, LaPointe DA (2009) Introduced avian diseases, climate change, and the future of Hawaiian honeycreepers. J Avian Med Surg 23:53–63

Atkinson CT, Van Riper C III (1991) Pathogenicity and epizootiology of avian haematozoa: *Plasmodium*, *Leucocytozoon*, and *Haemoproteus*. In: Loye JE, Zuk M (eds) Bird-parasite interactions: ecology, evolution, and behaviour. Oxford University Press, New York, pp 19–48

Atkinson CT, Dusek RJ, Lease JK (2001) Serological responses and immunity to superinfection with avian malaria in experimentally-infected Hawaii Amakihi. J Wildl Dis 37:20–27

Atkinson CT, Thomas NJ, Hunter DB (eds) (2009) Parasitic diseases of wild birds. Wiley, New York

Baccari GC, Pinelli C, Santillo A, Minucci S, Rastogi RK (2011) Mast cells in nonmammalian vertebrates: an overview. Int Rev Cell Mol Biol 290:1–53

Bairlein F, Totzke U (1992) New aspects on migratory physiology of trans-Saharan passerine migrants. Ornis Scand 23:244–250

Bańbura J, Bańbura M, Glądalski M, Kaliński A, Markowski M, Michalski M, Nadolski J, Skwarska J, Zieliński P (2011) Body condition parameters of nestling Great Tits *Parus major* in relation to experimental food supplementation. Acta Ornithol 46:207–212

Banerjee V, Banerjee M (1977) Variations of erythrocyte number and haemoglobin content of a migratory bird: *Philomachus pugnax* (Linnaeus). Zool Anz 199:261–264

Barnas GM, Rautenberg W (1990) Shivering and cardiorespiratory responses during normocapnic hypoxia in the pigeon. J Appl Physiol 68:84–87

Barrientos S, Brem H, Stojadinovic O, Tomic-Canic M (2014) Clinical application of growth factors and cytokines in wound healing. Wound Repair Regen 22:569–578

Barske JK (2013) Sexual selection and adaptations for performance of an elaborate courtship display by the Golden-collared Manakin (*Manacus vitellinus*). Ph.D. diss, Univ California, Los Angeles, CA

Barske J, Eghbali M, Kosarussavadi S, Choi E, Schlinger BA (2019) The heart of an acrobatic bird. Comp Biochem Physiol A 228:9–17

Barve S, Dhondt AA, Mathur VB, Cheviron ZA (2016) Life-history characteristics influence physiological strategies to cope with hypoxia in Himalayan birds. Proc R Soc B 283:20162201

Bauchinger U, McWilliams SR (2010) Extent of phenotypic flexibility during long-distance flight is determined by tissue-specific turnover rates: a new hypothesis. J Avian Biol 41:603–608

Beadell JS, Atkins C, Cashion E, Jonker M, Fleischer RC (2007) Immunological change in a parasite-impoverished environment: divergent signals from four island taxa. PLoS ONE 2:e896

Bean AGD, Lowenthal JW (2022) Avian cytokines and their receptors. In: Kaspers B, Schat KA, Göbel TW, Vervelde L (eds) Avian immunology, 3rd edn. Academic Press, London, pp 249–276

Beaufrére H, Ammersbach M, Tully TN Jr (2013) Complete blood cell count in Psittaciformes by using high-throughput image cytometry: a pilot study. J Avian Med Surg 27:211–217

Bensch S, Hellgren O, Pérez-Tris JR (2009) MalAvi: a public database of malaria parasites and related haemosporidians in avian hosts based on mitochondrial cytochrome B lineages. Mol Ecol Resour 9: 1353–1358

Bernstein MH (1989) Temperature and oxygen supply in the avian brain. In: Wood SC (ed) Comparative pulmonary physiology. Dekker, New York, pp 343–368

Bevan RM, Butler PJ, Woakes AJ, Boyd IL (2002) The energetics of Gentoo Penguins, *Pygoscelis papua*, during the breeding season. Funct Ecol 16:175–190

Birkhead TR, Fletcher F, Pellatt EJ (1998) Sexual selection in the Zebra Finch *Taeniopygia guttata*: condition, sex traits and immune capacity. Behav Ecol Sociobiol 44:179–191

Bischoff SC (2009) Physiological and pathophysiological functions of intestinal mast cells. Semin Immunopathol 31:185–205

Bishop CM (1997) Heart mass and the maximum cardiac output of birds and mammals: implications for estimating the maximum aerobic power input of flying animals. Philos Trans R Soc B 352:447–456

Bishop CM, Butler PJ (1995) Physiological modeling of oxygen consumption in birds during flight. J Exp Biol 198:2153–2163

Black CP, Tenney SM (1980) Oxygen transport during progressive hypoxia in high-altitude and sea-level waterfowl. Respir Physiol 39:217–239

Blount JD, Houston DC, Møller AP, Wright J (2003) Do individual branches of immune defence correlate? A comparative case study of scavenging and non-scavenging birds. Oikos 102:340–350

Boggs DF, Butler PJ, Wallace SE (1998) Differential air sac pressures in diving Tufted Ducks (*Aythya fuligula*). J Exp Biol 201:2665–2668

Boggs DF, Baudinette RV, Frappell P, Butler PJ (2001) The influence of locomotion on air sac pressures in Little Penguins. J Exp Biol 204:3582–3586

Bonadiman SF, Stratievsky GC, Machado JA, Albernaz AP, Rabelo GR, DaMatta RA (2009) Leukocyte ultrastructure, hematological and serum biochemical profiles of Ostriches (*Struthio camelus*). Poult Sci 88: 2298–2306

Bonneaud C, Mazuc J, Gonzalez G, Haussy C, Chastel O, Faivre B, Sorci G (2003) Assessing the cost of mounting an immune response. Am Nat 161:367–379

Bordel R, Haase E (1993) Effects of flight on blood parameters in homing pigeons. J Comp Physiol B 163:219–224

Borras A, Cabrera J, Senar JC (2010) Hematocrit variation in response to altitude changes in wild birds: a repeated-measures design. Condor 112:622–626

Boström JE, Dimitrova M, Canton C, Håstad O, Qvarnström A, Ödeen A (2016) Ultra-rapid vision in birds. PLoS ONE 11:e0151099

Boukens BJ, Kristensen DL, Filogonio R, Carreira LB, Sartori MR, Abe AS, Currie S, Joyce W, Conner J, Opthof T, Crossley DA II, Wang T, Jensen B (2019) The electrocardiogram of vertebrates: evolutionary changes from ectothermy to endothermy. Prog Biophys Mol Biol 144:16–29

Bowers EK, Hodges CJ, Forsman AM, Vogel AL, Masters BS, Johnson BGP, Johnson LS, Thompson CF, Sakaluk SK (2014) Neonatal body condition, immune responsiveness, and hematocrit predict longevity in a wild bird population. Ecology 95:3027–3034

Broughton JM (1994) Size of the bursa of Fabricius in relation to gonad size and age in Laysan and Black-footed albatrosses. Condor 96:203–207

Brown CR, Brown MB (2002) Spleen volume varies with colony size and parasite load in a colonial bird. Proc R Soc B 269:1367–1373

Brown TJ, Hammers M, Taylor M, Dugdale HL, Komdeur J, Richardson DS (2021) Hematocrit, age, and survival in a wild vertebrate population. Ecol Evol 11:214–226

Burggren WW (2013) Cardiovascular development and angiogenesis in the early vertebrate embryo. Cardiovasc Eng Technol 4:234–245

Burmeister AR, Marriott I (2018) The interleukin-10 family of cytokines and their role in the CNS. Front Cell Neurosci 12:458

Burmester T, Hankeln T (2004) Neuroglobin: a respiratory protein of the nervous system. News Physiol Sci 19: 110–113

Burmester T, Hankein T (2009) What is the function of neuroglobin? J Exp Biol 212:1423–1428

Burmester T, Hankeln T (2014) Function and evolution of vertebrate globins. Acta Physiol 211:501–514

Butler PJ (2001) Diving beyond the limits. News Physiol Sci 16:222–227

Butler PJ (2004) Metabolic regulation in diving birds and mammals. Respir Physiol Neurobiol 141:297–315

Butler PJ (2016) The physiological basis of bird flight. Philos Trans R Soc B 371:20150384

Butler PJ, Jones DR (1982) The comparative physiology of diving. Adv Comp Physiol Biochem 8:179–364

Butler PJ, Jones DR (1997) Physiology of diving birds and mammals. Physiol Rev 77:837–899

Butler PJ, Turner DL, Al-Wassia A, Bevan RM (1988) Regional distribution of blood flow during swimming in the Tufted Duck (*Aythya fuligula*). J Exp Biol 135: 461–472

Buzala M, Slomka A, Janicki B, Ponczek MB, Żekanowska E (2017) Review: the mechanism of blood coagulation, its disorders and measurement in poultry. Livest Sci 195:1–8

Campbell TW (1988) Avian hematology and cytology, 1st edn. Iowa State University Press, Ames, IA

Campbell TW (1994) Hematology. In: Ritchie BW, Harrison GJ, Harrison LR (eds) Avian medicine: principles and application. Wingers Publishing, Lake Worth, FL, pp 176–198

Campbell TW (2015) Avian hematology and cytology, 4th edn. Wiley, Hoboken, NJ

Campbell TW, Ellis CK (2007) Avian and exotic animal hematology and cytology, 3rd edn. Blackwell Publications, Ames, IA

Carey C, Morton ML (1976) Aspects of circulatory physiology of montane and lowland birds. Comp Biochem Physiol 54A:61–74

Carlson HC, Sweeny PR, Tokaryk JM (1968) Demonstration of phagocytic and trephocytic activities of chicken thrombocytes by microscopy and vital staining techniques. Avian Dis 12:700–715

Carvalho AR Jr, Diniz RM, Suarez MAM, Figueiredo CSS, Zagmignan A, Grisotto MAG, Fernandes ES, da Silva LCN (2018) Use of some asteraceae plants for the treatment of wounds: from ethnopharmacological studies to scientific evidences. Front Pharmacol 9:784

Carvalho-Paulo D, Neto JBT, de Oliveira TCG, de Sousa AA, dos Reis RR, dos Santos ZA, de Lima CM, de Oliveira MA, Said NM, Freitas SF, Sosthenes MCK (2021) Microglial morphology across distantly related species: phylogenetic, environmental and age influences on microglia reactivity and surveillance states. Front Immunol 12:2330

Casadevall A, Dadachova E, Pirofski L-A (2004) Passive antibody therapy for infectious diseases. Nat Rev Microbiol 2:695–703

Chawla A (2010) Control of macrophage activation and function by PPARs. Circ Res 106:1559–1569

Chico V, Nombela I, Puente-Marín S, del Mar Ortega-Villaizan M (2018) Nucleated red blood cells contribute to the host immune response against pathogens. IntechOpen. https://doi.org/10.5772/intechopen.80545

Cichon M, Sendecka J, Gustafsson L (2003) Age-related decline in humoral immune function in Collared Flycatchers. J Evol Biol 16:1205–1210

Clark NJ (2018) Phylogenetic uniqueness, not latitude, explains the diversity of avian blood parasite communities worldwide. Glob Ecol Biogeogr 27: 744–755

Clark P, Boardman WSJ, Raidal SR (2009) Atlas of clinical avian hematology. Wiley-Blackwell, Chichester, UK

Coke RL, West GD, Hoover JP (2004) Hematology and plasma biochemistry of captive Puna Ibis (*Plegadis ridgewayi*). J Wildl Dis 40:141–144

Conklin JR (2011) Extreme migration and the annual cycle: individual strategies in New Zealand Bar-tailed Godwits. Ph.D. dissertation, Massey University, Palmerston North, New Zealand

Cornell A, Gibson KF, Williams TD (2017) Physiological maturity at a critical life-history transition and flight ability at fledging. Funct Ecol 31:662–670

Cornick JL, Jensen J (1992) Anesthetic management of Ostriches. J Am Vet Med Assoc 200:1661–1666

Cotter PF (2022) A microscopic study on morphology of reactive thrombocytes in duckling. J World's Poult Res 12:142–150

Crossin GT, Phillips RA, Wynne-Edwards KE, Williams TD (2013) Postmigratory body condition and ovarian steroid production predict breeding decisions by female Gray-headed Albatrosses. Physiol Biochem Zool 86:761–768

Cuervo JJ, Møller AP, De Lope F (2007) Haematocrit is weakly related to condition in nestling Barn Swallows *Hirundo rustica*. Ibis 149:128–134

Cuervo JJ, Soler JJ, Avilés JM, Pérez-Contreras T, Navarro C (2011) Experimental feeding affects the relationship between hematocrit and body mass in Spotless Starling (*Sturnus unicolor*) nestlings. J Ornithol 152:201–206

Cuperus T, Coorens M, van Dijk A, Haagsman HP (2013) Avian host defense peptides. Dev Comp Immunol 41: 352–369

Dantzler WH (1989) Comparative physiology of the vertebrate kidney. Springer, New York

Davis AK, Maney DL, Maerz JC (2008) The use of leukocyte profiles to measure stress in vertebrates: a review for ecologists. Funct Ecol 22:760–772

Dawson RD, Bortolotti GR (1997) Are avian hematocrits indicative of condition? American Kestrels as a model. J Wildl Manag 61:1297–1306

de Angeli Dutra D, Fecchio A, Braga ÉM, Poulin R (2021) Migratory birds have higher prevalence and richness of avian haemosporidian parasites than residents. Int J Parasitol 51:877–882

de Carvalho LR, Macêdo Farias L, Nicoli JR, Silva MCF, Corsino ATSM, de Lima LA, Redondo RAF, Ferreira PCP, Pinto MEBM (2003) Dominant culturable bacterial microbiota in the digestive tract of the American Black Vulture (*Coragyps atratus*) and search for antagonistic substances. Braz J Microbiol 34:218–224

de Geus ED, Jansen CA, Vervelde L (2012) Uptake of particulate antigens in a nonmammalian lung: phenotypic and functional characterization of avian respiratory phagocytes using bacterial or viral antigens. J Immunol 188:4516–4526

de Kok-Mercado F, Habib M, Phelps T, Gregg L, Gailloud P (2012) National Science Foundation – visualization challenge. http://www.nsf.gov/news/special_reports/scivis/winners_2012.jsp

deGraw WA, Kern MD, King JR (1979) Seasonal changes in the blood composition of captive and free-living White-crowned Sparrows. J Comp Physiol 129:151–162

DeGroote LW, Rodewald PG (2010) Blood parasites in migrating wood-warblers (Parulidae): effects on refueling, energetic condition, and migration timing. J Avian Biol 41:147–153

Diamond JM, Karasov WH, Phan D, Carpenter FL (1986) Digestive physiology is a determinant of foraging bout frequency in hummingbirds. Nature 320:62–63

Dimitrov D, Palinauskas V, Iezhova TA, Bernotienė R, Ilgūnas M, Bukauskaitė D, Zehtindjiev P, Ilieva M, Shapoval AP, Bolshakov CV, Markovets MY (2015) *Plasmodium* spp.: an experimental study on vertebrate host susceptibility to avian malaria. Exp Parasitol 148: 1–16

Diniz CG, Magalhaes NGM, Sousa AA, Santos C, Diniz DG, Lima CM, Oliveira MA, Paulo DC, Pereira PDC, Sherry DF, Picanco-Diniz CW (2016) Microglia and neurons in the hippocampus of migratory sandpipers. Braz J Med Biol Res 49:20155005

Donnelly RE, Sullivan KA (1998) Foraging proficiency and body condition of juvenile American Dippers. Condor 100:385–388

Dora D, Arciero E, Hotta R, Barad C, Bhave S, Kovacs T, Balic A, Goldstein AM, Nagy N (2018) Intraganglionic macrophages: a new population of cells in the enteric ganglia. J Anat 233:401–410

Dragon S, Baumann R (2003) Hypoxia, hormones, and red blood cell function in chick embryos. News Physiol Sci 18:77–82

Dufva R (1996) Blood parasites, health, reproductive success, and egg volume in female Great Tits *Parus major*. J Avian Biol 27:83–87

Dussan FAO (2011) Modelling of the avian immune system for the optimization of specific antibody production. Ph.D. dissertation, Universidad Nacional de Colombia, Bogotá, Colombia

Dzialowski EM, Crossley DA (2015) The cardiovascular system. In: Scanes CG (ed) Sturkie's avian physiology. Academic Press, Cambridge, MA, pp 193–283

Ekino S, Arakawa H, Sonoda K, Noguchi K, Inui S, Yokoyama H, Kodama Y (2012) The origin of IgG-containing cells in the bursa of Fabricius. Cell Tissue Res 348:537–550

Elkins N (1983) Weather and bird behaviour. T & AD Poyser Ltd, Calton

Ellerby DJ, Henry HT, Carr JA, Buchanan CI, Marsh RL (2005) Blood flow in guinea fowl *Numida meleagris* as an indicator of energy expenditure by individual muscles during walking and running. J Physiol 564: 631–648

Emmenegger T, Bensch S, Hahn S, Kishkinev D, Procházka P, Zehtindjiev P, Bauer S (2021) Effects of blood parasite infections on spatiotemporal migration patterns and activity budgets in a long-distance migratory passerine. Ecol Evol 11:753–762

Esbaugh AJ, Tufts BL (2006) The structure and function of carbonic anhydrase isozymes in the respiratory system of vertebrates. Respir Physiol Neurobiol 154:185–198

Fair J, Whitaker S, Pearson B (2007) Sources of variation in haematocrit in birds. Ibis 149:535–552

Faraci FM (1986) Circulation during hypoxia in birds. Comp Biochem Physiol 85:613–620

Faraci FM (1991) Adaptations to hypoxia in birds: how to fly high. Annu Rev Physiol 53:59–70

Faraci FM, Kilgore DL, Fedde MR (1984) Oxygen delivery to the heart and brain during hypoxia: Pekin Duck vs. Bar-headed Goose. Am J Physiol 247:R69–R74

Favilla AB, Costa DP (2020) Thermoregulatory strategies of diving air-breathing marine vertebrates: a review. Front Ecol Evol 8:555509

Fecchio A, Clark NJ, Bell JA, Skeen HR, Lutz HL, De La Torre GM, Vaughan JA et al (2021) Global drivers of avian haemosporidian infections vary across zoogeographical regions. Glob Ecol Biogeogr 30:2393–2406

Fedde MR, Orr JA, Shams H, Scheid P (1989) Cardiopulmonary function in exercising Bar-headed Geese during normoxia and hypoxia. Respir Physiol 77:239–252

Ferdous F, Scott TR (2015) A comparative examination of thrombocyte/platelet immunity. Immunol Lett 163:32–39

Flajnik MF (2002) Comparative analyses of immunoglobulin genes: surprises and portents. Nat Rev Immunol 2:688

Folstad I, Karter AJ (1992) Parasites, bright males, and the immunocompetence handicap. Am Nat 139:603–622

Foo YZ, Nakagawa S, Rhodes G, Simmons LW (2017) The effects of sex hormones on immune function: a meta-analysis. Biol Rev 92:551–571

Fox AD, Kahlert J (2005) Changes in body mass and organ size during wing moult in non-breeding Greylag Geese *Anser anser*. J Avian Biol 36:538–548

Froget G, Butler PJ, Woakes AJ, Fahlman A, Kuntz G, Le Maho Y, Handrich Y (2004) Heart rate and energetics of free-ranging King Penguins (*Aptenodytes patagonicus*). J Exp Biol 207:3917–3926

Fuxjager MJ, Schuppe ER (2018) Androgenic signaling systems and their role in behavioral evolution. J Steroid Biochem Mol Biol 184:47–56

Gabanyi I, Muller PA, Feighery L, Oliveira TY, Costa-Pinto FA, Mucida D (2016) Neuro-immune interactions drive tissue programming in intestinal macrophages. Cell 164:378–391

Gaidica M, Dantzer B (2020) Quantifying the autonomic response to stressors—one way to expand the definition of "stress" in animals. Integr Comp Biol 60:113–125

Gallardo MH, Bickham J, Honeycutt RL, Ojeda RA, Kohler N (1999) Discovery of tetraploidy in a mammal. Nature 401:341

Gallo SSM, Ederli NB, Bôa-Morte MO, Oliveira FCR (2015) Hematological, morphological and morphometric characteristics of blood cells from rhea, *Rhea americana* (Struthioniformes: Rheidae): a standard for Brazilian birds. Braz J Biol 75:953–962

Gannes LZ (2001) Comparative fuel use of migrating passerines: effects of fat stores, migration distance, and diet. Auk 118:665–677

Gayathri KL, Hegde SN (2006) Alteration in haematocrit values and plasma protein fractions during the breeding cycle of female pigeons, *Columba livia*. Anim Reprod Sci 91:133–141

Genovese KJ, He H, Swaggerty CL, Kogut MH (2013) The avian heterophil. Dev Comp Immunol 41:334–340

Ghoddusi M, Kelly WR (2004) Ultrastructure of in situ perfusion-fixed avian liver, with special reference to structure of the sinusoids. Microsc Res Tech 65:101–111

Gleeson DJ, Blows MW, Owens IPF (2005) Genetic covariance between indices of body condition and immunocompetence in a passerine bird. BMC Evol Biol 5:61

Glomski CA, Pica A (2011) The avian erythrocyte: its phylogenetic odyssey. Science Publishers, Enfield, NH

Goossens KE, Ward AC, Lowenthal JW, Bean AGD (2013) Chicken interferons, their receptors and interferon-stimulated genes. Dev Comp Immunol 41:370–376

Graham AC, Temple RM, Obar JJ (2015) Mast cells and influenza a virus: association with allergic responses and beyond. Front Immunol 6:238

Green JA, Butler PJ, Woakes AJ, Boyd IL (2003) Energetics of diving in Macaroni Penguins. J Exp Biol 206:43–57

Gregory TR (2001) The bigger the C-value, the larger the cell genome size and red blood cell size in vertebrates. Blood Cell Mol Dis 27:830–843

Grilo ML, Vanstreels RET, Wallace R, García-Párraga D, Braga ÉM, Chitty J, Catão-Dias JL, Madeira de Carvalho LM (2016) Malaria in penguins–current perceptions. Avian Pathol 45:393–407

Grogan KB, Fernández RJ, Barranón FJR, Espinosa JG (2007) A brief review: avian immune system. Merial, Gainesville, GA

Grubb BR (1983) Allometric relations of cardiovascular function in birds. Am J Physiol 245:H567–H572

Grubb B, Jorgensen DD, Conner M (1983) Cardiovascular changes in the exercising Emu. J Exp Zool 104:193–201

Guglielmo CG, Williams TD (2003) Phenotypic flexibility of body composition in relation to migratory state, age, and sex in the Western Sandpiper (*Calidris mauri*). Physiol Biochem Zool 76:84–98

Guillaumet A, Paxton EH (2019) Evaluating community-level response to management actions across a diverse Hawaiian forest bird community. Ecol Appl 29:e01953

Gupta P, Vishnudas CK, Robin VV, Dharmarajan G (2020) Host phylogeny matters: examining sources of variation in infection risk by blood parasites across a tropical montane bird community in India. Parasit Vectors 13:1–13

Hahn S, Bauer S, Dimitrov D, Emmenegger T, Ivanova K, Zehtindjiev P, Buttemer WA (2018) Low intensity blood parasite infections do not reduce the aerobic performance of migratory birds. Proc R Soc B 28:20172307

Halsey LG, Butler PJ, Blackburn TM (2006) A phylogenetic analysis of the allometry of diving. Am Nat 167:276–287

Hamilton WD, Zuk M (1982) Heritable true fitness and bright birds: a role for parasites? Science 218:384–387

Handrich Y, Bevan RM, Charrassin J-B, Butler PJ, Ptz K, Woakes AJ, Lage J, Le Maho Y (1997) Hypothermia in foraging King Penguins. Nature 388:64–67

Harash G, Richardson KC, Alshamy Z, Hünigen H, Hafez HM, Plendl J, Al Masri S (2019) Heart ventricular histology and microvasculature together with aortic histology and elastic lamellar structure: a comparison of a novel dual-purpose to a broiler chicken line. PLoS ONE 14:e0214158

Harr KE (2002) Clinical chemistry of companion avian species: a review. Vet Clin Pathol 31:140–151

Harris BS, O'Brien TX, Gourdie RG (2002) Coronary arteriogenesis and differentiation of periarterial Purkinje fibers. Tex Heart Inst J 29:262–270

Hart B (1988) Biological basis of the behavior of sick animals. Neurosci Biobehav Rev 12:123–137

Härtle S, Magor KE, Göbel TW, Davison F, Kaspers B (2014) Structure and evolution of avian immunoglobulins. In: Schat K, Kaspers B, Kaiser P (eds) Avian immunology. Elsevier, Amsterdam, The Netherlands, pp 103–120

Härtle S, Magor KE, Göbel TW, Davison F, Kaspers B (2022) Structure and evolution of avian immunoglobulins. In: Kaspers B, Schat KA, Göbel TW, Vervelde L (eds) Avian immunology, 3rd edn. Academic Press, London, pp 101–119

Hartman FA (1961) Locomotor mechanisms of birds. Smithson Misc Collect 143:1–91

Harvey TC, Raichle ME, Winterborn MH, Jensen J, Lassen NA (1988) Effect of carbon dioxide in acute mountain sickness: a rediscovery. Lancet 8612:639–641

Hassanpour H, Shamsabadi MG, Dehkordi IN, Dehkordi MM (2014) Normal electrocardiogram of the Laughing Dove (*Spilopelia senegalensis*). J Zoo Wildl Med 45: 41–46

Hatch MI, Smith RJ, Owen JC (2010) Arrival timing and hematological parameters in Gray Catbirds (*Dumetella carolinensis*). J Ornithol 151:545–552

Hawkes LA, Batbayar N, Bishop CM, Butler PJ, Frappell PB, Meir JU, Milsom WK, Natsagdorj T, Scott GS (2017) Goose migration over the Himalayas: physiological adaptations. In: Prins HHT, Namgail T (eds) Bird migration across the Himalayas: wetland functioning amidst mountains and glaciers. Cambridge University Press, Cambridge, UK, pp 241–253

Hellgren O, Atkinson CT, Bensch S, Albayrak T, Dimitrov D, Ewen JG, Kim KS, Lima MR, Martin L, Palinauskas V, Ricklefs R (2015) Global phylogeography of the avian malaria pathogen *Plasmodium relictum* based on MSP1 allelic diversity. Ecography 388:842–850

Hempleman SC, Powell FL, Prisk GK (1992) Avian arterial chemoreceptor responses to steps of $CO_2$ and $O_2$. Respir Physiol 90:325–340

Hillman SS, Hedrick MS (2015) A meta-analysis of in vivo vertebrate cardiac performance: implications for cardiovascular support in the evolution of endothermy. J Exp Biol 218:1143–1150

Hirsowitz LA, Fell K, Torrance JD (1977) Oxygen-affinity of avian blood. Respir Physiol 31:51–62

Hodges RD (1979) The blood cells. In: King AS, McLelland J (eds) Form and function in birds, vol 1. Academic Press, New York, pp 361–379

Holz PH (1999) The reptilian renal portal system – a review. Bull Assoc Reptil Amphib Vet 9:4–9

Hong JM, Kim TJ, Shin DH, Lee JS, Joo IS (2013) Cardiovascular autonomic function in lateral medullary infarction. Neurol Sci 34:1963–1969

Hõrak P, Jenni-Eiermann S, Ots I, Tegelmann L (1998) Health and reproduction: the sex-specific clinical profile of Great Tits (*Parus major*) in relation to breeding. Can J Zool 76:2235–2244

Huang L, Appleton JA (2016) Eosinophils in helminth infection: defenders and dupes. Trends Parasitol 32: 798–807

Iannello A, Debbeche O, Martin E, Attalah LH, Samarani S, Ahmad A (2006) Viral strategies for evading antiviral cellular immune responses of the host. J Leukoc Biol 79:16–35

Ilgūnas M, Bukauskaitė D, Palinauskas V, Iezhova TA, Dinhopl N, Nedorost N, Weissenbacher-Lang C, Weissenböck H, Valkiūnas G (2016) Mortality and pathology in birds due to *Plasmodium* (*Giovannolaia*) *homocircumflexum* infection, with emphasis on the exoerythrocytic development of avian malaria parasites. Malar J 15:256

Ishtiaq F (2021) Ecology and evolution of avian malaria: implications of land use changes and climate change on disease dynamics. J Indian Inst Sci 101:213–225

Jamaati H, Mortaz E, Pajouhi Z, Folkerts G, Movassaghi M, Moloudizargari M, Adcock IM, Garssen J (2017) Nitric oxide in the pathogenesis and treatment of tuberculosis. Front Microbiol 8:2008

Janeway CA, Travers P, Walport M, Capra JD (1999) Immunobiology: the immune system in health and disease. Current Biology Publications, London

Jenni-Eiermann S, Jenni L (1991) Metabolic responses to flight and fasting in night migrating passerines. J Comp Physiol B 161:465–474

Jenni-Eiermann S, Jenni L (1998) What can plasma metabolites tell us about the metabolism, physiological state and condition of individual birds? An overview. Biologia e Conservazione Della Fauna 102:312–319

Jenni-Eiermann S, Jenni L, Kvist A, Lindstrom A, Piersma T, Visser GH (2002) Fuel use and metabolic response to endurance exercise: a wind tunnel study of a long-distance migrant shorebird. J Exp Biol 205: 2453–2460

Jensen B, Wang T, Christoffels VM, Moorman AFM (2013a) Evolution and development of the building plan of the vertebrate heart. Biochim Biophys Acta 1833:783–794

Jensen KH, Kim W, Holbrook NM, Bush JWM (2013b) Optimal concentrations in transport systems. J R Soc Interface 10:20130138

Johansen K, Berger M, Bicudo JEPW, Ruschi A, de Almeida PJ (1987) Respiratory properties of blood and myoglobin in hummingbirds. Physiol Zool 60: 2689–2278

Jones PJ (1983) Haematocrit values of breeding Red-billed Queleas *Quelea quelea* (Aves: Ploceidae) in relation to body condition and thymus activity. J Zool 201:217–222

Josephson B (1980) Aging and sexing Snowy Owls. J Field Ornithol 51:149–160

Joyce W (2018) Flying is not for the small-hearted. J Exp Biol 221:jeb169912

Jubril AJ, Olushola G, Adekola AA, Antia R (2021) Haematological profile of naturally infected haemoparasite positive and negative Japanese Quails (*Coturnix coturnix japonica*). Sahel J Vet Sci 18:21–26

Kalmbach E, Griffiths R, Crane JE, Furness RW (2004) Effects of experimentally increased egg production and laying dates in the Great Skua *Stercorarius skua*. J Avian Biol 35:501–514

Kharin SN (2004) Depolarisation and repolarization sequences of ventricular epicardium in chickens (*Gallus gallus domesticus*). Comp Biochem Physiol A 137:237–244

Kilgas P, Tilgar V, Mänd R (2006) Hematological health state indices predict local survival in a small passerine bird, the Great Tit (*Parus major*). Physiol Biochem Zool 79:565–572

Kilgore DL, Bernstein MH, Schmidt-Nielsen K (1973) Brain temperature in a large bird, the rhea. Am J Phys 225:739–742

Kilpatrick AM, LaPointe DA, Atkinson CT, Woodworth BL, Lease JK, Reiter ME, Gross K (2006) Effects of chronic avian malaria (*Plasmodium relictum*) infection on reproductive success of Hawaii Amakihi (*Hemignathus virens*). Auk 123:764–774

Kilpimaa J, Alatalo RV, Siitari H (2004) Trade-offs between sexual advertisement and immune function in the Pied Flycatcher (*Ficedula hypoleuca*). Proc R Soc B 271:245–250

Klasing KC (1998) Avian macrophages: regulators of local and systemic immune responses. Poult Sci 77:983–989

Klika E, Scheuermann DW, De Groodt-Lasseel MHA, Bazantova I, Switka A (1996) Pulmonary macrophages in birds (Barn Owl, *Tyto tyto alba*), domestic fowl (*Gallus gallus f. domestica*), quail (*Coturnix coturnix*), and pigeons (*Columba livia*). Anat Rec 246:87–97

Knower Stockard T, Heil J, Meir JU, Sato K, Ponganis KV, Ponganis PJ (2005) Air sac $P_{O2}$ and oxygen depletion during dives of Emperor Penguins. J Exp Biol 208:2973–2980

Kooyman GL, Ponganis PJ (1998) The physiological basis of diving to depth: birds and mammals. Annu Rev Physiol 60:19–32

Kozłowski J, Czarnołęski M, François-Krassowska A, Maciak S, Pis T (2010) Cell size is positively correlated between different tissues in passerine birds and amphibians, but not necessarily in mammals. Biol Lett 6:792–796

Kozlu T, Sari EK, Bozkurt YA, Altunay H (2011) A comparative study on the histological structure of the spleen in the Ostrich (*Struthio camelus*), the kestrel (*Falco tinnunculus*) and the Osprey (*Pandion haliaetus*). Acta Biol Hung 62:113–121

Kranen RW, Van Kuppevelt TH, Goedhart HA, Veerkamp CH, Lambooy E, Veerkamp JH (1999) Hemoglobin and myoglobin content in muscles of broiler chickens. Poult Sci 78:467–476

Kroneman J, Faber J, Wolschrijn C, Christoffels V, Jensen B (2019) Comparative analysis of avian hearts provides little evidence for variation among species with acquired endothermy. J Morphol 280:395–410

Laaksonen T, Fargallo JA, Korpimäki E, Lyytinen S, Valkama J, Pöyri V (2004) Year-and sex-dependent effects of experimental brood sex ratio manipulation on fledging condition of Eurasian Kestrels. J Anim Ecol 73:342–352

Laguë SL (2017) High-altitude champions: birds that live and migrate at altitude. J Appl Physiol 123:942–950

Laguë SL, Chua B, Alza L, Scott GR, Frappell PB, Zhong Y, Farrell AP, McCracken KG, Wang Y, Milsom WK (2017) Divergent respiratory and cardiovascular responses to hypoxia in Bar-headed Geese and Andean birds. J Exp Biol 220:4186–4194

Landys-Ciannelli MM, Jukema J, Piersma T (2002) Blood parameter changes during stopover in a long-distance migratory shorebird, the Bar-tailed Godwit *Limosa lapponica taymyrensis*. J Avian Biol 33:451–455

Lansford R, Rugonyi S (2020) Follow me! a tale of avian heart development with comparisons to mammal heart development. J Cardiovasc Dev Dis 7:8

Larson TA (2018) Sex steroids, adult neurogenesis, and inflammation in CNS homeostasis, degeneration, and repair. Front Endocrinol 9:205

Lasiewski RC, Calder WA (1971) A preliminary allometric analysis of respiratory variables in resting birds. Respir Physiol 11:152–166

Lee-Cruz L, Cunningham AA, Martínez P, Cruz M, Goodman SJ, Hamer KC (2016) Prevalence of *Haemoproteus* sp. in Galápagos Blue-footed Boobies: effects on health and reproduction. Parasitol Open 2:e1

Lefranc MP, Lefranc G (2020) Immunoglobulins or antibodies: IMGT® bridging genes, structures and functions. Biomedicines 8:319

Lenfant C, Johansen K, Torrance JD (1970) Gas transport and oxygen storage capacity in some pinnipeds and the sea otter. Respir Physiol 9:277–286

Leon-Velarde F, Sanchez J, Bigard AX, Brunet A, Lesty C, Monge C (1993) High altitude tissue adaptation in Andean Coots: capillarity, fibre area, fibre type and enzymatic activities of skeletal muscle. J Comp Physiol B Biochem Syst Environ Physiol 164:52–58

Li MO, Wan YY, Sanjabi S, Robertson A-KL, Flavell RA (2006) Transforming growth factor-β regulation of immune responses. Annu Rev Immunol 24:99–146

Liknes ET, Swanson DL (2011) Phenotypic flexibility of body composition associated with seasonal acclimatization in passerine birds. J Therm Biol 36:363–370

Lill A, Rajchl K, Yachou-Wos L, Johnstone CP (2013a) Are haematocrit and haemoglobin concentration reliable body condition indicators in nestlings: the Welcome Swallow as a case study. Avian Biol Res 6:57–66

Lill A, Yachou-Wos L, Johnstone CP (2013b) Are haematocrit and haemoglobin concentration reliable indicators of body condition in adult Welcome Swallows. Emu 113:93–98

Linde A, Ross CR, Davis EG, Dib L, Blecha F, Melgarejo T (2008) Innate immunity and host-defense peptides in veterinary medicine. J Vet Intern Med 22:247–265

Lindstrom A, Kvist A, Piersma T, Dekinga A, Dietz MW (2000) Avian pectoral muscle size rapidly tracks body mass changes during flight. J Exp Biol 203:913–919

Lindström KM, Foufopoulos J, Pärn H, Wikelski M (2004) Immunological investments reflect parasite abundance in island populations of Darwin's finches. Proc R Soc B 271:1513–1519

Lobato DN, Braga EM, Belo NDO, Antonini Y (2011) Hematological and parasitological health conditions of the Pale-breasted Thrush (*Turdus leucomelas*) (Passeriformes: Turdidae) in southeastern Brazil. Zoologia (Curitiba) 28:771–776

Lownie TJR, Jubinville I, Williams TD, Phillips RA, Crossin GT (2022) Varying aerobic capacity in relation to breeding stage and reproductive success in giant petrels (*Macronectes* spp.). Comp Biochem Physiol A 266:111155

Lu Y, James TN, Bootsma M, Terasaki F (1993) Histological organization of the right and left atrioventricular valves of the chicken heart and their relationship to the atrioventricular Purkinje ring and the middle bundle branch. Anat Rec 235:74–86

Lu M, Lee Y, Lillehoj HS (2023) Evolution of developmental and comparative immunology in poultry: the regulators and the regulated. Dev Comp Immunol 138: 104525

Lutfullah G, Abid Ali S, Abbasi A (2005) Molecular mechanism of high altitude respiration: primary structure of a minor hemoglobin component from Tufted Duck (*Aythya fuligula*). Biochem Biophys Res Commun 326:123–130

Lutz PL (1980) On the oxygen affinity of bird blood. Am Zool 20:187–198

Machida N, Aohagi Y (2001) Electrocardiography, heart rates, and heart weights of free-living birds. J Zoo Wildl Med 32:47–54

Madej JP, Chrząstek K, Piasecki T, Wieliczko A (2013) New insight into the structure, development, functions and popular disorders of bursa Fabricii. Anat Histol Embryol 42:321–331

Maginniss LA, Bernstein MH, Deitch MA, Pinshow B (1997) Effects of chronic hypobaric hypoxia on blood oxygen binding in pigeons. J Exp Zool 277:293–300

Maher MA (2019) Descriptive anatomy of hepatic and portal veins with special reference to biliary duct system in broiler chickens (*Gallus gallus domesticus*): a recent illustration. Braz J Poult Sci 21:1–12

Maina JN (2015) The design of the avian respiratory system: development, morphology and function. J Ornithol 156:41–63

Makanya AN (2017) Development of the airways and the vasculature in the lungs of birds. In: Maina JN (ed) The biology of the avian respiratory system. Springer, Cham, Switzerland, pp 147–178

Martin LB (2009) Stress and immunity in wild vertebrates: timing is everything. Gen Comp Endocrinol 163:70–76

Martínez-de la Puente J, Santiago-Alarcon D, Palinauskas V, Bensch S (2021) Parasite of the month: *Plasmodium relictum*. Trends Parasitol 37: 355–356

Marzal A, De Lope F, Navarro C, Møller AP (2005) Malarial parasites decrease reproductive success: an experimental study in a passerine bird. Oecologia 142:541–545

Mathieu-Costello O, Agey PJ (1997) Chronic hypoxia affects capillary density and geometry in pigeon pectoralis muscle. Respir Physiol 109:39–52

Matthieu-Costello O, Suarez RK, Hochachka PW (1992) Capillary-to-fiber geometry and mitochondrial density in hummingbird flight muscle. Respir Physiol 89:113–132

Maxwell MH, Robertson GW (1995) The avian basophilic leukocyte: a review. Worlds Poult Sci J 51:307–325

McKibben JS, Harrison GJ (1986) Clinical anatomy with emphasis on the Amazon Parrot. In: Harrison GJ, Harrison LR (eds) Clinical avian medicine and surgery. W. B. Saunders, Philadelphia, PA, pp 31–66

Medeiros MCI, Hamer GL, Ricklefs RE (2013) Host compatibility rather than vector–host-encounter rate determines the host range of avian *Plasmodium* parasites. Proc R Soc B 280:20122947

Meir JU, Milsom WK (2013) High thermal sensitivity of blood enhances oxygen delivery in the high-flying Bar-headed Goose. J Exp Biol 216:2172–2175

Meir JU, Stockard TK, Williams CL, Ponganis KV, Ponganis PJ (2008) Heart rate regulation and extreme bradycardia in diving Emperor Penguins. J Exp Biol 211:1169–1179

Meir JU, York JM, Chua BA, Jardine W, Hawkes LA, Milsom WK (2019) Reduced metabolism supports hypoxic flight in the high-flying Bar-headed Goose (*Anser indicus*). eLife 8:e44986

Michel CC, Curry FE (1999) Microvascular permeability. Physiol Rev 79:703–761

Midtgård U (1982) Patterns in the blood vascular system in the pelvic limb of birds. J Zool 196:545–567

Miguel RDS, Reinaldo RCPS, Fraga KB, Galvão APO, Silva JAC, Xavier JE, Magalhães CP (2014) Morphological and morphometric analysis from the heart of *Caracara plancus* (Carcará). J Morphol Sci 31:225–227

Milenkaya O, Weinstein N, Legge S, Walters JR (2013) Variation in body condition indices of Crimson Finches by sex, breeding stage, age, time of day, and year. Conserv Physiol 1:cot020

Milsom WK, Johansen K, Millard RW (1973) Blood respiratory properties in some Antarctic birds. Condor 75:472–474

Minias P (2014) High glucose concentrations are associated with symptoms of mild anaemia in Whiskered Terns: consequences for assessing physiological quality in birds. J Ornithol 155:1067–1070

Minias P (2015) The use of haemoglobin concentrations to assess physiological condition in birds: a review. Cons Physiol 3:cov007
Minias P (2020) Ecology and evolution of blood oxygen-carrying capacity in birds. Am Nat 195:788–801
Minias P, Kaczmarek K, Wlodarczyk R, Janiszewski T (2013) Hemoglobin concentrations in waders vary with their strategies of migration: a comparative analysis. Comp Biochem Physiol 165:7–12
Minias P, Wlodarczyk R, Plasecka A, Kaczmarek K, Janiszewski (2014) Ecological, physiological, and morphological correlates of blood hemoglobin concentration in a migratory shorebird. Physiol Biochem Zool 87:771–781
Møller AP (1990) Parasites and sexual selection: current status of the Hamilton and Zuk hypothesis. J Evol Biol 3:319–328
Møller AP, Erritzoe J (1996) Parasite virulence and host immune defense: host immune response is related to nest reuse in birds. Evolution 50:2066–2072
Møller AP, Erritzoe J (2000) Predation against birds with low immunocompetence. Oecologia 122:500–504
Monaghan P, Nager RG, Houston DC (1998) The price of eggs: increased investment in egg production reduces the offspring rearing capacity of parents. Proc R Soc B 265:1731–1735
Moreno J, Sanz JJ, Arriero E (1999) Reproductive effort and T-lymphocyte cell-mediated immunocompetence in female Pied Flycatchers *Ficedula hypoleuca*. Proc R Soc B 266:1105–1109
Moreno J, Yorio P, Garcia-Borboroglu P, Potti J, Villar S (2002) Health state and reproductive output in Magellanic Penguins (*Spheniscus magellanicus*). Ethol Ecol Evol 14:19–28
Morrison ES, Ardia DR, Clotfelter ED (2009) Cross-fostering reveals sources of variation in innate immunity and hematocrit in nestling Tree Swallows *Tachycineta bicolor*. J Avian Biol 40:573–578
Morton ML (1994) Hematocrits in montane sparrows in relation to reproductive schedule. Condor 96:119–126
Müller MS, Vyssotski AL, Yamamoto M, Yoda K (2017) Heart rate variability reveals that a decrease in parasympathetic ('rest-and-digest') activity dominates autonomic stress responses in a free-living seabird. Comp Biochem Physiol A 212:117–126
Murrant CL, Sarelius IH (2015) Local control of blood flow during active hyperaemia: what kinds of integration are important? J Physiol 593:4699–4711
Mutua MP, Muya S, Gicheru MM (2016) Protective roles of free avian respiratory macrophages in captive birds. Biol Res 49:29
Nager RG, Monaghan P, Houston DC (2001) The cost of egg production: increased egg production reduces future fitness in gulls. J Avian Biol 32:159–166
Narkkong N-A, Aengwanich W, Tanomthong A (2010) Morphological observations of the thrombocyte of White-bellied Sea Eagle *Haliaeetus leucogaster*. Comp Clin Pathol 19:263–267
Natarajan C, Jendroszek A, Kumar A, Weber RE, Tame JR, Fago A, Storz JF (2018) Molecular basis of hemoglobin adaptation in the high-flying Bar-headed Goose. PLoS Genet 14:e1007331
Navarro J, Gonzalez-Solis J, Viscor G (2007) Nutritional and feeding ecology in Cory's Shearwater *Calonectris diomedea* during breeding. Mar Ecol Prog Ser 351:261–271
Navarro J, Votier SC, Phillips RA (2014) Diving capabilities of diving petrels. Polar Biol 37:897–901
Navia A, Tejos R, Yañez R, Cuadra A, Dagnino B (2021) Optimizing the chicken wing anatomy: nomenclature review and description for microsurgery and supermicrosurgery training. J Reconstr Microsurg 38: e1–e2
Nespolo RF, González-Lagos C, Solano-Iguaran JJ, Elfwing M, Garitano-Zavala A, Mañosa S, Alonso JC, Altimiras J (2018) Aerobic power and flight capacity in birds: a phylogenetic test of the heart-size hypothesis. J Exp Biol 221:jeb162693
Nganvongpanit K, Kaewkumpai P, Kochagul V, Pringproa K, Punyapornwithaya V, Mekchay S (2020) Distribution of melanin pigmentation in 33 organs of Thai black-bone chickens (*Gallus gallus domesticus*). Animals 10:777
Norte AC, Ramos JA, Sampaio HL, Sousa JP, Sheldon BC (2010) Physiological condition and breeding performance of the Great Tit. Condor 112:79–86
O'Dwyer TW, Buttemer WA, Priddel DM (2007) Differential rates of offspring provisioning in Gould's Petrels: are better feeders better breeders? Aust J Zool 55:155–160
O'Toole ET, Hantgan RR, Lewis JC (1994) Localization of fibrinogen during aggregation of avian thrombocytes. Exp Mol Pathol 61:175–190
Odum EP (1945) The heart rate of small birds. Science 101:153–154
Oglesbee BL, Hamlin RL, Klingaman H, Cianciola J, Hartman SP (2001) Electrocardiographic reference values for macaws (*Ara* species) and cockatoos (*Cacatua* species). J Avian Med Surg 15:17–22
Oppliger A, Christe P, Richner H (1996) Clutch size and malaria resistance. Nature 381:565
Ordway GA, Garry DJ (2004) Myoglobin: an essential hemoprotein in striated muscle. J Exp Biol 207:3441–3446
Owen JC (2011) Collecting, processing, and storing avian blood: a review. J Field Ornithol 82:339–354
Owen JC, Moore FR (2006) Seasonal differences in immunological condition of three species of thrushes. Condor 108:389–398
Palinauskas V, Valkiūnas G, Bolshakov CV, Bensch S (2008) *Plasmodium relictum* (lineage P-SGS1): effects on experimentally infected passerine birds. Exp Parasitol 120:372–380
Palm EC, Newman SH, Prosser DJ, Xiao X, Ze L, Batbayar N, Balachandran S, Takekawa JY (2015) Mapping migratory flyways in Asia using dynamic Brownian bridge movement models. Mov Ecol 3:3

Paul S, Lal G (2017) The molecular mechanism of natural killer cells function and its importance in cancer immunotherapy. Front Immunol 8:1124

Peguero JG, Presti SL, Perez J, Issa O, Brenes JC, Tolentino A (2017) Electrocardiographic criteria for the diagnosis of left ventricular hypertrophy. J Am Coll Cardiol 69:1694–1703

Pendl H, Tizard I (2016) Immunology. In: Speer BL (ed) Current therapy in avian medicine and surgery. Elsevier, St. Louis, MO, pp 400–432

Pennycuick CJ (1998) Computer simulation of fat and muscle burn in long-distance bird migration. J Theor Biol 191:47–61

Perni S, Iyer VR, Franzini-Armstrong C (2012) Ultrastructure of cardiac muscle in reptiles and birds: optimizing and/or reducing the probability of transmission between calcium release units. J Muscle Res Cell Motil 33:145–152

Peters GW, Steiner DA, Rigoni JA, Mascilli AD, Schnepp RW, Thomas SP (2005) Cardiorespiratory adjustments of homing pigeons to steady wind tunnel flight. J Exp Biol 208:3109–3120

Petit M, Clavijo-Baquet S, Vézina F (2017) Increasing winter maximal metabolic rate improves intrawinter survival in small birds. Physiol Biochem Zool 90: 166–177

Piersma T (1997) Do global patterns of habitat use and migration strategies co-evolve with relative investments in immunocompetence due to spatial variation in parasite pressure? Oikos 80:623–631

Piersma T, Gudmundsson GA, Lilliendahl K (1999) Rapid changes in the size of different functional organ and muscle groups during refueling in a long-distance migrating shorebird. Physiol Biochem Zool 72:405–415

Piersma T, van der Velde M (2012) Dutch House Martins *Delichon urbicum* gain blood parasite infections over their lifetime, but do not seem to suffer. J Ornithol 153: 907–912

Piersma T, Everaarts JM, Jukema J (1996) Build-up of red blood cells in refueling Bar-tailed Godwits in relation to individual migratory quality. Condor 98:363–370

Ponganis PJ (2015) Diving physiology of marine mammals and seabirds. Cambridge University Press, Cambridge, UK

Ponganis PJ, Stockard TK, Meir JU, Williams CL, Ponganis KV, Van Dam RP, Howard R (2007) Returning on empty: extreme blood $O_2$ depletion underlies dive capacity of Emperor Penguins. J Exp Biol 210:4279–4285

Ponganis PJ, Meir JU, Williams CL (2010) Oxygen store depletion and the aerobic dive limit in Emperor Penguins. Aquat Biol 8:237–245

Potti J (2007) Variation in the hematocrit of a passerine bird across life stages is mainly of environmental origin. J Avian Biol 38:726–730

Potti J, Moreno J, Merino S, Frías O, Rodríguez R (1999) Environmental and genetic variation in the haematocrit of fledgling Pied Flycatchers *Ficedula hypoleuca*. Oecologia 120:1–8

Powell C, Lill A, Johnstone CP (2013) Body condition and chronic stress in urban and rural Noisy Miners. Open Ornithol J 6:25–31

Pretorius E, Vieira WA, Oberholzer HM, Auer REJ (2009) Comparative scanning electron microscopy of platelets and fibrin networks of human and different animals. Int J Morphol 27:69–76

Prosheva VI, Kaseva NN (2016) Location and functional characterization of the right atrioventricular pacemaker ring in the adult avian heart. J Morphol 277:363–369

Protasi F, Sun X-H, Franzini-Armstrong C (1996) Formation and maturation of the calcium release apparatus in developing and adult avian myocardium. Dev Biol 173:265–278

Pryke SR, Rollins LA (2012) Mothers adjust offspring sex to match the quality of the rearing environment. Proc R Soc B 279:4051–4057

Pryke SR, Rollins LA, Griffith SC (2011) Context-dependent sex allocation: constraints on the expression and evolution of maternal effects. Evolution 65:2792–2799

Pryke SR, Astheimer LB, Griffith SC, Buttemer WA (2012) Covariation in life-history traits: differential effects of diet on condition, hormones, behavior, and reproduction in genetic finch morphs. Am Nat 179: 375–390

Qureshi MA, Heggen CL, Hussain I (2000) Avian macrophage: effector functions in health and disease. Dev Comp Immunol 24:103–119

Ramírez-Plascencia O, Acosta-Hernández ME, García-García F (2019) The role of neuroglobin in brain function and sleep–wake cycle. In: Murillo-Rodriguez E (ed) The behavioral, molecular, pharmacological, and clinical basis of the sleep-wake cycle. Academic Press, New York, NY, pp 85–102

Ramiro AR, Nussenzweig MC (2004) Aid for AID. Nature 430:980–981

Ramsis MN, El-Gammal SM, Abo-El-Sooud K, Swielim GA (2021) Anatomical studies on the arterial blood supply of the pelvic limb of geese. Adv Anim Vet Sci 9:604–614

Ratcliffe MJH, Härtle S (2014) B cells, the bursa of Fabricius and the generation of antibody repertoires. In: Schat KA, Kaspers B, Kaiser P (eds) Avian immunology. Academic Press, Paris, France, pp 65–89

Rehman ZU, Meng C, Umar S, Mahrose KM, Ding C, Munir M (2017) Mast cells and innate immunity: master troupes of the avian immune system. Worlds Poult Sci J 73:621–632

Resano-Mayor J, Hernández-Matías A, Real J, Parés F, Moleón M, Mateo R, Ortiz-Santaliestra ME (2016) The influence of diet on nestling body condition of an apex predator: a multi-biomarker approach. J Comp Physiol B 186:343–362

Ricklefs RE (1992) Embryonic development period and the prevalence of avian blood parasites. Proc Natl Acad Sci USA 89:4722–4725

Rodman GP, Ebaugh FG Jr, Spivey Fox MR, Chambers DM (1957) The life span of the red blood cell and the red blood cell volume in the chicken, pigeon and duck as estimated by the use of $Na_2Cr_{51}O_4$. Blood 12:355–366

Romero LM, Gormally BMG (2019) How truly conserved is the "well-conserved" vertebrate stress response? Integr Comp Biol 59:273–281

Rothenberg ME, Hogan SP (2006) The eosinophil. Annu Rev Immunol 24:147–174

Ruiz G, Rosenmann M, Fernando Novoa F, Sabat P (2002) Hematological parameters and stress index in Rufous-collared Sparrows dwelling in urban environments. Condor 104:162–166

Sakaluk SK, Wilson AJ, Bowers EK, Johnson LS, Masters BS, Johnson BGP, Vogel LA, Forsman AM, Thompson CF (2014) Genetic and environmental variation in condition, cutaneous immunity, and haematocrit in House Wrens. BMC Evol Biol 14:242

Santhakumar D, Rubbenstroth D, Martinez-Sobrido L, Munir M (2017) Avian interferons and their antiviral effectors. Front Immunol 8:49

Santiago-Alarcon D, Palinauskas V, Schaefer HM (2012) Diptera vectors of avian Haemosporidian parasites: untangling parasite life cycles and their taxonomy. Biol Rev 87:928–964

Scanes CG (2015) Blood. In: Sturkie's avian physiology, 6th edn. Elsevier, New York

Schepelmann K (2005) Erythropoietic bone marrow in the pigeon: development of its distribution and volume during growth and pneumatization of bones. J Morphol 203:21–34

Scheuerlein A, Ricklefs RE (2004) Prevalence of blood parasites in European passeriform birds. Proc R Soc B 271:1363–1370

Schoenle LA, Kernbach M, Haussmann MF, Bonier F, Moore IT (2017) An experimental test of the physiological consequences of avian malaria infection. J Anim Ecol 86:1483–1496

Scholander PF (1940) Experimental investigations on the respiratory function in diving mammals and birds. Hvalradets Skrifter 22:1–131

Scott MG, Hancock RE (2000) Cationic antimicrobial peptides and their multifunctional role in the immune system. Crit Rev Immunol 20:407–431

Scoville SA, Doherty JP (2017) Baseline hematocrit values in birds: an important tool for monitoring exposure to polycyclic aromatic hydrocarbons. Environ Pollut Protect 2:145–152

Sedlmayr JC, Witmer LM (2002) Rapid technique for imaging the blood vascular system using stereoangiography. Anat Rec 267:330–336

Seymour RS, Blaylock AJ (2000) The principle of Laplace and scaling of ventricular wall stress and blood pressure in mammals and birds. Physiol Chem Zool 73: 389–405

Sheridan JA, Beissinger SR, Hughes CR (2004) Weak association between measures of health and reproductive success in Green-rumped Parrotlets (*Forpus passerinus*) in Venezuela. Auk 121:717–725

Shiels HA (2022) Avian cardiomyocyte architecture and what it reveals about the evolution of the vertebrate heart. Philos Trans R Soc B 377:20210332

Shiels HA, Galli GLJ (2014) The sarcoplasmic reticulum and the evolution of the vertebrate heart. Physiology 29:456–469

Signore AV, Tift MS, Hoffmann FG, Schmitt TL, Moriyama H, Storz JF (2021) Evolved increases in hemoglobin-oxygen affinity and Bohr effect coincided with the aquatic specialization of penguins. Proc Natl Acad Sci USA 118:e2023936118

Sirkiä PM, Qvarnström A (2021) Adaptive coloration in Pied Flycatchers (*Ficedula hypoleuca*)—the devil is in the detail. Ecol Evol 11:1501–1525

Smith FM (1994) Blood pressure regulation by aortic baroreceptors in birds. Physiol Zool 67:1402–1425

Smith KD, Barber CA (2012) Hematocrit does not indicate condition in nestling or adult European Starlings. Wilson J Ornithol 124:788–792

Smith KG, Hunt JL (2004) On the use of spleen mass as a measure of avian immune system strength. Oecologia 138:28–31

Smith FM, West NH, Jones DR (2000) The cardiovascular system. In: Whittow GC (ed) Sturkie's avian physiology, 5th edn. Academic Press, San Diego, CA, pp 41–231

Snyder GK, Sheafor BA (1999) Red blood cells: centerpiece in the evolution of the vertebrate circulatory system. Am Zool 39:189–198

Sol D, Jovani R, Torres J (2003) Parasite mediated mortality and host immune response explain age-related differences in blood parasitism in birds. Oecologia 135(542):547

Šolc D (2007) The heart and heart conducting system in the kingdom of animals: a comparative approach to its evolution. Exp Clin Cardiol 12:113

Sommer JR (1995) Comparative anatomy: in praise of a powerful approach to elucidate mechanisms translating cardiac excitation into purposeful contraction. J Mol Cell Cardiol 27:19–35

Sommer JR, Johnson EA (1969) Cardiac muscle: a comparative ultrastructural study with special references to frog and chicken hearts. Zeit Zellforsch Micro Anat 98:437–468

Soulsbury CD, Dobson J, Deeming DC, Minias P (2022) Energetic lifestyle drives size and shape of avian erythroctyes. Integr Comp Biol 62:71–80

Spronk HMH, Govers-Riemslag JWP, ten Cate H (2003) The blood coagulation system as a molecular machine. BioEssays 25:1220–1228

Stephenson R, Evans BK, Jones DR (1996) Physiological mechanisms for underwater endurance: Canada Goose (*Branta canadensis*) versus Pekin Duck (*Anas platyrhynchos*). J Comp Physiol B 166:46–54

Stephenson A, Adams JW, Vaccarezza M (2017) The vertebrate heart: an evolutionary perspective. J Anat 231:787–797

Stockard TK, Heil J, Meir JU, Sato K, Ponganis KV, Ponganis PJ (2005) Air sac $PO_2$ and oxygen depletion during dives of Emperor Penguins. J Exp Biol 208: 2973–2980

Storz JF (2007) Hemoglobin function and physiological adaptation to hypoxia in high-altitude mammals. J Mammal 88:24–31

Sutcliffe JF (1986) Black fly host location: a review. Can J Zool 64:1041–1053

Svoboda O, Bartunek P (2015) Origins of the vertebrate erythro/megakaryocytic system. Biomed Res Int 2015: 632171

Swanson DL (1990) Seasonal variation of vascular oxygen transport in the Dark-eyed Junco. Condor 92:62–66

Swanson DL, Vézina F (2015) Environmental, ecological and mechanistic drivers of avian seasonal metabolic flexibility in response to cold winters. J Ornithol 156: 377–388

Tadjalli M, Ghazi SR, Parto P (2009) Gross anatomy of the heart in Ostrich (*Struthio camelus*). Iran J Vet Res 10: 21–27

Taha AA, Abdel-Magied EM, King AS (1983) Ultrastructure of aortic and pulmonary baroreceptors in the Domestic Fowl. J Anat 137:197–207

Takekawa JY, Heath SR, Douglas DC, Perry WM, Javed S, Newman SH, Suwal RN, Rahmani AR, Choudhury BC, Prosser DJ, Yan B, Hou Y, Batbayar N, Natsagdorj T, Bishop CM, Butler PJ, Frappell PB, Milsom WK, Scott GR, Hawkes LA, Wikelski M (2009) Geographic variation in Bar-headed Geese *Anser indicus*: connectivity of wintering areas and breeding grounds across a broad front. Wildfowl 59:100–123

Tavernier P, Sagesse M, Van Wettere A, Redig P (2005) Malaria in an Eastern Screech Owl (*Otus asio*). Avian Dis 49:433–435

Tizard I (2002) The avian antibody response. Semin Avian Exotic Pet Med 11:2–14

Townsend AK, Wheeler SS, Freund D, Sehgal RNM, Boyce WM (2018) Links between blood parasites, blood chemistry, and the survival of nestling American Crows. Ecol Evol 8:8779–8790

Trinchieri G (2003) Interleukin-12 and the regulation of innate resistance and adaptive immunity. Nat Rev Immunol 3:133–146

Valkiūnas G (2005) Avian malaria parasites and other Haemosporidia. CRC Press, Boca Raton, FL

Valkiūnas G, Iezhova TA (2018) Keys to the avian malaria parasites. Malar J 17:212

van Oers K, Richardson DS, Sæther SA, Komdeur J (2010) Reduced blood parasite prevalence with age in the Seychelles Warbler: selective mortality or suppression of infection? J Ornithol 151:69–77

Vervelde L, Matthijs MGR, Van Haarlem DA, de Wit JJ, Jansen CA (2013) Rapid NK-cell activation in chicken after infection with infectious bronchitis virus M41. Vet Immunol Immunopathol 151:337–341

Vinkler M, Schnitzer J, Munclinger P, Votýpka J, Albrecht T (2010) Haematological health assessment in a passerine with extremely high proportion of basophils in peripheral blood. J Ornithol 151:841–849

Wagner EC, Stables CA, Williams TD (2008) Hematological changes associated with egg production: direct evidence for changes in erythropoiesis but a lack of resource dependence? J Exp Biol 211:2960–2968

Warr GW, Magor KE, Higgins DA (1995) IgY: clues to the origins of modern antibodies. Immunol Tod 16: 392–398

Warrington R, Watson W, Kim HL, Antonetti FR (2011) An introduction to immunology and immunopathology. Allergy Asthma Clin Immunol 7:S1

Weber RE, Jessen T-H, Malte H, Tame J (1993) Mutant hemoglobins alpha 119-Ala and beta 55-Ser: functions related to high-altitude respiration in geese. J Appl Physiol 75:2646–2655

West NH, Langille BL, Jones DR (1981) Cardiovascular system. In: King AS, McLelland J (eds) Form and function in birds. Academic Press, New York, pp 235–339

Westerdahl H, Waldenstrom J, Hansson B, Hasselquist D, von Schantz T, Bensch S (2005) Associations between malaria and MHC genes in a migratory songbird. Proc R Soc B 272:1511–1518

Westneat DF, Birkhead TR (1998) Alternative hypotheses linking the immune system and mate choice for good genes. Proc R Soc B 265:1065–1073

Wigley P, Kaiser P (2003) Avian cytokines in health and disease. Braz J Poult Sci 5:1–14

Wigley P, Hulme SD, Barrow PA (1999) Phagocytic and oxidative burst activity of chicken thrombocytes to *Salmonella*, *Escherichia coli* and other bacteria. Avian Pathol 28:567–572

Williams TD (2012) Physiological adaptations for breeding in birds. Princeton Univ Press, Princeton, NJ

Williams TD, Challenger WO, Christians JK, Evanson M, Love O, Vezina F (2004) What causes the decrease in haematocrit during egg production? Funct Ecol 18: 330–336

Wong AD, Ye M, Levy AF, Rothstein JD, Bergles DE, Searson PC (2013) The blood-brain barrier: an engineering perspective. Front Neuroeng 6:7

Woodbury RA, Hamilton WF (1937) Blood pressure studies in small animals. Am J Physiol 119:663–674

Wright TJ (2014) Myoglobin adaptation in terrestrial and diving birds and mammals. Ph.D. dissertation, Texas A&M University, College Station, TX

Wright AK, Ponganis KV, McDonald BI, Ponganis PJ (2014) Heart rates of Emperor Penguins diving at sea: implications for oxygen store management. Mar Ecol Prog Ser 496:85–98

Wu L, Sun Y, Li M, Li Y, Yao Y, Liu X, Hao Y, Li D, Wu Y (2016) Molecular cloning and 3D structure prediction of myoglobin and cytoglobin in Eurasian Tree Sparrow *Passer montanus*. J Ornithol 157:493–504

Wynn TA, Chawla A, Pollard JW (2013) Macrophage biology in development, homeostasis and disease. Nature 496:445–455

Yamamoto M, Kato A, Ropert-Coudert Y, Kuwahara M, Hayama S, Naito Y (2009) Evidence of dominant parasympathetic nervous activity of Great Cormorants (*Phalacrocorax carbo*). J Comp Physiol A 195:365–373

Yang D, Wan Y (2019) Molecular determinants for the polarization of macrophage and osteoclast. Semin Immunopathol 41:551–563

Yap KN, Tsai OHI, Williams TD (2019) Haematological traits co-vary with migratory status, altitude and energy expenditure: a phylogenetic, comparative analysis. Sci Rep 9:6351

Yatime L, Bajic G, Schatz-Jakobsen JA, Andersen GR (2016) Complement regulators and inhibitors in health and disease: a structural perspective. In: Howard KA, Vorup-Jensen T, Peer D (eds) Nanomedicine. Springer, New York, pp 13–42

Yogamaya B, Prafulla KM (2016) Haematology of Grey Heron (*Ardea cinerea*) and Black Crowned Night Heron (*Nycticorax nycticorax*) oi Chilika Wetland. Indian J Biol 3:151–156

Zaefarian F, Abdollahi MR, Cowieson A, Ravindran V (2019) Avian liver: the forgotten organ. Animals 9:63

Zepeda Mendoza ML, Roggenbuck M, Vargas KM, Hansen LH, Brunak S, Gilbert MTP, Sicheritz-Pontén T (2018) Protective role of the vulture facial skin and gut microbiomes aid adaptation to scavenging. Acta Vet Scand 60:61

Zhang X, Calvert RA, Sutton BJ, Doré KA (2017) IgY: a key isotype in antibody evolution. Biol Rev 92:2144–2156

# Respiration

# 7

## Contents



**Abstract**

Birds have high metabolic rates and so require an efficient means of delivering oxygen to their cells and eliminating carbon dioxide, and the avian respiratory system is exceptionally efficient. In this chapter, I describe in detail the anatomy of the avian respiratory system, from the nostrils and oral cavity, past the pharynx and larynx, trachea, and bronchi and to the lungs and air sacs. The process of ventilation, the mechanism by which air moves into, through, and out of the respiratory system, is described. The metabolic cost of breathing and how breathing is altered when birds walk, run, and fly are explained. The role of air sacs in the unidirectional flow of air through the lungs is described, as are other features of the avian lungs that make the exchange of gases between blood and air capillaries so efficient. Finally, the process by which the avian nervous system controls respiration is explained.

## 7.1 Introduction

Although life on earth began in an anaerobic environment, most organisms alive today are


G. Ritchison, *In a Class of Their Own*, Fascinating Life Sciences,
https://doi.org/10.1007/978-3-031-14852-1_7

dependent on a constant supply of oxygen. This "switch" was critical because aerobic respiration is far more efficient than anaerobic respiration. This enhanced efficiency was critical to the evolution of more complex and more active organisms, and birds are certainly both complex and active. The process whereby oxygen is delivered to and used by cells is called respiration. In other words, respiration really involves two processes: the exchange of oxygen (and carbon dioxide) between an organism and its external environment, and the use of oxygen (and production of carbon dioxide) by body cells. The former process is called external respiration; the latter, internal respiration. The focus of this chapter will be external respiration; the process whereby birds exchange oxygen and carbon dioxide (a waste product of ATP production by cells) with the external environment.

An organism's metabolic rate, the rate of transformation of energy and materials within an organism, is fueled by oxygen, and those with higher metabolic rates require more oxygen. Endotherms, of course, have higher metabolic rates than ectotherms. Metabolic rates of mammals and birds measured in the field are about 12 and 20 times higher, respectively, than those of similar-sized, but ectothermic, reptiles and birds (Nagy 2005). The high metabolic rates of birds, coupled with, for most birds, the energetic demands of flight require efficient delivery of oxygen to body cells, and birds have the most efficient, and most complex, respiratory system of any air-breathing vertebrate (Maina 1998). The oxygen extraction efficiency ($EO_2$) of avian lungs is greater than that of the lungs of other terrestrial vertebrates, with $EO_2$s for some birds as high as 60–70% (Brent et al. 1984) compared to typical values of about 20–30% for nonflying mammals (e.g., Larcombe 2002) and 35–45% for bats (Chappell and Roverud 1990; Thomas et al. 1995).

Vertebrate respiratory systems must have two components: a gas exchanger and a pump. The exchanger is where gases diffuse between the respiratory system and the blood, and the pump is the mechanism for moving air into and out of the exchanger. In most vertebrates, the lungs are where exchange takes place and, although pumping mechanisms vary, all involve the contraction of muscles. The respiratory systems of most terrestrial vertebrates other than birds have the same basic design, with atmospheric air moving through tubes into lungs that, to varying degrees, are subdivided into small terminal gas exchange units. The lungs of reptiles generally have more such units than those of amphibians, and the lungs of mammals more than those of reptiles. More gas exchange units mean more surface area and more efficient exchange of gases between the lungs and the blood. In human lungs, for example, the combined surface area of all small gas exchange units (alveoli) is about 150 $m^3$, a surface area about the size of a tennis court (Gehr et al. 1978). Although impressive, for birds and mammals of similar size, there is even more surface area available for gas exchange in avian lungs. An important limitation of the lungs of mammals is that gas exchange takes place in terminal sac-like structures called alveoli (Fig. 7.1)—air moves into and out of these sacs during each respiratory cycle. Smaller sacs would provide more surface area, but surface tension places a limit on how small these sacs can be.

Very small sacs coated with thin films of water are susceptible to collapse and more difficult to inflate because of surface tension. Cells called Type II cells in the lungs do produce a substance (surfactant) that reduces surface tension, but there is, nonetheless, a limit to how small these terminal sacs can be. Among mammals, the smallest reported alveoli (about 30 μm) are found in bats (Tenney and Remmers 1963). One reason why avian respiration is so efficient is that gas exchange takes place between air capillaries, i.e., "rather rotund interconnected structures" (Maina 2015a) and blood capillaries rather than between terminal sac-like structures and blood capillaries. The diameter of the air capillaries

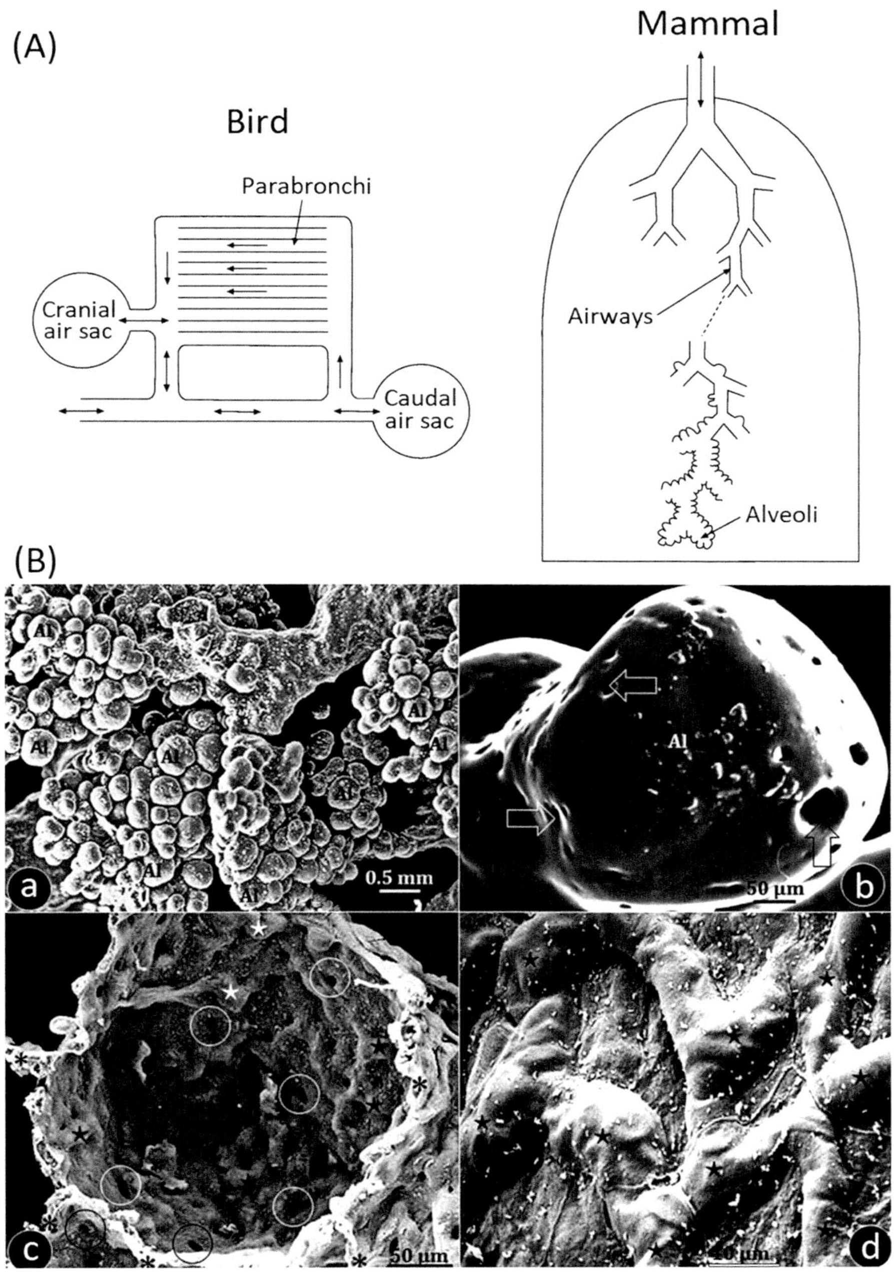


**Fig. 7.1** (**A**) Bird and mammal lungs, showing the flow-through arrangement for the gas-exchanging tissue of birds, and the reciprocating, or back-and-forth, pattern of mammals. (**B**) Latex cast of structures in the lung of a vervet monkey (*Cercopithecus aethiops*) showing (a) a cluster of alveoli (Al). (b) An alveolus. Arrows indicate interalveolar pores. (c) An alveolus with its surface lined by blood capillaries (stars). Circles indicate interalveolar pores. (d) Close-up of blood capillaries on the surface of an alveolus (stars). (Figure **A** from West et al. 2007a, b; Reproduced with permission of the © ERS 2020: European Respiratory Journal 29 (1) 11–17. DOI: 10.1183/09031936.00133306 Published 29 December 2006. Figure **B** from Maina 2015b; © 2015 Wiley Periodicals, Inc., used with permission)

does not change during the respiratory cycle and, therefore, surface tension is not as important. As a result, air capillaries can be very small in diameter (3–10 μm; Duncker 1972; Abdalla 1989; Maina and Woodward 2009) and that translates into more surface area for gas exchange. There is, of course, more to efficient respiration than surface area, and understanding how and why the avian respiratory system is so efficient first requires some knowledge of its anatomy.

Before providing the details, a quick overview may be helpful. In addition to the small diameter of the air capillaries where the exchange of gases takes place in the avian lung, another important feature of the bird respiratory system is that gas exchange and ventilation (movement of air into, through, and out of the respiratory system) are handled independently. In other words, mammal lungs serve to both exchange gases and ventilate; the lungs must expand and contract during each respiratory cycle to ensure the movement of gases into and out of the alveoli where gases are exchanged. In birds, thin-walled air sacs that extend from the lungs act as bellows to move air through the lungs (Powell and Hopkins 2004). Air moves into and out of the air sacs as they expand and contract and, because of their locations relative to the lungs, air simultaneously moves through the lungs and, specifically, through the air sacs. Air moves constantly and unidirectionally through the avian lung. This unique system ensures that there is no "dead air" in the air capillaries (as there is in mammalian alveoli) and so the oxygen content is higher. This, combined with the tremendous amount of surface area provided by all the air capillaries and their very thin walls, translates into the most efficient respiratory system known, with oxygen diffusing rapidly and efficiently into the blood capillaries that lay adjacent to the air capillaries. A more detailed look at the avian respiratory system will reveal more clearly how and why air moves through the system as it does and how gas exchange can be so efficient.

## 7.2 The Avian Respiratory System

Most birds, like mammals, can breathe through either their nares or mouth. Nostrils are typically located at the base of the beak and vary among taxa in shape and size (Figs. 7.2 and 7.3). Little is known about the possible importance of such variation, but, among scavengers (New and Old World vultures and caracaras), those more dependent on the use of odors to locate carcasses, i.e., Turkey Vultures (*Cathartes aura*) and Greater (*C. melambrotus*) and Lesser (*C. burrovianus*) Yellow-headed vultures, have larger nostrils than other scavengers that depend more on sight (Guangdi et al. 2015). Exceptionally, kiwis (*Apteryx* spp.) have nostrils located at the tip of their beak, allowing them to detect earthworms and other invertebrates by olfaction as they probe the soil with their beak. In addition, the nostrils of jacanas are located closer to the bill tip than those of most other species (Fig. 7.4) and, when approached by a human, a young Wattled Jacana (*Jacana jacana*) was found "... completely immobile while entirely submerged with just the beak [and nostrils] protruding vertically above the water" (Bosque and Herrera 1999). These authors suggested that the location of the nostrils of jacanas is an anatomical adaptation that allows "snorkeling" by young jacanas to avoid predation and aggressive conspecifics. In several species of birds (in 16 different orders, Persons and Currie 2015), special hair-like feathers (bristles) lie over the nares (Fig. 7.5). These bristles help prevent small particles, such as those caused by excavation of wood by woodpeckers or the capture of flying insects by flycatchers, from entering the nasal cavities. In some aquatic birds, including gannets, boobies, pelicans, and some penguins, the nostrils are permanently closed by the growth of cornified cells (King and McLelland 1984; Figs. 7.6 and 7.7). This capping of the nostrils ensures that water cannot enter the nasal passages when diving or swimming underwater. To aid in respiration, however, MacDonald (1960)

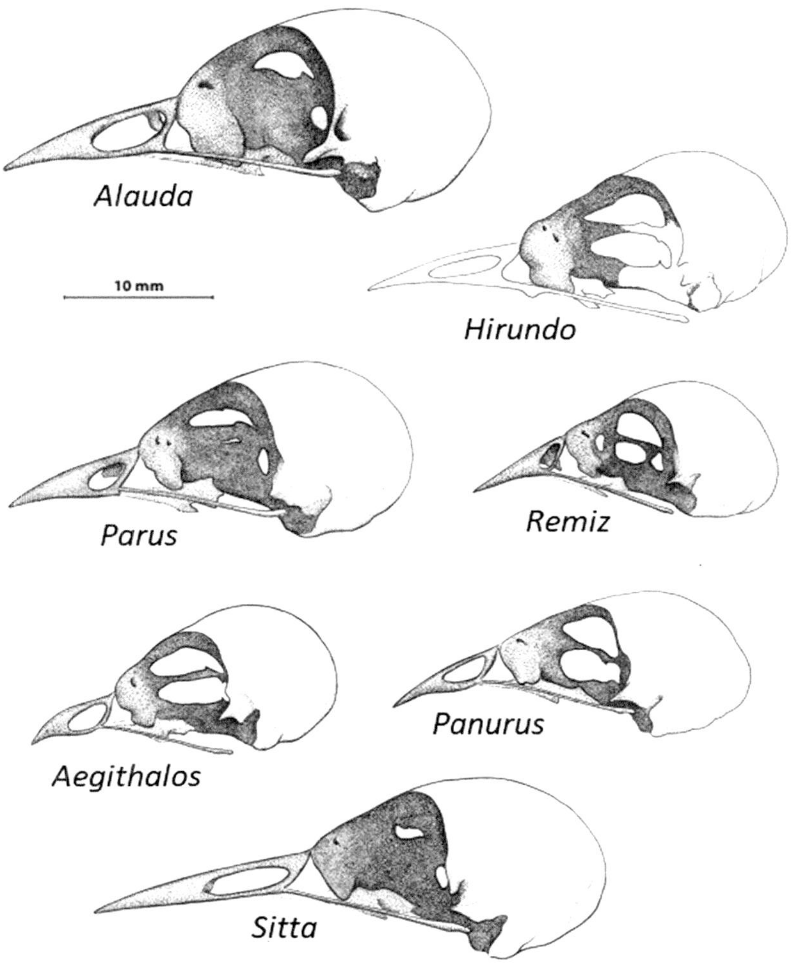


**Fig. 7.2** Variation in the shape and size of nostrils in seven genera of songbirds. (Figure from Ujhelyi 2016; published in Ornis Hungarica, an open-access journal, CC BY 4.0, https://creativecommons.org/licenses/by/4.0/)

discovered a moveable flap of rhamphotheca covering the upper jaw at the posterior corner of the mouth of gannets and boobies. This flap, called the jugal operculum, projects laterally to create a gap between the upper and lower jaws at the corner of the mouth that appears to serve as secondary oral external nares. Of course, breathing via the oral cavity means that inhaled air bypasses the olfactory epithelium in the nasal cavity, potentially depriving birds of the olfactory sense. These birds may, however, retain some sense of smell because air entering the oral cavity can pass through the internal nares (also called choana) and into the nasal cavity where the olfactory epithelium is located (Fig. 7.8).

Falcons have a central bony tubercle in their circular nostrils that is an extension of the septum of the nasal cavity. The function of this tubercle remains to be determined, but preliminary experiments suggest that it enhances air flow into the nasal passageways (Kesel et al. 2017, 2019; Fig. 7.9). Finally, the nostrils of some seabirds (order Procellariiformes) are at the end of "tubes" and these birds are sometimes called "tubenoses" (Box 7.1 Procellariiform Birds: The "Tube Knows" Air Speed?).

**Fig. 7.3** Nostrils of four raptors and two vultures. (**a**) Bald Eagle, *Haliaaetus leucocephalus*. (**b**) Aplomado Falcon, *Falco femoralis*. (**c**) Bearded Vulture, *Gypaetus barbatus* (note that the nostrils are covered by feathers). (**d**) Yellow-billed Kite, *Milvus aegyptius*. (**e**) Eastern Imperial Eagle, *Aquila heliaca*. (**f**) Turkey Vulture, *Cathartes aura*, with nostrils that are not divided by a septum. (Figure from Potier 2020; open-access article distributed under the terms of the Creative Commons CC BY license, https://creativecommons.org/licenses/by/4.0/)

**Fig. 7.4** The nostrils of jacanas, including this African Jacana (*Actophilornis africanus*) chick, are located closer to the tip of the bill than in many other species of birds and may be an adaptation to allow "snorkeling" by young jacanas as a way to avoid predation. (Photo from Wikipedia, CC BY 2.0, https://creativecommons.org/licenses/by/2.0/)

**Fig. 7.5** Bristles extend over the nostrils in several species of birds, including those in the family Corvidae like this Common Raven (*Corvus corax*), and likely function to prevent small particles from entering the nostrils. (Photo from pxhere.com, CC0 Public Domain)

**Fig. 7.6** The nostrils (left nostril location indicated by the arrow) of some aquatic birds, like this Northern Gannet (*Morus bassanus*), are covered by cornified cells. (Photo from pxhere.com, CC0 Public Domain)

**Fig. 7.7** The nostrils of some penguins, like the (**a**) Yellow-eyed Penguin (*Megadyptes antipodes*), are capped, whereas other penguins, including (**b**) Little Penguins (Eudyptula minor), have external nostrils (circled). Interestingly, only the species of penguins that nest in either burrows or crevices (*Spheniscus* spp. and Little Penguins) have external nostrils (Schreiber and Burger 2002; Bertelli and Giannini 2005). One possible explanation for this is that selection has favored external nostrils among species nesting in confined spaces, especially burrows, because oxygen levels might be lower and carbon dioxide levels higher in such locations and external nostrils might allow more efficient movement of air into and out of the respiratory system. (Photo of Yellow-eyed Penguin from pxhere.com, CC0 Public Domain. Photo of Little Penguin from pxhere.com, CC0 Public Domain, https://creativecommons.org/publicdomain/zero/1.0/)

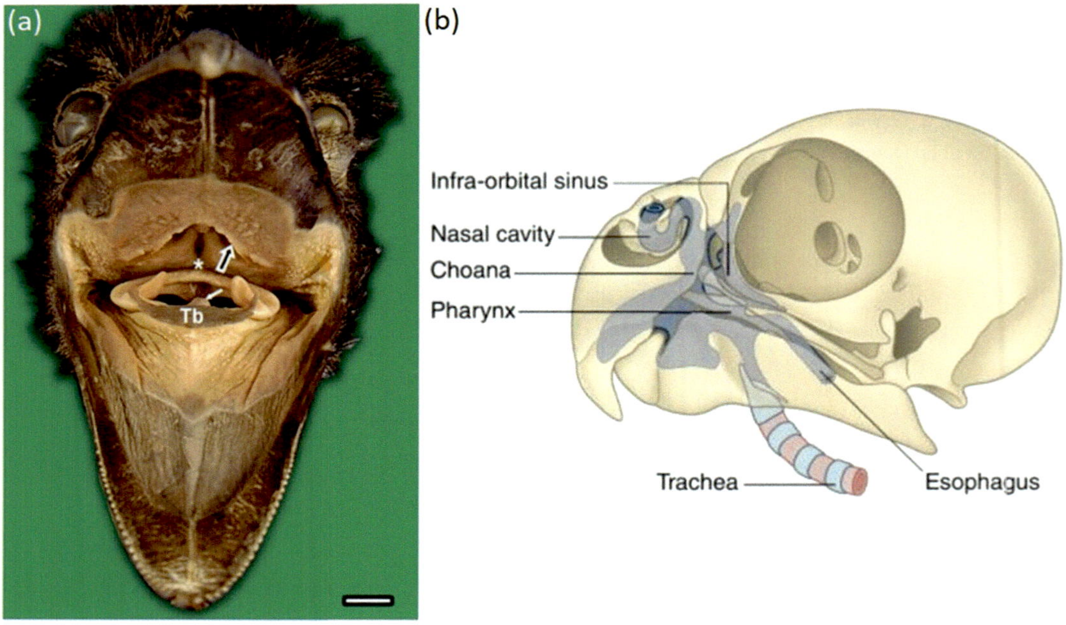


**Fig. 7.8** (**a**) Oral cavity of an Emu (*Dromaius novaehollandiae*) showing the tongue and internal nares (left internal naris is indicated by the arrow). Tb, body of tongue; *, tongue frenulum; white arrow, root of tongue. (**b**) For most birds, like Budgerigars (*Melopsittacus undulatus*) shown here, air entering the nasal passageways passes in the pharynx via the choana (internal nares) and then, from the pharynx, through the trachea and to the air sacs and lungs. Among birds with capped nostrils, some air can move from the oral cavity, through the internal nares, and into the nasal cavity where the olfactory epithelium is located. (Figure **a** from Crole and Soley 2010; © 2010 Blackwell Verlag GmbH, used with permission. Figure **b** from Evans 2016; © 2016 John Wiley & Sons, used with permission)

**Fig. 7.9** A Peregrine Falcon (*Falco peregrinus*) showing the tubercle in its nostril. This appears to act as a damper, reducing air velocity, increasing air pressure, and enhancing air flow into the nasal passageways. Air velocity increases and, therefore, air pressure outside of the nostrils declines during a high-speed stoop, making inhalation potentially more difficult (because, normally, air moves from an area of higher pressure outside of the nostrils to an area of lower pressure inside the avian respiratory system during inhalation). By reducing air velocity (and increasing pressure), the tubercles, therefore, may make it easier for a Peregrine Falcon to inhale during stoops (Kesel et al. 2017, 2019). If this hypothesis is eventually confirmed, the question of why all species of falcons have nostril tubercles, even though most do not attack prey using stoops like Peregrine Falcons, remains to be addressed. (Photo of Peregrine Falcon from pxhere.com, CC0 Public Domain, https://creativecommons.org/publicdomain/zero/1.0/)

**Box 7.1 Procellariiform Birds: The "Tube Knows" Air Speed?**

Albatrosses and smaller Procellariiform birds like petrels and shearwaters can travel long distances over the ocean by dynamic soaring (see Chap. 11). This method of soaring requires that these birds be able to detect variation in wind speed at various distances above the ocean surface. Pennycuick (2002) proposed that albatrosses and their relatives can use their tubenoses as a pitot tube to accurately determine air speed.

(continued)

**Box 7.1** (continued)

A "tubenosed" Northern Fulmar (*Fulmarus glacialis*). (Photo by Philipp Weigell, Wikipedia, CC BY 3.0, https://creativecommons.org/licenses/by/3.0/)

Pitot tubes on airplanes have two holes, one pointing forward (in the direction the plane is flying) to measure what is called the stagnation (or pitot) pressure, and a side hole that measures static pressure (the ambient pressure of the surrounding air). The difference between the stagnation pressure and static pressure is called the dynamic pressure, which can be used to determine a plane's air speed. For example, in the diagram below, as a plane increases its airspeed, the pressure generated at the stagnation point will increase relative to the static pressure and that increase will be detected by the pressure transducer. Pitot tubes are calibrated so that the pressure readings are translated into air speed readings.

(continued)

**Box 7.1** (continued)

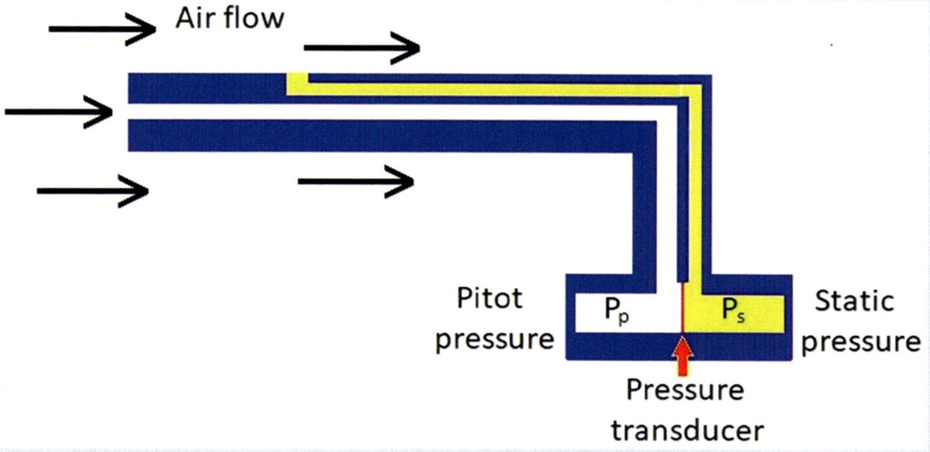


Pitot tubes have a front hole facing the air flow plus a side hole. The side hole measures static pressure, whereas the front hole measures the stagnation, or pitot, pressure. Differences in the two pressures are measured by a pressure transducer that converts the difference into air speed (shown in yellow). (Figure modified from NASA, CC0 Public Domain).

The tubenoses of albatrosses and other Procellariiform birds resemble pitot tubes and may function in the same way. The forward-pointing nostrils or tubenoses (pointing in the direction the birds fly), that could measure stagnation or pitot pressure, lead into nasal chambers also connected to the mouth, or oral, cavity, where pressure would correspond to static pressure. Mangold (1946) identified an expandable pocket or capsule on either side of the nasal septum of petrels and proposed that these were sense organs (comparable to the pressure transducers of pitot tubes) that could measure dynamic pressure, the difference between stagnation and static pressure, and provide the birds with information about their air speed. Information about air speed would be very useful because air speed often differs from ground speed, e.g., an albatross flying at an air speed at 15 km/h into a headwind of 14 km/h would have a ground speed of only 1 km/h. Additional study is needed to test Pennycuick's (2002) and Mangold's (1946) hypotheses.

(continued)

**Box 7.1** (continued)

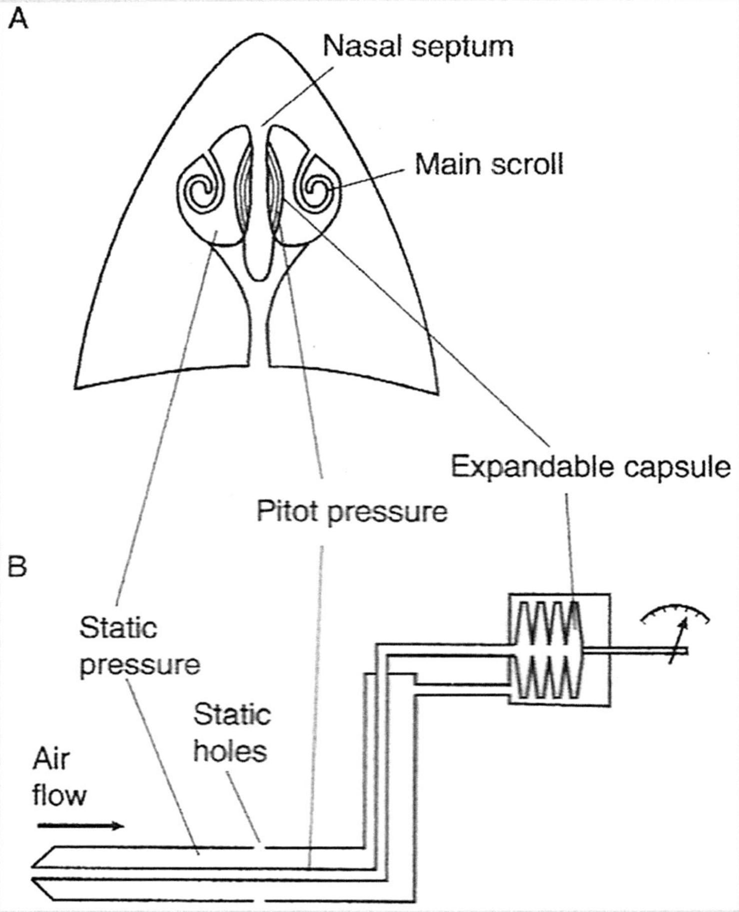


(**a**) Transverse section through a fulmar's skull showing nasal chambers with expandable pockets or capsules. The inside of the capsules is connected to the forward-pointing nostrils (possibly detecting stagnation or pitot pressure), whereas the chamber outside the capsules is connected to the oral or mouth cavity (possibly detecting static pressure). (**b**) Diagram of a typical pitot tube used to measure airspeed of airplanes. The expandable capsule registers the difference between the stagnation or pitot pressure coming from the forward-pointing tube and the static pressure coming from holes on the sides of the tube. (Figure from Pennycuick 2008; © 2008 Elsevier Inc., used with permission).

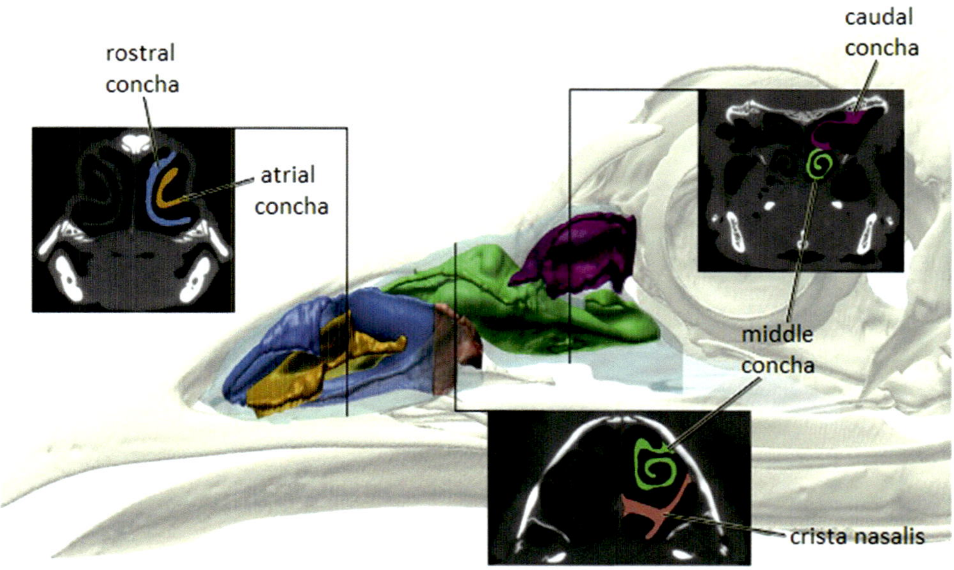


**Fig. 7.10** Cross sections showing the location of false-colored conchae and crista nasalis in the nasal capsule of a Wild Turkey (*Meleagris gallopavo*). (Figure modified from Bourke and Witmer 2016; © 2016 Elsevier B.V., used with permission)

### 7.2.1 Nasal Cavities

Upon inhalation, air passes into either or both the nasal cavities and mouth. The two nasal cavities of most birds are separated by a complete nasal septum. In others, the septum is either not present (e.g., Turkey Vultures, *Cathartes aura*) or perforated. These cavities are partitioned and subdivided to varying degrees by very thin walls of cartilage or bone. These walls are collectively referred to as the nasal conchae and serve to increase the exposed surface area with the nasal cavities (Figs. 7.10 and 7.11). During inhalation, air passing through the nasal cavities is warmed and humidified and the mucous membranes capture particles in the air (Fedde 1998). In addition, the surfaces of the caudal nasal conchae include olfactory receptors. The nasal cavities of birds with a well-developed sense of smell, such as many seabirds and Turkey Vultures (*Cathartes aura*), have more "partitions" and greater surface area with more olfactory receptors (Nevitt 1999; see Chap. 4 for more details).

From the nasal or oral cavities, air passes through the pharynx and larynx and into the trachea (Fig. 7.12). However, air can also enter and exit the respiratory system via the oral cavity and glottis. Importantly, the glottis must be closed when a bird swallows so that no food or water enters the trachea. In mammals, the larynx has an epiglottis that automatically covers the glottis when they swallow. The avian larynx, however, has no epiglottis. When Emus (*Dromaius novaehollandiae*) swallow, the lips of the glottis come together (closed by the intrinsic constrictor muscle, opened by the intrinsic dilatator muscle) and the tongue is retracted so that its root covers the rostral portion of the glottis. In combination, these events close the glottis (Crole and Soley 2012a; Fig. 7.13). The glottis of other species of birds may close in a similar manner (Fig. 7.14).

### 7.2.2 Larynx

The larynx of birds consists of a ring- or trough-shaped cricoid, an unpaired procricoid, and a pair

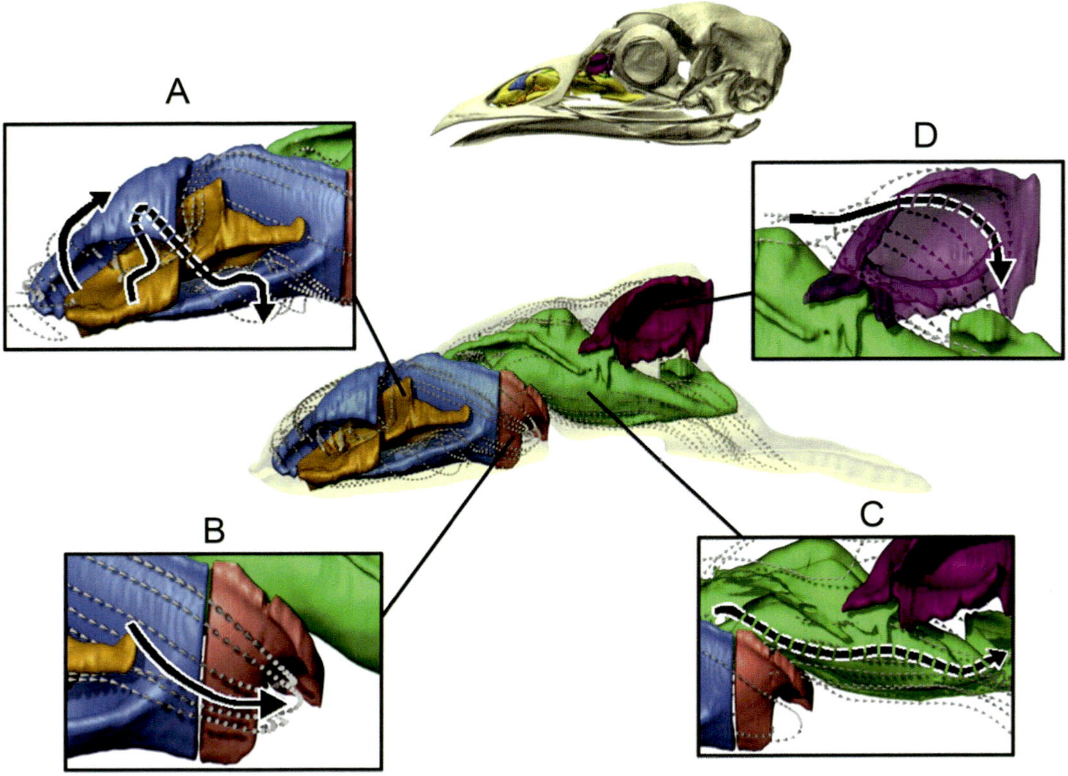


**Fig. 7.11** Airflow, as indicated by the black lines, through the nasal conchae of a Wild Turkey (*Meleagris gallopavo*). Dashed lines indicate air moving through the inside of conchae. (**a**) Air enters the nostril and the atrial concha splits the air field into a lateral channel and a longer medial channel. (**b**) Air moves caudally around the base of crista nasalis and into the middle nasal concha. (**c**) Airflow through the middle concha. (**d**) Airflow through the caudal concha where olfactory receptors are located. (Figure modified from Bourke and Witmer 2016; © 2016 Elsevier B.V., used with permission)

of arytenoids. The procricoid and arytenoids and the intrinsic laryngeal muscles exhibit little variation among birds, but the cricoid and the extrinsic laryngeal muscles do vary. Homberger (2017) placed the larynges of birds into three categories: (1) the akinetic larynx, the most common type (e.g., chickens, pigeons, and herons), consists of a trough-shaped cricoid (Fig. 7.15), (2) the psittaciform kinetic larynx has a flat ring-shaped cricoid (Fig. 7.16) and associated muscles that can change the shape and size of the laryngeal chamber, and (3) the passeriform kinetic larynx has a trough-shaped ventral cricoid forming the floor and sides of the laryngeal chamber and, as with the psittaciform larynx, associated muscles can change the shape and size of the laryngeal chamber. The ability of parrots and songbirds to change the shape of their larynges contributes to their ability to produce complex vocalizations and the resonating properties of their laryngeal chambers (Homberger 1999, 2017).

### 7.2.3 Trachea

The trachea is a tube supported by tracheal, or cartilaginous, rings through which air passes into and out of the avian respiratory system (Figs. 7.17 and 7.18). The number of cartilaginous rings (also called c-rings) varies with neck length, ranging

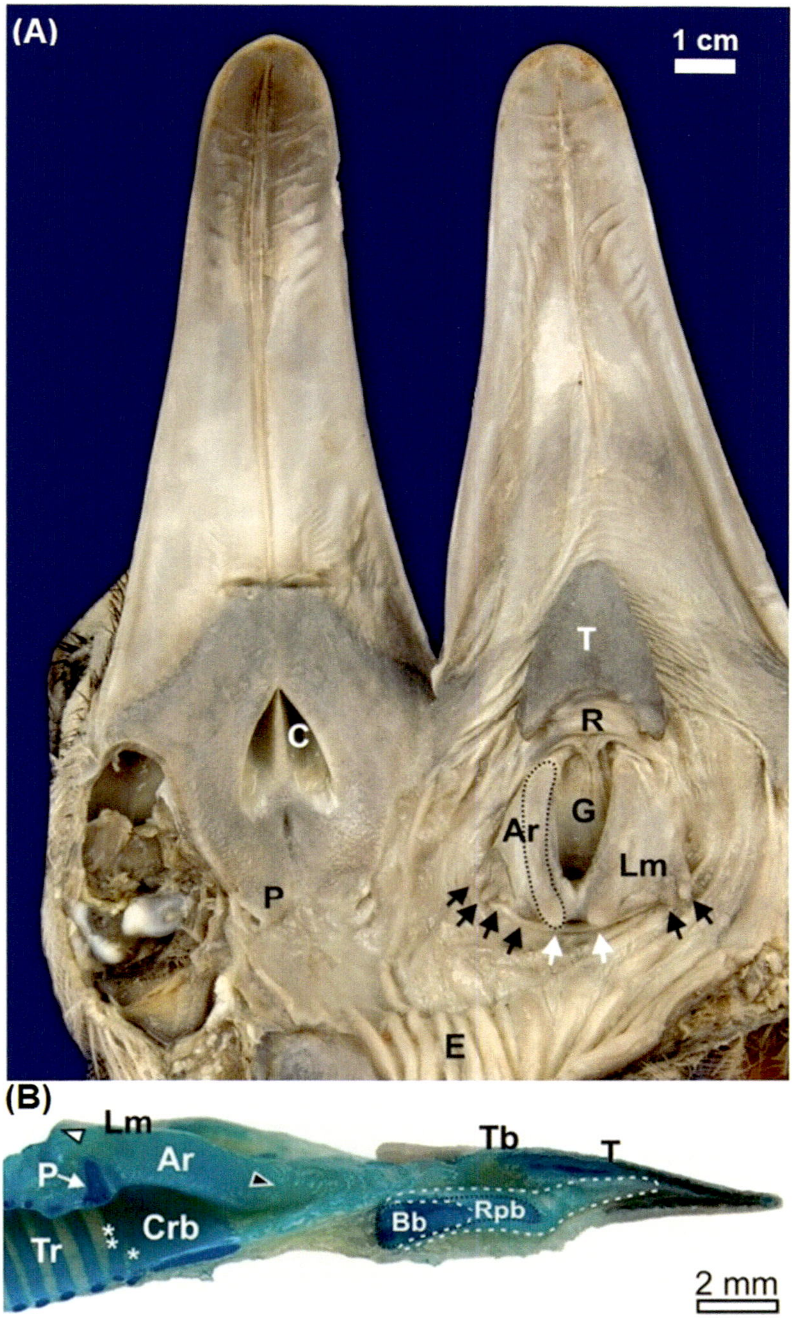


**Fig. 7.12** (**a**) Oral cavity and pharynx of a Greater Rhea (*Rhea americana*) showing the choana (C), tongue (T), tongue root (R), glottis (G), arytenoid cartilage (Ar), laryngeal mound (Lm), pharyngeal folds (P), and esophagus (E). Pharyngeal papillae are indicated by the black arrows, and medial pharyngeal papillae by the white arrows. (**b**) Longitudinal section of the tongue (T), larynx, and trachea (Tr) of a young Greater Rhea. Tb, tongue base; Rpb, rostral projection of the basihyal cartilage; Bb, body of the basihyal cartilage; Ar, arytenoid cartilage; Crb, body of the cricoid cartilage; Lm, largyngeal mound; P, procricoid cartilage. (Figures from Crole and Soley 2012b; © 2012 Springer-Verlag, used with permission)

from about 30 in small songbirds to about 350 in long-necked birds like cranes. The inner lining of the trachea consists of three types of cells: mucous (or goblet) cells, basal cells (progenitors of other tracheal epithelial cells), and ciliated cells (Fig. 7.19). Mucous cells produce mucous, a mixture of water (about 95%) and mucins (about 5%; a glycoprotein), adding moisture to the incoming

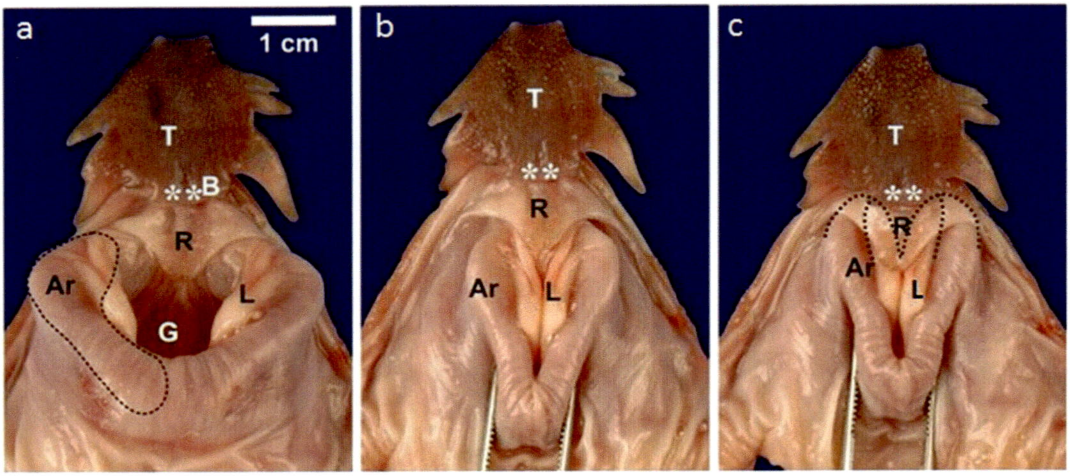


**Fig. 7.13** Closing of the glottis in an Emu (*Dromaius novaehollandiae*). **a**. Glottis (G), the entrance to the trachea, is open. The tongue body (T), base (B), and root (R), as well as caudal lingual papilla (white **), are in front of the open glottis. Structures of the larynx include the arytenoid cartilage (Ar, dotted outline) and the lips of the glottis (L). **b**. Glottis in a closed position with the lips together and the rostral portion of the glottis covered by the retracted and caudal papilla. **c**. The tongue root covers the rostral portion of the lips of the glottis (dotted outline) ensuring that no food or liquid enters the trachea. (Figure modified from Crole and Soley 2012a; open-access article distributed under the terms of the Creative Commons Attribution License, https://creativecommons.org/licenses/by/4.0/)

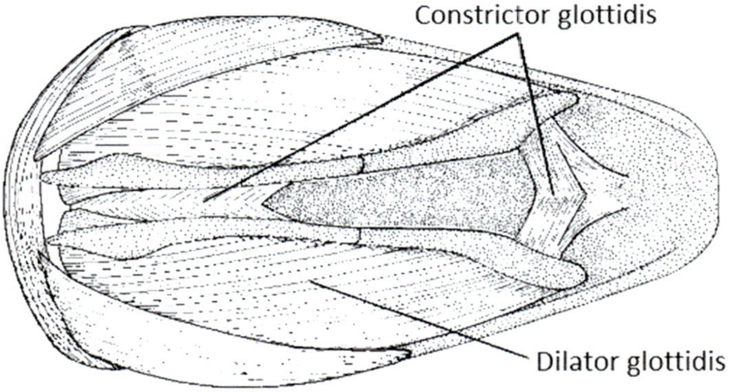


**Fig. 7.14** Muscles associated with the larynx of an American Crow (*Corvus brachyrhynchos*). Contraction of the dilator glottidis muscles opens the glottis, allowing air to pass into and out of the respiratory system. Contraction of the constrictor glottidis helps close the glottis and prevents food items or liquids from entering the trachea when swallowing. (Figure modified from Bock 1978; © 1978 Wilson Ornithological Society, used with permission)

air and preventing drying of the air sacs and lungs. The mucous also coats the ciliated epithelium that lines the inner walls of the trachea, making the cilia sticky and able to trap small particles in the incoming air (Fig. 7.20). Wave-like movements of the cilia then propel the particles back up the trachea to the larynx where they can be swallowed, moving down the

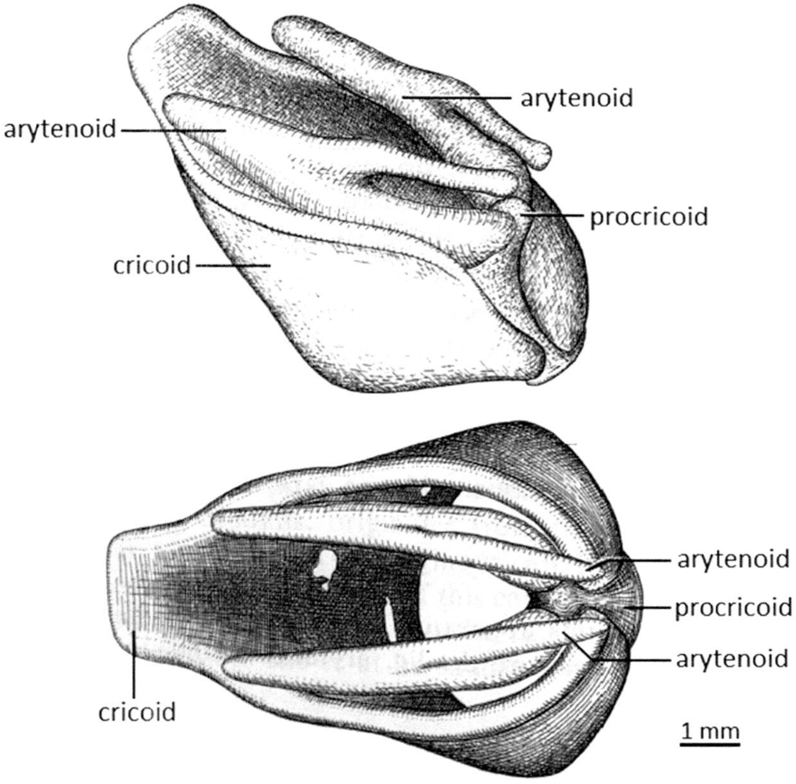


**Fig. 7.15** Dorsocaudolateral (top) and dorsal (bottom) views of the larynx of a Rock Pigeon (*Columba livia*). (Figure modified from Zweers et al. 1981; © 1981 Springer-Verlag, used with permission)

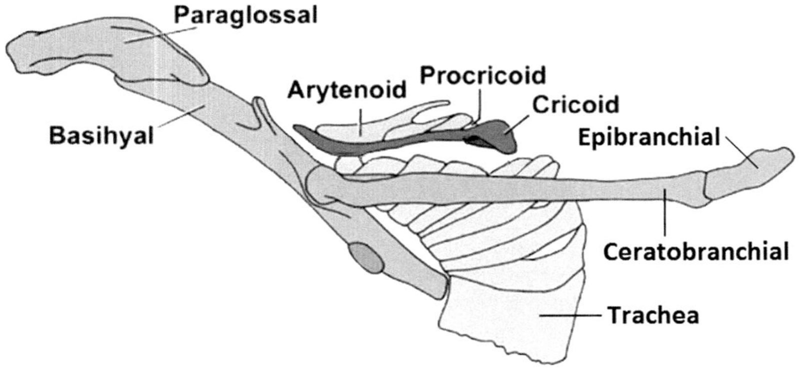


**Fig. 7.16** Kinetic larynx and hyoid of a Gray Parrot (*Psittacus erithacus*). The cricoid is ring-shaped, with the procricoid located between the two caudal ends of the cricoid and the arytenoids articulating with the procricoid and connected to the internal margins of the cricoid by ligaments. The extrinsic muscles of the larynx of parrots are more complex than those of other birds and allow a diversity of configurational changes that can alter sounds generated by the syrinx. (Figure from Homberger 1999, and used with permission of D. G. Homberger)

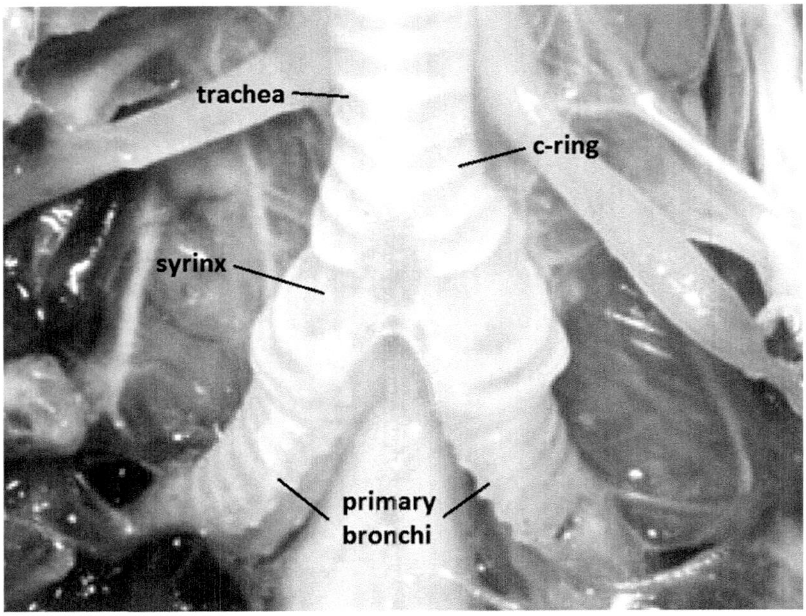


**Fig. 7.17** Trachea, cartilaginous rings (c-rings), syrinx, and primary bronchi of a Common Quail (*Coturnix coturnix*). (Figure modified from Çevik-Demirkan et al. 2007; © Blackwell Verlag GmbH, used with permission)

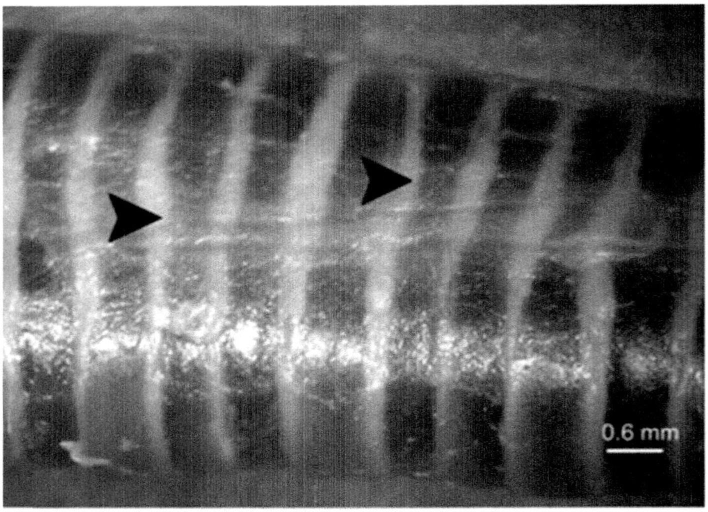


**Fig. 7.18** Dorsal view of a portion of the trachea of a Long-legged Buzzard (*Buteo rufinus*) showing the cartilaginous rings (darker areas). The cartilaginous rings completely encircle the trachea, but become slightly narrower both dorsally (arrows) and ventrally. (Figure from Kabak et al. 2007; © Blackwell Verlag GmbH, used with permission)

esophagus and into the acidic environment of the proventriculus. The mucous and cilia thus prevent the accumulation of particulate matter in the avian respiratory system. Interestingly, Burn et al. (2022) compared mucociliary transport speeds in the tracheas of four species of mammals (cow, pig, sheep, and rabbit) to two species of birds (Domestic Turkey [*Meleagris gallopavo domesticus*] and Common Ostrich [*Struthio camelus*]) and found that mucociliary clearance averaged about 3 to 7 times faster (range = about 100–300 μm/s) for the birds than the mammals (range = about 10–100 μm/s).

The tracheas of most birds are straight tubes that, reinforced by rings of cartilage to prevent collapse, extend from the larynx to the syrinx (where the trachea splits into two primary bronchi). However, several groups and species of

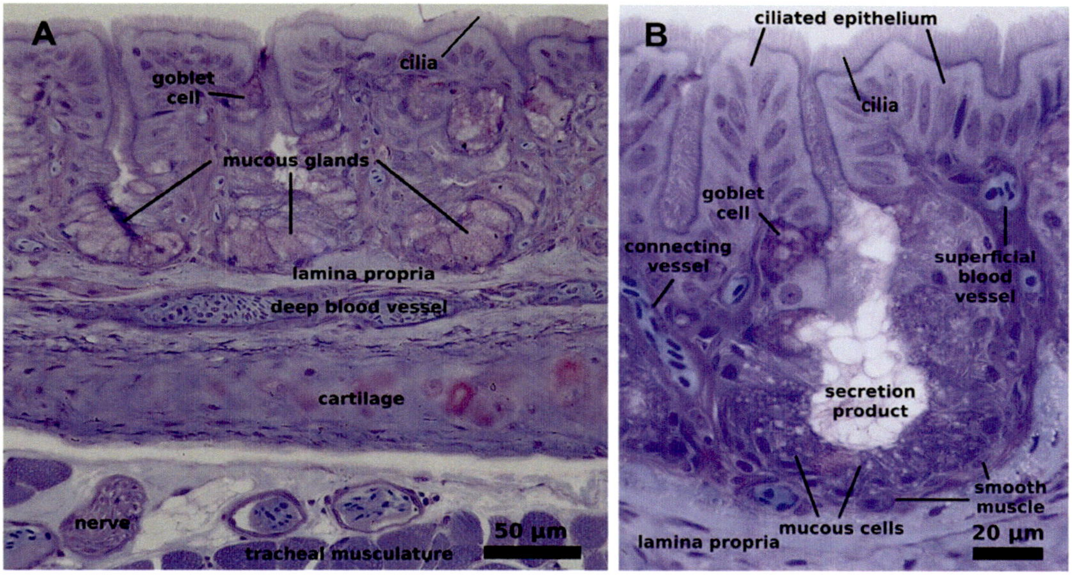


**Fig. 7.19** (**a**) Cross section through the wall of an avian trachea showing a portion of a cartilaginous ring that provides structural support, mucous glands that secrete mucous that adds moisture to incoming air, and mucous-coated cilia that trap particulates in inhaled air and transport them out of the trachea to the larynx where they are swallowed. (**b**) Close-up view of mucous cells and ciliated epithelial cells. (Figure from Sverdlova et al. 2012; © Sverdlova et al., open-access article distributed under the terms of the Creative Commons Attribution License, https://creativecommons.org/licenses/by/4.0/)

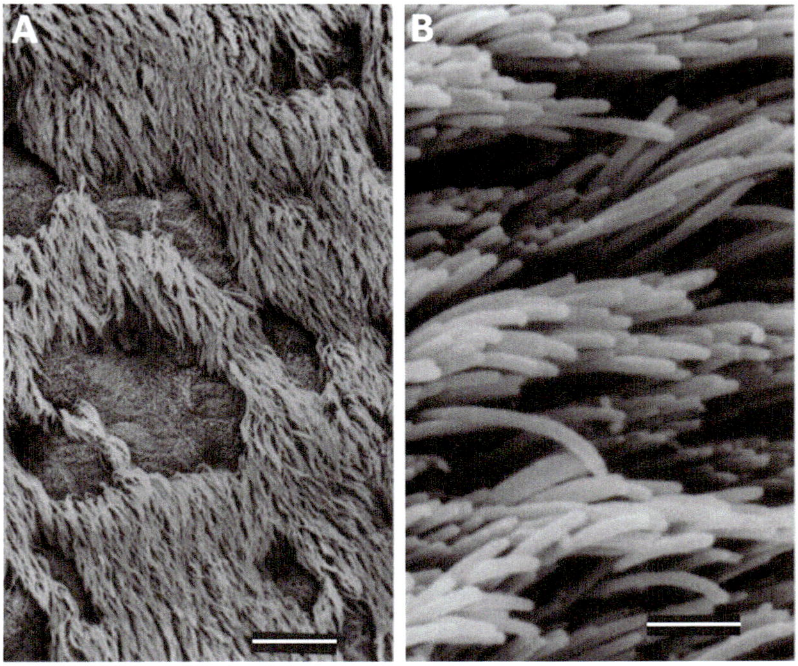


**Fig. 7.20** Electron micrographs of ciliated epithelium lining the trachea of a one-day-old Domestic Chicken (*Gallus g. domesticus*). (**a**) Embedding among the ciliated cells are non-ciliate mucous cells; magnified 1000× (Scale bar = 10 µm). (**b**) Cilia magnified 11,000× (Scale bar = 1 µm). (Figure from Sander et al. 1995; © America Association of Avian Pathologists, Inc., used with permission)

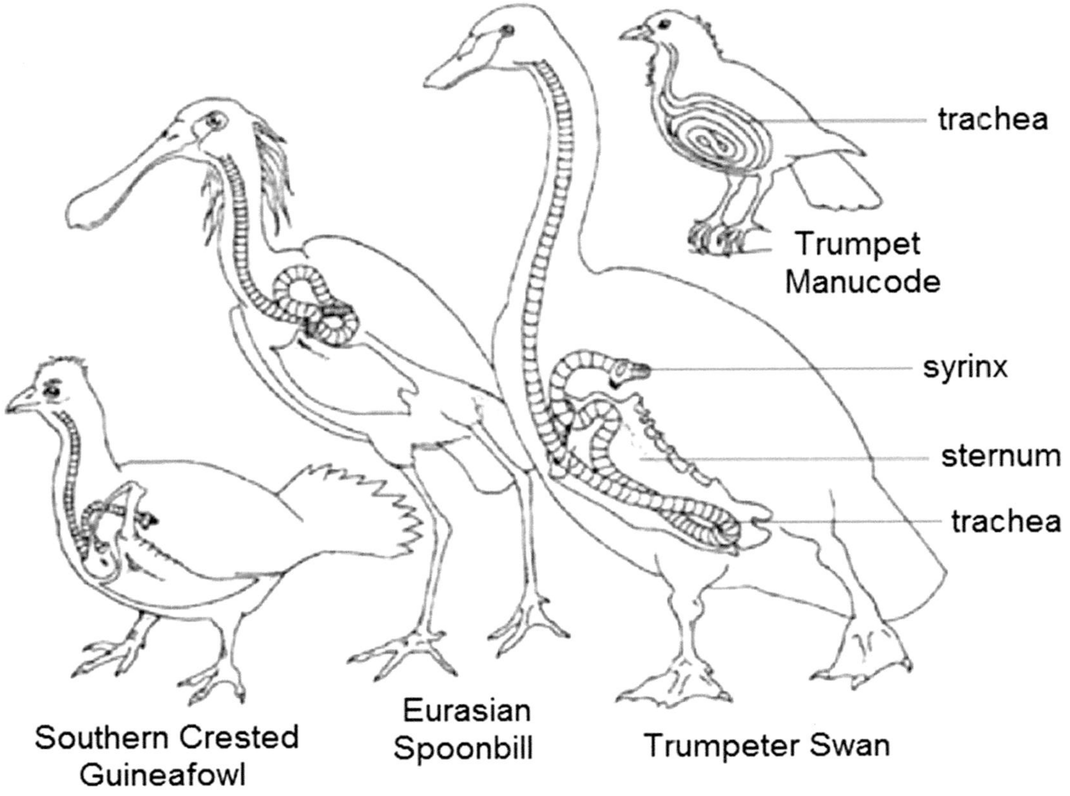


**Fig. 7.21** Examples of elongated, coiled avian tracheas. Intraclavicular coiling in Southern Crested Guineafowl (*Guttera edouardii*), intrathoracic coiling in Eurasian Spoonbills (*Platalea leucorodia*), intrasternal coiling in Trumpeter Swans (*Cygnus buccinators*), and subdermal coiling in Trumpet Manucodes (*Phonygammus keraudrenii*). (Figure adapted with permission from Cambridge University Press from Fitch [1999, Fig. 1] by Düring and Elemans 2016; © 2016 Springer International Publishing Switzerland, used with permission)

birds (at least 60 species, Fitch 1999), including swans, cranes, curassows, and others, have much longer trachea arranged into permanent loops or coils (Fig. 7.21). Exceptional examples of tracheal coiling can be found in some birds of paradise. Trumpet Manucodes (*Phonygammus keraudrenii*), about 30 cm in length, have tracheas up to 80 cm long (Clench 1978; Fig. 7.22). Tracheal loops are often found only in males or, if present in both sexes, are more complex in males (McLelland 1989a, b). One suggested function of coiled tracheas is that the increased surface area permits greater evaporative cooling and allows birds to pant without respiratory alkalosis. When panting, birds (and mammals) may breathe rapidly and eliminate carbon dioxide from the body faster than it forms and this, in turn, causes the pH of the blood to become more alkaline. Possible consequences include interference with eggshell formation in female bird and inhibition of respiratory control centers that could, by slowing respiration rate, interfere with a bird's ability to lose heat during hot weather (Barnas and Rautenberg 1987). However, individuals of four species of cranes were subjected to acute heat stress and their longer tracheas, although helpful, were not found to be essential for preventing respiratory alkalosis

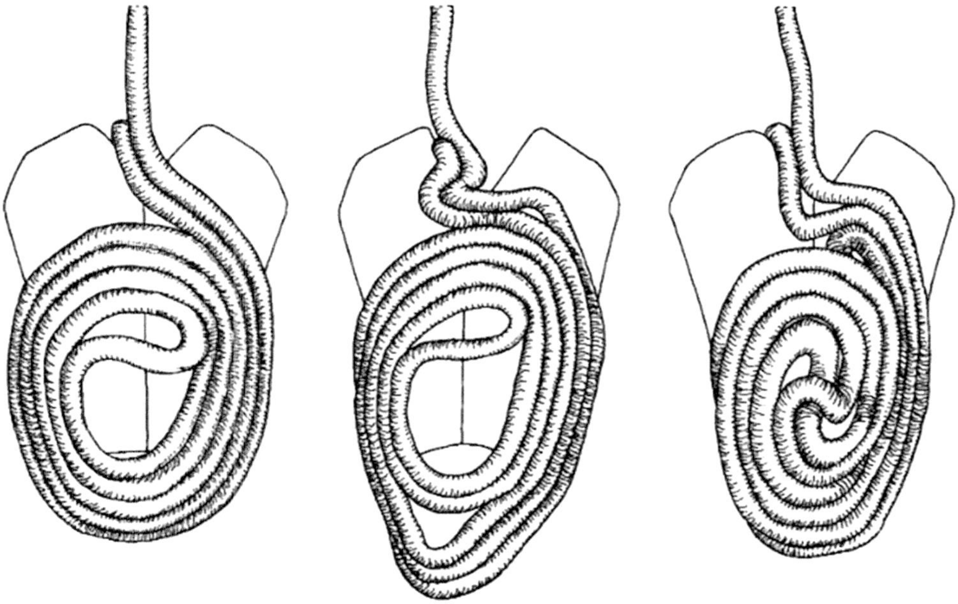

**Fig. 7.22** Elongated and coiled tracheas of three male Trumpet Manucodes (*Phonygammus keraudrenii*). Tracheas are not elongated in young birds, but increase in length as they age; the coiled traches of females are shorter than those of males. The longest trachea recorded is 828 mm. Concerning the vocalizations of Trumpet Manucodes, "The males utter a very loud and deep guttural note, unlike that of any other bird I am acquainted with, and it astonished me that a comparatively small bird could make so much noise" (a quote from Thorpe in Gilliard 1969, p. 105). (Figure from Clench 1978; © 1978 Oxford University Press, used with permission)

(Prange et al. 1985). One possible disadvantage of long, convoluted tracheas is the increased airflow turbulence and dead spaces where pathogens can settle and increase the likelihood of infection (Schwarz et al. 2016).

Available evidence indicates that tracheal loops help birds produce louder, lower frequency calls (Jones and Witt 2014). In support of this hypothesis, shortening the trachea of cranes was found to reduce call volume, and the degree of reduction was proportional to the degree of shortening (Gaunt et al. 1987). Tracheal elongation, by increasing the volume and lowering the frequency of calls, may also serve to exaggerate the apparent size of a vocalizing bird and aid males in obtaining mates and defending territories (the "size exaggeration" hypothesis; Fitch 1999). The greater tendency for males to exhibit tracheal coils supports the idea that sexual selection has played a role in their evolution (Frith 1994).

In further support of the "size exaggeration" hypothesis, Jones and Witt (2014) found a negative relationship between body size and migration among 15 species of cranes, with selection apparently favoring a reduction in size to minimize the energetic cost of migration. However, these investigators also found a positive relationship between migration distance and the proportional length of tracheas, suggesting that, as natural selection favored smaller size, sexual selection favored longer tracheas because of the benefits of perceived larger body size in intraspecific competition for territories and mates (Fig. 7.23).

How does tracheal coiling alter the characteristics of calls? One possibility is that long tracheas have resonant properties like any elongated tube and, therefore, are analogous to a wind instrument like a trombone. Another hypothesis is that, because they are embedded in or closely associate with the sternum, the coils

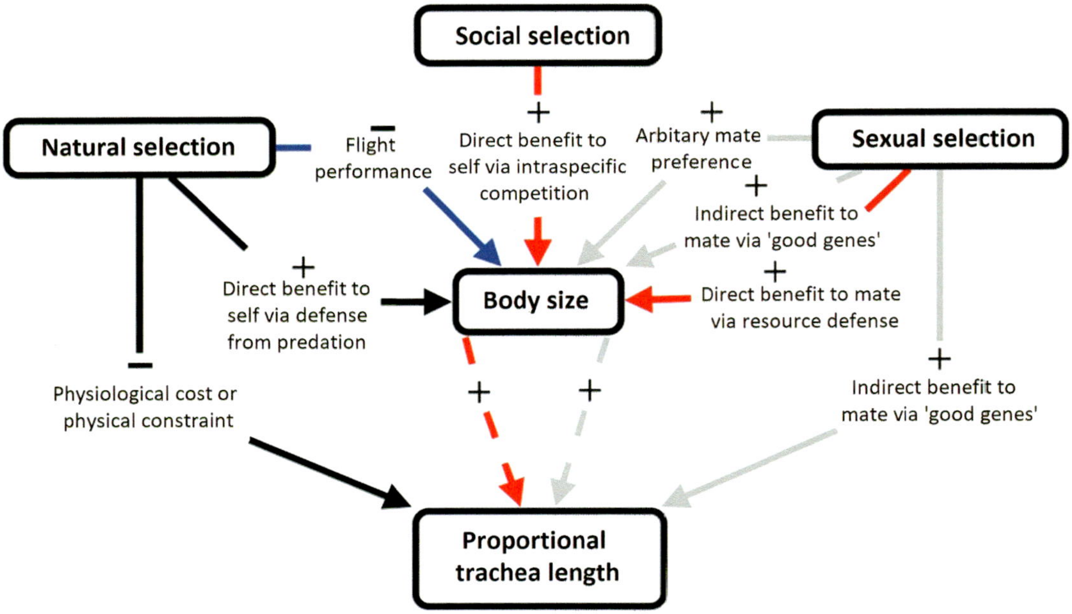


**Fig. 7.23** Potential mechanisms of selection leading to elongation of tracheas in cranes. Red lines indicate the factors most likely to contribute to the evolution of longer tracheas. Gray lines are alternative factors that seem less likely. Dashed lines indicate indirect selection on proportional trachea length as a result of direct selection on body size. Migration selects for improved flight performance favoring reduced body size (blue solid line), which increases the strength of the sexual and social selection on body size (red solid lines) and indirectly on trachea length (red dashed lines). (Figure from Jones and Witt 2014; © 2014 The Authors. Journal of Evolutionary Biology © 2014 European Society For Evolutionary Biology, used with permission)

serve a function analogous to the bridge of a stringed instrument, transmitting the vibrations of a tiny sound source to a large radiating surface, the sternum. The sternum then vibrates against the large internal air reservoir of the avian air-sac system. If correct, elongated, coiled tracheas, in association with the sternum and air sacs, function more like a violin than a trombone (Gaunt et al. 1987).

Two species of birds, Emus (*Dromaius novaehollandiae*) and Ruddy Ducks (*Oxyura jamaicensis*), are known to have inflatable sac-like outpocketings or diverticula of the trachea called tracheal sacs (McLelland 1989a, b). Emus have a slit in the ventral side of the trachea (just before the syrinx) that is covered with a thin membrane (Fig. 7.24). In adult males and females, this becomes an expandable pouch that, when inflated, is about 30 cm in diameter (Cho et al. 1984). Movement of air into this pouch or sac produces a loud booming call. Only male Ruddy Ducks have tracheal sacs and inflation of these sacs is an important component of their courtship display (Johnsgard 1995; Fig. 7.25).

Finally, some penguins and petrels have tracheas with a median septum (also called the septum trachealis medialis) that divides the trachea into two channels. The septum in the trachea of African Penguins (*Spheniscus demersus*) extends to just short of the larynx (King and McLelland 1979; Fig. 7.26), whereas the septum of Yellow-eyed Penguins (*Megadyptes antipodes*) is limited to the lower third of the trachea (Hocken 2002). The septum of King Penguins (*Aptenodytes patagonicus*) begins 6 cm below the larynx and extends to where the trachea splits to form bronchi (Kriesell et al. 2020; Fig. 7.27). The tracheas of still other penguins, including Emperor Penguins (*Aptenodytes forsteri*), have no septums (King and McLelland 1979). The

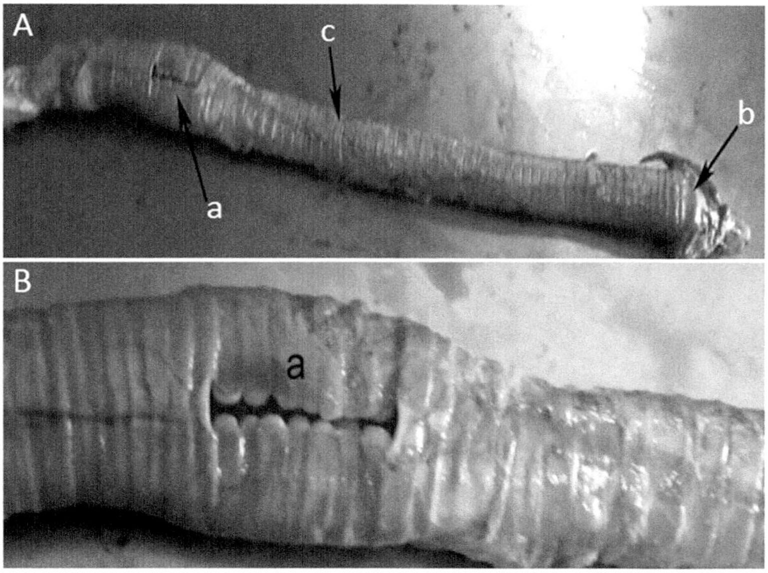


**Fig. 7.24** (**A**) Photo of the ventral surface of an Emu's (*Dromaius novaehollandiae*) trachea showing the (a) tracheal pouch, (b) larynx, and (c) trachea. (**B**) Close-up photo of the tracheal pouch. In live birds, the tracheal pouch is covered by a thin membrane that is, in turn, covered with loose skin that expands when an Emu vocalizes. The pouch is located about one-third of the way along the entire length of the trachea. (Figures from Jayachitra and Balasundaram 2015, published in the Indian Journal of Veterinary Anatomy, an open-access journal)

**Fig. 7.25** Drawing of the tracheal pouch of a male Ruddy Duck (*Oxyura jamaicensis*). The pouch "... serves as a "sounding board" for the bill-drumming, or "Bubble," display of the male, and is deflated to produce sound at the end of each display" (Johnsgard 1961, p. 66). (Figure from Johnsgard 2017; used with permission of Paul Johnsgard)

significance and possible function of the "double tracheas" and partial "double tracheas" of some birds remain unknown (McLelland 1989a, b). One possible function of the septum is to prevent collapse of the trachea during dives, which, in turn, could prevent pulmonary barotrauma (damage to the lungs caused by rapid changes in pressure) (Kriesell et al. 2020). In addition, Davenport et al. (2014) reported that the tracheas of leatherback turtles (*Dermochelys coriacea*) also have a partial septum and suggested that the septum prevents collapse of the trachea and

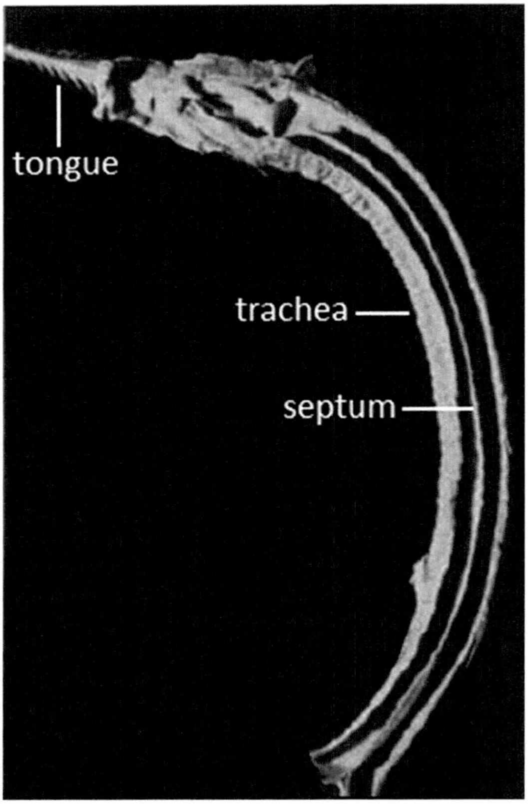


**Fig. 7.26** Trachea of an African Penguin (*Spheniscus demersus*) showing the median septum that divides the traches into two channels. (Figure modified from Zeek 1951; © 1951 Wiley-Liss, Inc., used with permission)

continued breathing when the adjacent esophagus is dilated by large prey items and presses on the trachea. Finally, the syrinx of some penguins, including King Penguins, is known to produce "two voice" vocalizations, i.e., simultaneous production of two sounds at different frequencies, and the tracheal septum might be important in keeping those two sounds separated until reaching the cranial end of the trachea (Kriesell et al. 2020).

Because birds have relatively long necks compared to similar-sized mammals, their tracheas, even with no coils and convolutions, are also relatively long. A typical bird has a trachea about 2.7 times longer than that of a mammal of similar size (Hinds and Calder 1971). The avian trachea is also a bit larger in diameter, averaging about 1.29 times that of a similar-sized mammal (Brown et al. 1997; Fig. 7.28). As a result, tracheal volume averages about 4.5 greater for birds (Hinds and Calder 1971). Birds compensate for the larger tracheal dead space by having relatively larger tidal volumes (air inhaled and exhaled with each respiratory cycle) and slower respiratory rates, about one third that of mammals. These two factors lessen the impact of the larger tracheal dead space volume on ventilation, with minute tracheal ventilation only about 1.5 to 1.9 times that of mammals (Ludders 2001). Among both birds and mammals, respiratory rate varies with body mass; smaller birds and mammals have higher respiratory rates (Fig. 7.29). For both birds and mammals, aquatic species tend to have lower breathing rates than terrestrial species and, among birds, the breathing rates of larger aquatic species are much lower than those of larger terrestrial species (Fig. 7.29). One possible contributing factor to this difference is that aquatic species of birds tend to have larger lung volumes and tidal volumes (Fig. 7.30). The greater lung volumes likely provide greater surface area for the exchange of gases, which, in turn, may increase levels of oxygen in the blood leaving the lungs and thus require fewer breaths per minute to maintain adequate blood oxygen levels.

The trachea bifurcates (or splits) into two primary bronchi at the syrinx (Fig. 7.31). The syrinx is unique to birds and is their "voicebox" (in mammals, sounds are produced in the larynx). The process by which the syrinx produces sounds is described in Chap. 12. Each primary bronchus then enters its corresponding lung. To this point, the avian respiratory system is not very different from the mammalian respiratory system. However, the lungs of birds, the tubular network within, and the associated air sacs are completely unique and incredibly efficient at gas exchange.

### 7.2.4 Lungs and Air Sacs

Avian lungs are relatively small and practically rigid (Maina 2005) and, among birds, lung

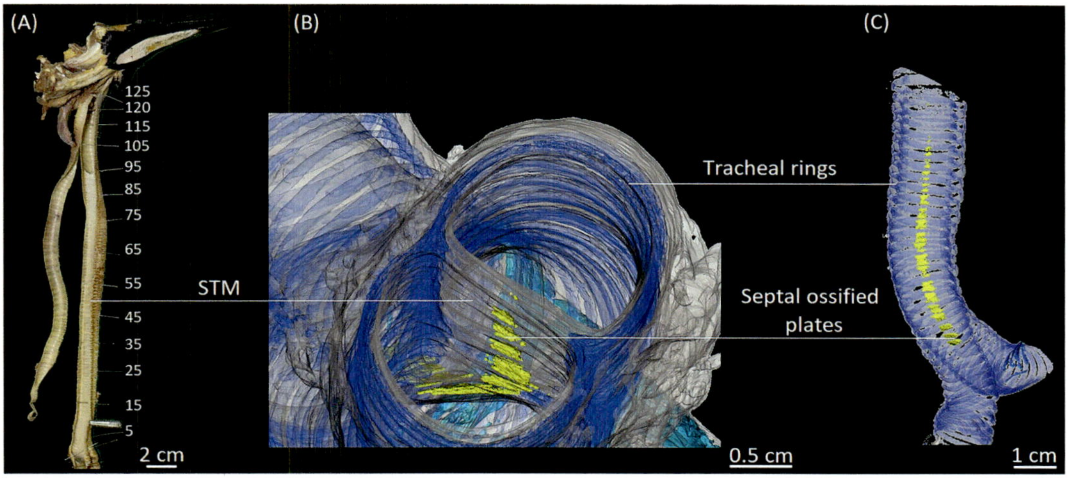


**Fig. 7.27** Trachea and the septum trachealis medialis (STM) separating the trachea of King Penguins (*Aptenodytes patagonicus*) into two lumina. (**a**) Upper vocal tract of a male King Penguin, with the right side of the trachea removed to expose the medial septum. The STM originates where the trachea splits into the two primary bronchi and extends to the 94th/96th tracheal rings, 6 cm below the larynx. Tracheal rings are numbered 1–125. (**b**, **c**) 3D reconstructions of trachea, syrinx, and bronchi showing different densities of the trachea and STM. Blue indicates ossified tracheal and bronchial rings, and yellow indicates ossified plates in the connective tissue (gray) of the tracheal septum. (**b**) Trachea showing the STM separating the two tracheal tubes and the irregular ossified plates (yellow) in the septum. (**c**) Lateral view of the trachea and ossified plates of the STM. (Figure from Kriesell et al. 2020; open-access article distributed under the terms of the Creative Commons CC BY license, https://creativecommons.org/licenses/by/4.0/)

volume is generally correlated with size (mass; Fig. 7.32). However, some species or populations of birds that have long existed at high altitudes have higher lung volumes per unit body mass than those of comparable body mass at lower altitudes, an adaptation for maintaining an adequate oxygen supply in the relatively hypoxic environments found at higher altitudes (Bakkeren et al. 2020; Fig. 7.33). Per unit body mass, the lungs of birds average about 27% smaller than those of mammals (Maina et al. 1989) and, during respiration, avian lung volume changes by only about 1.4% (Jones et al. 1985). The lungs of birds insert deeply into the vertebral ribs (Fig. 7.34) and are firmly attached to the thoracic wall. This firm attachment coupled with the lung's rigidity means that, as described above, the air capillaries do not "inflate" and, with surface tension less of a factor, can be, and are, very narrow. As a result, even though birds have smaller lungs than comparably sized mammals, the respiratory surface area is about 15% greater (Maina et al. 1989).

The exchange of gases between the lungs and the blood occurs by simple diffusion, so maintaining pressure gradients is essential (i.e., more oxygen in the lungs than in the blood so that oxygen will diffuse into the blood). Maintaining such gradients requires the continuous movement of "new" air into and out of the lungs, a process called ventilation. In mammals, the lungs play an essential role in ventilation, expanding as the inspiratory muscles contract (lowering the air pressure and drawing air into the lungs) and contracting during expiration (increasing the air pressure and forcing air out of the lungs). In contrast to the lungs of mammals, avian lungs are rigid and do not change in volume during respiration. In birds, ventilation involves a series of air sacs connected to the lungs via ostia (Figs. 7.35, 7.36, and 7.37).

Most birds have nine air sacs, including one interclavicular sac, two cervical sacs, two anterior thoracic sacs, two posterior thoracic sacs, and two abdominal sacs. However, loons (Gaviidae) and

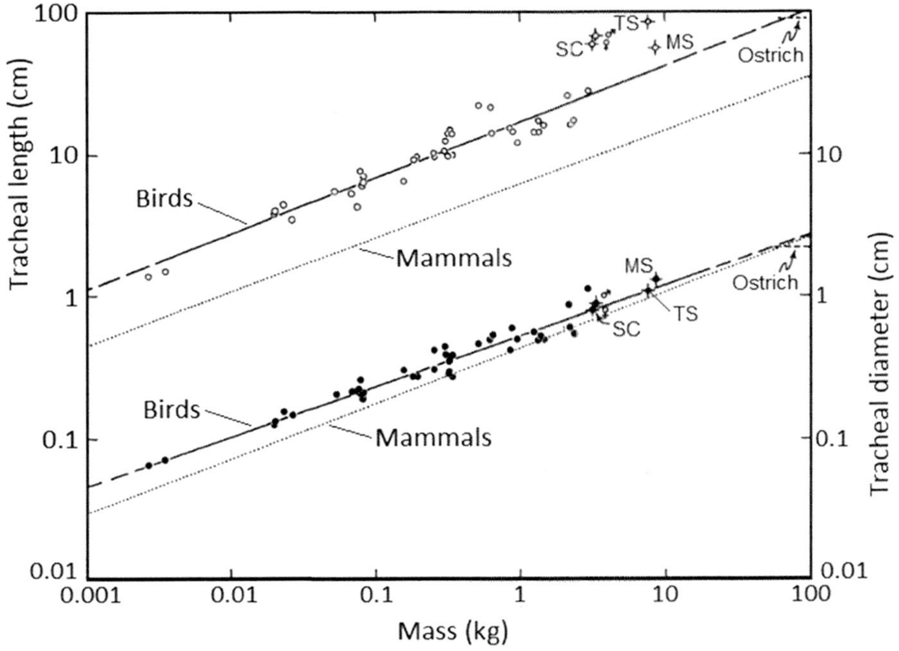


**Fig. 7.28** Relationships between tracheal dimensions and body mass of birds and mammals. Values for Tundra (*Cygnus columbianus*, TS) and Mute (*Cygnus olor*, MS) swans, Sandhill Cranes (*Antigone canadensis*, SC), and the Common Ostrich (*Struthio camelus*) were not used in generating regressions. (Figure modified from Hinds and Calder 1971; © 1971 The Society for the Study of Evolution, used with permission)

**Fig. 7.29** Breathing rates at rest of several species of terrestrial ($N = 35$ representing 27 species) and aquatic ($N = 48$ representing 22 species) birds that differ in mass. Data are from Mortola and Limoges (2006). (Figure modified from Mortola and Seguin 2009; © 2009 The Authors. Journal Compilation © 2009 The Zoological Society of London, used with permission)

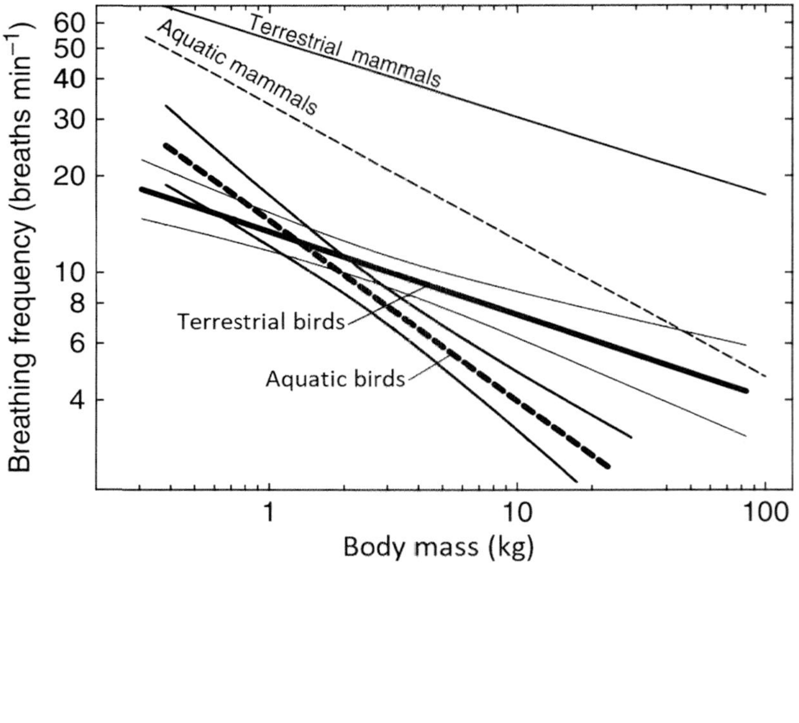

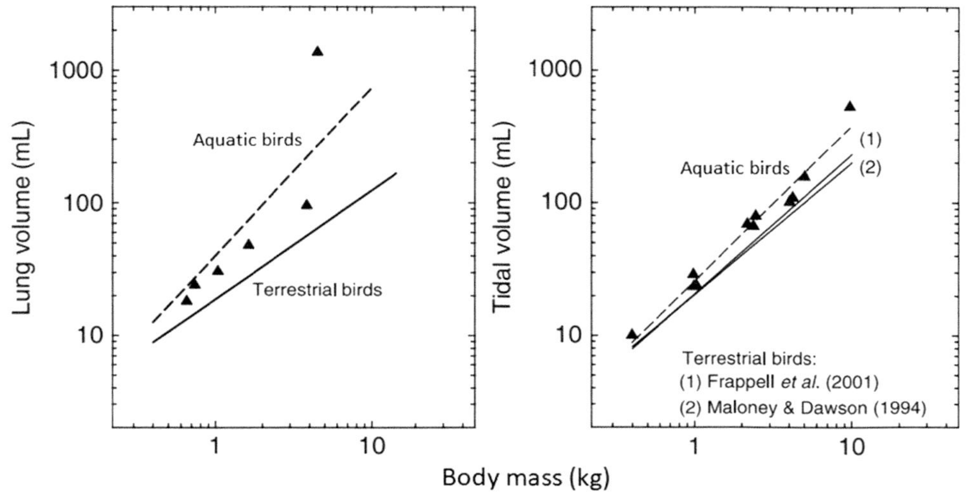


**Fig. 7.30** (Left) Lung volume of several species of aquatic (triangles) and terrestrial birds relative to body mass. Data from Maina (2005). (Right) Tidal volume of several species of aquatic (triangles) and terrestrial birds relative to body mass. Data for terrestrial birds are from Frappell et al. (2001) and Maloney and Dawson (1994). (Figure from Mortola and Seguin 2009; © 2009 The Authors. Journal compilation © 2009 The Zoological Society of London, used with permission)

grebes (Podicipedidae) do not have cervical air sacs and, in some species including Rock Pigeons (*Columba livia*) and House Sparrows (*Passer domesticus*), the interclavicular and cervical air sacs join to form a single cervicoclavicular air sac (Wetherbee 1951). Functionally, as will be explained later, air sacs can be divided into anterior sacs (interclavicular, cervicals, and anterior thoracics) and posterior sacs (posterior thoracics and abdominals). Air sacs have very thin walls lined primarily by simple squamous epithelium supported by a thin layer of connective and few blood vessels, so they do not play a direct role in gas exchange (Fig. 7.38). Rather, they act as "bellows" to ventilate the lungs (Powell 2000). In most species of birds, these thin-walled air sacs extend into postcranial bones, i.e., postcranial skeletal pneumaticity, but the extent of this pneumaticity varies among species (e.g., Atterholt and Wedel 2023; Fig. 7.39).

The extent of postcranial skeletal pneumaticity ranges from no pneumaticity (apneumatic) to "hyperpneumaticity." For many, generally small, birds, the only pneumatic bones are in the vertebral column (Fig. 7.40b). Examples include shorebirds, some waterfowl, and larids (gulls, terns, and skimmers) (O'Connor 2004, 2009). Other species have many more pneumatic bones, including the sternum, ribs, pectoral, and pelvic girdles, and, in some cases, proximal limb bones (humerus and femur) (Fig. 7.40c). Species with this "expanded" degree of postcranial skeletal pneumaticity are generally medium-sized and include non-diving ducks, geese, galliforms, columbiforms, and many hawks, falcons, storks, and herons (O'Connor 2004, 2009). These same bones are pneumatic in the largest species in these taxa, but, in addition, distal bones of the forelimbs and, less often, hindlimbs, are also pneumatic ("hyperpneumaticity," Fig. 7.40d). In contrast, species that pursue prey underwater (pursuit-specialists) are completely apneumatic, having no pneumatic bones (Fig. 7.40a). Apneumatic taxa include species in the families Spheniscidae (penguins), Gaviidae (loons), Podicipedidae (grebes), Pelecanoididae (diving petrels), Anhingidae (anhingas), and Phalacrocoracidae (cormorants and shags) (Figs. 7.41 and 7.42).

Some investigators have suggested that postcranial skeletal pneumaticity varies with bird

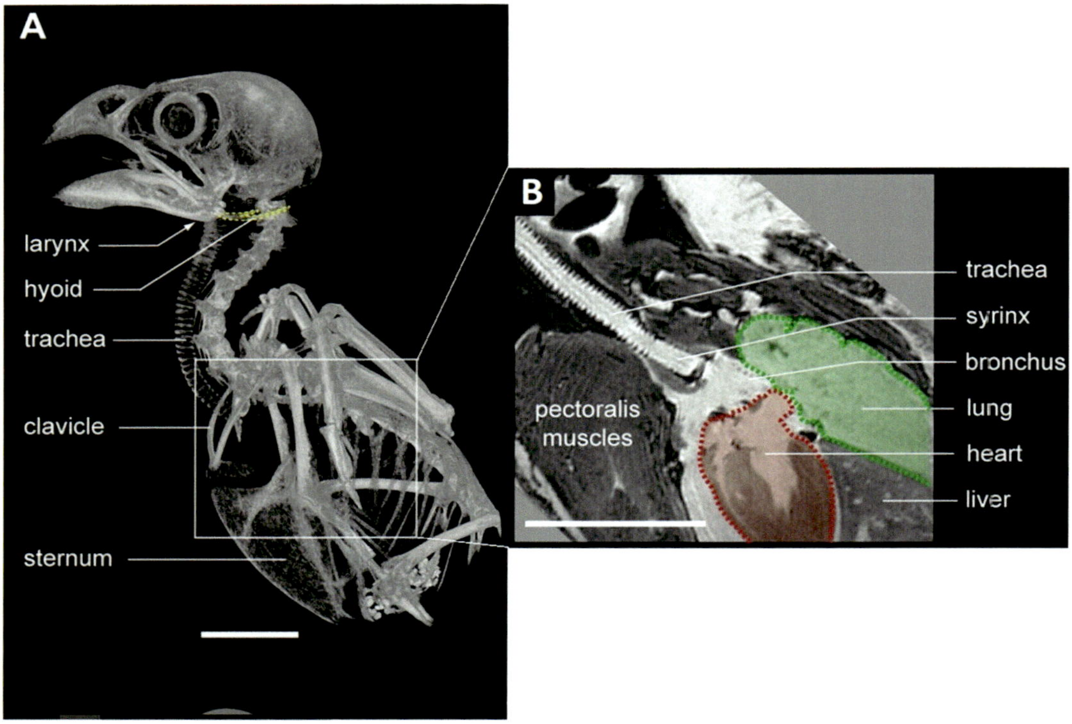


**Fig. 7.31** (**a**) Skeleton of a Zebra Finch (*Taeniopygia guttata*) showing the trachea extending down toward the syrinx. (**b**) Close-up of the trachea, syrinx, and a bronchus. The syrinx is located where the trachea splits into the two bronchi that then enters the lungs (shaded green). (Figure modified from Düring et al. 2013; © 2012 Düring et al. 2013, licensee BioMed Central Ltd., open-access article distributed under the terms of the Creative Commons Attribution License, http://creativecommons.org/licenses/by/4.0)

**Fig. 7.32** Relationship between body mass and lung volume for 29 species of birds. (Figure from Maina et al. 2017; © 2017 Maina et al., open-access article distributed under the terms of the Creative Commons Attribution License, https://creativecommons.org/licenses/by/4.0/)

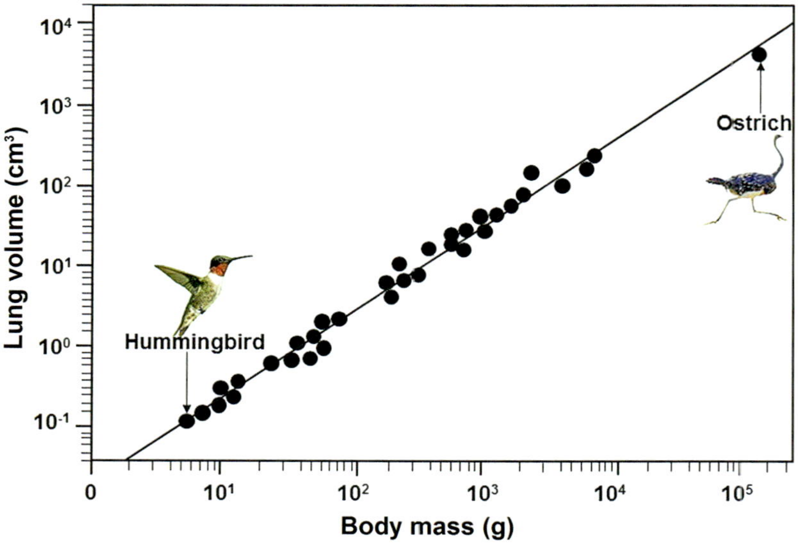

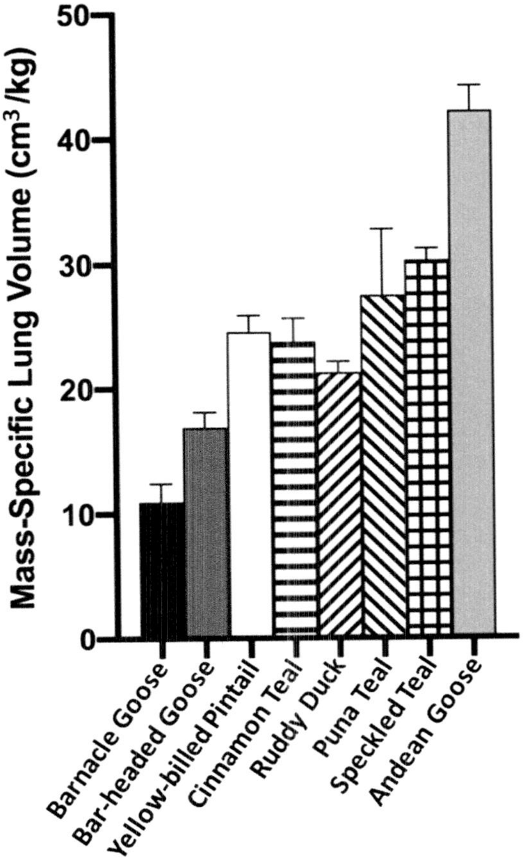


**Fig. 7.33** Mass-specific lung volumes for eight species of waterfowl. Barnacle Geese and Bar-headed Geese are found at lower altitudes (although Bar-headed Geese fly at high altitudes during migration), whereas the other six species are residents at high altitudes in the Andes Mountains of South America and have much greater lung volumes relative to their mass. Barnacle Goose, *Branta leucopsis*; Bar-headed Goose, *Anser indicus*; Yellow-billed Pintail, *Anas georgica*; Cinnamon Teal, *Spatula cyanoptera*; Ruddy Duck, *Oxyura jamaicensis*; Puna Teal, *Spatula puna*; Speckled (Yellow-billed) Teal, *Anas flavirostris*; Andean Goose, *Oressochen melanopterus*. (Figure modified from Bakkeren et al. 2020; © 2020 Anatomical Society, used with permission)

body mass, generally increasing with increasing body size (Currey and Alexander 1985; McLelland 1989a, b). However, other investigators have reported that this relationship is no longer significant when phylogenetic comparative methods are used (Gutherz and O'Connor 2022). Although body mass "... may account from some proportion of the variation in pneumaticity expression ...," factors likely more important for variation among taxa of birds in the degree, or absence, of skeletal pneumaticity than body mass include foraging ecology (e.g., underwater diving) and flight style (e.g., soaring flight) (Gutherz and O'Connor 2022). Among pursuit-diving birds, selection has likely favored apneumatic bones because it makes them less buoyant. The "hyperpneumatic" skeletons of soaring birds reduce their body mass and reduce the energetic requirements of flight, whereas the apneumatic skeletons of pursuit divers make them heavier, reducing the energy required for diving (O'Connor 2004, 2009; Smith 2012).

Gutherz and O'Connor (2021, 2022) conducted detailed analyses of two clades of non-aquatic birds, Cuculidae and Accipitrimorphae, and found limited variability in levels of pneumaticity among species in each clade. Species in these clades that did exhibit variation have "locomotor specializations" (Gutherz and O'Connor 2022). Specifically, among cuculids, roadrunners are far more cursorial than other cuculids and exhibit femoral pneumaticity (i.e., lighter hindlimbs aid in cursorial locomotion) and, among accipitrimorphs, vultures have greater distal forelimb pneumaticity that, with lighter distal forelimbs, likely aids in long-distance soaring.

Avian lungs are complex structures, with a series of tubes, or bronchi, moving air into, out of, and through the lungs. As the trachea approaches the lungs, it splits into two primary bronchi (singular, bronchus) that each enter one of the lungs. The primary bronchus of birds has only two clusters of secondary bronchi. Four (most birds), five (e.g., cassowaries), or even six (e.g., pelicans) branches (called ventrobronchi, or medioventral secondary bronchi) arise from the primary bronchus close to where it enters the lung and, from the caudal section of the primary bronchus, extend about 7 to 10 branches (called dorsobronchi, or mediodorsal secondary bronchi). A few lateral branches (laterobronchi) originate from the lateral and ventral sides of the primary bronchus and supply the small amount of lung below the primary bronchus (Brown et al. 1997; Duncker 2004; Fig. 7.43). Parabronchi extend from the dorsobronchi and laterobronchi

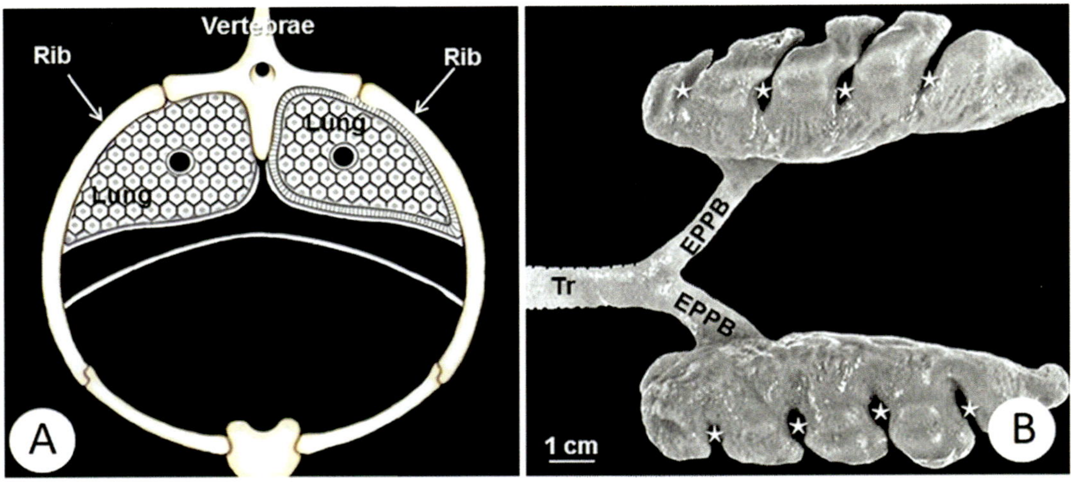


**Fig. 7.34** (**a**) Avian lungs are located between the vertebrae and ribs. (**b**) Trachea (Tr), primary bronchi (EPPB, or extrapulmonary primary bronchi), and lungs of an Ostrich (*Struthio camelus*) showing the deep costal sulci (*). (Figure modified from Maina 2015a; © 2015 Springer Nature, used with permission)

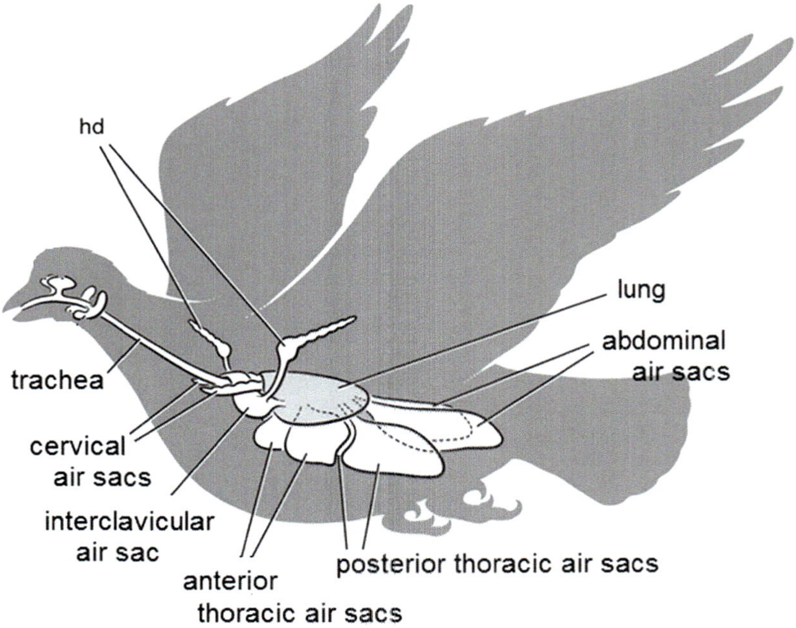


**Fig. 7.35** Avian respiratory system (hd, humeral diverticulum of the clavicular air sac). (Figure modified from Sereno et al. 2008; © 2008 Sereno et al., open-access article distributed under the terms of the Creative Commons Attribution License, https://creativecommons.org/licenses/by/4.0/)

to the ventrobronchi (Figs. 7.44 and 7.45). The primary bronchus continues out of the lung and opens into the abdominal air sac. One laterobronchus leaves the lung and opens into the posterior thoracic air sac. The cranial air sacs, including the anterior thoracic air sacs, the unpaired interclavicular air sac, and the cervical air sacs, are connected to branches of the ventrobronchi (Duncker 2004).

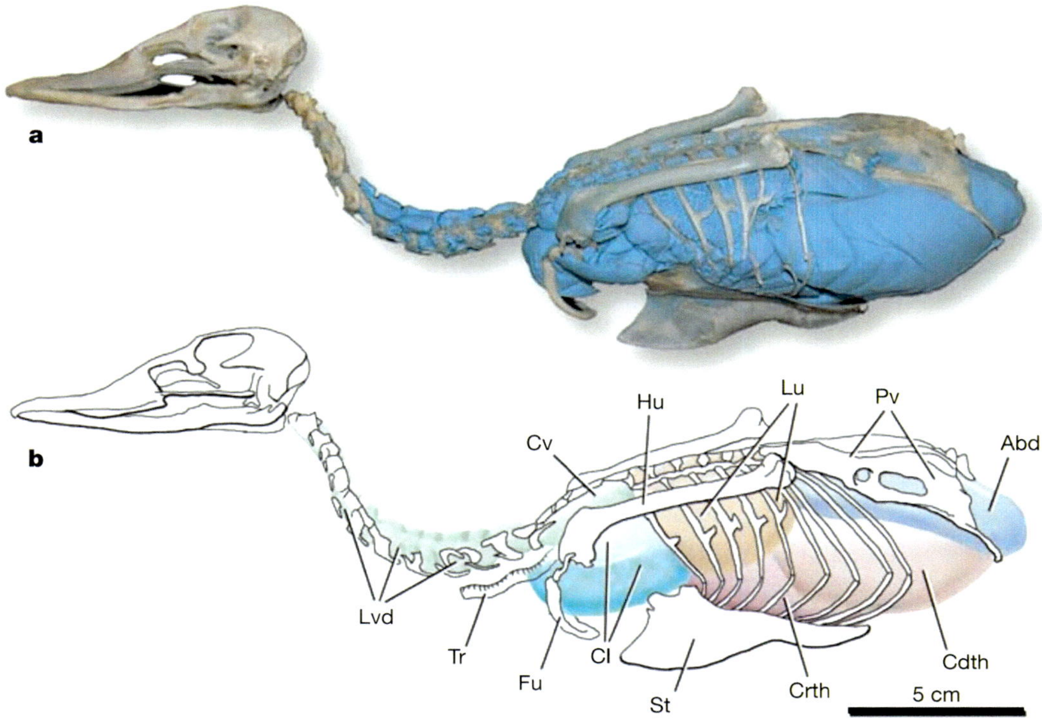


**Fig. 7.36** Pulmonary air-sac system of a Green-winged Teal (*Anas crecca*). **a**. Latex injection (blue) highlighting the location of air sacs. **b**. Main components of the avian flow-through system. Abd, abdominal air sac; Cdth, caudal thoracic air sac; Cl, clavicular air sac; Crth, cranial thoracic air sac; Cv, cervical air sac; Fu, furcula; Hu, humerus; Lu, lung; Lvd, lateral vertebral diverticula; Pv, pelvis; and Tr, trachea. (Figure from O'Connor and Claessens 2005; © 2005 Springer Nature, used with permission)

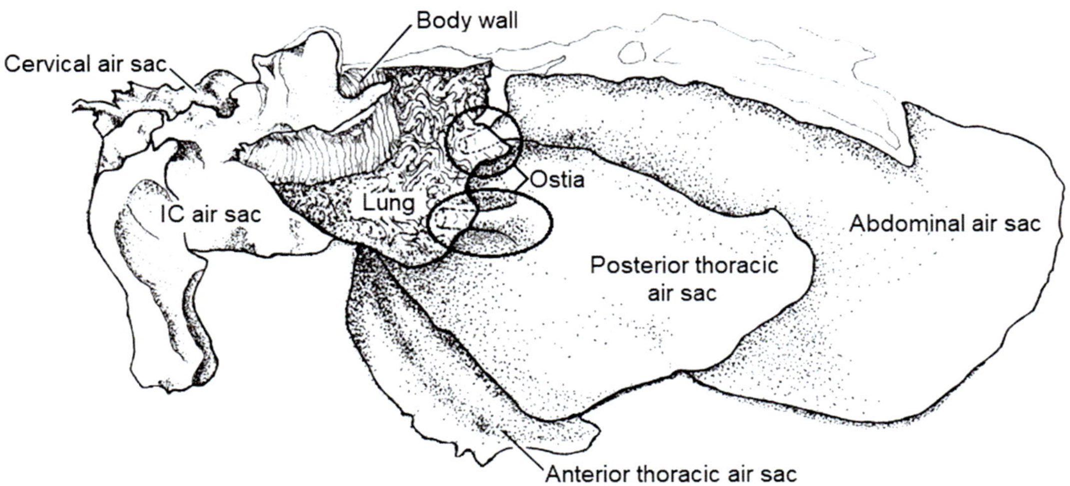


**Fig. 7.37** Drawing of a lung, air sacs, and two ostia that branch from the lung to allow air to move into and out of the posterior thoracic and abdominal air sacs. Ostia for the other air sacs are not shown. IC, interclavicular. (Figure modified from Hamlet and Fisher 1967; © 1967 Oxford University Press, used with permission)

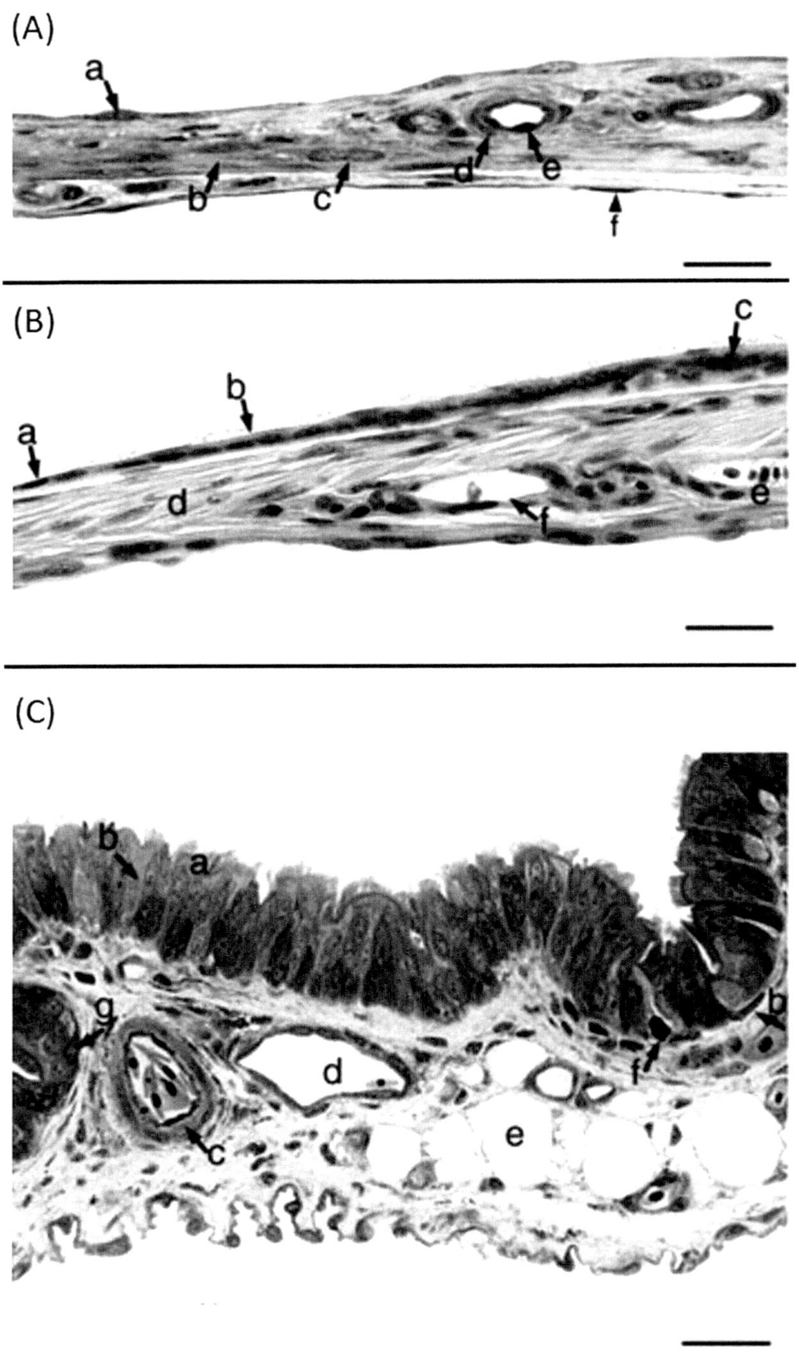


**Fig. 7.38** Epithelial lining of a thoracic air sac of a Domestic Chicken (*Gallus g. domesticus*). Most of the lining consisted of simple squamous epithelium, but there were also localized areas of cuboidal and columnar ciliated epithelium. (**A**) Squamous epithelium (a), connective tissue (b), nucleus of connective tissue cell, smooth muscle (d) in the wall of an arteriole (e), and squamous epithelium on the outer surface of the air sac (f). Scale bar = 18 μm. (**B**) Squamous epithelium (a), squamous ciliated epithelium (b), cuboidal ciliated epithelium (c), connective tissue (d), arteriole (e), and venule (f). Scale bar = 24 μm. (**C**) Columnar ciliated epithelium (a), goblet cell (secretes mucin, b), arteriole (c), venule (d), lymph vessel (e), leucocyte (f), and basal cell (cell that produces new epithelial cells, g). Scale bar = 19 μm. (Figure from Bezuidenhout 2005; © Blackwell Verlag GmbH, used with permission)

Birds exhibit some variation in lung structure and, specifically, in the arrangement of parabronchi. Most birds have two sets of parabronchi, the paleopulmonic ("ancient lung") and neopulmonic ("new lung") parabronchi. The paleopulmonic (or just paleopulmo) consists of

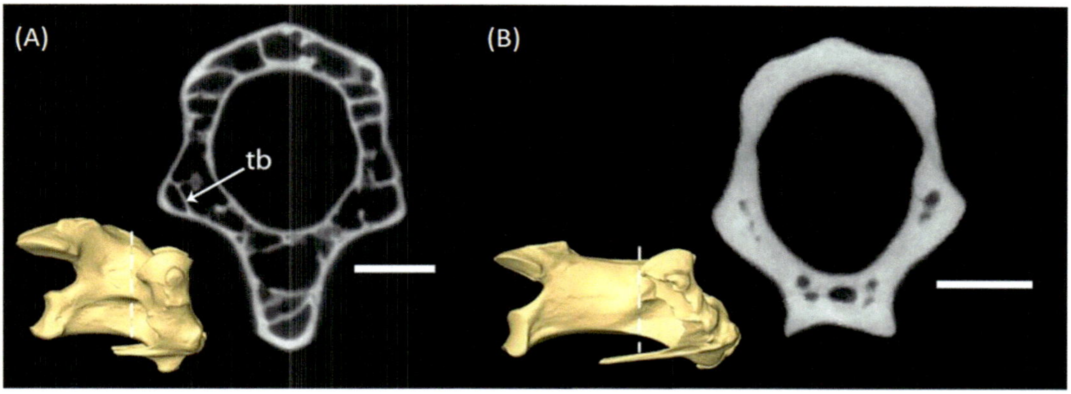


**Fig. 7.39** (**a**) Pneumatic cervical vertebra of a Great Skua (*Stercorarius skua*), and (**b**) apneumatic cervical vertebra of a Double-crested Cormorant (*Nannopterum auritus*). tb, trabecular bone. Scale = 0.25 cm. (Figure from Gutzwiller et al. 2013; © 2013 Wiley Periodicals, Inc., used with permission)

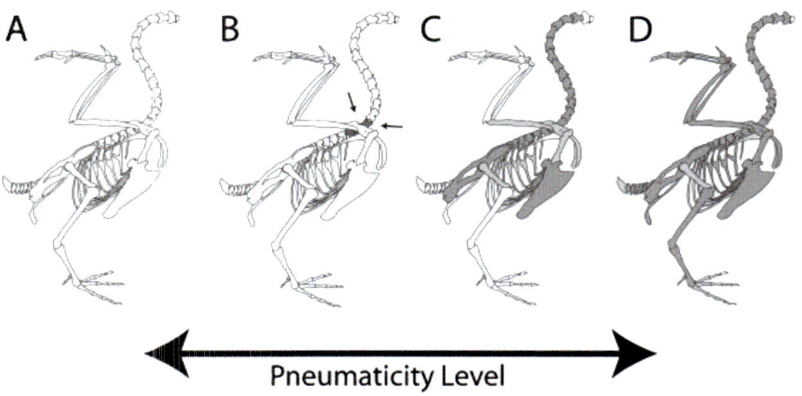


**Fig. 7.40** Variation among birds in the extent of skeletal pneumaticity. (**a**) No postcranial pneumaticity (e.g., diving birds like penguins, grebes, and loons), (**b**) pneumaticity is limited to portions of the cervicothoracic vertebral column (indicated by the arrows; e.g., shorebirds), (**c**) increased pneumaticity that extends through the postcranial axial skeleton, pectoral and pelvic girdles, and proximal limb bones (e.g., geese), and (**d**) hyperpneumaticity with pneumaticity extending through the postcranial axial and appendicular skeletons and distal bones of the limbs (e.g., pelicans and vultures). Pneumatic portions of the skeleton are indicated by dark gray shading. The arrow at the bottom is meant to indicate that there are continuous gradations in the degree of pneumaticity among birds. (Figure from O'Connor 2009; © 2009 Wiley-Liss, Inc., A Wiley Company, used with permission)

parallel stacks of parabronchi that occasionally anastomose (Maina 2015a). However, the neopulmonic region is absent in some birds (e.g., penguins and kiwis) and poorly developed in others (e.g., storks [Ciconiidae] and ducks [Anatidae]). In songbirds (Passeriformes), pigeons (Columbiformes), and gallinaceous birds (Galliformes), the neopulmonic region of the lung is well-developed (Maina 2008; Figs. 7.46, 7.47, and 7.48). In these latter groups, the neopulmonic parabronchi contain about 15 to 20% of the gas exchange surface of the lungs

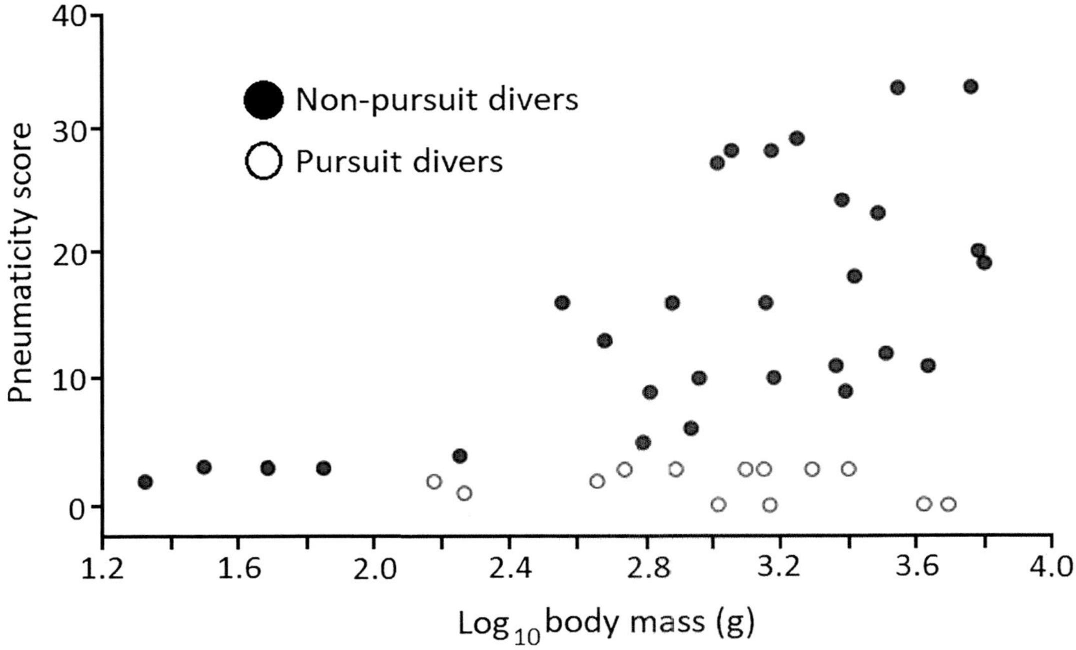


**Fig. 7.41** Relationship between body mass and pneumaticity score for several species of waterbirds. Pneumaticity scores were based on scoring of 50 pneumatic features of bones. Species of pursuit divers have lower pneumaticity scores than non-pursuit divers. In addition, pneumaticity scores tend to increase with increasing body mass. Species names are abbreviated. Pursuit divers include species in the families Spheniscidae (penguins), Gaviidae (loons), Podicipedidae (grebes), Pelecanoididae (diving petrels), Anhingidae (anhingas), and Phalacrocoracidae (cormorants and shags). Non-pursuit divers include species in the families Phaethontidae (tropicbirds), Diomedeidae (albatrosses), Procellariidae (petrels, prions, and shearwaters), Hydrobatidae (storm petrels), Phoenicopteridae (flamingoes), Ciconidae (storks), Ardeidae (herons, egrets, and bitterns), Threskiornithidae (ibises and spoonbills), Scopidae (Hamerkop), Balaenicepidae (Shoebill), Pelecanidae (pelicans), Fregatidae (frigatebirds), and Sulidae (gannets and boobies). (Figure modified from Smith 2012; © 2011 The Author. Evolution © 2011 The Society for the Study of Evolution, used with permission)

(Fedde 1998). Whereas airflow through the paleopulmonic parabronchi is unidirectional, airflow through the neopulmonic parabronchi is bidirectional (Scheid 1979; Fedde 1980). Duncker (1972) suggested that the neopulmo may be the main site of gas exchange when birds are resting and that gas exchange in the paleopulmo only occurs during period of activity. However, Holle et al. (1978) found that that blood flow through the neopulmo and paleopulmo was similar in resting Muscovy Ducks (*Cairina moschata*), suggesting that, for birds at rest, both areas contribute to gas exchange. Parabronchi can be several millimeters long and 0.5–2.0 mm in diameter (depending on the size of the bird; Maina 1989) and their walls contain hundreds of tiny, branching, and anastomosing air capillaries surrounded by a profuse network of blood capillaries. It is within these "air capillaries" that the exchange of gases (oxygen and carbon dioxide) between the lungs and the blood occurs.

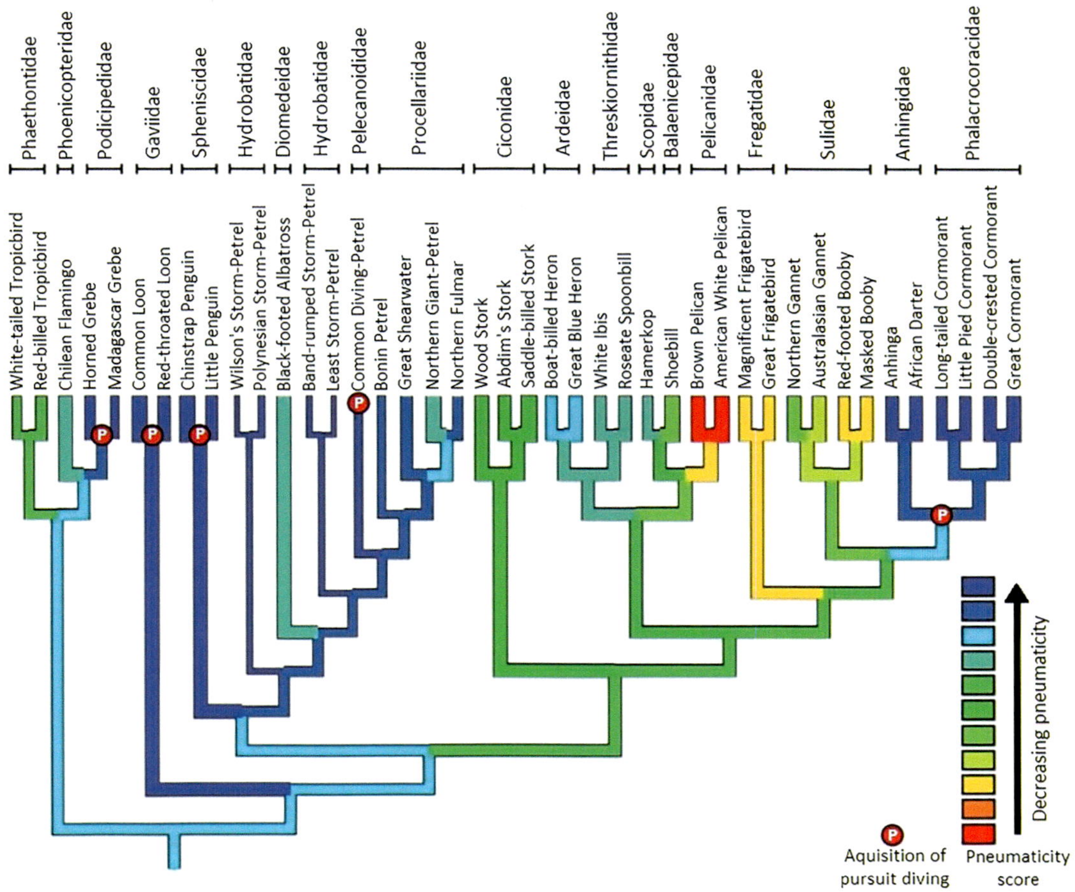


**Fig. 7.42** Variation in body mass and extent of postcranial skeletal pneumaticity in several taxa of waterbirds (phylogeny based on Hackett et al. 2008). Branch colors indicate pneumaticity score, branch thickness indicates $\log_{10}$ body mass, and red circles at nodes indicate the acquisition of pursuit diving. The bones of taxa that evolved to become pursuit divers became much less pneumatic. Waterbird families are indicated at top. White-tailed Tropicbird, *Phaethon lepturus*; Red-billed Tropicbird, *Phaethon aethereus*; Chilean Flamingo, *Pheonicopterus chilensis*; Horned Grebe, *Podiceps auritus*; Madagascar Grebe, *Tachybaptus pelzelnii*; Common Loon, *Gavia immer*; Red-throated Loon, *Gavia stellata*; Chinstrap Penguin, *Pygoscelis antarcticus*; Little Penguin, *Eudyptula minor*; Wilson's Storm-Petrel, *Oceanites oceanicus*; Polynesian Storm-Petrel, *Nesofregetta fuliginosa*; Black-footed Albatross, *Phoebastria nigripes*; Band-rumped Storm-Petrel, *Hydrobates castro*; Least Storm-Petrel, *Hydrobates microsoma*; Common Diving-Petrel, *Pelecanoides urinatrix*; Bonin Petrel, *Pterodroma hypoleuca*; Great Shearwater, *Ardenna gravis*; Northern Giant-Petrel, *Macronectes halli*; Northern Fulmar, *Fulmarus glacialis*; Wood Stork, *Mycteria americana*; Abdim's Stork, *Ciconia abdimii*; Saddle-billed Stork, *Ephippiorhynchus senegalensis*; Boat-billed Heron, *Cochlearius cochlearius*; Great Blue Heron, *Ardea herodias*; White Ibis, *Eudocimus albus*; Roseate Spoonbill, *Platalea ajaja*; Hamerkop, *Scopus umbretta*; Shoebill, *Balaeniceps rex*; Brown Pelican, *Pelecanus occidentalis*; American White Pelican, *Pelecanus erythrorhynchos*; Magnificent Frigatebird, *Fregata magnificens*; Great Frigatebird, *Fregata magnificens*; Northern Gannet, *Morus bassanus*; Australasian Gannet, *Morus serrator*; Red-footed Booby, *Sula sula*; Masked Booby, *Sula dactylatra*; Anhinga, *Anhinga anhinga*; African Darter, *Anhinga rufa*; Long-tailed Cormorant, *Microcarbo africanus*; Little Pied Cormorant, *Microcarba melanoleucos*; Double-crested Cormorant, *Nannopterum auritum*; Great Cormorant, *Phalacrocorax carbo*. (Figure modified from Smith 2012; © 2011 The Author. Evolution © 2011 The Society for the Study of Evolution, used with permission)

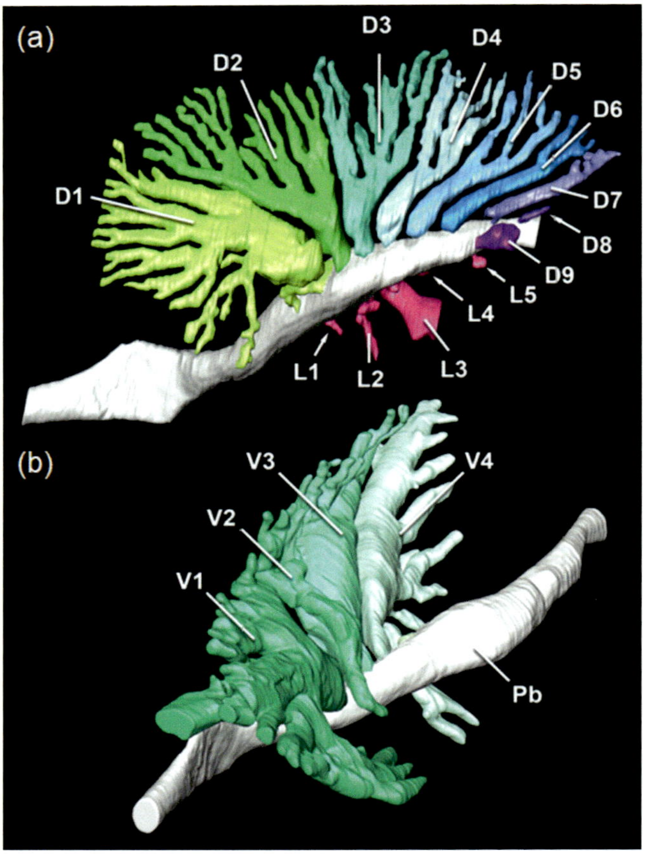


**Fig. 7.43** (**a**) Left lateral view of the "bronchial tree" of a Gray Parrot (*Psittacus erithacus*) showing nine dorsobronchi (D1–D9) and five laterobronch (L1–L5). (**b**) Left craniodorsal view of four ventrobronchi (V1–V4) branching from the primary bronchus (Pb) of an African Gray Parrot. (Figure from Lawson et al. 2021; © 2021 Wiley Periodicals LLC, used with permission)

## 7.3 Ventilation

During each respiratory cycle (inspiration plus expiration), air moves through the avian respiratory system in response to changes in air pressure. Birds that are perched or standing vary pressure in their respiratory system and, specifically, in their air sacs via muscular contraction that moves the sternum and rib cage (Fig. 7.49). During inspiration, the sternum moves slightly forward and downward, with movement of the caudal end of the sternum most pronounced, while the vertebral ribs move cranially to expand the sternal ribs and the thoracoabdominal cavity (Fig. 7.50). This expands the posterior and anterior air sacs in the subpulmonary cavity and lowers the pressure, causing air to move into those air sacs. Air moves through the trachea and bronchi into the posterior air sacs and, simultaneously, air from the lungs moves into the anterior air sacs.

When perched or standing, a bird's sternum can move freely. However, birds sometimes rest on their sternum, e.g., when incubating eggs, limiting movement of the sternum and potentially interfering with ventilation. Muscles associated with the ribs and, specifically, the uncinate processes of the ribs appear to be important under such conditions (Figs. 7.51 and 7.52; Box 7.3 Uncinate Processes and Avian Metabolic Rates). Codd et al. (2005) found that the activity of the appendicocostal muscle increased when sternal movements were restricted, with contraction of these muscles causing a lateral flaring of the rib cage and an increase in volume of the subpulmonary cavity and air sacs during

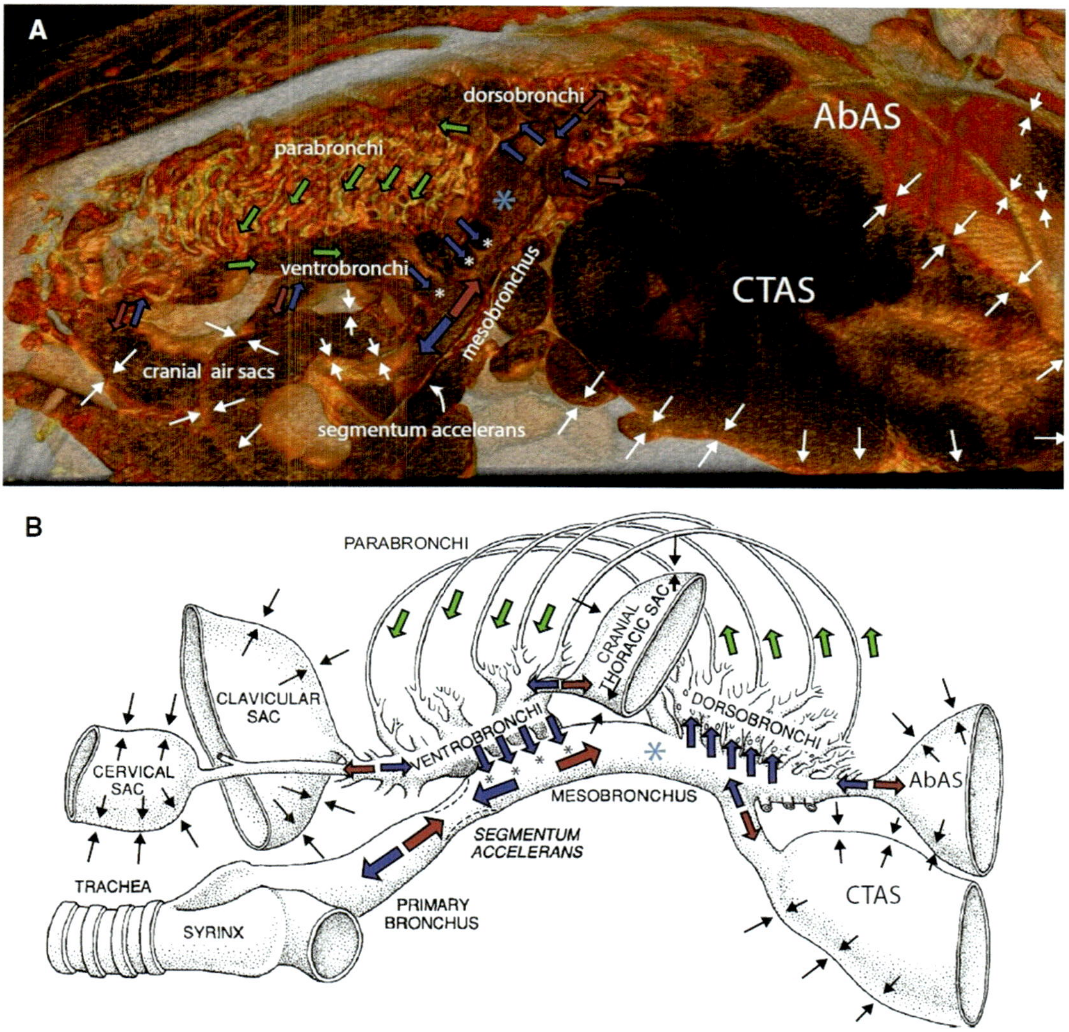


**Fig. 7.44** Illustration of unidirectional airflow in the avian respiratory system showing the bronchi located inside the lungs. Dorsobronchi and ventrobronchi branch off of the mesobronchus; parabronchi extend from the dorsobronchi to the ventrobronchi. Arrows indicate the direction of air flow. **a**. Lung and posterior air sacs of a parrot. **b**. Air flows through the lung as the air sacs expand and contract (small white arrows in **a** and black arrows in **b**). Red arrows show air flow during inspiration, blue arrows during expiration, and green arrows show air flow during both phases of respiration. The inspiratory aerodynamic valves are white and black asterisks, and the expiratory aerodynamic valve is a blue asterisk. The aerodynamic valves are discussed later in this chapter. AbAS, Abdominal air sac; CTAS, caudal thoracic air sac. (Figure A from Cieri and Farmer 2016; © 2016 Springer-Verlag Berlin Heidelberg, used with permission; Figure **b** adapted from Wang et al. 1992; © 1992 Wiley-Liss, Inc., A Wiley Company, used with permission)

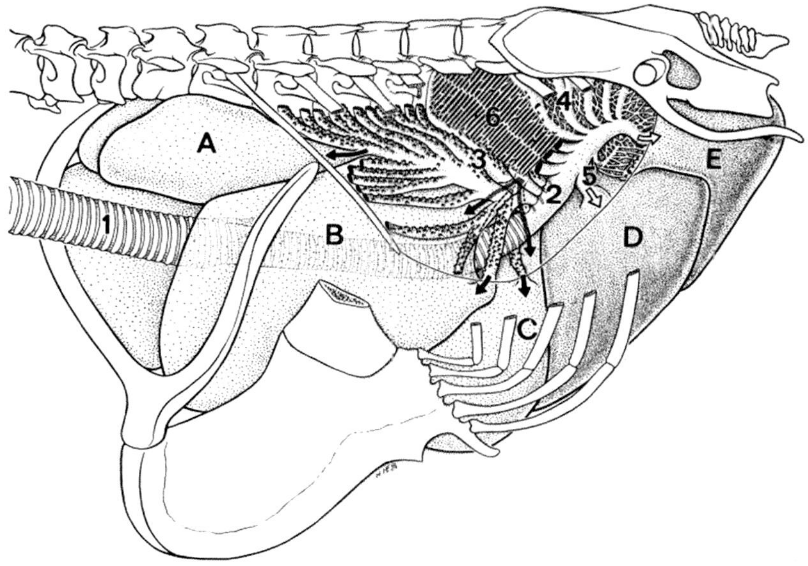


**Fig. 7.45** Drawing of the lung-air sac system of a large bird. The cranial half of the dorsobronchi (4) and the parabronchi (6) has been removed. 1, trachea; 2, primary bronchus; 3, ventrobronchi with the connections into (A) cervical, (B) interclavicular, and (C) cranial thoracic air sacs; 5, laterobronchus and caudal primary bronchus open into (D) posterior thoracic and (E) abdominal air sacs. (Figure from Duncker 1972; © 1972 Published by Elsevier B.V., used with permission)

inspiration. During expiration, the external oblique muscle contracts, moving the sternum dorsally (upward) and decreasing the volume of the subpulmonary cavity and air sacs (Fig. 7.53).

Baumel et al. (1990) studied anesthetized birds resting on their sternums and found that the longissimus dorsi muscle contracted during inspiration, elevating the pelvis and increasing the volume of the thoracoabdominal cavity. This would reduce the pressure in the thoracic and abdominal air sacs as air flowed into those air sacs during inspiration. During expiration, contraction of the infrapubic abdominal muscles caused depression of the pelvis, decreasing the volume of the thoracoabdominal cavity and increasing pressure in the thoracic and abdominal air sacs, which, in turn, would move air out of those air sacs and into the lungs. These movements of the pelvis that change air pressure in the thoracic and abdominal air sacs may be especially important when a bird is resting on, and impeding the movements of, its sternum (Baumel et al. 1990; Ruben et al. 2003; Fig. 7.54).

Boggs et al. (1997) suggested that pelvic movements may also play a role in lung ventilation when birds are flying. These investigators found that the synsacrum of some Black-billed Magpies (*Pica hudsonia*) rotated ventrally with each wing downstroke, compressing the thoracoabdominal cavity, and then rotated dorsally during the upstroke, expanding the thoracoabdominal cavity. This depression and elevation of the synsacrum and pelvis may facilitate ventilation during flight.

## 7.4 Metabolic Cost of Breathing

Some investigators have suggested that the energetic cost of breathing would be higher for birds than mammals because of the greater mass of muscle attached to the sternum (Markley and Carrier 2010). However, the energetic cost of breathing by Helmeted Guineafowl (*Numida meleagris*) was estimated to be between 1 and 2% of the basal metabolic rate (BMR; Ellerby et al. 2005; Markley and Carrier 2010). In addition, York et al. (2017) estimated that the metabolic cost of breathing of 11 species of waterfowl was less than 1% of BMR. Similarly, York et al. (2018) estimated the metabolic cost of breathing to be just 0.38% of BMR for Bar-headed Geese (*Anser indicus*), 0.68% of BMR for Andean Geese (*Oressochen melanoptera*), and 0.78% of BMR for Barnacle Geese (*Branta leucopsis*). However, experiments conducted by York et al.

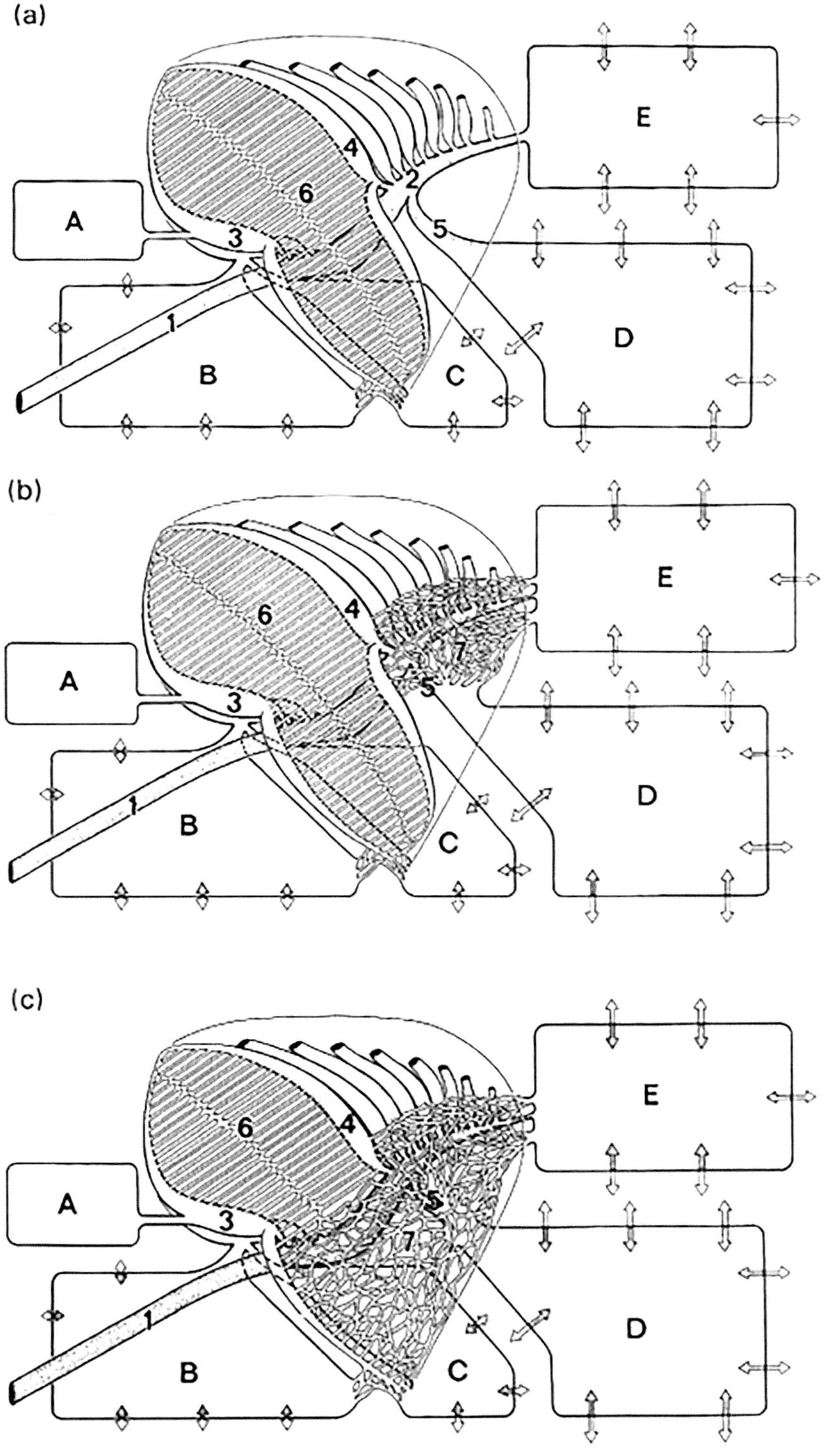


**Fig. 7.46** Differences among different birds in the development of the neopulmonic region of the lung. (**a**) Penguin lungs are entirely paleopulmonic. (**b**) Some birds, such as ducks, have a relatively small neopulmonic region. (**c**) Songbirds have a well-developed neopulmonic region. 1, trachea; 2, primary bronchus; 3, ventrobronchus; 4, dorsobronchus; 5, lateral bronchus; 6, paleopulmonic parabronchi; 7, neopulmonic parabronchi; A, cervical air

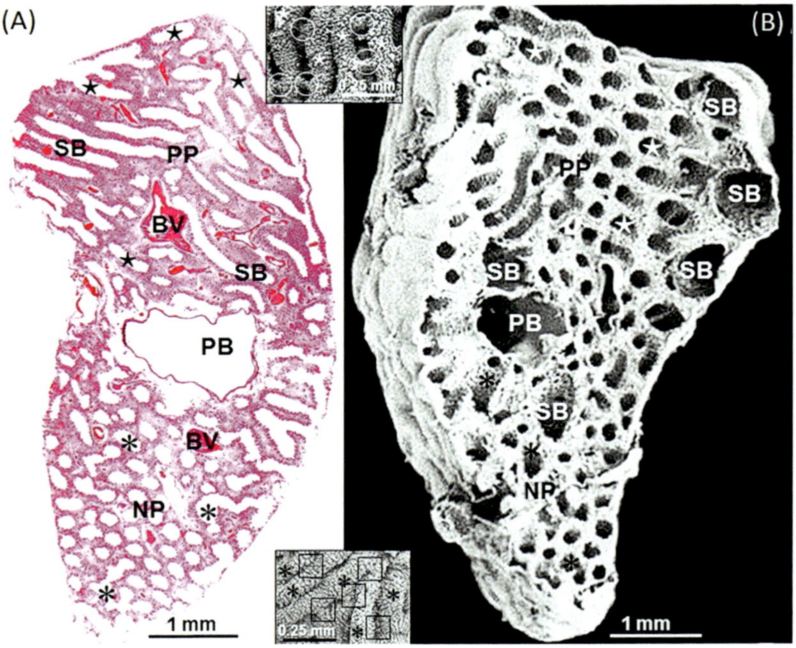


**Fig. 7.47** The paleomomic (PP) and the neopulmonic regions (NP) of the lung of a Domestic Chicken (*Gallus g. domesticus*). The paleopulmo is dorsal to the primary bronchus (PB) and the neopulmo ventral to it. Stars = parabronchi of the paleopulmo, asterisks = parabronchi of the neopulmo, SB = secondary bronchi, BV = blood vessel. Inserts: top, the paleopumo comprises stacks of parabronchi (stars) which interconnect sporadically (circles); bottom, the parabronchi of the neopulmo (asterisks) anastomose profusely (squares). (**a**) Stained section. (**b**) Electron micrograph. Inserts are electron micrographs of latex casts. (Figure from Maina 2015c; © 2015, Dt. Ornithologen-Gesellschaft e.V., used with permission)

(2017, 2018) were with waterfowl in a prone position (sternal recumbancy), which might result in underestimating the energetic cost of breathing compared to a standing bird. When standing or perched, the energetic cost of breathing might be higher because of the need to overcome the gravitational pull on the large flight muscles attached to the sternum (York et al. 2017).

## 7.5 Ventilation During Locomotion

Movements of the sternum are important in avian respiration. However, the sternum is also the site of origin of the two largest flight muscles, the pectoralis and supracoracoideus. The forces generated by these muscles during flight, therefore, might have some effect on respiration.

**Fig. 7.46** (continued) sac; B, interclavicular air sac; C, cranial thoracic air sac; D, caudal thoracic air sac; E, abdominal air sac. The white arrows indicate changes in volume of the air sacs during the respiratory cycle. (Figure from McLelland 1989a, b; © 1989 Elsevier, Inc., used with permission)

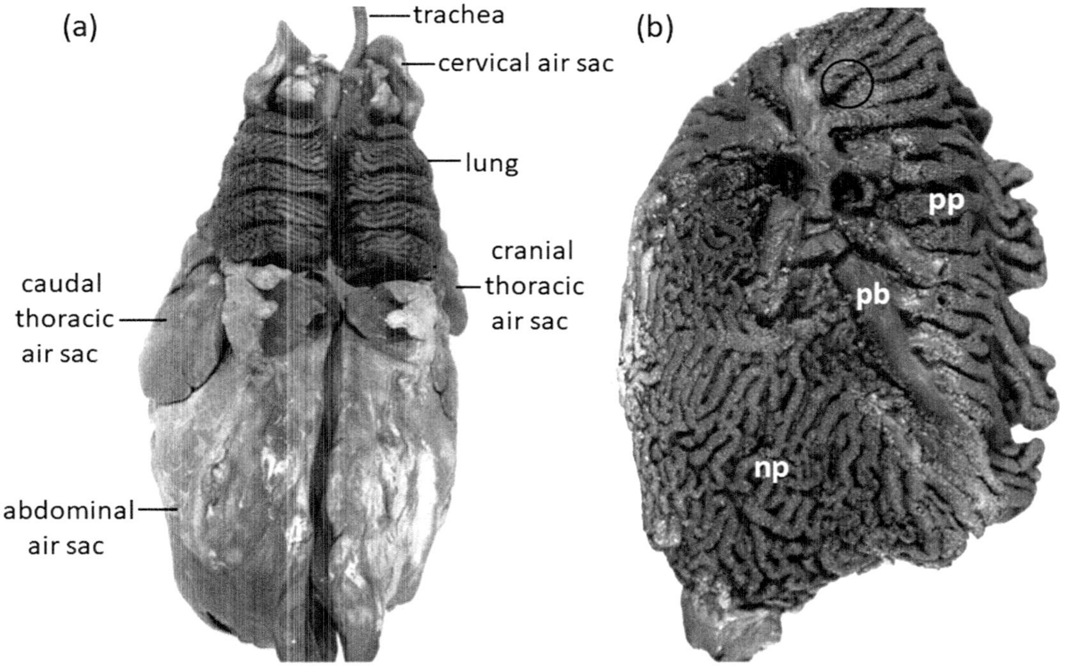


**Fig. 7.48** (**a**) Lungs and air sacs of a Domestic Chicken. (**b**) Medial view of a plastic cast of a lung of a Domestic Chicken. pb, primary bronchus; pp, paleopumonic parabronchi,; np, neopulmonic parabronchi; circle indicates a secondary bronchus. (Figure from Maina 2000; © 2000 Wiley-Liss, Inc., used with permission)

**Box 7.2 Evolution of Avian Air Sacs**

Like birds, air moves unidirectionally through the lungs of squamates and crocodilians. This means that unidirectional air flow could have been present in basal archosaurs, the common ancestor of crocodilians, dinosaurs, and birds. However, unidirectional pulmonary air flow is present in animals that do not have air sacs, so such air flow provides no insight concerning the evolution of air sacs. Pneumatic foramina and bones do provide evidence of air sacs and these have been found in many sauropods and non-avian theropods (Wedel 2003, 2006). However, in the early evolution of air sacs, they may not have penetrated into bones. As such, the early evolutionary history of air sacs remains, as noted by Lambertz et al. (2018), "enigmatic."

(continued)

**Box 7.2** (continued)

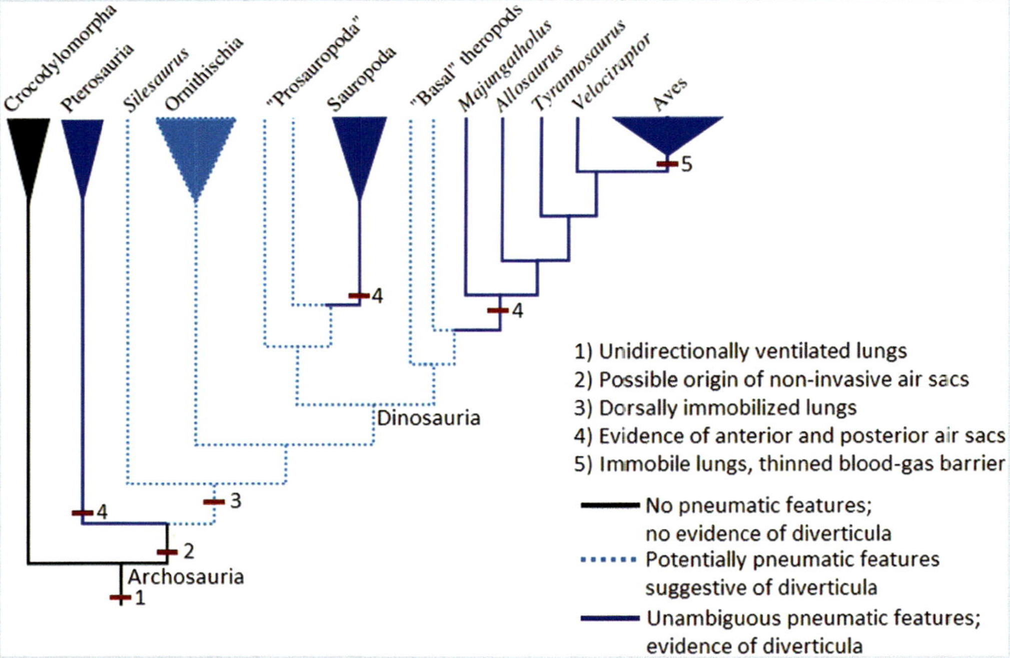


Possible evolution of the archosaur respiratory system. A phylogeny of Archosauria, showing key modifications to the respiratory system associated with the evolution of the avian lung-air sac system. (Figure from Brocklehurst et al. 2018; © 2018 The Authors. Published by the Royal Society under the terms of the Creative Commons Attribution License, http://creativecommons.org/licenses/by/4.0/).

(continued)

**Box 7.2** (continued)

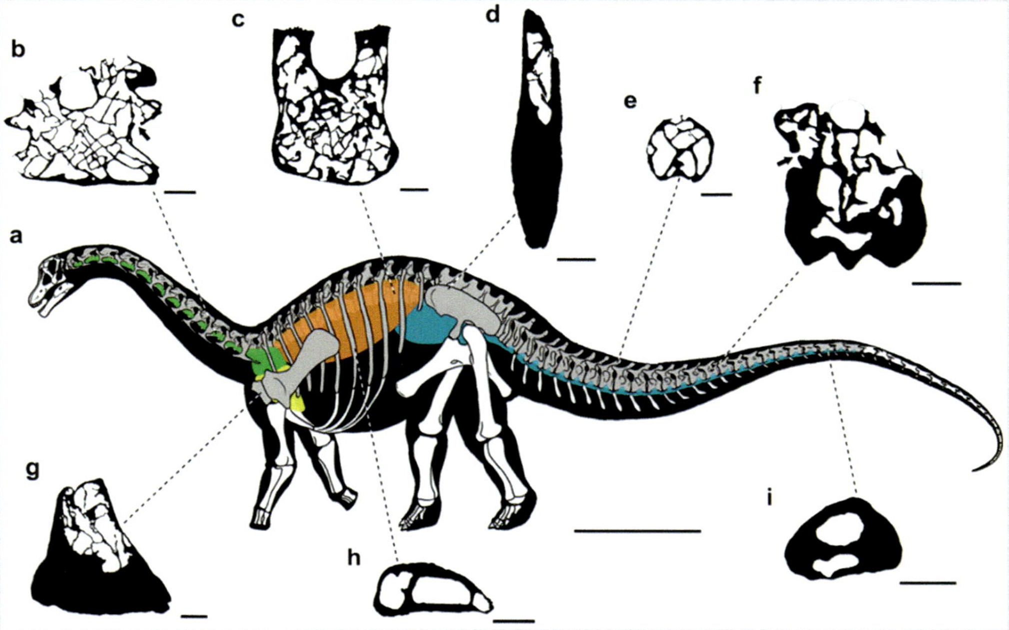


Reconstruction of the pulmonary air sacs of the sauropod, *Saltasaurus loricatus*. (**a**) Lateral view showing pneumatized postcranial bones (gray). Components of the pulmonary air-sac system are based on pneumaticity of the postcranial skeleton and include the cervical system (green), lung (orange), clavicular system (yellow), and abdominal system (blue). (**b**–**i**) Cross section of different bones showing internal cavities. (**b**) Incomplete cervical vertebra, (**c**) vertebral centrum, (**d**) process of a left ilium, (**e**) neural spine of a caudal vertebra, (**f**) caudal vertebra centrum, (**g**) coracoid, (**h**) thoracic rib, and (**i**) caudal vertebra centrum. Arrowheads indicate pneumatic foramina. Scale bars = 100 cm (**a**) and 2 cm (**b**–**i**). (Figure from Cerda et al. 2012; © 2012 Springer Nature, used with permission).

Lambertz et al. (2018) examined the bones of several present-day birds, some invaded by air sacs and some not invaded, and found that bone lined by air sacs differed histologically from bones not lined by air sacs. This unique tissue, referred to as pneumosteal tissue, was also found in the pneumatic caudal vertebrae of some theropods. The potential importance of the discovery of pneumosteal tissue is that it could be used to better understand the evolution of air sacs because its presence in or on apneumatic bones of avian ancestors may allow detection of the presence of air sacs prior to the evolution of pneumatic bones.

(continued)

**Box 7.2** (continued)

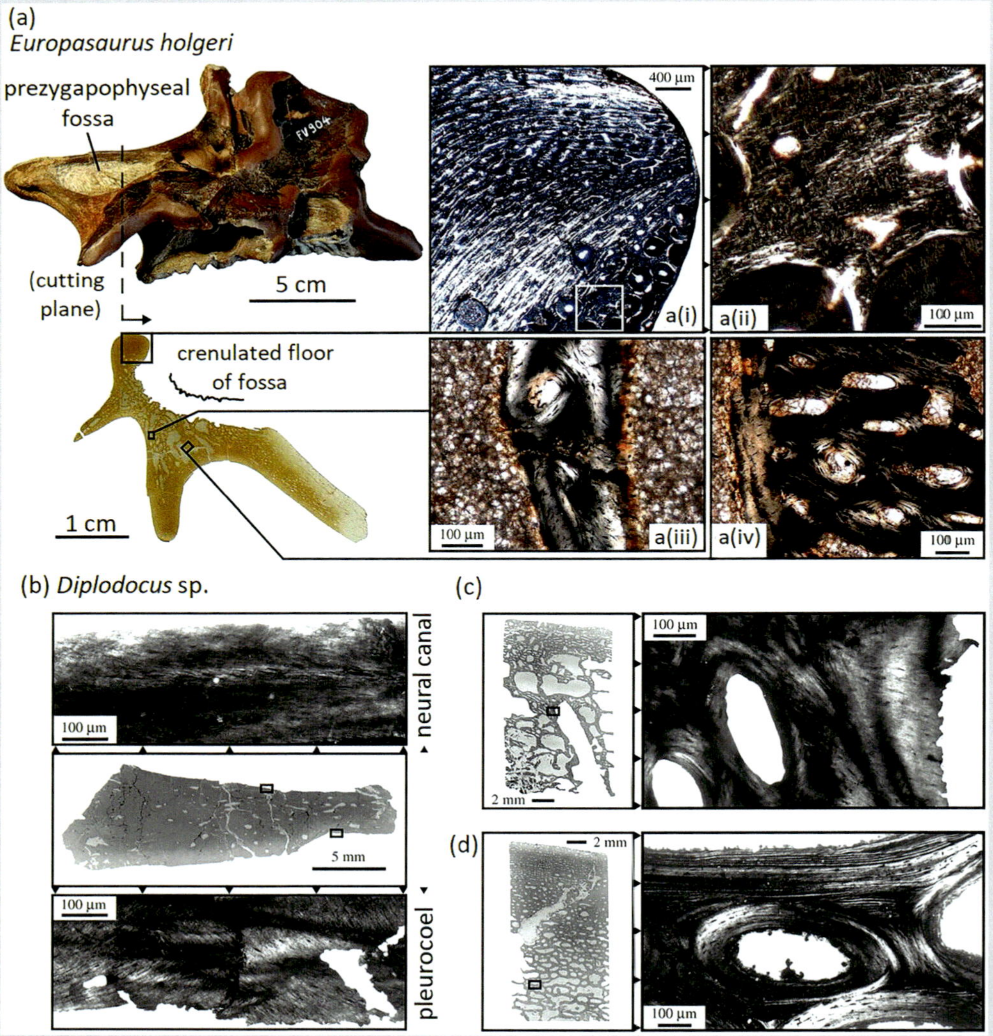


Pneumosteal tissue in sauropods. (**a**) Cross section of a prezygapophysis of a cervical vertebra of *Europasaurus holgeri*, the smallest known sauropod dinosaur (about 6.2 m long as an adult). The lamina connecting the neural spine and prezygapophysis consists of laminar bone dorsally (a (i)) and metaplastic bone (formed in connective tissue) ventrally (a(ii)). The surface of the lateral fossa is composed of trabecular bone with densely packed fine fibers: pneumosteal bone (a(iii–iv)). The appearance of pneumosteal fibers differs from fibers where tendons insert on bones a(ii). (**b**) Pneumatized vertebrae of *Diplodocus* sp. with pneumosteal tissue facing the pleurocoel (cavity in the side of a vertebra), but parallel-fibered bone lining the neural canal. (**c**) Pneumatized caudal vertebrae with pneumosteum in the trabeculae. (**d**) Apneumatic caudal vertebrae with no pneumosteum in their trabeculae. (Figure from Lambertz et al. 2018; ).

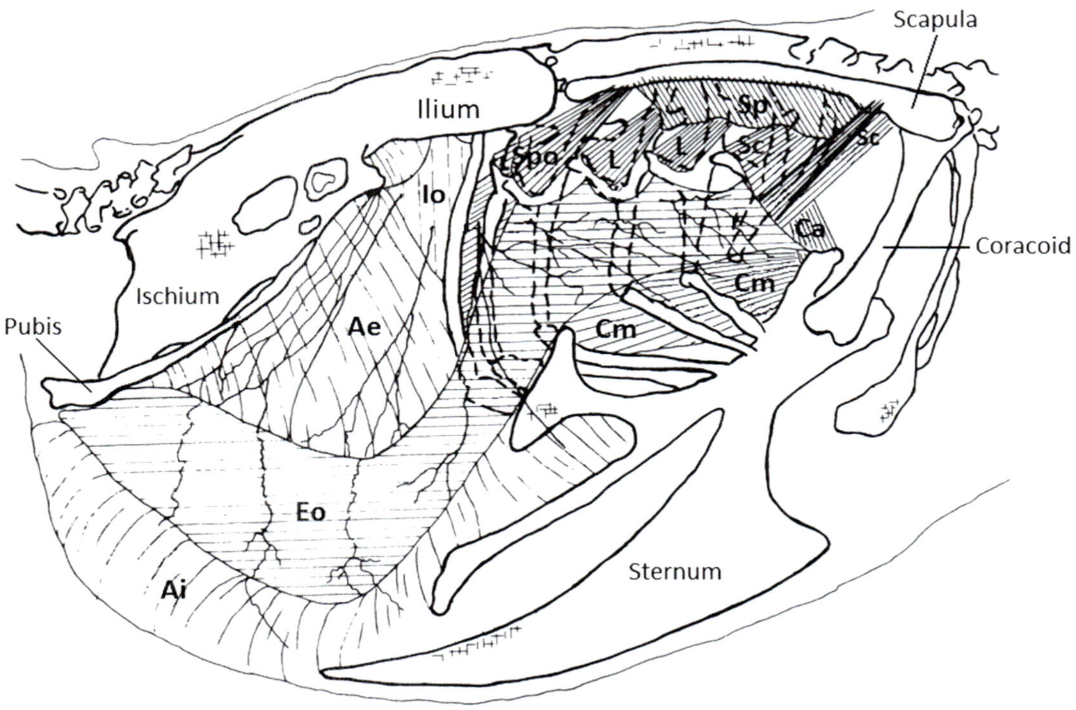


**Fig. 7.49** Muscles of inspiration include the scalenus (Sc), costosternalis major (Cm), levator costae (L), and serratus profundus (Sp). Muscles that play a role in expiration include the costosternalis minor (Ca), external oblique (Eo), internal oblique (Io), and serratus superficialis (Spo). Ae, aponeurosis of the insertion of the external oblique; Ai, aponeurosis of the origin of internal oblique. (Figure from DeWet et al. 1967; © 1967 Wiley-Liss, Inc., used with permission)

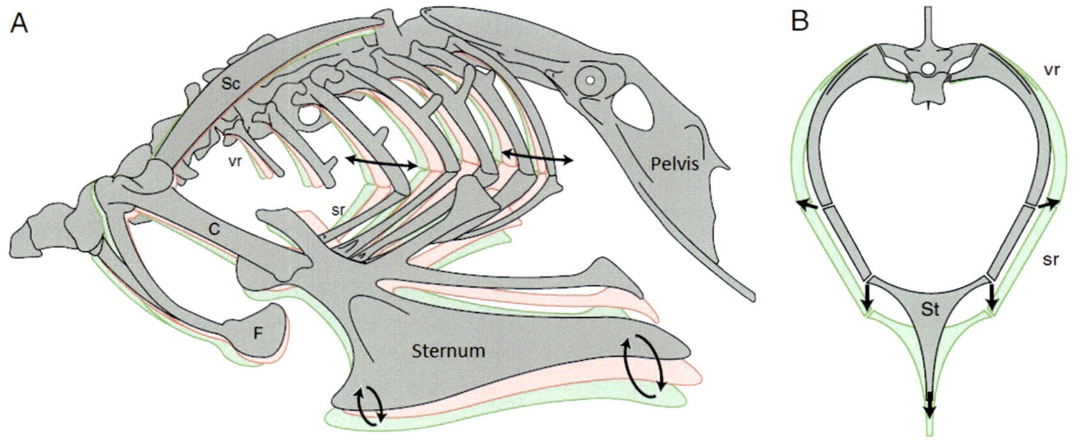


**Fig. 7.50** Model of skeletal movements involved in the avian aspiration pump, with elliptical sternal movement, elevation of the distal ends of the sternal ribs during inspiration, and furcular and coracosternal movement. Lateral (**a**) and transverse (**b**) views of the movements of the postcranial skeleton during lung ventilation. Gray = end of expiration, pink = mid-inspiration, and green = peak inspiration. Linear movement of the vertebral ribs results in elliptical sternal displacement (**a**). The vertebral ribs move forward, upward (**a**), and outward (**b**) during inspiration. Abbreviations: F, furcular; C, coracoid; Sc, scapula; vr, vertebral rib; sr, sternal rib; St, sternum. (Figure from Claessens 2009; © 2008 Wiley-Liss, Inc., A Wiley Company, used with permission)

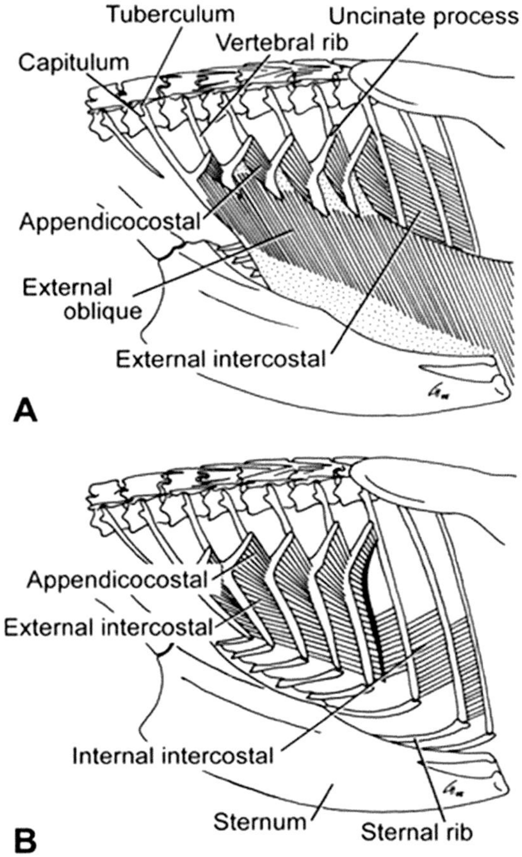


**Fig. 7.51** During inspiration, contraction of the appendicocostal muscles moves the vertebral ribs cranially and this, in turn, moves the sternum downward. Other muscles involved in inspiration, but not shown in this figure, include the costosternalis, scaleneus, and serratus profundus muscles (Fedde 1987). The appendicocostal muscles contract with greater force when a bird is sitting and movement of the sternum is restricted. These stronger contractions cause the lateral flaring or outward movement of the ribcage, expanding the air sacs during the process of inspiration (Codd et al. 2005). The external oblique muscle contracts during expiration, moving the sternum and ribcage upward. Other muscles involved in expiration include the internal intercostal muscles and several not shown in this figure, including the internal and external obliques, transverses, rhomboideus, and latissimus dorsi muscles (Fedde 1987). (Figure from Perry et al. 2005; used with permission of Steven Perry)

Studies of Rock Pigeons (*Columba livia*) and Black-billed Magpies (*Pica hudsonia*) flying in a wind tunnel have revealed changes in air-sac pressures as the pectoralis and supracoracoideus muscles contract and relax (Fig. 7.55). Contraction of the pectoralis (downstroke) has a compressive effect, whereas contraction of the supracoracoideum (upstroke) has an expansive effect on the air sacs (Boggs 1997; Boggs et al. 1997). These alternating compressive and expansive effects suggest "a mechanical assistance of downstroke to expiration and upstroke to inspiration" (Boggs 2002, p. 281), with changes (and differences) in pressures in the posterior and anterior air sacs enhancing movement of air through the lungs. However, both within and between species the ratio between wingbeats and breathing can vary (Table 7.1). Rock Pigeons have been found to take one breath per wingbeat cycle (downstroke, upstroke), but, for other species, this ratio has been found to vary from two wingbeat cycles per breath to five wingbeat cycles per breath (Table 7.1). This ratio has also been found to exhibit little variation with changes in flight speed in some species, but to change with changes in flight speed for other species. For example, the ratio of wingbeats to breaths exhibits little variation with flight speed for Rock Pigeons (Tobalske and Dial 1996), but this ratio changes for Black-billed Magpies shifting from slow, low amplitude flapping flight to rapid, high amplitude flight (Boggs 1997). Although the ratio changes for magpies, wing upstrokes still occur during the transition into inspiration, and downstrokes during the transition into expiration (Boggs 1997).

For birds that spend time under water, either wing-propelled as in penguins or foot-propelled as in many waterbirds, movements of the wings and feet may also alter pressures in the air sacs and enhance the movement of air through the lungs. Boggs et al. (2001) examined muscle activity and air-sac pressures of Little Penguins (*Eudyptula minor*) in the lab and found that fluctuations and differences in air pressure in the posterior and anterior air sacs caused by contraction and relaxation of the "flight" muscles caused

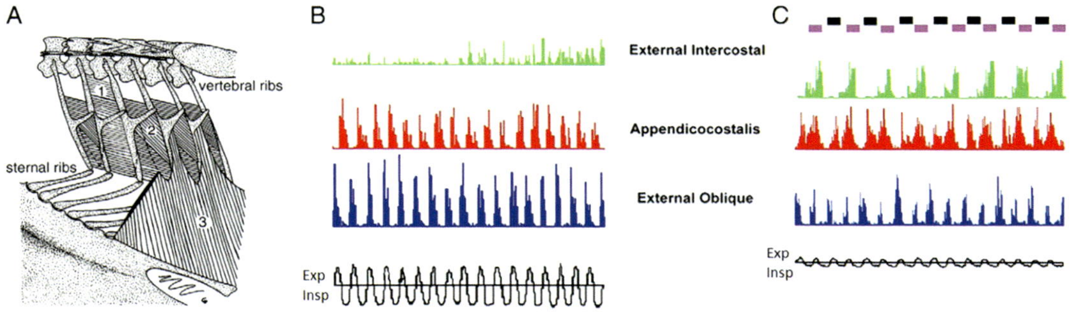


**Fig. 7.52** (**a**) Muscles associated with the uncinate processes of birds. (1) External intercostal, (2) Appendicocostalis, and (3) External oblique. Anterior is to the left. Electromyogram traces showing activity of (1) external intercostal muscle, green trace, no respiratory function during standing (**b**), but active during contralateral limb support (pink squares) (**c**), appendicocostalis, red trace, active during inspiration when standing (**b**) and ipsilateral limb support (black squares) during locomotion (**c**), and (3) blue trace, M. Externus obliquus, active during expiration when standing and walking. Exp, expiration, Insp, inspiration. (Figure from Codd 2010; © 2009 Elsevier Inc., All rights reserved, used with permission)

**Box 7.3 Uncinate Processes and Avian Metabolic Rates**

Uncinate processes that project caudally from the ribs of birds act as levers that improve the mechanical advantage for forward rotation of the dorsal ribs and the ventral movement of the sternum during inspiration (Tickle et al. 2007). The length of uncinate processes varies among birds that use different forms of locomotion, with those of diving birds longer and those of "walking" birds shorter than the intermediate-length uncinate processes of flying and swimming birds (Tickle et al. 2007, 2009). Diving birds also have relatively long sternums (see photos below), so longer uncinate processes are likely related to the need for a greater mechanical advantage to move the sternum during respiration (Tickle et al. 2007). In contrast, walking species of birds (i.e., flightless or not capable of sustained flight) typically have relatively small sternums and reduced flight muscle mass, thus requiring a relatively small mechanical advantage to move the sternum and ribs during respiration (Tickle et al. 2007).

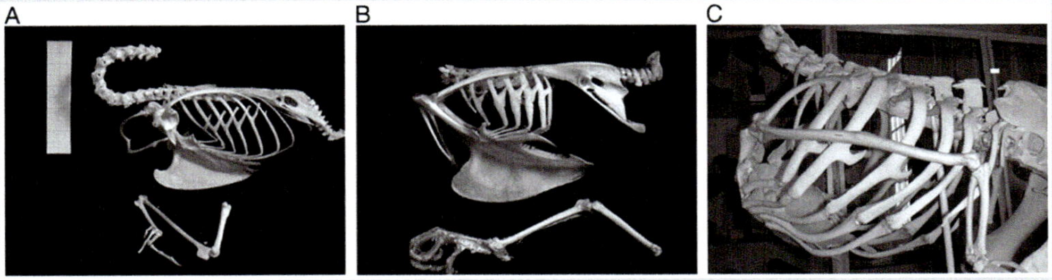


Birds with different forms of locomotion exhibit morphological differences in their rib cages. Here are three examples: (**a**) Common Tern (*Sterna hirundo*), an aquatic, diving species, (**b**) Rock Pigeon (*Columba livia*), a nonspecialist flying bird, and (**c**) Common Ostrich (*Struthio camelus*), a terrestrial, walking/running bird. Scale bar on left = 5 cm. Uncinate processes are

(continued)

**Box 7.3** (continued)

short in walking species, intermediate in length in typical birds, and relatively long in diving species. Muscles attached to uncinate processes (appendicocostales muscles) help rotate the ribs forwards, pushing the sternum down and inflating the air sacs during inspiration. Another muscle (external oblique) attached to uncinate processes pulls the ribs backward, moving the sternum upward during expiration. The longer uncinate processes of diving birds are probably related to the greater length of the sternum and the lower angle of the ribs to the backbone and sternum. The insertion of the appendicocostales muscles near the end of the uncinate processes may provide a mechanical advantage for moving the elongated ribs during breathing. (Figure from Codd 2010; © 2009 Elsevier Inc., used with permission).

Interestingly, Tickle et al. (2009) found that resting metabolic rates also varied with the length of avian uncinate processes, both among and within groups of birds with different forms of locomotion. These authors suggested that longer uncinate processes facilitate increases in metabolic rate by making respiration more efficient, i.e., using less effort to move more air and oxygen through the respiratory system. However, Tickle et al. (2009) suggested that longer uncinate processes may also come with a cost, possibly based on the need to change the shape of the associated musculature.

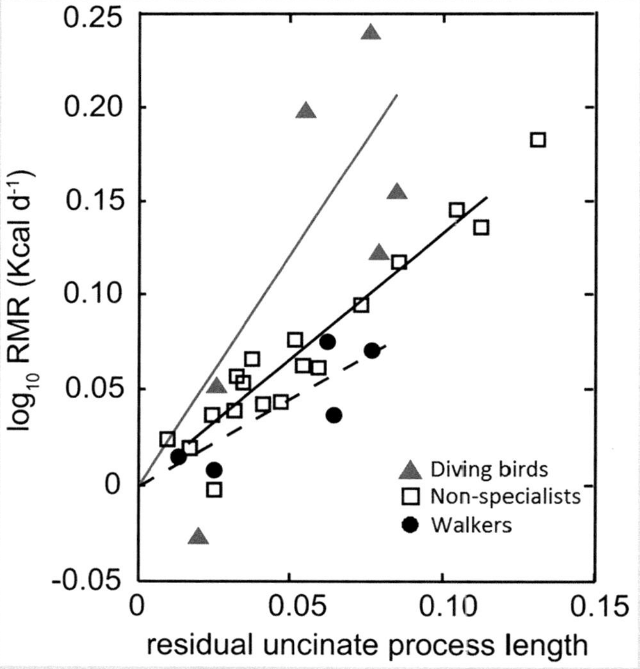


Relationship between resting metabolic rate and length of uncinate processes. Diving birds have relatively longer uncinated processes than most other birds, whereas "walkers," flightless birds and birds not capable of sustained flight have relatively shorter uncinate processes than most other birds. (Figure from Tickle et al. 2009; © 2009 Tickle et al., open-access article distributed under the terms of the Creative Commons Attribution License, https://creativecommons.org/licenses/by/4.0/).

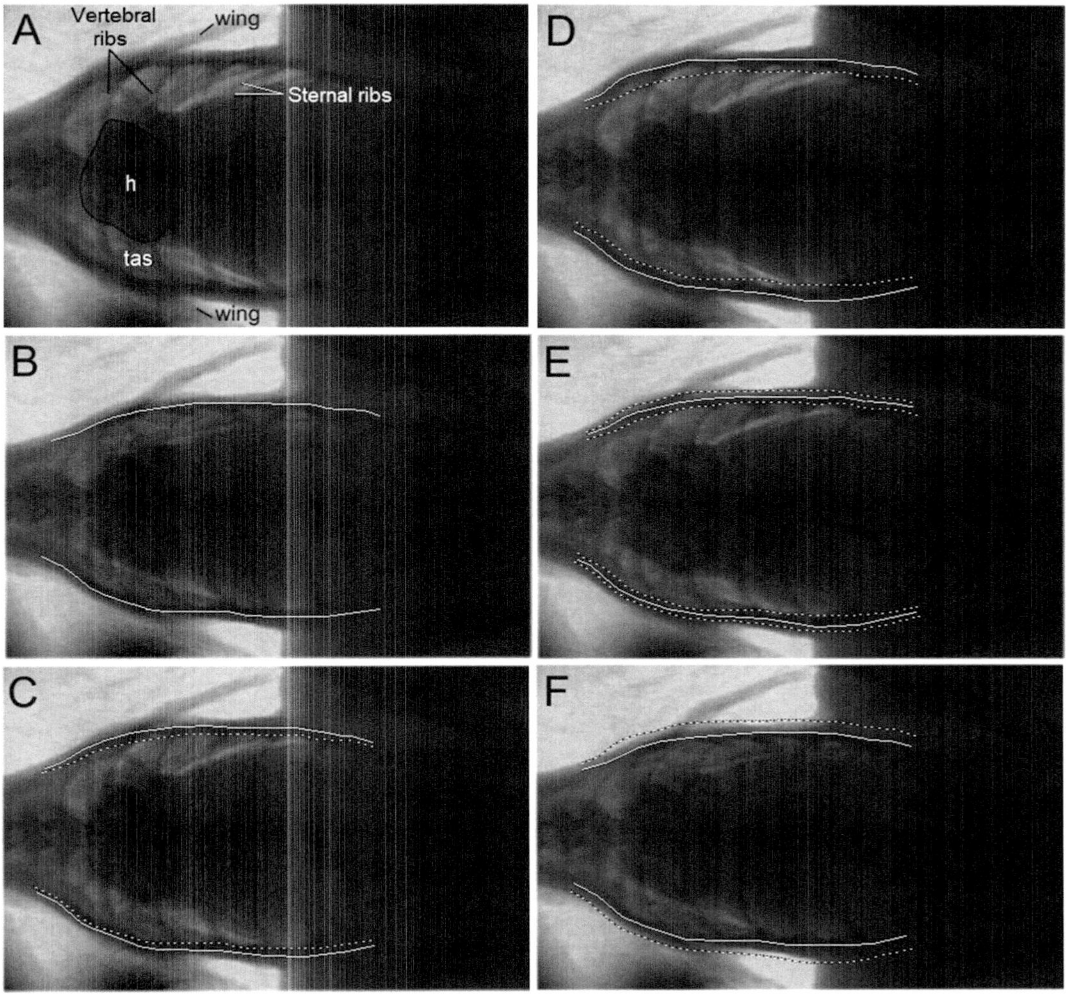


**Fig. 7.53** Illustration of the transverse displacement of the body wall during lung ventilation of a young Emu (*Dromaius novaeahollandiae*), and similar displacement of the ribs. (**a**)–(**c**) Inspiration, and (**d**)–(**f**) Expiration. Dashed outlines indicate the maximum expansion and contraction of the lateral margin of the rib cage, respectively. Solid lines indicate the position of the thoracic body wall in each image. Inflation and deflation of the thoracic air sacs can be seen in frames (**a**)–(**f**). h, heart; sr, sternal rib; tas, thoracic air sacs (superimposed); vr, vertebral rib. (Figure modified from Claessens 2009; © 2008 Wiley-Liss, Inc., A Wiley Company, used with permission)

air to move through the lungs, helping to optimize use of available oxygen when underwater. Similarly, movements of the hindlimbs of foot-propelled swimmers and divers also cause fluctuations and differences in air pressure in the posterior and anterior air sacs, with the resulting movement of air through the lungs, again optimizing use of available oxygen when underwater (Fig. 7.56).

Investigators have also found relationships between stride rates and breaths for several species of walking and running birds, with reported ratios of 2:1, 3:1, 1:2, 2:3, and 1.5:1 (Table 7.2; Fig. 7.57). However, birds may not always coordinate strides with breaths. For example, Funk

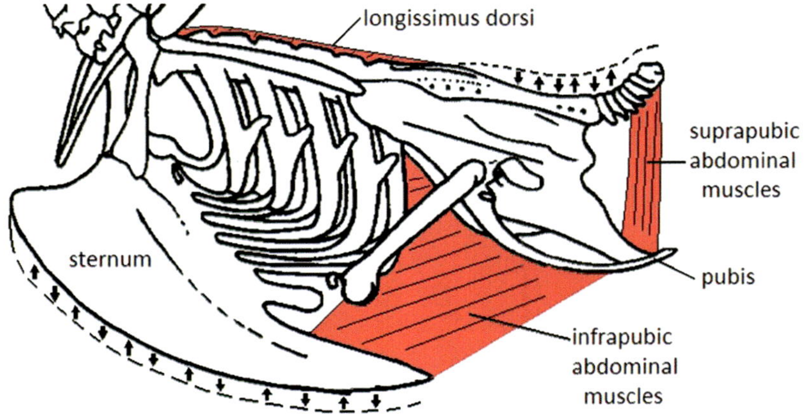


**Fig. 7.54** During inspiration, the sternum rotates ventrally and the longissimus dorsi muscles elevate the pelvic girdle. During expiration, the supra- and infrapubic musculature may rotate the pelvis and tail, changing the volume and air pressures in the posterior air sacs to aid in ventilation. This may be particularly important when a bird is resting on its sternum, limiting its movement (Skeleton from Wikipedia, CC BY 2.5, https://creativecommons.org/licenses/by/2.5/)

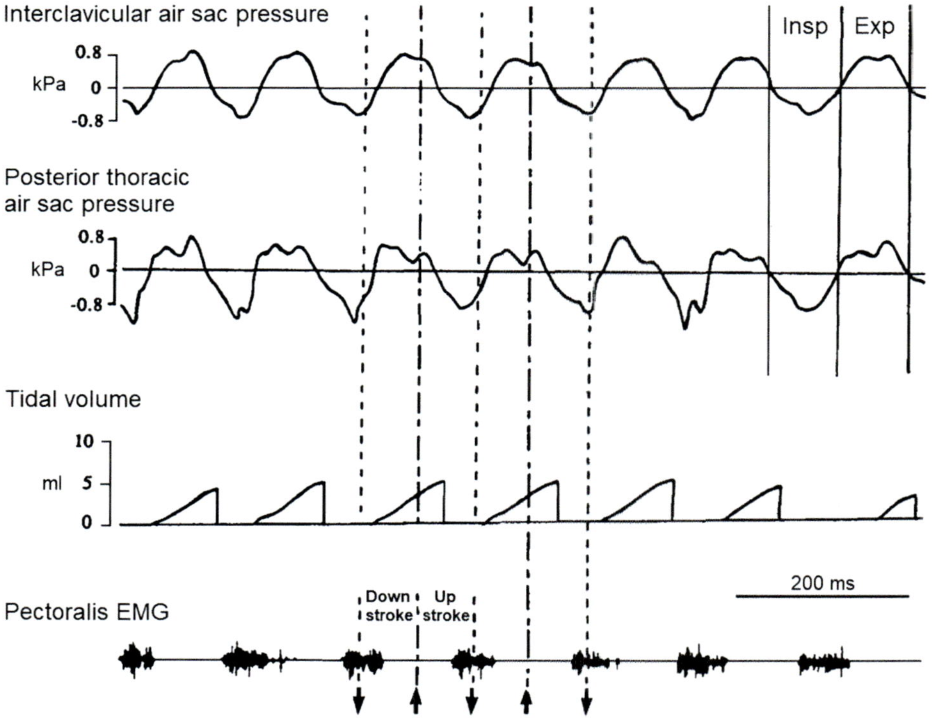


**Fig. 7.55** Pressure in anterior (interclavicular) and posterior (posterior thoracic) air sacs (relative to atmospheric pressure = 0) while a Rock Pigeon (*Columba livia*) is flying. The pectoralis EMG (electromyograph) indicates when the pectoralis muscle is contraction (downstroke) during the wingbeat cycle. Downstrokes begin at the dotted lines and upstrokes at the dashed lines. Note that Rock Pigeons take one breath per wingbeat cycle, with wing upstroke occurring at the transition from expiration to inspiration and wing downstroke at the transition from inspiration to expiration. (Figure from Boggs 1997; © 2015 Oxford University Press, used with permission)

**Table 7.1** Ratios of wingbeats to breaths for several species of birds. All species from Berger et al. (1970) except the last three (Table from Berger et al. 1970; © 1970 Springer-Verlag, used with permission)

| Species | Scientific name | Ratio of wingbeats to breaths |
|---|---|---|
| Ring-billed Gull | *Larus delawarensis* | 2:1, 5:1, 3:1 |
| American Crow | *Corvus brachyrhynchos* | 1:1, 3:2 |
| Black-capped Chickadee | *Poecile atricapillus* | 3:1, 7::2 |
| Rock Pigeon | *Columba livia* | 1:1 |
| Black-rumped Waxbill | *Estrilda troglodytes* | 3:1, 4:1 |
| Evening Grosbeak | *Coccothraustes vesperinus* | 3:1, 7:2, 4:1 |
| Wood Duck | *Aix sponsa* | 3:1, 4:1 |
| American Black Duck | *Anas rubripes* | 3:1, 4:1, 5:1 |
| Ring-necked Pheasant | *Phasianus colchicus* | 3:1, 4:1, 5:1 |
| Northern Bobwhite | *Colinus virginianus* | 5:1 |
| European Starling | *Sturnus vulgaris* | 5:1, 4:1 |
| Barnacle Goose[a] | *Branta leucopsis* | 3:1, 4:1, 2:1 |
| Canada Goose[b] | *Branta canadensis* | 3:1, 2:1 |
| Black-billed Magpie[c] | *Pica husonia* | 3:1, 5:2, 2:1 |

[a] From Butler and Woakes (1980)
[b] From Funk et al. (1993)
[c] From Boggs et al. (1997)

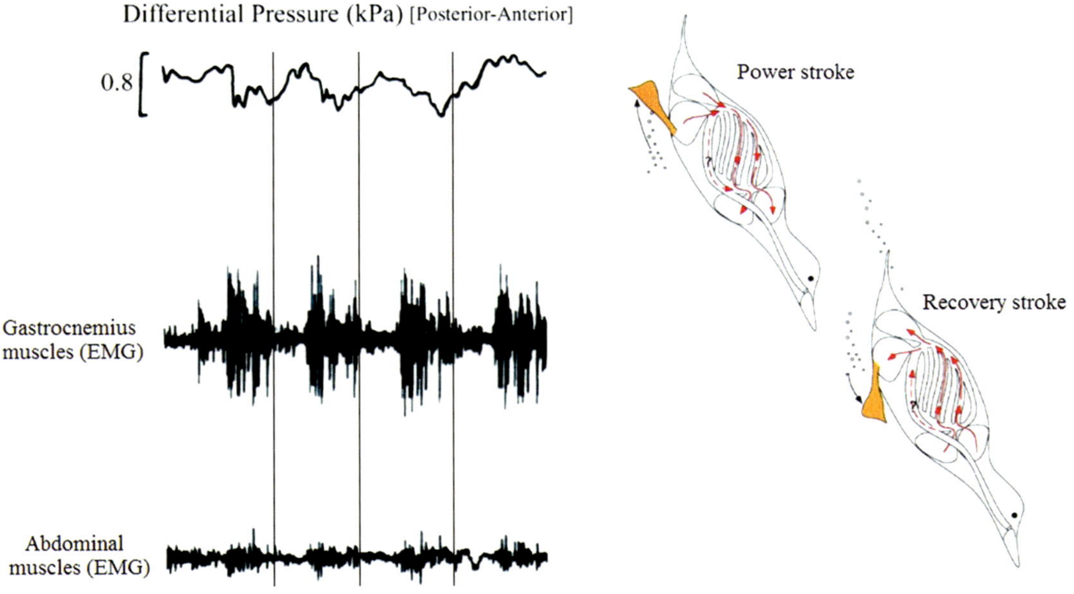


**Fig. 7.56** During a dive, differences between air pressure in the posterior and anterior air sacs oscillate. During the power stroke, the gastrocnemius muscle and abdominal muscles contract and the backward movement of the legs compresses the posterior air sacs. The increase in air pressure relative to that in the anterior air sacs may then cause air to move from those sacs through the lungs and into the anterior air sacs. When those muscles relax, air might then flow in the opposite direction. Such movement of air would allow more effective use of the oxygen stored in the respiratory system during a dive. (Figure modified from Boggs 2002; © 2002 Elsevier Science Inc., used with permission)

**Table 7.2** Ratios of strides to breaths for several species of walking or running birds (Table from Boggs 2002; © 2002 Elsevier Science Inc., used with permission)

| Species[a] | Strides:Breaths[b] | Speed (m/s) | Reference |
|---|---|---|---|
| Chukar | 2:1 | 1 | Boggs (unpublished) |
| Tufted Duck | 2:1 | – | Boggs (unpublished) |
| Little Penguin | 2:1 | 0.4 | Boggs (unpublished) |
| Domestic Chicken | 3:1, 1:2 | 1.4 | Brackenbury and Avery (1980) |
| Helmeted Guineafowl | 2:3 | 4.15 | |
| | 1:2 | 2.8–3.3 | Nassar et al. (2001) |
| Canada Goose | 1.5:1, 2:1 | 0.4 | Funk et al. (1989) |

[a] Scientific names: Chukar, *Alectoris chukar*; Tufted Duck, *Aythya fuligula*; Little Penguin, *Eudyptula minor*; Domestic Chicken, *Gallus g. domesticus*; Helmeted Guineafowl, *Numida meleagris*; Canada Goose, *Branta canadensis*

[b] Stride = one full cycle of footfall to foot fall for one foot or two bilateral steps

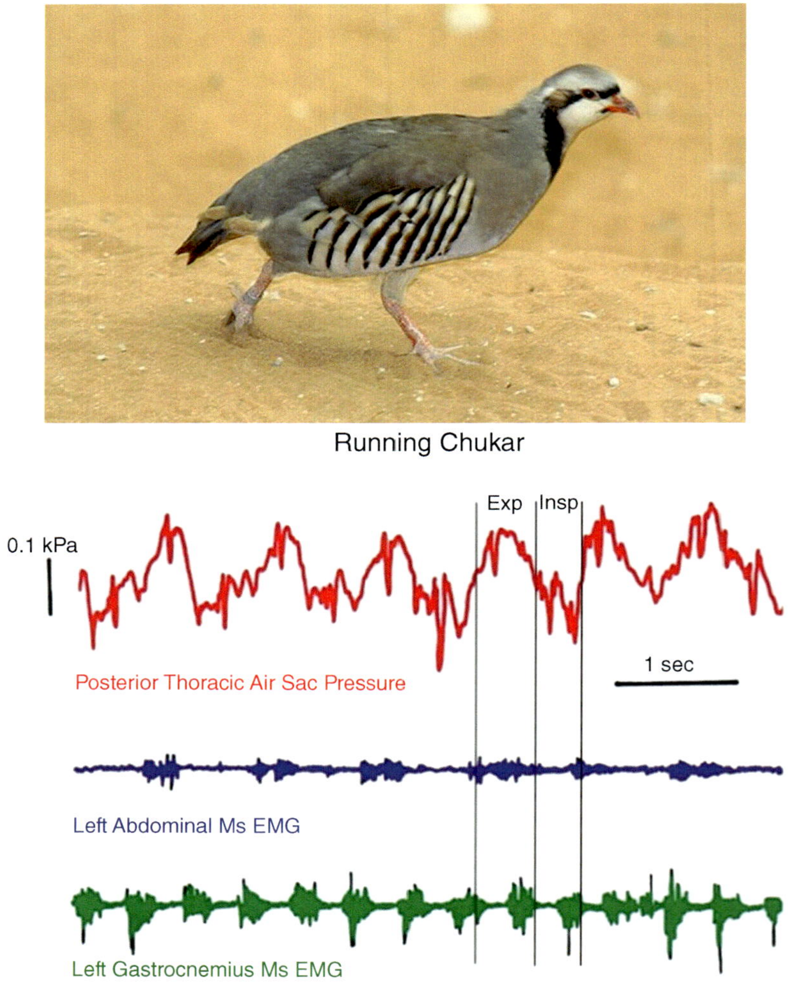


**Fig. 7.57** For a running Chukar (*Alectoris chukar*), abdominal muscle contraction coincides with expiration and the ratio of strides to breaths equals two, as indicated by successive contractions of the left gastrocnemius muscle and changes in air-sac pressure (Chukar photo by Shankar S., Wikipedia, CC BY 2.0, https://creativecommons.org/licenses/by/2.0/; Figure from Boggs 2002; © 2002 Elsevier Science Inc. All rights reserved, used with permission)

et al. (1989) found that strides and breaths were only coordinated during 41% of strides of running Canada Geese. Coordinating strides and breaths may be advantageous. Intercostal and abdominal muscles contract each time a running bird takes a step (Boggs et al. 2001; Boggs 2002; Codd et al. 2005), possibly to stabilize the body (Markley 2010), and these muscles can also play a role in expiration. As such, coordinating the contraction of these muscles with expiration may help maximize expiratory air flow (Markley 2010).

## 7.6 Movement of Air Through the Avian Respiratory System

In the avian respiratory system, most inhaled air enters the caudal air sacs (caudal thoracic and abdominal air sacs) during inspiration, with some air also moving into the lungs via the dorsobronchi. Simultaneously, air is moving through the parabronchi and ventrobronchi and into the anterior air sacs (cranial thoracic, cervical, and interclavicular air sacs) (Fig. 7.58). During expiration, air from the caudal air sacs passes through the dorsobronchi and into the parabronchi. At the same time, air is moving through the ventrobronchi and from the anterior air sacs into the primary bronchi and trachea and is exhaled (Fig. 7.59). As explained previously, this movement of air is driven by changes in air pressure in the air sacs that results from movements of the rib cage and sternum. Importantly, air flow through the paleopumonic parabronchi is unidirectional and continuous during both inspiration and expiration. Because of this unidirectional flow, air moving through the parabronchi contains high

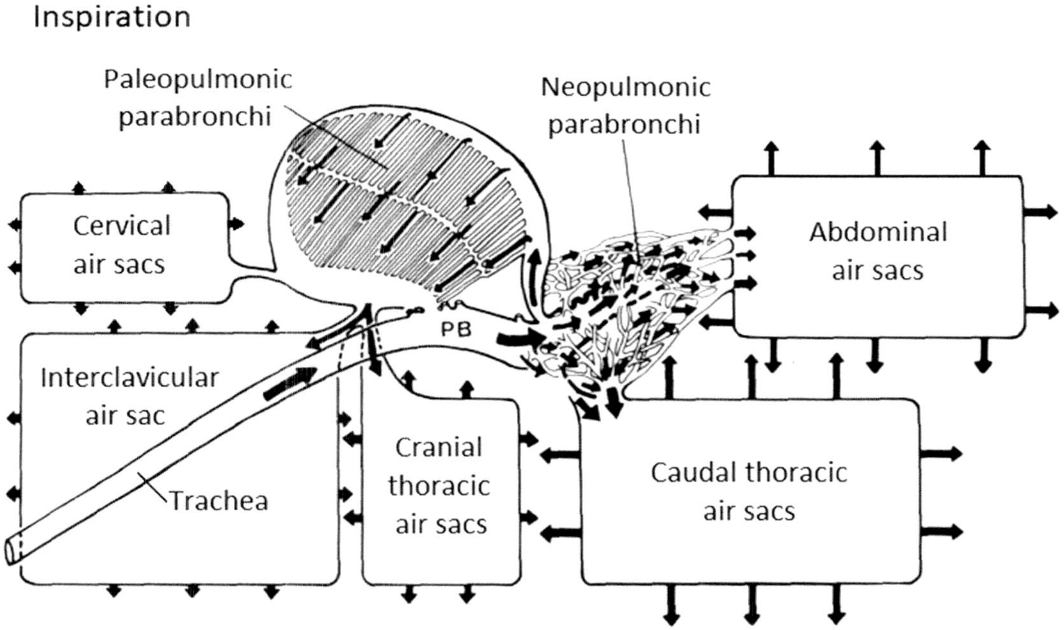


**Fig. 7.58** Pathway of air flow through the avian respiratory system. During inspiration, the air sacs expand as the inspiratory muscles contract, lowering the air pressure in the air sacs. As a result, air moves down the trachea through the neopulmonic parabronchi (for species that have them) and into the posterior air sacs, with some air also passing through the dorsobronchi and into the lungs (paleopulmonic parabronchi). Simultaneously, the anterior air sacs expand, lowering the air pressure, and air moves through the lungs (paleopulmonic parabronchi) ventrobronchi and into the anterior air sacs. PB, primary bronchus. (Figure from Fedde 1980; © 1980 Poultry Science Association Inc. Published by Elsevier Inc., used with permission)

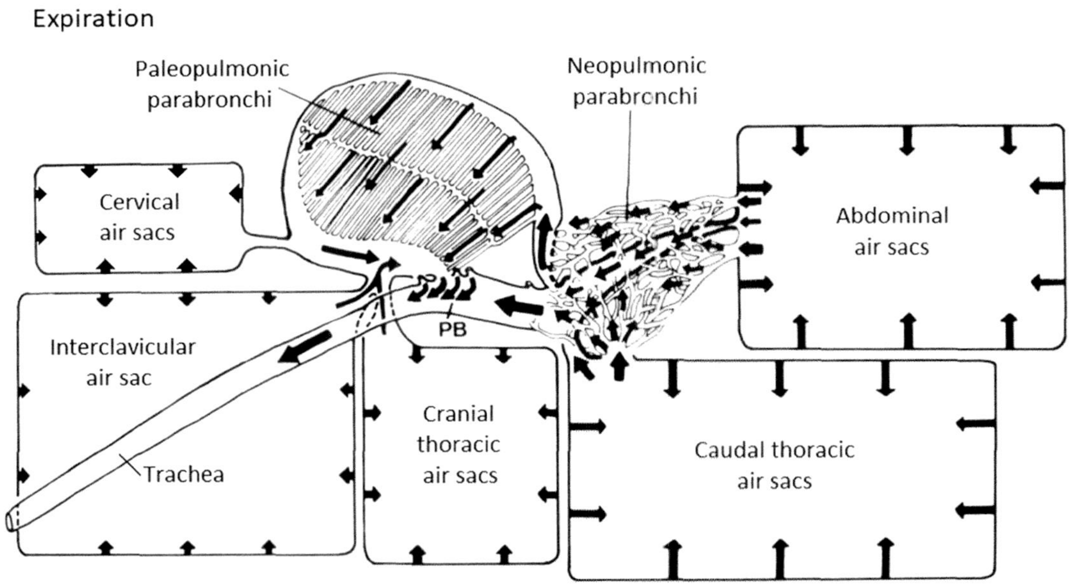


**Fig. 7.59** Pathway of air flow through the avian respiratory system. During expiration, the air sacs contract as the inspiratory muscles relax and expiratory muscles contract, increasing the air pressure in the air sacs. As a result, air moves from the posterior air sacs through the neopulmonic parabronchi (for species that have them) and through the dorsobronchi into and through the lungs (paleopulmonic parabronchi). Simultaneously, the anterior air sacs contract, increasing the air pressure, and air moves out of the anterior air sacs through the ventrobronchi and out of the respiratory system through the trachea. PB, primary bronchus. (Figure from Fedde 1980; © 1980 Poultry Science Association Inc. Published by Elsevier Inc., used with permission)

levels of oxygen (higher than in the lungs of other vertebrates) because the avian respiratory system has little dead air space (areas where air has less oxygen because they contain air that has passed through the lungs and given up some of its oxygen). In birds, dead air space includes the trachea and primary bronchi. So, when a bird inhales, the fresh air coming in mixes with the air moved into the trachea and primary bronchi during the preceding exhalation. That "mixing" of fresh and old air reduces the overall oxygen content of the air that will eventually pass through the lungs. In contrast to birds, the bidirectional respiratory system of mammals is completely dead air space; with each inhalation, fresh air is mixed with lots of "old air" (i.e., air already in the respiratory system that has less oxygen and more carbon dioxide than atmospheric air) so the air that reaches the alveoli (small sacs where the exchange of gases occurs between the blood and lungs in mammals) is mixed with old air and has a lower oxygen content than the air that passes through the parabronchi of birds.

In the avian lung, inspired air passing through the primary bronchus moves past and completely bypasses the openings of the ventrobronchi (or MVSB) and moves primarily into the caudal air sacs, a process termed inspiratory aerodynamic valving (IAV, Fig. 7.60; Banzett et al. 1987). The mechanism by which IAV occurs has been much debated (Maina 2005). Early investigators suspected that sphincters or valves helped direct air past the ventrobronchi, but none have been found (Maina 2006). As such, the inspiratory valving is now considered more aerodynamic than mechanical, something first suggested by Dotterweich (1936). With sufficient momentum (density times velocity squared; Butler et al. 1988), inspired air will simply continue past the entrances to the ventrobronchi and

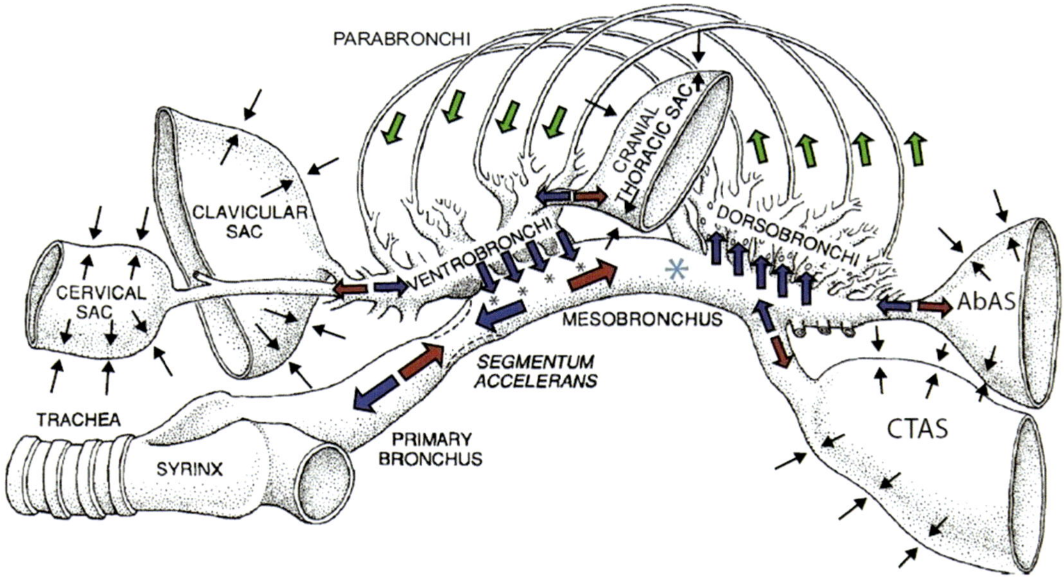


**Fig. 7.60** During inspiration, air moving through the primary bronchus must move past the openings into the ventrobronchi (also referred to as the medioventral secondary bronchi and most species of birds have four) and mostly pass into the posterior air sacs (with some entering the dorsobronchi, also referred to as the mediodorsal secondary bronchi). A narrowing of the primary bronchus called the segmentum accelerans causes air to move faster (Bernoulli effect) and paste the ventrobronchi. (Figure from Cieri and Farmer 2016; © 2016 Springer-Verlag Berlin Heidelberg, used with permission, and adapted from Wang et al. 1992; © 1992 Wiley-Liss, Inc., A Wiley Company, used with permission)

enter the dorsobronchi and posterior air sacs (Butler et al. 1988; Fig. 7.61, Box 7.4 Loopy Network Model of Bird Lungs). Potentially important in this process is a narrowing of the primary bronchi that accelerates the inspired air. This constriction is called the segmentum accelerans, a structure that may consist of erectile tissue because it contains numerous blood vessels (Maina 2002a, b). A segmentum accelerans has been observed in Domestic Chickens (*Gallus gallus*) as well as Rock Pigeons (*Columba livia*), and Graylag Geese (*Anser anser*) (Wang et al. 1992; Maina and Africa 2000), and Wang et al. (1992) reported that it dilated in response to increased levels of $CO_2$ in Graylag Geese. This suggests that constriction or dilation of the segmentum accelerans could affect the efficacy of the inspiratory valve. This aerodynamic valving has been found experimentally to be 95–100% effective (i.e., the percentage of the tidal volume that passes through the parabronchi) (Banzett et al. 1987, 1991).

The movement of much of the air from the caudal air sacs (abdominal and caudal thoracic air sacs) into the dorsobronchi, and less moving back through the mesobronchi (also called the intrapulmonary bronchi) and primary bronchi, has been referred to as expiratory aerodynamic valving (Brown et al. 1995). Hazelhoff (1943) suggested that the movement of air out of the ventrobronchi increases pressure in the mesobronchi and primary bronchi and this, in turn, directs air from the caudal air sacs into the dorsobronchi. Another possibility is that expiratory aerodynamic valving is due to compression of the mesobronchi (intrapulmonary bronchi), with pressure in the air sacs greater than the pressure in the mesobronchi. The pressure exerted by the air sacs may cause constriction of the mesobronchi, increasing resistance to air flow and directing air from the caudal air sacs into the dorsobronchi (Ludders 2015). More recently, Nguyen et al. (2021) suggested that the generation of vortices

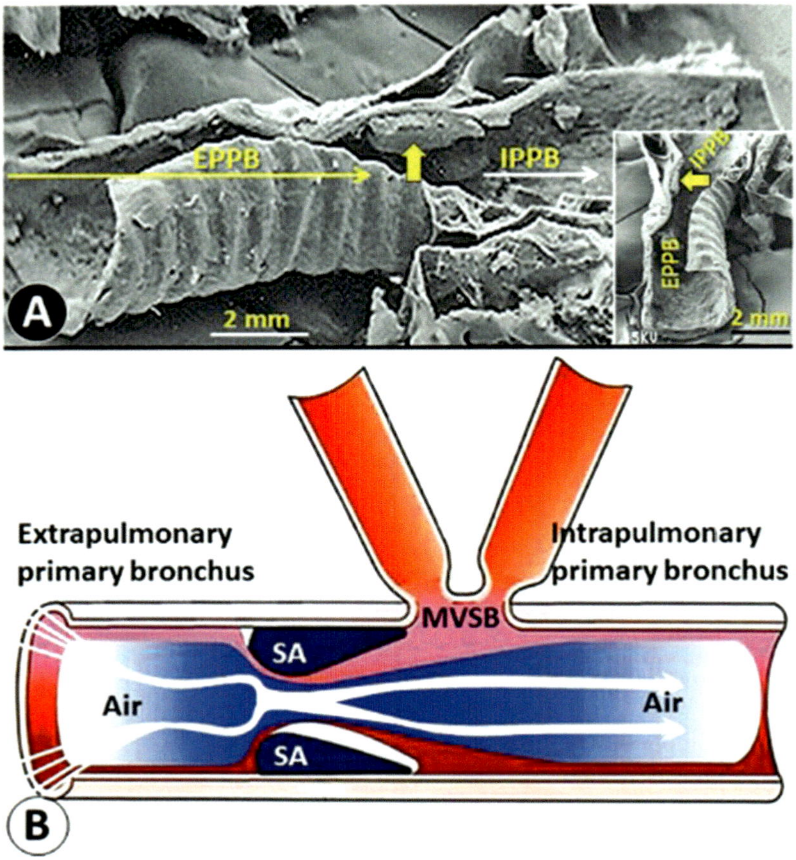


**Fig. 7.61** (**a**) Lateral view of an extrapulmonary primary bronchus (EPPB, i.e., section of the primary bronchus located outside of the lung) in the lung of a Domestic Chicken (*Gallus g. domesticus*) showing the segmentum accerelans (arrow) in the distal part of the airway; the insert on the lower right shows a dorsal view. IPPB, intrapulmonary primary bronchus (section of the primary bronchus located inside the lung). (**b**) Diagram illustrating inspiratory aerodynamic valving, with the inspired air accelerated by constriction of the primary bronchus by the segmentum accelerans (SA). The added momentum generated by this narrowing causes the air to be shunted past the entrances to the medioventral secondary bronchi (MVSB). (Figure from Maina 2017a; © Cambridge Philosophical Society, used with permission)

**Box 7.4 Loopy Network Model of Bird Lungs**

Within the avian respiratory system is a loop, with air moving from the primary bronchus through the mesobronchus, dorsobronchi, parabronchi, ventrobronchi, and back into the primary bronchus. This air flow results from the alternating contraction and expansion of the various air sacs, but there are no valves that direct the flow of air as it moves through the primary bronchus and past the entrance to the ventrobronchi into the mesobronchus or direct some air from the mesobronchus into the dorsobronchi as most of the air moves into the posterior air sacs. The segmentun accelerans may play a role in moving air past the entrance to the ventrobronchi during inhalation, but questions still remain concerning how air travels through the "loop" in the avian respiratory system.

Nguyen et al. (2021) found that, in a system of two loops with T-junctions and filled with water and flow-tracing particles and water driven by an oscillator serving as a piston, flow through one loop (lower one in c below) alternated, moving back and forth, whereas flow in the upper loop was unidirectional.

(continued)

**Box 7.4** (continued)

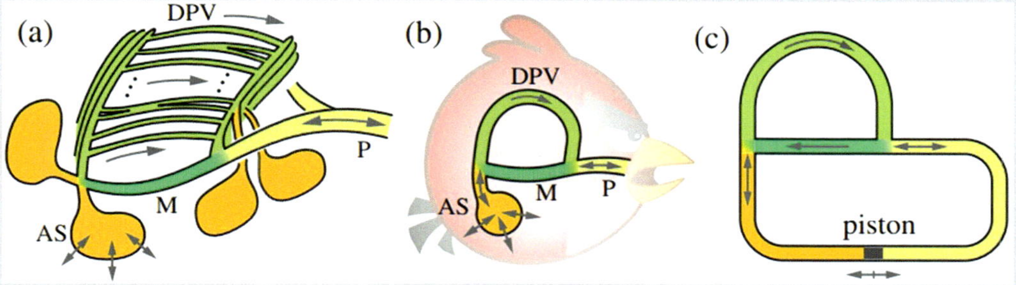


Alternating expansion and contraction of air sacs (AS) drives oscillatory inhalation and exhalation flow of air through the primary bronchus (P). Air flows unidirectionally through the dorso-, para-, and ventrobronchi (DPV) via the mesobronchus (M) and air sacs. (**b**) A "spherical bird" simplification with a single air sac (AS) and DPV. (**c**) A closed network formed by connecting the primary bronchus (P) to the air sac (AS), with the air sac replaced by a reciprocating piston. (Figure from Nguyen et al. 2021; Reprinted with permission of Nguyen et al., Physical Review Letters; © 2021 American Physical Society, used with permission).

Comparing those loops to the avian respiration system, during inhalation, air passing through the primary bronchi largely bypasses the entrance to the ventrobronchi (because inertia of the flow tends to maintain a straight course) and flows into the mesobronchus, with most then entering the posterior air sacs and some entering the dorsobronchi. Most air bypasses the dorsobronchi because a vortex forms at the entrance and acts as a plug. During exhalation, air moves from the posterior air sacs into the dorsobronchi, through the parabronchi, and flows from the ventrobronchi and mesobronchus into the primary bronchus. The converging currents of air from the ventrobronchi and mesobronchus generate a vortex that results in temporary flow separation and allows air to efficiently exit the ventrobronchi. Thus, based on Nguyen et al.'s (2021) model, vortices and flow separation combine to act as "valves" that regulate the movement of air, and specifically the unidirectional movement of air, through the dorso-, para-, and ventrobronchi in the avian respiratory system. These investigators also noted that "the T shape [junction] . . . is not critical, and the complexities of junction flows invite optimization over parameters such as branching angles." It seems likely that, during the evolution of the avian respiratory system, selection has favored branching angles that direct air flow in ways that help to maximize the efficiency of gas exchange.

(continued)

**Box 7.4** (continued)

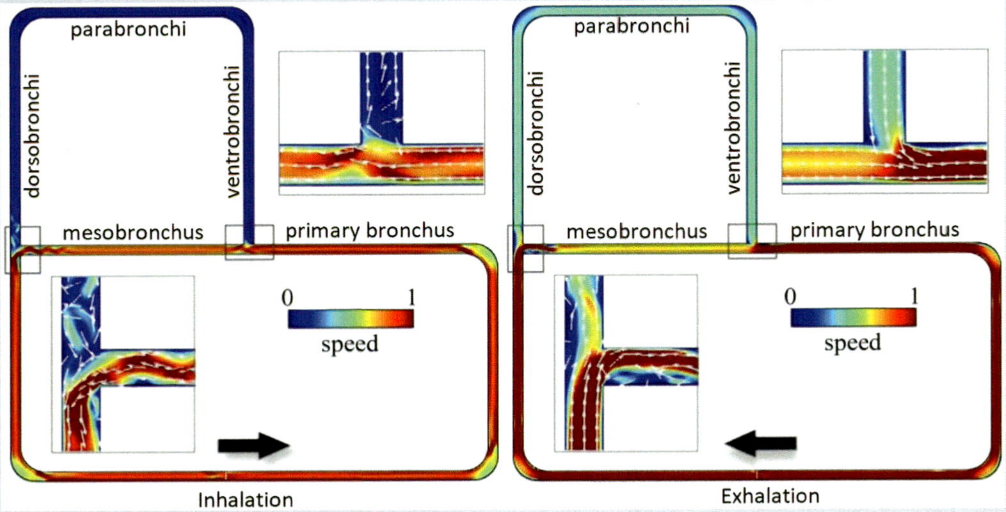


Airflow direction is indicated by arrows and air speed by color. During inhalation, note that air largely bypasses the entrance to the ventrobronchi and most, but not all, air bypasses the entrance to the dorsobronchi. During exhalation, air enters the dorsobronchi and exits the ventrobronchi. Although air exits and enters air sacs, air flows in one direction through the dorsobronchi, parabronchi, and ventrobronchi. (Figure made from two screenshots from a video made by Nguyen et al. 2021, https://www.youtube.com/watch?v=Bc-MDLlAMAA; Reprinted with permission of Nguyen et al., Physical Review Letters; © 2021 American Physical Society, used with permission).

that act as valves helps direct air flow from the posterior air sacs into the dorsobronchi (Box 7.4 Loopy Network Model of Bird Lungs). The efficacy of expiratory valving has been found to vary with species and with experimental conditions, e.g., 76–90% in chickens (Piiper et al. 1970; Bouverot and Dejours 1971), 88% in ducks (Powell et al. 1981), and 95% in geese (Brown et al. 1995), and the valving has been found to be more effective at higher flow rates, e.g., when birds are active (Harvey and Ben-Tal 2016).

## 7.7 Exchange of Gases

The paleopulmonic parabronchi are arranged in a parallel series of several hundred that connect the dorsobronchi and ventrobronchi (Figs. 7.62 and 7.63). Most birds also have neopulmonic parabronchi that exhibit irregular branching patterns. Air passing through the parabronchi (0.5 to 2 mm in diameter) moves into outpocketings of the parabronchi called atria and infundibulae and then into tiny air capillaries (Fig. 7.64). Air capillaries are very small in diameter, with typical diameters ranging from about 3 μm in songbirds to 10 μm in larger birds such as swans (Anatidae), and up to 20 μm in Common Ostriches (*Struthio camelus*; Duncker 1972). The exchange of gases between the avian lung and the blood occurs in the air capillaries, with oxygen diffusing from the air capillaries into the blood and carbon dioxide from the blood into the air capillaries.

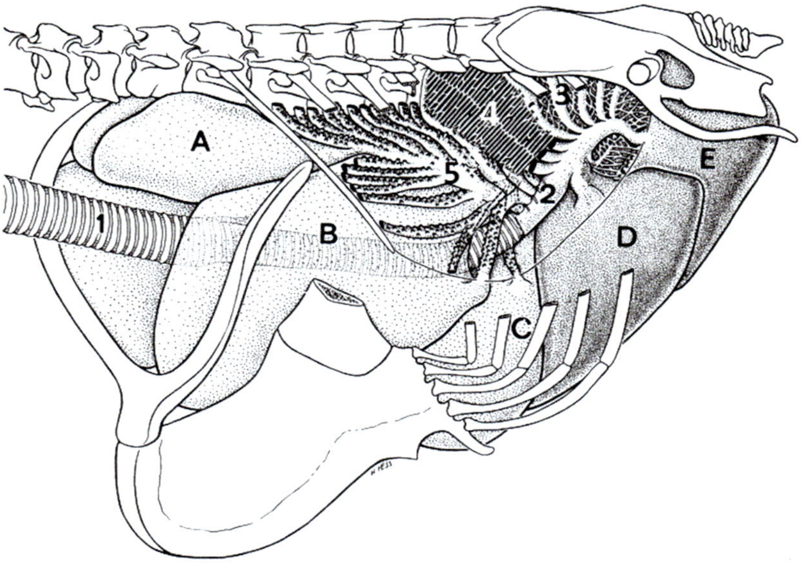


**Fig. 7.62** Air passes into the paleopulmonic parabronchi (4) from the dorsobronchi (3) and out of the paleopulmonic parabronchi into the ventrobronchi (5). (1) trachea, (2) primary bronchus, (A) cervical air sac, (B) interclavicular air sac, (C) anterior thoracic air sac, (D) posterior thoracic air sac, and (E) abdominal air sac. (Figure from Duncker 1972; © 1972 Published by Elsevier B.V., used with permission)

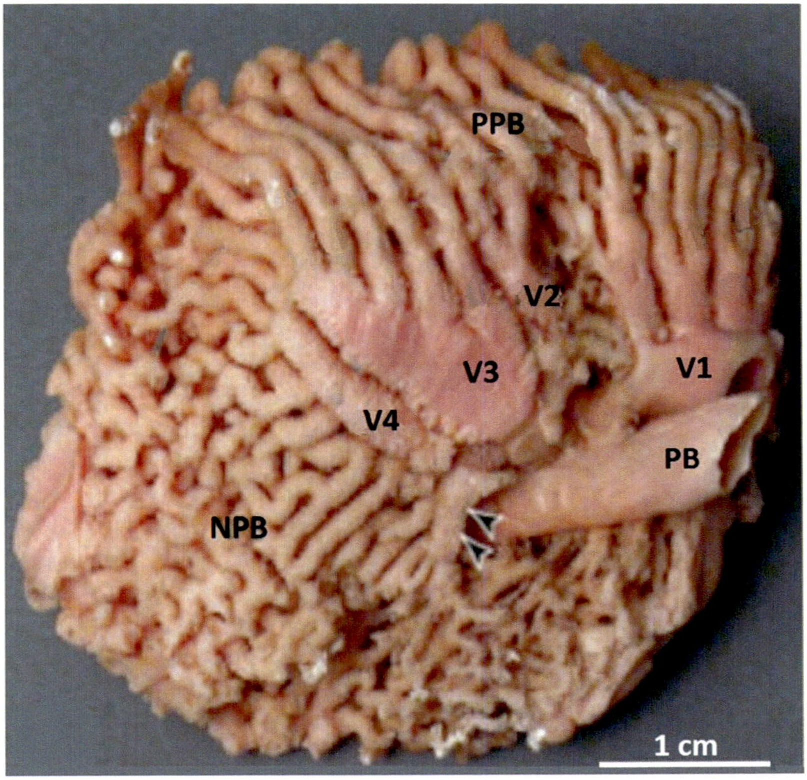


**Fig. 7.63** Medioventral section of a lung of a Domestic Chicken (*Gallus g. domesticus*) showing the primary bronchus (PB), ventrobronchi (V1–V4), paleopulmonic parabronchi (PPB), neopulmonic parabronchi (NPB), and lateroventral bronchus (arrows). In the avian lung, air moves from the parabronchi into the ventrobronchi and into the anterior air sacs. (Figure modified from Makanya and Djonov 2008; © 2008 Wiley-Liss, Inc., used with permission)

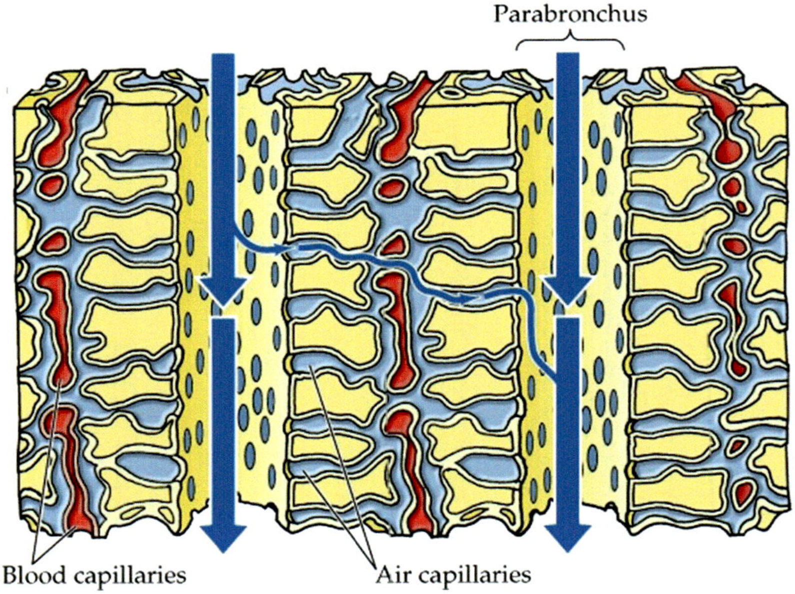


**Fig. 7.64** Parabronchi in the avian lung have numerous air capillaries that are closely associated with blood capillaries. Air moves unidirectionally through the parabronchi. Note the movement of air from the parabronchus on the left, through air capillaries, and into the parabronchus on the right. Oxygen diffuses from the air capillaries into the blood capillaries and carbon dioxide from the blood capillaries to the air capillaries. (Figure from Sadava et al. 2008; © 2007 Oxford University Press, used with permission)

This exchange is very efficient in birds for a number of reasons. First, the complex arrangement of blood and air capillaries (Figs. 7.65 and 7.66) and the small diameter of air capillaries in the avian lung create a substantial surface area through which gases can diffuse (Fig. 7.67). Air capillaries in bird lungs are about one-tenth the diameter of the alveoli in mammalian lungs (Maina 1982; Maina and Nathaniel 2001). The surface area available for exchange (SAE) varies with bird size. For example, the SAE is about 0.17 $m^2$ for House Sparrows (about 30 g; *Passer domesticus*), 0.9 $m^2$ for Rock Pigeons (about 350 g; *Columba livia*), 3.0 $m^2$ for a Mallard (about 1150 g; *Anas platyrhynchos*), and 8.9 $m^2$ for a male Graylag Goose (about 3.7 kg; *Anser anser*) (Maina 2008). However, smaller birds have a greater SAE per unit mass than do larger birds. For example, the SAE is about 90 $cm^2/g$ for Sparkling Violetears (*Colibri coruscans*; Dubach 1981), about 26 $cm^2/g$ for Mallards, and about 5.4 $cm^2/g$ for Emus (*Dromaius novaehollandiae*; Maina and King 1989). Among mammals, there is also a negative relationship between SAE and body size, with smaller mammals like shrews having a greater SAE per unit mass than larger mammals. However, for birds and mammals of similar size, the SAE of birds is generally about 15% greater (Maina et al. 1989).

A second reason why gas exchange in avian lungs is so efficient is that the blood-gas barrier through which gases diffuse is extremely thin. Blood-gas barriers consist of three thin layers, including a thin epithelial cell separated from an endothelial cell by an extracellular matrix space or a basement lamina (Figs. 7.68, 7.69, and 7.70). The thickness of this barrier is important because the amount of gas diffusing across this barrier is inversely proportional to its thickness. Among terrestrial vertebrates, the blood-gas barrier is thinnest in birds (Figs. 7.71 and 7.72). Natural selection has favored thinner blood-gas barriers in birds and mammals because endotherms use oxygen at higher rates than ectotherms like amphibians and reptiles. Among birds, the thickness of the blood-gas barrier varies, with smaller birds generally having thinner blood-gas barriers than larger birds. For example, the blood-gas barrier is 0.099-μm thick in Sparkling Violetear (*Colibri coruscans*) hummingbirds and 0.56-μm thick in Common Ostriches (*Struthio camelus*; West 2009).

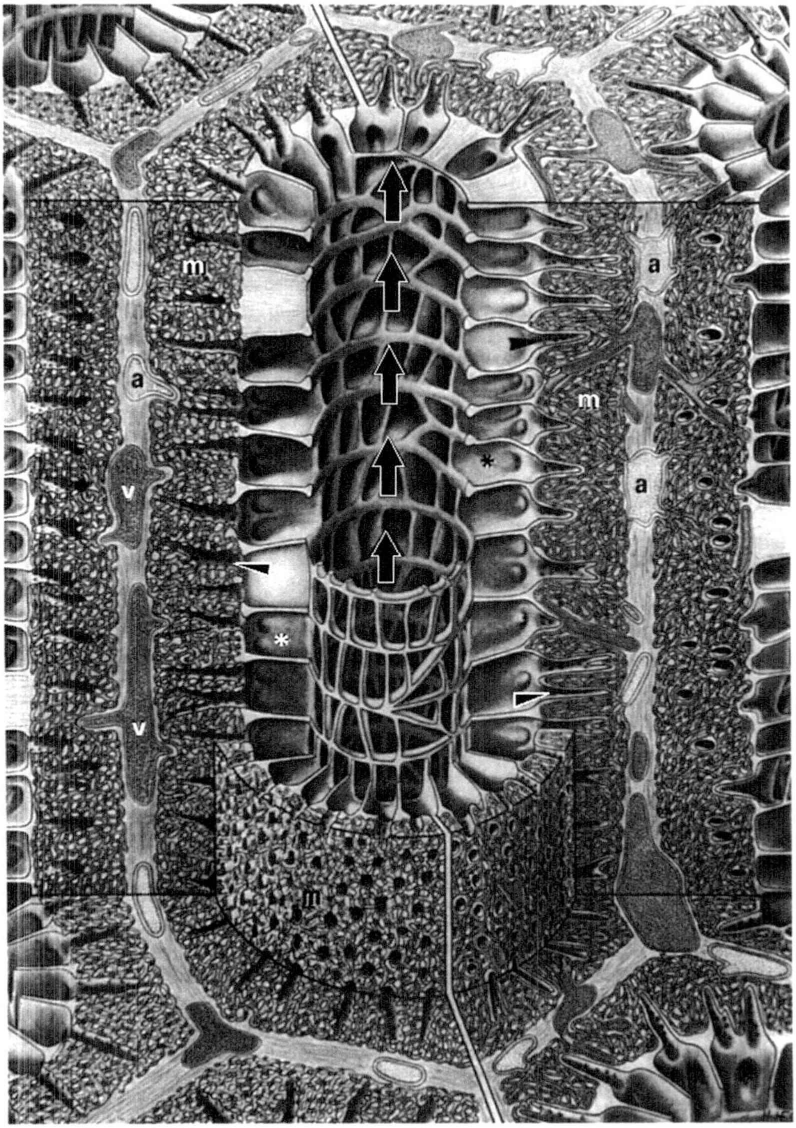


**Fig. 7.65** Anatomy of a parabronchus, the gas-exchange region of the bird's lung-air-sac respiratory system. The central parabronchial lumen, through which gas flows unidirectionally during both inspiration and expiration (large arrows), is surrounded by a mantle (m) of gas-exchange tissue consisting of a network of blood and air capillaries. The lumen of the parabronchus leads into multiple chambers called atria (*) that, in turn, lead into smaller chambers called infundibulae (arrowheads). Branching from the infundibulae are numerous air capillaries. Arteries (a) lead into the capillaries that are closely associated with the air capillaries and this is where (i.e., air and blood capillaries) oxygen and carbon dioxide are exchanged. After flowing through the capillaries, blood then moves into the veins (v) that will take the blood out of the lungs. (Figure from Duncker 1971; © 1971 Springer-Verlag Berlin Heidelberg, used with permission)

Also contributing to the efficiency of gas exchange in avian lungs is a process called cross-current exchange. Air passing through air capillaries and blood moving through blood capillaries generally travel at right angles to each other in what is called cross-current flow (Figs. 7.72 and 7.73; Makanya and Djonov 2008). Superimposed on this cross-current flow is a counter-current system, with the outward flow of air from the parabronchi into the air capillaries and the inward flow of blood in the blood capillaries from the periphery of the parabronchi (Hsia et al. 2013; Maina 2015c; Fig. 7.74). As a result, oxygen diffuses from the air capillaries into the blood across the entire length of the parabronchi, resulting in a greater concentration of oxygen (i.e., higher partial pressures) in the blood leaving the lungs than is possible in the alveolar lungs of mammals (Fig. 7.75).

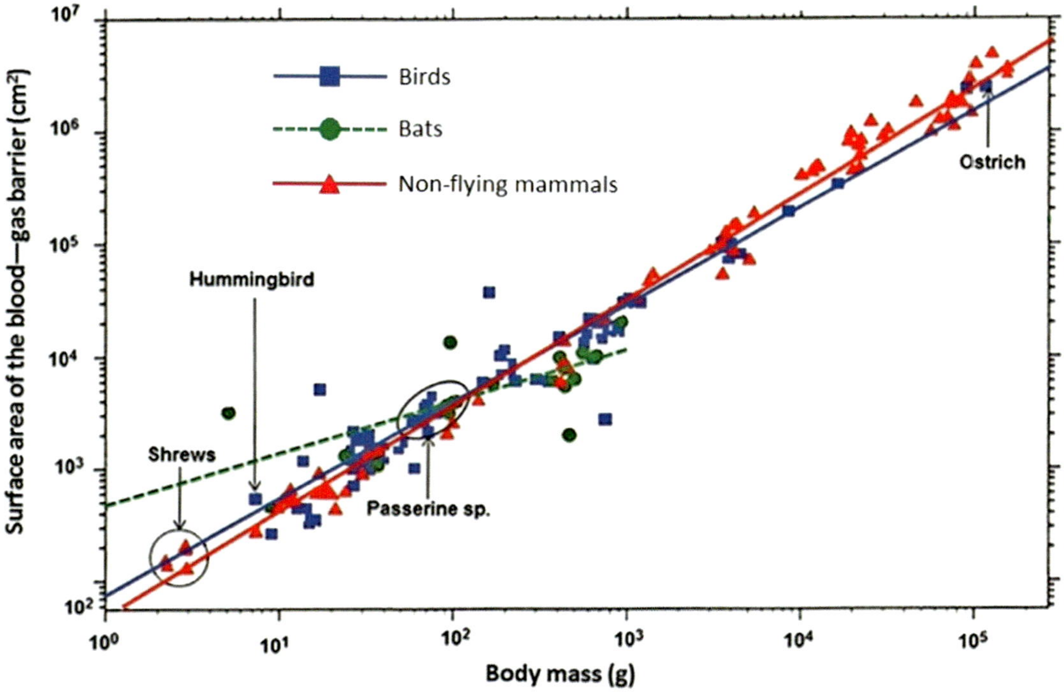


**Fig. 7.66** Regression lines for the surface area of blood-gas barriers relative to body mass in the lungs of birds, bats, and nonflying mammals. For animals of comparable body mass, bats have more respiratory surface area than nonflying mammals and birds. Small birds have a more respiratory surface area than similar-sized nonflying mammals, but the relationship is reversed for larger birds and mammals. Overall, however, the mean respiratory surface area of birds is 15% greater than that of nonflying mammals. (Figure from Maina 2017a; © 2016 Cambridge Philosophical Society, used with permission)

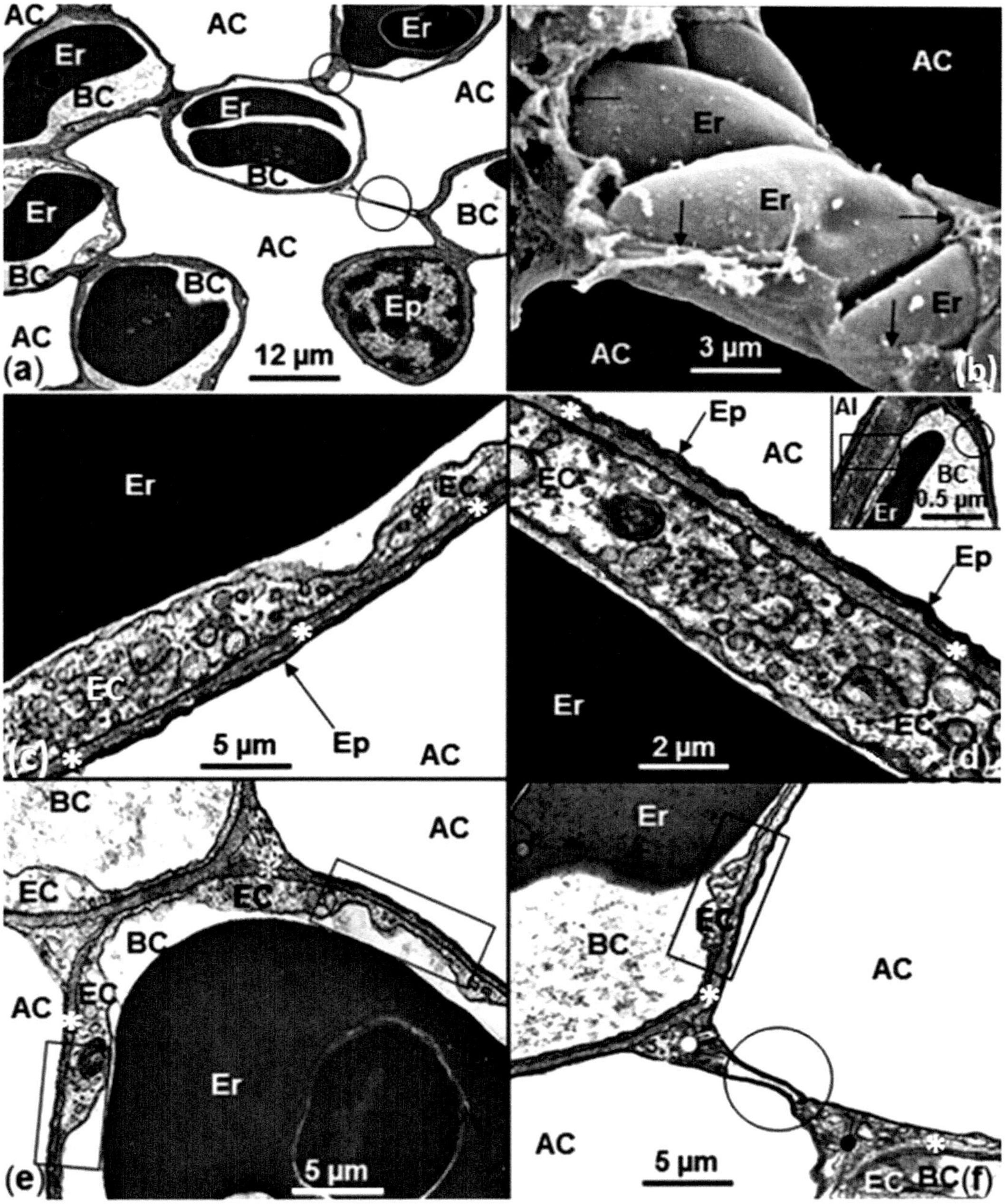


**Fig. 7.67** The thin blood-gas barrier of birds allows for the efficient exchange of oxygen and carbon dioxide between the air capillaries and the blood capillaries. (**a**) Micrograph of the exchange tissue in the lung of a Domestic Chicken showing air capillaries (AC), blood capillaries (BC), and red blood cells (erythrocytes, Er). Circles indicate epithelial cell-epithelial cell connections. (**b**) Micrograph from the lung of a Domestic Chicken showing erythrocytes (Er) in a blood capillary (black arrows are pointing to the walls of the capillary that separate the erythrocytes from air in the air capillaries (AC)). (**c**) Micrograph of the blood-gas barrier from the lung of a Black-headed Gull (*Chroicocephalus ridibundus*) showing an epithelial cell (Ep), endothelial cell (EC), basement membrane (*), erythrocyte (Er), and air capillary (AC). (**d**) Micrograph with greater magnification than in (**c**) of the blood-gas barrier of a Black-headed Gull showing an epithelial cell (Ep), basement membrane (*),

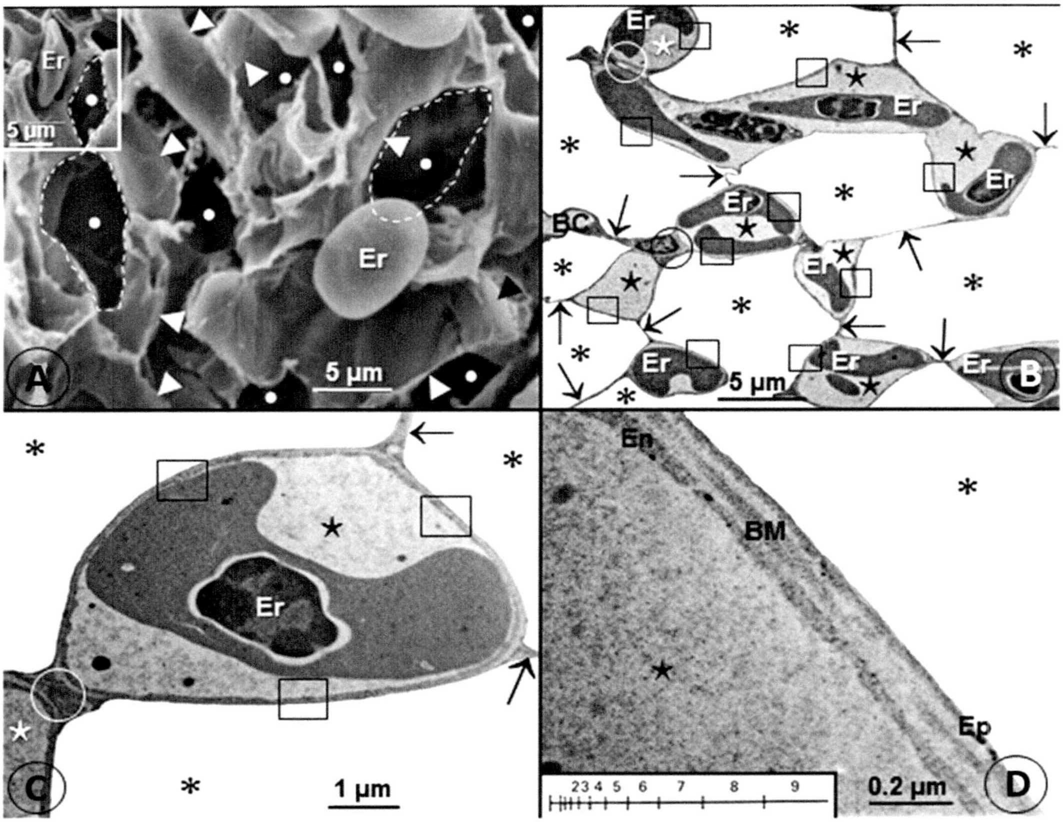


**Fig. 7.68** Air capillaries, blood capillaries, and the extremely thin blood-gas (tissue) barrier of the lung of the Andean Goose (*Oressochen melanopterus*). (**a**) Electron micrograph showing extravasated (i.e., outside of a capillary) erythrocytes (Er) that are similar in diameter to that of air capillaries (•). Some air capillaries are outlined with dashed lines. • = air capillaries; ▶ = blood capillaries. In the insert, an erythrocyte (Er) is located in an air capillary. Low- (**b**) and high (**c**) magnification electron micrographs showing air capillaries (*) and blood capillaries (★). In C, a blood capillary (★) is almost completely surrounded by air in the air capillaries (*). Arrows point to epithelial-epithelial cell connections; □ = blood-gas (tissue) barrier; Er, erythrocytes; circle (O), blood capillary-blood capillary connection. (**d**) High magnification electron micrograph of a blood-gas (tissue) barrier that consists of an epithelial cell (Ep), an endothelial cell (En), and a basement membrane (BM). ★, blood capillary; *, air capillary. The insert shows the logarithmic scale used to determine the thickness of the blood-gas (tissue) barrier. (Figure from Maina et al. 2017; © 2017 Maina et al., open-access article distributed under the terms of the Creative Commons Attribution License, https://creativecommons.org/licenses/by/4.0/)

**Fig. 7.67** (continued) endothelial cell (EC), erythrocyte (Er), and air capillary (AC). Insert shows the thicker blood-gas barriers from the lung of a lesser bushbaby (*Galago senegalensis*). Al, alveolus; BC, blood capillary; Er, erythrocyte. (**e**, **f**). Micrographs of the lung of a Domestic Chicken showing the periodic attenuation of the blood-gas barrier (boxed areas). BC, blood capillaries; AC, air capillaries; Er, erythrocytes; EC, endothelial cell; asterisk, basement membrane; dots in (**f**) indicate epithelial cell-epithelial cell connections. (Figure from Maina 2017b; © 2017 Springer International Publishing AG, used with permission)

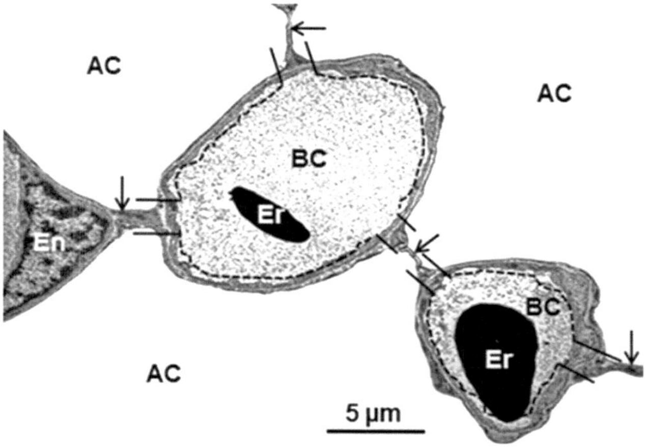


**Fig. 7.69** Exchange tissue of the avian lung showing air capillaries (AC) and blood capillaries (BC). Practically, all of the blood-gas barrier (dashed lines) is involved in gas exchange, while only a small part, the areas where blood capillaries interconnect (areas between the lines), is not involved in gas exchange. Arrows, epithelial–epithelial cell connections; En, endothelial cell; Er, erythrocyte. (Figure from Maina 2015a; © 2015 Springer International Publishing Switzerland, used with permission)

**Fig. 7.70** The blood-gas barrier of birds is significantly thinner than that of mammals, reptiles, and amphibians. Means are based on data obtained from 34 species of birds, 37 species of mammals, 16 species of reptiles, and 10 species of amphibians. (Data from Maina and West 2005)

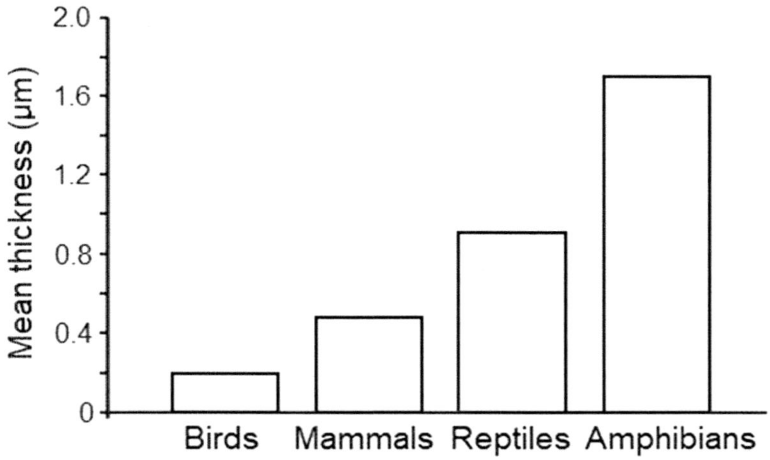

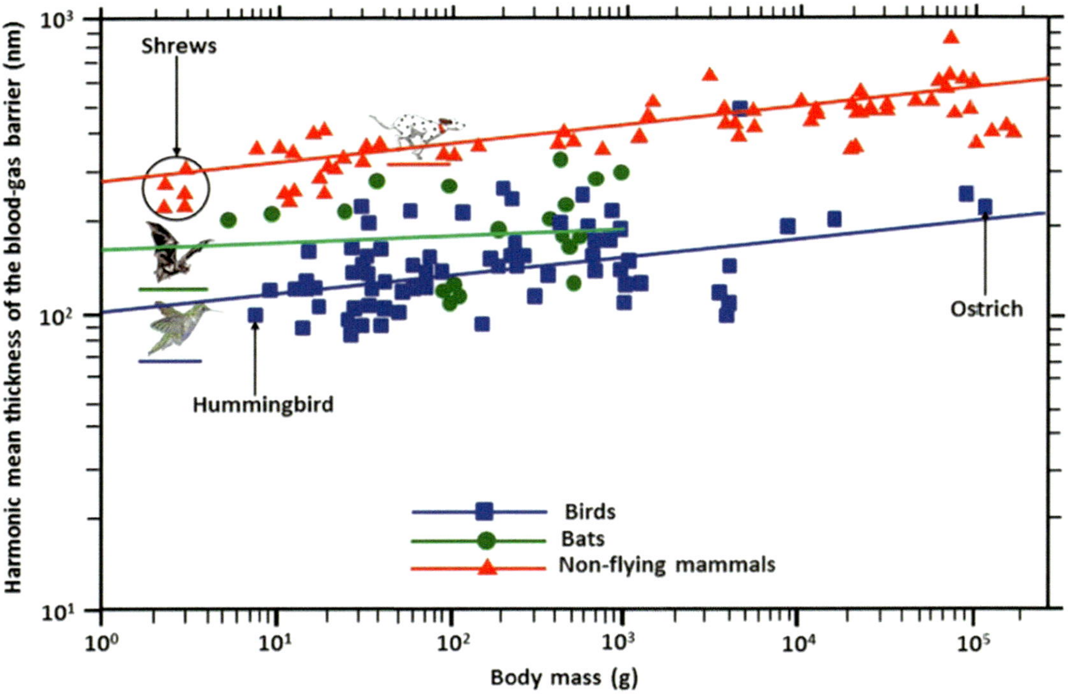


**Fig. 7.71** Comparison of the harmonic mean thickness of the blood-gas barriers relative to body mass in the lungs of birds, bats, and nonflying mammals. Birds have thinner barriers than bats or nonflying mammals. (Figure from Maina 2017b; © 2017 Springer International Publishing AG, used with permission)

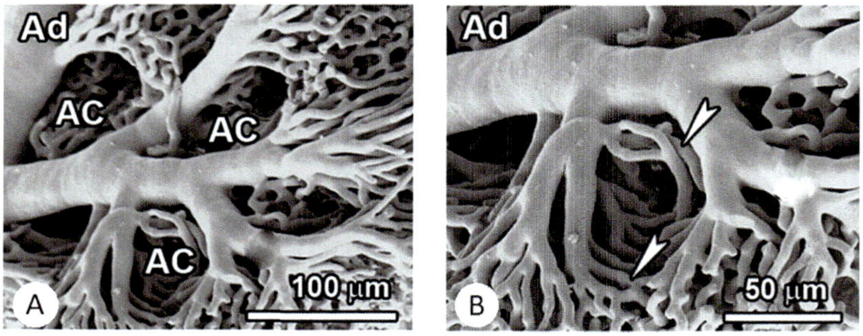


**Fig. 7.72** (**a**, **b**) Vascular casts of air capillaries (ACs) of an adult (Ad) Domestic Chicken show blood capillaries as smooth tubular structures (arrowheads in B) running perpendicular to the long axes of the ACs and surrounding the ACs to create a cross-current system. The numerous blood capillaries surrounding the air capillaries create an extensive surface area that enhances the exchange of gases. (Figure from Makanya et al. 2011; © 2010 Wiley-Liss, Inc., used with permission)

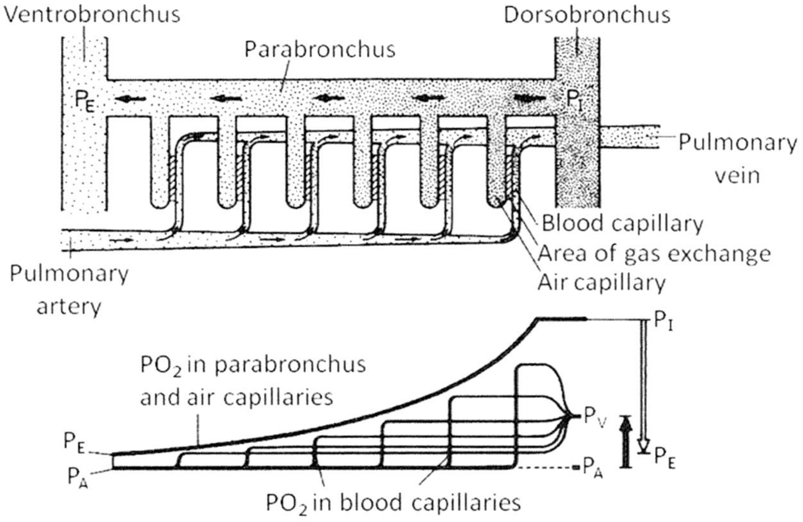


**Fig. 7.73** Top: Air flow (large arrows) and blood flow (small arrows) illustrating the cross-current gas-exchange mechanism operating in the avian lung (between the blood capillaries and air capillaries). Note the serial arrangement of blood capillaries running from the periphery to the lumen of the parabronchus and the air capillaries radially extending from the parabronchial lumen. The exchange of gases (simple diffusion of $O_2$ and $CO_2$) occurs only between blood capillaries and air capillaries. As air moves through a parabronchus and each successive air capillary, the partial pressure of oxygen ($P_{O_2}$) declines (as indicated by the decreased density of the stippling) because oxygen is diffusing into the blood capillaries associated with each air capillary. As a result of this diffusion, the partial pressure of oxygen in the blood leaving the lungs (pulmonary vein) is higher than that in blood entering the lungs (pulmonary artery) (as indicated by the increased density of the stippling). Bottom: Relative partial pressures of $O_2$ and $CO_2$ (1) for air entering a parabronchus (initial-parabronchial, PI) and air leaving a parabronchus (end-parabronchial, PE), and (2) for blood before entering blood capillaries in the lungs (pulmonary artery, PA) and for blood after leaving the blood capillaries in the lungs (pulmonary vein, PV). The partial pressure of oxygen ($P_{O_2}$ ) of venous blood (PV) is derived from a mixture of all serial air capillary-blood capillary units. Because of this cross-current exchange, the partial pressure of oxygen in avian pulmonary veins (PV) is greater than that of the air leaving the parabronchus (PE); air that will be exhaled. In mammals, the partial pressure of oxygen in veins leaving the lungs cannot exceed that of exhaled air (end-expiratory gas, or PE). (Figure adapted from Scheid and Piiper 1987). Importantly, the partial pressure of oxygen in blood leaving the avian lung is the result of "mixing"; blood from a series of capillaries associated with successive air capillaries along the length of a parabronchus is mixed as the blood leaves the capillaries and enters small veins. As a result, the direction of air flow through a parabronchus does not affect the efficiency of the cross-current exchange (because gases are only exchanged between blood capillaries and air capillaries, not between the parabronchus and the blood). So, in above diagram, reversing the direction of air flow would obviously mean that the air capillary on the far left would have the highest partial pressure of oxygen rather than the air capillary on the far right (so the stippling pattern that indicates the amount of oxygen in each air capillary would be reversed). However, because of the "mixing" of blood just mentioned, this reversal would not affect the PV, the partial pressure of oxygen in blood leaving via pulmonary veins. This is important because most birds have neopulmonic parabronchi as well as paleopulmonic parabronchi and, although air flow through paleopulmonic parabronchi is unidirectional, air flow through neopulmonic parabronchi is bidirectional. (Figure modified from Scheid 1979; © 1979 Springer-Verlag, used with permission)

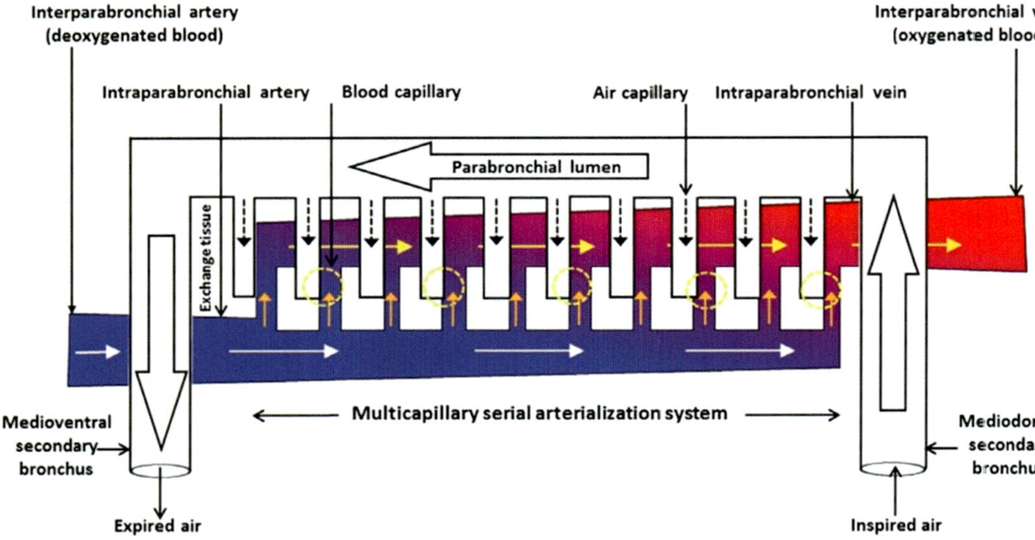


**Fig. 7.74** The exchange of gases in a bird lung involves both counter-current and cross-current flow of air and blood. Blood flows through the intraparabronchial arteries and veins in opposite directions (counter-current flow) and, in addition, the air in the air capillaries and blood in blood capillaries flow at right angle to each other. The atria and infundibula are not shown to simplify the diagram. The increasing red shading of the blood in the intraparabronchial vein indicates the increasing oxygenation of the blood as it passes through the gas exchange tissue of a parabronchus. (Figure from Maina 2017a; © 2016 Cambridge Philosophical Society, used with permission)

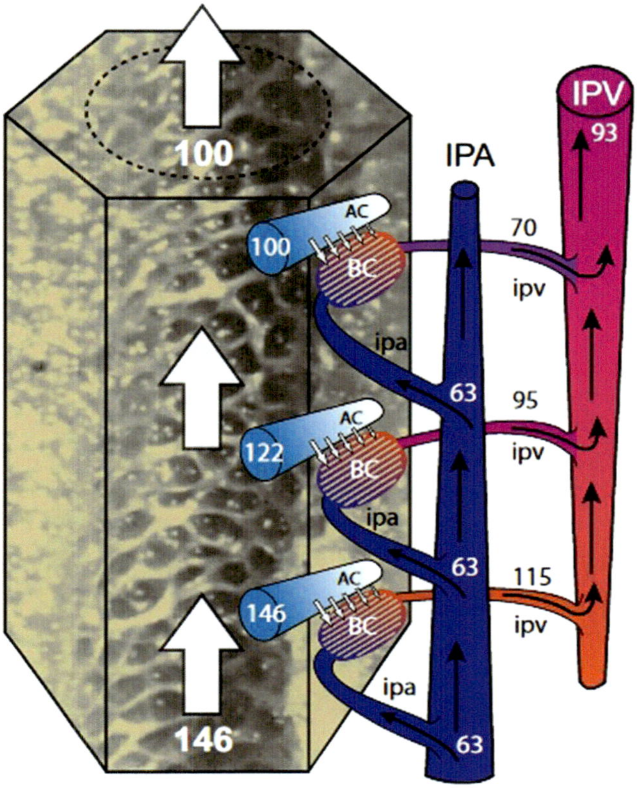


**Fig. 7.75** Air travels unidirectionally through the parabronchi (white arrows) and the exchange of oxygen and carbon dioxide occurs between a series of air capillaries (AC), and blood capillaries (BC) connected to the interparabronchial veins (IPV) and arteries (IPA) through intraparabronchial veins (ipv) and arteries (ipa). The airflow in the parabronchi and blood flow in the intrapulmonary vessels is perpendicular or cross-current and results in high rates of oxygen extraction. Oxygen levels in the blood in the ipa are low because this is the blood that has circulated through the bird's body and given up much of its oxygen. As air moves through the parabronchi, oxygen continues to diffuse from the air capillaries to the blood capillaries, so oxygen levels in the air moving through the parabronchi decline (e.g., from 146 to 122 to 100 in this example). However, oxygen levels in the air capillaries remain higher than those in the blood capillaries, so oxygen continues to diffuse into the blood capillaries. In other words, the blood–gas oxygen gradient is such that oxygen continues to diffuse from the air capillaries into the blood capillaries (and into the ipv) even as oxygen levels in the parabronchi and air capillaries decline as air moves from the beginning to the end of the parabronchi. The lung air and blood oxygen partial pressures ($P_{O_2}$) are those from a domestic duck (*Anas* sp.) from Jones and Holeton (1972) and are indicated in torr, which is approximately equal to partial pressure that is measured in terms of millimeters of mercury (ipv values are hypothetical). (Figure from Cieri and Farmer 2016; © 2016 Springer-Verlag Berlin Heidelberg, used with permission)

## 7.8 Control of Ventilation

Ventilation and respiratory rates are regulated to meet the demands imposed by changes in metabolic activity (e.g., rest and flight) as well as other sensory inputs (e.g., heat and cold). As in mammals, the central control areas for respiration in birds are located in the brainstem and are called the nuclei retroambigualis (RAm) and parambigualis (PAm) (Figs. 7.76 and 7.77). Respiratory frequency and inspiratory and expiratory duration also depend on feedback from receptors in the lungs as well as on extrapulmonary and intrapulmonary chemoreceptors and, possibly, mechanoreceptors (Ludders 2001; Scott and Milsom 2009; Schmidt and Wild 2014).

Hypoxia causes a decline in the partial pressure of oxygen ($PaO_2$) in arteries, which stimulates an increase in ventilation (both in breathing frequency and, to a lesser extent, in tidal volume; e.g., Powell et al. 1998; Fig. 7.78). This reflex is thought to be primarily initiated by the carotid bodies, peripheral chemoreceptors sensitive to changes in arterial $P_{O_2}$ and $P_{CO_2}$/pH that are located near the common carotid arteries that carry blood to the head and brain of birds (Gonzalez et al. 1994). The carotid body is innervated by the vagus nerve and is also located near the thyroid, parathyroid, and ultimobranchial glands (Kameda 1999; Fig. 7.79).

Central chemoreceptors affect ventilation in response to changes in arterial $P_{CO_2}$ and hydrogen ion concentration. Peripheral extrapulmonary chemoreceptors, specifically the carotid bodies (located in the carotid arteries), are influenced by $P_{O_2}$ and increase their discharge rate as $P_{O_2}$ decreases, thus increasing ventilation; they

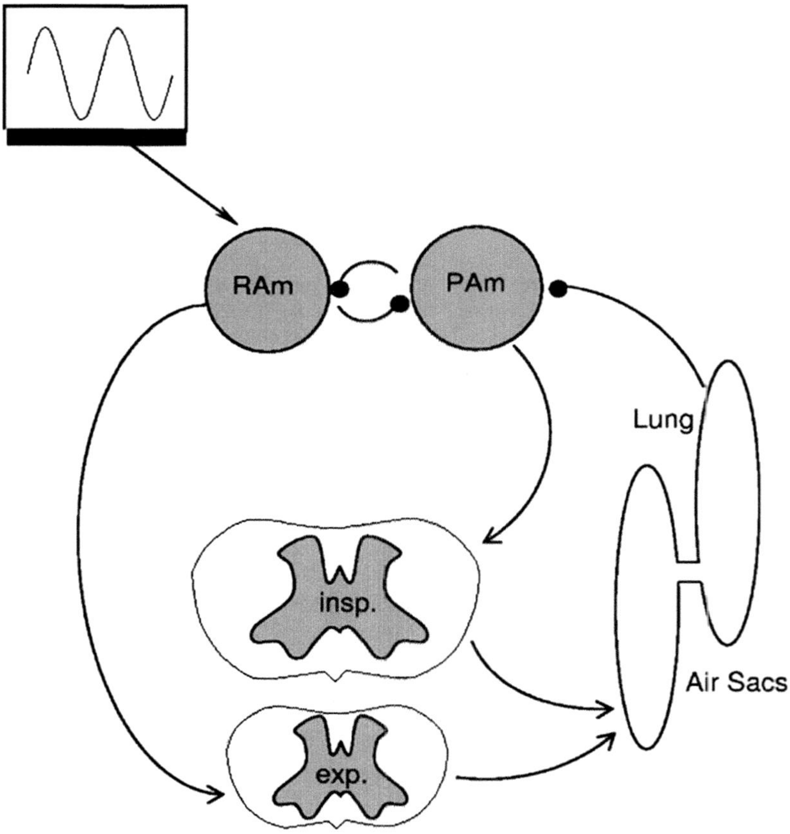


**Fig. 7.76** In the avian respiratory system in birds, nuclei parambigualis (PAm) and retroambigualis (RAm) control inspiration and expiration, respectively. The input is modeled by a rhythm generated by neurons in the telencephalon. Excitatory connections are indicated by arrows and inhibitory connections by filled circles. (Figure from Mendez et al. 2006; Reprinted with permission of Physical Review E, © 2006 American Physical Society)

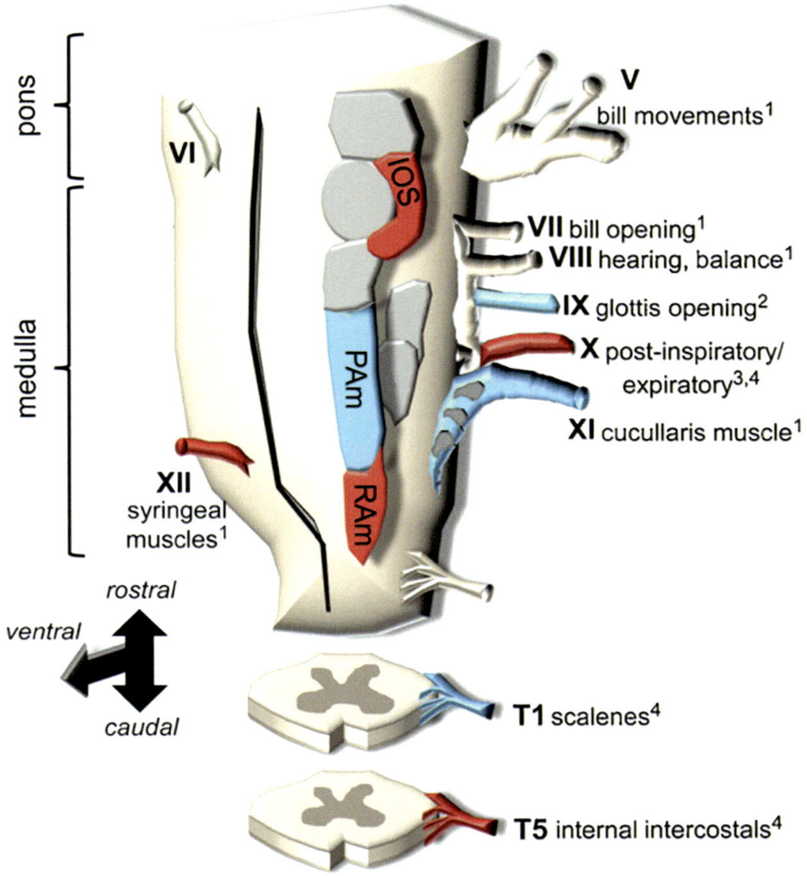


**Fig. 7.77** Areas in the pons and medulla that regulate respiratory rhythm, pattern, and control in birds. Red- and blue-shaded areas and cranial nerves indicate inspiratory- and expiratory-related areas, respectively. In birds, inspiration coincides with neural activity in an area in the medulla called the parambiguus (PAm) (and, possibly, the infra-olivaris superior, or IOS), and expiration coincides with neural activity in the retroambigualis (RAm). The color of the cranial nerve roots (indicated by Roman numerals) indicates inspiratory (blue) and expiratory (red) activity during normal breathing. IOS, nucleus infra-olivarus superior. T1 and T5 are spinal nerves in the thoracic region of the spinal cord. (Figure from Whitaker-Fornek et al. 2019; © 2019 Published by Elsevier B.V., used with permission)

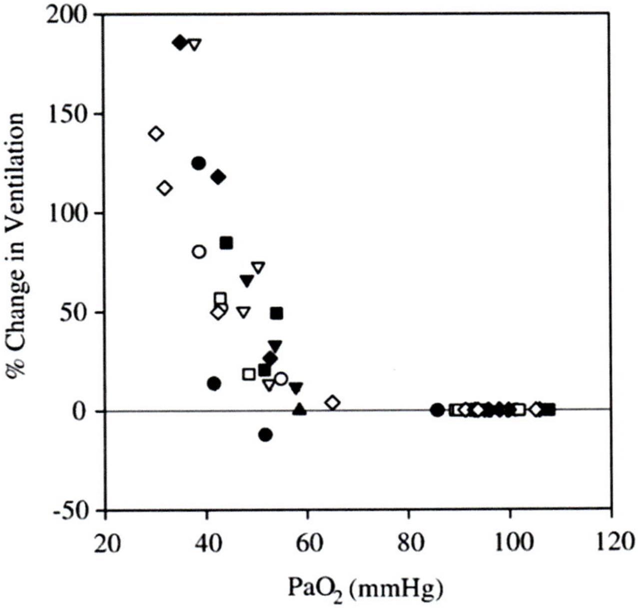


**Fig. 7.78** Relationship between the percent change in ventilation of Burrowing Owls (*Athene cunicularia*) breathing air or hypoxic gas mixtures (expressed as a percent change above normal values) and the partial pressure of oxygen in their blood ($PaO_2$). (Figure from Kilgore et al. 2008; © 2007 Elsevier Inc. All rights reserved, used with permission)

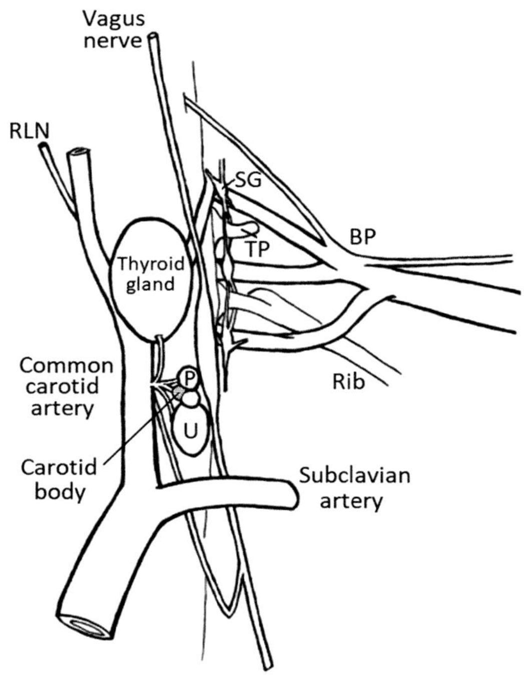


**Fig. 7.79** Location of the avian carotid body relative to nearby endocrine glands, arteries, and nerves in the cervicothoracic region. Ventral view on the left side. P, parathyroid; U, ultimobranchial gland; BP, brachial plexus; RLN, recurrent laryngeal nerve; SG, sympathetic ganglion; TP, transverse process of a cervical vertebra. (Figure from Kameda 1999; © 1999 Springer-Verlag, used with permission)

decrease their rate of discharge as $Po_2$ increases or $Pco_2$ decreases. These responses are the same as those observed in mammals. Unlike mammals, birds also have a unique group of peripheral receptors located in the lung called intrapulmonary chemoreceptors (IPC) that are acutely sensitive to carbon dioxide and insensitive to hypoxia. IPCs are distributed throughout the lungs of birds, including along parabronchi, but are more abundant in the caudal regions of the lungs and are innervated by the vagal nerve (Scheid et al. 1974; Nye and Burger 1978). Because of their locations, IPCs are able to respond to both venous (pulmonary arterial) and airway carbon dioxide (Scheid et al. 1978; Hempleman and Bebout 1994). IPCs are stimulated by low levels of lung $Pco_2$, inhibited by high levels of lung $Pco_2$, and their afferent neurons are responsible for reflex adjustments of breathing depth and rate. High levels of $CO_2$ in the blood (called hypercapnia) stimulate inspiration by reducing the inhibitory effects of IPCs on breathing (Fig. 7.80). IPCs affect the rate and volume of breathing on a breath-to-breath basis via an inspiratory-inhibitory reflex that is sensitive to the timing, rate, and extent of $CO_2$ washout from the lung during inspiration (Ludders 2001).

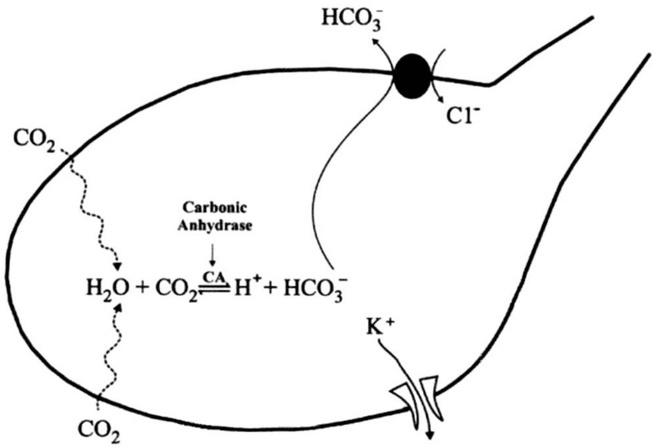


**Fig. 7.80** Avian intrapulmonary chemoreceptors respond to an inward diffusion of $CO_2$ from either airways or pulmonary blood that increases intracellular acidification via the reaction between $CO_2$ and $H_2O$ that, via the action of the enzyme carbonic anhydrase, produces hydrogen ions and $HCO_3^-$. This acidification causes potassium ($K^+$) channels to open and the outward diffusion of position ions hyperpolarizes the chemoreceptor membrane which, in turn, results in the chemoreceptors sending fewer impulses to respiratory control centers in the brainstem. The reduces the inhibitory effect of the IPCs which results in an increase in ventilation. (Figure modified from Scott and Milsom 2009; © 2009 Springer-Verlag Berlin Heidelberg, used with permission)

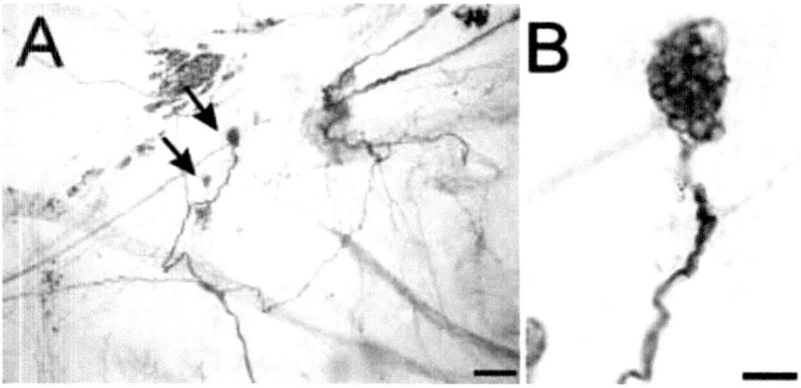


**Fig. 7.81** (**a**) Nerve endings (vagal terminations) form a dense network in air-sac membranes. (**b**) Higher magnification image of a single nerve ending. Scale bars: **a**, 100 μm; **b**, 20 μm. (Figure from Kubke et al. 2004; © Anatomical Society, used with permission)

The avian respiratory system also contains mechanoreceptors (or stretch receptors) that are stimulated by inflation of the respiratory system, but are not sensitive to hypoxia or hypercapnia. They fire throughout inspiration and their specific location is unknown, but some evidence suggests that they are located in the air sacs (Kubke et al. 2004; Fig. 7.81). The role of the mechanoreceptors in the respiratory function of birds is unclear.

## References


Abdalla MA (1989) The blood supply to the lung. In: King AS, McLelland J (eds) Form and function in birds, vol 4. Academic Press, London, pp 281–306

Atterholt J, Wedel MJ (2023) A computed tomography-based survey of paramedullary diverticula in extant Aves. Anat Rec 306:29–50

Bakkeren C, Smith E, York JM, Chua B, McCracken KG, Milsom WK (2020) A morphometric analysis of the lungs of high-altitude ducks and geese. J Anat 237: 188–196

Banzett RB, Nations CS, Barna JL, Lehr JL, Jones JH (1987) Inspiratory aerodynamic valving in goose lungs depends on gas density and velocity. Respir Physiol 70:287–300

Banzett RB, Nations CS, Wang N, Fredberg JJ, Butler JP (1991) Pressure profiles show features essential to aerodynamic valving in geese. Respir Physiol 84: 295–309

Barnas GM, Rautenberg W (1987) Temperature control. In: Seller TJ (ed) Bird respiration, vol 1. CRC Press, Inc., Boca Raton, pp 131–153

Baumel JJ, Wilson JA, Bergren DR (1990) The ventilatory movements of the avian pelvis and tail: function of the muscles of the tail region of the pigeon (*Columba livia*). J Exp Biol 151:263–277

Berger M, Roy OZ, Hart JS (1970) The coordination between respiration and wing beats in birds. Zeitschrift für Vergleichende Physiologie 66:190–200

Bertelli S, Giannini NP (2005) A phylogeny of extant penguins (Aves: Sphenisciformes) combining morphology and mitochondrial sequences. Cladistics 21: 209–239

Bezuidenhout AJ (2005) Light and electron microscopic study of the thoracic respiratory air sacs of the fowl. Anatomia, Histologia, Embryologia 34:185–191

Bock WJ (1978) Morphology of the larynx of *Corvus brachyrhynchos* (Passeriformes: Corvidae). Wilson Bull 90:553–565

Boggs DF (1997) Coordinated control of respiratory pattern during locomotion in birds. Am Zool 37:41–53

Boggs DF (2002) Interactions between locomotion and ventilation in tetrapods. Comp Biochem Physiol A 133:269–288

Boggs DF, Jenkins FA Jr, Dial KP (1997) The effects of the wingbeat cycle on respiration in Black-billed Magpies (*Pica pica*). J Exp Biol 200:1403–1412

Boggs DF, Baudinette RV, Frappell PB, Butler PJ (2001) The influence of locomotion on air-sac pressures in Little Penguins. J Exp Biol 204:3581–3586

Bosque C, Herrera EA (1999) "Snorkeling" by the chicks of the Wattled Jacana. Wilson Bull 111:262–265

Bourke JM, Witmer LM (2016) Nasal conchae function as aerodynamic baffles: experimental computational fluid dynamic analysis in a turkey nose (Aves: Galliformes). Respir Physiol Neurobiol 234:32–46

Bouverot P, Dejours P (1971) Pathway of respired gas in the air sacs-lung apparatus of fowl and ducks. Respir Physiol 13:330–342

Brackenbury JH, Avery P (1980) Energy consumption and ventilatory mechanisms in the exercising fowl. Comp Biochem Physiol A 66:439–445

Brent R, Rasmussen JG, Bech C, Martini S (1984) Lung ventilation and temperature regulation in the European Coot, *Fulica atra*. Physiol Zool 57:19–25

Brocklehurst RJ, Schachner ER, Sellers WI (2018) Vertebral morphometrics and lung structure in non-avian dinosaurs. R Soc Open Sci 5:180983

Brown RE, Kovacs CE, Butler JP, Wang N, Lehr J, Banzett RB (1995) The avian lung: is there an aerodynamic expiratory valve? J Exp Biol 198:2349–2357
Brown RE, Brain JD, Wang N (1997) The avian respiratory system: a unique model for studies of respiratory toxicosis and for monitoring air quality. Environ Health Perspect 105:188–200
Burn A, Schneiter M, Ryser M, Gehr P, Ricka J, Frenz M (2022) A quantitative interspecies comparison of the respiratory mucociliary clearance mechanism. European Biophys J 51:51–65
Butler PJ, Woakes AJ (1980) Heart rate, respiratory frequency and wing beat frequency of free flying Barnacle Geese, *Branta leucopsis*. J Exp Biol 85:213–226
Butler JP, Banzett RB, Fredberg JJ (1988) Inspiratory valving in avian bronchi: aerodynamic considerations. Respir Physiol 72:241–255
Cerda IA, Salgado L, Powell JE (2012) Extreme postcranial pneumaticity in sauropod dinosaurs from South America. Paläontologische Zeitschrift 86:441–449
Çevik-Demirkan A, Haziroğlu RM, Kürtül I (2007) Gross morphological and histological features of larynx, trachea and syrinx in Japanese Quail. Anatomia, Histologia, Embryologia 36:215–219
Chappell MA, Roverud RC (1990) Temperature effects on metabolism, ventilation, and oxygen extraction in Neotropical bat. Respir Physiol 81:401–412
Cho P, Brown R, Anderson M (1984) Comparative gross anatomy of ratites. Zoo Biol 3:133–144
Cieri RL, Farmer CG (2016) Unidirectional pulmonary airflow in vertebrates: a review of structure, function, and evolution. J Comp Physiol B 186:541–552
Claessens LPAM (2009) The skeletal kinematics of lung ventilation in three basal bird taxa (Emu, tinamou, and Guinea Fowl). J Exp Zool 311A:586–599
Clench M (1978) Tracheal elongation in birds-of-paradise. Condor 80:423–430
Codd JR (2010) Uncinate processes in birds: morphology, physiology and function. Comp Biochem Physiol Part A: Mol Integr Physiol 156:303–308
Codd JR, Boggs DF, Perry SF, Carrier DR (2005) Activity of three muscles associated with the uncinate processes of the Giant Canada Goose *Branta canadensis maximus*. J Exp Biol 208:849–857
Crole MR, Soley JT (2010) Gross morphology of the intraoral rhamphotheca, oropharynx and proximal oesophagus of the Emu (*Dromaius novaehollandiae*). Anatomia, Histologia, Embryologia 39:207–218
Crole MR, Soley JT (2012a) What prevents *Struthio camelus* and *Dromaius novaehollandiae* (Palaeognathae) from choking? A novel anatomical mechanism in ratites, the linguo-laryngeal apparatus. Front Zool 9:11
Crole M, Soley J (2012b) Gross anatomical features of the tongue, lingual skeleton and laryngeal mound of *Rhea americana* (Palaeognathae, Aves): morpho-functional considerations. Zoomorphology 131:265–273
Currey JD, Alexander RM (1985) The thickness of the walls of tubular bones. J Zool (London) 206:453–468
Davenport J, Jones TT, Work TM, Balazs GH (2014) Unique characteristics of the trachea of the juvenile leatherback turtle facilitate feeding, diving and endothermy. J Exp Mar Biol Ecol 450:40–46
DeWet PD, Fedde MR, Kitchell RL (1967) Innervation of the respiratory muscles of *Gallus domesticus*. J Morphol 123:17–34
Dotterweich H (1936) Die atmung der vogel. Zeitschrift für Vergleichende Physiologie 23:744–770
Dubach M (1981) Quantitative analysis of the respiratory system of the House Sparrow, Budgerigar, and Violet-eared Hummingbird. Respir Physiol 46:43–60
Duncker H-R (1971) The lung air sac system of birds. Adv Anat Embryol Cell Biol 45:1–171
Duncker HR (1972) Structure of avian lungs. Respir Physiol 14:44–63
Duncker H-R (2004) Vertebrate lungs: structure, topography and mechanics: a comparative perspective of the progressive integration of respiratory system, locomotor apparatus and ontogenetic development. Respir Physiol Neurobiol 144:111–124
Düring DN, Ziegler A, Thompson CK, Ziegler A, Faber C, Müller J, Scharff C, Elemans CPH (2013) The songbird syrinx morphome: a three-dimensional, high-resolution, interactive morphological map of the Zebra Finch vocal organ. BMC Biol 11:1
Evans HE (2016) Avian anatomy. In: Lovette IJ, Fitzpatrick JW (eds) Handbook of bird biology. John Wiley & Sons, Hoboken, pp 169–214
Fedde MR (1980) The structure and gas-flow pattern in the avian lung. Poult Sci 59:2642–2653
Fedde MR (1987) Respiratory muscles. In: Seller TJ (ed) Bird respiration, vol 1. CRC Press, Inc., Boca Raton, pp 3–37
Fedde MR (1998) Relationship of structure and function of the avian respiratory system to disease susceptibility. Poult Sci 77:1130–1138
Fitch WT (1999) Acoustic exaggeration of size in birds via tracheal elongation: comparative and theoretical analyses. J Zool 248:31–48
Frappell PB, Hinds DS, Boggs DF (2001) Scaling of respiratory variables and the breathing pattern in birds: an allometric and phylogenetic approach. Physiol Biochem Zool 74:75–89
Frith CB (1994) Adaptive significance of tracheal elongation in manucodes (Paradisaeidae). Condor 96:552–555
Funk GD, Milsom WK, Sholomenko GN, Steeves JD (1989) Role of the telencephalon in the synchronization of locomotor and respiratory frequencies during walking in Canada Geese. J Exp Biol 145:283–301
Funk GD, Sholomenko GN, Valenzuela IJ, Steves JD, Milsom WK (1993) Coordination of wing beat and respiration in Canada Geese during free flight. J Exp Biol 175:317–323
Gaunt AS, Gaunt SLL, Prange HD, Wasser JS (1987) The effects of tracheal coiling on the vocalizations of cranes (Aves; Gruidae). J Comp Physiol A 161:43–58

Gehr P, Bachofen M, Weibel ER (1978) The normal human lung: ultrastructure and morphometric estimation of diffusion capacity. Respir Physiol 32:121–140

Gilliard ET (1969) Birds of paradise and bower birds. Natural History Press, New York

Gonzalez C, Almaraz L, Obeso A, Rigual R (1994) Carotid body chemoreceptors: from natural stimuli to sensory discharges. Physiol Rev 74:829–898

Guangdi SI, Dong Y, Ma Y, Zhang Z (2015) Shape similarities and differences in the skulls of scavenging raptors. Zool Sci 32:171–177

Gutherz SB, O'Connor PM (2021) Postcranial skeletal pneumaticity in Cuculidae. Zoology 146:1–8

Gutherz SB, O'Connor PM (2022) Postcranial skeletal pneumaticity in non-aquatic neoavians: insights from accipitrimorphae. J Anat 241:1387–1398

Gutzwiller SC, Su A, O'Connor PM (2013) Postcranial penumaticity and bone structure in two clades of neognath birds. Anat Rec 296:867–876

Hackett SJ, Kimball RT, Reddy S, Bowie RCK, Braun EL, Braun MJ, Chojnonwski JL, Cox WA, Han KL, Harshman J, Huddleston CJ (2008) A phylogenomic study of birds reveals their evolutionary history. Science 320:1763–1768

Hamlet MP, Fisher HI (1967) Air sacs of respiratory origin in some Procellariiform birds. Condor 69:586–595

Harvey EP, Ben-Tal A (2016) Robust unidirectional airflow through avian lungs: new insights from a piecewise linear mathematical model. PLoS Comput Biol 12:e1004637

Hazelhoff EH (1943) Bouw en functic van de Vogellong. Versl Gewone Vergad Afd Naturk K ned Akad Wet 52:391–400

Hempleman SC, Bebout DE (1994) Increased venous $Pco_2$ enhances dynamic responses of avian intrapulmonary chemoreceptors. Am J Physiol 266:R15–R19

Hinds DS, Calder WA (1971) Tracheal dead space in the respiration of birds. Evolution 25:429–440

Hocken AG (2002) Post-mortem examination of penguins. DOC Science Internal Series 65, Department of Conservation, Wellington, New Zealand

Holle JP, Heisler N, Scheid P (1978) Blood flow distribution in the duck lung and its control by respiratory gases. Am J Physiol-Regul Integr Comp Physiol 234: R146–R154

Homberger DG (1999) The avian tongue and larynx: multiple functions in nutrition and vocalisation. In: Adams NJ, Slotow RH (eds) Proceedings of the 22nd International Ornithological Congress. BirdLife, Johannesburg, pp 94–113

Homberger DG (2017) The avian lingual and laryngeal apparatus within the context of the head and jaw apparatus, with comparisons to the mammalian condition: functional morphology and biomechanics of evaporative cooling, feeding, drinking, and vocalization. In: Maina JN (ed) The biology of the avian respiratory system. Springer, Cham, pp 27–98

Hsia CW, Schimtz A, Lambertz M, Perry SF, Maina JN (2013) Evolution of air breathing: oxygen homeostasis and the transitions from water-to-land and sky. Comp Physiol 3:849–915

Jayachitra S, Balasundaram K (2015) Gross morphological studies on the trachea of adult Emu. Indian J Vet Anat 27:63–64

Johnsgard PA (1961) Tracheal anatomy of the Anatidae and its taxonomic significance. Wildfowl 12:58–69

Johnsgard PA (1995) Daffy but not dapper: a serious approach to the comical-looking Ruddy Duck. Birder's World 1995:48–51

Johnsgard PA (2017) The North American whistling-ducks, pochards, and stifftails. Zea Books, Lincoln

Jones DR, Holeton GF (1972) Cardiovascular and respiratory responses of ducks to progressive hypocapnic hypoxia. J Exp Biol 56:657–666

Jones MR, Witt CC (2014) Migrate small, sound big: functional constraints on body size promote tracheal elongation in cranes. J Evol Biol 27:1256–1264

Jones JH, Effmann EL, Schmidt-Nielsen K (1985) Lung volume changes during respiration in ducks. Respir Physiol 59:15–25

Kabak M, Orhan IO, Haziroglu RM (2007) The gross anatomy of larynx, trachea and syrinx in the Long-legged Buzzard (*Buteo rufinus*). Anatomia, Histologia, Embryologia 36:27–32

Kameda Y (1999) VIP-, galanin-, and neuropeptide-Y-immunoreactive fibers in the chicken carotid bodies after various types of denervation. Cell Tissue Res 298:437–447

Kesel AB, Hoffmann F, Baars A, Danter L (2017) Do bony tubercle inside the falcon nostril effect breathing during high speed diving? Poster presentation, Annual Meeting of the Society for Experimental Biology, Gothenburg, Sweden

Kesel AB, Hoffmann F, Baars A, Danter L (2019) Flow phenomena in the falcon nose – an interpretation approach. Conference: Bionik: Patente aus der Natur. Bremen, Germany, pp 224–227

Kilgore DL Jr, Boggs DF, Kilgore TJ, Colby C, Williams BR Jr, Bavis RW (2008) Ventilatory and metabolic responses of Burrowing Owls, *Athene cunicularia*, to moderate and extreme hypoxia: analysis of the hypoxic ventilatory threshold vs. hemoglobin oxygen affinity relationship in birds. Comp Biochem Physiol A 150: 247–257

King AS, McLelland J (eds) (1979) Form and function in birds, vol 4. Academic Press, New York

King AS, McLelland J (1984) Birds: their structure and function, 2nd edn. Bailliere Tindal, London

Kriesell HJ, Le Bohec C, Cerwenka AF, Hertel M, Robin J-P, Ruthensteiner B, Gahr M, Aubin T, Düring DN (2020) Vocal tract anatomy of King Penguins: morphological traits of two-voiced sound production. Front Zool 17:5

Kubke MF, Ross JM, Wild JM (2004) Vagal innervation of the air sacs in a songbird, *Taenopygia guttata*. J Anat 204:283–292

Lambertz M, Bertozzo F, Sander PM (2018) Bone histological correlates for air sacs and their implications for understanding the origin of the dinosaurian respiratory system. Biol Lett 14:20170514

Larcombe A (2002) Effects of temperature on metabolism, ventilation, and oxygen extraction in the southern brown bandicoot *Isoodon obesulus* (Marsupialia: Peramelidae). Physiol Biochem Zool 75:405–411

Lawson AB, Hedrick BP, Echols S, Schachner ER (2021) Anatomy, variation, and asymmetry of the bronchial tree in the African Grey Parrot (*Psittacus erithacus*). J Morphol 282:701–719

Ludders JW (2001) Inhaled anesthesia for birds. In: Gleed RD, Ludders JW (eds) Recent advances in veterinary anesthesia and analgesia: companion animals. International Veterinary Information Service, Ithaca

Ludders JW (2015) Respiration in birds. In: Reece WO (ed) Dukes' physiology of domestic animals. Wiley & Sons, Hoboken, pp 245–262

MacDonald JD (1960) Secondary external nares of the gannet. Proc Zool Soc Lond 135:357–363

Maina JN (1982) A scanning electron microscopic study of the air and blood capillaries of the lung of the Domestic Fowl (*Gallus domesticus*). Experientia 35: 614–616

Maina JN (1989) The morphometry of the avian lung. In: King AS, McLelland J (eds) Form and function in birds. Academic Press, London, pp 307–368

Maina JN (1998) The gas exchangers: structure, function, and evolution of the respiratory processes. Springer-Verlag, Berlin

Maina JN (2000) Comparative respiratory physiology: themes and principles in the design and construction of gas exchangers. Anat Rec 261:25–44

Maina JN (2002a) Structure, function and evolution of the gas exchangers: comparative perspectives. J Anat 201: 281–304

Maina JN (2002b) Functional morphology of the vertebrate respiratory systems. Science Publishers, Inc., Enfield

Maina JN (2005) The lung air-sac system of birds: development, structure and function. Springer, Heidelberg

Maina JN (2006) Development, structure, and function of a novel respiratory organ, the lung-air sac system of birds: to go where no other vertebrate has gone. Biol Rev 81:545–579

Maina JN (2008) Functional morphology of the avian respiratory system, the lung-air system: efficiency built on complexity. Ostrich 79:117–132

Maina JN (2015a) Morphological and morphometric properties of the blood-gas barrier: comparative perspectives. In: Makanya AN (ed) The vertebrate blood-gas barrier in health and disease. Springer, Cham, pp 15–38

Maina JN (2015b) Structural and biomechanical properties of the exchange tissue of the avian lung. Anat Rec 298: 1673–1688

Maina JN (2015c) The design of the avian respiratory system: development, morphology and function. J Ornithol 156:41–63

Maina JN (2017a) Pivotal debates and controversies on the structure and function of the avian respiratory system: setting the record straight. Biol Rev 92:1475–1504

Maina JN (2017b) Functional design of the mature avian respiratory system. In: Maina JN (ed) The biology of the avian respiratory system: evolution, development, structure and function. Springer International Publishing AG, Cham, pp 191–218

Maina JN, Africa M (2000) Inspiratory aerodynamic valving in the avian lung: functional morphometry of the extrapulmonary primary bronchus. J Exp Biol 203: 2865–2876

Maina JN, King AS (1989) The lung of the Emu, *Dromaius novaehollandiae*: a microscopic and morphometric study. J Anat 163:67–74

Maina JN, Nathaniel C (2001) A qualitative and quantitative study of the lung of an Ostrich, *Struthio camelus*. J Exp Biol 204:2313–2330

Maina JN, West JB (2005) Thin and strong! The bioengineering dilemma in the structural and functional design of the blood-gas barrier: comparative and evolutionary perspectives. Physiol Rev 85:811–844

Maina JN, Woodward JD (2009) Three-dimensional serial section computer reconstruction of the arrangement of the structural components of the parabronchus of the Ostrich, *Struthio camelus* lung. Anat Rec 292:1685–1698

Maina JN, King AS, Settle G (1989) An allometric study of the pulmonary morphometric parameters in birds, with mammalian comparison. Philos Trans R Soc B 326:1–57

Maina JN, McCracken KG, Chua B, York JM, Milsom WK (2017) Morphological and morphometric specializations of the lung of the Andean Goose, *Chloephaga melanoptera*: a lifelong high-altitude resident. PLoS ONE 12:e0174395

Makanya AN, Djonov V (2008) Development and spatial organization of the air conduits in the lung of the Domestic Fowl, *Gallus gallus* variant *domesticus*. Microsc Res Tech 71:689–702

Makanya AN, El-Darawish Y, Kavoi BM, Djonov V (2011) Spatial and functional relationships between air conduits and blood capillaries in the pulmonary gas exchange tissue of adult and developing chickens. Microsc Res Tech 74:159–169

Maloney SK, Dawson TJ (1994) Ventilatory accommodation of oxygen demand and respiratory water loss in a large bird, the Emu (*Dromaius novaehollandiae*) and a reexamination of ventilatory allometry for birds. J Comp Physiol B 164:473–481

Mangold O (1946) Die nase der segelnden Vögel ein Organ des Strömungssinnes? Naturwissenschaften 33: 19–23

Markley JS (2010) Avian ventilator physiology: the coordination of running and breathing and the cost of ventilation. Ph.D. dissertation, University of Utah, Salt Lake City, UT

Markley JS, Carrier DR (2010) The cost of ventilation in birds measured via unidirectional artificial ventilation. Comp Biochem Physiol A 155:146–153

McLelland J (1989a) Larynx and trachea. In: King AS, McLelland J (eds) Form and function in birds, vol 4. Academic Press, San Diego, pp 69–103

McLelland J (1989b) Anatomy of the lungs and air sacs. In: King AS, McLelland J (eds) Form and function in birds, vol 4. Academic Press, San Diego, pp 221–279

Mendez JM, Alliende JA, Amador A, Mindlin GB (2006) Dynamical systems techniques reveal the sexual dimorphic nature of motor patterns in birdsong. Phys Rev E 74:041917

Mortola JP, Limoges M-J (2006) Resting breathing frequency in aquatic mammals: a comparative analysis with terrestrial species. Respir Physiol Neurobiol 154: 500–514

Mortola JP, Seguin J (2009) Resting breathing frequency in aquatic birds: a comparative analysis with terrestrial species. J Zool 279:210–218

Nagy KA (2005) Field metabolic rate and body size. J Exp Biol 208:1621–1625

Nassar PN, Jackson AC, Carrier DR (2001) Entraining the natural frequencies of running and breathing in Guinea fowl (*Numida meleagris*). J Exp Biol 204:1641–1651

Nevitt G (1999) Foraging by seabirds on an olfactory landscape. Am Sci 87:46–54

Nguyen QM, Oza AU, Abouezzi J, Sun G, Childress S, Frederick C, Ristroph L (2021) Flow rectification in loopy network models of bird lungs. Phys Rev Lett 126:114501

Nye PC, Burger RE (1978) Chicken intrapulmonary chemoreceptors: discharge at static levels of intrapulmonary carbon dioxide and their location. Respir Physiol 33:299–322

O'Connor PM (2004) Pulmonary pneumaticity in the postcranial skeleton of extant aves: a case study examining Anseriformes. J Morphol 261:141–161

O'Connor PM (2009) Evolution of archosaurian body plans: skeletal adaptations of an air-sac-based breathing apparatus in birds and other archosaurs. J Exp Zool Part A: Ecol Genet Physiol 311:629–646

O'Connor PM, Claessens LP (2005) Basic avian pulmonary design and flow-through ventilation in non-avian theropod dinosaurs. Nature 436:253–256

Pennycuick CJ (2002) Gust soaring as a basis for the flight of petrels and albatrosses (Procellariiformes). Avian Sci 2:1–12

Pennycuick CJ (2008) Information systems for flying animals. In: Pennycuick CJ (ed) Theoretical Ecology Series, vol. 5. Modelling the flying bird. Elsevier Inc., Amsterdam, pp 305–331

Perry SF, Codd JR, Klein W (2005) Evolutionary biology of aspiration breathing and origin of the mammalian diaphragm. Revue des Maladies Respiratoires 22: 2S19–2S38

Persons WS IV, Currie PJ (2015) Bristles before down: a new perspective on the functional origin of feathers. Evolution 69:857–862

Piiper J, Dres F, Scheid P (1970) Gas exchange in the domestic fowl during spontaneous breathing and artificial ventilation. Respir Physiol 9:234–245

Potier S (2020) Olfaction in raptors. Zool J Linn Soc 189: 713–721

Powell FL (2000) Respiration. In: Whittow GC (ed) Avian physiology, 5th edn. Academic Press, New York, pp 233–264

Powell FL, Hopkins SR (2004) Comparative physiology of lung complexity: implications for gas exchange. News Physiol Sci 19:55–60

Powell FL, Geiser J, Gratz RK, Scheid P (1981) Airflow in the avian respiratory tract: variations in $O_2$ and $CO_2$ concentrations in the bronchi of the duck. Respir Physiol 44:195–213

Powell FL, Milsom WK, Mitchell GS (1998) Time domains of the hypoxic ventilator response. Respir Physiol 112:123–134

Prange HD, Wasser JS, Gaunt AS, Gaunt SL (1985) Respiratory responses to acute heat stress in cranes (Gruidae): the effects of tracheal coiling. Respir Physiol 62:95–103

Ruben JA, Jones TD, Geist NR (2003) Respiratory and reproductive paleophysiology of dinosaurs and early birds. Physiol Biochem Zool 76:141–164

Sadava DE, Heller HC, Purves WK, Orians GH, Hillis DM (2008) Life: the science of biology. Macmillan, New York

Sander JE, Wilson JL, Rowland GN, Middendorf PJ (1995) Formaldehyde vaporization in the hatcher and the effect on tracheal epithelium of the chick. Avian Dis 39:152–157

Scheid P (1979) Mechanisms of gas exchange in bird lungs. In: Reviews of physiology, biochemistry and pharmacology, vol 86. Springer, Berlin, Heidelberg, pp 137–186

Scheid P, Piiper J (1987) In: Seller TJ (ed) Gas exchange and transport. CRC Press, Boca Raton, FL, pp 97–129

Scheid P, Slama H, Gatz RN, Fedde MR (1974) Intrapulmonary $CO_2$ receptors in the duck. III. Functional localization. Respir Physiol 22:123–136

Scheid P, Gratz RK, Powell FL, Fedde MR (1978) Ventilatory response to $CO_2$ in birds. II. Contribution by intra-pulmonary $CO_2$ receptors. Respir Physiol 35: 361–372

Schmidt MF, Wild JM (2014) The respiratory-vocal system of songbirds: anatomy, physiology, and neural control. Prog Brain Res 212:297–335

Schreiber EA, Burger J (2002) Biology of marine birds. CRC Press, Boca Raton

Schwarz T, Kelley C, Pinkerton ME, Hartup BK (2016) Computed tomographic anatomy and characteristics of respiratory aspergillosis in juvenile Whooping Cranes. Vet Radiol Ultrasound 57:16–23

Scott GR, Milsom WK (2009) Control of breathing in birds: implications for high-altitude flight. In:

Glass M, Wood S (eds) Cardio-respiratory control in vertebrates. Springer, Berlin, Heidelberg, pp 429–448
Sereno PC, Martinez RN, Wilson JA, Varricchio DJ, Alcober OA, Larsson HCE (2008) Evidence for avian intrathoracic air sacs in a new predatory dinosaur from Argentina. PLoS One 3:e3303
Smith ND (2012) Body mass and foraging ecology predict evolutionary patterns of skeletal pneumaticity in the diverse "waterbird" clade. Evolution 66:1059–1078
Sverdlova NS, Lambertz M, Witzel U, Perry SF (2012) Boundary conditions for heat transfer and evaporative cooling in the trachea and air sac system of the domestic fowl: a two-dimensional CFD analysis. PLoS ONE 7:e45315
Tenney SM, Remmers JB (1963) Comparative morphology of the lung: diffusing area. Nature 197:54–56
Thomas SP, Follette DB, Thomas GS (1995) Metabolic and ventilatory adjustments and tolerance of the bat *Pteropus poliocephalus* to acute hypoxic stress. Comp Biochem Physiol 112A:43–54
Tickle PG, Ennos AR, Lennox LE, Perry SF, Codd JR (2007) Functional significance of the uncinate processes in birds. J Exp Biol 210:3955–3961
Tickle P, Nudds R, Codd J (2009) Uncinate process length in birds scales with resting metabolic rate. PLoS ONE 4:e5667
Tobalske BW, Dial KP (1996) Flight kinematics of Black-billed Magpies and pigeons over a wide range of speeds. J Exp Biol 199:263–280
Ujhelyi P (2016) Cranial morphology of European passerine bird families (Aves, Passeriformes). Ornis Hungarica 24:54–77
Wang N, Banzett RB, Nations CS, Jenkins FA (1992) An aerodynamic valve in the avian primary bronchus. J Exp Biol 262:441–445
Wedel MJ (2003) Vertebral pneumaticity, air sacs, and the physiology of sauropod dinosaurs. Paleobiology 29: 243–255
Wedel MJ (2006) Origin of postcranial skeletal pneumaticity in dinosaurs. Integr Zool 1:80–85
West JB (2009) Comparative physiology of the pulmonary blood-gas barrier: the unique avian solution. Am J Physiol - Regul Integr Comp Physiol 297:R1625–R1634
West JB, Watson RR, Fu Z (2007a) Major differences in the pulmonary circulation between birds and mammals. Respir Physiol Neurobiol 157:382–390
West JB, Watson RR, Fu Z (2007b) The human lung: did evolution get it wrong? Eur Respir J 29:11–17
Wetherbee DK (1951) Air sacs in the English Sparrow. Auk 68:242–244
Whitaker-Fornek JR, Nelson JK, Lybbert CW, Pilarski JQ (2019) Development and regulation of breathing rhythms in embryonic and hatchling birds. Respir Physiol Neurobiol 269:103246
York JM, Chua BA, Ivy CM, Alza L, Cheek R, Scott GR, McCracken KG, Frappell PB, Dawson NJ, Laguë SL, Milsom WK (2017) Respiratory mechanics of eleven avian species resident at high and low altitude. J Exp Biol 220:1079–1089
York JM, Scadeng M, McCracken KG, Milsom WK (2018) Respiratory mechanics and morphology of Tibetan and Andean high-altitude geese with divergent life histories. J Exp Biol 221:jeb170738
Zeek PM (1951) Double trachea in penguins and sea lions. Anat Rec 111:327–343
Zweers GA, van Pelt HC, Beckers A (1981) Morphology and mechanics of the larynx of the pigeon (*Columba livia* L.): a drill-chuck system (Aves). Zoomorphology 99:37–69

# Endocrine System

# 8

## Contents



**Abstract**

The avian endocrine system includes the hypothalamus, pituitary gland, thyroid gland, ultimobranchial glands, adrenal glands, endocrine cells in the digestive system, endocrine cells in the pancreas, the testes of males, and the ovaries of females. In this chapter, I discuss the respective functions of hormones produced by these glands and cells. The hypothalamus produces and secretes hormones that help regulate the actions of the anterior pituitary and produces hormones that are stored in and released from the posterior pituitary. The anterior pituitary produces and secretes hormones that target specific organs, such as the gonads,


G. Ritchison, *In a Class of Their Own*, Fascinating Life Sciences,
https://doi.org/10.1007/978-3-031-14852-1_8

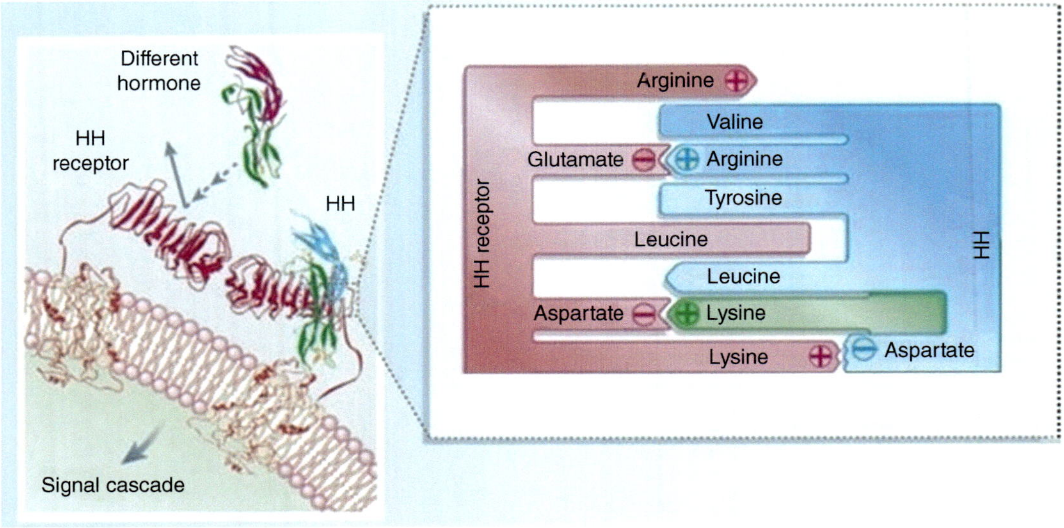


**Fig. 8.1** Hypothetical example of a hormone molecule binding to its receptor in the membrane of a target cell. The hypothetical peptide/protein hormone (HH) binds to its receptor in the membranes of target. Different hormones with different three-dimensional shapes are unable to bind to FSHR. The inset on the right shows how amino acids in the HH and HH receptor "interlock" to allow binding. (Figure modified from Dias 2005; © 2005 Nature Publishing Group, used with permission)

or have body-wide effects. The specific functions of these hormones are described in this chapter, as are the hormones released by the posterior pituitary. Also discussed in this chapter are the functions of hormones produced and secreted by the thyroid gland, parathyroid gland, ultimobranchial glands, adrenal glands, and the pancreas. The specific roles of thyroid hormones in reproduction, embryonic development and hatching, and migration are also discussed. The respective functions of hormones that play important roles in regulating bird appetites and digestion are also explained.

## 8.1 Endocrine Gland Hormones

Glands in the avian endocrine system produce and release hormones into the blood, with hormones representing "messages" sent to target organs or tissues. Hormone molecules can target specific organs and cells because only certain cells contain receptors for those molecules in their cell membranes or, for lipid-based hormones that can easily pass through cell and nuclear membranes, receptors in the nucleus. For example, follicle-stimulating hormone (FSH) is produced by the anterior pituitary and targets cells in the gonads that have follicle-stimulating hormone receptors (FSHR; Fig. 8.1). The binding of FSH to FSHR initiates a "signal cascade" in the target cells, stimulating the maturation of eggs in ovaries and the production of sperm in testes. All hormones similarly target cells that have receptors for those hormones in either their cell membranes or nuclei (Figs. 8.2 and 8.3).

The avian endocrine system includes the hypothalamus, pituitary gland, thyroid gland, ultimobranchial glands, adrenal glands, endocrine cells in the digestive system, endocrine cells in the pancreas, the testes of males, and the ovary or ovaries of females (Fig. 8.4). Cells in the endocrine system produce hormones and are typically categorized as monoamines, peptide/protein hormones, or lipid-derived (or steroid) hormones. Monoamines are hormones derived from amino acids and include epinephrine, norepinephrine,

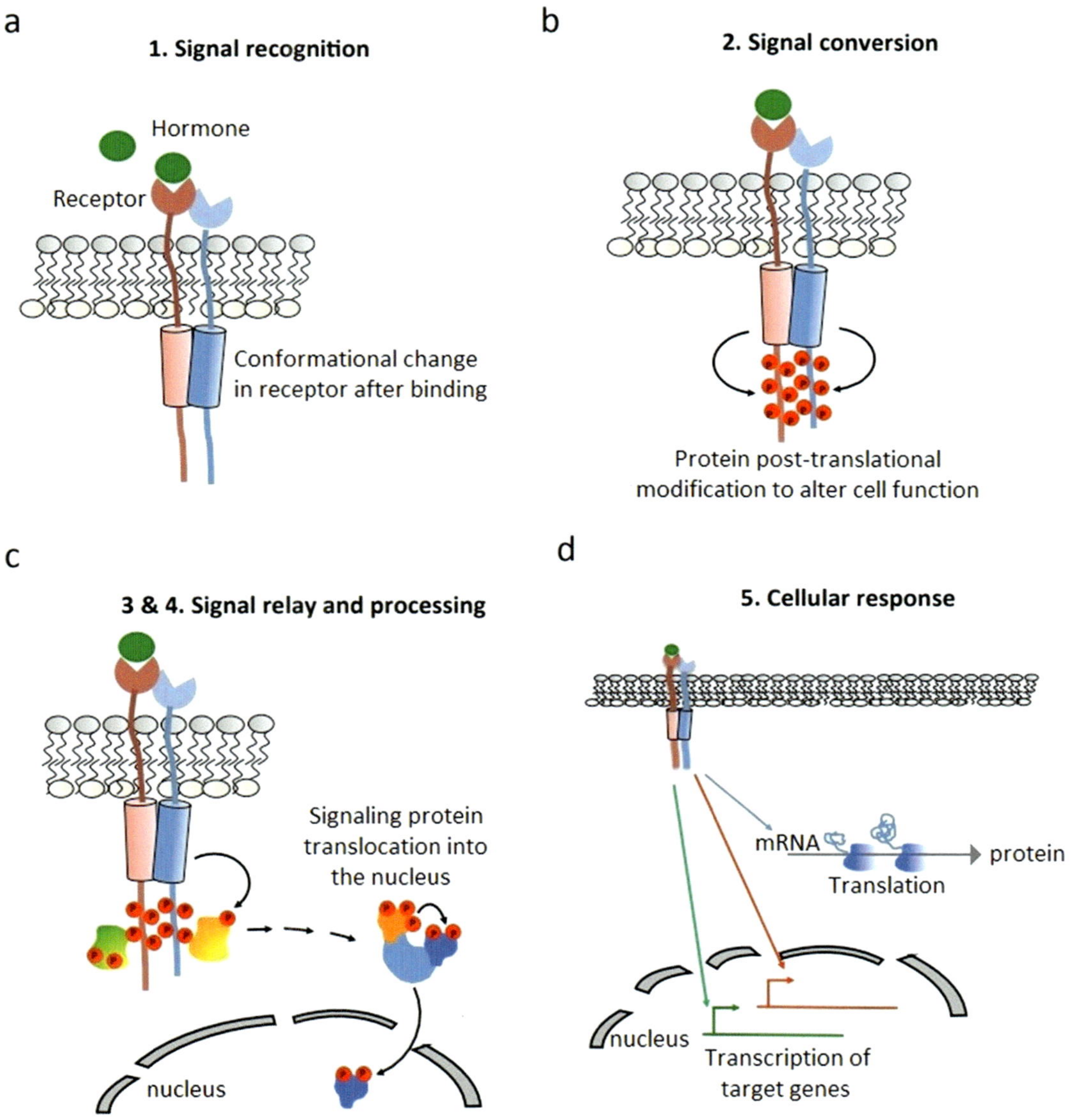


**Fig. 8.2** In general, signal transduction includes several steps to convert a hormonal "message" (i.e., hormone binding to its receptor) into a change in cell activity or function. In this example, the hormone molecule binds to its receptor (step 1). The receptor undergoes a transformation change and can modify proteins (enzymes) in a variety of different ways (step 2). Signal processing can also involve the formation of protein complexes that are translocated into the nucleus (steps 3 and 4) and initiate transcription of target genes and their translation into proteins (step 5) that alter cell function. (Figure from Yao et al. 2015; © 2015 Nature Publishing Group, a division of Macmillan Publishers Limited, used with permission)

and melatonin; monoamines also serve as neurotransmitters in the nervous system. Most hormones are peptide/protein hormones and include hormones produced by the anterior pituitary gland (e.g., adrenocorticotropic hormone, follicle-stimulating hormone, prolactin, and thyroid-stimulating hormone), hypothalamus (arginine vasotocin and mesotocin), thyroid gland

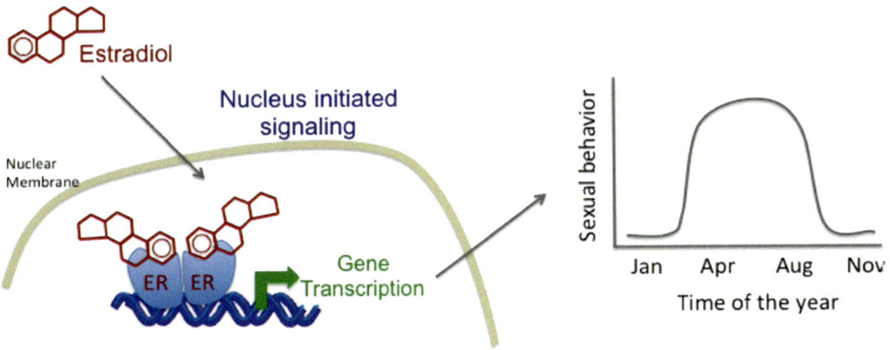


**Fig. 8.3** Lipid-based hormones like estradiol can pass through membranes and bind to, in this example, estradiol receptors (ER) in the nucleus. This alters gene expression to, for example, cause the cell to start producing estradiol that then triggers sexual behavior by female birds. (Figure from Rudolph et al. 2016; open-access article distrusted under the terms of the Creative Commons Attribution 4.0 International License (CC-BY), https://creativecommons.org/licenses/by/4.0/)

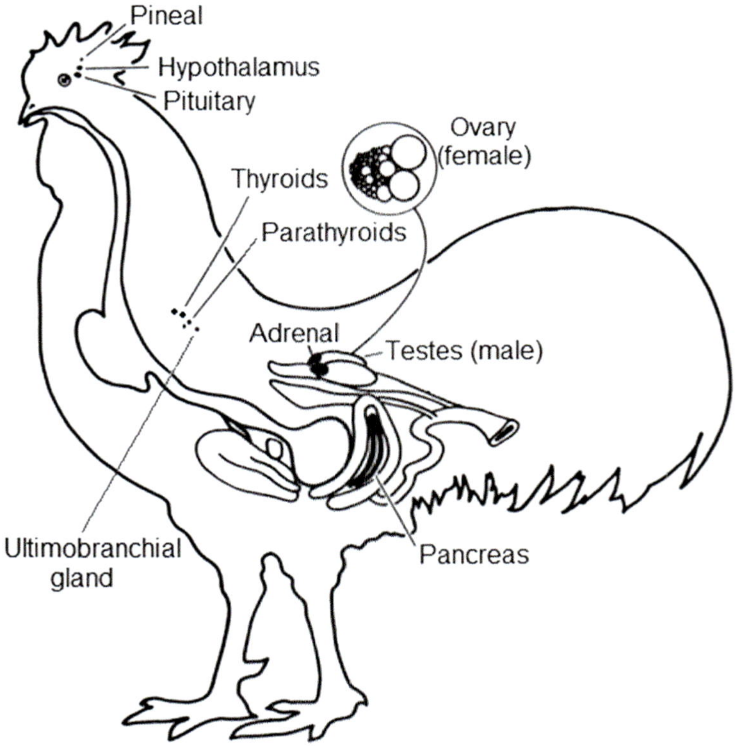


**Fig. 8.4** Endocrine glands of birds. (Figure modified from Parkhurst and Mountney 1988; © 1988 Van Nostrand Reinhold, used with permission)

(thyroxine, triiodothyronine, and parathyroid hormone), ultimobranchial glands (calcitonin), and pancreas (glucagon, insulin, pancreatic polypeptide, and somatostatin), and gastrointestinal hormones (e.g., gastrin, ghrelin, and secretin). Lipid (steroid)-based hormones are produced by the testes (testosterone), ovaries (estrogen, testosterone, and progesterone), and adrenal cortex (aldosterone and corticosterone).

## 8.2 Hypothalamus and the Pituitary Gland

The avian hypothalamus consists of several groups of cells that serve a variety of different functions, including temperature regulation, maintenance of diurnal rhythms, and control of the autonomic nervous system. The hypothalamus, located just above the pituitary gland (Fig. 8.5), is also a key component of the endocrine system, producing and secreting hormones that help regulate the actions of the anterior pituitary and producing hormones that are stored in and released from the posterior pituitary (Figs. 8.6 and 8.7). Hormones that regulate the actions of the anterior pituitary include gonadotropin-releasing hormone, gonadotropin-inhibitory hormone, corticotropin-releasing hormone, thyrotropin-releasing hormone, vasoactive intestinal polypeptide, and growth hormone-releasing hormone. Hormones produced by the hypothalamus that are stored and released from the posterior pituitary include arginine vasotocin and mesotocin (Fig. 8.6). Regulation of the secretion of hormones by the hypothalamus and pituitary is achieved by way of a feedback mechanism, with the amount of hormone produced and released into the blood by endocrine glands influencing the production and release of hormones by the hypothalamus and pituitary (Fig. 8.8). This feedback mechanism ensures that hormone levels in the blood are maintained at levels suitable for a particular time of year or situation.

### 8.2.1 Gonadotropin-Releasing Hormone

Two hormones produced by the hypothalamus are gonadotropin-releasing hormone (GnRH) and gonadotropin-inhibiting hormone (GnIH). These hormones either inhibit or stimulate reproduction by birds and their release can be influenced by both photic (light) and non-photic (e.g., rainfall) factors (Fig. 8.9). This topic is covered in more detail in Chap. 16, but, briefly, GnRH secretion triggers the release of follicle-stimulating hormone (FSH) and luteinizing hormone (LH) by the gonads. These hormones, in turn, stimulate the maturation of eggs and release of hormones (estrogen and progesterone) by

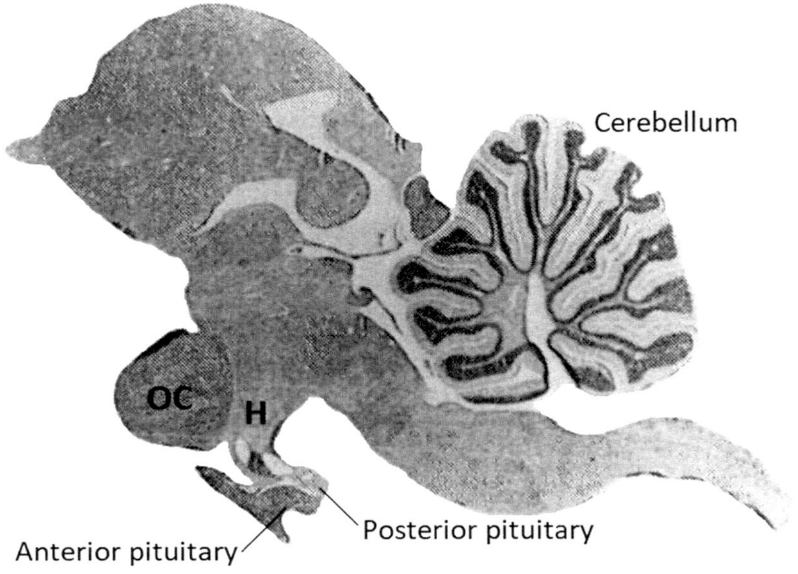


**Fig. 8.5** Section of the brain of a Domestic Chicken (*Gallus g. domesticus*) showing the location of the hypothalamus (H) and anterior and posterior pituitary. OC, optic chiasma. (Figure from Opel 1979; © 1979 Elsevier Inc., used with permission)

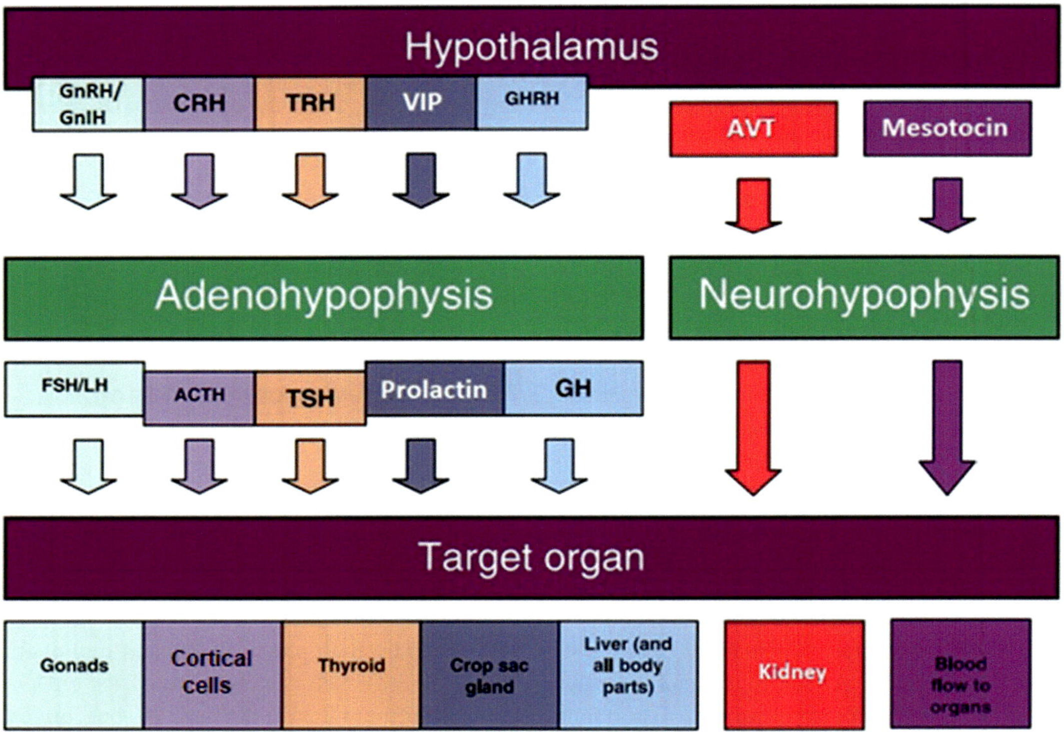


**Fig. 8.6** Hormones produced by the hypothalamus and pituitary (also called the hypophysis) of birds and their target organs. GnRH, Gonadotropin-Releasing Hormone; GnIH, Gonadotropin-Inhibitory Hormone; CRH, Corticotropin-Releasing Hormone; TRH, Thyrotropin-Releasing Hormone; VIP, Vasoactive Intestinal Polypeptide; GHRH, Growth Hormone-Releasing Hormone; AVT, arginine vasotocin; FSH, follicle-stimulating hormone; LH, luteinizing hormone; ACTH, adrenocorticotropic hormone; TSH, thyroid-stimulating hormone; GH, growth hormone. (Figure from Ritchie 2014; © 2008 Elsevier Inc., used with permission)

female ovaries and the production of sperm and release of testosterone by male testes (Figs. 8.9, 8.10, and 8.11; Box 8.1 Testosterone Levels in Sex-Role Reversed Species of Birds). Increasing plasma levels of estrogen and testosterone also stimulate various sexual behaviors by females (e.g., copulation solicitation and nest building) and males (e.g., singing), respectively (see Chap. 15 for additional detail).

### 8.2.2 Vasoactive Intestinal Peptide

Another hormone produced by the hypothalamus is vasoactive intestinal peptide (VIP). VIP is so-named because it was first isolated from the duodenum of pigs (Said and Mutt 1972); it serves no function in the avian digestive system. Rather, VIP stimulates the production and release of prolactin by the anterior pituitary (Maney et al. 1999). In temperate-zone birds, longer days activate the VIP/prolactin system, whereas, at lower latitudes, the ovarian steroids, estrogen and progesterone, may trigger activation (Kosonsiriluk et al. 2008). The number of VIP neurons in the hypothalamus increases during egg laying and peaks during incubation by female Domestic Chickens (*Gallus g. domesticus*; Figs. 8.12 and 8.13). As these numbers increase, plasma levels of prolactin increase, causing reduced secretion of ovarian steroids (Zadworny et al. 1988), inhibition of egg laying (Zadworny et al. 1988), and stimulating nest protective behavior (Zadworny and Etches 1987). After eggs hatch, the number

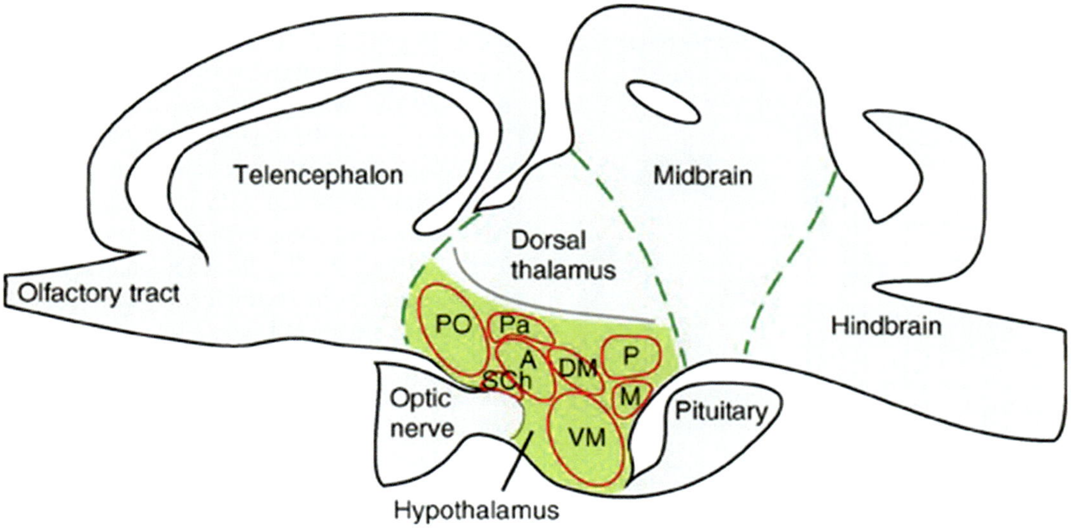


**Fig. 8.7** Control of pituitary hormone release by the hypothalamus is regulated by neurons from the paraventricular nucleus (Pa) and the suprachiasmatic nucleus (SCh), with neurons that project into the neurohypophysis to release arginine vasotocin and mesotocin, and neurons in the preoptic nucleus, paraventricular nucleus, and other nuclei produce and release the various releasing and inhibitory hormones plus vasoactive intestinal polypeptide. PO, preoptic area; Pa, paraventricular nucleus; SCh, suprachiasmatic nucleus; VM, ventromedial nucleus; M, mammillary body; A, anterior nucleus; DM, dorsomedial nucleus; P, posterior nucleus. (Figure from Bruce 2009; © 2009 Springer-Verlag GmbH Berlin Heidelberg, used with permission)

of VIP neurons decreases as do plasma levels of prolactin (Fig. 8.12).

Kingsbury et al. (2015) found that VIP expression was positively correlated with nest building behavior and time spent in nest cups by male and female Zebra Finches (*Taeniopygia guttata*). In songbirds, VIP appears to stimulate aggressive behavior, influencing singing behavior by territorial males during the breeding season (Goodson 1998a, b), but reduce aggressive behavior of males in colonial-breeding Zebra Finches (Goodson and Adkins-Regan 1999). In addition, Brown (1973) found that electrical stimulation of an area of the anterior hypothalamus with VIP-secreting cells evoked aggressive behavior by Steller's Jays (*Cyanocitta stelleri*).

Beyond its role in reproduction, VIP has also been found to affect gregariousness, with species that regularly occur in flocks having more neurons with VIP receptors in their brains than species that are less gregarious and more territorial (e.g., Goodson et al. 2006; Kingsbury and Wilson 2016). For example, two species of sparrows (Field Sparrows, *Spizella pusilla*, and Dark-eyed Juncos, *Junco hyemalis*) that form flocks during the nonbreeding season had higher concentrations of VIP neurons in an area of the brain important in regulating social behavior than two species of songbirds (Song Sparrows, *Melospiza melodia*, and Eastern Towhees, *Pipilo erythropthalmus*) that do not form flocks (Fig. 8.14). In further support of the role of VIP in flocking behavior, Kingsbury (2015) found that injection of a VIP antagonist (thus blocking the action of VIP) significantly reduced the amount of time that Zebra Finches spent in social contact with conspecifics.

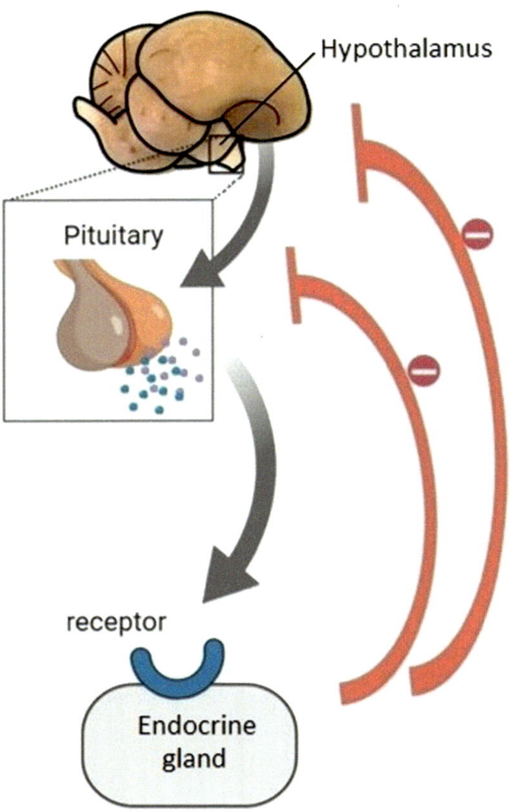


**Fig. 8.8** The hypothalamic-pituitary-endocrine gland feedback system. In this example, the hypothalamus releases hormones called releasing factors that stimulate the pituitary gland to release specific hormones into the blood. Hormone molecules bind to receptors in the cell membranes or nuclei of endocrine glands, causing the production and release of hormones into the blood. These hormones can then, depending on their concentration in the blood, influence the amount of releasing factor (hypothalamus) and hormone (pituitary) released which can then, in turn, influence the amount of hormone produced and released by the endocrine gland. (Figure modified from Loveland et al. 2021; © 2020 Elsevier Inc. All rights reserved, used with permission)

### 8.2.3 Prolactin

In birds, plasma levels of prolactin are low during nonbreeding periods, and then increase significantly during the incubation and post-hatching periods (e.g., Vleck and Vleck 2011; Smiley and Adkins-Regan 2016b; Fig. 8.15). High plasma levels of prolactin are correlated with the expression of parental care behaviors (e.g., Wang et al. 2020), reproductive success (e.g., Smiley and Adkins-Regan 2016a, b, Wang et al. 2020), and, for some species of birds, production of crop milk (Box 8.2 Crop Milk). For example, Wang et al. (2020) found that the time spent incubating eggs by male and female Kentish Plovers (*Charadrius alexandrinus*) increased with increasing plasma levels of prolactin (Fig. 8.16), as did daily nest survival (Fig. 8.17). In addition, prolactin concentrations were positively related to egg temperatures in Wood Ducks (*Aix sponsa*; Hope et al. 2020). In species where only females incubate eggs, females have higher prolactin levels than males (e.g., Dawson and Goldsmith 1982; Wingfield et al. 1989), but the opposite is true for sex-role reversed species like Red-necked Phalaropes (*Phalaropus lobatus*) where only males incubate eggs (Gratto-Trevor et al. 1990b).

The timing of prolactin secretion during incubation varies among different species of birds. For species with precocial young, plasma levels typically increase immediately after the egg-laying period, then decline shortly after eggs hatch (Fig. 8.18). In species with precocial young, parental care after eggs hatch is generally limited to leading young to good foraging areas and protecting young from predators. Among altricial species of birds where parental care continues through nestling and post-fledging periods, prolactin levels for those that regularly attend nests decline either when parental care ceases (e.g., during the mid- to late post-fledging period) or nests fail (e.g., due to predation). For altricial species where adults make extended trips to forage during the nestling period, e.g., some species of seabirds, prolactin levels may remain high for variable periods even after nest failure (Fig. 8.18). Once prolactin levels increase, levels of prolactin remain high for species that regularly attend nests through visual and tactile contact with nests, eggs, and young (e.g., Sharp et al. 1998). For species that leave nests for extended period to forage, an endogenously timed mechanism maintains the high levels of prolactin (Vleck and Vleck 2011). The behavior of helpers (i.e., nonbreeding individuals that help raise young of a breeding pair) in cooperatively breeding species of birds is an example of the importance of the

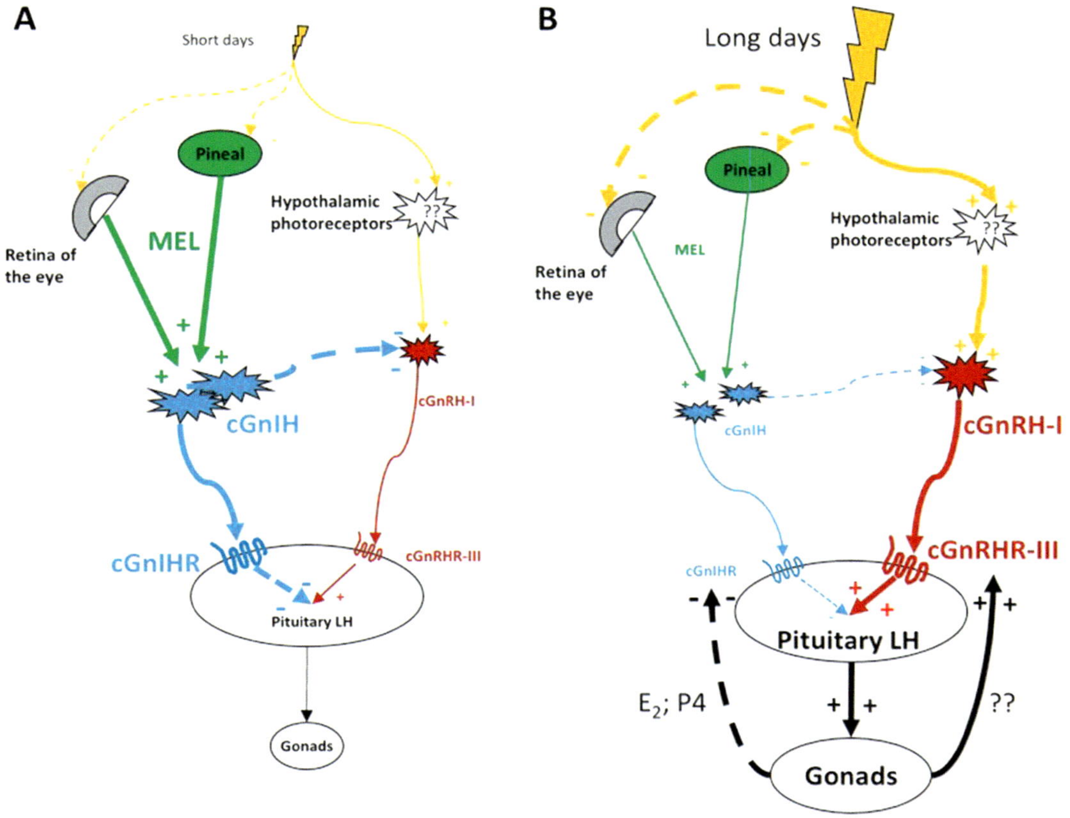


**Fig. 8.9** Summary of the dual stimulatory/inhibitory control of reproduction in Domestic Chickens. (**a**) Under short days, the pineal gland and retina of the eyes both increase melatonin (MEL) production, which in turn activates the hypothalamic synthesis and release of chicken gonadotropin-inhibitory hormone (cGnIH). Increased cGnIH activity directly inhibits chicken gonadotropin-releasing hormone I (cGnRH-I) neurons in the hypothalamus and reduces the release of luteinizing hormone (LH) by the anterior pituitary gland. At this stage, levels of cGnIH receptor (cGnIHR) and cGnRH receptor III (cGnRHR-III) in the pituitary gland are highest and lowest, respectively. Due to the lack of stimulation, the reproductive tract (including the gonads) remains immature; (**b**) Upon photostimulation with long days, a reduction in MEL production by the retina of the eye and the pineal gland reduces cGnIH release, thus lifting the inhibition. Simultaneously, the increase in photoperiod activates deep brain (hypothalamic) photoreceptors, which indirectly stimulate the synthesis and release of cGnRH-I. In turn, this switch from inhibitory to stimulatory control results in increased production of LH, triggering the maturation of the gonads and consequently the release of sex steroid hormones (estradiol E2 and progesterone P4). As the bird matures sexually, E2 and P4 inhibit the expression of the cGnIHR in the pituitary gland, while an increase in cGnRHR-III expression is observed. At this stage, the hypothalamus is primarily under stimulatory control and the anterior pituitary gland is sensitive primarily to cGnRH. (Figure from Bedecarrats 2015; © 2015 Poultry Science Association Inc. Published by Elsevier Inc., used with permission)

presence of nestlings in secretion of prolactin and providing parental care. Among helpers, plasma levels of prolactin increase when they begin provisioning young and elevated plasma levels of prolactin have been found to be positively associated with provisioning rates of helpers in Red-cockaded Woodpeckers (*Picoides borealis*; Khan et al. 2001), Florida Scrub-Jays (*Apheloma coerulesens*; Schoech et al. 1996), and Harris' Hawks (*Parabuteo unicinctus*; Vleck et al. 1991).

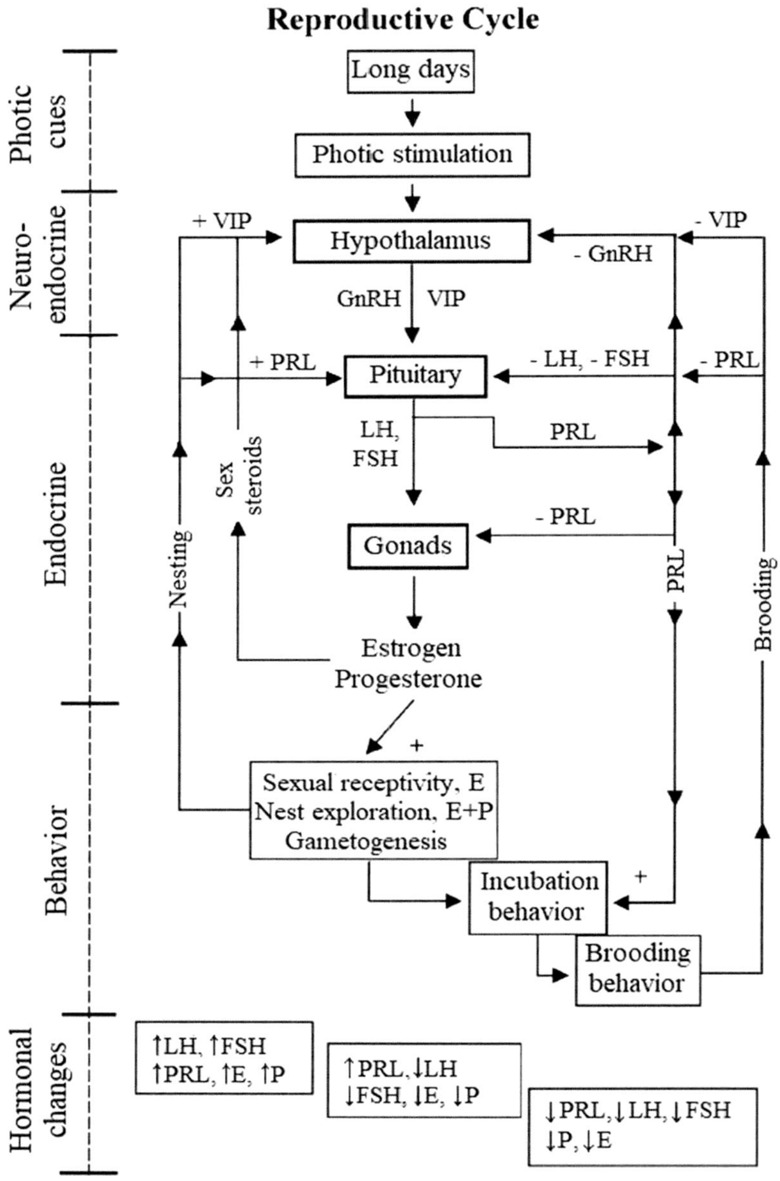


**Fig. 8.10** Flow chart showing the interrelationships between external and internal stimuli and the neuroendocrine control of the reproductive cycle of female Wild Turkeys (*Meleagris gallopavo*). + = stimulation/positive feedback; - = inhibition/negative feedback; ↑ = increase; ↓ = decrease; GnRH = gonadotropin-releasing hormone; VIP = vasoactive intestinal peptide; LH = luteinizing hormone; FSH = follicle-stimulating hormone; PRL = prolactin; P = progesterone; E = estrogen. (Figure modified from Chaiseha and El Halawani 2005; used with permission of the Japan Poultry Science Association)

### 8.2.4 Gonadotropin-Inhibitory Hormone

Gonadotropin-inhibitory hormone (GnIH) was first identified in the hypothalamus of Japanese Quail (*Coturnix japonica*) and was so-named because it can inhibit the release of gonadotropin-releasing hormone from the hypothalamus (Ubuka et al. 2008) and gonadotropins by the anterior pituitary (Tsutsui et al. 2000). Chowdhury et al. (2010) found that GnIH neurons in the hypothalamus have melatonin receptors, that melatonin administration increased the amount of GnIH released from the hypothalamus, and that the release of GnIH from the hypothalamus increased during periods with short days. Similarly, Kumar and Follett (1993) reported that melatonin secretion in Japanese

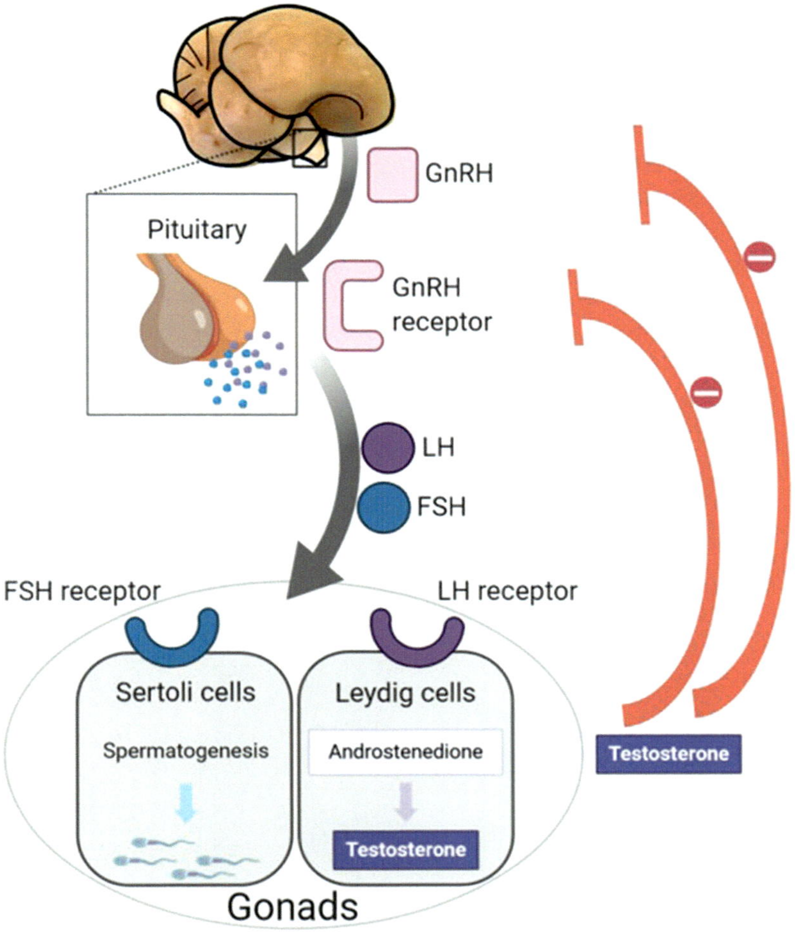


**Fig. 8.11** The hypothalamic-pituitary-gonadal (HPG) axis of male birds. Gonadotropin-releasing hormone is secreted by the hypothalamus and travels through the blood to the anterior pituitary. In response, the anterior pituitary releases luteinizing hormone (LH) and follicle-stimulating hormone (FSH) into the blood. In the testes, FSH binds to receptors in the membranes of Sertoli cells to stimulate spermatogenesis, and LH binds to receptors in the membranes of Leydig cells to stimulate the synthesis and release of androstenedione that is then converted to testosterone. Testosterone levels in the blood, in turn, regulate GnRH, LH, and FSH production by way of a negative feedback loop. (Figure from Loveland et al. 2021; © 2020 Elsevier Inc. All rights reserved, used with permission)

Quail varied with the length of nights, with secretion beginning soon after nightfall. These results indicate that GnIH transduces photoperiodic information via changes in the melatonin signal and, therefore, influences the reproductive axis of birds. More specifically, secretion of GnIH by the hypothalamus suppresses or inhibits reproduction.

The release of GnIH by the hypothalamus can also be influenced by environmental information and stress (Calisi et al. 2008; Tobari and Tsutsui 2019). Calisi et al. (2011) found that social status, as determined by having a nest box or not, and breeding status influenced brain levels of GnIH in European Starlings (*Sturnus vulgaris*), with those outcompeting others for nest boxes having fewer GnIH-producing cells than those without nest boxes. The effect of stress on GnIH production was revealed in a study where captive male House Sparrows (*Passer domesticus*) were found to have more GnIH-secreting cells than free-living males (DuRant et al. 2020). Thus, initiation of breeding by birds can be influenced by a bird's social environment and stress levels as well as by daylength (Fig. 8.19).

GnIH has other functions beyond its role in reproduction. For example, injection of GnIH into the cerebrospinal fluid of young Domestic Chickens stimulates food intake (Tachibana et al. 2005). Food availability is important for successful reproduction, so food scarcity during the breeding season can inhibit breeding. The

**Box 8.1 Testosterone Levels in Sex-Role Reversed Species of Birds**

Among species of birds that exhibit sex-role reversal, females are the more aggressive in sex and males incubate eggs and care for their precocial young. Given that, in most species of birds, males have higher plasma levels of testosterone than females, females in species that exhibit sex-role reversal might be expected to have higher plasma levels of testosterone than males. Lipshutz and Rosvall (2020b) conducted a meta-analysis on sex differences in plasma levels of testosterone in sex-role reversed species and showed that males in sex-role reversed species had significantly higher plasma levels of testosterone during the courtship period, but, when males were caring for young, testosterone levels of males and females did not differ. Such results indicate that sex-role reversal may select for lower testosterone levels in males, higher levels in females, or both.

In sex-role reversed species like Wilson's Phalaropes (*Phalaropus tricolor*), females (on the right) are more aggressive and colorful than males (on the left). (Photo by Tom Koerner/USFWS, Wikipedia, CC BY 2.0, https://creativecommons.org/licenses/by/2.0/)

Although females in sex-role reversed species must compete for mates, they still solicit copulations from males and lay eggs and, although males do not compete for mates, they still have higher plasma levels of testosterone than females during the courtship period. One possible explanation for this is that the sexes in sex-role reversed species may differ from non-sex-role reversed species in their sensitivity to the effects of circulating testosterone (Lipshutz and Rosvall 2020a, b).

(continued)

**Box 8.1** (continued)

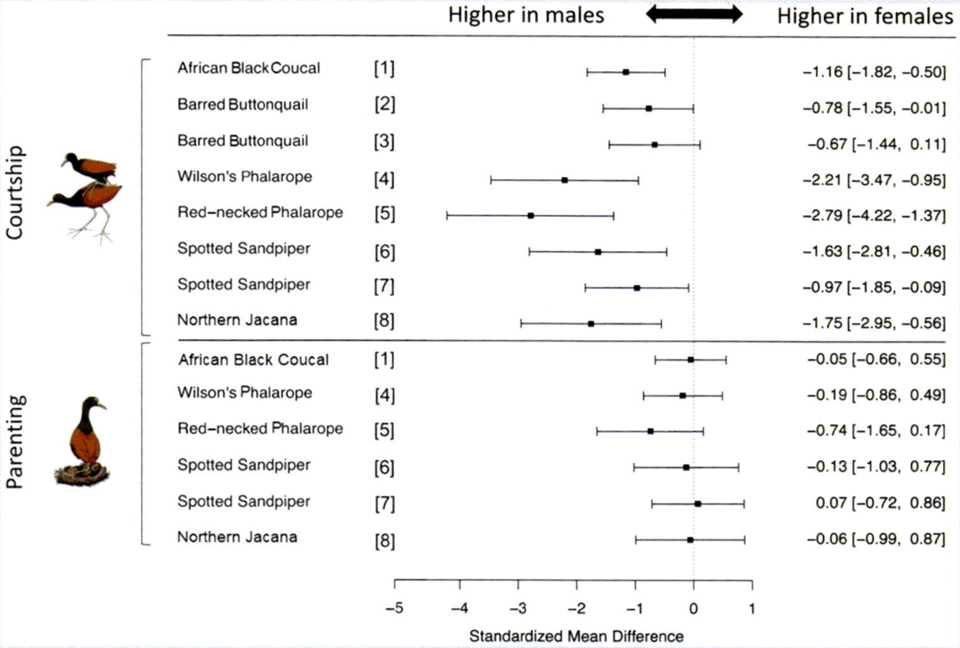


Mean effect sizes with (with 95% confidence intervals) for differences between males and females in plasma levels of testosterone during the courtship period and when caring for young after eggs hatch (Parenting). Numbers refer to references: [1] Goymann and Wingfield (2004), [2] Voigt (2016), [3] Muck and Goymann (2011), [4] Fivizzani et al. (1986), [5] Gratto-Trevor et al. (1990a), [6] Oring et al. (1986), [7] Rissman and Wingfield (1984), [8] Lipshutz and Rosvall (2020a, b). All studies were conducted with free-living individuals, except for Barred Buttonquails [2, 3]. African Black Coucal, *Centropus grillii*; Barred Buttonquail, *Turnix suscitator*; Wilson's Phalarope, *Phalaropus tricolor*; Red-necked Phalarope, *P. lobatus*; Spotted Sandpiper, *Actitis macularius*; Northern Jacana, *Jacana spinosa*. (Figure from Lipshutz and Rosvall 2020a, b; © 2020 Oxford University Press, used with permission)

stress of food scarcity can increase production of GnIH by the hypothalamus, which, in turn, could delay or inhibit breeding and, at the same time, stimulate feeding behavior (Kriegsfeld et al. 2015). GnIH also influences reproductive behavior by increasing the synthesis of neuroestrogen in the hypothalamus (Fig. 8.20). Although production of estrogens is often thought to occur only in the ovaries of females, estrogens and other steroid hormones are also synthesized in the brain (Balthazart et al. 2018) and can impact many sociosexual behaviors. For example, GnIH may inhibit female copulation solicitation displays by inhibiting GnRH2 neurons (Bentley et al. 2006).

### 8.2.5 Growth Hormone

Growth hormone helps regulate cellular differentiation, proliferation, and metabolism. Although cells in the anterior pituitary gland produce and secrete growth hormone (GH), embryonic birds do not have functional pituitary glands until later in their development. For example, GH-secreting cells in the pituitary of embryonic Domestic

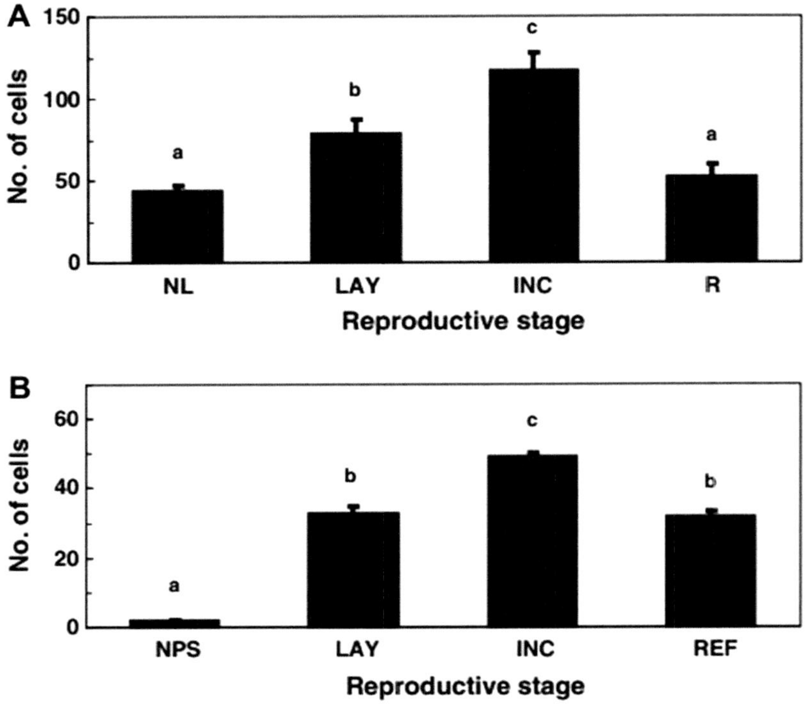


**Fig. 8.12** Changes in the number of VIP neurons in the hypothalamus of (**a**) Domestic Chickens and (**b**) Domestic Turkeys (*Meleagris gallopavo domesticus*) during different breeding stages. NL, non-laying; NPS, non-photostimulated; LAY, laying eggs; INC, incubating eggs; R, rearing chicks; REF, eggs removed and females provided with chicks. Values are presented as means ± SE. Significant differences between means are indicated by different letters ($P < 0.05$). (Figure from Kosonsiriluk et al. 2008; © 2008 Elsevier Inc. Published by Elsevier Inc., used with permission.)

Chickens are not responsive to growth hormone-releasing hormone (GHRH) being secreted by the hypothalamus until embryonic day 16 (hatch day is typically day 21; Porter et al. 1995). Prior to that, embryonic growth occurs without pituitary growth hormone in the blood (Harvey et al. 1998) and depends on extrapituitary GH. In Domestic Chicken embryos, GH mRNA and GH-immunoreactivity is apparently less than two days after incubation begins (ED 2; Harvey 2013) and is widespread in embryos by four days after incubation begins (ED 4; Fig. 8.21). In Domestic Chickens on embryonic day 4 (i.e., 4 days since the beginning of incubation; ED 4), extrapituitary GH is particularly detectable in the neural tube, notochord, somites, limb buds, heart, liver, and mesonephros (Harvey et al. 2000). By ED 7, GH is increasingly detectable in the limb buds and the brain (Murphy and Harvey 2001). Thus, growth hormone is produced by many extrapituitary tissues, including those in the nervous, immune, respiratory, and reproductive systems (Harvey 2010; Luna et al. 2014), and plays an important role as a local growth factor in the autocrine/paracrine regulation of cellular differentiation during embryonic development (Waters et al. 1999; Sanders and Harvey 2004).

Increasing plasma levels of growth hormone also stimulate the production and release of insulin-like growth factors (IGF-1 and IGF-2) by the liver and other tissues, and additional IGF-1 is produced by other tissues, including the brain, heart, and reproductive organs, both with and without stimulation by GH even before the anterior pituitary begins secreting GH (McMurtry et al. 1997; Fig. 8.22). Kikuchi et al. (1991) found that IGF-1 was detectable to the plasma of embryonic Domestic Chickens as early as ED9. GH and IGF levels in the blood are regulated via negative feedback by the hypothalamus (Fig. 8.22). By stimulating faster cell growth, differentiation, and survival, IGF-1 increases skeletal muscle mass, stimulates bone growth, and enhances neural and cardiac functions (e.g., Ren et al. 1999; Xiao et al. 2017). Other investigators have reported that

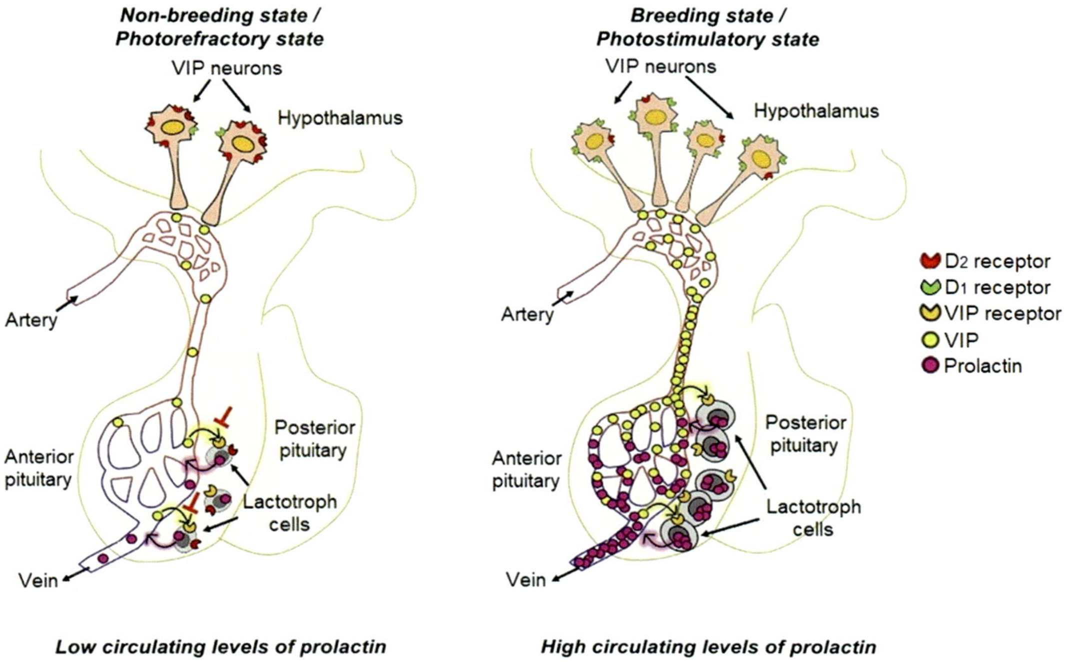


**Fig. 8.13** In female Domestic Chickens that are not laying eggs, dopamine binds primarily to D2 receptors (that are antagonistic to the actions of VIP) of the vasoactive intestinal peptide (VIP) neurons in the hypothalamus and cells in the anterior pituitary to inhibit the action of VIP (that normally stimulate the release of prolactin), so levels of prolactin in the blood are low. In females that are incubating eggs, dopamine binds primarily to D1 receptors (that stimulate VIP neurons) in the hypothalamus, releasing VIP that stimulates lactotroph cells in the anterior pituitary to release prolactin into the blood. During the breeding season, lactotroph cells increase in both size and number to increase the production and release of prolactin into the blood. (Figure modified from Smiley 2019; © 2019 Elsevier Inc., used with permission)

IGF-1 simulates growth and/or proliferation of several types of cells, including adipocytes, fibroblasts, lymphocytes, chondrocytes, osteoblasts, epithelial cells, and smooth and skeletal muscle cells (Baxter and Twigg, 2009; Duan et al. 2010; Kawai and Rosen 2010), and can even affect avian plumage (Box 8.3 IGF-1 and Plumage Traits).

In embryonic Domestic Chickens, plasma levels of IGF-1 were found to increase until embryonic (incubation) day (ED) 14.5 and then decline until hatching, after which levels again increased (Lu et al. 2007; Fig. 8.23). Different studies provide conflicting reports concerning plasma levels of IGF-2 in developing Domestic Chickens. Scanes et al. (1997) reported that plasma levels of IGF-2 were highest during the mid-embryonic period (ED 9.5–10.5) before declining, whereas Lu et al. (2007) reported increasing levels until ED 14, followed by fluctuating levels until hatching and declining levels thereafter (Fig. 8.24). The results of these two studies also differed in terms of the concentrations of IGF-1 and IGF-2, with Lu et al. (2007) reporting much higher plasma levels for both than Scanes et al. (1997). Together, however, the results of these studies suggest that both IGF-1 and IGF-2 are important for the growth and development of embryonic Domestic Chickens and that IGF-1 is important for the growth of Domestic Chickens after hatching.

Although almost all studies of IGF-1 have been conducted with Domestic Chickens and, to a lesser degree, Domestic Turkeys, Lodjak et al.

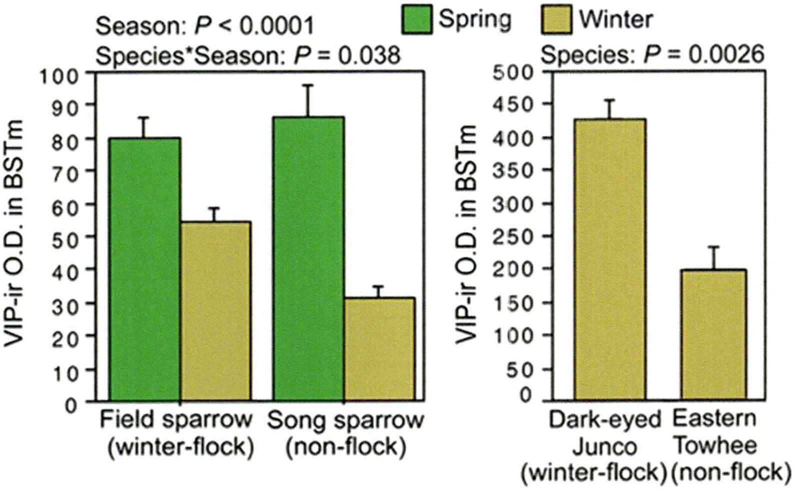


**Fig. 8.14** Density of VIP neurons (VIP immunoreactive (-ir) neurons) in the medial bed nucleus of the stria terminalis (BSTm; an area of the brain important in regulating social behavior) is greater in Field Sparrows (*Spizella pusilla*) and Dark-eyed Juncos (*Junco hyemalis*) that form flocks during the winter than in Song Sparrows (*Melospiza melodia*) and Eastern Towhees (*Pipilo erythrophthalmus*) that do not form flocks during the winter. Density of VIP neurons in the BSTm of Field Sparrows and Song Sparrows is higher during the spring when they form pair bonds during the breeding season. (Figure from Kingsbury and Wilson 2016; © 2016 Oxford University Press, used with permission)

(2014) found that nestling Great Tits (*Parus major*) in experimentally reduced broods had increased rates of postnatal growth rate, were in better body condition, and had higher plasma levels of IGF-1 prior to fledging than those in control and enlarged broods. These authors also found that plasma levels of IGF-1 were highest during the middle of the nestling period when nestling Great Tits were growing at the highest rates. In addition, plasma IGF-1 levels of nestling Great Tits were higher when nutritional conditions were better, resulting in higher growth rates and better body condition prior to fledging (Lodjak et al. 2014). In a study of European Pied Flycatchers (*Ficedula hypoleuca*), Lodjak et al. (2017) injected some nestlings with IGF-1 and found that growth rates, body mass, and tarsus length were all significantly greater in injected nestlings than control nestlings (Fig. 8.25).

Lodjak et al. (2018) examined possible associations between plasma IGF-1 levels and body size, mean life span, clutch size, and egg weight for 63 species of songbirds with slow and fast paces-of-life. Species with higher body masses and longer periods of parental care were considered to have a slower pace-of-life, and those with lower body mass and shorter periods of parental to have a faster pace-of-life. Analysis revealed that body mass was positively associated with plasma levels of IGF-1, but negatively associated with relative levels of IGF-1, mean life span was negatively associated with body mass for species with a fast pace-of-life, but not for species with a slow pace-of-life, and clutch size and egg weight were not associated with plasma levels of IGF-1 (Fig. 8.26). However, clutch size was positively associated with plasma levels of IGF-1 for species with a faster pace-of-life and negatively correlated with plasma levels of IGF-1 for those with a slower pace-of-life (Fig. 8.26c). In addition, egg weight was positively associated with plasma levels of IGF-1 for species with a slower pace-of-life and negatively associated with plasma levels of IGF-1 for those

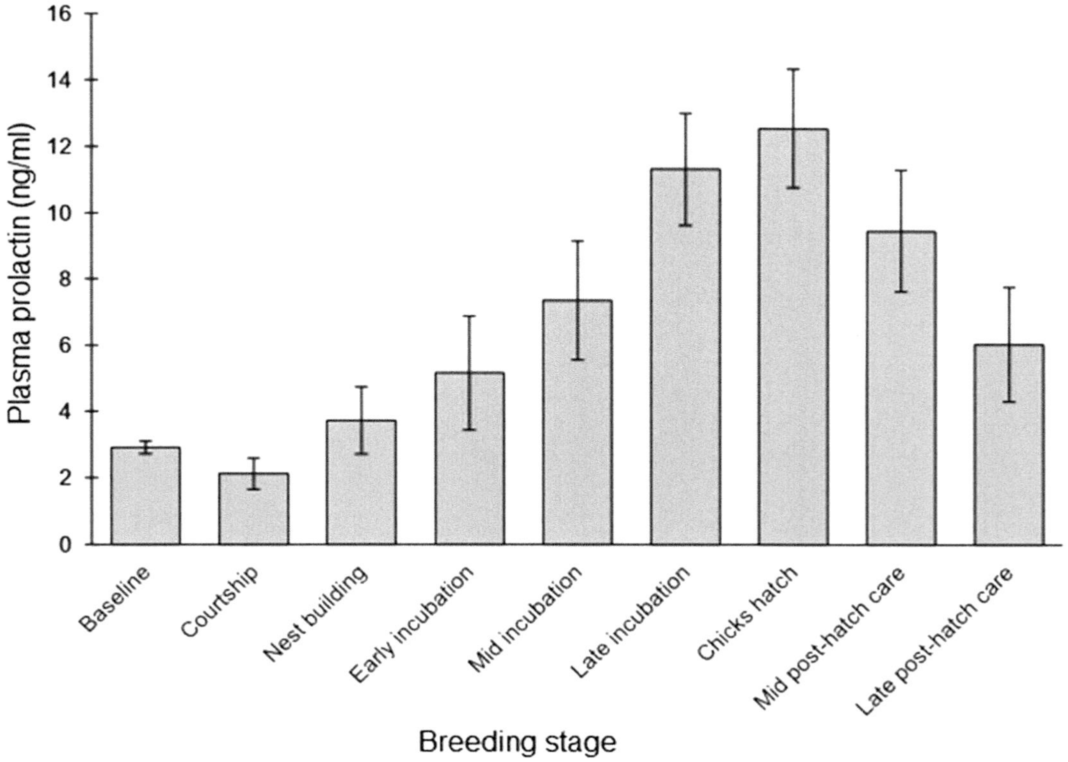


**Fig. 8.15** Mean (± SE) prolactin concentrations in the plasma of male and female Zebra Finches (*Taeniopygia guttata*) during different breeding stages. (Figure modified slightly from Smiley and Adkins-Regan 2016b; © 2016 Elsevier Inc., used with permission)

with a faster pace-of-life (Fig. 8.26d). Based on these results, Lodjak et al. (2018) concluded that IGF-1 appears to increase investment in reproduction and growth, and shorten lifespans, for species of songbirds with a faster pace-of-life, and favor longer lifespans and self-maintenance over greater annual fecundity for species with a slower pace-of-life. Whether similar results might be found in other orders of birds remains to be determined.

### 8.2.6 Mesotocin and Arginine Vasotocin

Mesotocin is the avian homolog of mammalian oxytocin, and arginine vasotocin (AVT) is the avian homolog to mammalian arginine vasopressin. Both mesotocin and AVT (also referred to as nonapeptides) are produced by the hypothalamus and released from the posterior pituitary (neurohypophysis) (Fig. 8.27). Neurons expressing these peptide hormones are most abundant in areas of the social behavior network in the brain (see Chap. 4 for more information about the avian social behavior network), suggesting an important role in the control of social behaviors (Goodson 2005), including cooperative breeding (Soares et al. 2010; Potticary and Duckworth 2021).

For example, injection of mesotocin has been found to increase preferences for larger over smaller social groups (Goodson et al. 2009) and increase prosocial preferences (Duque et al.

**Box 8.2 Crop Milk**

Adult pigeons and doves (Order Columbiformes), flamingoes, and Emperor Penguins (*Aptenodytes forsteri*) feed their young "crop milk," a semi-solid substance that consists of protein- and lipid-rich epithelial cells that slough-off from the lining of the crop (Columbiformes) or esophagus (flamingoes and Emperor Penguins). Crop milk is rich in both crude protein and fat based on dry weight (Campbell and Lack 2011; Hu et al. 2016) and also contains some vitamins and minerals, free amino acids, and carotenoids (Ward et al. 2001; Eraud et al. 2008). Crop milk is produced as plasma levels of prolactin increase during incubation and the post-hatch period (e.g., Smiley and Adkins-Regan 2016a, b). Prolactin activates the JAK-STAT signaling pathway in epithelial cells lining the crop or esophagus, stimulating protein and lipid synthesis and the proliferation of cells that will then slough-off to produce crop milk (Wan et al. 2019; Chen et al. 2020). The results of some studies suggest that the insulin and insulin-like growth factor 1 (IGF-1) may also be involved in the production of crop milk, but additional study is still needed to determine their possible roles (Hu et al. 2016; Xie et al. 2018).

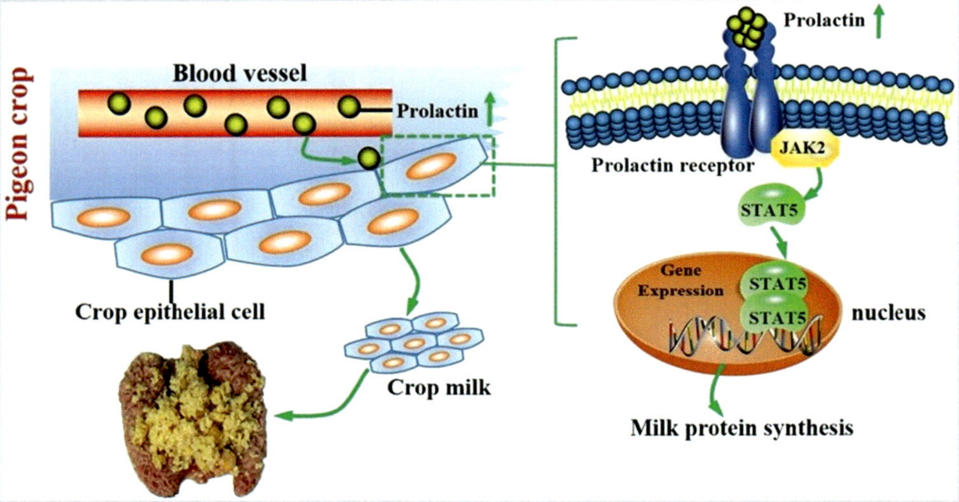


Prolactin binds to a prolactin receptor in the membrane of crop epithelial cells and initiates a signal cascade that promotes expression of the crop milk protein gene. JAK2, Janus kinase-signal transducer; STAT5, signal transducer and activator of transcription 5. (Figure modified from Chen et al. 2020; © 2020 The Royal Society of Chemistry, used with permission. Photo of pigeon crop from Hu et al. 2016, Rights managed by Taylor & Francis, used with permission)

(continued)

**Box 8.2** (continued)

An adult Emperor Penguin feeding its young. (Photo by Hannes Grobe, Wikipedia, CC BY 3.0, https://creativecommons.org/licenses/by/3.0/)

2018), whereas administering an antagonist disrupts pair-bond formation (Pedersen and Tomaszycki 2012). Mesotocin also plays a role in stimulating avian maternal care (Chokchaloemwong et al. 2013), pair bond formation (Pedersen and Tomaszycki, 2012; Klatt and Goodson 2013), flocking behavior by Zebra Finches (*Taeniopygia guttata*; Goodson et al. 2009), and prosociality (Duque et al. 2018). In addition, Hall et al. (2015) found that vasotocinergic and mesotocinergic neurons were active during nest building in Zebra Finches, and Kelly and Adkins-Regan (2020) reported that nestling Zebra Finches whose parents were injected with mesotocin-receptor antagonists had significantly lower growth rates and higher mortality rates than those with control parents that were not treated with mesotocin-blockers.

Estrildid finches that are gregarious year-round have greater densities of nonapeptide binding sites in the hypothalamus than do territorial estrildids (Goodson et al. 2006, 2009). Goodson et al. (2012) found that the density of mesotocin receptors increases in the lateral septum

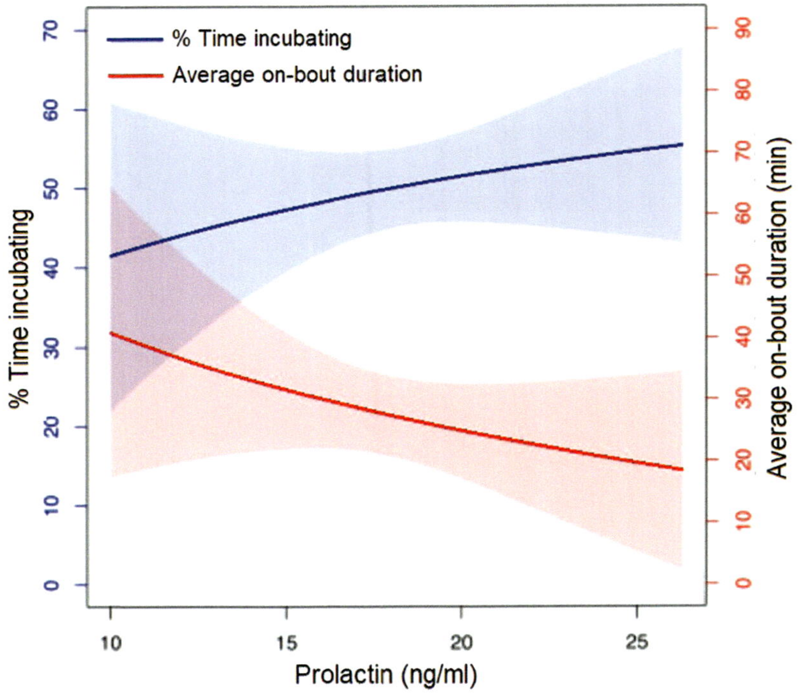


**Fig. 8.16** Relationship between plasma prolactin levels and incubation effort by male and female Kentish Plovers (*Charadrius alexandrinus*). (Figure from Wang et al. 2020; )

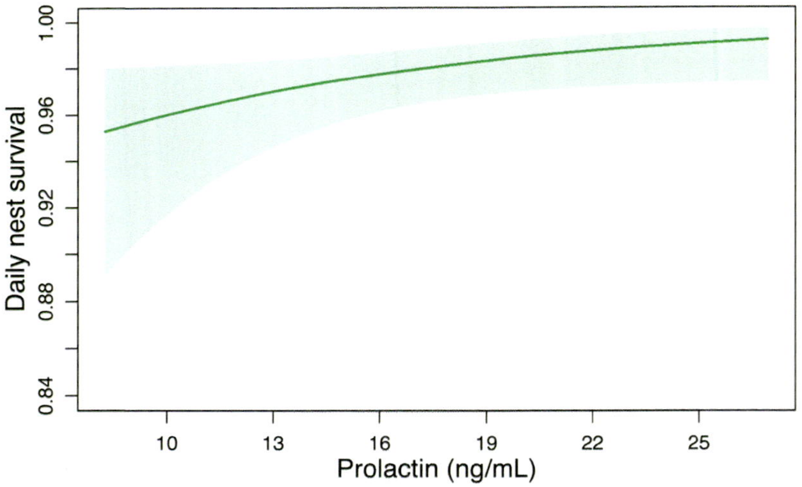


**Fig. 8.17** Relationship between plasma prolactin levels and daily nest survival of male and female Kentish Plovers (*Charadrius alexandrinus*). (Figure from Wang et al. 2020; )

(Fig. 8.28), an area involved in the regulation of social behavior, especially aggression, of Field Sparrows (*Spizella pusilla*) during the winter when they form flocks (Goodson et al. 1998). However, no such increase occurs in the lateral septum of Song Sparrows (*Melospiza melodia*) that are territorial year-round and do not form flocks (Goodson et al. 1998; Fig. 8.28). Similarly, flocking Dark-eyed Juncos (*Junco hyemalis*) have a higher density of mesotocin receptors in the

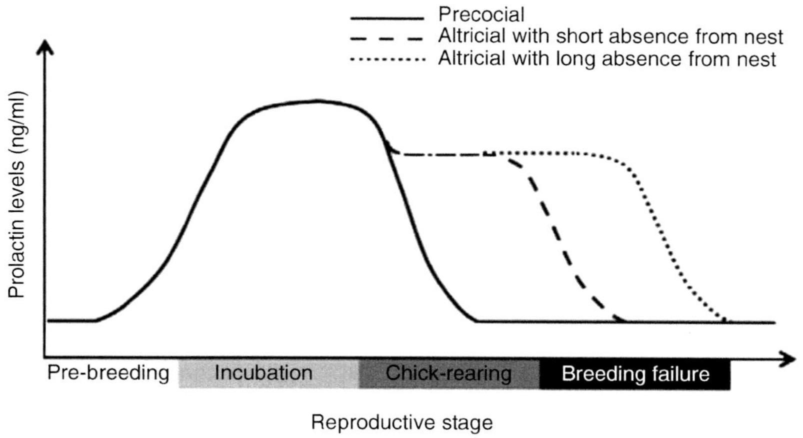


**Fig. 8.18** A schematic representation of the prolactin cycle of breeding birds with different reproductive modes, including precocial (e.g., ducks and geese), altricial with short absences from nests (e.g., gannets, starlings, and doves), and altricial with long absences from nests (e.g., penguins, albatrosses, and petrels). (Figure from Angelier et al. 2016; © 2015 Elsevier Inc. All rights reserved, used with permission)

lateral septum during winter when they form flocks, but non-flocking Eastern Towhees (*Pipilo erythrophthalmus*) do not (Goodson et al. 1998).

In another example of the role of mesotocin in social behavior, Kelly (2019) found that control and oxytocin-antagonist-treated Zebra Finches did not differ in the amount of time spent perched side by side, but treated male and female Zebra Finches spent significantly less time spent following their partner. Treated females also spent less time allopreening and treated males directed fewer singing bouts at their partners than control males. These results suggest that the mesotocin plays a role in regulating some, but not all, pair maintenance behaviors in Zebra Finches.

Arginine vasotocin (AVT) controls water balance by facilitating water reabsorption by the kidneys (see Chap. 9 for additional detail), but also has other functions. For example, during egg laying, AVT causes smooth muscle in the uterus of the oviduct to contract, presumably by binding to its receptors in muscle cells (Takahashi and Kawashima 2003; Fig. 8.29). When, as a result, an egg moves from the uterus to the vagina, rhythmic contractions of abdominal muscles then cause expulsion of the egg (Sturkie et al. 1962)

As with mesotocin, AVT also influences social interactions. For example, after intracranial injections of an AVT-antagonist, hatchling Zebra Finches (days 2–8 post-hatching) showed less attachment to their parents, and males failed to show normal affiliative interests in females when they reached sexual maturity and exhibited only modest levels of both courtship and affiliation in their interactions with females (Baran et al. 2016a). In contrast, hatchlings injected with AVT exhibited sustained interest in parents, and males showed increased interest in females as they matured (Baran et al. 2016a). Injection of AVT in adult male Zebra Finches increased their affiliative behavior with females, but caused a decline in singing rates (Baran et al. 2016b).

AVT also appears to play a role in song development. Baran et al. (2017) injected some hatchling male Zebra Finches with an AVT-antagonist on days 2–8 post-hatching and, after song development, recorded their songs. Songs of these treated males matched those of a tutor to a much lesser degree than songs of control males

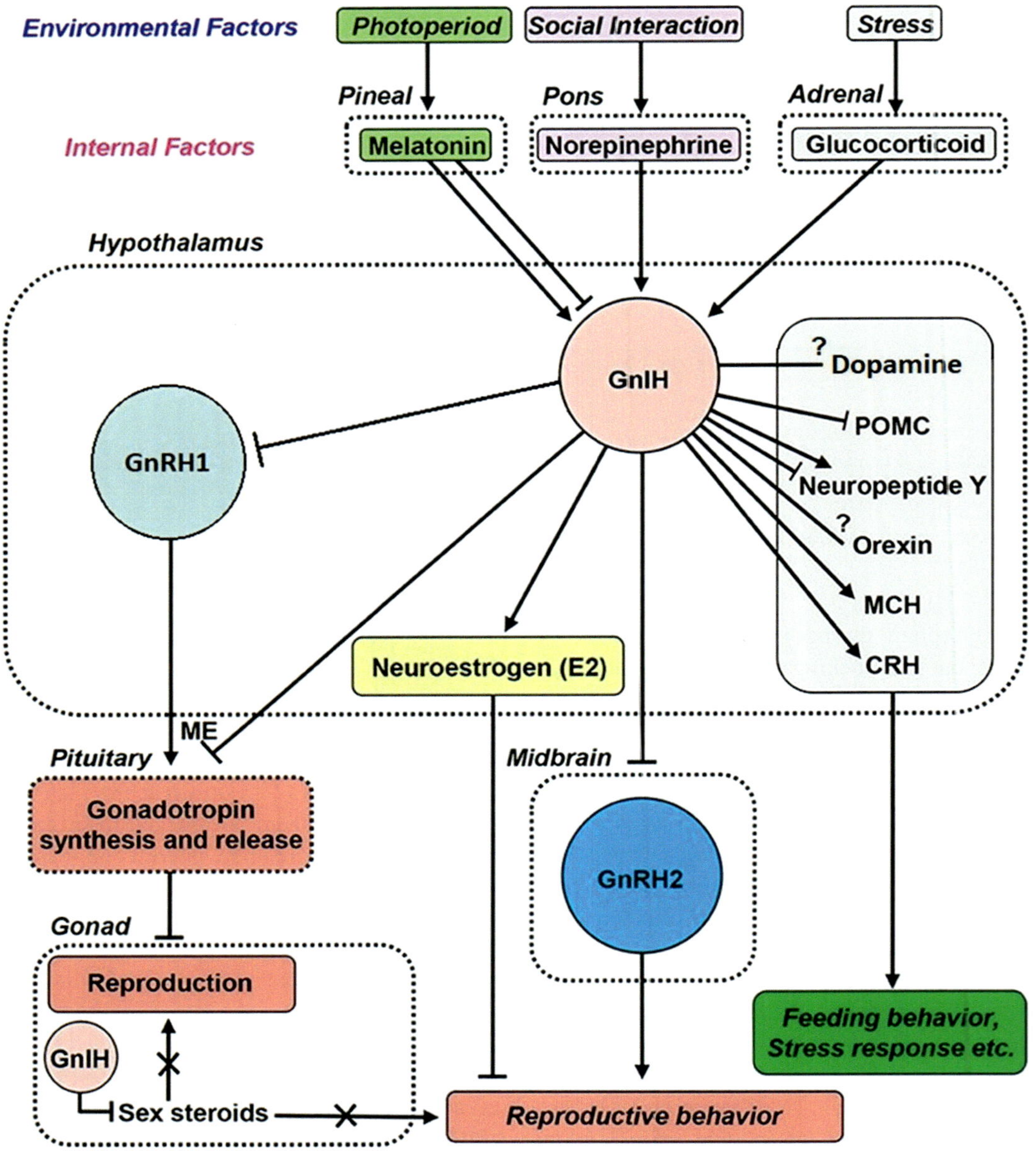


**Fig. 8.19** Factors influencing synthesis of GnIH. GnIH inhibits the synthesis and release of gonadotropins by the pituitary by inhibiting the activity of GnRH1 neurons in the hypothalamus. This prevents the production of sex hormone and germ cells in the gonads. GnIH also suppresses breeding behavior through effects in the brain and by stimulating synthesis of neuroestrogen (E2) in the hypothalamus. GnIH neurons also project to several other areas in the brain and so may have other effects. Environmental factors, like photoperiod, social interactions, and stress, and internal factors, like melatonin, glucocorticoid, and norepinephrine, are also important in regulating breeding physiology and behavior. Melatonin increases GnIH expression in Japanese Quail, stress increases GnIH expression via glucocorticoids, and the social environment influences the GnIH expression via the release of norepinephrine. Stimulatory effects are indicated by arrows, and inhibitory effects by blunt end lines. Question marks indicate morphological evidence, but no evidence of physiological actions. GnRH2 is secreted by cells in the midbrain and can influence reproductive behaviors such as female copulation solicitation displays. ME, median eminence; POMC, pro-opiomelanocortin; MCH, melanin-concentrating hormone; CRH, corticotropin-releasing hormone. (Figure modified from Tsutsui and Ubuka 2018; © 2018 Tsutsui and Ubuka, open-access article distributed under the terms of the Creative Commons Attribution License (CC BY), https://creativecommons.org/licenses/by/4.0/)

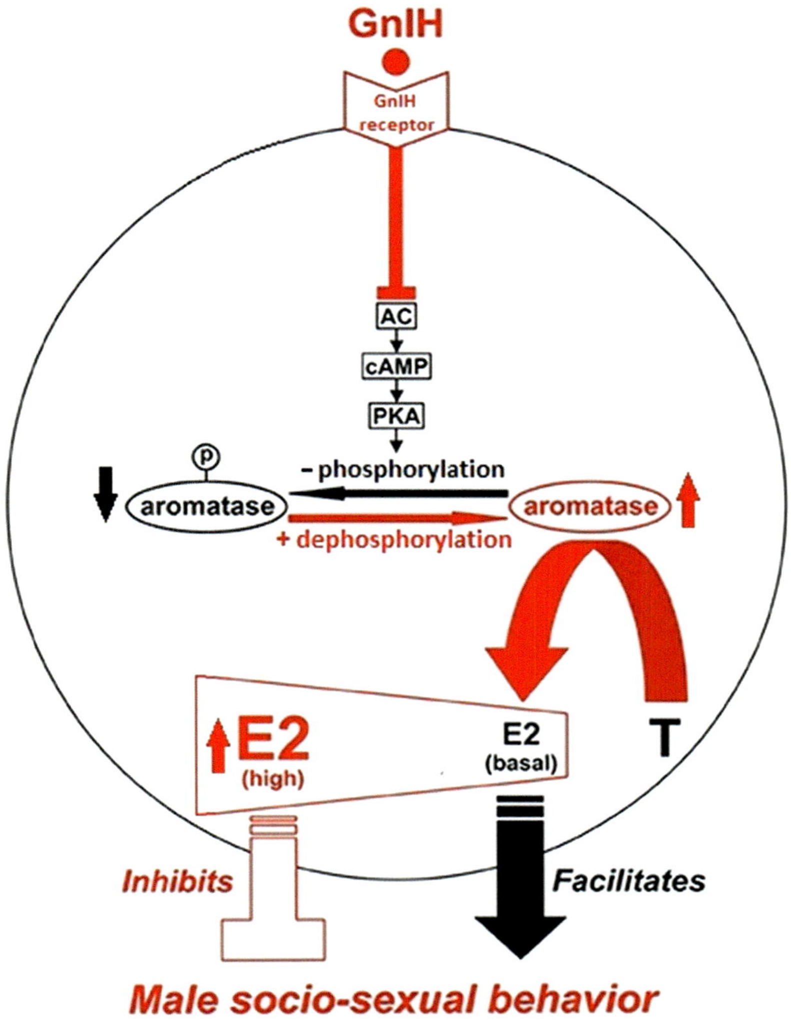


**Fig. 8.20** Possible mechanism by which GnIH may influence sociosexual behavior by regulating synthesis of neuroestrogen (E2). Neuroestrogen (E2) is synthesized from androgens like testosterone (T) by aromatase in the brain and inhibits male sociosexual behavior. GnIH inhibits the activity of protein kinase A (PKA) via adenylate cyclase (AC) and cAMP. This may decrease production of phosphorylated aromatase and increase production of dephosphorylated aromatase. Aromatase activity is inhibited by phosphorylation, and this inhibition is blocked by PKA, so increased production of GnIH may increase production of aromatase and neuroestrogen by increasing levels of phosphorylated aromatase. (Figure modified from Ubuka and Tsutsui 2014; © 2014 Ubuka and Tsutsui, open-access article distributed under the terms of the Creative Commons Attribution License (CC BY), https://creativecommons.org/licenses/by/4.0/)

(Fig. 8.30). In contrast, hatchling males injected with AVT were significantly better at learning songs from tutors than control males.

Finally, AVT-containing neurons in the bed nucleus of the stria terminalis (BSTm; an area of the brain important in regulation of social behavior) have been found to increase gregariousness (i.e., preferences for a large versus a small flock) and reduce anxiety behavior (Kelly et al. 2011). In addition, Ondrasek et al. (2018) found dense

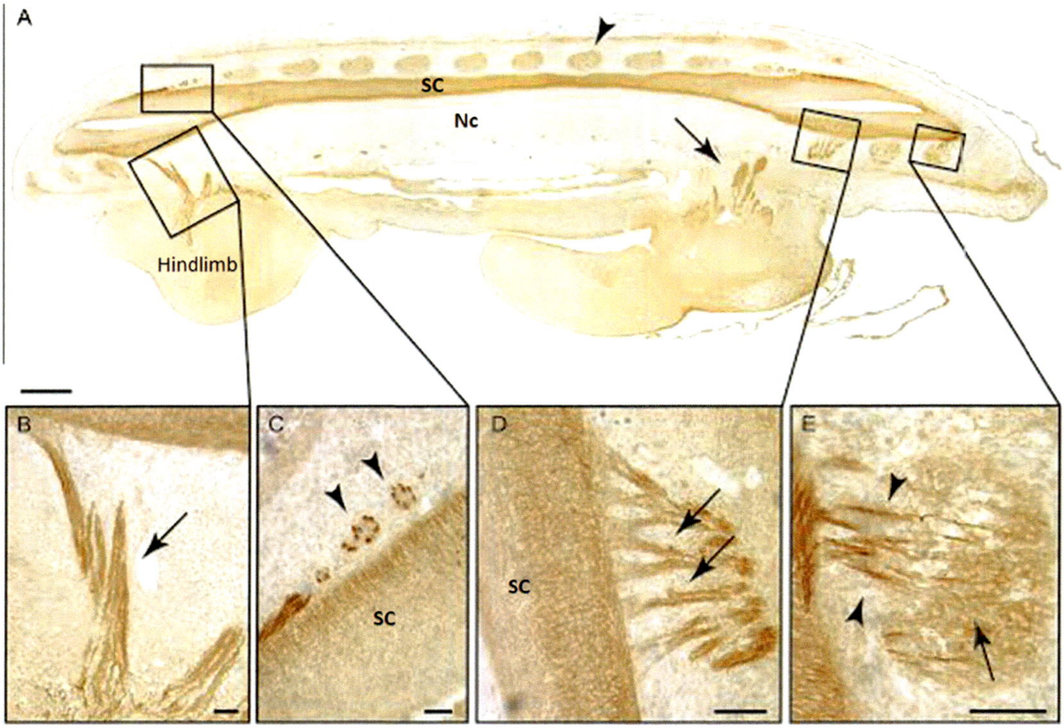


**Fig. 8.21** Growth hormone receptor (GHR) immunoreactivity in a cross section of an embryonic Domestic Chicken (ED 4). (**a**) Low magnification image with arrow (spinal nerves) and arrowhead (developing vertebra) pointing to areas with GHR immunoreactivity, (**b**–**e**) Higher magnification images of (**b**) spinal nerves innervating the hindlimb, (**c**) sensory neuron cell bodies in dorsal root ganglia, (**d**) spinal nerves innervating a developing vertebra (arrows), and (**e**) spinal nerves (arrowheads) and cytoplasm of cells of a developing vertebra (arrow). Nc, notochord; SC, spinal cord. Scale bars: (**a**) = 1 mm, (**b**–**e**) = 100 μM. (Figure from Harvey and Baudet 2014; © 2014 Elsevier Inc., used with permission)

concentrations of mesotocin- and vasotocin-like receptors in the hypothalamus of female European Starlings (*Sturnus vulgaris*) captured during the nonbreeding season when they form large flocks.

## 8.3 Thyroid Gland and Hormones

The avian thyroid gland is located above the heart (Fig. 8.31), and the cells of the thyroid produce and secrete thyroxine and triiodothyronine. The chemical formula for thyroxine is 3',5',3,5-L-tetra-iodothyronine and is typically referred to as T4, whereas the formula for triiodothyronine is 3',5,3-L-triiodothyronine and is referred to as T3. Control of thyroid hormone production in birds is regulated by the hypothalamus and pituitary gland, with the hypothalamus secreting thyrotropin-releasing hormone (TRH) in response to stress, hypothermia, or reduced levels of thyroid hormones in the blood (Merryman and Buckles 1998). TRH stimulates the release of thyroid-stimulating hormone (TSH) from the anterior pituitary and TSH then stimulates the production and release of thyroid hormones by the thyroid gland (Fig. 8.32). T4 is the most abundant thyroid hormone in the blood and is converted into the more biologically active T3 by enzymes in cells. The actions of thyroid hormones are mediated via thyroid hormone receptors in the nuclei of cells, with T3 binding

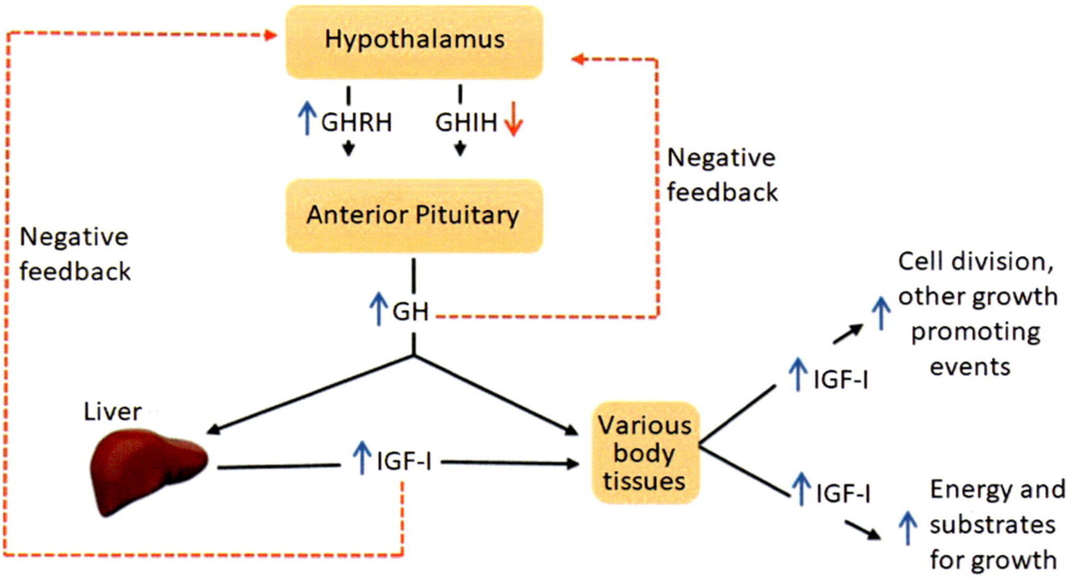


**Fig. 8.22** Growth hormone releasing hormone (GHRH) from the hypothalamus stimulates the release of growth hormone (GH) by the anterior pituitary. GH then stimulates the liver to release insulin-like growth factor 1 (IGF-1), and IGF-1 in the blood stimulates additional production and release of IGF-1 by other cells and tissue. GH and IGF-1 promote growth of many body tissues. (Figure from Angove and Forder 2020; Rights managed by Taylor & Francis, used with permission)

to receptors and promoting selective transcription of genes and translation to form proteins that alter cell function. This process is termed genomic action, but thyroid hormones can also affect cells via non-genomic actions by binding to receptors in cell membranes

Thyroid hormones serve a variety of functions, including helping to regulate basal metabolic rates (Fig. 8.33). This relationship between thyroid hormones and basal metabolic rate (i.e., the minimal energetic cost of living) has been understood since at least 1895 when Magnus-Levy (1895) showed that secretions from the thyroid gland caused an increase in human metabolic rates. Thyroid hormones are also required for the growth of birds and play important roles in brain development, initiation of gonadal development prior to breeding, molt (Box 8.4 Thyroid Hormones and Molt), and egg production by females (McNabb 2007). For embryonic Domestic Chickens, Lu et al. (2007) found that plasma levels of T3 were constant through mid-incubation and peaked the day before hatching. T4 levels were highest during the period of amnion resorption (ED 17–20 for chickens), then decreased after hatching. Elevated plasma levels of T3 play an important role in stimulating the developmental and metabolic processes needed for hatching, including the transition to lung respiration. In addition, both T3 and T4 were found to be positively correlated with the body mass of chicken embryos just prior to hatching (Lu et al. 2007).

Among vertebrates, plasma levels of T4 tend to increase from amphibians to reptiles, to birds, and to mammals, and plasma levels of T3 tend to increase from amphibians, to reptiles, to mammals, and to birds (Hulbert 2000; Fig. 8.34). However, these levels are largely determined by the presence of thyroid-hormone-binding plasma proteins (and thyroid hormones are only effective when not bound to plasma proteins) (Hulbert 2000). Levels of free thyroid hormones in the blood plasma of birds and mammals are generally similar (Fig. 8.34).

**Box 8.3 IGF-1 and Plumage Traits**

Androgens are known to influence the development of secondary sexual characteristics in birds (e.g., Peters et al. 2006; Laucht and Dale 2012), but expression of those characteristics may have costs, such as increased oxidative stress, that may require trade-offs between the degree of expression and self-maintenance (e.g., Flatt and Heyland 2011). IGF-1 plays an important role in regulating development and growth, and blood levels of IGF-1 can be influenced by nutritional status and factors such as stress levels (Dantzer and Swanson 2012; Tóth et al. 2018).

Male Bearded Reedlings (*Panurus biarmicus*) have melanin-based black "beards" that represent honest signals of condition (Hoi and Griggio 2008), as well as tails that vary in length and are also sexually selected traits (Romero-Pujante et al. 2002; Hoi and Griggio 2008). In addition, males have feathers on the head, sides, and chin that reflect light in ultraviolet (UV) wavelengths. Mahr et al. (2020) captured males in mist-nets, collected blood samples (so they could later determine plasma IGF-1 levels), measured beard and tail length, and calculated UV chroma, the proportion of reflectance in the UV ranges relative to total reflectance (320 – 700 nm, the color vision spectrum of songbirds).

Male Bearded Reedling. (Photo by Stefanie Leuker, Wikipedia, CC0 Public Domain)

Analysis revealed significant relationships between plasma levels of IGF-1 and both tail length and overall UV chroma. These results suggest that IGF-1 influences the development of secondary sexual characteristics important in both intra- and intersexual (mate choice) interactions. Plasma levels of IGF-1 can be influenced by a bird's nutritional status and condition (Regan et al. 2019) and, because IGF-1 stimulates protein synthesis in tissues, plasma levels might influence feather characteristics and growth when birds are molting. As such, expression

(continued)

**Box 8.3** (continued)

of traits influenced by IGF-1 levels in the blood can provide conspecifics with honest information about individual condition and quality.

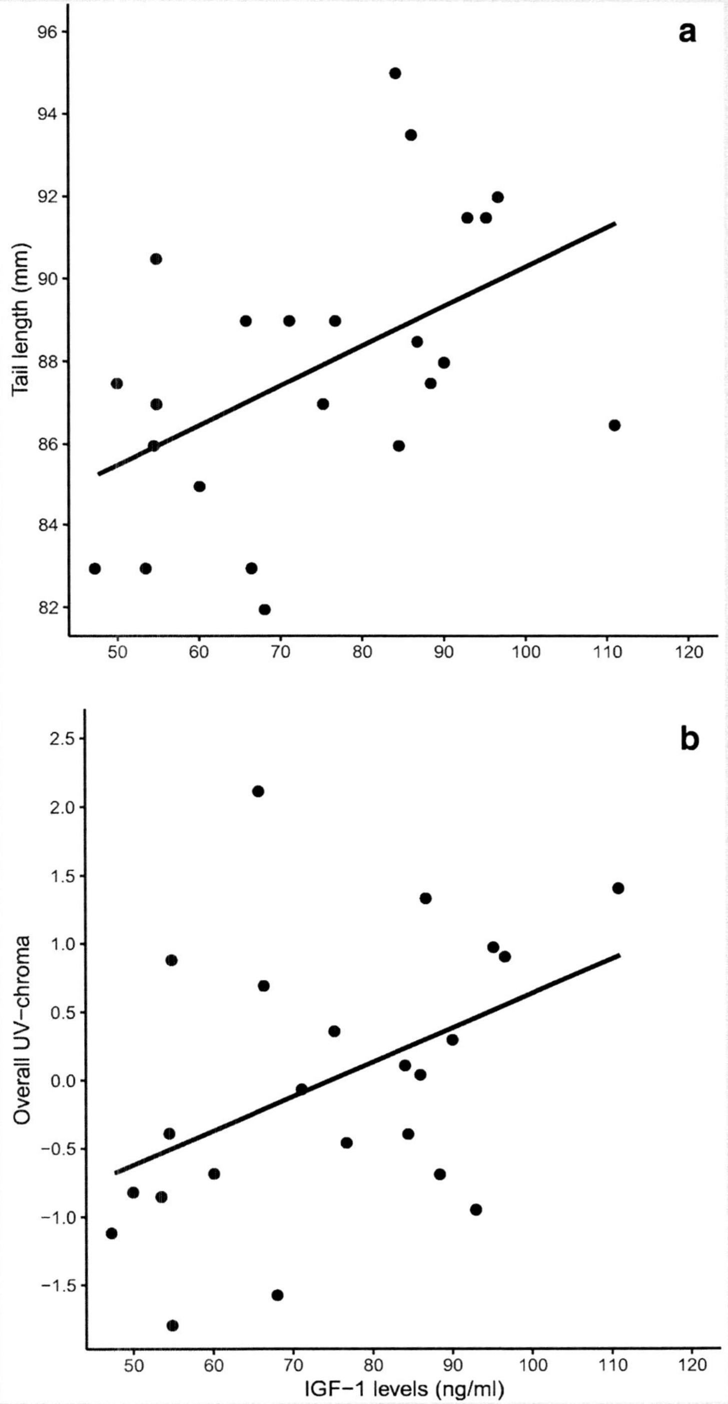


(continued)

**Box 8.3** (continued)

Relationship between baseline IGF-1 levels during molt and sexually selected ornaments of male Bearded Reedlings. Males with higher IGF-1 levels (**a**) had longer tails and (**b**) plumage with higher reflectance in ultraviolet (UV) wavelengths. (Figure from Mahr et al. 2020; open-access article licensed under a Creative Commons Attribution 4.0 International License, https://creativecommons.org/licenses/by/4.0/)

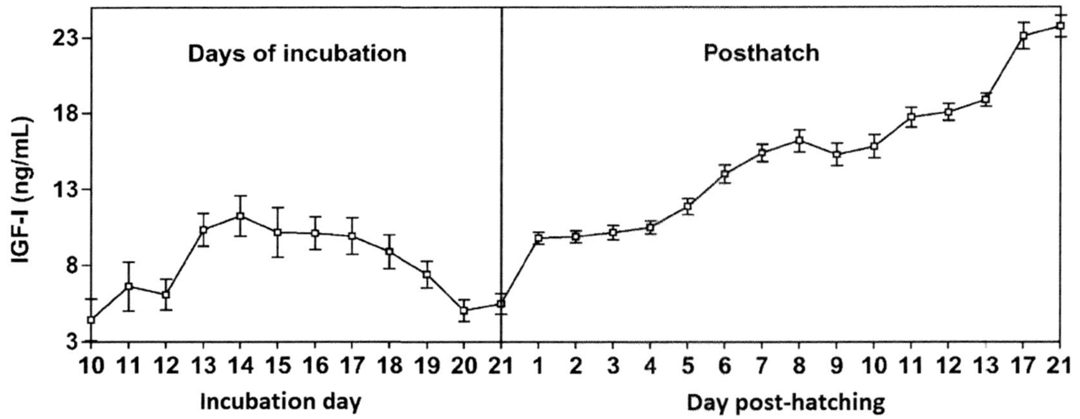


**Fig. 8.23** Plasma levels of IGF-1 of Domestic Chickens prior to and after hatching. Prior to hatching, levels increase until ED 14 then decline, but continue to increase after hatching. (Figure from Lu et al. 2007; © 2007 Poultry Science Association Inc. Published by Elsevier Inc., used with permission)

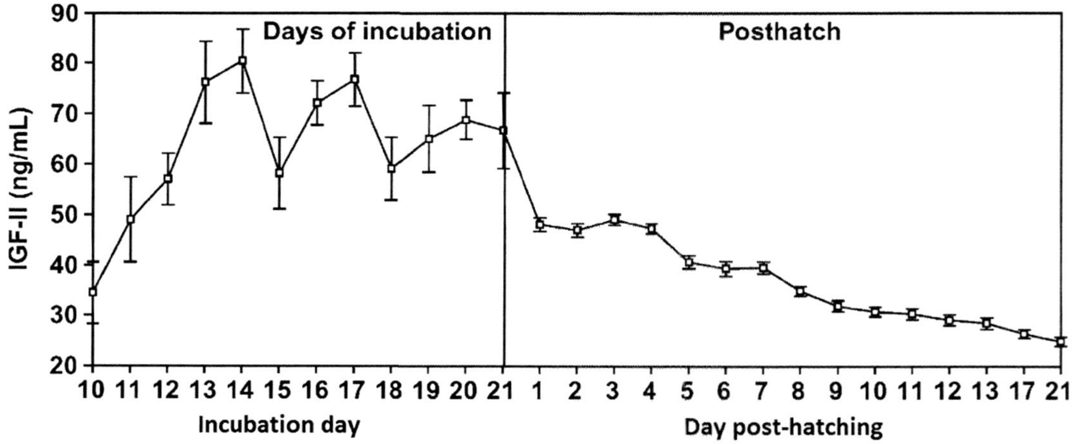


**Fig. 8.24** Plasma levels of IGF-II of Domestic Chickens prior to and after hatching. Prior to hatching, levels increase until ED 14 then decline, but continue to increase after hatching. (Figure from Lu et al. 2007; © 2007 Poultry Science Association Inc. Published by Elsevier Inc., used with permission)

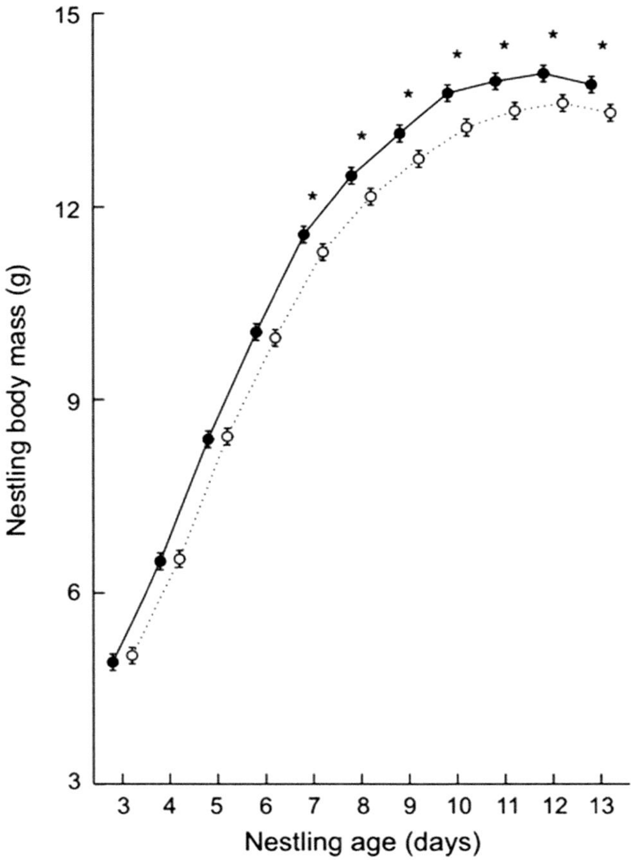


**Fig. 8.25** Changes in mean (±SE) body mass of nestling European Pied Flycatchers (*Ficedula hypoleuca*) with increasing age. Nestlings injected with IGF-1 are indicated by the filled circles; control nestlings by the open circles. Asterisks above whiskers indicate significant differences ($P < 0.05$, Tukey test). (Figure from Lodjak et al. 2017; © 2016 The Authors. Functional Ecology © 2016 British Ecological Society, used with permission)

Thyroid hormones, especially T3, are important in the regulation of many physiological processes, including cellular metabolic rates, functions of the heart and digestive system, muscle function, brain development, and maintenance of bone (Huang et al. 2008; Fig. 8.35). Thyroid hormones are important for normal bone growth, development, and repair (Fig. 8.36) and also influence myogenesis and metabolism of skeletal muscle (Fig. 8.37). In the heart, thyroid hormones enhance protein synthesis and lower vascular resistance, which increases cardiac output (Zhang et al. 2017a; Fig. 8.38). Thyroid hormones also influence liver function by regulating the basal metabolic rates of liver cells (hepatocytes), lipid metabolism, and cholesterol levels in the blood, with cholesterol being an important precursor of the steroid hormones (e.g., testosterone and estrogen) (Cicatiello et al. 2018). Thyroid hormones are also important in brain development of avian embryos by stimulating proliferation and differentiation of neurons (Dezonne et al. 2015).

Thyroid hormones are also important in thermogenesis and can increase energy expenditure and stimulate heat production by increasing muscle mass for shivering thermogenesis and increasing heat production in cells by increasing ATP production and altering ion gradients across membranes that then increase heat production by requiring additional ATP production and breakdown to help restore those gradients (Mullur et al. 2014; Fig. 8.39).

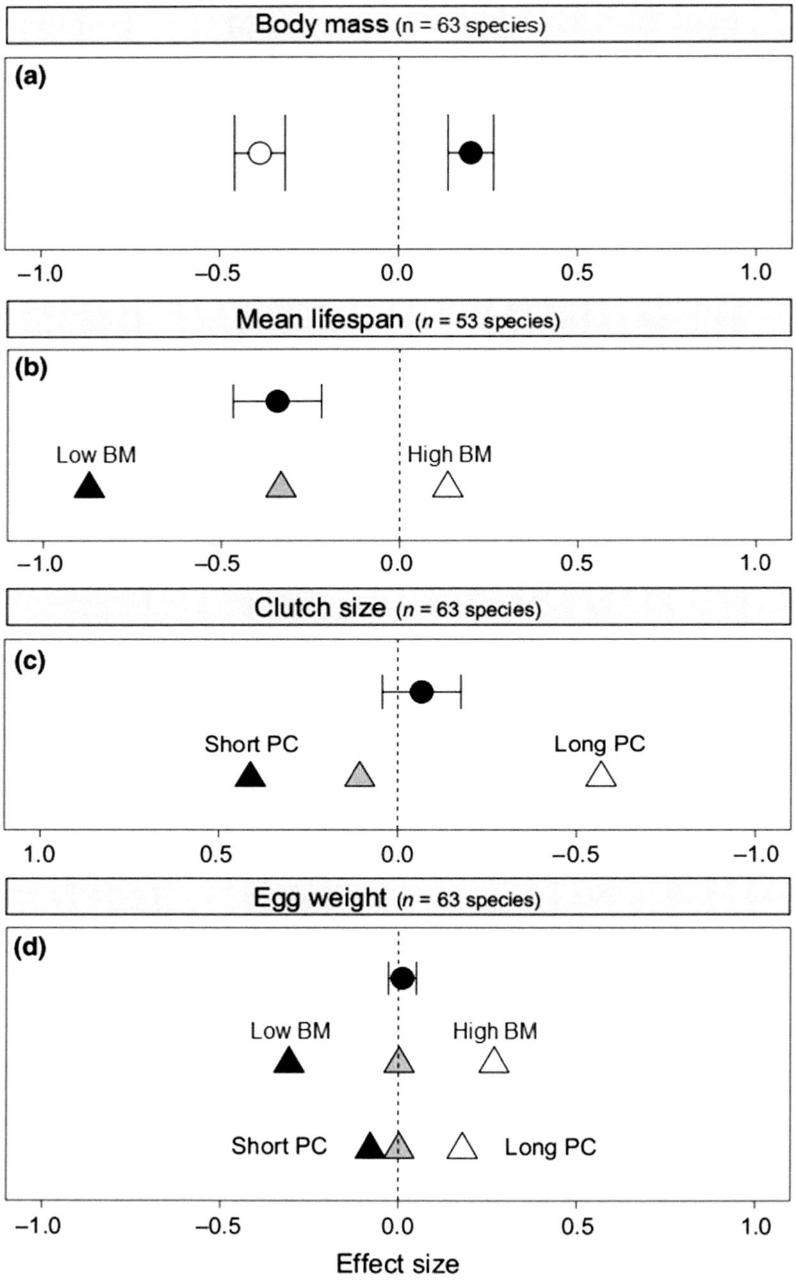


**Fig. 8.26** Effect sizes of the associations between plasma levels of IGF-1 and several life-history traits, including (**a**) body mass, (**b**) mean life span, (**c**) clutch size, and (**d**) egg weight. For body mass (**a**), plasma levels of IGF-1 were first included in the model as absolute values (filled circle) and then as relative values (per gram body mass; empty circle). For mean life span (**b**), clutch size (**c**), and egg weight (**d**), filled circles correspond to the main effect terms between traits and IGF-1 levels, triangles correspond to the interaction terms between traits and the continuous– continuous interaction between duration of parental care (PC) and plasma levels of IGF-1 (**c**, **d**), and the respective trait, and the continuous– continuous interaction between body mass (BM) and plasma levels of IGF-1 (**b**, **d**). A continuous–continuous interaction means that the slope between two traits changes depending on the

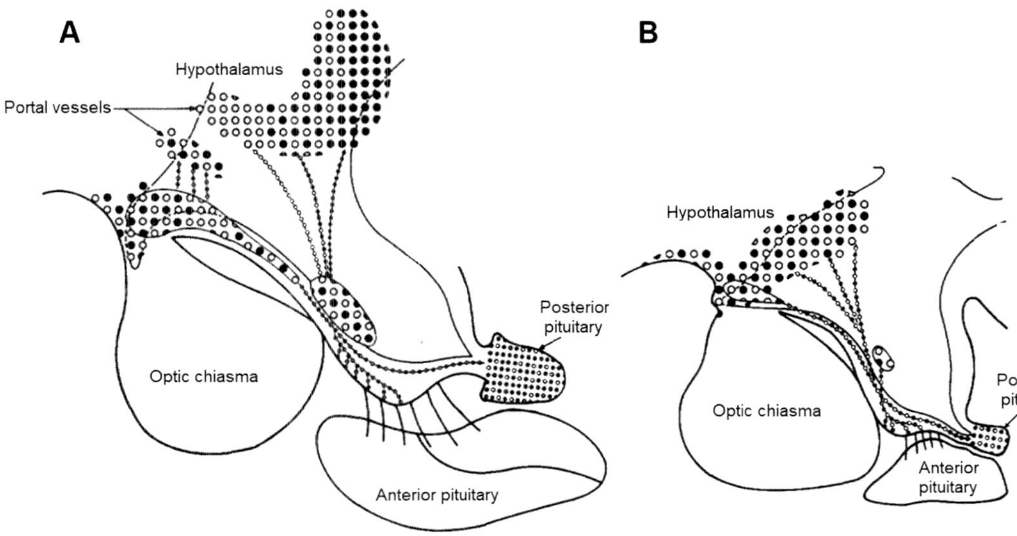


**Fig. 8.27** Drawings of the neurosecretory systems for mesotocin (filled circles) and arginine vasotocin (open circles) in the hypothalamus and posterior pituitary of a (**a**) Domestic Mallard (*Anas platyrhynchos*) and (**b**) Japanese Quail (*Coturnix japonica*). (Figure from Bons 1980; 1980 Springer-Verlag, used with permission)

## 8.4 Thyroid Hormones and Reproduction

Avian reproduction is regulated by the hypothalamic-pituitary-gonadal (HPG) axis, with the hypothalamus releasing gonadotropin-releasing hormone (GnRH) that, in turn, stimulates the release of gonadotropins by the anterior pituitary (luteinizing hormone [LH] and follicle-stimulating hormone [FSH]). LH and FSH then stimulate the production of testosterone and sperm by the testes. After breeding, the hypothalamus ceases production of GnRH, shutting down the HPG axis. A critical step in the activation of the HPG axis is the release of GnRH by the hypothalamus and triiodothyronine has been found to play an important role in that process.

For birds that initiate breeding in response to increasing day length, photoreceptor cells in the hypothalamus appear to trigger an uptake of thyroid hormones from the blood (hormones produced and released into the blood by stimulation of the thyroid gland by thyroid-stimulating hormone [TSH] produced by a specialized group of pituitary cells in the pars tuberalis; Wyse and Hazlerigg 2009) and also activate an enzyme called DIO2 (type 2 deiodinase) that converts

**Fig. 8.26** (continued) value of the third trait. To illustrate these interactions, effect sizes between the reproductive traits and IGF-1 levels are plotted separately in three quantiles (Q1 = black triangle, Q2 = gray triangle, and Q3 = white triangle) of duration of parental care (**c**, **d**) and body weight (**b**, **d**). Effect sizes on the left side of the panels represent strategies that are more common for species with a faster pace-of-life (lower body mass and weight and shorter periods of parental care); effect size on the right side represents species with a slower pace-of-life (greater body mass and longer periods of parental care). Effect sizes are standardized regression coefficients throughout. Whiskers denote 0.95 interaction intervals for the main effects. Note that the x-axis is reversed in panel c. (Figure from Lodjak et al. 2018; © 2017 The Authors. Functional Ecology © 2017 British Ecological Society, used with permission)

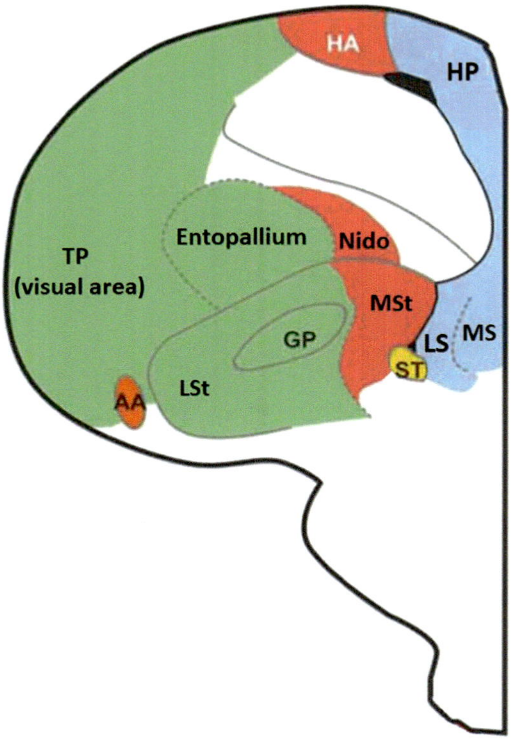


**Fig. 8.28** Section through the telencephalon of a Rock Pigeon (*Columba livia*). The lateral septum (LS) is strongly connected to the hippocampus and with other structures in the brain's social behavior system, e.g., medial amygdala, preoptic area, and hypothalamus. HP, hippocampus; HA, apical hyperpallium; GP, globus pallidus; MS, medial striatum; Nido, nidopallium; MSt, medial striatum; LSt, lateral striatum; TP, temporoparietal area; ST, stria terminalis; AA, anterior arcopallium. (Figure modified from Shanahan et al. 2013; © 2013 Shanahan, Bingman, Shimizu, Wild and Güntürkün, open-access article distributed under the terms of the Creative Commons Attribution License, https://creativecommons.org/licenses/by/4.0/)

thyroxine (T4) into triiodothyronine (T3) (Tamai and Yoshimura 2017). As a result, concentrations of T3 in the hypothalamus (and, specifically, in the mediobasal hypothalamus, or MBH) increase by 10-fold (Yoshimura et al. 2003). In the MBN, T3 causes morphological changes in the GnRH neurons that result in the release of GnRH and activation of the HPG axis (Shinomiya et al. 2014; Fig. 8.40).

## 8.5 Thyroid Hormones: Embryonic Development and Hatching

Thyroid hormones are needed during the development of bird embryos to coordinate the expression of numerous genes important in cell proliferation, differentiation, and maturation. However, the thyroid glands of embryos do not mature until about the middle of incubation in precocial species and about the time of hatching in altricial embryos. As a result, thyroid hormones deposited into the egg yolk by females play a critical role in embryonic development (Darras 2019).

Female birds deposit both thyroxine (T4) and triiodothyronine (T3) in egg yolk, with the amounts of T4 higher than T3 (Table 8.1). As embryos develop, the thyroid hormones are transferred from the egg yolk to the embryo (Fig. 8.41). Thyroid hormones are important in the differentiation and maturation of embryonic central nervous systems, skeletal systems, the heart, and skeletal muscles (McNabb 2007). More specifically, thyroid hormones are needed for the development of synaptic connections between neurons in the developing brain (McNabb 1992), bone elongation and fusion of ossification centers, differentiation of muscle cells and normal growth of cardiac and skeletal muscles, and maturation of glucose transport processes and enzyme production in the developing digestive system (in conjunction with glucocorticoid hormones) (McNabb 2007). Stier et al. (2020) examined the effect of experimentally increased thyroid hormone levels by injecting thyroid hormones into eggs of Collared Flycatchers (*Ficedula albicollis*) and found that, after hatching, nestlings from injected eggs had longer telomeres than those from control eggs. Telomere length can reflect phenotypic quality because it has been found to predict lifespan in Zebra Finches (Heidinger et al. 2012) and both lifespan and lifetime reproductive success in Purple-crowned Fairywrens (*Malurus coronatus*; Eastwood et al. 2019). In addition, the results of a meta-analysis of 27 studies revealed that individuals with longer telomeres live longer (Wilbourn et al. 2018). However, Stier et al. (2020) also noted that, if the

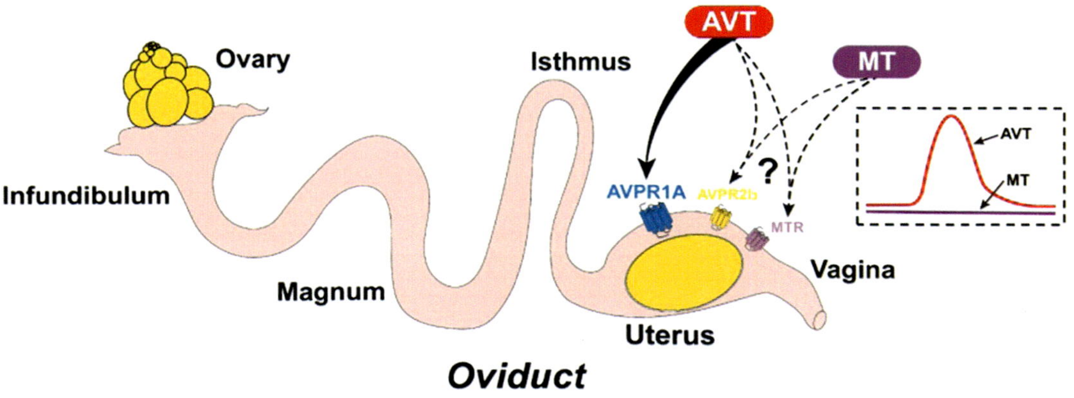


**Fig. 8.29** Proposed model of the role of AVT during oviposition. A surge of plasma AVT (but not mesotocin [MT]; shown in the box) may induce oviposition through receptors (AVPR1A) located in the membranes of smooth muscle cells in the uterus, with MT and other receptors having less important roles during oviposition. Wu et al. (2019) noted that this hypothesis needs further verification. (Figure from Wu et al. 2019; © 2019 Published by Elsevier Inc., used with permission)

effects of elevated levels of thyroid hormones in eggs were so beneficial in terms of increasing lifespans and enhancing reproductive fitness, natural selection would have favored females that transferred more thyroid hormones into their eggs. As such, they suggested that elevated levels of thyroid hormones in eggs must have some cost (s) (e.g., reduced immunity) that they were unable to detect. Additional study of the effects on birds of elevated levels of thyroid hormones in eggs will be needed to identify any possible negative effect (s).

In precocial species of birds, levels of thyroid hormones are highest in the days just before, during, and after hatching (Fig. 8.42). These hormones play important roles in muscle development, lung maturation, and switching from chorioallantoic to pulmonary respiration, retraction of the yolk sac, development of the digestive system, and, just prior to hatching, the growth and development of the musculus complexus, also referred to as the hatching muscle (De Groef et al. 2013; Fig. 8.43). In contrast, plasma levels of thyroid hormones in altricial species of birds do not increase until after hatching (Fig. 8.42), with the highest levels occurring as young become endothermic in European Starlings (*Sturnus vulgaris*; De Groef et al. 2013).

For embryos of precocial species, thyroid hormones have also been found to be important in brain development and, after hatching, learning behavior. Yamaguchi et al. (2012) reported that a surge in brain levels of triiodothyronine shortly after hatching is what initiates the start of the sensitive period for imprinting for young Domestic Chickens and this hormone also plays a critical role in learning later in life. For post-hatch Zebra Finches, levels of thyroid hormones in the brain continued to increase after hatching, with peak levels reached about 3 to 5 weeks after hatching (Yamaguchi et al. 2017; Fig. 8.44). Increasing levels of thyroid hormones in the brains of altricial young after hatching are likely important in brain development (Fig. 8.45) and also enhance their locomotor activities around the time of fledging (Yamaguchi et al. 2017). For young males in some species of birds, thyroid hormones may also play an important role in the process of song learning (Raymaekers and Darras 2017; Fig. 8.46).

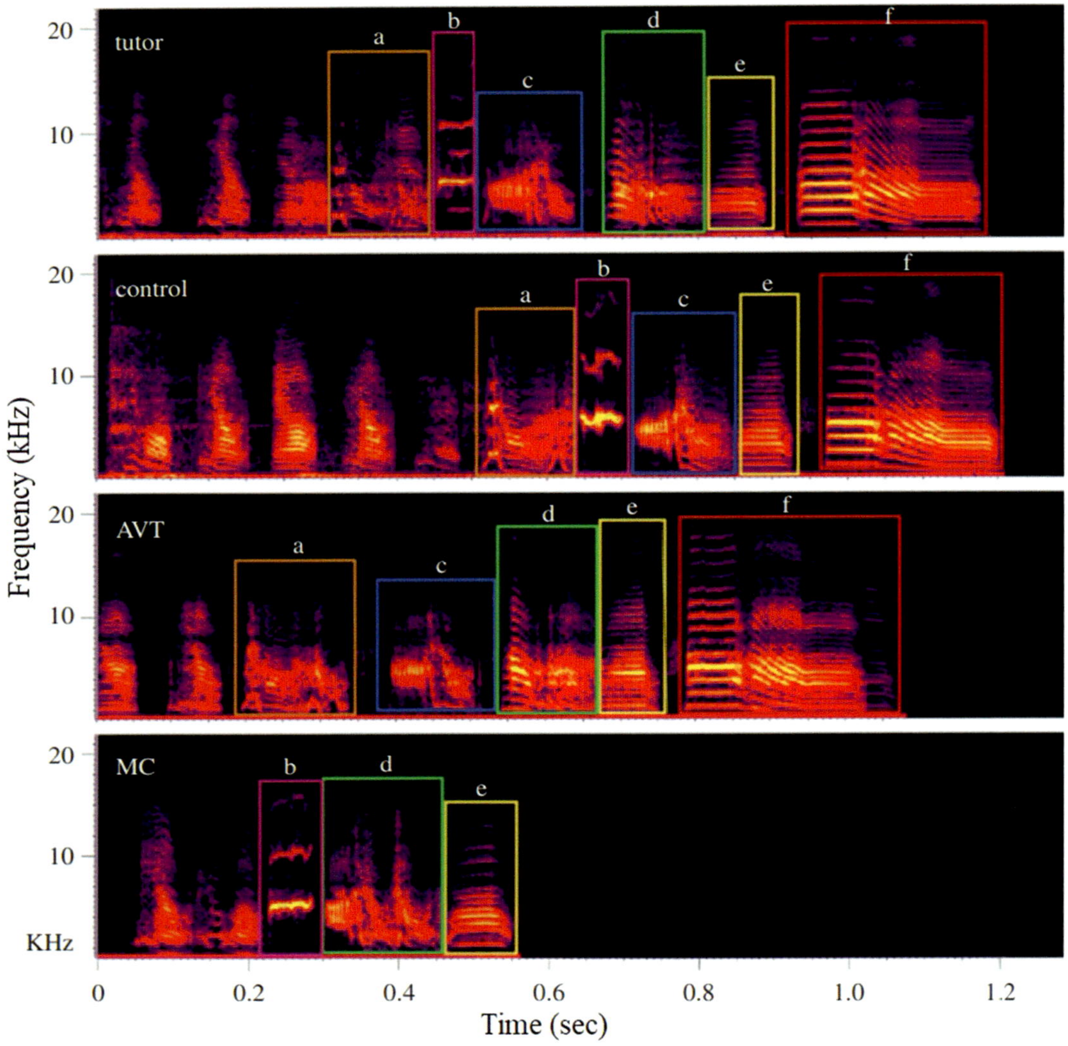


**Fig. 8.30** Examples of spectrograms of the songs of a tutor male Zebra Finch (*Taeniopygia guttata*) and males from each treatment group. Boxes outline individual song syllables. Letters indicate syllables that correspond between those of the tutor and males in each treatment group. Songs of MC males (treated with an AVT-antagonist) were significantly less similar to those of the tutor than those of AVT-treated and control males. Songs of AVT-treated and control males did not differ in similarity to songs of the tutor. (Figure from Baran et al. 2017; © 2017 The Authors. Published by the Royal Society. All rights reserved, used with permission)

## 8.6 Thyroid Hormones and Migration

Thyroid hormones appear to be needed for the expression of spring migration. Increased plasma levels of thyroid hormones prior to spring migration have been reported in several species of birds, including Canada Geese (*Branta canadensis*; John and George 1978), White-crowned Sparrows (*Zonotrichia leucophrys*; Smith 1982), and Rosy Starlings (*Sturnus roseus*; Pathak and Chandola 1984). In addition, Pant and

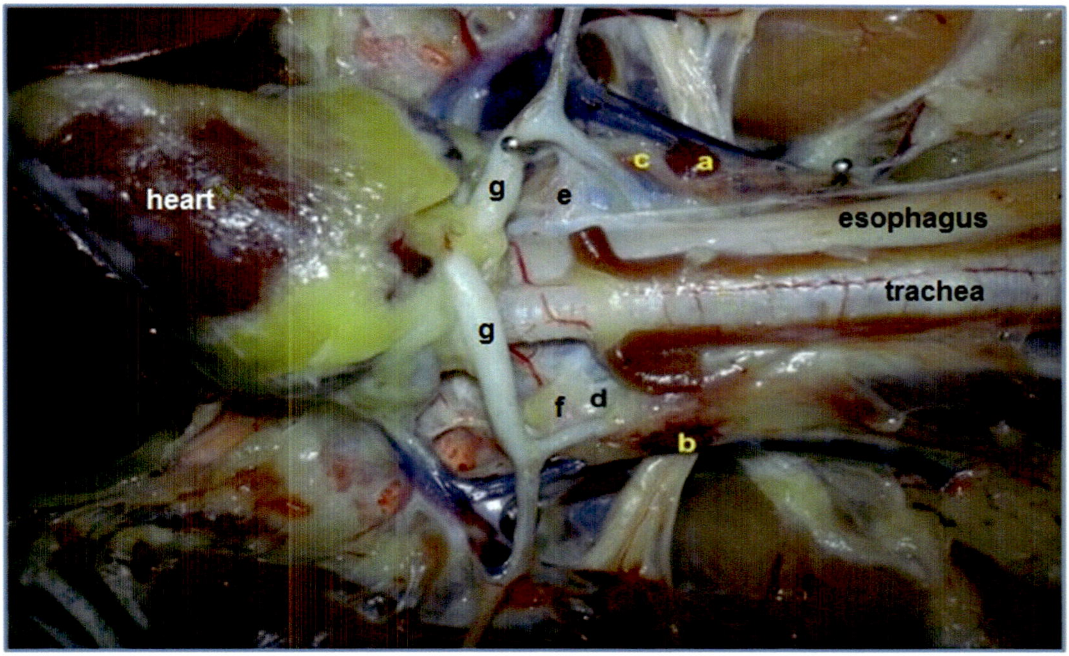


**Fig. 8.31** Photo of the right (**a**) and left (**b**) thyroid glands, right (**c**) and left (**d**) parathyroid glands, right (**e**) and left (**f**) ultimobranchial glands, the brachiocephalic arteries (**g**), heart, esophagus, and trachea of a Domestic Turkey (*Meleagris gallopavo domestica*). (Figure modified from Ali and Mirhish 2015; used with permission)

Chandola-Saklani (1993) found that Red-headed Buntings (*Emberiza bruniceps*) failed to gain mass or express nocturnal restlessness after thryroidectomies, and Ieromnimon (1978) reported that European Robins (*Erithacus rubecula*) treated with a thyroid-hormone blocker exhibited less nocturnal migratory restlessness, and injections of thyrotropin-releasing hormone stimulated migratory restlessness.

Pérez et al. (2016) found that chemical thyroid inhibition stopped all behavioral and morphological changes associated with spring migration by White-crowned Sparrows, including mass gain, fattening, muscle hypertrophy, and nocturnal restlessness (zugunruhe). Similar results have been reported for Red-headed Buntings (*Emberiza bruniceps*) (Nair et al. 1994). Sharma et al. (2018) examined plasma levels of triiodothyronine during the annual cycle of Red-headed Buntings and found that levels were significantly higher prior to fall migration than spring migration (Fig. 8.47). These authors suggested that plasma levels were higher prior to fall migration because, after the breeding season, the buntings had little or no subcutaneous fat and needed elevated metabolic rates to stimulate fat deposition.

## 8.7 Calcium Homeostasis: Parathyroid and Ultimobranchial Glands

Calcium is an essential nutrient that plays important roles in metabolism, blood clotting,

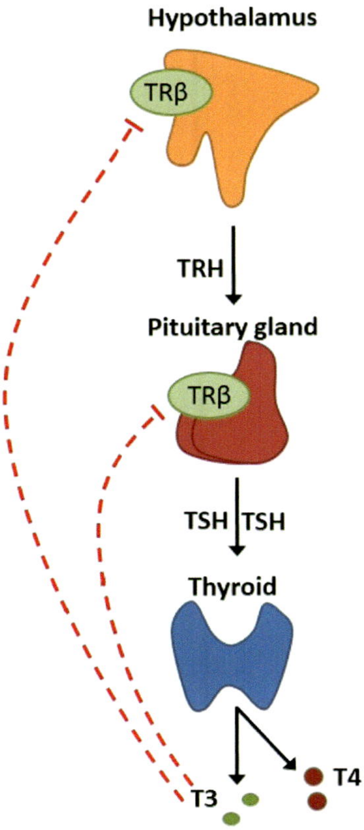


**Fig. 8.32** The hypothalamic-pituitary-thyroid (HPT) axis of birds. Thyrotropin-releasing hormone (TRH) released by the hypothalamus stimulates the secretion of thyroid-stimulating hormone (TSH) by the pituitary. Thyroid-stimulating hormone stimulates cells in the thyroid to increase secretion of the thyroid hormones triiodothyronine (T3) and thyroxine (T4). Levels of those hormones in the blood, in turn, regulate production of TRH by the hypothalamus. This feedback mechanism helps maintain appropriate levels of thyroid hormones in the blood. (Figure modified from Apostu et al. 2020; © 2020 by the authors. Licensee MDPI, Basel, Switzerland, open-access article distributed under the terms and conditions of the Creative Commons Attribution License (CC BY), https://creativecommons.org/licenses/by/4.0/)

neuromuscular function, muscle contraction, and intracellular signaling, and is, of course, also an essential ingredient of bones and eggshells. Calcium levels in the blood of birds are controlled via feedback mechanisms involving parathyroid hormone, vitamin D, and calcitonin. The ultimobranchial glands are located below the thyroid gland and secrete the hormone calcitonin (Figs. 8.31 and 8.48). Interestingly, mammals do not have ultimobranchial glands because the cells that secrete calcitonin are incorporated into the thyroid gland during development. Parathyroid hormone is secreted by the parathyroid glands.

Studies of the role of calcitonin in avian calcium metabolism have provided mixed results. For example, some investigators have reported no effect of calcitonin on plasma levels of calcium ions (e.g., Candlish and Taylor 1970; Dacke 1979), However, the results of other studies indicate that administration of calcitonin reduces plasma calcium levels (hypocalcemia), e.g., Rock Pigeon (*Columba livia*; Yadav and Srivastav 2009), a parrot (Swarup et al. 1980), and Japanese Quail (*Coturnix japonica*; Dhande et al. 1997). In addition, other investigators have found that plasma levels of calcitonin increase after calcium administration (Dacke 2000; de Matos 2008). Such results suggest that calcitonin is important in regulating plasma levels of calcium in birds. When secreted into the blood, calcitonin reduces plasma calcium levels by inhibiting bone resorption (Belanger and Copp 1972) and increasing the excretion of calcium by the kidneys (Sommerville and Fox 1987) (Fig. 8.49).

Parathyroid glands produce parathyroid hormone (PTH) in response to low plasma levels of calcium (hypocalcemia; Fig. 8.50). PTH increases plasma levels of calcium by stimulating bone resorption by osteoclasts and increasing calcium reabsorption by the kidneys. Low plasma levels of calcium also enhance expression of renal 1-α-hydroxylase (1αOHase) enzyme that converts 25-hydroxyvitamin D (25D) into 1,25-dihydroxyvitamin D (1,25D) that increases the absorption of calcium in the small intestine by acting on the intestinal vitamin D receptors. PTH is particularly important for female birds when producing and laying eggs, and calcium is needed to form eggshells (Fig. 8.51).

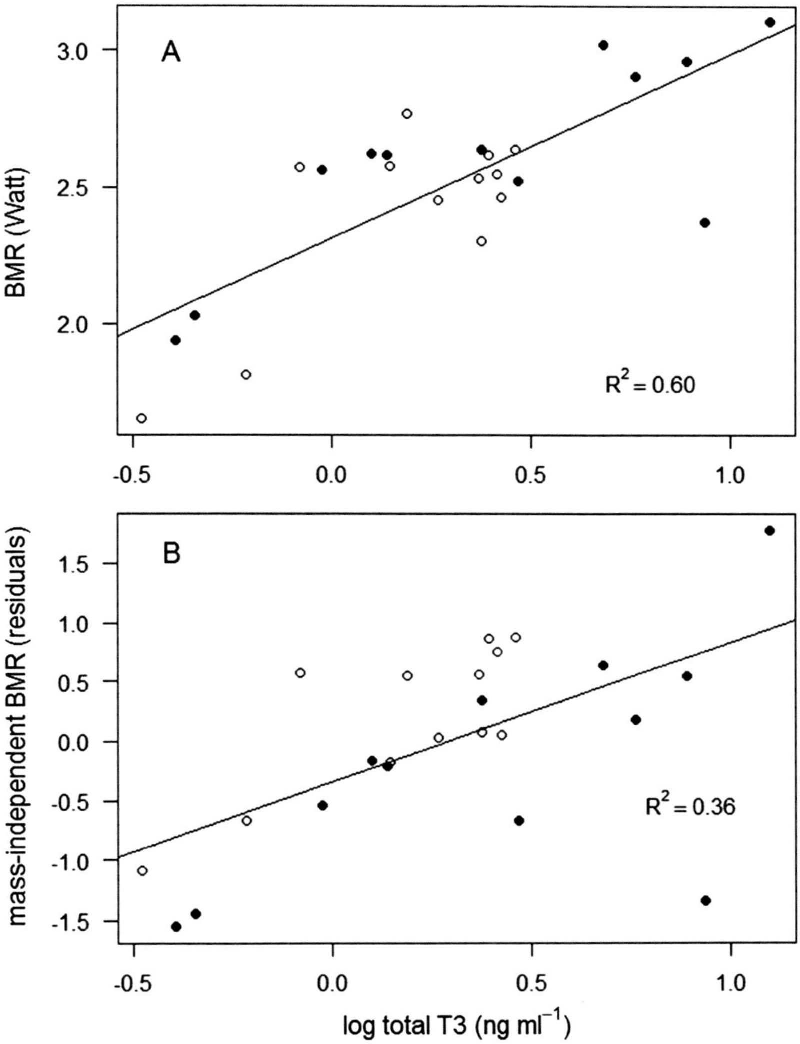


**Fig. 8.33** (**a**) Relationship between plasma levels of total triiodothyronine (T3) and basal metabolic rates (BMR), and (**b**) mass-independent BMR (residuals of BMR regressed on body mass) in breeding Black-legged Kittiwakes (*Rissa tridactyla*). (Figure from Welcker et al. 2013; © 2013 Welcker et al., open-access article distributed under the terms of the Creative Commons Attribution License, https://creativecommons.org/licenses/by/4.0/)

## 8.8 Adrenal Glands

The adrenal glands of birds are located above the kidneys (Fig. 8.52) and are composed of two types of cells, cortical cells and medullary cells. Cortical cells produce and secrete two categories of corticosteroids—glucocorticoids and mineralocorticoids. The primary mineralocorticoid produced by corticol cells is aldosterone, whereas the primary glucocorticoid produced by cortical cells is corticosterone; only small amounts of cortisol are produced (Carsia 2015). The medullary cells of the adrenal gland produce and secrete epinephrine and norepinephrine (Humayun et al. 2012). In mammals, the adrenal gland consists of an outer adrenal cortex with cells that synthesize and release stress hormones (cortisol) and an inner area called the adrenal medulla with cells that synthesize and release epinephrine and norepinephrine. In the avian adrenal gland, however, the cells that secrete the different hormones are not separated into different areas, but, rather, are intermixed throughout the adrenal gland (Fig. 8.53).

Glucocorticoids (GCs) are important in metabolic regulation, helping birds respond to their

**Box 8.4 Thyroid Hormones and Molt**

Birds at temperate latitudes time events like breeding and prenuptial molt (i.e., molt prior to the beginning of the breeding season) based on changing day length. Thyroid hormones are known to play an important role in the initiation of breeding by regulating the release of gonadotropin-releasing hormone by the hypothalamus and are also thought to be involved in the timing of prenuptial molt. To examine how thyroid hormones might affect the prenuptial molt of a songbird, Pérez et al. (2018) treated some White-crowned Sparrows (*Zonotrichia leucophrys*) with a substance that inhibits production of thyroid hormone (methimazole). Sparrows treated with methimazole did not molt, and birds treated with methimazole, but also treated with either T3 (triiodothyronine) or T4 (thyroxine), did molt. These results indicate that thyroid hormones are necessary for the expression of prenuptial molt in White-crowned Sparrows.

Based on an experiment where an area of the hypothalamus (ventral medial hypothalamus) was lesioned, Kuenzel (1974) suggested that this area was the site that controlled the prenuptial molt of White-throated Sparrows (*Zonotrichia albicollis*). However, Pérez et al. (2018) suggested that thyroid hormones control molt at the level of feather follicles rather than the hypothalamus because White-crowned Sparrows with only circulating T3 or T4 molted. This suggests that the hypothalamus was not involved because the results of previous studies indicate that T3 cannot cross the blood brain barrier and affect cells in the hypothalamus (e.g., Yasuo et al. 2005).

White-crowned Sparrow. (Photo by Ingrid Taylar, Wikipedia, CC BY 2.0, https://creativecommons.org/licenses/by/2.0/)

(continued)

**Box 8.4** (continued)

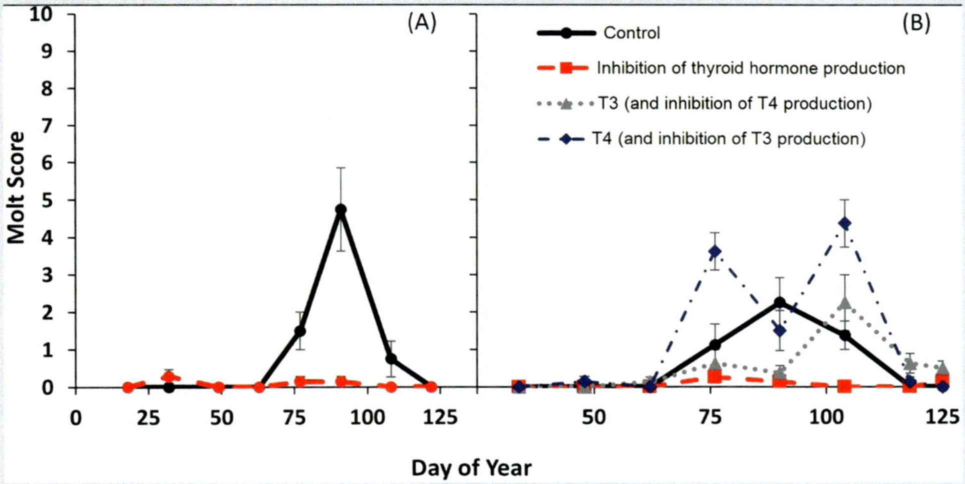


(**a**) Body molt scores for control White-crowned Sparrows and those treated with a chemical (methimazole) that inhibited production of thyroid hormones. (**b**) Body molt scores for control White-crowned Sparrows and sparrows treated with methimazole, T3 (triiodothyronine) and methimazole, and T4 (thyroxine) and methimazole. Values are means ± standard error. (Figure from Pérez et al. 2018; open-access article distributed under the terms of the Creative Commons Attribution License, https://creativecommons.org/licenses/by/4.0/)

unpredictable environments and, as such, have important effects on fitness (Table 8.2, Fig. 8.54). Baseline levels of GCs reflect variation in energy expenditure, i.e., metabolic rate (Fig. 8.55). However, when birds are faced with environmental and behavioral situations that are potentially harmful, i.e., stressful, the hypothalamo-pituitary-adrenal (HPA) axis is activated and GCs (primarily corticosterone) are released into the blood. Definitions of stress vary, but Cockrem (2007) defined stress as "a state when the hypothalamo-pituitary-adrenal (HPA) axis is activated with increased secretion of glucocorticoids in response."

Corticosterone (CORT) is secreted by the adrenal glands after the anterior pituitary releases adrenocorticotropic hormone (ACTH) into the blood, with the pituitary, in turn, releasing ACTH in response to the secretion of corticotropin-releasing hormone (CRH) and arginine vasotocin (AVT) by the hypothalamus. Kuenzel et al. (2020) examined the effects of both CRH and AVT on the synthesis and release of ACTH by the anterior pituitary of Domestic Chickens and found that, when injected into the cerebrospinal fluid, CRH was more effective in stimulating the release of ACTH than AVT, but, when injected into the blood, AVT was more effective than CRH. These authors suggested that the simultaneous release of both CRH and AVT may result in a "pronounced super-additive simulation of the glucocorticoid release."

Cells responsive to glucocorticoids have mineralocorticoid receptors, glucocorticoid receptors, or both types of receptors. After glucocorticoids bind to receptors, cell function is altered over variable time periods, sometimes within minutes, due to transcriptional effects. Mineralocorticoid receptors have a greater affinity for

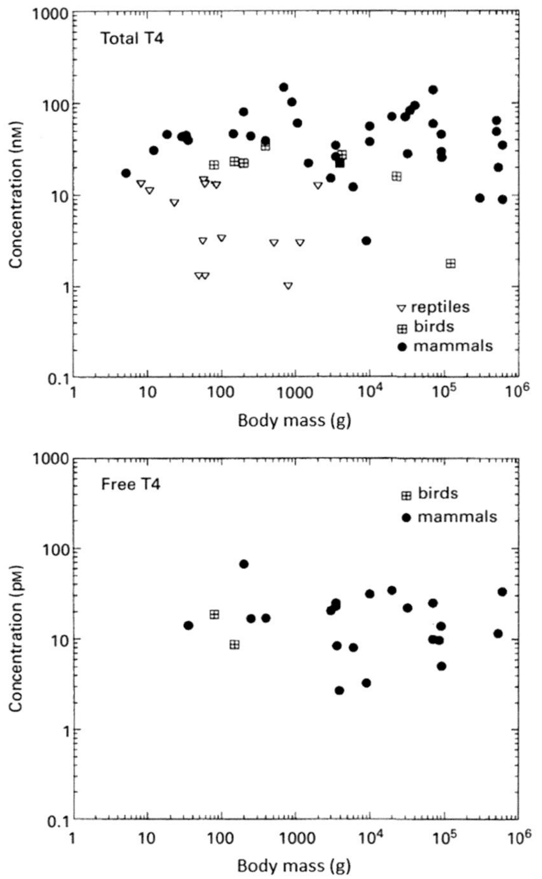


**Fig. 8.34** Concentrations of total (bound to plasma proteins) and free thyroxine (T4) and triiodothyronine (T3) in the plasma of adult reptiles, birds, and mammals. Note that, among birds, total plasma concentrations of T4 are generally between that of reptiles and mammals, whereas plasma concentrations of T3 are generally higher than or similar to that of mammals and higher than that of reptiles. Plasma concentrations of free T4 and T3 of birds and mammals are similar. (Figure modified from Hulbert 2000; © Cambridge Philosophical Society, used with permission)

glucocorticoids (GC) and so are more likely to bind with GC when plasma levels are lower; glucocorticoid receptors bind with GC when plasma levels of glucocorticoids are elevated and most mineralocorticoid receptors are occupied, e.g., during periods of stress (Lattin et al. 2012). Baseline (i.e., non-stressed) levels of GC in the blood help regulate glucose availability, maintain body mass, and mediate foraging behavior and other behaviors (Hau et al. 2016; Fig. 8.56). Administration of CORT, for example, has been found to stimulate food intake in several species of birds, e.g., Ring Doves (*Streptopelia risoria*, Koch et al. 2002), Japanese Quail (*Coturnix japonica*; Wall and Cockrem 2009), and Macaroni Penguins (*Eudyptes chrysolophus*, Crossin et al. 2012). In addition, Dallman et al. (1993) reported that daily food intake was typically highest when plasma levels of glucocorticoids were highest. However, Landys et al. (2006) noted that administration of CORT did not alter food consumption in some species of birds, e.g., Dark-eyed Juncos (*Junco hyemalis*) and Mountain Chickadees (*Poecile gambeli*),

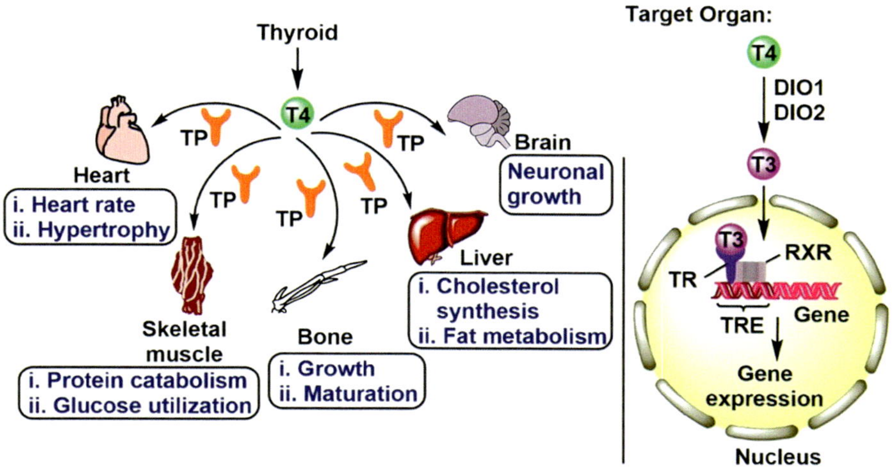


**Fig. 8.35** Examples of the effects of thyroid hormones on different organs and tissues. (Figure modified from Mondal et al. 2020; © 2019 Indian Institute of Science, used with permission)

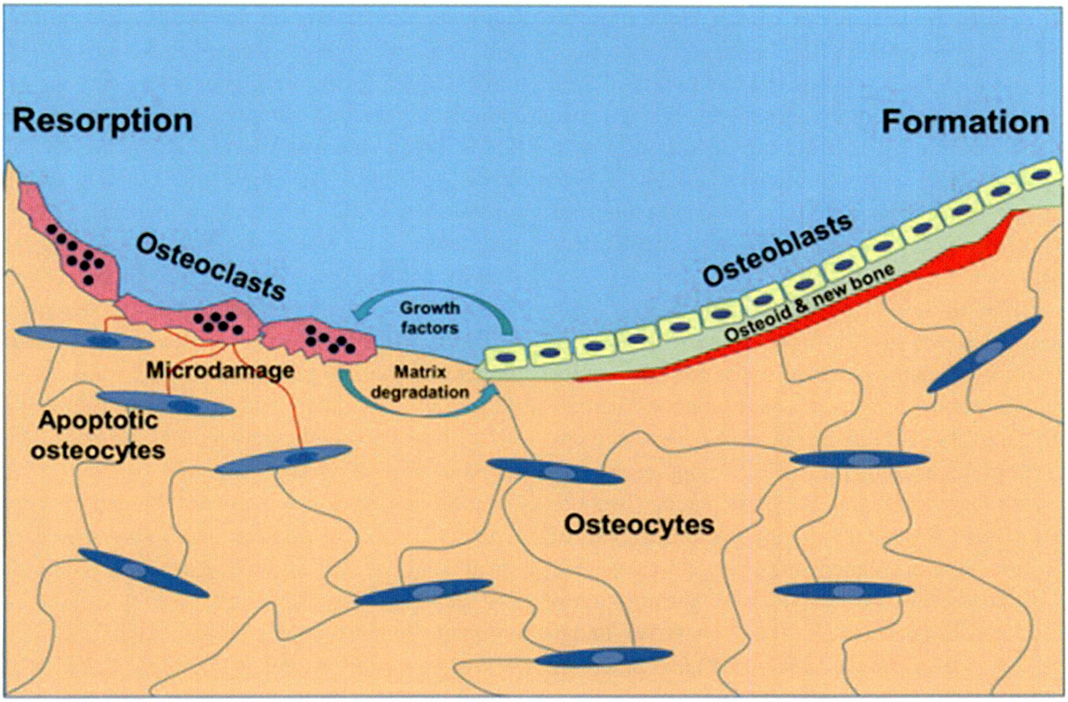


**Fig. 8.36** Bone remodelling. Sites in bones that have suffered minor damage are sensed by osteocytes that are embedded in bones and respond by undergoing apoptosis. Osteocytes resorb damaged bone and signal osteoblasts by releasing degraded bone matrix proteins and growth factors during bone resorption. Osteoblasts respond by forming new bone that repairs the damaged area. Triiodothyronine (T3) stimulates osteoblast activity and helps regulate osteoblast proliferation and differentiation. (Figure from Williams 2013; © 2012 Karger Publishers, used with permission)

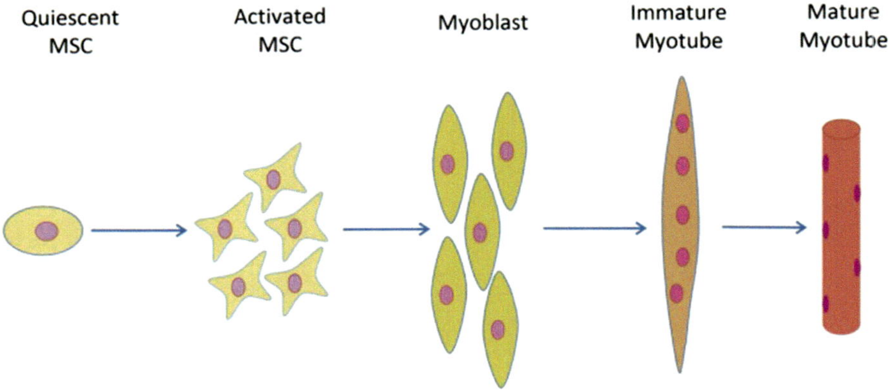


**Fig. 8.37** Triiodothyronine promotes differentiation of muscle satellite cells and maturation to form myotubes that will fuse to form muscle cells. (Figure from Salvatore et al. 2014; © 2013 Nature Publishing Group, a division of Macmillan Publishers Limited, used with permission)

and suggested that the effect of CORT on feeding behavior may differ among life-history stages, e.g., when birds need to store energy prior to and during migration. Baseline levels of corticosterone also help regulate plasma levels of free fatty acids and, possibly, plasma levels of glucose (Deviche et al. 2017). During periods of stress, plasma CORT either stimulates, inhibits (-), or has no effect (N) on glycolysis, stimulates or inhibits lipolysis, and stimulates proteolysis (Deviche et al. 2017; Fig. 8.57).

Baseline levels of GC also correlate with periods of activity during the daily cycle, with levels increasing just before periods of activity and remaining elevated for several hours thereafter (Schwabl et al. 2016; Fig. 8.58). This increase likely prepares birds for the upcoming increase in locomotor activity and energy demands (Breuner et al. 1999). Seasonal variation in plasma levels of GC are also correlated with periods associated with greater levels of activity, e.g., during migration (e.g., Eikenaar et al. 2020; Fig. 8.59), fledging from nests (Heath 1997), and natal dispersal (Belthoff and Dufty 1998; Fig. 8.60). DeSimone et al. (2020) developed a model that predicts how plasma levels of CORT and triglycerides (energy source) vary along with body mass for birds at stopover sites during migration (Fig. 8.61). Birds must refuel during stopover, increasing their body mass and plasma levels of triglycerides (TRIG), so they can depart on the next stage of their migration. Upon arrival, plasma CORT levels of birds vary depending on their physical condition and available energy stores. However, for all birds, CORT levels are higher when they arrive, stimulating feeding activity. As body mass and TRIG levels increase during stopover, CORT levels decline. As the time of departure approaches, CORT levels again increase, preparing birds for the upcoming increased locomotor activity and energetic demands (DeSimone et al. 2020). These authors tested their model with migrating Gray Catbirds (*Dumetella carolinensis*) and found that, as predicted, CORT levels were negatively correlated with body condition and TRIG levels for birds arriving and refueling at a stopover site. However, they did not detect an increase in CORT levels toward the end of stopover and prior to departure, possibly due to a small sample size and because Gray Catbirds that departed may have left their study site to complete their refueling phase at another location. A previous study revealed that Gray Catbirds in "migratory condition" had higher plasma CORT levels than those in "pre-migratory condition," suggesting that they likely do have higher CORT levels just prior to initiating longer migratory flights (Holberton et al. 1996), as also reported in studies of other species of migratory birds (e.g., Eikenaar et al. 2020).

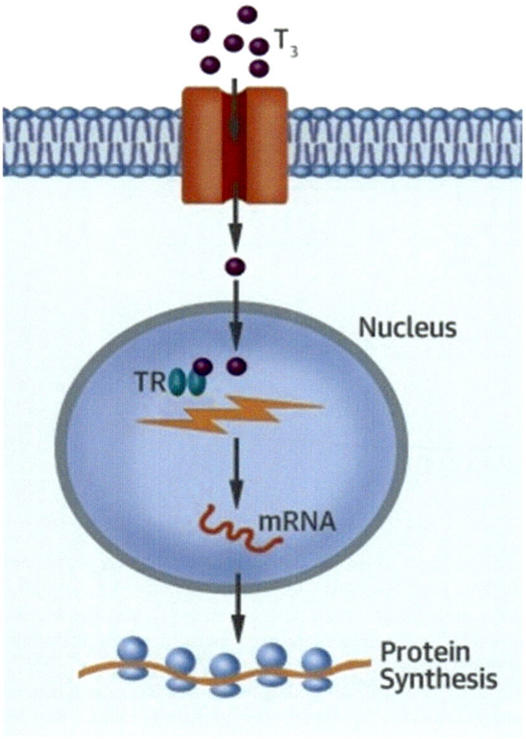


**Fig. 8.38** T3 has both genomic and non-genomic effects on cardiac muscle cells. Genomic effects involve the transport of T3 into a muscle cell and binding to a thyroid receptor (TR) and regulating the transcription and translation of specific genes. Non-genomic actions include regulation of ion movements through the cell membrane which can, for example, cause an increase in heart rate. (Figure modified from Razvi et al. 2018; © 2018 by the American College of Cardiology Foundation. Published by Elsevier, used with permission)

The CORT-fitness hypothesis posits that elevated baseline (non-stressed) plasma concentrations of GCs indicate that individuals are in poor condition (Bonier et al. 2009) and, as a result, may invest less in reproduction (Fig. 8.62). Among some species of seabirds, investigators have found that those with higher levels of CORT are less attentive to nests and more likely to abandon nests (e.g., Spée et al. 2010, 2011; Thierry et al. 2013). In further support of this hypothesis, Schoenle et al. (2017) found that female Red-winged Blackbirds (*Agelaius phoeniceus*) with higher baseline levels of CORT spent less time incubating eggs and had lower clutch masses than females with lower baseline levels of CORT. In addition, Great Tits (*Parus major*) with higher levels of CORT were more likely to abandon nests and had nestlings with lower average masses than those with lower levels (Ouyang et al. 2012). Studies of other species have also provided support for the CORT-fitness hypothesis (e.g., Saino et al. 2005).

However, the results of a number of studies suggest that relationships between baseline CORT levels can vary among life-history stages (e.g., Ouyang et al. 2013), environmental conditions (e.g., Riechert et al. 2014), and with reproductive strategy (Lancaster et al. 2008). As such, Bonier et al. (2009) developed a CORT-adaptation hypothesis that proposes that increased baseline levels of GCs can enhance fitness by helping individuals successfully respond to energetically challenging conditions and still devote resources to self-maintenance and reproduction. In support of the CORT-adaptation hypothesis, Riechert et al. (2014) reported that baseline CORT levels were positively related to breeding success in Common Terns (*Sterna hirundo*; Fig. 8.63), possibly because higher levels of CORT contributed to increased foraging activity and parental feeding of young. Other studies have also reported data that support the CORT-adaptation hypothesis (e.g., Bókony et al. 2009; Burtka et al. 2016; Fischer et al. 2020).

As the above examples indicate, relationships between baseline GC levels, breeding behavior, and fitness are complicated and can be negative, positive, have no apparent effect at all, or even differ for males and females of the same species. As an example of the latter case, Angelier et al. (2010) found a negative relationship between baseline CORT levels and quality for male Black-browed Albatrosses (*Thalassarche melanophris*), but no relationship between baseline CORT and quality for females. Relationships between baseline CORT levels and fitness likely depend on the particular environmental and energetic conditions faced by individual birds (Bonier et al. 2009; Fig. 8.64) and may also vary with life-history stage and species (Lattin et al. 2016; Vitousek et al. 2018). In addition, CORT levels are plastic and can vary during a breeding season

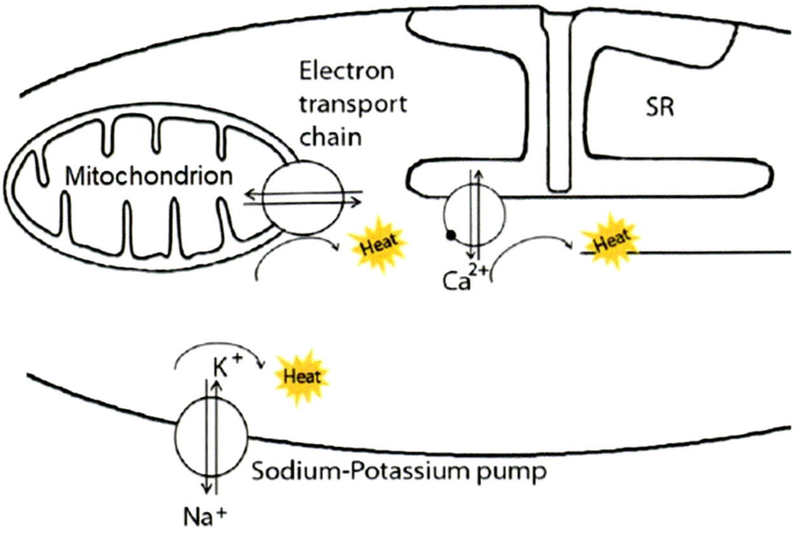


**Fig. 8.39** Thyroid hormones can increase basal metabolic rate by stimulating an increase in ATP production for metabolic processes and by generating and maintaining ion gradients across cell membranes. Thyroid hormones can also stimulate fat, glucose, and protein catabolism and anabolism. Two ion gradients influenced by thyroid hormones are the sodium (Na)-potassium (K) gradient across the cell membrane and the calcium ion gradient between the cytoplasm and sarcoplasmic reticulum of muscle cells. Thyroid hormones can alter the levels of sodium ions in cells and potassium outside of cells, which then requires increased activity of sodium-potassium pumps (i.e., active transport) and the breakdown of ATP (and release of energy) to maintain the gradient. (Figure modified from Nowack et al. 2017; © 2017 Nowack, Giroud, Arnold and Ruf, open-access article distributed under the terms of the Creative Commons Attribution License (CC BY), https://creativecommons.org/licenses/by/4.0/)

(Angelier et al. 2007), so the timing of blood sampling may influence perceived relationships between CORT levels and fitness. Investigators have also noted that the relationship between CORT levels and fitness can vary with reproductive strategies (Jaatinen et al. 2013).

In response to stressful events or conditions, the HPA axis and the sympathetic division of the Autonomic Nervous System are activated. The sympathetic division is often referred to as the "fight-or-flight" division because, when activated, its effects rapidly prepare organisms to become physically active, and that might mean fighting, e.g., a male bird responding aggressively to a conspecific male intruding into its breeding territory, or escaping, e.g., a bird taking flight and attempting to escape from a raptor. The effects of sympathetic stimulation include (1) increased heart rate and cardiac output, (2) changes in blood flow with more blood going to skeletal muscles, the heart, and central nervous system and less to the digestive system, (3) increased respiratory rate, (4) release of epinephrine and norepinephrine from the adrenal glands (which further enhances the effects of sympathetic stimulation), and (5) the release of glucose into the blood by the liver. All of these actions prepare birds for intense physical activity.

With the detection of stress, areas of the brain (e.g., amygdala and hippocampus) send nervous impulses to the hypothalamus and this activates (via the anterior pituitary and adrenal glands) the HPA axis and results in the release of glucocorticoids, primarily CORT, into the blood (Fig. 8.65). This response is not as rapid as the sympathetic response, but increasing levels of CORT in the blood can generally be detected within about three minutes and peak levels are typically reached in 15 to 30 min (Fig. 8.66).

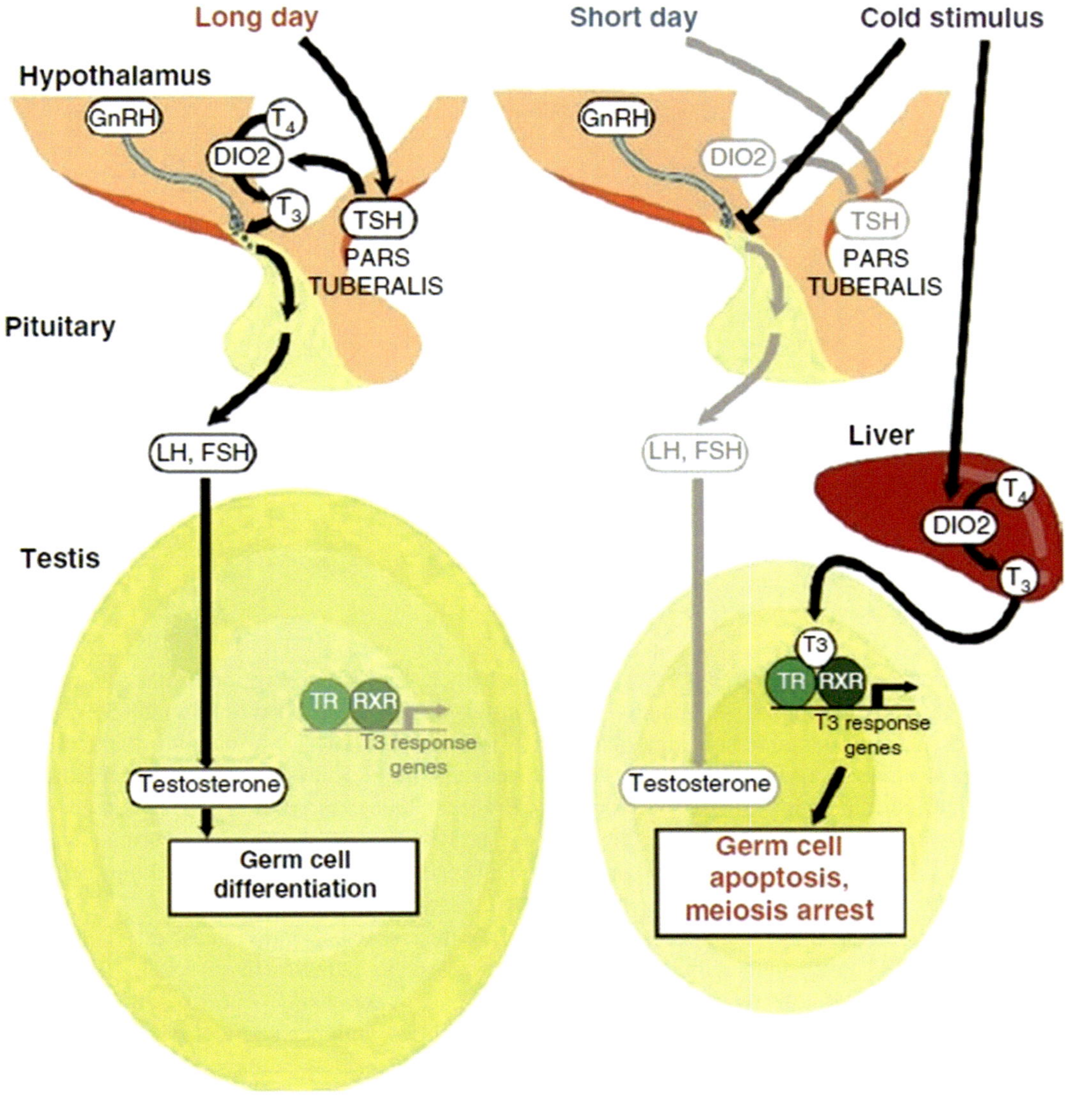


**Fig. 8.40** Role of thyroid hormones in regulation of the development and regression of testes before and after the breeding season. As days get longer, thyroid hormone stimulates secretion of gonadotropin-releasing hormone (GnRH) by the hypothalamus, which in turn stimulates the secretion of luteinizing hormone (LH) and follicle-stimulating hormone (FSH) by the anterior pituitary. LH stimulates the production of testosterone by the testes, whereas FSH stimulates sperm production. As days shorten and temperatures decline, uptake of thyroid hormones declines and DIO2, the enzyme that converts T4 to T3, is no longer produced, which blocks the release of GnRH from the hypothalamus. (Figure modified from Tamai and Yoshimura 2017; © 2017, Springer Nature Singapore Pte Ltd., used with permission)

CORT binds to receptors in target cells and tissues and activated receptors act as transcription factors that stimulate or inhibit production of different proteins. In the liver, CORT stimulates gluconeogenesis (production of glucose from noncarbohydrates) to increase blood glucose

**Table 8.1** Thyroid hormones deposited in egg yolk by females in different species (Table from Darras 2019; )

| Species | T4 in ng/g[a] | T3 in ng/g[a] | T4/T3[b] | References |
|---|---|---|---|---|
| Precocial Birds | | | | |
| Chickens | | | | |
| (*Gallus gallus*) | | | | |
| – Shaver strain (layer) | 3.80 ± 0.81 | 1.50 ± 0.39 | 2.13 | Prati et al. (1992) |
| – Ross strain (broiler) | 14.60 ± 2.24 | 1.23 ± 0.20 | 9.93 | Van Herck et al. (2013) |
| – Hy-line strain (layer) | 30.4 | 1.0 | 25.4 | Too et al. (2017) |
| – White leghorn strain (layer) | – | 1.05 ± 0.36 | – | Ho et al. (2011) |
| – Cornish rocks strain (layer) | – | 0.46 ± 0.38 | – | Ho et al. (2011) |
| Japanese Quail | 9.74 ± 2.20 | 3.44 ± 0.88 | 2.38 | Ruuskanen et al. (2018) |
| (*Coturnix japonica*) | 6 | 2.5 | 2.01 | Wilson and McNabb (1997) |
| Altricial birds | | | | |
| Rock Pigeon (*Columba livia*) | 3.06 ± 0.99 | 1.10 ± 0.21 | 2.33 | Hsu et al. (2016) |
| Great Tit (*Parus major*) | 1.15 ± 0.42 | 0.14 ± 0.07 | 6.73 | Ruuskanen et al. (2018) |
| Collared Flycatcher (*Ficedula albicollis*) | 7.21 ± 0.99 | 1.97 ± 0.48 | 3.06 | Ruuskanen et al. (2018) |
| Pied Flycatcher (*Ficedula hypoleuca*) | 5.72 ± 1.42 | 1.86 ± 0.56 | 2.57 | Ruuskanen et al. (2018) |

[a]Average concentration ± SD (when available)
[b]Ratio calculated on a molar basis

levels and provide cells with an important source of energy. CORT also reduces the uptake of glucose by tissues and organs not important in responding to stress (e.g., the digestive system), so those that are important have access to more glucose (Dallman et al. 1993). Thus, elevated plasma levels of CORT "facilitate a shift in energy balance to facilitate coping with a stressor" (MacDougall-Shackleton et al. 2019).

Increased CORT levels in the blood also block secretion of growth hormone by the anterior pituitary and have an inhibitory effect on reproduction by blocking the release of GnRH from the hypothalamus and limiting the effect of luteinizing hormone on ovaries and testes (Wingfield and Romero 2001). CORT also has an inhibitory effect on many aspects of the immune system, especially if CORT levels remain elevated for an extended period, e.g., limiting the activation and proliferation of B- and T-cells (Spencer et al. 2001). Stressors (and elevated CORT levels) lasting just minutes or a few days, e.g., aggressive encounters, tend to enhance immune responses; longer-term stressors lasting several days or weeks, e.g., inclement weather, suppress immune responses (e.g., Martin 2009; Vagasi et al. 2018). Finally, CORT can also alter behavior, e.g., increasing food intake and foraging behavior (e.g., Pravosudov 2003; Crossin et al. 2012), alter the begging behavior of nestlings (e.g., Kitaysky et al. 2003; Wada and Breuner 2008), increase aggressive tendencies (e.g., Kitaysky et al. 2003), and trigger greater exploratory behavior and risk-taking behavior (e.g., Martins et al. 2007).

Within species, individuals differ in their responses to stress (Fig. 8.67), and such differences are likely due to differences in personalities, with personalities defined as "suites of correlated behaviors that are expressed across different situations" (Carere and Eens 2005). Personalities vary along a continuum, but individuals are often categorized as either proactive (active) or reactive (passive) (Koolhaas et al. 1999; Table 8.3). In general, reactive individuals are expected to exhibit stronger HPA axis responses to stressors than proactive individuals (Table 8.3). In additional, proactive and reactive individuals are expected to differ in their behavioral responses to environmental stimuli and in relative fitness in predictable and unpredictable conditions (Table 8.3).

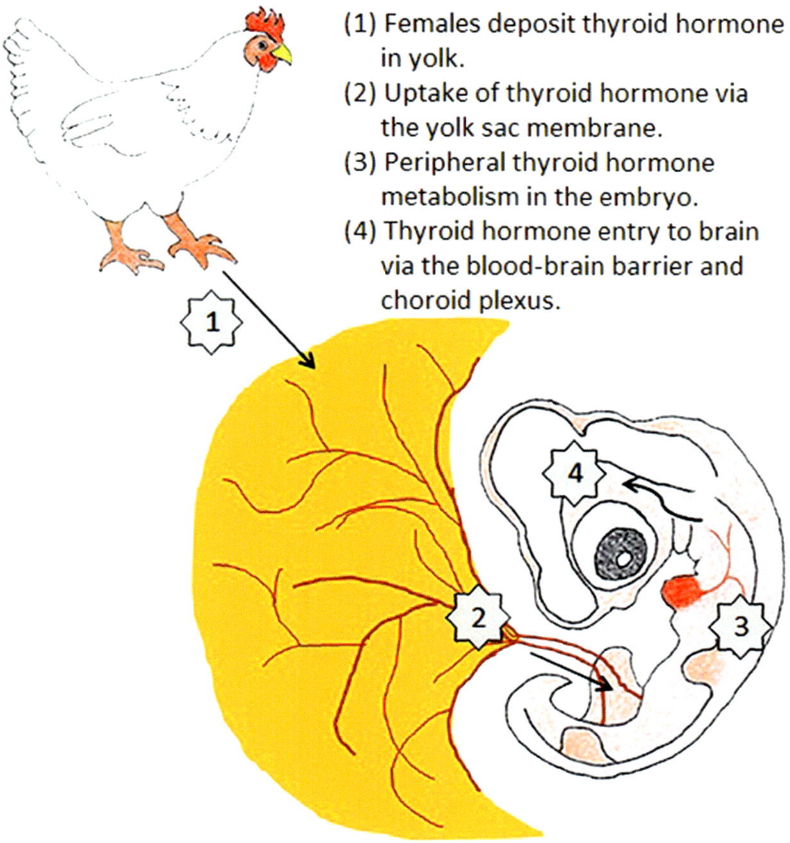


**Fig. 8.41** Thyroid hormones deposited by females in the yolk are transported from the yolk sac to the embryo, including the brain of the developing embryo. (Figure from Darras 2019; © 2019 Darras, open-access article distributed under the terms of the Creative Common Attribution License (CC BY), https://creativecommons.org/licenses/by/4.0/)

## 8.9 Hormones Important in Avian Feeding Behavior

For birds, especially those that fly, maintaining their body mass at a level that minimizes the likelihood of starvation, e.g., during cold winter nights at high latitudes or altitudes, and maximizes their chances of surviving, e.g., being able to escape from approaching predators, is typically, but not always, of critical importance. The exception is when birds must store more body fat, e.g., when preparing for a long-distance migration flight. Given the importance of body mass for birds, they are generally able to maintain their body mass within a very narrow range for extended periods (Fig. 8.68). Doing so requires complex interactions between the digestive system, the brain, and a variety of hormones and regulatory proteins (neuropeptides; Fig. 8.69). In the brain, the hypothalamus has an important role in regulating feeding, with different areas in the hypothalamus referred to as the satiety (ventromedial nucleus) and feeding (lateral hypothalamic area) centers. In peripheral tissues, and specifically in gastrointestinal endocrine cells (referred to as enteroendocrine cells; Fig. 8.70), several peptide hormones affect avian feeding behavior, including ghrelin, CCK, glucagon-like peptide (GLP)-1, and peptide YY (PYY) (Table 8.4).

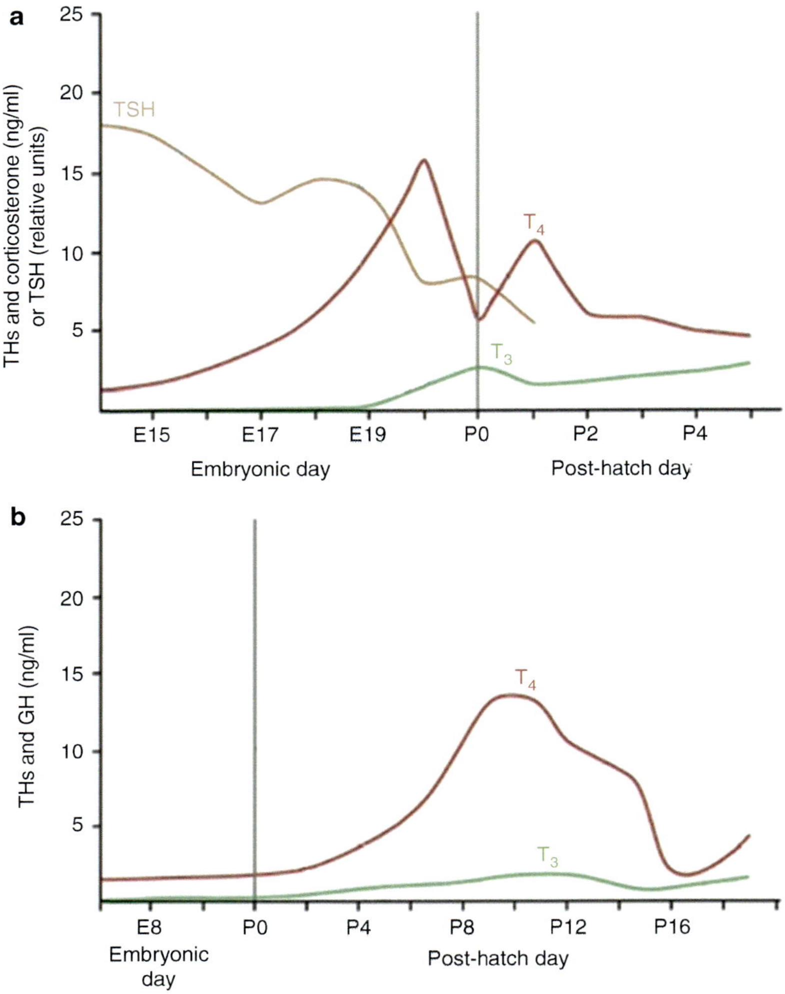


**Fig. 8.42** General patterns of the levels of hormones in blood plasma for (**a**) a precocial bird (Domestic Chicken), and (**b**) an altricial bird (European Starling, *Sturnus vulgaris*). Vertical lines indicate hatching. Plasma levels of thyroxine (T4) peak during the day prior to hatching and a secondary peak occurs the day after hatching; Plasma levels of triiodothyronine (T3) begin to increase two days prior to hatching and levels remain similar for several days post-hatching. THs, thyroid hormones. (Figure from De Groef et al. 2013; © 2013 De Groef, Grommen and Darras, open-access article distributed under the terms of the Creative Commons Attribution License, https://creativecommons.org/licenses/by/4.0/)

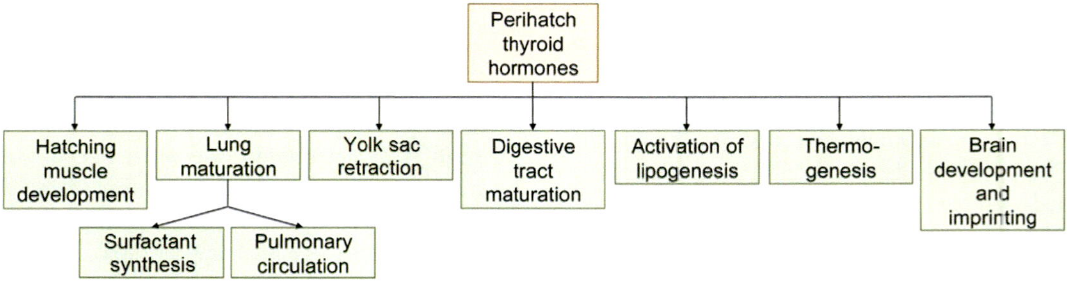


**Fig. 8.43** Overview of the roles of thyroid hormones in processes associated with hatching of precocial birds (Domestic Chicken). (Figure from De Groef et al. 2013; © 2013 De Groef, Grommen and Darras, open-access article distributed under the terms of the Creative Commons Attribution License, https://creativecommons.org/licenses/by/4.0/)

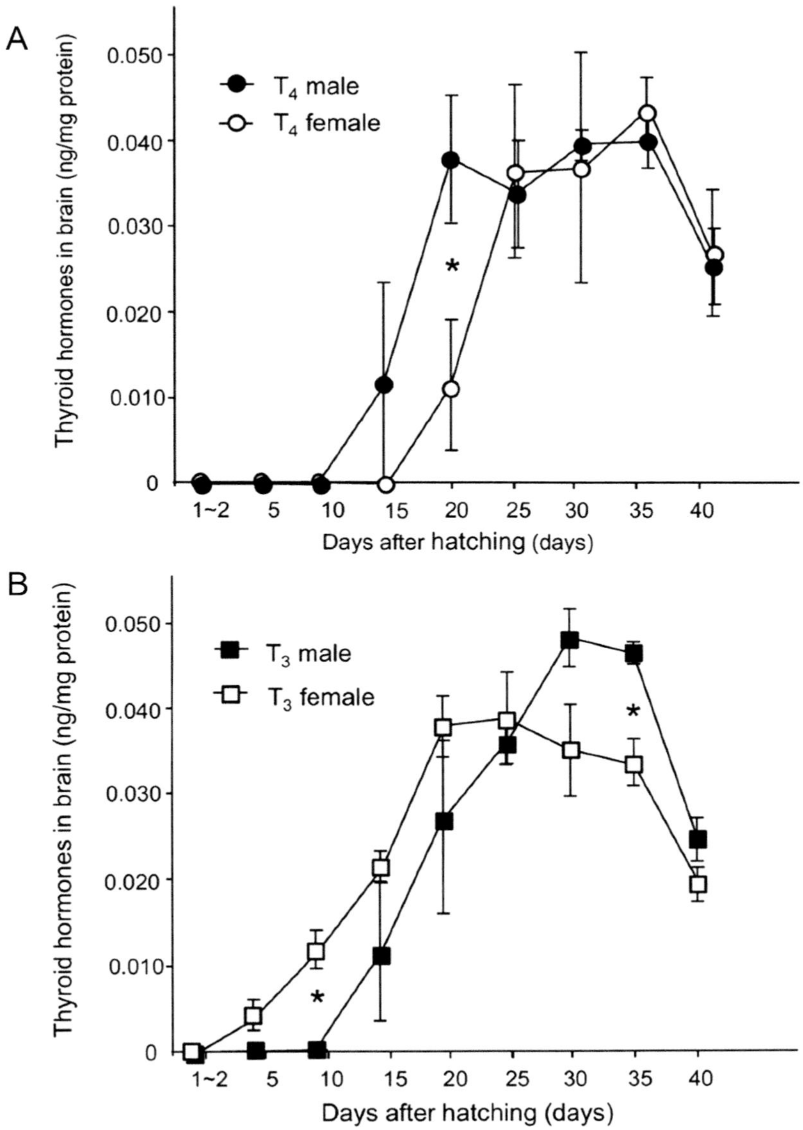


**Fig. 8.44** Plasma levels of thyroid hormones (T3 an T4) of Zebra Finches (*Taeniopygia guttata*) increase after hatching and peak at about 3 to 5 weeks post-hatch. Levels of thyroid hormones increase and peak faster for young males than young females. Young male Zebra Finches begin their critical song-learning period from about 25 to 60 days post-hatching and may start "singing" as early as 30-days post-hatching, so the earlier increase in brain thyroid hormones of males may be related to the early development of song learning and control areas in their brains. (Figure from Yamaguchi et al. 2017; © 2017 Yamaguchi et al., open-access article distributed under the terms of the Creative Commons Attribution License, https://creativecommons.org/licenses/by/4.0/)

Ghrelin, a peptide hormone, is secreted primarily by the proventriculus, with lesser amounts secreted by the gizzard (ventriculus) and the small intestine (Honda et al. 2017). Ghrelin stimulates feeding and food hoarding behavior in mammals (e.g., Keen-Rhinehart and Bartness 2005), but, when injected into either the cerebrospinal fluid or the blood, was found to inhibit feeding behavior in Domestic Chickens and Japanese Quail (*Coturnix japonica*; Shousha et al. 2005; Geelissen et al. 2006). In a more recent study, Goymann et al. (2017) studied Garden Warblers (*Sylvia borin*) at a stopover site during spring migration and found that injection of ghrelin reduced food intake by birds with lower body conditions, but had no effect on birds in good condition. In addition, Garden Warblers with more body fat had higher plasma levels of ghrelin. In contrast, Eikenaar et al. (2018) studied Eurasian Blackbirds (*Turdus merula*) at a

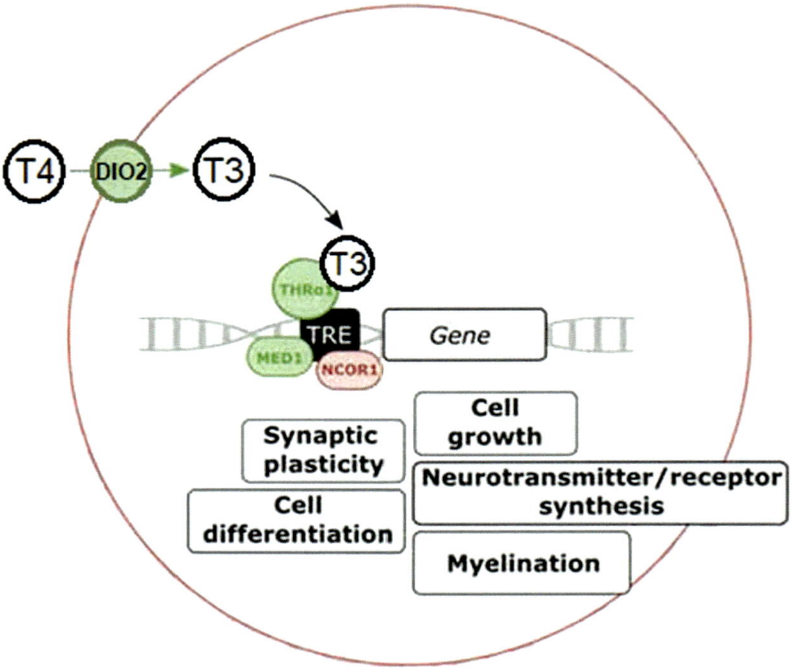


**Fig. 8.45** In the brain of young birds, T4 (thyroxine) is converted into T3 (triiodothyronine), which is more bioactive than T4, by DIO2, and T3 then activates signaling pathways that stimulate cell growth and differentiation, synthesis of neurotransmitters and myelin (myelination), and the formation of synaptic connections in the developing brain. (Figure from Batista and Hensch 2019; 2019 Batista and Hensch, open-access article distributed under the terms of the Creative Commons Attribution License (CC BY), https://creativecommons.org/licenses/by/4.0/)

stopover site during fall migration and found no relationship between plasma levels of ghrelin and lipid stores. However, these authors suggested that variation in the extent of food intake by migrants at stopover sites (especially birds like Eurasian Blackbirds that often consume large prey items) likely exhibits considerable variation that, in turn, might mask any relationship between lipid stores and plasma levels of ghrelin. In combination, the results of these studies suggest that artificially increasing plasma levels of ghrelin does inhibit the feeding behavior of birds, but that plasma levels in free-living birds fluctuate with condition and with time since birds last ate and how much they ate. Birds in poorer condition and/or those that have not eaten much recently will have lower ghrelin levels, which will stimulate feeding behavior, and those in good condition and/or those that have eaten recently will have high plasma levels of ghrelin, which will tend to inhibit feeding behavior. In other words, ghrelin "fine-tunes" food intake for birds (Goymann et al. 2017) and, for birds that fly, such fine-tuning is likely critical in terms of flight efficiency.

In addition to food consumption, ghrelin has also been found to influence avian caching behavior. Henderson et al. (2018) injected Coal Tits (*Periparus ater*), resident birds that cache food for later consumption, with high or low doses of ghrelin or saline (control) and found that low doses had no effect on caching behavior (as did injection of saline), but high doses of ghrelin caused Coal Tits to cache fewer items (sunflower seeds and peanut halves) (Fig. 8.71).

Leptin plays a critical role in regulating appetite and energy metabolism in mammals and other vertebrates, and administration of exogenous leptin has been found to reduce food intake in a number of mammals, reptiles, and fish (Londraville et al. 2014). In mammals, leptin is secreted primarily by adipose tissue and conveys information about the extent of lipid stores to the hypothalamus which, in turn, triggers behaviors that will either tend to decrease or increase those

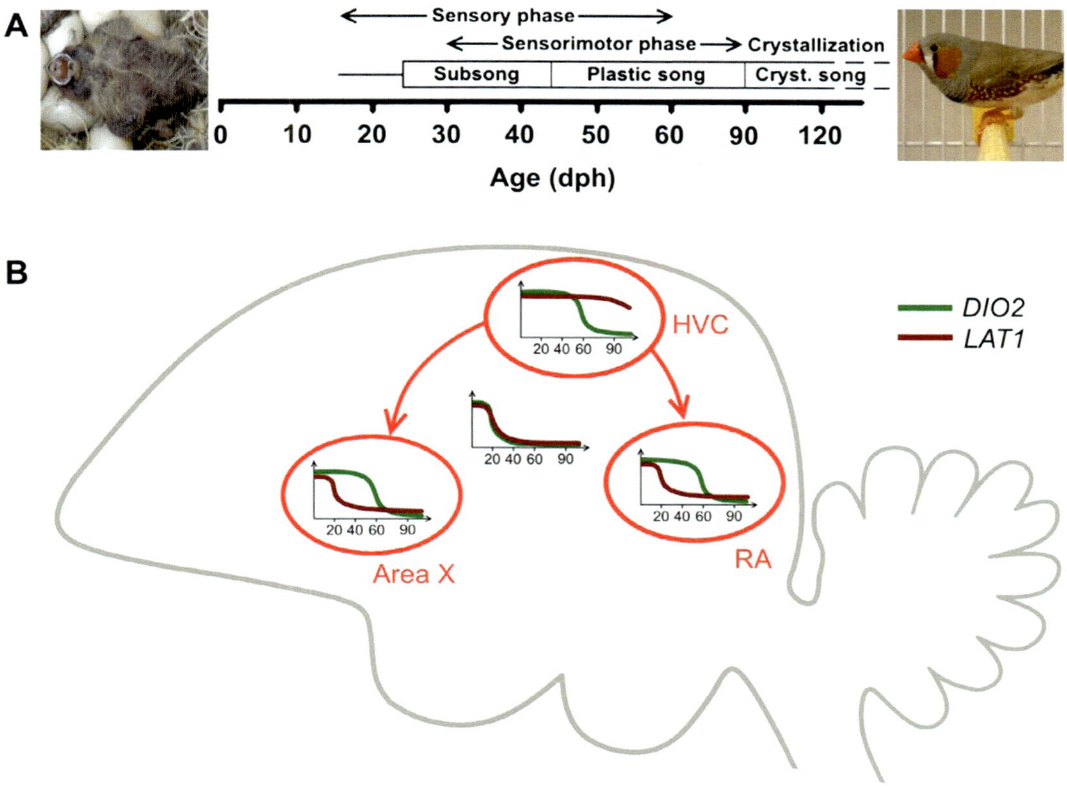


**Fig. 8.46** Levels of DIO2 and LAT1 expression in the song control nuclei of young male Zebra Finches (*Taeniopygia guttata*) during the song-learning period. LAT1 is found in cell membranes and transports thyroid hormones into cells. DIO2 converts T4 (thyroxine) into T3 (triiodothyronine); T3 is more bioactive than T4. (A) Timescale for song learning by male Zebra Finches. Sensory phase = critical period during which young males must be exposed to the songs of Zebra Finches. Males not exposed to songs during this period never develop typical Zebra Finch songs. Sensorimotor phase = young males begin to sing, trying to match aspects of songs that they have heard. Crystallization = song learning has ended. Males will now sing songs based on what they learned and practiced for the rest of their lives. Subsong = males sing, often at low volume; a period when males are just learning how to sing, using songs they have heard as a guide. Plastic song = songs become more adult-like, but are still variable because males are still in the process of developing the final version of their songs. (B) Graphs showing relative levels of DIO2 and LAT1 mRNA expression in song control nuclei in the brain, including Area X, HVC, and RA, compared to areas in the rest of the brain (represented by the central graph). Note that levels of expression remain higher for longer periods of time in the song control nuclei than in other parts of the brain, suggesting an important role for T3 in the song-learning process for male Zebra Finches. (Figure from Raymaekers and Darras 2017; © 2017 Elsevier Inc. All rights reserved, used with permission)

lipid stores (Friedman and Halaas 1998; Fig. 8.72).

Leptin was not discovered in birds until about two decades after its discovery in mammals (Huang et al. 2014), and the sequence of amino acids in avian leptin was found to differ substantially from that of mammalian (about 30% similar) and reptilian (about 35% similar) leptin (Friedman-Einat and Seroussi 2019). In addition, avian leptins also exhibit much interspecific variation, e.g., the amino-acid sequences of chicken and duck leptins are only 45% identical (Friedman-Einat and Seroussi 2019). Also, in contrast to mammals, avian leptin is not produced by adipose tissue, but is secreted primarily by the anterior pituitary (Fig. 8.73) and so does not

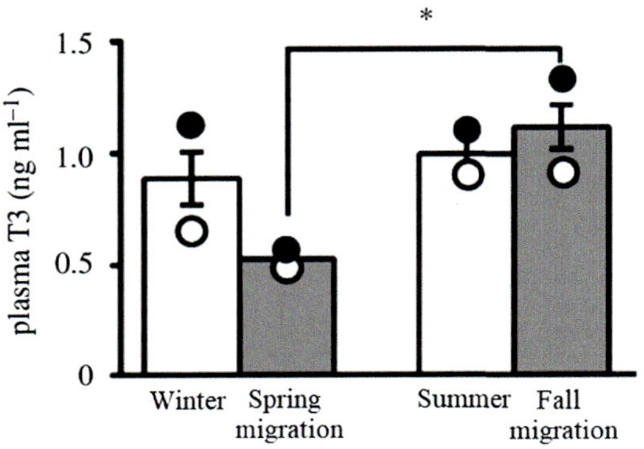


**Fig. 8.47** Plasma levels of triiodothyronine (T3) of Red-headed Buntings (*Emberiza bruniceps*) during winter and summer nonmigratory periods and during spring and fall migration. Closed and open circles indicate mean plasma levels during the night and day, respectively. (Figure from Sharma et al. 2018; © 2018 The Authors. Published by the Royal Society. All rights reserved, used with permission)

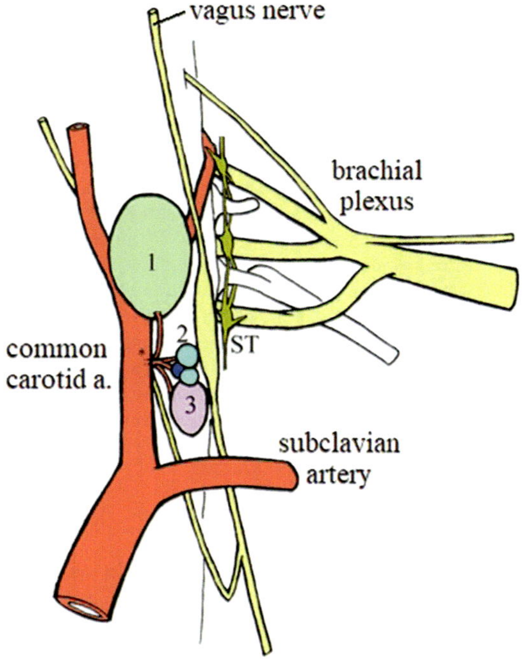


**Fig. 8.48** Drawing of the (1) thyroid, (2) parathyroid, and (3) ultimobranchial glands of a Domestic Chicken. ST, sympathetic trunk of the Autonomic Nervous System. (Figure modified from Kameda 2017; © 2017 Wiley Periodicals, Inc., used with permission)

provide birds with information about body stores of lipids as it does in mammals. However, the results of some studies suggest that leptin administration suppresses the feeding behavior of birds, e.g., Domestic Chickens (Denbow et al. 2000), Great Tits (*Parus major*; Lõhmus et al. 2003), Asian Blue Quail (*Coturnix chinensis*; Lõhmus et al. 2006), and White-throated Sparrows (*Zonotrichia albicollis*; Cerasale et al. 2011). Other investigators have reported no effect of leptin on food intake or body mass in Domestic Chickens (Sims et al. 2017), and Gertler et al. (2014) found that food intake by Domestic Chickens was not affected by treatment with a leptin-receptor blocker. In addition, Boswell and Dunn (2015) reviewed the literature on avian leptin and concluded that leptin ". . . may not even be involved in regulating energy balance" in birds. Also, Estienne et al. (2020) hypothesized that the level of avian leptin receptor expression in the hypothalamic-pituitary-gonadal and -adrenal axes suggests that leptin plays some role in regulating reproduction and avian stress responses rather than appetite and called for additional studies to investigate the possibilities. Interestingly, Lõhmus and Björklund (2009) found that treatment with leptin had a significant effect on the breeding behavior of female Great Tits, with five of 15 treated females and no untreated, control females ($N = 14$) starting second clutches within a week after treatment. These authors suggested that leptin treatment caused the females to "believe" they had sufficient resources to initiate a second nesting attempt. Although avian leptin does not have the same function (s) as mammalian leptin (i.e., leptin secreted by adipose tissue provides mammals with information about lipid stores), what that function or those functions might be remain to be determined.

Cholecystokinin (CCK) is produced and released from cells in the intestine, and the digestive products of fats, amino acids, and proteins stimulate the release of CCK by those cells (Mabayo et al. 1992; Furuse 1999). Cells with CCK receptors are located in the vagal nerve and pancreas. In mammals, CCK stimulates secretion by exocrine cells in the pancreas, but, although plasma CCK levels increase in

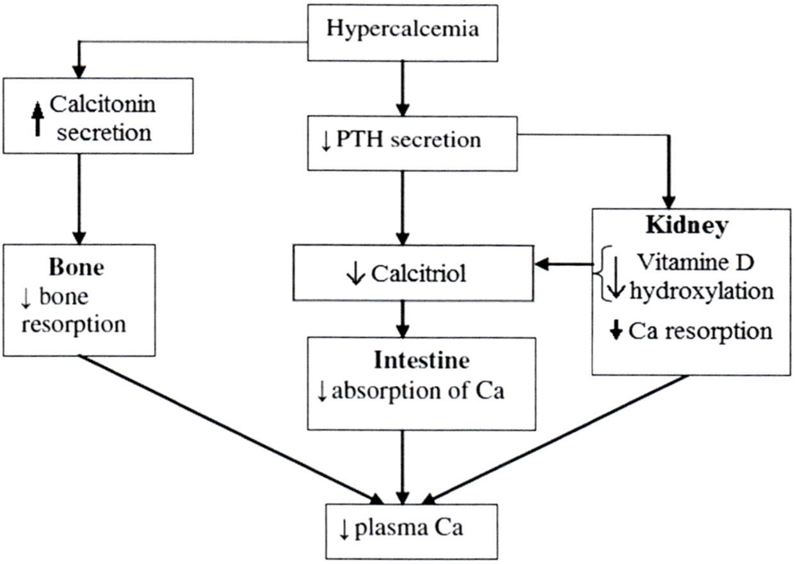


**Fig. 8.49** Role of calcitonin is the control of hypercalcemia. Calcitonin reduces plasma levels of calcium by stimulating bone deposition by osteoblasts. Hypercalcemia also inhibits the release of PTH by the parathyroid glands. Calcitriol = vitamin D; PTH, parathyroid hormone. (Figure modified from de Matos 2008; © 2008 Elsevier Inc., used with permission)

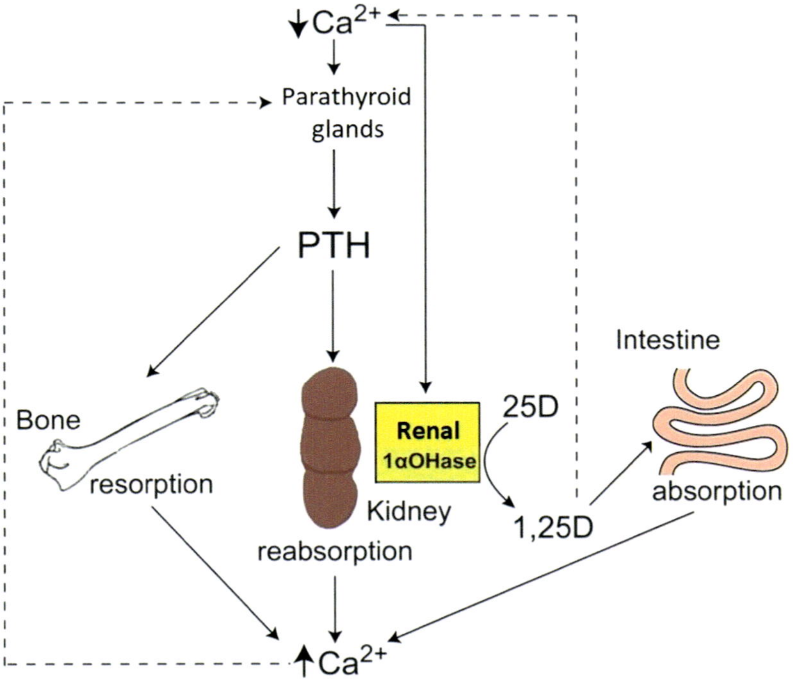


**Fig. 8.50** Role of parathyroid hormone (PTH) in the control of hypocalcemia. In response to low plasma levels of calcium, the parathyroid glands release PTH into the blood. PTH stimulates bone resorption, which releases calcium into the blood, and increased reabsorption of calcium by the kidneys. Low plasma levels of calcium also enhance expression of renal 1-α-hydroxylase (1αOHase) enzyme that converts 25-hydroxyvitamin D (25D) into biologically active 1,25-dihydroxyvitamin D (1,25D). Increased levels of 1,25D increase absorption of calcium in the small intestine by acting on the intestinal vitamin D receptor. (Figure modified from Hannan et al. 2019; © 2018 Springer Nature Limited, used with permission)

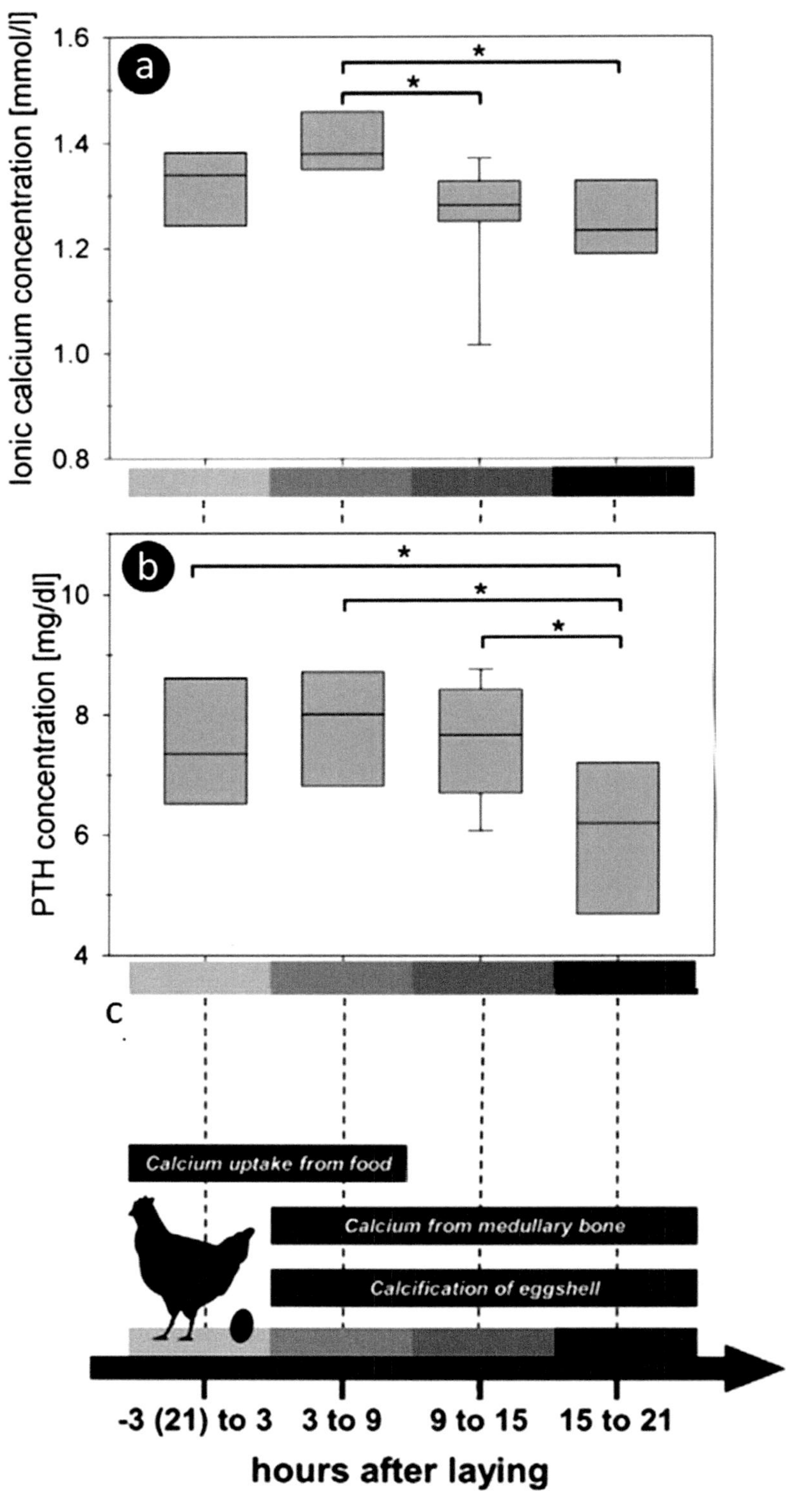


**Fig. 8.51** Calcium and parathyroid hormone levels in the blood of a female Domestic Chicken during the oviposition cycle. (**a**) Concentration of calcium ions increases during the early stages (until 9 h) after laying an egg as a result of consuming calcium-rich food. During later stages (9–21 h after laying), blood calcium levels decrease significantly as the eggshell calcifies. (**b**) Levels of parathyroid hormone (PTH) in the blood decline significantly during the later stages of the cycle as the need for calcium for the eggshell declines. (c) Sources of calcium and timing of eggshell calcification for a female Domestic Chicken during the oviposition cycle. * indicates statistically significant differences ($P < 0.05$). (Figures from Kerschnitzki et al. 2014; © 2014 Published by Elsevier Inc., used with permission)

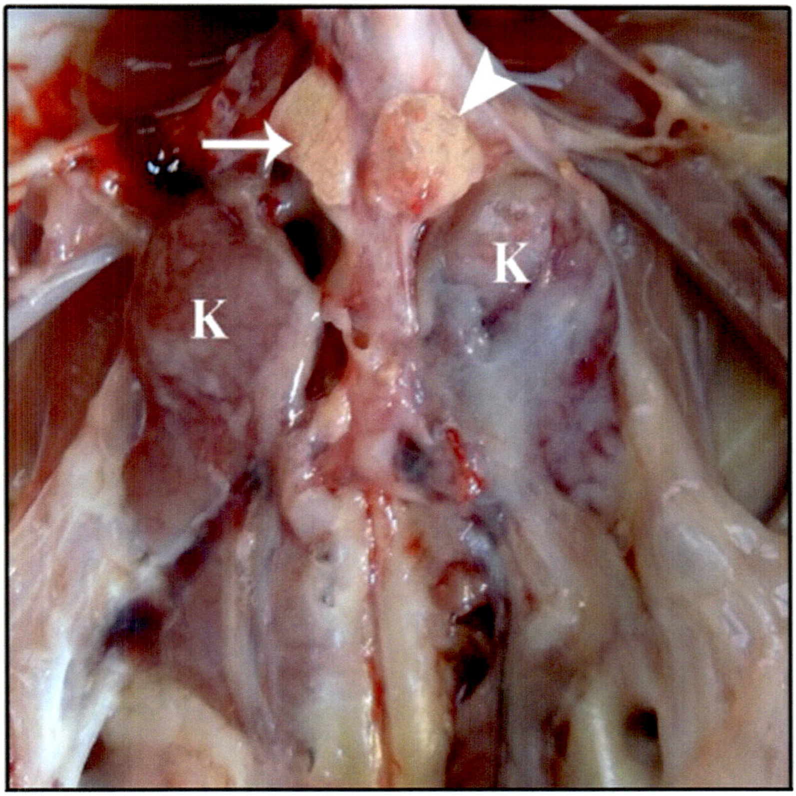


**Fig. 8.52** Adrenal glands of a male Domestic Chicken (at the arrow and arrowhead) are located just above the kidneys (K). (Figure from Moawad and Randa 2017; © 2017 Beni-Suef University. Production and hosting by Elsevier B.V., used with permission)

Domestic Chickens after ingestion of fats and proteins, levels do not increase sufficiently to stimulate the release of enzymes by pancreas exocrine cells (Wang and Cui 2007). Other investigators have reported similar results (e.g., Murai et al. 2000), so CCK does not appear to influence pancreatic secretion of enzymes in Domestic Chickens and, perhaps, birds in general. CCK has also been found to inhibit contraction of smooth muscle in the walls of the gizzard, but stimulate contractions at the gastroduodenal junction of Domestic Chickens (Martinez et al. 1993). In mammals, CCK suppresses feeding behavior, but Reid and Dunn (2018) found no difference between fed and fasted Domestic Chickens in plasma levels of CCK, suggesting that CCK neither suppresses nor stimulates their feeding behavior. Other investigators, however, have reported that CCK has an anorexigenic effect on Domestic Chickens (e.g., Tachibana et al. 2012) and White-crowned Sparrows (*Zonotrichia leucophyrs*; Richardson et al. 1993). Savory (1987) suggested that CCK-induced suppression of food intake in Domestic Chickens was an indirect effect due to malaise, with conditioned avoidance tests revealing that CCK caused a mild aversion. Given these mixed results, the effect of CCK on food intake by birds remains unclear. Finally, CCK stimulates the flow of bile from the gallbladder of Domestic Chickens which is important in the digestion and absorption of lipids in the intestine (DeGolier et al. 2013a).

Glucagon-like peptide (GLP)-1 is released from cells in the proventriculus and duodenum called L cells of Domestic Chickens after food ingestion (Richards and McMurtry 2008; Watanabe et al. 2018). In Domestic Chickens, Tachibana et al. (2003a) reported that GLP-1 regulated emptying of the crop, and Monir et al.

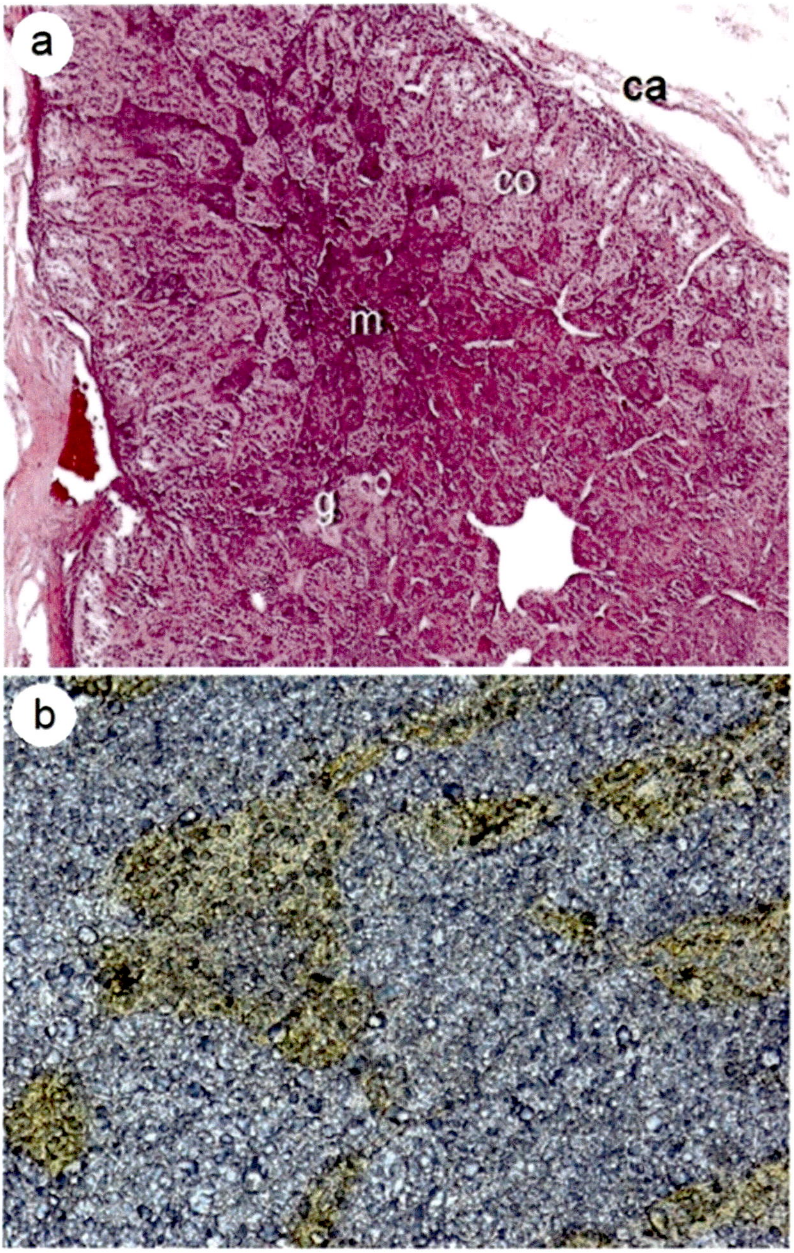


**Fig. 8.53** (**a**) Cross section of the adrenal gland of a Helmeted Guineafowl (*Numida meleagris*). The capsule consists of dense connective tissue. The peripheral portion of the adrenal consists largely of cortical cells with a few medullary cells. (**b**) Away from the periphery, medullary cells (yellow) are scattered among the cortical cells. ca, capsule; co, cortex; m, medulla; g, ganglion cell. (Figure from Moghadam and Mohammadpour 2017; © 2017 Springer-Verlag London Ltd., used with permission)

(2014) suggested that "GLP-1 secreted at the distal small intestine enhances the absorption of nutrients at the proximal portion by inhibiting the intestinal motility in the chicken alimentary tract." In addition, ICV injection of GLP-1 was found to reduce food intake by neonatal Domestic Chickens (Furuse et al. 1997) and Japanese Quail (Shousha et al. 2007).

Watanabe et al. (2018) also found that GLP-1 stimulated the secretion of somatostatin by the pancreas of Domestic Chickens, Northern Bob-white (*Colinus virginianus*), and Common

**Table 8.2** Hypotheses concerning how baseline and stress-induced levels of glucocorticoids might affect a bird's fitness (Table from Schoenle et al. 2018; © 2018 Oxford University Press, used with permission)

| Hypothesis | Relevant HPA axis component | Description | Source(s) |
|---|---|---|---|
| Non-adaptive stress | Baseline and stress-induced GCs | Maintaining high GCs for extended periods impairs reproduction and increases susceptibility to disease | Selye (1955), Christian (1980), Boonstra and Boag (1992) |
| Adaptive stress | Baseline and stress-induced GCs | GCs catalyze energy stores, reducing the need to forage, allowing animals to invest more time in reproduction at the cost of survival | Lee and Cockburn (1985), Boonstra and Boag (1992) |
| Alternate coping strategies | Baseline and stress-induced GCs | Selection maintains proactive and reacting coping styles associated with different HPA axix reactivities. The coping style positively associated with fitness depends on environmental conditions. | Koolhaas et al. (1999), Korte et al. (2005) |
| CORT-trade-Off | Stress-induced GCs | Stress-induced GCs mediate a life-history trade-off between survival and reproduction in favor of survival. This hypothesis was initially applied to stress-induced GCs, but has also been proposed for baseline GCs. | Wingfield and Sapolsky (2003) |
| Allostasis model | Baseline and stress-induced GCs | Environmental and life-history changes can increase the work required to maintain physiological constancy ("allostatic load"). If the energy demand required to maintain homeostasis exceeds available energy, an emergency life-history stage is initiated. If the allostatic load is too high for too long, pathological damage can result. | McEwen and Wingfield (2003) |
| CORT-fitness | Baseline GCs | Individuals or populations with elevated baseline GCs are facing greater challenges and/or are in worse condition and, therefore, will have lower fitness than those with lower baseline GCs. | Bonier et al. (2009) |
| CORT-adaptation (extension of CORT-fitness hypothesis) | Baseline GCs | When individuals face challenges associated with reproduction (e.g., feeding offspring), elevated GCs support increased reproductive effort. Thus, individuals with higher or intermediate GC concentrations will have greater reproductive success than those with lower GCs. | Bonier et al. (2009) |
| Reactive scope model | Baseline and stress-induced GCs, negative feedback | Physiological mediators (like GCs) associated with the stress response have a normal range of activity called the reactive scope. The reactive scope encompasses baseline circulating levels and stress-induced levels in the range that the mediator does not cause damage to the animal. If hormone concentrations go above or below the reactive scope, the animal | Romero et al. (2009) |

(continued)

**Table 8.2** (continued)

| Hypothesis | Relevant HPA axis component | Description | Source(s) |
|---|---|---|---|
| | | accumulates mediator-caused damage that could reduce fitness. | |
| Adaptive calibration model | Total physiological response to stressors (sympathetic and parasympathetic nervous systems and the HPA axis) | Organisms have evolved stress response systems that can be modified during development and throughout life to match their physical and social environments. Thus, individual variation in stress responsiveness is the result of an adaptive process and is the best possible response for an organism for their previous experiences. | Del Giudice et al. (2011) |
| Scope of flexibility hypothesis | Baseline and stress-induced GCs | If individuals who have a greater scope of flexibility (larger range of potential hormone concentrations) are better able to match the optimal endocrine phenotype across diverse conditions, they will have higher fitness in variable environments. | Hau et al. (2016), Taff and Vitousek (2016) |
| Speed of flexibility hypothesis | Baseline and stress-induced GCs, negative feedback, total vs. free GCs | Changes in environmental conditions will result in a mismatch between the optimal and expressed endocrine phenotypes. In moderately dynamic environments, individuals who can adjust their endocrine phenotype faster will generally have higher fitness. | Taff and Vitousek (2016) |

Ostriches (*Struthio camelus*) by binding to receptors in the membranes of pancreatic D-cells (Fig. 8.74) and suggested that somatostatin played an important role in the regulation of blood glucose levels. More specifically, somatostatin stimulates the uptake of glucose by the small intestine (Abalde et al. 1984) and inhibits glucagon-stimulated lipolysis in the adipose tissues of chickens (Di Scala-Guenot et al. 1985). In Ostriches, somatostatin is produced by cells in the mucosal layers of the proventriculus, duodenum, jejunum, and ileum (Chen et al. 2013). Tachibana et al. (2009) found that somatostatin stimulated the feeding behavior of young Domestic Chickens, and Yousefvand et al. (2019) reported similar results.

GLP-2, like GLP-1, is secreted by enteroendocrine L cells in the small and large intestine (Burrin et al. 2003) and was found to stimulate growth of Domestic Chickens after intraperitoneal injection (Hu et al. 2010). In addition, intraperitoneal injection of GLP-2 in Domestic Chickens did not influence feed intake, but caused an increase in height of villi in the small intestine, increased expression of a sodium and glucose cotransporter, and increased expression of a peptide transporter in the duodenum and jejunum (Hu et al. 2010).

Peptide tyrosine-tyrosine (PYY) is produced and secreted by cells in the avian small intestine and, primarily, the pancreas (Reid et al. 2017) and, in an early study to determine the possible effect of PYY of feeding behavior, Kuenzel et al. (1987) reported that injection of PYY increased food intake by Domestic Chickens. However, more recent studies have revealed that injection of PYY suppresses food intake by Domestic Chickens (Aoki et al. 2017). Melo-Duran et al. (2019) found that PYY levels in fasted Domestic Chickens declined over a 12-hour period, whereas PPY levels for those fed *ad libitum* remained constant, suggesting that higher plasma

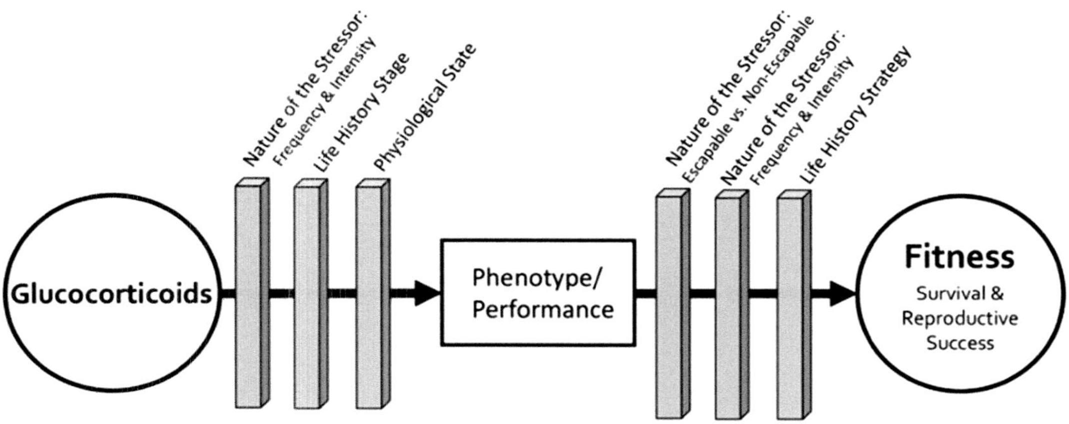


**Fig. 8.54** The relationship between glucocorticoids (GCs) and fitness will depend on (1) factors that affect how GCs influence phenotype and performance and (2) factors that influence the relationship between GC-mediated traits and fitness. The effects of GCs on phenotype and performance can vary depending on the frequency and intensity of stressors, the current life-history stage (e.g., juvenile vs. adult, breeding vs. nonbreeding), and the animal's physiological state (e.g., starvation, presence of parasite infection). The effect of GC-mediated phenotypes/performance on fitness will depend on whether a stressor is escapable or not, the overall frequency and intensity of the stressors, and life-history strategy (e.g., lifespan and value of a reproductive bout). (Figure from Schoenle et al. 2018; © 2018 Oxford University Press, used with permission)

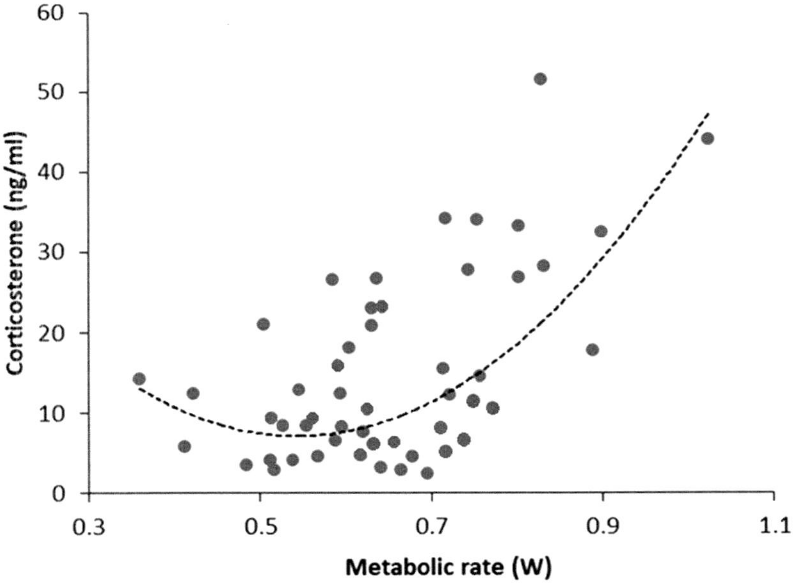


**Fig. 8.55** Relationship between plasma concentration of corticosterone and metabolic rate of Zebra Finches (*Taeniopygia guttata*). Variation in levels of corticosterone is explained by variation in metabolic rate. (Figure modified from Jimeno et al. 2018; open-access article licensed under a Creative Commons Attribution 4.0 International License, https://creativecommons.org/licenses/by/4.0/)

levels of PYY suppress feeding behavior and that low levels may stimulate feeding. Similarly, Reid et al. (2017) suggested that PYY acts as a short-term satiety signal in Domestic Chickens.

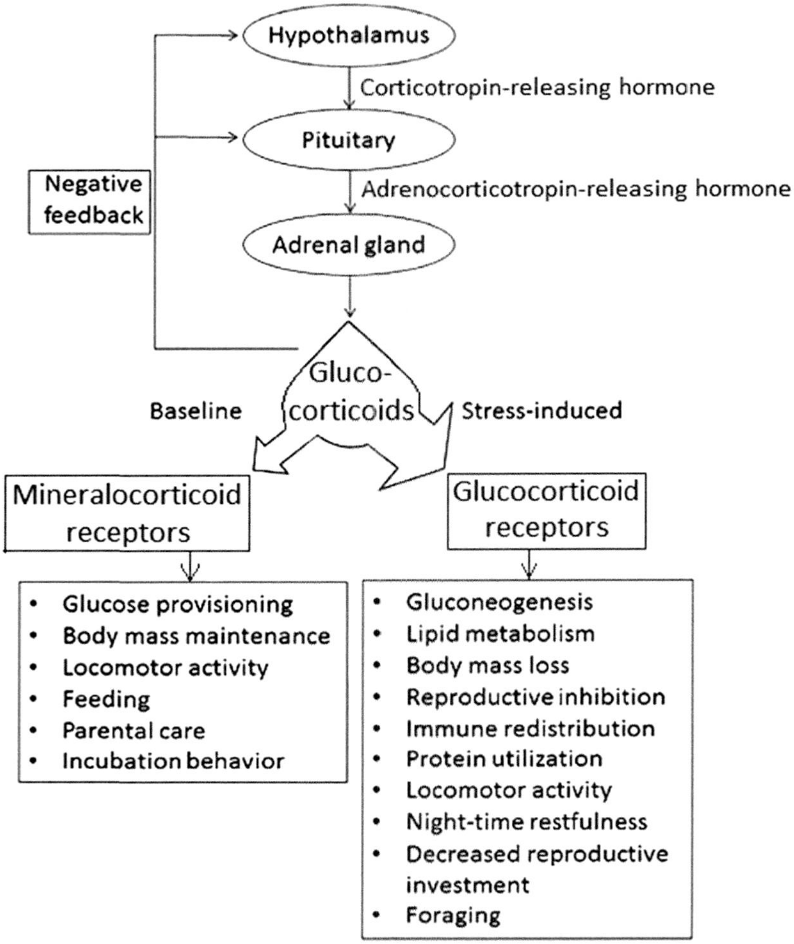


**Fig. 8.56** Glucocorticoid production pathways that result from activating the hypothalamic-pituitary-adrenal axis and the effects of glucocorticoids. Negative feedback can occur via both receptors. (Figure modified from Hau et al. 2016; © 2016 Elsevier Inc., used with permission)

Neuropeptide Y (NPY), a 36-amino acid peptide, is widely distributed in the central nervous systems of several species of mammals and birds (Kuenzel et al. 1999), as well as in skeletal muscle where it may play a role in the modulation of mitochondrial function (Damad et al. 2021). Injection of NPY has been found to stimulate feeding behavior by White-crowned Sparrows (*Zonotrichia leucophrys*; Richardson et al. 1995), Ringed Turtle-Doves (*Streptopelia risoria*; Strader and Buntin 2001), and Japanese Quail (*Coturnix japonica*; McConn et al. 2018), and stimulate food intake and locomotor activity by young Domestic Chickens (Chen et al. 2016). In addition, Zhou et al. (2005) reported that fasting increased concentrations of NPY in the hypothalamus of young Domestic Chickens, and Lees et al. (2017) reported similar results for Red Junglefowl (*Gallus gallus*; Fig. 8.75). Ramakrishnan et al. (2007) found that numbers of NPY cells and NPY mRNA levels in the hypothalamus of male and female Ring-necked Doves (*Streptopelia capicola*) were significantly higher during a period of parental hyperphagia after eggs hatched than during the incubation and hatching periods when food intake did not change (Fig. 8.76).

Agouti-related peptide (AgRP) is another hormone that stimulates food intake by birds. Food restriction has been found to increase AgRP expression in the hypothalamus of Domestic Chickens (Dunn et al. 2013) and Red Junglefowl

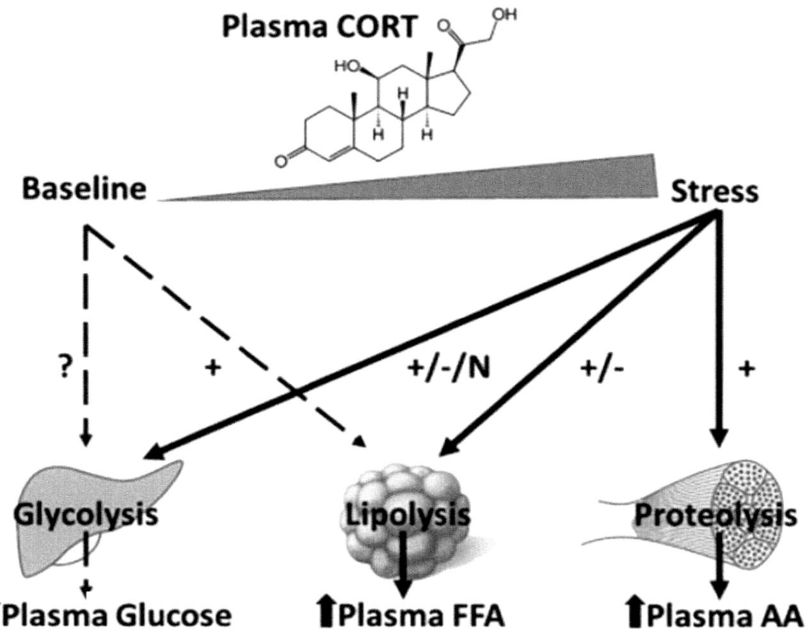


**Fig. 8.57** Role of corticosterone (CORT) in regulating avian energy metabolism under normal conditions (baseline) and during periods of stress. Arrows indicate possible pathways regulated by CORT, with solid arrows indicating substantial evidence and dashed arrows indicating weak evidence. Baseline plasma CORT stimulates (+) lipolysis, has no or an unknown (?) effect on glycolysis, and does not influence proteolysis. At elevated (stress) levels, plasma CORT either stimulates, inhibits (-), or has no effect (N) on glycolysis, stimulates or inhibits lipolysis, and stimulates proteolysis. (Figure from Deviche et al. 2017; © 2017 Oxford University Press, used with permission)

(Lees et al. 2017; Fig. 8.75a) and, after a period of food restriction, two days of ad libitum feeding was found to decrease AgRP expression in the hypothalamus by about 75% (Lees et al. 2017). The time needed to reach levels of AgRP expression in the hypothalamus after periods of food restriction for Red Junglefowl "appears to be proportional to the duration of prior restriction," suggesting that AgRP expression indicates "a bird's deviation from a desired physiological setpoint, most likely in terms of peripheral energy stores" (Lees et al. 2017).

Pro-opiomelanocortin (POMC) gene expression tends to decrease during periods of food restriction. For example, Hen et al. (2006) reported a significant decrease in POMC mRNA after female Domestic Chickens were food-restricted for seven days (Fig. 8.75b). Similarly, Higgins et al. (2010) observed a significant decrease in mRNA after food deprivation and other investigators have reported nonsignificant decreases (Phillips-Singh et al. 2003; Song et al. 2012). Boswell and Dunn (2015) moved Domestic Chickens from individual housing in cages to social housing in pens to increase their stress levels and, after six days, found both a 70% reduction in food intake and a significant increase in POMC gene expression. Such results suggest that POMC may inhibit food intake during periods of stress (Boswell and Dunn 2015), and Liu et al. (2007) hypothesized that this may also be the case in mammals.

POMC may also play a role in regulating food intake during periods when birds have limited, or no, time to feed, e.g., during long-distance migratory flights, when establishing and defending territories, and when incubating eggs or brooding young. For example, Dunn et al. (2015) compared

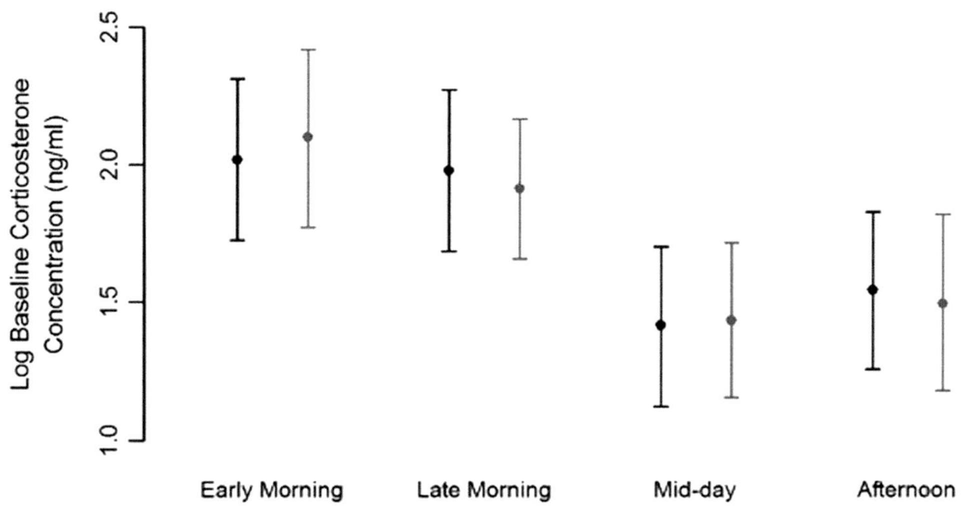


**Fig. 8.58** Mean concentrations (± 95% credible intervals) of baseline corticosterone levels in blood samples of birds captured at different times of day. Black points denote time intervals selected to partition data into sets of equal sample size ($n$ = 23, 23, 23 and 24, respectively). Gray points sort data into sets of equal length in time (i.e., 4 sequential intervals of 158 min each; $n$ = 19, 30, 25, and 19). (Figure from Schwabl et al. 2016; open-access article distributed under the terms of the Creative Commons CC BY license, https://creativecommons.org/licenses/by/4.0/)

neuropeptide gene expression in female Domestic Chicken that either did or did not begin to incubate eggs. Females initiating incubation exhibited a 60-70% reduction in food intake (Fig. 8.77), as well as a significant increase in the gene expression of POMC. These authors also found that incubating females exhibited increased gene expression for the orexigenic peptide agouti-related peptide (AgRP), but, given their reduced food intake, the effects of AgRP were apparently overridden by the anorexigenic effects of POMC.

Orexin neuropeptides were first identified in the brain and found to induce appetite and regulator feeding behavior and energy homeostasis in mammals (Li et al. 2014). In birds, however, mammalian orexins do not stimulate food intake, e.g., Rock Pigeons (*Columba livia*; da Silva et al. 2008) and young Domestic Chickens (Furuse et al. 1999a). Because avian orexins may differ in structure from those of mammals, additional studies are needed to determine what effect, if any, avian orexins might have on avian feeding behavior (Wong et al. 2011). Orexins have been found in the hypothalamus, gonads, the digestive system, liver, and muscle of birds (Lassiter et al. 2015). In the liver, Greene et al. (2020) found that orexins induce lipogenesis in the liver cells (hepatocytes). In avian skeletal muscles, orexin increases glucose metabolism and their sensitivity to insulin (Shiuchi et al. 2009). In mammals, orexins also help regulate sleep/wake cycles (e.g., Willie et al. 2003) and may play a similar role in birds because injection of mammalian orexins causes increased alertness in Domestic Chickens (Katayama et al. 2010) and increased exploratory movement and wing-flapping in Rock Pigeons (da Silva et al. 2008).

Serotonin is produced in the brain where it serves as a neurotransmitter, and by cells in the digestive system, particularly the duodenum, upper ileum, and rectum (Salvi et al. 1996). Serotonin plays an important role in the movement of food through the alimentary canal of Domestic Chickens by causing contraction of smooth muscle in the walls of the esophagus, proventriculus, ileum, and rectum (Mishra and Raviprakash 1983; Gwee et al. 1995; Kitazawa et al. 2006). In addition, several investigators have found that

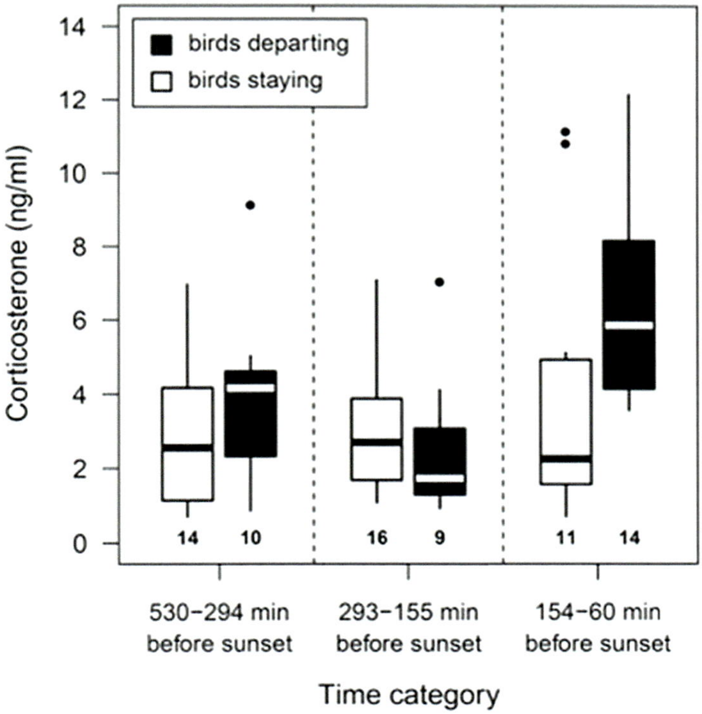


**Fig. 8.59** Top, Male Northern Wheatear (*Oenanthe oenanthe*). Bottom, Plasma concentration of corticosterone of Northern Wheatears sampled at different times prior to sunset when they would initiate their migratory flight. Corticosterone levels of individuals that left the stopover site the first night after corticosterone sampling had significantly higher plasma levels than those that did not depart. The increased levels of corticosterone for birds initiating migratory flight prepare them for the upcoming increase in locomotor activity and energy demands. These results also suggest that decisions about whether to initiate migratory flights are made in the hours just prior to the possible departure time when information about important cues that could affect the success of a migratory flight, such as energy stores and weather conditions, are available. Plotted are the median (black and white horizontal bars), 5th, 25th, 75th, and 95th percentiles. Black dots indicate values beyond the 5th or 95th percentiles. Numbers below boxes indicate the sample sizes. (Photo by Ron Knight, Wikipedia, CC BY 2.0, https://creativecommons.org/licenses/by/2.0/. Figure from Eikenaar et al. 2020; © 2020 Elsevier Inc., used with permission)

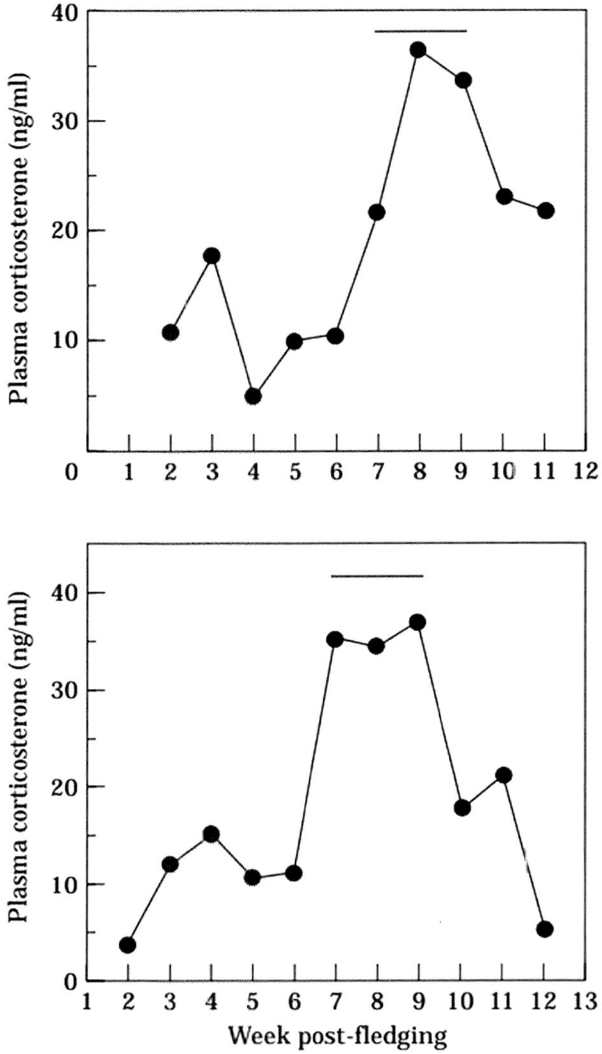


**Fig. 8.60** Concentrations of corticosterone in the plasma of two captive Western Screech-Owls (*Megascops kennicottii*) during the post-fledging period. Bars indicate the 3-week periods when owls exhibited peak locomotor activity levels (measured by backpack pedometers). Young, free-living Western Screech-Owls disperse from natal territories an average of eight weeks post-fledging. Increasing plasma levels of corticosterone may trigger "dispersal restlessness" and dispersal from natal territories. (Figure from Belthoff and Dufty 1998; © 1998 The Association for the Study of Animal Behaviour. Published by Elsevier Ltd., used with permission)

serotonin-treated Domestic Chickens reduced their food intake (e.g., Zhang et al. 2017b; Zendehdel et al. 2017). In contrast, Yousefi et al. (2019) found that introcerebroventricular injection of serotonin had no effect on food intake by five-day old Domestic Chickens. One possible explanation for such conflicting results is that serotonin-treatment causes a decrease in the activity of chickens and an increase in resting behavior, so the possible anorexigenic effects of serotonin may be a side effect, with less active chickens tending to eat less (Zhang et al. 2017a, b).

Gastrin is produced by cells in the small intestine, especially in the duodenum (Fig. 8.78), as well as by cells in the proventriculus and gizzard (Saito et al. 1989; Rawdon and Andrew 1999; Aksoy and Cinar 2009), and is released in response to the presence of amino acids and fatty acids (Furuse and Dockray 1995). Intracerebroventricularly injected gastrin has been found to inhibit food intake by young Domestic Chickens (Furuse et al. 1999b; Bohler et al. 2019). In addition to being anorexigenic, gastrin stimulates secretion of hydrochloric acid and pepsinogen in the proventriculus of birds

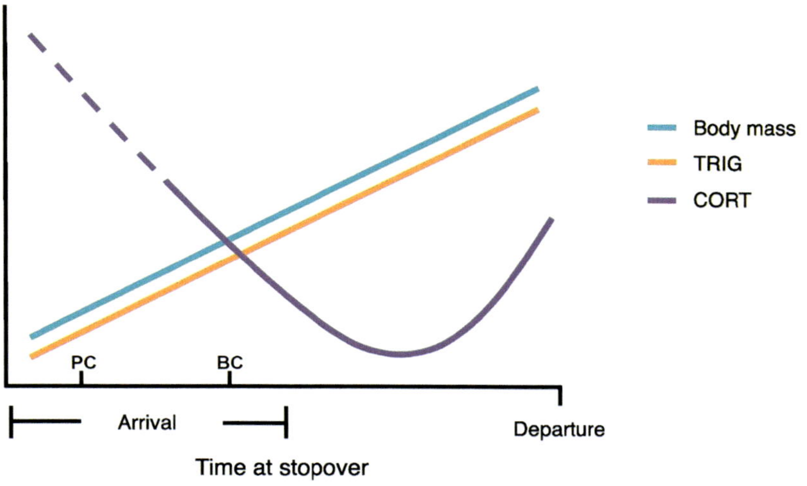


**Fig. 8.61** Predicted relationships among body mass, plasma levels of triglycerides (TRIG, i.e., the energy source), and plasma levels of CORT at a stopover site, from arrival at the stopover site on the left to departure on the right. Birds that arrive in better condition (BC) exhibit the same trajectory for body condition and TRIG and CORT levels, but begin that process further along (further to the right along the x-axis) than birds arriving in poorer condition (PC). CORT can reach stress-induced levels when birds arrive at a stopover site if they are in very poor energetic condition, as indicated by the dashed part of the CORT curve; CORT levels of birds that arrive in better energetic condition may be lower (solid line). CORT levels decrease with increasing body mass and energy stores (TRIG), but then increase as the time of departure approaches. (Figure from DeSimone et al. 2020; © 2020 Elsevier Inc., used with permission)

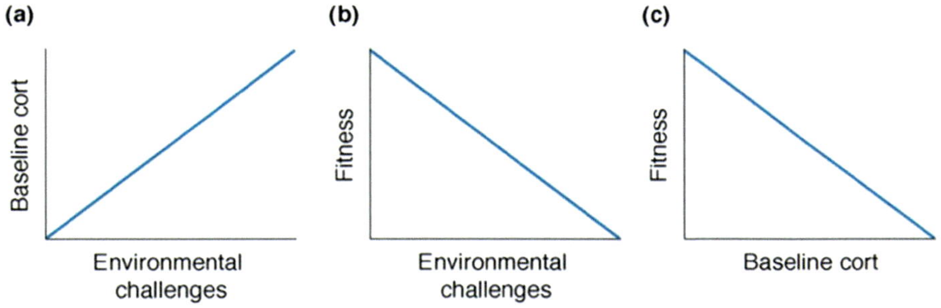


**Fig. 8.62** CORT–Fitness Hypothesis. (**a**) Baseline plasma levels of CORT are predicted to increase with increasing environmental challenges. (**b**) Increasing environmental challenges reduce fitness because available resources must be used to cope with those challenges rather than for either reproduction or self-maintenance. (**c**) Given a and b, the CORT–Fitness Hypothesis predicts a negative relationship between baseline CORT and fitness. (Figure from Bonier et al. 2009; © 2009 Elsevier Ltd., used with permission)

(Rawdon and Andrew 1981; Campbell et al. 1994).

Other anorexigenic neuropeptides that can suppress food intake include cocaine- and amphetamine-regulated transcript (CART) in young Domestic Chickens (Tachibana et al. 2003b), α- and β-melanocyte-stimulating-hormone in Domestic Chickens (Saneyasu et al. 2011; Cao et al. 2020), growth hormone-releasing hormone in Domestic Chickens (Tachibana et al.

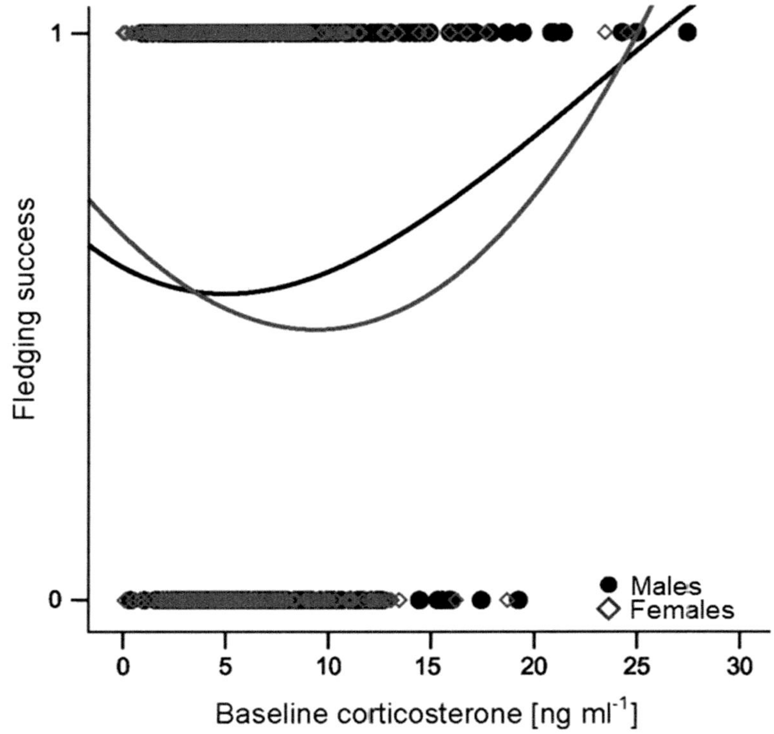


**Fig. 8.63** Relationship between baseline concentrations of CORT and fledging success (0 = no fledglings and 1 = at least one fledgling) of male and female Common Terns (*Sterna hirundo*). Curves (black, males; gray, females) show the best fit to the data. Black dots are not always visible through gray diamonds. (Figure from Riechert et al. 2014; © 2014 Springer-Verlag Berlin Heidelberg, used with permission)

2015), corticotropin-releasing factor in Domestic Chickens (Wang et al. 2019), Substance P in Japanese Quail (*Coturnix japonica*; Pauliukonis et al. 2020), oxyntomodulin in Domestic Chickens (Honda et al. 2014), vasoactive intestinal peptide in young Domestic Chickens (Khan et al. 2013) and Japanese Quail (Bohler et al. 2021), and neuromedin U/S in Japanese Quail and Domestic Chickens (Wan et al. 2018). In addition, Ukena et al. (2014) identified a small secretory protein that they called neurosecretory protein GL (NPGL) and found that subcutaneous infusion of NPGL caused young Domestic Chickens to gain body mass without affecting food intake. More recently, Shikano et al. (2019) reported that intracerebroventricular infusion of NPGL resulted in an increase in body mass of young Domestic Chickens due to increased mass of the liver, abdominal fat, and subcutaneous fat and, in contrast to the results reported by Ukena et al. (2014), this was accompanied by an increase in food intake. Finally, intracerebroventricular injection of galanin, a peptide found in the hypothalamus, was found to increase food intake by young Domestic Chickens (Tachibana et al. 2008).

As is clear from the preceding paragraphs, a number of neuropeptide hormones appear to affect the feeding behavior of birds. However, most studies of these hormones have involved injection of particular hormones and most have been conducted with neonatal or growing young chickens. As such, many questions remain about how these hormones actually interact to influence avian feeding behavior and how these interactions might vary with the age or sex of birds or with season. Much additional research is clearly needed.

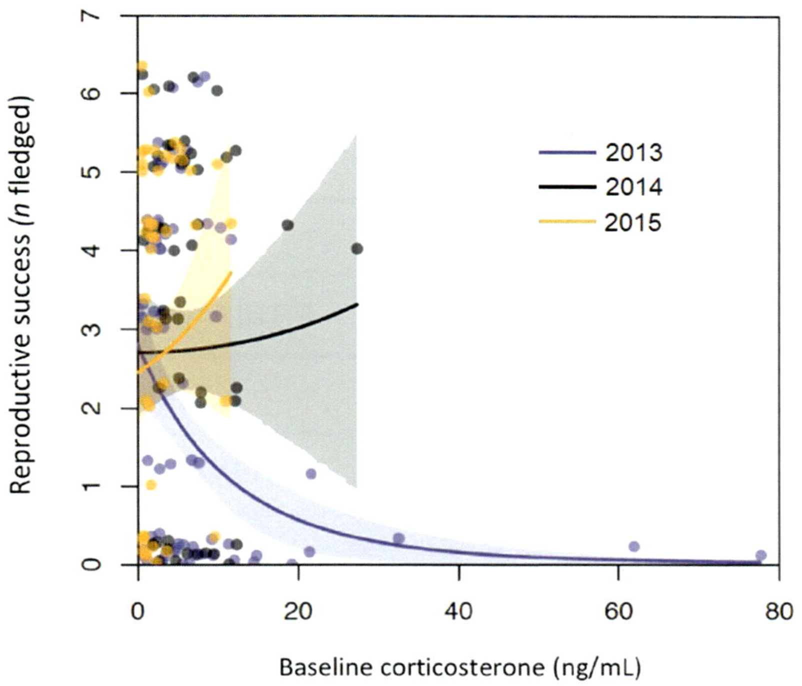


**Fig. 8.64** Baseline CORT levels and reproductive success of female Tree Swallows (*Tachycineta bicolor*) over three breeding seasons. Baseline levels were negatively associated with reproductive success (number of young fledged) in 2013, but not in either 2014 or 2015. During the 2013 breeding season, periods of cold weather reduced availability of insect prey and fledging success and females with higher baseline CORT levels during incubation fledged fewer young; this relationship was not apparent in either 2014 or 2015. (Figure from Vitousek et al. 2018; © 2018 Vitousek, Taff, Hallinger, Zimmer and Winkler, open-access article distributed under the terms of the Creative Commons Attribution License (CC BY), https://creativecommons.org/licenses/by/4.0/)

## 8.10 Gastrointestinal Hormones and Digestion

Some hormones produced by cells in the avian digestive tract do not affect feeding behavior, but do play a role in the process of digestion. One such hormone is neurotensin that is produced by enteroendocrine cells located along the entire intestinal tract of Domestic Chickens, but particularly in the large intestine (Numao et al. 2011). Secretion of neurotensin is stimulated by the presence of lipids, particularly oleic acid (DeGolier et al. 2013b). Neurotensin has been found to stimulate contraction of smooth muscle in the walls of the crop of Domestic Chickens (Denac and Scharrer 1987), inhibit gastric motility and pepsin secretion (DeGolier et al. 1997), increase activity of pancreatic lipase (DeGolier et al. 1999), and stimulate the production of bile (Gui et al. 2000) and its release into the small intestine from the gallbladder (DeGolier et al. 2013a). Bile emulsifies lipids, breaking them down into smaller particles with more surface area that enhances the action of lipolytic enzymes.

Motilin, a 22-amino-acid peptide, has been identified in Domestic Chickens, Japanese Quail, and Ring-necked Pheasants (*Phasianus colchicus*) and is produced primarily by cells in the walls of the duodenum and, by helping to regulate motility of the gastrointestinal tract, plays an important role in digestion and the absorption of nutrients (Kitazawa and Kaiya 2019). Motilin causes contraction of smooth muscle in the walls of all three sections of the small

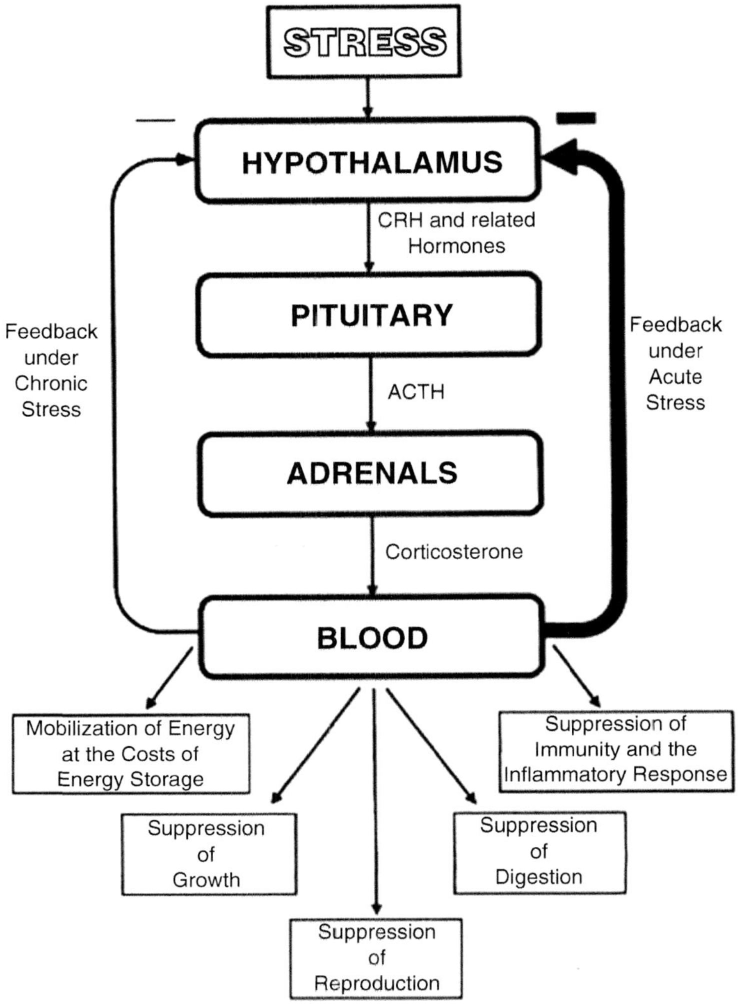


**Fig. 8.65** The hypothalamic-pituitary-adrenal (HPA) axis, the negative feedback response of birds to an acute stressor compared to those under chronic stress, and the major impacts stressors have on bodily processes. Also, note the negative feedback system, as levels of CORT in the blood increase, the hypothalamus responds by releasing less corticotropin-releasing hormone (CRH) which, in turn, reduces secretion of ACTH by the pituitary and CORT by the adrenal glands. (Figure from Sheriff et al. 2011; © 2011 Springer-Verlag, used with permission)

intestine and, to a lesser extent, the proventriculus, but smooth muscle in the walls of the crop and large intestine (colon) is less sensitive to motilin (Kitazawa and Kaiya 2019; Zhang et al. 2020; Fig. 8.79). Ghrelin, an anorexigenic hormone, also plays a role in regulating motility of the gastrointestinal tract of Domestic Chickens and Japanese Quail. Ghrelin stimulates contraction of smooth muscle in the walls of the crop, proventriculus, and large intestine, but not the small intestine (Kitazawa and Kaiya 2019). In addition, galanin has been found to stimulate contractions of longitudinal smooth muscle in the walls of ceca and the large intestine (DeGolier et al. 1999), and vasoactive intestinal peptide (among its other functions discussed earlier in this chapter), was found to stimulate secretion of pancreatic juice in Domestic Turkeys (Dimaline and Dockray 1979).

In Domestic Chickens, secretin receptors are found in the proventriculus, gizzard, duodenum, jejunum, ileum, and colon and, to a lesser degree, in the pancreas (Wang et al. 2012). Secretin plays a minor role in stimulating secretions of the pancreas in Domestic Chickens (Duke et al. 1987). However, Burhol (1974) found that intravenous injection of secretin caused a significant increase in both $H^+$ and pepsin in the proventriculus of

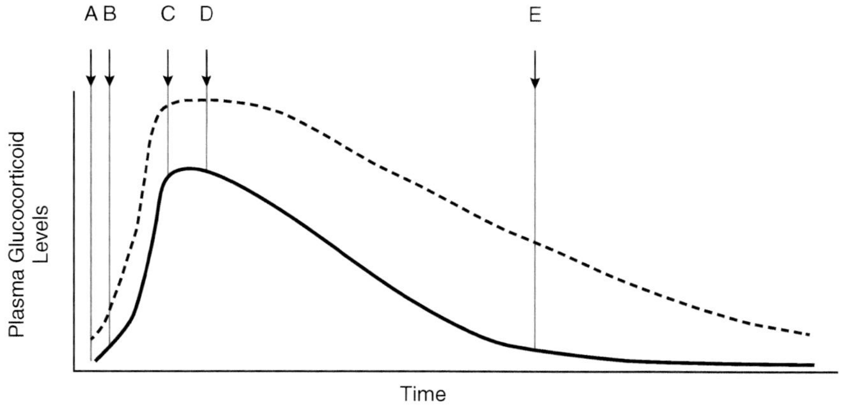


**Fig. 8.66** Timeline for a stress response. Numerous factors cause stress responses to vary (solid vs. dashed lines), including extrinsic factors, e.g., time of day, season, food availability, predation risk, social status, and habitat, and intrinsic factors, e.g., species, sex, condition, and age, but a greater response (dashed line) often indicates a greater level of stress. A—exposed to acute stressor, B—increased glucocorticoid levels (CORT) can be detected with three minutes, C—maximum CORT levels are reached in 15 to 30 minutes, D—CORT levels begin to decline sometime after peak levels were reached, and E—prestress levels may be reached in 60 to 90 minutes. However, with chronic stress, CORT levels may be higher and may be maintained for longer periods of time. (Figure from Sheriff et al. 2011; © 2011 Springer-Verlag, used with permission)

chickens, possibly via direct action on the mucosa. Although little studied in birds, secretin has multiple effects in the mammalian digestive system, including inhibiting the release of acid in the stomach, stimulating secretion of pancreatic juice, and inhibiting contraction of smooth muscle in the walls of the small and large intestines (Tam et al. 2014).

## 8.11 Adipokines

Adipokines are cytokine hormones produced by adipose tissue. Important adipokines in mammals include leptin, adiponectin, visfatin, and chemerin, and their functions include regulation of glucose and energy metabolism as well as playing an important role in reproduction (Kershaw and Flier 2004; Reverchon et al. 2014; Kiezun et al. 2014). In birds, leptin is not produced by adipose tissue, but, rather, is produced and secreted primarily by the anterior pituitary (Friedman-Einat and Seroussi 2019), so was discussed earlier in this chapter. However, avian adipose tissue does produce adiponectin, visfatin, and chemerin.

Adiponectin is an adipokine hormone secreted primarily by adipose tissue that appears to be important in regulating fat deposition and body mass (Fruebis et al. 2001). In Domestic Chickens, adiponectin limits lipid deposition in adipose tissue and induces food intake (Mellouk et al. 2018; Fig. 8.80). Similarly, Yan et al. (2014) reported a negative correlation between blood plasma levels of adiponectin and fat deposition, particularly the amount of abdominal fat and amount of lipid in the liver of Domestic Chickens.

Visfatin expression is positively correlated with chicken body mass and its functions including regulating growth of skeletal muscle (Krzysik-Walker et al. 2011), increasing food intake (Cline et al. 2008), and serving several reproductive functions (e.g., Ocón-Grove et al. 2010). Intracerebroventricular injection of visfatin caused an increase in food intake by young Domestic Chickens (Fig. 8.81) and reduced blood plasma levels of triglycerides, low-density lipoproteins, suggesting the visfatin

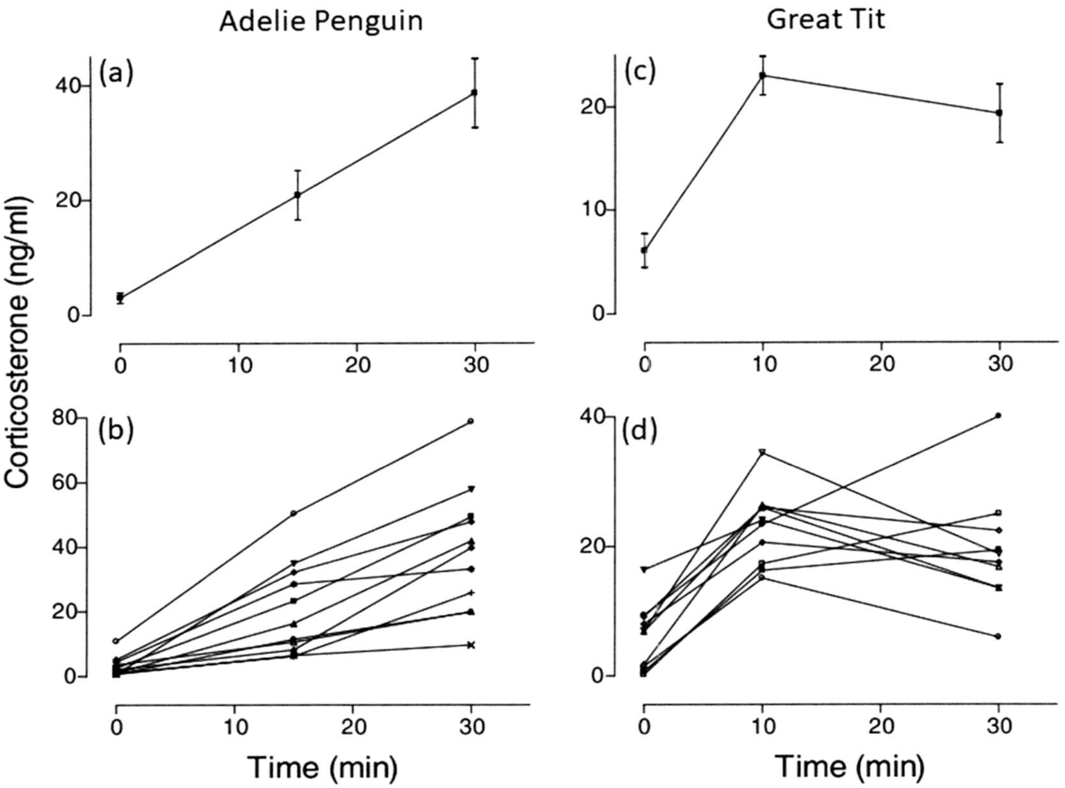


**Fig. 8.67** (**a**, **c**) Top, Mean (± SE) and (**b**, **d**) individual plasma corticosterone responses to handling for (**a**) Adelie Penguins (*Pygoscelis adeliae*) and (**b**) Great Tits (*Parus major*). Adelie Penguins were captured on their nests and held in mesh bags in darkened boxes after each blood sample was collected. Free-living Great Tits were captured and held individually in cages for 10 days before sampling. Great Tits were kept in cloth bags after each time blood samples were obtained. Blood samples were taken at time 0 (when first captured for Adelie Penguins and when first taken from cages for Great Tits) and again after 15 and 30 minutes, respectively. Data for Great Tits are from Cockrem and Silverin (2002). (Figure from Cockrem 2007; © 2007 Dt. Ornithologen-Gesellschaft e.V., used with permission)

plays an important role in lipid metabolism (Li et al. 2018). Visfatin injection also resulted in the downregulation of pro-opiomelanocortin (POMC) and corticotropin-releasing factor (CRF), suggesting that the orexigenic effect of visfatin might result from inhibition of these two hypothalamic anorexigenic hormones (Li et al. 2018).

In Domestic Chickens, chemerin and chemerin receptors are found in adipose tissue, as well as the liver, skeletal muscle, and the ovary, and levels of chemerin in the blood plasma are negatively correlated with levels of fat deposits (Mellouk et al. 2018). Estienne et al. (2021) found that intraperitoneally injected chemerin reduced food intake (Fig. 8.82) and body mass, increased plasma levels of insulin, and reduced plasma levels of glucose in Domestic Chickens. Estienne et al. (2021) also showed the chemerin and chemerin receptors were expressed in the hypothalamus of chickens, and administration of chemerin resulted in the reduced mRNA expression of orexigenic neuropeptide Y (Npy) and agouti-related peptide (Agrp). Such results

**Table 8.3** Possible characteristics of birds with proactive and reactive personalities (Table from Cockrem 2007; © 2007 Dt. Ornithologen-Gesellschaft e.V., used with permission)

| Proactive personality | Reactive personality |
|---|---|
| Behavior | |
| Bold and aggressive | Shy or cautious and not aggressive |
| Fast and superficial explorers | Slow and thorough explorers |
| Less fearful | More fearful |
| Less flexible and less sensitive to environmental stimuli | More flexible and more sensitive to environmental stimuli |
| Physiology | |
| Lower corticosterone responses to stressors | Higher corticosterone responses to stressors |
| Relatively high sympathetic nervous system reactivity | Relatively low sympathetic nervous system reactivity |
| Relatively low parasympathetic nervous system reactivity | Relatively high parasympathetic nervous system reactivity |
| Fitness | |
| More successful in constant or predictable conditions | More successful in changing or unpredictable conditions |

suggest that chemerin may help regulate appetite via modulation of hypothalamic peptides (Estienne et al. 2021).

## 8.12 Pancreas

The avian pancreas is attached to each side of the duodenal loop, with the duodenum being the first section of the small intestine (Fig. 8.83). The avian pancreas is both an exocrine and endocrine gland. As an exocrine gland, the pancreas produces enzymes and a bicarbonate solution that are secreted into the duodenum of the small intestine and aid in digestion. As an endocrine gland, the pancreas is important in regulating glucose levels in the blood, and the endocrine portion of the avian pancreas consists of large-diameter alpha (α) islets, a greater number of beta (β) islets, and a few or no mixed islets that contain D- and F-cells (Ku et al. 2000; Gülmez et al. 2004; Bayrakdar et al. 2011). Glucagon, insulin, somatostatin, and pancreatic polypeptide are produced and secreted by alpha cells, beta cells, D-cells, and F (or PP) cells, respectively (Braun and Sweazea 2008; Figs. 8.84, 8.85, and 8.86).

A typical concentration of blood glucose for mammals is about 7 – 8 millimoles per liter (or about 125–145 milligram per deciliter). Typical blood glucose levels of birds compared to mammals of similar mass are two to three times higher (Braun and Sweazea 2008; Polakof et al. 2011; Figs. 8.87 and 8.88), and birds are able to maintain these high blood glucose concentrations within tight limits even during periods of food restriction such as long-distance migration (Braun and Sweazea 2008; Scanes and Braun 2013). In addition, birds are less insulin-sensitive than mammals, with blood glucose levels less affected by exposure to insulin (Sweazea and Braun 2005; Sweazea et al. 2006), and extremely high glucose concentrations (30-40 mM) are needed to stimulate secretion of insulin by an avian pancreas (Honey et al. 1980; Colca and Hazelwood 1981). Mello and Lovell (2018) found that some genes known to influence energy homeostasis in mammals are not present in birds, potentially explaining the low sensitivity of birds to insulin and the high plasma levels of glucose. Another possible explanation for the difference between mammals and birds in the function of insulin is that birds lack a glucose-transport protein (called GLUT4) that is targeted by insulin in mammals (Braun and Sweazea 2008; Polakof et al. 2011; Fig. 8.89). However, birds are not totally insensitive to insulin; insulin-sensitive glucose transporters, although still less sensitive than GLUT4 in mammals, are located in the avian liver (possibly GLUT2) (Ji et al. 2020) and skeletal muscle (possibly GLUT12) (Tokushima et al. 2005; Sweazea and Braun 2005; Coudert et al. 2015).

In humans, plasma glucose levels as high as those of birds would cause serious neurological

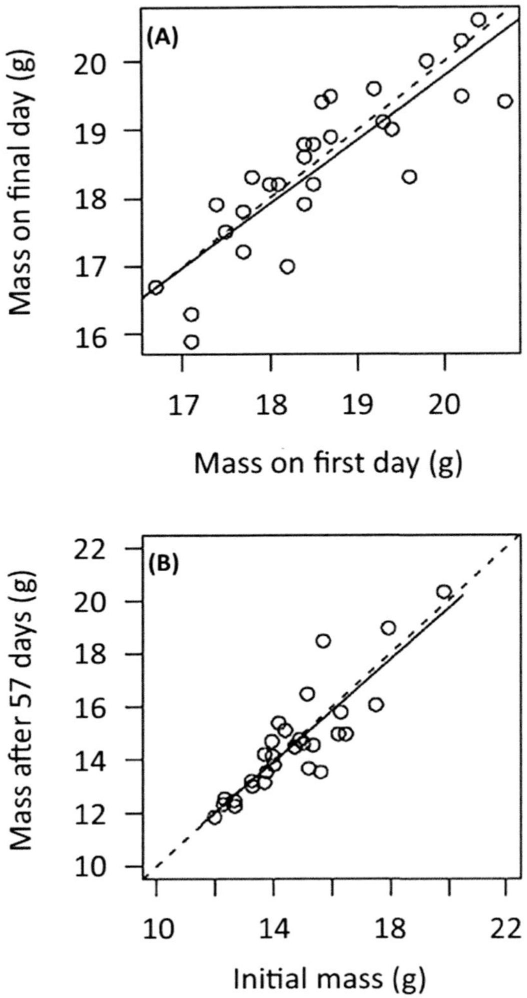


**Fig. 8.68** (**a**) Great Tits (*Parus major*) were automatically weighed when they visited bird feeders in their natural habitat. Their body masses when last determined were very similar to those when first determined (with as many as 40 days between measurements). (**b**) Captive Zebra Finches (*Taeniopygia guttata*) kept in cages with food always available maintained their body mass for 57 days. In both cases, best-fit lines (solid lines) are similar to lines of unity (dashed line). (Data for Great Tits from Moiron et al. (2018) and data for Zebra Finches from Yap et al. 2017; Figure from Halsey 2018; © 2018 Published by Elsevier Ltd., used with permission)

and microvascular pathologies, with glucose reacting with amino groups of proteins, nucleic acids, and lipids to form advanced glycation end products that accumulate in tissues and damage the retina and kidneys (Del Rio and Gutiérrez-Guerrero 2020). Birds may, in part, be able to tolerate such high levels of plasma glucose because an abundant plasma protein, albumin, is apparently able to resist glycation because avian albumen has a slightly different amino-acid composition than human albumin (Anthony-Regnitz et al. 2020). However, excessively high levels of blood glucose can still negatively impact birds. For example, adult Collared Flycatchers (*Ficedula albicollis*) with high levels of glycated hemoglobin (due to high levels of glucose in the blood) had lower survival rates than those with lower levels (Récapet et al. 2016). Similarly, Montoya et al. (2018) found that Zebra Finches (*Taeniopygia guttata*) with higher baseline levels of blood glucose had higher mortality rates than those with lower baseline levels.

Insulin and glucagon function differently in birds than in mammals, with birds having much lower plasma levels of insulin than mammals (about 10 times lower) and higher glucagon levels (8-10 times higher) (Hazelwood 1973). In mammals, insulin lowers and glucagon increases blood glucose levels. Insulin levels in the blood of birds are low relative to those of glucagon, which probably contributes to the high blood glucose levels of birds (Polakof et al. 2011). During periods of food deprivation, the ratio of glucagon to insulin in the blood of birds increases, increasing blood glucose levels and rates of gluconeogenesis (the synthesis of glucose from noncarbohydrates such as amino acids, glycerol, and pyruvate) (Cherel et al. 1988; Sartori et al. 1995).

In response to declining levels of glucose in the blood, glucagon is released by alpha cells in the pancreas. This triggers glycogenolysis in liver cells, i.e., the enzymatic breakdown of glycogen, a polysaccharide, into individual glucose molecules that are released into the blood to increase blood glucose levels (Hazelwood 1973). In addition, glucagon stimulates gluconeogenesis and lipolysis and suppresses hepatic lipogenesis (Braun and Sweazea 2008). Glucagon also appears to be important in maintaining normoglycemia when birds fast, e.g., during long flights (Vaillancourt and Weber 2015).

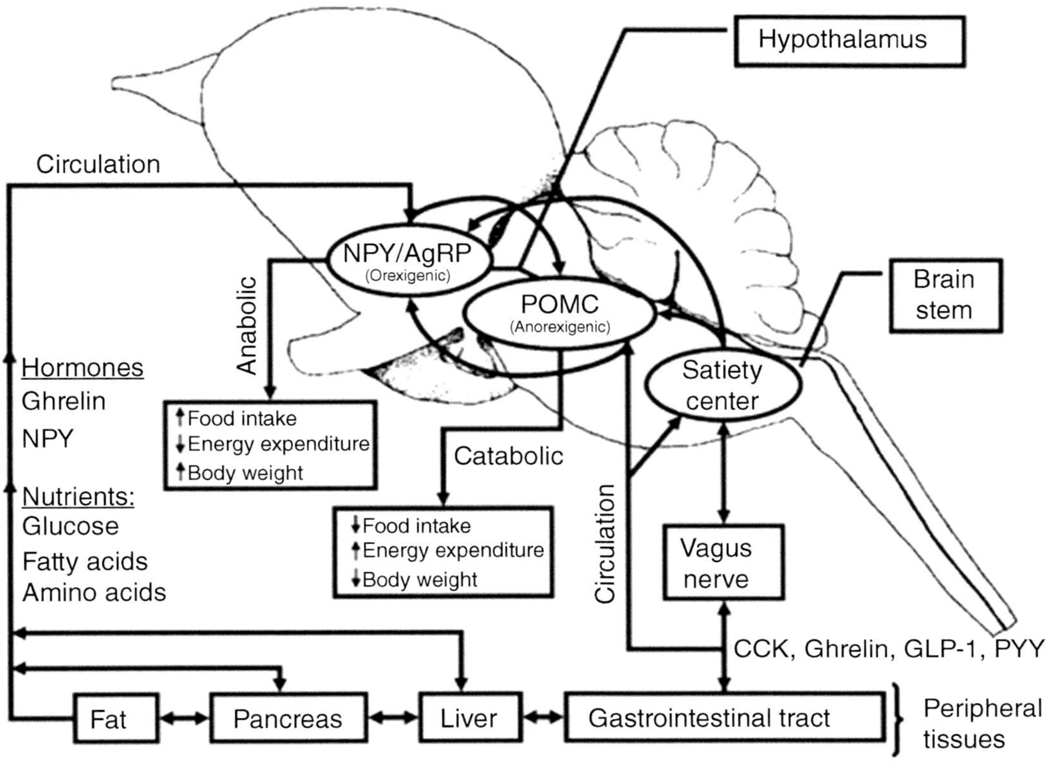


**Fig. 8.69** Proposed model explaining the long-term regulation of appetite and energy balance in birds and showing the roles of peripheral tissue and central nervous system circuits regulated by hormonal, neural, neuroendocrine, and nutrient signaling mechanisms. Signals from peripheral tissues include peptide hormones from fat (adipose tissue), the pancreas and liver, and gastrointestinal tract. The brainstem processes impulses from the vagus nerve and sends impulses back to the gastrointestinal tract that regulate peripheral tissue functions and generate a sense of satiety. Cholecystokinin (CCK) stimulates gastric emptying and the release of pancreatic enzymes. Stimulation of NPY/AgRP neurons leads to increased food intake, whereas stimulation of POMC neurons has the opposite effect. NPY = neuropeptide Y; AgRP = agouti-related peptide; POMC = pro-opiomelanocortin; CCK = cholecystokinin; GLP-1 = glucagon-like peptide 1; PYY = Peptide YY. (Figure from Richards and Proszkowiec-Weglarz 2007; © 2007 Elsevier, used with permission)

Beyond its role in regulating blood glucose levels, insulin may also influence food intake by birds. Intracerebroventricular injection of insulin was found to significantly reduce food intake by young Domestic Chickens (Shiraishi et al. 2008). Injection of insulin also caused a significant increase in expression of anorexigenic pro-opiomelanocortin (POMC) mRNA and a significant decrease in expression of orexigenic neuropeptide Y (NPY) mRNA, suggesting that insulin suppresses appetite via its effects on these two hypothalamic neuropeptides (Shiraishi et al. 2008).

Insulin may also be important in the formation of hypothalamic circuits during the early stages of avian embryonic development. Yacawych et al. (2019) reported evidence of insulin receptor expression in the hypothalamic nuclei important in energy homeostasis in embryonic Domestic Chickens, suggesting that insulin signaling affects the development of neural circuits in these areas of the hypothalamus. Additional study is needed to better understand the neural subpopulations affected and how insulin actually influences neuronal connections in the hypothalamus.

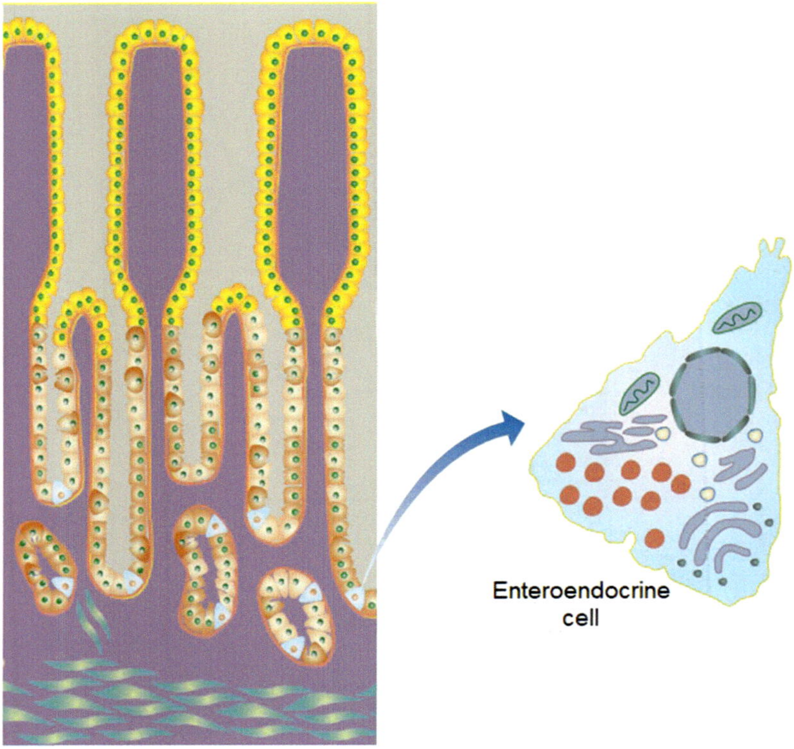


**Fig. 8.70** Left, cells in the gastrointestinal tract showing enteroendocrine cells (light blue) scattered among other epithelial cells. Right, an enterendocrine cell showing organelles involved in the production of peptide hormones that play important roles in regulating food intake by birds. (Figure from Skipper and Lewis 2000; © 1969 Nature America, Inc., used with permission)

**Table 8.4** Peptide hormones that influence food intake and energy expenditure by birds. Little is known about the specific effects of some of these hormones on avian feeding behavior. Those whose functions are better known are discussed in the text (Table modified from Richards and Proszkowiec-Weglarz 2007; © 2007 Poultry Science Association Inc. Published by Elsevier Inc., used with permission)

| Orexigenic | Anorexigenic |
|---|---|
| Neuropeptide Y[a] | α-/β-melanocyte stimulating hormone |
| Agouti-related peptide[a] | Growth hormone releasing hormone[a] |
| Peptide YY | Corticotropin-releasing factor[a] |
| Pancreatic polypeptide | Cocaine- and amphetamine-regulated transcript |
| Neurosecretory protein GL | Leptin (?)[b] |
| Galanin[b] | Ghrelin |
| Somatostatin | Glucagon-like peptide ½ |
| | Cholecystokinin (CCK) |
| | Pro-opiomelanocortin (POMC)[a] |
| | Gastrin |
| | Substance P |
| | Neuromedin U/S |
| | Oxyntomodulin |
| | Vasoactive intestinal peptide (VIP) |
| | Serotonin (?) |

[a]Brain peptides; all others are produced in peripheral tissues
[b]Produced in the brain and in peripheral tissues

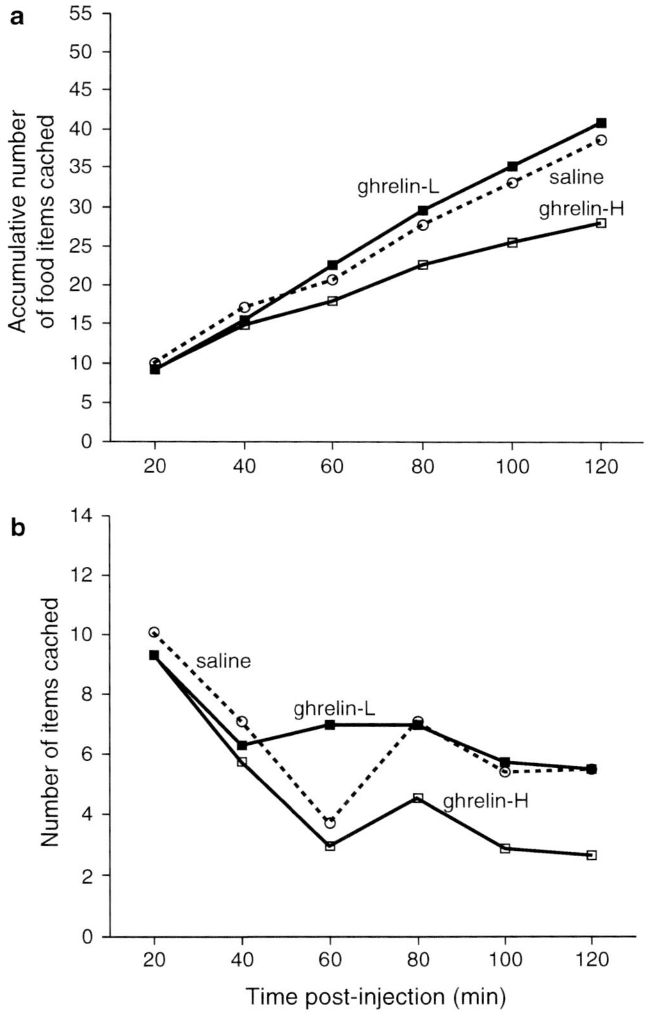


**Fig. 8.71** Effects of ghrelin on caching behavior of Coal Tits (*Periparus ater*). (**a**) Accumulative number of food items cached, and (**b**) the number of food items cached every 20 min, 2 h after injection with ghrelin-H, ghrelin-L, leptin or saline control. (Figure from Henderson et al. 2018; © 2018 The Authors, Published by the Royal Society, used with permission)

Insulin also has stimulatory effects on protein synthesis and growth in the liver and skeletal muscles of Domestic Chickens (Dupont et al. 2009). Tissue proteins and muscle mass are regulated via the processes of protein synthesis and proteolysis. A number of variables influence these processes, including physiological, genetic, and environmental factors such as diet and hormones, with insulin being of particular importance (Tesseraud et al. 2007). Insulin stimulates protein synthesis by maintaining levels of ribosomes and stimulating transcription and translation of ribosomal proteins (Proud and Denton 1997). Insulin influences protein synthesis and growth via its binding to membrane receptors. This binding activates a complex signaling pathway that results in an increase in protein synthesis, growth, and cellular differentiation (Fig. 8.90).

Pancreatic polypeptide is a 36-amino acid peptide that was first located in the avian pancreas in 1974 (Kimmel et al. 1975; Larsson et al. 1975) and is produced by F-cells. El-Salhy et al. (1982) reported that pancreatic polypeptide-immunoreactive cells were located in the duodenum jejunum, ileum, and large intestine (colon) of Domestic Chickens. Pancreatic polypeptide is a satiety hormone, with plasma levels increasing

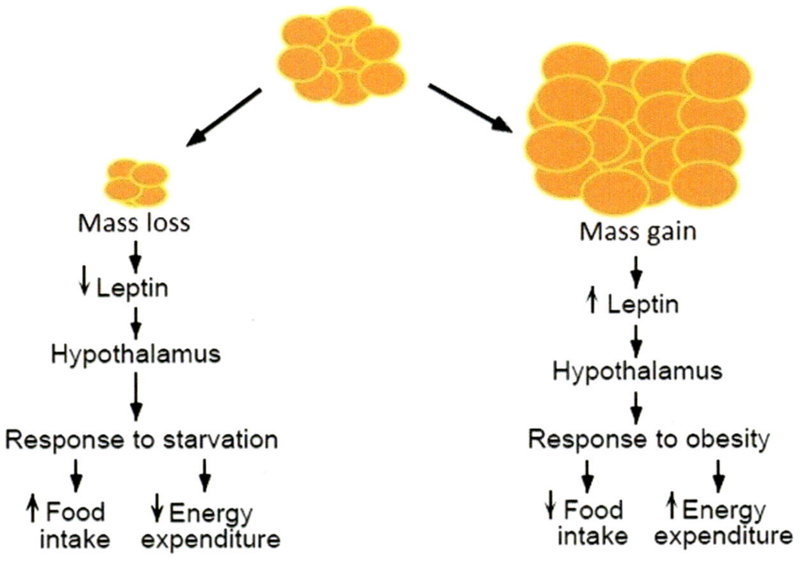


**Fig. 8.72** In mammals, leptin helps to maintain constant stores of fat. When the amount of body fat declines (insufficient food intake or starvation), adipose tissue releases less leptin into the blood which then stimulates the need to increase food intake and reduce energy expenditure, so the amount of body fat can be increased. Conversely, with increasing amounts of body fat, adipose tissue releases more leptin into the blood, reducing appetite and food intake and stimulating an increase in energy expenditure and leading to a decrease in the amount of body fat. (Figure modified from Friedman and Halaas 1998; © 1998 Macmillan Magazines Ltd., used with permission)

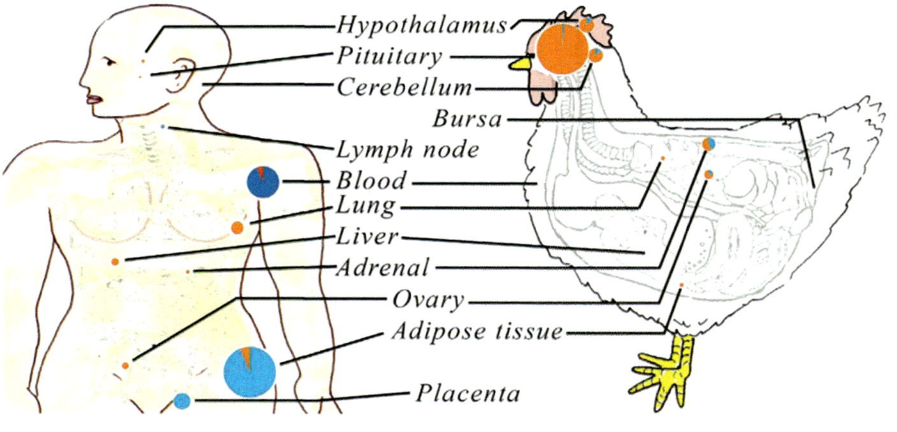


**Fig. 8.73** Pie charts indicate expression of mammalian leptin (light blue) and leptin-receptor (orange) mRNAs. Data for avian mRNA expression (color codes same as for mammals) are from Red Junglefowl (*Gallus gallus*) and Domestic Chickens. Due to much lower expression levels, pies representing avian mRNA are at a 12-fold larger scale than those of mammals. Leptin is expressed primarily in the adipose tissue of mammals and secreted into the blood. In birds, leptin is expressed primarily in the pituitary and hypothalamus, to a lesser extent in the adrenal glands and gonads, and not in the blood. Differences in expression patterns between the two clades indicate evolutionary divergence in the role of leptin. (Figure from Friedman-Einat and Seroussi 2019; © 2019 Elsevier Ltd., used with permission)

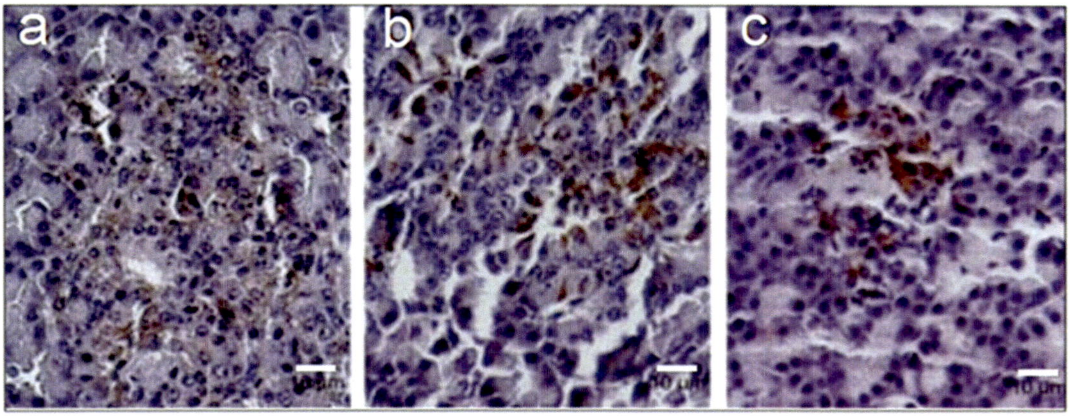


**Fig. 8.74** Somatostatin secreting cells in the (**a**) alpha, (**b**) beta, and (**c**) mixed islets in the pancreas of a Eurasian Sparrowhawk (*Accipiter nisus*). Scale bars = 10 μm. (Figure from Kara et al. 2014; Published in Kafkas Univ Vet Fak Derg, used with permission)

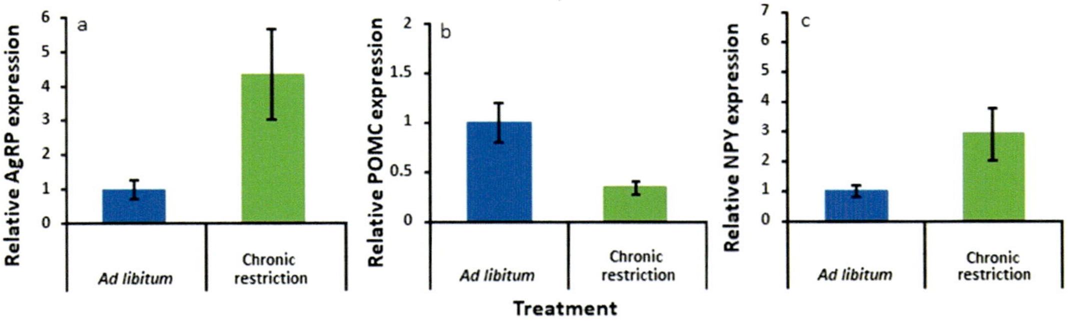


**Fig. 8.75** Gene expression of agouti-related peptide (AgRP) (**a**), pro-opiomelanocortin (POMC) (**b**), and neuropeptide Y (NPY) (**c**) in the arcuate nuclei of Red Junglefowl (*Gallus gallus*) chicks reared under *ad libitum* feeding (blue) and chronic feed restriction (green). Birds in the chronic restriction group were fed 60% of the amount of food provided to those in the ad libitum group for 11 days, then 75% of the amount fed to the ad libitum group for 12 more days. Error bars represent standard deviations of the mean. (Figure modified from Lees et al. 2017; open-access article licensed under a Creative Commons Attribution 4.0 International License, https://creativecommons.org/licenses/by/4.0/)

within minutes to hours after feeding (Johnson and Hazelwood 1982). Duke (1992) noted pancreatic polypeptide was released in response to the presence of ingesta in the upper small intestine of Domestic Turkeys and reduced gastroduodenal motility, slowing gastric emptying to provide more time for digestion of food in the small intestine. Similar results have been reported for Domestic Chickens (Duke et al. 1985). However, infusion of pancreatic polypeptide had no effect on gastric secretion and motility in Red-tailed Hawks (*Buteo jamaicensis*), suggesting that the effects of pancreatic polypeptide may differ among species of birds with different food habits (i.e., carnivore vs. granivore) (Duke et al. 1988). Finally, Fu and Gao (2000) provided one-day-old Domestic Chickens with either plain water (control) or water containing

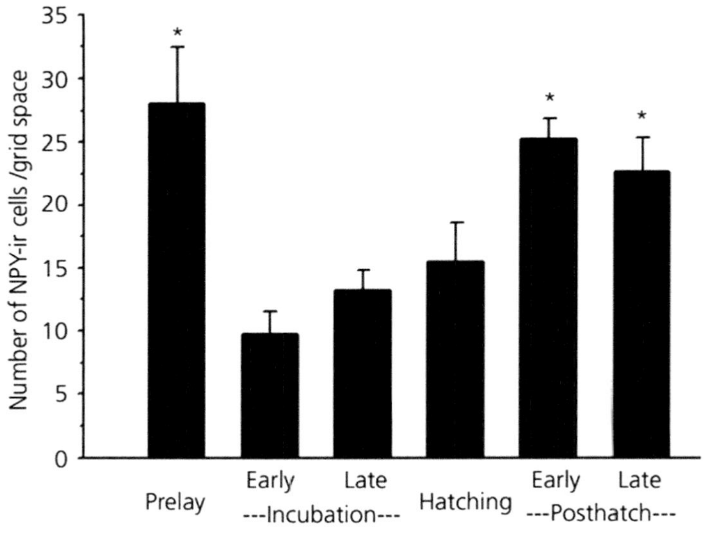


**Fig. 8.76** Ringed Turtle-Doves (*Streptopelia risoria*) had significantly more NPY-secreting cells in the hypothalamus during the pre-laying period and during the early and late post-hatch periods than during the incubation and hatching periods (Data are means ± SE). (Figure from Ramakrishnan et al. 2007; © British Society for Neuroendocrinology, used with permission)

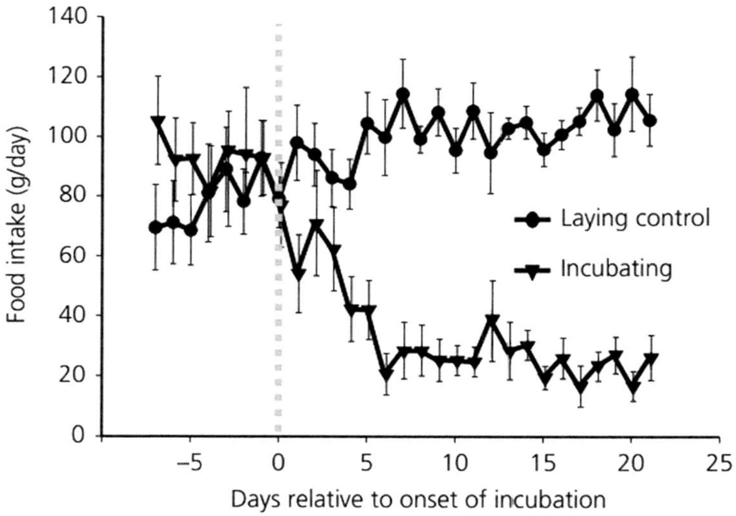


**Fig. 8.77** Mean (± SE) daily food intake of female Domestic Chickens (*Gallus g. domesticus*) transitioning, or not, to incubation behavior. The gray dotted line indicates where females in the incubating group began incubation. Laying control females were not provided with a nest box and so were not able to initiate incubation. (Figure from Dunn et al. 2015; open-access article distributed under the terms of the Creative Commons CC BY license, https://creativecommons.org/licenses/by/4.0/)

pancreatic polypeptide (treated) on a daily basis for 84 days and found that treated birds weighed more than control bird and had higher blood plasma concentrations of growth hormone, triiodothyronine, and thyroxine, suggesting that pancreatic polypeptide promotes the growth of young chickens.

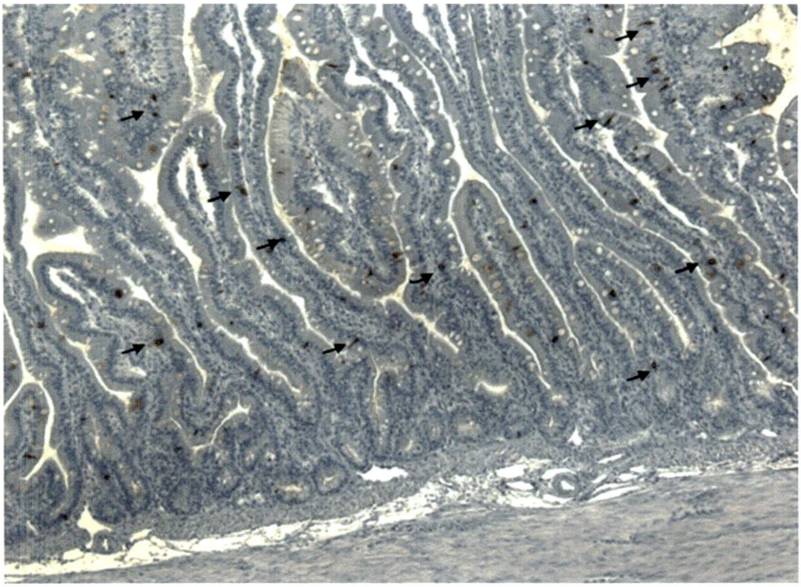

**Fig. 8.78** Gastrin-secreting enteroendocrine cells (indicated by the arrows) in the mucosa of the duodenum of a 30-day-old Common Ostrich (*Struthio camelus*). Magnification, 100×. (Figure from Dūrītis et al. 2021; © 2021 Wiley-VCH GmbH, used with permission)

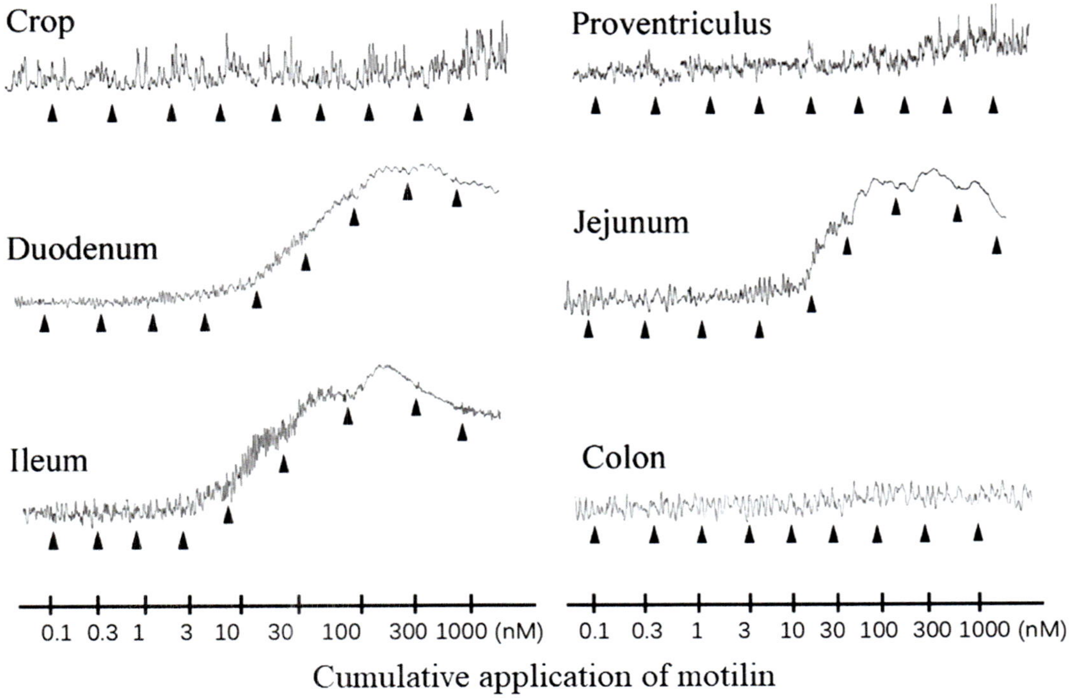


**Fig. 8.79** Contractile responses to chicken motilin in different areas of the gastrointestinal tract of Ring-necked Pheasants (*Phasianus colchicus*). Motilin was applied cumulatively at the times indicated by the arrowheads (about 2 min apart). The amplitude of contraction was relative to contraction after application of acetylcholine (100 μM). (Figure from Zhang et al. 2020; © 2019 Elsevier Inc. All rights reserved, used with permission)

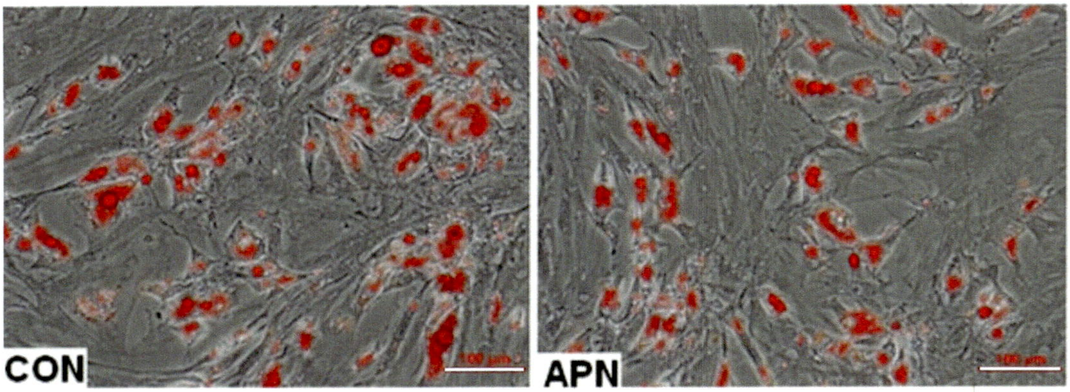


**Fig. 8.80** After nine days of exposure, chicken pre-adipocytes cultured in a medium containing 10 µg/ml of adiponectin (APN) contained less lipid than pre-adipocytes in a medium without adiponectin (control, CON). Scale bar = 100 µm. (Figure from Yan et al. 2014; Journal of Animal Physiology and Animal Nutrition © 2013 Blackwell Verlag GmbH, used with permission)

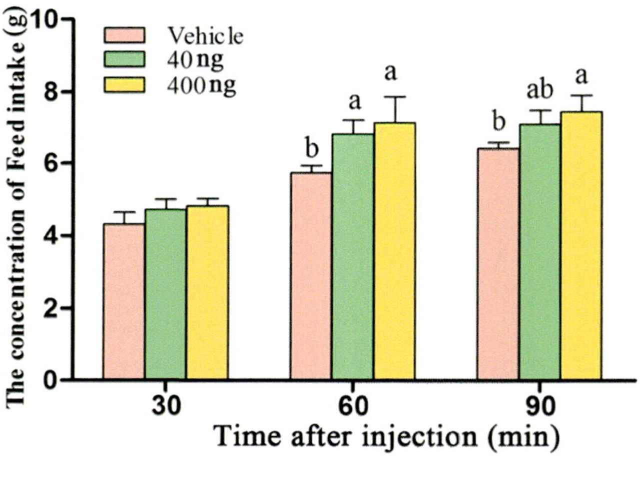


**Fig. 8.81** Intracerebroventricular injection of 40 ng and 400 ng of human visfatin caused an increase in food intake by 10-day-old Domestic Chickens. Visfatin was dissolved in artificial cerebrospinal fluid that also served as the control (vehicle). Values represent means ± SE, Different letters indicate a statistically significant difference ($P < 0.05$). (Figure from Li et al. 2018; © 2018 The Authors, open-access article distributed under the terms of the Creative Commons Attribution 4.0 International License, https://creativecommons.org/licenses/by/4.0/)

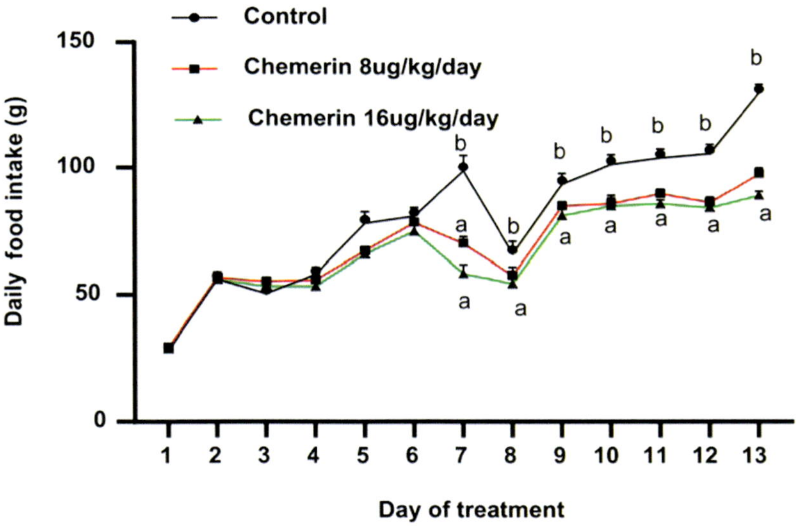


**Fig. 8.82** Two-week-old Domestic Chickens that were fed standard diets and injected intraperitoneally with saline plus different concentrations of chemerin ate significantly less food after seven days of treatment than chickens fed a standard diet and injected with just saline (control). Values represent means ± SE. Different letters indicate statistically significant differences ($P < 0.05$). (Figure from Estienne et al. 2021; © 2021 Elsevier Inc. All rights reserved, used with permission)

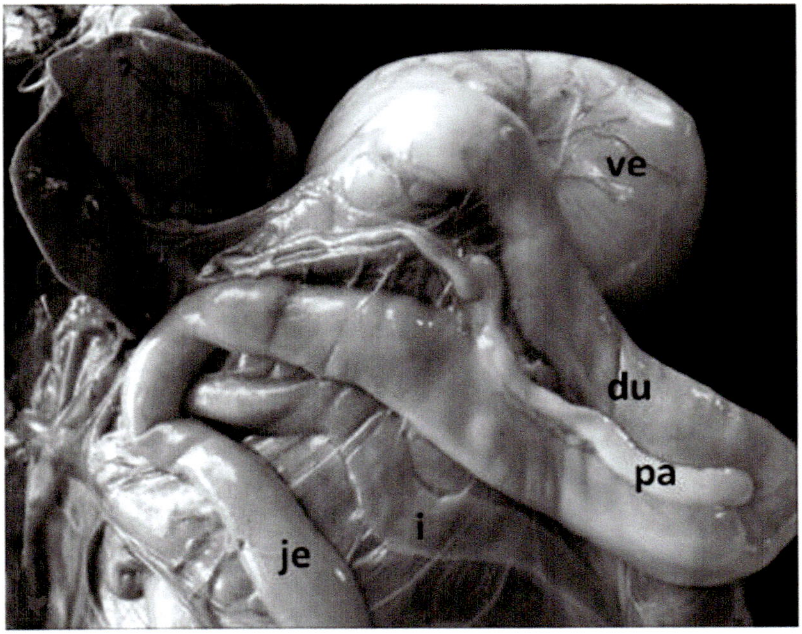


**Fig. 8.83** Pancreas (pa) of a Greater Rhea (*Rhea americana*). The pancreas in located between sections of the duodenum (du). ve, ventriculus (gizzard); je, jejunum of the small intestine; i, ileum of the small intestine. (Figure from Rodrigues et al. 2012; used with permission)

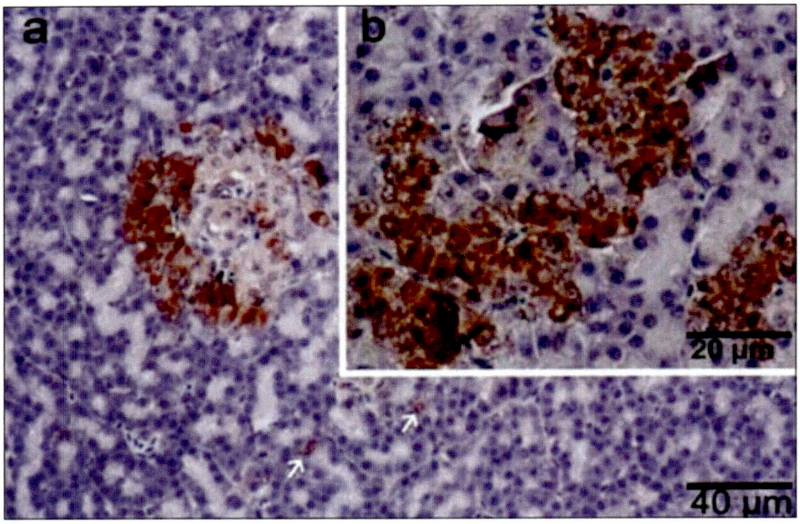


**Fig. 8.84** Glucagon secreting cells in (**a**) a mixed islet and (**b**) an alpha islet in the pancreas of a Eurasian Sparrowhawk (*Accipiter nisus*). (Figure from Kara et al. 2014; Published in Kafkas Univ Vet Fak Derg, used with permission)

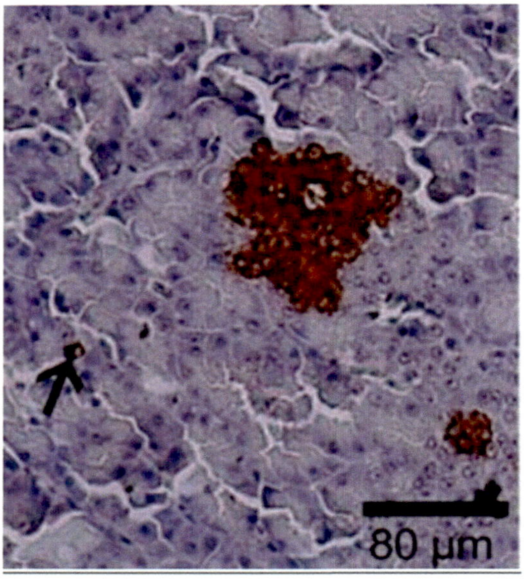


**Fig. 8.85** Insulin-secreting cells in a mixed islet in the pancreas of a Eurasian Sparrowhawk. Insulin secreting cells are also found in beta islets. (Figure from Kara et al. 2014; Published in Kafkas Univ Vet Fak Derg, used with permission)

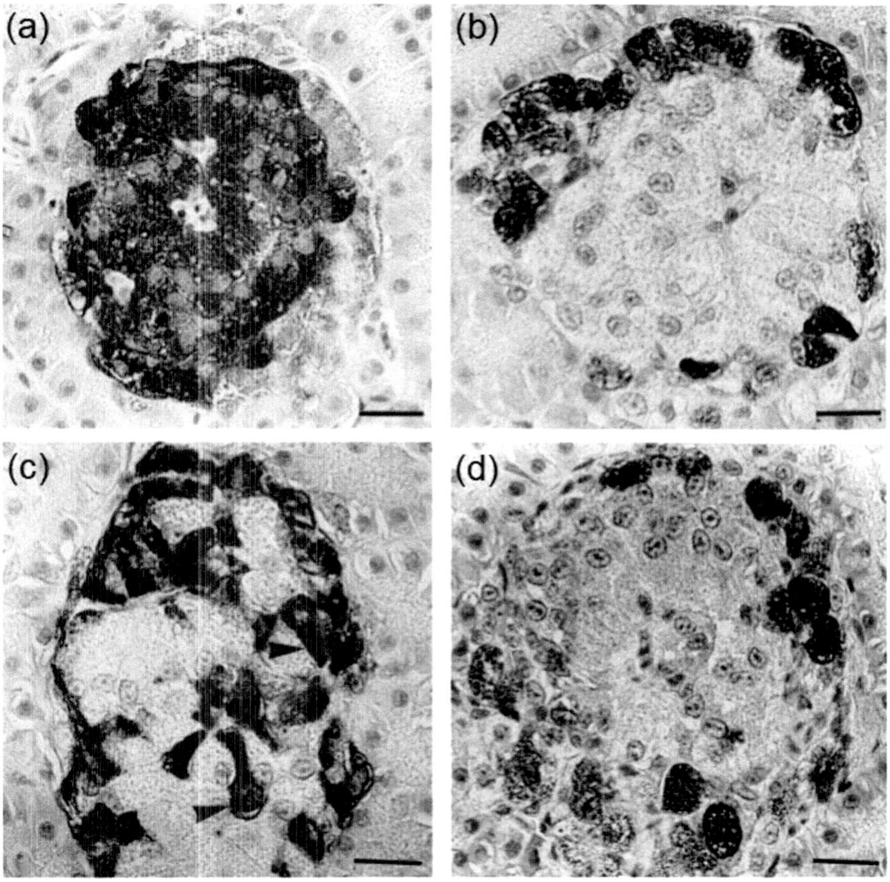


**Fig. 8.86** An alpha islet in the pancreas of a Houbara Bustard (*Chlamydotis undulata*) showing cells that produce and secrete (**a**) insulin, (**b**) glucagon, (**c**) somatostatin, and (**d**) pancreatic polypeptide. Cells immunoreactive to insulin are located in the center and all others are located on the periphery. Neurons immunoreactive to somatostatin are indicated by arrowheads. Scale bars = 6 μm. (Figure from Mensah-Brown et al. 2000; © 2000 Anatomical Society of Great Britain and Ireland, used with permission)

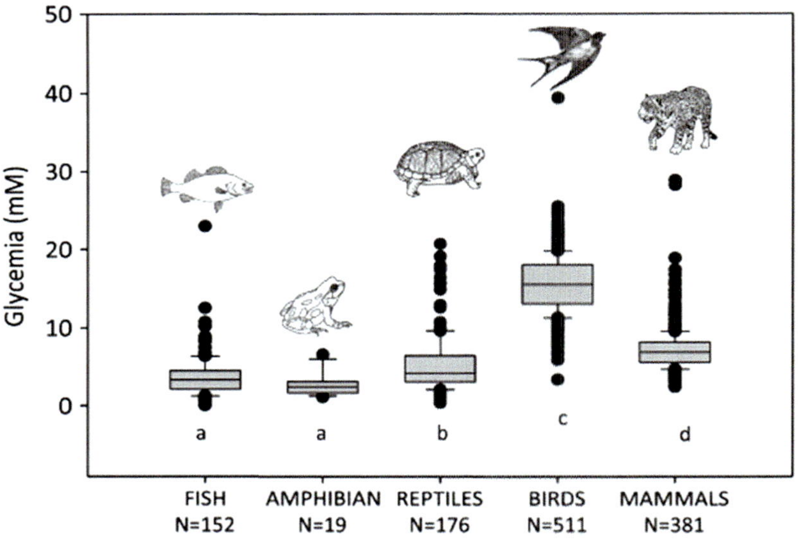


**Fig. 8.87** Individual (dots) and median (horizontal lines in boxes) glycemia (plasma glucose levels) values of several species ($N$ = sample size) of fish, amphibians, reptiles, birds, and mammals. Birds have significantly higher plasma glucose levels than the other classes of vertebrates. Different letters at the bottom indicate significant differences among taxa. The boundary of boxes closest to zero indicates the 25th percentile, the line in boxes is the median, and the boundary of boxes farthest from zero represents the 75th percentile. Whiskers above and below boxes represent the 10th and 90th percentiles. (Figure from Polakof et al. 2011; © 2011 Elsevier Inc. All rights reserved, used with permission)

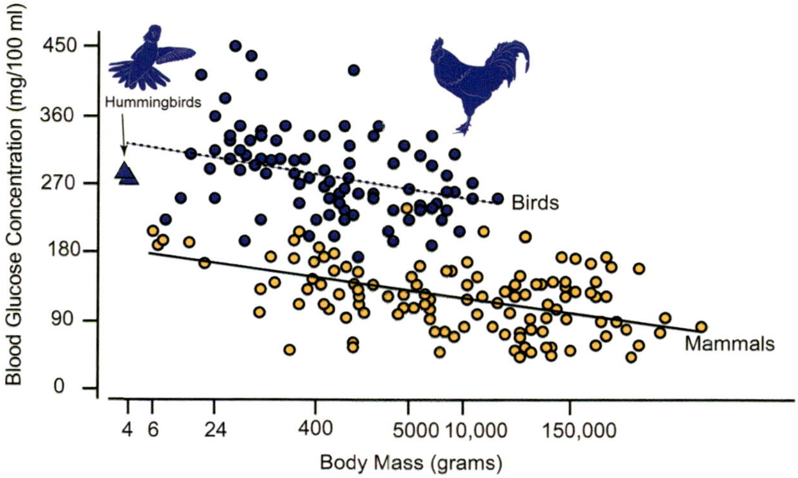


**Fig. 8.88** Blood glucose levels of birds are about twice those of mammals and, for both birds and mammals, levels decrease with increasing body mass. (Figure from Del Rio and Gutiérrez-Guerrero 2020; © 2020 Springer Science Business Media, LLC, part of Springer Nature, used with permission)

**Fig. 8.89** (a) Birds lack GLUT4. (b) In mammals, muscle contractile activity and insulin signaling independently, or in combination, enhance sugar uptake by promoting translocation of GLUT4 from intracellular vesicles to the fiber surface. Low transport capacity through GLUT5 and low affinity of hexokinase II (HK) for fructose largely limit the ability of mammalian muscle fibers to use this sugar directly as a fuel during exercise. (Figure modified from Welch and Chen 2014; © 2014 Springer-Verlag Berlin Heidelberg, used with permission)

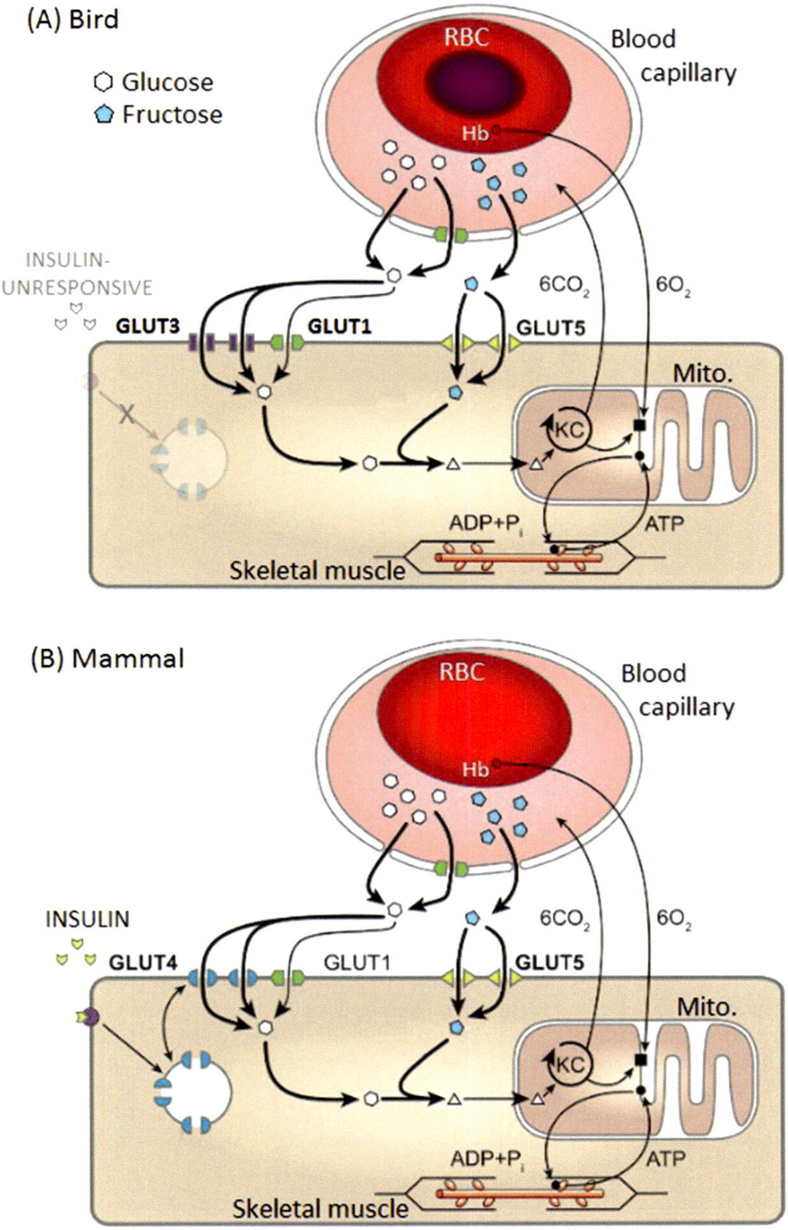

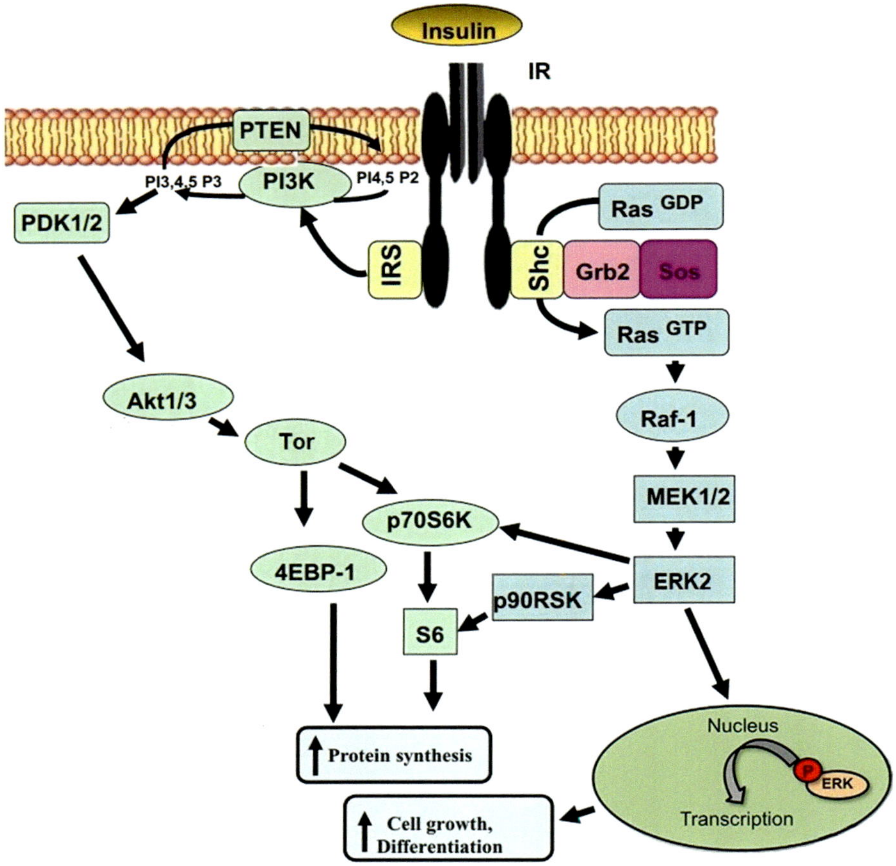


**Fig. 8.90** Insulin signaling pathway focusing on insulin's role in protein turnover. IR, insulin receptor; IRS, insulin receptor substrate; Shc, Src homology and collagen protein; PI3K, phosphatidylinositol-3 kinase; PTEN, phosphatase and tensin homolog; PDK1/2, phosphoinositide-dependent kinase 1/2; AKT1/3, protein kinase B 1/3; Tor, target of rapamycin; p70S6K, 70 kDA ribosomal protein S6 kinase; 4E-BP1, eukaryotic initiation factor 4E binding protein; S6, ribosomal protein S6; Grb2, growth factor receptor-bound protein 2; Sos, son of sevenless homolog 1; Raf, mitogen-activated protein kinase kinase kinase; ERK, extracellular signal-regulated protein kinase; MEK, Map/Erk kinase. Ras is a protein that transmits signals in cells; Ras GDP is the inactive form, and Ras GTP is the activated form. (Figure modified from Dupont et al. 2009; © 2008 Elsevier Inc. All rights reserved, used with permission. Nucleus image from Kirouac et al. 2017, open-access article distributed under the terms of the Creative Commons Attribution 4.0 International license, https://creativecommons.org/licenses/by/4.0/)

## References

Abalde M, Taboada C, Fernández P (1984) Somatostatin and 3-oxy-methyl-D-glucose (3-OMG) uptake in isolated chicken intestinal epithelial cells. Experientia 40:823–824

Aksoy A, Cinar K (2009) Distribution and ontogeny of gastrin-and serotonin-immunoreactive cells in the proventriculus of developing chick, *Gallus gallus domestica*. J Vet Sci 10:9

Ali HK, Mirhish SM (2015) Anatomical and histological study of thyroid, parathyroid and ultimobranchial glands in Iraqi local breed turkey "*Meleagris gallopavo*". Iraqi J Vet Med 39:40–48

Angelier F, Shaffer SA, Weimerskirch H, Trouvé C, Chastel O (2007) Corticosterone and foraging behavior in a pelagic seabird. Physiol Biochem Zool 80:283–293

Angelier F, Wingfield JC, Weimerskirch H, Chastel O (2010) Hormonal correlates of individual quality in a long-lived bird: a test of the 'corticosterone–fitness hypothesis'. Biol Lett 6:846–849

Angelier F, Wingfield JC, Tartu S, Chastel O (2016) Does prolactin mediate parental and life-history decisions in response to environmental conditions in birds? A review. Horm Behav 77:18–29

Angove JL, Forder REA (2020) The avian maternal environment: exploring the physiological mechanisms driving progeny performance. Worlds Poult Sci J 76: 100–118

Anthony-Regnitz C, Wilson AE, Sweazea KL, Braun EJ (2020) Fewer exposed lysine residues may explain relative resistance of chicken serum albumin to in vitro glycation in comparison to bovine serum albumin. J Mol Evol 88:653–661

Aoki K, Kondo M, Okuda M, Saneyasu T, Honda K, Kamisoyama H (2017) Identification, expression analysis, and functional characterization of peptide YY in chickens (*Gallus gallus domesticus*). Gen Comp Endocrinol 242:11–17

Apostu D, Lucaciu O, Oltean-Dan D, Mureşan AD, Moisescu-Pop C, Maxim A, Benea H (2020) The influence of thyroid pathology on osteoporosis and fracture risk: a review. Diagnostics 10:149

Balthazart J, Choleris E, Remage-Healey L (2018) Steroids and the brain: 50 years of research, conceptual shifts and the ascent of non-classical and membrane-initiated actions. Horm Behav 99:1–8

Baran NM, Sklar NC, Adkins-Regan E (2016a) Developmental effects of vasotocin and nonapeptide receptors on early social attachment and affiliative behavior in the Zebra Finch. Horm Behav 78:20–31

Baran NM, Tomaszycki ML, Adkins-Regan E (2016b) Early life manipulations of the nonapeptide system alter pair maintenance behaviors and neural activity in adult male Zebra Finches. Front Behav Neurosci 10:58

Baran NM, Peck SC, Kim TH, Goldstein MH, Adkins-Regan E (2017) Early life manipulations of vasopressin-family peptides alter vocal learning. Proc R Soc B 284:20171114

Batista G, Hensch TK (2019) Critical period regulation by thyroid hormones: potential mechanisms and sex-specific aspects. Front Mol Neurosci 12:77

Baxter RC, Twigg SM (2009) Actions of IGF binding proteins and related proteins in adipose tissue. Trends Endocrinol Metab 20:499–505

Bayrakdar A, Yaman M, Atalar O, Gencer Tarakci B, Ceribasi S (2011) Distribution of neuropeptides in endocrine and exocrine pancreas of Long-legged Buzzard (*Buteo rufinus*): an immunohistochemical study. Regul Pept 166:121–127

Bedecarrats GY (2015) Control of the reproductive axis: balancing act between stimulatory and inhibitory input. Poult Sci 94:810–815

Belanger LF, Copp DH (1972) The skeletal effect of prolonged calcitonin administration in birds under various conditions. In: Talmage RV, Munson PL (eds) Calcium, parathyroid hormone and the calcitonins. Excerpta Medica, Amsterdam, The Netherlands, pp 41–50

Belthoff JR, Dufty AM Jr (1998) Corticosterone, body condition and locomotor activity: a model for dispersal in screech-owls. Anim Behav 55:405–415

Bentley GE, Jensen JP, Kaur GJ, Wacker DW, Tsutsui K, Wingfield JC (2006) Rapid inhibition of female sexual behavior by gonadotropin-inhibitory hormone (GnIH). Horm Behav 49:550–555

Bohler M, Dougherty C, Tachibana T, Gilbert ER, Cline MA (2019) Gastrin releasing peptide-induced satiety is associated with hypothalamic and brainstem changes in chicks. Neurosci Lett 713:134529

Bohler M, Gilbert ER, Cline MA (2021) The anorexigenic effect of vasoactive intestinal polypeptide in Japanese Quail is associated with molecular changes in the arcuate and dorsomedial hypothalamic nuclei. Domest Anim Endocrinol 74:106499

Bókony V, Lendvai AZ, Liker A, Angelier F, Wingfield JC, Chastel O (2009) Stress response and the value of reproduction: are birds prudent parents? Am Nat 173: 589–598

Bonier F, Martin PR, Moore IT, Wingfield JC (2009) Do baseline glucocorticoids predict fitness? Trends Ecol Evol 24:634–642

Bons N (1980) The topography of mesotocin and vasotocin systems in the brain of the Domestic Mallard and Japanese Quail: immunocytochemical identification. Cell Tissue Res 213:37–51

Boonstra R, Boag PT (1992) Spring declines in *Microtus pennsylvanicus* and the role of steroid hormones. J Anim Ecol 61:339–352

Boswell T, Dunn IC (2015) Regulation of the avian central melanocortin system and the role of leptin. Gen Comp Endocrinol 221:278–283

Braun EJ, Sweazea KL (2008) Glucose regulation in birds. Comp Biochem Physiol B Biochem Mol Biol 151:1–9

Breuner CW, Wingfield JC, Romero LM (1999) Diel rhythms of basal and stress-induced corticosterone in a wild, seasonal vertebrate, Gambel's White-crowned Sparrow. J Exp Zool 284:334–342

Brown JL (1973) Behavior elicited by electrical stimulation of the brain of the Steller's Jay. Condor 75:1–16

Bruce LL (2009) Evolution of the hypothalamus in amniotes. In: Binder MD, Hirokawa N, Windhorst U (eds) Encyclopedia of neuroscience. Springer, Berlin, pp 1363–1367

Burhol PG (1974) Gastric stimulation by intravenous injection of cholecystokinin and secretin in fistula chickens. Scand J Gastroenterol 9:49–53

Burrin DG, Stoll B, Guan X (2003) Glucagon-like peptide 2 function in domestic animals. Domest Anim Endocrinol 24:103–122

Burtka JL, Lovern MB, Grindstaff JL (2016) Baseline hormone levels are linked to reproductive success but not parental care behaviors. Gen Comp Endocrinol 229:92–99

Calisi RM, Rizzo NO, Bentley GE (2008) Seasonal differences in hypothalamic EGR-1 and GnIH expression following capture-handling stress in House

Sparrows (*Passer domesticus*). Gen Comp Endocrinol 157:283–287

Calisi RM, Díaz-Muñoz SL, Wingfield JC, Bentley GE (2011) Social and breeding status are associated with the expression of GnIH. Genes Brain Behav 10:557–564

Campbell B, Lack E (2011) A dictionary of birds. T & AD Poyser, Calton, UK

Campbell B, Garner A, Dockray GJ, Hughes J, Dimaline R (1994) The mechanism of action of gastrin releasing peptide (GRP) in stimulating avian gastric acid secretion. Regul Pept 49:249–255

Candlish JK, Taylor TG (1970) The response-time to the parathyroid hormone in the laying fowl. J Endocrinol 48:143–144

Cao C, Gilbert ER, Cline MA (2020) The anorexigenic effect of beta-melanocyte-stimulating hormone involves corticotrophin-releasing factor and mesotocin in birds. Neurosci Lett 736:135282

Carere C, Eens M (2005) Unravelling animal personalities: how and why individuals consistently differ. Behaviour 142:1149–1157

Carsia RV (2015) Adrenals. In: Scanes CG (ed) Sturkie's avian physiology, 6th edn. Academic Press, New York, pp 577–611

Cerasale DJ, Zajac DM, Guglielmo CG (2011) Behavioral and physiological effects of photoperiod-induced migratory state and leptin on a migratory bird, *Zonotrichia albicollis*: I. Anorectic effects of leptin administration. Gen Comp Endocrinol 174:276–286

Chaiseha Y, El Halawani ME (2005) Neuroendocrinology of the female turkey reproductive cycle. J Poult Sci 42: 87–100

Chen M, He M, Peng K, Liu T, Jin C, Cao W, Wang L, Xiao K (2013) An immunohistochemical study of somatostatin in the stomach and the small intestine of the African Ostrich (*Struthio camelus*). Tissue Cell 45: 363–366

Chen G, Yang F, Wu T, Jiang J, Zhou W (2016) The stimulatory effect of cerebral intraventricular injection of cNPY on precocial feeding behavior in neonatal chicks (*Gallus domesticus*). PLoS One 11:e0153342

Chen M-J, Pan N-X, Wang X-Q, Yan H-C, Gao C-Q (2020) Methionine promotes crop milk protein synthesis through the JAK2-STAT5 signaling during lactation of Domestic Pigeons (*Columba livia*). Food Funct 11:10786

Cherel Y, Robin JP, Walch O, Karmann H, Netchitailo P, Le Maho Y (1988) Fasting in King Penguin. I. Hormonal and metabolic changes during breeding. Am J Phys Regul Integr Comp Phys 254: R170–R177

Chokchaloemwong D, Prakobsaeng N, Sartsoongnoen N, Kosonsiriluk S, El Halawani M, Chaiseha Y (2013) Mesotocin and maternal care of chicks in native Thai hens (*Gallus domesticus*). Horm Behav 64:53–69

Chowdhury VS, Yamamoto K, Ubuka T, Bentley GE, Hattori A, Tsutsui K (2010) Melatonin stimulates the release of gonadotropin-inhibitory hormone by the avian hypothalamus. Endocrinology 151:271–280

Christian JJ (1980) Endocrine factors in population regulation. In: Cohen MN, Malpass RS, Klein HG (eds) Biosocial mechanisms of population regulation. Yale University Press, New Haven, CT, pp 55–115

Cicatiello AG, Di Girolamo D, Dentice M (2018) Metabolic effects of the intracellular regulation of thyroid hormone: old players, new concepts. Front Endocrinol 9:474

Cline M, Nandar W, Prall B, Bowden C, Denbow D (2008) Central visfatin causes orexigenic effects in chicks. Behav Brain Res 186:293–297

Cockrem JF (2007) Stress, corticosterone responses and avian personalities. J Ornithol 148(Suppl 2):S169–S178

Cockrem JF, Silverin B (2002) Variation within and between birds in corticosterone responses of Great Tits (*Parus major*). Gen Comp Endocrinol 125:197–206

Colca JR, Hazelwood RL (1981) Insulin, pancreatic polypeptide, and glucagon release from the chicken pancreas in vitro: responses to changes in medium glucose and free fatty acid content. Gen Comp Endocrinol 45: 482–490

Coudert E, Pascal G, Dupont J, Simon J, Cailleau-Audouin E, Crochet S, Duclos MJ, Tesseraud S, Métayer-Coustard S (2015) Phylogenesis and biological characterization of a new glucose transporter in the chicken (*Gallus gallus*), GLUT12. PLoS ONE 10:e0139517

Crossin GT, Trathan PN, Phillips RA, Gorman KB, Dawson A, Sakamoto KQ, Williams TD (2012) Corticosterone predicts foraging behavior and parental care in Macaroni Penguins. Am Nat 180:E31–E41

da Silva ES, dos Santos TV, Hoeller AA, dos Santos TS, Pereira GV, Meneghelli C, Penzlin AI, dos Santos MM, Faria MS, Paschoalini MA, Marino-Neto J (2008) Behavioral and metabolic effects of central injections of orexins/hypocretins in pigeons (*Columba livia*). Regul Pept 147:9–18

Dacke GG (1979) Calcium regulation in sub-mammalian vertebrates. Academic Press, New York

Dacke CG (2000) The parathyroids, calcitonin, and vitamin D. In: Wittow GC (ed) Sturkie's avian physiology, 5th edn. Academic Press, San Diego, CA, pp 473–488

Dallman MF, Strack AM, Akana SF, Bradbury MJ, Hanson ES, Scribner KA, Smith M (1993) Feast and famine: critical role of glucocorticoids with insulin in daily energy flow. Front Neuroendocrinol 14:303–347

Damad A, Zampiga M, Greene ES, Sirri F, Dridi S (2021) Neuropeptide Y and its receptors are expressed in chicken skeletal muscle and regulate mitochondrial function. Gen Comp Endocrinol

Dantzer B, Swanson EM (2012) Mediation of vertebrate life histories *via* insulin-like growth factor-1. Biol Rev 87:414–429

Darras VM (2019) The role of maternal thyroid hormones in avian embryonic development. Front Endocrinol 10: 66

Dawson A, Goldsmith AR (1982) Prolactin and gonadotrophin secretion in wild starlings (*Sturnus vulgaris*) during the annual cycle and in relation to nesting, incubation, and rearing young. Gen Comp Endocrinol 48:213–221

De Groef B, Grommen SV, Darras VM (2013) Hatching the cleidoic egg: the role of thyroid hormones. Front Endocrinol 4:63

de Matos R (2008) Calcium metabolism in birds. Vet Clin North Am Exot Anim Pract 11:59–82

DeGolier TF, Duke GE, Carraway RE (1997) Neurotensin decreases pepsin output and gastrointestinal motility in chickens. Poult Sci 76:1435–1439

DeGolier TF, Place AR, Duke GE, Carraway RE (1999) Neurotensin modulates the composition of pancreatic exocrine secretions in chickens. J Exp Zool 283:455–462

DeGolier TF, Brown DR, Duke GE, Palmer MM, Swenson JR, Carraway RE (2013a) Neurotensin and cholecystokinin contract gallbladder circular muscle in chickens. Poult Sci 92:2156–2162

DeGolier TF, Carraway RE, Duke GE (2013b) Release of avian neurotensin in response to intraluminal contents in the duodenum of chickens. Poult Sci 92:418–423

Del Giudice M, Ellis BJ, Shirtcliff EA (2011) The adaptive calibration model of stress responsivity. Neurosci Biobehav Rev 35:1562–1592

Del Rio CM, Gutiérrez-Guerrero YT (2020) An evolutionary remedy for an abominable physiological mystery: benign hyperglycemia in birds. J Mol Evol 88:715–719

Denac M, Scharrer E (1987) Effect of neurotensin on the smooth muscle of the chicken crop. Comp Biochem Physiol C 87:325–327

Denbow DM, Meade S, Robertson A, McMurtry JP, Richards M, Ashwell C (2000) Leptin-induced decrease in food intake in chickens. Physiol Behav 69:359–362

DeSimone JG, Ramirez MG, Elowe CR, Griego MS, Breuner CW, Gerson AR (2020) Developing a stopover-CORT hypothesis: corticosterone predicts body composition and refueling rate in Gray Catbirds during migratory stopover. Horm Behav 124:104776

Deviche P, Bittner S, Gao S, Valle S (2017) Roles and mechanistic bases of glucocorticoid regulation of avian reproduction. Integr Comp Biol 57:1184–1193

Dezonne RS, Lima FRS, Trentin AG, Gomes FC (2015) Thyroid hormone and astroglia: endocrine control of the neural environment. J Neuroendocrinol 27:435–445

Dhande RR, Suryawanshi SA, Pandey AK (1997) Effect of salmon calcitonin administration on plasma calcium and inorganic phosphate levels of the intact and parathyroidectomized Grey Quail, *Coturnix coturnix coturnix* Linnaeus. Indian J Anim Sci 67:906–909

Di Scala-Guenot D, Strosser M-T, Mialhe P (1985) The biological activity of duck 'big'somatostatin on chicken adipose tissue. Biochimica et Biophysica Acta (BBA)-Molecular. Cell Res 845:261–264

Dias JA (2005) Fertility hormone in repose. Nature 433: 203–204

Dimaline R, Dockray GJ (1979) Potent stimulation of the avian exocrine pancreas by porcine and chicken vasoactive intestinal peptide. J Physiol 294:153–163

Duan C, Ren H, Gao S (2010) Insulin-like growth factors (IGFs), IGF receptors, and IGF-binding proteins: roles in skeletal muscle growth and differentiation. Gen Comp Endocrinol 167:344–351

Duke GE (1992) Recent studies on regulation of gastric motility in turkeys. Poult Sci 71:1–8

Duke GE, Kimmel JR, Hunt H, Pollock HG (1985) The influence of avian pancreatic polypeptide on gastric secretion and motility in laying hens. Poult Sci 64: 1231–1235

Duke GE, Larntz K, Hunt H (1987) The influence of cholecystokinin, vasoactive intestinal peptide and secretin on pancreatic and bilary secretion in laying hens. Comp Biochem Physiol C 86:97–102

Duke GE, Chaplin SB, Hunt H, Wiard-Bauer LK (1988) The influence of avian pancreatic polypeptide on gastric secretion and motility in Red-tailed Hawks (*Buteo jamaicensis*). Comp Biochem Physiol C 90:231–235

Dunn IC, Wilson PW, Smulders TV, Sandilands V, D'Eath RB, Boswell T (2013) Hypothalamic agouti-related protein expression is affected by both acute and chronic experience of food restriction and re-feeding in chickens. J Neuroendocrinol 25:920–928

Dunn IC, Wilson PW, D'Eath RB, Boswell T (2015) Hypothalamic agouti-related peptide mRNA is elevated during natural and stress-induced anorexia. J Neuroendocrinol 27:681–691

Dupont J, Tesseraud S, Simon J (2009) Insulin signaling in chicken liver and muscle. Gen Comp Endocrinol 163: 52–57

Duque JF, Leichner W, Ahmann H, Stevens JR (2018) Mesotocin influences Pinyon Jay prosociality. Biol Lett 14:20180105

DuRant S, Love AC, Belin B, Tamayo-Sanchez D, Pacheco MS, Dickens MJ, Calisi RM (2020) Captivity alters neuroendocrine regulators of stress and reproduction in the hypothalamus in response to acute stress. Gen Comp Endocrinol 295:113519

Dūrītis I, Hussar P, Mugurēvičs A (2021) Gastrin and somatostatin enteroendocrine cells in the small intestines of Ostrich (*Struthio camelus* var. *domesticus*) during pre-and post-hatching period. Anat Histol Embryol 50:550–555

Eastwood JR, Hall ML, Teunissen N, Kingma SA, Hidalgo Aranzamendi N, Fan M, Roast M, Verhulst S, Peters A (2019) Early-life telomere length predicts lifespan and lifetime reproductive success in a wild bird. Mol Ecol 28:1127–1137

Eikenaar C, Hessler S, Ballstaedt E, Schmaljohann H, Kaiya H (2018) Ghrelin, corticosterone and the resumption of migration from stopover, an automated telemetry study. Physiol Behav 194:450–455

Eikenaar C, Schäfer J, Hessler S, Packmor F, Schmaljohann H (2020) Diel variation in corticosterone and departure decision making in migrating birds. Horm Behav 122:104746
El-Salhy M, Wilander E, Grimelius L, Terenius L, Lundberg LM, Tatemoto K (1982) The distribution of polypeptide YY (PYY)-and pancreatic polypeptide (PP)-immunoreactive cells in the domestic fowl. Histochemistry 75:25–30
Eraud C, Dorie A, Jacquet A, Faivre B (2008) The crop milk: a potential new route for carotenoid-mediated parental effects. J Avian Biol 39:247–251
Estienne A, Brossaud A, Reverchon M, Ramé C, Froment P, Dupont J (2020) Adipokines expression and effects in oocyte maturation, fertilization and early embryo development: lessons from mammals and birds. Int J Mol Sci 21:3581
Estienne A, Ramé C, Ganier P, Chahnamian M, Barbe A, Grandhaye J, Dubois J-P, Batailler M, Migaud M, Lecompte F, Adriaensen H (2021) Chemerin impairs food intake and body weight in chicken: focus on hypothalamic neuropeptides gene expression and AMPK signaling pathway. Gen Comp Endocrinol 304:113721
Fischer D, Marrotte RR, Chin EH, Coulson S, Burness G (2020) Maternal glucocorticoid levels during incubation predict breeding success, but not reproductive investment, in a free-ranging bird. Biol Open 9: bio045898
Fivizzani AJ, Colwell MA, Oring LW (1986) Plasma steroid hormone levels in free-living Wilson's Phalaropes, *Phalaropus tricolor*. Gen Comp Endocrinol 62:137–144
Flatt T, Heyland A (eds) (2011) Mechanisms of life history evolution: the genetics and physiology of life history traits and trade-offs. Oxford University Press, Oxford, UK
Friedman JM, Halaas JL (1998) Leptin and the regulation of body weight in mammals. Nature 395:763–770
Friedman-Einat M, Seroussi E (2019) Avian leptin: bird's-eye view of the evolution of vertebrate energy-balance control. Trends Endocrinol Metab 30:819–832
Fruebis J, Tsao TS, Javorschi S, Ebbets-Reed D, Erickson MRS, Yen FT, Bihain BE, Lodish HF (2001) Proteolytic cleavage product of 30-kDa adipocyte complement-related protein increases fatty acid oxidation in muscle and causes weight loss in mice. Proc Natl Acad Sci USA 98:2005–2010
Fu W, Gao P (2000) Effects of crude avian pancreatic polypeptide on broiler growth, growth hormone and thyroid hormone levels in blood. Acta Zoonutrimenta Sin 12:39–42
Furuse M (1999) Release and endogenous actions of the gastrin/cholecystokinin (CCK) family in the chicken. J Exp Zool 283:448–454
Furuse M, Dockray GJ (1995) The regulation of gastrin secretion in the chicken. Regul Pept 55:253–259
Furuse M, Matsumoto M, Okumura J-I, Sugahara K, Hasegawa S (1997) Intracerebroventricular injection of mammalian and chicken glucagon-like peptide-1 inhibits food intake of the neonatal chick. Brain Res 755:167–169
Furuse M, Ando R, Bungo T, Shimojo M, Masuda Y (1999a) Intracerebroventricular injection of orexins does not stimulate food intake in neonatal chicks. Br Poult Sci 40:698–700
Furuse M, Ao R, Bungo T, Ando R, Shimojo M, Masuda Y, Saito N (1999b) Central gastrin inhibits feeding behavior and food passage in neonatal chicks. Life Sci 65:305–311
Geelissen SME, Swennen Q, Van der Geyten S, Kühn ER, Kaiya H, Kangawa K, Decuypere E, Buyse J, Darras VM (2006) Peripheral ghrelin reduces food intake and respiratory quotient in chicken. Domest Anim Endocrinol 30:108–116
Gertler A, Shinder D, Yosefi S, Shpilman M, Rosenblum CI, Ruzal M, Seroussi E, Friedman-Einat M (2014) Pegylated leptin antagonist with strong orexigenic activity in mice is not effective in chickens. J Exp Biol 217:180–184
Goodson JL (1998a) Territorial aggression and dawn song are modulated by septal vasotocin and vasoactive intestinal polypeptide in male Field Sparrows (*Spizella pusilla*). Horm Behav 34:67–77
Goodson JL (1998b) Vasotocin and vasoactive intestinal polypeptide modulate aggression in a territorial songbird, the Violet-eared Waxbill (Estrildidae: *Uraeginthus granatina*). Gen Comp Endocrinol 111: 233–244
Goodson JL (2005) The vertebrate social behavior network: evolutionary themes and variations. Horm Behav 48:11–22
Goodson JL, Adkins-Regan E (1999) Effect of intraseptal vasotocin and vasoactive intestinal polypeptide infusions on courtship song and aggression in the male Zebra Finch (*Taeniopygia guttata*). J Neuroendocrinol 11:19–26
Goodson JL, Eibach R, Sakata J, Adkins-Regan E (1998) Effect of septal lesions on male song and aggression in the colonial Zebra Finch (*Taeniopygia guttata*) and the territorial Field Sparrow (*Spizella pusilla*). Behav Brain Res 98:167–180
Goodson JL, Evans AK, Wang Y (2006) Neuropeptide binding reflects convergent and divergent evolution in species-typical group sizes. Horm Behav 50:223–236
Goodson JL, Schrock SE, Klatt JD, Kabelik D, Kingsbury MA (2009) Mesotocin and nonapeptide receptors promote estrildid flocking behavior. Science 325:862–866
Goodson JL, Kelly AM, Kingsbury MA, Thompson RR (2012) An aggression-specific cell type in the anterior hypothalamus of finches. Proc Natl Acad Sci USA 109:13847–13852
Goymann W, Wingfield JC (2004) Competing females and caring males. Sex steroids in African Black Coucals, *Centropus grillii*. Anim Behav 68:733–740
Goymann W, Lupi S, Kaiya H, Cardinale M, Fusani L (2017) Ghrelin affects stopover decisions and food

intake in a long-distance migrant. Proc Natl Acad Sci USA 114:1946–1951
Gratto-Trevor CL, Fivizzani AJ, Oring LW, Cooke F (1990a) Seasonal changes in gonadal steroids of a monogamous versus a polyandrous shorebird. Gen Comp Endocrinol 80:407–418
Gratto-Trevor CL, Oring LW, Fivizzani AJ, El Halawani ME, Cooke F (1990b) The role of prolactin in parental care in a monogamous and a polyandrous shorebird. Auk 107:718–729
Greene ES, Zampiga M, Sirri F, Ohkubo T, Dridi S (2020) Orexin system is expressed in avian liver and regulates hepatic lipogenesis via ERK1/2 activation. Sci Rep 10: 19191
Gui X, DeGolier TF, Duke GE, Carraway RE (2000) Neurotensin elevates hepatic bile acid secretion in chickens by a mechanism requiring an intact enterohepatic circulation. Comp Biochem Physiol C 127:61–70
Gülmez N, Kocamiş H, Aslan Ş, Nazli M (2004) Immunohistochemical distribution of cells containing insulin, glucagon and somatostatin in the goose (*Anser anser*) pancreas. Turk J Vet Anim Sci 28:403–407
Gwee MC, Gopalakrishnakone P, Cheah LS, Bowman WC (1995) L-arginine-nitric oxide pathway involvement in the nerve-evoked relaxant responses of the 5-HT precontracted chick isolated upper oesophagus. J Auton Pharmacol 15:151–158
Hall ZJ, Healy SD, Meddle SL (2015) A role for nonapeptides and dopamine in nest-building behaviour. J Neuroendocrinol 27:158–165
Halsey LG (2018) Keeping slim when food is abundant: what energy mechanisms could be at play? Trends Ecol Evol 33:745–753
Hannan FM, Kallay E, Chang W, Brandi ML, Thakker RV (2019) The calcium-sensing receptor in physiology and in calcitropic and noncalcitropic diseases. Nat Rev Endocrinol 15:33–51
Harvey S (2010) Extrapituitary growth hormone. Endocrine 38:335–359
Harvey S (2013) Growth hormone and growth? Gen Comp Endocrinol 190:3–9
Harvey S, Baudet M-L (2014) Extrapituitary growth hormone and growth? Gen Comp Endocrinol 205:55–61
Harvey S, Johnson CDM, Sharma P, Sanders EJ, Hull KL (1998) Growth hormone: a paracrine growth factor in embryonic development? Comp Biochem Physiol C: Pharmacol Toxicol Endocrinol 119:305–315
Harvey S, Johnson CD, Sanders EJ (2000) Extrapituitary growth hormone in peripheral tissues of early chick embryos. J Endocrinol 166:489–502
Hau M, Casagrande S, Ouyang JQ, Baugh AT (2016) Glucocorticoid-mediated phenotypes in vertebrates: multilevel variation and evolution. Adv Study Behav 48:41–115
Hazelwood RL (1973) The avian endocrine pancreas. Am Zool 13:699–709
Heath J (1997) Corticosterone levels during nest departure of juvenile American Kestrels. Condor 99:806–811
Heidinger BJ, Blount JD, Boner W, Griffiths K, Metcalfe NB, Monaghan P (2012) Telomere length in early life predicts lifespan. Proc Natl Acad Sci USA 109:1743–1748
Hen G, Yosefi G, Simchaev V, Shinder D, Hruby VJ, Friedman-Einat M (2006) The melanocortin circuit in obese and lean strains of chicks. J Endocrinol 190: 527–535
Henderson LJ, Cockcroft RC, Kaiya H, Boswell T, Smulders TV (2018) Peripherally injected ghrelin and leptin reduce food hoarding and mass gain in the Coal Tit (*Periparus ater*). Proc R Soc B 285:20180417
Higgins SE, Ellestad LE, Trakooljul N, McCarthy F, Saliba J, Cogburn LA, Porter TE (2010) Transcriptional and pathway analysis in the hypothalamus of newly hatched chicks during fasting and delayed feeding. BMC Genomics 11:162
Ho DH, Reed WL, Burggren WW (2011) Egg yolk environment differentially influences physiological and morphological development of broiler and layer chicken embryos. J Exp Biol 214:619–628
Hoi H, Griggio M (2008) Dual utility of a melanin-based ornament in Bearded Tits. Ethology 114:1094–1100
Holberton RL, Parrish JD, Wingfield JC (1996) Modulation of the adrenocortical stress response in Neotropical migrants during autumn migration. Auk 113:558–564
Honda K, Saneyasu T, Yamaguchi T, Shimatani T, Aoki K, Nakanishi K, Kamisoyama H (2014) Intracerebroventricular administration of chicken oxyntomodulin suppresses food intake and increases plasma glucose and corticosterone concentrations in chicks. Neurosci Lett 564:57–61
Honda K, Saneyasu T, Kamisoyama H (2017) Gut hormones and regulation of food intake in birds. J Poult Sci 54:103–110
Honey RN, Schwarz JA, Mathe CJ, Weir GC (1980) Insulin, glucagon, and somatostatin secretion from isolated perfused rat and chicken pancreas-duodenum. Am J Physiol Endocrinol Metab 238:E150–E156
Hope SF, DuRant SE, Angelier F, Hallagan JJ, Moore IT, Parenteau C, Kennamer RA, Hopkins WA (2020) Prolactin is related to incubation constancy and egg temperature following a disturbance in a precocial bird. Gen Comp Endocrinol 295:113489
Hsu BY, Dijkstra C, Darras VM, de Vries B, Groothuis TG (2016) Maternal adjustment or constraint: differential effects of food availability on maternal deposition of macro-nutrients, steroids and thyroid hormones in Rock Pigeon eggs. Ecol Evol 6:397–411
Hu XF, Guo YM, Huang BY, Bun S, Zhang LB, Li JH, Liu D, Long FY, Yang X, Jiao P (2010) The effect of glucagon-like peptide 2 injection on performance, small intestinal morphology, and nutrient transporter expression of stressed broiler chickens. Poult Sci 89: 1967–1974
Hu X-C, Gao C-Q, Wang X-H, Yan H-C, Chen Z-S, Wang X-Q (2016) Crop milk protein is synthesised following activation of the IRS1/Akt/TOR signalling pathway in

the Domestic Pigeon (*Columba livia*). Br Poult Sci 57: 855–862
Huang Y-H, Tsai M-M, Lin K-H (2008) Thyroid hormone dependent regulation of target genes and their physiological significance. Chang Gung Med J 31:325–334
Huang G, Li J, Wang H, Lan X, Wang Y (2014) Discovery of a novel functional leptin protein (LEP) in Zebra Finches: evidence for the existence of an authentic avian leptin gene predominantly expressed in the brain and pituitary. Endocrinology 155:3385–3396
Hulbert AJ (2000) Thyroid hormones and their effects: a new perspective. Biol Rev 75:519–631
Humayun KAKM, Aoyama M, Sugita S (2012) Morphological and histological studies on the adrenal gland of the chicken (*Gallus domesticus*). J Poult Sci 49:39–45
Ieromnimon V (1978) Beobachtungen über die wirkung von gormonen auf das zugverhalten bei Rotkehlchen (*Erithacus rubecula*). II. Die wirkung von pharmaka. Die Vogelwarte 29:221–230
Jaatinen K, Seltmann MW, Hollmén T, Atkinson S, Mashburn K, Öst M (2013) Context dependency of baseline glucocorticoids as indicators of individual quality in a capital breeder. Gen Comp Endocrinol 191:231–238
Ji J, Tao Y, Zhang X, Pan J, Zhu X, Wang H, Du P, Zhu Y, Huang Y, Chen W (2020) Dynamic changes of blood glucose, serum biochemical parameters and gene expression in response to exogenous insulin in Arbor Acres broilers and Silky fowls. Sci Rep 10:6697
Jimeno B, Hau M, Verhulst S (2018) Corticosterone levels reflect variation in metabolic rate, independent of 'stress'. Sci Rep 8:13020
John TM, George JC (1978) Circulating levels of thyroxine ($T_4$) and triiodothyronine ($T_3$) in the migratory Canada Goose. Physiol Zool 51:361–370
Johnson EM, Hazelwood RL (1982) Avian pancreatic polypeptide (APP) levels in fasted-refed chickens: locus of postprandial trigger? Proc Soc Exp Biol Med 169:176–182
Kameda Y (2017) Morphological and molecular evolution of the ultimobranchial gland of nonmammalian vertebrates, with special reference to the chicken C cells. Dev Dyn 246:719–739
Kara A, Tekiner D, Şimşek N, Balkaya H, Özüdoğru Z (2014) Distribution and location of endocrine cells in the pancreas of the sparrowhawk, *Accipiter nisus*. Kafkas Universitesi Veteriner Fakultesi Dergisi 20: 307–312
Katayama S, Hamasu K, Shigemi K, Cline MA, Furuse M (2010) Intracerebroventricular injection of orexin-A, but not orexin-B, induces arousal of layer-type neonatal chicks. Comp Biochem Physiol A Mol Integr Physiol 157:132–135
Kawai M, Rosen CJ (2010) The IGF-I regulatory system and its impact on skeletal and energy homeostasis. J Cell Biochem 111:14–19
Keen-Rhinehart E, Bartness TJ (2005) Peripheral ghrelin injections stimulate food intake, foraging, and food hoarding in Siberian hamsters. Am J Phys Regul Integr Comp Phys 288:R716–R722
Kelly EM (2019) Committed for the long haul: do nonapeptides regulate long-term pair maintenance in Zebra Finches? Gen Comp Endocrinol 276:86–92
Kelly EM, Adkins-Regan E (2020) Do nonapeptides regulate parental care depending on experience in Zebra Finches? Horm Behav 117:104603
Kelly AM, Kingsbury MA, Hoffbuhr K, Schrock SE, Waxman B, Kabelik D, Thompson RR, Goodson JL (2011) Vasotocin neurons and septal V1a-like receptors potently modulate songbird flocking and responses to novelty. Horm Behav 60:12–21
Kerschnitzki M, Zander T, Zaslansky P, Fratzl P, Shahar R, Wagermaier W (2014) Rapid alterations of avian medullary bone material during the daily egg-laying cycle. Bone 69:109–117
Kershaw EE, Flier JS (2004) Adipose tissue as an endocrine organ. J Clin Endocrinol Metab 89:2548–2556
Khan MZ, McNabb FMA, Walters JR, Sharp PJ (2001) Patterns of testosterone and prolactin concentrations and reproductive behavior of helpers and breeders in the cooperatively breeding Red-cockaded Woodpecker (*Picoides borealis*). Horm Behav 40:1–13
Khan MSI, Cline MA, Aramaki T, Ueda H, Tachibana T (2013) Feeding response following central administration of chicken vasoactive intestinal peptide in chicks. Gen Comp Endocrinol 184:61–66
Kiezun M, Smolinska N, Maleszka A, Dobrzyn K, Szeszko K, Kaminski T (2014) Adiponectin expression in the porcine pituitary during the estrous cycle and its effect on LH and FSH secretion. Am J Physiol Endocrinol Metab 307:E1038–E1046
Kikuchi K, Buonomo FC, Kajimoto Y, Rotwein P (1991) Expression of insulin-like growth factor-I during chicken development. Endocrinology 128:1323–1328
Kimmel JR, Hayden LJ, Pollock HG (1975) Isolation and characterization of a new pancreatic polypeptide hormone. J Biol Chem 250:9369–9376
Kingsbury MA (2015) New perspectives on vasoactive intestinal polypeptide as a widespread modulator of social behavior. Curr Opin Behav Sci 6:139–147
Kingsbury MA, Wilson LC (2016) The role of VIP in social behavior: neural hotspots for the modulation of affiliation, aggression, and parental care. Integr Comp Biol 56:1238–1249
Kingsbury MA, Jan N, Klatt JD, Goodson JL (2015) Nesting behavior is associated with VIP expression and VIP–Fos colocalization in a network-wide manner. Horm Behav 69:68–81
Kirouac L, Rajic AJ, Cribbs DH, Padmanabhan J (2017) Activation of Ras-ERK signaling and GSK-3 by amyloid precursor protein and amyloid beta facilitates neurodegeneration in Alzheimer's disease. Eneuro 4: e0149–e0116
Kitaysky AS, Kitaiskaia EV, Piatt JF, Wingfield JC (2003) Benefits and costs of increased levels of corticosterone in seabird chicks. Horm Behav 43:140–149

Kitazawa T, Kaiya H (2019) Regulation of gastrointestinal motility by motilin and ghrelin in vertebrates. Front Endocrinol 10:278

Kitazawa T, Ukai H, Komori S, Taneike T (2006) Pharmacological characterization of 5-hydroxytryptamine-induced contraction in the chicken gastrointestinal tract. Auton Autacoid Pharmacol 26:157–168

Klatt JD, Goodson JL (2013) Oxytocin-like receptors mediate pair bonding in a socially monogamous songbird. Proc R Soc B 280:20122396

Koch KA, Wingfield JC, Buntin JD (2002) Glucocorticoids and parental hyperphagia in Ring Doves (*Streptopelia risoria*). Horm Behav 41:9–21

Koolhaas JM, Korte SM, De Boer SF, Van Der Vegt BJ, Van Reenen CG, Hopster H, De Jong IC, Ruis MAW, Blokhuis HJ (1999) Coping styles in animals: current status in behavior and stress-physiology. Neurosci Biobehav Rev 23:925–935

Korte SM, Koolhaas JM, Wingfield JC, McEwen BS (2005) The Darwinian concept of stress: benefits of allostasis and costs of allostatic load and the trade-offs in health and disease. Neurosci Biobehav Rev 29:3–38

Kosonsiriluk S, Sartsoongnoen N, Chaiyachet O-A, Prakobsaeng N, Songserm T, Rozenboim I, El Halawani M, Chaiseha Y (2008) Vasoactive intestinal peptide and its role in continuous and seasonal reproduction in birds. Gen Comp Endocrinol 159:88–97

Kriegsfeld LJ, Ubuka T, Bentley GE, Tsutsui K (2015) Seasonal control of gonadotropin-inhibitory hormone (GnIH) in birds and mammals. Front Neuroendocrinol 37:65–75

Krzysik-Walker SM, Hadley JA, Pesall JE, McFarland DC, Vasilatos-Younken R, Ramachandran R (2011) Nampt/visfatin/PBEF affects expression of myogenic regulatory factors and is regulated by interleukin-6 in chicken skeletal muscle cells. Comp Biochem Physiol A 159:413–421

Ku S-K, Lee J-H, Lee H-S (2000) An immunohistochemical study of the insulin-, glucagon-and somatostatin-immunoreactive cells in the developing pancreas of the chicken embryo. Tissue Cell 32:58–65

Kuenzel WJ (1974) Multiple effects of ventromedial hypothalamic lesions in the White-throated Sparrow, *Zonotrichia albicollis*. J Comp Physiol 90:169–182

Kuenzel WJ, Douglass LW, Davison BA (1987) Robust feeding following central administration of neuropeptide Y or peptide YY in chicks, *Gallus domesticus*. Peptides 8:823–828

Kuenzel WJ, Beck MM, Teruyama R (1999) Neural sites and pathways regulating food intake in birds: a comparative analysis to mammalian systems. J Exp Zool 283:348–364

Kuenzel WJ, Kang SW, Jurkevich A (2020) The vasotocinergic system and its role in the regulation of stress in birds. Vitam Horm 113:183–216

Kumar V, Follett BK (1993) The circadian nature of melatonin secretion in Japanese Quail (*Coturnix coturnix japonica*). J Pineal Res 14:192–200

Lancaster LT, Hazard LC, Clobert J, Sinervo BR (2008) Corticosterone manipulation reveals differences in hierarchical organization of multidimensional reproductive trade-offs in r-strategist and K-strategist females. J Evol Biol 21:556–565

Landys MM, Ramenofsky M, Wingfield JC (2006) Actions of glucocorticoids at a seasonal baseline as compared to stress-related levels in the regulation of periodic life processes. Gen Comp Endocrinol 148: 132–149

Londraville RL, Macotela Y, Duff RJ, Easterling MR, Liu Q, Crespi EJ (2014) Comparative endocrinology of leptin: assessing function in a phylogenetic context. Gen Comp Endocrinol 203:146–157

Larsson L-I, Sundler F, Håkanson R (1975) Immunohistochemical localization of human pancreatic polypeptide (HPP) to a population of islet cells. Cell Tissue Res 156:167–171

Lassiter K, Greene E, Piekarski A, Faulkner OB, Hargis BM, Bottje W, Dridi S (2015) Orexin system is expressed in avian muscle cells and regulates mitochondrial dynamics. Am J Phys Regul Integr Comp Phys 308:R173–R187

Lattin CR, Bauer CM, de Bruijn R, Romero LM (2012) Hypothalamus-pituitary-adrenal axis activity and the subsequent response to chronic stress differ depending upon life history stage. Gen Comp Endocrinol 178: 494–501

Lattin CR, Breuner CW, Romero LM (2016) Does corticosterone regulate the onset of breeding in free-living birds?: The CORT-flexibility hypothesis and six potential mechanisms for priming corticosteroid function. Horm Behav 78:107–120

Laucht S, Dale J (2012) Correlations of condition, testosterone, and age with multiple ornaments in male House Sparrows: patterns and implications. Condor 114:865–873

Lee AK, Cockburn A (1985) Evolutionary ecology of marsupials. Cambridge University Press, Cambridge, UK

Lees JJ, Lindholm C, Batakis P, Busscher M, Altimiras J (2017) The physiological and neuroendocrine correlates of hunger in the Red Junglefowl (*Gallus gallus*). Sci Rep 7:17984

Li J, Hu Z, de Lecea L (2014) The hypocretins/orexins: integrators of multiple physiological functions. Br J Pharmacol 171:332–350

Li Z, Liu X, Zhang P, Han R, Sun G, Jiang R, Wangm Y, Liu X, Li W, Kang X, Tian Y (2018) Comparative transcriptome analysis of hypothalamus-regulated feed intake induced by exogenous visfatin in chicks. BMC Genomics 19:249

Lipshutz SE, Rosvall KA (2020a) Testosterone secretion varies in a sex- and stage-specific manner: insights on the regulation of competitive traits from a sex-role reversed species. Gen Comp Endocrinol 292:113444

Lipshutz SE, Rosvall KA (2020b) Neuroendocrinology of sex-role reversal. Integr Comp Biol 60:692–702

Liu J, Garza JC, Truong HV, Henschel J, Zhang W, Lu ZY (2007) The melanocortinergic pathway is rapidly recruited by emotional stress and contributes to stress-induced anorexia and anxiety-like behavior. Endocrinology 148:5531–5540

Lodjak J, Mägi M, Tilgar V (2014) Insulin-like growth factor 1 and growth rate in nestlings of a wild passerine bird. Funct Ecol 28:159–166

Lodjak J, Mägi M, Sild E, Mänd R (2017) Causal link between insulin-like growth factor 1 and growth in nestlings of a wild passerine bird. Funct Ecol 31:184–191

Lodjak J, Mänd R, Mägi M (2018) Insulin-like growth factor 1 and life-history evolution of passerine birds. Funct Ecol 32:313–323

Lõhmus M, Björklund M (2009) Leptin affects life history decisions in a passerine bird: a field experiment. PLoS ONE 4:e4602

Lõhmus M, Sundström LF, El Halawani M, Silverin B (2003) Leptin depresses food intake in Great Tits (*Parus major*). Gen Comp Endocrinol 131:57–61

Lõhmus M, Sundström LF, Silverin B (2006) Chronic administration of leptin in Asian Blue Quail. J Exp Zool A Comp Exp Biol 305:13–22

Loveland JL, Giraldo-Deck LM, Lank DB, Goymann W, Gahr M, Küpper C (2021) Functional differences in the hypothalamic-pituitary-gonadal axis are associated with alternative reproductive tactics based on an inversion polymorphism. Horm Behav 127:104877

Lu JW, McMurtry JP, Coon CN (2007) Developmental changes of plasma insulin, glucagon, insulin-like growth factors, thyroid hormones, and glucose concentrations in chick embryos and hatched chicks. Poult Sci 86:673–683

Luna M, Martínez-Moreno CG, Ahumada-Solórzano MS, Harvey S, Carranza M, Arámburo C (2014) Extrapituitary growth hormone in the chicken reproductive system. Gen Comp Endocrinol 203:60–68

Mabayo RT, Furuse M, Yang S-I, Okumura J-I (1992) Medium-chain triacylglycerols enhance release of cholecystokinin in chicks. J Nutr 122:1702–1705

MacDougall-Shackleton SA, Bonier F, Romero LM, Moore IT (2019) Glucocorticoids and "stress" are not synonymous. Integr Org Biol 1:obz017

Magnus-Levy A (1895) Über den respiratorischen Gaswechsel unter dem Einfluss der thyroidea sowie unter verschiedenen pathalogischen Zuständen. Berliner Klinische Wochenschrift 32:650

Mahr K, Vincze O, Tóth Z, Hoi H, Lendvai ÁZ (2020) Insulin-like growth factor 1 is related to the expression of plumage traits in a passerine species. Behav Ecol Sociobiol 74:39

Maney DL, Schoech SJ, Sharp PJ, Wingfield JC (1999) Effects of vasoactive intestinal peptide on plasma prolactin in passerines. Gen Comp Endocrinol 113:323–330

Martin LB (2009) Stress and immunity in wild vertebrates: timing is everything. Gen Comp Endocrinol 163:70–76

Martins TL, Roberts ML, Giblin I, Huxham R, Evans MR (2007) Speed of exploration and risk-taking behavior are linked to corticosterone titres in Zebra Finches. Horm Behav 52:445–453

Martinez V, Rodriguez-Membrilla A, Jimenez M, Gonalons E, Vergara P (1993) Immunohistochemical differentiation of gastrin and cholecystokinin in gastrointestinal tract of chickens. Poult Sci 72:2328–2336

McConn BR, Gilbert ER, Cline MA (2018) Appetite-associated responses to central neuropeptide Y injection in quail. Neuropeptides 69:9–18

McEwen BS, Wingfield JC (2003) The concept of allostasis in biology and biomedicine. Horm Behav 43:2–15

McMurtry JP, Francis GL, Upton Z (1997) Insulin-like growth factors in poultry. Domest Anim Endocrinol 14:199–229

McNabb FMA (1992) Thyroid hormones. Prentice Hall, Englewood Cliffs, NJ

McNabb FMA (2007) The hypothalamic-pituitary-thyroid (HPT) axis in birds and its role in bird development and reproduction. Crit Rev Toxicol 37:163–193

Mello CV, Lovell PV (2018) Avian genomics lends insights into endocrine function in birds. Gen Comp Endocrinol 256:123–129

Melo-Duran D, Gonzalez-Ortiz G, Sola-Oriol D, Martinez-Mora M, Perez JF, Bedford MR (2019) Relationship between peptide YY, cholecystokinin and fermentation products in fasted, re-fed and ad libitum fed broiler chickens. Anim Feed Sci Technol 247 141–148

Mellouk N, Rame C, Barbe A, Grandhaye J, Froment P, Dupont J (2018) Chicken is a useful model to investigate the role of adipokines in metabolic and reproductive diseases. Int J Endocrinol 2018:4579734

Mensah-Brown EPK, Bailey TA, Pallot DJ, Garner A (2000) Peptidergic hormones and neuropeptides, and aminergic neurotransmitters of the pancreatic islets of the Houbara Bustard (*Chlamydotis undulata*). J Anat 196:233–241

Merryman JI, Buckles EL (1998) The avian thyroid gland. Part one: a review of the anatomy and physiology. J Avian Med Surg 12:234–237

Mishra SK, Raviprakash V (1983) Mechanism of 5-hydroxytryptamine-induced biphasic responses of fowl rectum. Arch Int Pharmacodyn Ther 265:29–35

Moawad UK, Randa MH (2017) Histocytological and histochemical features of the adrenal gland of adult Egyptian native breeds of chicken (*Gallus gallus domesticus*). Beni-Suef Univ J Basic Appl Sci 6:199–208

Moghadam PD, Mohammadpour AA (2017) Histomorphological and stereological study on the adrenal glands of adult female Guinea fowl (*Numida meleagris*). Comp Clin Pathol 26:1227–1231

Moiron M, Mathot KJ, Dingemanse NJ (2018) To eat and not be eaten: diurnal mass gain and foraging strategies in wintering Great Tits. Proc R Soc B 285:20172868

Mondal S, Giri D, Mugesh G (2020) Halogen bonding in the molecular recognition of thyroid hormones and their metabolites by transport proteins and thyroid hormone receptors. J Indian Inst Sci 100:231–247

Monir MM, Hiramatsu K, Yamasaki A, Nishimura K, Watanabe T (2014) The influence of restricted feeding on glucagon-like peptide-1 (GLP-1)-containing cells in the chicken small intestine. Anat Histol Embryol 43: 153–158

Montoya B, Briga M, Jimeno B, Moonen S, Verhulst S (2018) Baseline glucose level is an individual trait that is negatively associated with lifespan and increases due to adverse environmental conditions during development and adulthood. J Comp Physiol B 188:517–526

Muck C, Goymann W (2011) Throat patch size and darkness covaries with testosterone in females of a sex-role reversed species. Behav Ecol 22:1312–1319

Mullur R, Liu Y-Y, Brent GA (2014) Thyroid hormone regulation of metabolism. Physiol Rev 94:355–382

Murai A, Satoh S, Okumura JI, Furuse M (2000) Factors regulating amylase secretion from chicken pancreatic acini in vitro. Life Sci 66:585–591

Murphy AE, Harvey S (2001) Extrapituitary βTSH and GH in early chick embryos. Mol Cell Endocrinol 185: 161–171

Nair NG, Pant K, Chandola-Saklani A (1994) Environmental and hormonal control of vernal migration in Redheaded Bunting (*Emberiza bruniceps*). J Biosci 19:453–466

Nowack J, Giroud S, Arnold W, Ruf T (2017) Muscle non-shivering thermogenesis and its role in the evolution of endothermy. Front Physiol 8:00889

Numao M, Sudo H, Yamamoto I, Nakao N, Kaiya H, Miyazato M, Tsushima N, Tanaka M (2011) Molecular characterization of structure and tissue distribution of chicken neurotensin receptor. Gen Comp Endocrinol 171:33–38

Ocón-Grove OM, Krzysik-Walker SM, Maddineni SR, Hendricks GL III, Ramachandran R (2010) NAMPT (visfatin) in the chicken testis: influence of sexual maturation on cellular localization, plasma levels and gene and protein expression. Reproduction 139:217–226

Ondrasek NR, Freeman SM, Bales KL, Calisi RM (2018) Nonapeptide receptor distributions in promising avian models for the neuroecology of flocking. Front Neurosci 12:713

Opel H (1979) The hypothalamus and reproduction in the female. Poult Sci 58:1607–1618

Oring LW, Fivizzani AJ, El Halawani ME, Goldsmith A (1986) Seasonal changes in prolactin and luteinizing hormone in the polyandrous Spotted Sandpiper, *Actitis macularia*. Gen Comp Endocrinol 62:394–403

Ouyang JQ, Quetting M, Hau M (2012) Corticosterone and brood abandonment in a passerine bird. Anim Behav 84:261–268

Ouyang JQ, Sharp P, Quetting M, Hau M (2013) Endocrine phenotype, reproductive success and survival in the Great Tit, *Parus major*. J Evol Biol 26:1988–1998

Pant K, Chandola-Saklani A (1993) A role for thyroid hormones in the development of premigratory disposition in Redheaded Bunting, *Emberiza bruniceps*. J Comp Physiol B 163:389–394

Parkhurst CR, Mountney GJ (1988) Physiology and reproduction in poultry. In: Poultry meat and egg production. Springer, Boston, MA, pp 31–48

Pathak VK, Chandola A (1984) Variations in circulating thyroxine and triiodothyronine concentration in relation to spring migration in Rosy Pastor, *Sturnus roseus*. Horm Behav 18:111–116

Pauliukonis AC, Buenaventura C, Cline MA, Gilbert ER (2020) Anorexigenic effects of substance P in *Coturnix japonica*. Neuropeptides 81:102045

Pedersen A, Tomaszycki M (2012) Oxytocin antagonist treatments alter the formation of pair relationships in Zebra Finches of both sexes. Horm Behav 62:113–119

Pérez JH, Furlow JD, Wingfield JC, Ramenofsky M (2016) Regulation of vernal migration in Gambel's White-crowned Sparrows: role of thyroxine and triiodothyronine. Horm Behav 84:50–56

Pérez JH, Meddle SL, Wingfield JC, Ramenofsky M (2018) Effects of thyroid hormone manipulation on pre-nuptial molt, luteinizing hormone and testicular growth in male White-crowned Sparrows (*Zonotrichia leuchophrys gambelii*). Gen Comp Endocrinol 255:12–18

Peters A, Delhey K, Goymann W, Kempenaers B (2006) Age-dependent association between testosterone and crown UV coloration in male Blue Tits (*Parus caeruleus*). Behav Ecol Sociobiol 59:666–673

Phillips-Singh D, Li Q, Takeuchi S, Ohkubo T, Sharp PJ, Boswell T (2003) Fasting differentially regulates expression of agouti-related peptide, pro-opiomelanocortin, prepro-orexin, and vasoactive intestinal polypeptide mRNAs in the hypothalamus of Japanese Quail. Cell Tissue Res 313:217–225

Polakof S, Mommsen TP, Soengas JL (2011) Glucosensing and glucose homeostasis: from fish to mammals. Comp Biochem Physiol B Biochem Mol Biol 160:123–149

Porter TE, Couger GS, Dean CE, Hargis BM (1995) Ontogeny of growth hormone (GH)-secreting cells during chicken embryonic development: initial somatotrophs are responsive to GH-releasing hormone. Endocrinology 136:1850–1856

Potticary AL, Duckworth RA (2021) A neuroendocrine perspective on the origin and evolution of cooperative breeding. Ornithology 138:ukab036

Prati M, Calvo R, Morreale G, Morreale de Escobar G (1992) L-thyroxine and 3,5,3′-triiodothyronine concentrations in the chicken egg and in the embryo before and after the onset of thyroid function. Endocrinology 130:2651–2659

Pravosudov VV (2003) Long-term moderate elevation of corticosterone facilitates avian food-caching behaviour and enhances spatial memory. Proc R Soc B 270: 2599–2604

Proud CG, Denton RM (1997) Molecular mechanisms for the control of translation by insulin. Biochem J 328: 329–341

Ramakrishnan S, Strader AD, Wimpee B, Chen P, Smith MS, Buntin JD (2007) Evidence for increased neuropeptide Y synthesis in mediobasal hypothalamus in relation to parental hyperphagia and gonadal activation in breeding Ring Doves. J Neuroendocrinol 19:163–171

Rawdon BB, Andrew A (1981) An immunocytochemical survey of endocrine cells in the gastrointestinal tract of chicks at hatching. Cell Tissue Res 220:279–292

Rawdon BB, Andrew A (1999) Gut endocrine cells in birds: an overview, with particular reference to the chemistry of gut peptides and the distribution, ontogeny, embryonic origin and differentiation of the endocrine cells. Prog Histochem Cytochem 34:3–82

Raymaekers SR, Darras VM (2017) Thyroid hormones and learning-associated neuroplasticity. Gen Comp Endocrinol 247:26–33

Razvi S, Jabbar A, Pingitore A, Danzi S, Biondi B, Klein I, Peeters R, Zaman A, Iervasi G (2018) Thyroid hormones and cardiovascular function and diseases. J Am Coll Cardiol 71:1781–1796

Récapet C, Sibeaux A, Cauchard L, Doligez B, Bize P (2016) Selective disappearance of individuals with high levels of glycated haemoglobin in a free-living bird. Biol Lett 12:20160243

Regan JC, Froy H, Walling CA, Moatt JP, Nussey DH (2019) Dietary restriction and insulin-like signalling pathways as adaptive plasticity: a synthesis and re-evaluation. Funct Ecol 34:107–128

Reid AM, Dunn IC (2018) Gastrointestinal distribution of chicken gastrin-cholecystokinin family transcript expression and response to short-term nutritive state. Gen Comp Endocrinol 255:64–70

Reid AMA, Wilson PW, Caughey SD, Dixon LM, D'Eath RB, Sandilands V, Boswell T, Dunn IC (2017) Pancreatic PYY but not PPY expression is responsive to short-term nutritional state and the pancreas constitutes the major site of PYY mRNA expression in chickens. Gen Comp Endocrinol 252:226–235

Ren J, Samson WK, Sowers JR (1999) Insulin-like growth factor I as cardiac hormone: physiological and pathophysiological implications in heart disease. J Mol Cell Cardiol 31:2049–2061

Reverchon M, Ramé C, Bertoldo M, Dupont J (2014) Adipokines and the female reproductive tract. Int J Endocrinol 2014:232454

Richards MP, McMurtry JP (2008) Expression of proglucagon and proglucagon-derived peptide hormone receptor genes in the chicken. Gen Comp Endocrinol 156:323–338

Richards MP, Proszkowiec-Weglarz M (2007) Mechanisms regulating feed intake, energy expenditure, and body weight in poultry. Poult Sci 86:1478–1490

Richardson RD, Boswell TI, Weatherford SC, Wingfield JC, Woods SC (1993) Cholecystokinin octapeptide decreases food intake in White-crowned Sparrows. Am J Phys Regul Integr Comp Phys 264:R852–R856

Richardson RD, Boswell T, Raffety BD, Seeley RJ, Wingfield JC, Woods SC (1995) NPY increases food intake in White-crowned Sparrows: effect in short and long photoperiods. Am J Phys Regul Integr Comp Phys 268:R1418–R1422

Riechert J, Becker PH, Chastel O (2014) Predicting reproductive success from hormone concentrations in the Common Tern (*Sterna hirundo*) while considering food abundance. Oecologia 176:715–727

Rissman EF, Wingfield JC (1984) Hormonal correlates of polyandry in the Spotted Sandpiper, *Actitis macularia*. Gen Comp Endocrinol 56:401–405

Ritchie M (2014) Neuroanatomy and physiology of the avian hypothalamic/pituitary axis: clinical aspects. Vet Clin Exotic Anim Pract 17:13–22

Rodrigues MN, de Oliviera GB, da Silva RSB, de Albuquerque JFG, Miglino MA, de Oliveira MF (2012) Morfologia e topografia do fígado e pâncreas de emas *Rhea americana*. Ciência Rural 42:474–479

Romero LM, Dickens MJ, Cyr NE (2009) The reactive scope model—a new model integrating homeostasis, allostasis, and stress. Horm Behav 55:375–389

Romero-Pujante M, Hoi H, Blomqvist D, Valera F (2002) Tail length and mutual mate choice in Bearded Tits (*Panurus biarmicus*). Ethology 108:885–895

Rudolph LM, Cornil CA, Mittelman-Smith MA, Rainville JR, Remage-Healey L, Sinchak K, Micevych PE (2016) Actions of steroids: new neurotransmitters. J Neurosci 36:11449–11458

Ruuskanen S, Hsu BY, Heinonen A, Vainio M, Darras VM, Sarraude T, Rokka A (2018) A new method for measuring thyroid hormones using nano-LC-MS/MS. J Chromatogr B 1093:24–30

Said SI, Mutt V (1972) Isolation from porcine-intestinal wall of a vasoactive octacosapeptide related to secretin and to glucagon. Eur J Biochem 28:199–204

Saino N, Romano M, Ferrari RP, Martinelli R, Møller AP (2005) Stressed mothers lay eggs with high corticosterone levels which produce low-quality offspring. J Exp Zool A Comp Exp Biol 303:998–1006

Saito T, Yamada J, Kitamura N, Yamashita T (1989) An immunohistochemical study on the distribution of endocrine cells in the gastrointestinal tract of domestic pigeon (*Columba livia var domestica*). Z Mikrosk Anat Forsch 103:237–246

Salvatore D, Simonides WS, Dentice M, Zavacki AM, Larsen PR (2014) Thyroid hormones and skeletal muscle - new insights and potential implications. Nature Rev Endocrin 10:206–214

Salvi E, Buffa R, Renda TG (1996) Ontogeny, distribution and amine/peptide content of chromogranin A-and B-immunoreactive endocrine cells in the small and large intestine of the chicken. Anat Embryol 194:89–98

Sanders EJ, Harvey S (2004) Growth hormone as an early embryonic growth and differentiation factor. Anat Embryol 209:1–9

Saneyasu T, Honda K, Kamisoyama H, Nakayama Y, Ikegami K, Hasegawa S (2011) Alpha-melanocyte stimulating hormone plays an important role in the regulation of food intake by the central melanocortin system in chicks. Peptides 32:996–1000

Sartori DRS, Migliorini RH, Veiga JAS, Moura JL, Kettelhut IC, Linder C (1995) Metabolic adaptations induced by long-term fasting in quails. Comp Biochem Physiol A 111:487–493

Savory CJ (1987) An alternative explanation for apparent satiating properties of peripherally administered bombesin and cholecystokinin in domestic fowls. Physiol Behav 39:191–202

Scanes CG, Braun E (2013) Avian metabolism: its control and evolution. Front Biol 8:134–159

Scanes CG, Thommes RC, Radecki SV, Buonomo FC, Woods JE (1997) Ontogenic changes in the circulating concentrations of insulin-like growth factor (IGF)-I, IGF-II, and IGF-binding proteins in the chicken embryo. Gen Comp Endocrinol 106:265–270

Schoech SJ, Mumme RL, Wingfield JC (1996) Prolactin and helping behaviour in the cooperatively breeding Florida Scrub-Jay, *Aphelocoma c coerulescens*. Anim Behav 52:445–456

Schoenle LA, Dudek AM, Moore IT, Bonier F (2017) Red-winged Blackbirds (*Agelaius phoeniceus*) with higher baseline glucocorticoids also invest less in incubation and clutch mass. Horm Behav 90:1–7

Schoenle LA, Zimmer C, Vitousek MN (2018) Understanding context dependence in glucocorticoid-fitness relationships: the role of the nature of the challenge, the intensity and frequency of stressors, and life history. Integr Comp Biol 58:777–789

Schwabl P, Bonaccorso E, Goymann W (2016) Diurnal variation in corticosterone release among wild tropical forest birds. Front Zool 13:19

Selye H (1955) Stress and disease. Laryngoscope 65:500–514

Shanahan M, Bingman VP, Shimizu T, Wild M, Güntürkün O (2013) Large-scale network organization in the avian forebrain: a connectivity matrix and theoretical analysis. Front Comput Neurosci 7:89

Sharma A, Singh D, Malik S, Gupta NJ, Rani S, Kumar V (2018) Difference in control between spring and autumn migration in birds: insight from seasonal changes in hypothalamic gene expression in captive buntings. Proc R Soc B 285:20181531

Sharp PJ, Dawson A, Lea RW (1998) Control of luteinizing hormone and prolactin secretion in birds. Comp Biochem Physiol C: Pharmacol Toxicol Endocrinol 119:275–282

Sheriff MJ, Dantzer B, Delehanty B, Palme R, Boonstra R (2011) Measuring stress in wildlife: techniques for quantifying glucocorticoids. Oecologia 166:869–887

Shikano K, Iwakoshi-Ukena E, Kato M, Furumitsu M, Bentley GE, Kriegsfeld LJ, Ukena K (2019) Neurosecretory protein GL induces fat accumulation in chicks. Front Endocrinol 10:392

Shinomiya A, Shimmura T, Nishiwaki-Ohkawa T, Yoshimura T (2014) Regulation of seasonal reproduction by hypothalamic activation of thyroid hormone. Front Endocrinol 5:fendo.2014.00012

Shiraishi JI, Yanagita K, Fujita M, Bungo T (2008) Central insulin suppresses feeding behavior via melanocortins in chicks. Domest Anim Endocrinol 34:223–228

Shiuchi T, Haque MS, Okamoto S, Inoue T, Kageyama H, Lee S, Toda C, Suzuki A, Bachman ES, Kim YB, Sakurai T (2009) Hypothalamic orexin stimulates feeding-associated glucose utilization in skeletal muscle via sympathetic nervous system. Cell Metab 10: 466–480

Shousha S, Nakahara K, Kojima M, Miyazato M, Hosoda H, Kangawa K, Murakami N (2005) Different effects of peripheral and central ghrelin on regulation of food intake in the Japanese Quail. Gen Comp Endocrinol 141:178–183

Shousha S, Nakahara K, Nasu T, Sakamoto T, Murakami N (2007) Effect of glucagon-like peptide-1 and -2 on regulation of food intake, body temperature and locomotor activity in the Japanese Quail. Neurosci Lett 415:102–107

Sims W, Yi J, Cline MA, Gilbert ER (2017) Central injection of a synthetic chicken partial leptin peptide does not affect food intake in chicks. Neurosci Lett 656:165–168

Skipper M, Lewis J (2000) Getting to the guts of enteroendocrine differentiation. Nat Genet 24:3–4

Smiley KO (2019) Prolactin and avian parental care: new insights and unanswered questions. Horm Behav 111: 114–130

Smiley KO, Adkins-Regan E (2016a) Prolactin is related to individual differences in parental behavior and reproductive success in a biparental passerine, the Zebra Finch (*Taeniopygia guttata*). Gen Comp Endocrinol 234:88–94

Smiley KO, Adkins-Regan E (2016b) Relationship between prolactin, reproductive experience, and parental care in a biparental songbird, the Zebra Finch (*Taeniopygia guttata*). Gen Comp Endocrinol 232: 17–24

Smith JP (1982) Changes in blood levels of thyroid hormones in two species of passerine birds. Condor 84:160–167

Soares MC, Bshary R, Fusani L, Goymann W, Hau M, Hirschenhauser K, Oliveira RF (2010) Hormonal mechanisms of cooperative behaviour. Philos Trans R Soc B 365:2737–2750

Sommerville BA, Fox J (1987) Changes in renal function of the chicken associated with calcitonin and parathyroid hormone. Gen Comp Endocrinol 66:381–386

Song Z, Liu L, Yue Y, Jiao H, Lin H, Sheikhahmadi A, Everaert N, Decuypere EE, Buyse J (2012) Fasting alters protein expression of AMP-activated protein kinase in the hypothalamus of broiler chicks (*Gallus gallus domesticus*). Gen Comp Endocrinol 178:546–555

Spée M, Beaulieu M, Dervaux A, Chastel O, Le Maho Y, Raclot T (2010) Should I stay or should I go?

Hormonal control of nest abandonment in a long-lived bird, the Adélie Penguin. Horm Behav 58:762–768
Spée M, Marchal L, Lazin D, Le Maho Y, Chastel O, Beaulieu M, Raclot T (2011) Exogenous corticosterone and nest abandonment: a study in a long-lived bird, the Adélie Penguin. Horm Behav 60:362–370
Spencer RL, Kalman BA, Dhabhar FS (2001) Role of endogenous glucocorticoids in immune system function: regulation and counterregulation. In: McEwen B, Goodman HM (eds) Handbook of physiology; section 7: the endocrine system; volume IV: coping with the environment: neural and endocrine mechanisms. Oxford University Press, New York, pp 381–423
Stier A, Hsu B-Y, Marciau C, Doligez B, Gustafsson L, Bize P, Ruuskanen S (2020) Born to be young? Prenatal thyroid hormones increase early-life telomere length in wild Collared Flycatchers. Biol Lett 16: 20200364
Strader AD, Buntin JD (2001) Neuropeptide-Y: a possible mediator of prolactin-induced feeding and regulator of energy balance in the Ring Dove (*Streptopelia risoria*). J Neuroendocrinol 13:386–392
Sturkie PD, Joiner P, Freedman SL (1962) Role of the "bearing down" reflex on oviposition in the chicken. Endocrinology 70:221–222
Swarup K, Tewari NP, Srivastav AK (1980) Effect of salmon calcitonin on serum calcium and inorganic phosphorus levels in parrot, *Psittacula psittacula*. Arch Biol 91:15–20
Sweazea KL, Braun EJ (2005) Glucose transport by English Sparrow (*Passer domesticus*) skeletal muscle: have we been chirping up the wrong tree? J Exp Zool A 303A:143–153
Sweazea KL, McMurtry JP, Braun EJ (2006) Inhibition of lipolysis does not affect insulin sensitivity to glucose uptake in the Mourning Dove. Comp Biochem Physiol B Biochem Mol Biol 144:387–394
Tachibana T, Matsumoto M, Furuse M, Hasegawa S, Yoshizawa F, Sugahara K (2003a) Central, but not peripheral, glucagon-like peptide-1 inhibits crop emptying in chicks. Comp Biochem Physiol A 134:777–781
Tachibana T, Takagi T, Tomonaga S, Ohgushi A, Ando R, Denbow DM, Furuse M (2003b) Central administration of cocaine-and amphetamine-regulated transcript inhibits food intake in chicks. Neurosci Lett 337:131–134
Tachibana T, Sato M, Takahashi H, Ukena K, Tsutsui K, Furuse M (2005) Gonadotropin-inhibiting hormone stimulates feeding behavior in chicks. Brain Res 1050:94–100
Tachibana T, Mori M, Khan MSI, Ueda H, Sugahara K, Hiramatsu K (2008) Central administration of galanin stimulates feeding behavior in chicks. Comp Biochem Physiol A 151:637–640
Tachibana T, Cline MA, Sugahara K, Ueda H, Hiramatsu K (2009) Central administration of somatostatin stimulates feeding behavior in chicks. Gen Comp Endocrinol 161:354–359
Tachibana T, Matsuda K, Kawamura M, Ueda H, Khan MSI, Cline MA (2012) Feeding-suppressive mechanism of sulfated cholecystokinin (26–33) in chicks. Comp Biochem Physiol A 161:372–378
Tachibana T, Sugimoto I, Ogino M, Khan MSI, Masuda K, Ukena K, Wang Y (2015) Central administration of chicken growth hormone-releasing hormone decreases food intake in chicks. Physiol Behav 139: 195–201
Taff CC, Vitousek MN (2016) Endocrine flexibility: optimizing phenotypes in a dynamic world? Trends Ecol Evol 31:476–488
Takahashi T, Kawashima M (2003) Arginine vasotocin induces bearing down for oviposition in the hen. Poult Sci 82:345–346
Tam JKV, Lee LTO, Jin J, Chow BKC (2014) Secretin/secretin receptors. J Mol Endocrinol 52:T1–T14
Tamai TK, Yoshimura T (2017) Molecular and neuroendocrine mechanisms of avian seasonal reproduction. In: Sasanami T (ed) Avian reproduction: from behavior to molecules. Singapore, Springer Nature, pp 125–136
Tesseraud S, Métayer S, Duchêne S, Bigot K, Grizard J, Dupont J (2007) Regulation of protein metabolism by insulin: value of different approaches and animal models. Domest Anim Endocrinol 33:123–142
Thierry A-M, Massemin S, Handrich Y, Raclot T (2013) Elevated corticosterone levels and severe weather conditions decrease parental investment of incubating Adélie Penguins. Horm Behav 63:475–483
Tobari Y, Tsutsui K (2019) Effects of social information on the release and expression of gonadotropin-inhibitory hormone in birds. Front Endocrinol 10:243
Tokushima Y, Takahashi K, Sato K, Akiba Y (2005) Glucose uptake in vivo in skeletal muscles of insulin-injected chicks. Comp Biochem Physiol B 141:43–48
Too HC, Shibata M, Yayota M, Darras VM, Iwasawa A (2017) Expression of thyroid hormone regulator genes in the yolk sac membrane of the developing chicken embryo. J Reprod Dev 63:463–472
Tóth Z, Ouyang JQ, Lendvai ÁZ (2018) Exploring the mechanistic link between corticosterone and insulin-like growth factor-1 in a wild passerine bird. PeerJ 6: e593
Tsutsui K, Ubuka T (2018) How to contribute to the progress of neuroendocrinology: discovery of GnIH and progress of GnIH research. Front Endocrinol 9:662
Tsutsui K, Saigoh E, Ukena K, Teranishi H, Fujisawa Y, Kikuchi M, Ishii S, Sharp PJ (2000) A novel avian hypothalamic peptide inhibiting gonadotropin release. Biochem Biophys Res Commun 275:661–667
Ubuka T, Tsutsui K (2014) Neuroestrogen regulation of socio-sexual behavior of males. Front Neurosci 8:323
Ubuka T, Kim S, Huang YC, Reid J, Jiang J, Osugi T, Chowdhury VS, Tsutsui K, Bentley GE (2008) Gonadotropin-inhibitory hormone neurons interact directly with gonadotropin-releasing hormone-I and -II neurons in European Starling brain. Endocrinology 149:268–278

Ukena K, Iwakoshi-Ukena E, Taniuchi S, Bessho Y, Maejima S, Masuda K, Shikano K, Kondo K, Furumitsu M, Tachibana T (2014) Identification of a cDNA encoding a novel small secretory protein, neurosecretory protein GL, in the chicken hypothalamic infundibulum. Biochem Biophys Res Commun 446:298–303

Vagasi CI, Pătraş L, Pap PL, Vincze O, Mureşan C, Nemeth J, Lendvai AZ (2018) Experimental increase in baseline corticosterone level reduces oxidative damage and enhances innate immune response. PLoS ONE 13:e0192701

Vaillancourt E, Weber JM (2015) Fuel metabolism in Canada Geese: effects of glucagon on glucose kinetics. Am J Phys Regul Integr Comp Phys 309:R535–R543

Van Herck SL, Geysens S, Bald E, Chwatko G, Delezie E, Dianati E, Ahmed RG, Darras VM (2013) Maternal transfer of methimazole and effects on thyroid hormone availability in embryonic tissues. J Endocrinol 218:105–115

Vitousek MN, Taff CC, Hallinger KK, Zimmer C, Winkler DW (2018) Hormones and fitness: evidence for trade-offs in glucocorticoid regulation across contexts. Front Ecol Evol 6:42

Vleck CM, Vleck D (2011) Hormones and regulation of parental behavior in birds. In: Norris D, Lopez KH (eds) Hormones and reproduction of vertebrates. Academic Press, New York, pp 181–203

Vleck CM, Mays NA, Dawson JW, Goldsmith AR (1991) Hormonal correlates of parental and helping behavior in cooperatively breeding Harris' Hawks (*Parabuteo unicinctus*). Auk 108:638–648

Voigt C (2016) Neuroendocrine correlates of sex-role reversal in Barred Buttonquails. Proc R Soc B 283: 20161969

Wada H, Breuner CW (2008) Transient elevation of corticosterone alters begging behavior and growth of White-crowned Sparrow nestlings. J Exp Biol 211: 1696–1703

Wall JP, Cockrem JF (2009) Effects of corticosterone treatment on responses to fasting in Japanese Quail. Comp Biochem Physiol A Mol Integr Physiol 154: 211–215

Wan Y, Zhang J, Fang C, Chen J, Li J, Li J, Wu C, Wang Y (2018) Characterization of neuromedin U (NMU), neuromedin S (NMS) and their receptors (NMUR1, NMUR2) in chickens. Peptides 101:69–81

Wan XP, Xie P, Bu Z, Zou XT, Gong DQ (2019) Prolactin induces lipid synthesis of organ-cultured pigeon crops. Poult Sci 98:1842–1853

Wang BJ, Cui ZJ (2007) How does cholecystokinin stimulate exocrine pancreatic secretion? From birds, rodents, to humans. Am J Phys Regul Integr Comp Phys 292:R666–R678

Wang Y, Huang G, Li J, Meng F, He X, Leung FC (2012) Characterization of chicken secretin (SCT) and secretin receptor (SCTR) genes: a novel secretin-like peptide (SCT-LP) and secretin encoded in a single gene. Mol Cell Endocrinol 348:270–280

Wang J, Matias J, Gilbert ER, Tachibana T, Cline MA (2019) Hypothalamic mechanisms associated with corticotropin-releasing factor-induced anorexia in chicks. Neuropeptides 74:95–102

Wang Y, Székely T, Zhang Z, Que P (2020) Prolactin concentrations predict parental investment and nest survival in a free-living shorebird. Horm Behav 119: 104633

Ward AM, Hunt A, Maslanka M, Brown C (2001) Nutrient composition of American Flamingo crop milk. Proceedings of the American Zoo and Aquarium Association (AZA) Nutrition Advisory Group, Lake Buena Vista, pp 189–194

Watanabe T, Hiramatsu K, Nishimura K, Ono T (2018) Glucagon-like peptide-1 receptor expression in the pancreatic D cells of three avian species; White Leghorn Chickens, Northern Bobwhites, and Common Ostriches. J Poult Sci 55:199–203

Waters MJ, Shang CA, Behncken SN, Tam SP, Li H, Shen B, Lobie PE (1999) Growth hormone as a cytokine. Clin Exp Pharmacol Physiol 26:760–764

Welch KC, Chen CC (2014) Sugar flux through the flight muscles of hovering vertebrate nectarivores: a review. J Comp Physiol B 184:945–959

Welcker J, Chastel O, Gabrielsen GW, Guillaumin J, Kitaysky AS, Speakman JR, Tremblay Y, Bech C (2013) Thyroid hormones correlate with basal metabolic rate but not field metabolic rate in a wild bird species. PLoS One 8:e56229

Wilbourn RV, Moatt JP, Froy H, Walling CA, Nussey DH, Boonekamp JJ (2018) The relationship between telomere length and mortality risk in non-model vertebrate systems: a meta-analysis. Philos Trans R Soc B 373: 20160447

Williams GR (2013) Thyroid hormone actions in cartilage and bone. Eur Thyroid J 2:3–13

Willie JT, Chemelli RM, Sinton CM, Tokita S, Williams SC, Kisanuki YY, Marcus JN, Lee C, Elmquist JK, Kohlmeier KA, Leonard CS, Richardson JA, Hammer RE, Yanagisawa M (2003) Distinct narcolepsy syndromes in orexin receptor-2 and orexin null mice: molecular genetic dissection of non-REM and REM sleep regulatory processes. Neuron 38:715–730

Wilson CM, McNabb FM (1997) Maternal thyroid hormones in Japanese Quail eggs and their influence on embryonic development. Gen Comp Endocrinol 107:153–165

Wingfield JC, Romero LM (2001) Adrenocortical responses to stress and their modulation in free-living vertebrates. In: McEwen BS, Goodman HM (eds) Handbook of physiology; section 7: The endocrine system; volume IV: Coping with the environment: neural and endocrine mechanisms. Oxford University Press, New York, pp 211–234

Wingfield JC, Sapolsky RM (2003) Reproduction and resistance to stress: when and how. J Neuroendocrinol 15:711–724

Wingfield JC, Ronchi E, Goldsmith AR, Marler C (1989) Interactions of sex steroid hormones and prolactin in male and female Song Sparrows, *Melospiza melodia*. Physiol Zool 62:11–24

Wong KKY, Ng SY, Lee LT, Ng HK, Chow BK (2011) Orexins and their receptors from fish to mammals: a comparative approach. Gen Comp Endocrinol 171:124–130

Wu C, Lv C, Wan Y, Li X, Zhang J, Li J, Wang Y (2019) Arginine vasotocin (AVT)/mesotocin (MT) receptors in chickens: evidence for the possible involvement of AVT-AVPR1 signaling in the regulation of oviposition and pituitary prolactin expression. Gen Comp Endocrinol 281:91–104

Wyse C, Hazlerigg D (2009) Seasonal biology: avian photoreception goes deep. Curr Biol 19:R685–R687

Xiao Y, Wu C, Li K, Gui G, Zhang G, Yang H (2017) Association of growth rate with hormone levels and myogenic gene expression profile in broilers. J Anim Sci Biotechnol 8:43

Xie P, Wan XP, Bu Z, Diao EJ, Gong DQ, Zou XT (2018) Changes in hormone profiles, growth factors, and mRNA expression of the related receptors in crop tissue, relative organ weight, and serum biochemical parameters in the Domestic Pigeon (*Columba livia*) during incubation and chick-rearing periods under artificial farming conditions. Poult Sci 97:2189–2202

Yacawych WT, Palmer AL, Doczi MA (2019) Insulin receptor localization in the embryonic avian hypothalamus. Neurosci Lett 698:126–132

Yadav S, Srivastav AK (2009) Influence of calcitonin administration on ultimobranchial and parathyroid glands of pigeon, *Columba livia*. Microsc Res Tech 72:380–384

Yamaguchi S, Aoki N, Kitajima T, Iikubo E, Katagiri S, Matsushima T, Homma KJ (2012) Thyroid hormone determines the start of the sensitive period of imprinting and primes later learning. Nat Commun 3:1081

Yamaguchi S, Hayase S, Aoki N, Takehara A, Ishigohoka J, Matsushima T, Wada K, Homma KJ (2017) Sex differences in brain thyroid hormone levels during early post-hatching development in Zebra Finch (*Taeniopygia guttata*). PLoS ONE 12:e0169643

Yan J, Yang H, Gan L, Sun C (2014) Adiponectin-impaired adipocyte differentiation negatively regulates fat deposition in chicken. J Anim Physiol Anim Nutr 98:530–537

Yao Z, Petschnigg J, Ketteler R, Stagljar I (2015) Application guide for omics approaches to cell signaling. Nat Chem Biol 11:387

Yap KN, Kim OR, Harris KC, Williams TD (2017) Physiological effects of increased foraging effort in a small passerine. J Exp Biol 220:4282–4291

Yasuo S, Watanabe M, Nakao N, Takagi T, Follett BK, Ebihara S, Yoshimura T (2005) The reciprocal switching of two thyroid hormone-activating and-inactivating enzyme genes is involved in the photoperiodic gonadal response of Japanese Quail. Endocrinology 146:2551–2554

Yoshimura T, Yasuo S, Watanabe M, Iigo M, Yamamura T, Hirunagi K, Ebihara S (2003) Light-induced hormone conversion of $T_4$ to $T_3$ regulates photoperiodic response of gonads in birds. Nature 426:178–181

Yousefi AR, Shojaei M, Zendehdel M (2019) Evaluation the role of central serotonin and 5HT2c serotonin receptor on feed intake in female layer-type Bovans chicken by intracerebroventricular (ICV) injection of para-chlorophenylalanine and SB242084. Vet Res Biol Prod 32:55–62

Yousefvand S, Hamidi F, Zendehdel M, Parham A (2019) Interaction of neuropeptide Y receptors (NPY1, NPY2 and NPY5) with somatostatin on somatostatin-induced feeding behaviour in neonatal chicken. Br Poult Sci 60: 71–78

Zadworny D, Etches RJ (1987) Effects of ovariectomy or force feeding on the plasma concentrations of prolactin and luteinizing hormone in incubating turkey hens. Biol Reprod 36:81–88

Zadworny D, Shimada K, Ishida H, Sumi C, Sato K (1988) Changes in plasma levels of prolactin and estradiol, nutrient intake, and time spent nesting during the incubation phase of broodiness in the Chabo hen (Japanese bantam). Gen Comp Endocrinol 71:406–412

Zendehdel M, Sardari F, Hassanpour S, Rahnema M, Adeli A, Ghashghayi E (2017) Serotonin-induced hypophagia is mediated via α2 and β2 adrenergic receptors in neonatal layer-type chickens. Br Poult Sci 58:298–304

Zhang D, Li Y, Liu S, Wang YC, Guo F, Zhai Q, Jiang J, Ying H (2017a) microRNA and thyroid hormone signaling in cardiac and skeletal muscle. Cell Biosci 7:14

Zhang W, Didehvar D, Wang G, Yi J, Gilbert ER, Cline MA (2017b) Anorexigenic effect of serotonin is associated with changes in hypothalamic nuclei activity in an avian model. Gen Comp Endocrinol 246:81–87

Zhang S, Okuhara Y, Iijima M, Takemi S, Sakata I, Kaiya H, Teraoka H, Kitazawa T (2020) Identification of pheasant ghrelin and motilin and their actions on contractility of the isolated gastrointestinal tract. Gen Comp Endocrinol 285:113294

Zhou W, Murakami M, Hasegawa S, Yoshizawa F, Sugahara K (2005) Neuropeptide Y content in the hypothalamic paraventricular nucleus responds to fasting and refeeding in broiler chickens. Comp Biochem Physiol A Mol Integr Physiol 141:146–152

# Urinary System, Salt Glands, and Osmoregulation

# 9

## Contents




**Abstract**

Osmoregulation refers to the mechanism by which birds regulate water and electrolyte levels in their bodies. In this chapter, I explain the various ways that birds obtain water, including drinking water, water in foods they ingest, and water produced by metabolic process. Birds also continuously lose water, and those processes, including respiration, cutaneous water loss, and excretion, are also explained. Birds can alter the amount of water lost in those different ways, and I explain how that occurs. The anatomy of the avian kidney is described in detail, including how avian kidneys differ from mammalian kidneys and the anatomy of the functional unit of the kidney—the nephron. The various functions of the kidneys are described, including filtration, excretion, secretion, and absorption. The process by which avian kidneys can increase or decrease the amount of water lost as urine is explained, including the roles of arginine vasotocin and proteins called aquaporins. Also explained are the roles of



G. Ritchison, *In a Class of Their Own*, Fascinating Life Sciences,
https://doi.org/10.1007/978-3-031-14852-1_9

the cloaca and lower intestine in avian osmoregulation. Nectar-feeding birds ingest large amounts of water, and the ways by which they eliminate excess water are described. How and why birds excrete nitrogenous wastes as uric acid is explained, and the chapter closes with a description of the structure and function of avian salt glands.

## 9.1 Introduction

Osmoregulation refers to the various mechanisms by which birds regulate water and electrolyte levels in their bodies. Body fluids, most importantly extracellular osmolality and blood volume, are regulated within narrow limits. The "osmo-" in osmoregulation refers to osmosis, the process by which water passes through semipermeable membranes (such as cell membranes) in response to differences in solute concentrations. Of course, living cells can only survive if the concentrations of solutes, or electrolytes, remain very similar inside (intracellular fluid) and outside (extracellular fluid) of cells. Ensuring that this is the case requires that both water and electrolyte levels in the body to be regulated very precisely (Fig. 9.1). In mammals, the kidneys play the critical role in this process. In birds, on the other hand, the kidneys, lower intestine, and, in some species, salt glands all play important roles in osmoregulation.

## 9.2 Intake of Water and Solutes

The body water content of birds generally ranges from about 63 to 66% (Ellis and Jehl 1991; Speakman 1997), but, for birds in arid habitats, values as high as 70–80% have been reported (Ambrose et al. 1996; Anava et al. 2000). To maintain those levels, water input must match water output. The importance of water input is apparent in the results of studies that have revealed that water availability can influence bird diversity directly, i.e., not indirectly via plant productivity. For example, Hagger et al. (2019) determined that water availability limited bird species richness in plant-restoration sites in Australia (Fig. 9.2). Similar results, with bird diversity influenced by water availability, have also been reported by other investigators (e.g., Hawkins et al. 2005; Shih 2018; de Camargo et al. 2020). However, other investigators have found that the presence of some bird assemblages such as those in more arid environments where birds are better adapted to drier conditions is not influenced by water availability (e.g., Pavey and Nano 2009). In addition, diet influences the importance of availability of surface water, with species able to obtain more water in their diet (e.g., nectarivores, frugivores, and insectivores) less dependent on surface water than the species that obtain less water in their diets (e.g., granivores; Fig. 9.3). With increasing temperatures, however, even species typically less dependent on surface water must visit sources of water more frequently to allow loss of heat via evaporative cooling (Fig. 9.4). More generally for birds in arid environments, those species that drink water on a regular basis are better able to tolerate higher ambient temperatures than those that do not because they are better able to elevate their rates of evaporative water loss (Czenze et al. 2020).

Most species of birds can obtain water directly by drinking. Many birds drink by what is called "dipping and tipping" (Figs. 9.5, 9.6, and 9.7) or by suction drinking. Within each of these two main categories, however, there are several kinds of mechanisms. For example, water is taken into the mouth by scooping water with the lower beak by cockatoos (Cacatuinae) and Greater Rheas (*Rhea americana*; Fig. 9.8). Other birds drink using a "tongue scoop" (Fig. 9.9). Mallards (*Anas platyrhynchos*) drink using a complex interplay of capillary action and pressure changes in different areas of the mouth (Fig. 9.10). Among the suction drinkers, parakeets ladle water with the tip of the tongue, and some parrots drink with a suctioning action. Pigeons and doves, on the other hand, use a "double-suction mechanism" where capillary action is responsible for bringing water between

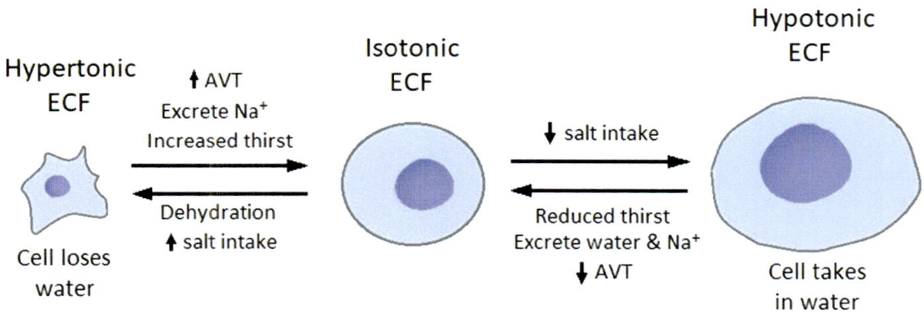


**Fig. 9.1** Processes involved in avian osmoregulation. Ideally, the extracellular fluid (ECF) is isotonic (the solute concentration equals that of the body cells). However, for birds drinking saltwater and/or consuming foods with high salt content or birds with reduced access to water (dehydration), the extracellular fluid can become hypertonic (solute concentration higher than that of body cells). If this occurs, water will tend to move by osmosis out of the cells (hence the "shriveled" cell on the left). When this happens to osmoreceptor cells (cells in the body responsible for monitoring body-wide water/solute balance), the pituitary will release more arginine vasotocin (AVT) into the blood, and this signals the kidneys to retain more water. In addition, more sodium ($Na^+$) is excreted from the body, and birds become thirsty. All of these in combination will tend to make the ECF more isotonic. In contrast, if a bird overhydrates (takes in more water than needed) or consumes food with limited amounts of salt, then the ECF becomes hypotonic (solute concentration lower than body cells), and water will tend to move by osmosis into cells (hence the "swollen" cell on the right). In response, the pituitary releases less AVT into the blood, which signals the kidneys to excrete more water and retain more $Na^+$, and birds become less thirsty. A more detailed explanation of these various processes is provided later in the chapter

the slightly gaping tips of the beak, and then the tongue acts as a piston to pump the water into the pharyngeal cavity. Aerial insectivores such as swifts, swallows, and caprimulgids obtain water on the wing by flying just above the surface of the water and scooping up water with their lower mandibles (Fig. 9.11).

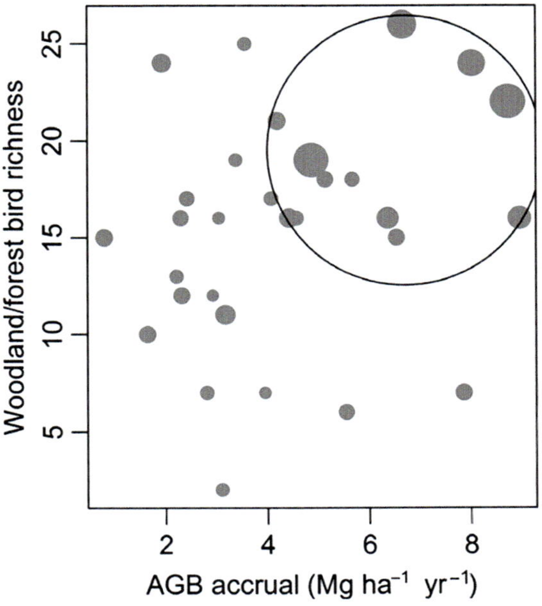


**Fig. 9.2** Relationship between above-ground biomass (AGB) and richness of woodland and forest birds at plant-restoration sites in Australia. The size of the circles corresponds to water availability (larger circles = greater water availability), and the cluster of wetter sites with both high AGB and bird richness is circled. (Figure from Hagger et al. 2019; open-access article distributed under the terms of the Creative Commons Attribution License, https://creativecommons.org/licenses/by/4.0/)

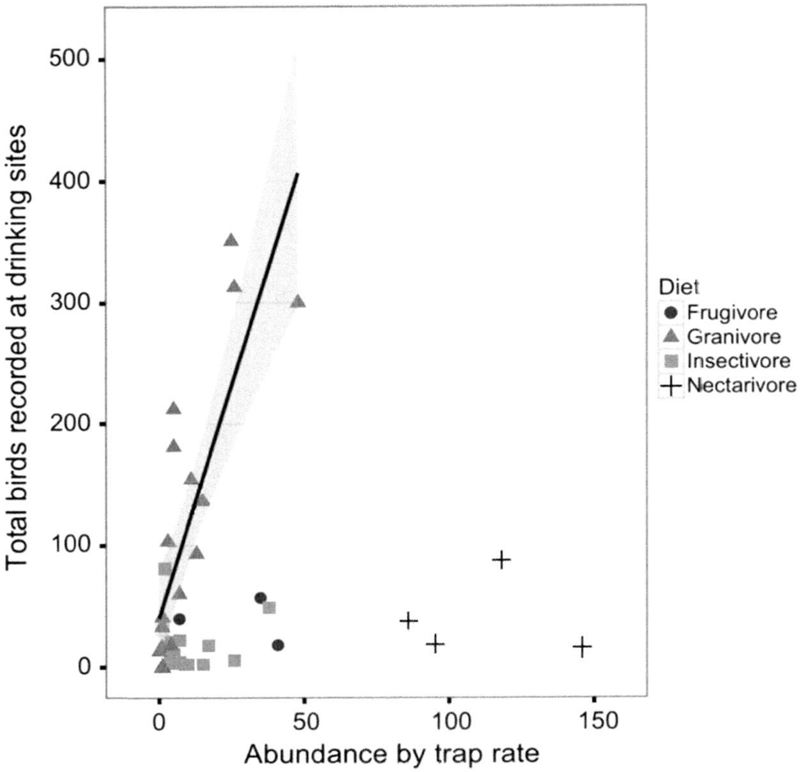


**Fig. 9.3** Visits by species of birds with different diets to sources of water relative to their abundance (as determined by capture rates in mist-nets). A linear regression line is plotted for granivores (with 95% confidence interval). (Figure from Lee et al. 2017; © 2017 John Wiley & Sons Ltd., used with permission)

Birds can also obtain water via the foods they ingest. For example, carnivores ingest animal tissue that is primarily water, frugivores consume fruits containing water, and the nectar consumed by nectarivores is, of course, largely water. In the Sonoran Desert of the southwestern United States and northwestern Mexico, the fruits of a cactus plant (saguaro, *Carnegiea gigantea*) appear to be an important source of water for several species of birds (Wolf et al. 2002; Wolf and Martinez del Rio 2003). Even foods that seemingly contain little water can serve as a water source because water is produced as a by-product of cellular metabolism (Fig. 9.12), with the two main sources being catabolism of fat and catabolism of protein (Jenni-Eiermann and Jenni 2012; Table 9.1). The importance of metabolic water in meeting the water requirement of birds varies considerably among different species of birds, ranging from less than 10% for some hummingbirds, about 14% for desert-adapted Ostriches (*Struthio camelus*), and up to 80% for some species of songbirds (MacMillen 1990; Williams et al. 1993). Metabolic water can be of particular importance to birds in arid environments (Williams 1996). For example, metabolic water represents about 14% of the total water needs of Common Ostriches (*Struthio camelus*) in the Namib Desert (Williams et al. 1993). Among birds in the Kalahari Desert in southern Africa, Smit et al. (2019) found that, with the exception of granivores and a few omnivores, most species were not dependent on surface water, likely getting sufficient water from their food plus metabolic water. For arid-adapted species of birds, the extent to which metabolic water contributes to daily water requirements may vary with ambient temperature. For example, Cooper et al. (2019) reported that metabolic water met 20% of the water requirements of Zebra Finches (*Taeniopygia guttata*) during hot days (temperature range = 40–44.5°C) and 32%

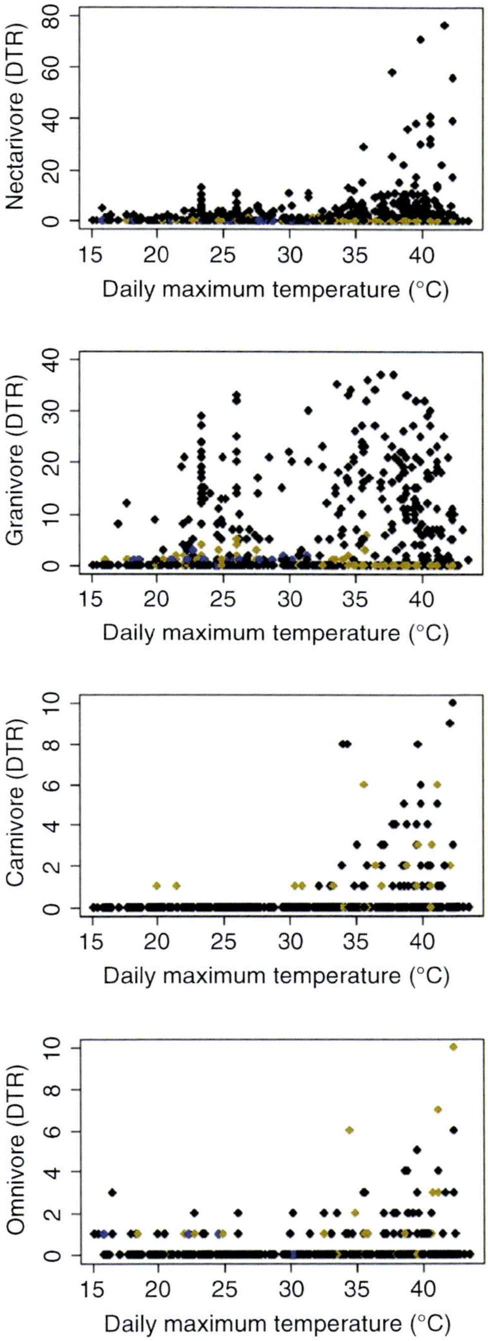


**Fig. 9.4** Daily trapping rates (DTR, or number of times per day that birds were observed at sources of water in photographs taken with cameras activated by bird presence) for birds categorized by diet in relation to daily maximum temperatures. Granivores, with dry seed diets, visited sources of water more often than birds in other diet groups, but even birds that obtain more water in their diets (nectarivores, carnivores, and omnivores) visited sources of water more frequently at extremely high ambient temperatures. Symbols with different colors indicate different sources of water (i.e., different springs and pools in the study area). (Figure from Votto et al. 2020; © 2020 Elsevier Ltd., used with permission)

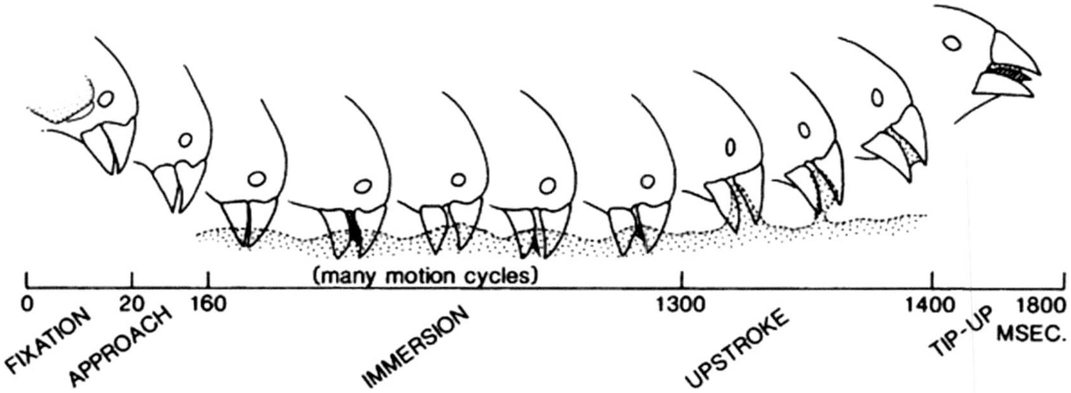


**Fig. 9.5** Drinking by a Zebra Finch (*Taeniopygia guttata*). During the immersion phase, the tongue rapidly protracts and retracts and, with each motion cycle, the space between the tongue and upper bill fills with water via capillary action. (Figure from Heidweiller and Zweers 1990; © 1990 Oxford University Press, used with permission)

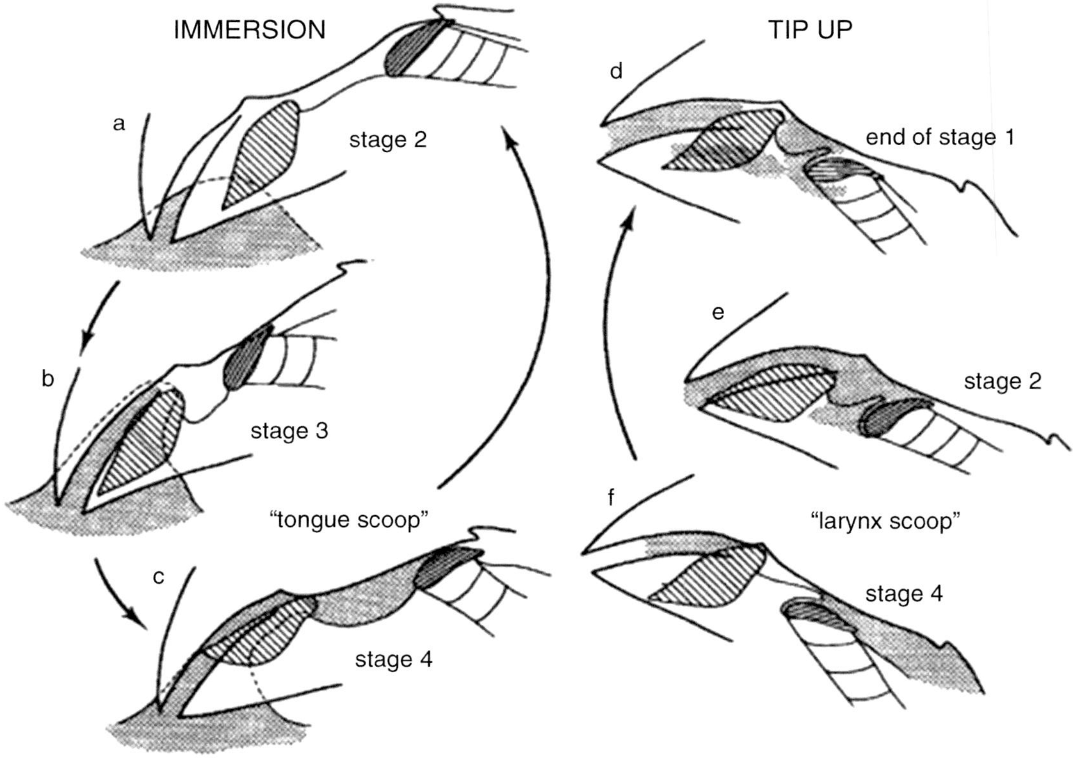


**Fig. 9.6** Drinking by a White-rumped Munia (*Lonchura striata*). Water runs between the beak tips as a result of adhesion and capillary action. The bill then tips up and that, along with tongue movements, moves water into the pharynx. (Figure from Heidweiller and Zweers 1990; © 1990 Oxford University Press, used with permission)

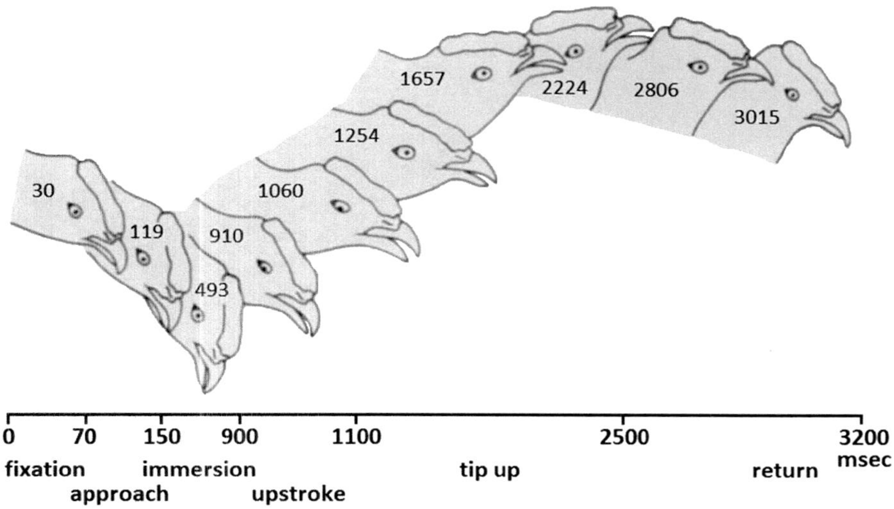


**Fig. 9.7** Drinking by a Domestic Chicken (*Gallus g. domesticus*). Each drink is divided into six phases. The immersion and tip up phases include a series of repetitive cycles of jaw, tongue, and pharynx floor motions. Numbers represent times in milliseconds. (Figure modified from Heidweiller et al. 1992; © 1992 Springer-Verlag, used with permission)

on cooler days (range = 27–27.6°C). This difference was likely due to the effect of ambient temperature on metabolic rates (Sabat et al. 2021).

The amount of water formed during fuel oxidation depends on a variety of factors, including metabolic rate and fuel type (carbohydrates, fats, or proteins; Schmidt-Nielsen 1997). Catabolism of fat alone produces about 27 µl of water per kilojoule, whereas catabolizing a mixture of 70%

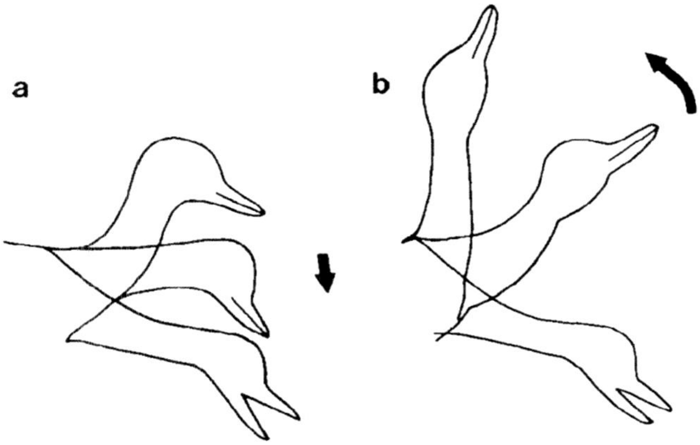


**Fig. 9.8** (**a**) When drinking water, Greater Rheas (*Rhea americana*) opens their bill, insert it into the water and, with a forward scooping motion of the head, water fills the lower bill. (**b**) The bill is then closed, the head is elevated, and the water is transported into the esophagus by elevation of the bill tip and retraction of the tongue. (Figure from Beaver 1978; © 1978 Published by Elsevier B.V., used with permission)

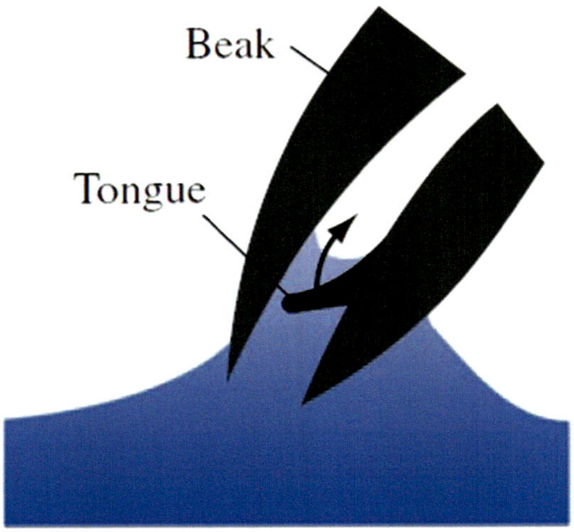


**Fig. 9.9** Close-up view of the "tongue scoop" of a finch drinking water. (Figure from Kim and Bush 2012; © Cambridge University Press 2012, used with permission)

fat and 30% protein yields about 34 μl of water per kilojoule (Klaassen 1996).

Most birds have regular access to water and drink more than needed on a regular basis. For example, Steller's Jays (*Cyanocitta stelleri*) and Dark-eyed Juncos (*Junco hyemalis*) were observed to visit a spring for water up to six times a day, and California Towhees (*Melozone crissalis*) typically came to the spring for water either in the morning or evening (Williams and Koenig 1980). Fisher et al. (1972) monitored the drinking behavior of several species of birds at natural and artificial sources of water in arid habitats in Australia and found that some species tended to visit water sources early and late in the day, whereas others visited water sources more regularly throughout the day (Fig. 9.13). In such cases, there is likely no fluid deficit or hormonal signal involved. However, when water loss exceeds water intake, osmoreceptor cells in the hypothalamus of the brain detect cellular dehydration caused by an increase in the solute concentration of the extracellular fluid (Thornton 1986; Orosz and Echols 2020). This, in turn, stimulates behaviors and physiological responses that will lead to increased intake and conservation of water. In addition, dehydration also causes the release of a hormone (angiotensin II) that is a powerful dipsogen (a substance that stimulates thirst; Fitzsimmons 1998; McKinley and Johnson 2004; Fig. 9.14).

Although many birds have regular access to water and drink as needed, other birds are faced with the potential problems of having either too little or too much water (Box 9.1 Water Availability at Migratory Stopover Sites). Birds in arid environments must often deal with limited supplies of water. For example, some resident birds in the Kalahari Desert of southern Africa, such as Namaqua Sandgrouse (*Pterocles namaqua*), travel long distances daily to drink water at isolated water sources (Willoughby and Cade 1967). However, birds in arid environments vary in their dependence on water for drinking, with some species drinking on a regular basis, some just occasionally, and others apparently never drinking (Maclean 1996; Smit et al. 2019). Granivorous birds that feed primarily on seeds with a low water content are more dependent on water sources than frugivores, carnivores, or omnivores (Smit et al. 2019; Votto et al. 2020).

Less obviously, birds that occupy saltwater environments, including pelagic species and those in coastal environments, may drink salt water and consume prey that contain high concentrations of salt. For these birds, water becomes a potential problem because, as will be discussed later, bird kidneys are not as efficient as those of mammals at producing concentrated urine (with high concentrations of salt), and excreting salt in relatively dilute urine would also mean excreting large amounts of water. At the other extreme are nectarivorous birds that have the potential problem of consuming too much water because nectar is typically 75–80% water (Nicolson and Fleming 2003).

## 9.3 Water Loss

Water balance, of course, is not just a matter of intake. Birds constantly lose water through respiration, cutaneous evaporation, and excretion (Fig. 9.15). The amount of water lost via excretion varies depending on a bird's water balance. In general, however, birds are able to produce relatively concentrated urine and can minimize

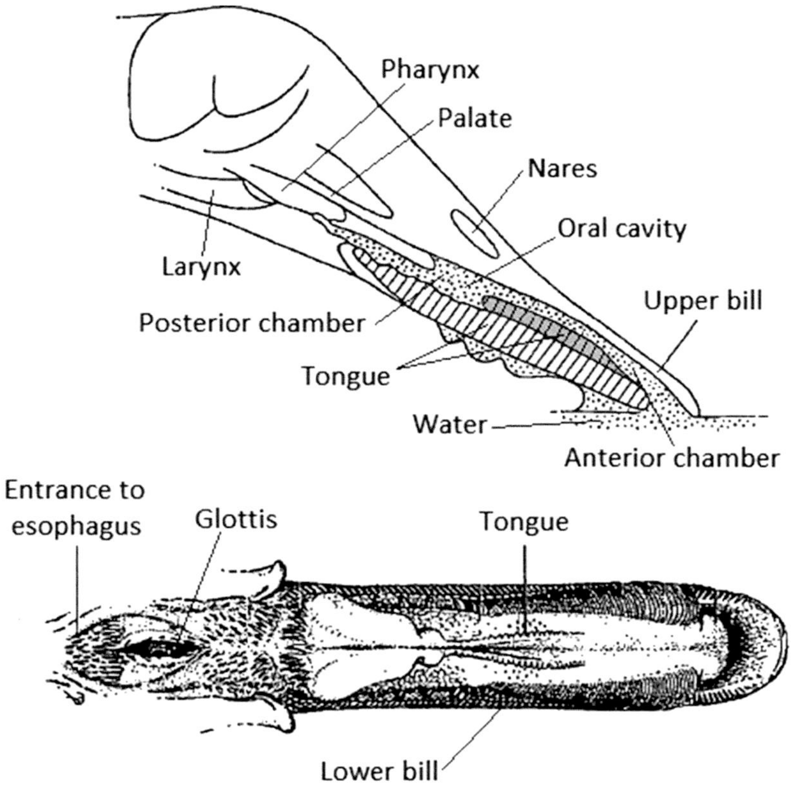


**Fig. 9.10** Drinking by a Mallard (*Anas platyrhynchos*). Water is transported into the pharynx in a several step process. First, the bill opens and water runs into the anterior chamber of the oral cavity (which is separated from the posterior chamber by a "bulge" on the top of the tongue (shaded area in top diagram). The bill then closes and the tongue is protracted, increasing the pressure in the anterior chamber of the oral cavity and forcing the water into the posterior chamber. This process is repeated several times, filling the posterior chamber with water. The bill then again closes, and the tongue is elevated, increasing the pressure in the posterior chamber and forcing the water into the pharynx and the larynx where it can then be swallowed. (Figure modified from Zweers 1991; © 2003 CCC Republication, used with permission)

water lost via excretion. Evaporative water loss through the skin and during respiration is how most water is lost. Small birds can lose up to five times more water by evaporation than by excretion (Dawson 1982). Total evaporative water loss (EWL) occurs via insensible EWL, the unavoidable loss of water given the permeability of the skin and respiratory surfaces to water vapor, but can also include thermoregulatory EWL, increased water loss by birds at high ambient temperatures to dissipate heat (Eto et al. 2017). At a thermally neutral temperature (25°C), total evaporative water loss is highly correlated with avian body mass (Fig. 9.16). In addition, total evaporative water loss is higher for passerines than non-passerines, with the higher basal metabolic rates of passerines requiring more evaporative water loss to maintain their necessary heat balance (Gavrilov 2017; Fig. 9.16). Because smaller birds have higher metabolic rates and more surface area per unit body mass, total evaporative water loss per day is higher for smaller birds than larger birds (McKechnie et al. 2021; Fig. 9.17).

### 9.3.1 Cutaneous Water Loss

As with other terrestrial vertebrates, the primary function of avian skin is to create a permeability

**Fig. 9.11** Aerial insectivores such as swallows obtain water by scooping up water with their lower mandibles as they fly just above the water's surface. This Tree Swallow (*Tachycineta bicolor*) appears to be slowing down (using its down-turned tail as a brake) as it approaches the water's surface. (Photo by Andy Reago and Chrissy McClarren, Wikipedia, CC BY 2.0, https://creativecommons.org/licenses/by/2.0/)

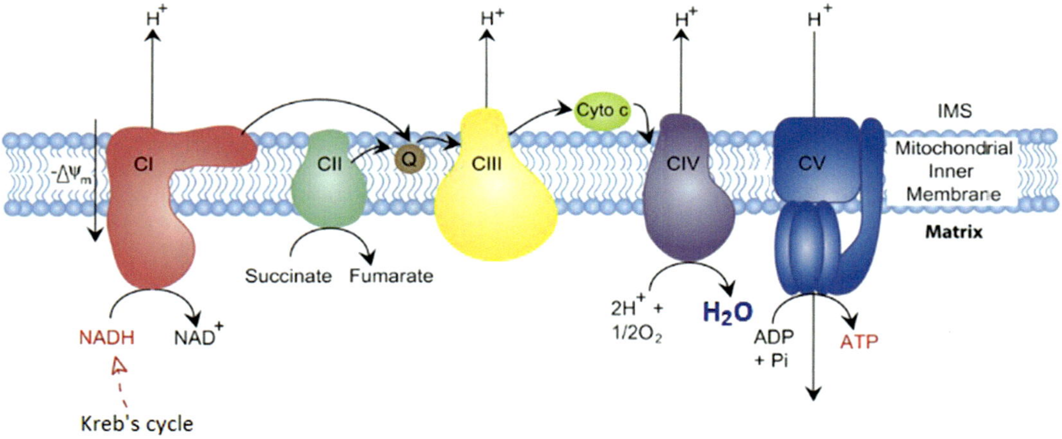


**Fig. 9.12** Metabolic water production via the electron transport chain takes place in the inner membrane of mitochondria. (Figure from Osellame et al. 2012; open-access article available under the terms of the Creative Commons Attribution License CC BY, https://creativecommons.org/licenses/by/4.0/)

**Table 9.1** Water content, metabolic water production, and total water production of carbohydrates (glycogen), lipids, and proteins (Table modified from Jenni-Eiermann and Jenni 2012; © 2012 Springer-Verlag Berlin Heidelberg, used with permission).

| | Glycogen | Lipids in adipose tissue | Protein in skeletal muscle or fat-free mass |
|---|---|---|---|
| Water content (%) | 75–80 | 5 | 70 |
| Metabolic water production (g $H_2O$ per g dry matter) | 0.556 | 1.05 | 0.39 |
| Total water production (g $H_2O$ per g wet tissue) | 0.89–0.91 | 1.10 | 0.82 |

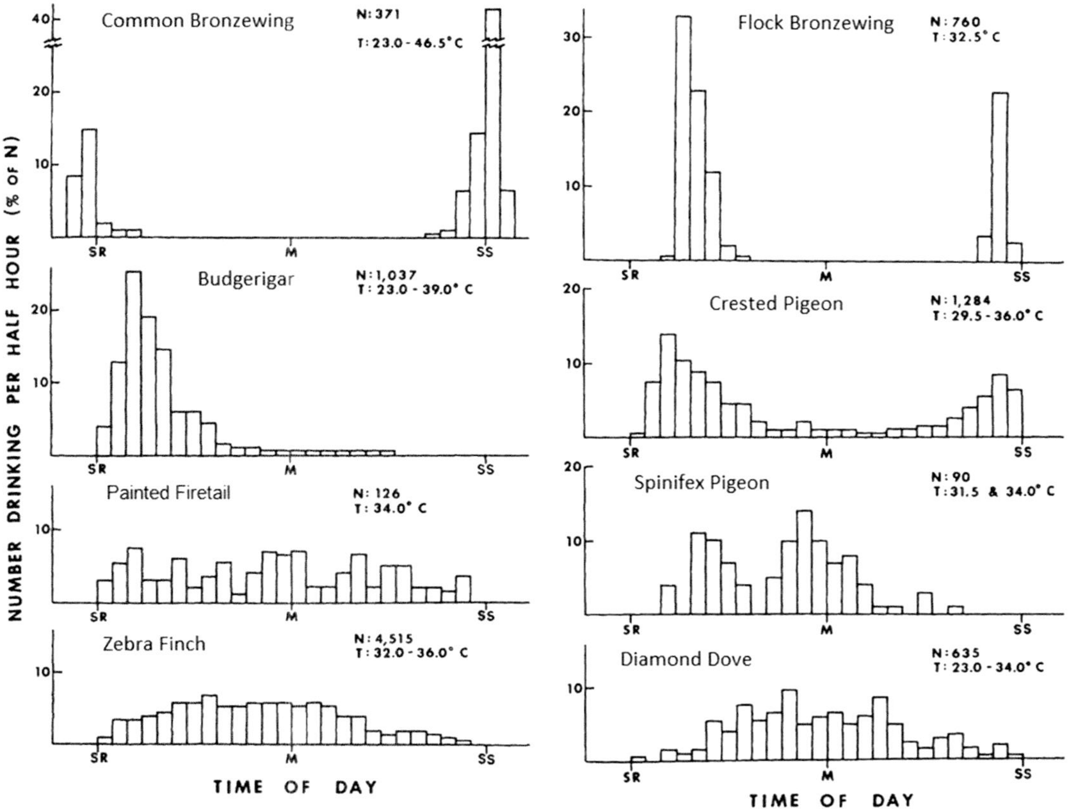


**Fig. 9.13** Daily patterns of drinking by several species of birds found in arid habitats in Australia. N = total number of birds drinking, T = maximum daily temperatures. Time of day: SR, sunrise; M, mid-day; SS, sunset. The pattern for Budgerigars was based on only 1% of the total number of birds drinking during every 30-min period. Common Bronzewing, *Phaps chalcoptera*; Flock Bronzewing, *Phaps histrionica*; Budgerigar, *Melopsittacus undulatus*; Crested Pigeon, *Ocyphaps lophotes*; Painted Firetail, *Emblema pictum*; Spinifex Pigeon, *Geophaps plumifera*; Zebra Finch, *Taeniopygia guttata*; Diamond Dove, *Geopelia cuneata*. (Figure from Fisher et al. 1972; © 1972 Oxford University Press, used with permission).

barrier that will prevent excessive evaporative water loss and death by dehydration. That barrier consists primarily of lipids (fats) present in the skin, and the reptilian ancestors of birds apparently first developed this lipid barrier about 300 million years ago (Matoltsy and Bereiter-Hahn 1986). The skin of birds has been modified by natural selection to facilitate evaporative cooling when needed while still providing waterproofing. Compared to mammals, cutaneous water loss is typically much higher in birds, suggesting a rather poor permeability barrier. However, as described in more detail in Chap. 10, this is beneficial for birds that are largely dependent on evaporative cooling for thermoregulation. The higher basal body temperature, increased heat production when flying, insulation by plumage, and the lack of sweat glands all require a higher rate of evaporative cooling through avian skin (Menon and Menon 2000).

For mammals with sweat glands, like humans, the loss of water through the skin can be obvious. Birds have no sweat glands, but water constantly passes directly through their skin and evaporates. In fact, cutaneous water loss has been found to account for about 65% of total evaporative water

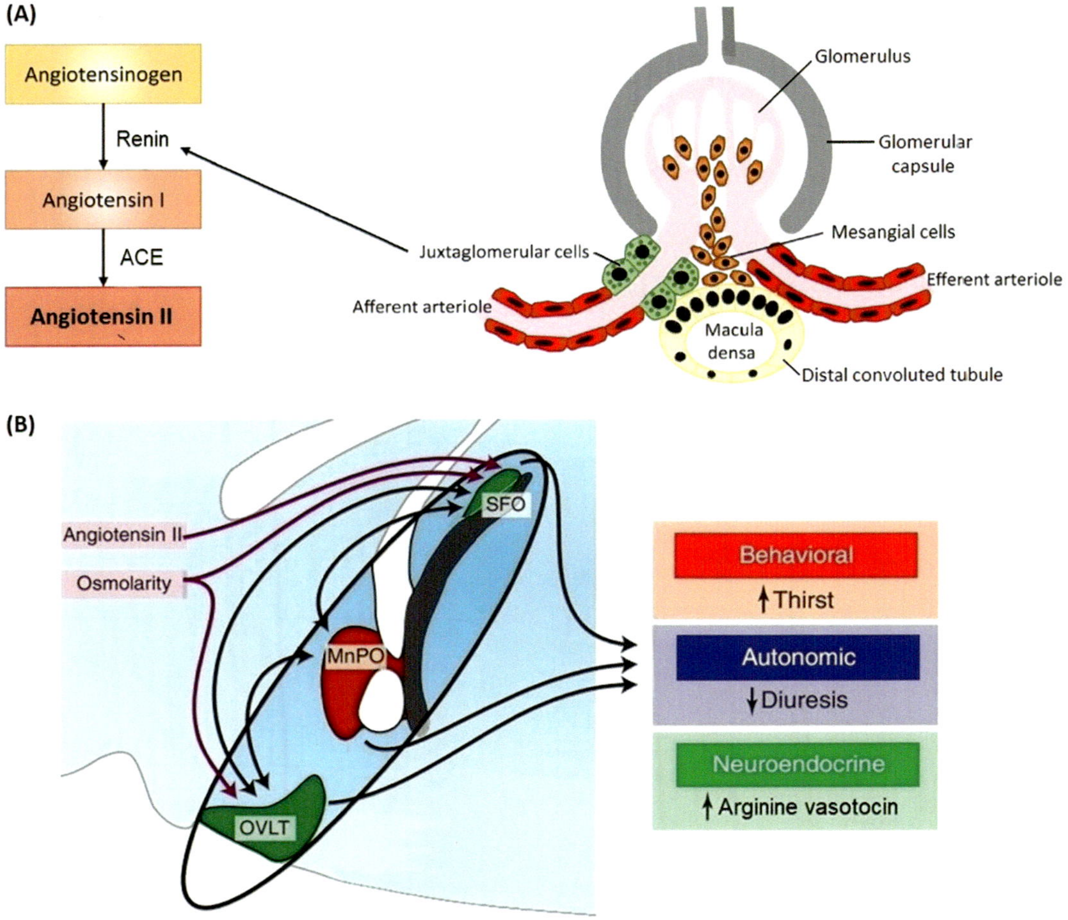


**Fig. 9.14** Responses to dehydration by cells in the avian kidney and brain (hypothalamus) that help restore water balance. (**a**) Dehydration reduces blood volume and lowers blood pressure. In response, cells of the macula densa in the distal convoluted tubules in the kidney stimulate juxtaglomerular cells that release renin. Renin converts angiotensinogen (produced by the liver and found in the blood plasma) into angiotensin I, which, in turn, is converted into angiotensin II by angiotensin-converting enzyme (also in the blood plasma and produced by cells in the lung, kidney, small intestine, and other organs) (Lipke and Olson 1988; Stevens et al. 1988). Macula densa cells are taller, darker, and denser than adjacent cells, hence the name macula densa (Orosz and Echols 2020). Mesangial cells provide structural support for the glomerular capillaries and help regulate glomerular capillary flow and filtration (Schlöndorff and Banas 2009). (**b**) Several areas in the hypothalamus in the brain monitor the osmolarity of the blood and help maintain water balance in the body. If a bird is dehydrated, the subfornical organ (SFO), also stimulated by increased production of angiotensin II, and the organum vasculosum (OVLT), along with the median preoptic nucleus (MnPO), trigger several responses, including increased thirst, decreased diuresis (decreased production of urine), and increased release of arginine vasotocin that increases reabsorption of water by the kidneys. (Figure **a** modified from Sequeira-Lopez and Gomez 2021; © 2021 Wolters Kluwer Health, used with permission. Figure **b** modified from Leib et al. 2016; © 2016 Elsevier Ltd., used with permission).

loss in birds at thermoneutral temperatures (Tieleman and Williams 2002; Ro and Williams 2010) so the rate of cutaneous water loss plays an important role in determining water (and heat) balance in birds (Champagne et al. 2012). Water is lost through the skin (cutaneous water loss, or CWL) of all terrestrial vertebrates, and the extent of this loss varies with the characteristics of the skin. Resistance of the skin to water loss is lowest in some amphibians, highest in desert reptiles, and somewhere in between in birds (Lillywhite 2006). Among birds, resistance to CWL varies about 10-fold and is influenced by ambient temperature and the number and position of feathers (Williams and Tieleman 2005; Fig. 9.18). A bird's plumage can reduce the passage of water vapor from the skin to the atmosphere by about 5–20%, with short-term variation due to feather position (elevated, or "fluffed," feathers are more effective at reducing water loss) and long-term variation due to seasonal differences in the number of feathers (Lillywhite 2006).

The main determinant of variation in resistance to CWL among different species of birds is the lipid (fat) composition of the epidermis (the outer, thin, nonvascular layer of the skin). Differences in the molecular structure of different lipids cause differences in the permeability of the skin to water. Evidence to date indicates that the epidermis of birds in more arid environments where minimizing CWL is critical contains lipids that impede the diffusion of water, whereas birds in areas with abundant water and high humidity have epidermal lipids that are less resistant to water diffusion (Haugen et al. 2003; Williams and Tieleman 2005; Muñoz-Garcia and Williams 2005, 2007; Champagne et al. 2012). In some species, epidermal lipids can be altered to increase or decrease the permeability of the skin to water (Figs. 9.18 and 9.19). For example, Zebra Finches (*Taeniopygia guttata*) maintained under a water deficit are capable of facultative waterproofing (Table 9.2). Menon et al. (1996) showed that within 16 h of water deprivation, adult Zebra Finches reduced cutaneous water loss by more than 50% by altering the arrangement and type of lipids in the epidermis.

Cutaneous water loss is also influenced by temperature, with cutaneous evaporation typically increasing with increasing air temperature (Fig. 9.20; Box 9.2 Feather Reflectivity and Evaporative Water Loss). Changes in blood flow are largely responsible for this relationship. At high ambient temperatures, birds must increase heat loss to maintain their body temperatures, and an important part of this process is the vasodilation of arterioles in the skin. This increased blood flow warms the skin, and the increased blood pressure also increases the outflow of water from skin capillaries. In combination, these changes increase cutaneous water loss and help reduce heat stress (Lillywhite 2006).

### 9.3.2 Respiratory Water Loss

The loss of water due to evaporation across mucous membranes during respiration is called

**Box 9.1 Water Availability at Migratory Stopover Sites**

During long migratory flights, birds use fat and protein from organs (particularly those in the digestive system) and tissues (particularly flight muscle [pectoralis muscle] tissue) to provide the energy (and water) needed to sustain flight. At stopover sites, birds must then rebuild their digestive systems, then other organs and tissues, including their flight muscles. Importantly, investigators have found that an important factor in the ability of birds to recover and refuel at stopover sites is the availability of water (e.g., Sapir et al. 2004; Tsurim et al. 2008). Mizrahi et al. (2011) conducted an experiment with Eurasian Blackcaps (*Sylvia atricapilla*) captured at a stopover site in Israel during spring migration to examine the effect of water availability on tissue renewal. Blackcaps were divided into two groups, both received food, but one group had unlimited access to water and the other group had access to water only for 30 min twice a day.

(continued)

**Box 9.1** (continued)

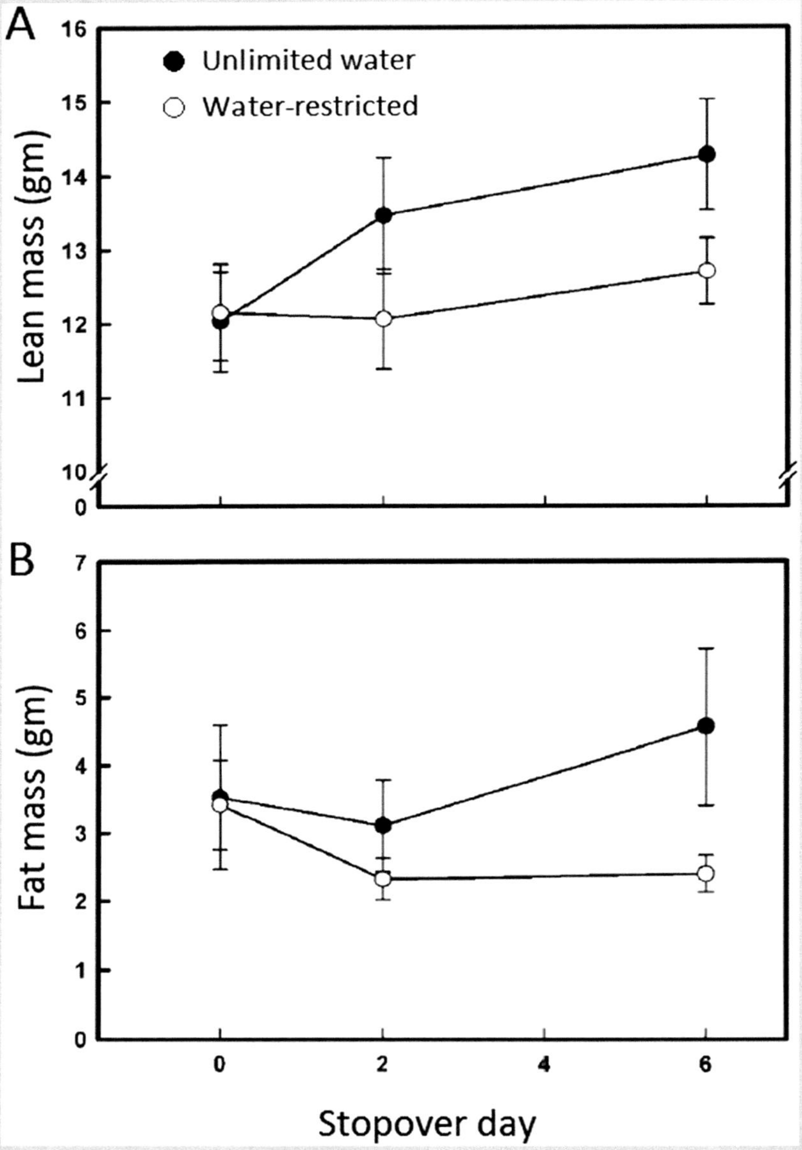


Mizrahi et al. (2011) found that the Eurasian Blackcaps provided with unlimited water gained significantly more lean (**a**) and fat mass (**b**) than those in the water-restricted group. These results suggest that birds refuel faster when sufficient water is available at stopover sites than when it is not. (Figure from Mizrahi et al. 2011; © 2011 Oxford University Press, used with permission).

(continued)

**Box 9.1** (continued)

Male Eurasian Blackcap. (Photo by Ron Knight, Wikipedia, CC BY 2.0, https://creativecommons.org/licenses/by/2.0/)

Domer et al. (2019) examined how water availability influenced fuel deposition rates of migrating Eurasian Blackcaps (*Sylvia atricapilla*). Because Eurasian Blackcaps (and other insectivorous songbirds) become more frugivorous during migration, blackcaps captured at a stopover site in Israel were subjected to different treatments with temporal separation in availability of water and fruit. Only blackcaps provided with both water and fruit at the same time increased their rate of fruit consumption and fuel deposition rates. These results suggest that drinking water "enhances lubrication or dissolution of nutrients within the digestive system or facilitates the penetration of digestive enzymes and acids into food particles, allowing birds to consume more fruits" and, consequently, increase fuel deposition rates. More generally, these results provide support for the hypothesis that availability of drinking water influences diet selection and hyperphagia of Eurasian Blackcaps and, perhaps, other migratory species of songbirds.

Availability of water at stopover sites may also influence how much time birds spend at stopover sites. Skrip et al. (2015) captured Garden Warblers (*Sylvia borin*) at a stopover site during spring migration, divided them into two groups (one with water and one without water), and examined their nocturnal restlessness (zugunruhe). For both groups, Garden Warblers with more stored fat exhibited the most nocturnal restlessness, indicating a greater tendency to continue migration.

(continued)

**Box 9.1** (continued)

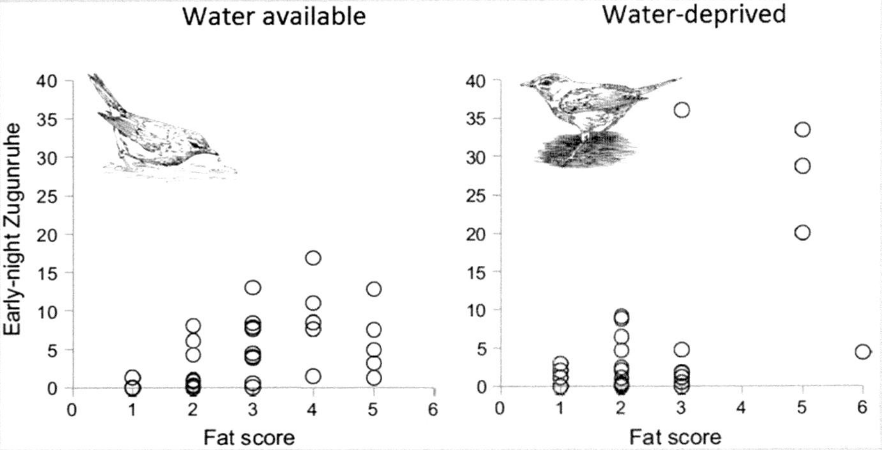


However, comparing warblers with comparably high fat scores in the two groups, levels of nocturnal restlessness were significantly higher for warblers that were water-deprived. These results suggest that warblers with adequate fat stores will spend less time at "dry" stopover sites than at sites with more water and provide further evidence that water availability plays an important role in the stopover behavior of migrating birds. (Figure from Skrip et al. 2015; © 2015 Dt. Ornithologen-Gesellschaft e.V., used with permission)

What remains to be determined is how water improves digestive efficiency. Possible explanations include: (1) a lack of water at stopover sites results in slower "rebuilding" of digestive organs, (2) less water increases viscosity of gut contents, slowing movement and lowering digestive efficiency, and (3) reduced access to water may reduce levels of foraging activity to minimize water loss (Mizrahi et al. 2011).

respiratory water loss, and the greater the volume of air exchanged during respiration, the greater the amount of water lost (Fig. 9.21). The importance of respiratory water loss (in terms of percentage of total evaporative water loss) has been found to increase at higher ambient temperatures for songbirds (e.g., Wolf and Walsberg 1996; Tieleman and Williams 2002; Box 9.3 Climate Change and Desert Songbirds) and likely for species in other taxa as well, e.g., sandgrouse (order Pterocliformes; McKechnie et al. 2016), nightjars (order Caprimulgiformes; Talbot et al. 2017), and owls (order Strigiformes; Talbot et al. 2018), but not for species in the order Columbiformes (e.g., McKechnie and Wolf 2004; Fig. 9.18). For example, cutaneous evaporative water loss accounted for about 60% of the total evaporative water loss by Zebra Finches (*Taeniopygia guttata*) at an ambient temperature of 25°C, but only about 20% at 40°C (Wojciechowski et al. 2021). For White-winged Doves (*Zenaida asiatica mearnsii*), in contrast, cutaneous evaporative water loss accounted for 64% of total evaporative water loss at an ambient temperature of 35°C, and this increased to 78% of total evaporative water loss at 45°C (McKechnie and Wolf 2004).

Regulation of respiratory water loss by birds involves adjustment of the volume of air expired (determined by breathing rate and the amount of

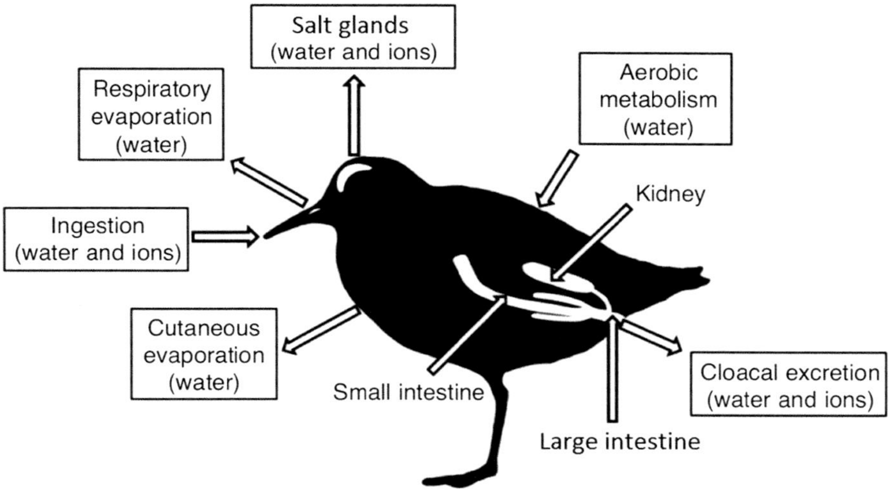


**Fig. 9.15** Pathways of water and ion influx and efflux, and the principal osmoregulatory organs of birds. Water and ions can be ingested, and water is produced in the process of aerobic metabolism (metabolic water produced during ATP production by cells), and both water and ions are excreted by the salt glands (in those species with salt glands) and via cloacal excretion. Water is lost via respiratory and cutaneous evaporation. (Figure from Gutiérrez 2014; © Gutiérrez, J. S., Ardeola, used with permission)

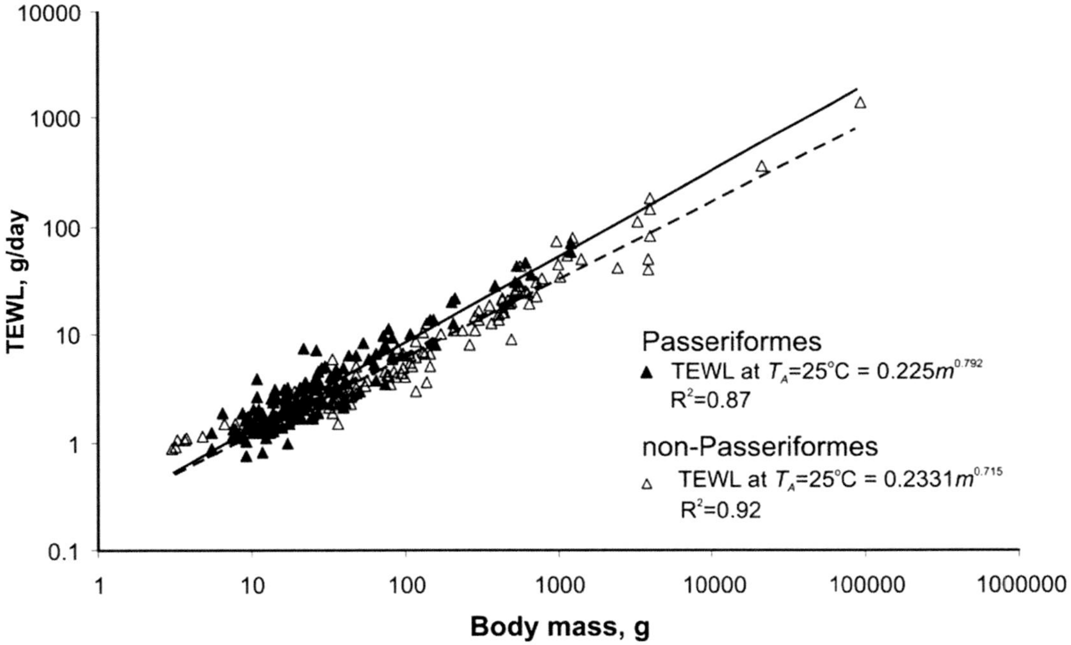


**Fig. 9.16** Total evaporative water loss (TEWL) at an ambient temperature of 25°C as a function of body mass for 152 species of passerines and 107 species of non-passerines. (Data from Williams 1996 and Gavrilov 2017; Figure from Gavrilov 2017; © 2017 Zoological Studies, open-access article distributed under the terms of the Creative Commons Attribution license, https://creativecommons.org/licenses/by/4.0/)

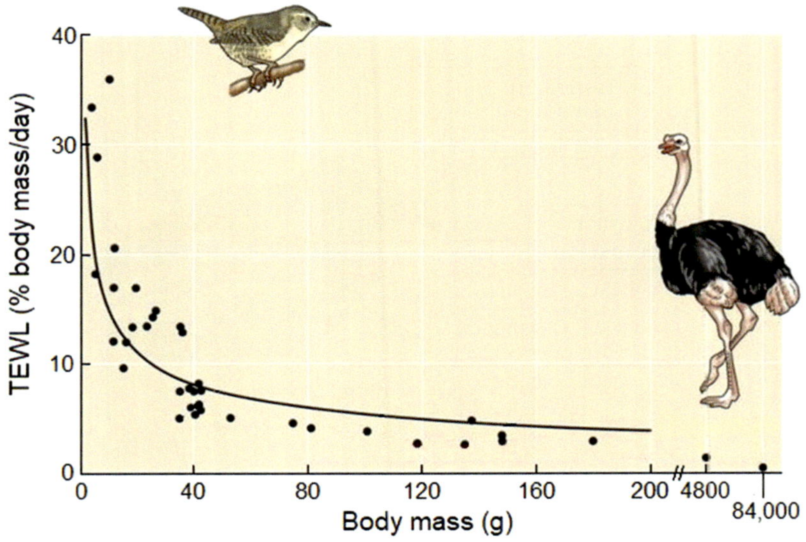


**Fig. 9.17** At a thermally neutral temperature, smaller birds have higher total evaporative water loss (TEWL) per day (as a percentage of body mass) than larger birds. (Figure modified from Hill 2012; © 2012 Oxford Publishing Limited, used with permission)

air expired with each exhalation) and the water content of that air. Expired air is usually saturated with water vapor in birds, and the higher the temperature of that air, the greater its water content. In addition, the greater the difference between the water content of inspired and expired air, the greater will be the respiratory water loss (Michaeli and Pinshow 2001). That is, the drier the air, the greater the respiratory water loss (Fig. 9.22). In addition, respiratory water loss tends to increase with increasing ambient temperature. For example, respiratory water loss for Verdins (*Auriparus flaviceps*) exposed to increasingly high air temperatures (from 30 to 50°C) increased more than 30-fold (Wolf and Walsberg 1996). Most of this increase occurred at temperatures above 39°C when the birds began using respiratory water loss to dissipate heat via panting or gular flutter (Fig. 9.23).

Birds can also reduce respiratory water loss via a counter-current mechanism involving the respiratory turbinates in their nasal cavities (Figs. 9.24 and 9.25) and, in some species, the trachea. During inspiration, air moves through the trachea and nasal passageways and over the moist surfaces of the tracheal epithelium and respiratory turbinates, both warming and humidifying the air and, at the same time, cooling the epithelium. Then, during exhalation, the outgoing air is cooled by the tracheal and turbinate epithelium, which causes water vapor in the air to condense on the surfaces of the respiratory turbinates (Fig. 9.26) and trachea. This mechanism allows birds to reduce both heat and water loss (Geist 2000). Owerkowicz et al. (2021) noted that, although the respiratory turbinate surface area of mammals averages almost three times that of similar-sized birds, the tracheas of birds average about 2.7 times longer and tracheal surface areas average almost 3.5 times that of similar-sized mammals. Given that longer tracheas mean that air spends more time in tracheas, Owerkowicz et al. (2021) suggested that, for some species of birds, tracheas may, along with the respiratory turbinates, play a role in reducing heat and water loss.

The nasal conchae of birds typically consist of two or three paired structures, with most having rostral and middle conchae (Fig. 9.25) and some (e.g., Wild Turkeys, *Meleagris gallopavo*; Fig. 9.24) also having caudal conchae (King 1993; Yokosuka et al. 2009a, b; Danner et al. 2017). The rostral and middle conchae appear to play a primary role in counter-current exchange, whereas the caudal conchae, roof of the nasal cavity, and, in some taxa, the middle conchae are more important for olfaction (Murrish 1973; Danner et al. 2017). In a comparison to two subspecies of Song Sparrows (*Melospiza*

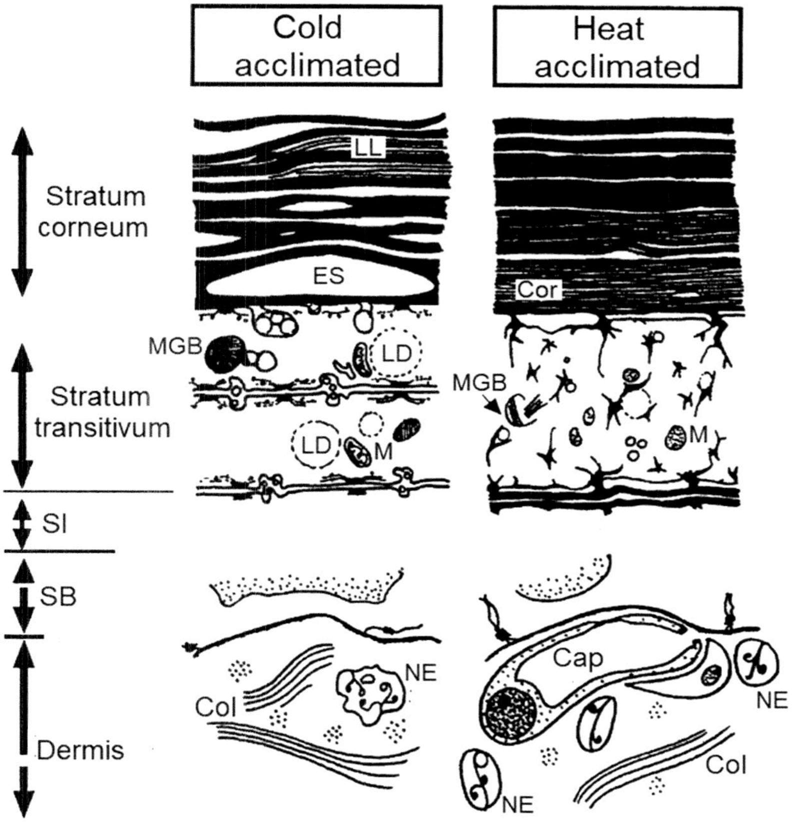


**Fig. 9.18** Differences in the epidermis and dermis of cold- and heat-acclimated Rock Pigeons (*Columba livia*; order Columbiformes). The skin of cold-acclimated pigeons contains more lipids, including extracellular laminar lipids (LL) in the stratum corneum, free lipid droplets (LD) in the stratum transitivum, and lipids secreted in the intercellular matrix by structures called multigranular bodies (MGB). These lipids help minimize cutaneous water loss. In contrast, the skin of heat-acclimated pigeons contains much less lipid in both the stratum corneum and stratum transitivum, plus MGBs no longer secrete lipids. Reducing the amount of lipid in the skin makes the skin more permeable to water vapor and allows greater cutaneous water loss. In addition, the dermis of heat-acclimated pigeons is more vascular, with more capillaries (Cap) that are more permeable to water. Thus, allowing further increases in cutaneous water loss as needed to regulate body temperature. Col, collagen bundles; Cor, corneocytes; ES, extracellular space; M, mitochondria; NE, nerve endings; SI, stratum intermedium; SB, stratum basale. (Figure modified from Arieli et al. 2002; © 2002 Elsevier Science Inc., used with permission)

*melodia*) found in habitats that differ in water availability, Danner et al. (2017) found greater conchae surface area in the subspecies that occupied a much drier habitat and suggested that the greater surface area might allow more efficient recovery of water during exhalation (Fig. 9.27).

Respiratory water loss also increases during flight. For European Starlings (*Sturnus vulgaris*), respiratory water loss during flight (average = 0.74 ml/h) was significantly higher than at rest (average = 0.18 ml/h) and increased with ambient temperature (Engel et al. 2006b). Similarly, respiratory water loss by flying Rock Pigeons (*Columba livia*) increased nearly eight times above that when at rest (Michaeli and Pinshow 2001). More generally, Engel et al. (2006a, b) estimated that respiratory evaporative water loss accounted for up to 88% of total

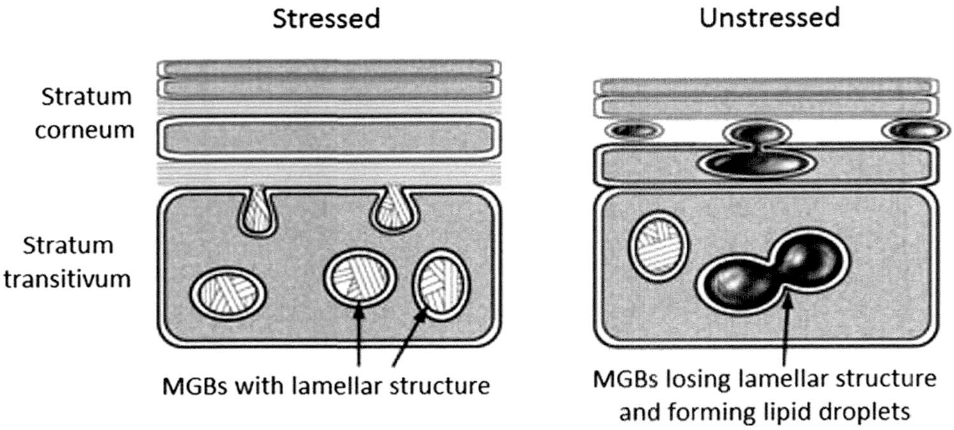


**Fig. 9.19** When not water-stressed, avian multigranular bodies (MGBs) form lipid droplets rather than layers, and the stratum corneum is then more permeable to water (with spaces between droplets where water is lost). However, when water-stressed, the MGBs form lipid layers that, after release, form layers between the corneocytes of the stratum corneum and reduce cutaneous evaporative water loss. (Figure from Menon and Dumbacher 2014; © 2014 John Wiley and Sons A/S. Published by John Wiley and Sons Ltd., used with permission)

**Table 9.2** Cutaneous water loss ($g/m^2/h$) by Zebra Finches. Note the decline in water loss after water was removed, followed by an increase (especially after 96 h) after water was provided. (Table from Menon et al. 1996; © 1996 Wiley-Liss, Inc., used with permission)

| | | | Post water deprivation | | | | | Replenishment of water | | |
|---|---|---|---|---|---|---|---|---|---|---|
| Bird number | Basal level | | 16 h | 24 h | 48 h | 72 h | | 24 h | 48 h | 96 h |
| 1 | 24 | Water removed | | 6 | 3 | 3 | Water provided | 5 | 5 | 28 |
| 2 | 22 | | | 7 | 7 | 5 | | 5 | 12 | 50 |
| 3 | 20 | | 9 | | | | | | | |
| 4 | 24 | | 7 | | | | | | | |

evaporative water loss during flight for European Starlings and Rose-colored Starlings (*Sturnus roseus*). Contributing to these increases is the increased respiratory rate of flying (or hovering) birds (Fig. 9.28).

## 9.4 Water Balance and Long-Distance Flight

During short flights, increased water loss is not a problem for birds. However, during long-distance flights, such as during migration, the continuing loss of water may limit a bird's flight range. Of course, when flying over land, birds always have the option of landing at a stopover site to replenish both energy and water reserves; that is not the case, however, when flying over deserts or oceans. Even for those birds flying over hospitable areas, the energy required to reach cruising altitude after a stopover may be an important consideration (Hedenström and Alerstam 1992).

Migrating birds will lose some water via excretion. However, Gerson and Guglielmo (2013) measured the glomerular filtration rates and fractional water reabsorption of Swainson's Thrushes (*Catharus ustulatus*) either resting or flying in a wind tunnel and found no difference between them in glomerular filtration rates, but fractional water reabsorption was greater for flying birds than resting birds. These results suggest that birds in flight minimize water loss by

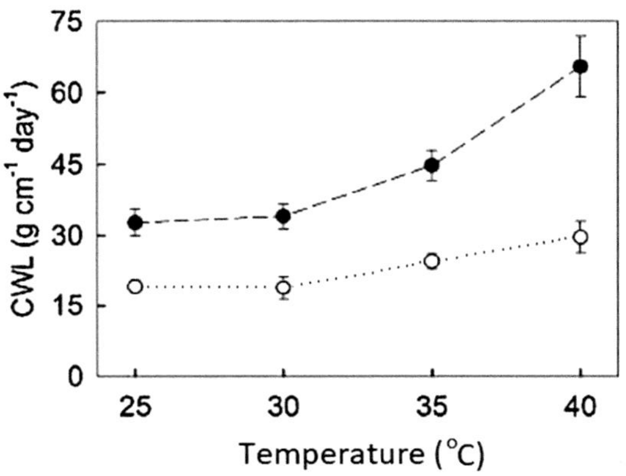


**Fig. 9.20** Mean (± SE) cutaneous water loss by House Sparrows (*Passer domesticus*) at different temperatures during summer (filled circles, dashed line) and winter (open circles, dotted line). Cutaneous water loss was greater during summer, possibly reflecting a mechanism by which birds use this water loss in warm, but not water-limited, environments. During winter, when colder and drier, minimizing cutaneous water loss likely represents a water- and heat-conserving mechanism. (Figure from Champagne et al. 2016; © 2015 Elsevier Ireland Ltd., used with permission)

increasing the amount of water reabsorbed by the kidney. Thus, the more important sources of water loss by birds in flight are respiratory and cutaneous water loss. One factor that influences respiratory water loss is flight altitude because humidity decreases with increasing altitude and the difference in the water content of inhaled and exhaled increases. However, a bird's body temperature increases during flight and flying at higher altitudes, where temperatures are lower, may reduce the need for evaporative cooling. For example, Black-tailed Godwits (*Limosa limosa*) migrating from the Netherlands to sub-Saharan West Africa spent 21% of their time at altitudes above 5000 m, likely because the cooler temperatures at those altitudes minimize the possibility of hyperthermia and water loss (Senner et al. 2018). Similarly, non-migrating Common Swifts (*Apus apus*) appear to select roosting altitudes (where they spend the night) based on ambient temperature, "roosting" at higher altitudes on warmer nights. One possible reason for such behavior is that swifts are selecting roosting altitudes that enhance maintenance of water balance (Backman and Alerstam 2001). Of course, migrating birds must consider factors other than water balance. For example, songbirds migrating at night likely select flight altitudes with supporting winds, with other factors such as temperature and water balance less important (Liechti et al. 2000; Schmaljohann et al. 2009; Horton et al. 2016).

Regardless of altitude, birds lose water as they fly, but, as fats and proteins are metabolized, they also benefit from metabolic water production. Given the simultaneous loss and production of water, how far can migrating birds fly without becoming dehydrated? Experiments with Rose-colored Starlings (*Sturnus roseus*) in a wind tunnel revealed a negative water balance during all flights, with water loss averaging 1.29 g/h and water gain (metabolic water) averaging 0.98 g/h (Engel et al. 2006a). Assuming fats were the only source of energy (and metabolic water production) and storing about 10 g of fat (mean body mass was 72 g), it was estimated that Rose-colored Starlings could fly for about 14 h before becoming dehydrated (Engel et al. 2006a). That is likely a conservative estimate, and other investigators have predicted that small birds, under the right conditions, could potentially fly for nearly 40 h or more hours before becoming dehydrated (Carmi et al. 1992). Coincidentally, based on the typical flight speeds of small migrants, it would take about 40 h for a small bird migrating between Europe and Africa to fly across the Sahara Desert. Recent evidence suggests that most songbirds migrating over the Sahara do not make non-stop flights (Schmaljohann et al. 2007). Because passerines stopping to rest in the desert typically do not feed, excessive water loss caused by flying through hot dry air is a likely reason why most birds do not fly non-stop. Birds flying over more humid areas (e.g., oceans), with favorable winds, can fly greater distances and for longer periods. For example, current evidence supports the hypothesis that most migrating Blackpoll Warblers (*Setophaga striata*) make an overwater flight from the northeastern United States and Canada's Maritime Provinces to northern South America (DeLuca et al. 2019). This route varies

**Box 9.2 Feather Reflectivity and Evaporative Water Loss**

About 45% of the energy in sunlight is in the ultraviolet (UV; 300–400 nm) and human-visible (400–700 nm) wavelength, but the remaining 55% is in near-infrared wavelengths (NIR; 700–2500 nm) that cannot be seen by birds and mammals (Luo et al. 2011). Medina et al. (2018) measured reflectance (300–2100 nm) of feathers of 616 individual birds representing 90 species, 34 families, and 14 orders of Australian birds and found that NIR reflectivity varied considerably, with much of variation due to differences in the shape of barbs and density of barbules (Stuart-Fox et al. 2018). In addition, species of birds found in hotter, drier environments had plumage with greater NIR reflectivity than birds in cooler, wetter environments. For small birds (10 g) in hot, dry environments, the higher NIR reflectivity reduces evaporative water loss by an estimated 2.1% of body mass per hour (Medina et al. 2018). For larger birds (100 g, 1 kg), increased reflectivity translates into a minimal reduction in evaporative water loss.

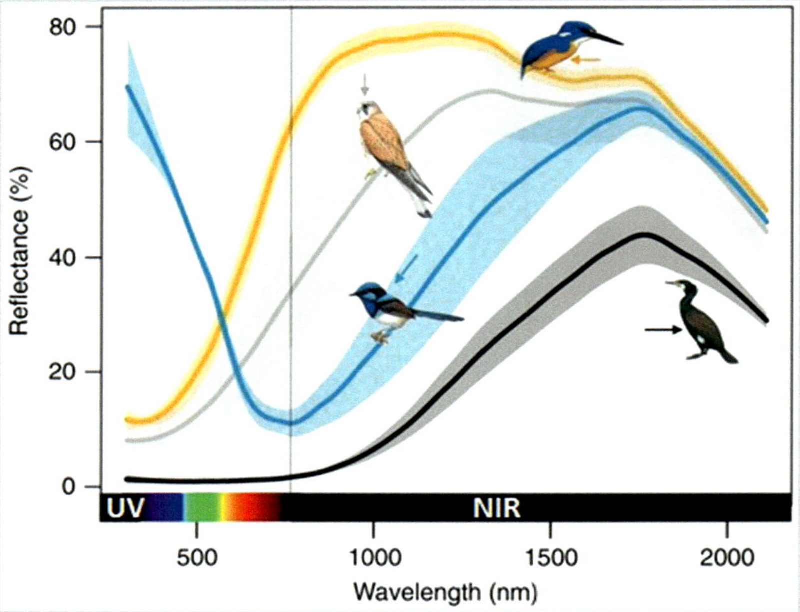


Variation in near-infrared reflectance in four species of Australian birds, with relatively high reflectivity for the gray crown of Nankeen Kestrels (*Falco cenchroides*) and orange belly of Azure Kingfishers (*Alcedo azurea*), and lower reflectivity for the blue mantle of male Superb Fairy wrens (*Malurus cyaneus*) and black belly of Great Cormorants (*Phalacrocorax carbo*). (Figure from Medina et al. 2018; open-access article licensed under a Creative Commons Attribution 4.0 International License, https://creativecommons.org/licenses/by/4.0/)

(continued)

**Box 9.2** (continued)

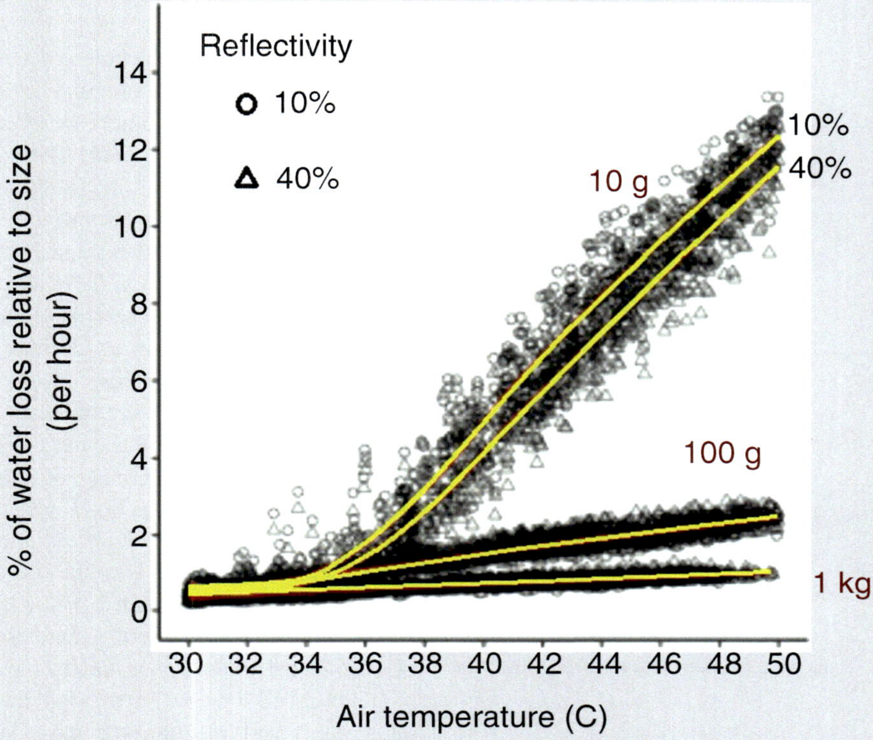


Relationship between air temperature and water loss according to body mass and total feather reflectivity. With high air temperatures, the increasing NIR reflectivity of feathers (e.g., 40% rather than 10%) reduces evaporative water loss for small birds and helps prevent extreme dehydration that could prove fatal. (Figure modified from Supplemental Fig. 6 from Medina et al. 2018; open-access article licensed under a Creative Commons Attribution 4.0 International License, https://creativecommons.org/licenses/by/4.0/)

The greater benefit of increased NIR reflectivity in reducing evaporative water loss for smaller birds may be due to having a thinner layer of insulating feathers and a higher surface-area-to-volume ratio (and greater potential cutaneous water loss) than larger birds. For small birds in hot, dry environments like the arid deserts of Australia, the high NIR reflectivity of their feathers and resulting reduced evaporative water loss at high temperatures may be critically important in preventing dehydration levels that could prove fatal (Medina et al. 2018).

from 2270 to 2770 km and is estimated to take from 52 to 72 h (DeLuca et al. 2015). Similarly, migrating Bar-tailed Godwits (*Limosa lapponica*) that breed in Alaska fly non-stop for up to 9 or more days (and over 11,000 km) to wintering areas in New Zealand (Gill et al. 2009) and those that winter in Africa and breed in Scandinavia and Siberia fly non-stop from Africa to stopover sites in northern Europe during spring migration (4300 km; Landys et al. 2000) and apparently exhibit no signs of dehydration upon arrival. To do so, godwits likely fly at altitudes (about 3000 m) where temperatures are cooler

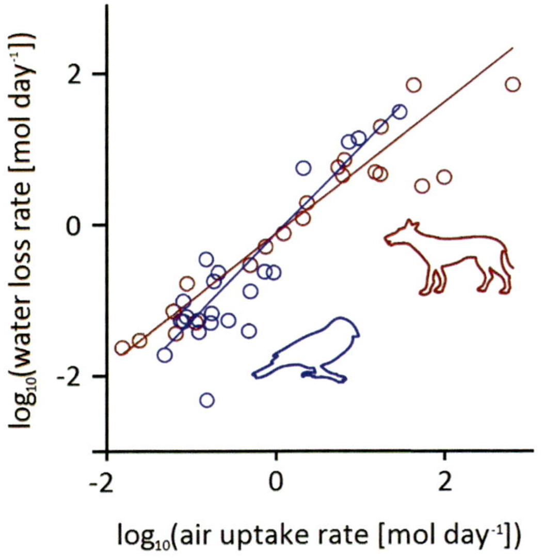


**Fig. 9.21** Relationship between volume of air exchanged during respiration and respiratory water loss for birds (blue circles; $N = 22$ species) and mammals (brown circles; $N = 21$ species). For both taxa, water loss scaled with air exchange with slopes of about 1, i.e., higher rates of gas exchange led to higher rates of respiratory water loss. (Figure modified from Woods and Smith 2010, used with permission of the United States National Academy of Sciences)

and evaporative water loss is minimized (Landys et al. 2000).

## 9.5 Kidney Structure and Function

To understand how birds cope with arid environments, salt water, and excess water, some understanding of the roles of the kidneys and lower intestine in osmoregulation is needed. Like the kidneys of other vertebrates, the functions of bird kidneys include filtration, excretion, secretion, and absorption. The kidneys filter water and some substances from blood, such as waste products of metabolism and ions that are then voided in the urine. Kidneys also play an important role in conserving water and reabsorbing needed substances (such as glucose).

The urinary system of birds consists of paired kidneys (Fig. 9.29) and ureters that transport urine to the urodeum of the cloaca (Fig. 9.30). The size or mass of the kidneys of songbirds is generally correlated with body size or mass (Fig. 9.31). Johnson (1968) noted that kidney mass is generally about 2% of the body mass of small birds (<100 g), about 1% for birds with masses between 100 and 1000 g, and less than 1% for even larger birds. Avian kidneys are divided into units called lobules. Each lobule has a cortex (outer area) and medulla (or medullary cone; Figs. 9.32 and 9.33). By volume, the avian kidney is primarily cortex (71–81%), plus a relatively small medulla (range = 5–15%) and blood vessels larger than capillaries (range 10–13%). The functional unit of the kidney is the nephron (Fig. 9.34). Nephrons filter the blood plasma to eliminate waste products, but, in doing so, must

**Box 9.3 Climate Change and Desert Songbirds**

When air temperatures exceed a bird's upper critical temperature (UCT, temperature at which heat must be lost via evaporative cooling and metabolic rates increase), evaporative heat loss is the only way to avoid further increases in body temperature. As air temperatures increase above the UCT, rates of evaporative water loss (EWL) continue to increase, and this loss of water can potentially result in dehydration and, at extremely high temperatures, even death. Albright et al. (2017) examined the effect of increasing heat on EWL and survival of five species of songbirds occupying desert habitat in the southwestern United States. These authors found that smaller songbirds such as Lesser Goldfinches (*Spinus psaltria*) exhibit higher rates of mass-specific EWL than larger songbirds such as Curve-billed Thrashers (*Toxostoma curvirostre*) and so are more likely to experience lethal dehydration at high temperatures.

(continued)

**Box 9.3** (continued)

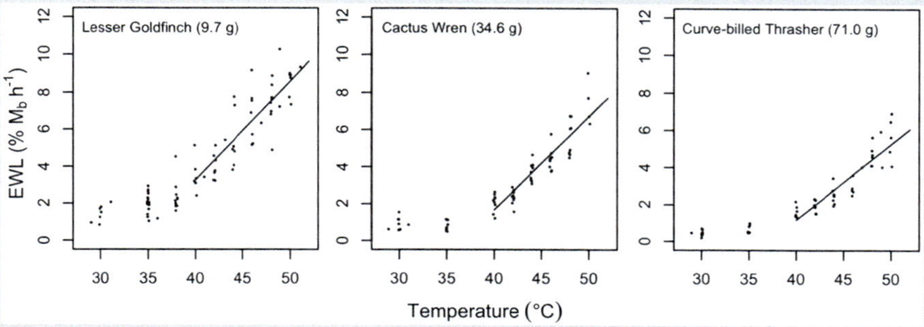


Evaporative water loss (EWL) as a percentage of body mass (Mb) for three species of songbirds across a range of elevated air temperatures. Mean body masses (Mb) are indicated, and a linear fit line ($P < 0.001$) has been added for air temperatures above the upper critical temperature of 40°C. Lesser Goldfinch, *Spinus psaltria*; Cactus Wren, *Campylorhynchus brunneicapillus*; Curve-billed Thrasher, *Toxostoma curvirostre*. (Figure from Albright et al. 2017; used with permission of the U. S. National Academy of Sciences)

Models developed by Albright et al. (2017) suggest that climate change will mean higher temperatures and an increasing frequency of heat waves in the future for the southwestern United States. These changes will likely expand the areas in the ranges of songbirds where they will be at risk of lethal dehydration, especially for smaller songbirds. Their model predicts, for example, that, by the year 2100, Lesser Goldfinches will be exposed to 60–120 days per year with potentially lethal temperatures over large regions in their current range. As such, in the future, small songbirds such as Lesser Goldfinches will likely be at increasing risk of extirpation in areas they currently occupy.

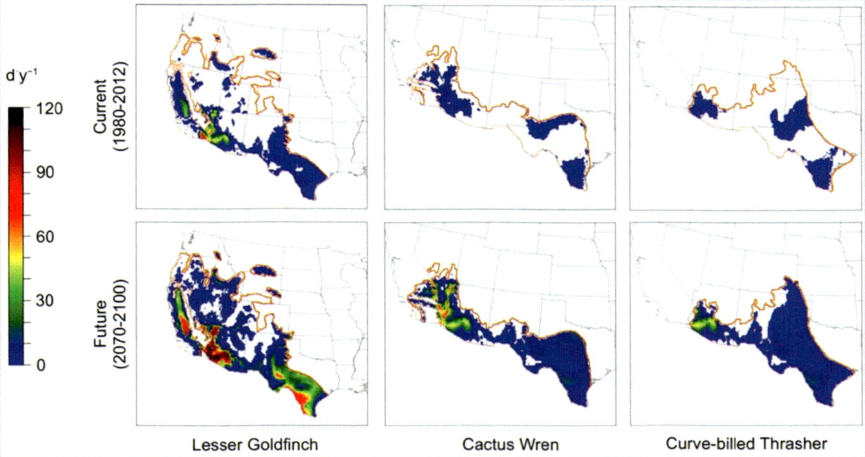


Current (1980–2012) and future (2070–2100) number of days per year between April 1 and September 30 with survival times ≤5 h across the ranges (orange lines) of three species of songbirds in the southwestern United States, arranged in order of increasing body mass. (Figure from Albright et al. 2017; used with permission of the U. S. National Academy of Sciences)

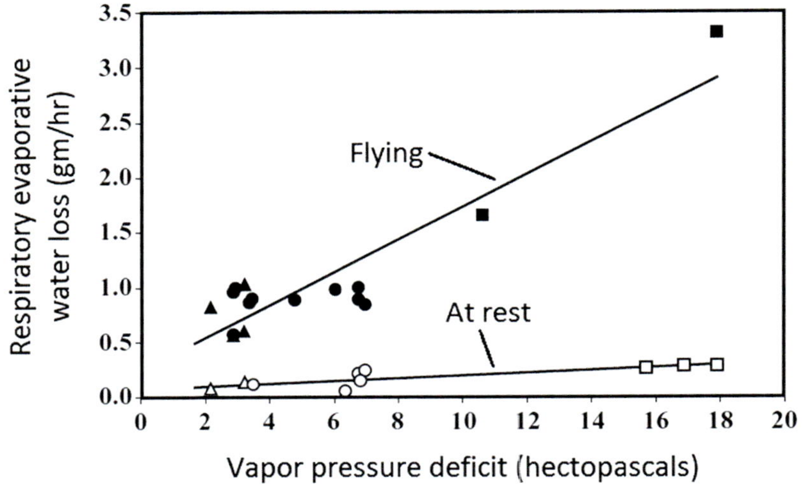


**Fig. 9.22** Respiratory evaporative water loss of European Starlings (*Sturnus vulgaris*) in flight (in a wind tunnel) and at rest in relation to the vapor pressure deficit of ambient air. Triangles are measurements made at 6°C, circles at 15°C, and squares at 25°C. Vapor pressure deficit is a measure of how much moisture is in the air, with high numbers indicating less moisture in the air. Especially when flying (when respiratory rate is higher), respiratory evaporative water loss increases with increasingly dry air. (Figure modified from Engel et al. 2006b; © 2006 Elsevier Inc., used with permission)

**Fig. 9.23** Thermal image of pelicans as they use gular flutter to increase respiratory evaporative water (and heat) loss. The bright areas along the pouches and the bare skin around the eyes indicate areas that are radiating much more heat than the areas covered by feathers. Gular flutter involves the rapid contraction of neck muscles with the bill open to increase air flow, water evaporation, and heat loss. (Photo credit: Witmer Lab/Ohio University and Glenn Tattersall/Brock University, used with permission)

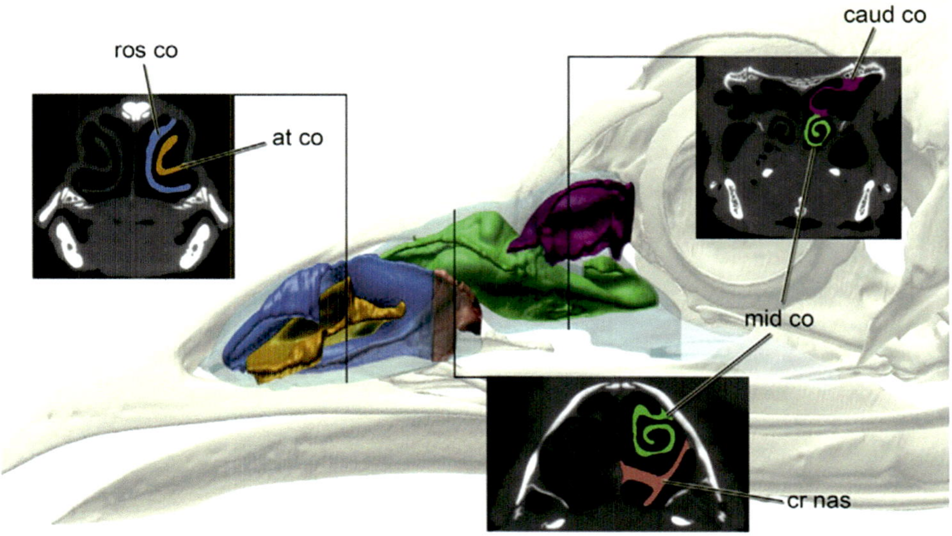


**Fig. 9.24** Nasal conchae in the nasal cavities of a Wild Turkey (*Meleagris gallopavo*). Conchae are projections of the nasal mucosa in the nasal passages formed by a cartilaginous skeleton, i.e., turbinates. ros co, rostral concha; at co, atrial concha; caud co, caudal concha; mid co, middle concha; and cr nas, crista nasalis. (Figure from Bourke and Witmer 2016; )

not lose either needed materials (such as glucose) or too much water.

Blood enters nephrons via small arteries called afferent arterioles (Fig. 9.35). This blood enters the glomerulus (a collection of capillaries; Figs. 9.35 and 9.36) under high pressure and "filters" through the walls of the capillaries and the walls of a surrounding structure called a glomerular capsule. The process of filtration depends on the glomerular hydrostatic pressure, and the glomerular filtration barrier consists of the endothelial cells of the glomerular capillaries, a basement membrane, and structures called pedicels, and materials are filtered based on their molecular

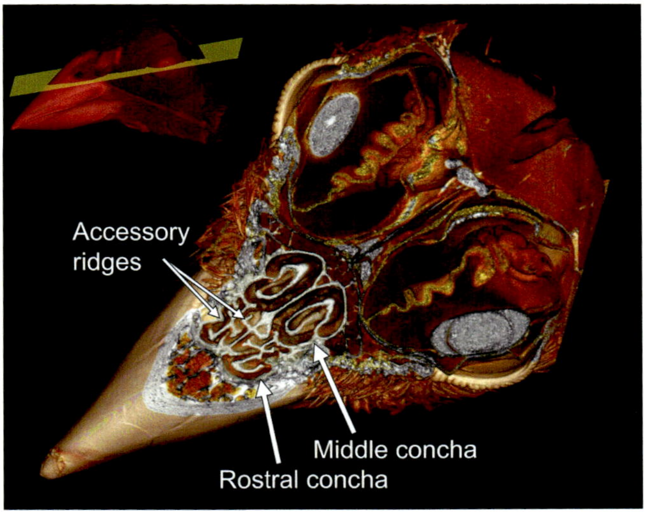


**Fig. 9.25** Transverse section of the head of a Song Sparrow (*Melospiza melodia*) head (dorsal view) showing the scroll-like middle concha and the complex branching rostral concha with small accessory ridges projecting from the nasal septum and lateral wall of the nasal cavity. The inset figure (upper left) shows the external surface of the bird with the location of the plane. (Figure from Danner et al. 2017; )

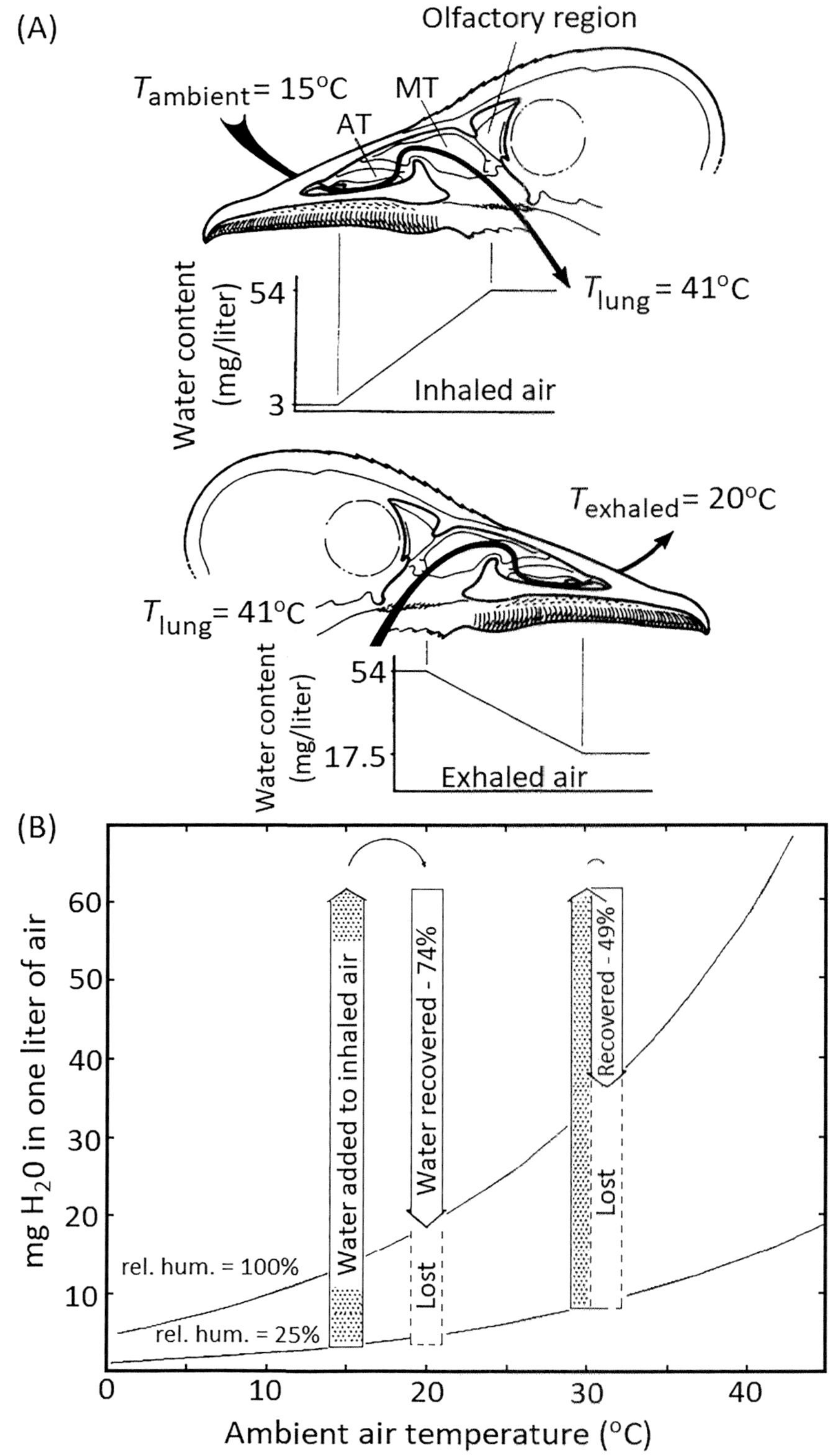


**Fig. 9.26** (**a**) Respiratory turbinate function in birds. During inhalation (top), moist surfaces of the respiratory turbinates warm and humidify incoming ambient air, and the incoming air cools the surfaces of the respiratory

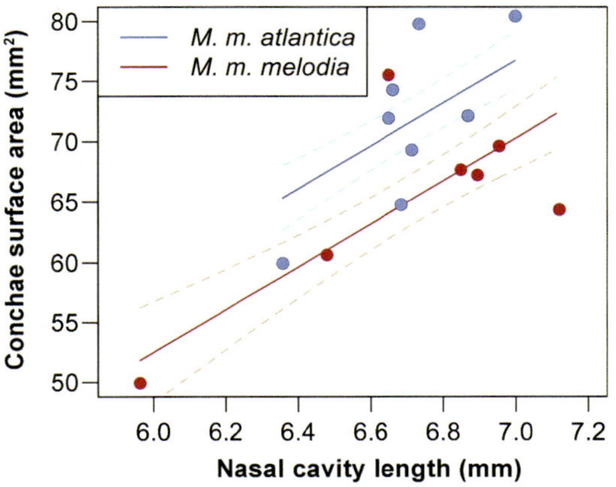


**Fig. 9.27** Surface area of nasal conchae relative to the length of nasal cavities for two subspecies of Song Sparrows, one (*Melospiza melodia melodia*) found in moister habitats throughout much of the eastern United States and the other (*M. m. atlantica*) endemic to the dry sand dunes of the Atlantic coast between North Carolina and New Jersey in the United States. The greater conchae surface area of the *atlantica* subspecies may allow more efficient recovery of water during exhalation. (Figure from Danner et al. 2017; © 2016 Oxford University Press, used with permission)

size and charge (Figs. 9.37 and 9.38). The filtrate that moves from the glomerular capsule into the proximal tubules is basically blood plasma without much protein (most protein molecules are too large to pass through). That filtrate, therefore, contains many important substances. In the proximal convoluted tubules, those needed substances, such as glucose, vitamins, and ions such as sodium, are reabsorbed into the blood (Fig. 9.39). Even sunbirds, nectar-feeders with a diet rich in glucose, reabsorb almost all (98%) of the glucose that filters into the kidney tubules (McWhorter et al. 2004). Even for Domestic Chickens, Morgan and Braun (2001) estimated that the ability of the renal tubules to reabsorb glucose was about 4–5 times that of human kidneys.

Avian nephrons are also involved in maintaining water balance, and this process involves the loops of Henle in the medulla of the kidney. Avian kidneys are unique among vertebrates in having two kinds of nephrons. The reptilian types, with no loops of Henle, are located in the cortex, and the mammalian types, with long or intermediate-length loops (Figs. 9.34 and 9.40), extend from the cortex and into the medullary cones of the medulla. In mammals, all nephrons have a loop of Henle. However, in birds, only 10–30% of nephrons have a loop of Henle, with the percentage varying among species (Goldstein 2006; Bolin and Burggren 2013).

Other than mammals, birds are the only vertebrates that conserve body water by producing urine osmotically more concentrated (with a higher concentration of solutes) than the blood plasma. However, the ability of birds to concentrate urine is limited compared to mammals. Typically, water-deprived birds produce urine that is 1.4–3 times more concentrated than plasma (Table 9.3), whereas most mammals produce urine about 5–10 times more concentrated than plasma, and some desert-adapted mammals may produce urine 20–25 times more concentrated than plasma (Casotti and Braun 2000). This difference is due to the smaller percentage of nephrons in the avian kidney with loops of

**Fig. 9.26** (continued) turbinates. During exhalation (bottom), this process is reversed. Warm, moist exhaled air passes over the cooler surfaces of the respiratory turbinates, reducing the temperature of the outgoing air and causing water vapor to condense on the surfaces of the respiratory turbinates. This reduces both heat loss and water loss. *T* temperature, AT anterior respiratory turbinate, MT middle respiratory turbinate. (**b**) Graph showing the amount of water vapor added to inhaled air (shaded bars) and the amount recovered during exhalation (open bars) by Cactus Wrens (*Campylorhynchus brunneicapillus*). The temperature of the inhaled air is indicated by the location of the shaded and open bars. Note that more water is lost at higher temperatures. (Figure **a** from Ruben et al. 1996; © 1996 American Association for the Advancement of Science, used with permission; Figure **b** from Schmidt-Nielsen et al. 1970; © 1970 Published by Elsevier B.V., used with permission)

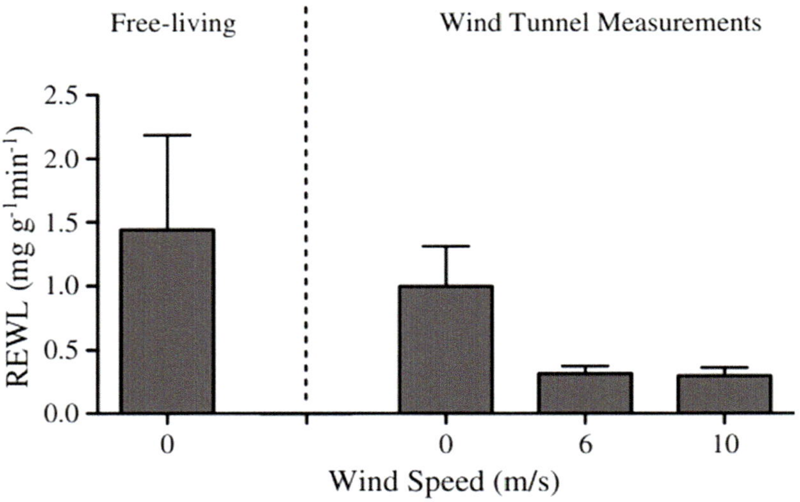


**Fig. 9.28** Mean (± SD) respiratory water loss (REWL) by Rufous Hummingbirds (*Selasphorus rufus*) when hovering (wind speed = 0) and during forward flight (wind speeds = 6 and 10 m/s). REWL was higher when hovering both in the field (free-living) and in a wind tunnel than during forward flight. REWL was higher when hovering because metabolic and respiration rates are higher than during hovering flight than during forward flight. (Figure from Powers et al. 2012; © 2011 Elsevier Inc. All rights reserved, used with permission)

Henle. In loops of Henle, solutes (sodium chloride) are actively transported out of the ascending limb (Fig. 9.41) and become concentrated in the interstitial fluid of the medulla of the kidney (medullary cone; Fig. 9.32). With more loops of Henle (as in mammals), solute concentrations become higher. When urine passes through the collecting ducts that pass through the medulla, water leaves the tubules by osmosis and the urine becomes concentrated. Because only the

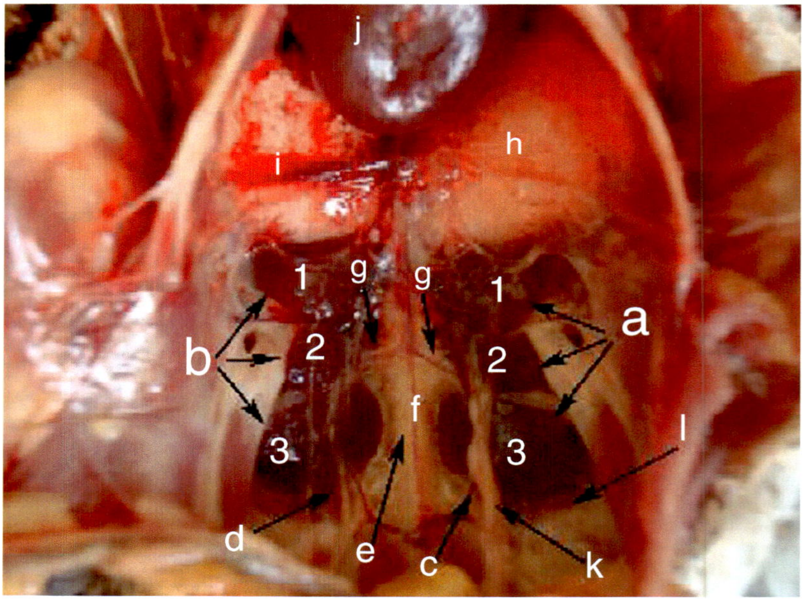


**Fig. 9.29** Kidneys and associated structures of a Rock Pigeon (*Columba livia*). (**a**) left kidney, (**b**) right kidney, (1, 2, 3) lobes of each kidney, (**c**) left ureter, (**d**) right ureter, (**e**) synsacrum, (**f**) aorta, (**g**) left and right ischiadic arteries, (**h**) left lung, (**i**) right lung, (**j**) heart, (**k**) oviduct, and (**l**) renal fossae. (Figure from Al-Ajeely and Mohammed 2012; © International Journal of Science and Nature, reproduction of the published material in any form is permitted with due acknowledgment of the source)

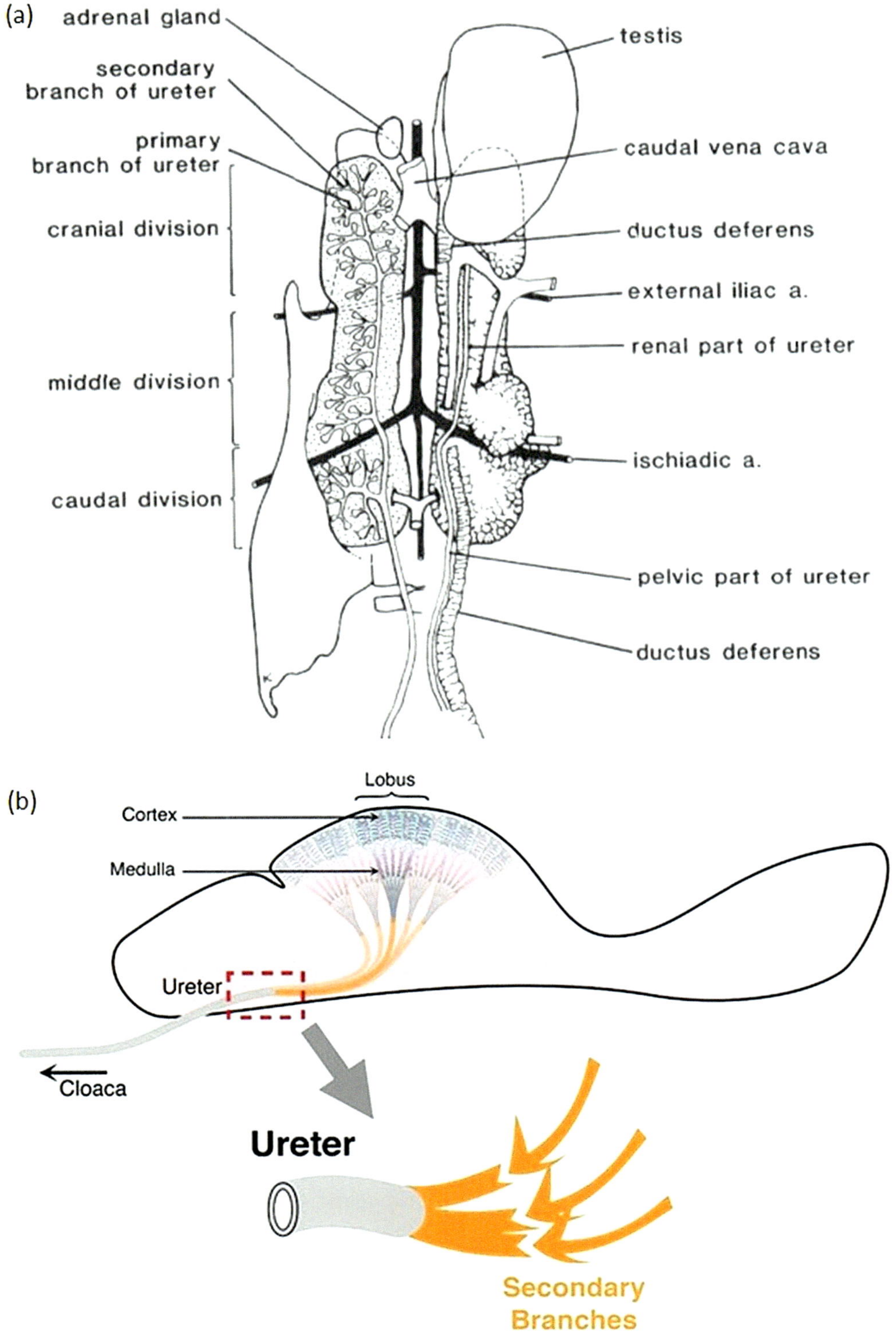


**Fig. 9.30** (**a**) Avian kidney (plus testis and ductus deferens). The kidney's three divisions (cranial, middle, and caudal) are indicated on the left. a., artery (**a**, **b**) The kidneys are drained by the ureters. (Figure **a** from Orosz

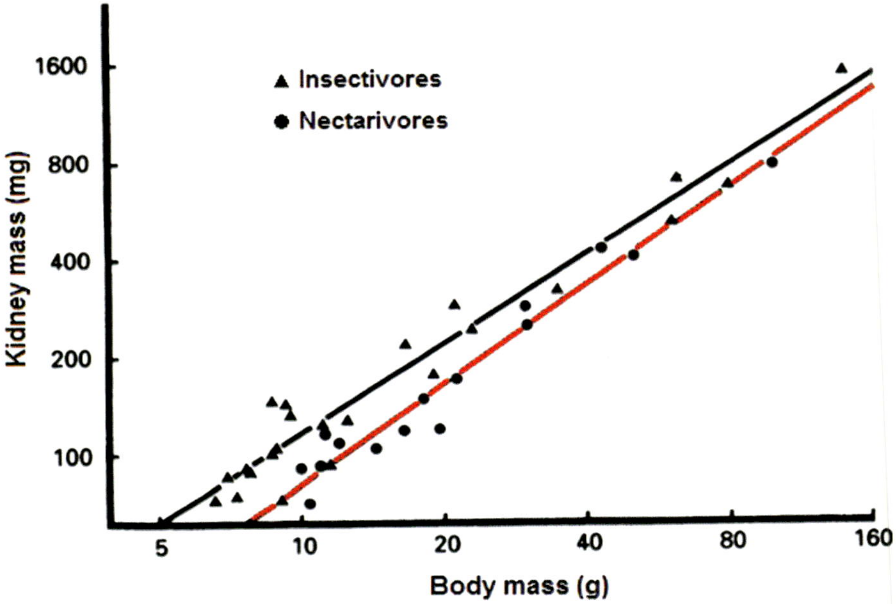


**Fig. 9.31** Log-log relationships between kidney mass and body mass for 23 species of Australian songbird insectivores and 15 species of Australian songbird honeyeaters (nectarivores). The upper line is the relationship for insectivores; the lower line for the honeyeaters. Honeyeaters, as nectarivores, tend to have smaller kidneys than insectivores, likely because they ingest so much water that they have little need for larger kidneys to help conserve water. (Figure modified from Richardson et al. 1991; © 1991 Anatomical Society of Great Britain and Ireland, used with permission)

looped nephrons contribute to the intramedullary osmotic gradient, the presence of loopless nephrons limits the ability of the kidneys to produce hyperosmotic (concentrated) urine. However, birds excrete non-water soluble uric acid as the primary end product of nitrogen metabolism, and mammals excrete water-soluble urea so birds can conserve water without producing urine as concentrated as that of mammals (Dantzler 2016).

This reduced capacity of avian kidneys to concentrate urine (compared to mammals) means that more water accompanies the solutes that travel from the kidneys through ureters to the cloaca. Water-deprived birds do have a mechanism for reducing the amount of water leaving the kidneys. In response to dehydration, the pituitary (via the hypothalamus) releases more of a hormone called arginine vasotocin (AVT) into the blood (Fig. 9.41). In the kidneys, AVT causes a reduction in the glomerular filtration rate (the rate at which plasma filters from the glomeruli into the glomerular capsule) so less water moves from the blood into the kidney tubules. In addition, AVT increases the permeability of the walls of collecting ducts to water by opening protein water channels called aquaporins (Fig. 9.42). As the collecting ducts become more permeable, more water moves by osmosis out of the collecting ducts (because of the higher solute concentration in the medullary cones) and can be reabsorbed by kidney capillaries. Studies to date suggest that the extent to which AVT can reduce urine production (and water loss) varies

**Fig. 9.30** (continued) and Echols 2020; © 2019 Elsevier Inc., used with permission. Figure **b** from André et al. 2015; open-access article distributed under the Creative Commons Attribution License, https://creativecommons.org/licenses/by/4.0/)

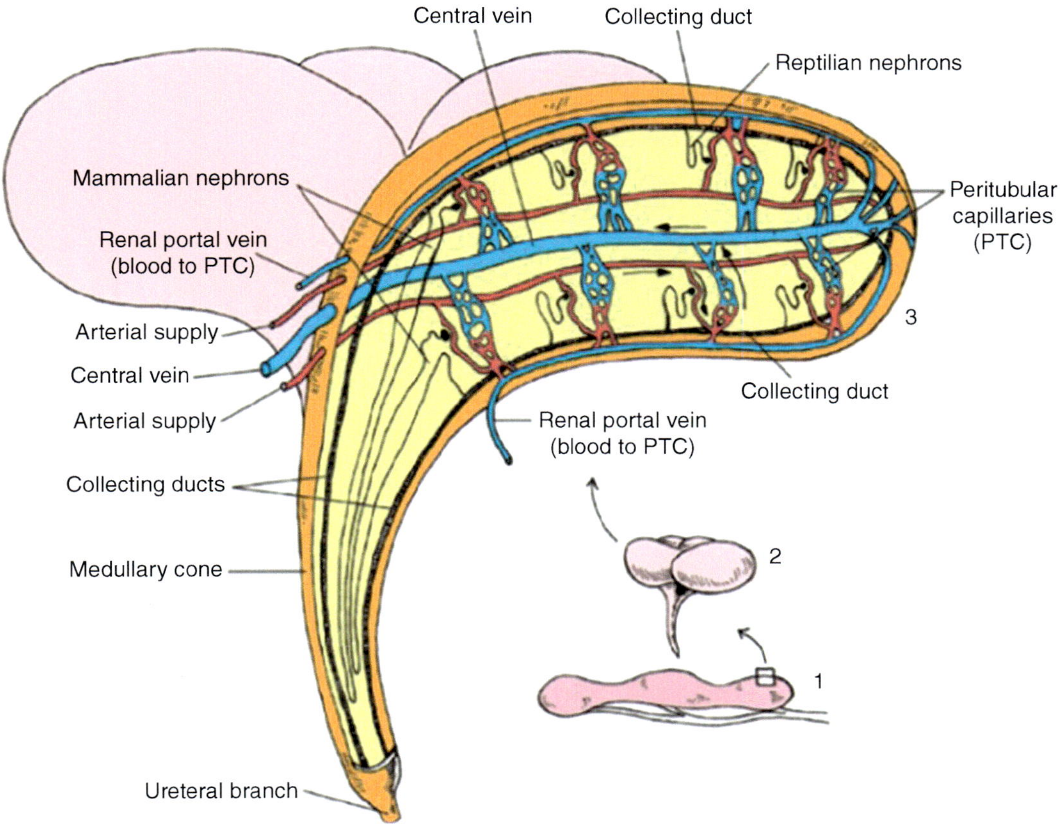


**Fig. 9.32** (1) Avian kidney with three lobes, (2) a lobule from one lobe, and (3) the inner anatomy of a lobule. The loops of Henle of mammalian-type nephrons extend into the medullary cone, which is cone-shaped because the number of loops of Henle decreases toward the apex of the medullary cones. Reptilian-type nephrons do not have loops of Henle. Fluid from both types of nephrons enters the collecting ducts where it is exposed to the high solute concentration of the interstitial fluid in the cone. Urine leaves the kidney via the ureteral branch. PTC, peritubular capillaries. (Figure from Reece 2015; © 2015 John Wiley and Sons, used with permission)

among species (Fig. 9.43), but, in general, AVT is less effective in conserving water than the mammalian equivalent (antidiuretic hormone, or ADH; Nishimura and Fan 2003). Therefore, water-deprived birds tend to produce more urine and lose more water from the kidneys than would a similar-sized water-deprived mammal.

## 9.6 Role of Cloaca and Lower Intestine in Osmoregulation

This loss of this water would clearly be a problem for birds with limited access to water or those that must excrete lots of solutes (e.g., salt). However, in contrast to mammals, urine that enters the cloaca of most birds moves by retrograde peristalsis into the lower intestine and cecae (Fig. 9.44). Fluid from the upper gastrointestinal tract also enters the cloaca. Therefore, the cloaca receives water from the kidneys and the small intestine.

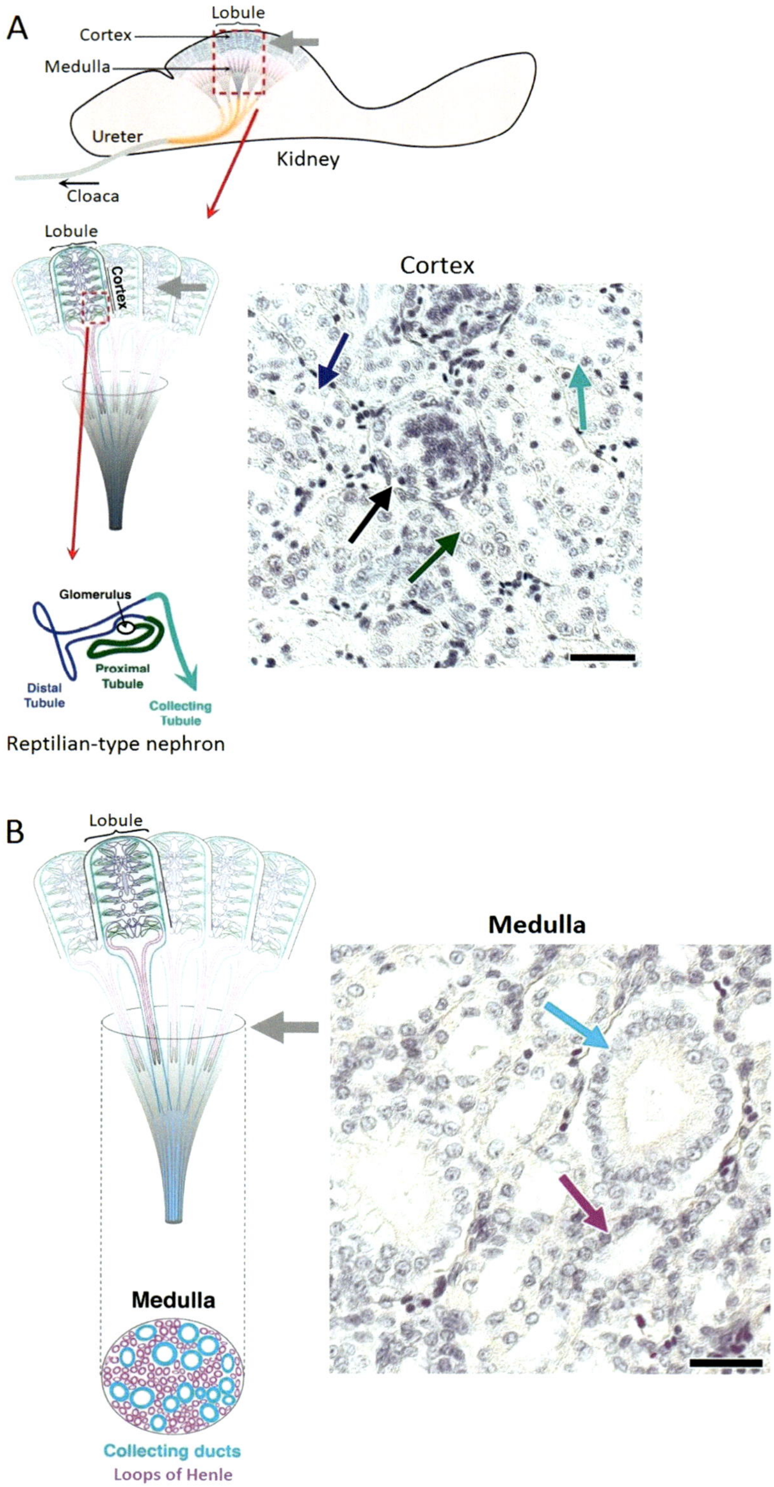


**Fig. 9.33** (**a**) Avian kidney, lobus, and a reptilian-type nephron. In the photomicrograph of the cortex, the arrows point to the structures that correspond to the colors of the structures illustrated in the reptilian-type nephron. Scale bar = 25 μm. (**b**) Avian kidney lobules showing the structure of the medullary cone, with large numbers of collecting ducts and Loops of Henle. In the photomicrograph of the medulla, the arrows point to a collecting duct (light blue arrow) and a Loop of Henle (pink arrow). Scale bar = 25 μm. (Figure modified from André et al. 2015; open-access article distributed under the Creative Commons Attribution License, https://creativecommons.org/licenses/by/4.0/)

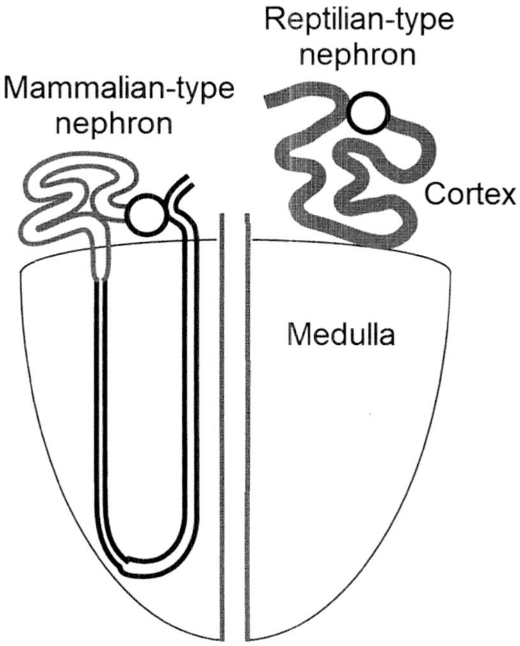


**Fig. 9.34** Avian nephrons. Left, mammalian-type nephron with loop of Henle that extends down into the medulla. Right, reptilian-type nephron located in the cortex and with no loop of Henle. (Figure modified from Liu et al. 2001; © 2001 International Society of Nephrology. Published by Elsevier Inc., used with permission)

The influx of water into the cloaca can be reabsorbed through the epithelium of the lower intestinal tract to maintain hydration. In the lower intestine and cecae, water and sodium chloride are reclaimed by the process of sodium-linked water reabsorption. In other words, positively charged sodium ions are actively transported out of the intestine and negatively charged chloride ions follow. Water then follows by osmosis (Hildebrandt 2001). As a result, birds can, if necessary, excrete urine as a relatively dry mixture along with feces.

Avian ceca have been well studied in several species of non-passerines and have been shown to absorb substrates and water. However, the ceca of passerine birds are smaller and, as result, may not have similar functions. The ceca of House Sparrows (*Passer domesticus*) have a central channel with a large number of side channels and all of the channels are lined by epithelial cells with a very dense microvillus brush border as well as a region densely packed with mitochondria just below the brush border (with the mitochondria possibly providing ATP to power substrate transport; Fig. 9.45). Although the importance of small ceca for fluid homeostasis remains to be determined, the structure of House Sparrow ceca suggests that even the small ceca of passerines birds may be important in water and electrolyte (e.g., sodium) balance (Reyes and Braun 2005).

As just described, the colon and ceca of birds can substantially modify the urine that initially enters the cloaca (coprodeum). The lower section of the bird gastrointestinal tract, by the same mechanisms, alters "feces" coming from the small intestine. As a result, the composition of bird droppings can differ dramatically from the composition of the urine and feces that initially entered the colon or large intestine. Depending on diet and access to water, droppings may have considerably less water, solutes (sodium and potassium), and uric acid (or urate) than the original urine and feces. In fact, Crouch et al. (2020) determined the chemical composition of the urine component of excreta of six species of bird using X-ray diffraction techniques and found that none of the samples contained uric acid (Fig. 9.46). Uric acid may not be present in excreted avian urine due to conditions and the microbiome in the ceca. For example, the pH in ceca is generally between 6 and 8 (Clench and Mathias 1995; Echols 2005) so can be slightly alkaline (Herd and Dawson 1984). With an alkaline pH, uric acid is converted to a urate ion (Crouch et al. 2020). In addition, bacteria in the ceca can degrade uric acid to volatile fatty acids and ammonia (Braun and Campbell 1989). As a result, Crouch et al. (2020) found that the urine component of avian excreta consisted of a variety of compounds including ammonium urate, struvite (magnesium ammonium phosphate), and two unknown compounds. The amount of these substances in the urine of birds would vary with diet, e.g., diets with more protein would increase uric acid levels in the blood and that, in turn, would increase the amount of urate in the excreta.

The coordinated action of the kidneys, lower GI tract, and the salt glands (see below) in the

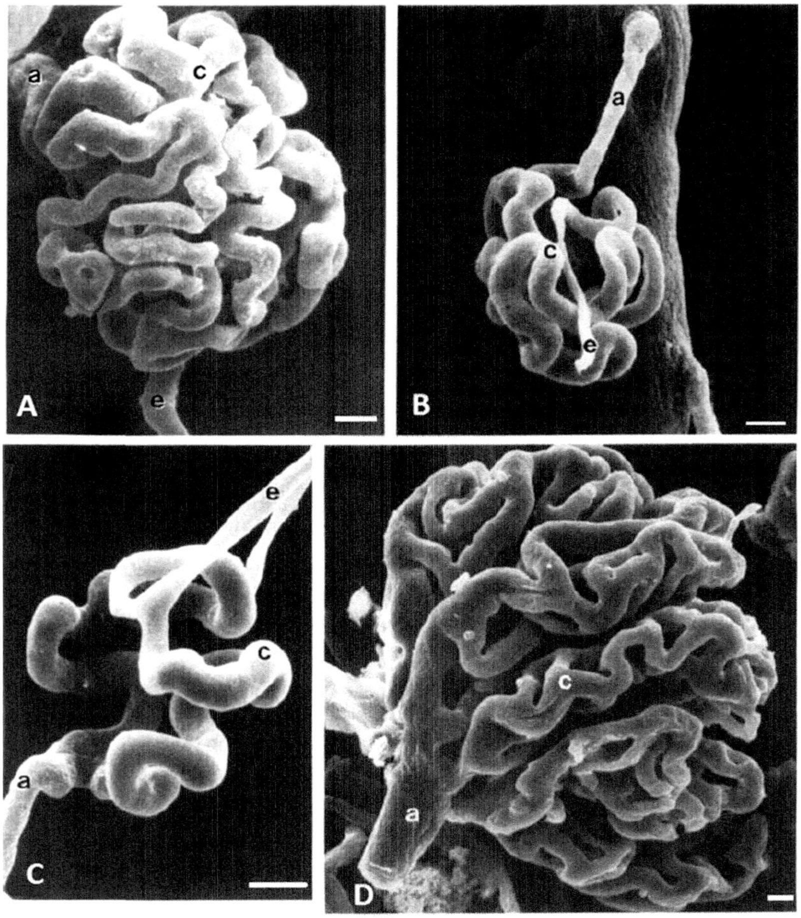


**Fig. 9.35** Scanning electron micrographs of the glomeruli of a Domestic Chicken (**a–c**; *Gallus gallus domesticus*) and a Norway rat (**d**; *Rattus norvegicus*) showing that the glomeruli of mammals are more convoluted and complex than those of birds. (**a**) Glomerulus of a mammalian-type nephron. (**b** and **c**) Glomeruli of reptilian-type nephrons. (**d**) Glomerulus of a Norway rat. a, afferent arteriole; c, capillaries; e, efferent arteriole. Scale bars = 10 μm. (Figure from Casotti and Braun 1995; © 1995 Wiley-Liss, Inc., used with permission)

regulation of fluid and ion balance is a classic example of the integration of organs required to maintain homeostatic balance. No single organ appears to have an outstanding capacity to conserve ions and water, but instead they all function in concert to maintain total body fluid homeostasis to allow birds to inhabit a wide range of environments (Braun 1996).

## 9.7 Nectar-Feeding Birds

For some birds, too much water is a potential problem. For example, nectar-feeding birds ingest large amounts of water (up to 5–6 times their own body mass every day) because nectars of most bird-pollinated flowers are 75–80% water (Nicolson and Fleming 2003). High intake rates mean there must be equally high excretion rates. Simultaneously, nectar-feeding birds must be able to retain the essential nutrients, such as amino acids, that are found in very low concentrations in nectar. In other terrestrial vertebrates, ingesting such prodigious amounts of water could lead to "water intoxication" and such negative effects as blood plasma dilution and subsequent swelling of red blood cells (Martinez del Rio et al. 2001).

Among the best-known nectar-feeding birds are the hummingbirds and sunbirds, and their mechanisms for dealing with excessive water intake differ. Hummingbirds absorb all of the water they ingest, but their kidneys are able to filter large amounts of blood plasma and,

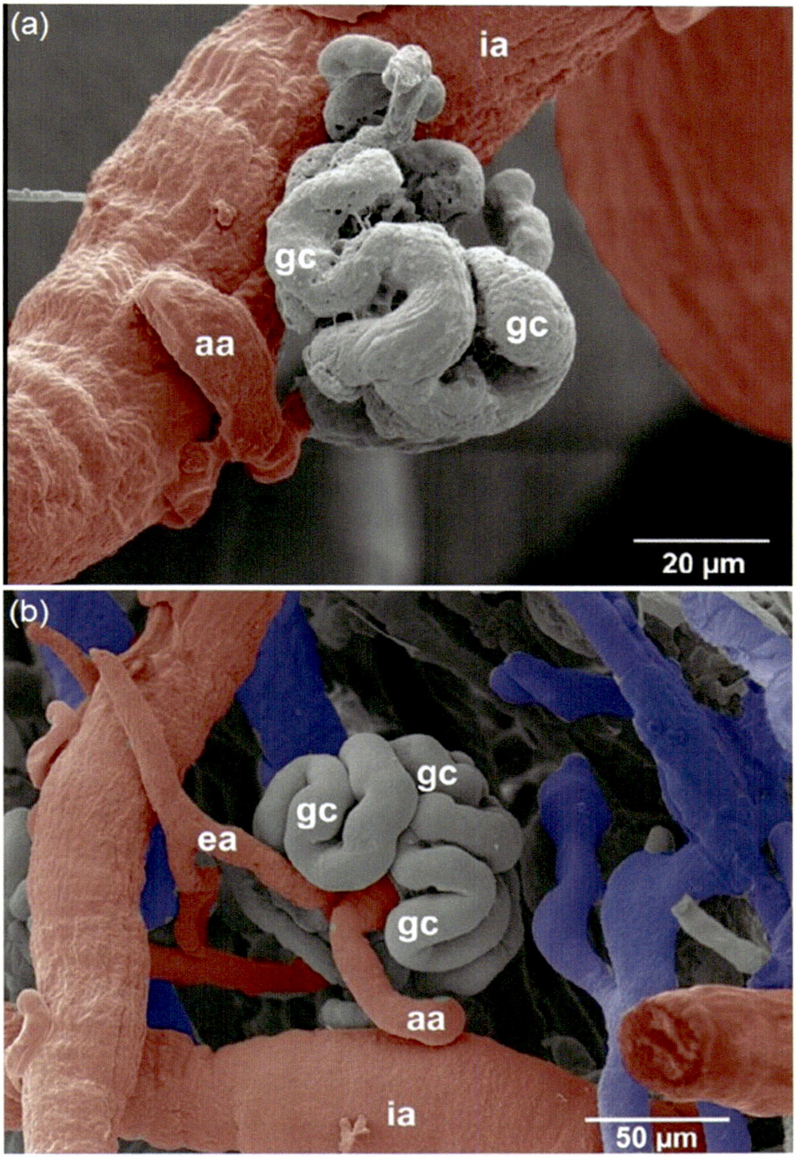


**Fig. 9.36** Glomeruli in the cortex of the kidney of a Domestic Chicken. (**a**) Glomerulus of a reptilian-type nephron. (**b**) Glomerulus of a mammalian-type nephron. Mammalian-type glomeruli are significantly larger than reptilian-type glomeruli. In addition, mammalian-type glomeruli form a complex capillary network with numerous loops, whereas reptilian-type glomeruli consist of a single capillary with few loops. Because of these differences, filtration rates of mammalian-type nephrons are more than double that of reptilian-type nephrons. aa, afferent arteriole (conducting blood into the glomerulus); ef, efferent arteriole (conducting blood out of the glomerulus); ia, intralobular artery; gc, glomerular capillary. (Figure from Hadžiomerović et al. 2021; © 2021 Wiley-VCH GmbH, used with permission)

depending on the amount of water ingested, can vary the amount of water reabsorbed. Excess water is eliminated by reducing the amount of water reabsorbed by the kidneys (Fig. 9.47). Contributing to the efficiency of this process, hummingbird kidneys, for their size, have about twice the number of glomeruli as a typical bird kidney (Beuchat et al. 1999). The structure of hummingbird kidneys permits the filtration of large amounts of plasma and the elimination of large amounts of water, but does not permit the production of concentrated urine (Goldstein and Skadhauge 2000). Another important feature of hummingbird kidneys is that the descending limbs of the few loops of Henle that are present have relatively thick walls that may facilitate the recovery of important electrolytes such as sodium and potassium (Beuchat et al. 1999).

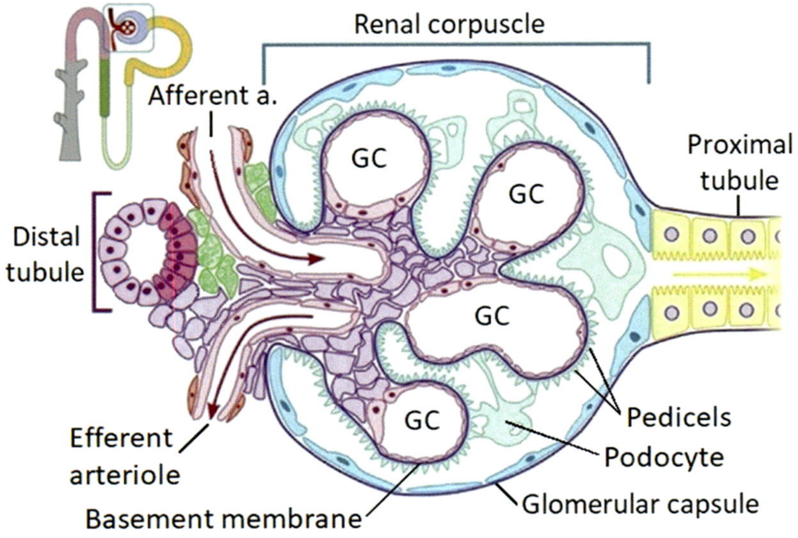


**Fig. 9.37** A renal corpuscle consists of the glomerular capillaries (GC) plus the glomerular capsule. Blood enters the glomerulus via an afferent arteriole and leaves via an efferent arteriole. Filtrate that enters the glomerular capsule then moves into the proximal convoluted tubule. Podocytes wrap around glomerular capillaries, secrete and maintain the basement membrane, and form processes called pedicels. The endothelial cells of the glomerular capillaries, the basement membrane, and pedicels together form the glomerular filtration barrier that filters materials based on molecular size and charge. (Image modified from M. Komorniczak, CC BY 3.0, https://creativecommons.org/licenses/by/3.0/)

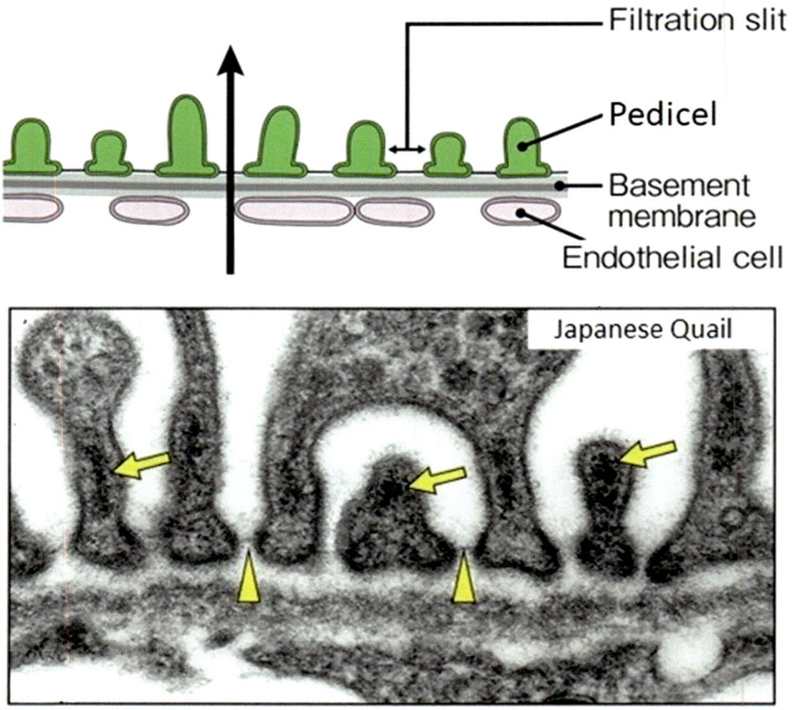


**Fig. 9.38** The glomerular filtration barrier consists of the endothelial cells of the glomerular capillaries, a basement membrane, and filtration slits located between pedicels. In the bottom image, the arrows are pointing at pedicles and the arrowheads at filtration slits. (Figure from Ichimura and Sakai 2018; © 2018 The Juntendo Medical Society, open-access article distributed under the terms of Creative Commons Attribution License (CC BY), https://creativecommons.org/licenses/by/2.0/)

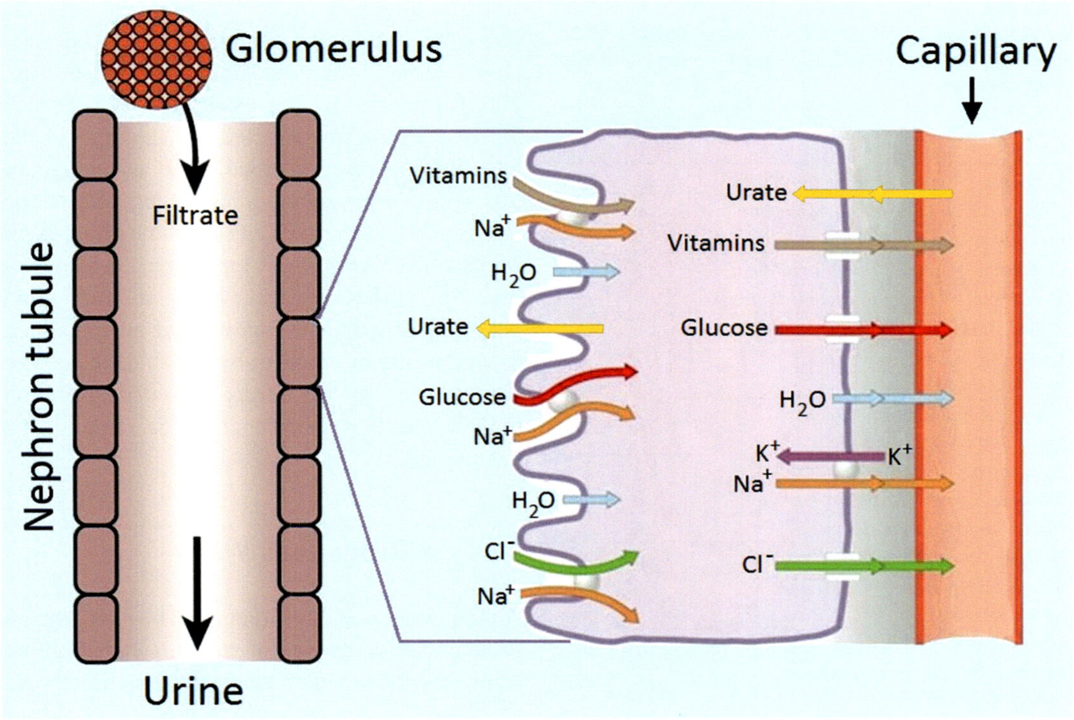


**Fig. 9.39** Reabsorption of materials, including glucose, vitamins, water, and important ions like chloride ions ($Cl^-$) and sodium ions ($Na^+$), from the proximal convoluted tubule back into the blood. Some materials, such as excess urate (uric acid) and excess potassium ($K^+$), can also be secreted from the blood into the nephron tubules for excretion in the urine. (Figure modified from Hediger 2002; © 2002 Springer Nature, used with permission)

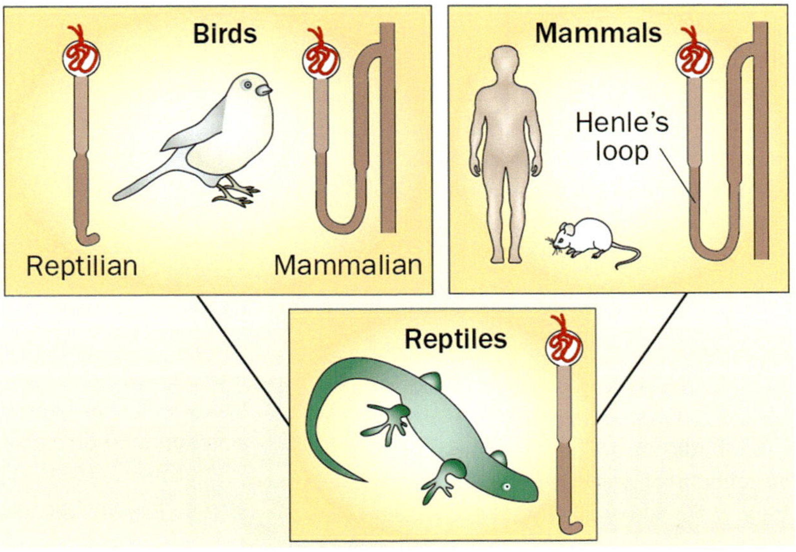


**Fig. 9.40** Nephrons of birds, mammals, and reptiles. Birds have both reptilian-type (without loops of Henle; 70–90% of nephrons) and mammalian-type (with loops of Henle; 10–30%) nephrons. (Figure from Romagnani et al. 2013; © 2013 Springer Nature, used with permission)

**Table 9.3** Maximum osmolal urine-to-plasma ratio (U/P) for several species of birds (Table from Dantzler 2016; © 2016 The American Physiological Society, used with permission)

| Species | Osmolal U/P |
|---|---|
| Domestic Chicken (*Gallus domesticus*) | 2.1[a] |
| Northern Bobwhite (*Colinus virginianus*) | 1.6[a] |
| California Quail (*Callipepla californica*) | 1.7[a] |
| Gambel's Quail (*Callipepla gambelii*) | 2.5[a] |
| House Finch (*Haemorhous mexicanus*) | 2.3[a] |
| Zebra Finch (*Taeniopygia guttata*)[b] | 2.8[c] |
| Laughing Dove (*Streptopelia senagalensis*) | 1.7[c] |
| Laughing Kookaburra (*Dacelo novaeguineae*) | 2.7[c] |
| Emu (*Dromaius novaehollandiae*)[b] | 1.4[c] |
| Crested Pigeon (*Ocyphaps lophotes*)[b] | 1.8[c] |
| Galah (*Eolophus roseicapilla*)[b] | 2.6[c] |
| Budgerigar (*Melopsittacus undulatus*)[b] | 3.0[d] |
| Savannah Sparrow (*Passerculus sandwichensis*) | 1.6[e] |

[a]Reviewed in Dantzler (1970)
[b]Birds from xeric environments
[c] Skadhauge (1974)
[d] Krag and Skadhauge (1972)
[e] Goldstein et al. (1990)

The efficiency with which hummingbird kidneys excrete water is impressive, but could potentially create problems when hummingbirds are not feeding on nectar. Their inability to concentrate urine could lead to dehydration during periods, such as during long migration flights and at night, when they are not feeding. The strategy used by Broad-tailed Hummingbirds (*Selasphorus platycercus*), as well as other hummingbirds and, more generally, other nectivorous birds such as sunbirds, is to almost stop filtering plasma during such periods (Bakken et al. 2004; Fleming et al. 2004; Figs. 9.48 and 9.49). The reduction in urine production is thought to be due to the vasoconstriction of blood vessels that supply the glomeruli where filtration occurs (Dantzler 1989), a process mediated by arginine vasotocin (Goecke and Goldstein 1997).

Sunbirds, in contrast to hummingbirds, vary the amount of water absorbed in their intestine (Figs. 9.50 and 9.51). With very dilute nectar, sunbirds eliminate up to two-thirds of ingested water by simply not absorbing the water as it passes through their intestine (McWhorter et al. 2003). Importantly, the absorption of sugars in the intestine is not affected by variation in the amount of water absorbed. One advantage of this ability to regulate water absorption in the intestine is that sunbirds can consume dilute floral nectars. The physiological mechanisms responsible for differences in how sunbirds and hummingbirds handle ingestion of excess water, as well as the mechanism by which sunbirds and honeyeaters vary the degree of water absorption in their intestines (Price et al. 2015), remain to be determined.

## 9.8 Nitrogenous Waste

An important part of the diet of all birds is protein. Proteins are composed of subunits called amino acids, and those amino acids are sometimes used as a source of energy or are converted into fats or carbohydrates. When amino acids are used for energy or converted to fats or carbohydrates, the amine ($NH_2$) group must be removed. These amine groups are toxic and must be eliminated, and most animals excrete these nitrogenous wastes as ammonia (ammonotelic), urea (ureotelic), or uric acid (uricotelic) (Figs. 9.52 and 9.53; Box 9.4 Are Birds Strictly Urocotelic?). Only aquatic animals, such as protochordates and fish, excrete these nitrogenous wastes primarily as ammonia because it takes lots of water to excrete highly water-soluble ammonia (about 400 ml of water for 1 g of ammonia; Wright 1995). Because water conservation is important for most terrestrial animals, natural selection favored mechanisms for concentrating nitrogenous waste and excreting less water. That involves the conversion of ammonia to either urea or uric acid; urea requires about 10 times less water than ammonia for excretion and uric acid about 50 times less water (Wright 1995).

Needham (1931) suggested that the evolution of the cleidoic egg (requiring only the exchange

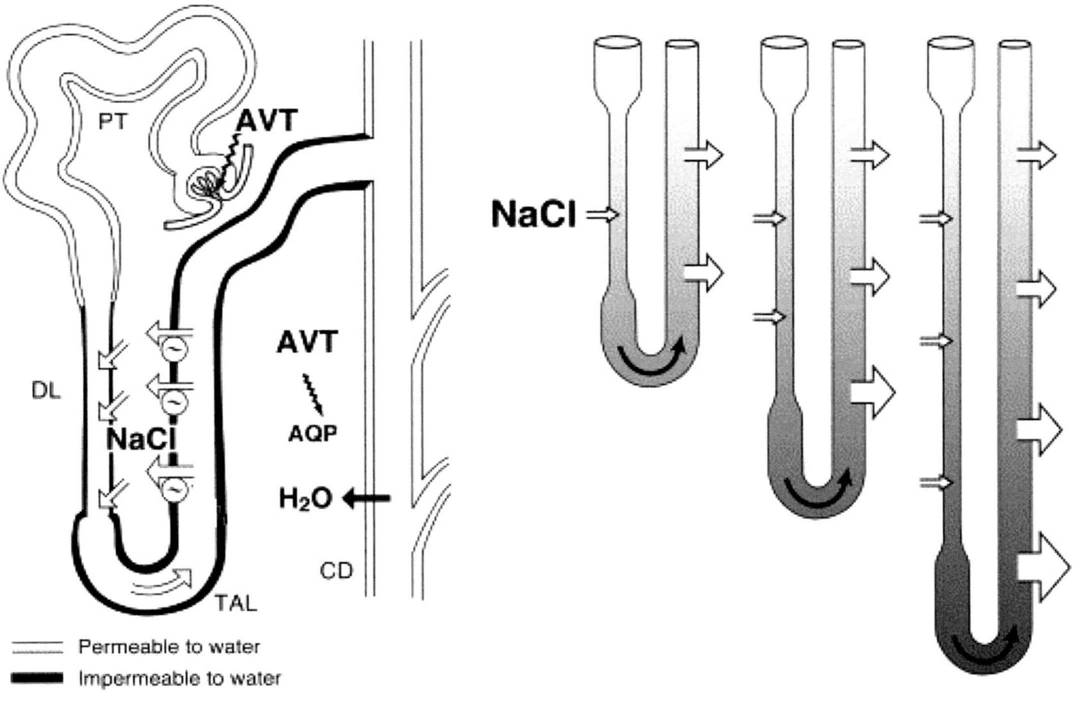


**Fig. 9.41** A proposed model for urine concentration in the mammalian-type nephron of birds showing transport of NaCl and water in various nephron segments. Left, NaCl is actively exported out of the thick ascending limb (TAL) and some reenters the descending limbs (DL) by simple diffusion (but, despite the resulting high solute concentration, water does not follow because the membrane is not permeable to water). PT, proximal tubule; CD, collecting duct. Right, loops of Henle exhibit some variation in length and more NaCl is transported out of the longer loops of Henle so the osmotic gradient is greatest near the end of the medulla. AVT (arginine vasotocin) is a hormone that helps birds conserve water by reducing the glomerular filtration rate (the rate at which plasma moves from glomeruli in the glomerular capsules) and by increasing the permeability of collecting ducts to water (by opening protein water channels called aquaporins; AQP). (Figure from Nishimura and Fan 2003; © 2003 Elsevier Science Inc., used with permission)

of gases with its environment) was closely linked to the evolution of uricoteley. Reasons for this linkage include the assumptions that ammonia and urea are so soluble in water that these wastes would diffuse from the allantoic sac (where wastes are stored in an egg) and kill the developing embryo or storing these wastes in an egg would require too much water (both in terms of the needs of the developing embryo and available space within an egg). However, avian embryos do produce some urea and ammonia during development. At hatching, for example, 71% of the nitrogen waste that has accumulated in the eggs of Red-winged Blackbirds (*Agelaius phoeniceus*) is in the form of uric acid, 26% in urea, and 3% in ammonia (Packard and Packard 1984). However, more importantly, little egg protein is catabolized during avian development in the egg (2–5%; Romanoff 1967) because lipids are the primary energy source for bird embryos. Thus, generating potentially toxic levels of water-soluble urea in a developing cleidoic egg may not have been a problem for ancient birds and, perhaps, their immediate ancestors (theropod dinosaurs).

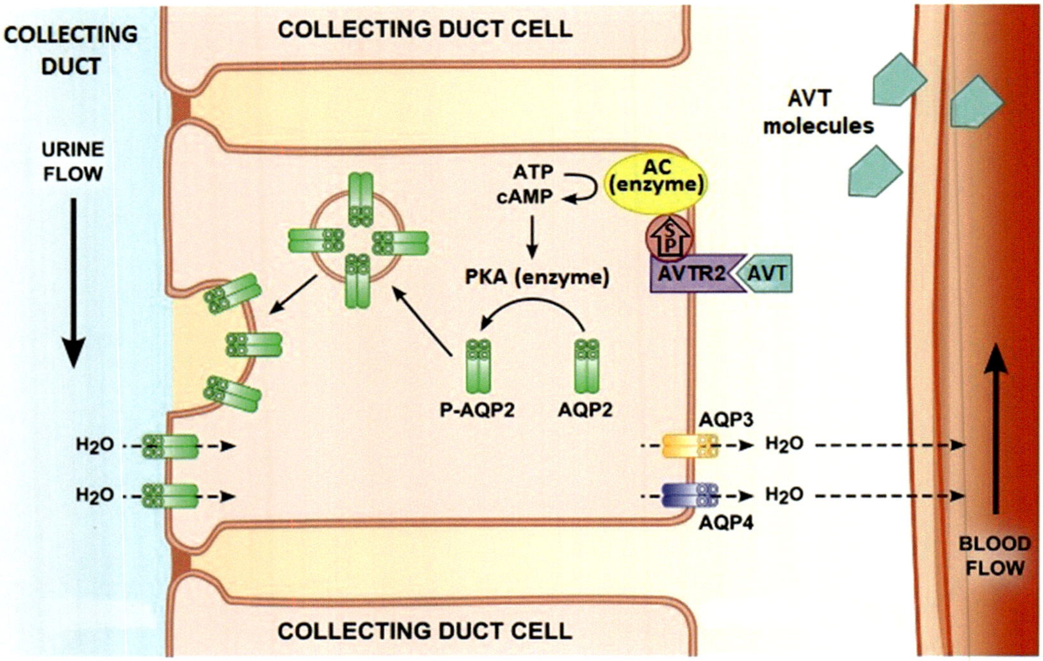


**Fig. 9.42** Aquaporins (AQP) are proteins important in water conservation in the kidneys. These proteins serve as channels for water, allowing water from collecting ducts to diffuse back into the blood so it is not lost in the urine. In response to dehydration, the anterior pituitary releases AVT (arginine vasotocin) into the blood. AVT binds to AVPR2 in the membrane of collecting duct cells which, in turn, activates a signaling protein (SP) that actives an enzyme (AC, adenylyl cyclase) that increases production of cAMP (cyclic AMP) from ATP. The cAMP activates another enzyme (PKA) that phosphorylates AQP2 (P-AQP2). P-AQP2 causes preexisting AQP2 water channels found in intracellular vesicles to move to the membrane of collecting duct cells on the side where the collecting duct is located. Once inserted, water molecules can then be reabsorbed, diffusing from the collected duct and into kidney capillaries via the collecting duct cells. The extent to which this process occurs depends on how much AVT is released by the anterior pituitary. (Figure modified from Qureshi et al. 2014; © 2014 Oxford University Press, used with permission)

One disadvantage of excreting nitrogenous wastes as uric acid is that, as a more complex molecule than either ammonia or urea, synthesizing it requires more energy. If producing urea would not be toxic for avian embryos and requires less energy, then other factors may be responsible for the evolution of uricoteley. The most likely such factor is water conservation. Excreting nitrogenous wastes as uric acid permits the excretion of a solid or semisolid paste with minimal water content (Dantzler and Braun 1980). Archosaurs that eventually gave rise to birds may have endured hot, dry conditions at the beginning of the Triassic that favored the evolution of uricoteley (Campbell et al. 1987). In support of this hypothesis, alligators are ammono-uricotelic rather than ureotelic, suggesting that the transition to uricoteley may have occurred early in the archosaurian lineage (Campbell et al. 1987).

Uric acid requires little water for excretion because it is insoluble in water, but it does require a significant amount of protein to maintain it in a colloidal suspension in the urine (i.e., urine spheres; Fig. 9.54). Uric acid is filtered from the blood through renal corpuscles (glomerulus and glomerular capsule) and secreted from the blood in the proximal tubules. As the concentration of

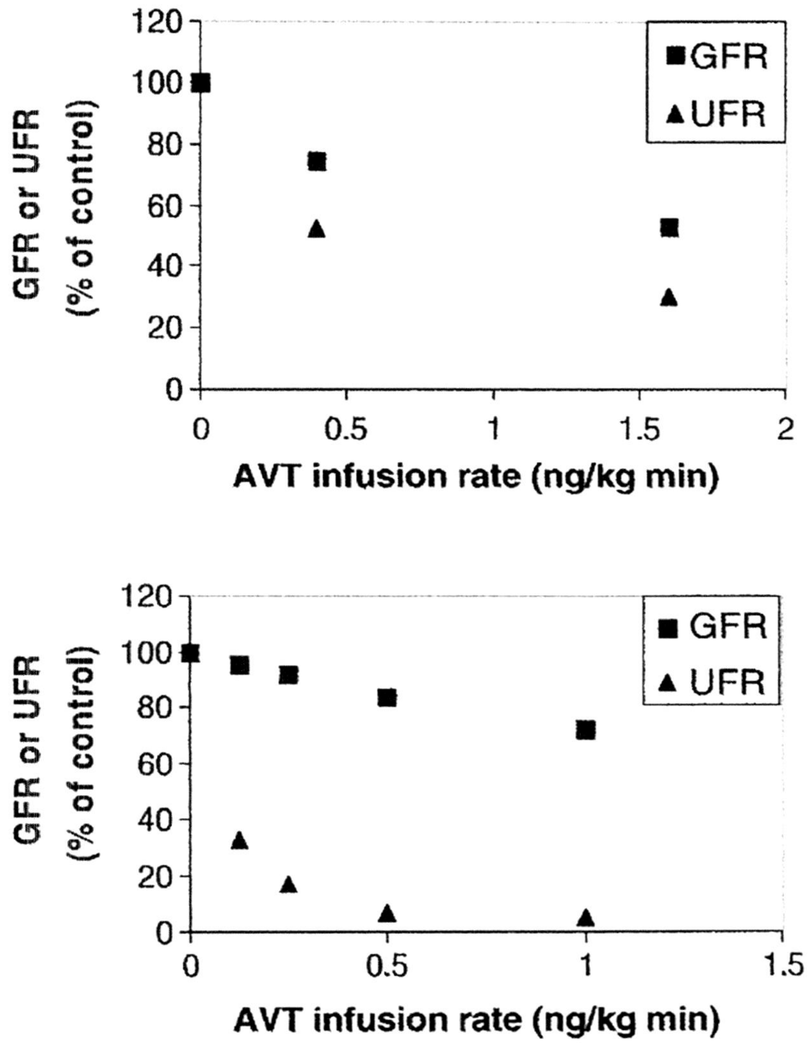


**Fig. 9.43** House Sparrows (*Passer domesticus*, top) and Domestic Chickens (*Gallus g. domesticus*, bottom) treated with different levels of arginine vasotocin (AVT) responded with a reduction in GFR (glomerular filtration rate) and UFR (urine flow rate, or urine production). This indicates that AVT can influence water retention in birds, but, given the differences in the two graphs, to varying degrees in different species, e.g., the effect on AVT on UFR was much more pronounced in chickens than House Sparrows. (Figure from Goldstein 2006; © 2005 Elsevier Inc., used with permission)

uric acid increases in the renal tubules, it forms small spheres together with some protein as well as a number of other ions filtered through renal corpuscles (Casotti and Braun 2004; Fig. 9.55). However, this protein is not lost because it is broken down when the urine enters the large intestine (Goldstein et al. 1999). In the large intestine, the composition of the urine is altered in several ways. The urine spheres are broken down, as is the protein that aided the formation of those spheres. Much of this degradation is accomplished by bacteria (e.g., Gasaway 1976). The amino acids or peptides that result are either used by the bacteria or absorbed by the epithelium of the large intestine and much, but not always all, of the uric acid is then excreted (Braun 1999). Some uric acid may be reabsorbed because uric acid is an important antioxidant for birds (Tsahar et al. 2005).

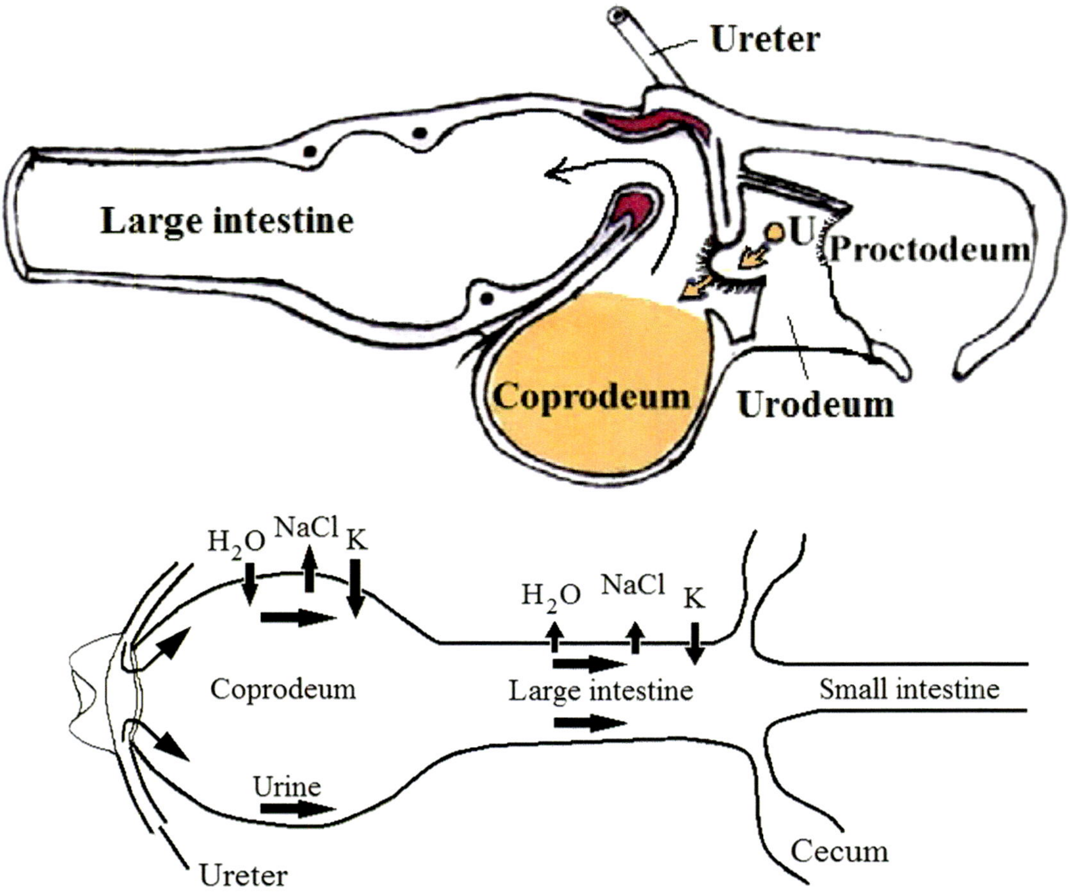


**Fig. 9.44** Top, urine enters the urodeum of the cloaca via the ureters and then moves into the coprodeum and, via reverse peristalsis, into the large intestine. U = where a ureter empties into the urodeum. Bottom, arrows indicate the net movements of water and electrolytes (NaCl and $K^+$) as well as the retrograde movement of ureteral urine (urine transported from the kidneys to the cloaca through the ureters). Initially, the high concentration of solutes in the urine causes water to move by osmosis out of the surrounding tissues and into the coprodeum (the section of the cloaca adjacent to the large intestine). After being transported by peristalsis into the large intestine, however, NaCl is transported out of the large intestine, and water follows the concentration gradient (osmosis) and is reabsorbed. (Figure modified from Laverty and Skadhauge 2008; © 2008 Elsevier Inc., used with permission)

## 9.9 Salt Gland Structure and Function

Because the kidneys of birds cannot produce a hypertonic urine (with lots of ions like sodium), the excretion of excess salt is a potential problem. Even quicker than humans, birds would be severely dehydrated after drinking saltwater and ingesting salty food. However, many species of birds, especially marine birds and shorebirds, can drink seawater as their only source of water. This is possible because these birds have another way (other than the kidneys) to eliminate excess salt–salt glands.

The salt glands of birds likely evolved from nasal glands of reptiles, and evidence of supraorbital salt glands in birds has been found in fossils from the Early Cretaceous (Fig. 9.56). These glands are immediately under the skin in supraorbital depressions of the frontal bone in the skull of

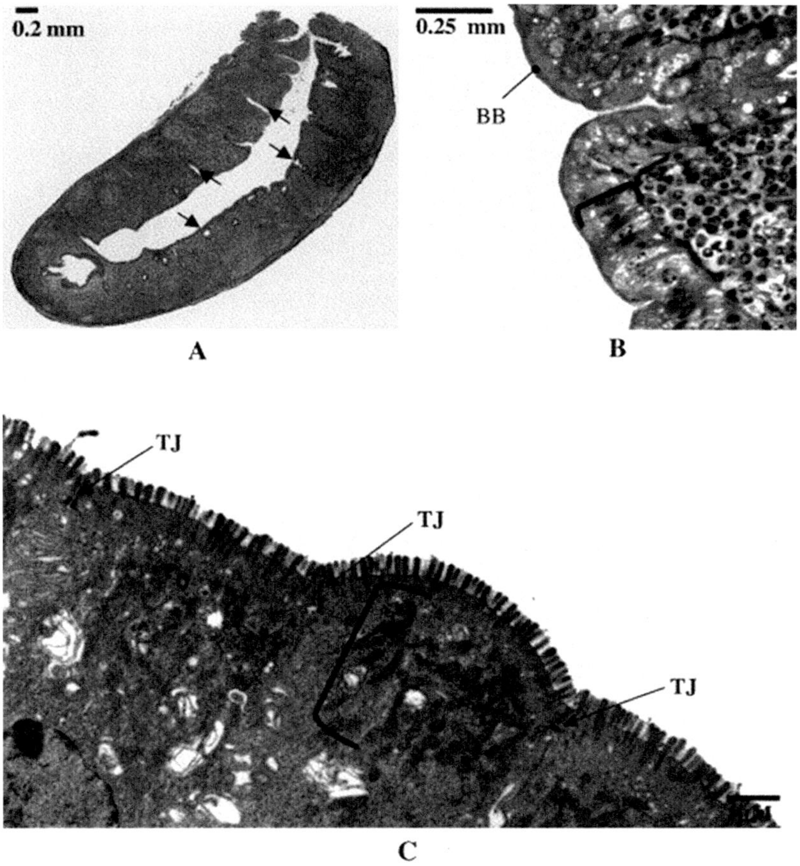


**Fig. 9.45** Light micrograph and electron micrographs of a House Sparrow (*Passer domesticus*) cecum. (**a**) A light micrograph of a cecum cut in sagittal section. The image shows one central channel with a proximal opening to the colon (upper right of section). Side channels (arrows) can be seen branching from the central channel. (**b**) Higher power light micrograph of a cecum. The image illustrates the columnar cells that line the channels (bracket) with their well-developed brush border (arrow). (**c**) Electron micrograph of a House Sparrow cecum (7400×). The image shows a portion of the epithelial cells that line channels of the ceca. Notice the dense microvillus brush border on the surface of the cells and tight junctions (TJ) between the cells. Just below the brush border is a dense layer of mitochondria (bracket). (Figure from Reyes and Braun 2005; © 2005 Elsevier Inc., used with permission)

Charadriiform birds (Figs. 9.57 and 9.58), but in other groups they may be located above the palate or within the orbit of the eye. Skulls of fossil birds, *Ichthyornis* and *Hesperornis*, have similar depressions in the skull, suggesting that these birds lived in marine habitats. The salt glands of marine birds (and some falconiform and desert birds) secrete excess NaCl via the salt glands using less water than is consumed, thus generating free water (Hughes 2003).

Salt glands have been identified in at least 40 families of birds, representing nearly all orders of birds except Passeriformes (Babonis and Brischoux 2012). Even though most studies of osmoregulation in birds have been conducted with marine taxa, nasal secretions are not to be restricted to these species. The presence of functional salt glands has been documented in several terrestrial orders. For example, Greater Roadrunners (*Geococcyx californianus*) and

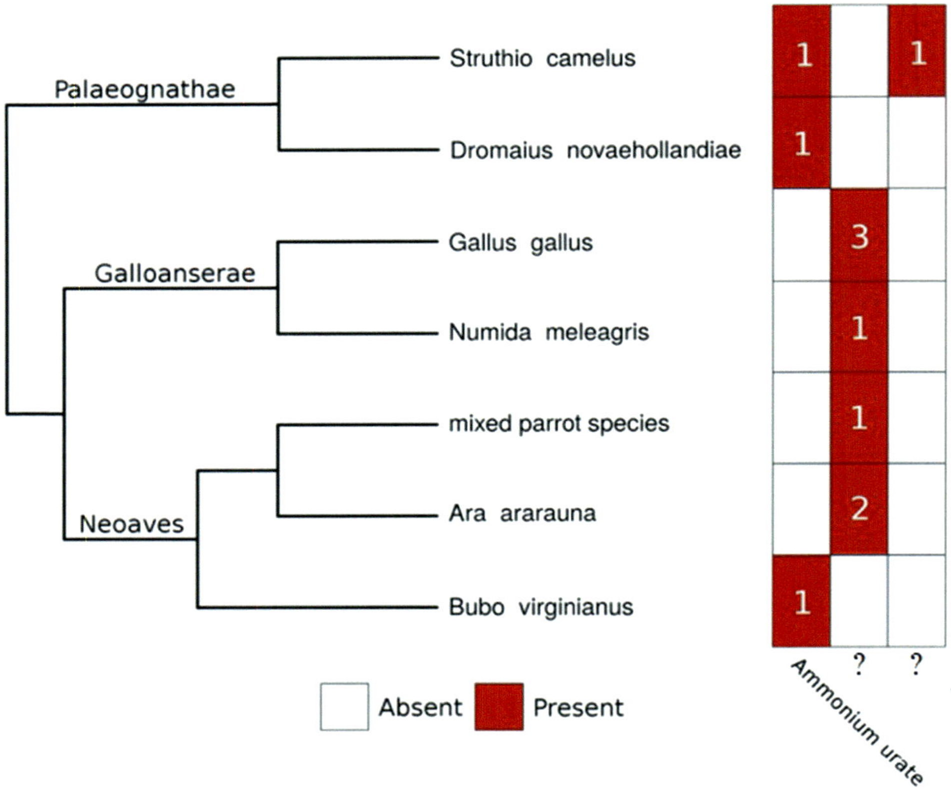


**Fig. 9.46** Analysis of the urine of several species of birds revealed the presence of no uric acid. The urine of three species included ammonium nitrate and the urine of several other species revealed the presence of unknown substances. Numbers indicated the number of samples of that species where the substances were identified. *Struthio camelus*, Common Ostrich; *Dromaius novaehollandiae*, Emu; *Gallus gallus*, Domestic Chicken; *Numida meleagris*, Helmeted Guineafowl; *Ara ararauna*, Blue-and-yellow Macaw; *Bubo virginianus*, Great Horned Owl. (Figure from Crouch et al. 2020; © 2019 Deutsche Ornithologen-Gesellschaft e.V., used with permission)

Savanna Hawks (*Buteogallus meridionalis*) have active salt glands and can produce hypertonic secretions in response to their protein-rich diets. Although these species are not stressed by high saline load, the active secretions of salt glands allow them to minimize water loss. Other desert birds, such as the Sand Partridge (*Ammoperdix heyi*) and the Common Ostrich (*Struthio camelus*), have functional salt glands that are stimulated in response to high temperature. Thus, salt glands are not restricted to birds that live in saline or maritime habitats, but are also present in some terrestrial forms that consume little water (Sabat 2000).

Salt glands have a system of countercurrent blood flow to remove and concentrate salt ions from the blood. The paired, crescent-shaped glands each contain several longitudinal lobes approximately 1 mm in diameter, and each lobe contains a central duct from which radiate thousands of tubules enmeshed in blood capillaries. These tiny capillaries carry blood along the tubules of the gland, which have walls just one cell thick and form a simple barrier between the salty fluid within the tubules and the bloodstream. It is here that salt excretion occurs (Fig. 9.59).

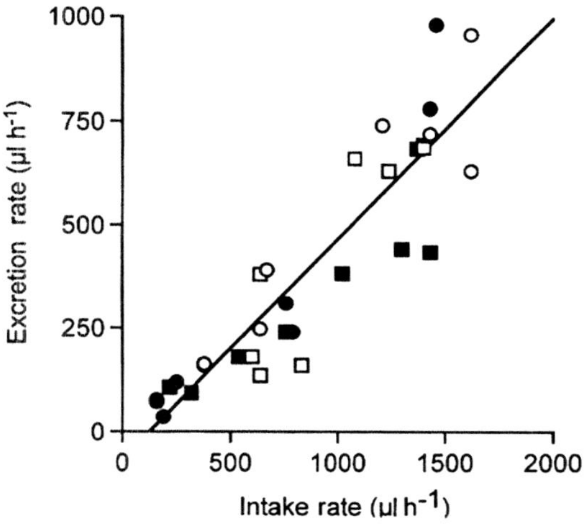


**Fig. 9.47** Relationship between nectar intake rate and excretion rate for Rufous Hummingbirds (*Selasphorus rufus*). Note that excretion rates increase linearly with intake rates. Different symbols indicate different individuals. (Figure from Lotz and Martinez del Rio 2004; © Journal of Avian Biology, used with permission)

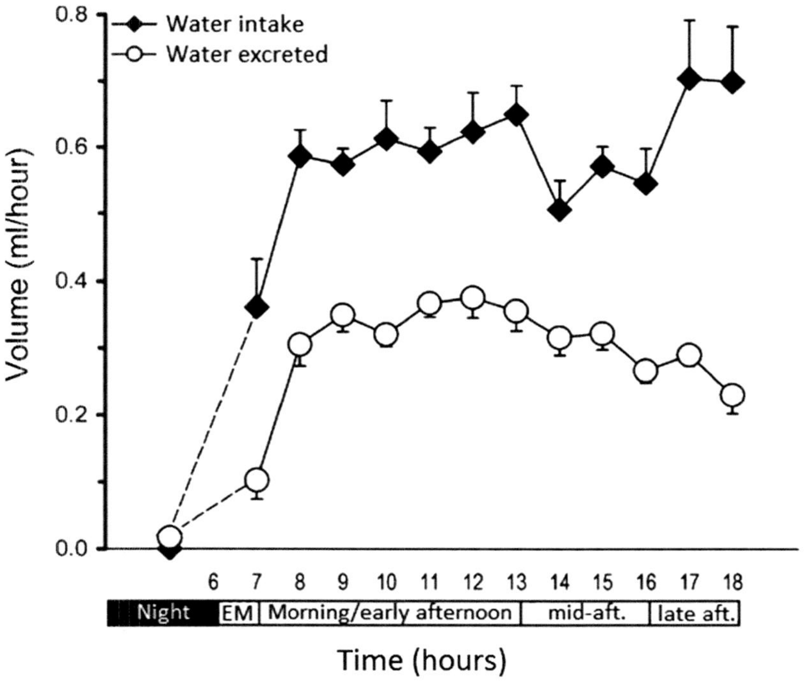


**Fig. 9.48** Variation over time in rates of water intake and water excretion mean by White-bellied Sunbirds (*Nectarinia talatala*). Filtering of water by the kidneys stops at night to help conserve as much water as possible. Filtration rates remain low in the early morning to conserve water and increase body supplies of water after losing water during the night. *EM* early morning. Values are means ± SE. (Figure modified from Fleming et al. 2004; © 2004 Springer-Verlag, used with permission)

When a bird becomes osmotically stressed, sodium enters the blood plasma from the intestine, and the solute concentration of the blood plasma increases. This causes water to move out of cells (osmosis) and increase the extracellular fluid volume. The increases in blood plasma solute concentration and the extracellular fluid volume are detected by osmoreceptors and volume

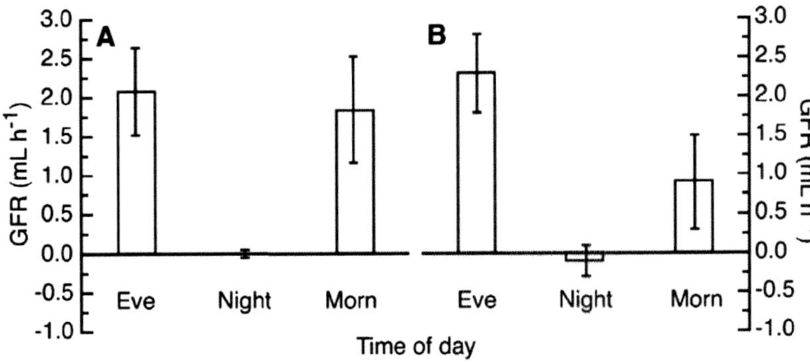


**Fig. 9.49** The glomerular filtration rates (GFRs) of hummingbirds exhibit diet variation, with filtering stopping at night. (**a**) Mean GFRs (± SD) of Green-backed Firecrowns (*Sephanoides sephanoides*) during the evening (eve), night, and morning (morn). (**b**) Mean GFRs of Broad-tailed Hummingbirds (*Selasphorus platycercus*) during the evening, night, and morning. By stopping renal filtration at night, hummingbirds prevent dehydration. (Figure from Bakken and Sabat 2008; © The Neotropical Ornithological Society, used with permission)

receptors in the brain and other peripheral locations that send impulses to the hypothalamus. The hypothalamus then sends impulses along parasympathetic neurons to the salt glands (Hanwell et al. 1971). Synapses on the surface of the secretory epithelial cells have two types of neurotransmitters, acetylcholine and vasoactive intestinal peptide or VIP (Hootman and Ernst 1981). Once released, these neurotransmitters fill receptor sites in the membranes of the secretory cells that stimulate the cells to secrete sodium chloride (Figs. 9.60 and 9.61).

The size and salt-excreting ability of avian salt glands vary among and within species as well as for individual birds. Variation in the salinity of habitats and the salt content of diets explains much of this variation, with species and individuals occupying habitats with higher levels of salinity and/or with diets with greater salt content having larger salt glands (Gutiérrez et al. 2012; Fig. 9.62). Some species also tend to avoid high-salinity environments or food items. For example, juvenile Common Eiders (*Somateria mollissima*) may selectively prey on smaller mussels with lower salt content in areas with high salinity levels (Nyström et al. 1991). Purdue and Haines (1977) found that Snowy Plovers (*Charadrius alexandrinus*) maintained water balance in salt-plain breeding areas by eating insects with a high water content and not drinking salt water. California Gulls (*Larus californicus*) breeding at Mono Lake, California, a lake with high concentrations of salt as well as sulfates and carbonates, avoided salt loading by taking in very little water when feeding on brine shrimp (*Artemia* spp.) and brine flies (family Ephydridae) and visiting sources of freshwater near the shore (Mahoney and Jehl 1985). Young birds with smaller or incompletely developed salt glands, including several species of gulls (Ensor and Phillips 1972; Dosch 1997) and Common Eiders (*Somateria mollissima*; DeVink et al. 2005), may be less able to tolerate prey items with high salt content so adults may provide them with prey having a high water content. For example, adult White Ibises (*Eudocimus albus*) feeding nestlings flew up to 25 km to forage for prey in freshwater environments (Bildstein et al. 1990). Among other species of birds, including some species of penguins, young have functional salt glands immediately after hatching (Janes 1997; Kim et al. 2020).

Among individual birds, the size and salt-excreting ability of salt glands vary with the amount of salt ingested in food and water (Fig. 9.63). However, these changes in salt gland size can take several days, so birds foraging in new locations (e.g., during migration) and/or on prey that vary in salt load may need to adjust their behavior accordingly. For example, in an

**Fig. 9.50** Nectar specialists all absorb the sugars present in nectar rapidly, but have different ways of dealing with ingestion of large amounts of water. Sunbirds and honeyeaters can vary the amount of water absorbed in their intestines, allowing excess water to simply pass through to be voided. The intestines of hummingbirds, however, absorb almost all water ingested, but their kidneys vary the amount of water absorbed so excess water can pass through the kidneys to be voided. (All photographs, CC0 Public Domain from pxhere.com)

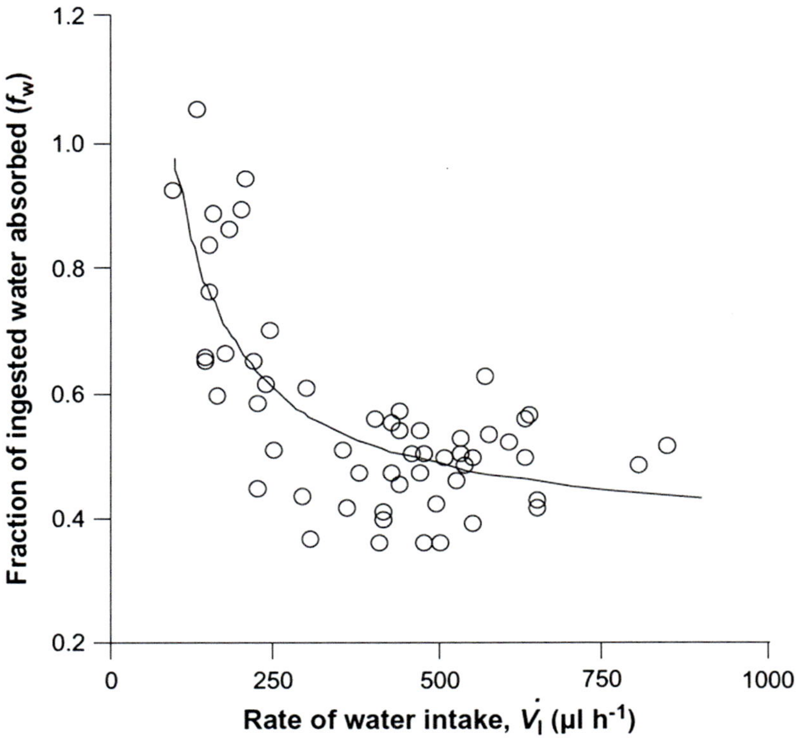


**Fig. 9.51** Amount of water absorbed in the small intestine of Palestine Sunbirds (*Cinnyris osea*) declines as water intake increases, with up to 64% of water not absorbed when water intake rates are high. The mechanism by which sunbirds vary water absorption in the intestine is unclear, but the permeability of their intestines to water may vary with the concentration of sugar in ingested water. (Figure from McWhorter et al. 2009; © 2009 The Authors. Journal compilation © 2009 Cambridge Philosophical Society, used with permission)

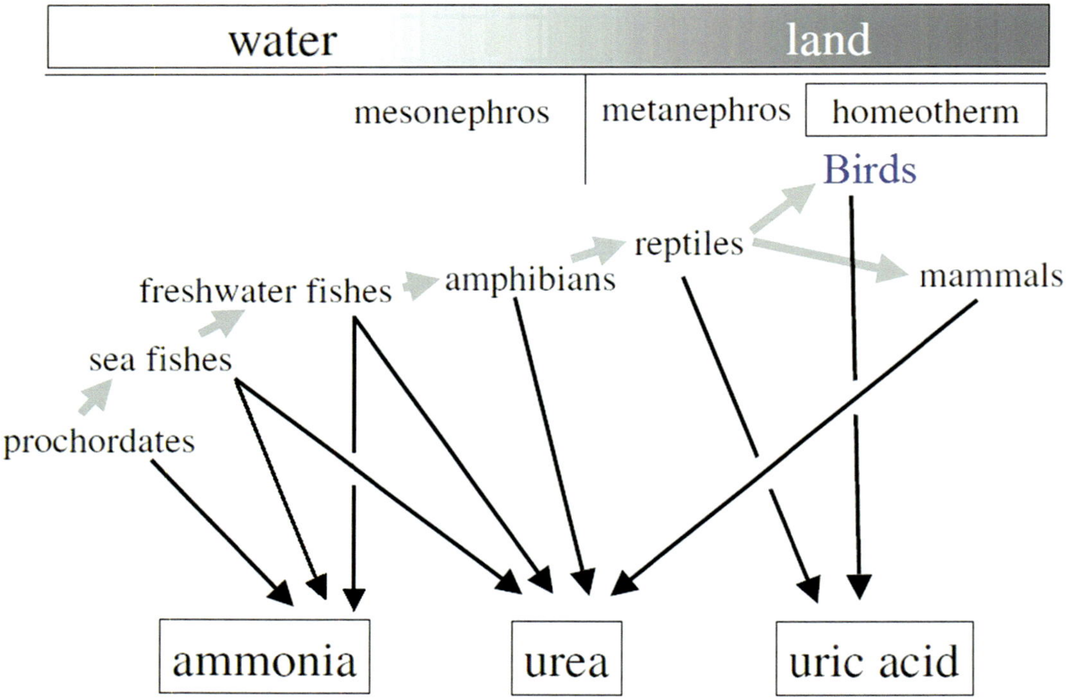


**Fig. 9.52** Excretion of nitrogenous wastes by prochordates and vertebrates. Only birds and reptiles excrete nitrogenous wastes primarily as uric acid. (Figure from Kondo et al. 2006; © 2006 Springer-Verlag, used with permission)

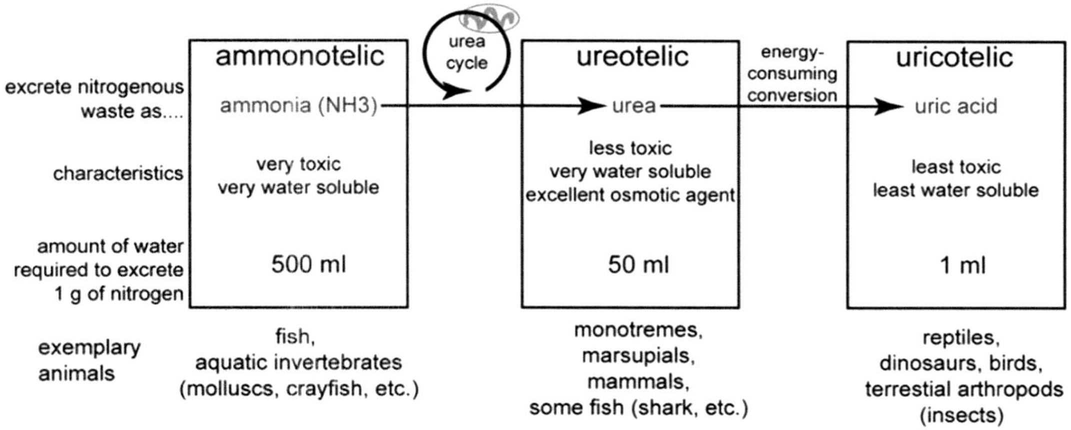


**Fig. 9.53** Excretion of nitrogenous waste. Aquatic animals excrete nitrogenous waste as ammonia. Although toxic, ammonia is quickly diluted by surrounding water. Mammals and some fish convert ammonia to less toxic water-soluble urea using the "urea cycle" (which partly takes place in mitochondria), but must excrete a minimal amount of water to eliminate the urea (in humans 0.5 L/ day). Birds, reptiles, and some insects convert nitrogenous waste to less toxic, non-water-soluble uric acid that is excreted as a white paste. (Figure from Schulte et al. 2015; © 2014 Oxford University Press, used with permission)

experiment with captive (temporarily) Red Knots (*Calidris canutus*), those with relatively small salt glands given a choice of different prey initially ate more and spent more time foraging on snails with low salt content (Gutiérrez and Piersma 2015). However, over a period of several days,

**Box 9.4 Are Birds Strictly Urocotelic?**

Urocotelic means that nitrogenous wastes are excreted primarily as uric acid, not necessarily only as uric acid. Birds do excrete nitrogen wastes primarily as uric acid and, therefore, are referred to as uricotelic. However, birds can also excrete some nitrogenous wastes as urea and ammonia (Aldea and Sabat 2007) and, under some circumstances in some species of birds, even primarily as ammonia. In contrast to uric acid, ammonia is highly soluble in water and requires less energy to synthesize, but is fairly toxic. However, nitrogenous waste may be excreted as ammonia if birds have high rates of water turnover and/or if there is a limited amount of protein in the diet (Tsahar et al. 2005).

(continued)

**Box 9.4** (continued)

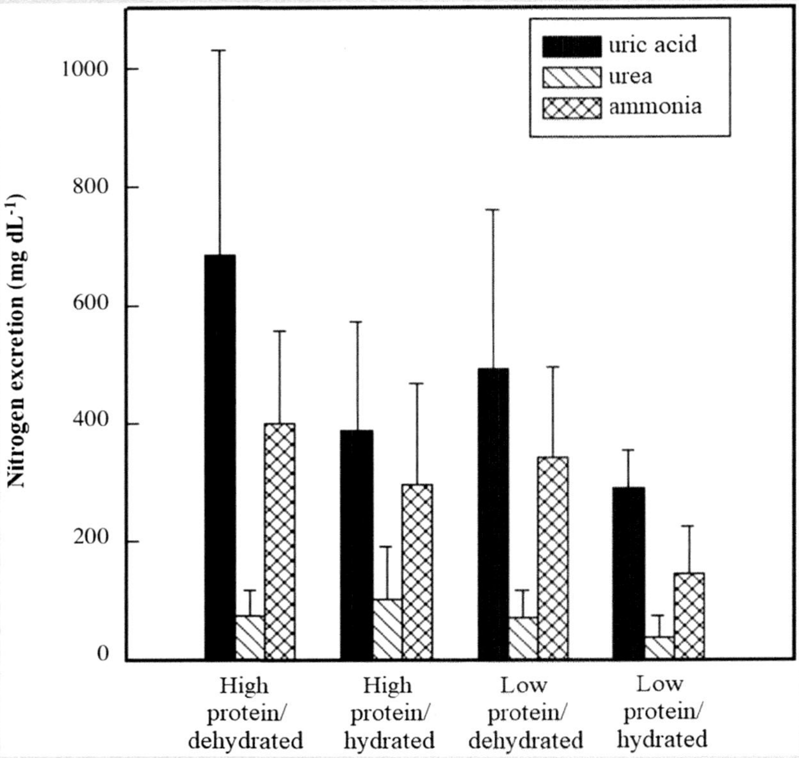


Changes in the concentration of different types of nitrogenous waste for Rufous-collared Sparrows (*Zonotrichia capensis*) fed high-protein (HP) and low-protein (LP) diets and provided with different amount of water (hydrated or dehydrated, with dehydrated birds given just 50% of the water they would normally consume). Sparrows in all four treatments excreted primarily uric acid, but also some urea and ammonia. Uric acid and urea concentrations were higher in dehydrated birds than in hydrated birds, but the proportions of nitrogen excreted as ammonia, uric acid, and urea in the urine did not change among treatments, suggesting that the Rufous-collared Sparrows cannot change nitrogen excretion pathways when their diets include different amounts of protein and their water intake varies. (Figure modified from Aldea and Sabat 2007; open-access article published under the Creative Commons Attribution License CC BY)

Preest and Beuchat (1997) suggested that it might be advantageous for birds that ingest large amounts of dilute, protein-poor nectar to shift from uricotely to ammonotely because ammonia can be voided rapidly, and the costs of synthesizing uric acid can be reduced. For example, with high water intake, Anna's Hummingbirds (*Calypte anna*) were found to excrete more than 50% of their nitrogen wastes as ammonia (i.e., they became facultatively ammonotelic), whereas with lower water intake they were uricotelic. Roxburgh and Pinshow (2002) found that Palestine Sunbirds also switched from uric acid to ammonia excretion under some conditions. However, Roxburgh and Pinshow (2002) noted that, for sunbirds with high water intake, the concentration of uric acid (also called urate) was higher in the ureters (the tubes that carry urine from the kidneys to the cloaca) than in excreta. Ammonotely in Palestine Sunbirds was, therefore, only "apparent" because it was not a result of excessive excretion of ammonia, but rather the result of

(continued)

**Box 9.4** (continued)

a reduction in excreted urate resulting from post-renal modification of urine. Tsahar et al. (2005) found that Yellow-vented Bulbuls (*Pycnonotus goiavier*), a frugivorous species, also appeared to switch from uricotely to ammonotely when they ingested large amounts of water and when their protein intake was low. As with Palestine Sunbirds, however, the amount of urate was lower in excreta than in ureteral urine, and Tsahar et al. (2005) hypothesized that some of the protein associated with urate spheres was digested in the lower intestine and recovered. Why would it be advantageous for birds to recover a nitrogenous metabolic waste? Although considered primarily a nitrogenous waste, uric acid is also a powerful antioxidant in both birds and mammals (Tsahar et al. 2005).

Anna's Hummingbird. (Photo by Becky Matsubara, Wikipedia, CC BY 2.0, https://creativecommons.org/licenses/by/2.0/)

So, can birds be ammonotelic? Apparently, yes. In some cases, as with Anna's Hummingbirds, ammonotely may be a response to the ingestion of lots of water (facultative ammonotely). In other cases, ammonotely may simply be "apparent," with urine produced by the kidneys being urotelic, but the actual excreta being ammonotelic because of reabsorption of uric acid in the lower intestine or colon prior to defecation.

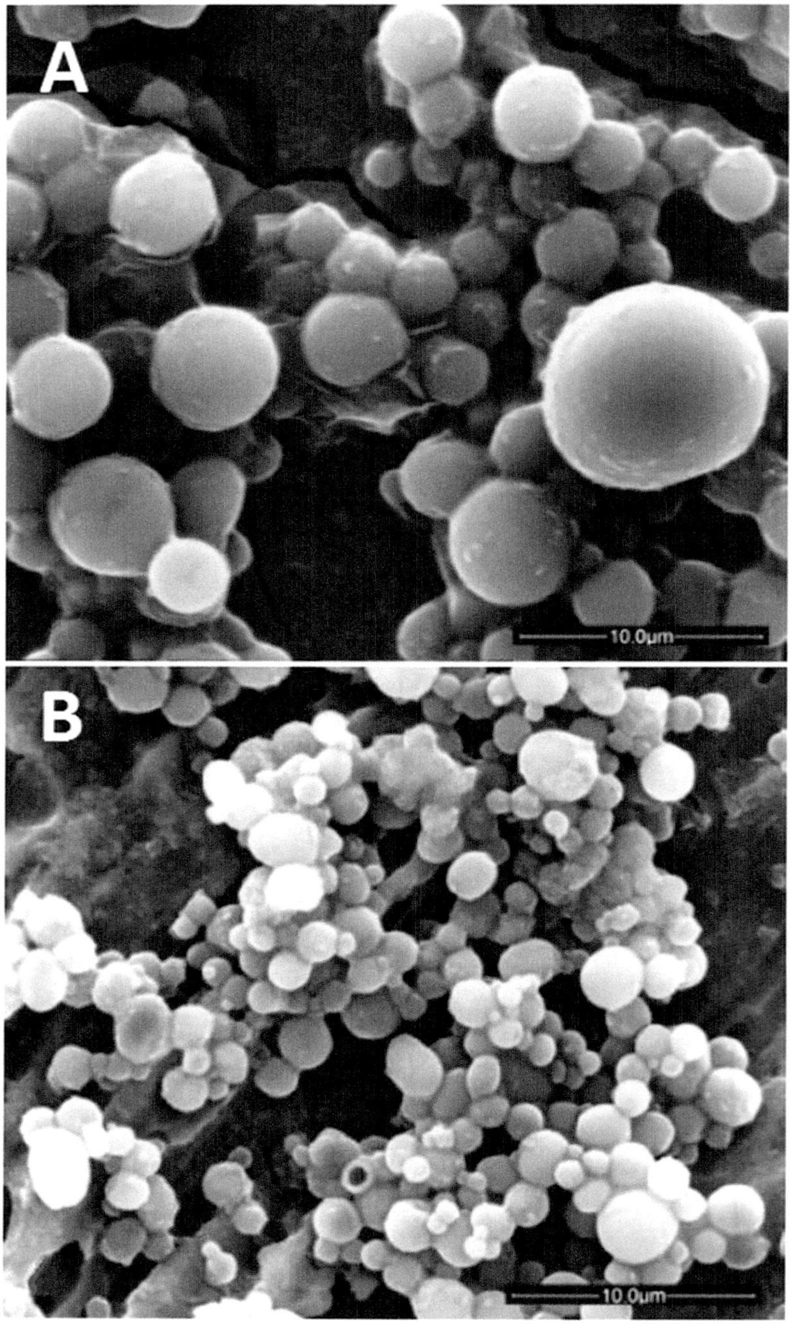


**Fig. 9.54** Urine spheres showing differences in size. (**a**) Domestic Chicken (*Gallus g. domesticus*), and (**b**) Wood Thrush (*Hylocichla mustelina*). In most species, urine spheres are between 0.5 and 5 µm in diameter, but those of Domestic Chickens can be up to 10 µm in diameter. (Figure from Casotti and Braun 2004; © 2004 Wiley-Liss, Inc., used with permission)

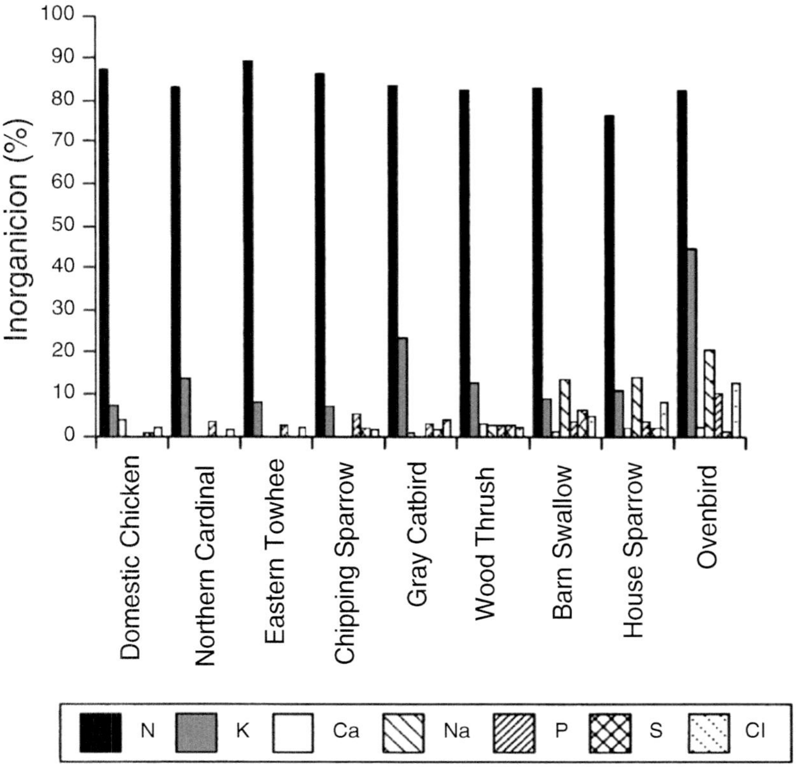


**Fig. 9.55** Composition of urine spheres as percent of total ions in the spheres of several species of birds. Domestic Chicken, *Gallus g. domesticus*; Northern Cardinal, *Cardinalis cardinalis*; Eastern Towhee, *Pipilo erythrophthalmus*; Chipping Sparrow, *Spizella passerina*; Gray Catbird, *Dumetella carolinensis*; Wood Thrush, *Hylocichla mustelina*; Barn Swallow, *Hirundo rustica*; House Sparrow, *Passer domesticus*; Ovenbird, *Seiurus aurocapilla*. (Figure modified from Casotti and Braun 2004; © 2004 Wiley-Liss, Inc., used with permission)

**Fig. 9.56** Photograph (top) and photomicrograph (bottom) of the skull of an ornithurine bird from the Early Cretaceous (about 125 mya) showing the supraorbital salt gland fossa (sgf). This fossil bird was associated with lake deposits so the presence of a salt gland suggests that lakes in the area at that time may have had relatively high salt concentrations. Other anatomical abbreviations: de, dentary; fp, frontal process; fr, frontal; h, hyoid; ju, jugal; la, lacrimal; ma, maxilla; na, nasal; pa, parietal; pm, premaxilla; q, quadrate; t, tooth/teeth. (Figure from Wang et al. 2018; open-access article licensed under a Creative Commons Attribution 4.0 International License, https://creativecommons.org/licenses/by/4.0/)

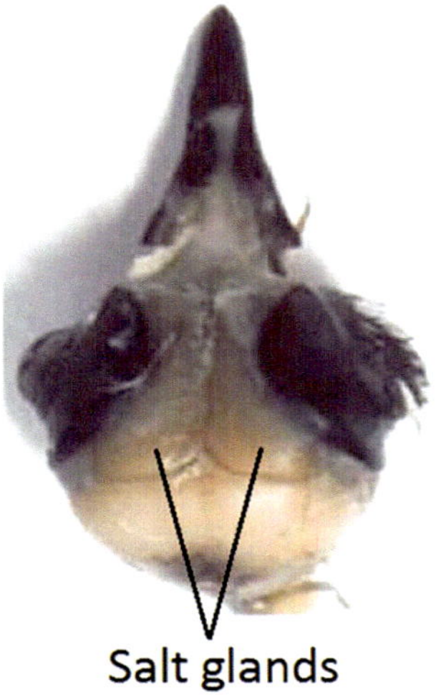


**Fig. 9.57** Salt glands of a Kentish Plover (*Charadrius alexandrinus*). (Figure from Jarrar 2009; open-access article published under the Creative Commons Attribution License CC BY, https://creativecommons.org/licenses/by/4.0/)

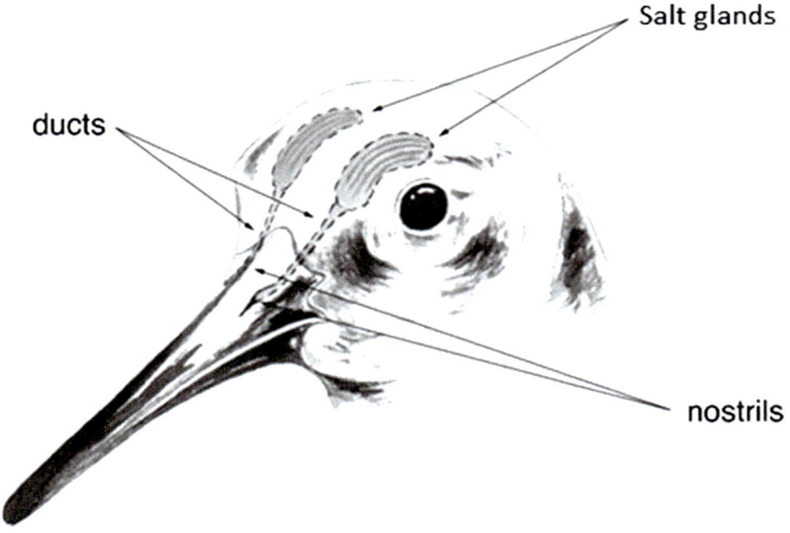


**Fig. 9.58** Supraorbital salt glands of a shorebird. Note that the ducts (outlined by dashed lines) pass through the beak and empty into the anterior nasal cavity so that the secretion flows out through the nares. (Figure from Gutiérrez 2014; © J. S. Gutiérrez–Ardeola; used with permission)

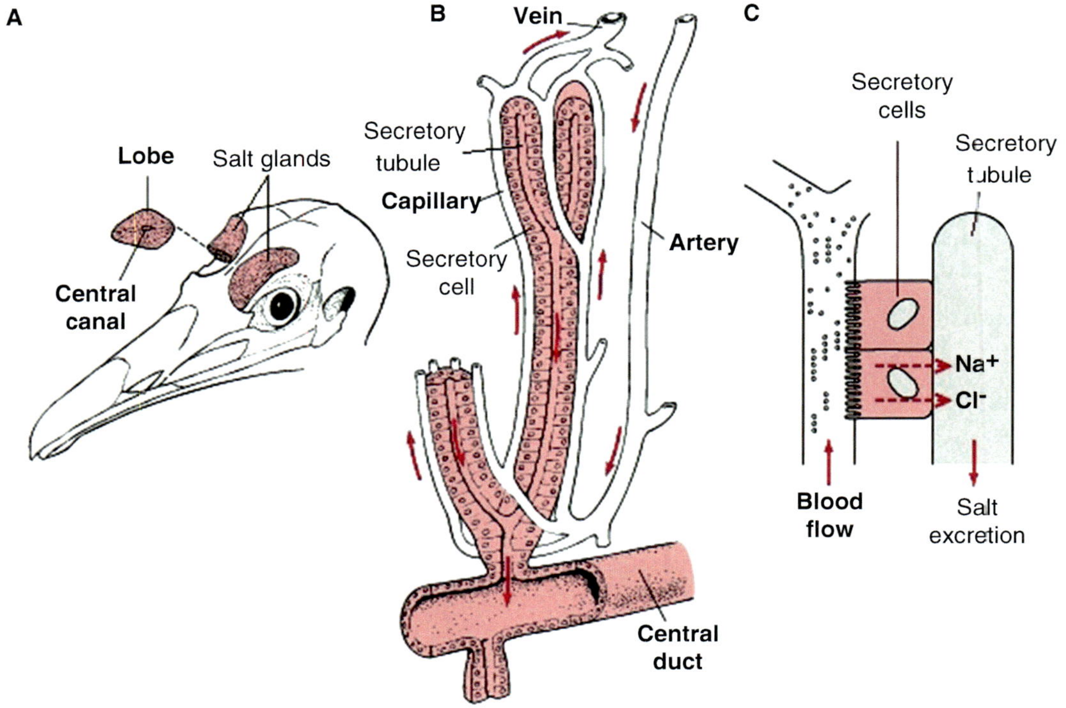


**Fig. 9.59** (**a**) Typical avian salt glands, each salt gland has two main ducts with lateral branches along their entire length. (**b**) Circulation showing flow of blood and filtrate in opposite directions. (**c**) Positively charged sodium ions are actively secreted into secretory tubules and negatively charged chloride ions passively follow. (Figure from Xiao and Liu 2009; © 2008 Springer Nature, used with permission)

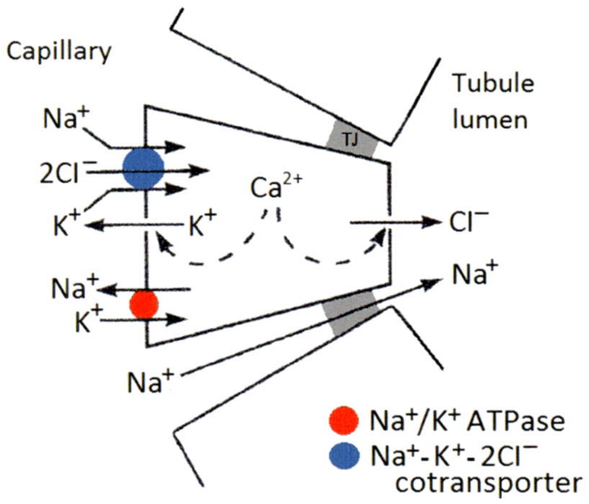


**Fig. 9.60** Possible mechanism for secretion of salt by secretory cells in avian salt glands. Positively charged $Na^+$ and $K^+$ ions are actively transported from the blood into secretory cells, and the negatively charged $Cl^-$ ions passively follow. The enzyme $Na^+$-$K^+$-ATPase provides energy (by breaking down ATP) for the co-transport of $Na^+$ and $K^+$. $Na^+$ flows between the principal cells, through the tight junctions and into the tubular lumen. High levels of $Ca^{2+}$ cause K+ and $Cl^-$ channels in the cell membrane to open so, given their high concentrations in the cell because they are being transported from the blood into the cell, K+ moves by simple diffusion into the blood and $Cl^-$ by simple diffusion into the tubule lumen. The high levels of negatively charged $Cl^-$ in the fluid of the tubular lumen cause the passible efflux of Na + through tight junctions into the tubular lumen. Small amounts of water may follow the solutes via the cellular or paracellular route (between cells through tight junctions), resulting in a hypertonic fluid in the tubular lumen composed primarily of NaCl together with smaller amounts of $K^+$ and traces of other ions. TJ, tight junction between cells. (Figure modified from Shuttleworth and Hildebrandt 1999; © 1999 Wiley-Liss, Inc., used with permission).

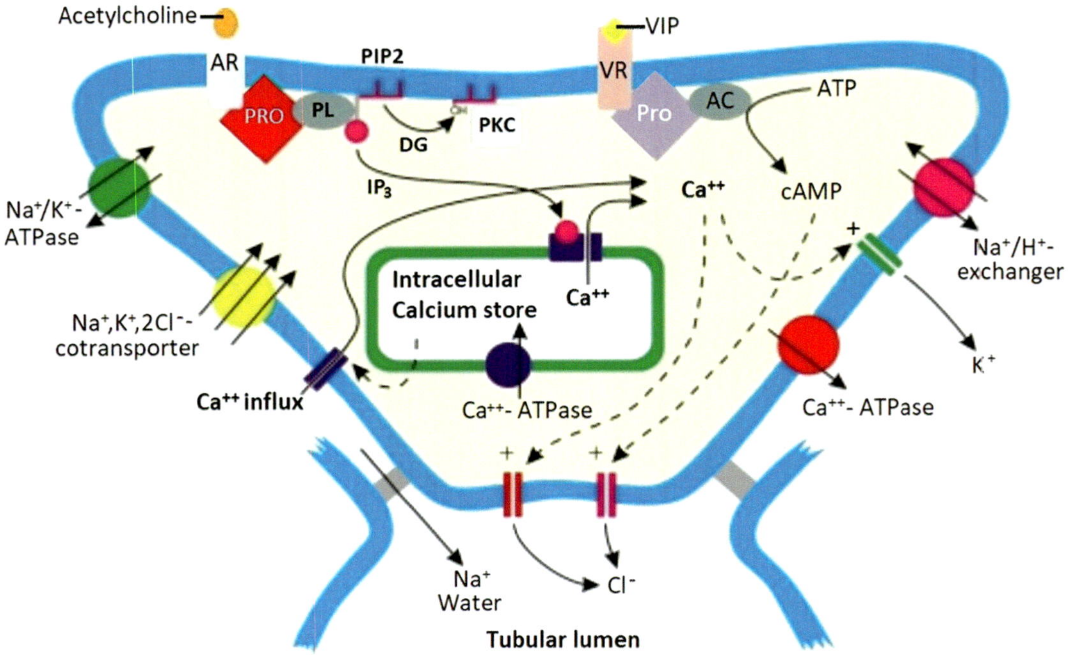


**Fig. 9.61** Possible mechanism of salt gland function. Acetylcholine binds to an acetylcholine receptor (AR) that activates a protein (PRO) combined with a phospholipase molecule (PL). Activated PL cleaves a membrane-bound lipid (PIP2) to generate two "second messengers," water-soluble IP3 and the lipophilic DG that activates protein kinase C (PKC). IP3 binds to specific receptors at the surface of intracellular calcium stores and opens calcium channels. Release of calcium from the stores causes an increase in the calcium concentration ($Ca^{++}$) in the cytoplasm that, in turn, causes calcium channels in the cell membrane to open, allowing calcium to move from the extracellular space into the cell. The increased concentration of calcium ions opens potassium and chloride channels. As those ions diffuse out of the cell, they are replaced by a $Na^+/K^+/2Cl^-$ cotransporter (shown on the left side). The negatively charged chloride ions cause the passive efflux of sodium ions (plus water) through the tight junctions (shown in gray) into the tubular lumen. Sodium ions that accumulate in the cell due to activation of the cotransporter and activation of the Na+/H+ exchanger (shown on the right) are transported out of the cell and exchanged with potassium ions by Na+/K +--ATPase (shown on the left). Secretion stops when calcium ions are removed from the cytoplasm by $Ca^{2+}$-ATPases in the cell membrane and the membrane of intracellular calcium stores. Receptors for another protein (VIP) bind to VIP membrane receptors (VP) to activate another protein (Pro). Activated adenylyl cyclase (AC) uses adenosine trisphosphate (ATP) to generate cyclic adenosine monophosphate (cAMP), which serves as an signaling molecule, opening another chloride channel and allowing the movement of chloride ions out of the cell. (Figure from Hildebrandt 2001; )

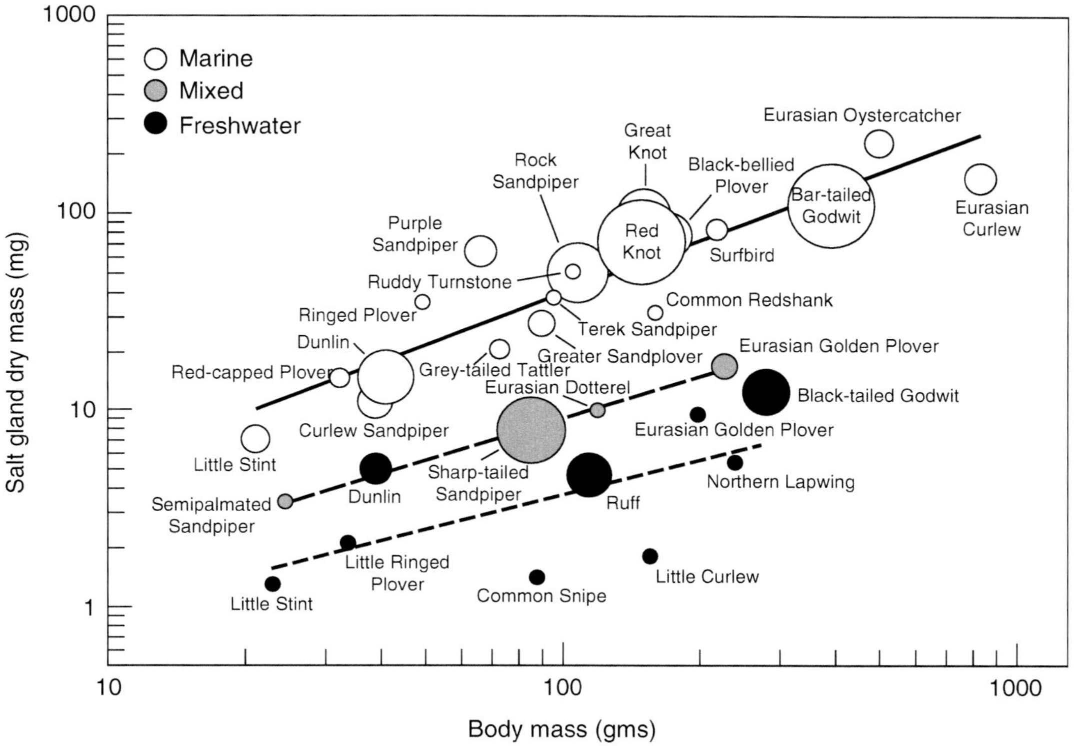


**Fig. 9.62** Relationship between body mass and salt gland dry mass for 29 marine, freshwater, and mixed (use both marine and freshwater habitats) species of shorebirds. Noted that shorebirds found in marine habitats have larger salt glands than those found in mixed or freshwater habitats. Symbols are scaled relative to sample sizes. Note that data for Little Stints, Dunlins, and Eurasian Golden Plovers are presented for more than one habitat (i.e., some populations occupy different habitats). Little Stint, *Calidris minuta*; Little Ringed Plover, *Charadrius dubius*; Common Snipe, *Gallinago gallinago*; Little Curlew, *Numenius minutus*; Semipalmated Sandpiper, *Calidris pusilla*; Dunlin, *Calidris alpina*; Sharp-tailed Sandpiper, *Calidris acuminata*; Ruff, *Calidris pugnax*; Northern Lapwing, *Vanellus vanellus*; Curlew Sandpiper, *Calidris ferruginea*; Eurasian Dotterel, *Charadrius morinellus*; Eurasian Golden-Plover, *Pluvialis apricaria*; Gray-tailed Tattler, *Tringa brevipes*; Greater Sand-Plover, *Charadrius leschenaultii*; Black-tailed Godwit, *Limosa limosa*; Red-capped Plover, *Charadrius ruficapillus*; Common Ringed Plover, *Charadrius hiaticula*; Terek Sandpiper, *Xenus cinereus*; Common Redshank, *Tringa totanus*; Ruddy Turnstone, *Arenaria interpres*; Red Knot, *Calidris canutus*; Surfbird, *Calidris virgata*; Purple Sandpiper, *Calidris maritima*; Rock Sandpiper, *Calidris ptilocnemis*; Great Knot, *Calidris tenuirostris*; Black-bellied Plover, *Pluvialis squatarola*; Bar-tailed Godwit, *Limosa lapponica*; Eurasian Oystercatcher, *Haematopus ostralegus*; Eurasian Curlew, *Numenius arquata*. (Figure from Gutiérrez et al. 2012; © 2011 The Authors. Functional Ecology © 2011 British Ecological Society, used with permission)

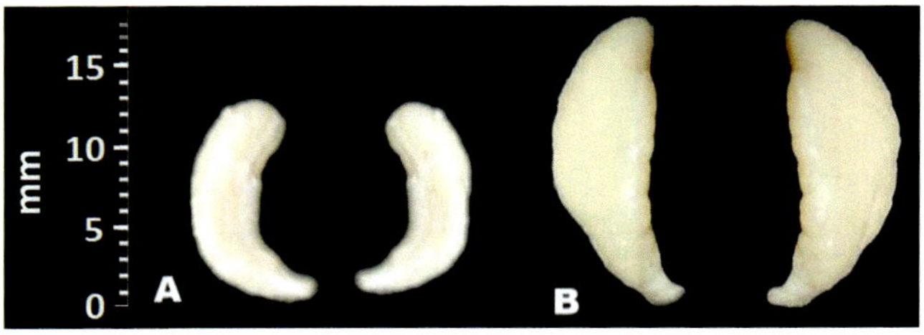


**Fig. 9.63** Salt glands of a Mallard (*Anas platyrhynchos*) after drinking (**a**) fresh water and (**b**) after drinking a 1% saline solution for 2 weeks. (Figure modified from El-Gohary et al. 2013; licensed under a Creative Commons 2.0 license (CC BY 2.0), https://creativecommons.org/licenses/by/2.0/)

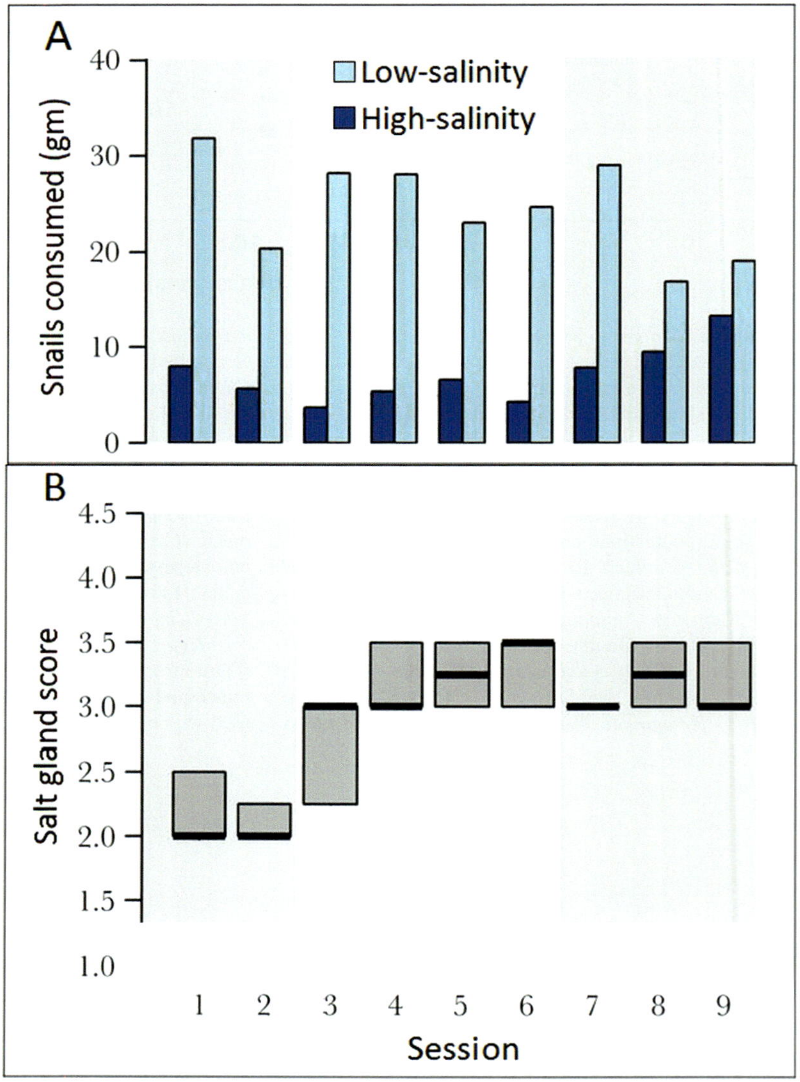


**Fig. 9.64** (**a**) Mean amount (in grams) of high- and low-salinity snails eaten by Red Knots (*Calidris canutus*) during nine 3-h experimental sessions. (**b**) Salt gland scores during the experiment. The shaded areas depict access to freshwater prior to the experimental session. Each bird had one session every other day. (Figure from Gutiérrez and Piersma 2015; © 2015 Oxford University Press, used with permission)

as their salt glands adapted to increasing amounts of salt in their diet by increasing in size and ability of excrete salt, the preference of Red Knots for low-salt prey disappeared (Fig. 9.64). Such results suggest that prey selection by birds with salt glands may be influenced by the relative size of their salt glands and the time needed for salt glands to adapt to environments and prey with different salt loads.

## References


Al-Ajeely RA, Mohammed FS (2012) Morpho-histological study on the development of kidney and ureter in hatching and adulthood racing pigeon (*Columba livia domestica*). Int J Sci Nat 3:665–677

Albright TP, Mutiibwa D, Gerson AR, Smith EK, Talbot WA, O'Neill JJ, McKechnie AE, Wolf BO (2017) Mapping evaporative water loss in desert passerines reveals an expanding threat of lethal dehydration. Proc Natl Acad Sci USA 114:2283–2288

Aldea P, Sabat P (2007) Osmoregulatory responses to dietary protein and water intake in the granivorous *Zonotrichia capensis* (Passerine, Emberizidae). Rev Chil Hist Nat 80:447–454

Ambrose SJ, Bradshaw SD, Withers PC, Murphy DP (1996) Water and energy balance of captive and free-ranging Spinifexbirds (*Eremiornis carteri*) north (Aves: Sylviidae) on Barrow Island, Western Australia. Aust J Zool 44:107–117

Anava A, Kam M, Shkolnik A, Degen AA (2000) Seasonal field metabolic rate and dietary intake in Arabian Babblers (*Turdoides squamiceps*) inhabiting extreme deserts. Funct Ecol 1:607–613

André S, Kaltner H, Manning JC, Murphy PV, Gabius H-J (2015) Lectins: getting familiar with translators of the sugar code. Molecules 20:1788–1823

Arieli Y, Peltonen L, Ophir E (2002) Cooling by cutaneous water evaporation in the heat-acclimated Rock Pigeon (*Columba livia*). Comp Biochem Physiol 131: 497–504

Babonis LS, Brischoux F (2012) Perspectives on the convergent evolution of tetrapod salt glands. Integr Comp Biol 52:245–256

Backman J, Alerstam T (2001) Confronting the winds: orientation and flight behaviour of roosting swifts. Proc R Soc B 268:1081–1087

Bakken BH, Sabat P (2008) The mechanisms and ecology of water balance in hummingbirds. Ornitol Neotrop 19 (Suppl):501–509

Bakken BH, McWhorter TJ, Tsahar E, Martinez del Rio C (2004) Hummingbirds arrest their kidneys at night: diel variation in glomerular filtration rate in *Selasphorus platycercus*. J Exp Biol 207:4383–4391

Beaver PW (1978) Ontogeny of feeding and drinking in the Greater Rhea (*Rhea americana*). Behav Process 3: 283–292

Beuchat CA, Preest MR, Braun EJ (1999) Glomerular and medullary architecture in the kidney of Anna's Hummingbird. J Morphol 240:95–100

Bildstein KL, Post W, Johnston J, Frederick P (1990) Freshwater wetlands, rainfall, and the breeding ecology of White Ibises in coastal South Carolina. Wilson Bull 102:84–98

Bolin G, Burggren WW (2013) Metanephric kidney development in the chicken embryo: glomerular numbers, characteristics and perfusion. Comp Biochem Physiol 166:343–350

Bourke JM, Witmer LM (2016) Nasal conchae function as aerodynamic baffles: experimental computational fluid dynamic analysis in a turkey nose (Aves: Galliformes). Respir Physiol Neurobiol 234:32–46

Braun J (1996) An overview of osmoregulation in birds. In: Abstract – VI international symposium of avian endocrinology, Chateau Lake Louise, Alberta, Canada

Braun EJ (1999) Integration of organ systems in avian osmoregulation. J Exp Zool 283:702–707

Braun EJ, Campbell CE (1989) Uric acid decomposition in the lower gastrointestinal tract. J Exp Zool 252:70–74

Campbell JW, Vorhaben JE, Smith DD Jr (1987) Uricoteley: its nature and origin during the evolution of tetrapod vertebrates. J Exp Zool 243:349–363

Carmi N, Pinshow B, Porter WP, Jaeger J (1992) Water and energy limitation on flight duration in small migrating birds. Auk 109:268–276

Casotti G, Braun EJ (1995) Structure of the glomerular capillaries of the Domestic Chicken and Desert Quail. J Morphol 224:57–63

Casotti G, Braun EJ (2000) Renal anatomy in sparrows from different environments. J Morphol 243:283–291

Casotti G, Braun EJ (2004) Protein location and elemental composition of urine spheres in different avian species. J Exp Zool 301A:579–587

Champagne AM, Munoz-Garcia A, Shtayyeh T, Tieleman BL, Hegemann A, Clement ME, Williams JB (2012) Lipid composition of the stratum corneum and cutaneous water loss in birds along an aridity gradient. J Exp Biol 215:4299–4307

Champagne AM, Allen HC, Bautista-Jimenez RC, Williams JB (2016) Organization of lipids in avian stratum corneum: changes with temperature and hydration. Chem Phys Lipids 195:47–57

Clench MH, Mathias JR (1995) The avian cecum: a review. Wilson Bull 107:93–121

Cooper CE, Withers PC, Hurley LL, Griffith SC (2019) The field metabolic rate, water turnover, and feeding and drinking behavior of a small avian desert granivore during a summer heatwave. Front Physiol 10:1405

Crouch NMA, Lynch VM, Clarke JA (2020) A re-evaluation of the chemical composition of avian urinary excreta. J Ornithol 161:17–24

Czenze ZJ, Kemp R, Van Jaarsveld B, Freeman MT, Smit B, Wolf BO, McKechnie AE (2020) Regularly

drinking desert birds have greater evaporative cooling capacity and higher heat tolerance limits than non-drinking species. Funct Ecol 34:1589–1600
Danner RM, Gulson-Castillo ER, James HF, Dzielski SA, Frank DC III, Sibbald ET, Winkler DW (2017) Habitat-specific divergence of air conditioning structures in bird bills. Auk 134:65–75
Dantzler WH (1970) Kidney function in desert vertebrates. In: Benson GK, Phillips JG (eds) Memoirs of the society of endrocrinology. 18. Hormones and the environment. Cambridge University Press, London, UK, pp 157–190
Dantzler WH, Braun EJ (1980) Comparative nephron function in reptiles, birds, and mammals. Am J Phys 239:R197–R213
Dantzler WH (1989) Comparative physiology of the vertebrate kidney. Springer, Berlin, Germany
Dantzler WH (2016) Comparative physiology of the vertebrate kidney, 2nd edn. Springer, New York, NY
Dawson WR (1982) Evaporative losses of water by birds. Comp Biochem Physiol 71:495–509
de Camargo B, Vanessa K, Rodewald AD, Ribeiro MC, Jahn AE (2020) Noise level and water distance drive resident and migratory bird species richness within a Neotropical megacity. Landsc Urban Plan 197:103769
DeLuca WV, Woodworth BK, Rimmer CC, Marra PP, Taylor PD, McFarland KP, Mackenzie SA, Norris DR (2015) Transoceanic migration by a 12 g songbird. Biol Lett 11:20141045
DeLuca WV, Woodworth BK, Mackenzie SA, Newman AEM, Cooke HA, Phillips LM, Freeman NE, Sutton AO, Tauzer L, McIntyre C, Stenhouse IJ, Weidensaul S, Taylor PD, Norris DR (2019) A boreal songbird's 20,000 km migration across North America and the Atlantic Ocean. Ecology 100:e02651
DeVink J-MA, Gilchrist HG, Diamond AW (2005) Effects of water salinity on growth and survival of Common Eider (*Somateria mollissima*) ducklings. Auk 122: 523–529
Domer A, Shochat E, Ovadia O, Sapir N (2019) Fruit consumption in migratory passerines is limited by water ingestion rather than by body water balance. J Avian Biol 50:e02021
Dosch JJ (1997) Salt tolerance of nestling Laughing Gulls: an experimental field investigation. Col Waterbirds 20: 449–457
Echols MS (2005) Evaluating and treating the kidneys. In: Harrison GJ, Lightfoot TL (eds) Clinical avian medicine. Palm Beach, FL, Spix, pp 451–492
El-Gohary ZM, El-Sayad FI, Hassan HA, Hamoda AM (2013) Structural adaptations associated with osmotic stress of the salt gland in birds. Egypt J Hosp Med 51: 332–345
Ellis HI, Jehl JR (1991) Total body water and body composition in phalaropes and other birds. Physiol Zool 64:973–984
Engel S, Biebach H, Visser GH (2006a) Water and heat balance during flight in the Rose-colored Starling (*Sturnus roseus*). Physiol Biochem Zool 79:763–774
Engel S, Suthers RA, Biebach H, Visser GH (2006b) Respiratory water loss during rest and flight in European Starlings (*Sturnus vulgaris*). Comp Biochem Physiol 145:423–432
Ensor D, Phillips J (1972) The effect of age and environment on extrarenal salt excretion in juvenile gulls (*Larus argentatus* and *L. fuscus*). J Zool 168:119–126
Eto EC, Withers PC, Cooper CE (2017) Can birds do it too? Evidence for convergence in evaporative water loss regulation for birds and mammals. Proc R Soc B 284:20171478
Fisher CD, Lindgren E, Dawson WR (1972) Drinking patterns and behavior of Australian desert birds in relation to their ecology and abundance. Condor 74: 111–136
Fitzsimmons JT (1998) Angiotensin, thirst, and sodium appetite. Physiol Rev 78:583–686
Fleming PA, Gray DA, Nicolson SW (2004) Circadian rhythm of water balance and aldosterone excretion in the Whitebellied Sunbird *Nectarinia talatala*. J Comp Physiol B 174:341–346
Gasaway WC (1976) Volatile fatty acids and metabolizable energy derived from cecal fermentation in the Willow Ptarmigan. Comp Biochem Physiol 53A:115–121
Gavrilov VM (2017) Total evaporative water loss in birds at different ambient temperatures: allometric and stoichiometric approaches. Zool Stud 56:e37
Geist NR (2000) Nasal respiratory turbinate function in birds. Physiol Biochem Zool 73:581–589
Gerson AR, Guglielmo CG (2013) Measurement of glomerular filtration rate during flight in a migratory bird using a single bolus injection of FITC-inulin. Am J Physiol-Ren Physiol 305:F823–F829
Gill RE Jr, Tibbitts TL, Douglas DC, Handel CM, Mulcahy DM, Gottschalck JC, Warnock N, McCaffery BJ, Battley PF, Piersma T (2009) Extreme endurance flights by landbirds crossing the Pacific Ocean: ecological corridor rather than barrier? Proc R Soc B 276: 447–457
Goecke CS, Goldstein DL (1997) Renal glomerular and tubular effects of antidiuretic hormone and two antidiuretic hormone analogues in House Sparrows (*Passer domesticus*). Physiol Zool 70:283–291
Goldstein DL (2006) Regulation of the avian kidney by arginine vasotocin. Gen Comp Endocrinol 147:78–84
Goldstein DL, Skadhauge E (2000) Renal and extrarenal regulation of body fluid composition. In: Whittow GC (ed) Sturkie's avian physiology. Academic Press, New York, NY, pp 265–298
Goldstein DL, Williams JB, Braun EJ (1990) Osmoregulation in the field by salt-marsh Savannah Sparrows *Passerculus sandwichensis beldingi*. Physiol Zool 63: 669–682
Goldstein DL, Reddy V, Plaga K (1999) Second messenger production in avian medullary nephron segments in response to peptide hormones. Am J Physiol Regul Integr Comp Physiol 276:R847–R854

Gutiérrez JS (2014) Living in environments with contrasting salinities: a review of physiological and behavioural responses in waterbirds. Ardeola 61:233–256

Gutiérrez JS, Piersma T (2015) Ecological context determines the choice between prey of different salinities. Behav Ecol 27:530–537

Gutiérrez JS, Dietz MW, Masero JA, Gill RE Jr, Dekinga A, Battley PF, Sánchez-Guzmán JM, Piersma T (2012) Functional ecology of saltglands in shorebirds: flexible responses to variable environmental conditions. Funct Ecol 26:236–244

Hadžiomerović N, Avdić R, Minnich B, Erlbacher K, Mlaćo N, Tandir F, Bejdić P (2021) 3D morphometric study of Domestic Fowl glomerulus – a scanning electron microscope study of vascular corrosion casts. Anat Histol Embryol 50:678–682

Hagger V, Wilson K, England JR, Dwyer JM (2019) Water availability drives aboveground biomass and bird richness in forest restoration plantings to achieve carbon and biodiversity cobenefits. Ecol Evol 9: 14379–14393

Hanwell A, Linzell JL, Peaker M (1971) Salt-gland secretion and blood flow in the goose. J Physiol 213:373–387

Haugen MJ, Tieleman BI, Williams JB (2003) Phenotypic flexibility in cutaneous water loss and lipids of the stratum corneum. J Exp Biol 206:3581–3588

Hawkins BA, Diniz-Filho JAF, Soeller SA (2005) Water links the historical and contemporary components of the Australian bird diversity gradient. J Biogeogr 32: 1035–1042

Hedenström A, Alerstam T (1992) Climbing performance of migrating birds as a basis for estimating limits for fuel-carrying capacity and muscle work. J Exp Biol 164:19–38

Hediger MA (2002) Kidney function: gateway to a long life? Nature 417:393–395

Heidweiller J, Zweers GA (1990) Drinking mechanisms in the Zebra Finch and the Bengalese Finch. Condor 92: 1–28

Heidweiller J, van Loon JA, Zweers GA (1992) Flexibility of the drinking mechanism in adult chickens (*Gallus gallus*) (Aves). Zoomorphology 111:141–159

Herd RM, Dawson TJ (1984) Fiber digestion in the Emu, *Dromaius novaehollandiae*, a large bird with a simple gut and high rates of passage. Physiol Zool 57:70–84

Hildebrandt J-P (2001) Coping with excess salt: adaptive functions of extrarenal osmoregulatory organs in vertebrates. Zoology 104:209–220

Hill RW (2012) Animal physiology, 3rd edn. Oxford Publishing, Oxford, UK

Hootman SR, Ernst SA (1981) Characterization of muscarinic acetylcholine receptors in the avian salt gland. J Cell Biol 91:781–789

Horton KG, Van Doren BM, Stepanian PM, Farnsworth A, Kelly JF (2016) Where in the air? Aerial habitat use of nocturnally migrating birds. Biol Lett 12: 20160591

Hughes MR (2003) Regulation of salt gland, gut and kidney interactions. Comp Biochem Physiol A 136:507–524

Ichimura K, Sakai T (2018) Acquisition and evolution of the renal glomerulus in vertebrates. Juntendo Med J 64: 37–45

Janes DN (1997) Osmoregulation by Adélie Penguin chicks on the Antarctic peninsula. Auk 114:488–495

Jarrar BM (2009) Microanatomy and histochemistry of the salt glands of the Kentish Plover, *Charadrius alexandrines* (Aves, Charadrridae). Online J Biol Sci 9:75–80

Jenni-Eiermann S, Jenni L (2012) Fasting in birds: general patterns and the special case of endurance flight. In: McCue MD (ed) Comparative physiology of fasting, starvation, and food limitation. Springer, Berlin Heidelberg, pp 171–192

Johnson DW (1968) Some morphological features of avian kidneys. Auk 85:216–228

Kim W, Bush JWM (2012) Natural drinking strategies. J Fluid Mech 705:7–25

Kim Y, Jeong M-S, Seo H-M, Kim H, Lee W-S, Choi C-Y (2020) Growth-related changes in salt gland mass in Gentoo and Chinstrap penguin chicks. Polar Res 39: 702

King AS (1993) Apparatus respiratorius (systema respiratorium). Handbook of avian anatomy: nomina anatomica avium. Nuttall Ornithological Club, Cambridge, MA

Klaassen M (1996) Metabolic constraints on long-distance migration in birds. J Exp Biol 199:57–64

Kondo Y, Morimoto T, Nishio T, Aslanova UF, Nishino M, Farajov EI, Sugawara N, Kumagai N, Ohsaga A, Maruyama Y, Takahashi S (2006) Phylogenetic, ontogenetic, and pathological aspects of the urine-concentrating mechanism. Clin Exp Nephrol 10:165–174

Krag B, Skadhauge E (1972) Renal salt and water excretion in the Budgerygah (*Melopsittacus undulatus*). Comp Biochem Physiol 41A:667–683

Landys MM, Piersma T, Visser GH, Jukema J, Wijker A (2000) Water balance during real and simulated long-distance migratory flight in the Bar-tailed Godwit. Condor 102:645–652

Laverty G, Skadhauge E (2008) Adaptive strategies for post-renal handling of urine in birds. Comp Biochem Physiol A 149:246–254

Lee ATK, Wright D, Barnard P (2017) Hot bird drinking patterns: drivers of water visitation in a fynbos bird community. Afr J Ecol 55:541–553

Leib DE, Zimmerman CA, Knight ZA (2016) Thirst. Curr Biol 26:R1260–R1265

Liechti F, Klaassen M, Bruderer B (2000) Predicting migratory flight altitudes by physiological migration models. Auk 117:205–214

Lillywhite HB (2006) Water relations of tetrapod integument. J Exp Biol 209:202–226

Lipke DW, Olson KR (1988) Distribution of angiotensin-converting enzyme-like activity in vertebrate tissues. Physiol Zool 61:420–428

Liu W, Morimoto T, Kondo Y, Iinuma K, Uchida S, Imai M (2001) "Avian-type" renal medullary tubule organization causes immaturity of urine-concentrating ability in neonates. Kidney Int 60:680–693

Luo D-G, Yue WWS, Ala-Laurila P, Yau K-W (2011) Activation of visual pigments by light and heat. Science 332:1307–1312

Maclean GL (1996) Ecophysiology of desert birds. Springer, Berlin, Germany

MacMillen RE (1990) Water economy of granivorous birds: a predictive model. Condor 92:379–392

Mahoney SA, Jehl JR Jr (1985) Physiological ecology and salt loading of California Gulls at an alkaline, hypersaline lake. Physiol Zool 58:553–563

Martinez del Rio C, Schondube JE, McWhorter TJ, Herrera LG (2001) Intake responses in nectar feeding birds: digestive and metabolic causes, osmoregulatory consequences, and coevolutionary effects. Am Zool 41:902–915

Matoltsy AG, Bereiter-Hahn J (1986) Introduction. In: Bereiter-Hahn J, Matoltsy AG, Richards KS (eds) Biology of the integument. 2. Vertebrates. Springer, Berlin, Germany, pp 1–7

McKechnie AE, Wolf BO (2004) Partitioning of evaporative water loss in White-winged Doves: plasticity in response to short-term thermal acclimation. J Exp Biol 207:203–210

McKechnie AE, Smit B, Whitfield MC, Noakes MJ, Talbot WA, Garcia M, Gerson AR, Wolf BO (2016) Avian thermoregulation in the heat: evaporative cooling capacity in an archetypal desert specialist, Burchell's Sandgrouse (*Pterocles burchelli*). J Exp Biol 219:2137–2144

McKechnie AE, Gerson AR, Wolf BO (2021) Thermoregulation in desert birds: scaling and phylogenetic variation in heat tolerance and evaporative cooling. J Exp Biol 224:jeb229211

McKinley MJ, Johnson AK (2004) The physiological regulation of thirst and fluid intake. Physiology 19:1–6

McWhorter TJ, Martínez del Rio C, Pinshow B (2003) Modulation of ingested water absorption by Palestine Sunbirds: evidence for adaptive regulation. J Exp Biol 206:659–666

McWhorter TJ, Martínez del Rio C, Pinshow B, Roxburgh L (2004) Renal function in Palestine Sunbirds: elimination of excess water does not constrain energy intake. J Exp Biol 207:3391–3398

McWhorter TJ, Caviedes-Vidal E, Karasov WH (2009) The integration of digestion and osmoregulation in the avian gut. Biol Rev 84:533–565

Medina I, Newton E, Kearney MR, Mulder RA, Porter WP, Stuart-Fox D (2018) Reflection of near-infrared light confers thermal protection in birds. Nat Commun 9:3610

Menon GK, Dumbacher JP (2014) A 'toxin mantle' as defensive barrier in a tropical bird: evolutionary exploitation of the basic permeability barrier forming organelles. Exp Dermatol 23:288–290

Menon GK, Menon J (2000) Avian epidermal lipids: functional considerations and relationship to feathering. Am Zool 40:540–552

Menon GK, Maderson PFA, Drewes RC, Baptista LF, Price LF, Elias PM (1996) Ultrastructural organization of avian stratum corneum lipids as the basis for facultative cutaneous waterproofing. J Morphol 227:1–13

Michaeli G, Pinshow B (2001) Respiratory water loss in free-flying pigeons. J Exp Biol 204:3803–3814

Mizrahi O, Bauchinger U, Aamidor SE, McWilliams SR, Pinshow B (2011) Availability of water affects renewal of tissues in migratory Blackcaps during stopover. Integr Comp Biol 51:374–382

Morgan C, Braun EJ (2001) Glucose handling by the kidney of the domestic fowl. FASEB J 15:A854. (abstract)

Muñoz-Garcia A, Williams JB (2005) Cutaneous water loss and lipids of the stratum corneum in House Sparrows *Passer domesticus* from arid and mesic environments. J Exp Biol 208:3689–3700

Muñoz-Garcia A, Williams JB (2007) Cutaneous water loss and lipids of the stratum corneum in Dusky Antbirds, a lowland tropical bird. Condor 109:59–66

Murrish DE (1973) Respiratory heat and water exchange in penguins. Respir Physiol 19:262–270

Needham J (1931) Chemical embryology. Cambridge University Press, Cambridge

Nicolson SW, Fleming PA (2003) Nectar as food for birds: the physiological consequences of drinking dilute sugar solutions. Plant Syst Evol 238:139–153

Nishimura H, Fan Z (2003) Regulation of water movement across vertebrate renal tubules. Comp Biochem Physiol 136:479–498

Nyström KGK, Pehrsson O, Broman D (1991) Food of juvenile Common Eiders (*Somateria mollissima*) in areas of high and low salinity. Auk 108:250–256

Orosz SE, Echols MS (2020) The urinary and osmoregulatory systems of birds. The Veterinary Clinics of North America. Exotic Anim Pract 23:1–19

Osellame LD, Blacker TS, Duchen MR (2012) Cellular and molecular mechanisms of mitochondrial function. Best Pract Res Clin Endocrinol Metab 26:711–723

Owerkowicz T, Musinsky C, Middleton KM, Crompton AW (2021) Respiratory turbinates and the evolution of endothermy in mammals and birds. In: Dial KP, Shubin N, Brainerd EL (eds) Great transformations in vertebrate evolution. University of Chicago Press, Chicago, IL, pp 143–166

Packard GC, Packard MJ (1984) Nitrogen excretion by embryos of an altricial bird, the Redwing Blackbird (*Agelaius phoeniceus*). J Exp Zool 231:363–366

Pavey CR, Nano CEM (2009) Bird assemblages of arid Australia: vegetation patterns have a greater effect than disturbance and resource pulses. J Arid Environ 73: 634–642

Powers DR, Getsinger PW, Tobalske BW, Wethington SM, Powers SD, Warrick DR (2012) Respiratory evaporative water loss during hovering and forward flight in

hummingbirds. Comp Biochem Physiol A Mol Integr Physiol 161:279–285
Preest MR, Beuchat CA (1997) Ammonia excretion by hummingbirds. Nature 386:561–562
Price ER, Brun A, Caviedes-Vidal E, Karasov WH (2015) Digestive adaptations of aerial lifestyles. Physiology 30:69–78
Purdue JR, Haines J (1977) Salt water tolerance and water turnover in the Snowy Plover. Auk 94:248–255
Qureshi S, Galiveeti S, Bichet DG, Roth J (2014) Diabetes insipidus: celebrating a century of vasopressin therapy. Endocrinology 155:4605–4621
Reece WO (2015) Kidney function in birds. In: Reece WO (ed) Dukes' physiology of domestic animals, 13th edn. Hoboken, NJ, Wiley-Blackwell, pp 193–202
Reyes L, Braun EJ (2005) The functional morphology of the English Sparrow cecum. Comp Biochem Physiol A 141:292–297
Richardson KC, Wooller RD, Casotti G (1991) The relative sizes and asymmetry of kidneys in passerine birds from Australia and North America. J Anat 175:181–185
Ro J, Williams JB (2010) Respiratory and cutaneous water loss of temperate-zone passerine birds. Comp Biochem Physiol A 156:237–246
Romagnani P, Lasagni L, Remuzzi G (2013) Renal progenitors: an evolutionary conserved strategy for kidney regeneration. Nat Rev Nephrol 9:137–146
Romanoff AL (1967) Biochemistry of the avian embryo. Macmillan, New York, NY
Roxburgh L, Pinshow B (2002) Ammonotely in a passerine nectarivore: the influence of renal and post-renal modification on nitrogenous waste product excretion. J Exp Biol 205:1735–1745
Ruben JA, Hillenius WJ, Geist NR, Leitch A, Jones TD, Currie PJ, Horner JR, Espe G III (1996) The metabolic status of some Late Cretaceous dinosaurs. Science 273: 1204–1207
Sabat P (2000) Birds in marine and saline environments: living in dry habitats. Rev Chil Hist Nat 73:401–410
Sabat P, Newsome SD, Pinochet S, Nespolo R, Sanchez-Hernandez JC, Maldonado K, Gerson AR, Sharp ZD, Whiteman JP (2021) Triple oxygen isotope measurements ($\Delta'^{17}O$) of body water reflect water intake, metabolism, and $\delta^{18}O$ of ingested water in passerines. Front Physiol 12:710026
Sapir N, Abramsky Z, Shochat E, Izhaki I (2004) Scale-dependent habitat selection in migratory frugivorous passerines. Naturwissenschaften 91:544–547
Schlöndorff D, Banas B (2009) The mesangial cell revisited: no cell is an Island. J Am Soc Nephrol 20: 1179–1187
Schmaljohann H, Liechti F, Bruderer B (2007) Songbird migration across the Sahara: the non-stop hypothesis rejected! Proc R Soc B 274:735–739
Schmaljohann H, Liechti F, Bruderer B (2009) Trans-Sahara migrants select flight altitudes to minimize energy costs rather than water loss. Behav Ecol Sociobiol 63:1609–1619
Schmidt-Nielsen K (1997) Animal physiology: adaptation and environment, 5th edn. Cambridge University Press, Cambridge, UK
Schmidt-Nielsen K, Hainsworth FR, Murrish DE (1970) Counter-current heat exchange in the respiratory passages: effect on water and heat balance. Respir Physiol 9:263–276
Schulte K, Kunter U, Moeller MJ (2015) The evolution of blood pressure and the rise of mankind. Nephrol Dial Transplant 30:713–723
Senner NR, Stager M, Verhoeven MA, Cheviron ZA, Piersma T, Bouten W (2018) High-altitude shorebird migration in the absence of topographical barriers: avoiding high air temperatures and searching for profitable winds. Proc R Soc B 285:20180569
Sequeira-Lopez MLS, Gomez RA (2021) Renin cells, the kidney, and hypertension. Circ Res 128:887–907
Shih W-Y (2018) Bird diversity of greenspaces in the densely developed city Centre of Taipei. Urban Ecosyst 21:379–393
Shuttleworth TJ, Hildebrandt J-P (1999) Vertebrate salt glands: short- and long-term regulation of function. J Exp Zool 283:689–701
Skadhauge E (1974) Renal concentrating ability in selected west Australian birds. J Exp Biol 61:269–276
Skrip MM, Bauchinger U, Goymann W, Fusani L, McWilliams SR (2015) Access to water affects the condition dependency of nocturnal restlessness in Garden Warblers on a Mediterranean Island stopover. J Ornithol 156:425–432
Smit B, Woodborne S, Wolf BO, McKechnie AE (2019) Differences in the use of surface water resources by desert birds are revealed using isotopic tracers. Auk 136:uky005
Speakman J (1997) Doubly labelled water: theory and practice. Chapman and Hall, London, UK
Stevens BR, Fernandez A, del Rio Martinez C (1988) Angiotensin converting enzyme in brush-border membranes of avian small intestine. J Exp Biol 135: 1–8
Stuart-Fox D, Newton E, Mulder RA, D'Alba L, Shawkey MD, Igic B (2018) The microstructure of white feathers predicts their visible and near-infrared reflectance properties. PLoS One 13:e0199129
Talbot WA, McWhorter TJ, Gerson AR, McKechnie AE, Wolf BO (2017) Avian thermoregulation in the heat: evaporative cooling capacity of arid-zone Caprimulgiformes from two continents. J Exp Biol 220:3488–3498
Talbot WA, Gerson AR, Smith EK, McKechnie AE, Wolf BO (2018) Avian thermoregulation in the heat: metabolism, evaporative cooling and gular flutter in two small owls. J Exp Biol 221:jeb171108
Thornton SN (1986) Osmoreceptor localization in the brain of the pigeon (*Columba livia*). Brain Res 377: 96–104
Tieleman BI, Williams JB (2002) Cutaneous and respiratory water loss in larks from arid and mesic environments. Physiol Biochem Zool 75:590–599

Tsahar E, Martínez del Rio C, Izhaki I, Arad Z (2005) Can birds be ammonotelic? Nitrogen balance and excretion in two frugivores. J Exp Biol 208:1025–1034
Tsurim I, Sapir N, Belmaker J, Shanni I, Izhaki I, Wojciechowski MS, Karasov WH, Pinshow B (2008) Drinking water boosts food intake rate, body mass increase and fat accumulation in migratory Blackcaps (*Sylvia atricapilla*). Oecologia 156:21–30
Votto SE, Dyer FJ, Caron V, Davis JA (2020) Thermally-driven thresholds in terrestrial avifauna waterhole visitation indicate vulnerability to a warming climate. J Arid Environ 181:104217
Wang X, Huang J, Hu Y, Liu X, Peteya J, Clarke JA (2018) The earliest evidence for a supraorbital salt gland in dinosaurs in new Early Cretaceous ornithurines. Sci Rep 8:3969
Williams JB (1996) A phylogenetic perspective of evaporative water loss in birds. Auk 113:457–472
Williams PL, Koenig WD (1980) Water dependence of birds in a temperate oak woodland. Auk 97:339–350
Williams JB, Tieleman BI (2005) Physiological adaptation in desert birds. Bioscience 55:416–425
Williams JB, Siegfried WR, Milton SJ, Adams NJ, Dean WRJ, du Plessis MA, Jackson S (1993) Field metabolism, water requirements, and foraging behavior of wild Ostriches in the Namib. Ecology 74:390–404
Willoughby ES, Cade CJ (1967) Drinking habits of birds in the central Namib Desert of south west Africa, vol 31. Scientific Papers of the Namib Desert Research Station, pp 1–35
Wojciechowski MS, Kowalczewska A, Colominas-Ciuró R, Jefimow M (2021) Phenotypic flexibility in heat production and heat loss in response to thermal and hydric acclimation in the Zebra Finch, a small arid-zone passerine. J Comp Physiol B 191:225–239
Wolf BO, Martinez del Rio C (2003) How important are columnar cacti as sources of water and nutrients for desert consumers? A review. Isot Environ Health Stud 39:53–67
Wolf BO, Walsberg GE (1996) Respiratory and cutaneous evaporative water loss at high environmental temperatures in a small bird. J Exp Biol 199:451–457
Wolf BO, Martínez del Rio C, Babson J (2002) Stable isotopes reveal that saguaro fruit provides different resources to two desert dove species. Ecology 83: 1286–1293
Woods HA, Smith JN (2010) Universal model for water costs of gas exchange by animals and plants. Proc Natl Acad Sci USA 107:8469–8474
Wright PA (1995) Nitrogen excretion: three end products, many physiological roles. J Exp Biol 198:273–281
Xiao J, Liu J (2009) Thermo-fluidic devices and materials inspired from mass and energy transport phenomena in biological system. Front Energy Power Eng China 3: 47–59
Yokosuka M, Hagiwara A, Saito TR, Aoyama M, Ichikawa M, Sugita S (2009a) Morphological and histochemical study of the nasal cavity and fused olfactory bulb of the Brown-eared Bulbul, *Hysipetes amaurotis*. Zool Sci 26:713–721
Yokosuka M, Hagiwara A, Saito TR, Tsukahara N, Aoyama M, Wakabayashi Y, Sugita S, Ichikawa M (2009b) Histological properties of the nasal cavity and olfactory bulb of the Japanese Jungle Crow *Corvus macrorhynchos*. Chem Senses 34:581–593
Zweers GA (1991) Behavioural mechanisms of avian drinking. Neth J Zool 42:60–84

# Energy Balance and Thermoregulation 10

## Contents



**Abstract**

Birds are endotherms and maintain their body temperatures with heat produced by cell metabolism. The energy consumed at rest, or the basal metabolic rate (BMR), varies within and among species, with rates increasing with decreasing body mass. In this chapter, I discuss factors beyond body mass that can influence BMR, such as latitude, as well as seasonal variation in BMR. Also discussed are the ways in which birds maintain their body temperatures when ambient temperatures fall below or rise above thermoneutral zones, ambient temperatures at which no additional


G. Ritchison, *In a Class of Their Own*, Fascinating Life Sciences,
https://doi.org/10.1007/978-3-031-14852-1_10

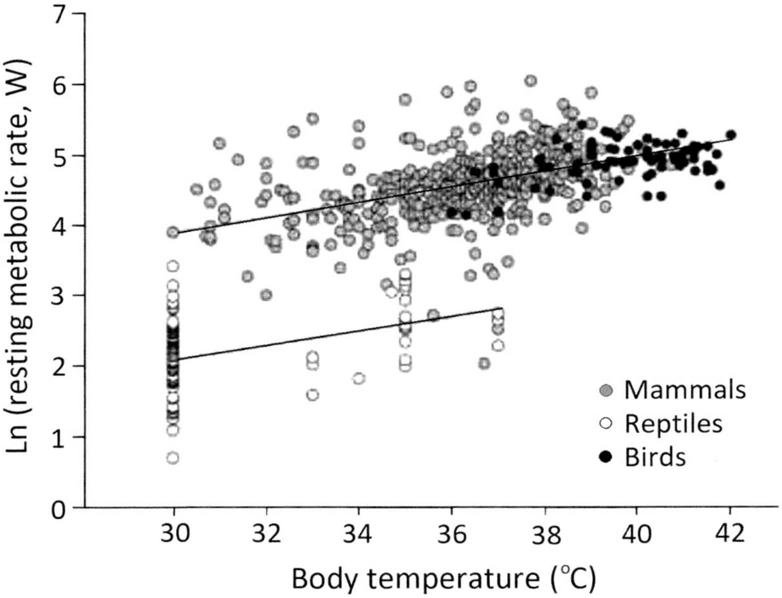


**Fig. 10.1** Resting (basal) metabolic rates of representative species of birds, mammals, and reptiles as a function of body temperature. Resting metabolic rate of a typical bird exceeds that of a typical reptile by a factor of about 15 (for a typical mammal, the factor is about 12). Body temperatures vary among species in all three taxa, and body temperatures of reptiles are, of course, influenced by environmental temperatures. Note that metabolic rates are on a logarithmic scale. (Figure from Clarke and Pörtner 2010; © 2010 The Authors. Biological Reviews © 2010 Cambridge Philosophical Society, used with permission)

energy is expended to maintain body temperatures. Some birds allow their body temperature to either increase (hyperthermia) or decrease (hypothermia), and how and why they do so is explained. Some species of birds allow their body temperatures to drop so low that they become unresponsive, a condition referred to as torpor, and I also explain how and why birds do this.

## 10.1 Evolution of Endothermy

Many vertebrates can produce and retain heat in ways that allow them to maintain relatively high body temperatures over a wide range of ambient temperatures. Birds and mammals differ, however, from other groups (insects, fish, amphibians, and reptiles) because heat is generated, even when resting, by internal organs (e.g., brain, heart, liver, and organs of the digestive system) and so is not dependent on contraction of skeletal muscles (Hulbert and Else 2004). This difference between birds and mammals and, for example, present-day reptiles is apparent when comparing their body temperatures relative to resting (or basal) metabolic rates (Fig. 10.1). Endothermy provides many benefits, such as allowing birds and mammals to quickly engage in and maintain high levels of activity regardless of the ambient temperature, but also comes with an important cost—much more energy (food) is required to fuel the higher temperatures and levels of activity. However, the ecological success of birds and mammals, as indicated by their broad geographic ranges and ability to succeed in a wide variety of habitats, is clear evidence that the benefits of endothermy exceed the costs. What is not so clear is when and how endothermy developed in birds.

Present-day birds diverged from a lineage of theropod dinosaurs during the Jurassic (about

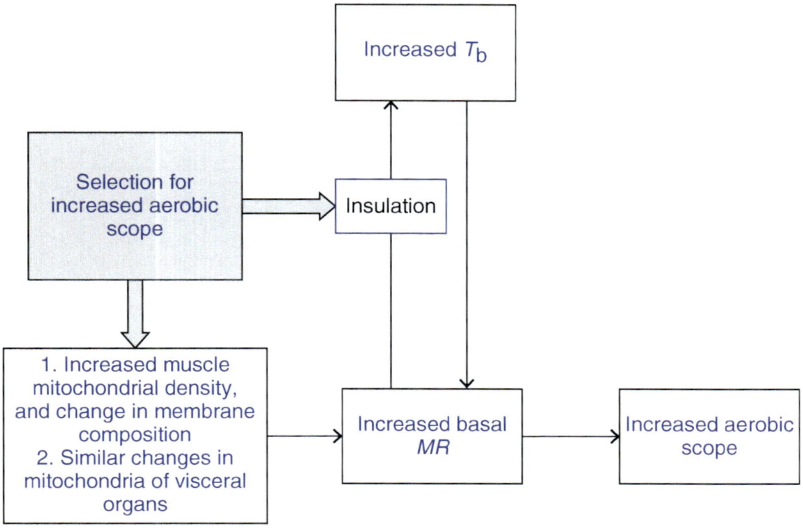


**Fig. 10.2** A conceptual model of the evolution of endothermy, emphasizing that selection for increased aerobic scope (= how far an organism can increase its metabolic rate above the basal metabolic rate) drove changes in both mitochondrial and muscle performance (as well as the number and distribution of mitochondria in different organs) and in insulation to retain metabolic heat. Changes in just one of these two features would have been insufficient to drive the evolution of true endothermy. $T_b$, body temperature, *MR*, metabolic rate. (Figure from Clarke and Pörtner 2010; © 2010 The Authors. Biological Reviews © 2010 Cambridge Philosophical Society, used with permission)

160 million years ago), and fossil evidence suggests that the ancestors of birds were likely active foragers (rather than sit-and-wait predators). Thus, one possible scenario for the evolution of endothermy in birds, based on Clarke and Pörtner (2010), is that natural selection favored the evolution of skeletal muscles in these ancestors that were capable of sustaining activity and high levels of metabolism for longer periods (i.e., aerobic scope; Fig. 10.2), an adaptation that would require muscle cells with greater numbers of, and more efficient (i.e., membranes more efficient at allowing movement of materials needed for synthesis of ATP) mitochondria for more sustained ATP (and, as a by-product, heat) production (Fig. 10.3). Fossil evidence has also revealed that the ancestors of birds had feathers, providing insulation important in retention of the heat generated by the skeletal muscles (Fig. 10.2). A higher body temperature, not just when active, but when resting as well, can also provide benefits. For example, with a higher and more stable resting body temperature, there is less or no need to spend time basking in the sun (to increase body temperature) or seeking shade (to reduce body temperature), providing more time for other activities. In addition, a higher body temperature allows animals to be active during cooler periods of the day, permitting occupation of crepuscular and nocturnal niches (Crompton et al. 1978), and can reduce vulnerability to predators, allowing rapid responses and movement even when environmental temperatures are low. Such advantages may have favored changes in smooth muscle similar to those in skeletal muscle, thus allowing increased heat production by visceral organs and the ability to maintain a higher body temperature even when skeletal muscles were not contracting (i.e., endothermy). The resulting increase in basal or resting metabolic rate would, in turn, allow a further increase in aerobic scope (maximum metabolic rate).

Basal metabolic rate (BMR) is defined as the rate of energy expenditure of a bird that is resting and not expending energy for digestion, growth,

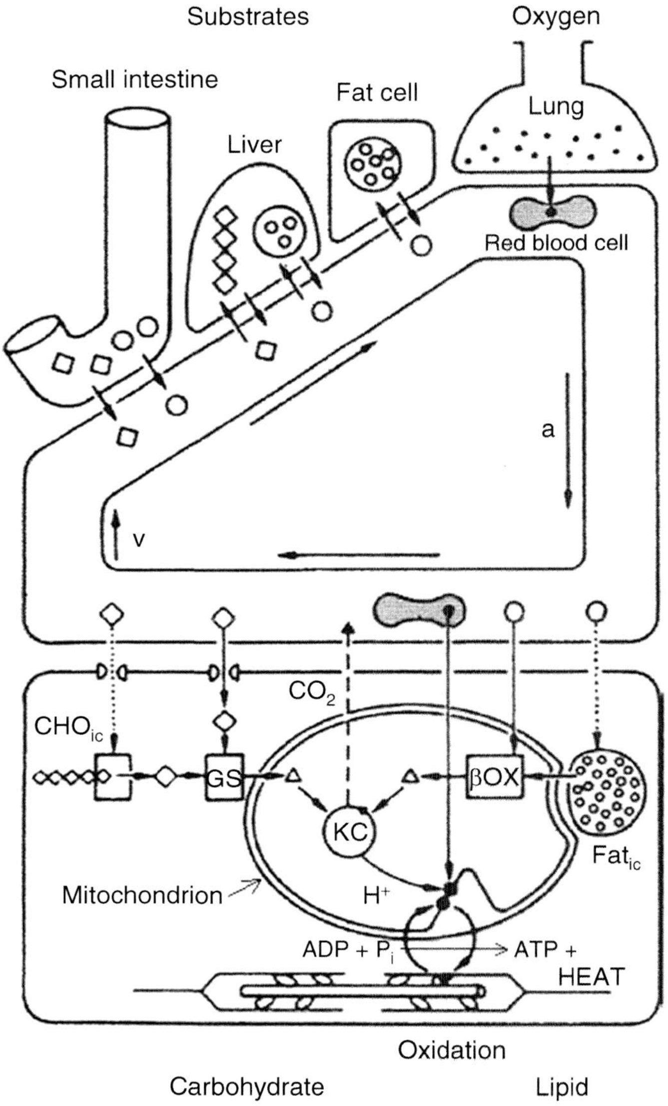


**Fig. 10.3** Illustration of how birds use different substrates as sources of energy and how those substrates are used to generate ATP and heat. Substrates enter the blood via the gut (small intestine) or the liver and fat cells where carbohydrates and fats are stored. Oxygen needed for ATP production enters the blood via the lungs. Substrates and oxygen (transported by hemoglobin in red blood cells) travel through arteries (a) to capillaries and are transported or diffuse into body cells (the rectangular structure at the bottom with the mitochondrion inside), including skeletal muscle cells (represented by the drawing of a sarcomere at the bottom of the figure). ATP is synthesized via glycolysis (GS) in the cytoplasm of body cells and, primarily, via the Kreb's cycle (KC) in each cell's mitochondria. Most ATP is made using carbohydrates and fats (lipids), but, if needed, proteins can also be metabolized and used to make ATP. A by-product of these chemical reactions is heat that helps endotherms like birds maintain their high body temperatures. From the capillaries, blood is transported via veins (v) back to the heart. CHO, carbohydrates; Fat, lipids; βOX, beta oxidation (enzymatic breakdown of lipids); open circles, lipids; open squares, carbohydrates. (Figure from Swanson 2010; ©2010 Springer New York, used with permission)

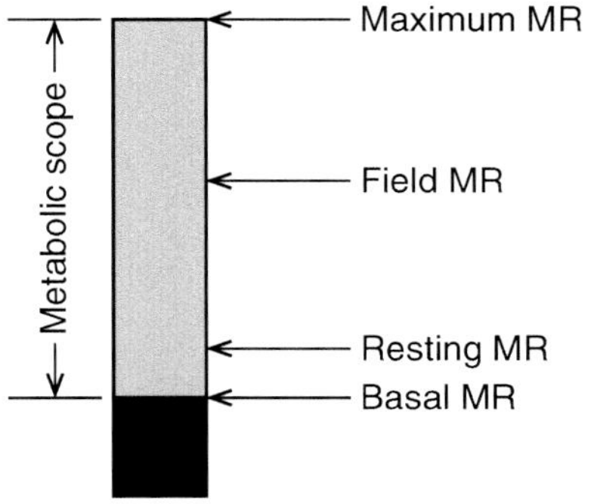


**Fig. 10.4** Relationships between basal metabolic rate, resting metabolic rate, field metabolic rates (during normal daily activity), summit metabolic rate, and maximum metabolic rate. The difference between basal metabolic rate and maximum metabolic rate is a bird's absolute metabolic or aerobic scope. (Figure modified from Clarke and Fraser 2004; © 2004 John Wiley and Sons, used with permission)

reproduction, or thermoregulation. BMR, usually determined by measuring the rate of oxygen consumption, represents a major portion of total energy expenditure of a bird (usually about 30%, Ricklefs et al. 1996). Another measure of metabolic rate is referred to as the resting, or field, metabolic rate (RMR or FMR). RMR is less restrictive than BMR and does not require that a bird is not expending energy for digestion. RMR and BMR, although not exactly the same, are sometimes considered similar enough by some investigators to be used interchangeably (as will be the case in this chapter; Fig. 10.4). A bird's maximal metabolic rate (MMR) represents its maximum rate of energy consumption when active, e.g., during sustained flight. However, MMR may be difficult to measure because of the difficulty of determining if a bird is actually performing at maximum levels. As a result, a more commonly used measure of "maximum" metabolic activity is a bird's maximum metabolic rate during cold exposure, also referred to as the summit metabolic rate (SMR). Both MMR and SMR represent maximum metabolic outputs (from intense physical activity and shivering, respectively), but MMR usually exceeds SMR. Among birds, MMR is generally 7–12 times higher than BMR, whereas SMR is about 4–8 times higher than BMR (Swanson 2010; Piersma 2011; Fig. 10.4). The relationship between BMR and MMR remains unclear, with the results of some studies suggesting that these values are positively correlated in some species of birds (e.g., Rezende et al. 2002), but not others (e.g., Swanson et al. 2012). What is known is that variation in the rate of energy use by birds can reflect different patterns of allocation of resources, time, and energy to growth and body maintenance as well as to reproduction and other activities.

## 10.2 Intraspecific Variation in BMR

Within species, BMR can exhibit significant inter-individual variation. For example, Versteegh et al. (2008) determined the BMRs of 18 European Stonechats (*Saxicola rubicola*) and found extensive individual variation (Fig. 10.5). Some of these variations are likely due to genotypic factors because available evidence suggests that, to varying degrees among species, BMR is heritable. Nilsson et al. (2009) examined the heritability of BMR in a free-living population of Eurasian Blue Tits (*Cyanistes caeruleus*) and estimated a heritability value (i.e., a measure of how much of variation in BMR is due to variation in genetic factors) of 0.59. In free-living European Pied Flycatchers (*Ficedula hypoleuca*), BMR heritability was estimated at 0.43 (Bushuev et al. 2010). Ronning et al. (2007) examined the heritability of BMR in a captive population of Zebra Finches (*Taeniopygia guttata*) and reported a heritability of 0.29. Additional indirect evidence for the heritability of BMR comes from a study that revealed differences in BMR among populations of European Stonechats (Wikelski et al. 2003a). The potential for selective responses to BMR could clearly enhance the fitness of individuals in species where factors that might favor different BMRs (e.g., ambient temperatures) vary spatially (e.g., for different populations of the same species as with European Stonechats; Wikelski et al. 2003a) and/or over time (Nilsson et al. 2009).

Differences in mass explain much of the variation among species in BMR and also explain

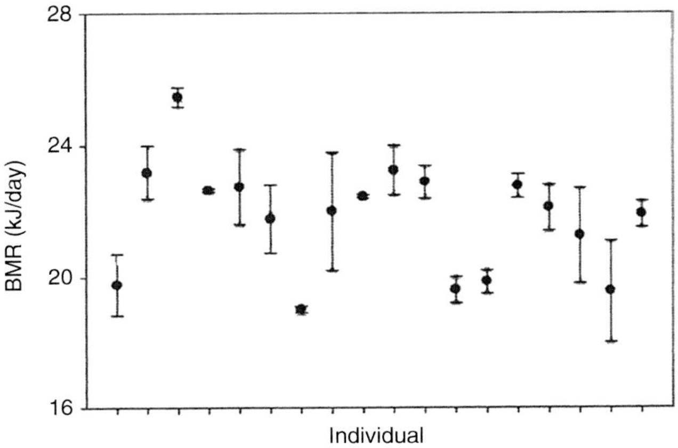


**Fig. 10.5** Individual variation in the basal metabolic rates (BMR) of 18 captive European Stonechats (*Saxicola rubicola*). Means are shown ± standard errors. Measurements were made during the winter, and individual birds were measured two or three times (with a mean time between first and last measurements of 25 days). (Figure from Versteegh et al. 2008; © 2008 Elsevier Inc., used with permission)

much of the variation within species. Significant relationships between body mass and BMR have been reported in several species of birds (e.g., Lindström and Rosén 2002; Broggi et al. 2007, 2009; Versteegh et al. 2008; Sabat et al. 2009). The results of some studies suggest that food habits can also influence BMR. McNab (1986) originally proposed the food habits hypothesis, suggesting that different components of diet, including food quality, food availability, and food predictability, produce individual variation in BMR. Studies testing this hypothesis have provided mixed results, with the results of several studies, but not all, providing support for the hypothesis (Table 10.1). Given the limited number of studies and the lack of any consensus in the results of those studies, additional studies of other species are needed. However, the results of some studies do suggest that changes in diet quality or reductions in food availability may have an effect on BMR. Sabat et al. (2009), for example, observed a decline in the BMR of Rufous-collared Sparrows (*Zonotrichia capensis*) switched from a diet of seeds (high digestibility) to one of mealworms (lower digestibility for seed-eating birds such as sparrows). In addition, the BMRs of Mallards (*Anas platyrhynchos*) were found to decline when placed on a restricted diet (Moe et al. 2005). Such results suggest that, at least for some species of birds, less food or feeding on lower-quality food reduces energy intake and that, in turn, may cause a decrease in BMR, perhaps via changes in organ mass.

Variation in pre-hatching and post-hatching environments may result in interindividual differences in adult phenotypes (a process referred to as developmental programming; McMillen and Robinson 2005), including adult BMRs (Fig. 10.5). For example, Broggi et al. (2022) found that ambient temperatures during the incubation period influenced winter BMRs of Great Tits (*Parus major*), with embryos that experienced lower ambient temperatures having higher BMRs during their first winters. One possible explanation for such results is that lower temperatures may result in increased production of thyroid hormones that, in turn, increase BMRs (Zheng et al. 2014; Ruuskanen et al. 2021; Broggi et al. 2022). Prenatal environments of young birds can also be influenced via maternal effects, with females able to influence deposition of sex hormones and other substances in egg yolk during egg formation (Schwabl 1993). Nilsson et al. (2011b) found that nestling Zebra Finches (*Taeniopygia guttata*) from eggs treated with 500 picograms of testosterone had higher BMRs

**Table 10.1** Summary of studies where investigators tested the food habits hypothesis (FHH) (Table from Bozinovic and Sabat 2010; © 2010 Oxford University Press, used with permission)[a]

| Taxa for FHH | BMR change | Approach | Support | Reference |
|---|---|---|---|---|
| Rufous-collared Sparrow | Increased BMR with increased trophic level | Field and interpopulational data | Yes | Sabat et al. (2009) |
| Rufous-collared Sparrow | Higher BMR with seed diet than with insect diet in some, but not all populations | Experimental studies and interpopulational data | Yes | Maldonado et al. (2009) |
| Pigeons and doves | No change in BMR with diet | Interspecies comparisons; phylogenetic approach | No | Schleucher and Withers (2002) |
| Multi-species (537) | BMR: nectar/pollen eaters > omnivores > frugivores/vertebrate eaters | Interspecies comparisons | Yes | McNab (2009) |
| Birds-of-paradise (Paradisaeidae) | BMR: omnivores and insectivores > frugivores | Interspecies comparisons | Yes | McNab (2003) |
| European Starling | No change in BMR with insect versus fruit diets | Experimental studies | No | Geluso and Hayes (1999) |
| Red Knot | Reduction in BMR when fed mussels compared to trout chow | Experimental studies | Yes | Piersma et al. (2004) |
| Zebra Finch | No change in BMR with diet quality | Experimental studies | No | Bech et al. (2004) |
| Mallard | Decrease in BMR after diet restriction | Experimental studies | Yes | Moe (2005) |
| Multi-species (139) | BMR and FMR were reduced in desert birds compared with Mesic species | Interspecies comparisons; phylogenetic approach | Yes | Tieleman and Williams (2000) |
| Alaudidae (12 species) | BMR decreased along a gradient of increased aridity | Interspecies comparisons; phylogenetic approach | Yes | Tieleman et al. (2003a, b) |
| Zebra Finch | Low quality diet in growing birds produced an elevation of RMR in adults | Experimental study | Yes | Criscuolo et al. (2008) |

[a]Scientific names: Rufous-collared Sparrow, *Zonotrichia capensis*; European Starling, *Sturnus vulgaris*; Red Knot, *Calidris canutus*; Zebra Finch, *Taeniopygia guttata*; Mallard, *Anas platyrhynchos*

than nestlings from untreated eggs and, as adults, these differences remained. The testosterone doses used by Nilsson et al. (2011a, b) were similar to the differences found naturally in eggs laid by female Zebra Finches. In another study of Zebra Finches, Verhulst et al. (2006) found that, when tested as 1-year-old birds, BMRs of young that had been reared in large broods (five or six nestlings) were significantly higher than those of young reared in small broods (two or three nestlings). Other investigators have found that nutritional stress during the nestling period can also result in increased BMRs when birds become adults (Criscuolo et al. 2008; Schmidt et al. 2012). Although based on a limited number of studies, such results suggest that maternal effects and conditions during the nestling period can have long-term effects on metabolic efficiency independent of mass (Verhulst et al. 2006). The long-term effects of such increases in BMR are unclear, but possible impacts are discussed below in the section entitled "Fitness-related effects of individual variation in BMR."

In many species of birds, males, and to a lesser extent, females, display conspicuous plumage or ornamentation, and investigators have, for example, found that indices of immunity vary with the quality or strength of those ornaments (e.g., Dunn et al. 2010). However, little is known about the extent to which variation in the degree of ornamental expression might correspond to variation in either physiological condition or BMR. Among the few studies to date, Kelly et al. (2012) examined this possible relationship in American Goldfinches (*Spinus tristis*) and found that, for both males and females, plumage luminence or brightness was positively related to BMR. Similarly, Viblanc et al. (2016) found that the

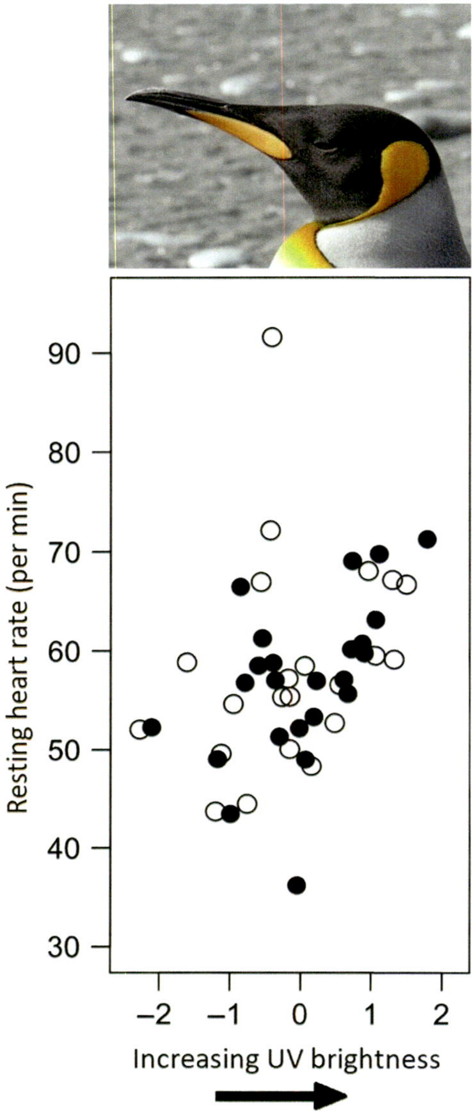


**Fig. 10.6** Relationship between UV brightness of beak spots and resting heart rates of male (filled circles) and female (open circles) King Penguins (*Aptenodytes patagonicus*). Males and females with brighter beak spots had higher resting heart rates, with resting heart rate serving as a qualitative measure of metabolic rate. (King Penguin photo from pxhere.com, CC0 Public Domain; Figure from Viblanc et al. 2016; © 2016 Oxford University Press, used with permission)

ultraviolet brightness of the beak spots of male and female King Penguins (*Aptenodytes patagonicus*) was positively correlated with resting heart rate, with heart rate serving as a measure of metabolic rate (Fig. 10.6). Individuals with higher BMRs tend to have higher activity levels and so may be better able to engage in activities contributing to reproductive success such as competing for territories and mates and provisioning young. If so, then American Goldfinches and King Penguins may benefit by choosing more brightly ornamented individuals as mates. The extent to which similar relationships between plumage quality and BMR may occur in other species of birds remains to be determined.

## 10.3 Relationships Between BMR, Age, and Survival

Studies of the effect of age on the BMR of birds have provided mixed results. BMR was found to decline significantly with age in captive Zebra Finches (*Taeniopygia guttata*; Moe et al. 2009; Rønning et al. 2014), free-living Great Tits (*Parus major*; Broggi et al. 2007; Fig. 10.7), and Thick-billed Murres (*Uria lomvia*; Elliott et al. 2015), increase with age in free-living European Stonechats (*Saxicola rubicola*; Versteegh et al. 2008; Fig. 10.8), and did not change significantly with age in Snow Petrels (*Pagodroma nivea*; Moe et al. 2007). For Thick-billed Murres, Elliott et al. (2015) found that thyroid hormone levels decreased with increasing age, suggesting that a possible reason for declining BMRs with age was hypothyroidism. Elliott et al. (2015) also suggested that a reduction in mitochondrial density in cells with increasing age might contribute to the age-related decline in BMR. Moe et al. (2007) suggested that long-lived species of birds, such as Snow Petrels, may have evolved mechanisms that allow a high degree of somatic maintenance and repair and prevent or delay age-related cellular damage that might lead to declines in BMR.

To explain the increase in BMR with age in European Stonechats, Versteegh et al. (2008) suggested selective mortality, with birds with lower BMRs tending to die when younger. In

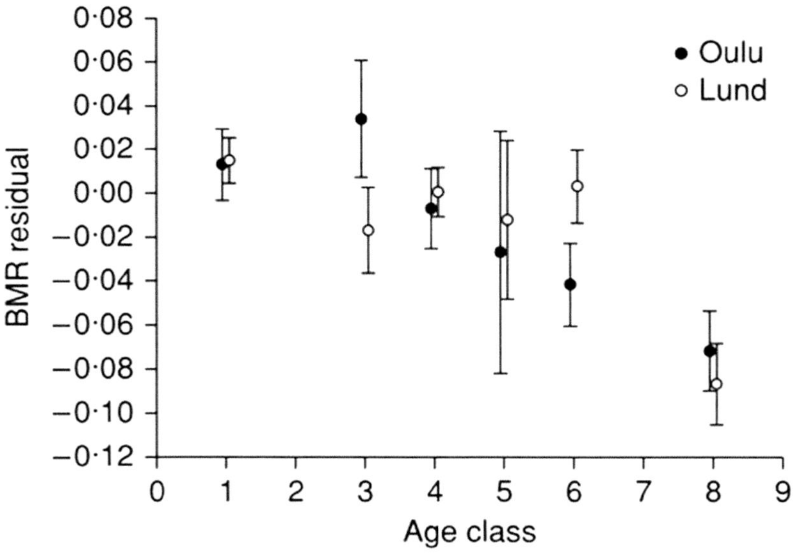


**Fig. 10.7** Mean basal metabolic rates (± SD) of Great Tits (*Parus major*) during the winter in two populations, one in Finland (Oulu) and one in Sweden (Lund), showing a decline with increasing age. Age classes: 1 = first winter, 2 = first winter or older, 3 = second winter, 4 = second winter or older, and so on). Residuals were obtained from each population after regressing all significant predictors except age on BMR values. (Figure from Broggi et al. 2007; © 2007 John Wiley and Sons, used with permission)

contrast, Scholer et al. (2019) examined the relationship between BMR and apparent annual survival for 37 species of tropical songbirds in Peru and found a negative relationship between BMR and apparent survival, i.e., individuals with higher BMRs had lower apparent annual survival. For tropical songbirds, selection appears to favor a slower "pace-of-life," perhaps explaining the greater apparent annual survival for birds with lower BMRs (Scholer et al. 2019). However, Boyce et al. (2020) measured metabolic rates and determined annual survival rates for 46 species of songbirds and found similar data for 147 species of songbirds from a range of latitudes in the literature. For all 193 species combined, these authors found that tropical species had lower metabolic rates than north-temperate species and, in addition, that species with higher metabolic rates had lower survival rates. However, most (but not all) of the variation in survival probability among latitudes was independent of metabolic rates. Such results suggest that the slower "pace-of-life" of songbirds at tropical latitudes may provide at least a partial explanation for the generally longer lifespans of tropical songbirds, but that extrinsic sources of mortality

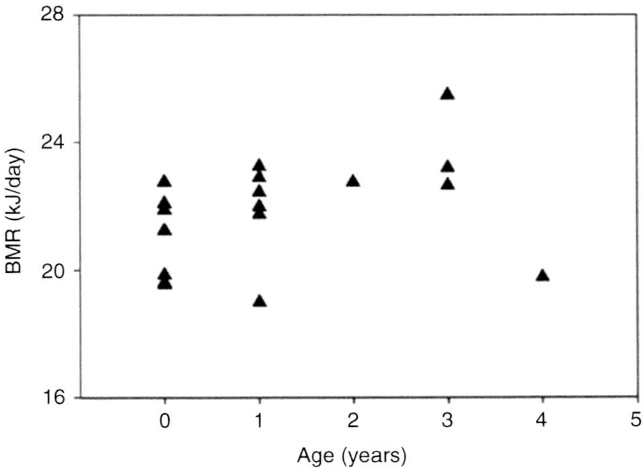


**Fig. 10.8** Basal metabolic rates of European Stonechats (*Saxicola rubicola*) tended to increase with age, perhaps because of the age-related death of individuals with lower metabolic rates. (Figure from Versteegh et al. 2008; © 2008 Elsevier Inc., used with permission)

may be the primary reason for latitudinal differences in survival (Boyce et al. 2020). Concerning the negative relationship between BMR and longevity, one possible explanation is that higher metabolic rates may cause greater cellular oxidative damage via production of reactive oxygen and nitrogen species and result in decreased longevity (e.g., Monaghan et al. 2009; Metcalfe and Alonso-Alvarez 2010; Sies et al. 2017).

## 10.4 Fitness-Related Effects of Individual Variation in BMR

Hypotheses concerning the relationship between BMR and fitness include the compensation hypothesis and the increased intake hypothesis. The compensation hypothesis posits that individuals with low BMRs will have greater fitness because their self-maintenance costs are lower, and they can devote more energy to growth and reproduction (Fig. 10.9). In contrast, the increased intake hypothesis proposes that individuals with high BMRs will have greater fitness because they are able to acquire and assimilate more energy for growth and reproduction (McNab 1980).

These two hypotheses were tested by Blackmer et al. (2005) in a study of Leach's Storm-Petrels (*Hydrobates leucorhoa*). They measured the BMRs and reproductive performance of Leach's Storm-Petrels breeding on Kent Island, New Brunswick, Canada, and, after controlling for internal (body mass, breeding age, and sex) and external (year, date, and time of day) effects on BMR, found that the eggs of males with relatively low BMRs hatched earlier in the season, and their chicks had faster wing growth rates than chicks of males with higher BMRs (Fig. 10.10). So, for male Leach's Storm-Petrels, these results support the compensation hypothesis; males with lower BMRs and lower self-maintenance costs had better reproductive performance than birds with higher BMRs, possibly because they can expend more energy on reproduction. BMR was not related to reproductive performance for female Leach's Storm-Petrels, perhaps because the results of previous studies of species in the same order (Procellariiformes) have revealed that males tend to contribute more effort to reproduction than females (e.g., more

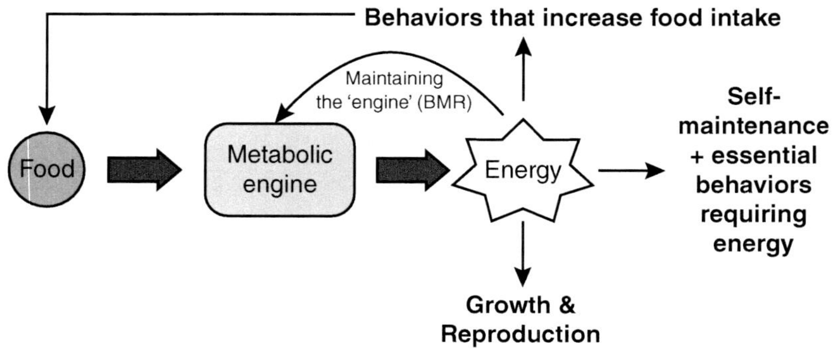


**Fig. 10.9** Relationships between a bird's "metabolic engine," energy, self-maintenance, behavior, and growth and reproduction. Some behaviors such as foraging or defending a territory where foraging takes place can increase food intake that allow a bird's "metabolic engine" (including all organelles, cells, organs, and systems responsible for digesting and metabolizing food) to generate energy; self-maintenance (e.g., preening and molting), other essential behaviors such as courtship simply require energy, and maintaining the "metabolic engine" simply requires energy. Once these latter energetic needs are met, the remaining energy available can be devoted to growth and reproduction. Individual variation in BMR translates into individual variation in the "size" of the "metabolic engine," and individuals with lower BMRs can devote more energy to growth and reproduction. (Figure modified from Biro and Stamps 2010; © 2010 Elsevier Ltd., used with permission)

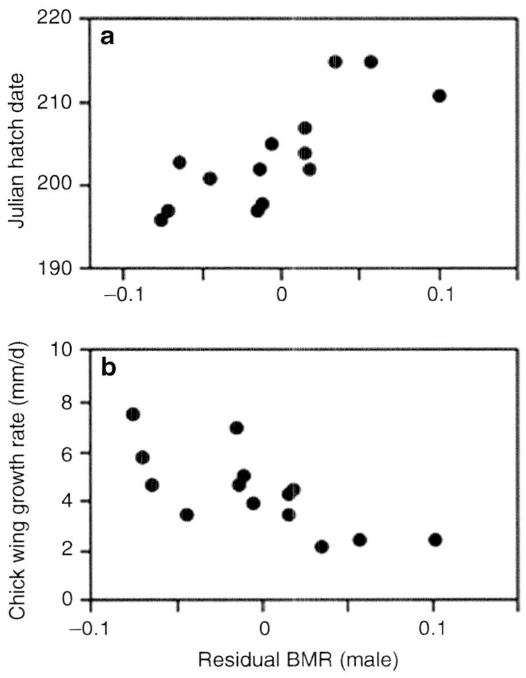


**Fig. 10.10** Relationship between residual BMR (0 = population mean) and (**a**) ordinal hatching date (i.e., day 1 = January 1, . . . day 190 = July 9, and so on), and (**b**) rates of nestling wing growth. The eggs of male Leach's Storm-Petrels (*Hydrobates leucorhous*) with lower BMRs hatched earlier and their nestlings grew faster. (Figure from Blackmer et al. 2005; © 2005 Oxford University Press, used with permission)

time incubating and provisioning chicks at higher rates). As a result, for breeding procellariiforms, male quality and reproductive effort may be more important in determining the reproductive success of pairs (Hatch 1990).

Chastel et al. (2003) examined the relationships between BMR, thyroid hormone, and timing of reproduction in multi-brooded House Sparrows (*Passer domesticus*) and found that, via a positive relationship between BMR and plasma levels of thyroid hormone, males and females with higher BMRs starting breeding earlier and, during the breeding season, raised an average of 2.3 more young than those with lower BMRs. These results support the increased intake hypothesis. In contrast, however, Rønning et al. (2016) found that female House Sparrows with lower BMRs produced more recruits than those with higher BMRs, results that support the compensation hypothesis. Given the conflicting results and the limited number of studies, no conclusions are possible concerning possible relationships between BMR and fitness. However, given that environmental factors such as temperature and food availability can vary during the breeding season and from year to year in most environments and habitats, relationships between BMR and reproductive success (fitness) are also likely to vary. In other words, a single optimal BMR may not exist for most species of birds. Individuals with higher BMRs may have higher reproductive success when conditions (e.g., food availability) are favorable, but those with lower BMRs may do better when conditions are less favorable (Burton et al. 2011).

## 10.5 Interspecific Variation in Basal Metabolic Rates

The greatest source of interspecific variation in basal metabolic rates of birds is body mass. Larger birds use more energy per day than smaller birds (Fig. 10.11). However, metabolic rates per unit mass decrease with increasing body mass (Fig. 10.12). Many investigators have attempted to formalize (and make universal for all birds) this relationship using the equation, $BMR = a \times M^b$, where M is mass, a is the y-intercept, and b is the scaling exponent. Values that investigators have assigned for this scaling exponent have varied, generally ranging between 0.64 and 0.75. The reason for this variation is that there is no single exponent that explains the relationship between the mass and basal metabolic rates for all birds (e.g., White et al. 2007b). This means that other variables must be considered when comparing the metabolic rates of different birds. What the variable scaling components do indicate, however, is that avian metabolic rates increase with decreasing body mass, but not in direct proportion to mass (if in direct proportion, the value of b would be 1.0). In other words, smaller birds use energy at disproportionally higher rates than larger birds.

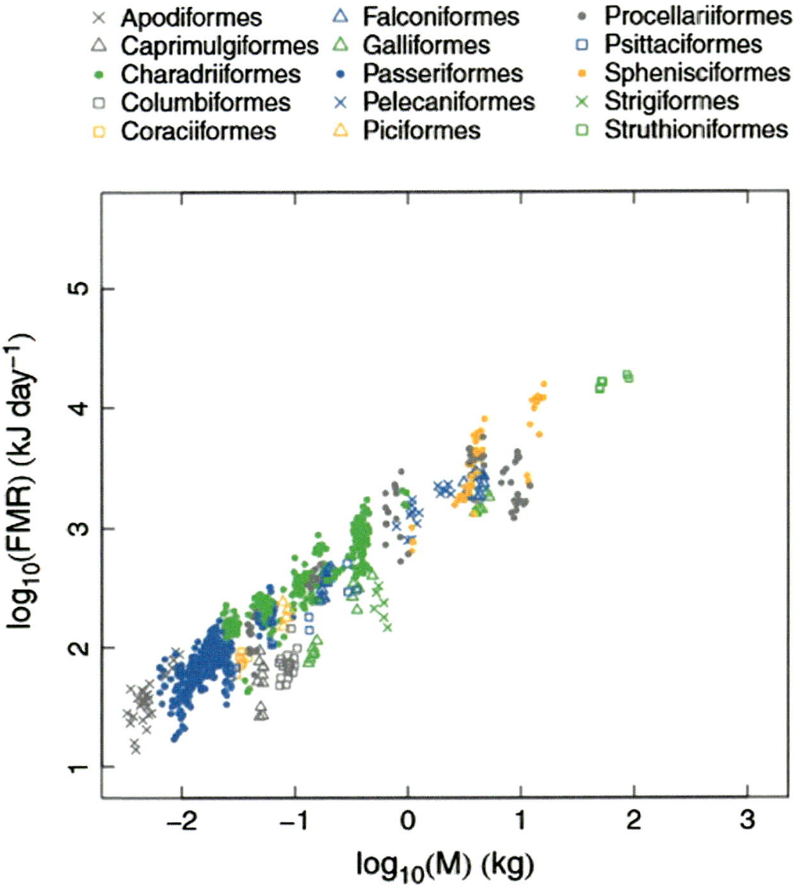


**Fig. 10.11** Field metabolic rates (FMR), in terms of energy use per day, increase with increasing bird mass. Each point is for an individual bird. (Figure from Hudson et al. 2013; © 2013 The Authors. Journal of Animal Ecology © 2013 British Ecological Society, used with permission)

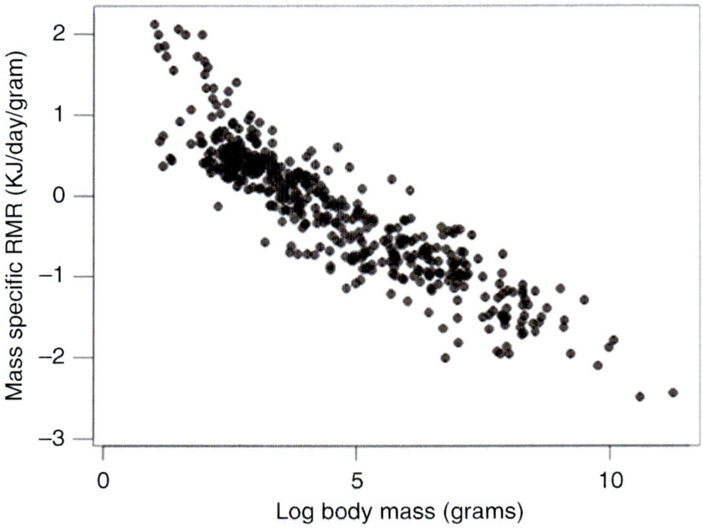


**Fig. 10.12** Relationship between resting metabolic rate (RMR), in terms of energy used per day per gram of body mass, and the body mass of birds (*N* = 459 species). Very small birds, such as hummingbirds (about 2–4 g), have much higher metabolic rates per unit mass than the largest birds, such as Ostriches (≥100 kg). (Figure from Furness and Speakman 2008; © 2008 Springer-Verlag Berlin Heidelberg, used with permission)

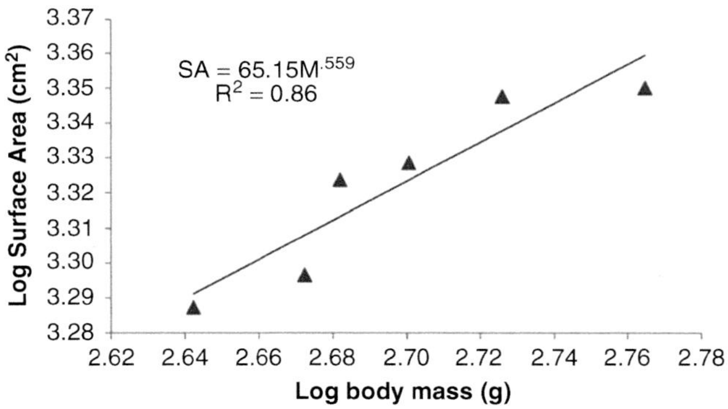


**Fig. 10.13** Relationship between skin surface area (SA) and the body mass for six homing pigeons (*Columba livia domestica*). Even within a species, surface area increased with body mass. (Figure from Perez et al. 2014; open-access article distributed under the terms of the Creative Commons Attribution License, http://creativecommons.org/licenses/by/4.0)

Mass has the greatest effect on avian basal metabolic rates because of the relationship between mass and surface area. A bird's surface is where the exchange of materials involved in metabolism, such as heat, takes place. The scaling exponent that explains the relationship between volume (or mass) and surface area is 0.67 (Walsberg and King 1978; Fig. 10.13), which means that small birds have disproportionately more surface area than larger birds. With more surface area per unit mass, small birds therefore will tend to lose heat faster than larger birds and, as a result, must generate more heat (via metabolism) to maintain their body temperatures.

Although the relationship between body mass and BMR is well established, body mass does not explain all interspecific differences in BMR (Fig. 10.14). BMR is known to be influenced by a number of factors, including the intensity and frequency of flight. Birds that fly more frequently and rely almost exclusively on flapping flight, e.g., ducks, parrots, pigeons, and passerines, generally have higher BMRs than birds that fly less frequently or with less intensity, e.g., owls and soaring hawks (McNab 2019). Other factors that can influence BMRs include latitude, migratory status (migratory or not), and habitat. However, what remains unclear are the underlying causes of

**Fig. 10.14** Relationship between body mass and BMRs of 533 species of birds (log scales). Note that, for particular body masses (e.g., 2.0), there is still much variation among different species in their BMRs, indicating that the body mass alone does not explain all interspecific variation in BMRs. (Figure from McNab 2009; © 2008 Elsevier Inc., used with permission)

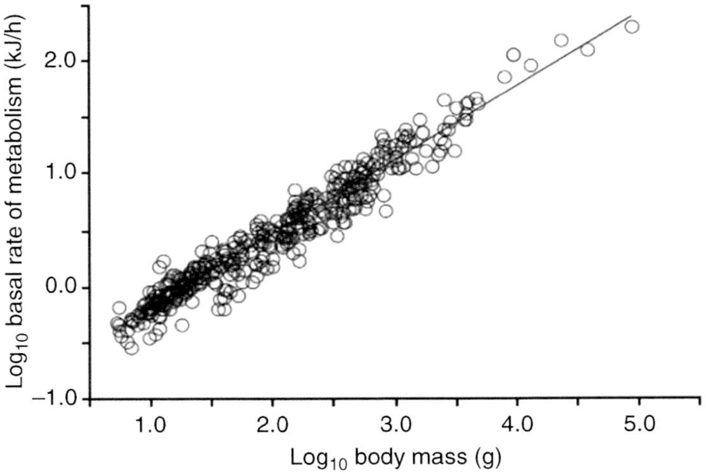

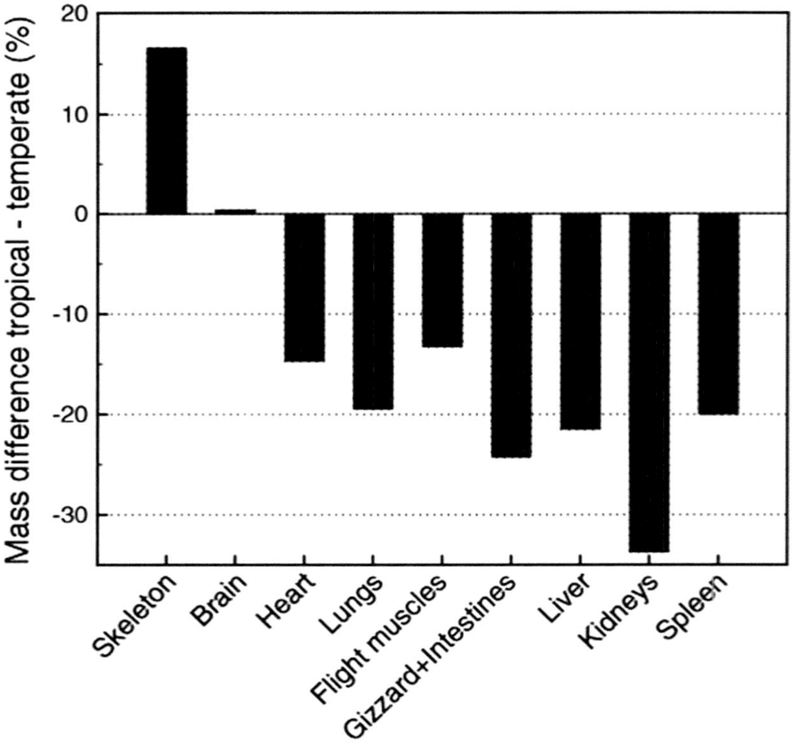


**Fig. 10.15** Relationship between differences in organ masses of tropical and temperate birds with equal body masses. With the exception of the skeleton and brain, organ masses of tropical birds were significantly lower than those of temperate birds. (Figure from Jimenez et al. 2014; © 2014 Springer-Verlag Berlin Heidelberg, used with permission)

variation in BMR, what possible differences among species and populations are there at the organ, tissue, cellular, and/or molecular levels that actually generate interspecific differences in BMR? One possibility is that variation in the mass of internal organs (such as the liver, kidneys, and heart) and the brain is the reason for variation in BMR because those organs are responsible for most metabolic activity when a bird is resting (Hulbert and Else 2000; Nespolo et al. 2011). Although some studies have revealed no evidence that differences in organ mass can explain differences in BMR (Tieleman et al. 2003b), other studies have revealed strong correlations between organ mass and BMR (Daan et al. 1990; Hammond et al. 2000; Williams et al. 2010). Analyses based on data available from previous studies showed that the heart, flight muscles, liver, pancreas, and kidneys were smaller in tropical species than similar-sized temperate species and, in addition, direct measurements of 49 species revealed smaller heart, lungs, liver, kidneys, ovaries, and testes in tropical species (Wiersma et al. 2012; Fig. 10.15). Selective pressures contributing to these differences might include day length (days are shorter in the tropics than during the breeding seasons in temperate areas so tropical species may be active for shorter periods of time) and temperature (warmer temperatures at lower latitudes may lower energy demands and favor downregulation of organ systems) (Wiersma et al. 2012), but, regardless of the selective pressures involved (and they may differ from those influenced by latitude, e.g., habitat or food habit differences), these results suggest a possible link between BMRs and internal organ mass that may help explain differences in the BMRs of similar-sized species.

Differences at the cellular level might also help explain interspecific differences in BMR. Jimenez et al. (2014a, b) cultured cells from 17 phylogenetically paired species of tropical and temperate birds of similar size and found

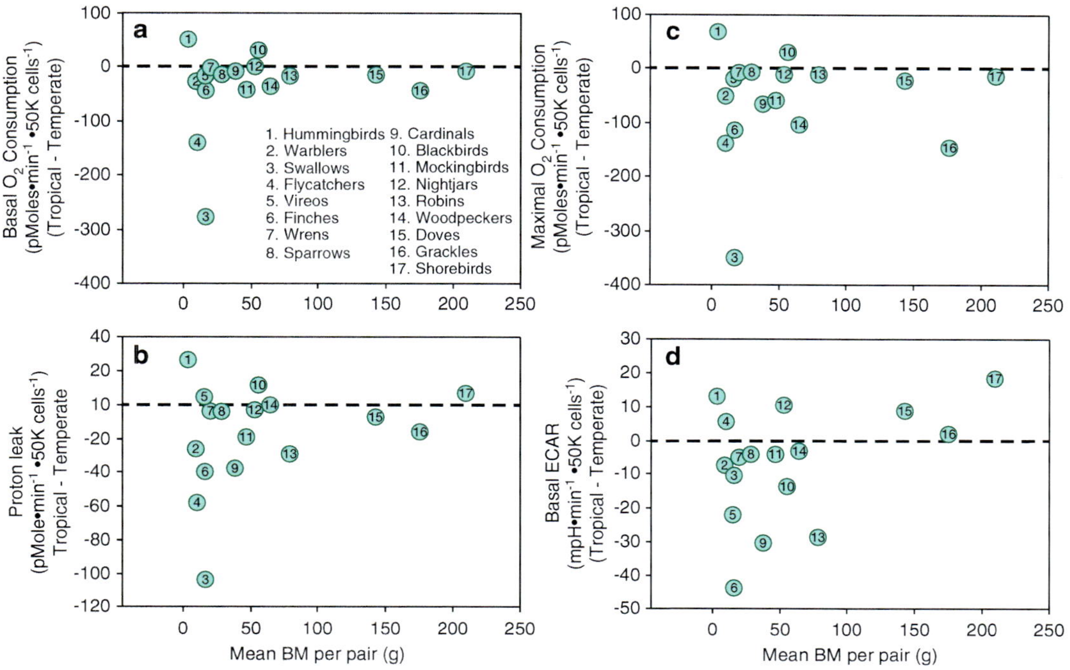


**Fig. 10.16** Differences between mean basal and maximal respiratory rate in tropical and temperate species of birds relative to mean body mass. (**a**) Basal oxygen consumption, (**b**) proton leak rates, (**c**) maximal oxygen consumption, and (**d**) basal glycolytic rates, where the dashed line indicates no difference between tropical and temperate species. Differences between rates were predominantly negative for most species, indicating that tropical values were generally lower than temperate values. Proton leak occurs in mitochondria and may play a role in thermogenesis. ECAR (extracellular acidification rate) provides an estimate of the rate of anaerobic glycolysis in cells. (Figure from Jimenez et al. 2014b; © 2014 Jimenez et al., open-access article distributed under the terms of the Creative Commons Attribution License, https://creativecommons.org/licenses/by/4.0/)

that several measures of cellular metabolic rates were significantly lower for the tropical species (Fig. 10.16). As with organ mass, the selective pressures contributing to these differences may be driven by latitudinal differences in day length, temperature, and other variables, but, more generally, these differences suggest that differences in cell metabolic rates may help explain interspecific differences in BMR of similar-sized birds that might be driven by other factors such as habitat or food habits.

As the preceding paragraphs clearly indicate, avian (and mammal) BMRs exhibit a latitudinal gradient, tending to increase with increasing latitude (Figs. 10.17, 10.18 and 10.19). Wiersma et al. (2007) compared the BMRs of several species of birds in Panama (9°7′N) to those of several species in Ohio (39°57′N) and found a clear trend, with those from Ohio (temperate) generally having higher BMRs than those in Panama (tropical) (Fig. 10.18). A bird's metabolic rate, or rate of energy consumption, influences biological processes such as growth rates, reproductive rates, and even life span. As such, the lower BMRs of tropical birds suggest a slower "pace of life," with slower growth, lower reproductive rates, and longer life spans than birds at higher latitudes. In fact, available evidence indicates that tropical birds do exhibit a slower "pace of life" than birds at higher latitudes. One possible contributing factor to this latitudinal difference is that tropical birds tend to live longer than temperate birds (Møller 2007), and this can influence evolutionary trade-offs between investment

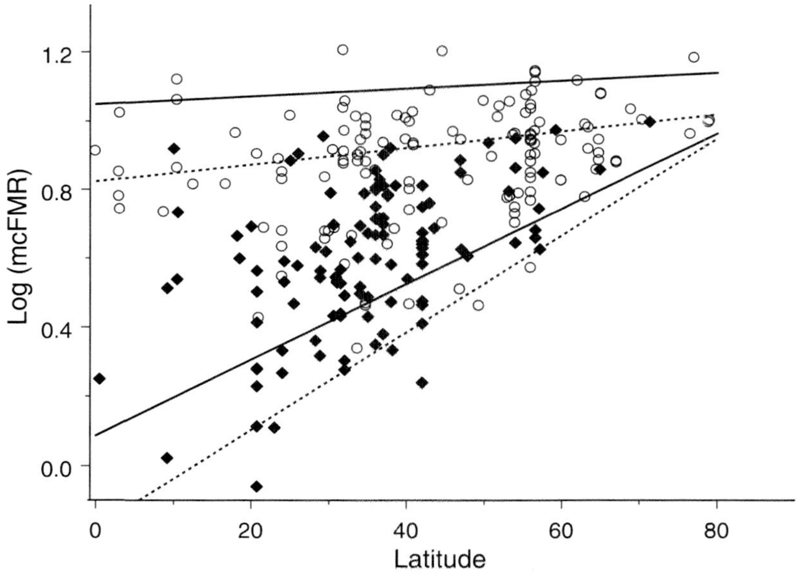


**Fig. 10.17** Latitudinal gradient in mass-corrected field metabolic rates (log) of birds (open circles) and mammals (solid diamonds) (mcFMR = FMR/$M^{3/4}$). Each symbol (circle or diamond) represents a different species. Lines (solid: birds, dashed: mammals) indicate the 0.05th and 0.95th regression estimates of the slope of the relationship. Data from Northern and Southern hemispheres are combined. (Figure from Anderson and Jetz 2005; © John Wiley & Sons Ltd./ CNRS, used with permission)

in each reproductive event and investment in self maintenance (Ricklefs 2000). With lower mortality rates, tropical species can invest less time and

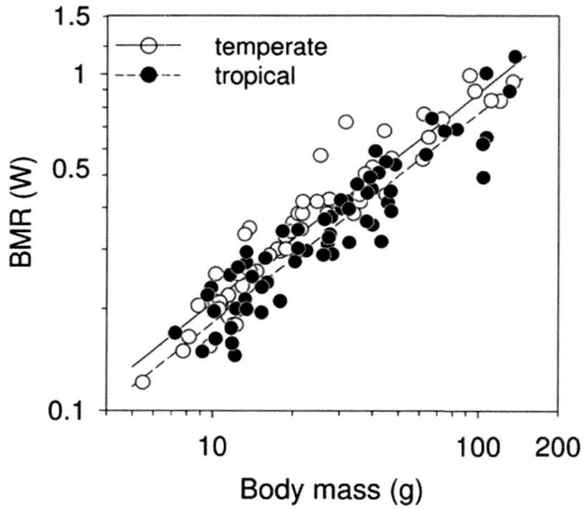


**Fig. 10.18** Basal metabolic rates ($\log_{10}$, and expressed in terms of watts, W) as a function of body mass ($\log_{10}$) for species of tropical ($N = 69$) and temperate ($N = 59$) birds. Each circle represents a single species. Note that BMRs are not expressed as raw values, but in terms of energy per unit mass. Larger birds do, of course, use more energy than smaller birds (i.e., large birds consume more food than smaller birds), but, as noted previously, smaller birds do use more energy per unit mass. (Figure from Wiersma et al. 2007; Copyright (2007) National Academy of Sciences, U.S.A, used with permission)

energy to reproduction (e.g., smaller clutches) because they are more likely to have additional breeding opportunities (and additional opportunities to maximize lifetime fitness) than temperate species. The low basal metabolic rates of tropical species could also be due to the low demands for thermogenesis in warm stable climates (McNab 2016; Bushuev et al. 2018).

Phylogeny, migratory status (migrant or not), habitat, and whether birds fly or are flightless are additional factors that can influence avian BMRs (McNab 2009). Comparisons of the BMRs of birds in different orders are limited by the availability of relevant data. For example, McNab (2009) compared the BMRs of birds in 26 different orders, but found data for relatively few species in many of those orders, e.g., only one species in the orders Podicipediformes (grebes) and Pteroclidiformes (sandgrouse). However, using available data, McNab (2009) found significant variation among orders in BMRs. Species in the order Passeriformes had among the highest BMRs (Fig. 10.20), but species in the orders Anseriformes, Charadriiformes, Pelecaniformes, and Procellariiformes (and not considering orders represented by a single species) had BMRs that did not differ significantly from those of passerines. Species with the lowest BMRs were

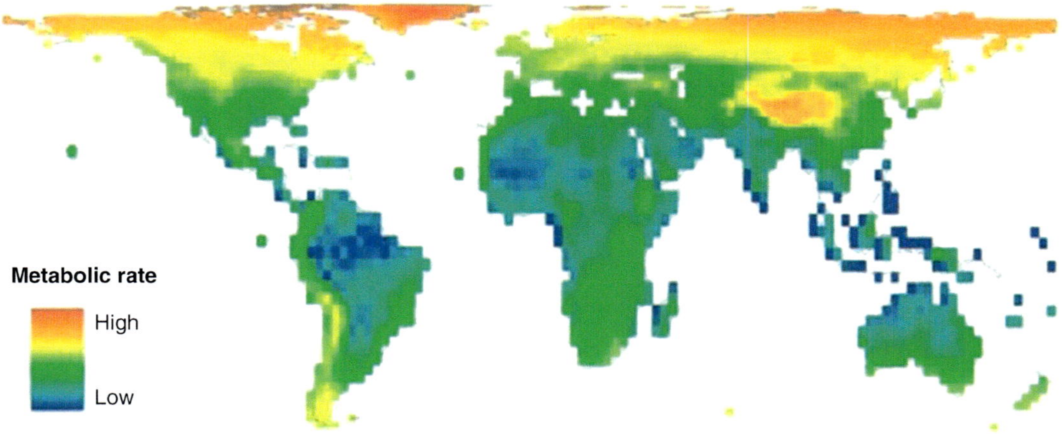


**Fig. 10.19** Variation in the basal metabolic rates of non-migratory birds with latitude. Metabolic rates of birds decrease with decreasing latitude, an indication of the slower "pace-of-life" of tropical and subtropical birds. (Figure from Buckley et al. 2012; © 2012 Blackwell Publishing Ltd., used with permission)

in the orders Caprimulgiformes (frogmouths and goatsuckers) and Coliiformes (mousebirds). Species in orders with BMRs comparable to passerines are all aquatic and high mobile and breed at polar or temperate latitudes, characteristics generally associated with higher BMRs (McNab 2009). Species in the order Caprimulgiformes are relatively large aerial insectivores that forage primarily during dawn, dusk, and at night. Feeding on aerial insects,

**Fig. 10.20** Passerines generally have higher basal metabolic rates than non-passerines. For example, the basal metabolic rates of most passerine frugivores (open circles, $N = 18$) are higher than those of non-passerine frugivores (filled circles, $N = 27$). (Figure from McNab 2016; © 2015 Published by Elsevier Inc., used with permission)

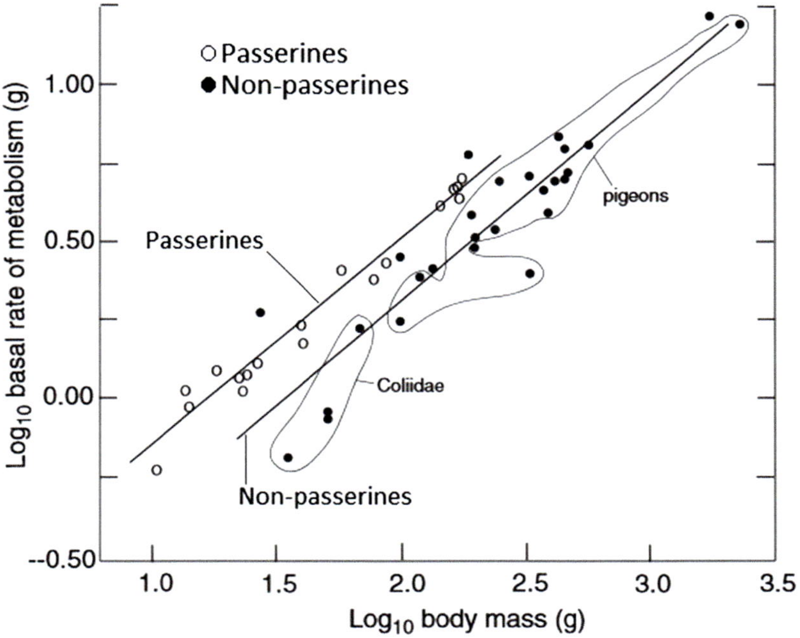

**Fig. 10.21** Red-faced Mousebird (*Urocolius indicus*). (Photo by Derek Keats, Wikipedia, CC BY 2.0, https://creativecommons.org/licenses/by/2.0/)

especially during periods of the day when temperatures are typically lowest, increases the likelihood of occasional food shortages because availability of insects is influenced by ambient temperature. Given their body size, the BMRs of caprimulgids are much lower than expected. For example, the BMR of Eastern Whip-poor-wills (*Antrostomus vociferus*) was found to be just 59% of that predicted based on their mass (Lane et al. 2004). With BMRs more typical of other birds their size, caprimulgids would likely, at times, be unable to meet their energetic needs. As such, the lower metabolic rates of caprimulgids and their ability to use daily hypothermia to conserve energy are evolutionary adaptations needed to occupy their unique niche.

Mousebirds (Coliiformes; Fig. 10.21) are a phylogenetically ancient order of birds found only in sub-Sahara Africa. The six species in this order are all highly social and exhibit huddling, or clustering, behavior. When clustering, groups of up to 26 individuals form a tight cluster on perches (Fry et al. 1988). Mousebirds are found in arid environments where temperatures can fluctuate widely, and their diet consists of fruits plus a wide variety of more difficult to digest plant materials, including leaves. Digestion of plant materials can be energetically costly, requiring fermentation and associated bacterial flora for breaking down cellulose, and takes time, meaning that energy is released slowly (Downs et al. 2000). Given their arid environment with low productivity and fluctuating temperatures, a diet of plant materials, and relatively small size (about 50 g), mousebirds are apparently unable, at times, to generate sufficient energy (and heat) to maintain a body temperature typical of birds their size (about 40°C). So, mousebirds have lower BMRs than most birds and are heterothermic, meaning that their body temperatures can fluctuate between about 28 and 38°C during day and night in response to variation in food availability and ambient temperatures (McKechnie et al. 2006). Importantly, however,

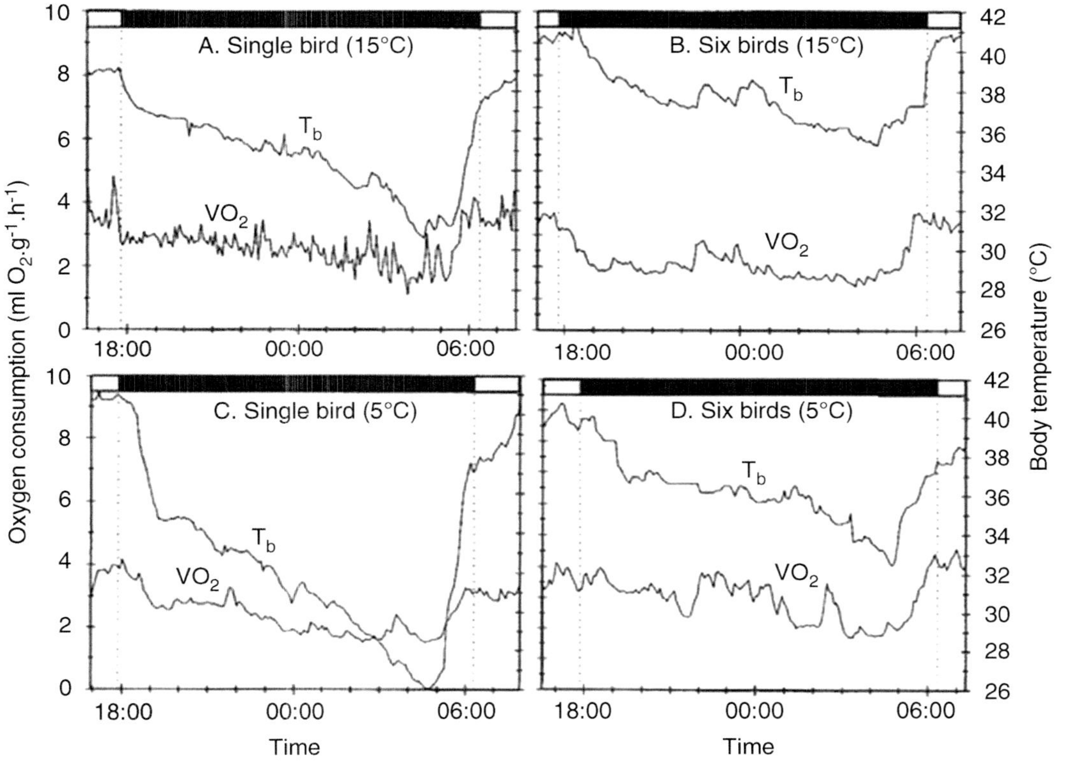


**Fig. 10.22** Oxygen consumption ($VO_2$) and body temperatures ($T_b$) of single and clustering (six birds) White-backed Mousebirds (*Colius colius*) at ambient temperatures of 15°C (**a** and **b**) and 5°C (**c** and **d**). Clustering birds had lower rates of oxygen consumption and maintained higher body temperatures than single birds. (Figure from McKechnie and Lovegrove 2001; © 1999 CCC Republication, used with permission)

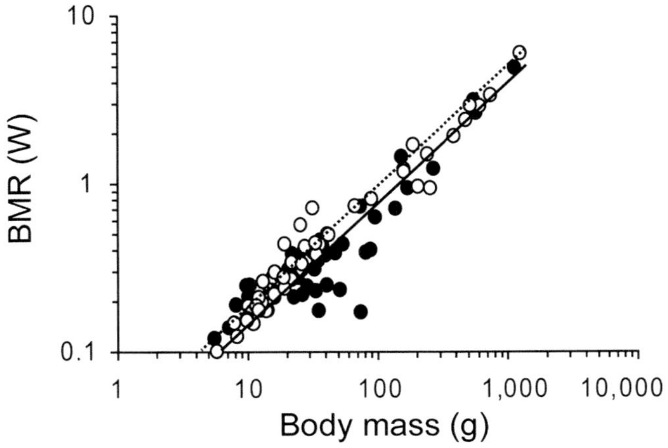


**Fig. 10.23** Basal metabolic rates (expressed in watts) of migratory species of birds (open circles, dotted line) are higher than those of nonmigratory species (closed circles, sold line). (Figure from Jetz et al. 2008; © 2008 Jetz et al., open-access article distributed under the terms of the Creative Commons Attribution License, https://creativecommons.org/licenses/by/4.0/)

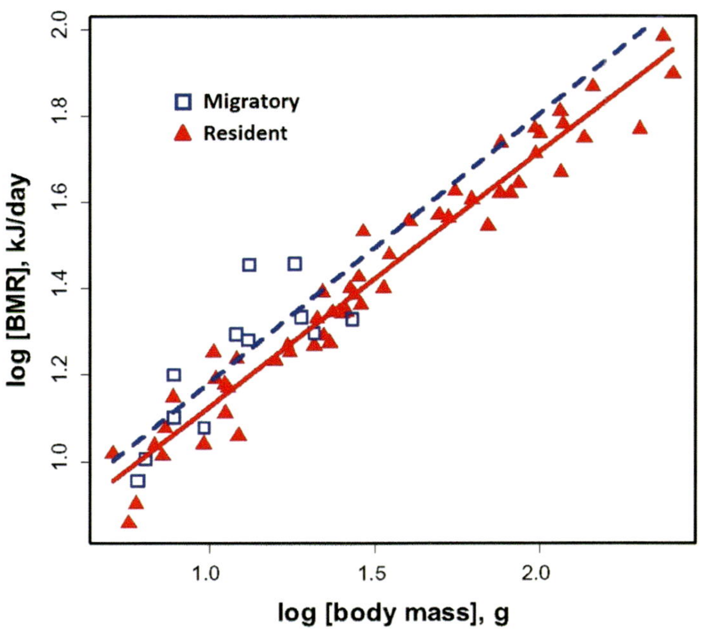


**Fig. 10.24** Relationship between BMR and body mass in resident (red line) and migratory (blue dashed line) birds in southern Vietnam. (Figure modified from Bushuev et al. 2018; © 2017 Oxford University Press, used with permission)

by clustering, mousebirds use less energy (as measured by oxygen consumption) and maintain higher body temperatures (McKechnie and Lovegrove 2001; Fig. 10.22).

Migratory status has also been found to influence avian BMRs, with migrants having higher BMRs than nonmigrants (Figs. 10.23 and 10.24). A possible explanation for this is that migration requires elevated metabolism and, given evidence that maximum metabolic rates (MMRs) may, at least for some species of birds, be correlated with BMRs (Lindström and Kvist 1995), selection for higher MMRs has caused a corresponding increase in BMRs (Jetz et al. 2008). Another possibility is that migrants typically breed at higher latitudes where temperatures are cooler and where short breeding seasons can be demanding energetically. These selective pressures may in turn favor higher BMRs and MMRs (Jetz et al. 2008).

Habitat, or environmental, characteristics can also influence avian BMRs. Species found in hot arid environments, for example, typically have lower BMRs than species in non-arid environments, with selection favoring a reduction in energy and water requirements with increasing aridity and decreasing primary productivity (Tieleman et al. 2003a). White et al. (2007a) examined relationships between BMRs of 139 species of birds and net primary productivity, precipitation, intra-annual variation in precipitation, mean annual ambient temperature, and annual temperature range, and found that BMR was negatively associated with ambient temperature and annual temperature range and positively associated with intra-annual variation in precipitation. Similarly, using data from 135 species of birds, Jetz et al. (2008) found that average ambient temperature was the most important single environmental predictor of BMR, with BMR significantly lower in warmer environments (Fig. 10.25). In warmer environments, body temperatures are more similar to ambient temperatures and, as a result, lower BMRs reduce both energy expenditure and endogenous heat load (McNab and Morrison 1963).

The results of several studies have also revealed differences in the basal metabolic rates of birds in marine or coastal habitats and birds in inland habitats, with marine and coastal species generally having higher basal metabolic rates (Vander Haegen et al. 1994; Duriez et al. 2010).

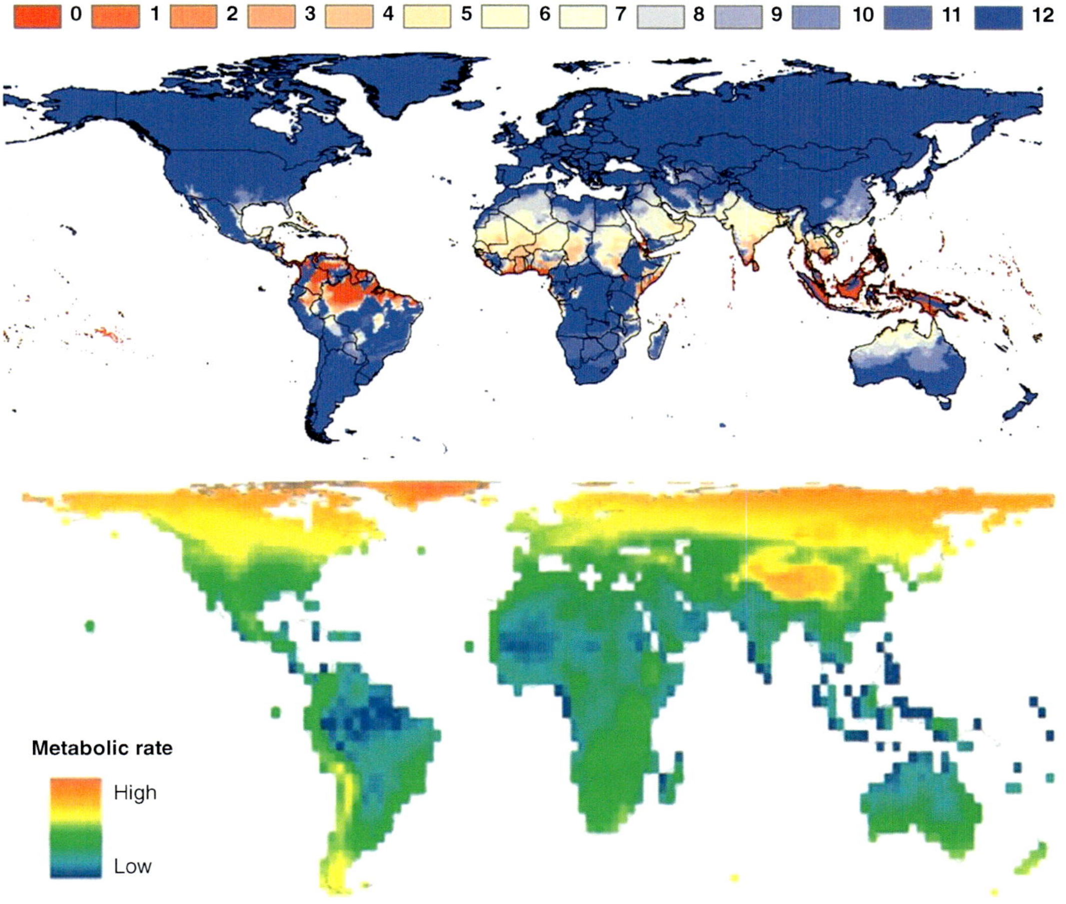


**Fig. 10.25** Top, number of months per year in different areas of the world where the average monthly minimum ambient temperature is less than the predicted lower critical temperature of a modal-sized bird (13 g, Blackburn and Gaston 1994). (Figure from Humphries and Careau 2011). Bottom, geographical variation in the basal metabolic rates of nonmigratory birds. Note the general correspondence between ambient temperatures and basal metabolic rates, with rates higher in areas with lower temperatures and lower in areas with warmer temperatures. (Figure from Buckley et al. 2012; © 2012 Blackwell Publishing Ltd., used with permission)

For example, Gutierrez et al. (2012) compared the basal metabolic rates of shorebirds in temperate inland freshwater habitats with those in coastal habitats and found that inland shorebirds generally had lower basal metabolic rates that those in coastal habitats (Fig. 10.26). Possible explanations for the lower basal metabolic rates of inland shorebirds include a difference in diet, with inland shorebirds ingesting less salt than coastal shorebirds. Coastal shorebirds must invest more in osmoregulation than inland shorebirds, and that investment can increase basal metabolic rates. Gutiérrez et al. (2011), for example, showed that the basal metabolic rates of Dunlins (*Calidris alpina*) increased by 17% when switching from foraging in freshwater habitats to foraging in a coastal environment. Another possible explanation for the lower basal metabolic rates of inland shorebirds is that they occupy less productive habitats than those in coastal habitats and, as a

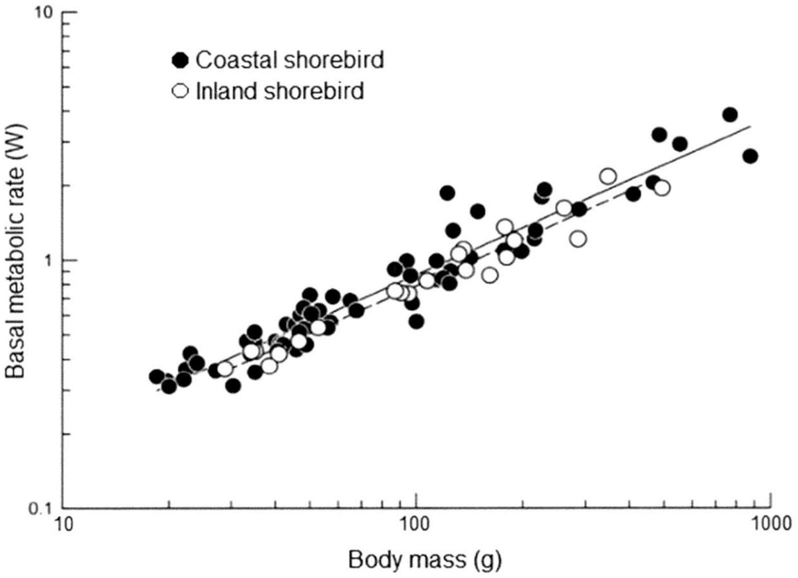


**Fig. 10.26** Basal metabolic rates of shorebirds in coastal habitats ($N = 70$) were found to be higher than those of shorebirds ($N = 22$) in inland habitats. (Figure from Gutierrez et al. 2012; © Gutiérrez et al., open-access article distributed under the terms of the Creative Commons Attribution License, https://creativecommons.org/licenses/by/4.0/)

result, have a slower "pace of living" than coastal shorebirds with access to abundant food (Gutierrez et al. 2012).

Whether or not birds fly also influences BMR. McNab (2009) compared the BMRs of 22 species of flightless birds with those of over 500 species of flying birds and found that flightless birds, with few exceptions, had lower BMRs than flying birds (Fig. 10.27). On average, basal metabolic rates of flightless birds averaged about 74% that of flying birds (McNab 2009). Flying is metabolically costly and, for species of birds that no longer must expend that energy, natural selection has favored a reduction in BMR.

**Fig. 10.27** Basal metabolic rates of flightless birds relative to those of other birds. On average, basal metabolic rates of flightless birds averaged about 74% that of flying birds. (Figure modified from McNab 2009; © 2008 Elsevier Inc., used with permission)

## 10.6 Latitude, Altitude, and Body Size

Ambient temperatures are influenced by a number of factors, with latitude and altitude being the most important. Moving from the equator toward the poles, mean annual temperatures drop by about 1°C every 145 km (Fig. 10.28). This variation in ambient temperatures has influenced the distribution of birds, with birds tending to be larger with increasing latitude (Fig. 10.29). This size gradient is even more pronounced during the winter in the Northern Hemisphere (North America) (Fig. 10.29). The effect of temperature on the size of birds was first noted by Bergmann in 1847 (Bergmann 1847) and this phenomenon is now referred to as Bergmann's rule. This pattern, with bird size increasing with latitude, is driven primarily by winter temperatures that generate selection pressures on birds to become larger. Larger birds have relatively less surface area per unit mass and, therefore, are better able to retain heat when temperatures are low. The body size gradient is less apparent during the summer because many smaller species migrate into the temperate zone to breed. The gradient is also less apparent in the Southern Hemisphere (South America; Fig. 10.29) because temperatures are not as low at higher latitudes as they are in North America, particularly during the winter. The mean January temperature in lowland northern Canada is −32 to −36°C, whereas mean July temperature in lowland southern Argentina is 1–3°C (Leemans and Cramer 1991).

Surprisingly, birds tend to be smaller at higher altitudes, particularly in tropical mountains such as the Andes in South America (Ramirez et al. 2007). Two factors may help explain this altitudinal gradient. First, even at relatively high altitudes, temperatures in tropical mountains remain higher than temperatures at high latitudes (Fig. 10.30). As such, there is little selection pressure for increasing size. In addition, avian species richness is very high in the Andes Mountains of low-latitude South America, with 650 to as many as 845 species in areas along the Andean arc (Fig. 10.31). Species may tend to be smaller in areas of high species richness because of factors related to community assembly (Brown and Nicoletto 1991; Meiri and Thomas 2007), rather than by direct adaptation to lower temperatures. Species richness influences body-size patterns

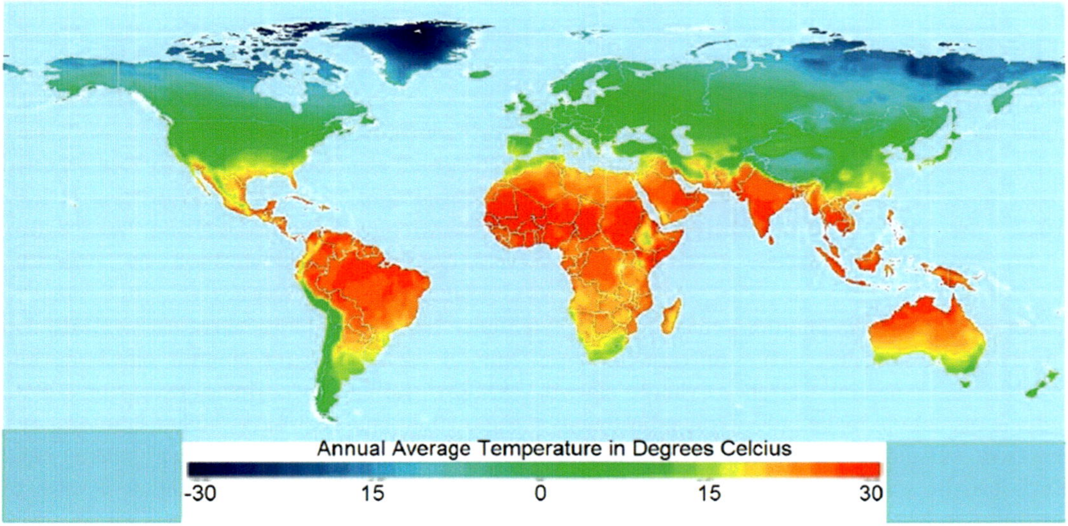


**Fig. 10.28** Variation in mean annual temperature with latitude. (Figure from Ooi and Noguchi 2017; © 2017 by the Authors. Licensee MDPI, Basel, Switzerland, open-access article distributed under the terms of the Creative Commons Attribution license, https://creativecommons.org/licenses/by/4.0/)

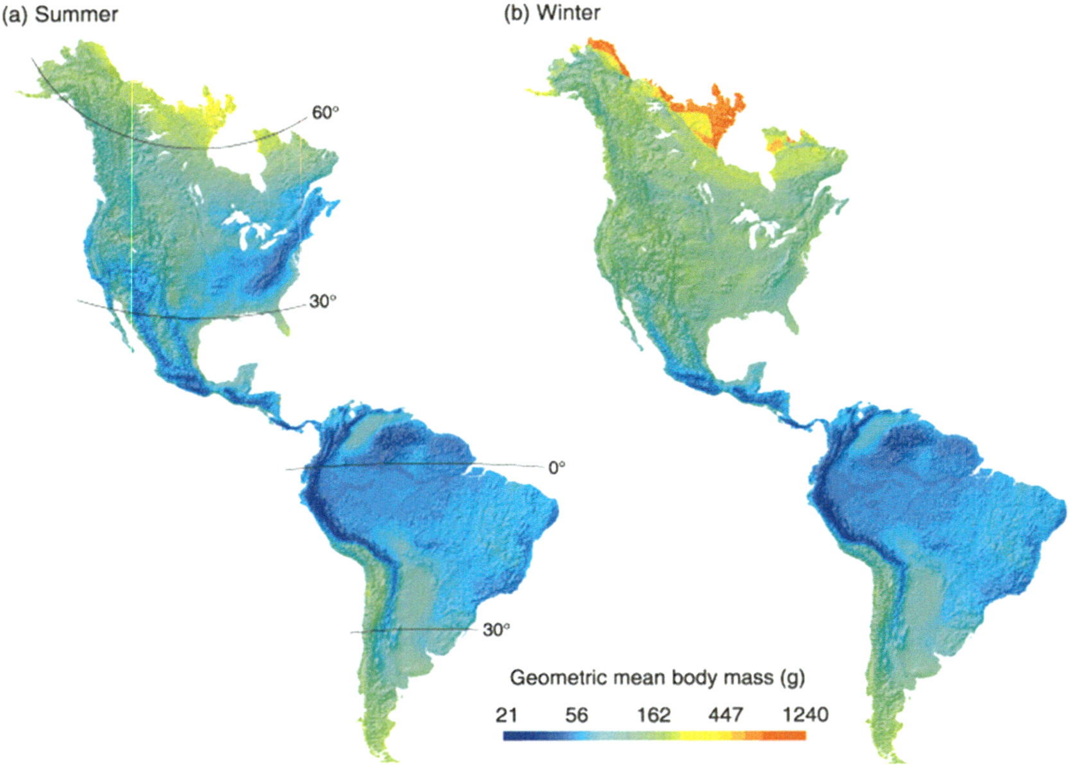


**Fig. 10.29** Mean masses of birds ($N = 3852$ species) during the (**a**) breeding and (**b**) nonbreeding seasons. Birds are larger at higher latitudes, with the gradient stronger during the winter. However, birds are smaller at higher altitudes, especially in tropical mountains such as the Andes Mountains in western South America. (Figure from Ramirez et al. 2007; © John Wiley & Sons Ltd., used with permission)

because species richness increases mainly by addition of small species (Cardillo 2002).

## 10.7 Phenotypic Flexibility in Avian Metabolic Rates

When environmental conditions change over relatively short timescales, such as different seasons of the year, natural selection should favor individuals able to make phenotypic adjustments that will enhance their likelihood of survival. Birds have been found to exhibit such phenotypic flexibility in behavior, morphology, and physiology, including BMR and summit metabolism. Much of the research focusing on flexibility in avian BMR has involved seasonal acclimatization and the physiological adjustments associated with long-distance migration (Box 10.1 Acclimatization vs. Acclimation).

Seasonal changes in BMR have been reported in several species of birds representing at least seven different orders (McKechnie 2008) (Box 10.2 Seasonal Acclimatization by American Goldfinches). Many species of birds found at higher latitudes exhibit increases in BMR and summit metabolism (or summit metabolic rate [SMR]) during the winter (Swanson 2010; Figs. 10.32 and 10.33). At higher latitudes, evidence suggests that birds can make rapid adjustments (i.e., within as little as 7 days or less) to BMR and summit metabolism, with adjustments made as needed to meet current conditions (Swanson and Olmstead 1999; Broggi et al. 2007). Metabolic rates typically increase with decreasing temperatures and vice versa,

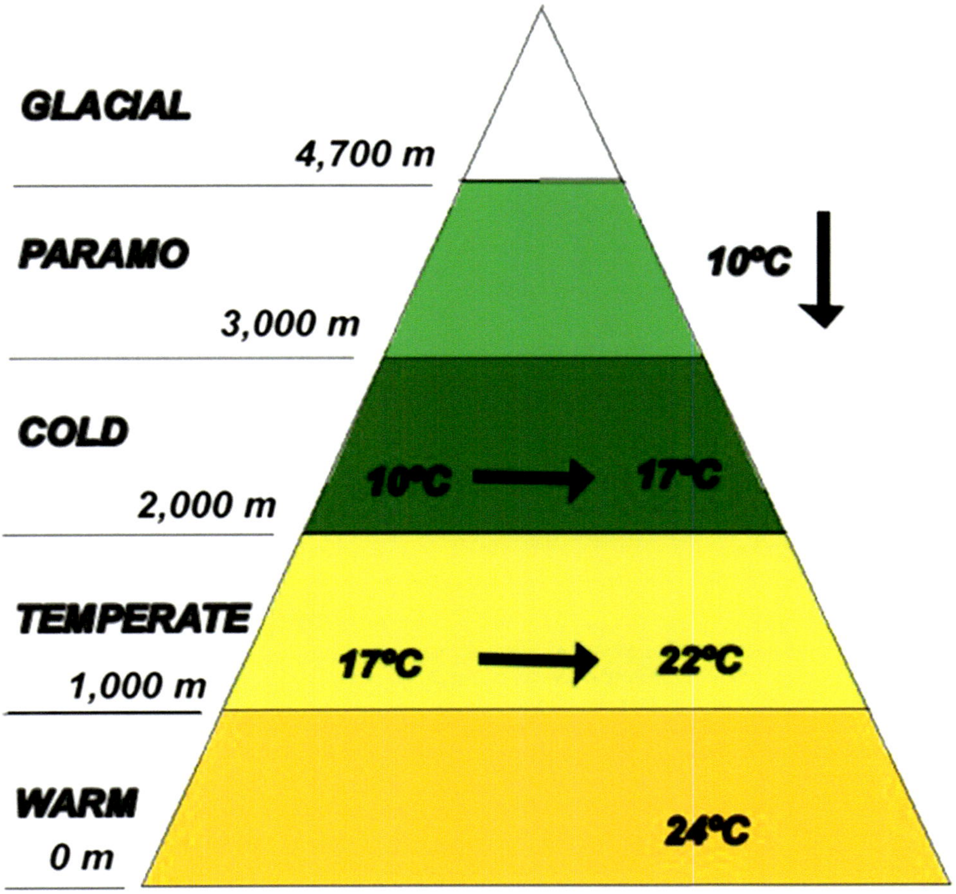


**Fig. 10.30** Typical temperatures with increasing elevation in the Andes Mountains of tropical South America. Even at altitudes as high as 3000 m, temperatures generally remain above freezing. Cloud forests are found at cold elevations; these forests consist of small trees and a diversity of bryophytes, lichens, ferns, bromeliads, and orchids (Hamilton et al. 1995). The Neotropical paramo is above the forest line and the primary vegetation in giant rosette plants, shrubs, and grasses (Luteyn 1999). (Figure licensed under the Creative Commons Attribution 3.0 Unported license, https://creativecommons.org/licenses/by/3.0/)

and available evidence, although based on just 17 species of birds, suggests that this may occur regardless of a bird's size or mass (Fig. 10.32). Winter increases in summit metabolic rates typically range from about 10 to 50% for small birds (Swanson 2010) and, as a result, birds have improved cold tolerance and enhanced capacities for thermogenesis (heat production).

Physiological mechanisms of seasonal acclimatization to colder temperatures vary among different species of birds. However, seasonal acclimatization generally involves some combination of the following: (1) an increase in food intake, with a corresponding increase in body mass and, often, an increase in stored fat, (2) an increase in the mass of internal organs (i.e., heart, intestine, liver spleen, and kidneys), (3) an increase in pectoralis muscle mass, (4) adjustments in transport capacities for oxygen and metabolic substrates, including higher concentrations of red blood cells (for more efficient transport of oxygen; Petit and Vézina 2014) and enhanced cell membrane transport of lipids (an important source of energy for bird muscles),

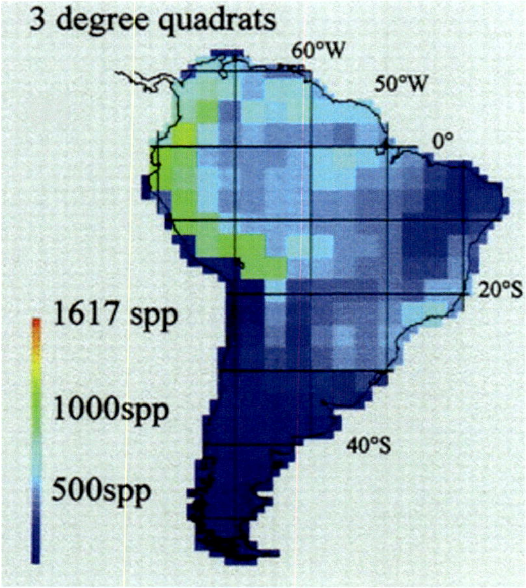


**Fig. 10.31** Spatial variation in species richness of 2869 breeding land and freshwater birds in South America (in blocks 3° × 3°). Note that the highest diversity is found along the Andean arc in northwestern South America. The diversity of different habitats and niches along altitudinal gradients has contributed to this impressive diversity. (Figure from Rahbek and Graves 2001; used with permission of the United States National Academy of Sciences)

and (5), possibly, upregulation of mitochondrial respiration rates and mitochondrial volume in red blood cells (Nord et al. 2021) (Box 10.3 Heat Production by Red Blood Cell Mitochondria). Seasonal increases in BMR are largely due to the increased mass of the internal organs, whereas increases in summit metabolism are primarily due to skeletal muscle activity, especially activity of the pectoralis muscle. Increased food intake provides the energy needed to support the increased mass of birds during winter, and the increased mass of internal organs likely enhances the digestion of and subsequent transport of nutrients throughout the body. The increased mass of the internal organs drives increases in BMR, whereas the increased mass of skeletal muscles and/or the enhanced transport of oxygen and the primary energy sources of skeletal muscles (lipids) drive increases in summit metabolism. The increased mass of the pectoralis muscles and/or enhanced transport of oxygen and lipids allows for increased heat production (thermogenic capacity) via shivering. Availability of lipids is critical for shivering because they supply nearly all the energy necessary to support shivering skeletal muscles, particularly at colder temperatures (Vaillancourt et al. 2005).

Seasonal acclimatization during the winter at higher latitudes also involves increased tolerance to cold, with cold tolerance defined as the length of time over which a bird can maintain its body temperature by thermogenesis (primarily by shivering) at a given temperature before becoming hypothermic (Swanson 2010). Increased cold tolerance during winter can, to some extent, be explained by a denser plumage and vasomotor (or blood flow) changes, but, to a much greater extent, the increased tolerance is a by-product of the increase in BMR and summit metabolism (Swanson 2010).

Most studies of seasonal acclimatization have focused on birds at higher latitudes where temperatures are typically much lower during the winter than the summer. Although less is known about seasonal variation in BMR of birds at subtropical and tropical latitudes, the results of some studies suggest that BMR of birds at these latitudes tends to decline during the nonbreeding, winter period. Wells and Schaeffer (2012) measured the summit metabolism of one temperate species (Tufted Titmouse, *Baeolophus bicolor*) in Ohio (39°N) and seven tropical species in Panama (9° N) during winter (February) and summer (June) and found that the summit metabolism of the temperate species was, as expected, higher during the winter. The tropical species, however, had higher summit metabolism during the summer, likely a result of the greater metabolic demands of breeding. At a subtropical latitude (25°S) in Africa (with the subtropics defined as the areas between about 23°N and S and 35°N and S), Smit and McKechnie (2010) studied five resident species and found that BMRs of all five were lower in winter than summer. During austral winter (June–August), temperatures in subtropical Africa are higher than during winter at higher latitudes and, in addition, birds are generally not breeding. Without the need

**Box 10.1 Acclimatization vs. Acclimation**

Avian physiologists studying the responses of birds to changes in environmental conditions sometimes use the term "acclimatization" and, at other times, the term "acclimation." Although both terms refer to how birds are responding to an environment, the term "acclimation" refers to responses to artificial environments, whereas "acclimatization" refers to responses to natural environments (Piersma and Drent 2003). So, for example, Maldonado et al. (2009) studied physiological responses of Rufous-collared Sparrows (*Zonotrichia capensis*) to thermal *acclimation* and seasonal *acclimatization*. They studied seasonal acclimatization by capturing birds in the wild (10 during summer and 10 during winter), measuring their BMRs at 30°C in the lab the next day, and then released them. To study thermal acclimation, they captured 19 birds, kept them in cages for 30 days at two different acclimation temperatures (10 birds at 15°C [cold-acclimated]; nine birds at 30°C [warm-acclimated]), then measured their BMRs at 30°C. Wild birds were, therefore, tested for seasonal acclimatization, and birds captive for 30 days to temperature acclimation.

Rufous-collared Sparrow. (Photo by Thomas Quine, Wikipedia, CC BY 2.0, https://creativecommons.org/licenses/by/2.0/)

to improve cold tolerance, birds in subtropical Africa are able to reduce BMRs during the winter.

More recently, McKechnie et al. (2015) compiled available data for several species of temperate and tropical/subtropical (between 38°N and 38°S) birds and found that seasonal adjustments in BMR were generally related to temperature and latitude, but adjustements among tropical and subtropical species were much more variable than those of temperate-latitude species (Fig. 10.34). For tropical/subtropical species, winter mass-specific BMRs ranged from 66% of

**Box 10.2 Seasonal Acclimatization by American Goldfinches**

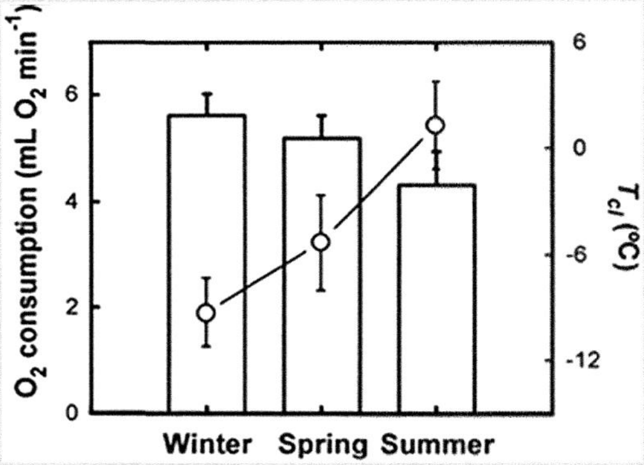


Seasonal variation in summit metabolism and cold tolerance for American Goldfinches (*Spinus tristis*). Bars represent the mean (± SD) summit metabolism achieved by acclimatized American Goldfinches in winter (January–February), spring, and summer (June–August). Mean ± SD temperature at cold-limit (Tcl), the bath temperature at which goldfinches became hypothermic, is represented by circles connected by a line. Note the positive relationship between summit metabolism and cold tolerance in this species. (Figure from Liknes et al. 2002; © 2002 Oxford University Press, used with permission)

Liknes et al. (2002) evaluated seasonal changes in cold tolerance, basal metabolic rate (BMR), and summit metabolic rate (maximum rate of metabolism in response to cold exposure) for American Goldfinches from South Dakota (USA) to determine if goldfinches differ in pattern of metabolic acclimatization from other species of small birds. Goldfinches were captured in winter (January and February), spring (April), and summer (June through August) and tested on the day of capture. Cold exposure tests involved subjecting birds to a decreasing series of temperatures. The temperature eliciting hypothermia was designated the cold limit. Whole-animal metabolic rates were analyzed. During winter, goldfinches had significantly higher BMRs (46%) and summit metabolic rates (31%) and significantly lower cold limits (−9.5 vs. 1.3°C) than during the summer. In the spring, goldfinches also had higher summit metabolic rates (21%) and lower cold limits (−5.3°C) than summer birds. Winter birds had higher BMRs (23%) and summit metabolic rates (8%) than spring birds. In winter birds, the cold limit was also significantly lower than in spring birds. These data support the view that increases in summit metabolic rates and BMR are components of winter acclimatization in American Goldfinches in South Dakota and that seasonal changes in metabolic rates of goldfinches are similar to those of other small wintering birds at temperate latitudes.

(continued)

**Box 10.2** (continued)

American Goldfinch. (Photo by Andy Morffew, CC0 Public Domain, pxhere.com)

summer BMR (Fork-tailed Drongo, *Dicrurus adsimilis*; Smit and McKechnie 2010) to 163% of summer BMR (Amethyst Sunbird, *Chalcomitra amethystine*; Lindsay et al. 2009a, b). One possible explanation from this variation among tropical/subtropical species is the greater diversity of metabolic niches at lower latitudes (McKechnie et al. 2015), ranging from arid deserts to wet rainforests (Lovegrove 2000). In contrast, at higher-latitudes, higher BMRs during the winter are a response to the relatively predictable decline in temperature.

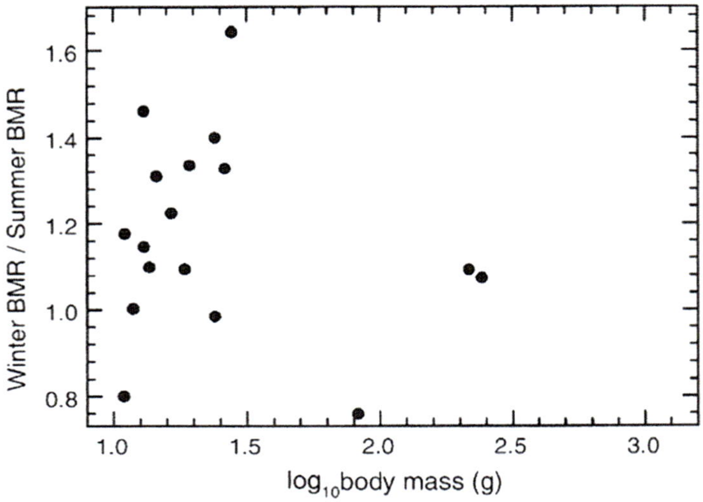


**Fig. 10.32** Relationship between body mass and the ratio of winter to summer BMR for 17 species of birds. Ratios above 1.0 indicate that BMR is higher during winter than summer. (Figure from McKechnie 2008; © 2007 Springer-Verlag, used with permission)

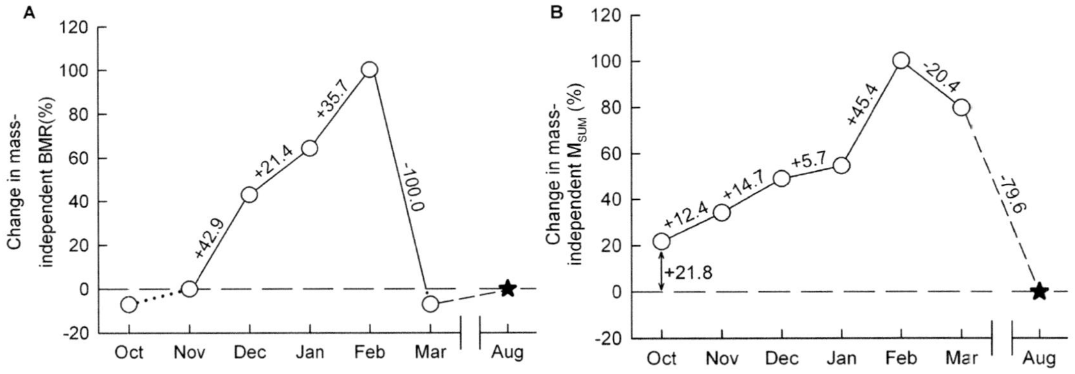


**Fig. 10.33** Changes in the (**a**) BMR and (**b**) summit metabolism of Black-capped Chickadees (*Poecile atricapillus*) during the fall and winter in Canada. Data obtained in August were used as summer reference points (black stars and dashed lines). Monthly changes in metabolic rates (values above lines) indicate percentages of the total difference between summer and maximum winter rates. (Figure from Petit et al. 2013; © 2013 Petit et al., open-access article distributed under the terms of the Creative Commons Attribution License, https://creativecommons.org/licenses/by/4.0/)

## 10.8 Metabolic Rates and Migration

The longest migrations of any animals on earth are made by birds, both in terms of total distances traveled and distances traveled per unit time (see Chap. 13—Migration). Migration, especially long-distance migration, can be energetically expensive. Evidence to date suggests that BMRs of migrants increase during the period of spring migration in the northern hemisphere. For many species of birds, fall migration occurs over a longer period of time and so may not be as energetically costly as spring migration (Nilsson et al. 2011a, 2013; Fig. 10.35) when birds typically attempt to reach breeding areas as soon as possible. For example, the body mass and BMRs of captive Red Knots (*Calidris canuta*) that typically migrate from wintering areas in western Europe to breeding areas in the Arctic were found to increase during the period of spring migration and peak when they would have been initiating breeding, whereas no such peak was apparent

**Box 10.3 Heat Production by Red Blood Cell Mitochondria**

Resident birds at higher latitudes must adapt to seasonal changes in temperature, photoperiod, and food availability. Lower temperatures during the winter require birds to expend more energy to maintain their body temperature. Nord et al. (2021) examined changes in the mitochondrial function of red blood cells of three resident species of birds in western Scotland (56°7′3″N) from autumn to winter, including Coal Tits (*Periparus ater*), Eurasian Blue Tits (*Cyanistes caeruleus*), and Great Tits (*Parus major*). Analysis revealed that bird blood was more thermogenic during winter than in the autumn, but at the possible cost of a reduction in ATP production per oxygen molecule consumed. These authors found that mitochondrial volumes in red blood cells (as measured by citrate synthase activity) increased during winter, but to a lesser degree for Coal Tits than Blue or Great tits.

(continued)

**Box 10.3** (continued)

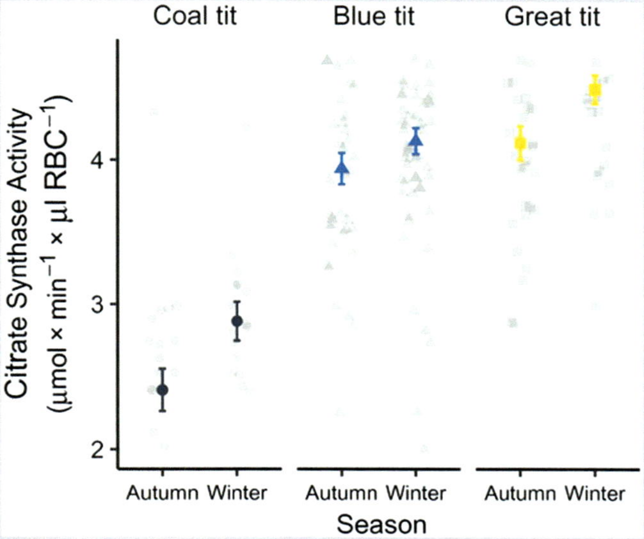


Seasonal variation in citrate synthase activity (a proxy for mitochondrial volume). This figure shows mean (± SE) citrate synthase activity in red blood cells of sympatric Coal, Blue, and Great tits in the beginning of autumn and in late winter in western Scotland. Gray symbols show raw data. (Figure from Nord et al. 2021; © 2021 John Wiley and Sons, used with permission)

Nord et al. (2021) also found that oxygen consumption during oxidative phosphorylation in red blood cells, as well as oxygen consumption devoted to ATP production, was generally higher in winter than in autumn (with the single exception of oxygen consumption devoted to ATP production for Eurasian Blue Tits, panels A and B below). In addition, for two of the three species, mitochondrial respiration per RBC had significantly higher uncoupled respiratory capacity (i.e., ETS was higher, panel C below) during winter than autumn. Finally, LEAK respiration in red blood cell mitochondria (oxygen consumption uncoupled from ATP production) was higher during winter than autumn for all three species of tits (panel D below). Although these seasonal changes might be the result of other factors, Nord et al. (2021) noted that it was "tempting to speculate that this was indeed causally related to seasonally declining temperatures (either experienced or anticipated). Cold-induced increases in proton conductance have been found in skeletal muscle, liver, and brain tissue in other studies on endotherms and are often interpreted as a mechanism for augmenting local thermogenesis." Additional studies are needed with other species over broad geographic ranges to improve our understanding of possible seasonal variation in avian red blood cell respiration.

(continued)

**Box 10.3** (continued)

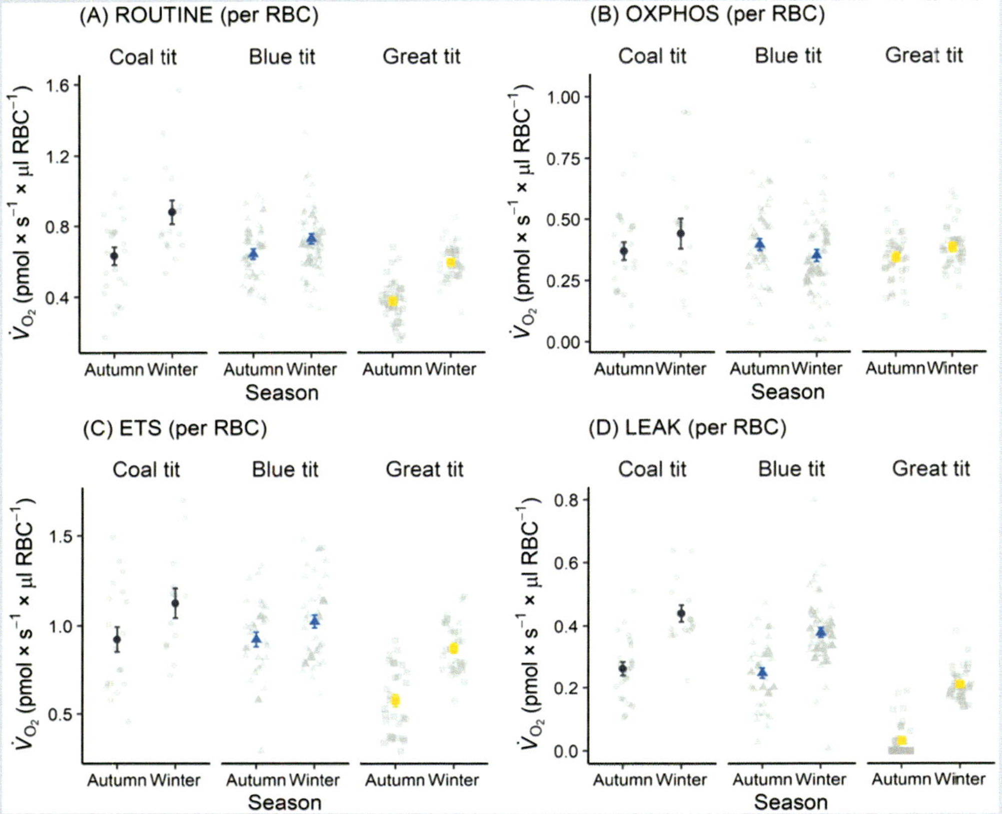


Seasonal variation in mitochondrial respiration per red blood cell in Coal, Blue, and Great tits in western Scotland. Shown is the mean (± SE) mitochondrial respiration rate per red blood cell for four mitochondrial respiration states. (**a**) Baseline ("ROUTINE") oxygen consumption during oxidative phosphorylation (i.e., ATP is being produced), (**b**) oxygen consumption devoted to ATP production alone ("OXPHOS"), (**c**) maximum working capacity of the electron transport system when uncoupled from ATP production ("ETS"), and (**d**) the part of baseline oxygen consumption (ROUTINE) attributed to offsetting the leak of protons across the inner mitochondrial membrane and, therefore, uncoupled from ATP production ("LEAK"). (Figure from Nord et al. 2021; © 2021 John Wiley and Sons, used with permission)

during the period of fall migration (Piersma et al. 1995; Fig. 10.36).

To meet the energetic demands of long-distance migration, birds preparing to migrate increase food intake by eating more (hyperphagia), and some have been found to select foods based in part on lipid content. Among some small species of migratory birds, pre-migration lipid reserves may represent as much as 50% of total body mass (Blem 1990; Figs. 10.37 and 10.38). Lipids are the best way to store energy because, when metabolized by cells, they yield up to 8–10 times more energy per unit gram than carbohydrates or proteins (Jenni and Jenni-Eiermann 1998; Box 10.4 Fats and Fat Metabolism). In addition, when metabolized, lipids also generate metabolic water

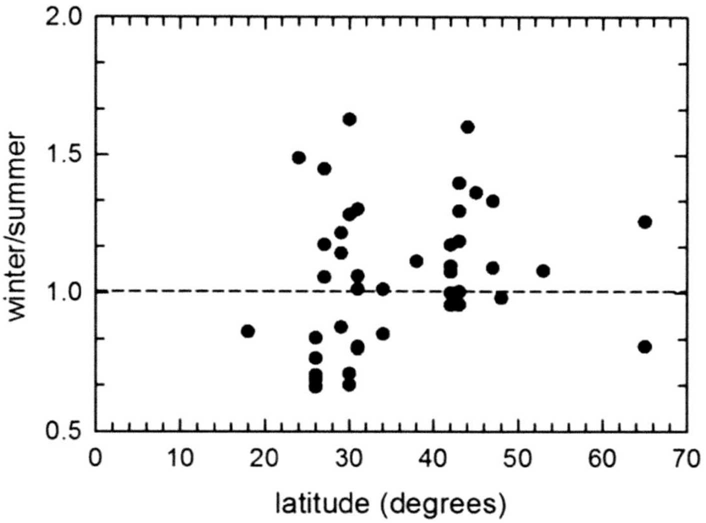


**Fig. 10.34** Winter/summer ratios of mass-specific basal metabolic rates for species of birds at different latitudes. Ratios generally increase with decreasing winter temperatures and increasing latitude, but species at lower latitudes (tropical/subtropical) exhibit greater variation. (Figure from McKechnie et al. 2015; © 2015 Springer Nature, used with permission)

(water produced as a by-product of ATP synthesis). For terrestrial birds making long-distance migrations across oceans, this metabolic water can be important in reducing the risk of dehydration (Gerson and Guglielmo 2011). During migration, most songbirds alternate between migrating at night and stopping over at different sites during the day to refuel.

As an example of diet switching, many songbirds in eastern North America switch from primarily insectivorous diets in their breeding areas to mainly frugivorous diets prior to and during fall migration. Insectivorous birds may also switch to preying on insects rich in lipids or carbohydrates (Jenni and Schaub 2003). However, the extent to which migrants switch diets

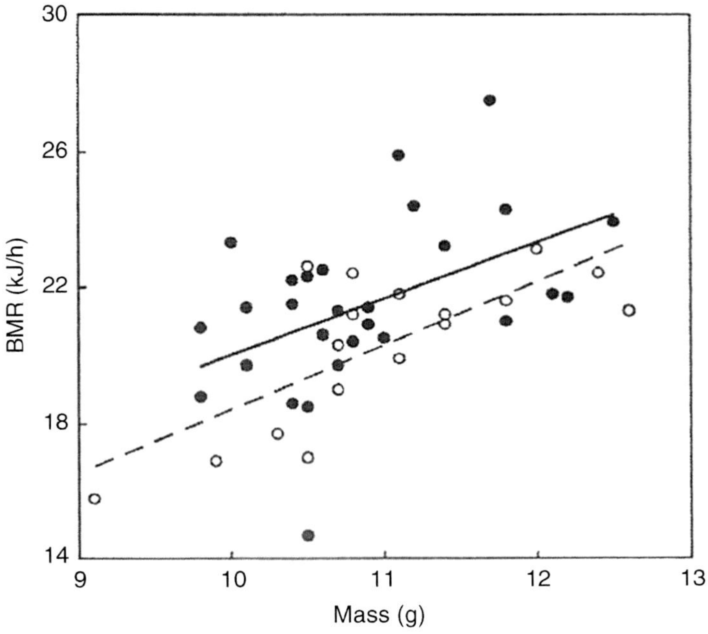


**Fig. 10.35** As with other species of birds, BMRs of Eurasian Blue Tits (*Cyanistes caeruleus*) vary with body mass. However, the BMRs of Blue Tits migrating during the fall (open circles and dashed line) were found to be lower than those of resident Blue Tits (closed circles and solid line). Migrating Blue Tits travel very slowly in the fall to warmer climates, whereas residents remain in areas with colder climates. To cope with colder temperatures, resident Blue Tits have higher BMRs. (Figure from Nilsson et al. 2011; © 2011 The Authors, used with permission)

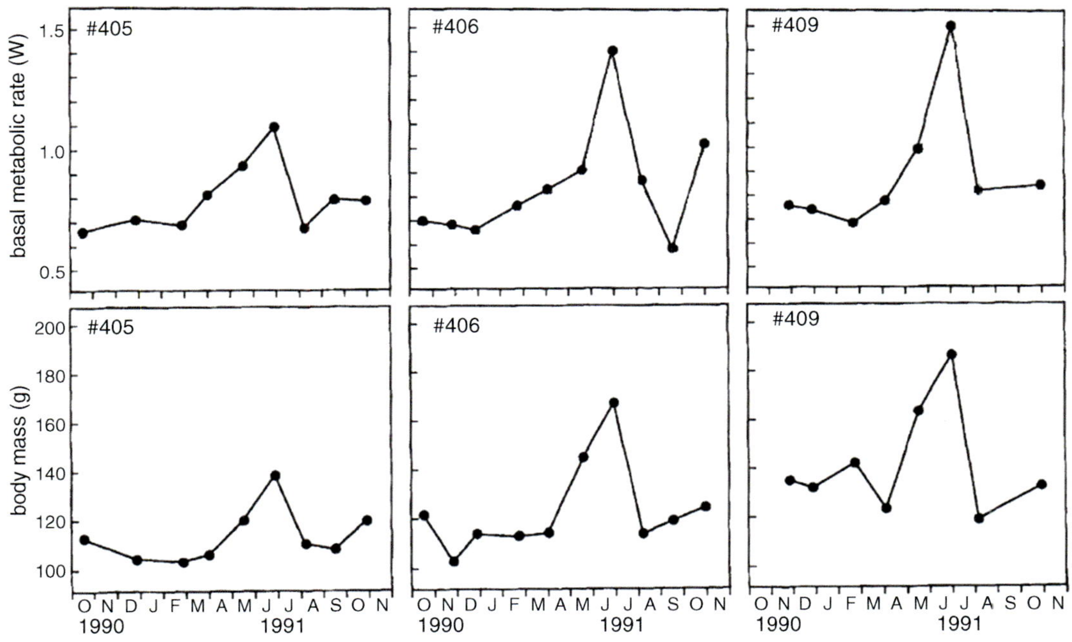


**Fig. 10.36** Yearly variation in body mass and BMR of three captive Red Knots (*Calidris canutus*) (O = October, N = November, and so on). Both mass and BMR increased during the period of spring migration and peaked in late June and early July when they would first begin breeding in their Arctic breeding areas. (Figure modified from Piersma et al. 1995; © 1995 Springer-Verlag, used with permission)

varies among species. For example, Parrish (1997) examined the food habits of resident and migrant songbirds along the east coast of the United States during fall migration and found that the diet of some migrants was ≥80% fruit (e.g., families Turdidae, Mimidae, and Cardinalidae; Fig. 10.39; Box 10.5 Fruit as a Source of Antioxidants). Even birds that are almost strictly insectivorous when not migrating, i.e., flycatchers (Tyrannidae), had diets consisting of about 60% fruit. In contrast, resident birds such as Carolina Wrens

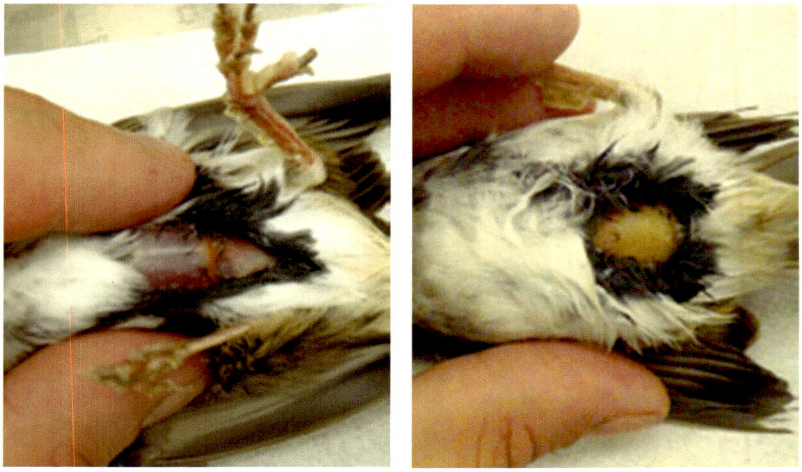

**Fig. 10.37** Examination of abdominal regions of a nonmigratory (left) and a pre-migratory (right) songbird reveals a substantial difference in the amount of fat or lipid (yellow in color) present. (Photos by Paul Bartell in Bartell and Moore 2013; used with permission of Paul Bartell).

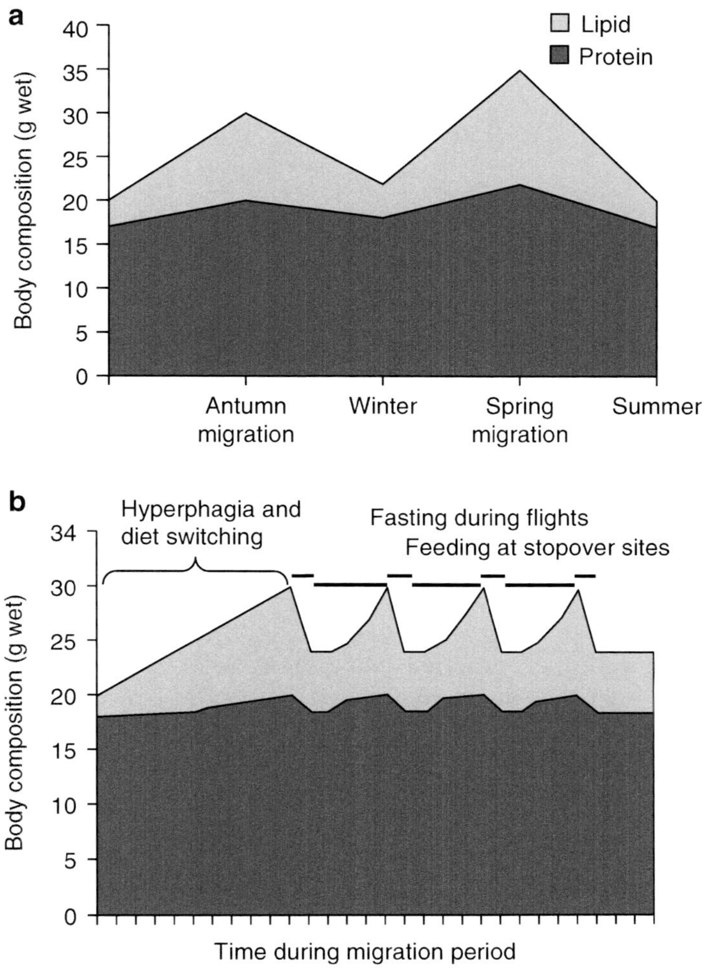


**Fig. 10.38** Change in body composition (**a**) throughout the year and (**b**) prior to and during migration in a typical temperate-zone small migratory songbird. (**a**) As the time of migration approaches, body mass increases from about 20 g to about 30 g for the longer duration fall migration and to about 35 g for the more rapid spring migration, with most additional mass being stored as lipids. (**b**) Preparing for migration, migrant songbirds increase food intake (hyperphagia) and may switch diets to more energy-rich foods. During migration, most songbirds migrate at night, then stopover during the day to refuel for the next night's flight. (Figure from Pierce and McWilliams 2005; © 2005 Oxford University Press, used with permission)

(*Thryothorus ludovicianus*; Troglodytidae) and White-breasted Nuthatches (*Sitta carolinensis*; Sittidae) continued to consume primarily insects (Fig. 10.39). By consuming more fruit, birds can use a food source that is often abundant during late summer and fall in temperate areas and, in addition, many fruits have a high fat or sugar content (Smith et al. 2007) that likely aids in migratory fattening.

Prior to migration, the body mass of long-distance migrants increases, largely, but not entirely, due to fat deposition; protein mass also increases (e.g., flight muscles increase in mass). Birds accumulate fat in several areas. McGreal and Farner (1956) identified 15 areas where fat deposits were located in White-crowned Sparrows (*Zonotrichia leucophrys*). Initially, fat is stored subcutaneously under the various feather tracts. With additional accumulation, more subcutaneous fat is stored in the furcular region (claviculocoracoid fat organ). In the fattest birds, subcutaneous deposits consist of extensive masses of fat, and the interfurcular region and the abdominal cavity are filled with fat (Fig. 10.40). In such birds, the lipid content of most areas of the body increases, with the exception of the heart (Blem 1976).

In addition to eating more, some birds may be able to increase their body mass and lipid stores by using less energy and increasing the efficiency of digestion and absorption. Eating more would seemingly require more time and effort spent foraging and, therefore, more energy. However, investigators have found that the daily energy

**Box 10.4 Fats and Fat Metabolism**

Birds store lots of lipids to power long migratory flights, and those lipids are stored as molecules called triglycerides or triacylglycerols. Triglycerides consist of three fatty acids attached to a glycerol molecule.

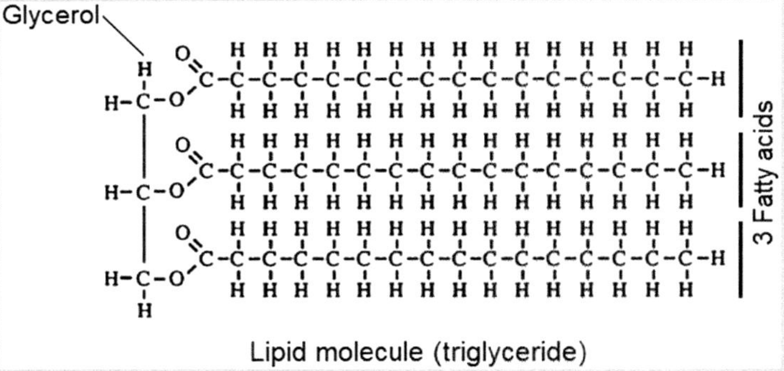


Different types of fatty acids contain different numbers of carbon atoms; glycerol always contains three carbon atoms. (Figure modified from Ong et al. 2011; © 2011 Elsevier Ltd., used with permission

When birds use triglycerides to power migratory flights, flight muscle cells use enzymes to break apart the molecules into individual fatty acids plus the glycerol. In the cytoplasm of the cells, glycerol is broken down by glycolytic enzymes into pyruvate, which, in turn, is broken down into Acetyl CoA (a two-carbon molecule) that can be used by mitochondria to make ATP (via the citric acid, or Kreb's, cycle). Enzymes break the long carbon chains of the fatty acids down into smaller molecules of Acetyl CoA (a process referred to as beta-oxidation) that, again, go through the citric acid cycle to produce lots of ATP. Triglycerides produce more ATP per unit mass than carbohydrates and proteins because, when metabolized, they produce many Acetyl CoA molecules that can be transformed by mitochondria into ATP.

expenditure (DEE) of some birds does not increase during the period of pre-migratory fattening. For example, female Common Eiders (*Somateria mollissima*) implanted with data loggers to monitor their activities and heart rates were found to spend three times more time foraging before than after migration, but their DEE during those periods (as measured by heart rates) did not differ (Fig. 10.41; Guillemette et al. 2012). Other investigators have also found that increased workloads (e.g., provisioning nestlings) may not always result in an increase in the DEE of birds (Deerenberg et al. 1998; Moreno et al. 1999). Female Common Eiders appear to use a change in behavior (less time flying) to keep DEE from increasing when their foraging time increases prior to migration (Fig. 10.41). However, other birds may compensate physiologically by, for example, becoming hypothermic at night (Butler and Woakes 2001; Wojciechowski and Pinshow 2009).

Increases in assimilation efficiency may also contribute to the ability of some migrants to rapidly increase lipid reserves. For example, Barlein (1985) reported that increases in the body mass of pre-migratory Garden Warblers (*Sylvia borin*) involve hyperphagia (~40% increase in food intake), but also involve an approximately 20% increase in assimilation efficiency. One potential explanation for this increased ability to absorb nutrients is the increased consumption of lipids, and perhaps, of lipids consisting of certain types

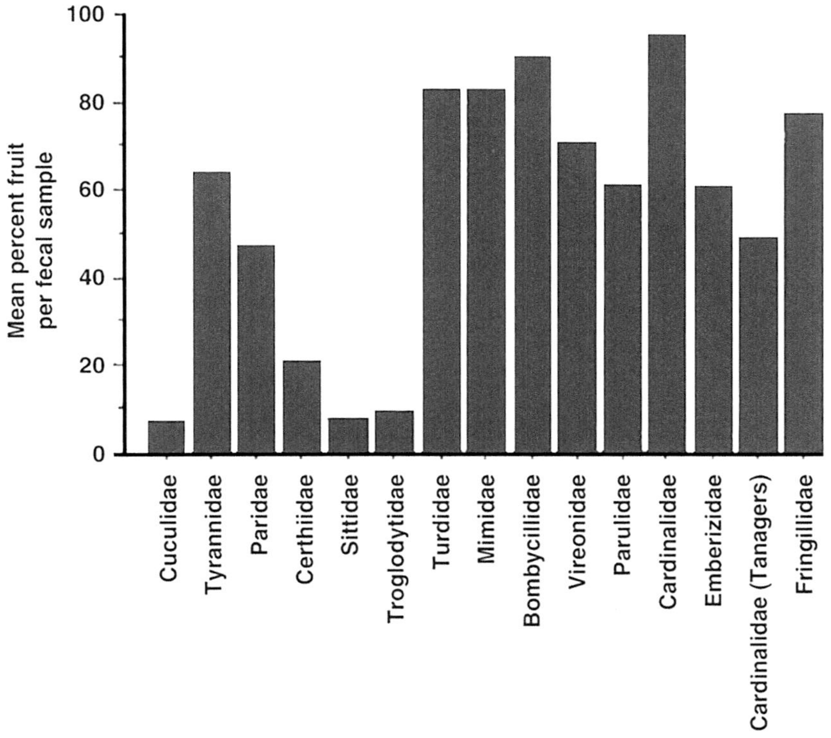


**Fig. 10.39** Percentage of fruit in the diet of Neotropical migrants in Rhode Island (USA) during fall migration expressed as the mean percent of fruit volume per fecal sample for each family of birds. Error bars = one standard deviation. (Figure from Parrish 1997; © 1997 Oxford University Press, used with permission)

of fatty acids because some fatty acids are known to be absorbed more efficiently than others (Leeson and Summers 2001).

Preparation for long-distance flight requires energy, but also requires phenotypic flexibility or, in other words, changes in key organs and muscles. As noted above, the body mass of long-distance migrants increases as they store fats to fuel their flight. In addition, however, the mass of many internal organs and muscles also changes prior to, during, and after migration. Prior to migration, the mass of many muscles and organs increases, including the mass of flight (pectoralis) and leg muscles, the heart, and

**Box 10.5 Fruit as a Source of Antioxidants**

Many migrating songbirds eat fruits prior to and during migration, even birds that are primarily insectivorous at other times of year. One reason for this is that fruits can be an excellent source of energy, rich in lipids and carbohydrates. However, another possible reason that birds preparing for migration or migrating eat fruits is that they can be an excellent source of dietary antioxidants.

As living cells generate ATP and other complex molecules, free radicals (atom or molecule with unpaired valence electrons or an open electron shell) are produced. Free radicals are highly reactive and, therefore, can react with and damage membranes, lipoproteins, and even DNA via

(continued)

**Box 10.5** (continued)

oxidation. This oxidative stress can have negative impacts on cell function and, over time, even lead to cell death.

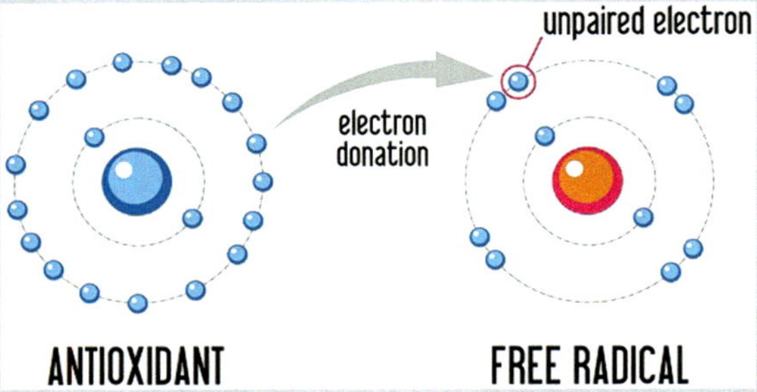


Antioxidants help prevent oxidative stress by reacting with free radicals (donate electrons) so they are no longer reactive (Figure from Wikipedia, CC0 Public Domain)

At a stopover site in Rhode Island, Bolser et al. (2013) found that, during fall migration, birds consumed fruits of arrowwood spp. more rapidly than those on other fruiting trees and shrubs, e.g., oriental bittersweet (*Celastrus orbiculatus*), multiflora rose (*Rosa multiflora*), and winterberry (*Ilex verticillata*), and the fruits of arrowwood spp. also had the highest concentrations of antioxidants. These authors suggested that birds may actively select fruits with more antioxidants during autumn migration "to protect themselves against the potentially damaging effects of oxidative stress caused by long-distance fasting flight."

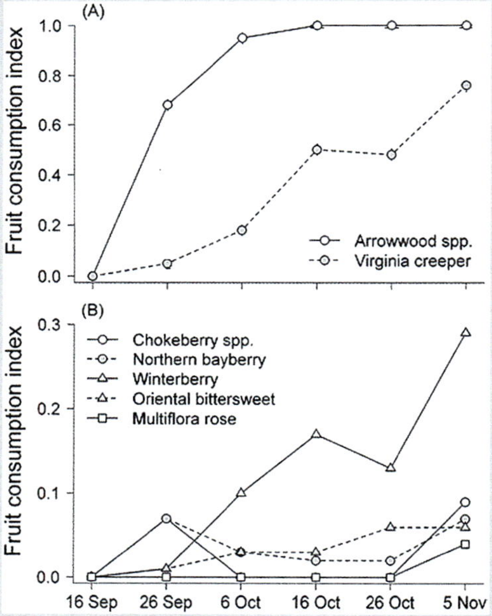


Fruit consumption indices for seven species of fruiting shrubs during autumn 2008 on Block Island, Rhode Island. (**a**) One species (arrowwood spp.) was consumed more than (**b**) northern bayberry (*Myrica pensylvanica*), oriental bittersweet, chokeberry spp., multiflora rose, and winterberry. A higher consumption index indicates a larger proportion of available fruit consumed. (Figure from Bolser et al. 2013; © 2013 Wilson Ornithological Society, used with permission)

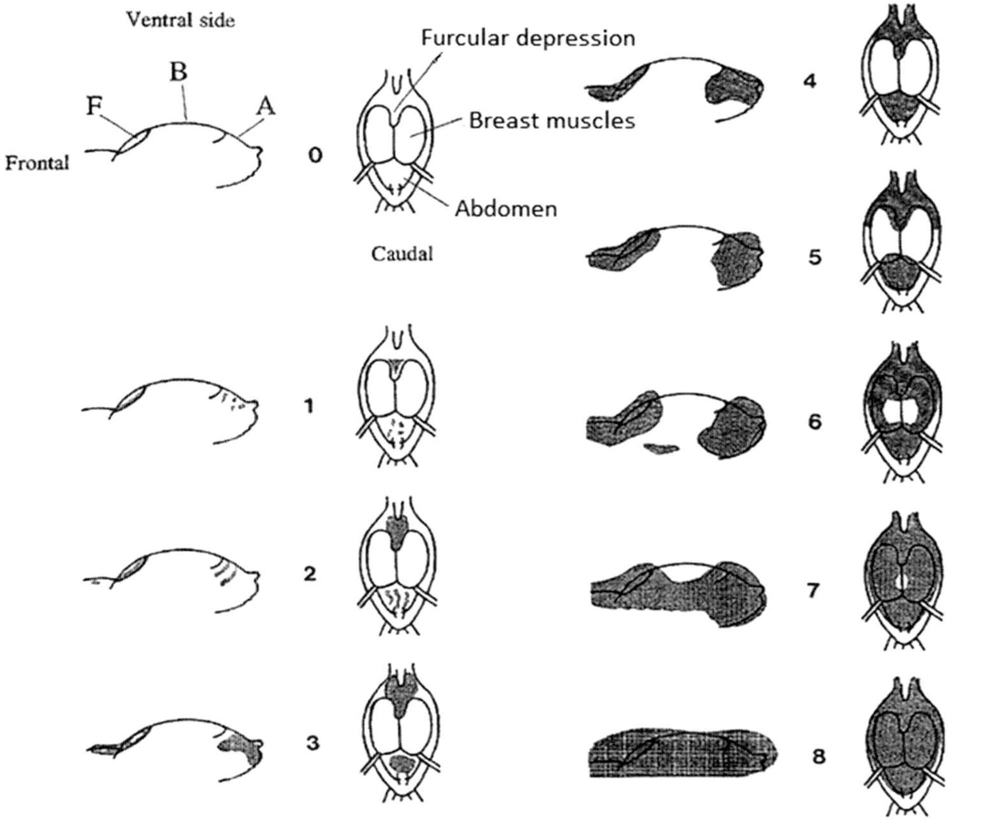


**Fig. 10.40** To quantify the extent of avian fat deposits, systems have been developed for "scoring" the amount of fat present. In this system, a score of "1" indicates the presence of little or no fat, whereas a score of "8" indicates an extremely fat bird. As birds store increasing amounts of fat prior to long-distance migration, this scoring system shows that fat is deposited. Initially, subcutaneous fat is stored in the furcular (interclavicular) depression (F), then increasingly in the abdominal region (A), and then in the breast region (B). (Figure from Kaiser 1993; © 1993 Association of Field Ornithologists, used with permission)

several organs in the digestive system (Fig. 10.42). These pre-migratory changes allow birds to efficiently digest food as they store fats and prepare the flight muscles for the upcoming flight. However, during long migratory flights (e.g., from Ethiopia to Egypt in Fig. 10.42), the mass of many organs and muscles declines as tissues are catabolized to provide energy (and metabolic water; Gerson and Guglielmo 2011) for flight muscles and other metabolically active organs (e.g., heart). The rate and extent to which organs and muscles lose mass vary, with the small intestine and liver losing the most mass, followed by the kidneys, gizzard, and heart, and, finally, the flight and leg muscles (Fig. 10.43). Several hypotheses have been proposed to explain this variation. For example, the "use-disuse" hypothesis proposes that digestive organs lose more mass than organs such as the heart or the flight muscles because they are energetically expensive to maintain and, during long migratory flights, are not being used anyway. Another hypothesis is that differences in the timing and rate of mass loss of different organs and muscles are due to differences in metabolic activity or rates (as measured by oxygen consumption). Bauchinger and McWilliams (2010) suggested that variation in mass loss among different organs

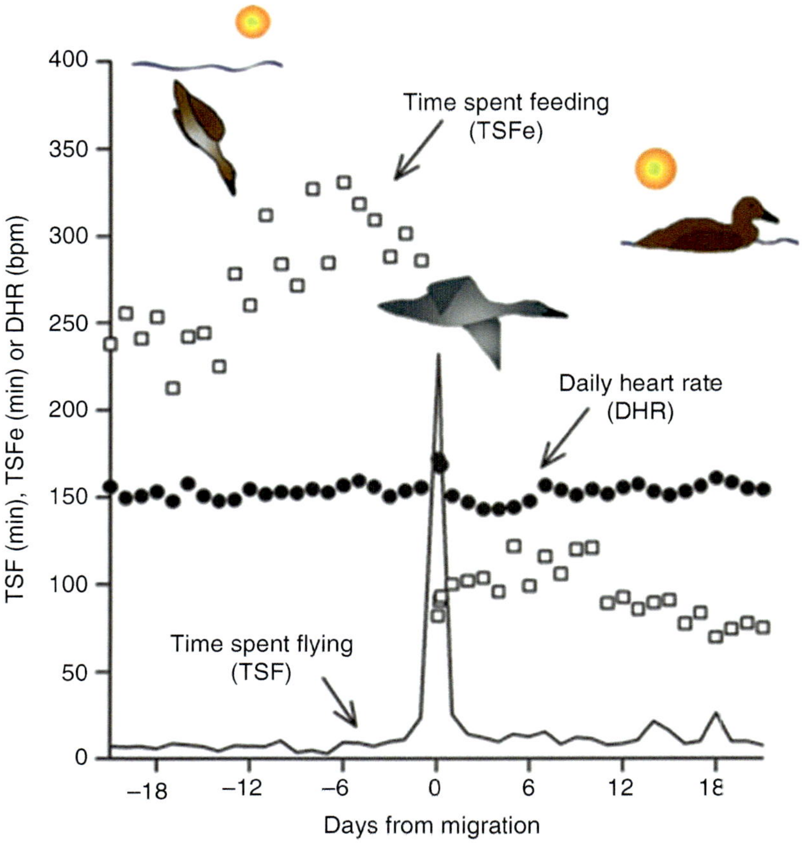


**Fig. 10.41** Mean values of time spent flying (TSF), time spent feeding (TSFe), and daily heart rate (DHR) for female Common Eiders (*Somateria mollissima*; $N = 16$) 3 weeks before and after molt migration. Data are from implanted data loggers that recorded heart beats and the time each female spent diving and flying. (Figure from Guillemette et al. 2012; © 2012 The Authors. Functional Ecology © 2012 British Ecological Society, used with permission)

and muscles is simply a result of differences in inter-organ differences in their biochemical pathways that result in differences in rates of protein turnover and the macromolecular composition of tissues. Reasons for these differences among organs and muscles remain unclear, but the result is that 80–90% of the energy required for flight is derived from fat, and the remaining 10–20% is from the breakdown and oxidation of proteins that occurs simultaneously with fat metabolism (Jenni and Jenni-Eiermann 1998; Gerson and Guglielmo 2011, 2013). Catabolizing lipids and proteins during migratory flights not only provides needed energy, but this loss of mass also reduces the energetic cost of flight (Pennycuick 1975).

The extent to which migrating birds lose body and organ mass varies with the distance and duration of migratory flights. Klaassen et al. (2000) trained a Thrush Nightingale (*Luscinia luscinia*) to fly in a wind tunnel for periods up to 16 h, allowing them to monitor changes in body mass and, by analysis of feces, what tissues were being catabolized. The rate of mass loss was about 50% higher during the first 2 h of flight than during the last 6 h (Fig. 10.44), likely because of the rapid utilization of glycogen stores and the emptying of the digestive tract and cloaca early in flight. The amount of nitrogen excreted in feces was used to estimate the proportion of protein in the tissues being catabolized during flight, and analysis revealed that the proportion averaged 2.1% (Fig. 10.44). As noted previously, when catabolized, proteins from smooth muscle in the digestive system and from skeletal muscle in flight and leg muscles provide a migrating bird with both energy and water.

During migration, most birds alternate flights of varying duration with stopovers where they can replenish energy stores and prepare for the

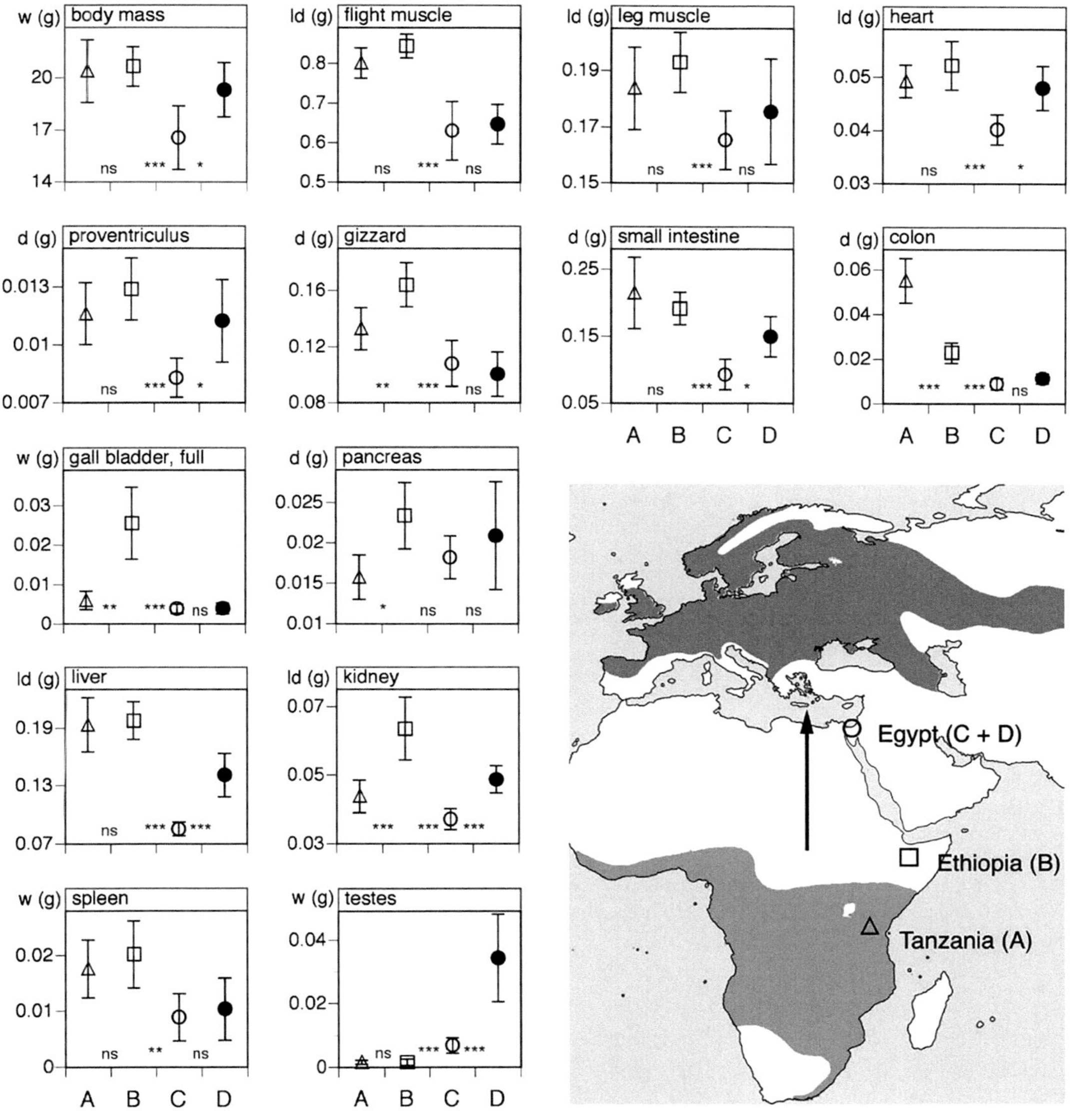


**Fig. 10.42** Changes of mean body mass and mean masses of 13 organs of Garden Warblers (*Sylvia borin*) during spring migration from their wintering range in Africa (lightly shaded area), across the Sahara Desert in northern Africa, and toward their breeding range in Eurasia (A, B, and C + D). Symbols on the map and in graphs represent sampling sites: triangle = pre-migration birds in Tanzania, square = birds after stopover in Ethiopia, open circle = migrating birds just after arrival in Egypt, and filled circle = migrating birds held captive for 9 days and provided with food and water. On the upper left of graphs, w = wet mass, d = dry mass, and ld = lean dry mass (calculated by subtracting fat mass from dry tissue mass). (Figure from Bauchinger et al. 2005; © 2005 Elsevier GmbH, used with permission)

next flight. For most migrating birds, more time can be spent at stopover sites than in flight. Hedenström and Alerstam (1998) estimated that some birds may spend up to 90% of their migration period at stopover sites. Analysis of data from Purple Martins (*Progne subis*) and Wood Thrushes (*Hylocichla mustelina*) fitted with geolocators revealed that 64% of their time

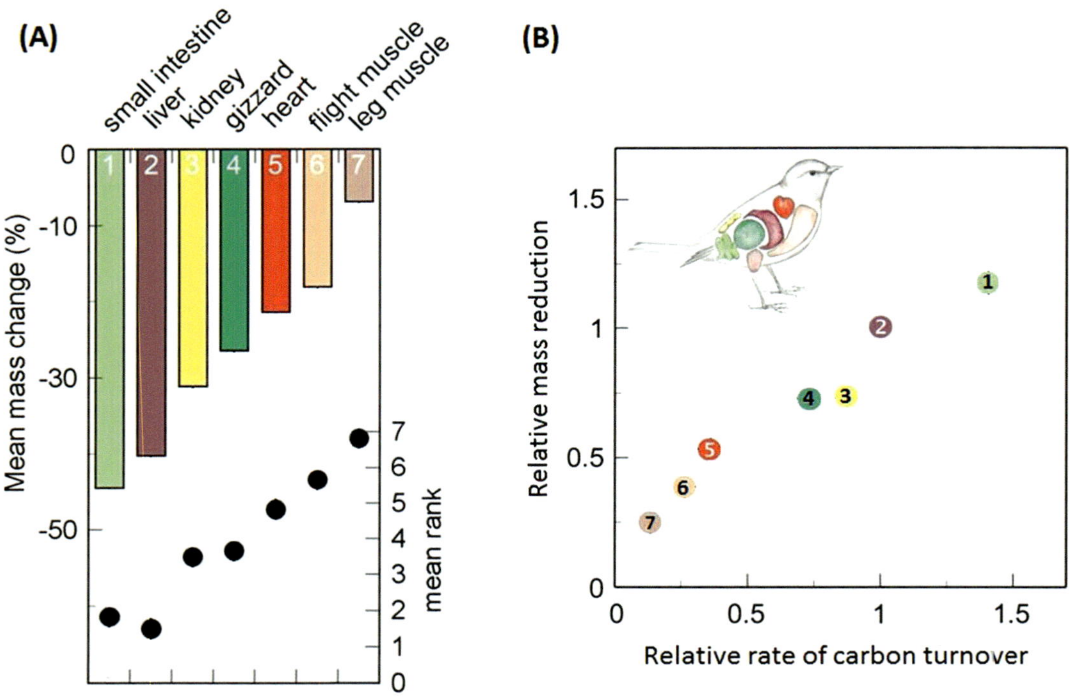


**Fig. 10.43** (**a**) Mass loss and relative rank order for seven organs/muscles from five species of migratory birds, including Great Knots (*Calidris tenuirostris*), Eurasian Blackcaps (*Sylvia atricapilla*), Garden Warblers (*Sylvia borin*), European Pied Flycatchers (*Ficedula hypoleuca*), and Willow Warblers (*Phyloscopus trochilus*). Birds had either fasted during actual (wild birds) or simulated (captive birds) migratory flights. (**b**) Organs with higher rates of carbon turnover also lost more mass. Numbers in circles correspond to the numbers of organs and muscles at the top of (**a**). Rate of carbon turnover by cells in organs and muscles was determined by the rate at which a carbon isotope was incorporated in laboratory studies. (Figure from Bauchinger and McWilliams 2010; © 2010 The Authors, used with permission)

was spent at stopovers during fall migration, but this declined to 24% during spring migration (Stutchbury et al. 2009; Box 10.6 How Do Geolocators Work?). At stopover sites, migrating birds may expend more energy than when flying. Wikelski et al. (2003a, b) estimated that migrating thrushes (*Catharus* spp.) expended more than twice the energy at stopover sites than in flight, with much of the energy expended during stopover needed for thermoregulation (Box 10.7 Conserving Energy at Stopover Sites). With warmer temperatures, less energy would be expended at stopover sites. Because of catabolism of protein during long migratory flights, the mass of the gastrointestinal tract (proventriculus, gizzard, and small and large intestines) of small birds, plus accessory organs such as the liver, can be reduced by as much as 40% (Muñoz-Garcia et al. 2012). Birds at stopover sites, therefore, not only expend energy foraging and for thermoregulation, but may face the problem of digesting food, absorbing nutrients, and replenishing energy reserves with a partially catabolized digestive system. Birds can meet this challenge in several ways, including increasing food intake rates, selecting stopover sites with high-quality food, and increasing digesta retention time during the first day of stopover, but shortening retention time in subsequent days (Karasov and Pinshow 2000; Bauchinger et al. 2009). By increasing digesta retention time on the first day at stopover sites, Eurasian Blackcaps (*Sylvia atricapilla*) maintain digestive efficiency while rebuilding their digestive systems, hence allowing for shorter digesta retention times on the following days and

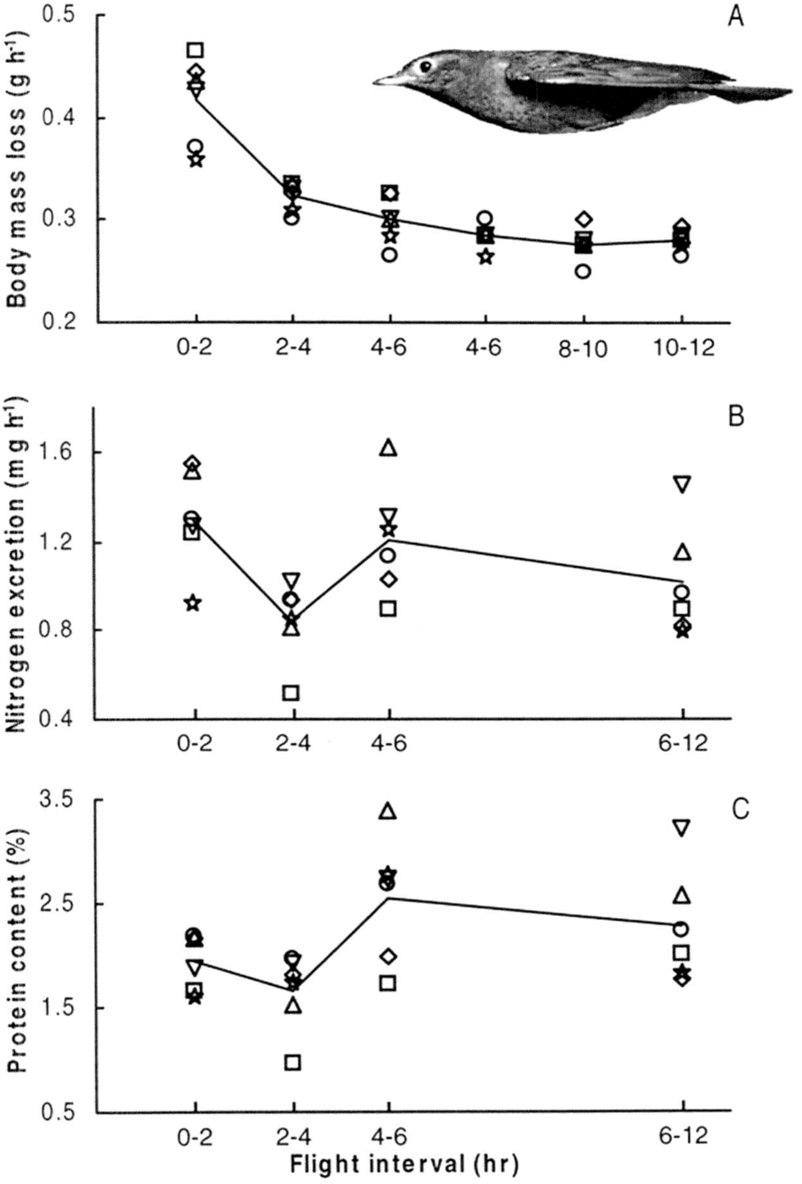


**Fig. 10.44** (**a**) Rate of mass loss and (**b**) the amount of nitrogen lost in feces, which was used to estimate (**c**) the proportion of protein in tissues being catabolized during six 12-h flights (each flight is represented by a different symbol) by a Thrush Nightingale (*Luscinia luscinia*) in a wind tunnel. The bird's mass at the start of flights varied between 26.7 and 29 g, and, during flights, it lost an average of 3.8 g (13.7% of initial mass). (Figure from Klaassen et al. 2000; © 2000 Oxford University Press, used with permission)

increased energy intake rates (Bauchinger et al. 2009).

Rather than alternating flights with stopovers to replenish energy reserves, some migrating birds, such as seabirds, raptors, and aerial insectivores, can instead use a fly-and-forage strategy. With this strategy, birds need not carry heavy lipid reserves that increase mass and increase the energetic cost of flight. Indeed, for some birds, such as raptors, accumulating the fat reserves needed for long-distance migrations would likely be difficult, if not impossible, given that many are sit-and-wait predators and many of their attacks are unsuccessful. As such, these birds migrate during the day and, when needed and the opportunity arises, spend time foraging. By traveling during the day, some birds can also exploit free energy from the atmosphere by using soaring or gliding flight, which is

**Box 10.6 How Do Geolocators Work?**

Geolocators use a light sensor to obtain and store light-level data at regular intervals. The location of a bird with a geolocator can be estimated based on the fact that day length varies with latitude and time of solar noon varies with longitude. By measuring these variables, a bird's general position can be determined. At sunrise and sunset, light measurements can be used to determine the sun's position because of the way light intensity varies with the angle of the sun. So low sun angles can be used to determine the time of sunrise and sunset, and solar noon is midway between those two times. Locations can then be estimated based on the length of the day (latitude) and the time of solar noon (longitude). Although this seems straightforward, collecting and analyzing data can be challenging because geolocators, once attached to a bird, have to be recovered to obtain the data. In other words, migratory birds captured and fitted with geolocators in their breeding areas must survive until the next breeding season, must return to the same general location where they were initially captured, and then must be recaptured. In addition, interpretation of data can be difficult because light levels can be influenced by environmental factors such as cloud cover and vegetative cover (i.e., shade). Despite these challenges, data from geolocators have provided and continue to provide new and exciting information about the movements and wintering areas of many species of migratory birds.

Example of a geolocator with a 15-mm "stalk" with the light sensor. (Photo by Alex Jahn, from Bridge et al. 2013; © 2013 Association of Field Ornithologists, used with permission)

much less energy-demanding than sustained flapping flight (Fig. 10.45).

For seabirds, energetics can play a critical role in determining migratory pathways. For those species with a fly-and-forage migration strategy, ensuring sufficient availability of food to meet to costs of migration may mean avoiding oligotrophic waters with limited food availability. For species that that do not feed during migration, minimizing the energetic costs of migration means seeking favorable wind conditions. For other species, migratory pathways may be determined by both of these factors. Studies of migrating Manx (*Puffinus puffinua*), Cory's (*Calonectris diomedia*), and Cape Verde (*C. edwardsii*) shearwaters fitted with geolocators revealed that they did not take the shortest route, but, rather, took routes that were 26–52% longer. Interestingly, shearwaters did not take routes that appeared to have the most favorable wind

**Box 10.7 Conserving Energy at Stopover Sites**

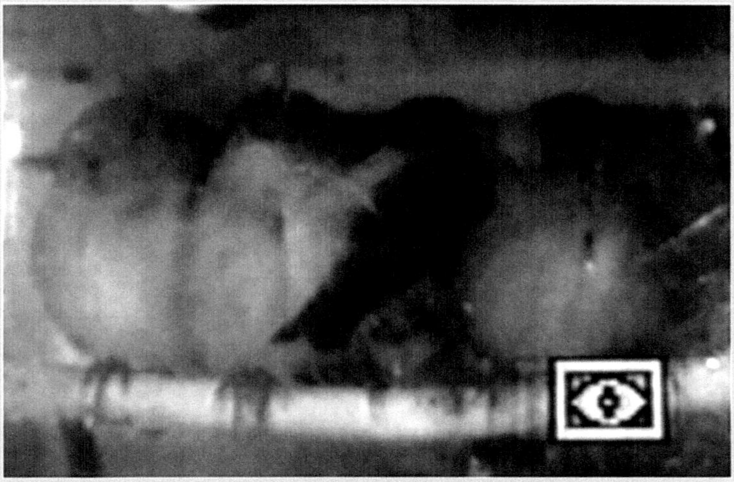

Four Eurasian Blackcaps (*Sylvia atricapilla*) huddling in a respiratory chamber. (Figure from Wojciechowski et al. 2011; © 2011 Oxford University Press, used with permission)

For small songbirds, successful migration requires rapid refueling at stopover sites. Efficient refueling requires efficient foraging, but minimizing energy use is also important. Wojciechowski et al. (2008) observed migrating Eurasian Blackcaps (*Sylvia atricapilla*) huddling at a stopover site and suggested that such behavior might reduce their energy expenditure. In the lab, Wojciechowski et al. (2011) found that huddling was beneficial for Blackcaps both in terms of energy and thermoregulation. Metabolic rates and body temperatures of Blackcaps placed in respirometry chambers overnight were measured both when alone and in groups of three or four at temperatures of 5, 10, and 15°C. Huddling Blackcaps maintained higher body temperatures than did solitary birds and also had metabolic rates about 30% lower than those of solitary birds. Huddling may, therefore, be an important way to save energy for small songbirds at migratory stopovers.

(continued)

**Box 10.7** (continued)

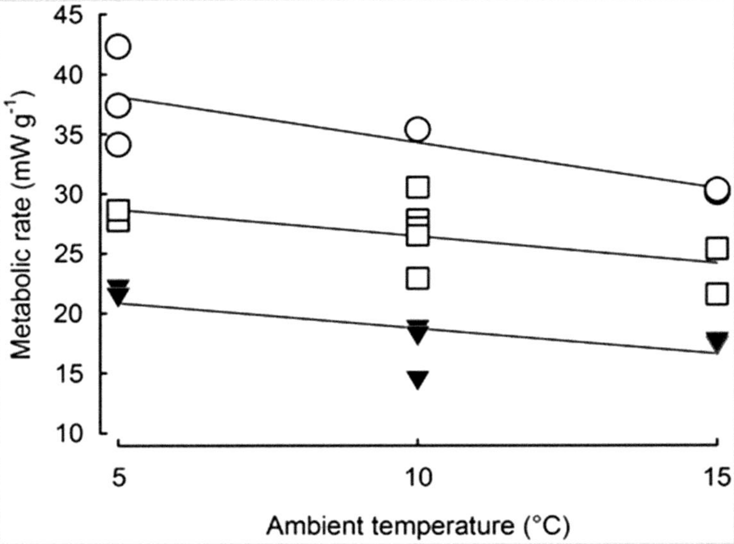


Mass-specific metabolic rates of solitary active (open circles), solitary resting (open squares), and huddling (inverted filled triangles) Eurasian Blackcaps at temperatures of 5, 10, and 15°C. Each symbol represents a single measurement. (Figure from Wojciechowski et al. 2011; © 2011 Oxford University Press, used with permission)

For migrating birds, nocturnal hypothermia can also help save energy, but it can also speed up fat accumulation at stopover sites. Wojciechowski and Pinshow (2009) monitored the body temperatures of eight Eurasian Blackcaps by radio-telemetry and found that their mean body temperature was 42.5°C during the day, but between 33 and 40°C at night. Measurements of their metabolic rates revealed that such hypothermia reduced energy expenditure by up to 30%. By conserving energy at night, Eurasian Blackcaps accelerate their rate of fuel accumulation during stopover.

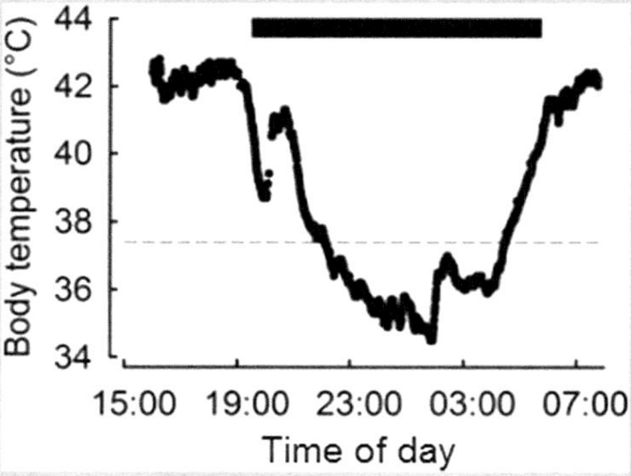


Body temperature of a Eurasian Blackcap during a night at a stopover site during migration. The black horizontal bar at the top of the figure indicates hours of darkness. The dashed horizontal line indicates the lower limit of normal, resting body temperature of 37.4°C. (Figure from Wojciechowski et al. 2011; © 2011 Oxford University Press, used with permission)

(continued)

**Box 10.7** (continued)

Yet another way for birds to conserve energy during stopovers is to sleep more and sleep with their head turned and tucked in the back feathers. Ferretti et al. (2019) examined sleep patterns of Garden Warblers (*Sylvia borin*) at a stopover site on an island in the Mediterranean Sea during spring migration. Warblers were captured, examined to determine their body condition (e.g., body mass, fat stores, and muscle mass), and placed in cages so their behavior could be recorded with infrared cameras. Warblers in better condition slept more during the day, but less during the night because they exhibited nocturnal restlessness (an indication of migratory disposition). In addition, when they slept they assumed an "untucked" posture with their head facing forward. In contrast, birds in poorer condition slept more at night (exhibiting less nocturnal restlessness) and assumed a "tucked" posture, with their heads turned backward and tucked in the scapular feathers.

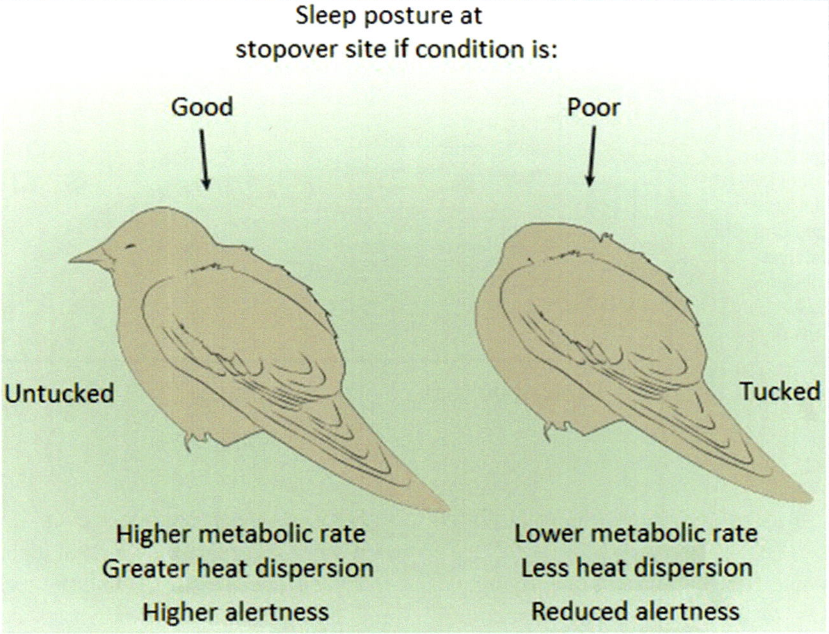


By sleeping more, birds in poorer condition conserved energy and sleeping in a "tucked" posture further reduced energy consumption, likely by reducing heat loss from the bill and eye region. However, birds assuming the "tucked" posture are likely at greater risk of predation because they sleep more deeply than birds sleeping with an "untucked" posture and, with their eyes tucked in their feathers, they would be less likely to detect approaching predators (Ferretti et al. 2019). Thus, Garden Warblers in poorer condition at stopover sites assume a greater risk of predation, but benefit by conserving energy. (Figure modified from Ferretti et al. 2019; © 2019 The Author(s). Published by Elsevier Ltd., used with permission)

conditions, rather their strategy was apparently to avoid unfavorable wind conditions when possible, but not at the expense of extending the duration of migration too much. Thus, the "compromise" routes likely required a bit more energy, but allowed the shearwaters to reach their

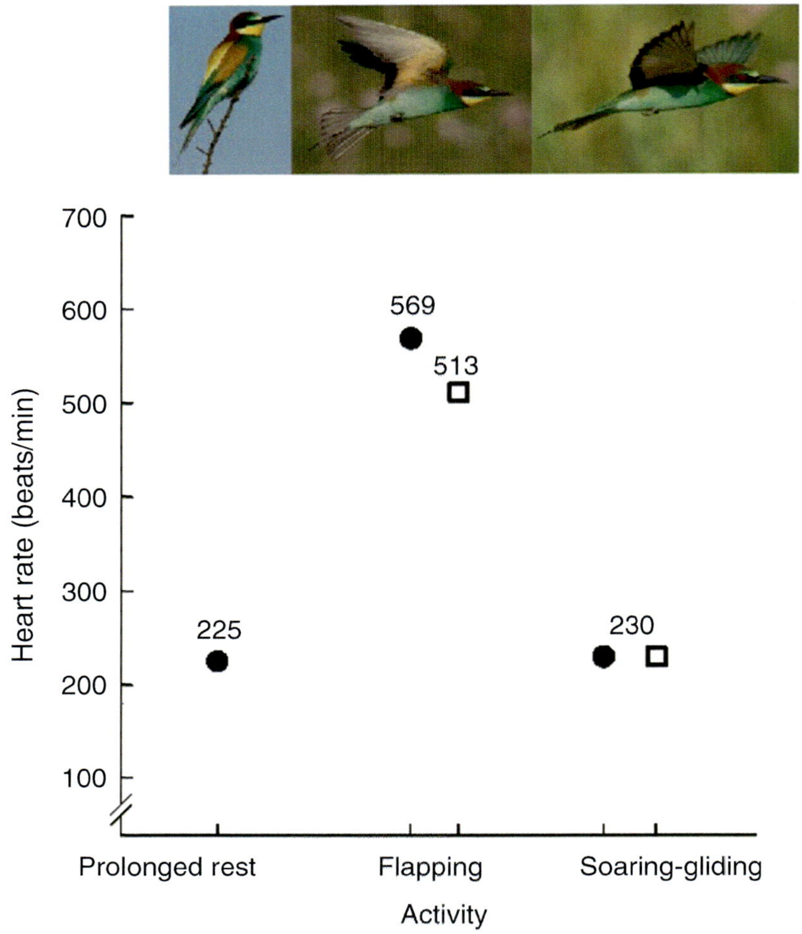


**Fig. 10.45** Heart rates of migrating European Bee-eaters (*Merops apiaster*) when resting and when using flapping flight and soaring-gliding flight. Bee-eaters during migratory flights are indicated by the open squares; filled circles represent bee-eaters at stopover sites. Numbers above symbols indicate averages. Assuming a correlation between heart rate and metabolic rate, energy use by European Bee-eaters when using soaring-gliding flight is similar to that when at rest. Use of soaring-gliding flight during migration, therefore, allows bee-eaters to use less energy during migration. (Figure from Sapir et al. 2010; © 2010 Sapir et al., open-access article distributed under the terms of the Creative Commons Attribution License, https://creativecommons.org/licenses/by/4.0/)

destinations sooner (González-Solís et al. 2009; Fig. 10.46).

When fly-and-forage migrants are flying, an important way to minimize energy expenditure is to choose flight altitudes with optimal wind speeds and direction. Using radar to track migrating birds crossing the Strait of Messina from Sicily to Italy's mainland, Mateos-Rodríguez and Liechti (2012) found that most hawks, harriers, kites, falcons, swifts, and herons flew at altitudes below 1135 m and that all species flew at altitudes where tailwinds were better than those at lower and higher altitudes (Fig. 10.47). Birds did not, however, necessarily fly at the altitude with the most favorable tailwind. Rather, birds continued to climb in altitude until tailwinds were most favorable relative to those at somewhat higher and lower altitudes, even if there were even more favorable tailwinds at higher altitudes.

The distances birds fly, and the speed at which they fly, during migration varies considerably. However, some birds, especially when migrating to breeding areas, cover impressive distances in rather short periods of time. Studies where investigators have used satellite transmitters or geolocators to track migration have revealed migration speeds ranging from 16 km/day to a phenomenal 2310 km/day (Table 10.2). Most migrants fly for only part of the day, e.g., most diurnal migrants fly during the day and rest at night and most nocturnal migrants fly only at night. However, some birds migrating across oceans have been found to fly continuously for multiple days. For example, during fall migration,

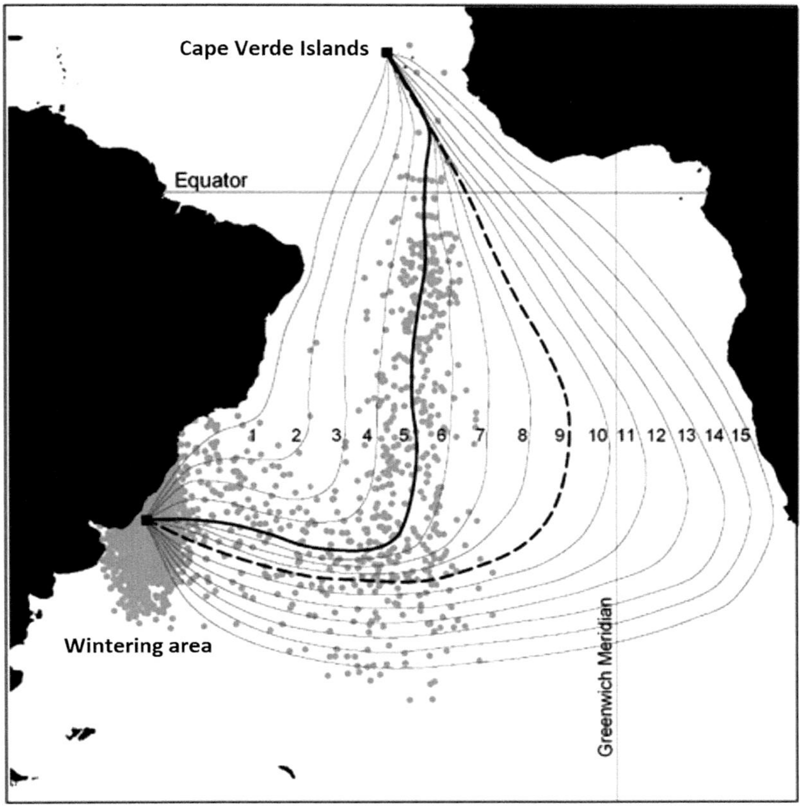


**Fig. 10.46** During migration to their breeding site on the Cape Verde Islands, Cape Verde Shearwaters (*Calonectris edwardsii*) must balance the need to minimize flight costs by seeking favorable winds with the need to get to the breeding site as soon as possible. In doing so, the route taken by shearwaters (dark solid line) is clearly not the shortest and fastest route, but neither do they take the route with the most favorable winds (dashed line) or other possible routes (lighter solid lines). (Figure from González-Solís et al. 2009; © Inter-Research 2009, used with permission)

some Blackpoll Warblers (*Setophaga striata*) fly continuously for as long as an estimated 3 days in a trans-oceanic flight from eastern Canada and the northeastern United States to northern South America (Williams et al. 1978; Fig. 10.48). In addition, migrating Pacific Golden Plovers (*Pluvialis fulva*) have been found to fly continuously for 5 days or more (Johnson et al. 2012) and, even more impressive, some migrating Bar-tailed Godwits (*Limosa lapponica*) have been found to fly non-stop for as long as 9.4 days (Gill et al. 2009; Fig. 10.49).

A number of factors influence the flight ranges, or the maximum distance traveled in a single flight, of birds that migrate using flapping flight (Fig. 10.50). Most important is the departure fuel load at departure, primarily fat, but also protein. Fats provide most of the energy needed for migration (85–95%), with proteins providing the rest (5–15%) (McWilliams et al. 2004; Weber 2009; Fig. 10.51). Birds also lose water as they fly (especially when ambient temperatures are high and/or humidity is low), but metabolic water production (produced as cells produce ATP) helps prevent dehydration (Gerson and Guglielmo 2011), as does flying at higher, cooler altitudes and flying at night, and so lipid and protein deposits are generally the primary determinant of flight ranges. Although these fuels are essential, the amount stored is limited because of the increased cost of flying as a bird's mass increases and the increased drag caused by the subcutaneous fat stores (and the corresponding increased frontal area). As fat stores and mass increase, therefore, a point of diminishing returns is reached, where flight range would begin to decline because of the additional mass and drag. Given this limit, a bird's flight range depends on its speed and the rate at which that stored energy is used at that speed. Of course, a bird's speed and the rate of energy use are influenced by their wing and body shape, with higher aspect ratios and

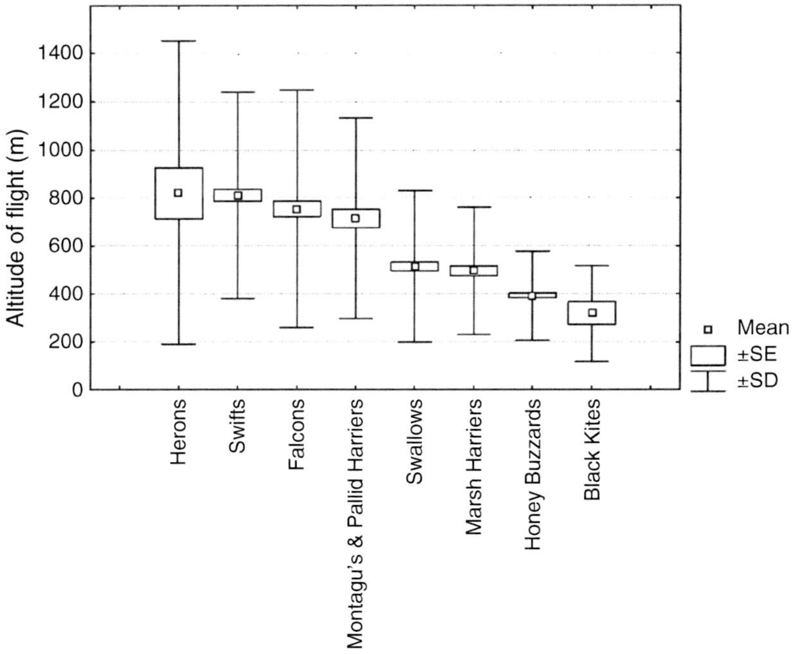


**Fig. 10.47** Mean flight altitudes of birds migrating across the Strait of Messina between Sicily and mainland Italy. Birds were tracked using radar, and wind speed and direction were monitored using balloons with aluminum reflectors that could also be tracked by radar. Flight altitudes ranged from 10 to 2495 m above ground. All species flew at altitudes with favorable tailwinds, and differences among species in mean altitude were due to differences in wind conditions during their spring migrations. Values are presented as means, standard errors (SE), and standard deviations (SD). Montagu's Harrier, *Circus pygargus*; Pallid Harrier, *Circus macrourus*; Eurasian Marsh-Harrier, *Circus aeruginosus*; European Honey-buzzard, *Pernis apivorus*; Black Kite, *Milvus migrans*. (Figure from Mateos-Rodríguez and Liechti 2012; © 2011 Oxford University Press, used with permission)

lower wing loading generally meaning faster speeds and a lower rate of energy use. As a result, those species of birds with the longest flight ranges are small to medium-sized birds with relatively high aspect ratios and relatively low wing loading (Table 10.3).

Birds with the longest flight ranges use their fuel at low rates. Although fuel consumption rates have been determined for few species of birds, the lowest value obtained thus far is for Bar-tailed Godwits (Table 10.3). These godwits make the longest known non-stop flights of any bird and are able to do so in part because of the low rate at which they use their energy stores. One factor contributing to this low rate is that Bar-tailed Godwits only initiate these flights when wind conditions are best (i.e., tailwinds or, at minimum, not headwinds) and are apparently able to do so by detection of changes in barometric pressure that signal the likelihood of favorable winds for the next several days (Gill et al. 2014).

## 10.9 Thermoregulation

Among birds, mean body temperatures range from about 36 to 45°C (Prinzinger et al. 1991). To maintain body temperature, the physiological and metabolic reactions that produce heat must be balanced against the heat being lost. Birds lose heat via conduction, convection, radiation, and evaporative water loss (Fig. 10.52):

**Table 10.2** Distances flown per day (excluding stopover days) by several species of migrating birds tracked using either geolocators or satellite transmitters

| Order | Common name | Scientific name | Distance flown per day (km/day) | Reference |
|---|---|---|---|---|
| Columbiformes | European Turtle-Dove | *Streptopelia turtur* | 240–812 | Eraud et al. (2013) |
| Apodiformes | Common Swift | *Apus apus* | 336 | Åkesson et al. (2012) |
| Charadriiformes | Little Ringed Plover | *Charadrius dubius* | 151–222 | Hedenström et al. (2013) |
| | Pacific Golden Plover[a] | *Pluvialis fulva* | 1438–1807 | Johnson et al. (2012) |
| | Long-billed Curlew | *Numenius americanus* | 358–479 | Page et al. (2014) |
| | Bar-tailed Godwit[b] | *Limosa lapponica* | 1400 | Battley et al. (2012) |
| | Red Knot | *Calidris canutus* | 1333 | Niles et al. (2010) |
| | Great Snipe[c] | *Gallinago media* | 1723–2310 | Klaassen et al. (2011) |
| | Long-tailed Jaeger | *Stercorarius longicaudus* | 800–900 | Sittler et al. (2010) |
| | Lesser Black-backed Gull | *Larus fuscus* | 177 | Klaassen et al. (2012) |
| | Sabine's Gull | *Larus sabini* | 813 | Stenhouse et al. (2012) |
| | Common Tern | *Sterna hirundo* | 520–720 | Nisbet et al. (2011) |
| | Arctic Tern | *Sterna paradisaea* | 390–670 | Egevang et al. (2010) |
| Procellariformes | Manx Shearwater | *Puffinus puffinus* | 1320 | Guilford et al. (2009) |
| | Westland Petrel | *Procellaria westlandica* | 1350 | Landers et al. (2011) |
| Bucerotiformes | Eurasian Hoopoe | *Upupa epops* | 120 | Bächler et al. (2010) |
| Accipitriformes | Osprey | *Pandion haliaetus* | 16–804 | Monti et al. (2018) |
| | European Honey-buzzard | *Pernis apivorus* | 103–319 | Vansteelant et al. (2015) |
| Falconiformes | Lesser Kestrel | *Falco naumanni* | 300–850 | Catry et al. (2011) |
| Passeriformes | Scissor-tailed Flycatcher | *Tyrannus forficatus* | 419 | Jahn et al. (2013) |
| | Veery | *Catharus fuscescens* | 350–570 | Heckscher et al. (2011) |
| | Wood Thrush | *Hylocichla mustelina* | 242 | Stutchbury et al. (2009) |
| | Great Reed Warbler | *Acrocephalus arundinaceus* | 294–584 | Lemke et al. (2013) |
| | Purple Martin | *Progne subis* | 833 | Stutchbury et al. (2009) |
| | Red-eyed Vireo | *Vireo olivaceus* | 508 | Callo et al. (2013) |
| | Tree Swallow | *Tachycineta bicolor* | 973–1359 | Bradley et al. (2014) |
| | Blackpoll Warbler | *Setophaga striata* | >900 | Williams et al. (1978), DeLuca et al. (2015) |
| | Golden-winged Warbler | *Vermivora chrysoptera* | 117 | Larkin et al. (2017) |

[a]Some Pacific Golden Plovers flew at speeds as high as 112 km/h, the fastest recorded ground speed for any shorebird
[b]Continuous flight for 8 days
[c]Great Snipes flew at speeds averaging 80–97 km/h

**Fig. 10.48** Route taken by a Blackpoll Warbler (*Setophaga striata*) during fall migration. Many of those that breed in eastern Canada make long transoceanic flights from eastern Canada to northern South America. (Figure from DeLuca et al. 2015; © 2015 The Authors. Published by the Royal Society, All rights reserved, used with permission)

1. Conduction occurs as warmer, faster-moving molecules transfer some of their energy (heat) to adjacent molecules, e.g., air or water.
2. Convective heat loss occurs as warmer, faster-moving molecules (e.g., air or water) transfer some of their energy (heat) to adjacent molecules.
3. Radiation is the movement of energy (heat in the form of infrared radiation) away from a bird's body, movement that does not require a medium (e.g., air or water).
4. Evaporative heat loss occurs when liquid water transitions to water vapor, a change that requires a bird's body heat.

Except within a range of ambient temperatures called the thermoneutral zone, maintaining a constant body temperature makes a steady demand either on the biochemical processes of heat production or the physical mechanisms for heat loss. Thermoneutral zones of birds range in extent from just a few degrees in small birds to as

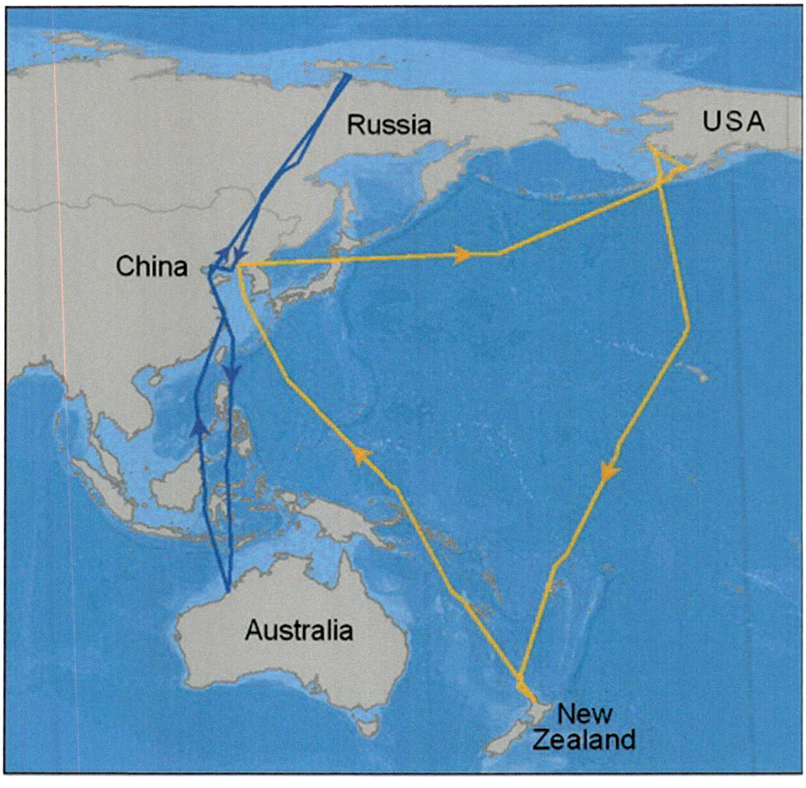


**Fig. 10.49** Round-trip migrations of two Bar-tailed Godwits (*Limosa lapponica*). One godwit (orange track) flew 29,280 km including three main migratory flights of 10,270 km (in 7 days) from New Zealand to China, 6510 km (4.9 days) from China to Alaska, and 11,690 km (8.1 days) from Alaska to New Zealand. The other godwit (blue track) flew 21,210 km in four main flights, including 5620 km (4.2 days) from Australia to China, 3990 km (2.1 days) from China to Siberia (Russia), 4090 km (2.6 days) from Siberia to China, and 6270 km (5.3 days) from China to Australia. (Figure from Battley et al. 2012; © 2012 The Authors, used with permission)

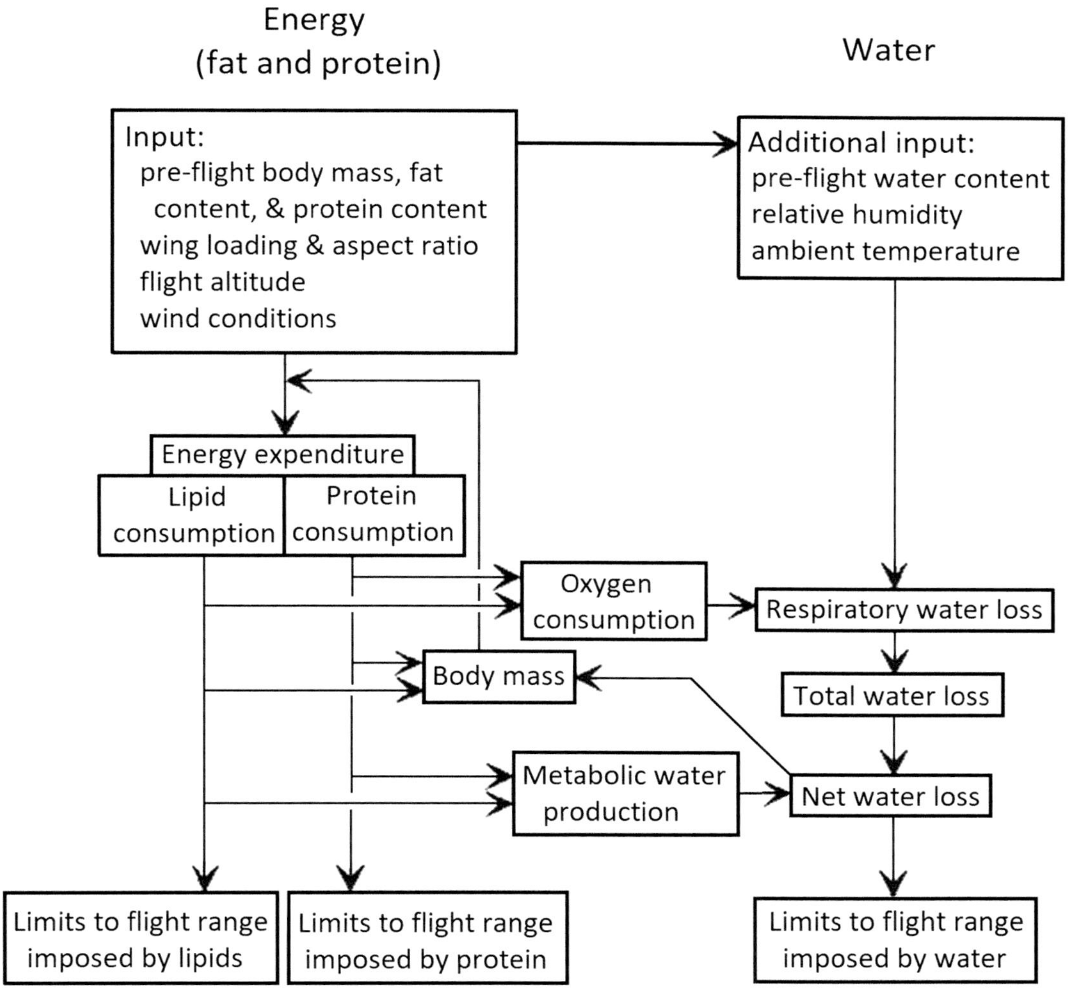


**Fig. 10.50** Variables that influence the flight ranges of migrating birds. Factors that are most important are pre-flight fat content and, to a lesser degree, pre-flight protein content. A bird's wing loading and aspect ratio as well as flying conditions (e.g., wind speed and direction) also impact flight ranges. Birds lose water via respiratory water loss, but metabolic water production and flying during conditions or at altitudes with higher humidity and lower temperatures can reduce the likelihood of dehydration during long flights. As birds lose fat mass and become lighter, the energetic cost of flight is reduced, allowing them to fly greater distances. (Figure modified from Jenni and Jenni-Eiermann 1999; used with permission of Susanne Jenni-Eiermann)

much as 30°C in larger birds (Dawson and Whittow 2000) and, within this range of temperatures, metabolic rates are at the minimum level (i.e., basal metabolic rate, or BMR). When ambient temperatures fall below the thermoneutral zone (lower critical temperature or LCT), heat production must increase and/or heat loss must decrease. Above the other end of the thermoneutral zone (upper critical temperature or UCT), heat loss must increase (Fig. 10.53). Thermoneutral zones, and the corresponding upper and lower critical temperatures, vary among species and with latitude. For example, examination of the upper and lower critical temperatures of birds at different latitudes reveals that both tend to decrease with

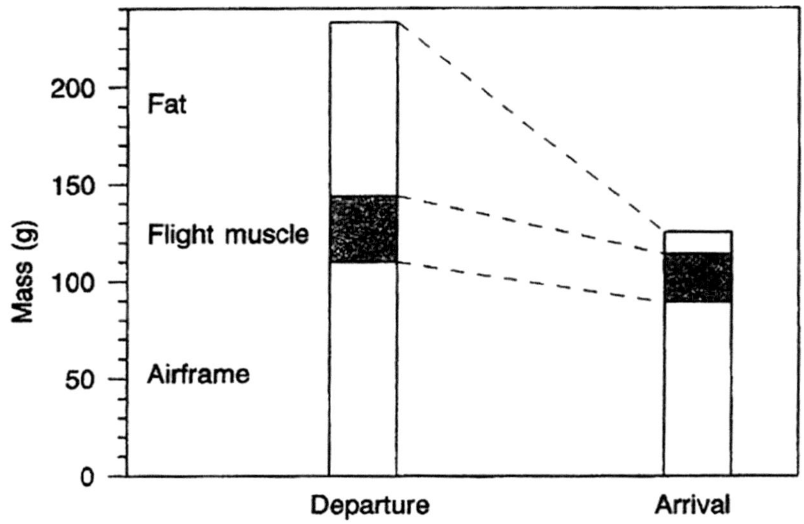


**Fig. 10.51** Proportions of a migrating Great Knot's (*Calidris tenuirostris*) body mass consisting of fat, flight muscle, and airframe (everything except fat and flight muscle) at departure and arrival after a non-stop flight of 5420 km from Australia to China. Fat provided most of the energy for the long flight, but mass was also lost from flight muscle and the bird's airframe (e.g., digestive system). (Figure from Pennycuick and Battley 2003; © Oikos. Published by John Wiley & Sons Ltd., used with permission)

increasing latitude, especially lower critical temperatures (Fig. 10.55).

Lower critical temperatures tend to exhibit more variation among species than upper critical temperatures (Araújo et al. 2013; Figs. 10.54 and 10.55). One possible explanation for this is that Earth's climate has largely been warm throughout its history (and throughout the period birds have existed), with occasional periods of colder climates (Ruddiman 2001). In addition, the equatorial region has remained consistently warm even during those periods with colder climates. For the ancestors of present-day species that evolved in areas with relatively warm temperatures, maximum performance temperatures would likely be closer to their upper critical temperatures (thermal maxima) than their lower critical temperatures (thermal minima) (Asbury and Angilletta 2010) due to enzyme inactivation at higher temperatures (Kingsolver 2009). Subsequent evolution and speciation have allowed birds to occupy cooler thermal niches, requiring lower thermal minima, at higher latitudes and altitudes (Araújo et al. 2013). A second possible explanation for the reduced interspecific variation among birds (and other organisms; Araújo et al. 2013) in tolerance to heat is that it is a consequence of limited interspecific variation in mechanisms for countering the destabilizing effects of high temperatures on cell membranes and protein structure (Angilletta 2009; Araújo et al. 2013).

In addition to ambient temperature, a bird's ability to, and the energy required to, thermoregulate can be influenced by wind velocity. The effect of wind on a bird's surface temperature is clearly illustrated by using thermal imaging (infrared thermography), where thermal radiation being emitted is converted to different colors (Fig. 10.56a). Zerba et al. (1999) used thermal imaging to examine the effect of wind on the surface temperatures of House Finches and found that even a light wind increased convective heat loss, particularly in the head region. The higher temperature of the head likely reflects a combination of greater blood flow and reduced insulation, particularly around the eyes. Similarly, Bakken et al. (2002) found that wind was a

**Table 10.3** Estimates of the proportion of body mass used to power flight during migration for species that make long-distance, long-duration flights (Data from Hedenström 2010; © 2010 Anders Hedenström, open-access article distributed under the terms of the Creative Commons Attribution License)

| Species | Scientific name | Mass (g) at beginning of flight | Distance flown (km) | Percent of body mass used per hour | Reference |
|---|---|---|---|---|---|
| Thrush Nightingale[a] | *Luscinia luscinia* | 25 | [b] | 1.00 | Kvist et al. (1998) |
| Bar-tailed Godwit | *Limosa lapponica* | 166 | 11,000 | 0.42 | Gill et al. (2009) |
| Great Knot | *Calidris tenuirostris* | 143 | 5400 | 0.52 | Pennycuick and Battley (2003) |
| Red Knot | *Calidris canutus* | 126 | 4800 | 0.77 | Piersma et al. (2005) |
| Ruddy Turnstone | *Arenaria interpres* | 115 | 3700 | 0.48 | Thompson (1974) |
| Ruby-throated Hummingbird | *Archilochus colubris* | 4.4 | 1100 | 2.0 | Lasiewski (1962) |
| Swainson's Thrush[a] | *Catharus ustulatus* | 96–100 | [c] | 0.51–0.66[d] | Gerson and Guglielmo (2011), Mack and Yong (2020) |
| American Robin[a] | *Turdus migratorius* | 75–85 | [e] | 4.3–4.9[f] | Gerson and Guglielmo (2013), Vanderhoff et al. (2020) |

[a]Study conducted in a wind tunnel
[b]Thrush Nightingales are long-distance migrants (>8000 km, Europe–Africa; Stach et al. 2013), but the distance of their longest non-stop flights during migration is unknown
[c]Swainson's Thrushes are long-distance migrants that typically fly 105–375 km/night (Mack and Yong 2020)
[d]Estimated using measurements of mass during flight by Gerson and Guglielmo (2011) and body masses provided by Mack and Yong (2020)
[e]Most American Robins migrate 500–1200 km (Vanderhoff et al. 2020), but not in a single flight
[f]Estimated using measurements of mass during flight by Gerson and Guglielmo (2013) and body masses provided by Mack and Yong (2020)

significant factor in the thermoregulation of young shorebirds in the Arctic tundra, with heat loss and metabolic rates increasing by as much as 30–50% during periods of high wind (30–50 km/h). Other investigators have also reported increases in bird metabolic rates with increasing wind velocity (e.g., Hayes and Gessaman 1980; Goldstein 1983; Wolf and Walsberg 1996; Fig. 10.56b).

Because of the high thermal conductivity of water (25 times that of air), swimming and diving birds face an energetic challenge when in cold water (Box 10.8 Metabolic Rates and Lower Critical Temperatures on Land and Water). Birds can potentially lose heat rapidly, resulting in high thermoregulatory costs. For example, metabolic rates of Little Penguins (*Eudyptula minor*) increase significantly with decreasing water temperatures (Stahel and Nicol 1982; Fig. 10.57). Similar results have been reported for other species of aquatic birds (e.g., Enstipp et al. 2005, 2006).

As the previous paragraphs indicate, variation in ambient temperatures, wind velocities, and, for aquatic birds, water temperatures all influence a bird's body temperature and can, at times, cause body temperatures to exceed or fall below the upper and lower critical temperatures. When that happens, birds must increase their metabolic rates in an attempt to maintain normal body temperatures. To avoid these increased metabolic costs, birds can use a variety of behavioral strategies to help maintain their body temperatures within their thermoneutral zones (Tables 10.4 and 10.5; Figs. 10.58, 10.59, 10.60, 10.61, 10.62, 10.63, 10.64, 10.65, 10.66 and 10.67).

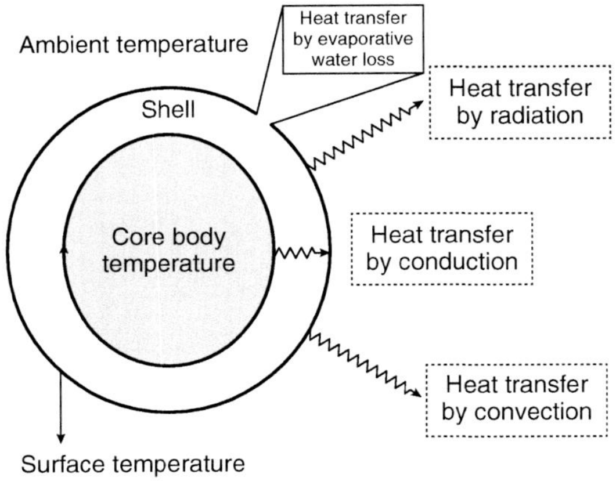


**Fig. 10.52** The central core of a bird's body is normally maintained within a range of just a few degrees C. The "shell" includes subcutaneous fat and a bird's feathers and has low thermal conductivity. The loss of heat from the core through the shell depends on a bird's surface area, the thickness and thermal conductivity of the shell, and the differences between core temperature, surface temperature, and ambient temperature. (Figure modified from Speakman and Król 2010; © 2010 The Authors. Journal compilation © 2010 British Ecological Society, used with permission)

Birds must sometimes choose between the relative importance of using particular behavioral strategies and other factors, such as the risk of predation. For example, wintering Dark-eyed Juncos (*Junco hyemalis*) can conserve energy during periods of cold weather by partially or completely covering their legs and feet with feathers (Fig. 10.68). However, Carr and Lima (2012a) found that ground-foraging juncos that assumed the most energy-conserving postures required more time to take off when startled, as they would be, for example, at the approach of a predator. This additional time could, at the approach of a fast-flying aerial predator like a Sharp-shinned Hawk (*Accipiter striatus*), mean the difference between escaping and being captured. Viewed from the perspective of evolutionary fitness, being killed by a predator is clearly worse than having to use some additional energy to try and maintain body temperature. As a result, juncos only used the most energy-conserving postures when most needed, i.e., when ambient temperatures fell below −10°C and well below their lower critical temperature of +20°C.

In another example of the trade-off needed when trying to minimize both thermoregulatory costs and predation risk, Carrascal et al. (2001) found that wintering Short-toed Treecreepers (*Certhia brachydactyla*) foraged in shaded patches of forest more than expected based on

**Fig. 10.53** The effect of changing ambient temperature on avian metabolic rates. When ambient temperatures fall below (lower critical temperature) or rise above (upper critical temperature) the thermoneutral zone, birds must expend energy to try and maintain their body temperature. (Figure modified from Brychta and Chen 2017; © 2016 Springer Nature, used with permission)

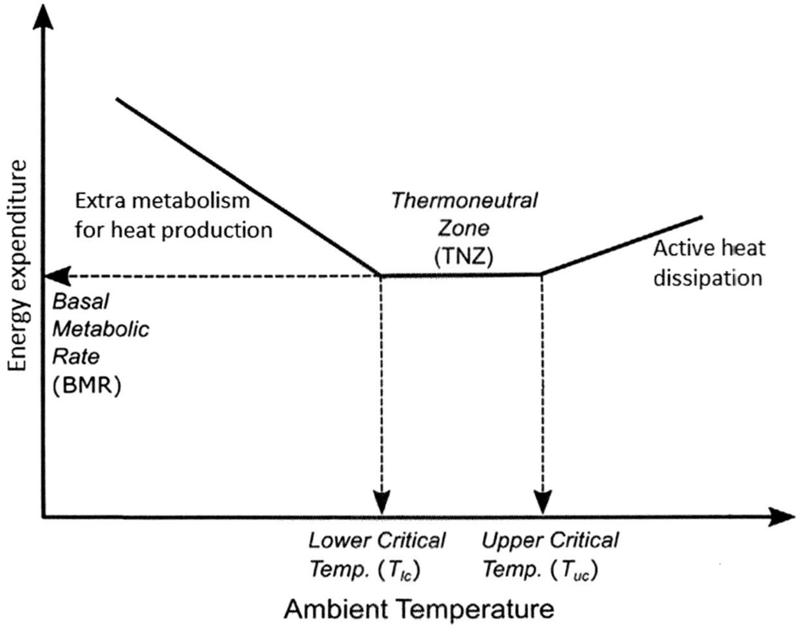

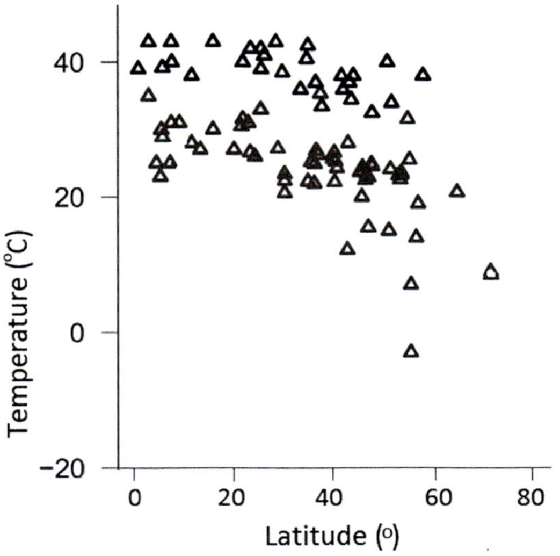


**Fig. 10.54** Latitudinal patterns in lower critical temperatures (gray triangles) and upper critical temperatures (black triangles) for several species of birds. Both upper and, especially, lower critical temperatures tend to decrease with increasing latitude. (Figure modified from Buckley et al. 2012; © 2012 Blackwell Publishing Ltd., used with permission)

availability (Fig. 10.69). Treecreepers might be expected to forage more in sunlit patches of forest when ambient temperatures fall below their lower critical temperature of 21.5°C because the solar radiation would help minimize thermoregulatory costs. Instead, however, treecreepers only began to forage in sunlit patches more than expected when ambient temperatures fell below about 8° C (Fig. 10.70). Short-toed Treecreepers are easier to detect (by humans and perhaps their predators as well) when foraging on sunlit trees (Fig. 10.72) and, therefore, likely at greater risk of predation. Short-toed Treecreepers, like the previously mentioned Dark-eyed Juncos, choose to minimize predation risk until ambient temperatures are well below their lower critical temperatures and the energetic cost of thermoregulation likely increases dramatically.

## 10.10 Responses to Temperatures Above and Below Avian Thermoneutral Zones

If, despite changes in location, posture, or behavior, birds are still unable to maintain their body temperatures, then they must either expend energy to increase heat loss (ambient temperature above upper critical temperatures) or increase heat production (ambient temperatures below the lower critical temperatures) (Box 10.10 Potential Roles of Activity and Eating on Reducing Thermoregulatory Costs). Alternatively, birds can allow their body temperatures to either increase (hyperthermia) or decrease (hypothermia).

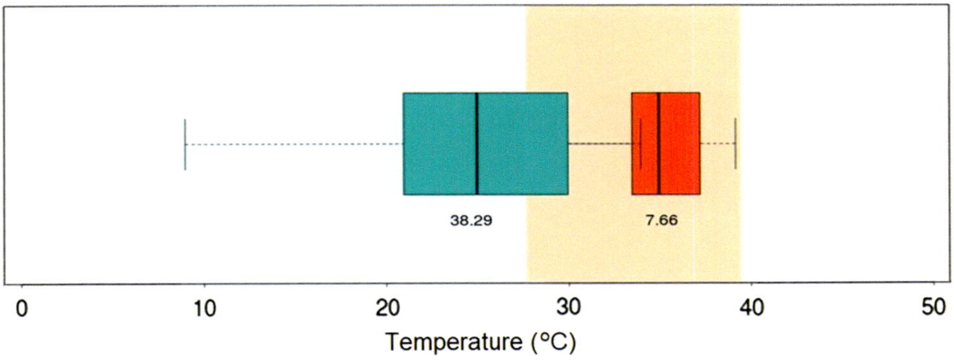


**Fig. 10.55** Box plot showing variation in the lower (blue) and upper (orange) critical temperatures of 39 species of birds. Birds exhibit much more interspecific variation in lower critical temperatures than upper critical temperatures. Thick vertical lines are the medians, boxes the interquartile ranges, and thin vertical lines the minimum and maximum values. (Figure modified from Araújo et al. 2016; open-access article available under the terms of the Creative Commons Attribution License (CC BY), https://creativecommons.org/licenses/by/4.0/)

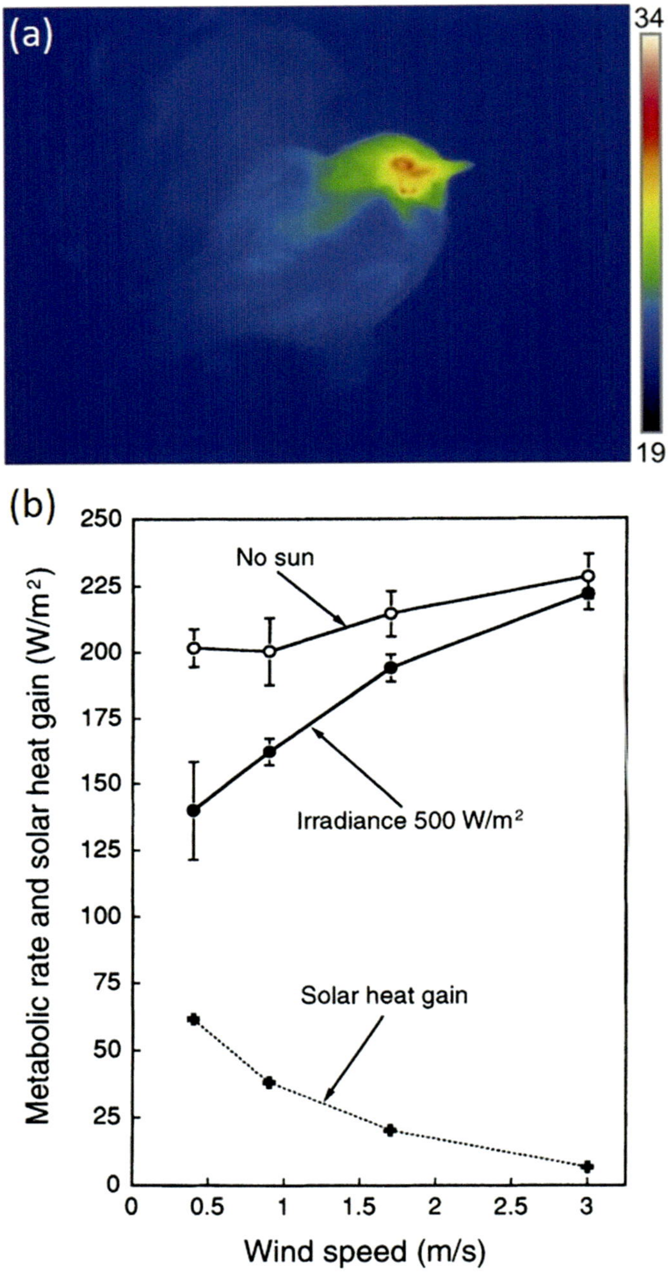


**Fig. 10.56** (**a**) Thermal image of a resting Garden Warbler (*Sylvia borin*) showing that most heat is being lost from the head region (t, temperature). (**b**) Mean metabolic rates (±95% confidence intervals) of Verdins (*Auriparus flaviceps*) at different wind speeds and in the presence and absence of simulated solar radiation (Irradiance) at an air temperature of 15°C. (Figure a from Ferretti et al. 2019; © 2019 The Authors. Published by Elsevier Ltd., used with permission. Figure b from Wolf and Walsberg 1996; © 1996 by the Ecological Society of America, used with permission)

Birds generate heat by shivering, the rapid, involuntary contraction and relaxation of skeletal muscles. The flight muscles, pectoralis and supracoracoideus, contribute most to shivering thermogenesis, but some leg muscles may also contribute (e.g., gastrocnemius and tibialis anterior; Carey et al. 1989). Shivering muscles contract about 10–20 times per second, with antagonistic muscle motor units activated by the nervous system so there is very little movement. However, shivering skeletal muscles, particularly the large pectoralis muscles, use substantial

**Box 10.8 Metabolic Rates and Lower Critical Temperatures on Land and Water**

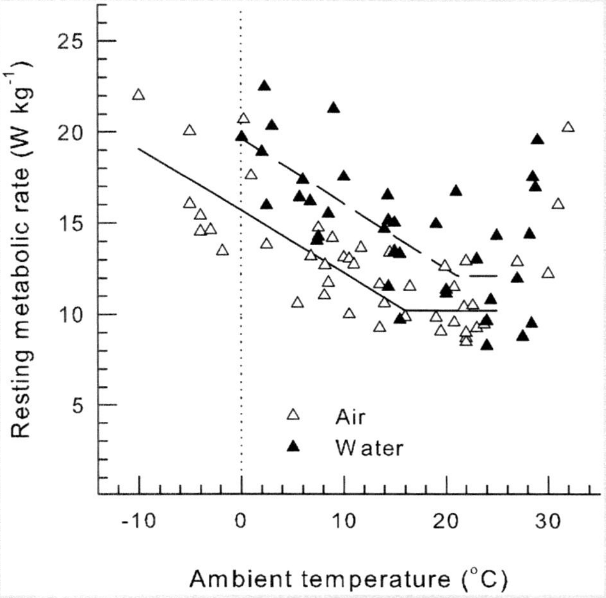


Resting metabolic rates and lower critical temperatures of Cassin's Auklets in water and on land. (Figure from Richman and Lovvorn 2011; © 1999 CCC Republication, used with permission)

For Cassin's Auklets (*Ptychoramphus aleuticus*) and most other seabirds (except large diving ducks that are apparently more cold-adapted, with exceptionally thick plumage), resting metabolic rates are higher when resting (floating) on water than when resting on land. Birds lose heat more rapidly when in water because thermal conductivity of water is 25 times higher than that of air. Richman and Lovvorn (2011) found that the lower critical temperature of Cassin's Auklets (~165 g) was higher in water (21°C) than in air (16°C). In addition, at lower temperatures, the resting metabolic rates of the auklets averaged 25% higher in water than in air. Because diving birds spend most of their time resting on water, the resulting high metabolic costs mean that they require large amounts of food (e.g., up to 67% of their body mass per day for Cassin's Auklets; Hodum et al. 1998), and that requirement may restrict them to areas of high-quality habitat with ample supplies of food (Grémillet et al. 1999).

(continued)

**Box 10.8** (continued)

Cassin's Auklet. (Photo from National Park Service, CC0 Public Domain)

amounts of oxygen and ATP (Fig. 10.71) and, as ATP is produced and subsequently converted to ADP, considerable heat is generated and transported throughout the avian body via the circulatory system. Shivering can generate a two- to fivefold increase in metabolic heat production over a period of just a few minutes (Tattersall et al. 2012).

To examine the possible effects of time of day and hunger levels on shivering by captive Rock Pigeons (*Columba livia*) maintained at a temperature below (21°C) their lower critical temperature, Hohtola et al. (1998) monitored their body temperatures, oxygen consumption, and timing and extent of shivering by their pectoralis muscles over a 10-day period. Rock Pigeons shivered much more during the light periods than the dark periods and also allowed their body temperatures to decline during the dark periods. Similar "nocturnal" hypothermia has been reported in many other species of birds. As expected, oxygen consumption (or energy used) increased during periods of shivering and shivering generated heat that increased pigeon body temperatures. When starved, pigeons conserved energy by allowing their body temperatures to decline more during dark periods, with "nocturnal" body temperatures declining more each dark period as the period of starvation continued. With lower "nocturnal" body temperatures, pigeons had to shiver more vigorously each "morning" to again increase their body temperature. Although conducted in the lab, the results of this study with Rock Pigeons demonstrate how effectively shivering can generate heat, although at a high energetic cost.

Most mammals can also produce heat via a process called non-shivering thermogenesis, where the mitochondria of some fat cells (brown adipose tissue or BAT) generate heat by transporting protons normally used by the mitochondria of other cells to produce adenosine triphosphate or ATP (Box 10.11 Non-shivering Thermogenesis in Mammals). Birds do not have brown adipose tissue and have lost the gene needed for UCP so cannot generate heat by non-shivering thermogenesis as mammals do. However, birds do have a gene for UCP3

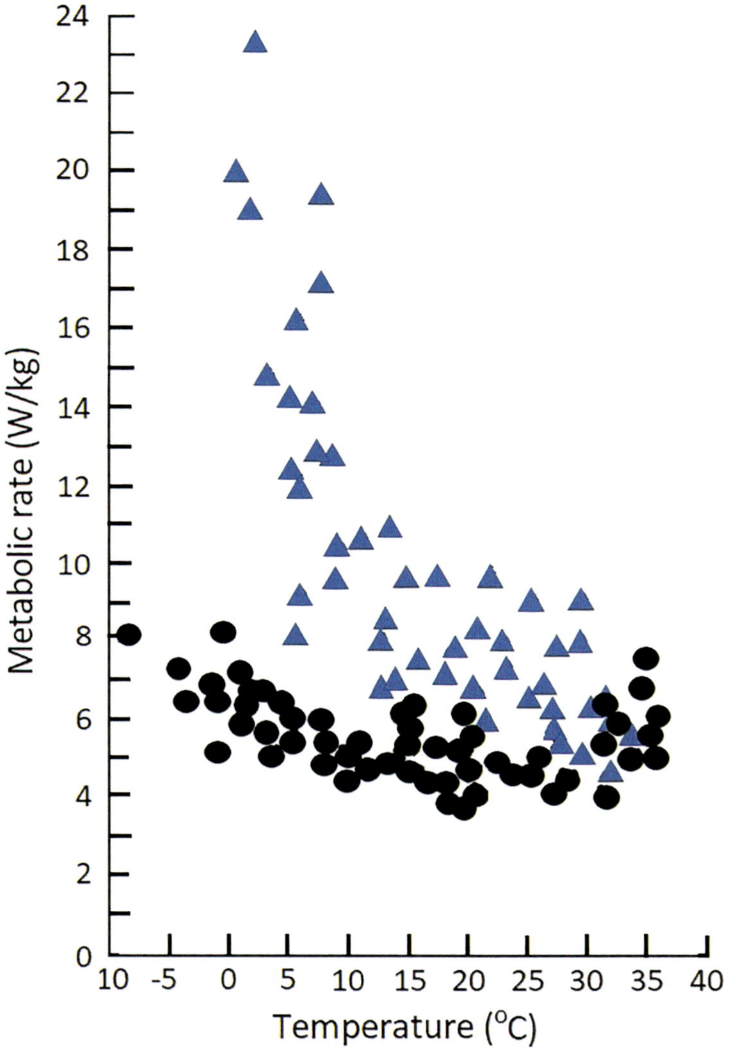


**Fig. 10.57** Effect of air (filled circles) and water (triangles) temperature on the mass-specific metabolic rates of Little Penguins (*Eudyptula minor*) when resting. With decreasing air and, especially, water temperatures, metabolic rates increased significantly. (Figure from Stahel and Nicol 1982; © 1982 Springer-Verlag, used with permission)

(also referred to as avUCP) that is expressed mainly in the skeletal muscles of cold-acclimatized birds (Teulier et al. 2010) where it appears to initiate non-shivering thermogenesis. However, the mechanism by which avUCP does so remains to be determined.

## 10.11 Regulating Heat Gain and Loss

Avian plumage helps birds retain heat, but also affects the amount of radiative heat they acquire from solar radiation. Heat is transferred through plumage by (1) conduction and convection through air, (2) conduction along the feathers themselves, and (3) radiation (Wolf and Walsberg 2000). About 95% of the heat transfer between a bird and its environment is via, and evenly divided between, the first two of these processes, whereas radiation accounts for only about 5% of the heat transfer (Wolf and Walsberg 2000). Conduction (i.e., the transfer of heat from faster-moving molecules to adjacent slower-moving molecules) plays a critical role in heat transfer and can be reduced with increased insulation. As such, birds in areas with colder temperatures

**Table 10.4** Behavioral responses that increase heat loss when ambient temperatures are high

| Behavior | Taxa or species | Example(s) |
|---|---|---|
| Panting | Many birds | Smit et al. (2016) |
| Vocal panting | Zebra Finches (*Taeniopygia guttata*) | Pessato et al. (2020) |
| Gular flutter | Many birds | O'Connor et al. (2017), Talbot et al. (2018) |
| Yawning | Budgerigars (*Melopsittacus undulates*) | Gallup et al. (2017) |
| Elevate scapulars when perched or incubating and/or extend head and neck | Many birds | Bartholomew and Dawson (1979) |
| Foraging or roosting in the shade | Many desert birds | Williams and Tieleman (2001), Ruth et al. (2020) |
| Spend less time foraging and more time resting during hottest parts of day | Many desert birds | Pattinson and Smit (2017), Funghi et al. (2019) |
| Press body against cool surface of shaded soil, tree bark, or plants to conduct heat away from body | Many desert birds | Williams et al. (1999) |
| Perch on stones or vegetation above ground and in shade and hold wings away from body to expose thinly feathered areas under wings | Many desert birds | Shobrak (1998) |
| Press ventral parts of body against wet and/or cool substrates to conduct heat away from body | Many birds | Wolf et al. (1996), Ryeland et al. (2021) |
| Use burrows of lizards or mammals as thermal refugia | Several desert species | Williams et al. (1999) |
| Roost in trees on north-facing slopes where temperatures are cooler | Spotted Owl (*Strix occidentalis*) | Barrows (1981) |
| Lifting webbed foot off ground in the shade of their bodies to increase convective cooling | Some albatrosses | Whittow (1980) |
| Extend legs when flying rather than bringing them up into the plumage | Many species of birds | Bryant (1983), Grant (1993), Neumann (2016) |
| Wing drooping (Figs. 10.65 and 10.66) | Many species of birds | Kahl (1971), Smit et al. (2016) |
| Defecate on legs, or urihidrosis (evaporation of moisture on legs enhances heat loss; Fig. 10.67) | Vultures and storks | Kahl (1963), Cabello-Vergel et al. (2021) |
| Foot-wetting and belly-soaking (Fig. 10.68) | Order Charadriiformes | Grant (1978) |
| Diving into shallow ponds then perching in the shade and spreading wings to enhance evaporative cooling | European Bee-eater (*Merops apiaster*) | Yosef (2010) |
| Perching in relief-air vents of air-conditioned buildings | Rosy-faced Lovebirds (*Agapornis roseicollis*) | Mills and McGraw (2021) |

typically have more insulation. For example, based on a phylogenetic comparative analysis of 152 species of birds, Osváth et al. (2018) found that feather mass and the density of downy feathers are higher for species in colder environments. In a study of high-elevation Himalayan songbirds, smaller birds, with greater surface area relative to volume than larger birds and, therefore, higher conductance, were found to have relatively longer feathers (and thus a more insulative plumage) than larger birds (Barve et al. 2021). Still air is the best insulating material available for birds (and other animals) and, importantly, feathers are more effective at trapping air than either the hair (fur) of mammals or fat (Tattersall et al. 2012; Fig. 10.72). In addition, birds can vary the amount of air trapped in their plumage and the thickness of their insulating layer by erecting their feathers (ptiloerection).

Depending on the ambient temperature, solar radiation can be an important source of heat for birds. For example, Verdins (*Auriparus*

**Table 10.5** Behavioral responses of birds to cool or cold ambient temperatures

| Behavior | Taxa or species | Example(s) |
|---|---|---|
| Sunbathing, sometimes with wings extended or drooping | Many birds, including Anhingas and cormorants (Fig. 10.64) | Hennemann (1982) |
| Forage or roost in sunlit areas with more exposure to solar radiation | Several species of birds | Huertas and Díaz (2001), Carrascal et al. (2001), Villén-Pérez et al. (2014) |
| Place beak and/or head under wing or wing scapulars when roosting or sleeping | Many species of birds, especially those found in aquatic habits (Fig. 10.60) | Bartholomew and Dawson (1979) |
| Pulling one leg up into plumage and stand on the other leg or squat so legs and feet are covered by ventral feathers (Fig. 10.60) | Many birds, especially those found in aquatic habitats and those with longer legs | Carr and Lima (2012a), Ryeland et al. (2019) |
| Leg-tuck, concealing legs and feet under their bodies (Fig. 10.59) | Many birds | Yorzinski et al. (2018) |
| Huddling or clustering (Fig. 10.69, Box 10.9 Huddling by Emperor Penguins) | Many birds | McKechnie and Lovegrove (2001), Gilbert et al. (2006) |
| Roosting in dense foliage or tree cavities | Many birds | Cooper (1999) |
| Forage and roost on warm beach wrack | Some shorebirds | Davis and Keppel (2021) |
| Roosting in snow burrows (Fig. 10.62) | Grouse and ptarmigans | Andreev (1999), Shipley et al. (2019, 2020) |
| Roost in fully enclosed nests with a small side entrance during the winter | Verdins (*Auriparus flaviceps*) | Buttemer et al. (1987) |
| Rest on blacktop (e.g., asphalt) roads that absorb and store heat | Many species of birds | Whitford (1985) |

*flaviceps*), small insectivores that are year-round residents of deserts in the southwestern United States and northern Mexico, can reduce their metabolic rates by as much as 50% by moving from a shaded, windy site to a sunny site protected from the wind (Wolf and Walsberg 1996). More generally, in montane forest habitat in central Spain, Huertas and Díaz (2001) found that wintering birds were more abundant than expected in sunlit patches of forest and suggested that, by foraging in these areas, birds could reduce the metabolic costs of thermoregulation. Of course, too much solar radiation can cause birds to become hyperthermic so, on sunny days with high ambient temperatures, birds alter their behavior to minimize heat gain. Cactus Wrens (*Campylorhynchus brunneicapillus*), for example, spend more time in shaded areas and reduce overall levels of activity when temperatures exceed 35°C (Ricklefs and Hainsworth 1968). The reduced heat load for Verdins moving from sunlit to shaded sites when ambient temperatures are high is thermally equivalent to a 12°C reduction in air temperature (Wolf and Walsberg 1996).

The amount of solar radiation acquired by birds from solar radiation can be influenced by plumage color. Dark-colored plumage is generally assumed to absorb more short-wave radiation (heat) than light-colored plumage. However, the amount of heat that actually reaches a bird's skin depends not just on plumage color, but on feather structure and wind speed as well. Contour feathers with denser and longer down barbs reduce heat transfer between a bird's skin and its environment (Fig. 10.73). As a result, darker feathers with these characteristics can better reduce heat penetration to a bird's skin than lighter-colored feathers. In addition, the position of bird feathers (degree to which they are depressed against the skin or erected) and wind speed can greatly impact heat transfer through plumage. The effect of wind on thermal penetration is nicely illustrated by two subspecies of Willow Ptarmigan (*Lagopus lagopus*) that differ in the color of their winter plumage. When exposed to solar radiation in still air, the plumage and skin of the reddish-brown subspecies are warmed more than those of the white subspecies.

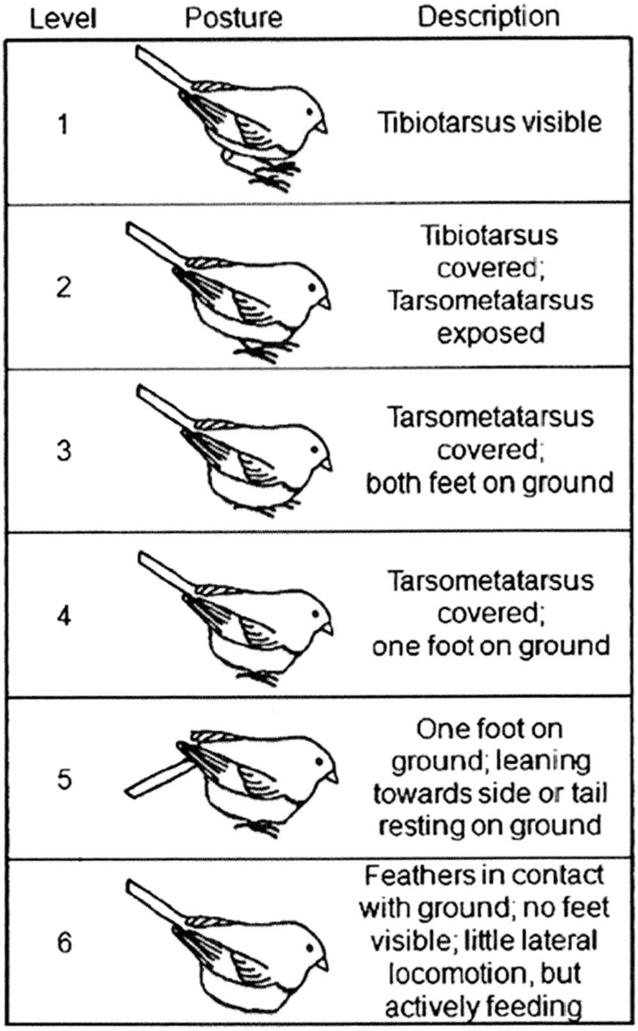


**Fig. 10.58** Postures sometimes used by small birds when roosting, sleeping, or even foraging to partially or totally cover legs and feet to reduce heat loss when ambient temperatures are low. (Figure from Carr and Lima 2012a; © 2011 Oxford University Press, used with permission)

However, at a wind speed of 11 km/h, there was no difference between the dark and light plumages (Ward et al. 2007, Fig. 10.74). Similarly, Wolf and Walsberg (2000) also found the feathers and skin of Rock Pigeons (*Columba livia*) with dark plumage acquire a greater heat load from solar radiation than white plumage in still air (Fig. 10.75). However, the difference between white and black plumage in heat load declined with ptiloerection of feathers (plumage depth of pigeons can vary from about 8 mm when depressed to about 31 mm with maximum ptiloerection) and with increasing wind speed (Fig. 10.75). In fact, at wind speeds above 6 m/s (or 21.6 km/h), the heat load from solar radiation for white plumage exceeded that of black plumage (Fig. 10.76). By increasing the amount of air trapped within the plumage, ptiloerection increases thermal resistance by reducing convective and radiative heat flow to (and from) a bird's body.

With cold temperatures below their thermoneutral zone, birds need to retain heat and, as with birds trying to minimize their heat load, use ptiloerection to reduce loss of heat from the body via convective and radiative heat flow (Fig. 10.76; Box 10.12 Ptiloerection, Shivering, and Body Temperature). For some birds at high

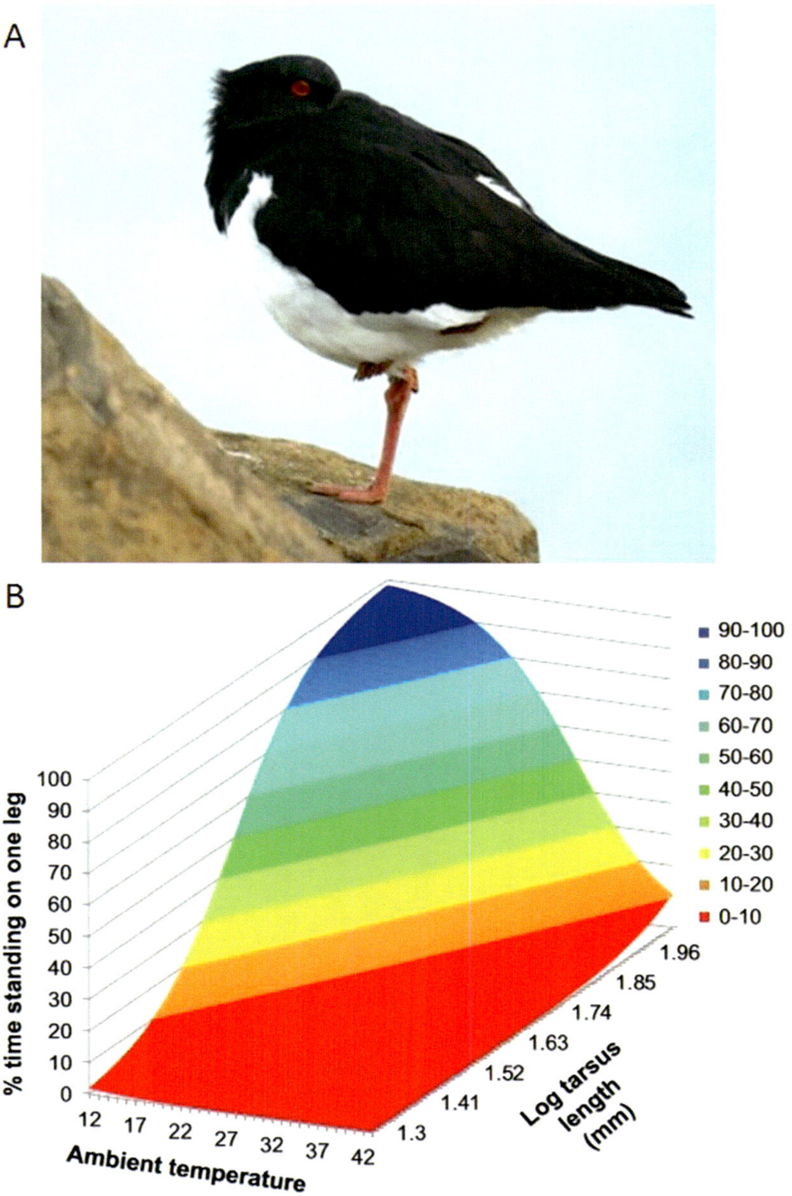


**Fig. 10.59** (**a**) Pied Oystercatcher (*Haemotopus longirostris*) standing on one leg and with the other leg tucking into the plumage to conserve heat. (Photo by M. A. Weston, Deakin University, from Pavlovic et al. 2019, used with permission of Michael Weston). (**b**) Predicted relationship between time spent standing on one leg while roosting in relation to ambient temperature (°C) and tarsus length. Birds with longer legs are predicted to spend more time standing on one leg, especially with declining ambient temperatures. (Figure from Ryeland et al. 2019; © 2019 The Authors. Published by John Wiley and Sons, used with permission)

latitudes and altitudes, their plumage becomes even more effective at reducing heat loss during the winter because they have more feathers than during the warmer summer months. Examples of this include American Goldfinches (*Carduelis tristis*; Middleton 1986), Dark-eyed Juncos (*Junco hyemalis*; Swanson 1991), Superb Fairywrens (*Malurus cyaneus*; Lill et al. 2006), Juniper Titmice (*Baeolophus griseus*; Cooper 2002), Mountain Chickadees (*Poecile gambeli*; Cooper 2002), House Sparrows (*Passer domesticus*; Barnett 1970), and others (Dawson and Carey 1976; Carey et al. 1978; Dawson et al. 1983).

Birds can also regulate body temperature by varying the amount of evaporative water, or heat, loss from the skin and respiratory membranes. Gas (or water vapor) molecules have more energy than liquid (or water) molecules because they

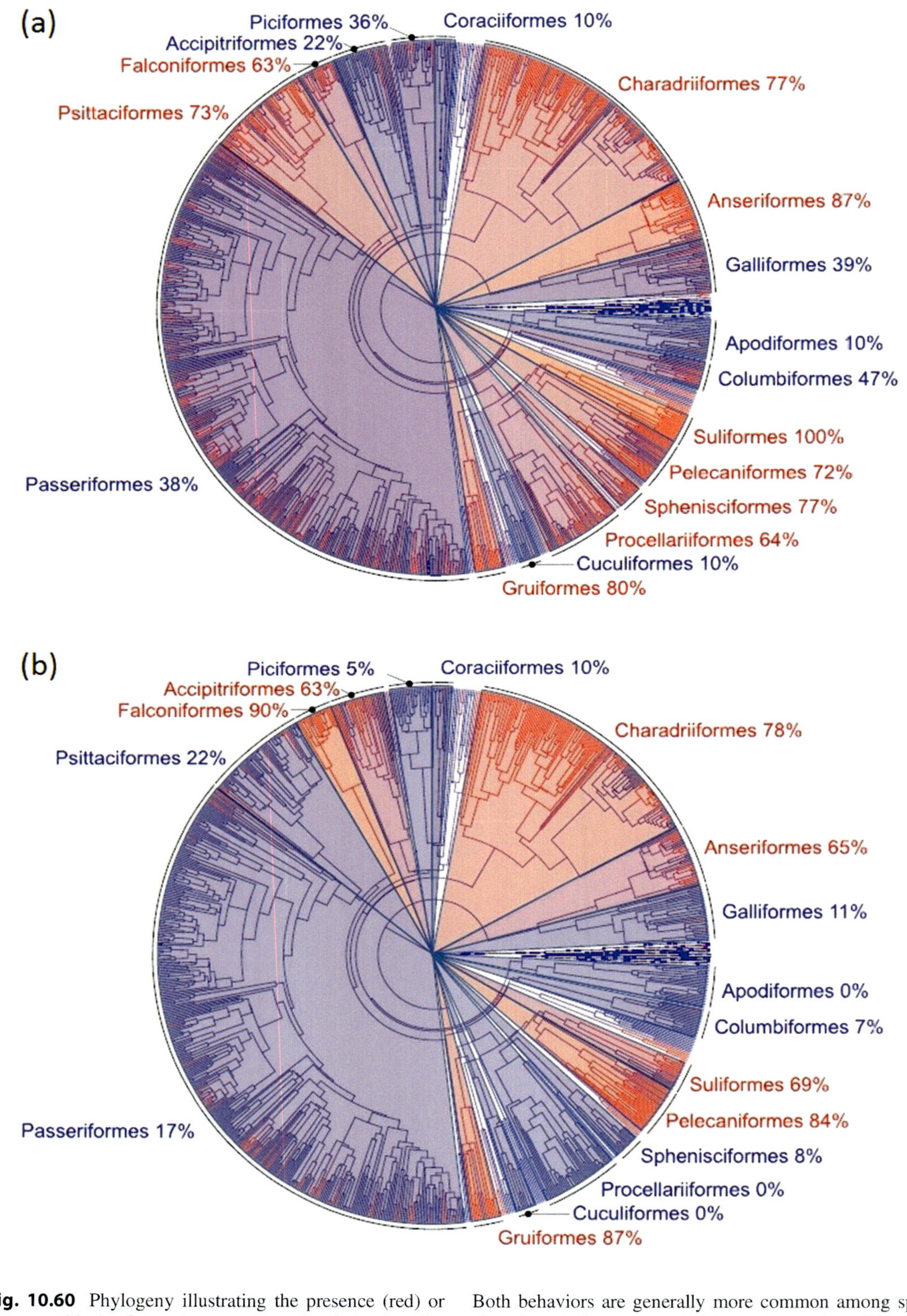


**Fig. 10.60** Phylogeny illustrating the presence (red) or absence (blue) of two heat conservation behaviors for 825 species of birds. (**a**) Placing bill in plumage, and (**b**) standing on one leg with the other tucked in the plumage. Both behaviors are generally more common among species found in aquatic habitats (e.g., Anseriformes, Charadriiformes, Suliformes, and Pelecaniformes), perhaps because body heat is lost faster in water than in air

move much faster (more kinetic energy). For a molecule of water to enter the gas phase, energy in the form of heat is needed from the surrounding environment. So, when heat from a bird's skin or respiratory membranes causes water on those surfaces to evaporate, that heat is lost and the skin or respiratory membrane becomes correspondingly cooler. The relative thermoregulatory importance of cutaneous (skin) and respiratory evaporative water loss (EWL) for birds exposed to ambient temperatures above their upper critical temperature varies among species. Based on data from 56 arid-zone species of birds, McKechnie et al. (2021) determined that ambient temperatures above which EWL increased varied by more than 10°C and, for most species, ranged between 36 and 43°C. Respiratory EWL becomes increasingly important with increasing ambient temperature. Williams and Tieleman (2001, 2005) examined the relationship between ambient temperature and total evaporative water loss (TEWL) for four species of songbirds and found that, at moderate temperatures, cutaneous water loss represented 50–75% of TEWL, but, at high temperatures, respiratory water loss dominated (Fig. 10.77). Studies of other species have revealed an increasing dependence on respiratory EWL for heat loss with increasing ambient temperatures (Bouverot et al. 1974; Richards 1976), especially among caprimulgids that depend primarily on gular flutter (Talbot et al. 2018). Species in the order Columbiformes, in contrast, rely more on cutaneous EWL to lose heat as ambient temperatures increase, which allows pigeons and doves to dissipate large heat loads with little increase in metabolism at ambient temperatures as high as 60°C (Marder and Arieli 1988; Smit et al. 2016; McKechnie et al. 2016). Despite this variation among taxa, McKechnie et al. (2021) reported a high degree of overlap in evaporative scope (ratio of maximum EWL to minimum thermoneutral EWL) among species of birds in different taxa and suggested that "the primary avenue of evaporation per se (i.e., panting, gular flutter or cutaneous evaporation) is not the major determinant of the maximum rate of EWL a bird can achieve." Rather, the primary determinant of the upper limit of evaporative cooling capacity and heat tolerance among different taxa of birds is the metabolic cost of the various heat dissipation pathways (McKechnie et al. 2021). However, questions remain about the metabolic costs of panting, cutaneous evaporation, and gular flutter. McKechnie et al. (2021) suggested that panting may be more metabolically costly so taxa that depend more on panting such as passerines and parrots (Psittaciformes) may be less able to tolerate high ambient temperatures than taxa depending more on cutaneous evaporation (Columbiformes) and gular flutter (e.g., Caprimulgiformes) that are less metabolically costly. However, Smit et al. (2018) found that Lilac-breasted Rollers (*Coracias caudatus*; Coraciiformes) that use panting as their primary means of evaporative heat loss exhibited heat tolerance comparable to that of columbids that depend primarily on cutaneous evaporation, and that the heat tolerance of African Cuckoos (*Cuculus gularis*; Cuculiformes) that depend primarily on gular flutter was similar to that of Burchell's Starlings (*Lamprotornis australis*; Passeriformes) that depend primarily on panting. Smit et al. (2018) concluded that additional sampling of more taxa is needed to better understand avian thermoregulatory responses to heat and the respective roles of panting, gular flutter, and cutaneous evaporation in those responses.

Differences in skin structure may help explain the differing use of cutaneous EWL among

**Fig. 10.60** (continued) of the same temperature. In addition, species of birds with larger bills (e.g., toucans, pelicans, albatrosses, and storks) are more likely to tuck them into their plumage, and birds with longer legs are more likely to stand on one leg to conserve heat. (Figure from Pavlovic et al. 2019; © 2018 The Authors. Functional Ecology © 2018 British Ecological Society, used with permission)

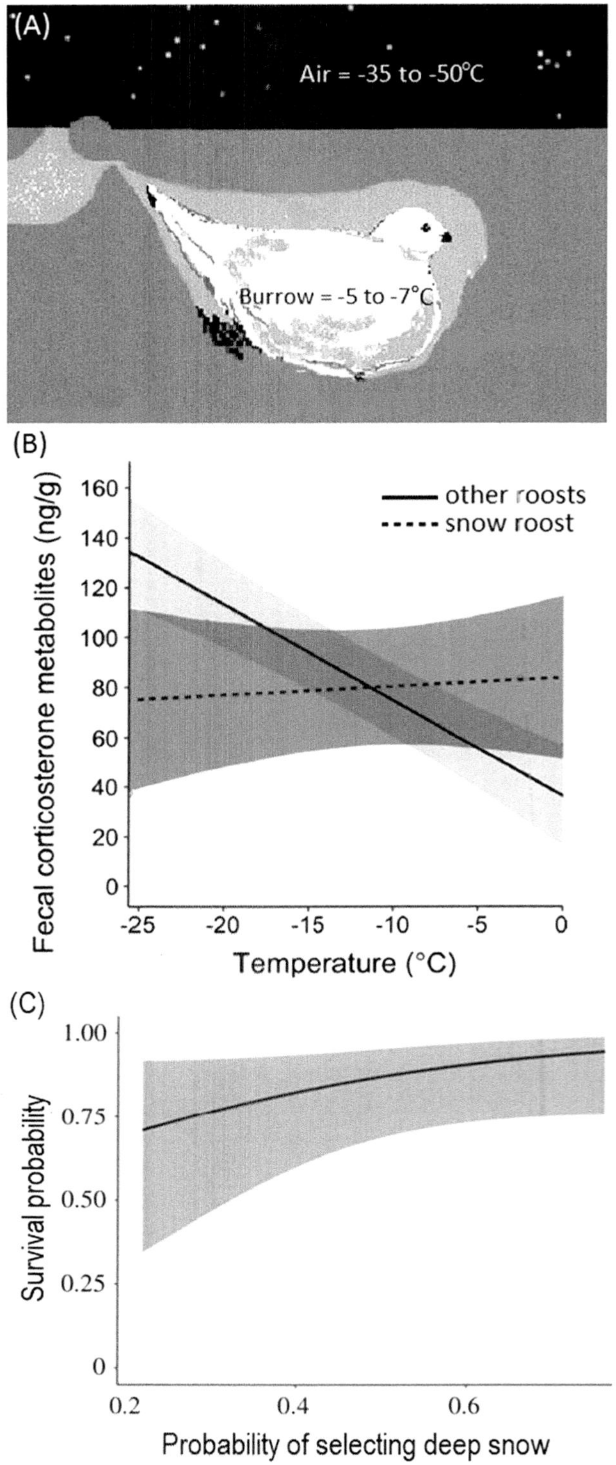


**Fig. 10.61** (**a**) Drawing of a Willow Ptarmigan (*Lagopus lagopus*) in a snow burrow in Russia. The insulating capacity of snow helps maintain burrow temperatures well above ambient temperatures. (Figure from Andreev 1999; © Taylor & Francis, used with permission). (**b**) Roosting at night in snow burrows not only aids in thermoregulation, but is also less stressful for Ruffed Grouse (*Bonasa umbellus*) when ambient temperatures are especially low, as indicated by the reduced levels of corticosterone (a hormone associated with increasing levels of stress) in the feces of grouse roosting in snow vs. not roosting in snow as temperatures fall below about −12°C (**c**) During periods with deep snow (≥15 cm), Ruffed Grouse in snow burrows in deeper snow were more likely to survive than those in burrows closer to the snow's surface (Shipley et al. 2020). (Figure **b** from Shipley et al. 2019; © 2019 Springer-Verlag GmbH Germany, part of Springer Nature, used with permission. Figure **c** from Shipley et al. 2020; © 2020 The Authors. Published by the Royal Society, used with permission)

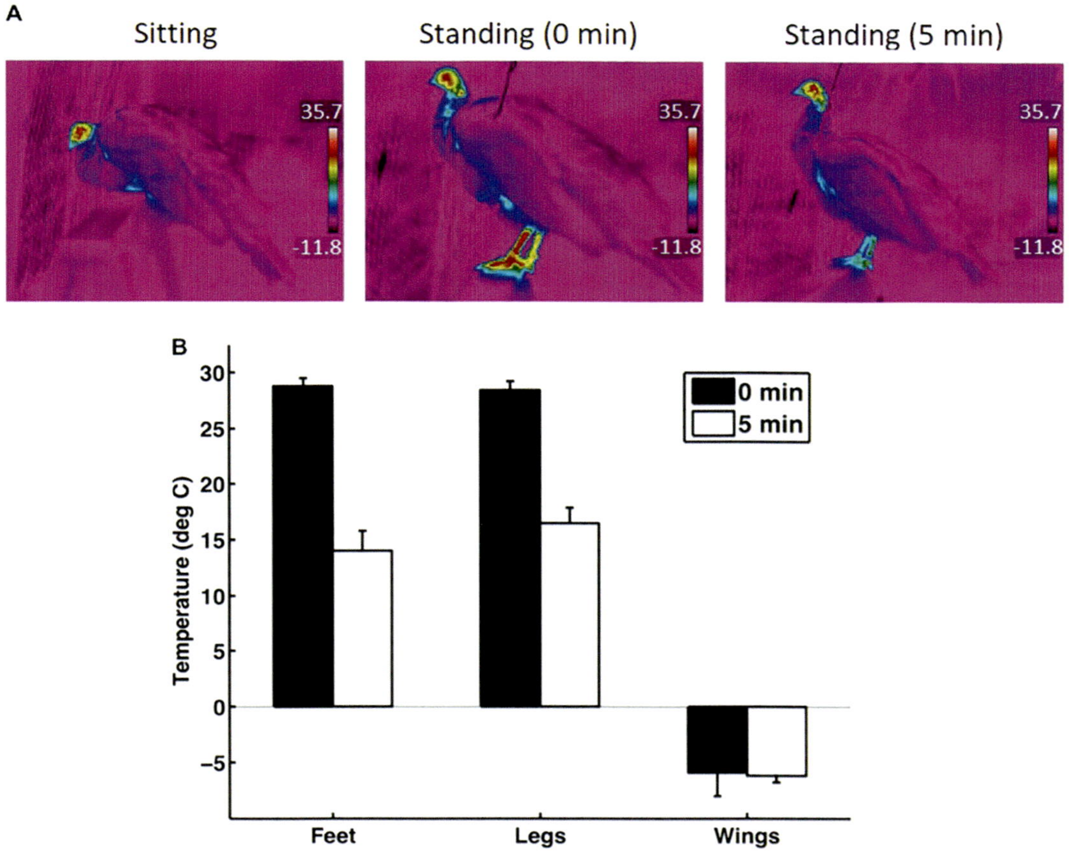


**Fig. 10.62** Effect of the leg-tuck posture on heat conservation by an Indian Peafowl (*Pavo cristatus*). (**a**) Left, thermal image of a peafowl sitting in a leg-tuck posture, with legs and feet covered by feathers. Middle, thermal image immediately after the peafowl stands up, with the legs and feet close to the normal body temperature. Right, thermal image taken 5 min after the peafowl stood up. The legs and feet are now much colder. (**b**) Mean temperatures (± SE) of feet, legs, and wings immediately after the peafowl stood up (0 min), and after standing for 5 min. The leg-tuck posture warms the legs and feet and, because they are covered by feathers, little heat is lost. However, after standing, the legs and feet cool rapidly to minimize heat loss to the surrounding air. (Figure from Yorzinski et al. 2018; open-access article distributed under the terms of the Creative Commons Attribution License, https://creativecommons.org/licenses/by/4.0/)

species at high environmental temperatures. Cutaneous EWL is known to be influenced by skin ultrastructure, specifically differences in the lipid composition of the epidermis. For example, Champagne et al. (2012) examined the relationship between cutaneous water loss and skin ultrastructure in 20 species of birds, including birds from both mesic and desert environments. Lipids in the skin of mesic and desert bird differed, and this difference contributed to reduced cutaneous water loss in desert birds. Differences between species in blood flow to the skin could also contribute to differences in cutaneous evaporative water loss. Ophir et al. (2002) suggested that changes in blood flood in the skin might allow these birds to regulate that heat loss. More precisely, variation in the extent of arteriole versus venous vasoconstriction (resulting from reflex responses of the hypothalamus and the autonomic nervous system to decreasing or increasing body

**Fig. 10.63** Typical spread-wing posture of anhingas and cormorants. (Photo from pxhere.com, CC0 Public Domain)

temperature) could alter the flow of water from skin capillaries that, in turn, would decrease or increase cutaneous EWL.

Birds can alter respiratory EWL and increase heat loss by gular flutter and panting. Panting is a controlled increase in respiratory rate accompanied by a decrease in tidal volume (the amount of air moving in and out of the respiratory system during a respiratory cycle) that increases ventilation of the upper respiratory tract, maintains air flow through the lungs, and increases evaporative heat loss. When panting, respiration rates can increase by as much as 15–25 times normal resting rates while tidal volume decreases (Fig. 10.78). The reduction in tidal volume means that ventilation is mainly limited to respiratory dead air space (trachea and bronchi), allowing a substantial increase in evaporative water loss without having much of an effect on blood $CO_2$ and acid–base balance (Box 10.13 Hypocapnia and Respiratory Alkalosis). Panting birds may occasionally resume more normal respiration for brief periods, taking slower, deeper breaths to ventilate the air sacs and lungs (Fig. 10.79).

Some heat-stressed birds also increase respiratory evaporative water loss and heat loss via gular flutter, which is the rapid fluttering or oscillation of the gular (throat) area (Fig. 10.80). Gular fluttering has been observed in many birds, including pelicans, cormorants, vultures, roadrunners, quail, and goatsuckers (nighthawks and poor-wills) (Fig. 10.81). During gular fluttering, birds can increase evaporative water loss by increasing respiratory rate (i.e., panting), increasing the flutter or oscillation rate, and/or increasing the amplitude of gular movement (McNab 2002). For most birds, panting and/or gular fluttering are the most effective means of losing heat and preventing increases in body temperatures. For example, by gular fluttering, Common Poorwills (*Phalaenoptilus nuttallii*) are able to maintain their body temperature well

**Fig. 10.64** Southern Pied Babblers (*Turdoides bicolor*) panting to aid in evaporative cooling and wing drooping to increase their surface area to aid in the passive loss of heat. Wing drooping could be useful for both warming and cooling birds. Drooping wings and facing the sun could warm a bird; drooping wings and facing away from the sun could cool a bird by exposing thinly feathered areas to the air and wind. Based on observations of birds in the Kalahari Desert of South Africa, Arizona in the southwest United States, and western Australia, Pattinson et al. (2020) found that larger birds used wing drooping to dissipate heat at lower ambient temperatures than smaller birds. These authors also found that heavier birds tended to be less active at temperatures between 35 and 40°C. The use of heat dissipating behaviors and reduction in activity levels by larger birds at lower ambient temperatures than for smaller birds are likely because larger birds have lower rates of mass-specific evaporative water loss (Weathers 1981; Whitfield et al. 2015; McKechnie et al. 2017). (Photo by Nicholas Pattinson, used with permission)

below ambient temperatures for long periods of time (Lasiewski and Bartholomew 1966; Fig. 10.82).

Blood flowing from the body core to arteries on the periphery of a bird's body, including the head, wings, legs, feet, and, for some species, even the talons (Fig. 10.83), can carry heat that can be readily lost through the skin (Box 10.14 Why Do Some Birds Have Bald Heads?). When ambient temperatures are high, this loss of heat is beneficial; when ambient temperatures are low, loss of heat is not beneficial. Birds regulate heat loss from these peripheral areas using counter-current heat exchangers (Fig. 10.84). For example, Song Sparrows (*Melospiza melodia*) alter blood flow to these peripheral areas to minimize heat loss when ambient temperatures are low and then maximize heat loss when ambient temperatures are high (Fig. 10.85). Birds, of course, do not consciously alter blood flow. Rather, their "thermostat," the hypothalamus, monitors body temperature and, via the Autonomic Nervous System, alters blood flow as needed to help maintain body temperature.

Counter-current heat exchangers are areas where blood vessels (arteries going in and veins coming out) are located in close proximity to allow, if necessary, heat to be recaptured and saved. In a counter-current exchanger, flow in two adjacent tubes (like blood vessels) is in opposite directions (Fig. 10.84). The counter-current heat exchangers of birds can be found in the head, the legs, and the wings. The number, complexity, and efficiency of these heat exchangers vary among different species of birds.

Many, but not all, birds have counter-current heat exchangers in their legs (Fig. 10.86). These heat exchangers consist of intermingled networks of arteries and veins and range in complexity from just a few arteries and veins to elaborate

**Fig. 10.65** Maribou Stork (*Leptoptilos crumenifer*) with excreta on its lower legs to increase evaporative water and heat loss. Cabello-Vergel et al. (2021) examined the use of urohidrosis by 19 species of storks and determined that, across species, urohidrosis was used for thermoregulation during periods with high temperature, humidity and solar radiation, and low wind speed. Species that typically forage in more open habitats were also found to use urohidrosis more than species that forage primarily in water. These authors also found that the ambient temperatures at which 50% of storks used urihidrosis ranged from 27.7 to 33.5°C, which was lower than other 50% thresholds for dissipating heat including panting (33.9–46.1°C) and wing drooping (35.3–44.6°C). Strokes likely use urihidrosis at lower ambient temperatures because it is relatively cheap in terms of metabolic cost and water loss. (Photo by Desidor, CC BY 3.0, https://creativecommons.org/licenses/by/3.0/)

networks consisting of as many as 60 arteries and 40 veins (flamingos). Importantly, birds also have superficial veins in their legs so that blood returning to the body can largely bypass the heat exchanger if a bird needs to lose heat (Fig. 10.87). However, when ambient temperatures fall below a bird's thermoneutral zone, these counter-current heat exchangers can greatly minimize heat loss via the legs and feet by transferring heat from arteries supplying the legs and feet to veins draining the legs and feet (Figs. 10.88, 10.89 and 10.90). When the ambient temperature falls below 0°C, intermittent pulses of increased blood flow to the legs and other extremities prevent freezing and tissue damage (Kilgore and Schmidt-Nielsen 1975). Some birds also have counter-current heat exchangers in their wings (or flippers; Fig. 10.91). Arteriovenous heat exchangers have been reported in the wings of penguins (e.g., Thomas et al. 2011) and vultures (Arad et al. 1989), and additional research may reveal their presence in other species of birds as well. Among penguins, the number of arteries involved in heat exchange with veins varies with size and water temperature. Larger penguins have flippers with more surface area, and so natural selection has favored more efficient heat exchangers that have more arteries (Fig. 10.92). Species of penguins found at higher latitudes where water temperatures are colder also have more arteries to more efficiently cool their flippers and minimize heat loss (Fig. 10.92).

The buccal (oral) cavity, bill, eyes, and nasal cavities of birds are important for thermoregulation because birds can also vary the amount of heat lost from these areas. There is a double shunt mechanism that allows both venous (via the ophthalmica vein and maxillaris vein) and arterial (via the cerebral carotid artery) blood to bypass the countercurrent heat exchanger (rete mirabile ophthalmicum, or ophthalmic rete; Fig. 10.93). As a result, birds can alter the amount of heat lost from the head, bill, and eyes by varying the proportion of arterial blood that passes through the rete ophthalmicum (Midtgård 1983). The amount of heat lost from bird bills has been found to be surprisingly important for thermoregulation in several species of birds (Fig. 10.94; Box 10.15 Summer Temperature and Bill Surface Area of Sparrows). Tattersall et al. (2017) used data from 214 species of birds in eight different taxonomic groups and found significant relationships between bill length and minimum monthly temperatures, with species in areas with colder climates having shorter bills (Fig. 10.95). The only exception was the finches (Estrildidae),

**Fig. 10.66** Female Kentish Plover (*Charadrius alexandrinus*) incubating eggs with wet belly feathers. (Figure from AlRashidi et al. 2011; © 2011 The Association for the Study of Animal Behaviour. Published by Elsevier Ltd., used with permission)

with no significant relationship between bill length and latitude (Fig. 10.95). Possible explanations for these results are that (1) selection favors shorter bills that reduce heat loss in areas with colder climates, (2) selection favors longer bills in areas with warmer climates to enhance heat loss, or (3) both of these. Regardless, these results suggest that thermoregulatory constraints have been an important selective force in the evolution of bill morphology (Tattersall et al. 2017).

**Fig. 10.67** A wintering group of about 2500 Emperor Penguins (*Aptenodytes forsteri*). (Figure from Gilbert et al. 2006; © 2006 Elsevier Inc., used with permission)

**Box 10.9 Huddling by Emperor Penguins**

Circular huddle of Emperor Penguins with individual penguins moving slowly in a counter-clockwise direction. (Figure from Gerum et al. 2013; open-access article licensed under the terms of the Creative Commons Attribution 3.0 license, https://creativecommons.org/licenses/by/3.0/)

Huddling helps male Emperor Penguins (*Aptenodytes forsteri*) save energy as they fast for more than 100 days during the pairing and incubation periods. Gilbert et al. (2006) attached data loggers glued to eight Emperor Penguins to examine their thermoregulatory behavior and to estimate the time spent huddling during the breeding season. They found that huddling episodes were discontinuous and of variable duration, lasting an average of 1.6 h. Despite heterogeneous huddling groups, birds had equal access to the warmth of the huddles because the penguins move in a highly coordinated manner to ensure mobility while keeping the huddle packed (Zitterbart et al. 2011). Every 30–60 s, all penguins make small steps that travel as a wave through an entire huddle. These small steps help achieve the highest packing density. As new penguins join the huddle at the periphery, the small steps compact the huddle. Throughout the breeding season, Gilbert et al. (2006) found that male Emperor Penguins huddled for an average of 38% of their time. Although varying with the "tightness" of huddles, huddling males may be able to reduce their metabolic rates by more than 50%, with about two-thirds of the reduction due to reduced exposure of their body surface and one-third to the milder microclimate created within huddles (Gilbert et al. 2008).

The cloaca may also play a role in avian thermoregulation. Hoffman et al. (2007) presented the first experimental evidence that enough water can evaporate from a bird's cloaca to have an important role in thermoregulation. They measured rates of evaporation occurring from the mouth, skin, and cloaca of Inca Doves (*Columbina inca*) and Common Quail (*Coturnix coturnix*). Cloacal evaporation in doves was negligible at ambient temperatures of 30, 35, and 40°C. However, at 42°C, the apportionment of total evaporation in doves was 53.4% cutaneous, 25.4% buccopharyngeal, and 21.2% cloacal (Fig. 10.96). In contrast, the evaporative apportionment in quail at 32°C (the highest ambient temperature tolerated by this species) was 58.2% cutaneous, 35.4% buccopharyngeal, and 6.4% cloacal. These results suggest that, for some birds, cloacal evaporation can be controlled and could play an important role in thermoregulation at high ambient temperatures.

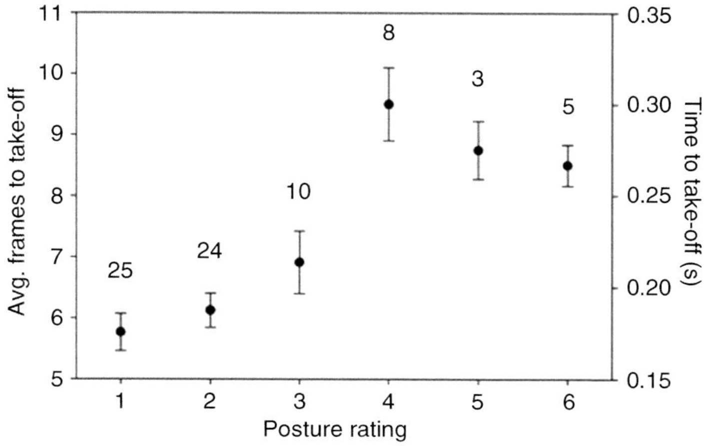


**Fig. 10.68** Average time needed (± standard error) for a Dark-eyed Junco (*Junco hyemalis*) to take off from the ground when startled (by a loud noise). Posture rating (number) refers to the thermoregulatory postures explained in Fig. 10.58. (Figure from Carr and Lima 2012a; © 2011 Oxford University Press, used with permission)

## 10.12 Avian Hyperthermia

During periods when ambient temperatures are well above their upper critical temperatures, birds can allow their body temperatures to increase above normal temperatures (hyperthermia). Body temperatures above about 46–48°C are generally lethal for birds (Dawson 1954; Arad and Marder 1982), but heat tolerance varies among species and orders of birds and may also vary with latitude (Fig. 10.97). In addition, Freeman et al. (2022) compared the maximum tolerable body temperatures of 53 species of birds and found differences among species that occupied different climates (i.e., different maximum air temperatures and humidity) in South Africa. Surprisingly, these authors found that birds in arid zones had lower maximum tolerable body temperatures than birds in humid lowland and mesic montane habitats. Although

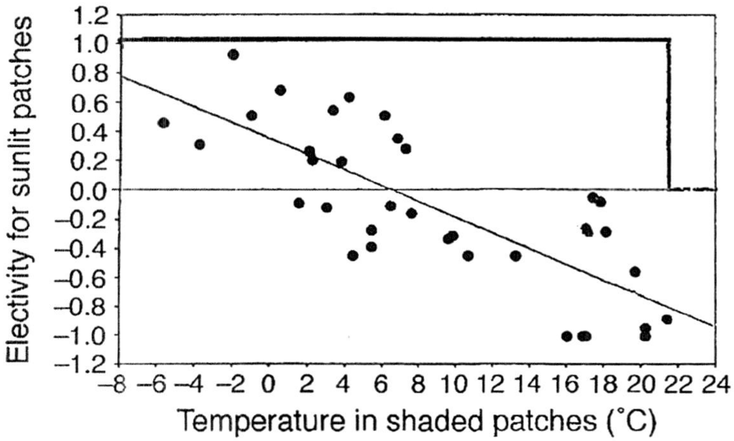


**Fig. 10.69** Choice of foraging sites by Short-toed Treecreepers (*Certhia brachydactyla*) during the winter in a Spain forest. The lower critical temperature of these treecreepers is 21.5°C so, for thermoregulatory purposes, they might be expected to forage in sunlit patches of forest when ambient temperatures fall below 21.5°C (thick line). However, foraging treecreepers did not exhibit a preference for sunlit patches until ambient temperatures were well below their lower critical temperature (about 8°C). Electivity was calculated as the proportion of time a treecreeper spent foraging in the sun relative to the proportion of available tree exposed to full sun. (Figure from Carrascal et al. 2001; © 2001 by the Ecological Society of America, used with permission)

**Fig. 10.70** Example of relative visibility of a Short-toed Treecreeper (*Certhia brachydactyla*) on a sunlit tree and a shaded tree. The arrows point to the location of a Short-toed Treecreeper in the sun (above) and in the shade (below). (Figure from Carrascal et al. 2001; © 2001 Ecological Society of America, used with permission)

arid-zone birds would experience higher air temperatures and reduced availability of water (with, therefore, strong selection pressure for water conservation) than birds in the other two climates, Freeman et al. (2022) suggested that "...synthesis of heat shock proteins in response to high temperatures are blunted in desert birds on account of the energetic costs involved," i.e., the relatively low maximum body temperatures of arid-zone birds, ". . . despite the adaptive value of hyperthermia for water conservation, suggests that extreme (body temperatures > 45°C) hyperthermia has substantial costs." For some species, however, the benefits of extreme hyperthermia must exceed those possible costs (Box 10.16 Extreme Hyperthermia Tolerance of Red-Billed Queleas).

During periods of hyperthermia, bird body temperatures typically increase by about 2–5°C (Fig. 10.98). McKechnie et al. (2021) quantified variation in the thermoregulatory performance of 56 arid-zone species of birds and determined that the difference between normothermic and maximum body temperature (with maximums associated with loss of balance or capacity for coordinated locomotion) ranged from 0.9°C for Mourning Doves (*Zenaida macroura*) to as high as 6.2°C for Australian Owlet-nightjars (*Aegotheles cristatus*) and Little Swifts (*Apus affinis*). A potential advantage of hyperthermia, particularly for birds in dry environments where availability of water may be limited, is that, as noted above, it can reduce evaporative water loss.

**Box 10.10 Potential Roles of Activity and Eating on Reducing Thermoregulatory Costs**

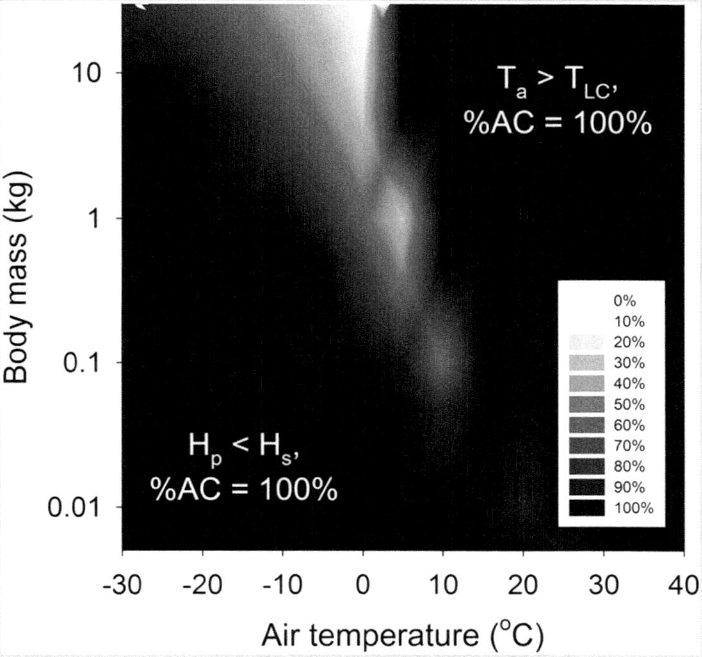


Predicted energetic cost of flying for birds as a function of body size and $T_a$ (ambient temperature). The cost of flight is expressed as a percentage of the cost of flying when ambient temperatures are within a bird's thermoneutral zone (%) and is calculated based on by how much $T_a < T_{LC}$ and $H_p > H_s$, with $T_{LC}$ being the lower critical temperature, $H_p$ being heat production, and $H_s$ being heat dissipation. The extent to which the energetic cost of flight is reduced by activity-thermoregulatory heat substitution is predicted to be greatest for larger birds. (Figure from Humphries and Careau 2011; © 2011 Oxford University Press, used with permission)

When ambient temperatures are lower than a bird's lower critical temperature (LCT), maintaining a constant body temperature requires an increase in heat production. When active (e.g., flying, foraging, preening, or vocalizing), however, birds generate heat via skeletal muscle contraction, and this heat can be used for thermoregulation. If heat produced by being active equals or exceeds that needed to maintain body temperature, then is activity at temperatures below a bird's LCT free (in terms of energetic cost)? This possibility is referred to as the "activity-thermoregulatory heat substitution" (ATHS) hypothesis. Humphries and Careau (2011) noted that ATHS requires ambient temperature to be lower than a bird's lower critical temperature ($T_a > T_{LC}$) and activity that generates heat ($H_p$) in excess of heat loss ($H_s$). The greater the difference between $H_P$ and $H_S$, the greater the potential for heat substitution and, in addition, because the energetic cost of activity (e.g., flight) and the amount of surface area relative to volume both decrease with increasing body size, larger birds should have proportionately less $H_p$ and less $H_s$ than smaller birds. Using available estimates of the energetic cost of bird flight and the effect of body size on heat loss, Humphries and Careau (2011) found that the extent of ATHS increases with body size and with increasing $T_a$ (up to $T_{LC}$). They estimated that flying birds ≤35 g at −10°C or ≤ 15 g at 0°C would not generate enough heat ($H_p$) to exceed heat loss ($H_s$), but that hovering birds of all body sizes would be capable of activity substitution when $T_a$ is

(continued)

**Box 10.10** (continued)

between 0°C and $T_{LC}$. More generally, as indicated in the figure above, Humphries and Careau (2011) predicted that activity would always be more costly at high temperatures because $T_a > T_{LC}$ and always costly for small birds at low $T_a$ because $H_S > H_P$. In contrast, for flying birds, the highest degree of substitution is predicted only for large-sized birds experiencing cool to cold $T_a$. Humphries and Careau's (2011) models plus the results of studies of captive birds suggest that ATHS could be important for some species of birds in the wild, but confirmation of this will require additional study of birds both in the lab and field.

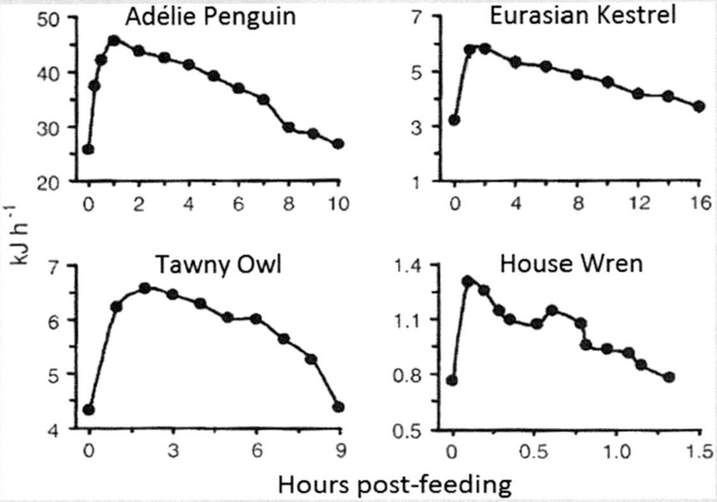


Metabolic rates of four species of birds after feeding. A bird's body mass, meal type, and meal mass can all influence the extent to which eating increases metabolic rates (metabolic rates are expressed in terms of kilojoules per hour). Adelie Penguin, *Pygoscelis adeliae*; Eurasian Kestrel, *Falco tinnunculus*; Tawny Owl, *Strix aluco*; House Wren, *Troglodytes aedon*. (Figure modified from Secor 2009; © 2008 Springer-Verlag, used with permission)

The process of digesting food also produces heat and is referred to either as the heat increment of feeding (HIF) or specific dynamic action. As with heat generated by activity, HIF can potentially reduce a bird's thermoregulatory costs. Although peristalsis, enhanced blood flow to the digestive system, enzyme synthesis, and active transport play a role, HIF appears to result primarily from intermediary metabolism (i.e., all anabolic and catabolic reactions that take place in cells), especially the metabolism of protein (Lovvorn 2007). The extent to which HIF can substitute for regulatory thermogenesis (e.g., energy used for shivering) has been found to vary among species. For example, HIF accounted for >20% of thermoregulatory costs of diving Lesser Scaup (*Aythya affinis*; Kaseloo and Lovvorn 2006). For Eurasian Kestrels (*Falco tinnunculus*) and Tawny Owls (*Strix aluco*), substitution from HIF was ~50 and > 90%, respectively, when resting at ambient temperatures below their lower critical temperatures (Masman et al. 1989; Bech and Præsteng 2004). In contrast, little evidence of thermal substitution was found for young Arctic Terns (*Sterna paradisaea*, Klaassen et al. 1989) or adult Mallards (*Anas platyrhynchos*; Kaseloo and Lovvorn 2003). Whether such interspecific differences are due to methodological differences among studies or represent actual physiological differences remains unclear. Additional studies of a variety of species are needed to better clarify the possible role of HIF in reducing the thermoregulatory costs of birds exposed to cold temperatures (Enstipp et al. 2008).

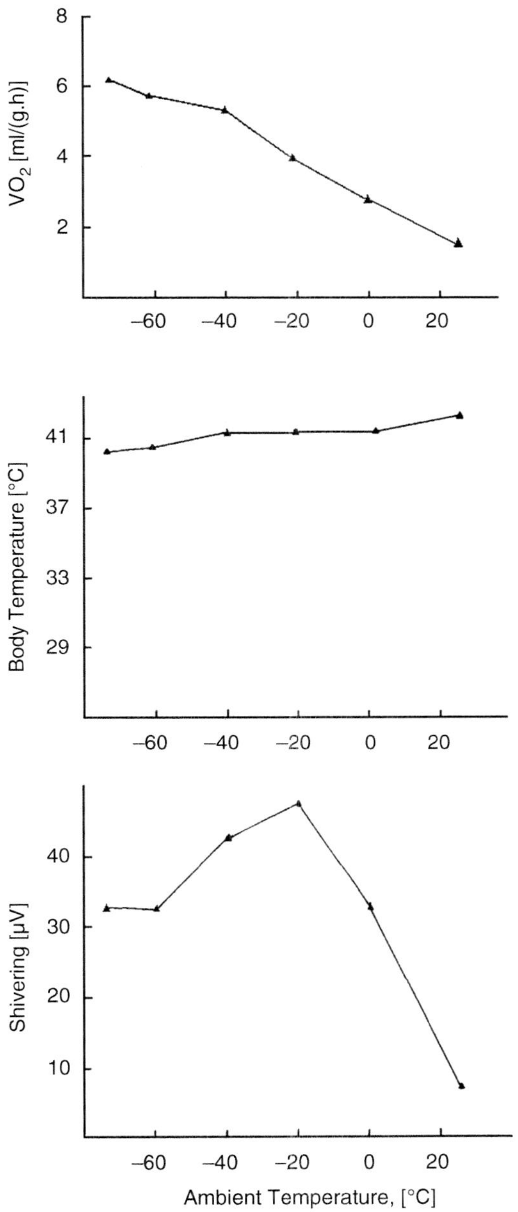


**Fig. 10.71** Mean oxygen consumption, body temperature, and shivering (microvolts, electromyogram) by Japanese Quail (*Coturnix japonica*) at different ambient temperatures. The quail maintained their body temperatures by shivering, and shivering caused a substantial increase in oxygen consumption. (Figure modified from Saarela and Heldmaier 1987; © 1987 Springer-Verlag, used with permission)

This reduction is possible because hyperthermia can potentially improve the thermal gradient between body and air temperature that drives non-evaporative heat loss (conduction and radiation), and heat stored in tissues can be dissipated later by these non-evaporative means when ambient temperatures decline. However, the relative importance of thermal gradients and heat storage for hyperthermic birds varies with body size and phylogeny (Williams and Tieleman 2005; Gerson et al. 2019; Fig. 10.99). Assuming an ambient temperature of 45°C and birds with a normal body temperature of 41°C whose body temperatures increase to 44°C, Tieleman and Williams (1999) calculated that different-sized birds hyperthermic for 1 h would all save the equivalent of about 50% of their total evaporative water loss. However, the improved thermal gradient was the most important contributing factor for smaller birds (with more surface area relative to body mass), whereas heat storage was most important for the largest birds.

The water-saving advantages of hyperthermia for birds of different sizes may also depend on its duration. Based on the same assumptions just noted, Tieleman and Williams (1999) calculated that smaller birds hyperthermic for 5 h still save water, but larger birds hyperthermic for that long do not. The reason for this difference is that, in contrast to water saved via the improved thermal gradient, heat storage is independent of time so water saved via heat storage is averaged over the duration of the hyperthermic bout. Available evidence suggests that this calculated time-dependent effect on the water-saving advantage of hyperthermia for large birds is correct, with Common Ostriches (*Struthio camelus*; about 100 kg) maintaining their normal body temperature (rather than becoming hyperthermic) over a period of 7.5 h at an ambient temperature of 51°C (Crawford and Schmidt-Nielsen 1967).

**Box 10.11 Non-shivering Thermogenesis in Mammals**

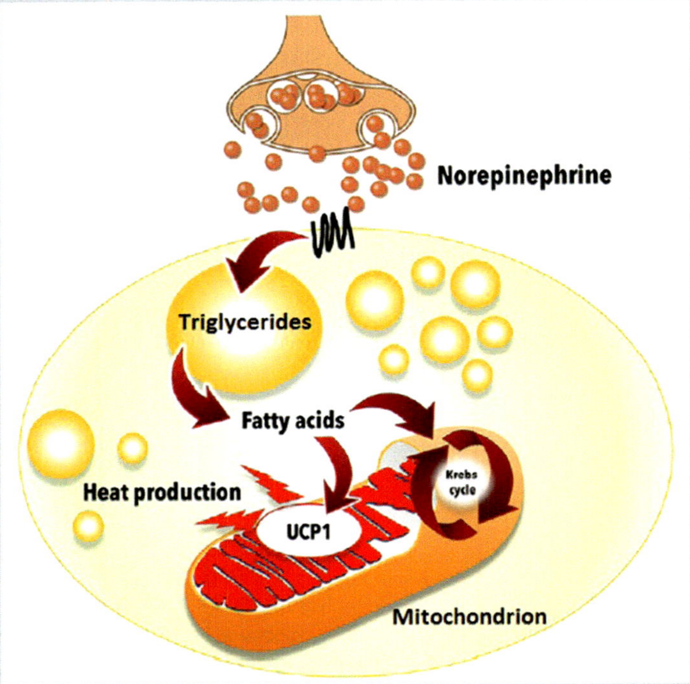


In cold-stressed mammals, norepinephrine (also called noradrenalin) stimulates brown fat cells (also called brown adipose tissue, or BAT) to break down triglycerides into free fatty acids (FFA) that, in turn, are metabolized via beta oxidation (breakdown of fatty acids) and the Kreb's, or citric acid, cycle. Products generated during these reactions interact with the enzymes of the electron transport chain in mitochondria, resulting in the transport of hydrogen ions (protons) through the inner mitochondrial membrane into the intermembrane space. This creates a substantial concentration gradient, with a much higher concentration of protons in the intermembrane space than within the mitochondrial matrix (inside the inner mitochondrial membrane). In most mitochondria, protons following this concentration gradient pass through molecules of ATP synthase and help generate ATP. However, in brown fat cells, an uncoupling protein (UCP1) transports protons back into the mitochondrial matrix without making ATP. The electron-driven proton pumping by the respiratory chain to again transport the protons back into the intermembrane space leads to oxygen consumption and heat production. (Figure modified from Carpentier et al. 2018; © 2018 Carpentier, Blondin, Virtanen, Richard, Haman and Turcotte, open-access article distributed under the terms of the Creative Commons Attribution License (CC BY), https://creativecommons.org/licenses/by/4.0/)

Unlike birds, most mammals have brown fat tissue that differs from white fat tissue by having very high densities of mitochondria. The function of brown fat tissue is to produce heat by simply oxidizing fat without making ATP. In mitochondria, protons (hydrogen ions) are constantly transported into the intermembrane space via the electron transport, or respiratory, chain, creating a concentration gradient. In most mitochondria, ATP is then produced as protons, following their concentration gradient, pass through molecules of ATP synthase located within the mitochondrial inner membrane. However, brown fat cell mitochondria have "uncoupling proteins" (UCP), so-called because they "uncouple" electron transport from ATP production, in their inner membranes. UCP transports protons back into the mitochondrial matrix (inside the

(continued)

**Box 10.11** (continued)

inner membrane) without synthesizing ATP. This transport reduces the proton concentration gradient that then must be restored by additional electron-driven proton pumping by the respiratory chain leading to oxygen consumption and heat production. This process of generating heat is referred to as non-shivering thermogenesis.

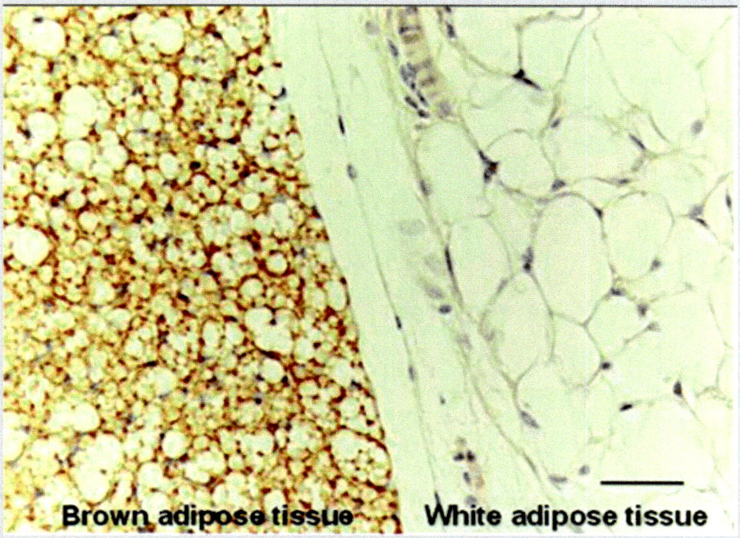


Histology of mammalian brown and white adipose tissue. White adipose cells have a single lipid droplet, whereas brown adipocytes contain numerous small droplets. The high concentration of mitochondria in brown adipose cells gives them their brown color. (Photo from Tam et al. 2012; © 2012 Wolters Kluwer Health, used with permission)

One possible explanation for why selection has not favored BAT-like tissue in birds is that the skeletal muscles of birds can achieve metabolic rates at least twice as high as in small mammals and can be as high as 8–18 times a bird's basal metabolic rate (Butler et al. 1998; Videler 2005). The reason for this difference between birds and mammals is that bird skeletal muscles are much more efficient at using plasma fatty acids (Jenni-Eiermann 2017). To generate heat, therefore, birds can import fuel in the form of fatty acids into skeletal muscles like the pectoralis, with the fatty acids then used by skeletal muscle cells to generate heat via ATP production by mitochondria and sodium-potassium pumps, i.e., active transport of sodium and potassium ions across muscle cell membranes to allow conduction of impulses and contraction of myofibrils (Nowack et al. 2017).

(continued)

**Box 10.11** (continued)

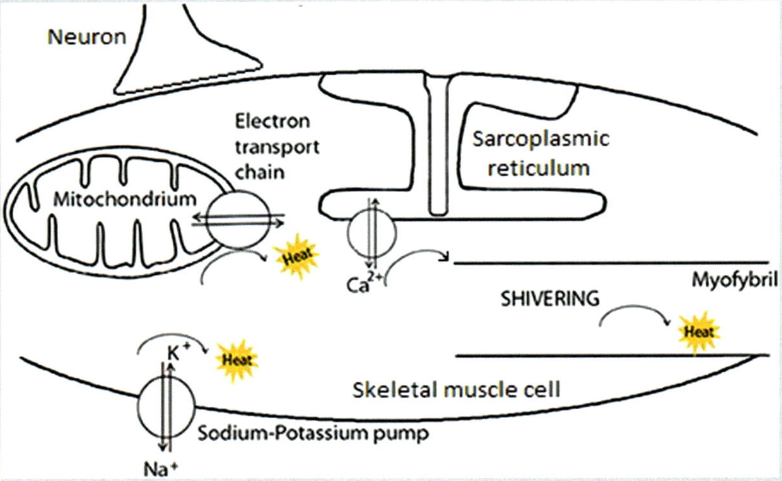


A possible mechanism for heat production by avian skeletal muscle cells. Fatty acids transported into muscle cells from the blood are converted into acetyl CoA in the mitochondria and via the Kreb's cycle and electron transport, ATP and heat are produced. Also producing heat are the numerous sodium-potassium pumps in the membranes of skeletal muscle cells and the contraction of myofibrils. (Figure modified from Nowack et al. 2017; © 2017 Nowack, Giroud, Arnold and Ruf, open-access article distributed under the terms of the Creative Commons Attribution License (CC BY), https://creativecommons.org/licenses/by/4.0/)

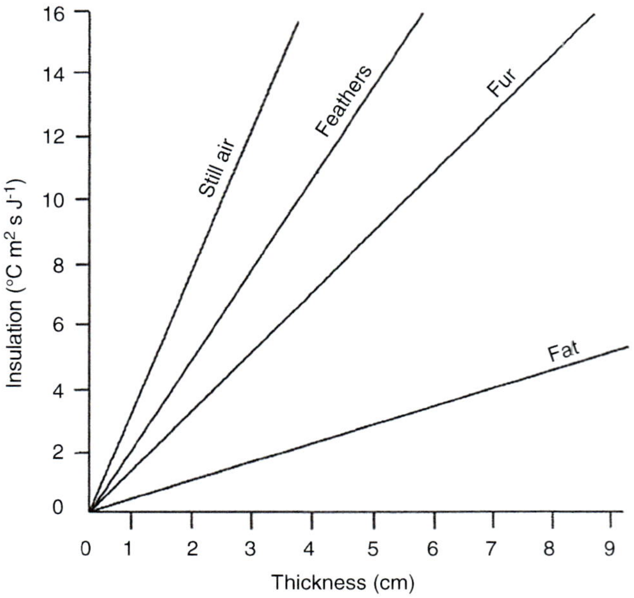


**Fig. 10.72** Effect of insulation thickness on thermal conductance of still air, feathers, fur, and fat. Still air is the best insulating material, and bird feathers and mammal hair are efficient at trapping still air close to the body, with feathers more effective at trapping air than fur. (Figure from Tattersall et al. 2012; © 2012 American Physiological Society, used with permission)

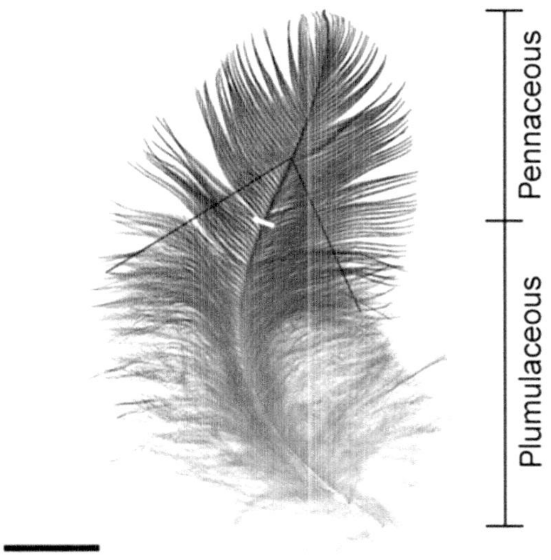


**Fig. 10.73** The structure of contour feathers can influence heat flux through plumage, including the relative length of the downy part of feathers, the length of the downy barbs, and barb density (number of barbs per unit length of the rachis). Longer downy parts and downy barbs and higher barb densities decrease heat flux from the skin to the environment and vice versa. White stripe delimits the two sections of the feather, and the black lines indicate the boundary between the pennaceous and plumulaceous sections of the vane. (Figure from Pap et al. 2017; © 2016 The Authors. Functional Ecology © 2016 British Ecological Society, used with permission)

## 10.13 Avian Hypothermia

Species in 16 different orders and 31 families of birds (Fig. 10.100) have been found to become hypothermic to varying degrees and for variable lengths of time (McKecknie and Mzilikazi 2011). By reducing body temperatures below normal levels, birds can reduce energy and water consumption during periods of environmental stress, such as cold temperatures and/or reduced availability of food. Controlled rest-phase hypothermia has been reported in birds ranging in body mass from 2.7 g (Booted Racket-tail, *Ocreatus underwoodii*) to 6500 g (Eurasian Griffon, *Gyps fulvus*) (McKechnie and Lovegrove 2002), but minimum body temperatures are lower for smaller birds than larger birds (Fig. 10.101). The word controlled in this context means that regulation of body temperature continues, but at a lower set point. In other words, a minimum body temperature is set by a bird's thermostat (hypothalamus), and if that minimal temperature is reached, metabolic heat production (shivering or non-shivering thermogenesis) is used to maintain body temperature at or above that minimum (Geiser 2004).

Hypothermia means that a bird's body temperature drops below its normal body temperature and, for some species, that drop is such that birds enter into a state referred to as torpor. The difference between being hypothermic and torpid depends on how torpor is defined, and not all investigators agree on that definition (e.g., Boyles et al. 2011; Brigham et al. 2011). For example, torpor can be defined as a physiological state associated with a controlled reduction in metabolism and body temperature resulting in energy savings (Geiser 2004; Brigham et al. 2011). However, the International Union of Physiological Sciences defined torpor as "a state of inactivity and reduced responsiveness" (IUPS 2003). Both of these definitions are correct, but not very specific. Perhaps that is because, as far as avian hypothermia and torpor, "one size does not fit all." In other words, the ability of birds to reduce their body temperatures varies among species such that there is no specific temperature that clearly separates hypothermia from torpor (Fig. 10.102; although some investigators have suggested that torpor be defined as having a skin temperature less than 30°C, Brigham et al. 2011). In addition, as body temperatures decline, the efficiency of physiological processes (e.g., speed of metabolic reactions or muscle contraction) and responsiveness to external stimuli (or, in other words, performance; Fig. 10.103) gradually decline. As such, no specific temperature or specific definition of responsiveness can be used to define torpor so that it would apply to all species of birds. Numerous studies of a wide variety of species of birds have revealed a continuum, with interspecific differences in the extent to which body temperatures decline during periods of hypothermia (Figs. 10.100, 10.101, and 10.102) as well as differences in the extent to which

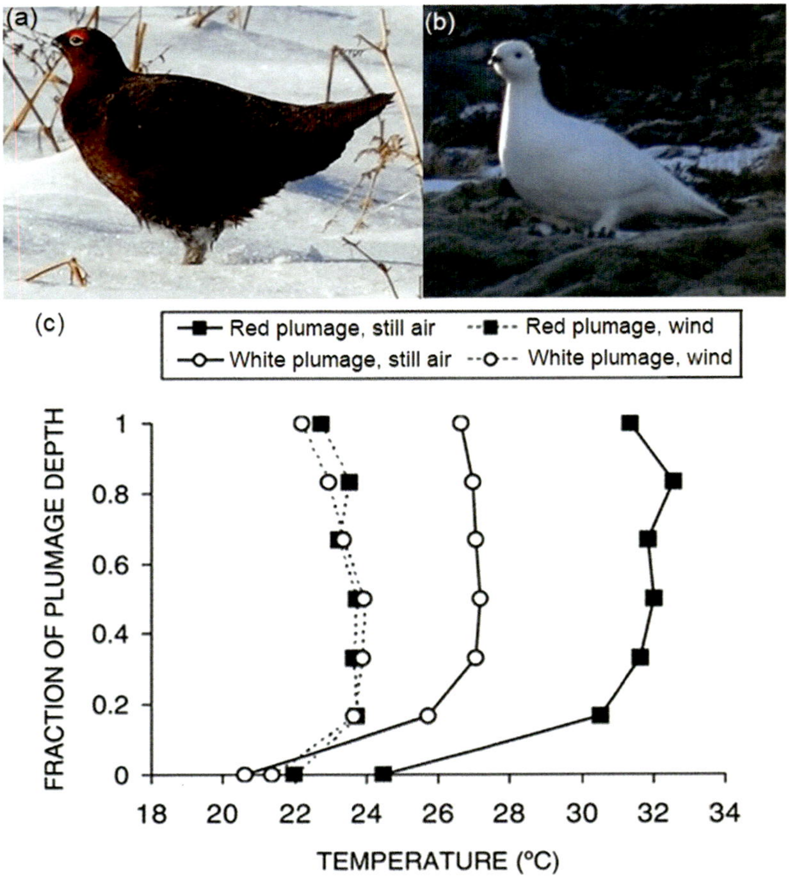


**Fig. 10.74** Two subspecies of Willow Ptarmigan (*Lagopus lagopus*), (**a**) one that remains reddish-brown during the winter and (**b**) another that has white winter plumage. (**c**) Temperature at various depths in the plumage after 30-min exposure to simulated solar radiation in still air and wind. Darker plumage (with higher absorptivity) acquires a greater heat load from solar radiation than lighter plumage in still air. However, under windy conditions, heat loads of dark and light plumage do not differ. Plumage depth of 0 = skin surface. (Figure **c** from Ward et al. 2007; open-access article published under a CC-BY license, https://creativecommons.org/licenses/by/4.0/; (**a**, **b**) Grouse photos from Wikipedia, CC BY 2.0, https://creativecommons.org/licenses/by/2.0/)

individual birds allow their body temperatures to drop below ambient temperatures (Box 10.17 A Cost of Being Cool). As such, for discussion of these topics, torpor will be defined as a pronounced drop in body temperature such that a bird is lethargic and capable of only minimal responsiveness to stimuli (e.g., limited movement of wings, legs, or rectrices; Bartholomew et al. 1957) or is unresponsive (Bligh and Johnson 1973; McKechnie and Lovegrove 2002). In most published studies of avian hypothermia, authors do not report data concerning levels of responsiveness. However, such limited responsiveness, i.e., torpor, has been reported for species in the avian families Columbidae (doves; Schleucher 2001), Trochilidae (hummingbirds), Apodidae (swifts), Caprimulgidae (nightjars), Todidae (todies; Merola-Zwartjes and Ligon 2000), Coliidae (mousebirds; Bartholomew and Trost 1970), and Nectariniidae (sunbirds; Downs and Brown 2002). Available evidence also suggests that Australian Owlet-nightjars (*Aegotheles cristatus*; Aegothelidae: Aegotheliformes) may occasionally become torpid, with body temperatures sometimes decreasing to as low as 19.4°C. With such low body temperatures and among the slowest rewarming rates of any species studied (McKechnie and Wolf 2004), these owlet-nightjars would likely be unable to arouse quickly if threatened by a predator (Doucette et al. 2012). In addition, body temperatures of food-deprived nestling Fork-tailed Storm-Petrels (*Oceanodroma furcate*, Procellariiformes) were found to range from about 13 to 25°C during the period from 6 to 28 days post-hatching (typical body temperature at those ages = 36–39°C; Boersma 1986). However, several of these nestlings subsequently died, and these declines in body temperature

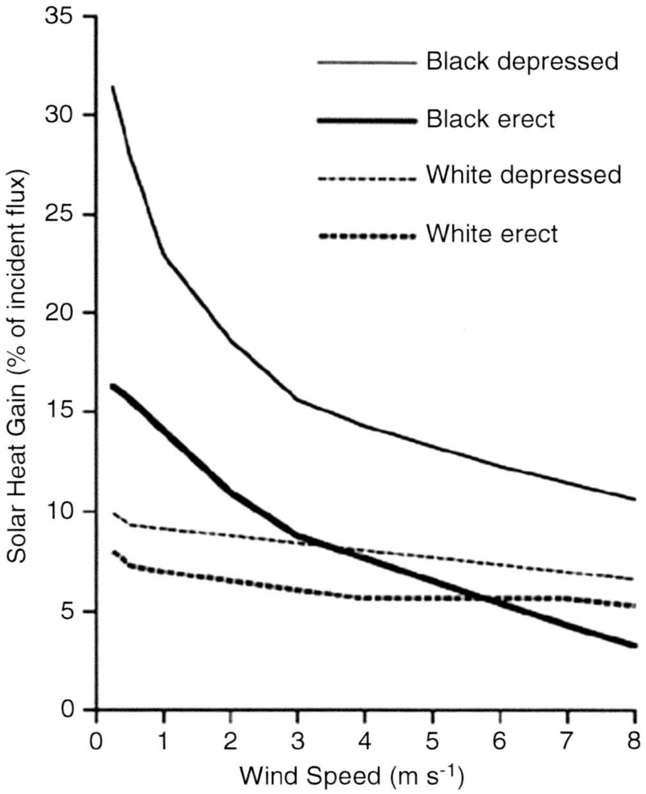


**Fig. 10.75** Effect of ptiloerection and wind speed on the solar heat gain acquired by the skin of Rock Pigeons (*Columba livia*) with black and white plumage. Darker plumage is warmed more by solar radiation than lighter plumage. However, with dark plumage, especially at low wind speeds, ptiloerection greatly decreases solar heat loads to the skin. In contrast, ptiloerection results in only small decreases in solar heat loads to the skin of pigeons with white plumage. (Figure from Wolf and Walsberg 2000; © 2015 Oxford University Press, used with permission)

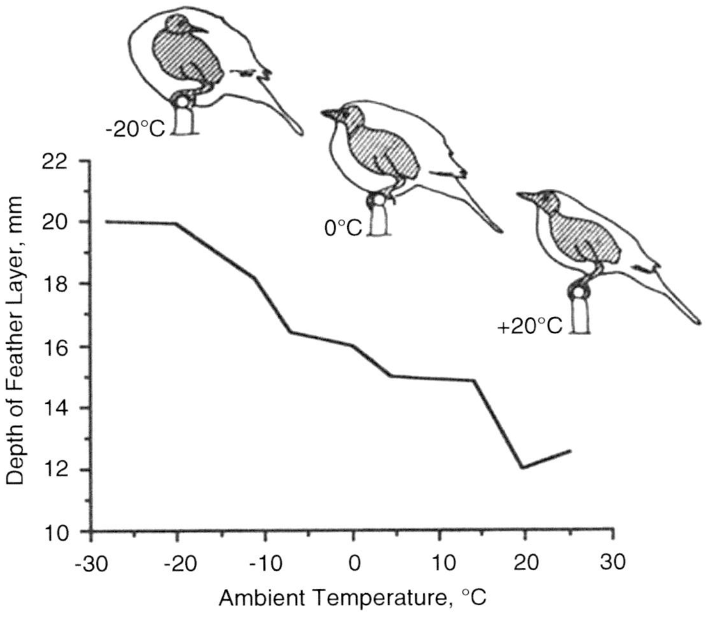


**Fig. 10.76** To reduce heat loss, the mean depth of the plumage of a Eurasian Blackbird (*Turdus merula*) increased via increasing ptiloerection with decreasing ambient temperatures. (Figure from Marsh and Dawson 1989; © 1989 Springer-Verlag Berlin Heidelberg, used with permission)

**Box 10.12 Ptiloerection, Shivering, and Body Temperature**

Ptiloerection traps warm air between the plumage and a bird's skin and can greatly reduce heat loss. As an example, Saarela et al. (1984) exposed summer-acclimated Rock Pigeons (*Columba livia*) to a temperature of 5°C and simultaneously measured the extent of ptiloerection and shivering and their body and foot temperatures over a 30-min period. At time 0 (first exposure to the colder temperature), the pigeons began to exhibit some ptiloerection to conserve heat while simultaneously beginning to shiver to generate heat. As the cold exposure continued, ptiloerection became more pronounced and the pigeons continued to shiver at low rates. By simultaneously exhibiting these heat-conservation and heat-generation measures (and reducing blood flow to the feet to reduce heat loss), the pigeons were able to minimize the decline in their body temperature (averaging only about 0.5°C).

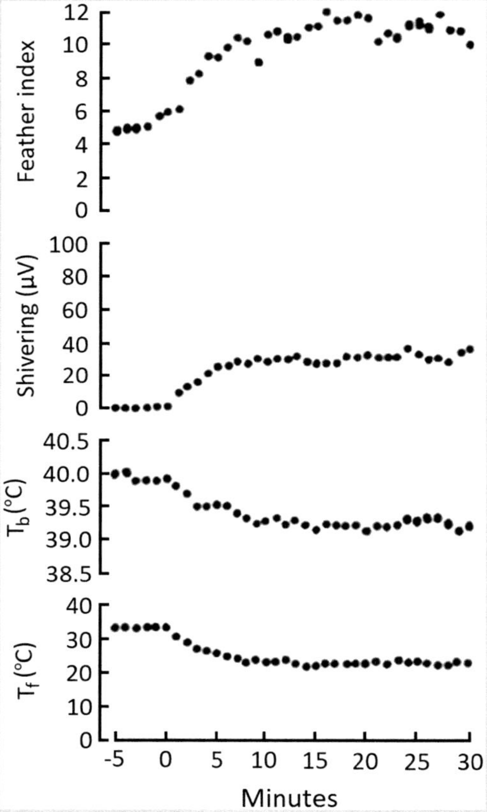


Feather indices (a measure of ptiloerection with higher numbers indicating greater ptiloerection), shivering intensity of pectoral muscles (as measured in microvolts, μV), and body ($T_b$) and foot ($T_f$) temperatures of pigeons exposed to a temperature of 5°C. (Figure modified from Saarela et al. 1984; © 1984 Springer-Verlag, used with permission)

**Fig. 10.77** Effect of ambient temperature on the percent of total evaporative water loss (TEWL) due to cutaneous water loss (CWL) for four species of songbirds. Dunn's Lark, *Eremalauda dunni* (top left); Greater Hoopoe Lark, *Alaemon alaudipes*; Eurasian Skylark, *Alauda arvensis*; Woodlark, *Lullula arborea* (top right). (Figure from Williams and Tieleman 2005; © 2005 Oxford University Press, used with permission. Photo of Dunn's Lark by Opisska, Wikipedia, CC0 Public Domain; Photo of Woodlark by Andrej Chudy, Wikipedia, CC BY 2.0, https://creativecommons.org/licenses/by/2.0/)

occurred primarily when nestlings were just 6–18 days old, suggesting that their ability to thermoregulate may not have been completely developed.

Because of differences among investigators in the definition of torpor, some rather larger birds, including Tawny Frogmouths (*Podargus strigoides*, about 500 g) and Laughing Kookaburras (*Dacelo novaeguineae*, about 360 g), have been reported to become torpid or exhibit "shallow torpor" (Körtner et al. 2000; Cooper et al. 2008). However, the lowest body temperatures recorded for these larger birds were about 29°C, or about 8–10°C below their normal body temperatures (Körtner et al. 2000; Cooper et al. 2008). Although the extent to which Tawny Frogmouths and Laughing Kookaburras are responsive at these low body temperatures has not been reported, Common Poorwills (*Phalaenoptilus nuttallii*; about 45 g) are capable of flight at body temperatures as low as 27.4°C (Austin and Bradley 1969). In addition, the minimal body temperatures of Tawny Frogmouths and Laughing Kookaburras are several degrees higher than those of smaller birds (from 4.3 to 23.5°C) that, when torpid, are unresponsive (see section entitled "Controlled, facultative hypothermia—torpor"). Among species that do not become torpid, birds remain alert and able to respond (though perhaps less effectively) during periods of hypothermia. For example, hypothermic Red-capped Manakins (*Ceratopipra mentalis*) were "lethargic but responsive" and, when disturbed, "they fluttered to the ground, maintained normal posture, and attempted to avoid capture" (Bartholomew et al. 1983:373–374). Although additional study is needed to determine the extent to which these larger birds are responsive during periods of below normal body temperatures, current evidence suggests that Tawny Frogmouths

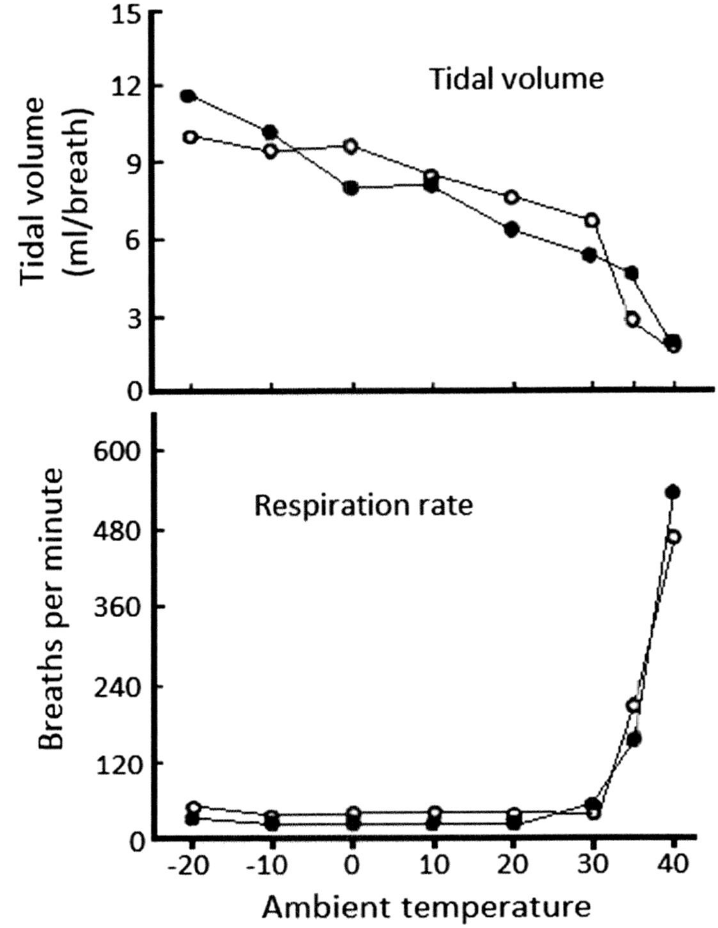


**Fig. 10.78** Effect of ambient temperate (°C) on the respiratory rates and tidal volumes of Chukars (*Alectoris chukar*) at different altitudes (open circles = 3800 m; closed circles = 340 m). At high ambient temperatures, Chukars began panting, with respiration rates increasing up to 23 times the rates at lower temperatures and tidal volumes declining. (Figure modified from Chappell and Bucher 1987; © 1986 Springer-Verlag, used with permission)

and Laughing Kookaburras do not exhibit true torpor, but, rather, like many other species of birds, use hypothermia to conserve energy.

Hypothermia is more common in smaller species of birds, but has also been reported in some larger species such as Laughing Kookaburras (*Dacelo navaeguineae*, 360 g; Dacelonidae: Coraciiformes; Cooper et al. 2008), Tawny Frogmouths (*Podargus strigoides*, 500 g; Podargidae, Caprimulgiformes; Körtner et al. 2000), Snowy Owls (*Nyctea scandiaca*, >2000 g; Strigidae, Strigiformes; Gessaman and Folk 1969), and Eurasian Griffons (*Gyps fulvus*, 6500 g; Accipitridae, Accipitriformes; Bahat et al. 1998). Larger birds are less likely to become hypothermic because they can store more energy, have lower metabolic rates, and, with less surface area per unit mass, tend to lose heat at lower rates than small birds. As a result, energy constraints are less likely for large species of birds (Geiser 2004). In addition, with their greater mass, rewarming rates for larger birds are slower than for small birds (McKechnie and Wolf 2004) and, as a result, given similar reductions in body temperature, larger birds would take longer to become normothermic than smaller birds—meaning less time for other activities and more time with increased risk of predation. Minimum body temperatures for hypothermic birds range from 26.8°C (11.1°C below normal body temperature, Golden-collared Manakin, *Manacus vitellinus*; Bartholomew et al. 1983) to about 40.5°C (about 2.5°C below normal body temperature, Mourning Dove, *Zenaida macroura*; Carr and

**Box 10.13 Hypocapnia and Respiratory Alkalosis**

When breathing rapidly, as when panting, birds could potentially lower levels of $CO_2$ in the blood. The same thing can occur when a human hyperventilates. Rapid breathing can lower levels of $CO_2$ in the lungs and that, in turn, causes more $CO_2$ to diffuse out of the blood and into the lungs. A lowering of blood $CO_2$ levels below normal levels is referred to as hypercapnia. Much of the $CO_2$ in the blood of birds (and humans) reacts with $H_2O$ to form carbonic acid ($H_2CO_3$) which, in turn, dissociates to form bicarbonate ($HCO_3^-$) and release a hydrogen ion ($H^+$). Having "normal" levels of $CO_2$ in the blood is important because the extent to which this reaction occurs helps maintain the blood's acid–base balance (more hydrogen ions = more acidic, fewer hydrogen ions = more basic or alkaline). When blood $CO_2$ levels drop, the balance between levels of $CO_2$ dissolved in the plasma and levels of hydrogen ions and bicarbonate is lost and the reaction reverses to compensate. $H^+$ in the blood plasma reacts with $HCO_3^-$ to generate more $CO_2$, and the resulting decline in blood $H^+$ levels causes the blood to become basic or alkaline, a condition referred to as respiratory alkalosis. Prolonged respiratory alkalosis can have a variety of negative effects on birds, including reduced food intake, endocrine dysfunction, reduced antioxidant capacity, and reducing calcium levels in the blood, which can, during the breeding season, affect eggshell quality (Ma et al. 2014). However, because tidal volume is generally reduced in panting birds and ventilation is largely limited to the trachea and bronchi, hypocapnia and respiratory alkalosis would likely only occur if birds are exposed to very high ambient temperatures for an extended period of time.

Lima 2012b). At these lower body temperatures, birds are still responsive, but less so than normal and potentially at greater risk of predation (Carr and Lima 2012b).

By lowering their body temperatures and metabolic rates, birds conserve energy and water. Although more energy can be conserved with lower body temperatures and metabolic rates, birds must balance the benefits of such savings with the increasing risk of predation resulting from reduced responsiveness as well as with the need to also expend energy (and time) to return body temperature back to normal. As an example of this trade-off, Laurila and Hohtola (2005) increased the perceived risk of predation for captive Rock Pigeons (*Columba livia*) by placing taxidermy mounts of Northern Goshawks (*Accipiter gentilis*) either next to (perched goshawk) or above (flying goshawk) their aviary. Because they were not fed during the experiments, pigeons had to balance the need to conserve energy with the need to minimize predation risk. As the risk of starvation increased (during the second day of the experiment without food), the pigeons further reduced their body temperatures during dark periods in both trials (Fig. 10.104). However, the pigeons also weighed the respective threats posed by perched versus "flying" goshawks against the potential need to be sufficiently responsive to avoid predation. With the "flying" goshawk perceived as the greater threat, the pigeons still became mildly hypothermic to conserve some energy, but maintained higher body temperatures so they would be better able to respond if threatened with predation (Fig. 10.104).

Most birds become hypothermic during periods of inactivity, at night for diurnal species and during the day for nocturnal species. However, some birds also become hypothermic when active. For example, some Black-capped Chickadees (*Poecile atricapillus*) were found to be hypothermic on cold winter days even when foraging (Lewden et al. 2014; Fig. 10.105). These

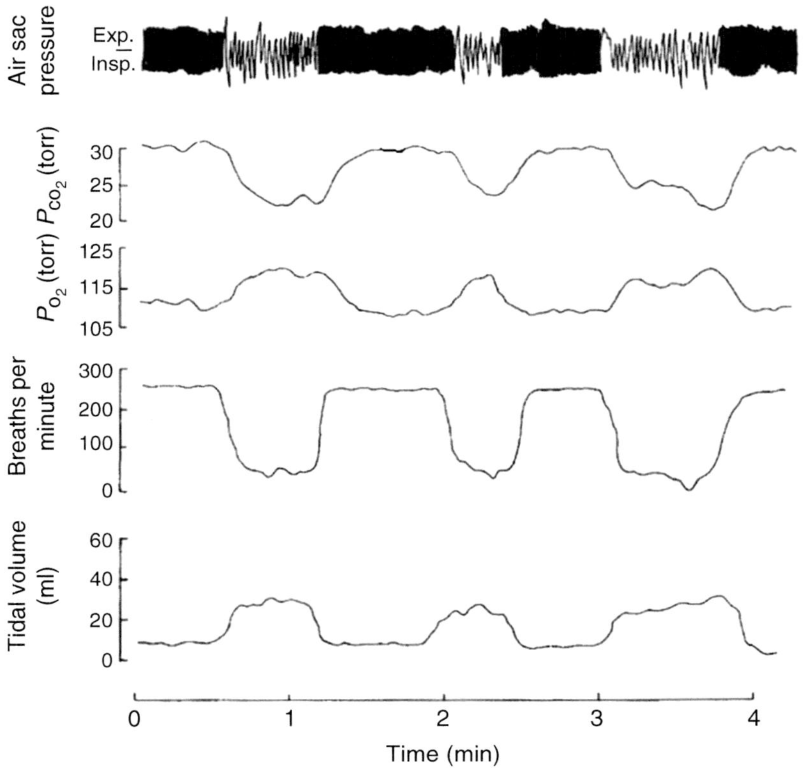


**Fig. 10.79** Variation in air-sac pressure (abdominal air sac), breathing rate, tidal volume, and partial pressures of $O_2$ and $CO_2$ in an abdominal air sac for Domestic Chickens (*Gallus g. domesticus*) maintained at an ambient temperature of 33°C. Partial pressure represents the amount or pressure of a gas in a mixture of gases; overall air pressure at sea level is about 760 mmHg. The chickens alternated periods of panting (with breathing rates averaging 271 per minute) with periods of slower, deeper breathing. The deeper breaths help eliminate $CO_2$ and increase $O_2$ levels in the air sacs and lungs. Exp, expiration; Insp, inspiration. Torr is essentially the same as millimeters of mercury. (Figure modified from Gleeson and Brackenbury 1983; © 1983 The Physiological Society, Published by John Wiley and Sons, used with permission)

chickadees also exhibit nocturnal hypothermia, when their body temperatures can drop lower than during the day (as low as 33.8°C; Chaplin 1976). By maintaining higher body temperatures when active, response times of mildly hypothermic chickadees are faster, enhancing their foraging ability and ability to avoid predation (Lewden et al. 2014). Interestingly, body temperatures of chickadees varied among individuals, with some remaining normothermic and others becoming hypothermic to varying degrees (Fig. 10.105). Reasons for such variation are unclear, but could be related to variation in fat reserves; individuals with less fat may become hypothermic to conserve energy so they can accumulate the fat reserves needed to survive the coming night (Lewden et al. 2014).

Daytime hypothermia has also been reported in two species of migratory songbirds, Garden Warblers (*Sylvia borin*) and Icterine Warblers (*Hippolais icterina*). Body temperatures of these songbirds were found to decline during the day at a stopover site during spring migration (and to decline even more during the night) (Carere et al. 2010). Hypothermia at stopover sites, both during the day and at night (e.g., Wojciechowski and Pinshow 2009), during migration may allow birds to conserve energy, thus allowing them to

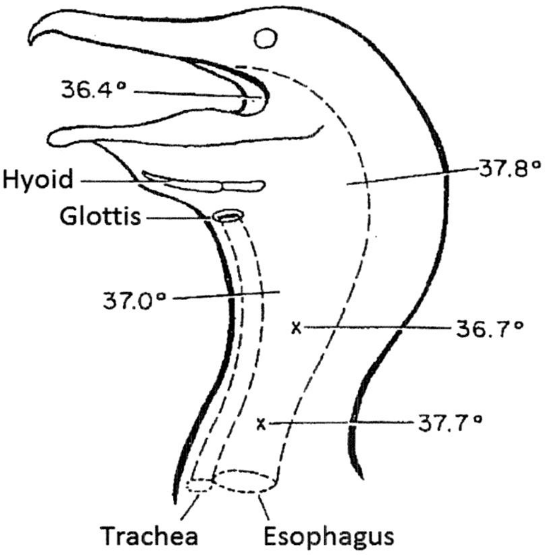


**Fig. 10.80** Temperatures in the head and neck regions of a cormorant during gular flutter. The increased evaporative water loss increases heat loss and helps birds maintain their body temperature. In this study, the ambient temperature was the same as the cormorant's body temperature (41.5°C). (Figure from Lasiewski and Snyder 1969; © 1969 Oxford University Press, used with permission)

restore energy reserves (fat) more quickly and continue their migratory flights. For example, a reduction in body temperature of about 4°C may allow birds to lower their metabolic rate and energy consumption by more than 50% (McKechnie and Lovegrove 2002).

Nocturnal hypothermia has been reported in a wide variety of birds and is particularly common among species that winter at higher latitudes and altitudes. During periods with low ambient temperatures, fewer hours of daylight for foraging, and the presence of snow and ice that may make foraging more difficult, birds, especially small birds with high surface-area-to-volume ratios, can be energetically challenged. To survive the long, often cold winter nights at higher latitudes, birds need to spend daylight hours securing sufficient food to ensure they have sufficient fat reserves. Haftorn (1992), for example, estimated that titmice (*Baeolophus* spp.; <20 g) needed to store about 10% of their body mass each day to survive the long, cold nights, and Krams et al. (2013) found that Great Tits (*Parus major*) that survived extremely cold nights (as low as −43°C) lost an average of 12.8% of their body mass. Six Great Tits that did not survive those cold nights lost an average of 17.5% of their body mass (range = 13.0–23.8%; Krams et al. 2013). Ambient temperature can clearly influence the fat-storing strategies of birds, but other factors can also play an important role (Box 10.18 Fat Accumulation in Wintering Songbirds).

A bird's ability to survive cold winter nights is influenced both by the amount of fat they accumulate during the day and the rate at which that fat is used during the night. Nocturnal hypothermia is an important factor in determining that rate, but other factors can also play a role.

Because roost site characteristics can greatly impact rates of heat loss, birds can benefit by selecting sites that better reduce radiative and convective heat loss (Table 10.5). Some birds reduce heat loss and the rate of fat utilization by huddling during cold nights, e.g., Bushtits (*Psaltriparus minimus*, Chaplin 1982), Golden-crowned Kinglets (*Regulus satrapa*, Heinrich 2003), Western Bluebirds (*Sialia mexicana*), and many others (see review by Gilbert et al. 2010). Huddling can result in substantial energy savings, potentially reducing energy expenditure by anywhere from 6 to 53% (Gilbert et al. 2010).

To the extent that body temperature can be lowered (i.e., as determined by physiological limitations and increasing mortality risk), controlled hypothermia can permit birds to balance the need to reduce body temperature with the amount of stored fat. For example, wintering Eurasian Blue Tits (*Cyanistes caeruleus*) with less fat were found to become more hypothermic than those with more fat (Fig. 10.106), suggesting an ability to regulate body temperature as needed based on available energy (Nord et al. 2011). Singing (*Gavicalis virescens*) and Brown (*Lichmera indistincta*) honeyeaters deprived of nectar prior to dusk had lower nocturnal metabolic rates and body temperatures than non-deprived birds (Collins and Briffa 1984). Similar results, with an inverse relationship between fat stores and nocturnal body temperatures, have been reported in Willow Tits (*Parus montanus*), Great Tits (*Parus major*), and Common Redpolls (*Acanthis flammea*;

**Fig. 10.81** (**a**) Bank Cormorant (*Phalacrocorax neglectus*) nesting colony. Note that some cormorants are sitting on nests incubating eggs or brooding young (upper left), whereas others are standing or crouching at their nests. Crouching and standing increase the vulnerability of eggs or nestlings of suboptimal temperatures, but adults may have to trade-off the need to thermoregulate against offspring survival. (**b**) Relationship between temperature and the proportion of time spent gular fluttering by Bank Cormorants. Observed data (points), predicted means (solid lines) and 95% credible intervals (shaded areas) for birds that were sitting (Si, blue) and those that were crouching or standing (Cr/St, yellow). Crouching or standing birds spent about 30% less time gular fluttering at an air temperature of 22°C than sitting birds. Colored lines that intersect the x-axis indicate the mean predicted temperature when birds would have to spend 50% of their time gular fluttering (dashed gray line). (Figures from Cook et al. 2020; © The Authors 2020. Published by Oxford University Press and the Society for Experimental Biology, open-access article distributed under the terms of the Creative Commons Attribution License, https://creativecommons.org/licenses/by/4.0/)

Reinertsen and Haftorn 1984, 1986), and Juniper Titmice (*Baeolophus ridgwayi*; Cooper and Gessaman 2005). In contrast, Cooper and Gessaman (2005) found that body mass did not influence the depth of hypothermia in Mountain Chickadees (*Poecile gambeli*). However, chickadees were provided with unlimited food during the study and, if food-restricted, results might differ (Cooper and Gessaman 2005).

Hypothermic birds can lower their body temperatures to conserve energy, but death can result if body temperatures get too low. For example, Brown Honeyeaters rarely survive if their body temperature drops below about 28°C (Collins and Briffa 1984), and European Starlings (*Sturnus vulgaris*) are unlikely to survive if their body temperature drops below 30°C (Brenner 1965). At these low body temperatures, death

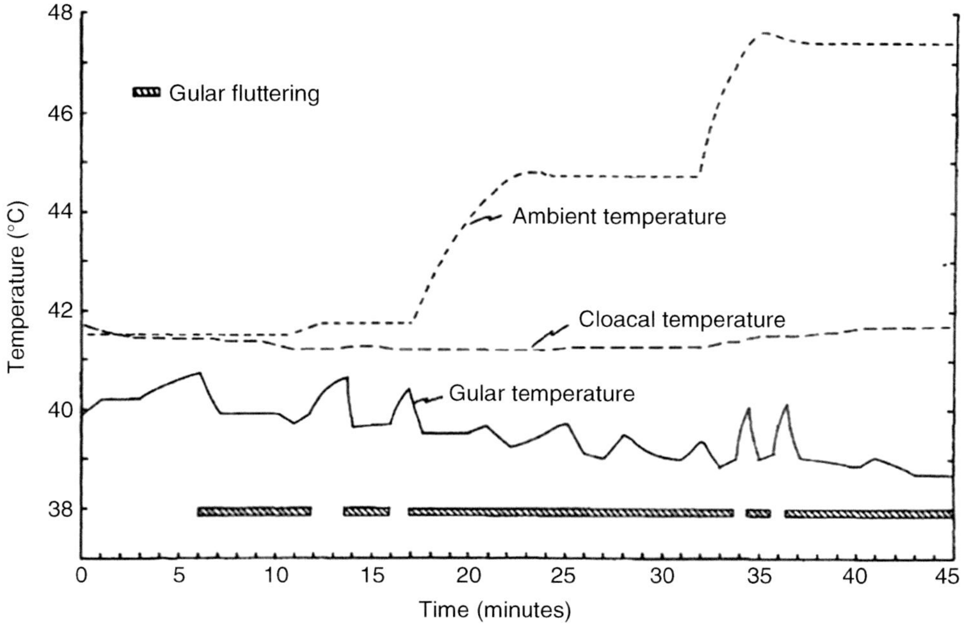


**Fig. 10.82** Effect of ambient temperature on time spent gular fluttering and gular and cloacal temperatures. Heat lost via gular fluttering allows Common Poorwills (*Phalaenoptilus nuttallii*) to maintain cloacal and body temperatures even at ambient temperatures approaching 48°C. Note the rapid increase in gular temperatures during brief periods when gular fluttering stopped. (Figure modified from Lasiewski and Bartholomew 1966; © 1899 CCC Republication, used with permission)

presumably results from reduced efficiency of cell or metabolic functions or an inability to generate sufficient heat to increase body temperature (Hainsworth and Wolf 1978).

Although hypothermia is a common strategy for conserving energy for small birds during the winter, some birds become hypothermic at other times of year and for reasons in addition to surviving cold nights (Fig. 10.107). Some birds become hypothermic on a regular basis throughout the year to reduce the energy needed for thermoregulation (e.g., Cooper and Gessaman 2005), and many birds become hypothermic in response to food deprivation, regardless of the ambient temperature. Hypothermia during periods of reduced food availability has been reported in at least 29 families of birds (Hohtola 2012). Birds are sometimes faced with both cold temperatures and reduced food availability and, in such cases, hypothermia is even more important for conserving energy (Fig. 10.108).

## 10.14 Controlled, Facultative Hypothermia: Torpor

As a bird's body temperature drops, its ability to respond to stimuli declines until, at some point, the bird becomes lethargic and capable of only minimal responsiveness to stimuli or is completely unresponsive. At that point, a bird has become torpid. Reported minimum body temperatures of torpid birds range from 23.5°C (Puerto Rican Tody, *Todus mexicanus*; Merola-Zwartjes and Ligon 2000) to as low as 4.3°C (nightjars, McKechnie and Lovegrove 2001) and 3.26°C (Black Metaltail hummingbird, *Metallura phoebe*; Wolf et al. 2020). Other reported minimum body temperatures include 23.3°C for sunbirds (McKechnie and Lovegrove 2001), 19.4°C for owlet-nightjars (Doucette et al. 2012), 18.2°C for mousebirds, 17°C for swifts, and 6.5°C for Andean Hillstar (*Oreotrochilus*

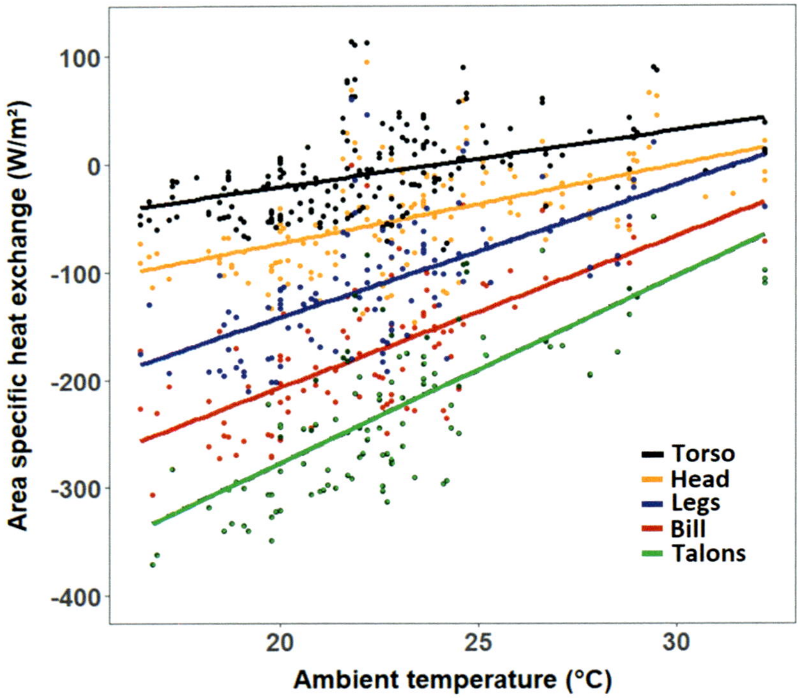


**Fig. 10.83** Area-specific heat exchange ($W/m^2$) for the different body parts of an Osprey's (*Pandion haliaetus*) body. Heat exchange was determined based on surface temperature measurements at ambient temperatures ranging from 16.5 to 32.2°C, and a wind speed of 0.1 m/s. Heat exchange <0 indicates a loss of heat and > 0 indicates heat gain. For Ospreys, most heat was dissipated through their talons, followed by the bill, legs, head, and torso. The morphology of raptor talons is likely determined more by prey type and size (Tsang et al. 2019) so the relatively long legs and talons of Osprey that prey on fish represent thermoregulatory exaptations (Rogalla et al. 2021). (Figure from Rogalla et al. 2021; © 2021 Elsevier Ltd., used with permission)

*estella*) hummingbirds and Broad-tailed Hummingbirds (*Selasphorus platycercus*; McKechnie and Lovegrove 2001).

During periods of torpor, body temperatures and energy use drop well below normothermic levels, with the extent of the declines influenced by the amount of stored fat and ambient temperature. Metabolic rates of birds that become torpid can decline by as much as about 95% (Table 10.6) and heart rates also decline (Table 10.7). Breathing rates also decline. For example, the breathing rates of torpid Allen's (*Selasphorus sasin*) and Anna's (*Calypte anna*) hummingbirds drop from a normal resting rate of about 265 breaths per minute to as low as six breaths per minute (Withers 1977).

As with hypothermia, but to a greater degree, torpor allows birds to conserve energy during periods of cold weather, food deprivation (e.g., Bize et al. 2007; Wellbrock et al. 2022), or both (Fig. 10.109). The need to conserve energy varies with size because larger birds can store more energy than small birds. In general, then, smaller birds become torpid more often than larger birds (Spence and Tingley 2021; Fig. 10.110). As an example, some small hummingbirds that weigh less than 5 g may become torpid almost every night, but a large hummingbird such as the Giant Hummingbird (*Patagona gigas*) of South America (about 19 g) rarely becomes torpid (Lasiewski et al. 1967). The relationship between food availability and torpor is nicely illustrated by Freckled Nightjars (*Caprimulgus tristigma*). For

**Box 10.14 Why Do Some Birds Have Bald Heads?**

Some birds have entirely or partially bare heads, including vultures and condors, some storks, several species of ibis, some spoonbills, some galliforms, a few passerines, (e.g., Bare-headed Laughingthrush, *Garrulax calvus*), and others. Bare or partially bare heads may serve a variety of functions. For example, the skin of unfeathered areas of some birds is brightly colored and may be important in inter- and intrasexual interactions. For vultures and condors, bare heads have often been assumed to be adaptations for scavenging, i.e., head feathers would become soiled and matted when feeding inside carcasses. However, another possible function is that these bare areas play a role in thermoregulation. Studies of Turkey Vultures (*Cathartes aura*), Griffon Vultures (*Gyps fulvus*), and Wild Turkeys (*Meleagris gallopavo*) have revealed that their bare heads do play important thermoregulatory roles. Ward et al. (2008) found that Griffon Vultures would be able to increase or decrease heat loss from their bare head and neck, depending on the ambient temperature, by simply changing their posture. In an "expanded posture," with head and neck extended, the vultures can expose more skin and increase heat loss whereas, in a "covered posture," vultures can greatly reduce the amount of exposed skin and dramatically reduce heat loss. Similarly, Arad et al. (1989) found that the unfeathered skin on the head of Turkey Vultures had numerous thin-walled veins near the skin's surface (epidermis), suggesting a role in heat dissipation.

Ward et al. (2008) found that Griffon Vultures would be able to increase or decrease heat loss from their bare head and neck, depending on the ambient temperature, by simply changing their posture. (**a**) In an "expanded posture," with head and neck extended, the vultures can expose more skin and increase heat loss, whereas (**b**) in a "covered posture," vultures can greatly reduce the amount of exposed skin and dramatically reduce heat loss.

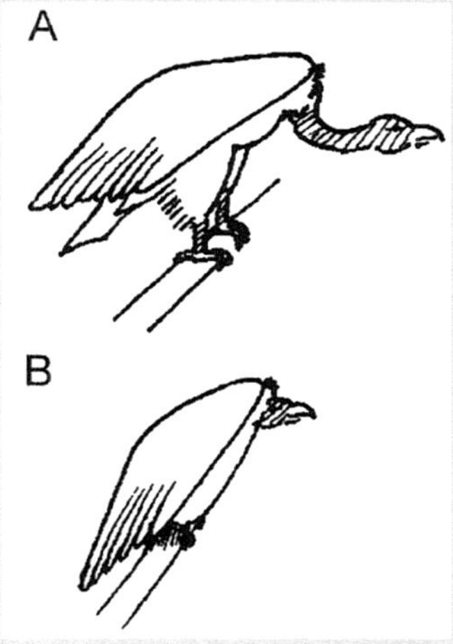


(Figure modified from Ward et al. 2008; © 2008 Elsevier Ltd., used with permission)

(continued)

**Box 10.14** (continued)

Turkey Vulture. (Photo from pxhere.com, CC0 Public Domain)

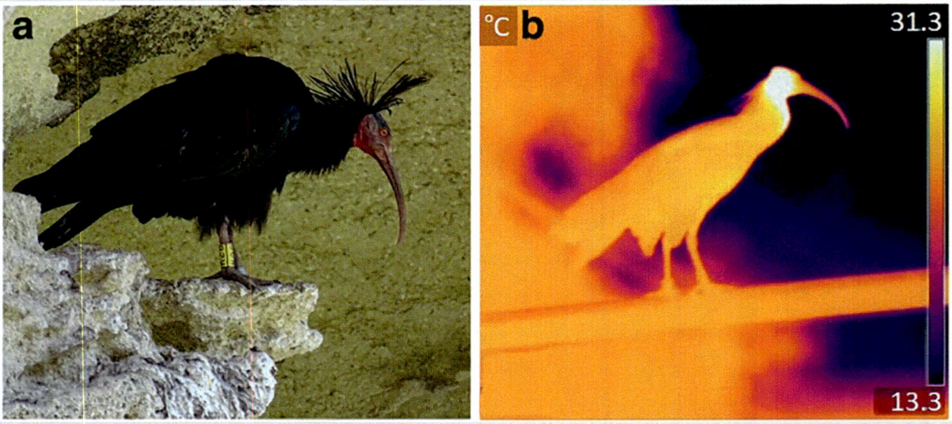


(**a**) A Northern Bald Ibis (*Geronticus eremita*). (Photo credit: J. H. Negro). (**b**) Thermal image of a Northern Bald Ibis illustrating the effectiveness of the bare head in dissipating heat. (Figure from Galván et al. 2017, © The Authors 2017, open-access article is distributed under the terms of the Creative Commons Attribution 4.0 International License, http://creativecommons.org/licenses/by/4.0/)

Buchholz (1996) examined the role of the unfeathered head and neck of Wild Turkeys and found that, at high ambient temperatures (35°C), heat dissipation from these unfeathered areas played a critical role in helping them maintain thermiostasis. In fact, the results of this study suggest that, if a Wild Turkey's head was feathered, it would not be able to dissipate sufficient

(continued)

**Box 10.14** (continued)

heat to maintain thermiostasis. In addition, as with Griffon Vultures, Wild Turkeys can, at low ambient temperatures, reduce the amount of exposed skin and reduce heat loss by using integumentary muscles to draw the feathered skin at the base of the neck up and over much of the usually bare areas on the back of the neck (Buchholz 1996).

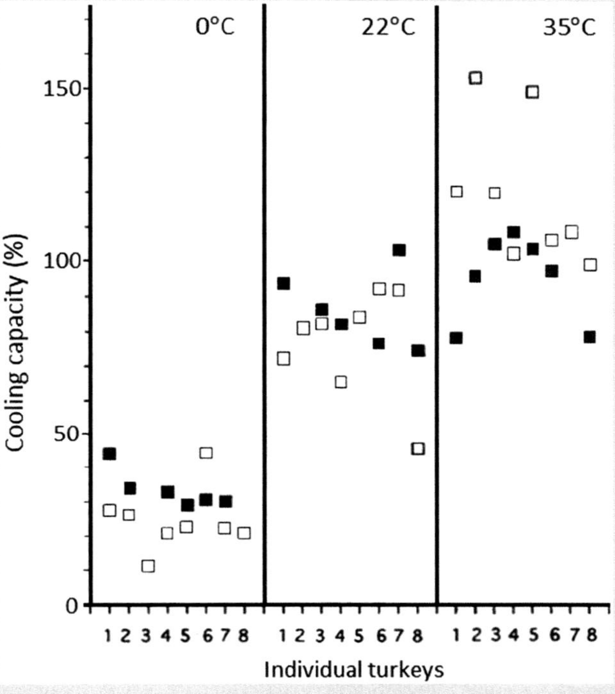


Cooling capacities of eight Wild Turkeys with heads bare (open squares) and with heads insulated (filled squares). "Insulated" turkeys had their head and neck covered with a sock. At an ambient temperature of 35°C, Wild Turkeys with bare heads were significantly better at dissipating excess heat by evaporation than those that were "insulated." At lower ambient temperatures, cooling capacities of bare-headed and insulated turkeys did not differ. Cooling capacity is a bird's ability to dissipate heat by evaporation and is expressed as a percentage (heat lost by evaporation divided by heat produced by metabolism times 100%). (Figure modified from Buchholz 1996; © 1996 Oxford University Press, used with permission)

crepuscular or nocturnal birds that depend on vision to capture prey such as most nightjars, food availability (foraging success) can vary with light levels. As a result, in response to low light levels when they spend much less time foraging (Fig. 10.111), Freckled Nightjars increasingly become hypothermic or torpid to conserve energy (Smit et al. 2011).

Torpor can be divided into three phases: (1) entry, where metabolic rate and body temperature drop below normothermic rest-phase levels, (2) maintenance, where body temperature either matches the ambient temperature or is thermogenically regulated at some point about the ambient temperature (i.e., the hypothalamic set point), and (3) rewarming, where body

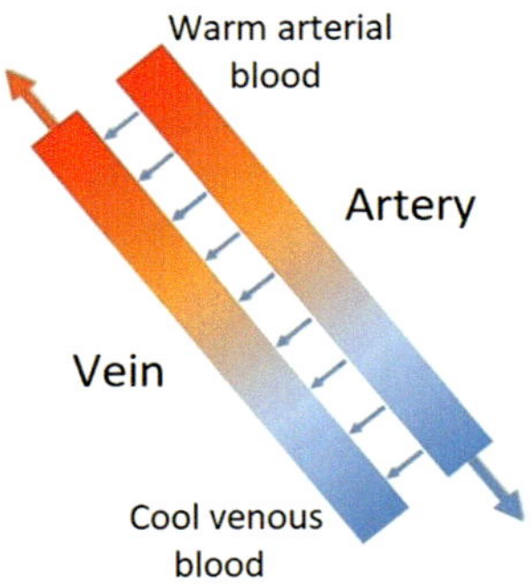


**Fig. 10.84** Example of the counter-current transfer of heat from an artery to a vein in the leg of a bird, e.g., a Mallard (*Anas platyrhynchos*) walking on ice. Heat is transferred from the warmer blood in the artery to the cooler blood in the vein. Because of this transfer of heat, the feet remain much cooler and, therefore, much less heat is lost through the feet even when walking on ice. (Photo of Mallard from pxhere.com, CC0 Public Domain; Figure of artery and vein modified from Gad-Briggs et al. 2017; © 2017 by the authors. Licensee MDPI, Basel, Switzerland, open-access article distributed under the terms and conditions of the Creative Commons Attribution (CC BY) license, http://creativecommons.org/licenses/by/4.0/)

temperature returns to normothermic levels by endogenous heat production (shivering or non-shivering thermogenesis), exogenous heat sources (e.g., solar radiation), or both (Fig. 10.112). Bouts of torpor generally continue throughout the night (for diurnal birds) and the day (for nocturnal birds), typically 10–12 h (Geiser and Ruf 1995). However, depending on a bird's energy reserves or, perhaps, in response to some external stimulus (e.g., sounds that might indicate the possible presence of a predator), bouts of torpor may be limited to only part of the night (or day) (Bech et al. 1997; Fig. 10.113). During the entry phase of torpor, metabolic rates and heat production (thermogenesis) drop, and the results of previous studies suggest that these declines likely result from an overall metabolic depression along with a reduction in rates of oxidative phosphorylation in mitochondria (Staples and Brown 2008). During the maintenance phase, body temperature continues to be controlled via the hypothalamus (i.e., the thermostat) and autonomic nervous system. Physiological functions such as heart rate, respiration rate, blood flow, and, when needed, thermogenesis via shivering (and, perhaps, non-shivering thermogenesis) are maintained at levels needed to maintain body temperature at or above ambient temperature or, if ambient temperature falls below a point that could be fatal for a bird, the hypothalamic set point temperature. The timing of arousal after a period of torpor appears to be primarily influenced by a bird's circadian rhythm (Körtner and Geiser 2000), with arousal typically (but not always) occurring either around dawn for diurnal species and sunset for nocturnal species.

The time needed to become torpid and to arouse from torpor varies with body mass (Figs. 10.114 and 10.115). Smaller birds have more surface area relative to their mass and so tend to lose heat at a faster rate than larger birds. Smaller birds, with higher metabolic rates per unit mass and less mass to warm, can also increase their body temperatures faster when arousing from torpor. Cooling and warming rates as birds become torpid and arouse from torpor also vary with ambient temperature, with cooling rates faster and warming rates slower with lower ambient temperatures. For birds that varied in size at different ambient temperatures, cooling rates were found to vary from 1.8 to 17.1°C per hour (Brigham 1992). The energy required to re-warm also varies with body mass and ambient temperature, with increasing amounts of energy needed as body mass increases and body temperature during

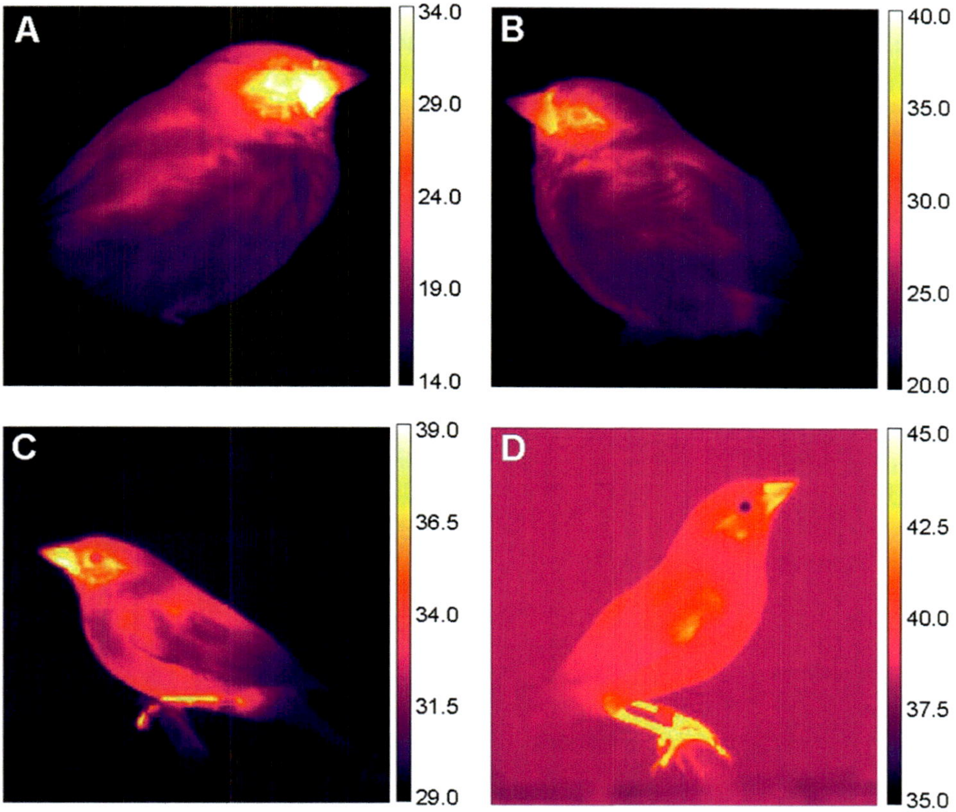


**Fig. 10.85** Thermal images of Song Sparrows (*Melospiza melodia*) at ambient temperatures of (**a**) 15°C, (**b**) 21°C, (**c**) 29°C, and (**d**) 37°C. With cooler ambient temperatures, heat is only being lost from about the eyes and base of the bill, but, as ambient temperatures increase, heat loss from the bill, legs, feet, and, to a lesser extent, the wings increases dramatically. Differences in the temperatures of a bird's head, bill, wings, legs, and feet can vary because of vascular countercurrent heat exchange. (Figure from Greenberg et al. 2012a; open-access article made available under the Creative Commons CC0 Public Domain)

torpor decreases (Fig. 10.116). Because of the time and energy needed to rewarm after a bout of torpor, with the exception of Caprimulgids, large birds (>30 g) do not use torpor to conserve energy.

Several factors likely explain why caprimulgids (e.g., nightjars and owlet-nightjars) are the exception to the rule that larger birds do not become torpid (Brigham et al. 2012; Figs. 10.117 and 10.118). Most importantly, their food habits and foraging behavior make it very likely that they will need to endure energy deficits. Most species are nocturnal aerial insectivores, and the availability of insect prey can decline dramatically with falling temperatures and/or rainfall. In addition, caprimulgids are visual predators so their foraging success generally declines during periods with less moonlight. Finally, caprimulgids have lower basal metabolic rates than other species of birds that are similar in size (Fig. 10.118).

Torpor allows birds to save considerable energy, but successful reproduction requires a considerable amount of energy to produce sperm or eggs, incubate and brood young, and provision nestlings. In addition, normal development of embryos and nestlings requires the transfer of heat from adults. Faced with this potential trade-off between conserving and expending energy, birds may become mildly hypothermic at times,

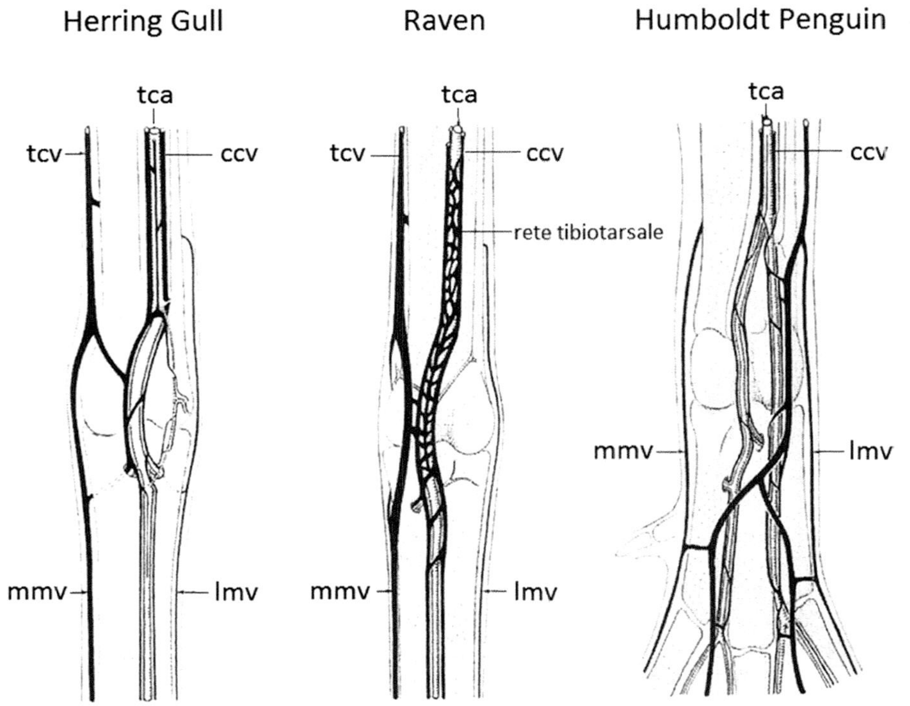


**Fig. 10.86** Anterior views of the left intertarsal joint of a Herring Gull (*Larus argentatus*), Common Raven (*Corvus corax*), and Humboldt Penguin (*Spheniscus humboldti*). Note the close association between the tibialis cranialis arteries and the counter-current veins, especially in the Common Raven where the veins form a complex network called the rete tibiotarsale. At this point, heat can be transferred from the artery to the veins to minimize heat loss from the lower leg and feet. With warm temperatures, more heat can be lost via the lower legs and feet by shunting more blood from the counter-current veins to the tibialis caudalis vein or lateral metatarsal vein. tca, tibialis cranialis artery; ccv, tibialis cranialis counter-current vein, tcv, tibialis caudalis vein; mmv, medial metatarsal vein; lmv, lateral metatarsal vein. (Figure modified from Midtgård 1981; © The Royal Swedish Academy of Sciences, used with permission)

but become torpid only under extreme conditions. For example, a female Anna's Hummingbird (*Calypte anna*) monitored during the incubation period only became torpid once for 4.5 h during a cold night (10°C) with almost constant rain (Vleck 1981). Similarly, Eberts et al. (2023) recorded nightly time-lapse thermal images of 14 nesting female Anna's Hummingbirds and found that females usually did not enter torpor. One female entered deep torpor on two nights (2% of nights monitored), and two other females appeared to enter shallow torpor on three nights (3% of nights monitored). Continuous recording of nest temperatures of female Broad-tailed Hummingbirds (*Selasphorus platycercus*) revealed only one brief bout (3.5 h) of torpor during 1 of 161 nights of incubation by several females over a 2-year period (Calder and Booser 1973). Female hummingbirds are largely able to avoid the use of torpor during incubation by foraging frequently during the day (Fig. 10.119) and by constructing well-insulated nests at sites that minimize heat loss (Calder 1974).

Caprimulgids also greatly restrict use of torpor during the breeding season. For example, during the incubation and brooding periods, male and female Common Poorwills (*Phalaenoptilus nuttallii*) became torpid during only 2 of 195 nights. In contrast, the same individuals became torpid during 27 of 44 nights during the nonbreeding period (Csada and Brigham 1994).

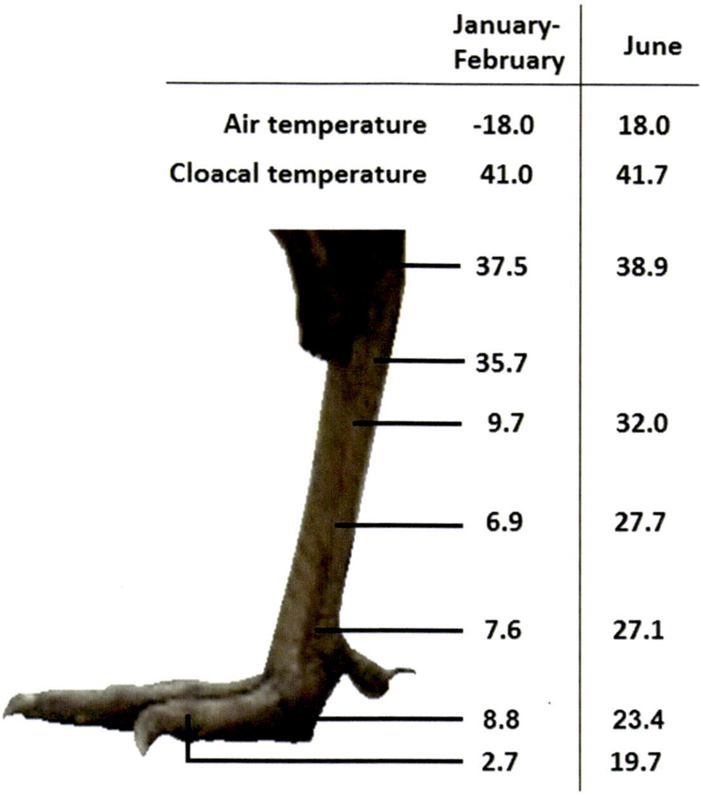


**Fig. 10.87** Leg temperatures (°C) of captive Ring-necked Pheasants (*Phasianus colchicus*; $N = 6$) kept in an outdoor enclosure in North Dakota and measured during the winter (January–February) and spring (June). The counter-current heat exchanger located just above the "heel" ensures that, when air temperatures are low, the temperature of the arterial blood passing into the unfeathered foot is greatly reduced, thereby reducing heat loss to the surrounding cold air. When ambient temperatures are higher, more blood bypasses the counter-current heat exchanger so warmer arterial blood flows into the feet (Temperatures from Ederstrom and Brumleve 1964; Photo of pheasant leg from pxhere.com, CC0 Public Domain)

## 10.15 Controlled, Facultative Hypothermia: Hibernation

Periods of torpor are limited to part of the day and generally no longer than 10–12 h. The only known exception is Common Poorwills (*Phalaenoptilus nuttallii*; Fig. 10.120), where individuals are known to remain torpid for multiple days. In other words, Common Poorwills are the only birds known to hibernate. These small (50 g) nocturnal insectivores are found in arid and semiarid habitats throughout western North

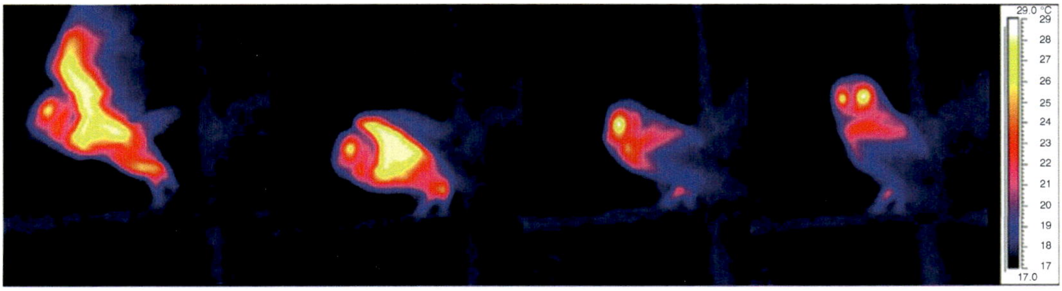


**Fig. 10.88** Barn Owl (*Tyto alba*) taking flight. Heat is being lost from the face and wings, but, because of the countercurrent arteries and veins, not from the legs and feet. (Figure from McCafferty 2013; © 2012 British Ornithologists' Union, used with permission)

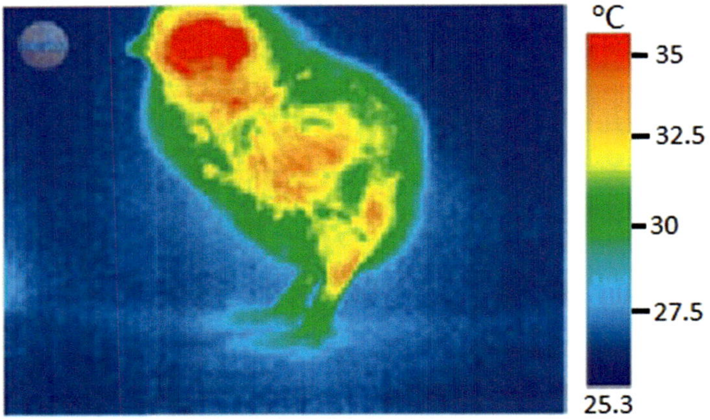


**Fig. 10.89** A young Domestic Chicken (*Gallus g. domesticus*) at an ambient temperature of about 9°C. Heat loss is reduced because the temperatures of the legs and feet are kept cooler by the counter-current exchange of heat in the legs. (Figure from Ferreira et al. 2011; open-access article under a Creative Commons Attribution License, https://creativecommons.org/licenses/by/4.0/)

America and may hibernate for periods as long as 25 days by free-ranging birds, 45 days by experimentally shaded birds, but typically exhibit multiday torpor for shorter periods (e.g., 5–10 days) (Woods and Brigham 2004; Woods et al. 2019). Common Poorwills hibernate frequently during the nonbreeding period (November–April), but do not do so during the breeding season (and only rarely become torpid during the breeding season). In a study conducted during the nonbreeding season in the southwestern United States (Arizona), Common Poorwills typically became torpid or hibernated when ambient temperatures approached or dropped below 10°C, a temperature at which few or no flying insects are available and the metabolic costs of thermoregulation increase. When hibernating, the respiration rates of Common Poorwills drop from normal levels (about 34 breaths per minute) to as low as four breaths per minute, and energy consumption is reduced by as much as 94% (based on the reduction in oxygen consumption; Withers 1977).

Why are Common Poorwills the only known species of bird capable of hibernating? Woods and Brigham (2004) hypothesized that hibernation is an adaptation of Common Poorwills to a diet of flying insects in an arid environment with limited productivity where they must also compete for food with other species of insectivores. In addition, areas where Common Poorwills are found exhibit considerable variation in temperature, with extended periods of low temperatures during the winter further limiting availability of flying insects. The need to minimize energy use during these sometimes extended periods with limited food availability has likely been the selective force favoring the evolution of multiday torpor bouts (hibernation) by Common Poorwills.

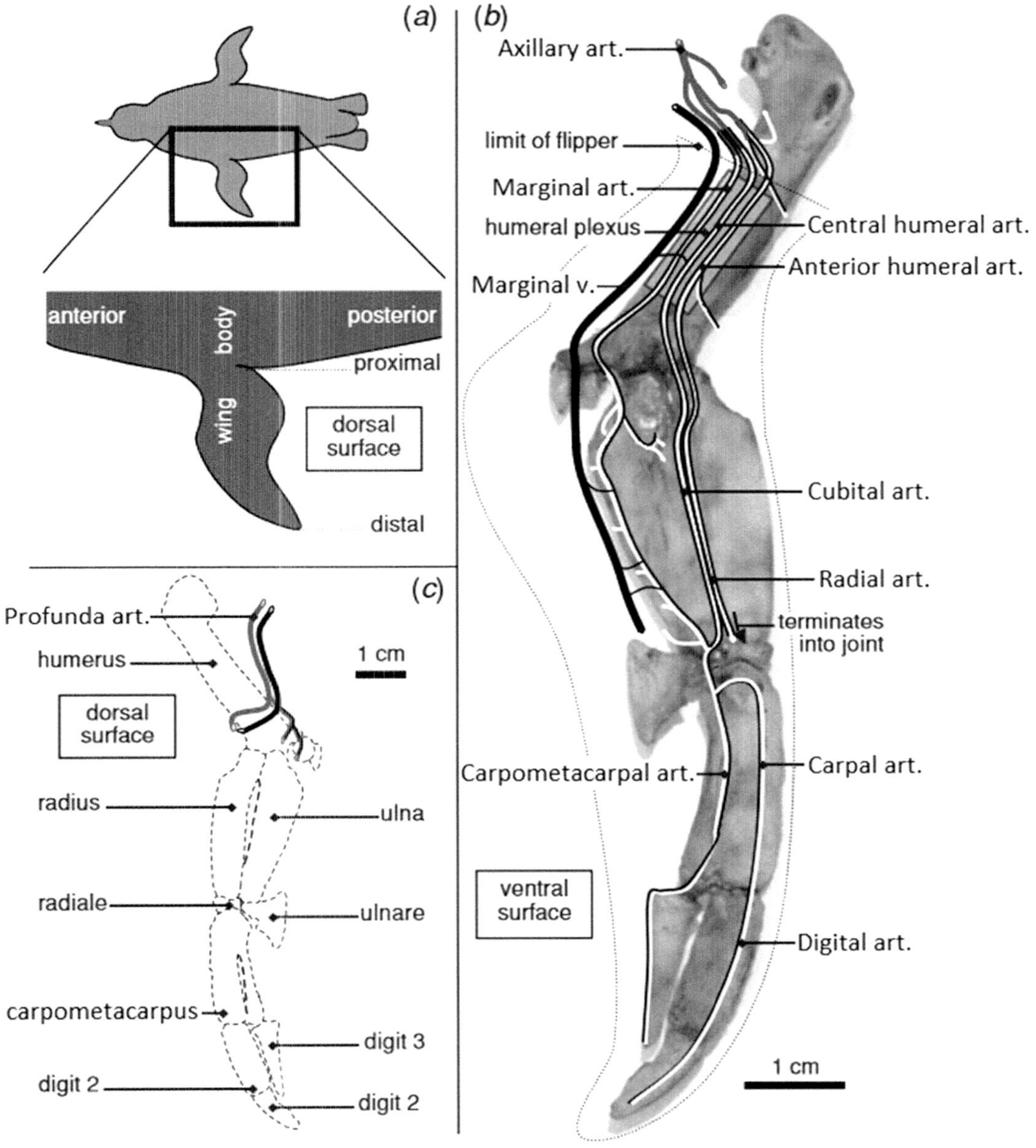


**Fig. 10.90** Counter-current heat retention in the penguin flipper. (**a**) Orientation terminology. (**b**) Arteries (white or gray) and veins (black) in the flipper of a Little Penguin (*Eudyptula minor*). Each branch of the axillary artery is associated with at least two veins so heat from the arteries is transferred to the adjacent veins, cooling the blood as it passes into the flipper and, as a result, limiting the amount of heat lost from the flippers. (**c**) Arteries (gray) and veins (black) on the dorsal surface of the flipper. (Figure from Thomas and Fordyce 2007; used with permission of CSIRO Publishing)

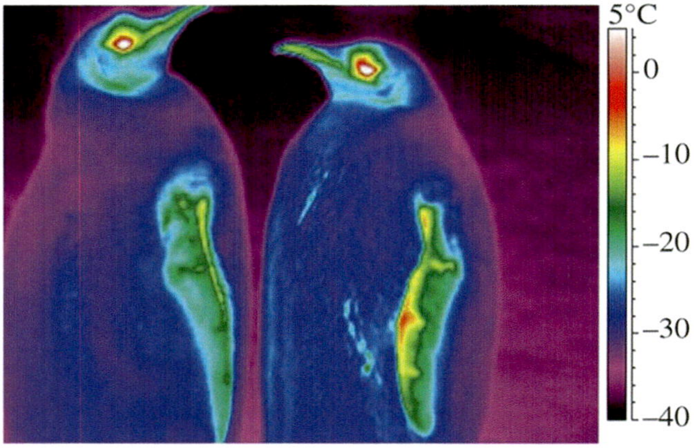


**Fig. 10.91** Thermal image of Emperor Penguins (*Aptenodytes forsteri*) at a breeding colony in Antarctica where the air temperature was −21°C. Note that the temperature of most of the penguin surfaces is lower than air temperature and well below freezing. This results from extreme radiative cooling of the surface that removes so much heat that the surface temperature is less than that of the air. Heat is lost only from unfeathered regions (feet, eye, and bill) and sparsely feathered flippers and is minimized by countercurrent heat exchange systems through arteriovenous networks in the head, flippers, and legs. Emperor Penguins also have small bills relative to their body size that help minimize heat loss. (Figure from McCafferty et al. 2013; © 2013 The Authors. Published by the Royal Society, used with permission)

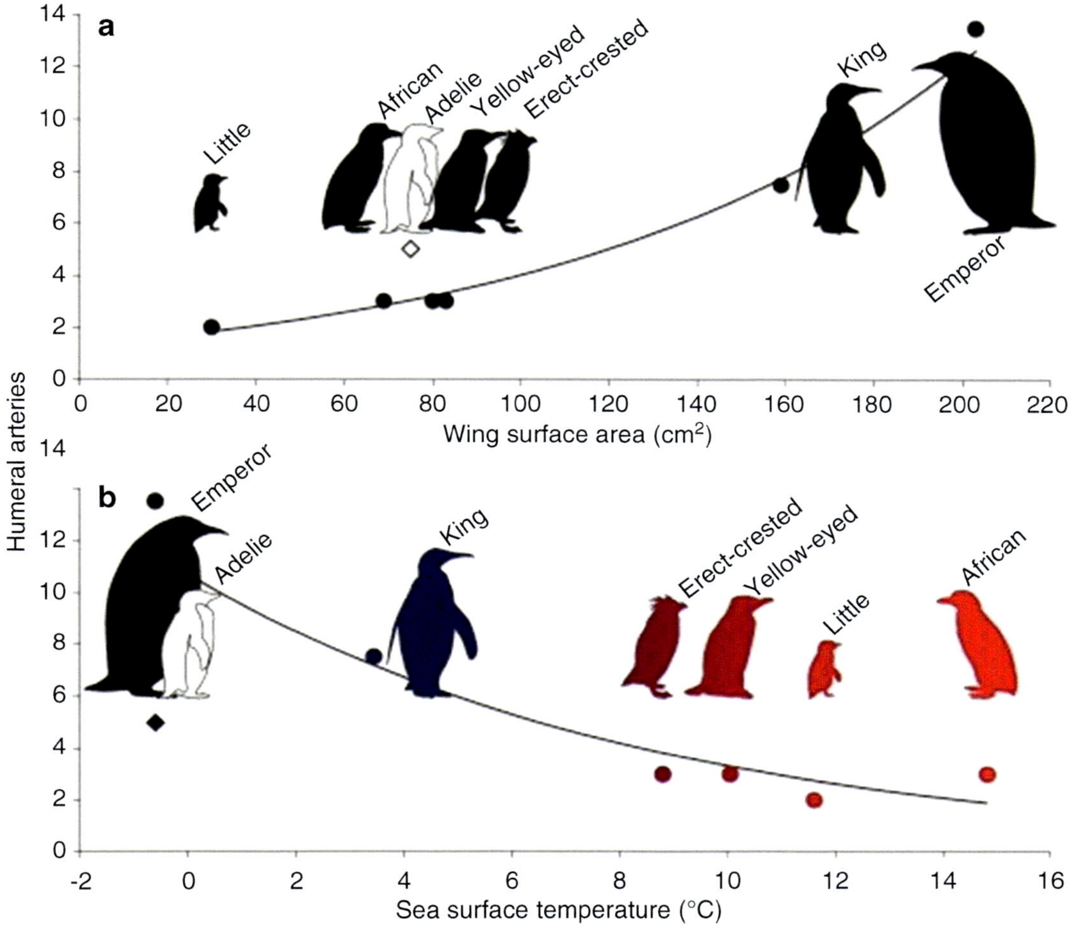


**Fig. 10.92** Relationships between (**a**) the number of humeral arteries in penguin wings and wing surface area and (**b**) artery count and mean sea surface temperature at breeding sites. Penguins with greater wing surface area and those that breed in areas with colder water have more arteries for more efficient counter-current exchange of heat. An exception to these trends is the Adélie Penguin (*Pygoscelis adeliae*). The large gradient between body core (38.5°C; Prinzinger et al. 1991) and the water (−0.6°C) temperatures where they forage may have favored the evolution of a more efficient mechanism of heat retention (i.e., more arteries) in Adélie Penguins than similar-sized penguins found at more temperate latitudes. In addition, Adélie Penguins have much smaller flippers, with less surface area, than Emperor Penguins (*Aptenodytes forsteri*), so, compared to Emperor Penguins, fewer arteries may be necessary to retain heat (Trawa 1970). Surface area and temperature measurements are from Stonehouse (1967). King Penguin, *Aptenodytes patagonicus*; Yellow-eyed Penguin, *Megadyptes antipodes*; Little Penguin, *Eudyptula minor*; African Penguin, *Spheniscus demersus*. (Figure from Thomas and Fordyce 2012; © 2011 Wiley Periodicals, Inc., used with permission)

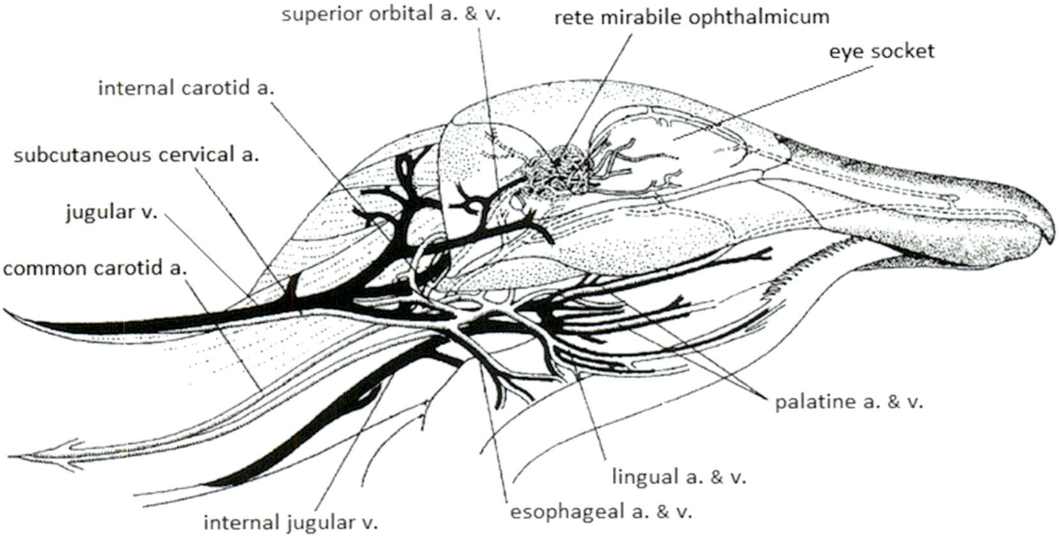


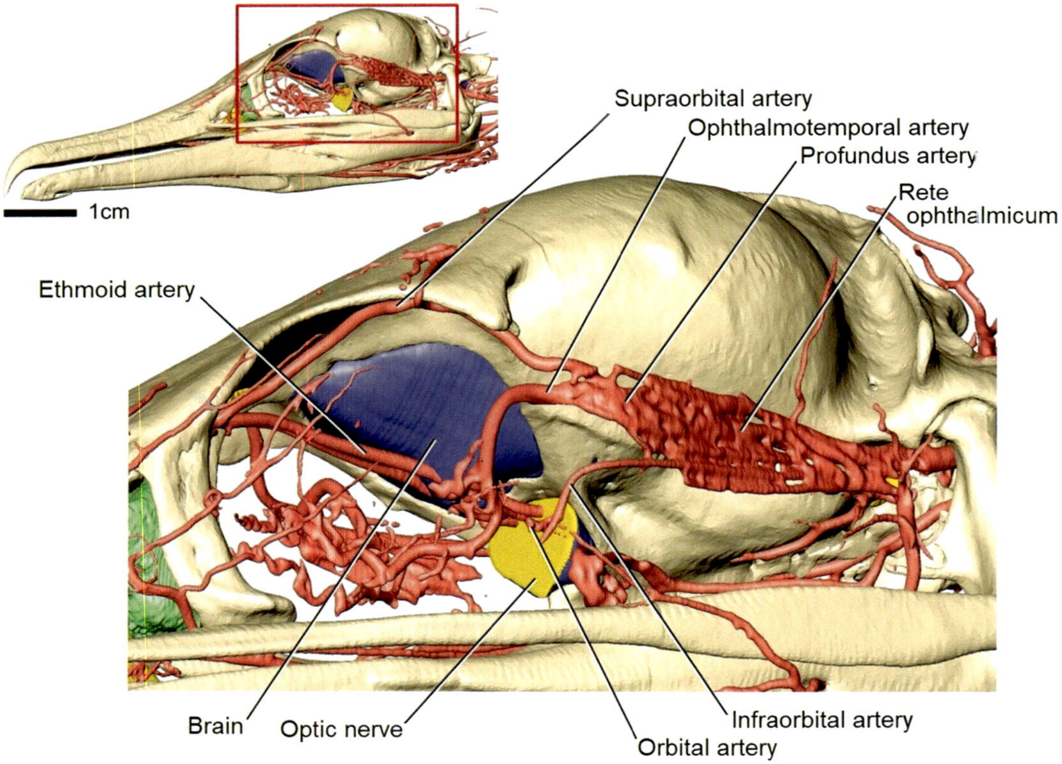


**Fig. 10.93** Top, blood vessels in the head of an African Penguin (*Spheniscus demersus*) showing the rete mirabile ophthalmicum. The tongue and associated muscles have been removed to provide a better view of the blood vessels. Note the numerous close associations between arteries and veins in the head. These associations, along with the rete mirabile, facilitate heat exchange. The very close association between small arteries and veins in the rete mirabile allows for more efficient exchange of heat. By varying the rate and amount of blood flow through these vessels, the amount of heat lost can be modified. Bottom, Rete ophthalmicum and other arteries and

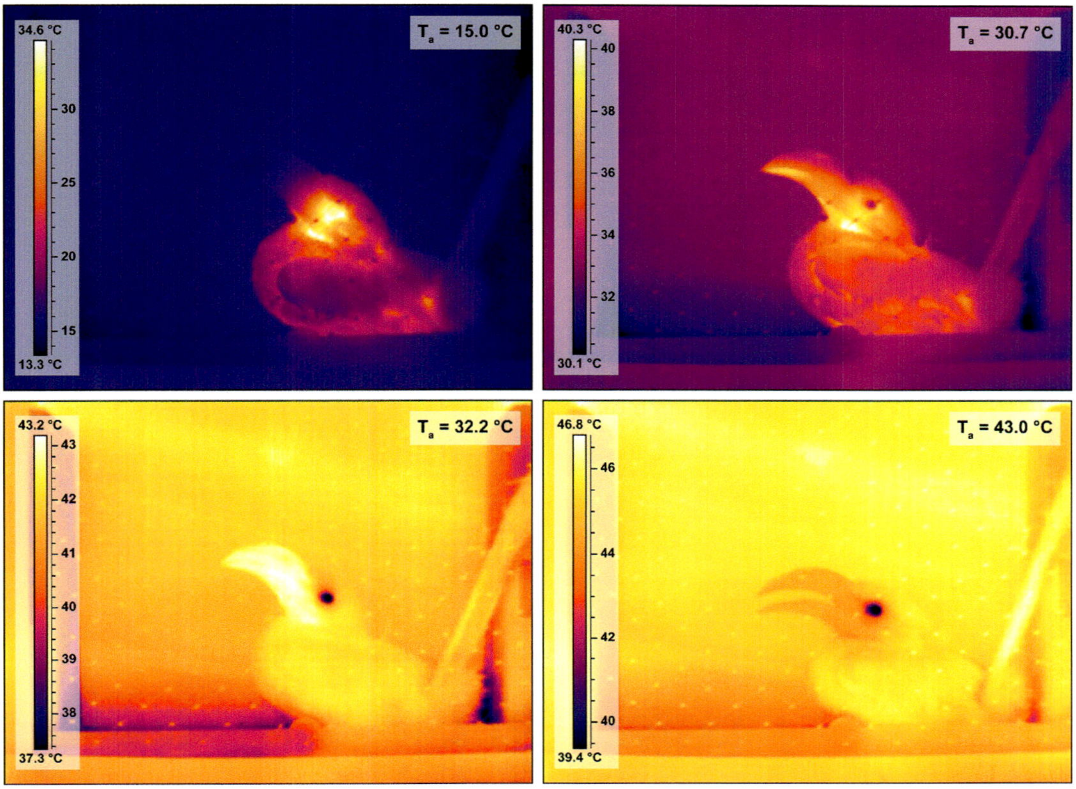


**Fig. 10.94** Thermal images of a female Southern Yellow-billed Hornbill (*Tockus leucomelas*) at different air temperatures. Surface temperatures (°C) are indicated by scale bar on the left side of each image. Top left: air temperature ($T_a$) is 15°C and the beak surface temperature matches that of the background temperature. Top right: air temperature is 30.7°C, and beak temperature is increasing, lower mandible first. Bottom left: air temperature is 32.2°C, and the beak temperature is much higher than that of the rest of the body and the environment so heat is being radiated from the beak. Bottom right: air temperature is 43°C and higher than the hornbill's body temperature. The beak is cooler than the ambient temperature and the hornbill is panting (note the open bill), using evaporative water loss to lose heat. (Figure from van de Ven et al. 2016; © 2016 van de Ven et al., open-access article distributed under the terms of the Creative Commons Attribution License, https://creativecommons.org/licenses/by/4.0/)

**Fig. 10.93** (continued) structures in the skull of a Double-crested Cormorant (*Phalacrocorax auritus*). The rete plays an important role in thermoregulation, allowing heat loss when necessary, but may also prevent excessive heat loss from the eyes when exposed to colder temperatures. Studies of other taxa suggest that visual acuity may be influenced by retinal temperatures (Parver 1991; Fritsches et al. 2005) (Top figure modified from Frost et al. 1975; © 2009 John Wiley and Sons, used with permission. Bottom figure modified from Porter and Witmer 2016; © 2016 Wiley Periodicals, Inc., used with permission)

**Box 10.15 Summer Temperature and Bill Surface Area of Sparrows**

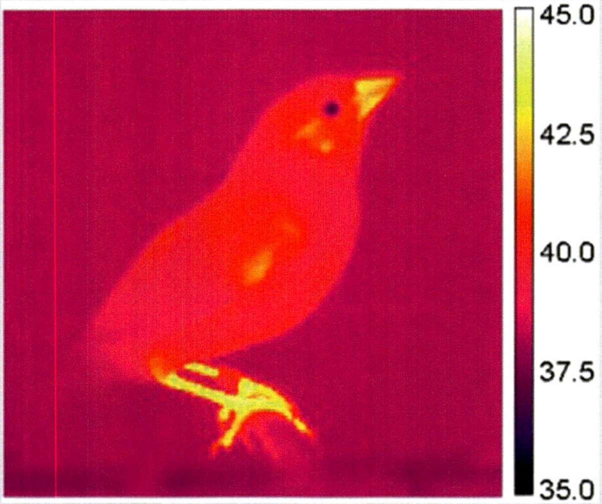


Infrared image of a Song Sparrow (*Melospiza melodia*) where the ambient temperature was 37°C. Note the loss of heat from the bill (and the legs). (Figure from Greenberg et al. 2012a; open-access article distributed under a Creative Commons CC0 license)

Physiological factors are rarely proposed to account for variation in the morphology of feeding structures. However, bird bills have been found to be important convective and radiant heat sinks. Larger bills with greater surface area could serve as more effective thermoregulatory organs under hot conditions. This heat-radiating function of bills would likely be more important in open habitats with little shade and stronger winds. As a way of dumping heat without losing water through evaporation, bills might be particularly important for birds that need to lose heat in areas where freshwater is limited. To examine the possible relationships between thermoregulation and bill size, Greenberg et al. (2012b) plotted the bill size of 10 species or subspecies against maximum summer and minimum winter temperatures. The surface area of sparrow bills was found to increase with increasing summer temperature, suggesting that bill morphology may be influenced by the need to efficiently lose heat in warmer environments.

(continued)

**Box 10.15** (continued)

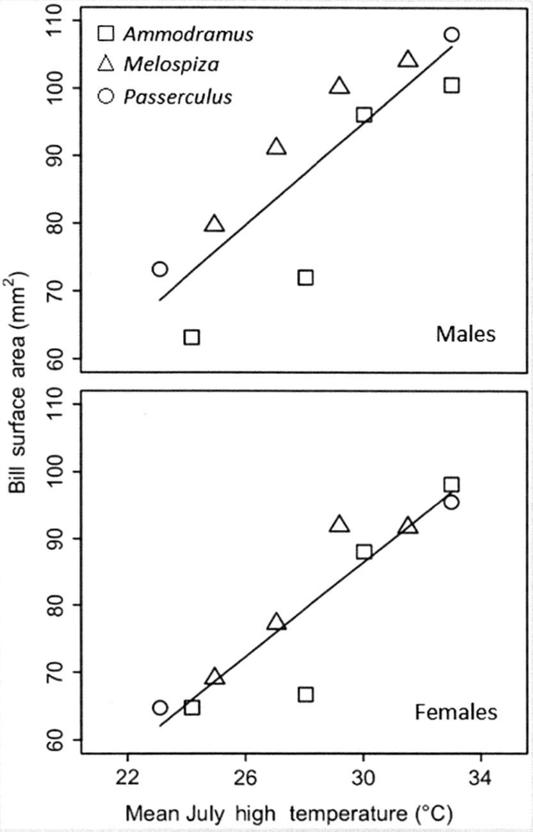


Surface area of bills of several species and subspecies of sparrows in three different genera tends to increase with increasingly high mean July temperatures. (Figure modified from Greenberg et al. 2012; © 2011 The Authors, used with permission)

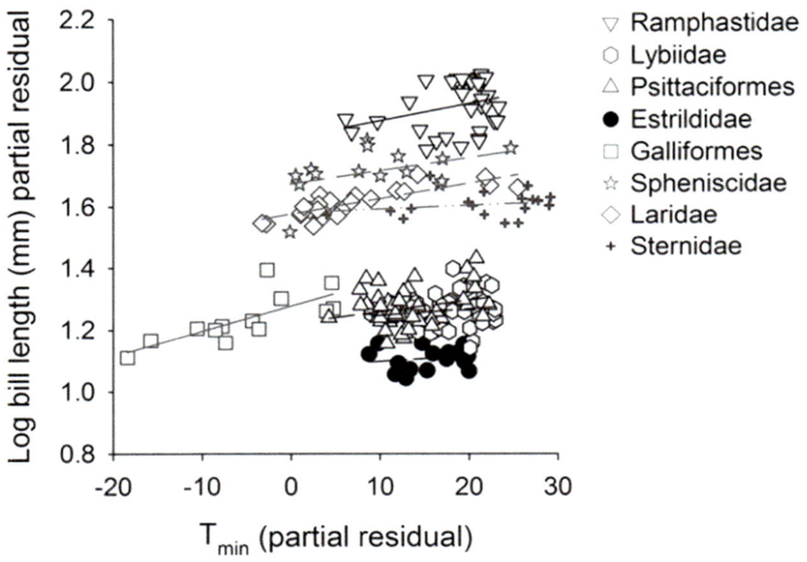


**Fig. 10.95** Relationship between bill length and minimum monthly temperature ($T_{min}$) for several taxa of birds. Best-fit lines are provided for each taxonomic group. For all groups except Estrildidae, bill length is significantly shorter where minimum monthly temperatures are lower. Ramphastidae—toucans; Lybiidae—African barbets and tinkerbirds; Psittaciformes—parrots and cockatoos; Estrildidae—grass finches; Galliformes—grouse, quail, ptarmigans, and Wild Turkey; Spheniscidae—penguins; Laridae—gulls; Sternidae—terns. (Figure from Tattersall et al. 2017; © 2016 Cambridge Philosophical Society, used with permission)

**Fig. 10.96** Average apportionment of total evaporation in Inca Doves (*Columbina inca*) at an ambient temperature of 42°C. (Data from Hoffman et al. 2007; Photo by Scott Loarie, CC BY 2.0, https://creativecommons.org/licenses/by/2.0/)

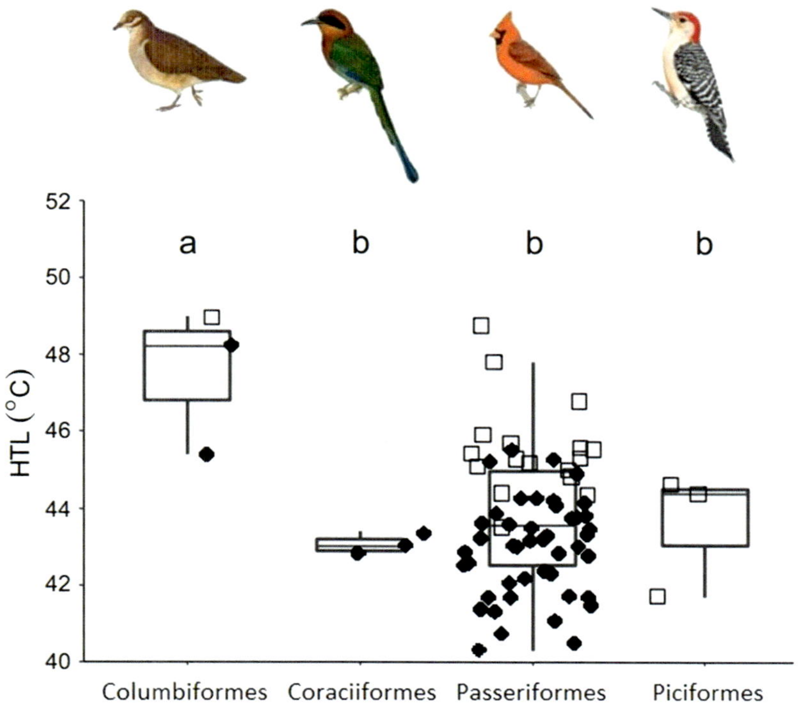


**Fig. 10.97** Box plots of heat tolerance limits (HTL) of four orders of birds, including Columbiformes (3 species), Coraciiformes (3 species), Passeriformes (71 species), and Piciformes (4 species). Species in the order Columbiformes exhibited greater heat tolerance limits than species in the other three orders, and heat tolerance limits of temperate species of birds (white squares) were greater than those of tropical species (black diamonds) Different letters indicate significant differences ($\alpha = 0.05$). Top, exemplar species from left to right: Ruddy Quail-dove (*Geotrygon montana*; Columbiformes), Rufous Motmot (*Baryphthengus martii*; Coraciiformes); Northern Cardinal (*Cardinalis cardinalis*; Passeriformes), and Red-bellied Woodpecker (*Melanerpes carolinus*; Piciformes). Data were collected from just one temperate location and one tropical location, so studies conducted in other habitats and locations are needed to evaluate the generality of these results. (Figure from Pollock et al. 2021; © 2020 British Ecological Society, used with permission)

**Box 10.16 Extreme Hyperthermia Tolerance of Red-Billed Queleas**

The maximum body temperatures that birds can tolerate rarely exceed 46°C, with this limit thought to be limited by the sensitivity of cell macromolecules to extreme heat (Laszlo 1992; Roti Roti 2008) or declining oxygen levels due to excessive oxygen demand at high temperatures (Pörtner 2001). Few examples of birds able to tolerate body temperatures above 46°C have been reported. In one such case, Weathers (1997) found that Variable Seed-eaters (*Sporophila aurita*) in Panama were able to tolerate body temperatures just above 46°C (46.8–47.0°C). More recently, Freeman et al. (2020) found that Red-billed Queleas (*Quelea quelea*) were able to tolerate body temperatures as high as 49.1°C, which they noted is "unprecedented among birds and mammals."

Red-billed Queleas are found in semi-arid habitats in sub-Saharan Africa and are considered one of the most abundant wild bird species in the world (ca. 1,500,000,000 individuals) (Craig 2020). In the lab, Freeman et al. (2020) exposed 20 Red-billed Queleas to air temperatures ranging from 28 to 52°C and monitored their body temperatures using temperature-sensitive passive integrated transponder (PIT) tags. The mean normothermic body temperature of the queleas was 40.9°C; individual maximum body temperatures ranged from 46.4 to 49.1°C, with 15 birds (75%) reaching body temperatures ≥48.0°C.

Male Red-billed Quelea in breeding plumage. (Photo by Pawel Ryszawa, Wikipedia, CC BY 3.0, https://creativecommons.org/licenses/by/3.0/)

(continued)

**Box 10.16** (continued)

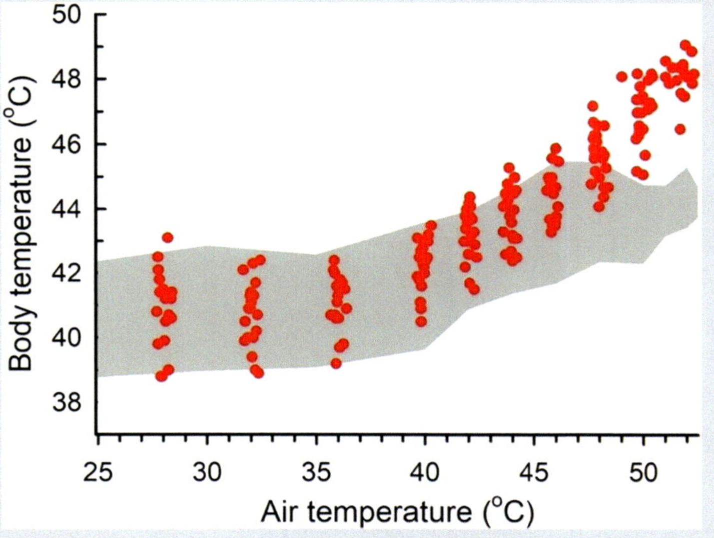


When exposed to extremely high air temperatures, body temperatures of Red-billed Queleas (red circles) generally remain within the range reported for other songbirds at air temperatures below 45°C. However, quelea body temperature increased well above those reported previously at air temperatures above 45°C. Gray shading indicates the range of body temperature for five species of Australian songbirds and three species of South African songbirds using the same experimental protocol. (Figure from Freeman et al. 2020; open-access article licensed under a Creative Commons Attribution 4.0 International License, https://creativecommons.org/licenses/by/4.0/)

The critical thermal maximum temperature of Red-billed Queleas was 2–3°C higher than the previously known range for birds, and the temperatures they were able to tolerate exceeded known lethal values for songbirds (Dawson 1954). The ability of these queleas to tolerate such high body temperatures is likely the result of several anatomical and molecular mechanisms, including a rete opthlmicum that helps them maintain cooler brain temperatures and "pronounced heat shock protein expression" (Freeman et al. 2020). Heat shock proteins function as "molecular chaperones," interacting with other proteins to reduce the likelihood that they will interact inappropriately with each other, reducing the likelihood of protein denaturation at high temperatures, clearing misfolded proteins from cells, repairing DNA damage, and maintaining cellular structures such as the cytoskeleton (Richter et al. 2010).

(continued)

**Box 10.16** (continued)

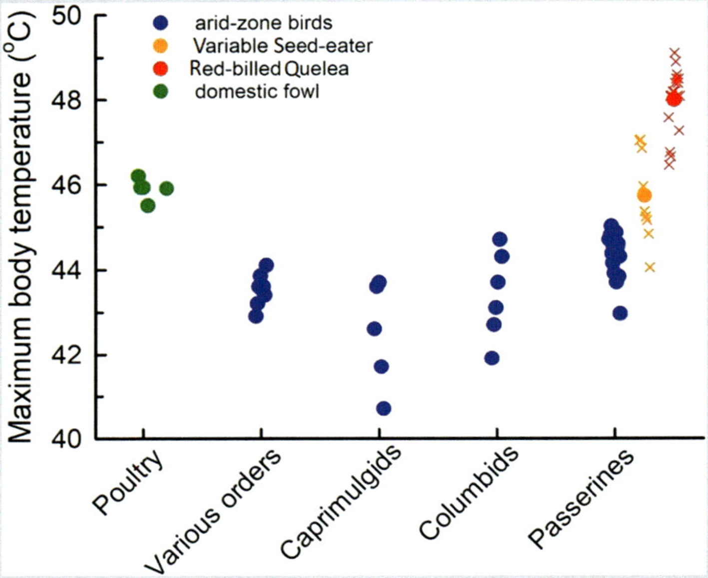


Maximum body temperatures attained of Red-billed Queleas exposed to high temperatures were much higher than those reported previously for other birds. For Variable Seedeaters (*Sporophila corvina*) and Red-billed Queleas, both species averages (filled circles) and values for individual birds (X) are shown. (Figure from Freeman et al. 2020; open-access article licensed under a Creative Commons Attribution 4.0 International License, https://creativecommons.org/licenses/by/4.0/)

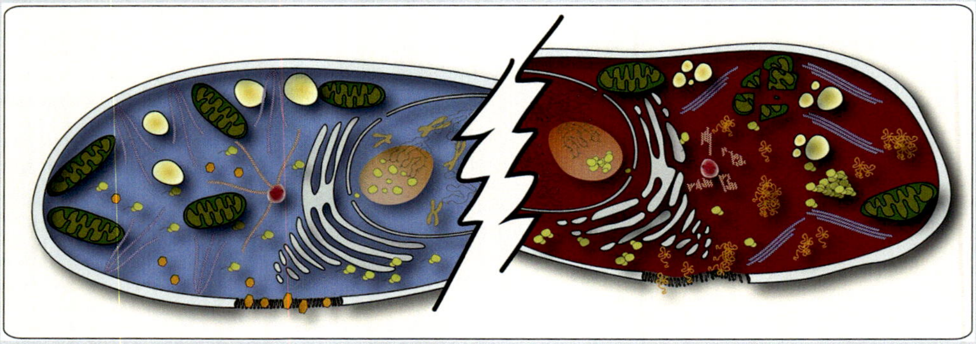

Possible effects of heat shock on eukaryotic cells. An unstressed eukaryotic cell (left) is compared to a heat-stressed cell (right). Heat stress can damage the cytoskeleton, including actin filaments (blue) and microtubules (red). Organelles such as the Golgi apparatus and the endoplasmic reticulum (white) fragment, mitochondria (green) fragment, and the number of lysosomes (yellow-white gradient) decrease, nucleoli (where ribosomes are assembled in the nucleus) swell and ribosomal proteins become visible, protein aggregates form in the cytoplasm (hexagonal versus spaghetti style, orange), and cell membrane morphology changes. (Figure from Richter et al. 2010; © 2010 Elsevier Inc., used with permission)

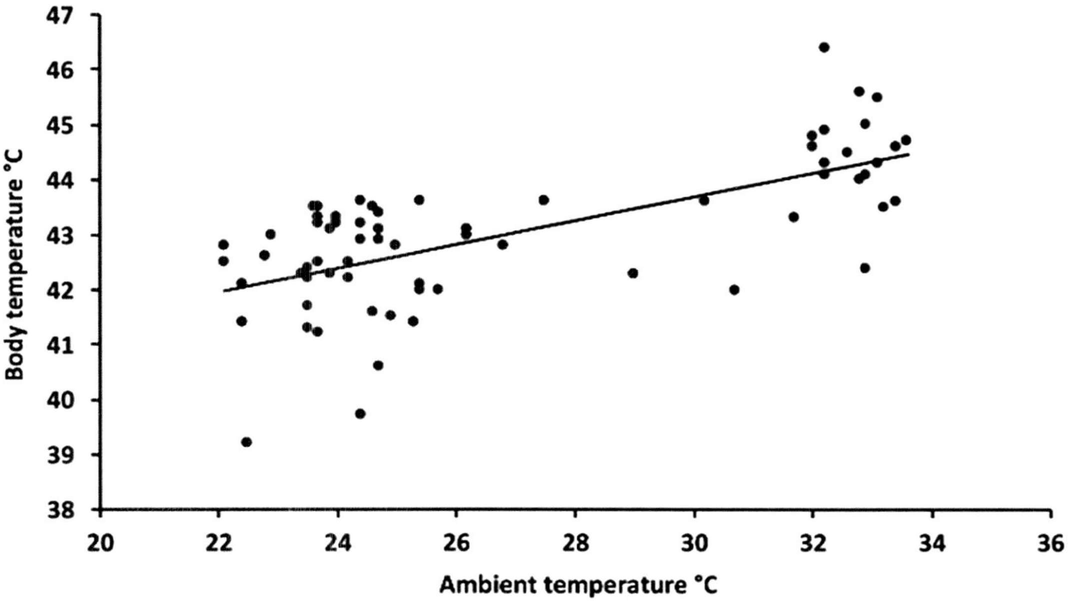


**Fig. 10.98** Relationship between ambient temperature in the shade and body temperature for 69 individuals representing 13 different species of birds in central Nigeria. With increasing ambient temperatures, birds allowed their body temperatures to increase. Data were collected during the hot, dry season with few sources of water available. (Figure from Nilsson et al. 2016; © 2016 Nilsson et al., open-access article distributed under the terms of the Creative Commons Attribution License, https://creativecommons.org/licenses/by/4.0/)

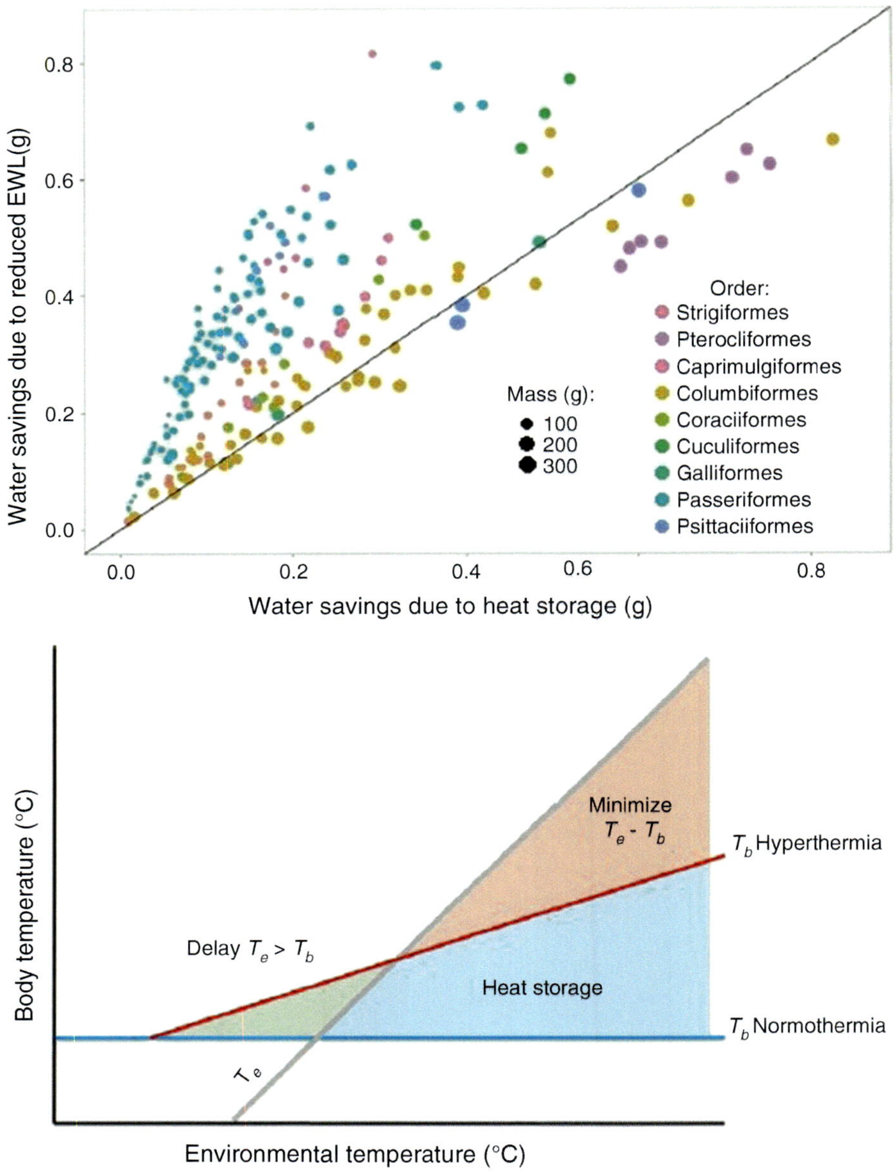


**Fig. 10.99** (Top) Cumulative water savings due to a reduction in evaporative water loss (EWL) relative to the total savings from heat storage for a 1-h heat exposure at 45°C. Most birds, regardless of size, but especially songbirds (Passeriformes) can save water by being hyperthermic and reducing EWL, but larger birds benefit more from water savings due to heat storage. (Bottom) Schematic showing possible benefits of hyperthermia. As environmental temperatures increase, keeping body temperature ($T_b$) above the environmental temperature via hyperthermia ($T_b > T_e$) allows heat loss by conduction, convection, and radiation rather than by evaporative water loss. However, when the environmental temperature exceeds body temperature ($T_e > T_b$), typically at about 45°C, further increases in body temperature help reduce the difference between $T_e$ and $T_b$ and so reduce the amount of evaporative water loss. In addition, heat stored in body tissues does not have to be lost via evaporative water loss, thus saving more water. (Figure modified from Gerson et al. 2019; © 2019 The Authors. Functional Ecology © 2019 British Ecological Society, used with permission)

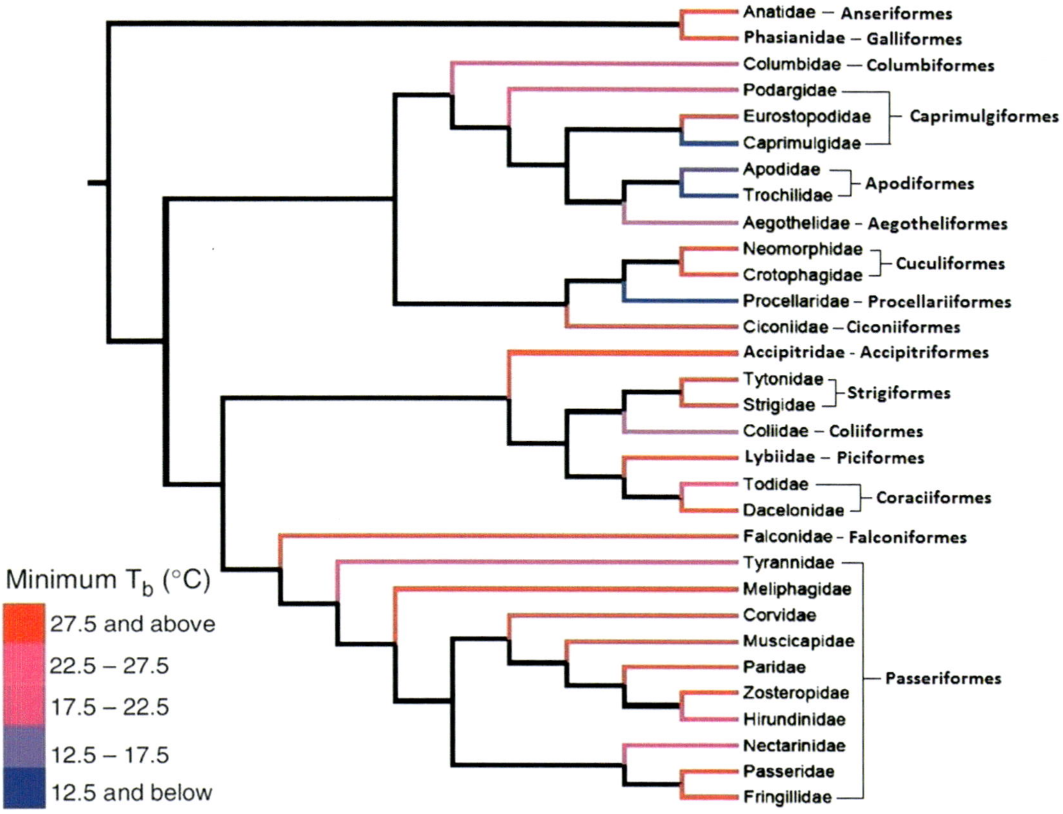


**Fig. 10.100** Minimum body temperatures of birds in orders and families of birds known to become hypothermic. Birds in the order Apodiformes (hummingbirds and swifts) and the families Caprimulgidae (nightjars) and Procellariidae (nestling petrels and shearwaters) are able to lower their body temperatures to a greater degree than other birds and also exhibit torpor. (Phylogeny based on Hackett et al. 2008; Figure modified from McKecknie and Mzilikazi 2011; © 2011 Oxford University Press, used with permission)

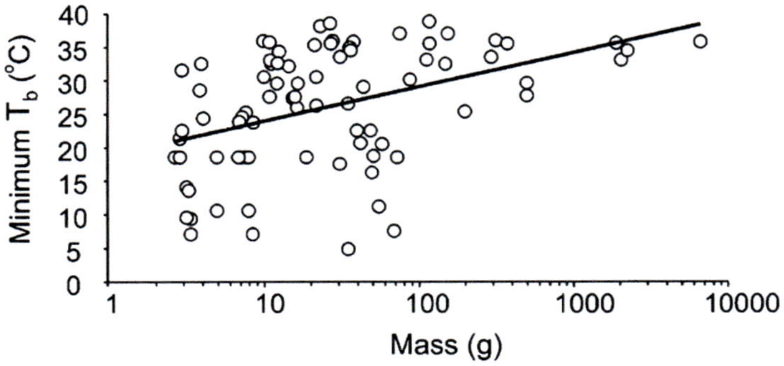


**Fig. 10.101** Relationship between the body mass of birds and their minimum body temperatures. Minimum body temperatures of smaller birds are much lower than those of larger birds. Each circle represents a single species. (Figure from Cooper et al. 2008, based on data from McKechnie and Lovegrove 2002; © 2008 Oxford University Press, used with permission)

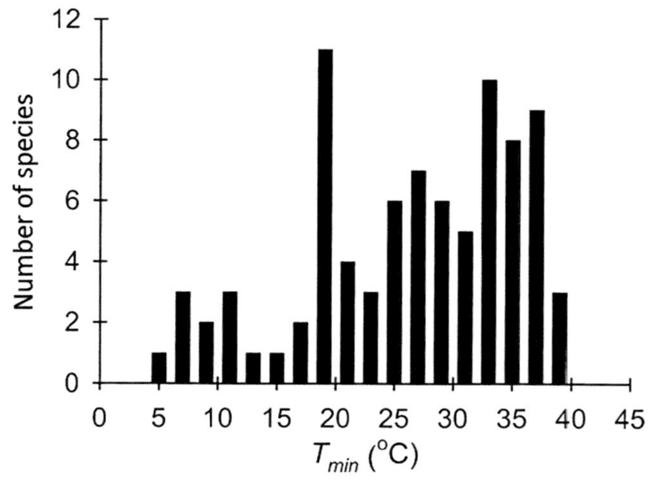


**Fig. 10.102** Distribution of minimum body temperatures of 85 species of birds. Available data clearly indicate that minimum body temperatures form a continuum, suggesting that facultative hypothermia and torpor do not represent two discrete physiological states. (Figure from McKechnie and Lovegrove 2002; © 2002 Oxford University Press, used with permission)

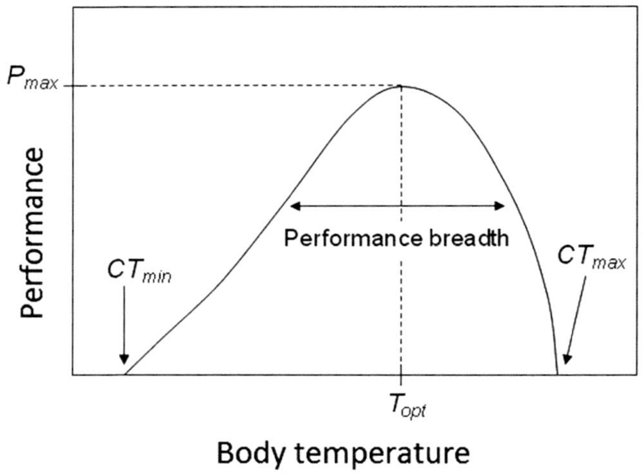


**Fig. 10.103** Hypothetical performance curve illustrating the relationship between a bird's body temperature and its performance, i.e., efficiency of metabolic process and ability to respond to external stimuli. With increasingly low and high body temperatures, performance declines. Colder temperatures slow down metabolic reactions and physiological responses; higher temperatures can alter protein structure and also slow down metabolic reactions and physiological responses. $CT_{min}$ and $CT_{max}$ represent the critical thermal limits below and above a bird cannot survive. (Figure from Angilletta et al. 2010; educational use does not require permission)

**Box 10.17 A Cost of Being Cool**

Hypothermia allows birds to conserve energy, but this conservation may come at a cost. In the lab, Carr and Lima (2012b) monitored the body temperatures of food-deprived Mourning Doves (*Zenaida macroura*) and conducted nocturnal flight tests to see if the doves could still fly while they were hypothermic. Prior to flight tests, a backpack weight (15% of starting body mass) was attached, based on the assumption that a bird able to fly with this extra weight would also be able to fly with the speed and maneuverability necessary to escape after a nocturnal attack by a predator. With each day of food deprivation, the body temperatures of the doves continued to decline further during periods of nocturnal hypothermia, dropping by an average of 4.1, 5.2, and 6.1°C during the first, second, and third nights of food deprivation, respectively. Eight of 21 doves were unable to fly upon release during the first night of food deprivation, four of 10 could not fly on the second night, and one of two tested could not fly on the third night (sample sizes decreased because birds apparently affected by food-deprivation were fed and, therefore, not tested on subsequent nights). These results suggest that birds conserving additional energy by allowing greater nocturnal declines in body temperatures are also increasingly at risk of predation.

Low body temperatures might also negatively impact temperature-dependent physiological processes such as immune responses. To simulate infection by a pathogen, Nord et al. (2013) injected some wintering Great Tits (*Parus major*) with small amounts of a bacterial endotoxin (a substance from the cell walls of bacteria that have no pathogenic effects) to trigger an immune response and found that injected birds maintained higher nocturnal body temperatures than control (non-injected) birds. Similar results have been reported for Song Sparrows (*Melospiza melodia*; Adelman et al. 2010). This suggests that hypothermic birds that become infected face a trade-off between the need to maintain a body temperature high enough for an effective immune response and the need to conserve energy by lowering their body temperature. Infections could, therefore, increase the risk of starvation for birds that depend on hypothermia to conserve energy during long, cold winter nights at high latitudes.

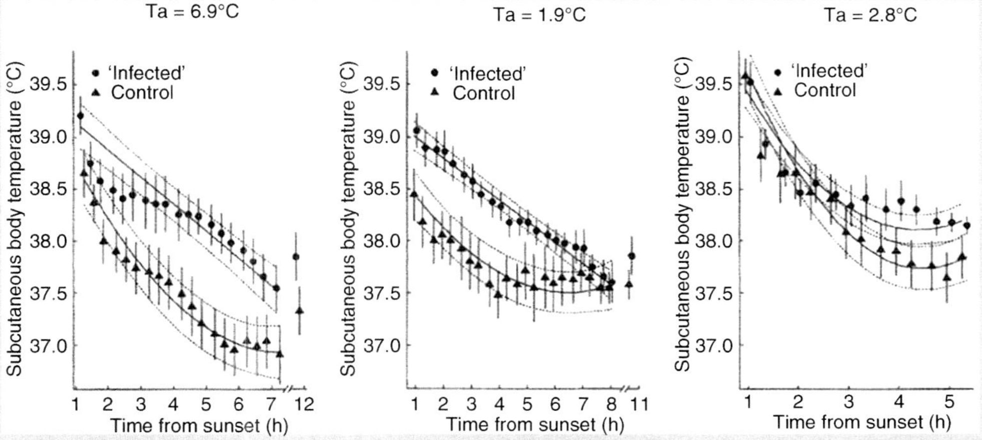


Mean body temperatures (± standard error) of control and "infected" Great Tits (*Parus major*) during winter nights with different ambient temperatures. Effective immune responses of "infected" birds required higher body temperatures, with the effect most pronounced when ambient temperatures (Ta) were lowest. (Figure from Nord et al. 2013; © 2012 The Authors. Functional Ecology © 2012 British Ecological Society, used with permission)

(continued)

**Box 10.17** (continued)

Great Tit. (Photo from pxhere.com, CC0 Public Domain)

Another possible cost of hypothermia is how it affects the quality of sleep. Hypothermic doves spent less time in rapid eye-movement (REM) sleep (Walker et al. 1983) and, although questions remain concerning the function(s) of sleep in birds and mammals, available evidence suggests that REM sleep may play an important role in learning and memory (Roth et al. 2010). If so, then hypothermic birds may need to try and balance the need to conserve energy with the need for REM sleep (e.g., to remember the locations of high-quality foraging sites or, for some species, where seeds have been cached). Additional study is needed to better understand the functions of sleep in birds and the possible impacts of lower quality sleep by hypothermic birds.

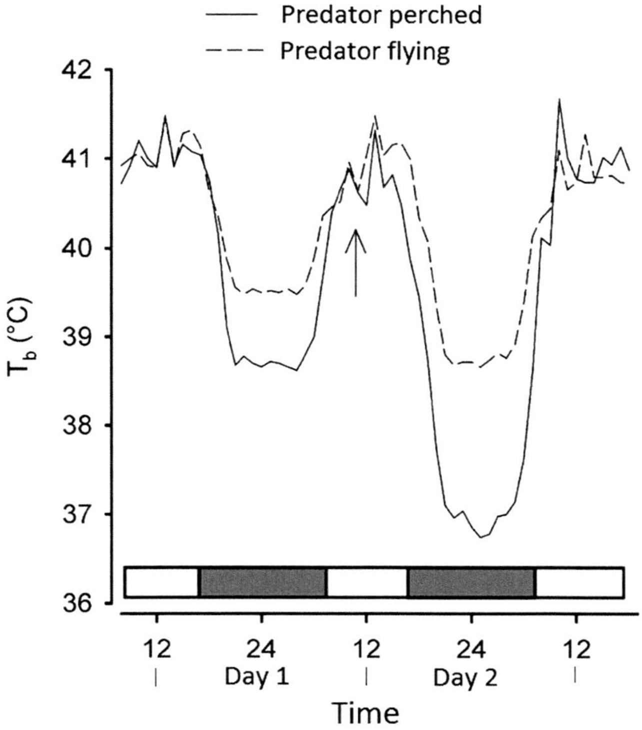


**Fig. 10.104** Mean body temperatures of starved Rock Pigeons (*Columba livia*) exposed to taxidermic mounts of a perched and flying (wings extended and hanging from a wire over the aviary) Northern Goshawk (*Accipiter gentilis*). To conserve energy, the pigeons became hypothermic during the "night" (dark periods indicated gray bars at the bottom of the figure), with lower body temperatures ($T_b$) during the second night without food. However, when exposed to a flying predator, hypothermic pigeons perceived a greater risk of predation and maintained higher body temperatures. The arrow indicates when the simulated predator was brought into the aviary during the light periods. (Figure modified from Laurila and Hohtola 2005; © 2005 Elsevier Ltd., used with permission)

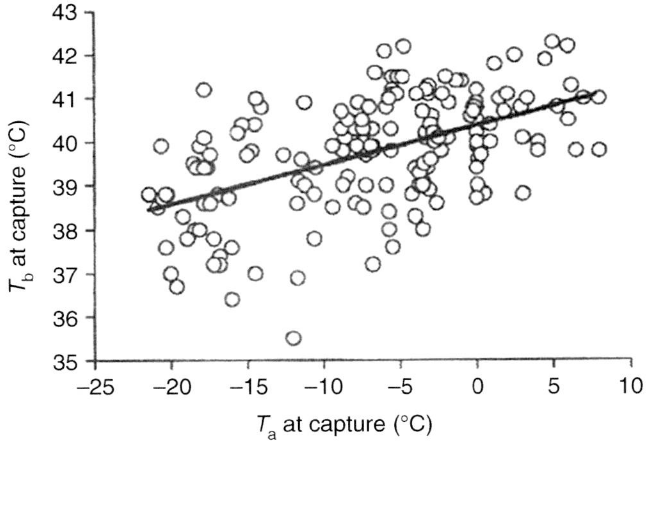


**Fig. 10.105** Relationship between ambient temperature ($T_a$) and the body temperatures ($T_b$) of Black-capped Chickadees (*Poecile atricapillus*) captured while foraging during winter in Canada. The normal body temperature of these chickadees is 41°C. (Figure from Lewden et al. 2014; © 2014 British Ornithologists' Union, used with permission)

**Box 10.18 Fat Accumulation in Wintering Songbirds**

Every day, small birds wintering at high latitudes need to accumulate sufficient fat to ensure survival during the coming night. Strategies used by these birds in accumulating this fat can be influenced by day length, ambient temperature, food availability, predation risk, and, in flocking species, dominance status. With shorter days, birds may forage and accumulate fat throughout the day to ensure sufficient fat reserves; with longer days, birds may forage more intensively and accumulate most of their fat later in the day. Adding most fat reserves later in the day reduces flight costs earlier in the day, and lighter birds are also more maneuverable to better able to escape from predators. However, the effect of day length on foraging strategies is influenced by temperature, predation risk, and food availability. With colder temperatures (more fat needed), reduced predation risk, and/or reduced food availability (increased chance of not finding sufficient food), birds may accumulate fat earlier in the day even if days are longer.

With increased food availability and/or a more predictable food supply, birds may accumulate fat throughout the day or wait until later in the day. For example, Koivula et al. (2002) provided some flocks of Willow Tits (*Parus montanus*) with supplemental food and found that the birds in supplemented flocks accumulated fat at similar rates during the morning and afternoon. (Figure below). However, birds in non-supplemented (control) flocks and less certain of finding sufficient food foraged more intensely and accumulated more fat during the morning, quickly ensuring they have sufficient fat reserves to survive the coming night.

Willow Tit. (Photo by Estormiz, Wikipedia, CC0 Public Domain)

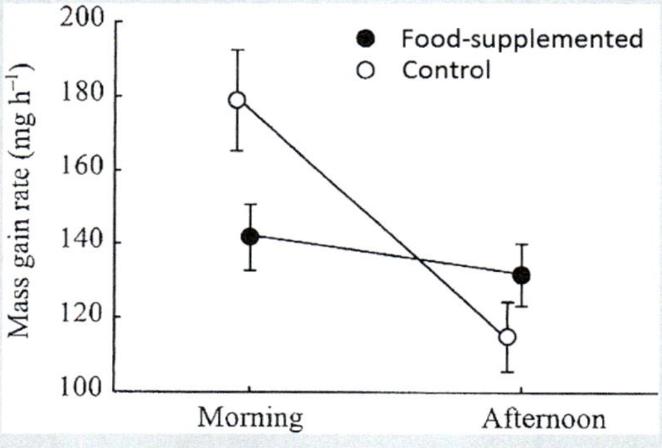


(continued)

**Box 10.18** (continued)

Mean (± standard error) rates of mass gain at different times of day by food-supplemented and control Willow Tits. (Figure from Koivula et al. 2002; © 2002 John Wiley and Sons, used with permission)

In flocking species, a bird's dominance status can influence its fat accumulation strategy. In wintering flocks of Great Tits (*Parus major*), dominant birds store less fat than subordinates (Krams et al. 2013). Given that dominant birds have priority access to food, they might be expected to store more fat. However, lighter birds are more maneuverable, and so, by storing less fat, dominant birds likely reduce their risk of predation. Although storing less fat could increase their risk of starvation during cold winter nights, foraging success of dominant birds is more predictable because of their priority access to food. After a cold night, therefore, hungry dominant birds have first access to food, reducing their likelihood of starvation. In contrast, access to food for subordinate birds is less predictable and becomes increasingly unpredictable with lower status and reduced access to food. The best strategy for subordinates, therefore, is to try and balance the need to store fat with the likelihood of gaining access to food. In other words, the lower a bird's status and likelihood of acquiring food, the more they should try to store fat as insurance against starving. In doing so, subordinate birds reduce the risk of starvation, but, as a trade-off, reduce their maneuverability in flight and likely increase their risk of predation.

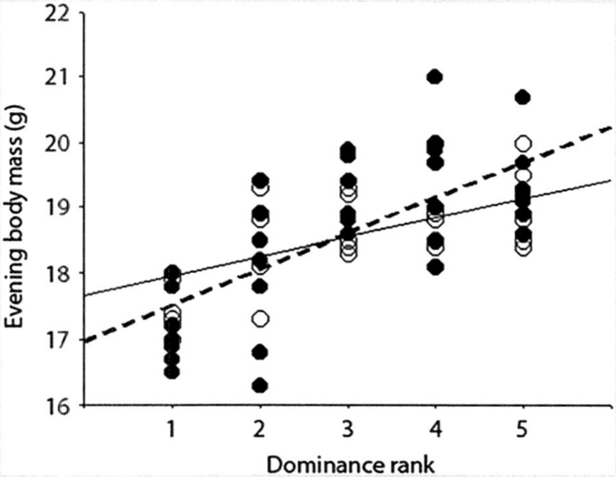


Relationship between evening body mass and dominance rank in Great Tits under conditions of mild ambient temperature (open circles, continuous line) and extremely low temperature (filled circles, dotted line). With colder temperatures and increased risk of starvation for subordinate birds, the different strategies of dominant and subordinate birds become even more apparent. (Figure from Krams et al. 2013; © 2012 Springer-Verlag Berlin Heidelberg, used with permission)

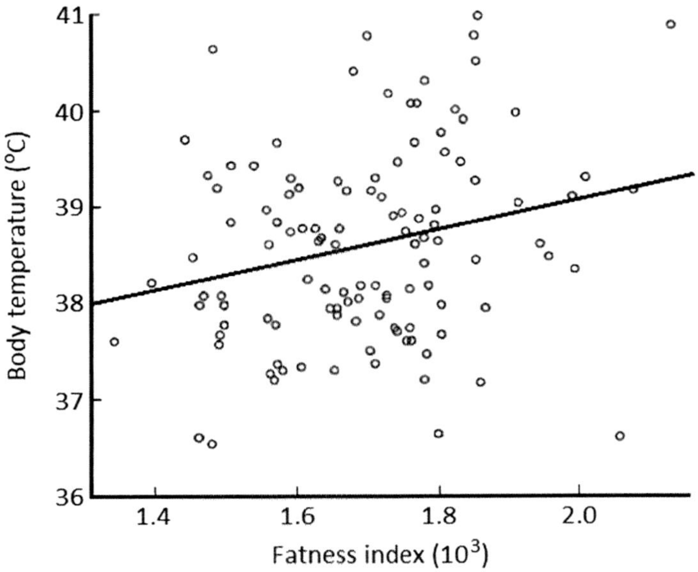


**Fig. 10.106** Effect of fat reserves on nocturnal body temperatures of roosting Eurasian Blue Tits (*Cyanistes caeruleus*). Birds with less fat tended to maintain lower body temperatures than birds with more fat. Ambient temperatures ranged from −6 to +3°C. Fatness index = mass/tarsus length$^3$. (Figure from Nord et al. 2011; © 2011 Springer-Verlag, used with permission)

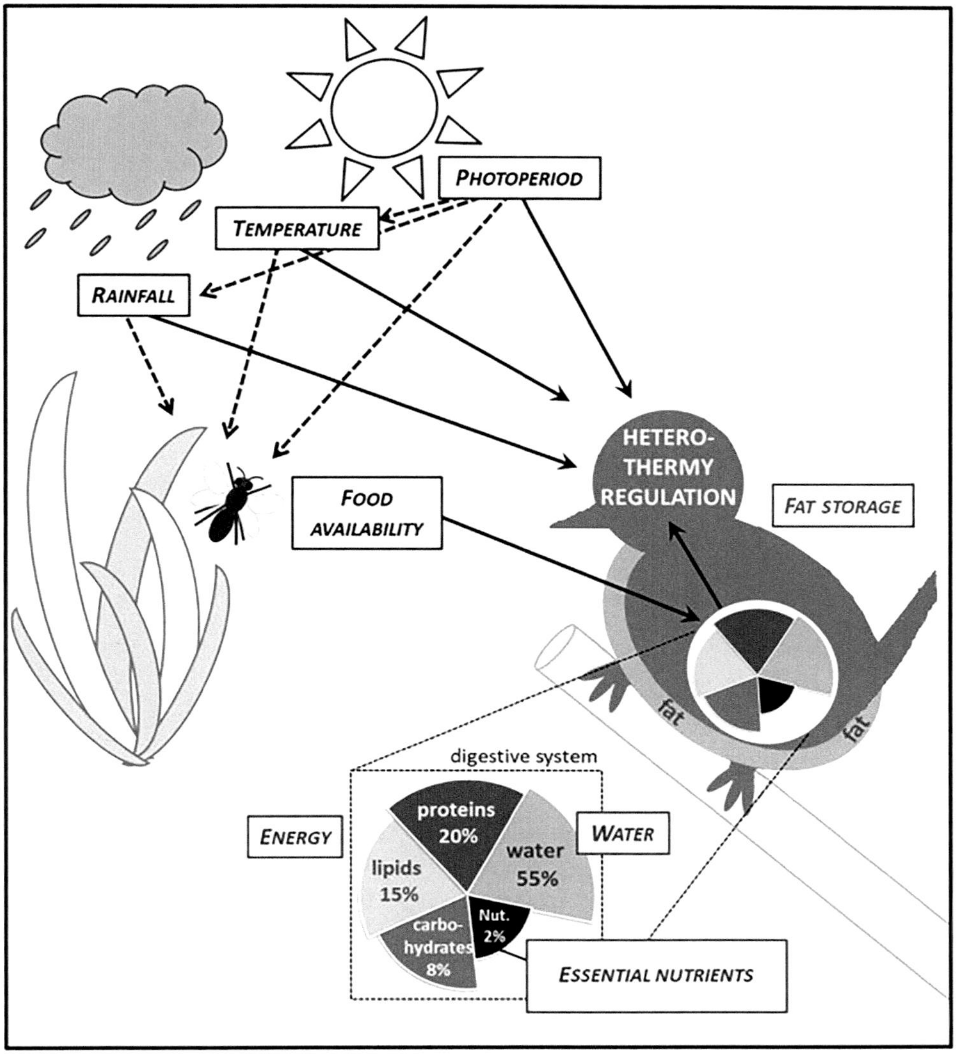


**Fig. 10.107** Many variables can influence the hypothermic responses of birds. Factors that have a direct influence on those responses are indicated with solid lines; factors with an indirect influence (via food availability) are indicated with dashed lines. Temperature and food availability generally have the greatest effects on hypothermic responses, but rainfall (e.g., by wetting feathers and reducing their insulating capacity) and photoperiod (i.e., season) can also have an effect. The effect of photoperiod is by way of daylength, e.g., at high latitudes, nights are longer in mid-winter, and birds may need to maintain lower nighttime body temperatures to ensure that energy stores last until morning. Daylength (season), temperature, and rainfall clearly impact food availability. (Figure modified from Vuarin and Henry 2014; © 2014 Springer-Verlag Berlin Heidelberg, used with permission)

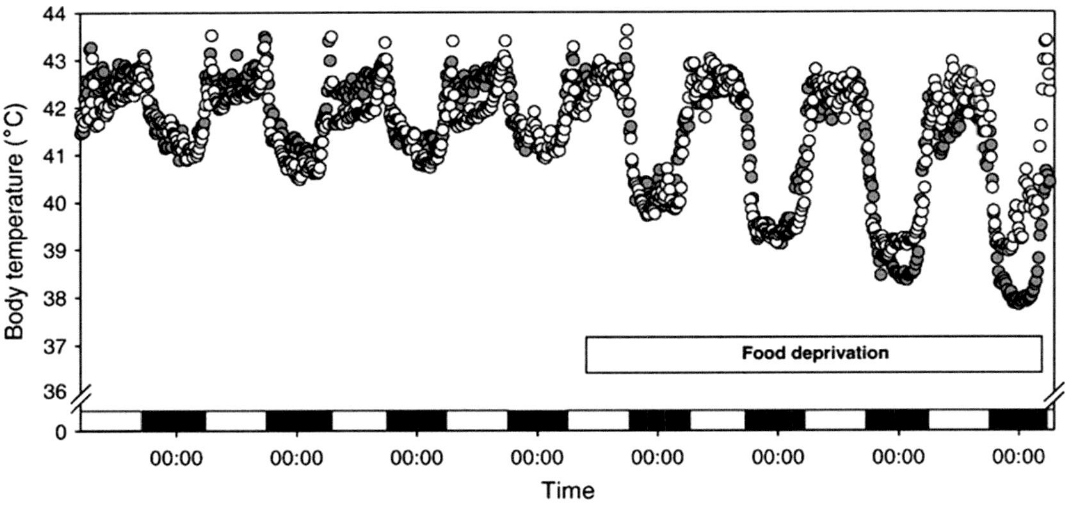


**Fig. 10.108** Mean body temperatures of Japanese Quail (*Coturnix japonica*) provided with food for 4 days, then deprived of food for the next 4 days. Periods of light and dark are indicated by the line at the bottom. Empty circles are body temperatures of quail kept at a thermoneutral temperature (33°C); gray circles are body temperatures of quail kept at a temperature below their lower critical temperature (13°C). In response to food deprivation, all quail become hypothermic and increasingly so with successive days without food. However, body temperatures of quail kept at the colder temperature decreased more than those of quail kept at a thermoneutral temperature. (Figure from Ben-Hamo et al. 2010; © 2009 Elsevier Inc., used with permission)

**Table 10.6** Examples of energy savings by torpid birds (relative to energy consumption if normal body temperature was maintained) from a variety of taxa

| Common name | Scientific name | Percent reduction in metabolic rate[a] | Reference |
|---|---|---|---|
| Speckled Mousebird | *Colius striatus* | 86.8 | McKechnie and Lovegrove (2001) |
| Red-backed Mousebird | *Colius castanotus* | 91.7 | Prinzinger et al. (1981) |
| Puerto Rican Tody[b] | *Todus mexicanus* | 69.3 | Merola-Zwartjes and Ligon (2000) |
| Costa's Hummingbird | *Calypte costae* | 87 | Lasiewski (1963, 1964) |
| Anna's Hummingbird | *Calypte anna* | 95.6 | Bartholomew et al. (1957), Lasiewski (1963) |
| Rufous Hummingbird | *Selasphorus rufus* | 87.2 | Lasiewski (1963), Hiebert (1990) |
| Broad-tailed Hummingbird | *Selasphorus platycercus* | 80 | Bucher and Chappell (1992, 1997) |
| Rivoli's Hummingbird | *Eugenes fulgens* | 84.4 | Dawson and Hudson (1970), Wolf and Hainsworth (1972) |
| Giant Hummingbird[c] | *Patagonia gigas* | 90 | Lasiewski et al. (1967) |
| Spotted Nightjar | *Eurostopodus argus* | 51.8 | Dawson and Fisher (1969) |
| Whip-poor-will | *Caprimulgus vociferus* | 73 | Lane (2002) |
| Common Nighthawk[d] | *Chordeiles minor* | 81.8 | Lasiewski and Dawson (1964) |
| Cloven-feathered Dove | *Drepanoptila holosericea* | 62.2 | Schleucher (2001) |
| Malachite Sunbird | *Nectarinia famosa* | 60 | Downs and Brown (2002) |

[a]Based on reduction in oxygen consumption
[b]Only female Puerto Rican Todies became torpid; females may be more energetically stressed than males due to the energetic demands of egg formation. High testosterone levels in males may also prevent them from becoming torpid (Merola-Zwartjes and Ligon 2000)
[c]Reduction estimated based on data in Fig. 1 in Lasiewski et al. (1967)
[d]Reduction estimated based on data in Fig. 1 in Lasiewski and Dawson (1964)

**Table 10.7** Examples of how heart rates of birds decline when they become torpid

| Common name | Scientific name | Resting heart rate (either maximum and minimum or average beats per minute) | Torpid heart rate (beats per minute) | Reference |
|---|---|---|---|---|
| Blue-naped Mousebird | *Urocolius macrourus* | 240 | 90 | Schaub and Prinzinger (1999) |
| Blue-throated Hummingbird | *Lampornis clemenciae* | 480–1260 | 140 at $T_a = 24°C$; 36 at $T_a = 15°C$ | Lasiewski and Lasiewski (1967) |
| Rivoli's Hummingbird | *Eugenes fulgens* | 420–1200 | 107 at $T_a = 20.5°C$; 55 at $T_a = 15°C$ | Lasiewski and Lasiewski (1967) |
| Black-chinned Hummingbird[a] | *Archilochus alexandri* | 600–1050 | 40 at 14°C | Lasiewski (1964) |
| Costa's Hummingbird[a] | *Calypte costae* | 500–900 | 40 at 7°C | Lasiewski (1964) |
| Giant Hummingbird | *Patagona gigas* | 300–1020 | 60 | Lasiewski et al. (1967) |
| Common Poorwill[b] | *Phalaenoptilus nuttallii* | 505 | 18 at 5°C | Bartholomew et al. (1962) |
| Common Nighthawk[c] | *Chordeiles minor* | 130–330 | 45–100 | Lasiewski and Dawson (1964) |

[a]Values estimated from data in Fig. 2 of Lasiewski (1964)
[b]It took 7.5 h for the heart rate to drop to 18 beats per minute. In addition, Common Poorwills are the only bird that hibernates (i.e., can remain torpid for several days)
[c]Values estimated from data in Fig. 2 in Lasiewski and Dawson (1964)

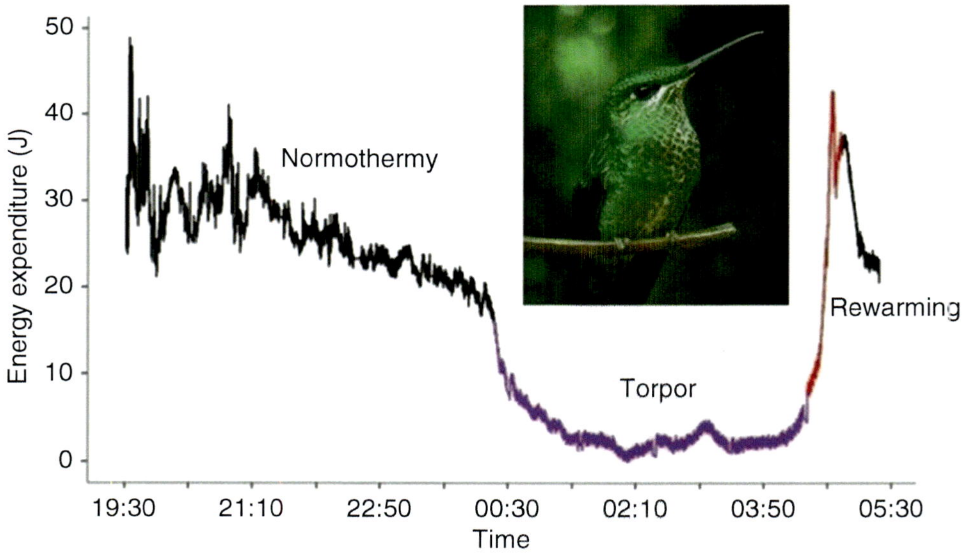


**Fig. 10.109** Energy expenditure (Joules) of a Green-crowned Brilliant (*Heliodoxa jacula*) prior to, during, and after a period of torpor over the course of a night. Note that energy expenditure declined from just over 30 J prior to torpor to less than 5 J during mid-torpor. (Figure from supplementary material from Shankar et al. 2020; © 2020 Nordic Society Oikos. Published by John Wiley & Sons Ltd., used with permission. Photo by Naomi Ibuki, Wikipedia, CC BY 2.0, https://creativecommons.org/licenses/by/2.0/)

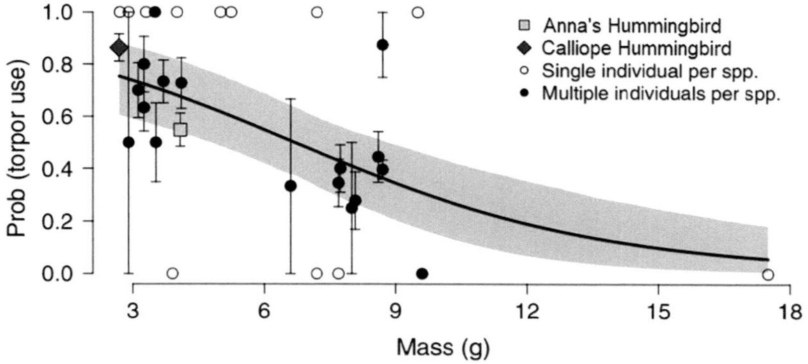


**Fig. 10.110** Effect of hummingbird body mass on the frequency of torpor use. Larger hummingbirds are less likely to use torpor than smaller hummingbirds (based on from 31 species). Points show mean frequency (± SE; gray = 95% credible interval) of torpor use per species (combined across studies, when applicable). Open circles represent observations of a single individual of a species; filled circles represent observations of multiple individuals per species. Anna's (*Calypte anna*) and Calliope (*Selasphorus calliope*) hummingbirds were the focus of the cited study. (Figure from Spence and Tingley 2021; © 2021 British Ecological Society, used with permission)

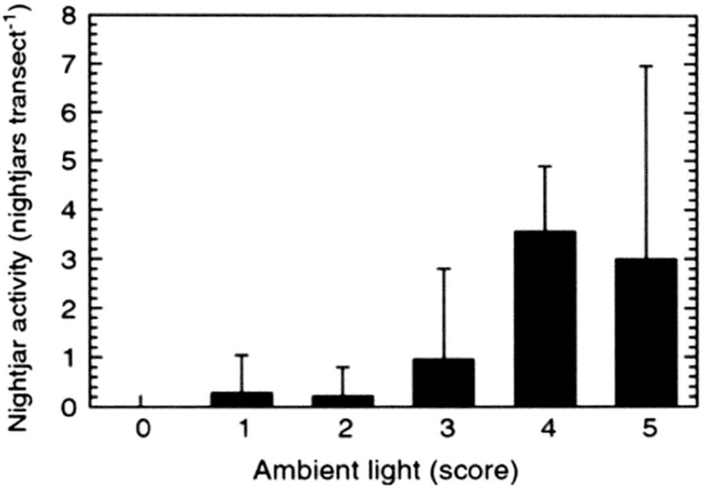


**Fig. 10.111** Freckled Nightjars (*Caprimulgus tristigma*) in a reserve in South Africa were much more active when light levels were higher. Light levels were scored on a scale of 0–5, with 0 being complete darkness with 100% cloud cover and 5 being a clear night with a full moon. (Figure from Ashdown and McKechnie 2008; © 2008 Dt. Ornithologen-Gesellschaft e.V., used with permission)

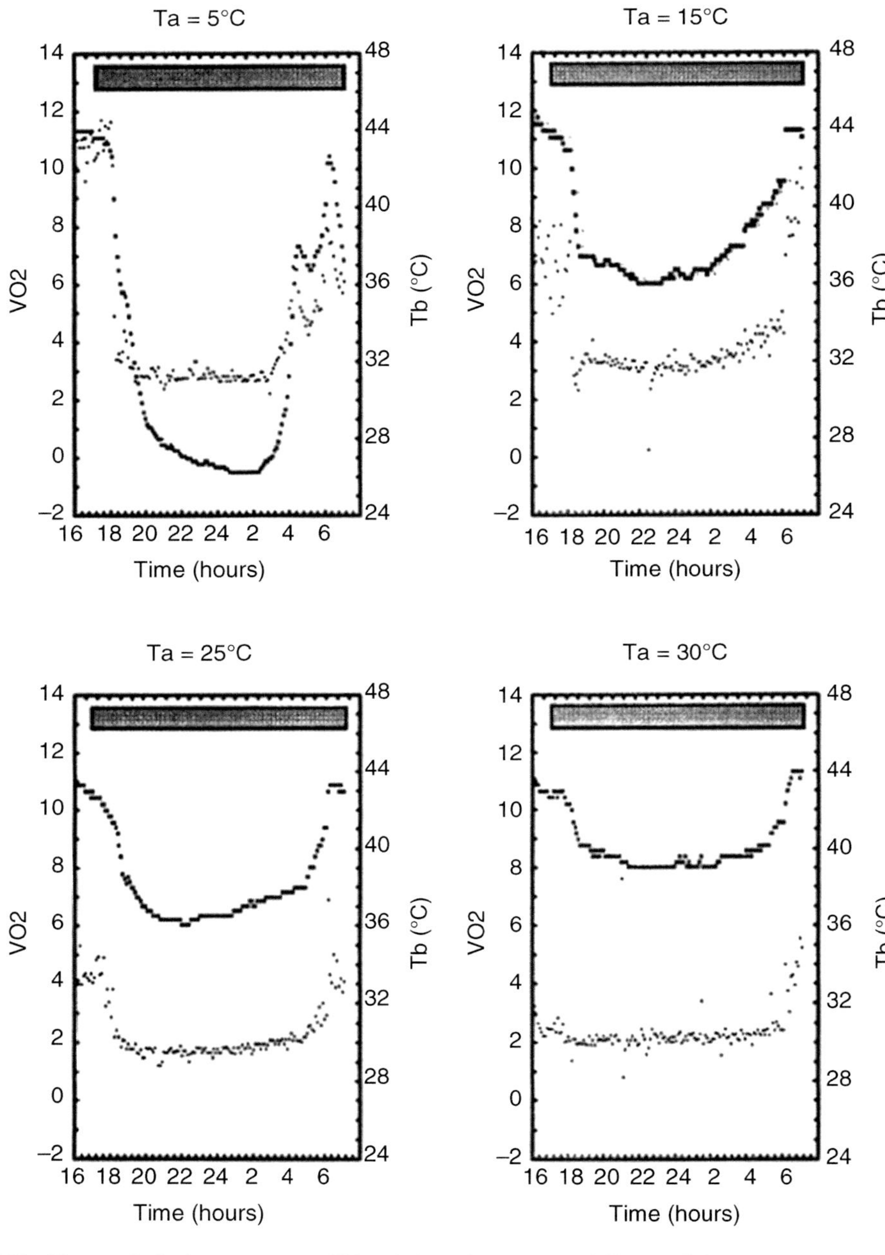


**Fig. 10.112** Changes in body temperature ($T_b$) and $O_2$ consumption ($VO_2$, mL $O_2$ $g^{-1}$ $h^{-1}$) during the night for a captive Malachite Sunbird (*Nectarinia famosa*) at ambient temperatures ($T_a$) ranging from 5 to 30°C. Small dots (lighter lines) = $VO_2$ and squares (darker lines) = $T_b$. With colder ambient temperatures, body temperatures are lower (especially at 5°C) and oxygen consumption increases as the sunbird must use additional energy to keep its body temperature from dropping further. (Figure from Downs and Brown 2002; © 2002 Oxford University Press, used with permission)

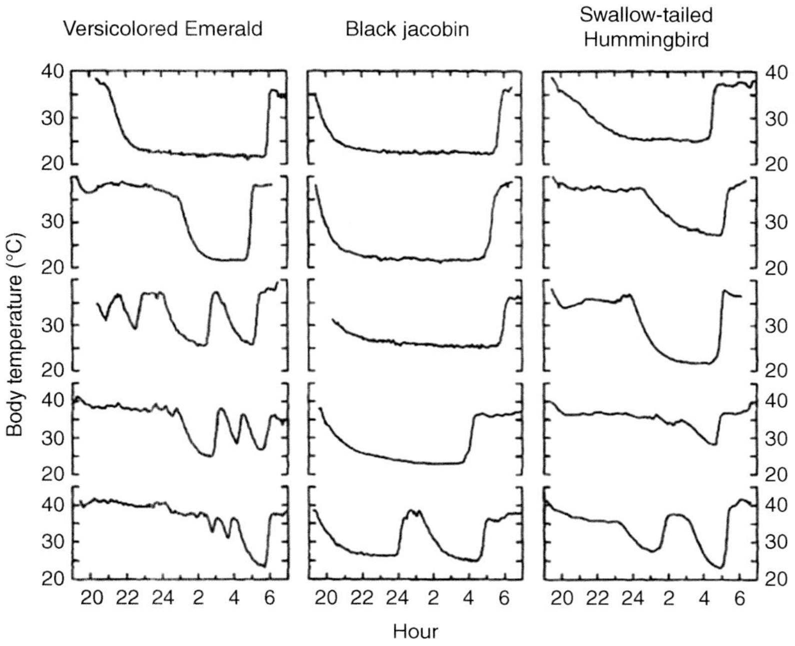


**Fig. 10.113** Examples of body temperatures of five individuals (each line represents a different individual) of three species of hummingbirds throughout the night. The duration of torpor bouts ranged from 2 h to all night (11 h). During torpor, body temperatures of individuals of all three species dropped from normothermic (about 37°C) to about 24°C (just above the ambient temperature of about 23°C). Versicolored Emerald, *Amazilia versicolor*; Black Jacobin, *Florisuga fusca*; Swallow-tailed Hummingbird, *Eupetomena macroura*. (Figure from Bech et al. 1997; © 1997 Oxford University Press, used with permission)

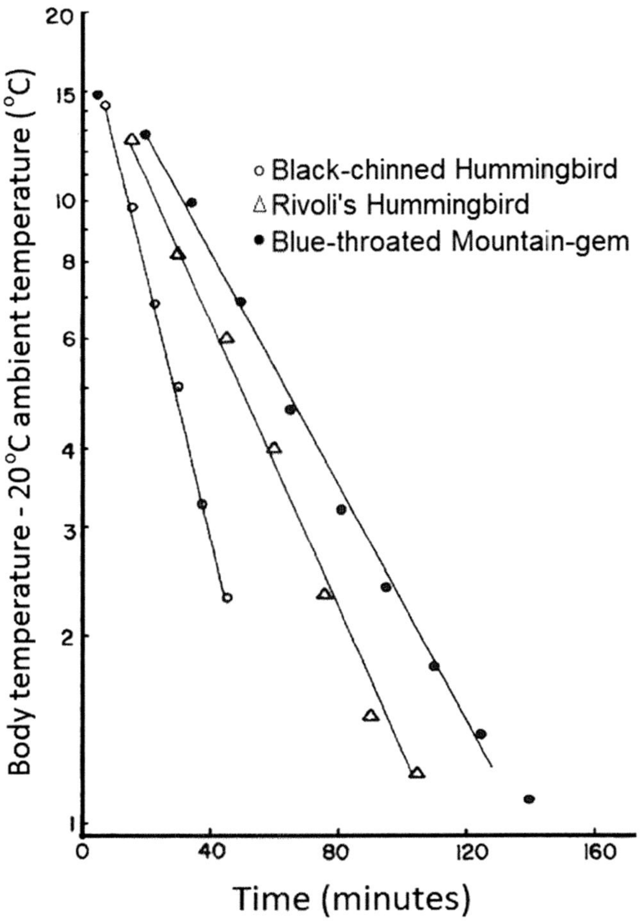


**Fig. 10.114** Time needed for three species of hummingbirds to become fully torpid at an ambient temperature of 20°C. Time needed was influenced by body mass, with the body temperature of a Black-chinned Hummingbird (*Archilochus alexandri*; 4 g) dropping faster than body temperatures of Rivoli's (*Eugenes fulgens*; 6.8 g) and Blue-throated Mountain-gem (*Lampornis clemenciae*; 8.5 g) hummingbirds. (Figure modified from Lasiewski and Lasiewski 1967; © 1967 Oxford University Press, used with permission)

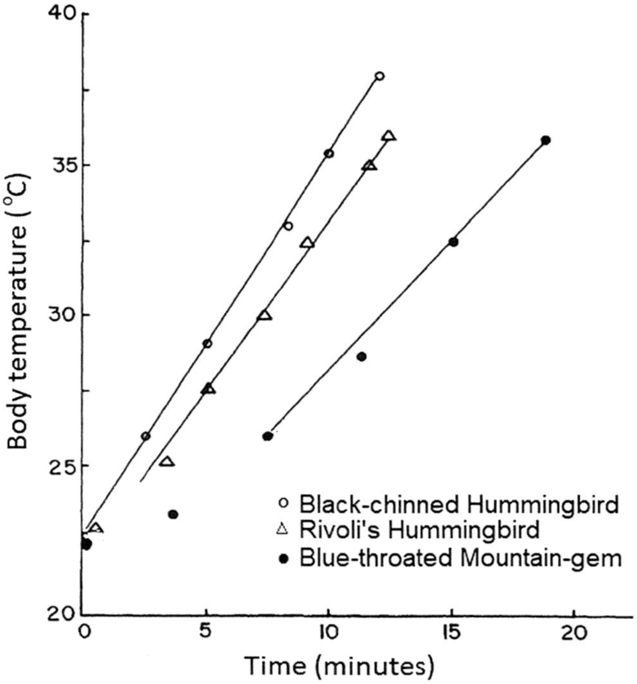


**Fig. 10.115** Time needed for three species of hummingbirds to become normothermic (or nearly so) after being torpid at an ambient temperature of 20°C. Time needed was influenced by body mass, with the body temperatures of a Black-chinned (3.3 g) and Rivoli's (6.8 g) increasing faster than the body temperature of a Blue-throated Mountain-gem (8.5 g) hummingbird. (Figure modified from Lasiewski and Lasiewski 1967; © 1967 Oxford University Press, used with permission)

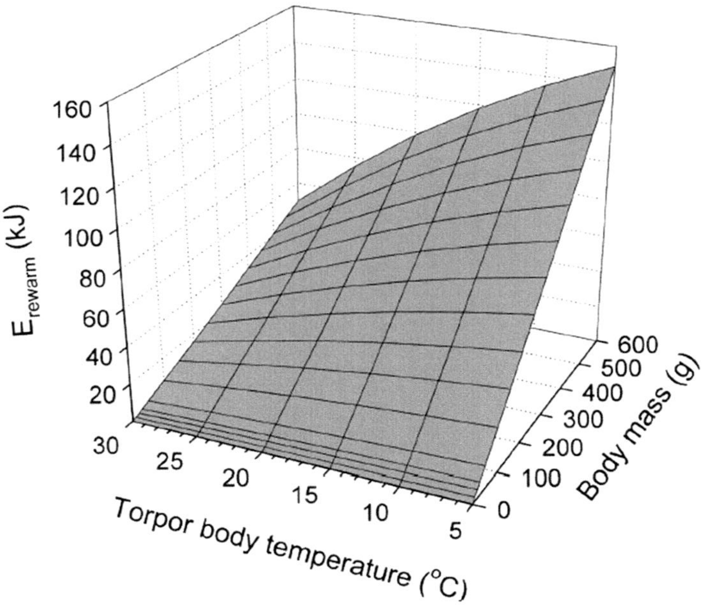


**Fig. 10.116** Predicted relationship between the energetic cost of rewarming ($E_{\text{rewarm}}$, in KJ), torpor body temperature, and body mass at an ambient temperature of 0°C. Based on this model, for a bird maintaining its body temperature at 18°C during a 12-h bout of torpor at an ambient temperature of 0°C, the energy required to re-warm to normal body temperature would represent 11% of the energy expended during torpor for a 5-g bird, but 73% of the energy expended during torpor for a 500-g bird. 1 Kilojoule = 0.24 Kilocalories. (Figure from McKechnie and Wolf 2004; © 2004 Institute of Arctic Biology, University of Alaska Fairbanks, used with permission)

**Fig. 10.117** Tawny Frogmouths (*Podargus strigoides*), caprimulgids found in Australia, are the largest known birds (290–680 g; Holyoak 2001) to use torpor. (Figure from Geiser 2013; © 2013 Elsevier Ltd., used with permission)

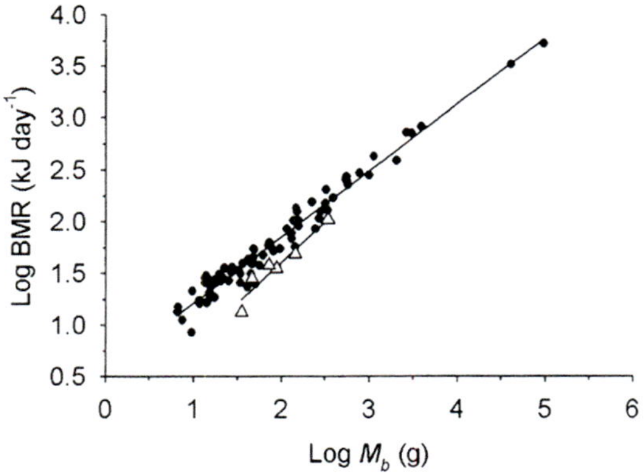


**Fig. 10.118** Log body mass ($M_b$) and log basal metabolic rate (BMR) for 82 bird species (filled circles) and six caprimulgids (unfilled triangles). For their body mass, caprimulgids have lower basal metabolic rates than similar-sized birds in other taxa. (Figure from Lane et al. 2004; © 2004 Oxford University Press, used with permission)

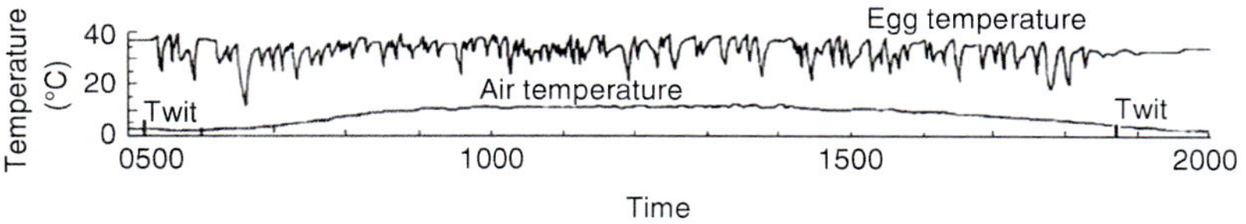


**Fig. 10.119** Egg and air temperatures during 1 day at a nest of a female Anna's Hummingbird (*Calypte anna*) located in a canyon in the Santa Monica Mountains in California. The incubating female maintained egg temperature well above air temperature, but egg temperature fluctuated dramatically and often during the day as the female alternated periods of incubation with periods of foraging. Amazingly, this female hummingbird was able to maintain a mean egg temperature of 34.9°C during incubation while still spending 25% of daylight hours foraging. Twit (twilight) indicates periods of increasing and decreasing light levels at dawn and dusk. (Figure modified from Vleck 1981; © 1981 Springer-Verlag, used with permission)

**Fig. 10.120** Common Poorwill. (Photo by Alan Schmierer, CC0 Public Domain)

## References

Adelman JS, Córdoba-Córdoba S, Spoelstra K, Wikelski M, Hau M (2010) Radiotelemetry reveals variation in fever and sickness behaviours with latitude in a free-living passerine. Funct Ecol 24:813–823

Åkesson S, Klaassen R, Holmgren J, Fox JW, Hedenström A (2012) Migration routes and strategies in a highly aerial migrant, the Common Swift *Apus apus*, revealed by light-level geolocators. PLoS ONE 7:e41195

AlRashidi M, Kosztolányi A, Shobrak M, Küpper C, Székely T (2011) Parental cooperation in an extreme hot environment: natural behaviour and experimental evidence. Anim Behav 82:235–243

Anderson KJ, Jetz W (2005) The broad-scale ecology of energy expenditure of endotherms. Ecol Lett 8:310–318

Andreev AV (1999) Energetics and survival of birds in extreme environments. In: Adams NJ, Slotow RH (eds) Proceedings of the 22nd International Ornithological Congress, vol 70, Ostrich, pp 13–22

Angilletta MJ (2009) Thermal adaptation: a theoretical and empirical synthesis. Oxford University Press, Oxford, UK

Angilletta MJ Jr, Cooper BS, Schuler MS, Boyles JG (2010) The evolution of thermal physiology in endotherms. Front Biosci E 2:861–881

Arad Z, Marder J (1982) Strain differences in heat resistance to acute heat stress, between the Bedouin Desert Fowl, the white leghorn and their crossbreeds. Comp Biochem Physiol 72:191–193

Arad Z, Midtgård U, Bernstein MH (1989) Thermoregulation in Turkey Vultures: vascular anatomy, arteriovenous heat exchange, and behavior. Condor 91:505–514

Araújo MB, Ferri-Yáñez F, Bozinovic F, Marquet PA, Valladares F, Chown SL (2013) Heat freezes niche evolution. Ecol Lett 16:1206–1219

Araújo MB, Ferri-Yáñez F, Bozinovic F, Chown SL, Marquet PA (2016) Erratum to Araújo et al. (2013). Ecol Lett 19:591–592

Asbury DA, Angilletta MJ (2010) Thermodynamic effects on the evolution of performance curves. Am Nat 176: E40–E49

Ashdown RAM, McKechnie AE (2008) Environmental correlates of Freckled Nightjar (*Caprimulgus tristigma*) activity in a seasonal, subtropical habitat. J Ornithol 149:615–619

Austin GT, Bradley WG (1969) Additional responses of the poor-will to low temperatures. Auk 86:717–725

Bächler E, Hahn S, Schaub M, Arlettaz R, Jenni L, Fox JW, Afanasyev V, Liechti F (2010) Year-round tracking of small trans-Saharan migrants using light-level geolocators. PLoS ONE 5:e9566

Bahat O, Choshniak I, Houston DC (1998) Nocturnal variation in body temperature of Griffon Vultures. Condor 100:168–171

Bakken GS, Williams JB, Ricklefs RE (2002) Metabolic response to wind of downy chicks of Arctic-breeding shorebirds (Scolopacidae). J Exp Biol 205:3435–3443

Barlein F (1985) Efficiency of food utilization during fat deposition in the long-distance migratory Garden Warbler, *Sylvia borin*. Oecologia 68:118–125

Barnett LB (1970) Seasonal changes in temperature acclimatization of the House Sparrow, *Passer domesticus*. Comp Biochem Physiol 33:559–578

Barrows CW (1981) Roost selection by Spotted Owls: an adaptation to heat stress. Condor 83:302–309

Bartell P, Moore A (2013) Avian migration: the ultimate red-eye flight. Am Sci 101:46–55

Bartholomew GA, Dawson WR (1979) Thermoregulatory behavior during incubation in Heermann's Gulls. Physiol Zool 52:422–437

Bartholomew GA, Trost CH (1970) Temperature regulation in the Speckled Mousebird, *Colius striatus*. Condor 72:141–146

Bartholomew GA, Howell TR, Cade TJ (1957) Torpidity in the White-throated Swift, Anna Hummingbird, and Poor-will. Condor 59:145–155

Bartholomew GA, Hudson JW, Howell TR (1962) Body temperature, oxygen consumption, evaporative water loss, and heart rate in the Poor-will. Condor 64:117–125

Bartholomew GA, Vleck CM, Bucher TL (1983) Energy metabolism and nocturnal hypothermia in two tropical passerine frugivores, *Manacus vitellinus* and *Pipra mentalis*. Physiol Zool 56:370–379

Barve S, Ramesh V, Dotterer TM, Dove CJ (2021) Elevation and body size drive convergent variation in thermo-insulative feather structure of Himalayan birds. Ecography 44:680–689

Battley PF, Warnock N, Tibbitts TL, Gill RE Jr, Piersma T, Hassell CJ, Douglas DC, Mulcahy DM, Gartell BD, Schuckard R, Melville DS, Riegen AC (2012) Contrasting extreme long-distance migration patterns in Bar-tailed Godwits *Limosa lapponica*. J Avian Biol 43:1–12

Bauchinger U, McWilliams SR (2010) Extent of phenotypic flexibility during long-distance flight is determined by tissue-specific turnover rates: a new hypothesis. J Avian Biol 41:603–608

Bauchinger U, Wohlmann A, Biebach H (2005) Flexible remodeling of organ size during spring migration of the Garden Warbler (*Sylvia borin*). Zoology 108:97–106

Bauchinger U, Kolb H, Afik D, Pinshow B, Biebach H (2009) Blackcap Warblers maintain digestive efficiency by increasing digesta retention time on the first day of migratory stopover. Physiol Biochem Zool 82:5410548

Bech C, Præsteng KE (2004) Thermoregulatory use of the heat increment of feeding in the Tawny Owl (*Strix aluco*). J Therm Biol 29:649–654

Bech C, Abe AS, Steffensen JF, Berger M, Bicudo JEPW (1997) Torpor in three species of Brazilian hummingbirds under semi-natural conditions. Condor 99:780–788

Bech C, Ronning B, Moe B (2004) Individual variation in the basal metabolism of Zebra Finches *Taeniopygia guttata:* no effect of food quality during early development. Int Congr Ser 1275:206–312

Ben-Hamo M, Pinshow B, McCue MD, McWilliams SR, Bauchinger L (2010) Fasting triggers hypothermia, and ambient temperature modulates its depth in Japanese Quail *Coturnix japonica*. Comp Biochem Physiol 156: 84–91

Bergmann C (1847) Über die verhältnisse der wärmeökonomie der thiere zu ihrer grösse. Gottinger Studien 3:595–708

Biro PA, Stamps JA (2010) Do consistent individual differences in metabolic rate promote consistent individual differences in behavior? Trends Ecol Evol 25: 653–659

Bize P, Klopfenstein A, Jeanneret C, Roulin A (2007) Intra-individual variation in body temperature and pectoral muscle size in nestling Alpine Swifts *Apus melba* in response to an episode of inclement weather. J Ornithol 148:387–393

Blackburn TM, Gaston KJ (1994) The distribution of body sizes of the world's bird species. Oikos 70:127–130

Blackmer AL, Mauck RA, Ackerman JT, Huntington CE, Nevitt GA, Williams JB (2005) Exploring individual quality: basal metabolic rate and reproductive performance in storm-petrels. Behav Ecol 16:906–913

Blem CR (1976) Patterns of lipid storage and utilization in birds. Am Zool 16:671–684

Blem CR (1990) Avian energy storage. Curr Ornithol 7: 59–113

Bligh J, Johnson KG (1973) Glossary of terms for thermal physiology. J Appl Physiol 35:941–961

Boersma PD (1986) Body temperature, torpor, and growth in chicks of Fork-tailed Storm-Petrels (*Oceanodroma furcata*). Physiol Zool 59:10–19

Bolser JA, Alan RR, Smith AD, Li L, Seeram NP, McWilliams SR (2013) Birds select fruits with more anthocyanins and phenolic compounds during autumn migration. Wilson J Ornithol 125:97–108

Bouverot P, Hildwein G, Le Goff D (1974) Evaporative water loss, respiratory pattern, gas exchange and acid-base balance during thermal panting in Pekin ducks exposed to moderate heat. Respir Physiol 21:255–269

Boyce AJ, Mouton JC, Lloyd P, Wolf BO, Martin TE (2020) Metabolic rate is negatively linked to adult survival but does not explain latitudinal differences in songbirds. Ecol Lett 23:642–652

Boyles JG, Smit B, McKechnie AE (2011) Does use of the torpor cut-off method to analyze variation in body temperature cause more problems than it solves? J Therm Biol 36:373–379

Bozinovic F, Sabat P (2010) On the intraspecific variability in basal metabolism and the food habits hypothesis in birds. Curr Zool 56:759–766

Bradley DW, Clark RG, Dunn PO, Laughlin AJ, Taylor CM, Vleck C, Whittingham LA, Winkler DW, Norris DR (2014) Trans-Gulf of Mexico loop migration of Tree Swallows revealed by solar geolocation. Curr Zool 60:653–659

Brenner FJ (1965) Metabolism and survival time of grouped Starlings at various temperatures. Wilson Bull 77:388–395

Bridge ES, Kelly JF, Contina A, Gabrielson RM, MacCurdy RB, Winkler DW (2013) Advances in tracking small migratory birds: a technical review of light-level geolocation. J Field Ornithol 84:121–137

Brigham RM (1992) Daily torpor in a free-ranging goatsucker, the Common Poorwill (*Phalaenoptilus nuttallii*). Physiol Zool 65:457–472

Brigham RM, Willis CKR, Geiser F, Mzilikazi N (2011) Baby in the bathwater: should we abandon the use of body temperature thresholds to quantify expression of torpor? J Therm Biol 36:373–379

Brigham RM, McKechnie AE, Doucette LI, Geiser F (2012) Heterothermy in Caprimulgid birds: a review of inter- and intraspecific variation in free-ranging populations. In: Ruf T, Bieber C, Arnold W, Millesi E (eds) Living in a seasonal world. Berlin, Germany, Springer, pp 175–187

Broggi J, Hohtola E, Koivula K, Orell M, Thomson RL, Nilsson J-Å (2007) Sources of variation in winter basal metabolic rate in the Great Tit. Funct Ecol 21:528–533

Broggi J, Hohtola E, Koivula K, Orell M, Nilsson J-Å (2009) Long-term repeatability of winter basal metabolic rate and mass in a wild passerine. Funct Ecol 23: 768–773

Broggi J, Hohtola E, Koivula K, Rytkönen S, Nilsson J-Å (2022) Prehatching temperatures drive inter-annual cohort differences in Great Tit metabolism. Oecologia 198:619–627

Brown JH, Nicoletto PF (1991) Spatial scaling of species composition: body masses of North American land mammals. Am Nat 138:1478–1512

Bryant DM (1983) Heat stress in tropical birds: behavioural thermoregulation during flight. Ibis 125: 313–323

Brychta RJ, Chen KY (2017) Cold-induced thermogenesis in humans. Eur J Clin Nutr 71:345

Bucher TL, Chappell MA (1992) Ventilatory and metabolic dynamics during entry into and arousal from torpor in *Selasphorus* hummingbirds. Physiol Zool 65:978–993

Bucher TL, Chappell MA (1997) Respiratory exchange and ventilation during nocturnal torpor in hummingbirds. Physiol Zool 70:45–52

Buchholz R (1996) Thermoregulatory role of the unfeathered head and neck in male Wild Turkeys. Auk 113:310–318

Buckley LB, Hurlbert AH, Jetz W (2012) Broad-scale ecological implications of ectothermy and endothermy in changing environments. Glob Ecol Biogeogr 21: 873–885

Burton T, Killen SS, Armstrong JD, Metcalfe NB (2011) What causes intraspecific variation in resting metabolic rate and what are its ecological consequences? Proc R Soc B 278:3465–3473

Bushuev AV, Kerimov AB, Ivankina EV (2010) Estimation of heritability and repeatability of resting metabolic rate in birds by the example of free-living Pied Flycatchers *Ficedula hypoleuca* (Aves: Passeriformes). Biol Bull Rev 71:402–424

Bushuev A, Tolstenkov O, Zubkova E, Solovyeva E, Kerimov A (2018) Basal metabolic rate in free-living tropical birds: the influence of phylogenetic, behavioral, and ecological factors. Curr Zool 64:33–43

Butler PJ, Woakes AJ (2001) Seasonal hypothermia in a large migrating bird: saving energy for fat deposition? J Exp Biol 204:1361–1367

Butler PJ, Woakes AJ, Bishop CM (1998) Behaviour and physiology of Svalbard Barnacle Geese *Branta leucopsis* during their autumn migration. J Avian Biol 29:536–545

Buttemer WA, Astheimer LB, Weathers WW, Hayworth AM (1987) Energy savings attending winter-nest use by Verdins (*Auriparus flaviceps*). Auk 104:531–535

Cabello-Vergel J, Soriano-Redondo A, Villegas A, Masero JA, Guzmán JMS, Gutiérrez JS (2021) Urohidrosis as an overlooked cooling mechanism in long-legged birds. Sci Rep 11:20018

Calder WA (1974) The thermal and radiant environment of a winter hummingbird nest. Condor 76:268–273

Calder WA, Booser J (1973) Hypothermia of Broad-tailed Hummingbirds during incubation in nature with ecological correlations. Science 180:751–753

Callo PA, Morton ES, Stutchbury BJM (2013) Prolonged spring migration in the Red-eyed Vireo (*Vireo olivaceus*). Auk 130:240–246

Cardillo M (2002) Body size and latitudinal gradients in regional diversity of New World birds. Glob Ecol Biogeogr 11:59–65

Carere C, Costantini D, Fusani L, Alleva E, Cardinale M (2010) Hypothermic abilities of migratory songbirds at a stopover site. Rendiconti Fisiche Accademia Lincei 21:323–334

Carey C, Dawson WR, Maxwell LC, Faulkner JA (1978) Seasonal acclimatization to temperature in Cardueline finches. II. Changes in body composition and mass in

relation to season and acute cold stress. J Comp Physiol B 125:101–113
Carey C, Johnston RM, Bekoff A (1989) Thermal thresholds for recruitment of muscles during shivering in winter acclimatized House Finches. In: Mercer JB (ed) Thermal physiology. Elsevier, Amsterdam, pp 685–690
Carpentier AC, Blondin DP, Virtanen KA, Richard D, Haman F, Turcotte ÉE (2018) Brown adipose tissue energy metabolism in humans. Front Endocrinol 9:447
Carr JM, Lima SL (2012a) Heat-conserving postures hinder escape: a thermoregulation-predation trade-off in wintering birds. Behav Ecol 23:434–441
Carr JM, Lima SL (2012b) Nocturnal hypothermia impairs flight ability in birds: a cost of being cool. Proc R Soc B 280:20131846
Carrascal LM, Díaz JA, Huertas DL, Mozetich I (2001) Behavioral thermoregulation by treecreepers: trade-off between saving energy and reducing crypsis. Ecology 82:1642–1654
Catry I, Dias MP, Catry T, Afanasyev V, Fox J, Franco AMA, Sutherland WJ (2011) Individual variation in migratory movements and winter behaviour of Iberian Lesser Kestrels *Falco naumanni* revealed by geolocators. Ibis 153:154–164
Champagne AM, Muñoz-Garcia A, Shtayyeh T, Tieleman BI, Hegemann A, Clement ME, Williams JB (2012) Lipid composition of the stratum corneum and cutaneous water loss in birds along an aridity gradient. J Exp Biol 215:4299–4307
Chaplin SB (1976) The physiology of hypothermia in the Black-capped Chickadee, *Parus atricapillus*. J Comp Physiol 112:335–344
Chaplin SB (1982) The energetic significance of huddling behavior in Common Bushtits (*Psaltriparus minimus*). Auk 99:424–430
Chappell MA, Bucher TL (1987) Effects of temperature and altitude on ventilation and gas exchange in Chukars (*Alectoris chukar*). J Comp Physiol B 157: 129–136
Chastel O, Lacroix A, Kersten M (2003) Pre-breeding energy requirements: thyroid hormone, metabolism and the timing of reproduction in House Sparrows *Passer domesticus*. J Avian Biol 34:298–306
Clarke A, Fraser KPP (2004) Why does metabolism scale with temperature? Funct Ecol 18:243–251
Clarke A, Pörtner H-O (2010) Temperature, metabolic power and the evolution of endothermy. Biol Rev 85: 703–727
Collins BG, Briffa P (1984) Nocturnal energy expenditure by honeyeaters experiencing food shortage and low environmental temperatures. Comp Biochem Physiol 78:77–81
Cook TR, Martin R, Roberts J, Häkkinen H, Botha P, Meyer C, Sparks E, Underhill LG, Ryan PG, Sherley RB (2020) Parenting in a warming world: thermoregulatory responses to heat stress in an endangered seabird. Conserv Physiol 8:coz109
Cooper SJ (1999) The thermal and energetic significance of cavity roosting in Mountain Chickadees and Juniper Titmice. Condor 101:863–866
Cooper SJ (2002) Seasonal metabolic acclimatization in Mountain Chickadees and Juniper Titmice. Physiol Biochem Zool 75:386–395
Cooper SJ, Gessaman JA (2005) Nocturnal hypothermia in seasonally acclimatized Mountain Chickadees and Juniper Titmice. Condor 107:151–155
Cooper CE, Kortner G, Brigham M, Geiser F (2008) Body temperature and activity patterns of free-living Laughing Kookaburras: the largest kingfisher is heterothermic. Condor 110:110–115
Craig A (2020) Red-billed Quelea (*Quelea quelea*), version 1.0. In: del Hoyo J, Elliott A, Sargatal J, Christie DA, de Juana E (eds) Birds of the World. Cornell Lab of Ornithology, Ithaca, NY
Crawford EC, Schmidt-Nielsen K (1967) Temperature regulation and evaporative cooling in the Ostrich. Am J Physiol 212:347–353
Criscuolo F, Monaghan P, Nasir L, Metcalfe NB (2008) Early nutrition and phenotypic development: 'catch-up' growth leads to elevated metabolic rate in adulthood. Proc R Soc B 275:1565–1570
Crompton AW, Taylor CR, Jagger JA (1978) Evolution of endothermy in mammals. Nature 272:333–336
Csada RD, Brigham RM (1994) Reproduction constrains the use of daily torpor by free-ranging Common Poorwills (*Phalaenoptilus nuttallii*) (Aves: Caprimulgidae). J Zool 234:209–216
Daan S, Masman D, Groenewold A (1990) Avian basal metabolic rates: their association with body composition and energy expenditure in nature. Am J Physiol 259:R333–R340
Davis TJ, Keppel G (2021) Fine-scale environmental heterogeneity and conservation management: beach-cast wrack creates microhabitats for thermoregulation in shorebirds. J Appl Ecol 58:1291–1301
Dawson WR (1954) Temperature regulation and water requirements of the Brown and Abert towhees, *Pipilo fuscus* and *Pipilo aberti*. In: Bartholomew GA, Crescitelli F, Bullock TH, Furgason WH, Schechtman AM (eds) University of California Publications in Zoology, vol 59. University of California Press, Berkeley, CA, pp 81–123
Dawson WR, Carey C (1976) Seasonal acclimatization to temperature in Cardueline finches. I. Insulative and metabolic adjustments. J Comp Physiol B 112:317–333
Dawson WR, Fisher CD (1969) Responses to temperature by the Spotted Nightjar (*Eurostopodus guttatus*). Condor 71:49–53
Dawson WR, Hudson JW (1970) Birds. In: Whittow GC (ed) Comparative physiology of thermoregulation. Academic Press, New York, NY, pp 223–310
Dawson WR, Whittow GC (2000) Regulation of body temperature. In: Whittow GC (ed) Avian physiology, 5th edn. Academic Press, New York, NY, pp 343–390

Dawson WR, Marsh RL, Buttemer WA, Carey C (1983) Seasonal and geographic variation of cold resistance in House Finches *Carpodacus mexicanus*. Physiol Zool 56:353–369

Deerenberg C, Overkamp GJF, Visser GH, Daan S (1998) Compensation in resting metabolism for experimentally increased activity. J Comp Physiol B 168:507–512

DeLuca WV, Woodworth BK, Rimmer CC, Marra PP, Taylor PD, McFarland KP, Mackenzie SA, Norris DR (2015) Transoceanic migration by a 12 g songbird. Biol Lett 11:20141045

Doucette LI, Brigham RM, Pavey CR, Geiser F (2012) Prey availability affects daily torpor by free-ranging Australian Owlet-nightjars (*Aegotheles cristatus*). Oecologia 169:361–372

Downs CT, Brown M (2002) Nocturnal heterothermy and torpor in the Malachite Sunbird (*Nectarinia famosa*). Auk 119:251–260

Downs CT, Wirminghaus JO, Lawes MJ (2000) Anatomical and nutritional adaptations of the Speckled Mousebird (*Colius striatus*). Auk 117:791–794

Dunn PO, Garvin JC, Whittingham LA, Freeman-Gallant CR, Hasselquist D (2010) Carotenoid and melanin-based ornaments signal similar aspects of male quality in two populations of Common Yellowthroat. Funct Ecol 24:149–158

Duriez O, Eraud C, Bretagnolle V (2010) First measurements of metabolic rates in the Stone-Curlew, a nocturnal inland wader. Water Study Group Bull 117: 119–122

Eberts ER, Tattersall GJ, Auger PJ, Curley M, Morado MI, Strauss EG, Powers DR, Camacho NM, Tobalske BW, Shankar A (2023) Free-living Allen's Hummingbirds (*Selasphorus sasin*) rarely use torpor while nesting. J Thermal Biol 112:103391

Ederstrom HE, Brumleve SJ (1964) Temperature gradients in the legs of cold-acclimatized pheasants. Am J Physiol 207:457–459

Egevang C, Stenhouse IJ, Phillips RA, Petersen A, Fox JW, Silk JRD (2010) Tracking of Arctic Terns *Sterna paradisaea* reveals longest animal migration. Proceed Natl Acad Sci USA 107:2078–2081

Elliott KH, Hare JF, Le Vaillant M, Gaston AJ, Ropert-Coudert Y, Anderson WG (2015) Ageing gracefully: physiology but not behaviour declines with age in a diving seabird. Funct Ecol 29:219–228

Enstipp MR, Grémillet D, Lorentsen S-H (2005) Energetic costs of diving and thermal status in European Shags (*Phalacrocorax aristotelis*). J Exp Biol 208:3451–3461

Enstipp MR, Grémillet D, Jones DR (2006) The effects of depth, temperature and food ingestion on the foraging energetics of a diving endotherm, the Double-crested Cormorant (*Phalacrocorax auritus*). J Exp Biol 209: 845–859

Enstipp MR, Grémillet D, Jones DR (2008) Heat increment of feeding in Double-crested Cormorants (*Phalacrocorax auritus*) and its potential for thermal substitution. J Exp Biol 211:49–57

Eraud C, Rivière M, Lormée H, Fox JW, Ducamp J-J, Boutin J-M (2013) Migration routes and staging areas of trans-Saharan Turtle Doves appraised from light-level geolocators. PLoS ONE 8:e59396

Ferreira VMOS, Francisco NS, Belloni M, Aguirre GMZ, Caldara FR, Nääs IA, Garcia RG, Almeida Paz ICL, Polycarpo GV (2011) Infrared thermography applied to the evaluation of metabolic heat loss of chicks fed with different energy densities. Revista Brasileira de Ciência Avícola 13:2

Ferretti A, Rattenborg NC, Ruf T, McWilliams SR, Cardinale M, Fusani L (2019) Sleeping unsafely tucked in to conserve energy in a nocturnal migratory songbird. Curr Biol 29:2766–2772

Freeman MT, Czenze ZJ, Schoeman K, McKechnie AE (2020) Extreme hyperthermia tolerance in the world's most abundant wild bird. Sci Rep 10:13098

Freeman MT, Czenze ZJ, Schoeman K, McKechnie AE (2022) Adaptive variation in the upper limits of avian body temperature. Proc Natl Acad Sci 119: e2116645119

Fritsches KA, Brill RW, Warrant EJ (2005) Warm eyes provide superior vision in swordfishes. Curr Biol 15: 55–58

Frost PGH, Siegfried WR, Greenwood PJ (1975) Arterio-venous heat exchange systems in the Jackass Penguin *Spheniscus demersus*. J Zool 175:231–241

Fry CH, Keith S, Urban EK (1988) The birds of Africa, vol 3. Academic Press, London, UK

Funghi C, McCowan LSC, Schuett W, Griffiith SC (2019) High air temperatures induce temporal, spatial and social changes in the foraging behaviour of wild Zebra Finches. Anim Behav 149:33–43

Furness LJ, Speakman JR (2008) Energetics and longevity in birds. Age 30:75

Gad-Briggs A, Nikolaidis T, Pilidis P (2017) Analyses of the effect of cycle inlet temperature on the precooler and plant efficiency of the simple and intercooled helium gas turbine cycles for generation IV nuclear power plants. Appl Sci 7:319

Gallup AC, Herron E, Militello J, Swartwood L, Cortes C, Eguibar JR (2017) Thermal imaging reveals sizable shifts in facial temperature surrounding yawning in Budgerigars (*Melopsittacus undulatus*). Temperature 4:429–435

Galván I, Palacios D, Negro JJ (2017) The bare head of the Northern Bald Ibis (*Geronticus eremita*) fulfills a thermoregulatory function. Front Zool 14:15

Geiser F (2004) Metabolic rate and body temperature reduction during hibernation and daily torpor. Annu Rev Physiol 66:239–274

Geiser F (2013) Hibernation. Curr Biol 23:R188–R193

Geiser F, Ruf T (1995) Hibernation versus daily torpor in mammals and birds: physiological variables and classification of torpor patterns. Physiol Zool 68:935–966

Geluso K, Hayes JP (1999) Effects of dietary quality on basal metabolic rate and internal morphology of

European Starlings *Sturnus vulgaris*. Physiol Biochem Zool 72:189–197
Gerson AR, Guglielmo CG (2011) Flight at low ambient humidity increases protein catabolism in migratory birds. Science 333:1434–1436
Gerson AR, Guglielmo CG (2013) Energetics and metabolite profiles during early flight in American Robins (*Turdus migratorius*). J Comp Physiol B 183:983–991
Gerson AR, McKechnie AE, Smit B, Whitfield MC, Smith EK, Talbot WA, McWhorter TJ, Wolf BO (2019) The functional significance of facultative hyperthermia varies with body size and phylogeny in birds. Funct Ecol 33:597–607
Gerum RC, Fabry B, Metzner C, Beaulieu M, Ancel A, Zitterbart DP (2013) The origin of traveling waves in an Emperor Penguin huddle. New J Phys 15:125022
Gessaman JA, Folk CE (1969) Body temperature and thermal conductance of the Snowy Owl. Physiologist 12:234
Gilbert C, Robertson G, Le Maho Y, Naito Y, Ancel A (2006) Huddling behavior in Emperor Penguins: dynamics of huddling. Physiol Behav 88:479–488
Gilbert C, Blanc S, Le Maho Y, Ancel A (2008) Energy saving processes in huddling Emperor Penguins: from experiments to theory. J Exp Biol 211:1–8
Gilbert C, McCafferty D, Le Maho Y, Martrette J-M, Giroud S, Blanc S, Ancel A (2010) One for all and all for one: the energetic benefits of huddling in endotherms. Biol Rev 85:545–569
Gill RE Jr, Tibbitts TL, Douglas DC, Handel CM, Mulcahy DM, Gottschalck JC, Warnock N, McCaffery BJ, Battley PF, Piersma T (2009) Extreme endurance flights by landbirds crossing the Pacific Ocean: ecological corridor rather than barrier? Proc R Soc B 276: 447–457
Gill RE Jr, Douglas DC, Handel CM, Tibbitts TL, Hufford G, Piersma T (2014) Hemispheric-scale wind selection facilitates Bar-tailed Godwit circum-migration of the Pacific. Anim Behav 90:117–130
Gleeson M, Brackenbury JH (1983) Ventilation, gaseous exchange and air sac gases during moderate thermal panting in Domestic Fowl. Q J Exp Physiol 68:591–602
Goldstein DL (1983) Effect of wind on avian metabolic rate with particular reference to Gambel's Quail. Physiol Zool 56:485–492
González-Solís J, Felicísimo A, Fox JW, Afanasyev V, Kolbeinsson Y, Muñoz J (2009) Influence of sea surface winds on shearwater migration detours. Mar Ecol Prog Ser 391:221–230
Grant GS (1978) Foot-wetting and belly-soaking by incubating Gull-billed Terns and Black Skimmers. J Bombay Nat Hist Soc 75:148–152
Grant GS (1993) Thermoregulatory extension of feet in flying Sooty Terns. Micron 26:217–219
Greenberg R, Cadena V, Danner RM, Tattersall GJ (2012a) Heat loss may explain bill size differences between birds occupying different habitats. PLoS ONE 7:e40933
Greenberg R, Danner R, Olsen B, Luther D (2012b) High summer temperature explains bill size variation in salt marsh sparrows. Ecography 35:146–152
Grémillet D, Wilson RP, Wanless S, Peters G (1999) A tropical bird in the Arctic (the cormorant paradox). Mar Ecol Prog Ser 188:305–309
Guilford T, Meade J, Phillips RA, Boyle D, Roberts S, Collett M, Freeman R, Perrins CM (2009) Migration and stopover in a small pelagic seabird, the Manx Shearwater *Puffinus puffinus*: insights from machine learning. Proc R Soc B 276:1215–1223
Guillemette M, Richman SE, Portugal SJ, Butler PJ (2012) Behavioural compensation reduces energy expenditure during migration hyperphagia in a large bird. Funct Ecol 26:876–883
Gutiérrez JS, Masero JA, Abad-Gómez JM, Villegas A, Sánchez-Gúzman JM (2011) Understanding the energetic costs of living in saline environments: effects of salinity on basal metabolic rate, body mass and daily energy consumption of a long-distance migratory shorebird. J Exp Biol 214:829–835
Gutierrez JS, Abad-Gómez JM, Sánchez-Guzmán JM, Navedo JG, Masero JA (2012) Avian BMR in marine and non-marine habitats: a test using shorebirds. PLoS ONE 7:e42206
Hackett SJ, Kimball RT, Reddy S, Bowie RCK, Braun EL, Braun MJ, Chojnowski JL, Cox WA, Han K-L, Harshman J, Huddleston CJ, Marks BD, Miglia KJ, Moore WS, Sheldon FH, Steadman DW, Witt CC, Yuri T (2008) A phylogenomic study of birds reveals their evolutionary history. Science 320:1763–1768
Haftorn S (1992) Ontogeny of food storing in titmice *Parus* spp. Ibis 134:69–71
Hainsworth FR, Wolf LL (1978) The economics of temperature regulation and torpor in nonmammalian organisms. In: Wang LCH, Hudson JW (eds) Strategies in cold: natural torpidity and thermogenesis. Academic Press, New York, NY, pp 147–184
Hamilton LS, Juvik JO, Scatena FN (eds) (1995) Tropical montane cloud forests. Springer, Berlin, Germany
Hammond KA, Chappell MA, Cardullo FA, Lin R-S, Johnsen TS (2000) The mechanistic basis of aerobic performance variation in Red Junglefowl. J Exp Biol 203:2053–2064
Hatch SA (1990) Incubation rhythm in the fulmar *Fulmarus glacialis*: annual variation and sex roles. Ibis 132:515–524
Hayes SR, Gessaman JA (1980) The combined effects of air temperature, wind and radiation on the resting metabolism of avian raptors. J Therm Biol 5:119–125
Heckscher CM, Taylor SM, Fox JW, Afanasyev V (2011) Veery (*Catharus fuscescens*) wintering locations, migratory connectivity, and a revision of its winter range using geolocator technology. Auk 128:531–542
Hedenström A (2010) Extreme endurance migration: what is the limit to non-stop flight? PLoS Biol 8:e1000362
Hedenström A, Alerstam T (1998) How fast can birds migrate? J Avian Biol 29:424–432

Hedenström A, Klaassen RHG, Åkesson S (2013) Migration of the Little Ringed Plover *Charadrius dubius* breeding in south Sweden by geolocators. Bird Study 60:466–474

Heinrich B (2003) Overnighting of Golden-crowned Kinglets during winter. Wilson Bull 115:113–114

Hennemann WW (1982) Energetics and spread-winged behavior of Anhingas in Florida. Condor 84:91–96

Hiebert SM (1990) Energy costs and temporal organization of torpor in the Rufous Hummingbird (*Selasphorus rufus*). Physiol Zool 63:1082–1097

Hodum PJ, Sydeman WJ, Visser GH, Weathers WW (1998) Energy expenditure and food requirement of Cassin's Auklets provisioning nestlings. Condor 100: 546–550

Hoffman TC, Walsberg GE, Denardo DF (2007) Cloacal evaporation: an important and previously undescribed mechanism for avian thermoregulation. J Exp Biol 210:741–749

Hohtola E (2012) Thermoregulatory adaptations to starvation in birds. In: McCue MD (ed) Comparative physiology of fasting, starvation, and food limitation. Springer, Berlin, Germany, pp 155–170

Hohtola E, Henderson RP, Rashotte ME (1998) Shivering thermogenesis in the pigeon: the effects of activity, diurnal factors, and feeding state. Am J Physiol 275: R1553–R1562

Holyoak DT (2001) Nightjars and their allies. Oxford University Press, Oxford, UK

Hudson LN, Isaac NJ, Reuman DC (2013) The relationship between body mass and field metabolic rate among individual birds and mammals. J Anim Ecol 82:1009–1020

Huertas DL, Díaz JA (2001) Winter habitat selection by a montane forest bird assemblage: the effects of solar radiation. Can J Zool 79:279–284

Hulbert AJ, Else PO (2000) Mechanisms underlying the cost of living in animals. Annu Rev Physiol 62:207–235

Hulbert AJ, Else PL (2004) Basal metabolic rate: history, composition, regulation and usefulness. Physiol Biochem Zool 77:869–876

Humphries MM, Careau V (2011) Heat for nothing or activity for free? Evidence and implications for activity-thermoregulatory heat substitution. Integr Comp Biol 51:419–431

IUPS (International Union of Physiological Sciences) Commission for Thermal Physiology (2003) Glossary of terms for thermal physiology, third ed. J Therm Biol 28:75–106

Jahn AE, Cueto VR, Fox JW, Husak MS, Kim DH, Landoll DV, Ledezma JP, Lepage HK, Levey DJ, Murphy MT, Renfrew RB (2013) Migration timing and wintering areas of three species of flycatchers (*Tyrannus*) breeding in the Great Plains of North America. Auk 130:247–257

Jenni L, Jenni-Eiermann S (1998) Fuel supply and metabolic constraints in migrating birds. J Avian Biol 29: 521–528

Jenni L, Jenni-Eiermann S (1999) Fat and protein utilisation during migratory flight. In: Adams NJ, Slotow RH (eds) Proceedings of the 22nd International Ornithological Congress. BirdLife South Africa, Johannesburg, pp 1437–1449

Jenni L, Schaub M (2003) Behavioural and physiological reactions to environmental variation in bird migration: a review. In: Berthold P, Gwinner E, Sonnenschein E (eds) Avian migration. Springer, Berlin, Germany, pp 155–171

Jenni-Eiermann S (2017) Energy metabolism during endurance flight and the post-flight recovery phase. J Comp Physiol A 203:431–438

Jetz W, Freckleton RP, McKechnie AE (2008) Environment, migratory tendency, phylogeny and basal metabolic rate in birds. PLoS ONE 3:e3261

Jimenez AG, Cooper-Mullin C, Calhoon EA, Williams JB (2014a) Physiological underpinnings associated with differences in pace of life and metabolic rate in north temperate and Neotropical birds. J Comp Physiol B 184:545–561

Jimenez AG, van Brocklyn J, Wortman M, Williams JB (2014b) Cellular metabolic rate is influenced by life-history traits in tropical and temperate birds. PLoS ONE 9:e87349

Johnson OW, Fielding L, Fisher JP, Gold RS, Goodwill RH, Bruner AE, Furey JF, Brusseau PA, Brusseau NH, Johnson PM, Jukema J, Prince LL, Tenney MJ, Fox JW (2012) New insight concerning transoceanic migratory pathways of Pacific Golden-Plovers (*Pluvialis fulva*): the Japan stopover and other linkages as revealed by geolocators. Wader Study Group Bull 119:1–8

Kaiser A (1993) A new multi-category classification of subcutaneous fat deposits of songbirds. J Field Ornithol 64:246–255

Kahl MP (1971) Spread-wing postures and their possible functions in the Ciconiidae. Auk 88:715–722

Kahl MP Jr (1963) Thermoregulation in the Wood Stork, with special reference to the role of the legs. Physiol Zool 36:141–151

Karasov WH, Pinshow B (2000) Test for physiological limitation to nutrient assimilation in a long-distance passerine migrant at a springtime stopover site. Physiol Biochem Zool 73:335–343

Kaseloo PA, Lovvorn JR (2003) Heat increment of feeding and thermal substitution in Mallard ducks feeding voluntarily on grain. J Comp Physiol B 173:207–213

Kaseloo PA, Lovvorn JR (2006) Substitution of heat from exercise and digestion by ducks diving for mussels at varying depths and temperatures. J Comp Physiol B 176:265–275

Kelly RJ, Murphy TG, Tarvin KA, Burness G (2012) Carotenoid-based ornaments of female and male American Goldfinches (*Spinus tristis*) show sex-specific correlations with immune function and metabolic rate. Physiol Biochem Zool 85:348–363

Kilgore DL, Schmidt-Nielsen K (1975) Heat loss from ducks' feet immersed in cold water. Condor 77:475–517

Kingsolver JG (2009) The well-temperatured biologist. Am Nat 174:755–768

Klaassen M, Bech C, Slagsvold G (1989) Basal metabolic rate and thermal conductance in Arctic Tern chicks and the effect of heat increment of feeding on thermoregulatory expenses. Ardea 77:193–200

Klaassen M, Kvist A, Lindström Å (2000) Flight costs and fuel composition of a bird migrating in a wind tunnel. Condor 102:444–451

Klaassen RHG, Alerstam T, Carlsson P, Fox JW, Lindström Å (2011) Great flights by Great Snipes: long and fast non-stop migration over benign habitats. Biol Lett 7:833–835

Klaassen RHG, Ens BJ, Shamoun-Baranes J, Exo K-M, Bairlein F (2012) Migration strategy of a flight generalist, the Lesser Black-backed Gull *Larus fuscus*. Behav Ecol 23:58–68

Koivula K, Orell M, Lahti K (2002) Plastic daily fattening routines in Willow Tits. J Anim Ecol 71:816–823

Körtner G, Geiser F (2000) The temporal organization of daily torpor and hibernation: circadian and circannual rhythms. Chronobiol Int 17:103–128

Körtner G, Brigham RM, Geiser F (2000) Winter torpor in a large bird. Nature 407:318

Krams I, Cirule D, Vrublevska J, Nord A, Rantala MJ, Krama T (2013) Nocturnal loss of body reserves reveals high survival risk for subordinate Great Tits wintering at extremely low ambient temperatures. Oecologia 172:339–346

Kvist A, Klaassen M, Lindström A (1998) Energy expenditure in relation to flight speed: what is the power of mass loss rate estimates? J Avian Biol 29:485–498

Landers TJ, Rayner MJ, Phillips RA, Hauber ME (2011) Dynamics of seasonal movements by a trans-Pacific migrant, the Westland Petrel. Condor 113:71–79

Lane JE (2002) Thermoregulation of Whip-poor-wills (*Caprimulgus vociferus*) under natural and laboratory conditions. M.S. thesis, University of Regina, Regina, SK, Canada

Lane JE, Swanson DL, Brigham RM, McKechnie AE (2004) Physiological responses to temperature by Whip-poor-wills: more evidence for the evolution of low metabolic rates in Caprimulgiformes. Condor 106: 921–925

Larkin JL, Raybuck D, Roth A, Chavarría-Duriaux L, Duriaux G, Siles M, Smalling C (2017) Geolocators reveal migratory connectivity between wintering and breeding areas of Golden-winged Warblers. J Field Ornithol 88:288–298

Lasiewski RC (1962) The energetics of migrating hummingbirds. Condor 64:324

Lasiewski RC (1963) Oxygen consumption of torpid, resting, and flying hummingbirds. Physiol Zool 36:122–140

Lasiewski RC (1964) Body temperatures, heart and breathing rate, and evaporative water loss in hummingbirds. Physiol Zool 37:212–223

Lasiewski RC, Bartholomew GA (1966) Evaporative cooling in the poor-will and the Tawny Frogmouth. Condor 68:253–262

Lasiewski RC, Dawson WR (1964) Physiological responses to temperature in the Common Nighthawk. Condor 66:477–490

Lasiewski RC, Lasiewski RJ (1967) Physiological responses of the Blue-throated and Rivoli's hummingbirds. Auk 84:34–48

Lasiewski RC, Snyder GK (1969) Responses to high temperature in nestling Double-crested and Pelagic cormorants. Auk 86:529–540

Lasiewski RC, Weathers WW, Bernstein MH (1967) Physiological responses of the Giant Hummingbird, *Patagonia gigas*. Comp Biochem Physiol 23:797–813

Laszlo A (1992) The effects of hyperthermia on mammalian cell structure and function. Cell Prolif 25:59–87

Laurila M, Hohtola E (2005) The effect of ambient temperature and simulated predation risk on fasting-induced nocturnal hypothermia of pigeons in outdoor conditions. J Therm Biol 30:392–399

Leemans R, Cramer WP (1991) The IIASA database for mean monthly values of temperature, precipitation and cloudiness of a global terrestrial grid, RR-91-18. IIASA, Laxenburg, Austria

Leeson S, Summers JD (2001) Nutrition of the chicken, 4th edn. University Books, Guelph, ON

Lemke HW, Tarka M, Klaassen RHG, Åkesson M, Bensch S, Hasselquist D, Hansson B (2013) Annual cycle and migration strategies of a trans-Saharan migratory songbird: a geolocator study in the Great Reed Warbler. PLoS ONE 8:e79209

Lewden A, Petit M, Milbergue M, Orio S, Vézina F (2014) Evidence of facultative daytime hypothermia in a small passerine wintering at northern latitudes. Ibis 156:321–329

Liknes ET, Scott SM, Swanson DL (2002) Seasonal acclimatization in the American Goldfinch revisited: to what extent do metabolic rates vary seasonally? Condor 104:548–557

Lill A, Box J, Baldwin J (2006) Do metabolism and contour plumage insulation vary in response to seasonal energy bottlenecks in Superb Fairy-wrens? Aust J Zool 54:23–30

Lindsay CV, Downs CT, Brown M (2009a) Physiological variation in Amethyst Sunbirds (*Chalcomitra amethystina*) over an altitudinal gradient in summer. J Therm Biol 34:190–199

Lindsay CV, Downs CT, Brown M (2009b) Physiological variation in Amethyst Sunbirds (*Chalcomitra amethystina*) over an altitudinal gradient in winter. J Exp Biol 212:483–493

Lindström Å, Kvist A (1995) Maximum energy intake rate is proportional to basal metabolic rate in passerine birds. Proc R Soc B 261:337–343

Lindström Å, Rosén M (2002) The cost of avian winter stores: intra-individual variation in basal metabolic rate of a wintering passerine, the Greenfinch *Carduelis chloris*. Avian Science 2:1–5

Lovegrove BG (2000) The zoogeography of mammalian basal metabolic rate. Am Nat 156:201–219

Lovvorn JR (2007) Thermal substitution and aerobic efficiency: measuring and predicting effects of heat balance on endotherm diving energetics. Philos Trans R Soc B 362:2079–2093

Luteyn JL (1999) Páramos: a checklist of plant diversity, geographical distribution, and botanical literature. New York Botanical Garden Press, Bronx, NY

Ma X, Lin Y, Zhang H, Chen W, Wang S, Ruan D, Jiang Z (2014) Heat stress impairs the nutritional metabolism and reduces the productivity of egg-laying hens. Anim Reprod Sci 145:182–190

Mack DE, Yong W (2020) Swainson's Thrush (*Catharus ustulatus*), version 1.0. In: Poole AF, Gill FB (eds) Birds of the World. Cornell Lab of Ornithology, Ithaca, NY

Maldonado K, Cavieres G, Veloso C, Canals M, Sabat P (2009) Physiological responses in Rufous-collared Sparrows to thermal acclimation and seasonal acclimatization. J Comp Physiol B 179:335–343

Marder J, Arieli Y (1988) Heat balance of acclimated pigeons (*Columba livia*) exposed to temperatures up to 60°C Ta. Comp Biochem Physiol 91:165–170

Marsh RL, Dawson WR (1989) Avian adjustments to cold. In: Wang LCH (ed) Advances in comparative and environmental physiology, vol 4. Springer, Berlin Heidelberg, pp 205–253

Masman D, Daan S, Dietz M (1989) Heat increment of feeding in the kestrel, *Falco tinnunculus*, and its natural seasonal variation. In: Bech C, Reinertsen RE (eds) Physiology of cold adaptation in birds. Plenum Press, New York, NY, pp 123–136

Mateos-Rodríguez M, Liechti F (2012) How do diurnal long-distance migrants select flight altitude in relation to wind? Behav Ecol 23:403–409

McCafferty DJ (2013) Applications of thermal imaging in avian science. Ibis 155:4–15

McCafferty DJ, Gilbert C, Thierry A-M, Currie J, Le Maho Y, Ancel A (2013) Emperor Penguin body surfaces cool below air temperature. Biol Lett 9: 20121192

McGreal RD, Farner DS (1956) Premigratory fat deposition in the Gambel White-crowned Sparrow: some morphological and chemical observations. Northwest Sci 30:12–23

McKechnie AE (2008) Phenotypic flexibility in basal metabolic rate and the changing view of avian physiological diversity: a review. J Comp Physiol B 178:235–247

McKechnie AE, Lovegrove BG (2001) Thermoregulation and the energetic significance of clustering behavior in the White-backed Mousebird (*Colius colius*). Physiol Biochem Zool 74:238–249

McKechnie AE, Lovegrove BG (2002) Avian facultative hypothermic responses: a review. Condor 104:705–724

McKechnie AE, Wolf BO (2004) The energetics of the rewarming phase of avian torpor. In: Barnes BM, Carey HV (eds) Life in the cold: mechanisms, adaptation, and application. Biological Papers of the University of Alaska, no. 27, Fairbanks, AK, pp 265–273

McKechnie AE, Körtner G, Lovegrove BG (2006) Thermoregulation under semi-natural conditions in Speckled Mousebirds: the role of communal roosting. Afr Zool 41:155–163

McKechnie AE, Noakes MJ, Smit B (2015) Global patterns of seasonal acclimatization in avian resting metabolic rates. J Ornithol 156:367–376

McKechnie AE, Whitfield MC, Smit B, Gerson AR, Smith EK, Talbot WA, McWhorter TJ, Wolf BO (2016) Avian thermoregulation in the heat: efficient evaporative cooling allows for extreme heat tolerance in four Southern Hemisphere columbids. J Exp Biol 219: 2145–2155

McKechnie AE, Gerson AR, McWhorter TJ, Smith EK, Talbot WA, Wolf BO (2017) Avian thermoregulation in the heat: evaporative cooling in five Australian passerines reveals within-order biogeographic variation in heat tolerance. J Exp Biol 220:2436–2444

McKechnie AE, Gerson AR, Wolf BO (2021) Thermoregulation in desert birds: scaling and phylogenetic variation in heat tolerance and evaporative cooling. J Exp Biol 224:jeb229211

McKecknie AE, Mzilikazi N (2011) Heterothermy in Afrotropical mammals and birds: a review. Integr Comp Biol 51:349–363

McMillen IC, Robinson JS (2005) Developmental origins of the metabolic syndrome: prediction, plasticity, and programming. Physiol Rev 85:571–633

McNab BK (1980) Food habits, energetics, and the population biology of mammals. Am Nat 116:106–124

McNab BK (1986) The influence of food habits on the energetics of eutherian mammals. Ecol Monogr 56:1–19

McNab BK (2002) The physiological ecology of vertebrates. Cornell University Press, Ithaca, NY

McNab BK (2003) Metabolism: ecology shapes bird bioenergetics. Nature 426:620–621

McNab BK (2009) Ecological factors affect the level and scaling of avian BMR. Comp Biochem Physiol 152: 22–45

McNab BK (2016) Avian energetics: the passerine/non-passerine dichotomy. Comp Biochem Physiol 191: 152–155

McNab BK (2019) What determines the basal rate of metabolism? J Exp Biol 222:jeb205591

McNab BK, Morrison P (1963) Body temperature and metabolism in subspecies of *Peromyscus* from arid and mesic environments. Ecol Monogr 33:62–82

McWilliams SR, Guglielmo C, Pierce B, Klaassen M (2004) Flying, fasting, and feeding in birds during

migration: a nutritional and physiological ecology perspective. J Avian Biol 35:377–393
Meiri S, Thomas GH (2007) The geography of body size – challenges of the interspecific approach. Glob Ecol Biogeogr 16:689–693
Merola-Zwartjes M, Ligon JD (2000) Ecological energetics of the Puerto Rican Tody: heterothermy, torpor, and intra-Island variation. Ecology 81:990–1003
Metcalfe NB, Alonso-Alvarez C (2010) Oxidative stress as a life-history constraint: the role of reactive oxygen species in shaping phenotypes from conception to death. Funct Ecol 24:984–996
Middleton ALA (1986) Seasonal changes in plumage structure and body composition of the American Goldfinch, *Carduelis tristis*. Can Field-Nat 100:545–549
Midtgård U (1981) The rete tibiotarsale and arterio-venous association in the hind limb of birds: a comparative morphological study on counter-current heat exchange systems. Acta Zool 62:67–87
Midtgård U (1983) Scaling of the brain and the eye cooling system in birds: a morphometric analysis of the rete ophthalmicum. J Exp Zool 225:197–207
Mills R, McGraw KJ (2021) Cool birds: facultative use by an introduced species of mechanical air conditioning systems during extremely hot outdoor conditions. Biol Lett 17:20200813
Moe B, Stolevik E, Bech C (2005) Ducklings exhibit substantial energy-saving mechanisms as a response to short-term food shortage. Physiol Biochem Zool 78:90–104
Moe B, Angelier F, Bech C, Chastel O (2007) Is basal metabolic rate influenced by age in a long-lived seabird, the Snow Petrel? J Exp Biol 210:3407–3414
Moe B, Rønning B, Verhulst S, Bech C (2009) Metabolic ageing in individual Zebra Finches. Biol Lett 5:86–89
Møller AP (2007) Senescence in relation to latitude and migration in birds. J Evol Biol 20:750–757
Monaghan P, Metcalfe NB, Torres R (2009) Oxidative stress as a mediator of life history trade-offs: mechanisms, measurements and interpretation. Ecol Lett 12:75–92
Monti F, Grémillet D, Sforzi A, Dominici JM, Bagur RT, Navarro AM, Fusani L, Klaassen RHG, Alerstam T, Duriez O (2018) Migration distance affects stopover use but not travel speed: contrasting patterns between long-and short-distance migrating Ospreys. J Avian Biol 49:e01839
Moreno J, Merino S, Potti J, de León A, Rodriguez R (1999) Maternal energy expenditure does not change with flight costs or food availability in the Pied Flycatcher (*Ficedula hypoleuca*): costs and benefits for nestlings. Behav Ecol Sociobiol 46:244–251
Muñoz-Garcia A, Aamidor SE, McCue MD, McWilliams SR, Pinshow B (2012) Allocation of endogenous and dietary protein in the reconstitution of the gastrointestinal tract of migratory Blackcaps at stopover sites. J Exp Biol 215:1069–1075
Nespolo RF, Baciagalupe LD, Figueroa CC, Koteja P, Opazo JC (2011) Using new tools to solve an old problem: the evolution of endothermy in vertebrates. Trends Ecol Evol 26:414–423
Neumann C (2016) Behavioural thermoregulation in the Common Swift during flight. Br Birds 109:10–13
Niles LJ, Burger J, Porter RR, Dey AD, Minton CDT, Gonzalez PM, Baker AJ, Fox JW, Gordon C (2010) First results using light level geolocators to track Red Knots in the Western Hemisphere show rapid and long intercontinental flights and new details of migration pathways. Wader Study Group Bull 117:123–130
Nilsson J-Å, Åkesson M, Nilsson JF (2009) Heritability of resting metabolic rate in a wild population of Blue Tits. J Evol Biol 22:1867–1874
Nilsson ALK, Nilsson J-Å, Alerstam T (2011a) Basal metabolic rate and energetic cost of thermoregulation among migratory and resident Blue Tits. Oikos 120: 1784–1789
Nilsson JF, Tobler M, Nilsson J-Å, Sandell MI (2011b) Long-lasting consequences of elevated yolk testosterone for metabolism in the Zebra Finch. Physiol Biochem Zool 84:287–291
Nilsson C, Klaassen RHG, Alerstam T (2013) Differences in speed and duration of bird migration between spring and autumn. Am Nat 181:837–845
Nilsson JÅ, Molokwu MN, Olsson O (2016) Body temperature regulation in hot environments. PLoS ONE 11:e0161481
Nisbet ICT, Mostello CS, Veit RRR, Fox JW, Afanasyev V (2011) Migrations and winter quarters of five Common Terns tracked using geolocators. Waterbirds 34: 32–39
Nord A, Nilsson JF, Nilsson J-Å (2011) Nocturnal body temperature in wintering Blue Tits is affected by roost-site temperature and body reserves. Oecologia 167:21–25
Nord A, Chiriac S, Hasselquist D, Nilsson J-Å (2013) Endotoxin injection attenuates rest-phase hypothermia in wintering Great Tits through the onset of fever. Funct Ecol 27:236–244
Nord A, Metcalfe NB, Page JL, Huxtable A, McCafferty DJ, Dawson NJ (2021) Avian red blood cell mitochondria produce more heat in winter than in autumn. FASEB J 35:e21490
Nowack J, Giroud S, Arnold W, Ruf T (2017) Muscle non-shivering thermogenesis and its role in the evolution of endothermy. Front Physiol 8:889
O'Connor RS, Wolf BO, Brigham RM, McKechnie AE (2017) Avian thermoregulation in the heat: efficient evaporative cooling in two southern African nightjars. J Comp Physiol B 187:477–491
Ong HC, Mahlia TMI, Masjuki HH, Norhasyima RS (2011) Comparison of palm oil, *Jatropha curcas* and *Calophyllum inophyllum* for biodiesel: a review. Renew Sust Energ Rev 15:3501–3515
Ooi KB, Noguchi M (2017) Verification of the performance of a vertical ground heat exchanger applied to a test house in Melbourne, Australia. Energies 10:1558
Ophir E, Arieli Y, Marder J, Horowitz M (2002) Cutaneous blood flow in the pigeon *Columba livia*: its

possible relevance to cutaneous water evaporation. J Exp Biol 205:2627–2636

Osváth G, Daubner T, Dyke G, Fuisz TI, Nord A, Pénzes J, Vargancsik D, Vágási CI, Vincze O, Pap PL (2018) How feathered are birds? Environment predicts both the mass and density of body feathers. Funct Ecol 32:701–712

Page GW, Warnock N, Tibbitts TL, Jorgensen D, Hartman CA, Stenzel LE (2014) Annual migratory patterns of Long-billed Curlews in the American west. Condor 116:50–61

Pap PL, Vincze O, Wekerle B, Daubner T, Vágási CI, Nudds RL, Dyke GJ, Osváth G (2017) A phylogenetic comparative analysis reveals correlations between body feather structure and habitat. Funct Ecol 31: 1241–1251

Parrish JD (1997) Patterns of frugivory and energetic condition in Nearctic landbirds during autumn migration. Condor 99:681–697

Parver LM (1991) Temperature modulating action of choroidal blood flow. Eye 5:181–185

Pattinson NB, Smit B (2017) Seasonal behavioral responses of an arid-zone passerine in a hot environment. Physiol Behav 179:268–275

Pattinson NB, Thompson ML, Griego M, Russell G, Mitchell NJ, Martin RO, Wolf BO, Smit B, Cunningham SJ, McKechnie AE, Hockey PA (2020) Heat dissipation behaviour of birds in seasonally hot arid-zones: are there global patterns? J Avian Biol 51: jav02350

Pavlovic G, Weston MA, Symonds MR (2019) Morphology and geography predict the use of heat conservation behaviours across birds. Funct Ecol 33:286–296

Pennycuick CJ (1975) Mechanics of flight. In: Farner DS, King JR (eds) Avian biology, vol 5. Academic Press, New York, NY, pp 1–75

Pennycuick CJ, Battley PF (2003) Burning the engine: a time-marching computation of fat and protein consumption in a 5420-km non-stop flight by Great Knots, *Calidris tenuirostris*. Oikos 103:323–332

Perez CR, Moye JK, Pritsos CA (2014) Estimating the surface area of birds: using the homing pigeon (*Columba livia*) as a model. Biol Open 3:486–488

Pessato A, McKechnie AE, Buchanan KL, Mariette MM (2020) Vocal panting: a novel thermoregulatory mechanism for enhancing heat tolerance in a desert-adapted bird. Sci Rep 10:18914

Petit M, Vézina F (2014) Phenotypic manipulations confirm the role of pectoral muscles and haematocrit in avian maximal thermogenic capacity. J Exp Biol 217: 824–830

Petit M, Lewden A, Vézina F (2013) Intra-seasonal flexibility in avian metabolic performance highlights the uncoupling of basal metabolic rate and thermogenic capacity. PLoS ONE 8:e68292

Pierce BJ, McWilliams SR (2005) Seasonal changes in composition of lipid stores in migratory birds: causes and consequences. Condor 107:269–279

Piersma T (2011) Why marathon migrants get away with high metabolic ceilings: towards an ecology of physiological restraint. J Exp Biol 214:295–302

Piersma T, Drent J (2003) Phenotypic flexibility and the evolution of organismal design. Trends Ecol Evol 18: 228–233

Piersma T, Cadée N, Daan S (1995) Seasonality in basal metabolic rate and thermal conductance in a long-distance migrant shorebird, the knot (*Calidris canutus*). J Comp Physiol B 165:37–45

Piersma T, Gessaman JA, Dekinga A, Visser GH (2004) Gizzard and other lean mass components increase, yet basal metabolic rates decrease, when Red Knots *Calidris canutus* are shifted from soft to hard-shelled food. J Avian Biol 35:99–104

Piersma T, Pérez-Tris J, Mouritsen H, Bauchinger U, Bairlein F (2005) Is there a "migratory syndrome" common to all migrant birds? Ann N Y Acad Sci 1046:282–293

Pollock HS, Brawn JD, Cheviron ZA (2021) Heat tolerances of temperate and tropical birds and their implications for susceptibility to climate warming. Funct Ecol 35:93–104

Porter WMR, Witmer LM (2016) Avian cephalic vascular anatomy, sites of thermal exchange, and the rete ophthalmicum. Anat Rec 299:1461–1486

Pörtner H (2001) Climate change and temperature-dependent biogeography: oxygen limitation of thermal tolerance in animals. Naturwissenschaften 88:137–146

Prinzinger R, Goppel R, Lorenz A, Kulzer E (1981) Body temperature and metabolism in the Red-backed Mousebird (*Colius castanotus*) during fasting and torpor. Comp Biochem Physiol A 69:689–692

Prinzinger R, Pressmar A, Schleucher E (1991) Body temperature in birds. Comp Biochem Physiol 99:499–506

Rahbek C, Graves GR (2001) Multiscale assessment of patterns of avian species richness. Proc Natl Acad Sci USA 98:4534–4539

Ramirez L, Diniz-Filho JAF, Hawkins BA (2007) Partitioning phylogenetic and adaptive components of the geographical body-size pattern of New World birds. Glob Ecol Biogeogr 17:100–110

Reinertsen RE, Haftorn S (1984) The effect of short-time fasting on metabolism and nocturnal hypothermia in the Willow Tit *Parus montanus*. J Comp Physiol B 154:23–28

Reinertsen RE, Haftorn S (1986) Different metabolic strategies of northern birds for nocturnal survival. J Comp Physiol B 156:655–663

Rezende EL, Swanson DL, Novoa FF, Bozinovic F (2002) Passerines versus non-passerines: so far, no statistical differences in the scaling of avian energetics. J Exp Biol 205:101–107

Richards SA (1976) Evaporative water loss in Domestic Fowls and its partition in relation to ambient temperature. J Agric Sci 87:527–532

Richman SE, Lovvorn JR (2011) Effects of air and water temperatures on resting metabolism of auklets and other diving birds. Physiol Biochem Zool 84:316–332
Richter K, Haslbeck M, Buchner J (2010) The heat shock response: life on the verge of death. Mol Cell 40:253–266
Ricklefs RE (2000) Density dependence, evolutionary optimization, and the diversification of avian life histories. Condor 103:9–22
Ricklefs RE, Hainsworth FR (1968) Temperature dependent behavior of the Cactus Wren. Ecology 49:227–233
Ricklefs RE, Konarzewski M, Daan S (1996) The relationship between basal metabolic rate and daily energy expenditure in birds and mammals. Am Nat 147: 1047–1071
Rogalla S, Shawkey MD, Vanthournout B, D'Alba L (2021) Thermoregulation and heat exchange in Ospreys (*Pandion haliaetus*). J Therm Biol 96:102857
Rønning B, Moe B, Berntsen HH, Noreen E, Bech C (2014) Is the rate of metabolic ageing and survival determined by basal metabolic rate in the Zebra Finch? PLoS ONE 9:e108675
Rønning B, Broggi J, Bech C, Moe B, Ringsby TH, Pärn H, Hagen IJ, Sæther BE, Jensen H (2016) Is basal metabolic rate associated with recruit production and survival in free-living House Sparrows? Funct Ecol 30:1140–1148
Ronning B, Jensen H, Moe B, Bech C (2007) Basal metabolic rate: heritability and genetic correlations with morphological traits in the Zebra Finch. J Evol Biol 20:1815–1822
Roth TC II, Rattenborg NC, Pravosudov VV (2010) The ecological relevance of sleep: trade-off between sleep, memory and energy conservation. Philos Trans R Soc B 365:945–959
Roti Roti JL (2008) Cellular responses to hyperthermia (40–46°C): cell killing and molecular events. Int J Hyperth 24:3–15
Ruddiman WF (2001) Earth's climate: past and future. Macmillan
Ruth JM, Talbot WA, Smith EK (2020) Behavioral response to high temperatures in a desert grassland bird: use of shrubs as thermal refugia. West N Am Nat 80:265–275
Ruuskanen S, Hsu BY, Nord A (2021) Endocrinology of thermoregulation in birds in a changing climate. Mol Cell Endocrinol 519:111088
Ryeland J, Weston MA, Symonds MRE (2019) Leg length and temperature determine the use of unipedal roosting in birds. J Avian Biol 50:e02008
Ryeland J, Weston MA, Symonds MRE (2021) The importance of wetland margin microhabitat mosaics; the case of shorebirds and thermoregulation. J Appl Ecol 58:382–391
Saarela S, Heldmaier G (1987) Effect of photoperiod and melatonin on cold resistance, thermoregulation and shivering/nonshivering thermogenesis in Japanese Quail. J Comp Physiol B 157:625–633
Saarela S, Rintamäki H, Saaarela M (1984) Seasonal variation in the dynamics of ptiloerection and shivering correlated changes in the metabolic rate and body temperature of the pigeon. J Comp Physiol B 154:47–53
Sabat P, Cavieres G, Veloso C, Canals M, Bozinovic F (2009) Intraspecific basal metabolic rate varies with trophic level in Rufous-collared Sparrows. Comp Biochem Physiol 154:502–507
Sapir N, Wikelski M, McCue MD, Pinshow B, Nathan R (2010) Flight modes in migrating European Bee-eaters: heart rate may indicate low metabolic rate during soaring and gliding. PLoS ONE 5:e13956
Schaub R, Prinzinger R (1999) Long-term telemetry of heart rates and energy metabolic rate during the diurnal cycle in normothermic and torpid African Blue-naped Mousebirds (*Urocolius macrourus*). Comp Biochem Physiol 124:439–445
Schleucher E (2001) Heterothermia in pigeons and doves reduces energetic costs. J Therm Biol 26:287–293
Schleucher EP, Withers C (2002) Metabolic and thermal physiology of pigeons and doves. Physiol Biochem Zool 75:439–450
Schmidt KL, MacDougall-Shackleton EA, MacDougall-Shackleton SA (2012) Developmental stress has sex-specific effects on nestling growth and adult metabolic rates but no effect on adult body size or body composition in Song Sparrows. J Exp Biol 215:3207–3217
Scholer MN, Arcese P, Puterman ML, Londoño GA, Jankowski JE (2019) Survival is negatively related to basal metabolic rate in tropical Andean birds. Funct Ecol 33:1436–1445
Schwabl H (1993) Yolk is a source of maternal testosterone for developing birds. Proc Natl Acad Sci USA 90: 11446–11450
Secor SM (2009) Specific dynamic action: a review of postprandial metabolic response. J Comp Physiol B 179:1–56
Shipley AA, Sheriff MJ, Pauli JN, Zuckerberg B (2019) Snow roosting reduces temperature-associated stress in a wintering bird. Oecologia 190:309–321
Shipley AA, Cruz J, Zuckerberg B (2020) Personality differences in the selection of dynamic refugia have demographic consequences for a winter-adapted bird. Proc R Soc B 287:20200609
Shobrak M (1998) Notes on the breeding and cooling behaviour of Hoopoe Larks *Alaemon alaudipes* in central Saudi Arabia. Sandgrouse 20:53–55
Sies H, Berndt C, Jones DP (2017) Oxidative stress. Annu Rev Biochem 86:715–748
Sittler B, Aebischer A, Gilg O (2010) Post-breeding migration of four Long-tailed Skuas (*Stercorarius longicaudus*) from north and east Greenland to west Africa. J Ornithol 152:375–381
Smit B, McKechnie AE (2010) Avian seasonal metabolic variation in a subtropical desert: basal metabolic rates are lower in winter than in summer. Funct Ecol 24: 330–339

Smit B, Boyles JG, Brigham RM, McKechnie AE (2011) Torpor in dark times: patterns of heterothermy are associated with the lunar cycle in a nocturnal bird. J Biol Rhythm 26:241–248

Smit B, Zietsman G, Martin RO, Cunningham SJ, McKechnie AE, Hockey PAR (2016) Behavioural responses to heat in desert birds: implications for predicting vulnerability to climate warming. Clim Change Resp 3:9

Smit B, Whitfield MC, Talbot WA, Gerson AR, McKechnie AE, Wolf BO (2018) Avian thermoregulation in the heat: phylogenetic variation among avian orders in evaporative cooling capacity and heat tolerance. J Exp Biol 221:jeb174870

Smith SB, McPherson KH, Backer JM, Pierce BJ, Podlesak DW, McWilliams SR (2007) Fruit quality and consumption by songbirds during autumn migration. Wilson J Ornithol 119:419–428

Speakman JR, Król E (2010) Maximal heat dissipation capacity and hyperthermia risk: neglected key factors in the ecology of endotherms. J Anim Ecol 79:726–746

Spence AR, Tingley MW (2021) Body size and environment influence both intraspecific and interspecific variation in daily torpor use across hummingbirds. Funct Ecol 35:870–883

Stach C, Jakobsson S, Kullberg C, Fransson T (2013) Geolocators reveal three consecutive wintering areas in the Thrush Nightingale. Anim Migr 1:1–7

Stahel CD, Nicol SC (1982) Temperature regulation in the Little Penguin, *Eudyptula minor*, in air and water. J Comp Physiol 148:93–100

Staples JF, Brown JCL (2008) Mitochondrial metabolism in hibernation and daily torpor: a review. J Comp Physiol B 178:811–827

Stenhouse IJ, Egevang C, Phillips RA (2012) Trans-equatorial migration, staging sites and wintering areas of Sabine's Gulls *Larus sabini* in the Atlantic Ocean. Ibis 154:42–51

Stutchbury BJM, Tarof SA, Done T, Gow E, Kramer PM, Tautin J, Fox JW, Afanasyev V (2009) Tracking long-distance songbird migration by using geolocators. Science 323:896

Swanson DL (1991) Seasonal adjustments in metabolism and insulation in the Dark-eyed Junco. Condor 93: 538–545

Swanson DL (2010) Seasonal metabolic variation in birds: functional and mechanistic correlates. Curr Ornithol 17:75–129

Swanson DL, Olmstead KL (1999) Evidence for a proximate influence of winter temperatures on metabolism in passerine birds. Physiol Biochem Zool 72:566–575

Swanson DL, Thomas NE, Liknes ET, Cooper SJ (2012) Intraspecific correlations of basal and maximal metabolic rates in birds and the aerobic capacity model for the evolution of endothermy. PLoS ONE 7:e34271

Talbot WA, Gerson AR, Smith EK, McKechnie AE, Wolf BO (2018) Avian thermoregulation in the heat: metabolism, evaporative cooling and gular flutter in two small owls. J Exp Biol 221:jeb171108

Tam CS, Lecoultre V, Ravussin E (2012) Brown adipose tissue: mechanisms and potential therapeutic targets. Circulation 125:2782–2791

Tattersall GJ, Sinclair BJ, Withers PC, Fields PA, Seebacher F, Cooper CE, Maloney SK (2012) Coping with thermal challenges: physiological adaptations to environmental temperatures. Comp Physiol 2:2151–2202

Tattersall GJ, Arnaout B, Symonds MRE (2017) The evolution of the avian bill as a thermoregulatory organ. Biol Rev 92:1630–1656

Teulier L, Rouanet J-L, Letexier D, Romestaing C, Belouze M, Rey B, Duchamp C, Roussel D (2010) Cold-acclimation-induced nonshivering thermogenesis in birds is associated with upregulation of avian UCP but not with innate uncoupling or altered ATP efficiency. J Exp Biol 213:2476–2482

Thomas DB, Fordyce RE (2007) The heterothermic loophole exploited by penguins. Aust J Zool 55:317–321

Thomas DB, Fordyce RE (2012) Biological plasticity in penguin heat-retention structures. Anat Rec 295:249–256

Thomas DB, Ksepka DT, Fordyce RE (2011) Penguin heat-retention structures evolved in a greenhouse Earth. Biol Lett 7:461–464

Thompson MG (1974) Migratory patterns of Ruddy Turnstones in the central Pacific region. Living Bird 12:5–23

Tieleman BI, Williams JB (2000) The adjustments of avian metabolic rates and water fluxes to desert environments. Physiol Biochem Zool 73:461–479

Tieleman BI, Williams JB, Bloomer P (2003a) Adaptation of metabolism and evaporative water loss along an aridity gradient. Proc R Soc B 270:207–214

Tieleman BI, Williams JB, Buschur ME, Brown CR (2003b) Phenotypic variation of larks along an aridity gradient: are desert birds more flexible? Ecology 84: 1800–1815

Trawa G (1970) Note preliminaire sur la vascularisation des membres des Spheniscides de terre Adelie. L'Oiseau et la Revue Français d'Ornithologie 40:142–156

Tsang LR, Wilson LAB, Ledogar J, Wroe S, Attard M, Sansalone G (2019) Raptor talon shape and biomechanical performance are controlled by relative prey size but not by allometry. Sci Rep 9:7076

Vaillancourt E, Prud'Homme S, Haman F, Guglielmo CG, Weber J-M (2005) Energetics of a long-distance migrant shorebird (*Philomachus pugnax*) during cold exposure and running. J Exp Biol 208:317–325

van de Ven TMFN, Martin RO, Vink TJF, McKechnie AE, Cunningham SJ (2016) Regulation of heat exchange across the hornbill beak: functional similarities with toucans? PLoS ONE 11:e0154768

Vander Haegen WM, Owen RB, Krohn WB (1994) Metabolic rate of American Woodcock. Wilson Bull 106: 338–342

Vanderhoff N, Pyle P, Patten MA, Sallabanks R, James FC (2020) American Robin (*Turdus migratorius*), version

1.0. In: Rodewald PG (ed) Birds of the World. Cornell Lab of Ornithology, Ithaca, NY
Vansteelant WMG, Bouten W, Klaassen RHG, Koks BJ, Schlaich AE, Van Diermen J, Van Loon EE, Shamoun-Baranes J (2015) Regional and seasonal flight speeds of soaring migrants and the role of weather conditions at hourly and daily scales. J Avian Biol 46:25–39
Verhulst S, Holveck M-J, Riebel K (2006) Long-term effects of manipulated natal brood size on metabolic rate in Zebra Finches. Biol Lett 2:478–480
Versteegh MA, Helm B, Dingemanse NJ, Tieleman BI (2008) Repeatability and individual correlates of basal metabolic rate and total evaporative water loss in birds: a case study in European Stonechats. Comp Biochem Physiol 150:452–457
Viblanc VA, Dobson FS, Stier A, Schull Q, Saraux C, Gineste B, Pardonnet S, Kauffmann M, Robin J-P, Bize P (2016) Mutually honest? Physiological 'qualities' signalled by colour ornaments in monomorphic King Penguins. Biol J Linn Soc 118:200–214
Videler JJ (2005) Avian flight. Oxford University Press, Oxford, UK
Villén-Pérez S, Carrascal LM, Gordo O (2014) Wintering forest birds roost in areas of higher sun radiation. Eur J Wildl Res 60:59–67
Vleck CM (1981) Hummingbird incubation: female attentiveness and egg temperature. Oecologia 51:199–205
Vuarin P, Henry P-Y (2014) Field evidence for a proximate role of food shortage in the regulation of hibernation and daily torpor: a review. J Comp Physiol B 184: 683–697
Walker LE, Walker JM, Palca JW, Berger RJ (1983) A continuum of sleep and shallow torpor in fasting doves. Science 221:194–195
Walsberg GE, King JR (1978) The relationship of the external surface area of birds to skin surface area and body mass. J Exp Biol 76:185–189
Ward JM, Ruxton GD, Houston DC, McCafferty DJ (2007) Thermal consequences of turning white in winter: a comparative study of Red Grouse *Lagopus lagopus scoticus* and Scandanavian Willow Grouse *L. l. lagopus*. Wildl Biol 13:120–129
Ward J, McCafferty DJ, Houston DC, Ruxton GD (2008) Why do vultures have bald heads? The role of postural adjustment and bare skin areas in thermoregulation. J Therm Biol 33:168–173
Weathers WW (1981) Physiological thermoregulation in heatstressed birds: consequences of body size. Physiol Zool 54:345–361
Weathers WW (1997) Energetics and thermoregulation by small passerines of the humid, lowland tropics. Auk 114:341–353
Weber J-M (2009) The physiology of long-distance migration: extending the limits of endurance metabolism. J Exp Biol 212:593–597
Wellbrock AHJ, Eckhardt LRH, Kelsey NA, Heldmaier G, Rozman J, Witte K (2022) Cool birds: first evidence of energy-saving nocturnal torpor in free-living Common Swifts *Apus apus* resting in their nests. Biol Lett 18: 20210675
Wells ME, Schaeffer PJ (2012) Seasonality of peak metabolic rate in non-migrant tropical birds. J Avian Biol 43:481–485
White CR, Blackburn TM, Martin GR, Butler PJ (2007a) Basal metabolic rate of birds is associated with habitat temperature and precipitation, not primary productivity. Proc R Soc B 274:287–293
White CR, Cassey P, Blackburn TM (2007b) Allometric exponents do not support a universal metabolic allometry. Ecology 88:315–323
Whitfield MC, Smit B, McKechnie AE, Wolf BO (2015) Avian thermoregulation in the heat: scaling of heat tolerance and evaporative cooling capacity in three southern African arid-zone passerines. J Exp Biol 218:1705–1714
Whitford PC (1985) Bird behavior in response to the warmth of blacktop roads. Trans Wis Acad Sci 73: 135–143
Whittow GC (1980) Thermoregulatory behavior of the Laysan and Black-footed albatrosses. Elepaio 40:97–98
Wiersma P, Muñoz-Garcia A, Walker A, Williams JB (2007) Tropical birds have a slow pace of life. Proceed Natl Acad Sci USA 104:9340–9345
Wiersma P, Nowak B, Williams JB (2012) Small organ size contributes to the slow pace of life in tropical birds. J Exp Biol 215:1662–1669
Wikelski M, Spinney L, Schelsky W, Scheuerlein A, Gwinner E (2003a) Slow pace of life in tropical sedentary birds: a common-garden experiment on four Stonechat populations from different latitudes. Proc R Soc B 270:2383–2388
Wikelski M, Tarlow EM, Raim A, Diehl RH, Larkin RP, Visser GH (2003b) Avian metabolism: costs of migration in free-flying songbirds. Nature 423:704
Williams JB, Tieleman BI (2001) Physiological ecology and behavior of desert birds. Curr Ornithol 16:299–353
Williams JB, Tieleman BI (2005) Physiological adaptation in desert birds. Bioscience 55:416–425
Williams TC, Williams JM, Ireland LC, Teal JM (1978) Estimated flight time for transatlantic autumnal migrants. Am Birds 32:275–280
Williams JB, Tieleman BI, Shobrak M (1999) Lizard burrows provide thermal refugia for larks in the Arabian desert. Condor 101:714–717
Williams JB, Miller RA, Harper JM, Wiersma P (2010) Functional linkages for the pace of life, life-history, and environment in birds. Integr Comp Biol 50:855–868
Withers PC (1977) Respiration, metabolism, and heat exchange of euthermic and torpid poorwills and hummingbirds. Physiol Zool 50:43–52
Wojciechowski MS, Pinshow B (2009) Heterothermy in small, migrating passerine birds during stopover: use of hypothermia at rest accelerates fuel accumulation. J Exp Biol 212:3068–3075

Wojciechowski MS, Jefimow M, Pinshow B (2008) Blackcaps huddle at night at migratory stopovers. In: Lovegrove BG, McKechnie AE (eds) Hypometabolism in animals: torpor, hibernation and cryobiology. Interpak Books, University of KwaZulu-Natal, Pietermaritzburg, pp 405–414

Wojciechowski MS, Jefimow M, Pinshow B (2011) Heterothermy, and the energetic consequences of huddling in small migrating passerine birds. Integr Comp Biol 51:409–418

Wolf LW, Hainsworth FR (1972) Environmental influence on regulated body temperature in torpid hummingbirds. Comp Biochem Physiol 41A:167–173

Wolf BO, Walsberg GE (1996) Thermal effects of radiation and wind on a small bird and implications for microsite selection. Ecology 77:2228–2236

Wolf BO, Walsberg GE (2000) The role of the plumage in heat transfer processes of birds. Am Zool 40:575–584

Wolf BO, Wooden DM, Walsberg GE (1996) The use of thermal refugia by two small desert birds. Condor 98: 424–428

Wolf BO, McKechnie AE, Schmitt CJ, Czenze ZJ, Johnson AB, Witt CC (2020) Extreme and variable torpor among high-elevation Andean hummingbird species. Biol Lett 16:20200428

Woods CP, Brigham RM (2004) The avian enigma: "hibernation" by Common Poorwills (*Phalaenoptilus nuttallii*). In: Barnes BM, Carey HV (eds) Life in the cold: evolution, mechanisms, adaptation, and application. University of Alaska, Fairbanks, AK, pp 231–240. Biological Papers of the University of Alaska, no. 27

Woods CP, Czenze ZJ, Brigham RM (2019) The avian 'hibernation' enigma: thermoregulatory patterns and roost choice of the Common Poorwill. Oecologia 189:47–53

Yorzinski JL, Lam J, Schultz R, Davis M (2018) Thermoregulatory postures limit antipredator responses in peafowl. Biology Open 7:bio031005

Yosef R (2010) Thermoregulatory behavior in migratory European Bee-eaters (*Merops apiaster*). Wilson J Ornithol 122:378–380

Zerba E, Dana AN, Lucia MA (1999) The influence of wind and locomotor activity on surface temperature and energy expenditure of the eastern House Finch (*Carpodacus mexicanus*) during cold stress. Physiol Biochem Zool 72:265–276

Zheng W-H, Liu J-S, Swanson DL (2014) Seasonal phenotypic flexibility of body mass, organ masses, and tissue oxidative capacity and their relationship to resting metabolic rate in Chinese Bulbuls. Physiol Biochem Zool 87:432–444

Zitterbart DP, Wienecke B, Butler JP, Fabry B (2011) Coordinated movements prevent jamming in an Emperor Penguin huddle. PLoS ONE 6:e20260

# Avian Locomotion: Flying, Running, Walking, Climbing, Swimming, and Diving

# 11

## Contents




G. Ritchison, *In a Class of Their Own*, Fascinating Life Sciences,
https://doi.org/10.1007/978-3-031-14852-1_11

**Abstract**

Most birds can fly, but can also, to varying degrees depending on the species and their habitats, walk, run, climb, swim, and dive. With a focus on flight, this chapter covers the evolution of flight and provides a detailed explanation of how birds fly and factors that have contributed to the evolution of different wing shapes and variation in wing loading. The different ways that birds fly, from gliding and soaring to flapping flight to hovering, are discussed, and how birds take off and land is explained. The metabolic cost of flight is also explained. Finally, the various ways that birds move along the ground, in trees and other substrates, and on and in the water are also explained.

## 11.1 Introduction

One reason why birds are so popular is that, to many people, they represent freedom. The ability to fly means that birds can conceivably travel just about anywhere at any time. Clearly, flight does provide birds with impressive mobility that, in turn, gives them the ability to travel farther in search of food, mates, nest sites, and other resources than nonflying vertebrates of similar size, and also gives them the ability to efficiently elude many predators, particularly land-based predators. These are important advantages of flight and, perhaps not surprisingly then, powered flight has evolved independently in three vertebrate classes: Reptilia, Mammalia, and, of course, Aves (Fig. 11.1).

## 11.2 Evolution of Flight

It might be useful, before focusing on birds, to examine what is known about the evolution of flight in reptiles and mammals. The earliest known bats in the fossil record are from the Eocene (49–53 mya) and these bats were already capable of flapping flight. Many authors have suggested that the hypothetical ancestor of these early bats would have been nocturnal, insectivorous, arboreal, and a glider. An initial gliding phase would have involved extension of the digits and further development of membranes between the digits to provide additional lift. Gliding would then have been gradually replaced by powered flight, providing early bats with additional mobility and reduced predation risk (Speakman 2001).

Pterosaurs are thought to be derived from a bipedal, cursorial archosaur in the late Triassic period (about 225 million years ago). As with bats, the origin and evolution of pterosaurs, and flight, is not well understood because of the poor fossil record. Available fossil evidence indicates that pterosaurs ranged in size from sparrow-like forms to the impressively large *Quetzalcoatlus*, with a wingspan of 11–12 m. Small pterosaurs apparently had sufficient available power for taking off from the ground, larger pterodactyls like *Quetzalcoatlus* were likely limited in their take-off capabilities, possibly depending on brief periods of high anaerobic power generation for a running take-off with a moderate headwind and thermal updrafts. For large pterosaurs, taking off from a high cliff may have been more cost effective than a ground take-off. What about the earliest pterosaurs? Recent evidence suggests that pterosaurs did not descend from dinosaur-

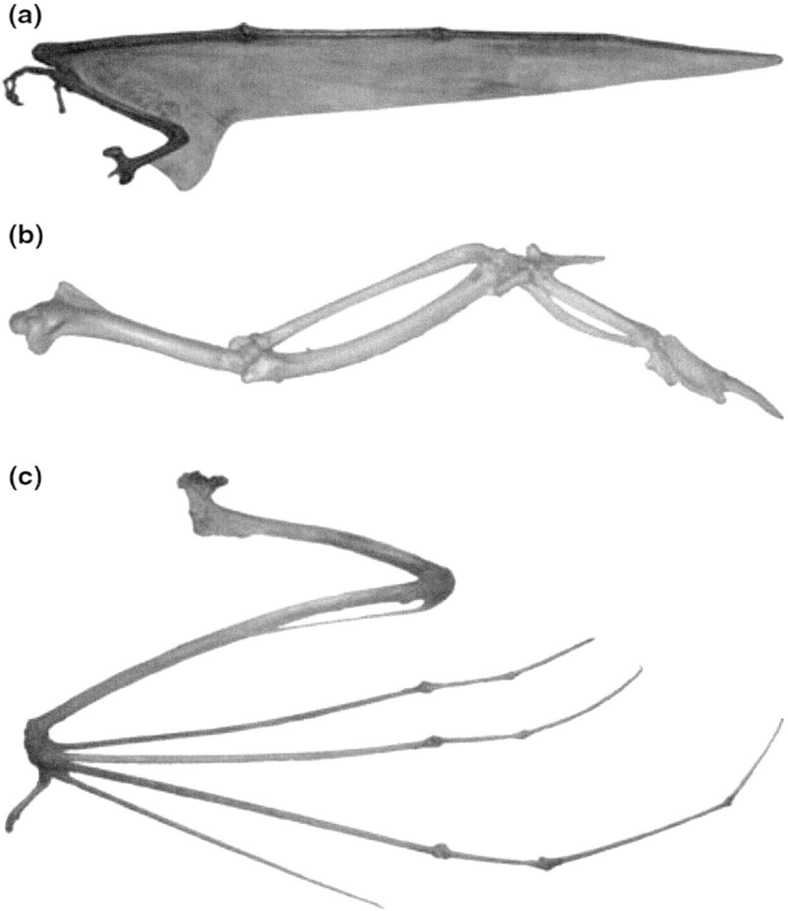


**Fig. 11.1** Forelimbs of flying vertebrates including a (**a**) pterosaur, (**b**) bird, and (**c**) bat. The wing membrane of pterosaurs was supported by an elongated fourth digit, whereas the wing membrane of bats is supported by four elongated digits. The bird wing, of course, consists of long feathers anchored by the humerus, ulna, and manus (carpometacarpus and digits). (Figure from Bell et al. 2011; © 2011 The Authors. *Journal of Evolutionary Biology* © 2011 European Society for Evolutionary Biology, used with permission)

like cursorial bipeds, but from more basal quadrupedal, climbing reptiles. Basal pterosaurs exhibit a dorsoventrally shallow body and sharply curved hand and foot claws. Such features indicate tree climbing ability and suggest that pterosaurs had arboreal origins (Naish and Martill 2003; Witton 2013).

Available evidence suggests, then, that flight may have evolved from the trees down in both bats and pterosaurs. What about birds? For many years, the choices have been limited to three main hypotheses. Primary among them are the arboreal hypothesis (e.g., Feduccia 1996), with the ancestors of *Archaeopteryx* living in trees (or at least climbing into trees on a regular basis) and initially gliding before developing flapping flight, and the cursorial hypothesis (e.g., Burgers and Chiappe 1999; Dececchi and Larsson 2011), with these ancestors using long, powerful legs to run fast with their arms (wings) outstretched perhaps to aid in maneuvering and capturing prey and, eventually, developing sufficient lift to take flight (Fig. 11.2).

An additional hypothesis, a derived version of the cursorial hypothesis, is the WAIR (wing-assisted incline running) hypothesis. Dial (2003a, b) noticed the ability of young Chukars (a partridge; *Alectoris chukar*) to escape danger by scrambling up inclined surfaces. The chicks first run very fast, flapping their rather small, partially feathered wings to creating enough momentum to run up an inclined surface to safety. The ancestors of birds may have using proto-wings in a similar fashion, with wings eventually

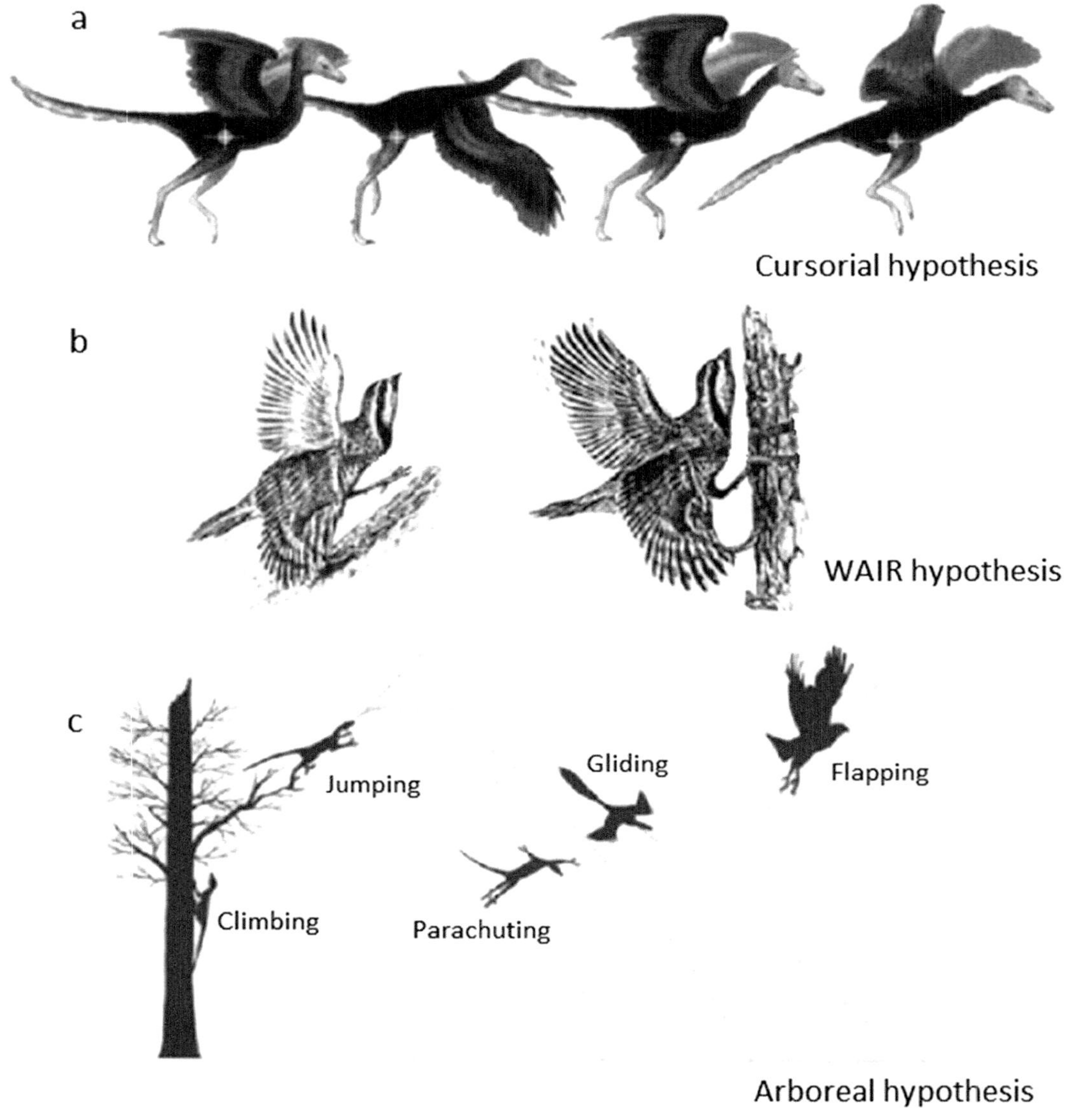


**Fig. 11.2** Hypotheses for the evolution of avian flight. (**a**) Cursorial hypothesis, with the feathered ancestors of birds running along the ground with sufficient speed to generate lift. (**b**) Wing-assisted incline running (WAIR) hypothesis, with the feathered ancestors first using their "wings" to climb trees, with the further development of wings then allowing gliding and, eventually, powered flapping flight. (**c**) Arboreal hypothesis, with flying ability transitioning from simple parachuting, to gliding, to powered flapping flight. (Figure modified from Chatterjee and Templin 2012; © 2012 Springer Science Business Media B.V., used with permission)

evolving to the point of permitting not only running up inclined surfaces but, for an animal running across the ground, flight. However, the idea that flight evolved along a particular pathway in a linear, straightforwardly way "... is being replaced by an emerging picture of a rapid diversification of volant and near-volant pennaraptorans characterised by high levels of experimentation, homoplasy, and exaptation with regard to the aerodynamic apparatus" (Sullivan et al. 2017: 214).

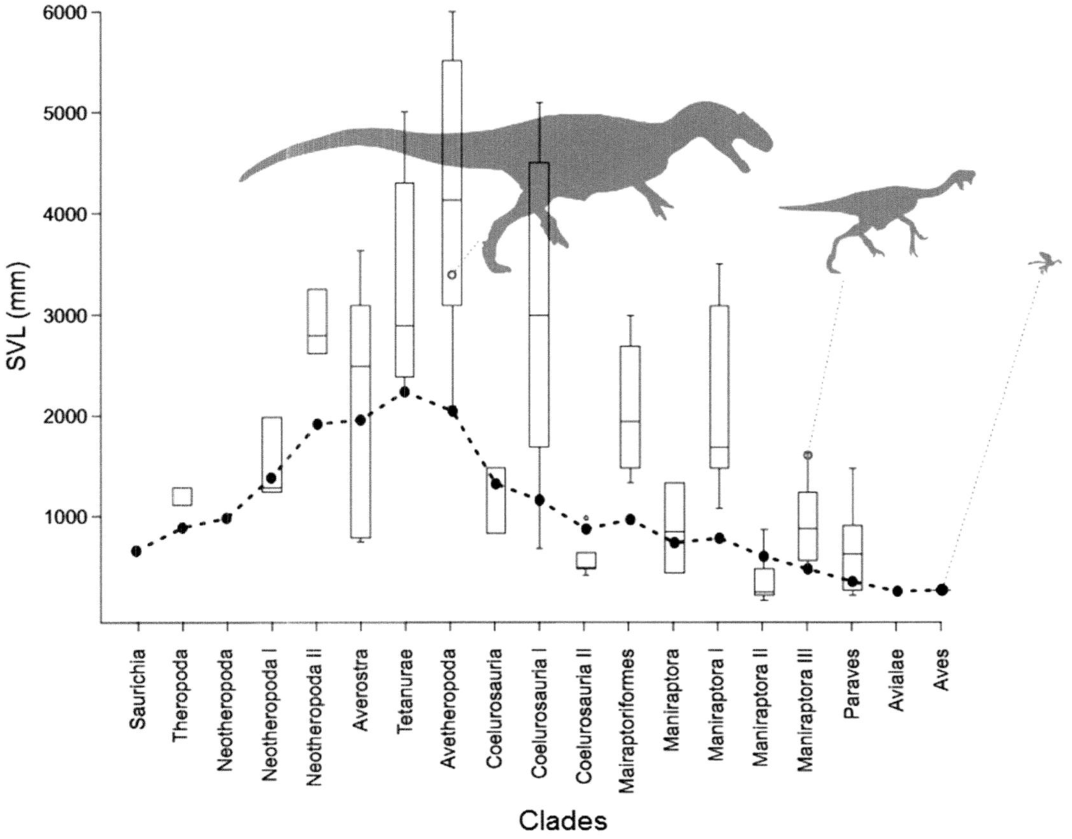


**Fig. 11.3** Evolution of body size of theropods to the time of *Archaeopteryx*. Snout-vent lengths (SVLs) are provided for different theropod lineages (boxplots) and nodes (closed circles and dashed line). Nodes represent the common ancestors of lineages. (Figure from Dececchi and Larsson 2013; © 2013 The Authors. Evolution © 2013 The Society for the Study of Evolution, used with permission)

Before further addressing the question of origins, consider the characteristics needed to make flight possible. One characteristic needed for powered flight is a reasonably high metabolic rate because powered flight requires significant amounts of energy. Bats are mammals, mammals are endotherms with high metabolic rates, and so were the immediate ancestors of bats. For pterosaurs, the picture is less clear, but many, if not all, pterosaurs had hair, suggesting an elevated metabolism (Naish and Martill 2003). The evolution of flight in birds must have also been preceded by the evolution of increased metabolic rates. Rezende et al. (2020) suggested that metabolic rates of theropods began increasing about 180 to 170 million years ago when many of them already had protofeathers and feathers and, further that, paravians may have been fully endothermic. In addition, sustained flapping flight also requires a high aerobic capacity, with a cardiovascular system and, specifically, a heart capable of generating stroke volumes and cardiac output sufficient to provide flight muscles with the oxygen needed for flapping flight (Altimiras et al. 2017).

Another characteristic important for the evolution of flight is relatively small size. There have certainly been some very large flying birds, pterosaurs, and, to a lesser degree, bats. However, the evolution of flight is more likely for smaller organisms because less lift is needed to achieve flight. Over a period of about 50 million years, the theropod ancestors of birds decreased in size (Dececchi and Larsson 2013; Fig. 11.3). A variety of factors likely contributed to this "miniaturization," including selection for an increasingly horizontal (and biomechanically demanding) orientation of the femur, a stiff tail, and greater agility and cursoriality (Lee et al. 2014).

Flight, of course, also requires wings, and the wings of bats and pterosaurs consist of elongated digits and membranes. An essential component of bird wings is, of course, feathers. For birds, then, the evolution of flight necessarily had to be preceded by the evolution of feathers. The first feathers were relatively simple filamentous structure, e.g., the compsognathid *Sinosauropteryx* (Chen et al. 1998), but, over time, pennaceous feathers, like those of the oviraptorosaurs *Caudipteryx* and *Protarcheopteryx* (Ji et al. 1998), evolved. However, given their relatively large body size, short forearms, and resulting high wing loading, most non-avian theropods would not have been able to fly (Fig. 11.4). Rather, these non-avian theropods may have used their feathered forelimbs for display, brooding, or for high-speed maneuvering and braking and balancing when pursuing prey (Zelenitsky et al. 2012; Li et al. 2012; Fowler et al. 2011; Dececchi et al. 2016).

Flight requires longer forelimbs and pennaceous feathers to transform wings into increasingly effective aerodynamic structures (Fig. 11.5), but what selective factors might have favored increasingly longer forelimbs and pennaceous feathers? If small, feathered forelimbs were beneficial in terms of nonflying behaviors such as high-speed maneuvering, leaping, and rapid braking and turning and those benefits would be enhanced if forelimbs and feathers were longer, then selection would seemingly favor such forelimbs and feathers (Heers et al. 2014). As suggested by Dececchi et al. (2016), "Enhancements of even a few percent may have had tremendous advantages to these animals, particularly if we compare the small margins of performance differences of extant predator-prey interactions." Further, Pei et al. (2020) noted that such advantages may have provided these "agile" non-avian theropods with opportunities to fill a diversity of ecological niches (Box 11.1 Evolution of the Avian Wing).

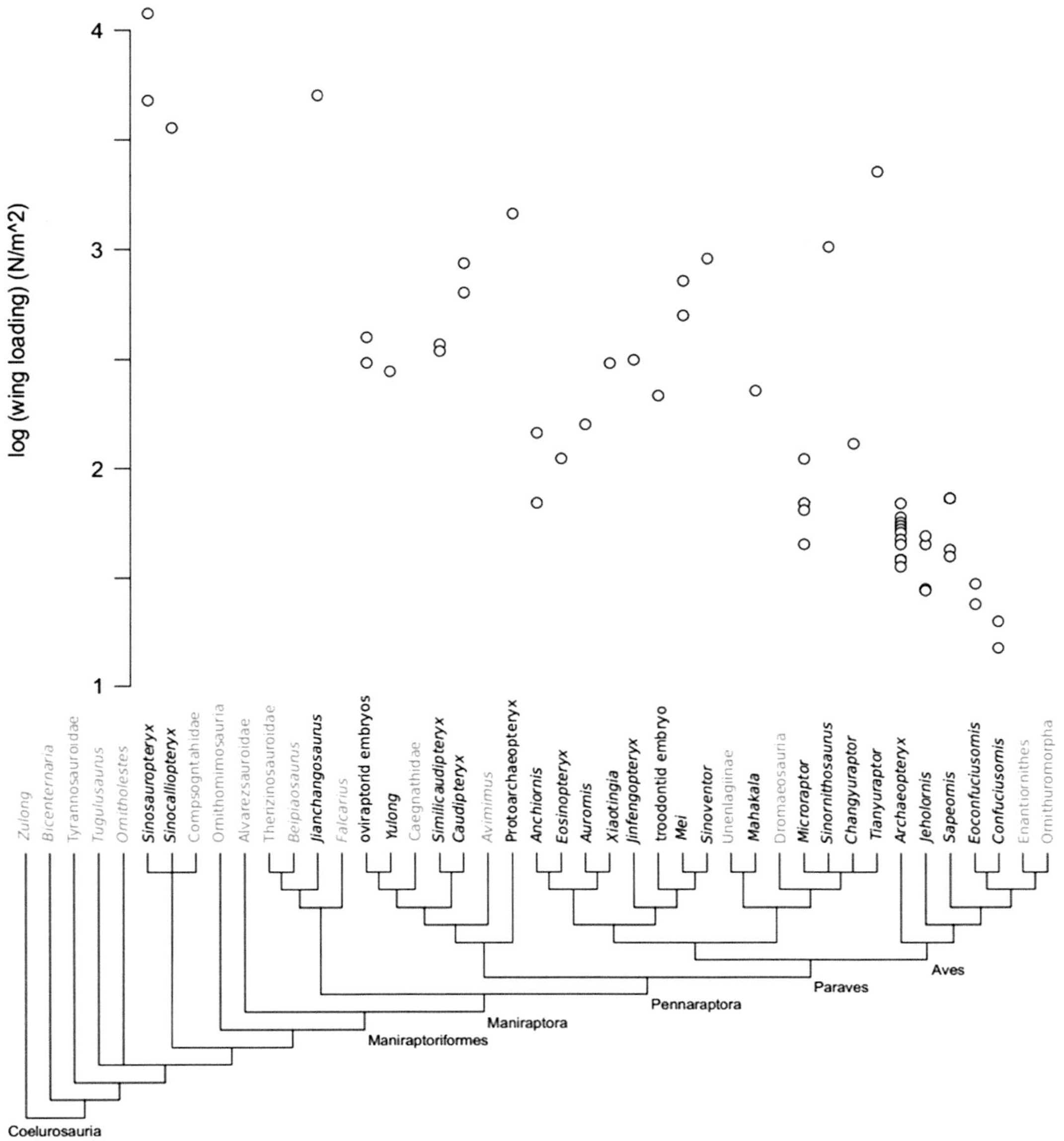


**Fig. 11.4** Wing loading values of non-avian theropods and birds (Aves). For taxa with multiple specimens, values for each specimen are provided. Note that few non-avian theropods have wing loading values comparable to those of *Archaeopteryx*. (Figure modified from Dececchi et al. 2016; open-access article distributed under the terms of the Creative Commons Attribution License, https://creativecommons.org/licenses/by/4.0/)

**Fig. 11.5** Incipient wings in the theropod-avian lineage. Note the increasing wing surface area from *Caudipteryx* to *Archaeopteryx* and *Microraptor* and to present-day Chukars (*Alectoris chukar*). (Figure modified from Heers et al. 2014; © The Paleontological Society, used with permission)

**Box 11.1 Evolution of the Avian Wing**

Pennaceous feathers have been found on the wings of several theropods that were clearly unable to fly. These non-avian theropods may have used their feathered forelimbs for display and/or to aid in changing direction and braking when pursuing prey (Zelenitsky et al. 2012; Li et al. 2012; Fowler et al. 2011; Dececchi et al. 2016). Over millions of years of evolution feathered wings became efficient airfoils.

(continued)

**Box 11.1** (continued)

Pennaceous feathers with symmetrical vanes on the wing of *Caudipteryx zoui*, a Maniraptoran theropod. Scale bar, 1 cm (Image from Ji et al. 1998; © 1998 Springer Nature, used with permission)

However, the feathered wings of the earliest gliders or flyers differed from those of present-day birds. For example, the wings of *Anchiornis*, a non-avian theropod, consisted of feathers that were "unspecialized and undifferentiated," with remiges that had thin rachises and were relatively short and symmetrical (Longrich et al. 2012). In addition, the wing coverts were oriented in different directions and formed multiple layers. The remiges in the wings of *Archaeopteryx* were longer and had asymmetrical vanes, but the rachises were still relatively thin and the wing coverts were relatively long. Longrich et al. (2012) suggested that the multiple layers of coverts in the wings of *Anchiornis* and *Archaeopteryx* may have provided the reinforcement needed to allow the wings to function as airfoils. In contrast to *Anchiornis* and *Archaeopteryx*, the wings of *Confuciusornis sanctus*, a pygostylian, were much more similar to appearance to those of present-day birds that fly, with long asymmetrical primaries that had thicker rachises and much shorter wing coverts. However, the outer primaries that formed the leading edge of the wing were shorter than those of present-day birds and the wings of *Confuciusornis* still lacked an alula (Longrich et al. 2012). Wings more closely resembling those of present-day birds, with longer outer primaries and an alula, were first observed in an enantiornithine bird from the early Cretaceous, *Eoalulavis hoyasi* (Sanz et al. 1996).

(continued)

**Box 11.1** (continued)

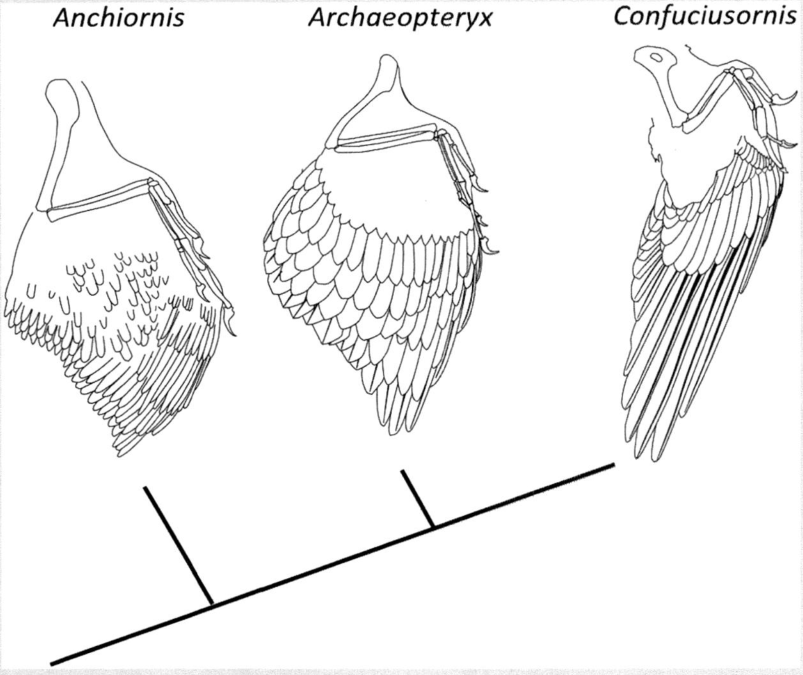


*Anchiornis*—Ancestral wing morphology where the wing is composed of slender, symmetrical, and poorly differentiated flight feathers. *Archaeopteryx*—Wing of early Avialae where the remiges are elongated, broad, and asymmetrical. *Confuciusornis* - Wing of pygostylia where the primary remiges are further elongated and the coverts are shortened. (Figure from Longrich et al. 2012; © 2012 Elsevier Ltd., used with permission)

Flight also requires muscles and supporting skeletal structures. Extant birds that fly have large pectoralis muscles and supporting skeletal structures like the sternal keel that provide the power and support needed for sustained flapping flight. The mass of the pectoralis muscles of extant birds generally ranges between 10 and 20% of total body mass (Hartman 1961; Greenewalt 1975) whereas the pectoral muscles of *Archaeopteryx* have been estimated to represent only 0.5% of its body mass (Bock 2013). The pectoral girdle is also critical, with the glenoid cavity particularly important because it influences the movements of the humerus. In present-day birds that fly, the glenoid cavity is oriented in more of a latero-dorsal direction and allows a wide range of dorsoventral (i.e., up and down), but limited anteroposterior (i.e., back and forth), movements of the humerus. In contrast, basal avialans like *Archaeopteryx*, *Bambiraptor*, and *Buitreraptor* as well as flightless ratites like Ostriches and rheas have glenoid cavities that are more vertically oriented (Fig. 11.6). As a result, wing movements of ratites are, and those of basal avialans likely were, largely oriented in an anteroposterior direction rather than the dorsoventral direction of present-day birds that fly. If correct, then basal avialans would likely not have been capable of powered flight (Agnolin et al. 2019). However, Voeten and Cubo (2018) suggested that "*Archaeopteryx*'s large coracoids and robust, flattened and more dorsally positioned

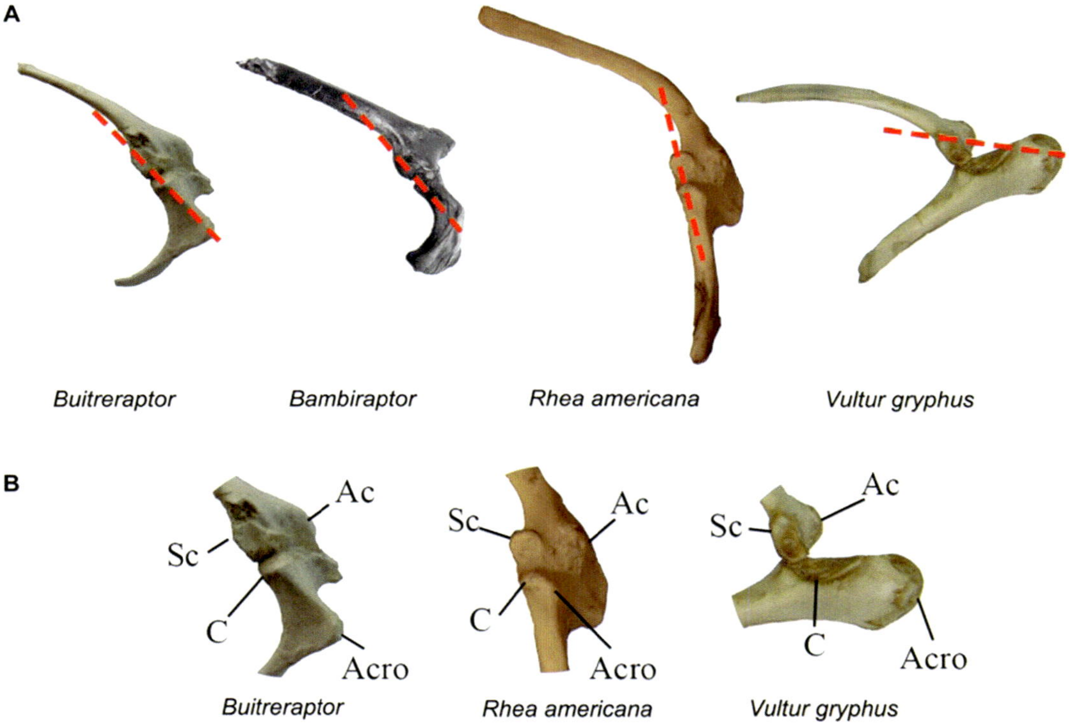


**Fig. 11.6** (**a**) Lateral view of the right scapulocoracoid of selected basal and derived paravians. The dotted lines indicate the inclination of the major axis of the glenoid cavity. (**b**) Detail of the glenoid of the right scapulocoracoid of selected paravians in lateral view. Not to scale. C, coracoidal portion of the glenoid; Sc, scapular portion of the glenoid; Ac, acromion; Acro, acrocoracoid or biceps tubercle. *Rhea americana*, Greater Rhea; *Vultur gryphus*, Andean Condor. (Figure from Agnolin et al. 2019; © 2019 Agnolin, Motta, Brissón Egli, Lo Coco and Novas, open-access article distributed under the terms of the Creative Commons Attribution License (CC BY), https://creativecommons.org/licenses/by/4.0/)

furcula lacking hypocleidial communication with the sternum could have provided support for an anterodorsally posteroventrally oriented flight stroke cycle that was morphologically closer to the "grabbing" motion of maniraptorans and did not or hardly extend over the dorsum." In addition, Norberg (2007) noted that the lateral-facing glenoid of *Archaeopteryx* would have allowed lifting of its "... wings well above the horizontal plane through the shoulder, preparatory to a lift-producing downstroke."

Numerous authors have examined available evidence concerning the potential gliding or flying ability of non-avian theropods and the first birds (Table 11.1). Because pennaceous feathers are a critical requirement for either gliding or flying, this narrows the list of potentially volant taxa to members of the clade Pennaraptora (Fig. 11.7). Included in this clade are the oviraptors, none of which were volant, including *Similicaudipteryx* (Dececchi et al. 2016), *Protoarchaeopteryx* (Padian 2001), and *Caudipteryx* (Talori et al. 2018).

Paravian theropods that preceded *Archaeopteryx* in the fossil record such as *Anchiornis huxleyi*, *Xiaotingia zhengi*, and *Eosinopteryx brevipenna* had pennaceous feathers and both *Anchiornis* and *Xiaotingia* have been referred to a "four-winged theropods" because they also had "hindwings," i.e., long tibial and metatarsal feathers. These non-avian theropods occurred in what is now China and lived about 155–160 million years ago. In contrast, *Eosinopteryx brevipenna* had relatively short forelimbs with

**Table 11.1** Conclusions reached in studies of the flight capabilities of *Archaeopteryx* and other Mesozoic non-avian theropods and paravians

| Taxa | Conclusion | Reasoning | Source |
|---|---|---|---|
| *Archaeopteryx* | Occasional, opportunistic flyer | "... *Archaeopteryx* is envisioned as an opportunistic flier in shoreline habitats exposed to steady trade winds all year round. By exploiting those winds *Archaeopteryx* could eliminate the 'velocity gap' between its maximal running speed (currently estimated at about 2–3 m s) and its stalling speed (or minimal flying speed, estimated at 5–6 m s). On windless days, *Archeopteryx* may have been unable to fly and would have subsisted in the role of a small theropod dinosaur adapted for hunting and foraging along the shore. Breezes (>4 m s) might have allowed *Archaeopteryx* to take off with a running or leaping start. While stronger winds (>6 m s) might have presented the option of virtually effortless take-off. Brief episodes of wind-assisted lift-off and airborne drift, controlled by inertial drag of the long feathered tail and by intermittent use of the wings, might have permitted *Archaeopteryx* to extend its range of foraging at minimal energetic cost." | Thulborn (2003) |
| *Archaeopteryx* | Limited flight capacity | In *Archaeopteryx*, the forelimb scaling to snout-vent length is not as pronounced as in present-day birds and forelimb proportions were similar to those of deinonychosaurians such as *Microraptor*. "This may be linked to the limited flight capacity, if any, of *Archaeopteryx* compared to more derived basal avians such as *Sapeornis*." | Dececchi and Larsson (2013) |
| *Microraptor* | Effective gliding flight | "The gliding flight of *Microraptor*, which our experimental results show was remarkably effective over medium distances starting from moderate heights, did not require highly derived lifting surfaces and can be regarded as representing a highly successful experiment in theropod adaptation to aerial behaviour, consistent with at least some evidence for climbing in this taxon." | Dyke et al. (2013) |

(continued)

**Table 11.1** (continued)

| Taxa | Conclusion | Reasoning | Source |
|---|---|---|---|
| *Eosinopteryx* (basal troodontid) | Not capable of flying or gliding | "With a shorter humerus and manus and a reduced plumage, *Eosinopteryx* had a much shorter wing span than other feathered paravians, including *Archaeopteryx*, *Wellnhoferia* and *Anchiornis*. The straight and closely aligned ulna-radius . . . also means that pronation/ supination of the manus with respect to the upper arm would have been limited; combined with the absence of a bony sternum and weakly developed proximal humerus, these attributes suggest that *Eosinopteryx* had little or no ability to oscillate the arms to produce a wing-beat. The phalanges of the third toe of *Eosinopteryx* decrease proximodistally and its ungual phalanges are not very recurved and particularly short, suggesting a ground dwelling, 'cursorial' mode-of-life . . ." | Godefroit et al. (2013) |
| *Archaeopteryx* | Capable of flight | "Recent work focusing on neurological adaptations for flight has identified a surficial, neornithine-like Wulst in *Archaeopteryx* that has been inferred as a neurological indicator of volancy . . . Limitations in the study of fossils alone, however, present the opportunity to integrate our first ever direct examination of Wulst activity . . . with inferences from the fossil record. We concur with the interpretations of *Archaeopteryx*." | Gold et al. (2016) |
| *Changyuraptor* | Likely capable of gliding and perhaps other semi-aerial locomotion | "The morphology of *Changyuraptor yangi* suggests that aerial abilities (be it gliding, powered flight, incline running or other semi-aerial locomotion) were not necessarily limited to small-bodied animals but were also present among more sizable animals." | Han et al. (2014) |
| *Anchiornis, Microraptor, Archaeopteryx, Jeholornis, Sapeornis, Zhongjianornis,* and *Confuciousornis* | Capable of controlled aerial behaviors | ". . . the capacity for maneuvering characterized the early stages of flight evolution, before forewings with a power stroke fully evolved . . . overall gross movement of the large tail of early paravians yielded high aerodynamic control | Evangelista et al. (2014) |

(continued)

**Table 11.1** (continued)

| Taxa | Conclusion | Reasoning | Source |
|---|---|---|---|
| | | effectiveness and the body possessed some degree of stability. Combined with likely dynamic forces and torques generated by either tail whipping at the root . . . or mild asymmetric . . . or symmetric forewing flapping (flapping limited by less robust skeletal and feather morphology or porosity), this suggests that early avialans and their ancestors were still capable of controlled aerial behaviors at high angles of attack. Gradual evolution of improved maneuvering ability. . . via consistent aerodynamic mechanisms is consistent with a continuum of aerial behaviors ranging to full aerial. The staggered acquisition of certain morphological characters (e.g., sternum ossification; pygostyle) is consistent with aerial maneuvering evolving in an incremental and continuous manner. Subsequent shifts in control would be consistent with more shallow glides facilitated by incipient wing flapping, which may have served initially in control but then ultimately became the power stroke characteristic of modern birds." | |
| Mesozoic paravians | Capable of gliding, not modern high-powered flight | "The presence of the plesiomorphic Mesozoic flight feather morphology (small cutting-edge and small trailling vane barb angles) on the hindlimbs of *Microraptor* indicates that this feather morphology was functionally associated with passive gliding, at least in that taxon . . . The observation that this plesiomorphic morphology is shared with *Archaeopteryx*, *Confuciusornis* and *Sapeornis*, raises the prospect that aerial locomotion in these taxa also consisted primarily of passive gliding; . . . our data are congruent with the idea that Mesozoic paravians stemward of Enantiornithes may have been incapable of fully modern high-powered flight." | Feo et al. (2015) |

(continued)

**Table 11.1** (continued)

| Taxa | Conclusion | Reasoning | Source |
|---|---|---|---|
| Non-avian theropods, *Archaeopteryx*, and *Microraptor* | Possibly capable of wing-assisted incline running (WAIR) and estimated to have the potential for ground-based take-off | "Biomechanical mathematical models based on known aerodynamic principals and in vivo experiments and ground truthed using extant avians" [were used to determine] ". . . if an incipient flight stroke may have contributed sufficient force to permit flap running, WAIR, or leaping takeoff along the phylogenetic lineage from Coelurosauria to birds . . . We find no support for widespread prevalence of WAIR in non-avian theropods, but cannot reject its presence in large winged, small-bodied taxa like *Microraptor* and *Archaeopteryx*." [In addition], "*Microraptor gui and Archaeopteryx* . . . were estimated to have had the potential for ground based takeoff at both sprint speeds and leaping takeoff values." | Dececchi et al. (2016) |
| *Archaeopteryx* and basal avians | Not capable of sustained flapping flight | "Complementary anatomical evidence from the position of the glenoid facet . . . and the small cross-section of the pectoralis muscle . . . indicates that *Archaeopteryx* and other phylogenetically basal avians were not capable of sustained flapping flight and complicated aerial maneuvers." | Mayr (2017) |
| Stem birds, including *Archaeopteryx* | Could have flown using flight modes of modern birds | "Our estimates of AR [aspect ratio] and WL [wing loading] show that stem birds could have flown using most of the flight modes known for modern birds, although we have found no evidence for either dynamic soarers or flighted divers (hovering was not analyzed)." | Serrano et al. (2017) |
| *Archaeopteryx* and *Microraptor* | Downstrokes provided access to trees and expanded foraging opportunities | ". . . adding a single downstroke generating only 30% bodyweight support would increase *Archaeopteryx* and *Microraptor* long jump ranges by ~20%—by investing more energy into locomotion, these proto-fliers could have expanded their foraging volumes in trees and gained critical advantages over competitors . . . extending long jumps with proto-wingbeats to increase foraging gain provides a self-reinforcing, gradual path through which protobirds could have honed their flight skills." | Chin and Lentink (2017) |

(continued)

**Table 11.1** (continued)

| Taxa | Conclusion | Reasoning | Source |
|---|---|---|---|
| *Archaeopteryx* | Actively employed wing flapping | "Our analyses reveal that the architecture of *Archaeopteryx*'s wing bones consistently exhibits a combination of cross-sectional geometric properties uniquely shared with volant birds, particularly those occasionally utilising short-distance flapping. We therefore interpret that *Archaeopteryx* actively employed wing flapping to take to the air through a more anterodorsally posteroventrally oriented flight stroke than used by modern birds." | Voeten and Cubo (2018) |
| *Archaeopteryx* and *Confuciousornis* | Capable of gliding | ". . . paravian dinosaurs and early birds progressively developed longer, stiffer, and more aerodynamic fore- and hind-limb feathers, so that the lift–drag ratio of these animals progressively became higher, allowing for more efficient gliding. The earliest birds, as exemplified by *Archaeopteryx* and *Confuciusornis*, already possessed fully developed asymmetrical flight feathers, allowing for longer distance glides, compared with the more basal paravian *Anchiornis* . . . upslope wind in mountain areas . . . or strong winds in the plain grounds provided the . . . conditions for the gliding of avian ancestors from the ground to the trees . . . smaller, feathered paravians were able to generate enough lift to glide down to the trees from the mountain slopes and even perhaps to glide up to high trees in the plain areas in strong airflow." | Shahid et al. (2019) |
| Basal avialans and paravians | Not capable of powered flight | "In basal avialans . . . and basal paravians . . ., the glenoid facet is laterally oriented. [Also,] the acrocoracoid process is represented by a small bump or process called the biceps tubercle, [and] this tubercle largely dictates the course of the tendon of the main humerus protractor . . ., the *M. supracoracoideus*. [Based on these anatomical traits, the main forelimb movements of basal avialans and paravians such as | Agnolin et al. (2019) |

(continued)

**Table 11.1** (continued)

| Taxa | Conclusion | Reasoning | Source |
|---|---|---|---|
| | | *Archaeopteryx* and *Buitreraptor* were more front to back rather than up and down so *Archaeopteryx* was] very probably not capable of powered flight." | |
| *Alcmonavis* | Primitive form of flapping flight | "The development of a pronounced tubercle for the insertion of the *m. biceps brachii* on the radius in volant basal avialans is thus consistent with the idea that these animals used a primitive form of flapping flight . . . probably starting as burst fliers . . . Thus, although a potential flight performance of the most basal avialans like *Anchiornis* is controversial . . . active flapping flight might have originated early within the lineage." | Rauhut et al. (2019) |
| Avialans, dromaeosaurids, and one troodontid | Capable of powered flight | "Using . . . phylogenetic results and available data for vaned feathered paravians, maximum and minimum estimates of wing loading and specific lift calculated using ancestral state reconstruction analysis . . . as proxies for the potential for powered flight . . ." all avialans (including *Archaeopteryx*), six dromaeosaurids (*Bambiraptor, Buitreraptor, Changyuraptor, Mahakala, Microraptor* and *Rahonavis*) and one troodontid (*Jinfengopteryx*) among the vane-feathered paravians sampled have wing loading estimates at or below the 2.5 g $cm^{-2}$ threshold for modern flapping flyers. | Pei et al. (2020) (bioRxiv) |
| *Archaeopteryx* | Flap gliding likely possible, flapping flight, if possible, was rudimentary | ". . . the primitive morphology of the glenoid, the unossified sternum, and the lack of a supracoracoideus pulley (acrocoracoid process) . . . suggest that flapping flight, if it was possible, was rudimentary. In contrast, flap gliding . . . with glides interspersed with low amplitude flapping flight . . . was likely possible. Sustained level flight at high speeds, again using low-amplitude flapping . . . may have been possible, given that flight is less expensive and flight kinematics are simpler . . . Low-speed flight, requiring more elaborate wing kinematics and high-power output, seems less likely." | Longrich et al. (2020) |

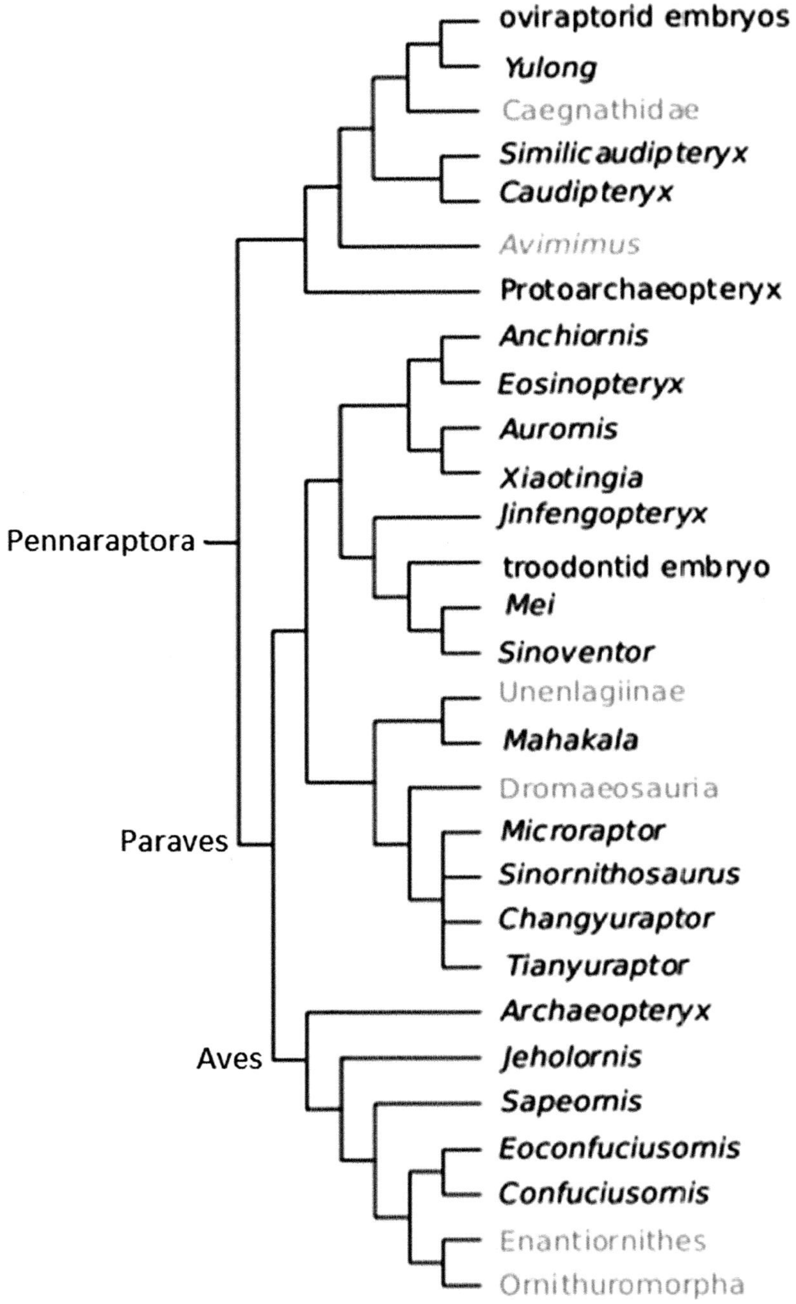


**Fig. 11.7** Taxa in the clade Pennaraptora. (Figure modified from Dececchi et al. 2016; open-access article distributed under the terms of the Creative Commons Attribution License, https://creativecommons.org/licenses/by/4.0/)

relatively few, symmetrically vaned pennaceous feathers, suggesting they were non-volant ground-dwelling theropods (Godefroit et al. 2013). Forelimb and hindlimb feathers of *Anchiornis* and *Xiaotingia* were more numerous and longer than those of *Eosinopteryx*, but were also symmetrically vaned (Hu et al. 2009) and shorter than those of more derived avialans (Hu et al. 2009). Foth et al. (2014) suggested that *Anchiornis* was probably non-volant and

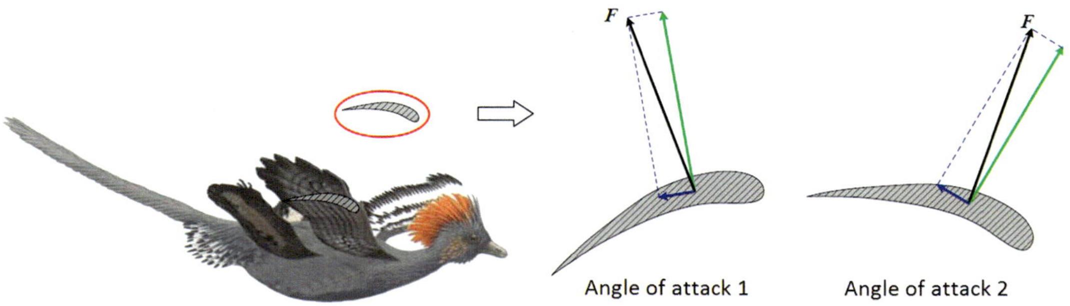


**Fig. 11.8** "Four-winged" *Anchiornis* during gliding flight. Angle of attack 1 could increase lift (green arrow) and allow *Anchiornis* to glide upward, whereas angle of attack 2 would permit longer glides. Blue arrow = drag, F = resultant force. (Figure a supplemental figure from Shahid et al. 2019; © 2019 By the Authors, open-access article distributed under the terms and conditions of the Creative Commons Attribution (CC BY) license, https://creativecommons.org/licenses/by/4.0/)

used their hindwing feathers for display, breeding, or other functions. Pei et al. (2017) also concluded that the forelimb and hindlimb feathers of *Anchiornis* probably did not serve "aerodynamic functions," and Pei et al. (2020) concluded that *Xiaotingia* would not have been capable of powered flight. Pei et al. (2020) further suggested that *Anchiornis* might have been capable of powered flight based on wing loading values and lift estimates, but also noted features that might have prevented such flight, including symmetrical feathers, relatively short ulnae and humerii compared to more derived avialans, apparently limited pectoral musculature, and lack of a bony sternum. The arrangement of feathers on the wings of *Anchiornis* also differed from that of more derived avialans, but Longrich et al. (2012) suggested that their wings, with multiple layers of wing coverts, could have still functioned as airfoils and allowed high-speed gliding or flapping flight. Similarly, noting that the forelimb pennaceous feathers of *Anchiornis* may have been open-vaned, lacking functional barbicels (Saitta et al. 2018), Lefèvre et al. (2020) hypothesized that the forelimbs of *Anchiornis* were used primarily as visual displays, but did not rule out the possibility that their wings may have allowed gliding from tree to tree. Clearly, opinions differ concerning the possibility that *Anchiornis* was volant, but they were likely not capable of powered flight and, perhaps at best, might have been able to climb trees and glide from tree to tree. Shahid et al. (2019) suggested that *Anchiornis* may have taken advantage of favorable wind conditions to glide from tree to ground or tree to tree (Figs. 11.8 and 11.9).

Additional "four-winged" non-avian theropods in the families Dromaeosauridae and Troodontidae have also been described, with the first being the dromaeosaurid *Microraptor gui* (Xu et al. 2003). Others include the dromaeosaurid *Changyuraptor yangi* (Han et al. 2014), and the troodontid *Jianianhualong tengi* (Xu et al. 2017; Fig. 11.10). Still other dromaeosaurids had large wings with pennaceous feathers, but did not appear to have vaned feathers on their legs, e.g., *Zhenyuanlong suni* (Lü and Brusatte 2015). Of these, *Microraptor* and, possibly, *Changyuraptor*, were capable of at least gliding flight (Box 11.2 Four-Winged Dromaeosaurids). However, these paravians lived tens of millions of years after birds split off from other dinosaurs so provide limited insight concerning the origin of flight in birds.

**Fig. 11.9** Arboreal gliding by *Anchiornis* and other "four-winged" theropods. (**a**) Parachuting from a branch of a tree to the ground, (**b**) Attempting unsuccessfully to glide from one tree to another, (**c**) Gliding from the higher position in one tree to the lower position in another tree, and (**d**) Changing trajectory by adjusting the angle of attack of its wing to gain lift and glide to a more distant tree. (Figure modified from Shahid et al. 2019; © 2019 By the Authors, open-access article distributed under the terms and conditions of the Creative Commons Attribution (CC BY) license, https://creativecommons.org/licenses/by/4.0/)

**Fig. 11.10** Life construction of *Jianianhualong tengi*. (Drawing by Julius T. Csotonyi, Wikipedia, licensed under the Commons Attribution 4.0 International license, https://creativecommons.org/licenses/by/4.0/)

**Box 11.2 Four-Winged Dromaeosaurids**

*Microraptor gui*, a four-winged dromaeosaur from the early Cretaceous of China, provides evidence for an arboreal-gliding origin of avian flight, with asymmetric flight feathers on the forearm as well as the legs (Xu et al. 2003). Using a model of *Microraptor gui*, Dyke et al. (2013) examined its aerodynamic performance in a wind tunnel and concluded that a "legs-down" gliding posture provided stability in flight and allowed microraptors to glide estimated distances of about 70–100 m when leaping from a perch 30 m above ground (the likely common height of trees during the late Jurassic). These authors suggested that microraptors likely spent some time foraging on the ground as well as in trees, and only occasionally used their gliding ability. More recently, Pei et al. (2020) concluded that *Microraptor* (and *Rahonavis*, another dromaeosaurid) were capable of true flapping flight. These dromaeosaurids did have a sternum, but no triosseal canal, so the deltoid (shoulder) muscles likely helped power the upstroke during flapping flight (Pittman et al. 2022).

(continued)

**Box 11.2** (continued)

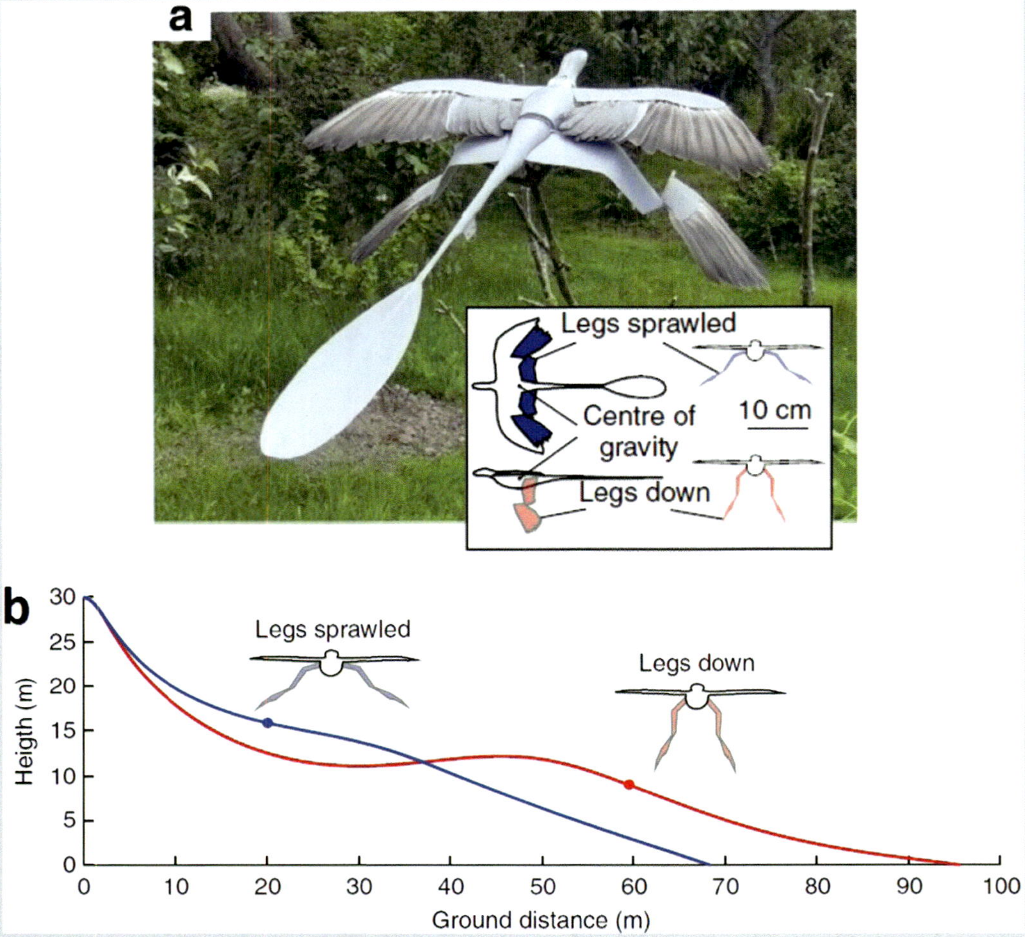


(**a**) The model of *Microraptor gui* illustrating the legs sprawled posture. (**b**) Based on data collected using a wind tunnel, the legs-down posture allowed *Microraptor* to glide greater distances. With that posture, *Microraptor* would likely have lost altitude faster after leaping, and the greater speed generated as they dropped allowed them to then glide further. (Figures from Dyke et al. 2013; © 2013 Springer Nature, used with permission)

Han et al. (2014) described a much larger dromaeosaurid theropod with four wings, *Changyuraptor yangi*. Its morphology also suggests aerial ability, e.g., gliding, powered flight, incline running, or other semi-aerial locomotion, with a long tail that would likely help compensate for small adjustments in pitch when in flight (Han et al. 2014).

(continued)

**Box 11.2** (continued)

*Changyuraptor yangi*, a four-winged predatory dromaeosaurid theropod with tail feathers nearly 30-cm long. (Figure from Han et al. 2014; © 2014 Springer Nature, used with permission)

With the exception that *Anchiornis* (Paraves) might have been capable of gliding flight, fossils have not yet revealed evidence of flight before *Archaeopteryx* (Fig. 11.11). There is no clear consensus concerning the aerial abilities of *Archaeopteryx*, but multiple investigators have concluded that *Archaeopteryx* was likely able to glide (Table 11.1; Figs. 11.12 and 11.13). Concerning the possibility of powered flight, a primary concern raised by some investigators is the orientation of the glenoid cavity that would likely have limited the amplitude of *Archaeopteryx's* wing stroke (e.g., Mayr 2017). However, others have suggested that flap gliding, with glides interspersed with low amplitude flapping flight, was likely possible (e.g., Longrich et al. 2020; Table 11.1). Gliding flight would likely have required *Archaeopteryx* to access elevated perches, possibly low conifers or, in some locations, taller conifers (Box 11.3 Solnhofen and

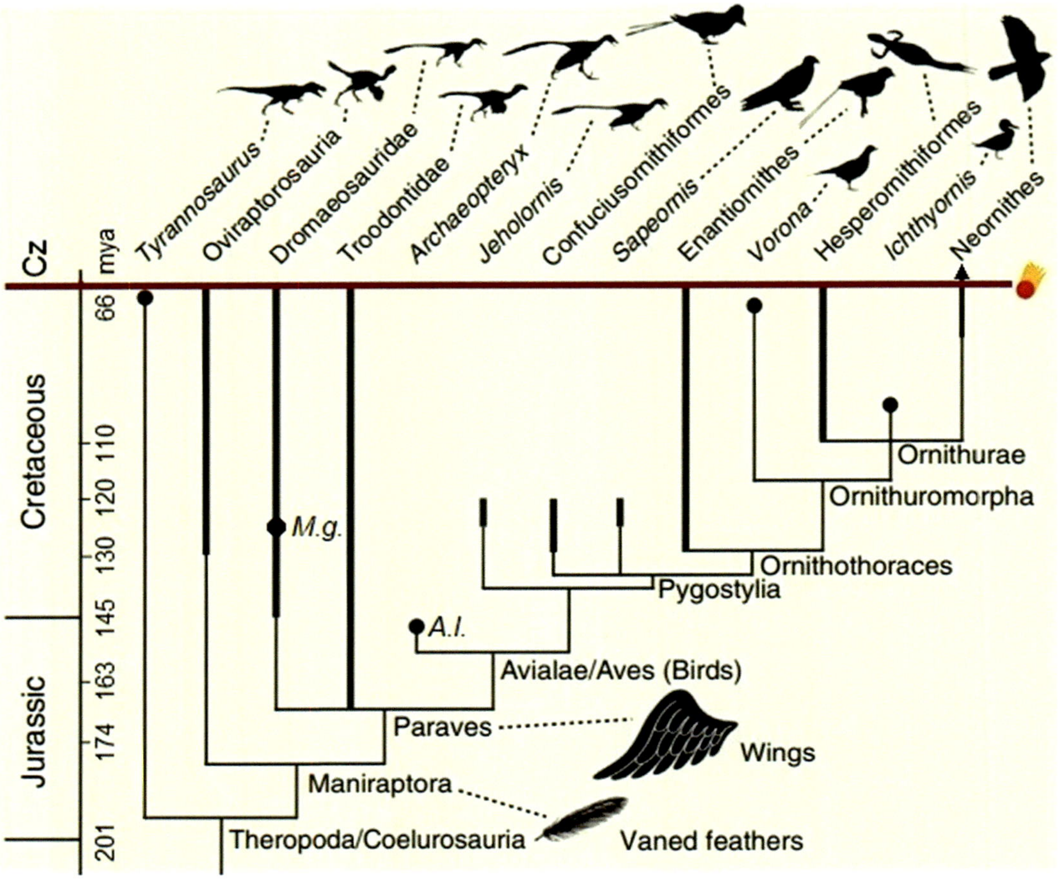


**Fig. 11.11** Phylogeny of Theropoda/Coelurosauria. *Microraptor gui* (*M.g.*) and other four-winged non-avian theropods in the families Dromaeosauridae and Troodontidae lived about 25 million years after *Archaeopteryx lithographica* (*A.l.*). The mass extinction at the Cretaceous–Paleogene boundary was caused by asteroid impact (denoted by fireball on the right). Only Neornithes survived the extinction event. Circles represent species known from a particular point in time. Cz, Cenozoic interval after the end-Cretaceous extinction. (Figure from Brusatte et al. (2015); © 2015 Elsevier Ltd., used with permission)

Jehol Paleoenvironments). *Archaeopteryx* had claws on both its forelimbs and hindlimbs, potentially contributing to an ability to climb (Abourachid et al. 2019). Based on morphology of hindlimb claws, Cobb and Sellers (2020) suggested that *Archaeopteryx* may have been largely arboreal and, based on phalangeal proportions, Fowler et al. (2011) suggested that *Archaeopteryx* may have been able to grasp elevated perches. In addition, *Archaeopteryx* may have been able to reach elevated perches using wing-assisted incline running (WAIR; Dececchi et al. 2016), although other investigators have noted that the wingbeat frequencies required for WAIR may not have been possible for *Archaeopteryx* (Nudds and Dyke 2009).

In trees, *Archaeopteryx* could have used its wings to efficiently move from branch to branch (Box 11.4 Proto-Wingbeat Long Jumping). Chin and Lentink (2017) hypothesized that, with a single downstroke, *Archaeopteryx* could increase the length of jumps by as much as 20%, providing them with access to more branches and an ability to forage more efficiently. This ability to jump

**Fig. 11.12** Reconstruction of *Archaeopteryx*. The orientation of the glenoid cavity of *Archaeopteryx* would likely have limited the amplitude of their wing strokes, but gliding flight would still have been possible. (Drawing from DataBase Center for Life Science, Wikipedia, CC BY 4.0, https://creativecommons.org/licenses/by/4.0/)

greater distances would have also allowed *Archaeopteryx* to move more efficiently to higher branches in trees, which, in turn, might have allowed them to glide greater distances.

Could *Archaeopteryx* also have taken off from the ground? A ground take-off would require a running speed equal to the minimum speed that could support *Archaeopteryx's* mass in the air. Based on examination of bone histology, Erickson et al. (2009) found that *Archaeopteryx* grew rather slowly and may have taken about 900 days after hatching to reach their maximum mass (about 900 g; Fig. 11.14). This extended period of growth explains the variation in the estimated mass of different *Archaeopteryx* specimens (range = 110–480 g; Table 11.2). Regardless of this variation in mass, estimates of wing loading for all *Archaeopteryx* specimens are within the range of those of present-day birds that use either continuous flapping flight or flap-gliding flight (Fig. 11.15). Of course, this does not necessarily mean that *Archaeopteryx* was able to take off from the ground because, as noted by Mayr (2017), their flight stroke would likely have been limited in both range (i.e., the position of the glenoid facet) and power (i.e., the small cross section of the pectoralis muscle). However, in addition to favorable wing loading, other factors also suggest that *Archaeopteryx* may have been capable of at least limited powered flight and, perhaps, ground take-offs. For example, *Archaeopteryx* had wing bones similar to those of volant birds, particularly birds that often use short-distance flapping flights (Voeten and Cubo 2018; Rauhut et al. 2019), a long tail with asymmetrical pennaceous feathers that may have provided additional lift (Gatesy and Dial 1996), and a neurological indicator of volancy (i.e., a neornithine-like Wulst; Gold et al. 2016). *Archaeopteryx* also had an estimated running speed of 3.0–3.4 m per second (Dececchi et al. 2020; Fig. 11.16) that could have aided in generating lift. In addition, *Archaeopteryx* appears to have occupied shore-line habitats with steady winds that could have further aided in generating lift (Thulborn 2003). When taking off from the ground, *Archaeopteryx* could have also benefited from ground effect that would have reduced the amount of lift required (O'Farrell et al. 2002 additional information about ground effect is provided later in this chapter).

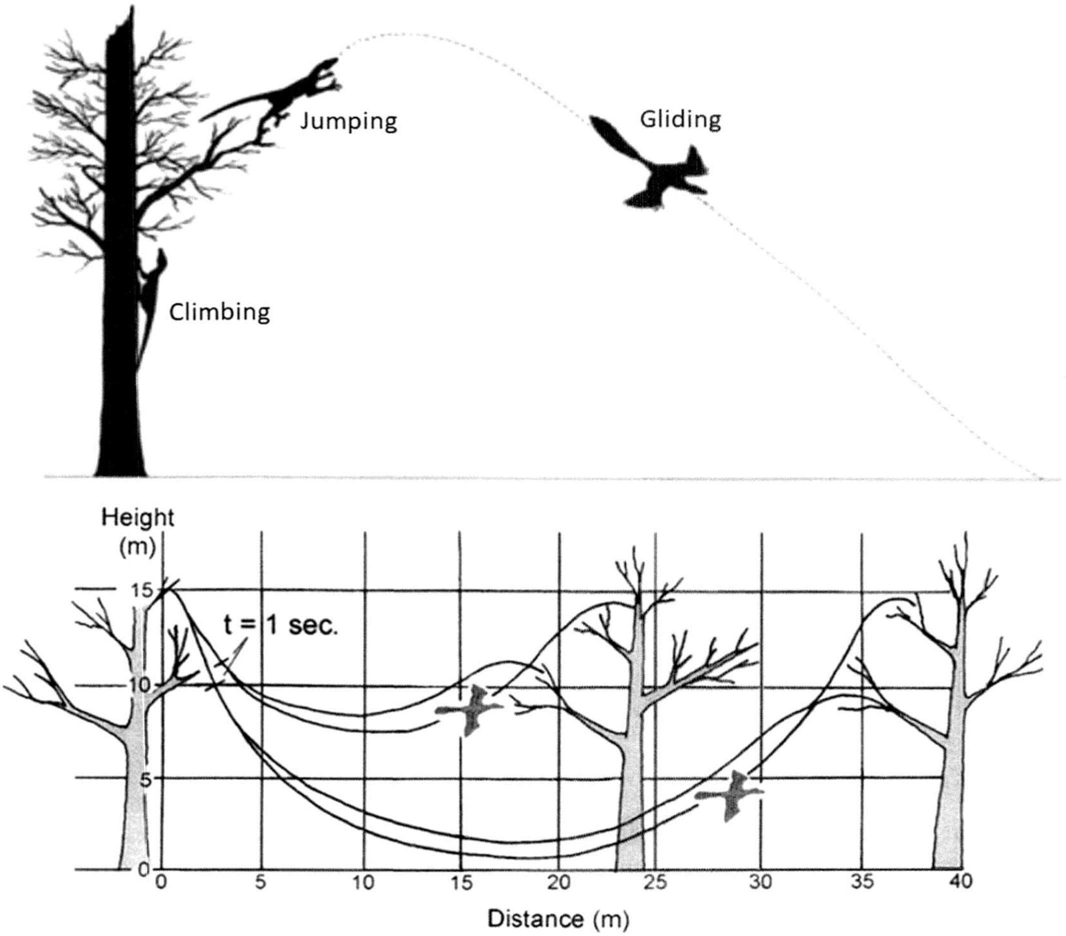


**Fig. 11.13** Top, *Archaeopteryx* climbing, jumping, then gliding. Bottom, *Archaeopteryx* may have been able to glide varying distances depending on starting height and angle of descent, with steeper angles generating more lift and allowing longer glides. (Figures from Chatterjee and Templin 2012; © 2012 Springer Science Business Media B.V., used with permission)

The possibility exists, therefore, that *Archaeopteryx* could have taken flight from trees or other elevated perches ("arboreal") and from the ground ("cursorial"), and was capable of gliding and short-distance powered flights. Beyond *Archaeopteryx*, however, the diversity of pennaraptorans and the habitats they occupied make it most unlikely that there was a single pathway to powered flight. Rather, flight almost certainly arose independently in multiple lineages, with some lineages being more arboreal and others more terrestrial. As such, the arboreal vs. cursorial debate concerning the origin of flight in non-avian theropods and birds almost certainly oversimplifies the question. The more likely answer would seem to be "all of the above," with some lineages becoming volant from the trees down, others from the ground up, and still others occupying both arboreal and terrestrial habitats and potentially using wing-assisted incline running (WAIR) to access the trees and eventually become volant. Regardless of how flight originated, after initially becoming volant, selection in some lineages would have

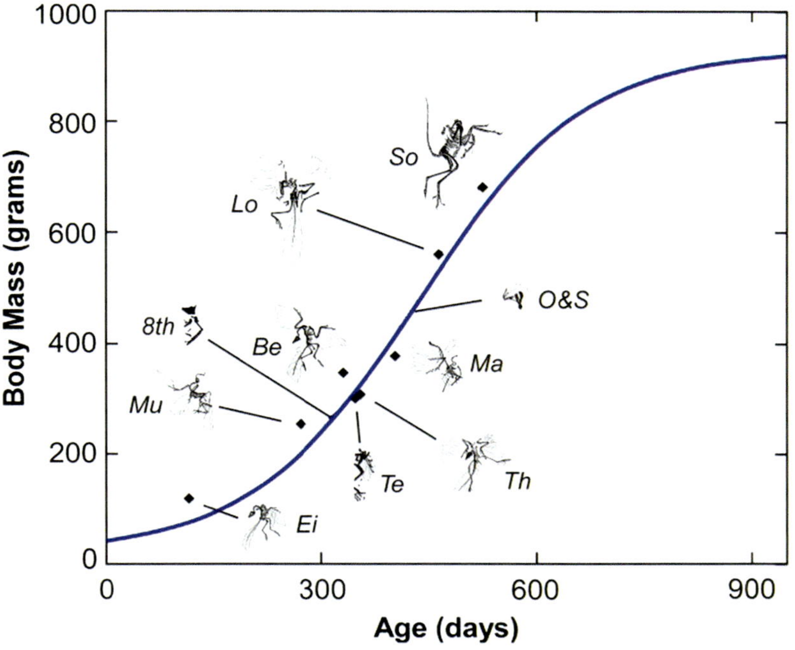


**Fig. 11.14** Size and estimated age for 10 specimens of *Archaeopteryx*. The growth curve is based on the age and size estimates (diamonds) for the eight specimens where femoral length is known. (Figure from Erickson et al. 2009; © 2009 Erickson et al., open-access article distributed under the terms of the Creative Commons Attribution License, https://creativecommons.org/licenses/by/4.0/)

favored improved flying ability, resulting in larger pectoralis and supracoracoideus muscles, associated skeletal modifications, and the other characteristics we now associate with the flight of present-day birds.

**Table 11.2** Wing measurements, body mass, wing area, and wing loading estimates for several specimens of *Archaeopteryx*. *Archaeopteryx* with bold font are specimens without preserved forelimb remiges so feather lengths were estimated based on the other specimens. For body mass estimates, "Liu" indicates estimates based on Liu et al. (2012), and "Fe" indicates estimates based on Field et al. (2013) (Table from Dececchi et al. 2016; © 2016 Dececchi et al., open-access article distributed under Creative Commons CC BY 4.0, https://creativecommons.org/licenses/by/4.0/)

| Taxa | References | Wing length (m) | Span (m) | Mass (kg) Liu | Mass (kg) FE | Wing area ($m^2$) | Wing loading ($N/m^2$) |
|---|---|---|---|---|---|---|---|
| ***Archaeopteryx*** | Foth et al. (2014) | **0.31** | **0.65** | **0.24** | – | **0.06** | 38 |
| *Archaeopteryx* | Foth et al. (2014) | 0.31 | 0.65 | – | 0.36 | 0.06 | 57 |
| ***Archaeopteryx*** | Mayr et al. (2007) | **0.29** | **0.61** | **0.23** | – | **0.06** | 38 |
| *Archaeopteryx* | Mayr et al. (2007) | 0.29 | 0.61 | – | 0.32 | 0.06 | 55 |
| ***Archaeopteryx*** | Elzanowski (2002) | **0.33** | **0.69** | **0.31** | – | **0.07** | 45 |
| *Archaeopteryx* | Elzanowski (2002) | 0.33 | 0.69 | – | 0.48 | 0.07 | 70 |
| ***Archaeopteryx*** | Mayr et al. (2007), Nudds and Dyke (2010) | **0.26** | **0.55** | **0.18** | – | **0.05** | 38 |
| *Archaeopteryx* | Mayr et al. (2007), Nudds and Dyke (2010) | 0.26 | 0.55 | – | 0.25 | 0.05 | 53 |
| ***Archaeopteryx*** | Mayr et al. (2007) | **0.27** | **0.57** | **0.19** | – | **0.05** | 36 |
| *Archaeopteryx* | Mayr et al. (2007) | 0.27 | 0.57 | – | 0.27 | 0.05 | 51 |
| ***Archaeopteryx*** | Mayr et al. (2007) | **0.19** | **0.39** | **0.11** | – | **0.02** | 47 |
| *Archaeopteryx* | Mayr et al. (2007) | 0.19 | 0.39 | – | 0.14 | 0.02 | 60 |

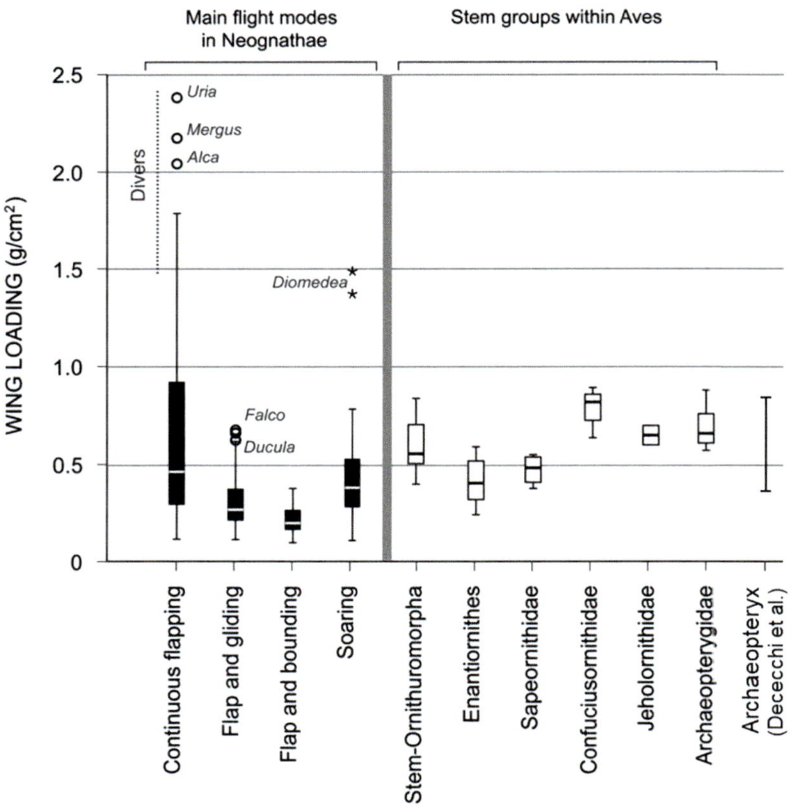


**Fig. 11.15** Wing loading values for present-day birds (according to main flight mode) and stem avian taxa. Values for *Archeopteryx* on the far right are from Dececchi et al. (2016). Box length indicates the interquartile range (25th and 75th percentiles), whisker lengths indicate the 5–95% confidence limits, open circles represent individuals out of this range, and asterisks indicate outliers. (Figure modified from Serrano et al. 2017; © 2016 The Paleontological Society, used with permission)

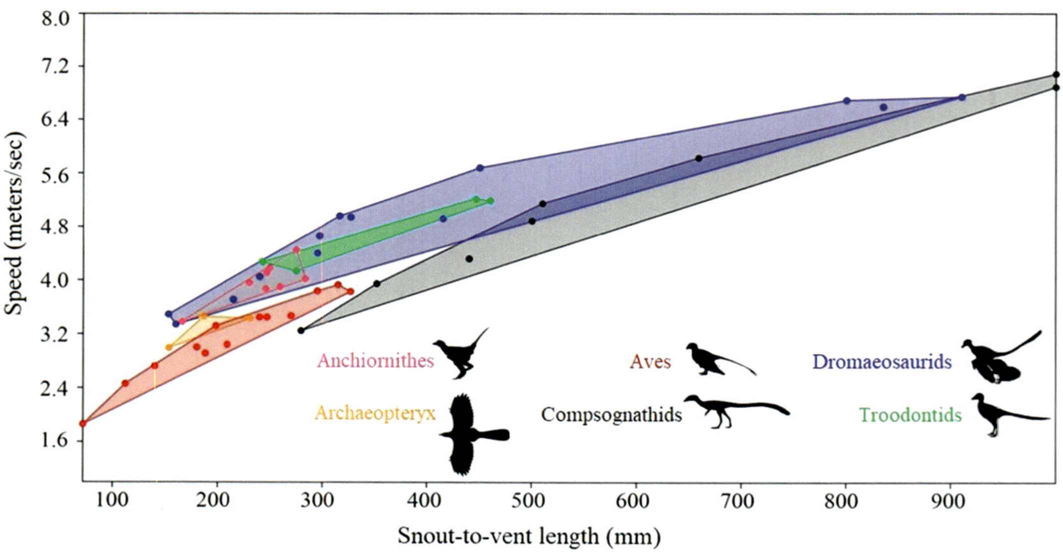


**Fig. 11.16** Comparison of the estimated maximum speeds of several clades of small coelurosaurs. Note the maximum speed of *Archaeopteryx* is estimated to be about 3.0 to 3.4 m per second (or 10.8 to 12.2 km per hour). (Figure modified from Dececchi et al. 2020; open-access article made available under the Creative Commons CC0 Public Domain dedication)

**Box 11.3 Solnhofen and Jehol Paleoenvironments**

At the time of *Archaeopteryx*, most of present-day Europe was an inland sea with two islands roughly similar in size to present-day Madagascar and a series of smaller islands, called the Solnhofen archipelago, extending west of the two larger islands. This archipelago is where *Archaeopteryx* lived. Selden and Nudds (2012) suggested that these islands were covered with shrubby vegetation consisting of low conifers adapted to a salty soil, but with no tall trees. Viohl (1985) suggested that the islands were likely arid, relatively flat, and consisted of open areas with patches of vegetation consisting of small conifers no more than 3 m in height. Fossil evidence indicates the presence of terrestrial insects and lizards that likely served as prey for *Archaeopteryx* (Rauhut et al. 2012).

Interestingly, however, the most recent specimen of *Archaeopteryx* (12th) was found about 30 km east (near present-day Schamhaupten, Germany) of where other specimens were located and, based on fossil evidence, this area was likely more terrestrial than where other specimens of *Archaeopteryx* occurred during the Jurassic. Thus suggests the possibility that some areas within the range of *Archaeopteryx* had more diverse and taller vegetation than previously suggested (Foth and Rauhut 2017). For example, Burnham (2007) suggested that at least some of the areas occupied by *Archaeopteryx* had a taller coniferous canopy of woody gymnosperms (Araucaria, Palaeocyparis, and Brachyphyllum).

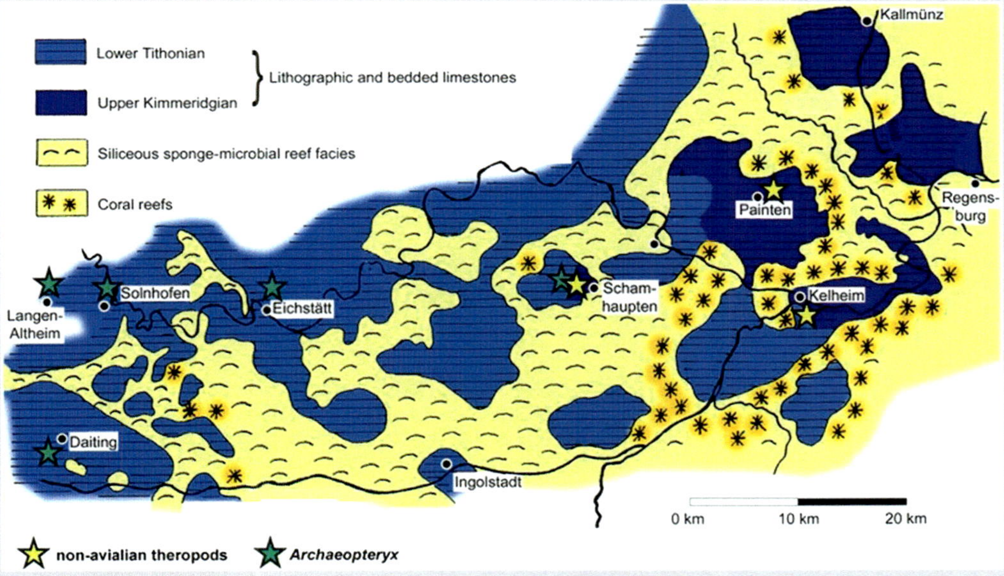


The most recent specimen of *Archaeopteryx* was located about 30 km east (near Schamhaupten) of where the other specimens were located. Based on other fossils found in the same general area, this more eastern location appears to have been more terrestrial, as indicated by abundant plant remains, lepidosaurs, and non-avian theropod dinosaurs, including *Compsognathus*, *Juravenator*, and *Sciurumimus* (Foth and Rauhut 2017; open-access article distributed under the terms of the Creative Commons Attribution 4.0 International License, https://creativecommons.org/licenses/by/4.0/)

The Jehol Biota has been defined as "... the organisms that lived in Early Cretaceous volcanic-influenced environments of northeastern China, and were buried in lacustrine and,

(continued)

**Box 11.3** (continued)

rarely, fluvial sediments, where most turned into exceptionally preserved fossils" (Pan et al. 2013). More precisely, the Jehol Biota existed from 135 to 120 Ma, and this period has been divided into early (Fuxin formation), middle (Jiufotang formation), and late (Yixian formation) periods (Yang et al. 2020). This biota occurred in what is now eastern Asia and in an ecosystem that is thought to have been largely forested because a distinctive feature of the Jehol vertebrate assemblage is the presence of several arboreal (or scansorial) and herbivorous forms including birds, pterosaurs, dinosaurs, mammals, and lizards (Zhou 2004). The oldest enantiornithine and ornithuromorph birds in the Jehol Biota first appeared about 131–129 Ma (Yang et al. 2020), Plant fossils suggest that much of the Jehol ecosystem consisted of sloping, open fields and forests with numerous volcanogenic and large lakes (Xu et al. 2020). In addition, fossils of mountain-dwelling insects indicate the presence of relatively high-altitude mountains (Xu et al. 2020), and the presence of conifers and wood fossils suggest there were seasonal climatic changes (Tian et al. 2015). Approximately 75% of the birds identified in the Jehol assemblage were arboreal, whereas fewer than 25% (mainly ornithurines) were terrestrial or lake-shore dwellers (Zhonghe and Yuan 2010).

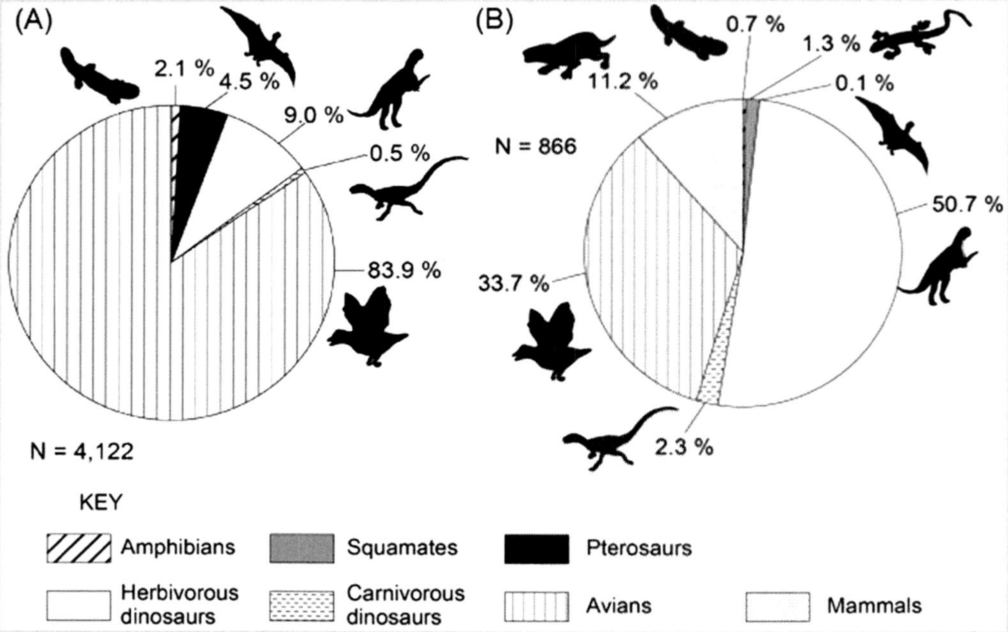


Ratios of the vertebrate faunas based on estimated population densities during the (**a**) middle (Jiufotang formation) and (**b**) late (Yixian formation) periods of the Jehol Biota. Note that birds dominated during the middle period whereas herbivorous dinosaurs became more prevalent during the late period. (Figure modified from Matsukawa et al. 2014; © 2014 Oxford University Press, used with permission)

**Box 11.4 Proto-Wingbeat Long Jumping**

Birds foraging in trees and shrubs often hop or jump from branch to branch, and sometimes use a wingbeat or two (or more) to increase the distance of these jumps. For short jumps, the legs and leg muscles can generate the momentum (or bodyweight impulse, which is a product of a force and the time over which the force is applied) needed to make it to the next branch. With increases in the distance of jumps, however, the wings must contribute.

(continued)

**Box 11.4** (continued)

A Pacific Parrotlet (*Forpus coelestis*) using a single wingbeat to increase the distance of a "long jump" from one branch to another. (Figure assembled from screen-captures from a video by Diana Chin; used with permission of Diana Chin).

By increasing the distance they could jump, either in trees or on the ground, the proto-wings of the early ancestors of birds and the better-developed wings of somewhat later ancestors such as *Microraptor* and *Archaeopteryx* may have improved foraging efficiency (Chin and Lentink 2017). In trees, the ability to jump further would allow a forager, especially smaller ones, to access more branches when foraging and, if a prey item was spotted on another branch, to potentially capture more prey. On the ground, longer jumps might have increased the likelihood of capturing prey that were being pursued. These potential advantages would have increased as the ancestors of birds became smaller and/or wing surface area increased (Chin and Lentink 2017). Such changes, and a continuing improvement in "jumping" ability could have also, over the long-term, contributed to the development of powered flight.

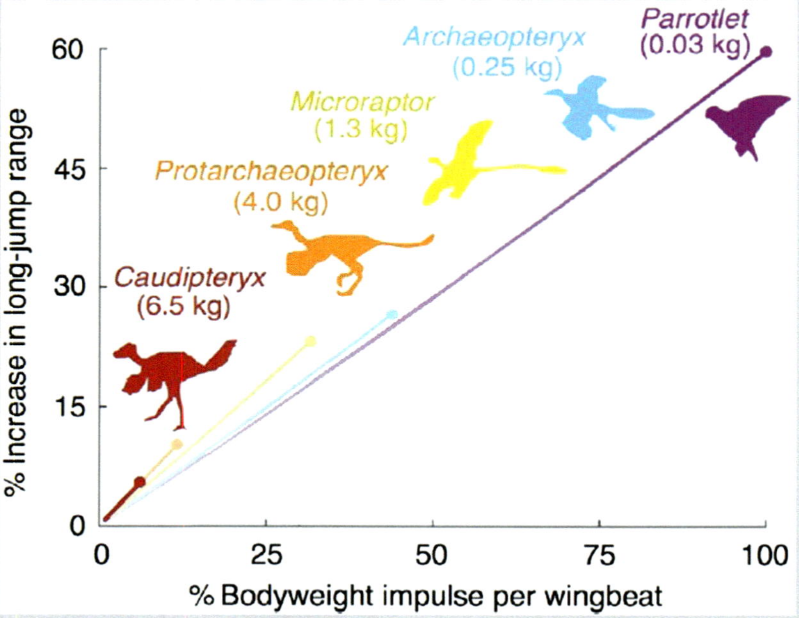


Even one wingbeat or, in the case of the early ancestors of birds such as *Caudipteryx* and *Protoarchaeopteryx*, one beat of a proto-wing can increase the distance traveled when jumping. Bodyweight impulse is a product of a force (in this case, generated by the wings) and the time over which the force is applied. (Figure modified from Chin and Lentink 2017; © American Association for the Advancement of Science, used with permission)

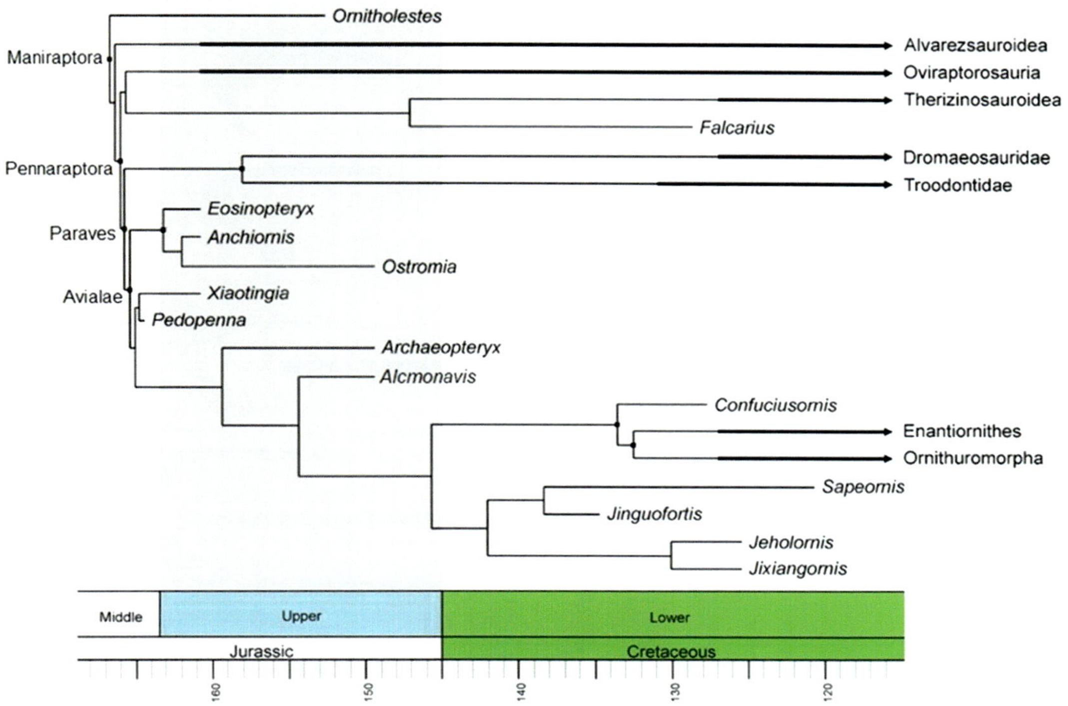


**Fig. 11.17** Time-calibrated, simplified reduced consensus tree proposed by Rauhut et al. (2019). Based on characteristics of the radius, Rauhut et al. (2019) proposed that *Alcmonavis* was capable of "a primitive form of flapping flight." (Figure modified from Rauhut et al. 2019; © Rauhut et al., open-access article distributed under the terms of the Creative Commons Attribution License, https://creativecommons.org/licenses/by/4.0/)

An example of how evolution began gradually improving the flying ability of birds is *Alcmonavis poeschi*, an avialan theropod that lived several million years after *Archaeopteryx* and whose fossilized right wing was discovered in southern Germany in 2017 (Fig. 11.17). Analysis revealed more pronounced attachment sites (compared to *Archaeopteryx*) for the pectoralis muscle on the humerus and for the biceps brachii muscle on the radius, suggesting "an early improvement in flapping flight capabilities already in the Late Jurassic" (Rauhut et al. 2019). *Alcomonavis* also had more robust metacarpals and phalanges that serve as attachment sites for the primary feathers (Rauhut et al. 2019).

Over the next several million years, the flying abilities of birds continued to improve. For example, the angle between the scapula and coracoid decreased and altered the orientation of the glenoid cavity to allow a greater range of motion of the wing, especially in the dorso-ventral plane (Fig. 11.18). The sternum developed a keel, providing an attachment site for the primary flight muscles (pectoralis and supracoracoideus; Fig. 11.19). Several taxa of birds coexisted during the Cretaceous, including confuciusornithiforms, sapeornithiformes, jeholornithiformes, and enantiornithines, and varied in the degree to which they could fly. Enantiornithines were especially adept flyers, matching the abilities of present-day birds. For more information about these birds and their flying abilities, see Chap. 1.

## 11.3 Flying

Birds in flight must generate lift (upward force) and thrust (forward force) to overcome gravity and drag (friction). Perhaps the most fascinating of these forces is lift.

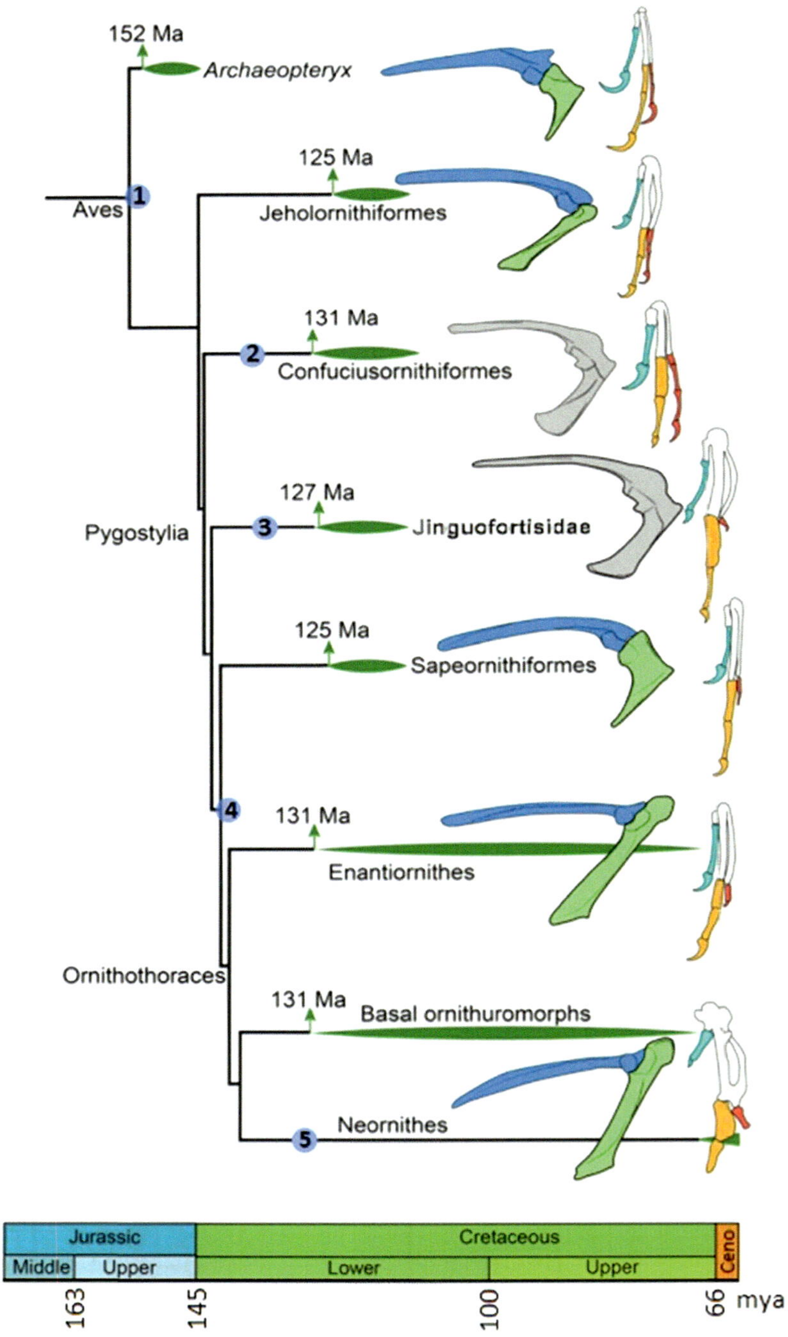


**Fig. 11.18** Cladogram of basal Aves showing changes to the pectoral girdle and hand. Thick green lines indicate the temporal range of each group. Changes to the pectoral girdle and hand include (1) scapula and coracoid are separate and form an angle of ~90°, (2,3) fusion to form scapulocoracoid, (4) scapula and coracoid no longer fused and form an acute angle, and (5) fusion of carpals and metacarpals to form carpometacarpus; loss of some phalanges. (Figure modified from Wang et al. 2018a; used with permission of the US National Academy of Sciences)

### 11.3.1 Lift

Lift is generated primarily by a bird's wings (with, in some cases, some contribution from the tail), and wing morphology varies widely (Fig. 11.20). To generate lift while minimizing friction (or drag), a wing shaped like an airfoil (in cross section) is needed. An airfoil is a tear-

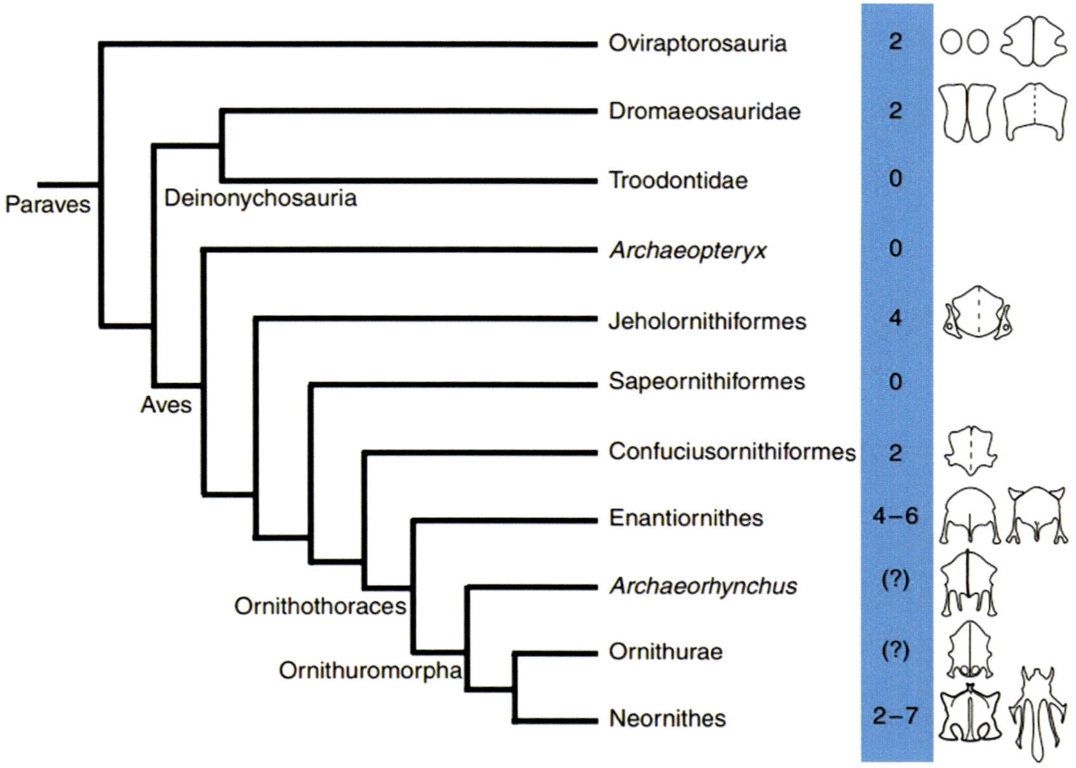


**Fig. 11.19** Evolution of sternal morphology in Paraves. Numbers in blue column represent the number of ossification centers, and dashed lines indicate where regions of ossification fused. (Figure from Zheng et al. (2012); © 2012 Springer Nature, used with permission)

drop-shaped structure that, from the front (leading edge) to the back (trailing edge), becomes progressively thinner (Fig. 11.21). Air moving past an airfoil will generate lift if the leading edge is higher than the trailing edge or, in other words, if the angle of attack (Fig. 11.22) is greater than 0°.

Many explanations of lift focus on the Bernoulli effect (or principle) which states that an increase in the speed of a fluid or air causes a reduction in the static pressure. However, one misconception is that, with a cambered wing (Fig. 11.23), air flows faster over the top because of the longer path along the top of the wing than the bottom of the wing (i.e., air molecules on top move faster to keep up with molecules flowing below) and this, in turn, means that pressure is greater under the wing than on top and lift is generated. This, however, is not correct. Air does flow faster over the top of a cambered wing than along the bottom, but not because the air molecules on top are trying to keep up with those flowing along the bottom of the wing! Rather, air flows faster over the top of a cambered wing because the flow area above the wing becomes narrower than the flow area under the wing and, when the area through which air narrows, the air flows faster (Fig. 11.24). Because mass can neither be created nor destroyed (conservation of mass principle), when the area through which air passes narrows or widens, the air must either speed up or slow down to maintain a constant amount of air moving through the area.

Lift is also generated when wings direct air downwards. When the leading edge of a wing is higher than its trailing edge, air is deflected downward and, because of Newton's Third Law of

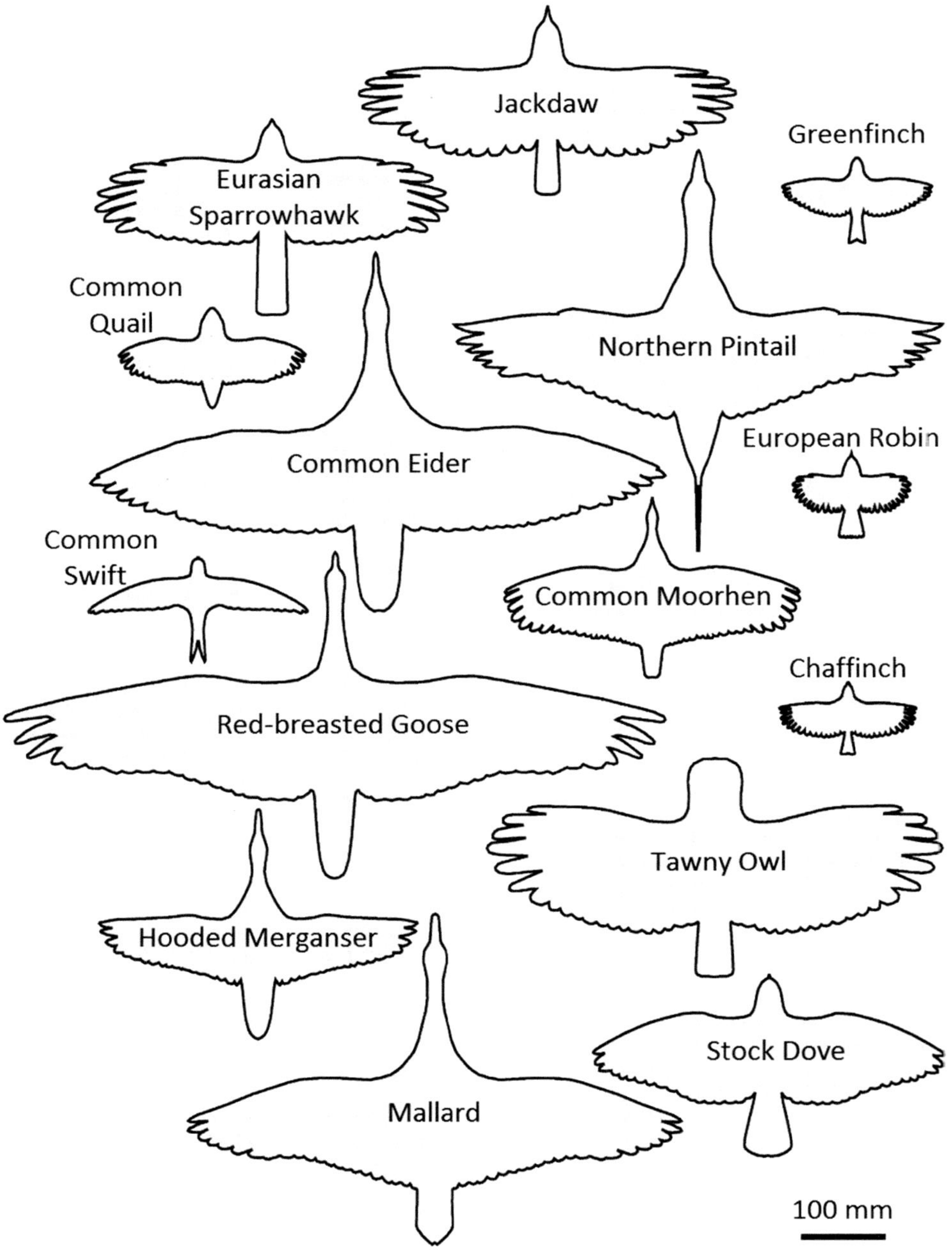


**Fig. 11.20** Examples of variation in wing morphology of different species of birds. Some wings are more pointed at the tips, others more rounded, some are broad relative to their length, and others are narrow relative to their length. All, however, must generate lift. Scientific names: Eurasian Jackdaw, *Corvus monedula*; Eurasian Sparrowhawk, *Accipiter nisus*; Common Quail, *Coturnix coturnix*; Common Eider, *Somateria mollissima*; Common Swift, *Apus apus*; Red-breasted Goose, *Branta ruficollis*; Hooded Merganser, *Mergus cucullatus*; Mallard, *Anas platyrhynchos*; Stock Dove, *Columba oenas*; Tawny Owl, *Strix aluco*; Common Chaffinch, *Fringilla coelebs*; Common Moorhen, *Gallinula chloropus*; European Robin, *Erithacus rubecula*; Northern Pintail, *Anas acuta*; European Greenfinch, *Chloris chloris*. (Figure modified from Thomas and Taylor 2001; © 2001 Academic Press, used with permission)

**Fig. 11.21** An airfoil in a wind tunnel with lines of smoke showing movement of air past the airfoil. The shape of the airfoil minimizes friction (drag) because the air flows smoothly past both the top and bottom. However, this airfoil is generating no lift because the streams of air are moving at equal speeds across the top and bottom (so there is no difference in air pressure) and, with its horizontal position (i.e., angle of attack [as determined by the angle of a line from the front of the airfoil to the back] = 0°), the air is not being deflected downward (which would create the equal and opposite reaction of "pushing" the airfoil upward). (Figure is a screenshot from a video posted on YouTube by H. Babinski at the Dept. of Engineering at the University of Cambridge in 2008; https://www.youtube.com/watch?v=6UlsArvbTeo; used with permission of Holger Babinsky)

Motion (for every action there is an equal and opposite reaction), the wing is deflected upward. Both the upper and lower surfaces of the wing deflect the air. The upper surface deflects air down because the airflow "sticks" to the wing surface and follows the tilted wing (the Coanda effect; Fig. 11.25).

Because of inertia, the faster-moving air over the top of the wing tends to keep moving in a straight line while, simultaneously, the Coanda effect tends to keep the air against the top of the wing. That inertia, however, keeps the air moving over the wing from "pushing" against the top of the wing with as much force as it would if the wing was not moving. This creates an area of

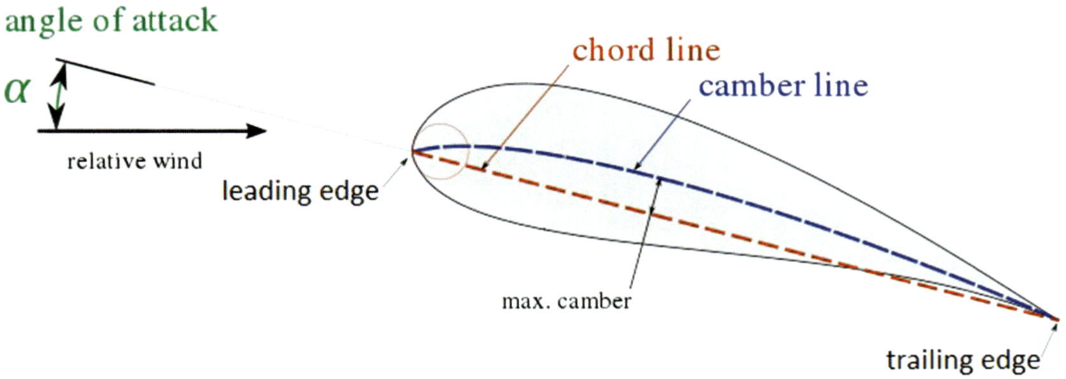


**Fig. 11.22** The angle of attack of an airfoil is the angle between the chord line of a cambered airfoil (formed by a line between the leading edge and trailing edge) and the oncoming flow of air. (Image by Olivier Cleynen, Wikipedia, CC0 Public Domain)

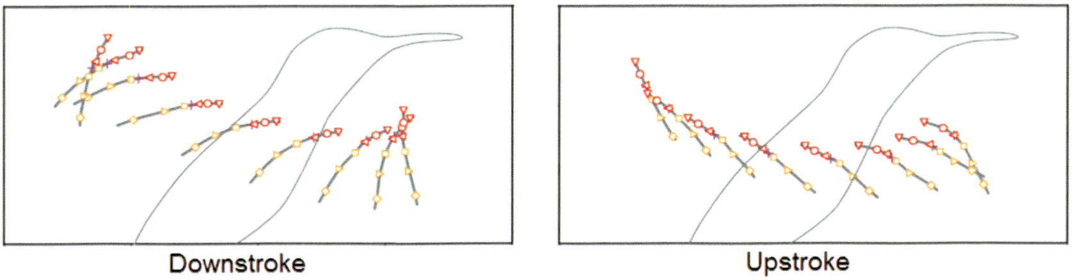


**Fig. 11.23** Drawings of the wing cross section of an Amazilia Hummingbird (*Amazilia amazilia*) in hovering flight showing the cambered shape of the wing during both the downstroke and upstroke. Symbols represent different wing feathers. (Figure from Maeda et al. 2017; © 2017 The Authors. Published by the Royal Society under the terms of the Creative Commons Attribution License 4.0, https://creativecommons.org/licenses/by/4.0/)

lower pressure above the wing. Because air tends to move from areas of high pressure to areas of low pressure, air tends to move from the high pressure area below and ahead of the wing to the lower pressure area above and behind the wing. This air moves, therefore, toward the trailing edge of the wing, or the same direction as the airflow created by the wing's motion. As a result, air flows faster over the top of the wing. Because air under the wing is dragged slightly in the direction of travel, it moves slower than does the air moving over the top of the wing (Fig. 11.26). The faster-moving air going over the top of the wing exerts less pressure than the slower moving air under the wing and, as a result, the wing is pushed upwards by the difference in

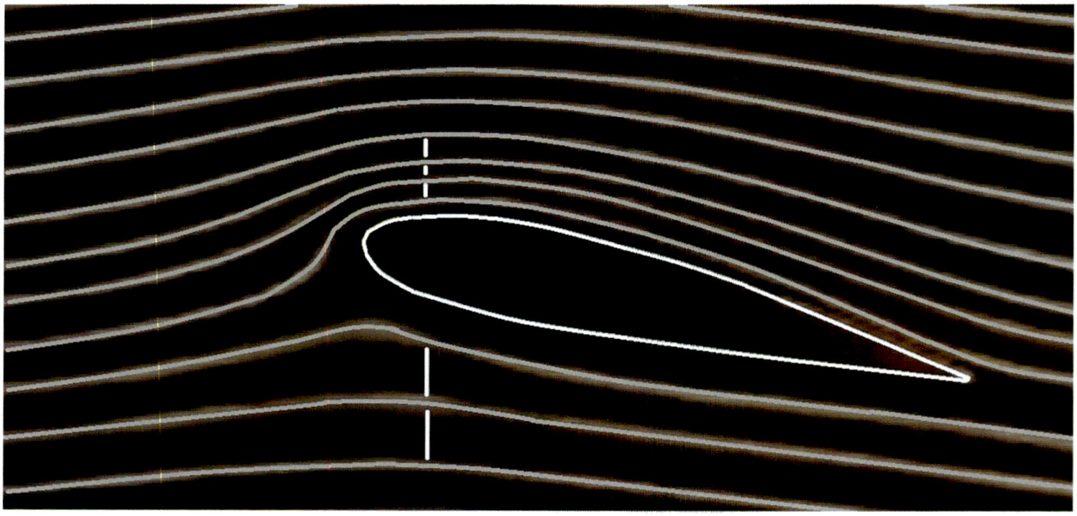

**Fig. 11.24** As air hits the leading edge of a wing, the flow area above the wing becomes narrower than the flow area under the wing and, when the area through which air narrows, the air flows faster. (Figure is a screenshot from a video posted on YouTube by H. Babinski at the Dept. of Engineering at the University of Cambridge in 2008; https://www.youtube.com/watch?v=6UlsArvbTeo; used with permission of Holger Babinski)

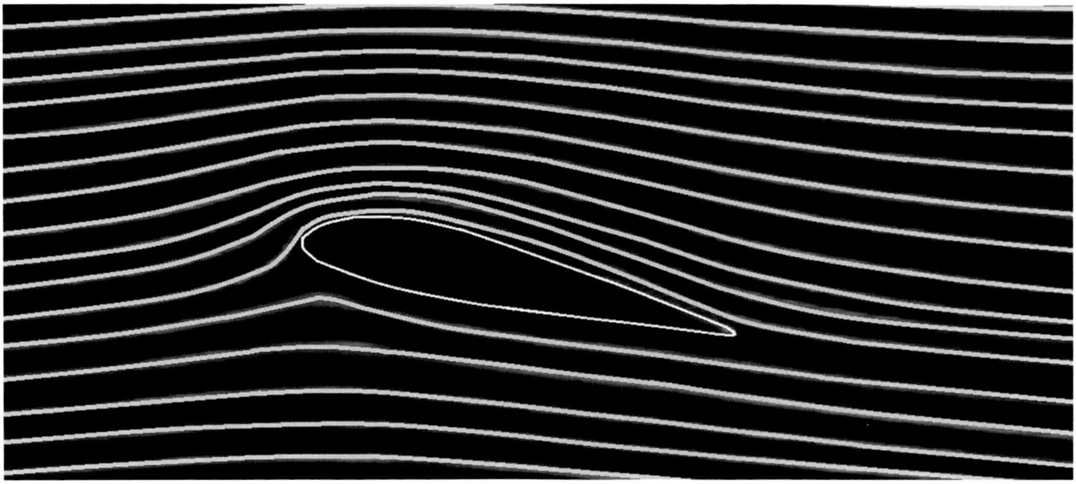

**Fig. 11.25** Smoke lines moving over and under an airfoil with an angle of attack of about 5°. Note how the air (smoke line) hits the leading edge of the airfoil, is deflected up, but then moves smoothly along the top of the airfoil. This illustrates the Coanda effect, where water or air molecules tend to be attracted to a nearby surface (this effect is named after Henri Marie Coandă, a Romanian scientist who recognized its importance in aerodynamics). (Figure is a screenshot from a video posted on YouTube by H. Babinski at the Dept. of Engineering at the University of Cambridge in 2008; https://www.youtube.com/watch?v=6UlsArvbTeo; used with permission of Holger Babinski)

pressure between the top and the bottom (the Bernoulli effect). So, both the development of low pressure above the wing (Bernoulli's principle) and the wing's reaction to the downward-deflected air below it (Newton's third Law) generate forces that create lift.

Several factors influence the amount of lift a wing generates, including angle of attack, the amount of camber, air speed, and, of course, wing morphology or shape (Box 11.5 Wing (and Feather) Morphing). Increasing the angle of attack increases the force at which air is pushed downward and, therefore, the force at which the wing is pushed upward. For birds, the optimum angle of attack is likely similar to that for airplane wings, or about 5–15°. If the angle of attack exceeds that, air flow tends to separate from the wing, less lift is generated, and stalling may result (Fig. 11.27). Birds also tend to stall at low speeds because slower moving air may not move smoothly over the wing. At low speeds (such as during take-off and landing) where the angle of attack is high, maintaining smooth air flow is critical to maintaining lift. A structure at the leading edge of bird wings, called the alula, helps maintain smooth air flow over the wing (Box 11.6 Alulas in the Fossil Record). The alula is formed by feathers (usually 3 or 4, but ranging from 2 to 6) attached to the first digit. When these feathers are elevated, they keep air moving smoothly over the wing and help birds maintain lift (Fig. 11.28, Box 11.7 Interspecific Differences in Alulae). Early investigators suggested that the alula maintained airflow over the wing like the leading-edge, retractable slats on aircraft wings that are used during take-off and landing (i.e., at lower air speeds; e.g., Alvarez et al. 2001). However, Lee et al. (2015) suggested that the alula maintains airflow over the wing by generating vortices that help prevent flow separation at high angles of attack (Fig. 11.29).

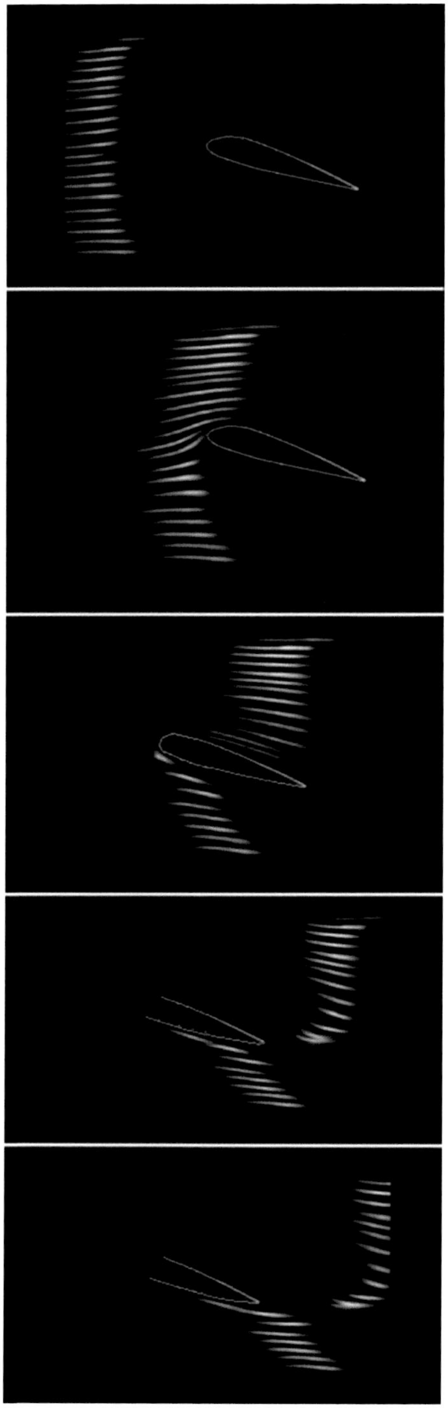

**Fig. 11.26** Smoke lines moving over and under an airfoil with an angle of attack of about 15°. From top to bottom, note how the air (pulsed smoke lines) moves faster over the top of the wings than under the bottom, generating lift via the Bernoulli effect. Friction generated as the air hits the bottom of the wing slows the air. Also note how the air tends to flow smoothly over the top of the wing despite the 15° angle of attack. Air molecules tend to stick to the wing surface because of the Coanda effect. Finally, note that the air moving past the trailing edge of the airfoil is directed only slightly downward because this airfoil is not cambered. (Figure generated from screenshots of a video posted on YouTube by H. Babinski at the Dept. of Engineering at the University of Cambridge in 2008; https://www.youtube.com/watch?v=6UlsArvbTeo; used with permission of Holger Babinsky)

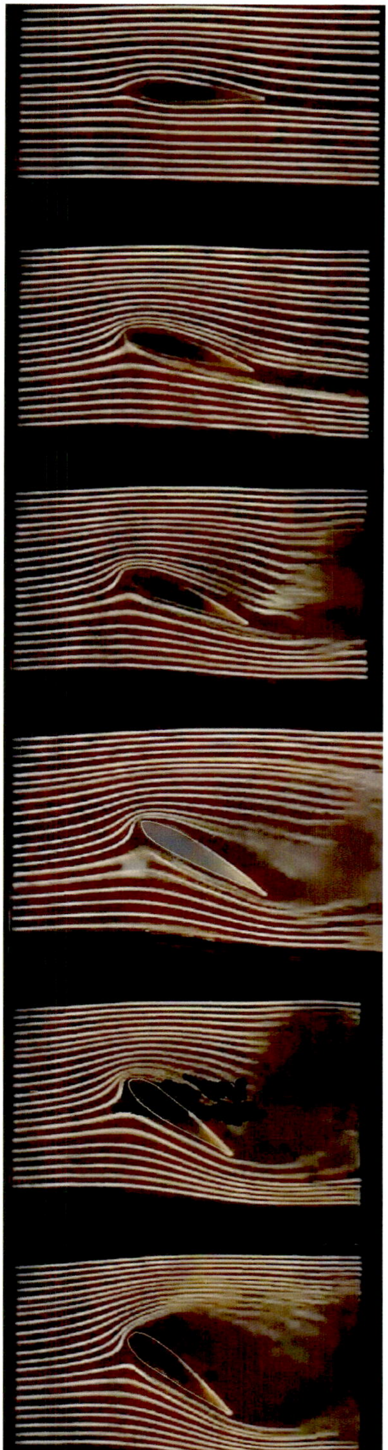

**Fig. 11.27** If the angle of attack is too great, air flow becomes detached from the top of the wing and more turbulent and less lift is generated. (Figure generated from screenshots of a video posted on YouTube by H. Babinski at the Department of Engineering at the University of Cambridge in 2008; https://www.youtube.com/watch?v=6UlsArvbTeo; used with permission of Holger Babinsky)

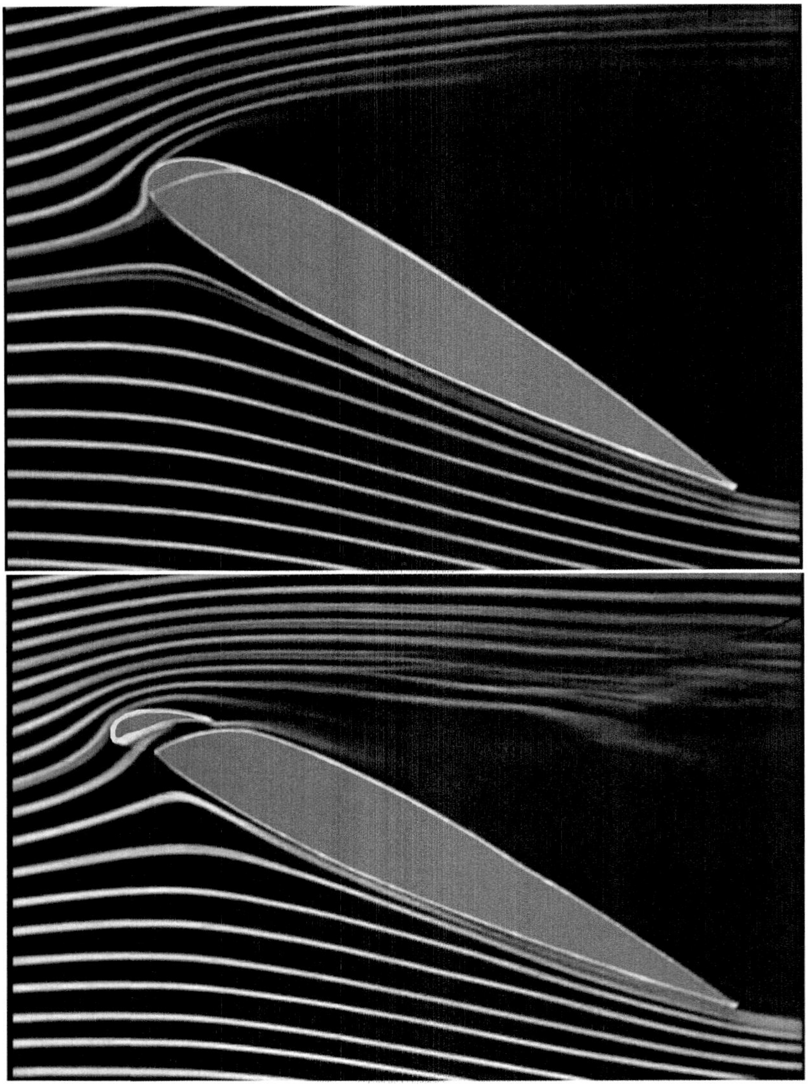

**Fig. 11.28** At high angles of attack (e.g., when taking off and landing), birds elevate their alulas to help maintain smooth air flow of the top of the wing which, in turn, helps maintain lift. Compare the top image with the "alula" (a leading-edge slot) not elevated to the bottom image with the alula elevated. In the top image, no lift would be generated, i.e., stalling, because the air flow has separated from the top of the airfoil. In the bottom image, some lift would still be generated because air is still moving across the top of the airfoil. (Images are screen-captures from "1930s test conducted at NASA Langley Research Center's Transonic Tunnel during its NACA era," CC0 Public Domain)

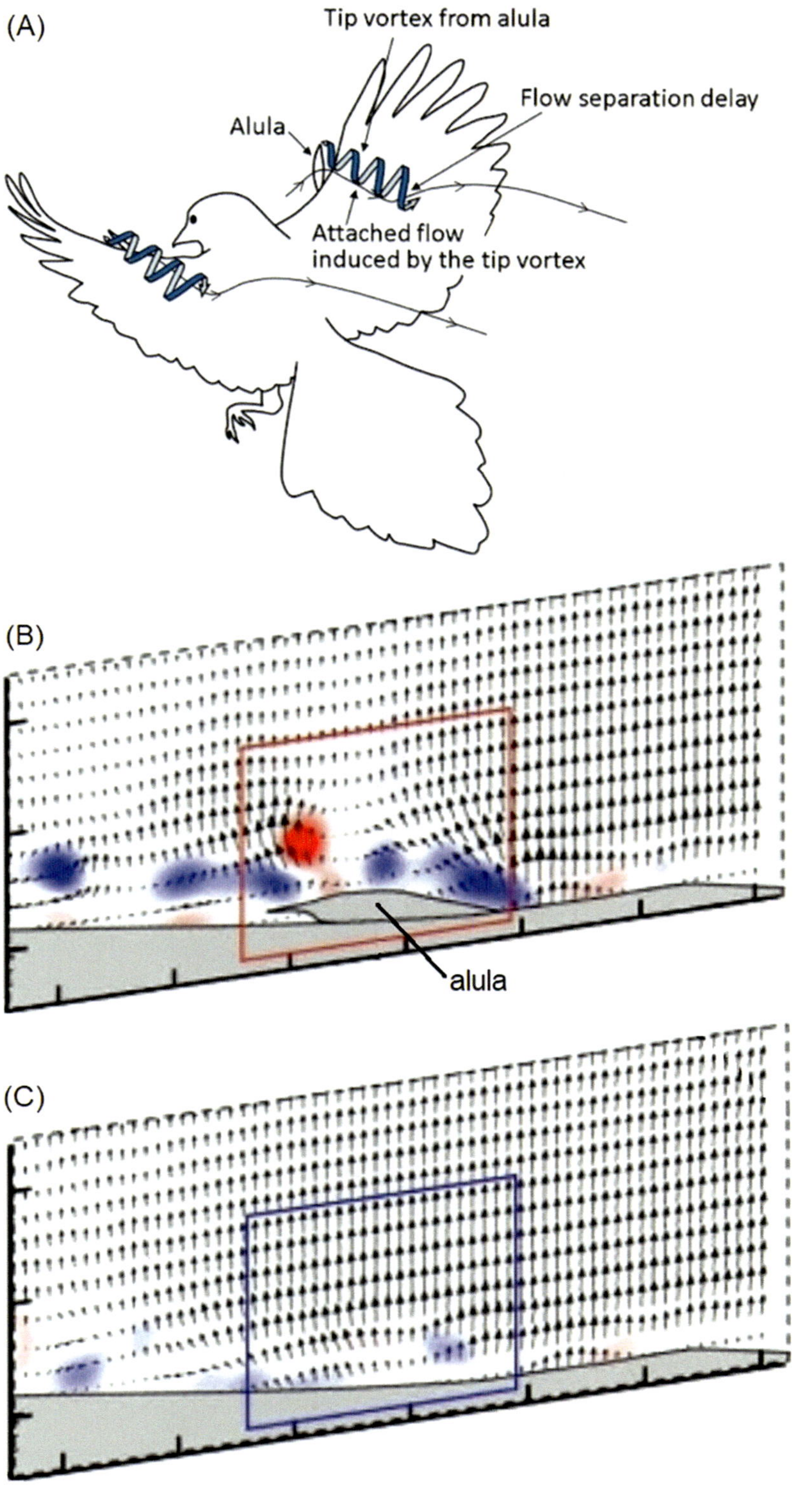


**Fig. 11.29** (**a**) Alula creating tip vortices that help maintain airflow over the wing at high angles of attack. (**b**) Tip vortex created by the alula that directs airflow down toward the wing to prevent flow separation. Note the direction of the arrows with the blue areas indicating airflow beginning to rotate down toward the wing. (**c**) In the absence of an alula, no vortex is created, air flow separates (as indicated by the arrows) from the wing. Angle of attack in B and C is 24° and air is moving from left to right. (Figure modified from Lee et al. 2015; open-access article licensed under a Creative Commons Attribution 4.0 International License, https://creativecommons.org/licenses/by/4.0/)

**Box 11.5 Wing (and Feather) Morphing**

Birds can change the shape of their wings because wings have joints, many muscles, and spreadable feathers. Entire wings can fold, extend, and twist, and, at the wrist joint, primaries can be flexed or extended.

Wing morphing by a gull in flight. (Figure from Harvey et al. 2019; © 2019 The Authors. Published by the Royal Society. All rights reserved, used with permission)

During flapping flight, especially at slower speeds, birds typically fold their wings during the upstroke by flexing the elbow and adducting the wrist. This folding reduces wing area which, in turn, reduces the effort and energy needed to move the wing upward. For soaring birds, wing morphing allows them to change their glide speed and glide angle. For example, a Harris' Hawk (*Parabuteo unicinctus*) gliding in a tilted wind tunnel folded its wings as air speed increased, reducing its wing area and reducing lift and drag to maintain a constant speed and sink rate (Tucker and Heine 1990). Folding also varies with flight speed. For example, the extended wings of swifts generate more lift and so are better at lower speeds whereas folding, or swept, wings are better at high speeds and for sharp turns (Lentink et al. 2007). Peregrine Falcons (*Falco peregrinus*) fold their wings when stooping at prey to minimize drag and maximize their speed (Gowree et al. 2018). Wing folding can also help maintain flight stability in turbulent air (Reynolds et al. 2014) and is useful when flying through cluttered habitats like forests (Williams and Biewener 2015).

(continued)

**Box 11.5** (continued)

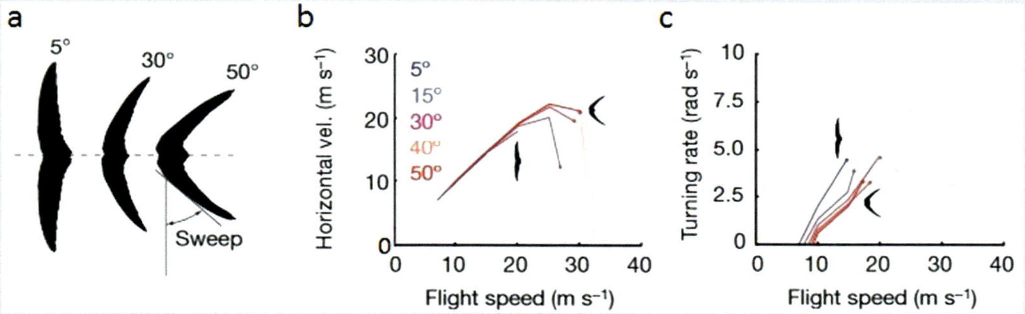


(**a**) Swifts can change the sweep angle of their wings to alter lift and drag. (**b**) Extended wings are best at low flight speeds and swept wings are best at higher speeds. (**c**) Swept wings also allow sharper turning angles. Solid lines, glide angle <45°; dot, glide angle = 45°; no dot, wing fails before 45° is reached; faint lines, glide angle >45°. (Figures from Lentink et al. 2007; © 2007 Springer Nature, used with permission)

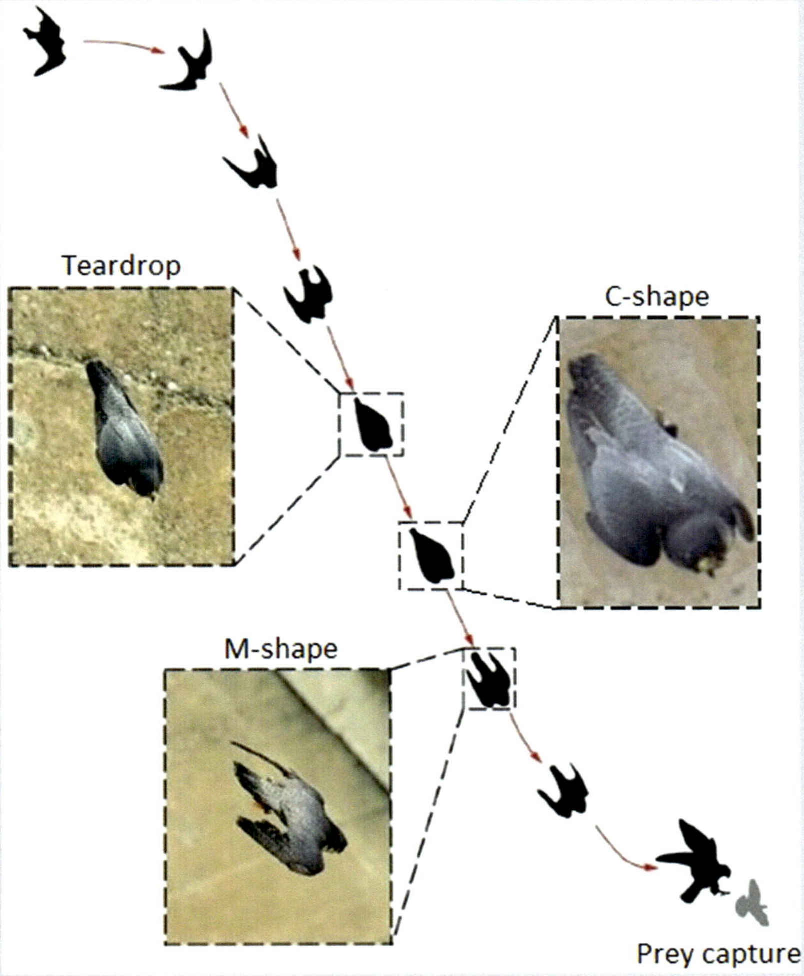


(continued)

**Box 11.5** (continued)

Flight path of a stooping Peregrine Falcon (*Falco peregrinus*). With extended wings that maximize lift, Peregrine Falcons ascend to high altitude. As the stoop is initiated, they adopt a "teardrop" shape with wings folded to minimize drag and maximize speed. As they begin to pull out of the stoop, they adopt a cupped-wing shape, or "C-shape," with wings slightly unfolded to create a space between the wings and body. Shortly thereafter, they adopt an "M-shape" with the wings unfolding a bit more. This increases drag and slightly reduces their speed, but increases maneuverability and increases the likelihood of a successful attack. (Figure modified from Gowree et al. 2018; open-access article under a Creative Commons Attribution 4.0 International License, https://creativecommons.org/licenses/by/4.0/; C-shape image from Ponitz et al. 2014; © 2014 Ponitz et al., open-access article distributed under the terms of the Creative Commons Attribution License, https://creativecommons.org/licenses/by/4.0/).

As wings morph, flight feathers must slide past each other, but still maintain contact to prevent the formation of gaps between feathers. Matloff et al. (2020) found that underlapping flight feathers of birds have numerous lobate cilia that lock and unlock with hooked rami of overlapping feathers as they slide past each other. Importantly, the frictional forces between these microstructures, referred by Matloff et al. (2020) as "directional Velcro," are sufficient to maintain contact between overlapping flight feathers, prevent gaps, and make morphing wings resistant to turbulence, but not sufficient to greatly increase the energetic cost of wing morphing (i.e., the force of muscle contraction required to move the feathers).

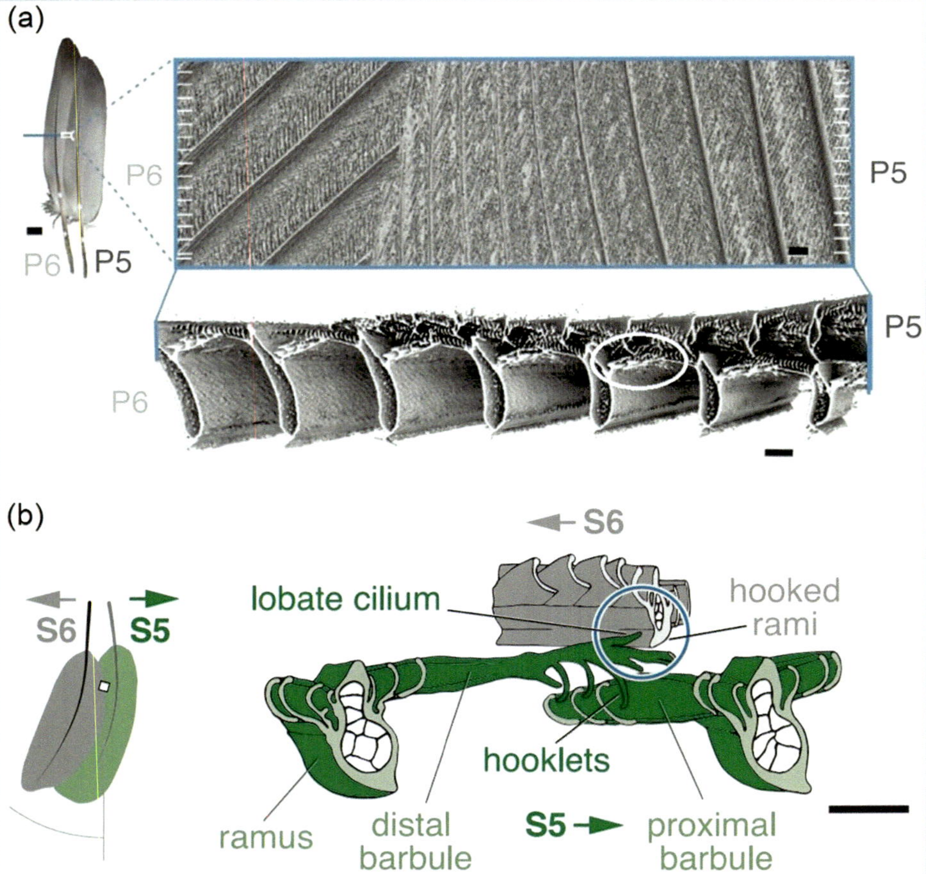


(continued)

**Box 11.5** (continued)

(**a**) Micro-CT scans of two overlapping feathers (primaries 5 and 6) to show how their surfaces engage. Scale bars are 10 mm (left) and 100 mm (top right and bottom right). Micro-CT uses X-rays to see inside objects. The white oval indicates where the lobate cilia and hooked rami are located. (**b**) Drawing showing the interaction between a single lobate cilium (on secondary feather number 5, S5) and hooked ramus (on secondary feather 6, S6). Scale bar, 50 mm. (Figure from Matloff et al. (2020); © 2020 The Authors, some rights reserved; exclusive licensee American Association for the Advancement of Science, used with permission)

Wing twisting is defined as a change in the angle of incidence along the length of the wing. Active wing twisting is achieved through pronating and supinating the wrist. Because the angle of attack tends to increase toward the tip of wings, birds twist their wings to control the angle of attack at the wing tips (Altshuler et al. 2015).

Birds can bend their wings by flexing and extending the wrist dorsoventrally, i.e., perpendicular to the plane of the wing. Birds often flex and fold their wings simultaneously during the upstroke, folding to reduce wing surface area and reduce the effort and energy needed to move the wing upward (Altshuler et al. 2015).

The extent to which birds can morph their wings does vary among taxa and, more specifically, with body mass and flight behavior. Based on data from 61 species of birds, Baliga et al. (2019) found that smaller birds exhibit a greater range of wing motion (i.e., flexing, extending, bending, and twisting) than larger, heavier birds. They also found that the wings of birds that dynamically soar or glide tend to be more rigid, and that birds that used their wings to swim underwater exhibit little ability to bend and twist their wings. In contrast, the wings of birds that use flapping and bounding flight exhibit a much greater ability to flex, extend, bend, and twist their wings.

Of course, individual feathers can also bend and twist. Klaassen van Oorschot et al. (2020) measured the deflection of individual, slotted primary feathers of seven species of raptors in a wind tunnel and found that the feathers deflected passively in response to changes in air speed and angle of attack. Compared to a rigid airfoil, feather twisting was found to delay the onset of stall by reducing the angle of attack at the feather tip. For five raptors (American Kestrel [*Falco sparverius*], Merlin [*Falco columbarius*], Cooper's Hawk [*Accipiter cooperii*], Red-tailed Hawk [*Buteo jamaicensis*], and Peregrine Falcon [*Falco peregrinus*]), their feathers bent upward as wind velocity increased, perhaps working like the dihedral wings (wings angled upward) of aircraft, i.e., dihedral wings increase lateral stability by orienting lift forces over the center of mass. The feathers of Great Horned Owls (*Bubo virginianus*) and Osprey (*Pandion haliaetus*) exhibited anhedral shapes at lower wind velocities (8 m/s) and bent only slightly upward at higher wind velocities (12 m/s), and anhedral wings (wings angled downward) are known to enhance maneuverability (Thomas and Taylor 2001).

**Box 11.6 Alulas in the Fossil Record**

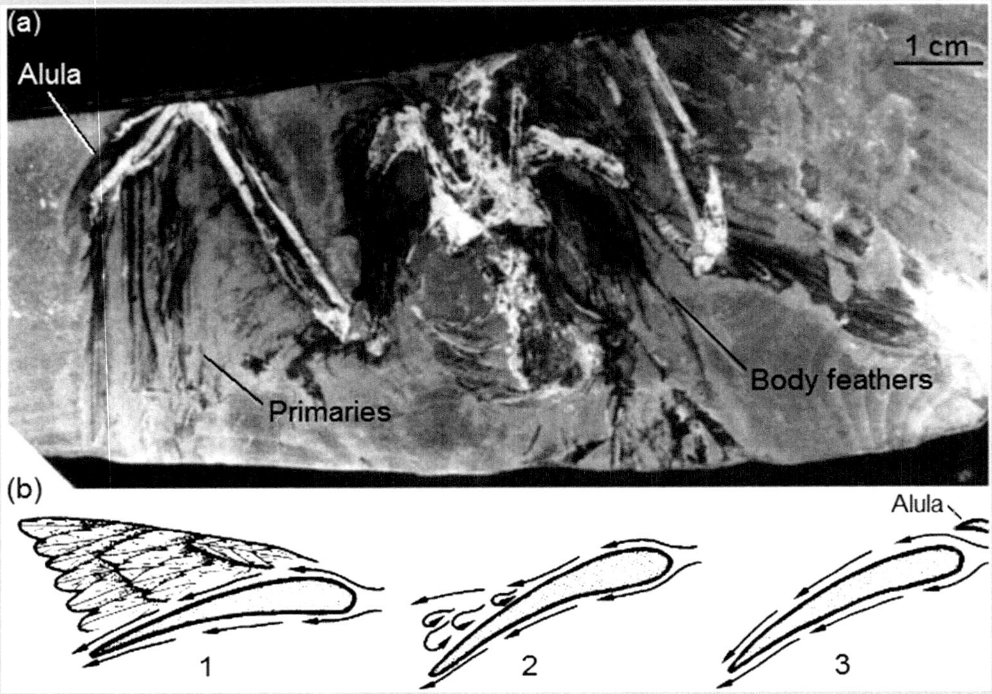


Fossil of *Eoalulavis hoyasi*. (**a**) Fluorescence-induced ultraviolet photo of the specimen showing an alula (**b**) Drawings of a cross section of a wing showing (1) a low angle of attack, (2) a greater angle of attack causing some turbulence at the trailing edge of the wing, and (3) a large angle of attack with an extended alula maintaining smooth airflow over the wing. (Figure from Sanz et al. 1996; © 1996 Springer Nature, used with permission)

The fossilized remains of a tiny bird provide evidence that birds flew as nimbly 115 million years ago as their descendants do today. The fossilized bird, *Eoalulavis hoyasi*, was found in a limestone quarry in Spain (Sanz et al. 1996). About the size of a goldfinch, the bird had an alula that would have helped it maintain lift at slow speeds. *Eoalulavis* is the most primitive bird known with an alula. *Archaeopteryx* probably flapped and glided, but did not have an alula. *Eoalulavis* provides evidence that by 30 million years after *Archaeopteryx*, at least one group of early birds had developed the alula. *Eoalulavis hoyasi*, which means "dawn bird with a bastard wing from Las Hoyas," was discovered at a site where a freshwater lake existed millions of years ago.

**Box 11.7 Interspecific Differences in Alulae**

The position of the alulae at the leading edge of bird wings is controlled by several small muscles and, when elevated, the alula helps maintain smooth airflow over the wings at high angles of attack. Like all contour feathers, alulae have a calamus that anchors the feathers (in this case, to the first digit) and a rachis. The rachis consists of an outer cortex and an interior consisting of foam-like material. Differences among species in the thickness and shape of the cortex and the properties of the foam-like material inside contribute to variation in the mechanical properties of contour feathers, including alulae. Schmitz et al. (2015) compared the alulae of three raptors and Rock Pigeons (*Columba livia*) and found that those of Eurasian Sparrowhawks (*Accipiter nisus*) resisted bending better than those of the other species. A likely explanation for this is that Eurasian Sparrowhawks prey primarily on small birds in cluttered habitats like forests. As such, when pursuing prey, they must be able to quickly change speeds and direction as they maneuver in and around trees, folding wings in when necessary to avoid trunks and branches, and then brake quickly when attacking their prey. Frequent changes in flight speed mean frequent changes in the angle of attack of their wings that, in turn, require frequent use of the alulae. For Eurasian Sparrowhawks, then, natural selection has favored the evolution of alulae with a rachis very resistant to bending and able to maintain their shape and, in doing so, maintain lift regardless of wing shape and speed. The flight styles of Peregrine Falcons (*Falco peregrinus*), Rock Pigeons, and Eurasian Kestrels (*Falco tinnunculus*) differ from that of Eurasian Sparrowhawks and, as a result, these species apparently do not require alulae as resistant to bending.

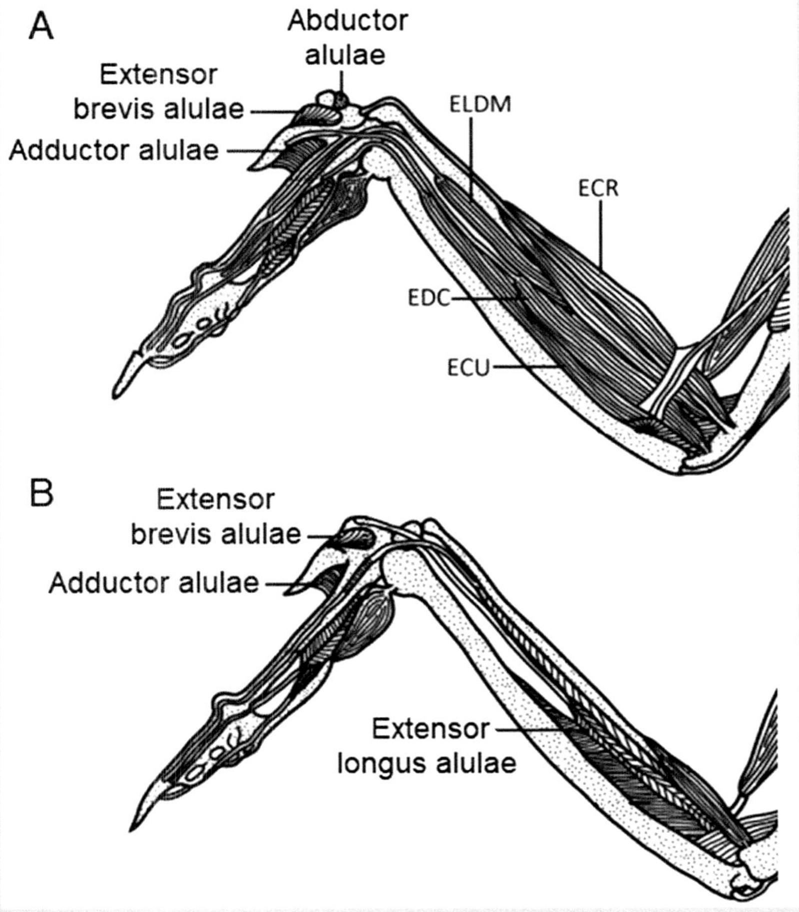


(continued)

**Box 11.7** (continued)

Dorsal view of (**a**) superficial and (**b**) deep muscles of the wing of a Eurasian Sparrowhawk (*Accipiter nisus*). Muscles that control the position of the alula include the abductor alulae, adductor alulae, extensor brevis alulae, and the extensor longus alulae. On the ventral side of the wing, the flexor alulae muscle can also alter the position of the alula. Other muscles: ELDM, extensor longus digiti majoris; ECR, extensor carpi radialis; EDC extensor digitorum communis, ECU, extensor carpi ulnaris. (Figure modified from Bribiesca-Contreras et al. 2019; © 2019 American Association for Anatomy, used with permission)

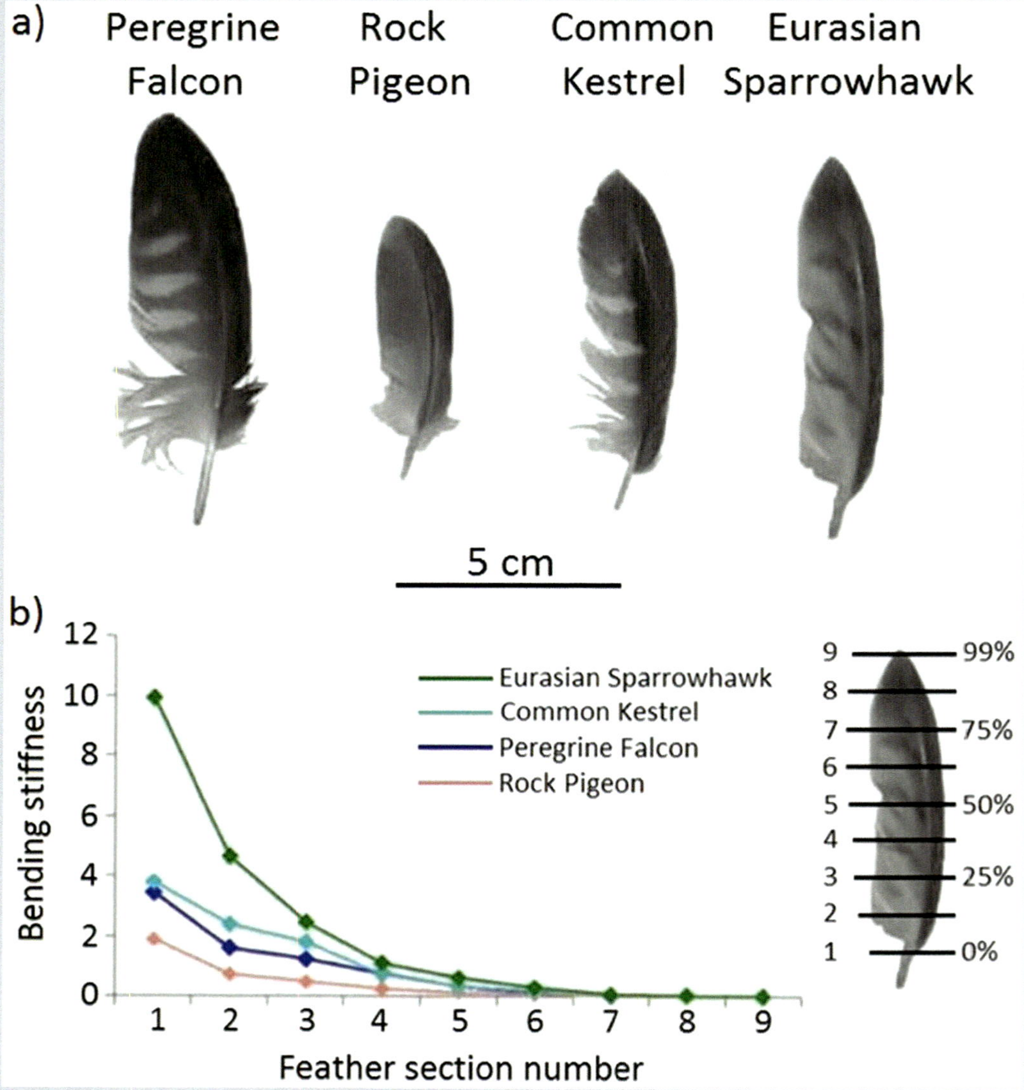


(**a**) Outermost alulae of four species of birds. (**b**) The bending stiffness of alulae was measured at nine points along the feather shaft and the numbers along the X-axis correspond to those in the alula feather to the right. Note that the outermost alula of Eurasian Sparrowhawks is stiffer and more resistant to bending, particularly near the proximal end of the rachis, than those of the other three species. Eurasian Sparrowhawk, *Accipiter nisus*; Common Kestrel, *Falco tinnunculus*; Peregrine Falcon, *Falco peregrinus*; Rock Pigeon, *Columba livia*. (Figure modified from Schmitz et al. 2015; © 2014 Wiley Periodicals, Inc., used with permission)

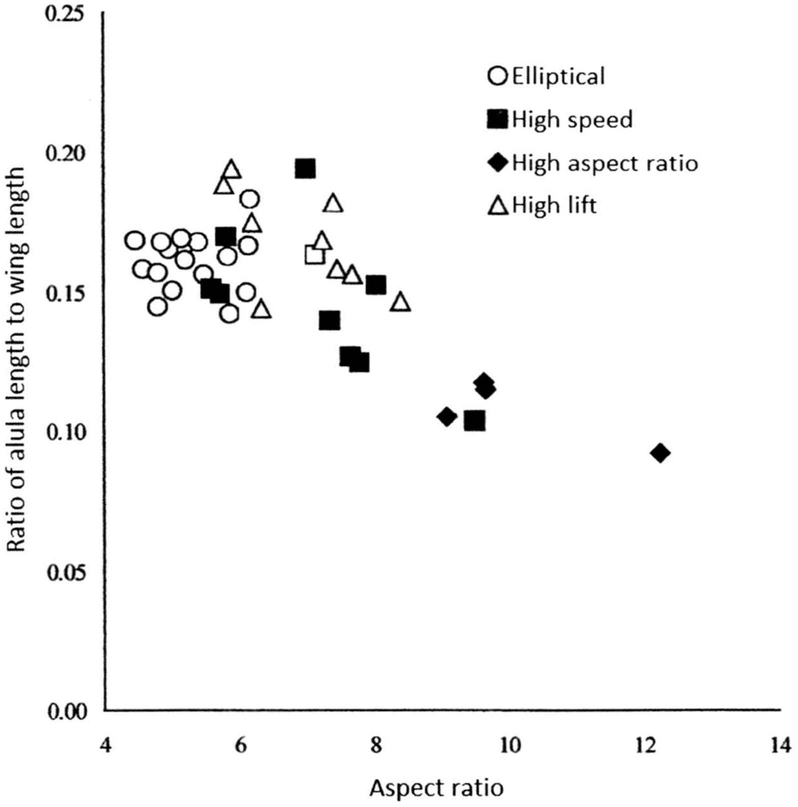


**Fig. 11.30** Variation in the length of alulas relative to wing type. Birds with elliptical wings (like most songbirds) and high-lift wings (like soaring hawks and vultures) have relative long alulas. Birds with high-aspect ratio wings (like albatrosses and other seabirds) have relatively short alulas, and birds with high-speed wings (like falcons, swallows, and swifts) exhibit more variation in relative alula length. (Figure from Alvarez et al. 2001; © Álvarez, J. C. et al., Ardeola, used with permission)

The position of the alula can be controlled by several muscles (Bribiesca-Contreras et al. 2019; Box 11.7 Interspecific Differences in Alulae), but, because of its location at the leading edge of the wing where suction develops as the angle of attack increases, the alula also tends to elevate naturally with an increasing angle of attack (Meseguer et al. 2003). As it does so, air moving between the leading edge and the alula is deflected downward and flows smoothly over the upper surface of the wing.

The relative length of alulas varies among different birds (specifically the aspect ratio relative to the ratio of alula length to wingspan; Fig. 11.30), with those having elliptical wings (like most songbirds) and high-lift wings (like soaring hawks and vultures) also having relative long alulas. Alulas are important for birds with elliptical wings because many of these birds take off and land frequently and many occupy habitats like forests where maneuverability is important (i.e., rapid maneuvers that might generate rapid changes in wing position and angle of attack). Birds with high-lift wings often fly at relatively slow speeds (when soaring), making longer alulae important. In contrast, birds with high-aspect ratio wings (like albatrosses and other seabirds) have relatively short alulas. These birds tend to take off and land less frequently and, when landing, typically do so in water that minimizes the impact of harder landings and reduces the need for long alulae. Birds with high-speed wings (like falcons, swallows, and swifts) exhibit more variability in relative alula length, perhaps because some spend more time in flight (and less time taking off and landing, e.g., swallows) and would benefit less from long alulae, whereas others do take off and land more frequently (e.g., some falcons) or might require more control when they to make infrequent landings (e.g., swifts entering hollow trees, narrow caves, or chimneys).

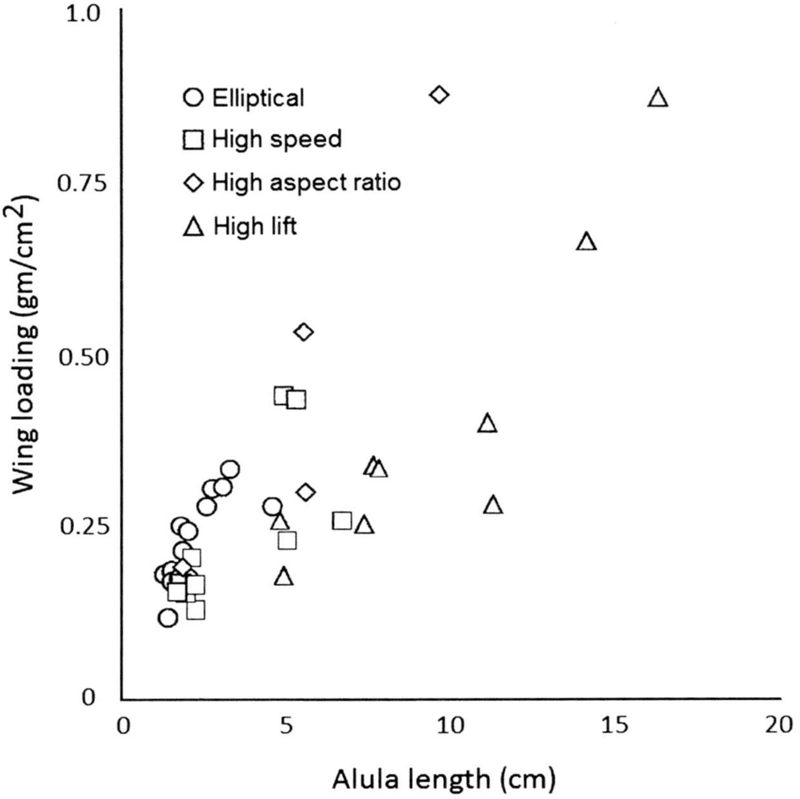


**Fig. 11.31** Variation in the length of alulas relative to wing type. Alula length generally increases with wing loading regardless of wing type, but birds with high-lift wings (like soaring hawks and vultures) have relatively longer alulas than birds with other types of wings. (From modified from Alvarez et al. 2001; © Álvarez, J. C., et al., Ardeola, used with permission)

Relative alula length also varies with wing loading (a bird's mass relative to its wing area). In general, relative alula length increases with increased wing loading (Fig. 11.31), regardless of wing type (elliptical, high speed, high-aspect ratio, or high lift). However, birds with light lift wings (e.g., soaring hawks, vultures, cranes, and storks) have relatively longer alula relative to wing loading than do birds with other types of wings. Apparently, the tendency of these birds to fly at relatively slow speeds that would tend to generate eddies has selected for relatively long alulas that help maintain smooth air flow over their wings.

Along with the alulae, upper wing coverts can also help maintain lift at slow speeds and high angles of attack. At increasing angles of attack, an eddy starts to propagate from the trailing edge toward the leading edge of the wing. As a result, air flowing over the top of the wing separates from the upper surface and lift is lost. However, when coverts are lifted upward by the eddy (Fig. 11.32), they prevent the spread of the eddy and work as "eddy-flaps" (Fig. 11.33). The "covert eddy-flaps," by preventing the spread of the eddy toward the leading edge of the wing, help maintain lift (i.e., prevent stalling) at high angles of attack, e.g., when taking off or landing.

Bird wings can also create lift by generating leading-edge vortices (LEVs). LEVs are typically produced by wings at relatively high angles of attack (>20°) and have been reported to develop at the leading edge of the wings of birds ranging in size from hummingbirds to geese. The air pressure at the core of LEVs is low, creating a suction that provides additional lift. In addition, the vortices direct air downward over the top and across the wing to help generate lift (Figs. 11.34 and 11.35).

**Fig. 11.32** Wing coverts act like eddy-flaps, preventing the spread of turbulence across the top of the wing to help maintain lift. (Figure from Bhushan 2009; used with permission of The Royal Society, from Biomimetics: lessons from nature—an overview, © 2009; permission conveyed through Copyright Clearance Center, Inc.)

### 11.3.2 Thrust

In addition to lift, bird wings must generate forward thrust. Flapping flight involves up-and-down movement of the wings and, during such flight, the entire wing, and especially the proximal portion (Fig. 11.36), produces lift, but the distal part of the wing generates most of the thrust that propels a bird forward. During the downstroke (power stroke), a wing moves downward and forward. However, the proximal portion of the wing moves less than the distal portion and the resulting forces also differ. The proximal wing remains more-or-less parallel to the axis of the body, the velocity of the air moving over the proximal wing (closest to the body) is the same as the bird's forward velocity (assuming calm air), and, therefore, the resulting force (lift) is largely upward or vertical. However, the distal portion of the wing is angled downward (with the leading edge lower than the trailing edge) and air moving past the distal wing is moving faster, and at a

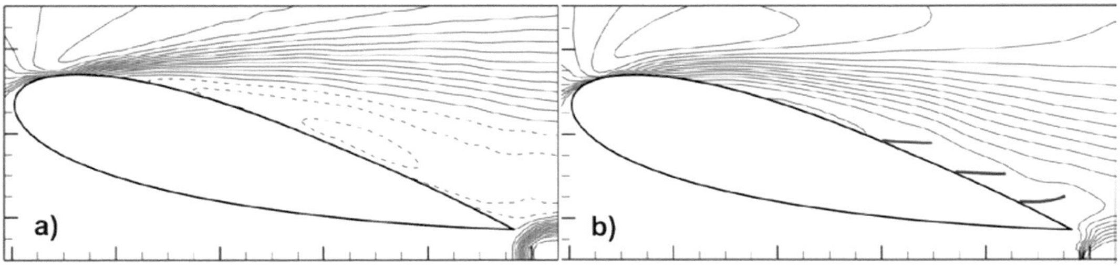


**Fig. 11.33** When vortices begin to form near the trailing edge of a wing, (**a**) they can spread toward the leading edge and result in flow separation, i.e., air flow "separates" from the surface of the wing and lift is lost. (**b**) Wing coverts, however, are lifted up by the vortices and help maintain airflow over at least a portion of the wing and, by doing so, help maintain lift. (Figure modified from Brücker and Weidner 2013; used with permission of C. Brücker)

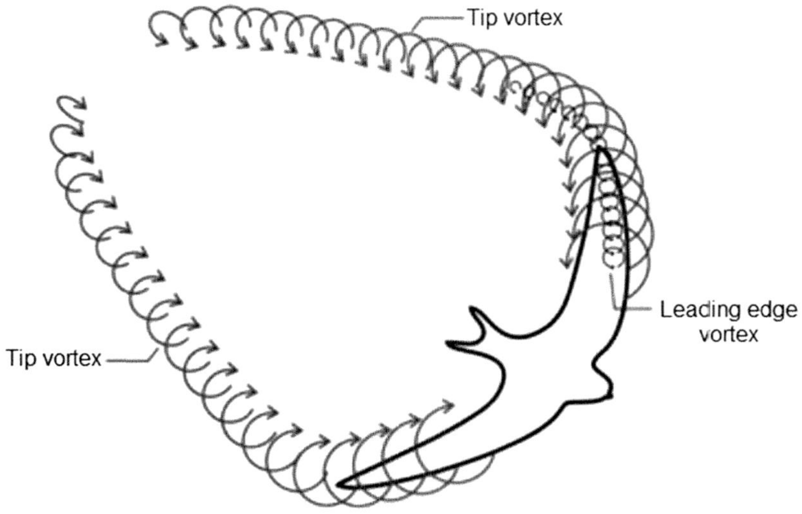


**Fig. 11.34** Illustration of the conical leading-edge vortices (LEVs) generated by the wing of a swift in gliding flight. The oncoming air is deflected downward by the attached LEV to generate lift. LEV separation starts at the wrists and the LEVs then start to go upward and behind the wingtip. (Figure from Jenni-Eiermann and Srygley 2017; Copyright © 2017, Springer International Publishing AG, part of Springer Nature, used with permission)

different angle, because of the wing's flapping motion. As a result, the resulting force is angled forward and this produces the thrust that moves the bird forward (Fig. 11.37).

During the upstroke, especially when flying at slower speeds, the wings flex at the wrist (reducing the wing area) and the angle of attack is reduced so that little or no force (thrust) is generated. Because of the wing's angle during the upstroke (leading edge higher than the trailing edge), some reverse thrust can be generated, but the forward thrust generated by the downstroke is more than sufficient to compensate (Alexander 2002). Particularly at slower speeds and for birds with high-lift and elliptical wings, the tips of the primaries separate during the upstroke and the "slots" between the primaries reduce drag as the wing comes up (Fig. 11.38).

### 11.3.3 Drag

A wing moving through the air is opposed by friction, or drag. In general, drag is inversely related to the degree of streamlining, with a more streamlined shape generating less drag. The three types of drag important for a bird in flight are parasitic drag, profile drag, and induced drag. Parasitic drag is due to friction generated as air moves across the surface of a bird's body (excluding the wings) and can be expressed as:

$$D_{\text{par}} = \tfrac{1}{2}\,\rho V^2 S_{\text{par}} C_{\text{Dpar}},$$

where $V$ is air speed, $\rho$ is air density, $S_{\text{par}}$ is the minimum frontal area of the body, and $C_{\text{Dpar}}$ is the body drag coefficient and represents the degree of streamlining. Parasitic drag is so-named because the bird's body does not generate lift and so depends on the wings and tail to take flight (i.e., like a parasite depends on a host to survive). Parasitic drag occurs because a bird's body causes an increase in pressure in front of it and tends to decelerate the oncoming flow of air. A large frontal area causes more drag, whereas a slim, streamlined body minimizes drag (Hedenström 2002). The frontal area of birds (responsible for parasitic drag) increases in a predictable pattern with body size so, for example, smaller birds are not more streamlined than larger birds (Nudds and Rayner 2006). However, the bodies of larger birds tend to be more dorsoventrally flattened than those of smaller birds and the "flatter" bodies of large birds may provide more body lift during flight (Maybury

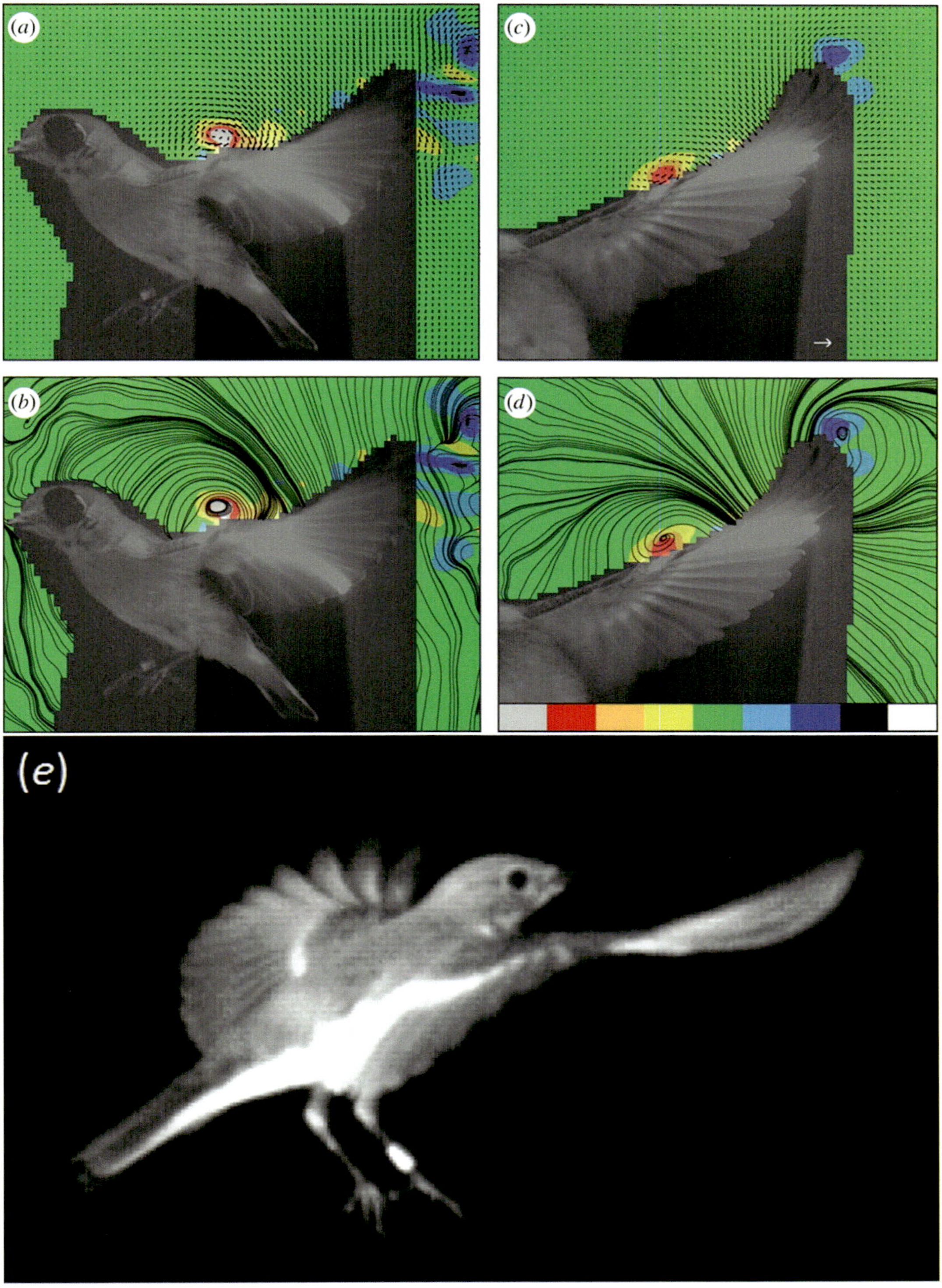


**Fig. 11.35** Air movement during the downstroke of a European Pied Flycatcher (*Ficedula hypoleuca*) flying at a speed of 1 m/s in a wind tunnel. (**a**, **b**). Proximal section of wing. (**c**, **d**) Distal section of wing. All panels (**a**–**d**) show the vorticity fields. Images **b** and **d** show streamlines and **a** and **c** show velocity vectors. In contrast to swifts where the strength of the leading-edge vortex (LEV) increases toward the wingtip, the LEV of Pied Flycatchers

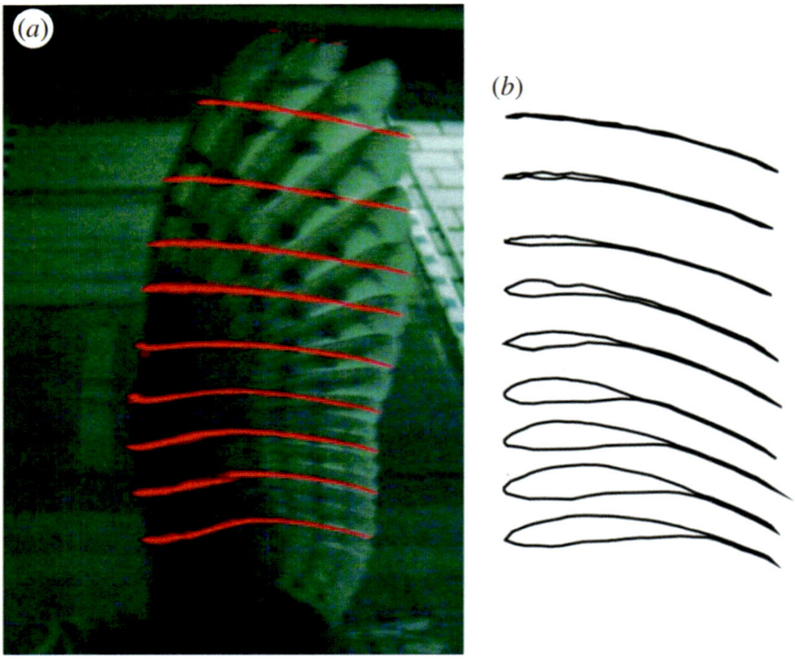


**Fig. 11.36** (**a**, **b**) A bird's wing is more cambered closer to the body where the secondary flight feathers are located. As a result, more lift is generated by the proximal portion of wings than the distal portion. (Figure from Wagner et al. 2017; © 2016 The Authors. Published by the Royal Society, used with permission)

2000), a possible energy-saving adaptation. Parasitic drag varies with the degree of "fairing" so that any projecting structures, including the head or feet, increase drag (Pennycuick et al. 1996; Fig. 11.39). Birds that spend more time in flight and depend on flight to capture food, such as swifts, swallows, and falcons, are more streamlined, whereas birds that tend to fly more slowly and use flight more for "transportation" between locations than feeding, such as herons and egrets, are less streamlined, with prominent heads and dangling legs.

**Fig. 11.35** (continued) decreased in strength toward the wingtip (i.e., compare **a** and **c** and **b** and **d**). This reduction in strength in the LEV toward the wingtip occurs because the primary feathers passively bend up near the wingtip during a downstroke (as shown in **e**, a high-speed video image of a Pied Flycatcher flying in a wind tunnel). (Figure from Muijres et al. 2012b; © 2012 The Royal Society, used with permission)

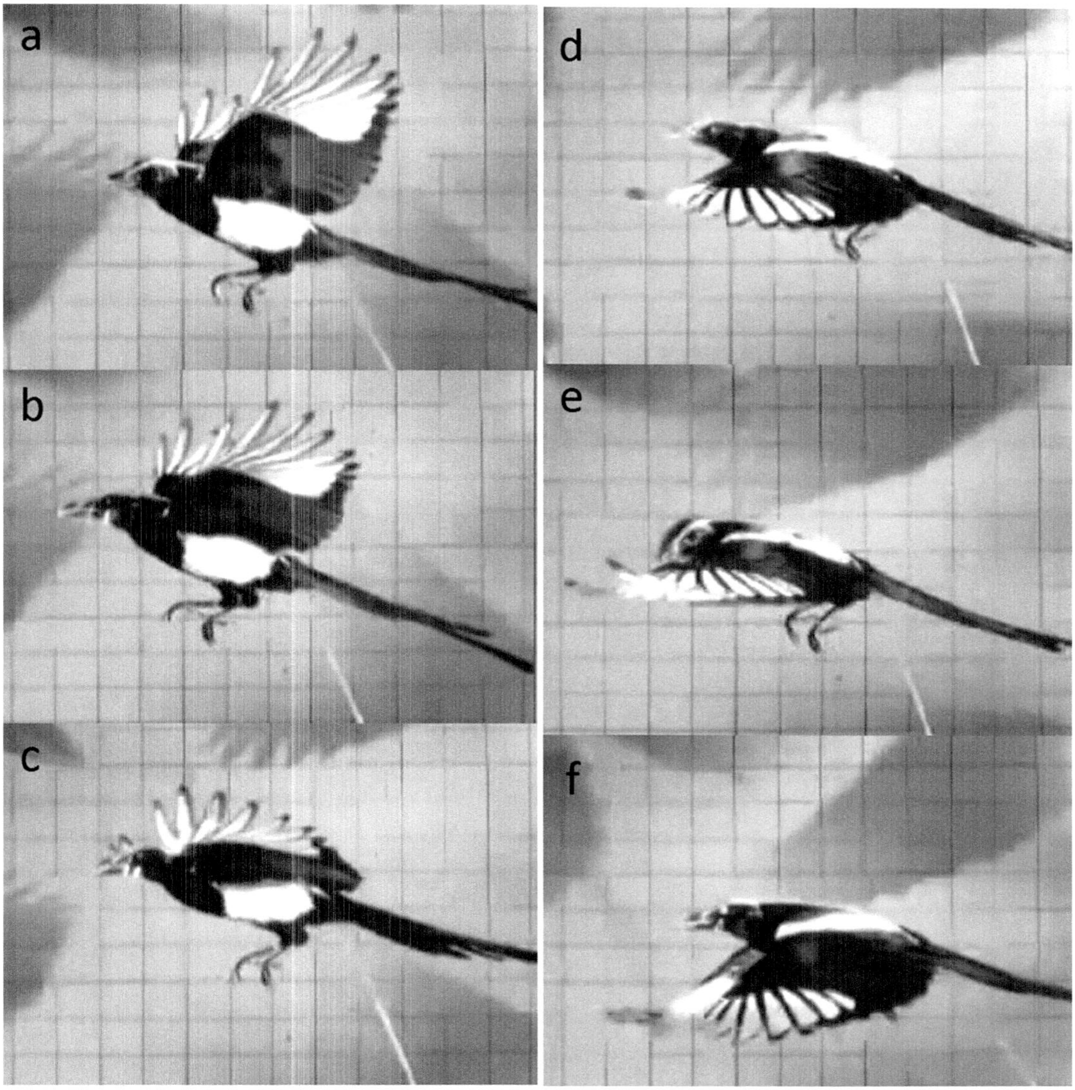


**Fig. 11.37** (**a**–**f**) Downstroke of a Black-billed Magpie (*Pica hudsonia*) flying at a speed of 4 m/s in a wind tunnel. Note how the wing moves both downward and forward and the concave shape; with the leading edge of the wing lower than the trailing edge (especially in images **a** and **b**), the wing, as it moves downward, "pushes" air downward and backward. That "backward push" generates an opposite and equal reaction (Newton's 3rd law), which propels the magpie forward. In other words, the downstroke is not only providing lift (especially the proximal portion of the wing that is moving less than the distal end), but forward thrust as well. (Images are screenshots from a slow-motion video from the University of Montana Flight lab on YouTube; https://www.youtube.com/watch?v=8vARD17nQjg, used with permission)

**Fig. 11.38** Upstroke of a Black-billed Magpie (*Pica hudsonia*) flying at a speed of 4 m/s in a wind tunnel. Note how the wing moved both upward and backward. Note also how the wing is slightly folded and the tips of the primaries separate; the "slots" between the primaries reduce drag as the wing comes up. (Images are screenshots from a slow-motion video from the University of Montana Flight lab on YouTube; https://www.youtube.com/watch?v=8vARD17nQjg, used with permission)

**Fig. 11.39** Body, or parasitic, drag is low for birds like (**a**) Sand Martins (*Riparia riparia*) where the head and body plumage make a smooth transition and where the feet can be pulled up into the plumage. When the feet of birds like (**b**) Red-wattled Lapwings (*Vanellus indicus*) or the head and feet of birds like (**c**) Great Egrets (*Ardea alba*) and (**d**) Eurasian Griffons (*Gyps fulvus*) project from the body, the amount of parasitic, or body, drag increases. (Photo of Sand Martin from Wikipedia, CC BY 2.0, https://creativecommons.org/licenses/by/2.0/; Masked Lapwing photo from Wikipedia, CC BY 2.0, https://creativecommons.org/licenses/by/4.0/; Great Egret photo from pxhere.com, CC0 Public Domain; Eurasian Griffon photo by Frank Wouters, Wikipedia, CC BY 2.0, https://creativecommons.org/licenses/by/2.0/)

**Box 11.8 Wing Color, Drag, and Flight Performance**

Albatrosses and many other birds that soar have dark wings, particularly on the upper surface. One potential advantage of dark wings is that they absorb solar radiation and convert it to heat. As this heat is released, the air just above the wing gets warmer and, as a result, is less dense and viscous. That makes it easier for air to pass across the top of the wing (an area referred to as the boundary layer), and the result may be a reduction in drag (Hassanalian et al. 2017; Rogalla et al. 2019). Rogalla et al. (2021) conducted wind-tunnel experiments and showed that radiative heating of bird wings improved flight efficiency, suggesting that seabirds, and perhaps other birds that use soaring and gliding flight, may have evolved wing pigmentation because it enhances flight performance.

(continued)

**Box 11.8** (continued)

Two views of an albatross showing the lighter-colored underwing (left) and the darker upper surface of the wings. (Figure from Hassanalian et al. 2017; © 2017 Elsevier Ltd., used with permission)

When sunlight strikes the dark upper surface of a wing, the dark feathers absorb the solar radiation and convert it into heat. The heat in turn warms the molecules of air just above the surface and warmer, faster-moving molecules are less dense and viscous. (Figure modified from Hassanalian et al. 2017; © 2017 Elsevier Ltd., used with permission)

(continued)

**Box 11.8** (continued)

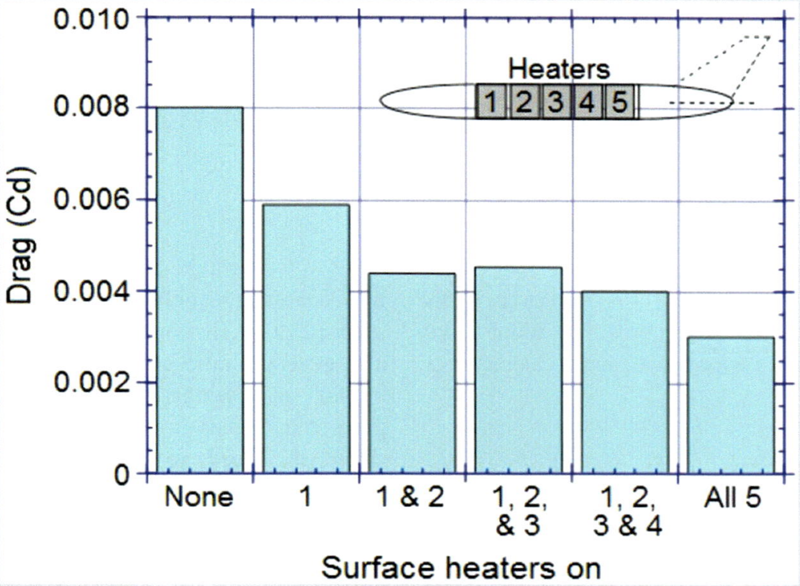


Effect of heating the surface of a model airplane in a wind tunnel on the amount of drag. When more of the surface of the model airplane was heated (all 5 surface heaters turned on), the result was a substantial reduction in drag. (Figure modified from Kramer et al. 1999; reprinted by permission of the American Institute of Aeronautics and Astronautics, Inc.)

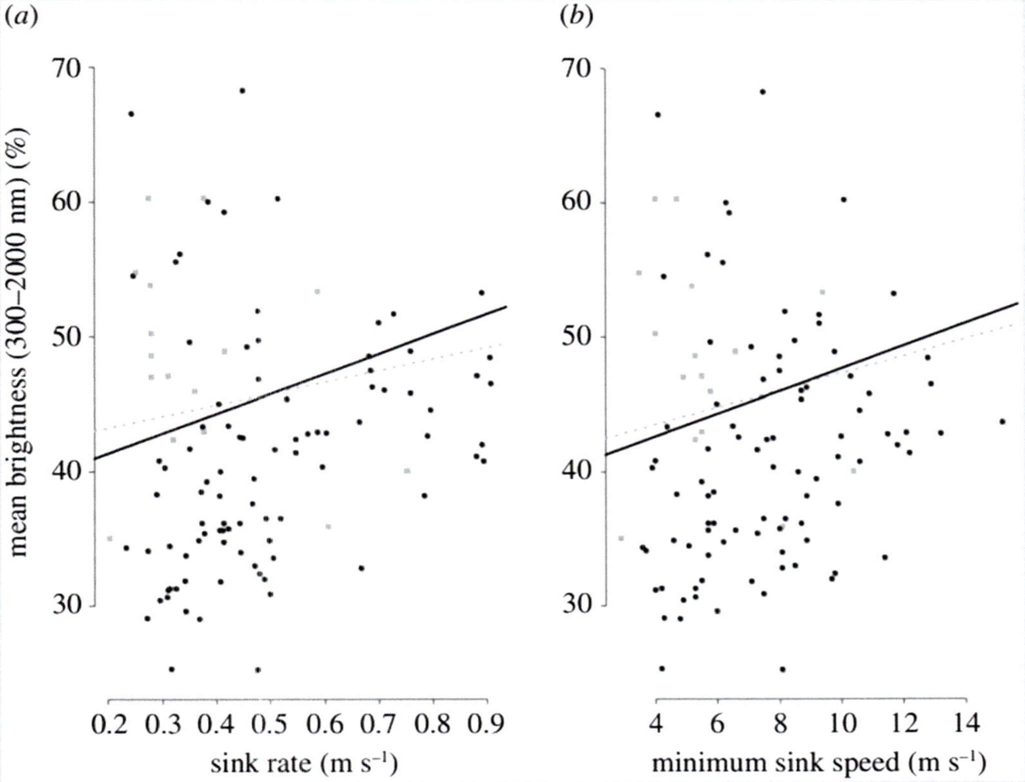


(continued)

**Box 11.8** (continued)

Wing mean brightness as a function of glide performance for 90 species of oceanic seabirds (black dots, solid line; orders Procellariiformes, Suliformes, Pelecaniformes, Phaethontiformes, and Charadriiformes) and 16 additional species of seabirds (gray dots, dashed line). (**a**) Sink rate. (**b**) Minimum sink speed. Sink rate and minimum sink speed increased with increasingly bright wings, i.e., birds with darker wings that heat up more than brighter wings had lower sink rates and lower minimum sink speeds. Birds other than seabirds, e.g., vultures, condors, and other soaring birds, might also benefit from having darker wings. (Figure modified from Rogalla et al. 2021; © 2021 The Authors. Published by the Royal Society, used with permission)

Profile drag is generated as air moves across the surface of a bird's wings (Box 11.8 Wing Color, Drag, and Flight Performance) and is calculated as:

$$D_{\text{pro}} = \tfrac{1}{2}\,\rho V^2 S_{\text{pro}} C_{D\text{pro}},$$

where $\rho$ is the air density, $V$ is the air speed, $S$ is the projected area of the wings, and $C_{D\text{pro}}$ is the profile drag coefficient. When birds flex their wings, wing area varies approximately linearly with the wingspan, which reduces profile drag. This is likely why birds flex their wings at higher flight speeds (Tucker 1987).

Induced drag, also called lift-induced drag, occurs when a moving object redirects air flow and can be expressed as:

$$D_{\text{ind}} = \frac{L^2}{\frac{1}{2}\rho V^2 \pi b^2},$$

where $L$ is lift, $\rho$ is air density, $V$ is air speed, and $b$ is wingspan (i.e., aspect ratio). Induced drag occurs because a wing moving through the air pushes air downward, creating lift, but this deflection of air also generates drag that tilts the net lift vector backward (Fig. 11.40). Any time wings produce lift, they also produce some induced drag and, as the angle of attack increases, the vector of the resultant force tilts back with it. Another way of looking at it is that wings act to accelerate air downward, but wings also accelerate air slightly forward and, as the angle of attack increases, wings produce more forward movement and less downward movement of air.

Wings create both a lift force and a drag force and the force resulting from the interaction of these two forces is the backward component of lift. As a wing's angle of attack increases so does induced drag. However, higher angles of attack also generate more lift. Because the speed of air moving past wings affects the amount of lift generated, higher angles of attack are needed at slower air speeds to generate the necessary lift even though this results in an increase in induced drag. At higher speeds, lower angles of attack will generate the same amount of lift while also reducing induced drag. However, at any given angle of attack, increasing air speed reduces the induced drag (i.e., more air is directed downward rather than forward) which, in turn, will increase lift (Fig. 11.41). This is also apparent from the induced drag equation provided above, induced drag decreases with increasing flight speed.

At wing tips, air tends to move from the area of high pressure below the wing to the area of lower pressure above the wing. This movement creates wingtip vortices that disrupts the smooth flow of air over the wing and reduce lift. One way to compensate for this is to increase the angle of attack, which in turn can increase induced drag. In other words, wingtip vortices can cause an increase in induced drag, but the effect is indirect rather than direct.

Induced drag is also affected by a wing's aspect ratio, i.e., length relative to width. Wings with higher aspect ratios (longer relative to their width) generate less induced drag than wings with lower aspect ratios (shorter relative to their width). Wings with higher aspect ratios generate more lift at a given angle of attack than wings with lower aspect ratios. As with increasing air speed, higher aspect ratio wings reduce induced drag because they increase lift forces to a greater

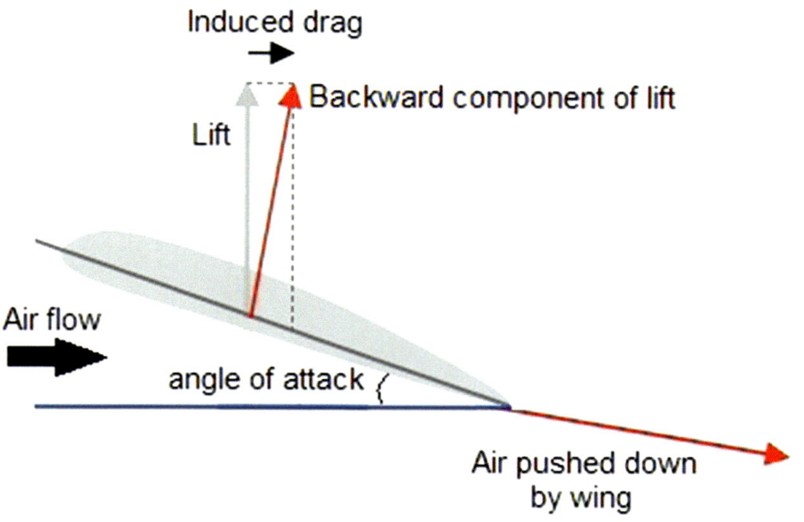


**Fig. 11.40** As a wing moves through the air, air is pushed downward. However, this deflection of air tilts the net lift vector backward or, in other words, creates drag (induced drag) by impeding forward movement (Image from NASA.gov, CC0 Public Domain)

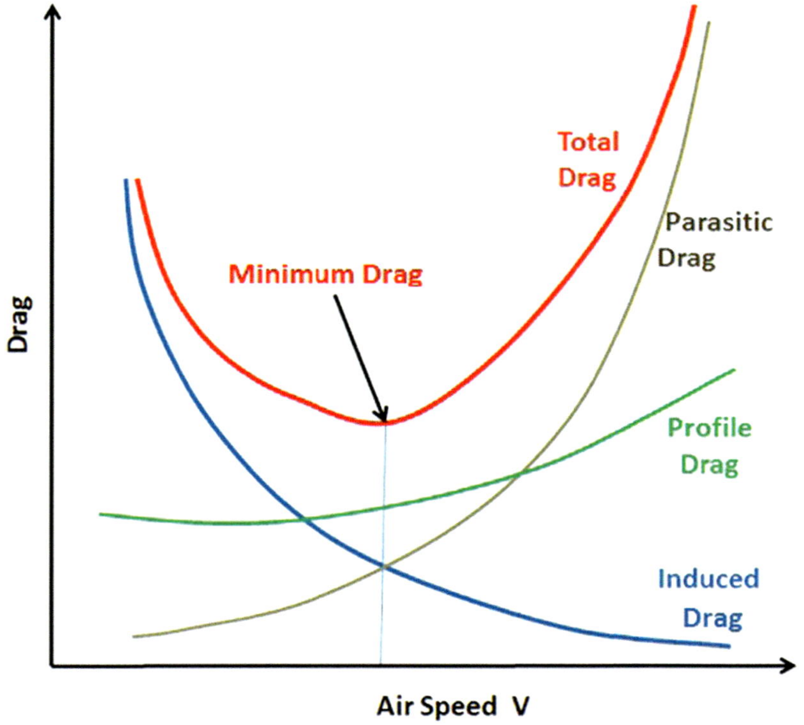


**Fig. 11.41** Parasitic drag is created both by a bird's body (called parasitic drag) and by friction resulting from air moving across a bird's wings (called profile drag). Parasitic drag increases with the square of velocity. Induced drag occurs at wing tips as air moves from the area of high pressure (under the wing) to the area of low pressure (top of the wing) and is inversely proportional to the square of velocity. Total drag is the sum of parasitic drag and induced drag. Minimum drag occurs at the air speed where the lift-to-drag ratio is at its maximum. (Figure source: http://www.mpoweruk.com//flight_theory.htm, © Woodbank Communications Ltd 2005; use permitted by acknowledging Electropaedia as the source)

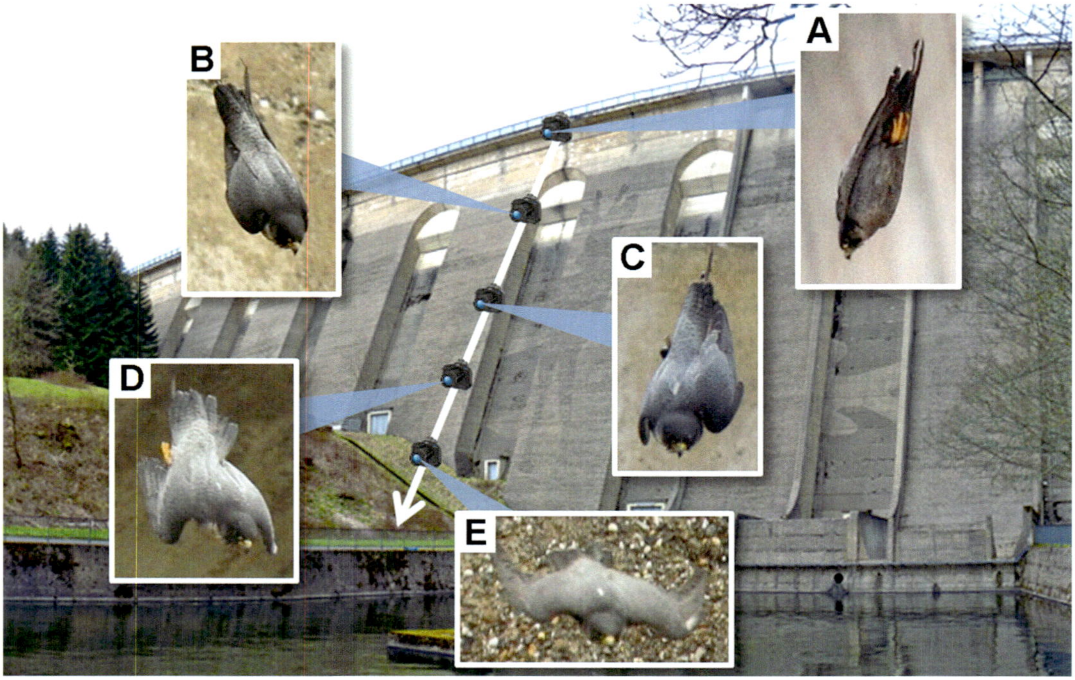


**Fig. 11.42** Photos of a trained Peregrine Falcon (*Falco peregrinus*) launching into a stoop from the top of a dam. Cameras positioned at different height were used to photograph the falcon during the stoop. For most of the stoop (**a**–**e**), the Peregrine Falcon's wings are pulled close to the body, almost eliminating induced and profile drag so that almost all remaining drag is parasitic drag. (Figure from Ponitz et al. 2014; © 2014 Ponitz et al., open-access article distributed under the terms of the Creative Commons Attribution License, https://creativecommons.org/licenses/by/4.0/)

degree than induced drag forces. The induced drag equation also clearly shows that increases in aspect ratio cause corresponding decreases in induced drag.

Because induced drag is highest at low speeds and profile drag increases with increasing speed, the relationship between a bird's speed and drag can be expressed as a U-shaped curve (Fig. 11.41). All birds must fly at slow speeds when taking off and landing, but, of course, birds differ in their typical and maximum speeds. Because drag increases with increasing speed, birds that typically fly faster have more adaptations to minimize drag at those speeds than do slower-flying birds, including more streamlining and narrower wing tips. Streamlining reduces parasite and profile drag, whereas narrow wing tips minimize induced drag. At the extreme, a bird like a Peregrine Falcon (*Falco peregrinus*), when stooping at prey, can pull in its wings to greatly reduce induced and profile drag (Fig. 11.42).

Another way to reduce induced drag is to fly near the surface of the ground or water and take advantage of the ground effect. When flying near a surface, birds experience a reduction in drag because the surface (of either ground or water) interferes with the formation of wingtip vortices responsible for induced drag (Figs. 11.43 and 11.44). A reduction in induced drag means there is a corresponding increase in lift. The change in airflow across the wing when flying near the surface of the ground or water also increases the air pressure under the wing, generating more lift. As a result, birds can save considerable energy when using ground effect, especially when gliding (Rayner 1991). Many birds take advantage of ground effect on a regular basis, including many large aquatic birds like pelicans and cormorants and even some rather small birds like swallows (Fig. 11.45).

**Fig. 11.43** A Brown Pelican gliding near the surface of the water taking advantage of ground effect. (Photo by Gary Ritchison)

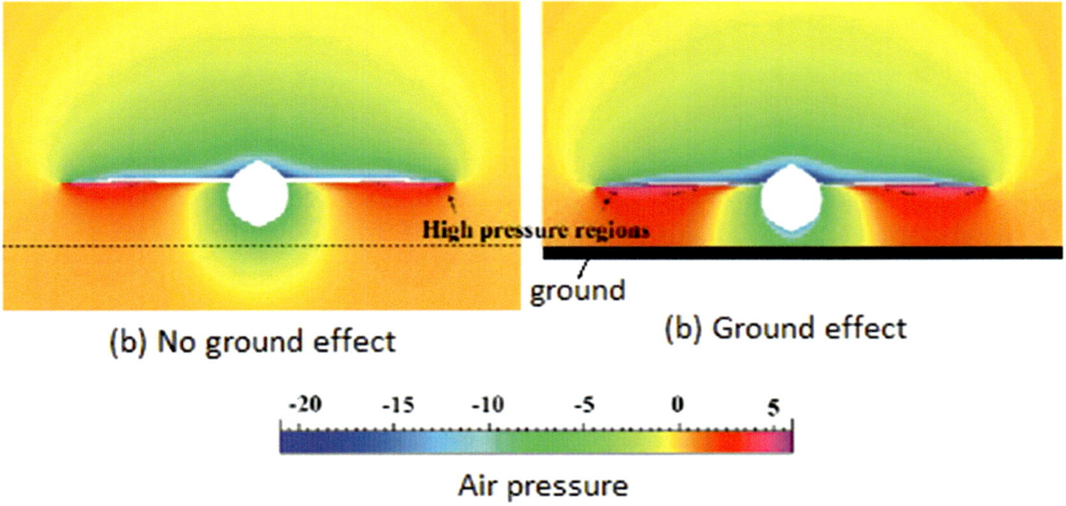


**Fig. 11.44** When flying in ground effect, the stagnation point (the point on the leading edge of an airfoil where the airflow separates, with some going over and some under the surface of the airfoil) shifts slightly to the lower side of wing so more air is diverted over the wing. As a result, the air speed under the wing is slightly reduced, increasing the air pressure and generating more lift. (Figure modified from Tang et al. 2013; © The Society of Theoretical and Applied Mechanics, R.O.C. 2013, used with permission)

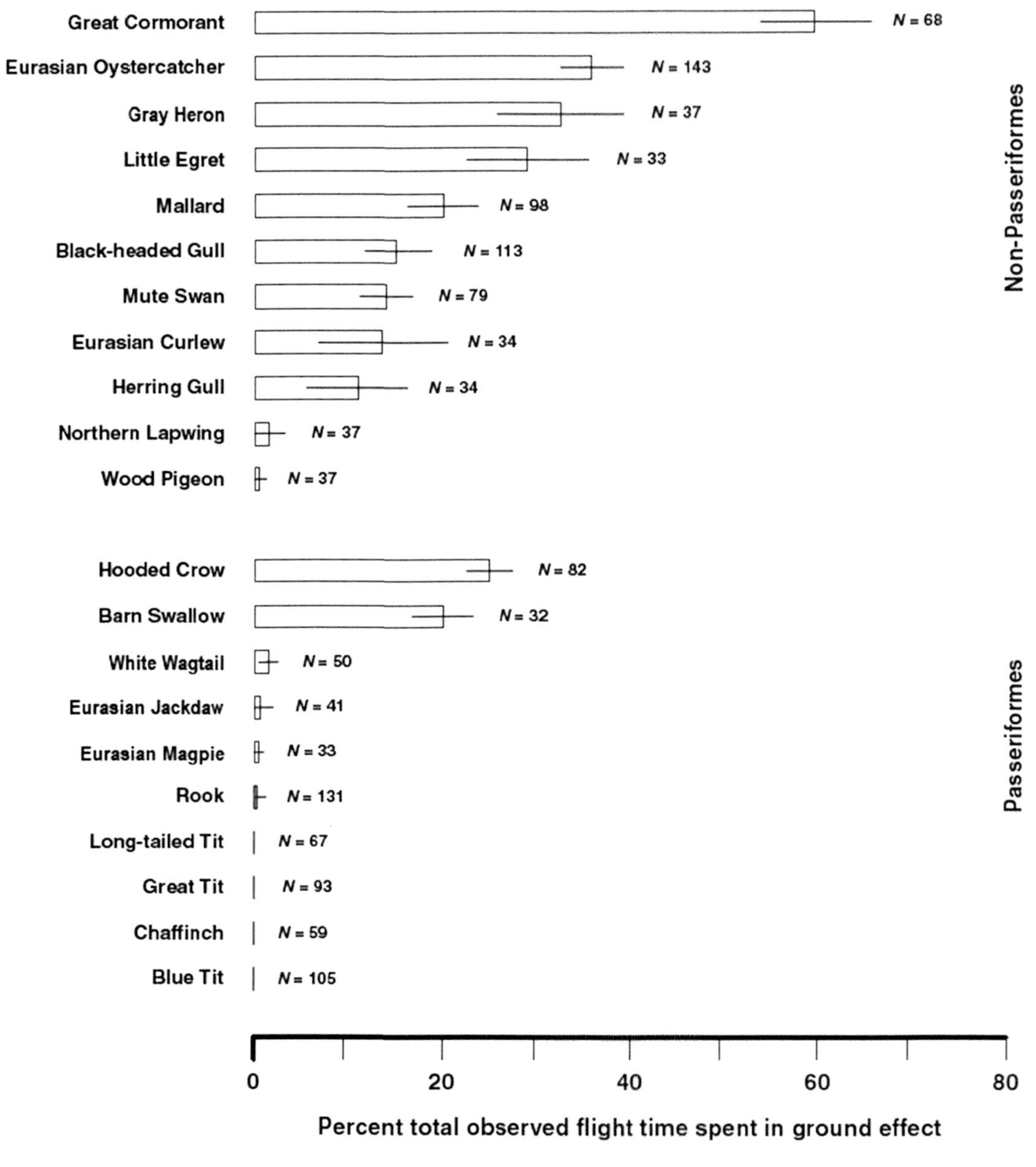


**Fig. 11.45** Percent time (mean ± SE) spent using ground effect by several species of passerines and non-passerines in Ireland. Small passerines, especially those found in habitat where vegetation and other obstacles prevent birds from flying close to the ground, rarely take advantage of ground effect. However, large aquatic birds often do use ground effect. $N$ = number of individuals observed. Great Cormorant, *Phalacrocorax carbo*; Eurasian Oystercatcher, *Haematopus ostralegus*; Gray Heron, *Ardea cinerea*; Little Egret, *Egretta garzetta*; Mallard, *Anas platyrhynchos*; Black-headed Gull, *Chroicocephalus ridibundus*; Mute Swan, *Cygnus olor*; Eurasian Curlew, *Numenius tenuirostris*; Herring Gull, *Larus argentatus*; Northern Lapwing, *Vanellus vanellus*; Common Wood Pigeon, *Columba palumbus*; Hooded Crow, *Corvus cornix*; Barn Swallow, *Hirundo rustica*; White Wagtail, *Motacilla alba*; Eurasian Jackdaw, *Corvus monedula*; Eurasian Magpie, *Pica pica*; Rook, *Corvus frugilegus*; Long-tailed Tit, *Aegithalos caudatus*; Great Tit, *Parus major*; Common Chaffinch, *Fringilla coelebs*; Eurasian Blue Tit, *Cyanistes caeruleus*. (Figure from Finn et al. 2012; © Association of Field Ornithologists, used with permission)

Feathers play a critical role in minimizing drag, with the coat of feathers transforming the avian body, with its globular to oval trunk, relatively long thin neck, and bulbous thighs, into a fusiform shape (Homberger and de Silva 2000). This fusiform body shape is especially important for reducing drag in small birds because of their relatively larger surface area, and small birds are generally more fusiform than larger ones (Pennycuick et al. 1996). The surface of the feather coat, however, is not completely smooth because of the longitudinal ribbing created by the rachis of feathers that project slightly above the surface of the vanes. This ribbing has been hypothesized to reduce drag in a manner similar to that of the ribbing of scales in fast-swimming sharks (Ball 1999, Homberger and de Silva 2000; but see van Bokhorst et al. 2015).

In flight, then, a bird's body generates drag that tends to reduce speed and, by flapping its wings (or, if gliding, converting potential energy into work), a bird produces both lift and thrust (discussed later) to balance the pull of gravity and drag (Hedenstrom and Liechti 2001). An important parameter, of course, is the lift-to-drag ratio, an indicator of the efficiency of a wing that determines, for example, how far a bird can glide before having to flap its wings or land. Drag in this case is total drag, or the sum of induced drag and parasitic drag. The higher the lift-to-drag ratio, the less thrust needed to attain the required lift, and long, thin wings tend to have the highest lift-to-drag ratios. Although a high lift-to-drag ratio is clearly beneficial, natural selection, acting upon birds that occupy different habitats and with different foraging strategies and different needs in terms of flight performance, including maneuverability, acceleration, distance, and speed, has, for most birds, balanced the need to minimize thrust with these other factors. As a result, the wings of birds vary dramatically in shape.

### 11.3.4 Wing Shape

One way of expressing variation in wing shape is the aspect ratio, or the ratio of wingspan squared divided by wing area (i.e., aspect ratio $= b^2/S$, where $b =$ wingspan and $S =$ wing surface area). Aspect ratios of present-day birds range from about 4–15, and differences result from differences in the wing skeleton and the wing feathers (secondaries and, especially, the primaries). Birds with high-aspect ratio wings, such as albatrosses, have relatively long proximal wing bones (humerus, radius, and ulna; e.g., Fig. 11.46), but most variation in wing shape is due to differences in the shape of the flight feathers.

Wing tips also vary in shape based on the relative lengths and shapes of the primary flight feathers, and range from pointed to rounded (Fig. 11.47). Small birds tend to have more rounded wings (Lockwood et al. 1998), but, among larger birds, wingtip shape can differ among wings with similar aspect ratios (Fig. 11.47). Aspect ratio and wingtip shape are important because of their relationship to induced drag (tip vortex). Long, narrow wings (high-aspect ratio) generate less induced drag than short, broad wings (elliptical wings). For birds that regularly fly long distances, such as species like albatrosses that travel long distances in search of prey, very high-aspect ratio wings are beneficial because such wings minimize induced drag and, therefore, the costs of long-distance flight. Albatrosses have the highest aspect ratios of any birds not only because many must travel long distances, but also because the absence of barriers in the open ocean eliminates a selective factor that would tend to limit wing length.

In addition to high-aspect ratio wings, bird wings are typically categorized as high-speed, high-lift, and elliptical. High-speed wings, like those of falcons, swallows, and swifts, have relatively high-aspect ratios. These narrow, tapering wings can be flapped rapidly to generate lots of speed with minimal drag (because, again, the pointed tip minimizes induced drag). High-lift wings have lower aspect ratios and there are spaces between the emarginate primary feathers at the end of the wing (Box 11.9 Emarginate Outer Primary Feathers). These slots reduce drag at slow speeds because the separated tip feathers act as "winglets" and spread vorticity both

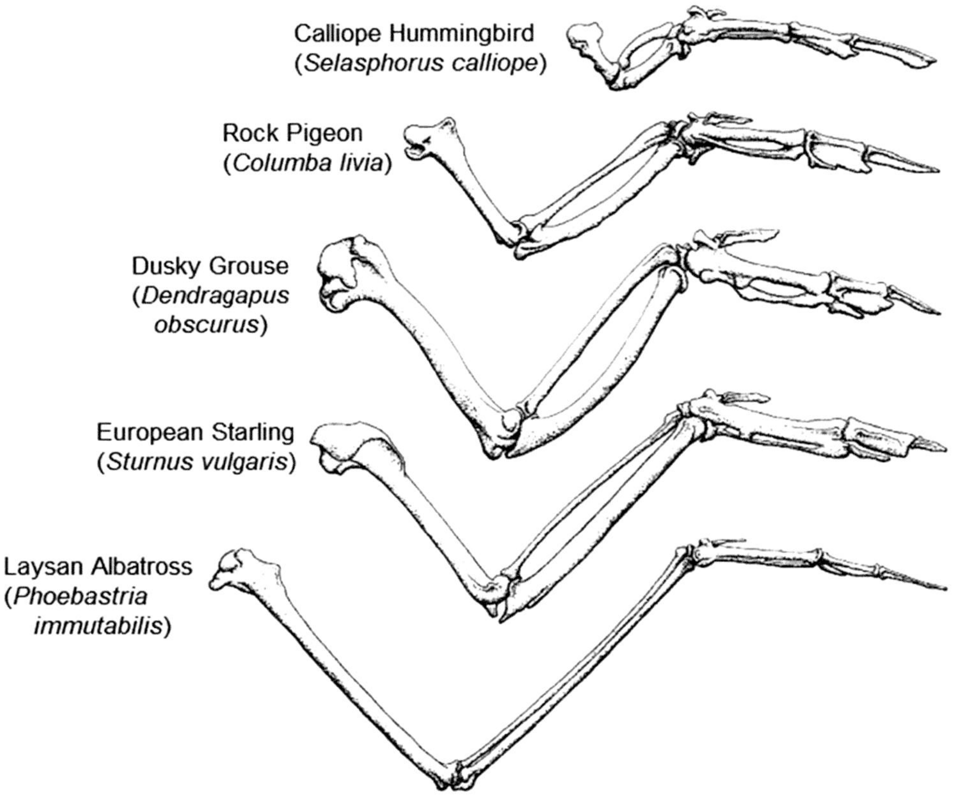


**Fig. 11.46** Bones of the wings of five species of birds scaled so that the carpometacarpi are equal in length. For hovering flight, hummingbirds have evolved a relatively short humerus. The relative stoutness of the radius and ulna and the amount of lateral bowing of these two bones are related to the muscle mass attached. Birds that need to change wing shape as they maneuver through and around vegetation, like hummingbirds, doves, grouse, and songbirds like European Starlings, have more wing (forearm) musculature. Birds that fly in open areas and need not be so maneuverable, like albatrosses, need less musculature and have more slender, less bowed forearm bones. (Figure from Dial 1992; © 1992 Oxford University Press, used with permission)

horizontally and vertically. Slotted tips allow short wings to have the same induced drag as longer wings without slots and shorter wings are advantageous because they permit increased maneuverability that is particularly important in cluttered habitats and when circling within thermals to take advantage of rising air. Although slotted tips increase the profile drag of the wings, such drag is likely minimal at low speeds (Tucker 1993). Wings with low aspect ratios (elliptical wings), like those of many songbirds, woodpeckers, pheasants, and quail, permit sharp turns when flying among trees and shrubs.

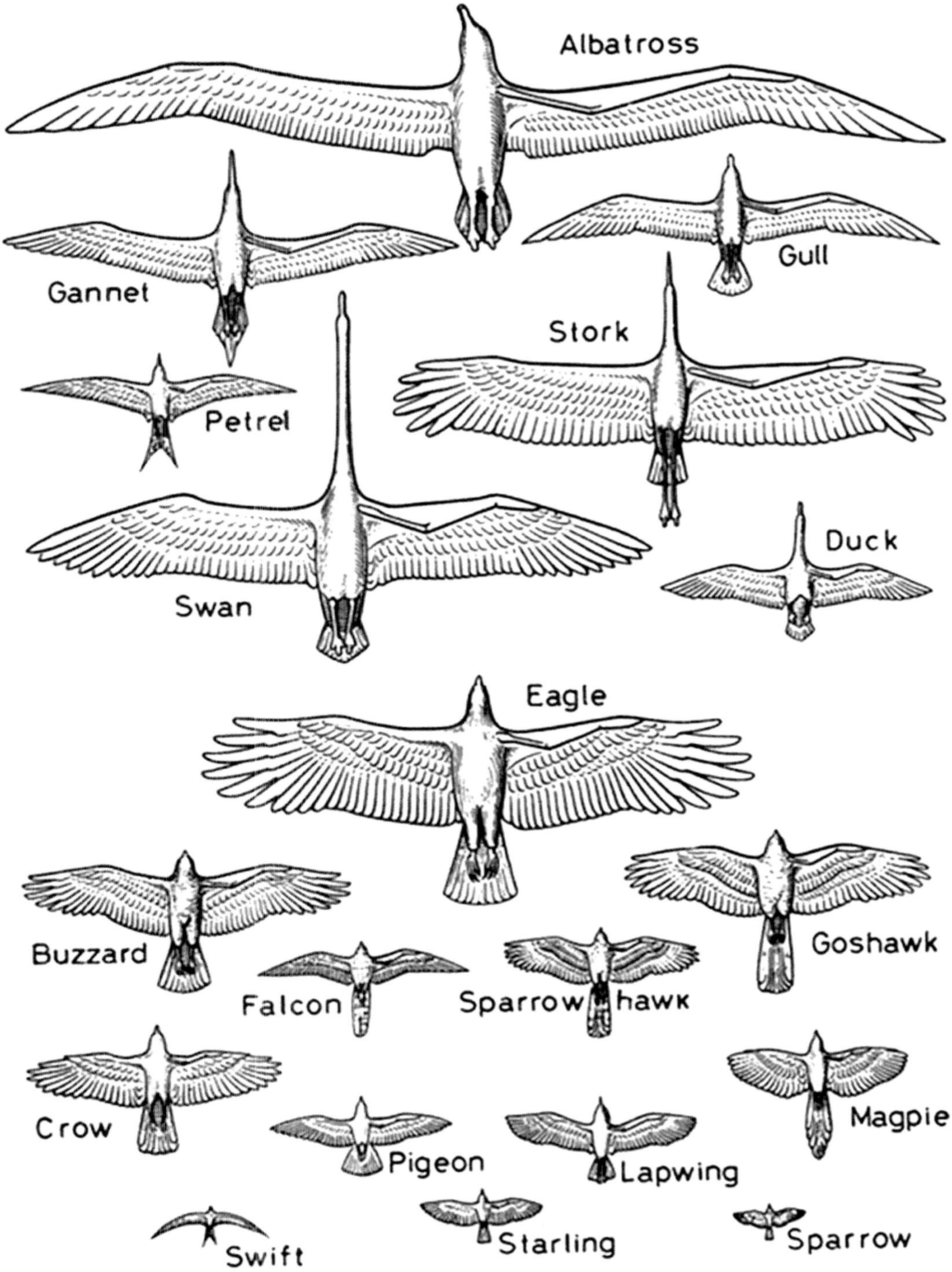


**Fig. 11.47** Wing shapes of birds. (Figure from Norberg 1990; © 1990 Springer-Verlag Berlin Heidelberg, used with permission)

**Box 11.9 Emarginate Outer Primary Feathers**

The outermost primaries of some species of birds are, to varying degrees, emarginated or "notched," with the distal ends noticeably narrower than the proximal ends. The outer primaries of pelagic species, like gulls and albatrosses, are not emarginated, those of freshwater species like ducks have some emargination, and those of large terrestrial species, like hawks, eagles, and owls, have a high degree of emargination.

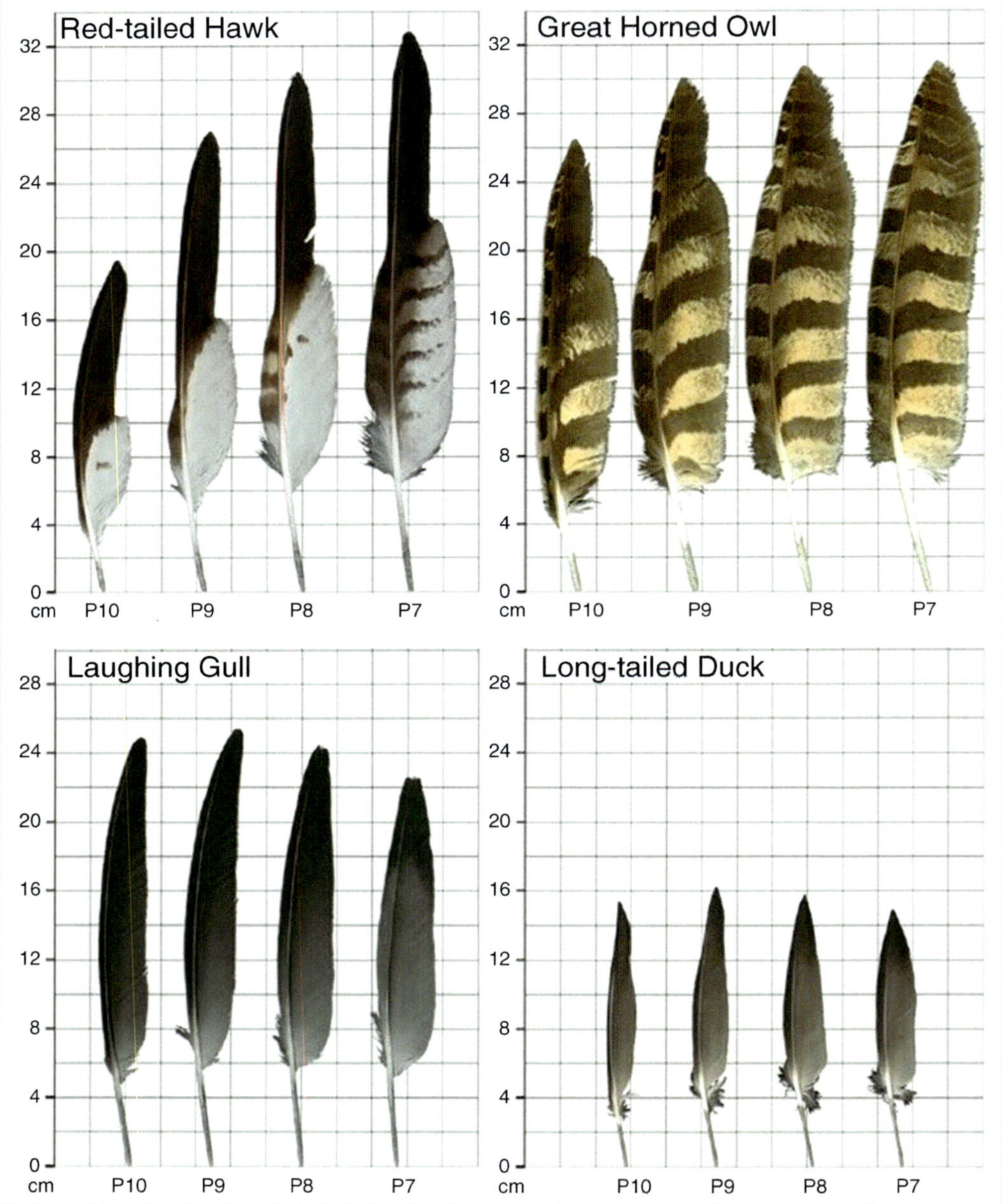


Distal primary feathers (P7–P10) of four species of birds. The distal primaries of Red-tailed Hawks (*Buteo jamaicensis*) and Great Horned Owls (*Bubo virginianus*) are very emarginated, those of Long-tailed Ducks

(continued)

**Box 11.9** (continued)

(*Clangula hyemalis*) are slightly emarginated, and those of Laughing Gulls (*Leucophaeus atricilla*) show no emargination. (Figure from Klaassen van Oorschot et al. 2017; © 2017 Wiley Periodicals, Inc., used with permission)

Drawings of a White Stork (*Ciconia ciconia*) from above (left) and from behind (right) showing how the slotted primaries separate to form individual winglets. (Figure from Eder et al. 2015; © 2015 Springer Nature, used with permission)

Emarginated outer primaries can potentially serve a number of important functions, including a reduction in induced drag. Slotted wings generate several smaller vortices whereas wings without slots generate one larger wingtip vortex. Spreading these smaller vortices both horizontally and vertically reduces the downwash force on wings and so, in other words, reduces induced drag. Less drag means less energy is expended.

(continued)

**Box 11.9** (continued)

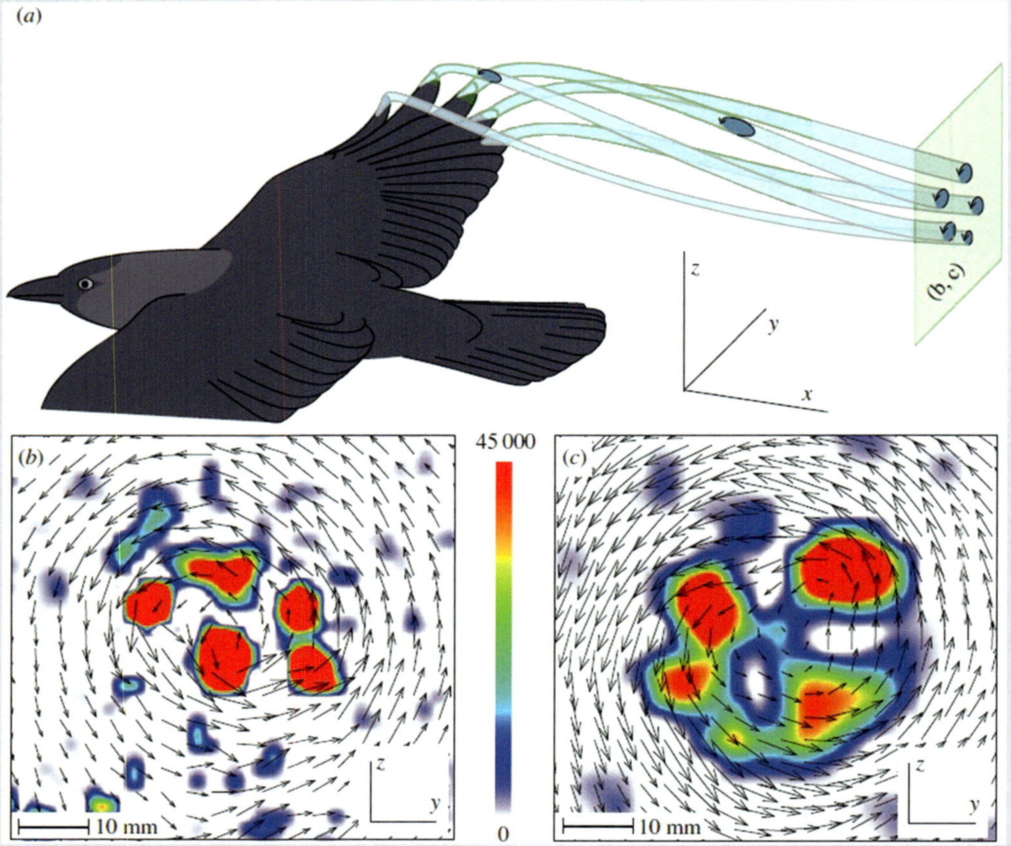


(**a**) Drawing that shows the wingtip vortices generated by the slotting wing of a Eurasian Jackdaw (*Corvus monedula*). (**b**) Location of the wingtip vortices during gliding flight and (**c**) during the mid-downstroke of flapping flight. Slotting wings generate smaller vortices, spread out the vorticity, and reduce the induced downwash on the wing and so reduce induced drag. (Figure modified from KleinHeerenbrink et al. 2017; © 2017 The Royal Society, used with permission)

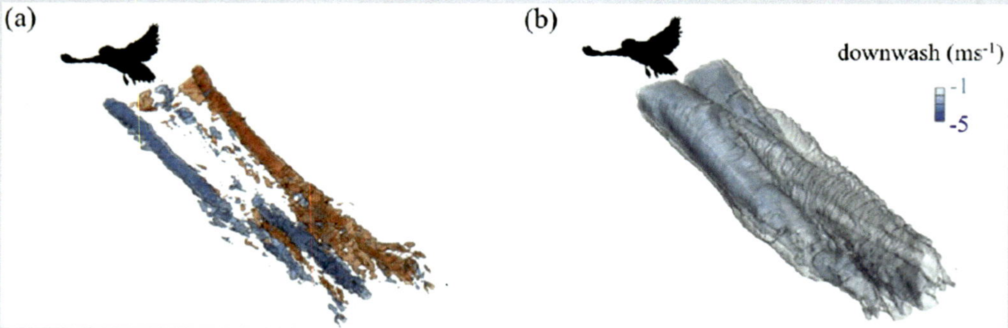


(continued)

**Box 11.9** (continued)

Wingtip vortices generated by the non-slotted wings of a Pacific Parrotlet (*Forpus coelestis*). Each wing generates one relatively large wingtip vortex with corresponding downwash and induced drag. (Figure modified from Gutierrez et al. 2017; open-access article licensed under a Creative Commons CC BY 3.0 license, https://creativecommons.org/licenses/by/3.0/)

An additional advantage of emarginate or slotted wings is that they reduce a wing's surface area and, as a result, reduce the energetic cost of vigorous flapping at slow flight speeds such as when taking off and landing (Klaassen van Oorschot et al. 2017). This is especially the case for larger birds, as indicated by the relationship between body mass and degree of emargination for terrestrial and coastal and freshwater species of birds.

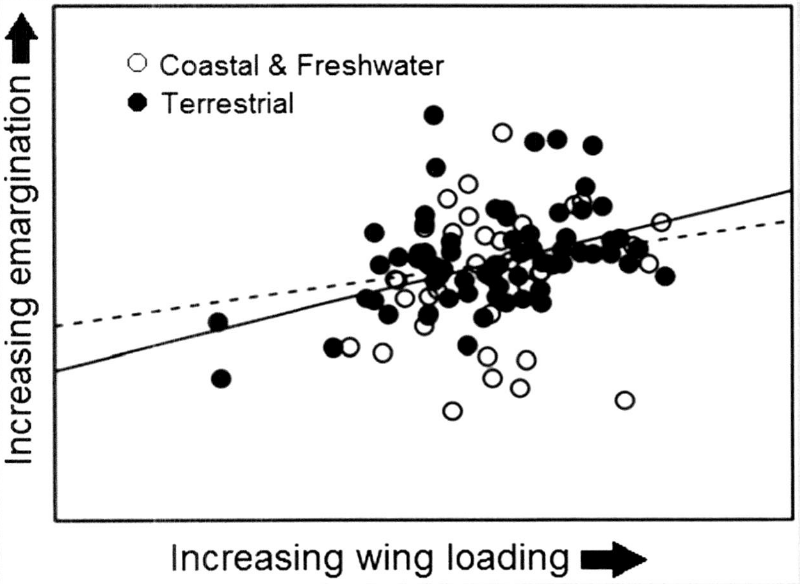


Relationship between extent to which outer primaries are emarginated and wing loading for coastal and freshwater species of birds (dashed line) and terrestrial species (solid line). For both groups, emargination increased with increasing wing loading, suggesting that selection has favored a greater reduction in surface area at the wing tips of heavier birds to reduce the energetic cost of the vigorous flapping required for birds with greater wing loading when taking off and landing. (Figure modified from Klaassen van Oorschot et al. 2017; © Wiley Periodicals, Inc., used with permission)

(continued)

**Box 11.9** (continued)

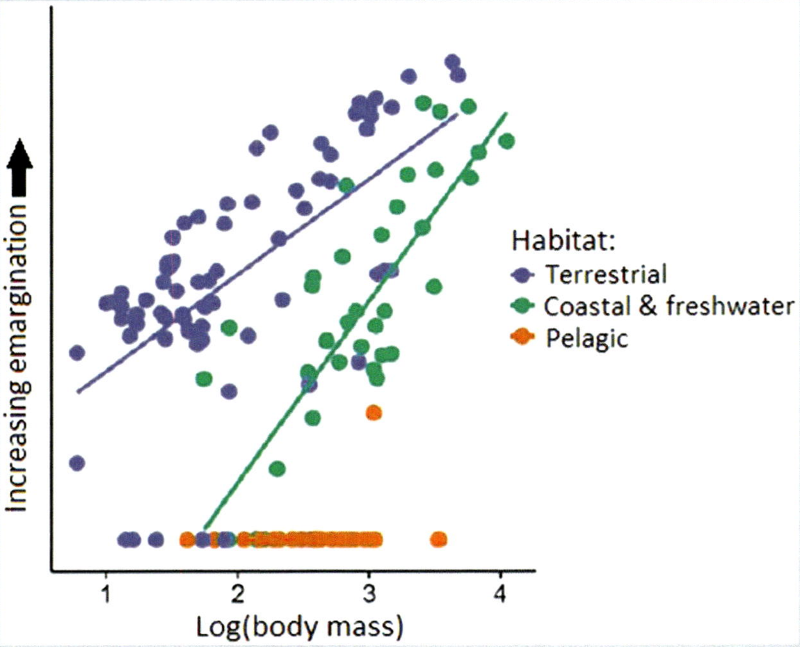


Emargination of primary feathers is influenced by mass in terrestrial species and coastal/freshwater species of birds, but does not change in pelagic species. In addition, coastal/freshwater and terrestrial species of birds have significantly more feather emargination than pelagic species. (Figure from Klaassen van Oorschot et al. 2017; © 2017 Wiley Periodicals, Inc., used with permission)

Klaassen van Oorschot et al. (2017) also found that the degree of emargination of primary feathers was influenced by habitat. Dynamic-soaring species occupy a habitat with persistent wind and, by taking off and landing into headwinds, such species "... are freed from the constraints of slow flight during take-off and landing." In contrast, terrestrial soaring species often take off from the ground and fly slowly when soaring, so emarginated primary feathers "may enhance lift and reduce induced drag costs which dominate at slow speeds." In addition, terrestrial birds may need to quickly gain height to avoid obstacles such as shrubs and trees, further enhancing the benefits of emarginated primary feathers. Coastal and freshwater species generally have no such obstacles and occupy more open habitats where headwinds may be present, perhaps explaining why they generally have primaries less emarginated than those of terrestrial species.

### 11.3.5 Wing Loading

Birds move through the air in various ways, including gliding, soaring, flapping, and hovering. All birds that fly use flapping flight, but do so to varying degrees. Some birds flap continuously, or nearly so, as they fly, whereas other birds prefer to glide and soar and only use flapping flight when needed (e.g., when taking off). Foraging hummingbirds frequently hover. Different "styles" of flight require different types of wings. However, a bird's flight style is also determined by its mass, more specifically mass relative to wing surface area, and the ability of its flight muscles to generate power.

A bird's mass divided by the area of its wings is defined as its wing loading and, among birds, wing loading values range from about 0.12 (Coal Tit) to 2.13 g/cm$^2$ (Arctic Loon; Alerstam et al. 2007). Although wing loading is important because it influences maneuverability, or turning radius (Pennycuick 1975), low wing loading is not required for high maneuverability; most birds can affect a turn of very small radius (facultative maneuverability; Warrick et al. 1998, 2002; Figs. 11.48 and 11.49). For example, when landing, many birds exhibit, for a brief period, impressive maneuverability. However, birds with low wing loadings are more efficient at maneuvering because they can make turns of small radii without expending the energy involved in slowing and flapping (Hails 1979). Therefore, for birds that spend much of their time maneuvering, such as frigatebirds and swallows, low wing loading is

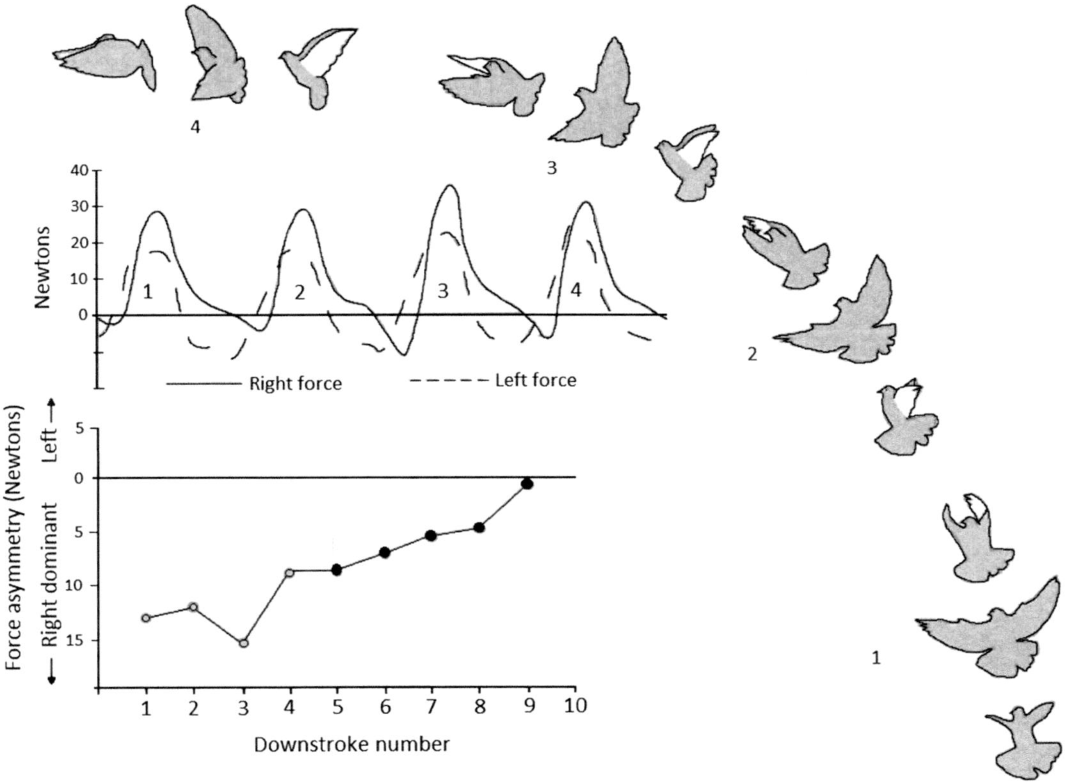


**Fig. 11.48** A Rock Pigeon (*Columba livia*) making an ascending left turn. In successive wingbeats, the wing on the outside of the turn produced greater peak force than the wing on the inside of the turn. This continuous production of asymmetrical downstroke force is likely important for birds making turns during slow, flapping flight. (Figure from Warrick et al. 1998; © 1998 Oxford University Press, used with permission)

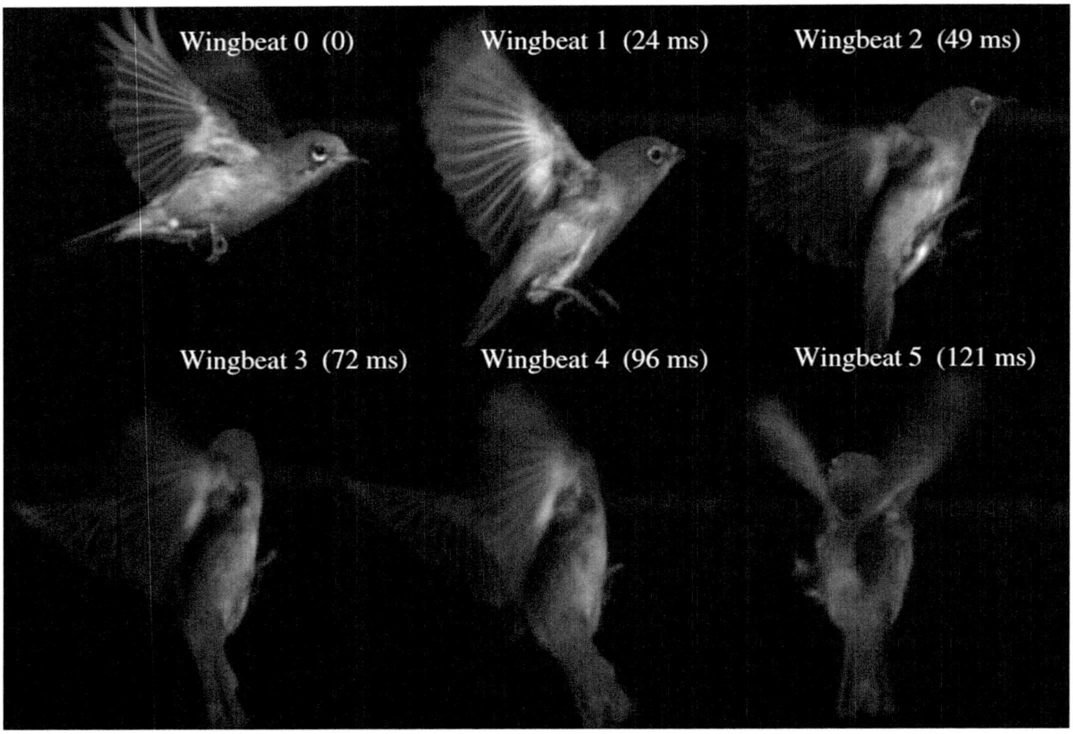


**Fig. 11.49** A Japanese White-eye (*Zosterops japonicus*) performing a 90° "hovering turn." Wingbeat 0 corresponds to the beginning of the turn and wingbeat 5 the end of the turn. This radius of this hovering turn was just 0.015 m, and the turn took only 0.121 s. Note that the body position of the Japanese White-eye quickly changed from nearly horizontal to nearly vertical in just 49 ms; a vertical orientation is better for making sharp turns when flying very slowly. (Figure from Su et al. 2012; © 2011 Society for Experimental Mechanics. Published by Springer Nature, used with permission)

clearly advantageous (Norberg 1986b, 1990; Warrick 1998). Conversely, if maneuvering is restricted to a few short, but critical, moments (take-off and landing), efficiency of maneuvering becomes unimportant. Thus, wing loading can be used a measure of time spent maneuvering, but not as a measure of absolute maneuverability (Warrick et al. 2002).

### 11.3.6 Banking

Birds can also turn by banking to the left or right. Banking begins with a bird increasing the angle of attack of the outside wing relative to the inside wing, which produces a roll into the turn (Fig. 11.50). Roll into the turn is stopped and maintained by decreasing the angle of attack of the outside wing to equal that of the inside wing. Roll back into a level flight begins with an increase in the angle of attack of the inside wing relative to the outside wing, and is again stopped when the angles of attack of each wing match. As these changes in wing angle of attack occur, the bank of the tail relative to the body also changes during rolls into and out of turns. During banking rolls, the banking angle of the tail tends to be greater than that of the body (Fig. 11.50; Carruthers et al. 2009). This increased banking of the tail may help prevent adverse yaw (Carruthers et al. 2009), which is the tendency for a bird to move (yaw) in the direction opposite of a roll because of differences in induced drag between the wings (greater on the outside wing).

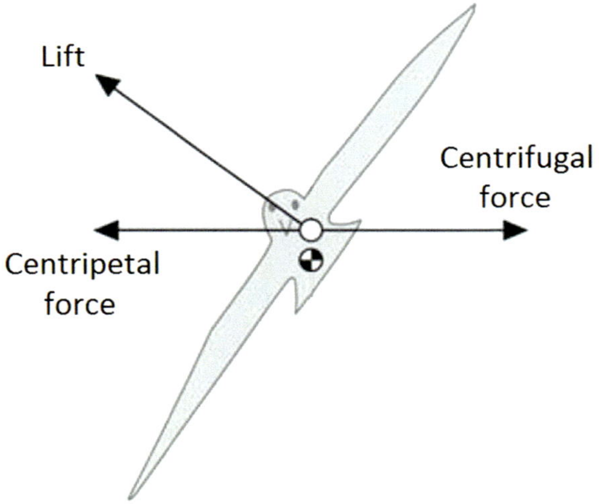


**Fig. 11.50** (Top) A Peregine Falcon (*Falco peregrinus*) banking to the right. (Photo by Don Endicott, National Park Service; used with permission). (Bottom) A banked turn to the right by a bird in non-flapping flight (gliding or soaring). Centripetal force is the force that makes a body follow a curved path and its direction is always at a right angle (orthogonal) to the velocity of the body and toward the fixed point of the center of curvature of the flight path. Shown in the top photo, but not in the bottom figure, is the position of the tail during a banked turn. Note that the banking of the tail is in the same direction as the bird's body, but to a greater degree. (Bottom figure modified from Hemelrijk and Hildenbrandt 2011; © 2011 Hemelrijk, Hildenbrandt, open-access article distributed under the terms of the Creative Commons Attribution License, https://creativecommons.org/licenses/by/4.0/)

## 11.4 Flying in Cluttered Habitats

Birds, particularly those that occupy cluttered habitats like forests, not only have to sometimes make sharp turns, but also need to, at times, fly through narrow spaces like small gaps between trees or branches. Williams and Biewener (2015) examined the strategies of Rock Pigeons (*Columba livia*) flying through a series of vertical poles and found that they used two different strategies (Fig. 11.51), with a more efficient wings-paused posture used when the gap between poles was wider and a less-efficient wings-folded posture when the gap was narrower and the risk of collision greater. Raptors and other birds in forest habitats use similar postures when flying between trees and branches (Fig. 11.52). Of course, if possible, birds likely try to avoid flying through narrow gaps that require use of either of these strategies. To do so, Bhagavatula et al. (2011) found that birds can negotiate cluttered habitats by balancing the speeds of image motion experienced by the two eyes (Fig. 11.53). When flying through, for example, a forest, a bird detects its

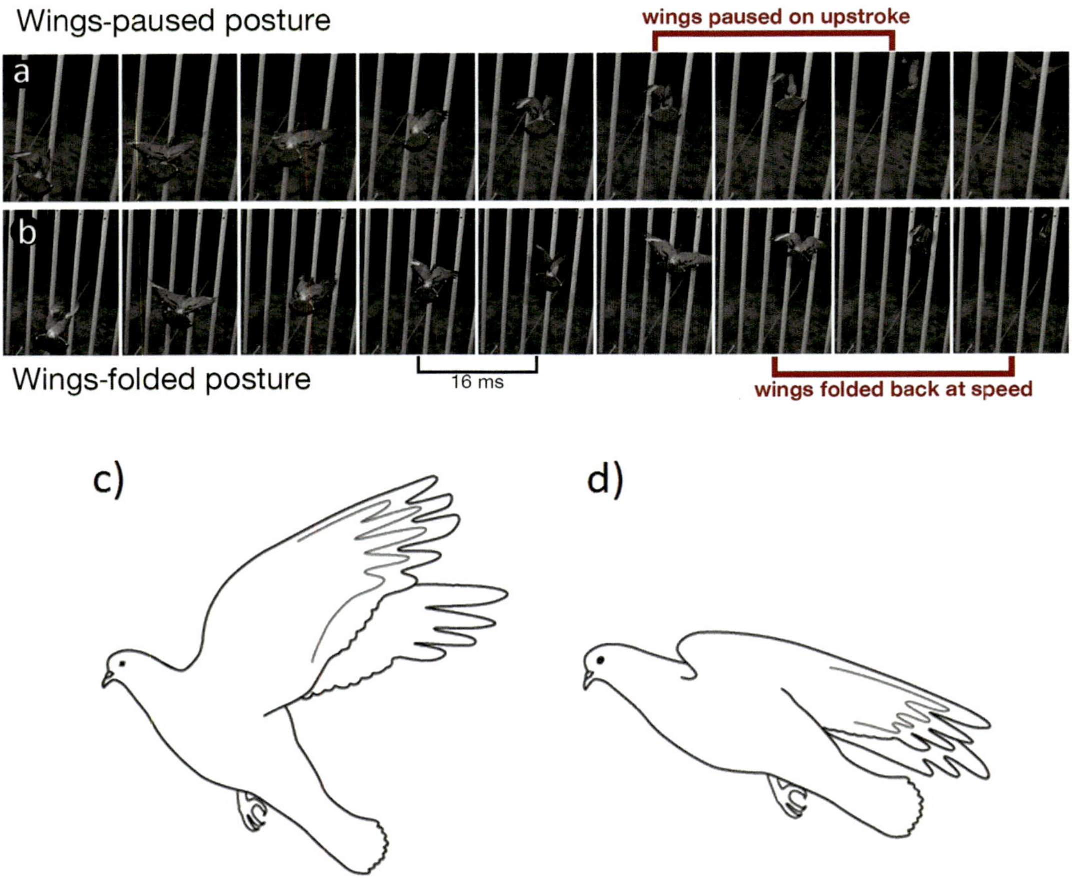


**Fig. 11.51** When flying through an array of vertical poles, Rock Pigeons (*Columba livia*) used two different strategies, either (**a**, **c**) a very brief pause in the normal wingbeat cycle, or (**b**, **d**) a wings-folded posture. The wings-folded posture takes longer to assume and causes a slight loss in altitude so is less efficient than the brief-pause strategy. Pigeons used the more efficient wings-paused strategy when the gap was wider (**a**), but switched to the less-efficient wings-folded posture when the gap was narrower (**b**) and there was a greater chance of colliding with a pole. (Figures from Williams and Biewener 2015; used with permission of the United States National Academy of Sciences)

relative motion by observing the apparent speed with which trees move past its two eyes. Of course, when objects like trees are closer, they appear to move past the eye faster. So, to safely fly through groups of trees, a bird attempts to balance the speed with which trees on each side move past its two eyes, i.e., attempting to maintain a relatively equal distance from the trees on each side.

## 11.5 Flight Styles

Not surprisingly, given the substantial variation among birds in aspect ratios, wing loading, and power available for flight, the flight styles of birds are incredibly diverse and difficult to categorize. Traditionally, wing shape has been used to categorize flight style (Savile 1957), with (1) elliptical

**Fig. 11.52** A Cooper's Hawk (*Accipiter gentilis*) in pursuit of prey will have its wings extended as it approaches a narrow opening between trees or branches, then folds its wings to pass through narrow openings before again extending its wings after passing through the narrow gap. (Photo from http://www.tug44.org/; CC0 Public Domain)

wings for low-speed, maneuverable flight in forested or shrubby habitats, (2) high-speed wings characteristic of aerial foragers and birds that make long-distance migrations, (3) high-aspect ratio wings for soaring over the ocean, and (4) slotted high-lift wings characteristic of terrestrial soaring birds (Welty and Baptista 1988). However, based on differences in aspect ratios and wing loading (Rayner 1988; Fig. 11.54), flight styles can also be categorized as either specialized or nonspecialized. The nonspecialists have average aspect ratios and average wing loading and are excellent flyers (capable of long flights and with good maneuverability) that typically use flapping flight. The nonspecialists can be further subdivided, based on aspect ratio and speed, as slow nonspecialists and fast nonspecialists. In the slow category would be most passerines (Passeriformes), pelicans (Pelicaniformes), herons, egrets, ibises, and storks (Ciconiiformes), pigeons and doves (Columbiformes), cuckoos (Cuculiformes), most owls (Strigiformes), trogons (Trogoniformes), most birds in the order Gruiformes (e.g., gallinules, rails, and bustards), mousebirds (Coliiformes), woodpeckers (Piciformes), and parrots (Psittaciformes). Fast nonspecialists include many falcons (Falconidae), gulls (Larinae), and storm-petrels (Hydrobatidae).

Birds with morphological attributes (aspect ratio and wing loading) that differ (beyond one or two standard deviations) from those of "typical" birds exhibit specialized flight styles (Rayner 1988). Among these specialized styles are:

1. Marine soarers are birds that fly for long periods over the open ocean and have very high-aspect ratio wings and average or low wing loading that reduce the energetic cost of flight. Birds in this category include the albatrosses (Procellariiformes).
2. Aquatic divers/swimmers are birds with medium to high-aspect ratios and high wing loading, including murres, loons, grebes, scoters, mergansers, ducks, and swans. These birds fly rapidly, but with limited maneuverability, characteristics useful for birds that often fly long distances (e.g., during migration) and take off and land on water where precise maneuverability is not as important.

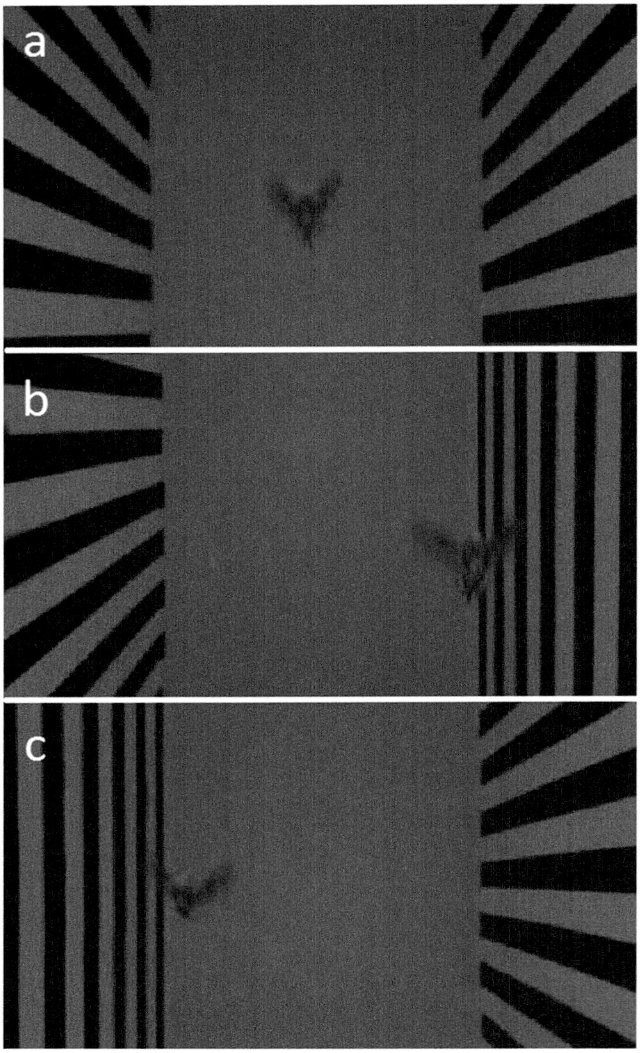


**Fig. 11.53** Budgerigars (*Melopsittacus undulatus*) were trained to fly down a corridor with vertical and/or horizontal lines on the walls. (**a**) When both wall had vertical lines, Budgerigars flew down the center of the corridor, but, when one wall had vertical lines and the other had horizontal lines (**b** and **c**), Budgerigars were unable to accurately judge their distance from the wall with horizontal lines (because the horizontal lines always look the same; there is no image motion) and so flew much closer to that wall than to the wall with vertical lines, sometimes even hitting the wall with horizontal lines. (Figure made from screenshots of supplemental videos provided online by Bhagavatula et al. 2011; © 2011 Elsevier Ltd., used with permission)

3. Aerial hunters are birds with high-aspect ratio wings and low wing loading, a combination permitting rapid flight and excellent maneuverability. Aerial hunters include swallows and martins (Passeriformes), swifts (Apodiformes), nightjars (Caprimulgiformes), Swallow-tailed Kites (*Elanoides forficatus*; Accipitriformes), frigatebirds (Fregatidae), terns (Sterninae), some falcons (e.g., hobbies and Eleonora's Falcon [*Falco eleonorae*]), and tropicbirds (Phaethontidae).
4. Soarers/coursers include birds with low aspect ratios and low wing loading, characteristics that allow relatively large birds to either soar or fly just above the vegetation in open habitats in search of prey. Birds in the soaring category include hawks and eagles (Accipitriformes), vultures and condors (Cathartiformes), and cranes (Gruiformes). Coursing birds include some owls, e.g., Barn Owl (*Tyto alba*) and Short-eared Owl (*Asio flammeus*; Strigiformes) and harriers (Accipitriformes).

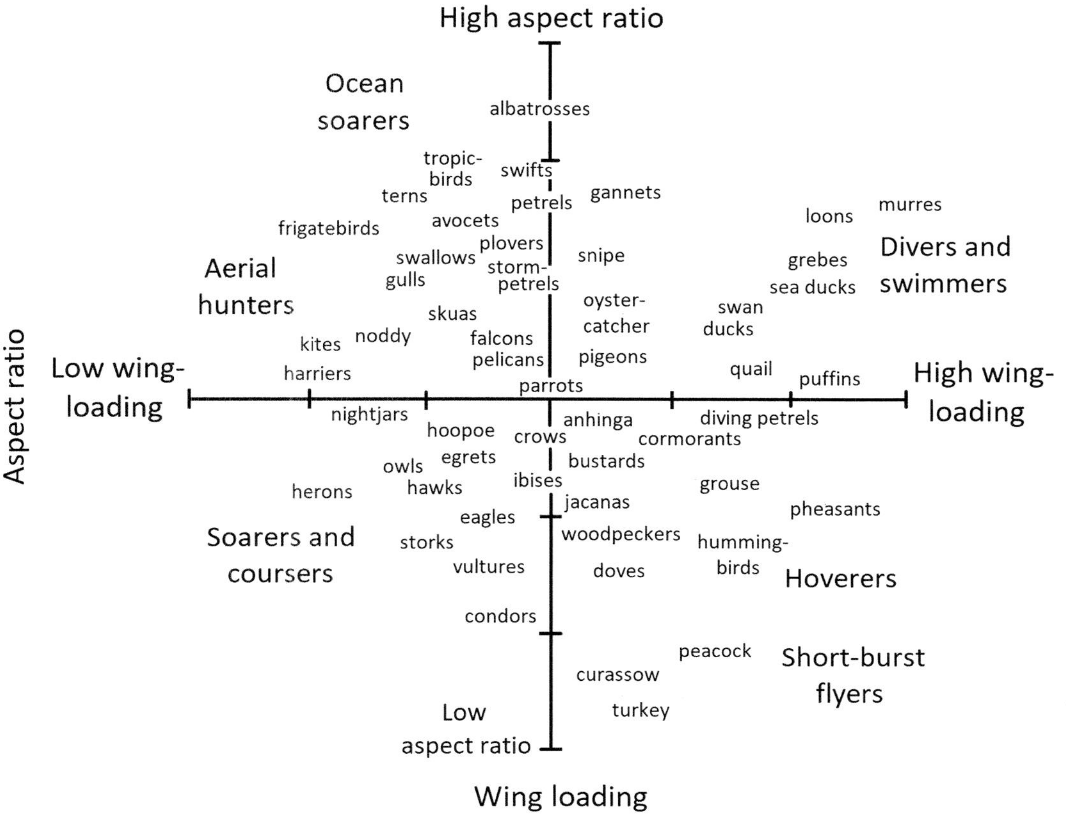


**Fig. 11.54** Aspect ratios relative to wing loading for some major groups of birds and for the six flight styles described by Rayner (1988). (Figure modified from Rayner 1988; © 1988 Plenum Press, New York, used with permission)

5. Short-burst fliers are birds with low aspect ratios and high wing loading that fly infrequently and only for short distances (Fig. 11.55). Birds in this category include those in the orders Galliformes (e.g., turkeys, pheasants, quail, grouse, and megapodes) and Tinamiformes (tinamous).
6. Hoverers are birds capable of flying in one position without wind and have high-aspect ratios and, surprisingly, high wing loading. The high-aspect ratio reduces the energetic cost of flight, whereas the high wing loading permits relatively fast, agile flight (Rayner 1988). The only true hoverers are the hummingbirds (Apodiformes).

### 11.5.1 Gliding and Soaring

Whether specialized for certain styles of flight or not, most birds do use different types of flight. For example, albatrosses spend most of their time soaring over the open ocean, but must use flapping flight when taking off from land (e.g., nest sites) and water. Of the various types of flight, gliding is the simplest. Gliding is relatively inexpensive because a bird covers the aerodynamic cost by losing potential energy. On the negative side, however, gliding is a relatively slow type of flight. A gliding bird uses its weight (mass) to overcome air resistance to its forward motion. To do this effectively, of course, requires a certain

**Fig. 11.55** Short-burst flyers like Ring-necked Pheasants (*Phasianus colchicus*) "burst" into flight, perhaps in response to an approaching predator, flap continuously for a few seconds (or more depending on how far it wants to fly), then use the momentum generated to glide to their landing spot. (Photo by Tom Koerner/USFWS, Wikipedia, CC BY 2.0, https://creativecommons.org/licenses/by/2.0/)

mass and, as a result, only relatively large birds, such as vultures, glide on a regular basis. When gliding, birds lose altitude at a sinking rate ($V_z$) while traveling forward at some flight speed ($V_{xy}$; Fig. 11.56). A bird's glide ratio equals $V_{xy}/V_z$ (the distance traveled forward per unit of altitude lost). The best gliders are the albatrosses, with glide ratios as high as 23:1 (Anderson and Eberhardt 2001). By comparison, White-backed Vultures (*Gyps africanus*) have a glide ratio of 15:1, typical airliners 16:1, and the best man-made gliders 60:1 (Videler 2006). Of course, gliding birds can store potential energy by soaring in updrafts or thermals, by using powered flight to increase altitude, or by dynamic soaring (discussed below; Rosén and Hedenström 2001). By doing so, a gliding bird can maintain or increase its altitude without flapping its wings.

Updrafts can be generated when a steady wind strikes an obstruction, such as a hill, cliff, or building, and by the uneven heating of air near the earth's surface (Fig. 11.57). Such uneven heating creates thermals. For example, one side of a hill may be heated by the sun while the other side is not. Regardless of what causes the differential heating, warmer air has a lower density and tends to rise, creating an updraft. Air in a thermal is typically about 1–2°C warmer than the surrounding air and rises at a rate of about 2–3 m per second (Pennycuick 2003). As the air rises, typically to altitudes of 2–3 km (Pennycuick 2003), it gradually cools until it eventually descends back toward the ground. Thermals are typically columnar in shape and the location of such thermals is often apparent from the location of cumulous clouds because the cooling air inside

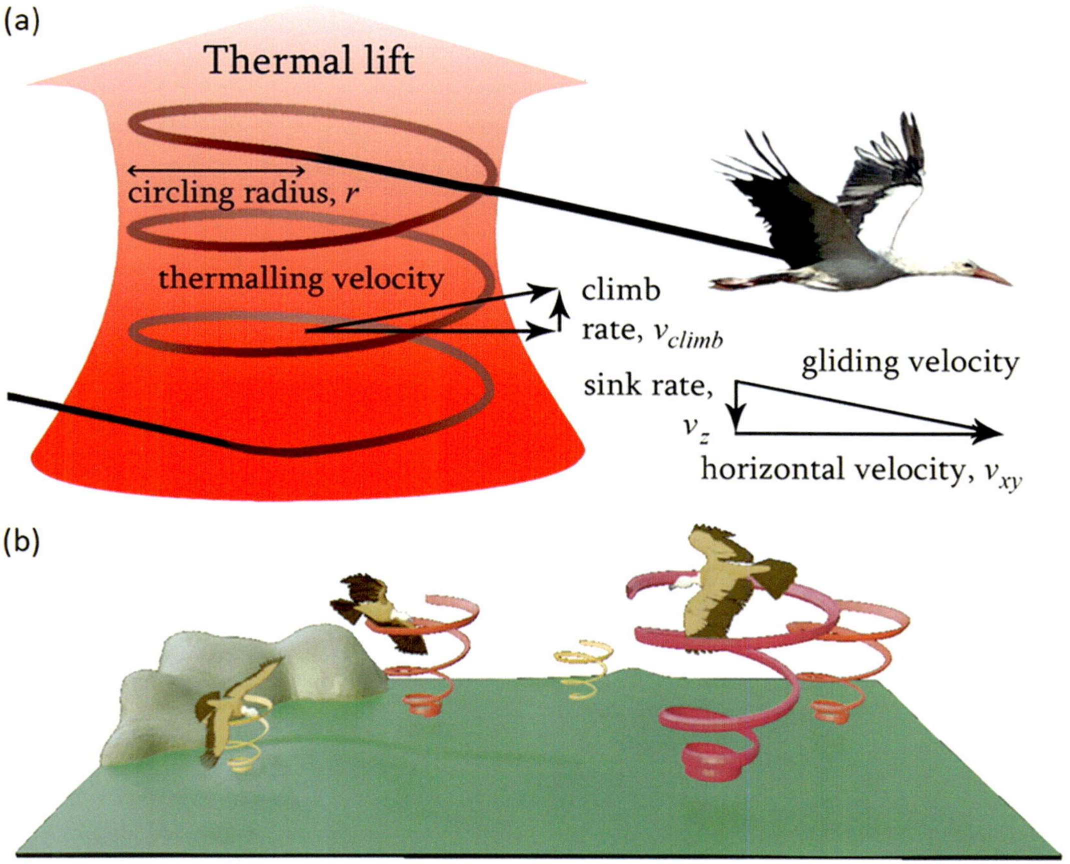


**Fig. 11.56** (**a**) Soaring birds traveling long distances can use thermals to gain altitude and then glide while searching for another thermal. As birds glide, altitude is lost, but can be regained if another area of rising air is located before too much altitude is lost. (**b**) The location of thermals may, to varying degrees depending on the landscape, be predictable in time and space, but soaring and gliding birds may also use social information, i.e., the location of other soaring birds, to identify the locations of thermals. (Figure **a** from Ákos et al. 2008; Copyright (2008) National Academy of Sciences, USA, used with permission. Figure **b** from Williams and Safi 2021; © 2021 The Authors. Published by Elsevier Ltd., used with permission)

a rising thermal sometimes causes water vapor in the air mass to condense. Once created, thermals move with the wind (as is apparent by the movement of cumulous clouds). Because of the columnar shape of most thermals, birds using thermals for lift typically fly in circles (to stay in the area of rising air). However, the circling radius of a bird climbing in a thermal is proportional to its wing loading (mass relative to wing area); birds with lower wing loading (lighter relative to the surface area of their wings) have a smaller circling radius in thermals, and improved climbing capacity than birds with higher wing loading (Alerstam 1994; Fig. 11.58, Box 11.10 Wind-Drift Circling Soaring by Great Frigatebirds). Typically, thermals begin to develop in the morning as ground temperatures rise. The development of thermals typically peaks at midday, then declines in the late afternoon. Soaring birds enter thermals where lift is strong enough to promote soaring; birds climb in a thermal by circling, and then glide between thermals (Fig. 11.59).

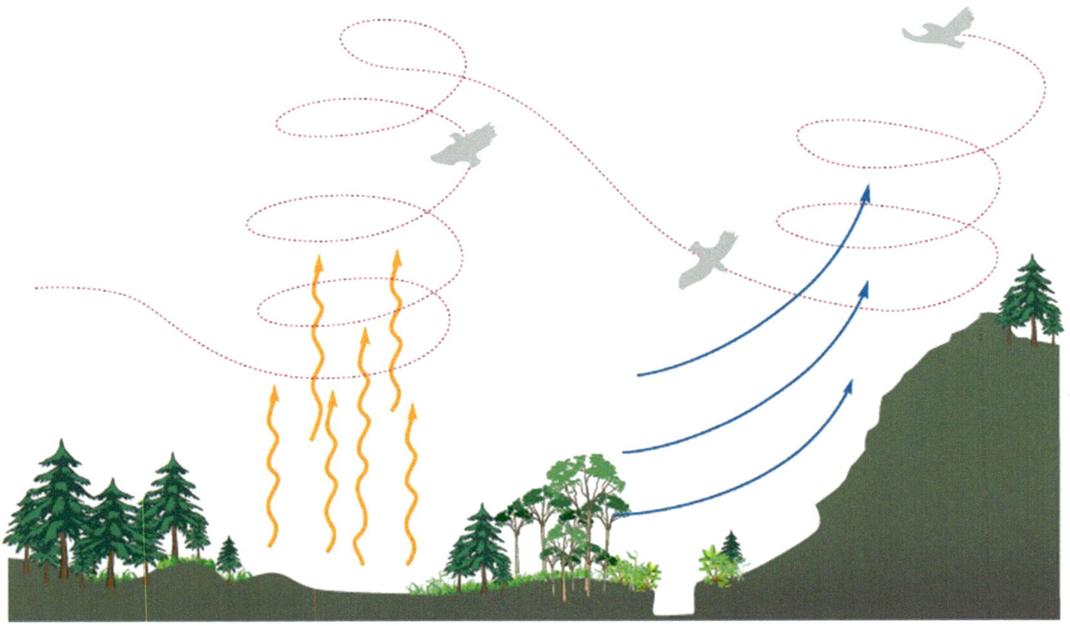

**Fig. 11.57** Soaring birds can take advantage of rising air generated by uneven heating of the earth's surface (thermals; left) or by uneven topography (obstruction or orographic lift; right). (Figure from Guerrero 2009; used with permission of Joel Guerrero)

Soaring is a very energy-efficient way to fly. For example, Duriez et al. (2014) fitted Eurasian Griffons (*Gyps fulvus*) and Himalayan Griffons (*Gyps himalayensis*) with data-loggers that

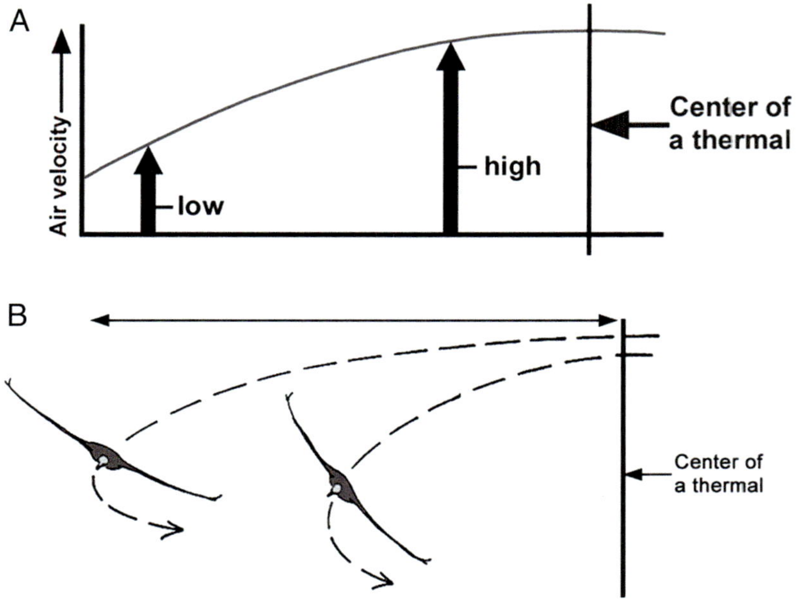


**Fig. 11.58** The velocity of rising air and, therefore, the amount of lift are greater closer to the center of thermals. (**a**) As a result, the smaller the circling radius in a thermal, the greater the lift. (**b**) Smaller birds have smaller circling radii than larger ones, but even large soaring birds can bank at greater angles to reduce their circling radius and get more lift. (Figure modified from Chatterjee et al. 2007; used with permission of the US National Academy of Sciences)

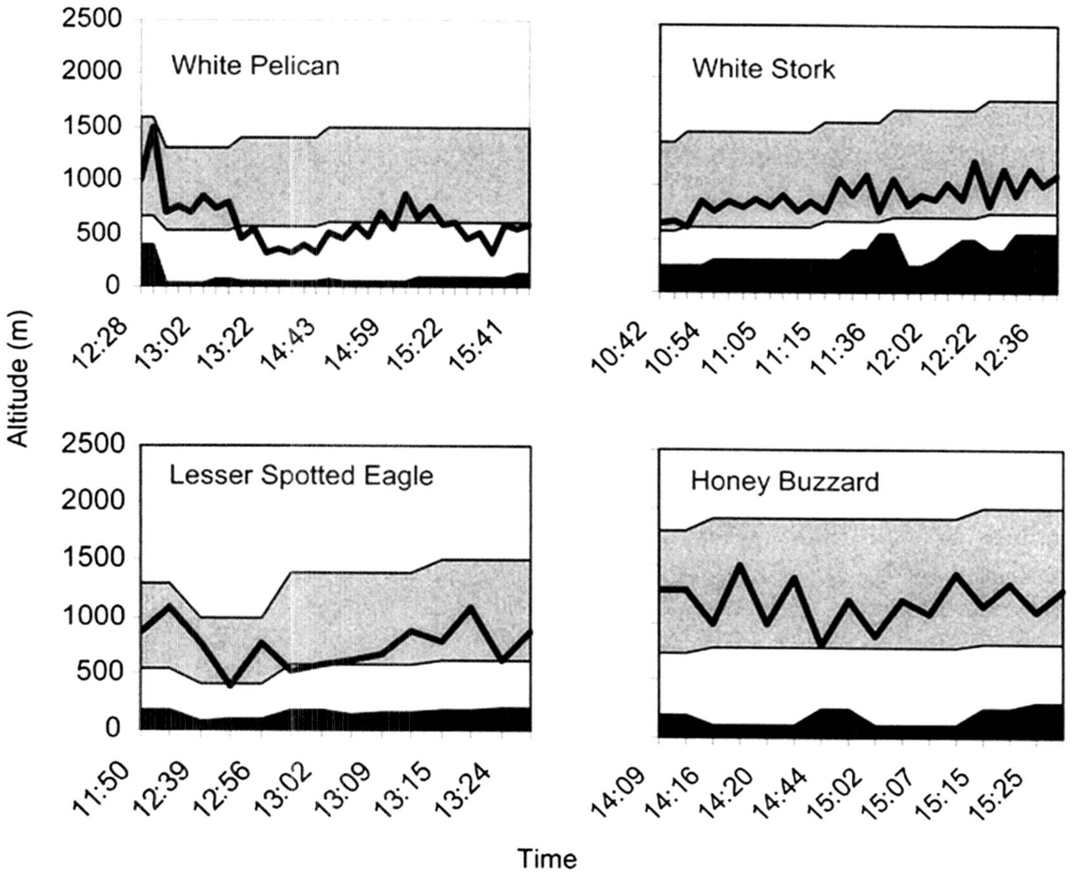


**Fig. 11.59** Examples of soaring paths, and time spent taking advantages of the rising air in thermals, taken by flocks of four species during migration over central Israel. Upper lines indicate the altitude at the top of thermals. Solid lines indicate the flight altitude of one flock during one day: Top left, American White Pelicans (*Pelecanus erythrorhynchos*), Top right, White Storks (*Ciconia ciconia*), Bottom left, Lesser Spotted Eagles (*Clanga pomarina*), and Bottom right, European Honey-buzzards (*Pernis apivorus*). All four species during this study spent most of their time soaring at altitudes between 41% and 100% of the maximum predicted thermal depth (gray area, representing, 75%, 78%, 96%, and 74%, respectively, of all observations of each species). Filled area indicates topography. (Figure from Shamoun-Baranes et al. (2003); © 2003 Oxford University Press, used with permission)

**Box 11.10 Wind-Drift Circling Soaring by Great Frigatebirds**

There are five species of frigatebirds, all in the genus *Frigata*. The genus name is derived from the French name for these birds *la frégate*, referring to a frigate or a fast warship (Jobling 2010). Frigatebirds can fly fast, but are best known for their ability to remain airborne for long periods at sea. Because their feathers are not waterproof, frigatebirds cannot land in the water, obtaining food by picking up items at the water's surface (or, sometimes, by chasing other birds and forcing them to drop prey items that are then retrieved by the frigatebird).

(continued)

**Box 11.10** (continued)

Great Frigatebird (*Fregata minor*) picking a fish off the water's surface. (Photo by Duncan Wright, Wikipedia, CC0 Public Domain)

Frigatebirds have the lowest wing loading of any bird and so are specialized for soaring flight. On land, soaring birds often use thermals to gain altitude and remain airborne while expending minimal energy. Weimerskirch et al. (2016) discovered that Great Frigatebirds also use rising air to remain airborne for long periods over the Indian Ocean. More specifically, Great Frigatebirds take advantage of favorable winds and thermals created by convection from the ocean's surface (rising of warmer, lighter air) to soar in a manner referred to as wind-drift circling soaring. Updrafts of rising air "drift" along with the trade winds and so the frigatebirds, to remain in the column of rising air, must also "drift." Using a combination of soaring (to gain altitude) and gliding flight, Great Frigatebirds are able to minimize the energetic cost of flight and move over considerable distances, typically at altitudes ranging from 50 to 600 m where they can scan the ocean's surface looking for foraging opportunities, e.g., prey driven toward the surface by foraging schools of tuna (*Katsuwonus* and *Euthynnus* spp.), other predatory fishes, or dolphins (*Stenella*, *Delphinus*, and *Steno* spp.) (Au and Pitman 1986).

Great Frigatebirds were sometimes found to attain altitudes as high as 4000 m by soaring inside cumulus clouds. Beginning at such altitudes, they can then glide considerable distances (>60 km) searching for favorable "wind-drift circling soaring" conditions and even sleeping for short periods (~ 2–12 min) (Weimerskirch et al. 2016; Rattenborg et al. 2016).

(continued)

**Box 11.10** (continued)

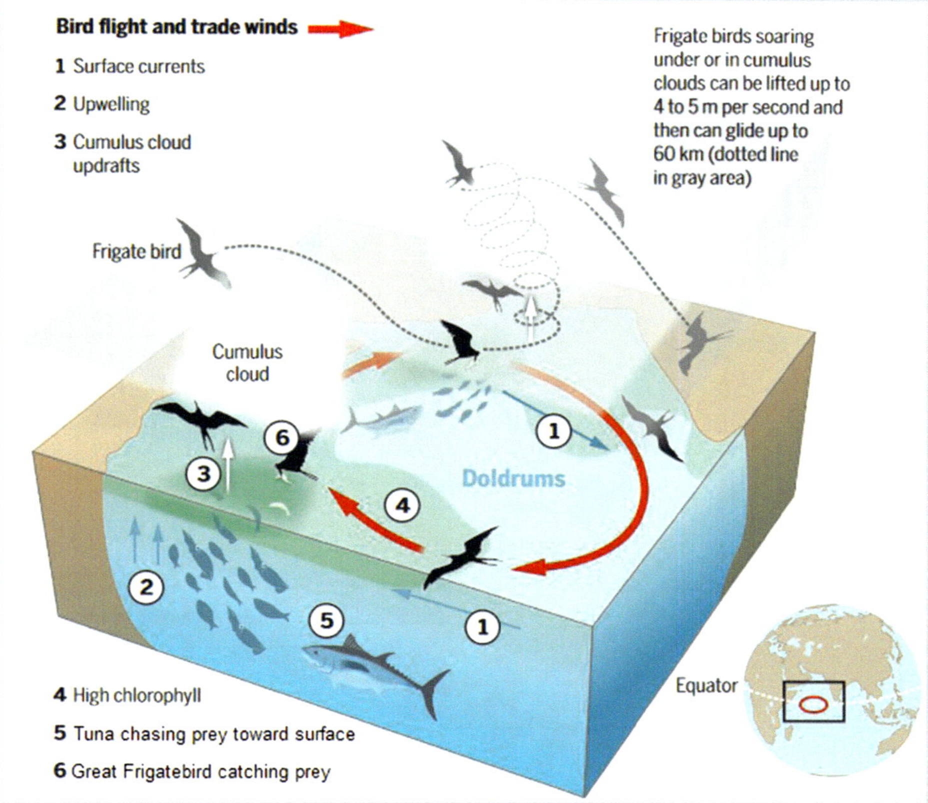


Weimerskirch et al. (2016) used satellite technology to track the flights of Great Frigatebirds around the Indian Ocean. In the Indian Ocean, trade winds blow clockwise around an area of calm air (called the doldrums). Great Frigatebirds take advantage of convection from the ocean surface to soar and glide in search of foraging opportunities. The tradewinds generate surface currents and create upwellings that bring nutrients to the surface and create areas of high primary productivity (chlorophyll a). Tuna and other predators chase fish and squid to the surface in these areas and make them accessible to Great Frigatebirds. (Figure modified from Huey and Deutsch 2016, Illustration by K. Sutliff/*Science;* © 2015 AAAS, used with permission)

measured location (GPS), relative movement (accelerometer), and heart rate (as a proxy for metabolic rate) and found that, when soaring, their heart rates (and metabolic rates) were comparable to those at baseline or resting levels (Fig. 11.60). Heart rates of the vultures increased dramatically at take-off when flapping flight was required, but just as quickly declined as they began soaring and gliding. Given the low energetic cost of soaring and gliding, vultures and other large soaring birds like Broad-winged Hawks (*Buteo platypterus*) can fly for long periods and over large distances, e.g., during migration (Fig. 11.61), with minimal energetic cost. Because some vultures can fly more than 200 km per day searching for carrion, the low

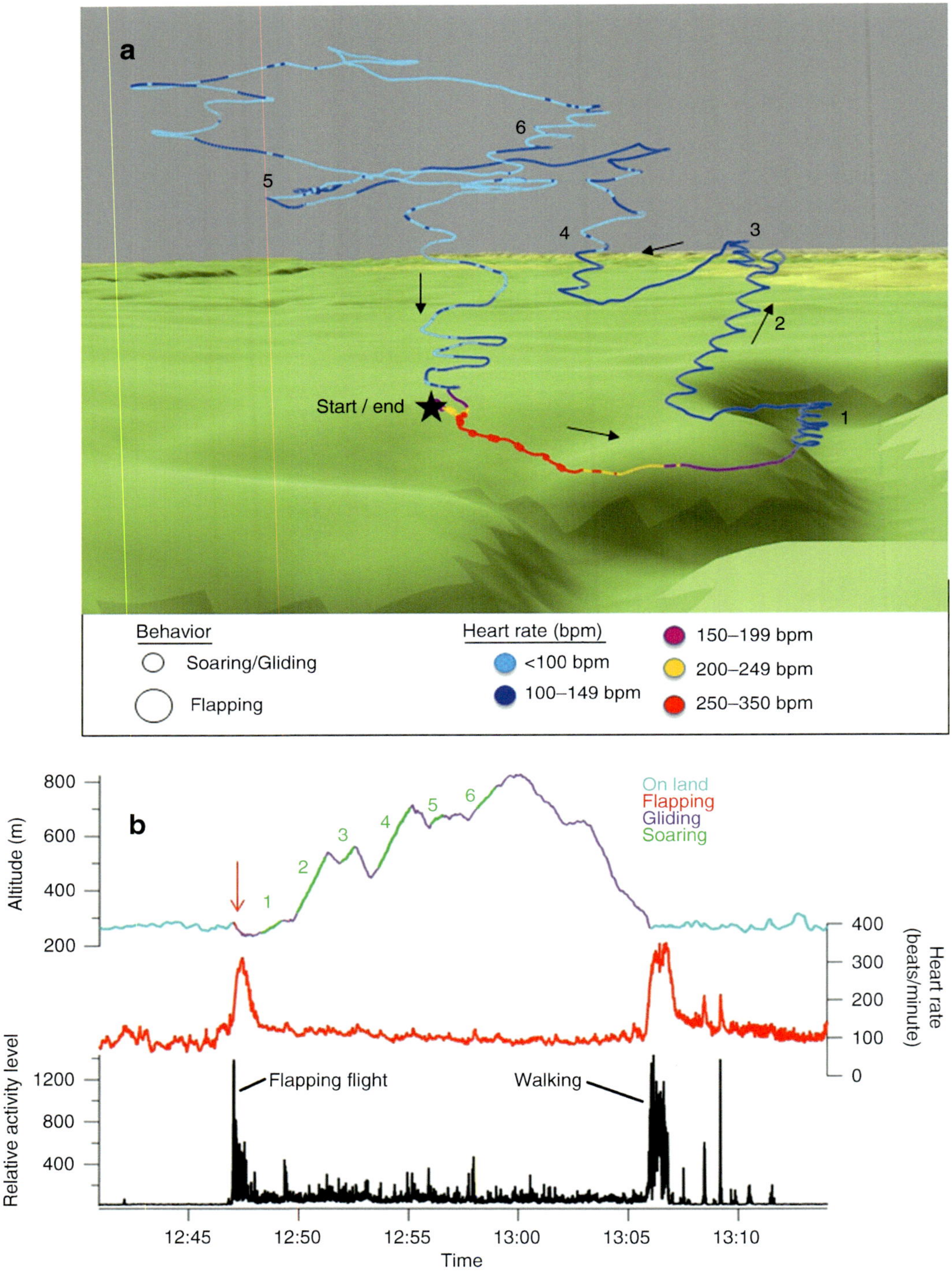


**Fig. 11.60** Example of flight track, heart rate and behavior of a Himalayan Griffon (*Gyps himalayensis*). (**a**) The color of the flight path indicates the vulture's heart rate, ranging from red (250–350 beats per minute) down to light blue (less than 100 beats per minute). During most of the flight, the vulture was soaring or gliding (indicated by a narrow line); only briefly at take-off did the vulture use flapping flight (thicker lines to the right of the star indicating the take-off and landing points). The black arrows indicate the flight direction and periods of soaring are numbered and correspond with those in **b** indicating the vulture's altitude. Also provided in **b** are the heart-rate

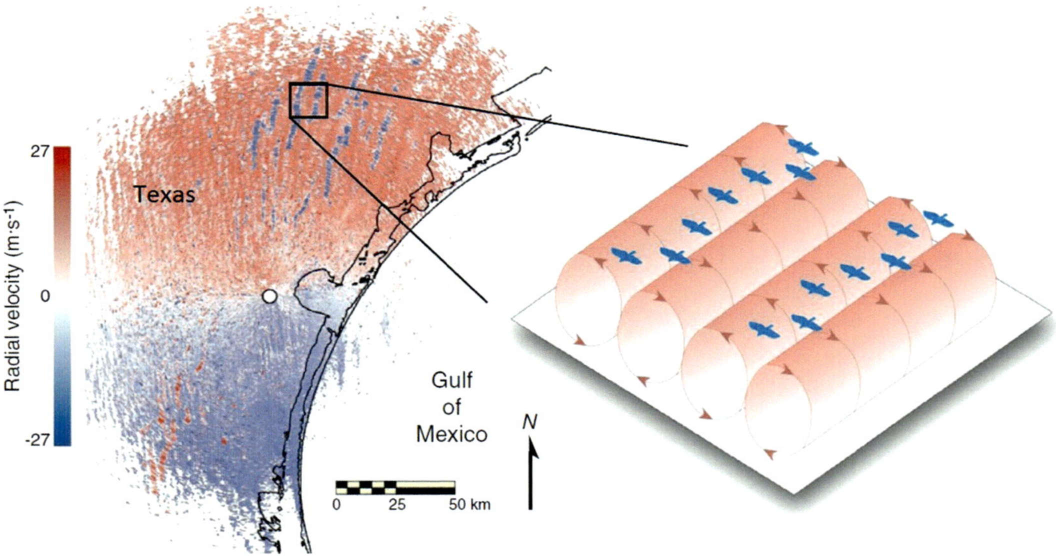


**Fig. 11.61** Soaring birds can also take advantage of horizontal bands of rising air generated by horizontal roll vortices. These vortices appear as striations in radar images and, in this radar image, blue indicates movement toward the radar (white dot) and red indicates movement away from the radar. The blue streaks within the red area are large numbers of migrating (and soaring) Broad-winged Hawks (*Buteo platypterus*) flying north and following bands of rising air between the roll vortices. Surveys conducted near the radar station revealed that nearly 100,000 raptors migrated through the area on the day when this image was generated. Horizontal roll vortices can form over land or water and require rising air (generated by warm land or water). Conditions conducive to their formation often include variable wind velocities, with faster winds at the convective boundary layer (altitude at which rising warmed air is sufficiently cooled to stop rising, usually about 1–2 km above ground) and slower winds at lower altitudes. As the warm air rises, water condenses and forms horizontal bands of clouds between the vortices. (Figure from Diehl 2013; Published by Elsevier Ltd., used with permission)

energetic cost of soaring and gliding is critically important (Duriez et al. 2014). As an extreme example, Williams et al. (2020) equipped eight Andean Condors (*Vultur gryphus*) with data-loggers that included accelerometers so wingbeats could be monitored and determined that, in 235 h of flight time, the condors used flapping flight only 1.3% of the time and more than 75% of that flapping flight occurred when the condors were taking off.

### 11.5.2 Dynamic Soaring

Over the open ocean, large birds like Wandering Albatrosses (*Diomedea exulans*) take advantage of wind velocity gradients in a type of soaring called dynamic soaring (Fig. 11.62). Wind velocity typically varies with height above the water's surface because swells and waves impede air flow close to the surface. So, with increasing height above the surface, wind velocity tends to increase. Dynamic soaring requires a minimum

**Fig. 11.60** (continued) data and accelerometer data. Note the correspondence between times with more activity (flapping flight and walking) and rapid increases in heart rate. For altitude, the color of the line changes with behavior, and the beginning of the brief bout of flapping flight is indicated by the red arrow; the green numbers indicate periods of soaring corresponding to the numbers in **a**. (Figure modified from Duriez et al. 2014; © 2014 Duriez et al., open-access article distributed under the terms of the Creative Commons Attribution License, https://creativecommons.org/licenses/by/4.0/)

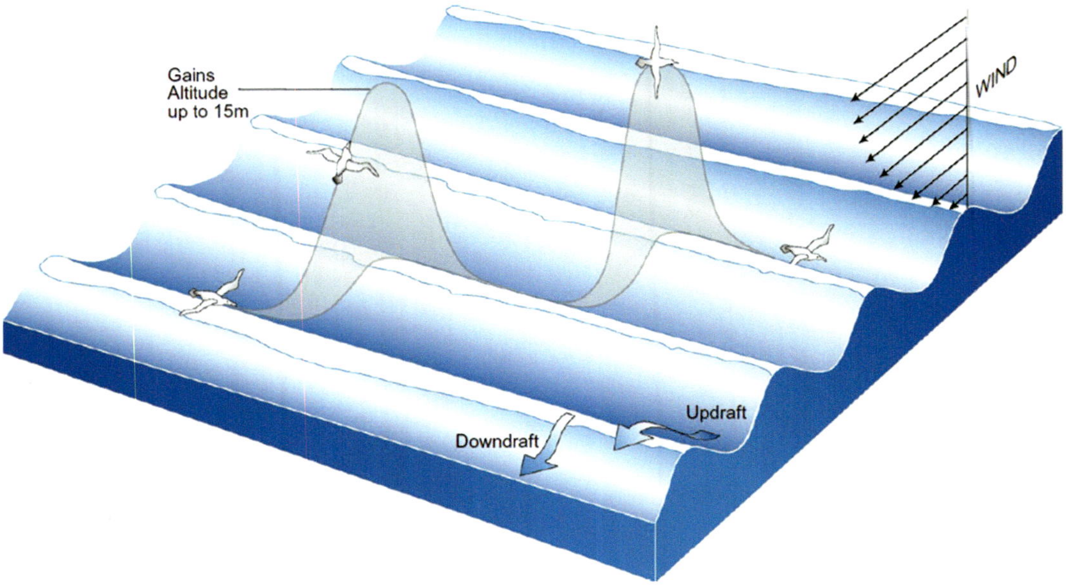


**Fig. 11.62** Example of an albatross dynamic soaring. The length of the arrows on the right indicates relative wind speed, with longer arrows indicating higher wind speeds. A dynamic-soarer, such as an albatross, flies into the wind and the increased speed of the air moving over its wings generates lift, and the bird then turns downwind and gains momentum as it moves down toward the slower moving air in the troughs between swells or waves. Although the downwind gliding appears parallel to the waves, the movement is actually downwind because the waves are moving. After gliding some distance and starting to lose momentum, the bird again turns into the wind to gain lift and repeat the dynamic-soaring cycle. (Figure from Richardson 2011; used with permission of Philip L. Richardson)

wind velocity of about 5 m/s close to the water's surface and a minimum velocity of about 9 m/s about 10–30 m above the surface (Sachs 2005). Dynamic soaring consists of periodically repeated cycles with four phases: (1) climb, or windward flight, to a height of about 10–30 m above the ocean surface, (2) upper curve, or change in flight direction to leeward (with the wind), (3) descent, or leeward flight, and (4) lower curve, or change of flight direction to windward (into the wind; Fig. 11.62). For example, assume an albatross gliding with the wind (15 m/s) at an altitude of about 20 m. As it glides, the albatross picks up speed and momentum while simultaneously losing altitude. When just above the water, the albatross swings around into the wind. As it does so, the velocity of the wind over its wings increases because wind velocity will equal the wind speed plus the albatross's air speed. Assuming an air speed a bit below wind speed as the albatross turns into the wind, the effective wind velocity over its wings when the albatross reaches the fastest moving air would be 12 m/s plus 15 m/s, or 27 m/s. As a result, the amount of lift generated remains high even as the albatross begins to slow down relative to the ocean surface. As its velocity relative to the surface approaches zero, the albatross then wheels around and flies with the wind to again pick up momentum as it begins another glide downward toward the ocean surface. By repeating this process (with each cycle taking about 15 s; Sachs et al. 2012), albatrosses and other large birds of the open ocean can travel long distances without flapping and, therefore, at minimum energetic cost (Fig. 11.63). Helping to minimize the cost of dynamic soaring by albatrosses is a specialization called a shoulder-lock, a sheet of tendon that locks the wing when fully extended to help keep

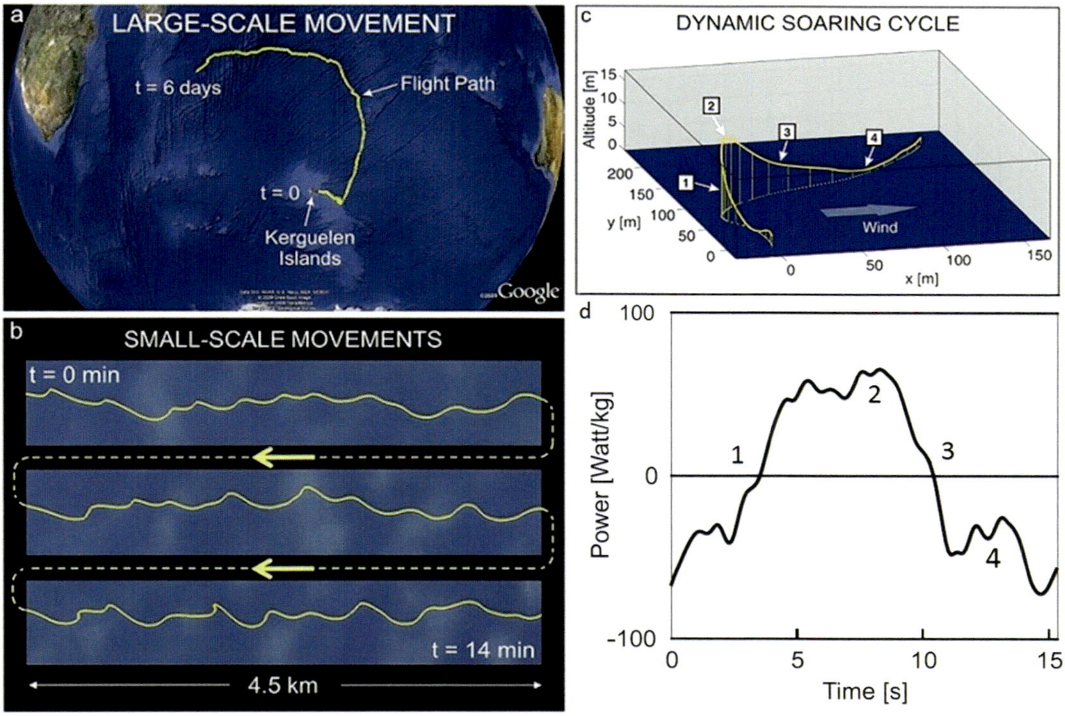


**Fig. 11.63** Movements of a dynamic-soaring Wandering Albatross (*Diomedea exulans*) at different scales. (**a**) Path of a 6-day, 4950-km flight of a Wandering Albatross in the southern Indian Ocean. (**b**) A 14-min portion of the long flight showing that the flight path consisted of a repeating series of small turns. (**c**) Each small turn in (**b**) represents a dynamic-soaring cycle, with each cycle consisting of (1) a windward climb to an altitude of about 12 m, with the increasing wind velocity at higher altitudes moving across the bird's wings and creating lift, (2) a curve that takes the albatross in a downwind direction, (3) a descent closer to the surface of the water where wind velocity is lower, and (4) a curve that will take the albatross into another windward climb. (**d**) During a windward climb, albatrosses gaining power or energy from the wind. During the leeward descent and curve from lee- to windward, energy is lost. However, the overall positive power, or energy gained, exceeds the negative power and this power surplus is sufficient to compensate for drag. (Figures modified from Sachs et al. 2012; © Sachs et al., open-access article distributed under the terms of the Creative Commons Attribution License, https://creativecommons.org/licenses/by/4.0/)

the wings extended (Pennycuick 1982; Meyers and Stakebake 2005).

Wandering Albatrosses have been recorded making foraging trips as long as 15,200 km and making flights around the world in as little as 46 days, flying at speeds of 80 km/h or higher and flying as far as 900 km in a single day (Jouventin and Weimerskirch 1990; Croxall et al. 2005; Fig. 11.63). When dynamic soaring, albatrosses gain energy from the wind and, as a result, fly long distances without expending any more energy than when they are resting on the water. As an example, Weimerskirch et al. (2000) fitted a Wandering Albatross with a data-logger that allowed them to monitor its heart rate (which is an indirect measure of energy expenditure). Although the albatrosses' heart rate (and energy expenditure) increased when walking and wing flapping (e.g., when taking off), its heart rate when dynamic soaring is just above that when it was resting on the water (Fig. 11.64). Similarly, Sakamoto et al. (2013) used data-loggers to monitor activity patterns and heart rates of Black-browed Albatrosses (*Thalassarche melanophris)* and found that heart rates averaged 144 beats/min when dynamic soaring, similar to the mean heart

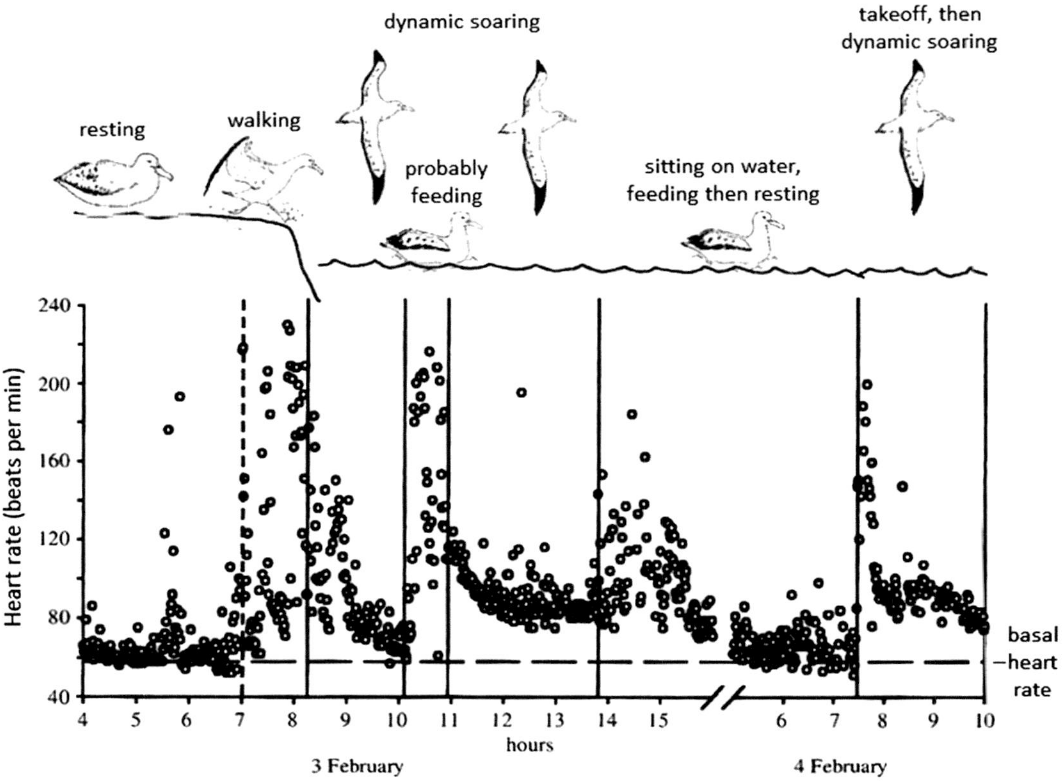


**Fig. 11.64** Heart rate of a male Wandering Albatross (*Diomedea exulans*) over 15 hours as recorded by a data-logger. The albatross engaged in several activities (top of figure), ranging from resting to flying. Vertical lines indicate take-offs and landings, and the dashed horizontal line indicates the albatrosses heart rate when resting on land. Heart rates, and therefore energy consumption, increased dramatically at take-off and to a lesser extent when the albatross was likely feeding. When dynamic soaring, heart rates were nearly as low as when the albatross was resting on land, indicating that dynamic soaring requires surprisingly little energy. (Figure from Weimerskirch et al. 2000; © 2000 The Royal Society, used with permission)

rate when sitting on the water (150 beats/min). In contrast, heart rates were much higher when albatrosses were taking off or landing (about 200 beats/min).

Dynamic soaring by birds may also be possible over land in areas with wind-shear layers. Wind-shear layers can develop in areas where obstructions such as mountain ridges (Fig. 11.65) or other orthographic features create a gradient of wind speeds, with greater velocities further above the ground. For example, when there was sufficient wind velocity (≥2.8 m/s), Schijf (2008) observed gulls dynamic soaring near an airport in the Netherlands. In addition, Sachs (2013) reported that albatrosses sometimes use dynamic soaring over land.

### 11.5.3 Sea-Anchor Soaring

Yet another type of soaring is sea-anchor soaring (Sugimoto 1998; Pennycuick 2008). Foraging storm-petrels (family Hydrobatidae) have a unique method for locating and picking up food from the water's surface. They "soar" into the wind with their body clear of the surface, but with their webbed feet in the water, helping to "anchor" them. Parasitic drag and hydrodynamic

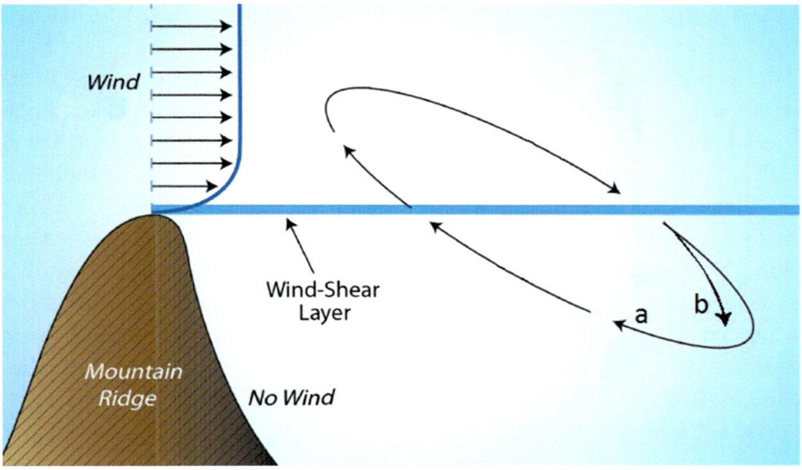


**Fig. 11.65** Dynamic soaring by birds may also be possible over land by taking advantage of variable wind speeds downwind from, for example, a mountain ridge. Starting at point a where there is little or no wind, a soaring bird could fly up into the faster-moving air and turn downwind to gain speed. With increased momentum, the bird could then move back down below the wind-shear layer and turn more sharply (b) back up into the faster-moving air to gain additional altitude. By repeating this process several times, a dynamic-soaring bird could gain both altitude and speed, then turn downwind and glide to potentially find another location with similar conditions. (Figure modified from Richardson 2012; used with permission of Philip L. Richardson)

drag (or resistance) tend to impede forward movement (Fig. 11.66). When lift and forward propulsion forces generated by the wings match, or nearly match, drag, then storm-petrels remain in place (or nearly in place), suspended above the water's surface at a height that allow them to pick small food items from the surface.

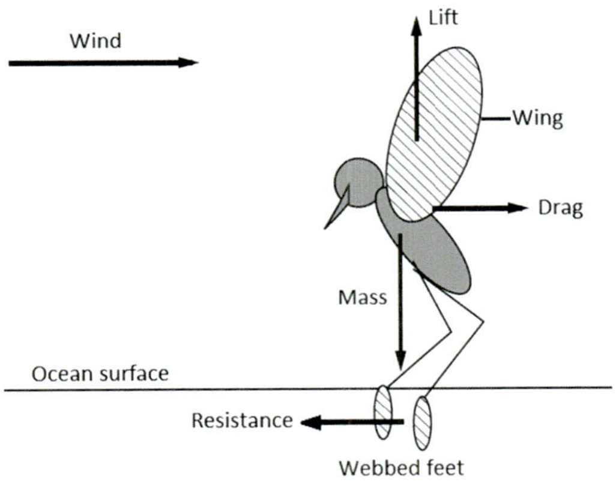

**Fig. 11.66** Left, Forces acting on a sea-anchor-soaring petrel. Wings generate lift as a petrel's flight speed matches, or closely matches, that of the wind. Parasitic drag and hydrodynamic drag or resistance (the latter caused by the petrel's webbed feet in the water) tend to impede forward movement. When these forces are balanced or nearly so, a petrel can remain stationary or "soar" slightly forward or even backward. Right, a sea-anchor-soaring Cape Petrel (*Daption capense*). (Left, Figure modified from Sugimoto 1998; © 1998 Academic Press, used with permission; right, Photo by Ed Dunens, Wikipedia, CC BY 2.0, https://creativecommons.org/licenses/by/2.0/)

### 11.5.4 Wave-Meandering Wing-Sailing

Some seabirds have another way of taking advantage of the variation in wind speeds about the ocean surface. In what Spivey et al. (2014) called a wave-meandering wing-sailing flight strategy (Fig. 11.66), Manx Shearwaters (*Puffinus puffinus*) alternate brief bursts of flapping flight with longer gliding phases just above the water's surface where wind velocities are lower. Manx Shearwaters using this flight strategy flap their wings when necessary to maintain their airspeed, e.g., during some ascending flights into particularly strong wings, but gliding phases help them conserve energy. When using this flight strategy, gliding shearwaters turn into the wind above the waves and the increased air speed over the wings, along with updrafts generated by wave crests, generates lift. With the increased altitude, shearwaters can then descend closer to the water's surface and the momentum generated with that drop in altitude powers them into another glide (Fig. 11.67).

### 11.5.5 Wave-Slope Soaring

Soaring birds take advantage of rising air, and some birds appear to take advantage of air deflected upward by ocean waves. Based on observations made near South Georgia in the early 1920s, Idrac (1925) first suggested that albatrosses could use wave-induced updrafts to generate lift and save energy. More recently, Stokes and Lucas (2021) studied similar behavior by Brown Pelicans (*Pelecanus occidentalis*), and referred to this mode of flight as wave-slope soaring (Fig. 11.68). In wave-slope soaring, the updrafts are created by traveling ocean waves pushing against the air and forcing it upwards rather than winding hitting a fixed object (Fig. 11.69). When wave-slope soaring, Stokes and Lucas (2021) estimated possible reductions in the cost of flight of about 70% for a pelican flying at a height of 0.5 m above a 2-m high wave, and greater reductions with lower flight heights and taller waves. However, wave-slope soaring is likely only possible during periods with low wind

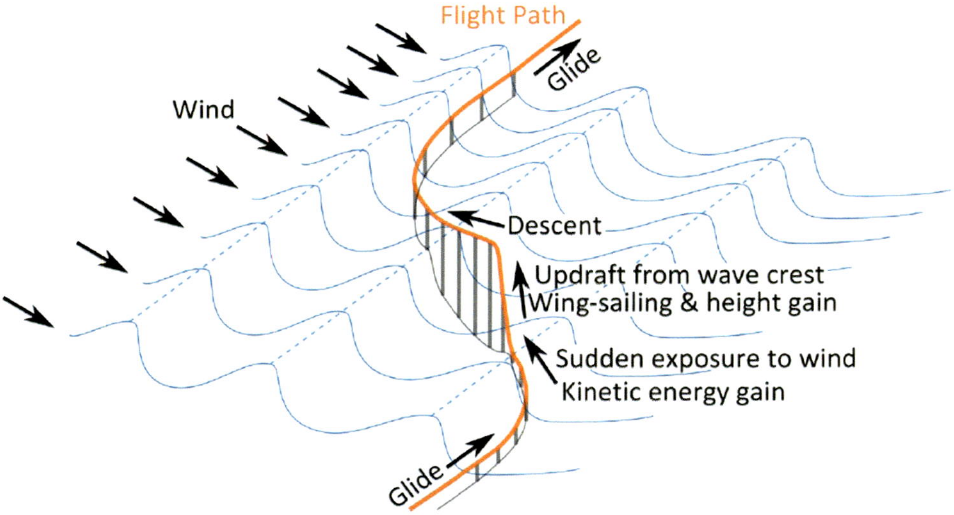


**Fig. 11.67** A single cycle of a wave-meandering wing-sailing flight strategy. A gliding Manx Shearwater (*Puffinus puffinus*) turns into the wing and gains height as the faster wind generates lift which is aided by updrafts from wave crests. Depending on wind speed, shearwaters may have to flap their wings during this ascent into the wind. After gaining sufficient altitude, shearwaters can then descend to just above the water's surface between wave crests when the wind velocity is much lower. The momentum generated as shearwaters descend toward the water allows them to glide variable distances (depending on wind speed and the momentum generated in their descent) before repeating the cycle. (Figure from Spivey et al. 2014; © 2014 Elsevier Ltd., used with permission)

**Fig. 11.68** Wave-slope soaring Brown Pelicans (*Pelecanus occidentalis*). (Photo by Gil Eckrich; used with permission of Gil Eckrich)

speeds because, with increasing wind speed, updrafts produced by traveling waves would be reduced in magnitude or lose coherence (Stokes and Lucas 2021).

### 11.5.6 Flapping Flight

Most birds flap their wings when they fly and do so at different rates, with flapping rate generally declining as size increases. Of course, flapping

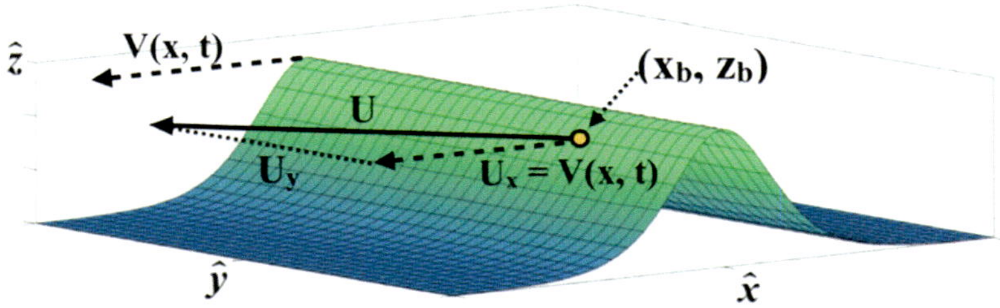


**Fig. 11.69** Trajectories of a Brown Pelican (*Pelecanus occidentalis*) wave-slope soaring on a solitary wave. $x$^ is the direction of wave propagation, $\hat{y}$ is parallel to the wave front, and }\hat{z}{ is in the vertical direction. $(x_b, z_b)$ is the pelican's location, $V(x, t)$ is the velocity of the wave, and $U$ is the horizontal component of the wave-induced wind. To benefit from the updraft created by the wave, the pelican's groundspeed in the direction of the wave must match the velocity of the wave. (Figure from Stokes and Lucas 2021; © 2021 The Authors. Published by the Royal Society under the terms of the Creative Commons Attribution License, https://creativecommons.org/licenses/by/4.0/)

rate also varies with speed. However, at cruising speeds, the ratio of wing stroke frequency (wing beats per second) and amplitude (vertical distance traveled by the tip of the wing during the flapping stroke) to forward speed (distance traveled per second) differs little among birds (and for other flying and swimming animals). This dimensionless ratio is the Strouhal number and Taylor et al. (2003) found that, for 42 species of birds, bats, and flying insects, this number typically ranges from 0.2 to 0.4. For example, as noted by Corum (2003), Eurasian Kestrels (*Falco tinnunculus*) cruise with a frequency of 5.61 strokes per second, an amplitude of 0.339 m per stroke, and a speed of 8.1 m per second. Because the Strouhal number is the ratio of the frequency ($f$) of wing strokes times the amplitude ($A$) of the wing, divided by the animal's forward speed ($U$), or:

$$\frac{fA}{U},$$

for the kestrel:

$$\frac{(5.61\ \text{strokes/second}) \times (0.339\,\text{m/stroke})}{(8.1\,\text{m/s})}$$

The strokes cancel each other out, leaving:

$$\frac{1.9(\text{m/s})}{8.1(\text{m/s})}$$

or the ratio of the vertical distance traveled by the tip of the wing during the flapping stroke ($f\,A$) over the speed of the animal ($U$).

The meters per second also cancel each other out, leaving:

$$\frac{1.9}{8.1}$$

or a Strouhal number of **0.235** for Eurasian Kestrels.

Such similarity among species suggests that locomotion systems are finely tuned by natural selection to a narrow range of Strouhal numbers because this maximizes propulsion efficiency. Birds flying below cruising speed or hovering do not have Strouhal numbers in the range 0.2–0.4 because as speed decreases the Strouhal number increases. In the case of hovering, speed will be zero and the Strouhal number will approach infinity (Hedenström 2004).

Birds fly at different speeds, varying from extremely slow (e.g., during take-off and landing) to very fast (e.g., 31.1 m/s by a Common Swift, *Apus apus*; Henningsson et al. 2010). However, birds typically employ two, and maybe three, types of flight, or gaits, distinguished by the aerodynamic function of the upstroke and the pattern of movement of the wings. During slow-speed flight (about 7 m/s or less), the upstroke is aerodynamically passive (no lift is generated), and the wings flex and move close to the body to minimize inertia and drag. At faster speeds, the wings generate lift during both the downstroke and upstroke, and the wings exhibit much less flexure at the wrist. At even higher speeds, birds may retract their wings to an increasing degree to reduce profile drag. The slow gait is referred to as the vortex ring gait, the fast gait is called the continuous-vortex gait, and, at the highest speeds, a ladder-wake gait, where birds use an inflexible wing motion and vortex structures are connected via the tip vortices, may be generated (Hedrick et al. 2002; Fig. 11.70). Although the concept of "gaits" in flight is useful in distinguishing between slow and fast flight, the transition between gaits as a function of speed, for most birds, is likely to be continuous, with the aerodynamic role of the upstroke gradually increasing with increasing flight speed (Usherwood et al. 2003).

As described previously, wings generating lift also generate a system of trailing vortices behind the wingtips (as air flows from the area of high pressure below the wing to the area of lower pressure above the wing). During slow flight, wings generate a doughnut-shaped vortex ring with each downstroke so, for example, a series of vortex rings is produced as a bird takes off (Fig. 11.71). With increasing speed, each wingtip generates a separate vortex trail during both the upstroke and downstroke and lift is generated during both the upstroke and downstroke (Fig. 11.71). This shift from a vortex-ring to a continuous-vortex gait requires sufficient forward velocity to provide airflow over the wing during the upstroke that is similar to that during the

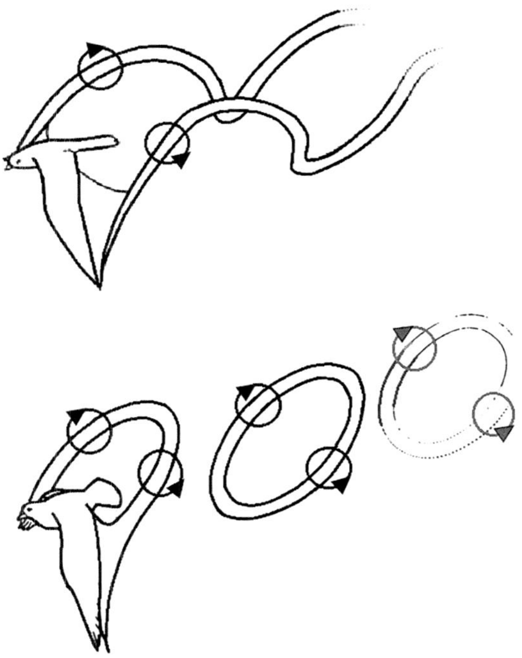

**Fig. 11.70** Top, continuous gait of a bird in cruising flight. In the continuous-vortex gait, lift is produced during both the upstroke and the downstroke. The downstroke produces a positive upward force and forward thrust; the upstroke produces a positive upward force and rearward thrust. Partial flexion of the wing during the upstroke reduces the magnitude of the rearward thrust to less than that of the forward thrust produced during the downstroke, providing net positive thrust per wingbeat. Bottom, vortex ring gait of a slow-flying bird (Rayner 2001; © 2001 John Wiley & Sons, Ltd., used with permission)

downstroke, reflecting the need to minimize drag and produce forward thrust to fly at high speed (Hedrick et al. 2002).

Two characteristics of flapping flight may explain the shift away from the continuous-vortex gait at very fast flight speeds: (1) use of the distal wing as the primary source of thrust, and (2) negative thrust production during a lifting upstroke. Most of the thrust in flapping flight is produced by the distal wing during the downstroke. Therefore, increasing thrust production requires the distal-wing-bound circulation to increase. Consequently, if the proximal circulation during the upstroke does not increase to match the distal downstroke circulation, a bird will necessarily shift into a discontinuous vortex-ring or ladder-wake gait. The difficulty for a bird flying at high speed and attempting to sustain circulation and lift during the upstroke is that upstroke circulation necessarily generates negative thrust (i.e., lift has a rearward component). Therefore, increasing proximal wing circulation during the upstroke to match distal downstroke circulation and maintain a continuous-vortex gait would actually diminish the bird's forward acceleration by increasing negative thrust production. Observations of Cockatiels (*Nymphicus hollandicus*) and Ringed Turtle-Doves (*Streptopelia risoria*) in a flight tunnel during periods of very fast flight confirmed this, showing that, although estimated upstroke circulation did not cease, it does not increase to match the elevated downstroke circulation, resulting in the loss of continuous circulation and a shift to a ladder-wake gait (Hedrick et al. 2002).

The generation of continuous vortices during fast flight explains why some birds, particularly larger birds, fly in formation. Birds in a flock can fly more cheaply in V-formation, saving energy by taking advantage of the uplift created by the bird in front of them (Fig. 11.72). Great White Pelicans (*Pelecanus onocrotalus*) have lower

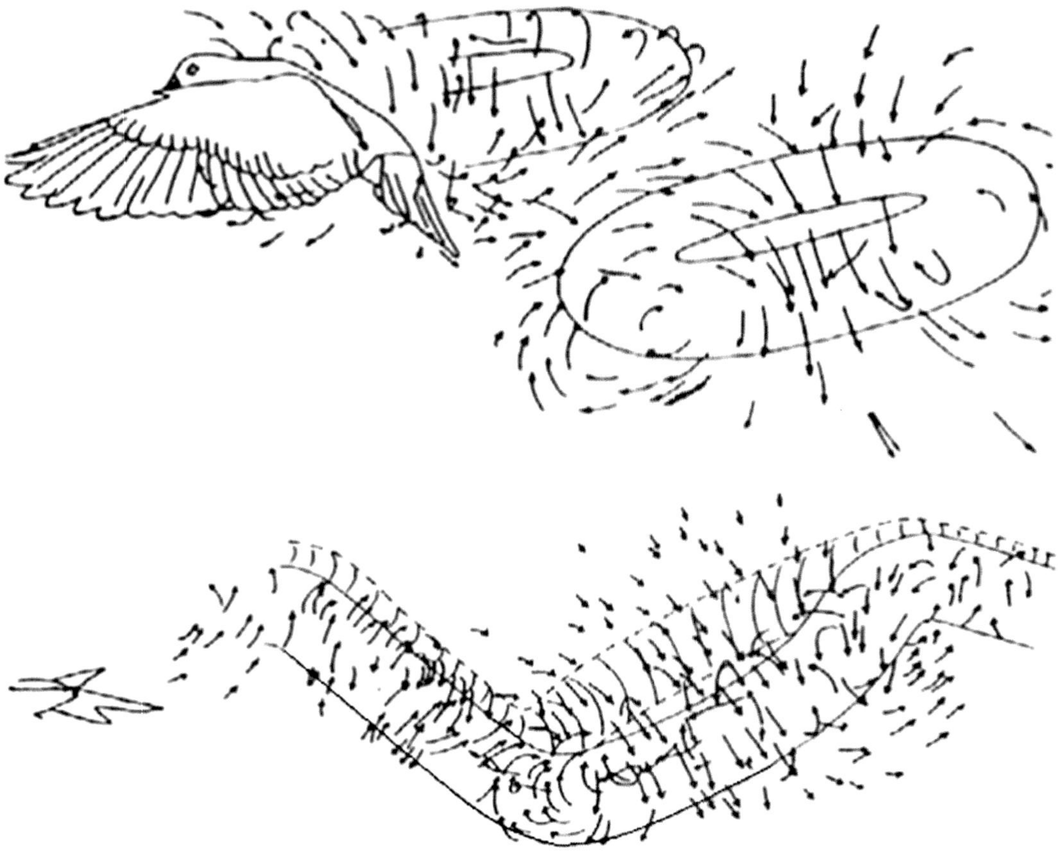

**Fig. 11.71** Top, the vortex-ring gait of a Rock Pigeon (*Columba livia*) in slow flight (~3 m/s) and, Bottom, the continuous-vortex gait of a Eurasian Kestrel (*Falco tinnunculus*) flying at ~7 m/s. (Figures from Rayner 1988; © 1988 Plenum Press, New York, Published by Springer Nature, used with permission)

heart and wingbeat rates, glide more often, and save as much as 11–14% energy when flying in the vortex wake of another bird (Weimerskirch et al. 2001). Birds probably gain the greatest energetic benefits if wingtips overlap laterally by about one-tenth of a wingspan relative to that of the bird in front (Figs. 11.73 and 11.74). If members of a flock differ in size, energy savings will be particularly high for smaller, lighter individuals, e.g., juveniles, or females in species that exhibit sexual size dimorphism (Hummel 1995). Not all birds that fly or migrate in flocks use formation flight. For small birds with relatively short wings, rapid, large-amplitude wing strokes create complex wake patterns with less predictable patterns of rising and falling air that makes the use of formation flight to conserve energy impossible (Hummel 1995).

### 11.5.7 Flap-Bounding and Flap-Gliding

Many birds do not flap their wings continuously when flying, but, instead, alternate periods of thrust generated by their flapping wings with periods of gliding or bounding (Rayner et al. 2001; Fig. 11.75). During flap-bounding flight, birds alternate flapping phases with flexed-wing bounding phases (Figs. 11.76 and 11.77).

**Fig. 11.72** Northern Bald Ibises (*Geronticus eremita*) flying in V-formation. (Image from Portugal et al. 2014; © Springer Nature, used with permission)

Bounding flight is confined to small birds, up to a maximum mass of about 250 g. Bounding is the characteristic flight mode of most small passerines (except longer-winged species such as swallows [Hirundinidae] and wood-swallows [Artamidae]) that use brief intermittent glides, smaller owls (Strigiformes), parrots (Psitaciiformes), woodpeckers and flickers (the largest birds that flap-bound; Piciformes), and even hummingbirds. Because birds lose mechanical energy to the air when bounding and replenish energy during the flapping phase, bounding flight paths involve cyclic variation in altitude (Rayner et al. 2001). For birds that flap-bound, the percentage of time spent flapping tends to decrease with increasing speed. For example, Zebra Finches (*Taeniopygia guttata*) in a wind tunnel flapped their wings about 84% of the time when flying at a speed of 2 m/s, and this decreased to about 56% of the time when flying at a speed of 14 m/s (Tobalske et al. 1999). Because the wings are drawn in against the body during the bounding phase, profile and induced drag is reduced and the cost of flight is reduced. Profile drag on a bird's wings increases with increasing speed so flexing the wings during bounds means less energy is required than would be the case with continuous flapping at fast flight speeds (Tobalske 2001). Flap-bounding may also maximize horizontal velocity by transferring the vertical acceleration during the descending (bounding) phase into forward velocity (Wang et al. 2018b).

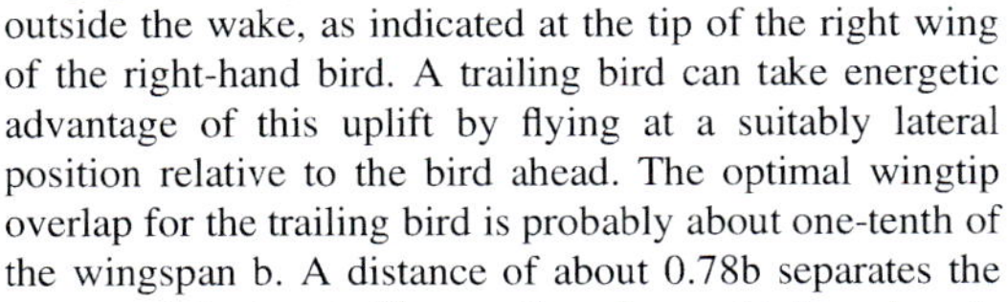


**Fig. 11.73** A vortex is formed in the wake of each wingtip, creating downflow behind the wing and uplift outside the wake, as indicated at the tip of the right wing of the right-hand bird. A trailing bird can take energetic advantage of this uplift by flying at a suitably lateral position relative to the bird ahead. The optimal wingtip overlap for the trailing bird is probably about one-tenth of the wingspan b. A distance of about 0.78b separates the centers of the two trailing vortices from a bird or aircraft. (Figure from Andersson and Wallander 2004; © 2004 Oxford University Press, used with permission)

Undulating flight, or flap-gliding, is much more common among birds, with the alteration of glides and bursts of flapping a characteristic flight mode of many birds of intermediate size (Fig. 11.78). It is prevalent in species with

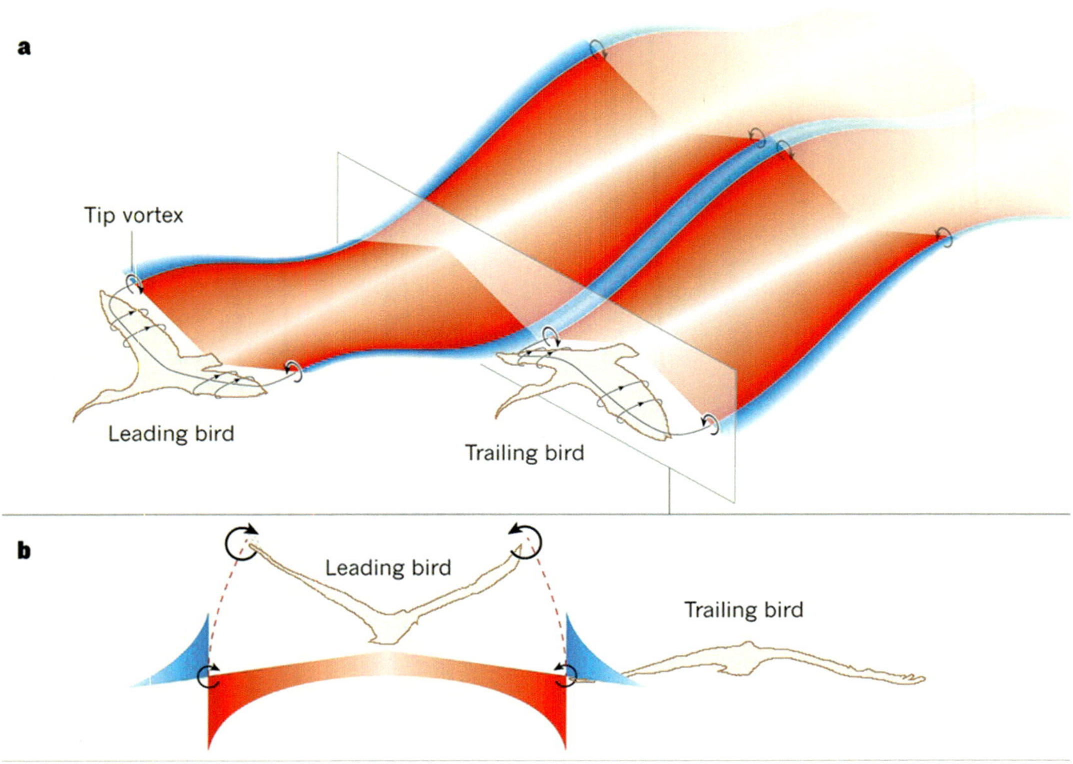


**Fig. 11.74** (**a**) (side view) and (**b**) (rear view). Air moving across the wings is forced downward (red area behind the birds; creating lift via Newton's third law) except at the wing tips where the air with higher pressure under the wing flows around the wingtip to the area of lower pressure above the wing, i.e., the wingtip vortices responsible for induced drag (blue area behind the wing tips). A bird flying behind another bird can be affected by both the downward moving air and the upward moving air. However, by synchronizing their wing movements while flying in a V-formation and flying with their right wingtip in the wingtip vortex generated by the bird in front, the trailing bird benefits from the lift generated by the rising air and saves some energy. (Figure modified from Muijres and Dickinson 2014; © 2014 Springer Nature, used with permission)

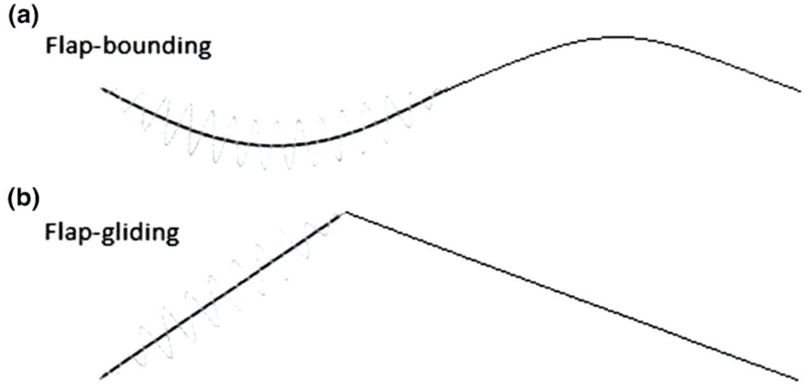


**Fig. 11.75** Intermittent flight patterns. (**a**) Flap-bounding and (**b**) flap-gliding flight. (Figure from Rayner et al. 2001; © 2015 Oxford University Press, used with permission)

**Fig. 11.76** (**a**, **b**) Screenshots from a video of a flap-bounding White Wagtail (*Motacilla alba*). (https://jimusherwoodresearch.com/flapping-gaits-bounding-and-flap-gliding/; used with permission)

relatively long wings, including some petrels, gulls, and corvids. Flap-gliding is thought to require less energy at most flight speeds and, particularly, during slow flight (Rayner 1985). Flap-gliding flight may save energy by lowering muscle mechanical power requirements and allowing the pectoralis to use more economical isometric muscle contractions rather than relying solely on concentric contractions (Tobalske and Dial 1994).

Some birds can both flap-bound and flap-glide; flap-gliding to save energy when flying slowly and shifting to flap-bounding, and saving even more energy, at faster speeds (Tobalske and Dial

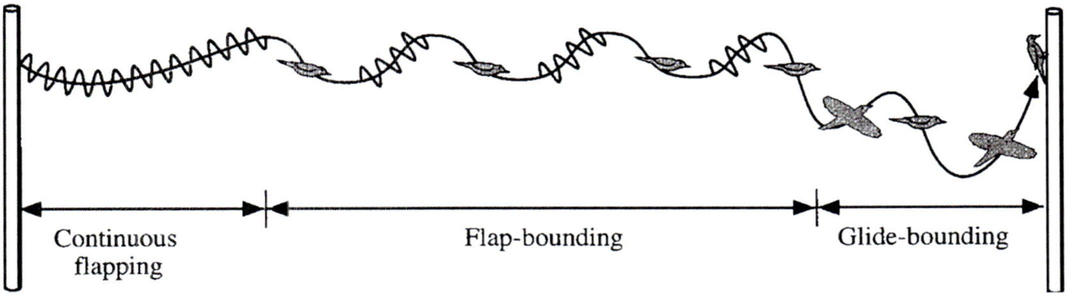


**Fig. 11.77** Continuous flapping, flap-bounding, and glide-bounding flight of a Lewis' Woodpecker (*Melanerpes lewis*). (Figure from Tobalske 1996; © 1996 Oxford University Press, used with permission)

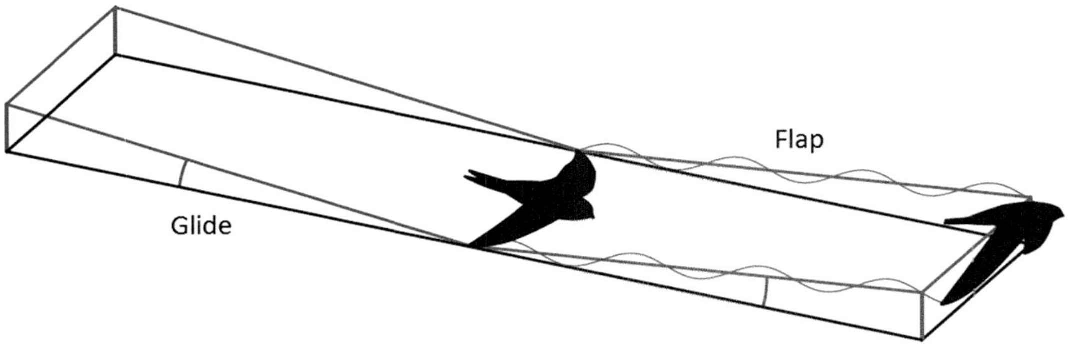


**Fig. 11.78** Flap-gliding birds tend to descend during the glide phase, then maintain altitude or climb during the flapping phase. (Figure modified from Muijres et al. 2012a, b; © 2012 Elsevier Ltd Published by Elsevier Ltd., used with permission)

1994). For example, as flight speed increases, Budgerigars (*Melopsittacus undulatus*), European Starlings (*Sturnus vulgaris*), and Black-billed Magpies (*Pica hudsonia*) spend less time flapping (Fig. 11.79), then flap-glide more at intermediate speeds and both flap-glide and flap-bound more at their highest speeds (Tobalske 1995, 2001; Tobalske and Dial 1996).

## 11.6 Flight Speed

Cruising speeds of birds vary considerably. Alerstam et al. (2007) found that the mean air-speed of 138 species of birds ranged between 8 and 23 m/sec (or between 18 and 51 miles/h; Fig. 11.80). Birds of prey, songbirds, gulls, terns, and herons had flight speeds in the lower part of this range, whereas pigeons, some shorebirds,

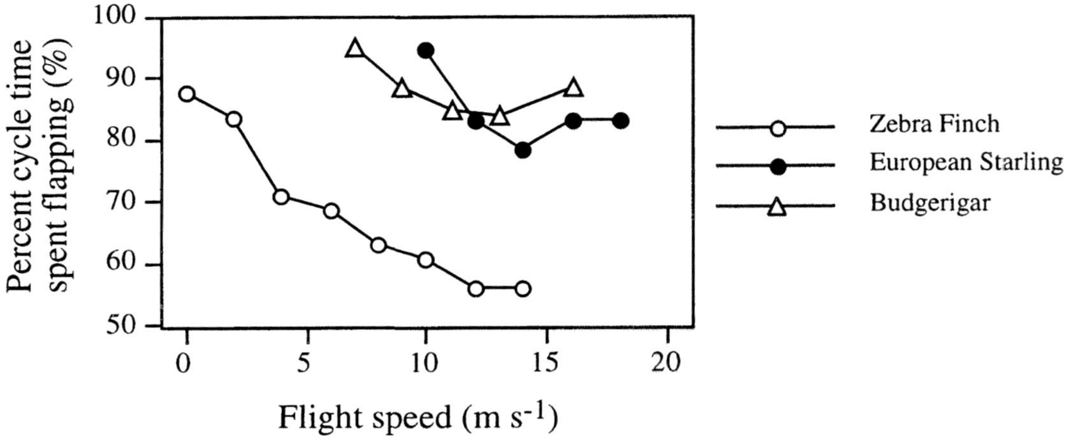


**Fig. 11.79** Effect of flight speed on the percent time spent flapping during intermittent flight by Zebra Finches (*Taenopygia guttata*), Budgerigars (*Melopsittacus undulatus*), and European Starlings (*Sturnus vulgaris*). (Figure from Tobalske 2001; © 2015 Oxford University Press, used with permission)

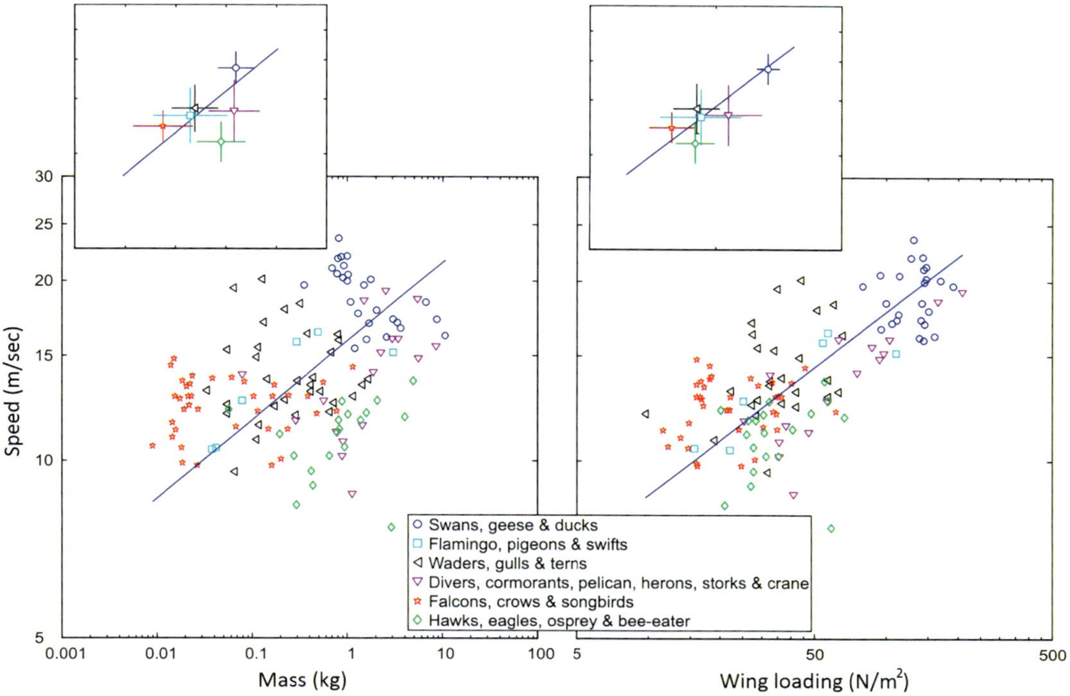


**Fig. 11.80** Bird flight speeds (m/s) relative to body mass (kg) and wing loading ($N/m^2$) for 138 species of birds in six monophyletic groups of birds. Inserts above show means (± standard deviations) for the six groups. Species in the same group tend to fly at similar speeds. N, Newtons (1 Newton is the force needed to accelerate 1 kg of mass at the rate of 1 m/s). (Figure from Alerstam et al. 2007; © 2007 Alerstam et al., open-access article distributed under the terms of the Creative Commons Attribution License, https://creativecommons.org/licenses/by/4.0/)

loons, swans, geese, and ducks were fast flyers in the range of 15–20 m/s. Cormorants, cranes, and skuas flew at intermediate speeds, about 15 m/s. Diving ducks had the fastest mean speeds, with several species having mean speeds of 20–23 m/s. Variation in cruising speed among species was largely due to phylogeny (evolutionary history) and wing loading. Depending on their ecological lifestyle and foraging behavior, birds are adapted to different aspects of flight performance, such as speed, agility, lift generation, escape, take-off, and energetic cost of flight. For example, species flying at comparatively slow cruising speeds frequently use thermal soaring (raptors and storks), are adapted for hunting and load carrying (raptors), or for take-off and landing in dense vegetation (herons). Functional differences in flight apparatus and musculature among birds of different life and flight styles (differences often associated with evolutionary origin) have a significant influence on a bird's performance and speed in sustained cruising flight (Alerstam et al. 2007).

## 11.7 Hovering

A hovering bird generates only lift rather than lift and thrust and this requires a lot of muscle power and energy. Hovering becomes more difficult with increasing size so, among birds, the largest species capable of true hovering is the Pied Kingfisher (*Ceryle rudis*) (but some larger birds can hover for brief periods; Videler 2006). Larger birds, however, can remain in place by flying into the wind at a speed equal to that of the wind, a type of flight called wind-hovering (Videler et al. 1983). By flying into the wind at

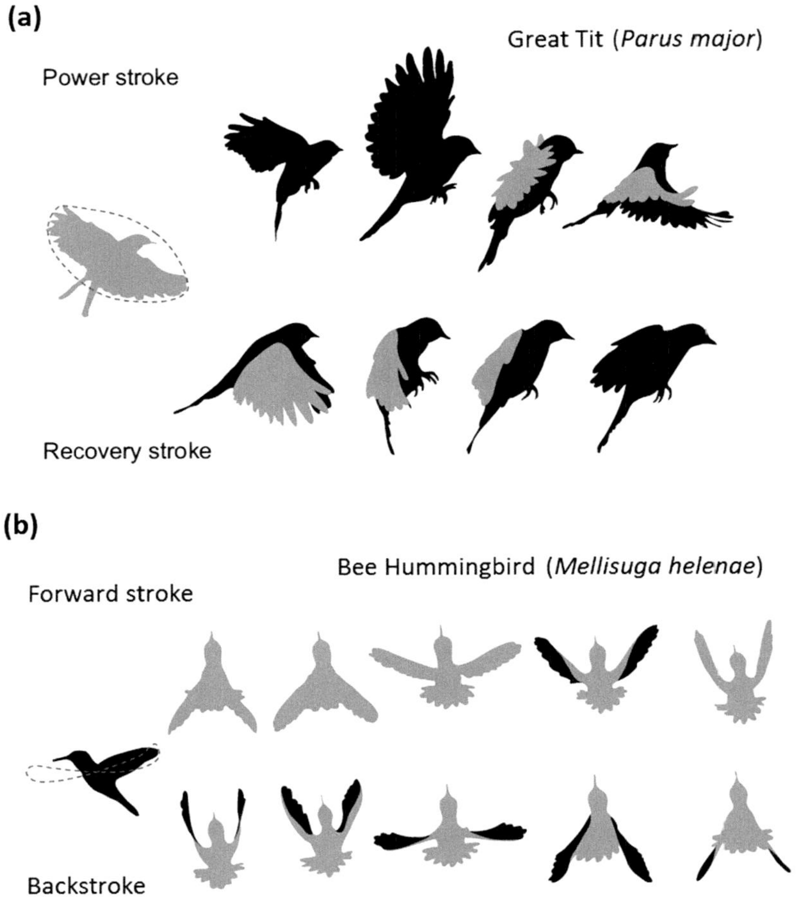


**Fig. 11.81** (**a**) Hovering by small birds, like Great Tits, involves generating maximum lift on the downstroke then pulling the wing in and minimizing drag of the upstroke. (**b**) Hummingbirds, like Bee Hummingbirds, can rotate their wings for the "upstroke" (or backstroke when hovering) and so are able to generate lift during both the down- and upstroke. (Figure modified from Guerrero 2009; used with permission of Joel Guerrero)

the speed of the wind, predatory birds, such as raptors, terns, and kingfishers, can keep their eyes in a fixed position relative to the ground (or water) to better detect movements of their prey (Videler 2006).

Among birds, there are two ways to hover (Guerrero 2009). With the exception of hummingbirds, birds cannot rotate their wings during flapping flight so, when hovering, lift is only produced during the downstroke and, during the upstroke, the wings are flexed inward to reduce drag (Fig. 11.81a). Hummingbirds, on the other hand, can rotate their wings, so are "true" hoverers (Fig. 11.81b).

Hummingbirds can remain in the same place in still air as long as they wish—they are true hoverers. A hovering hummingbird keeps its body at about a 45° angle to the ground and moves its wings in more or less a figure-8 pattern, with the "8" lying on its side, and lift is generated during both the downstroke and upstroke (Fig. 11.82). Although the downstroke and upstroke of hummingbirds are almost symmetrical, studies of Rufous Hummingbirds (*Selasphorus rufus*) indicate that wing velocity is slightly greater during the downstroke than the upstroke and, as a result, the downstroke provides more of the total lift (75%) than the upstroke (Warrick et al. 2005). In all other birds, the downstroke provides 100% of the lift during slow flight and hovering (but not during flight at higher speeds). Compared to other birds, the shoulder joint of hummingbirds permits greater rotation of the humerus, and the relatively long primaries allow more of the wing to be inverted during the upstroke. However, the proximal

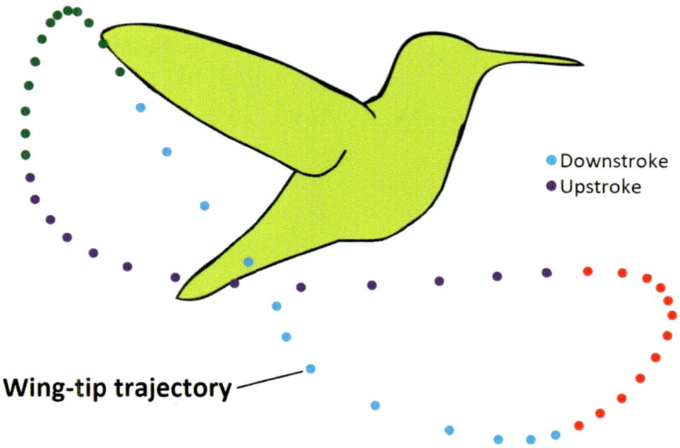


**Fig. 11.82** A single wingbeat of a Ruby-throated Hummingbird (*Archilochus colubris*). By focusing on the wingtip, the figure 8 movement of the wing becomes apparent, with the wingtip moving forward and down on the downstroke, flipping over, then moving backward and up on the upstroke. (Figure modified from Leys et al. 2016; open-access article distributed under the terms of the Creative Commons Attribution License, https://creativecommons.org/licenses/by/4.0/)

portion of the hummingbird wing is not as inverted as the distal portion and, with less camber, less lift is generated during the upstroke (Warrick et al. 2005). By comparison, hovering insects can totally reverse the camber of their wings and so 50% of the lift is provided by the downstroke and 50% by the upstroke (Willmott et al. 1997; Fig. 11.83).

Forward flight is also important to hummingbirds, and they increase flight velocity by simultaneously tilting their body angle and stroke-plane angle of the wings, without varying wingbeat frequency (Tobalske et al. 2007). The figure-8 path of the wing tips apparent at flight speeds of 0 and 2 m/s is similar to that exhibited during slow flight by other birds with pointed wings or wings of relatively high-aspect ratio, including Rock Pigeons (*Columba livia*) and Cockatiels (*Nymphicus hollandicus*; Tobalske et al. 2003). The wing motion of hummingbirds flying at faster speeds is also similar to that of other birds (Tobalske and Dial 1996; Fig. 11.84). As with other small birds, hummingbirds may also, based on studies in wind tunnels, use intermittent glides and bounds when flying at higher speeds (Tobalske et al. 2007).

## 11.8 The Role of Bird Tails in Flight

Avian tails can play an important role in flight and, in some species, serve as ornaments important in mate choice. For a bird in flight, the tail can help control stability, produce lift, improve maneuverability, and reduce body drag (Evans 2003). Tails reduce drag by acting as "splitter plates"; they "split" or keep separate the boundary (or shear) layers of the opposite sides of a bird's body. Boundary layers are very thin layers of air moving over the surface of a bird. Because air has viscosity, this layer of air tends to adhere to the bird's surface. However, as this layer moves toward the tail of a bird, it begins to lose speed due to friction and becomes turbulent (Fig. 11.85). By separating these layers of turbulent air and delaying their interaction, a bird's tail (along with the upper and lower tail coverts) reduces drag. Experiments with mounted specimens of European Starlings (*Sturnus vulgaris*) in a wind tunnel (and using thin streams of smoke to visualize air flow) revealed that removing the tail increased the total parasitic drag of the body and tail by between 25 and 55% (Maybury and Rayner 2001).

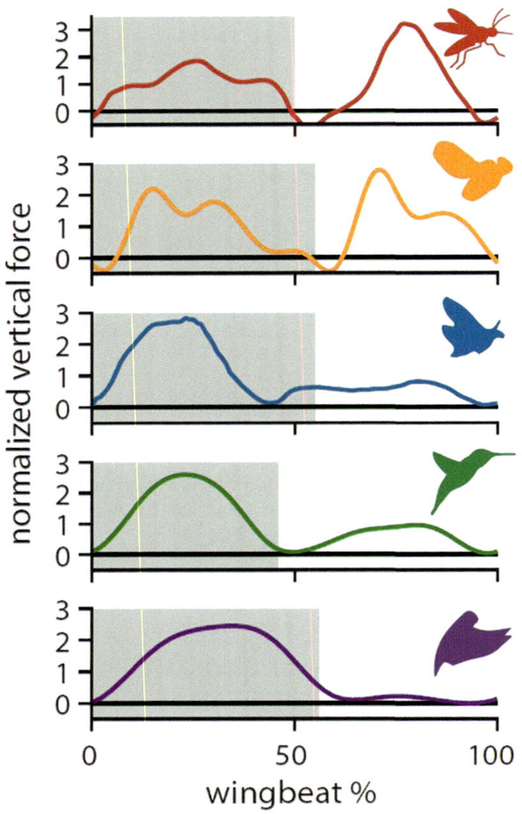


**Fig. 11.83** Aerodynamic weight support profiles for (from top to bottom) a mosquito, fly, hawkmoth, Anna's Hummingbird (*Calypte anna*), and Pacific Parrotlet (*Forpus coelestic*). Mosquitos and flies generate as much lift during the upstroke (50–100% of wingbeat) as the downstroke (0–50% of wingbeat). Hawkmoths and hummingbirds generate about 75% of their lift during the downstroke, whereas Pacific Parrotlets generate almost all of their lift during the downstroke. (Figure from Hightower et al. 2021; © 2021 Hightower et al., open-access article distributed under the terms of the Creative Commons Attribution License, https://creativecommons.org/licenses/by/4.0/)

Bird tails can also generate lift (Fig. 11.86). Wind-tunnel experiments indicate that the amount of lift generated can be substantial. For example, the maximum lift produced by the tails of a European Starling (*Sturnus vulgaris*) and House Sparrow (*Passer domesticus*) at a speed of 10 m/s and with tails spread was equivalent to 45% and 73% of their body weight, respectively (Evans 2003). These results were based on models with the wings set in a gliding posture and the amount of lift generated by tails might differ during flapping flight. However, the wakes generated by the wings of Eurasian Kestrels (*Falco tinnunculus*) during flapping and gliding flight were found to be similar (Spedding 1987a, b), suggesting that lift generated by tails may also be similar regardless of the type of flight (Evans 2003).

During flight, the muscles that control the position of a bird's tail are active (Gatesy and Dial 1993), and observations of birds in flight reveal movements of the tail such as spreading and twisting (Hedenström 2002). For example, swallows in flight use a variety of tail movements as they pursue flying insects (Warrick et al. 2002; Fig. 11.87). Such movements likely enhance maneuverability. For example, swallows and swifts use their tails as flaps (spread and depressed) to increase total lifting area (and reduce wing loading) and reduce turning radius when pursuing flying insects (Warrick et al. 2002). For rapid downward flight (called bunting), swallows elevate their tails to reduce lift.

Tails clearly serve important functions for flying birds. However, in several species, tails also serve as ornaments that can influence mate choice and reproductive success, with peafowl (*Pavo* spp.) being the classic examples. Modifications of tails that transform them into ornaments impinge on their ability to perform their naturally selected functions and, as a result, the benefits of enhanced ornaments must be balanced by the costs of impaired flying ability (Evans 2004). For example, elongated tails serve as ornaments in a wide variety of species, but they can also influence the costs of flight, the lift-to-drag ratio, and a bird's ability to maneuver. These potential costs place limits on tail length that vary among species. Species that spend most of their time on the ground are less affected by these costs and, therefore, tail ornaments are particularly well-developed in many gallinaceous birds, such as pheasants (Evans 2004). For birds that spend more time in flight, potential costs of tail elongation are smaller for larger birds because the size of the tail relative to the rest of the body tends to be smaller and, therefore, characteristics of the tail have less of an impact on flying ability. As a

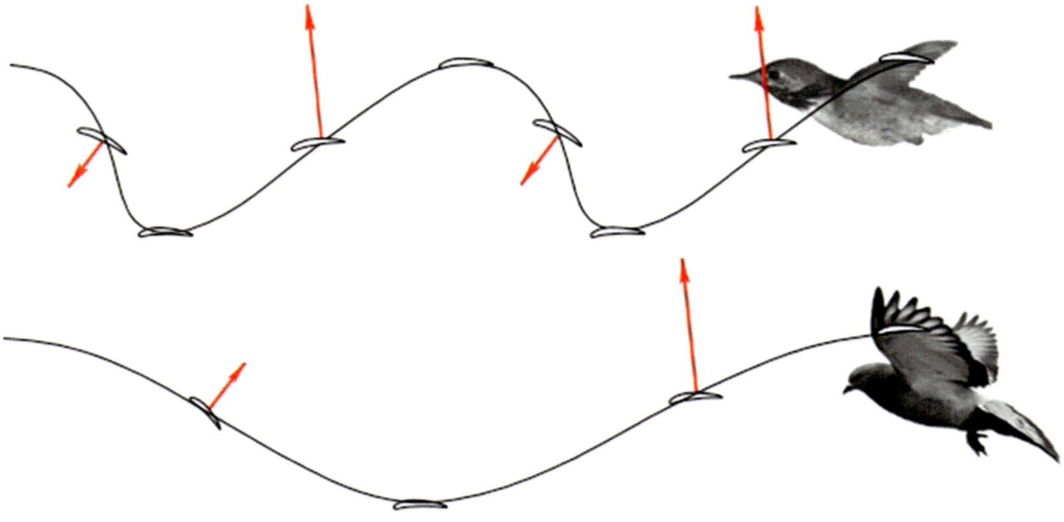

**Fig. 11.84** Similar wing trajectories and force production of a hummingbird and pigeon during forward flight. (Figure from Song et al. 2016; © 2016 The Authors. Published by the Royal Society under the terms of the Creative Commons Attribution License, https://creativecommons.org/licenses/by/4.0/)

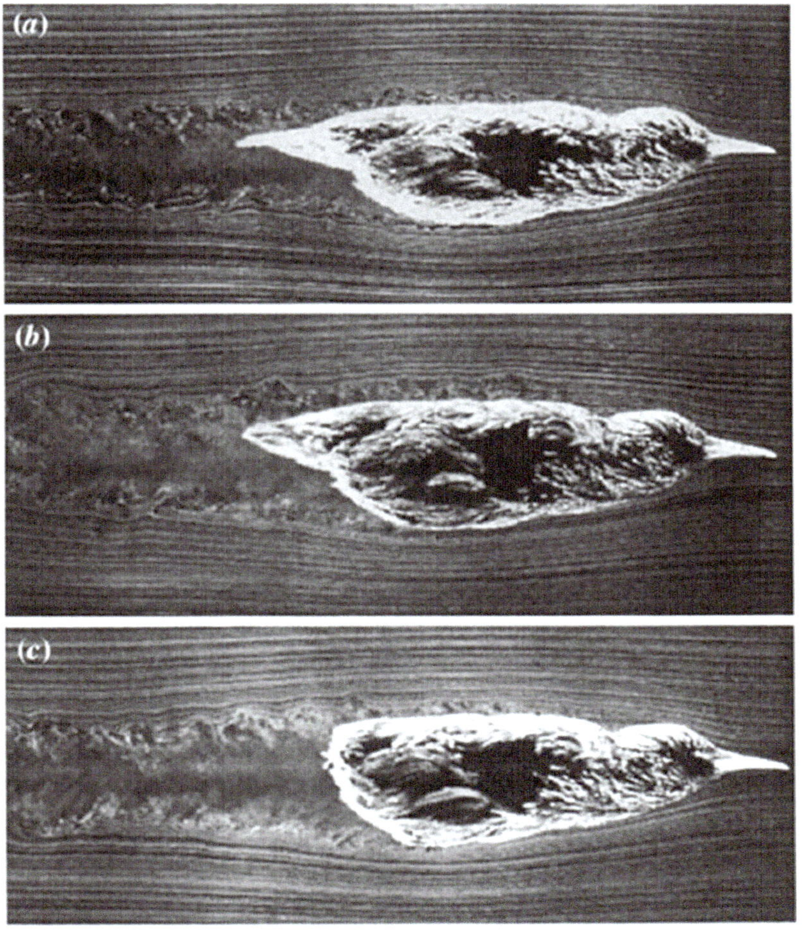


**Fig. 11.85** Flow visualization around mounted wingless body of a European Starling (*Sturnus vulgaris*) using the smoke-wire technique in a wind tunnel at 9 $ms^{-1}$. (**a**) The bird with intact tail and covert feathers; (**b**) tail feathers protruding beyond ventral coverts trimmed to the same length as coverts; (**c**) tail feathers and both upper and under-tail coverts removed. The height of the wake increases from (**a**) to (**c**). The dorsal boundary layer also becomes increasingly turbulent in (**b**) and (**c**) compared with the intact tail-body configuration in (**a**). (Figure from Hedenström (2002); © 2002 Elsevier Science Ltd., used with permission)

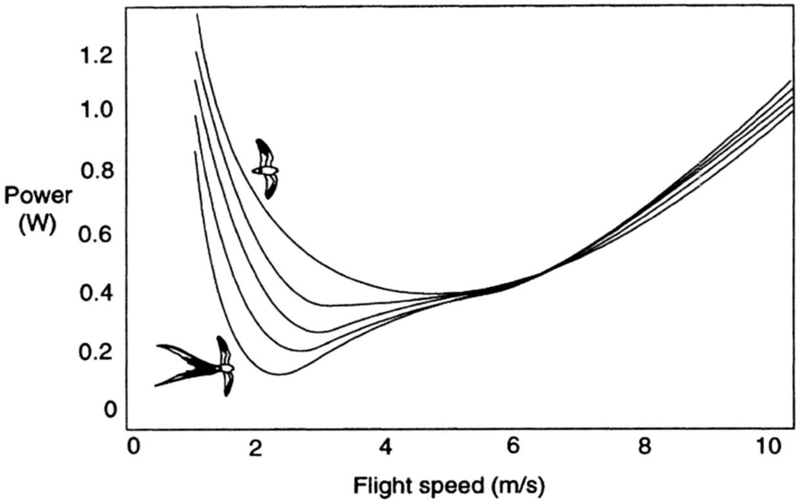


**Fig. 11.86** The energetic cost of flight (in terms of Watts) at low flight speeds varies with tail length of Lyre-tailed Nightjars (*Uropsalis lyra*). With decreasing tail length (successively higher curves in the graph, with the tail completely removed in the top curves), less lift is generated and the energetic cost of flight at low speeds increases. (Figure from Thomas (1997); © 1997 Oxford University Press, used with permission)

result, elongated tails are more common among "medium-sized" birds, such as Scissor-tailed Flycatchers (*Tyrannus forficatus*; about 40 grams). However, these long tails still come with a cost. For example, Norberg (1995) estimated that the long tails of male Long-tailed Widowbirds (*Euplectes progne*) increased parasitic drag five times and, as a result, they tend to fly relatively slowly and also have larger wings than other birds their size to compensate for the increased tail drag (Fig. 11.88).

Some smaller birds also have elongated tails, including Barn Swallows (*Hirundo rustica*; about 19 grams). However, to minimize the costs typically associated with long tails, natural selection has favored the development of forked tails so that only the outer rectrices are elongated.

## 11.9 Maneuverability

The flight of birds is sometimes straight and level. During migration, birds may fly considerable distances without changing direction or altitude. However, for most birds much of the time, flight involves changes in altitude and direction or, in

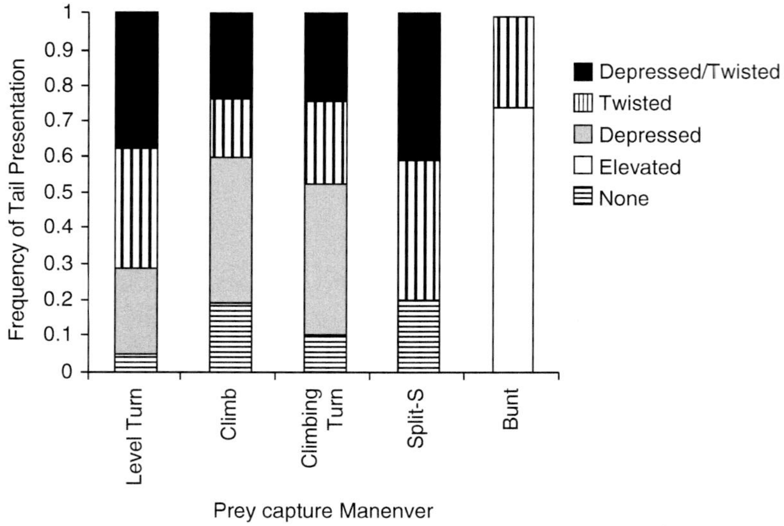


**Fig. 11.87** Observed tail use by swallows during capture of flying insects (Split-S = one-half roll in level flight followed by a one-half inside loop to level flight in the opposite direction; Bunt = pitching down). (Figure from Warrick et al. 2002; © 2002 Oxford University Press, used with permission)

**Fig. 11.88** The long tail of a male Long-tailed Widowbird (*Euplectes progne*). Although favored via sexual selection, the long tail increases parasitic drag when males fly, increasing the energetic cost of flight. (Photo by Derek Keats, Wikipedia, CC BY 2.0, https://creativecommons.org/licenses/by/2.0/)

other words, maneuverability. In some cases, such changes can be made in a leisurely fashion. For example, a soaring raptor or vulture must turn gradually to take advantage of a column of rising air in a thermal. To do so, they can slightly increase the angle of attack of one wing (increasing lift) and decrease it (and lift) for the other wing and this lift asymmetry rolls a bird into a banked turn (in the direction of the wing with the reduced angle of attack and lift). Gliding vultures that turn in this manner served as the inspiration for the wing-warping design used in the first aircraft constructed by the Wright brothers (Wright 1900; Warrick et al. 2002). By altering the angles of attack of the two wings to different degrees, gliding birds can alter the angle of their bank or turn.

To make a sharper turn, birds can quickly increase the angle of attack of one wing (the wing on the side in which they want to turn) to the point where lift is lost, stalling occurs, and drag on the wing increases dramatically. By doing so, a bird can turn (yaw) sharply to one side or the other. Terns have been observed using this maneuver, turning sharply to roll into a dive when foraging for fish (Warrick et al. 2002).

To turn during slow flapping flight, Rock Pigeons use asymmetries in downstroke force. By producing greater force on the wing on the outside of the turn (e.g., greater force on the right wing when turning left) throughout a turn, they roll, or bank, in the direction they want to turn. As they roll, pigeons can even adjust forces during a downstroke to prevent excessive rolling, producing a greater downstroke velocity on the outside wing during the first portion of a downstroke and a greater downstroke velocity on the inside wing during the last part of the downstroke (Warrick et al. 2002; Fig. 11.89).

Among the most maneuverable birds are the coursing insectivores, such as swallows. Swallows use a variety of maneuvers to capture flying insects (Fig. 11.90) and their maneuverability varies with flight speed. Turning radius is independent of flight speed during level flight, but not when turns have a vertical component. Blake et al. (1990) found that Barn Swallows (*Hirundo rustica*) tend to fly at higher speeds when flying at

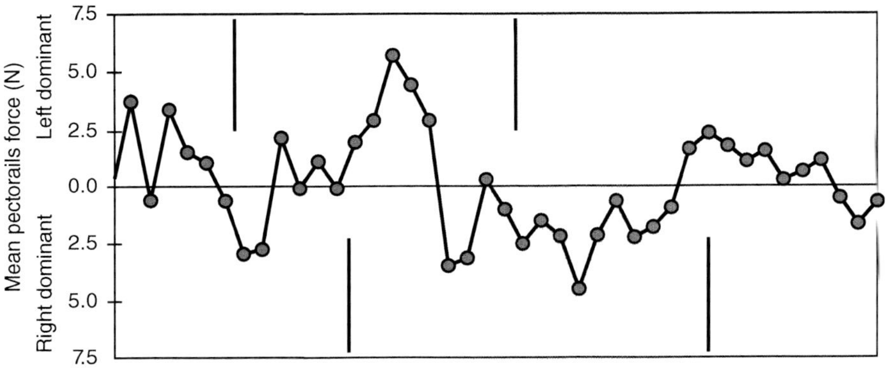


**Fig. 11.89** Force asymmetries of the Rock Pigeon (*Columba livia*) flying through an obstacle course (with the vertical lines representing the obstacles). To turn, the pigeon exerted force asymmetrically on its wings, with more force on the right wing to turn left and vice versa. Pectoralis force is indicated in Newtons (N), with one Newton defined as the force that accelerates an object with a mass of 1 kg one meter per second each second. (Figure modified from Warrick et al. 2002; © 2002 Oxford University Press, used with permission)

lower altitudes, and Warrick et al. (1998) suggested that they do so because swallows often capture prey from below (particularly when flying lower) and they can "pitch up" at a sharper angle when flying faster (Fig. 11.72). In addition, at greater speeds, wing asymmetries caused by flexing wings or altering angles of attack cause faster banking and rolling movements (Norberg and Rayner 1987) and such agility can improve foraging efficiency.

## 11.10 Take-Off and Landing

Although some birds spend considerable time in flight, including swifts that even roost and sleep in flight and Magnificent Frigatebirds (*Fregata magnificens*) that soar for long periods as they forage for prey near the ocean's surface, all birds that fly must take off and land. Birds take flight from a variety of substrates, including perches of various types (e.g., cliffs, trees, and a variety of man-made structures), the ground, and the water's surface. When taking off from an elevated perch, some birds use gravity to aid in accelerating to a velocity needed for level flight, and this can be particularly important for birds with high wing loading (relatively high mass relative to wing area). For example, Least (*Aethia pusilla*) and Crested (*A. cristatella*) auklets nest on cliffs and, when taking off from ledges, use gravity to different degrees by varying the angle of their downward trajectories and, as a result, their speed (Renner 2006). When threatened by an aerial predator, such as a Peregrine Falcon (*Falco peregrinus*) or Glaucous-winged Gull (*Larus glaucescens*), the auklets fly straight down the cliff face to maximize their speed. Unfortunately for the larger Crested Auklets, smaller Least Auklets can generate greater speed at take-off and, therefore, Glaucous-winged Gulls tend to preferentially chase Crested Auklets that are easier to catch (and represent a larger meal; Renner 2006).

Diving birds with high wing loading that do not nest on cliffs, such as grebes, loons, and swans, must take off from the water's surface and run, or skitter, along the surface for some distance before generating sufficient lift to take-off. At take-off, Red-throated Loons (*Gavia stellata*) have to skitter to obtain the necessary relative airspeed at the wings, with the feet contributing the extra vertical force necessary to keep them out of water while acceleration is provided by feet and wings. Loons typically skitter for about 15–40 m before reaching their take-off speed of about 36 km/h (Norberg and Norberg 1971). Aquatic birds with lower wing loadings,

**Fig. 11.90** Barn Swallows (*Hirundo rustica*) have a smaller turning radius when "pitching upward" at higher speeds. Although a complete loop is shown, swallows rarely if ever pitch up beyond a 90° angle when foraging for insects. The numbers around the loop correspond to numbers on the graph. (Photo by Derek Keats, Wikipedia, CC BY 2.0, https://creativecommons.org/licenses/by/2.0/. Figure from Warrick 1998; used with permission of Canadian Science Publishing © 1998; permission conveyed through Copyright Clearance Center, Inc.)

like Mallards, are able to take off directly from the water's surface (Fig. 11.91).

Birds taking off from perches and, especially, the ground maximize their initial flight velocity using leg thrust rather than wing flapping (Figs. 11.92 and 11.93; Box 11.11 Age-Specific Decline in Take-Off Performance). European Starlings (*Sturnus vulgaris*) taking off from the ground use their hindlimbs to produce 91% of the total vertical velocity of take-off, with the first downstroke of the wings adding 9% to the vertical velocity. Similarly, Common Quail (*Coturnix coturnix*) taking off from the ground generate 88% of the total vertical velocity with their hindlimbs (Earls 2000). As a general rule, birds taking off from a perch or the ground obtain 80–90% of their initial flight velocity using leg thrust (Box 11.12 Using Drag to Take Flight). An

**Fig. 11.91** Female Mallard (*Anas platyrhynchos*) taking off from water. Before taking off, legs are drawn up toward the body and wings moved cranially. During take-off, the wings oar through the water, moving downward and backward at the same time that legs are extending and the tail is depressed. The Mallard extends its neck as its body angle shifts from horizontal to almost vertical throughout the take-off. The actions of the wings, tail, and legs propel the Mallard out of the water and, once out of the water, the wings are elevated and begin flapping (Taylor-Burt and Biewener 2020). (Figure made from screenshots of a video made by Kari Taylor-Burt and used with her permission)

exception to that rule is hummingbirds. Hummingbirds have small hindlimbs unable to generate much leg thrust and depend more on their wings than other birds to generate thrust at take-off (Fig. 11.94). Tobalske et al. (2004) measured hindlimb forces on a perch mounted with strain gauges and found that hummingbird hindlimbs contributed 59% of initial flight velocity at take-off. Unlike other birds, hummingbirds beat their wings several times as they thrust using their hindlimbs.

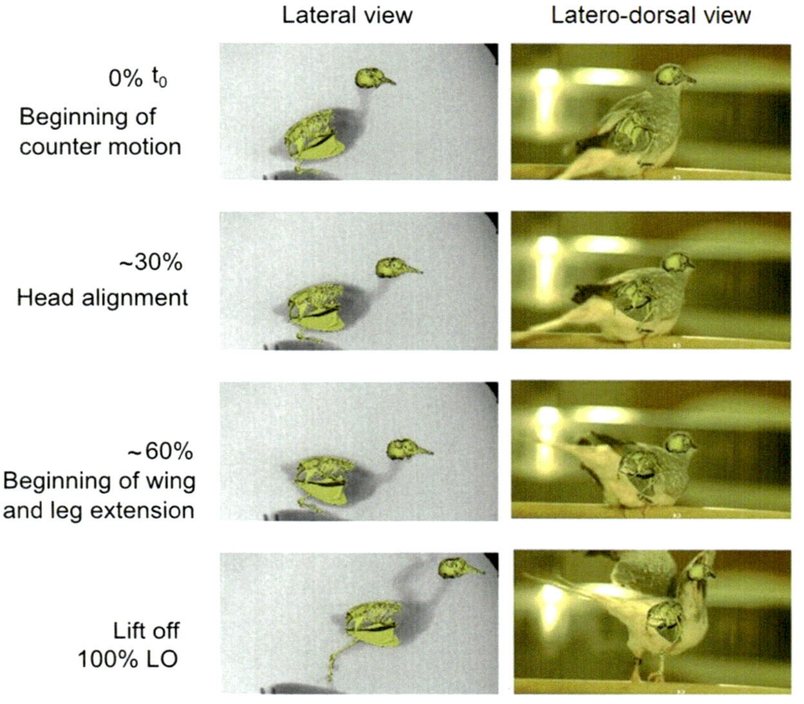


**Fig. 11.92** Two views of a Diamond Dove (*Geopelia cuneata*) taking off. Key events include the start of take-off when a bird crouches, the beginning of leg extension, and lift-off when a bird loses contact with the substrate. (Figure modified from Provini and Abourachid 2018; © 2018, Springer-Verlag GmbH Germany, part of Springer Nature, used with permission)

**Box 11.11 Age-Specific Decline in Take-Off Performance**

When detecting and attempting to evade predators, a bird's take-off speed can determine whether it survives an attack or not. The results of some studies have revealed that a bird's age can affect their behavior and "performance." For example, older Wandering Albatrosses (*Diomedea exulans*) were found to have lower foraging success than younger individuals (Catry et al. 2006). Because other aspects of "performance" could also be affected by a bird's age, Labocha et al. (2015) examined the possibility that a bird's take-off performance might vary with age. Using a captive population of Zebra Finches (*Taeniopygia guttata*; a species with a lifespan of up to 5 years in the wild and 8 years in captivity), Labocha et al. (2015) compared the ability of Zebra Finches in three age classes (0.5, 2, and 4.5 years old, respectively) to "escape" after being released in a vertical flight chamber. They found that older Zebra Finches were about 10% slower than the youngest ones. This difference could be due to age-specific deterioration in the power output of muscles (as reported in humans, Koopman and Van Loon 2009) and/or in feather quality (Swaddle et al. 1996). Regardless of the cause, given its potentially critical importance in the wild, the slower take-off speed of older Zebra Finches could contribute to higher mortality rates of older birds

(continued)

**Box 11.11** (continued)

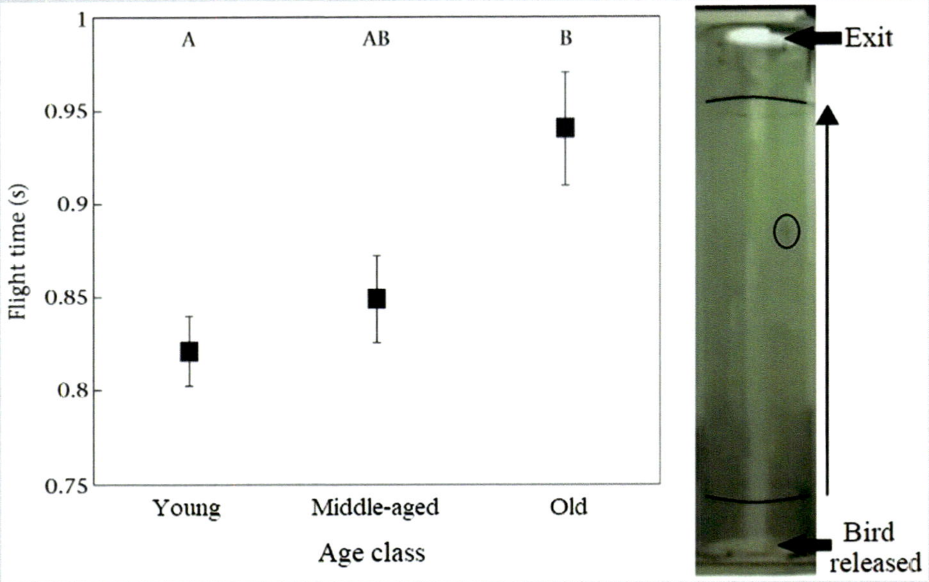


Mean (± SE) time needed for vertical flights of 140 cm by Zebra Finches in three age classes: young (0.5 years old), middle-aged (2 years old), and old (4.5 years old). Older birds took significantly longer to travel the 140 cm distance after taking off than did younger birds. Different letters denote significant differences between the means. The test apparatus is on the right, showing the 140-cm distance (between the dark lines) the birds flew. A bird flying in the apparatus is circled. (Figure from Labocha et al. (2015); © 2015 The Association for the Study of Animal Behaviour. Published by Elsevier Ltd., used with permission)

**Box 11.12 Using Drag to Take Flight**

With high angles of attack (up to 60°), Chin and Lentink (2019) found that Pacific Parrotlets (*Forpus coelestic*) can use drag to support up to half their body weight during take-off for slow, short, perch-to-perch flights. During the first wingbeats after take-off, drag is mostly directed up and slightly backward and lift is directed up and forward. With continued acceleration, wing strokes begin to level off. Using drag to support a bird's weight is energetically costly, but helps to maximize force generation and lift during take-off. High angles of attack have been observed for many other species of birds during take-off, suggesting that those species might also use drag to take flight (Chin and Lentink 2019).

(continued)

**Box 11.12** (continued)

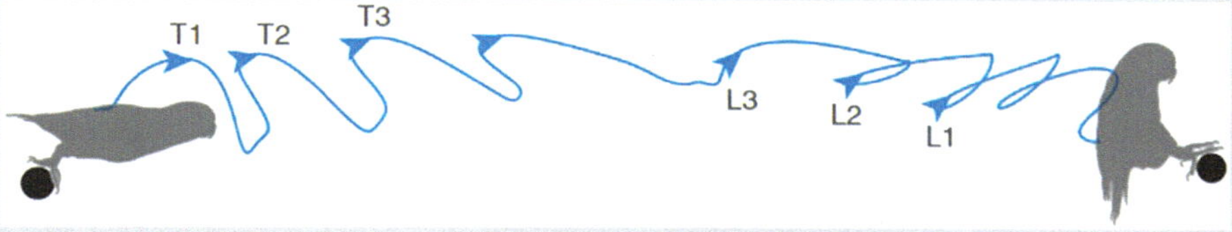


Birds initiate flights with inclined (i.e., high angle of attack) wing strokes (Take-off T1, T2, and T3) that level off, but then pitch backward before landing (arrows indicate the start of wingbeats. (Figure from Chin and Lentink 2019; open-access article distributed under the terms of the Creative Commons CC BY license, https://creativecommons.org/licenses/by/4.0/)

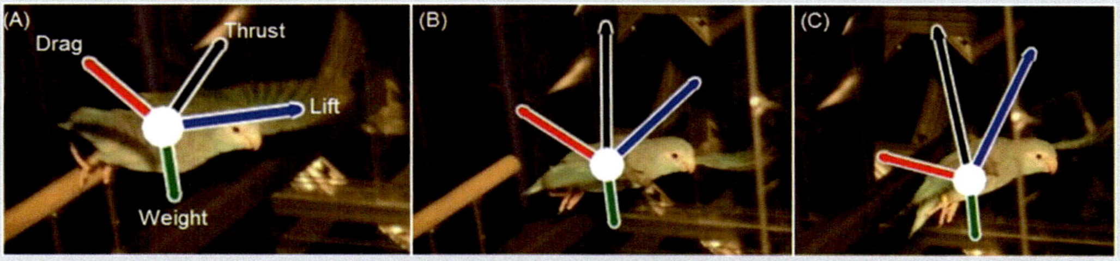


A Pacific Parrotlet taking off incline their wing strokes to orient lift forward to accelerate and orient drag upward to generate lift. A, B, and C are the first three wingbeats after leaving the perch. Note that, with successive wingbeats, lift becomes more vertical and drag becomes more backward oriented. (Figure made from screenshots from Supplementary Movie 1 from Chin and Lentink 2019; open-access article distributed under the terms of the Creative Commons CC BY license, https://creativecommons.org/licenses/by/4.0/)

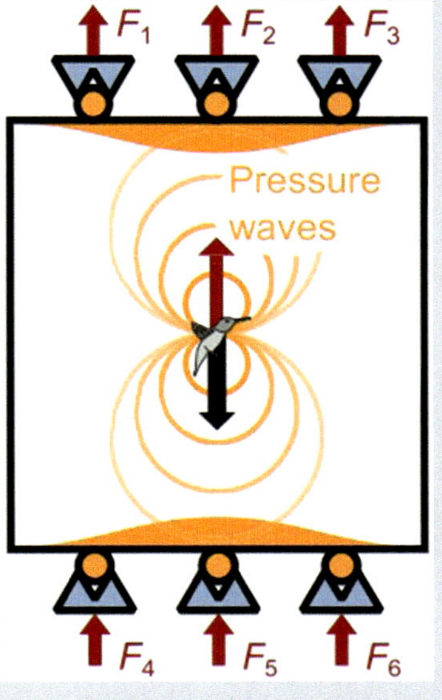


Forces were measured in an aerodynamic force platform. As a bird flew through the container, both downward and upward pressures could be measured. (Figure modified from Hightower et al. 2017; open-access article published under the Creative Commons Attribution 3.0 license, https://creativecommons.org/licenses/by/3.0/)

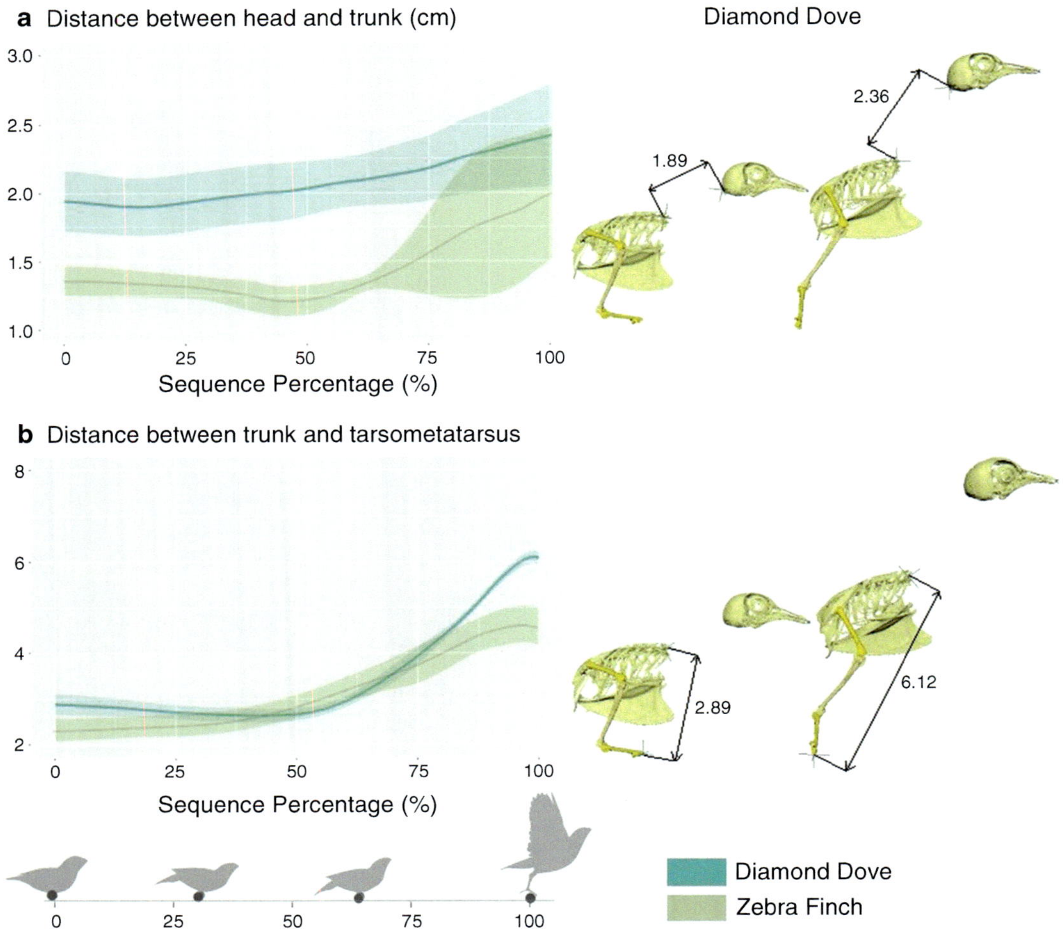


**Fig. 11.93** Body extension during take-off of Diamond Doves (*Geopelia cuneata*) and Zebra Finches (*Taeniopygia guttata*) shown as mean distances (± SD) between (**a**) the head and trunk and (**b**) the proximal part of the trunk and the distal end of the tarsometatarsus. (Figure from Provini and Abourachid 2018; © 2018, Springer-Verlag GmbH Germany, part of Springer Nature, used with permission)

Birds land from flight on many substrates, including the ground, the water's surface, vertical tree trunks, and perches such as branches or ledges. One way to land is to make a slow, hovering descent onto a perch that provides time for adjustments and minimizes the chance of mistakes. However, hovering is energetically expensive and only possible for small birds. Birds landing on a particular spot, e.g., a branch or ledge, typically follow a shallow descent trajectory and, when close to the perch, increase the angle of attack of their wings to decrease both speed and lift (Green and Cheng 1998; Figs. 11.95, 11.96, 11.97, 11.98, and 11.99; Box 11.13 Using Lift as a Brake). Birds also spread and lower their tails, with the tail increasing drag and acting like a brake (Figs. 11.98 and 11.99). Finally, legs and feet are extended for landing. Birds with low wing loading and, therefore, good maneuverability make landing appear effortless. However, birds with greater wing loading and less maneuverability make landing appear more challenging. For example, here is a description of a Black-crowned Night-Heron (*Nycticorax*

**Fig. 11.94** Female Booted Racket-tail (*Ocreatus underwoodii*). Note the short legs that are unable to generate much leg thrust to aid in taking off from a perch. (Photo by Andy Morffew, CC BY 2.0, https://creativecommons.org/licenses/by/2.0/)

*nycticorax*) landing in a tree: "At about 200 feet (61 m) from the landing site the glide begins with feet dropped and head fully retracted; the heron then extends its neck and starts flapping the wings approximately 30 feet (9 m) from the perch … The wing flapping frequency increases until the bird is over the perch. The perch is grasped by the feet and the wings continue to beat until equilibrium is established" (Maxwell and Putnam 1968).

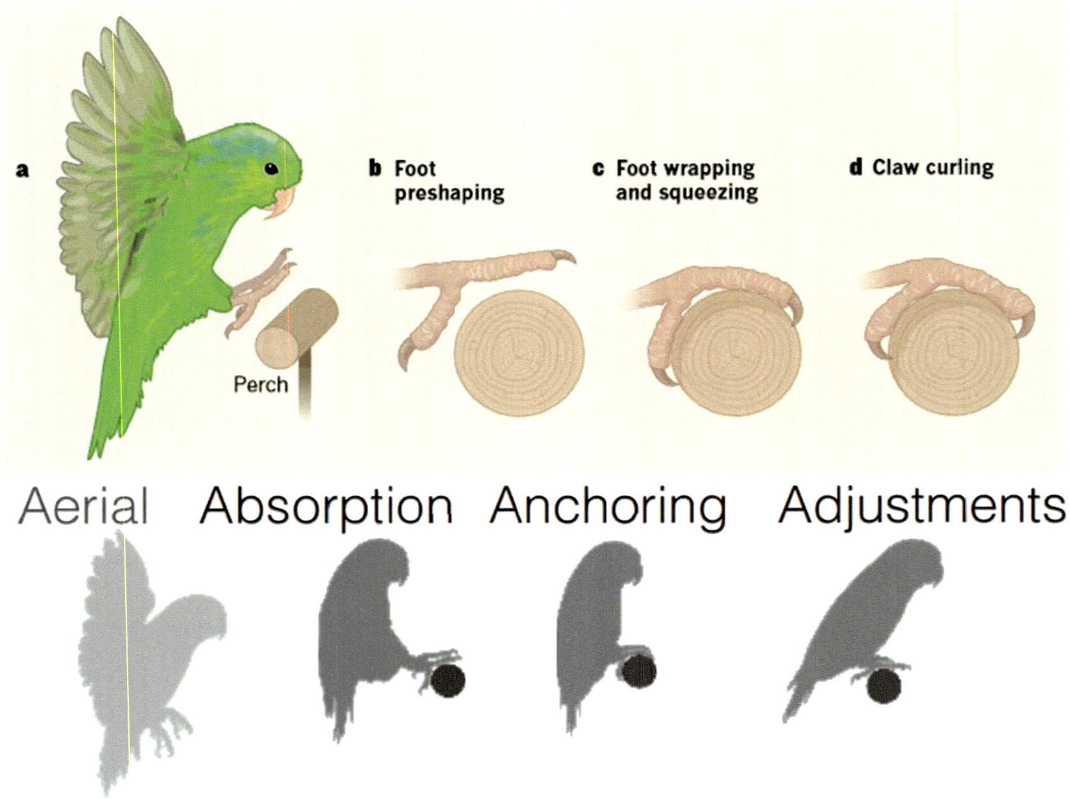


**Fig. 11.95** How a Pacific Parrotlet (*Forpus coelestis*) lands stably on a perch. (**a**) When about to land, the wings and tail are used for braking and the legs and feet are extended toward the perch. (**b**) Just before making contact, the toes begin to close (preshaping). (**c**) After making contact, the toes wrap around the perch and squeeze it, (**d**) followed by curling of the claws (Top figure from Biewener (2019); Nature 574: 180–181; Bottom figure from Roderick et al. (2019); © Roderick et al. article distributed under the terms of the Creative Commons Attribution License, https://creativecommons.org/licenses/by/4.0/)

**Fig. 11.96** (**a**) Positions of foot and claws when approaching and making contact with a perch. (**b**) Changing angles of the foot and claw when approaching and making contact with a perch. (Figure from Roderick et al. 2019; © Roderick et al., article distributed under the terms of the Creative Commons Attribution License, https://creativecommons.org/licenses/by/4.0/)

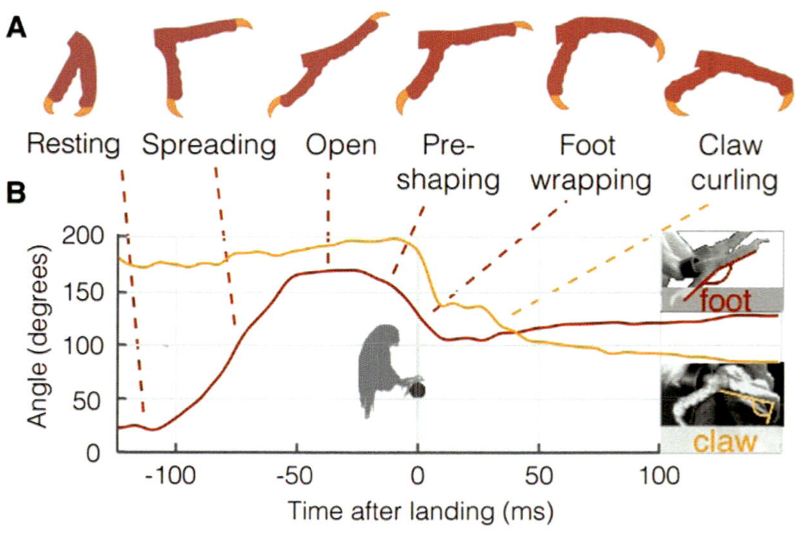

**Fig. 11.97** Eurasian Kestrel (*Falco tinnunculus*) landing. Note the increasing angle of attack that causes a reduction in lift, and spreading of the rectrices to aid in reducing speed as the kestrel approaches the perch. (Photo by Sachin Nihcas, Pexels, CC0 Public Domain)

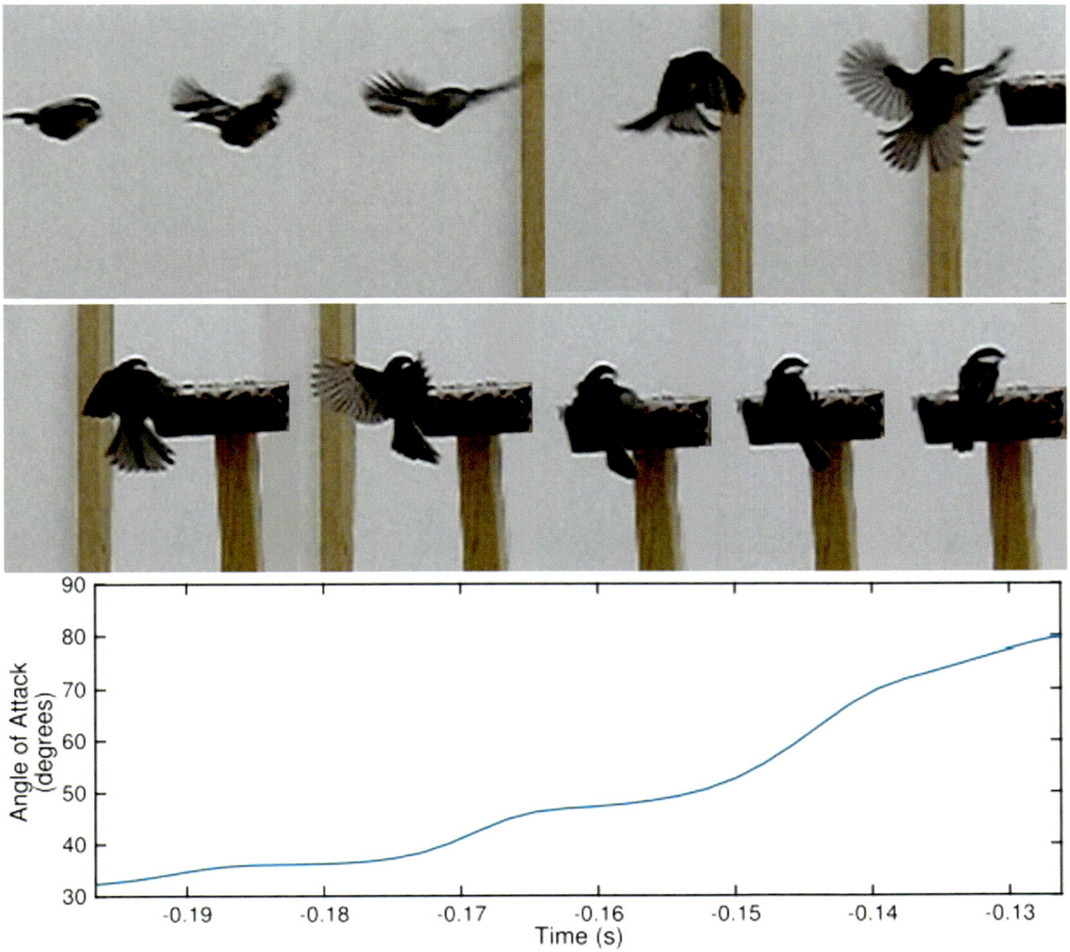


**Fig. 11.98** A Black-capped Chickadee (*Poecile atricapillus*) coming in for a landing. Bottom, note the rapid increase in angle of attack as the chickadee approaches the perch. (Images are screenshots from a Supplemental Video from Polet and Rival 2015 and bottom figure is also from Polet and Rival 2015; © 2015 IOP Publishing, used with permission)

**Fig. 11.99** Drawing that shows the flight of a Harris's Hawk (*Parabuteo unicinctus*) between two perches. A take-off jump was followed by a powered dive (yellow line). At the flight's lowest point (black dot), the hawk transitioned to an unpowered climb (blue line), then finished by pitching upward with its body nearly vertical and with its wings outstretched as its feet contacted the perch. This perching maneuver by larger birds eliminates the need to hover as they land on a perch (as smaller birds can do). By gliding up to a perch, large birds may also be better able to get a visual fix on the perch, which might be more difficult when flapping the wings (KleinHeerenbrink et al. 2022). (Figure modified from KleinHeerenbrink et al. 2022; open-access article licensed under a Creative Commons Attribution 4.0 International License, https://creativecommons.org/licenses/by/4.0/)

**Box 11.13 Using Lift as a Brake**

When birds take off, lift opposes drag and accelerates birds forward. However, Chin and Lentink (2019) found that, when Pacific Parrotlets (*Forpus coelestic*) land after short flights, lift is oriented backward, assisting drag in decelerating and contributing about one-quarter of their braking force. Parrotlets preparing to land on a perch greatly increase their angle of attack, increasing drag, and, during the last two wingbeats before landing, the angle of attack is so high that the lift force becomes a braking force.

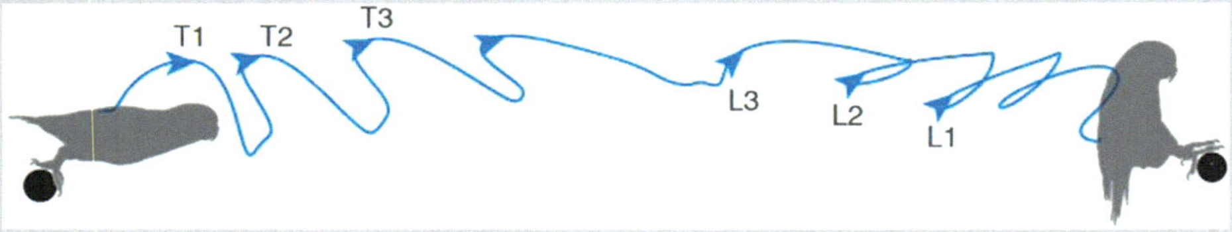


Birds initiate flights with an inclined wingstroke that gradually levels out and then angles backward before landing. (Figure from Chin and Lentink 2019; open-access article distributed under the terms of the Creative Commons CC BY license, https://creativecommons.org/licenses/by/4.0/)

(continued)

**Box 11.13** (continued)

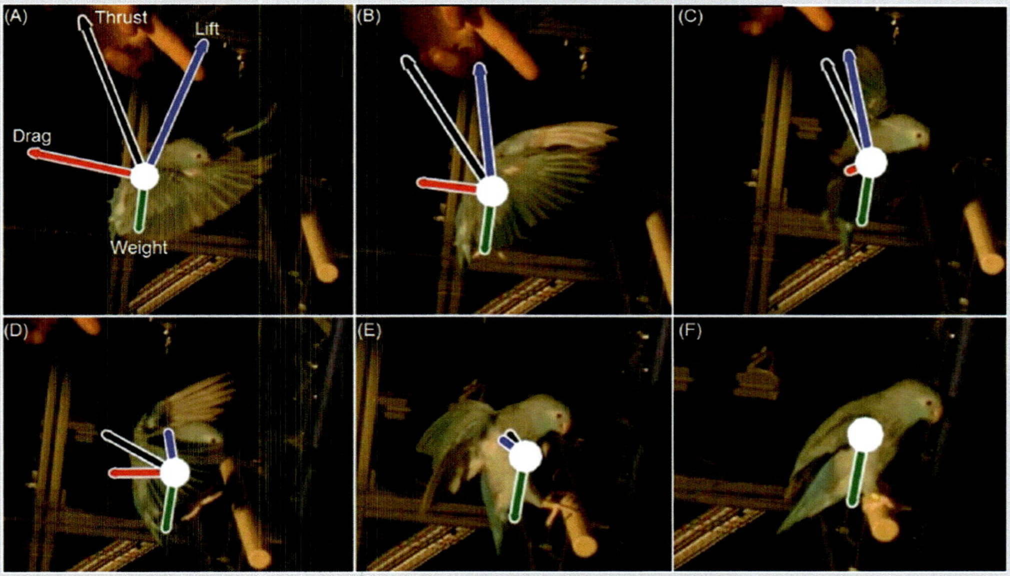


As a bird lands, the lift force changes from a vertical direction to a slightly backward direction, helping to reduce speed to zero at landing. (Figure made from screenshots from Supplementary Movie 1 from Chin and Lentink 2019; open-access article distributed under the terms of the Creative Commons CC BY license, https://creativecommons.org/licenses/by/4.0/)

Another strategy for landing, used particularly by larger birds, is to fly relatively fast to a point ahead of and below the landing point (KleinHeerenbrink et al. 2022). The bird then swoops upward, using its forward momentum, and gradually decelerates due to the force of gravity until, as its speed reaches zero, it arrives at the landing spot (Fig. 11.99). This is particularly useful means of landing for cliff-nesting birds. Birds landing on water, like those landing on a perch, also follow a shallow trajectory and reduce speed and lift using their wings and tails (and, sometimes, by whiffling; Box 11.14 Whiffling). However, many aquatic birds also lower their webbed feet to further increase drag and reduce speed. Then, just before landing, the webbed feet are extended forward, permitting birds to hydroplane to a halt (Fig. 11.100).

**Fig. 11.100** Swan landing with webbed feet extended forward and hydroplaning. (Photo from pxhere.com, CC0 Public Domain)

**Box 11.14 Whiffling**

Ducks and geese, and occasionally other species of birds and even bats (Norberg 1976), coming in for a landing sometimes twist their bodies (but not their heads) and even, briefly, may turn upside down, a behavior referred to as whiffling. The likely explanation for such behavior is that these movements, especially when upside down, cause them to lose lift as they approach their projected landing location. When upside down, whiffling ducks and geese can rapidly lose altitude because upside-down wings, rather than generating upward lift, generate downward "lift" or, in other words, a loss of altitude (the Bernoulli effect and Newton's third law act to force them downward; Norberg 1976). Birds with relatively high wing loading, like ducks and geese, must fly relatively fast to maintain lift so, when landing, one way to maintain relatively high speeds, but still lose altitude, is to whiffle.

(continued)

**Box 11.14** (continued)

Photo of a whiffling Graylag Goose (*Anser anser*). Note that, although flying upside down, the very flexible neck of birds allows the goose to maintain its head position (Copyright Brian MacFarlane/Albanpix, used with permission of Brian MacFarlane)

(continued)

**Box 11.14** (continued)

Barnacle Geese whiffling as they come in for a landing. Compare images a through j, and note that a whiffling goose (circled) loses both speed (falling behind other geese in the flock) and altitude (comparison of images g through j and noting the position of the goose relative to its background, clearly illustrates the loss of altitude caused by whiffling) (Images are screenshots from video by de Koning and van Eekhout 2011 (https://www.youtube.com/watch?v=Tyzl4mDM5GM; licensed under a Creative Commons Attribution license, https://creativecommons.org/licenses/by/2.0/).

## 11.11 Energetics of Flight

Flight requires muscle contraction and muscle contraction requires energy. Much of the needed energy is used for the contraction of muscle, primarily skeletal muscle but cardiac (heart) and smooth (e.g., blood vessels) muscle as well. Energy is defined as the ability to work, with work being the application of a force that moves an object. In a muscle, like the pectoralis (Fig. 11.101) muscle of birds, energy (ATP) is used to move (shorten) muscle cells (Fig. 11.102; Box 11.15 Flying Generates Heat). The collective action of many muscle cells causes the movements of the skeleton (and associated structures, including feathers) needed for sustained, flapping flight. Another way to look at the energy needed for flight is in terms of power. Power is defined as work divided by time, so faster muscles (contracting faster) generate more power. This is important for birds because larger muscles tend to contract more slowly than smaller muscles and, therefore, larger muscles generate relatively less power. For example, the flight muscles of a small warbler can contract faster and generate more power per unit mass than those of an eagle. Of course, the pectoralis muscle of an eagle is much larger than that of a warbler and so the total power that can be generated is much greater. As a result, even large birds (up to 14–16 kg in mass) can generate the minimum power necessary for flight despite having less "efficient" flight muscles.

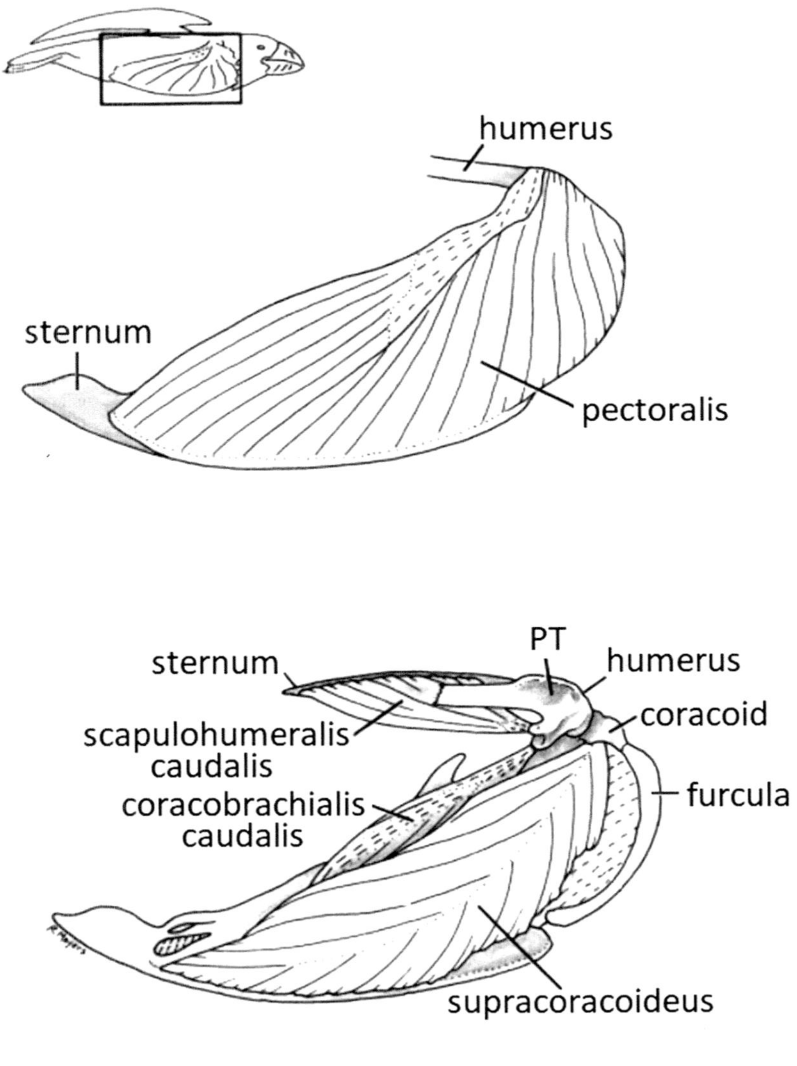


**Fig. 11.101** Flight muscles of an Atlantic Puffin (*Fratercula arctica*). The pectoralis is the primary downstroke muscle and the supracoracoideus the primary upstroke muscle. (Figure from Kovacs and Meyers (2000); © 2000 Wiley-Liss, Inc., used with permission)

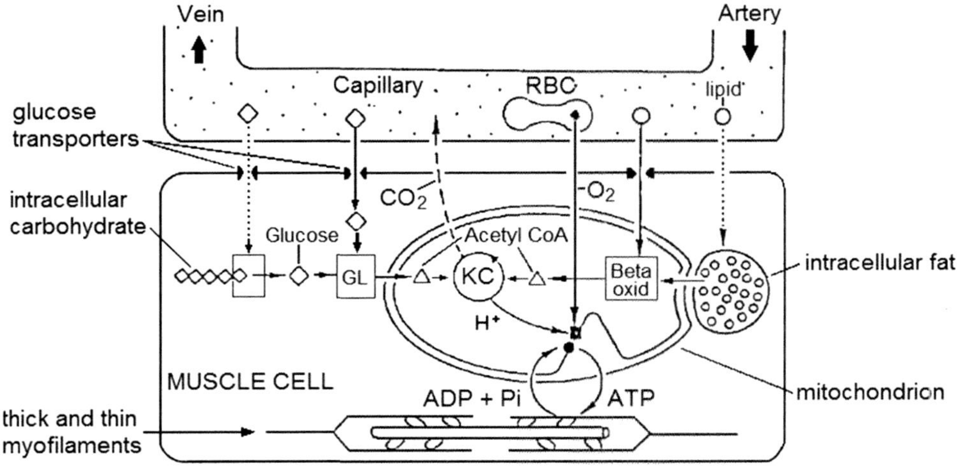


**Fig. 11.102** Nutrients (carbohydrates and lipids) and oxygen are transported to muscle cells. In those cells, acetyl CoA is generated from the breakdown of carbohydrates (glycolysis, GS) and fats (β–oxidation, Beta oxid) and acetyl CoA, in turn, is used to produce ATP via the Kreb's cycle (KC). ATP provides the energy needed for muscle contraction. (Figure modified from Hoppeler 1999; © 1999 Springer Nature, used with permission)

**Box 11.15 Flying Generates Heat**

An estimated 85–90% of the metabolic energy generated by flight muscles during flapping flight in birds (and bats) is dissipated as heat, with only 10–15% converted into mechanical power (Wells 1993; Speakman and Thomas 2003; Welch 2007). To maintain a normal body temperature, this heat must be dissipated. Ward et al. (1999) studied European Starling (*Sturnus vulgaris*) flying in a wind tunnel and, using infrared thermography, found that most heat was lost from the legs, head, and under the wings. To enhance heat loss, starlings lowered their legs into the air stream at higher air temperatures, with loss of heat apparently more important than minimizing drag. Birds like herons and storks that always have their legs in the air stream behind their bodies when flying regulate heat transfer from the legs either by via a rete system or by altering the flow of blood to the legs (Ward et al. 1999).

In a similar study, Powers et al. (2015) used infrared thermography to examine heat loss by Calliope Hummingbirds (*Selasphorus calliope*) hovering or flying in a wind tunnel. As with European Starlings, most heat was lost from the head (and specifically the eyes) and the area under the wings. In addition, the hummingbirds extended their legs and feet when hovering to enhance heat loss. Beating wings create air flow across the legs and feet, further enhancing heat loss when hovering (Powers et al. 2015).

(continued)

**Box 11.15** (continued)

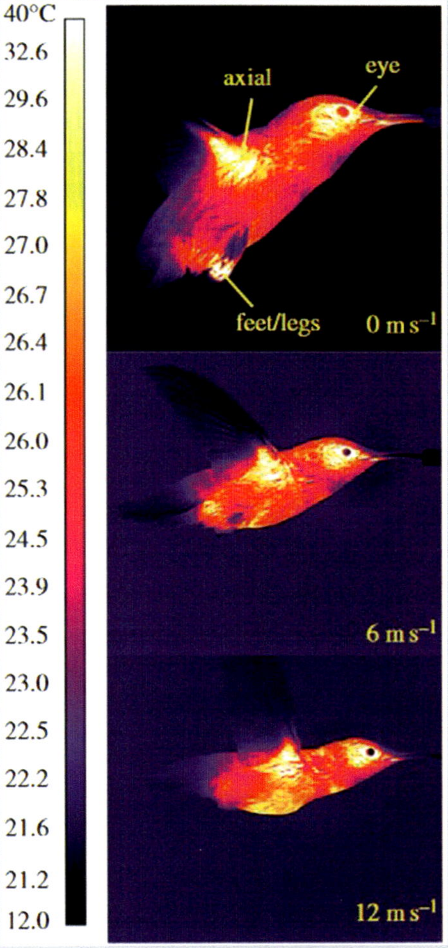


Infrared thermographic images of Calliope Hummingbirds (*Selasphorus calliope*) when hovering (top) and when flying at speeds of 6 (intermediate speed; middle) and 12 (high speed; bottom) meters per second. The eyes and areas under the wings were important areas of heat loss regardless of air speed. When hovering, the hummingbirds extended their feet and legs to enhance heat loss, but, during forward flight, the legs and feet were usually retracted to minimize drag. (Figure from Powers et al. (2015); © 2015 The Authors. Published by the Royal Society under the terms of the Creative Commons Attribution License, http://creativecommons.org/licenses/by/4.0/)

During long-duration flights by birds that must flap their wings continuously, e.g., during migration, the risk of hyperthermia likely increases. A strategy used by many migrants is to fly at higher altitudes where cooler air temperatures enhance heat loss. Alternating periods of flight with periods of time at stopover sites may be another means of avoiding hyperthermia. Migrating birds can replenish and restore energy reserves at stopover site, but these breaks between periods of flight may also help birds avoid the risks of hyperthermia.

Guillemette et al. (2017) captured female Common Eiders (*Somateria mollissima*) at a breeding colony in Denmark and implanted data-loggers in their body cavities that recorded

(continued)

**Box 11.15** (continued)

body temperature every 16 s. One year later, females that returned to the breeding area ($N = 19$) were recaptured, data-loggers were removed, and stored data were downloaded. Analysis revealed that body temperatures of the female eiders increased during migratory flights and, to avoid overheating, periods of flight were alternated with periods of cooling and resting at sea. This strategy, however, may be limited to birds adapted for diving because, when flying, their flight muscles likely use more energy and generate more heat because of their high wing loading. Migratory species of birds with lower wing loading values do not need stopovers to avoid overheating, with heat loss from the head, under the wings, and, perhaps, the legs and feet sufficient to avoid hyperthermia.

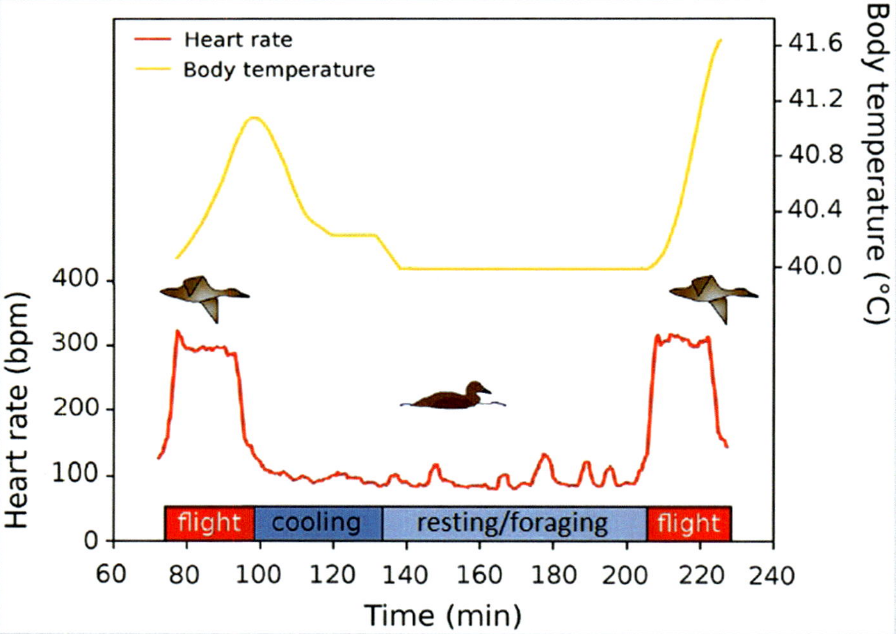


Flight cycle of Common Eiders during migration. Heart rate increases to and remains at about 300 beats per minute during periods of flight. However, body temperature continues to increase during periods of flight so, to avoid hyperthermia, migrating eiders must land to cool off before again taking flight and resuming migration. (Figure modified from Guillemette et al. 2017; © 2017 Guillemette, Polymeropoulos, Portugal and Pelletier, open-access article distributed under the terms of the Creative Commons Attribution License (CC BY), https://creativecommons.org/licenses/by/4.0/)

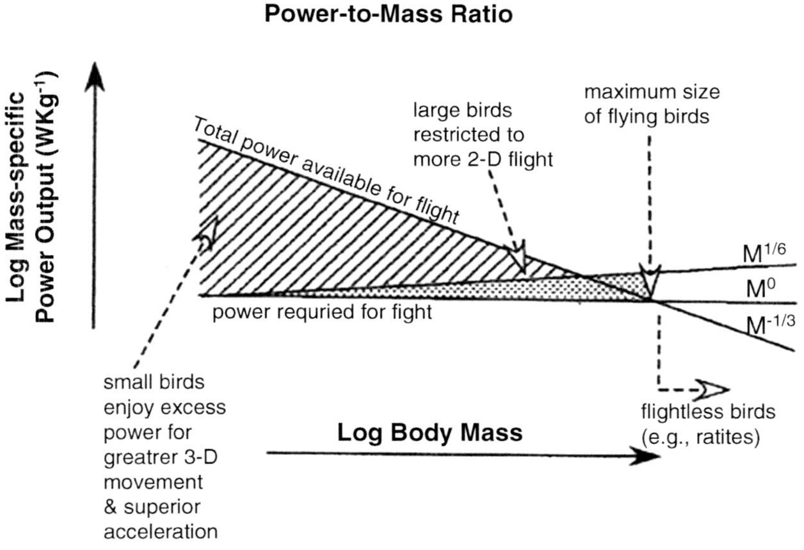


**Fig. 11.103** Theoretical relationship between body size and power output among birds. Power required is thought to scale independent ($M^0$) or slightly positive ($M^{1/6}$) among different-sized birds. Power available is thought to scale negatively ($M^{-1/3}$). Small birds have an excess of muscle power output relative to their body mass and, therefore, are able to enjoy a more three-dimensional world in time and space. (Figure from Dial (2003a, b); © 2003 Oxford University Press, used with permission)

Thus, the power needed for flight scales independently of body mass or, in other words, the power needed for flight is proportional to body mass. However, the flight muscles of larger birds cannot produce as much power per unit body mass as smaller ones. Larger birds flap their wings more slowly than smaller ones, and the general trend is for wing-flap frequency to scale negatively to the one-third body mass ($M^{-1/3}$). Because of the relationship between the speed of contraction and force generated, the mass-specific power available for flight likely also scales negatively to the one-third body mass (Ellington 1991; Dial 2003a, b; Fig. 11.103). The limited power available for flight limits the flight performance of larger birds, whereas the "excess" power available for smaller birds translates into greater flight capabilities (Dial 2003a, b). Most small species, therefore, inhabit more three-dimensional space and enjoy greater accelerative performance than larger birds (Dial 2003a, b).

Power is a measure of the rate at which energy is converted and the cost of bird flight is typically expressed in terms of Watts per unit mass (with a Watt equal to 1 Joule and a Joule being the amount of energy needed to exert a force of 1 Newton through a distance of 1 m). As an example, it takes about 10 Newton-meters of energy to lift a 1-kg mass to a height of one meter and, because 1 Newton-meter equals 1 Joule, that equals about 10 Joules or 10 Watts of power. What is important is the relative cost of flight (in terms of power, or Watts) compared to other forms of locomotion and how that cost varies with flight speed.

The power required for flapping flight for any bird or other flying animal is a function of flight speed, and the relationship between speed and power is typically expressed as a power curve. Aerodynamic theory predicts that mechanical power should vary as a function of forward velocity in a U-shaped curve (Fig. 11.104). In other words, theory predicts that more power is needed to fly slow and fast than to fly at intermediate speeds. This is because power is needed to overcome drag and drag is greater at slow and fast speeds. Remember, however, that there are two types of drag—induced and parasite drag—and they change in different ways as speed increases. Induced drag is very high at slow speeds, but declines as speed increases (Fig. 11.105). At low speeds, airflow over wings is relatively slow and, to maximize lift, birds must maximize the angle of attack and flap their wings fast to increase air speed and this requires lots of power. With increasing speed, induced drag declines as does the power required for flight. However, as speed continues to increase, so does parasite drag and

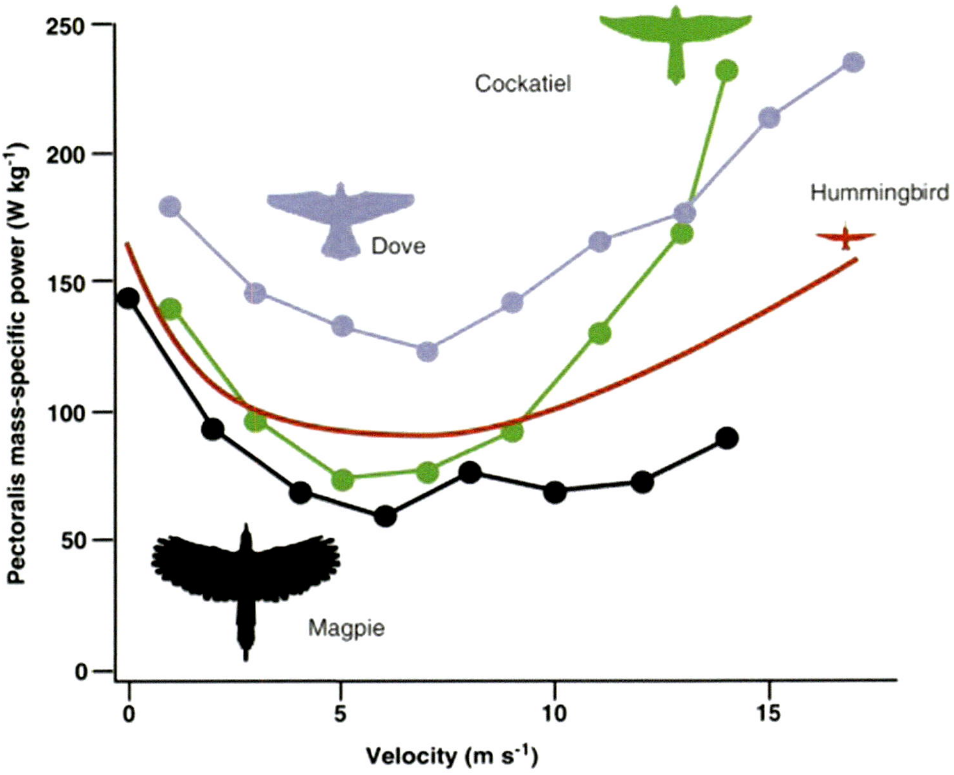


**Fig. 11.104** The mechanical power produced by birds in flight. Note that, for species ranging from hummingbirds to magpies, the relationship between the power required for flight and flight speed forms a U-shaped curve. (Figure from Warrick et al. (2012); © 2012 Elsevier Ltd., used with permission)

additional power is needed (Fig. 11.105). Although elevated respiration and blood circulation are partially responsible (Hedenström 2002), most of the additional power needed to overcome drag and fly at greater speeds is due to the increased activity of flight muscles (particularly the pectoralis muscles). Some birds beat their wings at greater rates at higher speeds. For example, wingbeat frequencies of both Cockatiels (*Nymphicus hollandicus*) and Ringed Turtle-doves (*Streptopelia risoria*) increase (along with pectoralis power) with increasing flight speed (Fig. 11.106; Tobalske 2007). Another means of modulating pectoralis power is to vary the proportion of motor units (the proportion of muscle cells) used, with fewer motor units used at slower speeds and increasing numbers used with increasing speed (Tobalske 2007).

With a U-shaped power curve, the lowest point represents the speed that requires the minimum power and is called the minimum-power speed ($V_{mp}$; Fig. 11.107). Energetically, the minimum-power speed is best and, under most circumstances, would be the flight speed of choice. However, for long-distance flights, such as during migration, flying a bit faster than the minimum-power speed is optimal because it allows a bird to maximize distance flown relative

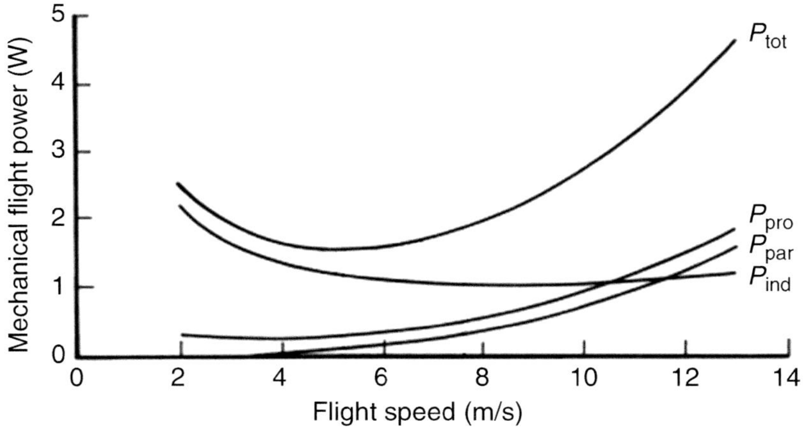


**Fig. 11.105** Mechanical power (in Watts) output ($P_{aero}$) of birds relative to flight speed during flapping flight. Power needed to overcome both profile ($P_{pro}$) and parasitic ($P_{par}$) drag increases with increasing flight speed, but power needed to overcome induced drag ($P_{ind}$) decreases with increasing flight speed. (Figure modified from Bishop and Butler 2015; © 2015 Elsevier Inc. All rights reserved, used with permission)

to the amount of energy used. By flying faster, at the maximum-range speed ($V_{mr}$; Fig. 11.107), a bird uses more energy, but the increased speed allows it to travel farther. In other words, the maximum-range speed maximizes airspeed relative to power and, as a result, maximizes distance relative to power (or energy required). Some evidence suggests that birds may adjust their speeds to fit the circumstances. For example, Eurasian Skylarks (*Alauda arvensis*) use minimum-power speed during song flights ($V_{mp}$), but fly at the maximum-range speed (or a bit faster) during migration ($\geq V_{mr}$; Hedenström and Alerstam 1996). Similarly, European Kestrels (*Falco tinnunculus*) wind-hover longer and spend more time hovering during periods when wind speeds are close to their $V_{mp}$ (Masman and Klaasen 1987). Other investigators have reported that birds may not always fly at what is considered (or calculated to be) the optimum speed. For example, to maximize energy delivered to nestlings, birds that must travel some distance from nest sites to foraging sites might be expected to fly faster than $V_{mr}$ (Norberg 1981) and, to reduce the risk of predation (e.g., from aerial predators), birds traveling from nest or roost sites to foraging sites may also fly faster than $V_{mr}$ (Elliott and Gaston 2005). In addition, however, even when data are available for body mass and wing measurements, the exact shape of power curves is poorly known for most species. Also, estimates of the flight speeds are typically made at low altitudes rather than the higher altitudes used by many migrating birds. Additional study is needed to determine if birds typically fly at their "optimum" speeds and, if not, to determine the factors that contribute to deviation from predicted values.

Although theory predicts that power should vary as a function of forward velocity in a U-shaped curve, power curves determined for several species of birds flying in wind tunnels are sometimes J-shaped (power for slow flight is not high), L-shaped (power does not rise as speed increases), or relatively flat. One factor that could contribute to this difference between theory and fact (or at least empirical data) is that birds may be reluctant or unable to fly at certain speeds

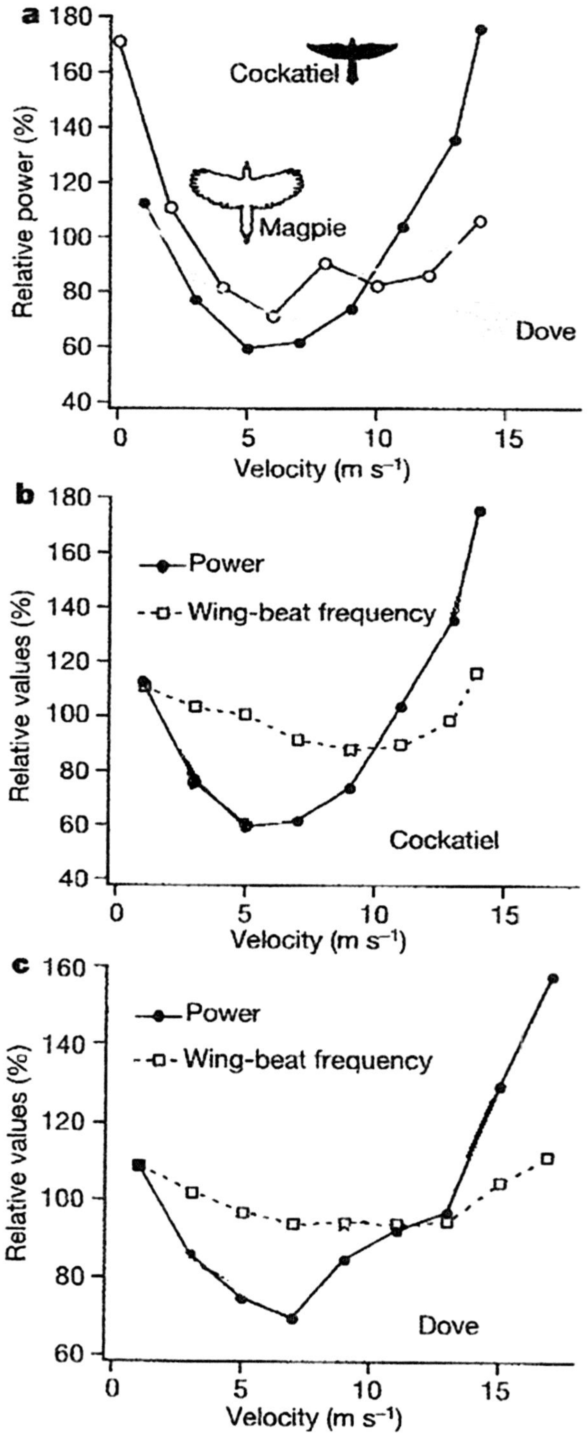


**Fig. 11.106** Pectoralis power and wingbeat frequency relative to flight speed (with values expressed as a percentage of the observed within-species mean for each variable). (**a**) Mean relative pectoralis power in Cockatiels (*Nymphicus hollandicus*), Black-billed Magpies (*Pica pica*), and Ringed Turtle-doves (*Streptopelia risoria*). (**b**, **c**) Mean relative pectoralis power and wingbeat frequency in Cockatiels (**b**) and Ringed Turtle-doves (**c**). (Figure modified from Tobalske et al. 2003b; © 2003 Springer Nature, used with permission)

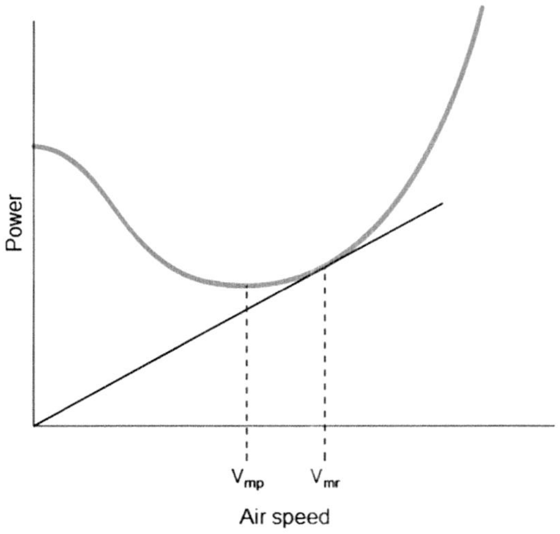


**Fig. 11.107** Relationship between air speed and mechanical power for a bird in forward flapping flight. The minimum-power speed ($V_{mp}$) is that speed at which the power required for flight is lowest; the maximum-range speed ($V_{mr}$) is the speed that allows a bird to travel the greatest distance with a given amount of fuel. (Figure from Hedenstrom and Lindstrom 2017; © 2017 The Authors, used with permission)

(in wind tunnels and in the wild) either because of the large expenditure of energy or mechanical/physiological limitations and, therefore, only a portion of the U-shaped curve is generated. With data collected only for a certain range of speeds, power curves could be relatively flat (no data available for relatively low or high speeds), J-shaped (no data for low speeds), or L-shaped (no data for high speeds; Rayner 1999).

Many birds use only flapping flight, but many other, particularly larger, terrestrial birds spend much of their time in the air soaring and gliding. Because soaring birds use rising air to generate lift, the energetic cost of flight where birds alternate between soaring and gliding can be much lower than that of continuous flapping flight (Figs. 11.108 and 11.109). Vultures may spend more than 50% of their daily time budgets in flight (Xirouchakis and Andreou 2009) so the low energetic cost of alternating soaring and gliding is likely essential for scavengers that may fly on average over 100 km a day in search of food (Monsarrat et al. 2013; Box 11.16 Contorted Soaring). The low energetic cost of soaring means that continuing to search for food may sometimes be a better strategy than landing, competing for access to a carcass, and then again taking off (Duriez et al. 2014).

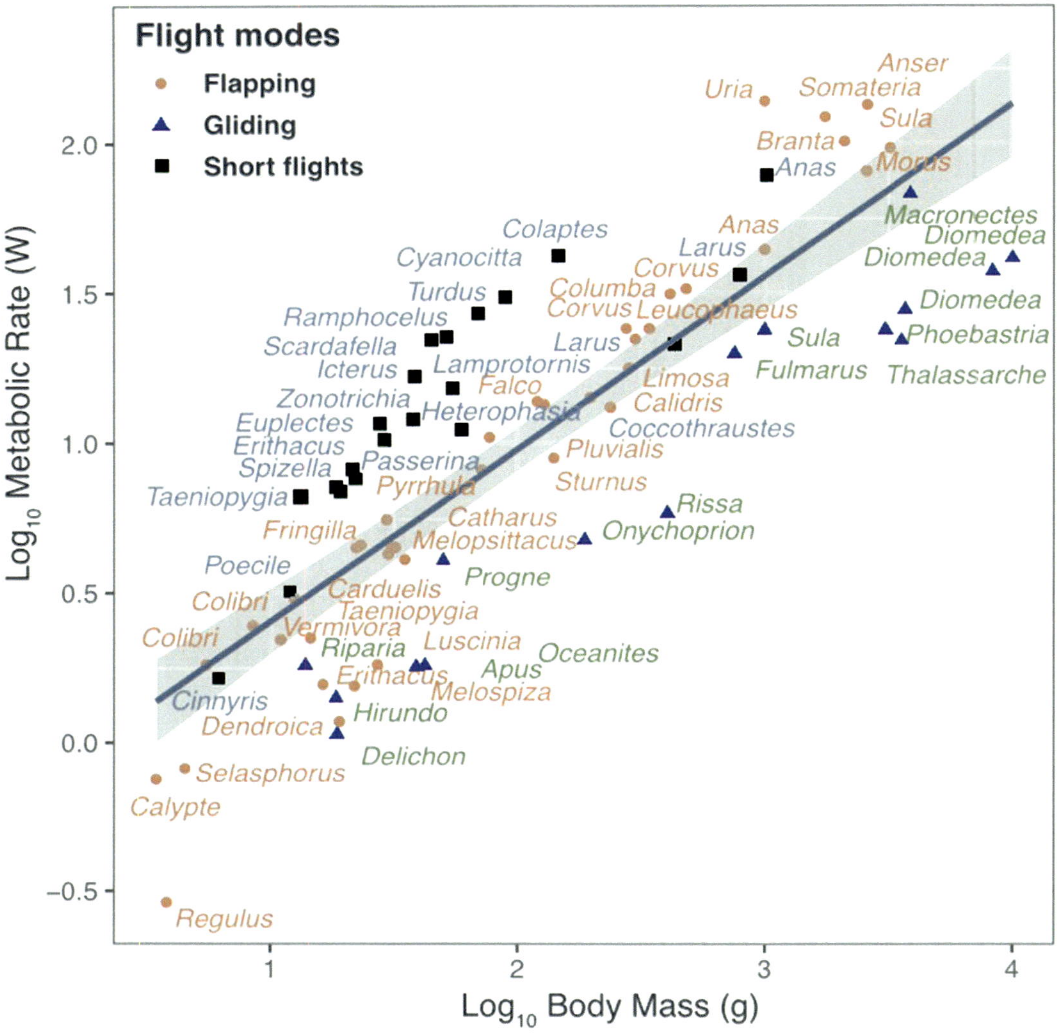


**Fig. 11.108** Variation in flying metabolic rate among different types of flight after correcting for body mass. Note that sustained flapping flight takes less energy than short flights, likely because of the cost associated with taking off and accelerating, and sustained flapping flight is more costly than sustained gliding flight. Data obtained from studies of birds in the wild, in wind tunnels, and in flight cages. (Figure from Guigueno et al. (2019); © 2019 Elsevier Inc., used with permission)

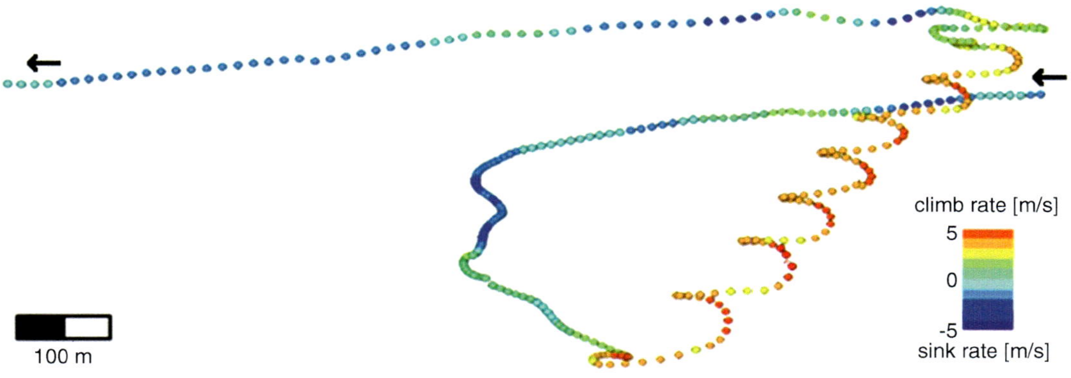


**Fig. 11.109** Example of a flight path of a Eurasian Griffon (*Gyps fulvus*) during a day favorable for the formation of thermals. Flight modes alternated between gliding (blue sections) and thermal soaring (ranging from yellow to red). (Figure from Harel et al. (2016); open-access article licensed under a Creative Commons Attribution 4.0 International License, https://creativecommons.org/licenses/by/4.0/)

**Box 11.16 Contorted Soaring**

Turkey Vultures (*Cathartes aura*) can be found throughout South and Central America as well as in Mexico, the United States, and southern Canada. These vultures are well known for their distinctive dihedral flight posture and style, holding their wings above the horizontal to form a shallow V and often teetering back and forth as they move through the air. Mallon et al. (2016) referred to this type of flight, consisting of numerous vertical and horizontal deviations from a straight-line path, as contorted soaring. Turkey Vultures tend to engage in contorted soaring on days that are cool and windy and when flying over forested areas. On warmer, calmer days, they are more likely to use thermals to gain altitude by soaring followed by gliding flight.

Illustration of a Turkey Vulture engaged in contorted soaring just above tree height. The flight path, ~10 s in duration, deviates horizontally and vertically from a linear flight path while maintaining both altitude and general direction. (Figure from Mallon et al. (2016); © 2015 Oxford University Press, used with permission)

(continued)

**Box 11.16** (continued)

On windy days in generally open habitats with scattered woodlots and riparian areas, as well as the occasional urban area, wind hitting an area of trees or buildings produces areas of uplift. However, because air flow becomes turbulent when striking these barriers, areas of uplift are interspersed with areas where the swirling air is moving in other directions. This means that, as Turkey Vultures are flying, one wing, say the left wing, but not the other, may be in an area of rising air. When this happens, a vulture will "teeter" to the right so that the left wing experiencing more lift tilts upward, generating less lift, while the right wing tilts downward. As they continue moving forward, the situation will reverse and the vulture will "teeter" back to the left. By controlling how much lift is generated by each wing, vultures are able maintain their preferred altitude as they search for carrion. Although other species of birds, including Black Vultures (*Coragyps atratus*), also engage in contorting soaring at times (Mallon et al. 2016), Turkey Vultures likely do so most often because, in contrast to other species of vultures, they use olfaction to detect and locate carrion. As a result, Turkey Vultures are better able to detect carrion in cluttered habitats like forests and even some urban areas and are, therefore, more likely to be found in these areas and able to take advantage of the patterns of air flow in these areas that make contorted soaring possible.

(continued)

**Box 11.16** (continued)

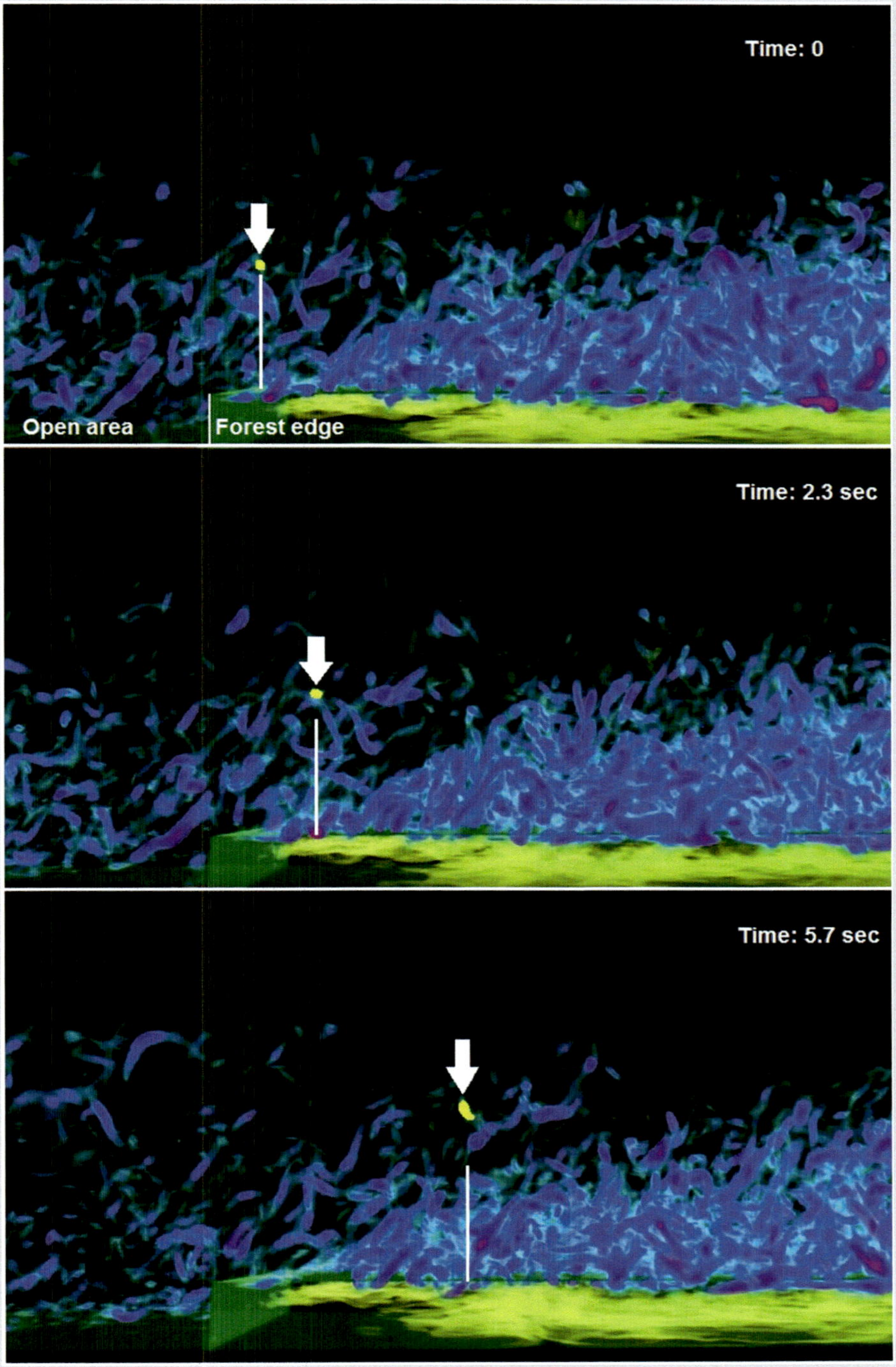


A series of images from an animation showing turbulence as wind hits the edge of a forested area. Within the area of turbulence are areas of rising air as illustrated by "packet" of air highlighted with yellow. The white lines below

(continued)

**Box 11.16** (continued)

the yellow areas are the same height, illustrating that as the wind travels along the top of the forest (from top image to the bottom image), this "packet" of air is rising. (Figure made from screen-captures from an animation on YouTube: https://www.youtube.com/watch?v=NClfGdfwWTA; Copyright 2014 Institut für Meteorologie und Klimatologie, Leibniz Universität Hannover (LUH), Authors: Farah Kanani, Björn Maronga, Helge Knoop, Siegfried Raasch; Animation based on Kanani-Sühring and Raasch 2015; used with permission)

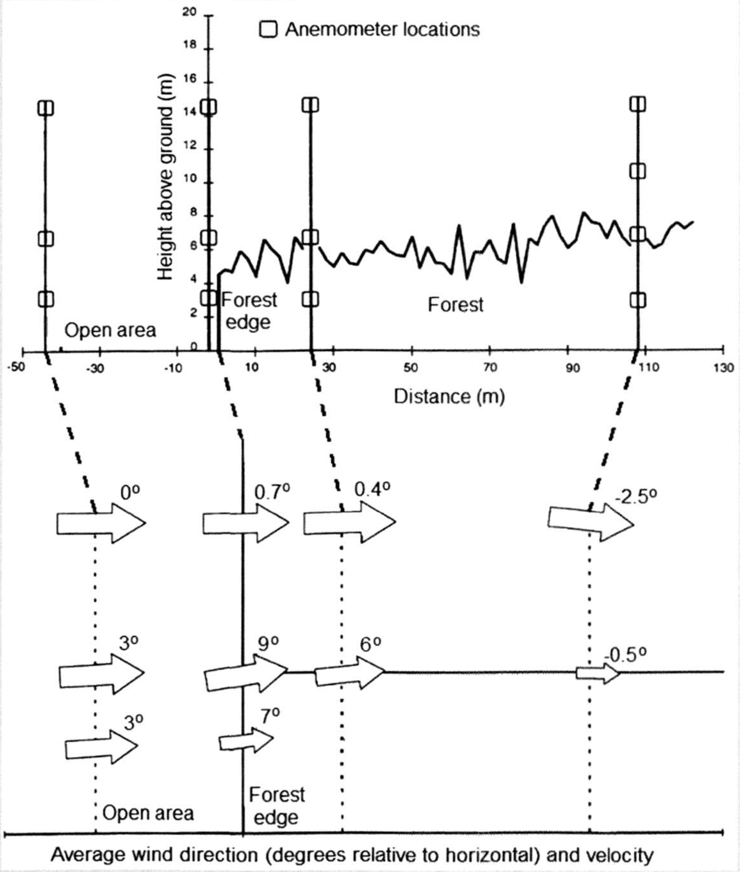


Effect of a forest edge on wind direction and velocity, with velocity measured using anemometers located at different heights and different distances relative to the forest edge. Turbulence is generated above the forest canopy, but the average wind direction above the canopy at the forest edge slightly upward at an angle of 9°. Arrow size represents relative wind velocity, with larger arrows indicated faster wind speeds. (Figure modified from Irvine et al. 1997; © 1997 Kluwer Academic Publishers, used with permission)

(continued)

**Box 11.16** (continued)

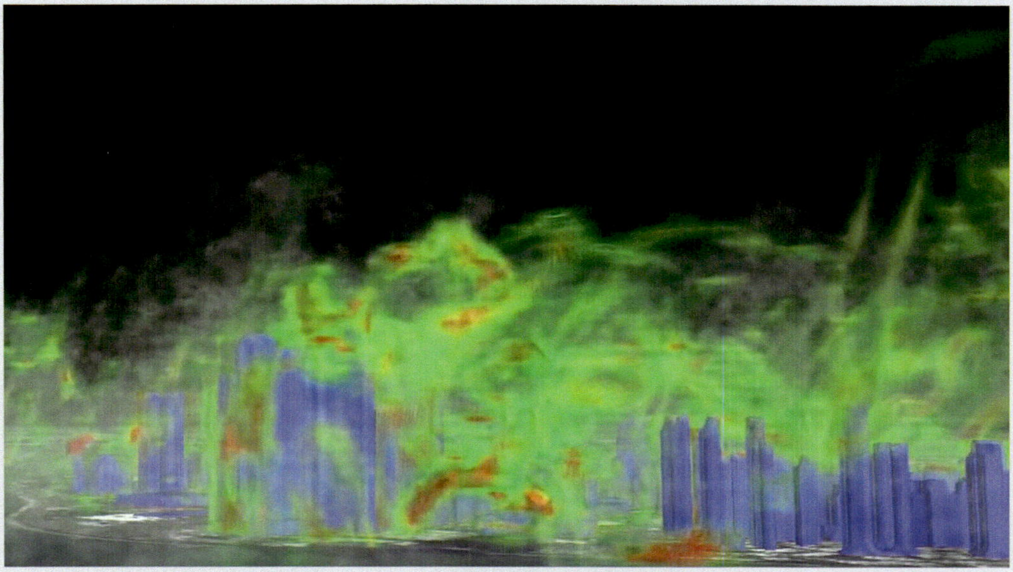

Animation showing turbulence generating rising air as wind hits a series of tall buildings in a city. (Figure is a screen-capture from an animation on YouTube: https://www.youtube.com/watch?v=y1sSRXFBN7k; Copyright 2014 Institut für Meteorologie und Klimatologie, Leibniz Universität Hannover (LUH), Authors: Helge Knoop, Marius Keck, Siegfried Raasch. Animation based on Kanani-Sühring and Raasch 2015, used with permission)

(continued)

**Box 11.16** (continued)

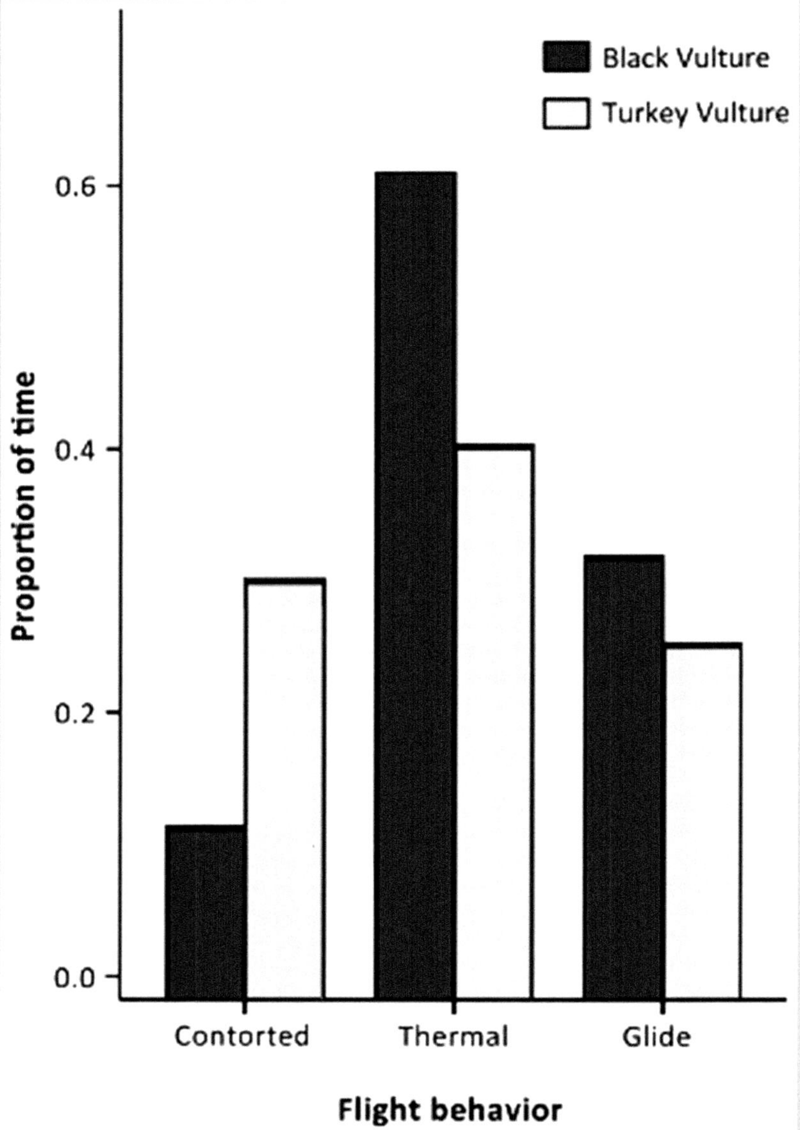


Proportion of time that Turkey (*Cathartes aura*) and Black (*Coragyps atratus*) vultures spent using contorted soaring, thermal soaring, and gliding flight. (Figure modified from Mallon et al. 2016; © 2015 Oxford University Press, used with permission)

## 11.12 Loss of Flight

Flying allows birds to travel widely across landscapes, continents, and oceans searching for food, mates, and other important resources and can also greatly increase their chances of escaping from potential predators. Despite these and other potential advantages of being able to fly, selection has favored the loss of flight in many species of birds. Sayol et al. (2020) conducted an extensive search of the literature and determined that at least 226 species of extinct ($N = 166$) and extant birds ($N = 60$) lost the ability to fly during the period from the Late Pleistocene to the present (Fig. 11.110). Many of these flightless birds are or were found in island archipelagoes, particularly Hawaii (23 species) and New Zealand (26 species). Before human-caused extinctions of many flightless birds, Sayol et al. (2020) determined that 23 of 39 orders of birds had at least one flightless species. Currently, the 60 extant species of flightless birds represent just nine orders, and the families Rallidae (rails) and Spheniscidae (penguins) account for 35 of those species. Most flightless species of birds evolved in areas where ground-based predators, especially mammals, were absent, so when humans arrived in these areas accompanied by other non-native mammals, they and their eggs and young were, and still are, often easy prey (e.g., Milberg and Tyrberg 1993; Duncan and Blackburn 2004).

As just noted, most species of birds that lost the ability to fly occurred in areas without mammalian predators and, in addition, in areas where flight was not required to access essential resources, especially food (e.g., McNab 1994, 2002). Given these conditions, selection may favor the loss of flight because, despite its many potential advantages, flying takes considerable energy. The transition of flightlessness involves the redistribution of muscle mass from the forelimbs to the hindlimbs, with great reductions in the mass of the primary flight muscles (pectoralis and supracoracoideus) and larger legs (e.g., Olson 1973; Livezey 2003; Wright et al. 2016). For example, the wing muscle mass of Greater Rheas (*Rhea americana*) represents only 0.89% of their total body mass (Le Coco et al. 2022). Much smaller flight muscles that are rarely used require much less energy and this substantial reduction in energy consumption has likely been an important factor in the loss of flight for many species of flightless birds (McNab 1994; McNab and Ellis 2006).

Penguins are flightless, but still have well-developed pectoralis and supracoracoideus muscles. Among seabirds that both fly and are wing-propelled divers, selection has favored smaller wings that reduce the energetic cost of "flying" underwater, but, as a consequence, increase the energetic cost of flying (Fig. 11.111a). This high cost of flight for wing-propelled diving birds like murres, plus increasing energetic costs of diving with increasing body mass for flying divers compared to nonflying divers (penguins; Fig. 11.111b), likely explains the evolution of flightlessness in large, wing-propelled divers, like penguins and extinct larger auks (Elliott et al. 2013).

## 11.13 Nonflying Modes of Locomotion

### 11.13.1 Walking, Running, Hopping, and Waddling

Most birds also use forms of locomotion other than flying. Exceptions include hummingbirds and swifts that have weakly developed legs and feet used only for perching or, in the case of swifts, for grasping vertical surfaces. Of course, some birds are flightless and depend entirely on walking, running, swimming, or diving to get from place to place. The extent to which birds use forms of locomotion other than flight varies, and such variation results in corresponding differences in the morphology of bird legs, feet, and wings.

Many birds walk, run, hop, or waddle with varying degrees of ability. In bipedal walking and running (Box 11.17 Body Mass and Bipedal Locomotion), the feet move alternately, half a cycle out of phase of each other. If the duty factor (the fraction of time that each foot is on the

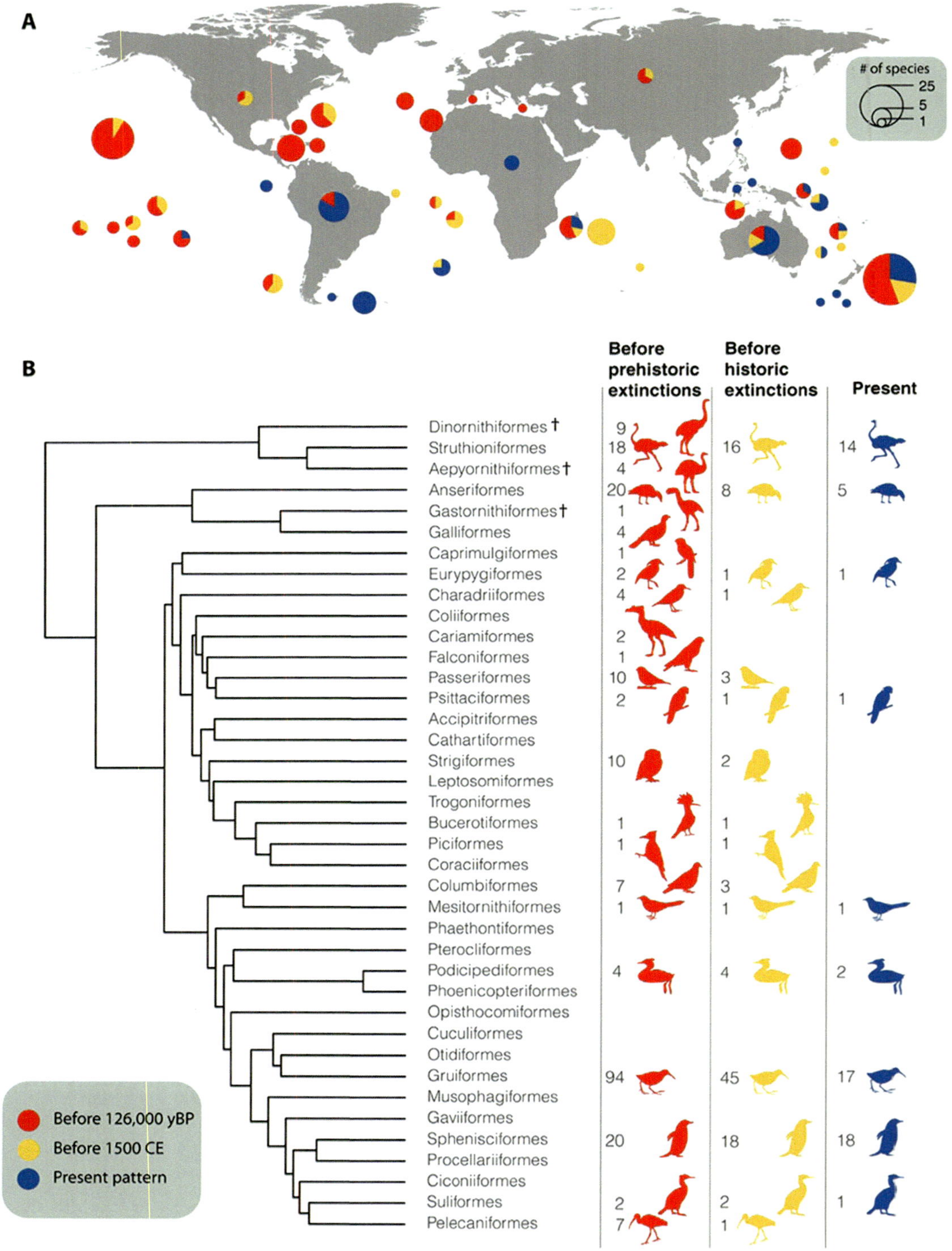


**Fig. 11.110** Geographic and phylogenetic distribution of flightless birds. (**a**) The global distribution of flightless species is indicated by the circles and the size of circles indicates the diversity of known flightless species. Numbers of extant species are show in blue, historical extinctions (i.e., after 1500 CE [common era]) in yellow, and prehistoric extinctions (i.e., Late Pleistocene and Holocene up to 1500 CE) in red. (**b**) The number of

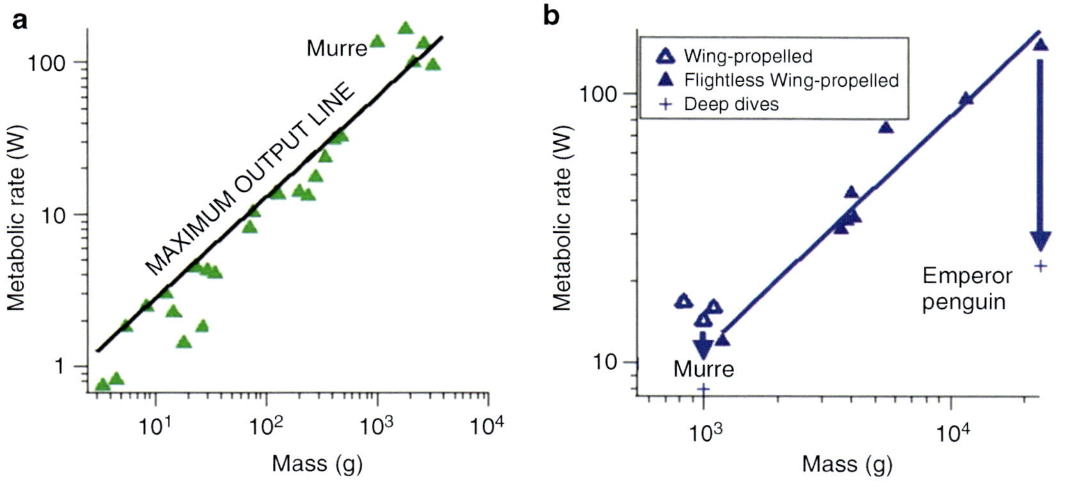


**Fig. 11.111** (**a**) Comparison of power output during flapping flight for different bird species. Note the relatively high energetic cost of flight for wing-propelled diving murres. The sustained maximum output limit as proposed by Videler (2006) is indicated by the solid line. (**b**) Dive costs as a function of body mass for several species of birds. Thick-billed murre diving costs averaged across dive times. Average dive costs during deep dives by murres and Emperor Penguins are also shown. Metabolic rates decrease during deep dives because buoyance, heart rate, and temperature decrease during such dives. (Figure from Elliott et al. (2013); used with permission of the US National Academy of Sciences)

ground) is greater than 0.5, a bird is walking (Figs. 11.112c and 11.113) and, if less than 0.5, it is running (Fig. 11.112a, b). When hopping, the feet move simultaneously (Alexander 2004; Fig. 11.114). Terrestrial species tend to walk and run, whereas those that are primarily arboreal typically hop (Clark Jr. 1975). Substrate may also play a role, with hopping more likely among rocks or uneven ground, and running and walking on more level terrain (Short 1971).

**Fig. 11.110** (continued) flightless species for each order is shown for each of the three time periods. Original silhouettes are deposited at phylopic.org under a public domain license. Orders where all species are extinct are indicated with †; yBP = years before the present. Original silhouettes are deposited at phylopic.org under a public domain license. (Figure from Sayol et al. (2020); © 2020 The Authors, exclusive licensee American Association for the Advancement of Science, used with permission)

**Fig. 11.112** X-ray images of a Northern Lapwing (*Vanellus vanellus*) on a treadmill. (**a**) Rapid running, (**b**) Slower running, and (**c**) walking. Stride length and rate increases with increasing speed. Note that when running rapidly, the Northern Lapwing has a brief aerial phase (**a**, far right image). (Figure from Nyakatura et al. (2012); © 2012 Wiley Periodicals, Inc., used with permission)

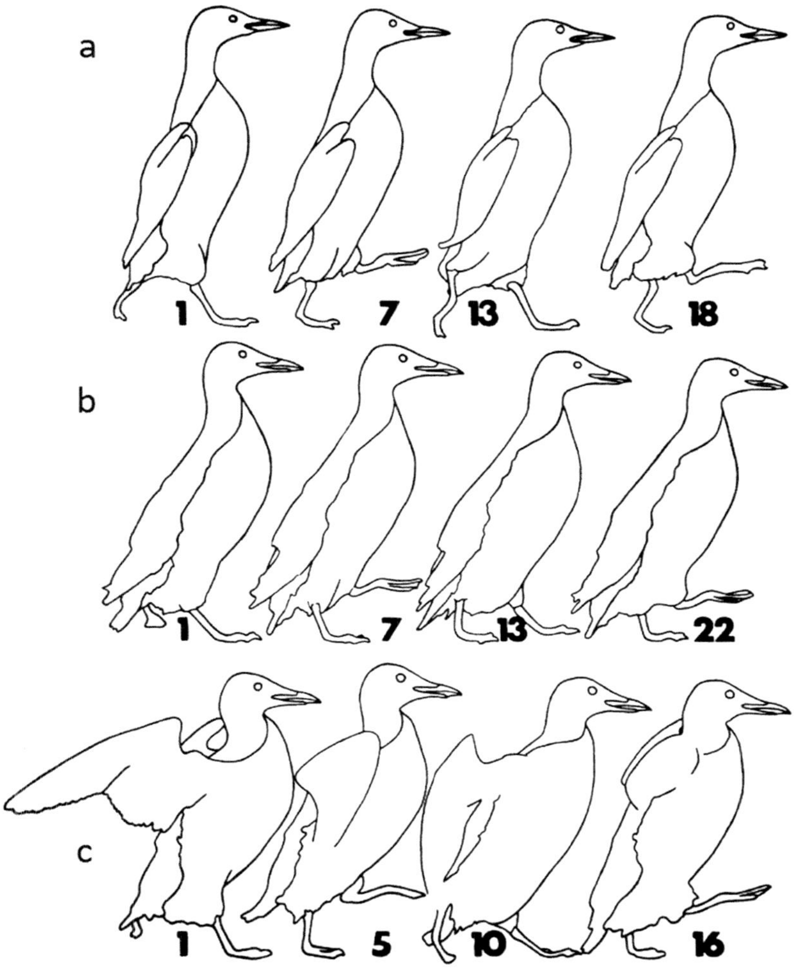


**Fig. 11.113** (**a**) A Common Murre (*Uria aalge*) walking, (**b**) a Thick-billed Murre (*Uria lomvia*) walking slowly, (**c**) a Thick-billed Murre walking fast. Numbers denote video frames in sequence. (Figure from Spring (1971); © 1971 Oxford University Press, used with permission)

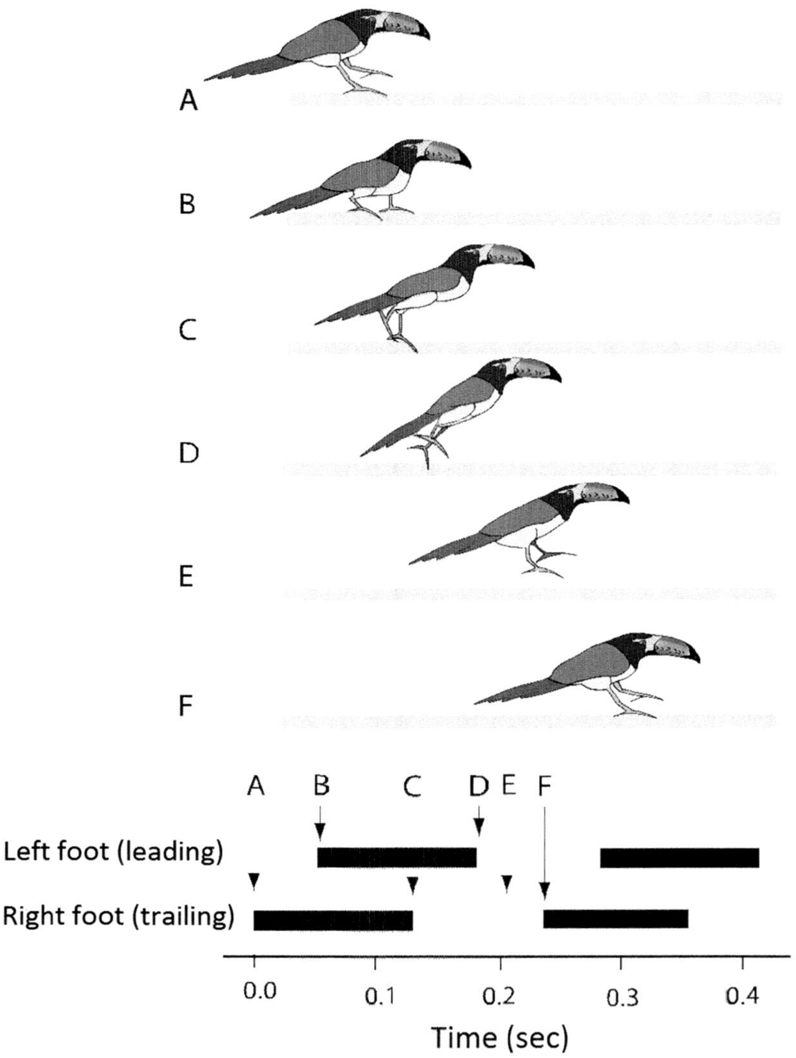


**Fig. 11.114** Hopping by a Lettered Aracari (*Pteroglossus inscriptus*) on a small branch. (**a**) touchdown of the right foot, (**b**) touchdown of the left foot, (**c**) right foot leaves the branch, (**d**) left foot leaves the branch, (**e**) no foot on the branch, (**f**) right foot touchdowns the branch again. Bottom: Gait diagram representing relative duration of the stance (dark lines) and swing (space between two dark lines) phases of each foot. Mean speed = 1.49 m per second. (Figure from Höfling et al. 2006; used with permission)

**Box 11.17 Body Mass and Bipedal Locomotion**

Extant birds exhibit a > 2500-fold range in body mass and corresponding diversity in the length of hindlimb bones (Gatesy and Middleton 1997; Zeffer et al. 2003). Despite this variation in leg length, the bones and joints in legs of different-sized birds move in a similar manner when they walk and run. However, Daley and Birn-Jeffery (2018) found that, at dynamically similar speeds, larger birds have relatively shorter stride lengths and higher stride frequencies than smaller birds. For example, a 0.1-kg bird must stride at a frequency three times higher than a 100-kg bird at a dynamically similar speed. Stability and agility are likely among the factors influencing these differences. Small birds may face more obstacles as they run, with vegetation

(continued)

**Box 11.17** (continued)

and other objects on the ground being larger relative to their leg length than for larger birds. As such, the longer and slower strides (i.e., more time between successive strides) of small birds might provide more stability and improve their ability to maneuver through relatively rough terrain (Daley and Birn-Jeffery 2018).

Small birds like use relatively longer strides and lower stride frequencies than larger birds at equivalent dimensionless running speeds (dimensionless speed is based on the idea that bipedal birds, and other bipedal animals, move in the same way when walking or running regardless of size (leg length). The formula for dimensionless speed (DS) is $DS = \frac{\text{speed}}{\sqrt{(\text{leg length} \times g)}}$, where $g$ equals acceleration due to gravity, or 9.8 m/s. Brown Tinamou, *Crypturellus obsoletus*; Red-winged Tinamou, *Rhynchotus refescens*; Solitary Tinamou, *Tinamus solitarius*; Black-legged Seriema, *Chunga burmeisteri*; Red-legged Seriema, *Cariama cristata.* (Figure from Abourachid et al. 2005; © The Neotropical Ornithological Society, used with permission)

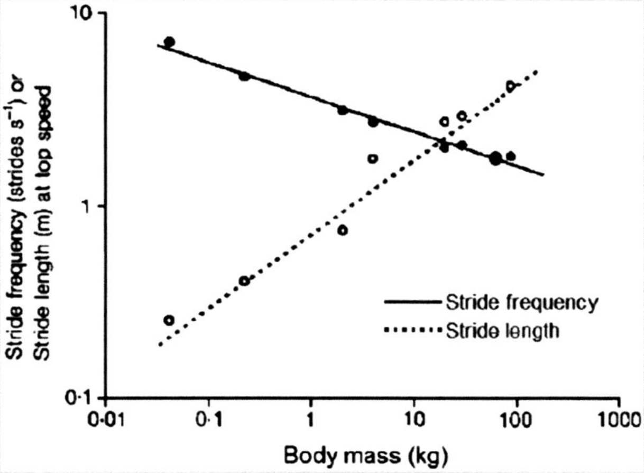


Stride length and stride frequency relative to body mass. Larger birds tend to have higher stride frequencies and shorter stride lengths than smaller birds at top speeds. (Figure modified from Gatesy and Biewener 1991; © 2009 John Wiley and Sons, used with permission)

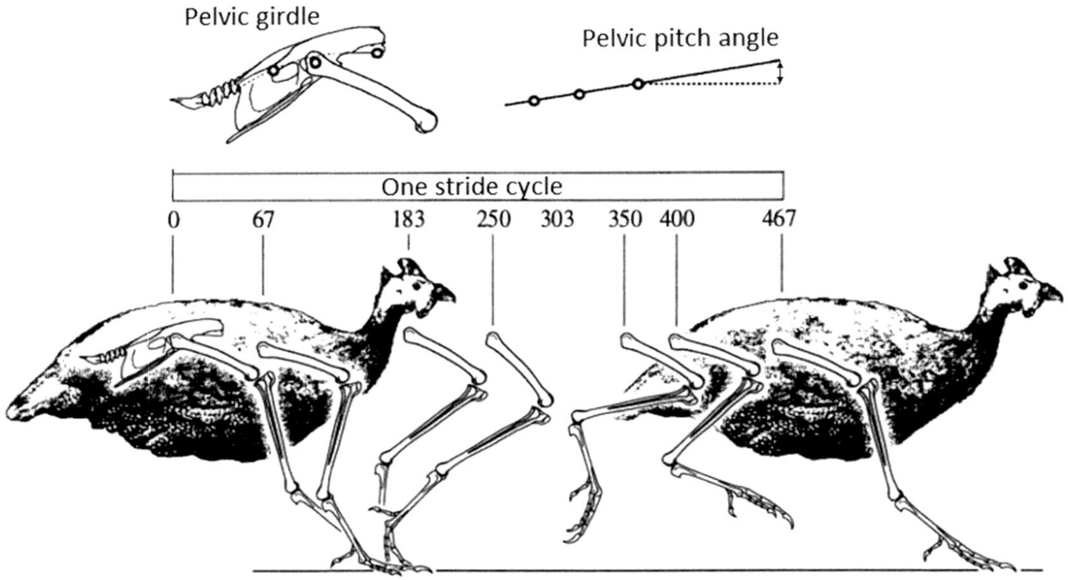


**Fig. 11.115** A Helmeted Guineafowl (*Numida meleagris*) running at 1.0 m/s. Note the low pelvic pitch angle. (Figure from Gatesy 1999; © 1999 Wiley-Liss, Inc., used with permission)

Duty factors of birds are similar to or a little smaller than those of humans when walking, but larger when running. The lowest duty factor recorded for a bird is 0.29 for a fast-running Common Ostrich (*Struthio camelus*; Alexander et al. 1979). Among humans, duty factors may be as low as 0.25. Penguins walk with their trunk vertical, but other birds keep their bodies much more horizontal when walking and running. For example, the pelvic girdle of a Helmeted Guineafowl (*Numida meleagris*) is at an angle of about 25° when during slow walking and drops to about 11° during fast running (Gatesy 1999; Fig. 11.115).

The metabolic energy used by most birds when walking and running appears to increase linearly with speed:

$$\text{Metabolic rate/body mass} = A + B\ (\text{speed}),$$

where $A$ and $B$ are constants for each species. This equation predicts that a bird will have a metabolic rate $A$ per unit body mass when stationary, and will use additional energy $B$ per unit distance traveled (so $B$ is the metabolic cost of walking or running). For many species of birds and mammals that walk and run, $B$ is generally close to the value given by this equation (Taylor et al. 1982):

$$\text{Metabolic cost of transport (J/kg m)} = 10.7 \times (\text{body mass, kg})^{-0.32}$$

For most birds, the metabolic costs of walking and running have not been determined. Among birds that have been studied, the cost of walking and running by gallinaceous birds is close to that predicted by this equation. However, walking and running by shorebirds is less costly than predicted, whereas, for waterfowl and penguins, it is more costly than predicted. The relatively high costs of terrestrial locomotion for waterfowl (ducks and geese) and penguins are due to their use of a gait involving lateral displacement (waddling; Fig. 11.116). This is more costly because energy is used not only for movement in the main direction, but from side-to-side as well. This high cost can be thought of as an evolutionary trade-off. The morphological adaptations of waterfowl and penguins for aquatic locomotion necessarily

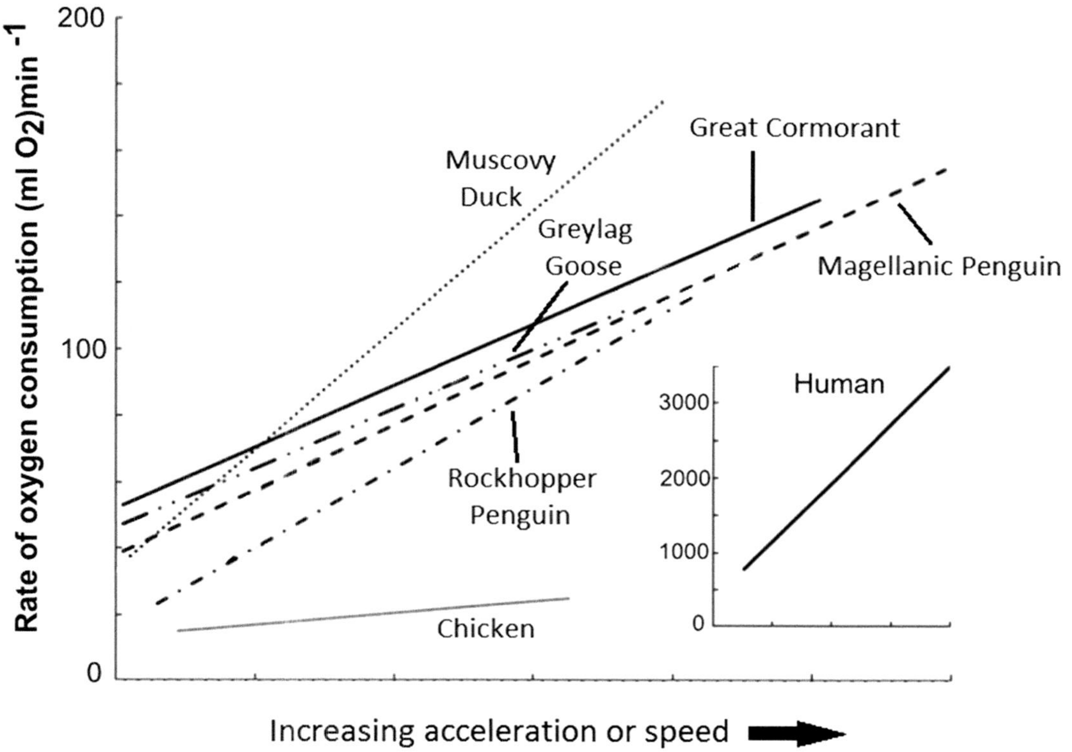


**Fig. 11.116** Energetic cost of walking or waddling by penguins (estimated based on rate of oxygen consumption) for a gallinaceous bird (Domestic Chicken) and five species of aquatic birds, including two species of penguins. Terrestrial locomotion is less energetically expensive for gallinaceous birds than for aquatic birds more adapted for swimming and diving. Data for humans are included for comparison (and note that humans expend far more energy when walking and running than birds do). Muscovy Duck, *Cairina moschata*; Great Cormorant, *Phalacrocorax carbo*; Graylag Goose, *Anser anser*; Magellanic Penguin, *Spheniscus magellanicus*; Southern Rockhopper Penguin, *Eudyptes chrysocome*; Domestic Chicken, *Gallus g. domesticus*. (Figure modified from Halsey et al. 2011; )

translate into less-efficient terrestrial locomotion (Bruinzeel et al. 1999).

Ratites, including Common Ostriches (*Struthio camelus*), Emus (*Dromaius novaehollandiae*), and rheas, are flightless and specialized for walking and running. Several features of the hindlimbs of ratites reflect their ability to run at speeds exceeding 50 km/h, including elongation of the hindlimb bones, a reduction in the number of toes (2 or 3 rather than the usual 4 of most birds), and increased development of the hindlimb musculature. The leg extensor muscles of Emus and Common Ostriches make up 22–29% of total body mass, allowing them to take long, powerful, and rapid strides (Hutchinson 2004). Among ratites, the contribution to total body mass of the hindlimb muscles is comparable to that of the flight muscles of flying birds.

Much of the energy lost by a human runner in the first half of a step is stored as elastic strain energy in stretched tendons and ligaments, and returned by elastic recoil in the second half. Similarly, running Common Ostriches, turkeys, and guineafowl save energy via elastic storage in the digital flexor tendons (Alexander et al. 1979; Daley and Biewener 2003). As a likely result of this energy-saving mechanism, the energetic cost of running for Common Ostriches, rheas, and Emus increases linearly with increasing speed,

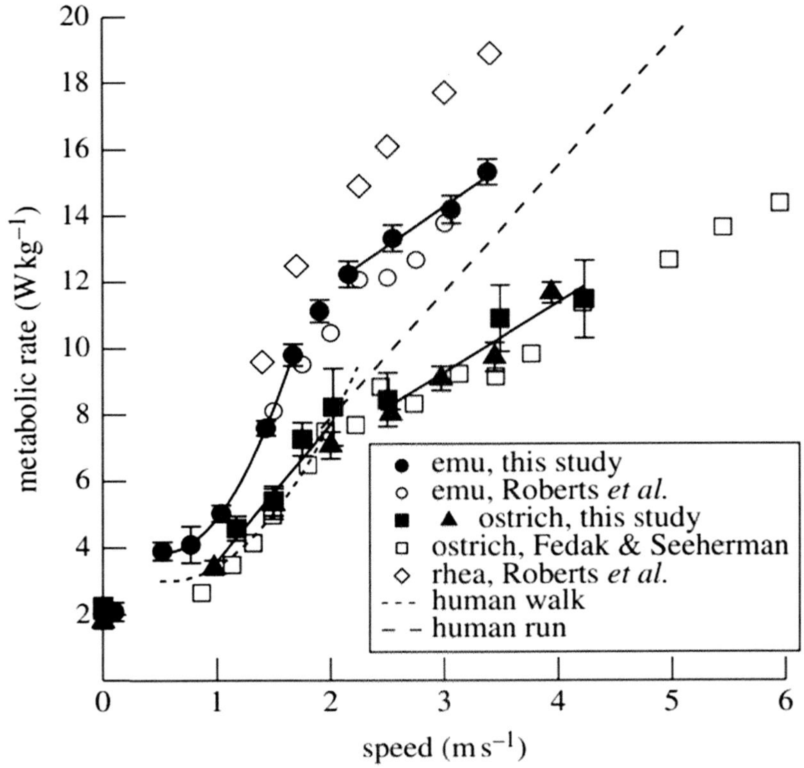


**Fig. 11.117** Relationship between running speed and mass-specific metabolic rate for Common Ostriches (*Struthio camelus*), Emus (*Dromaius novaehollandiae*), and Greater Rheas (*Rhea americana*) (and humans) as determined during three different studies. Values at zero speed are mean values for birds standing quietly on the treadmill. Cited studies: Roberts et al. (1998), Fedak and Seeherman (1979). (Figure from Watson et al. 2011; © 2010 The Royal Society, used with permission)

but only up to a point (Fig. 11.117). At speeds of about 1.7–2.0 m/s, the energetic cost of running by these birds continues to increase, but at a slower rate. As a result, running Common Ostriches, rheas, and Emus, for example, used about 20% less energy than predicted based on data from other birds and mammals and about 40% less energy than smaller non-ratite birds (galliformes), likely because of energy saved via elastic storage in their digital flexor tendons (Watson et al. 2011).

Compared to typical gallinaceous birds, like grouse, that also spend much of their time walking and running, the cost of running by similar-sized shorebirds is about 20% lower. This reduced cost is likely due to the relatively longer legs of many shorebirds that allow them to cover more distance with fewer steps (Box 11.18 Running Shorebirds). The energetic benefits of long legs may also help explain the rather long legs of the precocial young of shorebirds (Bruinzeel et al. 1999). The relationship between limb length and cost of terrestrial locomotion has been established for a wide variety of terrestrial animals, with longer limbs permitting longer strides and lower stride rates. Remarkably, limb length explains 98% of the observed variation in locomotor cost across a wide range of terrestrial species including mammals, birds, reptiles and arthropods (Pontzer 2007).

Many species of birds move their heads forward through a series of successive, fixed positions when walking, including those in the orders Galliformes, Gruiformes, Ciconiiformes, Pelecaniformes, and Columbiformes (Necker 2007; Fig. 11.118). This unique "head-bobbing" behavior stabilizes visual fields during body movement, preventing motion blur of the retinal image. Gaze stabilization could be required for successful visual search, particularly for moving objects, but the time available for stabilization varies with walking speed. No direct evidence has been published showing that birds favor the stabilization phase while foraging either for moving or immobile food. Cronin et al. (2005) examined head-bobbing by foraging Whooping

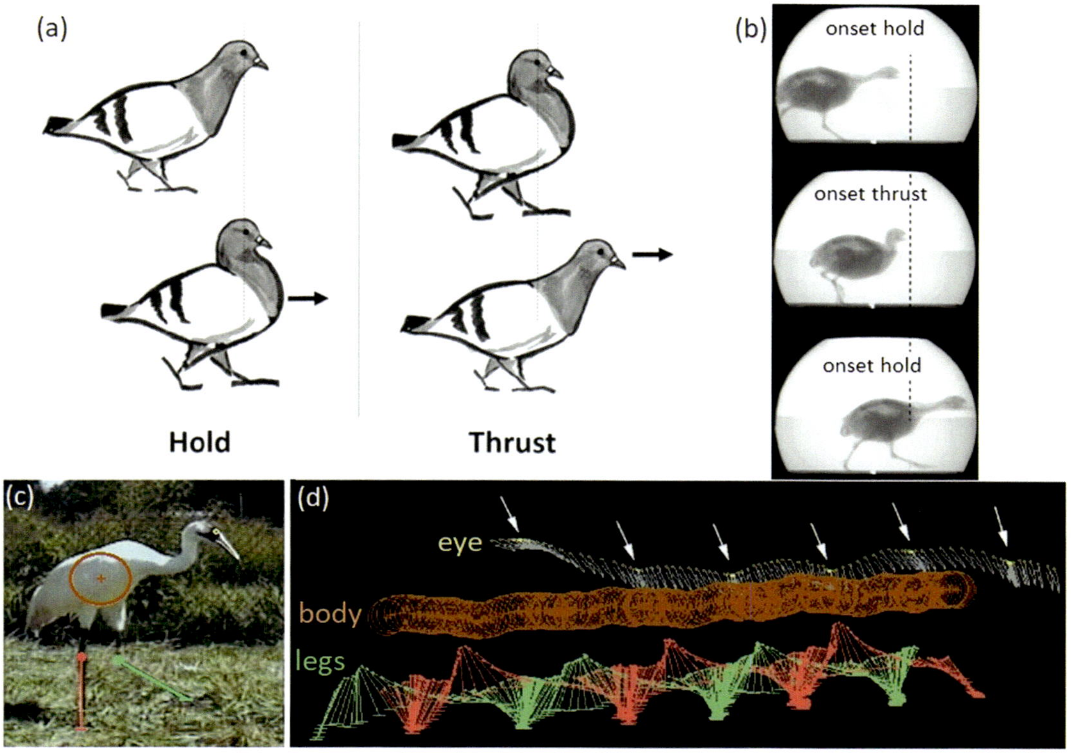


**Fig. 11.118** (**a**) Rock Pigeons (*Columba livia*) and several other species of birds bob their heads while moving, holding their head still while their body moves forward, then thrusting their head in front of their body. (**b**) Images from an X-ray sequence of a running Common Quail (*Coturnix coturnix*). During the hold phase, the quail holds its head in a fixed position while the rest of the body moves forward. During the thrust phase, the quail's head moves forward rapidly relative to the body. (**c**) A walking Whooping Crane (*Grus americana*) showing locations where movement of the head, body, and leg position was monitored. Head movements are indicated by a line from the eye to the end of the bill, body movements focus on the area in the circle that is centered over the pelvis, and leg movements are indicated by lines from heel to the foot (tarsometatarsus) (green, right leg; red, left leg). (**d**) Sequence of body locations at intervals of 33 ms for a foraging Whooping Crane making several steps. The crane walked at a speed of about 0.46 m $s^{-1}$. The right foot completed nearly three steps, and the left foot about 2.5 steps. The head was stabilized throughout most of each foot's step, with its positions at each of these times indicated by the arrows. (Figure **a** from Pritchard and Healy 2018; open-access article distributed under the terms of the Creative Commons Attribution 4.0 International License, https://creativecommons.org/licenses/by/4.0/. Figure **b** from Nyakatura and Andrada 2014; © 2014 Nyakatura and Andrada, licensee BioMed Central Ltd., open-access article distributed under the terms of the Creative Commons Attribution License, https://creativecommons.org/licenses/by/4.0/. Figures **c** and **d** from Cronin et al. 2005; © 2005 Elsevier Ltd. All rights reserved, used with permission)

Cranes (*Grus americana*) searching the ground for food, and found that they walked at speeds that allowed the head to be immobilized at least 50% of the time. The stable phase of bird head-bobbing movements is particularly interesting because the behavior, unique to birds, clearly contributes to visual gaze stabilization. Pigeons head-bob when landing, and herons stabilize their heads rigidly when walking or when their perch moves, almost certainly for visual function. Head movements nevertheless play essential roles in vision, giving visual cues for distances and relative locations of objects, providing an opportunity for changes in head angle, and permitting birds to fixate new objects of visual interest.

**Box 11.18 Running Shorebirds**

The hindlimb morphology of shorebirds is extremely variable, ranging from the short hindlimbs of several species of sandpipers to the very long hindlimbs of stilts, avocets, and godwits. Leg length in these species is correlated with water depth where they typically forage. For example, American Avocets (*Recurvirostra americana*) and Long-billed Dowitchers (*Limnodromus scolapaceus*) have much longer legs than Least (*Calidris minutilla*) and Western (*C. mauri*) sandpipers and, at stopover sites during migration, forage in significantly deeper water (Davis and Smith 2001).

Top, American Avocet (*Recurvirostra americana*) with chick. Bottom, Least Sandpiper (*Calidris minutilla*) (American Avocet photo by Ingrid Taylar, Wikipedia, CC BY 2.0, https://creativecommons.org/licenses/by/2.0/; Least Sandpiper photo by Alan Vernon, Wikipedia, CC BY 2.0, https://creativecommons.org/licenses/by/2.0/)

Among birds, including shorebirds, walking and running speeds are determined by stride length and stride rate; longer legs and strides plus faster stride rates equal greater speed. Stride length is limited by leg length so, as a result, to attain a particular speed, species with shorter legs must stride at higher rates than species with longer legs. For example, Northern Lapwings (*Vanellus vanellus*) and Eurasian Oystercatchers (*Haematopus ostralegus*) have shorter legs than Pied Avocets (*Recurvirostra avosetta*) and must, therefore, stride at faster rates to achieve the same speed as a Pied Avocet striding at lower rates.

(continued)

**Box 11.18** (continued)

Examples of differences in length of the hindlimbs of shorebirds (Data from Kilbourne et al. 2016; Northern Lapwing photo by Pawel Ryszawa, Wikipedia, CC BY 3.0, https://creativecommons.org/licenses/by/3.0/; Eurasian Oystercatcher photo from pxhere.com, CC0 Public Domain; Pied Avocet photo by Derek Keats, Wikipedia, CC BY 2.0, https://creativecommons.org/licenses/by/2.0/)

Leg movements of a running Northern Lapwing. (Figure from Nyakatura et al. (2012); © 2012 Wiley Periodicals, Inc., used with permission)

(continued)

**Box 11.18** (continued)

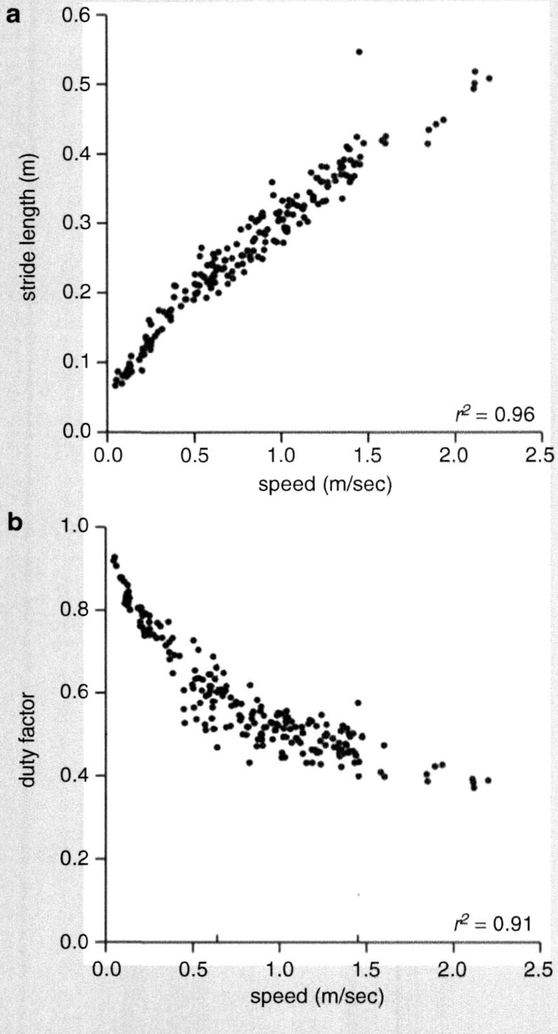


With increasing running speed, the stride length of Northern Lapwings increases and the duty factor (percent of stride cycle duration that hind limbs are in contact with the ground) decreases. (Figure from Nyakatura et al. (2012); © 2012 Wiley Periodicals, Inc., used with permission)

## 11.13.2 Climbing

Birds adapted for climbing vertical or nearly vertical surfaces, like woodpeckers, treecreepers, and nuthatches, have sharply recurved claws and toes, sometimes relatively long, that can be spread apart to help firmly grip the substrate (typically tree bark) (Leblanc et al. 2023). Other adaptations for climbing differ with foraging habits. Climbers that typically move up trees, like woodpeckers (Picidae; 240 species), treecreepers (Certhidae), and woodcreepers (Dendrocolaptidae; 57 species), have relatively short legs (particularly the tibotarsus) that keep their center of mass close to the substrate and a long, stiff tail that provides support against the force of gravity (Figs. 11.119 and 11.120). The robust leg bones of climbing birds (Carrascal et al. 1990; Norberg 2008a) likely represent an adaptation that may "... decrease the moment about the claw, subsequently reducing the force to keep the body close to the tree trunk (Zeffer et al. 2003)" (Leblanc et al. 2023). As woodpeckers move up a tree, they "hop" upward and inward (to counteract the force of gravity that tends to pull them away from a vertical tree trunk), moving both feet in unison. Tail support provides two advantages: (1) the long tail creates a long baseline between the points of attachment (feet and tail) and, the longer this baseline, the smaller the horizontal force between feet and bark against which the bird must work when pulling itself toward the trunk when hopping upward, and (2), when not moving, the tail, rather than the leg muscles, supports part or all of the bird's weight (Norberg 1981b).

Climbing birds that use their tail for support almost always move upwards when foraging (Fig. 11.121). Foraging woodpeckers, treecreepers, and woodcreepers approaching the top of one tree, typically fly downward to a lower position on another tree then again climb upwards

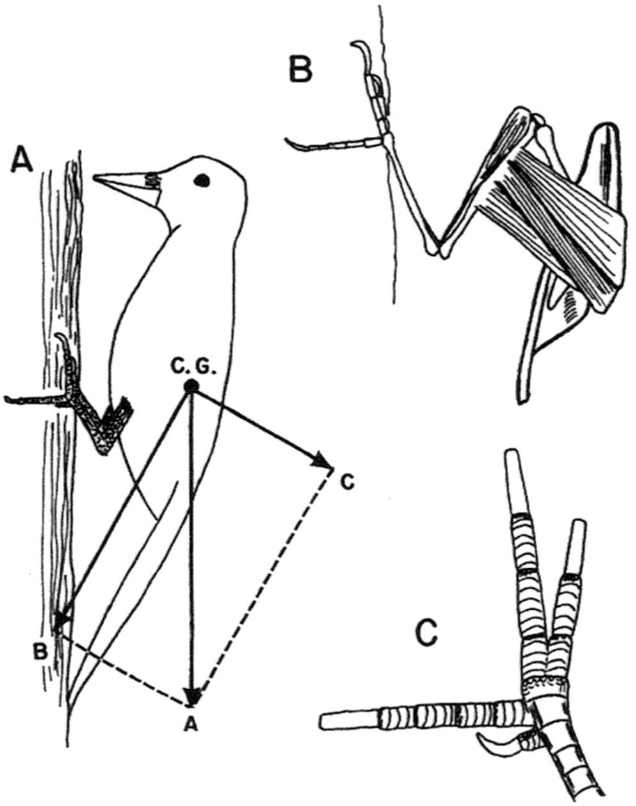


**Fig. 11.119** Features of climbing woodpeckers. (**a**) Forces acting on a climbing woodpecker. The pull of gravity, **a**, is divided into its two components: down and in, **b**, and outward, **c**. The forces originate from the center of gravity (C.G.). (**b**). The pelvic girdle and hind limb of a woodpecker showing muscles important in flexing the leg, including the flexors of the tibia (biceps femoris, semitendinosus, and semimembranosus) and the flexor of the tarsus (tibialis anterior). (**c**) Dorsal surface of a foot showing the typical spread position of the toes during climbing. (Figure from Bock and Miller 1959; used with permission of the American Museum of Natural History)

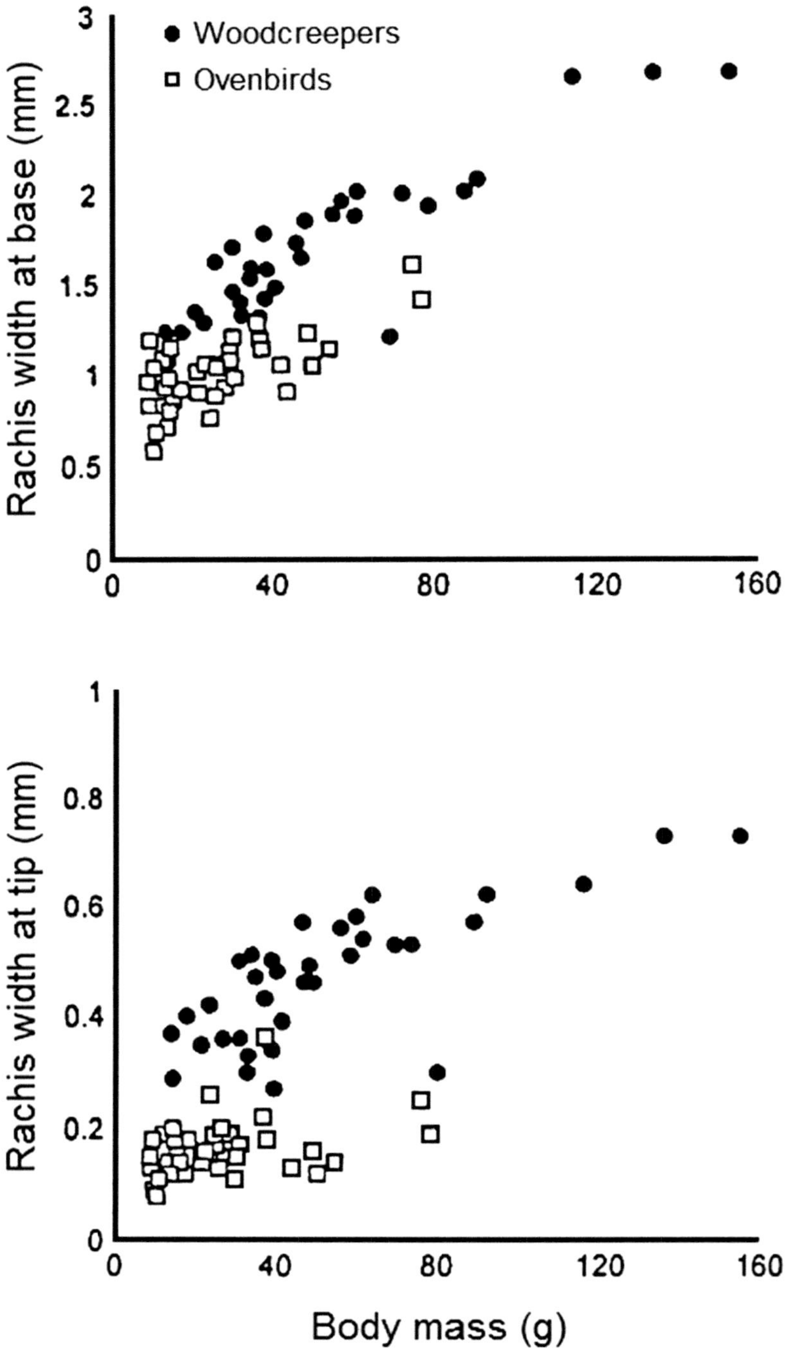


**Fig. 11.120** Compared to species of birds like Neotropical ovenbirds that do not use their tail for support against substrates like tree trunks, those species that do, like woodcreepers, have rectrices with wider rachises that provide greater structural support. (Figure modified from Tubaro et al. 2002; © 2002 Oxford University Press, used with permission)

and, when approaching the top, repeat the process. Not only does such a foraging strategy make sense energetically (because flying downward is less costly), but attempting to move downward when foraging would create at least three problems (Norberg 1981b): (1) difficulty in seeing where to grasp the bark after a hop, (2) the stiff tail could get caught on the irregular surface

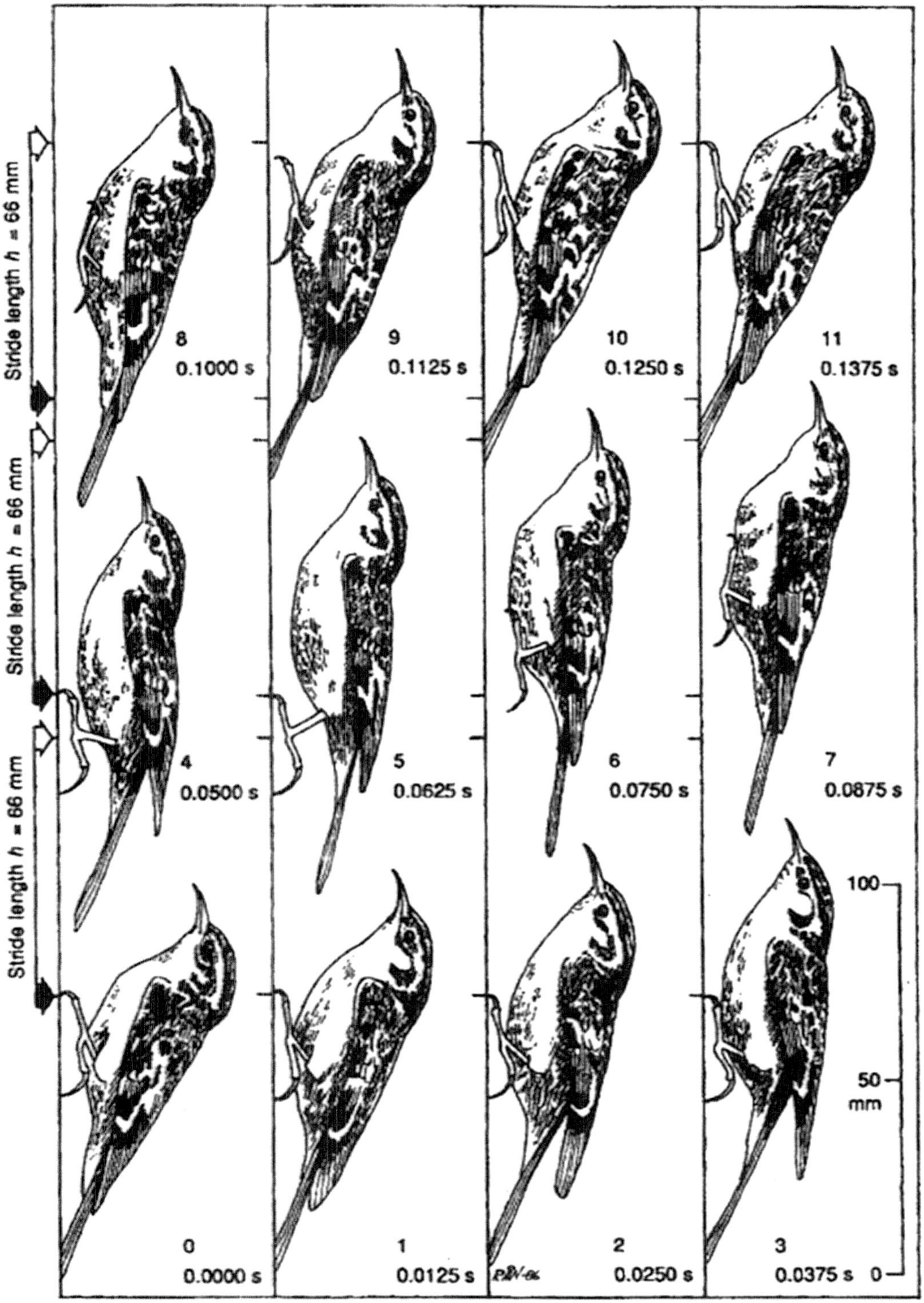


**Fig. 11.121** A complete movement cycle, or "hop," of a treecreeper climbing upwards on a vertical trunk. The sequence is from lower left to upper right, and numbers give picture number and time from the start in resting position. The bird is shown relative to positions of foreclaws on the trunk before and after the hop. (Norberg 1986a; © 1970 CCC Republication, used with permission)

**Fig. 11.122** A White-breasted Nuthatch (*Sitta carolinensis*). (Photo from pxhere.com, CC0 Public Domain)

of the bark, and (3) potential prey would be alerted to the presence of a possible predator before the bird could get in a position to capture them.

Nuthatches (Sittidae) are adapted for climbing downward as well as upwards (Fig. 11.122). Their relatively short tails are not used for support and, rather than hopping, nuthatches walk up-and-down tree trunks and branches with alternating leg movements. Nuthatches have relatively long legs (particularly the tibiotarsus), allowing a relatively long baseline between the feet and reducing the horizontal force between feet and bark and the energetic cost of locomotion (Norberg 1981b).

Many other birds spend much of their time foraging in vegetation for various types and, rather than climbing, use their hindlimbs to move from perch-to-perch and to position them so that they can detect and obtain food items such as insects, fruits, and seeds. Birds that typically forage in the outer or distal portions of trees and shrubs tend to have longer legs (particularly the tarsometatarsus; Carrascal et al. 1990). The advantage of longer legs for these birds is that they can search for and reach food items farther from their perches.

### 11.13.3 Aquatic Locomotion

Many birds have adapted to varying degrees to aquatic lifestyles and use a variety of ways to travel on and below the water's surface. One way to travel through water is to simply dive in, and many species of birds (about 130; Tyler and Younger 2022), including terns, Brown Pelicans (*Pelecanus occidentalis*), boobies, and gannets, attempt to capture fish and other aquatic prey by plunging into the water (Fig. 11.123). Most plunge divers are after prey at or near the water's surface, but some travel to depths approaching 10 m. Gannets are among the most impressive plunge divers, initiating plunges from heights up

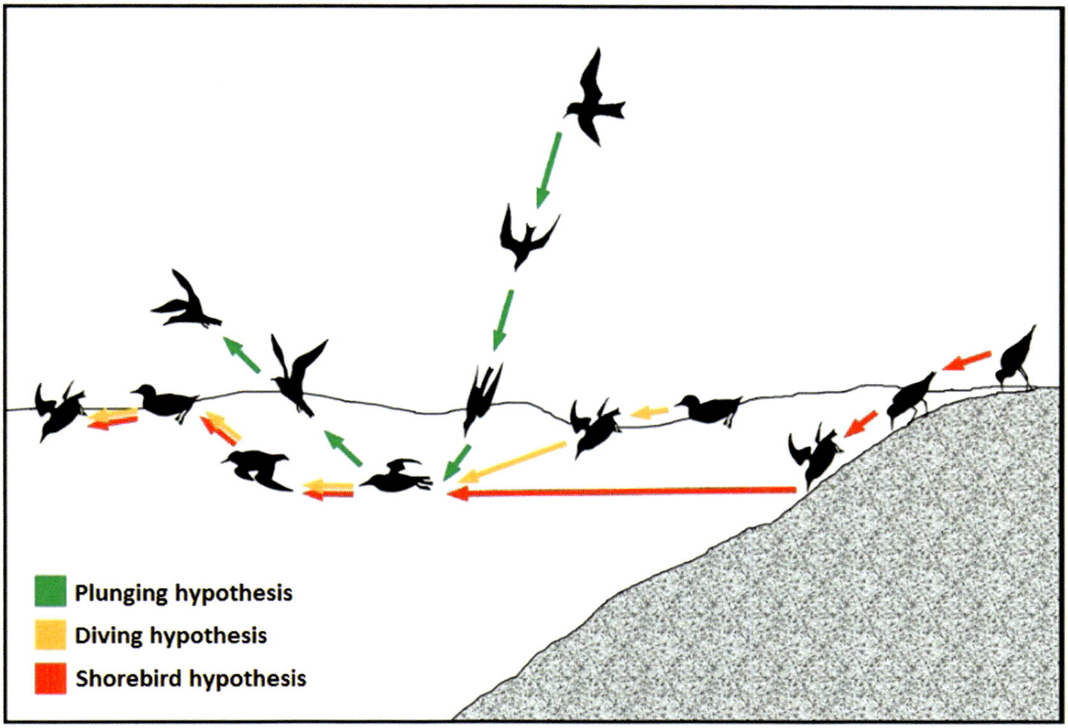


**Fig. 11.123** Only about 45 species of birds are capable of wing-propelled diving (Tyler and Younger 2022), including 33 species in the order Charadriiformes (e.g., dovekies, murres, murrelets, guillemots, and puffins). Hypotheses to explain the evolution of wing-propelled diving behavior in this group of birds include: (1) shorebirds foraging in shallow water began diving into deeper water as a way to escape from predators (shorebird hypothesis), (2) plunge-diving diving birds gradually became more adept at wing-propelled underwater location (plunging hypothesis), (3) surface and shallow-water feeding birds that initially used foot-propelled locomotion began seeking prey in deeper water and transitioned to wing-propelled diving (diving hypothesis) (Smith and Clarke 2014). Based on ancestral state reconstructions, Tyler and Younger (2022) found no evidence for hypothesis 2, i.e., that plunge diving eventually led to wing-propelled underwater locomotion. (Figure modified from Smith and Clarke 2014; © 2014 The Authors. Published by John Wiley and Sons, used with permission)

to 30 m above the water's surface and sometimes reaching depths of nearly 10 m before, if needed, moving even deeper by flapping their wings. Cape Gannets (*Morus capensis*) equipped with data-loggers were found to dive to a mean depth of about 3.5 m and remain underwater for about 5 s. After capturing or chasing prey, they glide passively back to the surface, using buoyancy rather than muscle power to complete their dive (Ropert-Coudert et al. 2004). Brown (*Sula leucogaster*) and Red-footed (*S. sula*) boobies were found to dive to less impressive depths, with mean dive depths of about 0.75–0.9 m in pursuit of their main prey (flying fish and squid; Lewis et al. 2005).

After a plunge dive, gannets often continue their pursuit of prey by flapping their wings. Most diving birds forego the plunge and pursue prey by propelling themselves through the water with their wings or their feet. Wing-propelled diving has evolved at least five times, including in the (1) penguins, (2) auks, (3) diving-petrels, (4) extinct Plotopteridae (related to gannets and boobies), and (5) some dippers (Raikow et al. 1988; Fig. 11.124). Of the five species of dippers, three are wing-propelled divers, including White-throated Dippers (*Cinclus cinclus*), Brown

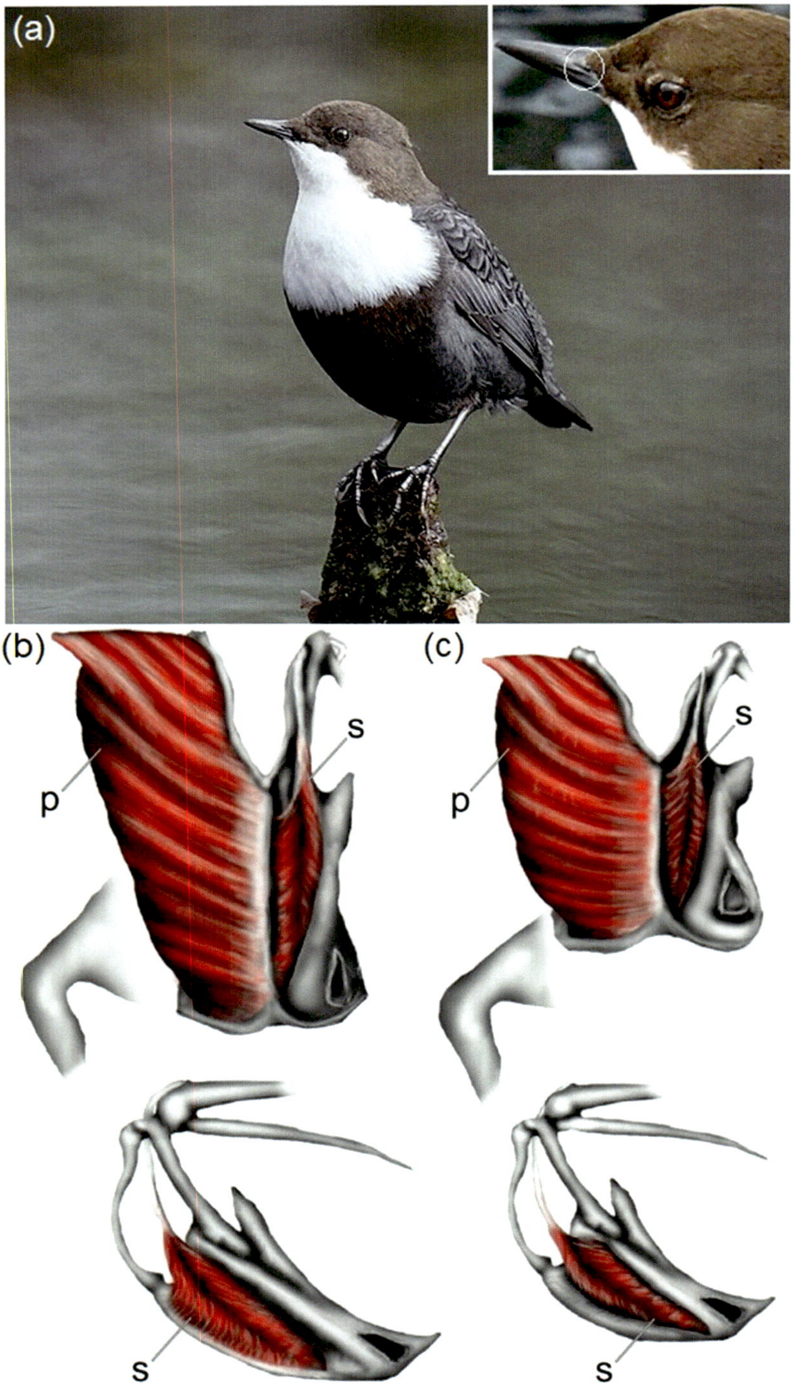


**Fig. 11.124** (**a**) White-throated Dipper (*Cinclus cinclus*) with elongated-slit nostrils and strongly curved claws. Inset, close-up view of a nostril. (**b**, **c**) Pectoralis (p) and supracoracoideus (s) muscles of (b) an American Dipper (*Cinclus mexicanus*) and (**c**) a Swainson's Thrush (*Catharus ustulatus*) Dippers (*Cinclus* spp.) share some characteristics with other wing-propelled divers, including nares reduced in size and covered by narial flaps and/or feathers (to prevent aspiration of water when diving), more robust flight muscles (pectoralis and supracoracoideus) than other songbirds of similar size (favored by selection because water is much denser than air), and no apteria

Dippers (*C. pallassi*), and American Dippers (*C. mexicanus*), whereas White-capped Dippers (*C. leucocephalus*) and Rufous-throated Dippers (*C. schulzi*) only dip their heads under water when foraging (Rijke and Jesser 2010). Penguins are flightless, as were the plotopterids, and the more recently extinct Great Auks (*Pinguinus impennis*), whereas auks, diving-petrels, and dippers can fly as well as use their wings for propulsion under water. However, because the densities of air and water are so different, the flap amplitudes and rates and stroke-plane angles also differ (Fig. 11.125). Because flying requires subtle and varied movements of the wings, the wings of auks, diving-petrels, and dipper exhibit greater joint mobility than those of penguins. Penguin wings are relatively small (with aspect ratios of 3.5–4.6; Hui 1988, Bannasch 1995) and rigid with limited joint mobility (Fig. 11.126), forming strong, compact flippers that allow penguins to "fly" through a medium (water) nearly 800 times denser than air. Penguin wings are much smaller than those of flying birds of similar mass, e.g., about 1/40th of the area of the wings of an eagle (Hui 1988). The wings of birds that both fly and swim represent a compromise in terms of wing area (Fig. 11.127). Common Murres (*Uria aalge*) fly and swim and have relatively small wings compared to birds of similar mass that only fly. As a result, wing loading for a murre is nearly four times that of a similar-sized Herring Gull (*Larus argentatus*; Pennycuick 1997).

An important difference between penguins and the auks and diving-petrels is that penguins have relatively larger upstroke muscles (supracoracoideus). Among auks and diving-petrels, the ratio of pectoralis (downstroke muscle) to supracoracoideus mass is typically about 3.3–3.5 (Mill and Baldwin 1983). For penguins, that ratio is typically about 1.8–2.3 (Osa 1994), indicating that the supracoracoideus is larger (relative to the pectoralis) than in auks and diving-petrels. This difference influences how these birds dive. Penguins generate forward propulsion during both the downstroke and upstroke, and can do so because of their relatively large upstroke muscles. By comparison, auks and diving-petrels generate forward propulsion during both the downstroke and upstroke only in relatively shallow water (<10 m; Fig. 11.128). In deeper water, most forward propulsion is generated by the downstroke. With increasing depth, diving birds become less buoyant because increasing water pressure compresses the air in their respiratory systems (mainly the air sacs) and plumage (air volume decreases by a factor of $1/n$, where $n$ is the number of 10-m depth increments). In shallow water, therefore, more propulsion must be generated to overcome buoyancy. In deeper water, with reduced buoyancy, auks and diving-petrels respond by slowing their upstroke and depending more on the downstroke (and their pectoralis muscle) to generate propulsion. Penguins continue to generate propulsion during both the downstroke and upstroke, regardless of depth and buoyancy (Watanuki et al. 2006). With slower upstrokes, downstrokes must be faster to maintain the same speed, resulting in increased drag and increased energy expenditure (Lovvorn 2001).

Penguins swimming near the surface sometimes exhibit "porpoising" behavior, leaping briefly out of the water as they swim

**Fig. 11.124** (continued) (a characteristic shared with penguins). Given the absence of apteria, dippers also have more contour feathers (with thicker and longer rachises and also more water resistant) and down feathers than other songbirds of comparable size. However, dippers do not exhibit the skeletal or neuroendocranial characteristics of other wing-propelled diving birds, likely because of the relative recency of their divergence from other songbirds (Rijke and Jesser 2010; Smith et al. 2022). (Photo of White-throated Dipper by Dirk-Jan van Roest, Wikipedia, CC BY 2.0, https://creativecommons.org/licenses/by/2.0/. Figures **b** and **c** from Smith et al. 2022; © 2021 The Authors. The Anatomical Record published by Wiley Periodicals LLC on behalf of American Association for Anatomy, used with permission)

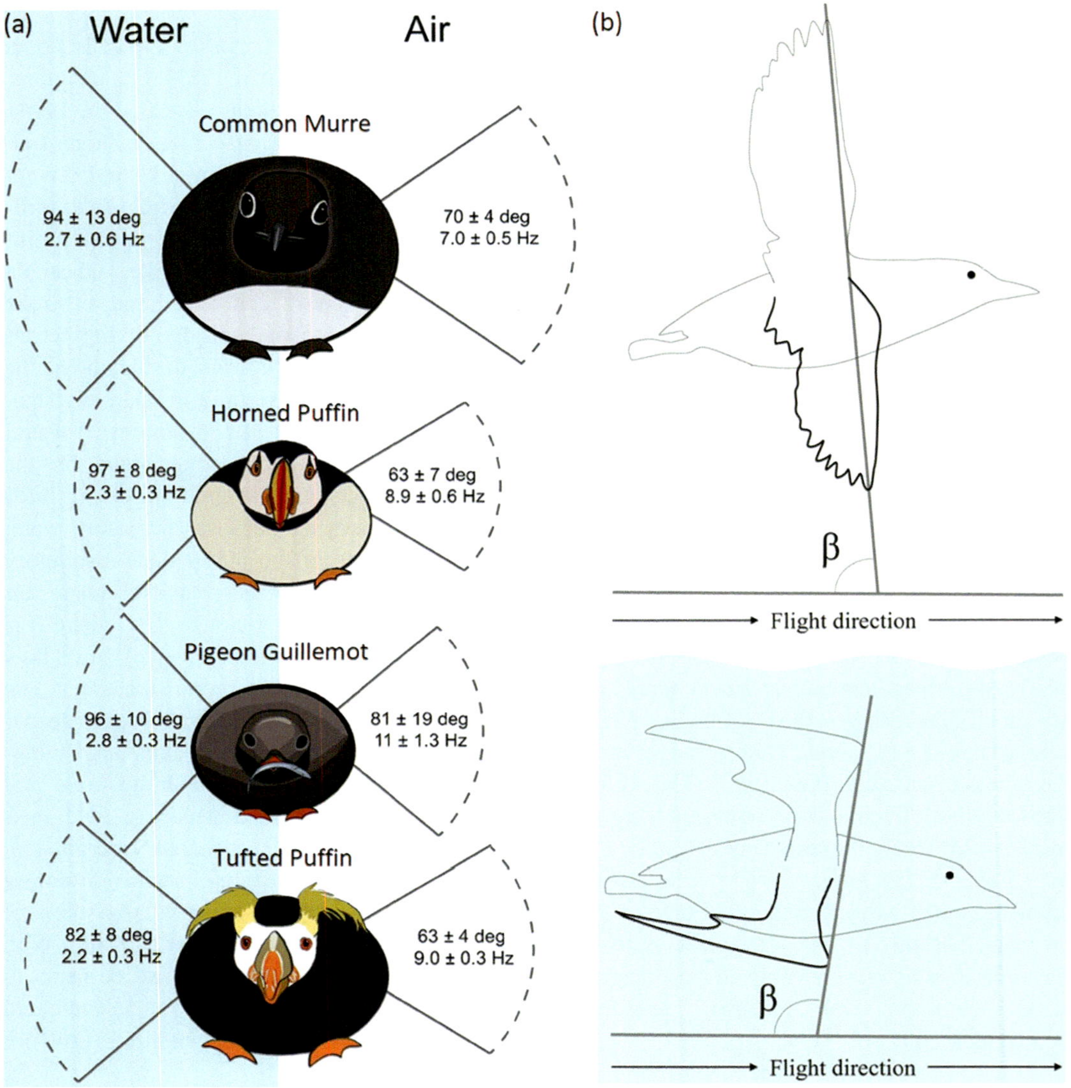


**Fig. 11.125** Wing-propelled diving birds that can also fly do so through two very different mediums, with air about 800 times less dense than water. In addition, in water, these birds must also counteract their buoyancy. As a result, different wingbeat amplitudes and frequencies are required in the two mediums. (**a**) Compared to aerial flight, when wing-propelled diving, these birds have higher wingbeat amplitudes (measures in degrees, deg) and lower wingbeat frequencies (measured in hertz, Hz). Flying in low-density air requires much faster and shallower wingbeats. (**b**) Stroke-plane angles (β) also differ in air versus water. This shift allows for both aerial and aquatic propulsion with the same wings. During aerial downstrokes, the wing also moves forward, creating the vertical force needed to counteract gravity. In water, birds are pulled upward by their buoyancy so, during the upstroke, the wing rotates cranially, shifting the net force production to counteract drag. Common Murre, *Uria aalge*; Horned Puffin, *Fratercula corniculata*; Pigeon Guillemot, *Cepphus columba*; Tufted Puffin, *Fratercula cirrhata*. (Figures **a** and **b** [with **b** modified] from Lapsansky et al. 2020; © 2020 Lapsansky et al., open-access article distributed under the terms of the Creative Commons Attribution License, https://creativecommons.org/licenses/by/4.0/. Artwork for Figure a by Emily Moore)

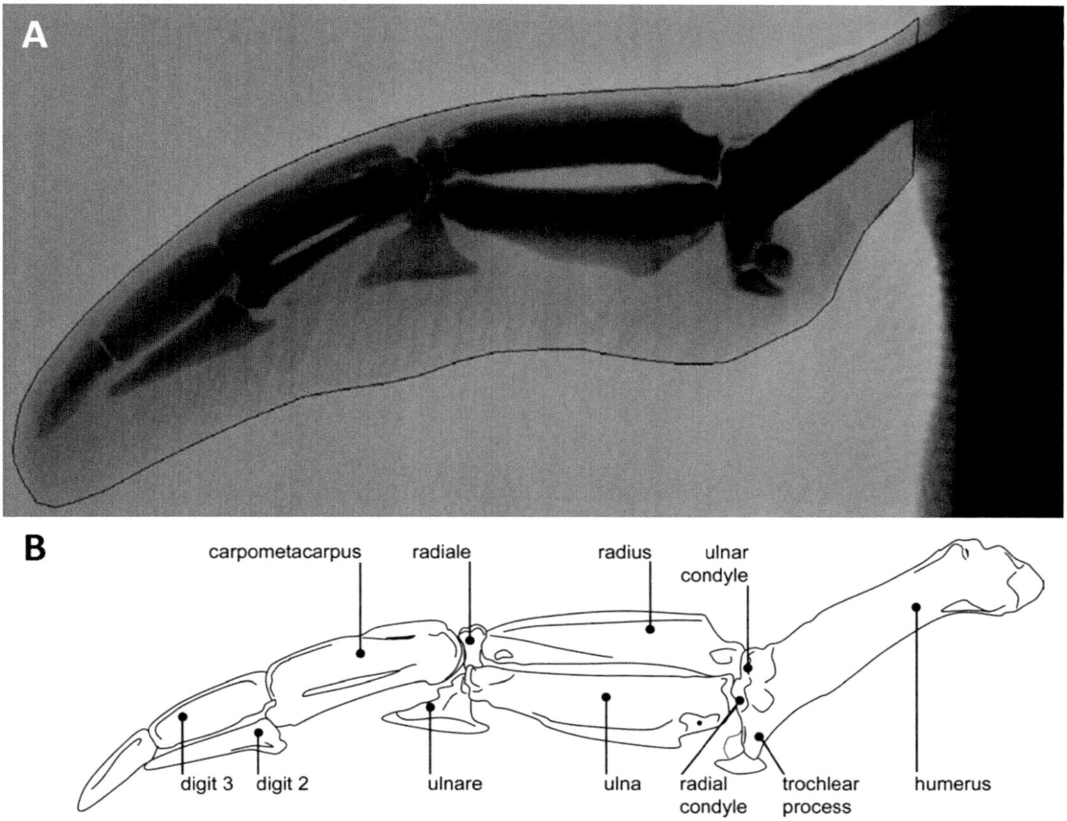


**Fig. 11.126** (**a**) X-ray of the wing of a Southern Rockhopper Penguin (*Eudyptes chrysocome*). (**b**) Drawing of a penguin wing, with bones and features labeled. For wing-propelled swimming underwater, the "paddle" must be highly mobile at the shoulder (humerus-pectoral girdle articulation), but the remaining joints need to be relatively fixed to minimize the muscle contraction needed to maintain the proper position. (Fig. **a** from DeBlois and Motani 2019, © 2019 Wiley Periodicals, Inc., used with permission; Fig. **b** from Thomas and Fordyce 2012, © 2011 Wiley Periodicals, Inc., used with permission)

(Fig. 11.129). Possible functions of "porpoising" include energy conservation, avoiding predation, allowing breathing without reducing swimming speed, observing potential landing sites, and enjoyment (Yoda et al. 1999). Because penguins do not porpoise on a regular basis and, when they do, distances between leaps may be rather long, it is doubtful that porpoising makes travel more energy-efficient. Porpoising could allow penguins to breath when swimming because the time spent out of the water is sufficient to permit inhalation (Hui 1987). In addition, porpoising could be a response to the threat of predation. Penguins have a variety of potential predators, including leopard seals (*Hydrurga leptonyx*) and killer whales (*Orcinus orca*), that sometimes congregate near the shores where penguins are nesting (Wilson 1995). Porpoising may allow penguins to rapidly cross these areas and reduce the risk of predation. In support of this hypothesis, *Spheniscus* penguins apparently only porpoise when alarmed by a potential predator (Wilson 1995).

Many diving birds used foot propulsion rather than wing propulsion and some use both types of propulsion. About 61 species of birds use foot propulsion (Tyler and Younger 2022), including several species of waterfowl, plus loons, grebes,

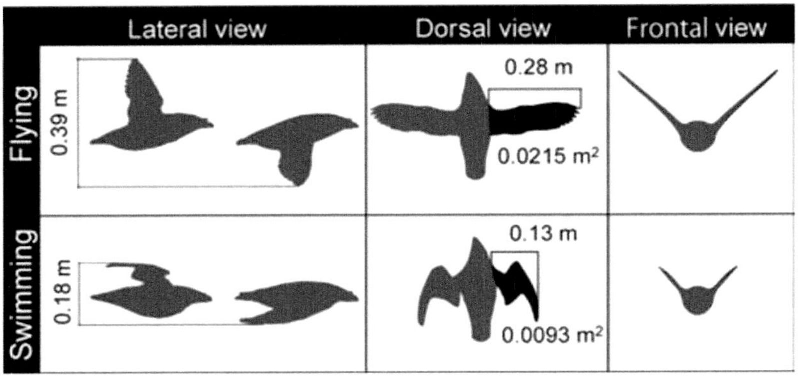


**Fig. 11.127** The wings of birds that both fly and swim represent an evolutionary compromise. Flying favors greater wing area to generate more lift in very-low-density air, but swimming favors smaller wings to reduce drag and allow more energetically efficient flapping in high-density water. As a result of these conflicting selection pressures on wing morphology, the energetic cost of both flying and swimming are higher for birds like Rhinocerous Auklets (*Cerorhinca monocerata*) that can fly and swim than for birds more specialized for either just flying or just swimming. However, to reduce the energetic cost of swimming, birds like auklets reduce their effective wing area by partially folding their wings when swimming. The wingbeat amplitude is shown in the lateral view, and wing length and wing surface area are shown in the dorsal view. (Figure modified from Kikuchi et al. 2015; © The Authors, used with permission)

anhingas, and cormorants; eiders use both foot and wing propulsion. Most foot-propelled diving birds have webbing between either three toes (palmate) or all four toes (totipalmate, e.g., cormorants). Grebes, however, have stiff skin lobes that increase toe surface area and only the proximal portions of the front three toes are fused by webbing. In addition, the lobing is asymmetrical, with the lateral lobes being smaller than the medial lobes. Foot propulsion is typically drag-based propulsion, with the webbed feet acting like oars or paddles. Cormorants, for example, propel themselves underwater by stroking (paddling) with both feet simultaneously (with webbed feet spread and wings-folded tightly against the body) to generate thrust, then glide with legs and feet extended, and finally move the feet back toward the body (with webs closed to reduce drag) in preparation for the next stroke—a technique referred to as burst-and-glide swimming (Ribak et al. 2005; Fig. 11.130).

Grebes, in contrast to the other foot-propelled divers, apparently rely on lift-based propulsion (Fig. 11.131). The difference between drag-based and lift-based propulsion is nicely illustrated by comparing oars and paddles (drag-based) to propellers (lift-based). The blades of a propeller act as rotating wings and produce force through both Bernoulli's principle and Newton's third law, generating pressure differences between the front and rear surfaces of the air-foil-shaped blades and by pushing water backward. The power stroke of swimming grebes is perpendicular to the swimming direction and the toes do not move in a direction opposite to the swimming direction (as is the case with drag-based propulsion), suggested that propulsion must be lift-based. Because a grebe's toes are separate, they may function like the separate blades of a propeller, maximizing lift and reducing drag as they fold and move forward during the recovery stroke (Johansson and Norberg 2001).

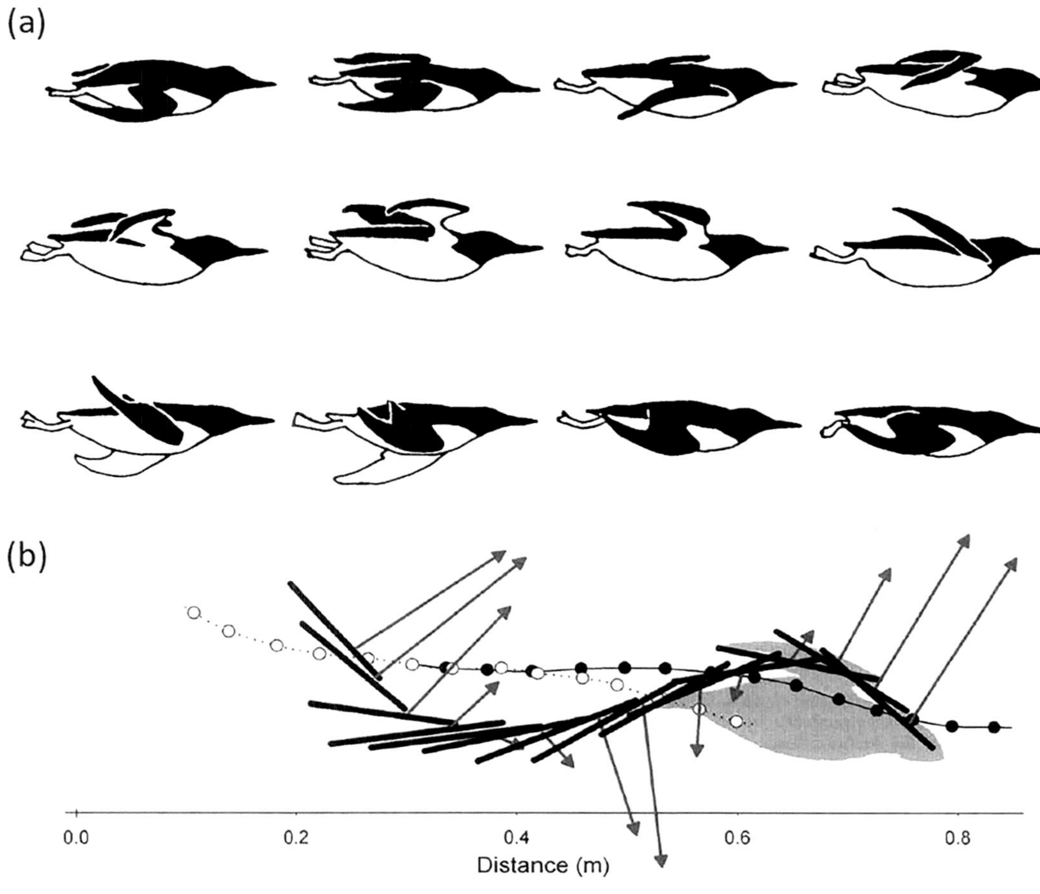


**Fig. 11.128** Top, wing positions of a swimming Common Murre (*Uria aalge*), with an upstroke (beginning at upper left) followed by a downstroke. Bottom, wings of a puffin generate a forward- and upward-directed force during the downstroke and a forward- and downward-directed force during the upstroke. (Top figure from Lovvorn 2001; © 2015 Oxford University Press, used with permission. Bottom figure from Johansson 2003; © 2003 Canadian Science Publishing, used with permission)

Lift-based wing propulsion is generally more efficient than drag-based foot propulsion. For horizontal swimming under water at a constant speed, the energetic cost of drag-based foot propulsion ranges from 26 to 40% higher than for lift-based wing propulsion (Lovvorn and Liggins 2002). Because wing propulsion can (especially for penguins) generate thrust on both downstroke and upstroke and drag-based foot propulsion can only generate thrust during the power stroke, a bird using foot propulsion must generate greater speed during the power stroke to maintain the same speed as a bird using wing propulsion. Because drag increases with increasing speed, greater speed during the power stroke causes greater friction drag and overcoming that drag requires more energy (Lovvorn and Liggins 2002). However, foot propulsion may allow greater maneuverability (lunging, braking, and turning) at lower speeds, and may be more

**Fig. 11.129** Adelie Penguins (*Pygoscelis adeliae*) porpoising in the Southern Ocean. (Photo from NSF.gov, CC0 Public Domain)

suitable for hovering over and probing into benthic substrates for food items such as mussels (Duffy et al. 1987).

Common Eiders (*Somateria mollissima*) use both wing and foot propulsion. When diving, these eiders time the power stroke of their feet so it occurs at the transition between the upstroke and downstroke of the wings when drag is high because of the increased angle of attack (Fig. 11.132). Such timing likely reduces drag, and the energetic cost of swimming, by helping to maintain a more constant velocity (Heath et al. 2006). Additional study may reveal other diving birds that also use both wing and foot propulsion.

Although diving birds use different methods to propel themselves through the water, including wing propulsion, drag-based foot propulsion, lift-based foot propulsion, and both wing and foot propulsion, typical swimming speeds of diving birds exhibit little variation. Studies of diving birds ranging from penguins to auks to cormorants (Box 11.19 Unusual Feathers of Cormorants) have revealed that typical speeds range from about 1.4–2.3 m/s (5–8 km/h, Sato et al. 2007), with larger birds tending to swim faster than smaller birds (Watanabe et al. 2011). Interestingly, most marine mammals tend to swim at similar speeds (Fig. 11.133). For example, the mean swimming speed of northern elephant seals (*Mirounga angustirostris*) is about 1.8 m/s and that of sperm whales (*Physeter macrocephalus*) is about 1.6 m/s (Sato et al. 2007). Mean stroke rates of flippers or fins are, however, correlated with mass, with smaller organisms exhibiting higher stroke rates than larger ones (e.g., mean stroke rate is 3.6 per second for Little Penguins and 0.2 per second for sperm whales). The predictable scaling of stroke rate and mass suggests that diving specialists among marine mammals and seabirds have evolved similar proportions of propulsive muscles and muscle contraction rates that translate into similar cruising speeds (Sato et al. 2007).

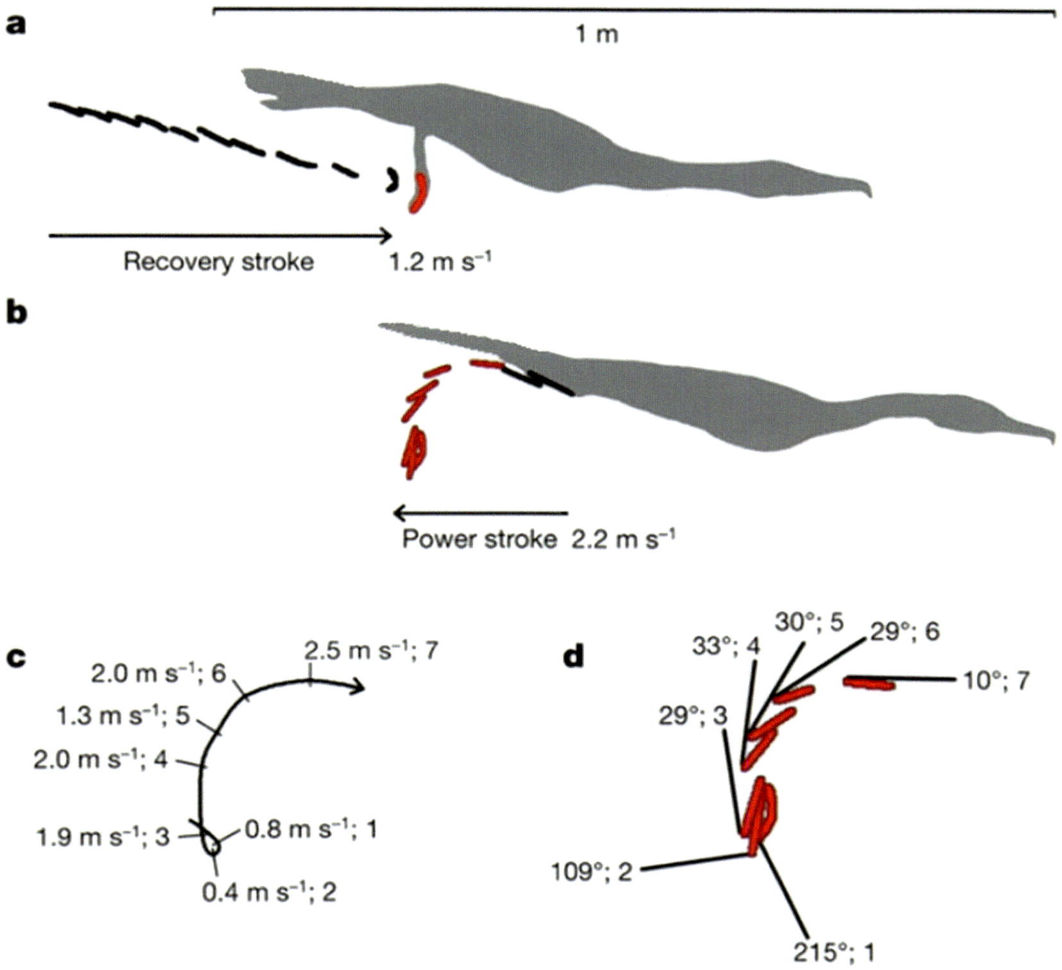


**Fig. 11.130** (**a**) The recovery stroke, with the foot positioned to begin the power stroke. (**b**) The power stroke acceleration of the body increases from 1.2 m $s^{-1}$ at the end of recovery stroke to 2.2 m $s^{-1}$ at the end of the power stroke. (**c**) The still-water speed of the trailing edge of the foot is indicated at seven positions along its trajectory. (**d**) The angle of attack at the trailing edge of the foot at the same seven foot positions. (Figure from Johansson and Norberg (2003a); © 2003 Springer Nature, used with permission)

**Fig. 11.131** (**a**) Stroboscopic pictures of the grebe's right foot as it is moved mechanically in a still-water tank, simulating the power stroke. The foot is moved from right (first part of the power stroke) to left (last part of power stroke) in the figure, which shows five different phases. Note the staggering and twisting of the toes in the picture on the left. (**b**) Dorsal view of the same foot as shown in (**a**). The asymmetry of the toes is most pronounced on the fourth toe, which is leading the movement during the power stroke, and least on the second. The second and third toes in particular show increasing asymmetry toward their tips. (Figure from Johansson and Norberg 2003b; © 2000 Springer Nature, used with permission)

**Fig. 11.132** Diving Common Eiders (*Somateria mollissima*). Wing strokes involve an upstroke and a downstroke; foot strokes involve the thrust, retract, and glide stages. During the wing stroke cycle, the feet are most often in the glide phase (legs fully extended and parallel to the body and toes closed to minimize drag). During foot retraction, the feet are pulled forward with the webbing closed. Foot retraction is followed by the power or thrust phase, with the legs quickly pushing backward with the webbing extended. (Heath et al. 2006; Image is a screenshot from the film *People of a Feather*, www.arcticeider.com/diving; used with permission)

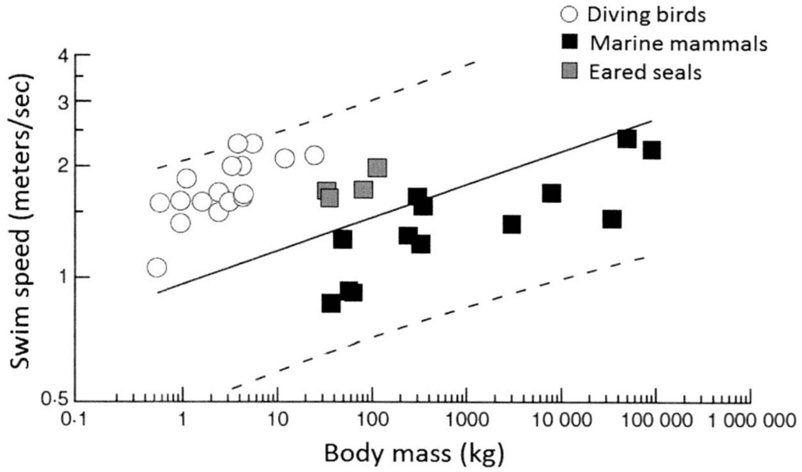


**Fig. 11.133** Swim speeds during propulsive swimming relative to body mass of 17 species of diving birds, 13 marine mammals (other than eared seals, family Otariidae), and four species of eared seals. Most birds, marine mammals, and eared seals swim at similar speeds, with a few outliers. The fastest swimmers among birds (top two open circles) are penguins. Slower-swimming marine mammals (the bottom three black squares) are some smaller marine mammals, including some earless seals (family Phodicae) and harbor porpoises (*Phocoena phocoena*), one of the smallest marine mammals. The solid and dashed lines show the scaling relationship (speed = $0.95 \times \text{mass}^{0.09}$) and 95% confidence intervals. (Figure modified from Watanabe et al. 2011; © 2010 The Authors. Journal compilation © 2010 British Ecological Society, used with permission)

**Box 11.19 Unusual Feathers of Cormorants**

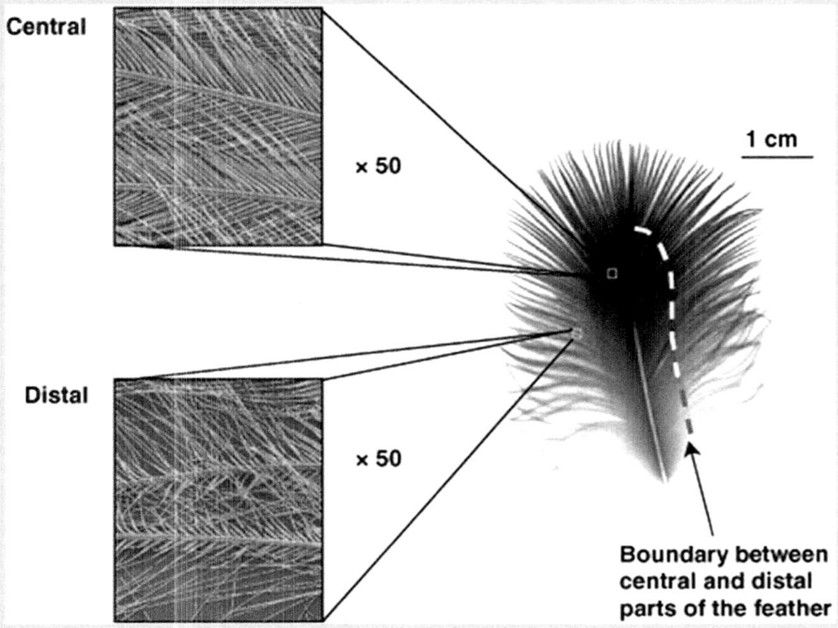


(continued)

**Box 11.19** (continued)

Great Cormorant feathers have a regular and highly waterproof central part whereas the distal region is irregular and wettable. A ventral feather is shown. (Figure from Grémillet et al. 2005; © Journal of Avian Biology, used with permission)

Because diving birds can potentially lose much heat in cold water, most have dense, waterproof plumage that traps an insulating layer of air. Anhingas and cormorants are exceptions, with plumage that is apparently penetrated by water during dives. With wettable plumage that would seemingly provide little insulation when diving, Great Cormorants (*Phalacrocorax carbo*) might be limited to areas with warmer water temperatures. However, these cormorants have an extensive breeding range that extends into cold-water areas. Grémillet et al. (2005) examined the insulative properties of body plumage of Great Cormorants ranging from tropical to polar regions. All subspecies retained an insulating air layer in their plumage, but the layer was thinner than for other species of diving birds. Examination of the plumage showed that cormorant body feathers have a loose, wettable outer section and a waterproof central portion. With partly wettable plumage, Great Cormorants can maintain a thin layer of air in their plumage. This plumage provides these cormorants with some insulation and, at the same time, absorbs some water, making the cormorants less buoyant and more energy-efficient divers.

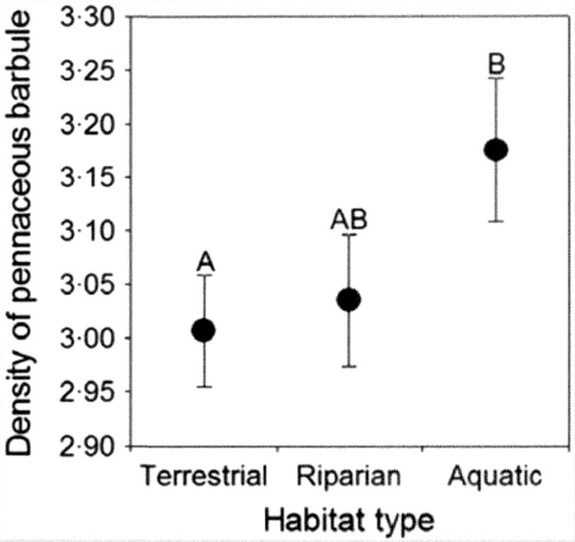


Most aquatic birds have higher densities of barbules (per mm) than birds in riparian or terrestrial habitats, making their feathers more water resistant and better at trapping an insulating layer of air. Data based on examination of feathers of 1043 individuals representing 194 species. Values are means ± SE, and different letters indicate a significant difference at $P < 0.05$. (Figure from Pap et al. 2017; © 2016 The Authors. Functional Ecology © 2016 British Ecological Society, used with permission)

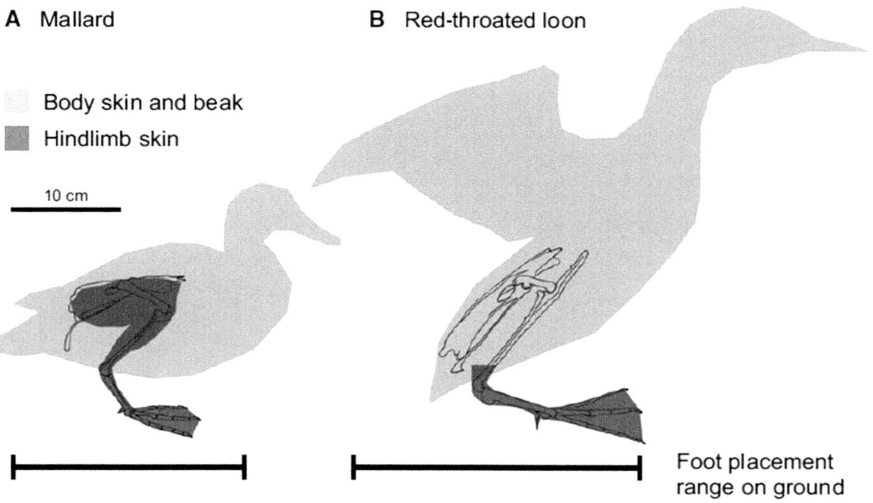


**Fig. 11.134** (**a**) Surface-swimming birds like Mallards (*Anas platyrhyncos)* can easily stand on land with hindlimbs that allow a large range of motion at the hip and knee. (**b**) Birds specialized for diving like Red-throated Loons (*Gavia stellata*) have more streamlined bodies, but find it difficult to stand on land. The medial surface of the thigh (femur) is not covered by skin, but is located next to the ribs, which restricts mobility at the hip and knee. Only the feet extend from the body. (Figure from Clifton et al. 2018; © 2017 Anatomical Society, used with permission)

Diving birds have adaptations that permit efficient movement through the water, such as the ability to alter their specific gravity, streamlined bodies with femurs tucked next to the ribcage (Fig. 11.134), and specialized wings or legs and feet. However, they also spend time on the water's surface and some aquatic birds, including many ducks, geese, and swans, do little or no diving and are instead adapted for traveling along the water's surface. Birds specialized for surface-swimming typically have low specific gravities (lightweight so they are very buoyant) and feathers with lots of barbules and hooklets that are less permeable to water. As a result, the plumage does not absorb water (as occurs in cormorants) and the birds remain "lighter." Both diving and surface-swimming birds also have either webbed feet (waterfowl) or lobed toes (grebes and coots) that provided propulsion when traveling along the water's surface (Fig. 11.135).

Dive costs increase rapidly with body mass for flying divers (diving birds that can also fly) compared to those of flightless divers (penguins), and this holds true for both flying wing-propelled divers and, especially, for foot-propelled divers (Fig. 11.136). Foot-propelled diving is energetically costly because feet, unlike wings during the upstroke, cannot generate thrust during retraction so, to attain a speed similar to that of a wing-propelled diver, additional thrust is needed during the power stroke (Lovvorn and Liggins 2002). That additional thrust (and speed), in turn, results in increased drag, further increasing energetic costs (Lovvorn 2001).

The metabolic cost of diving for flying wing-propelled divers, although generally lower than that of foot-propelled divers, is higher than that for penguins, the nonflying wing-propelled divers (Fig. 11.137). To be able to fly, flying wing-propelled divers need larger wings with more surface area than the nonflying wing-propelled divers. The larger wings create more drag that, in turn, increases the energetic cost of diving above that of penguins. However, this selective "compromise," with smaller wings than other birds their size to increase diving efficiency, but wings still large enough to fly, means that flying is more energetically costly (Fig. 11.137). The smaller wings of flying wing-propelled divers

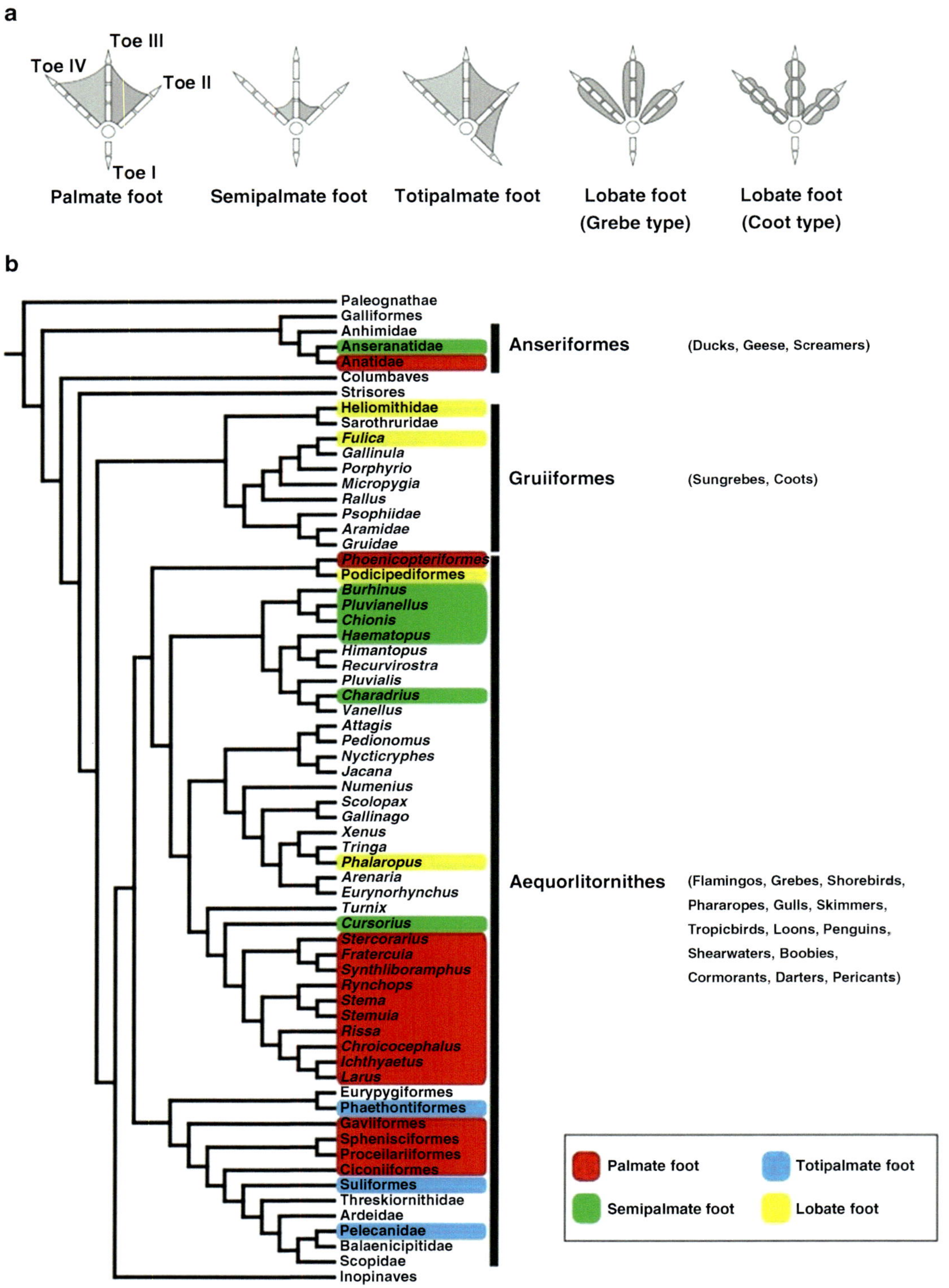


**Fig. 11.135** Diversity in the webbed feet of aquatic birds. (**a**) Different types of webbed feet in different taxa of birds. (**b**) Phylogenetic positions of waterbird taxa that have different types of webbed feet. (Figure from Tokita et al. 2020; open-access article licensed under a Creative Commons Attribution 4.0 International License, https://creativecommons.org/licenses/by/4.0/)

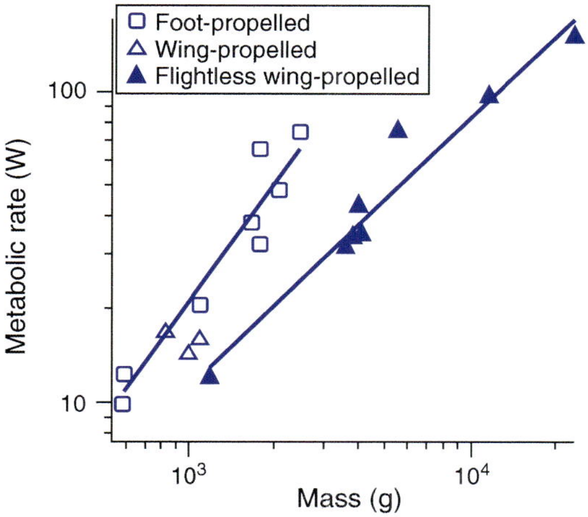


**Fig. 11.136** Energetic cost of foot-propelled diving versus wing-propelled diving as a function of body mass. Trend line for foot-propelled divers is significantly different from the trend line for penguins. (Figure from Elliott et al. 2013; used with permission of the US National Academy of Sciences)

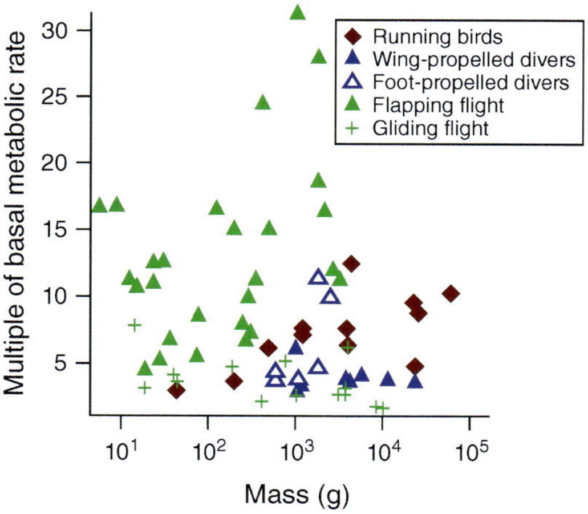


**Fig. 11.138** Energetic cost, as a multiple of basal metabolic rate, of running, diving, flapping flight, and gliding flight. The running metabolic rate is the maximum during sustained running, whereas metabolic costs of diving, flapping flight, and gliding flight are minimum rates. (Figure modified from Elliott et al. 2013; used with permission of the US National Academy of Sciences)

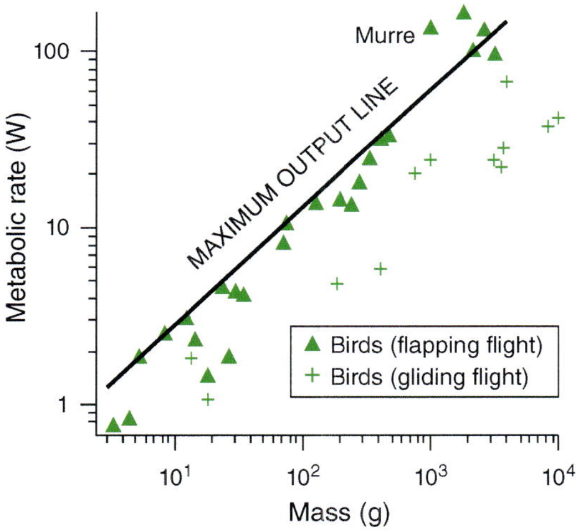


**Fig. 11.137** Metabolic cost of flight for several species of birds. The metabolic cost of flight for birds that fly, but also use their wings for propulsion under water, like Thick-billed Murres (*Uria lomvia*), is relatively high because natural selection has favored smaller wings with less surface area. The smaller wings make wing-propelled diving possible, but they also produce less lift and require a greater expenditure of energy when flying. The solid line represents the sustained maximum output limit for flying birds proposed by Videler (2006). (Figure modified from Elliott et al. 2013; used with permission of the US National Academy of Sciences)

generate less lift, thus requiring additional expenditure of energy by the flight muscles. Although penguins cannot fly, their smaller, stronger "wings," or flippers, minimize drag and generate propulsion on both the downstroke and upstroke, thus reducing the energetic cost of wing-propelled diving.

Both foot-propelled and wing-propelled diving require energy, but both types of locomotion remain, on average, less energetically costly than flapping flight and, for some species, even less costly than running. In general, the types of locomotion requiring the least energy are running by small birds and gliding flight (Fig. 11.138).

As birds swim along the water's surface, they move their legs and feet in a general back (tailward, or caudally) and forth (headward, or cranially) manner. As the backward, or power stroke, begins, the webbed feet or lobed toes are fully extended and, as they move back, push against the water like an oar and generate a force that moves the bird forward. Then, as the foot is moved forward in the recovery stroke, the toes come together and the web folds so there is less resistance to the water (Fig. 11.139). Most foot-propelled swimming birds, however, do not

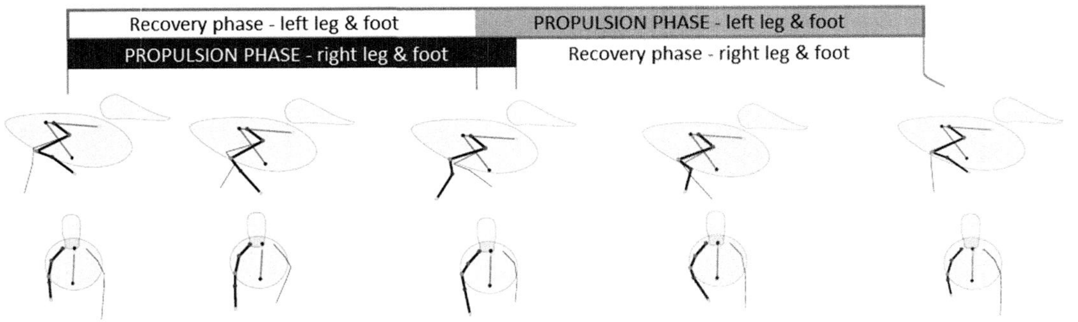


**Fig. 11.139** Leg and foot movements of a Ringed Teal (*Callonetta leucophrys*) swimming at the water's surface. During the propulsion phase, the toes and webbing are extended to increase surface area and a greater propulsive force. During the recovery phase, toes come together to minimize surface area and reduce resistance as the leg and foot move forward. (Figure modified from Provini et al. 2012; © 2012 Wiley Periodicals, Inc., used with permission)

sweep their webbed or lobed feet straight backward, but, rather, in a curved path. This is particularly true for diving specialists with elongated pelvic girdles and legs positioned further back on their bodies. As the power stroke begins, the foot moves backward in a line parallel to the bird's body and, as it does so, it is perpendicular, or near so, to the longitudinal axis of the body and the water's surface. This maximizes the surface area of the webbed or lobed foot pushing against the water and generates a substantial propulsive force. However, as the power stroke continues, the foot curves backward, inward toward the body, and upward in the direction of the water's surface so that the angle of the foot relative to the water's surface becomes increasingly smaller (i.e., from perpendicular to increasingly parallel to the surface; Fig. 11.140). As the foot moves, the propulsive force shifts from drag-based (when the foot acts like a paddle) to lift-based (Johansson and Norberg 2003a). Lift is generated as water flows past the top of the foot with greater velocity (due to the swimming speed combined with movement of the foot) than over the bottom of the foot (the side of the foot now closer to the surface). The slower moving water on the bottom of the webbed or lobed foot exerts greater pressure than the faster-moving water on the top (Bernoulli effect) and the resultant lift force generates a propulsive force forward (Fig. 11.140f). The relative contribution of this lift-based propulsive force compared to that generated during the "paddle" phase of the power stroke (drag-based) likely varies among species, and is probably more important for diving specialists like loons and cormorants than for surface-swimming specialists like dabbling ducks and coots. Lift-based propulsive force would also be more important as birds increase their swimming speed (and the amount of lift and the resulting propulsive force forward also tend to increase).

When swimming at the surface, aquatic birds have to overcome the drag created as their bodies move through the water (Box 11.20 Formation Swimming by Ducklings). In addition, when swimming at the surface (but not when diving), a bird's body creates both bow and stern waves and, at a certain speed, the wavelength of the bow wave will equal the length of the body in the water (Fish 2000). At this speed, referred to as the hull speed, the hull of the object is entrapped in a trough between the bow and stern waves and drag (Prange and Schmidt-Nielsen 1970; Figs. 11.141 and 11.142), and the energetic cost of swimming, increases dramatically (Fig. 11.143). Aquatic birds may tend to avoid swimming at speeds above the hull speed, e.g., Mallards (*Anas platyrhynchos*; Prange and Schmidt-Nielsen 1970) and American Black Ducks (*Ana rubripes*; Baudinette and Gill 1985). Ancel et al. (2000) reported that Brandt's Cormorants (*Phalacrocorax penicillatus*) being tested in the lab were reluctant to swim on the surface at speeds higher than 1 m/s (hull speed) and, instead, attempted to dive.

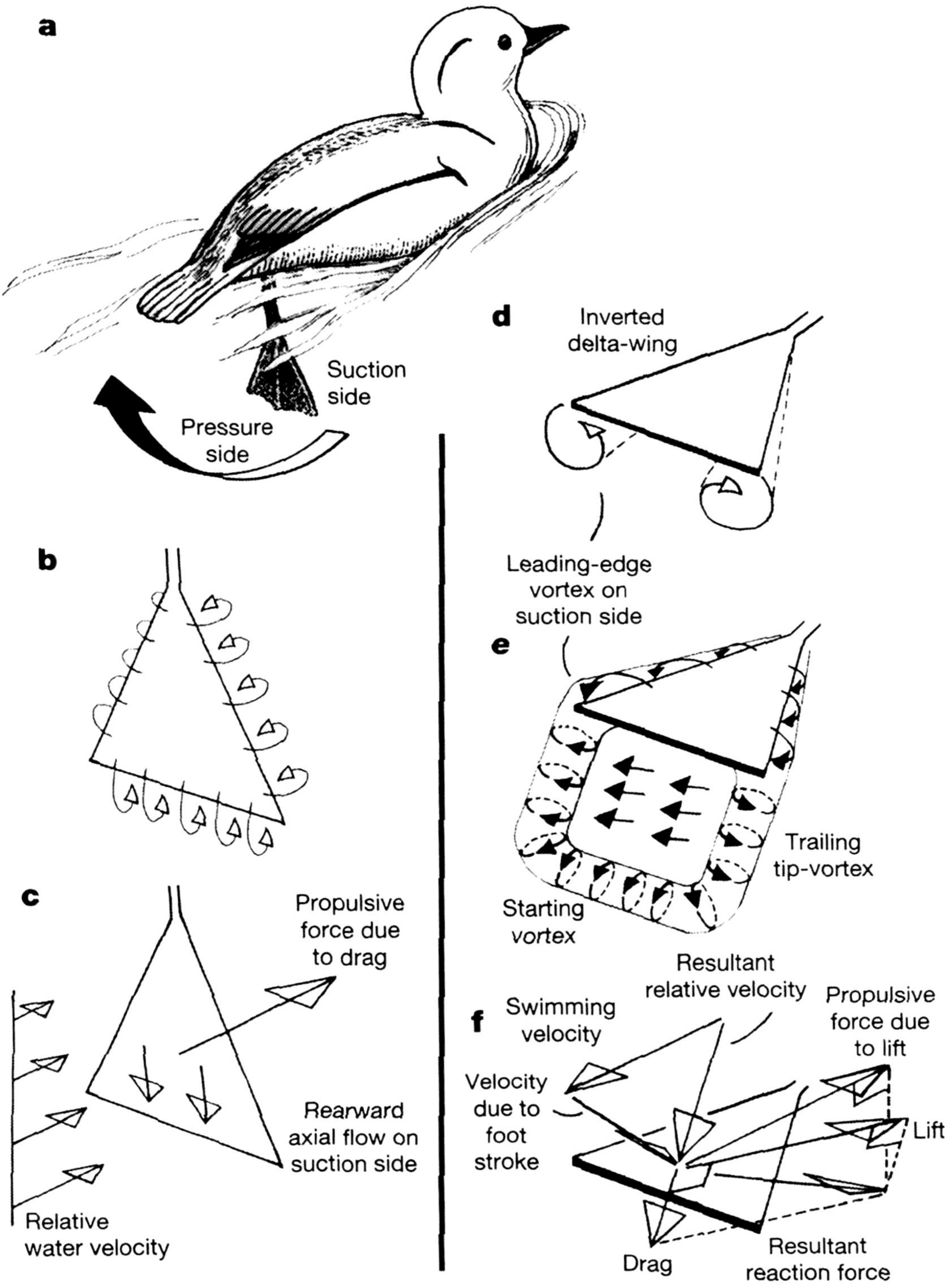


**Fig. 11.140** Movements of a webbed-foot and the resulting vortices and forces generated during a power stroke for a bird swimming relatively fast. **a**, early in the power stroke pressures are greater on the bottom of the webbed-foot (pressure side) than the top (suction side), creating a propulsive force forward (**b**, **c**). **d**–**f**, later in the power stroke, lift is generated due to different water velocities over the bottom and top of the webbed-foot and produces a propulsive force forward. (Figure from Johansson and Norberg 2003a; 

**Fig. 11.141** A Eurasian Coot (*Fulica atra*) swimming at or near hull speed. Note the bow wave in front of the coot and the stern wave at the back, with the coot in a trough in between. Swimming faster would take much more energy because the coot would need to give up the beneficial stern wave and swim up and through the bow wave. (Photo by Ian Barker and used with permission)

**Fig. 11.142** Top, A rubber duck being towed in a flow tank at a speed just under the hull speed with well-formed bow and stern waves. Bottom, A rubber duck being towed at a speed just above the hull speed. The beneficial stern wave is gone and the duck is cutting through the bow wave. For a real duck, swimming above the hull speed requires much more energy than swimming just below the hull speed. (Reproduced from Vogel 1998, with the permission of AIP Publishing)

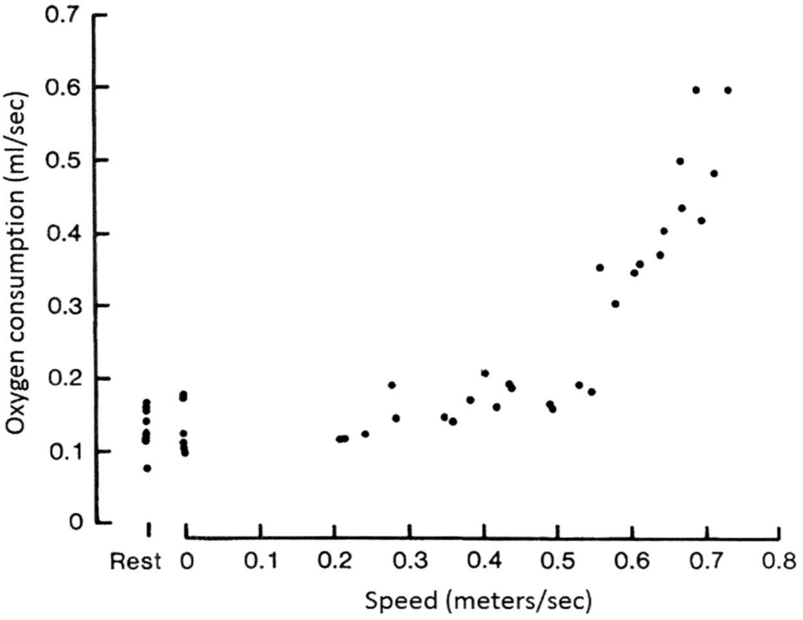


**Fig. 11.143** Rate of oxygen consumption (as a measure of energetic cost) of a male Tufted Duck (*Aythya fuligula*) swimming at different speeds in a variable speed water channel. When swimming, oxygen consumption changed little up to a speed of 0.5–0.6 m/s, i.e., the hull speed, when oxygen consumption increased dramatically. At a speed of 0.5 m/s, oxygen consumption was 32% above that when at rest. However, at their maximum sustainable speed (0.78 m/section), oxygen consumption was 3.8 times higher than when resting. (Data from Woakes and Butler 1983, Figure from Butler 2000; © 1999 CCC Republication, Published by University of Chicago Press, used with permission)

**Box 11.20 Formation Swimming by Ducklings**

Young ducks and geese often swim in lines behind adults, and energetic measurements indicate that ducklings can save up of 63% in metabolic effort by swimming in the wake created by a leading adult (Fish 1994, 1995). Yuan et al. (2021) determined that ducklings riding waves generated by a leading adult can be pushed forward, thus saving energy. Ducklings following an adult act as "wave passers," passing the energy of the wave to the duckling behind it so that all ducklings save energy. Interestingly, Yuan et al. (2021) suggested that this principle "... could be potentially applied to design modern freight carrying vessels, e.g. a water-train, to transport more cargoes without extra fuel cost."

(continued)

**Box 11.20** (continued)

Family of Canada Geese (*Branta canadensis*) swimming in a row. (Photo by Anchor Lee; free to use under Unsplash license)

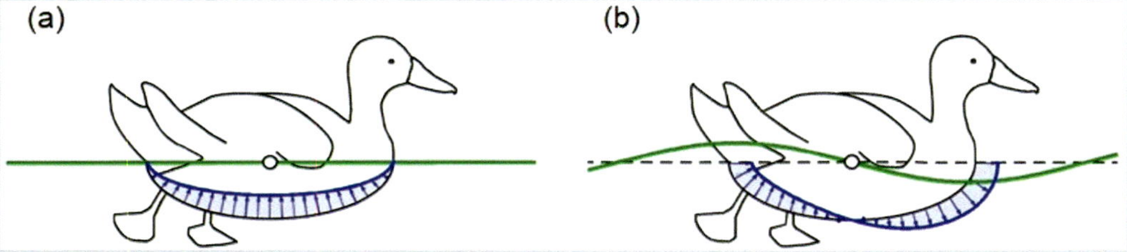


(**a**) In calm water, a duckling is subject to hydrostatic pressure without a horizontal component. (**b**) A duckling is riding a wave with its breast on a wave trough and abdomen on a wave crest so it will be propelled by the wave, reducing the amount of energy needed to move forward. This benefit is maintained as long as the position of the duckling relative to the wave does not change, i.e., a duckling's forward speed must equal the group velocity of the wave. (Figure from Yuan et al. 2021; © The Authors. Published by Cambridge University Press, open-access article distributed under the terms of the Creative Commons Attribution license, https://creativecommons.org/licenses/by/4.0/).

(continued)

**Box 11.20** (continued)

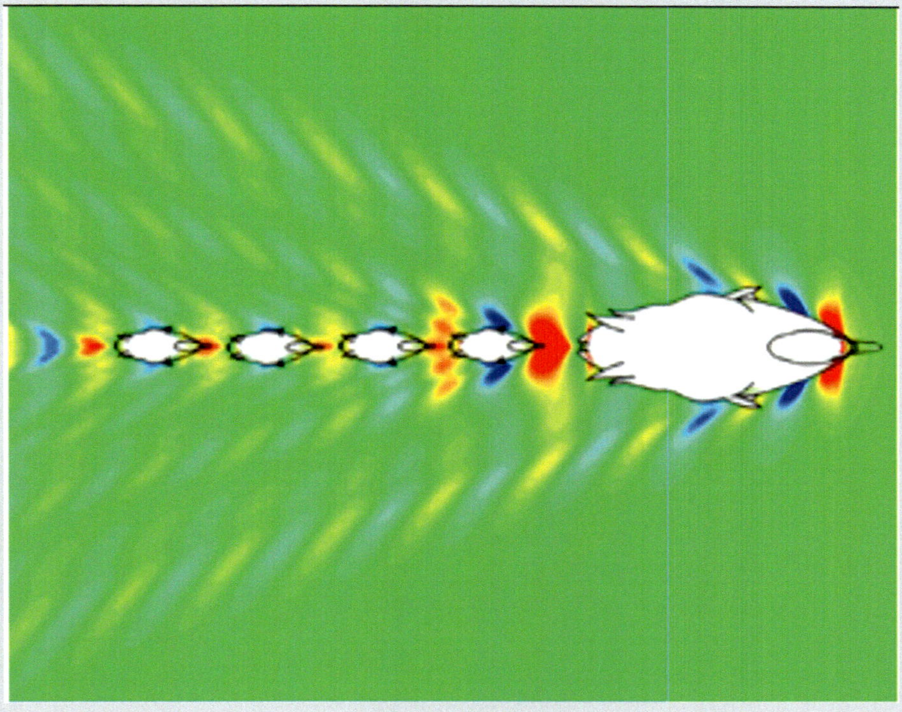

Waves produced by a leading adult and trailing ducklings. Red areas are troughs and blue areas are wave crests. Ducklings are propelled forward as the crest of waves propels them toward the troughs. The adult can also benefit energetically from the frontal wave produced by the duckling behind it. (Figure from Yuan et al. 2021; © The Authors. Published by Cambridge University Press, open-access article distributed under the terms of the Creative Commons Attribution license, https://creativecommons.org/licenses/by/4.0/).

Optimum swimming formation for ducklings following an adult duck. This wave-riding benefit can sustained by the rest of the ducklings in a single-file line formation. (Figure modified from Yuan et al. 2021; © The Authors. Published by Cambridge University Press, open-access article distributed under the terms of the Creative Commons Attribution license, https://creativecommons.org/licenses/by/4.0/).

**Fig. 11.144** Skittering Mallard (*Anas platyrhynchos*) ducklings. (Photo by Sherry Widmer, used with permission)

When necessary, aquatic birds can exceed the hull speed by "burst swimming" or "skittering" (Raikow 1973; Aigeldinger and Fish 1995). For example, young waterfowl are not able to fly for several weeks after hatching and may not be efficient divers so an important escape tactic when confronted with a potential predator is "burst swimming." Burst swimming is faster than normal paddle swimming, e.g., the hull speed of Mallard ducklings was found to be about 0.4 m/s, whereas their burst speed was 1.7 m/s, about four times faster (Aigeldinger and Fish 1995). To "burst" through the bow wave, young Mallards nearly double their paddling frequency. This generates both thrust and lift, raising their breasts above the water's surface and above the bow wave and allowing them to briefly hydroplane along the surface (Fig. 11.144).

Some waterfowl, including ducks and alcids, have other ways to exceed hull speed, including using their wings as "oars" in what is referred to as "skittering" or "steaming" (Gough et al. 2015). When "steaming" (so-named because movements of the wings mimic a steamboat's circular paddlewheel), birds use their wings (and likely their feet as well) to generate the thrust and lift needed to move up and over the bow wave and exceed the hull speed (Fig. 11.145). The speed of a steaming Common Eider (*Somateria mollissima*) is 5.5 times faster than their hull speed, averaging about 4 m/s (Gough et al. 2015). Although taking flight or diving (for diving duck) would seemingly be better strategies when threatened by a predator, birds that are molting flight feathers (e.g., waterfowl that molt all flight feathers at the same time) or with high

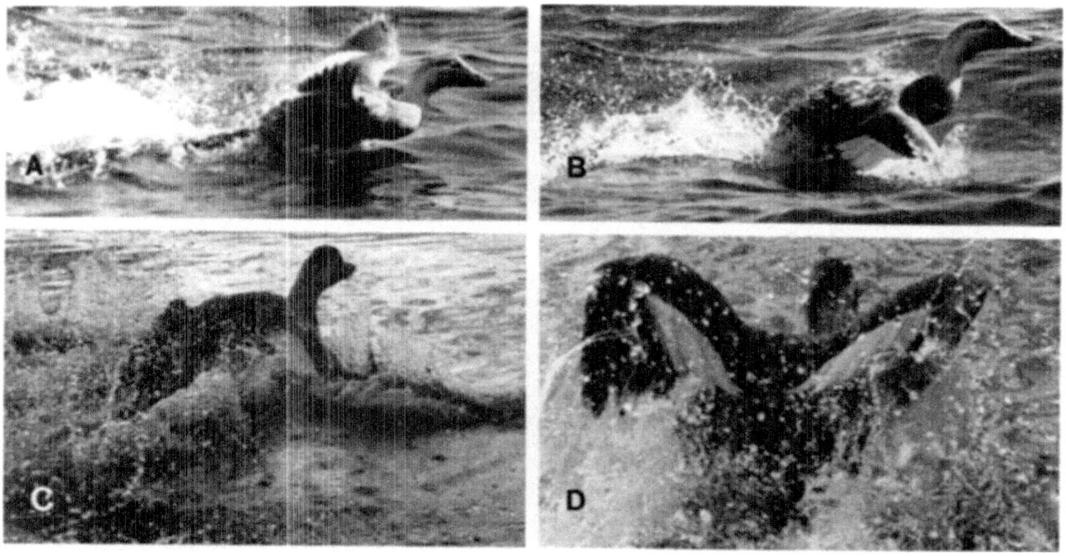


**Fig. 11.145** Steaming behavior of a White-headed Steamer Duck (*Tachyeres leucocephalus*). (**a**) During the downstroke the wings are fully extended and (**b**, **c**) come down to and below the water simultaneously. (**d**) Beginning of the recovery stroke, with wings moving upward. (Figure from Livezey and Humphrey 1983; © 1884 CCC Republication, used with permission)

wing loading (like eiders) must sometimes attempt to escape in other ways such as by steaming. Even for diving ducks, steaming may be preferable when threatened by aquatic predators such as seals (Guillemette and Ouellet 2005).

Yet another way to exceed hull speed is paddle-assisted flying, where aquatic birds flap their wings while their feet alternately paddle along the water's surface (Fig. 11.146). Many aquatic birds with high wing loading, such as albatrosses, shearwaters, geese, swans, and loons, always use paddle-assisted flying when taking off, but, when not taking off, it can also be an effective means of evading a potential predator. One advantage of paddle-assisted flying over steaming is the greater speed, with Common Eiders averaging nearly 5 m/s (versus 4 m/s when steaming).

**Fig. 11.146** Paddle-assisted take-off by a Common Eider (*Somateria mollissima*). (Photo by Ron Knight, Wikipedia, CC BY 2.0, https://creativecommons.org/licenses/by/2.0/)

## References


Abourachid A, Höfling E, Renous S (2005) Walking kinematics parameters in some paleognathous and neognathous Neotropical birds. Ornitol Neotrop 16: 471–479

Abourachid A, Herrel A, Decamps T, Pages F, Fabre A-C, Van Hoorebeke L, Adriaens D, Amado MAG (2019) Hoatzin nestling locomotion: acquisition of quadrupedal limb coordination in birds. Sci Adv 5:eaat0787

Agnolin FL, Motta MJ, Brissón Egli F, Lo Coco G, Novas FE (2019) Paravian phylogeny and the dinosaur-bird transition: an overview. Front Earth Sci 6:252

Aigeldinger TL, Fish FE (1995) Hydroplaning by ducklings: overcoming limitations to swimming at the water surface. J Exp Biol 198:1567–1574

Ákos Z, Nagy M, Vicsek T (2008) Comparing bird and human soaring strategies. Proc Natl Acad Sci USA 105:4139–4143

Alerstam T (1994) Bird migration. Cambridge University Press, Cambridge

Alerstam T, Rosén M, Bäckman J, Ericson PGP, Hellgren O (2007) Flight speeds among bird species: allometric and phylogenetic effects. PLoS Biol 5:e197

Alexander DE (2002) Nature's flyers: birds, insects, and the biomechanics of flight. Johns Hopkins University Press, Baltimore, MD

Alexander RM (2004) Bipedal animals, and their differences from humans. J Anat 204:321–330

Alexander RM, Maloiy GMO, Njau R, Jayes AS (1979) Mechanics of running of the Ostrich (*Struthio camelus*). J Zool 187:169–178

Altimiras J, Lindgren I, Giraldo-Deck LM, Matthei A, Garitano-Zavala Á (2017) Aerobic performance in tinamous is limited by their small heart. A novel hypothesis in the evolution of avian flight. Sci Rep 7: 1–15

Altshuler DL, Bahlman JW, Dakin R, Gaede AH, Goller B, Lentink D, Segre PS, Skandalis DA (2015) The biophysics of bird flight: functional relationships integrate aerodynamics, morphology, kinematics, muscles, and sensors. Can J Zool 93:961–975

Alvarez JC, Meseguer J, Meseguer E, Pérez A (2001) On the role of the alula in the steady flight of birds. Ardeola 48:161–173

Ancel A, Starke LN, Ponganis PJ, van Dam R, Kooyman GL (2000) Energetics of surface swimming in Brandt's Cormorants (*Phalacrocorax penicillatus* Brandt). J Exp Biol 203:3727–3731

Anderson D, Eberhardt S (2001) The Newtonian description of lift of a wing. Fermilab-Pub-01/036-E, Batavia, IL

Andersson M, Wallander J (2004) Kin selection and reciprocity in flight formation? Behav Ecol 15:158–162

Au DWK, Pitman RL (1986) Seabird interactions with dolphins and tunas in the eastern tropical Pacific. Condor 88:304–317

Baliga VB, Szabo I, Altshuler DL (2019) Range of motion in the avian wing is strongly associated with flight behavior and body mass. Sci Adv 5:eaaw6670

Ball P (1999) Shark skin and other solutions. Nature 400(507):509

Bannasch R (1995) Hydrodynamics of penguins – an experimental approach. In: Dann P, Norman I, Reilly P (eds) The penguins. Surrey Beatty and Sons, Chipping Norton, NSW, pp 141–176

Baudinette RV, Gill P (1985) The energetics of 'flying' and 'paddling' in water: locomotion in penguins and ducks. J Comp Physiol B 155:373–380

Bell E, Andres B, Goswami A (2011) Integration and dissociation of limb elements in flying vertebrates: a comparison of pterosaurs, birds and bats. J Evol Biol 24:2586–2599

Bhagavatula PS, Claudianos C, Ibbotson MR, Srinivasan MV (2011) Optic flow cues guide flight in birds. Curr Biol 21:1794–1799

Bhushan B (2009) Biomimetics: lessons from nature – an overview. Phil Trans R Soc A 373:2037

Biewener AA (2019) Getting to grips with how birds land stably on complex surfaces. Nature 574:180–181

Bishop CM, Butler PJ (2015) Flight. In: Scanes CG (ed) Sturkie's avian physiology, 6th edn. Academic Press, New York, NY, pp 919–974

Blake RW, Kolotylo R, de la Cueva H (1990) Flight speeds of the Barn Swallow, *Hirundo rustica*. Can J Zool 68:1–5

Bock WJ (2013) The furcular and the evolution of avian flight. Paleontol J 47:1236–1244

Bock WJ, Miller WD (1959) The scansorial foot of the woodpeckers, with comments on the evolution of perching and climbing feet in birds. Am Mus Novit 1931:1–45

Bribiesca-Contreras F, Parslew B, Sellers WI (2019) A quantitative and comparative analysis of the muscle architecture of the forelimb myology of diurnal birds of prey (Order Accipitriformes and Falconiformes). Anat Rec 302:1808–1823

Brücker C, Weidner C (2013) Separation control via self-adaptive hairy flaplet arrays. In: Unsteady separation in fluid-structure interaction. ERCOFTAC International Symposium, Mykonos

Bruinzeel LW, Piersma T, Kersten M (1999) Low costs of terrestrial locomotion in waders. Ardea 87:199–205

Brusatte SL, O'Connor JK, Jarvis ED (2015) The origin and diversification of birds. Curr Biol 25:R888–R898

Burgers P, Chiappe LM (1999) The wings of *Archaeopteryx* as a primary thrust generator. Nature 399:60–62

Burnham DA (2007) Paleoenvironment, paleoecology, and evolution of Maniraptoran 'dinosaurs.' Ph.D. dissertation, University of Kansas, Lawrence, KS

Butler PJ (2000) Energetic costs of surface swimming and diving of birds. Physiol Biochem Zool 73:699–705

Carrascal LM, Moreno E, Telleria JL (1990) Ecomorphological relationships in a group of insectivorous birds of temperate forests in winter. Holarct Ecol 13:105–111

Carruthers AC, Gillies J, Ozawa Y, Walker SM, Thomas ALR, Taylor GK (2009) Engineering applications of bird flight. European Office of Aerospace Research and Development, Ruislip

Catry P, Phillips RA, Phalan B, Croxall JP (2006) Senescence effects in an extremely long-lived bird: the Grey-headed Albatross *Thalassarche chrysostoma*. Proc R Soc B 273:1625–1630

Chatterjee S, Templin J (2012) Palaeoecology, aerodynamics, and the origin of avian flight. In: Talent JA (ed) Earth and life. New York, NY, Springer Science, pp 585–612

Chatterjee S, Templin RJ, Campbell KE Jr (2007) The aerodynamics of *Argentavis*, the world's largest flying bird from the Miocene of Argentina. Proc Natl Acad Sci USA 104:12398–12403

Chen P-J, Dong ZM, Zhen SN (1998) An exceptionally well-preserved theropod dinosaur from the Yixian Formation of China. Nature 391:147–152

Chin DD, Lentink D (2017) How birds direct impulse to minimize the energetic cost of foraging flight. Sci Adv 3:e1603041

Chin DD, Lentink D (2019) Birds repurpose the role of drag and lift to take off and land. Nat Commun 10: 5354

Clark GA Jr (1975) Additional records of passerine terrestrial gaits. Wilson Bull 87:384–389

Clifton GT, Carr JA, Biewener AA (2018) Comparative hindlimb myology of foot-propelled swimming birds. J Anat 232:105–123

Cobb SE, Sellers WI (2020) Inferring lifestyle for Aves and Theropoda: a model based on curvatures of extant avian ungual bones. PLoS ONE 15:e0211173

Corum J [online] (2003) The Strouhal number in cruising flight. http://www.style.org/strouhalflight/

Cronin TW, Kinloch MR, Olsen GH (2005) Head-bobbing behavior in foraging Whooping Cranes favors visual fixation. Curr Biol 15:R243–R244

Croxall JP, Silk JRD, Phillips RA, Afanasyev V, Briggs DR (2005) Global circumnavigations: tracking year-round ranges of nonbreeding albatrosses. Science 307: 249–250

Daley MA, Biewener AA (2003) Muscle force-length dynamics during level versus incline locomotion: a comparison of in vivo performance of two Guinea fowl ankle extensors. J Exp Biol 206:2941–2958

Daley MA, Birn-Jeffery A (2018) Scaling of avian bipedal locomotion reveals independent effects of body mass and leg posture on gait. J Exp Biol 221:jeb152538

Davis CA, Smith LM (2001) Foraging strategies and niche dynamics of coexisting shorebirds at stopover sites in the Southern Great Plains. Auk 118:484–495

de Koning H, van Eekhout L [online] (2011) Whiffling geese in slow motion. https://www.youtube.com/watch?v=Tyzl4mDM5GM

DeBlois MC, Motani R (2019) Flipper bone distribution reveals flexible trailing edge in underwater flying marine tetrapods. J Morphol 280:908–924

Dececchi TA, Larsson HCE (2011) Assessing arboreal adaptations of bird antecedents: testing the ecological setting of the origin of the avian flight stroke. PLoS ONE 6:e22292

Dececchi TA, Larsson HC (2013) Body and limb size dissociation at the origin of birds: uncoupling allometric constraints across a macroevolutionary transition. Evolution 67:2741–2752

Dececchi TA, Larsson HC, Habib MB (2016) The wings before the bird: an evaluation of flapping-based locomotory hypotheses in bird antecedents. PeerJ 4:e2159

Dececchi TA, Mloszewska AM, Holtz TR Jr, Habib MB, Larsson HCE (2020) The fast and the frugal: divergent locomotory strategies drive limb lengthening in theropod dinosaurs. PLoS ONE 15:e0223698

Dial KP (1992) Avian forelimb muscles and nonsteady flight: can birds fly without using the muscles in their wings? Auk 109:874–885

Dial KP (2003a) Wing-assisted incline running and the evolution of flight. Science 299:402–404

Dial KP (2003b) Evolution of avian locomotion: correlates of flight style, locomotor modules, nesting biology, body size, development, and the origin of flapping flight. Auk 120:941–952

Diehl RH (2013) The airspace is habitat. Trends Ecol Evol 28:377–379

Duffy DC, Todd FS, Siegfried WR (1987) Submarine foraging behavior of alcids in an artificial environment. Zoo Biol 6:373–378

Duncan RP, Blackburn TM (2004) Extinction and endemism in the New Zealand avifauna. Glob Ecol Biogeogr 13:509–517

Duriez O, Kato A, Tromp C, Dell'Omo G, Vyssotski AL, Sarrazin F, Ropert-Coudert Y (2014) How cheap is soaring flight in raptors? A preliminary investigation in freely-flying vultures. PLoS ONE 9:e84887

Dyke G, de Kat R, Palmer C, van der Kindere J, Naish D, Ganapathisubramani B (2013) Aerodynamic performance of the feathered dinosaur *Microraptor* and the evolution of feathered flight. Nat Commun 4:2489

Earls KD (2000) Kinematics and mechanics of ground take-off in the starling (*Sturnus vulgaris*) and the quail (*Coturnix coturnix*). J Exp Biol 203:725–739

Eder H, Fiedler W, Neuhäuser M (2015) Evaluation of aerodynamic parameters from infrared laser tracking of free-gliding White Storks. J Ornithol 156:667–677

Ellington CP (1991) Limitations on animal flight performance. J Exp Biol 160:71–91

Elliott KH, Gaston AJ (2005) Flight speeds of two seabirds: a test of Norberg's hypothesis. Ibis 147: 783–789

Elliott KH, Ricklefs RE, Gaston AJ, Hatch SA, Speakman JR, Davoren GK (2013) High flight costs, but low dive costs, in auks support the biomechanical hypothesis for flightlessness in penguins. Proc Natl Acad Sci USA 110:9380–9384

Elzanowski A (2002) Archaeopterygidae (Upper Jurassic of Germany). In: Chiappe LM, Witmer LM (eds) Mesozoic birds: above the heads of dinosaurs. Univ California Press, Berkeley, CA, pp 129–159

Erickson GM, Rauhut OWM, Zhou Z, Turner AH, Inouye BD, Hu D, Norell MA (2009) Was dinosaurian physiology inherited by birds? Reconciling slow growth in *Archaeopteryx*. PLoS ONE 4:e7390

Evangelista D, Cam S, Huynh T, Kwong A, Mehrabani H, Tse K, Dudley R (2014) Shifts in stability and control effectiveness during evolution of Paraves support aerial maneuvering hypotheses for flight origins. PeerJ 2:e632

Evans MR (2003) Birds' tails do act like delta wings but delta-wing theory does not always predict the forces they generate. Proc R Soc B 270:1379–1385

Evans MR (2004) Limits on the evolution of tail ornamentation in birds. Am Nat 163:341–357

Fedak MA, Seeherman HJ (1979) Reappraisal of energetics of locomotion shows identical cost in bipeds and quadrupeds including Ostrich and horse. Nature 282: 713–716

Feduccia A (1996) The origin and evolution of birds. Yale University Press, New Haven, CT

Feo TJ, Field DJ, Prum RO (2015) Barb geometry of asymmetrical feathers reveals a transitional morphology in the evolution of avian flight. Proc R Soc B 282: 20142864

Field DJ, Lynner C, Brown C, Darroch SAF (2013) Skeletal correlates for body mass estimation in modern and fossil flying birds. PLoS ONE 8:e82000

Finn J, Carlsson J, Kelly T, Davenport J (2012) Avoidance of headwinds or exploitation of ground effect – why do birds fly low? J Field Ornithol 83:192–202

Fish FE (1994) Energy conservation by formation swimming: metabolic evidence from ducklings. In: Maddock L, Bone Q, Rayner JMV (eds) Mechanics and physiology of animal swimming. Cambridge University Press, Cambridge, pp 193–204

Fish FE (1995) Kinematics of ducklings swimming in formation: consequences of position. J Exp Zool 273: 1–11

Fish FE (2000) Biomechanics and energetics in aquatic and semiaquatic mammals: platypus to whale. Physiol Biochem Zool 73:683–698

Foth C, Rauhut OW (2017) Re-evaluation of the Haarlem *Archaeopteryx* and the radiation of maniraptoran theropod dinosaurs. BMC Evol Biol 17:236

Foth C, Tischlinger H, Rauhut OWM (2014) New specimen of *Archaeopteryx* provides insights into the evolution of pennaceous feathers. Nature 511:79–82

Fowler DW, Freedman EA, Scannella JB, Kambic RE (2011) The predatory ecology of *Deinonychus* and the origin of flapping in birds. PLoS ONE 6:e28964

Gatesy SM (1999) Guineafowl hind limb function. I: Cinderadiographic analysis and speed effects. J Morphol 240:115–125

Gatesy SM, Biewener AA (1991) Bipedal locomotion: effects of speed, size and limb posture in birds and humans. J Zool 224:127–147

Gatesy SM, Dial KP (1993) Tail muscle activity patterns in walking and flying pigeons (*Columba livia*). J Exp Biol 176:55–76

Gatesy SM, Dial KP (1996) From frond to fan: *Archaeopteryx* and the evolution of short-tailed birds. Evolution 50:2037–2048

Gatesy SM, Middleton KM (1997) Bipedalism, flight, and the evolution of theropod locomotor diversity. J Vertebr Paleontol 17:308–329

Godefroit P, Demuynck H, Dyke G, Hu D, Escuillié F, Claeys P (2013) Reduced plumage and flight ability of a new Jurassic paravian theropod from China. Nat Commun 4:1394

Gold MEL, Schulz D, Budassi M, Gignac PM, Vaska P, Norell MA (2016) Flying starlings, PET and the evolution of volant dinosaurs. Curr Biol 26:R265–R267

Gough WT, Farina SC, Fish FE (2015) Aquatic burst locomotion by hydroplaning and paddling in Common Eiders (*Somateria mollissima*). J Exp Biol 218:1632–1638

Gowree ER, Jagadeesh C, Talboys E, Lagemann C, Brücker C (2018) Vortices enable the complex aerobatics of Peregrine Falcons. Commun Biol 1:27

Green PR, Cheng P (1998) Variation in kinematics and dynamics of the landing flights of pigeons on a novel perch. J Exp Biol 201:3309–3316

Greenewalt CH (1975) The flight of birds: the significant dimensions, their departure from the requirements for dimensional similarity, and the effect on flight aerodynamics of that departure. Trans Am Philos Soc 65:1–67

Grémillet D, Chauvin C, Wilson RP, Le Maho Y, Wanless S (2005) Unusual feather structure allows partial plumage wettability in diving Great Cormorants *Phalacrocorax carbo*. J Avian Biol 36:57–63

Guerrero J (2009) Numerical stimulation of the unsteady aerodynamics of flapping flight. Ph.D. dissertation, University of Genova, Genova

Guigueno MF, Shoji A, Elliott KH, Aris-Brosou S (2019) Flight costs in volant vertebrates: a phylogenetically-controlled meta-analysis of birds and bats. Comp Biochem Physiol A 235:193–201

Guillemette M, Ouellet J-F (2005) Temporary flightlessness as a potential cost of reproduction in pre-laying Common Eiders *Somateria mollissima*. Ibis 147:301–306

Guillemette M, Polymeropoulos ET, Portugal SJ, Pelletier D (2017) It takes time to be cool: on the relationship between hyperthermia and body cooling in a migrating seaduck. Front Physiol 8:532

Gutierrez E, Quinn DB, Chin DD, Lentink D (2017) Lift calculations based on accepted wake models for animal flight are inconsistent and sensitive to vortex dynamics. Bioinspir Biomim 12:016004

Hails CJ (1979) A comparison of flight energetics in hirundines and other birds. Comp Biochem Physiol 63A:581–585

Halsey LG, Shepard EL, Wilson RP (2011) Assessing the development and application of the accelerometry technique for estimating energy expenditure. Comp Biochem Physiol A Mol Integr Physiol 158:305–314

Han G, Chiappe LM, Ji S-A, Habib M, Turner AH, Chinsamy A, Liu X, Han L (2014) A new raptorial dinosaur with exceptionally long feathering provides insights into dromaeosaurid flight performance. Nature Comm 5:4382

Harel R, Horvitz N, Nathan R (2016) Adult vultures outperform juveniles in challenging thermal soaring conditions. Sci Rep 6:27865

Hartman FA (1961) Locomotor mechanisms of birds. Smithsonian Miscellaneous Collections, vol 143. Smithsonian Institution, Washington, DC

Harvey C, Baliga VB, Lavole P, Altshuler DL (2019) Wing morphing allows gulls to modulate static pitch stability during gliding. J Roy Soc Inter 16:20180641

Hassanalian M, Abdelmoula H, Ben Ayed S, Abdelkefi A (2017) Thermal impact of migrating birds' wing color on their flight performance: possibility of new generation of biologically inspired drones. J Therm Biol 66: 27–32

Heath JP, Gilchrist HG, Ydenberg RC (2006) Regulation of stroke pattern and swim speed across a range of current velocities: diving by Common Eiders wintering in polynyas in the Canadian Arctic. J Exp Biol 209: 3974–3983

Hedenström A (2002) Aerodynamics, evolution and ecology of avian flight. Trends Ecol Evol 17:415–422

Hedenström A (2004) A general law for animal locomotion? Trends Ecol Evol 19:217–219

Hedenström A, Alerstam T (1996) Skylark optimal flight speeds for flying nowhere and somewhere. Behav Ecol 7:121–126

Hedenstrom A, Liechti F (2001) Field estimates of body drag coefficient on the basis of dives in passerine birds. J Exp Biol 204:1167–1175

Hedenstrom A, Lindstrom A (2017) Wind tunnel as a tool in bird migration research. J Avian Biol 48:37–48

Hedrick TL, Tobalske BW, Biewener AA (2002) Estimates of circulation and gait change based on a three-dimensional kinematic analysis of flight in Cockatiels (*Nymphicus hollandicus*) and Ringed Turtle-doves (*Streptopelia risoria*). J Exp Biol 205:1389–1409

Heers AM, Dial KP, Tobalske BW (2014) From baby birds to feathered dinosaurs: incipient wings and the evolution of flight. Paleobiology 40:459–476

Hemelrijk CK, Hildenbrandt H (2011) Some causes of the variable shape of flocks of birds. PLoS ONE 6:e22479

Henningsson P, Johansson LC, Hedenström A (2010) How swift are swifts *Apus apus*? J Avian Biol 41:94–98
Hightower BJ, Ingersoll R, Chin DD, Lawhon C, Haselsteiner AF, Lentink D (2017) Design and analysis of aerodynamic force platforms for free flight studies. Bioinspir Biomim 12:064001
Hightower BJ, Wijnings PWA, Scholte R, Ingersoll R, Chin DD, Nguyen J, Shorr D, Lentink D (2021) How oscillating aerodynamic forces explain the timbre of the hummingbird's hum and other animals in flapping flight. eLife 10:e63107
Höfling E, Abourachid A, Renous S (2006) Locomotion behavior of the Lettered Aracari (*Pteroglossus inscriptus*) (Ramphastidae). Ornitol Neotrop 17:363–371
Homberger DG, de Silva KN (2000) Functional microanatomy of the feather-bearing integument: implications for the evolution of birds and avian flight. Am Zool 40:553–574
Hoppeler H (1999) Skeletal muscle substrate metabolism. Int J Obes 23:S7–S10
Hu DY, Hou LH, Zhang LJ, Xu X (2009) A pre-*Archaeopteryx* troodontid theropod from China with long feathers on the metatarsus. Nature 461: 640–643
Huey RB, Deutsch C (2016) How frigate birds soar around the doldrums. Science 353:26–27
Hui CA (1987) The porpoising of penguins: an energy-conserving behavior for respiratory ventilation? Can J Zool 65:209–211
Hui CA (1988) Penguin swimming. I. Hydrodynamics. Physiol Zool 61:333–343
Hummel D (1995) Formation flight as an energy-saving mechanism. Isr J Zool 41:261–278
Hutchinson JR (2004) Biomechanical modeling and sensitivity analysis of bipedal running ability. I. Extant taxa. J Morphol 262:421–440
Idrac P (1925) Experimental study of the "soaring" of albatrosses. Nature 115:532
Irvine MR, Gardiner BA, Hill MK (1997) The evolution of turbulence across a forest edge. Bound-Layer Meteorol 84:467–496
Jenni-Eiermann S, Srygley RB (2017) Physiological aeroecology: anatomical and physiological adaptations for flight. In: Chilson P, Frick W, Kelly J, Liechti F (eds) Aeroecology. Springer, Cham, pp 87–118
Ji Q, Currie PJ, Norell MA, Ji SA (1998) Two feathered dinosaurs from northeastern China. Nature 393:753–761
Jobling JA (2010) The Helm dictionary of scientific bird names. Christopher Helm, London
Johansson LC (2003) Indirect estimates of wing-propulsion forces in horizontally diving Atlantic Puffins (*Fratercula arctica* L.). Can J Zool 81:816–822
Johansson LC, Norberg UML (2001) Lift-based paddling in diving grebe. J Exp Biol 204:1687–1696
Johansson LC, Norberg RA (2003a) Delta-wing function of webbed feet gives hydrodynamic lift for swimming propulsion in birds. Nature 424:65–68
Johansson LC, Norberg UML (2003b) Biomechanics: asymmetric toes aid underwater swimming. Nature 407:582–583
Jouventin P, Weimerskirch H (1990) Satellite tracking of Wandering Albatrosses. Nature 343:746–748
Kanani-Sühring F, Raasch S (2015) Spatial variability of scalar concentrations and fluxes downstream of a clearing-to-forest transition: a large-eddy simulation study. Bound-Layer Meteorol 155:1–27
Kikuchi DM, Watanuki Y, Sato N, Hoshina K, Takahashi A, Watanabe YY (2015) Strouhal number for flying and swimming in Rhinoceros Auklets *Cerorhinca monocerata*. J Avian Biol 46:406–411
Kilbourne BM, Andrada E, Fischer MS, Nyakatura JA (2016) Morphology and motion: hindlimb proportions and swing phase kinematics in terrestrially locomoting Charadriiform birds. J Exp Biol 219:1405–1416
Klaassen van Oorschot BK, Tang HK, Tobalske BW (2017) Phylogenetics and ecomorphology of emarginated primary feathers. J Morphol 278:936–947
Klaassen van Oorschot B, Choroszucha R, Tobalske BW (2020) Passive aeroelastic deflection of avian primary feathers. Bioinspir Biomim 15:056008
KleinHeerenbrink M, Johansson LC, Hedenström A (2017) Multi-cored vortices support function of slotted wing tips of birds in gliding and flapping flight. J R Soc Interface 14:20170099
KleinHeerenbrink M, France LA, Brighton CH, Taylor GK (2022) Optimization of avian perching manoeuvres. Nature 607:91–96
Koopman R, Van Loon LJC (2009) Aging, exercise, and muscle protein metabolism. J Appl Physiol 105:2040–2048
Kovacs CE, Meyers RA (2000) Anatomy and histochemistry of flight muscles in a wing-propelled diving bird, the Atlantic Puffin, *Fratercula arctica*. J Morphol 244: 109–125
Kramer BR, Smith BC, Heid JP, Noffz GK, Richwine DM, Ng T [online] (1999) Drag reduction experiments using boundary layer heating. AIAA Paper 1999-0134, 37th AIAA Aerospace Sciences Meeting, Reno, NV. http://www.rollinghillsresearch.com/Aero_Research/Files/AIAA-1999-0134_BLHeat.pdf
Labocha MK, Rutkowska J, Cichoń M, Bauchinger U (2015) Age-specific decline in take-off flight performance in a small passerine. Anim Behav 105:103–107
Lapsansky AB, Zatz D, Tobalske BW (2020) Alcids 'fly' at efficient Strouhal numbers in both air and water but vary stroke velocity and angle. eLife 9:e55774
Le Coco GE, Motta MJ, Agnolín FL, Novas FE (2022) Wing osteology, myology, and function of *Rhea americana* (Aves, Rheidae). J Morphol 283:1015–1047
Leblanc K, Pintore R, Galvão A, Heitz E, Provini P (2023) Foot adaptation to climbing in ovenbirds and woodcreepers (Furnariidae). J Anat 242:607–626

Lee MSY, Cau A, Naish D, Dyke GJ (2014) Sustained miniaturization and anatomical innovation in the dinosaurian ancestors of birds. Science 345:6196

Lee S-I, Kim J, Park H, Jabłoński PG, Choi H (2015) The function of the alula in avian flight. Sci Rep 5:9914

Lefèvre U, Cau A, Hu D, Godefroit P (2020) Feather evolution in Pennaraptora. In: Foth C, Rauhut OWM (eds) The evolution of feathers: from their origin to the present. Cham, Springer Nature Switzerland AG, pp 103–118

Lentink D, Müller UK, Stamhuis EJ, De Kat R, Van Gestel W, Veldhuis LLM, Henningsson P, Hedenström A, Videler JJ, Van Leeuwen JL (2007) How swifts control their glide performance with morphing wings. Nature 446:1082–1085

Lewis S, Schreiber EA, Daunt F, Schenk GA, Orr K, Adams A, Wanless S, Hamer KC (2005) Sex-specific foraging behaviour in tropical boobies: does size matter? Ibis 147:408–414

Leys F, Reynaerts D, Vandepitte D (2016) Outperforming hummingbirds' load-lifting capability with a lightweight hummingbird-like flapping-wing mechanism. Biology Open 5:1052–1060

Li Q, Gao K-Q, Meng Q, Clarke JA, Shawkey MD, D'Alba L, Pei R, Ellison M, Norell MA, Vinther J (2012) Reconstruction of *Microraptor* and the evolution of iridescent plumage. Science 335:1215–1219

Liu D, Zhou ZH, Zhang YG (2012) Mass estimate and evolutionary trend in Chinese Mesozoic fossil birds. Vert PalAs 50:39–52

Livezey BC (2003) Evolution of flightlessness in rails (Gruiformes: Rallidae): phylogenetic, ecomorphological, and ontogenetic perspectives. Ornithol Monogr 53:1–654

Livezey BC, Humphrey PS (1983) Mechanics of steaming in steamer-ducks. Auk 100:485–488

Lockwood R, Swaddle JP, Rayner JMV (1998) Avian wingtip shape reconsidered: wingtip shape indices and morphological adaptations to migration. J Avian Biol 29:273–292

Longrich NR, Vinther J, Meng Q, Li Q, Russell AP (2012) Primitive wing feather arrangement in *Archaeopteryx lithographica* and *Anchiornis huxleyi*. Curr Biol 22: 2262–2267

Longrich NR, Tischlinger H, Foth C (2020) The feathers of the Jurassic Urvogel *Archaeopteryx*. In: Foth C, Rauhut OWM (eds) The evolution of feathers: from their origin to the present. Cham, Springer Nature Switzerland AG, pp 119–146

Lovvorn JR (2001) Upstroke thrust, drag effects, and stroke-glide cycles in wing-propelled swimming by birds. Am Zool 41:154–165

Lovvorn JR, Liggins GA (2002) Interactions of body shape, body size and stroke-acceleration patterns in costs of underwater swimming by birds. Funct Ecol 16:106–112

Lü J, Brusatte SL (2015) A large, short-armed, winged dromaeosaurid (Dinosauria: Theropoda) from the Early Cretaceous of China and its implications for feather evolution. Sci Rep 5:11775

Maeda M, Nakata T, Kitamura I, Tanaka H, Liu H (2017) Quantifying the dynamic wing morphing of hovering hummingbird. R Soc Open Sci 4:170307

Mallon JM, Bildstein KL, Katzner TE (2016) In-flight turbulence benefits soaring birds. Auk 133:79–85

Masman D, Klaasen M (1987) Energy expenditure during free flight in trained and free-living Eurasian Kestrels. Auk 104:603–616

Matloff LY, Chang E, Feo TJ, Jeffries L, Stowers AK, Thomson C, Lentink D (2020) How flight feathers stick together to form a continuous morphing wing. Science 367:293–297

Matsukawa M, Shibata K, Sato K, Xing X, Lockley MG (2014) The Early Cretaceous terrestrial ecosystems of the Jehol Biota based on food-web and energy-flow models. Biol J Linn Soc 113:836–853

Maxwell GR, Putnam LS (1968) The maintenance behavior of the Black-crowned Night Heron. Wilson Bull 80: 467–478

Maybury WJ (2000) The aerodynamics of bird bodies. Ph. D. dissertation, University of Bristol, Bristol

Maybury WJ, Rayner JMV (2001) The avian tail reduces body parasite drag by controlling flow separation and vortex shedding. Proc R Soc B 268:1405–1410

Mayr G (2017) Pectoral girdle morphology of Mesozoic birds and the evolution of the avian supracoracoideus muscle. J Ornithol 158:859–867

Mayr G, Pohl B, Hartman S, Peters DS (2007) The tenth skeletal specimen of *Archaeopteryx*. Zool J Linnean Soc 149:97–116

McNab BK (1994) Energy conservation and the evolution of flightlessness in birds. Am Nat 144:628–642

McNab BK (2002) Minimizing energy expenditure facilitates vertebrate persistence on oceanic islands. Ecol Lett 5:693–704

McNab BK, Ellis HI (2006) Flightless rails endemic to islands have lower energy expenditures and clutch sizes than flighted rails on islands and continents. Comp Biochem Physiol A 145:295–311

Meseguer J, Alvarez J, Meseguer E, Perez A (2003) The alula: a leading edge, high lift device of birds. IDR/UPM, Universidad Politecnica de Madrid, Madrid

Meyers RA, Stakebake EF (2005) Anatomy and histochemistry of spread-wing posture in birds. 3. Immunochemistry of flight muscles and the 'shoulder lock' in albatrosses. J Morphol 263:12–29

Milberg P, Tyrberg T (1993) Naive birds and noble savages: a review of man-caused prehistoric extinctions of island birds. Ecography 16:229–250

Mill GK, Baldwin J (1983) Biomechanical correlates of swimming and diving behavior in the Little Penguin, *Eudyptula minor*. Physiol Zool 56:242–254

Monsarrat S, Benhamou S, Sarrazin F, Bessa-Gomes C, Bouten W, Duriez O (2013) How predictability of feeding patches affects home range and foraging habitat selection in avian social scavengers? PLoS ONE 8: e53077

Muijres FT, Dickinson MH (2014) Bird flight: fly with a little flap from your friends. Nature 505:295–296
Muijres FT, Henningsson P, Stuiver M, Hedenström A (2012a) Aerodynamic flight performance in flap-gliding birds and bats. J Theor Biol 306:120–128
Muijres FT, Johansson LC, Hedenström A (2012b) Leading edge vortex in a slow-flying passerine. Biol Lett 8: 554–557
Naish D, Martill DM (2003) Pterosaurs – a successful invasion of prehistoric skies. Biologist 50:213–216
Necker R (2007) Head-bobbing of walking birds. J Comp Physiol A 193:1177
Norberg UM (1976) Some advanced flight manoeuvers of bats. J Exp Biol 64:489–495
Norberg RA (1981) Optimal flight speed in birds when feeding young. J Anim Ecol 50:473–477
Norberg RA (1981b) Why foraging birds in trees should climb and hop upwards rather than downwards. Ibis 123:281–288
Norberg RA (1986a) Treecreeper climbing; mechanics, energetics, and structural adaptations. Ornis Scand 17:191–209
Norberg UM (1986b) Evolutionary convergence in foraging niche and flight morphology in insectivorous aerial-hawking birds and bats. Ornis Scand 17:253–260
Norberg UM (1990) Vertebrate flight: Mechanics, physiology, morphology, ecology, and evolution. Springer, Berlin
Norberg RÅ (2008a) Why foraging birds in trees should climb and hop upwards rather than downwards. Ibis 123:281–288
Norberg RA, Norberg UM (1971) Take-off, landing, and flight speed during fishing flights of *Gavia stellata*. Ornis Scand 2:55–67
Norberg UM (1995) How a long tail and changes in mass and wing shape affect the cost of flight in animals. Funct Ecol 9:48–54
Norberg UM, Rayner JMV (1987) Ecological morphology and flight in bats: wing adaptations, flight performance, foraging strategy and echolocation. Philos Trans R Soc B 316:335–427
Norberg UML (2007) Evolution of flight in animals. In: Liebe R (ed) Flow phenomena in nature, vol. 1. WIT Press, Southhampton, UK, pp 36–54
Norberg UML (2008b) Evolution of flight in animals. In: Liebe R (ed) Flow phenomena in nature, vol 1. WIT Press, Ashurst, pp 36–54
Nudds RL, Dyke GJ (2009) Forelimb posture in dinosaurs and the evolution of the avian flapping flight-stroke. Evolution 63:994–1002
Nudds RL, Dyke GJ (2010) Narrow primary feather rachises in *Confuciusornis* and *Archaeopteryx* suggest poor flight ability. Science 328:887–889
Nudds RL, Rayner JMV (2006) Scaling of body frontal area and body width in birds. J Morphol 267:341–346
Nyakatura JA, Andrada E (2014) On vision in birds: coordination of head-bobbing and gait stabilises vertical head position in quail. Front Zool 11:27
Nyakatura JA, Andrada E, Grimm N, Weise H, Fischer MS (2012) Kinematics and center of mass mechanics during terrestrial locomotion in Northern Lapwings (*Vanellus vanellus*, Charadriiformes). J Exp Zool A 317:580–594
O'Farrell B, Davenport J, Kelly T (2002) Was *Archaeopteryx* a wing-in-ground effect flier? Ibis 144:686–688
Olson SL (1973) Evolution of the rails of the South Atlantic islands (Aves: Rallidae). Smithson Contrib Zool 152:1–53
Osa Y (1994) Functional morphology of diving and flying seabirds. Ph.D. dissertation, Tokyo Fisheries University, Tokyo
Padian K (2001) Cross-testing adaptive hypotheses: phylogenetic analysis and the origin of bird flight. Am Zool 41:598–607
Pan Y, Sha J, Zhou Z, Fürsich FT (2013) The Jehol Biota: definition and distribution of exceptionally preserved relicts of a continental Early Cretaceous ecosystem. Cretac Res 44:30–38
Pap PL, Vincze O, Wekerle B, Daubner T, Vágási CI, Nudds RL, Dyke GJ, Osváth G (2017) A phylogenetic comparative analysis reveals correlations between body feather structure and habitat. Funct Ecol 31: 1241–1251
Pei R, Li Q, Meng Q, Norell MA, Gao KQ (2017) New specimens of *Anchiornis huxleyi* (Theropoda: Paraves) from the Late Jurassic of northeastern China. Bull Am Mus Nat Hist 411:1–67
Pei R, Pittman M, Goloboff PA, Dececchi TA, Habib MB, Kaye TG, Larsson HCE, Norell MA, Brusatte SL, Xu X (2020) Potential for powered flight neared by most close avialan relatives, but few crossed its thresholds. Curr Biol 30:4033–4046
Pennycuick CJ (1975) Mechanics of flight. In: Farner DS, King JR (eds) Avian biology, vol 5. Academic, New York, NY, pp 1–75
Pennycuick CJ (1982) The flight of petrels and albatrosses (Procellariiformes), observed in South Georgia and its vicinity. Philos Trans R Soc B 300:75–106
Pennycuick CJ (1997) Actual and 'optimum' flight speeds: field data reassessed. J Exp Biol 200:2355–2361
Pennycuick CJ (2003) The concept of energy height in animal locomotion: separating mechanics from physiology. J Theor Biol 224:189–203
Pennycuick CJ (2008) Gliding flight and soaring. In: Pennycuick CJ (ed) Theoretical ecology series, vol. 5. Modelling the flying bird. Elsevier, Amsterdam, pp 271–304
Pennycuick CJ, Klaassen M, Kvist A, Lindström A (1996) Wingbeat frequency and the body drag anomaly: wind-tunnel observations on a Thrush Nightingale (*Luscinia luscinia*) and a teal (*Anas crecca*). J Exp Biol 199:2757–2765
Pittman M, Kaye TG, Wang X, Hartman SA (2022) Preserved soft anatomy confirms shoulder-powered upstroke of early theropod flyers, reveals enhanced early pygostylian upstroke, and explains early sternum loss. Proc Natl Acad Sci USA 119:e2205476119

Polet DT, Rival DE (2015) Rapid area change in pitch-up manoeuvres of small perching birds. Bioinspir Biomim 10:066004

Ponitz B, Schmitz A, Fischer D, Bleckmann H, Brücker C (2014) Diving-flight aerodynamics of a Peregrine Falcon (*Falco peregrinus*). PLoS ONE 9:e86506

Pontzer H (2007) Effective limb length and the scaling of locomotor cost in terrestrial animals. J Exp Biol 210: 1752–1761

Portugal SJ, Hubel TY, Fritz J, Heese S, Trobe D, Voelkl B, Hailes S, Wilson AM, Usherwood JR (2014) Upwash exploitation and downwash avoidance by flap phasing in ibis formation flight. Nature 505: 399–402

Powers DR, Tobalske BW, Wilson JK, Woods HA, Corder KR (2015) Heat dissipation during hovering and forward flight in hummingbirds. R Soc Open Sci 2:150598

Prange HD, Schmidt-Nielsen K (1970) The metabolic cost of swimming in ducks. J Exp Biol 53:763–777

Pritchard DJ, Healy SD (2018) Taking an insect-inspired approach to bird navigation. Learn Behav 46:7–22

Provini P, Abourachid A (2018) Whole-body 3D kinematics of bird take-off: key role of the legs to propel the trunk. Sci Nat 105:12

Provini P, Goupil P, Hugel V, Abourachid A (2012) Walking, paddling, waddling: 3D kinematics Anatidae locomotion (*Callonetta leucophrys*). J Exp Zool 317: 275–282

Raikow RJ (1973) Locomotor mechanisms in North American ducks. Wilson Bull 85:295–307

Raikow RJ, Bicanovsky L, Bledsoe AH (1988) Forelimb joint mobility and the evolution of wing-propelled diving in birds. Auk 105:446–451

Rattenborg NC, Voirin B, Cruz SM, Tisdale R, Dell'Omo G, Lipp HP, Wikelski M, Vyssotski AL (2016) Evidence that birds sleep in mid-flight. Nat Commun 7:12468

Rauhut OWM, Heyng AM, Lopez-Arbarello A, Hecker A (2012) A new Rhynchocephalian from the Late Jurassic of Germany with a dentition that is unique amongst tetrapods. PLoS ONE 7:e46839

Rauhut OWM, Tischlinger H, Foth C (2019) A non-archaeopterygid avialan theropod from the Late Jurassic of southern Germany. eLife 8:e43789

Rayner JMV (1985) Bounding and undulating flight in birds. J Theor Biol 117:47–77

Rayner JMV (1988) Form and function in avian flight. Curr Ornithol 5:1–66

Rayner JMV (1991) On the aerodynamics of animal flight in ground effect. Philos Trans R Soc B 334:119–128

Rayner JMV (1999) Estimating power curves of flying vertebrates. J Exp Biol 202:3449–3461

Rayner JM (2001) Mathematical modelling of the avian flight power curve. Math Methods Appl Sci 24:1485–1514

Rayner JMV, Viscardi PW, Ward S, Speakman JR (2001) Aerodynamics and energetics of intermittent flight in birds. Am Zool 41:188–204

Renner M (2006) Repeatable measures of take-off flight performance in auklets. J Zool 268:395–404

Reynolds KV, Thomas AL, Taylor GK (2014) Wing tucks are a response to atmospheric turbulence in the soaring flight of the Steppe Eagle *Aquila nipalensis*. J R Soc Interface 11:20140645

Rezende EL, Bacigalupe LD, Nespolo RF, Bozinovic F (2020) Shrinking dinosaurs and the evolution of endothermy in birds. Sci Adv 6:eaaw4486

Ribak G, Weihs D, Arad Z (2005) Submerged swimming of the Great Cormorant *Phalacrocorax carbo sinensis* is a variant of the burst-and-glide gait. J Exp Biol 208: 3835–3849

Richardson PL (2011) How do albatrosses fly around the world without flapping their wings? Prog Oceanogr 88: 46–58

Richardson PL (2012) High-speed dynamic soaring. RC Soar Dig 29:36–49

Rijke AM, Jesser WA (2010) The feather structure of dippers: water repellency and resistance to water penetration. Wilson J Ornithol 122:563–568

Roberts TJ, Kram R, Weyand PG, Taylor CR (1998) Energetics of bipedal running. I. Metabolic cost of generating force. J Exp Biol 201:2745–2751

Roderick WRT, Chin DD, Cutkosky MR, Lentink D (2019) Birds land reliably on complex surfaces by adapting their foot-surface interactions upon contact. eLife 8:e46415

Rogalla S, D'Alba L, Verdoodt A, Shawkey MD (2019) Hot wings: impacts of wing coloration on surface temperature during bird flight. J R Soc Interface 16: 20190032

Rogalla S, Nicolaï MPJ, Porchetta S, Glabeke G, Battistella C, D'Alba L, Gianneschi NC, van Beeck J, Shawkey MD (2021) The evolution of darker wings in seabirds in relation to temperature-dependent flight efficiency. J R Soc Interface 18:20210236

Ropert-Coudert Y, Grémillet D, Ryan P, Kato A, Naito Y, Le Maho Y (2004) Between air and water: the plunge dive of the Cape Gannet *Morus capensis*. Ibis 146: 281–290

Rosén M, Hedenström A (2001) Gliding flight in a Jackdaw: a wind tunnel study. J Exp Biol 204:1153–1166

Sachs G (2005) Minimum shear wind strength required for dynamic soaring of albatrosses. Ibis 147:1–10

Sachs G (2013) Dynamic soaring of albatrosses over land. Am Inst Aeronaut Astronaut 2013:4842

Sachs G, Traugott J, Nesterova AP, Dell'Omo G, Kümmeth F, Heidrich W, Vyssotski AL, Bonadonna F (2012) Flying at no mechanical energy cost: disclosing the secret of Wandering Albatrosses. PLoS ONE 7: e41449

Saitta ET, Gelernter R, Vinther J (2018) Additional information on the primitive contour and wing feathering of paravian dinosaurs. Palaeontology 61:273–288

Sakamoto KQ, Takahashi A, Iwata T, Yamamoto T, Yamamoto M, Trathan PN (2013) Heart rate and estimated energy expenditure of flapping and gliding in Black-browed Albatrosses. J Exp Biol 216:3175–3182

Sanz JL, Chiappe LM, Perez-Moreno P, Buscalioni AD, Moratalla JJ, Ortega F, Payata-Ariza FJ (1996) An early Cretaceous bird from Spain and its implications for the evolution of avian flight. Nature 382:442–445

Sato K, Watanuki Y, Takahashi A, Miller PJO, Tanaka H, Kawabe R, Ponganis PJ, Handrich Y, Akamatsu T, Watanabe Y, Mitani Y, Costa DP, Bost C-A, Aoki K, Amano M, Trathan P, Shapiro A, Naito Y (2007) Stroke frequency, but not swimming speed, is related to body size in free-ranging seabirds, pinnipeds and cetaceans. Proc R Soc B 274:471–477

Savile DBO (1957) Adaptive evolution in the avian wing. Evolution 11:212–224

Sayol F, Steinbauer MJ, Blackburn TM, Antonelli A, Faurby S (2020) Anthropogenic extinctions conceal widespread evolution of flightlessness in birds. Sci Adv 6:eabb6095

Schijf ML (2008) The gull's flight behavior over land explained with the atmospheric boundary layer stability: a study at Schiphol Airport in The Netherlands. M.S. thesis, University of Amsterdam, Amsterdam

Schmitz A, Ponitz B, Brücker C, Schmitz H, Herweg J, Bleckmann H (2015) Morphological properties of the last primaries, the tail feathers, and the alulae of *Accipiter nisus*, *Columba livia*, *Falco peregrinus*, and *Falco tinnunculus*. J Morphol 276:33–46

Selden PA, Nudds JR (2012) The Solnhofen limestone. In: Evolution of fossil ecosystems, 2nd edn. Elsevier, London, pp 156–167

Serrano FJ, Palmqvist P, Chiappe LM, Sanz JL (2017) Inferring flight parameters of Mesozoic avians through multivariate analyses of forelimb elements in their living relatives. Paleobiology 43:144–169

Shahid F, Zhao J, Godefroit P (2019) Aerodynamics from cursorial running to aerial gliding for avian flight evolution. Appl Sci 9:649

Shamoun-Baranes J, Leshem Y, Yom-Tov Y, Liechti O (2003) Differential use of thermal convections by soaring birds over central Israel. Condor 105:208–218

Short LL (1971) The evolution of terrestrial woodpeckers. Am Mus Novit 2467:1–23

Smith NA, Clarke JA (2014) Systematics and evolution of the Pan-Alcidae (Aves, Charadriiformes). J Avian Biol 45:1–16

Smith NA, Koeller KL, Clarke JA, Ksepka DT, Mitchell JS, Nabavizadeh A, Ridgley RC, Witmer LM (2022) Convergent evolution in dippers (Aves, Cinclidae): the only wing-propelled diving songbirds. Anat Rec 305:1563–1591

Song J, Tobalske BW, Powers DR, Hedrick TL, Luo H (2016) Three-dimensional simulation for fast forward flight of a Calliope Hummingbird. Royal Soc Open Sci 3:160230

Speakman JR (2001) The evolution of flight and echolocation in bats: another leap in the dark. Mammal Rev 31:111–130

Speakman JR, Thomas DW (2003) Physiological ecology and energetics of bats. In: Kunz TH, Fenton MB (eds) Physiological ecology and energetics of bats. University of Chicago Press, Chicago, IL, pp 430–490

Spedding GR (1987a) The wake of a kestrel (*Falco tinnunculus*) in gliding flight. J Exp Biol 127:45–57

Spedding GR (1987b) The wake of a kestrel (*Falco tinnunculus*) in flapping flight. J Exp Biol 127:59–78

Spivey RJ, Stansfield S, Bishop CM (2014) Analysing the intermittent flapping flight of a Manx Shearwater, *Puffinus puffinus*, and its sporadic use of a wave-meandering wing-sailing flight strategy. Prog Oceanogr 125:62–73

Spring L (1971) A comparison of functional and morphological adaptations in the Common Murre (*Uria aalge*) and Thick-billed Murre (*Uria lomvia*). Condor 73:1–27

Stokes IA, Lucas AJ (2021) Wave-slope soaring of the Brown Pelican. Mov Ecol 9:13

Su J-Y, Ting S-C, Yang J-T (2012) How a small bird executes a sharp turning maneuver: a mechanical perspective. Exp Mech 52:693–703

Sugimoto T (1998) A theoretical analysis of sea-anchor soaring. J Theor Biol 192:393–402

Sullivan C, Xu X, O'Connor JK (2017) Complexities and novelties in the early evolution of avian flight, as seen in the Mesozoic Yanliao and Jehol Biotas of Northeast China. Palaeoworld 26:212–229

Swaddle JP, Witter MS, Cuthill IC, Budden A, McCowen P (1996) Plumage condition affects flight performance in Common Starlings: implications for developmental homeostasis, abrasion and moult. J Avian Biol 27:103–111

Talori YS, Liu Y-F, Zhao J-S, Sullivan C, O'Connor JK, Li Z-H (2018) Winged forelimbs of the small theropod dinosaur *Caudipteryx* could have generated small aerodynamic forces during rapid terrestrial locomotion. Sci Rep 8:17854

Tang J-H, Su J-Y, Wang C-H, Yang J-T (2013) Numerical investigation of the ground effect for a small bird. J Mech 29:433–441

Taylor CR, Heglund NC, Maloiy GMO (1982) Energetics and mechanics of terrestrial locomotion. I. Metabolic energy consumption as a function of speed and body size in birds and mammals. J Exp Biol 97:1–21

Taylor GK, Nudds RL, Thomas ALR (2003) Flying and swimming animals cruise at a Strouhal number tuned for high power efficiency. Nature 425:707–711

Taylor-Burt KR, Biewener AA (2020) Aquatic and terrestrial takeoffs require different hindlimb kinematics and muscle function in Mallard ducks. J Exp Biol 223: jeb223743

Thomas ALR (1997) On the tails of birds. Bioscience 47: 215–225

Thomas ALR, Taylor GK (2001) Animal flight dynamics I. Stability in gliding flight. J Theor Biol 212:399–424

Thomas DB, Fordyce RE (2012) Biological plasticity in penguin heat-retention structures. Anat Rec 295:249–256

Thulborn RA (2003) Wind-assisted flight of *Archaeopteryx*. Neues Jahrb Geol Paläontol Abh 229:61–74

Tian N, Xie A, Wang Y, Jiang Z, Li L, Yin Y, Zhu Z, Wang J (2015) New records of Jurassic petrified wood in Jianchang of western Liaoning, China and their palaeoclimate implications. Sci China Earth Sci 58: 2154–2164

Tobalske BW (1995) Neuromuscular control and kinematics of intermittent flight in European Starlings (*Sturnus vulgaris*). J Exp Biol 198:1259–1273

Tobalske BW (1996) Scaling of muscle composition, wing morphology, and intermittent flight behavior in woodpeckers. Auk 113:151–177

Tobalske BW (2001) Morphology, velocity, and intermittent flight in birds. Am Zool 41:177–187

Tobalske BW (2007) Biomechanics of bird flight. J Exp Biol 210:3135–3146

Tobalske BW, Altshuler DL, Powers DR (2004) Take-off mechanics in hummingbirds (Trochilidae). J Exp Biol 207:1345–1352

Tobalske BW, Dial KP (1994) Neuromuscular control and kinematics of intermittent flight in Budgerigars (*Melopsittacus undulatus*). J Exp Biol 187:1–18

Tobalske BW, Dial KP (1996) Flight kinematics of Black-billed Magpies and pigeons over a wide range of speeds. J Exp Biol 199:263–280

Tobalske BW, Hedrick TL, Biewener AA (2003) Wing kinematics of avian flight across speeds. J Avian Biol 34:177–184

Tobalske BW, Hedrick TL, Dial KP, Biewener AA (2003b) Comparative power curves in bird flight. Nature 421:363–366

Tobalske BW, Peacock WL, Dial KP (1999) Kinematics of flap-bounding flight in the Zebra Finch over a wide range of speeds. J Exp Biol 202:1725–1739

Tobalske BW, Warrick DR, Clark CJ, Powers DR, Hedrick TL, Hyder GA, Biewener AA (2007) Three-dimensional kinematics of hummingbird flight. J Exp Biol 210:2368–2382

Tokita M, Matsushita H, Asakura Y (2020) Developmental mechanisms underlying webbed foot morphological diversity in waterbirds. Sci Rep 10:8028

Tubaro PL, Lijtmaer DA, Palacios MG, Kopuchian C (2002) Adaptive modification of tail structure in relation to body mass and buckling in woodcreepers. Condor 104:281–296

Tucker VA (1987) Gliding birds: the effect of variable wing span. J Exp Biol 133:33–58

Tucker VA (1993) Gliding birds: reduction of induced drag by wing tip slots between the primary feathers. J Exp Biol 180:285–310

Tucker VA, Heine C (1990) Aerodynamics of gliding flight in a Harris' Hawk, *Parabuteo unicinctus*. J Exp Biol 149:469–489

Tyler J, Younger JL (2022) Diving into a dead-end: asymmetric evolution of diving drives diversity and disparity shifts in waterbirds. Proc R Soc B 289:20222056

Usherwood JR, Hedrick TL, Biewener AA (2003) The aerodynamics of avian take-off from direct pressure measurements in Canada Geese (*Branta canadensis*). J Exp Biol 206:4051–4056

van Bokhorst E, de Kat R, Elsinga GE, Lentink D (2015) Feather roughness reduces flow separation during low Reynolds number glides of swifts. J Exp Biol 218: 3179–3191

Videler JJ (2006) Avian flight. Oxford University Press, Oxford

Videler JJ, Weihs D, Daan S (1983) Intermittent gliding in the hunting flight of the kestrel, *Falco tinnunculus*. J Exp Biol 102:1–12

Viohl G (1985) Geology of the Solnhofen lithographic limestone and the habitat of *Archaeopteryx*. In: Hecht MK, Ostrom JH, Viohl G, Wellnhofer P (eds) The beginnings of birds. Proceedings of the International *Archaeopteryx* Conference Eichstatt 1984. Freunde des Jura-Museums Eichstatt, Willibadsburg, pp 31–44

Voeten DF, Cubo J, De Margerie E, Röper M, Beyrand V, Bureš S, Tafforeau P, Sanchez S (2018) Wing bone geometry reveals active flight in *Archaeopteryx*. Nat Commun 9:923

Vogel S (1998) Exposing life's limits with dimensionless numbers. Phys Today 51:22–27

Wagner H, Weger M, Klaas M, Schröder W (2017) Features of owl wings that promote silent flight. Interface Focus 7:20160078

Wang M, Stidham TA, Zhou Z (2018a) A new clade of basal Early Cretaceous pygostylian birds and developmental plasticity of the avian shoulder girdle. Proc Natl Acad Sci USA 115:10708–10713

Wang Y, Tobalske BW, Cheng B, Deng X (2018b) Gravitation-enabled forward acceleration during flap-bounding flight in birds. J Bionic Eng 15:505–515

Ward S, Rayner JMV, Möller U, Jackson DM, Nachtigall W, Speakman JR (1999) Heat transfer from starlings *Sturnus vulgaris* during flight. J Exp Biol 202:1589–1602

Warrick DR (1998) The turning and linear maneuvering performance of birds: the cost of efficiency for coursing insectivores. Can J Zool 76:1063–1079

Warrick DR, Bundle MW, Dial KP (2002) Bird maneuvering flight: blurred bodies, clear heads. Integr Comp Biol 42:141–148

Warrick DR, Dial KP, Biewener AA (1998) Asymmetrical force production in the maneuvering flight of pigeons. Auk 115:916–928

Warrick D, Hedrick T, Fernández MJ, Tobalske B, Biewener A (2012) Hummingbird flight. Curr Biol 22:R472–R477

Warrick DR, Tobalske BW, Powers DR (2005) Aerodynamics of the hovering hummingbird. Nature 435: 1094–1097

Watanabe YY, Sato K, Watanuki Y, Takahashi A, Mitani Y, Amano M, Aoki K, Narazaki T, Iwata T, Minamikawa S, Miyasaki N (2011) Scaling of swim speed in breath-hold divers. J Anim Ecol 80:57–68

Watanuki Y, Wanless S, Harris M, Lovvorn JR, Miyazaki M, Tanaka H, Sato K (2006) Swim speeds and stroke patterns in wing-propelled divers: a

comparison among alcids and a penguin. J Exp Biol 209:1217–1230
Watson RR, Rubenson J, Coder L, Hoyt DF, Propert MWG, Marsh RL (2011) Gait-specific energetics contributes to economical walking and running in Emus and Ostriches. Proc R Soc B 278:2040–2046
Weimerskirch H, Bishop C, Jeanniard-du-Dot T, Prudor A, Sachs G (2016) Frigate birds track atmospheric conditions over months-long transoceanic flights. Science 353:74–78
Weimerskirch H, Guionnet T, Martin J, Shaffer SA, Costa DP (2000) Fast and fuel efficient? Optimal use of wind by flying albatrosses. Proc R Soc B 267:1869–1874
Weimerskirch H, Martin J, Clerquin Y, Alexandre P, Jiraskova S (2001) Energy saving in flight formation. Nature 413:697–698
Welch KC, Altshuler DL, Suarez RK (2007) Oxygen consumption rates in hovering hummingbirds reflect substrate-dependent differences in P/O ratios: carbohydrate as a 'premium fuel'. J Exp Biol 210:2146–2153
Wells DJ (1993) Muscle performance in hovering hummingbirds. J Exp Biol 178:39–57
Welty JC, Baptista L (1988) The life of birds. Saunders College Publishing, New York, NY
Williams CD, Biewener AA (2015) Pigeons trade efficiency for stability in response to level of challenge during confined flight. Proc Natl Acad Sci USA 112:3392–3396
Williams HJ, Safi K (2021) Certainty and integration of options in animal movement. Trends Ecol Evol 36: 990–999
Williams HJ, Shepard ELC, Holton MD, Alarcón PAE, Wilson RP, Lambertucci SA (2020) Physical limits of flight performance in the heaviest soaring bird. Proc Natl Acad Sci USA 117:17884–17890
Willmott AP, Ellington CP, Thomas ALR (1997) Flow visualization and unsteady aerodynamics in the flight of the hawkmoth, *Manduca sexta*. Philos Trans R Soc B 352:303–316
Wilson RP (1995) Foraging ecology. In: Williams TD (ed) The penguins. Oxford University Press, Oxford, pp 81–106
Witton MP (2013) Pterosaurs: natural history, evolution, anatomy. Princeton University Press, Princeton, NJ
Woakes AJ, Butler PJ (1983) Swimming and diving in Tufted Ducks, *Aythya fuligula*, with particular reference to heart rate and gas exchange. J Exp Biol 107: 311–329
Wright NA, Steadman DW, Witt CC (2016) Predictable evolution toward flightlessness in volant island birds. Proc Natl Acad Sci USA 113:4765–4770
Wright W (1900) Letter to Octave Chanute, May 13, 1900. In: McFarland MW (ed) The papers of Wilbur and Orville Wright, vol. 1, 1899–1905. McGraw-Hill, New York, NY, pp 15–19
Xirouchakis SM, Andreou G (2009) Foraging behaviour and flight characteristics of Eurasian Griffons *Gyps fulvus* in the island of Crete, Greece. Wildl Biol 15: 37–52
Xu X, Currie P, Pittman M, Xing L, Meng Q, Lü J, Hu D, Yu C (2017) Mosaic evolution in an asymmetrically feathered troodontid dinosaur with transitional features. Nature Comm 8:14972
Xu X, Zhou Z, Wang X, Kuang X, Zhang F, Du X (2003) Four-winged dinosaurs from China. Nature 421:335–340
Xu X, Zhou Z, Wang Y, Wang M (2020) Study on the Jehol Biota: recent advances and future prospects. Sci China Earth Sci 63:757–773
Yang S, He H, Jin F, Zhang F, Wu Z, Li Q, Wang M, O'Connor JK, Deng C, Zhu R (2020) The appearance and duration of the Jehol Biota: constraint from SIMS U-Pb zircon dating for the Huajiying Formation in northern China. Proc Natl Acad Sci 117:14299–14305
Yoda K, Sato K, Niizuma Y, Kurita M, Bost C-A, Le Maho Y, Naito Y (1999) Precise monitoring of porpoising behaviour of Adelie Penguins determined using acceleration data loggers. J Exp Biol 202:3121–3126
Yuan Z-M, Chen M, Jia L, Ji C, Incecik A (2021) Wave-riding and wave-passing by ducklings in formation swimming. J Fluid Mech 928:R2
Zeffer A, Johansson LC, Marmebro Å (2003) Functional correlation between habitat use and leg morphology in birds (Aves). Biol J Linn Soc 79:461–484
Zelenitsky DK, Therrien F, Erickson GM, DeBuhr CL, Kobayashi Y, Eberth DA, Hadfield F (2012) Feathered non-avian dinosaurs from North America provide insight into wing origins. Science 338:510–514
Zheng X, Wang X, O'Connor J, Zhou Z (2012) Insight into the early evolution of the avian sternum from juvenile enantiornithines. Nat Commun 3:1116
Zhonghe Z, Yuan W (2010) Major features of the vertebrate diversity of the Early Cretaceous Jehol Biota and their paleoecological implications. J Earth Sci 21 (Special Issue):228–230
Zhou ZH (2004) Vertebrate radiations of the Jehol Biota and their environmental background. Chin Sci Bull 49: 754–756

# Sound and Vocal Production and Function

# 12

## Contents





G. Ritchison, *In a Class of Their Own*, Fascinating Life Sciences,
https://doi.org/10.1007/978-3-031-14852-1_12

**Abstract**

Acoustic communication plays a critical role in the lives of most species of birds. The focus of this chapter is on the ways that birds produce both nonvocal and vocal sounds. The anatomy of the avian syrinx and how it varies among different taxa of birds is discussed. The mechanism by which the syrinx produces sounds is explained and the types and functions of those sounds (calls and songs) are discussed in detail. Different species of birds vary dramatically in the size of their vocal repertoires, including call and song types, and possible explanations for such variation are provided. Many aspects of the singing behavior of birds are discussed, including the mechanisms by which the central nervous system controls singing behavior and differences between the sexes in those mechanisms, variation within and among species in song structure, the various functions of bird song, intra- and interspecific variation in the size of song repertoires, and the process of song learning. Also discussed are vocal mimicry, duetting, group choruses, and the roles of vocalizations in male cooperative courtship.

## 12.1 Introduction

For most species of birds, acoustic signaling plays an important role in species and individual recognition and interactions with conspecifics. Acoustic communication involves a sender, receiver(s), and the environment through which the sound travels (Fig. 12.1). Particularly for long-distance acoustic communication, e.g., attracting mates and defending territories, several environmental factors can influence the quality and, therefore, the effectiveness (in terms of accurately transmitting information) of acoustic signals that reach receivers. Several factors can contribute to the

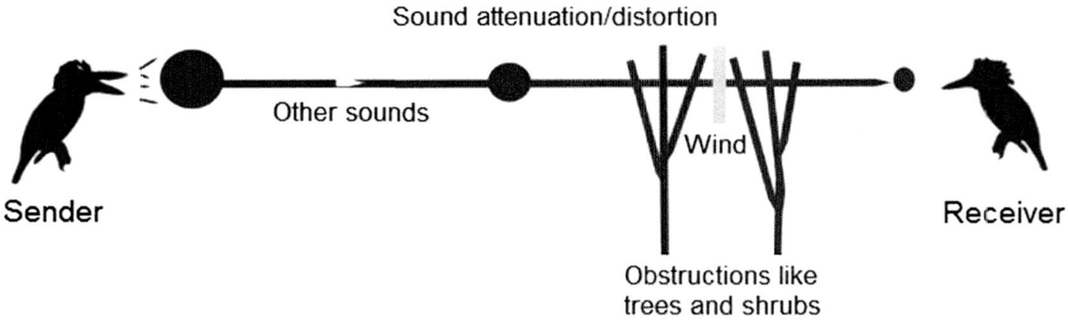


**Fig. 12.1** When sender or signaler vocalizes, the sound is transmitted toward one or more receivers. However, other overlapping sounds and attenuation (decreasing amplitude with increasing distance), as well as wind and obstructions that can distort sounds can make successful transmission of the sound more difficult. (Figure modified from Kaplan 2014; © 2014 John Wiley & Sons, Ltd., used with permission)

**Box 12.1 Terminology**

Acoustic communication involves the production of sounds. Many species of birds generate sounds by using substrates (woodpeckers), their wings or tails, or their bills (bill-clapping or striking their bills against hard substrates) (such nonvocal sounds are described in more detail below). Of course, birds also generate sounds via their vocal organ, the syrinx. These sounds are categorized as calls or songs, and can be visualized as sonograms (also called spectrograms or spectrographs). Sonograms are three-dimensional, showing the duration, frequency (or pitch), and amplitude of calls and songs (with amplitude indicated via variation in comparative darkness [if shown in black and white] or different colors). Three examples of sonograms of songs of three different species of birds are provided below. These songs are made up of subunits called **notes** or elements and are defined as continuous sounds (no silent intervals) separated in time from other such notes or elements. In some cases, two or more notes always occur together and these are called **syllables**. A series of very short-duration notes with very short internote intervals is called a **trill**. Consecutive sequences of two or more of the same syllables are called **phrases** (or motifs). A particular sequence of particular notes, syllables, phrases, and/or trills is called a **song type** (Brenowitz et al. 1997).

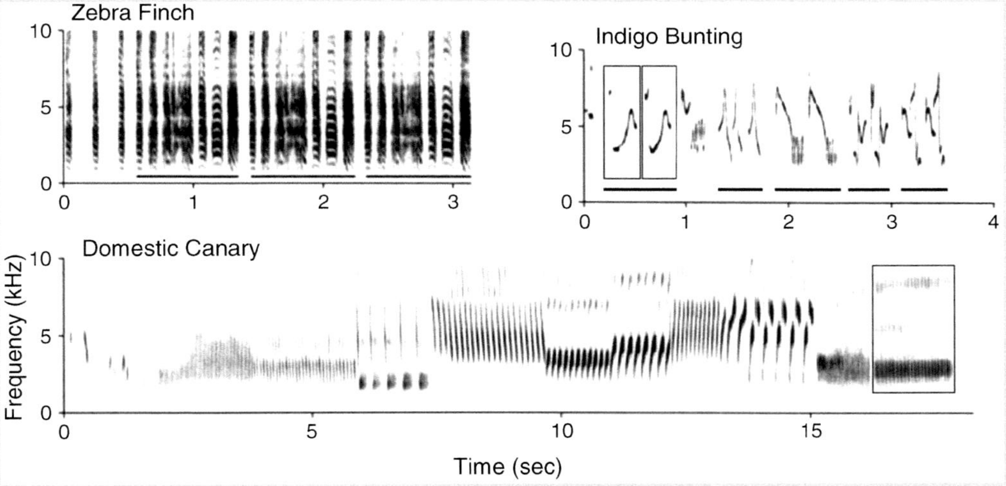


Examples of the songs of three species of birds. Each song begins with a few **notes**. In the song of the Indigo Bunting (*Passerina cyanea*), two short notes are followed by two **syllables** (consisting of two notes) that together make up a **phrase**. The songs of both the Zebra Finch (*Taeniopygia guttata*) and Indigo Bunting include multiple phrases (underlined). The song of the Domestic Canary (*Serinus canaria domestica*) includes several **trills**, with the intervals between notes so brief that the gaps between them are difficult to see. (Figure modified from Brenowitz et al. 1997; © 1997 John Wiley & Sons, Inc., used with permission)

degradation of sounds (Marten and Marler 1977; Wiley and Richards 1978, 1982; Brown and Handford 2000; Box 12.1 Terminology). Air turbulence, ground reflectance, and scatter through vegetation affect sound transmission in three ways: (1) attenuation, (2) irregular fluctuation in amplitude, and (3) reverberation. Attenuation is frequency-dependent, with lower-frequency sounds attenuating less than higher-frequency sounds. This is especially true in cluttered habitats like forests because lower-frequency sounds have longer wavelengths so

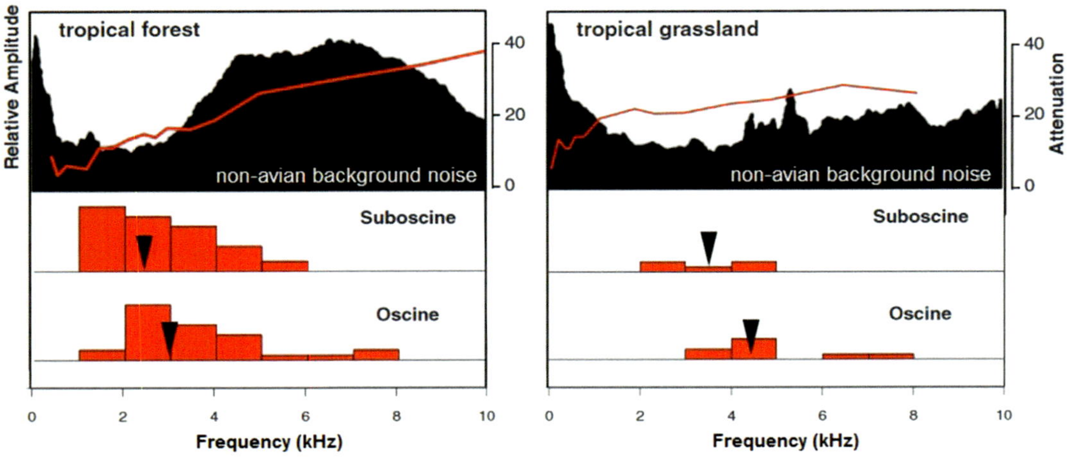


**Fig. 12.2** Frequency, or pitch, of songs (with means indicated by the black triangles) of species of New World oscine and suboscine birds at tropical (0° to 23°) latitudes. The black areas represent the relative amplitude of the non-avian background noise at different frequencies in tropical wet forests and grasslands (in Panama, data from Ryan and Brenowitz 1985). Red lines represent attenuation (dB/100 m) in tropical forests and grasslands. The frequency of songs of songbirds in tropical forest is generally lower than that of the songs of songbirds in tropical grasslands; the dense vegetation of tropical forests, as well as the greater amplitude of high-frequency background noise (primarily by insects), has favored lower-frequency songs. (Figure from Weir et al. (2012); © 2012 The Authors. Evolution © 2012 The Society for the Study of Evolution, used with permission)

they better diffract (bend or change direction) around obstacles than higher-frequency sounds (Wiley and Richards 1978). Irregular amplitude fluctuations are caused by air turbulence, and reverberation is caused by the reflection of sounds by structures in the environment, including the ground and vegetation. Reflected sounds must travel further than a direct signal so are perceived later by the receiver. Reverberations affect the temporal structure of sounds, and have a greater effect of sounds repeated at short intervals like trills, e.g., masking the silent intervals between successive sounds.

Given the ways that characteristics of the environment can affect sound transmission, Morton (1975) proposed that differences in vegetation in different habitats should impose different selective pressures on the acoustic signals of birds that would result in optimal transmission from senders to receivers. Concerning avian songs, this "Acoustic Adaptation Hypothesis" (AAH) predicts that songs with lower frequencies, reduced frequency range, less frequency modulation, longer notes, and longer internote intervals should be favored in habitats with dense vegetation, and songs with higher frequencies, greater frequency ranges, more frequency modulation, shorter notes, and shorter internote intervals should be favored in open habitats (Fig. 12.2). The AAH further predicts that other factors that can influence song characteristics, such as morphological constraints (e.g., bird and syrinx size) and sexual selection, are similar in different habitats. In addition to differences in vegetation, other investigators have found that differences among habitats in ambient noise, e.g., by insects or anthropogenic noise (Box 12.2 Anthropogenic Noise and Bird Vocalizations), can also influence the characteristics of songs (Slabbekoorn and Smith 2002; Weir et al. 2012).

The results of multiple studies have provided support for the AAH. For example, as predicted by the AAH, the songs of male Rufous-collared Sparrows (*Zonotrichia capensis*) found in open habitats include trills, whereas those in forest habitats where reverberation can mask short-internote intervals do not (Fig. 12.3). In southeastern Peru, Tobias et al. (2010) broadcast artificial sounds in bamboo forests and *terra firme*

**Box 12.2 Anthropogenic Noise and Bird Vocalizations**

Urban areas can be filled with sounds regarded as unwanted noise by most people. These sounds are generated by vehicles (e.g., motors and road noise) and stationary machines (e.g., air conditioners) and, as the volume of these sounds increases, the distance at which birds can hear the vocalizations of conspecifics decreases.

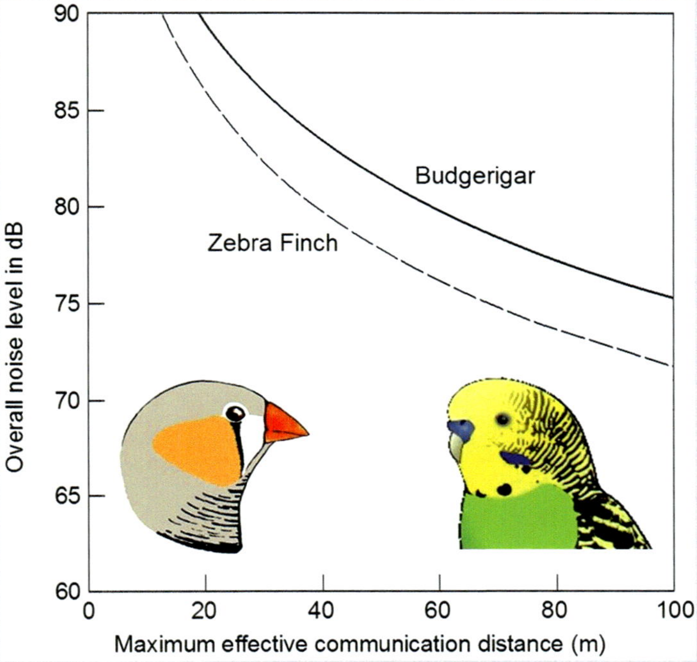


Maximum effective communication distances of the calls of Budgerigars (*Melopsittacus undulatus*) and Zebra Finches (*Taeniopygia guttata*) decrease with increasing noise levels. (Figure modified from Halfwerk et al. 2018; © 2018 Springer Science Business Media, LLC, part of Springer Nature, used with permission. Budgerigar drawing from Wikipedia, CC BY 3.0, https://creativecommons.org/licenses/by/3.0/)

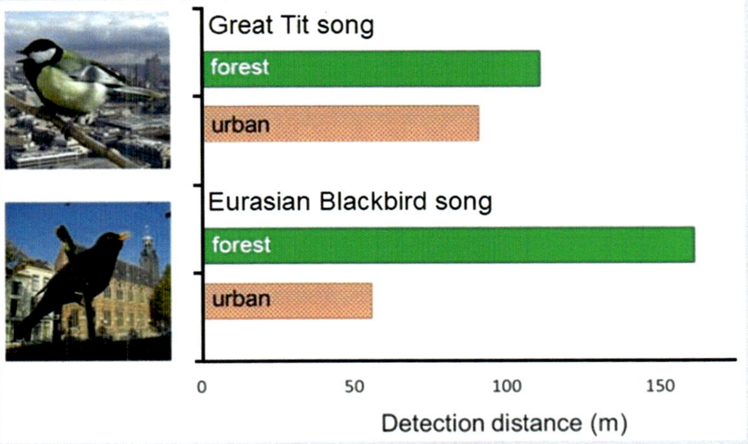


Detection distances for the songs of Great Tits (*Parus major*) and Eurasian Blackbirds (*Turdus merula*) are shorter in urban areas than in forest habitats. However, the songs of Great Tits are at a higher frequency than those of

(continued)

**Box 12.2** (continued)

Eurasian Blackbirds so their detection distances are greater than those of Eurasian Blackbirds in urban areas. (Figure from Halfwerk et al. (2018); © 2018 Springer Science Business Media, LLC, part of Springer Nature, used with permission)

Most of this noise is low frequency (less than 3 kHz). For birds with lower-frequency songs (<3 kHz), the overlapping anthropogenic noise may "mask" their songs, interfering with their ability to communicate with conspecifics. In urban areas, therefore, effective communication is possible only if birds with lower-frequency songs modify their songs and/or their behavior. Several species of songbirds have been found to respond to anthropogenic noise by shifting the frequency of the songs above that of noise. Roca et al. (2016) found that smaller species with lower-frequency songs were more likely to shift song frequencies than larger species. Larger birds generally do sing at lower frequencies (Ryan and Brenowitz 1985) and may be limited in their ability to produce higher-frequency songs (e.g., due to the structure of the vibratile membranes of the syrinx). However, larger birds may overcome acoustic masking in other ways. For example, larger birds may sing louder rather than shifting frequencies (Zollinger and Brumm 2015) or they may sing during less noisy periods (e.g., at night or during periods with less noise during the day; Fuller et al. 2007).

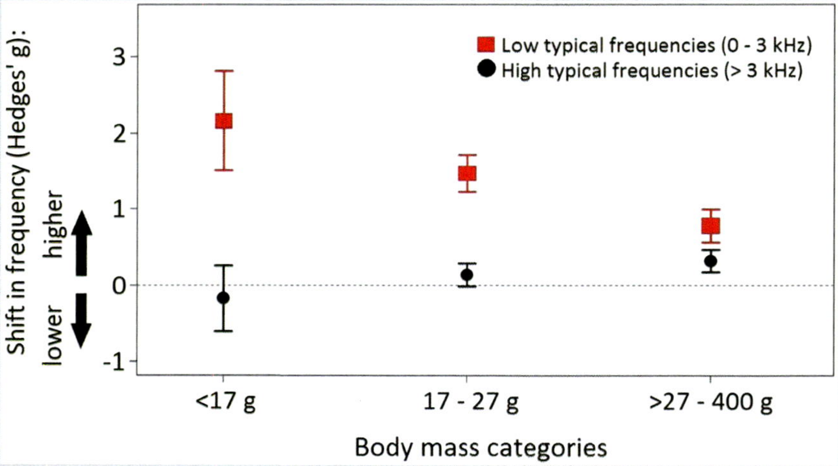


Relationship between mean (± SE) bird body mass ($N = 55$ species) and two typical frequency ranges (low and high). Smaller birds with songs that are normally below 3 hHz in frequency (the same frequency range as noise) generally shift their songs to higher frequencies in areas with anthropogenic noise. Larger birds do not. (Figure modified from Roca et al. 2016; © 2016 Oxford University Press, used with permission)

forests (tropical forests that do not flood) to quantify possible difference in sound transmission in the two habitats. They found that (1) higher-frequency sounds (4–5 kHz) were more attenuated in bamboo than in *terra firme* forest, (2) higher-frequency sounds were also more degraded by reverberation in bamboo forests than *terra firme* forests, and (3) signals produced at shorter intervals were more distorted in *terra firme* forests than bamboo forests. As predicted

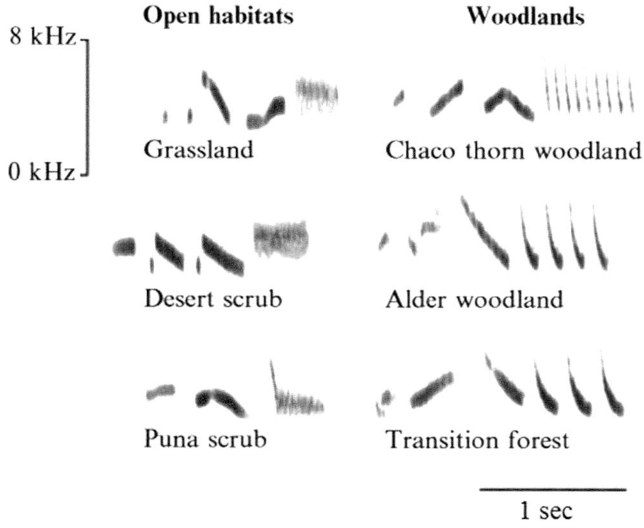


**Fig. 12.3** Variation in the characteristics of song of male Rufous-collared Sparrows (*Zonotrichia capensis*) in different habitats in northwestern Argentina. As predicted by the Acoustic Adaptation Hypothesis, these sparrows have shorter internote intervals (trills) in open habitats and longer internote intervals in forest habitats. (Figure from Brumm and Naguib (2009); © 2009 Elsevier Inc., used with permission)

by the AAH, the songs of songbirds in bamboo forests were lower in frequency than those in *terra firme* forests, and the songs of birds in *terra firme* forest were generally slower-paced (longer internote intervals) than those in bamboo forests (Fig. 12.4).

Although supported by a number of studies, other investigators have reported results that either do not support the AAH or provide only partial support (i.e., not all characteristics of songs can be explained by AAH). One possible explanation for this is that the frequency of songs is subject to physical constraints such as body mass or the size of vocal tracts (Ryan and Brenowitz 1985) so, for example, smaller species of birds may not be able to produce low- frequency songs. In addition, potential "eavesdropping" (i.e., obtaining information, including information about a signaler's location, from signaling interactions between others; McGregor and Dabelsteen 1996) by predators and, for some species, brood parasites may be a selective factor in the evolution of song characteristics. For example, the direction toward a sound source is determined by receivers by differences between their ears in the time of arrival, intensity, and phase (i.e., sound consists of pressure waves and, for longer-waved, low-frequency sounds, peak pressures may reach the two ears at slightly different times) of sound waves (Marler 1955) and, as a result, nonmodulated, high-frequency songs with narrow frequency ranges may be more difficult to locate than other sounds (Catchpole and Slater 2008). Therefore, for some species of birds, in contrast with the AAH predictions, the risk of eavesdropping by predators and/or brood parasites may select for higher-frequency songs in densely vegetated habitats or for nonmodulated songs in open habitats.

## 12.2 Sound Production: Nonvocal

Although most avian acoustic communication involves vocalizations, many species of birds generate sounds by using substrates (woodpeckers), wings or tails, special wings (manakins; Fig. 12.5), or their bills (bill-clapping; Fig. 12.6). Woodpeckers "tap" (strike their bills against hard substrates) when foraging, but also "drum," using resonant surfaces to increase volume and sound transmission distance, when seeking mates and advertising territories (Box 12.3 Woodpecker Drumming). Among birds, most wing- and tail-generated sounds, referred to as sonations, are thought to result from three mechanisms: (1) flutter, where airflow causes

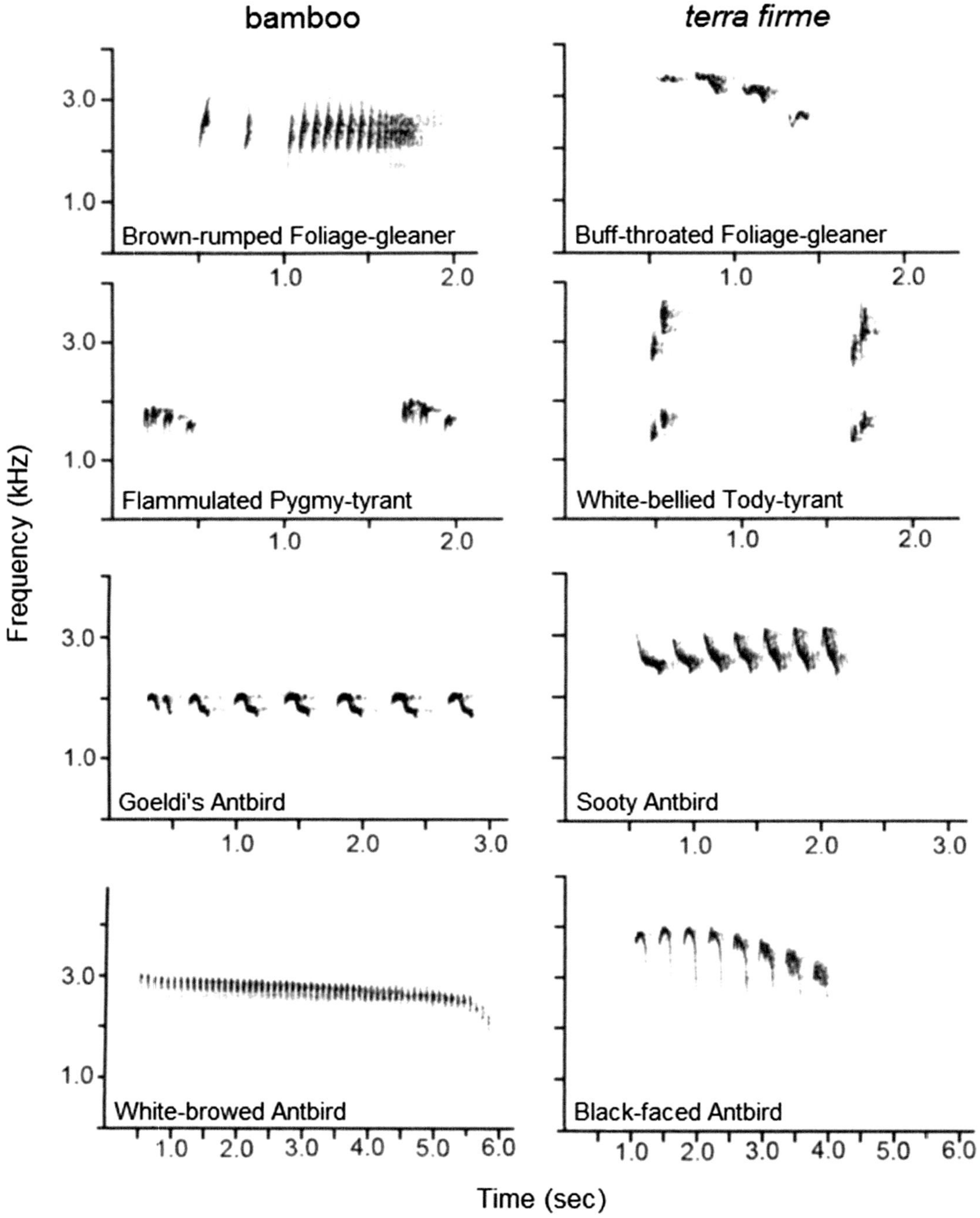


**Fig. 12.4** Sonograms of the songs of four species pairs, with songs of bamboo-specialist species on the left, and those of their closest relatives in *terra firme* forests (forests that are not inundated by flooded rivers) on the right. Brown-rumped Foliage-gleaner, *Automolus malanopezus*; Buff-throated Foliage-gleaner, *A. ochrolaemus*; Flammulated Pygmy-tyrant, *Hemitriccus flammulatus*; White-bellied Tody-tyrant, *H. griseipectus*; Goeldi's Antbird, *Myrmeciza goeldi*; Sooty Antbird, *M. fortis*; White-browed Antbird, *Myrmoborus leucophrys*; Black-faced Antbird, *M. myotherinus*. (Figure modified from Tobias et al. 2010; © 2010 The Authors. Journal compilation © 2010 The Society for the Study of Evolution, used with permission)

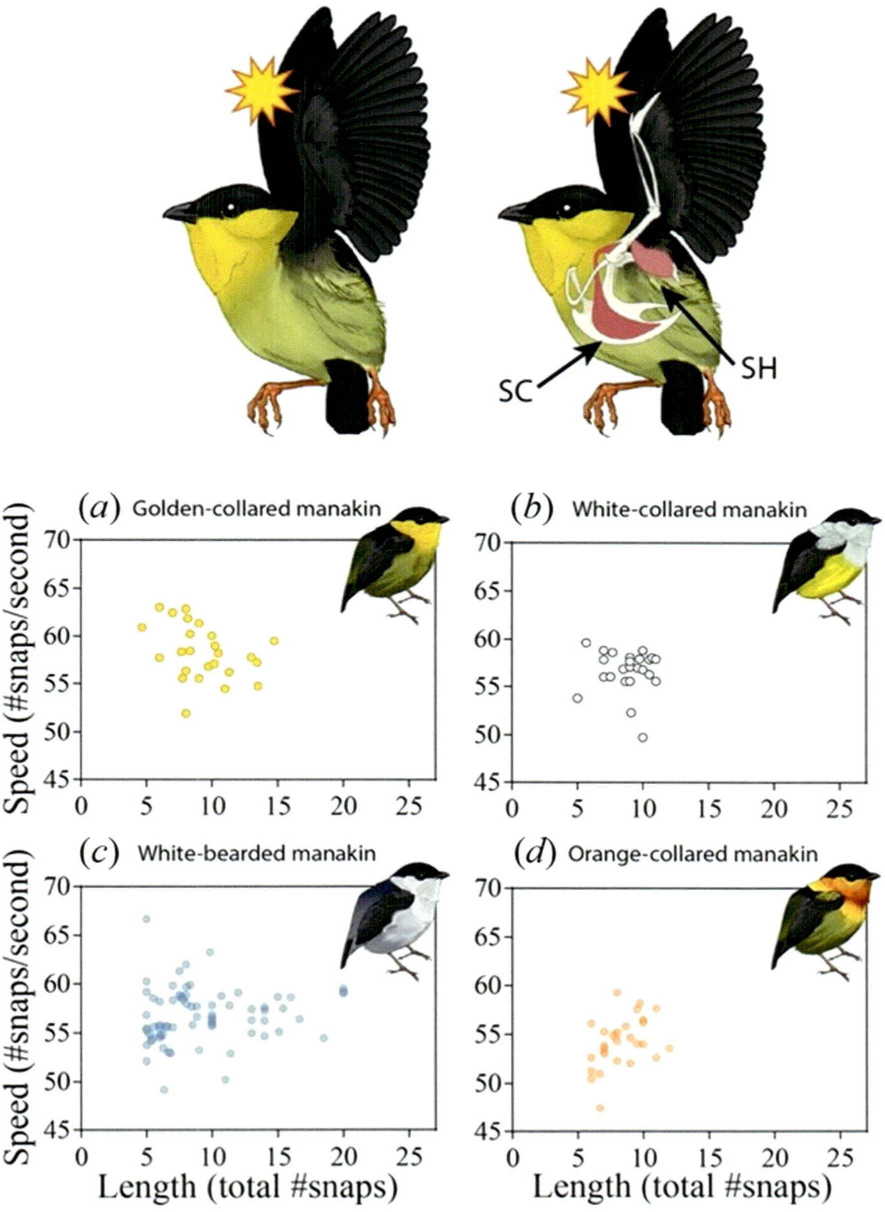


**Fig. 12.5** Male manakins in the genus *Manacus* have courtship displays called roll-snaps where they generate a trill-like sound by rapidly hitting their wings together over their back, and females tend to prefer males able to do at the fastest rate. (Top left) Male Golden-collared Manakin (*Manacus vitellinus*) producing a roll-snap. (Top right) The two main wing muscles involved in the roll-snap display are the supracoracoideus (SC) and scapulohumeralis caudalis (SH). The SC muscle lifts the wings and the SH snaps the wings together by retracting the elevated humeri. Top right, a male Orange-collared Manakin (*Manacus aurantiacus*). Below, mean snap rates of individuals of each species (indicated by colored or open circles) vary. For (**a**) Golden-collared Manakins and (**b**) White-collared Manakins (*M. candei*), snap rates tend to decline as the total number of snaps in a display increase, whereas the total number of snaps in a display have no significant effect on snap rates of (**c**) White-bearded Manakins (*M. manacus*) and (**d**) Orange-collared Manakins (*M. aurantiacus*). (Top left and right from Fuxjager and Schuppe 2018, © 2018 Elsevier Ltd., used with permission. Figures **a**–**d** from Miles et al. (2018); © Miles et al., open-access article distributed under the terms of the Creative Commons Attribution License, https://creativecommons.org/licenses/by/4.0/)

**Fig. 12.6** Nestling Great Horned Owl (*Bubo virginianus*). Owl's bill-snap, forcefully closing their bills to generate sound, when disturbed or stressed as a warning or aggressive signal (Artuso et al. 2020). For many other species of birds, bill-snapping conveys aggression toward conspecifics, predators, and brood parasites (e.g., Prendergast 1990; Hancock et al. 1992; Sproat and Ritchison 1994; Ficken and Popp 1996; Welbergen and Davies 2008). (Photo by Chris Reiss, National Park Service, CC0 Public Domain)

part of a feather to oscillate at audible frequencies, (2) percussion (snaps and claps caused by collisions between wings or between wings and another body part), and (3) wing whirring, sounds produced by rapidly flapping wings (Clark and Prum 2015). Other less common mechanisms of sound production include "whooshes" (produced by turbulent airflow generated by wingbeats), stridulation, drumming with wings, and "rustling" (Clark and Prum 2015). "Whoosh" sounds are produced during courtship displays by males in some species of manakins, with the sound apparently generated during rapid flight by air moving past modified primaries (Bostwick and Prum 2003). Several species of birds, including some manakins (Bostwick and Prum 2003), produce "wing-snapping" sounds (e.g., Hilty and Brown 1986; Bostwick and Prum 2003; Schulenberg et al. 2010; Eddowes and Lea 2021), but how those sounds are generated in many of those species remains unclear. One hypothesis is that such sounds are analogous to human hand-clapping, with sound produced when air is rapidly expelled from between primary feathers (Rico-Guevara et al. 2022).

The only species of bird known to generate sound by stridulation is the Club-winged Manakin (*Machaeropterus deliciosus*), with males generating sound by the resonant vibration of modified wing feathers. As described by Bostwick et al. (2010:1–2), a male flips "the dorsal surface of his wings craniodorsally, so the secondaries form an upright wall in a plane roughly perpendicular to the male's back. As the wings are being positioned, the male adducts the medial edges of the wings together such that the innermost pair of the modified feathers (seventh secondaries) collides across the back. Immediately following this collision, the wings are shivered laterally and medially, pulling these feathers just millimeters apart, only to be adducted approximately 9 ms later to produce another collision. The pair of feathers knock and rebound as the wing shivers, and the sonation tone is produced continuously throughout this process."

To attract mates and defend territories, male Ruffed Grouse (*Bonasa umbellus*) produce a very low-frequency sound known as "drumming" by beating their wings 40–50 times in rapid succession over 8–11 s while standing on a log (Garcia et al. 2012; Krilow and Iwaniuk 2015; Fig. 12.7). Other species of birds, including some species of nightjars [Caprimulgiformes; e.g., wing-boom display of Common Nighthawks (*Chordeiles minor*; Miller 1925; Knight et al. 2022)], generate similar sounds (Mengel et al. 1972) using

**Box 12.3 Woodpecker Drumming**

Woodpeckers (family Picidae) used a variety of calls to communicate with conspecifics, but they also communicate via physical displays called drums, a repetitive striking of their bills again a resonate surface. The results of a phylogenetic reconstruction suggest that drumming is an ancestral behavior in the family Picidae, extending back to an estimated 22.5 million years ago (Shakya et al. 2017). Drumming has been thought to be an innate behavior, but Schuppe et al. (2022a) examined gene expression in the brains of woodpeckers and, because their forebrain regions were anatomically similar to the pallial song nuclei of vocal learning birds, concluded that "... all aspects of drumming are learned early in life, much like in song learning birds" or, alternatively, "individuals learn specific features of the drum in a way that modifies an existing innate motor template for the behavior. Species-specific rhythm, for example, might be acquired through learning." Other investigators have also suggested that characteristics of drums are largely species-specific in local communities (e.g., Florentin et al. 2016; Garcia et al. 2020). However, other investigators have suggested that the characteristics of drums, such as duration, number of beats in a drum, and drum rate (beats per second), are sometimes too similar to permit species recognition (Stark et al. 1998). Woodpeckers generally drum early in their breeding seasons, suggesting that they are important in establishing and defending territories (e.g., Schuppe et al. 2016). Drumming may also be important in female mate choice (Miles et al. 2018).

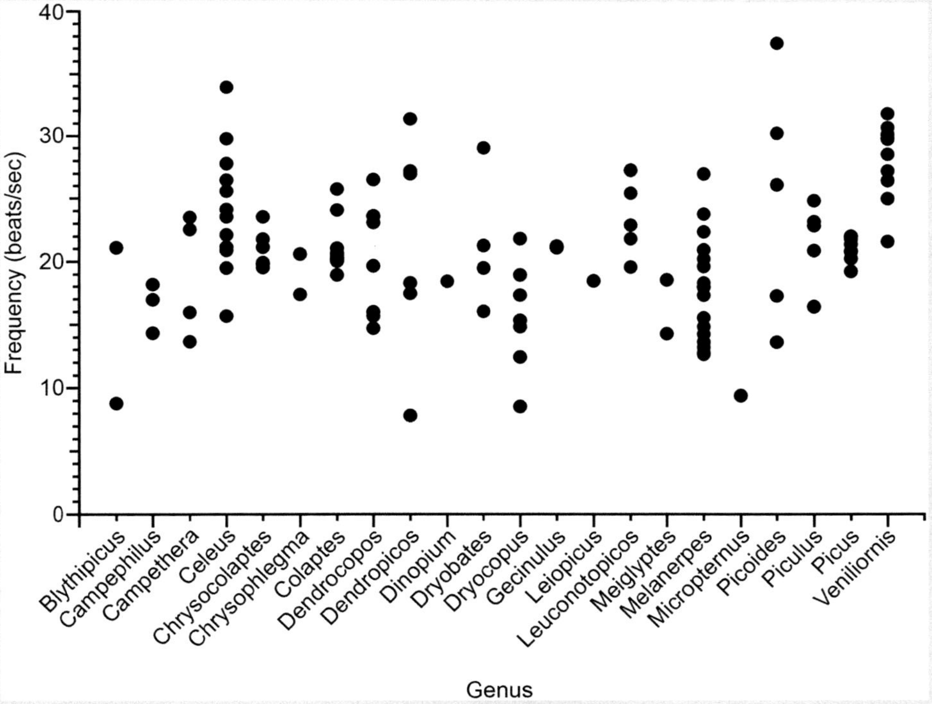


Variation in drumming frequency (beats/second) among and within 22 genera of woodpeckers. Each point is a mean frequency for a single species (Data from Miles et al. 2018. Figure from Schuppe et al. (2021); © 2021 Schuppe, Rutter, Roberts and Fuxjager, open-access article distributed under the terms of the Creative Commons Attribution License (CC BY), https://creativecommons.org/licenses/by/4.0/).

(continued)

**Box 12.3** (continued)

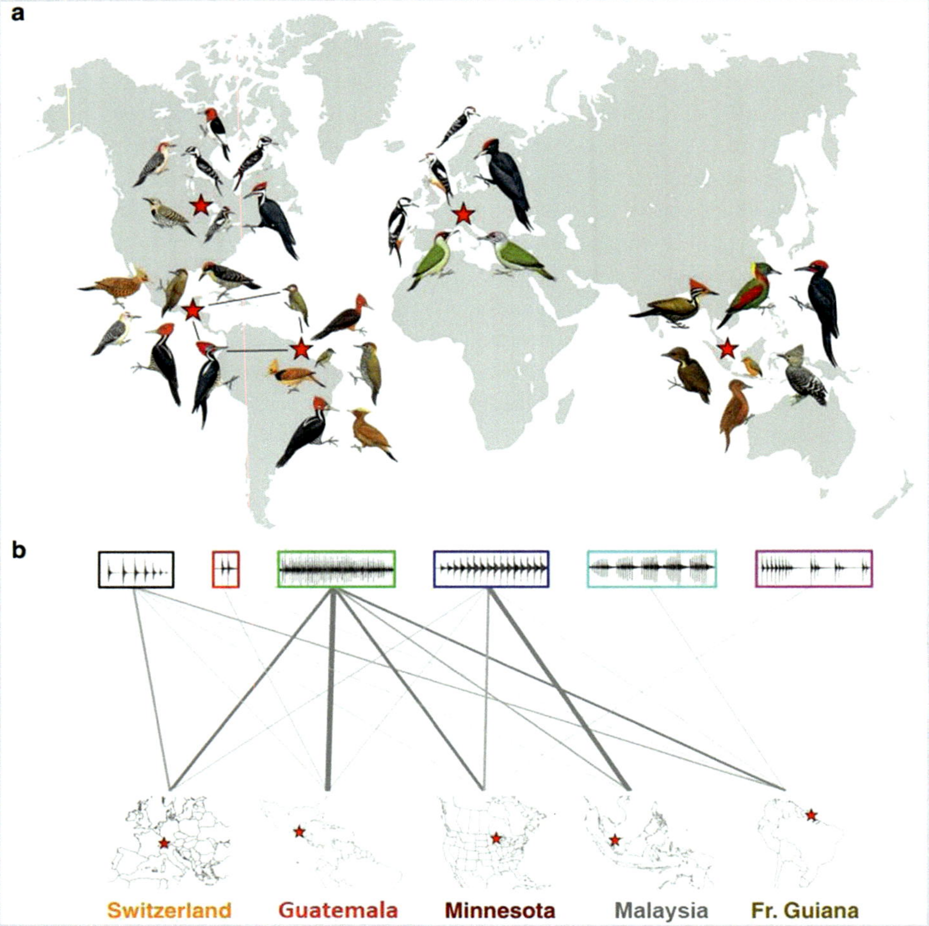


(**a**) Species composition of woodpecker communities in five different areas, including a Palearctic deciduous forest in Switzerland, a Neotropical forest in Guatemala, a Nearctic deciduous forest in Minnesota (USA), an Indomalayan rainforest in Malaysis, and a Neotropical rainforest in French Guiana. (**b**) Based on analysis of the drums of 92 species of woodpeckers, Garcia et al. (2020) identified six basic types of drums, including (left to right) accelerated drums, double knocks, steady fast drums, steady slow drums, regular sequence drums, and irregular sequence durms. Sympatric species of woodpeckers in different woodpecker communities use different drum types, which typically allows species recognition. Width of lines connecting locations and drum types is proportional to the number of species using each drum type. (Figure from Garcia et al. (2020); open-access article licensed under a Creative Commons Attribution 4.0 International License, https://creativecommons.org/licenses/by/4.0/).

Given the importance of drumming in intra- (territory defense) and intersexual (mate choice) interactions, selection may favor its "elaboration" in some species of woodpeckers, e.g., higher drum rates (Moody et al. 2022). For example, playback experiments revealed that higher drum rates elicited stronger agonistic responses from resident Downy Woodpeckers (*Dryobates pubescens*; Schuppe and Fuxjager 2018). In further support of the importance of drum rate in Downy Woodpeckers and perhaps other species of woodpeckers, Schuppe et al. (2018) found

(continued)

**Box 12.3** (continued)

that the muscle most important in drumming (longus colli ventralis) in Downy and Red-bellied (*Melanerpes carolinus*) woodpeckers exhibited greater expression of genes that influence $Ca^{++}$ handling than a muscle not involved in drumming (scapulohumeralis). This suggests the selection has favored the ability of Downy Woodpeckers to drum at higher rates because the efficiency of $Ca^{++}$ handling in muscle cells (diffusion out of the sarcoplasmic reticulum when a muscle contracts and active transport back into the sarcoplasmic reticulum when a muscle relaxes) influences the rate at which they can contract. In addition, Schuppe et al. (2022a, b) determined that the longus colli ventralis muscle of Downy Woodpeckers produces its own testosterone and, as a result, maintains higher levels of testosterone than muscles not involved in drumming. Given this ability to produce testosterone and assuming that androgens are needed in this muscle to allow drumming (Schuppe and Fuxjager 2019), Downy Woodpeckers are able to drum during the nonbreeding season when circulating levels of testosterone are very low (Ellison 1992; Soma et al. 2008; Jackson and Ouellet 2020). How testosterone affects the longus colli ventralis muscle is unclear, but testosterone may mediate muscular twitch speeds (Fuxjager et al. 2013, 2017; Tobiansky et al. 2020) and maintain muscular hypertrophy (Luine et al. 1980).

Male Downy Woodpecker. (Photo by Ken Thomas, Wikipedia, CC0 Public Domain)

What about drum rates of other species of woodpeckers? Miles et al. (2018) examined and compared drum rates of 164 species and found that drum rates tended to decline with increasing body size. Selection may favor higher drum rates, but body size places a limit on those rates. Larger species must rotate their heads and bills greater distances when drumming and are,

(continued)

**Box 12.3** (continued)

therefore, unable to drum as fast as smaller species. Interestingly, however, Miles et al. (2018) also found that body size had no effect on drum length (duration). These results suggest that when morphology, in this case body size, limits the exaggeration of one component of a display (drum rate), sexual selection may instead exaggerate a trait that is not limited by morphology (drum duration) (Moody et al. 2022).

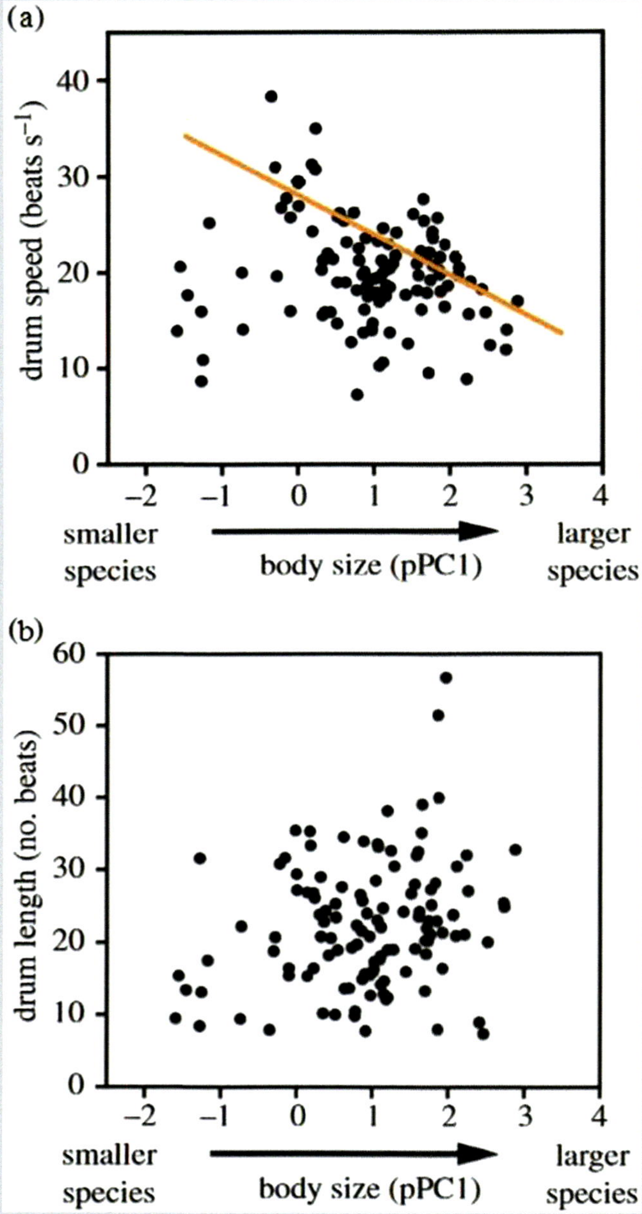


Relationship between body size and (**a**) drum speed (beats/second) and (**b**) drum length (total number of beats during drums). The relationship between body size and drum speed is statistically significant, with smaller species

(continued)

**Box 12.3** (continued)

having more beats per second than larger species. However, there is no significant relationship between woodpecker body size and the length of drums. (Figure modified from Miles et al. (2018); open-access article distributed under the terms of the Creative Commons CC BY 4.0 license, https://creativecommons.org/licenses/by/4.0/)

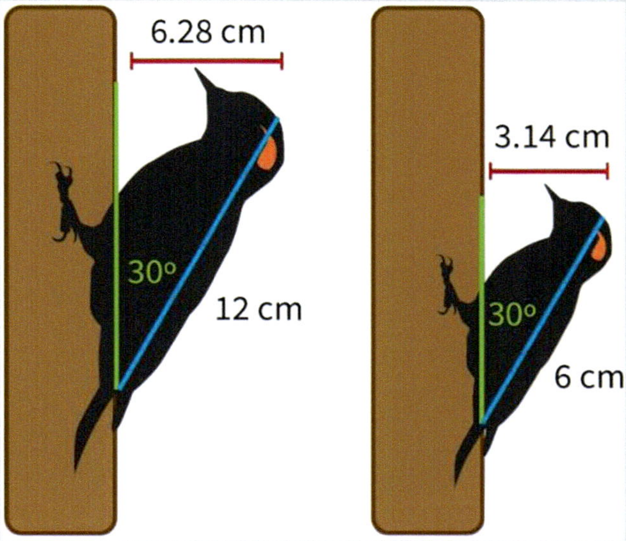


Woodpeckers drum by rotating their body so, assuming similar body angles, larger-bodied woodpeckers have to rotate greater distances to strike a resonant surface than smaller-bodied woodpeckers. This translates into slower drum rates for larger species of woodpeckers. (Figure from supplementary material from Miles et al. (2018); open-access article distributed under the terms of the Creative Commons CC BY 4.0 license, https://creativecommons.org/licenses/by/4.0/)

different mechanisms, including wing whirring [a series of sound pulses generated during wing downstrokes, possibly caused by fluttering feathers (Jordan and Areta 2020)] and aeroelastic flutter (Clark and Prum 2015). Many other species of birds also produce sounds by aeroelastic flutter, where feathers or portions of feathers vibrate at different frequencies at air speeds above a critical velocity (Clark and Prum 2015). In most species where aeroelastic flutter has been reported, the sounds and corresponding displays play a role in courtship (Clark and Prum 2015). For example, during the courtship display of male Anna's Hummingbirds, they fly at high speed toward perched females and very briefly spread their rectrices, with the outer rectrices vibrating and generating a chirp sound (Fig. 12.8). Among species in the genus *Columbina* (order Columbiformes), a modified primary feather (P7) has a section with elongated barbs that cause it to flutter and collide with the adjacent feather (P6) to produce a buzzing sound that may be used in sexual signaling or to indicate the presence of a predator (Niese et al. 2020; Amorim and Dias 2021). Crested Pigeons (*Ocyphaps lophotes*) also have modified flight feathers that produce louder whistles when alarmed birds take-off with faster-than-usual wing beats. Experiments revealed that conspecifics took flight only when these "alarmed" whistles were played back over speakers, suggesting that the whistles

**Fig. 12.7** A male Ruffed Grouse performing a drumming display. Rapid wing movements generate the drumming sounds. As described by Rusch et al. (2020), "While firmly perched on a log, stump, boulder, or other slightly raised platform, with tail braced, male spreads his wings and rotates wings forward, then quickly backward. This creates a sudden compression and release of air pressure, which produces the drum-like sound. Sound is actually the rushing of air into a momentary vacuum, creating a miniature sonic boom. Grouse begins with 2 or 3 slow beats, then gradually increases speed of wing beats, creating a drum-like roll, until drumming blends continuously together. In the 8–11 s it takes to complete a sequence, wings will have beaten up to 50 times." (Photo by Neal Herbert, Yellowstone National Park, CC0 Public Domain)

serve as a predator-detection signal (Hingee and Magrath 2009).

## 12.3 Sound Production: Vocal

In most vertebrates, the larynx produces sound via the vibration of vocal folds or cords. Birds, however, have a unique vocal organ called the syrinx (named after the chaste Greek nymph, Syrinx, who, after asking river nymphs for help when being pursued by the Greek God Pan, was transformed into water reeds that Pan then used to make a set of pan flutes). The syrinx consists of a cartilaginous framework, oscillatory soft tissues, and muscles for regulating airflow for respiration and for vocal control (Fig. 12.9). Although syringeal elements are typically referred to as cartilages, the syringes of some species include ossified or partially ossified elements as well (King 1989). The fundamental frequency of sound and fast temporal dynamics of songs are largely controlled by the syrinx (Suthers and Zollinger 2004), but characteristics of those sounds can be affected in various ways by upper vocal tract filtering (Ohms et al. 2012). The fine-temporal structure of sounds is under the direct motor control of syringeal muscles, with the (sub)millisecond temporal precision found in some species of birds controlled by superfast syringeal muscles (Elemans et al. 2008).

Birds have three major types of syringes. Most species of songbirds (and hummingbirds; Fig. 12.10) have a tracheobronchial syrinx

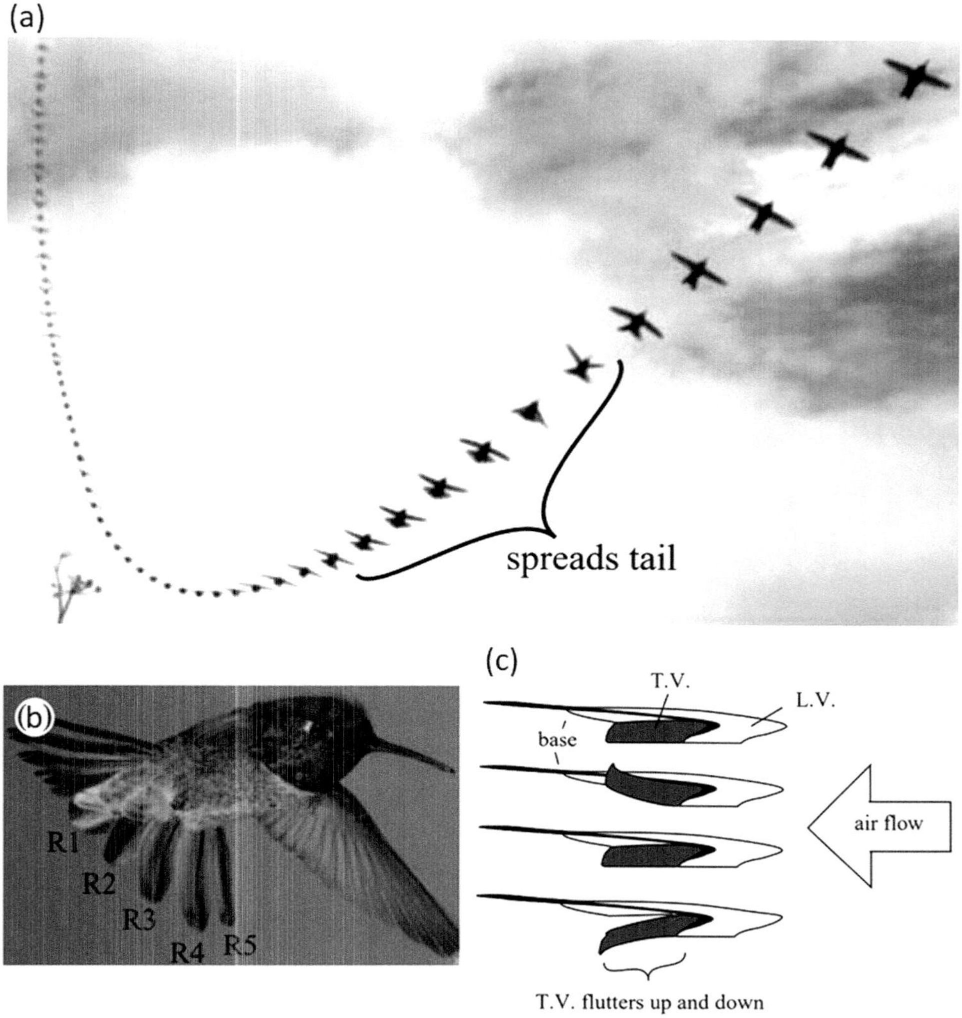


**Fig. 12.8** Tail chirp of a male Anna's Hummingbird (*Calypte anna*). (**a**) Composite image of a male diving toward a perched female, with consecutive images 0.01 s apart. During dives, males spread their tails for only 0.06 s. (**b**) Male Anna's Hummingbird with spread rectrices R1 through R5. (**c**) Drawings illustrating how the trailing vane (T.V.) of rectrix R5 flutters up and down to generate the tail chirp. L.V., leading vane. (Figures from Clark and Feo (2008); © 2008 The Royal Society, used with permission)

located at the junction of the trachea and bronchi. However, some species, notably those in the superfamily Furnarioidea (ovenbirds, woodcreepers, antbirds, antthrushes, gnateaters, and tapaculos) and some parrots, have a tracheal (or tracheophone) syrinx, with sound produced by vibration of membranes located in the trachea (Fig. 12.11). A third type of syrinx is the bronchial (or haploophone) syrinx found in cuckoos (Cuculiformes), nightjars (Caprimulgiformes), Oilbirds (*Steatornis caripensis*; Fig. 12.12), and some owls (Strigidae), where sounds are generated by vibration of membranes in the bronchi. Although most birds have one of these three main types of

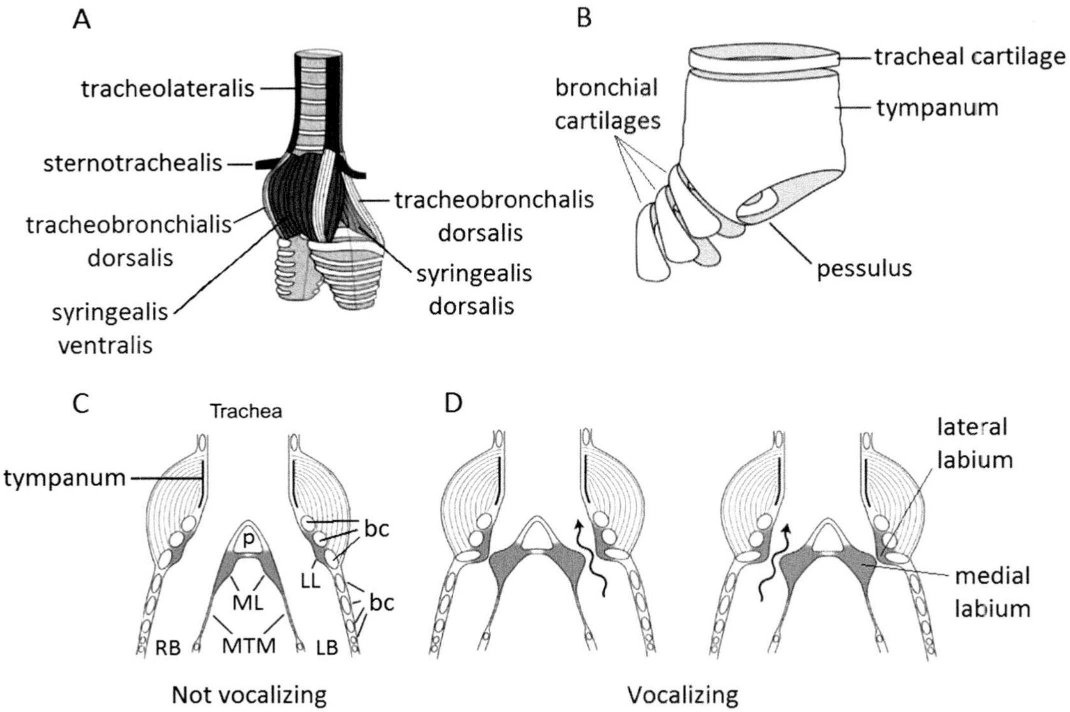


**Fig. 12.9** Anatomy of the tracheobronchial syrinx of a songbird. (**a**) Syringeal muscles. (**b**) Cartilage components of the syrinx include three tracheobronchial semi-rings plus the tympanum (consisting of four closely apposed or fused tracheosyringeal cartilages) with the pessulus at its caudal end. (**c**) Songbird syrinx when not vocalizing and (**d**) when vocalizing. When vocalizing, the medial and lateral labia vibrate in response to passing air. Contraction of the syringeal dorsalis and tracheobronchialis dorsalis muscles rotates the bronchial cartilages into the syringeal lumen, moving the lateral and medial labia into the streams of moving (expired) air and causing them to vibrate and produce sound. Sound can be produced either on one side of the syrinx (as shown) or by both sides simultaneously (not shown). Abbreviations in (**c**): ML, medial labium; LL, lateral labium; MTM, medial tympaniform membrane; p, pessulus; bc, bronchial cartilages; RB, right bronchus; LB, left bronchus. (Figure modified from Schmidt and Wild 2014; © 2014 Elsevier B.V., used with permission)

syringes, the syringes of some species are intermediate in structure (King 1989).

The skeleton of a tracheobronchial syrinx consists of tracheal rings and bronchial half-rings that are typically cartilaginous, but may partially or completely ossify with increasing age. In many species of birds, several tracheal and bronchial rings are partially or completely fused to form a structure called the tympanum, which is the insertion point for syringeal muscles that influence the movements of bronchial rings and, in turn, the position and tension of the sound-producing labia (Elemans 2014). In many species, the bottom of the tympanum creates a bridge-like structure called the pessulus that helps support membranes in the syrinx called the medial labia that vibrate to generate sound (Fig. 12.9).

Syringeal muscles are categorized as either intrinsic or extrinsic. Intrinsic muscles originate in the syrinx and insert on other structures in the syrinx, including bronchial rings, internal and external tympaniform membranes, and syringeal cartilage. Extrinsic muscles either originate or insert outside the syrinx (Larson and Goller 2002; Fig. 12.13). The musculature of the syrinx of some species of birds includes only two pair of extrinsic muscles (sternotrachealis and tracheolateralis), as in many nonpasserines (e.g.,

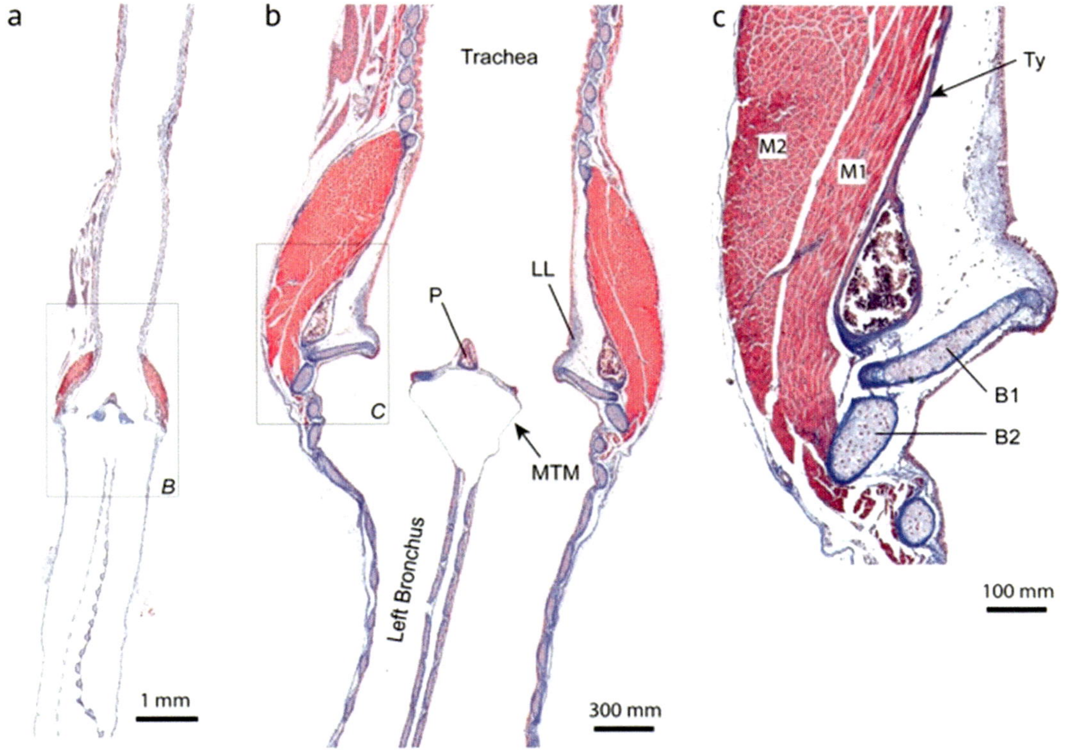


**Fig. 12.10** Syrinx of a Bee Hummingbird (*Mellisuga helenae*). (**a**) Lower magnification view of the syrinx. (**b**) Higher magnification view. P, pessulus; MTM, medial tympaniform membrane; LL, lateral labium. (**c**) Structures associated with the syrinx. Ty, tympanum; M1 and M2, muscles; B1 and B2, first and second accessory cartilages. The syrinx of hummingbirds appears to be similar to that of songbirds based on four features shared by songbirds and, at least, Bee Hummingbirds, including multiple sets of intrinsic muscles, a large tympanum, multiple accessory cartilages, and two sets of opposing soft tissue masses. (Figure from Riede and Olson 2020; open-access article distributed under the terms of the Creative Commons CC BY license, https://creativecommons.org/licenses/by/4.0/)

species in the orders Anseriformes and Columbiformes), whereas other species have two pairs of extrinsic muscles and anywhere from two to six pairs of intrinsic muscles (During et al. 2013). Muscles on each side of the syrinx are separately innervated by the tracheosyringeal branch of the hypoglossal nerve that relay motor impulses from song control nuclei in the brain (Brenowitz et al. 1997). The role of syringeal muscles in sound production is to move the labia or lateral tympaniform membranes into the airway, causing them to vibrate.

Contraction of syringeal muscles move the labia or lateral tympaniform membranes into the airway and tense the syrinx. When vocalizing, syringeal muscles can modulate the sounds generated by varying the degree of constriction and tension of the syrinx. For example, contraction of the syringeal dorsalis and tracheobronchialis dorsalis muscles causes inward movement of the medial and lateral labial, constricting the syringeal apertures. In contrast, contraction of the tracheobronchialis ventralis muscle causes movement of the lateral labium

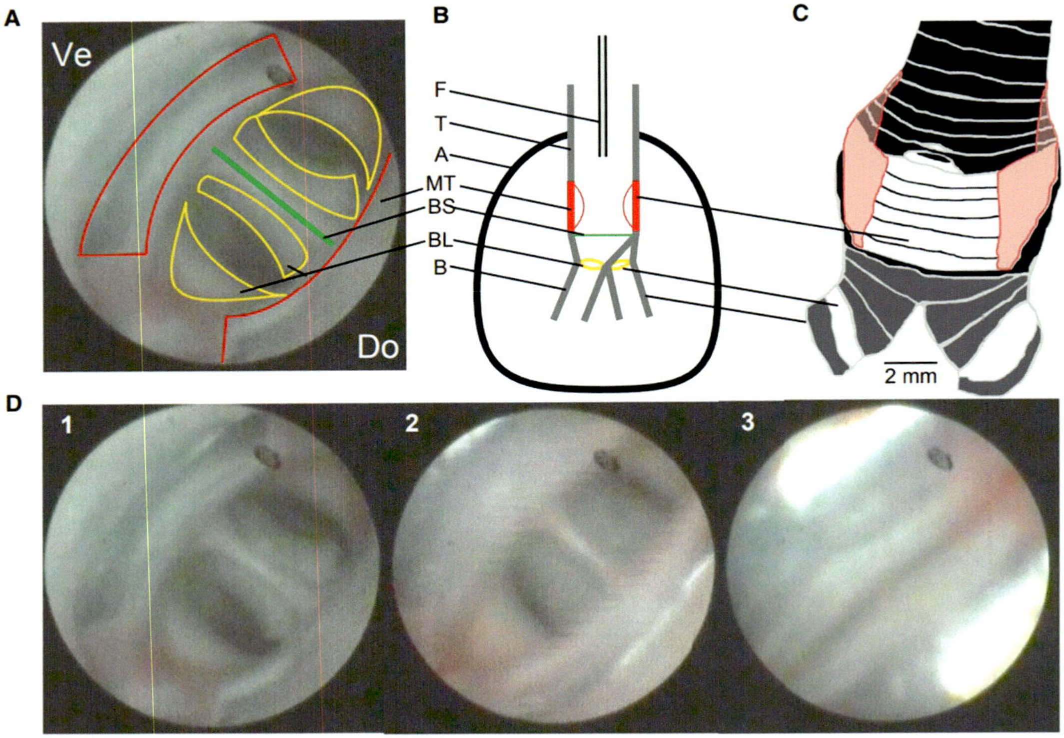


**Fig. 12.11** The tracheophone syrinx of a Dusky Antbird (*Furnarius rufus*). Sounds are generated by fibrations of the membrane trachealis in the trachea as well as the bronchial labia. (**a**) A fiberscopic view in the trachea and syrinx, with structures labeled as they appear during quiet respiration (**a**, corresponding to D1). Ve, ventral; Do, dorsal. (**b**) A drawing of a lateral view of the tracheobronchial junction (F, fiber optic cable; T, trachea; A, air sac; MT, membrana trachealis; BS, bronchial septum; BL, bronchial labia; B, bronchus), and (**c**) a drawing of a ventral view of the syrinx. (**d**) Still shots of the syrinx during quiet respiration (1), when the bronchial sound sources are visible and vibrating (2), and when the membrana trachealis (MT) are vibrating (3). (Figure from Garcia et al. 2017; © 2017 Elsevier Ltd., used with permission)

and enlarges the syringeal aperture (Larson and Goller 2002; Méndez and Goller 2020; Fig. 12.14). The general effects of specific syringeal muscles have been determined via electrical stimulation of those muscles and the resulting effects on the syrinx (e.g., Larson and Goller 2002). In addition, studies of songbirds have revealed correlations between syringeal muscle activity (as measured by electromyography) and the production of sound (e.g., Elemans et al. 2008). However, isolating the independent contributions of individual syringeal muscles is difficult because the actions of different muscles can be correlated (e.g., Goller and Suthers 1996), and individual syringeal muscles can influence different characteristics of sounds generated in songs and calls (Srivastava et al. 2015).

Most sounds generated by the syrinx are produced during expiration (Figs. 12.15 and 12.16, but see Fig. 12.17) and, when a bird is singing, the expiratory pressure increases at least 10–20 times above that during normal respiration. Respiratory pressure also affects the position of the labia because the syrinx is located in the interclavicular air sac. Without the pressure exerted by the interclavicular air sac, the labia cannot vibrate. The labia are soft connective tissues that consist of collagen and elastin fibers and hyaluronan

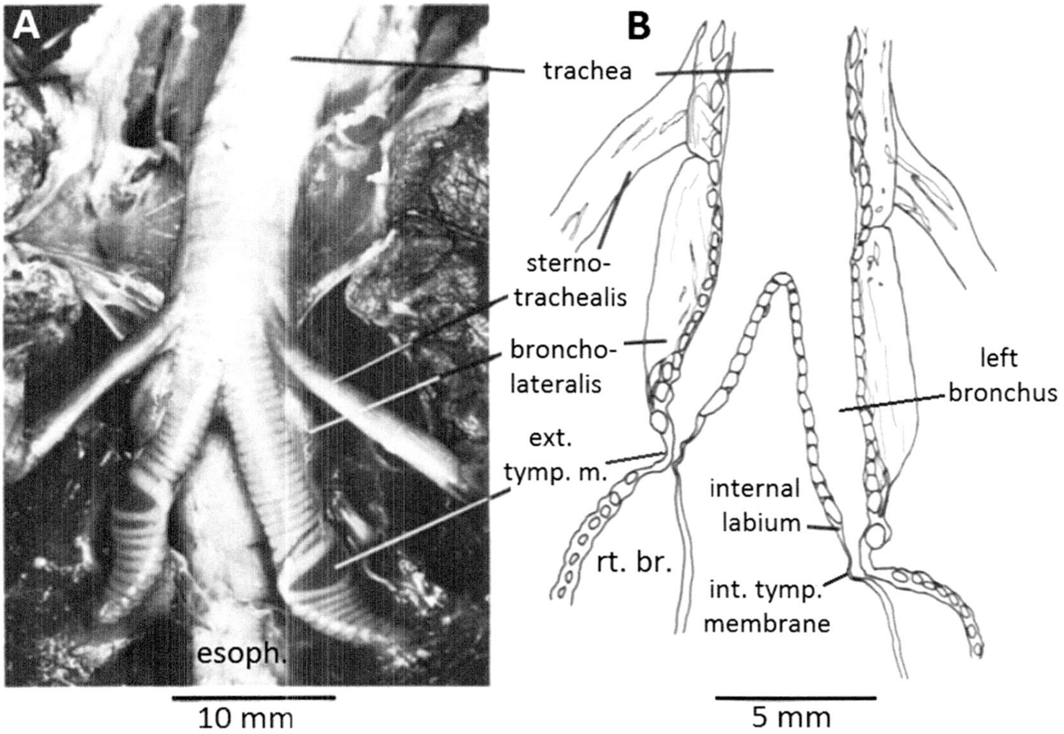


**Fig. 12.12** (**a**, **b**) Bronchial syrinx of an Oilbird (*Steatornis caripensis*). esoph., esophagus; ext. tymp. m., external tympaniform membrane; int. tymp. membrane, internal tympaniform membrane; rt. br., right bronchus. The sternotrachealis and broncholateralis are syringeal muscles. (Figure modified from Suthers and Hector 1985; © 1985 Springer-Verlag, used with permission)

(a polysaccharide) (Fig. 12.18). Vibrations of the lateral and medial labia are the primary means of sound production in a tracheobronchial syrinx (Fig. 12.19), whereas vibration of the lateral tympaniform membranes generates sound in birds with a tracheal syrinx (Fig. 12.20).

## 12.4 Two Sound Sources

Birds with a tracheobronchial syrinx have two sound sources and each is independently controlled by ipsilateral syringeal muscles innervated by the tracheosyringeal branch of the hypoglossal nerve (Goller and Suthers 1996; Fig. 12.21). The two sides of the syrinx are typically tuned to different frequencies, with the left side generating sounds with lower-average frequency than the right side (Suthers 1997). For example, the frequency range of the left side of the syrinx of European Starlings is 1–3.5 kHz whereas the range of the right side is 3–10 kHz (Uchida et al. 2010). Exceptions to this include juvenile Australian Magpies (*Gymnorhina tibicen*; Suthers et al. 2011) and Bengalese Finches (*Lonchura striata domestica*; Secora et al. 2012), where the right side of the syrinx produces lower frequency notes and the left side produces higher-frequency notes. Songbirds use these two sound sources to increase vocal complexity in several ways, including producing two separate sounds at the same time and switching between the two sides to produce abrupt changes in frequency between notes (Suthers and Zollinger 2004; Fig. 12.22).

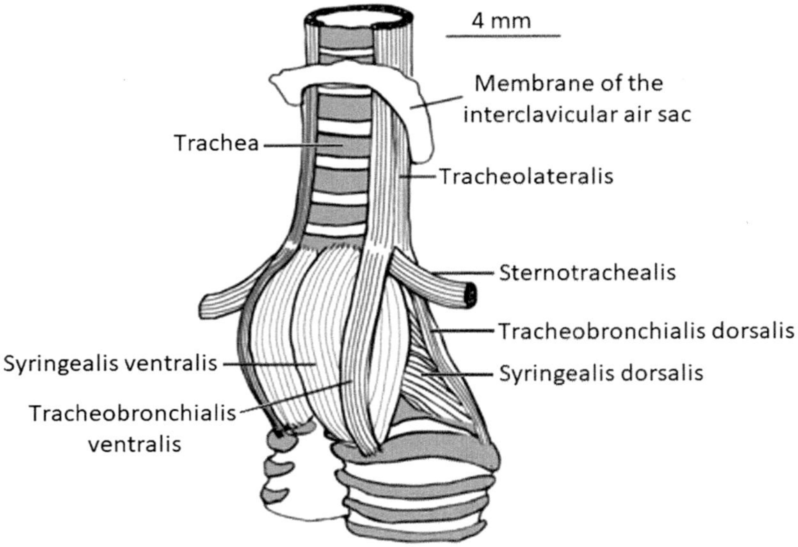


**Fig. 12.13** Ventrolateral view of the syrinx of a typical oscine songbird showing the syringeal muscles. The tracheobronchialis and syringealis muscles are intrinsic syringeal muscles, whereas the tracheolateralis and sternotrachealis are extrinsic syringeal muscles. The tracheobronchial muscles play an important role in controlling the flow of air through each side of the syrinx and, by moving the lateral labium in and out of the bronchial lumen, contribute to the onset of vibrations (Méndez and Goller 2020). (Figure modified from Suthers and Zollinger 2004; © 2006 John Wiley and Sons, used with permission)

## 12.5 Syringeal Membrane Oscillation

The myoelastic-dynamic properties of the vibrating membranes in the syrinx influence the oscillation frequency and, therefore, the frequency (or pitch) of the generated sound. For example, the arrangement of fibrous proteins in the labia into different numbers of distinct layers in different species of birds is correlated with the frequency range of their vocalizations (Riede and Goller 2014; Figs. 12.23 and 12.24). However, the rate of oscillation can also be influenced by the air pressure in the interclavicular air sac that surrounds the syrinx, changes in tension caused by syringeal muscle activity, and source-filter interactions (Goller and Suthers 1996; Alonso et al. 2014). Amador and Margoliash (2013) examined the effect of changes in air-sac pressure on the frequency of bird vocalizations by implanting Zebra Finches with valves that could be opened and closed to alter air pressure in the interclavicular air sac (Fig. 12.25a, b). When the valve was opened, causing a drop in interclavicular air pressure, the frequency of notes dropped (Fig. 12.25c). The effect of this drop in pressure on the vibratory characteristics of the labial membranes caused a reduction in their oscillation rate or, in other words, a decline in frequency.

## 12.6 Source-Filter Supra-Syringeal Structures

Source-filter interactions refer to the role of supra-syringeal structures, including the trachea, oropharyngeal–esophageal cavity (OEC), tongue, and beak, on the characteristics of sounds generated by the syrinx. By changing the shape of their vocal tract, birds can alter the spectral properties of their vocalizations.

Changes in tracheal length can potentially affect the spectral properties of bird vocalizations, but such changes appear to be minimal when

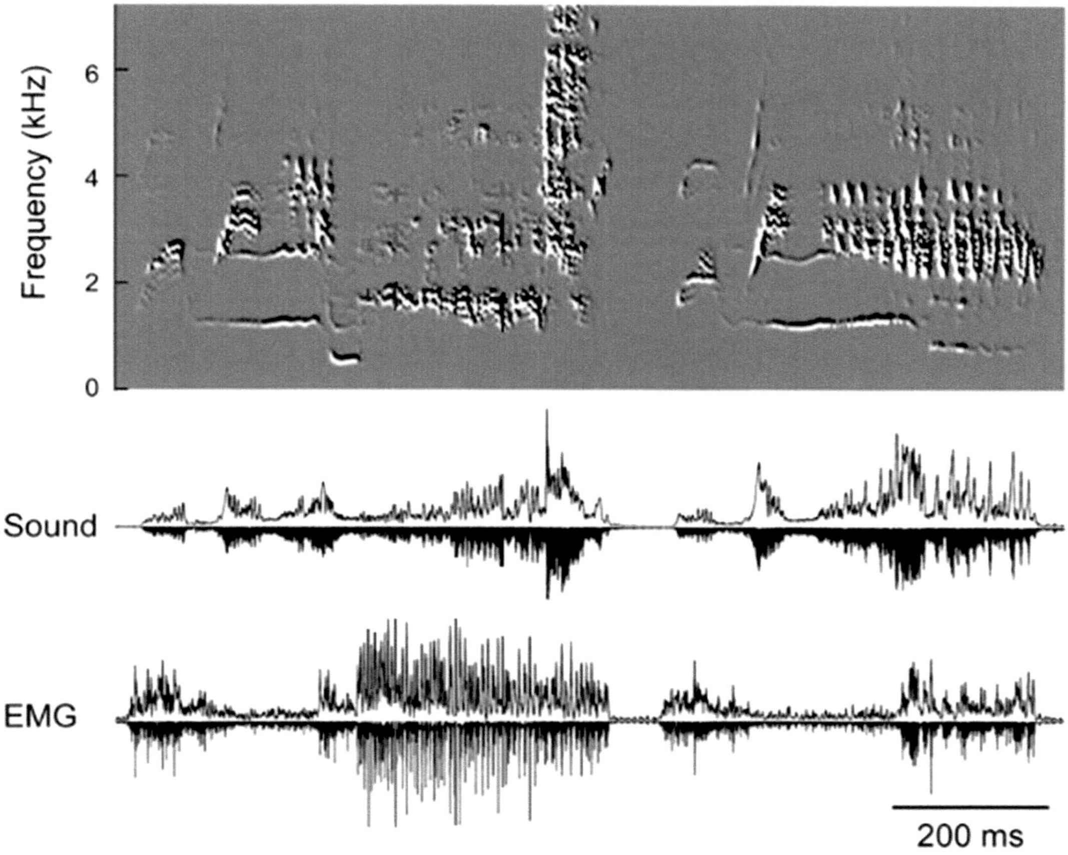


**Fig. 12.14** Song of a European Starling (*Sturnus vulgaris*) illustrating correlation between activity of the tracheobronchialis muscle (EMG), oscillogram of the sound wave pressure generated (Sound), and sonogram showing the various components of the song. (Figure modified from Elemans et al. 2008; open-access article distributed under the terms of the Creative Commons Attribution License, https://creativecommons.org/licenses/by/4.0/)

songbirds (Passeriformes) vocalize. For example, as Zebra Finches sing, changes in trachea length were found to be small (less than 0.2 mm), so likely have no effect on song characteristics (Daley and Goller 2004). However, studies of some species in the family Psittacidae have revealed tracheal shortening when calls are uttered. The length of the Monk Parakeet (*Myiopsitta monachus*) trachea was found to shorten in the range of 9% to 44% when vocalizing (Ohms et al. 2012), and similar tracheal shortening has been reported for Gray Parrots (*Psittacus erithacus*; Patterson et al. 1997). However, what is not yet known is how such changes in tracheal length influence the acoustic features of the calls of psittacids.

Among most species of birds, tracheas are relatively straight tubes a bit longer than a bird's neck. For a relatively small number of species, however, tracheas are coiled to varying degrees and so can be much longer than necks. One possible explanation for this is that longer, coiled tracheas alter the characteristics of calls in a way that calling birds "sound" larger than they

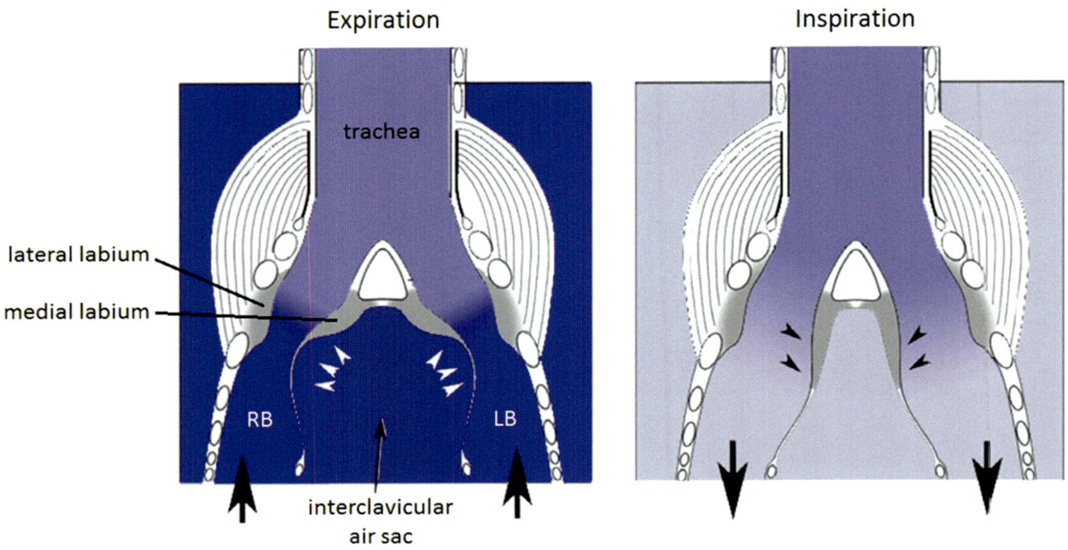


**Fig. 12.15** The syrinx is located in the interclavicular air sac and, during expiration, increasing pressure in the interclavicular air sac pushes the medial labia into the bronchial lumen (white arrow heads). During inspiration, the medial labia move back as the pressure in the interclavicular air sac declines. Dark blue indicates higher pressure than lighter blue. Arrows in the bronchi indicate the direction of airflow. RB, right bronchus; LB, left bronchus. (Figure modified from Goller and Riede 2013; © 2012 Elsevier Ltd., used with permission)

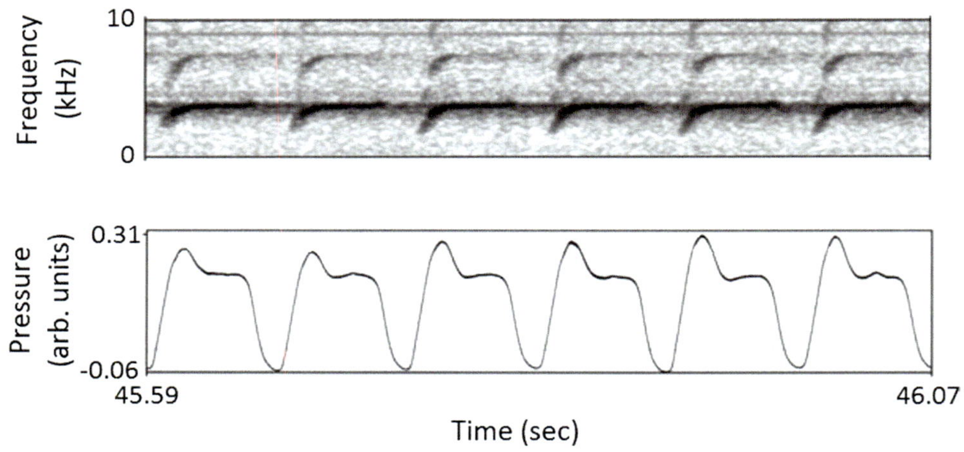


**Fig. 12.16** Simultaneous recording of interclavicular air-sac pressure (measured in arbitrary pressure units) and song of a Domestic Canary (*Serinus canaria*). (Figure modified from Alonso et al. 2014; © 2014 American Physical Society, used with permission)

actually are to conspecifics hearing those calls (Box 12.4 Tracheal Elongation).

In addition to the trachea, the oropharyngeal–esophageal cavity (OEC), tongue, and beak can also alter the characteristics of sounds generated

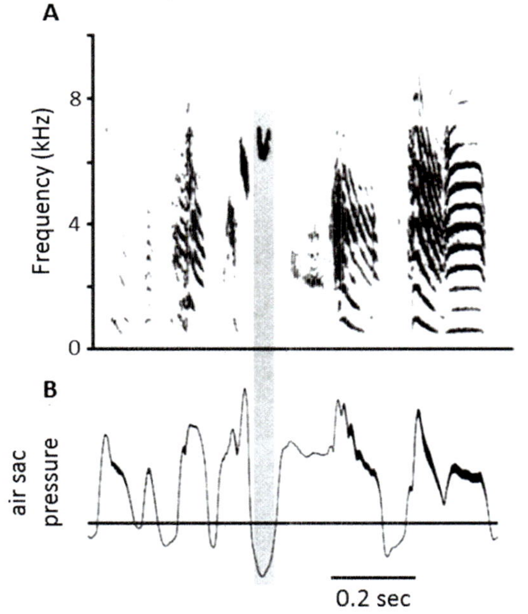


**Fig. 12.17** (**a**) Sonogram of a song of a Zebra Finch (*Taeniopygia guttata*). Most sounds or syllables are produced during expiration, (**b**) as indicated by increased pressure in the interclavicular air sac (horizontal line indicates ambient, or atmospheric, pressure). However, a single syllable (indicated by the gray bar) was generated during inspiration (as indicated by negative pressure in the air sac). (Figure from Goller and Daley 2001; © 2001 The Royal Society, used with permission)

by the syrinx. Studies of singing Zebra Finches (Ohms et al. 2010, Northern Cardinals (*Cardinalis cardinalis*; Riede et al. 2006), and White-throated Sparrows (*Zonotrichia albicollis*; Riede and Suthers 2009) have revealed that, as each note or syllable of their songs is uttered, movement of the hyoid skeleton causes an increase in the volume of the OEC (Fig. 12.26) which then acts as a resonance chamber that tracks the fundamental frequency of the notes or syllables (Fig. 12.27). As a resonance chamber, pressure waves generated by the syrinx cause the walls of the OEC to oscillate with greater amplitude at the fundamental frequencies, amplifying the fundamental frequency, suppressing overtones, and, for species with whistle-like notes and calls, increasing the tonal purity of notes or syllables (Riede et al. 2006). For species with notes and syllables with rich harmonic content like Zebra Finches, movements of the OEC generate resonance peaks over a broader frequency range (Riede et al. 2013).

As with expansion of the OEC, beak gape can also filter sounds produced by the syrinx. In general, beak gape is positively correlated with fundamental frequency, with wider beak gapes concentrating energy of higher-frequency sounds

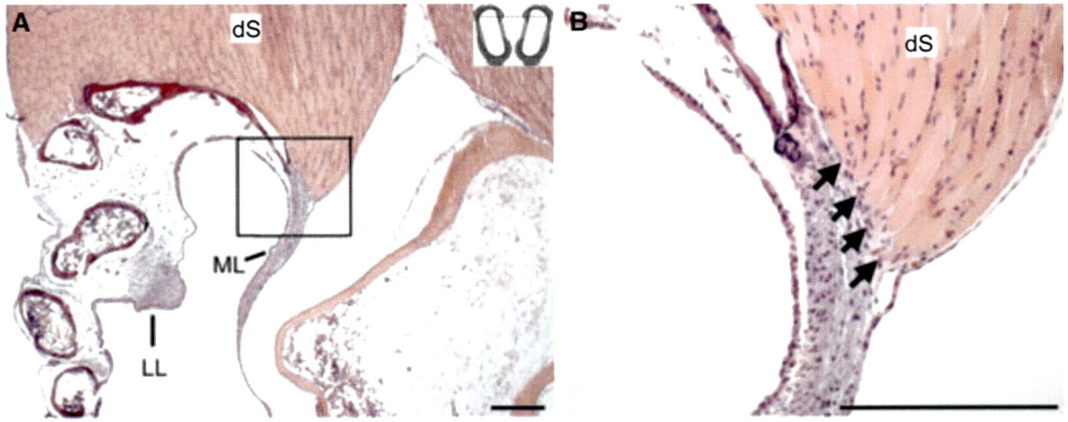


**Fig. 12.18** (**a**) Section through the dorsal section a syrinx of a White-crowned Sparrow (*Zonotrichia leucophrys*). The square is the area magnified in **b**. (**b**) Arrows are pointing to the point where the syringeal dorsalis muscle (dS) attaches to the medial labium. ML, medial labium; LL, lateral labium. (Figure from Riede and Goller 2010a; © 2009 The Authors. Journal compilation © 2009 Anatomical Society of Great Britain and Ireland, used with permission)

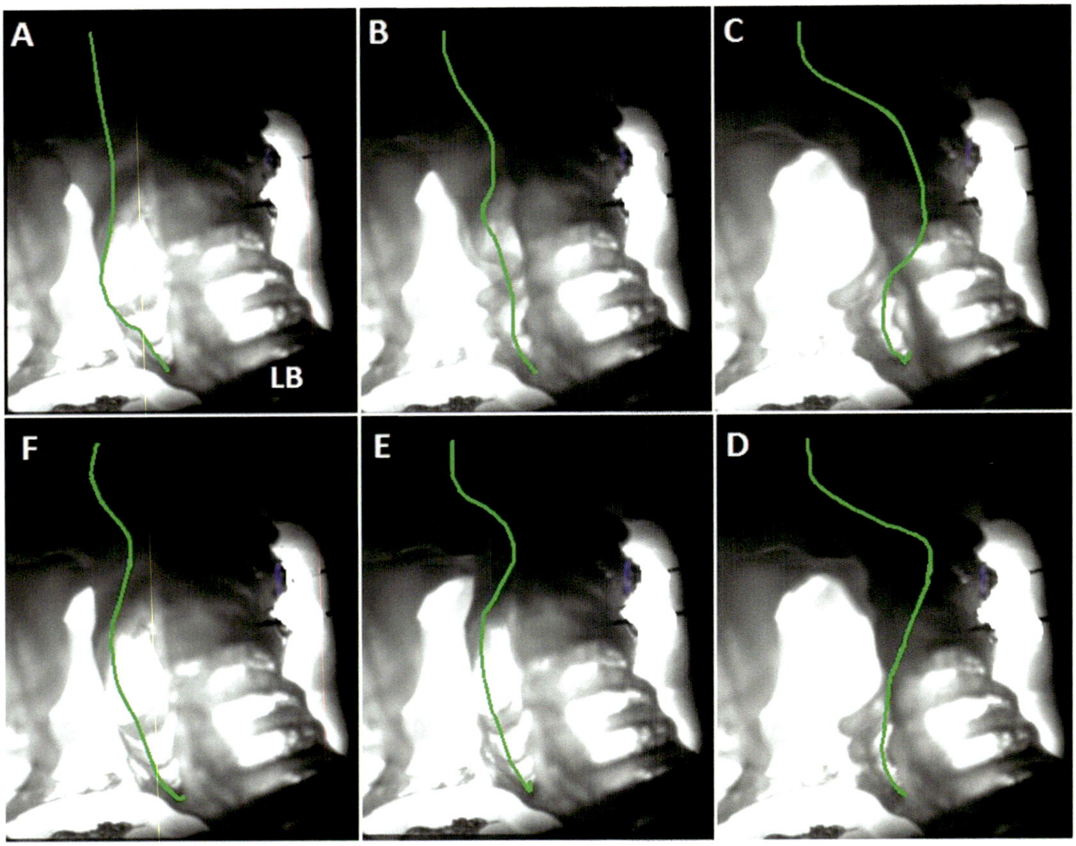


**Fig. 12.19** Screen-captures from a high-speed video (2000 frames per second) showing vibration of the left medial labium of a Zebra Finch (*Taeniopygia guttata*). LB, left bronchus. (Images from Supplementary Movie 6 from Elemans et al. (2015); open-access article distributed under the terms of the Creative Commons CC BY 4.0 license, https://creativecommons.org/licenses/by/4.0/)

and smaller beak gapes do so with lower frequency sounds. This concentration can improve the tonal quality or purity of calls, notes, and syllables (Nelson et al. 2005). For species producing complex notes or syllables with energy distributed over a range of frequencies, beak movements appear to serve other functions. For example, the songs of Zebra Finches include complex, broad frequency syllables, and wider beak gapes were found to increase the amplitude (volume) and peak frequency of syllables in those songs (Williams 2001). Similarly, wider beak gapes also corresponded with higher frequencies in Large Tree-Finches (*Camarhynchus psittacula*, Podos et al. 2004).

Finally, tongue movements can also filter sounds. Suthers et al. (2016) observed elevation of the tongue by Northern Cardinals when notes below 2 kHz in frequency were uttered, suggesting that tongue elevation lowers the resonance frequency below 2 kHz by reducing the area of the passage from the oral cavity into the beak. Ohms et al. (2012) found a positive correlation between beak displacement and sound amplitude for the chatter calls of Monk Parakeets. Similarly, Zebra Finches open their beaks when producing loud notes (Ohms et al. 2010). Monk Parakeets also move their tongues when vocalizing, but how those movements affect

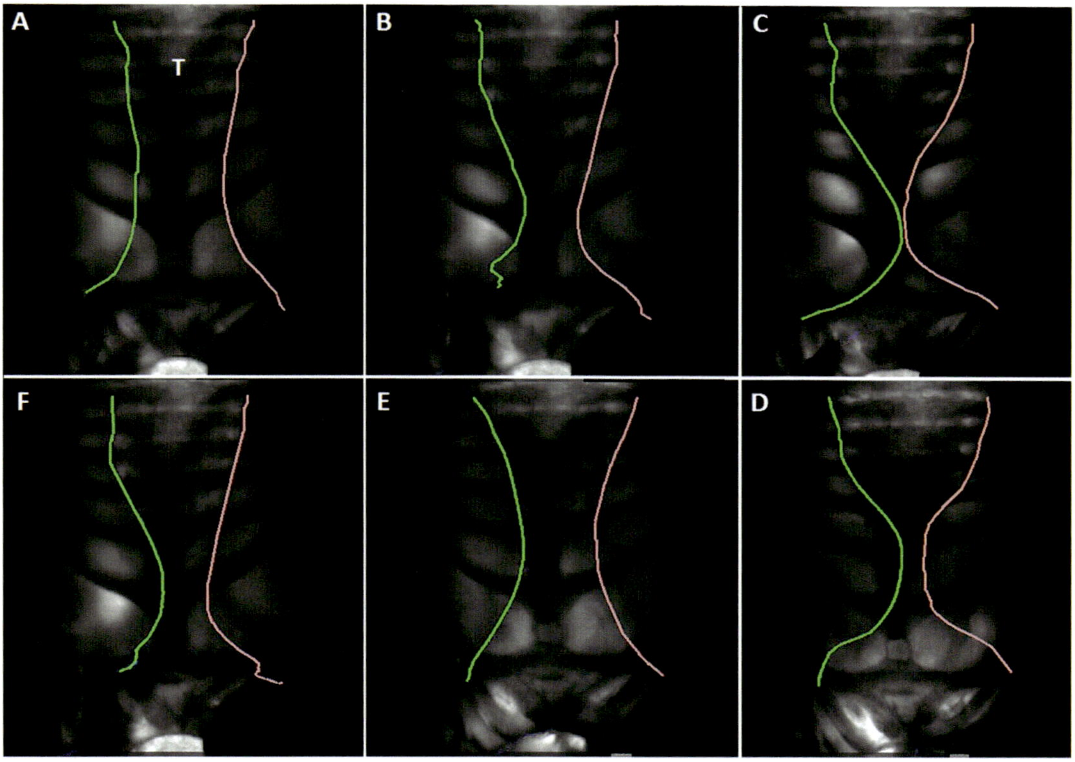


**Fig. 12.20** Screen-captures from a high-speed video (4000 frames per second) showing vibration of the lateral tympaniform membranes in the tracheal syrinx of a Rock Pigeon (*Columba livia*). As rapidly moving air is expired and moves past the membranes, a negative pressure is generated (Bernoulli pressure) that brings the membranes together (**a** through **c**). Air pressure then increases below the membranes and eventually is sufficient to force the membranes apart (**d** through **f**). The elastic nature of the membranes then causes these movements to continue and their rapid vibration generates sound, i.e., the calls and songs of birds. Notice in **f** that the membrane is beginning to move in a wave-like pattern upward so the membranes do not just oscillate back-and-forth, but in wave-like motions as well. T, trachea. (Images from Supplementary Movie 1 from Elemans et al. 2015; open-access article distributed under the terms of the Creative Commons CC BY 4.0 license, https://creativecommons.org/licenses/by/4.0/)

sound quality remains to be determined (Ohms et al. 2012).

## 12.7 Vocalizing and Breathing

When vocalizing, the duration of respiratory phases (expiration and inspiration) becomes more variable. During brief silent periods during songs, birds take short-duration inspirations called minibreaths. For example, the songs of Zebra Finches typically include a series of expiratory pulses, corresponding to the song notes or syllables that alternate with minibreaths (Plummer and Goller 2008). The volume of air inspired with minibreaths equals the volume required to produce the preceding note or syllable so birds can continue singing (Hartley and Suthers 1989). However, when uttering notes at high rates, e.g., trills, there is insufficient time for minibreaths between notes.

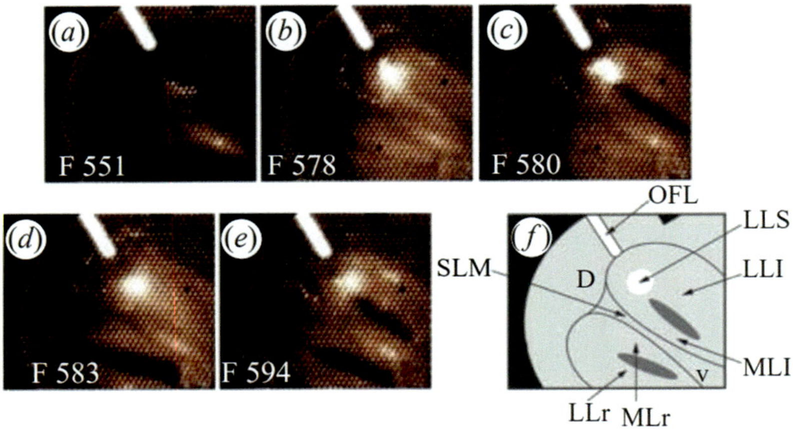


**Fig. 12.21** High-speed video (1000 frames per second) showing alternating opening of the left and right sides of the syrinx. Video frames (frame number in white) showing (**a**) both sides of the syrinx open (* = open) prior to sound production, (**b**) both sides closed, (**c**) left side open, (**d**) right side open, and (**e**) both sides open when vocalizing. (**f**) Drawing indicated the view in frames **a**–**e**. OFL, optical fiber layer; LLS, laser light spot; LLl and LLr, left and right lateral labium; MLl and MLr, left and right medial labium; D, dorsal; v, ventral; SLM, semilunar membrane. (Figure from Jensen et al. 2007; © 2007 The Royal Society, used with permission)

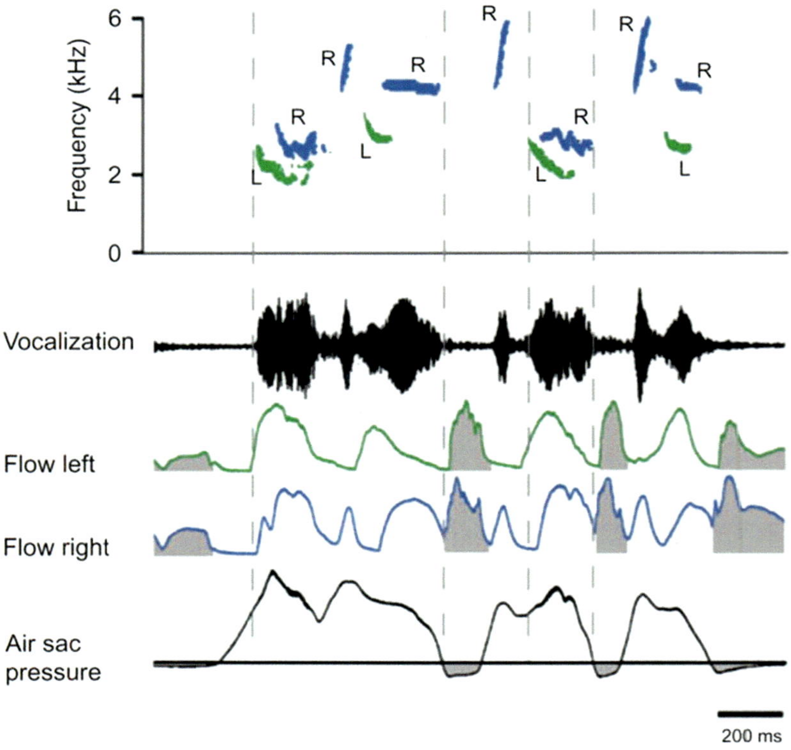


**Fig. 12.22** Portion of a song of a Brown Thrasher (*Toxostoma rufum*) illustrating the switching sounds generated by the left and right side of the syrinx, sometimes separately and sometimes simultaneously. Sounds generated by the right side (R) of the syrinx are shown in blue, those generated by the left side of the syrinx (L) in green, and inspiratory minibreaths are shaded in gray. (Figure from Schmidt and Wild 2014; Copyright © 2014 Elsevier B.V. All rights reserved, used with permission)

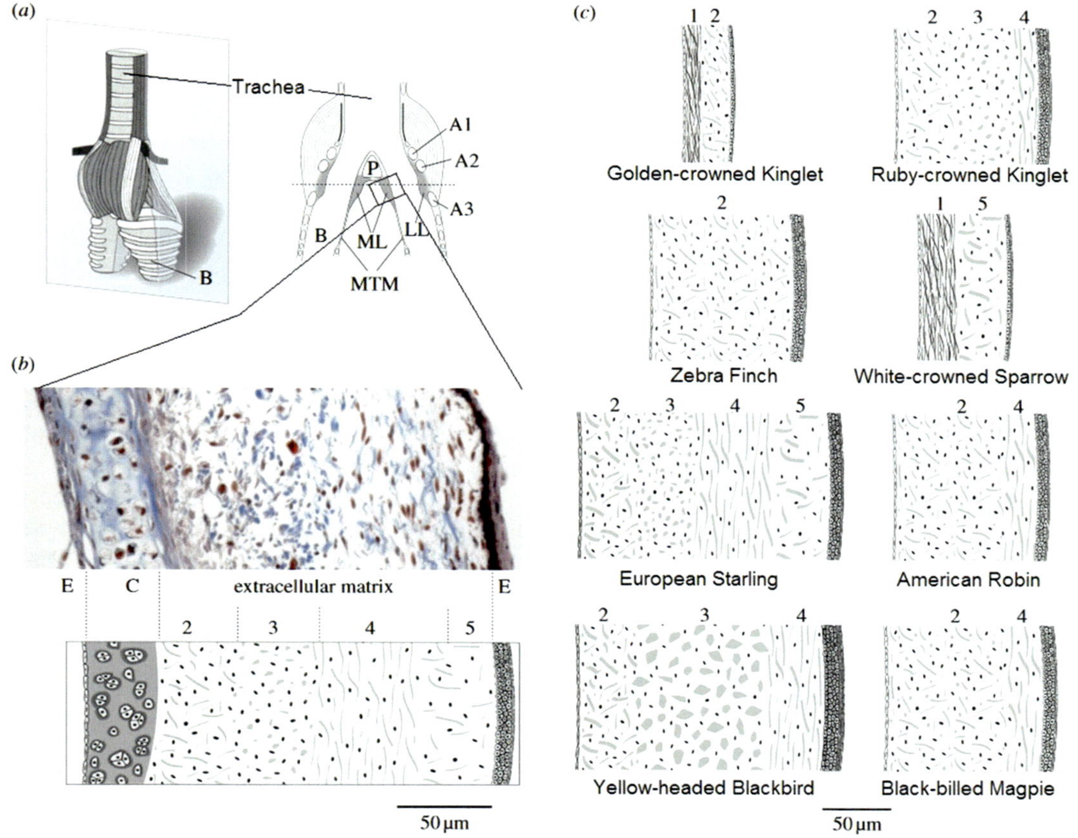


**Fig. 12.23** (**a**) The syrinx consists of a cartilaginous framework, soft tissues called labia that oscillate, and muscles that control airflow and labial tension. The syrinx of oscine songbirds has two sound sources, each consists of lateral (LL) and medial (ML) labia. A1–A3, syringeal cartilages; P, pessulus; B, bronchus; MTM, medial tympaniform membrane. The black box indicates the plane for sectioning. (**b**) Stained section of a medial labium of a European Starling showing its cellular and extracellular components. Thin epithelial tissue (E) on each side encompasses the extracellular matrix that consists of connective tissue consisting of collagen (blue) and elastin fibers (black). The layered structure of the extracellular matrix is shown in the drawing below, with layers identified and numbered based on the type and orientation of collagen fibers. Most collagen fibers are oriented randomly in layers 2 and 5, dorsoventrally in layer 3, and craniocaudally in layer 4. Numbers (2–4) identify similar layers in different species. European Starlings do not have a layer dominated by elastin (layer 1 in (**c**)). (**c**) Drawings showing the composition of the extracellular matrix and size differences of the medial labia in the eight species of songbirds, with layer thickness drawn to scale. Species: Golden-crowned Kinglet, *Regulus satrapa*; Ruby-crowned Kinglet, *Regulus calendula*; Zebra Finch; White-crowned Sparrow, *Zonotrichia leucophrys*; European Starling, *Sturnus vulgaris*; American Robin, *Turdus migratorius*; Yellow-headed Blackbird, *Xanthocephalus xanthocephalus*; and Black-billed Magpie, *Pica hudsonia.* (Figure from Riede and Goller 2014; © 2014 The Authors. Published by the Royal Society, used with permission)

## 12.8 Sexual Dimorphism in Syringeal Anatomy

In many (but not all; e.g., Miller et al. 2008) species of birds, the syringeal anatomy of males and females differs. For example, in the family Anatideae, only males have syringeal bulla (Johnsgard 1971; Fig. 12.28). In Zebra Finches, the syringeal muscles of males are larger than those of females and can also contract at higher rates (Wade and Buhlman 2000; Elemans et al.

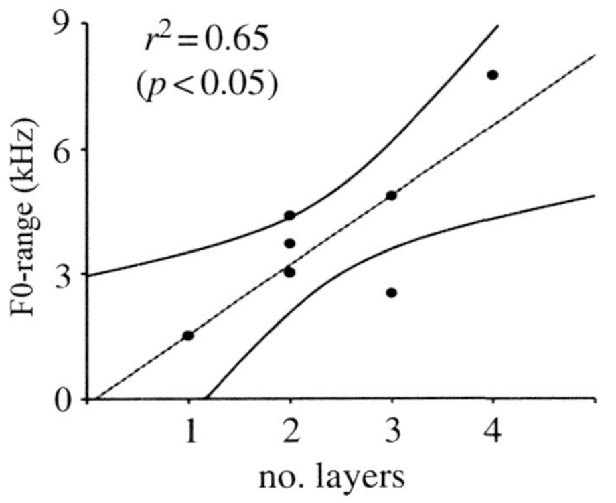


**Fig. 12.24** The number of distinct layers in the extracellular matrix of the medial labia is positively correlated with the fundamental frequency (F0) range of a species' vocalizations. (Figure from Riede and Goller 2014; © 2014 The Authors. Published by the Royal Society, used with permission)

2008; Fig. 12.29). Male Herring Gulls (*Larus argentatus*) have larger and thicker vibratile membranes (tympaniform membranes) than females (Hardouin et al. 2014). Differences in syringeal anatomy have also been reported in several other species and taxa of birds (e.g., Miller 1934; Ballintijn and ten Cate 1997; Bottino et al. 2006). With the exception of Herring Gulls (Hardouin et al. 2014), intersexual differences in syringeal anatomy have been found to contribute to intersexual differences of the characteristics of vocalizations. Generally, the calls of males are lower in frequency than the calls of females (e.g., Miller 1934; Ballintijn and ten Cate 1997; Volodin et al. 2005). In Zebra Finches, however, intersexual differences in syringeal anatomy allow males to produce higher-frequency songs and to produce a wider range of high fundamental frequencies (Riede et al. 2010). A potential advantage of intersexual differences in syringeal anatomy is that the corresponding intersexual differences in call structure permit vocal sex recognition. However, intersexual differences may also result from intersexual (i.e., female preference for certain call or song characteristics) and/or intrasexual (i.e., advantages of certain call or song characteristics in male-male competition) selection.

## 12.9 Vocal Sacs and Closed-Beak Vocalizations

Vocal sacs are specialized structures that can be inflated and deflated when vocalizing. Vocal sacs have flexible walls containing elastin, collagen, and muscle fibers and so collapse when not in use (Riede et al. 2008). Birds have three different types of vocal sacs, including esophageal, tracheal, and gular sacs. The inflated esophagus of Ring Doves (domesticated form of Barbary Dove, *Streptopelia risoria*) acts as an amplifier and resonant filter, eliminating higher-frequency harmonics to generate pure-tone coo calls (Fletcher et al. 2004; Fig. 12.30).

Most birds vocalize with their beaks open, accompanied by conspicuous and complex movements of the neck and beak. These movements are a dynamic component of upper-vocal-tract filtering (Riede and Goller 2010a, b). Another vocalization pattern is to seal off the pharynx from the oral and nasal cavities and inflate the esophagus or tracheal pouches in the ventral neck area, between the beak and thoracic aperture, e.g., Ringed Turtle-Doves (*Streptopelia risoria*; Riede et al. 2004) and Greater Sage-Grouse (*Centrocercus urophasianus*; Krakauer et al. 2009). Several species also use closed-beak vocalizations, vocalizing with a closed beak and the simultaneous inflation of the esophagus or tracheal pouches (Riede et al. 2016; Fig. 12.31). An inflated esophagus and radiation of calls through the skin of the neck have a strong filtering effect (Fletcher et al. 2004; Riede et al. 2004). The stretched skin acts as a band-pass filter eliminating higher and lower frequencies, thereby generating a pure tone (Fletcher et al. 2004).

Inflation of the esophagus or tracheal pouches is often associated with visual signals or displays, e.g., colorful skin or feathers in the throat area (frigatebirds and Greater Sage-Grouse), elaborate body movements (head bobbing by pigeons and doves), or stereotyped flight displays (sandpipers; Miller 1983). The increase in neck size in less colorful species (e.g., Pectoral Sandpipers, and pigeons and doves) may serve as important

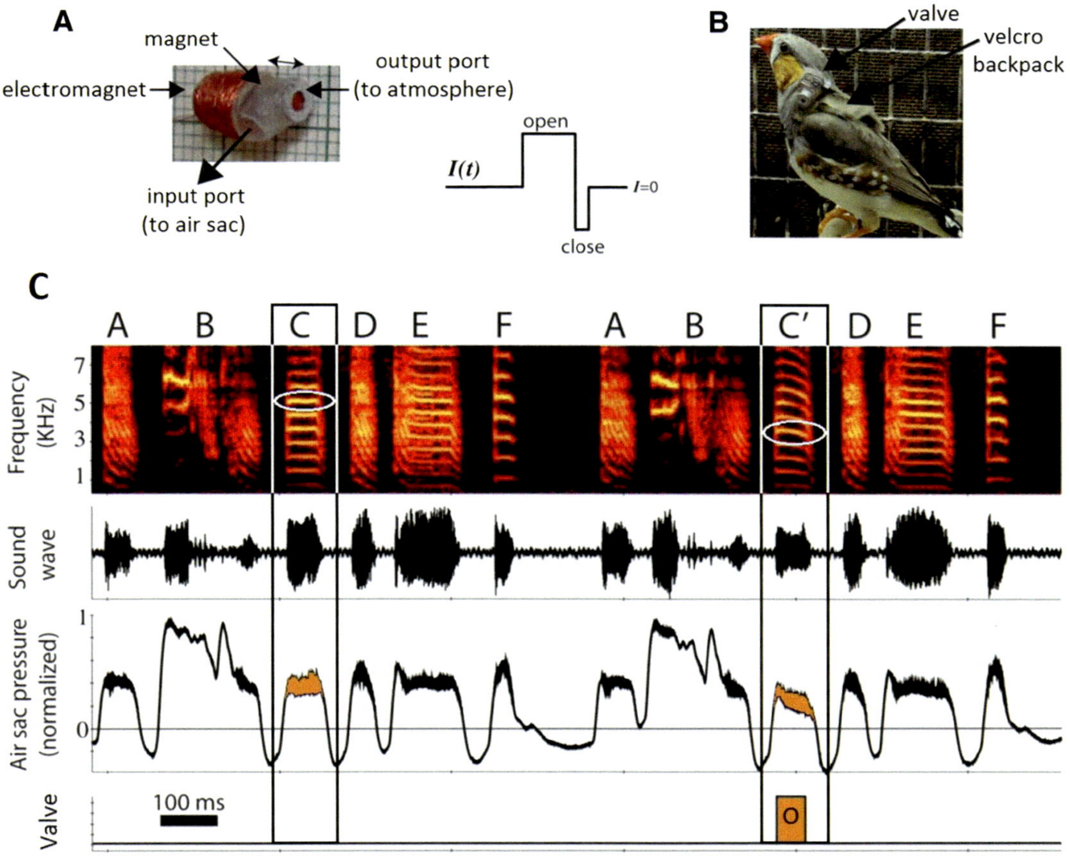


**Fig. 12.25** Method for manipulating vocalizations. (**a**) A miniature valve. Current through an electromagnet moves a magnet to open or close a hole that connects the air-sac system with the atmosphere. Activating the electromagnet generates current [$I(t)$] and causes movement of the magnet that opens the valve and reduces air-sac pressure; with no current ($I = 0$), the magnet moves back and closes the valve so the air-sac system is sealed. (**b**) Implanted bird carrying a backpack with the valve attached. (**c**) A decrease in the pressure in the interclavicular air sac when uttering the second note C (C′) resulted in a decrease in its fundamental frequency. A valve regulating air-sac pressure was briefly opened (O at bottom) to cause a drop in air-sac pressure. (Figure modified from Amador and Margoliash 2013; © Society for Neuroscience, used with permission)

**Box 12.4 Tracheal Elongation**

As sounds produced by the syrinx pass through the trachea, they vibrate at certain frequencies called the formant frequencies and they represent peaks of acoustic energy (Fant 1960). Formant spacing is the average distance (measured in hertz or kilohertz) between adjacent formants (Fitch 1997). This spacing is determined by the length of the trachea and can be calculated by the equation,

$$\text{Distance between formants} = c/2L,$$

(continued)

**Box 12.4** (continued)

where *c* is the speed of sound and *L* is the length of the sound-emitting tube (in meters; Fitch 1999). This equation shows that there is an inverse relationship between trachea length and formant spacing, as trachea length increases, formants are closer together. Because larger birds tend to have longer tracheas, formant spacing tends to decrease with increasing body size. Subtle changes in formant spacing can be perceived by some species of birds and so can provide information about the body size of vocalizing birds (Fitch 1999).

Among some species of birds, sexual selection tends to favor larger body size because size is generally associated with greater competitive abilities. However, increases in body size may be constrained by genetic or ontogenetic factors and/or because natural selection favors a smaller body size (Blanckenhorn 2000; Badyaev 2002). In some species of birds, these opposing factors have resulted in selection favoring the elongation of tracheas, with longer tracheas reducing the distance between formants and, therefore, increasing the "apparent" size of callers. This elongation has occurred via the looping and coiling of trachea and has been reported both in fossils (Li et al. 2018) and in at least 60 extant species of birds representing seven different orders, including Gruiformes (14 species), Charadriiformes (1 species), Ciconiiformes (1 species), Pelecaniformes (1 species), Anseriformes (6 species), Galliformes (32 species), and Passeriformes (5 species) (Fitch 1999).

In addition to this "size exaggeration" hypothesis, other investigators have proposed that selection has favored tracheal elongation to increase call volume or amplitude (e.g., long tracheas cause vibration of adjacent sternum and air sacs and increase energy transfer, and volume of calls, to the atmosphere; Gaunt et al. 1987) or because it provides some physiological benefit (e.g., longer trachea = more dead air space in respiratory system = increased levels of carbon dioxide in lungs and blood = chemoreceptors in aorta and carotid arteries stimulated = maintain breathing rate that matches wingbeat frequency of large birds like cranes in flight; Schmidt-Nielsen 1972). Fitch (1999) pointed out flaws in both of these alternative hypotheses, with the most important being that so few species of birds have elongated tracheas and, in those species, usually only one sex (typically males) has elongated traches. If tracheal elongation provided physiological benefits and/or increased call volume, selection would have certainly favored the evolution of longer tracheas in many more species of birds and in both sexes.

Given that sexual selection has favored increased size, most often in males, in many species of birds, selection favoring an increase in "apparent size" might also be expected in many species. One possible explanation for the limited number of species with elongated tracheas is that an increase in "apparent size" may only be beneficial in habitats or times of day with reduced visibility where visual contact is limited and/or nocturnal species where vocal cues are more important than visual cues in determining the body size of conspecifics. In fact, many species with elongated tracheas are found in habitats with dense vegetation and limited visibility, e.g., cracids, cranes, and manucodes, and/or frequently call at night, e.g., some cracids, Limpkins (*Aramus guarauna*), Greater Painted-Snipes (*Rostratula benghalensis*), Eurasian Spoonbill (*Platalea leucorodia*), and Freckled Ducks (*Stictonetta naevosa*) (Fitch 1999).

The size exaggeration hypothesis predicts that only males will have elongated tracheas in species where only males defend territories or where females prefer large-sounding males, but both sexes will have elongated tracheas in species where both sexes help defend territories (Fitch 1999). In support of this hypothesis, both sexes do have elongated tracheas in many species where both defend territories, including cranes, swans, and European Spoonbills, and only males

(continued)

**Box 12.4** (continued)

have elongated tracheas in many species where only males defend territories, e.g., most cracids and Australian Painted-Snipes (*Rostratula australis*), and species where females choose among males displaying and calling at leks (manucodes). In further support of the size exaggeration hypothesis, Greater Painted-Snipes are polyandrous and only females defend territories and have elongated tracheas (Fitch 1999).

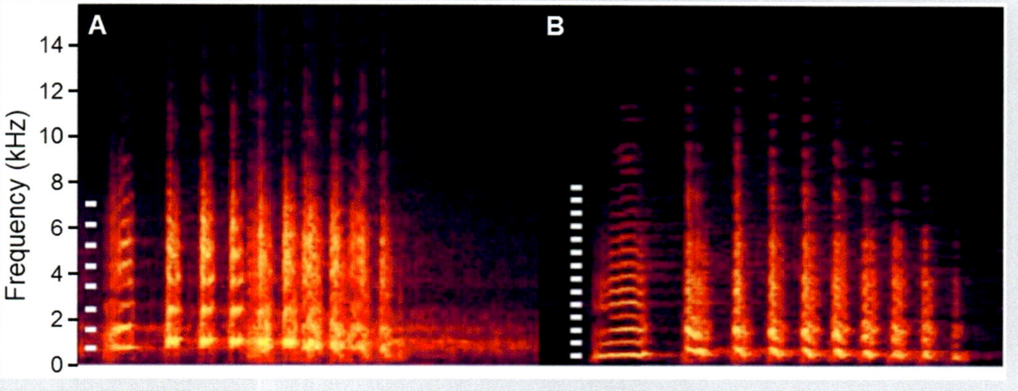


Sonograms of guard calls of two subspecies of Sandhill Cranes that differ in size. (**a**) Calls of the smaller *Grus canadensis canadensis*, and (**b**) calls of the larger *G. c. tabida*. The white horizontal bars on the left side of each sonogram indicate positions of the fundamental frequency and harmonics. (Figure modified from Jones and Witt 2012; © 2012 The Wildlife Society, used with permission)

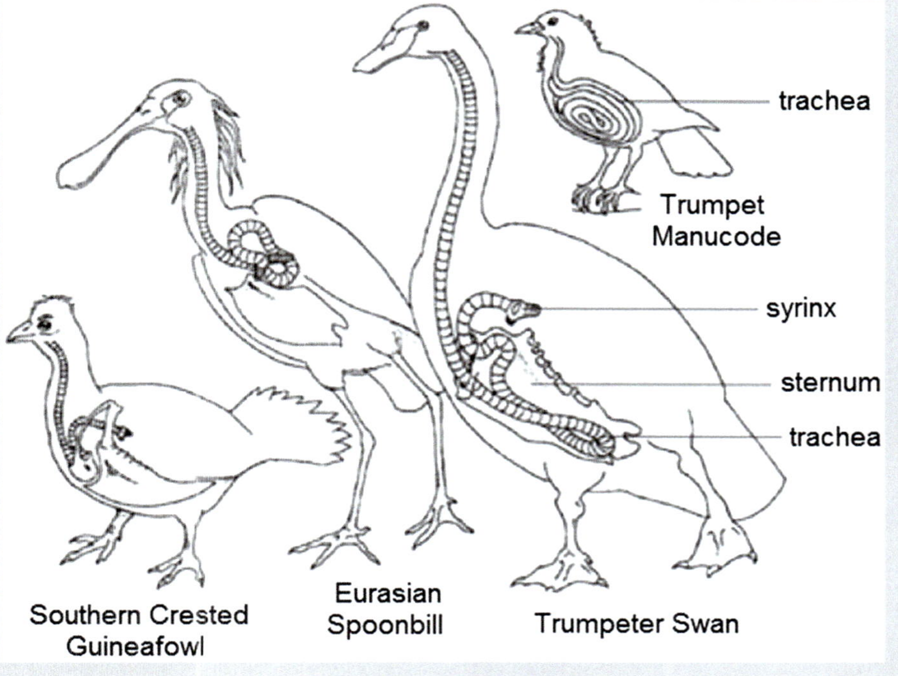


(continued)

**Box 12.4** (continued)

Examples of tracheal elongation in four species of birds. Southern Crested Guineafowl, *Guttera edouardi*; Eurasian Spoonbill, *Platalea leucorodia*; Trumpeter Swan, *Cygnus buccinator*; and Trumpet Manucode, *Phonygammus keraudrenii*. (Figure from Fitch and Hauser 2003; © 2003 Springer-Verlag New York, Inc., used with permission)

advertisement signals for females or other males (Riede et al. 2015; Fig. 12.32).

## 12.10 Central Motor Control of Song

Song production involves the coordinated actions of several motor systems: syringeal muscle control, upper respiratory tract control, and respiration (Suthers and Zollinger 2008). The motor system of the syrinx includes the HVC and the robust nucleus of the arcopallium, or RA. The HVC song nucleus is located in the forebrain and is known to be essential for singing. HVC neurons are known to transmit impulses to the song motor nucleus RA when birds sing, with different HVC neurons stimulated at different times during songs (Hahnloser et al. 2002). One hypothesis is that these neurons in the HVC control song timing and evidence in support of this hypothesis comes from experiments where bilateral cooling of the HVC (i.e., cooling the HVC nucleus in both hemispheres) results in a slowing of song timing (Andalman et al. 2011; Fig. 12.33).

The results of other studies, however, suggest that singing involves more complex interactions among different areas in the avian brain. The HVC, acting via the RA, is the source of motor impulses that control the acoustic and temporal

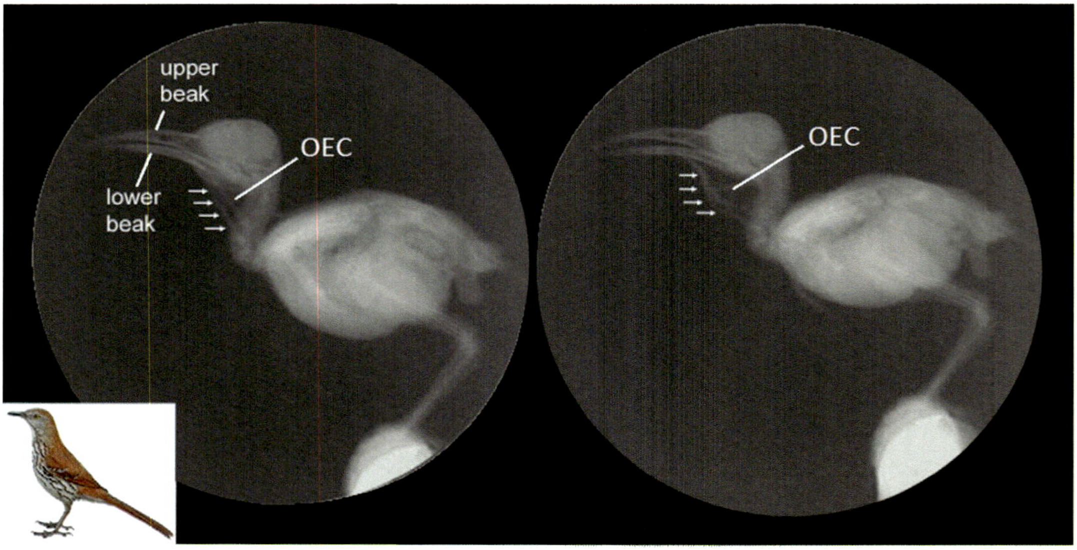


**Fig. 12.26** X-ray image of a Brown Thrasher (*Toxostoma rufum*) before (left) and during production of a song syllable (right). The arrows indicate the position of the trachea. When a note or syllable is uttered, the larynx and trachea move down and out. Extrinsic laryngeal muscles protract and retract the hyoid skeleton, larynx, and upper part of the trachea causing changes in the volume of the oropharyngeal-esophageal cavity (OEC) when a Brown Thrasher sings. (Figure modified from Riede and Goller 2010b; © 2009 Elsevier Inc., used with permission)

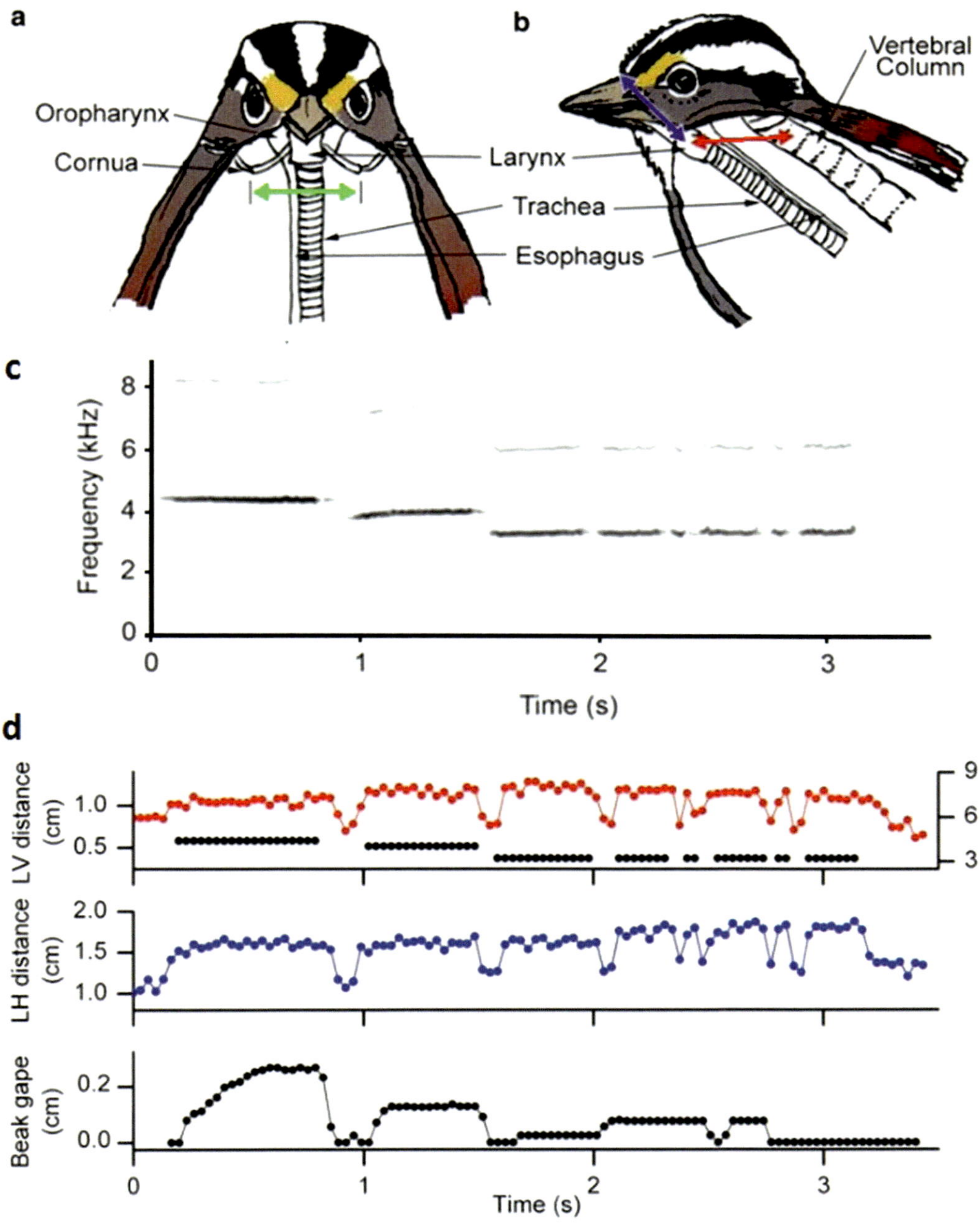


**Fig. 12.27** As White-throated Sparrows (*Zonotrichia albicollis*) sing, coordinated movements of the larynx and hyoid cornua increase the volume of the oropharyngeal–esophageal cavity (OEC) as they utter each note. (**a**, **b**) Drawings of a White-throated Sparrow showing (**a**) the hyoid cornua and the distance between them (green arrow), and (**b**) dorsoventral movement (LV, red arrow) of the larynx measured from the middle of the second cervical vertebra and craniocaudal movement (LH, blue arrow) of the larynx relative to the dorsal edge of the beak–skull transition. (**c**) Song of a White-throated Sparrow. (**d**) Movements of larynx, cornua, and beak during a song are indicated by changes in the distance between the anatomical structures that define the length of each of the vectors measured. LV, LH, and beak gape in successive frames of the X-ray movie (30 fps) during a descending song. The timing of each syllable and its fundamental frequency (f0) is plotted in black beneath LV. (Figure modified from Riede and Suthers 2009; © 2008 Government Employee: National Institutes of Health, USA, used with permission)

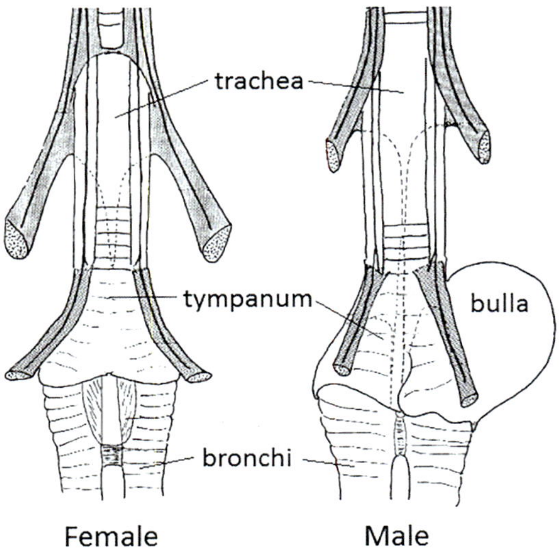


**Fig. 12.28** The syringes of female and male Mallards (*Anas platyrhynchos*). Note the large ossified bulla on the left side of the male's syrinx. (Figure modified from Lockner and Youngren 1976; © 1976 Oxford University Press, used with permission)

characteristics of songs (Amador et al. 2013). Motor impulses from the RA stimulate the dorsomedial nucleus (DM) in the midbrain, respiratory centers in the brainstem, and the hypoglossal motor nucleus (nXIIts), with the latter then sending motor impulses to the syringeal muscles (Fig. 12.34). The DM and respiratory centers control the temporal features of songs via regulation of respiration; in general, notes and syllables are uttered during expiration and gaps between them occur during inspiration (Schmidt and Wild 2014).

When a bird is singing, the HVC receives input from a nucleus in the thalamus called Uva (or Uvaeformis), which in turn receives input from the DM and respiratory centers (Figs. 12.35 and 12.36). This "respiratory-thalamic" pathway is thought to provide information about the timing of songs to the HVC and, more specifically, about the coordination of syringeal muscle control and breathing (Schmidt and Wild 2014). The production of songs, therefore, depends on a "circular" neural network, with the HVC receiving the input necessary to control the characteristics and timing of songs.

Birds have two cerebral hemispheres and, therefore, have two HVCs and two RAs. Mammals have a structure called the corpus callosum that provides a pathway for nervous impulses to travel directly between cerebral

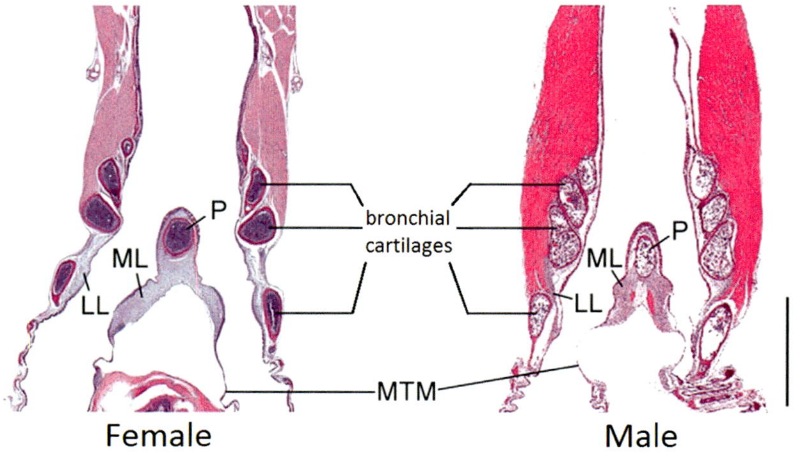


**Fig. 12.29** Sections through the syringes of a female and male Zebra Finch (*Taeniopygia guttatus*) illustrating sexual dimorphism in muscle mass and the structure of the bronchial cartilages. P, pessulus; ML, medial labium; LL, lateral labium; MTM, medial tympaniform membrane. (Figure modified from Riede et al. 2010; open-access article distributed under the terms of the Creative Commons Attribution License, https://creativecommons.org/licenses/by/4.0/)

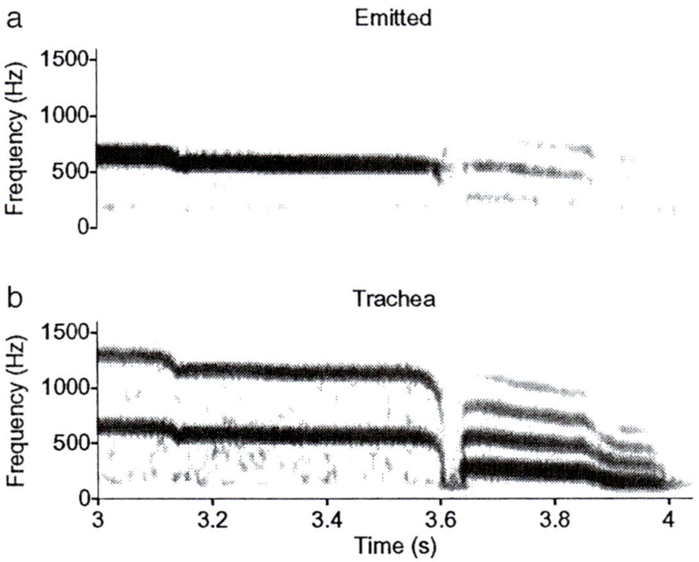


**Fig. 12.30** Coo call of a Ring Dove illustrating the filtering by the inflated esophagus. (**a**) Sonogram of a coo call of a Ring Dove. (**b**) Sonogram of a coo call as it would appear in the trachea prior to filtering. Note that the high-frequency harmonics have been eliminated, producing a low-frequency, pure-tone call. (Figure modified from Beckers et al. 2003a; © 2003 National Academy of Sciences USA, used with permission)

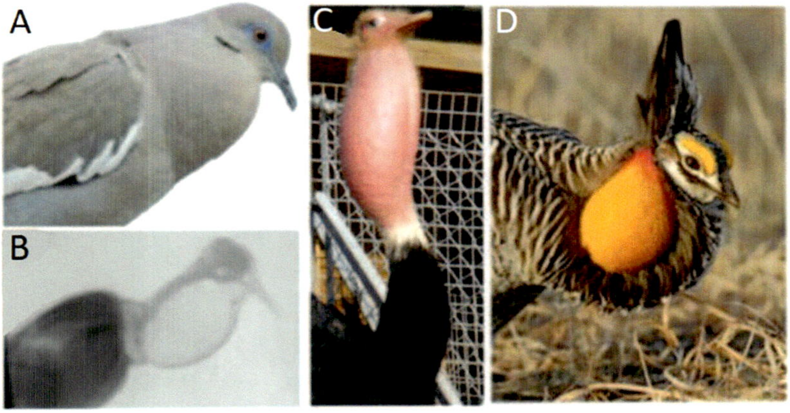


**Fig. 12.31** Examples of species of birds with closed-beak vocalizations. (**a**) Ring Dove and (**b**) Ring Dove X-ray image showing inflated esophagus, (**c**) Common Ostrich (*Struthio camelus*), and (**d**) Lesser Prairie-Chicken (*Tympanuchus pallidicinctus*). (Figure from Riede et al. 2016; © 2016 The Authors. Evolution © 2016 The Society for the Study of Evolution, used with permission)

hemispheres. Birds, however, do not. How then are the actions of the song control regions (HVC and RA) in the two hemispheres coordinated? Such coordination is needed because the song control regions in each hemisphere only control the muscles on one side of the syrinx, and the songs of some songbirds include rapidly alternating sequences of notes or syllables produced by the two sides of the syrinx. The respiratory-thalamic pathway provides a mechanism of coordination across cerebral hemispheres. The right and left Uvas in the thalamus receive input from respiratory centers and the DM of the midbrain, and the HVCs in each cerebral

**Fig. 12.32** Courtship display and vocalizations (low-frequency gurgling sounds) of a male Pectoral Sandpiper (*Calidris melanotos*), performed only when near a female. The calls are generated with a fully inflated throat (esophagus), and a closed beak (but open tip). (Figures from Riede et al. 2015; © 2014 Oxford University Press, used with permission)

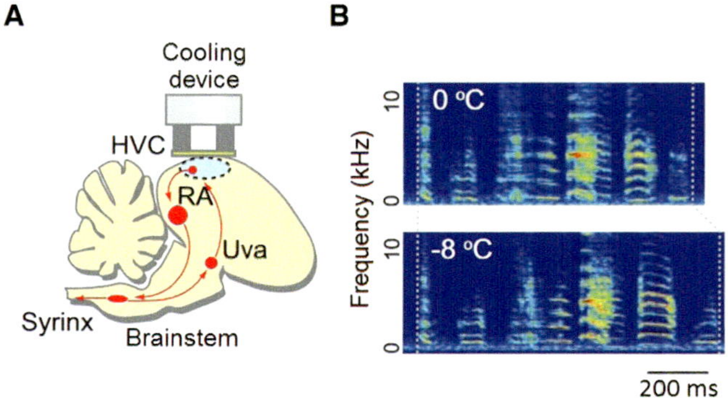


**Fig. 12.33** (**a**) Cooling of the neurons in the HVC causes a slowing in song timing, (**b**) with the duration of both syllables and gaps between syllables increasing. RA, robust nucleus of the arcopallium; Uva, relay nucleus in the thalamus. (Figure modified from Hamaguchi et al. 2016; © 2016 Elsevier Inc., used with permission)

hemisphere, in turn, receive input from the Uvas. This input is thought to synchronize the activity of the two HVCs and permit the precise timing of notes, syllables, and trills in the songs of songbirds (Schmidt and Wild 2014; Fig. 12.35).

In addition to songbirds, species in the order Psittaciformes and family Trochilidae (order Apodiformes; hummingbirds) are also vocal learners and have motor areas in the brain that control sound production via the syrinx. As with

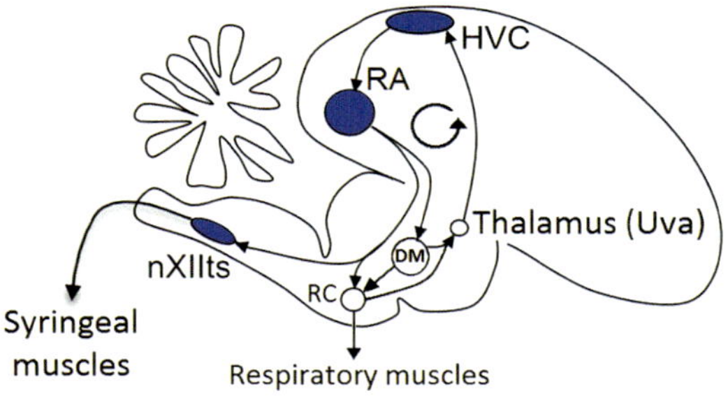


**Fig. 12.34** Drawing showing the song motor pathway. Motor impulses from the RA (robust nucleus of the arcopallium) stimulate the dorsomedial nucleus (DM) in the midbrain, respiratory centers in the brainstem (RC), and the hypoglossal motor nucleus (nXIIts), which then sends motor impulses to the syringeal muscles. (Figure modified from Andalman et al. 2011; © 2011 Andalman et al., open-access article distributed under the terms of the Creative Commons Attribution License, https://creativecommons.org/licenses/by/4.0/)

the songbird HVC, the primary vocal control areas are located in the nidopallium: the central nucleus of the lateral nidopallium (NLc) in the psittacids and the vocal nucleus of the lateral nidopallium (VLN) in hummingbirds (Fig. 12.36). Motor impulses from these areas stimulate neurons in the AAC (central nucleus of the anterior arcopallium) and VA (vocal nucleus of the arcopallium) in psittacids and hummingbirds, respectively. These impulses then, via the DM (dorsal medial nucleus of the midbrain), stimulate the hypoglossal nucleus that transmits impulses to the syringeal muscles. Among nonsong-learning species of birds, the vocal control area is located in the midbrain (the DM) (Fig. 12.37).

## 12.11 Seasonality of Song

Most species of birds breed and sing seasonally, with the timing typically influenced by environmental factors such as day length or precipitation. Seasonal changes in these factors trigger changes in gene expression in song control nuclei in the brain that, in turn, cause an increase in the number of neurons in, and the volume of, those nuclei (Fig. 12.38). These changes are thought to be regulated primarily by increased production of gonadal testosterone, with testosterone activating the neuronal gene needed to produce brain-derived neurotrophic factor (BDNF). BDNF, in turn, triggers neurogenesis and an increase in the number of neurons on the song control nuclei, the HVC, and robust nucleus of the arcopallium (RA) (Fig. 12.39). In addition, however, Li et al. (2000) found that the act of singing also appears to stimulate neurogenesis in the HVC of male Canaries (*Serinus canaria*) by enhancing production of BDNF (Fig. 12.40). Among male songbirds, the number of neurons in the HVC can increase dramatically. For example, the number of neurons in the HVC of male Song Sparrows (*Melospiza melodia*) increases from about 150,000 in the fall to 250,000 during the breeding season (Smith et al. 1997). The production of new neurons can also be surprisingly fast. The number of neurons in the HVC of captive White-crowned Sparrows (*Zonotrichia leucophrys*) treated with testosterone (via an implant) and exposed to 20-h days increased from 90,000 to 150,000 in just 7 days (Smith et al. 1997). This increase in the number of neurons in the HVC generally corresponds to an increase in singing rates (Alvarez-Borda and Nottebohm 2002; Fig. 12.41).

As mean volumes increase, males begin singing. As day length decreases after the breeding season, T levels drop, HVC neurons die, and males sing less or not at all (Top figure from Cohen et al. (2016); open-access article distributed under the terms of the Creative Commons Attribution License; bottom figure modified from Ball et al. (2004); © 2006 John Wiley and Sons, used with permission).

## 12.12 Sex Differences in the Song Control System

Among songbirds, differences in the singing behavior of males and females generally translate into differences in their song control systems,

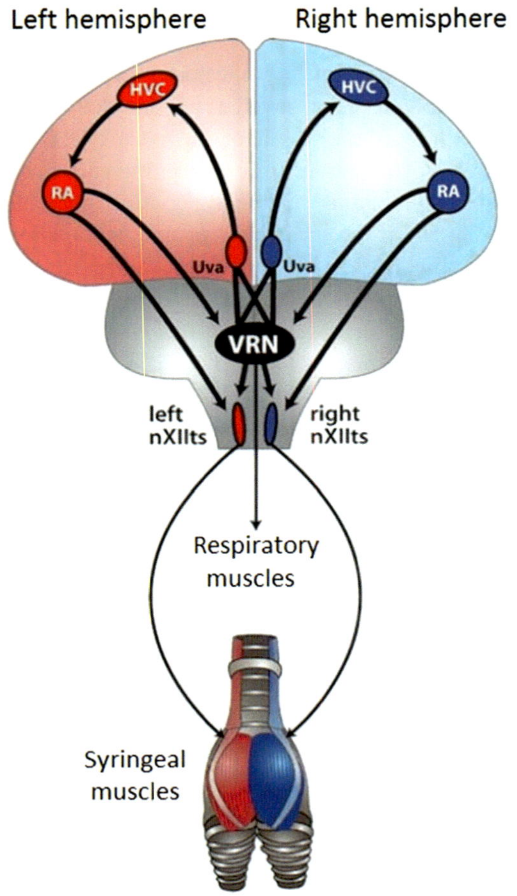


**Fig. 12.35** Motor portion of the avian song system illustrating the bilateral organization. Muscles on each side of the syrinx are innervated by motor neurons from the brainstem (ipsilateral nXIIts). These nuclei receive motor commands from the HVC and RA in the ipsilateral forebrain. As a result, motor impulses generated in the left hemisphere (red) stimulate muscles on the left side of the syrinx, and those on the right side (blue) stimulate syringeal muscles on the right side. Motor impulses from each hemisphere also stimulate brainstem nuclei that are part of the vocal-respiratory network (VRN), which is interconnected across the midline of the brainstem. The VRN, which includes the respiratory centers and the dorsomedial nucleus of the midbrain (DM; see Figure 12.34), also sends impulses to the HVC via the Uva. These impulses are thought to help synchronize the actions of both hemispheres and aid in the rapid hemispheric switching needed to generate sounds by alternating sides of the syrinx. (Figure modified from Schmidt 2008; © 2008 Marc F. Schmidt, open-access article distributed under the terms of the Creative Commons Attribution License, https://creativecommons.org/licenses/by/4.0/)

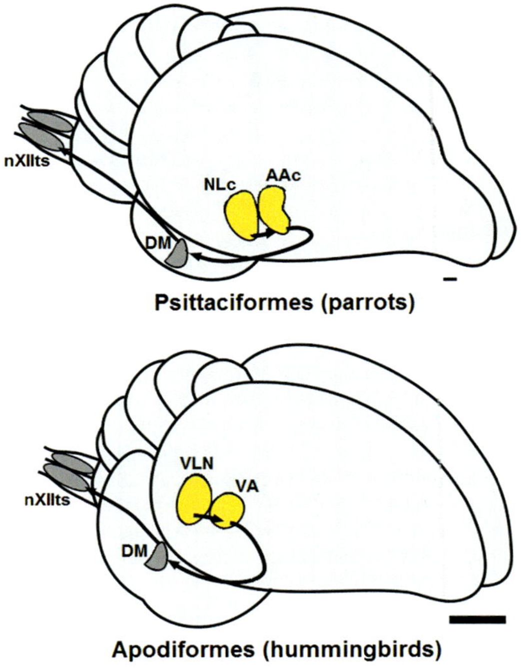


**Fig. 12.36** Motor pathways for vocalizing in parrots and hummingbirds. The primary vocal control areas are the NLc (central nucleus of the lateral nidopallium) for parrots and the VLN (vocal nucleus of the lateral nidopallium) for hummingbirds. AAc, Central nucleus of the anterior arcopallium; DM, dorsal medial nucleus of the midbrain; nXIIts, hypoglossal nucleus; VA, vocal nucleus of the arcopallium. Scale bars = 1 mm. (Figure modified from Chakraborty et al. 2015; © 2015 Chakraborty et al., open-access article distributed under the terms of the Creative Commons Attribution License, https://creativecommons.org/licenses/by/4.0/)

particularly the HVC. Species of songbirds that exhibit the greatest difference in singing behavior, i.e., males sing and females do not, exhibit the greatest sex differences in the volume of areas that make up the song control system. As differences between the sexes in singing behavior decrease, sex differences in the volume of their song control areas also tend to decrease (MacDougall-Shackleton and Ball 1999). However, even among species of songbirds where females sing as much, or even more, than males, the volume of the song control areas of males is greater than that of females. For example, females sing regularly in many species of songbirds found in the tropics and sub-tropics (Odom et al. 2014),

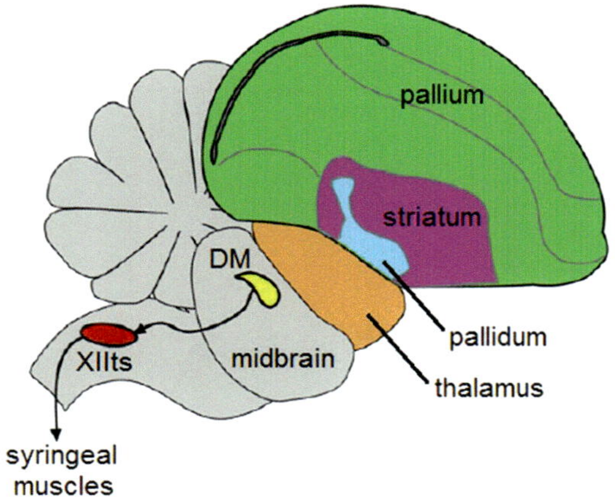


**Fig. 12.37** Species of birds that do not learn songs do not have forebrain vocal control areas. Rather, motor impulses to the syrinx originate in the dorsal medial nucleus of the midbrain (DM). XIIts, hypoglossal nucleus. (Figure modified from Arriaga and Jarvis 2013; © 2012 Elsevier Inc., used with permission)

but available evidence suggests that males in these species still have larger song control areas. Blue-capped Cordonbleus (*Uraeginthus cyanocephalus*) is one such species, with females singing at rates comparable to males and also singing complex songs. However, the ratio of female to male mean volumes of the HVC and RA was found to range between 0.5 and 0.7 (Lobato et al. 2015; Fig. 12.42). Similar results, where females sing as much as males, but have smaller song control areas, have been reported in Slate-colored Boubous (*Laniarius funebris*; Gahr et al. 1998), Buff-breasted Wrens (*Cantorchilus leucotis*; Brenowitz et al. 1985), Bay Wrens (*C. nigricapillus*; Brenowitz and Arnold 1986), and Dark-backed (Forest) Weavers (*Ploceus bicolor*; Gahr et al. 2008). Even in Streak-backed Orioles (*Icterus pustulatus*), where females sing at higher rates than males, the volume of male HVCs averaged 75% larger than those of females (Hall et al. 2010).

One possible explanation for sex differences in songbird song-control areas is that selection may favor male songbirds with song circuits able to produce songs attractive to females whereas, for females choosing mates, selection may favor more telencephalic space for perception, analysis, and comparison of those male songs (Ball and Balthazart 2008). Perhaps related to this hypothesis, Lobato et al. (2015) suggested that the song control areas of female songbirds, specifically the HVC and RA, function differently from those of males, enabling female to sing as much and as well as males, but with fewer neurons. If true, that difference remains to be determined.

**Fig. 12.38** Seasonal changes in the volume of two song-control areas in the brain of male Spotted Towhees (*Pipilo maculatus*). (**a**) HVC and (**b**) robust nucleus of the archistriatum (RA). Scale bars = 0.5 mm. (Figure from Tramontin and Brenowitz 2000; © 2000 Elsevier Science Ltd., used with permission)

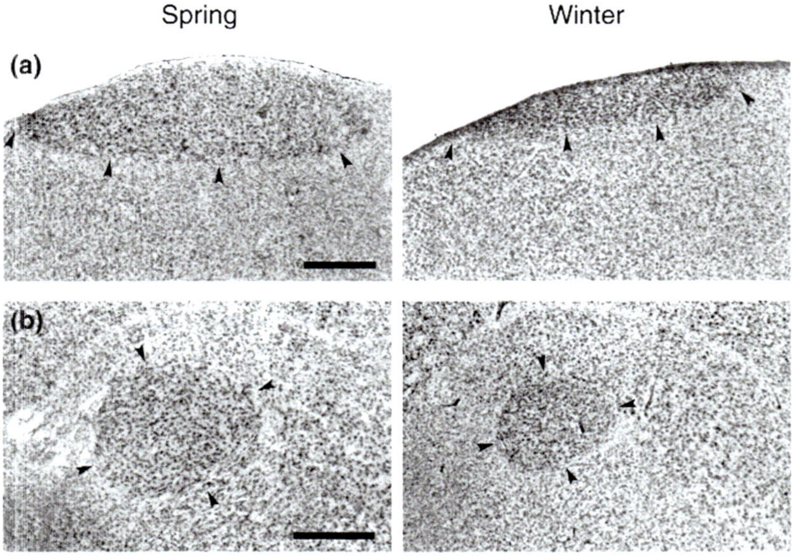

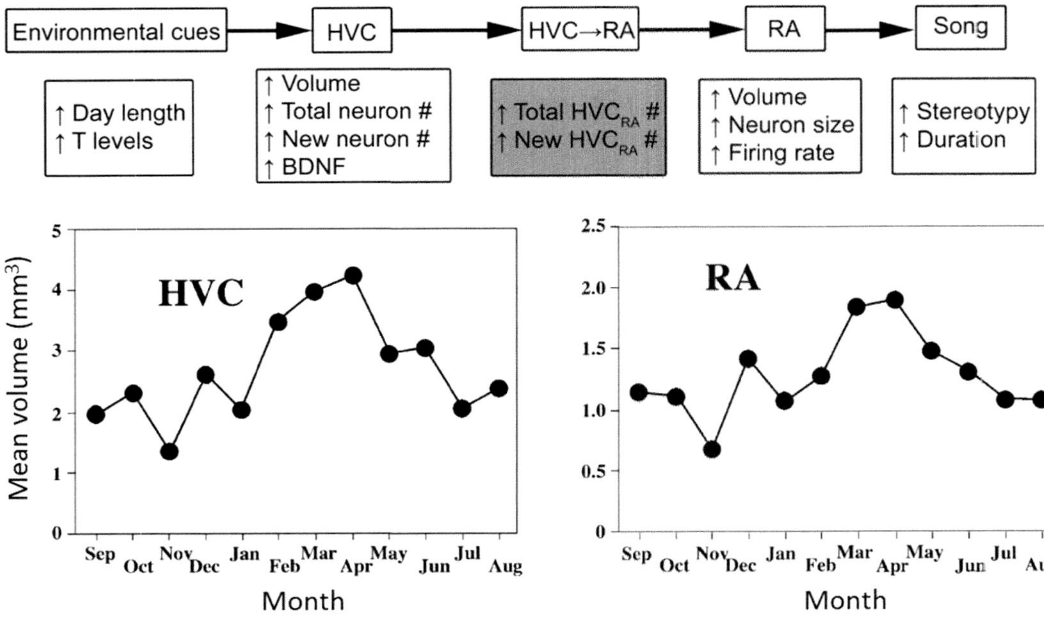


**Fig. 12.39** Seasonal changes in physiology, morphology, and behavior of songbirds. As day length increases at the onset of the breeding season, plasma testosterone (T) levels increase. The increase in T causes an increase in the expression of the gene that codes for brain-derived neurogenic factor (BDNF); BDNF triggers neurogenesis, increasing the number of neurons in the HVC and the robust nucleus of the arcopallium (RA). In male European Starlings in Belgium (50°N), mean volumes of the HVC and robust nucleus of the arcopallium (RA) are highest in late winter and early spring as day length increases and the breeding season begins (bottom). As mean volumes increase, males begin singing. As day length decreases after the breeding season, T levels drop, HVC neurons die, and males sing less or not at all. (Top figure from Cohen et al. 2016: open-access article distributed under the terms of the Creative Commons Attribution License; Bottom figure modified from Ball et al. 2004; copyright 2006 John Wiley and Sons, used with permission)

## 12.13 Classification of Vocalizations

The vocalizations of birds are typically divided into two categories: songs and calls. Calls are generally short in duration, most, but not all, are acoustically simple and uttered by both sexes throughout the year, and can serve a wide variety of different functions. Songs are typically longer in duration and more complex, and are important for defending territories and attracting mates (Catchpole 1982). At higher latitudes, singing is generally limited to the breeding season, but birds in the tropics and subtropics may sing throughout the year. The number of different calls in the vocal repertoires of birds varies widely, ranging from birds like pelicans and storks with very limited repertoires to those like some species of parrots and corvids with very large, complex repertoires (Appendix 1). Many species of birds have call repertoires ranging from 2 to 15 call types; fewer species have call repertoires greater than 15 call types (Appendix 1). Admittedly, however, determining the number of different calls in the repertoires of some species of birds is difficult because they may combine different calls or note types in their repertoires in different ways in different contexts or, for calls that consist of variable numbers of the same note types, they may vary the number of notes in these calls in different contexts.

By determining the context(s) in which calls are uttered and, in some cases, conducting experiments where certain calls are played back over speakers and the responses of birds are monitored, investigators have been able to identify a number of functional categories of calls (e.g., Clay et al. 2012; Rosa et al. 2016). These include:

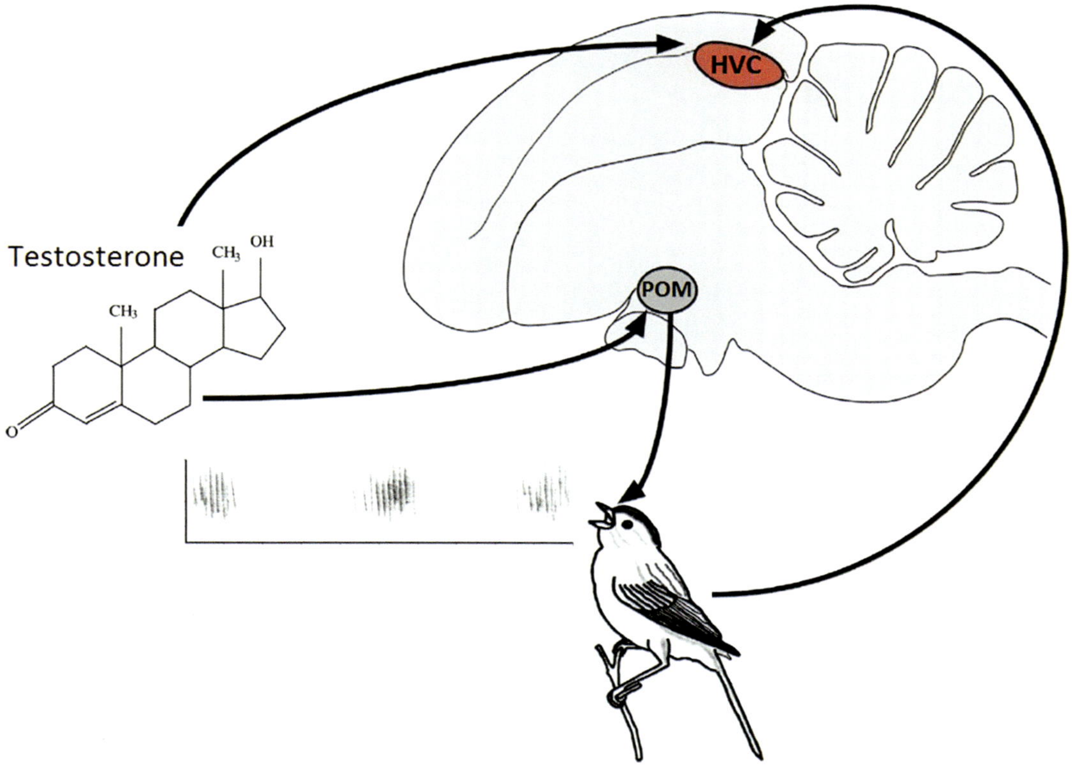


**Fig. 12.40** Alternative ways by which testosterone modulates neurogenesis in the HVC. The presence of androgen receptors in the HVC suggests that testosterone acts directly to modulate neurogenesis. However, stimulation of the preoptic nucleus (POM) by testosterone is known to increase singing rates of songbirds, and the act of singing can also cause an increase in neuronal recruitment in the HVC under certain conditions. Such song-induced increased neurogenesis in HVC is thought to be mediated by increased production of brain-derived neurotrophic factor (BDNF) in the HVC. (Figure modified from Balthazart and Ball 2016; © 2016 Published by Elsevier Inc., used with permission)

- **Contact calls** aid in maintaining contact with one or more conspecifics, but can also encode other information such as individual identity (Kondo and Watanabe 2009).
- **Feeding, or food-associated, calls** inform conspecifics that food has been located. For example, male Domestic Chickens (*Gallus g. domesticus*) utter "food calls" when they discover food, particularly when a female is present (Evans and Evans 1999). Other species of birds reported to utter food calls include House Sparrows (*Passer domesticus*), Cliff Swallows (*Petrocheldion pyrrhonota*), Carolina Chickadees (*Poecile carolinensis*), and Common Ravens (*Corvus corvax*) (Clay et al. 2012). Given that advertising the presence and location of food may lead to increased competition, food calls must also provide callers with some benefit. Possible reasons for uttering food calls include recruiting conspecifics to assist in scanning for predators while eating (e.g., House Sparrows) or to help track movements of swarms of flying insects (e.g., Cliff Swallows). Juvenile Common Ravens use food calls to attract conspecifics to food sources and make

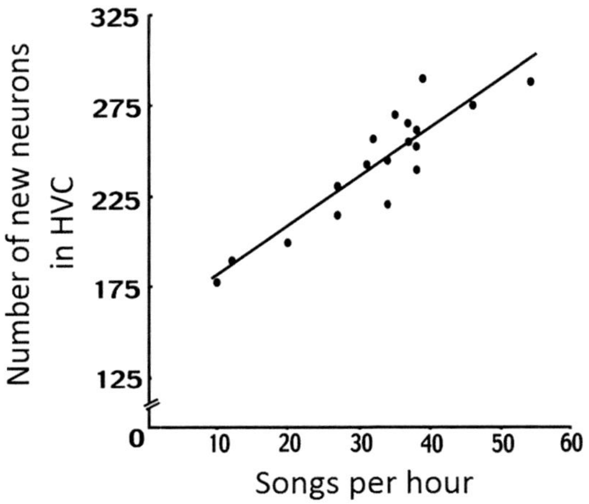


**Fig. 12.41** Relationship between singing rates of male Canaries and number of new neurons in the HVC. (Figure from Alvarez-Borda and Nottebohm 2002; © 2002 Society for Neuroscience, used with permission)

it more difficult for dominant individuals to prevent others from feeding.

- **Provisioning calls** are given by adults in some species when they approach or arrive at nests to feed nestlings. For example, adult Black-capped Chickadees (*Poecile atricapillus*) utter "squawk" calls at nests to stimulate gaping by young nestlings (Clemmons 1995). Provisioning calls uttered prior to arrival at nests may prepare nestlings for a food delivery and, after arrival, stimulate begging by nestlings. In either case, by reducing the time needed to feed nestlings, adults spend less time at nests (possibly reducing the likelihood of attracting the attention of predators) and have more time to forage.
- **Flight calls** convey an individual's intention to take flight and these calls also help maintain flock cohesion during flight. Some species of birds utter flight calls when migrating at night. These species-specific nocturnal flight calls may serve multiple functions during migration, including group cohesion and coordination of movements (Fig. 12.43). Several investigators have reported that nocturnal flight calls are given at higher rates in the hours before sunrise, possibly signaling to

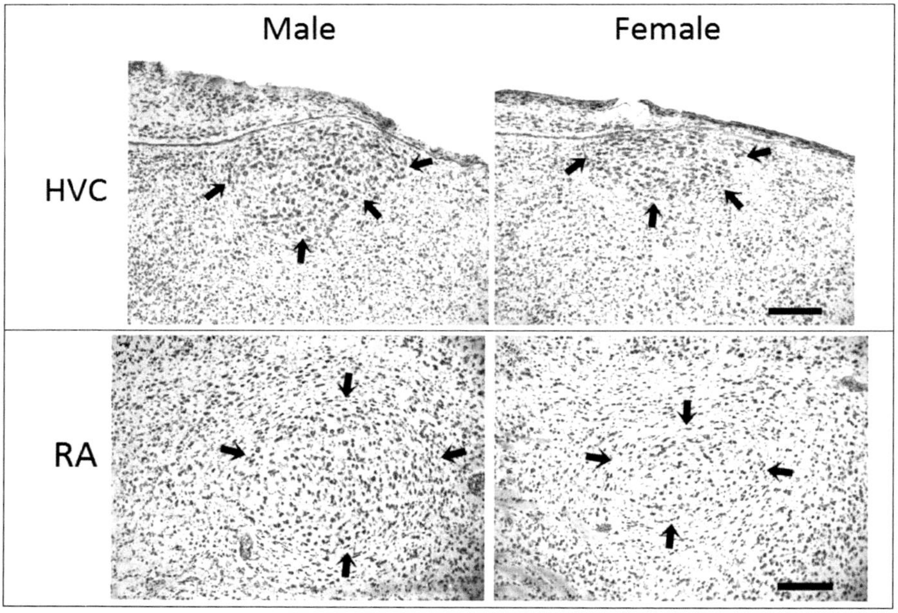


**Fig. 12.42** Song-control nuclei of male Blue-capped Cordonbleus (left; *Uraeginthus cyanocephalus*) are larger in volume than those of females (right). RA, robust nucleus of the arcopallium. Scale bars = 100 μm. (Figures modified from Lobato et al. 2015; open-access article distributed under the terms of the Creative Commons Attribution License, https://creativecommons.org/licenses/by/4.0/)

**Fig. 12.43** Examples of the species-specific nocturnal flight calls uttered during migration by several species of warblers. Ovenbird, *Seiurus aurocapilla*; Magnolia Warbler, *Setophaga magnolia*; American Redstart, *Setophaga ruticilla*; Black-and-white Warbler, *Mniotilta varia*; Tennessee Warbler, *Oreothlypis peregrina*; Virginia's Warbler, *Leiothlypis virginiae*; Black-throated Gray Warbler, *Setophaga nigrescens*. (Figure modified from Lanzone et al. 2009; © 2009 Oxford University Press, used with permission)

conspecifics the approaching end of the flight and search for a stopover site (e.g., Hüppop and Hilgerloh 2012; Fig. 12.44).

- **Flight coordination calls** aid in coordinating the movements of individuals in flocks of flying birds. For example, Blue-fronted (Turquoise-fronted) Amazons (*Amazona aestiva*) utter calls ("Grr-uíp") when changing positions in a flock or when maneuvering prior to landing in trees (Fernández et al. 1998). Arnold et al. (2022) recorded the vocal behavior and position of each bird in flocks of six Zebra Finches (*Taeniopygia guttata*) flying in a wind tunnel and found that they used both visual and acoustic cues for three-dimensional flock coordination. However, when visual information was lacking (e.g., when flying at or near the front

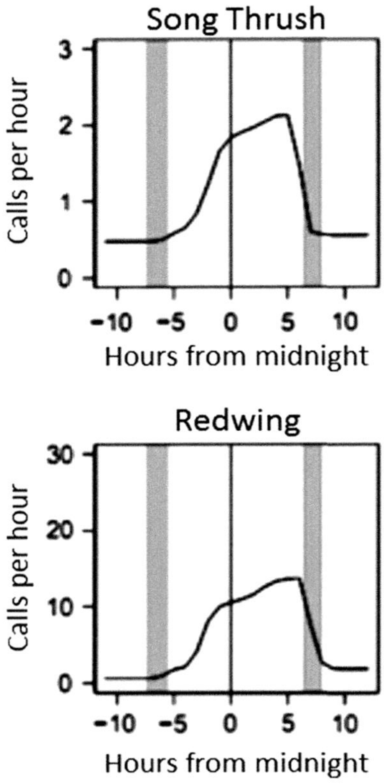


**Fig. 12.44** Some migrants, including Song Thrushes (*Turdus philomelos*) and Redwings (*Turdus iliacus*), utter nocturnal flight calls at higher rates prior to sunrise. Vertical gray bars indicate sunset and sunrise. (Figure modified from Hüppop and Hilgerloh 2012; © 2012 The Authors, used with permission)

of the flock), Zebra Finches increasingly used vocal communication ("stack" calls) to avoid collisions with flock mates.

- **Territorial calls** aid in maintaining territory boundaries. Among songbirds, territory defense is one important function of singing. However, songbirds may also use calls when defending territories as do many nonsongbirds, e.g., female Common Cuckoos (*Cuculus canorus*; Moskát and Hauber 2019).
- **Aggressive calls** convey aggressive intent to conspecifics, indicating an increasing likelihood of attack and serving to potentially intimidate opponents, e.g., mewing call of male Corn Crakes (*Crex crex*; Rek and Osiejuk 2011).
- **Submissive calls** signal submission to a dominant conspecific. For example, lower-ranking male Long-tailed Manakins (*Chiroxiphia linearis*) utter a specific call ("chitter") after being chased by, or involved in an agonistic interaction with, a dominant male (Trainer and McDonald 1993).
- **Alarm calls** alert conspecifics to the presence of a predator (Suzuki 2011; Gill and Bierema 2013). "Eavesdropping" heterospecifics sometimes also recognize that these calls indicate the presence of a predator and respond accordingly (Potvin et al. 2018). Potential benefits of eavesdropping include an increased likelihood of detecting predators, less time spent being vigilant, and, with the increased information about predator presence, being able to select safer breeding and foraging areas (Sullivan 1984; Schmidt et al. 2010). In some cases of eavesdropping, individuals are responding to heterospecific calls that are acoustically similar to their own. In many cases, however, eavesdropping likely requires learning (Figs. 12.45 and 12.46).
- **Aerial predator alarm calls** are more specific and alert conspecifics to the presence of an aerial predator. These calls often have characteristics that make it difficult for the predator to determine the location of the calling bird (e.g., Wood et al. 2000; Box 12.5 Locatability).
- **Terrestrial predator alarm calls** are also more specific and alert conspecifics to the presence of a nonflying predator (e.g., Regosin 2002).
- **Mobbing calls** are used to harass predators and attract other birds to aid in mobbing a predator, with mobbing defined as the movement of potential prey species toward a predator while uttering easily located mobbing calls to attract conspecifics and, often, heterospecific and, in some cases, flying at and even striking the predator (Curio 1978). The likely function of mobbing is to induce the predator to leave the area ("move-on" hypothesis; Curio 1978). The mobbing calls of some species of birds are functionally referential (see "Referential calls" section).

**Fig. 12.45** Example of learning to associate the alarm call of another species with the threat of predation. Magrath et al. (2015) conducted an experiment where Superb Fairywrens (*Malurus cyaneus*) were exposed to novel "alarm" calls (i.e., one the investigators synthesized and one from a species not found in their study area). Initially, the fairywrens did not respond when the novel alarm calls were played back over a speaker (Before training). The fairywrens were then exposed to both playback of the novel calls plus gliding models of two species of predators (Training). After two days of training, most fairywrens took cover in vegetation when just exposed to playback of the calls. (Figure modified from Magrath et al. 2015; © 2015 Elsevier Ltd., used with permission)

- **Distress calls** are given when a bird is being held by a human or is attacked by a predator (Box 12.6 Distress Calls).
- **Courtship, or copulation solicitation, calls** signal reproductive intent and, more specifically, intent to copulate. For example, male Rock Sparrows (*Petronia petronia*) utter two-note courtship calls, in combination with courtship displays, in the presence of females and, if females respond by assuming a copulation-soliciting display, the pair copulates (Matessi et al. 2005). Calls by females uttered prior to copulation, referred to as copulation solicitation calls, mating calls, or precopulatory calls, have been reported in numerous species of birds (e.g., Anderson 2009; Amy et al. 2015; Reichard and Welklin 2015).
- **Copulation calls** have been reported in several (but not most) species of birds. They are uttered when birds copulate and may serve different functions in different species. In some species, females may utter calls during copulation that are loud and are sometimes referred to as distress calls rather than copulation calls. These calls may attract other males

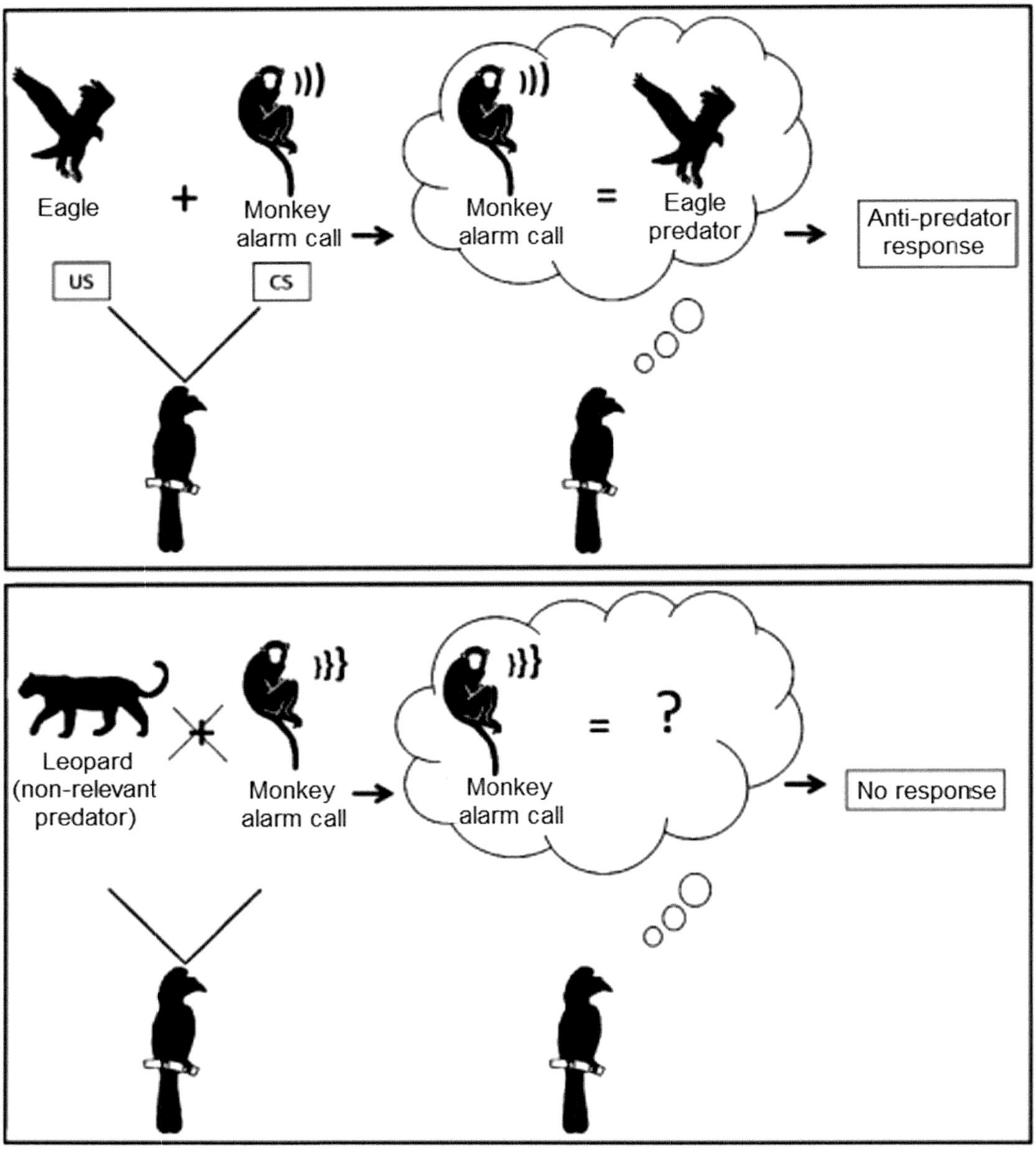


**Fig. 12.46** Heterospecific social communication via learning (Rainey et al. 2004). (Top) An association between an alarm call given by Diana monkeys (*Cercopithecus diana*) responding to the presence of an aerial predator that also represents a potential threat to birds results in Yellow-casqued Hornbills (*Ceratogymna elata*) learning that the call indicates the presence of an aerial predator. (Bottom) However, a terrestrial predator like a leopard (*Panthera pardus*) is not a threat to hornbills so the alarm call given by Diana monkeys responding to the presence of a leopard elicits no response by hornbills. (Figure modified from Avarguès-Weber et al. 2013; © 2013 The Zoological Society of London, used with permission)

**Box 12.5 Locatability**

In a paper published in 1955, Peter Marler proposed that animal sounds are "adapted in structure to the function they have to perform." For example, the vocal repertoire of many species of birds includes contact or location calls that are important in maintaining pair or group cohesion. As such, these calls should have characteristics that allow conspecifics, or even heterospecies among species that occur in mixed-species flocks, to accurately determine the location of the calling individual. Similarly, many species of birds also have mobbing calls that are given when a potential predator, such as an owl, is located. These calls attract conspecifics, and sometimes heterospecifics, which then, collectively, "mob" or harass the predator, perhaps to encourage the predator to leave the area. As with contact calls, to attract others to the location of the predator, the bird or birds uttering mobbing calls must be easy to locate. Characteristics that make calls locatable include an abrupt beginning and ending and extending over a wide range of frequencies, like the contact calls of Pied Bushchats (*Saxicola caprata*) and the mobbing calls of Brown Thornbills (*Acanthiza pusilla*), New Holland Honeyeaters (*Phylidonyris novaehollandiae*), and Superb Fairywrens (*Malurus cyaneus*).

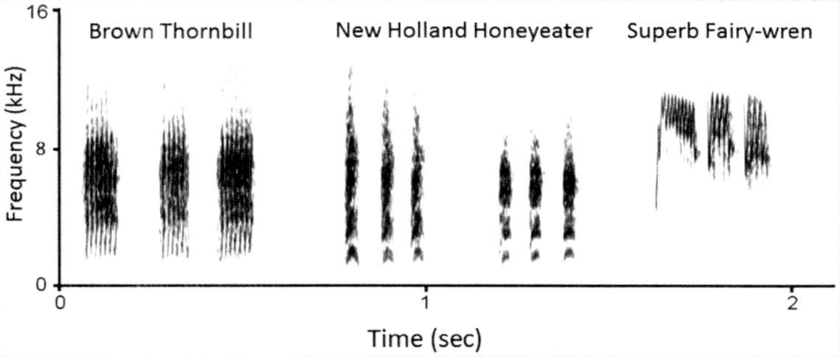


Mobbing calls of three songbirds: Brown Thornbill, New Holland Honeyeater, and Superb Fairywren. (Figure modified from Igic and Magrath 2014; © 2014 Oxford University Press, used with permission)

Marler (1955) also suggested that, in some cases, calling individuals may benefit by not providing listeners with information about their location. An obvious example is when a bird calls to inform conspecifics that an aerial predator has been spotted, and that call might also be heard by the predator. In this case, uttering a call that conveys the needed information to conspecifics, but does not help the predator determine the specific location of caller might be beneficial. The aerial predator alarm calls of several species of birds, including Common Chaffinches (*Fringilla coelebs*), Common Blackbirds (*Turdus merula*), and Great Tits (*Parus major*), are high-frequency, narrow-frequency-band calls that gradually increase and decrease in volume as they begin and end. Humans find it difficult to identify the source of calls (often referred to as "seet" calls) with these characteristics and, importantly for species of birds with these calls, laboratory experiments have revealed that at least some raptors have difficulty as well (Jurisevic and Sanderson 1998; Jones and Hill 2001).

(continued)

**Box 12.5** (continued)

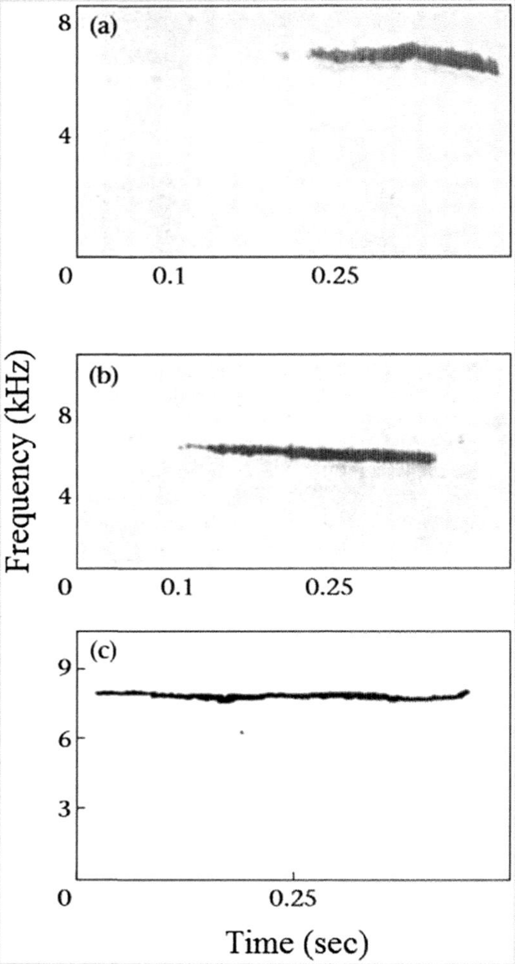


Examples of the aerial predator alarm (or "seet") calls of (**a**) Common Chaffinches, (**b**) Common Blackbirds, and (c) Great Tits. (Figure from Searcy and Yasukawa 2017; © 2016 The Association for the Study of Animal Behaviour. Published by Elsevier Ltd., used with permission)

Brown (1982) tested the ability of Red-tailed Hawks (*Buteo jamaicensis*) and Great Horned Owls (*Bubo virginianus*) to localize the mobbing calls of Red-winged Blackbirds (*Agelaius phoeniceus*) and "seet" calls of American Robins (*Turdus migratorius*) and found that both raptors were more than twice as good at localizing the source of mobbing calls than seet calls. Similarly, Jones and Hill (2001) tested the ability of eight species of raptors to localize the "seet" call of American Robins and the mobbing call of Tufted Titmice (*Baeolophus bicolor*) and found they were significantly better at localizing the source of mobbing calls than seet calls. These results lend support to Marler's (1955) hypothesis that some vocal signals can both convey information to conspecifics and minimize risk to the caller by "hiding" their location.

**Box 12.6 Distress Calls**

Distress calls (also referred to as distress screams or fear screams) are loud, harsh calls given by some species of birds when they are captured by a predator or handled by humans. Hypotheses proposed to explain the evolution and function of distress calls include: (1) calling for help from kin, (2) warning kin, (3) eliciting mobbing behavior, (4) startling the predator, and (5) distracting the predator by attracting other predators (Neudorf and Sealy 2002). Several investigators have tested these hypotheses. For example, during thousands of hours in the field over a period of 38 years, Møller and Nielsen (2010) observed that distress calls uttered by captured birds do sometimes attract predators (12 observations made when removing birds from mist-nets), and that distress calls uttered by birds being chased or captured by a predator can attract conspecifics (26 observations) that, by harassing the predator, sometimes help the calling bird escape (10 of the 26 observations, or 38%). These authors also noted that species of birds where individuals exhibited greater genetic similarity were more likely to utter distress calls than species with less genetic similarity, lending support for the kin-warning hypothesis.

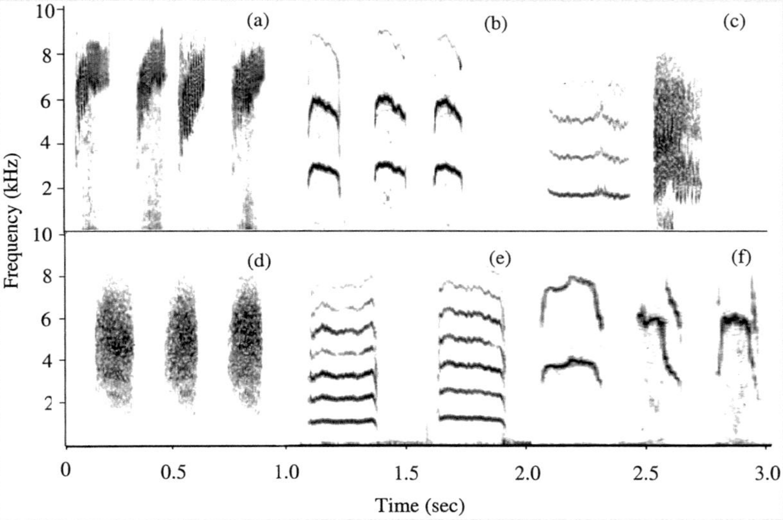


Sonograms of distress calls of six species of birds, including (**a**) Sooty-capped Chlorospingus (*Chlorospingus pileatus*), (**b**) Black-capped Flycatcher (*Empidonax atriceps*), (**c**) Mexican Violetear (*Colibri thalassinus*), (**d**) Gray-breasted Wood-Wren (*Henicorhina leucophrys*), (**e**) Streak-breasted Treehunter (*Thripadectes rufobrunneus*), and (**f**) Yellowish Flycatcher (*Empidonax flavescens*). Each sonogram represents 1 s of calling. (Figure from Neudorf and Sealy 2002; © 2002 Association for Tropical Biology and Conservation, used with permission)

Branch and Freeberg (2012) studied Tufted Titmice (*Baeolophus bicolor*) and found that their distress calls did not readily attract larger avian predators to the area, and suggested that the calls function either to confuse or startle predators or warn conspecifics and elicit their mobbing behavior. Lee et al. (2015) suggested that the distress calls of Vinous-throated Parrotbills (*Sinosuthora webbiana*) may function to attract conspecifics and other species, potentially disturbing the predator and giving the caller a chance to escape. In contrast, Koenig et al. (1991) found that the distress calls of Acorn Woodpeckers (*Melanerpes formicivorus*) did not attract conspecifics, but did sometimes attract large mammalian predators, lending support to the predator-attraction hypothesis. Finally, based on a study of the distress calls of Phainopeplas

(continued)

**Box 12.6** (continued)

(*Phainopepla nitens*), Chu (2001) concluded that distress calls elicit mobbing by other birds, but could also have other functions, including startling predators and attracting secondary predators. Collectively, then, the results of these studies suggest that multiple selective factors have favored the evolution of distress calls and these calls can simultaneously serve multiple functions that vary among species.

which, after disrupting the copulation, may then copulate with the female that gave the calls. The function of these female copulation calls may be to attract a higher-quality mating partner (Løvlie et al. 2014). In other species, males utter calls when copulating with females. For example, male Herring Gulls (*Larus argentatus*) utter a series of loud calls when copulating, perhaps to discourage other males from interfering (Weseloh et al. 2020). In yet other species, either or both the male and female call when copulating. For example, female White-breasted Nuthatches (*Sitta carolinensis*) utter a "phee-oo" call immediately after copulating (Ritchison 1983). Male and female Common Black Hawks (*Buteogallus anthracinus*) utter a series of notes for about 10 seconds while copulating (Schnell 2020). Female Swainson's Hawks (*Buteo swainsoni*) utter low volume, "weeee . . ." calls during copulation (Bechard et al. 2020). The possible function(s) of these often low-volume copulation calls remain(s) to be determined.

- **Recruitment calls** are used by some group-living species to attract conspecifics to the location of the calling bird. For example, Pale-winged Trumpeters (*Psophia leucoptera*) utter a recruitment call ("drums") when a potentially dangerous prey item that might require group-effort to kill, such as a snake, is observed (Seddon et al. 2002). In Pied Babblers (*Turdoides bicolor*), parents and helpers use a recruitment call to direct nutritionally independent, but inexperienced, young to particular food patches (Radford and Ridley 2006).
- **Female nest-departure calls** have been reported in several species of songbirds and may serve to elicit nest guarding by mates (e.g., Grunst et al. 2014).
- **Female nest-defense calls** given by females at nests may deter potential predators (e.g., Koosa and Tilgar 2016; Zub et al. 2017).
- **Adult begging calls** are uttered by adult females during the breeding season to solicit food from their mates (often referred to as courtship feeding or allofeeding) (e.g., Moore and Rohwer 2012; Boucaud et al. 2016).
- **Juvenile begging calls** are uttered by young birds to solicit feeding by adults (Budden and Wright 2001).
- **Embryo calls** are uttered in the days just prior to hatching, more commonly in precocial than altricial species of birds (Oppenheim 1972). These calls may signal to parents the impending hatching of young (Saino and Fasola 1996) and, in some cases, solicit incubation by parents when embryo temperatures decline (Brua et al. 1996). Embryos may also call to each other to better synchronize the time of hatching (Brua 2002; Rumpf and Tzschentke 2010). In addition, adult Red-capped Plovers (*Charadrius ruficapillus*) were more likely to engage in high-intensity antipredator displays when embryos were vocalizing than when embryos were not vocalizing (Kostoglou et al. 2020).

As just described, different calls in the repertoires of different species of birds serve different functions, but they can also provide information about individual identity (Stoddard 1996), kinship (McDonald and Wright 2011; Keen et al.

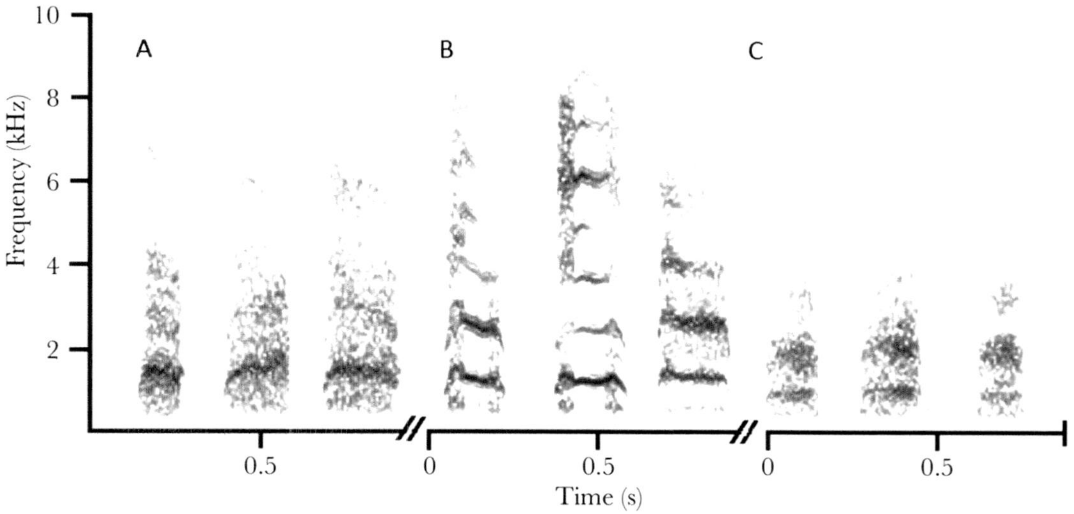


**Fig. 12.47** Contact calls of three different Apostlebirds (*Struthidea cinerea*; **a**, **b**, and **c**) in the same social group. These calls help Apostlebirds maintain contact with others in their groups, but differences in the characteristics of the calls of different individuals also allow Apostlebirds to identify the specific individual giving the calls. (Figure modified from Warrington et al. 2015; © 2014 Oxford University Press, used with permission)

2013), and group membership (Hopp et al. 2001; Keen et al. 2013) that can be perceived by conspecifics (Sharp et al. 2005). For example, Apostlebirds (*Struthidea cinerea*) live in social groups and can identify individuals in those groups based on individual variation in the characteristics of their contact calls (Warrington et al. 2015; Fig. 12.47). Similar individual variation in contact calls, as well as calls that serve other functions, has been reported in numerous other species (e.g., Sharp and Hatchwell 2005, Kennedy et al. 2009, Berg et al. 2011; Fig. 12.48). The ability to recognize the calls of different individuals is particularly common among group-living species of birds. For species with complex social systems, individual recognition can be important in a variety of contexts, including mate recognition, dominance hierarchies, and discriminating between group members and nongroup members (Dale et al. 2001).

Although the characteristics of some bird calls, such as contact calls and flight calls, exhibit little intra- and interindividual variation. The characteristics of other calls can vary with context, effectively increasing the size of the vocal repertoire of a species. Two functional categories of calls where such variation occurs in several species of birds are alarm calls and mobbing calls. For example, characteristics of the aerial alarm calls of White-browed Scrubwrens (*Sericornis frontalis*) vary with the distance from a predator, when predators get closer the number of elements in the call and the minimum frequency both increase (Leavesley and Magrath 2005). For American Crows (*Corvus brachyrhynchos*) responding to the presence of a predator, longer-duration mobbing calls uttered at a higher rate indicate a greater degree of danger (Yorzinski and Vehrencamp 2009; Fig. 12.49). Similar variation in the characteristics of alarm and mobbing calls has been reported in many other species of birds and this variation provides general information about the threat posed by a predator, but not specific information about the type of predator (referred to as referential signaling).

Among some species of birds, the characteristics of calls can vary as a means of more effectively communicating with

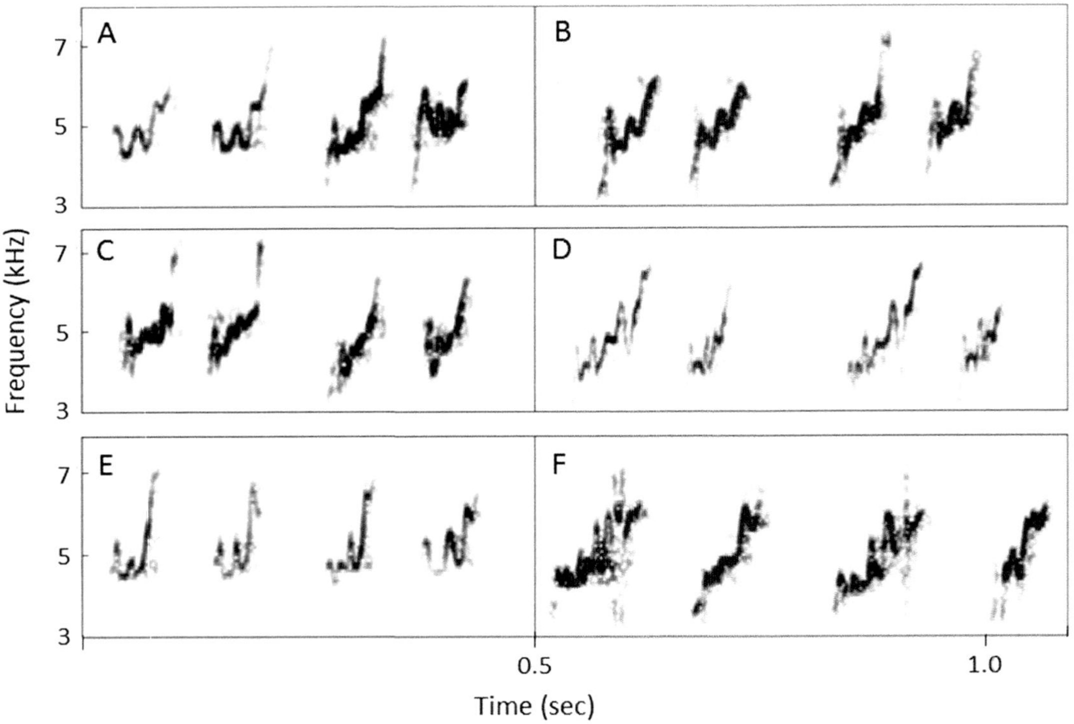


**Fig. 12.48** Variation in the contact calls of six Green-rumped Parrotlets (*Forpus passerinus*; **a–f**). This variation appears to be particularly important in the recognition of mates during the breeding season of these parrotlets. (Figure modified from Berg et al. 2011; © 2010 The Association for the Study of Animal Behaviour, used with permission)

conspecifics. For example, male Spotted Crakes (*Porzana porzana*) have a vocal repertoire of only five basic calls. However, by modifying the characteristics of those calls, territorial males actually have much larger vocal repertoires that allow them to convey more specific information about the level of their "aggressive motivation" (Ręk 2015; Fig. 12.50). Carolina Chickadees (*Poecile carolinensis*) have larger vocal repertoires than many other species of birds, and one call in their repertoire is the "chick-a-dee" call. This call consists of a series of different note types and the order and number of these notes in different "chick-a-dee" calls can vary, but the extent of that variation was not known until Krams et al. (2012) studied a captive flock of six individuals. Recording and analysis revealed an increasing number of unique calls (i.e., with a different order and number of different notes) with an increasing number of calls recorded. Even after recording 100 calls, the number of unique calls was still increasing (Fig. 12.51), indicating that their repertoires of different "chick-a-dee" calls are very large. The extent to which this extreme variation plays a role in chick-adee communication remains to be determined. Many other species of birds also modify the characteristics of calls, e.g., duration, frequency, and/or intervals between notes in calls, and, as such, the size of their "functional vocal repertoires" is often much larger than their "basic vocal repertoires" (Appendix 1).

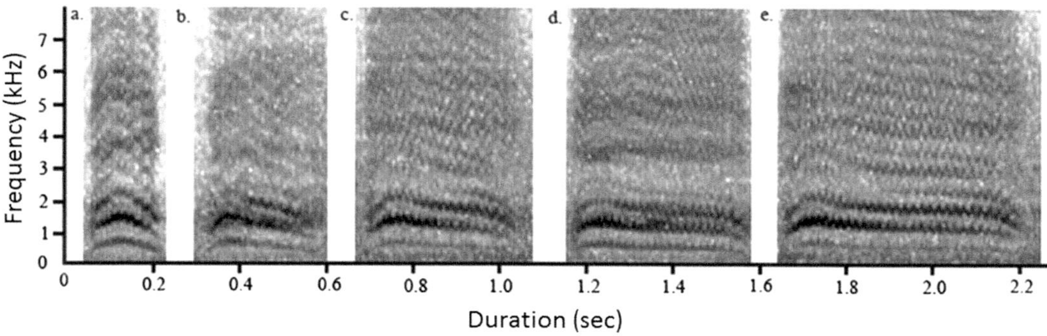


**Fig. 12.49** Sonograms of mobbing calls of American Crows (*Corvus brachyrhynchos*). Calls of increasing duration (**a**–**e**) indicate an increasing degree of threat posed by a predator. (Figure from Yorzinski and Vehrencamp 2009; © 2009, Oxford University Press, used with permission)

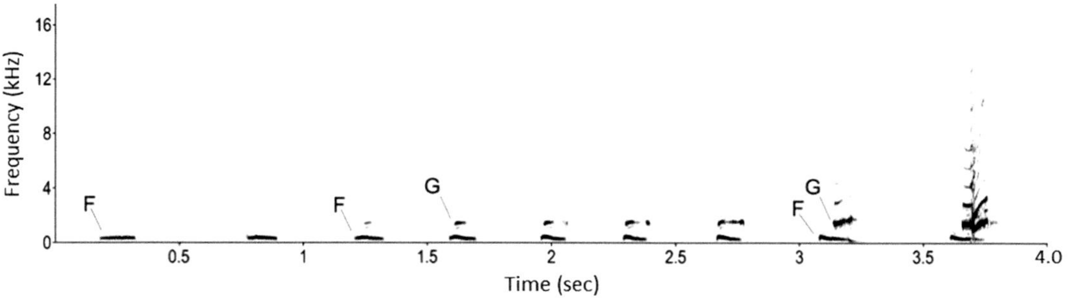


**Fig. 12.50** Sonograms of the calls of male Spotted Crakes (*Porzana porzana*) showing several different variants of a single call type. One call in their vocal repertoire is the "whitt" call that often consists of two notes (F and G). However, during interactions with conspecific males, males produce several variants of the "whitt" call to signal different levels of aggression. (Figure from Ręk 2015; © 2015 Oxford University Press, used with permission)

## 12.14 Referential Calls

Referential calls are calls or a call with particular characteristics that are given in responses to a particular stimulus and elicit appropriate, specific response from conspecifics that hear the call (Smith 2017). Investigators have reported several examples of referential calls given by several species of birds, and those calls have been found to provide information either about the location of food or about predators (Table 12.1). Food calls provide mates, young, or other conspecifics with information about the location of a food source. For example, subordinate Common Ravens (*Corvus corax*) utter a food call ("haa") at food sources being defended by more dominant individuals (Heinrich and Marzluff 1991; Bugnyar et al. 2001). Lower-ranking ravens may benefit by calling and attracting others to a food source because, with the help of other ravens, they may increase their chances of gaining access to the food despite the attempts of more dominant birds to deny them access (Marzluff and Heinrich 1991). More generally, food calls attract conspecifics to a food source. Although attracting additional foragers could potentially mean less food for the calling individual, such behavior likely has benefits as well. For example, increased foraging success of kin (e.g., newly fledged young) may enhance a bird's inclusive fitness. Even attracting nonkin

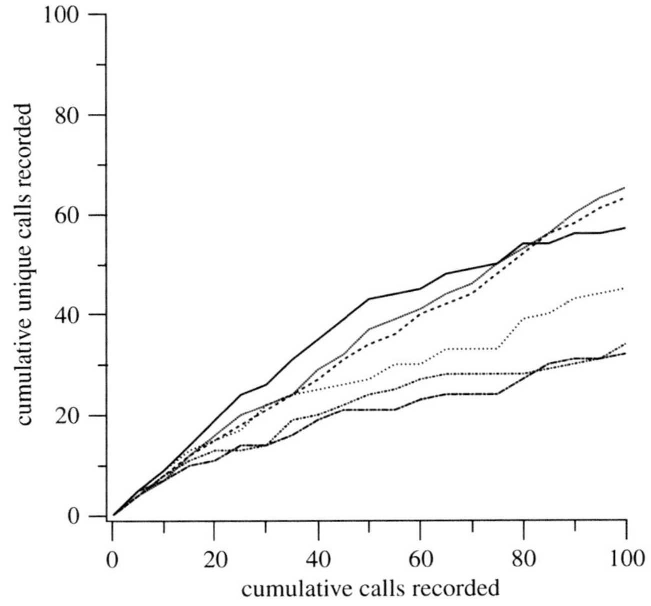


**Fig. 12.51** The cumulative number of novel "chick-a-dee" calls uttered by six Carolina Chickadees continued to increase even after recording 100 calls. Each line represents a different Carolina Chickadee. (Figure from Krams et al. 2012; open-access article distributed under the terms of the Creative Commons Attribution License, https://creativecommons.org/licenses/by/4.0/)

can be beneficial if increasing the number of individuals at a food source reduces predation risk via a dilution effect, increased vigilance, or both (Clay et al. 2012).

As noted previously, several species of birds have different alarm calls for aerial and terrestrial predators (Table 12.1), and conspecifics respond differently, and appropriately, when hearing those calls. As an example, when hearing the call indicating the presence of an aerial predator, Smooth-billed Anis (*Crotophaga ani*) seek cover in vegetation, but, when responding to the terrestrial predator alarm call, they take flight (Grieves et al. 2014). Cavity-nesting Japanese Tits (*Parus minor*) also have two types of alarm calls, but use them to warn their nestlings (Suzuki 2011). For older nestlings in a cavity, the threat posed by a predator depends on whether or not that predator can enter the nest cavity so the alarm calls of adult Japanese Tits provide that information. In response to the presence of a larger predator unable to enter a nest cavity, like Jungle Crows (*Corvus macrorhynchos*), nestlings respond to the adult alarm call ("chicka") by crouching down, making it less likely that the predator can reach them with their bill or feet (Fig. 12.52). However, in response to the call uttered by adults alerting nestlings to the presence of a predator able to enter a nest cavity, like a Japanese rat snake (*Elaphe climacophora*), nestlings immediately leave the cavity. Even if unable to fly, young that fledge prematurely are more likely to survive than young trapped in a nest cavity by a predator like a rat snake.

Three species of parids (family Paridae) have different alarm calls for aerial predators in flight ("seet" calls; described previously) and aerial predators that are perched ("chick-a-dee" calls). Additionally, however, these parids vary the number of "dee" notes in their "chick-a-dee" calls, with more "dee" notes in the calls when responding to the presence of smaller, more threatening aerial predators (Figs. 12.53 and 12.54). "Chick-a-dee" calls, therefore, not only alert conspecifics to the presence of a perched aerial predator, but also provide information about the threat posed by those aerial predators.

## 12.15 Learning Calls

Vocal learning is the ability to imitate the vocalizations of conspecifics and, in some cases, heterospecifics. Much of what is known about

**Table 12.1** Examples of functionally referential calls given by several species of birds

| Species | Scientific name | Information conveyed | Citation |
|---|---|---|---|
| Domestic Chicken | *Gallus domesticus* | Food | Evans and Evans (1999) |
| | | Aerial predator size, speed, & proximity | Wilson and Evans (2012) |
| Red Junglefowl | *Gallus gallus* | Food | Van Kampen and Hogan (2000) |
| White-tailed Ptarmigan | *Lagopus leucurus* | Predator type[a] | Ausmus and Clarke (2014) |
| Red-legged Partridge | *Alectoris rufa* | Predator type | Binazzi et al. (2011) |
| Smooth-billed Ani | *Crotophaga ani* | Predator type | Grieves et al. (2014) |
| Cliff Swallow | *Hirundo pyrrhonota* | Food | Brown et al. (1991) |
| Barn Swallow | *Hirundo rustica* | Cuckoo vs. hawk[b] | Yu et al. (2016) |
| Common Raven | *Corvus corax* | Food | Bugnyar et al. (2001) |
| Australian Magpie | *Gymnorhina tibicen* | Predator type | Kaplan and Rogers (2013) |
| Australian Magpie | *Gymnorhina tibicen* | Predator distance | Dutour et al. (2021) |
| Pied Babbler | *Turdoides bicolor* | Food | Raihani and Ridley (2007) |
| Arabian Babbler | *Turdoides squamiceps* | Predator type | Regosin (2002) |
| Noisy Miner | *Manorina melanocephala* | Predator type | Farrow et al. (2017) |
| White-browed Scrubwren | *Sericornis frontalis* | Predator type | Platzen and Magrath (2005) |
| Superb Fairywren | *Malurus cyaneus* | Predator type | Colombelli-Négrel et al. (2010) |
| New Holland Honeyeater | *Phylidonyris novaehollandiae* | Predator threat | McLachlan and Magrath (2020) |
| Southern House Wren | *Troglodytes musculus* | Predator type | Fasanella and Fernández (2009) |
| Northern Mockingbird | *Mimus polyglottos* | Predator type | Savage et al. (2020) |
| House Sparrow | *Passer domesticus* | Food | Elgar (1986) |
| Southern Lapwing | *Vanellus vanellus* | Aerial predator | Walters (1990) |
| Siberian Jay | *Perisoreus infaustus* | Aerial predator +[c] | Griesser (2008, 2009) |
| Japanese Great Tit | *Parus minor* | Predator category | Suzuki (2011) |
| Great Tit | *Parus major* | Predator threat | Kalb et al. (2019) |
| Willow Tit | *Poecile montanus* | Food | Suzuki (2012) |
| Black-capped Chickadee | *Poecile atricapilla* | Predator size/behavior[d] | Templeton et al. (2005) |
| Carolina Chickadee | *Poecile carolinensis* | Predator size/behavior | Soard and Ritchison (2009) |
| Tufted Titmouse | *Baeolophus bicolor* | Predator size/behavior | Courter and Ritchison (2010) |
| Yellow Warbler | *Setophaga petechia* | Presence of cowbird[e] | Gill and Sealy (2004) |

[a]Predator type = aerial predator versus terrestrial predator
[b]Cuckoo vs. hawk = Brood-parasite Common Cuckoo (*Cuculus canorus*) versus Eurasian Sparrowhawk (*Accipiter nisus*)
[c]Aerial predator + = calls indicate that presence of an aerial predator plus the predator's behavior (perched, searching for prey, or attacking)
[d]Predator size/behavior = different calls indicate where an aerial predator is in flight or perched and, if perched, variation in call characteristics provides information about the relative size and degree of threat posed by the aerial predator
[e]Presence of cowbird = presence of a brood-parasitic Brown-headed Cowbird (*Molothrus ater*) near a nest

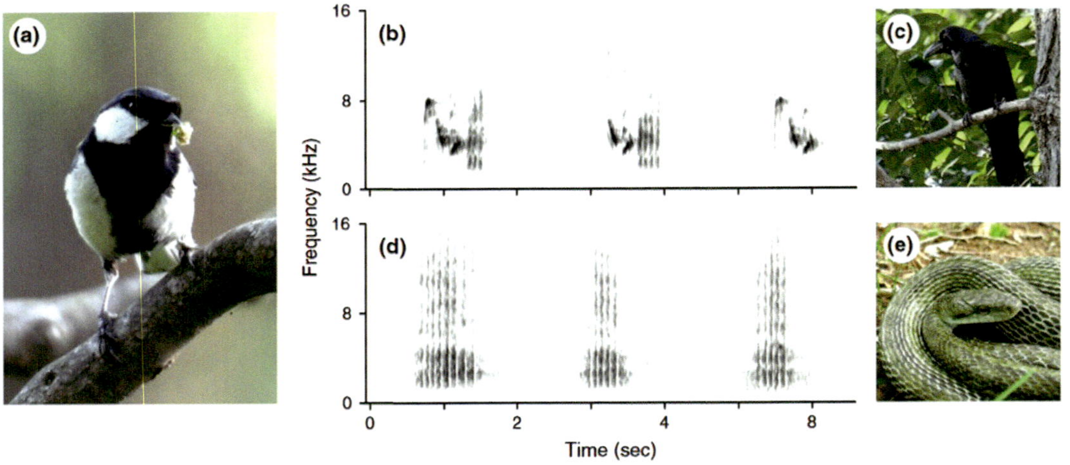


**Fig. 12.52** Japanese Tits (*Parus minor*), (**a**) have different alarm calls for different nest predators. (**b**) "Chicka" calls are given in response to the presence of Jungle Crows (*Corvus macrorhynchos*), (**c**) and (**d**) "jar" calls in response to the presence of Japanese rat snakes (*Elaphe climacophora*, **e**). (Figure from Suzuki 2016; open-access article distributed under the terms of the Creative Commons Attribution License, https://creativecommons.org/licenses/by/4.0/)

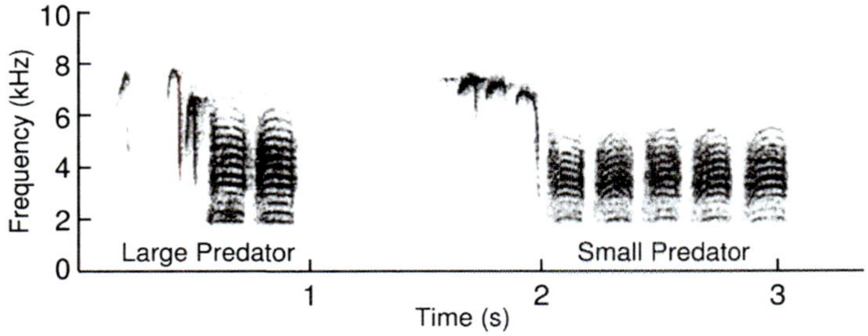


**Fig. 12.53** Sonograms of two "chick-a-dee" calls of Black-capped Chickadees (*Poecile atricapillus*), with two "dee" notes when responding to the presence of a large, nonthreatening predator and five "dee" notes when responding to a small, more threatening predator. (Figure from Templeton and Greene 2007; © 2007 National Academy of Sciences USA, used with permission)

vocal learning comes from studies of song learning in songbirds (Passeriformes). However, hummingbirds (Trochilidae) and species in the order Psittaciiformes (e.g., parrots and cockatoos) also exhibit vocal learning, and for some species in the orders Passeriformes and Psittaciiformes, that include learning calls and not just songs. For many species, learned calls are important in the formation and maintenance of pair bonds or to encode membership in a group or family (Table 12.2; Box 12.7 Geographic Variation in Parrot Contact Calls). Call learning has been reported in many species of parrots and cockatoos. For example, male Budgerigars (*Melopsittacus undulatus*) typically learn to imitate the contact calls of their mate during the process of pair formation (Hile et al. 2000, 2005; Fig. 12.55). Hile et al. (2005) hypothesized that mate choice by female Budgerigars involves several stages. When meeting a potential mate, females assess the potential ability of the male to learn their call

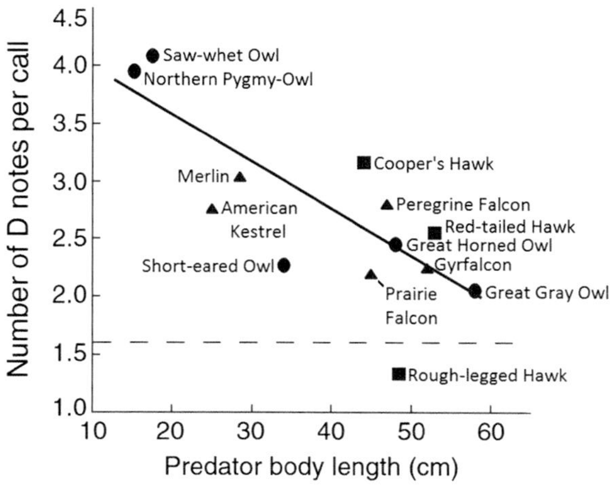


**Fig. 12.54** Mean number of "dee" notes per "chick-a-dee" call uttered by Black-capped Chickadees exposed to raptors that differ in size and, therefore, in the threat they pose to the chickadess. Larger raptors are unlikely to prey on birds as small as chickadees, but smaller raptors, particularly nocturnal raptors, do sometimes prey on chickadees. The dashed line indicates the mean number of "dee" notes per "chick-a-dee" call when no raptor was present. Northern Saw-whet Owl, *Aegolius acadicus*; Northern Pygmy-Owl, *Glaucidium gnoma*; Merlin, *Falco columbarius*; American Kestrel, *F. sparverius*; Short-eared Owl, *Asio flammeus*; Cooper's Hawk, *Accipiter cooperii*; Peregrine Falcon, *Falco peregrinus*; Red-tailed Hawk, *Buteo jamaicensis*; Great Horned Owl, *Bubo virginianus*; Gyrfalcon, *Falco rusticolus*; Prairie Falcon, *F. mexicanus*; Great Gray Owl, *Strix nebulosi*; Rough-legged Hawk, *Buteo lagopus*. (Figure modified from Templeton et al. 2005; © 2005 American Association for the Advancement of Science, used with permission)

by the presence or absence of similar calls in the vocal repertoire of the male. Then, as courtship proceeds to pair formation, females further assess males by how well they imitate the females' calls. This hypothesis assumes that precise imitation of female calls can be difficult for males, and that males that better imitate a particular female's call will be a better mate and, perhaps, parent. In support of this hypothesis, females paired with "imitation-impaired" males were more likely to seek interactions with other males (Hile et al. 2005).

Among other species of birds shared, learned calls encode group or family membership (the "badge" or "password" hypothesis, Feekes 1982). Such passwords may be beneficial to those in a social group or family group if being a part of the group involves cooperative interactions that may be costly and vulnerable to cheating (Sewall et al. 2016). Group- or family-specific calls likely also aid in coordinating movements and maintaining social or family bonds during interactions with conspecifics (Brown and Farabaugh 1997; Figs. 12.56 and 12.57).

In at least two species of fairywrens, Red-backed (*Malurus melanocephalus*) and Superb (*M. cyaneus*) fairywrens, females call to their young prior to hatching and, after hatching, females whose young then better imitate their "prehatching calls" when begging tend to provision their young at higher rates than females whose young produce less accurate imitations (Colombelli-Négrel et al. 2016). One possible explanation for such behavior is that young better able to imitate calls heard prior to hatching are higher-quality young and females should be willing to invest more in such young. Another possibility is that females in better condition call more prior to hatching, making it easier for young to imitate their calls, and females in better condition are also able to provision nestlings at higher rates

**Table 12.2** Examples of species where learning of conspecific contact calls has been documented and the social context in which those calls are used (Portion of table from Sewall et al. 2016; © 2016 The Association for the Study of Animal Behaviour, used with permission)

| Order | Species[a] | Level of social organization with shared call | Evidence of call learning | Social context | References |
|---|---|---|---|---|---|
| Psittaciformes | Budgerigar | Pair | Imitation over time | Mate choice Pair bond formation or mate choice | Moravec et al. (2006), Hile et al. (2000, 2005) |
| | | Group | Convergence over time | Social cohesion | Hile and Striedter (2000) |
| | | | Imitation and convergence over time | Mediating social associations | Dahlin et al. (2014) |
| | | | | Social cohesion | Farabaugh et al. (1994) |
| | Galah | Interaction | Rapid convergence during social interactions | Mediating social associations | Scarl and Bradbury (2009) |
| | Orange-fronted Conure | Interaction | Rapid convergence during social interactions | Mediating social associations | Vehrencamp et al. (2003), Balsby and Bradbury (2009) |
| | Yellow-naped Parrot | Populations | Imitation over time | Social integration | Salinas-Melgoza and Wright (2012) |
| Passeriformes | European Siskin | Pair | Imitation over time | Social integration | Mundinger (1970) |
| | Red Crossbill | Family | Imitation over time | Family cohesion | Sewall (2011) |
| | | Pair | Convergence over time | Social affiliation | Sewall (2009) |

[a]Scientific names: Budgerigar, *Melopsittacus undulatus*; Galah, *Eolophus roseicapilla*; Orange-fronted Conure, *Eupsittula canicularis*; Yellow-naped Parrot, *Amazona auropalliata*; European Siskin, *Spinus spinus*; Red Crossbill, *Loxia curvirostra*

(Colombelli-Négrel et al. 2016). The extent to which such communication between parents and prehatched young might influence parental behavior in other species of birds remains to be determined.

## 12.16 Variation Among Species in the Size of Call Repertoires

The number of different calls in the repertoires of different taxa and species of birds varies from as few as one to over 50 (Appendix 1). Several factors could potentially contribute to such variation. Cluttered and/or noisy habitats (e.g., dense forests or habitats that are often windy or those like tropical rain forests that might support an abundance of sound-producing insects) might favor development of new calls if, for example, vocalizing individuals need a different call to alert conspecifics before uttering a more "information-rich" vocalization (Ord and Garcia-Porta 2012). Additionally, cluttered habitats might favor larger vocal repertoires if reduced visibility interferes with transmission of visual displays, potentially leading to the development of multimodal displays that include both visual and vocal components (Freeberg et al. 2012). Leighton and Birmingham (2021) examined the vocal repertoires of 821 species of birds and determined that cooperative breeding was "a significant predictor of larger repertoire size in birds,"

**Box 12.7 Geographic Variation in Parrot Contact Calls**

Oscine songbirds and hummingbirds learn their songs. Parrots also learn, but learn calls rather than songs. Unlike many songbirds, vocal learning by parrots generally continues throughout their lives, allowing them to modify calls or learn new ones as they interact with conspecifics. Among parrots and other social species of birds, contact calls provide information about an individual's location as well as information about the identity of the caller. In addition, given their ability to learn, the characteristics of contact calls of individuals that regularly interact may converge and, thereby, provide information about group-level identity. One result of this is that the characteristics of contact calls of social groups in different geographical locations may differ, resulting in contact call "dialects." For example, such dialects have been found in Yellow-naped (*Amazona auropalliata*) and Yellow-crowned (*A. ochrocephala*) parrots (Wright and Wilkinson 2001) and in several other species of parrots (e.g., Bond and Diamond 2005; Cortopassi and Bradbury 2006; Ribot et al. 2012).

Because the contact calls of some species of parrots can be important in mate choice and mate recognition (e.g., Moravec et al. 2006), vocal dialects could potentially lead to reduced gene flow and genetic differences among populations with different vocal dialects. Although this was found to be the case with Crimson Rosellas (*Platycercus elegans*), with contact call dialects corresponding to two different subspecies, geographic variation in calls does not correspond to genetic variation in other species, but, rather, develops because individuals that disperse to different populations learn the characteristics of calls in their new population (e.g., Wright and Wilkinson 2001).

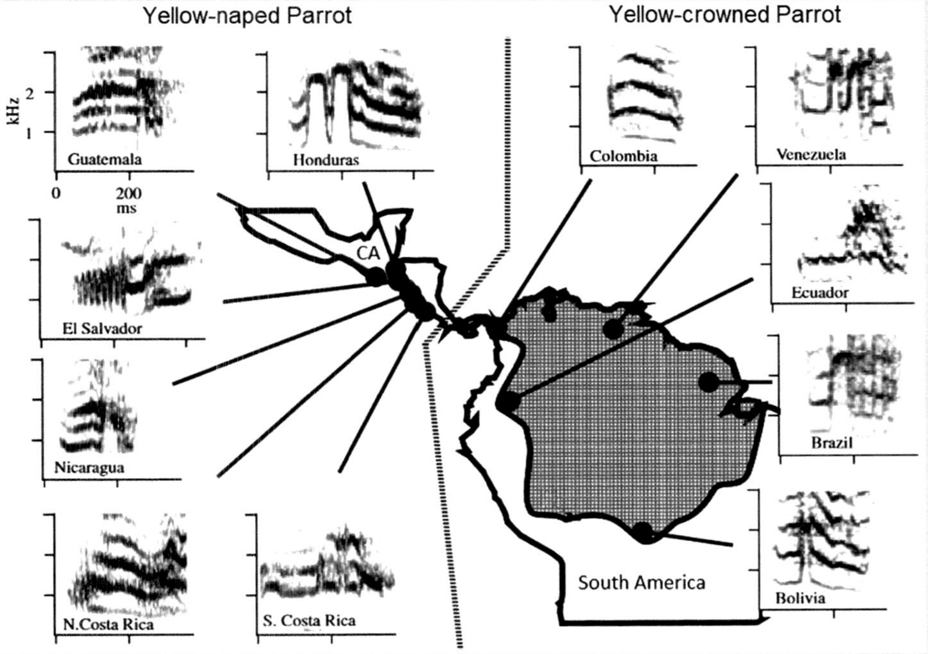


Geographical variation in the contact calls of Yellow-naped and Yellow-crowned parrots. Analysis revealed high levels of gene flow among populations so the persistence of these call dialects is likely due to postdispersal learning of population-specific contact calls. CA, Central America. (Figure from Wright and Wilkinson 2001; © 2001 The Royal Society, used with permission)

(continued)

**Box 12.7** (continued)

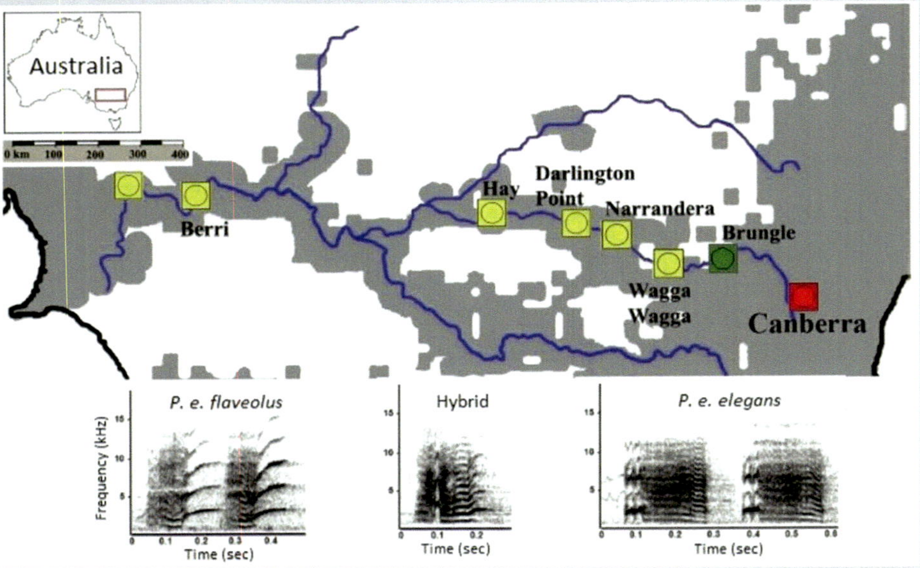


Crimson Rosellas (*Platycercus elegans*) are found throughout southeastern Australia (including the areas shaded gray) and the characteristics of their contact calls vary among populations as illustrated by the three sonograms. This vocal variation corresponds with genetic variation between two subspecies, *P. e. flaveolus* (populations indicated by yellow) and *P. e. elegans* (population indicated by red). The population at Brungle (green) consists of hybrids of the two subspecies. Inset (upper left) shows the area in southeastern Australia where these populations were located. (Figure modified from Ribot et al. 2012; © 2012 Ribot et al., open-access article distributed under the terms of the Creative Commons Attribution License, https://creativecommons.org/licenses/by/4.0/)

suggesting that "social complexity leads to communicative complexity." Other investigators have also noted the potential role of social complexity in explaining interspecific variation in the size of vocal repertoires. The social complexity hypothesis posits that species with more complex social systems require more complex systems of communication, e.g., more complex vocal repertoires, to more efficiently interact with conspecifics. What makes a social system complex? Complex social systems are those where more individuals interact more frequently with the same individuals in different contexts (Freeberg et al. 2012; Fig. 12.58). In such social systems, the increased number of and variability in the types of interactions with conspecifics may require a corresponding increase in the number of different calls in the vocal repertoire to convey more specific messages.

Many species of parrots, for example, have relatively large repertoires of note and call types, including Lilac-crowned Parrots (*Amazona finschi*; Montes-Medina et al. 2016), Brown-headed Parrots (*Poicephalus cryptoxanthus*; Taylor and Perrin 2005), Orange-winged (*Amazona amazonica*; de Moura et al. 2011), Turquoise-fronted Parrots (*Amazona aestiva*; Fernández et al. 1998), and Gray Parrots (*Psittacus erithacus*; May 2004), and also have complex social structures (Montes-Medina et al. 2016). Some parrots also migrate seasonally as they track food resources (e.g., Renton 2001), and this may require learning new vocalizations as

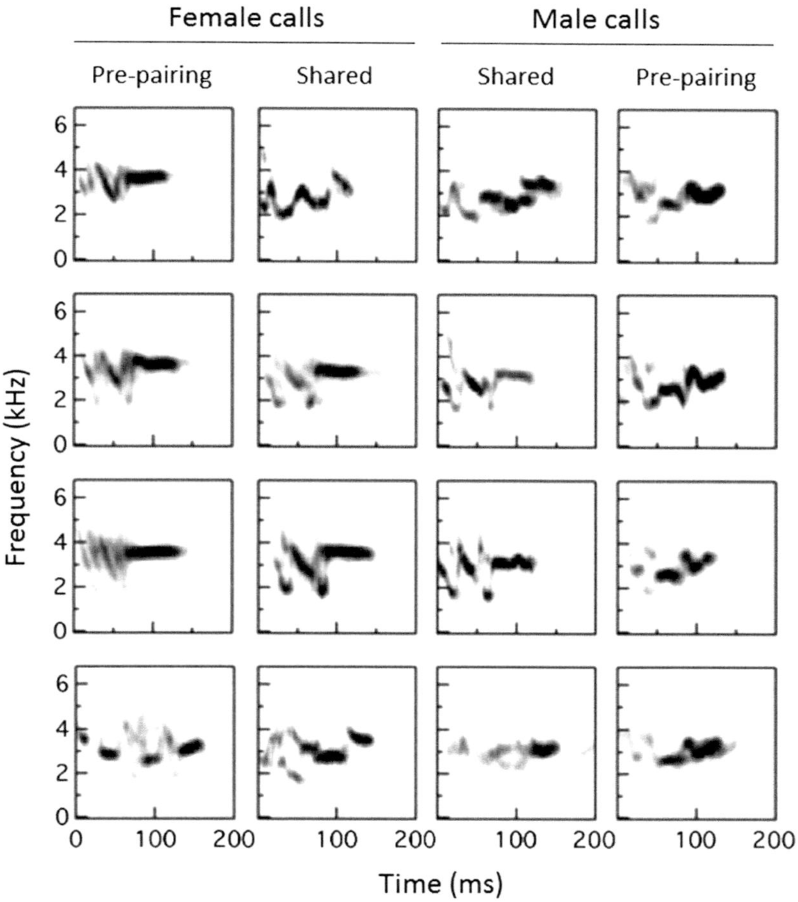


**Fig. 12.55** Sonograms of the calls of four pairs of Budgerigars (*Melopsittacus undulatus*) prior to and after pair formation. Note that, except for the pair at the bottom, the calls of males changed more to match those of females than vice versa. (Figure modified from Hile et al. 2005; © 2005 John Wiley and Sons, used with permission)

they interact with conspecifics in different locations (Salinas-Melgoza and Wright 2012).

## 12.17 Functions of Bird Song

As noted previously, compared to calls, the songs of birds are typically longer in duration and more complex, and are important for defending territories and attracting mates (Catchpole 1982). However, providing a more specific definition of song that would be acceptable to all ornithologists is probably not possible. For example, Spector (1994) reviewed the bird song literature and came up with more than 80 different definitions of song, with the most common defining criteria including structure (e.g., duration), physiological (e.g., under hormonal control), developmental (e.g., requires learning), and functional (e.g., attracting mates and/or defending territories). However, there are exceptions to all criteria used to define songs (Table 12.3) so, in this discussion of the functions of song, the focus will be on the "typical" songs of songbirds and hummingbirds, with some references to the exceptions.

Some known functions of song include:

- Identification
  Songs have characteristics that permit other birds to identify the species, sex (if both males and females sing), and individual identity of a singer. Characteristics important in permitting specific, sexual, and individual recognition vary among species, but may include song duration, the interval between song elements, frequency, syntax (order of elements

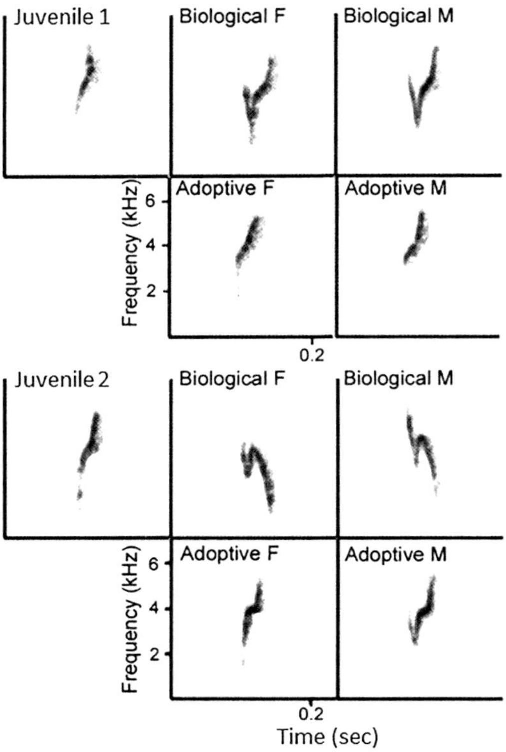


**Fig. 12.56** Examples of call learning by two juvenile Red Crossbills (*Loxia curvirostra*). Juveniles 1 and 2 were cross-fostered and so raised by "adoptive" parents. Both juveniles developed calls similar to those of their adoptive parents, showing the calls are learned rather than innate. (Figure modified from Sewall 2011; © 2010 Springer-Verlag, used with permission)

in songs), and the structure of elements, e.g., duration and frequency (Becker 1982).

- Attracting and stimulating females
  In many species of birds, the choice of mates by females is influenced by the songs and singing behavior of males. Studies of cavity-nesting species where females inspected or occupied boxes from which songs were played faster than boxes where songs were not played provide evidence that females are attracted to male song (e.g., Eriksson and Wallin 1986, Johnson and Searcy 1996). Other studies have revealed that females exposed to male songs may increase their nest-building activities or perform courtship displays (e.g., Kroodsma 1976; Searcy and Marler 1981; Mota 1999; Gong et al. 2020), initiate egg laying earlier when mates sing at higher rates (Wright and Cuthill 1992), or, if exposed to "attractive" song repertoires (larger repertoires and longer songs) increase the amount of testosterone in their eggs (Gil et al. 2004).

  In some species of songbirds, females have also been found to prefer to mate with males with larger song repertoires or more complex songs (Byers and Kroodsma 2009; Table 12.4). However, the results of studies focusing on the possible role of song repertoires have provided mixed results. The results of laboratory studies are more likely to have provided evidence that females prefer males with larger song repertoires than field studies and, for some species, different studies have provided conflicting results (Table 12.4). Based on their analysis, Byers and Kroodsma (2009) concluded that female preferences for larger song repertoires are "not widespread" and that selection has favored song repertoires in some species of songbirds because social conditions have favored "complex signaling systems." The results of other studies suggest that, in some species, females prefer males able to sing longer or more complex songs (e.g., Gentner and Hulse 2000; Caro et al. 2010; Dunning et al. 2020) or males that sing at higher rates (e.g., Nolan and Hill 2004).

- Territory establishment and defense
  Singing is used as an intrasexual threat signal to establish and defend territories against competitors in many species of birds. For example, speaker-replacement experiments (where males are temporarily removed from territories and replaced with speakers playing back their songs) show that male songs, even in the absence of males, can deter conspecific males from intruding (e.g., Searcy et al. 1998). McDonald (1989b) temporarily muted male Seaside Sparrows (*Ammospiza maritima*) by rupturing their interclavicular air sacs, leaving them unable to sing for about 2 weeks (but still able to produce normal calls). Male muted early in the breeding season acquired territories later than sham-operated or

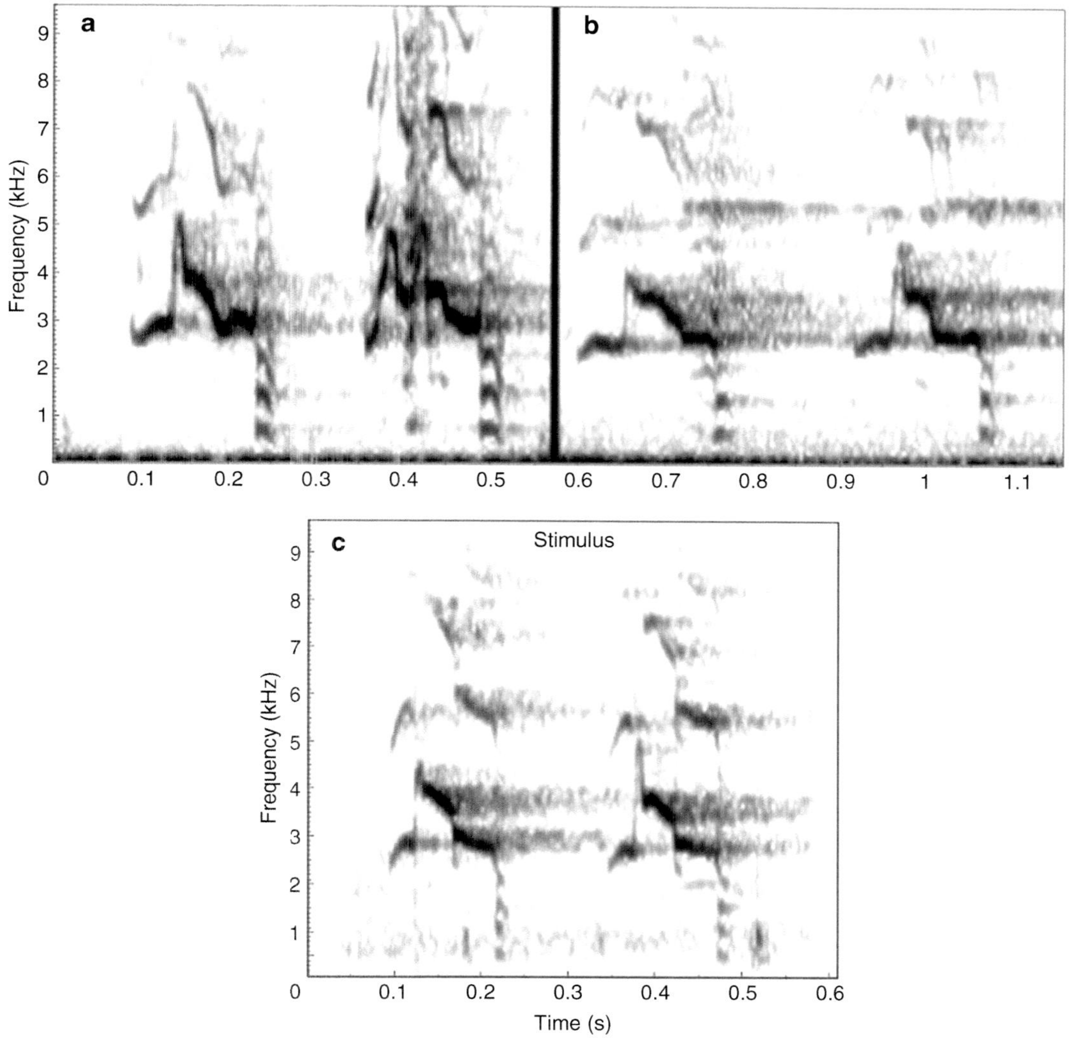


**Fig. 12.57** Call convergence by male Galahs (*Eolophus roseicapilla*). (**a**, **b**) Sonagrams of consecutive response calls given by one male galah in response to a male stimulus call (**c**). No other bird called within 5 s of either response call, suggesting that the bird produced these calls in response to the stimulus. The responding bird produced the second call (**b**) 6 s after the first call (**a**). (Figure from Scarl and Bradbury 2009; © 2009 The Association for the Study of Animal Behaviour, used with permission)

nonexperimental males, and males that had established territories and were muted later in the season either lost their territories or had them reduced in size. However, these latter males either established new territories or, if they retained smaller territories, were able to re-expand their territories.

- Motivation and fitness

Birds may provide information to conspecifics through variation in singing rates (e.g., increasing during aggressive encounters), song duration (e.g., may increase or decrease, depending on the species, during conflict situations), syllable rates (Funghi et al. 2015), song amplitude (or volume), e.g., may either decrease (Akçay et al. 2015; Ali and Anderson

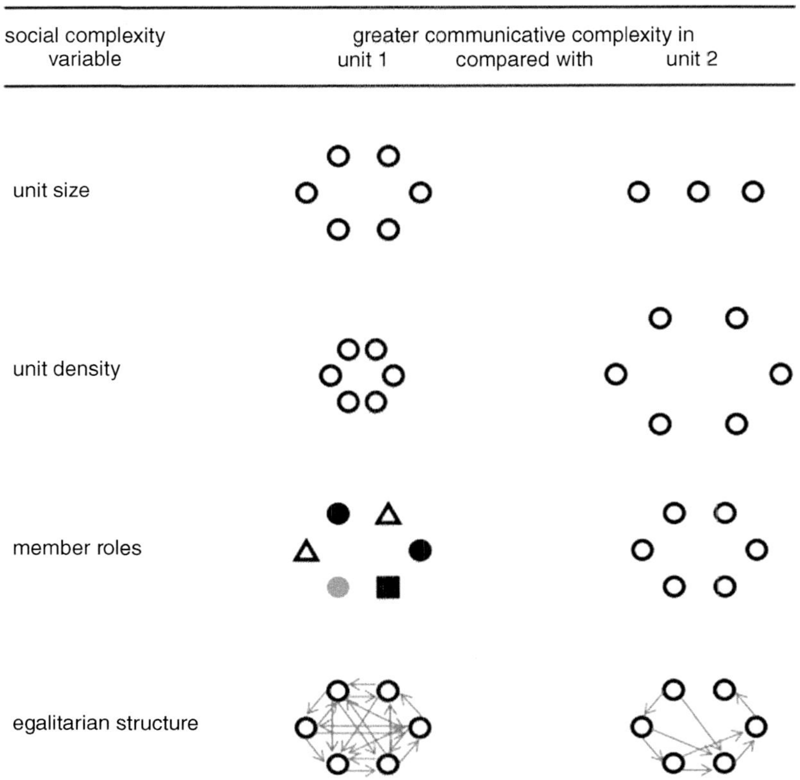


**Fig. 12.58** Several factors can influence social and vocal complexity, including group (or unite) size, group (or unit) density (with increasing density meaning that birds remain closer together and, therefore, interact more frequently), the number of different "roles" of members in a group (e.g., breeding adults, young of different ages, or helpers), and the egalitarian structure (i.e., social relationships such as group members all with similar social status and frequent reversals in the outcome of agonistic interactions versus a linear dominance hierarchy with few or no reversals in the outcome of agonistic interactions). (Figure from Freeberg et al. 2012; © 2012 The Royal Society, used with permission)

2018) or increase (Brumm and Ritschard 2011) to signal aggression, song pitch or frequency (e.g., may increase during conflict situations (Thompson 1972), song complexity, e.g., songs may consist of more or fewer elements during conflict situations, or song-type matching (Becker 1982).

- Distraction of potential predators or warning mates about the presence of potential predators
  Some studies have revealed that males may sing in an attempt to warn incubating or brood mates about the presence of potential predators or to distract potential predators in an attempt to direct them away from nests. For example, Dyrcz et al. (2011) found that singing rates of male Aquatic Warblers (*Acrocephalus paludicola*) increased sevenfold when a potential predator (human) was detected near (about 30 m) from nests. In addition, Ritchison (1991) suggested that the flight songs of male Common Yellowthroats (*Geothlypis trichas*) warned incubating or brooding mates and directed the attention of potential predators toward the males and away from nests. Similarly, Pereyra and Morton (2010) suggested that the flight songs of Dusky Flycatchers (*Empidonax oberholseri*) may alert mates and offspring about the presence of aerial predators or serve as a pursuit-deterrent signal.
- Coordination of activities
  In some species of birds, singing may help coordinate the actions of males and females (Leonard 2008). For example, Halkin (1997) suggested that nest-vicinity exchanges of song helped to coordinate parental care at nests. Similarly, Boucaud et al. (2016) suggested that male and female Zebra Finches use song to coordinate incubation duties.

**Table 12.3** Criteria sometimes used to determine if a bird vocalization is a song (Table modified from Spector 1994; 

| |
|---|
| **Song structure** |
| Duration, number of components (notes, syllables, and/or phrases) |
| Volume |
| **Song delivery** |
| Consistently repeated, species-specific pattern |
| Occurs in bouts (i.e., periods of singing interspersed with periods with no singing) |
| **Song delivery—who and when** |
| Only males sing in some species (particularly at higher latitudes), but both sexes sing in other species (particularly in the tropics) |
| Only during the breeding season in many species (particularly at higher latitudes), but all year in other species (particularly at tropical latitudes) |
| **Song neurobiology, physiology, and development** |
| Singing depends on a song control system in the brain |
| Singing often controlled by gonadal production of testosterone and its adrenergic and estrogenic neural metabolites[a] |
| Learned in oscine songbirds and hummingbirds, but not in suboscine songbirds or other taxa that are sometimes said to sing[b] |
| **Song function** |
| Often, but not always, used in long-distance communication[c] |
| Used to attract mates and defend territories (but with other functions as well) |
| **Singing taxa** |
| Songbirds (Order Passeriformes) sing, but so do some hummingbirds (Order Apodiformes) and the vocalizations of species in other taxa are also sometimes referred to as songs[b]. |

[a]Examples of exceptions to this include several species of tropical songbirds (e.g., Rödl et al. 2004, York et al. 2016)
[b]Examples of other taxa where species are sometimes referred to as being able to sing include todies (Order Coraciiformes; Pérez Mena and Mora 2011), doves (Order Columbiformes; Beckers et al. 2003b), and owls (Order Strigiformes; Herting and Belthoff 2001)
[c]As just one example, low volume songs, sometimes called soft or quiet songs are sometimes used in aggressive contexts (e.g., Akçay et al. 2015)

## 12.18 Structure and Function of Female Songs

Although more studies are now being conducted in other parts of the world, most studies of bird song published during the last several decades were conducted in North America and Europe. Because of this geographical bias, the general consensus has been that, among most species of songbirds (Passeriformes), males sing, but females do not. However, as our understanding of the vocal behavior of songbirds outside of Europe and North America, and especially in the tropics, has increased, it has become increasingly clear that singing by females is more common and important than previously thought. Odom et al. (2014) conducted an extensive survey of 1141 species of oscine songbirds, were able to determine unequivocally if females did or did not sing for 323 of these species, and found that females did sing in 229 of the 323 species (71%) and in 32 of 34 families (94%). Further analysis based on the phylogeny of oscine songbirds revealed a 92% probability that females sang in the ancestor of present-day songbirds (Fig. 12.59). Investigators have also reported singing by females in several species of suboscine songbirds (e.g., Tobias et al. 2011). Thus, available data suggest that singing by female songbirds is both relatively common and, because it is ancestral, selection has favored the loss of female song in many present-day species of songbirds (Odom et al. 2014).

Singing by female songbirds can be important for both intersexual and, more often, intrasexual communication. Females in a few species use song to attract males. For example, Alpine Accentors (*Prunella collaris*) are polygynandrous

**Table 12.4** Species studied more than once to determine the possible effect of repertoire size on female mate choice[a] (Table from Byers and Kroodsma 2009 and citations of studies referred to in this table are available in their paper; 

| Species | Laboratory | Field | |
|---|---|---|---|
| | | Social mating | Extrapair mating |
| Song Sparrow | √ √ √ | √ x | – |
| Great Reed Warbler | √ | x x x | √ |
| Sedge Warbler | √ | √ √ | x x |
| Great Tit | √ | x x | – |
| Pied Flycatcher | √ | √ | – |
| Red-winged Blackbird | √ | x | |
| European Starling | √ x | √ √ | – |
| House Finch | x | x | |
| Willow Warbler | – | x | x |
| Zebra Finch | √ √ x | – | – |
| Swamp Sparrow | √ x | – | – |
| Canary | √ √ | – | – |

[a]Checkmarks indicate a study that identified a repertoire size effect; each x indicates a study that tested for, but did not detect, such an effect. Song Sparrow, *Melospiza melodia*; Great Reed Warbler, *Acrocephalus arundinaceus*; Sedge Warbler, *A. schoenebaenus*; Great Tit, *Parus major*; Pied Flycatcher, *Ficedula hypoleuca*; Red-winged Blackbird, *Agelaius phoeniceus*; European Starling, *Sturnus vulgaris*; House Finch, *Haemorhous mexicanus*; Willow Warbler, *Phylloscopus trochilus*; Zebra Finch, *Taeniopygia guttata*; Swamp Sparrow, *Melospiza georgiana*; Canary, *Serinus canaria domesticus*

and, when fertile, females sing to attract males (Langmore et al. 1996). Female Red-cheeked Cordonbleus (*Uraeginthus bengalus*) sing most often before egg laying and males respond by engaging in courtship behavior (Gahr and Guttingery 1986). In addition, females in some species of birds have been found to either begin singing or increase singing rates after removal or loss of their mates (Drent 1983; Kempenaers et al. 1995; Morton 1996; Morton et al. 2000; Tobias et al. 2011).

Among many species of songbirds at tropical latitudes, paired males and females typically maintain long-term pair bonds and defend territories year-round (Langmore 1998, 2000; Slater and Mann 2004). Such behavior increases competition among females for mates and territories which, in turn, favors singing by females to aid in defense of their territories and mates (e.g., Langmore 1998, 2000). Both observational and experimental (e.g., song playback) studies have revealed the importance of female song in defending territories and mates (mate guarding) in numerous species, primarily among tropical species (e.g., Illes and Yunes-Jimenez 2009; van den Heuvel et al. 2014; Cain and Langmore 2015), but in some temperate species as well (e.g., Pavlova et al. 2007; Krieg and Getty 2016).

Female song may also function in maintaining contact and communicating with mates to coordinate or stimulate breeding activity (e.g., Kovach 2013; Hall et al. 2015). Singing may also aid males and females in coordinating and cooperating in parental care during the breeding seasons (e.g., Halkin 1997; Leonard 2008).

Given these various potential functions, why might selection favor the loss of female song? The most important factor is likely that most songbirds that breed at higher latitudes, including both resident species and migrants, do not maintain long-term pair bonds and defend territories year-round. These are the primary factors that favor female song at tropical latitudes so, in their absence in higher latitudes, female song is much less common. Interestingly, the loss of female song among species of songbirds at higher latitudes matches the tendency of female

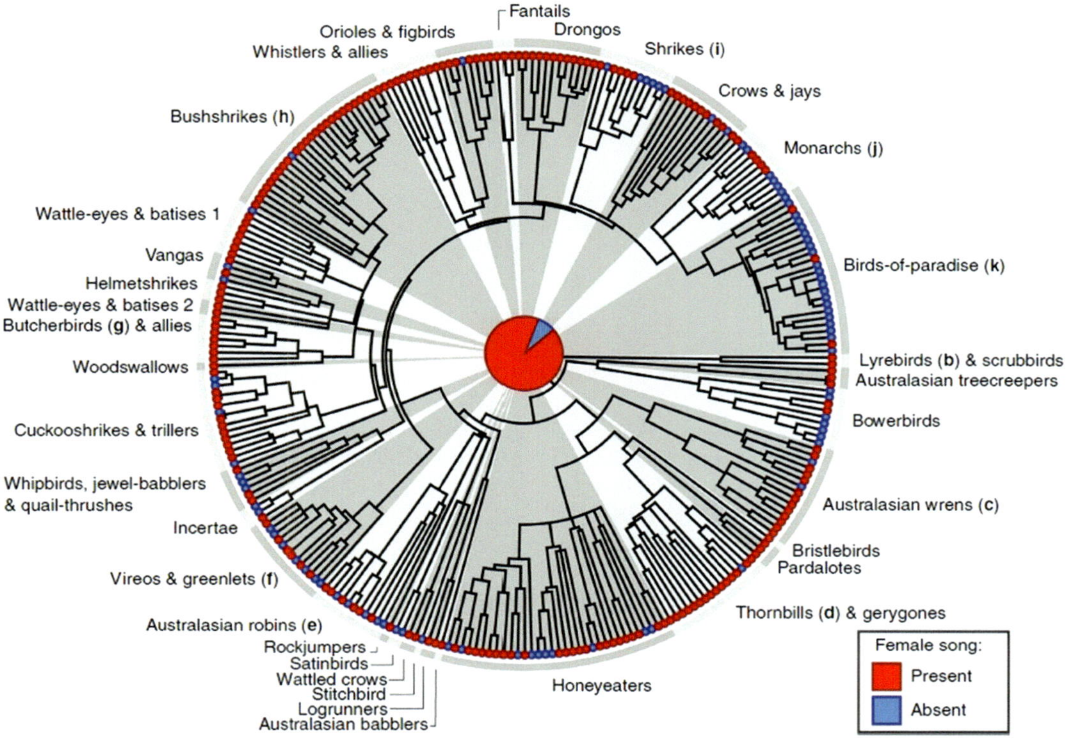


**Fig. 12.59** Prevalence of female song among oscine songbirds. Based on an analysis of 323 species where information was available, females sing in 229 species of birds (representing 32 families) and females do not sing in 94 species (19 families). The pie chart in the center shows that female song was likely present (red) in the common ancestor of modern songbirds. (Figure from Odom et al. 2014; © 2014, Springer Nature, used with permission)

**Fig. 12.60** Relationship between tendency of female plumage to be as elaborate as male plumage and the tendency for females to sing. Females in species with more elaborate plumage are also more likely to sing. (Figure modified from Webb et al. 2016; © 2016 Webb, Brunton, Aguirre, Thomas, Valcu and Dale, open-access article distributed under the terms of the Creative Commons Attribution License (CC BY), https://creativecommons.org/licenses/by/4.0/)

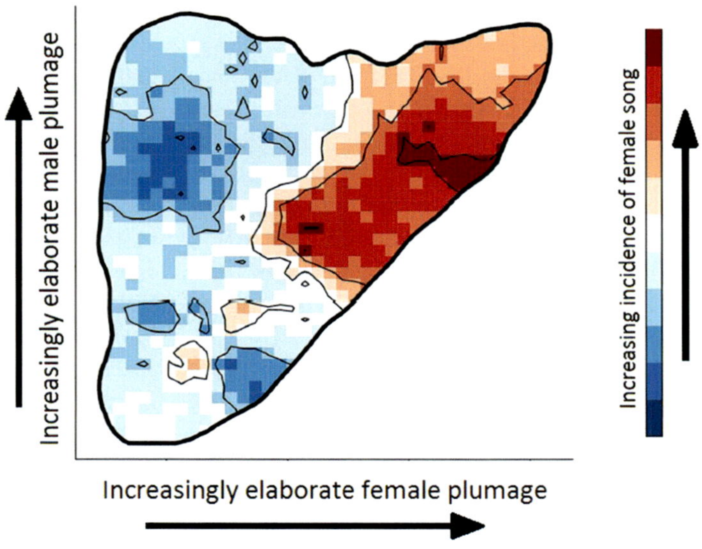

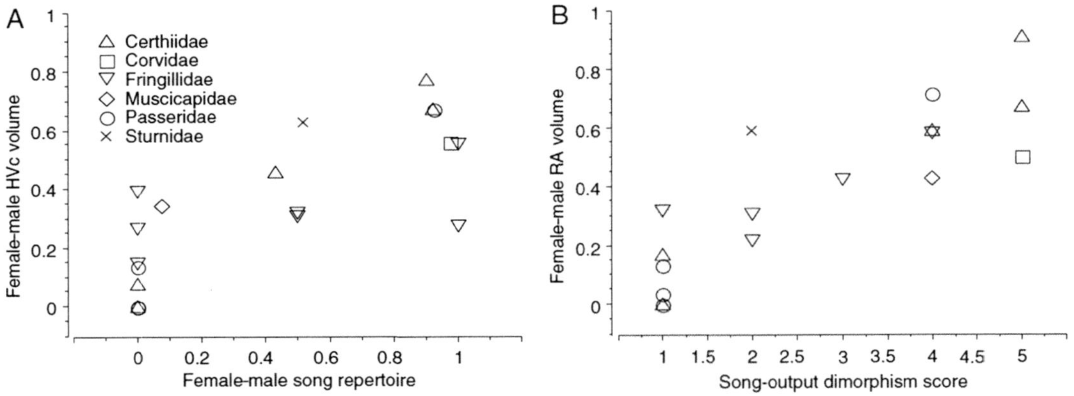


**Fig. 12.61** Relationships between differences in the volume of song control areas in the brains of males and females in several species of songbirds and sex differences in the size of song repertoires (**a**) and singing rates (**b**). RA, robust nucleus of the arcopallium. (Figure from MacDougall-Shackleton and Ball 1999; © 1999 Elsevier Science Ltd., used with permission)

songbirds at higher latitudes to be less colorful or ornamented than females at tropical latitudes (Fig. 12.60). Competition for resources is greater at tropical latitudes, making visual and vocal signaling more important than at higher latitudes. In addition, sexual selection tends to be more male-biased at higher latitudes, reducing the selective pressure for social signaling in females (Dale et al. 2015).

As with male songbirds, the HVC and RA are important song control regions in the brains of females, both for song development and production. However, among species where both sexes sing, these areas are smaller in females than in males. For example, the HVC and RA (robust nucleus of the arcopallium) of female Blue-capped Cordonbleus (*Uraeginthus cyanocephalus*) are about 55 and 40% smaller, respectively, than those of males, with about 50% fewer neurons in each area (Lobato et al. 2015). Similar differences between the sexes in these song control areas have been reported in many other species of songbirds (e.g., MacDougall-Shackleton and Ball 1999, Hall et al. 2010; Fig. 12.61), even in species where females sing as much or more than males and have songs that are equally complex (e.g., Gahr et al. 1998; Schwabl et al. 2015). Female songbirds, therefore, can sing as much and as well as males despite having smaller song control areas with correspondingly fewer neurons. One possible explanation for this is that the song control neural circuits of female songbirds function differently than those of males and allow females to learn and produce song despite the presence of fewer neurons (Lobato et al. 2015). If so, those differences have not yet been identified.

## 12.19 Song Learning by Females

Little is known about song learning by female songbirds, including when and how they learn and who they learn from (Riebel 2016). In one of the few studies of female song learning to date, Evans and Kleindorfer (2016) found that young male and female Superb Fairywrens acquired song elements from both their social fathers and mothers and had comparable song repertoires when 7–10 weeks old. Similarly, young female Blue-capped Cordonbleus (*Uraeginthus cyanocephalus*) learn songs from adults of both sexes and learning appears to begin between days 30 and 40 post-hatching and continues until sometime after day 250 post-hatching (Geberzahn and Gahr 2013; Lobato et al. 2015). During song development, the songs and syllable repertoires of female Blue-capped Cordonbleus are similar to

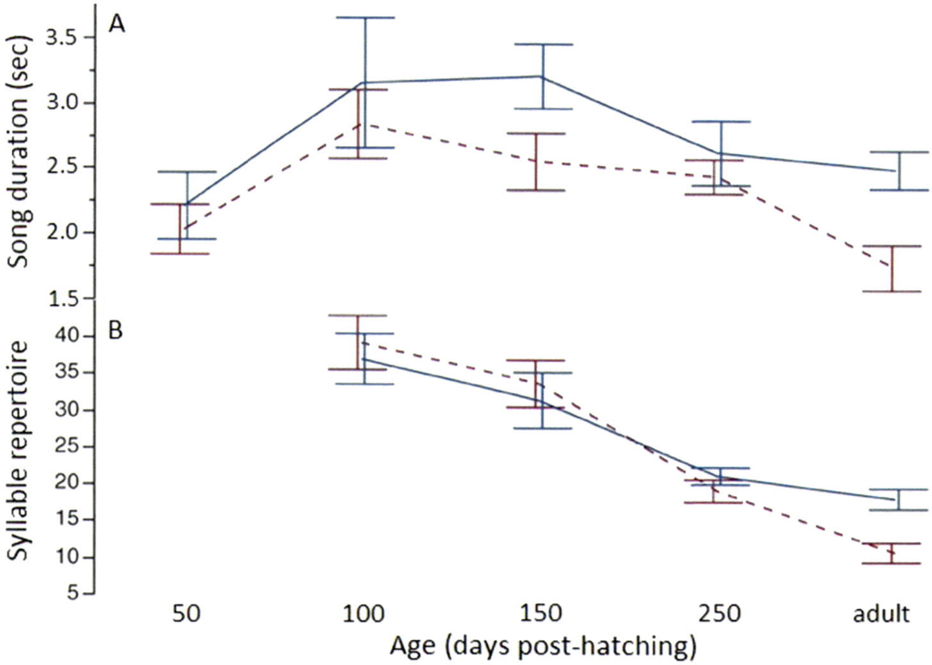


**Fig. 12.62** Development of song in male (solid blue line) and female (dashed red line) Blue-capped Cordonbleus (*Uraeginthus cyanocephalus*). (**a**) Song duration and (**b**) the size of syllable repertoires develop similarly in males and females until about 250 days post-hatching, but then become significantly different, with females having shorter songs and smaller syllable repertoires as adults. Values are means ± SE. (Figure modified from Lobato et al. 2015; open-access article distributed under the terms of the Creative Commons Attribution License, https://creativecommons.org/licenses/by/4.0/)

those of males. However, after the period of song learning ends, adult females have shorter songs and smaller syllable repertoires than males (Fig. 12.62). These results indicate that the process of song learning is similar in females and males; differences in the songs of adult male and female may be due sexual selection favoring longer songs and more complex syllable repertoires in males rather than any neuroanatomical differences between the sexes (Lobato et al. 2015).

## 12.20 Geographical Variation in Songs

Among oscine songbirds and hummingbirds, the characteristics of songs vary geographically (including latitudinally; Box 12.8 Latitudinal Variation in Bird Song). This variation may be clinal, with characteristics of songs more similar among birds in the same population and increasingly different with increasing distance from a population (Fig. 12.63), random, with no clear macrogeographic pattern (Fig. 12.64), or abrupt, with sharp boundaries between populations with different song characteristics (Fig. 12.65; Podos and Warren 2007). In the latter case, with little distance and sharp boundaries between different

**Box 12.8 Latitudinal Variation in Bird Song**

Some characteristics of the songs of songbirds tend to vary with latitude. Weir et al. (2012), for example, found that the songs of songbirds (both oscines and suboscines) tend to be lower in frequency in tropical forests than in temperate forests. Likely reasons for this difference are that there is more ambient noise (especially high-frequency insect noise) in tropical forests and high-frequency sounds degrade more rapidly in the dense vegetation of tropical forests. As a result, to enhance signal transmission, selection has favored lower-frequency songs (between 1 and 4 kHz) in tropical forests than in temperate forests.

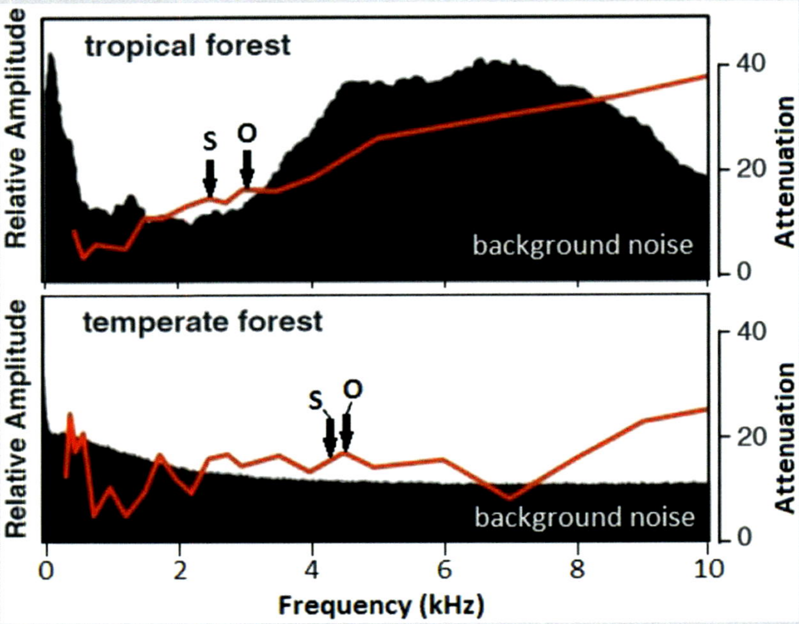


Mean frequency of songs (indicated by black arrows) of suboscine (S) and oscine (O) songbirds in forests located at tropical (0–23°) and temperate (>23°) latitudes. Black area indicates typical relative amplitude of background noise and red lines plot levels of sound attenuation (dB/100 m). The greater degradation or attenuation of sound in tropical forests than in temperate forests in combination with increased levels of background noise in tropical forests favors lower- frequency songs than in temperate forests. (Figure modified from Weir et al. 2012; © 2012 The Authors. Evolution; © 2012 The Society for the Study of Evolution, used with permission)

The songs of songbirds at higher latitudes, in addition to generally being higher in frequency, also tend to be more complex than the songs of songbirds at tropical latitudes. As one example, Kaluthota et al. (2016) examined the songs of male House Wrens (*Troglodytes aedon*) recorded at locations ranging from the southern tip of South America to southern Canada and found that songs of males at higher latitudes consisted of more elements or notes uttered at a faster rate than the songs of males at tropical latitudes.

(continued)

**Box 12.8** (continued)

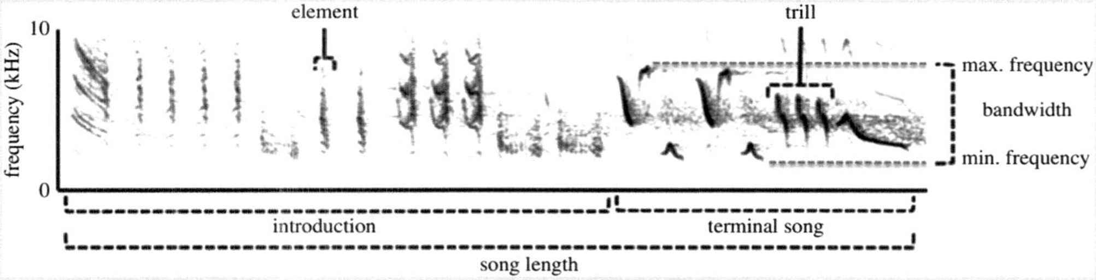


Sonogram of a typical House Wren song consisting of a series of elements or notes and trills. (Figure from Kaluthota et al. 2016; © 2016 The Authors. Published by the Royal Society, used with permission)

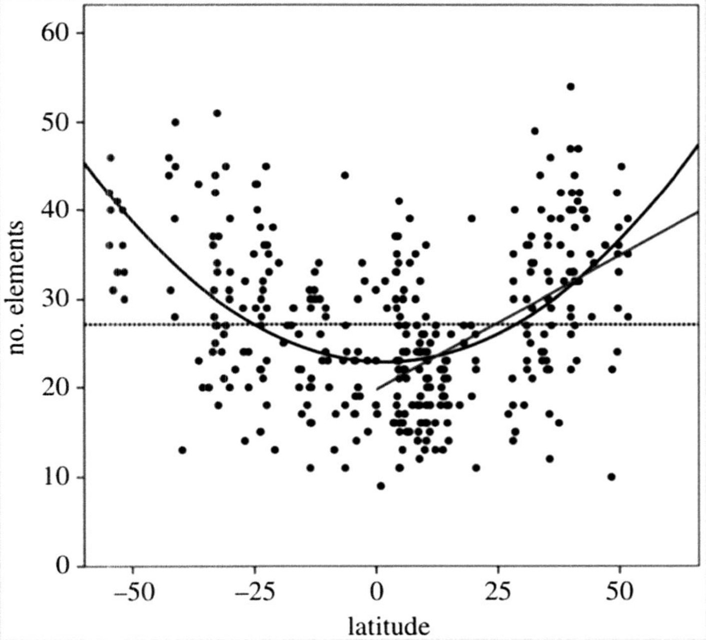


The songs of male House Wrens located at higher latitudes consisted of more elements or notes than the songs of males at tropical latitudes. In addition, males at higher latitudes uttered those notes at faster rates than did males at tropical latitudes. The U-shaped line represents the relationship between latitude and number of elements in House Wren songs. The horizontal line and the line on the right represent the nonsignificant linear relationship and significant quadratic relationship, respectively, between the two variables across both hemispheres. (Figure from Kaluthota et al. 2016; © 2016 The Authors. Published by the Royal Society, used with permission)

Other investigators have also found that songs of songbirds at higher latitudes tend to be more complex than those of songbirds at tropical latitudes (e.g., Weir and Wheatcroft 2011, Singh and Price 2015; but see Najar and Benedict 2019). One possible explanation for this latitudinal difference is that songbirds at higher latitudes have shorter breeding seasons with less time to establish territories and attract mates. This time constraint may increase male-male competition for territories (intrasexual selection) and drive selection for songs better at attracting mates

(continued)

**Box 12.8** (continued)

(intersexual selection). Thus, both competition for territories and mates may favor songs of increasing complexity. Alternatively, higher tree and foliage densities in tropical habitats may favor less complex songs to improve signal transmission. In contrast, the reduced density of foliage at higher latitudes, in combination with lower levels of ambient noise, may favor more complex songs (Singh and Price 2015; Kaluthota et al. 2016).

"song populations," the different songs are referred to as dialects. Although the characteristics of song vary geographically in most species of oscine songbirds, fewer species have song dialects. In a review of the literature, Podos and Warren (2007) identified just 31 species of songbirds in eight families, mostly in the family Fringillidae, with song dialects (Fig. 12.66).

One characteristic that appears to be essential for the development of dialects is learning (Podos and Warren 2007). Only oscine songbirds and hummingbirds, taxa where songs are learned, have song dialects and, in addition, many species of vocal-learning parrots have call dialects (see section on call dialects earlier in this chapter). Among oscine songbirds, the timing of song acquisition is not critical for the development of

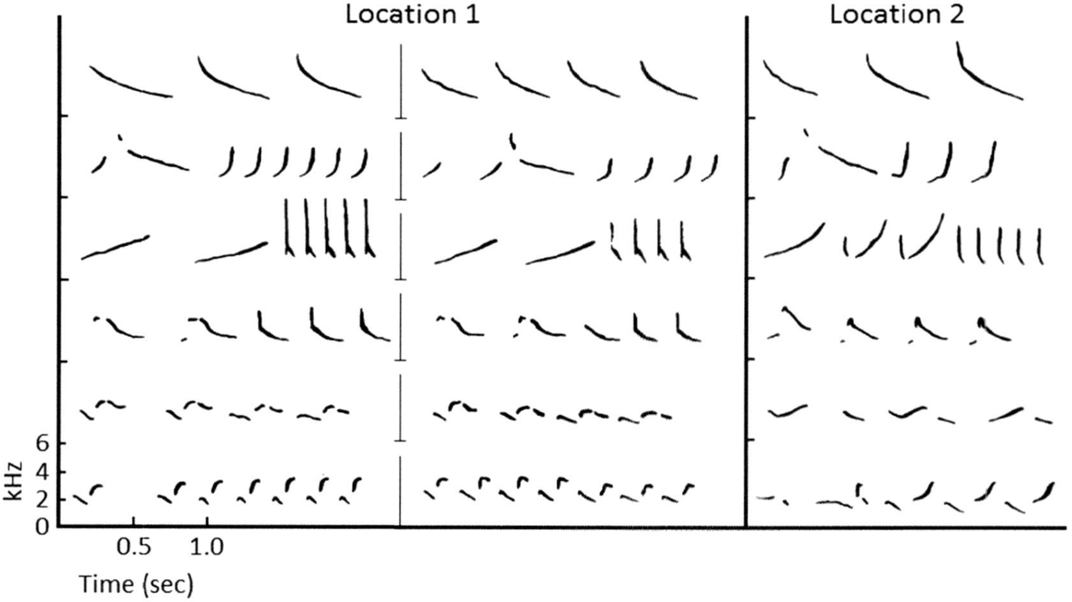


**Fig. 12.63** Example of clinal variation in song characteristics of male Northern Cardinals (*Cardinalis cardinalis*). The songs consist of a series of elements called syllables and the characteristics and use of these syllables exhibit clinal variation, with differences increasing with increasing distances. On the left are sonograms of six different songs of two males recorded at the same location and, on the right, are sonograms of six songs of male recorded at a location 35 km from Location 1. Note that the syllables and song types of males at Location 1 are very similar whereas the syllables and songs of the male at Location 2 differ to varying degrees from those of the males at Location 1. (Figure modified from Lemon 1975; © 1975 Oxford University Press, used with permission)

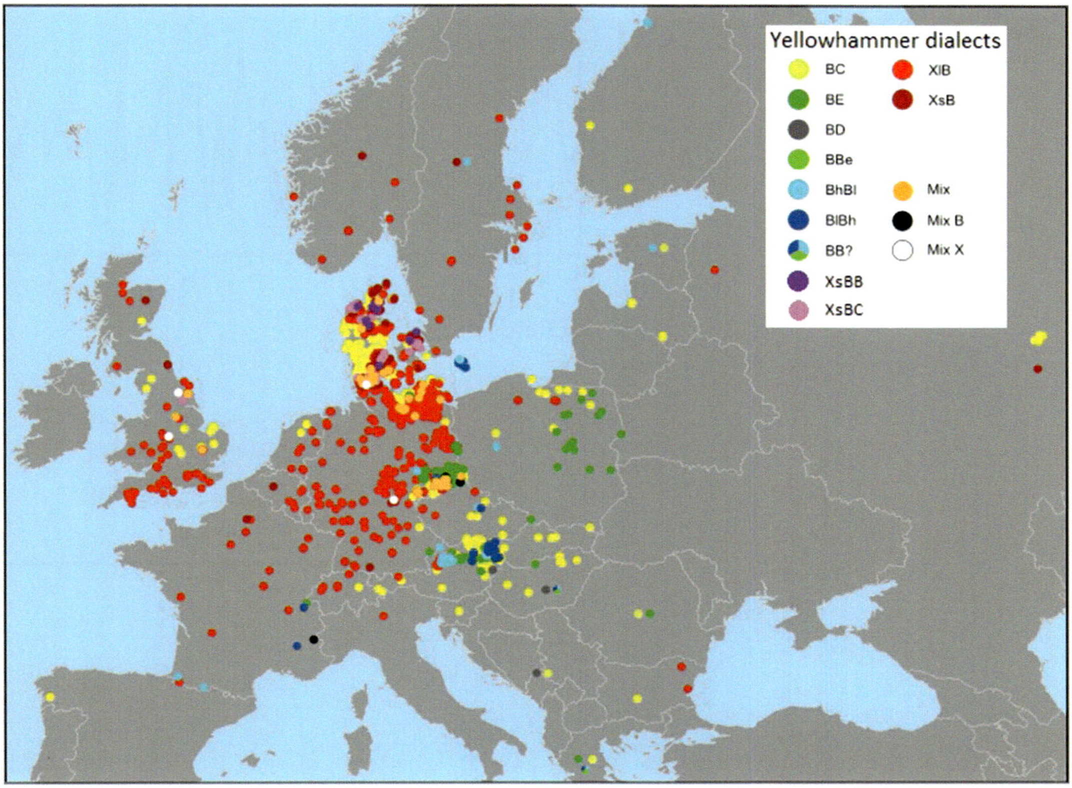


**Fig. 12.64** Random distribution of song dialects of Yellowhammers (*Emberiza citronella*) in Europe. (Figure modified from Petrusková et al. 2015; Copyright © 2014 Dt. Ornithologen-Gesellschaft e.V., used with permission)

dialects, with dialects reported both in species that acquire songs prior to and after natal dispersal (Podos and Warren 2007). The characteristics of songs in particular dialect areas, therefore, become similar either if males learn songs prior to dispersal, but disperse short distances, or if males disperse further, but learn songs from neighboring individuals after natal dispersal. This suggests that dialects may simply be incidental by-products, or as referred to by Andrew (1962), epiphenomena, resulting from different patterns of dispersal and song learning.

In contrast, the local adaptation hypothesis proposes that individuals are adapted to their local environment, which favors limited dispersal, and females develop a preference for the songs of local males, resulting in divergence in song characteristics at different locations or, in other words, song dialects (Nottebohm 1969). Along with limited dispersal, song learning should occur prior to natal dispersal and young males should settle in areas where conspecific males sing similar songs. As result, females can enhance their fitness by mating with local males because their young should be better adapted to the conditions of the local environment (Podos and Warren 2007). If populations become locally adapted in this manner and females prefer males that sing the local dialect, then different populations should also become genetically differentiated. Many investigators have examined the relationship between song and genetic

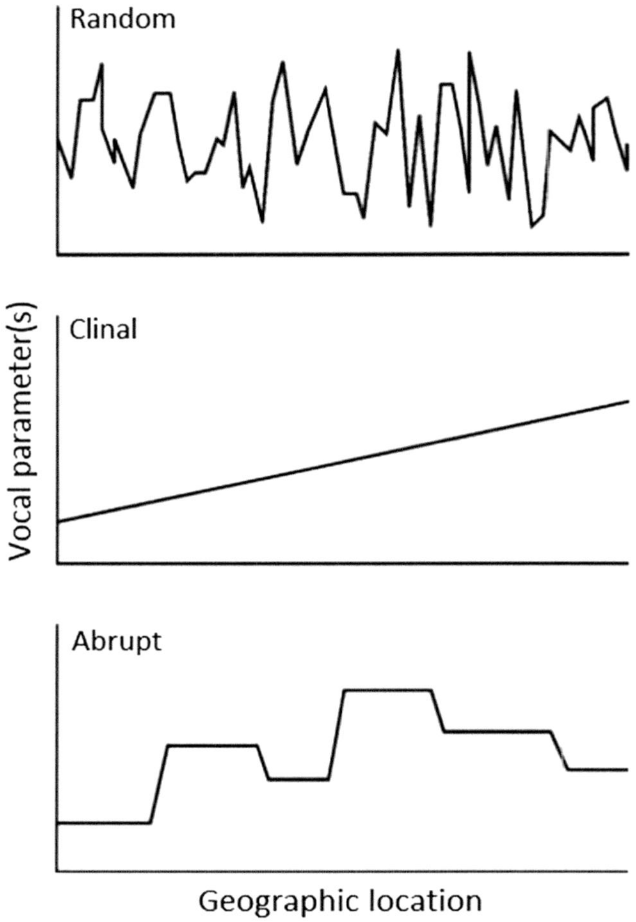


**Fig. 12.65** Different patterns of geographic variation in the characteristics of bird songs. (Figure modified from Podos and Warren 2007; © 2007 Elsevier Inc., used with permission)

differences in several species of songbirds and hummingbirds and have found no genetic differentiation among populations with different song dialects (e.g., Leader et al. 2008, Potvin et al. 2013, González and Ornelas 2014, Linossier et al. 2016; Fig. 12.67). In these species, young males may disperse from natal areas and then learn their songs from neighboring males in the postdispersal location. Such behavior results in gene flow between populations while maintaining local song dialects (González and Ornelas 2014).

The results of other studies have revealed genetic divergence between populations of songbirds with different song dialects, i.e., local adaptation. For example, Lipshutz et al. (2017) found genetic divergence in populations (subspecies) of White-crowned Sparrows (*Zonotrichia leucophrys*) with different song dialects (Fig. 12.68). Preliminary support for the local adaptation hypothesis has also been reported for Mountain Chickadees (*Poecile gambeli*), with males at different elevations having different songs (Branch and Pravosudov 2015, 2020). Although genetic analysis is still needed, altitudinal differences in environments, such as the need for better spatial memory for seed caching at higher elevations with longer winters, may favor local adaptation, with female fitness enhanced by

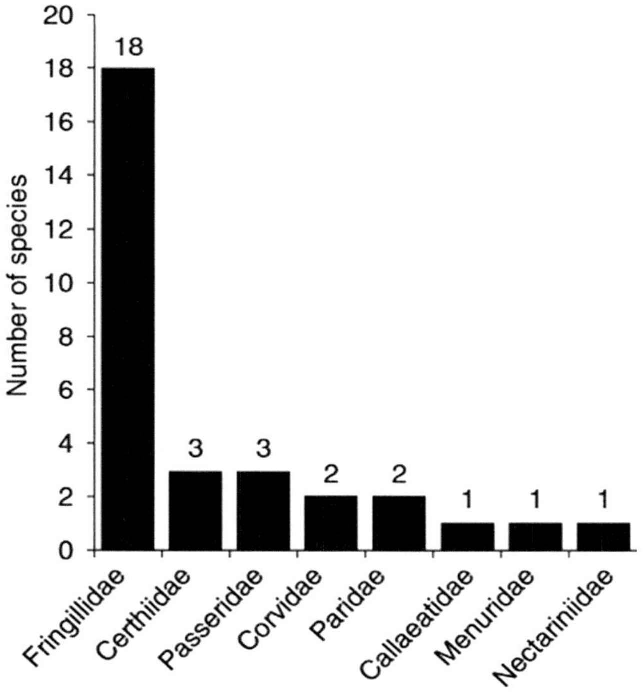


**Fig. 12.66** Families of birds (order Passeriformes) and numbers of species of birds in those families where vocal dialects have been described. (Figure from Podos and Warren 2007; © 2007 Elsevier Inc., used with permission)

preferring males with the local song type or dialect as mates (Branch and Pravosudov 2015). For White-crowned Sparrows, differences in vegetation density between areas with different song dialects appear to favor local adaptation (Derryberry 2009), with female fitness enhanced by not dispersing from the area with the local dialect and preferring as mates males that sing that dialect, having daughters that do the same, and having sons with limited natal dispersal distances that learn the local dialect and are subsequently preferred as mates by local females. Evidence in support of the local adaptation hypothesis has also been reported for Gray-breasted Wood-wrens (*Henicorhina leucophrys*; Dingle et al. 2008), House Finches (*Carpodacus mexicanus*; Badyaev et al. 2008), and Swamp Sparrows (*Melospiza georgiana*; Ballentine et al. 2013).

According to the social adaptation hypothesis, song dialects might also develop if males gain fitness advantages by singing songs similar to those of other males in their area or region (Payne 1982). This hypothesis requires that young males learn songs after natal dispersal and gene flow between populations would mean no genetic differentiation among individuals in areas with different song dialects. Based on the definition of song dialects provided above, i.e., with little distance and sharp boundaries between different song populations, no studies to date provide support for the social adaptation hypothesis. However, the social adaptation hypothesis does explain geographical variation in song for several species of birds where the variation is clinal or gradual rather than sharp, including Eurasian Skylarks (*Alauda arvensis*; Briefer et al. 2008), Common Nightingales (*Luscinia megarhynchos*; Kiefer et al. 2010), Canada Warblers (*Cardellina canadensis*; Demko et al. 2016), and several others. As predicted by the social adaptation hypothesis, males in these species disperse into an area then develop songs or song repertoires similar to those of other males in the population. Males may then benefit via increased reproductive success (e.g., Payne et al. 1988) and enhanced ability to interact with neighboring males (e.g., engaging in aggressive

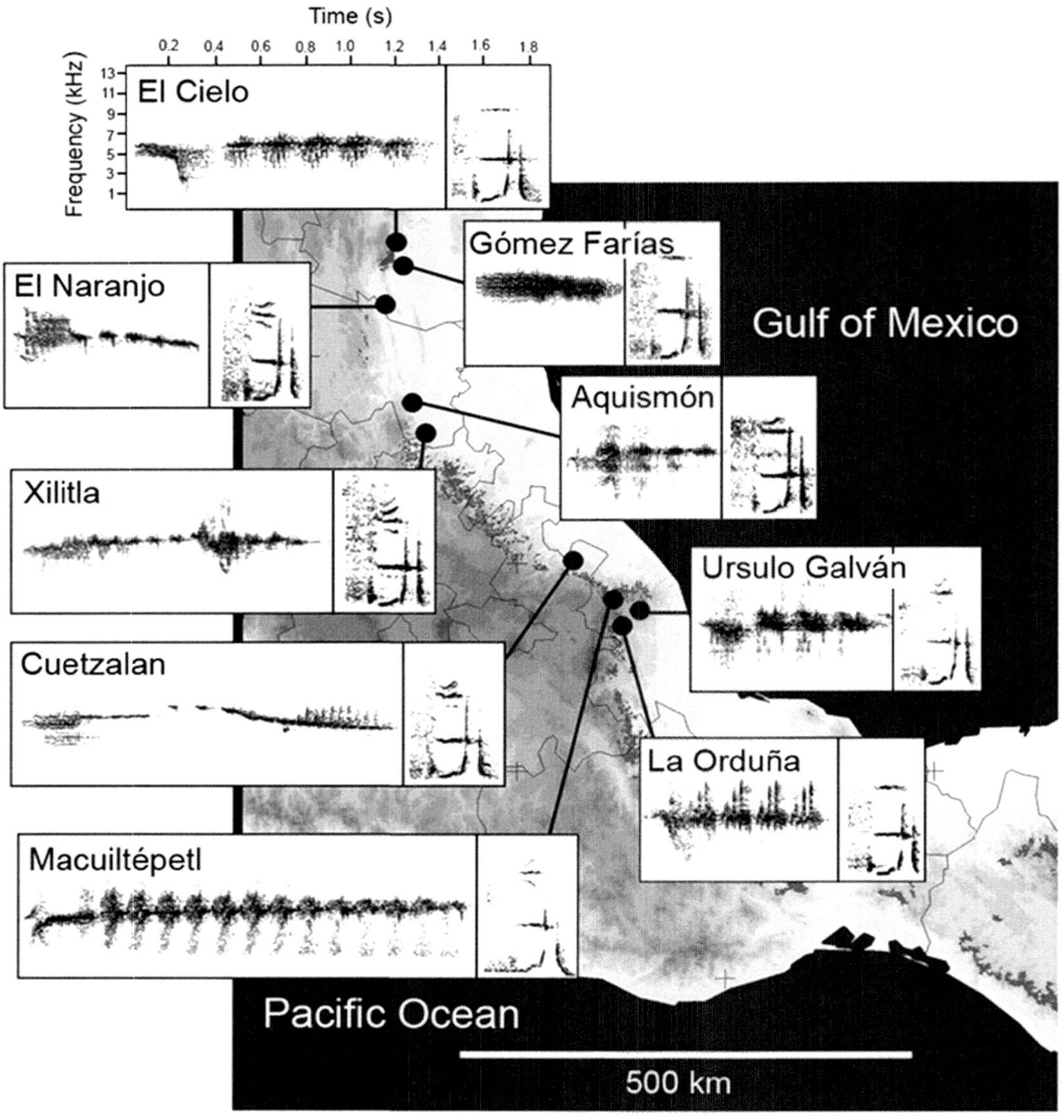


**Fig. 12.67** Song dialects in a hummingbird, the Wedge-tailed Sabrewing (*Campylopterus curvipennis*). Sonograms show the introductory (left) notes that vary among, and are unique to, different dialect areas and another note (right) that is similar among males in all dialect areas. The absence of any genetic differentiation among males in different populations suggests that male Wedge-tailed Sabrewings learn their vocalizations after dispersal, probably to facilitate territory establishment and/or access to breeding leks. (Figure from González and Ornelas 2014; © 2014 González, Ornelas, open-access article distributed under the terms of the Creative Commons Attribution License, https://creativecommons.org/licenses/by/4.0/)

encounters using shared songs; Vehrencamp 2001) and retain territories in subsequent breeding seasons (e.g., Beecher et al. 2000).

## 12.21 Song Repertoires

The songs of oscine songbirds can consist of variable numbers of notes, syllables, phrases,

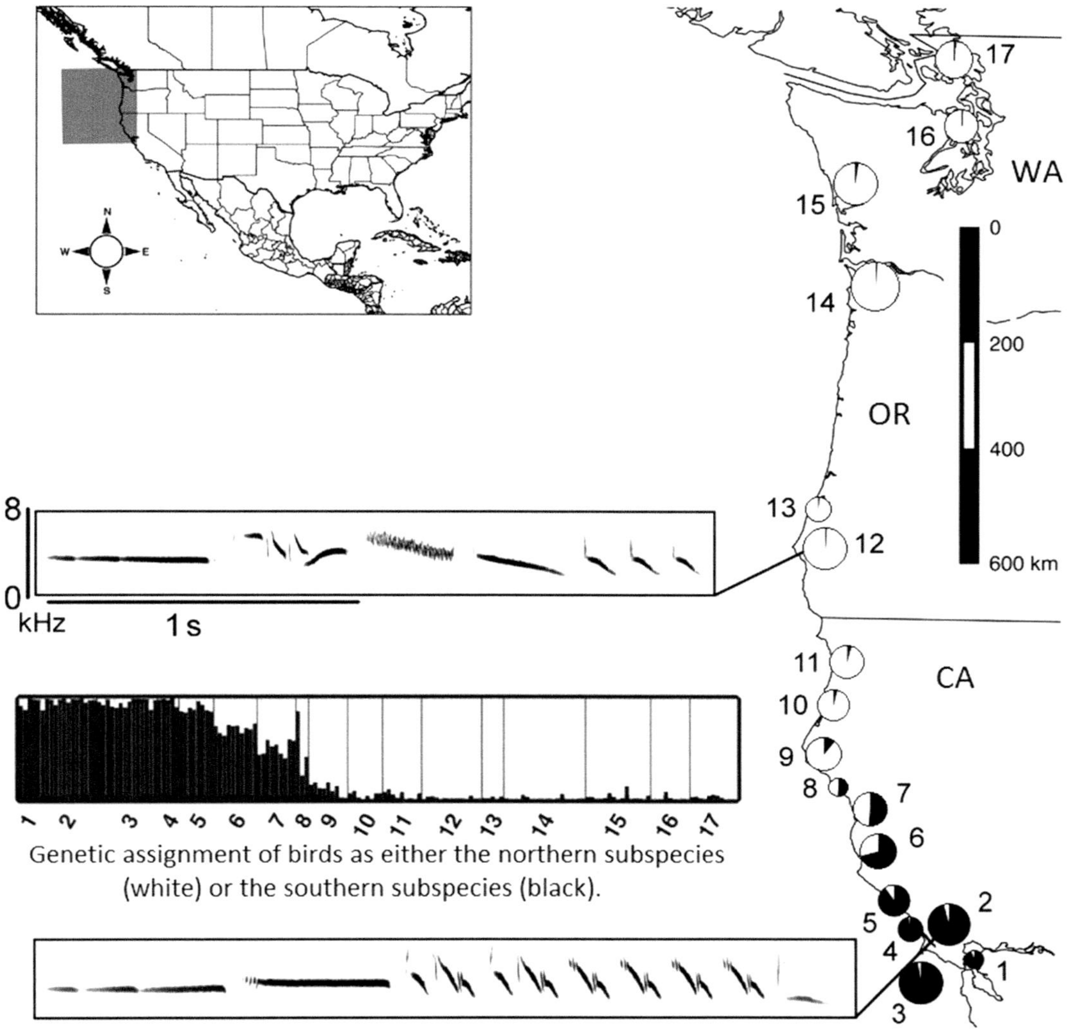


**Fig. 12.68** The songs and genetic structure of White-crowned Sparrows (*Zonotrichia leucophrys*) were sampled at 17 locations in the northwestern United States (WA, Washington; OR, Oregon; CA, California). Males have single-song repertoires and males at locations 9–17 largely shared the same song type or dialect (top sonogram). Similarly, males at locations 1–5 shared the same song dialect (bottom sonogram). Males in between, at locations 6, 7, and 8, had "intermediate" song types. The pie-charts and graph between the sonogram indicate the genetic assignment of birds at the different locations and clearly reveal genetic differentiation with a transition zone including locations 6, 7, and 8. (Figure modified from Lipshutz et al. 2017; © 2016 John Wiley & Sons Ltd., used with permission)

and/or trills and particular sequences of these subunits are called song types (Brenowitz et al. 1997). The number of song types in the repertoires of male songbirds varies considerably, ranging from one to a thousand or more. Males in about 30% of all species of oscine songbirds have single-song repertoires, about 50% have relatively small repertoires (≤5 song types), and the remaining 20% have repertoires consisting of six or more song types (Beecher and Brenowitz 2005). Most species of songbirds with large song repertoires are open-ended learners capable

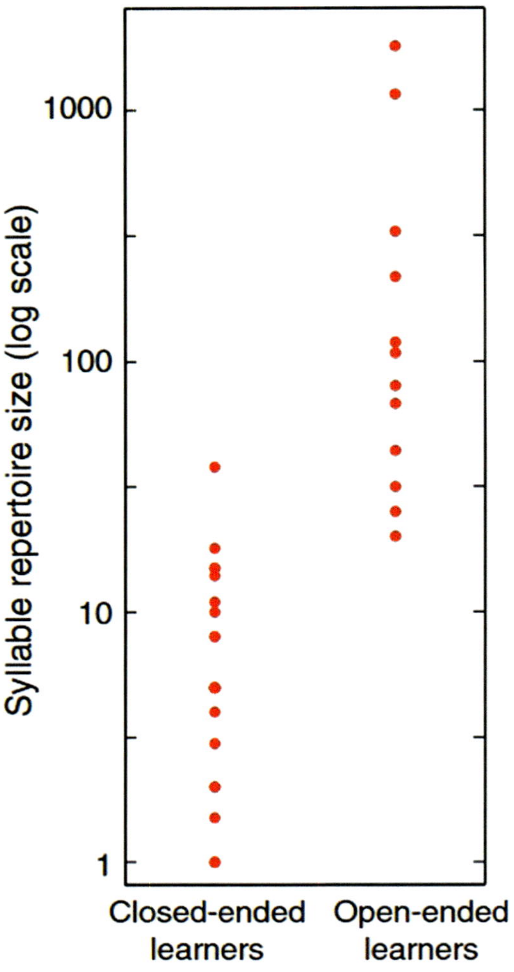


**Fig. 12.69** Size of syllable repertoires of several species of songbirds that are either closed-ended learners (song learning limited to a specific time period early in life) or open-ended learners (capable of learning new songs throughout their lives), with open-ended learners having, on average, significantly larger syllable repertoires. (Figure from Creanza et al. 2016; open-access article distributed under the terms of the Creative Commons Attribution License, https://creativecommons.org/licenses/by/4.0/)

of learning new songs or song notes and syllables throughout their lives (Fig. 12.69). Well-known examples of open-ended learners include Northern Mockingbirds (*Mimus polyglottos*), with 100–200 song types in their repertoires (Derrickson 1987), and Brown Thrashers (*Toxostoma rufum*), with song repertoires including more than 1100 song types (Boughey and Thompson 1981).

One possible reason why selection might have favored development of song repertoires is that different song types can be used to convey different information. For example, many species of North American wood-warblers (Parulidae) have two types of songs that differ in when and where they are used (Spector 1992). One type is typically sung at relatively low rates and primarily early in the breeding season when males are near females so is more important in intersexual communication. In contrast, the second type is more important in intrasexual communication and is typically more complex, sung at higher rates later in the breeding season, and during male-male interactions (e.g., Staicer 1989; Highsmith 1989; Wiley et al. 1994). Similar use of different song types in different contexts has been reported in a number of species, including Grasshopper Sparrows (*Ammodramus savannarum*; Proppe and Ritchison 2008), Field Sparrows (*Spizella pusilla*; Carey et al. 2020), Dusky Warblers (*Phylloscopus fuscatus*; Forstmeier and Balsby 2002), Common Yellowthroats (*Geothlypis trichas*; Guzy and Ritchison 2020), and several other species.

Two additional hypotheses proposed to explain the evolution of multisong repertoires are the Repertoire hypothesis and the Sharing hypothesis. The Repertoire hypothesis posits that the size of song repertoires is under strong directional sexual selection in some species of songbirds, with repertoire size driven by female preferences for larger repertoires (Searcy and Andersson 1986). Based on a review of the literature, and as noted previously, Byers and Kroodsma (2009: 13) concluded that female preferences for males with larger song repertoires "are not widespread." However, such preferences have been found in several species, including Song Sparrows (*Melospiza melodia*; Reid et al. 2004), European Starlings (*Sturnus vulgaris*; Mountjoy and Lemon 1996), European Pied Flycatchers (*Ficedula hypoleuca*; Lampe and Saetre 1995), and several others. An adaptive explanation for this female preference is that repertoire size reliably signals male quality, with

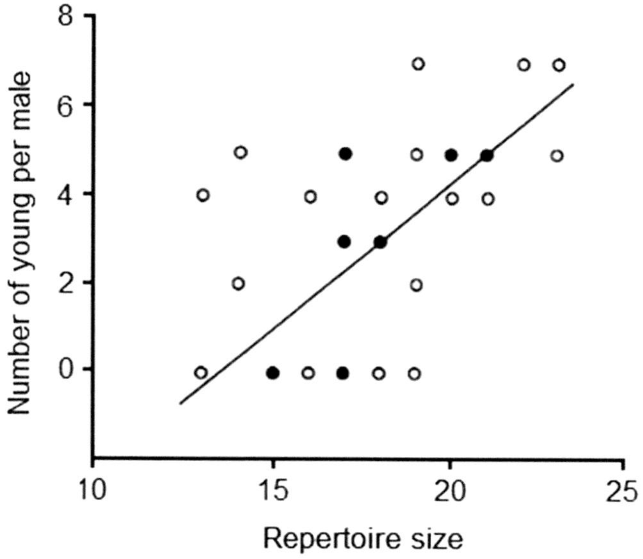


**Fig. 12.70** Correlation between song repertoire size of male Great Reed Warblers (*Acrocephalus arundinaceus*) and number of young that successfully fledged. Open circles = one male, filled circles = two or more males. (Figure modified from Catchpole 1986; © 1986 Springer-Verlag, used with permission)

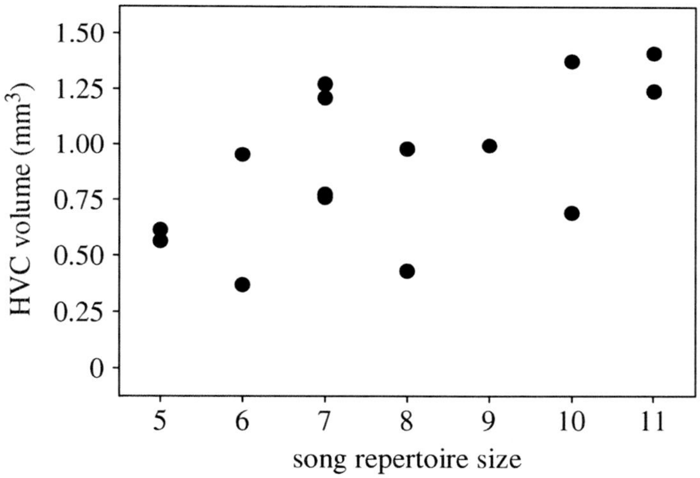


**Fig. 12.71** Relationship between song repertoire size and HVC volume for 16 adult male Song Sparrows (*Melospiza melodia*). Males with larger song repertoires also tended to have large HVCs and were in better body condition. As such, the size of male song repertoires provides conspecifics with information about male quality. (Figure from Pfaff et al. 2007; © 2007 The Royal Society, used with permission)

males having large repertoires providing direct benefits to females, such as superior parental care or territorial defense and/or indirect benefits in the form of good genes for their offspring (Searcy 1992; Bartsch et al. 2015). In support of this hypothesis, some investigators have found that song repertoire size is positively correlated with several traits important for survival and reproduction such as body size and mass (e.g., Lampe and Espmark 1994, Doutrelant et al. 2000, Kipper et al. 2006), and other studies have revealed a positive correlation between song repertoire size and breeding success (e.g., Catchpole 1986, Potvin et al. 2015; Fig. 12.70). In addition, males with larger song or syllable repertoires have been found to have larger song control areas in the brain, another potential indicator of male quality (Pfaff et al. 2007; Moore et al. 2011; Figs. 12.71 and 12.72).

The Sharing hypothesis predicts that sharing song types with neighboring conspecifics can be beneficial, potentially enhancing a male's fitness.

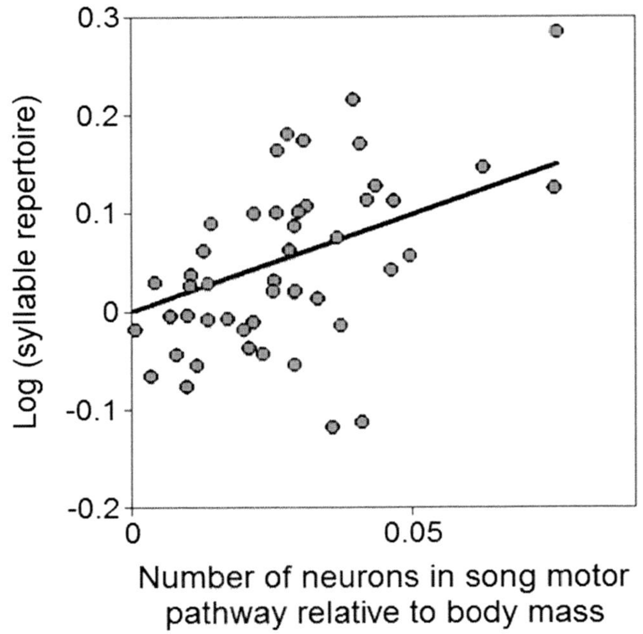


**Fig. 12.72** Relationship between the size of a species' syllable repertoire and relative number of neurons in the song motor pathway of the brain. Each point represents a different species. (Figure modified from Moore et al. 2011; used with permission of the National Academy of Sciences USA)

In contrast to the Repertoire hypothesis, the Sharing hypothesis does not require large song repertoires. In fact, the Sharing hypothesis might even favor smaller repertoires (so males need not remember, and be able to sing, as many different song types; Beecher and Brenowitz 2005, but see Moser-Purdy and Mennill 2016). Males that share song types with neighboring conspecifics can potentially benefit in several ways. First, birds are able to "range," i.e., use the extent of song degradation (as emitted songs travel through the environment to listeners) to estimate their distance from a singer. However, birds can range more accurately if they are familiar with a song type (e.g., McGregor and Krebs 1984) so having shared songs allows a territory holder to better estimate their distance from, and the location of, singing neighbors. This is important because, for example, a neighbor singing close to a shared territory boundary may be more of a threat than one singing further away.

Another potential advantage of song sharing is the ability to address a message to a particular individual. For example, if multiple neighbors are singing, but one is known to be very close to a shared territory boundary (via ranging) or, early in the breeding season before boundaries have been established, close to a disputed boundary, a territorial male can match the song type of that male to indicate that his "keep-out message" is intended for that male (King and McGregor 2016).

Among species of songbirds with multisong repertoires and song sharing, males generally share some, but not all, of their song types. This potentially allows males to use both shared and unshared song types during interactions with neighbors to convey more specific "messages." For example, male Song Sparrows (*Melospiza melodia*) have multisong repertoires, with some song types typically shared with other males in the same area and some not shared. When territory boundaries are being established, male Song Sparrows can either escalate or de-escalate encounters depending on which song type they sing (Fig. 12.73). The use of shared and unshared songs allows male Song Sparrows to better indicate their level of aggression by using songs that either indicate a willingness to escalate or a desire to de-escalate the intensity of an interaction (Beecher and Brenowitz 2005).

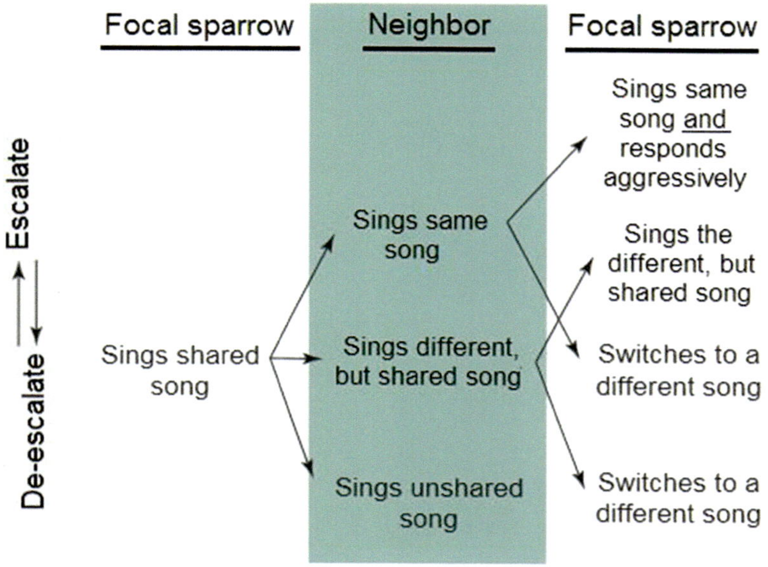

**Fig. 12.73** Song Sparrows typically interact with neighbors by singing one of the song types they share. When a Song Sparrow sings, its neighbor can (1) respond with low-threat signal, i.e., singing a song they share, but not the one neighbor just sang, (2) escalate the interaction by singing the same song type as the neighbor (song-matching), or (3) de-escalate the interaction by singing an unshared song type. If the neighbor responds with the same song type, the focal sparrow can either respond to the escalation by continuing to sing the same type and responding aggressively (e.g., moving closer, singing lower-volume songs, and/or performing a visual display that conveys aggression such as waving their wings) or de-escalate by switching to another song type and not responding aggressively. If the intensity of the interaction continues to escalate and a focal sparrow sings the same song and also responds with an additional signal of aggression (e.g., wing-waving), then actual fighting may soon follow. (Figure modified from Beecher and Brenowitz 2005; © 2005 Elsevier Ltd., used with permission; photo of wing-waving Song Sparrow by Çağlar Akçay and used with permission)

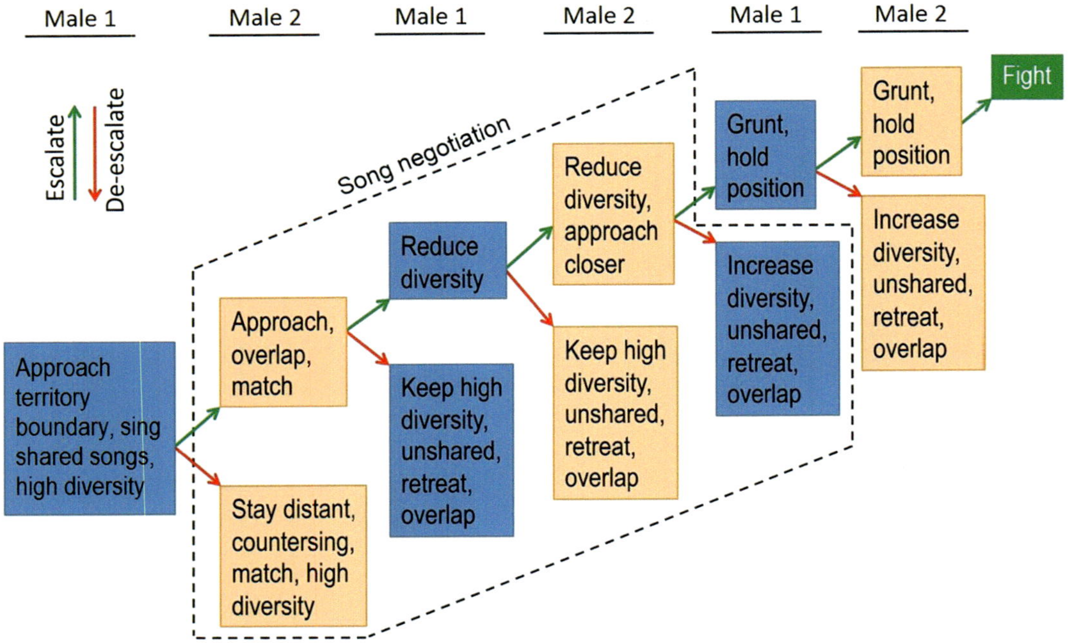


**Fig. 12.74** Flow chart showing how male Banded Wrens (*Thryophilus pleurostictus*) can either escalate or de-escalate the intensity of an interaction via a combination of song and distance from each other. In this example, two males alternate in making decisions whether to escalate or de-escalate an interaction. If both choose to approach, they "negotiate" by increasing or decreasing the number of different song types uttered (song-type diversity). Males choosing to de-escalate an encounter continue singing with high song-type diversity, but also use unshared song types and then overlap the other male's songs while retreating. If an interaction continues to escalate, both males may respond with grunt calls and, if so, a fight may ensue. (Figure modified from Vehrencamp et al. 2014; © 2014 Oxford University Press, used with permission)

As another example, male Banded Wrens (*Thryophilus pleurostictus*) in the process of establishing or attempting to maintain or move the location of shared territory boundaries use variation in song-type diversity and use of unshared song types to either escalate or de-escalate interactions with neighboring males (Fig. 12.74). Males intent on initiating an interaction (Male 1) approach a boundary and sing several different song types (high diversity) that are also in the repertoire of the other male. In response, that male (Male 2) can choose to either continue the interaction by moving closer to the aggressor or indicate an unwillingness to further interact by not approaching and singing with high diversity while also song matching. If the former, then each male, when responding to their neighbor, can choose to continue escalating the encounter or not. If escalation continues, then the males may eventually fight (Fig. 12.74).

Given the potential advantages of multisong repertoires, why do males in about 30% of all species of oscine songbirds have single-song repertoires? One possible explanation is that multisong repertoires have greater developmental and energetic costs because they require more neurons in areas of the brain involved in song learning and production (Garamszegi and Eens 2004; Moore et al. 2011). Given this constraint that may require trade-offs in brain space devoted to singing and to other functions (Garamszegi and Eens 2004), selection may favor single-song repertoires. In addition, selection may not favor multisong repertoires if sexual selection favors nonacoustic traits (Gontard-Danek and Møller 1999) or acoustic traits such as consistency or performance (e.g., Byers 2007; Podos 2017) rather than larger song repertoires.

Finally, selection may favor single-song repertoires because males with just a single song type may be able to convey information to conspecifics by varying the rate or volume at which the song is uttered, the duration of song bouts, by varying song characteristics, and by varying when and where they sing. For example, the duration of song bouts and/or song rate may reflect a male's condition and, as such, can be an honest advertisement of male quality (e.g., Otter et al. 1997; Grunst and Grunst 2014). Male Chipping Sparrows (*Spizella passerina*) have a single song type that consists of a rapid series of similar short notes, but, during the dawn chorus, they sing from low perches near territory boundaries and sing short songs at a rapid rate; these songs are likely directed at neighboring males and serve to maintain territory boundaries (Liu and Kroodsma 2007). In contrast, during the day, male Chipping Sparrows sing longer songs from high perches that likely function to attract females (Liu and Kroodsma 2007). Male Black-capped Chickadees (*Poecile atricapillus*) also have a single song type, but convey information by varying the frequency (pitch) of the two notes of their "fee-bee" song (Christie et al. 2004). More specifically, during the dawn chorus, males match the frequency of the songs of neighboring males, particularly males with similar dominance status. This matching may be important in assessing the quality and condition of neighbors (Foote et al. 2008).

Most species of oscine songbirds with single-song repertoires have songs that consist of more than one note or syllable type. Read and Weary (1992) surveyed the literature and noted the size of song repertoires of 113 species. Of those, 27 (24%) had single-song repertoires. However, 22 of those 27 species had multiple note or syllable types in their repertoires (mean = 4.8 syllables, range = 2–10), potentially allowing them to produce different songs by altering the order of those syllables or the number of syllables used. For example, male Indigo Buntings (*Passerina cyanea*) have a single song that consists of several different notes and syllables, but vary the number of notes and syllables used in songs depending on the context; males use shorter songs uttered at high rates to attract mates and longer songs to convey aggression during male-male interactions (Beckett and Ritchison 2010; Fig. 12.75). Even species where males have one song type that consists of just one note or syllable type can vary song duration, and potentially convey different information in different contexs, by varying the number of those notes or syllables in different songs (Fig. 12.76).

## 12.22 Energetic Cost of Singing

Studies of the energetic cost of singing have provided mixed results. Laboratory studies of singing birds with energetic costs determined by oxygen consumption have generally revealed rather small metabolic costs. For example, Ward et al. (2004) found that the metabolic cost of song production by male European Pied Flycatchers (*Ficedula hypoleuca*) was less than 1.12, the cost of just standing. Similarly, Ward and Slater (2005) used thermal images to estimate energetic costs and found only a 14% increase in heat transfer when Canaries (*Serinus canaria*) were singing. In contrast, using the doubly labeled water method to measure daily energy expenditure, Hasselquist and Bensch (2008) found a relatively high energetic cost of singing by free-living Great Reed Warblers (*Acrocephalus arundinaceus*) (Fig. 12.77). In addition, other investigators have provided indirect evidence that the energetic costs of singing may be higher than reported in lab-based studies. For example, male Common Nightingales (*Luscinia megarhynchos*) sing at night, but vary in their singing rates. Thomas et al. (2002) found that males that sang at higher rates lost more mass during the night than those singing at lower rates, suggesting that singing at higher rates required more energy. In addition, other investigators have found that restricting food availability can reduce singing rates (Ritschard and Brumm 2012) whereas providing supplemental food can increase singing rates (Grava et al. 2009; Barnett and Briskie 2011) and even song

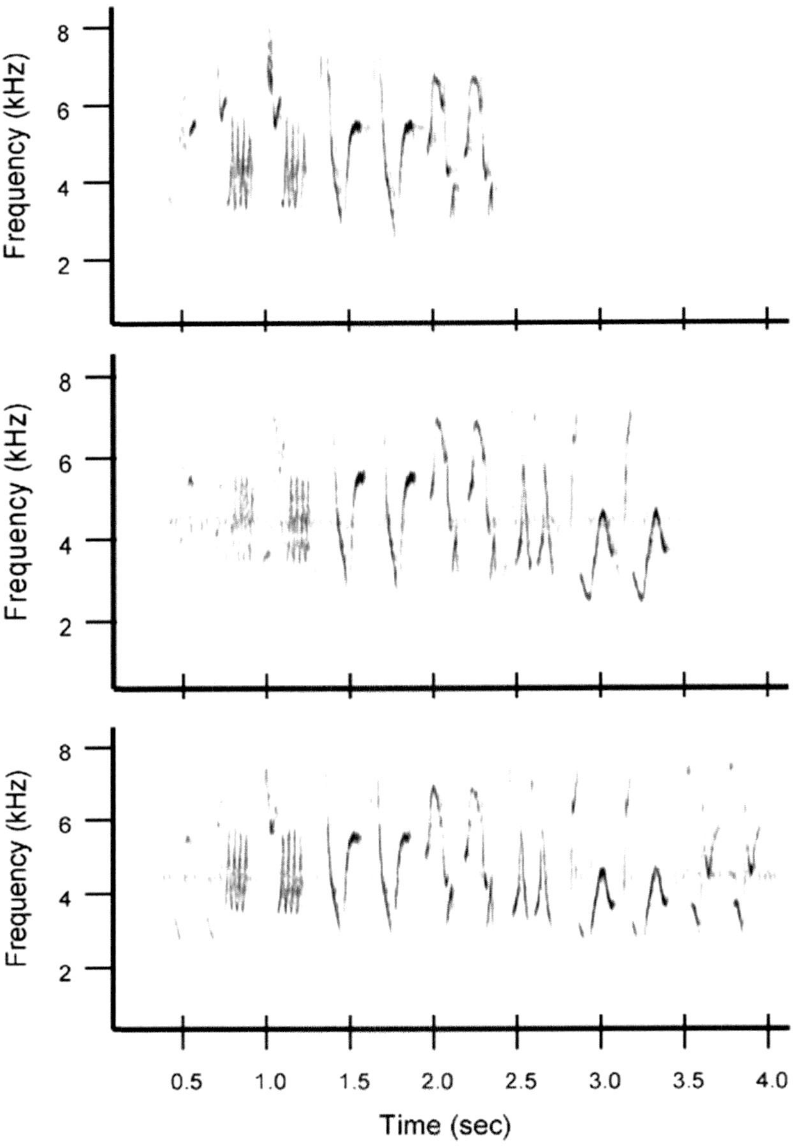


**Fig. 12.75** Example of how a male Indigo Bunting (*Passerina cyanea*) varies the duration of songs by varying the number of notes and syllables. (Figure from Beckett and Ritchison 2010; © 2010 Wilson Ornithological Society, used with permission)

quality (Yamada and Soma 2016: Fig. 12.78). Such results suggest that the energetic costs of singing may affect singing rates and, perhaps, reproductive success. The question of whether or not energetic costs can affect singing rates is an important one because, if so, then singing rates would be an honest signal of condition, with males capable of singing at higher rates advertising their quality and condition to male and female conspecifics.

Beyond energetic costs, another potential cost of singing is its potential effect on a bird's oxidative status. Oxidative status is determined by the relative concentrations of reactive oxygen species (ROS) and antioxidants in the blood. ROS are by-products of cellular metabolism (ATP production) that can react with and damage molecules in cells including proteins and nucleic acids. Antioxidants are molecules that interact with ROS to prevent such damage. Active brain

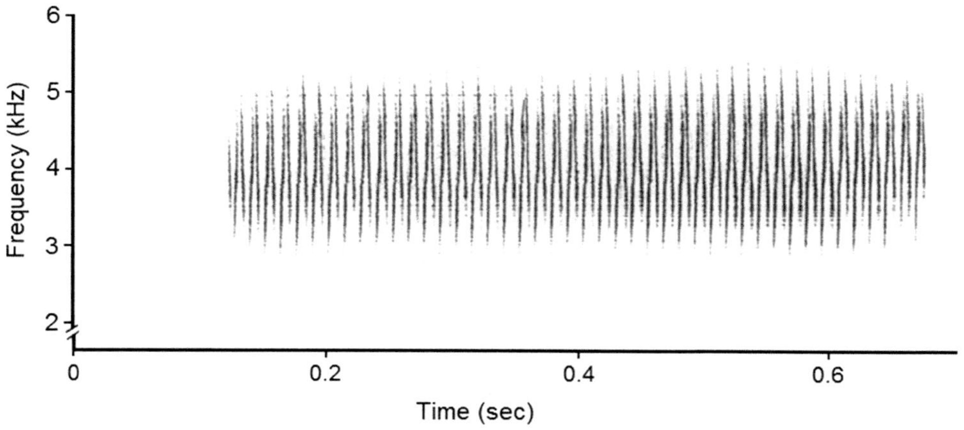


**Fig. 12.76** Male Clay-colored Sparrows (*Spizella pallida*) have one song type that consists of series of the same syllable type. Males are known to vary the duration of songs by varying the number of syllables in songs, but the possible function(s) of such variation remains to be determined. (Figure modified from Knapton 1982; © Brill Academic Publishers, used with permission)

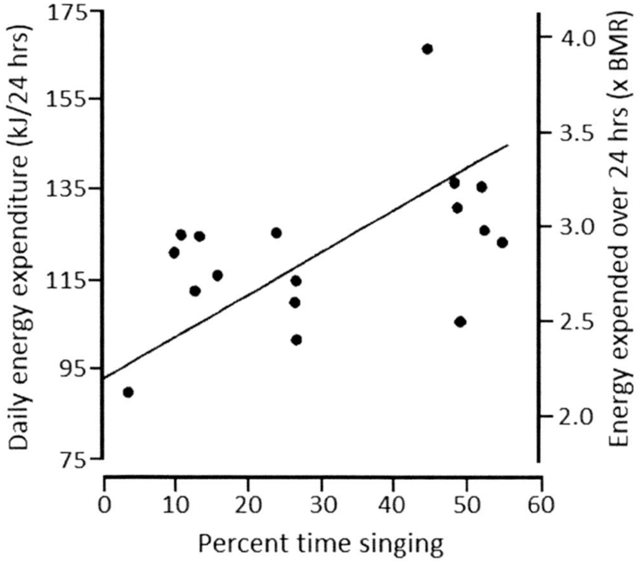


**Fig. 12.77** Energetic cost of singing by male Great Reed Warblers (*Acrocephalus arundinaceus*). Males typically utter songs during 50% of daylight hours (percent time singing does not include silent periods between songs). (Figure modified from Hasselquist and Bensch 2008; © 2008 The Authors, used with permission)

tissues, including areas of the avian brain important in song production, use much oxygen and, therefore, produce lots of ROS. If not balanced by sufficient availability of antioxidants, neuronal damage can occur.

Studies of European Starlings (*Sturnus vulgaris*) and Snow Buntings (*Plectrophenax nivalis*) have revealed that a male's "redox state" (i.e., ratio of ROS relative to antioxidants) can influence singing behavior (Fig. 12.79). Messina et al. (2017) found that experimental inhibition of an important antioxidant (glutathione) resulted in increased oxidative damage and reduced singing rates by subordinate male

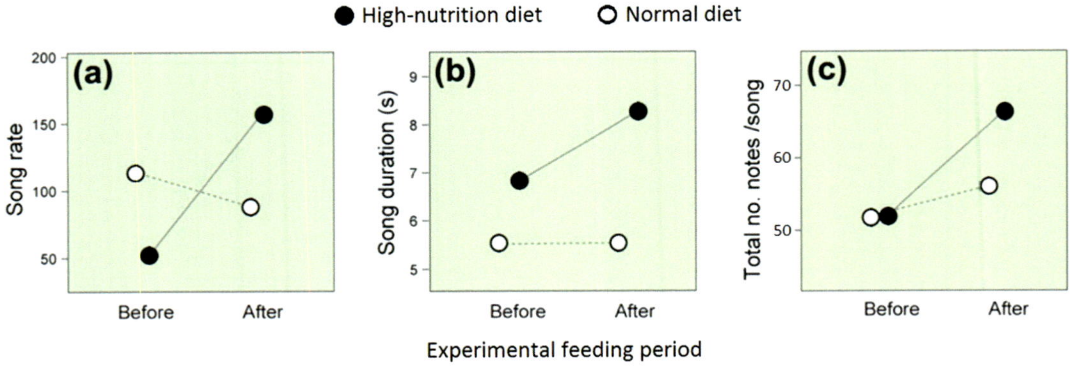


**Fig. 12.78** Changes in (**a**) song rate (number of songs during a 12-h day), (**b**) song duration (seconds), and (**c**) number of notes per song for male Bengalese Finches (*Lonchura striata domestica*) before and after an experimental feeding period. Males in both experimental groups were provided with the same food before the experimental feeding period. However, after that period some males (normal) were maintained on the same diet whereas other males (high nutrition) were provided with a more nutritious diet. Males with the more nutritious diet sang more and sang longer, more complex songs (with more notes) than males provided with the normal diet, suggesting that a male's nutritional status can potentially influence both song rate and quality. (Figure modified from Yamada and Soma 2016; © 2016 The Authors, used with permission)

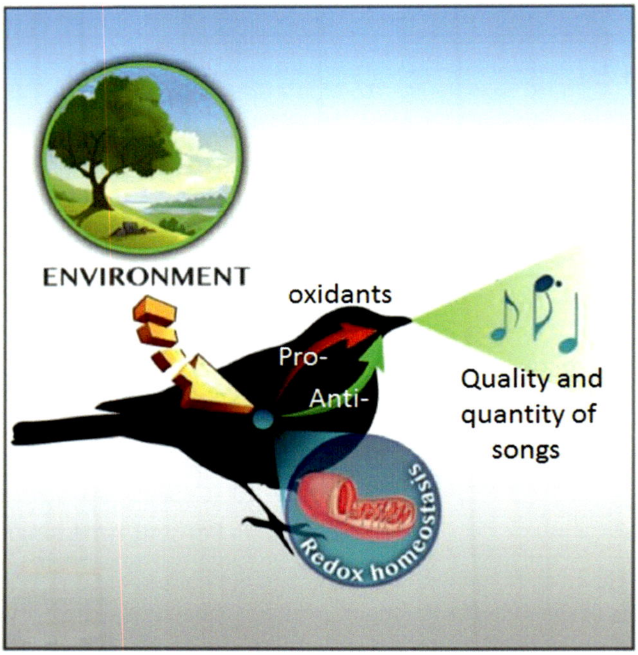


**Fig. 12.79** Singing may be an honest signal because it is costly and can result in oxidative damage. This hypothesis predicts that production of antioxidants is dependent on the quality of males and/or their territories. The ability to increase production of antioxidants (molecules able to react with and neutralize reactive oxygen metabolites produced by mitochondria) allows males in better condition to sing higher quality songs at higher rates. Of course, a male's environment also influences oxidative status. (Figure modified from Casagrande et al. 2016; open-access article distributed under the terms of the Creative Commons Attribution License, https://creativecommons.org/licenses/by/4.0/)

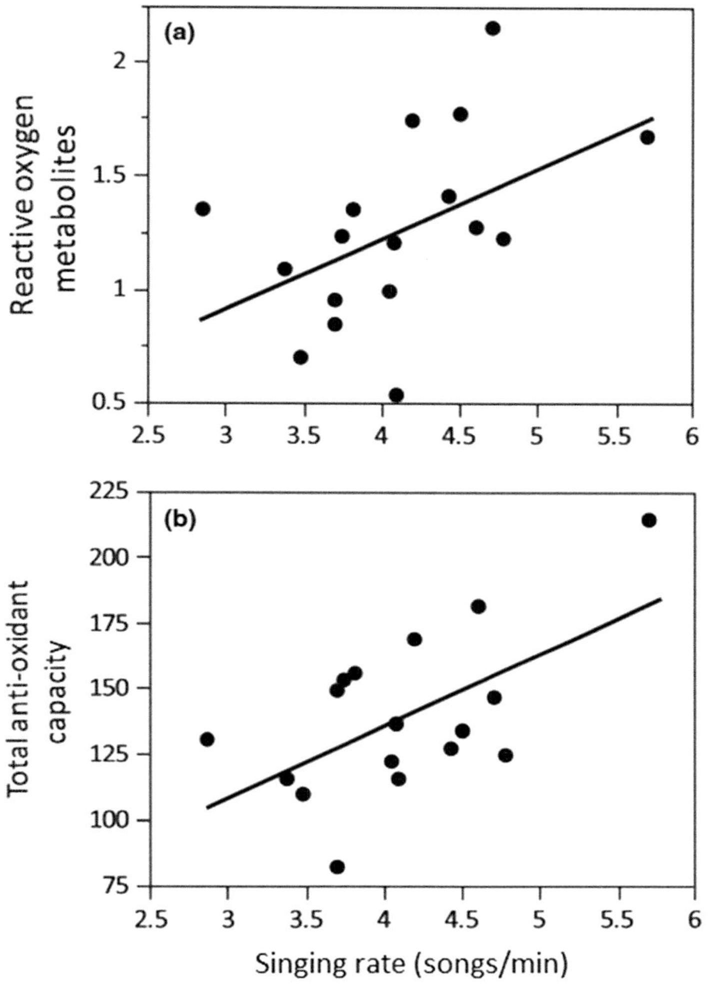


**Fig. 12.80** Male Snow Buntings (*Plectrophenax nivalis*) that sang at higher rates had both (**a**) higher levels of reactive oxygen species or metabolites (measured as mg/100 ml of blood) and (**b**) total antioxidant capacity (expressed as 1 mol of HOCl/ml neutralized). (Figure from Baldo et al. 2015; © 2015 Blackwell Verlag GmbH, used with permission)

European Starlings, but inhibition had no effect on dominant males. In a study of male Snow Buntings, Baldo et al. (2015) found that males singing at higher rates had higher levels of ROS, suggesting that the metabolic demands of singing can increase production of ROS. In addition, however, these authors found that some males were better than others at increasing their antioxidant capacity and, as a result, were able to sing at higher rates (Fig. 12.80). These results, although based on just two species of songbirds, suggest that singing can increase production of ROS and that singing rates can provide conspecifics with honest information about a male's oxidative status or redox state.

## 12.23 Song Learning by Males

The songs of birds can be innate, learned, or invented. In some species, like Brown-headed Cowbirds (*Molothrus ater*), males develop normal songs even when raised in acoustic isolation. However, for most species of oscine songbirds, at least some aspects of their singing behavior are learned. That learning, however, is guided by innate mechanisms. Vocal learning occurs in

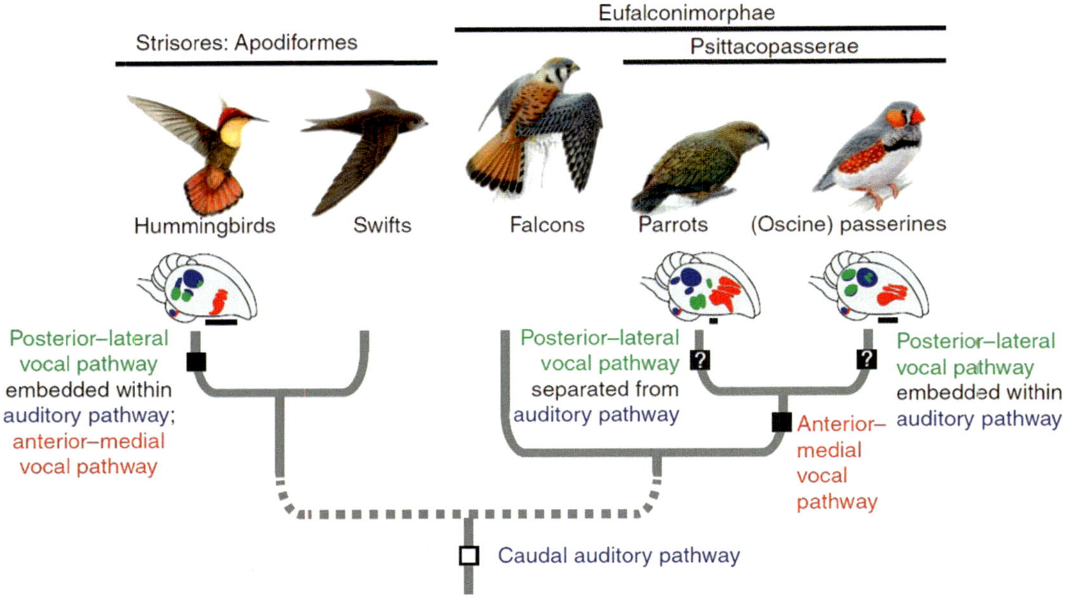


**Fig. 12.81** Evolution of vocal learning in birds. Schematic brain drawings of hummingbirds, parrots, and songbirds (oscine passerines). Some traits associated with vocal learning (for example, the anterior–medial vocal pathway (red)) are potentially homologous and thus evolved as an autapomorphic trait (unique to a given taxon, black square) in the Psittacopasseran ancestor, but also independently in the distantly related ancestor of hummingbirds (dashed lines indicate that several neoavian lineages are not shown). Neuroanatomical studies on early branching passerines (that is, Suboscines and Acanthisittidae) are necessary to infer that the posterior–lateral vocal pathway (green, located at different brain regions in parrots and oscine passerines) is either homologous or evolved independently in the two lineages. The caudal auditory pathway (blue) is a plesiomorphic trait (white square) and was probably inherited from a common avian ancestor. The putative location of the auditory pathway in falcons and swifts has not yet been investigated. Scale bars below brains, 2 mm. (Figure from Suh et al. 2011; © 2011 Springer Nature, used with permission)

three taxa of birds, the orders Passeriformes and Psittaciformes and the family Trochilidae (Apodiformes, hummingbirds). Several hypotheses have been proposed to explain the evolution of vocal learning. One is that vocal learning evolved independently in songbirds, parrots, and hummingbirds (Brenowitz 1997; Jarvis et al. 2000). Another hypothesis is that, because vocal centers in the brains of songbirds, parrots, and hummingbirds are similar, many taxa of birds that are "nonlearners" lost the ability to learn independently, multiple times (Wada et al. 2004). A third hypothesis is that vocal learning evolved twice, once in the ancestor of present-day parrots and songbirds and again in hummingbirds after the divergence of hummingbirds and swifts (Suh et al. 2011; Fig. 12.81). In support of their hypothesis, Suh et al. (2011) suggested that assuming that vocal learning evolved in the ancestor of parrots and songbirds "seems more parsimonious than hypothesizing four independent evolutions of vocal learning within Psittacopasserae," i.e., oscine songbirds, suboscine families Cotingidae and Pipridae with some evidence of vocal learning (Trainer et al. 2002; Saranathan et al. 2007), and parrots.

Among oscine songbirds, the learning of songs exhibits considerable variation. For example, the period during which songs can be learning can last for the first few months after hatching (e.g., White-crowned Sparrows, *Zonotrichia leucophrys*; Marler 1970a), the first year and into their second year of life (e.g., Indigo Buntings, *Passerina cyanea*; Payne and Payne

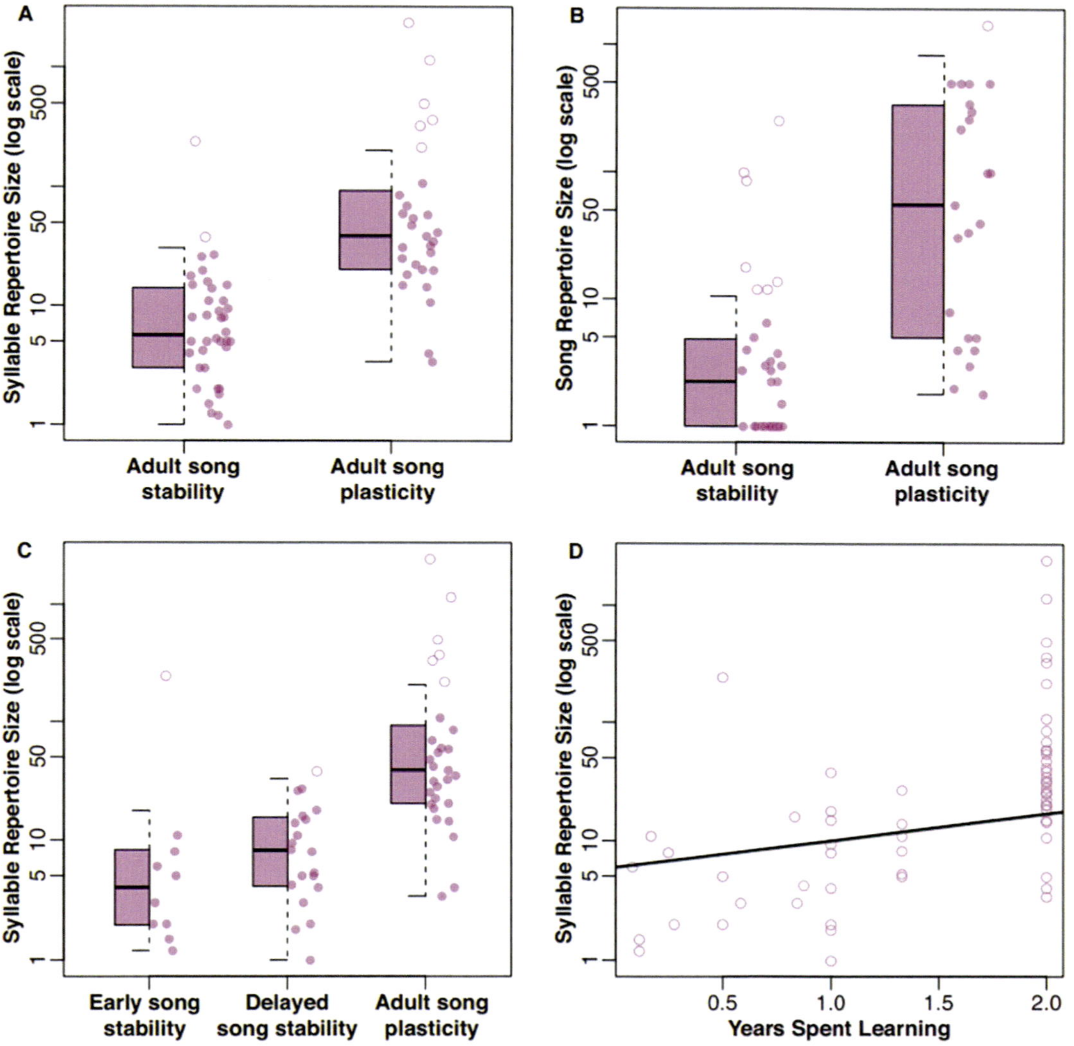


**Fig. 12.82** Song and syllable repertoire size for closed-ended song learners (i.e., song stability) and open-ended song learners (i.e., song plasticity). (**a**, **c**, and **d**) Open-ended learners have significantly larger syllable repertoires and (**b**) significantly larger song repertoires. (**c**) Syllable repertoires for species in three categories: early song stability (species where all learning occurs before their first breeding season), delayed song stability (species where learning can still learn songs during their first breeding season, but not after), and adult song plasticity (open-ended learners). Boxes indicate the 25th, 50th, and 75th percentiles. The lower whisker is either the minimum value or the 25th percentile minus 1.5 times the interquartile range, whichever was larger. The upper whisker is either the maximum value or the 75th percentile plus 1.5 times the interquartile range, whichever was smaller. Solid dots are within the range of the box and whiskers, and open dots are outliers. (Figure from Robinson et al. 2019; © Robinson et al., open-access article distributed under the terms of the Creative Commons Attribution License, https://creativecommons.org/licenses/by/4.0/)

1997), or for a bird's lifetime (e.g., Northern Mockingbirds, *Mimus polyglottus*; Derrickson 1987). Species where song learning is limited in time are referred to as closed-ended learners and those able to learn songs or song elements throughout their lives are referred to as open-

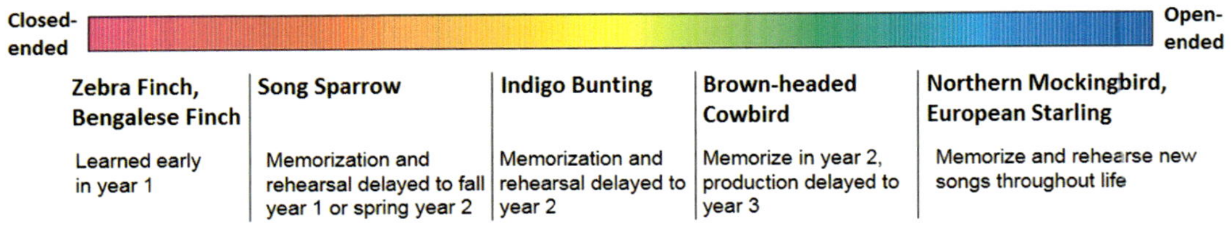


**Fig. 12.83** Among songbirds, song learning programs form a continuum between closed-ended and open-ended learning strategies. "Years" refers to calendar years. Zebra Finches (*Taeniopygia guttata*) and Bengalese Finches (*Lonchura striata*) are closed-ended learners. Once song is crystallized in the first year, males sing the same song for the rest of their life. Young male Song Sparrows (*Melospiza melodia*) learn song during the summer and fall of their first year, and during the following spring when they memorize songs of adult males in the area where they will establish their breeding territory. Indigo Buntings (*Passerina cyanea*) typically delay song learning until their first breeding season. Male Brown-headed Cowbirds (*Molothrus ater*) learn new songs during their first breeding season, but do not sing them until the next breeding season. At the opposite end of the song-learning spectrum are species like Northern Mockingbirds (*Mimus polyglottus*) and European Starlings (*Sturnus vulgaris*) that continue to learn new songs throughout their lives. (Figure modified from Brenowitz and Beecher 2005; Copyright © 2005 Elsevier Ltd., used with permission)

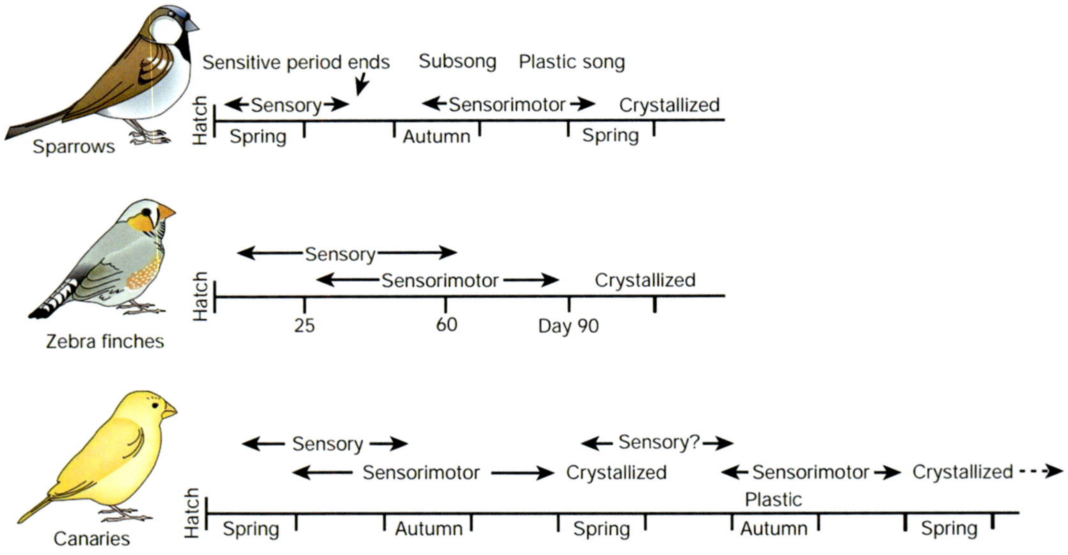


**Fig. 12.84** Variation in the process of song learning among oscine songbirds. (Top) For many species, like White-crowned Sparrows (*Zonotrichia leucophrys*), the sensory and sensorimotor phases of learning occur at different times. During the sensory phase, young birds listen to and memorize one or more tutor songs. During sensorimotor learning, young birds match their own song to the memorized model using auditory feedback. Sensorimotor learning includes an initial stage of babbling-like vocalizations called subsong, followed by more structured, but still variable, plastic song. Plastic song is progressively refined until the song or songs crystallize into their final form used as adults. (Middle) Young Zebra Finches (*Taeniopygia guttata*) first memorize the song of an adult "tutor" during a critical period for "sensory" learning of song and, during the overlapping sensorimotor learning period, they begin to sing and gradually match their initially immature vocalizations to the memorized song using auditory feedback. Song crystallization occurs between 90 and 120 days after hatching. (Bottom) Open learners like Canaries (*Serinus canaria domesticus*) continue to learn songs or components of song throughout their lives. (Figure from Brainard and Doupe 2002; © 2002 Nature Publishing Group, used with permission)

**Table 12.5** Definitions of terms used to describe the process of song learning (Table from Johnson and Clark (2020); Published by Elsevier Ltd on behalf of The Association for the Study of Animal Behaviour, used with permission)

| Term | Definition | Source |
|---|---|---|
| Sensory phase | Subject listens to and memorizes species-specific song(s) | Marler (1970a, b, Konishi (1985) |
| Sensorimotor phase | Subject begins singing, learning to use its vocal apparatus and match the motor output, i.e., the vocalizations, with the memory of the song(s) from the sensory phase | Marler and Peters (1987) |
| Subsong | The first vocalizations a vocal-learning bird produces. The vocalizations are usually short in duration, vary in frequency, and are acoustically and structurally dissimilar in sound and appearance from the adult song(s) | Thorpe (1958), Thorpe and Pilcher (1958) |
| Plastic song | Stage of song development where a bird practices its song(s) and starts to match its song(s) to its memorized song | Marler and Peters (1981, 1982) |
| Crystalized song | Final stage of song development, when the song(s) match the memorized song(s); song(s) is(are) stereotyped (i.e., low variability) and cease (s) changing | Marler et al. (1988) |

ended learners. Open-ended learners have larger song and syllable repertoires than closed-ended learners (Fig. 12.82), and Robinson et al. (2019) suggested that sexual selection for larger repertoires could have driven selection for open-ended learning. Among closed-ended learners, the time frame involved for learning varies among species (Fig. 12.83).

Among young oscine songbirds, song learning occurs in several stages or phases, the timing of which varies among species (Fig. 12.84, Table 12.5). First is the sensory phase during which young listen to and memorize the song (s) of one or more tutors. Depending on the timing of the sensory phase, a bird's tutors can be its neighbors during the breeding season it hatches, its neighbors during its first breeding season, or neighbors during subsequent breeding seasons. Next comes a sensorimotor phase during which young attempt to match their own song (s) with the memorized one(s) using auditory feedback. The sensorimotor phase includes an initial stage of "experimentation" called subsong, where vocalizations may have little similarity to the tutor song(s). During the subsong phase, birds may utter low volume, unstructured notes initially and, over time, sounds become more like the species-specific song of the tutor(s) (Fig. 12.85). The subsong phase is followed by a plastic song phase during which songs are more structured, but still variable. During the final phase of song learning called song crystallization, songs closely match those of the tutor(s) and will be the ones used during the rest of their lives. Although much less is known about the process of song learning in hummingbirds, song learning in hummingbirds is known to be culturally transmitted through imitation (Jarvis et al. 2000). In addition, the timing of song learning appears to vary in hummingbirds much as it does for oscine songbirds. For example, the results of a study of male Costa's Hummingbirds (*Calypte costae*) suggest that songs are crystallized at about 120 days post-hatching (Fig. 12.86). In contrast, variation in the songs of males in different leks suggests that males in some species of hummingbirds learn songs after dispersal, probably to facilitate territory establishment and access to leks (González and Ornelas 2014; Lara et al. 2015). In addition, Araya-Salas and Wright (2013) found that male Long-billed Hermits (*Phaethornis longirostris*) had crystallized single-song repertoires, but some males changed their songs after periods ranging from 186 to 547 days (mean = 307 days). Such changes suggest that male Long-billed Hermits are open-ended learners (Araya-Salas and Wright 2013). However, the extent to which other species of hummingbirds might be closed-ended or open-ended learners will require additional study.

Parrots (Order Psittaciformes) are also vocal learners. For most species of parrots, vocal

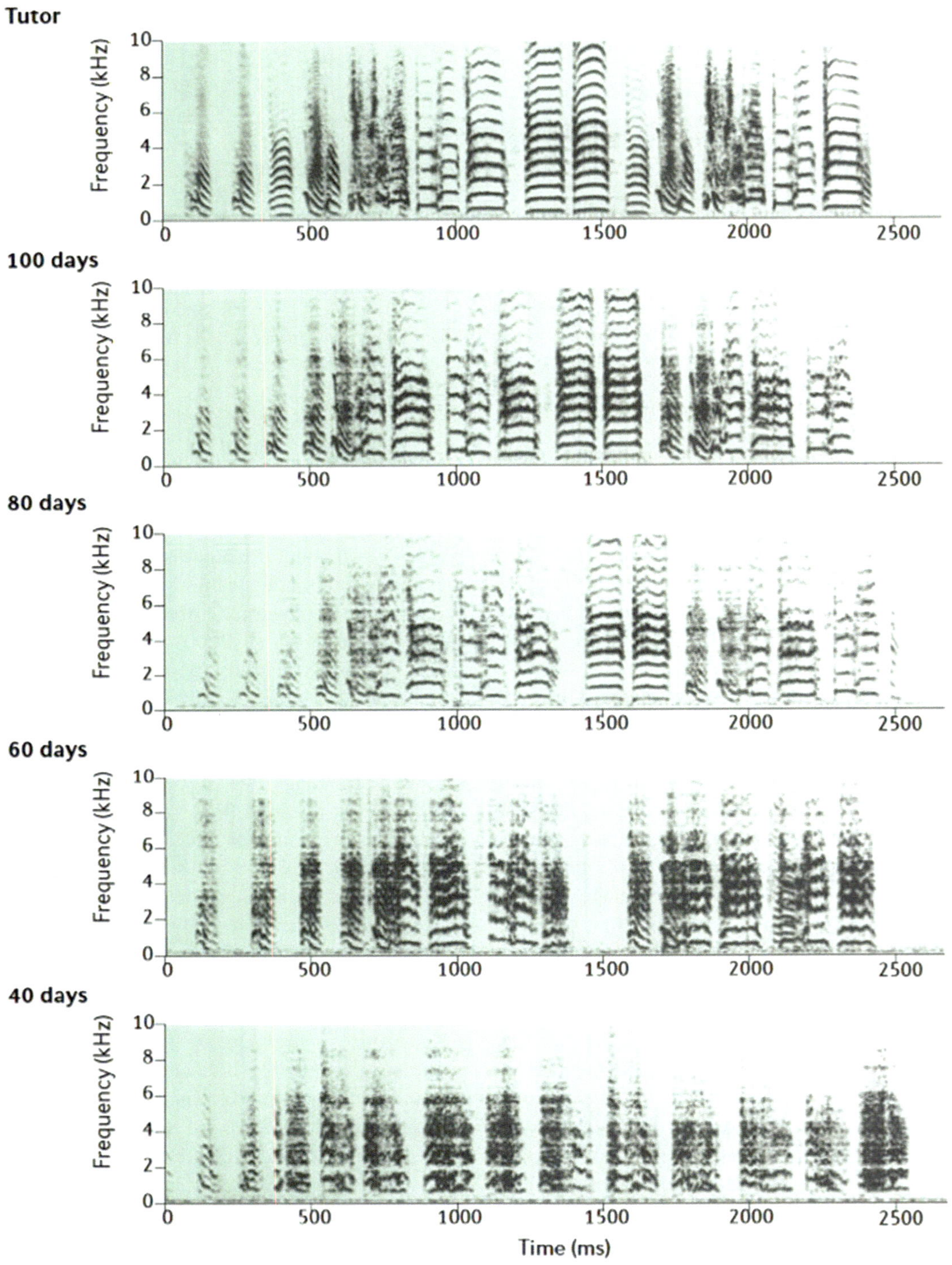


**Fig. 12.85** Sonograms of the song of an adult male Zebra Finch (the tutor; *Taeniopygia guttata*) and those of one of his male offspring at different stages of development. The first vocalizations (40 days post-hatching) are called subsong. At about 60–80 days post-hatching, the young male produces plastic song that resembles the tutor song more than subsong. At about day 100, the young male produces crystallized song that closely resembles the tutor's song. (Figure from Bolhuis and Gahr 2006; © 2006 Springer Nature, used with permission)

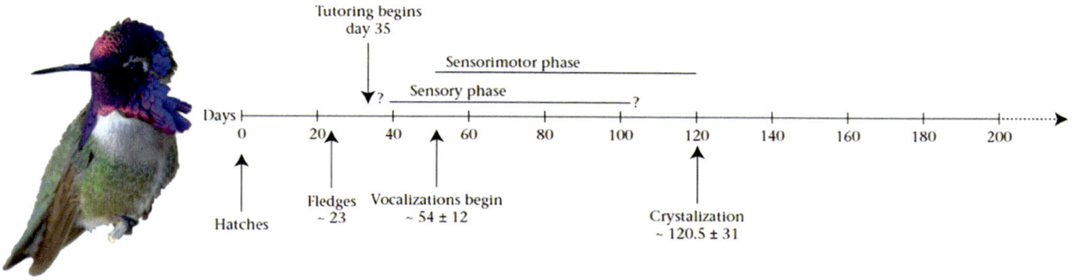


**Fig. 12.86** Possible timeline for song development by Costa's Hummingbirds (*Calypte costae*). (Figure from Johnson and Clark 2020; Published by Elsevier Ltd on behalf of The Association for the Study of Animal Behaviour, used with permission)

learning is similar for males and females, open-ended throughout their lives, and typically results in large vocal repertoires (Bradbury 2003; Colbert-White et al. 2014). Bradbury and Balsby (2016) identified two different uses of vocal learning in parrots: the importance of learning local calls or duets during defense of nest sites and acquiring contact calls to mediate social dynamics, with the acquisition of contact calls "linked to the unique diets of parrots and the attendant consequences for social foraging strategies."

Some oscine songbirds develop songs via improvisation. For example, Kroodsma et al. (1999) found that male Sedge Wrens (*Cistothorus platensis*) in North America exposed to different song types do not imitate those song types, but, rather, generate larger song repertoires largely by improvisations of the songs they heard. Examples of other species where males develop songs by improvisation rather than imitation include Grasshopper Sparrows (*Ammodramus savannarum*; Soha et al. 2016), Pied Butcherbirds (*Cracticus nigrogularis*; Taylor 2010), Common Nightingales (*Luscinia megarhynchos*; Hultsch and Kopp 1989), and Gray Catbirds (*Dumetella carolinensis*; Kroodsma et al. 1997). In these species, males appear to use information from songs of conspecific males, but do not imitate particular songs precisely. In some species, this type of song development may be due to selection for large song repertoires (Kroodsma 1996), e.g., Gray Catbirds and other mimicking species. In others such as Grasshopper Sparrows and Sedge Wrens, improvisation may be correlated with low breeding site fidelity (Kroodsma et al. 2002). Kroodsma et al. (1999) suggested that improvisation reduces variation across geographical areas and, as a result, facilitates species recognition regardless of where birds breed. Finally, Soha et al. (2009) suggested that improvisation might occur if "precise imitation of song is particularly difficult," and noted that the rapid buzz notes in the songs of male Grasshopper Sparrows might represent such a case.

For most species of oscine songbirds, an "auditory template," i.e., songs of tutors, is a critical step in forming a memory of those songs. This memory provides a way to evaluate how vocalizations match those of a tutor. Errors in a bird's vocal performance can be identified by comparing auditory feedback with the auditory template and to make changes as necessary (Keller and Hahnloser 2009). Neurons in the auditory cortex of the songbird brain enable songbirds to recognize conspecific songs and vocalizations (Moore and Woolley 2019) so neurons in that region have been thought to also serve as the auditory template, specifically an area called the caudomedial nidopallium (or NCM) (Bolhuis and Moorman 2015; Fig. 12.87). However, Canopoli et al. (2016) found that lesions (using injections of ibotenic acid) of the NCM of young Zebra Finches did not prevent them from learning to imitate the song of a tutor, suggesting that the memory guiding

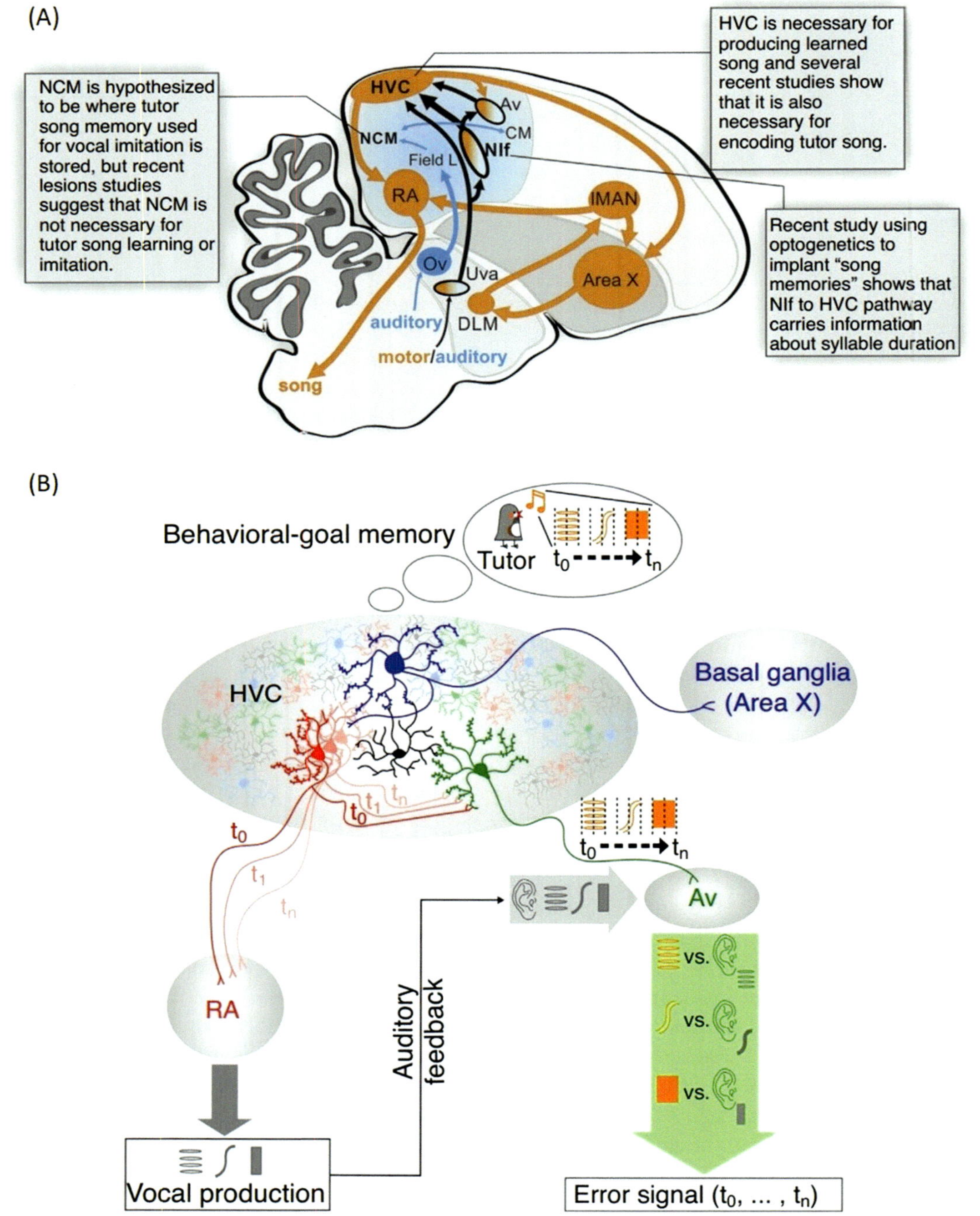


**Fig. 12.87** (**a**) Schematic showing the auditory (blue) and song-producing motor circuits (orange) in a songbird brain. The striato-pallidal basal ganglia nucleus Area X, nucleus avalanche (Av), caudal mesopallium (CM), dorsolateral thalamic nucleus (DLM), primary auditory forebrain (field L), premotor song nucleus HVC (proper name), lateral magnocellular nucleus of the anterior nidopallium (LMAN), caudomedial nidopallium (NCM), nucleus interface of the nidopallium (NIf), thalamic nucleus ovoidalis (Ov), nucleus Uvaeformis (Uva), and robust nucleus of the arcopallium (RA). (**b**) Hypothesized model for comparing memories and vocal production to evaluate songs during the learning process. Based on this hypothesis, the memory of tutor songs is represented by neural circuits in the HVC and is used to guide vocal learning. As a young bird vocalizes (using the sound

imitation is not located in the NCM. Ikeda et al. (2020) suggested that the NCM may be involved in recognizing or perceiving tutor songs, but that process could be independent of encoding a memory used for song imitation.

Other investigators have suggested that the HVC has a role in encoding memories of tutor songs (e.g., Roberts et al. 2010; Adret et al. 2012). In adult songbirds, activity of neurons in the HVC is the result of input from an area called the nucleus interface of the nidopallium (NIf) (e.g., Coleman and Mooney 2004, Hahnloser and Fee 2007; Fig. 12.87a). Zhao et al. (2019) found that the NIf contains projection neurons (i.e., neurons that transmit impulses with the central nervous system) that project to either the HVC or an area called the nucleus avalanche (Av) (Fig. 12.87a), and that lesions of only the HVC projection neurons block the ability of young Zebra Finches to store memories of tutor songs. However, when the lesions occur after young have already heard tutor songs, they can still imitate those songs, suggesting that the NIf to HVC neurons encode song memories. Ross et al. (2019) found that the process of song tutoring causes changes in neural circuits involving HVC neurons projecting to the robust nucleus of the arcopallium (RA) and to Area X (Fig. 12.87b). These changes may allow young songbirds to compare memories of tutor songs to those they produce as they practice, specifically via neural circuits between the HVC and the RA and nucleus avalanche (Av) and those between a young bird's ears and the Av (an auditory center in the brain) (Fig. 12.87b). By doing so, young birds can make changes to their song(s) as needed to better match their songs to those in their memory.

## 12.24 Why Learn Songs?

Among birds, vocal learning, where individuals imitate the vocalizations of others, has evolved in three different orders: Passeriformes, Apodiformes (hummingbirds), and Psittaciformes (parrots and allies). Investigators have proposed four different hypotheses to explain the evolution of vocal learning in birds (Fig. 12.88). One hypothesis is that learning promotes the development of local song dialects, with males learning the dialect in their natal area, females having a preference for, and selecting as mates, males with that dialect (Nottebohm 1972). This seems unlikely because males in many species do not learn songs in their natal areas, but, rather, after they have dispersed, and there is little evidence that dialect populations actually correspond to genetically differentiated populations indicative of adaptation to local environments (Podos and Warren 2007). In addition, the vocal dialect hypothesis only provides males with a selective advantage if dialects and female preferences already exist, so this hypothesis could explain the maintenance of vocal learning, but not its origin (Nowicki and Searcy 2014).

The sexual selection hypothesis posits that vocal learning was a selective response to female preferences for complex songs, with greater complexity possible when songs are learned rather than being innate (Nottebohm 1972). In support of this hypothesis, studies have demonstrated a female preference for more complex songs in several species of birds (Searcy and Yasukawa 1996). This preference would benefit females if song complexity is positively correlated with male quality. However, assuming an ancestral population where male song was innate, how might a female preference for greater vocal

**Fig. 12.87** (continued) production pathway that includes Area X), impulses along neurons from the HVC to the RA and an area called the nucleus avalanche (Av, an auditory area) convey information about a specific point in a song ($t_0$ neuron). The young bird hears this specific point in a song and this information is transmitted as impulses to the Av of its auditory system. By comparing the memory to the sound produced via auditory feedback, the young bird can evaluate its song performance and make changes as needed to better match performance and memory. (Figures from Ikeda et al. 2020; © 2019 Elsevier Ltd., used with permission)

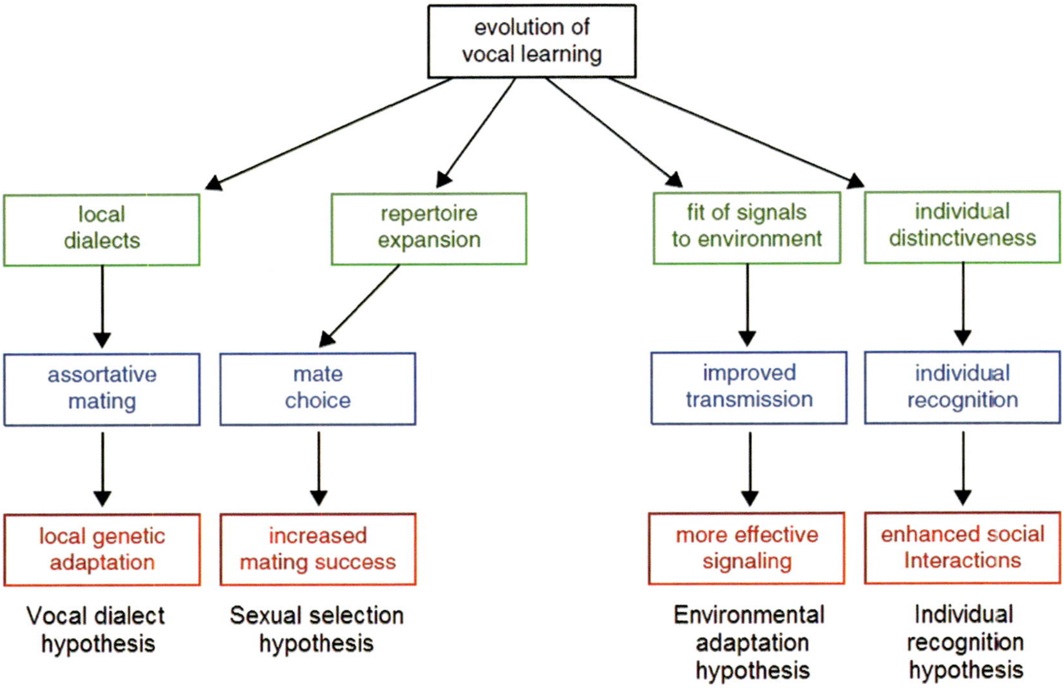


**Fig. 12.88** Four hypotheses on the selective benefits leading to the evolution of vocal learning. For each hypothesis, vocal learning has a proximate effect (in green) that translates by some mechanism (in blue) to a selective benefit (in red). (Figure modified from Nowicki and Searcy 2014; © 2014 Elsevier Ltd., used with permission)

complexity arise? Nowicki and Searcy (2014) suggested that "sensory bias" might have played a role. Such bias can occur because repeated exposure to the same stimulus might result in habituation, e.g., females might become less responsive to males repeatedly singing the same song or songs. If so, males with even slightly more variable songs might be preferred by females, providing such males with a selective advantage and potentially favoring the evolution of vocal learning (Nowicki and Searcy 2014).

The environmental adaptation hypothesis posits that the selective advantage of song learning is that it allows birds that occupy particular habitats to select those songs or song components with the acoustic characteristics that best transmit in that habitat (Hanson 1979; Moseley et al. 2018). Young birds, in other words, determine which songs or song components transmit best to them, learn them, and, as a result, their songs also transmit well. This would provide a selective advantage to those in a population able to learn and so could explain both the origin of vocal learning and its maintenance (Nowicki and Searcy 2014).

Finally, the individual recognition hypothesis proposes that vocal learning allows vocalizations to be more individually distinctive and enhance social cohesion (Nowicki and Searcy 2014). For example, Green-rumped Parrotlets (*Forpus passerinus*) and other species of parrots learn individually unique contact calls that allow identification of mates and other individuals in a population (Berg et al. 2011, 2012). Because learning a distinctive call and being recognized as a particular individual would likely be beneficial in a variety of social contexts, e.g., interactions with mates, even if others in a population had not learned such calls, the individual recognition hypothesis could explain both the origin and

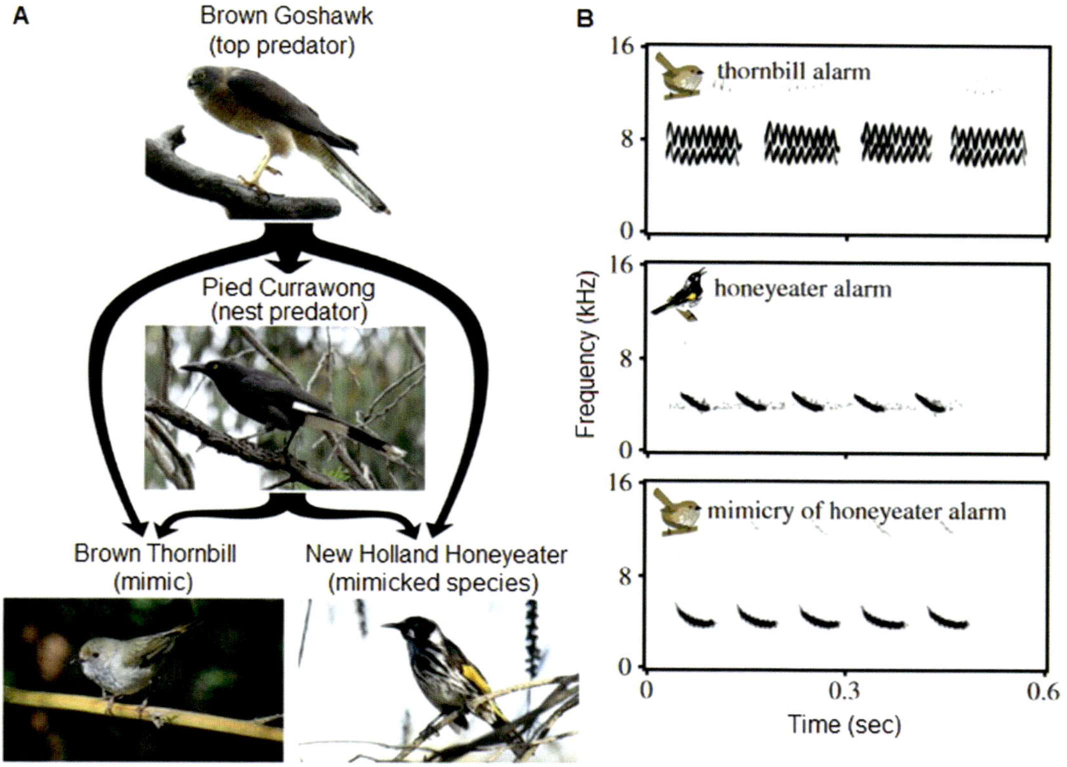


**Fig. 12.89** Predator–prey relationships between species, and aerial alarm vocalizations used in experiments. (**a**) Brown Goshawks (*Accipiter fasciatus*) are large, aerial predators, Pied Currawongs (*Strepera graculina*) are nest predators, and Brown Thornbills (*Acanthiza pusilla*) mimic the aerial-predator alarm calls of several other species of birds, including those of New Holland Honeyeaters (*Phylidonyris novaehollandiae*). (**b**) Sonograms of Brown Thornbill aerial-predator alarm calls, aerial-predator alarm calls of New Holland Honeyeaters, and the honeyeater aerial alarm calls mimicked by a Brown Thornbill. (Figure from Igic et al. 2015; © 2015 The Authors. Published by the Royal Society, used with permission)

maintenance of vocal learning (Nowicki and Searcy 2014). However, studies of nonvocal-learning species have revealed that individual recognition does not necessarily require learned vocalizations (e.g., Budka and Osiejuk 2013) and, in addition, Nottebohm (1972) noted that vocal learning could potentially increase the size of vocal repertoires and make individual recognition more difficult. On balance, therefore, the individual recognition hypothesis seems an unlikely explanation for the evolution of vocal learning in birds.

## 12.25 Vocal Mimicry

About 20% of all songbirds are vocal mimics, producing sounds resembling those of heterospecifics (Baylis 1982). Among oscine songbirds, Goller and Shizuka (2018) traced the evolutionary history of vocal mimicry and their analysis revealed that mimicry evolved independently at least 237 times and was lost at least 52 times. In some species, mimicry may serve a deceptive function. For example, Fork-tailed Drongos (*Dicrurus adsimilis*) sometimes mimic the alarm calls of other species of birds to "steal" food. When a "target" bird (or mammal; it

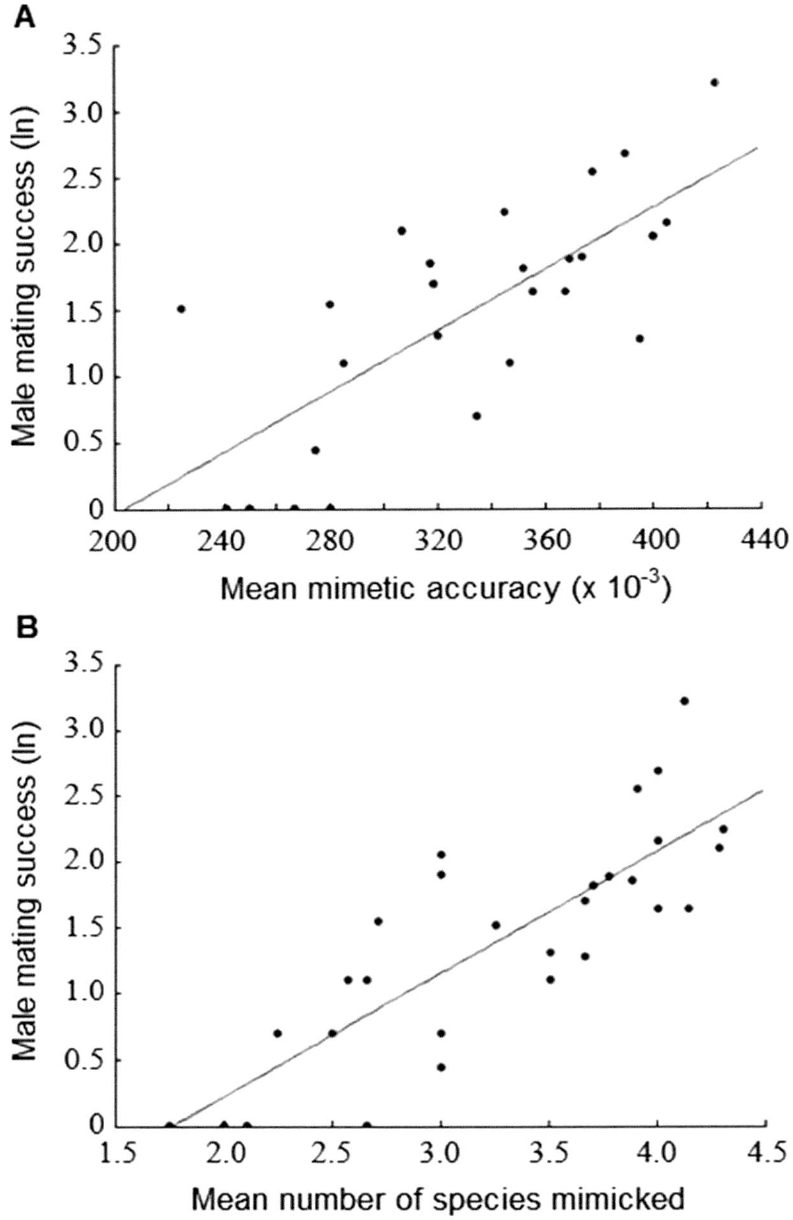


**Fig. 12.90** Effect of mimetic accuracy and number of heterospecifics mimicked on the mating success of male Satin Bowerbirds (*Ptilonorhynchus violaceus*). (Figure from Coleman et al. 2007; © 2007 The Royal Society, used with permission)

sometimes even works with meerkats [*Suricata suricatta*]) has food, drongos utter false alarm calls (no predator is nearby) and, as the target seeks cover, they may leave their food behind (Flower 2011). The target species continue to respond to these false alarm calls because drongos also utter "honest" alarm calls when a predator is nearby. Brown Thornbills (*Acanthiza pusilla*) are also deceptive mimics, uttering alarm calls and mimicking the alarm calls given in response to the presence of an aerial predator by other species of birds when Pied Currawongs (*Strepera graculina*), primary nest predators 40 times larger than thornbills, approach their nests (Igic et al. 2015). Experiments have demonstrated that Pied Currawongs respond to the mimicked calls just as they would to the "real" calls, either by scanning the sky for a predator or flying away to seek cover (Igic et al. 2015; Fig. 12.89).

In other examples of deceptive mimicry, Greater Racket-tailed Drongos (*Dicrurus paradiseus*) mimic the songs and calls of other species that participate in mixed-species foraging flocks to attract those species and, by "managing" flock size, increase their foraging efficiency (Goodale and Kotagama 2006a). In another context, Greater Racket-tailed Drongos benefit by mimicking the alarm and mobbing calls of other species when they detect predators, gaining the assistance of other species in monitoring the location of or mobbing potential predators (Goodale and Kotagama 2006b).

For other species of birds, mimicry may function to attract mates (Dobkin 1979). Although this hypothesis has rarely been tested, Coleman et al. (2007) found that male Satin Bowerbirds (*Ptilonorhynchus violaceus*) that mimicked more species and mimicked those species more accurately achieved the greatest mating success (i.e., mated with more females) (Fig. 12.90). The results of other studies suggest that mimicry may play a role in attracting or stimulating mates. For example, male Northern Mockingbirds (*Mimus polyglottos*) with larger song repertoires were found to have larger territories and also mated earlier in the breeding season (Howard 1974). In addition, Gammon (2014) found that mimetic frequency by male Northern Mockingbird was highest early in the breeding season, suggesting that mimicry might play a role in attracting or stimulating females. Superb Lyrebirds (*Menura novaehollandiae*) are excellent mimics, but adult males mimic more sounds and mimic sounds with greater accuracy than subadults. This suggests there is a cost involved in developing mimetic song and that selection, and females, may favor males that mimic more sounds and do so with greater accuracy (Zann and Dunstan 2008).

Among avian vocal mimics, what factors influence the choice of species, calls, or songs to mimic? Gammon (2013) proposed five hypotheses that might help explain the choices made by Northern Mockingbirds (*Mimus polyglottis*): (1) Proximity hypothesis: preferential imitation of species found in the same area as the mimic, (2) Aggression hypothesis: preferential imitation of species with which the mimic interacts aggressively, (3) Passive sampling hypothesis: preferential imitation of species and vocalizations heard more frequently by the mimic, (4) Alarm hypothesis: preferential imitation of alarm-associated vocalizations, and (5) Acoustic similarity hypothesis: preferential imitation of species and vocalizations most similar to the nonimitative songs and calls of the vocal mimic. Analysis of the songs of male Northern Mockingbirds revealed that the most-supported hypothesis was the Acoustic similarity hypothesis, with males more likely to copy heterospecific vocalizations most similar to their nonimitative vocalizations in terms of pitch or frequency, maximum duration, delivery rate, and whether or not notes or syllables were repeated. Studies of other avian vocal mimics, including Marsh Warblers (*Acrocephalus palustris*; Dowsett-Lemaire 1979), European Starlings (*Sturnus vulgaris*; Hausberger et al. 1991), and five species of robin-chats (*Cossypha* spp.; Ferguson et al. 2002), have also revealed preferential imitation of vocalizations with certain characteristics.

## 12.26 Duetting

Duetting occurs when two individuals, usually a male and female of a mated pair, combine their vocalizations in a temporally coordinated acoustic display (Hall 2009). Overall, duetting has been reported in an estimated 15.9% of all species of birds, including in 26 of 39 orders (66.7%) of birds (Tobias et al. 2016; Table 12.6). Based on a comparative analysis of over 10,000 species of birds, Tobias et al. (2016) concluded that duetting is strongly associated with year-round territory defense and long-term pair bonds; other factors, including latitude, migratory status, climate, and habitat, had little apparent effect on the evolution of duetting behavior. The apparent relationship between latitude and/or climate on the global distribution of duetting species (Fig. 12.91) is a result of two factors. First, there are more species

**Table 12.6** Percent of duetting species in 37 orders of birds. Orders in bold font have no duetting species (Data based on supplementary information provided by Tobias et al. 2016)

| Order | No. of species | No. of duetting species | % of duetting species |
|---|---|---|---|
| **Struthioniformes** | 2 | 0 | 0 |
| **Rheiformes** | 2 | 0 | 0 |
| Tinamiformes | 47 | 7 | 14.9 |
| **Casuariiformes** | 4 | 0 | 0 |
| Apterygiformes | 5 | 5 | 100.0 |
| Galliformes | 295 | 109 | 36.9 |
| Anseriformes | 168 | 3 | 1.8 |
| **Phoenicopteriformes** | 6 | 0 | 0 |
| Podicipediformes | 19 | 5 | 26.3 |
| Columbiformes | 320 | 17 | 5.3 |
| Mesitornithiformes | 3 | 3 | 100.0 |
| **Pteroclidiformes** | 16 | 0 | 0 |
| **Apodiformes** | 453 | 0 | 0 |
| Caprimulgiformes | 114 | 4 | 3.5 |
| Cuculiformes | 147 | 29 | 19.7 |
| Otidiformes | 27 | 6 | 22.2 |
| Musophagiformes | 23 | 8 | 34.8 |
| **Opisthocomiformes** | 1 | 0 | 0 |
| Gruiformes | 161 | 70 | 43.5 |
| Charadriiformes | 377 | 11 | 2.9 |
| Gaviiformes | 5 | 1 | 20.0 |
| Procellariiformes | 124 | 97 | 78.2 |
| **Sphenisciformes** | 17 | 0 | 0 |
| **Ciconiiformes** | 19 | 0 | 0 |
| Pelecaniformes | 168 | 7 | 4.2 |
| Eurypygiformes | 2 | 2 | 100.0 |
| **Phaethontiformes** | 3 | 0 | 0 |
| Accipitriformes | 258 | 4 | 1.6 |
| Strigiformes | 216 | 88 | 40.7 |
| **Coliiformes** | 6 | 0 | 0 |
| **Leptosomiformes** | 1 | 0 | 0 |
| **Trogoniformes** | 42 | 0 | 0 |
| Bucerotiformes | 73 | 21 | 28.8 |
| Coraciiformes | 153 | 48 | 31.4 |
| Piciformes | 426 | 90 | 21.1 |
| Cariamiformes | 2 | 2 | 100.0 |
| Falconiformes | 65 | 3 | 4.6 |
| Psittaciformes | 372 | 14 | 3.8 |
| Passeriformes | 6187 | 992 | 16.0 |
| **Total** | 10,329 | 1646 | 15.9 |

of birds at low latitudes (and, therefore, more duetting species as well) and, second, there are many more species that defend year-round territories at low latitudes than at higher latitudes. Selection has favored defense of year-round territories at tropical latitudes because: (1) the climate is more stable (Ghalambor et al. 2006) and, as a result, (2) defendable resources such as food are present throughout the year. In addition, (3) the lower mortality rates and longer lifespans of birds in the tropics mean that populations may be closer to carrying capacity, resulting in

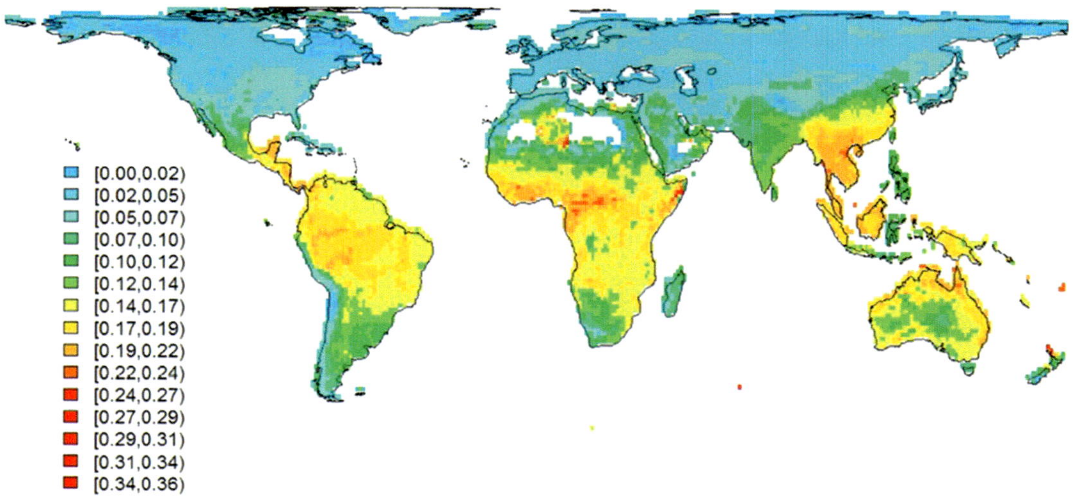


**Fig. 12.91** Global distribution of species of birds that duet, calculated as the proportion of total species occurring in 110 × 110 km grid cells. The greatest proportions of duetting species are found in western and central Africa, Indo-Malaya, northern Australia, and central America. Legend indicates lower and upper proportions represented by each color. (Figure from Tobias et al. 2016; open-access article distributed under the terms of the Creative Commons Attribution License, https://creativecommons.org/licenses/by/4.0/)

increased competition for territories (Wiersma et al. 2007; Williams et al. 2010). This competition, in turn, favors shared defense of territories, and cooperative signaling and singing, by paired males and females (Tobias et al. 2016).

Although joint territory defense is its primary function, duetting can also serve other functions. In Red-backed Fairywrens (*Malurus melanocephalus*), for example, males in territories where pairs were faster at initiating duets in response to simulated territorial intrusion (playback of conspecific songs) were cuckolded less often than slower pairs, suggesting that duetting serves to strengthen the pair bond and/or deter intruding males. If the latter, then, for males, duetting plays an important role in mate guarding (Baldassarre et al. 2016). Similarly, female Eastern Whipbirds (*Psophodes olivaceus*) were more likely to engage in duets with their mates in response to playback of female songs than male songs, suggesting that females use duets to "guard" their position in their partnership with their mate (Rogers et al. 2007).

For most species of birds, duetting likely serves multiple functions. For example, in different contexts, duets by Rufous-and-white Wrens (*Thryothorus rufalbus*) likely play a role in territory defense, mate guarding, simply maintaining contact, and establishing and maintaining pair bonds (Mennill and Vehrencamp 2008). Similarly, duets by Cabanis's Wrens (*Cantorchilus modestus*) appear to be important for both joint territory defense and mate guarding (Marshall-Ball et al. 2006). Dahlin and Benedict (2014) reviewed studies of 59 different duetting species of birds and found that duetting served multiple functions in most species, with joint resource defense the most commonly reported function (88% of all species). Other reported functions of duetting include mate guarding (44% of all species) and, less frequently, reproductive synchronization, maintaining contact with mates, and signaling commitment to a partnership (Benedict 2010; Dahlin and Benedict 2014).

The ways in which pairs of birds duet vary among species (Table 12.7). In many, but not all, species, duets exhibit temporal precision, with

**Table 12.7** Characteristics of the duets of 59 species of birds representing nine different orders[a] (Table from Dahlin and Benedict 2014; © 2013 Blackwell Verlag GmbH, used with permission)

| Structure characteristic | Number and percentage of species with data (%) | Number of orders | Percentages and ranges of duet structure characteristics |
|---|---|---|---|
| Sex-specific initiation | 35 (59) | 8 | Male or female: 80% Male only: 17% Female only: 3% |
| Sex-specific termination | 20 (34) | 6 | Male or female: 75% Male only: 0% Female only: 25% |
| Sex-specific notes | 48 (81) | 9 | Yes: 73% No: 27% |
| Individual or pair specific? | 20 (34) | 5 | Yes: 55% No: 45% |
| Temporal precision | 33 (56) | 5 | High: 69% Low: 27% Variable: 3% |
| Note timing | 51 (86) | 9 | Antiphonal: 29% Overlap: 31% Simultaneous: 2% Multiple ways: 37% |
| Female: number of note types | 22 (37) | 7 | Range: 1 to 20+ |
| Male: number of note types | 25 (42) | 7 | Range: 1 to 40+ |
| Duet duration (s) | 25 (42) | 8 | Range: 2 to 49+ |
| Amplitude | 19 (32) | 5 | High: 63% Low: 26% Variable: 1% |
| Sometimes associated with visual display? | 20 (34) | 5 | Yes: 84% No: 16% |

[a]Apterygiformes (1 species), Galliformes (3 species), Mesitornithiformes (1 species), Gruiformes (2 species), Psittaciformes (5 species), Cuculiformes (1 species), Strigiformes (2 species), Piciformes (11 species), and Passeriformes (33 species)

short response latencies and/or synchronous note timing (Dahlin and Benedict 2014). In most species, either sex can initiate duets, and notes uttered during duets are sex-specific, allowing nearby conspecifics to identify the respective contributions of males and females. In addition, the way that duets are organized varies among species, with notes sometimes uttered antiphonally, sometimes overlapping, and, rarely, uttered simultaneously (Fig. 12.92). Finally, among most duetting species of birds, vocal duets are accompanied by visual displays (Table 12.7), likely because vocal and visual signals combined transmit information more effectively than a single signal. Alternatively, the vocal component of duets may serve a territorial function, whereas the visual component might help pairs coordinate the vocal duets (Ręk and Magrath 2016).

## 12.27 Group Choruses

In some species of group-living birds, three or more members of a group vocalize at the same time. In some species, these vocal choruses are coordinated, with individuals contributing in particular sequences of notes or calls (Fig. 12.93), but, in other species, such as Green Woodhoopoes (*Phoeniculus purpureus*, Radford 2003), individuals in the group simply call at the same time. Such choruses may represent a cooperative signaling system serving as an important indicator of group identity and playing a role in maintaining group territories, e.g., Rufous-naped Wrens (*Campylorhynchus rufinucha;* Bradley and Mennill 2009; Fig. 12.94), Australian Magpies (*Cracticus tibicen*; Baker 2009), and Black-breasted Wood-Quail (*Odontophorus leucolaemus*; Hale 2006). Chorus volume and duration may also provide other groups with information about group size (e.g., Radford

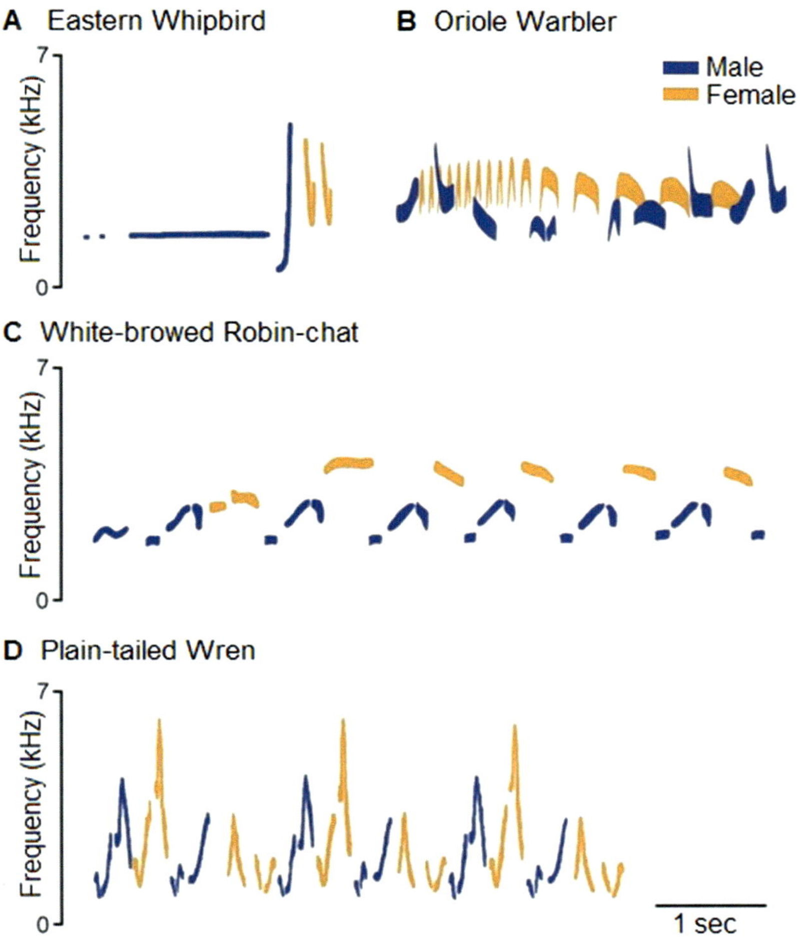


**Fig. 12.92** Structure of avian duets. (**a**) Stimulus-response, where a female responds to a male, e.g., Eastern Whipbirds (*Psophodes olivaceus*). (**b**) Song of a female overlaps that of a male, e.g., Oriole Warblers (*Hypergerus atriceps*). (**c**) Alternating duet, where the notes of males and females alternate, e.g., White-browed Robin-chats (*Cossypha heuglini*). (**d**) Coordinated duet, where the contributions of males and females fit precisely together, e.g., Plain-tailed Wrens (*Pheugopedius euophrys*). (Figure modified from Brumm and Slater 2007; © 2007 Elsevier Ltd., used with permission)

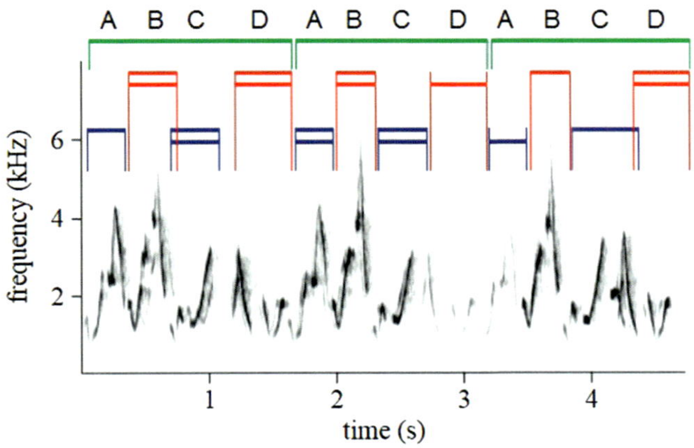


**Fig. 12.93** Highly synchronized chorus involving two male (blue) and two female (red) Plain-tailed Wrens (*Pheugopedius euophrys*). One bird of each sex is closer (upper colored line) and the other is more distant (lower-colored line). Double lines indicate the two birds are singing the same note or phrase in synchrony. (Figure from Mann et al. 2005; © 2005 The Royal Society, used with permission)

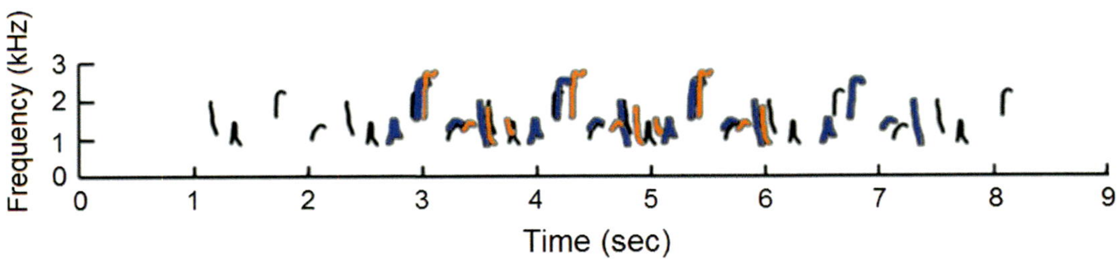


**Fig. 12.94** Chorus that involved three Rufous-naped Wrens (*Campylorhynchus rufinucha*) as indicated by different-colored notes. (Figure modified from Bradley and Mennill 2009; © 2009 Dt. Ornithologen-Gesellschaft e.V., used with permission)

2003). Group choruses may also aid in maintaining social bonds within groups and in maintaining social hierarchies among members of a group (e.g., Laughing Kookaburra, *Dacelo novaeguineae*; Reyer and Schmidl 1988).

## 12.28 Male Cooperative Courtship

Cooperative courtship involves the use of coordinated vocalizations and displays by pairs or groups of males to attract and solicit copulations from females. Coordinated male displays have been reported in just a few species of birds, primarily among manakins (family Pipridae: order Passeriformes). In most species of manakins, males defend individual display territories at leks. Cooperative displays have been reported in four species in the genus *Chiroxiphia*, four species in the genus *Pipra* (Castro-Astor et al. 2004; Díaz-Muñoz et al. 2014), and one species in the genus *Corapipo* (Jones et al. 2014). Among three species in the genus *Chiroxiphia* [Lance-tailed Manakin, *C. lanceolata* (Figs. 12.95 and 12.96); Blue Manakin, *C. caudata*; and Blue-backed Manakin, *C. pareola*], males often, but not always, cooperate (DuVal 2007a, b); in one species, Long-tailed Manakins (*Chiroxiphia linearis*), males always cooperate (McDonald 1989a; Figs. 12.97 and 12.98). In addition, males in all four species in the genus *Pipra* (Wire-tailed Manakin, *P. filicauda*; White-fronted Manakin, *P. serena*; Band-tailed Manakin, *P. fasciicauda*; and Red-headed Manakin, *P. rubrocapilla*), and a single species in the genus *Corapipo* (White-ruffed Manakin, *C. altera*; Jones et al. 2014), often, but not always, cooperate (Lukianchuk and Doucet 2014).

In all but one of the species just mentioned (Long-tailed Manakins), males sometimes display alone, and the relative frequency of sole versus cooperative displays varies. For example, about a third of all displays by Blue Manakins (Foster 1981) and about three-quarters of the displays of White-ruffed Manakins (Jones et al. 2014) are solo. In addition, female White-ruffed Manakins are equally likely to copulate with males after solo and cooperative displays (Fig. 12.99) and females in other species of manakins have also been observed copulating with males after solo displays (Snow 1963; Foster 1981). However, in most species of manakins with facultative cooperative displays, most displays are cooperative rather than solo. Cooperative displays might be preferred by females because such displays are more conspicuous than solo displays. For example, male Wire-tailed Manakins display at higher rates when coordinating with another male than when displaying alone (Fig. 12.100). In addition to being more conspicuous, cooperative displays may require greater cognitive ability and motor skills than solo displays and so may provide

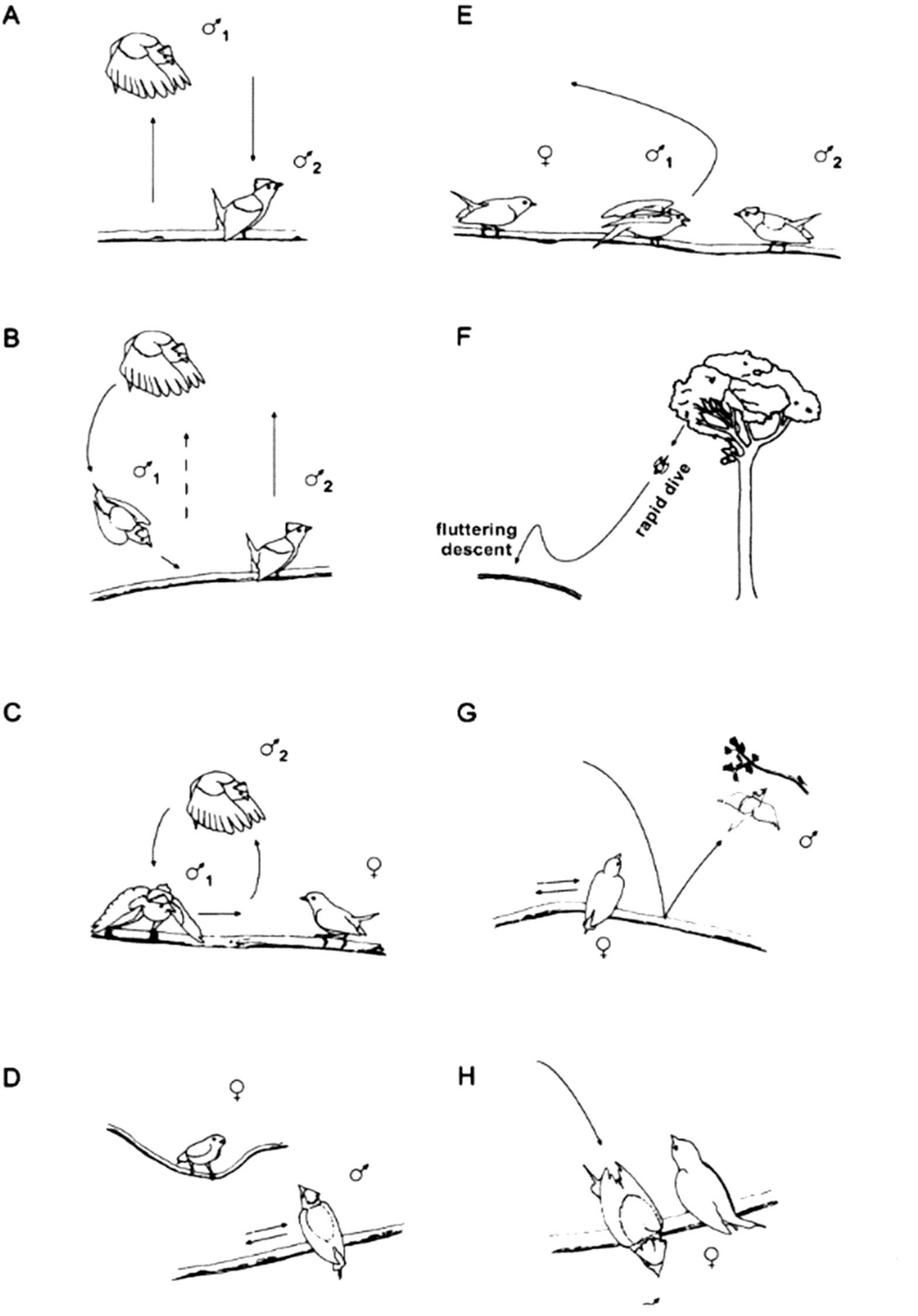


**Fig. 12.95** Cooperative displays of male Lance-tailed Manakins (*Chiroxiphia lanceolata*). (**a**) Up-and-down display, (**b**) quick-down variant of the up-and-down display, (**c**) leapfrog dance, (**d**) back-and-forth by male after female leaves the display perch, (**e**) *eek* display following a bout of leapfrog dancing, with the alpha male (male 1) giving the *eek* call and then leaving the display perch before the beta male (male 2) partner follows, (**f**) swoop, (**g**) bounce at the end of a courtship sequence, with back-and-forth by the female, and (**h**) a bow before copulation. (Figure from DuVal 2007a; © 2007 Oxford University Press, used with permission)

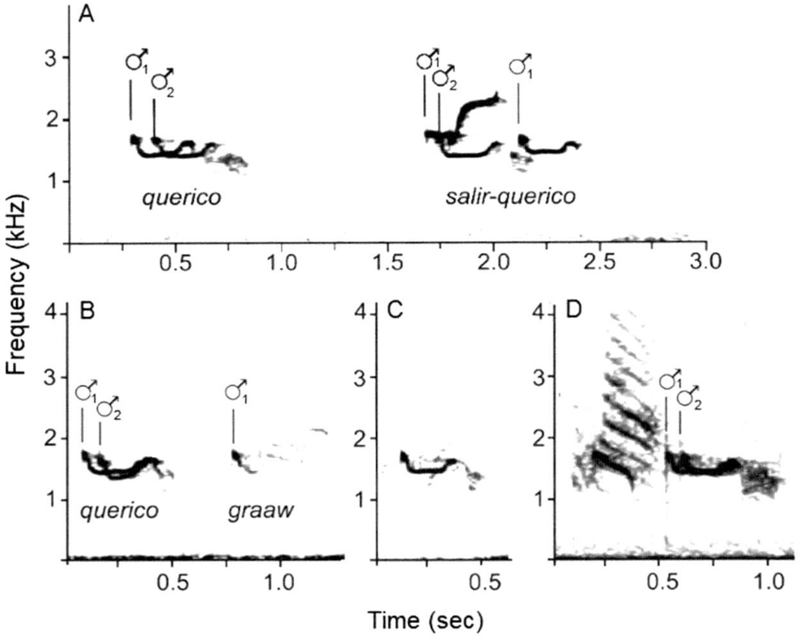


**Fig. 12.96** Vocalizations given during cooperative displays by male Lance-tailed Manakins (*Chiroxiphia lanceolata*). (**a**) Duet song, with the vertical lines indicating the start of each male's call. The male giving the *salir* call gives a *querico* immediately afterwards. (**b**) A *querico-graaw* call, with either male then giving the *graaw* call. (**c**) A solitary *querico* call. (**d**) A duet song preceded by coordinated *wah* calls. (Figure modified from DuVal 2007a; © 2007 Oxford University Press, used with permission)

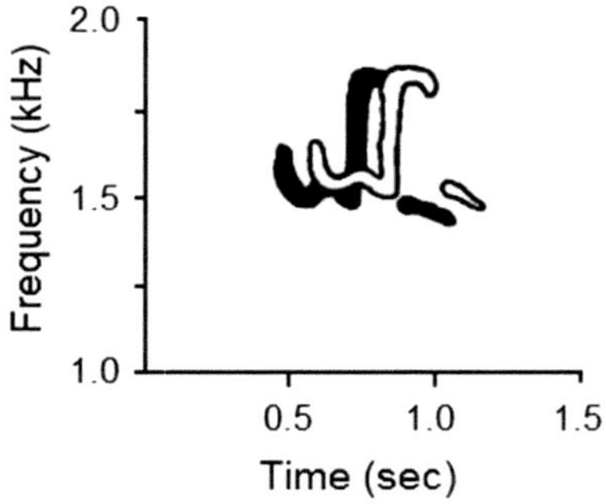


**Fig. 12.97** Sonogram of a "duet" of "toledo" calls by two male Long-tailed Manakins (*Chiroxiphia linearis*), with the contribution of one male solid and the other open. After the alpha male utters a "teeamoo" call to attract the beta male to the display area, they "duet," uttering a series of "toledo" call to attract females to the area. When a female arrives, the males begin their courtship display. (Figure modified from Trainer and McDonald 1995; © 1995 Springer-Verlag, used with permission)

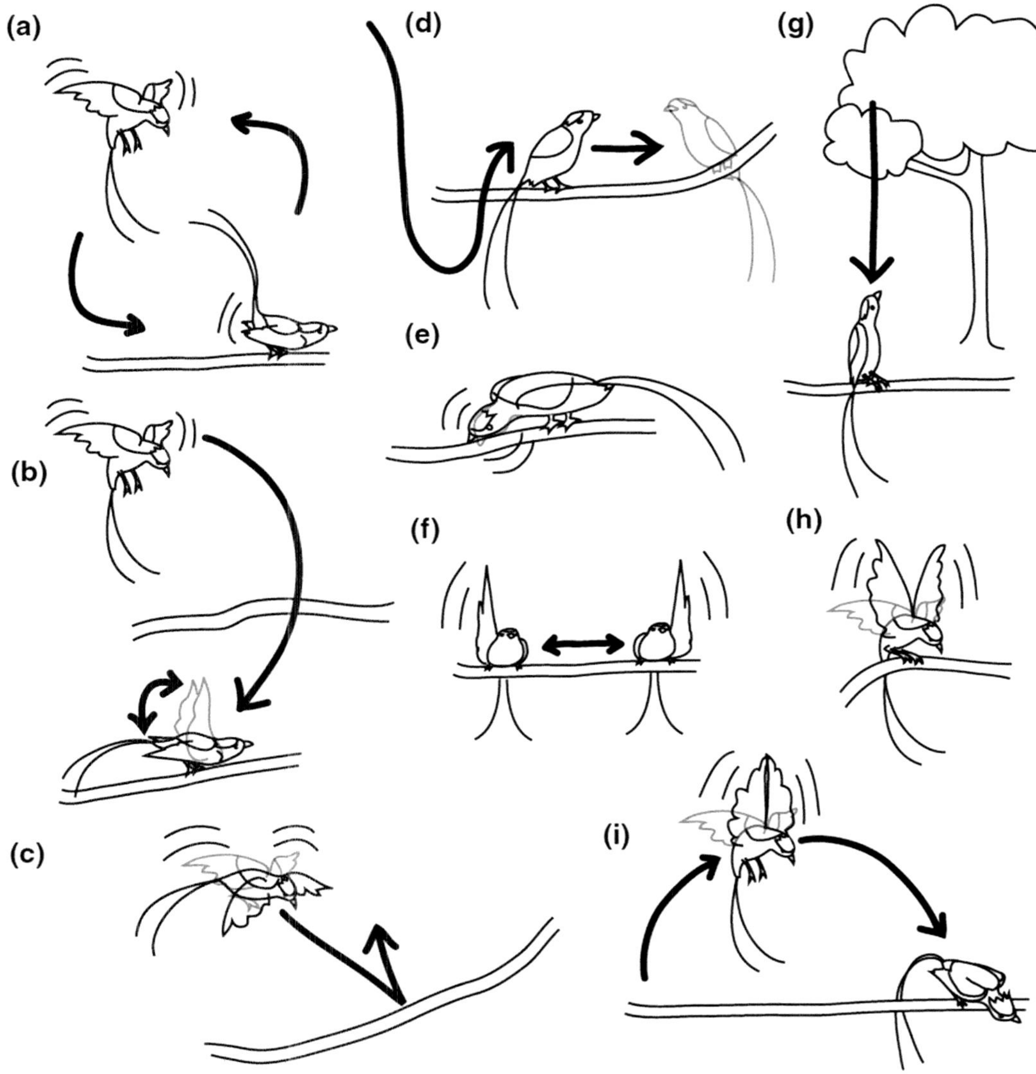


**Fig. 12.98** The coordinated courtship display of male Long-tailed Manakins (*Chiroxiphia linearis*). (**a**) Two males performing "leapfrogs" (males alternate hopping in the air while uttering short calls, with the male remaining on the perch moving forward and the leaping male then landing behind him), (**b**) male on the top perch performing a "buzz-weent" display (alpha male utters a "weent" call, then quickly flies down past the lower perch and swoops up to land on a nearby perch), followed by "tucked wing flicks" (alpha male rapidly and repeatedly flicks his wings behind his body so they almost meet), (**c**) male performing "butterfly flight" (short, slow flight with slow wing flaps), then jumping upward after landing, (**d**) arrow showing flight path of an "s-shaped dart" (short, rapid flights between the display perch and nearby perches, with few wing flaps), followed by an "about-face" (180° turn on the perch) after landing, (**e**) male bill wipes, (**f**) male "back-and-forth display" (one male moves rapidly from side to side, apparently using one wing to propel himself), (**g**) arrow showing flight path of a "lightning bolt" (flying upward, then making a rapid vertical descent to a display perch), then assuming an upright posture after landing, (**h**) male performing a "frenzied flutter," and (**i**) male performing an "angel flight" (a short flight of about 10–15 cm, with exaggerated wing flaps to the wings touch with each flap), then landing in a "bow" position (bowing with crest erected and spread, and head lowered toward the front of the female). (Illustrations by K. C. Lukianchuk; Figure from Lukianchuk and Doucet 2014; © 2014 Dt. Ornithologen-Gesellschaft e.V., used with permission)

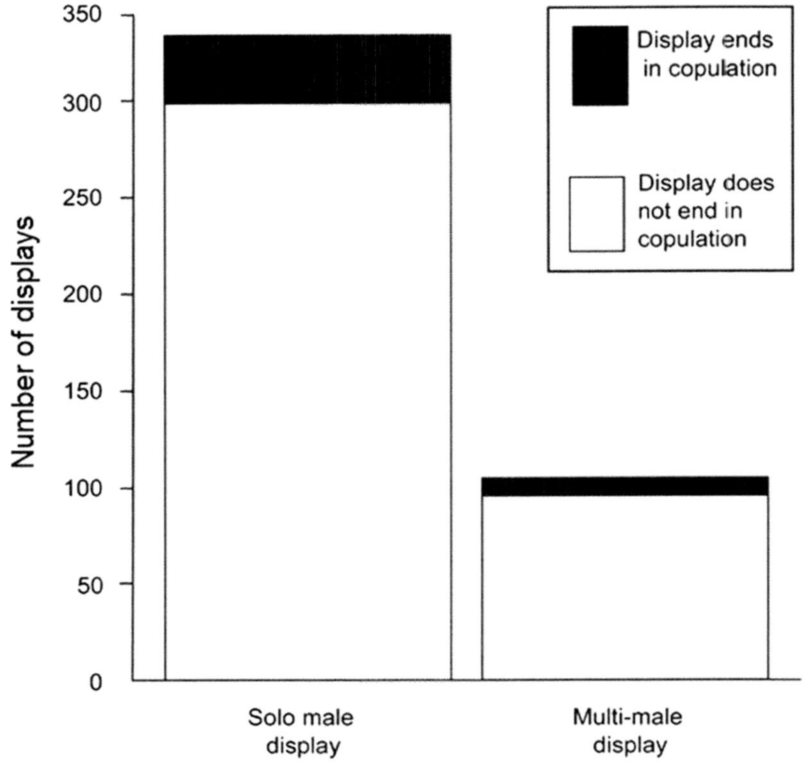


**Fig. 12.99** Male White-ruffed Manakins (*Corapipo altera*) usually display solo for females (76.3% of displays), but sometimes participate in cooperative displays with other males. The likelihood of a display ending with a copulation was similar for solo and multimale displays. (Figure from Jones et al. 2014; © 2014 Oxford University Press, used with permission)

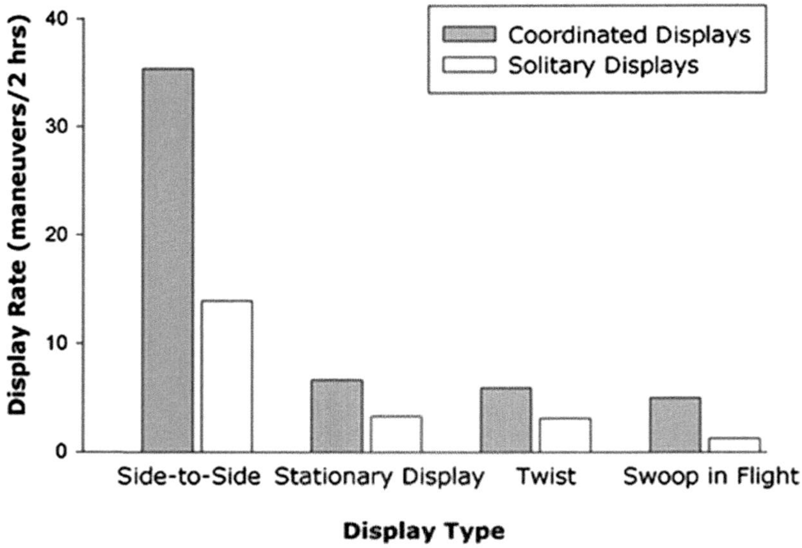


**Fig. 12.100** Male Wire-tailed Manakins (*Pipra filicauda*) displayed at higher rates when coordinating with another male than when displaying alone as shown by the rate at which four common display maneuvers were performed. (Figure modified from Ryder et al. 2011; © 2011 Oxford University Press, used with permission)

females with a better indication of male quality (Boogert et al. 2011; Vanderbilt et al. 2015).

Among the pairs or groups of cooperating male manakins, one male is always dominant and, with very few exceptions, these alpha males are the ones that copulate with females and sire young. Why then do the other males engage in these cooperative displays? Possible explanations include: (1) inclusive fitness benefits if males are related to the alpha male, and (2) delayed direct benefits if cooperation eventually results in participating males becoming alpha males. Genetic analysis has provided no support for the first explanation; cooperating males are not related (McDonald

| Order | Example | Nonsong vocal repertoire | Citations |
|---|---|---|---|
| Struthioniformes | Common Ostrich (*Struthio camelus*) | Two calls, with one given primarily by chicks | Aravinth and Selvan (2015) |
| Rheiformes | Greater Rhea (*Rhea americana*) | Five chick call-notes, but past 7 weeks old only infrequent "hissing" and a male "grunt" call | Beaver (1978) |
| Apterygiformes | Little Spotted Kiwi (*Apteryx owenii*) | Six vocalizations | Digby et al. (2013) |
| Casuariiformes | Southern Cassowary (*Casuarius casuarius*) | Six adult calls; one juvenile call | Biggs (2013) |
| Tinamiformes | Black Tinamou (*Tinamus osgoodi*) | Eight vocalizations | Negret and Laverde-R (2015) |
| Anseriformes | Mallard (*Anas platyrhynchos*) | Eight female calls; Four male calls | Abraham (1974) |
| | Wood Duck (*Aix sponsa*) | Four female calls; Three male calls | Armbruster (1982) |
| Galliformes | Willow Ptarmigan (*Lagopus lagopus*) | 11 calls | Martin et al. (1995) |
| | Great Curassow (*Crax rubra*) | Five call types | Baldo and Mennill (2011) |
| | Red Junglefowl (*Gallus gallus*) | 24 vocal signals; some intergrade so absolute size of repertoire unclear | Collias (1987) |
| Sphenisciformes | African Penguin (*Spheniscus demersus*) | Four adult calls; two juvenile begging calls | Favaro et al. (2014) |
| Gaviiformes | Yellow-billed Loon (*Gavia adamsii*) | Eight call types | Sjölander and Ågren (1976) |
| Procellariformes | Procellariiidae (Petrels) | 3–7 calls | Bretagnolle (1996) |
| Pelecaniformes | Brown Pelican (*Pelecanus occidentalis*) | One adult call plus bill snap | Shields (2020) |
| Ciconiiformes | Wood Stork (*Mycteria americana*) | One call plus bill claps | Coulter et al. (2020) |
| Suliformes | Australian Gannet (*Morus serrator*) | 13 distinct call elements | Krull et al. (2012) |
| Podicipediformes | Western Grebe (*Aechmophorus occidentalis*) | Nine calls | LaPorte et al. (2020) |
| | Great Grebe (*Podiceps major*) | Eight adult calls | Greenquist (1982) |
| Accipitriformes | Cooper's Hawk (*Accipiter cooperii*) | Four vocalizations | Rosenfield and Bielefeldt (1991) |
| | Osprey (*Pandion haliaetus*) | Eight acoustic signals | Bretagnolle and Thibault (1993) |
| Falconiformes | Gray Falcon (*Falco hypoleucos*) | Two adult calls; one juvenile call | Baylis et al. (2015) |

(continued)

| Order | Example | Nonsong vocal repertoire | Citations |
|---|---|---|---|
| Gruiformes | Pale-winged Trumpeter (*Psophia leucoptera*) | 11 close-range calls | Seddon et al. (2002) |
| | Pukeko (*Porphyrio melanotus*) | 25 call types in nine broad categories | Clapperton and Jenkins (1984) |
| | Gray Crowned Crane (*Balearica regulorum*) | 11 adult calls; three chick calls | Budde (1999) |
| Charadriiformes | Ancient Murrelet (*Synthliboramphus antiquus*) | Nine calls | Jones et al. (1989) |
| | Auklets (Five species) | 3–5 per species | Seneviratne et al. (2009) |
| | Piping Plover (*Charadrius melodus*) | 14 adult calls; three chick calls | Sung et al. (2005) |
| | Wilson's Plover (*Charadrius wilsonia*) | Eight adult calls; two chick calls | Bergstrom (1988) |
| | Eurasian Stone-Curlew (*Burhinus oedicnemus*) | 11 different call types plus some subtypes | Dragonetti et al. (2013) |
| | Long-billed Curlew (*Numenius americanus*) | Eight adult calls; four juvenile calls | Forsythe (1970) |
| | Brown Noddy (*Anous stolidus*) | 9 temporal Arrangements of click notes | Riska (1986) |
| Columbiformes | Mourning Dove (*Zenaida macroura*) | Three calls | Otis et al. (2020) |
| Psittaciformes | Kea (*Nestor notabilis*) | Seven distinct call types | Schwing et al. (2012) |
| | North Island Kaka (*Nestor meridionalis septentrionalis*) | Five call types | Van Horik et al. (2007) |
| | Carnaby's Black Cockatoo (*Calyptorhynchus latirostris*) | 15 different calls | Saunders (1983) |
| | Blue-fronted Amazon (*Amazona aestiva*) | Nine vocalizations | Fernández et al. (1998) |
| | Lilac-crowned Amazon (*Amazona finschi*) | 58 commonly used note types (but 101 total note types) | Montes-Medina et al. (2016) |
| | Gray Parrot (*Psittacus erithacus*) | 39 call types | May (2004) |
| | Palm Cockatoo (*Probosciger aterrimus*) | 27 distinct note types | Zdenek et al. (2015) |
| Cuculiformes | Smooth-billed Ani (*Crotophaga ani*) | 11 call types and one group vocalization | Grieves et al. (2015) |
| | Asian Koel (*Eudynamys scolopaceus*) | Six adults vocalizations | Maller and Jones (2001) |
| Strigiformes | Barred Owl (*Strix varia*) | 13 distinct calls | Odom and Mennill (2010) |
| Caprimulgiformes | Common Nighthawk (*Chordeiles minor*) | Five calls | Brigham et al. (2020) |
| | Chuck-will's-widow (*Antrostomus carolinensis*) | Seven calls | Straight and Cooper (2020) |
| Apodiformes | Sombre Hummingbird (*Aphantochroa cirrochloris*) | Six call types | Ferreira et al. (2006) |
| | Rufous-breasted Hermit (*Glaucis hirsutus*) | Three call types | Ferreira et al. (2006) |
| | Amethyst-throated Hummingbird (*Lampornis amethystinus*) | Three call types | Ornelas et al. (2002) |

(continued)

| Order | Example | Nonsong vocal repertoire | Citations |
|---|---|---|---|
| | Blue-throated Hummingbird (*Lampornis clemenciae*) | 14 acoustic signals | Ficken et al. (2002) |
| Trogoniformes | Malabar Trogon (*Harpactes fasciatus*) | Five adult calls; one juvenile call | Varghese (2013) |
| Coliiformes | Blue-naped Mousebird (*Urocolius macrourus*) | 21 calls | de Juana (2001) |
| Coraciiformes | Belted Kingfisher (*Megaceryle alcyon*) | Four structural groups of calls | Davis (1988) |
| Piciformes | Gila Woodpecker (*Melanerpes uropygialis*) | Two calls, but they vary in number of notes so may convey different information in different contexts | Brenowitz (1978) |
| | Red-bellied Woodpecker (*Melanerpes carolinus*) | Six calls, plus drumming | Wilkins (1996) |
| | Acorn Woodpecker (*Melanerpes formicivorus*) | Seven adult calls, four nestling/ juvenile calls, plus drumming | MacRoberts and MacRoberts (1976) |
| | Yellow-bellied Sapsucker (*Sphyrapicus varius*), Northern Flicker (*Colaptes auratus*), Hairy Woodpecker (*Dryobates villosus*), and Downy Woodpecker (*Dryobates pubescens*) | Drumming, tapping, plus six adult calls; one nestling call | Lawrence (1967) |
| | Red-cockaded Woodpecker (*Dryobates borealis*) | 20 calls, plus drumming and tapping | Jackson (2020) |
| | Pileated Woodpecker (*Dryocopus pileatus*) | Seven calls, plus drumming and tapping | Bull and Jackson (2020) |
| | Black-backed Woodpecker (*Picoides arcticus*) | Seven calls, plus drumming and tapping | Tremblay et al. (2020) |
| Passeriformes | | | |
| Acanthisittidae | Rock Wren (*Xenicus gilviventris*) | Two calls | Lloyd-Jones (2014) |
| Philepittidae | Velvet Asity (*Philepitta castanea*) | Three calls plus whirring sound produced by wings | Prum and Razafindratsita (1997) |
| Pittidae | Papuan Pitta (*Erythropitta macklotii*) | Four calls, plus song | del Hoyo et al. (2018) |
| Pipridae | Long-tailed Manakin (*Chiroxiphia linearis*) | 13 distinct calls | Trainer and McDonald (1993) |
| Cotingidae | Swallow-tailed Cotinga (*Phibalura flavirostris*) | One call | Bromfield et al. (2004) |
| Tyrannidae | Eastern Phoebe (*Sayornis phoebe*) | Eight vocalizations | Smith (1969) |
| | Great Crested Flycatcher (*Myiarchus crinitus*) | 12 different vocalizations | Smith and Smith (1996) |
| | Rufous Twistwing (*Cnipodectes superrufus*) | Two calls | Tobias et al. (2008) |
| Conopophagidae | Gnateaters | All species (N = 12) have a song and three or four calls | Whitney (2018) |
| Thamnophilidae | White-bibbed Antbird (*Myrmeciza loricata*) | Three calls | Amorim et al. (2014) |
| Rhinocryptidae | Rock Tapaculo (*Scytalopus petrophilus*) | Four distinct vocalizations | Whitney et al. (2010) |

(continued)

| Order | Example | Nonsong vocal repertoire | Citations |
|---|---|---|---|
| Formicariidae | Ground Antbirds | | |
| Furnariidae | Thorn-tailed Rayadito (*Aphrastura spinicauda*) | Five types of vocalizations | Ippi et al. (2011) |
| | Black-banded Woodcreeper | Four adult calls; one juvenile call | Willis (1982) |
| Maluridae | Spendid Fairywren (*Malurus splendens*) | 13 different vocalizations | Greig and Pruett-Jones (2008) |
| Meliphagidae | Bell Miner (*Manorina melanophrys*) | 18 adult calls; four juvenile calls | Heathcote (1989) |
| | Tui (*Prosthemadera novaeseelandiae*) | Two calls | Hill (2011) |
| | Noisy Miner (*Manorina melanocephalus*) | 13 calls | Holt et al. (2017) |
| Pomatostomidae | Chestnut-crowned Babbler (*Pomatostomus ruficeps*) | 18 vocalizations | Crane et al. (2016) |
| Acanthizidae | Australian Warblers | | |
| Campephagidae | Javan Cuckoo-shrike (*Caracina javensis*) | Two calls | Taylor (2016a) |
| | Western Wattled Cuckoo-shrike (*Lobotos lobatus*) | Two calls | Taylor (2016b) |
| Vireonidae | White-eyed Vireo (*Vireo griseus*) | Two calls | Bradley (1980) |
| Aegithinidae | Marshall's Iora (*Aegithina nigrolutea*) | Two basic calls | Siriwardhane (2007) |
| Vangidae | Helmet Vanga (*Euryceros prevostii*) | Three call types | La Morco and Thorstrom (2000) |
| Corcoracidae | Apostlebird (*Struthidea cinerea*) | 17 common vocalizations | Warrington et al. (2014) |
| Monarchidae | Hawaii Elepaio (*Chasiempis sandwichensis*) | Eight calls | Vanderwerf (2020) |
| Laniidae | Chinese Gray Shrike (*Lanius sphenocercus*) | Three calls | Opaev (2014) |
| Corvidae | American Crow (*Corvus brachyrhynchos*) | Seven adult calls; two juvenile calls | Tarter (2008) |
| | Common Raven (*Corvus corax*) | 12 calls/individual | Enggist-Dueblin and Pfister (2002) |
| | Common Raven | 18 call types | Conner (1985) |
| | Yucatan Jay (*Cyanocorax yucatanicus*) | 24 call types | Hardy (1979) |
| | Hawaiian Crow (or ʻAlalā) (*Corvus hawaiiensis*) | 24 call types | Tanimoto et al. (2017) |
| | White-throated Magpie-Jay (*Calocitta formosa*) | 14 functionally distinct call types; at least 150 structurally distinct types | Ellis (2008) |
| | Red-billed Chough (*Pyrrhocorax pyrrhocorax*) | Eight structurally different calls | Laiolo et al. (2000) |
| Petroicidae | Scarlet Robin (*Petroica boodang*) | Three adult calls; one fledgling call | Debus (2006) |
| Regulidae | Ruby-crowned Kinglet (*Regulus calendula*) | Three adult calls; one fledgling call | Swanson et al. (2020b) |
| | Golden-crowned Kinglet (*Regulus satrapa*) | Two adult calls; two nestling calls | Swanson et al. (2020a) |
| Turdidae | Veery (*Catharus fuscescens*) | Two structurally variable graded calls and four discrete call types | Heckscher (2007) |

(continued)

| Order | Example | Nonsong vocal repertoire | Citations |
|---|---|---|---|
| Paridae | Willow Tit (*Parus montanus*) | Seven main vocalizations | Haftorn (1993) |
| | Black-capped Chickadee (*Poecile atricapillus*) | 11 adult vocalizations; two juvenile vocalizations | Ficken et al. (1978) |
| Alaudidae | Horned Lark (*Eremophila alpestris*) | Three call types | Beason (2020) |
| Hirundinidae | Purple Martin (*Progne subis*) | Ten call types | Brown (1984) |
| | South African Cliff Swallow (*Petrochelidon spilodera*) | Seven distinctive calls | Earlé (1986) |
| | Barn Swallow (*Hirundo rustica*) | Eight adult calls; one nestling call | Samuel (1971) |
| | Cliff Swallow (*Petrochelidon pyrrhonota*) | Two adult calls; one nestling call | Samuel (1971) |
| Pyconotidae | Red-vented Bulbul (*Petrochelidon pyrrhonota*) | Seven call types | Kumar and Bhatt (2000) |
| | Sooty-headed Bulbul (*Pycnonotus aurigaster*) | Ten call types | Payakkhabut (2012) |
| Locustelliidae | Timor Bush Warbler (*Locustella timorensis*) | Two calls | Trainor et al. (2012) |
| Sylviidae | Arctic Warbler (*Phylloscopus borealis*) | Nine call types | Lowther and Sharbaugh (2020) |
| Zosteropidae | White-collared Yuhina (*Yuhina diademata*) | Nine types of calls | Ning et al. (2008) |
| Leiothrichidae | Jungle Babbler (*Turdoides striata*) | 11 vocalizations | Gaston (1977) |
| Dulidae | Palmchat (*Dulus dominicus*) | Number of calls in repertoire unknown; no true song | Temple (2018) |
| Hypocoliidae | Gray Hypocolius (*Hypocolius ampelinus*) | Three calls | Porter et al. (2018) |
| Bombycillidae | Cedar Waxwing (*Bombycilla cedrorum*) | Three calls | Witmer et al. (2020) |
| | Bohemian Waxwing (*Bombycilla garrulus*) | Seven different calls | Witmer (2020) |
| Ptiliogonatidae | Long-tailed Silky-flycatcher (*Ptiliogonys caudatus*) | Four different calls | Skutch (1965) |
| Sittidae | White-breasted Nuthatchl (*Sitta carolinensis*) | 13 adult calls; two juvenile calls | Ritchison (1983) |
| Certhiidae | Eurasian Treecreeper (*Certhia familiaris*) | Two call types | Tietze et al. (2008) |
| Mohouidae | New Zealand Brown Creeper (*Mohoua novaeseelandiae*) | 19 vocal displays | Cunningham (1985) |
| Polioptilidae | Black-tailed Gnatcatcher (*Polioptila melanura*) | Six call types | Farquhar and Ritchie (2020) |
| Troglodytidae | Cactus Wren (*Campylorhynchus brunneicapillus*) | Eight adult call-notes; eight fledgling and juvenile vocalizations | Wolford (1969) |
| Mimidae | Brown Thrasher (*Toxostoma rufum*) | Six calls | Cavitt and Haas (2020) |
| | Gray Catbird (*Dumetella carolinesis*) | Three call types | Smith et al. (2020) |
| Sturnidae | Indian Hill Mynah (*Gracula religiosa*) | 15 vocalizations (four categories of vocalizations, one of which is "calls," with 12 different calls described) | Bertram (1970) |
| Cinclidae | American Dipper (*Cinclus mexicanus*) | Three adult calls; two juvenile calls | Fite (1984) |

(continued)

| Order | Example | Nonsong vocal repertoire | Citations |
|---|---|---|---|
| Muscicapidae | Indian Chat (*Oenanthe fusca*) | Five adult calls; three chick calls | Sethi and Bhatt (2008) |
| | Pied Bush Chat (*Saxicola caprata*) | Two adult calls; three juvenile calls | Sethi et al. (2012) |
| Dicaeidae | Mistletoebird (*Dicaeum hirundinaceum*) | Ten calls, plus songs | Cheke and Mann (2018) |
| Nectariniidae | Shining Sunbird (*Cinnyris habessinicus*) | Three calls | Williams (1955) |
| Irenidae | Asian Fairy-bluebird (*Irena puella*) | 20 calls plus song | Wells et al. (2018) |
| Ploceidae | Village Weaver (*Ploceus cucullatus*) | 15 different calls | Collias (1963) |
| Estrildidae | Black-faced Quailfinch (*Ortygospiza atricollis*) | Six adult calls; two nestling calls | Nuttall (1993) |
| Fringillidae | Bullfinch (*Pyrrhula pyrrhula*) | 14–17 vocal categories | Wilkinson (1990) |
| | American Goldfinch (*Spinus tristis*) | Six groups of vocalizations | McGraw and Middleton (2020) |
| Passeridae | White-browed Sparrow-weaver (*Plocepasser mahali*) | 13 calls | Collias and Collias (2004) |
| | Gray-capped Social Weaver (*Pseudonigrita arnaudi*) | 13 calls | Collias and Collias (2004) |
| | Sociable Weaver (*Philetairus socius*) | 20 calls | Collias and Collias (2004) |
| Motacillidae | American Pipit (*Anthus rubescens*) | Nine calls | Hendricks and Verbeek (2020) |
| Cardinalidae | Northern Cardinal (*Cardinalis cardinalis*) | 16 different calls | Halkin et al. (2021) |
| | Pine Grosbeak (*Pinicola enucleator*) | Four different calls | Adkisson (1981) |
| Thraupidae | Summer Tanager (*Piranga rubra*) | Five adult calls; two juvenile calls | Robinson (2020) |
| Emberizidae | White-crowned Sparrow (*Zonotrichia leucophrys*) | Five calls uttered by both sexes; four only by males | Hill and Lein (1985) |
| Icteridae | Great-tailed Grackle (*Quiscalus mexicanus*) | 16 patterns of sound for males; five for females | Kok (1971) |
| Parulidae | Kentucky Warbler (*Geothlypis formosa*) | Two calls | McDonald (2020) |
| | Prairie Warbler (*Setophaga discolor*) | ≥ Ten calls for males; ≥ eight calls for females | Nolan (1978) |
| | Prothonotary Warbler (*Protonotaria citrea*) | Two calls | Petit (2020) |

and Potts 1994; Loiselle et al. 2007; DuVal 2007b). Rather, beta males that participate in coordinated displays can benefit if, by doing so, they enhance their chances of becoming alpha males in future breeding seasons. Coordinating with alpha males may increase the chances of beta males obtaining display areas and, as a further potential benefit, their experiences as beta males may allow them to improve the quality of their displays, potentially enhancing their chances of obtaining copulations if and when they do become alpha males (DuVal 2013).

## Appendix: Non-song Call Repertoires Among Several Different Taxa and Species of Birds

## References


Abraham RL (1974) Vocalizations of the Mallard (*Anas platyrhynchos*). Condor 76:401–420

Adkisson CS (1981) Geographic variation in vocalizations and evolution of North American Pine Grosbeaks. Condor 83:277–288

Adret P, Meliza CD, Margoliash D (2012) Song tutoring in presinging Zebra Finch juveniles biases a small population of higher-order song-selective neurons toward the tutor song. J Neurophysiol 108:1977–1987

Akçay Ç, Anderson RC, Nowicki S, Beecher MD, Searcy WA (2015) Quiet threats: soft song as an aggressive signal in birds. Anim Behav 105:267–274

Ali S, Anderson R (2018) Song and aggressive signaling in Bachman's Sparrow. Auk 135:521–533

Alonso R, Goller F, Mindlin GB (2014) Motor control of sound frequency in birdsong involves the interaction between air sac pressure and labial tension. Phys Rev. E 89:032706

Alonso RG, Trevisan MA, Amador A, Goller F, Mindlin GB (2015) A circular model for song motor control in *Serinus canaria*. Front Comput Neurosci 9:41

Alvarez-Borda B, Nottebohm F (2002) Gonads and singing play separate, additive roles in new neuron recruitment in adult Canary brain. J Neurosci 22:8684–8690

Amador A, Margoliash D (2013) A mechanism for frequency modulation in songbirds shared with humans. J Neurosci 33:11136–11144

Amador A, Perl YS, Mindlin GB, Margoliash D (2013) Elemental gesture dynamics are encoded by song premotor cortical neurons. Nature 495:59–64

Amo L, Avilés JM, Parejo D, Peña A, Rodríguez J, Tomás G (2012) Sex recognition by odour and variation in the uropygial gland secretion in starlings. J Anim Ecol 81: 605–613

Amorim PS, Dias RI (2021) A non-vocal alarm? Effects of wing trill playbacks on antipredator responses in the Scaled Dove. Acta Ethologica 24:119–126

Amorim TOS, Manhães MA, Azevedo AF, Andriolo A (2014) The vocal repertoire of *Myrmeciza loricata* (Lichtenstein, 1823) (Aves: Thamnophilidae). J Nat Hist 48:1181–1198

Amy M, Salvin P, Naguib M, Leboucher G (2015) Female signalling to male song in the Domestic Canary, *Serinus canaria*. R Soc Open Sci 2:140196

Andalman AS, Foerster JN, Fee MS (2011) Control of vocal and respiratory patterns in birdsong: dissection of forebrain and brainstem mechanisms using temperature. PLoS ONE 6:e25461

Anderson RC (2009) Operant conditioning and copulation solicitation display assays reveal a stable preference for local song by female Swamp Sparrows *Melospiza georgiana*. Behav Ecol Sociobiol 64:215–223

Andrew RJ (1962) Evolution of intelligence and vocal mimicking. Science 137:585–589

Aravinth A, Selvan ST (2015) General behavioural pattern in Ostrich a key for better management. Int J Sci Res 4: 672–674

Araya-Salas M, Wright T (2013) Open-ended song learning in a hummingbird. Biol Lett 9:20130625

Armbruster JS (1982) Wood Duck displays and pairing chronology. Auk 99:116–122

Arnold AP (1980) Sexual differences in the brain. Am Sci 68:165–173

Arnold AP, Nottebohm F, Pfaff DW (1976) Hormone concentrating cells in vocal control and other brain regions of the Zebra Finch (*Poephila guttata*). J Comp Neurol 165:487–512

Arnold F, Staniszewski MS, Pelzl L, Ramenda C, Gahr M, Hoffmann S (2022) Vision and vocal communication guide three-dimensional spatial coordination of Zebra Finches during wind-tunnel flights. Nat Ecol Evol 6:1221–1230

Arriaga G, Jarvis ED (2013) Mouse communication system: are ultrasounds learned or innate? Brain Lang 124:96–116

Artuso C, Houston CS, Smith DG, Rohner C (2020) Great Horned Owl (*Bubo virginianus*), version 1.0. In: Poole AF (ed) Birds of the World. Cornell Lab of Ornithology, Ithaca, NY

Ausmus DM, Clarke JA (2014) Mother knows best: functionally referential alarm calling in White-tailed Ptarmigan. Anim Cogn 17:671–679

Avarguès-Weber A, Dawson EH, Chittka L (2013) Mechanisms of social learning across species boundaries. J Zool 290:1–11

Badyaev AV (2002) Growing apart: an ontogenetic perspective on the evolution of sexual size dimorphism. Trends Ecol Evol 17:369–378

Badyaev AV, Young RL, Oh KP, Addison C (2008) Evolution on a local scale: developmental, functional, and genetic bases of divergence in bill form and associated changes in song structure between adjacent habitats. Evolution 62:1951–1964

Baker MC (2009) Information content in chorus songs of the group-living Australian Magpie (*Cracticus tibicen dorsalis*) in western Australia. Ethology 115:227–238

Baldassarre DT, Greig EI, Webster MS (2016) The couple that sings together stays together: duetting, aggression and extra-pair paternity in a promiscuous bird species. Biol Lett 12:20151025

Baldo S, Mennill DJ (2011) Vocal behavior of Great Curassows, a vulnerable Neotropical bird. J Field Ornithol 82:249–258

Baldo S, Mennill DJ, Guindre-Parker S, Gilchrist HG, Love OP (2015) The oxidative cost of acoustic signals: examining steroid versus aerobic activity hypotheses in a wild bird. Ethology 121:1081–1090

Ball GF, Balthazart J (2008) Individual variation and the endocrine regulation of behaviour and physiology in birds: a cellular/molecular perspective. Phil Trans R Soc B 363:1699–1710

Ball GF, Auger CJ, Bernard DJ, Charlier TD, Sartor JJ, Riters LV, Balthazart J (2004) Seasonal plasticity in the song control system: multiple brain sites of steroid hormone action and the importance of variation in song behavior. Ann N Y Acad Sci 1016:586–610

Ballentine B, Horton B, Brown ET, Greenberg R (2013) Divergent selection on bill morphology contributes to nonrandom mating between Swamp Sparrow subspecies. Anim Behav 86:467–473

Ballintijn MR, ten Cate C (1997) Sex differences in the vocalizations and syrinx of the Collared Dove (*Streptopelia decaocto*). Auk 114:22–39

Balsby T, Bradbury J (2009) Vocal matching by Orange-fronted Conures (*Aratinga canicularis*). Behav Process 82:133–139

Balthazart J, Ball GF (2016) Endocrine and social regulation of adult neurogenesis in songbirds. Front Neuroendocrinol 41:3–22

Barnett CA, Briskie JV (2011) Strategic regulation of body mass and singing behavior in New Zealand Robins. Ethology 117:28–36

Bartsch C, Weiss M, Kipper S (2015) Multiple song features are related to paternal effort in Common Nightingales. BMC Evol Biol 15:115

Baylis JR (1982) Avian vocal mimicry: its function and evolution. In: Kroodsma DE, Miller EH, Ouellet H (eds) Acoustic communication in birds, vol 2. Academic, New York, NY, pp 51–83

Baylis T, van Gessel FW, Debus SJS (2015) Some vocalisations of the Grey Falcon *Falco hypoleucos*. Corella 39:73–76

Beason RC (2020) Horned Lark (*Eremophila alpestris*), version 1.0. In: Billerman SM (ed) Birds of the World. Cornell Lab of Ornithology, Ithaca, NY

Beaver PW (1978) Ontogeny of vocalization in the Greater Rhea. Auk 95:382–388

Bechard MJ, Houston CS, Saransola JH, England AS (2020) Swainson's Hawk (*Buteo swainsoni*), version 1.0. In: Poole AF (ed) Birds of the World. Cornell Lab of Ornithology, Ithaca, NY

Becker PH (1982) The coding of species-specific characteristics in bird sounds. In: Kroodsma DE, Miller EH (eds) Acoustic communication in birds, vol 1. Academic, New York, NY, pp 213–252

Beckers GJL, Suthers RA, ten Cate C (2003a) Pure-tone birdsong by resonance filtering of harmonic overtones. Proc Natl Acad Sci USA 100:7372–7376

Beckers GJL, Suthers RA, ten Cate C (2003b) Mechanisms of frequency and amplitude modulation in Ring Dove song. J Exp Biol 206:1833–1843

Beckett MD, Ritchison G (2010) Effects of breeding stage and behavioral context on singing behavior of male Indigo Buntings. Wilson J Ornithol 122:655–665

Beecher MD, Brenowitz EA (2005) Functional aspects of song learning in songbirds. Trends Ecol Evol 20:143–149

Beecher MD, Campbell SE, Nordby JC (2000) Territory tenure in Song Sparrows is related to song sharing with neighbours, but not to repertoire size. Anim Behav 59:29–37

Benedict L (2010) California Towhee vocal duets are multi-functional signals for multiple receivers. Behaviour 147:953–978

Bentley GE, Van't Hof T, Ball GF (1999) Seasonal neuroplasticity in the songbird telecephalon: a novel role for melatonin. Proc Natl Acad Sci USA 96:4674–4679

Berg KS, Delgado S, Okawa R, Beissinger SR, Bradbury JW (2011) Contact calls are used for individual mate recognition in free-ranging Green-rumped Parrotlets, *Forpus passerinus*. Anim Behav 81:241–248

Berg KS, Delgado S, Cortopassi KA, Beissinger SR, Bradbury JW (2012) Vertical transmission of learned signatures in a wild parrot. Proc R Soc B 279:585–591

Bergstrom PW (1988) Breeding displays and vocalizations of Wilson's Plovers. Wilson Bulletin 100:36–49

Bertram B (1970) The vocal behaviour of the Indian Hill Mynah, *Gracula religiosa*. Anim Behav Monogr 3:79–192

Biggs JR (2013) Captive management guidelines for the Southern Cassowary *Casuarius casuarius johnsonii*. Cairns Tropical Zoo, Palm Cove

Binazzi R, Zaccaroni M, Nespoli A, Massolo A, Dessi-Fulgheri F (2011) Anti-predator behaviour of the Red-legged Partridge *Alectoris rufa* (Galliformes: Phasianidae) to simulated terrestrial and aerial predators. Italian J Zool 78:106–112

Blanckenhorn WU (2000) The evolution of body size: what keeps organisms small. Q Rev. Biol 75:385–402

Bloomfield LL, Phillmore LS, Weisman RG, Sturdy CB (2005) Note types and coding in parid vocalizations. III: the chick-a-dee call of the Carolina Chickadee (*Poecile carolinensis*). Can J Zool 83:820–833

Bolhuis JJ, Gahr M (2006) Neural mechanisms of birdsong memory. Nat Rev. Neurosci 7:347–357

Bolhuis JJ, Moorman S (2015) Birdsong memory and the brain: in search of the template. Neurosci Biobehav Rev 50:41–55

Bonadonna F, Sanz-Aguilar A (2012) Kin recognition and inbreeding avoidance in wild birds: the first evidence for individual kin-related odour recognition. Anim Behav 84:509–513

Bond AB, Diamond J (2005) Geographic and ontogenetic variation in the contact calls of the Kea (*Nestor notabilis*). Behaviour 142:1–20

Boogert NJ, Fawcett TW, Lefebvre L (2011) Mate choice for cognitive traits: a review of the evidence in nonhuman vertebrates. Behav Ecol 22:447–459

Bostwick KS, Prum RO (2003) High-speed video analysis of wing-snapping in two manakin clades (Pipridae: Aves). J Exp Biol 206:3693–3706

Bostwick KS, Elias DO, Mason A, Montealegre-Z F (2010) Resonating feathers produce courtship song. Proc R Soc B 277:835–841

Bottino F, Artoni SMB, Oliveira D, Franzo VS, Pacheco MR, Orsi AM, Amoroso L (2006) Influence of the sexual dimorphism on the morphology of the syrinx of Guinea Fowl (*Numida meleagris*). Ciência Rural 36: 1424–1428

Boucaud ICA, Mariette MM, Villain AS, Vignal C (2016) Vocal negotiation over parental care? Acoustic communication at the nest predicts partners' incubation share. Biol J Linn Soc 117:322–336

Boughey MJ, Thompson NS (1981) Song variety in the Brown Thrasher (*Toxostoma rufum*). Z Tierpsychol 56: 47–58

Bradbury JW (2003) Vocal communication in wild parrots. In: de Waal FBM, Tyack PL (eds) Animal social complexity: intelligence, culture and individualized societies. Harvard University Press, Cambridge, MA, pp 293–316

Bradbury JW, Balsby TJS (2016) The functions of vocal learning in parrots. Behav Ecol Sociobiol 70:293–312

Bradley RA (1980) Vocal and territorial behavior in the White-eyed Vireo. Wilson Bull 92:302–311

Bradley DW, Mennill DJ (2009) Solos, duets and choruses: vocal behavior of the Rufous-naped Wren (*Campylorhynchus rufinucha*), a cooperatively breeding Neotropical songbirds. J Ornithol 150:743–753

Brainard MS, Doupe AJ (2002) What songbirds teach us about learning. Nature 417:351–358

Branch CL, Freeberg TM (2012) Distress calls in Tufted Titmice (*Baeolophus bicolor*): are conspecifics or predators the target? Behav Ecol 23:854–862

Branch CL, Pravosudov VV (2015) Mountain Chickadees from different elevations sing different songs: acoustic adaptation, temporal drift or signal of local adaptation? R Soc Open Sci 2:150019

Branch CL, Pravosudov VV (2020) Variation in song structure along an elevational gradient in a resident songbird. Behav Ecol Sociobiol 74:9

Brenowitz GL (1978) An analysis of Gila Woodpecker vocalizations. Wilson Bulletin 90:451–455

Brenowitz EA (1997) Comparative approaches to the avian song system. J Neurobiol 33:517–531

Brenowitz EA, Arnold AP (1986) Interspecific comparisons of the size of neural song control regions and song complexity in duetting birds: evolutionary implications. J Neurosci 6:2875–2879

Brenowitz EA, Beecher MD (2005) Song learning in birds: diversity and plasticity, opportunities and challenges. Trends Neurosci 28:127–132

Brenowitz EA, Larson TA (2015) Neurogenesis in the adult avian song-control system. Cold Spring Harb Perspect Biol 7:a019000

Brenowitz EA, Lent K (2002) Act locally and think globally: intracerebral testosterone implants induce seasonal-like growth of adult avian song control circuits. Proc Natl Acad Sci USA 99:12421–12426

Brenowitz EA, Arnold AP, Levin RN (1985) Neural correlates of female song in tropical duetting birds. Brain Res 343:104–112

Brenowitz EA, Margoliash D, Nordeen CW (1997) An introduction to birdsong and the avian song system. J Neurobiol 33:495–500

Bretagnolle V (1996) Acoustic communication in a group of nonpasserine birds, the petrels. In: Kroodsma DE, Miller EH (eds) Ecology and evolution of acoustic communication in birds. Cornell University Press, Ithaca, NY, pp 160–177

Bretagnolle V, Thibault J-C (1993) Communicative behavior in breeding Ospreys (*Pandion haliaetus*): description and relationship of signals to life history. Auk 110:736–751

Briefer E, Aubin T, Lehongre K, Rybak F (2008) How to identify dear enemies: the group signature in the complex song of the Skylark *Alauda arvensis*. J Exp Biol 211:317–326

Brigham RM, Ng J, Poulin RG, Grindal SD (2020) Common Nighthawk (*Chordeiles minor*), version 1.0. In: Poole A (ed) Birds of the World. Cornell Lab of Ornithology, Ithaca, NY

Bromfield G, Ritchie WN, Bromfield V, Ritchie J, Hennessey AB (2004) New information on plumage, nesting, behaviour and vocalisations of the Bolivian Swallow-tailed Cotinga *Phibalura flavirostris boliviana* from the Apolo area of Madidi National Park, Bolivia. Cotinga 21:63–67

Brown CH (1982) Ventriloquial and locatable vocalizations in birds. Z Tierpsychol 59:338–350

Brown CR (1984) Vocalizations of the Purple Martin. Condor 86:433–442

Brown ED, Farabaugh SM (1997) What birds with complex social relationships can tell us about vocal learning: vocal sharing in avian groups. In: Snowdon CT, Hausberger M (eds) Social influences on vocal development. Cambridge University Press, New York, NY, pp 98–127

Brown TJ, Handford P (2000) Sound design for vocalizations: quality in the woods, consistency in the fields. Condor 102:81–92

Brown CR, Brown MB, Shaffer ML (1991) Food-sharing signals among socially foraging Cliff Swallows. Anim Behav 42:551–564

Brua RB (2002) Parent-embryo interactions. In: Deeming DC (ed) Avian incubation: behavior, environment, evolution. Oxford University Press, Oxford, pp 88–99

Brua RB, Nuechterlein GL, Buitron D (1996) Vocal response of Eared Grebe embryos to egg cooling and egg turning. Auk 113:525–533

Brumm H, Naguib M (2009) Environmental acoustics and the evolution of bird song. Adv Study Behav 40:1–33

Brumm H, Ritschard M (2011) Song amplitude affects territorial aggression of male receivers in Chaffinches. Behav Ecol 22:310–316

Brumm H, Slater P (2007) Animal communication: timing counts. Curr Biol 17:R521–R523

Budde C (1999) The vocal repertoire of the Grey crowned Crane *Balearica regulorum gibbericeps*. I: the tonal and the non-harmonic calls. Bioacoustics 10:161–173

Budden AE, Wright J (2001) Begging in nestling birds. In: Nolan V Jr, Kettersen E (eds) Current ornithology, vol 16. Springer, Boston, MA, pp 83–118

Budka M, Osiejuk TS (2013) Neighbour-stranger call discrimination in a nocturnal rail species, the Corncrake *Crex crex*. J Ornithol 154:685–694

Bugnyar T, Kijne M, Kotrschal K (2001) Food calling in ravens: are yells referential signals? Anim Behav 61:949–958

Bull EL, Jackson JA (2020) Pileated Woodpecker (*Dryocopus pileatus*), version 1.0. In: Poole AF (ed) Birds of the World. Cornell Lab of Ornithology, Ithaca, NY

Byers BE (2007) Extrapair paternity in Chestnut-sided Warblers is correlated with consistent vocal performance. Behav Ecol 18:139–136

Byers BE, Kroodsma DE (2009) Female mate choice and songbird song repertoires. Anim Behav 77:13–22

Cain KE, Langmore NE (2015) Female and male song rates across breeding stage: testing for sexual and nonsexual functions of female song. Anim Behav 109:65–71

Canopoli A, Zai AT, Hahnloser RH (2016) Lesions of a higher auditory brain area during a sensorimotor period do not impair birdsong learning. Matters 2: e201603000018

Carey M, Burhans DE, Nelson DA (2020) Field Sparrow (*Spizella pusilla*), version 1.0. In: Poole AF (ed) Birds of the World. Cornell Lab of Ornithology, Ithaca, NY

Caro SP, Sewall KB, Salvante KG, Sockman KW (2010) Female Lincoln's Sparrows modulate their behavior in response to variation in male song quality. Behav Ecol 21:562–569

Casagrande S, Pinxten R, Eens M (2016) Honest signaling and oxidative stress: the special case of avian acoustic communication. Front Ecol Evol 4:52

Castro-Astor IN, Alves MAS, Cavalcanti RB (2004) Display behavior and spatial distribution of the Red-headed Manakin in the Atlantic Forest of Brazil. Condor 106:320–335

Catchpole CK (1982) The evolution of bird sounds in relation to mating and spacing behavior. In: Kroodsma DE, Miller EH, Ouellet H (eds) Acoustic communication in birds, vol 1. Academic, New York, NY, pp 297–319

Catchpole CK (1986) Song repertoires and reproductive success in the Great Reed Warbler *Acrocephalus arundinaceus*. Behav Ecol Sociobiol 19:439–445

Catchpole CK, Slater PJB (2008) Bird song: biological themes and variations, 2nd edn. Cambridge University Press, New York, NY

Cavitt JF, Haas CA (2020) Brown Thrasher (*Toxostoma rufum*), version 1.0. In: Poole A (ed) Birds of the World. Cornell Lab of Ornithology, Ithaca, NY

Chakraborty M, Walløe S, Nedergaard S, Fridel EE, Dabelsteen T, Pakkenberg B, Bertelsen MF, Dorrestein GM, Brauth SE, Durand SE, Jarvis ED (2015) Core and shell song systems unique to the parrot brain. PLoS ONE 10:e0118496

Cheke R, Mann C (2018) Mistletoebird (*Dicaeum hirundinaceum*). In: del Hoyo J, Elliott A, Sargatal J, Christie DA, de Juana E (eds) Handbook of the Birds of the World Alive. Lynx Edicions, Barcelona

Chiale MC (2016) La glándula uropigia de aves de distintos ambientes: su estructura y función. Ph.D. dissertation, Universidad Nacional de La Plata, Buenos Aires

Christie PJ, Mennill DJ, Ratcliffe LM (2004) Pitch shifts and song structure indicate male quality in the dawn chorus of Black-capped Chickadees. Behav Ecol Sociobiol 55:341–348

Chu M (2001) Heterospecific responses to scream calls and vocal mimicry by Phainopeplas (*Phainopepla nitens*) in distress. Behaviour 138:775–787

Clapperton BK, Jenkins PF (1984) Vocal repertoire of the Pukeko (Aves: Rallidae). N Z J Zool 11:71–84

Clark CJ, Feo TJ (2008) The Anna's Hummingbird chirps with its tail: a new mechanism of sonation in birds. Proc R Soc B 275:955–962

Clark CJ, Prum RO (2015) Aeroelastic flutter of feathers, flight and the evolution of non-vocal communication in birds. J Exp Biol 218:3520–3527

Clay Z, Smith CL, Blumstein DT (2012) Food-associated vocalizations in mammals and birds: what do these calls really mean? Anim Behav 83:323–330

Clemmons JR (1995) Vocalizations and other stimuli that elicit gaping in nestling Black-capped Chickadees (*Parus atricapillus*). Auk 112:603–612

Cohen RE, Macedo-Lima M, Miller KE, Brenowitz EA (2016) Adult neurogenesis leads to the functional reconstruction of a telencephalic neural circuit. J Neurosci 36:8947–8956

Colbert-White EN, Coibalhs MC, Fragaszy DM (2014) Where apes and songbirds are left behind: a comparative assessment of the requisites for speech. Comp Cogn Behav Rev 9:1–28

Coleman MJ, Mooney R (2004) Synaptic transformations underlying highly selective auditory representations of learned birdsong. J Neurosci 24:7251–7265

Coleman SW, Patricelli GL, Coyle B, Siani J, Borgia G (2007) Female preferences drive the evolution of

mimetic accuracy in male sexual displays. Biol Lett 3: 463–466
Collias NE (1963) A spectrographic analysis of the vocal repertoire of the African Village Weaverbird. Condor 65:517–527
Collias NE (1987) The vocal repertoire of the Red Junglefowl: a spectrographic classification and the code of communication. Condor 89:510–524
Collias NE, Collias EC (2004) Comparison of vocal signals of three species of African finches. Behaviour 141:1151–1171
Colombelli-Négrel D, Robertson J, Sulloway FJ, Kleindorfer S (2010) Extended parental care of fledglings: parent birds adjust anti-predator response according to predator type and distance. Behaviour 147:853–870
Colombelli-Négrel D, Webster MS, Dowling JL, Hauber ME, Kleindorfer S (2016) Vocal imitation of mother's calls by begging Red-backed Fairywren nestlings increases parental provisioning. Auk 133:273–285
Conner RN (1985) Vocalizations of Common Ravens in Virginia. Condor 87:379–388
Cortopassi KA, Bradbury JW (2006) Contact call diversity in wild Orange-fronted Parakeet pairs, *Aratinga canicularis*. Anim Behav 71:1141–1154
Coulter MC, Rodgers JA, Ogden JC, Depkin FC (2020) Wood Stork (*Mycteria americana*), version 1.0. In: Poole AF, Gill FB (eds) Birds of the World. Cornell Lab of Ornithology, Ithaca, NY
Courter JR, Ritchison G (2010) Alarm calls of Tufted Titmice convey information about predator size and threat. Behav Ecol 21:936–942
Crane JMS, Savage JL, Russell AF (2016) Diversity and function of vocalisations in the cooperatively breeding Chestnut-crowned Babbler. Emu 116:241–253
Creanza N, Fogarty L, Feldman MW (2016) Cultural niche construction of repertoire size and learning strategies in songbirds. Evol Ecol 30:285–305
Cunningham JB (1985) Breeding ecology, social organization and communicatory behaviour of the Brown Creeper (*Finschia novaeseelandiae*). Ph.D. dissertation, University of Canterbury, Christchurch
Curio E (1978) The adaptive significance of avian mobbing. I. Teleonomic hypotheses and predictions. Z Tierpsychol 48:175–183
Dahlin CR, Benedict L (2014) Angry birds need not apply: a perspective on the flexible form and multifunctionality of avian vocal duets. Ethology 120:1–10
Dahlin CR, Young AM, Cordier B, Mundry R, Wright TF (2014) A test of multiple hypotheses for the function of call sharing in female Budgerigars, *Melopsittacus undulates*. Behav Ecol Sociobiol 68:145–161
Dale J, Lank DB, Reeve HK (2001) Signaling individual identity versus quality: a model and case studies with Ruffs, queleas, and House Finches. Am Nat 158:75–86
Dale J, Dey CJ, Delhey K, Kempenaers B, Valcu M (2015) The effects of life history and sexual selection on male and female plumage colouration. Nature 527:367–370
Daley M, Goller F (2004) Tracheal length changes during Zebra Finch song and their possible role in upper vocal tract filtering. J Neurobiol 59:319–330
Dalziell AH, Welbergen JA (2016) Elaborate mimetic vocal displays by female Superb Lyrebirds. Front Ecol Evol 4:34
Dave AS, Margoliash D (2000) Song replay during sleep and computational rules for sensorimotor vocal learning. Science 290:812–816
Davis WMJ (1988) Acoustic communication in the Belted Kingfisher: an example of temporal coding. Behaviour 106:1–23
de Juana E (2001) Family Coliidae (Mousebirds). In: del Hoyo J, Elliott A, Sargatal J (eds) Handbook of the Birds of the World, vol 6. Lynx Edicions, Barcelona
De Lawrence L, K. (1967) A comparative life-history study of four species of woodpeckers. Ornithol Monogr 5:1–156
de Moura LN, Da Silva ML, Vielliard J (2011) Vocal repertoire of wild breeding Orange-winged Parrots *Amazona amazonica* in Amazonia. Bioacoustics 20: 331–339
Debus SJS (2006) Breeding biology and behaviour of the Scarlet Robin *Petroica multicolor* and Eastern Yellow Robin *Eopsaltria australis* in remnant woodland near Armidale, New South Wales. Corella 30:59–65
del Hoyo J, Collar N, Kirwan GM (2018) Papuan Pitta (*Erythropitta macklotii*). In: del Hoyo J, Elliott A, Sargatal J, Christie DA, de Juana E (eds) Handbook of the Birds of the World Alive. Lynx Edicions, Barcelona
Demko AD, Reitsma LR, Staicer CA (2016) Repertoire structure, song sharing, reproductive success, and territory tenure in a population of Canada Warblers (*Cardellina canadensis*) in central New Hampshire. Can J Zool 94:283–290
Derrickson KC (1987) Yearly and situational changes in the estimate of repertoire size in Northern Mockingbirds (*Mimus polyglottos*). Auk 104:198–207
Derryberry EP (2009) Ecology shapes birdsong evolution: variation in morphology and habitat explains variation in White-crowned Sparrow song. Am Nat 174:24–33
Díaz-Muñoz SL, DuVal EH, Krakauer AH, Lacey KA (2014) Cooperating to compete: altruism, sexual selection and causes of male reproductive cooperation. Anim Behav 88:67–78
Digby A, Bell BD, Teal PD (2013) Vocal cooperation between the sexes in Little Spotted Kiwi *Apteryx owenii*. Ibis 155:229–245
Dingle C, Halfwerk W, Slabbekoorn H (2008) Habitat-dependent song divergence at subspecies level in the Grey-breasted Wood-wren. J Evol Biol 21:1079–1089
Dobkin DS (1979) Functional and evolutionary relationships of vocal copying phenomena in birds. Z Tierpsychol 50:348–363
Doutrelant C, Blondel J, Perret P, Lambrechts MM (2000) Blue Tit song repertoire size, male quality and interspecific competition. J Avian Biol 31:360–366

Dowsett-Lemaire F (1979) The imitative range of the song of the Marsh Warbler *Acrocephalus palustris*, with special reference to imitations of African birds. Ibis 121:453–468

Dragonetti M, Caccamo C, Corsi FO, Farsi F, Giovacchini P, Pollonara E, Giunchi D (2013) The vocal repertoire of the Eurasian Stone-Curlew (*Burhinus oedicnemus*). Wilson J Ornithol 125:34–49

Drent PJ (1983) The functional ethology of the Great Tit (*Parus major*). Ph.D. dissertation, University of Groningen, Groningen

Duffy DL, Ball GF (2002) Song predicts immunocompetence in male European Starlings (*Sturnus vulgaris*). Proc R Soc B 269:847–852

Dunning JL, Pant S, Murphy K, Prather JF (2020) Female finches prefer courtship signals indicating male vigor and neuromuscular ability. PLoS ONE 15:e0226580

Düring DN, Elemans CPH (2016) Embodied motor control of avian vocal production. In: Suthers RA, Fitch WT, Fay RR, Popper AN (eds) Vertebrate sound production and acoustic communication. Springer, Cham, pp 119–157

During DN, Ziegler A, Thompson CK, Ziegler A, Faber C, Muller J, Scharff C, Elemans CPH (2013) The songbird syrinx morphome: a three-dimensional, highresolution, interactive morphological map of the Zebra Finch vocal organ. BMC Biol 11:1–27

Dutour M, Walsh SL, Ridley AR (2021) Australian Magpies adjust their alarm calls according to predator distance. Bioacoustics 30:458–468

DuVal EH (2007a) Cooperative display and lekking behavior of the Lance-tailed Manakin (*Chiroxiphia lanceolata*). Auk 124:1168–1185

DuVal EH (2007b) Adaptive advantages of cooperative courtship for subordinate male Lance-tailed Manakins. Am Nat 169:423–432

DuVal EH (2013) Does cooperation increase helpers' later success as breeders? A test of the skills hypothesis in the cooperatively displaying Lance-tailed Manakin. J Anim Ecol 82:884–893

Dyrcz A, Zdunek W, Hagen KS (2011) Increased male singing in response to predator presence may represent reproductive investment in a promiscuous species, the Aquatic Warbler *Acrocephalus paludicola*. Acta Ornithol 46:93–96

Earlé RA (1986) Vocalizations of the South African Cliff Swallow *Hirundo spilodera*. S Afr J Zool 3:229–232

Eddowes MJ, Lea AJ (2021) A review of 'wing-clapping' in European Nightjars. British Birds 114:629–637

Elemans CPH (2014) The singer and song: the neuromechanics of avian sound production. Curr Opin Neurobiol 28:172–178

Elemans CPH, Spierts ILY, Muller UK, van Leeuwen JL, Goller F (2004) Superfast muscles control dove's trill. Nature 431:146

Elemans CPH, Mead AF, Rome LC, Goller F (2008) Superfast vocal muscles control sound production in songbirds. PLoS ONE 3:e2581

Elemans CPH, Rasmussen JH, Herbst CT, Düring DN, Zollinger SA, Brumm H, Srivastava K, Svane N, Ding M, Larsen ON, Sober SJ, Švec JG (2015) Universal mechanisms of sound production and control in birds and mammals. Nat Commun 6:8978

Elgar M (1986) House Sparrows establish foraging flocks by giving chirrup calls if the resources are divisible. Anim Behav 34:169–174

Ellis J (2008) The vocal repertoire of the White-throated Magpie-jay, *Calocitta formosa*. Ph.D. dissertation, Cornell University, Ithaca, NY

Ellison WG (1992) Different drummers: identifying the rhythms of northeastern woodpeckers. Birding 24: 350–355

Enggist-Dueblin P, Pfister U (2002) Cultural transmission of vocalizations in ravens, *Corvus corax*. Anim Behav 64:831–841

Eriksson D, Wallin L (1986) Male bird song attracts females—a field experiment. Behav Ecol Sociobiol 19:297–299

Evans CS, Evans L (1999) Chicken food calls are functionally referential. Anim Behav 58:307–319

Evans C, Kleindorfer S (2016) Superb Fairy-Wren (*Malurus cyaneus*) sons and daughters acquire song elements of mothers and social fathers. Front Ecol Evol 4:9

Fant G (1960) Acoustic theory of speech production. Mouton, The Hague

Farabaugh SM, Linzenbold A, Dooling RJ (1994) Vocal plasticity in Budgerigars (*Melopsittacus undulatus*): evidence for social factors in the learning of contact calls. J Comp Psychol 108:81–92

Farquhar CC, Ritchie KL (2020) Black-tailed Gnatcatcher (*Polioptila melanura*), version 1.0. In: Poole AF, Gill FB (eds) Birds of the World. Cornell Lab of Ornithology, Ithaca, NY

Farrow LF, Doohan SJ, McDonald PG (2017) Alarm calls of a cooperative bird are referential and elicit context-specific antipredator behavior. Behav Ecol 28:724–731

Fasanella M, Fernández GJ (2009) Alarm calls of the Southern House Wren *Troglodytes musculus*: variation with nesting stage and predator model. J Ornithol 150: 853–863

Favaro L, Ozella L, Pessani D (2014) The vocal repertoire of the African Penguin (*Spheniscus demersus*): structure and function of calls. PLoS ONE 9:e103460

Feekes F (1982) Song mimesis within colonies of *Cacicus c. cela* (Icteridae, Aves): a colonial password? Z Tierpsychol 58:119=152

Ferguson JWH, Van Zyl A, Delport K (2002) Vocal mimicry in African *Cossypha* robin chats. J Ornithol 143: 319–330

Fernández-Juricic E, Martella MB, Alvarez EV (1998) Vocalizations of the Blue-fronted Amazon (*Amazona aestiva*) in the Chancaní Reserve, Córdoba, Argentina. Wilson Bull 110:352–361

Ferreira AR, Smulders TV, Sameshima K, Mello CV, Jarvis ED (2006) Vocalizations and associated behaviors of the Sombre Hummingbird (*Aphantochroa cirrhochloris*) and the Rufous-breasted Hermit (*Glaucis hirsutus*). Auk 123:1129–1148

Ficken MS, Popp J (1996) A comparative analysis of passerine mobbing calls. Auk 113:370–380

Ficken MS, Ficken RW, Witkin SR (1978) Vocal repertoire of the Black-capped Chickadee. Auk 95:34–48

Ficken MS, Rusch KM, Taylor SJ, Powers DR (2002) Reproductive behavior and communication in Blue-throated Hummingbirds. Wilson Bull 114:197–209

Fitch WT (1997) Vocal tract length and formant frequency dispersion correlate with body size in rhesus macaques. J Acoust Soc Am 102:1213–1222

Fitch WT (1999) Acoustic exaggeration of size in birds via tracheal elongation: comparative and theoretical analyses. J Zool 248:31–48

Fitch WT, Hauser MD (2003) Unpacking 'honesty': vertebrate vocal production and the evolution of acoustic signals. In: Simmons A, Fay RR, Popper AN (eds) Acoustic communication. Springer, New York, NY, pp 65–137

Fite MK (1984) Vocal behavior and interactions among parents and offspring in the American Dipper. M.S. thesis, Utah State University, Logan, UT

Fletcher NH, Riede T, Beckers GJL, Suthers RA (2004) Vocal tract filtering and the "coo" of doves. J Acoust Soc Am 116:3750–3756

Florentin J, Dutoit T, Verlinden O (2016) Identification of European woodpecker species in audio recordings from their drumming rolls. Eco Inform 35:61–70

Flower T (2011) Fork-tailed Drongos use deceptive mimicked alarm calls to steal food. Proc R Soc B 278: 1548–1555

Foote JR, Fitzsimmons LP, Mennill DJ, Ratcliffe LM (2008) Male chickadees match neighbors interactively at dawn: support for the social dynamics hypothesis. Behav Ecol 19:1192–1199

Forstmeier W, Balsby TJS (2002) Why mated Dusky Warblers sing so much: territory guarding and male quality announcement. Behaviour 139:89–111

Forsythe DM (1970) Vocalizations of the Long-billed Curlew. Condor 72:213–224

Foster MS (1981) Cooperative behavior and social organization of the Swallow-tailed Manakin (*Chiroxiphia caudata*). Behav Ecol Sociobiol 9:167–177

Freeberg TM, Dunbar RIM, Ord TJ (2012) Social complexity as a proximate and ultimate factor in communicative complexity. Philos Trans R Soc B 367:1785–1801

Fuller RA, Warren PH, Gaston KJ (2007) Daytime noise predicts nocturnal singing in urban robins. Biol Lett 3: 368–370

Funghi C, Cardoso GC, Mota PG (2015) Increased syllable rate during aggressive singing in a bird with complex and fast song. J Avian Biol 46:283–288

Fuxjager MJ, Schuppe ER (2018) Androgenic signaling systems and their role in behavioral evolution. J Steroid Biochem Mol Biol 184:47–56

Fuxjager MJ, Longpre KM, Chew JG, Fusani L, Schlinger BA (2013) Peripheral androgen receptors sustain the acrobatics and fine motor skill of elaborate male courtship. Endocrinology 154:3168–3177

Fuxjager MJ, Miles MC, Goller F, Petersen J, Yancey J (2017) Androgens support male acrobatic courtship behavior by enhancing muscle speed and easing the severity of its tradeoff with force. Endocrinology 158: 4038–4046

Gahr M, Guttingery H-R (1986) Functional aspects of singing in male and female *Uraeginthus bengalus* (Estrilidae). Ethology 72:123–131

Gahr M, Sonnenschein E, Wickler W (1998) Sex difference in the size of the neural song control regions in a duetting songbird with similar song repertoire size of males and females. J Neurosci 18:1124–1131

Gahr M, Metzdorf R, Schmidl D, Wickler W (2008) Bi-directional sexual dimorphisms of the song control nucleus HVC in a songbird with unison song. PLoS ONE 3:e3073

Gammon DE (2013) How is model selection determined in a vocal mimic?: Tests of five hypotheses. Behaviour 150:1375–1397

Gammon DE (2014) Seasonal patterns of vocal mimicry in Northern Mockingbirds *Mimus polyglottos*. J Avian Biol 45:545–550

Garamszegi LZ, Eens M (2004) Brain space for a learned task: strong intraspecific evidence for neural correlates of singing behavior in songbirds. Brain Res Rev 44:187–193

Garamszegi LZ, Møller AP, Erritzøe J (2003) The evolution of immune defense and song complexity in birds. Evolution 57:905–912

Garamszegi LZ, Eens M, Török J (2008) Birds reveal their personality when singing. PLoS ONE 3:e2647

Garcia M, Charrier I, Rendall D, Iwaniuk AN (2012) Temporal and spectral analyses reveal individual variation in a non-vocal acoustic display: the drumming display of the Ruffed Grouse (*Bonasa umbellus*, L.). Ethology 118:292–301

Garcia SM, Kopuchian C, Mindlin GB, Fuxjager MJ, Tubaro PL, Goller F (2017) Evolution of vocal diversity through morphological adaptation without vocal learning or complex neural control. Curr Biol 27: 2677–2683

Garcia M, Theunissen F, Sèbe F, Clavel J, Ravignani A, Marin-Cudraz T, Fuchs J, Mathevon N (2020) Evolution of communication signals and information during species radiation. Nat Commun 11:4970

Gaston AJ (1977) Social behaviour within groups of Jungle Babblers (*Turdoides striatus*). Anim Behav 25: 828–848

Gaunt AS, Gaunt SLL, Prange HD, Wasser JS (1987) The effects of tracheal coiling on the vocalizations of cranes (Aves: Gruidae). J Comp Physiol A 161:43–59

Geberzahn N, Gahr M (2013) Song learning in male and female *Uraeginthus cyanocephalus*, a tropical songbird species. J Comp Psychol 127:352–364

Gentner TQ, Hulse SH (2000) Female European Starling preference and choice for variation in conspecific male song. Anim Behav 59:443–458

Ghalambor CK, Huey RB, Martin PR, Tewksbury JJ, Wang G (2006) Are mountain passes higher in the tropics? Janzen's hypothesis revisited. Integr Comp Biol 46:5–17

Gil D, Leboucher G, Lacroix A, Cue R, Kreutzer M (2004) Female canaries produce eggs with greater amounts of testosterone when exposed to preferred male song. Horm Behav 45:64–70

Gill FB (1995) Ornithology, 2nd edn. W.H. Freeman and Co., New York, NY

Gill SA, Bierema AM-K (2013) On the meaning of alarm calls: a review of functional reference in avian alarm calling. Ethology 119:449–461

Gill SA, Sealy SG (2004) Functional reference in an alarm signal given during nest defence: seet calls of Yellow Warblers denote brood-parasitic Brown-headed Cowbirds. Behav Ecol Sociobiol 56:71–80

Goller F, Daley MA (2001) Novel motor gestures for phonation during inspiration enhance the acoustical complexity of birdsong. Proc R Soc B 268:2301–2305

Goller F, Larsen ON (1997a) A new mechanism of sound generation in songbirds. Proc Natl Acad Sci USA 94:14787–14791

Goller F, Larsen ON (1997b) *In situ* biomechanics of the syrinx and sound generation in pigeons. J Exp Biol 200:2165–2176

Goller F, Larsen ON (1999) Role of syringeal vibrations in bird vocalizations. Proc R Soc B 266:1609–1615

Goller F, Riede T (2013) Integrative physiology of fundamental frequency control in birds. J Physiol Paris 107: 230–242

Goller M, Shizuka D (2018) Evolutionary origins of vocal mimicry in songbirds. Evol Lett 2:417–426

Goller F, Suthers RA (1996) Role of syringeal muscles in controlling the phonology of bird song. J Neurophysiol 76:287–300

Gong Y, Qin B, Qin Y, Sun W, Wang H (2020) Playback of male song advanced onset of nest building but not initial egg date or breeding synchrony in the Great Tit *Parus major*. J Ornithol 161:585–588

Gontard-Danek MC, Møller AP (1999) The strength of sexual selection: a meta-analysis of bird studies. Behav Ecol 10:476–486

González C, Ornelas JF (2014) Acoustic divergence with gene flow in a lekking hummingbird with complex songs. PLoS ONE 9:e109241

Goodale E, Kotagama SW (2006a) Vocal mimicry by a passerine bird attracts other species involved in mixed-species flocks. Anim Behav 72:471–477

Goodale E, Kotagama SW (2006b) Context-dependent vocal mimicry in a passerine bird. Proc R Soc B 273: 875–880

Grava T, Grava A, Otter KA (2009) Supplemental feeding and dawn singing in Black-capped Chickadees. Condor 111:560–564

Greenquist EA (1982) Displays, vocalizations and breeding biology of the Great Grebe (*Podiceps major*). Condor 84:370–380

Greig E, Pruett-Jones S (2008) Splendid songs: the vocal behaviour of Splendid Fairy-wrens (*Malurus splendens melanotus*). Emu 108:103–114

Griesser M (2008) Referential calls signal predator behavior in a group-living bird species. Curr Biol 18:69–73

Griesser M (2009) Mobbing calls signal predator behavior in a group-living species. Proc R Soc B 276:2887–2892

Grieves LA, Logue DM, Quinn JS (2014) Joint-nesting Smooth-billed Anis, *Crotophaga ani*, use a functionally referential alarm call system. Anim Behav 89:215–221

Grieves LA, Logue DM, Quinn JS (2015) Vocal repertoire of cooperatively breeding Smooth-billed Anis. J Field Ornithol 86:130–143

Grunst ML, Grunst AS (2014) Song complexity, song rate, and variation in the adrenocortical stress response in Song Sparrows (*Melospiza melodia*). Gen Comp Endocrinol 200:67–76

Grunst ML, Grunst AS, Rotenberry JT (2014) Phenotypic plasticity in nest departure calls: weighing costs and benefits. Anim Behav 90:31–39

Guzy MJ, Ritchison G (2020) Common Yellowthroat (*Geothlypis trichas*), version 1.0. In: Rodewald PG (ed) Birds of the World. Cornell Lab of Ornithology, Ithaca, NY

Haftorn S (1993) Ontogeny of the vocal repertoire in the Willow Tit *Parus montanus*. Ornis Scand 24:257–289

Hahnloser RH, Fee MS (2007) Sleep-related spike bursts in HVC are driven by the nucleus interface of the nidopallium. J Neurophysiol 97:423–435

Hahnloser RHR, Kozhevnikov AA, Fee MS (2002) An ultra-sparse code underlies the generation of neural sequences in a songbird. Nature 419:65–70

Hale AM (2006) The structure, context and functions of group singing in Black-breasted Wood-Quail. Behaviour 143:511–533

Halfwerk W, Lohr B, Slabbekoorn H (2018) Impact of man-made sound on birds and their songs. In: Slabbekoorn H, Dooling RJ, Popper AN, Fay RR (eds) Effects of anthropogenic noise on animals. Springer, New York, NY, pp 209–242

Halkin SL (1997) Nest-vicinity song exchanges may coordinate biparental care of Northern Cardinals. Anim Behav 54:189–198

Halkin SL, Shustack DP, DeVries MS, Jawor JM, Linville SU (2021) Northern Cardinal (*Cardinalis cardinalis*), version 2.0. In: Rodewald PG, Keeney BK (eds) Birds of the World. Cornell Lab of Ornithology, Ithaca, NY

Hall ML (2009) A review of vocal duetting in birds. Adv Study Behav 40:67–121

Hall ZJ, MacDougall-Shackleton SA, Osorio-Beristain M, Murphy TG (2010) Male bias in the song control system despite female bias in song rate in Streak-backed Orioles (*Icterus pustulatus*). Brain Behav Evol 76:168–175

Hall ML, Rittenbach MRD, Vehrencamp SL (2015) Female song and vocal interactions with males in a Neotropical wren. Front Ecol Evol 3:12

Hamaguchi K, Tanaka M, Mooney R (2016) A distributed recurrent network contributes to temporally precise vocalizations. Neuron 91:680–693

Hancock JA, Kushlan J, Kahl MP (1992) Storks, ibises, and spoonbills of the world. Academic Press, San Diego, CA

Handford P (2004) Song dialects, habitats, and genetics: the Rufous-collared Sparrow. In: Marler P, Slabbekoorn H (eds) Nature's music: the science of birdsong. Academic Press, San Diego, CA, p 201

Hanson P (1979) Vocal learning: its role in adapting sound structures in long-distance propogation, and a hypothesis on its evolution. Anim Behav 27:1270–1271

Hara E, Rivas MV, Ward JM, Okanoya K, Jarvis ED (2012) Convergent differential regulation of parvalbumin in the brains of vocal learners. PLoS ONE 7:e29457

Hardouin LA, Reby D, Bavoux C, Burneleau G, Bretagnolle V (2007) Communication of male quality in owl hoots. Am Nat 169:552–562

Hardouin LA, Thompson R, Stenning M, Reby D (2014) Anatomical bases of sex- and size-related acoustic variation in Herring Gull alarm calls. J Avian Biol 45:157–166

Hardy JW (1979) Vocal repertoire and its possible evolution in the black and blue jays (*Cissilopha*). Wilson Bull 91:187–201

Haribal M, Dhondt AA, Rosane D, Rodriguez E (2005) Chemistry of preen gland secretions of passerines: different pathways to same goal? Why? Chemoecology 15:251–260

Hartley RS, Suthers RA (1989) Airflow and pressure during canary song: direct evidence for mini-breaths. J Comp Physiol A165:15–26

Hasselquist D, Bensch S (2008) Daily energy expenditure of singing Great Reed Warblers *Acrocephalus arundinaceus*. J Avian Biol 39:384–388

Hausberger M, Jenkins PF, Keene J (1991) Species-specificity and mimicry in bird song: are they paradoxes? A reevaluation of song mimicry in the European Starling. Behaviour 117:53–81

Heathcote CF (1989) The acoustic repertoire of the Bell Miner, *Manorina melanophrys*. Ph.D. dissertation, University of Melbourne, Melbourne

Hebets EA, Barron AB, Balakrishnan CN, Hauber ME, Mason PH, Hoke KL (2016) A systems approach to animal communication. Proc R Soc B 283:20152889

Heckscher CM (2007) Use of the Veery (*Catharus fuscescens*) call repertoire in vocal communication. Ph. D. dissertation, University of Delaware, Newark, DE

Heinrich B, Marzluff J (1991) Do Common Ravens yell because they want to attract others? Behav Ecol Sociobiol 28:13–21

Heinrich B, Marzluff J (1995) Why ravens share. Am Sci 83:342–349

Hendricks P, Verbeek NA (2020) American Pipit (*Anthus rubescens*), version 1.0. In: Billerman SM (ed) Birds of the World. Cornell Lab of Ornithology, Ithaca, NY

Herting BL, Belthoff JR (2001) Bounce and double trill songs of male and female Western Screech-Owls: characterization and usefulness for classification of sex. Auk 118:1095–1101

Highsmith RT (1989) The singing behavior of Golden-winged Warblers. Wilson Bull 101:36–50

Hile AG, Striedter GF (2000) Call convergence within groups of female Budgerigars (*Melopsittacus undulatus*). Ethology 106:1105–1114

Hile AG, Plummer TK, Striedter GF (2000) Male vocal imitation produces call convergence during pair bonding in Budgerigars, *Melopsittacus undulates*. Anim Behav 59:1209–1218

Hile AG, Burley NT, Coopersmith CB, Foster VS, Striedter GF (2005) Effects of male vocal learning on female behavior in the Budgerigar, *Melopsittacus undulates*. Ethology 111:901–923

Hill SD (2011) The vocalisations of Tui (*Prosthemadera novaeseelandiae*). M.S. thesis, Massey University, Albany, New Zealand

Hill BG, Lein MR (1985) The non-song vocal repertoire of the White-crowned Sparrow, *Zonotrichia leucophrys*. Condor 87:327–335

Hilty SL, Brown WL (1986) A guide to the birds of Colombia. Princeton University Press, Princeton, NJ

Hingee M, Magrath RD (2009) Flights of fear: a mechanical wing whistle sounds the alarm in a flocking bird. Proc R Soc B 276:4173–4179

Holt J, Barati A, McDonald PG (2017) The complex acoustic repertoire of a highly social species, the Noisy Miner, *Manorina melanocephala*. Emu 117:19–30

Hopp SL, Jablonski P, Brown JL (2001) Recognition of group membership by voice in Mexican Jays, *Aphelocoma ultramarina*. Anim Behav 62:297–303

Howard RD (1974) Influence of sexual selection and interspecific competition on mockingbird song (*Mimus polyglottos*). Evolution 28:428–438

Hultsch H, Kopp ML (1989) Early auditory learning and song improvisation in Nightingales, *Luscinia megarhynchos*. Anim Behav 37:510–512

Hüppop O, Hilgerloh G (2012) Flight call rates of migrating thrushes: effects of wind conditions, humidity and time of day at an illuminated offshore platform. J Avian Biol 43:85–90

Igic B, Magrath RD (2014) A songbird mimics different heterospecific alarm calls in response to different types of threat. Behav Ecol 25:538–548

Igic B, McLachlan J, Lehtinen I, Magrath RD (2015) Crying wolf to a predator: deceptive vocal mimicry by a bird protecting young. Proc R Soc B 282: 20150798

Ihle M, Forstmeier W (2013) Revisiting the evidence for inbreeding avoidance in Zebra Finches. Behav Ecol 24:1356–1362

Ikeda MZ, Trusel M, Roberts TF (2020) Memory circuits for vocal imitation. Curr Opin Neurobiol 60:37–46

Illes E, Yunes-Jimenez L (2009) A female songbird out-sings male conspecifics during simulated territorial intrusions. Proc R Soc B 276:981–986

Ippi S, Vasquez RA, van Dongen WF, Lazzoni I (2011) Geographical variation in the vocalizations of the suboscine Thorn-tailed Rayadito *Aphrastura spinicauda*. Ibis 153:789–805

Jackson JA (2020) Red-cockaded Woodpecker (*Picoides borealis*), version 1.0. In: Poole AF, Gill FB (eds) Birds of the World. Cornell Lab of Ornithology, Ithaca, NY

Jackson JA, Ouellet HR (2020) Downy Woodpecker (*Dryobates pubescens*), version 1.0. In: Rodewald PG (ed) Birds of the World. Cornell Lab of Ornithology, Ithaca, NY

Jarvis ED, Ribeiro S, Vielliard J, Da Silva ML, Ventura D, Mello CV (2000) Behaviorally driven gene expression reveals song nuclei in hummingbird brain. Nature 406: 628–632

Jenni L (1981) Das skelettmuskelsystem des halses von buntspecht und mittelspecht *Dendrocopos major* und *medius*. J Ornithol 122:37–63

Jensen KK, Cooper BG, Larsen ON, Goller F (2007) Songbirds use pulse tone register in two voices to generate low-frequency sound. Proc R Soc B 274: 2703–2710

Johnsgard PA (1971) Observations on sound production in the Anatidae. Wild 22:46–59

Johnson KE, Clark CJ (2020) Ontogeny of vocal learning in a hummingbird. Anim Behav 167:139–150

Johnson LS, Searcy WA (1996) Female attraction to male song in House Wrens (*Troglodytes aedon*). Behaviour 133:357–366

Jones KJ, Hill WL (2001) Auditory perception of hawks and owls for passerine alarm calls. Ethology 107:717–726

Jones MR, Witt CC (2012) Utility of vocal formant spacing for monitoring Sandhill Crane subspecies. Wildl Soc Bull 36:47–53

Jones IL, Falls JB, Gaston AJ (1989) The vocal repertoire of the Ancient Murrelet. Condor 91:699–710

Jones MA, DuVal EH, Boyle WA (2014) Individual and temporal variability in the courtship behavior of White-ruffed Manakins (*Corapipo altera*), a species with facultative cooperative displays. Auk 131:727–742

Jordan EA, Areta JI (2020) Bisonic mechanical wing songs and complex kinematics in aerial displays of the Subtropical Doradito (*Pseudocolopteryx acutipennis*). Integr Comp Biol 60:1173–1187

Jurisevic MA, Sanderson KJ (1998) Acoustic discrimination of passerine anti-predator signals by Australian raptors. Aust J Zool 46:369–379

Kaiser M (1990) Untersuchungen zur biomechanik des Balztrommelns der Specte (Picidae). Beiträge zur Vogelkunde 36:129–159

Kalb N, Anger F, Randler C (2019) Subtle variations in mobbing calls are predator-specific in Great Tits (*Parus major*). Sci Rep 9:6572

Kaluthota C, Brinkman BE, dos Santos EB, Rendall D (2016) Transcontinental latitudinal variation in song performance and complexity in House Wrens (*Troglodytes aedon*). Proc R Soc B 283:20152765

Kaplan G (2014) Animal communication. Wiley Interdiscip Rev Cogn Sci 5:661–677

Kaplan G, Rogers LJ (2013) Stability of referential signalling across time and locations: testing alarm calls of Australian Magpies (*Gymnorhina tibicen*) in urban and rural Australia and in Fiji. PeerJ 1:e112

Keen SC, Meliza CD, Rubenstein DR (2013) Flight calls signal group and individual identity but not kinship in a cooperatively breeding bird. Behav Ecol 24:1279–1285

Keller GB, Hahnloser RHR (2009) Neural processing of auditory feedback during vocal practive in a songbird. Nature 457:187–190

Kempenaers B, Verheyen GR, Dhondt AA (1995) Mate guarding and copulation behaviour in monagamous and polygynous Blue Tits: do males follow a best-ofa-bad-job strategy? Behav Ecol Sociobiol 36:33–42

Kennedy RAW, Evans CS, McDonald PG (2009) Individual distinctiveness in the mobbing call of a cooperative bird, the Noisy Miner *Manorina melanocephala*. J Avian Biol 40:481–490

Kiefer S, Sommer C, Scharff C, Kipper S (2010) Singing the popular songs? Nightingales share more song types with their breeding population in their second season than in their first. Ethology 116:619–626

King AS (1989) Functional anatomy of the syrinx. In: King AS, McLelland J (eds) Form and function in birds, vol 4. Academic, San Diego, CA, pp 105–192

King SL, McGregor PK (2016) Vocal matching: the what, the why and the how. Biol Lett 12:20160666

Kipper S, Mundry R, Sommer C, Hultsch H, Todt D (2006) Song repertoire size is correlated with body measures and arrival date in Common Nightingales, *Luscinia megarhynchos*. Anim Behav 71:211–217

Knapton RW (1982) Geographic similarity and year-to-year retention of song in the Clay-colored Sparrow (*Spizella pallida*). Behaviour 79:189–200

Knight EC, Brigham RM, Bayne EM (2022) The Big Boom Theory: the Common Nighthawk wing-boom display delineates exclusive nesting territories. Ornithology 139:ukab066

Koenig WD, Stanback MT, Hooge PN, Mumme RL (1991) Distress calls in the Acorn Woodpecker. Condor 93:637–643

Kok OB (1971) Vocal behavior of the Great-tailed Grackle (*Quiscalus mexicanus prosopidicola*). Condor 73:348–363

Kondo N, Watanabe S (2009) Contact calls: information and social function. Jpn Psychol Res 51:197–208

Konishi M (1985) Birdsong: from behavior to neuron. Annu Rev Neurosci 8:125–170

Koosa K, Tilgar V (2016) Is hissing behavior of incubating Great Tits related to reproductive investment in the wild? Acta Ethol 19:173–180

Kostoglou KN, Van Dongen WFD, Weston MA (2020) Parental defence in shorebirds is mediated by embryonic calling, ambient temperature and predator latency. J Ornithol 161:1153–1165

Kovach KA (2013) Cooperative functions of duetting behavior in tropical wrens. M.S. thesis, University of Windsor, Windsor, ON, Canada

Krakauer AH, Tyrrell M, Lehmann K, Losin N, Goller F, Patricelli GL (2009) Vocal and anatomical evidence for two-voiced sound production in the Greater Sage-Grouse *Centrocercus urophasianus*. J Exp Biol 212:3719–3727

Krams I, Krama T, Freeberg TM, Kullberg C, Lucas JR (2012) Linking social complexity and vocal complexity: a parid perspective. Philos Trans R Soc B 367: 1879–1891

Krause ET, Krüger O, Kohlmeier P, Caspers BA (2012) Olfactory kin recognition in a songbird. Biol Lett 8: 327–329

Krieg CA, Getty T (2016) Not just for males: females use song against male and female rivals in a temperate zone songbird. Anim Behav 113:39–47

Krilow JM, Iwaniuk AN (2015) Seasonal variation in forebrain region sizes in male Ruffed Grouse (*Bonasa umbellus*). Brain Behav Evol 85:189–202

Kroodsma DE (1976) Reproductive development in a female songbird: differential stimulation by quality of male song. Science 192:574–575

Kroodsma DE (1996) Ecology of passerine song development. In: Kroodsma DE, Miller EH (eds) Ecology and evolution of acoustic communication in birds. Cornell University Press, Ithaca, NY, pp 3–19

Kroodsma DE, Houlihan PW, Fallon PA, Wells JA (1997) Song development by Grey Catbirds. Anim Behav 54: 457–464

Kroodsma DE, Liu WC, Goodwin E, Bedell PA (1999) The ecology of song improvisation as illustrated by North American Sedge Wrens. Auk 116:373–386

Kroodsma DE, Woods RW, Goodwin EA (2002) Falkland Island Sedge Wrens (*Cistothorus platensis*) imitate rather than improvise large song repertoires. Auk 119:523–528

Krull CR, Ranjard L, Landers TJ, Ismar SM, Matthews JL, Hauber ME (2012) Analyses of sex and individual differences in vocalizations of Australasian Gannets using a dynamic time warping algorithm. J Acoust Soc Am 132:1189–1198

Kumar A, Bhatt D (2000) Vocal signals in a tropical avian species, the Red-vented Bulbul *Pycnonotus cafer*: their characteristics and importance. J Biosci 25:387–396

La Morco G, Thorstrom R (2000) Breeding biology, diet and vocalization of the Helmet Vanga *Euryceros prevostii* on the Masoala Peninsula, Madagascar. Ostrich 71:400–403

Laiolo P, Palestrini C, Rolando A (2000) A study of Choughs' vocal repertoire: variability related to individuals, sexes and ages. J Ornithol 141:168–179

Laje R, Gardner T, Mindlin GB (2002) Neuromuscular control of vocalization in bird song: a model. Phys Rev E 65:051921

Lampe HM, Espmark YO (1994) Song structure reflects male quality in Pied Flycatchers, *Ficedula hypoleuca*. Anim Behav 47:869–876

Lampe HM, Saetre GP (1995) Female Pied Flycatchers prefer males with larger song repertoires. Proc R Soc B 262:163–167

Langmore NE (1998) Functions of duet and solo songs of female birds. Trends Ecol Evol 13:136–140

Langmore NE (2000) Why female birds sing. In: Espmark Y, Amundsen T, Rosenqvist G (eds) Animal signals: signalling and signal design in animal communication. Tapir Academic Press, Tronheim, Norway, pp 317–327

Langmore NE, Davies NB, Hatchwell BJ, Hartley IR (1996) Female song attracts males in the Alpine Accentor *Prunella collaris*. Proc R Soc B 263:141–146

Lanzone M, Deleon E, Grove L, Farnsworth A (2009) Revealing undocumented or poorly known flight calls of warblers (Parulidae) using a novel method of recording birds in captivity. Auk 126:511–519

LaPorte N, Storer RW, Nuechterlein GL (2020) Western Grebe (*Aechmophorus occidentalis*), version 1.0. In: Rodewald PG (ed) Birds of the World. Cornell Lab of Ornithology, Ithaca, NY

Lara C, Martínez-García V, Ornelas JF (2015) Microgeographical variation in song repertoire and structure between the leks of Green Violetears *Colibri thalassinus* in central Mexico. Acta Ornithol 50:23–32

Larsen C, Speed M, Harvey N, Noyes HA (2007) A molecular phylogeny of the nightjars (Aves: Caprimulgidae) suggests extensive conservation of primitive morphological traits across multiple lineages. Mol Phylogenet Evol 42:789–796

Larson ON, Goller F (2002) Direct observation of syringeal muscle function in songbirds and a parrot. J Exp Biol 205:25–35

Lawrence LDK (1967) A comparative life-history study of four species of woodpeckers. Ornithol Monogr 5:1–156

Leader N, Geffen E, Mokady O, Yom-Tov Y (2008) Song dialects do not restrict gene flow in an urban population of the Orange-tufted Sunbird, *Nectarinia osea*. Behav Ecol Sociobiol 62:1299–1305

Leavesley AJ, Magrath RD (2005) Communicating about danger: urgency alarm calling in a bird. Anim Behav 70:365–373

Lee JW, Lee JY, Noh HJ, Kim MS, Kim KH, Yoo JC (2015) Habitats matter: the incidence of and response to fear screams in a habitat generalist, the Vinous-throated Parrotbill *Paradoxornis webbianus*. Behav Ecol Sociobiol 69:1575–1584

Leighton GM, Birmingham T (2021) Multiple factors affect the evolution of repertoire size across birds. Behav Ecol 32:380–385

Lemon RE (1975) How birds develop song dialects. Condor 77:385–406

Leonard M (2008) An overview and comparative analysis of singing on the nest in North American birds. Can J Zool 86:1101–1110
Li X-C, Jarvis ED, Alvarez-Borda B, Lim DA, Nottebohm F (2000) A relationship between behavior, neurotrophin expression, and new neuron survival. Proc Natl Acad Sci USA 97:8584–8589
Li Z, Clarke JA, Eliason CM, Stidham TA, Deng T, Zhou Z (2018) Vocal specialization through tracheal elongation in an extinct Miocene pheasant from China. Sci Rep 8:8099
Linossier J, Zsebök S, Baudry E, Aubin T, Courvoisier H (2016) Acoustic but not genetic divergence in migratory and sedentary populations of Blackcaps, *Sylvia atricapilla*. Biol J Linn Soc 119:68–79
Lipshutz SE, Overcast IA, Hickerson MJ, Brumfield RT, Derryberry EP (2017) Behavioural response to song and genetic divergence in two subspecies of White-crowned Sparrows (*Zonotrichia leucophrys*). Mol Ecol 26:3011–3027
Liu W-C, Kroodsma DE (2007) Dawn and daytime singing behavior of Chipping Sparrows (*Spizella passerina*). Auk 124:44–52
Lloyd-Jones DJ (2014) Ultrasonic harmonics in the calls of Rock Wren (*Xenicus gilviventris*). Notornis 61:165–169
Lobato M, Vellema M, Gahr C, Leitão A, de Lima S, Geberzahn N, Gahr M (2015) Mismatch in sexual dimorphism of developing song and song control system in Blue-capped Cordon-bleus, a songbird species with singing females and males. Front Ecol Evol 3:117
Lockner FR, Youngren OM (1976) Functional syringeal anatomy of the Mallard. I. In situ electromyograms druing ESB elicited calling. Auk 93:324–342
Loiselle BA, Ryder TB, Duräes R, Tori W, Blake JC, Parker PG (2007) Kin selection does not explain male aggregation at leks of 4 manakin species. Behav Ecol 18:287–291
Løvlie H, Zidar J, Berneheim C (2014) A cry for help: female distress calling during copulation is context dependent. Anim Behav 92:151–157
Lowther PE, Sharbaugh S (2020) Arctic Warbler (*Phylloscopus borealis*), version 1.0. In: Billerman SM (ed) Birds of the World. Cornell Lab of Ornithology, Ithaca, NY
Luine V, Nottebohm F, Harding C, McEwen BS (1980) Androgen affects cholinergic enzymes in syringeal motor neurons and muscle. Brain Res 192:89–107
Lukianchuk KC, Doucet SM (2014) Cooperative courtship display in Long-tailed Manakins *Chiroxiphia linearis*: predictors of courtship success revealed through full characterization of display. J Ornithol 155:729–743
MacDougall-Shackleton SA, Ball GF (1999) Comparative studies of sex differences in the song-control system of songbirds. Trends Neurosci 22:432–436
Mack AL, Jones J (2003) Low-frequency vocalizations by cassowaries (*Casuarius* spp.). Auk 120:1062–1068
MacRoberts MH, MacRoberts BR (1976) Social organization and behavior of the Acorn Woodpecker in central coastal California. Ornithol Monogr 21:1–115
Magrath RD, Haff TM, McLachlan JR, Igic B (2015) Wild birds learn to eavesdrop on heterospecific alarm calls. Curr Biol 25:2047–2050
Maller CJ, Jones DN (2001) Vocal behaviour of the Common Koel, *Eudynamys scolopacea*, and implications for mating systems. Emu 101:105–112
Mann NI, Dingess KA, Slater PJB (2005) Antiphonal four-part synchronized chorusing in a Neotropical wren. Biol Lett 2:1–4
Marler P (1955) Characteristics of some animal calls. Nature 176:6–8
Marler P (1970a) A comparative approach to vocal learning: song development in White-crowned Sparrows. J Comp Physiol Psychol 71:1–25
Marler P (1970b) Birdsong and speech development: could there be parallels? Am Sci 58:669–673
Marler P, Peters S (1981) Sparrows learn adult song and more from memory. Science 213:780–782
Marler P, Peters S (1982) Long-term storage of learned birdsongs prior to production. Anim Behav 30:479–482
Marler P, Peters S (1987) A sensitive period for song acquisition in the Song Sparrow, *Melospiza melodia*: a case of age-limited learning. Ethology 76:89–100
Marler P, Peters S, Ball GF, Dufty AM Jr, Wingfield JC (1988) The role of sex steroids in the acquisition and production of birdsong. Nature 336:770–772
Marshall-Ball L, Mann N, Slater PJB (2006) Multiple functions to duet singing: hidden conflicts and apparent cooperation. Anim Behav 71:823–831
Marten K, Marler P (1977) Sound transmission and its significance for animal vocalizations. I. Temperate habitats. Behav Ecol Sociobiol 2:271–290
Martin K, Horn AG, Hannon SJ (1995) The calls and associated behavior of breeding Willow Ptarmigan in Canada. Wilson Bull 107:496–509
Marzluff J, Heinrich B (1991) Foraging by Common Ravens in the presence and absence of territory holders: an experimental analysis of social foraging. Anim Behav 42:755–770
Matessi G, McGregor PK, Peake TM, Dabelsteen T (2005) Do male birds intercept and use rival courtship calls to adjust paternity protection behaviours? Behaviour 142: 507–524
May DL (2004) Vocal repertoire of Grey Parrots (*Psittacus erithacus*) living in the Congo Basin. Ph.D. dissertation, University of Arizona, Tuscon, AZ
McDonald DB (1989a) Cooperation under sexual selection: age-graded changes in a lekking bird. Am Nat 134:709–730
McDonald MV (1989b) Function of song in Scott's Seaside Sparrow, *Ammodramus maritimus peninsulae*. Anim Behav 38:468–485
McDonald MV (2020) Kentucky Warbler (*Geothlypis formosa*), version 1.0. In: Poole AF (ed) Birds of the World. Cornell Lab of Ornithology, Ithaca, NY

McDonald DB, Potts WK (1994) Cooperative display and relatedness among males in a lek-mating bird. Science 266:1030–1032

McDonald PG, Wright J (2011) Bell Miner provisioning calls are more similar among relatives and are used by helpers at the nest to bias their effort towards kin. Proc R Soc B 278:3403–3411

McGraw KJ, Middleton AL (2020) American Goldfinch (*Spinus tristis*), version 1.0. In: Rodewald PG (ed) Birds of the World. Cornell Lab of Ornithology, Ithaca, NY

McGregor PK, Dabelsteen T (1996) Communication networks. In: Kroodsma DE, Miller EH (eds) Ecology and evolution of acoustic communication in birds. Cornell University Press, Ithaca, NY, pp 409–425

McGregor PK, Krebs JR (1984) Sound degradation as a distance cue in Great Tit (*Parus major*) song. Behav Ecol Sociobiol 16:49–56

McLachlan JR, Magrath RD (2020) Speedy revelations: how alarm calls can convey rapid, reliable information about urgent danger. Proc R Soc B 287:20192772

Méndez JM, Goller F (2020) Multifunctional bilateral muscle control of vocal output in the songbird syrinx. J Neurophysiol 124:1857–1874

Mengel RM, Sharpe RS, Woolfenden GE (1972) Wing clapping in territorial and courtship behavior of the Chuck-will's-widow and Poor-will (Caprimulgidae). Auk 89:440–444

Mennill DJ, Vehrencamp SL (2008) Context-dependent functions of avian duets revealed by microphone-array recordings and multispeaker playback. Curr Biol 18:1314–1319

Messina S, Eens M, Casasole G, AbdElgawad H, Asard H, Pinxten R, Costantini D (2017) Experimental inhibition of a key cellular antioxidant affects vocal communication. Funct Ecol 31:1101–1110

Mihailova M, Berg ML, Buchanan KL, Bennett ATD (2014) Odour-based discrimination of subspecies, species and sexes in an avian species complex, the Crimson Rosella. Anim Behav 95:155–164

Miles MC, Schuppe ER, Ligon RM, Fuxjager MJ (2018) Macroevolutionary patterning of woodpecker drums reveals how sexual selection elaborates signals under constraint. Proc R Soc B 285:20172628

Miller AH (1925) The boom-flight of the Pacific Nighthawk. Condor 27:141–143

Miller AH (1934) The vocal apparatus of some North American owls. Condor 36:204–213

Miller EH (1983) The structure of aerial displays in three species of *Calidridinae* (Scolopacidae). Auk 100:440–451

Miller EH, Seneviratne SS, Jones IL, Robertson GJ, Wilhelm SI (2008) Syringeal anatomy and allometry in murres (Alcidae: *Uria*). J Ornithol 149:545–554

Møller AP, Nielsen JT (2010) Fear screams and adaptation to avoid imminent death: effects of genetic variation and predation. Ethol Ecol Evol 22:183–202

Montes-Medina AC, Salinas-Melgoza A, Renton K (2016) Contextual flexibility in the vocal repertoire of an Amazon parrot. Front Zool 13:40

Moody NM, Vivlamore EK, Fuxjager MJ (2022) Woodpecker drum evolution: an analysis of covariation in elements of a multicomponent acoustic display among and within species. Evolution 76:1469–1480

Moore SD, Rohwer VG (2012) The functions of adult female begging during incubation in sub-Arctic breeding Yellow Warblers. Anim Behav 84:1213–1219

Moore JM, Woolley SMN (2019) Emergent tuning for learning vocalizations in auditory cortex. Nat Neurosci 22:1469–1476

Moore JM, Székely T, Büki J, DeVoogd TJ (2011) Motor pathway convergence predicts syllable repertoire size in oscine birds. Proc Natl Acad Sci USA 108:16440–16445

Moravec ML, Striedter GF, Burley NT (2006) Assortative pairing based on contact call similarity in Budgerigars, *Melopsittacus undulates*. Ethology 112:1108–1116

Morton ES (1975) Ecological sources of selection on avian sounds. Am Nat 109:17–34

Morton ES (1996) A comparison of vocal behavior among tropical and temperate passerine birds. In: Kroodsma DE, Miller EH (eds) Ecology and evolution of acoustic communication in birds. Cornell University Press, Ithaca, NY, pp 258–268

Morton ES, Derrickson KC, Stutchbury BJM (2000) Territory switching behavior in a sedentary tropical passerine, the Dusky Antbird (*Cercomacra tyrannina*). Behav Ecol 11:648–653

Moseley DL, Derryberry GE, Phillips JN, Danner JE, Danner RM, Luther DA, Derryberry EP (2018) Acoustic adaptation to city noise through vocal learning by a songbird. Proc R Soc B 285:20181356

Moser-Purdy C, Mennill DJ (2016) Large vocal repertoires do not constrain the dear enemy effect: a playback experiment and comparative study of songbirds. Anim Behav 118:55–64

Moskát C, Hauber ME (2019) Sex-specific responses to simulated territorial intrusions in the Common Cuckoo: a dual function of female acoustic signaling. Behav Ecol Sociobiol 73:1–10

Mota PG (1999) The functions of song in the Serin. Ethology 105:137–148

Mountjoy DJ, Lemon RE (1996) Female choice for complex song in the European Starling: a field experiment. Behav Ecol Sociobiol 38:65–71

Moyer BR, Rock AN, Clayton DH (2003) Experimental test of the importance of preen oil in Rock Doves (*Columba livia*). Auk 120:490–496

Mundinger PC (1970) Vocal imitation and individual recognition of finch calls. Science 168:480–482

Najar N, Benedict L (2019) The relationship between latitude, migration and the evolution of bird song complexity. Ibis 161:1–12

Negret PJ, Laverde-R O (2015) The enigmatic Black Tinamou: Do distribution, climate, and vocalizations reveal more than one species? Auk 132:132–139

Nelson BS, Beckers GJL, Suthers RA (2005) Vocal tract filtering and sound radiation in a songbird. J Exp Biol 208:297–308

Neudorf DL, Sealy SG (2002) Distress calls of birds in a Neotropical cloud forest. Biotropica 34:118–126

Niese RL, Clark CJ, Tobalske BW (2020) Specialized feathers produce sonations during flight in *Columbina* ground doves. Integr Comp Biol 60:1160–1172

Ning JZ, Guo YS, Han YL (2008) Preliminary study on vocalizations of captive White-collored Yuhina. J China West Normal Univ (Nat Sci) 3:011

Nolan PM, Hill GE (2004) Female choice for song characteristics in the House Finch. Anim Behav 67: 403–410

Nolan V Jr (1978) The ecology and behavior of the Prairie Warbler *Dendroica discolor*. Ornithol Monogr 26:1–595

Nomtes-Median AC, Salinas-Melgoza A, Renton K (2016) Contextual flexibility in the vocal repertoire of an Amazon parrot. Front Zool 13:40

Nottebohm F (1969) The song of the Chingolo, *Zonotrichia capensis*, in Argentina: description and evaluation of a system of dialects. Condor 71:299–315

Nottebohm F (1972) The origins of vocal learning. Am Nat 106:116–140

Nottebohm F (1980) Testosterone triggers growth of brain vocal control nuclei in adult female canaries. Brain Res 189:429–436

Nottebohm F (1981) A brain for all seasons: cyclical anatomical changes in song control nuclei of the canary brain. Science 214:1368–1370

Nottebohm F (1995) The neural basis of birdsong. PLoS Biol 3:e164

Nowicki S, Searcy WA (2014) The evolution of vocal learning. Curr Opin Neurobiol 28:48–53

Nowicki S, Hasselquist D, Bensch S, Peters S (2000) Nestling growth and song repertoire size in Great Reed Warblers: evidence for song learning as an indicator mechanism in mate choice. Proc R Soc B 267: 2419–2424

Nowicki S, Searcy WA, Peters S (2002) Quality of song learning affects female response to male bird song. Proc R Soc B 269:1949–1954

Nuttall RJ (1993) Vocal behaviour of the Quail Finch *Ortygospiza atricollis*. Ostrich 64:97–104

Odom KJ, Mennill DJ (2010) A quantitative description of the vocalizations and vocal activity of the Barred Owl. Condor 112:549–560

Odom KJ, Hall ML, Riebel K, Omland KE, Langmore NE (2014) Female song is widespread and ancestral in songbirds. Nat Commun 5:3379

Ohms VR, Snelderwaard PC, ten Cate C, Beckers GJL (2010) Vocal tract articulation in Zebra Finches. PLoS ONE 5:e11923

Ohms VR, Beckers GJL, ten Cate C, Suthers RA (2012) Vocal tract articulation revisited: the case of the Monk Parakeet. J Exp Biol 215:85–92

Opaev AS (2014) Behavior and vocalization of the Chinese Gray Shrike (*Lanius sphenocerus* Cabanis, 1873) at the beginning of its breeding cycle. Biol Bull 41: 879–884

Oppenheim RW (1972) Prehatching and hatching behaviour in birds: a comparative study of altricial and precocial species. Anim Behav 20:644–655

Ord TJ, Garcia-Porta J (2012) Is sociality required for the evolution of communicative complexity? Evidence weighed against alternative hypotheses in diverse taxonomic groups. Philos Trans R Soc B 367:1811–1828

Ornelas JF, González C, Uribe J (2002) Complex vocalizations and aerial displays of the Amethyst-throated Hummingbird (*Lampornis amethystinus*). Auk 119:1141–1149

Otis DL, Schulz JH, Miller D, Mirarchi RE, Baskett TS (2020) Mourning Dove (*Zenaida macroura*), version 1.0. In: Poole AF (ed) Birds of the World. Cornell Lab of Ornithology, Ithaca, NY

Otter K, Chruszcz B, Ratcliffe L (1997) Honest advertisement and song output during the dawn chorus of Black-capped Chickadees. Behav Ecol 8:167–173

Parejo D, Amo L, Rodríguez J, Avilés JM (2012) Rollers smell the fear of nestlings. Biol Lett 8:502–504

Patterson DK, Pepperberg IM, Story BH, Hoffman E (1997) How parrots talk: insights based on CT scans, image processing, and mathematical models. Proc SPIE 3033:14–24

Pavlova DZ, Pinxten R, Eens M (2007) Seasonal singing patterns and individual consistency in song activity in female European Starlings (*Sturnus vulgaris*). Behaviour 144:663–680

Payakkhabut S (2012) Vocal communication in Sooty-headed Bulbul, *Pycnonotus aurigaster*. Int J Biol Biomol Agric Food Biotechnol Eng 6:140–145

Payne RB (1982) Ecological consequences of song matching: breeding success and intraspecific song mimicry in Indigo Buntings. Ecology 63:401–411

Payne RB, Payne LL (1997) Field observations, experimental design, and the time and place of learning bird songs. In: Snowdon CT, Hausberger M (eds) Social influences on vocal development. Cambridge University Press, Cambridge, pp 57–84

Payne RB, Payne LL, Doehlert SM (1988) Biological and cultural success of song memes in Indigo Buntings. Ecology 69:104–117

Pereyra ME, Morton ML (2010) Flight songs of Dusky Flycatchers: a response to bird-hunting raptors? J Field Ornithol 81:42–48

Pérez Mena EE, Mora EC (2011) Geographic song variation in the non-oscine Cuban Tody (*Todus multicolor*). Wilson J Ornithol 123:76–84

Petit LJ (2020) Prothonotary Warbler (*Protonotaria citrea*), version 1.0. In: Poole AF, Gill FB (eds) Birds of the World. Cornell Lab of Ornithology, Ithaca, NY

Petrusková T, Diblíková L, Pipek P, Frauendorf E, Procházka P, Petrusek A (2015) A review of the distribution of Yellowhammer (*Emberiza citrinella*) dialects in Europe reveals the lack of a clear macrogeographic pattern. J Ornithol 156:263–273

Pfaff JA, Zanette L, MacDougall-Shackleton SA, MacDougall-Shackleton EA (2007) Song repertoire size varies with HVC volume and is indicative of male quality in Song Sparrows (*Melospiza melodia*). Proc R Soc B 274:2035–2040

Pickering SPC, Berrow SD (2001) Courtship behavior of the Wandering Albatross *Diomedea exulans* at Bird Island, South Georgia. Mar Ornithol 29:29–37

Platzen D, Magrath RD (2005) Adaptive differences in response to two types of parental alarm calls in altricial nestlings. Proc R Soc B 272:1101–1106

Plummer EM, Goller F (2008) Singing with reduced air sac volume causes uniform decrease in airflow and sound amplitude in the Zebra Finch. J Exp Biol 211:66–78

Podos J (2017) Birdsong performance studies: reports of their death have been greatly exaggerated. Anim Behav 125:e17–e24

Podos J, Warren PS (2007) The evolution of geographic variation in birdsong. Adv Study Behav 37:403–458

Podos J, Southall JA, Rossi-Santos MR (2004) Vocal mechanics in Darwin's finches: correlation of beak gape and song frequency. J Exp Biol 207:607–619

Porter R, Aspinall S, Kirwan GM (2018) Hypocolius (*Hypocolius ampelinus*). In: del Hoyo J, Elliott A, Sargatal J, Christie DA, de Juana E (eds) Handbook of the Birds of the World Alive. Lynx Edicions, Barcelona

Potvin DA, Parris KM, Mulder RA (2013) Limited genetic differentiation between acoustically divergent populations of urban and rural Silvereyes (*Zosterops lateralis*). Evol Ecol 27:381–391

Potvin DA, Crawford PW, MacDougall-Shackleton SA, MacDougall-Shackleton EA (2015) Song repertoire size, not territory location, predicts reproductive success and territory tenure in a migratory songbird. Can J Zool 93:627–633

Potvin DA, Ratnayake CP, Radford AN, Magrath RD (2018) Birds learn socially to recognize heterospecific alarm calls by acoustic association. Curr Biol 28:2632–2637

Prendergast HDV (1990) A second note on bill-snappping in Australian birds. Canberra Bird Notes 15:35–36

Proppe DS, Ritchison G (2008) Use and possible functions of the primary and sustained songs of male Grasshopper Sparrows. Am Midl Nat 160:1–6

Pröve E (1978) Courtship and testosterone in male Zebra Finches. Z Tierpsychol 48:47–67

Prum RO, Razafindratsita VR (1997) Lek behavior and natural history of the Velvet Asity (*Philepitta castanea*: Eurylaimidae). Wilson Bull 109:371–392

Radford AN (2003) Territorial vocal rallying in the Green Woodhoopoe: influence of rival group size and composition. Anim Behav 66:1035–1044

Radford AN, Ridley AR (2006) Recruitment calling: a novel form of extended parental care in an altricial species. Curr Biol 16:1700–1704

Raihani NJ, Ridley AR (2007) Adult vocalizations during provisioning: offspring response and postfledging benefits in wild Pied Babblers. Anim Behav 74:1303–1309

Rainey HJ, Zuberbühler K, Slater PJB (2004) Hornbills can distinguish between primate alarm calls. Proc R Soc B 27:755–759

Read AF, Weary DM (1992) The evolution of bird song: comparative analyses. Philos Trans R Soc B 338:165–187

Regosin JV (2002) Alarm calling and predator discrimination in the Arabian Babbler (*Turdoides squamiceps*). Bird Behav 15:11–19

Reichard DG, Welklin JF (2015) On the existence and potential functions of low-amplitude vocalizations in North American birds. Auk 132:156–166

Reid JM (2007) Secondary sexual ornamentation and non-additive genetic benefits of female mate choice. Proc R Soc B 274:1395–1402

Reid JM, Arcese P, Cassidy AL, Hiebert SM, Smith JNM, Stoddard PK, Marr AB, Keller LF (2004) Song repertoire size predicts initial mating success in male Song Sparrows, *Melospiza melodia*. Anim Behav 68:1055–1063

Ręk P (2015) High functional complexity despite an extremely small repertoire of calls in the Spotted Crake (*Porzana porzana*). Auk 132:613–623

Ręk P, Magrath RD (2016) Multimodal duetting in magpie-larks: how do vocal and visual components contribute to a cooperative signal's function? Anim Behav 117:35–42

Rek P, Osiejuk TS (2011) No male identity information loss during call propogation through dense vegetation: the case of the Corncrake. Behav Process 86:323–328

Renton K (2001) Lilac-crowned Parrot diet and food resource availability: resource tracking by a parrot seed predator. Condor 103:62–69

Reyer H-U, Schimdl D (1988) Helpers have little to laugh about: group structure and vocalization in the Laughing Kookaburra *Dacelo novaeguineae*. Emu 88:150–160

Reynolds S (2013) The anatomy and histomorphology of the uropygial gland in New Zealand endemic species. M.S. thesis, Massey University, Palmerston North, New Zealand

Ribot RFH, Buchanan KL, Endler JA, Joseph L, Bennett ATD, Berg ML (2012) Learned vocal variation is associated with abrupt cryptic genetic change in a parrot species complex. PLoS ONE 7:e50484

Rico-Guevara A, Echeverri-Mallarino L, Clark CJ (2022) Oh, snap! A within-wing sonation in Black-tailed Trainbearers. J Exp Biol 225:jeb243219

Riebel K (2016) Understanding sex differences in form and function of bird song: the importance of studying song learning processes. Front Ecol Evol 4:62

Riede T, Goller F (2010a) Functional morphology of the sound-generating labia in the syrinx of two songbird species. J Anat 216:23–36

Riede T, Goller F (2010b) Peripheral mechanisms for vocal production in birds – differences and similarities to human speech and singing. Brain Lang 115:69–80

Riede T, Goller F (2014) Morphological basis for the evolution of acoustic diversity in oscine songbirds. Proc R Soc B 281:20132306

Riede T, Olson CR (2020) The vocal organ of hummingbirds shows convergence with songbirds. Sci Rep 10:2007

Riede T, Suthers RA (2009) Vocal tract motor patterns and resonance during constant frequency song: the White-throated Sparrow. J Comp Physiol A 195:183–192

Riede T, Beckers GJL, Blevins W, Suthers RA (2004) Inflation of the esophagus and vocal tract filtering in Ring Doves. J Exp Biol 207:4025–4036

Riede T, Suthers RA, Fletcher NH, Blevins WE (2006) Songbirds tune their vocal tract to the fundamental frequency of their song. Proc Natl Acad Sci USA 103:5543–5548

Riede T, Tokuda IT, Munger JB, Thomson SL (2008) Mammalian laryngeal air sacs add variability to the vocal tract impedance: physical and computational modeling. J Acoust Soc Amer 124:634–647

Riede T, Fisher JH, Goller F (2010) Sexual dimorphism of the Zebra Finch syrinx indicates adaptation for high fundamental frequencies in males. PLoS ONE 5: e11368

Riede T, Schilling N, Goller F (2013) The acoustic effect of vocal tract adjustments in Zebra Finches. J Comp Physiol A 199:57–69

Riede T, Forstmeier W, Kempenaers B, Goller F (2015) The functional morphology of male courtship displays in the Pectoral Sandpiper (*Calidris melanotos*). Auk 132:65–77

Riede T, Eliason CM, Miller EH, Goller F, Clarke JA (2016) Coos, booms, and hoots: the evolution of closed-mouth vocal behavior in birds. Evolution 70: 1734–1746

Riska DE (1986) An analysis of vocal communication in the adult Brown Noddy (*Anous stolidus*). Auk 103: 359–369

Ritchison G (1983) Vocalizations of the White-breasted Nuthatch. Wilson Bull 95:440–451

Ritchison G (1991) The flight songs of Common Yellowthroats: description and causation. Condor 93: 12–18

Ritschard M, Brumm H (2012) Zebra Finch song reflects current food availability. Evol Ecol 26:801–812

Roberts TF, Tschida KA, Klein ME, Mooney R (2010) Rapid spine stabilization and synaptic enhancement at the onset of behavioural learning. Nature 463:948–952

Robinson WD (2020) Summer Tanager (*Piranga rubra*), version 1.0. In: Poole AF (ed) Birds of the World. Cornell Lab of Ornithology, Ithaca, NY

Robinson CM, Snyder KT, Creanza N (2019) Correlated evolution between repertoire size and song plasticity predicts that sexual selection on song promotes open-ended learning. eLife 8:e44454

Roca IT, Desrochers L, Giacomazzo M, Bertolo A, Bolduc P, Deschesnes R, Martin CA, Rainville V, Rheault G, Proulx R (2016) Shifting song frequencies in response to anthropogenic noise: a meta-analysis on birds and anurans. Behav Ecol 27:1269–1274

Rödl T, Goymann W, Schwabl I, Gwinner E (2004) Excremental androgen metabolite concentrations and gonad sizes in temperate zone vs. tropical Stonechats (*Saxicola torquata* ssp.). Gen Comp Endocrinol 139: 124–130

Rogers AC, Langmore NE, Mulder RA (2007) Function of pair duets in the Eastern Whipbird: cooperative defense or sexual conflict? Behav Ecol 18:182–188

Rosa GLM, Ellis JM, Bonaccorso E, dos Anjos L (2016) Friend or foe? Social system influences the allocation of signals across functional categories in the repertoires of the New World jays. Behaviour 153:467–524

Rosenfield RN, Bielefeldt J (1991) Vocalizations of Cooper's Hawks during the pre-incubation stage. Condor 93:659–665

Ross MT, Flores D, Bertram R, Johnson F, Wu W, Hyson RL (2019) Experience-dependent intrinsic plasticity during auditory learning. J Neurosci 39:1206–1221

Rumpf M, Tzschentke B (2010) Perinatal acoustic communication in birds: why do birds vocalize in the egg? Open Ornithol J 3:141–149

Rusch DH, Destefano S, Reynolds MC, Lauten D (2020) Ruffed Grouse (*Bonasa umbellus*), version 1.0. In: Poole AF, Gill FB (eds) Birds of the World. Cornell Lab of Ornithology, Ithaca, NY

Ryan MJ, Brenowitz EA (1985) The role of body size, phylogeny, and ambient noise in the evolution of bird song. Am Nat 126:87–100

Ryder TB, Blake JG, Parker PG, Loiselle BA (2011) The composition, stability, and kinship of reproductive coalitions in a lekking bird. Behav Ecol 22:282–290

Saino N, Fasola M (1996) The function of embryonic vocalization in the Little Tern (*Sterna albifrons*). Ethology 102:265–271

Salinas-Melgoza A, Wright TF (2012) Evidence for vocal learning and limited dispersal as dual mechanisms for dialect maintenance in a parrot. PLoS ONE 7:e48667

Samuel DE (1971) Vocal repertoires of sympatric Barn and Cliff swallows. Auk 88:839–855

Saranathan V, Hamilton D, Powell GVN, Kroodsma DE, Prum RE (2007) Genetic evidence supports song learning in the Three-wattled Bellbird *Procnias tricarunculata* (Cotingidae). Mol Ecol 16:3689–3702

Saunders DA (1983) Vocal repertoire and individual vocal recognition in the Short-billed White-tailed Black Cockatoo, *Calyptorhynchus funereus latirostris* Carnaby. Aust Wildl Res 10:527–536

Savage K, Sihm JS, Robinson SK, Stracey CM (2020) Alarm calls of nesting Northern Mockingbirds (*Mimus polyglottos*) are associated with predator type. Wilson J Ornithol 132:608–618

Scarl JC, Bradbury JW (2009) Rapid vocal convergence in an Australian cockatoo, the Galah *Eolophus roseicapillus*. Anim Behav 77:1019–1026

Schmidt MF (2008) Using both sides of your brain: the case for rapid interhemispheric switching. PLoS Biol 6:e269

Schmidt MF, Wild JM (2014) The respiratory-vocal system of songbirds: anatomy, physiology, and neural control. Prog Brain Res 212:297–335

Schmidt KA, Dall SRX, van Gils JA (2010) The ecology of information: an overview on the ecological significance of making informed decisions. Oikos 119:304–331

Schmidt MF, McLean J, Goller F (2012) Breathing and vocal control: the respiratory system as both a driver and a target of telencephalic vocal motor circuits in songbirds. Exp Physiol 97:455–461

Schmidt-Nielsen K (1972) How animals work. Cambridge University Press, New York, NY

Schnell JH (2020) Common Black Hawk (*Buteogallus anthracinus*), version 1.0. In: Poole AF, Gill FB (eds) Birds of the World. Cornell Lab of Ornithology, Ithaca, NY

Schulenberg TS, Stotz DF, Lane DF, O'Neill JP, Parker TA III (2010) Aves de Peru. Sirie Biodiversidad Corbidi, Lima

Schuppe ER, Fuxjager MJ (2018) High-speed displays encoding motor skill trigger elevated territorial aggression in Downy Woodpeckers. Funct Ecol 32:450–460

Schuppe ER, Fuxjager MJ (2019) Phenotypic variation reveals sites of evolutionary constraint in the androgenic signaling pathway. Horm Behav 115:104538

Schuppe ER, Sanin GD, Fuxjager MJ (2016) The social context of a territorial dispute differentially influences the way individuals in breeding pairs coordinate their aggressive tactics. Behav Ecol Sociobiol 70:673–682

Schuppe ER, Petersen JO, Fuxjager MJ (2018) Woodpecker drumming behavior is linked to elevated expression of genes that encode calcium handling proteins in the neck musculature. J Exp Biol 221: jeb180190

Schuppe ER, Rutter AR, Roberts TJ, Fuxjager MJ (2021) Evolutionary and biomechanical basis of drumming behavior in woodpeckers. Front Ecol Evol 9:649146

Schuppe ER, Cantin L, Chakraborty M, Biegler MT, Jarvis ER, Chen CC, Hara E, Bertelsen MF, Witt CC, Jarvis ED, Fuxjager MJ (2022a) Forebrain nuclei linked to woodpecker territorial drum displays mirror those that enable vocal learning in songbirds. PLoS Biol 20: e3001751

Schuppe ER, Tobiansky D, Goller F, Fuxjager MJ (2022b) Specialized androgen synthesis in skeletal muscles that actuate elaborate social displays. J Exp Biol 225: jeb243730

Schwabl H, Dowling J, Baldassarre DT, Gahr M, Lindsay WR, Webster MS (2015) Variation in song system anatomy and androgen levels does not correspond to song characteristics in a tropical songbird. Anim Behav 104:39–50

Schwing R, Parsons S, Nelson XJ (2012) Vocal repertoire of the New Zealand Kea Parrot, *Nestor notablis*. Curr Zool 58:727–740

Searcy WA (1992) Song repertoire and mate choice in birds. Am Zool 32:71–80

Searcy WA, Andersson M (1986) Sexual selection and the evolution of song. Annu Rev Ecol Syst 17:507–533

Searcy WA, Marler P (1981) A test for responsiveness to song structure and programming in female sparrows. Science 213:926–928

Searcy WA, Yasukawa K (1996) Song and female choice. In: Kroodsma DE, Miller EH (eds) Ecology and evolution of acoustic communication in birds. Cornell University Press, Ithaca, NY, pp 454–473

Searcy WA, Yasukawa K (2017) Eavesdropping and cue denial in avian acoustic signals. Anim Behav 124:273–282

Searcy W, Nowicki S, Hughes M (1998) The territory defense function of song in Song Sparrows: a test with the speaker occupation design. Behaviour 135: 615–628

Secora KR, Peterson JR, Urbano CM, Chung B, Okanoya K, Cooper BG (2012) Syringeal specialization of frequency control during song production in the Bengalese Finch (*Lonchura striata domestica*). PLoS ONE 7:e34135

Seddon N, Tobias JA, Alvarez A (2002) Vocal communication in the Pale-winged Trumpeter (*Psophia leucoptera*): repertoire, context and functional reference. Behaviour 139:1331–1359

Seneviratne SS, Jones IL, Miller EH (2009) Vocal repertoires of auklets (Alcidae: Aethiini): structural organization and categorization. Wilson J Ornithol 121:568–584

Sethi VK, Bhatt D (2008) Call repertoire of an endemic avian species, the Indian Chat *Cercomela fusca*. Curr Sci 94:1173–1179

Sethi VK, Bhatt D, Kumar A (2012) Characteristics and behavioral correlates of call types in a tropical bird, the Pied Bush Chat *Saxicola caprata*. Pak J Zool 44:1231–1238

Sewall KB (2009) Limited adult vocal learning maintains call dialects but permits pair-distinctive calls in Red Crossbills. Anim Behav 77:1303–1311

Sewall KB (2011) Early learning of discrete call variants in Red Crossbills: implications for reliable signaling. Behav Ecol Sociobiol 65:157–166

Sewall KB, Young AM, Wright TF (2016) Social calls provide novel insights into the evolution of vocal learning. Anim Behav 120:163–172

Shakya SB, Fuchs J, Pons J-M, Sheldon FH (2017) Tapping the woodpecker tree for evolutionary insight. Mol Phylogenet Evol 116:182–191

Sharp SP, Hatchwell BJ (2005) Individuality in the contact calls of cooperatively breeding Long-tailed Tits (*Aegithalos caudatus*). Behaviour 142:1559–1575

Sharp SP, McGowan A, Wood MJ, Hatchwell BJ (2005) Learned kin recognition cues in a social bird. Nature 434:1127–1130

Shawkey MD, Pillai SR, Hill GE (2003) Chemical warfare? Effects of uropygial oil on feather-degrading bacteria. J Avian Biol 34:345–349

Shields M (2020) Brown Pelican (*Pelecanus occidentalis*), version 1.0. In: Poole AF (ed) Birds of the World. Cornell Lab of Ornithology, Ithaca, NY

Singh P, Price TD (2015) Causes of the latitudinal gradient in birdsong complexity assessed from geographical

variation within two Himalayan warbler species. Ibis 157:511–527
Siriwardhane M (2007) Notes on the natural history of Marshall's Iora. Siyoth 2:19–24
Sjölander S, Ågren G (1976) Reproductive behavior of the Yellow-billed Loon, *Gavia adamsii*. Condor 78:454–463
Skutch AF (1965) Life history of the Long-tailed Silky-flycatcher, with notes on related species. Auk 82:375–426
Slabbekoorn H, den Boer-Visser A (2006) Cities change the songs of birds. Curr Biol 16:2326–2331
Slabbekoorn H, Smith TB (2002) Habitat-dependent song divergence in the Little Greenbul: an analysis of environmental selection pressures on acoustic signals. Evolution 56:1849–1858
Slater PJB, Mann NI (2004) Why do the females of many bird species sing in the tropics? J Avian Biol 35:289–294
Smith WJ (1969) Displays of *Sayornis phoebe* (Aves, Tyrannidae). Behaviour 33:283–321
Smith CL (2017) Referential signaling in birds: the past, present and future. Anim Behav 124:315–323
Smith WJ, Smith AM (1996) Vocal signalling of the Great Crested Flycatcher, *Myiarchus crinitus* (Aves, Tyrannidae). Ethology 102:705–723
Smith GT, Brenowitz EA, Beecher MD, Wingfield JC (1997) Seasonal changes in testosterone, neural attributes of song control nuclei, and song structure in wild songbirds. J Neurosci 17:6001–6010
Smith RJ, Hatch MI, Cimprich DA, Moore FR (2020) Gray Catbird (*Dumetella carolinensis*), version 1.0. In: Poole AF (ed) Birds of the World. Cornell Lab of Ornithology, Ithaca, NY
Snow DW (1963) The evolution of manakin displays. In: Sibley CG, Hickey JJ, Hickey MB (eds) Proc XII Int Ornithol Congr. American Ornithologists' Union, Baton Rouge, LA, pp 553–561
Soard CM, Ritchison G (2009) 'Chick-a-dee' calls of Carolina Chickadees convey information about degree of threat posed by avian predators. Anim Behav 78: 1447–1453
Soha JA, Nelson DA, Parker PG (2004) Genetic analysis of song dialect populations in Puget Sound White-crowned Sparrows. Behav Ecol 15:636–646
Soha JA, Lohr B, Gill DE (2009) Song development in the Grasshopper Sparrow, *Ammodramus savannarum*. Anim Behav 77:1479–1489
Soha JA, Poesel A, Nelson DA, Lohr B (2016) Non-salient geographic variation in birdsong in a species that learns by improvisation. Ethology 4:343–353
Soma KK, Scotti MAL, Newman AE, Charlier TD, Demas GE (2008) Novel mechanisms for neuroendocrine regulation of aggression. Front Neuroendocrinol 29:476–489
Spector DA (1992) Wood-warbler song systems: a review of Paruline singing behaviors. Curr Ornithol 9:199–238
Spector DA (1994) Definition in biology: the case of 'bird song'. J Theor Biol 168:373–381
Spencer KA, Buchanan KL, Leitner S, Goldsmith AR, Catchpole CK (2005) Parasites affect song complexity and neural development in a songbird. Proc R Soc B 272:2037–2043
Sproat TM, Ritchison G (1994) The antipredator vocalizations of adult Eastern Screech-Owls. J Raptor Res 28:93–99
Srivastava KH, Elemans CPH, Sober SJ (2015) Multifunctional and context-dependent control of vocal acoustics by individual muscles. J Neurosci 36:10–14
Staicer CA (1989) Characteristics, use, and significance of two singing behaviors in Grace's Warbler (*Dendroica graciae*). Auk 106:49–63
Stark RD (2002) An analysis of eastern Nearctic woodpecker drums. Ph.D. dissertation, Ohio State University, Columbus, OH
Stark RD, Dodenhoff DJ, Johnson EV (1998) A quantitative analysis of woodpecker drumming. Condor 100: 350–356
Stoddard PK (1996) Vocal recognition of neighbors by territorial passerines. In: Kroodsma DE, Miller EH (eds) Ecology and evolution of acoustic communication in birds. Cornell University Press, Ithaca, NY, pp 356–374
Straight CA, Cooper RJ (2020) Chuck-will's-widow (*Antrostomus carolinensis*), version 1.0. In: Poole AF (ed) Birds of the World. Cornell Lab of Ornithology, Ithaca, NY
Suh A, Paus M, Kiefmann M, Churakov G, Franke FA, Brosius J, Kriegs JO, Schmitz J (2011) Mesozoic retroposons reveal parrots as the closest living relatives of passerine birds. Nat Commun 2:443
Sullivan KA (1984) The advantages of social foraging in Downy Woodpeckers. Anim Behav 32:16–22
Sung H-C, Miller EH, Flemming SP (2005) Breeding vocalizations of the Piping Plover (*Charadrius melodus*): structure, diversity, and repertoire organization. Can J Zool 83:579–595
Suthers RA (1997) Peripheral control and lateralization of birdsong. J Neurbiol 33:632–652
Suthers RA, Hector DH (1985) The physiology of vocalization by the echolocating Oilbird, *Steatornis caripensis*. J Comp Physiol A 156:243–266
Suthers RA, Zollinger SA (2004) Producing song: the vocal apparatus. Ann N Y Acad Sci 1016:109–129
Suthers RA, Zollinger SA (2008) From brain to song: the vocal organ and vocal tract. In: Ziegler HP, Marler P (eds) Neuroscience of birdsong. Cambridge University Press, Cambridge, pp 78–98
Suthers RA, Wild JM, Kaplan G (2011) Mechanisms of song production in the Australian Magpie. J Comp Physiol A 197:45–59
Suthers RA, Rothgerber JR, Jensen KK (2016) Lingual articulation in songbirds. J Exp Biol 219:491–500
Suzuki TN (2011) Parental alarm calls warn nestlings about different predatory threats. Curr Biol 21:R15–R16

Suzuki TN (2012) Calling at a food source: context-dependent variation in note composition of combinatorial calls in Willow Tits. Ornithol Sci 11:103–107

Suzuki TN (2016) Semantic communication in birds: evidence from field research over the past two decades. Ecol Res 31:307–319

Swanson DL, Ingold JL, Galati R (2020a) Golden-crowned Kinglet (*Regulus satrapa*), version 1.0. In: Poole AF (ed) Birds of the World. Cornell Lab of Ornithology, Ithaca, NY

Swanson DL, Ingold JL, Wallace GE (2020b) Ruby-crowned Kinglet (*Regulus calendula*), version 1.0. In: Poole AF (ed) Birds of the World. Cornell Lab of Ornithology, Ithaca, NY

Tanimoto AM, Hart PJ, Pack AA, Switzer R (2017) Vocal repertoire and signal characteristics of 'Alalā, the Hawaiian Crow (*Corvus hawaiiensis*). Wilson J Ornithol 129:25–35

Tarter R (2008) The vocal behavior of the American Crow, *Corvus brachyrhynchos*. M.S. thesis, Ohio State University, Columbus, OH

Taylor H (2010) Blowin'in birdland: improvisation and the Australian Pied Butcherbird. Leonardo Music J 20: 79–83

Taylor B (2016a) Javan Cuckoo-shrike (*Coracina javensis*). In: del Hoyo J, Elliott A, Sargatal J, Christie DA, de Juana E (eds) Handbook of the Birds of the World Alive. Lynx Edicions, Barcelona

Taylor B (2016b) Western Wattled Cuckoo-shrike (*Lobotos lobatus*). In: del Hoyo J, Elliott A, Sargatal J, Christie DA, de Juana E (eds) Handbook of the Birds of the World Alive. Lynx Edicions, Barcelona

Taylor S, Perrin MR (2005) Vocalisations of the Brown-headed Parrot, *Poicephalus cryptoxanthus*: their general form and behavioural context. Ostrich 76:61–72

Temple H (2018) Palmchat (*Dulidae*). In: del Hoyo J, Elliott A, Sargatal J, Christie DA, de Juana E (eds) Handbook of the Birds of the World Alive. Lynx Edicions, Barcelona

Templeton CN, Greene E (2007) Nuthatches eavesdrop on variations in heterospecific chickadee mobbing alarm calls. Proc Natl Acad Sci USA 104:5479–5482

Templeton CN, Greene E, Davis K (2005) Allometry of alarm calls: Black-capped Chickadees encode information about predator size. Science 308:1934–1937

Thomas RJ (2002) The costs of singing in Nightingales. Anim Behav 63:959–966

Thomas RJ, Székely T, Cuthill IC, Harper DGC, Newson SE, Frayling TD, Wallis PD (2002) Eye size in birds and the timing of song at dawn. Proc R Soc B 269:831–837

Thompson WL (1972) Singing behavior of the Indigo Bunting. Z Tierpsychol 31:39–59

Thorpe WH (1958) The learning of song patterns by birds with especial reference to the song of the Chaffinch, *Fringilla coelebs*. Ibis 100:535–570

Thorpe WH, Pilcher PM (1958) The nature and characteristics of subsong. British Birds 51:509–514

Tietze DT, Martens J, Sun Y-H, Päckert M (2008) Evolutionary history of treecreeper vocalisations (Aves: *Certhia*). Org Divers Evol 8:306–324

Tobiansky DJ, Miles MC, Goller F, Fuxjager MJ (2020) Androgenic modulation of extraordinary muscle speed creates a performance trade-off with endurance. J Exp Biol 223:jeb222984

Tobias JA, Lebbin DJ, Aleixo A, Andersen MJ, Guilherme E, Hosner PA, Seddon N (2008) Distribution, behavior, and conservation status of the Rufous Twistwing (*Cnipodectes superrufus*). Wilson J Ornithol 120:38–49

Tobias JA, Aben J, Brumfield RT, Derryberry EP, Halfwerk W, Slabbekoorn H, Seddon N (2010) Song divergence by sensory drive in Amazonian birds. Evolution 64:2820–2839

Tobias JA, Gamarra-Toledo V, García-Olaechea D, Pulgarin PC, Seddon N (2011) Year-round resource defence and the evolution of male and female song in suboscine birds: social armaments are mutual ornaments. J Evol Biol 24:2118–2138

Tobias JA, Sheard C, Seddon N, Meade A, Cotton AJ, Nakagawa S (2016) Territoriality, social bonds, and the evolution of communal signaling in birds. Front Ecol Evol 4:74

Trainer JM, McDonald DB (1993) Vocal repertoire of the Long-tailed Manakin and its relation to male-male cooperation. Condor 95:769–781

Trainer JM, McDonald DB (1995) Singing performance, frequency matching and courtship success of Long-tailed Manakins (*Chiroxiphia linearis*). Behav Ecol Sociobiol 37:249–254

Trainer JM, McDonald DB, Learn WA (2002) The development of coordinated singing in cooperatively displaying Long-tailed Manakins. Behav Ecol 13:65–69

Trainor CR, Verbelen P, Hoste S (2012) Rediscovery of the Timor Bush Warbler *Locustella timorensis* on Alor and Timor, Wallacea: clarifying taxonomic affinities, defining habitat and survey recommendations. Bird Conserv Int 22:354–369

Tramontin AD, Brenowitz EA (2000) Seasonal plasticity in the adult brain. Trends Neurosci 23:251–258

Tremblay JA, Dixon RD, Saab VA, Pyle P, Patten MA (2020) Black-backed Woodpecker (*Picoides arcticus*), version 1.0. In: Rodewald PG (ed) Birds of the World. Cornell Lab of Ornithology, Ithaca, NY

Uchida AM, Meyers RA, Cooper BG, Goller F (2010) Fibre architecture and song activation rates of syringeal muscles are not lateralized in the European Starling. J Exp Biol 213:1069–1078

van den Berg J (1958) Myoelastic-aerodynamic theory of voice production. J Speech Hear Res 1:227–244

van den Heuvel IM, Cherry MI, Klump GM (2014) Land or lover? Territorial defence and mutual mate guarding in the Crimson-breasted Shrike. Behav Ecol Sociobiol 68:373–381

Van Horik J, Bell B, Burns KC (2007) Vocal ethology of the North Island Kaka (*Nestor meridionalis septentrionalis*). N Z J Zool 34:337–345

Van Kampen HS, Hogan JA (2000) Courtship food-calling in Burmese Red Junglefowl: III. Factors influencing the male's behavior. Behaviour 137: 1191–1209

Vanderbilt CC, Kelley JP, DuVal EH (2015) Variation in the performance of cross-contextual displays suggests selection on dual-male phenotypes in a lekking bird. Anim Behav 107:213–219

VanderWerf EA (2020) Hawaii Elepaio (*Chasiempis sandwichensis*), version 1.0. In: Rodewald PG (ed) Birds of the World. Cornell Lab of Ornithology, Ithaca, NY

Varghese AP (2013) Ecology and behaviour of Malabar Trogon *Harpactes fasciatus malabaricus*. M.S. thesis, Mahatma Gandhi University, Kerala, India

Vehrencamp SL (2001) Is song-type matching a conventional signal of aggressive intentions? Proc R Soc B 268:1637–1642

Vehrencamp SL, Ellis JM, Cropp BF, Koltz JM (2014) Negotiation of territorial boundaries in a songbird. Behav Ecol 25:1436–1450

Vehrencamp SL, Ritter AF, Keever M, Bradbury JW (2003) Responses to playback of local vs. distant contact calls in the Orange-fronted Conure *Aratinga canicularis*. Ethology 109:37–54

Volodin IA, Volodina EV, Klenova AV, Filatova OA (2005) Individual and sexual differences in the calls of the monomorphic White-faced Whistling Duck *Dendrocygna viduata*. Acta Ornithol 40:43–52

Wada K, Sakaguchi H, Jarvis ED, Hagiwara M (2004) Differential expression of glutamate receptors in avian neural pathways for learned vocalization. J Comp Neurol 476:44–64

Wade J, Buhlman L (2000) Lateralization and effects of adult androgen in a sexually dimorphic neuromuscular system controlling song in Zebra Finches. J Comp Neurol 426:154–164

Walters JR (1990) Anti-predator behavior of lapwings: field evidence of discriminative abilities. Wilson Bull 102:49–70

Ward S, Slater PJB (2005) Heat transfer and the energetic cost of singing by Canaries *Serinus canaria*. J Comp Physiol A Neuroethol Sens Neural Behav Physiol 191: 953–964

Ward S, Lampe HM, Slater PJB (2004) Singing is not energetically demanding for Pied Flycatchers, *Ficedula hypoleuca*. Behav Ecol 15:477–484

Warrington MH, McDonald PG, Sager AK, Griffith SC (2014) The vocal repertoire of the cooperatively breeding Apostlebird (*Struthidea cinerea*). Emu 114:206–221

Warrington MH, McDonald PC, Griffith SC (2015) Within-group vocal differentiation of individuals in the cooperatively breeding Apostlebird. Behav Ecol 26:493–501

Webb WH, Brunton DH, Aguirre JD, Thomas DB, Valcu M, Dale J (2016) Female song occurs in songbirds with more elaborate female coloration and reduced sexual dichromatism. Front Ecol Evol 4:22

Weir JT, Wheatcroft D (2011) A latitudinal gradient in rates of evolution of avian syllable diversity and song length. Proc R Soc B 278:1713–1720

Weir JT, Wheatcroft DJ, Price TD (2012) The role of ecological constraint in driving the evolution of avian song frequency across a latitudinal gradient. Evolution 66:2773–2783

Welbergen JA, Davies NB (2008) Reed warblers discriminate cuckoos from sparrowhawks with graded alarm signals that attract mates and neighbours. Anim Behav 76:811–822

Wells D, Elliott A, Kirwan GM (2018) Asian Fairy-bluebird (*Irena puella*). In: del Hoyo J, Elliott A, Sargatal J, Christie DA, de Juana E (eds) Handbook of the Birds of the World Alive. Lynx Edicions, Barcelona

Weseloh DV, Hebert CE, Mallory ML, Poole AF, Ellis JC, Pyle P, Patten MA (2020) Herring Gull (*Larus argentatus*), version 1.0. In: Billerman SM (ed) Birds of the World. Cornell Lab of Ornithology, Ithaca, NY

Whitney B (2018) Gnateaters (Conopophagidae). In: del Hoyo J, Elliott A, Sargatal J, Christie DA, de Juana E (eds) Handbook of the Birds of the World Alive. Lynx Edicions, Barcelona

Whitney BM, de Vasconcelos MF, Silveira LF, Pacheco JF (2010) *Scytalopus petrophilus* (Rock Tapaculo): a new species from Minas Gerais, Brazil. Rev Bras Ornitol 18:73–88

Whittaker DJ, Richmond KM, Miller AK, Kiley R, Burns CB, Atwell JW, Ketterson ED (2011) Intraspecific preen oil odor preferences in Dark-eyed Juncos (*Junco hyemalis*). Behav Ecol 22:1256–1263

Whittaker DJ, Gerlach NM, Soini HA, Novotny MV, Ketterson ED (2013) Bird odour predicts reproductive success. Anim Behav 86:697–703

Wiersma P, Muñoz-Garcia A, Walker A, Williams JB (2007) Tropical birds have a slow pace of life. Proc Natl Acad Sci USA 104:9340–9345

Wiley RH, Richards DG (1978) Physical constraints on acoustic communication in the atmosphere: implications for the evolution of animal vocalizations. Behav Ecol Sociobiol 3:69–94

Wiley RH, Richards DG (1982) Adaptations for acoustic communication in birds: sound transmission and signal detection. In: Kroodsma DE, Miller EH (eds) Acoustic communication in birds, vol I. Academic, New York, NY, pp 131–181

Wiley RH, Godard R, Thompson AD Jr (1994) Use of two singing modes by Hooded Warblers as adaptations for signalling. Behaviour 129:243–278

Wilkins HD (1996) The acoustic signals of male and female Red-bellied Woodpeckers: description and causation. M.S. thesis, Eastern Kentucky University, Richmond, KY

Wilkinson R (1990) Vocal behavior and call development in the Bullfinch (*Pyrrhulla pyrrhula*). Bioacoustics 3: 179–197

Williams JG (1955) A systematic revision and natural history of the Shining Sunbird of Africa. Condor 57: 249–262

Williams H (2001) Choreography of song, dance and beak movements in the Zebra Finch (*Taeniopygia guttata*). J Exp Biol 204:3497–3506

Williams JB, Miller RA, Harper JM, Wiersma P (2010) Functional linkages for the pace of life, life-history, and environment in birds. Integr Comp Biol 50:855–868

Willis EO (1982) The behavior of Black-banded Woodcreepers (*Dendrocolaptes picumnus*). Condor 84:272–285

Wilson DR, Evans CS (2012) Fowl communicate the size, speed and proximity of avian stimuli through graded structure in referential alarm calls. Anim Behav 83: 535–544

Wilson PL, Vehrencamp SL (2001) A test of the deceptive mimicry hypothesis in song-sharing Song Sparrows. Anim Behav 62:1197–1205

Witmer MC (2020) Bohemian Waxwing (*Bombycilla garrulus*), version 1.0. In: Billerman SM (ed) Birds of the World. Cornell Lab of Ornithology, Ithaca, NY

Witmer MC, Mountjoy DJ, Elliot L (2020) Cedar Waxwing (*Bombycilla cedrorum*), version 1.0. In: Poole AF (ed) Birds of the World. Cornell Lab of Ornithology, Ithaca, NY

Wolford MJ (1969) Vocal repertoire of the Cactus Wren (*Campylorhynchus brunneicapillus*). M.S. thesis, University of Arizona, Tucson, AZ

Wood SR, Sanderson KJ, Evans CS (2000) Perception of terrestrial and aerial alarm calls by honeyeaters and falcons. Aust J Zool 48:127–134

Wright J, Cuthill I (1992) Monogamy in the European Starling. Behaviour 120:262–285

Wright TF, Wilkinson GS (2001) Population genetic structure and vocal dialects in an Amazon parrot. Proc R Soc B 268:609–616

Yamada K, Soma M (2016) Diet and birdsong: short-term nutritional enrichment improves songs of adult Bengalese Finch males. J Avian Biol 47:865–870

Yamaguchi A (1998) A sexually dimorphic learned birdsong: the Northern Cardinal. Condor 100:504–511

Yamaguchi A (2001) Sex differences in vocal learning. Nature 411:257–258

York JE, Radford AN, de Vries B, Groothuis TG, Young AJ (2016) Dominance-related seasonal song production is unrelated to circulating testosterone in a subtropical songbird. Gen Comp Endocrinol 233:43–52

Yorzinski JL, Vehrencamp SL (2009) The effect of predator type and danger level on the mob calls of the American Crow. Condor 111:159–168

Yu J, Wang L, Xing X, Yang C, Ma J, Møller AP, Wang H, Liang W (2016) Barn Swallows (*Hirundo rustica*) differentiate between Common Cuckoo and sparrowhawk in China: alarm calls convey information on threat. Behav Ecol Sociobiol 70:171–178

Zann R, Dunstan E (2008) Mimetic song in Superb Lyrebirds: species mimicked and mimetic accuracy in different populations and age classes. Anim Behav 76: 1043–1054

Zdenek CN, Heinsohn R, Langmore NE (2015) Vocal complexity in the Palm Cockatoo (*Probosciger aterrimus*). Bioacoustics 24:253–267

Zhang Y-H, Du Y-F, Zhang J-X (2013) Uropygial gland volatiles facilitate species recognition between two sympatric sibling bird species. Behav Ecol 24:1271–1278

Zhao W, Garcia-Oscos F, Dinh D, Roberts TF (2019) Inception of memories that guide vocal learning in the songbird. Science 366:83–89

Zollinger SA, Brumm H (2015) Why birds sing loud songs and why they sometimes don't. Anim Behav 105:289–295

Zub K, Czeszczewik D, Ruczyński I, Kapusta A, Walankiewicz W (2017) Silence is not golden: the hissing calls of tits affect the behaviour of a nest predator. Behav Ecol Sociobiol 71:1–7

GPSR Compliance

*The European Union's (EU) General Product Safety Regulation (GPSR) is a set of rules that requires consumer products to be safe and our obligations to ensure this.*

*If you have any concerns about our products, you can contact us on ProductSafety@springernature.com*

In case Publisher is established outside the EU, the EU authorized representative is:

Springer Nature Customer Service Center GmbH
Europaplatz 3
69115 Heidelberg, Germany

**Batch number: 10756351**

Printed by Printforce, the Netherlands

Fascinating Life Sciences

This interdisciplinary series brings together the most essential and captivating topics in the life sciences. They range from the plant sciences to zoology, from the microbiome to macrobiome, and from basic biology to biotechnology. The series not only highlights fascinating research; it also discusses major challenges associated with the life sciences and related disciplines and outlines future research directions. Individual volumes provide in-depth information, are richly illustrated with photographs, illustrations, and maps, and feature suggestions for further reading or glossaries where appropriate. Interested researchers in all areas of the life sciences, as well as biology enthusiasts, will find the series' interdisciplinary focus and highly readable volumes especially appealing.

Gary Ritchison

# In a Class of Their Own

## A Detailed Examination of Avian Forms and Functions

Volume 3

Springer

Gary Ritchison
Richmond, KY, USA

ISSN 2509-6745 ISSN 2509-6753 (electronic)
Fascinating Life Sciences
ISBN 978-3-031-14851-4 ISBN 978-3-031-14852-1 (eBook)
https://doi.org/10.1007/978-3-031-14852-1



This Springer imprint is published by the registered company Springer Nature Switzerland AG
The registered company address is: Gewerbestrasse 11, 6330 Cham, Switzerland

Paper in this product is recyclable.

# Preface

This book represents my attempt to provide an up-to-date examination of what we know about the structure and function of birds and to do so in a way that is useful and understandable to a wide variety of readers. Readers will find that some chapters are more detailed than others, largely due to differences in the amount of relevant literature available for different topics and, to a much lesser degree perhaps, my interest in the different topics. Those who read multiple (or all) chapters in the book will find some overlap, with some topics covered in more than one chapter. In such cases, topics are covered in much more detail in particular chapters and to a lesser degree in others. I wanted each chapter to tell a particular story and, to do so, felt it is necessary to sometimes mention the same topic in different chapters (and, to emphasize important points, to sometimes mention the same topic more than once in individual chapters). In addition, with the e-version of the book, interested readers may only wish to read certain chapters. In such cases, they will learn something about topics that might be covered in more detail in chapters they do not read. In terms of organization, all chapters cover aspects of avian form and function, but the initial chapters (after a first chapter about the evolution of birds) focus more on structure (e.g., the skeleton, skeletal muscles, integument, and so on), whereas later chapters include more information about function. In many chapters, I also point out where I think additional study is needed. Each chapter includes multiple figures and photos that I hope complement and add to the information provided in the text. Finally, because scientific names are used more than common names in many parts of the world, and to aid those living in those countries, scientific names of species are often provided with common names more than once throughout each chapter.

Richmond, KY Gary Ritchison

# Acknowledgments

I sincerely thank the many reviewers for comments and suggestions that were incredibly helpful in many ways, including correcting errors (but, of course, any that remain are entirely my own and please, in case there might be another edition of this book sometime in the future, let me know of any such errors by email, i.e., ritchisong@gmail.com), providing useful references, suggesting ways to improve chapter organization, and improving the quality of the writing, among others. Thanks also to the many individuals who provided permission to use their figures and photos in the book. I also thank my editor, Lars Koerner, for providing assistance (and access to books published by Springer) whenever I asked and finding the many helpful reviewers. Writing this book has taken many years and so, most of all, I want to thank my family for their understanding and for listening to me when, during evening meals and on many other occasions, I felt the need to tell them something that I'd learned about birds during my research for the book. Although they often smiled, and sometimes even (briefly) laughed (oh no, not again!), they never once told me to please not continue (at least that's how I remember things). To be specific, thanks to my wonderful children, Brandon Tyler and Brianna Carol, for sharing my interest in birds and nature, for all the great times we had when you were kids, and for all the great times we've had since then. Some of my best memories include the many times our family spent in the woods and elsewhere hiking and enjoying nature. During the final stages of writing this book, Brandon's wife, Lisa, gave birth to their son, Carter James. Congratulations to the proud parents, and I look forward to introducing Carter to the wonderful world of birds (Brandon is already introducing him to the wonderful world of Star Trek)! Finally, a big, big thank you to my wife, Tameria. She never complained about the many times and many hours I was in the field studying birds, in the office writing about birds, or engaged in the many other things required for those in academics. After long days working on the book (and other assorted duties), I always looked forward to getting home to spend time with her. In addition to being a great wife, mother, and, more recently, grandmother, she also shares my love of birds and nature! Thanks Tam, you're the best! Finally, thanks to all of you, Tameria, Brandon, Brianna, Lisa, and Carter, for so many happy memories, and I look forward to making many more such memories in the future.

# Contents

# Contents for Volume 1

# Contents for Volume 2

# Migration

# 13

## Contents



**Abstract**

About half of the approximately 10,000 species of birds are classified as migrants. Ideas about the origins of avian migration are discussed in this chapter and, for present-day birds, the distances, routes, and heights that migrants fly are explained. In mountainous regions, many species of birds are altitudinal migrants and the reasons why birds exhibit such behavior are discussed. The reasons why some migrants follow loop and figure-eight migration routes are also explained. Also discussed is the importance of stopover sites for migrants. Other topics covered in this chapter include bird migration in the Neotropics and seasonal differences in the speed of migration. Many migratory species exhibit



G. Ritchison, *In a Class of Their Own*, Fascinating Life Sciences,
https://doi.org/10.1007/978-3-031-14852-1_13

protandry, with males arriving in breeding areas before females and factors contributing to such behavior are discussed. Reasons why birds migrate during the day versus at night are explained, and the chapter closes with a discussion of the possible effects of climate change on bird migration.

## 13.1 Introduction

Birds have likely been migrating for millions of years. Although anatomical evidence available from fossils suggests that birds were likely capable of migratory flights at least 100 million years ago, Steadman (2005) suggested that landbirds probably did not migrate long distances until a period of global cooling in the late Eocene and Oligocene (about 40–25 million years ago). After a warmer interval from 25–15 million years ago, earth began to cool and, because passerine diversity also increased during this period, it is possible that long-distance migration was not a widespread phenomenon until the mid-Miocene (15–11 million years ago; Steadman 2005).

Migration continues to be a widespread phenomenon, with more than half of the world's approximately 10,000 species of birds classified as migrants (Berthold 1998). However, the percentage of bird species that exhibit migratory behavior varies with latitude. For example, Newton (2003a, b) examined the migratory behavior of birds that breed in eastern North America and found a clear trend, with a higher percentage of birds that breed at higher latitudes being migratory species (Fig. 13.1). Similar trends are found in other areas of the Northern Hemisphere (Kuo et al. 2013; Fig. 13.2). Birds in the Southern Hemisphere also migrate (Fig. 13.3), but such migration (called austral migration) tends to be shorter in distance than for many birds in the Northern Hemisphere and also tends to exhibit more variation with respect to routes and patterns (Dingle 2008). In mountainous regions of the world, many species of birds exhibit altitudinal migration, breeding at higher altitudes and spending the nonbreeding period at lower altitudes (Boyle 2008b).

Before proceeding further, some definitions of terms are needed. Migration involves movement of course, but not all bird movements can be called migration. Migration is the regular, endogenously controlled, seasonal movement of birds between breeding and nonbreeding areas (Salewski and Bruderer 2007). Some birds make less predictable nonmigratory movements in response to proximate environmental factors, but such movements are generally termed irruptions or nomadic movements rather than migration (although also sometimes referred to as facultative migration, e.g., Watts et al. 2017). For example, whenever spruce seeds are scarce or absent over much of their normal breeding range in the boreal forests of the western Palearctic, large numbers of Red Crossbills (*Loxia curvirostra*) move long distances (often more than 1000 km and up to 3000 km) toward southern and western Europe (Newton 1972, 2006; Marquiss 2002). Such irruptions are irregular, but not infrequent; at least 40 such irruptions were known to have occurred during the 120-year period from 1881 to 2000 (Newton 2006). Other species of birds classified as irruptive species and whose movements are also triggered by declining food supplies include several that feed on seeds or fruit, such as Bramblings (*Fringilla montifringilla*), waxwings (*Bombycilla* spp.), and Pine Siskins (*Spinus pinus*; Strong et al. 2015), and some raptors, such as Snowy (*Bubo scandiacus*) and Great Gray (*Strix nebulosa*) owls.

Migration distance varies widely among different species of birds (Fig. 13.4). Many migratory birds are well known for their long-distance journeys. For example, many species of birds, such as Purple Martins (*Progne subis*), Common Nighthawks (*Chordeiles minor*), and Bobolinks (*Dolichonyx oryzivorus*), breed in North America and spend the winter in South America, with round-trip journeys of over 20,000 km. However, some migratory species of birds exhibit variation among populations or among individuals within populations in tendency to migrate and migration

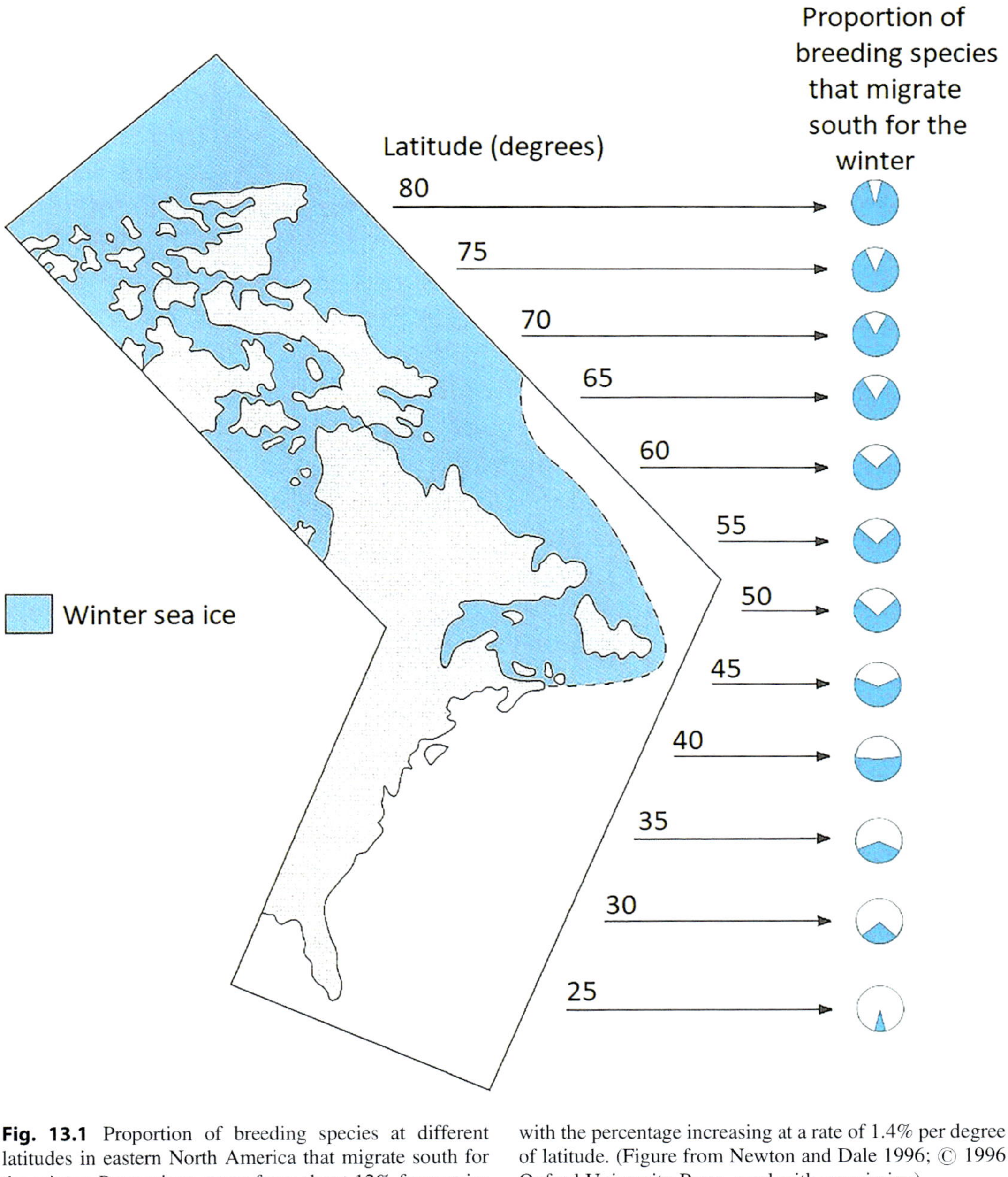


**Fig. 13.1** Proportion of breeding species at different latitudes in eastern North America that migrate south for the winter. Proportions range from about 12% for species that breed at 25°N to 87% for species breeding at 80°N, with the percentage increasing at a rate of 1.4% per degree of latitude. (Figure from Newton and Dale 1996; © 1996 Oxford University Press, used with permission)

distances (Fig. 13.5). For example, some species are partial migrants, with some individuals in a population migrating and others being sedentary. Other species exhibit differential migration, with differences among individuals in migration patterns based on age or sex. Still other species exhibit leap-frog migration, with populations or subspecies breeding at higher latitudes wintering further south (i.e., "leap-frogging") than those breeding at lower latitudes (Figs. 13.6 and 13.7).

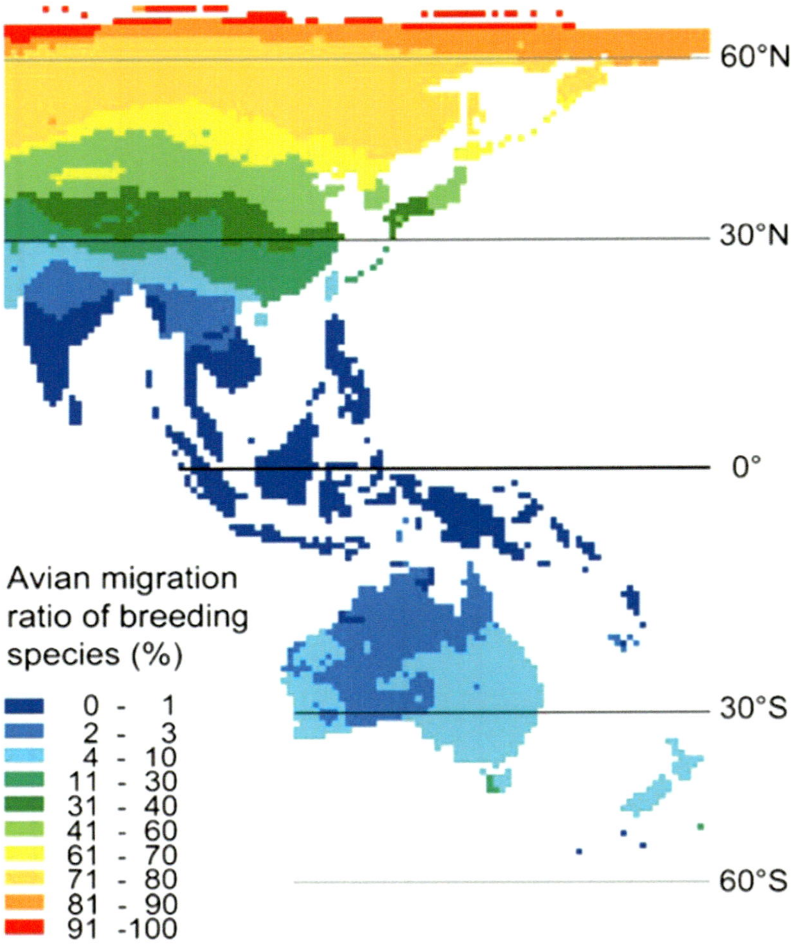


**Fig. 13.2** The percentage of birds that are migratory in east and southeast Asia increases with latitude in the Northern Hemisphere. Few species of birds in the Southern Hemisphere migrate, likely due to paleogeographical isolation and a stable and warm paleoclimate. (Figure from Kuo et al. 2013; © 2013 Springer-Verlag Berlin Heidelberg, used with permission)

Finally, some species take different routes during spring and fall migration, i.e., loop or figure-eight migrations (Figs. 13.8 and 13.9). Thus, birds exhibit a diversity of migratory behaviors and many investigators have addressed the question of how such behaviors evolved.

## 13.2 Origin of Avian Migration

Several lines of evidence suggest that many birds have an innate capacity to migrate. For example, several nonmigratory, resident species of birds, including African Stonechats (*Saxicola torquata*; Helm and Gwinner 2006), Silvereyes (*Zosterops lateralis*; Chan 1994), and White-crowned Sparrows (*Zonotrichia leucophrys*; Smith et al. 1969), have been found to exhibit migratory restlessness (or zugunruhe). In addition, comparative studies indicate that migratory behavior has evolved repeatedly in different avian lineages (Helbig 2003; Outlaw et al. 2003; Joseph 2005; Davis et al. 2006; Outlaw and Voelker 2006; Dufour et al. 2020).

Based on such observations and studies, some investigators have suggested that migration has been a common and widespread characteristic of birds for many millions of years and, if so, then many present-day birds may have inherited the capacity to migrate from their ancestors (Berthold 1999). Dufour et al. (2020) estimated that the ancestor of Passeriformes (~ 47 million years

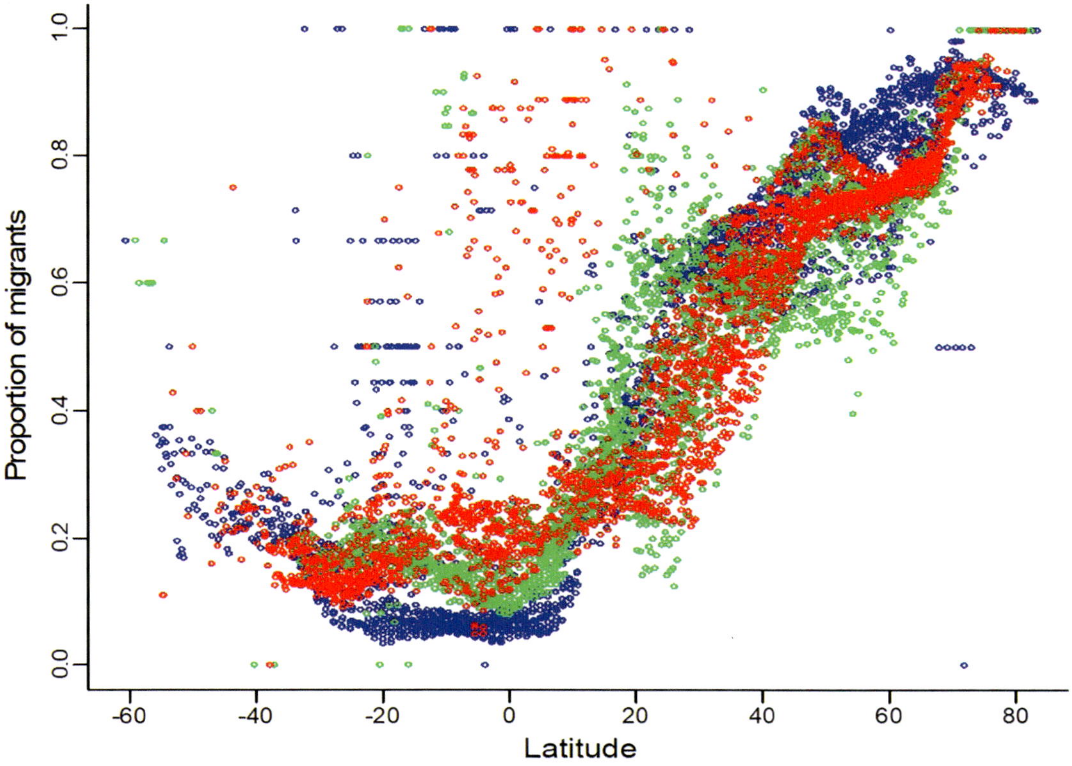


**Fig. 13.3** Proportion of species that breed at different latitudes that are migratory. Asymmetry in proportions between Northern and Southern Hemispheres is likely due, at least in part, to the fact that climate seasonality is more extreme in the Northern than in the Southern Hemisphere because of the relative amount of land versus ocean is much smaller in the Southern Hemisphere (i.e., the greater area of ocean has a buffering effect on climate). The Americas Flyway is in blue, the African-Eurasian Flyway is in green, and the East Asian-Australasian Flyway in red. (Figure from Somveille et al. 2013; © 2013 Somveille et al., open-access article distributed under the terms of the Creative Commons Attribution License, https://creativecommons.org/licenses/by/4.0/)

ago, Oliveros et al. 2019) was migratory, and that migratory behavior first occurred about 22.5 million years ago in the order Charadriiformes and 6.6 million years ago in the order Anseriformes (Box 13.1 Origin of Migratory Species and Populations). Hypotheses proposed to explain the origin of migration by birds include the "southern-home" and "northern-home" hypotheses (Cox 1968; Gauthreaux 1982; Levey and Stiles 1992; Rappole 1995; Bell 2000; Salewski and Bruderer 2007; Winger et al. 2012). The "southern-home" hypothesis posits that long-distance migrants evolved from resident species of birds in the tropics that shifted their breeding ranges poleward to avoid competition and gain access to resources available (and, perhaps, the longer periods of daylight available to access those resources; Sockman and Hurlbert 2020) during temperate latitude springs and summers. The "northern-home" hypothesis proposes that migratory behavior evolved as resident species at higher latitudes began wintering at lower latitudes to avoid the harsh conditions of high-latitude winters (Fig. 13.10). What these two hypotheses have in common is that migration likely evolved in response to the development of

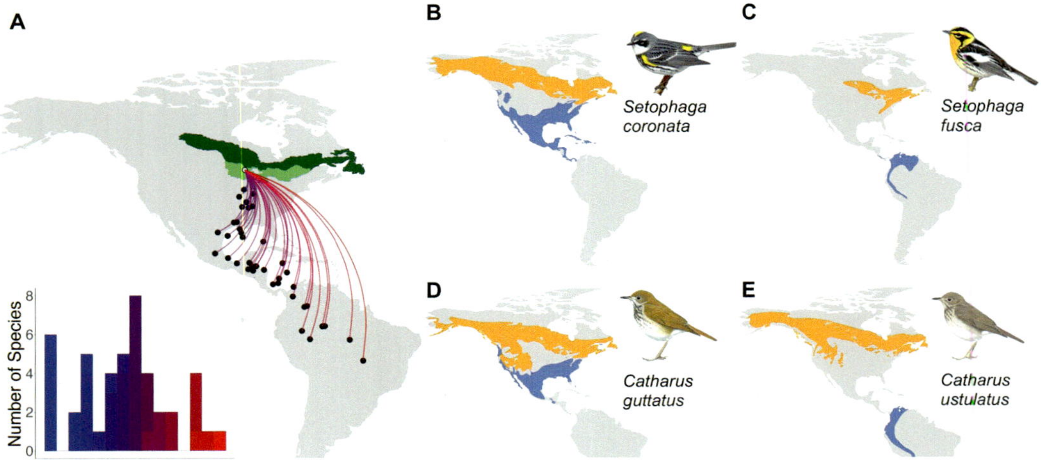


**Fig. 13.4** (**a**) Variation in migration distances among 45 species of birds that breed in the boreal forests of North America, but exhibit much variation in how far they migrate to nonbreeding locations. Dark green area is coniferous forest; light green area is coniferous-deciduous forest transition area. Hypothetical migration routes originate at a single breeding location in northern Minnesota (USA). Blue to red transition in the histogram indicates increasing migration distances. (**b–e**) Examples of species in the same genus with different migration distances, with **b** and **d** representing short-distance migrants and **c** and **e** representing long-distance migrants. *Setophaga coronata*, Yellow-rumped Warbler; *Setophaga fusca*, Blackburnian Warbler; *Catharus guttatus*, Hermit Thrush; *Catharus ustulatus*, Swainson's Thrush. (Figure from Winger and Pegan 2021; © 2021 Oxford University Press, used with permission)

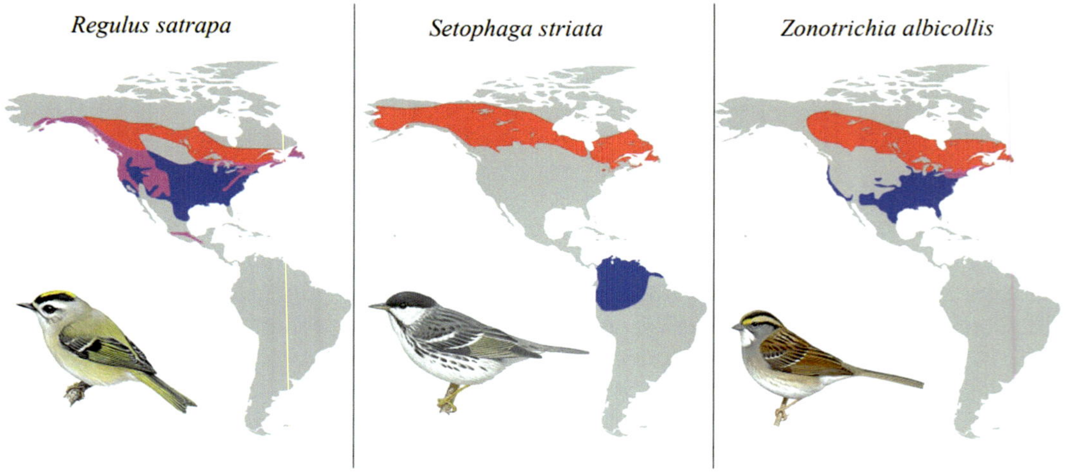


**Fig. 13.5** Examples of variation among species of birds in migratory behavior. Golden-crowned Kinglets (*Regulus satrapa*) breed in boreal forests in North America (red), with both resident populations (purple) and short-distance migrants (blue). Blackpoll Warblers (*Setophaga striata*) are long-distance migrants that breed in far northern North America (red) and winter in northern South America (blue). White-throated Sparrows (*Zonotrichia albicollis*) migrate short distances from breeding areas in Canada and the northern United States (red) to wintering areas in the United States. (Figure from Winger et al. 2019; © 2018 Cambridge Philosophical Society, used with permission)

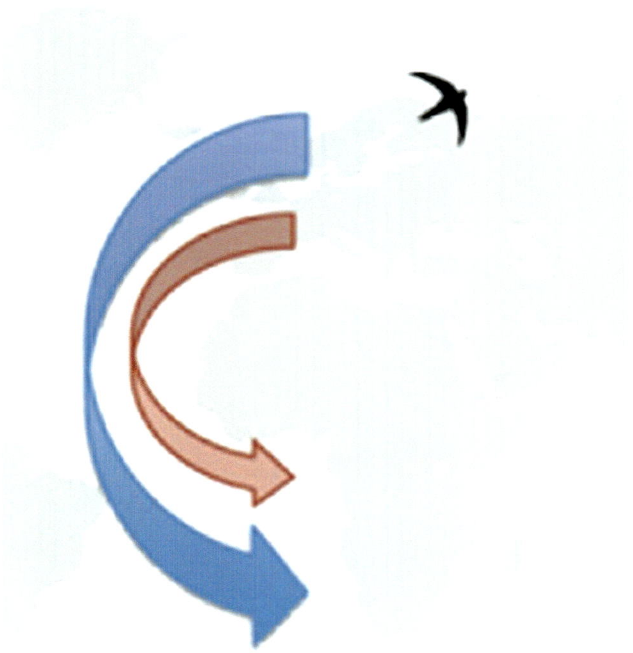

**Fig. 13.6** Example of leap-frog migration where all individuals in two breeding populations of Common Swifts (*Apus apus*) migrate, but those in the more northern population migrate farther and "leap-frog" individuals from the more southerly breeding population. Another type of leap-frog migration is where a migratory population located north of a sedentary population migrates to a location south of the sedentary population. (Figure from Åkesson et al. 2020; © 2020 The Authors. Evolution © 2020 The Society for the Study of Evolution, used with permission)

seasonal environments, with substantially different climates at different latitudes. An important difference is that the "southern-home" hypothesis focuses on competition and access to resources, whereas the "northern-home" hypothesis focuses on seasonality; the fitness benefits of moving from areas with unfavorable conditions to areas with more favorable conditions outweigh the costs of migration (Dokter et al. 2018; Winger et al. 2019).

For millions of years, natural selection has acted upon some or all of the physiological and behavioral components important in avian migration, maintaining a diversity of migratory behaviors, including, often, its suppression (i.e., nonmigratory behavior). For example, based on an ancestral state reconstruction of migratory behavior using a well-supported phylogeny of species in the family Parulidae, Winger et al. (2012) concluded that "the ancestral wood-warbler was migratory and that losses of migration have been at least as prevalent as gains throughout the history of Parulidae." However, by removal of the suppression of those genes controlling the various components of migration, migratory behavior can quickly reappear in a population or species. Among living birds, then, the expression of migratory activity is subject to selective pressures, with those pressures determining if birds are sedentary, short-distance migrants, or long-distance migrants. Although additional study can always alter current ideas or hypotheses, the importance of hypotheses concerning the origin of evolution may "mostly depend on the evolutionary history of the clades under consideration" (Winger et al. 2014, 2019). For example, the "northern-home" hypothesis appears to best explain the origin of migration for species in the superfamily Emberizoidea (Winger et al. 2014), whereas the "southern-home" hypothesis appears to best explain the origin of migration among bee hummingbirds (Tribe Mellisugini; Licona-Vera and Ornelas 2017).

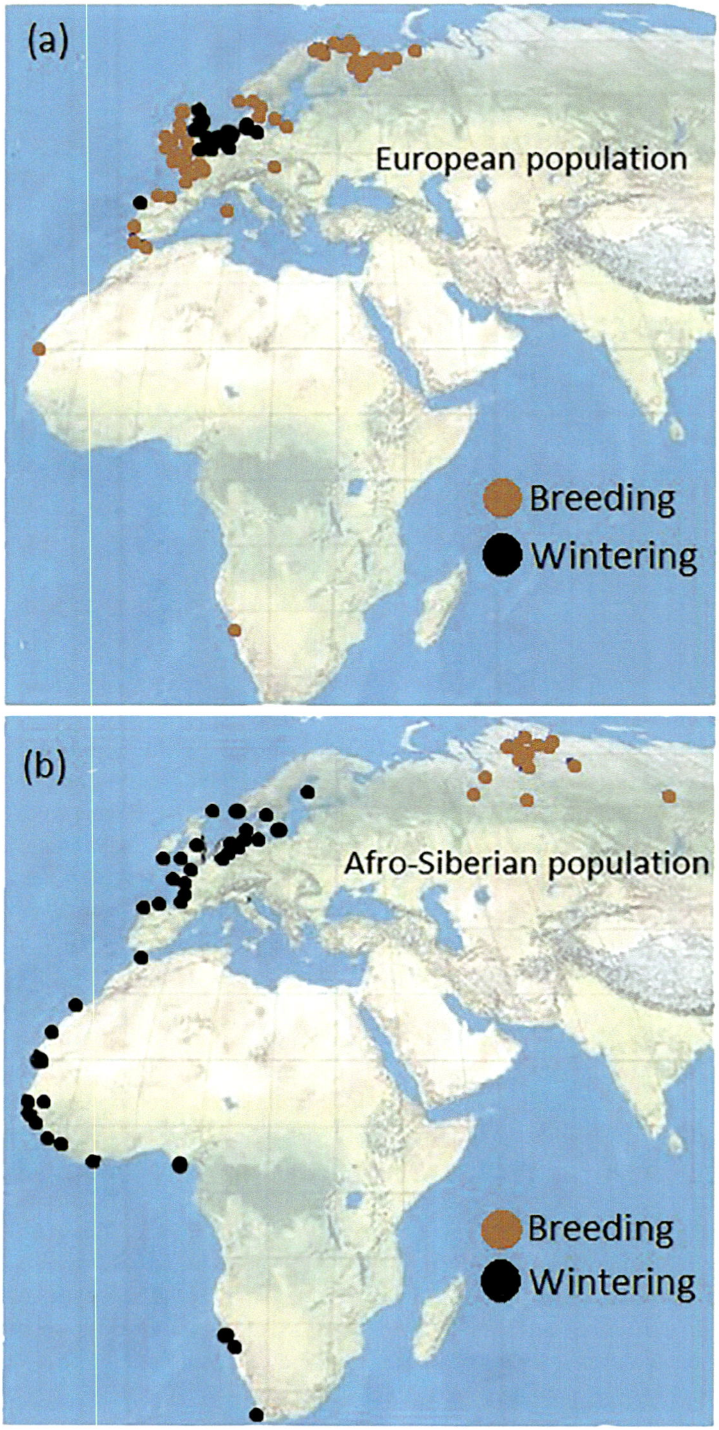


**Fig. 13.7** (**a**, **b**) Leap-frog migration by two populations of Bar-tailed Godwits (*Limosa lapponica*). The population that breeds in Siberia (Afro-Siberian population) winters much further south (west coast of Africa) than the population the largely breeds in Europe (European population). (Figure modified from Duijns et al. 2012; © Netherlands Ornithologists' Union, used with permission)

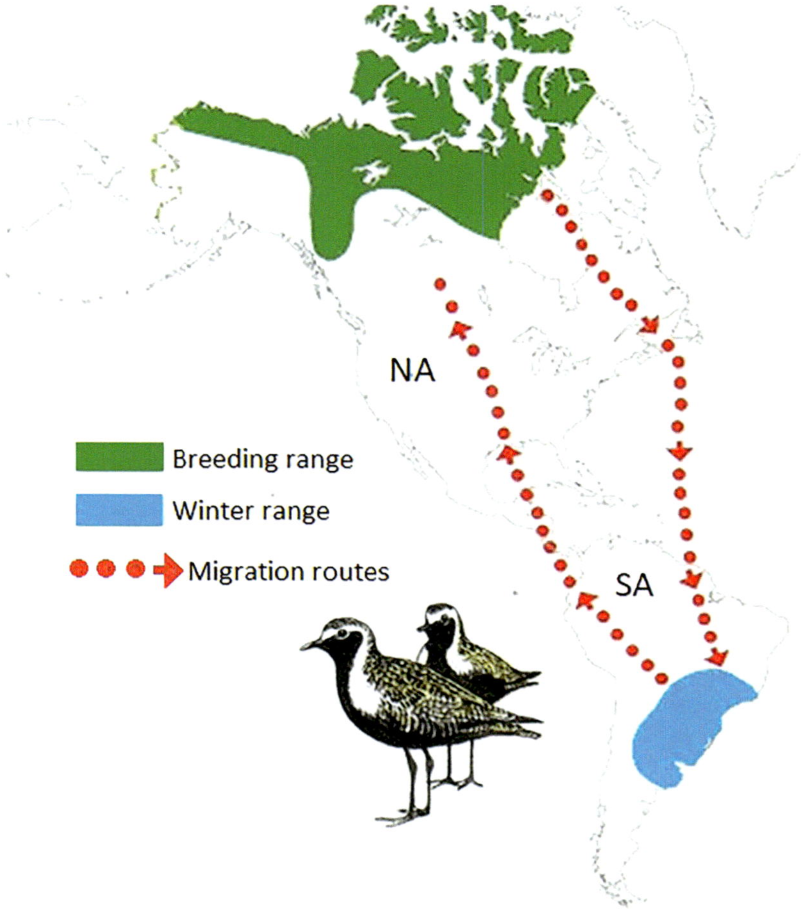


**Fig. 13.8** Migration routes of American Golden-Plovers (*Pluvialis dominica*). Many of these plovers take a nonstop route across the Atlantic Ocean during fall migration, whereas they migrate further west during spring migration, moving north through South and Central America, across the Gulf of Mexico and through the United States and Canada to their breeding areas. (Figure from U. S. Geological Survey, CC0 Public Domain)

## 13.3 Migratory and Sedentary Behaviors of Present-Day Birds

Migrating is costly in terms of energy and many birds die during migration. For example, Strandberg et al. (2010) used satellite telemetry to monitor four species of raptors crossing the Sahara Desert while attempting to migrate from Europe to Africa and found that 31% of all juveniles and 2% of adults died en route. Klaassen et al. (2014) tracked individuals of three species of raptors using satellite telemetry and determined that mortality rates were about six times higher during migration seasons than during stationary periods (Fig. 13.11). For Black-throated Blue Warblers (*Setophaga caerulescens*), apparent mortality rates were found to be 15 times higher during migration than during stationary periods of the year, and more than 85% of apparent annual mortality occurred during migration (Sillett and Holmes 2002). Similarly, an estimated 62% of the estimated annual mortality of Willow Flycatchers (*Empidonax traillii*) occurred during migration (Paxton et al. 2017). Possible causes of mortality during migration include increased risk of predation in unfamiliar habitats (Hope et al. 2014; Sabal et al. 2021), disease, exhaustion, navigation errors, and food shortages at stopover sites (Newton 2008). In addition, weather-related mass mortality events can also occur during migration. Newton (2007a) summarized previously documented cases of bird mortality during migration and examples include more than 10,000

**Fig. 13.9** (Top) Cory's Shearwater (*Calonectris diomedea*). (Bottom) Annual migrations of Cory's Shearwaters, with each line corresponding to one individual. Colored circles indicate locations of three different breeding populations. The white solid line on the left represents the approximate route and direction of most birds. Movements were dominated by prevailing trade winds and westerlies; prevailing winds circulate clockwise in the Northern Hemisphere and counter-clockwise in the Southern Hemisphere. Calm, oligotrophic areas were avoided. Wintering areas were associated with upwellings and high rates of primary production. (Photo from Wikipedia, CC0 Public Domain; Bottom figure from González-Solís et al. 2007; © The Ecological Society of America, used with permission)

Magnolia Warblers (*Setophaga magnolia*) and other warblers killed in a rainstorm off the Texas coast in 1981, more than 20,000 songbirds killed as a result of dense fog off the coast of Sweden in 1998, and an estimated 200,000 jays, thrushes, and warblers killed during a rainstorm over Lake Manitoba. Dionne et al. (2008) calculated a total reduction of about 50% in the breeding population of Chimney Swifts (*Chaetura pelagica*) in the province of Quebec, Canada, in the year after a hurricane developed in the Caribbean during the fall migration period. Other investigators have also reported mass mortalities of bird during migration (e.g., Diehl et al. 2014; Yang et al. 2021). For migratory species, the benefits of migrating must exceed these costs and, therefore, any explanation for the evolution of migration requires identification of those benefits.

As already noted, birds that migrate vary in the extent to which they migrate (e.g., all vs. part of the population) and in the distances traveled. With such variation, and given that thousands of different species of birds migrate, it is almost certainly the case that different selective factors influence the migratory behavior of present-day

**Box 13.1 Origin of Migratory Species and Populations**

Zink (2002, 2011) suggested that avian migratory behavior has likely developed and been lost many times throughout the evolutionary history of birds and, as a result, birds may possess an innate migratory program (i.e., "migratory syndrome") that, given certain environmental conditions, allows sedentary populations or species of birds to become migratory. Indeed, the results of experimental studies have revealed that migratory behavior can develop or be suppressed in just a few generations (Berthold et al. 1992; Mueller et al. 2011). As an example of how rapidly migratory behavior can develop, House Finches (*Haemorhous mexicanus*) from a largely sedentary population in southern California were introduced on Long Island, New York, in about 1940 and, by the early 1960s, many of these eastern House Finches had become migratory (Able and Belthoff 1998).

Male House Finch. (Photo by John Benson, Wikipedia, CC BY 2.0, https://creativecommons.org/licenses/by/2.0/)

(continued)

**Box 13.1** (continued)

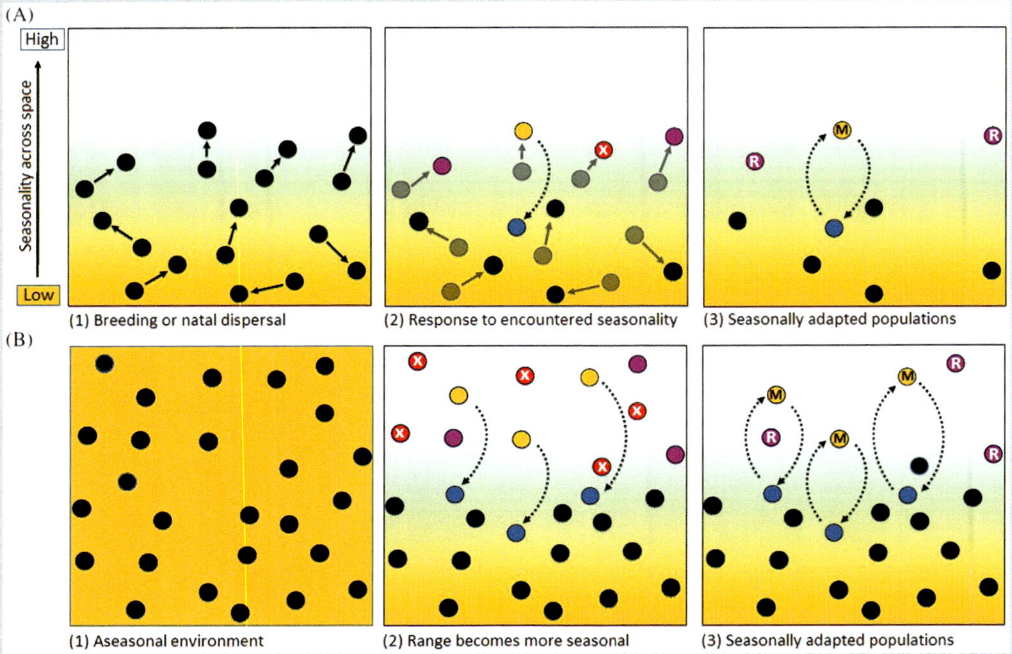


(**a**) Migratory populations of bird may be the result of dispersal of individuals into more seasonal environments. (1) Birds in relatively aseasonal environments may disperse, either as adults (breeding dispersal) or young (natal dispersal), into more seasonal environments and establish breeding territories. (2) As environmental conditions deteriorate in the more seasonal area, birds that move back to the more aseasonal area with better conditions during the nonbreeding period may be more likely to survive. Of those that remain, some may survive (purple) and others may not (red x). (3) Given the benefits of fidelity to breeding sites, birds that move back to their breeding territories in the more seasonal area (orange M) may be more likely to successfully reproduce, thus becoming migratory. Those that remained in the more seasonal environment and survived (purple R) may become residents and, over time, become better adapted to the more seasonal environment. (**b**) Migratory populations of birds could also be the result of an increase in seasonality over time. If aseasonal environments become more seasonal over time, then individuals may respond either by becoming migratory (orange M) or adapting to the changing environment and remaining residents (purple R). (Figure modified from Winger et al. 2019; © 2018 Cambridge Philosophical Society, used with permission)

birds. However, some factors are likely more important than others, and those most commonly linked to the origin of migration include seasonal variation in food availability, direct climatic effects on physiological condition, and the risk of nest predation (Fretwell 1980; Cox 1985; Alerstam 1990; Berthold 2001). Variation in food availability could favor migration by allowing exploitation of a seasonal peak in availability during the breeding season, forcing movement out of unproductive areas, or both. Arctic Terns (*Sterna paradisaea*), with the longest migration route of any bird (about 40,000 km; Hatch et al. 2020), represent an extreme example of this. These terns migrate between maximally productive areas (in terms of daily solar energy reaching the earth's surface), breeding in productive latitudes in the Northern

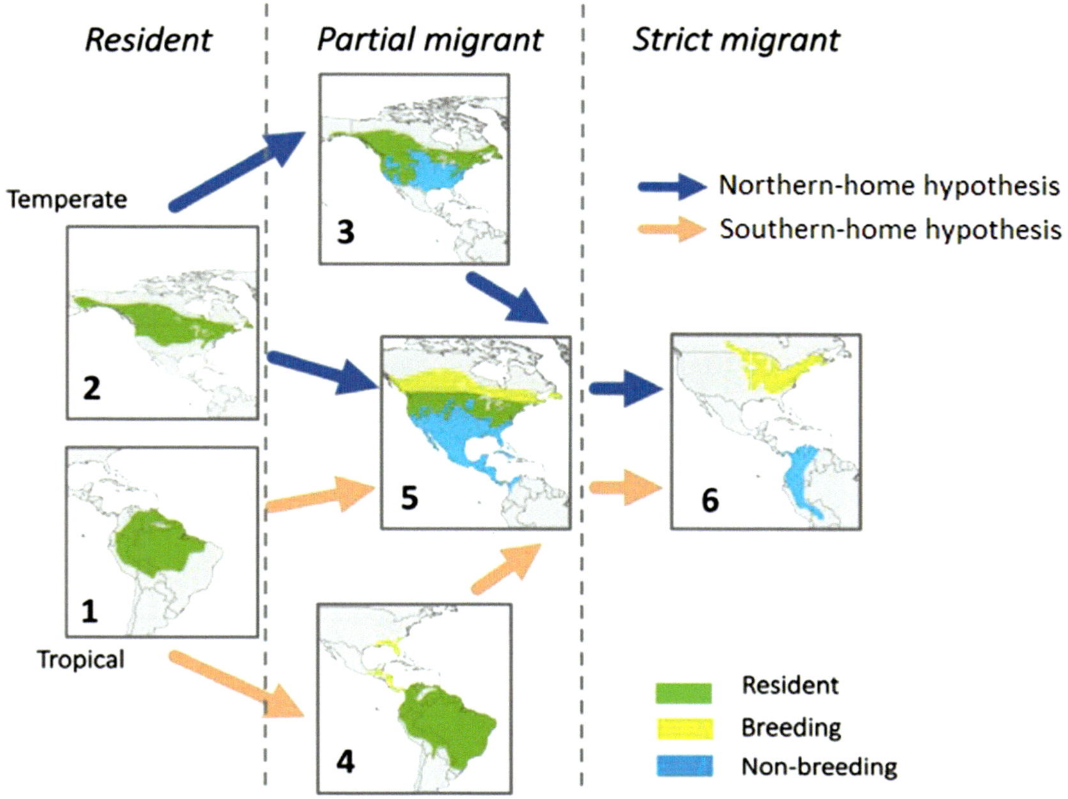


**Fig. 13.10** Hypotheses to explain the evolution of migratory behavior. For both hypotheses, resident birds may have first become partial migrants before eventually becoming strict migrants. Possible stages in the evolution of migration are shown in maps 1 to 6. (Figure from Dufour et al. 2020; © 2019 John Wiley & Sons Ltd., used with permission)

Hemisphere and wintering in equally productive latitudes in the Southern Hemisphere (Fig. 13.12). The results of an analysis based on the distribution of the world's terrestrial bird species also suggest that birds (1) migrate to breed in areas with a surplus of resources, preferring areas with harsh winters and, presumably, less competition from resident species of birds, and (2) migrate to nonbreeding areas with milder winters that reduce winter mortality (Somveille et al. 2015). The results of this analysis also suggest that migrant birds tend to winter in suitable areas that are as close to breeding areas as possible, which would explain why the highest concentration of migrants (most of which are found in the Northern Hemisphere) is in the southern latitudes of the Northern Hemisphere (Fig. 13.13). Climate could also contribute to the origin of migration if seasonal variation in temperatures affected breeding success or survival rates. Migration would also be selected for if risk of nest failure due to predation varied predictably with location, e.g., with latitude or altitude. Of course, these factors are not mutually exclusive and each, to varying degrees, might contribute to the origin of migratory behavior (Boyle and Conway 2007).

Levey and Stiles (1992) proposed that bird migration between the Nearctic and Neotropics arose from the tendency of some species to move in response to changes in resource abundance. Specifically, they noted that many short-distance

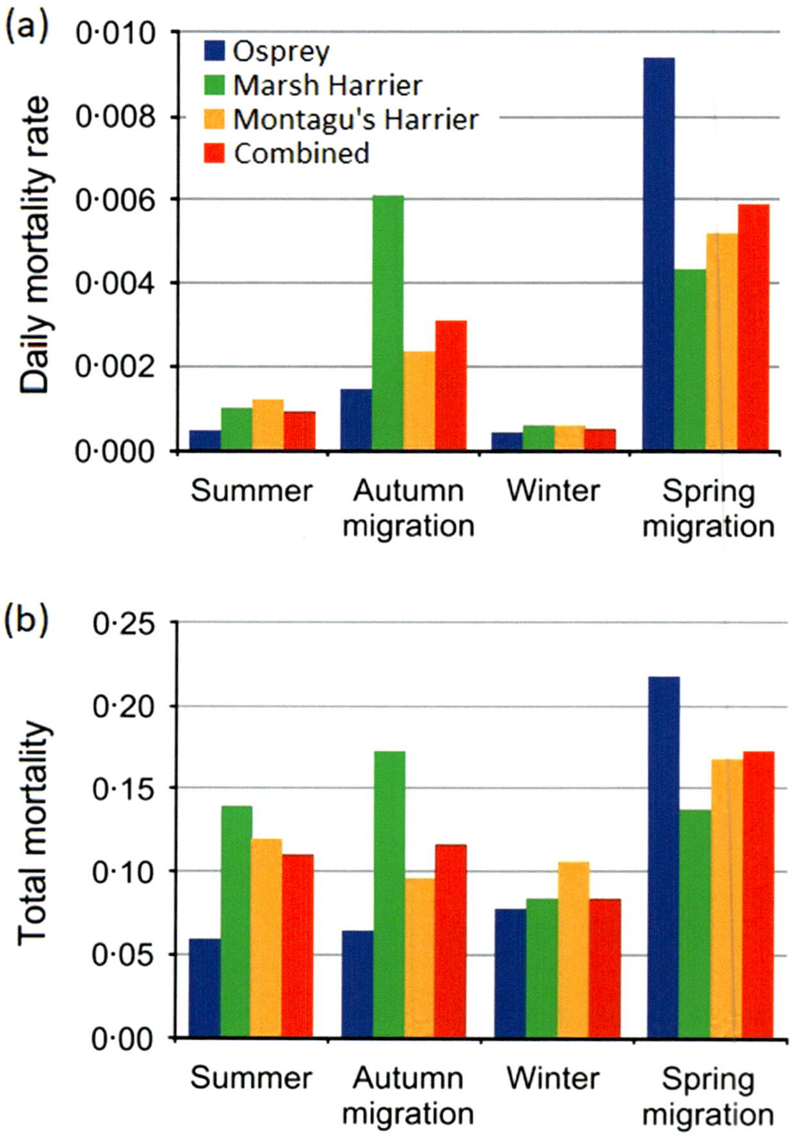


**Fig. 13.11** (**a**) Daily mortality rates for adults in three species of raptors were significantly higher during autumn and spring migration than during wintering and breeding (summer) periods. (**b**) Total mortality during the four periods of the year for adults in the three species of raptors. Osprey, *Pandion haliaetus*; Eurasian Marsh-Harrier, *Circus aeruginosus*; Montagu's Harrier, *Circus pygargus*. (Figure from Klaassen et al. 2014; © 2013 The Authors. Journal of Animal Ecology © 2013 British Ecological Society, used with permission)

Neotropical migrants are primarily frugivorous and occupy habitats, including open, nonforested areas, forest edge, and the forest canopy, that exhibit greater fluctuation in temperature and humidity than forest-interior habitat. Fluctuating conditions cause fluctuation in resource availability that favors species that migrate (Somveille et al. 2015; Fig. 13.12). Over time, some birds in lineages dependent on these "fluctuating" habitats and resources and with preexisting tendencies to migrate moved longer distances and became long-distance migrants. Based on this "evolutionary precursor" hypothesis, habitat characteristics contribute to variation in resource availability that favors the evolution of migratory behavior.

In a comparative study involving 379 species in the suborder Tyranni (flycatchers, manikins, cotingas, tityras, and becards), Boyle and Conway (2007) examined the possible

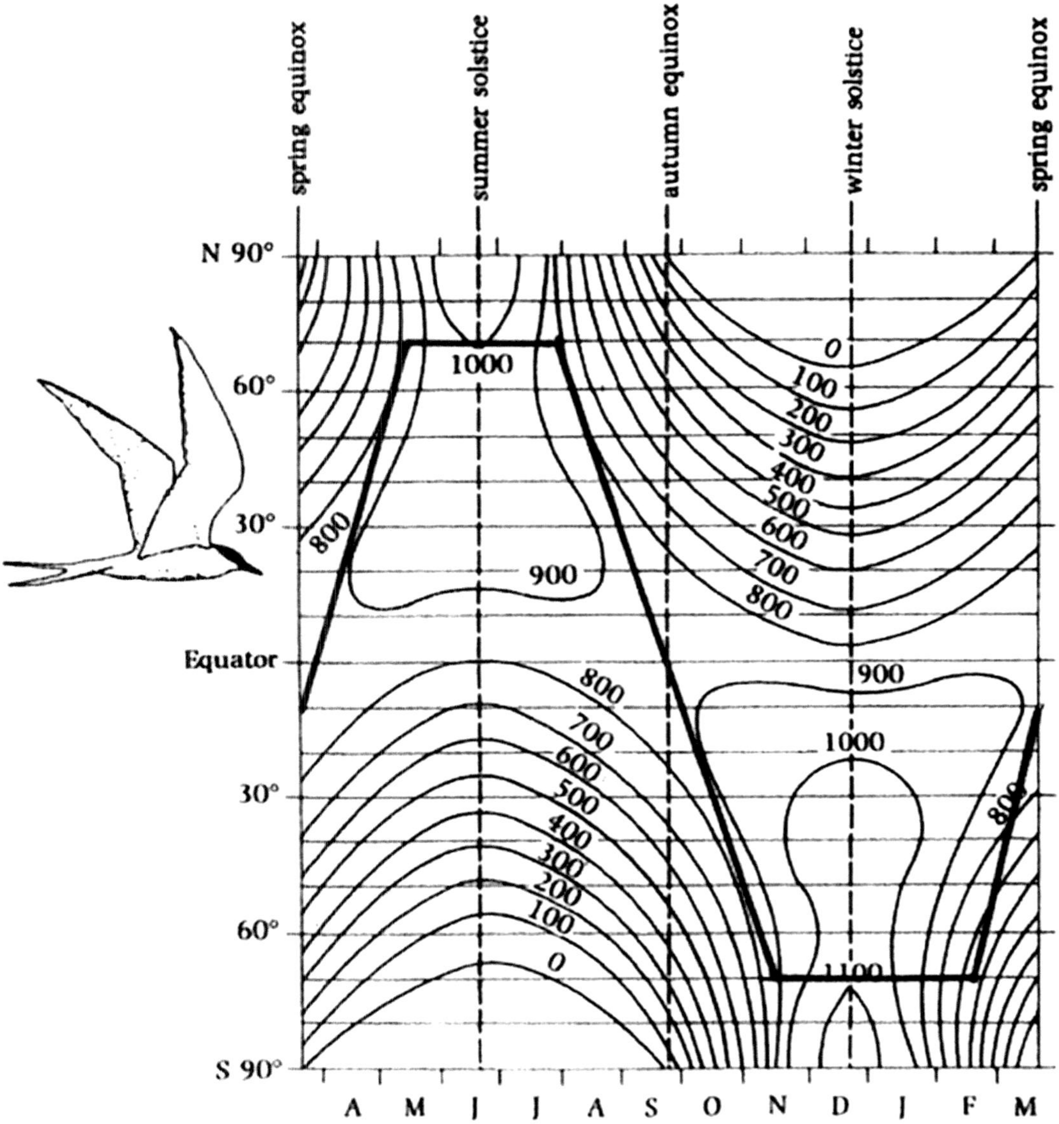


**Fig. 13.12** Daily solar energy (cal $cm^{-2}$) reaching the Earth at different latitudes and times of the year. The dark line shows the migratory pathway of Arctic Terns (*Sterna paradisaea*) through this "energy landscape" between high latitudes in the Northern Hemisphere where they breed and those in the Southern Hemisphere where they molt and spend part of the nonbreeding season. (Figure from Alerstam et al. 2003; © Oikos. Published by John Wiley and Sons, Inc., used with permission)

relationships between habitat, diet, behavior, and migration more broadly, considering both sedentary and migratory species, birds with a wider variety of food habits (frugivores and insectivores) and social organization (solitary vs. flocking species), and birds occupying a wider variety of habitats (ground/thickets, forest understory, forest midstory, canopy, open/arid, and disturbed). Their analysis revealed that habitat and diet influence the

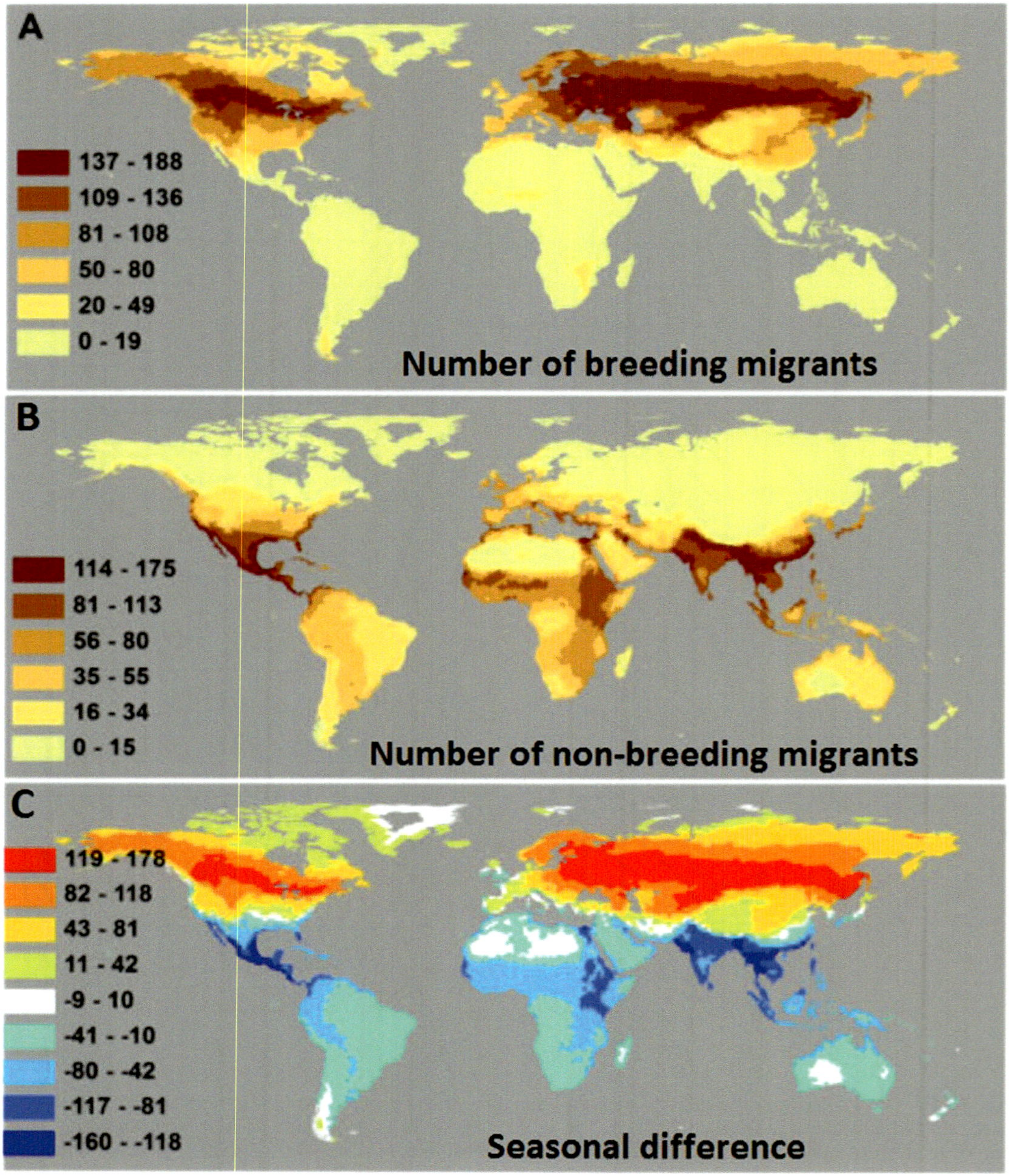


**Fig. 13.13** Global patterns of the diversity of migratory landbirds. (**a**) Number of bird species present only during the breeding season, (**b**) number of bird species present only during their nonbreeding season, and (**c**) number of breeding migrants minus the number of nonbreeding migrants. (Figures modified from Somveille et al. 2015; © 2015 John Wiley & Sons Ltd., used with permission)

tendency of birds to migrate, but in rather complex ways (Fig. 13.14). Species found in thickets and forest canopy and understory were more likely to migrate if frugivorous, but species in open/arid and disturbed habitats were more likely to migrate if insectivorous. Without better

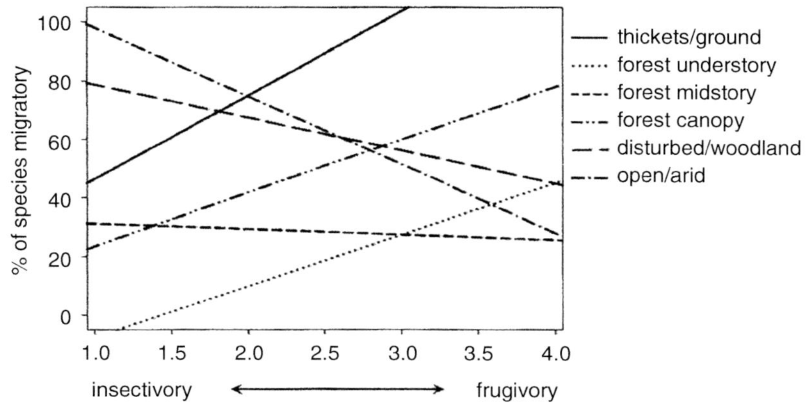


**Fig. 13.14** Percentage of species in the Tyranni that migrate varies with diet (highly insectivorous to highly frugivorous) and habitat use. The six lines illustrate how diet and habitat interact; birds of thickets, forest understory, and canopy are more likely to be migratory if they are frugivorous. In contrast, increasing frugivory is associated with a decreasing likelihood of being migratory for birds of disturbed and arid habitats. Boyle and Conway (2007) plotted linear regression lines for each habitat category based on the proportion of species that migrate for each level of diet along a scale from highly insectivorous (1.0) to highly frugivorous (4.0). (Figure from Boyle 2006, used with permission of Alice Boyle)

information concerning the extent to which food resources vary in different habitats, interpreting these results is difficult. However, one possibility is simply that, regardless of habitat, species depending on food resources that exhibit seasonal variation in availability are more likely to evolve migratory behavior than species depending on less seasonal resources (resource availability hypothesis; Boyle 2006; Boyle and Conway 2007).

In addition to resource availability, another factor that appears to influence migratory behavior is the size of foraging groups. Boyle and Conway (2007) found that species of birds that are solitary foragers are more likely to be migratory than species that forage in pairs or groups. A likely explanation for this relationship between foraging behavior and tendency to migrate is that foraging with conspecifics likely improves foraging efficiency. If so, then migration and foraging with conspecifics may simply represent two alternative strategies for coping with seasonal variation in resources, either leave or forage with others to increase foraging efficiency.

## 13.4 Migration Distance, Routes, and Heights

Migratory birds exhibit great variation in the distance of their migratory journeys, with some short-distance migrants moving just a few hundred kilometers and some long-distance migrants traveling several thousand kilometers. Several factors can contribute to this variation in migration distance among different species and populations. Watanabe (2016) examined the migratory flights of 196 species of birds ranging in mass from 12 to 10,350 grams and found that migration distance declined with increasing body mass for birds that used flapping flight. However, for birds that migrate primarily by soaring, body mass had no effect of migration distance (Fig. 13.15). For flapping birds, Watanabe (2016) also found that body mass had no effect on either the duration of migration or the number of flight legs (number of flights interspersed with periods at stopover sites) (Fig. 13.16). However, migration speed decreased with increasing body mass (Fig. 13.16). As body mass increases, the

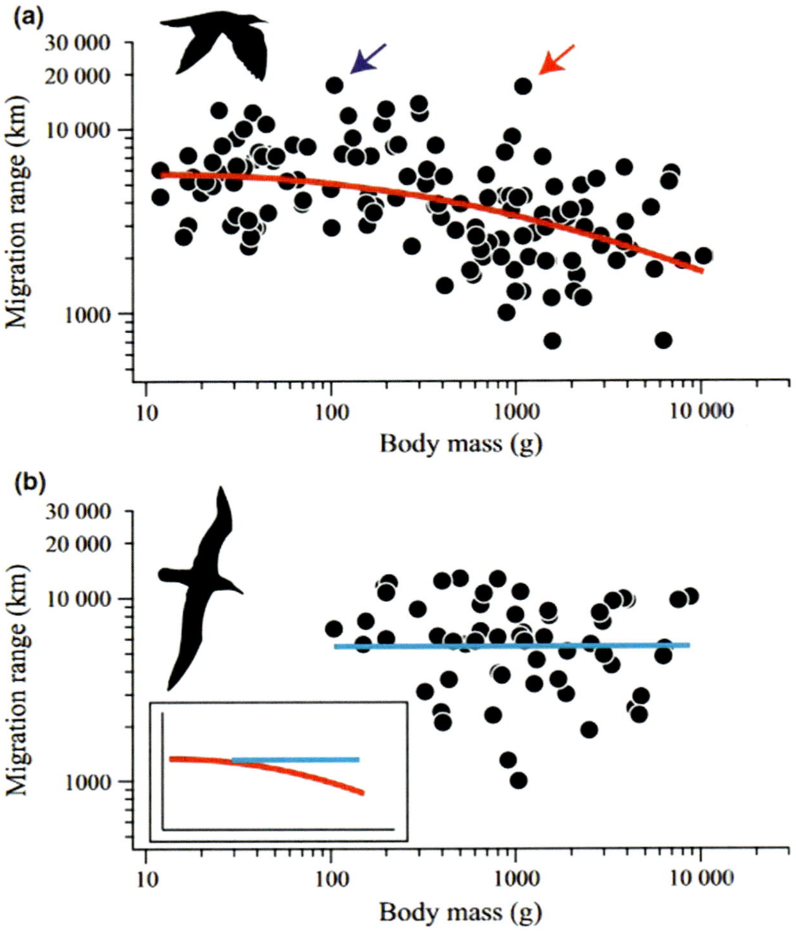


**Fig. 13.15** Migration range as a function of body mass for (**a**) flapping birds and (**b**) soaring birds. Red and blue arrows in (**a**) represent the South Polar Skua (*Stercorarius maccormicki*) and Arctic Tern (*Sterna paradisaea*), respectively. Inset in (**b**) shows the comparison between flapping and soaring birds. (Figure from Watanabe 2016; © 2016 John Wiley & Sons Ltd./CNRS, used with permission)

energy needed for flapping flight increases, likely explaining the shorter migration distances and slower migration speeds of heavier birds that use flapping flight. For birds that soar, energetic costs of flight are similar, regardless of a bird's mass. Other factors can, however, influence the migratory behavior of soaring birds, including the availability of rising air (e.g., thermals or topography like mountain ranges that generate updrafts) for terrestrial birds and, for dynamic soarers in the southern oceans, the size of the earth (Watanabe 2016).

Among birds in the suborder Tyranni, migration distance appears to be influenced by food habits, with insectivorous species tending to migrate longer distances than frugivorous species (Boyle and Conway 2007). Similarly, analysis of the relationship between migration distance and food habits among songbirds of the western Palearctic revealed that insectivorous species tend to migrate greater distances than granivorous (seed-eating) birds or those that fed on both seeds and insects (Fig. 13.17; Newton 1995, 2003b). Year-round insectivory is likely a consequence rather than the cause of long-distance migration in birds (Boyle and Conway 2007). That is, during the breeding season, fruit and seeds are not as abundant at high latitudes as insects and, therefore, fewer frugivorous and granivorous birds breed there. In addition, during the nonbreeding season, fruit and seeds are likely more abundant than insects at mid-latitudes. As a result of the availability of prey, therefore, many insectivorous species breed at higher latitudes and migrate

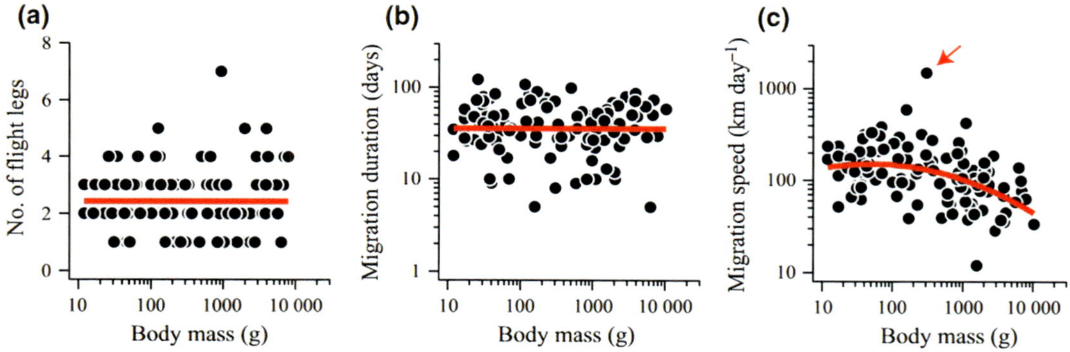


**Fig. 13.16** Relationships between body mass and (**a**) number of flight legs (i.e., number of flights interspersed with stopovers), (**b**) the duration of migration, and (**c**) migration speed (distance divided by duration). Body mass has no effect on either the number of flight legs (average = 2.4) or the duration of migration (average = 35.8 days). However, flight speed tends to decline with increasing body mass. In (**c**), the red arrow indicates a species with exceptional flight speed, the Bar-tailed Godwit (*Limosa lapponica*). They can fly nonstop from Alaska to New Zealand (Gill et al. 2009). (Figure from Watanabe 2016; © 2016 John Wiley & Sons Ltd./CNRS, used with permission)

relatively long distances to winter at low latitudes.

Among raptors, food habits can also influence migration distance. For example, among raptors of the western Palearctic, those that prey on endotherms (or warm-blooded prey; i.e., birds and mammals) tend to migrate shorter distances than those that prey on either ectotherms (or cold-blooded prey; i.e., insects, amphibians, and reptiles) or both ectotherms and endotherms (Fig. 13.18; Newton 2003a). Again, this difference is a result of food availability, raptors that breed at higher latitudes and tend to prey on ectotherms must migrate farther to wintering areas because their primary prey may not be available at high- to mid-latitudes during the non-breeding period.

Although food habits can clearly influence migration distance, for many populations and species of birds, other factors can also be important. Among shorebirds that breed at high latitudes in North America, the risk of nest predation has historically been found to decrease with increasing latitude (McKinnon et al. 2010,

**Fig. 13.17** Migration distance relative to breeding location (latitude) and diet for songbirds of the western Palearctic. Lines were calculated by linear regression analysis based on data for individual species. (Figure from Newton 2003b; © 2003 Springer-Verlag Berlin Heidelberg, used with permission)

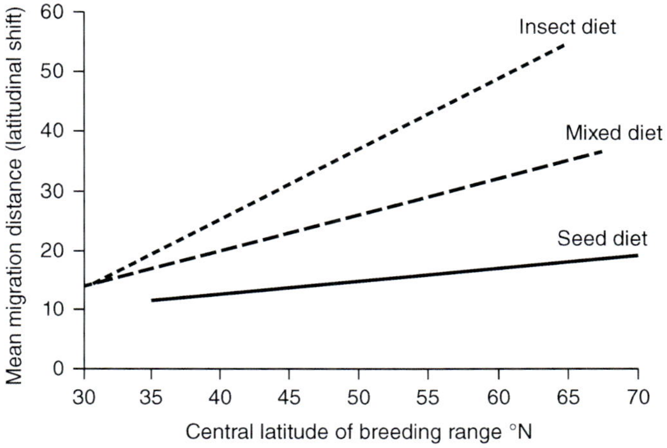

**Fig. 13.18** Migration distances of western Palearctic raptors relative to breeding latitude and diet. Lines calculated by linear regression analyses. (Figure from Newton 2003a; © 2003 Elsevier B.V., used with permission)

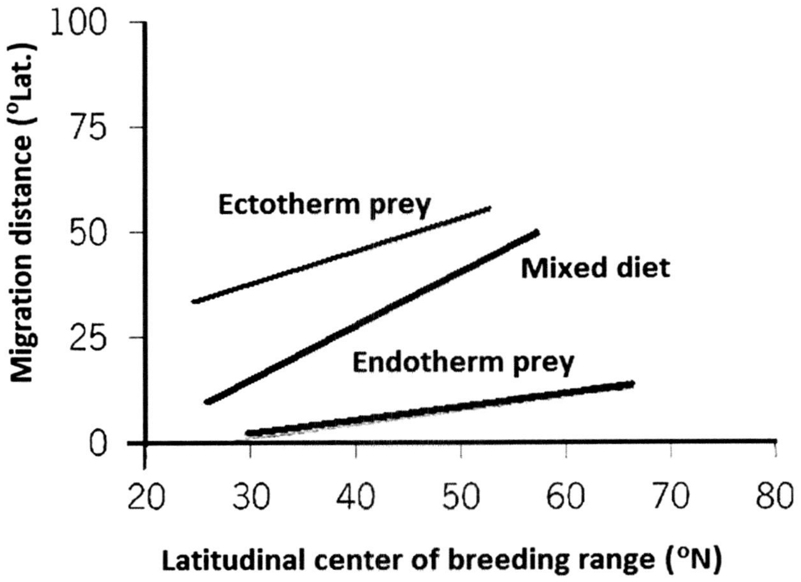


Fig. 13.19; Box 13.2 Effect of Climate Change on Shorebird Nest Predation). As a result, natural selection favored individuals that migrated further to nest at higher latitudes. Of course, migrating further north also entails a greater energetic cost and the need to withstand the harsher environmental conditions, but the benefit of lower rates of nest predation must have exceeded those costs and favored long-distance migration by many species of shorebirds.

Migration distances and specific routes taken can also be influenced by the presence of ecological or geographical barriers, such as large bodies of water, deserts, and mountain ranges (Box 13.3 Afro-Palearctic Landbird Migration). In most cases, the distance across these barriers is within the birds' potential flight range capacity and, in many cases, some, if not most, birds do cross them. For example, numerous Neotropical migrants cross the Gulf of Mexico during migration to and from breeding and wintering areas in North American and Central and South America, and many migrants in the East Asian-Australasian Flyway cross expanses of ocean to reach wintering areas on islands in southeast Asia (Fig. 13.20). However, other birds use longer migration routes to avoid crossing the barriers (Figs. 13.21 and 13.22). For some migrating birds, some barriers simply cannot be crossed. For example, most birds, such as Broad-winged Hawks (*Buteo platypterus*), that rely on soaring flight to travel long distances are dependent on rising air generated by thermals and, therefore, avoid migration routes that would take them over large bodies of water (Haines et al. 2003, Fig. 13.23; Box 13.4 Gliding Speed of Migrating Birds that Rely on Soaring). One apparent exception is Ospreys (*Pandion haliaetus*) that have been reported to use weak thermals when crossing the Mediterranean Sea (Fig. 13.24). For birds physiologically capable of crossing barriers, numerous factors likely influence the decision to either cross or avoid barriers located along their migration routes, including their physical condition, e.g., fat stores and weather conditions (Fig. 13.25), wind speed and direction (Fig. 13.26), and risk of predation (Alerstam 2001; Deppe et al. 2015).

Age and experience can also influence the choice of migratory routes and the timing and speed of migration. For example, older birds have been found to migrate faster than younger birds in some species (McKinnon et al. 2014; Crysler et al. 2016), but not in others (McKinnon et al. 2014; Deppe et al. 2015). Among some species, but not others (e.g., Schmaljohann et al. 2018), first-year birds tend to exhibit more variation in migratory directions than adults (e.g.,

**Box 13.2 Effect of Climate Change on Shorebird Nest Predation**

Historically, rates of nest predation for shorebirds decreased with increasing latitude in the Northern Hemisphere. However, since the year 2000, rates of nest predation at north-temperate latitudes, and especially the Arctic, have increased significantly (Kubelka et al. 2018).

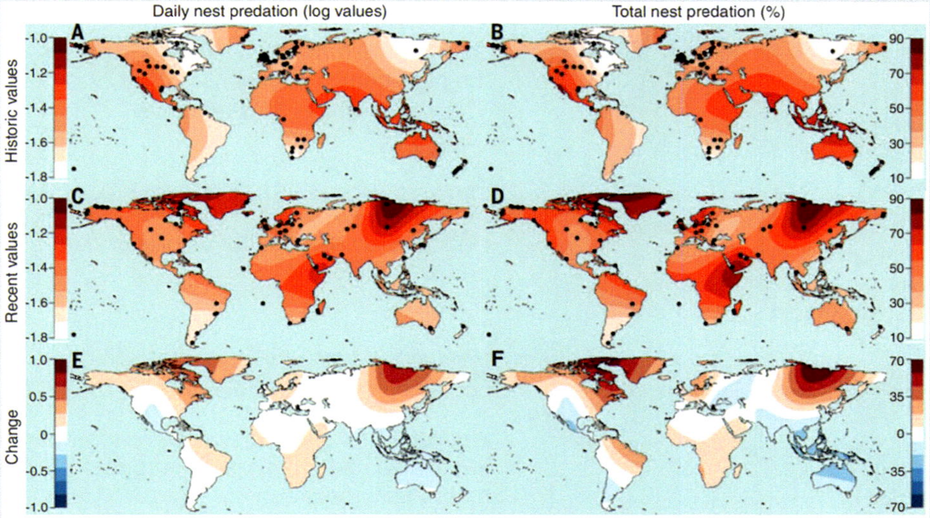


Nest predation in shorebirds. (**a** and **b**) Historic rates of nest predation (1944–1999, 145 populations). (**c** and **d**) Rates of nest predation from 2000 to 2016 (102 populations). (**e** and **f**) Changes between historic and more recent nest predation rates. Dots show study locations. (**a**), (**c**), and (**e**) Daily nest predation (log transformed), and (**b**), (**d**), and (**f**) Total nest predation (percentage). (Figure from Kubelka et al. 2018; © 2018 The Authors, some rights reserved; exclusive licensee American Association for the Advancement of Science, used with permission)

Compared to historic rates, rates of nest predation nearly doubled in the north-temperate zone and tripled in the Arctic during the period from 2000 to 2016 (Kubelka et al. 2018). These increased levels of nest predation correspond with increases in ambient temperatures and variation in temperature resulting from global climate change. Increasing temperatures may contribute to increased rates of nest predation by (1) reducing populations of lemmings and voles as a result of insufficient snow cover (Ims et al. 2008) and, as a result, altering predator-prey interactions and increasing predation pressure on alternative prey, including shorebird nests, (2) altering the behavior and/or distribution of nest predators like foxes (*Vulpes* spp.) in ways that increase their likelihood of preying on shorebird eggs (Blois et al. 2013), and (3) altering the structure of vegetation around shorebird nests and increasing their vulnerability to nest predators (Meltofte et al. 2007). Unfortunately, the increase in rates of nest predation at high latitudes suggests that the Arctic may now be an ecological trap (Schlaepfer et al. 2002) for migrating birds and that populations of Arctic birds will likely decline in the future, not only due to increased rates of nest predation, but also due to a trophic mismatch (Gilg et al. 2012), i.e., eggs may no longer hatch during the period of peak food abundance, a reduction in suitable habitat (Gilg et al. 2012), and reduce survival during migration (Studds et al. 2017).

(continued)

**Box 13.2** (continued)

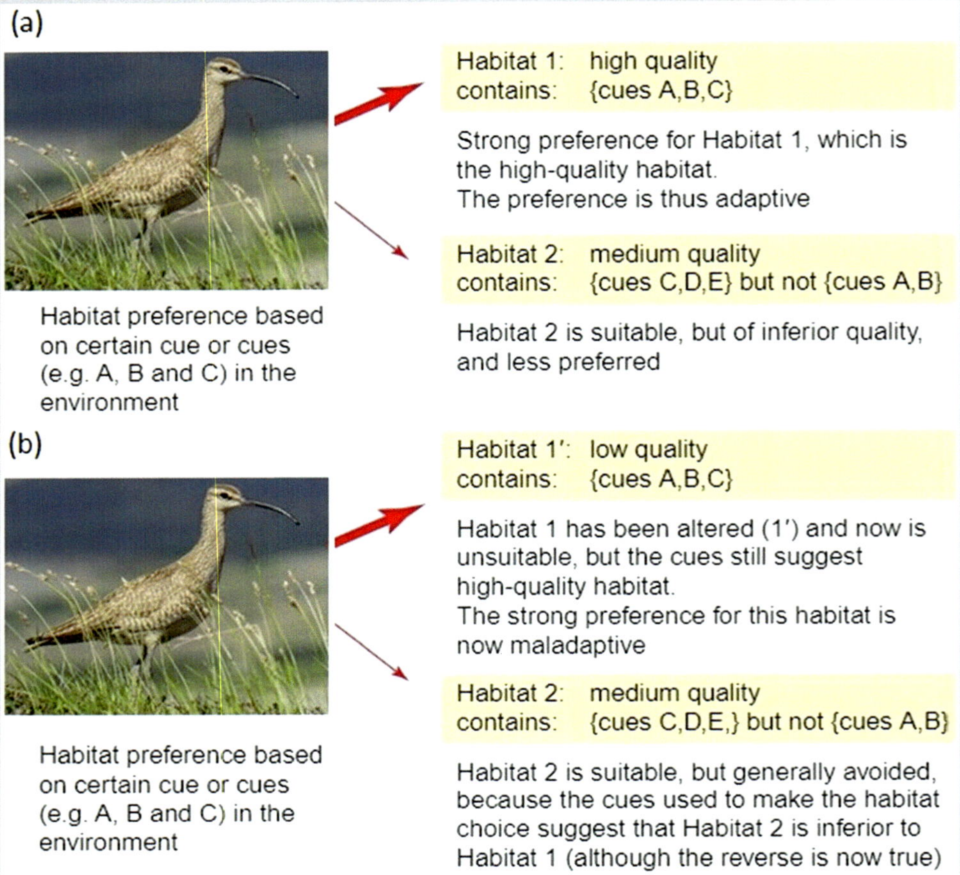


A bird's response to environmental cues in a normal environment (**a**) and an ecological trap (**b**). Each habitat contains numerous potential cues, some of which will be used to make decisions, e.g., cues A, B, and C are used for habitat and nest-site selection. The stronger the cues, the stronger the preference (indicated by the thickness of arrows). An ecological trap can occur when a habitat is degraded by human activities (e.g., climate change), but the original cues persist, misleading birds into behaving, in this example, as if the habitat is still high quality. (Figure modified from Schlaepfer et al. 2002; © 2002 Elsevier Science Ltd., used with permission; Photo of Whimbrel by Rachel Richardson, USGS, Alaska Science Center, CC0 Public Domain)

Helbig 1996; Åkesson et al. 2005). For species where the orientation of first-time migrants and adults is similar during migration, endogenous programs likely orient young birds in the same direction as adults (Åkesson et al. 2021). However, the migration routes of older and younger birds have been found to differ in several species of birds. For example, Crysler et al. (2016) examined the migratory behavior of Ipswich Sparrows (*Passerculus sandwichensis princeps*) that breed on Sable Island in Nova Scotia, Canada, and found that juveniles initiated fall migration about 24 days before adults, and flew at different orientations to the mainland that resulted in

**Fig. 13.19** Mean latitudinal decrease in nest predation risk and map of the shorebird breeding sites where artificial nests were monitored. The decrease in predation risk (3.6% per degree relative to the southernmost site, Akimiski Island) is indicated at 5° intervals on the latitudinal scale at right. (Figure from McKinnon et al. 2010; © 2010 American Association for the Advancement of Science, used with permission)

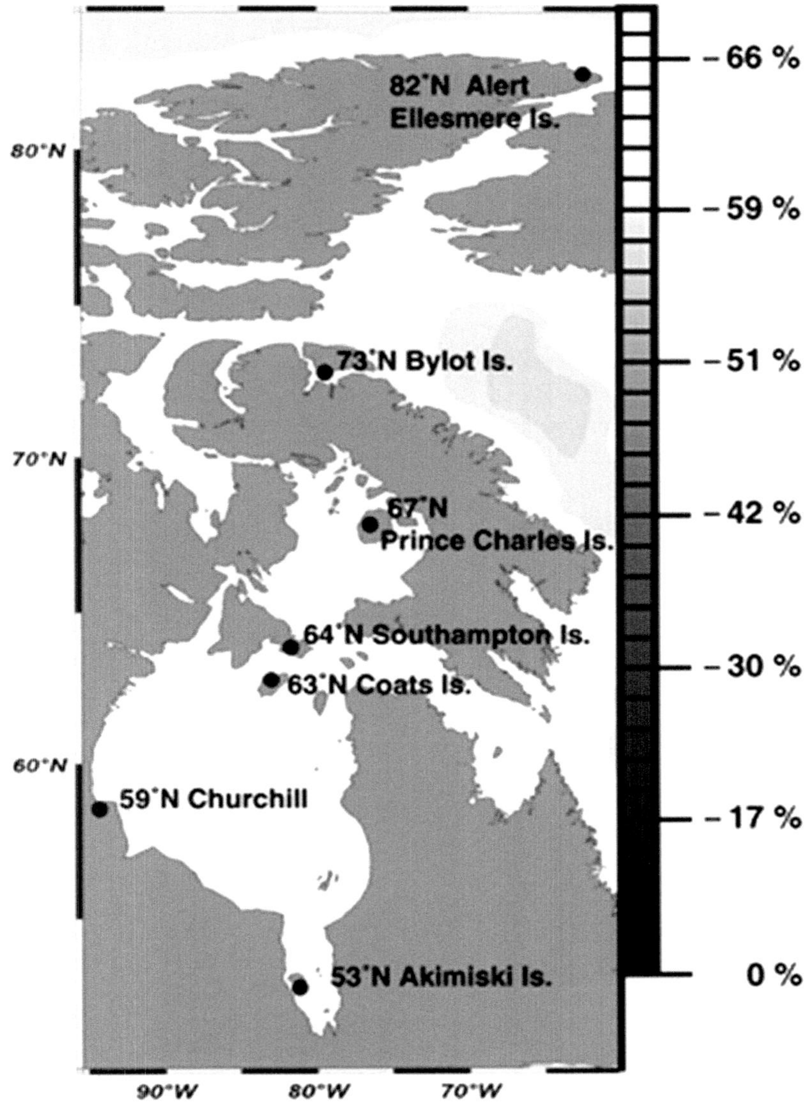


shorter ocean crossings and longer overall distances traveled than adults (Fig. 13.27). Hake et al. (2003) studied the fall migration of satellite-tagged European Honey-buzzards (*Pernis apivorus*) from Sweden to western Africa and found that all adults crossed the Mediterranean Sea at the Strait of Gibraltar whereas juveniles crossed the Mediterranean at a variety of locations. Miller et al. (2016) found that the migration routes of adult Golden Eagles (*Aquila chrysaetos*) in eastern North America were more direct than those of juveniles and subadults (Fig. 13.28). Investigators have reported differences in the migration routes of adults and juveniles in several other species of birds (e.g., Handel and Gill 2010; Mueller et al. 2013; Sergio et al. 2014). With increasing age, the migratory behavior of juveniles becomes similar to that of adults due to both experience and selective mortality (Sergio et al. 2014).

When migrating, birds can encounter a wide range of atmospheric conditions, conditions that can affect their ability to remain on course and schedule (Shamoun-Baranes et al. 2010). As such, birds might be expected to fly at altitudes with the most favorable atmospheric conditions. In general, investigators have found that migrating birds prefer to fly at relatively low altitudes (< 2100 m above ground level, Bruderer et al. 2018; Fig. 13.29), but fly at higher altitudes if wind conditions are more favorable (e.g., Schmaljohann et al. 2009; Kemp et al. 2013; Galtbalt et al. 2021; Fig. 13.30). Flying at lower altitudes can be beneficial because the higher air density increases lift, plus climbing to higher altitudes requires additional energy (Hedenström and Alerstam 1992, Pennycuick 2008). At higher altitudes, reduced air density does reduce friction (drag), but requires a higher flapping frequency that increases the cost of flight (Pennycuick 2008). Oxygen and humidity levels are also higher at lower altitudes; at higher altitudes, with less oxygen and generally lower humidities, increased rates of respiration require additional energy and increase water loss (Carmi et al. 1992; Klaassen 1996). In addition, flying at a lower altitude allows birds to better assess their ground speed and, for some species, provides a better view of landmarks that might be useful for navigational purposes (Galtbalt et al. 2021).

After the preference by most migrants to fly at low altitudes, favorable winds are next in importance. Favorable tailwinds increase flight speed, which can reduce the time needed to complete migration (e.g., Kemp et al. 2010; Åkesson and Bianco 2021; Fig. 13.31) and also reduce the energetic costs of migration (e.g., Butler et al. 1997, Schmaljohann et al. 2009; Fig. 13.32). Despite such benefits, migrating birds do not always fly at heights that maximize the use of favorable tailwinds (Fig. 13.33). As noted by Horton et al. (2016a), "non-aerodynamic constraints, such as costs associated with the time and energy to sample airspace, navigate and stop over…, may cause migrants to seek conditions sufficient, rather than optimal, for flight."

Among the other variables that can influence flight altitudes during migration is temperature. Senner et al. (2018) monitored the migratory flights of four Black-tailed Godwits (*Limosa limosa*) from Europe to Africa using GPS trackers and found that the godwits flew at altitudes above 5000 meters during 21% of their flights and reached maximum altitudes of almost 6000 meters. Flight altitudes appeared to be primarily

**Box 13.3 Afro-Palearctic Landbird Migration**

More than 100 species of birds migrate annually between the western Palearctic and Africa (Newton 2008). These migrants generally travel along two longitudinally parallel flyways between breeding areas and wintering areas, with birds that breed in western Europe typically wintering in western Africa and those breeding in eastern Europe wintering in eastern Africa. These two migration pathways are at opposite ends of two ecological barriers, the Mediterranean Sea and the Sahara Desert. By largely avoiding the central (and broadest) portions of these two barriers, migrants traveling along these two flyways reduce the distance they must travel over these two "inhospitable areas" (Adamik et al. 2016; Briedis et al. 2020). Interestingly, birds using the western flyway initiate fall migration an average of 5 to 7 days earlier than those using the eastern flyway. Briedis et al. (2020) suggested that this difference was likely driven by differences in phenology, with a "… strong SW-NE gradient in the timing of… autumn vegetation senescence across Europe that is likely to result from the degree of continentality, that is, by how much the climate of a region is influenced by the advection of air masses from the ocean and from the landmass (Franzén 1991)."

(continued)

**Box 13.3** (continued)

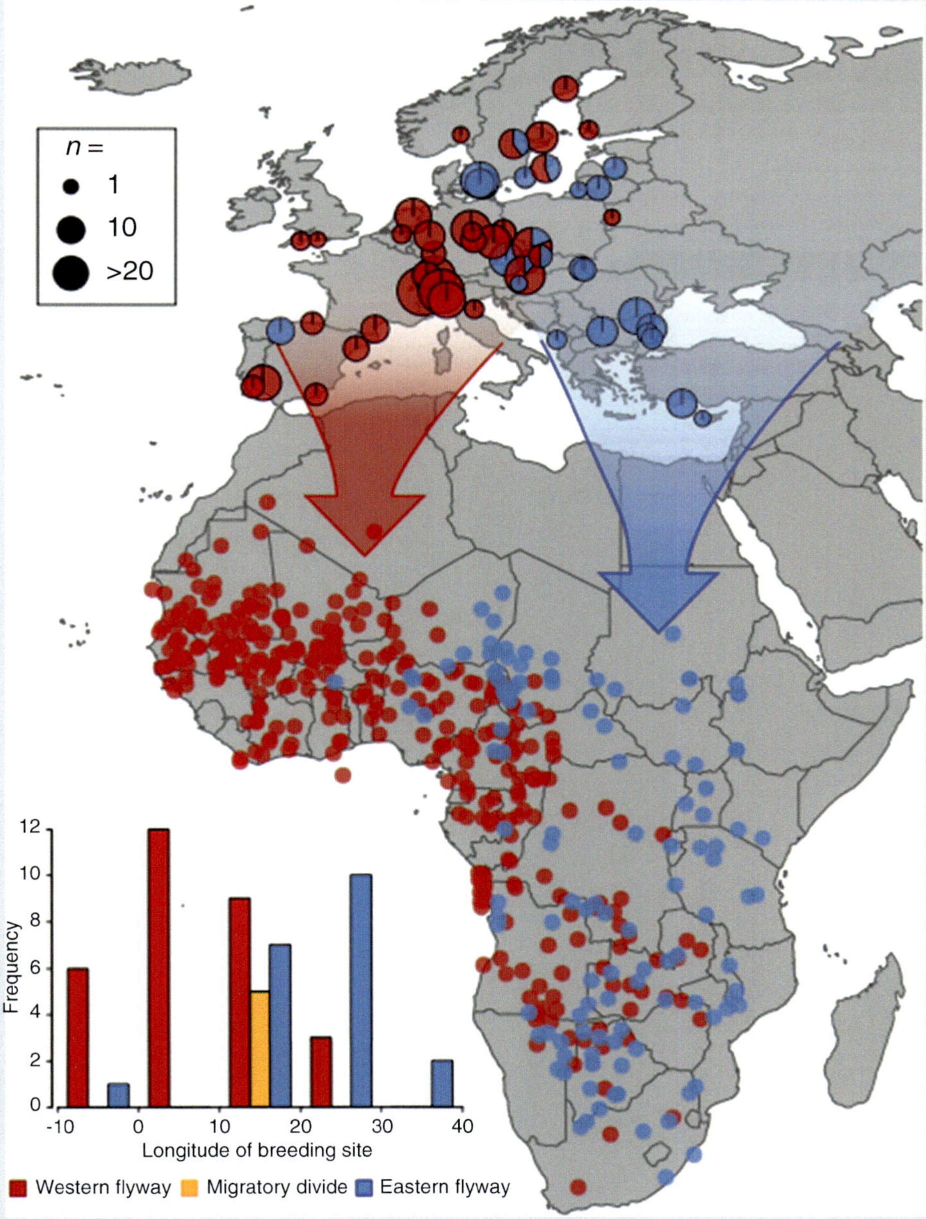


Routes used by long-distance migrants that breed in the western Palearctic and winter in Africa. Birds generally follow two flyways, a western flyway (red) and an eastern flyway (blue). The pie charts indicate the locations of breeding areas and their diameters indicate the sample sizes (n) and the proportion of birds using each flyway in each population. Nonbreeding locations in Africa are indicated by the red and blue dots. The inset (lower left) shows the frequency distribution of flyway use relative to the longitude of breeding locations, with values binned into 10° longitudinal bands. The orange bar indicates populations of birds that use both flyways (i.e., populations with a migratory divide). (Figure from Briedis et al. 2020; © 2020 John Wiley & Sons Ltd., used with permission)

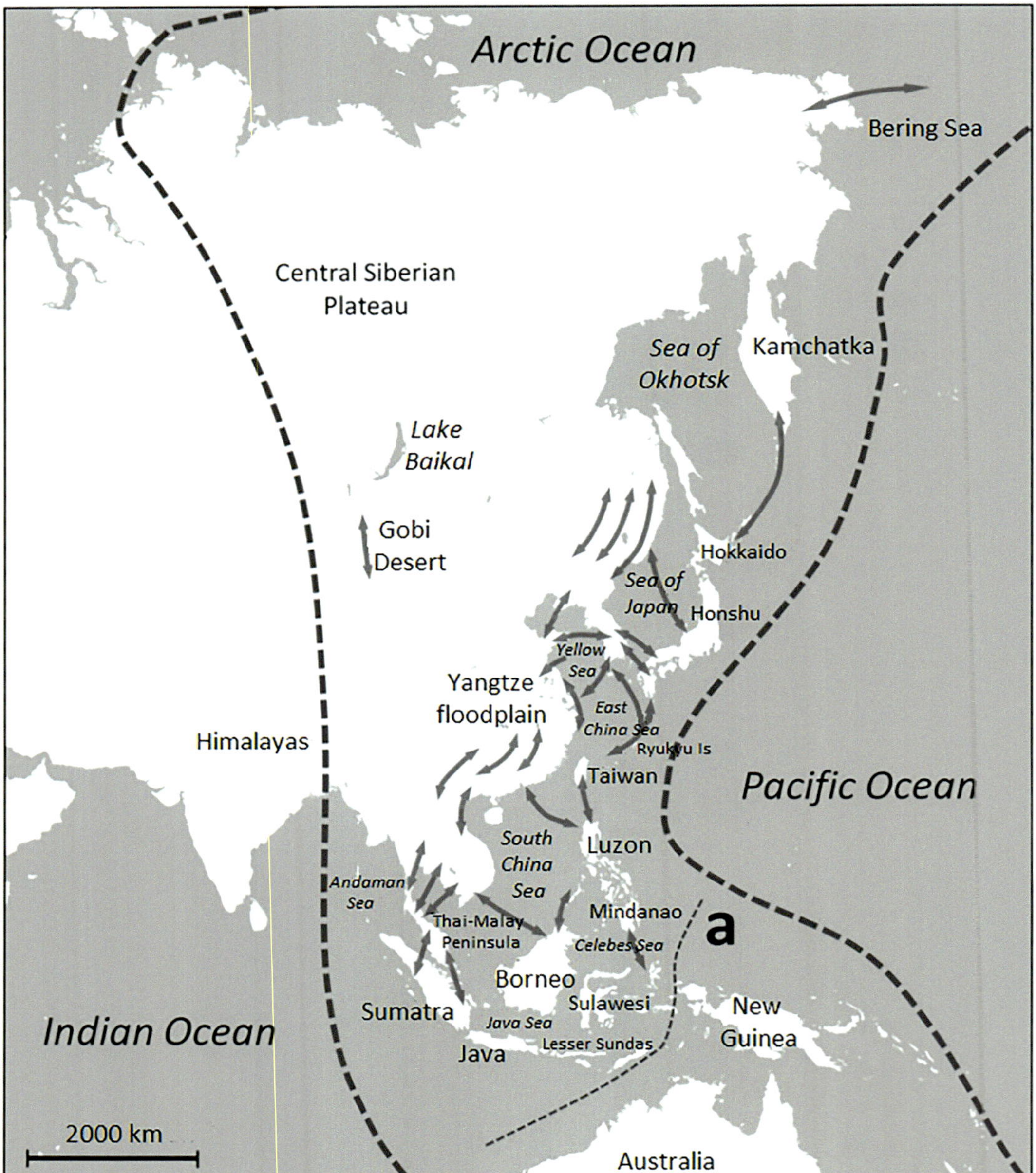


**Fig. 13.20** East Asian-Australasian Flyway showing locations (arrows) where large-scale movements of migrating songbirds have been observed. Dotted line (a) indicates the southeastern limit of the wintering ranges of most migratory songbirds (> 95%) in this flyway. (Figure from Yong et al. 2015; © BirdLife International 2015, used with permission)

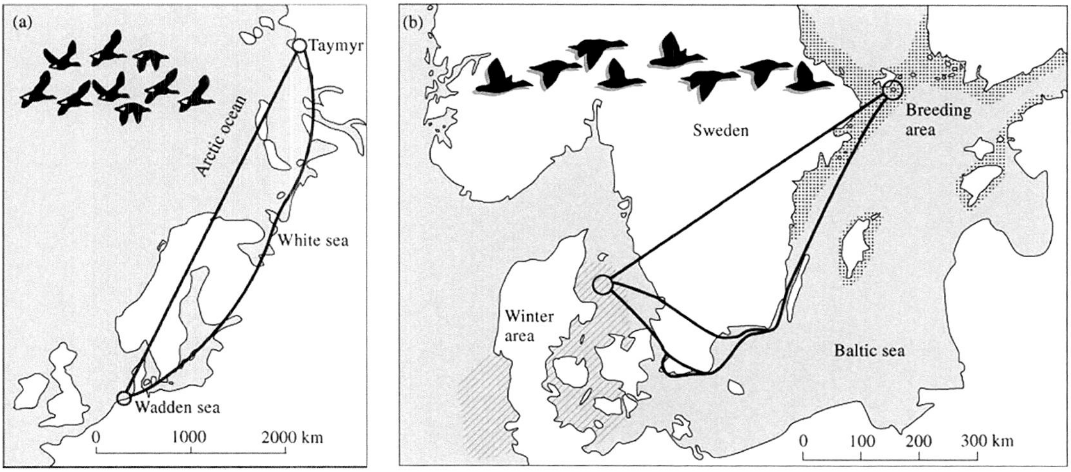


**Fig. 13.21** Detour migration by Brant Geese (*Branta bernicla*; left map, (**a**) and Common Eiders (*Somateria mollissima*; right map, (**b**). Departure and destination locations are indicated by small circles. The shortest distance migration routes are shown as straight lines; curved lines indicate the actual migration routes. (Figure from Alerstam 2001; © 2001 Academic Press, used with permission)

influenced by ground temperature and the presence of tailwinds (Fig. 13.34). When experiencing high ground temperatures, godwits tended to move to higher altitudes with lower-ambient temperatures, likely to avoid hyperthermia. Especially during the latter portions of their migratory flights, the godwits flew to higher altitudes with faster tailwinds. Because the godwits are lighter toward the end of long flights because they lose mass as they fly, the energetic cost of flying to higher altitudes declines. In addition, as energy stores are depleted during long migratory flights, the benefit of faster tailwinds (and faster ground speeds) may exceed the cost of flying to higher altitudes (Senner et al. 2018). Similar behavior has been reported for both Great Reed Warblers (*Acrocephalus arundinaceus*; Sjöbert et al. 2021) and Great Snipes (*Gallinago media*; Lindström et al. 2021) migrating between breeding areas in Europe and wintering areas in Africa. Migrating Great Reed Warblers flew at a mean altitude of 2394 m during nocturnal flight, but flew to a mean cruising altitude of 5367 m (and a maximum of 6267 meters) at dawn. Increasing their altitude to avoid increasing ambient temperatures closer to the ground is one possible explanation for this behavior. However, flying at higher altitudes during the day may also reduce the likelihood of predation by Eleonora's Falcons (*Falco eleonora*) that are known to prey on migrating songbirds and extend the vision of warblers under daylight conditions. These possible explanations are not, of course, mutually exclusive. Similarly, Lindström et al. (2021) suggested that the movement of Great Snipes to higher altitudes at dawn may provide an improved view for orientation by landmarks, reduce the likelihood of predation, and help them avoid hyperthermia.

## 13.5 Differential and Partial Migration

Among some species of migratory birds, all individuals in a population or species may share the same general breeding and wintering areas, with no spatial separation of adults and juveniles or males and females. However, for other species, particularly among short- and medium-distance migrants, wintering areas may vary with sex, age, or both. Some individuals may migrate

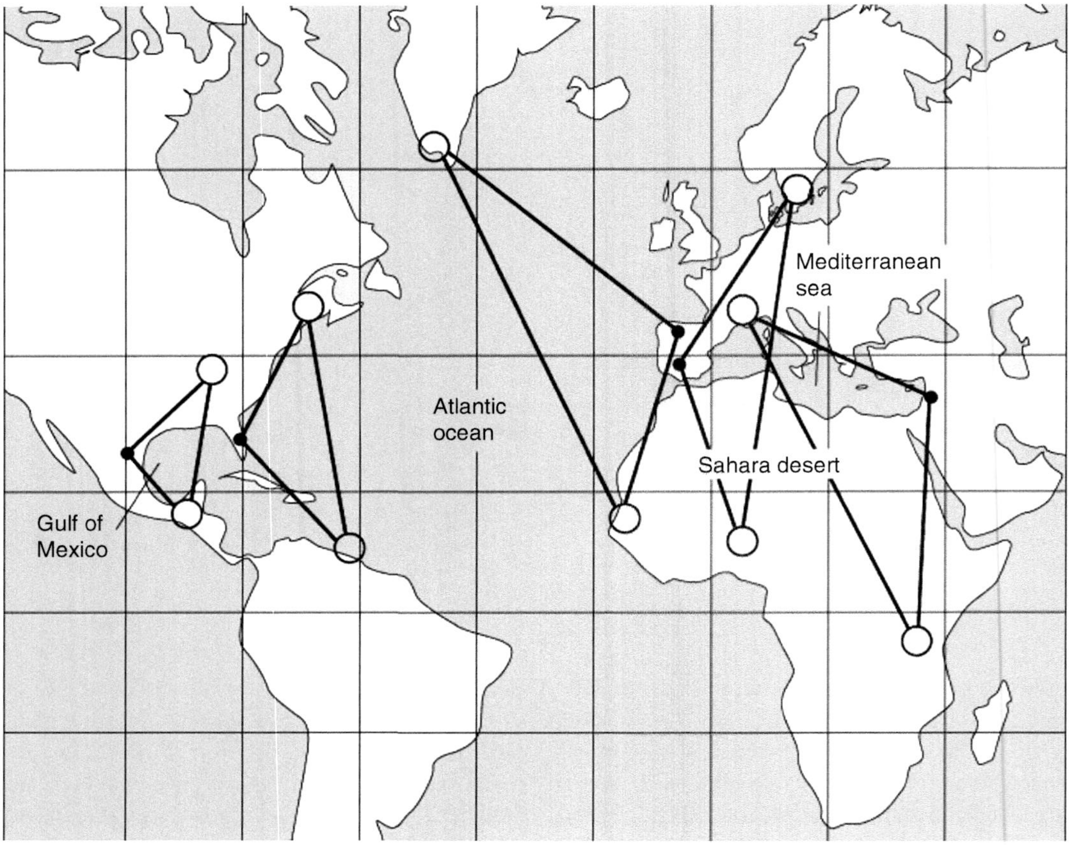


**Fig. 13.22** Examples of observed and potential detours in bird migration at ecological barriers like the Mediterranean Sea, Sahara Desert, Atlantic Ocean, and Gulf of Mexico. Shortest routes are indicated by straight lines between large open circles. Longer "detour" routes taken by some migrants are indicated by lines between connecting large open circles via the small filled circles. (Figure from Alerstam 2001; © 2001 Academic Press, used with permission)

whereas others do not (partial migration) or some individuals migrate greater distances than others (differential migration; Fig. 13.35).

Among partial and differential migrants, a number of factors can potentially influence either a bird's decision to migrate or not or how far to migrate, including age, sex, physical condition, size, and dominance status. In addition, the migratory behavior of partial migrants can either be obligate, i.e., genetically (innately) fixed at the individual level (Lundberg 1988), or facultative, with migratory decisions based on conditions that can change over time (Ketterson and Nolan 1983; Box 13.5 Obligate Versus Facultative Migration).

Several hypotheses have been proposed to explain partial and differential migration (Table 13.1). The Arrival Time hypothesis proposes that the individuals that establish breeding territories are less likely to migrate or, if they migrate, to migrate shorter distances because remaining in or near breeding areas makes it more likely that they will be able to acquire (or reacquire) high-quality territories (King et al. 1965). The Dominance hypothesis suggests that migratory decisions are based on dominance status, with subordinate individuals in a population more likely to migrate because, if they stay or migrate shorter distances, dominant individuals

**Fig. 13.23** (**a**) Broad-winged Hawk (*Buteo platypterus*). (**b**) Entirely land-based fall migration routes of Broad-winged Hawks that breed in Canada and the United States and winter in South America. During spring migration, the same general land-based routes are followed to return to breeding areas. (Photo a by Len Blumin, Wikipedia, CC BY 2.0, https://creativecommons.org/licenses/by/2.0/. Figure b from McCabe et al. 2020; © 2020 Wilson Ornithological Society, used with permission)

are likely to outcompete them for access to needed resources (Gauthreaux 1982). The Body Size hypothesis posits that larger individuals with smaller surface area-to-volume ratios are less likely to migrate or to migrate long distances because they are better able to withstand colder temperatures and food shortages (Ketterson and Nolan 1976). These hypotheses are all based on the results of studies conducted in north-temperate areas where there can be extreme seasonal differences in environmental conditions (e.g., temperature and day length) and food availability. In addition, all three are based on the assumption that staying further north can be costly due to adverse weather conditions, but can also be beneficial because of shorter migration distances. Testing these hypotheses is often difficult because, in many species of birds, males are larger, dominant, and establish breeding territories. In such species, all three hypotheses lead to the same prediction: larger, dominant males should winter further north. These hypotheses are also not mutually exclusive; multiple factors can contribute to the evolution of partial migration.

As an example of the difficulty in differentiating among these hypotheses, White-throated Sparrows (*Zonotrichia albicollis*) breed across most of eastern Canada and the northeastern United States and, during the nonbreeding season, migrate as far south as the Gulf of Mexico. These sparrows exhibit differential migration, with males tending to winter further north than females. Male White-throated Sparrows are larger than and dominant to females (Piper and Wiley 1989), and arrive in breeding areas one to two weeks earlier than females to establish territories (Falls and Kopachena 2020).

**Box 13.4 Gliding Speed of Migrating Birds that Rely on Soaring**

More than 300 species of birds use soaring and gliding flight when migrating (Del Hoyo et al. 1992). These birds typically soar, gaining altitude by circling in columns of rising warm air (thermals) generated by the uneven heating of the earth's surface. As a bird soars higher in a thermal, the warm air providing the lift gradually cools, producing less and less lift. Approaching the top of a thermal, soaring birds then continue toward their migratory destination by gliding in search of another thermal, using observations of circling birds in the distance or the presence of cumulus clouds (formed as moist air in a thermal that cools with increasing altitude and causing condensation of water). Gliding birds can adjust both the distance they travel and their speed, with speed increasing with sharper angles of descent.

Using tracking radar, Horvitz et al. (2014) collected data from 1346 migrating birds representing 12 different species that soar and glide during migration from Europe and Asia to Africa. Surprisingly, the mean gliding speeds of all 12 species converged within a relatively narrow range of 2.7 meters per second. A likely explanation for this convergence is that soaring and gliding birds attempt to minimize energy use rather than maximize migration speed. Energy use is minimized by birds gliding at speeds that maximize the chance of reaching a new thermal where they can then soar and minimize the chance that they will, instead, need to use some flapping flight to reach a new thermal.

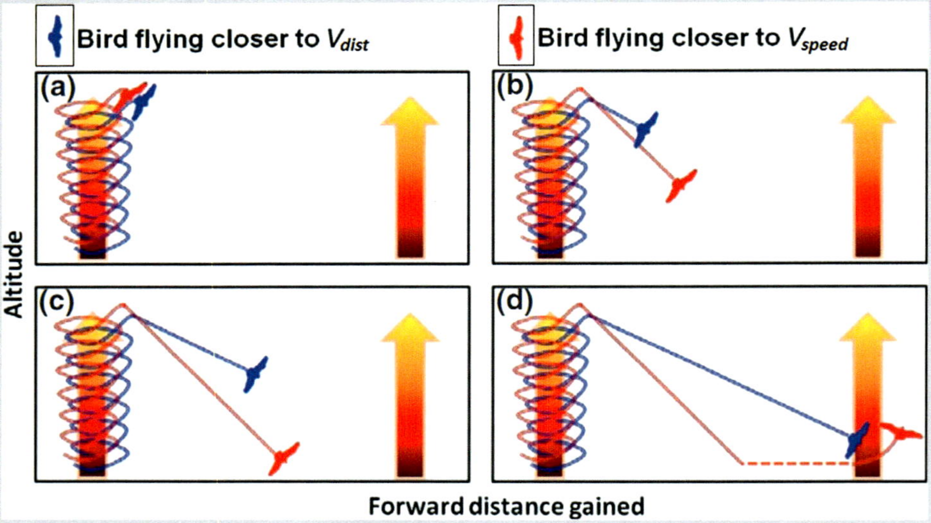


Birds that soar and glide during migration can glide at higher speeds or velocities ($V$) to maximize distance traveled ($V_{speed}$, red bird) or they can glide at slower speeds that allow them to glide further ($V_{dist}$, blue bird). In this example, the $V_{speed}$ bird glides faster because of its steeper angle of descent, but, by doing so, loses altitude faster and fails to reach the next thermal. To avoid landing, the $V_{speed}$ bird must use flapping flight (and much more energy) to reach the base of the thermal. By gliding more slowly at a shallower angle, the $V_{dist}$ bird glides further, reaches the next thermal, and saves energy because no flapping flight was required. (Figure from Horvitz et al. 2014; © 2014 John Wiley & Sons Ltd./CNRS, used with permission)

(continued)

**Box 13.4** (continued)

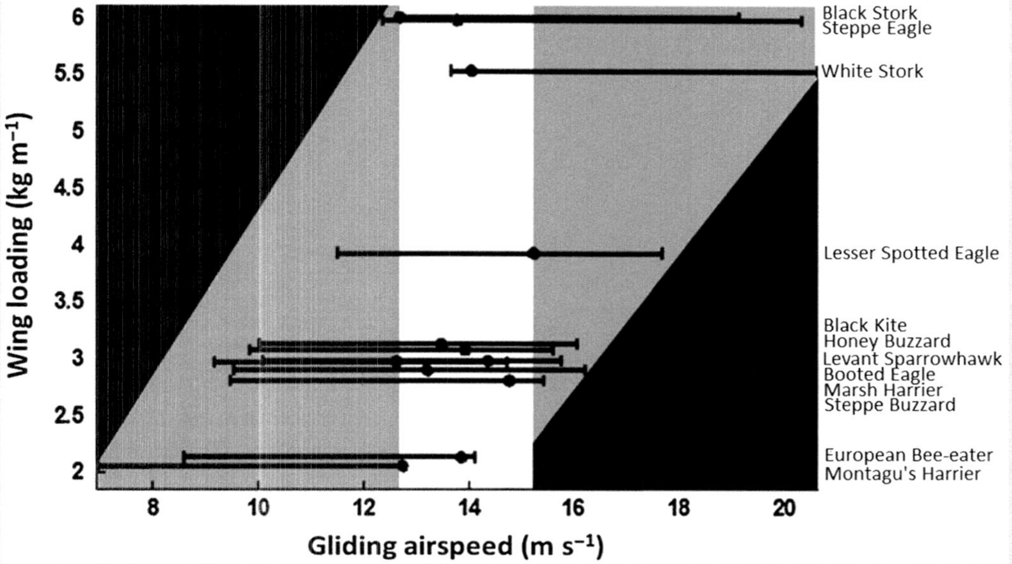


Species of birds that use soaring and gliding flight during migration can vary their gliding speeds by changing their angles of descent, with those having higher wing loading able to glide at faster speeds. However, Horvitz et al. (2014) found that several species of storks, raptors, and even a songbird (European Bee-eater) all tended to glide within a rather narrow range of speeds (2.7 m/sec; white area in the middle) within the larger range of possible speeds (13.7 m/sec; gray area). Mean gliding speeds for each species are indicated by the black dot. Mean gliding speeds for the larger birds with higher wing loading are well below their maximum gliding speeds, but that allows them to glide further and maximize the chance of reaching another thermal while gliding and minimize the chance of needing to use flapping flight. Black Stork, *Ciconia nigra*; Steppe Eagle, *Aquila nipalensis*; White Stork, *Ciconia ciconia*; Lesser Spotted Eagle, *Aquila pomarina*; Black Kite, *Milvus migrans*; European Honey-buzzard, *Pernis apivorus*; Levant Sparrowhawk, *Accipiter brevipes*; Booted Eagle, *Hieraaetus pennatus*; Eurasian Marsh-Harrier, *Circus aeruginosus*; Steppe Buzzard, *Buteo buteo vulpinus*; European Bee-eater, *Merops apiaster*; Montagu's Harrier, *Circus pygargus*. (Figure from Horvitz et al. 2014; © 2014 John Wiley & Sons Ltd./CNRS, used with permission)

(continued)

**Box 13.4** (continued)

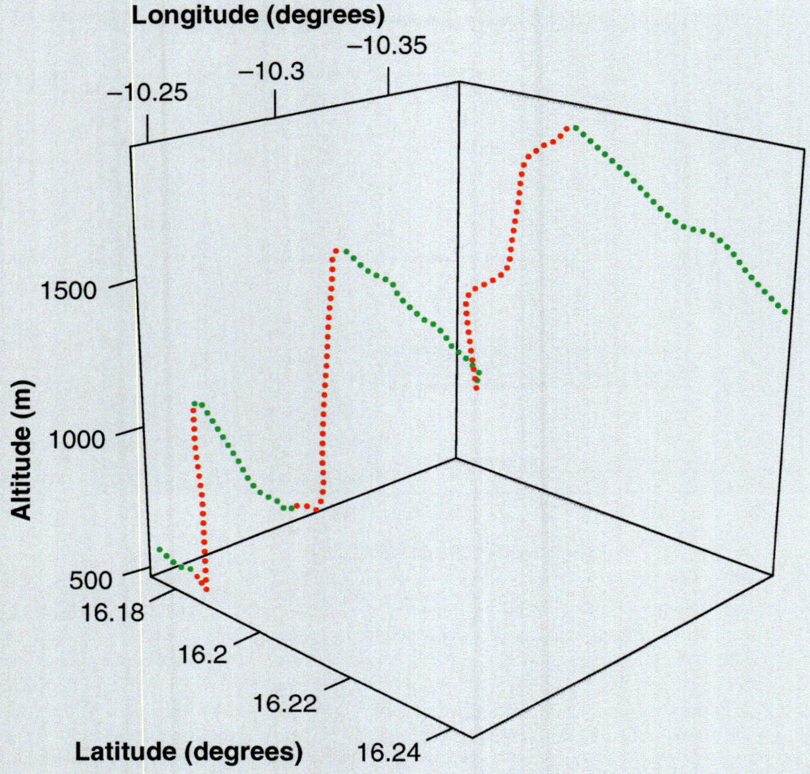


Example of a 30-minute period during which a migrating European Honey-buzzard (*Pernis apivorus*) equipped with a GPS-logger alternated soaring to gain altitude (red) with gliding to locate another thermal (green). (Figure from Vansteelant 2016; used with permission of Wouter Vansteelant)

(continued)

**Box 13.4** (continued)

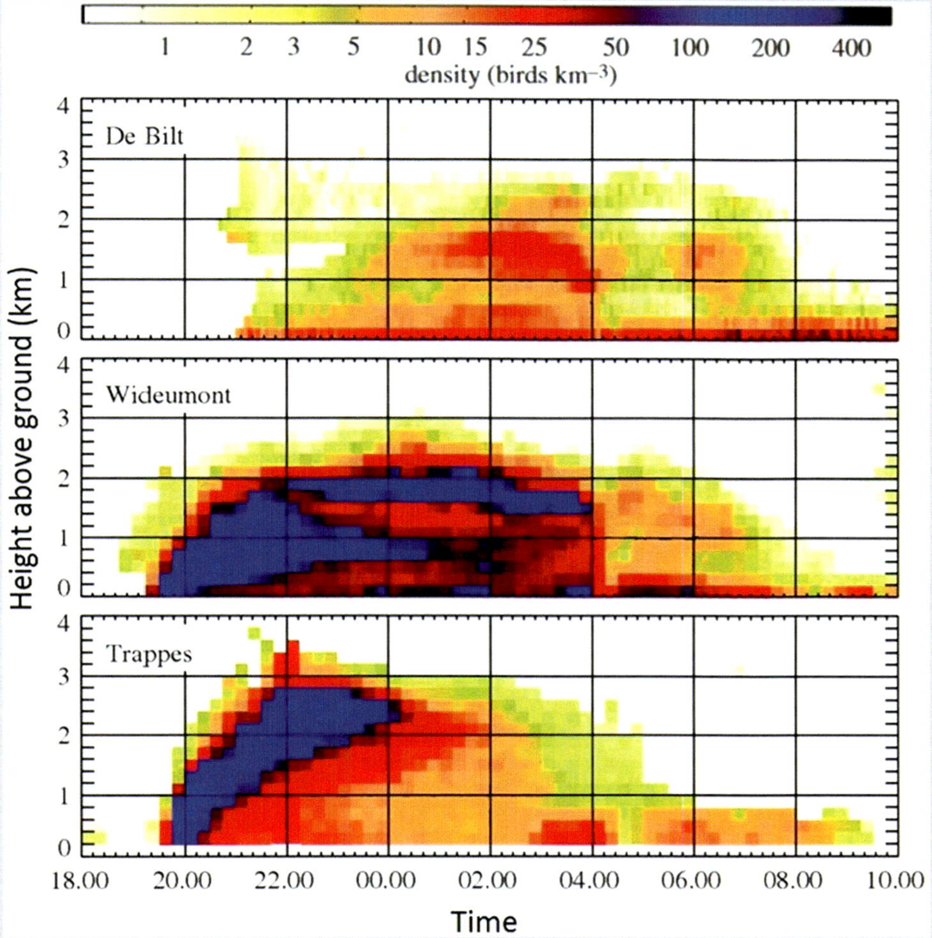


Decisions to migrate and migration altitude vary with weather conditions. Dokter et al. (2011) used weather radar to study bird migration in western Europe and found that wind conditions influenced the altitudes at which birds flew. At three different locations with differing wind conditions, birds exhibited different migration strategies. At Trappes, where there were favorable tail winds at higher altitudes (and unfavorable winds at lower altitudes), most birds quickly ascended to altitudes above 2 km. At Wideumont (280 km northeast of Trappes), birds flew at altitudes below 2 km because winds were more favorable there than at higher altitudes. At De Bilt, few birds departed shortly after sunset due to a weak occlusion front (where a cold front overtakes a warm front) that generated low clouds and precipitation a short distance to the south. Such results clearly show how birds monitor weather conditions and adjust migration strategies accordingly. (Figure modified from Dokter et al. 2011; 

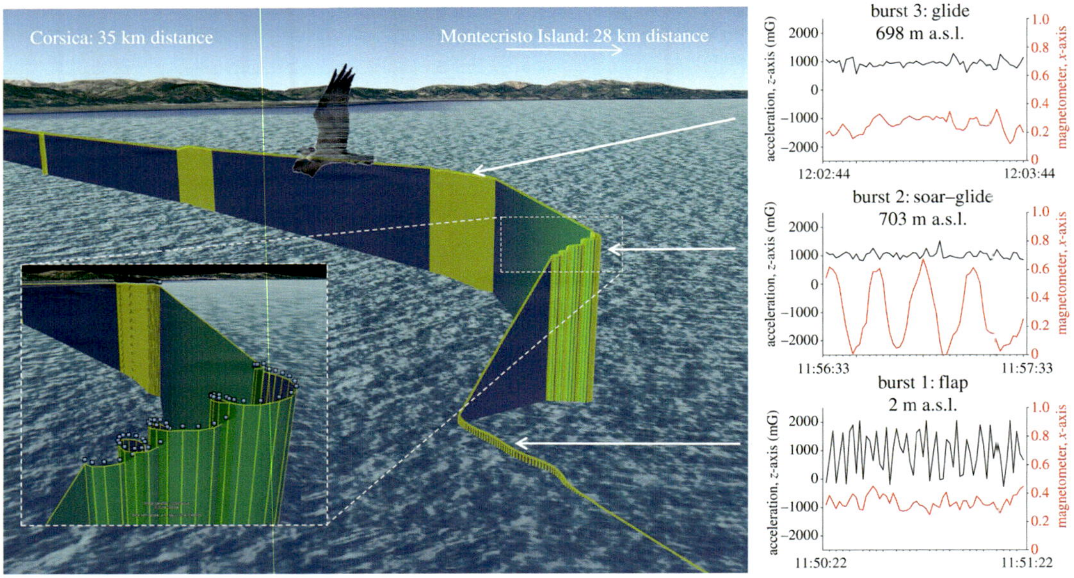


**Fig. 13.24** Three-dimensional view of a section of the migratory flight path of an Osprey (*Pandion haliaetus*) crossing the Mediterranean Sea between Italy (via Montecristo Island) and Corsica. The GPS tag was recording GPS positions for periods of 60 seconds with one position per second, followed by a pause of 5 min. The vertical yellow lines show the projection of the 3D track. The inset shows the detail of a thermal soaring bout, revealed by a GPS burst where the Osprey's position was recorded once per second. The spiral indicates that the thermal drifted to the north on the southerly wind. The panels on the right show acceleration (in black) and the magnetometer (in red) signals for three GPS bursts. Burst 1 signals (with rapid oscillation on the accelerometer) indicate rapid flapping flight, but burst 2 signals (with constant acceleration and oscillation of the magnetometer signal) indicate soaring flight without flapping. Burst 3 signals (relatively constant on both sensors) indicate gliding flight. (Figure from Duriez et al. 2018; © 2018 The Authors. Published by the Royal Society, used with permission)

Thus, any or all of the proposed hypotheses (Arrival Time, Dominance, or Body Size hypotheses) could explain differential migration by White-throated Sparrows.

Some species, however, have characteristics that make them suitable for testing these hypotheses, and studies have revealed interspecific differences in the factors that have led to the evolution of partial migration. For example, House Finches (*Haemorhous mexicanus*) in the eastern United States exhibit differential migration, with males tending to winter further north than females. This difference appears to be best explained by the Body Size hypothesis (Belthoff and Gauthreaux 1991) because male House Finches do not defend territories (and so, based on the Arrival Time hypothesis, they have no need to winter closer to breeding areas) and females are typically dominant to males (so, based on the Dominance hypothesis, females should winter further north). However, male House Finches are larger than females and, as predicted by the Body Size hypothesis, should winter further north because they can better cope with colder temperatures and reduced food availability.

Partial migration by Lesser Black-backed Gulls (*Larus fuscus*) appears to be best explained by the Arrival Time hypothesis (Marques et al. 2010). Older Lesser Black-backed Gulls tend to winter further north, closer to breeding areas, than younger gulls. These gulls exhibit minimal variation in body size so the Body Size hypothesis cannot explain the age-related difference in migration distance. The Dominance hypothesis predicts that dominant gulls (those 4 or more

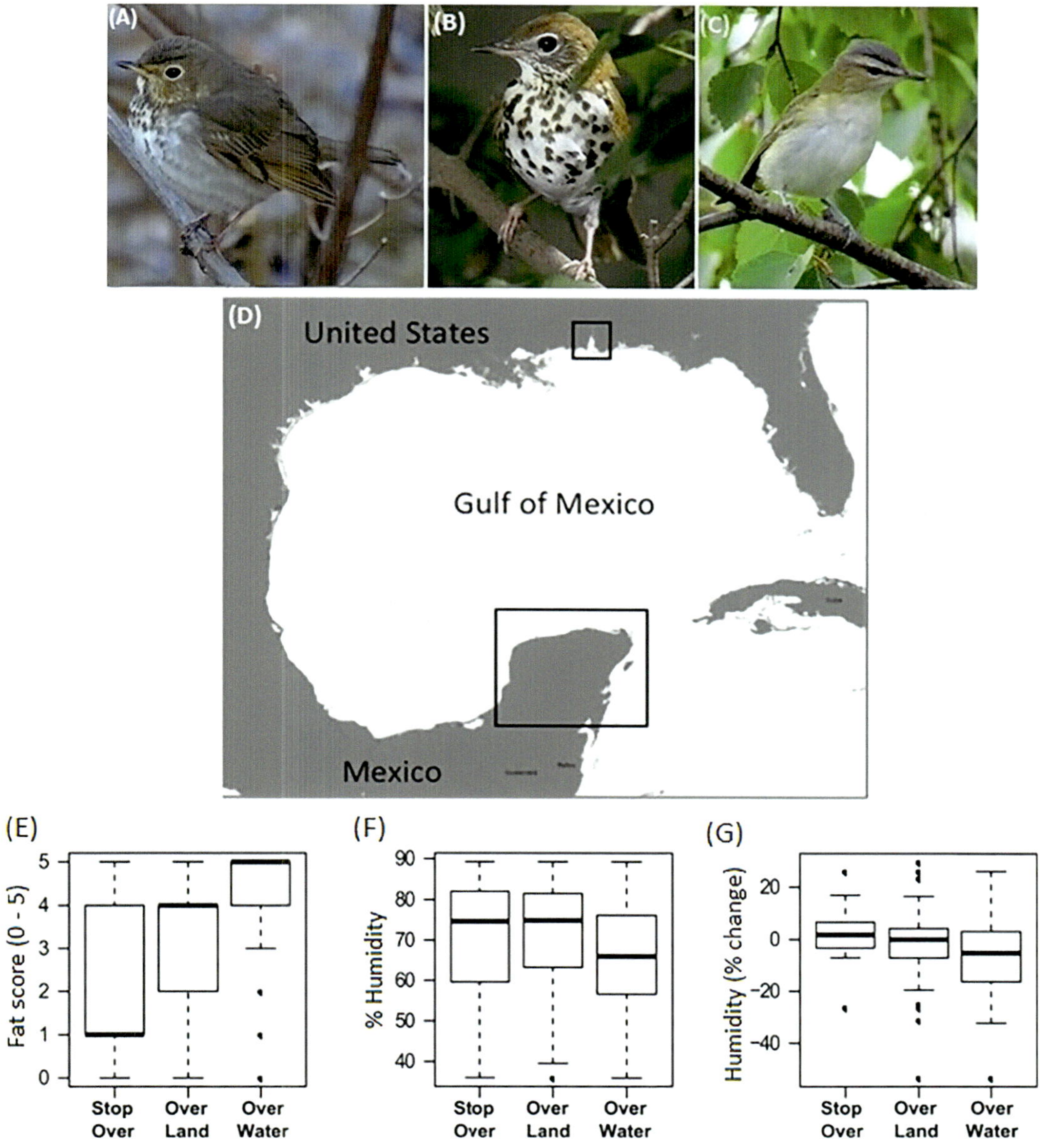


**Fig. 13.25** Decisions about whether to stopover, continue migration along a land route, or continue migration across the Gulf of Mexico from the Gulf Coast of the United States to the Yucatan Peninsula of Mexico during fall migration by (**a**) Swainson's Thrushes (*Catharus ustulatus*), (**b**) Wood Thrushes (*Hylocichla mustelina*), and (**c**) Red-eyed Vireos (*Vireo olivaceus*) are influenced by (**e**) fat scores, (**f**) humidity, and (**g**) change in humidity during the preceding 24 hours. The distance between the two regions ranges from 950 to 1040 km. Birds with the highest fat scores and, therefore, greatest energy reserves were more likely to migrate across the Gulf of Mexico. Birds were also more likely to migrate across the Gulf when humidity levels were low and declining, conditions associated with clear skies, cooler temperatures, and tailwinds. Box plots indicate median and quartiles for each variable. Negative values for change in humidity indicate a decline in humidity; positive values indicate an increase in humidity. ((**a**) Photo by Andy Reago and Chrissy McClarren, Wikipedia, CC BY 2.0, https://creativecommons.org/licenses/by/2.0/, (**b**) Photo by Steve Maslowski, CC0 Public Domain, (**c**) Photo by dfaulder, Wikipedia, CC BY 2.0, https://creativecommons.org/licenses/by/2.0/; Figs. **d–g** from Deppe et al. 2015; used with permission of the U. S. National Academy of Sciences)

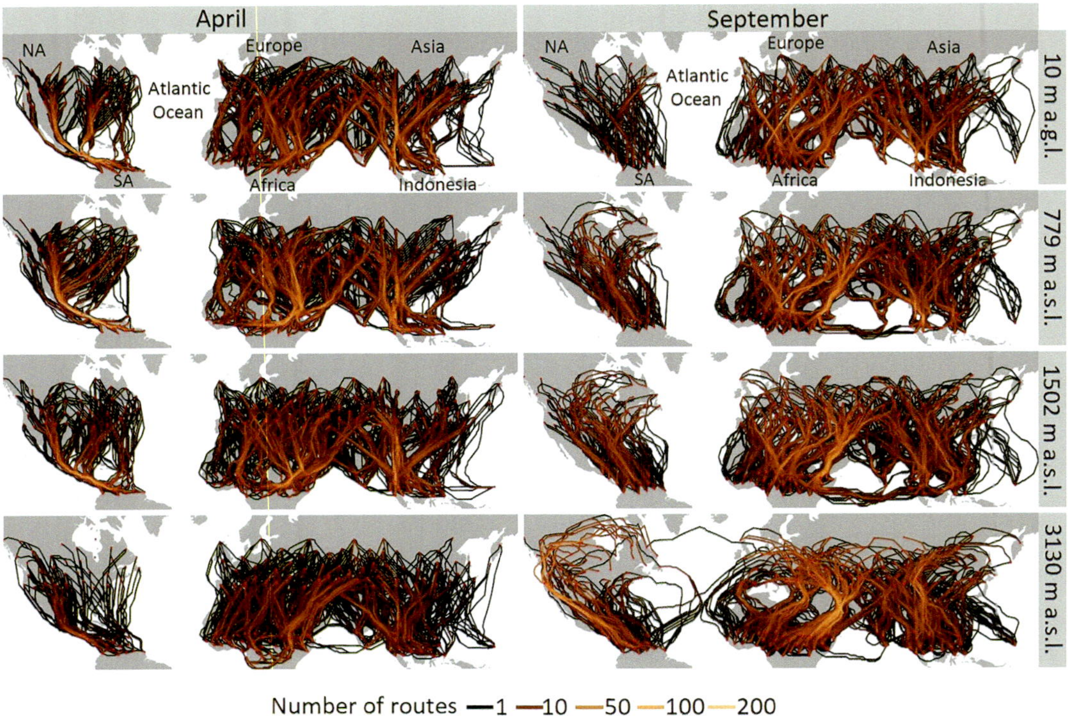


**Fig. 13.26** Migratory routes may be influenced by prevailing wind patterns and where, and at what altitudes, migrating birds can find the most favorable wind conditions. Birds are moving from south to north in April, and from north to south in September. In September, more routes between the Americas are over ocean, whereas, in April, more routes are closer to the continents. Routes between Africa and Europe tend to be farther east in September than in April. The number of routes in each area are indicated by different colors. (Figure from Kranstauber et al. 2015; © 2015 John Wiley & Sons Ltd./CNRS, used with permission)

years old) should winter closest to the breeding grounds. However, three-year old gulls that will be breeding for the first-time winter as close or even closer to breeding areas than many older, more dominant gulls, suggesting that begin close and arriving early in breeding areas best explains the winter distribution of Lesser Black-backed Gulls (Marques et al. 2010).

Few studies have provided support for the Dominance hypothesis. However, Kjellén (1994) found that, in several species of raptors, juveniles were more likely to migrate than adults and, in addition, females were less likely to migrate than males. Adult raptors are dominant to juveniles and exhibit reversed sexual dimorphism, with females larger than and dominant over males. These results, therefore, support the Dominance hypothesis, with dominant adults and larger, more dominant females tending to winter further north.

Few investigators have examined partial migration in the tropics where wet-dry cycles predominate. However, a recent study of Tropical Kingbirds (*Tyrannus melancholicus*) provided support for the Food Limitation hypothesis (Jahn et al. 2010). This hypothesis predicts that, among insectivorous species, larger individuals with greater energetic needs are more likely to migrate to wetter areas to find sufficient food. In contrast to the other three hypotheses where larger individuals, generally males, are predicted to be less likely to migrate, Jahn et al. (2010) found that the largest male Tropical Kingbirds that were typically older and dominant over younger individuals were most likely to migrate from breeding areas. Because Tropical Kingbirds feed

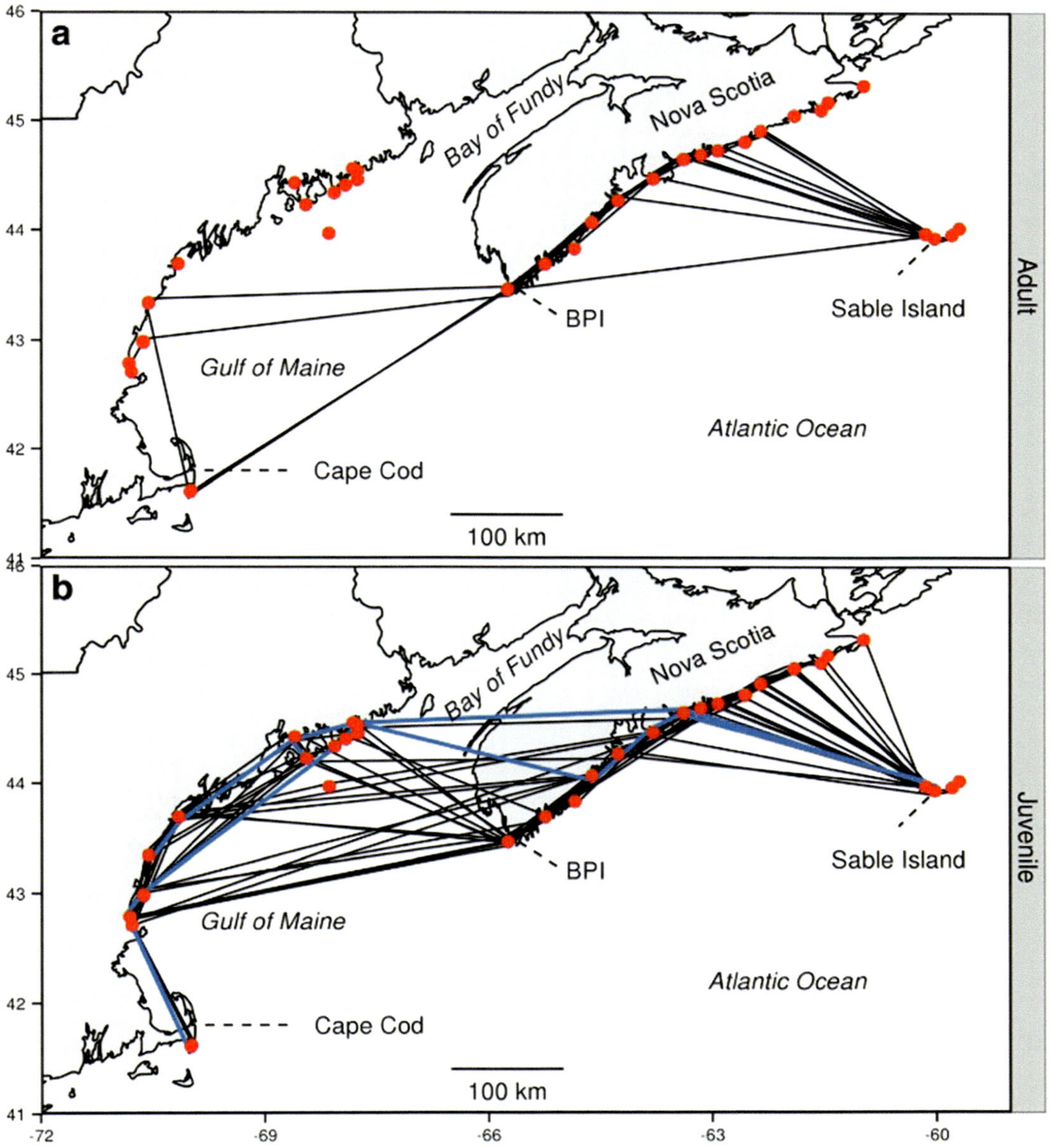


**Fig. 13.27** Migration routes from Sable Island to the mainland by (**a**) 11 adult and (**b**) 28 juvenile Ipswich Sparrows (*Passerculus sandwichensis princeps*). Juveniles took routes that resulted in shorter ocean crossings, but overall longer distances to the mainland. Two juveniles made direct flights across Nova Scotia (blue lines). Red circles indicate the locations of receiver stations. BPI, Bon Portage Island. (Figure from Crysler et al. 2016; © 2016 Crysler et al., open-access article distributed under the terms of the Creative Commons CC BY license, https://creativecommons.org/licenses/by/4.0/)

on flying insects and never forage in flocks, dominance status has less effect on their ability to access resources (compared to many granivores and omnivores that feed in flocks). What is more important is the abundance of flying insects. When insect availability drops during the dry season (coinciding with the nonbreeding period), larger males may be unable to meet their energetic

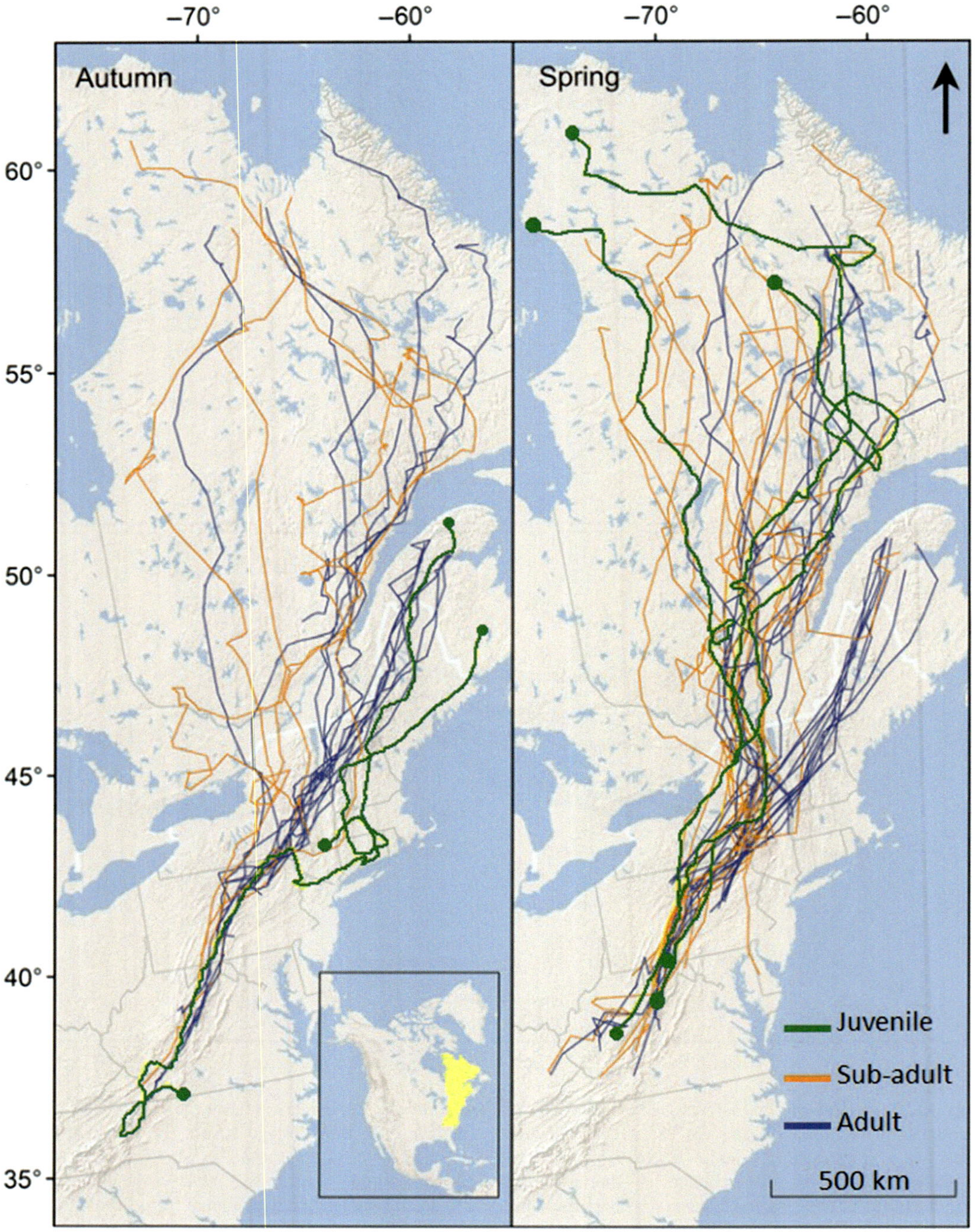


**Fig. 13.28** Migratory tracks of juvenile, subadult, and adult Golden Eagles (*Aquila chrysaetos*) in eastern North America during spring and autumn, 2007–2013. In general, especially in the spring, the migratory tracks of adults were more direct than those of subadults and juveniles. Starting and ending points of migratory tracks of juveniles are indicated by solid green circles. (Figure modified from Miller et al. 2016; © 2015 British Ornithologists' Union, used with permission)

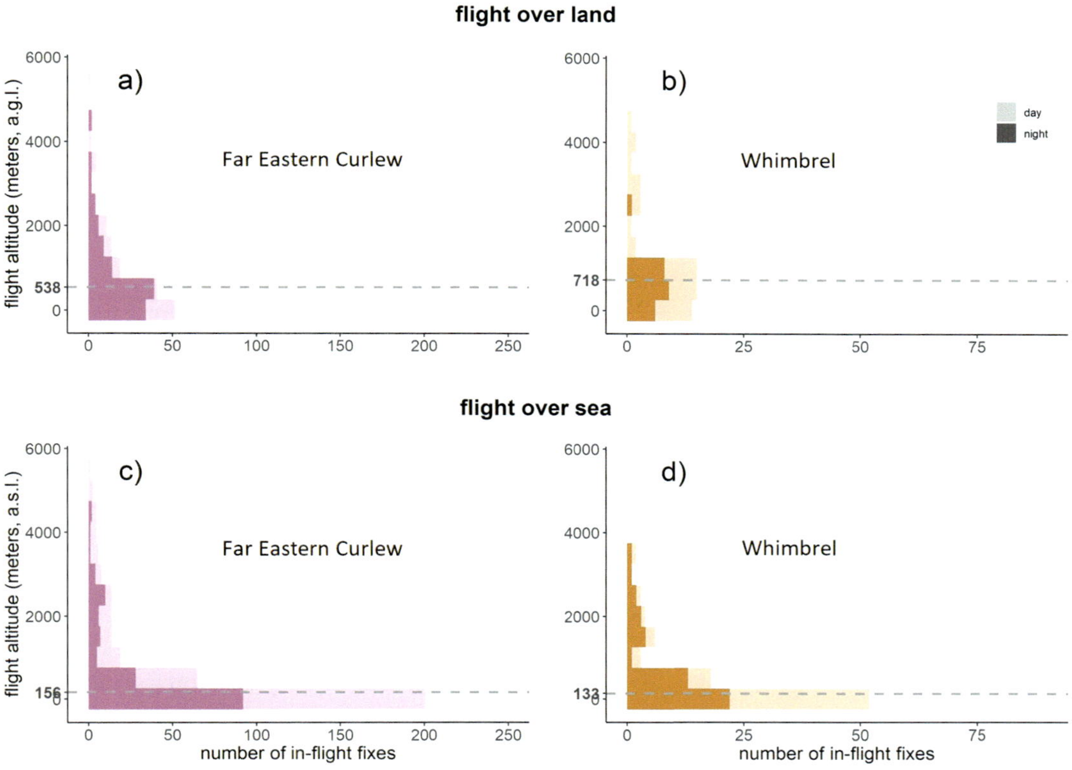


**Fig. 13.29** Histograms showing the vertical distribution of flight altitudes of Far Eastern Curlews (*Numenius madagascariensis*) and Whimbrels (*Numenius phaeopus*) migrating between breeding areas in Russia and wintering areas in Australia when flying over land (**a**, **b**) and over water (**c**, **d**). Both species flew primarily at relatively low altitudes, but at higher altitudes over land than water. One possible explanation for flying higher over land is that ground and vegetation levels vary so flying at a higher altitude may reduce the risk of encountering possible obstacles; water levels do not vary over water. Dashed lines indicate median altitudes. (Figure from Galtbalt et al. 2021; open-access article licensed under a Creative Commons Attribution 4.0 International License, https://creativecommons.org/licenses/by/4.0/)

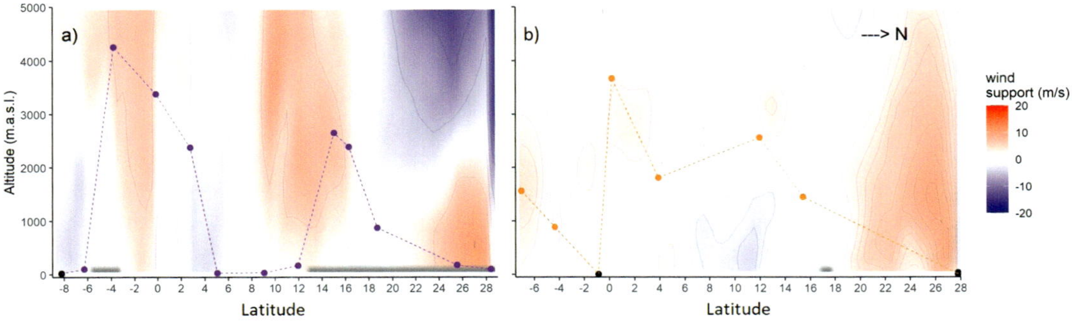


**Fig. 13.30** Effect of wind support on the altitude of a migrating (**a**) Far Eastern Curlew (*Numenius madagascariensis*) and (**b**) Whimbrel (*Numenius phaeopus*). Note that both birds altered their flight altitudes to find more favorable wind conditions. (Figure from Galtbalt et al. 2021; open-access article licensed under a Creative Commons Attribution 4.0 International License, https://creativecommons.org/licenses/by/4.0/)

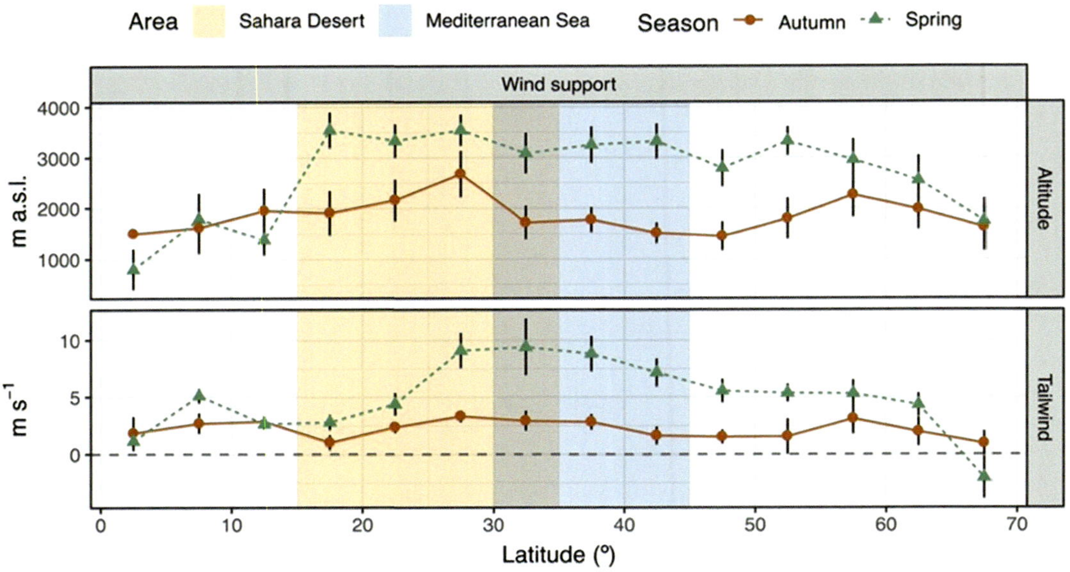


**Fig. 13.31** Common Swifts (*Apus apus*) migrating north from sub-Sahara Africa to breeding areas in Sweden at higher altitudes with more favorable tailwinds in the spring than during autumn migration. Tailwinds allowed the swifts to travel an average of about 8000 km in an average of 15 days (range = 9–25 days). Autumn migration included more stopovers and took an average of 42 days (range = 18–66 days). (Figure from Åkesson and Bianco 2021; open-access article distributed under the terms of the Creative Commons CCBY license, https://creativecommons.org/licenses/by/4.0/)

needs and must migrate to wetter areas with more insects. In contrast, smaller individuals require less energy and fewer insects and need not migrate.

## 13.6 Altitudinal Migration

During their annual cycles, many birds in mountainous areas move up and down in altitude, a phenomenon called altitudinal migration. For example, Barcante et al. (2017) conducted a literature survey and compiled a list of 1238 species of birds that were considered altitudinal migrant. Among songbirds, 12.6% of all species (830 of 6579) are considered altitudinal migrants (Pageau et al. 2020). By moving relatively short distances in altitude, birds gain the same climatic benefit as latitudinal migrants that travel hundreds or thousands of kilometers (Newton 2008). Short-distance altitudinal migration is particularly common in tropical regions. For example, on the Atlantic slope of Costa Rica, about 30% of breeding bird species exhibit altitudinal migration and most altitudinal migrants are primarily frugivores or nectarivores (Stiles 1983).

A number of different factors are associated with altitudinal migration by birds (Table 13.2). As with latitudinal migration, variation in food availability may be a factor in the altitudinal migration of some species. For example, Spotted Owls (*Strix occidentalis*) in the Sierra Nevada Mountains of California migrate to lower altitudes during the winter (Laymon 1989), likely because heavy snow at higher altitudes makes locating and capturing their prey (e.g., rodents) more difficult. In Costa Rica, Bare-necked Umbrellabirds (*Cephalopterus glabricollis*) appear to respond to variation in fruit abundance, breeding at higher elevations during the period of peak fruit abundance and moving to lowland areas where fruit abundance peaks during the nonbreeding season (Chaves-Campos et al. 2003). More generally, Pageau et al. (2020)

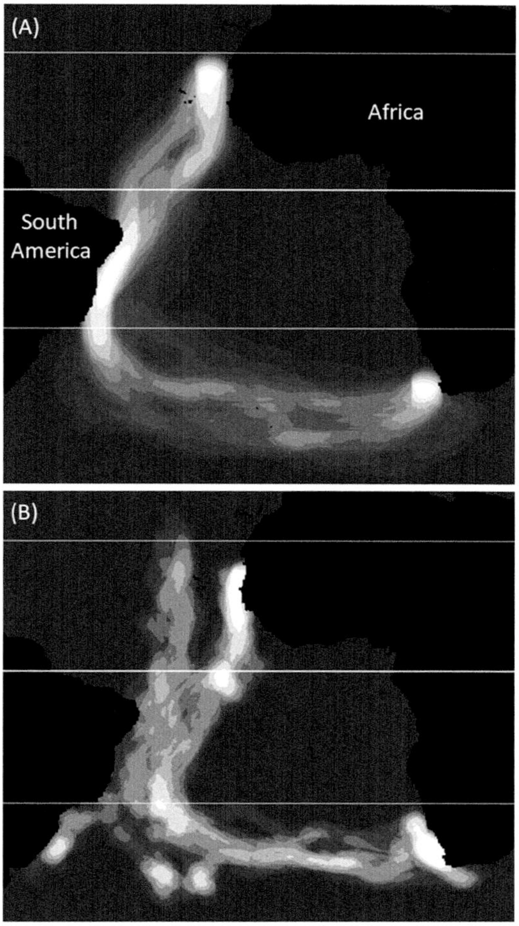


**Fig. 13.32** (**a**) Minimum cost wind corridor. Lighter areas represent areas with the most favorable winds; wind direction is from northwest coast of Africa toward South America, then across the Atlantic Ocean to the southwest coast of Africa. (**b**) Flight trajectories of 15 Cory's Shearwaters (*Calonectris diomedea*) migrating from islands off the northwest coast of Africa to wintering areas off the southwest coast of Africa. Lighter areas indicate a greater concentration of trajectories. Flight trajectories closely followed the "wind highways" that minimized the cost of migratory flight. (Figures from Felicísimo et al. 2008; © 2008 Felicísimo et al., open-access article distributed under the terms of the Creative Commons Attribution License, https://creativecommons.org/licenses/by/4.0/)

examined relationships between foraging guilds and altitudinal migration by songbirds and found that the guilds with the highest proportion of altitudinal migrants were those that fed on fruit, nectar, seeds, or plants, with these species more likely to exhibit altitudinal migration because of seasonal changes in the availability and abundance of those food items at different elevations (Fig. 13.36). Other investigators have also suggested that most altitudinal migrants are frugivores or nectarivores (e.g., Boyle 2010; Tsai et al. 2021). However, among all species and orders of birds, Barcante et al. (2017) found that, worldwide, most species of altitudinal migrants were invertivores (Fig. 13.37) and suggested that differences among studies concerning relationships between foraging guilds and altitudinal migration were likely due to differences in the geographic scope of studies. For example, most studies of altitudinal migration have been conducted in the Neotropics where most altitudinal migrants are either frugivores or nectarivores (e.g., Levey and Stiles 1992; Blake and Loiselle 2000).

The risk of nest predation may also influence distances moved by altitudinal migrants. Using artificial nests, each with two eggs (one infertile canary egg and one egg made of clay), placed along an altitudinal gradient in Costa Rica, Boyle (2008b) found that nest predation rates generally declined with increasing altitude (Fig. 13.38). This suggests that, for altitudinal migrants in the tropics, one potentially important factor in determining how high to migrate to breeding sites is predation risk, and some birds may migrate further and higher because of the benefits associated with lower rates of nest predation.

Altitudinal migration may also be an outcome of competition. For example, American Dippers (*Cinclus mexicanus*) are aquatic songbirds that breed along fast-flowing rivers and mountain streams and feed on freshwater invertebrates. Dippers construct domed nests close to water and prefer sites inaccessible to predators, protected from floods, and with a horizontal ledge or crevice for support (Kingery and Willson 2020). Dippers prefer to breed at lower elevations, and those that do so produce more offspring than those that breed at higher elevations (Gillis et al. 2008). However, the availability of territories with suitable nest sites is

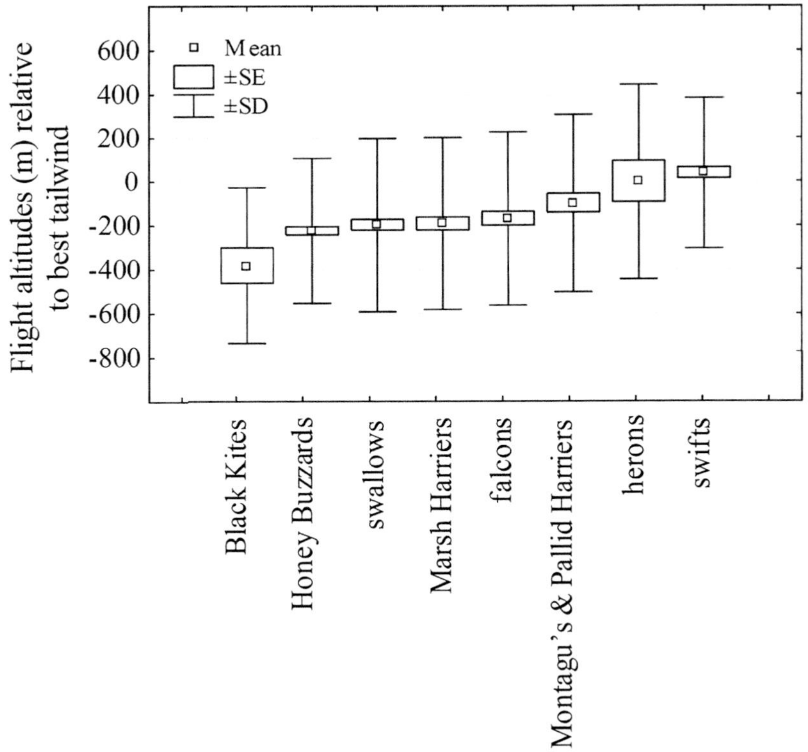


**Fig. 13.33** Flight altitudes of several species of birds were, with the exception of swifts, generally lower than the altitude with the most favorable tailwinds. Black Kite, *Milvus migrans*; European Honey-buzzard, *Pernis apivorus*; Eurasian Marsh-Harrier, *Circus aeruginosus*; Montagu's Harrier, *Circus pygargus*; Pallid Harrier, *Circus macrourus*. (Figure from Mateos-Rodríguez and Liechti 2012; © 2011 Oxford University Press, used with permission)

limited at lower elevations and most Dippers must either migrate to higher elevations to breed (Gillis et al. 2008, Mackas et al. 2010). Competition may also help explain why some altitudinal migrants move to higher elevations during the nonbreeding season (Tsai et al. 2021; Fig. 13.39). Studies conducted in Taiwan (Tsai et al. 2021) and North America (Boyle 2017) revealed that about 23% and 25% of altitudinal migrants, respectively, moved to higher elevations during the nonbreeding season. More than 180 species of long-distance migratory species of birds winter in Taiwan (CWBF 2017), and most occupy habitats at low to medium elevations (Shiu and Lee 2003). This influx of migrants might create greater competition for resources and, to avoid this competition, some resident species may move to higher elevations during the nonbreeding period (Tsai et al. 2021).

Molting is energetically expensive for birds and altitudinal migration in some species may reflect a strategy of moving to more productive areas at higher elevations after breeding to better meet the energetic and nutritive demands of molt. For example, Cassin's Vireos (*Vireo cassinii*) breed in low elevation coniferous forests in the Cascade Mountains, but, after breeding, move up-slope at least 300 m to molt in wetter, high-elevation Douglas-fir (*Pseudotsuga menziesii*) forests (Rohwer et al. 2008) where insect prey may be more abundant. Altitudinal molt migration has also been reported in several other species of birds, including Blue-tailed Hummingbirds (*Amazila cyanura*; Fraser et al. 2010), Orange-crowned Warblers (*Leiothlypis celata*; Steele and McCormick 1995), Western Tanagers (*Piranga ludoviciana*; Butler et al. 2002), and Wilson's Warblers (*Cardellina pusilla*; Wiegardt et al. 2017).

Weather can also lead to altitudinal migration. For example, White-ruffed Manakins (*Corapipo altera*) in Costa Rica migrate to higher altitudes to breed because fruit availability is greater at higher than at lower altitudes, but migration to lower

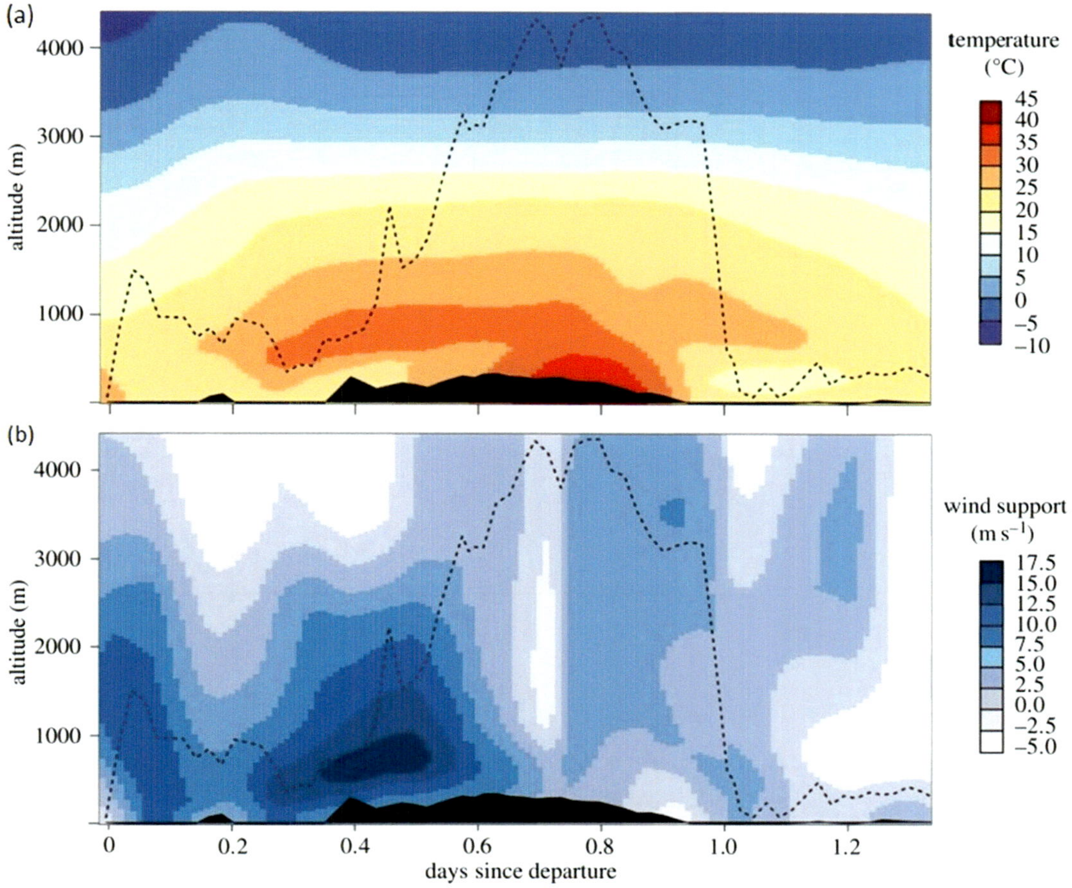


**Fig. 13.34** Flight conditions during a migratory flight by a female Black-tailed Godwit (*Limosa limosa*) with a GPS tracker. The flight lasted 31.74 hours, beginning in Spain and ending in Senegal. For each panel, black polygons denote the ground elevation and dotted lines depict the flight path. Colors indicate (**a**) temperatures and (**b**) wind support (tailwinds). (Figure from Senner et al. 2018; © 2018 The Authors. Published by the Royal Society, used with permission)

altitudes during the nonbreeding period appears to be in response to weather conditions (heavy rains at high altitudes; Boyle 2010; Boyle et al. 2010b).

## 13.7 Loop and Figure-Eight Migration Routes

Migrating birds sometimes follow different routes during spring and fall, and when one route is east or west of the other, it is referred to as loop migration and if the two migration routes intersect, it is referred to as figure-eight migration. The use of different migration routes in spring and fall occurs when conditions, typically wind direction, favor different routes in fall and spring. For example, American Golden Plovers (*Pluvialis dominica*), several other species of shorebirds, and even Blackpoll Warblers (*Setophaga striata*; Fig. 13.40), migrate from the northeastern United States across the Atlantic Ocean to the northeastern coast of South America during the fall. This long flight is made easier, and energetically less expensive, by prevailing winds that help carry birds off the coast of the United

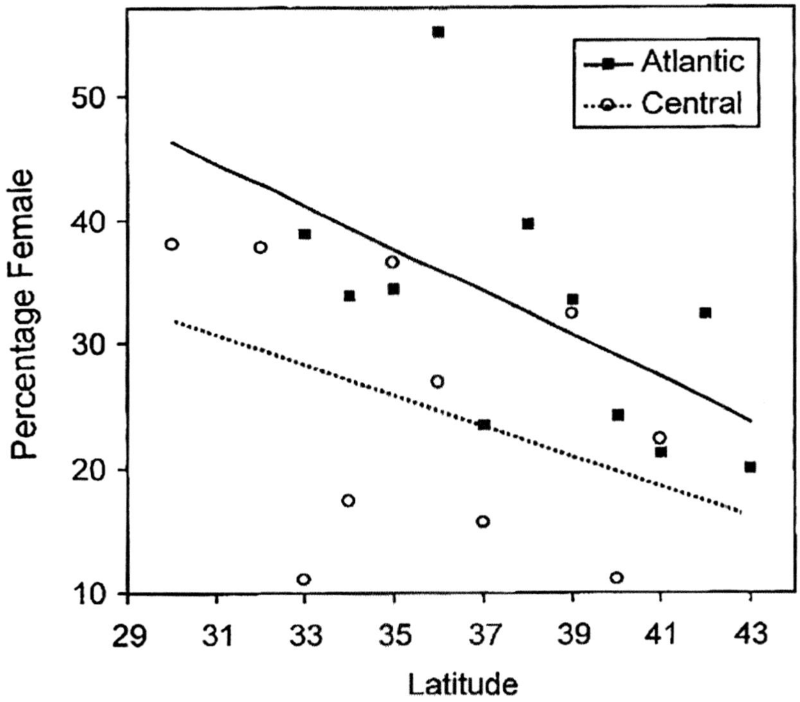


**Fig. 13.35** Relationship between latitude and percentage of females in the wintering populations of White-throated Sparrows (*Zonotrichia albicollis*) in the eastern (Atlantic) and central regions of the United States. Sex ratios are more biased toward females at more southern latitudes, indicating that females tend to winter further south than males. (Figure from Jenkins and Cristol 2002; © 2002 Oxford University Press, used with permission)

States toward the southeast and then, in the mid-Atlantic, winds that help carry birds southwest to the coast of South America (Fig. 13.41). However, during the spring, wind conditions over the Atlantic are no longer favorable (Fig. 13.39) and a northward migration further west and largely over land is the more favorable route. Examples of loop migration and figure-eight migrations have also reported for several species in the Afro-Palearctic bird migration system (e.g., Briedis et al. 2018; Pedersen et al. 2020; Fig. 13.42) and the East Asia-Australian migration system (e.g., Yamaguchi et al. 2021; Fig. 13.43).

## 13.8 Reverse Migration

During both fall and spring migration, birds are sometimes observed moving in the reverse direction of what would be expected (e.g., Richardson 1982; Åkesson 1999; Nilsson and Sjöberg 2016; Fig. 13.44). The percentage of birds that exhibit reverse migration varies among locations, weather conditions, and with methods used to detect it (e.g., band recoveries vs. radar), but has been found to range from about 2% to 20% (Åkesson 1999, Nilsson and Sjöberg 2016). Reverse migrating birds tend to fly slower and at lower altitudes than those migrating in the appropriate direction (Bruderer and Liechti 1998; Nilsson and Sjöberg 2016). A bird's decision to make a reverse movement is likely influenced by a combination of factors, with the bird's condition and weather conditions among the most important (Sandberg and Moore 1996; Schmaljohann and Naef-Daenzer 2011; Nilsson and Sjöberg 2016). Birds with less stored fat may reverse direction to locate suitable foraging habitat, especially if forward migration would require crossing a barrier such as a large body of water or desert (Åkesson et al. 1996). In some cases, reverse migration may simply be due to orientation errors (Komenda-Zehnder et al. 2002).

Weather conditions, particularly wind direction, can also be a factor in reverse migration (Pastorino et al. 2017). Flying against the wind can reduce flight speed and increase the energetic

**Box 13.5 Obligate Versus Facultative Migration**

Among some species of birds, individuals are genetically programmed to migrate, leaving breeding areas and returning at certain times of year. Other species are facultative migrants, with variable numbers of individuals in a population migrating in some years, but not others, and with individuals migrating different distances to different destinations. Still other species fall somewhere in between. For example, White Storks (*Ciconia ciconia*) migrate from Europe to Africa every year, but, after reaching Africa, their movements are facultative, with final destinations apparently influenced by rainfall and food supplies (Berthold et al. 2002, 2004).

The timing of migration by obligate migrants is based primarily on circannual clocks whereas facultative migrants are much more responsive to local environmental conditions. In addition, obligate migrants typically exhibit site fidelity, consistently breeding and wintering in the same areas, whereas facultative migrants exhibit limited or no fidelity to specific breeding and wintering areas (Newton 2012). Examples of facultative migrants include Snowy Owls (*Nyctea scandiaca*), siskins (*Carduelis* spp.), Common Redpolls (*Carduelis flammea*), Purple Finches (*Haemorhous purpureus*), Evening Grosbeaks (*Coccothraustes vespertinus*), and crossbills (*Loxia* spp.).

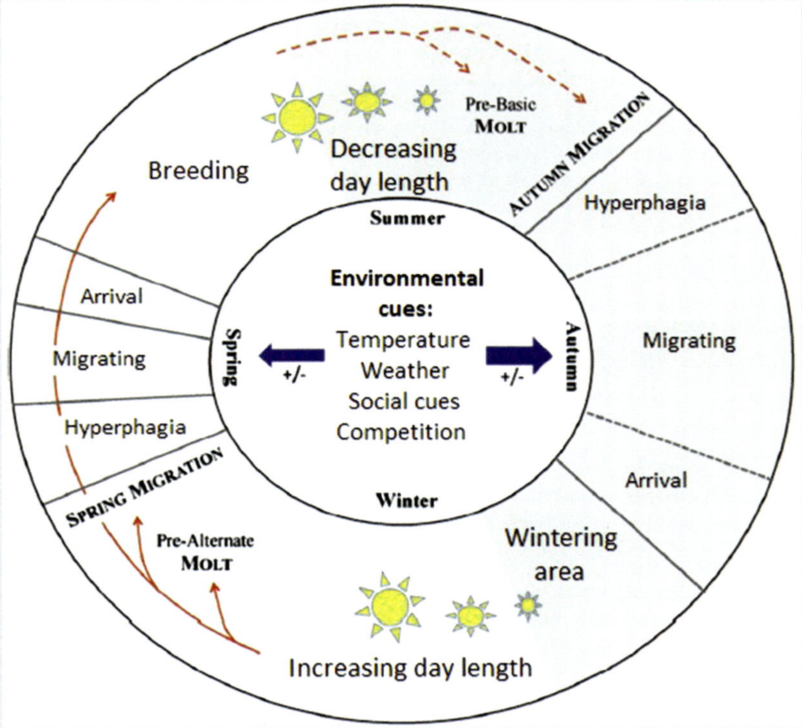


The timing of migration by obligate migrants is determined primarily by changing photoperiods, with some "fine-tuning" of the actual times of departure based on environmental cues. (Figure modified from Cornelius et al. 2013; © 2013 Elsevier Inc., used with permission)

(continued)

**Box 13.5** (continued)

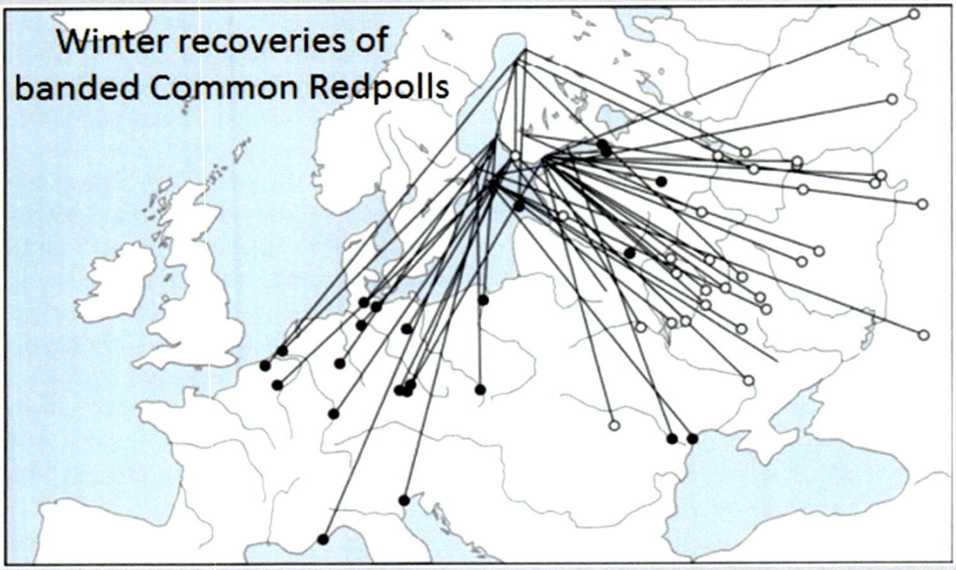


Example of the lack of winter-site fidelity by a facultative migrant. Circles indicate the locations where Common Redpolls banded in breeding areas in Finland were found during two different winters (filled vs. open circles). (Figure from Newton 2012, after Zink and Barlein 1995; © 2011 Dt. Ornithologen-Gesellschaft e.V., used with permission)

Movements by facultative migrants, e.g., in response to declining food resources, are thought to be driven by a stress response. The results of many studies have revealed a relationship between elevated plasma levels of the hormone corticosterone and changes in environmental conditions such as reduced food availability and inclement weather (Hahn et al. 2004; Ramenofsky et al. 2008). Reduced availability of food or changes in the feeding activity or number of flock mates could activate the hypothalamic-pituitary-adrenal (HPA) axis, with the release of corticosterone from the adrenal glands triggering behavioral and physiological changes, e.g., gluconeogenesis, increased plasma glucose levels, and increased locomotor activity, that could include migration (Wingfield 2003). For example, experimentally elevated levels of corticosterone increased the likelihood of facultative migration to lower altitudes during spring storms by White-crowned Sparrows (*Zonotrichia leucophrys*; Breuner and Hahn 2003), and plasma corticosterone levels increased when captive Red Crossbills (*Loxia curvirostra*), an irruptive migrant, were provided with less food (Cornelius et al. 2010).

(continued)

**Box 13.5** (continued)

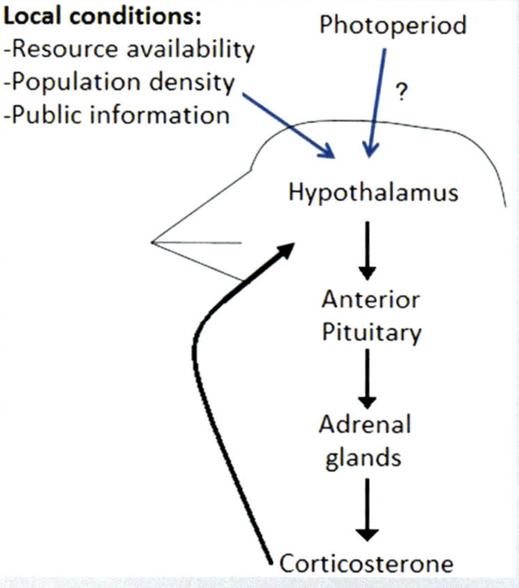


Possible mechanism whereby changes in local conditions might initiate a stress response that will trigger migration by facultative migrants. The possible role of changing photoperiods remains to be determined. (Figure modified from Ramenofsky et al. 2012; © 2012 Dt. Ornithologen-Gesellschaft e.V., used with permission)

Some characteristics of typical obligate and facultative migrants (Table from Newton 2012; © 2011 Dt. Ornithologen-Gesellschaft e.V., used with permission)

| Characteristic | Obligate migrants | Facultative migrants |
|---|---|---|
| Habitat/food | Predictable | Unpredictable |
| Breeding areas | Fixed | Variable |
| Wintering areas | Fixed | Variable |
| Site fidelity | High | Low |
| Fall migration: | | |
| Proportion migrating | Constant | Variable |
| Timing | Consistent | Variable |
| Distance | Consistent | Variable |
| Direction | Consistent | Variable |
| Main presumed ultimate stimulus | Food supply | Food supply |
| Main presumed proximate stimulus | Probably day length[a] | Food supply |

[a]The timing of fall migration is highly modified by other factors, such as the completion of previous events in the annual cycle (whether breeding or molt), which can delay departure. In both regular and irruptive migrants, flights are also dependent on appropriate weather conditions

**Table 13.1** Summary of hypotheses to explain partial migration by birds (based on Chapman et al. 2011; © 2011 The Authors, used with permission)

| Hypothesis | Reference(s) | Description | Empirical support |
|---|---|---|---|
| Arrival time | Ketterson and Nolan (1976) | Intrasexual competition for breeding territories favors residency | Willow Tits (Silverin et al. 1989) |
| Competitive release (dominance) | Gauthreaux (1982) | Dominant individuals remain resident; subdominants migrate (usually females and juveniles) | Blue Tits (Smith and Nilsson 1987), Eurasian Blackbirds (Lundberg 1985) |
| Fasting endurance (body size) | Boyle (2008a), Jahn et al. (2010) | Seasonal reduction in food availability drives migration. Individuals at greater risk of starvation (e.g., smaller, higher metabolic rates) are more likely to migrate | White-ruffed Manakin (Boyle et al. 2010a, Boyle 2011), Tropical Kingbirds (Jahn et al. 2010) |
| Thermal tolerance | Belthoff and Gauthreaux (1991), Able and Belthoff (1998) | Individuals less able to endure | House Finches (Belthoff and Gauthreaux 1991; Able and Belthoff 1998) |
| Trophic polymorphism | Chapman et al. (2011) | Individuals with more food-limited niches during nonbreeding periods are more likely to migrate colder temperatures (e.g., smaller individuals with greater surface area per unit mass) | Some evidence for Eurasian Kestrels (Aparicio 2000) |

Scientific names: Willow Tit, *Poecile montanus*; Eurasian Blue Tit, *Cyanistes caeruleus*; Eurasian Blackbird, *Turdus merula*; White-ruffed Manakin, *Corapipo altera*; Tropical Kingbird, *Tyrannus melancholicus*; House Finch, *Haemorhous mexicanus*; Eurasian Kestrel, *Falco tinnunculus*

cost of flight. Wind direction and velocity can be particularly important for birds faced with crossing a barrier and, at locations at the edge of such barriers, more birds may engage in reverse direction when wind conditions are unfavorable (Nilsson and Sjöberg 2016). At locations where there are no ecological barriers, continuing forward migration is less risky and, as a result, fewer birds make reverse movements (Åkesson 1999).

## 13.9 Stopover Sites

Migrating birds rely on stored energy and nutrients to fuel their flights, and many birds, especially small landbirds, cannot store enough energy to fly nonstop between breeding and wintering areas. So, for most birds, migration is divided into alternating periods of flight and stopover, with time at stopover sites spent foraging to deposit fuel for the subsequent flight (s) (Figs. 13.45 and 13.46). A stopover can be defined as "an interruption of migratory endurance flight to minimize immediate and/or delayed fitness costs" (Fig. 13.47; Schmaljohann et al. 2022). The time spent at stopover sites is influenced by a bird's condition when arriving at a site (Fig. 13.48) and by conditions, such as food availability and weather, at the site. If conditions are unfavorable at initial stopover sites, then birds may make "landscape movements" in search for better-quality sites (Fig. 13.49). The overall speed of migration is greatly influenced by the time spent at stopover sites, and this speed can be of critical importance because it determines when migrants arrive at breeding and wintering sites.

Experiments suggest that stopover duration is short if foraging success is poor and fuel deposition rates are low or negative (Biebach 1985; Yong and Moore 1993). More generally, Schaub et al. (2008) found that birds that accumulated fuel stores at medium rates remained at stopover sites longer than those that either lost fuel stores during their stopover or were able to increase their fuel stores quickly. However, the decision about when to leave a stopover site also appears to be influenced by the location of a site, with birds at sites located just before a large ecological

**Table 13.2** Variables that likely play important roles in altitudinal migration by birds, plus possible explanations and citations (Table from Barcante et al. 2017; © 2017 Association of Field Ornithologists, used with permission)

| Variable | Explanation |
|---|---|
| Food resources | During the breeding season, birds migrate to higher altitudes following food resources. This pattern has been reported primarily in the Neotropics (Levey 1988; Blake and Loiselle 1991, 2000, 2002; Loiselle and Blake 1991; Levey and Stiles 1992; Rosselli 1994; Solórzano et al. 2000; Kimura et al. 2001; Chaves-Campos et al. 2003; Chaves-Campos 2004, Boyle 2008a, b, 2010, 2011; Boyle et al. 2010a; Bridge et al. 2010; Faaborg et al. 2010) |
| | Migrant frugivorous species can explore food sources available in young and old second-growth forest in the Neotropics (Blake and Loiselle 2000) |
| | Some birds can experience limited opportunities for foraging in the highlands due to storm events. These limited opportunities for foraging are associated with poorer physical condition and, in the Neotropical realm, lead frugivorous males (that are at greater risk of mortality at high altitudes due to their smaller size) to leave breeding areas and migrate to lower elevations (Boyle 2008a, b, 2011, Boyle et al. 2010b) |
| | Birds migrate to higher altitudes, following food resources required for molting (Butler et al. 2002; Rohwer et al. 2008; Fraser et al. 2010) |
| Climatic conditions | Birds migrate in response to climatic conditions. Individuals in poorer physical condition will migrate to lower elevations where extreme conditions (frequency and duration of storms) are rarer. This patterns has only been reported in the Neotropics (Boyle 2008b, 2011; Boyle et al. 2010b, 2011). |
| Mating and nesting | Birds migrate to higher altitudes (1650–2780 m) where rates of nest predation are lower. This pattern was found in one study in the Neotropics (Boyle 2008a, b). opportunities |
| | Birds that are poor competitors migrate altitudinally to areas less suitable for breeding (Gillis et al. 2008; Mackas et al. 2010). |
| | Some large male Dark-eyed Juncos remain resident in breeding areas (higher altitudes) throughout the year and compete for food, driving subordinate individuals to lower elevations (Rabenold and Rabenold 1985). |
| | Some male White-ruffed Manakins migrate to lower elevations after the breeding season to survive, causing a reduction in their social status and mating success in the next breeding season (Boyle et al. 2011). |

Scientific names: Dark-eyed Junco, *Junco hyemalis*; White-ruffed Manakin, *Corapipo alatera*

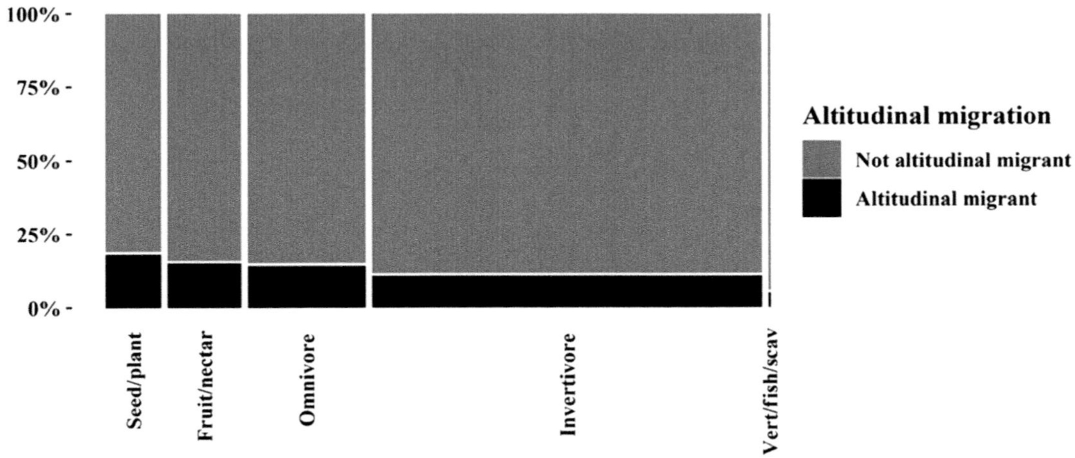


**Fig. 13.36** The proportion of songbird species that are altitudinal migrants (black) or not (gray) in each foraging guild. Note that species feeding on seeds/plants and fruit/nectar had a greater proportion of altitudinal migrations than the other guilds. The width of the bars of the x-axis indicates the proportion of species in each category. (Figure from Pageau et al. 2020; © 2020 The Authors, open-access article under the terms of the Creative Commons Attribution License, https://creativecommons.org/licenses/by/4.0/)

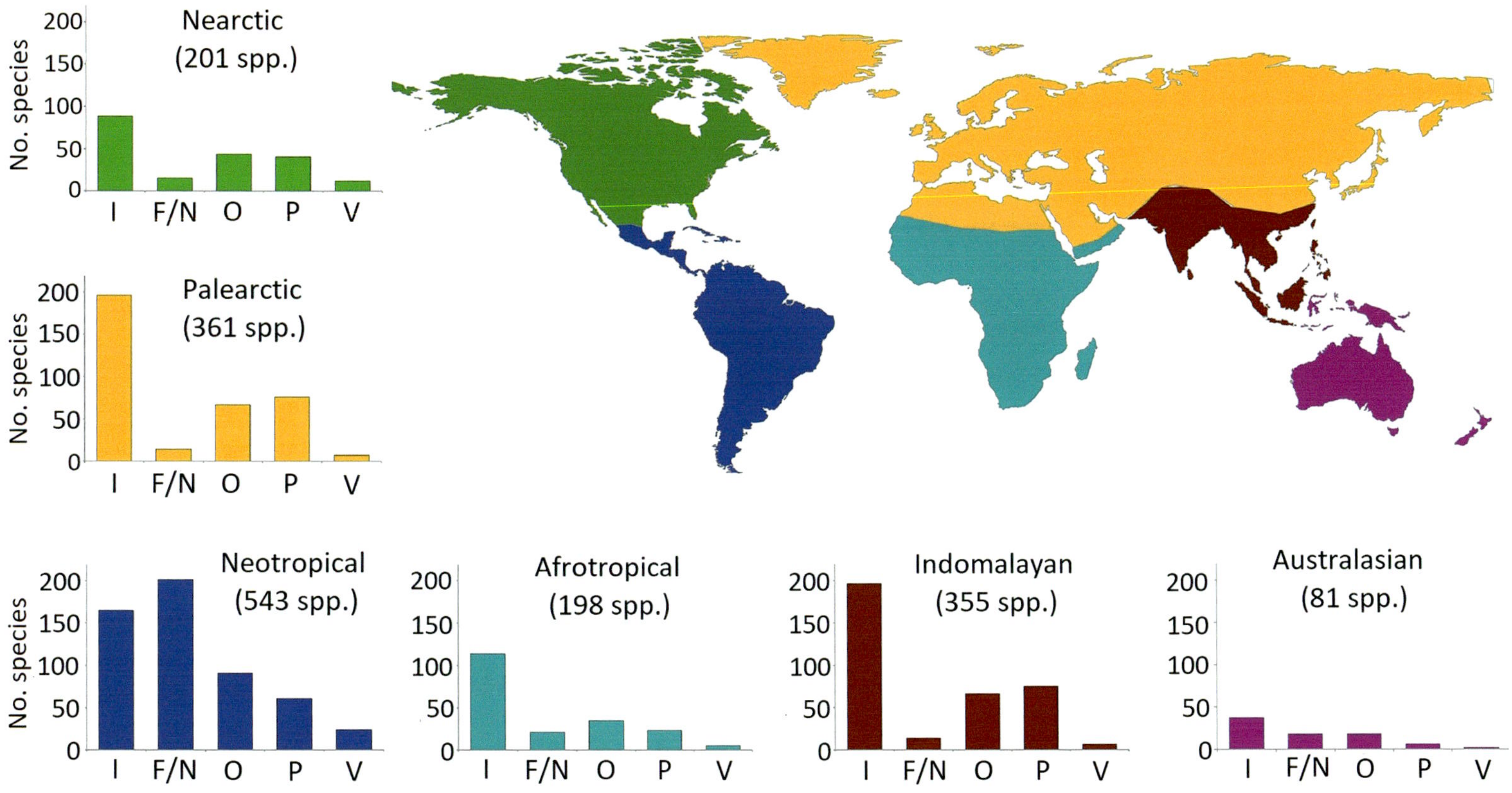


**Fig. 13.37** Foraging guilds of species of birds that are altitudinal migrants in the different zoogeographic realms. *I* invertivore birds, *F/N* frugivores/nectarivores, *O* omnivores, *P* plantivores, and *V* carnivores that consume mainly vertebrates and/or carrion. Note that most altitudinal migrants in the Neotropics are frugivores/nectarivores, but, in the other realms, most are invertivores. (Figure from Barcante et al. 2017; © 2017 Association of Field Ornithologists, used with permission)

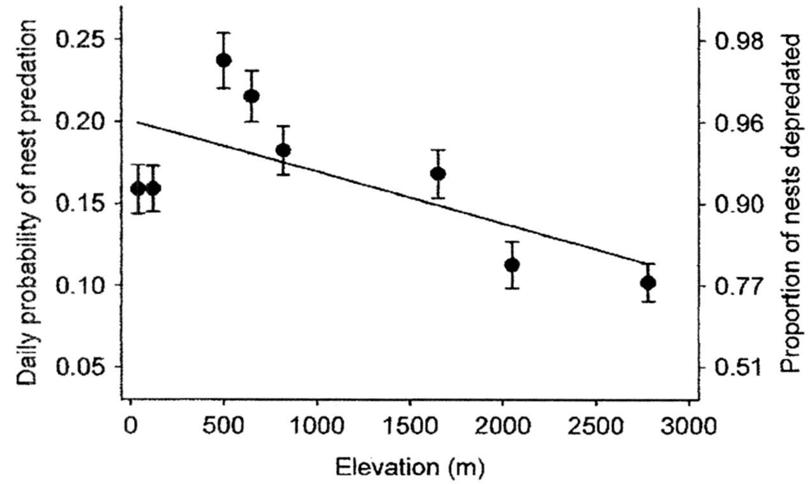


**Fig. 13.38** Relationship between elevation and probability of nest predation for 375 nests located at eight sites ranging in elevation from 40 to 2780 m on the Atlantic slope of Costa Rica. The axis on the right represents the proportion of nests predated at each site. The line is a regression line showing the linear relationship between elevation and predation. (Figure from Boyle 2008a, b; © 2007 Springer-Verlag, used with permission)

barrier (e.g., a desert) generally staying long enough to deposit sufficient fuel to cross the barrier (Schaub et al. 2008).

Yet another factor that can influence stopover duration is weather. Several studies have demonstrated that birds tend to leave stopover sites when winds are favorable (tailwinds; Richardson 1990; Liechti and Bruderer 1998). However, precipitation can be a complicating factor; rain, for example, can saturate a bird's plumage, increase wing loading, and increase rates of heat loss (Newton 2007a), so birds typically do not leave stopover sites during periods of precipitation. However, because early arrival times at breeding and wintering sites can be critically important, extended periods of harsh weather conditions (e.g., precipitation) can force birds to depart from a site even when weather conditions are not optimal, e.g., when there are headwinds (Erni et al. 2002; Jenni and Schaub 2003).

Although important for refueling, stopovers can also serve other functions for migrating birds (Schmaljohann et al. 2022). Migrants may use stopovers to recover from the physiological challenges of migratory flight, recovering from oxidative challenges, strengthening their immune responses, sleeping, and, for species with high wing loading like Common Eiders (*Somateria mollissima*), dissipating heat to prevent hyperthermia (Owen and Moore 2008; Skrip et al. 2015; Guillemette et al. 2017; Eikenaar et al. 2020; Linscott and Senner 2021). Migrating birds may also stopover when conditions become unsuitable for flying, e.g., heavy rain and headwinds, and wait until conditions improve (Schaub et al. 2004; Clipp et al. 2020). Conditions at stopover sites may also provide birds with information about the likely environmental conditions in breeding areas. As migrating birds get closer to breeding areas, such information becomes more accurate and increases the likelihood that birds will arrive in breeding areas when conditions are optimal (Winkler et al. 2014; Fig. 13.50). Although some species of birds have been found to alter the phenology of migration based on environmental cues (e.g., Marra et al. 2005; Kelly et al. 2016; Briedis et al. 2017; Fig. 13.51), other species apparently do not (e.g., Both and Visser 2001).

## 13.10 Migration in the Neotropics

Most of what is known about the migratory behavior of birds is based on studies of birds that breed in the Northern Hemisphere and migrate south during the nonbreeding season. Some birds that breed in the Southern Hemisphere also migrate, but less is known about their migratory behavior. In a review of the available literature, Jahn et al. (2020) noted that there

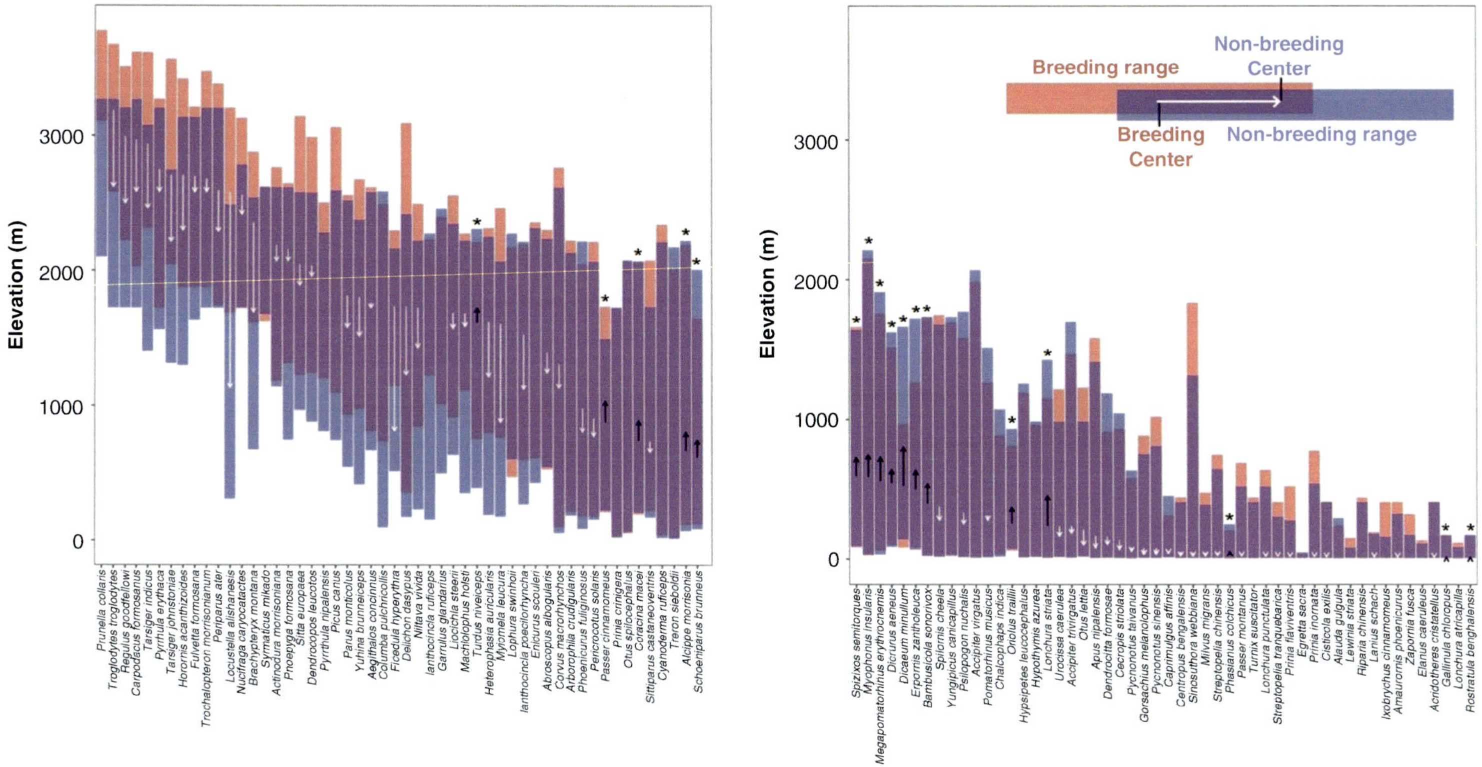


**Fig. 13.39** Seasonal shifts in the elevational distributions of 104 species of birds in Taiwan. Sixty of the 104 species (58%) exhibited seasonal shifts in the center of their distributions and/or their upper or lower boundaries. More species moved to lower elevations during the nonbreeding period, but 17 species (indicated by upward-pointing arrows and asterisks) either moved to higher elevations or shifted their upper boundary to higher elevations during the nonbreeding period. For each species, red and blue bars indicate the elevational ranges in the breeding and nonbreeding seasons, respectively, and the overlap between the bars (in purple) indicates the range where the species was recorded in both seasons. The red color on the top of bars indicates a postbreeding downhill shift in the lower boundary of the elevational distribution of a species. The blue color on the top of bars indicates a postbreeding uphill shift in the upper boundary. The arrow inside the bars shows the seasonal shift in the distributional center, with a white arrow indicating a downhill shift and a black one indicating an uphill shift. Arrows are shown only for species with a significant shift in their distributional center. (Figure modified from Tsai et al. 2021; © 2020 The Authors. Published by John Wiley & Sons Ltd. On behalf of Nordic Society Oikos, open-access article distributed under the terms of the Creative Commons Attribution License, https://creativecommons.org/licenses/by/4.0/)

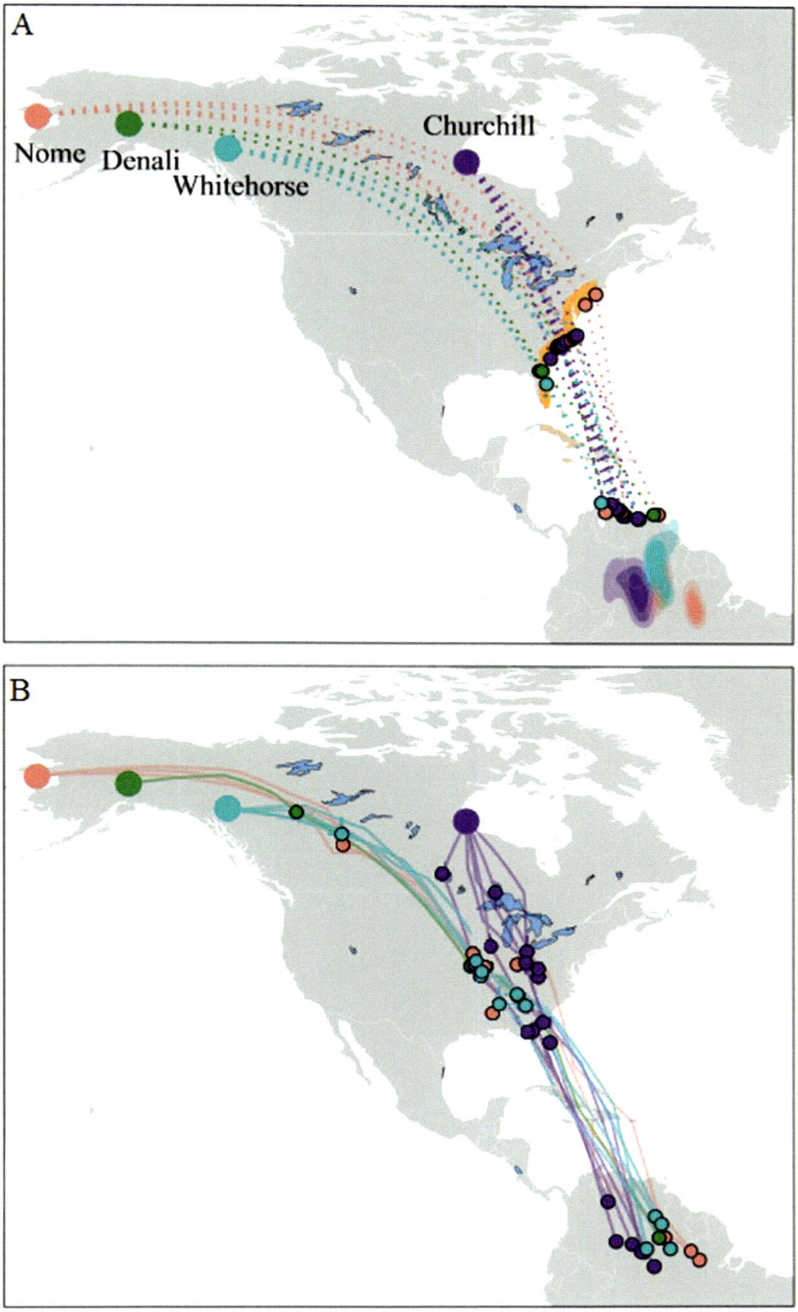


**Fig. 13.40** (**a**) Blackpoll Warblers (*Setophaga striata*) migrate over the Atlantic Ocean from the northeastern coast of the United States to the northeastern coast of South America in the fall, (**b**) but take a more westerly route over the Gulf of Mexico and the United States and Canada in the spring. (Figure from DeLuca et al. 2019; © 2019 by the Ecological Society of America, used with permission)

were at least four types of annual migratory movements by birds that breed in the Neotropics: intratropical, altitudinal, Neotropical austral, and longitudinal (Table 13.3). Intratropical migration can be altitudinal, latitudinal, or longitudinal in the area between the Tropic of Cancer and Tropic of Capricorn, altitudinal migration is a seasonal movement to areas at different altitudes, Neotropical austral migration is the movement of birds that breed at south-temperate latitudes to

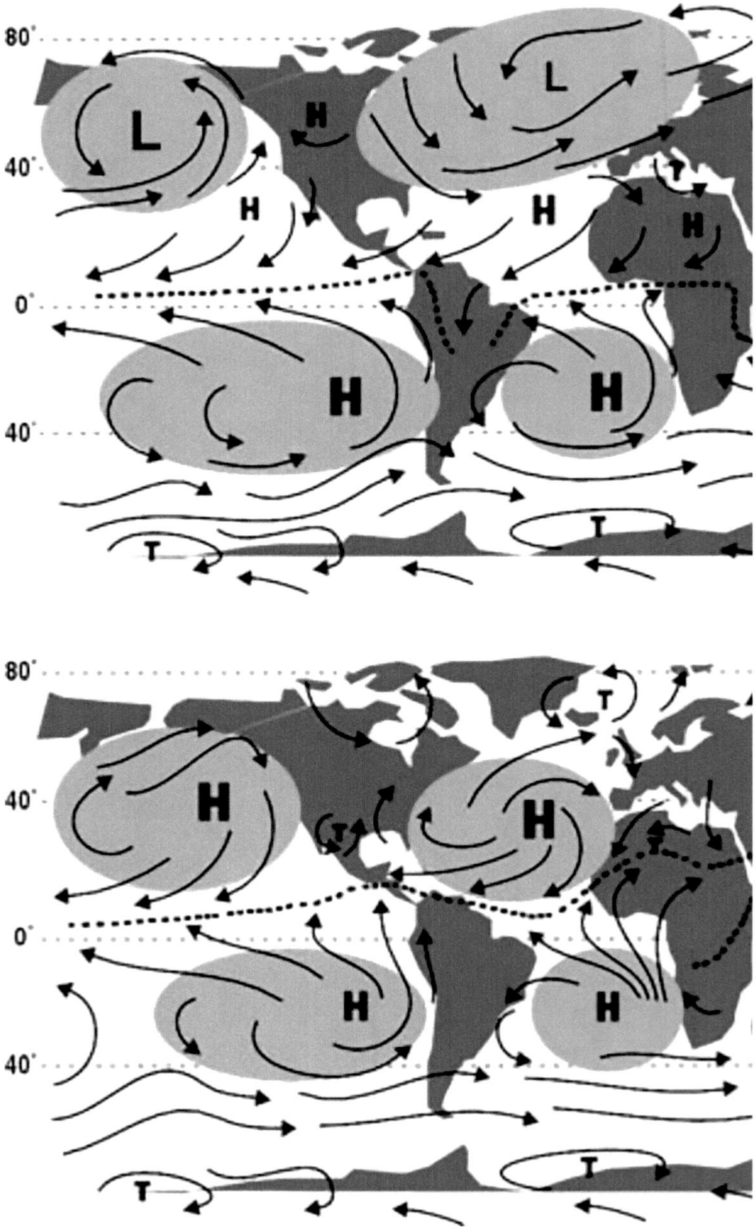


**Fig. 13.41** Distribution of average main pressure areas and wind patterns during the fall and winter (above) and spring and summer (below). Note that wind conditions are favorable for flights from the northeastern United States to South America over the Atlantic Ocean in the fall (winds generally to the southeast over the North Atlantic and switching to the southwest and carrying birds toward the northeastern coast of South America further south). In contrast, wind conditions are not favorable for a return flight over the Atlantic in the spring, with birds facing a headwind off the northeastern coast of South America. (Figure from Liechti 2006; © 2006, Dt. Ornithologen-Gesellschaft e.V., used with permission)

wintering areas closer to the equator, and longitudinal migration is the seasonal movement of birds to different longitudes (east-west) (Jahn et al. 2020). Factors contributing to migration in the Neotropics are food availability, temperature limitations, and nest-site competition or limitation (Table 13.3).

At least 220 species have been documented as Neotropical austral migrants (compared to about 340 species of Nearctic-Neotropical migrants; Rappole 1995), with about one-third of those species being tyrannid flycatchers (Chesser 1994; Stiles 2004; Jahn et al. 2019). That number is likely an underestimate because in Brazil alone,

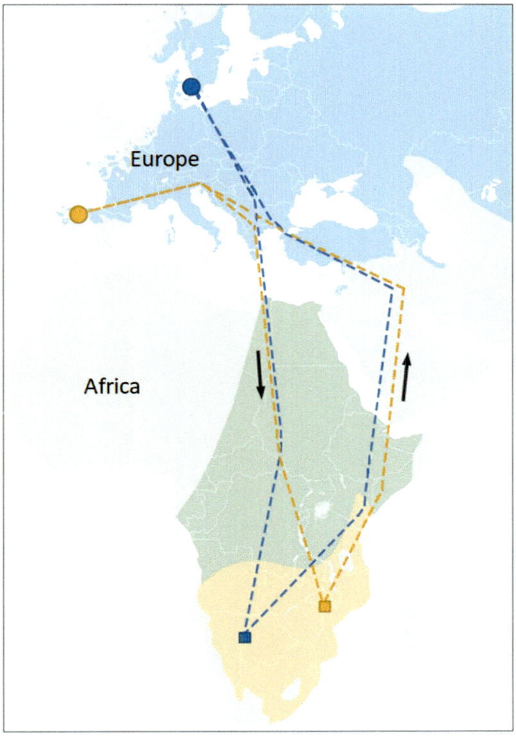


**Fig. 13.42** Loop migration routes of populations of Red-backed Shrikes (*Lanius collurio*) that breed in southern Scandinavia and Spain and winter in southern Africa. Dashed lines illustrate overall migration routes for the two populations and arrows indicate migratory direction. Estimates of winds along the migration routes show significant tailwind assistance during both fall and spring migration (Tøttrup et al. 2017). (Figure modified from Pedersen et al. 2020; © 2020 Nordic Society Oikos. Published by John Wiley & Sons Ltd., used with permission)

of 1919 species of birds, 198 (10.3%) have been found to exhibit migratory behavior (Somenzari et al. 2018). As noted by Jahn et al. (2020), "... detecting migration in New World birds that breed at tropical and south-temperate latitudes is often more difficult than in those that breed at north-temperate latitudes because most movements in the Neotropics occur over relatively short distances (e.g., hundreds, rather than thousands, of kilometers) ... and partial migration is common, such that resident/nonmigratory individuals may 'mask' movements of migratory individuals." However, there are fewer Neotropical austral migrants because southern South America is much smaller in area than northern North America (about five times smaller; Chesser 1994).

Compared to Nearctic-Neotropical migrants, Neotropical austral migrants tend to migrate shorter distances and, in some cases, breeding and wintering ranges overlap (Figs. 13.52 and 13.53). Only about 7% of Neotropical austral migrants have completely separate breeding and wintering areas (Stotz et al. 1996; Fig. 13.54). The average distance migrated by Neotropical austral migrants equals 9.2 degrees in latitude, whereas, for Nearctic-Neotropical migrants, the average distance migrated equals 22.5 degrees in latitude (Chesser 2005). Most of the land area in South America is located near the equator, so birds need not migrate as far. There are, however, exceptions. For example, White-crested Elaenias (*Elaenia albiceps chilensis*) are long-distance Neotropical austral migrants that breed in Patagonian forests in southern South America and overwinter in tropical South America. Some individuals tracked with light-level geolocators traveled more than 6000 km from breeding areas to their wintering areas (Bravo et al. 2017). Other species of Neotropical austral migrants that travel long distances between breeding and wintering areas include Barn Swallows (*Hirundo rustica*, Winkler et al. 2017; some with >14,000-km round-trips), Snail Kites (*Rostrhamus sociabilis*, Jahn et al. 2021; up to 4000 km), and Fork-tailed Flycatchers (*Tyrannus savanna*, Jahn et al. 2013b; some >4000 km).

Because of the distribution of land area in South American, there is far more wintering area available than breeding area. In North America, in contrast, there is far more breeding area available than wintering area. In addition to the shorter migration distances, most Neotropical austral migrants, in contrast to many Nearctic-Neotropical migrants that must cross the Gulf of Mexico, face no geographic barriers or obstacles. Possible exceptions include species that breed in open habitats and must migrate across the Amazonian forest to open habitats in their wintering areas in northern South America (e.g., Fork-tailed

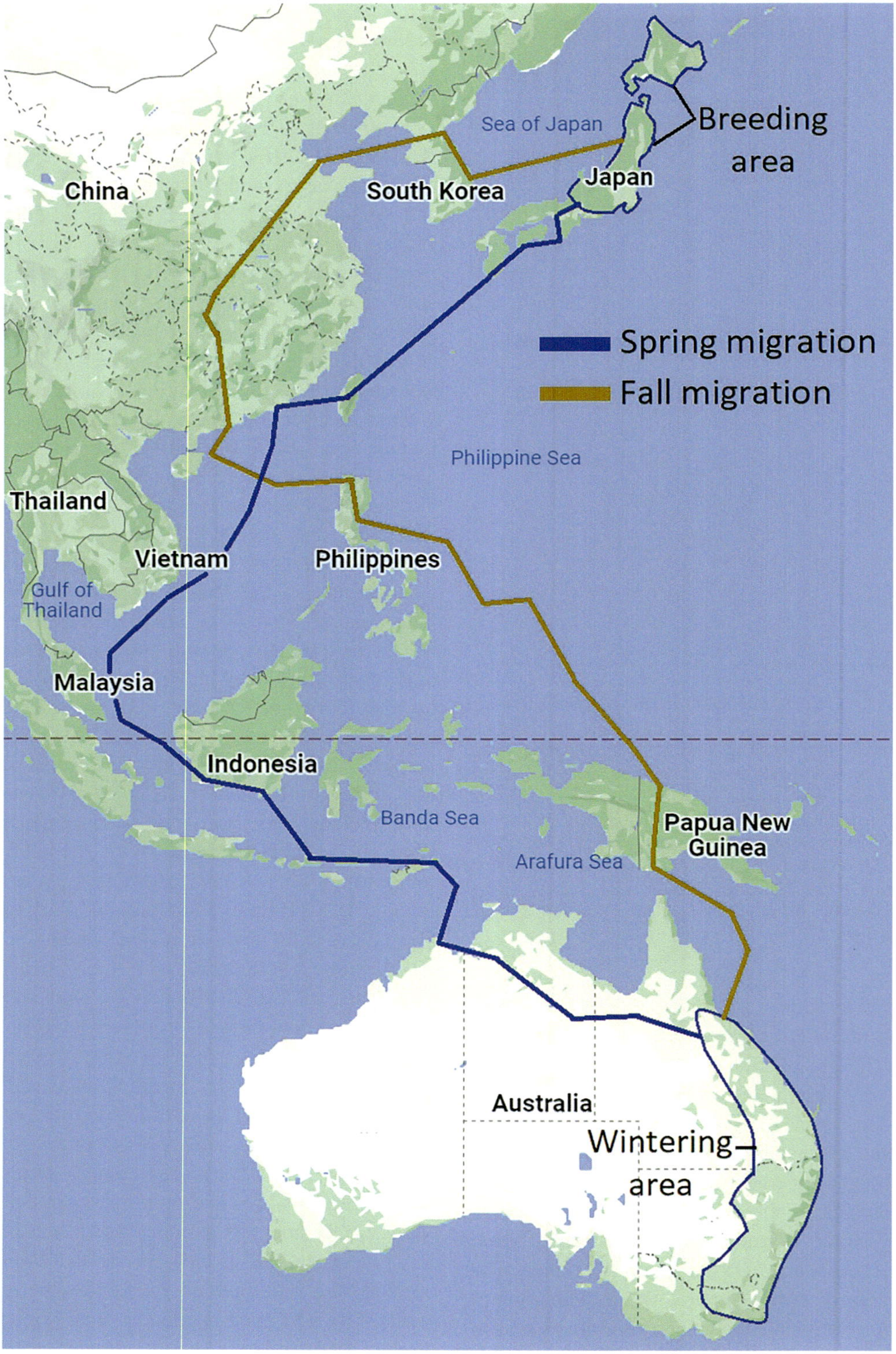


**Fig. 13.43** Figure-eight migration of a White-throated Needletail (*Hirundapus caudacutus*; family Apodidae) between its breeding area in Japan and wintering area in Australia. (Map provided by Google Maps; migration paths based on data from Yamaguchi et al. 2021; used with permission)

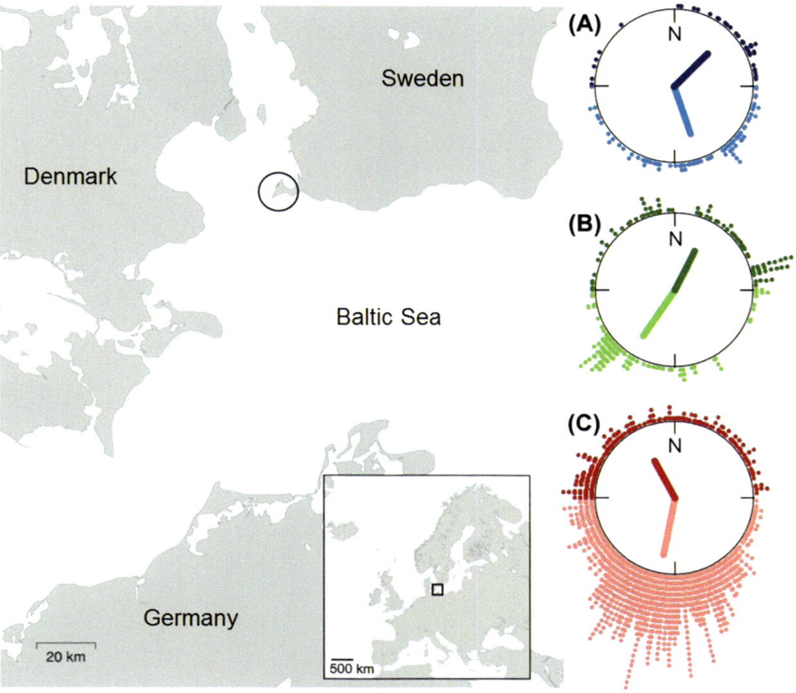


**Fig. 13.44** Departure directions of birds from Falsterbo, Sweden (black circle), during fall migration based on data collected from (**a**) radio telemetry (blue), (**b**) banded birds (green), and (**c**) radar tracking (red). Reverse migrants are indicated by dark colors and forward migrants in light colors. Mean directions are indicated inside each circle. Insert map shows the location of the region (black box) in Europe. (Figure from Nilsson and Sjöberg 2016; © The Authors. Published by John Wiley and Sons, used with permission)

Flycatchers, Jahn et al. 2013a, b). Interestingly, although many species of birds that breed in temperate areas of North America winter in temperate areas in South America (e.g., several species of shorebirds), there are no birds that breed in the south-temperate zone that winter in the north-temperate zone (Chesser 1994; Jahn et al. 2004). Why this is the case is unclear (Jahn et al. 2004). Ecologically, most Neotropical austral migrants breed in open habitats like grasslands, whereas most Nearctic-Neotropical migrants breed in forested habitats (Jahn et al. 2004). In both cases,

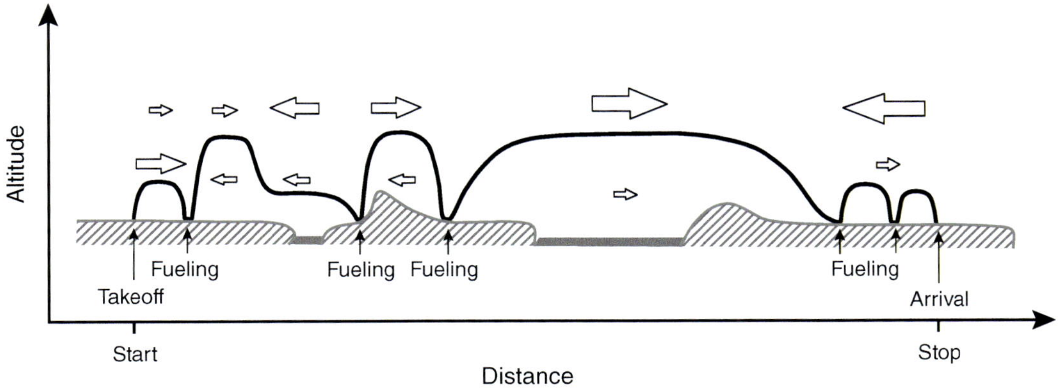


**Fig. 13.45** Example of the migration journey of a hypothetical bird, showing how distances flown and flight altitudes can vary due to barriers like mountains and bodies of water as well as wind direction and velocity (with variation indicated by the different-sized arrows; note that wind velocities are typically higher at higher altitudes). (Figure from Åkesson and Hedenström 2007; © 2007 Oxford University Press, used with permission)

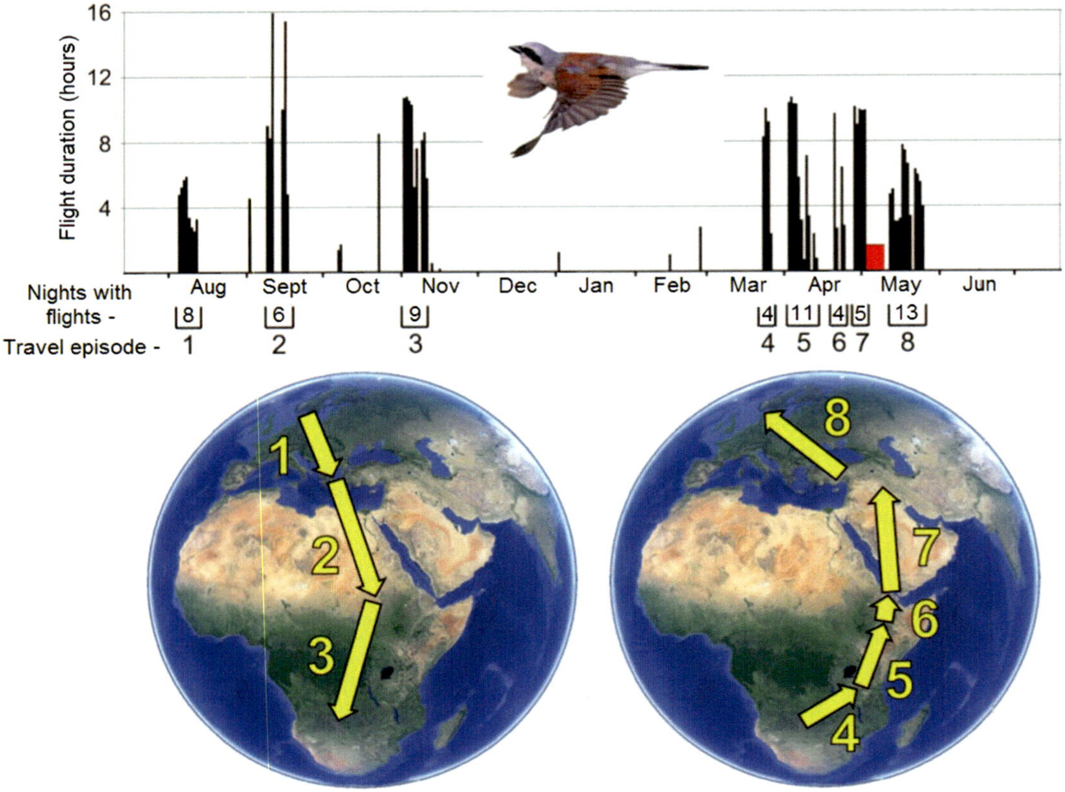


**Fig. 13.46** Migration of a Red-backed Shrike (*Lanius collurio*) from its breeding territory in Denmark to its wintering area in southern Africa and back again. Overall, this shrike made 69 flights during migration during three periods of travel during fall migration and five during spring migration. Data were collected using an accelerometer and data were missing during the period from 3 to 9 May (indicated in red in the graph). (Figure from Bäckman et al. 2017; Published by John Wiley and Sons, open-access article available under the terms of the Creative Commons Attribution License (CC BY), https://creativecommons.org/licenses/by/4.0/)

however, migrants tend to be habitat generalists during the nonbreeding season (Jahn et al. 2004).

Intratropical migration is the annual long-distant movement (≥ 100 km) of birds that occurs between the Tropic of Cancer and the Tropic of Capricorn (Hayes 1995; Heckscher et al. 2011), and has been reported to occur in species that both breed and winter in the tropics (e.g., Fork-tailed Flycatchers; Fig. 13.55) as well as species that breed at higher latitudes and winter in tropics, e.g., Bobolinks (*Dolichonyx oryzivorus*), Purple Martins (*Progne subis*), and Thrush Nightingales (*Luscinia luscinia*). Among tropical species, intratropical migration allows individuals to track seasonal shifts in the availability of food resources (Morton 1977; Levey and Stiles 1992). Although more common among frugivores and nectarivores (Levey and Stiles 1992), intratropical migration has also been documented in insectivores, e.g., Fork-tailed Flycatchers (Jahn et al. 2016) and Pallid Swifts (*Apus pallidus*, Norevik et al. 2019).

Intratropical migration by birds that migrate from higher latitudes to the tropics is defined as the prolonged occupation of two or more widely separated locations within the tropical wintering area of a species (Stutchbury et al. 2016). Intratropical migration has been observed in several species of long-distance latitudinal migrants (Table 13.4) and additional study will certainly

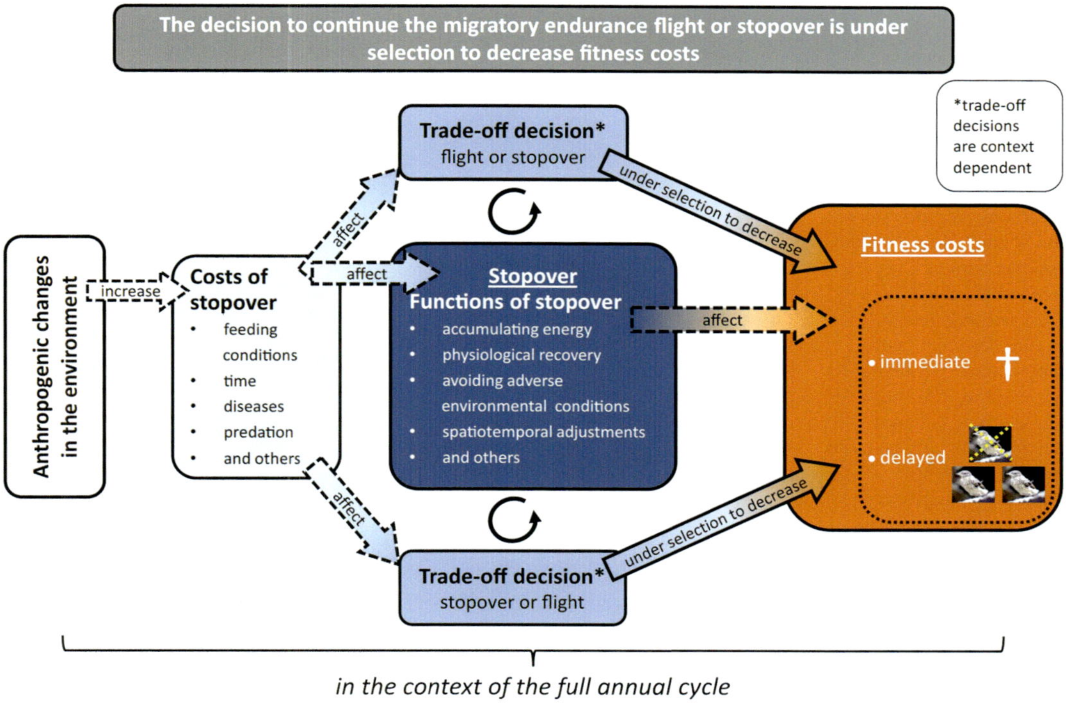


**Fig. 13.47** The decision to either continue migration or stopover is under selection to decrease immediate and/or delayed fitness costs in the context of a bird's full annual cycle. The various costs and functions of stopover influence this decision. Anthropogenic changes in the environment (e.g., habitat degradation and fragmentation) may increase the costs of stopover through, for example, habitat loss and reduced food availability. Spatiotemporal adjustments refer to the potential ability of birds to use condition at stopover sites to assess seasonal progression (e.g., ambient temperature and/or extent of vegetation growth) to better optimize arrival in breeding areas during spring migration. The immediate fitness cost is mortality, whereas the delayed fitness costs include reduced reproductive success and/or lower probability of survival in the future. (Figure from Schmaljohann et al. 2022; © 2022 The Authors. Biological Reviews published by John Wiley & Sons Ltd. on behalf of Cambridge Philosophical Society, used with permission)

add more species to that list. In most studies to date, the reason(s) for intratropical movements during the wintering period could not be determined and, in some species, not all individuals made such movements. For example, Stutchbury et al. (2016) found that 85 of 191 (44%) Purple Martins made at least one intratropical migration.

As with intratropical migration by tropical species, one possible explanation for intratropical migration by latitudinal migrants is that it allows individuals to track seasonal variation in food availability. For example, Stach et al. (2012) suggested that intratropical migration allowed Thrush Nightingales (*Luscinia luscinia*) to track seasonal variation in rain and vegetation (and, presumably, food resources; Fig. 13.56). Lemke et al. (2013) suggested that Great Reed Warblers (*Acrocephalus arundinaceus*) complete molt at their first wintering sites and only then move to their second wintering site (Fig. 13.57). Stutchbury et al. (2016) suggested that intratropical migration by Purple Martins was triggered by increasing densities of birds at roost sites, perhaps because roost sites may have a "carrying capacity" or increasing densities' mean increasing competition for food.

Some pelagic species may also track seasonal variation in food availability during their non-breeding periods. For example, Arctic Terns (*Sterna paradisaea*) travel considerable distances

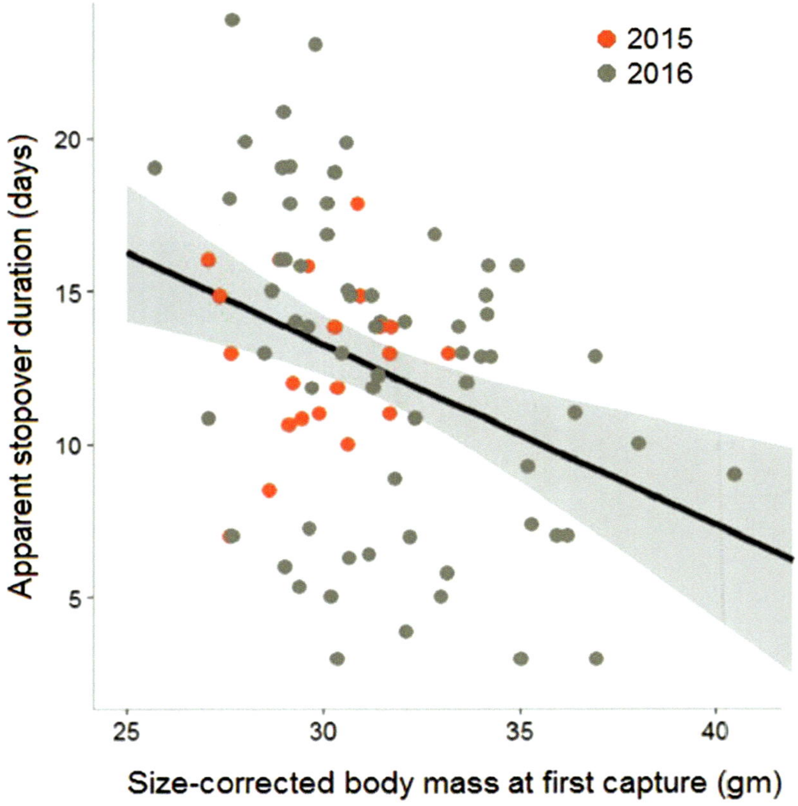


**Fig. 13.48** Relationship between body mass at first capture (assumed to correspond to time of arrival at a stopover site in northern Colombia) and apparent stopover duration ("apparent" because uncertainty about when a bird actually arrived at the stopover site) of Gray-cheeked Thrushes (*Catharus minimus*). (Figure modified from Supplementary Material for Gómez et al. 2017; open-access article licensed under a Creative Commons Attribution 4.0 International License, https://creativecommons.org/licenses/by/4.0/)

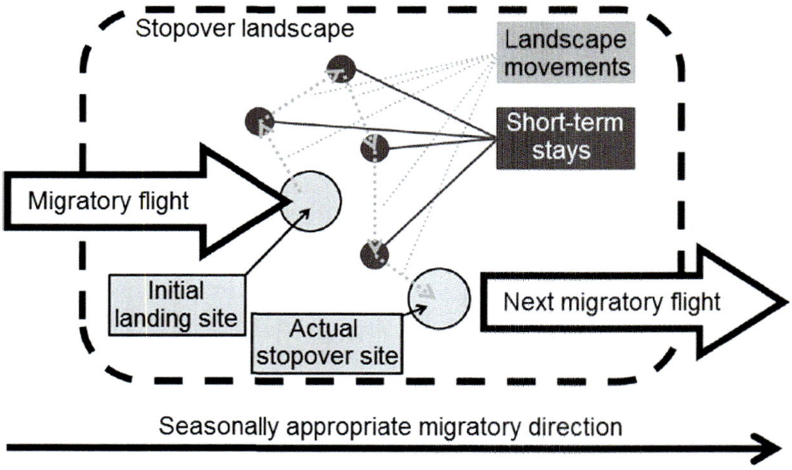


**Fig. 13.49** Migrating birds arriving at stopover sites may move from initial landing sites in search of better-quality sites, e.g., sites with greater food availability, fewer predators, or less competition for food resources. These landscape movements are not necessarily oriented in a particular direction. (Figure modified from Schmaljohann and Eikenaar 2017; © 2017 Springer-Verlag Berlin Heidelberg, used with permission)

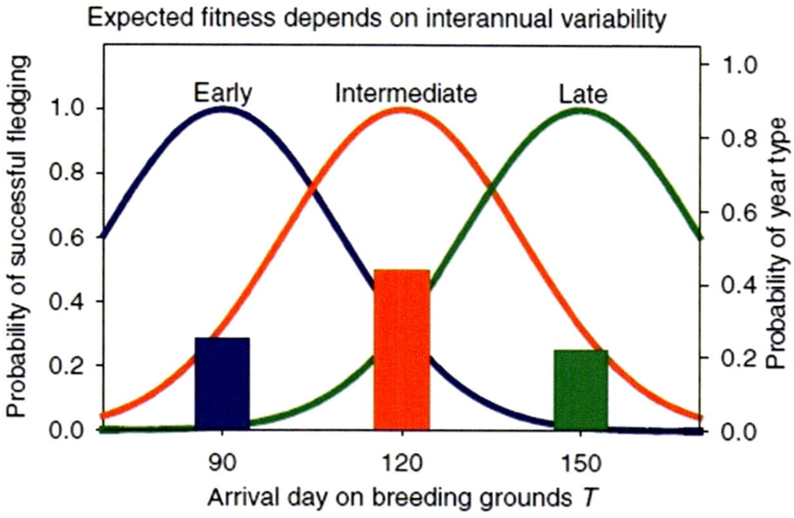


**Fig. 13.50** Information gathered at stopover sites can be used by migrating birds to determine the best time (in terms of fitness benefits) to arrive in breeding areas. Expected breeding success (probability of successfully fledging young) depends on timing, with the peak of the curves indicating optimal arrival times (ordinal date, T). The dates when environmental conditions are optimal for initiating breeding vary among years, sometimes early, most often intermediate, and sometimes late. The probability of each type of year is indicated by the blue, red, and green bars and the scale on the right. (Figure from Winkler et al. 2014; © 2014 Winkler et al., licensee BioMed Central Ltd., open-access article distributed under the terms of the Creative Commons Attribution License, https://creativecommons.org/licenses/by/4.0/)

in the Southern Hemisphere as they track variation in ocean productivity and food availability (Alerstam et al. 2019; Fig. 13.58). Other species of seabirds have also been found to use different areas during the nonbreeding period, presumably also tracking variation in productivity and food availability, e.g., Tristan Albatrosses (*Diomedea dabbenena*, Reid et al. 2013) and Brown Boobies (*Sula leucogaster*, Kohno et al. 2019). However, most studies of seabirds have revealed that individuals make consistent use of variably sized areas of ocean during the nonbreeding period (e.g., Phillips et al. 2007; Pinet et al. 2011; Becker et al. 2016; Delord et al. 2019; Goldstein et al. 2019; Grecian et al. 2019).

Longitudinal migration in the Neotropics involves seasonal movements of birds from either east to west or west to east. Examples of species the exhibit longitudinal migration include several species of waterfowl that breed in inland freshwater wetlands and winter along the Atlantic coast (e.g., Capllonch 2004; Capllonch et al. 2008). Longitudinal migration has also been reported in several species of songbirds, with these movements often from breeding areas in eastern South America (e.g., the Atlantic Forest) to more western locations during the nonbreeding season (e.g., Lees 2016; Fig. 13.59).

## 13.11 Seasonal Differences

For many species of birds, spring migration is shorter in duration than fall migration, with selection favoring a time-minimized strategy. Based on data from weather surveillance radar, Horton et al. (2016b) determined that groundspeeds and airspeeds of nocturnal migrants in the eastern United States were significantly higher in spring than fall and, more generally, birds made more rapid and precise flights in spring than fall. For migrants, early arrival in breeding areas can be important because, by arriving early, migrants may be more likely to gain access to resources

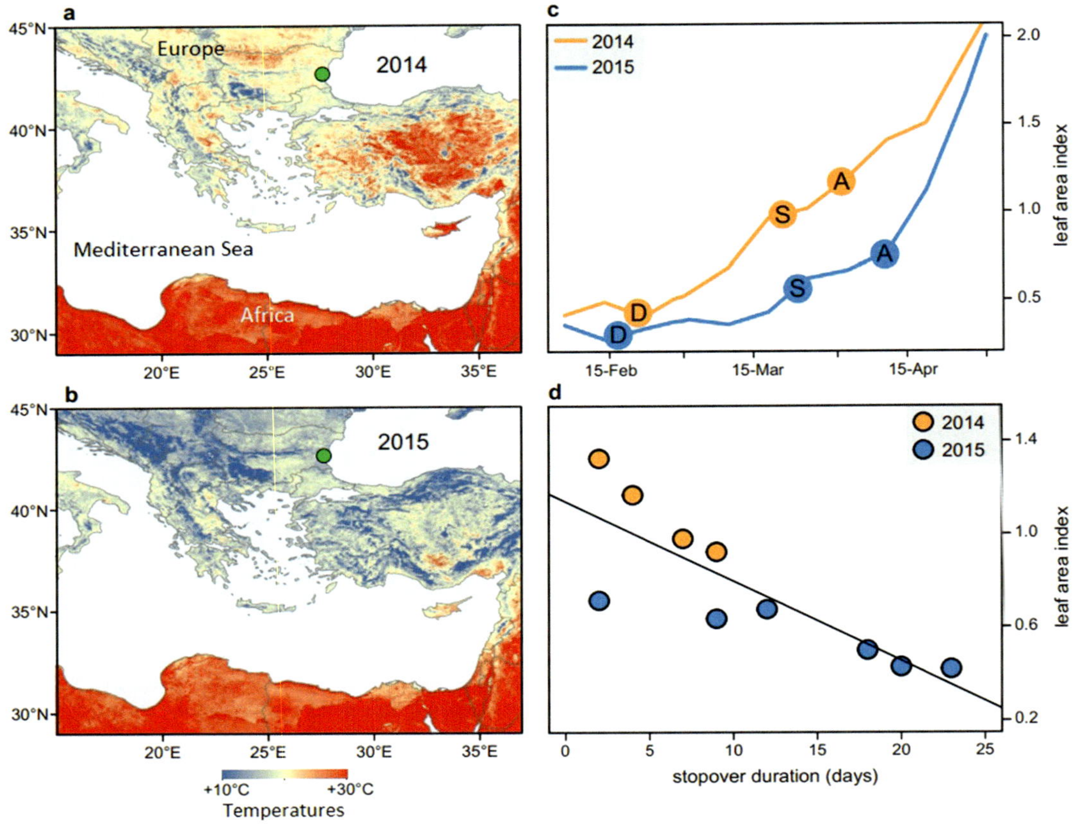


**Fig. 13.51** Effect of temperature and plant phenology on the phenology of spring migration of Semicollared Flycatchers (*Ficedula semitorquata*). During the period from 22 March to 7 April, temperatures in the migratory pathway were much higher in 2014 (**a**) than in 2015 (**b**). The average temperature across the northern Mediterranean Sea was 20.5 °C in 2014 and only 14.8 °C in 2015. (**c**) Plant phenology (as measured by the leaf area index, $m^2$ of leaf area per $m^2$ of ground area) at the breeding site was similar between years when the flycatchers departed (D) from their wintering area in eastern-central Africa (D), but was considerably delayed by the time flycatchers reached a stopover site (S) on the southern edge of the Mediterranean Sea and when they eventually arrived in their breeding area in eastern Bulgaria (green circle). (**d**) In response to the colder temperature and delayed plant phenology, the flycatchers remained at the stopover location an average of 10 days longer in 2015 compared to 2014. (Figure from Briedis et al. 2017; © The Authors 2017, open-access article distributed under the terms of the Creative Commons Attribution 4.0 International License, https://creativecommons.org/licenses/by/4.0/)

such as breeding territories and mates that can influence reproductive success. In addition, birds that arrive early can start breeding earlier, and early breeding can enhance reproductive success (van Noordwijk et al. 1995; Fig. 13.60). Early arrival in breeding areas also means more time to breed, perhaps improving a bird's chances of raising multiple broods and more time for young to develop skills needed for survival and to prepare for fall migration (McNamara et al. 1998).

Although multiple factors can influence the duration of migration, one important factor is flight speed and many species of birds have been found to fly faster during spring migration than fall migration (Figs. 13.61 and 13.62). For example, radar studies have revealed that nocturnal migrants fly on average at 12–16% higher speeds during spring migration than during fall migration (Karlsson et al. 2012; Nilsson et al. 2014), with higher speeds allowing migrants to

**Table 13.3** Examples of species exhibiting the four migratory systems of birds in the Neotropics (Table from Jahn et al. 2020; © 2020 Oxford University Press, used with permission)

| Species | Breeding site | Nonbreeding site | Potential driver(s) of migration | References |
|---|---|---|---|---|
| **Intratropical migrants** | | | | |
| Black Skimmer (*Rhynchops Niger*) | Southern Peru | Northern Bolivia, southwestern Brazil | Nest-site limitation due to flooding | Davenport et al. (2016) |
| Orinoco Goose (*Neochen jubata*) | Southern Peru | Northern Bolivia | Competition for nesting sites | Davenport et al. (2012) |
| Lesser Elaenia (*Elaenia chiriquensis*) | Central and southern Brazil | Central and western Amazonia | Arthropod (and fruit?) food availability | Marini and Cavalcanti (1990) |
| Rufous-thighed Kite (*Harpagus diodon*) | Southeastern Brazil | Amazonia | Food availability | Lees and Martin (2015) |
| **Altitudinal migrants** | | | | |
| Violet-throated Metaltail (*Metallura baroni*) | Higher altitudes of Ecuadorian Andes | Lower altitudes of Ecuadorian Andes | Nectar (and arthropod) food availability | Hobson et al. (2003) |
| Yellow-legged Thrush (*Turdus flavipes*) | Higher altitudes of the Atlantic rainforest | Lower altitudes of the Atlantic rainforest | Fruit and arthropod (other?) food availability | de Castro et al. (2012) |
| Three-wattled Bellbird (*Procnias tricaarunculatus*) | Higher altitudes in Central America | Lower altitudes in central America | Fruit food availability | Powell and Bjork (2004) |
| **Neotropical austral migrants** | | | | |
| **South American temperate-tropical migrants** | | | | |
| White-crested Elaenia (*Elaenia albiceps chilensis*) | Patagonian forests | Atlantic rainforest and Brazilian interior | Fruit and arthropod food availability | Bravo et al. (2017) |
| Barn Swallow (*Hirundo rustica*) | Central Argentina | Northern South America | Arthropod food availability | Winkler et al. (2017) |
| Creamy-bellied Thrush (*Turdus amaurochalinus*) | South-Central South America | Central South America (and other?) food availability | Fruit and arthropod | Capllonch et al. (2008) |
| Slaty Elaenia (*Elaenia strepera*) | Southern Bolivia and northwestern Argentina | Northern South America | Fruit and arthropod food availability | Marantz and Remsen (1991) |
| **South American cool-temperate migrants** | | | | |
| Chocolate-vented Tyrant (*Neoxolmis rufiventris*) | Patagonia | Northeastern Argentina and Uruguay | Arthropod food availability | Farnsworth and Langham (2018) |
| Austral Negrito (*Lessonia rufa*) | Patagonia | South-Central South America | Arthropod food availability | Farnsworth et al. (2018) |
| Ruddy-headed Goose (*Chloephaga rubidiceps*) | Patagonia | Pampas grasslands | Temperature limitation/food availability | Blanco et al. (2003, 2006) |
| **Longitudinal migrants** | | | | |
| Cinnamon Warbling-Finch (*Poospiza ornata*) | Western Argentina | Eastern Argentina | Seed food availability | Cueto et al. (2011) |
| Rosy-billed Pochard (*Netta peposaca*) | Central Argentina | Eastern Brazil | Temperature limitation/food availability | Capllonch (2004) |

(continued)

**Table 13.3** (continued)

| Species | Breeding site | Nonbreeding site | Potential driver(s) of migration | References |
|---|---|---|---|---|
| Neotropic Cormorant (*Phalacrocorax brasilianus*) | Central Argentina | Eastern Brazil | Temperature limitation/food availability | Capllonch (2004) |
| Ash-throated Casiornis (*Casiornis fuscus*) | Caatinga ecoregion | Amazonia | Arthropod (and fruit?) food availability | Lees (2016) |

travel further each day. Interestingly, based on data obtained from tracking 1937 radio-tagged individuals of four species of shorebirds, Duijns et al. (2019) determined that, whereas the overall duration of spring migration was shorter than that of fall migration due to difference in stopover duration, all four species flew at higher speeds during fall migration. One possible explanation for higher flight speeds in the fall is that the shorebirds fly faster to minimize predation risk from migrating raptors that reach peak abundance later in the fall (Fig. 13.63). However, other factors, including deteriorating weather conditions, competition, and declining food availability may also contribute to the faster airspeeds during fall migration (Duijns et al. 2019).

Another important factor that influences the speed of migration is the time spent at stopover sites (Fig. 13.64). Schmaljohann (2018) reviewed the results of 64 studies and found that migration speed was significantly faster during the spring than the fall for shorebirds (waders), gulls, swifts, and songbirds (Fig. 13.64) and, further, that the main driver of this difference was a seasonal difference in total stopover duration. With less time spent at stopover sites, spring migration of gulls, songbirds, and swifts averaged 250%, 171%, and 154% faster, respectively, than fall migration (Schmaljohann 2018).

Among some groups of birds, the speed of migration is similar during the spring and fall, e.g., European Nightjars (*Caprimulgus europaeus*; Jacobsen et al. 2017), Bar-headed Geese (*Anser indicus*; Köppen et al. 2010), and Light-bellied Brant Geese (*Branta berical hrota*; Vissing et al. 2020), and the speed of fall migration for some species, e.g., many waterfowl, is shorter than that of spring migration (Schmaljohann 2018; Figs. 13.65 and 13.66). Similarly, although little is known about the migratory behavior of owls, Linkhart et al. (2016) found that, for one species—Flammulated Owls (*Psiloscops flammeolus*) —fall migration took longer than spring migration. The duration of fall migration is also longer than spring

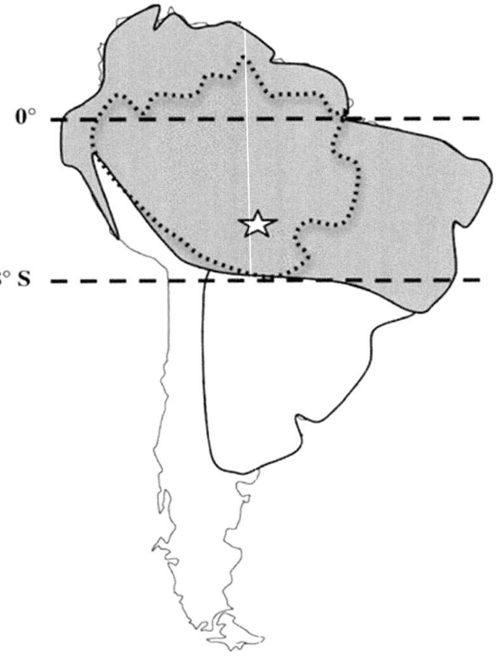


**Fig. 13.52** The breeding and winter ranges of Tropical Kingbirds (*Tyrannus melancholicus*) in South America overlap. The gray area mostly north of 18°S latitude represents the area where they are found year-round. The dashed line within the gray area indicates the Amazon Basin. The white area indicates where some Tropical Kingbirds breed during the austral summer. (Figure from Jahn et al. 2010; © 2010 The Authors. Journal compilation © 2010 British Ecological Society, used with permission)

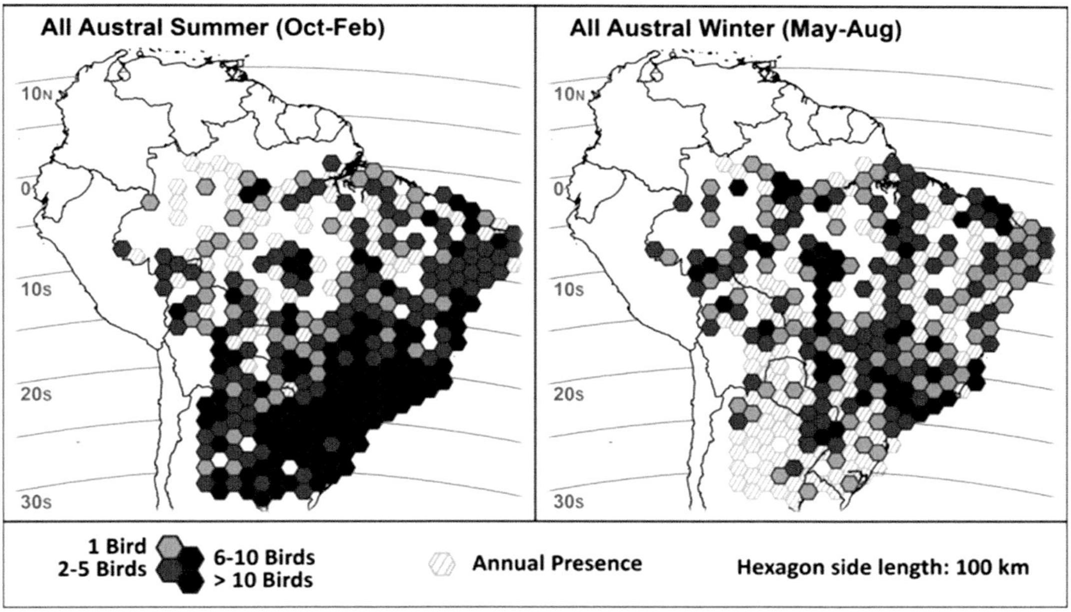


**Fig. 13.53** Distribution of Common Potoos (*Nyctibius griseus*) during the austral summer (October–February) and austral winter (May–August) based on data from both WikiAves and eBird. Note the migration of most individuals from breeding areas further south to wintering areas further north. (Figure from DeGroote et al. 2021; © 2020 British Ornithologists' Union, used with permission)

migration for other species of birds, e.g., Whimbrels (*Numenius phaeopus*; Carneiro et al. 2019) and Egyptian Vultures (*Neophron percnopterus*; Phipps et al. 2019). Although early arrival in breeding areas in the spring can be beneficial, arriving too early may not be beneficial, and even result in mortality, if conditions are unfavorable in breeding areas (Marcström and Mascher 1979). Schmaljohann (2018) suggested that starting spring migration early, but then adjusting the speed of migration based on environmental conditions encountered en route may be the best strategy for determining the optimal time to arrive in breeding areas. For example, Flammulated Owls breed at high elevations in the United States and Canada and those that arrive too early may encounter snow cover and low temperatures that decrease availability of their prey (Linkhart et al. 2016). To avoid such conditions, owls can use environmental conditions and prey availability at stopover sites as cues to adjust the time of arrival in their breeding areas. Waterfowl that breed in the Arctic may also encounter unfavorable conditions if they arrive too early and, therefore, may also benefit from a strategy of using conditions en route to adjust their time of arrival (Bety et al. 2004). Kölzsch et al. 2016) found that the duration of spring migration of Greater White-fronted Geese (*Anser albifrons*) that breed in the Arctic was about twice as long as fall migration. An important reason for this difference is that, during spring migration, geese need to spend time foraging at stopover sites to store the energy that will be needed to breed (i.e., capital breeding; Fig. 13.67). In contrast, in the fall, the geese can forage near breeding areas, then wait for favorable wind conditions and quickly fly to wintering areas.

## 13.12 Timing of Migration

The arrival times of migrating birds at their breeding and wintering areas can be of critical importance. In spring in the northern hemisphere, birds

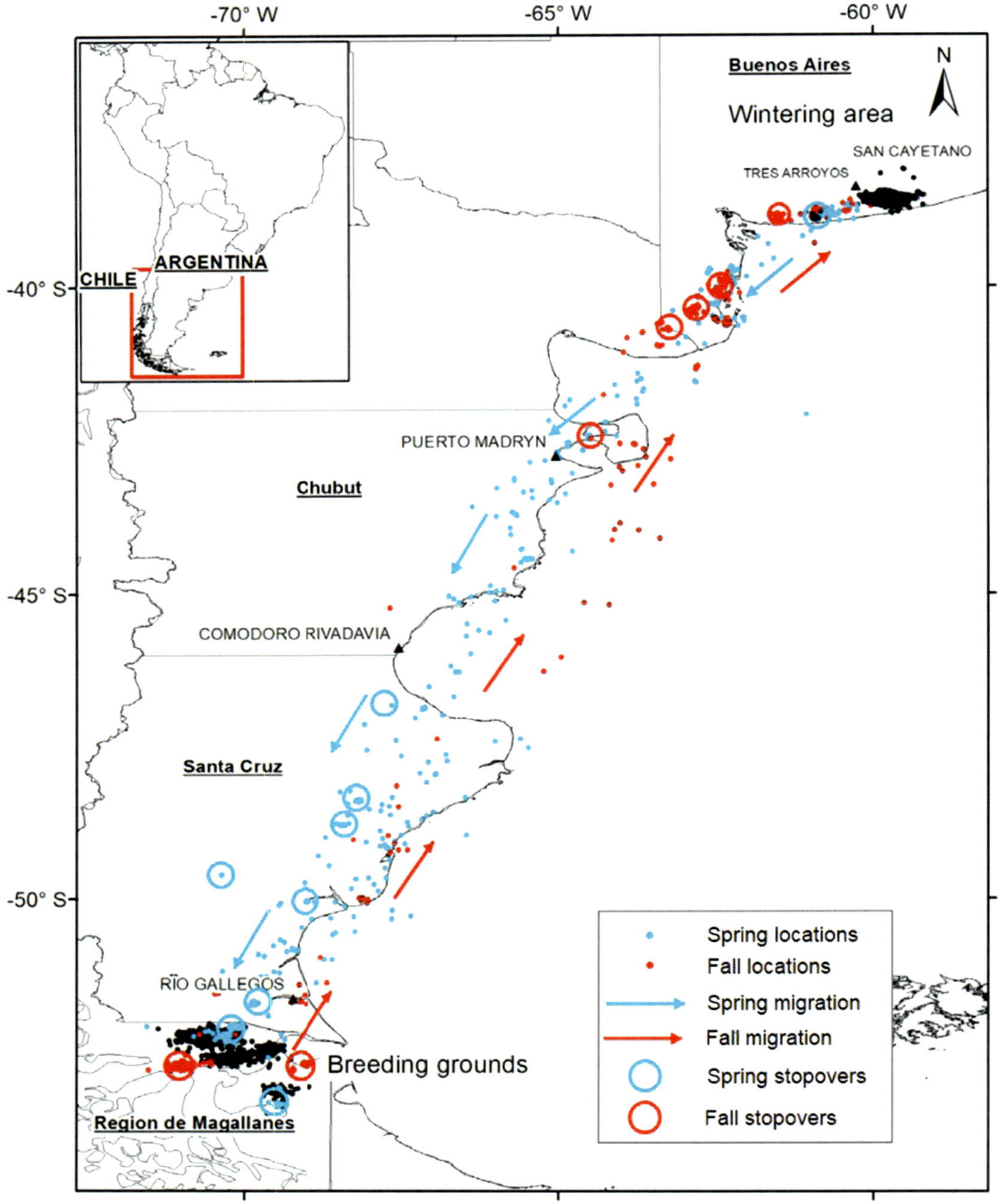


**Fig. 13.54** Migration routes of Ruddy-headed Geese (*Chloephaga rubidiceps*) tracked from 2015 to 2018. Blue indicates spring migration and red indicates fall migration. Circles show stopovers' locations during spring and fall migration. Black points in Buenos Aires province indicate the wintering area and black points in Patagonia indicate the breeding area. The geese left the wintering site in August and migrated 2395 km to the breeding area in the Patagonian region. (Figure from Pedrana et al. 2020; © The Authors 2020. Published by Cambridge University Press on behalf of Birdlife International, used with permission)

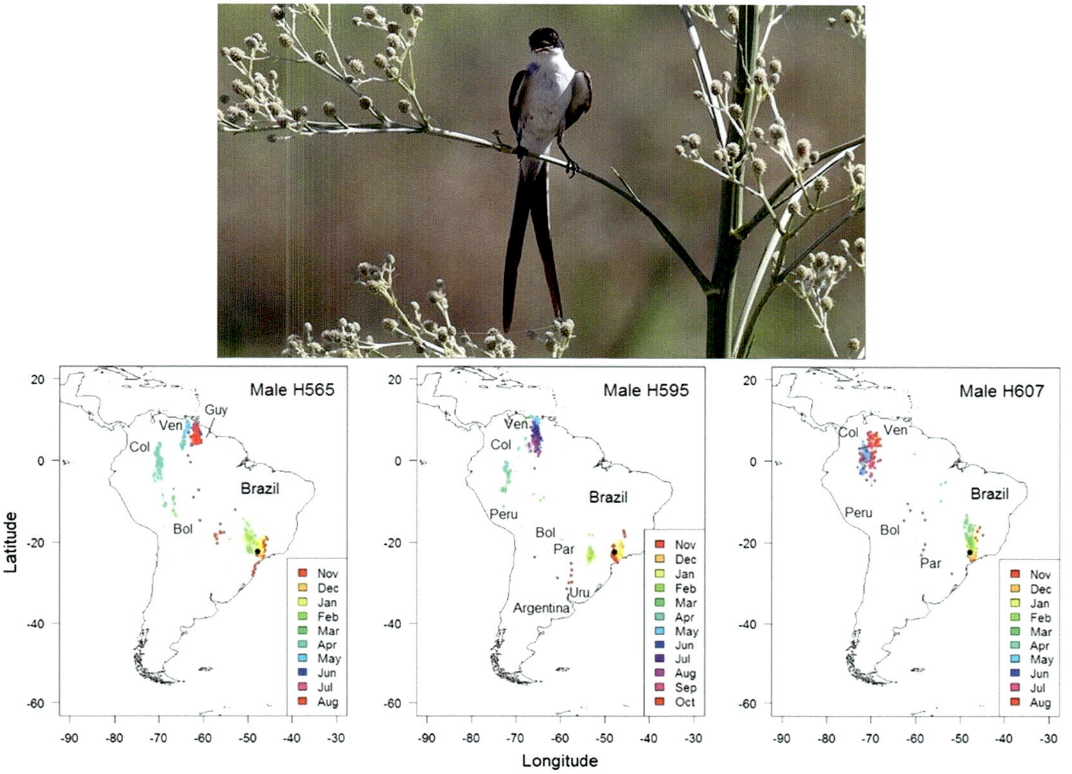


**Fig. 13.55** Wintering areas and migration routes of three male Fork-tailed Flycatchers (*Tyrannus savanna*) captured at a breeding site in southeastern Brazil. Fork-tailed Flycatchers in Brazil breed from late January to April and, after breeding, these three male Fork-tailed Flycatchers migrated to northern South America. This migration may be timed so that the Fork-tailed Flycatchers arrive in areas with abundant food resources when they undergo their annual molt (Col, Colombia; Ven, Venezuela; Guy, Guyana; Bol, Bolivia; Par, Paraguay; Uru, Uruguay. (Figure modified from Jahn et al. 2016; open-access article licensed under a Creative Commons Attribution 4.0 International License, https://creativecommons.org/licenses/by/4.0/. Photo by dfaulder, Wikipedia, CC BY 2.0, https://creativecommons.org/licenses/by/2.0/)

arriving too early may face unfavorable weather (especially at higher latitudes), but those arriving too late may have reduced breeding success (Kokko 1999; Vergara et al. 2007). Many factors can influence the timing of migration, including genetic factors. For example, differences in the timing of fall migration by two species of redstarts, Common (*Phoenicurus phoenicurus*) and Black (*P. ochruros*) redstarts, were found to have a genetic basis (Berthold 1998). In a comparative analysis of 18 species, Berthold (1990) found that the onset of migratory activity in captive birds (migratory restlessness, or zugunruhe; Fig. 13.68) maintained under controlled conditions was highly correlated with that of birds in the wild, suggesting that the timing of migration has a genetic, or innate, component. However, the extent of genetic control probably varies among species, with such control more likely in species, such as long-distance migrants, that breed in highly seasonal environments where environmental conditions are predictable within and across years (Ogonowski and Conway 2009) and when wintering and breeding areas are far apart and conditions in one location provide no evidence of conditions at the other location.

**Table 13.4** Species of long-distance latitudinal migrants known to exhibit intratropical migration

| Species | Scientific name | Wintering area | References |
|---|---|---|---|
| Common Ringed Plover | *Charadrius hiaticula* | Africa | Lislevand et al. (2017) |
| Common Swift | *Apus apus* | Africa | Wellbrock et al. (2017) |
| Pallid Swift | *Apus pallidus* | Africa | Norevik et al. (2019) |
| Common Cuckoo | *Cuculus canorus* | Africa | Thorup et al. (2017) |
| Fork-tailed Flycatcher | *Tyrannus savana* | South America | Jahn et al. (2016) |
| Eastern Kingbird | *Tyrannus tyrannus* | South America | Jahn et al. (2013b) |
| Lesser Elaenia | *Elaenia chiriquensis* | South America | De Paiva and Marini (2013) |
| White-crested Elaenia | *Elaenia albiceps* | South America | Marini and Cavalcanti (1990) |
| Red-backed Shrike | *Lanius collurio* | Africa | Thorup et al. (2017) |
| Great Reed Warbler | *Acrocephalus arundinaceus* | Western Africa | Lemke et al. (2013), Koleček et al. (2018) |
| Purple Martin | *Progne subis* | South America | Stutchbury et al. (2016) |
| Tree Swallow | *Tachycineta bicolor* | Southern United States, Mexico, and Caribbean islands | Knight et al. (2019) |
| Veery | *Catharus fuscescens* | South America | Heckscher et al. (2011) |
| Swainson's Thrush | *Catharus ustulatus* | South America | Delmore et al. (2012) |
| Thrush Nightingale | *Luscinia luscinia* | Eastern Africa | Stach et al. (2012) |
| Bobolink | *Dolichonyx oryzivorus* | South America | Renfrew et al. (2013) |
| Tawny Pipit | *Anthus campestris* | Western Africa | Briedis et al. (2016) |
| Red-eyed Vireo | *Vireo olivaceus* | South America | Callo et al. (2013) |
| Prothonotary Warbler | *Protonotaria citrea* | Caribbean, Central America | Wolfe and Johnson (2015) |

Among species where migratory behavior has a strong innate component, there must also be an internal clock or some internal mechanism for determining when to initiate migration. Such endogenous circannual clocks have been reported in more than 20 species of birds (Newton 2007b). The circannual clocks of birds have a period of about one year, but, for increased accuracy, are synchronized using environmental cues, such as day length, light intensity, and seasonal rainfall patterns (Wikelski et al. 2008). Because it is highly predictable and consistent between years, most birds use changes in day length to synchronize their internal clocks. Of course, variation in day length varies with latitude and, near the equator, day length changes little during the year. However, laboratory experiments indicate that even annual changes in day length of one hour (equivalent to the change at 9° north or south of the equator) are sufficient for precisely synchronizing the circannual clocks of European Starlings (*Sturnus vulgaris*; Dawson 2007). Even more impressively, experiments revealed that even changes in daylength as little as 17 minutes produced behavioral and physiological changes in Spotted Antbirds (*Hylophylax naevioides*; Hau et al. 1998). Additional study is needed, but these results suggest that even birds near the equator may be able to use changes in day length to synchronize circannual clocks.

Although the existence of circannual clocks has been well documented, how such clocks function remains unclear. One idea is that birds count days to derive an annual cycle (frequency de-multiplication hypothesis; Mrosovsky 1978; Gwinner 1986), but there is currently no experimental evidence to support this hypothesis. More recently, Wikelski et al. (2008) proposed an energy turnover hypothesis (ETH), where birds track the total amount of energy expended over a year and, by measuring or accounting for energy turnover (by some currently unknown

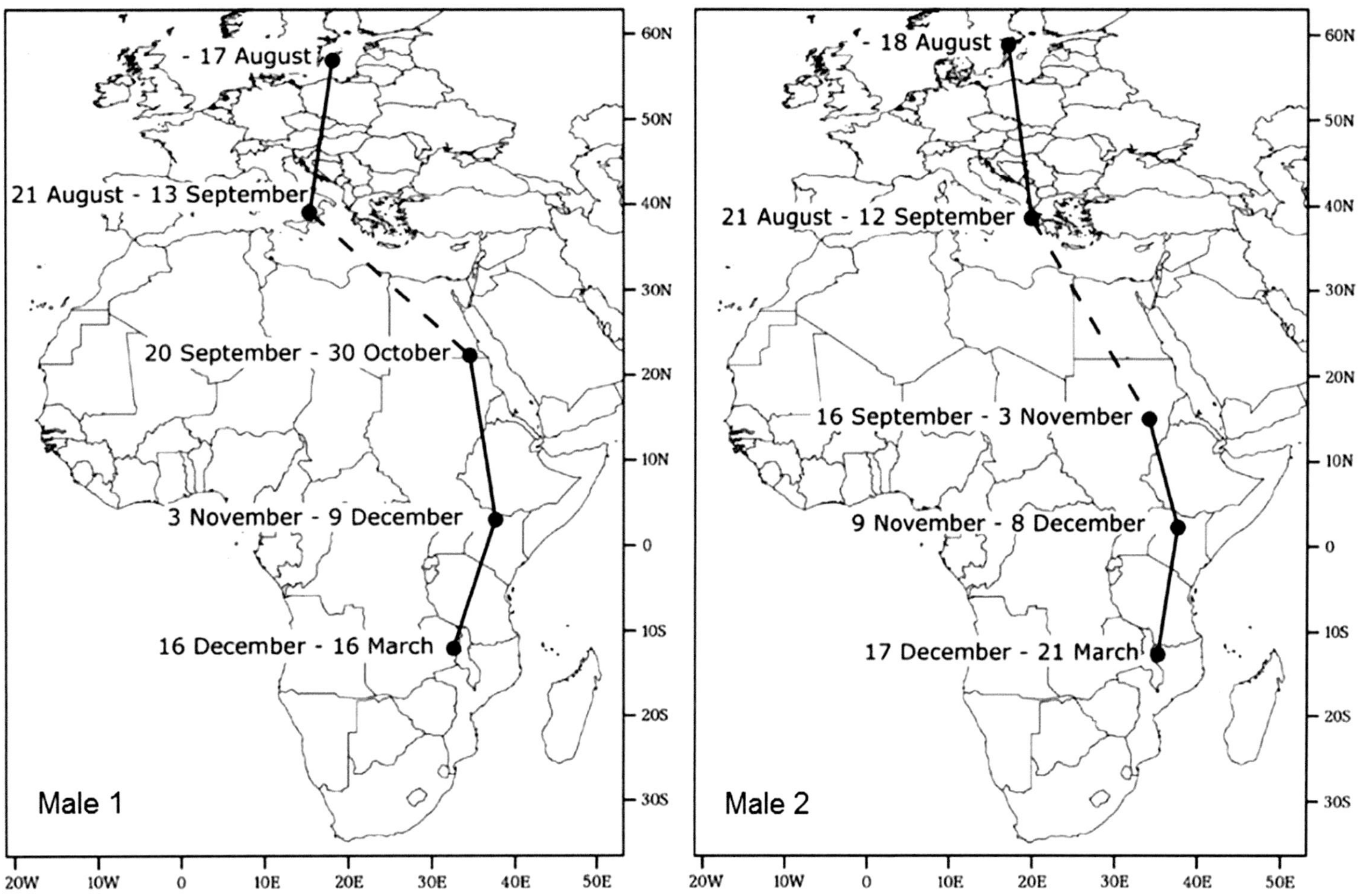


**Fig. 13.56** Intratropical migration by two male Thrush Nightingales (*Luscinia luscinia*). Both were captured in their breeding territories in Sweden and arrived in their first wintering areas in mid-September (Male 1 in southern Egypt and Male 2 in eastern Sudan). The second wintering areas for both males were in northern Kenya. The third wintering areas were in eastern Zambia for Male 1 and northwest Mozambique for Male 2. (Figure from Stach et al. 2012; open-access article distributed under the terms of the Creative Commons CC BY license, https://creativecommons.org/licenses/by/4.0/)

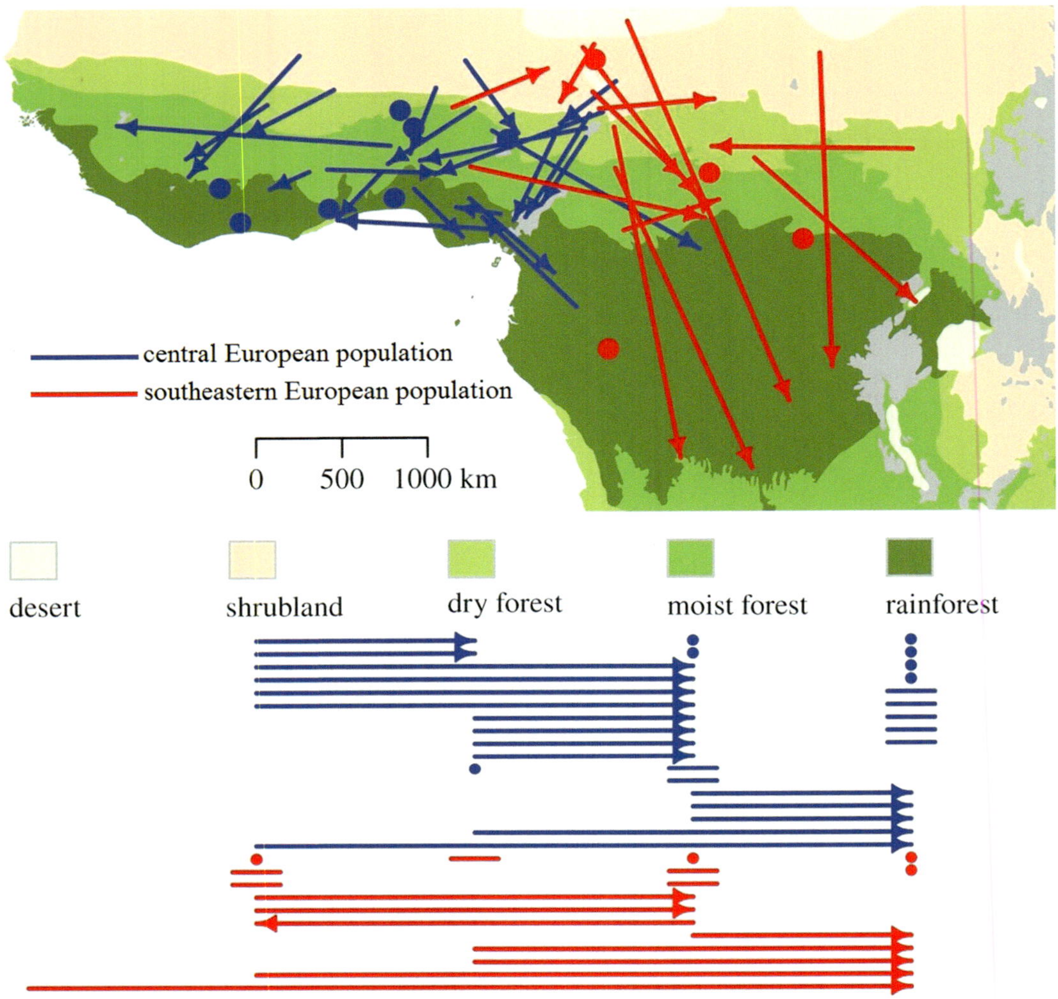


**Fig. 13.57** Intratropical movements between the first and the second sub-Saharan wintering sites of 22 Great Reed Warblers (*Acrocephalus arundinaceus*) from the central European population and 13 from the southeastern European population. Arrows connect individual locations and indicate the direction of movements and, at bottom, movements between different ecological zones. Movements between zones are indicated by arrows; movements within zones by lines; filled circles indicate birds that did not move. Gray color indicates mountains. (Figure from Koleček et al. 2018; © 2018 The Authors, open-access article published by the Royal Society under the terms of the Creative Commons Attribution License, https://creativecommons.org/licenses/by/4.0/)

mechanism), they can tell the time of year. This hypothesis remains to be tested. Regardless of how circannual clocks function, some birds clearly use such clocks to determine when to initiate migration.

For short-distance migrants, environmental factors have a greater influence on the timing of migration. When wintering and breeding areas are closer and where conditions in one location may provide some indication of expected conditions at the other location, environmental cues can be more useful. Environmental factors clearly influence the migratory behavior of irruptive species that do not migrate regularly. For example, decreasing food supplies trigger the migratory movements of Red Crossbills (*Loxia curvirostra*;

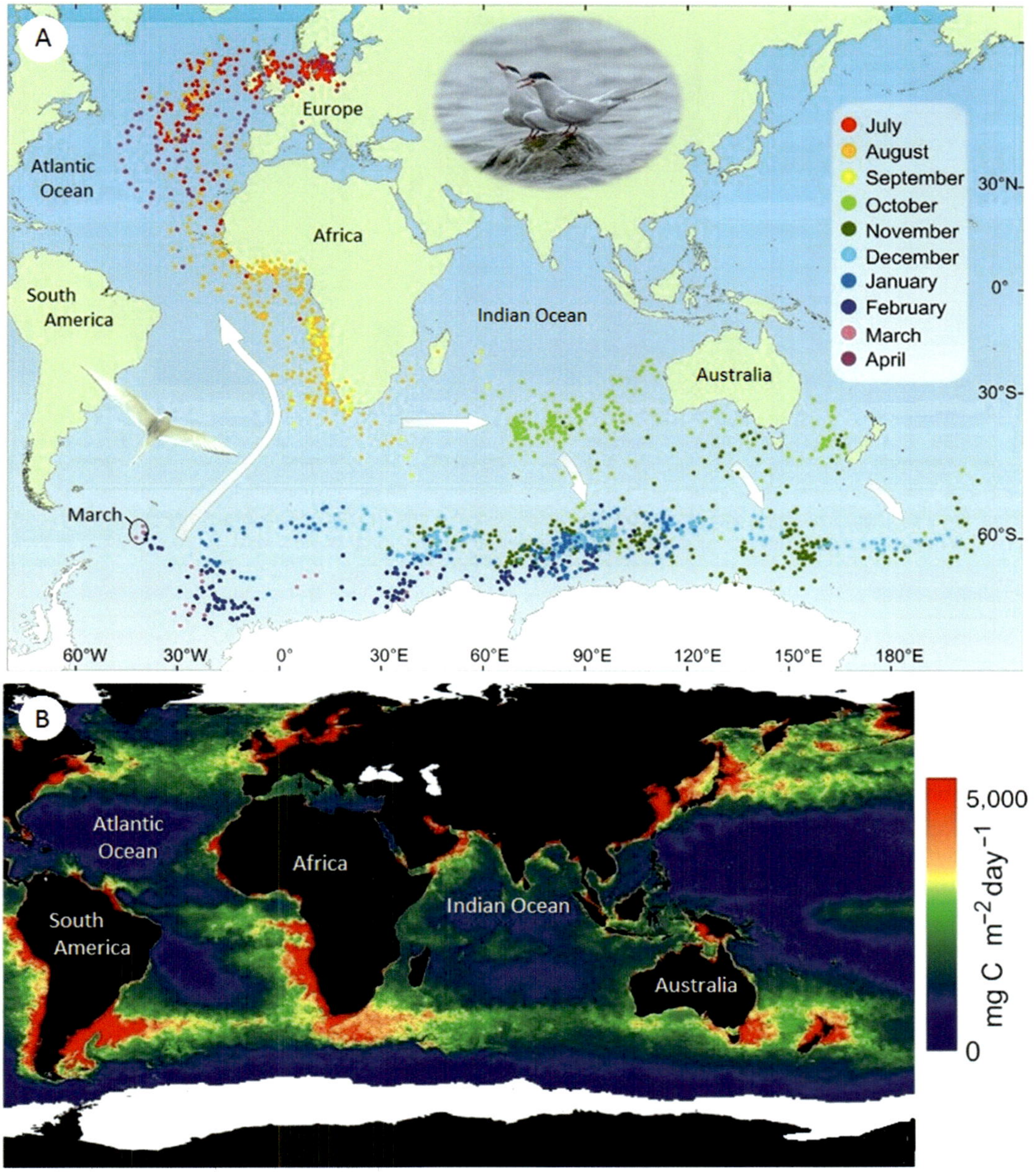


**Fig. 13.58** (**a**) Locations of Arctic Terns (*Sterna paradisaea*) from July through April. (**b**) Marine productivity (net primary production measured as milligrams of carbon per m$^2$ per day) in October. During September and October, many Arctic Terns are found in the highly productive waters off the west coast of Africa, then travel across the Indian Ocean to the Tasman Sea off the southeast coast of Australia. (Figure modified from Alerstam et al. 2019; Published by John Wiley and Sons, open-access article available under the terms of the Creative Commons Attribution License (CC BY), https://creativecommons.org/licenses/by/4.0/)

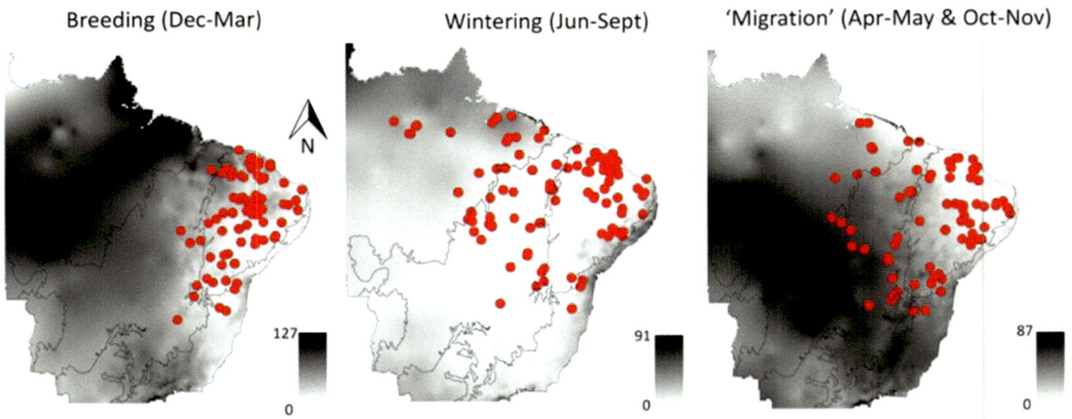


**Fig. 13.59** Example of longitudinal migration by Ash-throated Casiornises (*Casiornis fuscus*) in Brazil. Red circles indicate locations of Ash-throated Casiornises during the breeding, wintering, and migration periods. Dark shading indicates the mean precipitation (bars on the lower right of the maps) during the months of February (breeding), August (nonbreeding), and November (migration). These data suggest that not all Ash-throated Casiornises migrate from their breeding areas. The westward movement by some individuals in the population occurs during the dry season in Amazonia and the driest and hottest part of the year in their breeding areas. (Figure from Lees 2016; © 2016 Association of Field Ornithologists, used with permission)

Koenig and Knops 2001) and other irruptive species of birds.

Regardless of the degree to which the timing of migration might be under genetic control, evidence suggests that birds can respond to a variety of cues during migration to better optimize arrival times. So, even for species where the time when migration begins is largely under genetic control, conditions en route can influence the duration of the migratory journey. Of course, some migrants must cross extensive areas of unsuitable habitat, such as deserts or, for landbirds, extensive bodies of water, and flights over those areas may often be crossed or, in the case of large bodies of water for landbirds, must be crossed nonstop. After initiating such flights, birds sometimes, due to unfavorable winds or other factors (insufficient fat reserves), decide to return and reverse direction. For example, birds migrating from Europe to Africa and initiating flights across the Mediterranean Sea sometimes reverse course and return to land (reverse migration; Bruderer and Liechti 1998).

For long-distance migrants, migration typically consists of a series of flights, with those flights interspersed with time spent at stopover sites where energy reserves can be replenished. Birds can potentially use environmental cues at stopover sites to determine how long to remain and what distance to fly before again stopping, and such decisions ultimately determine the overall speed of migration and the time of arrival at the breeding grounds. For example, Pink-footed Geese (*Anser brachyrhynchus*) appear to use plant phenology to determine speed of migration and arrival times in breeding areas, with these herbivores adjusting migration speed to more closely follow a green wave of plant growth (Duriez et al. 2009). Tøttrup et al. (2010) examined the timing of spring migration of 12 songbirds in Europe and found that local temperature best predicted arrival times in breeding areas. Interestingly, however, this was true only for individuals that were first to arrive in breeding areas, usually adult males closely followed by adult females. Thus, experienced birds monitored environmental conditions and timed their migration accordingly; inexperienced, first-time migrants relied entirely on endogenous, or innate, cues.

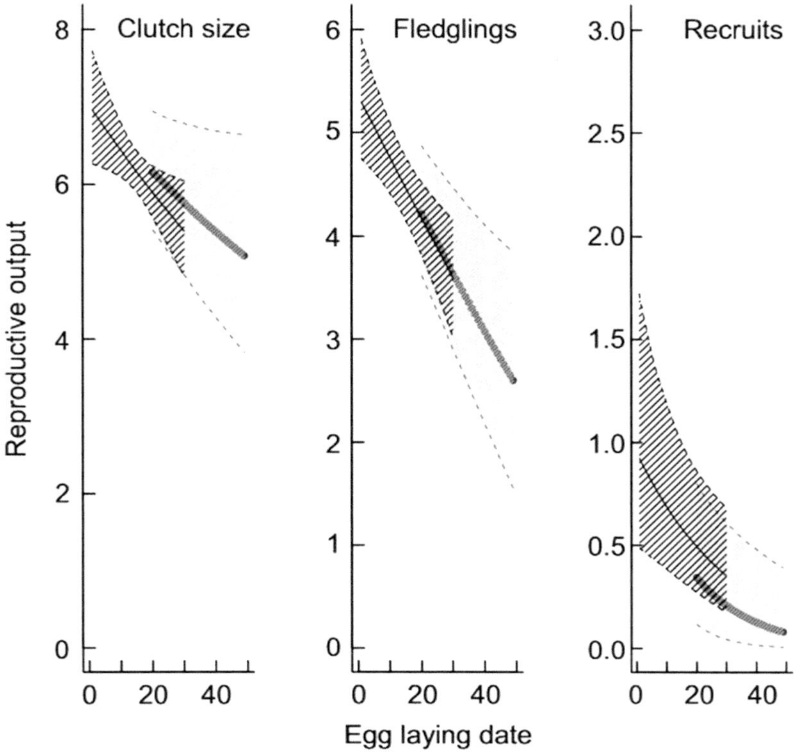


**Fig. 13.60** Effect of egg-laying date on the reproductive output of female Northern Wheatears (*Oenanthe oenanthe*). Females that initiated egg laying earlier in the breeding season had larger clutches, more fledglings, and more young surviving until the next breeding season (recruits). They also were able to initiate second clutches earlier and so also had greater reproductive output in their second breeding attempt. Egg-laying date 1 = 1 May. First breeding attempts are indicated by the thin black lines and second breeding attempts by the thicker gray line (± 95% credible intervals). (Figure from Low et al. 2015; Published by John Wiley and Sons, open-access article available under the terms of the Creative Commons Attribution License (CC BY), https://creativecommons.org/licenses/by/4.0/)

## 13.13 Protandry

Among many species of migratory birds, males tend to arrive in breeding areas before females, a phenomenon called protandry (Fig. 13.69). Protogyny, where females arrive in breeding areas before males, has been reported only for species that exhibit sex-role reversal (e.g., phalaropes; Oring and Lank 1982; Reynolds et al. 1986). Among the hypotheses proposed to explain the evolution of protandry (and protogyny for sex-role reversed species) are the rank-advantage hypothesis and the mate-opportunity hypothesis. The rank-advantage hypothesis suggests that early arriving males benefit by acquiring higher-quality territories (Ketterson and Nolan 1976; Morbey and Ydenberg 2001). The mate-opportunity hypothesis posits that early arriving males are more successful than later arriving males in acquiring mates; this would be particularly important for polygynous species and in populations with male-biased sex ratios where some males may not obtain mates (Morbey and Ydenberg 2001; Kokko et al. 2006). These hypotheses are, of course, not mutually exclusive because males

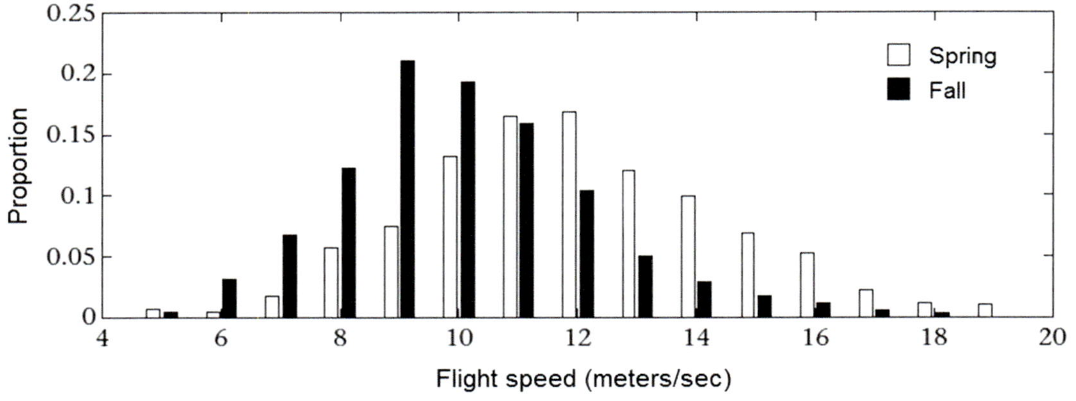


**Fig. 13.61** Flights speeds of nocturnal songbird migrants during spring and fall migration at a location in northern Sweden. Flight speeds averaged 21% faster during spring migration than fall migration. (Figure modified from Karlsson et al. 2012; © 2011 The Association for the Study of Animal Behaviour. Published by Elsevier Ltd., used with permission)

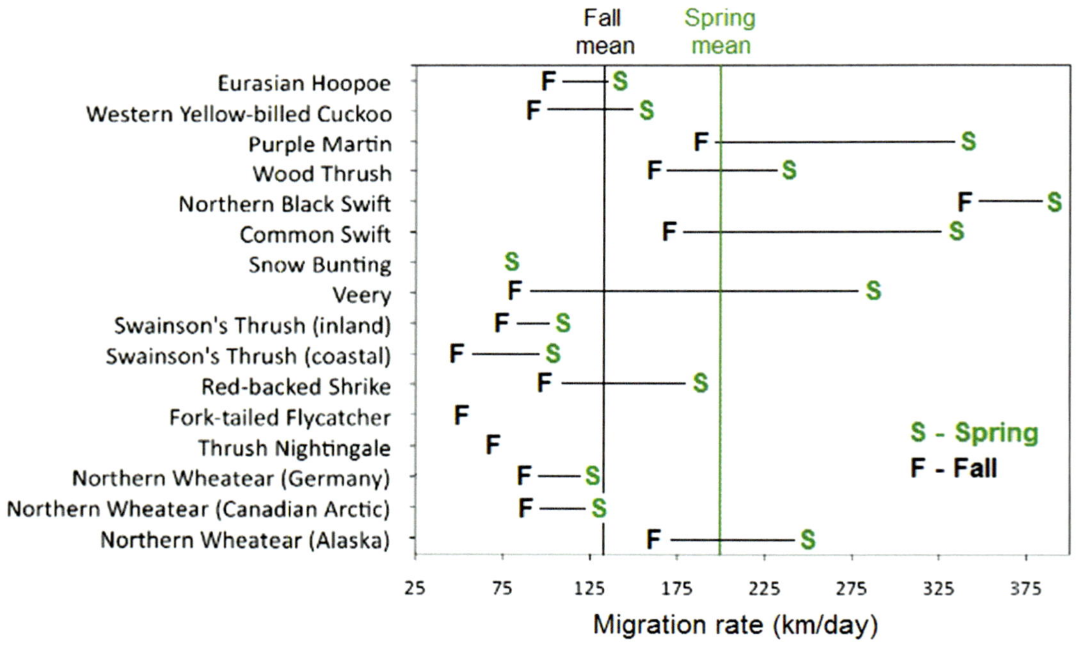


**Fig. 13.62** Migration rates during spring ($N = 13$ species) and fall ($N = 15$ species) for several species of birds. Mean migration speeds for all species combined were 135 km/day in the fall and 206 km/day in the spring. Eurasian Hoopoe, *Upupa epops*; Yellow-billed Cuckoo, *Coccyzus americanus*; Purple Martin, *Progne subis*; Wood Thrush, *Hylocichla mustelina*; Black Swift, *Cypseloides niger*; Common Swift, *Apus apus*; Snow Bunting, *Plectrophenax nivalis*; Veery, *Catharus fuscescens*; Swainson's Thrush, *Catharus ustulatus*; Red-backed Shrike, *Lanius collurio*; Fork-tailed Flycatcher, *Tyrannus savanna*; Thrush Nightingale, *Luscinia luscinia*; Northern Wheatear, *Oenanthe oenanthe*. (Figure modified from McKinnon et al. 2013; © 2013 Oxford University Press, used with permission)

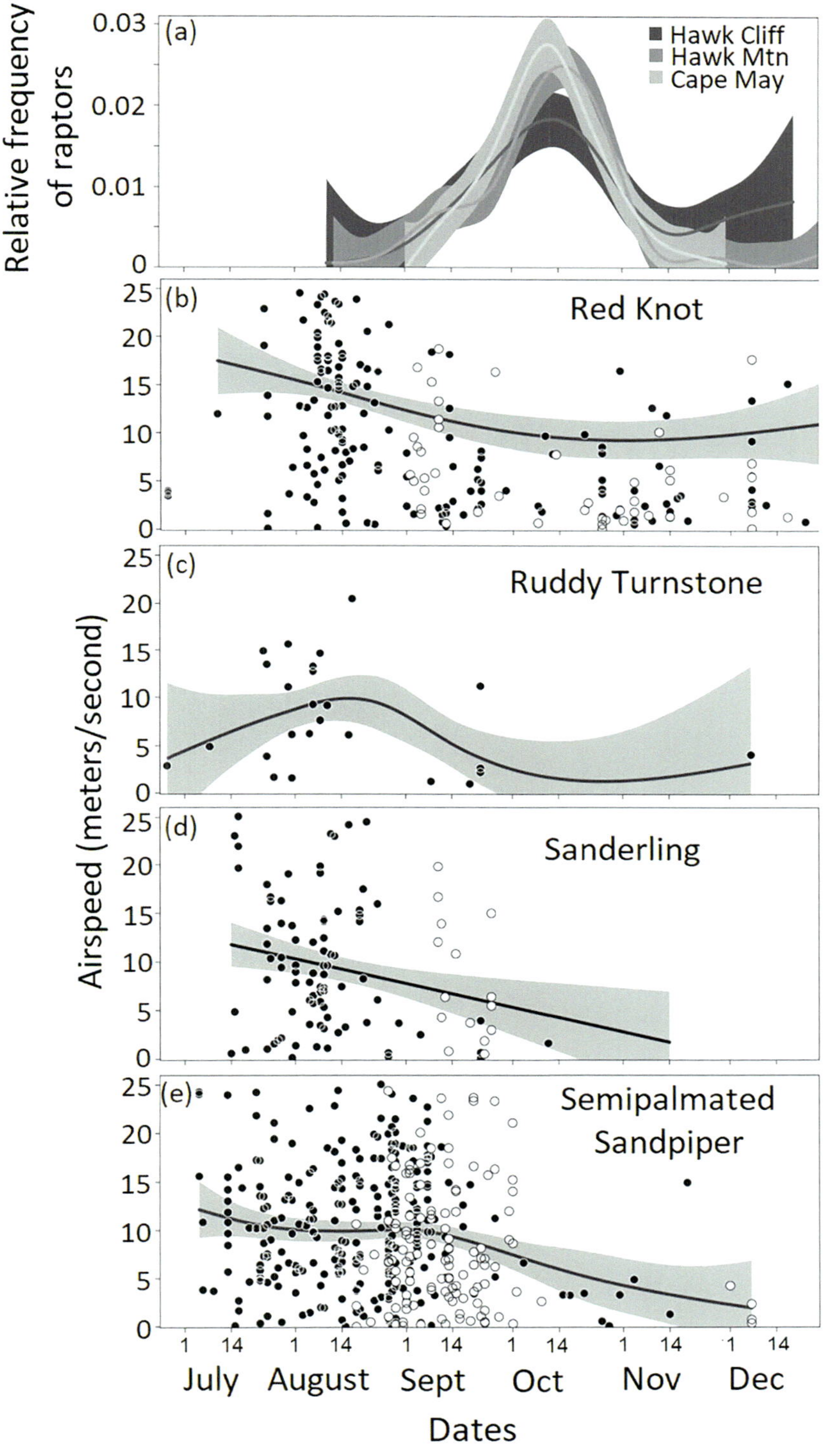


**Fig. 13.63** (**a**) Mean relative frequency (per day) of avian predators at three sites in North America where migrating raptors are observed in relatively high numbers. Predators considered as relevant predators for the four shorebird species considered during the study included Peregrine Falcons (*Falco peregrinus*), Merlins (*Falco columbarius*),

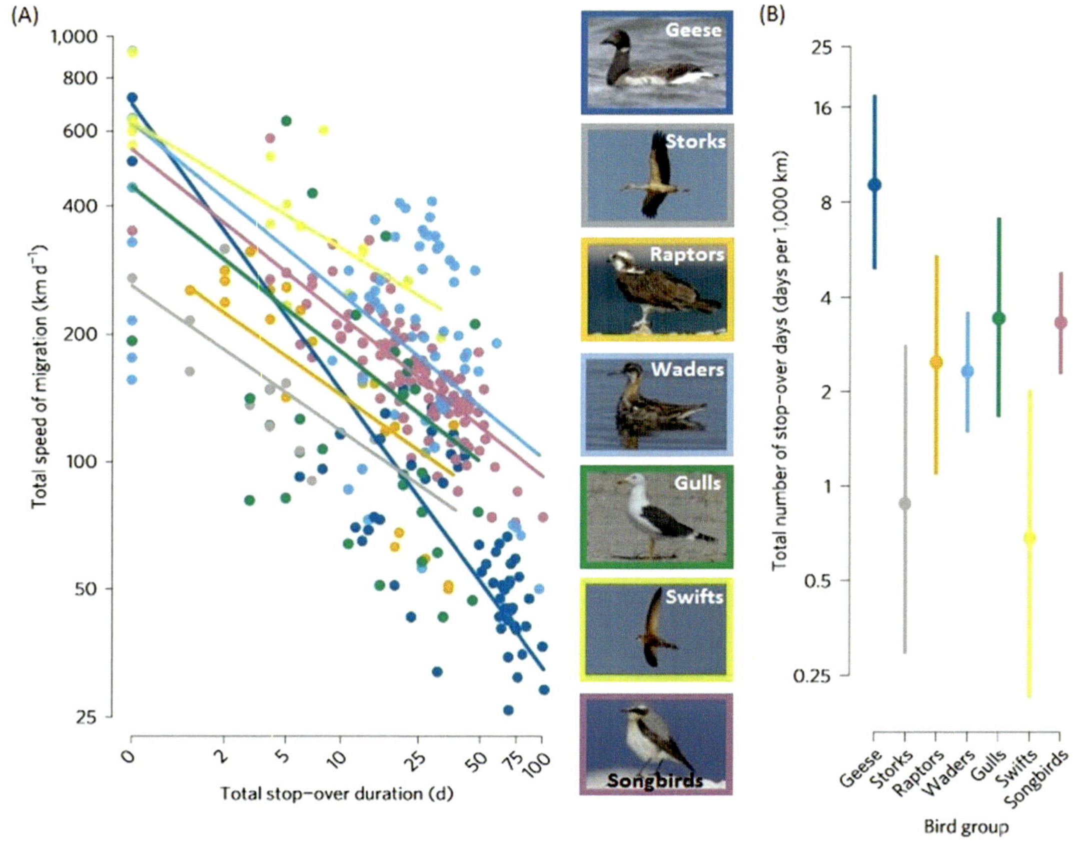


**Fig. 13.64** Total stopover duration explained variation in overall speed of migration in seven taxa of birds. (**a**) Total speed of migration plotted against individual total stopover duration for seven taxa of birds in spring. Total stopover duration and bird group explained 66% of variation in total migration speed. (**b**) Mean number (± 95% credible intervals) of stopover days for 1000-km migratory journeys. Waders = shorebirds. (Figure modified from Schmaljohann and Both 2017; © 2017 Springer Nature, used with permission)

(or females in sex-role reversed species) could benefit by both obtaining superior territories and being more likely to obtain mates.

If the outcome of male-male competition for the best territories influences male reproductive success and selects for protandry, then protandrous species might also be expected to exhibit sexual size dimorphism. This would be the case if larger males were more likely to outcompete smaller males. In addition, for species that breed at high latitudes where environmental conditions for early arriving migrants can be

**Fig. 13.63** (continued) and Cooper's Hawks (*Accipiter cooperii*). (**b–e**) Airspeeds of four species of shorebirds during the fall migration period. Note that airspeeds for all four species are higher prior to the peak in raptor abundance. Solid circles = adults, open circles = juveniles. The solid line represents the regression and the gray area indicates 95% confidence intervals. Red Knot, *Calidris canutus*; Ruddy Turnstone, *Arenaria interpres*; Sanderling, *Calidris alba*; Semipalmated Sandpiper, *Calidris pusilla*. (Figure modified from Duijns et al. 2019; open-access article licensed under a Creative Commons Attribution 4.0 International License, https://creativecommons.org/licenses/by/4.0/)

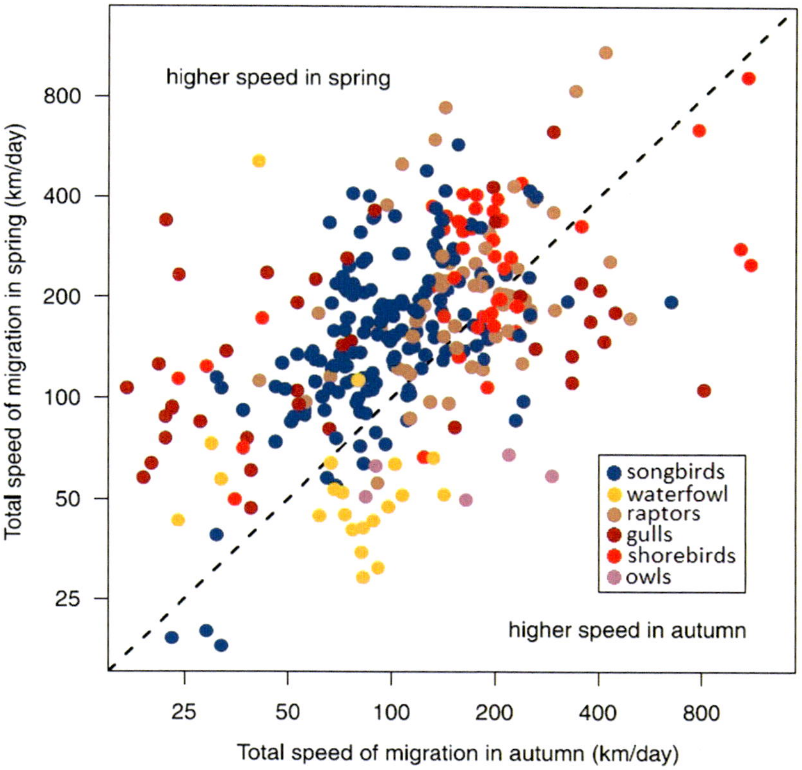


**Fig. 13.65** Total speed of spring migration plotted against the corresponding speed of fall migration for individual birds in six groups of birds. Dots above the dashed line indicate individual birds that migrated faster in spring than fall, and dots below the dashed line indicate birds that migrated faster in the fall than the spring. (Figure modified from Schmaljohann 2018; open-access article licensed under a Creative Commons Attribution 4.0 International License, https://creativecommons.org/licenses/by/4.0/)

harsh, selection may also favor larger males better able to compete for access to limited food resources and, in addition, with smaller surface area-to-volume ratios and, therefore, better able to withstand cold conditions (Rubolini et al. 2004).

Of course, sexual size dimorphism might also be expected if mate acquisition is driving the evolution of protandry because larger males may outcompete smaller ones for access to females and females may prefer larger males. In addition, however, among songbirds, intense intersexual selection typically also favors male ornamentation, i.e., sexual dichromatism.

Rubolini et al. (2004) examined the relationship between the degree of protandry (the difference between male and female arrival times) and the degree of sexual dichromatism (differences between the plumage of males and females) and sexual size dimorphism for 21 species of migratory songbirds. They found a strong, positive association between the degree of protandry and the degree of sexual dichromatism (Fig. 13.70), but no relationship between the degree of protandry and sexual size dimorphism. In other words, the most protandrous species (where males tended to arrive at breeding grounds the earliest relative to females) were more sexually dichromatic, but not more sexually size dimorphic. Such results are consistent with the predictions of the mate-opportunity hypothesis and suggest that increased success at acquiring mates may be a major driving force in the evolution of protandry. In further support of the mate-opportunity hypothesis, Kokko et al. (2006) generated models in an attempt to better understand the selective pressures favoring protandry and also found that the mate-opportunity hypothesis better explained the evolution of protandry than the rank-advantage hypothesis. Their models suggest that competition for territories alone is not sufficient to favor the evolution of protandry; early arrival must also improve mating opportunities.

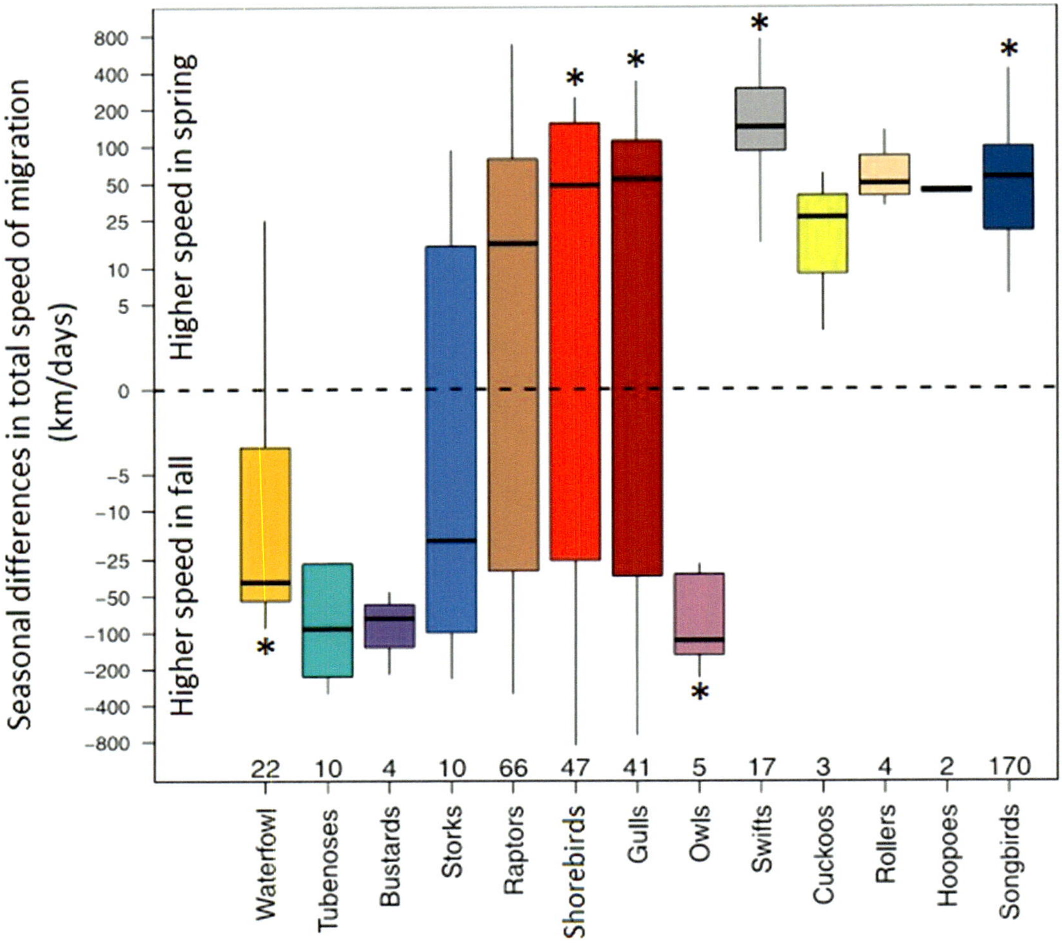


**Fig. 13.66** Seasonal differences in total speed of migration for different taxa of birds. Seasonal differences in total speed of migration equal the spring value minus the corresponding fall value for each group. Positive values indicate faster migration in the spring and negative values indicate faster migration in the fall. Asterisks indicate significant differences, with shorebirds, gulls, swifts, and shorebirds migrating significantly faster in the spring, and waterfowl and owls migrating significantly faster in the fall. Box plots show the median and 25th and 75th percentiles, whiskers indicate values within 1.5 times the interquartile range, and outliers are indicated by black dots. The number of individuals birds sampled for each group is indicated by the numbers just above the x-axis. The y-axis is on a logarithmic scale. (Figure modified from Schmaljohann 2018; open-access article licensed under a Creative Commons Attribution 4.0 International License, https://creativecommons.org/licenses/by/4.0/)

## 13.14 Diurnal Versus Nocturnal Migration

Many species of birds migrate at night, but some migrate during the day and still others are flexible and can migrate at any time of the day. Most waterfowl, including ducks and geese, are flexible and migrate at various times of day (Box 13.6 Hyperthermia and Flight Duration of a Short-Distance Migrant). However, most long-distance migratory songbirds and shorebirds migrate only at night, and some songbirds, including finches, swallows, and corvids, as well as pigeons and doves migrate only during the day (Alerstam 1990).

An important advantage of nocturnal migration is that, for birds that forage during the day,

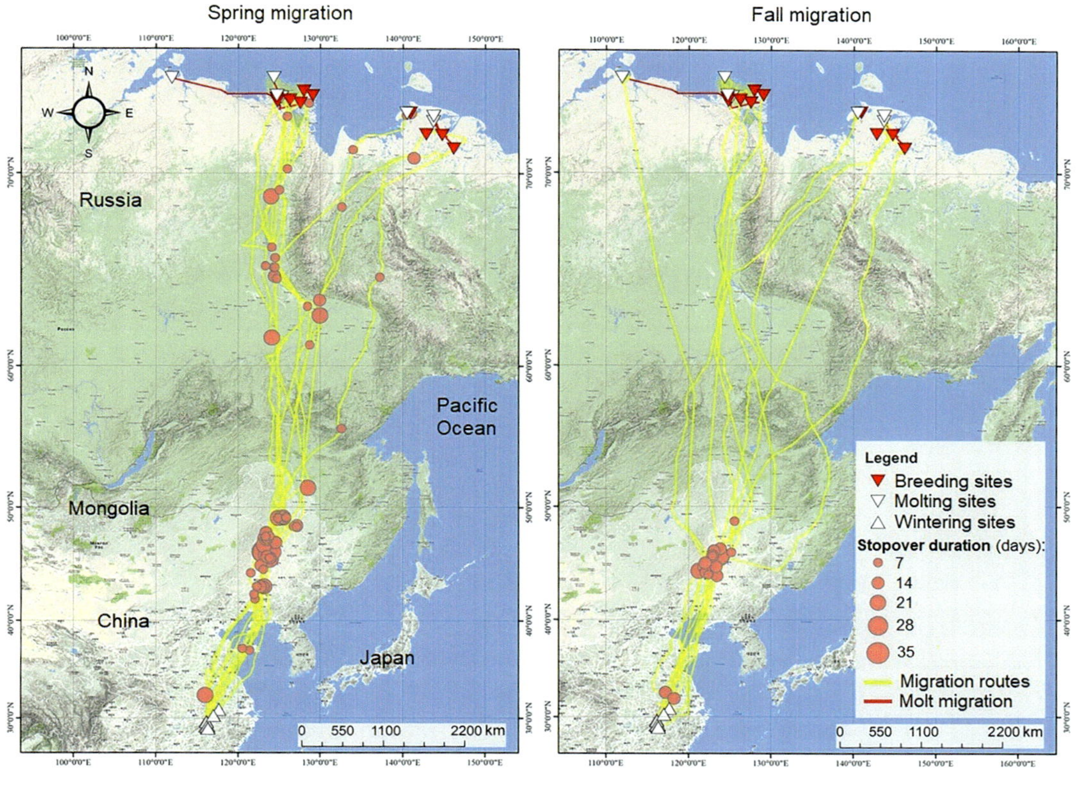


**Fig. 13.67** Routes taken by individual Greater White-fronted Geese (*Anser albifrons*) during spring and fall migration. The geese used more stopover sites (pink circles) during spring migration than fall migration. As a result, the mean duration of spring migration was 79 days, whereas the mean duration of fall migration was 35 days. Distances traveled during spring and fall migration did not differ. Greater White-fronted Geese are thought to be at least partial capital breeders (i.e., relying on stored energy reserves to meet the energy requirements of reproduction) so the many stopovers during spring migration may have allowed them to acquire the energy stores needed for reproduction. However, the timing of spring thaw may have also contributed to the longer duration and number stopovers during spring migration. Breeding individuals arrived to breeding sites (red triangles) and remained in the same areas to molt, but nonbreeding individuals made molt migration flights to remote molt sites (downward white triangles). (Figure from Deng et al. 2019; © The Authors 2019, open-access article distributed under the terms of the Creative Commons Attribution 4.0 International License, https://creativecommons.org/licenses/by/4.0/)

more time is available for foraging. Nocturnal migrants can initiate migration after a day spent foraging and storing energy, whereas diurnal migrants must balance flight time and foraging during the day and spend the night roosting and sleeping. As a result, nocturnal migrants can spend more time flying, resulting in faster migration (Alerstam 2009). The potential importance of maximizing foraging time is perhaps best illustrated by shorebirds. Lank (1989) found that shorebirds typically initiate migratory flights at dusk, after light levels made foraging difficult, but also noted that shorebirds sometimes initiated migratory flights during the day if foraging at a particular location is prevented (e.g., due to rising tides).

Beyond increased foraging and flight time, nocturnal migration may also be advantageous because flying conditions are typically better, with less atmospheric turbulence, less wind, and reduced evaporative water loss (because temperatures are cooler). Flying at night may also take less energy because, with cooler

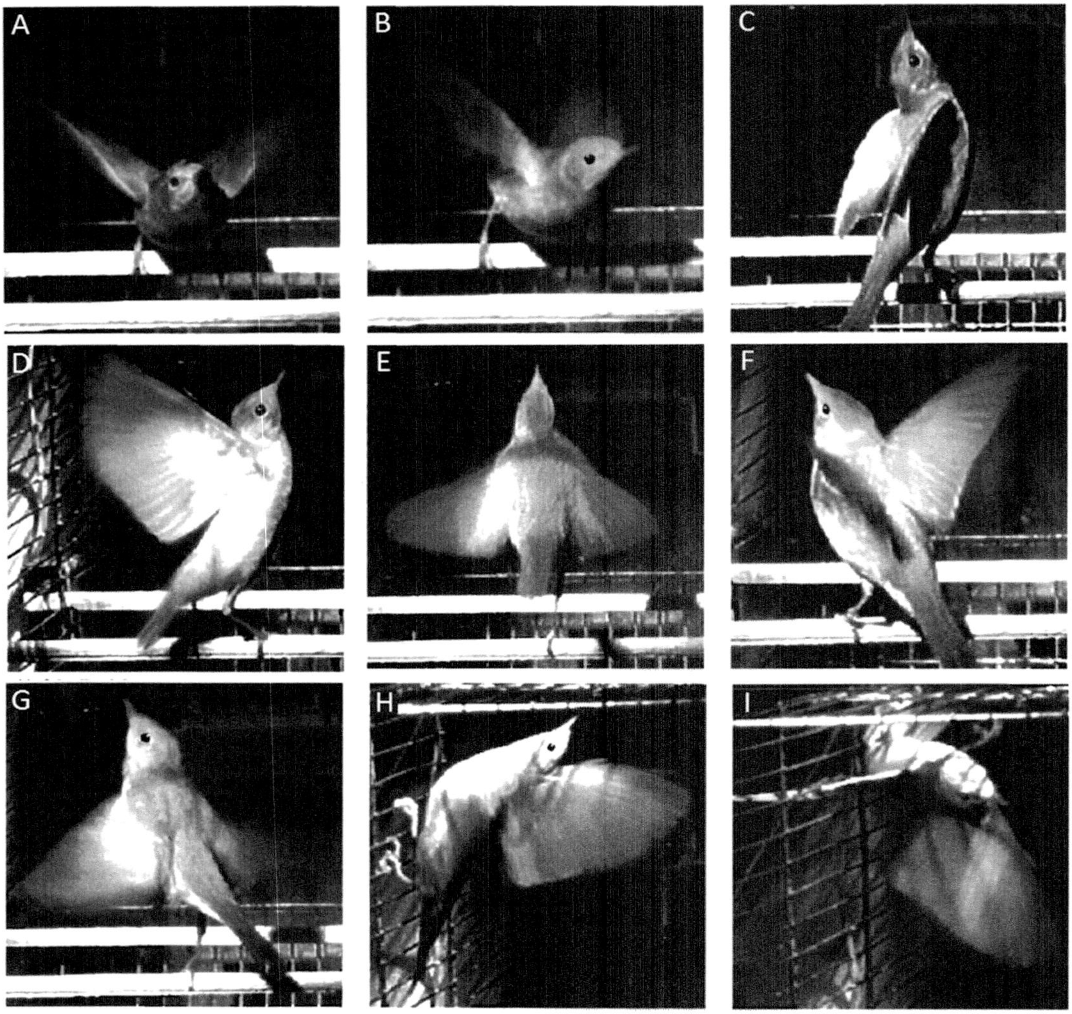


**Fig. 13.68** (a–i) A captive Eurasian Blackcap (*Sylvia atricapilla*) exhibiting zugunruhe or migratory restlessness. Zugunruhe is a German word with "Zug" meaning to move or migrate and "unruhe" meaning anxiety or restlessness. A prominent aspect of this behavior is "wing whirring," rapidly beating wings while perched. Migratory restlessness by caged birds also involves hopping, climbing, and flying. (Figures from Berthold et al. 2000; © 2000 Deutsche Ornithologen-Gesellschaft/ Blackwell Wissenschafts-Verlag, used with permission)

temperatures and higher humidity increasing air density, generating lift is easier (Kerlinger and Moore 1989; Alerstam 2009). In addition, with few or no aerial predators (except, in some cases, for bats; Box 13.7 Bats Preying on Migrating Birds), predation risk is lower for birds migrating at night. Finally, many birds use navigational cues that are only available at dusk or at night (e.g., stars; see Chap. 14).

Given the many advantages of nocturnal migration, why do some birds migrate during the day? In contrast to the continuous flapping flight of nocturnal migrants, some species, such as hawks, vultures, and storks, soar and glide during migration. Soaring requires less energy than flapping flight, but requires rising currents of air that are generated during the day, either by unequal heating of the earth's surface (thermals)

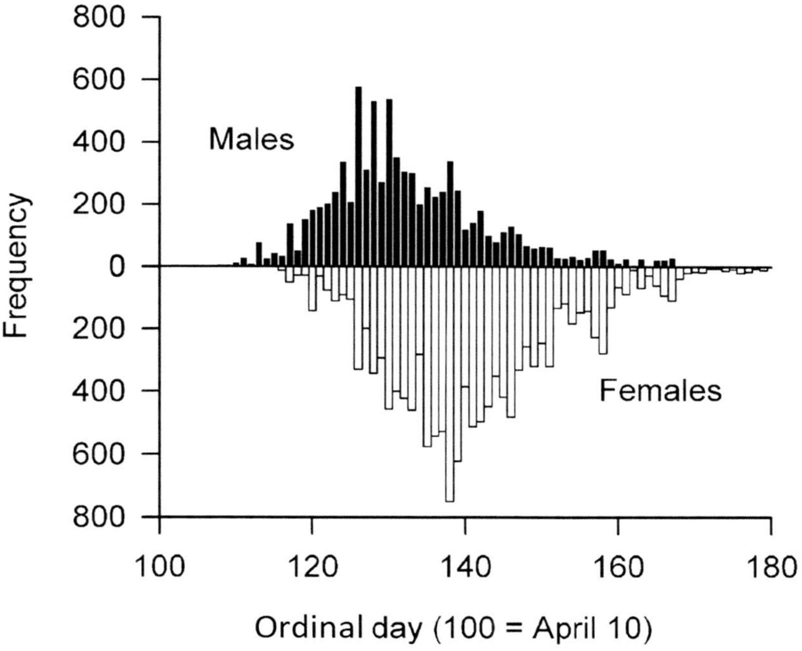


**Fig. 13.69** Protandry in Common Redstarts (*Phoenicurus phoenicurus*), with males beginning to arrive in breeding areas several days before the first females arrive. (Figure from Coppack and Pulido 2009; © 2009 Oxford University Press, used with permission)

or by wind directed upward by mountain ranges or other obstacles (obstruction lift; see Chap. 11). Of course, birds that migrate during the day can also forage, using a fly-and-forage strategy. For example, Hobbies (*Falco subbuteo*) monitored by satellite tracking were found to occasionally fly at lower speeds, especially during the afternoon, with reduced speed suggesting a temporary switch to focus on hunting (Strandberg et al. 2009). Aerial insectivores, such as swallows, may also use a fly-and-forage strategy, occasionally spending time foraging during their migration flights. Depending on the migration route, birds may be limited in their use of the fly-and-

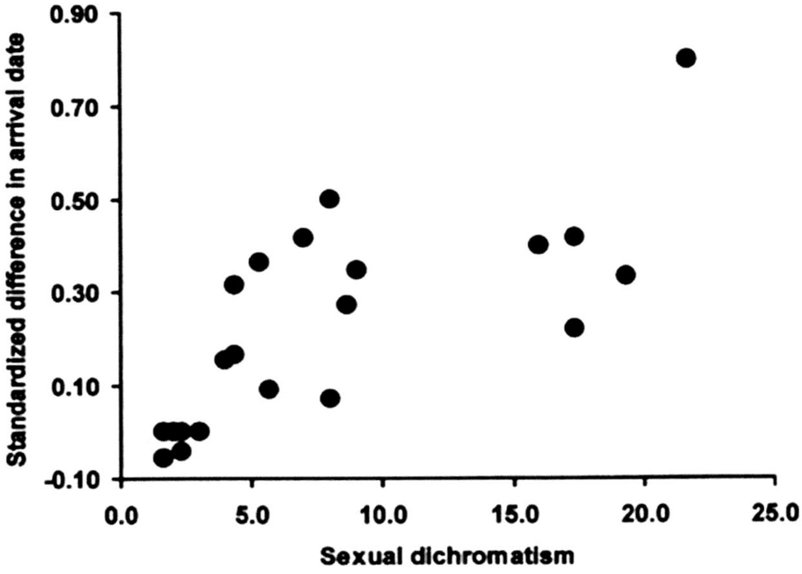


**Fig. 13.70** Relationship between standardized difference in migration dates between females and males (greater values indicate that males are migrating earlier than females) and degree of sexual dichromatism (scored on a 0–24 scale) for 21 species of long-distance migratory songbirds. (Figure from Rubolini et al. 2004; © 2004 Oxford University Press, used with permission)

**Box 13.6 Hyperthermia and Flight Duration of a Short-Distance Migrant**

Some species of long-distance migrants are capable of flights lasting many hours and, in some cases, even many days. Other species, however, use a stop-and-go migration strategy, making several short-distance flights interspersed with periods to rest or forage. One such species is the Common Eider (*Somateria mollissima*). These sea ducks are short-distance migrants and, when migrating, they make several short-duration flights (mean = 14 flights/day that last an average of 15.7 min). Guillemette et al. (2016) fitted 45 Common Eiders with data-loggers that recorded heart rate and body temperature. Using the heart-rate data, they were able to determine when the eiders were either flying (when heart rates were 3–4 times the resting level) or not (heart rate at the resting level).

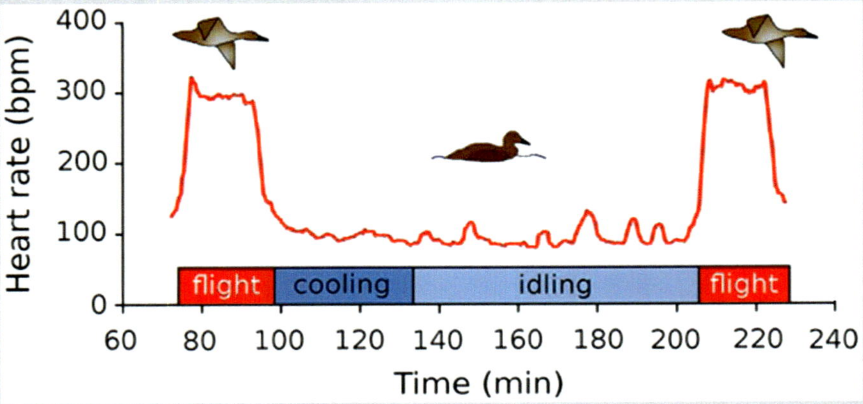


Flight cycle of Common Eiders during migration. Heart rates during and between flights were used to determine time intervals between flights. (Figure from Guillemette et al. 2017; © 2017 Guillemette, Polymeropoulos, Portugal, and Pelletier, open-access article distributed under the terms of the Creative Commons Attribution License, https://creativecommons.org/licenses/by/4.0/)

Data from the data-loggers revealed that the body temperatures of Common Eiders in flight increased and continued to do so as flight duration increased. In addition, the eiders usually did forage between flights (as indicated by heart rates remaining at the resting level rather than at levels associated with diving behavior). Based on these data, Guillemette et al. (2016) suggested that the flight durations of migrating Common Eiders were limited at least in part by hyperthermia, with increasing body temperatures causing the eiders to end their longer-duration flights (>20–40 min).

(continued)

**Box 13.6** (continued)

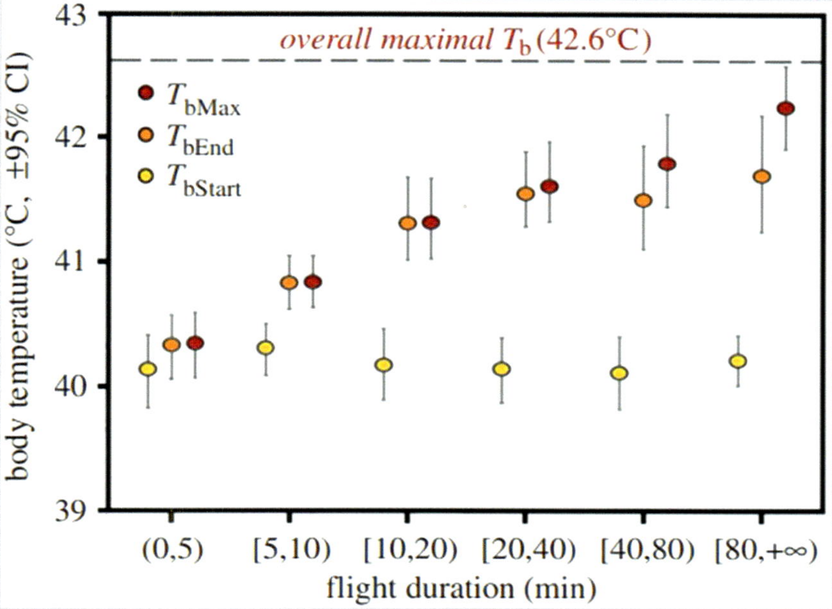


Mean body temperatures of Common Eiders at the start and end of flights of different durations, plus the maximum body temperatures recorded during those flights. (Figure from Guillemette et al. 2016; © 2016 The Authors. Published by the Royal Society. All rights reserved, used with permission)

Male Common Eider. (Photo by Ron Knight, Wikipedia, CC BY 2.0, https://creativecommons.org/licenses/by/2.0/)

(continued)

**Box 13.6** (continued)

For most species of birds that migrate, hyperthermia is unlikely to influence flight duration. In contrast to most other birds, diving ducks like Common Eiders have high wing loading and continuous fast-flapping when flying, factors that increase the cost of flight and, as a result, generate lots of heat. In addition, diving ducks are relatively large and larger birds have relatively less surface area than smaller birds, reducing the ability of larger birds to lose heat. Finally, when migrating, many species of birds have flight strategies that reduce the likelihood of hyperthermia, including flying at night when temperatures are cooler and flying at high altitudes where temperatures are lower (Guillemette et al. 2016).

forage strategy. For example, Barn Swallows (*Hirundo rustica*) migrating from breeding areas in Europe to wintering areas in Africa, must, like long-distance nocturnal migrants, build up fat stores before making long flights across the Mediterranean Sea and the Sahara Desert (Rubolini et al. 2002).

For nocturnal migrants, a variety of factors contribute to when migratory flights begin each evening (Box 13.8 Nocturnal Departure Times). Some evidence suggests that, for some species, the time of departure is endogenously controlled. For example, wild-caught migratory Common Redstarts (*Phoenicurus phoenicurus*) held captive for three days and kept under constant dim light without any photoperiodic cues, with a constant supply of food, and without access to environmental cues, still exhibited clear patterns of migratory activity, with the onset of migration each night coinciding with the time of local sunset (Coppack et al. 2008). This suggests that a circadian clock determines the time of departure. If Common Redstarts initiated nightly migratory flights based on factors such as habitat quality, physical condition, or experience, then they would not exhibit such consistent temporal patterns, regardless of age or point of origin (Coppack et al. 2008).

Another factor that may influence nocturnal departure time is the total migration distance, with species or populations that migrate greater distances tending to depart earlier than those

**Box 13.7 Bats Preying on Migrating Birds**

Although bats preying on birds has rarely been documented, examination of 14,000 fecal pellets revealed that greater noctule bats (*Nyctalus lasiopterus*) capture and eat large numbers of migrating songbirds in Spain. Ibáñez et al. (2001) reported that these bats likely capture and eat birds in flight, just as aerial-hawking bats normally do with insects. Bats can approach and surprise birds without being detected because their echolocation calls are well above the frequencies songbirds can detect. Greater noctule bats are one of the largest aerial-hawking bats in the world, with a mean mass of 48 grams and typical wingspan of 45 cm, so can easily overpower small songbirds flying at night. In addition to greater noctule bats, two other species of bats—bird-like noctules (*Nyctalus aviator*) in northeastern Asia and great evening bats (*Ia io*) in India—have also been reported to prey on migrating birds (Thabah et al. 2007, Fukui et al. 2013). Using DNA metabarcoding, Gong et al. (2021) found that great evening bats in China consumed 22 species of birds representing seven families of songbirds with body masses ranging from 6 to 19 grams.

(continued)

**Box 13.7** (continued)

Greater noctule bat. (Photo by Nicol Harper, Wikipedia, CC BY 2.5, https://creativecommons.org/licenses/by/2.5/)

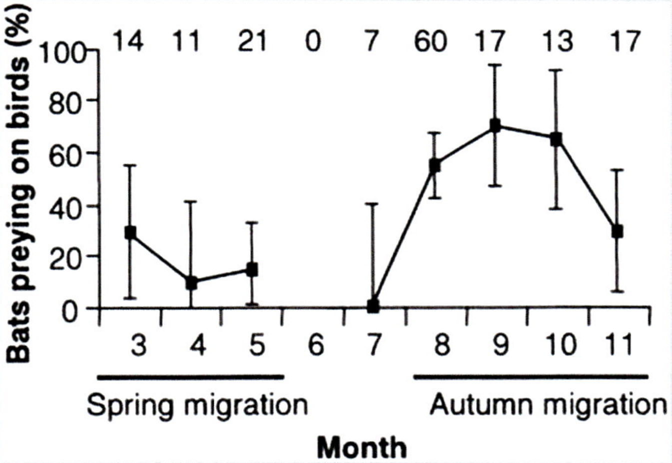


Percentage (± 95% confidence intervals) of greater noctule bats captured that (based on feathers in their fecal pellets) were preying on birds during spring and fall migration. The numbers above the lines represent the number of bats captured. (Figure from Ibáñez et al. 2001; Copyright (2001) National Academy of Sciences, U.S.A., used with permission)

(continued)

**Box 13.7** (continued)

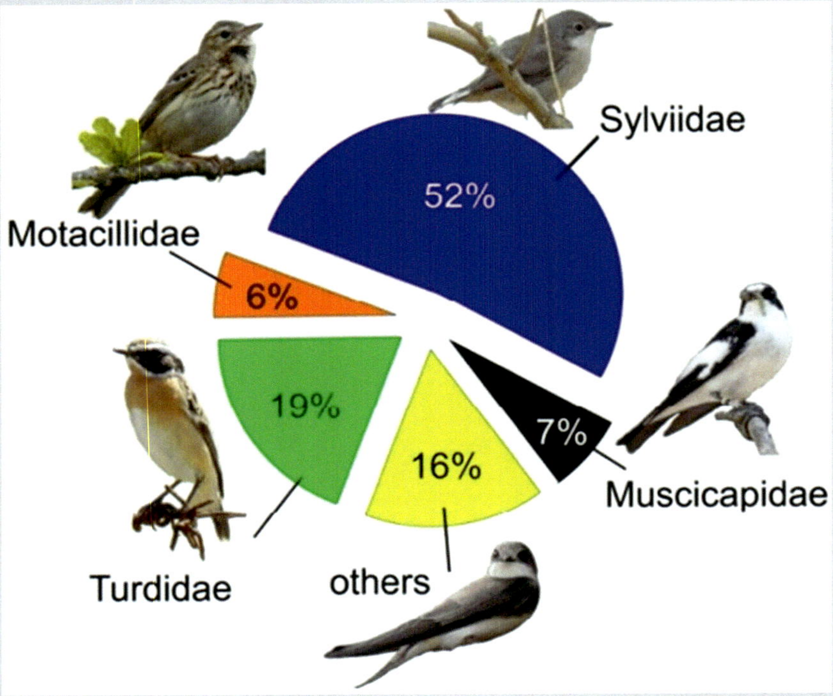


Families of birds preyed on by greater noctule bats during migration. (Figure from Ibáñez et al. 2016; © 2016 John Wiley & Sons Ltd., used with permission)

traveling shorter distances (Schmaljohann et al. 2013). For example, Bolshakov and Bulyuk (2001) monitored departure times of several species of nocturnally migrating songbirds using searchlights at a stopover site near the Baltic Sea and noted a trend for species migrating short- and middle-distances to depart later after sunset than long-distance migrants.

## 13.15 Bird Migration and Climate Change

For birds that breed at mid- to high latitudes, the time of arrival at their breeding areas has been selected over time so they largely miss the adverse weather conditions of late winter and early spring and arrive when food availability is increasing. Over the past several decades, however, global temperatures have increased due to anthropogenic climate change, with corresponding changes in the timing of seasonal climate conditions. These changes in temperature have not been uniform across the globe; temperatures in the Northern Hemisphere, particularly at higher latitudes, have increased more than those in the Southern Hemisphere, and temperatures in Europe and northern Asia have increased more than those in most of North America (excluding parts of Alaska; Fig. 13.71).

Not surprisingly, increasing temperatures have altered the timing or phenology of biological processes in many areas, but particularly at higher latitudes in the Northern Hemisphere. For

**Box 13.8 Nocturnal Departure Times**

Decisions made by migrating birds are influenced by both innate factors and environmental conditions. Innate factors include tendencies to migrate at certain times of the year and certain times of day (e.g., day or night), in certain directions, and for certain distances. Correlative evidence suggests that individuals with greater fuel loads and favorable wind conditions depart earlier, whereas individuals with little fuel and in areas with unfavorable winds vary in their departure times (Müller et al. 2016).

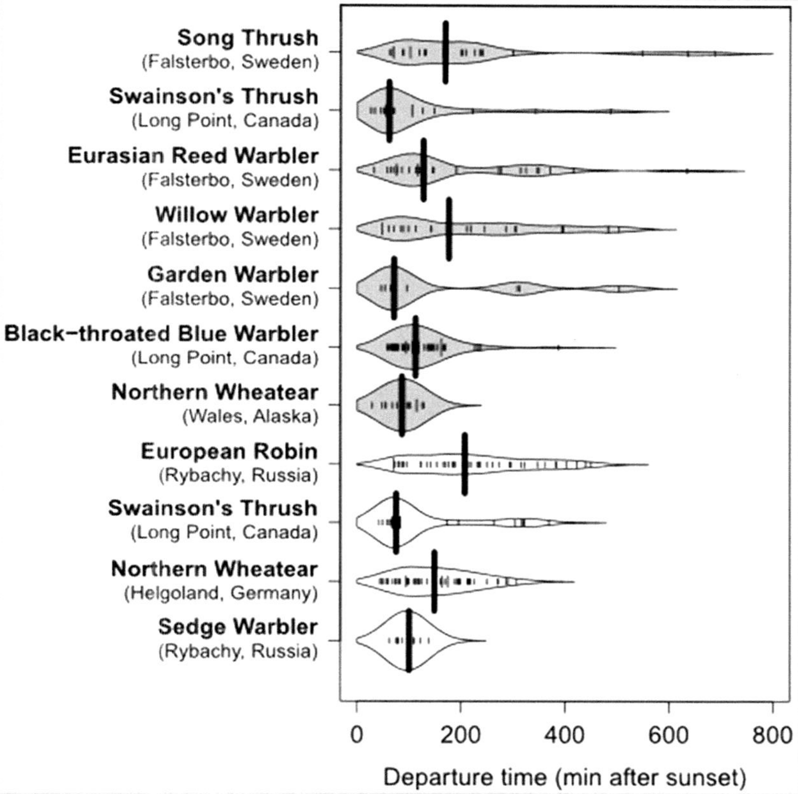


Variation in the nocturnal departure time of different songbird species as determined from several radio-tracking studies. Plots illustrate variation in nocturnal departure times relative to sunset, with the dark vertical lines indicating the median departure time. Shaded plots = fall migration; white plots = spring migration. Song Thrush, *Turdus philomelos*; Swainson's Thrush, *Catharus ustulatus*; Eurasian Reed Warbler, *Acrocephalus scirpaceus*; Willow Warbler, *Phylloscopus trochilus*; Garden Warbler, *Sylvia borin*; Black-throated Blue Warbler, *Setophaga caerulescens*; Northern Wheatear, *Oenanthe oenanthe*; European Robin, *Erithacus rubecula*; Sedge Warbler, *Acrocephalus schoenobaenus*. (Figure from Müller et al. 2016; © 2016 The Authors, open-access article distributed under the terms of the Creative Commons Attribution 4.0 International License, https://creativecommons.org/licenses/by/4.0/)

(continued)

**Box 13.8** (continued)

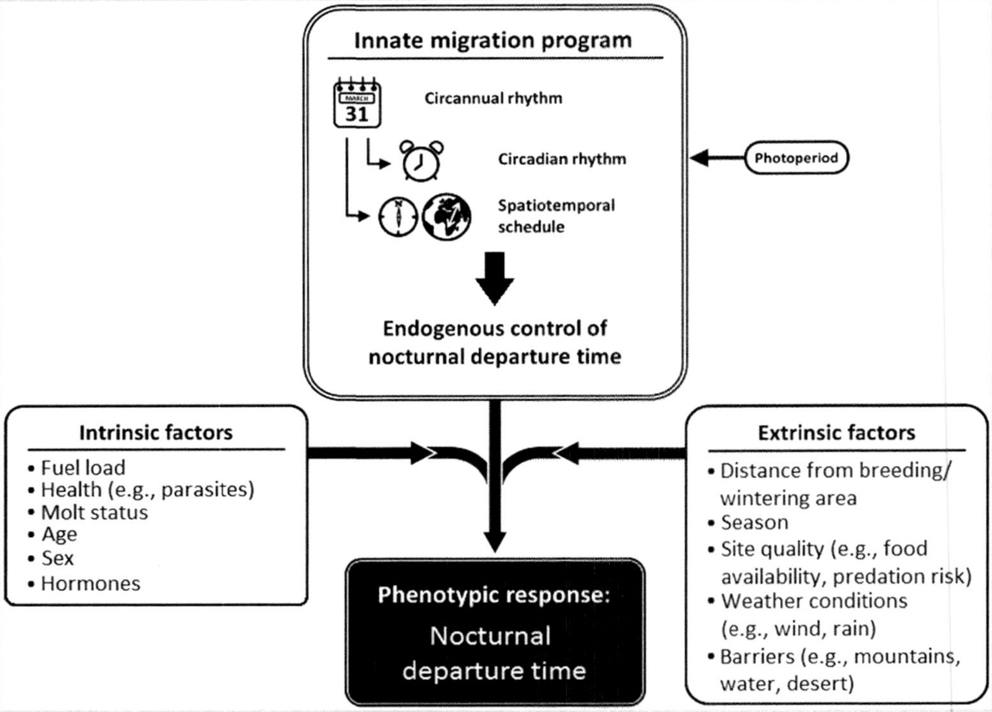


Factors influencing departure times of nocturnal songbird migrants. The innate program provides the circannual, circadian rhythms, and spatiotemporal schedule of migration. The photoperiod is used to calibrate or reset the innate migration program, but a number of intrinsic and extrinsic factors can modulate the endogenous stimuli. Birds with greater fuel loads typically depart earlier and fly for longer periods. Poor health may result in less available energy to sustain flight so departure times of individuals likely vary depending on their condition. Molting during migration likely means less energy is available for flight and, potentially, an increased cost of flight (depending on stage of molt and feathers being molted), but how these might affect departure times is unclear. Effects of age and sex on departure times are also unclear, although males may depart earlier if earlier arrival in breeding areas contributes to an increased likelihood of breeding success. Birds that are farther from breeding or wintering areas may tend to depart earlier than those that are closer, and birds that initiate migration later in a season likely have earlier departure times than those that initiated migration earlier in the season. Site quality can determine a bird's fuel load which, in turn, can influence departure times. Birds in areas with poor weather conditions (e.g., rain or unfavorable winds) would tend to depart later (or not at all on a given night). Birds needing to cross barriers where landing is not possible (e.g., large bodies of water) need to consider their fuel load and flying conditions when making decisions about departure times (or whether to depart at all) because the wrong decision could have lethal consequences. (Figure modified from Müller et al. 2016; © 2016 The Authors, open-access article distributed under the terms of the Creative Commons Attribution 4.0 International License, https://creativecommons.org/licenses/by/4.0/)

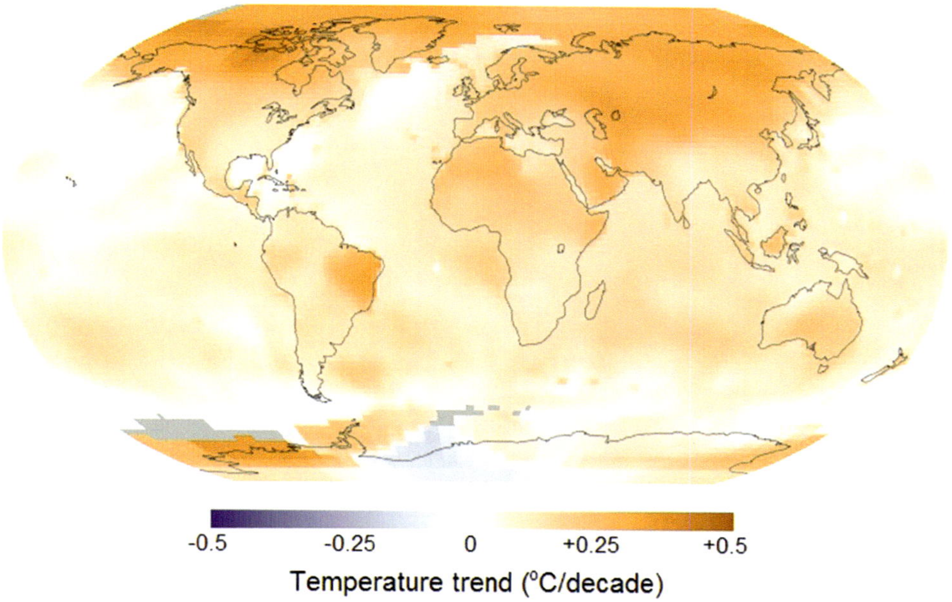


**Fig. 13.71** Temperature trend for the period from 1950 to 2014. The most extreme warming is taking place in the Arctic. Few areas had cooler temperatures. Gray areas over parts of the Southern Ocean are places where temperatures were not recorded. Note that temperatures have increased more in Europe and Asia than in North America. (Image source: https://www.giss.nasa.gov/research/news/20150116/, CC0 Public Domain)

example, in the Arctic, where annual average temperatures have increased at almost twice the rate of the rest of the world (Callaghan et al. 2005), warmer winters and earlier springs have advanced the date of peak abundance of arthropods by about 7 days (Tulp and Schekkerman 2008). Climate change has also altered the phenology of migration for many species of birds. For example, Végvári et al. (2010) determined first-arrival dates (the day on which the first individuals of a migratory species are observed in the spring) of migrants in eastern Hungary (47°N latitude) and found that, during the period from 1969 to 2007, the time of arrival was significantly earlier for 45 of 177 species (25.4%; Fig. 13.72). First-arrival dates of short-distance migrants (species wintering north of the Sahara Desert) were advanced more than those of long-distance migrants (species wintering south of the Sahara in tropical Africa) (Figs. 13.72 and 13.73). The shorter the migration distance, the greater the tendency for earlier arrival dates (Fig. 13.72). Similarly, Murphy-Klassen et al. (2005) examined a 63-year dataset (1939 to 2001) of first spring sightings for 96 species of birds in Manitoba (50° N latitude) and found that 25 species (26%) had significant trends for increasingly early arrival. As also found in Hungary, more short-distance migrants (33.2% of species) exhibited significant tendencies for earlier arrival dates than long-distance migrants (18.8% of species). Horton et al. (2020) used 24 years of data (1995–2018) collected in the United States using weather surveillance radar to examine variation in the date of peak spring migration. Analysis revealed a significant trend for earlier peak migration at 35, 40, and 45° N latitude, but not at 30° N latitude (Fig. 13.74).

Although studies have revealed that many species of migratory birds are now arriving on their breeding grounds earlier than in the past, this trend is by no means universal because many other species have exhibited no change in the timing of migration. Several factors might help explain differences among migratory species in their response to climate change, including

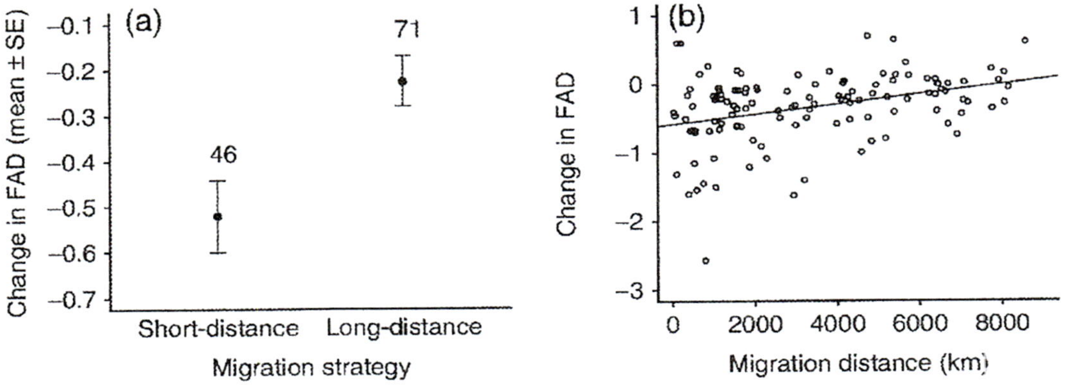


**Fig. 13.72** Annual change (days per year) in the date when the first individuals of migratory birds were observed in eastern Hungary relative to (**a**) migration strategy and (**b**) migration distance. Numbers above error bars in (**a**) indicate the number of species. (Figure from Végvári et al. 2010; © Blackwell Publishing Ltd., used with permission)

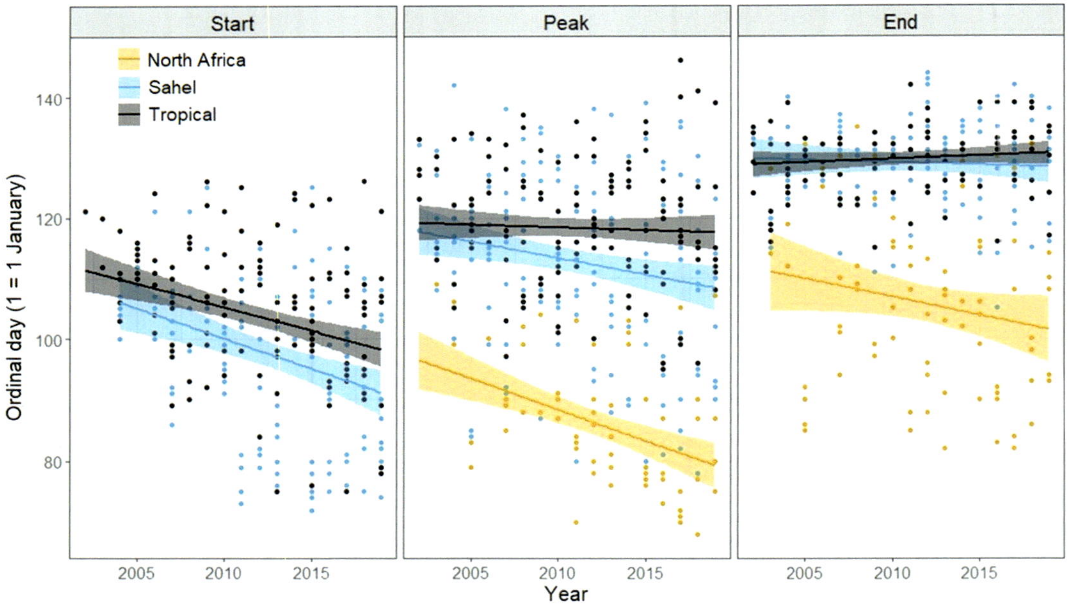


**Fig. 13.73** Migration phenology of 30 species of birds that winter in Africa and migrate in spring across the Mediterranean Sea to breeding areas in the Palearctic. Migrating birds were captured en route at a small island 50 km south of Italy. Wintering areas included North Africa (north of the Sahara Desert), the Sahel (just south of the Sahara Desert), and tropical Africa (savanna and tropical forests). Note that species wintering further south and further from breeding areas exhibited less change in migration phenology than those wintering closer to breeding areas (i.e., wintering in north Africa). (Figure from Maggini et al. 2020; © 2020 Maggini et al., open-access article distributed under the terms of the Creative Commons Attribution License, https://creativecommons.org/licenses/by/4.0/)

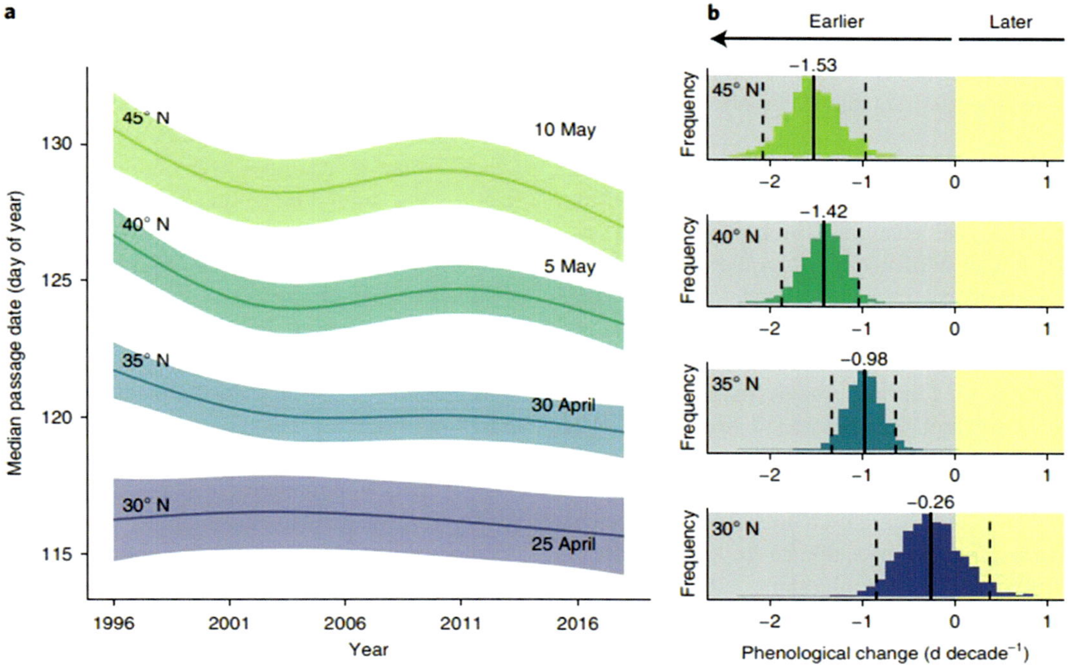


**Fig. 13.74** (**a**) Changes in time for peak spring migration dates at different latitudes in the United States as determined using weather surveillance radar. (**b**) Phenological change in peak spring migration (days per decade) at different latitudes. Solid lines indicate means and the dashed lines the fifth and 95th percentiles. Phenological changes are greater at higher latitudes. (Figure from Horton et al. 2020; © 2019 The Authors, under exclusive license to Springer Nature Limited, used with permission)

habitat, feeding ecology, breeding behavior, migratory strategy (e.g., distance between wintering and breeding areas), and structural traits such as body size (Lehikoinen and Sparks 2010). In a comparative study that included data from 117 species of birds, Végvári et al. (2010) found that advancement of first-arrival dates was greater in migratory species with more generalized diets, shorter migration distances, more broods per year, and less extensive pre-breeding molt.

As also noted in a number of studies, one important factor in predicting the response of a species to climate change is migration distance, with short-distance migrants advancing first-arrival dates more than long-distance migrants (Végvári et al. 2010; Usui et al. 2017). A likely explanation for this difference is that the timing of migration for long-distance migrants is under stronger endogenous control (e.g., responding to changes in day length) whereas short-distance migrants respond more to variation in climate. On their distant wintering areas, long-distance migrants may be less sensitive to weather conditions on their breeding grounds. For short-distance migrants, weather conditions in wintering areas are more likely to be related to conditions in breeding areas, allowing them to adjust the timing of migration accordingly. However, some long-distance migrants are also able to alter their migration phenology in response to climate change. For example, departure dates for Bar-tailed Godwits (*Limosa lapponica*) that migrate from New Zealand to Alaska during spring migration were found to have advanced by six days during the period from 2008 to 2020 (Conklin et al. 2021). Long-distance migrants with earlier departure dates may be responding to improving conditions in wintering areas

(Studds and Marra 2011; Haest et al. 2020) or advances in spring arrival dates may be an evolutionary response if migration timing is a trait that is sufficiently heritable (Schmaljohann and Both 2017).

The results of some studies suggest that changes in the climates of wintering areas may correlate with changes in breeding areas and, as a result, can influence the migration phenology of long-distance migrants. Jonzén et al. (2006) estimated trends in arrival times using data collected from 1980 to 2004 at four bird observatories in Scandinavia and a site in southern Italy and found that 17 species of long-distance migrants had advanced their arrival times at breeding areas in northern Europe. These authors suggested that this advancement in arrival dates was a climate-driven evolutionary change. Although migratory activity is under endogenous control, individuals exhibit phenotypic flexibility in their responses to the cues that can influence the location of wintering areas (e.g., "short-stopping" or wintering in areas closer to breeding areas; Elmberg et al. 2014; Potvin et al. 2016; Mendgen et al. 2023; Fig. 13.75) and/or the timing of the onset of migration (Coppack et al. 2003). As such, songbirds like those studied by Jonzén et al. (2006) with short generation times, breed when one year old, can potentially exhibit a rapid evolutionary response to environmental changes. Other investigators have also provided evidence that environmental conditions in breeding areas, e.g., warmer temperature early in the breeding season or abnormally high temperature later in the breeding season, can alter selection on phenological traits such as arrival times (Visser et al. 2015; Marrot et al. 2017; Fig. 13.76).

Diet, number of broods per year, and molting strategies also appear to influence migration strategies (Végvári et al. 2010). Birds with more general diets are likely better able to find sufficient food during migration and after arrival in breeding areas than those with specialized diets and, therefore, are better able to respond to climate change by advancing the timing of migration. Because changes in global temperatures have not been uniform and specific weather conditions vary between and within years, generalized diets would be beneficial for birds migrating earlier because the availability of certain types of food (e.g., arthropods, Tulp and Schekkerman 2008) may vary geographically (e.g., along migration routes) and temporally (e.g., between years). In contrast, birds with specialized diets would be less likely to survive if variable weather conditions reduced availability of their required foods.

The time of arrival on breeding grounds for multibrooded species has tended to advance more than that of species that typically produce fewer broods (Møller et al. 2008; Végvári et al. 2010). Species that can potentially raise multiple broods during a breeding season are under stronger selection for earlier arrival (providing more time for more broods), accelerating their response to a warming environment (Végvári et al. 2010).

First-arrival dates of species with no pre-breeding molt have advanced more than those species with a pre-breeding molt and those that molt flight feathers in wintering areas (Végvári et al. 2010). Birds with a pre-breeding molt are unable to initiate migration before completing molt because the energetic costs of replacing feathers while migrating would be too high. The need to complete molt before migrating, therefore, means that the timing of migration for these species is less flexible than for species with no pre-breeding molt.

Previous studies have revealed conflicting results concerning the possible effect of habitat on the responses of birds to climate change. Végvári et al. (2010) found no effect of habitat on first-arrival dates. However, Butler (2003) found that the first-arrival dates of grassland birds were more advanced than those of birds that breed in other habitats. Most grassland birds are omnivores and feed on seeds when other types of prey (e.g., insects) are less available. With warming temperatures, snow cover may melt sooner, exposing underlying vegetation and seeds and providing early arriving grassland birds with a source of food (Butler 2003). By contrast, arrival times of species that feed

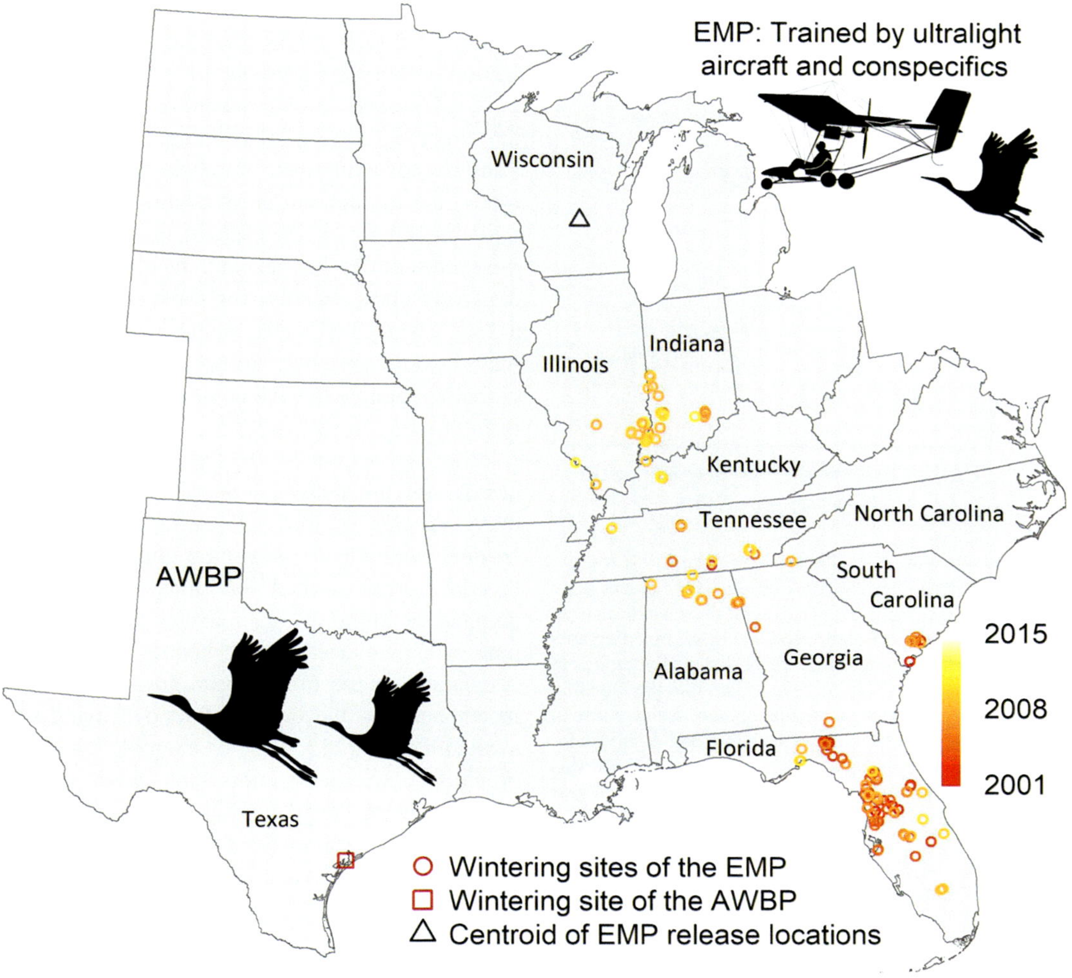


**Fig. 13.75** Traditional and more recent wintering sites of the two migrating populations of Whooping Cranes. Birds in the Aransas Wood Buffalo population (AWBP) continue to migrate from a breeding area in central Canada to their wintering area in the Aransas National Wildlife Refuge (NWR) on the coast of Texas. Whooping Cranes in the Eastern Migratory population (EMP) were initially trained to migrate (i.e., by following ultralight aircraft) from Wisconsin to wintering sites in Florida (and, in one year, Alabama). Since 2001, shortstopping behavior by cranes in the EMP has resulted in birds wintering in areas much farther north and correspondingly reduced migration distances to their release locations (breeding area) in Wisconsin. Cranes in the EMP have located wintering areas with suitable habitat in several states located well north of Florida. Suitable wintering habitat for the AWBP appears to be available north of Texas, but, in contrast to cranes in the EMP, juveniles in the AWBP follow their parents to the Aransas NWR during fall migration. This social learning reduces the likelihood of cranes in the AWBP wintering at other locations. (Figure modified from Mendgen et al. 2023; © 2023 The Authors. Published by Elsevier B.V., used with permission)

primarily on arthropods (e.g., many forest-dwelling birds) are more constrained by the time of emergence of their food source.

Although global warming has clearly affected the timing of spring migration of many species of birds, studies in both Europe and North America have generally revealed no consistent trends in the timing of fall migration. Sparks et al. (2007) examined departure dates of 23 species and a trend for slightly later departure, but few species

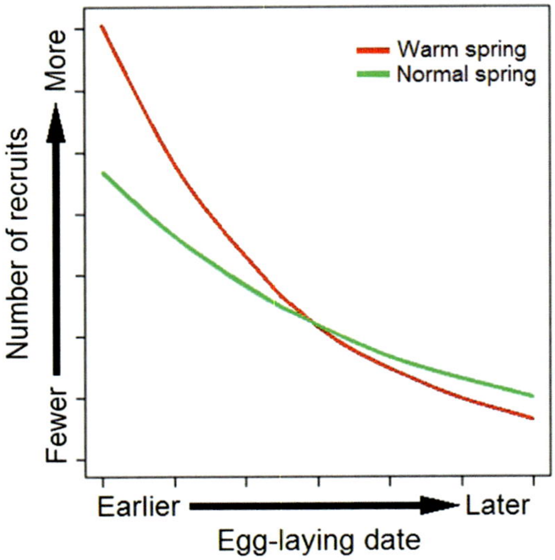


**Fig. 13.76** Among Pied Flycatchers (*Ficedula hypoleuca*), spring arrival dates of first-year females were found to be positively correlated with their hatch date. In addition, when temperatures at arrival in the breeding area were warmer than usual, these earlier-hatched young were more like to survive (i.e., to be recruited into the breeding population) than when temperatures were normal, suggesting a possible cost of arriving earlier in colder years. However, climate warming may have reduced this cost and, as a result, higher temperatures in the arrival year of recruits were found to be associated with stronger selection for earlier reproduction. (Figure modified from Visser et al. 2015; © 2015 Visser et al., open-access article distributed under the terms of the Creative Commons Attribution License, https://creativecommons.org/licenses/by/4.0/)

exhibited significant changes in departure dates. Similarly, Van Buskirk et al. (2009) reported conflicting results, with some species tending to depart later and some earlier; no variable (e.g., migration distance or number of broods) was significantly associated with departure date. The effect of global warming on the timing of fall migration by birds is less clear-cut because, in contrast to the clear relationship between fitness and early spring arrival on the breeding grounds (e.g., Møller 1994; Nystrom 1997), fitness considerations in the fall might favor earlier departure, later departure, or no change in departure time (Mills 2005). For example, if global warming allows earlier breeding (and the earlier completion of breeding), then species where individuals compete for winter territories might be expected to leave breeding areas earlier, e.g., Yellow Warblers (*Setophaga petechia*; Mills 2005). However, global warming may lengthen the breeding season, permitting, for example, more broods, additional time between broods, or more time to complete postbreeding molt before initiating migration. In each of these cases, fall migration might be delayed, resulting in later departure dates. Finally, for some species, such as long-distance migrants, the timing of migration may be endogenously fixed and, as a result, global warming will have no impact on the timing of fall migration (Mills 2005).

Although responses vary, global warming has clearly affected the timing of spring migration for many species of birds. Because selection has favored individuals that time their arrival in breeding areas so that the time of peak food demand (feeding young) matches the period of peak food availability, one potential consequence of changes in the timing of migration is a mismatch between the timing of food demand and availability (Fig. 13.77). For example, as noted previously, warmer winters and earlier springs have advanced the date of peak abundance of arthropods in some areas of the Arctic tundra by about 7 days (Tulp and Schekkerman 2008). If arrival dates of migrant birds are not similarly advanced, then the time of peak food demand may no longer match the time of peak availability and such a mismatch could have a negative impact on breeding success. Such a mismatch has apparently contributed to a decline in populations of long-distance migrants that breed in forested habitats in Europe (Both et al. 2010) and, more generally, migrant bird populations in both the Nearctic and Palearctic (Jones and Cresswell 2010). In general, species of birds that breed in habitats characterized by short bursts of food availability (e.g., temperate forests and tundra) and where the timing of migration is controlled by endogenous factors (long-distance migrants) are most likely to be negatively impacted by such mismatches (Both 2010; Both et al. 2010). Short-distance migrants are better able to adjust the timing of migration and avoid or minimize mismatches between food demand

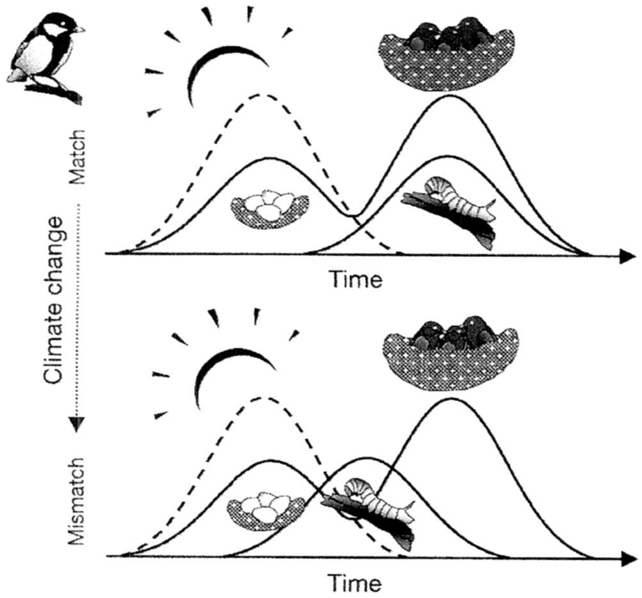


**Fig. 13.77** Example of a mismatch caused by global warming. Before global warming (above), environmental cues (dashed line) that triggered the onset of egg laying (daylength) resulted in nestlings being present when food availability (caterpillars) was highest. With global warming (below), food availability now peaks prior to peak food demand (nestlings present) because the cues that trigger egg laying (daylength) are no longer synchronized with the cues that trigger hatching and development of caterpillars (temperature). (Figure from Durant et al. 2007; © 2007 The Authors, used with permission)

and availability. Similarly, birds that breed in habitats characterized by longer peaks in food availability (e.g., marshes and coniferous forests) are less likely to be impacted by such mismatches (Both et al. 2010).

Available evidence indicates that a mismatching of food demand and availability is having a negative impact on populations of some species of migratory birds, but relationships between global warming, food availability, and timing of bird migration are complex. For example, species of birds breeding in the same habitat may experience different mismatches or be able to avoid such mismatches due to subtle differences in food habits (Jones and Cresswell 2010). For example, Wood Warblers (*Phylloscopus sibilatrix*) now breed on average a week earlier than they did in the early 1980s, timing that no longer corresponds to the time of peak caterpillar biomass. However, because of their generalist diet, there has been no decline in breeding success (Mallord et al. 2017). Climate change may cause shifts in or expansion of breeding and wintering ranges, affecting the degree to which food demand and availability might or might not mismatch. Clearly, additional study is needed to better understand how global warming is impacting the behavior, breeding success, and population status of migratory birds. Although increasing global temperatures have clearly altered the migratory behavior of many species of birds, the potential future impacts of global warming on birds are likely to be even more pronounced. If, as predicted by some, global mean temperatures increase by up to 8 degrees C by the year 2100 (Fig. 13.78), the resulting changes in climate and sea levels will undoubtedly have tragic consequences for many species of birds.

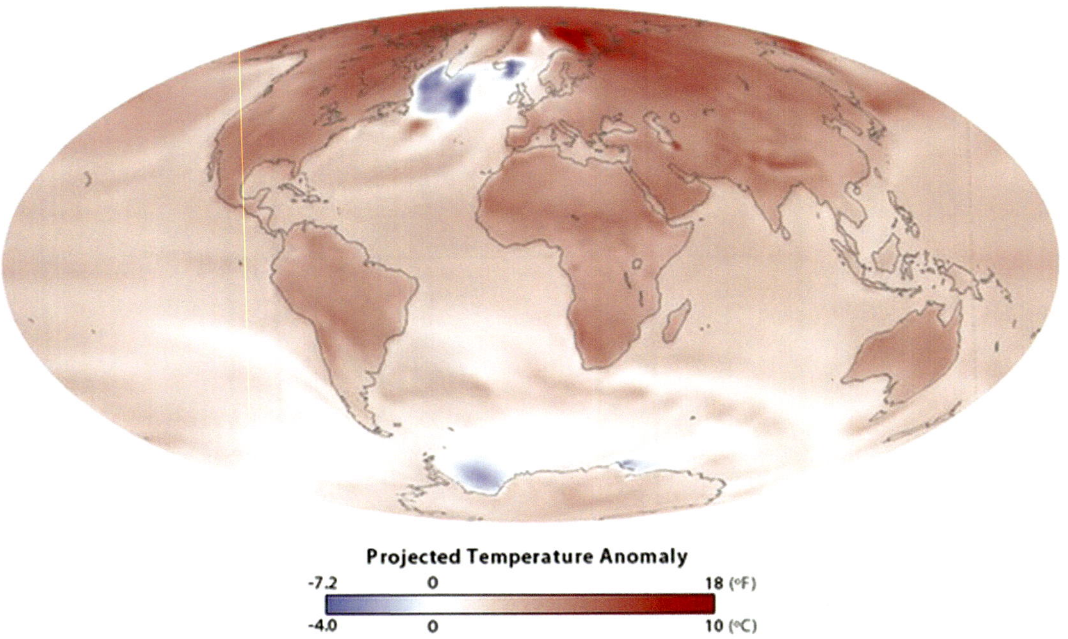


**Fig. 13.78** Predicted distribution of temperature changes by the end of the century due to global warming based on data from the NOAA Geophysical Fluid Dynamics Laboratory. (Image source: Climate.gov; CC0 Public Domain)

## References

Able KP, Belthoff JR (1998) Rapid 'evolution' of migratory behaviour in the introduced House Finch of eastern North America. Proc R Soc B 265:2063–2071

Adamik P, Emmenegger T, Briedis M, Gustafsson L, Henshaw I, Krist M, Laaksonen T, Liechti F, Prochazka P, Salewski V, Hahn S (2016) Barrier crossing in small avian migrants: individual tracking reveals prolonged nocturnal flights into the day as a common migratory strategy. Sci Reports 6:21560

Åkesson S (1999) Do passerine migrants captured at an inland site perform temporary reverse migration in autumn? Ardea 87:129–137

Åkesson S, Bianco G (2021) Wind-assisted sprint migration in Northern Swifts. iScience 24:102474

Åkesson S, Hedenström A (2007) How migrants get there: migratory performance and orientation. Bioscience 57:123–133

Åkesson S, Karlsson L, Walinder G, Alerstam T (1996) Bimodal orientation and the occurrence of temporary reverse bird migration during autumn in South Scandinavia. Behav Ecol Sociobiol 38:293–302

Åkesson S, Morin J, Muheim R, Ottosson U (2005) Dramatic orientation shift of White-crowned Sparrows displaced across longitudes in the high Arctic. Curr Biol 15:1591–1597

Åkesson S, Atkinson PW, Bermejo A, de la Puente J, Ferri M, Hewson CM, Holmgren J, Kaiser E, Kearsley L, Klaassen RH, Kolunen H (2020) Evolution of chain migration in an aerial insectivorous bird, the Common Swift *Apus apus*. Evolution 74:2377–2391

Åkesson S, Bakam H, Hernandez EM, Ilieva M, Bianco G (2021) Migratory orientation in inexperienced and experienced avian migrants. Ethol Ecol Evol 33:206–229

Alerstam T (1990) Bird migration. Cambridge University Press, Cambridge, UK

Alerstam T (2001) Detours in bird migration. J Theor Biol 209:319–331

Alerstam T (2009) Flight by night or day? Optimal daily timing of bird migration. J Theor Biol 258:530–536

Alerstam T, Hedenström A, Åkesson S (2003) Long-distance migration: evolution and determinants. Oikos 103:247–260

Alerstam T, Bäckman J, Grönroos J, Olofsson P, Strandberg R (2019) Hypotheses and tracking results about the longest migration: the case of the Arctic Tern. Ecol Evol 9:9511–9531

Aparicio JM (2000) Differences in the diets of resident and non-resident kestrels in Spain. Ornis Fennica 77:169–175

Bäckman J, Andersson A, Alerstam T, Pedersen L, Sjöberg S, Thorup K, Tøttrup AP (2017) Activity and migratory flights of individual free-flying songbirds

throughout the annual cycle: method and first case study. J Avian Biol 48:309–319
Barcante L, Vale MM, Alves MAS (2017) Altitudinal migration by birds: a review of the literature and a comprehensive list of species. J Field Ornithol 88: 321–335
Becker PH, Schmaljohann H, Riechert J, Wagenknecht G, Zajková Z, González-Solís J (2016) Common Terns on the East Atlantic flyway: temporal–spatial distribution during the non-breeding period. J Ornithol 157:927–940
Bell CP (2000) Process in the evolution of bird migration and pattern in avian ecogeography. J Avian Biol 31: 258–265
Belthoff JR, Gauthreaux SA Jr (1991) Partial migration and differential winter distribution of House Finches in the eastern United States. Condor 93:374–382
Berthold P (1990) Wegzugbeginn und einsetzen der zugunruhe bie 19 vogelpopulationen – eine vergleichende untersuchung. In: van den Elzen R, Schuchmann K-L, Schmidt-Koenig K (eds) Proceedings of the International Centennial Meeting of the DO-G, Current Topics in Avian Biology. Bonn, Germany, pp. 217–222
Berthold P (1998) Bird migration: genetic programs with high adaptability. Zoology 101:235–245
Berthold P (1999) A comprehensive theory for the evolution, control and adaptability of migration. Ostrich 70: 1–11
Berthold P (2001) Bird migration: a general survey. Oxford University Press, Oxford, UK
Berthold P, Helbig AJ, Mohr G, Querner U (1992) Rapid microevolution of migratory behaviour in a wild bird species. Nature 360:668–670
Berthold P, Fiedler W, Querner U (2000) Die zugunruhe bei vögeln – eine darstellung nach videoaufnahmen bie infrarotlichtbeleuchtung. J Ornithol 141:285–299
Berthold P, van den Bosch W, Jakubiec Z, Kaatz C, Kaatz M, Querner U (2002) Long-term satellite tracking sheds light on variable migration strategies of White Storks (*Ciconia ciconia*). J Ornithol 143:489–495
Berthold P, Kaatz M, Querner U (2004) Long-term satellite tracking of White Stork (*Ciconia ciconia*) migration: constancy versus variability. J Ornithol 145:356–359
Bety J, Giroux JF, Gauthier G (2004) Individual variation in timing of migration: causes and reproductive consequences in Greater Snow Geese (*Anser caerulescens atlanticus*). Behav Ecol Sociobiol 57:1–8
Biebach H (1985) Sahara stopover in migratory flycatchers: fat and food affect the time program. Experientia 41:695–697
Blake JG, Loiselle BA (1991) Variation in resource abundance affects capture rates of birds in three lowland habitats in Costa Rica. Auk 108:114–130
Blake JG, Loiselle BA (2000) Diversity of birds along an elevational gradient in the Cordillera Central, Costa Rica. Auk 117:663–686
Blake JG, Loiselle BA (2002) Manakins (Pipridae) in second-growth and old-growth forests: patterns of habitat use, movement, and survival. Auk 119:132–148
Blanco DE, Zalba SM, Belenguer CJ, Pugnali G, Goñi HR (2003) Status and conservation of the Ruddy-headed Goose *Chloephaga rubidiceps* Sclater (Aves, Anatidae) in its wintering grounds (province of Buenos Aires, Argentina). Rev Chil Hist Nat 76:47–55
Blanco DE, Matus R, Blank O, de la Balze VM, Zalba SM (2006) The Ruddy-headed Goose *Chloephaga rubidiceps* mainland population: a flyway perspective. In: Boere GC, Galbraith CA, Stroud DA (eds) Waterbirds around the world: a global overview of the conservation, management and research of the world's waterbird flyways. The Stationary Office, Edinburgh, pp 195–196
Blois JL, Zarnetske PL, Fitzpatrick MC, Finnegan S (2013) Climate change and the past, present, and future of biotic interactions. Science 341:499–504
Bolshakov CV, Bulyuk VN (2001) New comprehensive systematic data concerning the time of nocturnal departure in some passerine migrants in autumn. Ring 23: 131–137
Both C (2010) Food availability, mistiming, and climatic change. In: Møller AP, Fiedler W, Berthold P (eds) Effects of climate change on birds. Oxford University Press, New York, NY, pp 129–147
Both C, Visser ME (2001) Adjustment to climate change is constrained by arrival date in a long-distance migrant bird. Nature 411:296–298
Both C, Van Turnhout CAM, Bijlsma RG, Siepel H, Van Strien AJ, Foppen RPB (2010) Avian population consequences of climate change are most severe for long-distance migrants in seasonal habitats. Proc R Soc B 277:1259–1266
Boyle WA (2006) Why do birds migrate? The role of food, habitat, predation, and competition, Ph.D. dissertation. University of Arizona, Tucson, AZ
Boyle WA (2008a) Partial migration in birds: tests of three hypotheses in a tropical lekking frugivore. J Anim Ecol 77:1122–1128
Boyle WA (2008b) Can variation in risk of nest predation explain altitudinal migration in tropical birds? Oecologia 155:397–403
Boyle WA (2010) Does food abundance explain altitudinal migration in a tropical frugivorous bird? Can J Zool 88:204–213
Boyle WA (2011) Short-distance partial migration of Neotropical birds: a community-level test of the foraging limitation hypothesis. Oikos 120:1803–1816
Boyle WA (2017) Altitudinal bird migration in North America. Auk 134:443–465
Boyle WA, Conway CJ (2007) Why migrate? A test of the evolutionary precursor hypothesis. Am Nat 169:344–359
Boyle WA, Conway CJ, Bronstein JL (2010a) Why do some, but not all, tropical birds migrate? A comparative study of diet breadth and fruit preference. Evol Ecol 25:219–236

Boyle WA, Norris DR, Guglielmo CG (2010b) Storms drive altitudinal migration in a tropical bird. Proc R Soc B 277:2511–2519
Boyle WA, Guglielmo CG, Hobson KA, Norris DR (2011) Lekking birds in a tropical forest forego sex for migration. Biol Lett 7:661–663
Bravo SP, Cueto VR, Gorosito CA (2017) Migratory timing, rate, routes and wintering areas of White-crested Elaenia (*Elaenia albiceps chilensis*), a key seed disperser for Patagonian forest regeneration. PLoS One 12:e0170188
Breuner CW, Hahn TP (2003) Integrating stress physiology, environmental change, and behavior in free-living sparrows. Horm Behav 43:115–123
Bridge ES, Kelly JF, Bjornen PE, Curry CM, Crawford PHC, Paritte JM (2010) Effects of nutritional condition on spring migration: do migrants use resource availability to keep pace with a changing world? J Exp Biol 213:2424–2429
Briedis M, Beran V, Hahn S, Adamík P (2016) Annual cycle and migration strategies of a habitat specialist, the Tawny Pipit *Anthus campestris*, revealed by geolocators. J Ornithol 157:619–626
Briedis M, Hahn S, Adamík P (2017) Cold spell en route delays spring arrival and decreases apparent survival in a long-distance migratory songbird. BMC Ecol 17:11
Briedis M, Hahn S, Krist M, Adamík P (2018) Finish with a sprint: evidence for time-selected last leg migration in a long-distance migratory songbird. Ecol Evol 8: 6899–6908
Briedis M, Bauer S, Adamík P, Alves JA, Costa JS, Emmenegger T, Gustafsson L, Koleček J, Krist M, Liechti F, Lisovski S (2020) Broad-scale patterns of the Afro-Palaearctic landbird migration. Glob Ecol Biogeogr 29:722–735
Bruderer B, Liechti F (1998) Flight behaviour of nocturnally migrating birds in coastal areas: crossing or coasting. J Avian Biol 29:499–507
Bruderer B, Peter D, Korner-Nievergelt F (2018) Vertical distribution of bird migration between the Baltic Sea and the Sahara. J Ornithol 159:315–336
Butler CJ (2003) The disproportionate effect of global warming on the arrival dates of short-distance migratory birds in North America. Ibis 145:484–495
Butler RW, Williams TD, Warnock N, Bishop MA (1997) Wind assistance: a requirement for migration of shorebirds? Auk 114:456–466
Butler LK, Donahue MG, Rohwer S (2002) Molt-migration in Western Tanagers (*Piranga ludoviciana*): age effects, aerodynamics, and conservation implications. Auk 119:1010–1023
Callaghan TV, Bjorn LO, Chapin FS, Chernov Y, Christensen TR, Huntley B, Ims RA, Johansson M, Jolly D, Jonasson S, Matveyeva N, Oechel WC, Panikov N, Shaver GR, Elster J, Henttonen H, Jonsdottir IS, Laine K, Schaphoff S, Sitch S, Taulavuori E, Taulavuori K, Zockler C (2005) Arctic tundra and polar desert ecosystems. In: Arctic climate impact assessment. Cambridge Univ Press, Cambridge, pp 243–352
Callo PA, Morton ES, Stutchbury BJM (2013) Prolonged spring migration in the Red-eyed Vireo (*Vireo olivaceus*). Auk 130:240–246
Capllonch P (2004) Migraciones de aves en el litoral Argentino. Miscelánea 12:363–368
Capllonch P, Ortiz D, Soria K (2008) Migración del Zorzal Común *Turdus amaurochalinus* (Turdidae) en Argentina. Revista Brasileira de Ornitologia 16:12–22
Carmi N, Pinshow B, Porter WP, Jaeger J (1992) Water and energy limitations on flight duration in small migrating birds. Auk 109:268–276
Carneiro C, Gunnarsson TG, Alves JA (2019) Faster migration in autumn than in spring: seasonal migration patterns and nonbreeding distribution of Icelandic Whimbrels *Numenius phaeopus islandicus*. J Avian Biol 50:jav.01938
Chan K (1994) Nocturnal activity of caged resident and migrant Silvereyes (Zosteropidae: Aves). Ethology 96:313–321
Chapman BB, Brönmark C, Nilsson JÅ, Hansson LA (2011) The ecology and evolution of partial migration. Oikos 120:1764–1775
Chaves-Campos J (2004) Elevational movements of large frugivorous birds and temporal variation in abundance of fruits along an elevational gradient. Ornitología Neotropical 15:433–445
Chaves-Campos J, Arévalo JE, Araya M (2003) Altitudinal movements and conservation of Bare-necked Umbrellabird *Cephalopterus glabricollis* of the Tilarán Mountains, Costa Rica. Bird Conserv Int 13:45–58
Chesser RT (1994) Migration in South America, an overview of the austral system. Bird Conservation International 4:91–107
Chesser RT (2005) Seasonal distribution and ecology of South American austral migrant flycatchers. In: Greenberg R, Marra PP (eds) Birds of two worlds. Johns Hopkins University Press, Baltimore, MD, pp 168–181
Clipp HL, Cohen EB, Smolinsky JA, Horton KG, Farnsworth A, Buler JJ (2020) Broad-scale weather patterns encountered during flight influence landbird stopover distributions. Remote Sens 12:565
Conklin JR, Lisovski S, Battley PF (2021) Advancement in long-distance bird migration through individual plasticity in departure. Nat Commun 12:4780
Coppack T, Pulido F (2009) Proximate control and adaptive potential of protandrous migration in birds. Integr Comp Biol 49:493–506
Coppack T, Pulido F, Czisch M, Auer DP, Berthold P (2003) Photoperiodic response may facilitate adaptation to climatic change in long-distance migratory birds. Proc R Soc B (Suppl) 270:S43–S46
Coppack T, Becker SF, Becker PJJ (2008) Circadian flight schedules in night-migrating birds caught on migration. Biol Lett 4:619–622
Cornelius JM, Breuner CW, Hahn TP (2010) Under a neighbor's influence: public information affects stress hormones and behaviour of a songbird. Proc R Soc B 277:2399–2404

Cornelius JM, Boswell T, Jenni-Eiermann S, Breuner CW, Ramenofsky M (2013) Contributions of endocrinology to the migration life history of birds. Gen Comp Endocrinol 190:47–60

Cox GW (1968) The role of competition in the evolution of migration. Evolution 22:180–192

Cox GW (1985) The evolution of avian migration systems between temperate and tropical regions of the New World. Am Nat 126:451–474

Crysler ZJ, Ronconi RA, Taylor PD (2016) Differential fall migratory routes of adult and juvenile Ipswich Sparrows (*Passerculus sandwichensis princeps*). Movement Ecol 4:3

Cueto VR, Milesi FA, Sagario MC, De Casenave JL, Marone L (2011) Distribución geográfica y patrones de movimiento de la Monterita Canela (*Poospiza ornata*) y el Yal Carbonero (*Phrygilus carbonarius*) en Argentina. Ornitologia Neotropical 22:483–494

CWBF (Chinese Wild Bird Federation) [online]. (2017) Checklists of birds of Taiwan, version Year:2017. http://www.bird.org.tw/index.php/works/lists

Davenport LC, Nole Bazán I, Carlos Erazo N (2012) East with the night: longitudinal migration of the Orinoco Goose (*Neochen jubata*) between Manú National Park, Peru and the llanos de Moxos. Bolivia. PLoS ONE 7: e46886

Davenport LC, Goodenough KS, Haugaasen T (2016) Birds of two oceans? Trans-Andean and divergent migration of Black Skimmers (*Rynchops niger cinerascens*) from the Peruvian Amazon. PLoS One 11:e0144994

Davis LA, Roalson EH, Cornell KL, McClanahan KD, Webster MS (2006) Genetic divergence and migration patterns in a North American passerine bird: implications for evolution and conservation. Mol Ecol 15:2141–2152

Dawson A (2007) Seasonality in a temperate zone bird can be entrained by near equatorial photoperiods. Proc R Soc B 274:721–725

de Castro ER, Côrtes MC, Navarro L, Galetti M, Morellato LPC (2012) Temporal variation in the abundance of two species of thrushes in relation to fruiting phenology in the Atlantic rainforest. Emu 112:137–148

De Paiva LV, Marini MÅ (2013) Timing of migration and breeding of the Lesser Elaenia (*Elaenia chiriquensis*) in a Neotropical savanna. Wilson J Ornithol 125:116–120

DeGroote LW, Hingst-Zaher E, Moreira-Lima L, Whitacre JV, Slyder JB, Wenzel JW (2021) Citizen science data reveals the cryptic migration of the Common Potoo *Nyctibius griseus* in Brazil. Ibis 163:380–389

Del Hoyo J, Elliott A, Sargatal J (1992) Handbook of birds of the world. Lynx Edicions, Barcelona

Delmore KE, Fox JW, Irwin DE (2012) Dramatic intraspecific differences in migratory routes, stopover sites and wintering areas, revealed using light-level geolocators. Proc R Soc B 279:4582–4589

Delord K, Barbraud C, Pinaud D, Ruault S, Patrick SC, Weimerskirch H (2019) Individual consistency in the non-breeding behavior of a long-distance migrant seabird, the Grey Petrel *Procellaria cinerea*. Mar Ornithol 47:93–103

DeLuca WV, Woodworth BK, Mackenzie SA, Newman AEM, Cooke HA, Phillips LM, Freeman NE, Sutton AO, Tauzer L, McIntyre C, Stenhouse IJ (2019) A boreal songbird's 20,000 km migration across North America and the Atlantic Ocean. Ecology 100:e02651

Deng X, Zhao Q, Fang L, Xu Z, Wang X, He H, Cao L, Fox AD (2019) Spring migration duration exceeds that of autumn migration in Far East Asian Greater White-fronted Geese (*Anser albifrons*). Avian Res 10:9

Deppe JL, Ward MP, Bolus RT, Diehl RH, Celis-Murillo A, Zenzal TJ, Moore FR, Benson TJ, Smolinsky JA, Schofield LN, Enstrom DA, Paxton EH, Bohrer G, Beveroth TA, Raim A, Obringer RL, Delaney D, Cochran WW (2015) Fat, weather, and date affect migratory songbirds' departure decisions, routes, and time it takes to cross the Gulf of Mexico. Proc Natl Acad Sci U S A 112:E6331–E6338

Diehl RH, Bates JM, Willard DE, Gnoske TP (2014) Bird mortality during nocturnal migration over Lake Michigan: a case study. Wilson J Ornithol 126:19–29

Dingle H (2008) Bird migration in the southern hemisphere: a review comparing continents. Emu 108: 341–359

Dionne M, Maurice C, Gauthier J, Shaffer F (2008) Impact of Hurricane Wilma on migrating birds: the case of the Chimney Swift. Wilson J Ornithol 120:784–792

Dokter AM, Liechti F, Stark H, Delobbe L, Tabary P, Holleman I (2011) Bird migration flight altitudes studied by a network of operational weather radars. J R Soc Interface 8:30–43

Dokter AM, Farnsworth A, Fink D, Ruiz-Gutierrez V, Hochachka WM, La Sorte FA, Robinson OJ, Rosenberg KV, Kelling S (2018) Seasonal abundance and survival of North America's migratory avifauna determined by weather radar. Nature Ecol Evol 2:1603–1609

Dufour P, Descamps S, Chantepie S, Renaud J, Guéguen M, Schiffers K, Thuiller W, Lavergne S (2020) Reconstructing the geographic and climatic origins of long-distance bird migrations. J Biogeogr 47:155–166

Duijns S, Jukema J, Spaans B, van Horssen P, Piersma T (2012) Revisiting the proposed leap-frog migration of Bar-tailed Godwits along the East-Atlantic flyway. Ardea 100:37–43

Duijns S, Anderson AM, Aubry Y, Dey A, Flemming SA, Francis CM, Friis C, Gratto-Trevor C, Hamilton DJ, Holberton R, Koch S, McKellar AE, Mizrahi D, Morrissey CA, Neima SG, Newstead D, Niles L, Nol E, Paquet J, Rausch J, Tudor L, Turcotte Y, Smith PA (2019) Long-distance migratory shorebirds travel faster towards their breeding grounds, but fly faster post-breeding. Sci Rep 9:9420

Durant JM, Hjermann DØ, Ottersen G, Stenseth NC (2007) Climate and the match or mismatch between

predator requirements and resource availability. Clim Res 33:271–283
Duriez O, Bauer S, Destin A, Madsen J, Nolet BA, Stillman RA, Klaasen M (2009) What decision rules might Pink-footed Geese use to depart on migration? An individual-based model. Behav Ecol 20:560–569
Duriez O, Peron G, Gremillet D, Sforzi A, Monti F (2018) Migrating Ospreys use thermal uplift over the open sea. Biol Lett 14:20180687
Eikenaar C, Hessler S, Hegemann A (2020) Migrating birds rapidly increase constitutive immune function during stopover. R Soc Open Sci 7:192031
Elmberg J, Hessel R, Fox AD, Dalby L (2014) Interpreting seasonal range shifts in migratory birds: a critical assessment of 'short-stopping'and a suggested terminology. J Ornithol 155:571–579
Erni B, Liechti F, Underhill LG, Bruderer B (2002) Wind and rain govern the intensity of nocturnal bird migration in Central Europe - a log-linear regression analysis. Ardea 90:155–166
Faaborg J, Holmes RT, Anders AD, Bildstein KL, Dugger KM, Gauthreaux SA Jr, Heglund P, Hobson KA, Jahn AE, Johnson DH, Latta SC, Levey DJ, Marra PP, Merkord CL, Nol E, Rothstein SI, Sherry TW, Sillett TS, Thompson FR III, Warnock N (2010) Recent advances in understanding migration systems of New World land birds. Ecol Monogr 80:3–48
Falls JB, Kopachena JG (2020) White-throated Sparrow (*Zonotrichia albicollis*), version 1.0. In: Poole AF (ed) Birds of the world. Cornell Lab of Ornithology, Ithaca, NY
Farnsworth A, Langham G (2018) Chocolate-vented Tyrant (*Neoxolmis rufiventris*), version 1.0. In: del Hoyo J, Elliott A, Sargatal J, Christie DA, de Juana E (eds) Handbook of the birds of the world alive. Lynx Edicions, Barcelona
Farnsworth A, Lebbin D, Kirwan GM (2018) Austral Negrito (*Lessonia rufa*). In: del Hoyo J, Elliott A, Sargatal J, Christie DA, de Juana E (eds) Handbook of the birds of the world alive. Lynx Edicions, Barcelona
Felicísimo ÁM, Muñoz J, González-Solis J (2008) Ocean surface winds drive dynamics of transoceanic aerial movements. PLoS One 3:e2928
Franzén LG (1991) The changing frequency of gales on the Swedish west coast and its possible relation to the increased damage to coniferous forests of southern Sweden. Int J Climatol 11:769–793
Fraser KC, Diamond AW, Chavarría L (2010) Evidence of altitudinal moult-migration in a central American hummingbird, *Amazilia cyanura*. J Trop Ecol 26:645–648
Fretwell SD (1980) Evolution of migration in relation to factors regulating bird numbers. In: Keast A, Morton ES (eds) Migrant birds in the Neotropics. Smithsonian Institution, Washington, D.C., pp 517–527
Fukui D, Dewa H, Katsuta S, Sato A (2013) Bird predation by the birdlike noctule in Japan. J Mammal 94:657–661
Galtbalt B, Lilleyman A, Coleman JT, Cheng C, Ma Z, Rogers DI, Woodworth BK, Fuller RA, Garnett ST, Klaassen M (2021) Far Eastern Curlew and Whimbrel prefer flying low-wind support and good visibility appear only secondary factors in determining migratory flight altitude. Mov Ecol 9:32
Gauthreaux SA (1982) The ecology and evolution of avian migration systems. In: Farner DS, King JR (eds) Avian biology, vol 6. Academic Press, New York, NY, pp 93–168
Gilg O, Kovacs KM, Aars J, Fort J, Gauthier G, Grémillet D, Ims RA, Meltofte H, Moreau J, Post E, Schmidt NM (2012) Climate change and the ecology and evolution of Arctic vertebrates. Ann N Y Acad Sci 1249:166–190
Gill RE Jr, Tibbitts TL, Douglas DC, Handel CM, Mulcahy DM, Gottschalck JC, Warnock N, McCaffery BJ, Battley PF, Piersma T (2009) Extreme endurance flights by landbirds crossing the Pacific Ocean: ecological corridor rather than barrier? Proc R Soc B 276: 447–458
Gillis EA, Green DJ, Middleton HA, Morrissey CA (2008) Life history correlates of alternative migratory strategies in American Dippers. Ecology 89:1687–1695
Goldstein MI, Duffy DC, Oehlers S, Catterson N, Frederick J, Pyare S (2019) Interseasonal movements and non-breeding locations of Aleutian Terns *Onychoprion aleuticus*. Mar Ornithol 47:67–76
Gómez C, Bayly NJ, Norris DR, Mackenzie SA, Rosenberg KV, Taylor PD, Hobson KA, Cadena CD (2017) Fuel loads acquired at a stopover site influence the pace of intercontinental migration in a boreal songbird. Sci Rep 7:3405
Gong L, Shi B, Wu H, Feng J, Jiang T (2021) Who's for dinner? Bird prey diversity and choice in the great evening bat, *Ia io*. Ecol Evol 11:8400–8409
González-Solís J, Croxall JP, Oro D, Ruiz X (2007) Trans-equatorial migration and mixing in the wintering areas of a pelagic seabird. Front Ecol Environ 5:297–301
Grecian WJ, Williams HJ, Votier SC, Bearhop S, Cleasby IR, Grémillet D, Hamer KC, Le Nuz M, Lescroël A, Newton J, Patrick SC (2019) Individual spatial consistency and dietary flexibility in the migratory behavior of Northern Gannets wintering in the Northeast Atlantic. Front Ecol Evol 7:214
Guillemette M, Woakes AJ, Larochelle J, Polymeropoulos ET, Granbois J-M, Butler PJ, Pelletier D, Frappell PB, Portugal SJ (2016) Does hyperthermia constrain flight duration in a short-distance migrant? Philos Trans R Soc B 371:20150386
Guillemette M, Polymeropoulos ET, Portugal SJ, Pelletier D (2017) It takes time to be cool: on the relationship between hyperthermia and body cooling in a migrating seaduck. Front Physiol 8:532
Gwinner E (1986) Circannual rhythms. Springer Verlag, Heidelberg, Germany
Haest B, Hüppop O, Bairlein F (2020) Weather at the winter and stopover areas determines spring migration

onset, progress, and advancements in Afro-Palearctic migrant birds. Proc Natl Acad Sci 117:17056–17062

Hahn TP, Sockman KW, Breuner CW, Morton ML (2004) Facultative altitudinal movements by Mountain White-crowned Sparrows (*Zonotrichia leucophrys oriantha*) in the Sierra Nevada. Auk 121:1269–1281

Haines AM, McGrady MJ, Martell MS, Dayton BJ, Henke MB, Seegar WS (2003) Migration routes and wintering locations of Broad-winged Hawks tracked by satellite telemetry. Wilson Bull 115:166–169

Hake M, Kjellén N, Alerstam T (2003) Age-dependent migration strategy in Honey Buzzards *Pernis apivorus* tracked by satellite. Oikos 103:385–396

Handel CM, Gill RE Jr (2010) Wayward youth: trans-Beringian movement and differential southward migration by juvenile Sharp-tailed Sandpipers. Arctic 63:273–288

Hatch JJ, Gochfeld M, Burger J, Garcia EFJ (2020) Arctic Tern (*Sterna paradisaea*), version 1.0. In: Billerman SM (ed) Birds of the World. Cornell Lab of Ornithology, Ithaca, NY

Hau M, Wikelski M, Wingfield JC (1998) A Neotropical forest bird can measure the slight changes in tropical photoperiod. Proc R Soc B 265:89–95

Hayes FE (1995) Definition for migrant birds: what is a Neotropical migrant? Auk 112:521–523

Heckscher CM, Taylor SM, Fox JW, Afanasyev V (2011) Veery (*Catharus fuscescens*) wintering locations, migratory connectivity, and a revision of its winter range using geolocator technology. Auk 128:531–542

Hedenström A, Alerstam T (1992) Climbing performance of migrating birds as a basis for estimating limits for fuel-carrying capacity and muscle work. J Exp Biol 164:19–38

Helbig AJ (1996) Genetic basis, mode of inheritance and evolutionary changes of migratory directions in Palearctic warblers (Aves: Sylviidae). J Exp Biol 199:49–55

Helbig AJ (2003) Evolution of bird migration: a phylogenetic and biogeographic perspective. In: Berthold P, Gwinner E, Sonnenschein F (eds) Avian migration. Springer, Berlin, Germany, pp 3–20

Helm B, Gwinner E (2006) Migratory restlessness in an equatorial nonmigratory bird. PLoS Biol 4:e110

Hobson KA, Wassenaar LI, Milá B, Lovette I, Dingle C, Smith TB (2003) Stable isotopes as indicators of altitudinal distributions and movements in an Ecuadorean hummingbird community. Oecologia 136:302–308

Hope DD, Lank DB, Ydenberg RC (2014) Mortality-minimizing sandpipers vary stopover behavior dependent on age and geographic proximity to migrating predators. Behav Ecol Sociobiol 68:827–838

Horton KG, Van Doren BM, Stepanian PM, Farnsworth A, Kelly JF (2016a) Where in the air? Aerial habitat use of nocturnally migrating birds. Biol Lett 12:20160591

Horton KG, Van Doren BM, Stepanian PM, Farnsworth A, Kelly JF (2016b) Seasonal differences in landbird migration strategies. Auk 133:761–769

Horton KG, La Sorte FA, Sheldon D, Lin T-Y, Winner K, Bernstein G, Maji S, Hochachka WM, Farnsworth A (2020) Phenology of nocturnal avian migration has shifted at the continental scale. Nat Clim Chang 10:63–68

Horvitz N, Sapir N, Liechti F, Avissar R, Mahrer I, Nathan R (2014) The gliding speed of migrating birds: slow and safe or fast and risky? Ecol Lett 17:670–679

Ibáñez C, Juste J, Garcia-Mudarra JL, Agirre-Medi PT (2001) Bat predation on nocturnally migrating birds. Proc Natl Acad Sci USA 98:9700–9702

Ibáñez C, Popa-Lisseanu AG, Pastor-Beviá D, García-Mudarra JL, Juste J (2016) Concealed by darkness: interactions between predatory bats and nocturnally migrating songbirds illuminated by DNA sequencing. Mol Ecol 25:5254–5263

Ims RA, Henden J-A, Killengreen ST (2008) Collapsing population cycles. Trends Ecol Evol 23:79–86

Jacobsen LB, Jensen NO, Willemoes M, Hansen L, Desholm M, Fox AD, Tøttrup AP, Thorup K (2017) Annual spatiotemporal migration schedules in three larger insectivorous birds: European Nightjar, Common Swift and Common Cuckoo. Anim Biotelemetry 51:4

Jahn AE, Levey DJ, Smith KG (2004) Reflections across hemispheres: a system-wide approach to New World bird migration. Auk 121:1005–1013

Jahn AE, Levey DJ, Hostetler JA, Mamani AM (2010) Determinants of partial bird migration in the Amazon Basin. J Anim Ecol 79:983–992

Jahn AE, Cueto VR, Fox JM, Husak MS, Kim DH, Landoli DV, Ledezma JP, Lepage HK, Levey DJ, Murphy MT, Renfrew RB (2013a) Migration timing and wintering areas of three species of flycatchers (*Tyrannus*) breeding in the Great Plains of North America. Auk 130:247–257

Jahn AE, Levey DJ, Cueto VR, Ledezma JP, Tuero DT, Fox JW, Masson D (2013b) Long-distance bird migration within South America revealed by light-level geolocators. Auk 130:223–229

Jahn AE, Seavy NE, Bejarano V, Guzmán MB, Provinciato ICC, Pizo MA, MacPherson M (2016) Intra-tropical migration and wintering areas of Fork-tailed Flycatchers (*Tyrannus savana*) breeding in São Paulo, Brazil. Revista Brasileira de Ornitologia 24:116–121

Jahn AE, Cereghetti J, Cueto VR, Hallworth MT, Levey DJ, Marini MÂ, Masson D, Pizo MA, Sarasola JH, Tuero DT (2019) Breeding latitude predicts timing but not rate of spring migration in a widespread migratory bird in South America. Ecol Evol 9:5752–5765

Jahn AE, Cueto VR, Fontana CS, Guaraldo AC, Levey DJ, Marra PP, Ryder TB (2020) Bird migration within the Neotropics. Auk 137:ukaa033

Jahn AE, Buechley E, Bugoni L, Cereghetti J, Repenning M, Marra PP, Ryder TB (2021) Variable seasonal movement dynamics among individual Snail Kites (*Rostrhamus sociabilis*) in South America. J Raptor Res 55:151–168

Jenkins KD, Cristol DA (2002) Evidence of differential migration by sex in White-throated Sparrows (*Zonotrichia albicollis*). Auk 119:539–543

Jenni L, Schaub M (2003) Behavioural and physiological reactions to environmental variation in bird migration: a review. In: Berthold P, Gwinner E, Sonnenschein E (eds) Avian migration. Springer, Berlin, Heidelberg, pp 155–171

Jones T, Cresswell W (2010) The phenology mismatch hypothesis: are declines in migratory birds linked to uneven global climate change? J Anim Ecol 79: 98–108

Jonzén N, Lindén A, Ergon T, Knudsen E, Vik JO, Rubolini D, Piacentini D, Brinch C, Spina F, Karlsson L, Stervander M (2006) Rapid advance of spring arrival dates in long-distance migratory birds. Science 312:1959–1961

Joseph L (2005) Molecular approaches to the evolution and ecology of migration. In: Greenberg R, Marra PP (eds) Birds of two worlds: the ecology and evolution of migration. Johns Hopkins University Press, Baltimore, MD, pp 18–26

Karlsson H, Nilsson C, Bäckman J, Alerstam T (2012) Nocturnal passerine migrants fly faster in spring than in autumn: a test of the time minimization hypothesis. Anim Behav 83:87–93

Kelly JF, Horton KG, Stepanian PM, de Beurs KM, Fagin T, Bridge ES, Chilson PB (2016) Novel measures of continental-scale avian migration phenology related to proximate environmental cues. Ecosphere 7:e01434

Kemp MU, Shamoun-Baranes J, Van Gasteren H, Bouten W, Van Loon EE (2010) Can wind help explain seasonal differences in avian migration speed? J Avian Biol 41:672–677

Kemp MU, Shamoun-Baranes J, Dokter AM, van Loon E, Bouten W (2013) The influence of weather on the flight altitude of nocturnal migrants in mid-latitudes. Ibis 155:734–749

Kerlinger P, Moore FR (1989) Atmospheric structure and avian migration. Curr Ornithol 6:109–142

Ketterson ED, Nolan V Jr (1976) Geographic variation and its climatic correlates in the sex ratio of eastern-wintering Dark-eyed Juncos (*Junco hyemalis*). Ecology 57:679–693

Ketterson ED, Nolan V Jr (1983) The evolution of differential bird migration. Curr Ornithol 1:357–402

Kimura K, Yumoto T, Kikuzawa K (2001) Fruiting phenology of fleshy-fruited plants and seasonal dynamics of frugivorous birds in four vegetation zones on mt. Kinabalu. Borneo. J Trop Ecol 17:833–858

King JR, Farner DS, Mewaldt LR (1965) Seasonal sex and age ratios in populations of the White-crowned Sparrow of the race *gambelii*. Condor 67:489–504

Kingery HE, Willson MP (2020) American Dipper (*Cinclus mexicanus*), version 1.0. In: Rodewald PG (ed) Birds of the World. Cornell Lab of Ornithology, Ithaca, NY

Kjellén N (1994) Differences in age and sex ratio among migrating and wintering raptors in southern Sweden. Auk 111:274–284

Klaassen M (1996) Metabolic constraints on long-distance migration in birds. J Exp Biol 199:57–64

Klaassen RHG, Hake M, Strandberg R, Koks BJ, Trierweiler C, Exo K-M, Bairlein F, Alerstam T (2014) When and where does mortality occur in migratory birds? Direct evidence from long-term satellite tracking of raptors. J Anim Ecol 83:176–184

Knight SM, Gow EA, Bradley DW, Clark RG, Bélisle M, Berzins LL, Blake T, Bridge ES, Burke L, Dawson RD, Dunn PO, Garant D, Holroyd GL, Hussell DJT, Lansdorp O, Laughlin AJ, Leonard ML, Pelletier F, Shutler D, Siefferman L, Taylor CM, Trefry HE, Vleck CM, Vleck D, Whittingham LA, Winkler DW, Norris DR (2019) Nonbreeding season movements of a migratory songbird are related to declines in resource availability. Auk 136:1–13

Koenig WD, Knops JMH (2001) Seed-crop size and eruptions of North American boreal seed-eating birds. J Anim Ecol 70:609–620

Kohno H, Mizutani A, Yoda K, Yamamoto T (2019) Movements and activity characteristics of the Brown Booby *Sula leucogaster* during the non-breeding period. Mar Ornithol 47:169–174

Kokko H (1999) Competition for early arrival in migratory birds. J Anim Ecol 68:940–950

Kokko H, Gunnarsson TG, Morrell LJ, Gill JA (2006) Why do female migratory birds arrive later than males? J Anim Ecol 75:1293–1303

Koleček J, Hahn S, Emmenegger T, Procházka P (2018) Intra-tropical movements as a beneficial strategy for Palearctic migratory birds. R Soc Open Sci 5:171675

Kölzsch A, Müskens GJDM, Kruckenberg H, Glazov P, Weinzierl R, Nolet BA, Wikelski M (2016) Towards a new understanding of migration timing: slower spring than autumn migration in geese reflects different decision rules for stopover use and departure. Oikos 125: 1496–1507

Komenda-Zehnder S, Liechti F, Bruderer B (2002) Is reverse migration a common feature of nocturnal bird migration? An analysis of radar data from Israel. Ardea 90:325–334

Köppen U, Yakovlev AP, Barth R, Kaatz M, Berthold P (2010) Seasonal migrations of four individual Bar-headed Geese *Anser indicus* from Kyrgyzstan followed by satellite telemetry. J Ornithol 151:703–712

Kranstauber B, Weinzieri R, Wikelski M, Safi K (2015) Global aerial pathways allow efficient travelling. Ecol Lett 18:1338–1345

Kubelka V, Salek M, Tomkovich P, Vegvari Z, Freckleton RP, Szekely T (2018) Global pattern of nest predation is disrupted by climate change in shorebirds. Science 362:680–683

Kuo Y, Lin D-L, Chuang F-M, Lee P-F, Ding T-S (2013) Bird species migration ratio in East Asia, Australia, and surrounding islands. Naturwissenschaften 100: 729–738

Lank DB (1989) Why fly by night? Inferences from tidally-induced migratory departures of sandpipers. J Field Ornithol 60:154–161
Laymon SA (1989) Altitudinal migration movements of Spotted Owls in the Sierra Nevada, California. Condor 91:837–841
Lees AC (2016) Evidence for longitudinal migration by a "sedentary" Brazilian flycatcher, the Ash-throated Casiornis. J Field Ornithol 87:251–259
Lees AC, Martin RW (2015) Exposing hidden endemics in a Neotropical forest raptor using citizen science. Ibis 157:103–114
Lehikoinen E, Sparks TH (2010) Changes in migration. In: Møller AP, Fiedler W, Berthold P (eds) Effects of climate change on birds. Oxford University Press, New York, NY, pp 89–112
Lemke HW, Tarka M, Klaassen RHG, Åkesson M, Bensch S, Hasselquist D, Hansson B (2013) Annual cycle and migration strategies of a trans-Saharan migratory songbird: a geolocator study in the Great Reed Warbler. PLoS ONE 8:e79209
Levey DJ (1988) Spatial and temporal variation in Costa Rican fruit and fruit-eating bird abundance. Ecol Monogr 58:251–269
Levey DJ, Stiles FG (1992) Evolutionary precursors of long-distance migration: resource availability and movement patterns of Neotropical landbirds. Am Nat 140:447–476
Licona-Vera Y, Ornelas JF (2017) The conquering of North America: dated phylogenetic and biogeographic inference of migratory behavior in Bee Hummingbirds. BMC Evol Biol 17:126
Liechti F (2006) Birds: blowin' by the wind? J Ornithol 147:202–211
Liechti F, Bruderer B (1998) The relevance of wind for optimal migration theory. J Avian Biol 29:561–568
Lindström Å, Alerstam T, Andersson A, Bäckman J, Bahlenberg P, Bom R, Ekblom R, Klaassen RH, Korniluk M, Sjöberg S, Weber JK (2021) Extreme altitude changes between night and day during marathon flights of Great Snipes. Curr Biol 31:2021.05.047
Linkhart BD, Fox JW, Yanco SW (2016) Migration timing and routes, and wintering areas of Flammulated Owls. J Field Ornithol 87:42–54
Linscott JA, Senner NR (2021) Beyond refueling: investigating the diversity of functions of migratory stopover events. Ornithol Appl 123:duaa074
Lislevand T, Briedis M, Heggøy O, Hahn S (2017) Seasonal migration strategies of Common Ringed Plovers *Charadrius hiaticula*. Ibis 159:225–229
Loiselle BA, Blake JG (1991) Temporal variation in birds and fruits along an elevational gradient in Costa Rica. Ecology 72:180–193
Low M, Arlt D, Pärt T, Öberg M (2015) Delaying timing of breeding as a cost of reproduction. J Avian Biol 46: 325–331
Lundberg P (1985) Dominance behaviour, body weight and fat variations, and partial migration in European Blackbirds *Turdus merula*. Behav Ecol Sociobiol 17:185–189
Lundberg P (1988) The evolution of partial migration in birds. Trends Ecol Evol 3:172–175
Mackas RH, Green DJ, Whitehorne IBJ, Fairhurst EN, Middleton HA, Morrisey CA (2010) Altitudinal migration in American Dippers (*Cinclus mexicanus*): do migrants produce higher quality offspring? Can J Zool 88:369–377
Maggini I, Cardinale M, Sundberg JH, Spina F, Fusani L (2020) Recent phenological shifts of migratory birds at a Mediterranean spring stopover site: species wintering in the Sahel advance passage more than tropical winterers. PLoS ONE 15:e0239489
Mallord JW, Orsman CJ, Cristinacce A, Stowe TJ, Charman EC, Gregory RD (2017) Diet flexibility in a declining long-distance migrant may allow it to escape the consequences of phenological mismatch with its caterpillar food supply. Ibis 159:76–90
Marantz CA, Remsen JV Jr (1991) Seasonal distribution of the Slaty Elaenia, a little-known austral migrant of South America. J Field Ornithol 62:162–172
Marcström V, Mascher JW (1979) Weights and fat in lapwings *Vanellus vanellus* and oystercatchers *Haematopus ostralegus* starved to death during a cold spell in spring. Ornis Scand 10:235–240
Marini MÅ, Cavalcanti RB (1990) Migrações de *Elaenia albiceps chilensis* e *Elaenia chiriquensis albivertex* (Aves: Tyrannidae). Boletim do Museu Paraense Emílio Goeldi Zoologia 6:59–67
Marques PAM, Sowter D, Jorge PE (2010) Gulls can change their migratory behavior during lifetime. Oikos 119:946–951
Marquiss M (2002) Common Crossbill. In: Wernham CV, Toms MP, Marchant JH, Clark JA, Siriwardena GM, Baillie SR (eds) The migration atlas: movements of the birds of Britain and Ireland. T. & A. D. Poyser, London, pp 663–665
Marra PP, Francis CM, Mulvihill RS, Moore FR (2005) The influence of climate on the timing and rate of spring bird migration. Oecologia 142:307–315
Marrot P, Garant D, Charmantier A (2017) Multiple extreme climatic events strengthen selection for earlier breeding in a wild passerine. Philos Trans R Soc B 372:20160372
Mateos-Rodríguez M, Liechti F (2012) How do diurnal long-distance migrants select flight altitude in relation to wind? Behav Ecol 23:403–409
McCabe RA, Goodrich LJ, Barber DR, Master TL, Watson JL, Bayne EM, Harrison A-L, Marra PP, Bildstein KL (2020) Satellite tracking reveals age and origin differences in migration ecology of two populations of Broad-winged Hawks (*Buteo platypterus*). Wilson J Ornithol 132:1–14
McKinnon L, Smith PA, Nol E, Martin JL, Doyle FI, Abraham KF, Gilchrist HG, Morrison RIG, Bêty J (2010) Lower predation risk for migratory birds at high latitudes. Science 327:326–327

McKinnon EA, Fraser KC, Stutchbury BJ (2013) New discoveries in landbird migration using geolocators, and a flight plan for the future. Auk 130:211–222
McKinnon EA, Fraser KC, Stanley CQ, Stutchbury BJM (2014) Tracking from the tropics reveals behaviour of juvenile songbirds on their first spring migration. PLoS One 9:e105605
McNamara JM, Welham RK, Houston AI (1998) The timing of migration within the context of an annual routine. Oikos 29:416–423
Meltofte H, Piersma T, Boyd H, McCaffery B, Ganter B, Golovnyuk VV, Graham K, Gratto-Trevor CL, Morrison RIG, Nol E, Rösner H-U, Schamel D, Schekkerman H, Soloviev MY, Tomkovich PS, Tracy DM, Tulp I, Wennerberg L (2007) Effects of climate variation on the breeding ecology of Arctic shorebirds. Museum Tusculanum Press, Copenhagen, Denmark
Mendgen P, Converse SJ, Pearse AT, Teitelbaum CS, Mueller T (2023) Differential shortstopping behavior in Whooping Cranes: habitat or social learning? Global Ecol Cons 41:e02365
Miller TA, Brooks RP, Lanzone MJ, Brandes D, Cooper J, Tremblay JA, Wilhelm J, Duerr A, Katzner TE (2016) Limitations and mechanisms influencing the migratory performance of soaring birds. Ibis 158:116–134
Mills AM (2005) Changes in the timing of spring and autumn migration in North American migrant passerines during a period of global warming. Ibis 147:259–269
Møller AP (1994) Sexual selection and the Barn Swallow. Oxford University Press, Oxford
Møller AP, Rubolini D, Lehikoinen E (2008) Populations of migratory bird species that did not show a phonological response to climate change are declining. Proc Natl Acad Sci U S A 105:16195–16200
Morbey YE, Ydenberg RC (2001) Protandrous arrival timing to breeding areas: a review. Ecol Lett 4:663–673
Morton ES (1977) Intratropical migration in the Yellow-green Vireo and Piratic Flycatcher. Auk 94:97–106
Mrosovsky N (1978) Circannual cycles in hibernators. In: Wang LCH, Hudson JW (eds) Strategies in the cold: natural torpidity and thermogenesis. Academic Press, New York, NY, pp 21–65
Mueller JC, Pulido F, Kempenaers B (2011) Identification of a gene associated with avian migratory behaviour. Proc R Soc B 278:2848–2856
Mueller T, O'Hara RB, Converse SJ, Urbanek RP, Fagan WF (2013) Social learning of migratory performance. Science 341:999–1002
Müller F, Taylor PD, Sjöberg S, Muheim R, Tsvey A, Mackenzie SA, Schmaljohann H (2016) Towards a conceptual framework for explaining variation in nocturnal departure time of songbird migrants. Mov Ecol 4:24
Murphy-Klassen HM, Underwood TJ, Sealy SG, Czyrnyj AA (2005) Long-term trends in spring arrival dates of migrant birds at Delta Marsh, Manitoba, in relation to climate change. Auk 122:1130–1148
Newton I (1972) Finches. Collins, London
Newton I (1995) Relationship between breeding and wintering ranges in Palaearctic-African migrants. Ibis 137: 241–249
Newton I (2003a) The speciation and biogeography of birds. Academic Press, San Diego, CA
Newton I (2003b) Geographical patterns in bird migration. In: Berthold P, Gwinner E, Sonnenschein E (eds) Avian migration. Springer-Verlag, New York, NY, pp 211–224
Newton I (2006) Movement patterns of Common Crossbills *Loxia curvirostra* in Europe. Ibis 148:782–788
Newton I (2007a) Weather-related mass-mortality events in migrants. Ibis 149:453–467
Newton I (2007b) The ecology of bird migration. Academic Press, London, UK
Newton I (2008) The migration ecology of birds. Academic Press, London, UK
Newton I (2012) Obligate and facultative migration in birds: ecological aspects. J Ornithol 153(Suppl. 1): S171–S180
Newton I, Dale LC (1996) Bird migration at different latitudes in eastern North America. Auk 113:626–635
Nilsson C, Sjöberg S (2016) Causes and characteristics of reverse bird migration: an analysis based on radar, radio tracking and ringing at Falsterbo, Sweden. J Avian Biol 47:354–362
Nilsson C, Bäckman J, Alerstam T (2014) Seasonal modulation of flight speed among nocturnal passerine migrants: differences between short-and long-distance migrants. Behav Ecol Sociobiol 68:1799–1807
Norevik G, Boano G, Hedenström A, Lardelli R, Liechti F, Åkesson S (2019) Highly mobile insectivorous swifts perform multiple intra-tropical migrations to exploit an asynchronous African phenology Oikos 128:640–648
Nystrom KGK (1997) Food density, song rate, and body condition in territory-establishing Willow Warblers (*Phylloscopus trochilus*). Can J Zool 75:47–58
Ogonowski MS, Conway CJ (2009) Migratory decisions in birds: extent of genetic versus environmental control. Oecologia 161:199–207
Oliveros CH, Field DJ, Ksepka DT, Barker FK, Aleixo A, Andersen MJ, Alström P, Benz BW, Braun EL, Braun MJ, Bravo GA (2019) Earth history and the passerine superradiation. Proc Natl Acad Sci 116:7916–7925
Oring LW, Lank DB (1982) Sexual selection, arrival times, philopatry and sitefidelity in the polyandrous Spotted Sandpiper. Behav Ecol Sociobiol 10:185–191
Outlaw DC, Voelker G (2006) Phylogenetic tests of hypotheses for the evolution of avian migration: a case study using the Motacillidae. Auk 123:455–466
Outlaw DC, Voelker G, Mila B, Girman DJ (2003) Evolution of long-distance migration in and historical biogeography of *Catharus* thrushes: a molecular phylogenetic approach. Auk 120:299–310
Owen JC, Moore FR (2008) Relationship between energetic condition and indicators of immune function in

thrushes during spring migration. Can J Zool 86:638–647

Pageau C, Vale MM, de Menezes MA, Barçante L, Shaikh M, Alves MAS, Reudink MW (2020) Evolution of altitudinal migration in passerines is linked to diet. Ecol Evol 10:3338–3345

Pastorino A, Ramirez JR, Agostini N, Dell'Omo G, Panuccio M (2017) Fog and rain lead migrating White Storks *Ciconia ciconia* to perform reverse migration and to land. Avocetta 41:5–12

Paxton EH, Durst SL, Sogge MK, Koronkiewicz TJ, Paxton KL (2017) Survivorship across the annual cycle of a migratory passerine, the Willow Flycatcher. J Avian Biol 48:1126–1131

Pedersen L, Onrubia A, Vardanis Y, Barboutis C, Waasdorp S, van Helvert M, Geertsma M, Ekberg P, Willemoes M, Strandberg R, Matsyna E (2020) Remarkably similar migration patterns between different Red-backed Shrike populations suggest that migration rather than breeding area phenology determines the annual cycle. J Avian Biol 51:jav.02475

Pedrana J, Puetz K, Bernad L, Munoz S, Gorosabel A, Castresana G, Leiss A, Pon JPS (2020) Spatial and temporal variation in the migration of Ruddy-headed Goose in southern South America using satellite tagging. Bird Cons Int 30:634–648

Pennycuick CJ (2008) Modelling the flying bird. Academic Press, London, UK

Phillips RA, Catry P, Silk JR, Bearhop S, McGill R, Afanasyev V, Strange IJ (2007) Movements, winter distribution and activity patterns of Falkland and Brown skuas: insights from loggers and isotopes. Mar Ecol Prog Ser 345:281–291

Phipps WL, Lopez-Lopez P, Buechley ER, Oppel S, Alvarez E, Arkumarev V, Bekmansurov R, Berger-Tai O, Bermejo A, Bounas A, Alanis IC (2019) Spatial and temporal variability in migration of a soaring raptor across three continents. Front Ecol Evol 7:00323

Pinet P, Jaquemet S, Pinaud D, Weimerskirch H, Phillips RA, Le Corre M (2011) Migration, wintering distribution and habitat use of an endangered tropical seabird, Barau's Petrel *Pterodroma baraui*. Mar Ecol Prog Ser 423:291–302

Piper WH, Wiley RH (1989) Correlates of dominance in wintering White-throated Sparrows: age, sex, and location. Anim Behav 37:298–310

Potvin DA, Välimäki K, Lehikoinen A (2016) Differences in shifts of wintering and breeding ranges lead to changing migration distances in European birds. J Avian Biol 47:619–628

Powell GVN, Bjork RD (2004) Habitat linkages and the conservation of tropical biodiversity as indicated by seasonal migrations of Three-wattled Bellbirds. Conserv Biol 18:500–509

Rabenold KN, Rabenold PP (1985) Variation in altitudinal migration, winter segregation, and site tenacity in two subspecies of Dark-eyed Juncos in the southern Appalachians. Auk 102:805–819

Ramenofsky M, Agatsuma R, Ramfar T (2008) Environmental conditions affect the behavior of captive, migratory Whitecrowned Sparrows. Condor 110:658–671

Ramenofsky M, Cornelius JM, Helm B (2012) Physiological and behavioral responses of migrants to environmental cues. J Ornithol 153:181–191

Rappole JH (1995) The ecology of migrant birds, a Neotropical perspective. Smithsonian Institution Press, Washington, D.C.

Reid TA, Wanless RM, Hilton GM, Phillips RA, Ryan PG (2013) Foraging range and habitat associations of non-breeding Tristan Albatrosses: overlap with fisheries and implications for conservation. Endanger Species Res 22:39–49

Renfrew RB, Kim D, Perlut N, Smith J, Fox J, Marra PP (2013) Phenological matching across hemispheres in a long-distance migratory bird. Divers Distrib 19:1008–1019

Reynolds JD, Colwell MA, Cooke F (1986) Sexual selection and spring arrival times of Red-necked and Wilson's phalaropes. Behav Ecol Sociobiol 18:303–310

Richardson WJ (1982) Northeastward reverse migration of birds over Nova Scotia, Canada, in autumn. Behav Ecol Sociobiol 10:193–206

Richardson WJ (1990) Timing and amount of bird migration in relation to weather: updated review. In: Gwinner E (ed) Bird migration: physiology and ecophysiology. Springer-Verlag, Berlin, Germany, pp 78–101

Rohwer VG, Rohwer S, Barry JH (2008) Molt scheduling of western Neotropical migrants and up-slope movement of Cassin's Vireo. Condor 110:365–370

Rosselli L (1994) The annual cycle of the White-ruffed Manakin *Corapipo leucorrhoa*, a tropical frugivorous altitudinal migrant, and its food plants. Bird Cons Int 4:143–160

Rubolini D, Pastor AG, Pilastro A, Spina F (2002) Ecological barriers shaping fuel stores in Barn Swallows *Hirundo rustica* following the central and western Mediterranean flyways. J Avian Biol 33:15–22

Rubolini D, Spina F, Saino N (2004) Protandry and sexual dimorphism in trans-Saharan migratory birds. Behav Ecol 15:592–601

Sabal MC, Boyce MS, Charpentier CL, Furey NB, Luhring TM, Martin HW, Melnychuk MC, Srygley RB, Wagner CM, Wirsing AJ, Ydenberg RC (2021) Predation landscapes influence migratory prey ecology and evolution. Trends Ecol Evol 36:737–749

Salewski V, Bruderer B (2007) The evolution of bird migration – a synthesis. Naturwissenschaften 94:268–279

Sandberg R, Moore FR (1996) Migratory orientation of Red-eyed Vireos, *Vireo olivaceus*, in relation to energetic condition and ecological context. Behav Ecol Sociobiol 39:1–10

Schaub M, Liechti F, Jenni L (2004) Departure of migrating European Robins, *Erithacus rubecula*, from a stopover site in relation to wind and rain. Anim Behav 67:229–237

Schaub M, Jenni L, Bairlein F (2008) Fuel stores, fuel accumulation, and the decision to depart from a migration stopover site. Behav Ecol 19:657–666

Schlaepfer MA, Runge MC, Sherman PW (2002) Ecological and evolutionary traps. Trends Ecol Evol 17:474–480

Schmaljohann H (2018) Proximate mechanisms affecting seasonal differences in migration speed of avian species. Sci Rep 8:4106

Schmaljohann H, Both C (2017) The limits of modifying migration speed to adjust to climate change. Nat Clim Chang 7:573

Schmaljohann H, Eikenaar C (2017) How do energy stores and changes in these affect departure decisions by migratory birds? A critical view on stopover ecology studies and some future perspectives. J Comp Physiol A 203:411–429

Schmaljohann H, Naef-Daenzer B (2011) Body condition and wind support initiate the shift of migratory direction and timing of nocturnal departure in a songbird. J Anim Ecol 80:1115–1122

Schmaljohann H, Liechti F, Bruderer B (2009) TransSahara migrants select flight altitudes to minimize energy costs rather than water loss. Behav Ecol Sociobiol 63:1609–1619

Schmaljohann H, Korner-Nievergelt F, Naef-Daenzer B, Nagel R, Maggini I, Bulte M, Bairlein F (2013) Stopover optimization in a long-distance migrant: the role of fuel load and nocturnal take-off time in Alaskan Northern Wheatears (*Oenanthe oenanthe*). Front Zool 10:26

Schmaljohann H, Mueller F, Klinner T, Eikenaar C (2018) Potential age differences in the migratory behaviour of a nocturnal songbird migrant during autumn and spring. J Avian Biol 49:e01815

Schmaljohann H, Eikenaar C, Sapir N (2022) Understanding the ecological and evolutionary function of stopover in migrating birds. Biol Rev 97:1231–1252

Senner NR, Stager M, Verhoeven MA, Cheviron ZA, Piersma T, Bouten W (2018) High-altitude shorebird migration in the absence of topographical barriers: avoiding high air temperatures and searching for profitable winds. Proc R Soc B 285:20180569

Sergio F, Tanferna A, De Stephanis R, Jiménez LL, Blas J, Tavecchia G, Preatoni D, Hiraldo F (2014) Individual improvements and selective mortality shape lifelong migratory performance. Nature 515:410–413

Shamoun-Baranes J, Bouten W, van Loon EE (2010) Integrating meteorology into research on migration. Integr Comp Biol 50:280–292

Shiu H-J, Lee P-F (2003) Seasonal variation in bird species richness along elevational gradients in Taiwan. Acta Zoologica Taiwanica 14:1–21

Sillett TS, Holmes RT (2002) Variation in survivorship of a migratory songbird throughout its annual cycle. J Anim Ecol 71:296–308

Silverin B, Viebke PA, Westin J (1989) Hormonal correlates of migration and territorial behavior in juvenile Willow Tits during autumn. Gen Comp Endocrinol 75:148–156

Sjöberg S, Malmiga G, Nord A, Andersson A, Bäckman J, Tarka M, Willemoes M, Thorup K, Hansson B, Alerstam T, Hasselquist D (2021) Extreme altitudes during diurnal flights in a nocturnal songbird migrant. Science 372:646–648

Skrip MM, Bauchinger U, Goymann W, Fusani L, Cardinale M, Alan RR, McWilliams SR (2015) Migrating songbirds on stopover prepare for, and recover from, oxidative challenges posed by long-distance flight. Ecol Evol 5:3198–3209

Smith HG, Nilsson JA (1987) Intraspecific variation in migratory pattern of a partial migrant, the Blue Tit (*Parus caeruleus*): an evaluation of different hypotheses. Auk 104:109–115

Smith R, Brown I, Mewaldt R (1969) Annual activity patterns of caged non-migratory White-crowned Sparrows. Wilson Bull 81:419–440

Sockman KW, Hurlbert AH (2020) How the effects of latitude on daylight availability may have influenced the evolution of migration and photoperiodism. Funct Ecol 34:1752–1766

Solórzano S, Castillo S, Valverde T, Avila L (2000) Quetzal abundance in relation to fruit availability in a cloud forest in southeastern Mexico. Biotropica 32: 523–532

Somenzari M, do Amaral PP, Cueto VR, de Camargo Guaraldo A, Jahn AE, Lima DM, Lima PC et al (2018) An overview of migratory birds in Brazil. Papéis Avulsos de Zoologia 58:e20185803

Somveille M, Manica A, Butchart SHM, Rodrigues ASL (2013) Mapping global diversity patterns for migratory birds. PLoS One 8:e70907

Somveille M, Rodrigues ASL, Manica A (2015) Why do birds migrate? A macroecological perspective. Glob Ecol Biogeogr 24:664–674

Sparks TH, Huber K, Bland RL, Crick HQP, Croxton PJ, Flood J, Loxton RG, Mason CF, Newnham JA, Tryjanowski P (2007) How consistent are trends in arrival (and departure) dates of migrant birds in the UK? J Ornithol 148:503–511

Stach R, Jakobsson S, Kullberg C, Fransson T (2012) Geolocators reveal three consecutive wintering areas in the Thrush Nightingale. Anim Migr 1:1–7

Steadman DW (2005) The paleoecology and fossil history of migratory landbirds. In: Greenberg R, Marra PP (eds) Birds of two worlds: the ecology and evolution of migration. Johns Hopkins University Press, Baltimore, MD, pp 5–17

Steele J, McCormick J (1995) Partitioning of the summer grounds by Orange-crowned Warblers into a breeding grounds, adult molting grounds and juvenile staging areas. North American Bird Bander 20:152

Stiles FG (1983) Birds. In: Janzen DH (ed) Costa Rican natural history. University of Chicago Press, Chicago, IL, pp 502–530

Stiles FG (2004) Austral migration in Colombia: the state of knowledge, and suggestions for action. Ornitologia Neotropical 15(Suppl):349–355

Stotz DF, Fitzpatrick JW, Parker TA III, Moskovits DK (1996) Neotropical birds: ecology and conservation. University of Chicago Press, Chicago, IL

Strandberg R, Klaassen RHG, Olofsson P, Alerstam T (2009) Daily travel schedules of adult Eurasian Hobbies (*Falco subbuteo*) – variability in flight hours and migration speed along the route. Ardea 97:287–295

Strandberg R, KlaassFen RHG, Hake M, Alerstam T (2010) How hazardous is the Sahara Desert crossing for migratory birds? Indications from satellite tracking of raptors. Biol Lett 6:297–300

Strong C, Zuckerberg B, Betancourt JL, Koenig WD (2015) Climatic dipoles drive two principal modes of North American boreal bird irruption. Proc Natl Acad Sci 112:E2795–E2802

Studds CE, Marra PP (2011) Rainfall-induced changes in food availability modify the spring departure programme of a migratory bird. Proc R Soc B 278: 3437–3443

Studds CE, Kendall BE, Murray NJ, Wilson HB, Rogers DI, Clemens RS, Gosbell K, Hassell CJ, Jessop R, Melville DS, Milton DA (2017) Rapid population decline in migratory shorebirds relying on Yellow Sea tidal mudflats as stopover sites. Nat Commun 8: 14895

Stutchbury BJM, Siddiqui R, Applegate K, Hvenegaard GT, Mammenga P, Mickle N, Pearman M, Ray JD, Savage A, Shaheen T, Fraser KC (2016) Ecological causes and consequences of intratropical migration in temperate-breeding migratory birds. Am Nat 188:S28–S40

Thabah A, Li G, Wang Y, Liang B, Hu K, Zhang S, Jones G (2007) Diet, echolocation calls, and phylogenetic affinities of the great evening bat (*Ia io*; Vespertilionidae): another carnivorous bat. J Mammal 88:728–735

Thorup K, Tøttrup AP, Willemoes M, Klaassen RH, Strandberg R, Vega ML, Dasari HP, Araújo MB, Wikelski M, Rahbek C (2017) Resource tracking within and across continents in long-distance bird migrants. Sci Adv 3:e1601360

Tøttrup AP, Rainio K, Coppack T, Lehikoinen E, Rahbek C, Thorup K (2010) Local temperature fine-tunes the timing of spring migration in birds. Integr Comp Biol 50:293–304

Tøttrup AP, Pedersen L, Onrubia A, Klaassen RHG, Thorup K (2017) Migration of Red-backed Shrikes from the Iberian Peninsula: optimal or sub-optimal detour? J Avian Biol 48:149–154

Tsai P-Y, Ko C-J, Chia SY, Lu Y-J, Tuanmu M-N (2021) New insights into the patterns and drivers of avian altitudinal migration from a growing crowdsourcing data source. Ecography 44:75–86

Tulp I, Schekkerman H (2008) Has prey availability for Arctic birds advanced with climate change? Hindcasting the abundance of tundra arthropods using weather and seasonal variation. Arctic 61:48–60

Usui T, Butchart SHM, Phillimore AB (2017) Temporal shifts and temperature sensitivity of avian spring migratory phenology: a phylogenetic meta-analysis. J Anim Ecol 86:250–261

van Buskirk J, Mulvihill RS, Leberman RC (2009) Variable shifts in spring and autumn migration phenology in North American songbirds associated with climate change. Glob Chang Biol 15:760–771

van Noordwijk AJ, McCleery RH, Perrins CM (1995) Selection for the timing of Great Tit breeding in relation to caterpillar growth and temperature. J Anim Ecol 64:451–458

Vansteelant WMG (2016) From thermal to flyway: how weather shapes the soaring migration of European Honey Buzzards *Pernis apivorus* at multiple scales, Ph.D. dissertation. University of Amsterdam, Amsterdam

Végvári Z, Bókony V, Barta Z, Kovács G (2010) Life history predicts advancement of avian spring migration in response to climate change. Glob Chang Biol 16: 1–11

Vergara P, Aguirre JI, Fernández-Cruz M (2007) Arrival date, age and breeding success in White Stork *Ciconia ciconia*. J Avian Biol 38:573–579

Visser ME, Gienapp P, Husby A, Morrisey M, de la Hera I, Pulido F, Both C (2015) Effects of spring temperatures on the strength of selection on timing of reproduction in a long-distance migratory bird. PLoS Biol 13:e1002120

Vissing MS, Fox AD, Clausen P (2020) Non-stop autumn migrations of light-bellied Brent Geese *Branta bernicla hrota* tracked by satellite telemetry – racing for the first *Zostera* bite? Wild 70:76–93

Watanabe YY (2016) Flight mode affects allometry of migration range in birds. Ecol Lett 19:907–914

Watts HE, Robart AR, Chopra JK, Asinas CE, Hahn TP, Ramenofsky M (2017) Seasonal expression of migratory behavior in a facultative migrant, the Pine Siskin. Behav Ecol Sociobiol 71:1–12

Wellbrock AHJ, Bauch C, Rozman J, Witte K (2017) Same procedure as last year? Repeatedly tracked swifts show individual consistency in migration pattern in successive years. J Avian Biol 48:897–903

Wiegardt AK, Barton DC, Wolfe JD (2017) Post-breeding population dynamics indicate upslope molt-migration by Wilson's Warblers. J Field Ornithol 88:47–52

Wikelski M, Martin LB, Scheuerlein A, Robinson MT, Robinson ND, Helm B, Hau M, Gwinner E (2008) Avian circannual clocks: adaptive significance and possible involvement of energy turnover in their proximate control. Philos Trans R Soc B 363: 411–423

Winger BM, Pegan TM (2021) Migration distance is a fundamental axis of the slow-fast continuum of life history in boreal birds. Ornithology 138:ukab0043

Winger BM, Lovette IJ, Winkler DW (2012) Ancestry and evolution of seasonal migration in the Parulidae. Proc R Soc B 279:610–618

Winger BM, Barker FK, Ree RH (2014) Temperate origins of long-distance seasonal migration in New World songbirds. Proc Natl Acad Sci 111:12115–12120

Winger BM, Auteri GG, Pegan TM, Weeks BC (2019) A long winter for the Red Queen: rethinking the evolution of seasonal migration. Biol Rev 94:737–752

Wingfield JC (2003) Control of behavioural strategies for capricious environments. Anim Behav 66:807–816

Winkler DW, Jørgensen C, Both C, Houston AI, McNamara JM, Levey DJ, Partecke J, Fudickar A, Kacelnik A, Roshier D, Piersma T (2014) Cues, strategies, and outcomes: how migrating vertebrates track environmental change. Mov Ecol 2:10

Winkler DW, Gandoy FA, Areta JI, Iliff MJ, Rakhimberdiev E, Kardynal KJ, Hobson KA (2017) Long-distance range expansion and rapid adjustment of migration in a newly established population of Barn Swallows breeding in Argentina. Curr Biol 27:1080–1084

Wolfe JD, Johnson EI (2015) Geolocator reveals migratory and winter movements of a Prothonotary Warbler. J Field Ornithol 86:238–243

Yamaguchi NM, Mori S, Yonekawa H, Waga D, Higuchi H (2021) Light-level geolocators reveal that White-throated Needletails (*Hirundapus caudacutus*) follow a figure-eight migration route between Japan and Australia. Pac Sci 75:75–84

Yang D, Yang A, Yang J, Xu R, Qiu H (2021) Unprecedented migratory bird die-off: a citizen-based analysis on the spatiotemporal patterns of mass mortality events in the western United States. GeoHealth 5: e2021GH000395

Yong W, Moore FR (1993) Relation between migratory activity and energetic condition among thrushes following passage across the Gulf of Mexico. Condor 95:934–943

Yong DL, Liu Y, Low BW, Espanola CP, Choi C-Y, Kawakami K (2015) Migratory songbirds in the east Asian-Australasian flyway: a review from a conservation perspective. Bird Cons Int 25:1–37

Zink RM (2002) Towards a framework for understanding the evolution of avian migration. J Avian Biol 33:433–436

Zink RM (2011) The evolution of avian migration. Biol J Linnean Soc 104:237–250

Zink G, Barlein F (1995) Der zug Europäischer Singvögel. AULA-Verlag, Wiesbaden, Germany

# Navigation and Orientation 14

## Contents




**Abstract**

When moving over both short and long distances, birds that have a specific destination must be able to determine their location and the direction they need to travel. In this chapter, I discuss the various ways that birds navigate and orient. For orientation, identifying the compass direction a bird wants to travel, some birds are known to use a star compass, sun compass, polarized light, and magnetic cues. Each of the mechanisms is discussed in detail in this chapter. Also discussed is how birds might use path integration and beacon-based navigation. The ways by which birds might determine their location are also explained. Mechanisms potentially used by birds for long- and short-range navigation are discussed, as well as how some birds might use noncompass orientation, including topographical features, landmarks, olfaction, and infrasounds. The role of the avian hippocampus in navigation is also described.




G. Ritchison, *In a Class of Their Own*, Fascinating Life Sciences,
https://doi.org/10.1007/978-3-031-14852-1_14

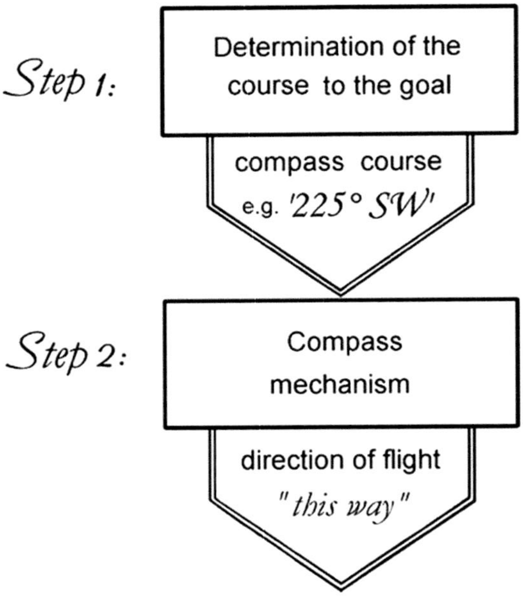


**Fig. 14.1** The two-step navigation system proposed by Kramer (1953, 1957). (Figure from Wiltschko and Wiltschko 2009; © 2009 Oxford University Press, used with permission)

## 14.1 Introduction

Birds are often faced with the need to return to a particular location, such as a nest or roost site, a source of food or water, or, for migratory species, a breeding territory or wintering area. Such directed movement is called navigation or, more precisely, true navigation and involves the ability of a bird to locate its position, whether in a familiar or unfamiliar area, with respect to where it wants to go. Orientation, on the other hand, is more simply the ability to move in a given compass direction. True navigation, as first described by Kramer (1953, 1957), is a two-step process: (1) determining the correct direction of travel, and (2) being able to correctly identify that direction. In other words, birds must determine the direction that will take them toward their goal, just as we would when using a map, then, as we might using a compass, locate or identify that direction (Fig. 14.1). Among many species of birds, true navigation is apparently accomplished only by "experienced" birds, i.e., birds that have become familiar, on a smaller scale, with a local area or, on a larger, migratory scale, have successfully completed a migratory journey at least once.

For some species, young, naïve birds appear to be capable only of vector navigation, or the ability to maintain a particular orientation for a specified time or distance (Fig. 14.2; Bingman and Cheng 2005). Differences between adult and juvenile birds have been revealed in experiments where they are captured during migration at the same location and then transported some distance from the normal migration route and released. Such experiments with European Starlings (*Sturnus vulgaris*; Perdeck 1958), Green-winged Teal (*Anas crecca*; Wolff 1970), and White-crowned Sparrows (*Zonotrichia leucophrys*; Thorup et al. 2007) during fall migration have all demonstrated that adults typically compensate for the displacement and fly in the direction that will take them to their normal wintering grounds, whereas juveniles continue to fly in the direction they were moving prior to displacement (Fig. 14.3). These results indicate that, for these species, first-time migrants reach their wintering areas by an innate mechanism that influences the direction and distance of movement or, in other words, vector navigation.

Results of other studies suggest that, in some species, juveniles may be able to compensate for displacement. Using captive birds, Åkesson et al. (2005) recorded the orientation of migratory White-crowned Sparrows after longitudinal displacement of distances ranging from 266 to 2862 km across high-Arctic North America and found that both adults and juveniles compensated for the displacement and altered their orientation in the manner that would take them back either to their normal wintering areas or to their normal migration route. Thorup and Rabøl (2007) reanalyzed the results of several displacement experiments and found that, in many cases, both adults and juveniles showed directional shifts after displacement.

Studies to date, then, indicate that, among migratory species, adults and, in some cases, juveniles are capable of true navigation. For juveniles, this ability must be innate. For species where adults are capable of true navigation and

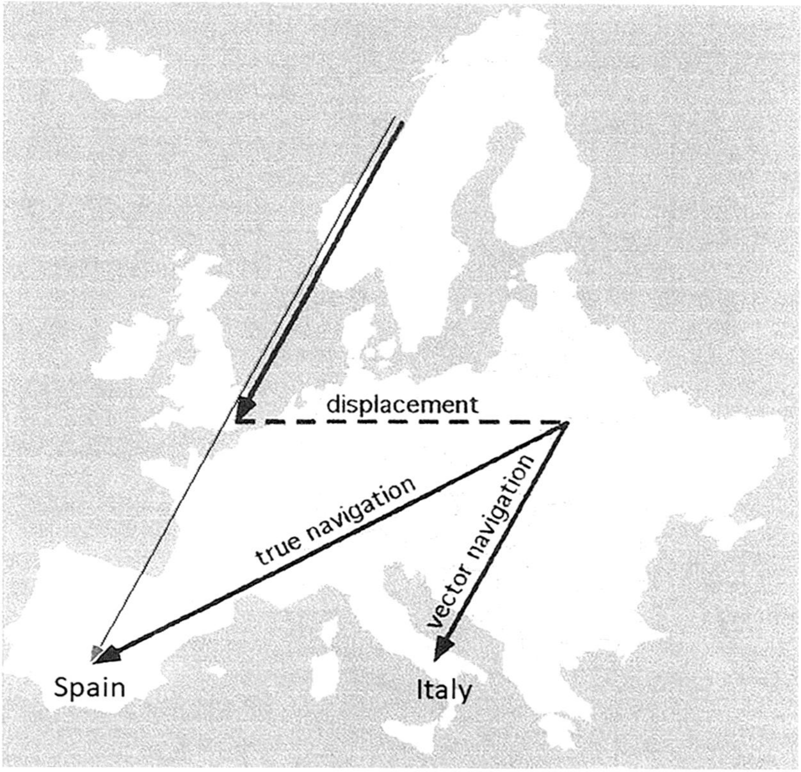


**Fig. 14.2** Example of the difference between true navigation and vector navigation by a hypothetical songbird migrant in Europe with a fall migration route from Norway to Spain (thin solid arrow). In this hypothetical experiment, migrants' en route is captured and displaced (dashed arrow) from their traditional migratory path to a distant, unfamiliar site in eastern Europe and then released. True navigation (the thick solid arrow to Spain) would require the ability to determine the "new" location and adjust the migratory route to compensate for the displacement and still end up at the over-wintering site in Spain. Because vector navigation (the thick solid arrow to Italy) is only the ability to continue moving in a particular direction for a certain distance or time, there would be no compensating to take into account the "new" location. So, migration would continue along the same orientation and for the same distance, but the hypothetical migrant would end up in Italy rather than Spain. (Figure from Bingman and Cheng 2005; Rights managed by Taylor & Francis, used with permission)

juveniles are not, learning takes place during the first migratory journal and birds acquire the information needed for true navigation. Regardless of whether innate or learned, true navigation requires the ability to discern one's location with respect to the ultimate destination or goal (Thorup and Holland 2009). For a bird returning to a particular location in a known or familiar area like a nest in a breeding territory, visual landmarks recognized from previous visits can be used to determine its location and the correct direction of travel. However, navigation from an unfamiliar area requires some other mechanism for determining one's location.

Navigation by humans is based on the latitude/longitude coordinate systems. True navigation by birds is likely based on a similar bi-coordinate, or grid-based, system, with cues varying along gradients that allow extrapolation of familiar gradients to unfamiliar areas (Thorup and Holland 2009; Fig. 14.4). Cues important in bird navigation are celestial, magnetic, and olfactory. Each of these same cues can also be used for orientation by birds and, before examining how

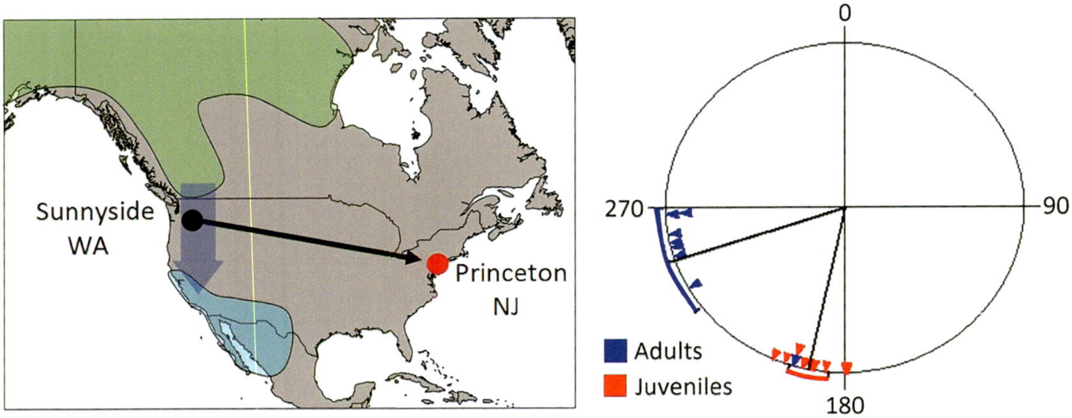


**Fig. 14.3** Thirty White-crowned Sparrows (*Zonotrichia leucophrys*; 15 adults and 15 juveniles) were displaced 3700 km from Seattle, Washington, USA, to Princeton, New Jersey, USA. The map on the left shows the displacement and breeding area (green), wintering area (cyan), and normal migration route (blue). The circle to the right shows the directions with the mean and confidence interval indicated where adults (blue) and juveniles (red) flew after being released in New Jersey. All but one adult flew in the direction toward their normal wintering grounds, whereas juveniles continued in their normal migration direction. (Figure modified from Thorup et al. 2007; used with permission of the U. S. National Academy of Sciences)

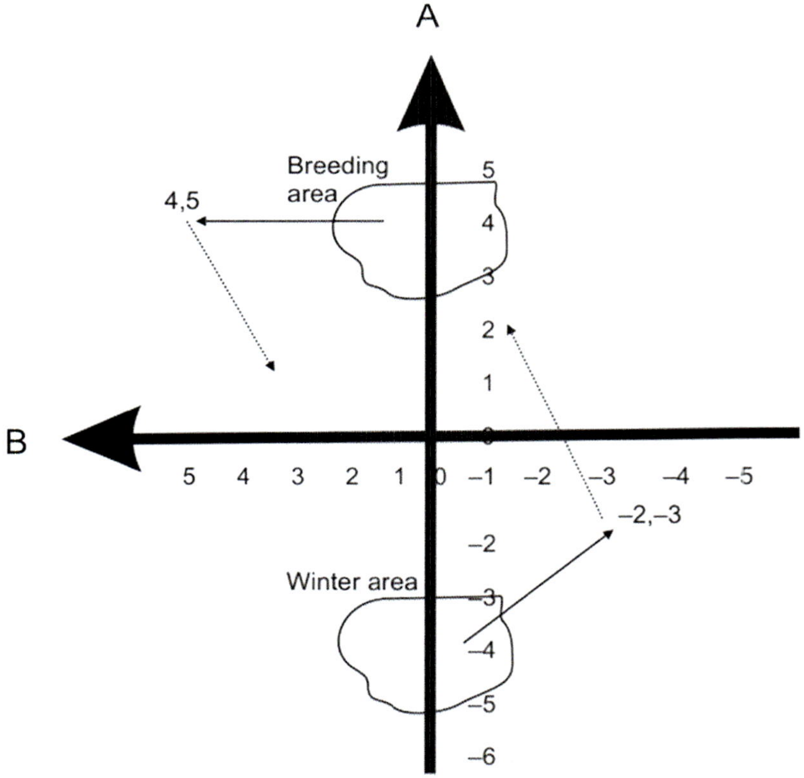


**Fig. 14.4** With a bi-coordinate gradient map, two gradients (arrows A and B) are learned by exploration. If the gradients increase or decrease predictably, then a displaced bird can compare values at displaced locations with those of its breeding or wintering areas to determine the needed direction of movement to return to its desired location. For long-distance migrants, these cues would need to vary consistently on a continental or global scale. (Figure from Holland 2014; © 2014 The Zoological Society of London, used with permission)

**Fig. 14.5** Long-term exposure of the Northern Hemisphere night sky. The star near the center of rotation is Polaris (North Star). (Image by SkyB, Wikipedia, CC BY 4.0, https://creativecommons.org/licenses/by/4.0)

birds might use them for navigation (to determine their location), an understanding of how they function is needed.

## 14.2 Compass Orientation: Star Compass

In a series of classic experiments, Emlen (1967a, b, 1969, 1970) showed that Indigo Buntings (*Passerina cyanea*) use the stars for orientation during migration. By manipulating the location and movements of stars and constellations in a planetarium, Emlen (1967b) found that buntings derived directional information from the pattern of constellations relative to each other and to the celestial pole. Specifically, young buntings learned the location of "north" by observing the rotation of constellations around the celestial pole (Fig. 14.5). In these planetarium experiments, Emlen (1966) determined the direction that buntings were attempting to fly, or their orientation, using funnels with ink pads at the bottom and lined with blotting paper (Emlen funnels; Fig. 14.6). Each time a bunting hopped upward (and, exhibiting nocturnal restlessness, or zugunruhe, they did so frequently), ink on the feet and plumage was left on the blotting paper, leaving a record of the direction the bunting was attempting to travel (Fig. 14.6).

Star-compass orientation has been demonstrated in several species of night-migrating songbirds, including Bobolinks (*Dolichonyx oryzivorus*; Beason 1987), Garden Warblers (*Sylvia borin*; Wiltschko et al. 1987), Savannah Sparrows (*Passerculus sandwichensis*; Able and Able 1996), European Robins (*Erithacus rubecula*; Pakhomov et al. 2017), and European Pied Flycatchers (*Ficedula hypoleuca*) and Eurasian Blackcaps (*Sylvia atricapilla*; Mouritsen and Larsen 2001). As first reported by Emlen (1967a, b), birds that use stellar cues

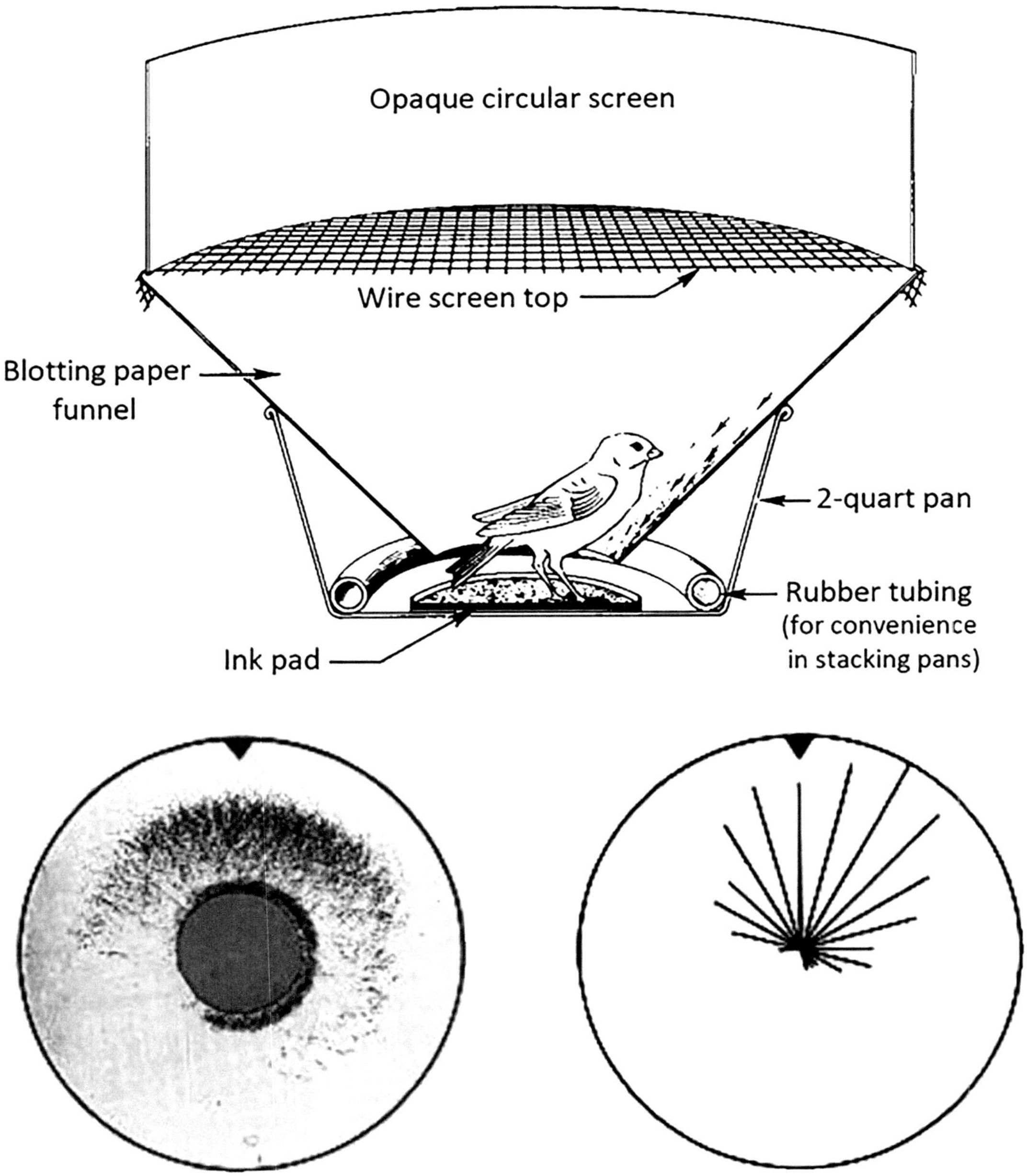


**Fig. 14.6** (Top) An "Emlen funnel." Birds could view the planetarium "sky" through the opaque circular screen. (Bottom) An example of a "footprint" record (left) and a resulting vector diagram (right) for quantifying the footprint records. In these examples, the bird exhibited a tendency to move in a northeasterly direction. (Figures from Emlen 1966; © 1966 Oxford University Press, used with permission)

for orientation do not do so innately, but must learn to use this "compass" by observation of stellar rotation. What is innate, however, among birds that use a star compass, is the comprehension that they must ". . . look for rotating light dots in the sky and to interpret the center of rotation as north" (Mouritsen et al. 2016). In the lab, Michalik et al. (2014) found that young

**Fig. 14.7** Long-term exposure of the Southern Hemisphere night sky as viewed from the Chile's Atacama Desert. There are no stars near the center of rotation like Polaris in the Northern Hemisphere, but night-migrating birds could still use the star compass for orientation. (Image by A. Duro, Jr., Wikipedia, CC BY 4.0, https://creativecommons.org/licenses/by/4.0/)

European Robins exposed to a rotating artificial star pattern for seven nights subsequently showed inappropriate or random orientation, suggesting that seven nights were not sufficient to establish a star compass. Other investigators have determined that birds require at least two weeks or more (Wiltschko et al. 1987; Able and Able 1990; Prinz and Wiltschko 1992). Young birds likely require several nights to develop a star compass because the earth's speed of rotation is so slow (0.0042 degrees/s; Mouritsen et al. 2016). Given the relatively slow rotational motion of stars, how are birds able to determine the center of rotation? Alert et al. (2015) suggest that, when learning how to use the stellar compass, young identify the center of rotation using a "snapshot strategy," i.e., by comparing star patterns with a memorized snapshot of the pattern relative to fixed local landmarks from some prior time. Once learned, birds no longer need to observe the rotation.

Experiments have also revealed that birds do not rely on specific stars like the north star (Polaris), but obtain directional information from the position of stars and constellations relative to each other and to the celestial pole (Wiltschko and Wiltschko 2009). This means that songbirds in the Southern Hemisphere (where there is no bright star directly or nearly directly above the south pole) could also use star-compass orientation (Fig. 14.7). For example, Bobolinks breed in temperate North America and winter in temperate South America, so cross the equator during both fall and spring migration. Bobolinks tested in a planetarium with fixed star patterns, and with magnetic fields incrementally changing during the night from an artificial Northern Hemisphere field to an artificial Southern Hemisphere field, maintained a constant heading throughout the experiment. In the wild, nocturnally migrating Bobolinks exhibited similar behavior that would have continued their flight across the equator into the Southern Hemisphere, and this ability to maintain an accurate heading after crossing the magnetic equator may be based on the use of visual cues such as the stars of the Southern Hemisphere night sky (Beason 1992).

Although it seems that many, if not most, night-migrating songbirds (Passeriformes) are capable of star-compass orientation, some

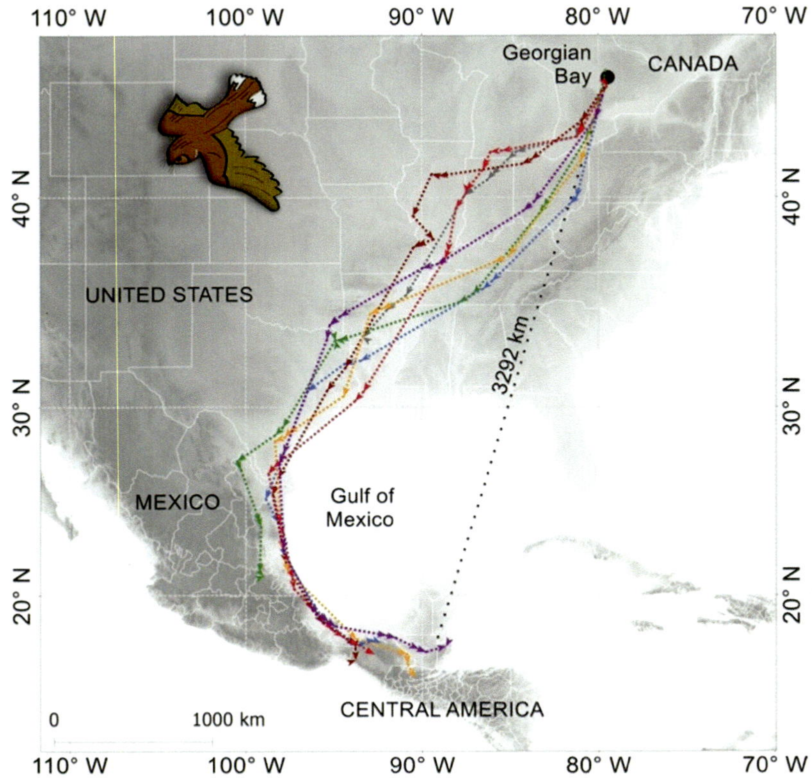


**Fig. 14.8** Locations and connecting track lines for seven Eastern Whip-poor-wills (*Antrostomus vociferus*) fitted with GPS receiver tags and migrating from breeding areas in eastern Canada to wintering areas in southern Mexico, northern Guatemala (one bird), and Belize (one bird). Connecting lines do not necessarily reflect exact paths. Mean shortest distance from deployment site to wintering site was 3292 km; actual average total distance flown was 4180 km. Although several species in the orders Caprimulgiforms and Strigiforms migrate, how they navigate during even relatively long-distance migrations like those of Eastern Whip-poor-wills remains to be determined. (Figure from Korpach et al. 2019; © 2019 Deutsche Ornithologen-Gesellschaft e.V., used with permission)

evidence suggests that use of a star compass by night-migrating songbirds is not universal. For example, Garden Warblers (*Sylvia borin*) exposed to a magnetic field that did not contain any compass information were unable to orient in the seasonally appropriate direction later in the night even if they had access to the correct stellar information (Pakhomov and Chernetsov 2020). Chernetsov (2015) noted that some species may not have a stellar compass or, if they do, may use it less efficiently than other species. Possible explanations for the absence of a stellar compass are that its use is both time-consuming and cognitively challenging (Chernetsov 2015). In addition, little is known about the possible use of stellar cues for orientation by taxa of birds other than songbirds. Of course, an ability to orient using stellar cues is not needed by birds that migrate during the day, such as raptors and waterfowl. However, some nonsongbirds, such as some owls, do migrate at night and studies examining the cues used for orientation by such species are needed. In addition, nothing is currently known about the possible ability of birds that are normally active at night, such as owls and caprimulgids, to orient using stellar (or other navigational) cues (Fig. 14.8). In one of the few studies (and perhaps only study) of owls to date,

Smith (2009) examined migratory restlessness and orientation of Flammulated (*Psiloscops flammeolus*) and Northern Saw-whet (*Aegolius acadicus*) owls using orientation cages. Flammulated Owls are long-distance migrants that breed in western North America and winter in southern Mexico and Central America (Linkhart and McCallum 2020), and Northern Saw-whet Owls are partial, short-distance migrants that breed in the northern and western United State and southern Canada, with some individuals migrating to the southern United States during the nonbreeding period (Rasmussen et al. 2020). Only four of 16 Flammulated Owls exhibited migratory restlessness and a significant directionality in orientation, whereas 59 of 97 (61%) Northern Saw-whet Owls did so. However, neither species had a consistent directionality of orientation, with orientations differing among individuals for both species. This lack of consistency suggests that different individuals have different migration strategies and different preferred wintering areas (Smith 2009). Interestingly, however, Northern Saw-whet Owls were less active in experiments where no celestial cues were available, suggesting the possibility that such cues, perhaps the stars, are needed for orientation (Smith 2009).

The stellar map of birds does not provide information about longitude so other cues are likely needed for long-distance navigation by birds. For example, Mouritsen and Larsen (2001) found that European Pied Flycatchers used stars only as a compass, orienting in the same direction all night in an experiment where stars were stationary. If using stars to navigate, Pied Flycatchers would have been expected to change direction toward the end of the night to avoid longitudinal displacement.

## 14.3 Compass Orientation: Sun Compass

The use of the sun for orientation by birds was first demonstrated by Kramer (1950). He used a test apparatus where mirrors altered the apparent position of the sun and found that European Starlings (*Sturnus vulgaris*) altered their directional preference accordingly. The avian sun compass is based on the sun's azimuth (the point on the horizon directly below the sun), with the sun's altitude being irrelevant. Because the sun moves across the sky during the day, the use of a sun compass requires an internal clock because it is the sun's azimuth at a particular time of day that provides directional information (Figs. 14.9 and 14.10).

The importance of a bird's internal clock for correctly using the sun compass can be demonstrated by "shifting" the clock. For example, Schmidt-Koenig (1958) found that the bearings of Rock Pigeons (*Columba livia*) that had been kept for several days in a room where the "artificial" photoperiod was shifted by 6 h either forward or backward differed from those of nonshifted pigeons by about 90°. How can this

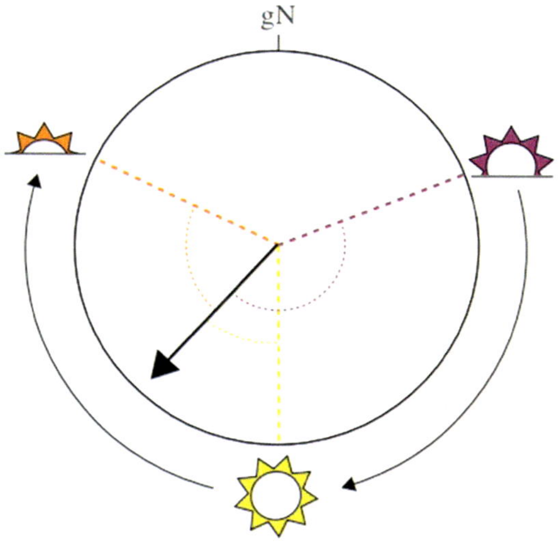


**Fig. 14.9** The avian sun compass is time compensated and must consider the movement of the sun during the day. Birds using a time-compensated sun compass must compensate for the azimuthal change of the sun during the day when determining the direction they wish to travel (dark arrow). In this example, three positions of the sun during a day are shown (sunrise on the right, noon at bottom, and sunset on the left) along with the changing angles (different colors) between the sun's position and the desired direction. To determine these angles correctly, birds must be able to precisely measure local time. *gN*, geographical north. (Figure from Muheim 2011; © 2011 The Royal Society, used with permission)

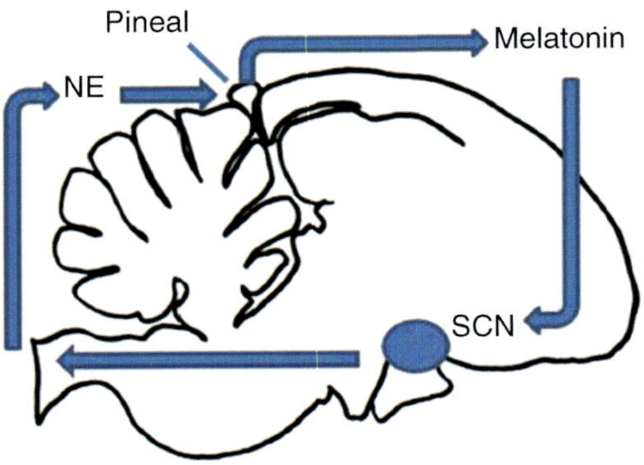


**Fig. 14.10** The avian circadian clock depends on daily rhythms in the secretion of the hormone melatonin. The pineal gland and the retina (not shown here) secrete melatonin during the night and inhibit nerve cells in two nuclei in the hypothalamus (called the medial suprachiasmatic nucleus and visual suprachiasmatic nucleus; collectively the SCN here). As day approaches, the SCN then inhibits the secretion of melatonin by the pineal gland via release of norepinephrine (NE) by the sympathetic nervous system. The clock is synchronized by the daily light cycle. (Figure from Cassone and Westneat 2012; open-access article distributed under the Creative Commons Attribution License (CC BY 4.0), https://creativecommons.org/licenses/by/4.0/)

be explained? As an example, assume two pigeons, one nonshifted and one shifted 6 h forward, are released at noon (12:00) on a clear day and both attempt to return to a home loft located 10 km to the south. Upon release, both pigeons see the sun, but the "clock-shifted" pigeon mistakenly interprets the "noon sun" as the 6:00 p.m. (18:00) sun. As a result, the nonshifted pigeon knowing the "noon sun" is located to the south will, correctly, fly south. However, the clock-shifted pigeon incorrectly interprets the "noon sun" as the "evening sun" to the west and, intending to fly south, flies east instead.

Birds using the sun for orientation must compensate for the sun's movement, a task made more difficult by variation in the speed of movement during the day. For example, shortly after sunrise and before sunset, the sun's altitude changes rapidly, but the azimuth changes slowly. At mid-day, in contrast, the altitude changes slowly, but the azimuth changes rapidly. In addition, the rate of change in the sun's azimuth varies with latitude and season (Chernetsov 2017). Despite this apparent complexity, birds that use the sun compass appear to have a precise understanding of how the sun's azimuth changes not only throughout a day, but with season and latitude (Wiltschko et al. 1998, 2000; Duff et al. 1998).

The avian solar compass as just described is time-dependent, but birds may also be able to determine directions using the sun in a nontime-dependent manner. If birds determine the point of sunset (either directly or based on the location of maximum polarization in the sky) and remember that point, then, at the same location, determine the point of sunrise, they could determine the bisector of the angle between the points of sunset and sunrise which is the north-south axis (Muheim et al. 2006a; Chernetsov 2017). The extent to which migrating birds might use this method to orient remains to be determined (Chernetsov 2017).

Birds must learn to use the sun compass and current evidence indicates that the area of the brain most important in this process is the left hippocampus (Gagliardo et al. 2005). Few investigators have examined the process by which birds learn to use the sun compass. However, studies of young homing pigeons indicate that they are first able to use the sun compass when about 12 weeks old. The specific time of development of the sun compass depends on flying experience, with learning taking place after a pigeon is confronted with the need to orient during flights around its home loft (Wiltschko and Wiltschko 1981).

The relative importance of the sun compass for migrating birds remains unclear, although Wiltschko and Wiltschko (2015) suggested that it "does not seem to play a large role in bird migration" and noted the results of studies where investigators found no evidence for use of the sun compass by migrating birds (e.g., Munro and Wiltschko 1993, 1995; Åkesson et al. 2006). Because use of a sun compass would require birds migrating north or south to continually compensate for changes in latitude, and birds moving east

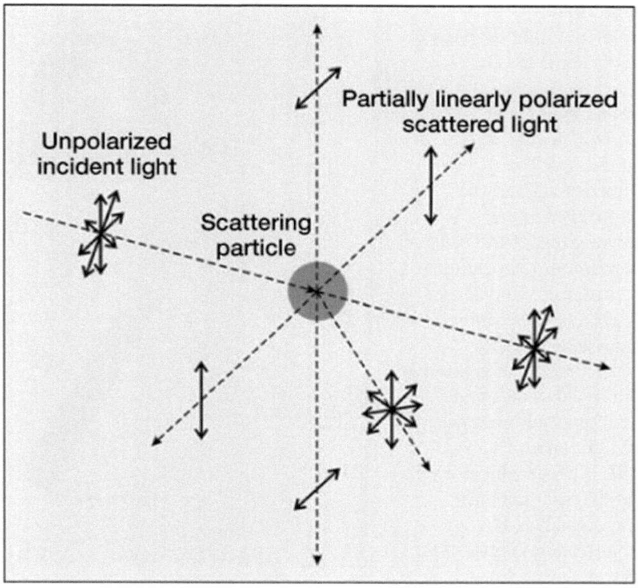


**Fig. 14.11** After scattering on a particle, unpolarized light, with an electric field vector (double-headed arrows) vibrating in all possible directions perpendicular to the direction of propagation (dashed arrows), becomes partially linearly polarized. (Figure from Horváth et al. 2009; © The Ecological Society of America, used with permission)

or west must readjust their internal clocks with changes in longitude, Wiltschko and Wiltschko (2015) suggested that "the sun compass does not seem to be the optimal mechanism." Chernetsov et al. (2017), however, suggested that the results of experiments where investigators found no evidence for use of a sun compass "may partially be due to difficulties of testing migratory birds in orientation cages in the daytime, with phototaxis and shade effects playing a role." Given these conflicting opinions, determining the extent to which migrating birds do or do not use the sun compass will obviously require additional study.

## 14.4 Compass Orientation: Polarized Light

Gas and water molecules in the atmosphere scatter light from the sun in all directions, an effect that is responsible for blue skies and a phenomenon called atmospheric polarization. When a photon from the sun strikes a gas molecule, the electric field from the photon induces a vibration and subsequent reradiation of polarized light from the molecule (Fig. 14.11). Light scattering of atoms and molecules in the atmosphere is unpolarized if the light keeps traveling in the same direction and is linearly polarized if it scatters in a direction perpendicular (either vertically or horizontally) to the way it was traveling. So, sunlight coming directly toward you is unpolarized. Light is more polarized in directions perpendicular to the sun's rays so, at noon, polarization would be most apparent along the horizon. However, at sunset, polarized light forms an image like a large bow-tie—located overhead at sunset—pointing north and south (Fig. 14.12). Importantly, polarization patterns are apparent even when skies are cloudy.

Many invertebrates and vertebrates can perceive polarized light and use that ability in activities such as foraging and orientation. It remains to be determined how many species of birds can detect polarized light and, for those that can, the mechanism involved (Greenwood et al. 2003). There is currently no widely accepted mechanism for detection of polarized light by birds, but such detection almost certainly involves the eyes. Some investigators have suggested that double cones in the retina of bird eyes are involved in detection of polarized light (Cameron

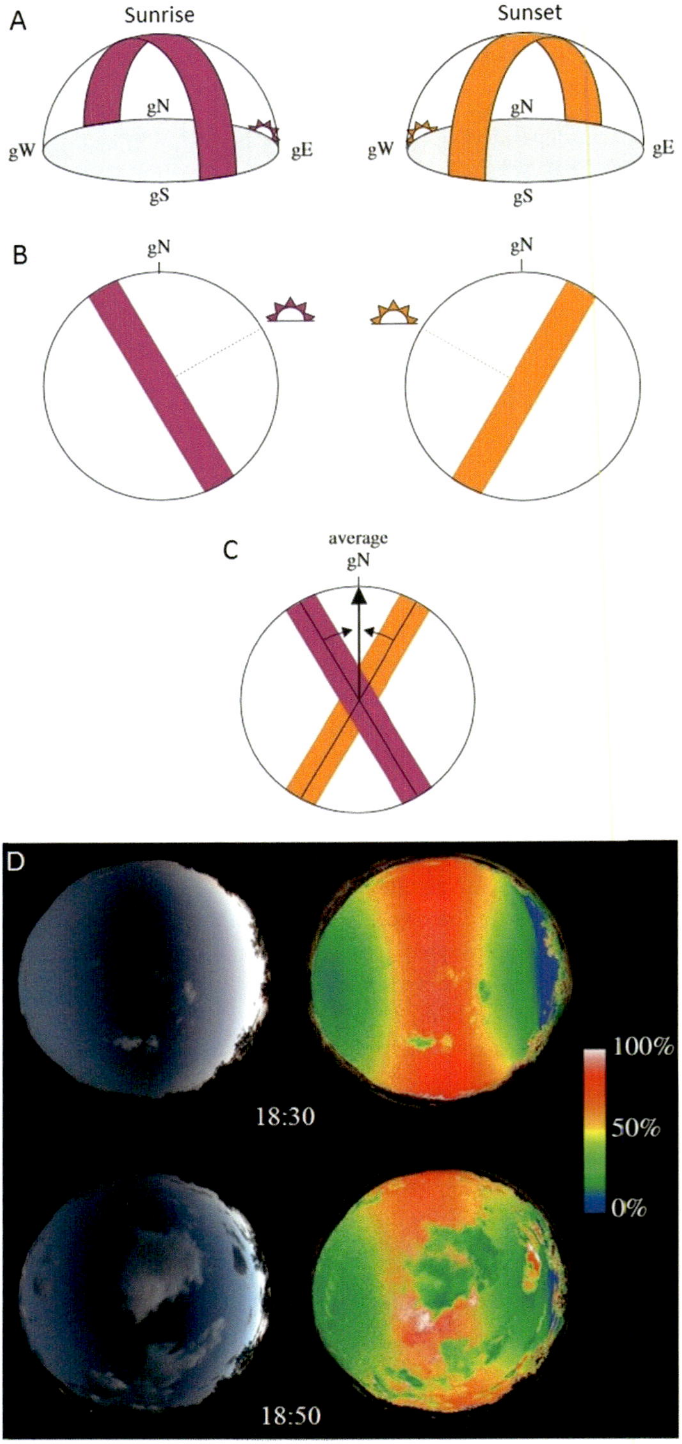


**Fig. 14.12** Use of polarized light by birds to navigate. Bands of maximum polarization vertically intersect the horizon at sunrise and sunset. (**a**) Three-dimensional and (**b**) two-dimensional illustrations of the band of maximum polarization at sunrise (left) and sunset (right). (**c**) Averaging sunrise and sunset polarization bands provides birds with an accurate geographical reference. gN, gS, gE, and gW = geographical north, south, east, and west, respectively. (**d**) Polarization visible in the sky at sunset. Images were made using a digital camera with a 180° "fisheye" lens, aimed at the zenith, and with a linear polarizer placed behind the lens. For each image, north is at the top and east to the left. Actual photos, with the polarizer oriented east/west, are shown on the left. Because the polarization band passing through the zenith is polarized perpendicular to the position of the sun below the horizon in the west, the dark band shows polarization oriented north/south. False-color images showing the degree of polarization are shown on the right. Sunset was at 18:14, so the polarization pattern remains until well after sunset (18:50). (Figures a-c from Muheim 2011; © 2011 The Royal Society, used with permission. Figure d from Cronin and Marshall 2011; © 2011 The Royal Society, used with permission)

and Pugh 1991). Young and Martin (1984) suggested that the oil droplets in the principal cone of double cones and an absence of screening pigment between the outer segments of double cones might be the basis for perception of polarized light by some species of birds. Experiments indicate that several species of migratory birds (but not all; Wiltschko et al. 2008, Melgar et al. 2015) can perceive polarized light and use it as a cue for orientation. Specifically, investigators have provided experimental evidence that migratory Savannah Sparrows (*Passerculus sandwichensis*; Muheim et al. 2006a, 2006b), Swainson's (*Catharus ustulatus*) and Gray-cheeked thrushes (*C. minimus*; Cochran et al. 2004), and White-throated Sparrows (*Zonotrichia albicollis*; Muheim et al. 2009) use polarized light cues at sunrise or sunset to calibrate their magnetic compass. Because the relationship between magnetic and geographic north changes with location, birds need to recalibrate the different compass systems with respect to each other on a regular basis to prevent navigational errors (Muheim et al. 2006b, 2007).

Not all species of birds use the same cues to navigate or to calibrate their compass systems. As such, and not surprisingly, experiments have revealed that not all species of birds use light polarization patterns to recalibrate any of their compass systems, e.g., Northern Wheatears (*Oenanthe oenanthe*; Schmaljohann et al. 2013), Song Thrushes (*Turdus philomelos*; Chernetsov et al. 2011), and Dunnocks (*Prunella modularis*), European Robins (*Erithacus rubecula*), and Sedge Warblers (*Acrocephalus schoenobaenus*) (Åkesson et al. 2014). One possible explanation for differences among species in the use (or not) of light polarization patterns for recalibration purposes is that the compass system hierarchy differs among different species of birds. For example, some species may calibrate their inclination compass using celestial cues, others may calibrate their celestial compass using magnetic cues, and still other may rely on one compass without calibrating during migration (Pakhomov and Chernetsov 2020),

## 14.5 Compass Orientation: Magnetic Cues

The hypothesis that migrating birds could use the earth's magnetic field for orientation was first proposed in 1859 (von Middendorff 1859), and experimental evidence to support that hypothesis was first provided in 1968 in a study of European Robins (Wiltschko 1968). Since then, the use of a magnetic compass has been demonstrated in several species of migratory birds (Wiltschko and Wiltschko 2009; Huttunen 2009; Mouritsen 2018) as well as nonmigratory species (e.g., Zebra Finch, *Taeniopygia guttata*, Voss et al. 2007; Domestic Chicken, *Gallus g. domesticus*, Freire et al. 2005). Because the lineages leading to present-day gallinaceous birds (order Galliformes) and songbirds (order Passeriformes) likely separated as early as the Cretaceous period (Cooper and Penny 1997), the magnetic compass of birds likely developed before then and may even have developed in the common ancestors of all modern birds (Wiltschko et al. 2007; Wiltschko and Wiltschko 2015).

Earth's magnetic field has three components: (1) a declination, (2) intensity, and (3) an inclination. Declination is the angle between the magnetic pole and the geographic pole. The earth's magnetic field, caused by electrical current generated by the rotating molten core, is slightly tilted relative to the spin axis so the two poles (magnetic north and south) are located several hundred kilometers from the geographic poles. The intensity of the magnetic field varies among geographic locations and latitude (e.g., ranging from about 25 microTesla at the equator to 65 microTesla at the poles, Skiles 1985; Fig. 14.13), and the inclination is the angle of the magnetic field relative to earth's surface, and varying from 90° at the magnetic poles to 0° at the magnetic equator (Fig. 14.14). Available data indicate that birds use a combination of these components, potentially in combination with stellar and/or olfactory, cues to determine their location and heading (Kishkinev et al. 2013). Based on studies to date, birds detect magnetic fields by

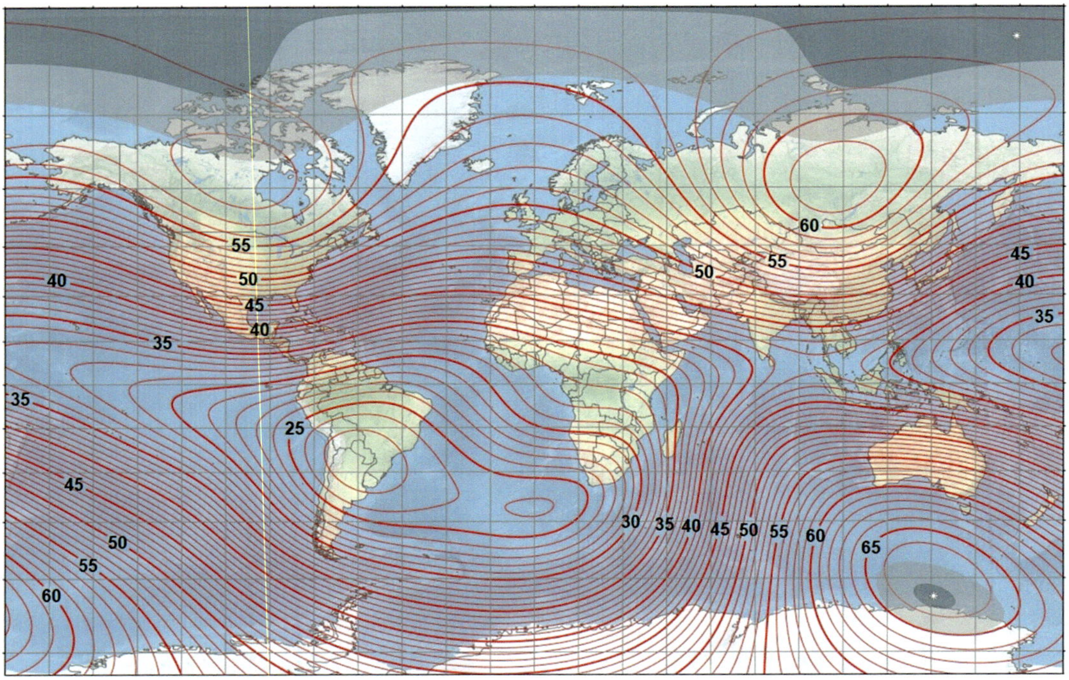


**Fig. 14.13** The intensity of earth's magnetic field varies geographically and with latitude, ranging from about 25–65 microTesla. (Figure from NOAA.gov; CC0 Public Domain)

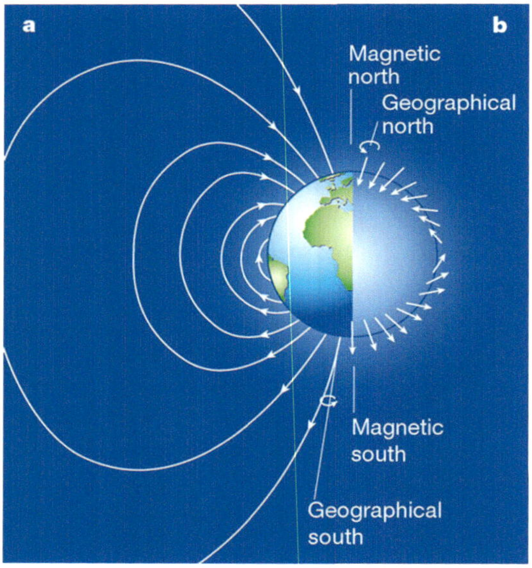


**Fig. 14.14** The earth's magnetic field showing how field lines (represented by arrows) intersect the earth's surface, and how inclination angle varies with latitude. (Figure from Alerstam 2003; © 2003 Springer Nature, used with permission)

formation of radical pairs in their retina, intracellular magnetite, and/or electromagnetic induction (Malkemper et al. 2019).

An inclination compass does not distinguish between north and south, but between the direction of the poles, where field lines are inclined increasingly steeper, and the direction of the equator, where the inclination in increasingly shallow. For birds, the north and south magnetic poles cannot be distinguished because their "compass" only distinguishes between the direction of a pole and the direction of the equator (Wiltschko and Wiltschko 2015; Fig. 14.15). Birds that cross the equator, then, must somehow switch from a "move toward the equator" strategy to a "move toward the pole" strategy. Experience with the horizontal magnetic field found at the magnetic equator appears to trigger this switch, possibly in combination with use of a star compass (Beason 1992).

The avian magnetic compass based on the formation of radical pairs in the retina is light

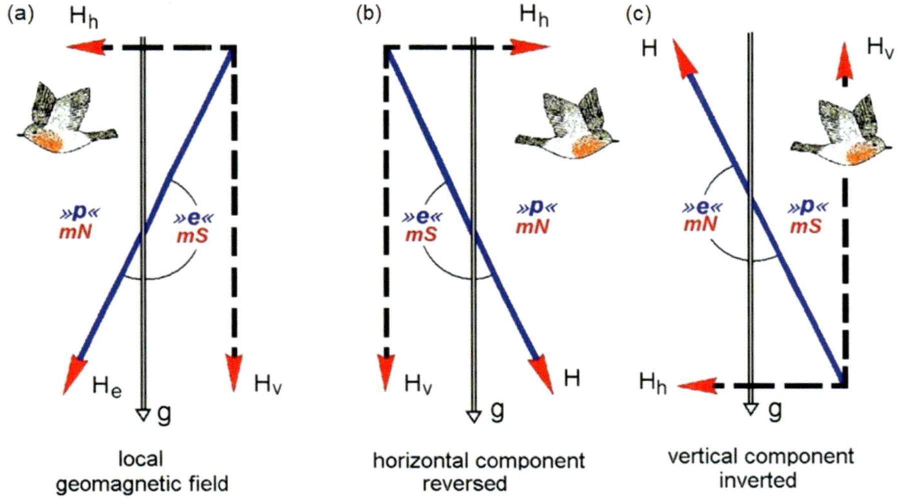


**Fig. 14.15** With the inclination compass of birds, the polarity of the field does not matter. Inclinations of the magnetic field ($H_e$, shown in blue) as seen from the west are shown. (**a**) A robin migrating toward its breeding area in the Northern Hemisphere uses the detected inclination of the magnetic field to fly poleward (»*p*«) in the direction of magnetic north (mN). The polarity of the magnetic field is indicated by the red arrow tips. (**b**) If the inclination of the magnetic field is reversed, the robin reverses its direction and flies south, believing that it is still, based on the inclination of the magnetic field, flying north. Note that the polarity of the field is reversed, but polarity plays no role in the avian magnetic (inclination) compass so is not detected by the robin. (**c**) A reversal in the polarity of the magnetic field is not detected by the robin, so, with the inclination the same as in (**b**), the robin continues flying in the same direction, which is now toward magnetic South (mS). $H_h$, $H_v$, horizontal and vertical components of the magnetic field; g, gravity vector indicating downward; »*p*«, poleward; »*e*«, equatorward. (Figure from Wiltschko and Wiltschko 2015; © 2015 Elsevier B.V., used with permission)

dependent. More specifically, short-wavelength light is required. For example, birds were found to be oriented with light with peak wavelengths ranging from 373 (ultraviolet) to 585 nm (yellow), but were disoriented with light with peak wavelengths ranging from 617 to 645 nm (red) (e.g., Wiltschko and Wiltschko 1995, 1999; Rappl et al. 2000; Muheim et al. 2002; Pinzon-Rodriguez and Muheim 2017). With light at those short wavelengths, birds actually "see" magnetic field lines because the receptors are located in the retina of the avian eye. Specifically, the magnetic sense appears to be based on light-photoreceptor proteins called cryptochromes found in the retina of the avian eye. The eyes of birds contain five types of cryptochromes, and some investigators have suggested that cryptochrome 4 is the one most likely involved in allowing birds to "see" magnetic fields (Bolte et al. 2021; Xu et al. 2021). Cryptochromes are found in photoreceptor cells and ganglion cells, with cryptochrome 4 found in double cones (Günther et al. 2018). Although questions remain concerning the mechanism, one possibility is that short-wavelength light excites photoreceptor cryptochrome molecules in a bird's retina (Fig. 14.16). The light excitation initiates a reaction involving the formation of a pair of radicals (molecules with a single, unpaired electron) and electron-transfer within cryptochrome molecules (Box 14.1 Radical-Pair Magnetoreception), and magnetic fields can influence this pathway by affecting the spin or rotation of the electrons. Because of the rounded shape of the retina, the magnetic field affects the cryptochrome molecules differently in different parts of the retina. As a result, the magnetic field is translated into a visual pattern transmitted to the brain from the retina (Mouritsen and Ritz 2005).

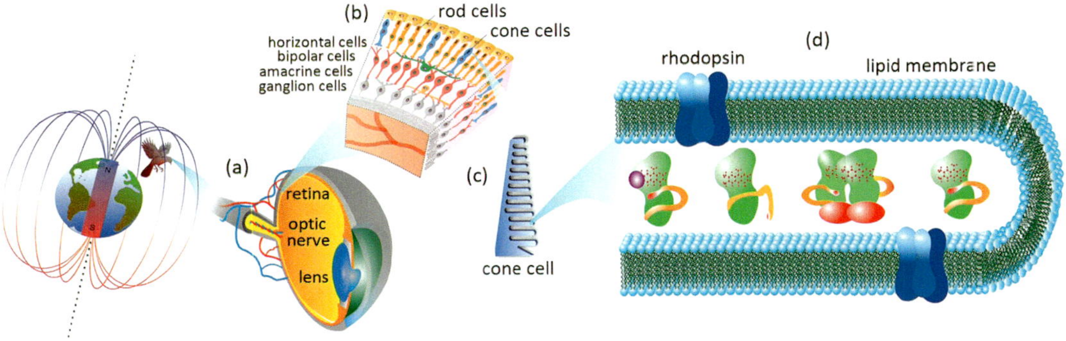


**Fig. 14.16** Illustration showing the components of a magnetic compass in a bird's eye. (**a**) The retina includes cells called rods and cones that convert light into nervous impulses that are conducted to the brain via the optic nerve. (**b**) Section of the retina showing the different types of cells arranged in layers. The primary signal is generated in the photoreceptor cone cells, passed to other cell layers, and transmitted to the brain by the ganglion cells. (**c**) Disc of the outer segments of a cone cell where the photo-magnetoreceptor cryptochrome is suggested to be located. (**d**) Section of a disc membrane of a cone cell. Rhodopsin proteins (shown in blue) transmit visual information. Cryptochrome proteins (shown in green) with different conformations or shapes may transmit magnetic information via formation of magnetically sensitive radical pair reactions. (Figure from Procopio and Ritz 2016, used with permission of Maria Procopio)

Assuming this reaction occurs best when the magnetic field is parallel to cryptochrome molecules, a bird looking in different directions relative to the magnetic field might "see" visual patterns that look something like that illustrated in Figs. 14.17 and 14.18 and, using this information, be able to determine its direction of movement relative to a magnetic pole (north or south) and the magnetic equator. In fact, migratory birds relying on their magnetic compass are known to look in several directions (head scans), apparently to determine the appropriate direction, before initiating activity (Mouritsen et al. 2004a, b).

When birds are using magnetic cues for orientation, an area in the forebrain called cluster N becomes active (Brodbeck et al. 2023). In contrast, neuronal activity in cluster N does not increase in nonmigratory birds during the night, and activity ceases when a bird's eyes are covered, indicating that there is a direct neuronal connection between the retina and cluster N. This connection, plus the location of cluster N adjacent to and part of a visual area of the brain called the visual Wulst, indicates that migratory birds use their visual system to "see" the geomagnetic field (Heyers et al. 2007). In support of this hypothesis, Zapka et al. (2009) showed that European Robins with lesions of cluster N were no longer able to orient using magnetic fields, but were still able to orient correctly using their star compass and sun compass (Fig. 14.19).

Some investigators have suggested that magnetosensitive materials might be located somewhere in the inner ear (Viguier 1882, Harada et al. 2001, Lauwers et al. 2013; but see Malkemper et al. 2019). Nimpf et al. (2019) proposed that pigeons detect magnetic fields via electromagnetic induction in their semicircular canals, showing that changing magnetic stimuli induced electric fields that could likely be detected (Figs. 14.20 and 14.21). Electromagnetic induction has previously been reported in elasmobranch fish (sharks, rays, and skates), with electrosensitive epithelia detecting voltages as the fish move through conductive seawater in the earth's magnetic field (Kalmijn 1971, 1982). Fluid in the semicircular canals of birds (endolymph), like conductive seawater, may similarly allow birds to detect magnetic fields, but

**Box 14.1 Radical-Pair Magnetoreception**

According to the radical pair mechanism, the earth's magnetic field affects chemical reactions in sensory cells in the avian retina by altering the spin state of electrons in "entangled" radical pairs. So, what exactly does that mean? Radical-pair magnetoreception is thought to involve light-sensitive proteins called cryptochromes that are found in the retinas of birds. When high-energy light (blue or green light) strikes these proteins, the energy is transferred to electrons that can now overcome the attractive force of their atomic nuclei and become part of another molecule to form a radical pair, i.e., molecules with unpaired electrons that are "entangled" such that each affects the other even though they are in different molecules.

All electrons have a property called spin and, for magnetoreception, what is important about spin is that there are just two different types or states, and magnetic fields cause the electrons in the radical pair to change their spin state with respect to each other (because they are "entangled"). The extent and rate of this spin change is thought to depend on how a bird is orientated in a magnetic field and these different spin states might translate into a signal that the bird can interpret, perhaps the production of different chemical products that would allow a bird to "see" in which direction it needs to fly.

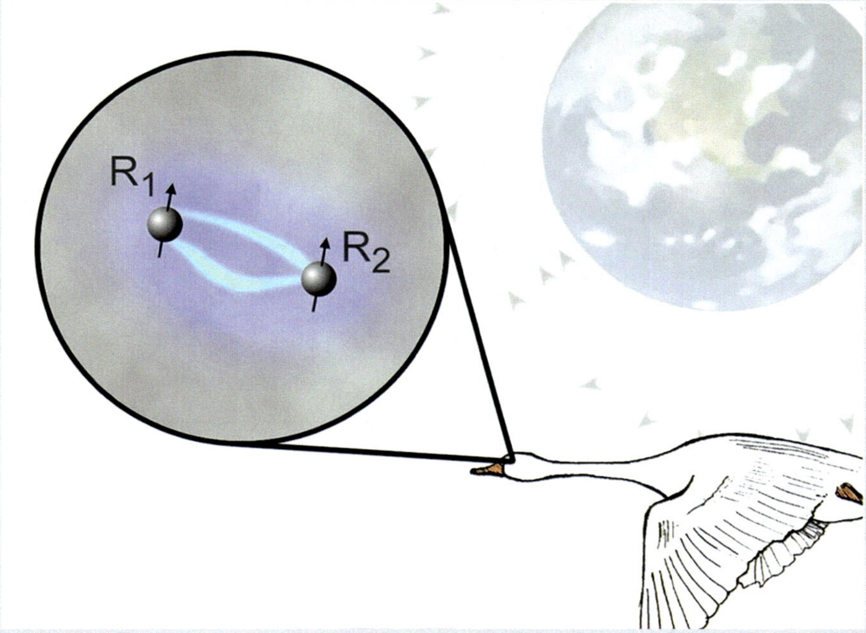


Magnetoreceptive molecules in the birds' eyes include a pair of radicals ($R_1$, $R_2$) that allow birds to sense the Earth's magnetic field. (Bottom) These radicals form when a molecule, or part of a molecule (e.g., cryptochrome) is photochemically excited by absorption of a photon. Radicals are magnetic because electrons have a property referred to as spin, which is analogous to the spins of larger objects (like a planet). Of course, the analogy is imperfect because electrons are wave-like (unless observed). However, because electrons have magnetic properties like those of larger charged, spinning particles (e.g., they are deflected when they move through magnetic fields), physicists used the term "spin" to describe those properties of electrons. The radical pair is in either a singlet (opposite spin) or triplet (same spin) state. In the singlet state, there is only a single way to have opposite spins, whereas, in the triplet state, there are three ways to have correlated spins. (Figure modified from Pedersen et al. 2016; open-access article licensed under a Creative Commons Attribution 4.0 International License)

(continued)

**Box 14.1** (continued)

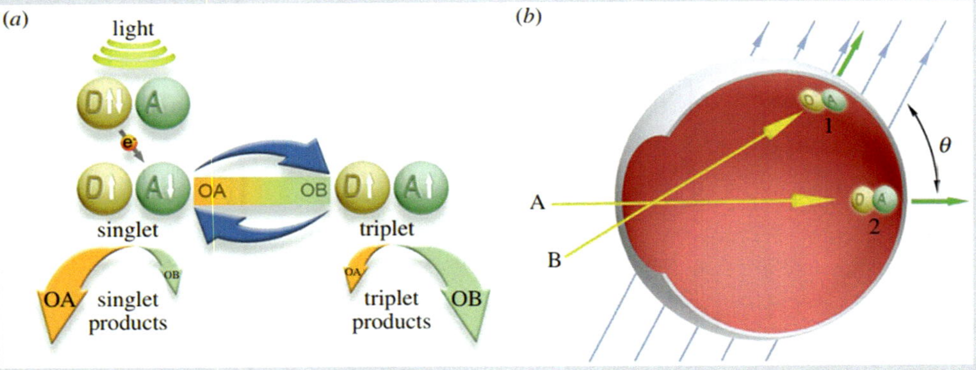


(**a**) Radical pairs oscillate and the interconversion between singlet and triplet states (blue arrows) in response to different magnetic fields can affect the extent and frequency of this process (e.g., at orientation A (OA) or B (OB)) that, in turn, changes the ratio of singlet to triplet products (denoted by the size of the bottom arrows). (**b**) Light entering the eye (e.g., A and B) drives radical pair formation in cryptochrome molecules oriented normal to the retina surface (green arrows) at sites 1 and 2 that are oriented at different angles (u) relative to the external magnetic field (blue lines). How singlet/triplet ratios might be transformed into a nervous impulse for transmission to the brain is not yet known (Wiltschko and Wiltschko 2019). However, differences in radical pair production across different areas of retina could potentially produce a superimposed impression of the magnetic field in the bird's sight (Figure from Shaw et al. 2015; © 2015 The Authors. Published by the Royal Society. All rights reserved, used with permission)

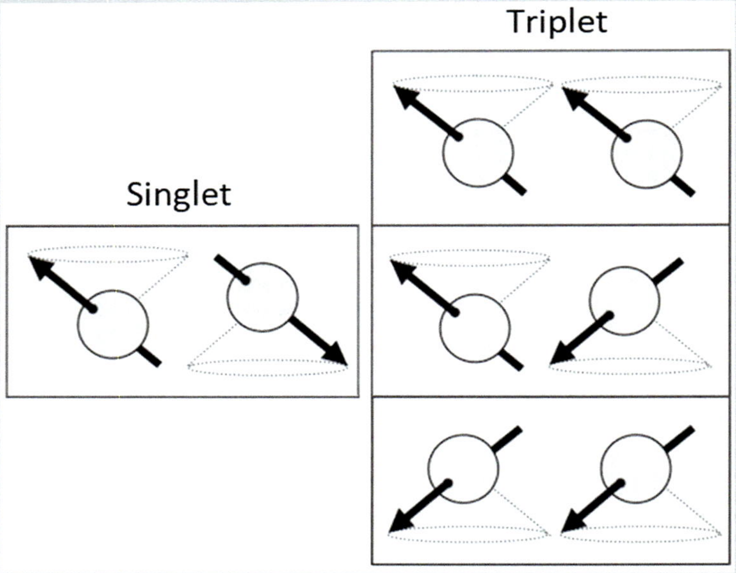


Radical pairs in the singlet state have only a single way to have opposite spins, whereas there are three ways to have correlated spins in the triplet state. (Figure modified from Nagashima and Velan 2013; © 2014 Wiley Periodicals, Inc., used with permission)

(continued)

**Box 14.1** (continued)

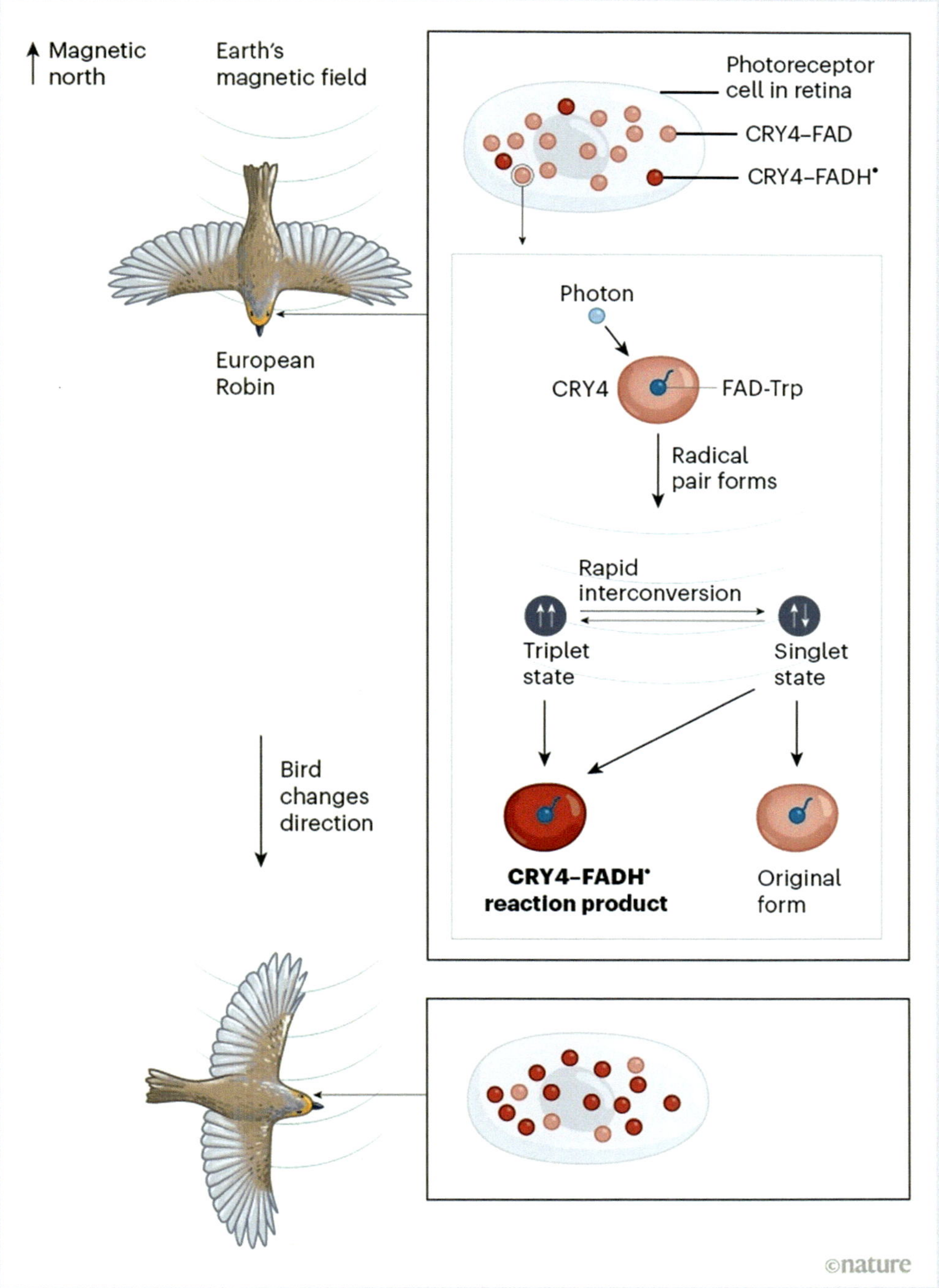


Photoreceptor cells (cones) in the retina of migratory European Robins (*Erithacus rubecula*) contain a protein called cryptochrome 4 (CRY4) that binds to a molecule called FAD. As explained by Warrant (2021), when this CRY4–FAD complex absorbs a photon of short-wavelength light, an electron is released due to an interaction

(continued)

**Box 14.1** (continued)

between FAD and an amino acid in the CRY4 molecule (tryptophan, or trp). Radical pairs (two radicals created simultaneously), so-called because they have an odd number of electrons, are then formed when FAD is reduced (gains the electron). The addition of the odd electron makes the radicals magnetic because electrons behave like tiny magnets with "spin." In molecules with even numbers of electrons, the spins of each pair of electrons are canceled out so the molecule is not magnetic. The radical pair oscillates rapidly (millions of times per second) between singlet and triplet states (indicated by the white arrows). These pairs can generate CRY4–FADH• (which is hypothesized to be the signaling molecule for magnetoreception), but the singlet state can also revert to its original form (CRY4–FAD) and reduce production of CRY4-FADH⁺. The direction of earth's magnetic field influences which outcome predominates. If the bird changes direction, the change in the relative orientation of the magnetic field drives a shift in the proportions of the singlet and triplet states, potentially altering the yield of CRY4–FADH•, and providing birds with information about the inclination of the magnetic field (Warrant 2021; © 2021 Springer Nature, used with permission)

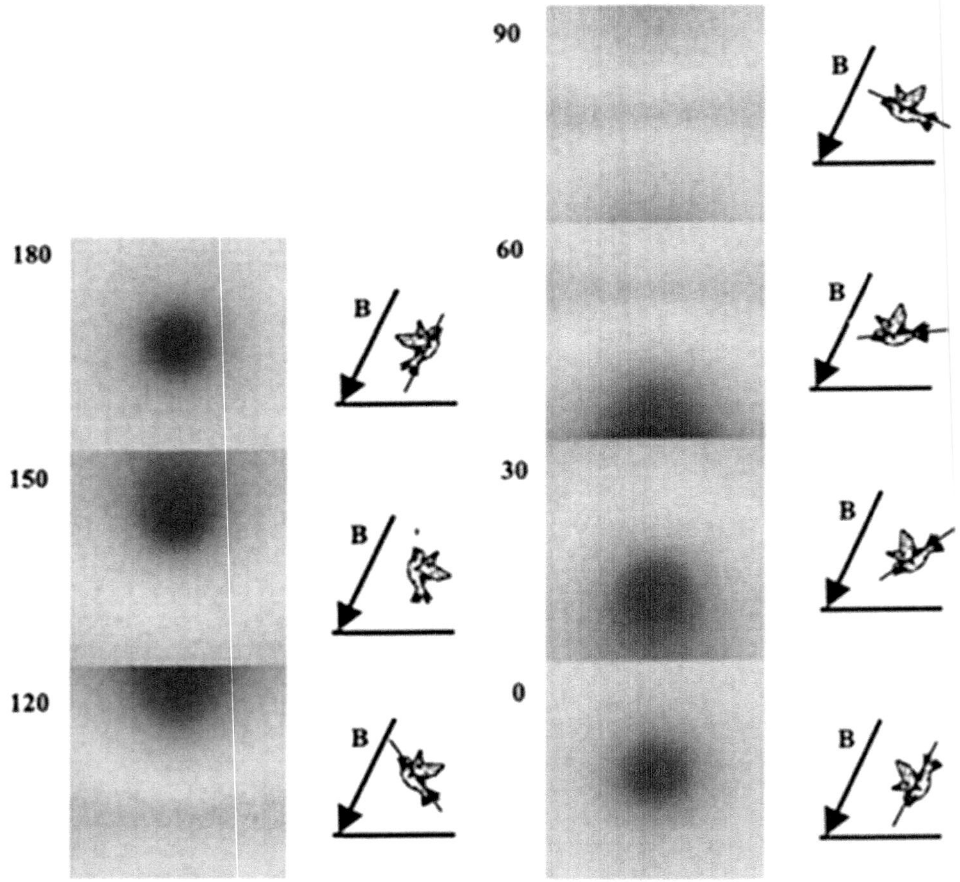


**Fig. 14.17** A bird's-eye view of the geomagnetic field as seen when looking in different directions. In this example, the geomagnetic field (arrow) has an inclination of 68°. (Figure from Ritz et al. 2000; © 2000 The Biophysical Society. Published by Elsevier Inc., used with permission)

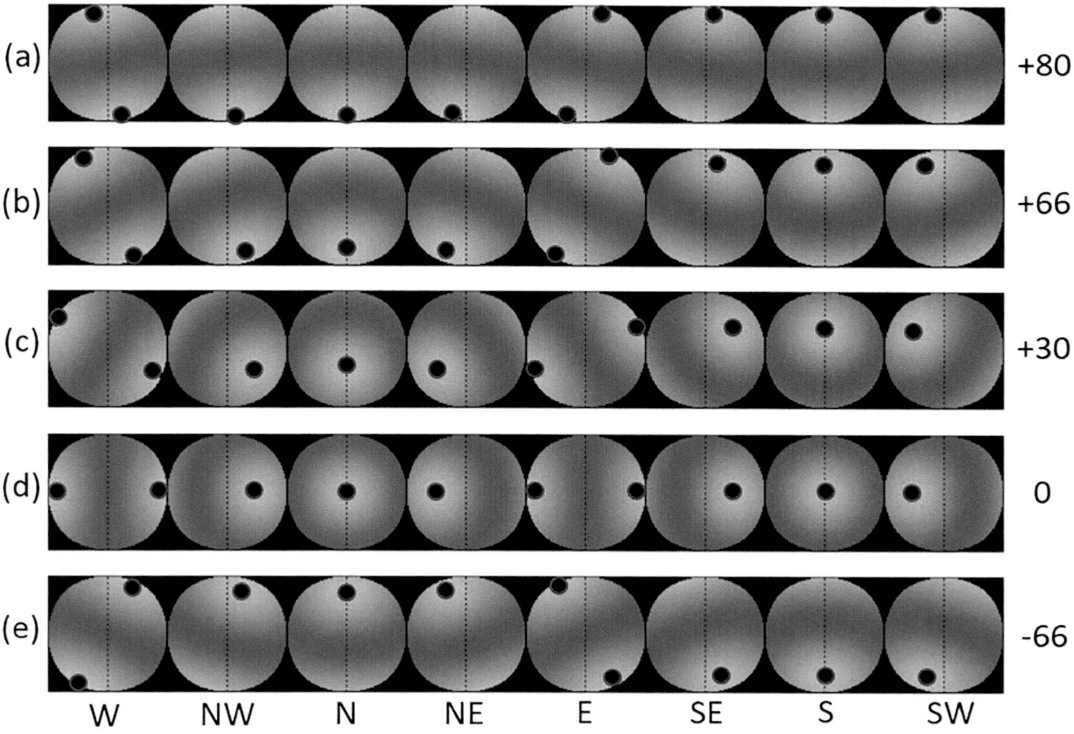


**Fig. 14.18** Another bird's-eye view of the magnetic field for a bird changing its viewing direction clockwise in 45° increments in magnetic fields of different inclinations. For easier recognition, the crosses mark the position of the brightest spot in the signal modulation patterns. For each inclination, the eight panels represent a full 360° sweep, showing all cardinal directions, from west (left panel) to southwest (right panel). Inclination angles are positive for the northern hemisphere with the magnetic field pointing downward. (**a**) +80° inclination, e.g., in northern Canada, (**b**) +66° inclination, e.g., in central Europe, (**c**) +30° inclination, e.g., in the Sahara Desert, (**d**) 0° inclination, at the equator, (**e**) −66° inclination, e.g., in southern Australia. (Figure from Wang et al. 2006; © 2006 IOP Publishing Ltd., used with permission)

additional studies and experiments are needed to test this "electromagnetic induction" hypothesis.

Night-migrating songbirds also have a magnetic sense involving the ophthalmic branch of the trigeminal nerve. Experiments where the orientation of spring-migrating Eurasian Reed Warblers (*Acrocephalus scirpaceus*) is determined before and after translocation have revealed that, after translocation, those with short sections of their ophthalmic branch removed are not able to correctly reorient toward their breeding areas. However, sham-operated Eurasian Reed Warblers with intact ophthalmic branches are able to successfully reorient toward their breeding areas after translocation (Kishkinev et al. 2013, Pakhomov et al. 2017; Fig. 14.22). These results indicate that the ophthalmic branch of the trigeminal nerve transmits information that is an essential component of their magnetic map. Further, because experienced Eurasian Reed Warblers are known to use magnetic declination, which requires both magnetic and celestial compass information to determine longitude, the results of these studies suggest that both retina-based and trigeminal-based magnetic information are needed for the magnetic map to work correctly (Pakhomov et al. 2017). What is not yet known, however, are the location and identification of the sensors that detect the magnetic information transmitted by the trigeminal nerve

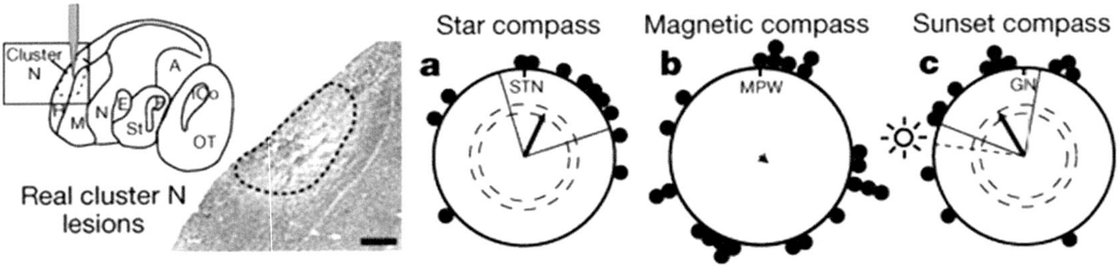


**Fig. 14.19** (**a**) European Robins (*Erithacus rubecula*) with a damaged cluster N that were tested in a planetarium simulating the local starry sky (STN, star north) oriented in the typical north-northeast spring migratory direction. (**b**) Birds with cluster N damage could not orient using their magnetic compass (MPW, magnetic pole). (**c**) Birds with cluster N damage could also orient during sunset using their sun compass. (Figure from Zapka et al. 2009; © 2009 Springer Nature, used with permission)

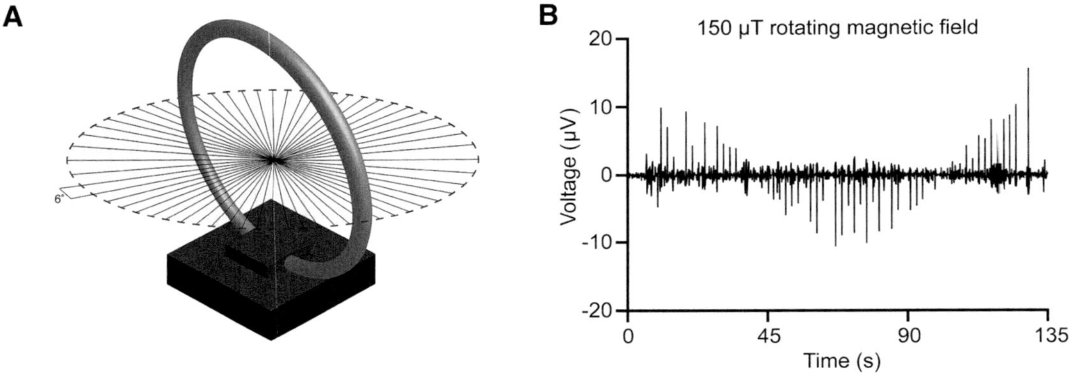


**Fig. 14.20** (**a**) A model of a Homing Pigeon's (*Columba livia*) semicircular canal (21 cm in diameter) that contained artificial endolymph was exposed to a rotating magnetic field (150 microTesla) and any induced voltage was measured using a nanovoltmeter. The stimulus consisted of six shifts every 2 s, taking 2 min to complete one rotation around the vertical axis. (**b**) Voltage spikes were correlated with changes in the magnetic field. Peak voltage (15.6 μV) was detected when the stimulus was directed 90° to the plane of the model semicircular canal and lowest (1.6 μV) when the vector was parallel to the plane of the semicircular canal. If pigeons, and perhaps other birds, detect magnetic fields via electromagnetic induction, they would likely use head scanning to alter the orientation of their semicircular canals to earth's magnetic field. (Figure from Nimpf et al. 2019; open-access article distributed under the terms of the Creative Commons CC-BY license, https://creativecommons.org/licenses/by/4.0/)

(Box 14.2 Source of the Ophthalmic Branch Magnetic Sense).

## 14.6 Navigation

Birds can potentially navigate using a process called path integration (sometimes referred to as dead reckoning). Path integration is a navigational, or homing, strategy used by many animals, ranging from arthropods to mammals. Using path integration, an animal is able to return to a specific location after traveling to any point some distance from it, even if the path taken is circuitous, by using information collected from the environment during the journey to determine a

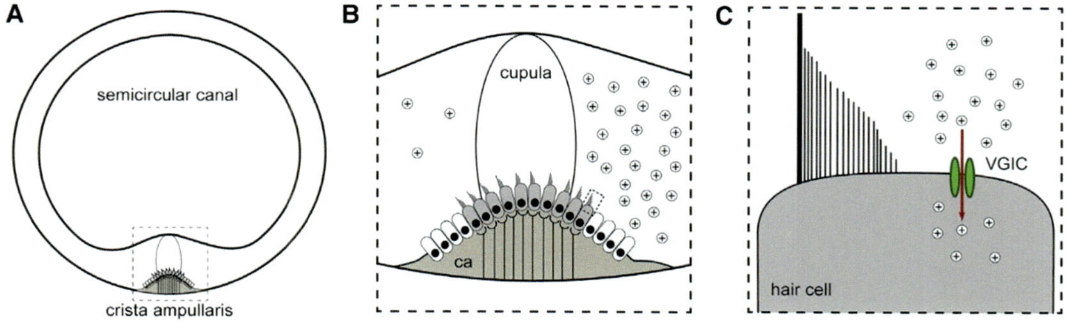


**Fig. 14.21** Possible mechanism for detection of magnetic fields in the inner ear of pigeons by electromagnetic induction (generating electric current, or nervous impulses, with a magnetic field), i.e., as birds move through earth's magnetic field, voltage is induced via changes in the distribution of ions in the fluid (endolymph) in semicircular canals. (**a**) Semicircular canal with sensory hair cells located at the base of the gelatinous cupula in the crista ampullaris (ca). (**b**) Rotation of the bird's head perpendicular to the plane of the semicircular canal and through a magnetic field causes a redistribution of charges across the cupula. (**c**) Voltage-gated ion channels (VGIC) located in hair cell membranes can respond to this redistribution of charges by allowing the inward diffusion of positive ions into hair cells that causes action potentials and nervous impulses perceived by the brain as magnetic information. (Figure from Nimpf et al. 2019; open-access article distributed under the terms of the Creative Commons CC-BY license, https://creativecommons.org/licenses/by/4.0/)

direct (straight-line) route back (Fig. 14.23). Using path integration also referred to a route-based navigation, an animal determines its position and the positions of other objects in the environment by integrating the distance and directions traveled during a journey (Fig. 14.24). Distance and direction information can potentially be obtained from a number of sources, including proprioceptive cues, vestibular or somatosensory cues, and solar and magnetic cues. Among birds, Wiltschko and Wiltschko et al. (1998) suggested that young pigeons use path integration (or, using their terminology, route reversal) when initially learning about their environment (until they are about 3 months old). However, Wallraff (2000) concluded that the hypothesis that "pigeons develop a very sophisticated path integration mechanism for use within only a few weeks after which they forget it lacks both plausibility and experimental support." Beyond homing pigeons, little is known about the use of path integration by birds. However, because the likelihood of errors increases with increasing distance (Able 2000), path integration, if used by birds, is likely of greater importance for short-distance navigation.

Beacon-based navigation involves the use of landmarks, or beacons, for short-range homing (returning to a familiar area or site; Fig. 14.25). However, how do birds located at greater distances from a target destination determine where they are located? As noted earlier, a grid-based navigational system seems most likely because "map-based" cues (those that a bird can memorize and use to navigate) are too local to be useful over long distances (e.g., visual landmarks). Cues other than visual landmarks, such as atmospheric odors (Zannoni et al. 2020) and infrasounds (sounds below 20 Hz generated by ocean waves, mountain winds, and large waterfalls; Hagstrum 2000, Bedard 2021), could potentially be used as features of a "map-based" navigational system.

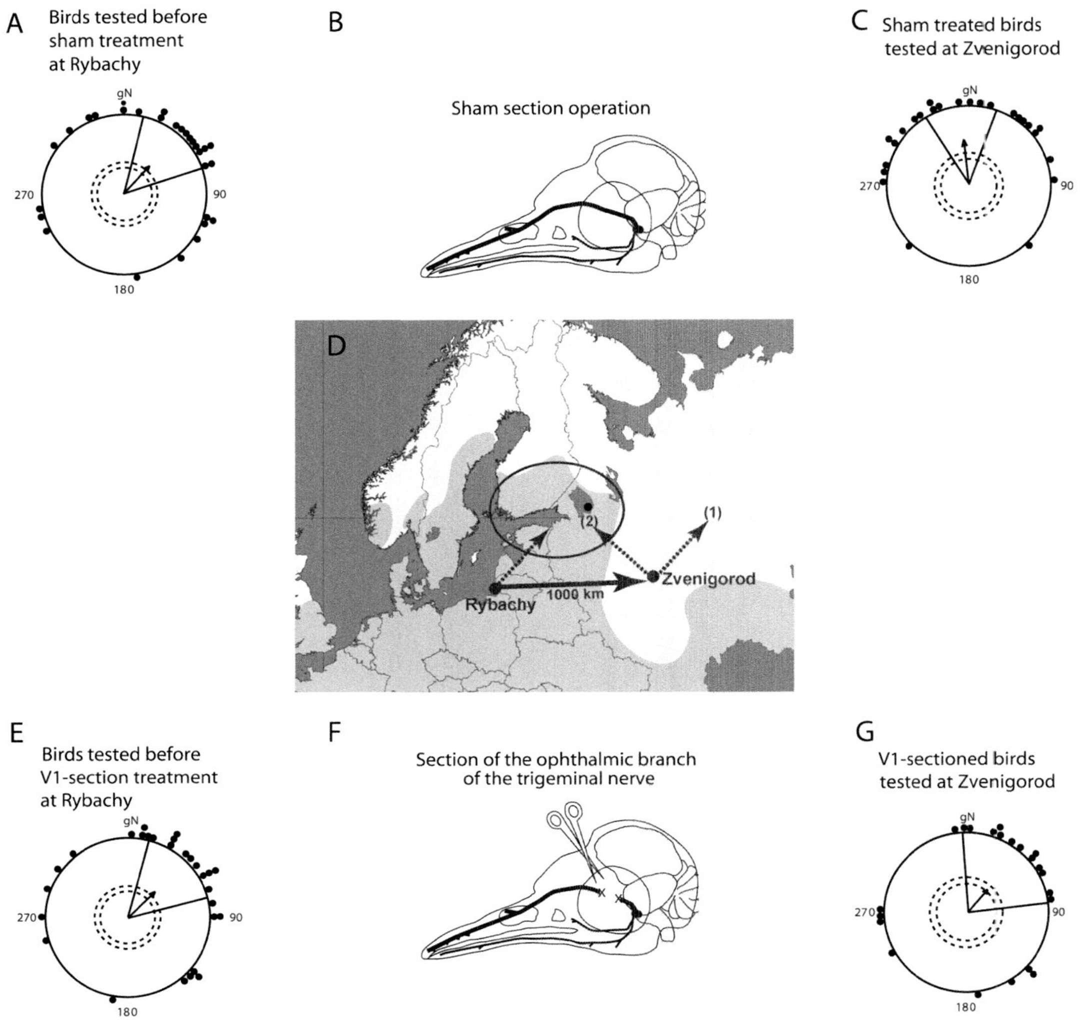


**Fig. 14.22** Orientation of Eurasian Reed Warblers (*Acrocephalus scirpaceus*) at their capture and displacement sites. (**a, c**) Results for sham-operated birds before surgery at the capture site (**a**) and after sham-surgery and translocation to the displacement site (**c**). (**e, g**) Results for nerve-sectioned birds before sectioning at the capture site (**e**) and after sectioning surgery and translocation to the displacement site (**g**). Each dot on the circular diagram periphery indicates the mean orientation of one bird. (**b, f**) Both sham-operated and nerve-sectioned birds underwent the same surgical procedure, except that a 3–5 mm section of the ophthalmic branch of the trigeminal nerve was removed from the nerve-sectioned birds. The ophthalmic branch is shown in bold. (Figure from Kishkinev et al. 2013; © 2013 Kishkinev et al., open-access article distributed under the terms of the Creative Commons Attribution License, https://creativecommons.org/licenses/by/4.0/)

**Box 14.2 Source of the Ophthalmic Branch Magnetic Sense**

The ophthalmic branch of the trigeminal nerve is known to transmit magnetic information to the brain of some species of birds, but where those nervous impulses originate remains to be determined. Kirschvink and Gould (1981) suggested that biogenic magnetite ($Fe_3O_4$) crystals might be the basis for magnetoreception in animals, and some investigators suggested that such crystals in the upper beak of homing pigeons might be a source of magnetic information (Beason and Nichols 1984; Fleissner et al. 2003). However, subsequent analysis revealed that the cells previously identified as magnetosensitive neuron were actually iron-rich macrophages and were, therefore, not the source of any magnetic information (Treiber et al. 2012). Treiber et al. (2012) ended the abstract of their paper by stating that, "Our work necessitates a renewed search for the true magnetite-dependent magnetoreceptor in birds." That search has not yet revealed the source of the ophthalmic-based magnetic information.

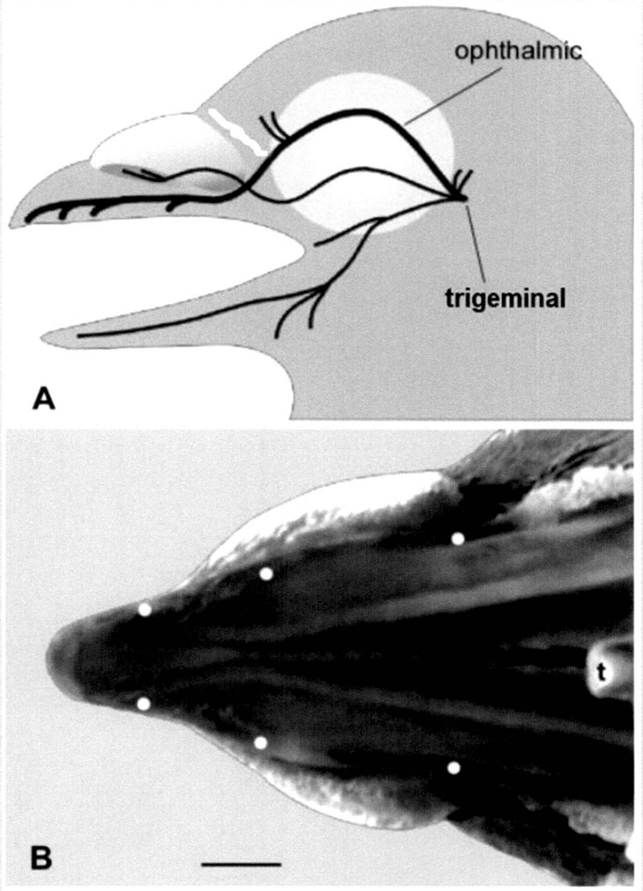


(**a**) Drawing of a Rock Pigeon (*Columba livia*) skull showing the ophthalmic branch of the trigeminal nerve. (**b**) Inside view showing the location of the putative magnetoreceptor nerve endings (white dots). t, tongue. (Figure from Fleissner et al. 2003; © 2003 Wiley-Liss, Inc., used with permission)

Other investigators have suggested that ferromagnetic particles in the lagena provide birds with information about the direction and magnitude of magnetic fields (Wu and Dickman 2011,

(continued)

**Box 14.2** (continued)

2012). However, Malkemper et al. (2019) conducted a high-sensitivity screening for iron in the lagena of pigeons and found no evidence of magnetic otoconia.

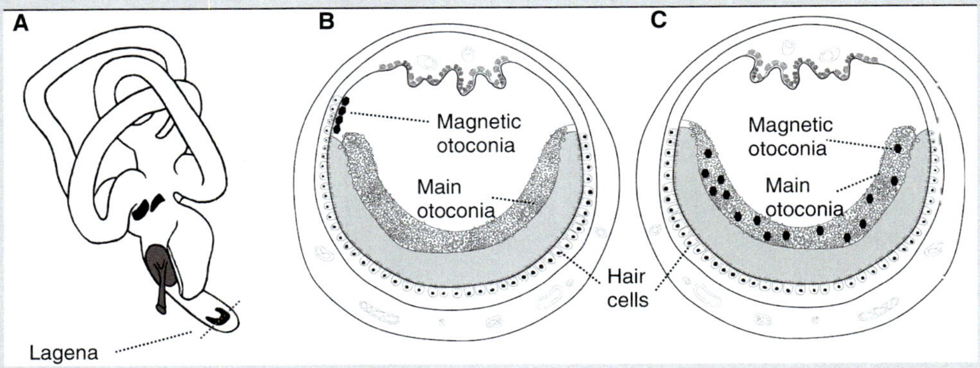


(**a**) Drawing of the inner ear of a pigeon showing the location of the lagena. The otoconia are shown in black (otoconia are small crystals of calcium carbonate). (**b**) The magnetic otoconia hypothesis posited that magnetic crystals (containing $Fe_3O_4$) responding to magnetic fields would stimulate hair cells that transduced magnetic information into nervous impulses. Drawing is a cross-section through the lagena at the level indicated by the dashed line in (**a**) magnetic otoconia could be linked to a certain section of hair cells separate from those associated with calcium-carbonate otoconia. (**c**) Alternative to the hypothesis presented in (**b**), magnetic structures could be interspersed with calcium carbonate otoconia in the lagena. (Figure from Malkemper et al. 2019; © 2018 Published by Elsevier Ltd., used with permission)

One untested hypothesis is that the source of magnetic information might be symbiotic magnetotactic bacteria (SMB). Natan et al. (2020) proposed that SMB residing in the lacrimal glands move across the avian cornea in response to magnetic fields and that movement could potentially be perceived by birds. Coincidentally, based on experiments with Nathusius' bats (*Pipistrellus nathusii*), Lindecke et al. (2021) suggested that the cornea might be a site of magnetoreception in bats. In support of that hypothesis, these investigators found that bats treated with corneal anesthesia in both eyes flew in random direction, whereas as bats treated in a single eye and control bats flew in the seasonally appropriate direction.

As proposed by Natan et al. (2020), magnetotactic bacteria moving across the cornea in response to magnetic fields may provide birds with information about the inclination of the earth's magnetic field. (Figure from Natan et al. 2020; © 2020 The Authors. Published by the Royal Society, used with permission)

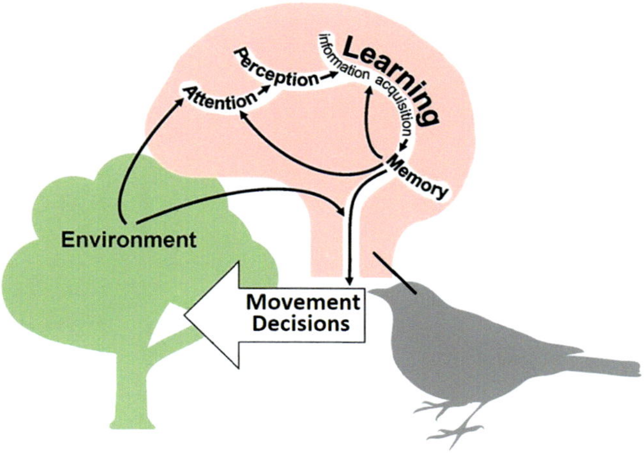


**Fig. 14.23** Schematic showing the process of learning by a bird prior to path navigation. The bird gathers information about landscape features in the environment while moving from a starting location. As currently understood, information gathering involves attention, perception, learning, and memory. The acquired information will then allow the bird to navigate, i.e., return to its starting point from its present location. (Figure modified from Lewis et al. 2021; © 2021 Lewis, Fagan, Auger-Méthé, Frair, Fryxell, Gros, Gurarie, Healy and Merkle, open-access article distributed under the terms of the Creative Commons Attribution License (CC BY), https://creativecommons.org/licenses/by/4.0/)

## 14.7 True Navigation

A bird is capable of true navigation if "following displacement, it selects the correct direction to return to its goal" (Putman 2021). Further, as noted by Kishkinev et al. (2021), true navigation is "the ability to return to a known destination after displacement to an unknown location without relying on familiar surroundings, cues that emanate from the destination, or information collected during the outward journey." Thus, true navigation requires that a bird be able to determine its current spatial location relative to some distant goal (typically called "map" information), it can then choose the appropriate direction of travel.

Humans navigate using a bi-coordinate system. Determine your latitude and longitude (easily accomplished with a GPS unit) and you know exactly where you are located. Based on studies to date, avian navigation is likely also based on a bi-coordinate system (but it could very well be multi-coordinate; Thorup and Holland 2009). If so, for true navigation, birds need some way to discern two coordinates (corresponding to our latitude and longitude). Birds can potentially determine their location along one coordinate (latitude) using two magnetic cues: inclination and, with experience, intensity (Gould 1998). Inclination provides latitudinal information because the inclination of the earth's magnetic field, as noted previously, varies predictable with latitude (Fig. 14.26), and many birds can "see" this inclination. Although exhibiting more variability than inclination, the intensity of earth's magnetic field also exhibits gradients that vary with latitude. To be useful for navigation, birds must be able to "sense" the intensity of the magnetic field and birds are known to have receptors that allow them to do so (Schiffner and Wiltschko 2011; Wiltschko and Wiltschko 2013). Using geomagnetic satellite data, Zein et al. (2022) determined that Greater White-fronted Geese (*Anser albifrons*) migrating from breeding areas in the Russian Arctic to wintering areas in

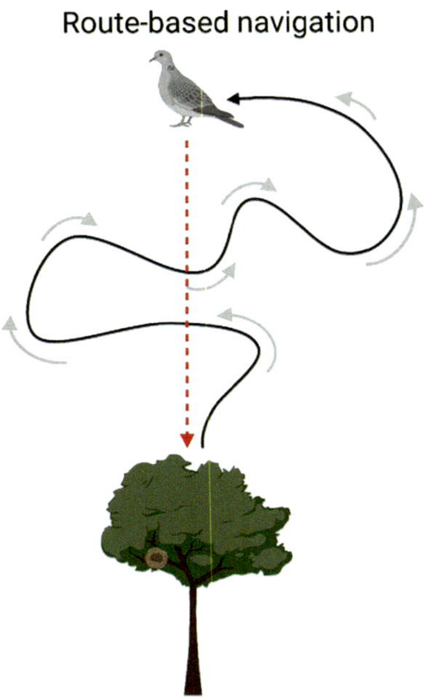


**Fig. 14.24** With route-based navigation, a bird remembers its movements relative to its starting location, summing the distance and direction of those movements to estimate its current location, then reversing course to return to its starting point. (Figure from Kashetsky et al. 2021; © 2021 Kashetsky, Avgar and Dukas, open-access article distributed under the terms of the Creative Commons Attribution License (CC BY), https://creativecommons.org/licenses/by/4.0/)

**Fig. 14.25** With beacon-based navigation, a bird identifies and remembers landmarks that, in sequence, can be used to return it to its goal. (Figure from Kashetsky et al. 2021; © 2021 Kashetsky, Avgar and Dukas, open-access article distributed under the terms of the Creative Commons Attribution License (CC BY), https://creativecommons.org/licenses/by/4.0/)

western Europe appeared to use variation in geomagnetic intensity, in possible combination with other cues (e.g., landmarks), for navigational purposes.

If their magnetic receptors provide information about the inclination (the light-based receptor), strength, and direction of magnetic fields, how might birds use that information to generate grid-based navigational maps? A grid-based navigational map requires that birds determine both their latitudinal and longitudinal position. Latitude can be determined using the inclination and, perhaps, the strength of the magnetic field. In some locations, lines of equal inclination and lines of equal strength are nearly parallel and oriented in an east-west direction, potentially allowing birds to determine latitude using either (or both) cue(s). However, in other locations, lines of equal inclination and lines of equal strength are not parallel and lines of equal strength are not oriented in an east-west direction. In such areas, the two sources of magnetic information might provide birds with information about both their latitude and longitude or, in other words, a bi-coordinate magnetic map.

One area where the lines of equal magnetic inclination and lines of equal magnetic strength or intensity vary in direction to form a grid-like pattern is northwestern Russia. Chernetsov et al. (2008) examined the navigational abilities of adult Eurasian Reed Warblers (*Acrocephalus scirpaceus*) during spring migration by displacing them about 1000 km to the east (Fig. 14.27). The warblers corrected for the displacement by shifting their orientation from the northeast at the capture site (that would take them to their breeding ground) to the northwest after the displacement (Fig. 14.19). Such results indicate that

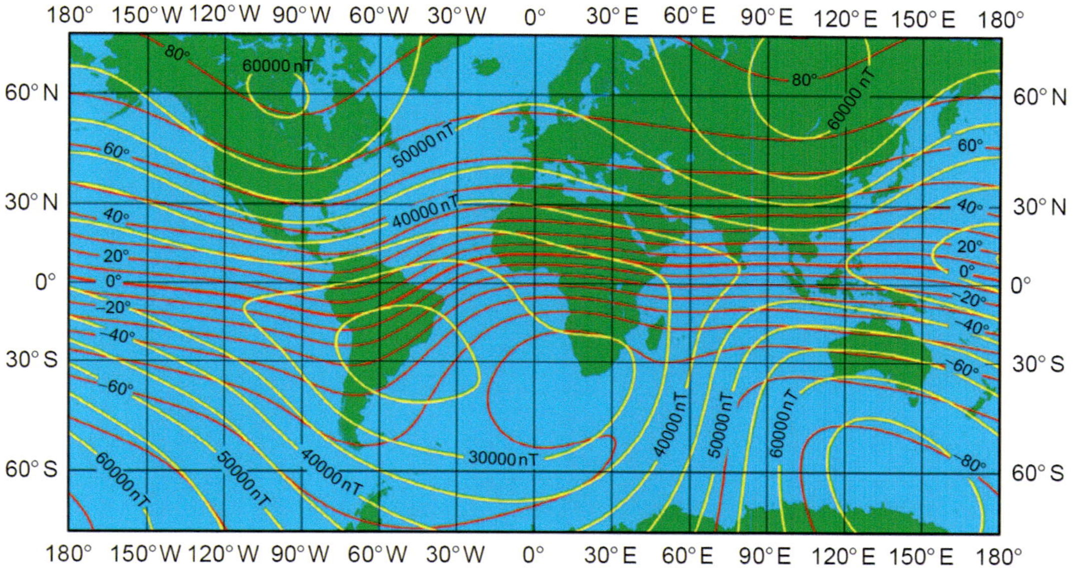


**Fig. 14.26** Map showing isodynamic lines of the total intensity of earth's magnetic field (yellow lines, in nano-Tesla) and isoclines of magnetic inclination (red lines, in degrees). Isodynamic lines and isoclines can form a grid (e.g., Siberia or the Indian Ocean) or may be close to parallel to each other (e.g., United States). (Figure from United States Geological Survey, CC0 Public Domain)

the warblers were somehow able to determine that a longitudinal shift had occurred and were able to correctly orient in the direction that would take them to their breeding areas. Because lines of equal inclination and strength form a grid-like pattern in the study area, the warblers may have used those cues to accurately determine their position and then orient in the direction that would take them to their breeding areas.

The results of a number of studies have revealed that birds have an innate ability to use their inclination compass; no experience is needed. However, using variation in magnetic strength to navigate requires experience because lines of equal magnetic strength vary in their orientation in different areas and, therefore, birds must learn the pattern of variation in the areas they occur. One possible illustration of this comes from the results of a study of White-crowned Sparrows (*Zonotrichia leucophrys*; Thorup et al. 2007; see Fig. 14.3). Adult White-crowned Sparrows displaced 3700 km from the west to the east coast of the United States were able to determine their "new" longitudinal position and correctly orient in the direction that would take them to their wintering areas; juveniles incorrectly continued to orient as if they were still on the west coast. One explanation for such results is that the adult sparrows had acquired the needed magnetic information during previous migratory journeys, whereas juveniles had not.

In some areas, such as in and around the Caribbean (Fig. 14.28), the lines of equal magnetic inclination and field strength are nearly parallel and do not form a grid-like pattern and, therefore, may not, in combination, be useful for navigation. As an alternative, birds might use some combination of other magnetic cues for navigation, including inclination, total intensity, gradients of intensity, horizontal field intensity, vertical field intensity, declination (difference

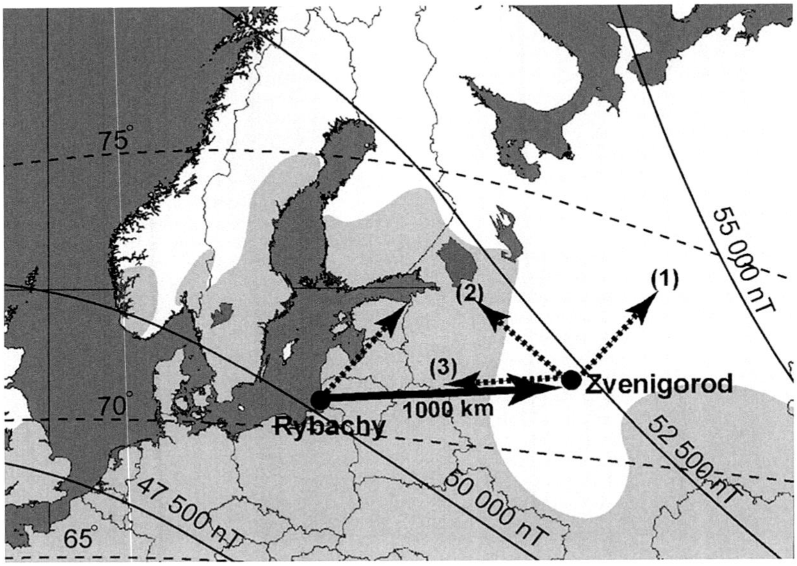


**Fig. 14.27** Isolines of magnetic intensity (solid thin lines) and inclination (dashed lines) relative to capture (Rybachy, Kaliningrad region) and displacement (Zvenigorod, Moscow region) sites and the breeding range of Eurasian Reed Warblers (*Acrocephalus scirpaceus*) in the region (shaded light gray). Solid arrow shows the displacement direction. The broken arrow at the capture site shows the mean migratory direction, and the broken arrows at the displacement sites show our working hypotheses: (1) no compensation, (2) compensation toward the breeding destinations, and (3) compensation toward the capture site. (Figure from Chernetsov et al. 2008; © 2008 Elsevier Ltd., used with permission)

between geographic north and magnetic north, Fig. 14.29; Box 14.3 Eurasian Reed Warblers Use Magnetic Declination), or, perhaps, even polarity (Solov'yov and Greiner 2009). For example, some investigators have noted that magnetic intensity exhibits gradients between the poles and the magnetic equator (Walker 1998), and may provide birds with something like "magnetic latitude" (Wiltschko et al. 2006).

Few investigators have examined the possibility that birds might use these gradients for navigational purposes. However, Dennis et al. (2007) released homing pigeons in areas in and around a natural magnetic anomaly (where the Earth's magnetic field is spatially distorted and caused by differences in the magnetization of the rocks in the Earth's crust). Using flight trajectories recorded by GPS-based tracking devices, they found that many of the pigeons released at unfamiliar sites initially flew, sometimes several kilometers, in directions that were either parallel or perpendicular to the bearing of the local intensity field (Fig. 14.30). Pigeons exhibited this behavior regardless of the bearing of their home loft and significantly more often than would be expected by chance. These results provide evidence that pigeons are able to detect spatial variation in the strength of the Earth's magnetic field. By aligning their flight paths either parallel or perpendicular to lines of magnetic field strength, pigeons might be "sampling" the strength and variability of the local magnetic field, information that could potentially be useful for navigation (Dennis et al. 2007).

Wiltschko et al. (2009) also released homing pigeons near a magnetic anomaly and found that

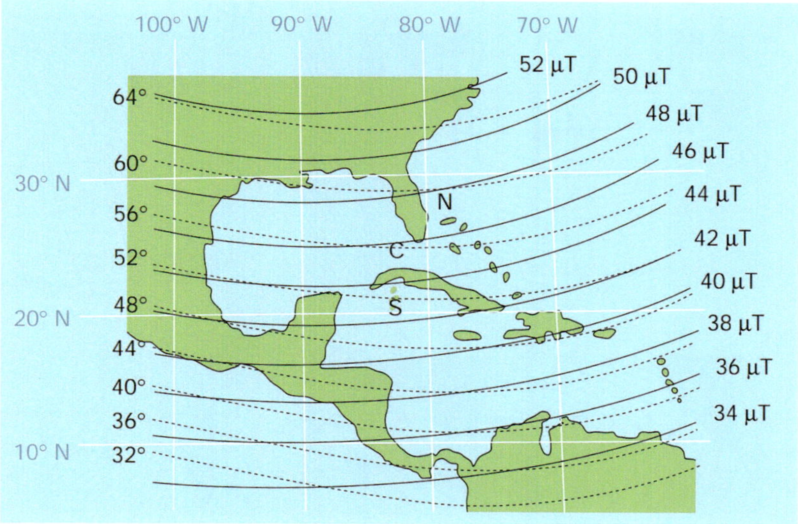


**Fig. 14.28** Magnetic inclination (degrees) and total field intensity (microTesla) in the Caribbean and adjacent areas. (Figure from Alerstam 2003; © 2003 Springer Nature, used with permission)

the irregular magnetic field in the area of the anomaly caused confusion. Pigeons took longer than normal to decide which direction to finally orient (termed the vanishing interval in homing pigeon research). Because other orientation cues are not affected by magnetic anomalies (i.e., the inclination compass and the sun compass), such results indicate that the pigeons detected and were apparently confused by the anomalous magnetic intensity. This, in turn, indicates that homing pigeons normally "record" and use information about magnetic intensity and its changes as part of their navigational process. In contrast to Dennis et al. (2007), Wiltschko et al. (2009) found no evidence that the pigeons in their study followed gradients of magnetic intensity. More recently, however, Kishkinev et al. (2021) exposed Eurasian Reed Warblers to geomagnetic cues that differed from those in their normal range and the warblers were able to accurately reorient toward their normal range. Because the "translocation" of the warblers was based on manipulation of magnetic cues in the lab, all other environmental cues were unchanged. Thus, a possible explanation for such results is that the warblers possessed a magnetic map that allowed them to extrapolate beyond their normal range, determine their location, and determine the direction needed to return to their normal range. Alternatively, Kishkinev et al. (2021) proposed that birds might use a simpler "rule of thumb" mechanism; the warblers determined their general "displaced" location by comparing the magnetic inclination and intensity to that of their original location and used differences in those cues to determine the direction they needed to fly to return to their original location (Fig. 14.31).

A study where White-crowned Sparrows (*Zonotrichia leucophrys*) were displaced longitudinally across distances ranging from 266 to 2869 km provides some additional possible clues about how magnetic cues might be used by birds (Åkesson et al. 2005). Both young and adult White-crowned Sparrows were captured in their breeding area in the Northwest Territories, Canada, toward the end of the breeding season and shortly before they would normally begin fall migration (15 July to 10 August). One group of sparrows (15 adults and 15 juveniles) was transported (on an icebreaker) to unfamiliar sites along a northeasterly route to the magnetic north pole (79.0° N, 105.1° W) and then further southeast (Fig. 14.32). A control group (5 adults and 39 juveniles) was transported a short distance west of the capture site. Using Emlen funnels (Fig. 14.6), the directional orientation of both

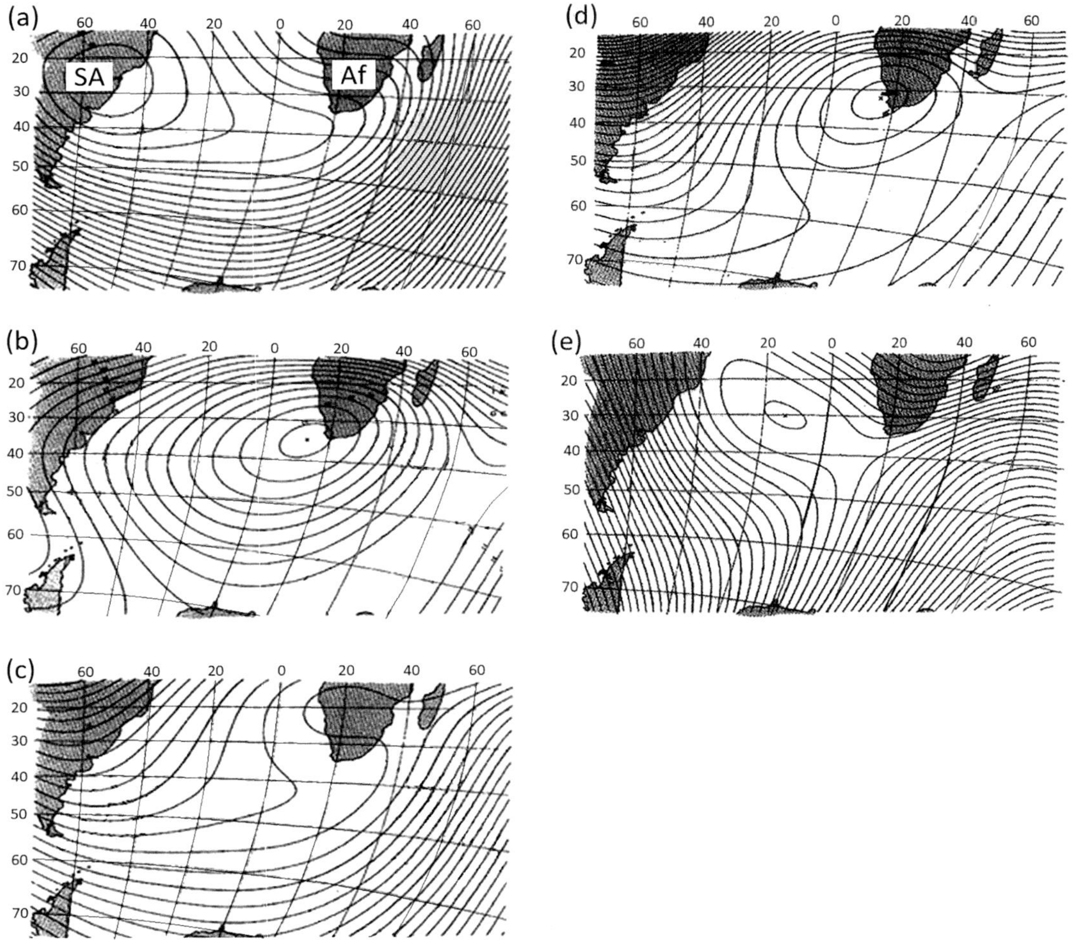


**Fig. 14.29** The isomagnetics for (**a**) total field intensity, (**b**) horizontal field intensity, (**c**) vertical field intensity, (**d**) inclination, and (**e**) declination in a portion of the Southern Hemisphere including parts of South America and southern Africa. Birds navigating in the area could potentially use two or more of these cues (particularly those that intersect at an angle to form grid-like patterns) to generate a magnetic navigational map. (Figure from Åkesson and Alerstam 1998; © 1998 John Wiley & Sons, Inc., used with permission)

**Box 14.3 Eurasian Reed Warblers Used Magnetic Declination to Determine Longitude**

Use of a bi-coordinate map requires information about both latitude and longitude. Birds can potentially determine their latitude using two magnetic cues: inclination and, with experience, intensity. Less is known about how birds might determine longitude, what Gould (1998) referred to as "the longitude problem." One possibility would be the use, at least in some areas, of nonparallel lines of equal inclination and lines of equal strength. Another possibility, and the focus of a study of Eurasian Reed Warblers (*Acrocephalus scirpaceus*), is that birds might use magnetic declination, the difference between true (geographic) north and magnetic north. Chernetsov et al. (2017) captured adult and juvenile Reed Warblers during fall migration at Rybachy (Russia) and, using Emlen funnels, determined their nighttime migratory orientations. Both adults and juveniles oriented, on average, in the appropriate southwesterly direction that would take them to their wintering areas. Using a system of coils that allowed manipulation of the magnetic field, the birds were then virtually displaced to Scotland and their migratory orientations were again determined.

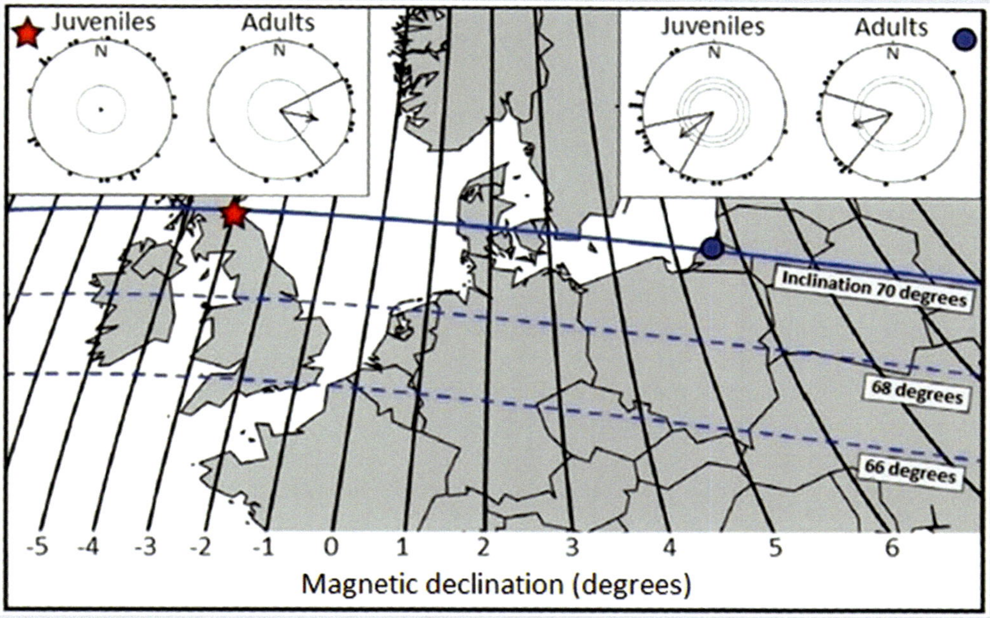


The average migratory orientation of juvenile and adult Eurasian Reed Warblers (upper right, indicated by arrows) in Rybachy (blue circle) was, as expected, in a southwesterly direction that would take them to their wintering areas in Africa. The birds were then exposed to a magnetic field that matched the natural magnetic field in Scotland (red star). Most adult Reed Warblers compensated for the apparent change in location by orienting in a direction that would return them to their correct migratory pathway, the migratory orientations of juveniles, however, were random (upper left). (Figure modified from Chernetsov et al. 2017; © 2017 Elsevier Ltd., used with permission)

(continued)

**Box 14.3** (continued)

Eurasian Reed Warbler (Photo by Kleine Kareklet, Wikipedia, CC BY 2.0, https://creativecommons.org/licenses/by/2.0/)

Adult Eurasian Reed Warblers compensated for the change in location and oriented in southeasterly direction that would return them to their typical migration route, but juveniles did not. Because all other potential cues (magnetic inclination and intensity, celestial compass information, and possible olfactory cues) were kept constant, these results suggest that adult Eurasian Reed Warblers used magnetic declination to determine their new position and the appropriate migratory orientation. The use of magnetic declination means that adults were able to accurately determine their new longitudinal position by comparing the direction of magnetic north (using their inclination compass) to that of geographic north (using celestial cues): at Rybachy, the difference was about 5.5° whereas, in Scotland, the difference was −3° (see Figure above). The random orientations of juvenile Eurasian Reed Warblers after their virtual displacement also suggest that they must learn to use magnetic declination to navigate (Chernetsov et al. 2017). In a subsequent study, Chernetsov et al. (2020) determined that adult and juvenile European Robins (*Erithacus rubecula*) and adult Garden Warblers (*Sylvia borin*) were unable to use magnetic declination in the same way as adult Eurasian Reed Warblers, suggesting that different species of birds differ in the map cues used for navigation.

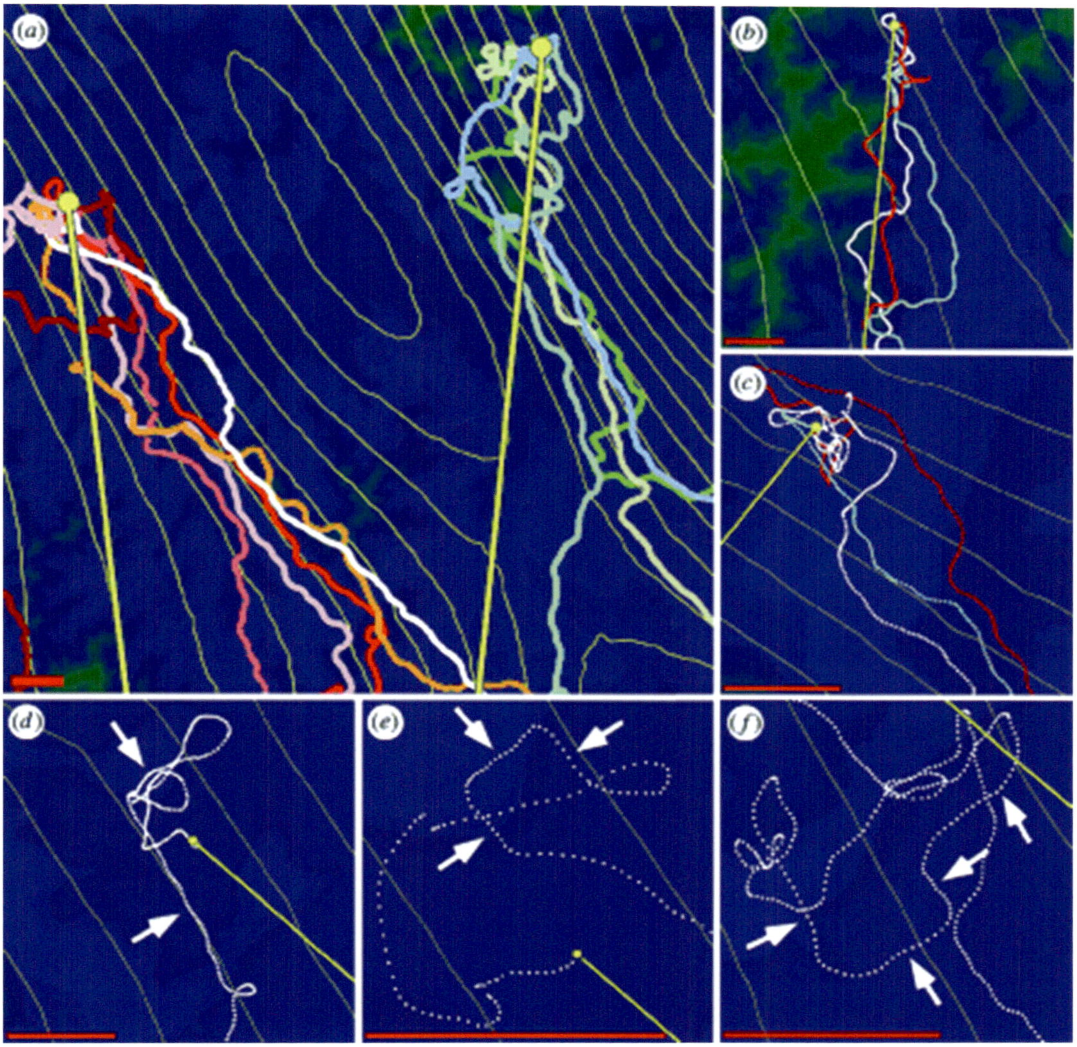


**Fig. 14.30** Examples of the orientation of flight trajectories of homing pigeons relative to isopleths of geomagnetic intensity. Single-color lines and points indicate flight trajectories and position fixes of individual pigeons. Yellow circles show the location of release sites, and yellow lines designate the straight-line direction to the home loft. Thin green lines are magnetic intensity isopleths (10 nT intervals). Background color depicts relative elevation (low-elevation areas are blue and high-elevation areas are green). Red scale bars are 500 m. Arrows indicate locations of alignment of individual birds: (**a**) Long-distance alignment along lines of similar magnetic intensity; (**b, c**) Parallel and/or perpendicular alignments at two other release sites; (**d–f**). Detailed views of examples of parallel and perpendicular alignments. (Figure from Dennis et al. 2007; © 2007 The Royal Society, used with permission)

groups of sparrows was determined, with the experimental group tested at nine different locations (including their breeding area; site 1 in Fig. 14.32). Sparrows in the control group generally oriented to the southeast (Fig. 14.33), the general direction of their presumed wintering area.

Sparrows being transported to the east, however, shifted their orientation from their normal migratory direction to a direction leading back to

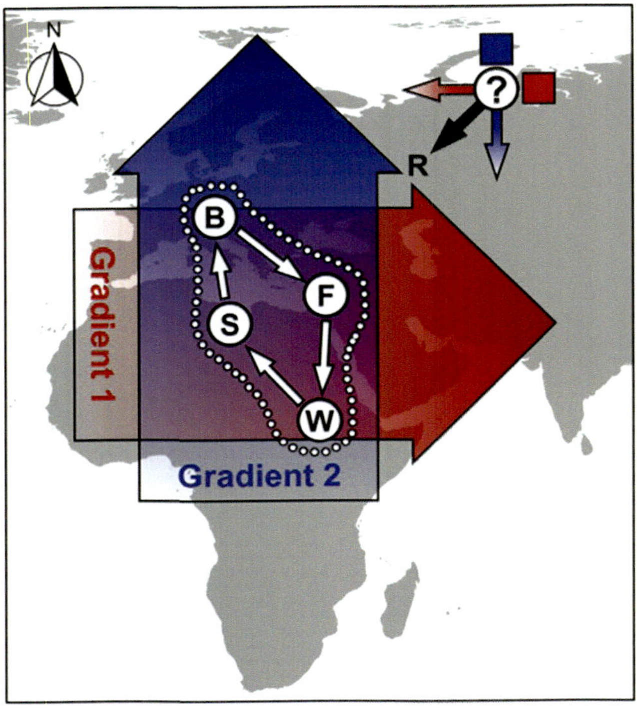


**Fig. 14.31** Potential use of a bi-coordinate map for navigational purposes. The map is formed by extrapolation from two gradients, e.g., magnetic inclination and intensity, learned during a year-round experience. The area within the dotted line represents the range of a hypothetical bird after fledging, including the breeding area (B), stopover site during fall migration (F), wintering range (W), and stopover site used during spring migration (S) as it returns to its breeding area. The two possible gradients increase from west to east (gradient 1, red) and from south to north (gradient 2, blue). A bird displaced to an unfamiliar site located northeast of its year-round range (? indicates an unfamiliar site) perceives changes in both magnetic inclination and intensity and determines that they exceed the maximum ranges of magnitude previously encountered in its year-round range. To determine its new location and the direction of movement needed to return to its year-round range, the bird would need to determine that, based on gradient 1 (e.g., magnetic intensity), its current location is farther east than most eastern familiar site, so it needs to move west. Based on gradient 2 (e.g., magnetic inclination), its current position is farther north than its breeding site so it needs to move south. Based on those two sources of information, the direction most likely to return the bird to its year-round range (R) would represent the average of those two sources of information, which would be to the southwest. (Figure from Kishkinev et al. 2021, © 2021 Elsevier Inc., used with permission)

the breeding area or their typical migration route, suggesting compensation for the west-to-east displacement using geomagnetic cues (perhaps in combination with solar cues). One possible cue used by the sparrows is geomagnetic declination (the angle formed by the difference between geographic and magnetic north). The White-crowned Sparrows could have used the stars to determine geographic north (i.e., rotation center of sky) and their inclination compass to determine magnetic north. Determining the exact angle of declination at high latitudes is likely difficult because of the steep geomagnetic field lines. However, recognizing a positive versus a negative declination would be easier. If so, the area near the point of zero declination could have been used as a longitudinal cue that, when passed, could have triggered a shift in orientation.

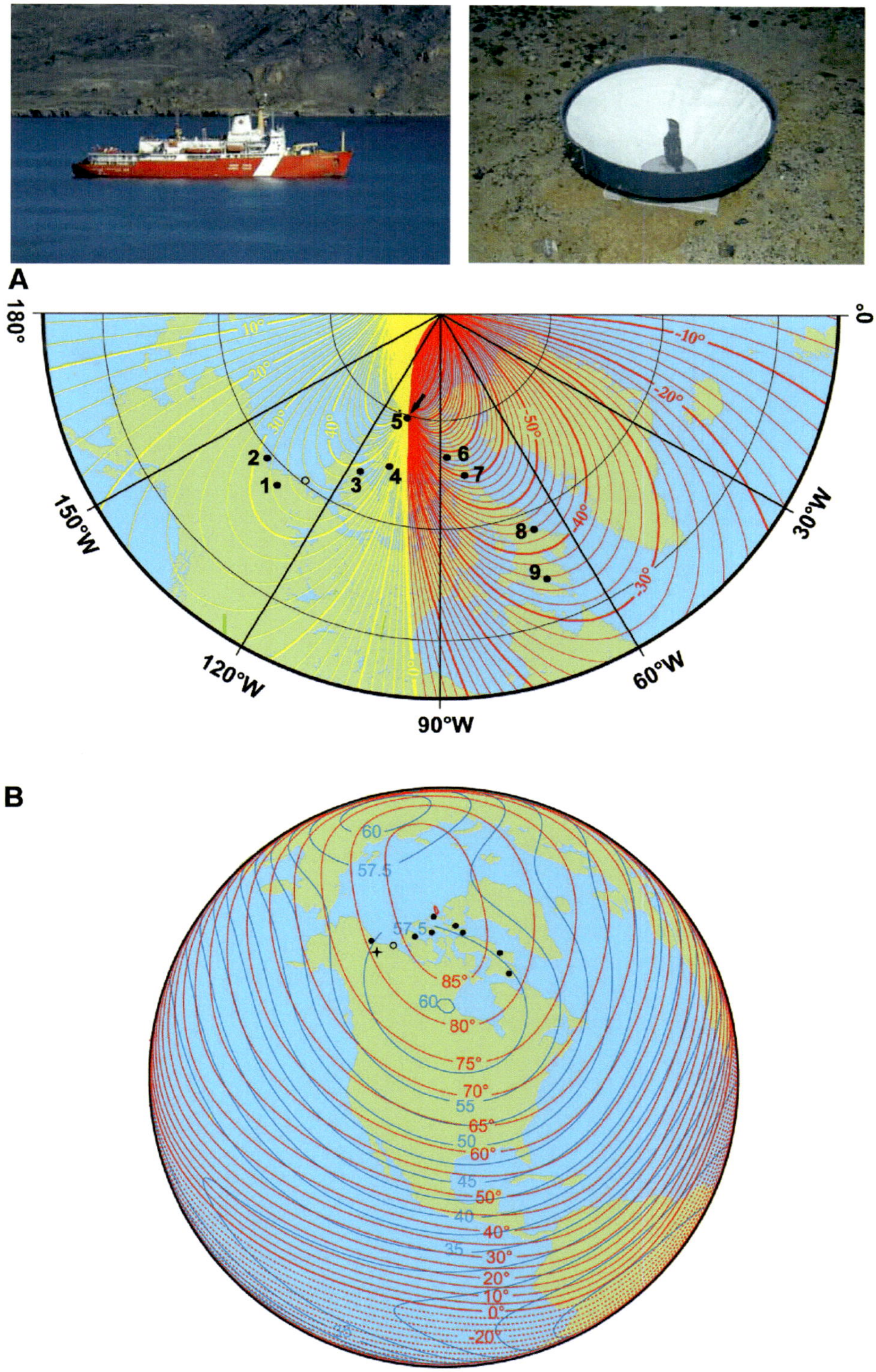


**Fig. 14.32** Sites where orientation cage experiments with White-crowned Sparrows (*Zonotrichia leucophrys*) were conducted in 1999 by Åkesson et al. (2005). Sparrows were captured in the breeding area (site 1), and their orientation was recorded in circular cages at sites 1–9. At one site (open circle), the birds were transported to the

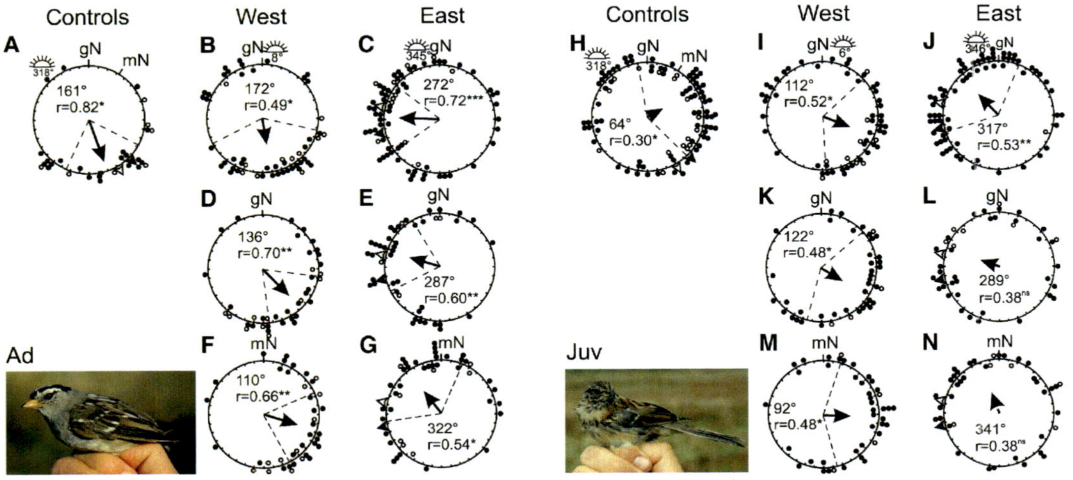


**Fig. 14.33** Orientation of displaced White-crowned Sparrows (*Zonotrichia leucophrys*). A-G, adults; H-N, juveniles. Control birds and those tested at a location west of the capture location on the breeding grounds typically oriented to the east or southeast in the general direction of their wintering area. Sparrows transported in an icebreaker east of the capture site tended to orient to the west or northwest, a direction that would lead them back to the breeding area or their typical migration route. (Figure from Åkesson et al. 2005; © 2005 Elsevier Ltd., used with permission)

## 14.8 Long- and Short-Range Navigation

Available evidence clearly indicates that many, if not most, birds have a variety of "compasses" that can be used for orientation and, although how they do it remains a question, are also able to determine their location. Thus, many birds can determine where they are and then select a course or direction that will take them to their goal. When selecting that direction, birds can potentially choose from among a variety of compass mechanisms, including celestial cues (e.g., star compass, sun compass, and polarized light) and magnetic cues (e.g., inclination compass). When moving closer to a goal, birds can also potentially use landmarks, olfactory cues, or even infrasounds. Given this array of potentially available cues, how do birds use or integrate the information from the various cues to find their way?

The availability of different compass mechanisms changes with time of day (e.g., sun

**Fig. 14.32** (continued) tundra, but no experiments were performed because of high speeds. In 1999, the geomagnetic North Pole was located at Ellef Ringnes Island (site 5, arrow in [A]). (**a**) The map showing magnetic declination. Yellow isolines indicate positive declination (deviations to the east of geographic north) and red negative (deviations to the west) values, respectively. (**b**) Map showing isoclinics (geomagnetic inclination) as red (broken) lines and isodynamics (total field intensity, μT) as blue (filled) lines. The star indicates the site of capture. (Figure from Åkesson et al. 2005; © 2005 Elsevier Ltd., used with permission) and star compasses) and weather conditions. When multiple cues are potentially available, birds may preferentially use one compass or may integrate the information from, for example, the inclination compass and the celestial compass. Birds may also use one compass cue to calibrate another cue (Chernetsov 2015; Box 14.4 Avian Compass Systems and Calibration). Such calibration may be critical for maintaining an accurate heading because cue availability changes with weather conditions, season, time of day, and latitude, and directional information between different compass systems sometimes diverge (e.g., because magnetic declination, the difference between magnetic and geographic north, varies with location). As a result, birds must calibrate

**Box 14.4 Avian Compass Systems and Calibration**

Migratory birds potentially have multiple ways to determine their direction, including, but not limited to, the sun/polarized light compass, stellar compass, and magnetic compass. However, availability of these cues may vary with time of day, season, and latitude. To compensate for this, birds must calibrate the different compasses relative to some reference on a regular basis (e.g., Muheim et al. 2006a, 2006b; Sjöberg and Muheim 2016). For some species of birds, Sjöberg and Muheim (2016) suggested that polarized light cues serve as the primary reference and are initially used to recalibrate the magnetic compass. Once stars become visible, birds then recalibrate their star (stellar) compass using the recalibrated magnetic compass. Of course, this proposed mechanism could only be sued by birds able to detect and use all of these cues—polarized light, magnetic fields, and a stellar compass. Some species of migratory birds may not possess all of these compass systems (Chernetsov 2015) and would need to recalibrate those they do have in a way different from that of birds that do possess them (Pakhomov and Chernetsov 2020).

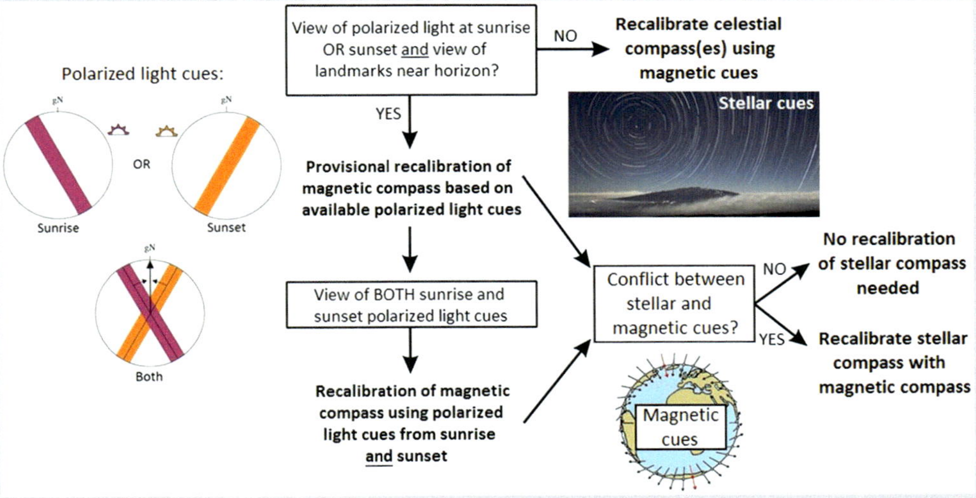


Possible "pathways" by which migrating birds calibrate various cues needed to accurately navigate. Sjöberg and Muheim (2016) proposed that polarized light cues are the key reference and are used to recalibrate the magnetic compass which in turn is used to recalibrate the stellar compass. (Figure modified from Sjöberg and Muheim 2016; 

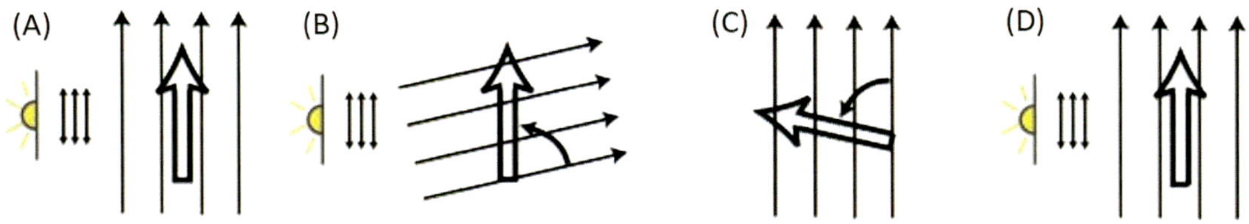


**Fig. 14.34** If the solar twilight azimuth is used to calibrate the magnetic compass on a daily basis, then (**a**) thrushes experiencing a normal magnetic field and observing sunset should orient north, (**b**) thrushes experiencing a magnetic field turned toward the east and observing sunset should determine that they need to orient 90° counterclockwise to head north, (**c**) thrushes who have experienced a magnetic field turned toward the east and observed sunset and determine that they need to orient 90° counterclockwise to head north should, if experiencing the normal magnetic field without an opportunity to view sunset, orient toward the west because they've "learned" to orient 90° counterclockwise relative to the magnetic field, and (**d**) on subsequent nights when exposed to both the normal magnetic field and sunset, they should return to their normal northerly migratory direction. (Figure from Cochran et al. 2004; © 2004 American Association for the Advancement of Science, used with permission)

the different compasses with respect to a common reference both before and during migration to avoid navigational errors.

As an example of preferential use of compass cues, clock-shift experiments indicate that homing pigeons use the sun as their preferred compass, using magnetic cues when the sun is not available (e.g., cloudy days; Walcott 2005). Experiments with night-migrating Gray-cheeked (*Catharus minimus*) and Swainson's (*C. ustulatus*) thrushes revealed that they depended primarily on sunset cues (the location of sunset or polarized light), using the solar twilight azimuth to calibrate their magnetic compass (Cochran et al. 2004). Temporarily caged thrushes exposed to an east-pointing magnetic field, but allowed to see the setting sun, apparently ignored the magnetic field and still oriented north. However, based on this experience, they learned that the direction "north" (as indicated by the setting sun) was rotated 90° counterclockwise from the magnetic field. When subsequently released after sunset (and, out of their cages, now experiencing the normal magnetic field) with radio-transmitters so they could be followed, they flew west, mistakenly still assuming that they needed to orient 90° counterclockwise from the magnetic field to head north (Fig. 14.34). Fortunately, the next evening, the radio-tagged thrushes used both cues (the normal magnetic field and observing sunset), determined which direction was north, and flew in the northerly direction that would take them to their breeding areas (Fig. 14.35). Interestingly, some experimental Gray-cheeked Thrushes, when released and perhaps uncertain about their proper heading because the magnetic and sunset cues conflicted, did not immediately resume migration (Fig. 14.35).

Migrating Swainson's and Gray-cheeked thrushes used one cue (sunset) to calibrate a second cue (magnetic north) and similar behavior has been reported by other birds. For example, migrating Savannah Sparrows (*Passerculus sandwichensis*) use polarized light cues to calibrate their magnetic (inclination) compass (Muheim et al. 2006a, 2006b, 2007). Other studies have revealed that some species of birds appear to use their magnetic compass to calibrate star and sunset cues (Sandberg et al. 2000). More generally, Sandberg et al. (2000) suggested that migrating birds may depend primarily on their magnetic (inclination) compass when faced with long-distance, nonstop flights across inhospitable terrain, such as deserts or large bodies of water. Migrating birds making shorter flights over more hospitable areas may preferentially use celestial cues calibrated using the magnetic field. Birds might exhibit such differences because, in contrast to celestial cues, the inclination compass is always available, regardless of time of day or weather conditions. Clearly, when traveling over

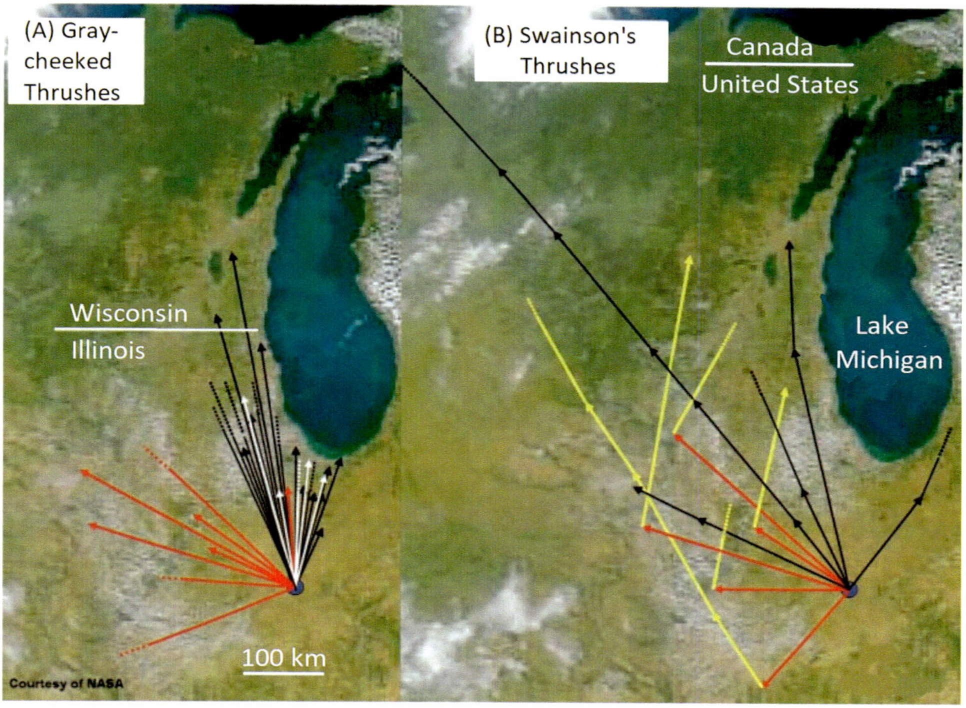


**Fig. 14.35** Tracks of free-flying (**a**) Gray-cheeked Thrushes (*Catharus minimus*) and (**b**) Swainson's Thrushes (*C. ustulatus*). Arrows indicate the direction and ground track of flights. Data are depicted differently in (**a**) and (**b**) because for Gray-cheeked Thrushes, experimental and control birds are different individuals, whereas for Swainson's Thrushes, the same experimental individuals were followed for at least two successive nocturnal migrations (because of the large spread in natural headings). Connected arrows show flights of the same individual during successive nights. Natural migratory flights are indicated by the black arrows; flights where thrushes were exposed to magnetic fields turned east before take-off (and not allowed to see sunset) are indicated by red arrows. For Swainson's Thrushes, flights of experimental birds on the second night after release are indicated by yellow arrows. For Gray-cheeked Thrushes, birds exposed to the "east" magnetic field that were released, did not migrate on the first night after release, did so 1–6 day later are indicated by white arrows. Broken lines indicate that birds were lost during tracking at the site where the broken lines start. (Figure from Cochran et al. 2004; © 2004 American Association for the Advancement of Science, used with permission)

large, inhospitable areas, a bird losing access to its primary compass because of changing weather conditions (e.g., increased cloud cover) and becoming disoriented may not survive.

In addition to preferentially using certain compass cues to calibrate others, for long-range navigation, migrating birds may use different compass cues at different times or different locations. For example, Bingman and Cheng (2005) suggested that, for long-distance navigation, birds might first determine their location relative to their "goal" in a mental, bi-coordinate map that could potentially be generated, as described previously, using magnetic cues (Fig. 14.36). Initial orientation toward that goal could involve multiple compass mechanisms, including the use of celestial (stars and sun) and magnetic (inclination compass) cues. A bird getting closer to its goal and entering more familiar areas may rely more on regional or local cues such as odors, major landmarks (such as mountains, rivers, coastlines, or lakes), or even infrasound. When nearing the

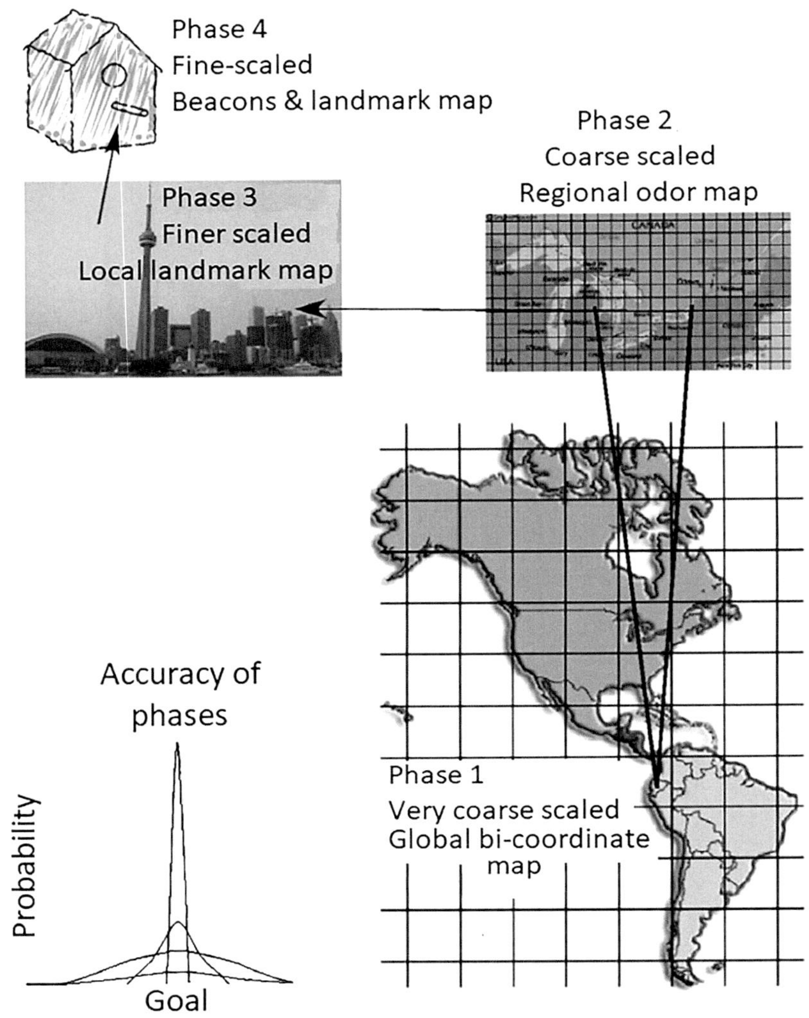


**Fig. 14.36** A global-navigation mechanism based on the use of multiple cues at different scalar resolutions (Bingman and Cheng 2005). As an example, an experienced migrant begins its spring migration in South America and heads in an approximate direction toward the breeding site in Ontario, Canada, exploiting low-resolution bi-coordinate, grid-based navigation relying on variations in the Earth's magnetic field (Phase 1). As the migrant approaches its breeding area, control of its navigational behavior switches to a higher-resolution, bi-coordinate, grid-based navigation based on variation in atmospheric odors (Phase 2). Getting closer to its breeding site, control of the migrant's navigational behavior again switches to even higher-resolution map-based navigation based on familiar visual landmarks (Phase 3). The migratory journey closes as the bird beacons in on its breeding territory of the previous year (Phase 4). The spatial resolution of the different navigational mechanisms can be interpreted using the inset to the lower left. The increasing accuracy of successive phases is indicated by narrower probability distributions around the target (goal) direction. Thus, the later phases are more likely to get the bird close to its goal than the earlier phases. (Figure from Bingman and Cheng 2005; © Dipartimento di Biologia, Università di Firenze, Italia, reprinted by permission of Informa UK Limited, trading as Taylor & Francis Group, www.tandfonline.com on behalf of Dipartimento di Biologia, Università di Firenze, Italia)

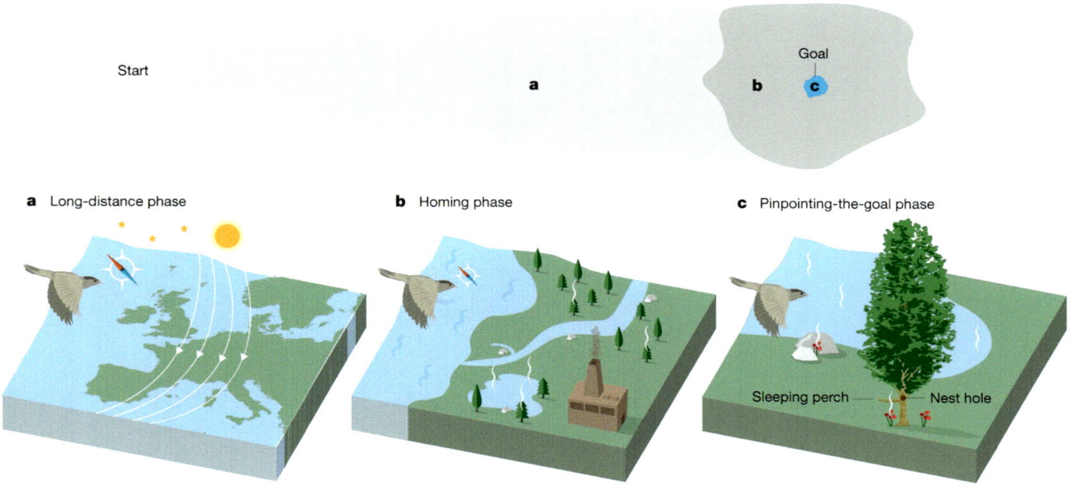


**Fig. 14.37** A succession of at least three stages is needed for long-distance navigation. (a) First is the long-distance stage likely involving the use of celestial and magnetic compass cues, although landmarks like coastlines or rivers could also be important. (b) Next, during the homing phase, compasses are usually still important, but regional map cues such as visual landmarks, olfactory gradients, or soundscapes can become important. (c) Finally, during the pinpointing-the-goal phase, cues like a specific trees or lakes are used to identify a specific location like a breeding territory. (Figure from Mouritsen 2018; © 2018 Macmillan Publishers Ltd., part of Springer Nature, used with permission)

location of its goal, a bird would likely use finer landscape features such as general topography, patterns generated by areas or patches of different habitat types, bodies of water, and, in some areas, man-made features such as roads. Finally, birds likely use specific features of or landmarks in their "goal" area, such as their breeding territory or wintering area, to determine that they have arrived at their "goal" or destination (Fig. 14.37).

This hypothetical "scenario" of how a migrating bird might use a variety of cues to reach its goal is certainly plausible. However, it remains to be determined how birds actually use the various compass mechanisms to successfully navigate and, specifically, how birds integrate and use the multiple sources of "compass information" available to them. Of course, using compass cues to reach a specific goal or destination is only possible if a bird knows its current position relative to that goal. Although some type of "magnetic bi-coordinate map" system seems at least possible, the relative importance of magnetic cues as well as celestial, olfactory, or other cues in allowing birds to determine their location remains unknown. In fact, it is possible that birds may not even use a bi-coordinate latitude/longitude map.

Until recently, most studies of avian navigation and orientation have been lab-based. Such studies have greatly improved, and will continue to improve, our understanding of compass cues used by and the navigational abilities of birds. However, improving technology is enhancing our ability to monitor the behavior and movements of birds in the wild and, by learning more about the routes actually taken by migrating birds and the cues potentially available to them at various locations, we will be able to better understand how birds are able to successfully navigate over, in some cases, incredible distances.

## 14.9 Noncompass Orientation

In some cases, birds may orient their flight relative to landmarks or topographic features. The results of numerous studies have revealed that Rock (homing) Pigeons (*Columba livia*) returning to home lofts after release at locations

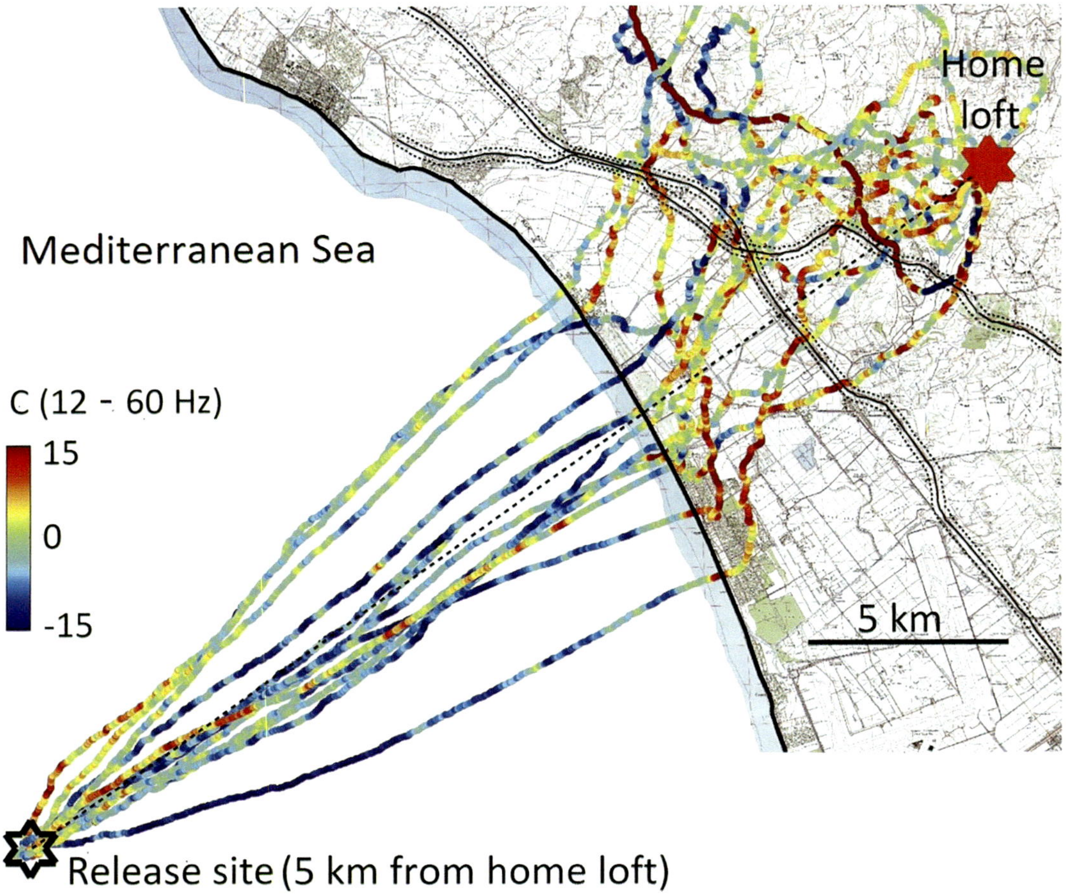


**Fig. 14.38** EEGs of pigeons flying from a release site over the Mediterranean Sea to their home loft several kilometers inland. Flight paths of the pigeons are color-coded to show EEG activity as they flew over the ocean and onto land toward their home loft, with yellow, orange, and red colors indicating increased EEG activity and recognition of landmarks. Pigeons generally exhibited minimal EEG activity over the featureless Mediterranean Sea, but increased EEG activity over familiar land areas as they moved closer to their home loft. Note the numerous orange and red sections of the flight paths as the pigeons observed familiar landmarks near their home loft. (Figure from Vyssotski et al. 2009; © 2009 Elsevier Ltd., used with permission)

several kilometers away use visual cues in areas where they have had previous flight experience (e.g., Gagliardo et al. 2020a, 2020b). Pigeons fitted with GPS loggers to indicate their locations and data-loggers to record their electroencephalograms (EEG) exhibited increased EEG activity as they approached their home loft, indicating visual perception and identification of familiar landmarks (Vyssotski et al. 2009; Fig. 14.38). On a smaller scale, several species of birds use landmarks to locate nest sites (Slagsvold and Wiebe 2021), foraging patches or locations (Shaw et al. 2019), the boundaries of their territories (Heap et al. 2012), or food caches. For example, Clark's Nutcrackers (*Nucifraga columbiana*) and other food-storing birds use multiple landmarks to navigate (e.g., LaDage et al. 2009), and Rufous Hummingbirds (*Selasphorus rufus*) were able to use visual landmarks to remember how to return to sources of nectar or sugar solutions (Pritchard et al. 2015, 2016). In addition, some brood-parasitic birds appear to rely on spatial memory to remember the location of multiple host nests during a breeding season (Lois-Milevicich et al. 2021).

## 14.10 Navigation and the Hippocampus

The avian hippocampal formation (HF), located in the telencephalon, is known to be important in spatial cognition so may play some role in avian navigation. In that context, some investigators have suggested that the avian HF may be involved in learning and then recalling familiar, landmark/landscape-based maps (Bingman et al. 2005; Herold et al. 2015). In support of this hypothesis, investigators have reported migratory species or populations that have relatively larger HFs than nonmigratory species or populations (Healy et al. 1996; Cristol et al. 2003; Pravosudov et al. 2006), and migratory species or populations where HFs have more or more densely packed neurons (Cristol et al. 2003, Pravosudov et al. 2006). In addition, old pigeons have more neurons in the HF (triangular region) than young pigeons, e.g., a 14-year-old pigeon had 1.9 million cells whereas one-month-old pigeons had an average of 740,000 nerve cells; this increase in number of nerve cells with age might be a result of greater navigational experience and, possibly, enhanced spatial memory (Meskenaite et al. 2016). Overall, the results of such studies suggest that the HF plays a role in the navigational abilities of birds.

The HF is known to be important for local landmark/landscape navigation by homing pigeons and likely for other species of birds as well (Bingman and MacDougall-Shackleton 2017). Gagliardo et al. (2014) determined that hippocampal-lesioned homing pigeons were able to establish their location relative to their home area after release from a location 40–50 km away and begin flying toward home, but were unable to locate their home loft. These results suggest that pigeons obtain navigational information upon release at distance locations, e.g., atmospheric odors, sun compass, and earth's magnetic field, but switch to the use of familiar landmarks as they approach their home loft (Gagliardo et al. 2015). Damage to the HF left pigeons unable to identify familiar landmarks near home lofts and, therefore, unable to locate their home loft. Similar results concerning the importance of the HF in forming "map-like memory representations of visual landmarks/landscape features" and allowing pigeons to locate home lofts in familiar locations have also been reported in other studies (Gagliardo et al. 2020a, 2020b, 2021). Migratory species of birds may also "rely on the hippocampus when navigating a familiar, local space, whether in proximity to a breeding site, stopover site or over-wintering site" (Bingman and MacDougall-Shackleton 2017).

Although important for navigation in familiar areas, there is no evidence that the HF is important for navigation at distances hundreds of kilometers from targeted destinations (i.e., mid-spatial resolution, Bingman and MacDougall-Shackleton 2017; Fig. 14.39). Few investigators have examined the possible role of the HF in long-range navigation (i.e., many hundreds to thousands of kilometers). However, magnetosensitive cells have been reported in the HF of Zebra Finches (*Taeniopygia guttata*) and homing pigeons (Vargas et al. 2006; Wu and Dickman 2011; Keary and Bischof 2012). In addition, Bingman et al. (2021) found that HF lesions prevented homing pigeons from detecting variation in the intensity of magnetic fields, although they were still able to detect variation in magnetic inclination. Based on those results, Bingman et al. (2021) suggested that the HF may allow pigeons to learn a map-like representation of landscape features based on landscape-level variation in magnetic intensity. The extent to which the HF plays a role in the long-range navigation of migrating birds, possibly based on magnetic cues, remains to be determined.

## 14.11 Topographical Features and Landmarks

During migration, migrants may orient their flight using coastlines, mountain ridges, or rivers (Fig. 14.40). Tracking studies have revealed that

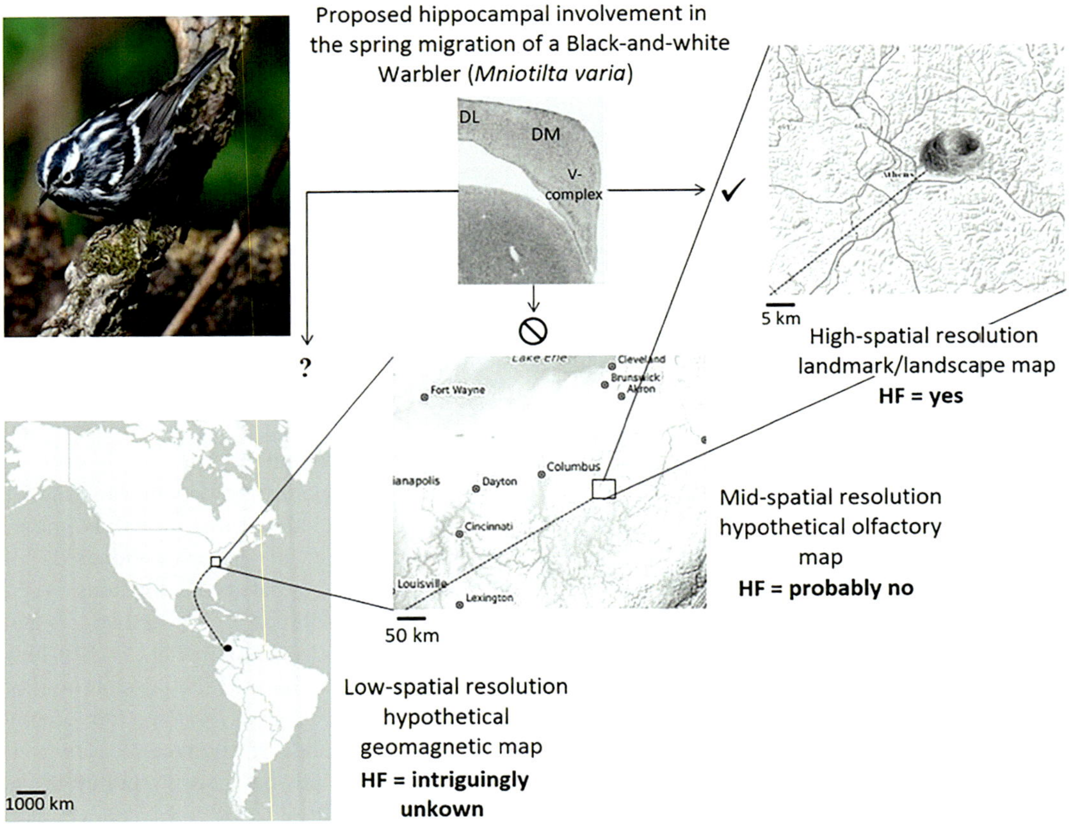


**Fig. 14.39** Illustration of how the hippocampal formation may aid in navigation by a Black-and-white Warbler (*Mniotilta varia*) migrating from its wintering area in Colombia to its breeding area in the United States (Ohio) using three hypothetical maps at different spatial scales as proposed by Bingman and Cheng (2005). For long-range navigation, migratory birds may rely more on geomagnetic cues, possibly switching the olfactory cues and mid-range distances (hundreds of kilometers), then to a hippocampus landmark/landscape map at shorter distances. Little is known about the possible relevance of the hippocampal formation when a migrant is navigating using a hypothetical long-range, low-resolution spatial map perhaps based on the variation in the earth's magnetic field. However, the hippocampus likely provides birds with a high-spatial resolution landmark/landscape map in familiar areas. Areas of the hippocampus formation include the V-complex and the dorsomedial (DM) and dorsolateral (DL) subdivisions. (Photo of a Black-and-white Warbler by Andy Reago and Chrissy McClarren, Wikipedia, CC BY 2.0, https://creativecommons.org/licenses/by/2.0/; Figure from Bingman and MacDougall-Shackleton 2017; © 2017 Springer-Verlag Berlin Heidelberg, used with permission)

migrating Whooping Cranes (*Grus americana*) appear to retain spatial memories of landscape features that aid in navigation (Kuyt 1992; Mueller et al. 2013). Kok et al. (2020) tracked the migration of a female Red Knot (*Calidris canutus*; referred to by the authors as Paula) across Greenland using a solar-powered satellite transmitter and concluded that, based on routes taken over multiple migrations and sudden course changes during migration, "... a broad overview of the landscape from the high altitude at which Paula may have been flying, together with the variable topography of the Icecap (reaching levels up to 3300 m asl), may contribute distinct visual

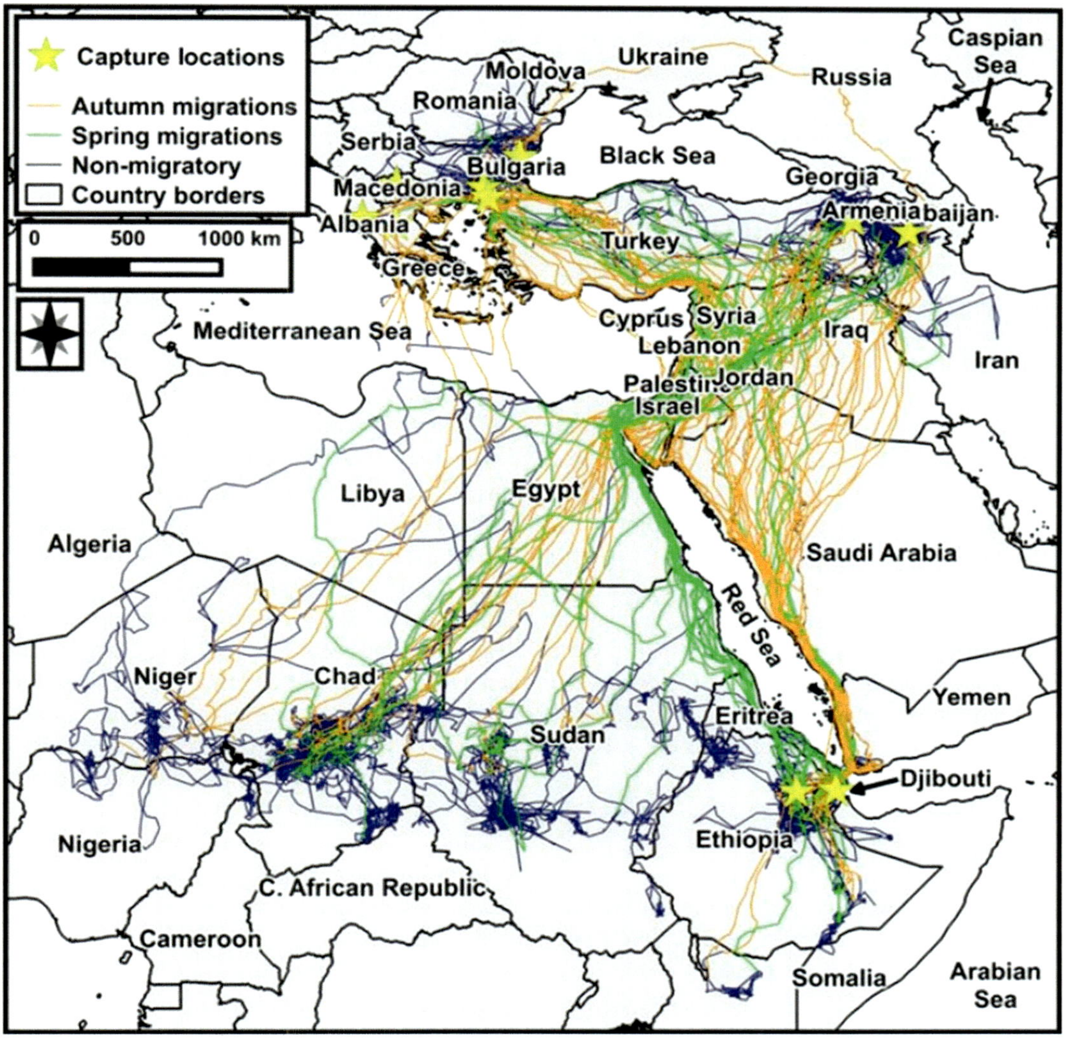


**Fig. 14.40** Migration routes of 45 Egyptian Vultures determined by satellite tracking. Egyptian Vultures are diurnal soaring migrants and birds from Eastern Europe and the Middle East leave their breeding areas and migrate to northern and east Africa. Note that migration routes of many of the vultures concentrated along the eastern Mediterranean and the eastern (during autumn migration) and western (during spring migration) coasts of the Red Sea. (Figure from Buechley et al. 2018; © 2018 The Authors. Published by John Wiley and Sons, used with permission)

cues to aid orientation." In Europe, many birds migrating south that encounter the Alps turn southwest and fly parallel to the mountains, with fewer flying through passes between the mountains (Liechti et al. 1996; Bruderer and Liechti 1999). Similar behavior has been reported for birds migrating along the northern Appalachian Mountains (Williams et al. 2001). Some nocturnal migrants in eastern New York appear to travel along routes parallel to the Hudson River (Bingman et al. 1982), and diurnally migrating raptors are well known for migrating along prominent topographic features such as coastlines and mountain ridges (Bildstein 2006). Of course, migrating birds that might appear to be using geographical features for orientation purposes may be using them for entirely different reasons. For example, migrating waterfowl may fly along rivers not for orientation purposes, but because they represent roosting or foraging habitat (O'Neal et al. 2015), and soaring birds may migrate along mountain ranges because updrafts facilitate soaring flight.

Other studies have revealed that some species do not use familiar landmarks when migrating. For example, flight paths of individual Ospreys (*Pandion haliaetus*) making repeated migrations between Europe and west Africa over several years were often 120 to 305 km apart (east-west separation), with a maximum of 1400 km for one male (Alerstam et al. 2006). Similar results have been reported for other species of birds (Berthold et al. 2004, Vardanis et al. 2011, Stanley et al. 2014, López-López et al. 2014; Fig. 14.41), with the use of different routes potentially due to variation in meteorological conditions, individual energetic condition, and, for species that migrate using a fly-and-forage strategy (Strandberg and Alerstam 2007), variation in food availability (Stanley et al. 2014; López-López et al. 2014).

## 14.12 Olfactory Navigation

The possible importance of olfactory cues in bird navigation was first reported by Hans Wallraff, with pigeons raised in enclosures with glass screens (thus able to view their surroundings, but not exposed to the outside air) less successful at orienting homeward than pigeons with shielded views (but exposed to the outside air) (Wallraff 1966, 1979). Similarly, Papi et al. (1971) found that pigeons with severed olfactory nerves and released at an unfamiliar location never returned to their home loft, whereas pigeons with intact olfactory nerves did return. Since then, studies of homing pigeons have revealed that experienced pigeons use an "olfactory map" (based on the spatial distribution of natural odor sources) to determine their position and navigate (Wallraff 2004) (Fig. 14.42). An olfactory map like that envisioned by Wallraff (2004) was created by Zannoni et al. (2020) by identifying and measuring volatile organic compounds at a home loft in Italy, the surrounding forests, and from an aircraft at 180 m above ground. Measurements revealed clear regional, horizontal, and vertical gradients in volatile organic compounds that could be the basis of an olfactory map suitable for homing pigeons. Importantly, experiments revealed that pigeons released at three different locations within the olfactory-map area were able to successfully return to the home loft.

The results of studies of other species of birds have also provided evidence of olfactory navigation. Holland et al. (2009) treated some individuals with zinc sulfate to produce anosmia and found that olfaction may play some role in determining the direction of migration by adult Gray Catbirds (*Dumetella carolinensis*), but how catbirds might use olfactory information relative to other cues remains to be determined. Wikelski et al. (2015) found that Lesser Black-backed Gulls (*Larus fuscus*) whose olfactory nerves were sectioned were less likely to orient toward and return to typical migration pathways after being longitudinally displaced by 1080 km than control gulls. Studies of other species have revealed no evidence that olfaction cues provide birds with information about their geographic position (e.g., Eurasian Reed Warblers, *Acrocephalus scirpaceus*, Kishkinev et al. 2020). Given the few species studied, the conflicting results of studies with the most-studied species (homing pigeons), and the improbability of navigation over distances beyond a few hundred kilometers using airborne olfactory cues (Bingman and Cheng 2005), long-range navigation using olfactory "maps" currently seems unlikely for most, but not all, birds.

Possible exceptions, however, include the tube-nosed seabirds (order Procellariiformes) and penguins (order Sphenisciformes). Tube-nosed seabirds have large olfactory bulbs (Zelenitsky et al. 2011) and a well-developed sense of smell. Studies have revealed that many species, including several storm-petrels (Nevitt et al. 1995) and shearwaters (Nevitt and Hunt 1996; Dell'Ariccia et al. 2014), Blue Petrels (*Halobaena caerulea*; Cunningham et al. 2003), and Antarctic Prions (*Pachyptila desolata*; Nevitt and Bonadonna 2005), can detect dimethyl

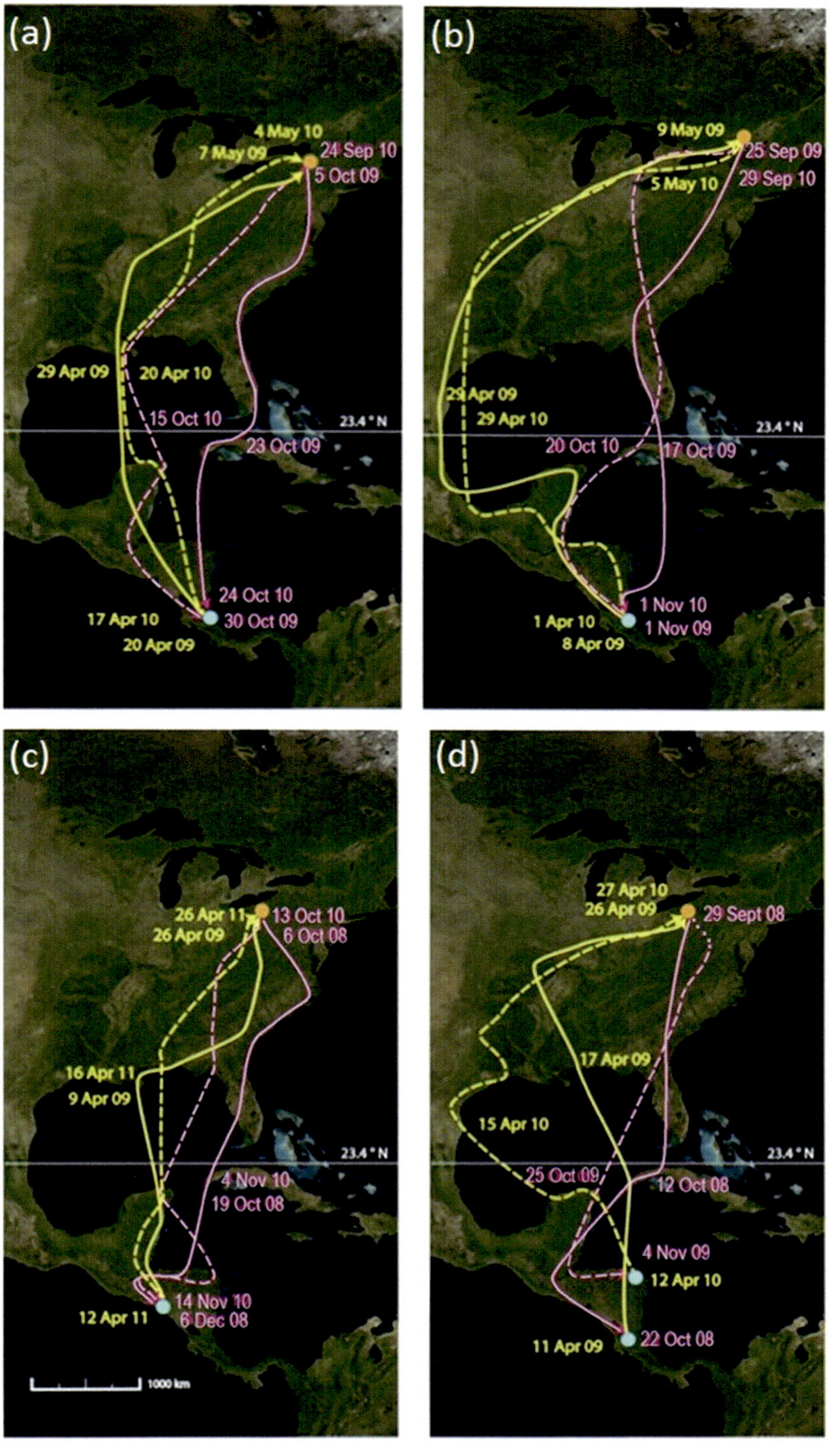


**Fig. 14.41** Migration routes of four different Wood Thrushes (*Hylocichla mustelina*) tracked in consecutive years using geolocators. Note the different routes taken in different years by each of the Wood Thrushes during both spring migration (yellow) and fall migration (pink). Solid and dashed lines indicate the different years/Orange circles = breeding sites, white circles = winter sites. Dotted lines in d indicate where the migration route was

sulfide (DMS) at very low concentrations and use it as a cue to identify foraging areas and for navigational purposes. Although penguins have relatively small olfactory bulbs (Lu et al. 2016), three species, including African Penguins (*Spheniscus demersus*; Cunningham et al. 2008), Chinstrap Penguins (*Pygoscelis antarctica*; Amo et al. 2013), and King Penguins (*Aptenodytes patagonicus*; Cunningham et al. 2016), have also demonstrated an attraction to DMS.

In the marine environment, DMS is associated with the abundance of phytoplankton that is in turn associated with predictable oceanic features (Nevitt 2000; Fig. 14.43). Marine phytoplankton produce dimethylsulfonioproprionate (DMSP) which is released into the water column via algal senescence and grazing by zooplankton (Dacey and Wakeham 1986). DMSP is then converted to DMS by algal and bacterial enzymes (DMSP lyases) (Kwint et al. 1996; Archer et al. 2003). DMS released into the atmosphere can then be used as an olfactory and navigational cue by a variety of seabirds (Abolaffio et al. 2018; Box 14.5 Olfactory Navigation by Seabirds).

## 14.13 Possible Use of Infrasounds

Infrasounds are sound waves with frequencies below about 20 Hz, the lower threshold of sound detection by humans. Higher-frequency sounds tend to be attenuated by the atmosphere, but infrasounds can travel many thousands of kilometers. The use of an "infrasound map" for navigation would require persistent sources of infrasounds that birds could hear. Persistent infrasounds are generated by mountain winds and ocean waves (Hagstrum 2000), and laboratory tests indicate that some birds, including pigeons (Kreithen and Quine 1979), Domestic Chickens (Hill et al. 2014), Helmeted Guineafowl

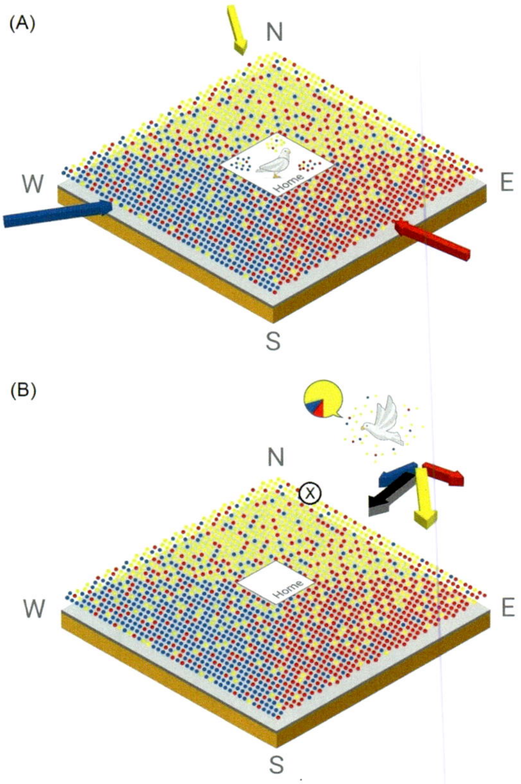


**Fig. 14.42** Over a period of months, fledgling pigeons learn to associate odors with certain wind directions and times of day. (**a**) Colored dots represent regional scale odor gradients and the pigeon's olfactory map. (**b**) Pigeons taken to an unfamiliar location and released (at the location indicated by the X) are exposed to the local odors. Pigeons compare the odors at the release site to those experienced at the home loft and orient accordingly homeward. For example, a pigeon at location X would orient generally toward the south because the yellow odor is more prevalent at the release point than at the home loft and the pigeon recalls that that odor comes from the north. By comparing all of the odors (indicated by the yellow, blue, and red dots) at the release site to those at the home loft, a course oriented toward the home loft (shown as the black arrow that results from evaluating the relative strength of the yellow, blue, and red odors). (Figure from Zannoni et al. 2020; open-access article licensed under a Creative Commons Attribution 4.0 International License, https://creativecommons.org/licenses/by/4.0/)

**Fig. 14.41** (continued) unknown due to poor-quality light data or failure of the geolocator battery. (Figure from Stanley et al. 2012; © 2012 Stanley et al., open-access article distributed under the terms of the Creative Commons Attribution License, https://creativecommons.org/licenses/by/4.0/)

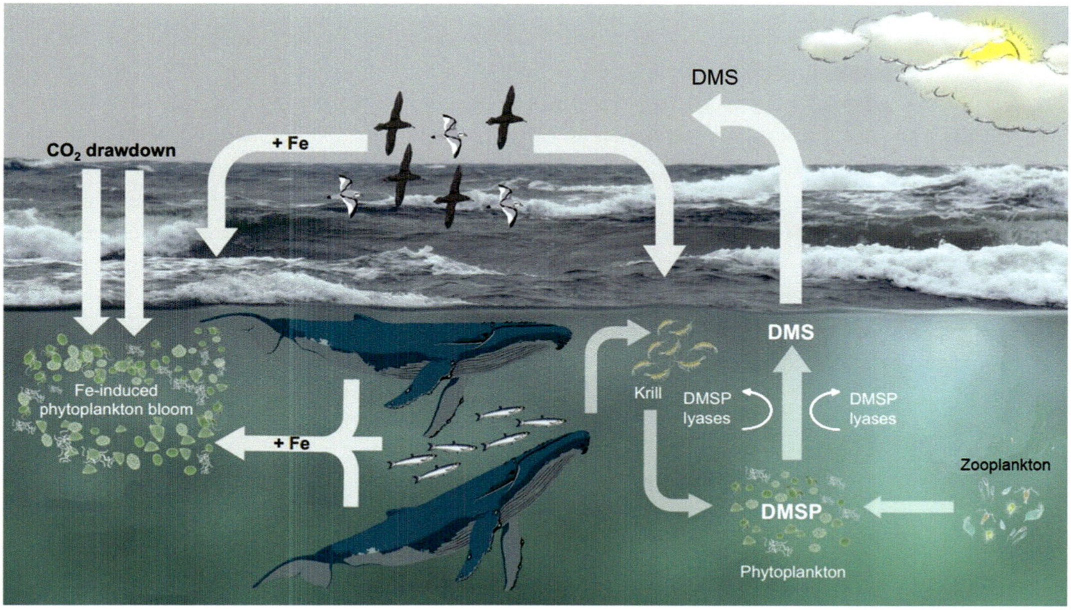


**Fig. 14.43** Dimethyl sulfide emissions increase over ocean waters where phytoplanktons are plentiful (e.g., above upwellings that bring nutrients into the water column) and are being actively grazed by zooplankton. Dimethylsulfonioproprionate (DMSP) is released into the water by phytoplanktons that die or are grazed upon by zooplankton and krill. DMSP is then converted to dimethylsulfide (DMS) by algal and bacterial enzymes (DMSP lyases). Such emissions are often predictably present at certain locations and these locations may provide tube-nosed seabirds with navigational guideposts. Seabirds and other ocean predators attracted to the area prey on krill, fish, and other prey, and release iron (Fe) into the ocean via their feces. Iron is a limiting factor in most ocean waters so the addition of iron stimulates increased production of phytoplankton. As phytoplankton photosynthesize, they use carbon dioxide in the water, resulting in the diffusion of atmospheric carbon dioxide into the ocean. (Figure from Savoca 2018; © 2018 Springer International Publishing AG, part of Springer Nature, used with permission)

**Box 14.5 Olfactory Navigation by Seabirds**

The open ocean would seem to be a featureless environment, providing few if any cues to seabirds. However, many seabirds travel considerable distances between foraging locations and between foraging areas and breeding colonies, but must somehow be able reach their specific destinations. One hypothesis to explain the navigational abilities of seabirds is that they use magnetic cues. However, experiments designed to test this hypothesis have revealed no supporting evidence (e.g., Bonadonna et al. 2003, 2005; Gagliardo et al. 2013). Another hypothesis is that seabirds might navigate by using olfactory cues (Benhamou et al. 2003), and experiments designed to test this hypothesis have provided evidence to support this hypothesis. For example, Gagliardo et al. (2013) conducted an experiment with Cory's Shearwaters (*Calonectris borealis*), capturing several at their breeding colony and then releasing them about 800 km away. The shearwaters were dividing into three treatment groups: control (no manipulation), magnetic treatment (magnets attached to the back of their heads to deprive them of any geomagnetic information), and anosmic treatment (deprived of their sense of smell by washing their olfactory mucosa with a weak solution of zinc sulfate). Control and treatment

(continued)

**Box 14.5** (continued)

birds took relatively direct paths back to the colony, returning in just a few days. However, the anosmic birds were unable to find their way back to the colony, wandering for many kilometers across the open ocean. Similarly, Padget et al. (2017) found that anosmic Scopoli's Shearwaters (*Calonectris borealis*) were not able to accurately orient toward their breeding colonies from distances beyond 40 km. Such results provide strong support for the hypothesis that at least some seabirds can navigate over long distances using olfactory cues. Cory's Shearwaters and other seabirds may acquire an "olfactory map" as they forage in different areas and then use that "map" to navigate. One possible olfactory cue used by seabirds is a substance called dimethyl sulfide, a substance released by phytoplankton and released in greater amounts in areas of the ocean that are more productive. Because some areas remain productive areas for long periods (e.g., in the vicinity of upwellings), they may generate an "olfactory landscape" that seabirds can use to navigate.

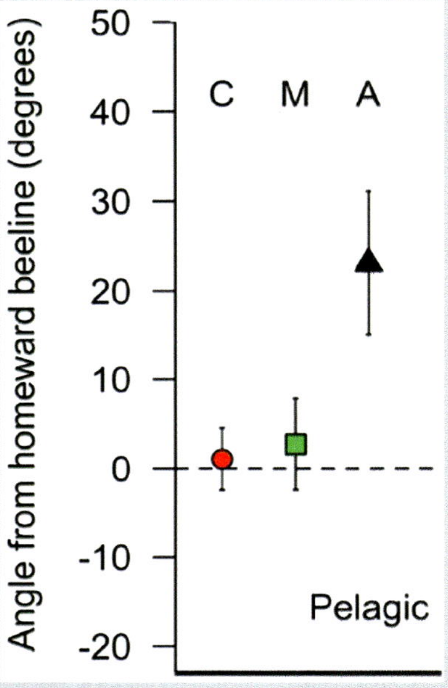


In contrast to control (C) and magnetically disrupted (M) Scopoli's Shearwaters, anosmic Scopoli's Shearwaters (A) were not able to accurately orient toward their breeding colony. The direction to the breeding colony was 0°. (Figure from Padget et al. 2017; open-access article distributed under the terms of the Creative Commons CC BY license, https://creativecommons.org/licenses/by/4.0/)

(continued)

**Box 14.5** (continued)

Cory's Shearwater (Wikipedia, CC0 Public Domain)

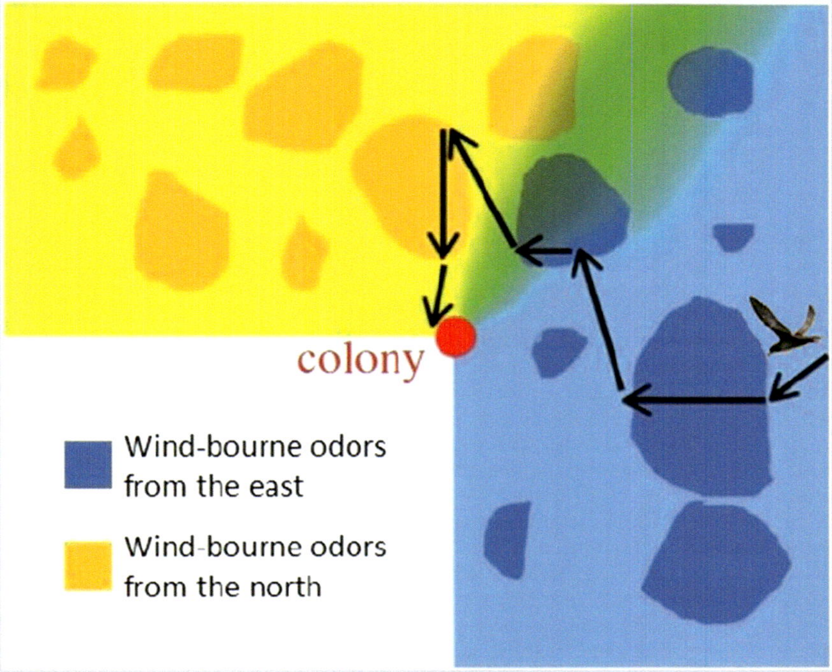


(continued)

**Box 14.5** (continued)

Possible mechanism of olfactory navigation by pelagic birds. Seabirds may associate odors with wind directions, with, for example, some odors (e.g., dimethyl sulfide originating in more productive areas of the ocean) consistently coming from an easterly direction (blue) and others from a northerly direction (yellow). In some areas, depending on wind conditions, these odors may mix (green area). In addition, because of atmospheric turbulence and variation in the concentration of odors in the air, the odors are not detectable everywhere (lighter blue and yellow areas). A bird foraging east of its breeding colony and wanting to return to the colony will fly until detecting the odor then head west knowing that the odor is being blown in that direction. If the bird flies into an area where the odor cannot be detected, it changes direction to try and relocate the odor. If the odor is again detected, the bird turns west, then loses the odor, but then detects the odor coming from the north and turns south. This takes the seabird closer to the breeding colony where they can then potentially use other sources of information such as landmarks, flight paths of other birds, and odors emanating from the colony to reach its destination. (Figure from Reynolds et al. 2015; © 2015 The Authors. Published by the Royal Society, used with permission)

(*Numida meleagris*; Theurich et al. 1984), and Indian Peafowl (*Pavo cristatus*; Heffner et al. 2020) can detect infrasounds (Box 14.6 Using Infrasounds to Avoid Tornadoes?). In addition, although untested, Patrick et al. (2021) suggested that seabirds could use sources of infrasound for navigation, with those sources potentially acting as beacons, landmarks, or gradients (Fig. 14.44).

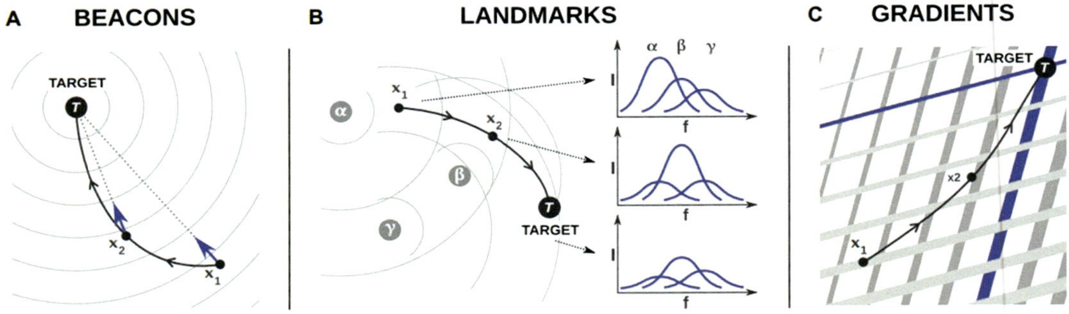


**Fig. 14.44** Possible ways that seabirds (and perhaps other birds) could use infrasound to navigate. In this figure, the flight path of a hypothetical bird (in black) is sampled at locations $x_1$ and $x_2$, and at its "target" at times 1, 2, and 3. In **a**, **b**, and **c**, blue elements indicate the relevant information used by the bird. (**a**) If infrasound serves as a beacon, the target location, e.g., the bird's colony is the infrasound source that the bird detects and moves toward. With this hypothesis, the bird either directly perceives the direction of the beacon (indicated by the blue arrows) or at least is able to determine if is approaching or moving away from the target. (**b**) As an example of the landmark hypothesis, there are three sources of infrasound (α, β, and γ, e.g., three islands). Infrasound from the three sources creates a unique acoustic landmark, i.e., an infrasound signature at the target (power spectrum in blue, I is the sound intensity, and f the frequency); the bird compares the power spectra from the infrasound sources to reassess its location. (**c**) Large-scale infrasonic amplitude gradient fields could provide a functional coordinate reference system. In this example, there are two gradient fields coming from two different sources (gradient isolines are represented by gray parallel lines, with directions orthogonal to the gradient directions, and widths proportional to the signal intensities). The bird could use the gradient fields as a coordinate system to navigate toward its target characterized by specific signal intensities (blue arrows). (Figure from Patrick et al. 2021, © 2021 Patrick, Assink, Basille, Clusella-Trullas, Clay, den Ouden, Joo, Zeyl, Benhamou, Christensen-Dalsgaard, Evers, Fayet, Köppl, Malkemper, Martín López, Padget, Phillips, Prior, Smets and van Loon. This is an open-access article distributed under the terms of the Creative Commons Attribution License (CC BY), https://creativecommons.org/licenses/by/4.0/)

**Box 14.6 Using Infrasounds to Avoid Tornadoes?**

During a study of Golden-winged Warblers (*Vermivora chrysoptera*) fit with geolocators, Streby et al. (2015) documented an unusual facultative migration by males leaving their breeding areas to avoid a severe tornadic storm. After migration about 5000 km from wintering areas in eastern Colombia, five male Golden-winged Warblers arrived in their breeding territories in eastern Tennessee between 13 and 27 April 2014. From 27 to 30 April, severe storms moved east through the central and eastern United States. During this period, there were 84 confirmed tornadoes, 35 human fatalities, and >1 billion dollars (USD) in property damage. On 26 and 17 April, just a day or two before the storm's arrival in their breeding area, the five male warblers left their territories. All took different routes, but all arrived in Florida by 29 April, with four located in the Florida panhandle ~700 km from their breeding area and one farther to the south. All five males returned to their breeding territories by 1 or 2 May, and were recaptured during the period from 3 to 9 May to retrieve the geolocators.

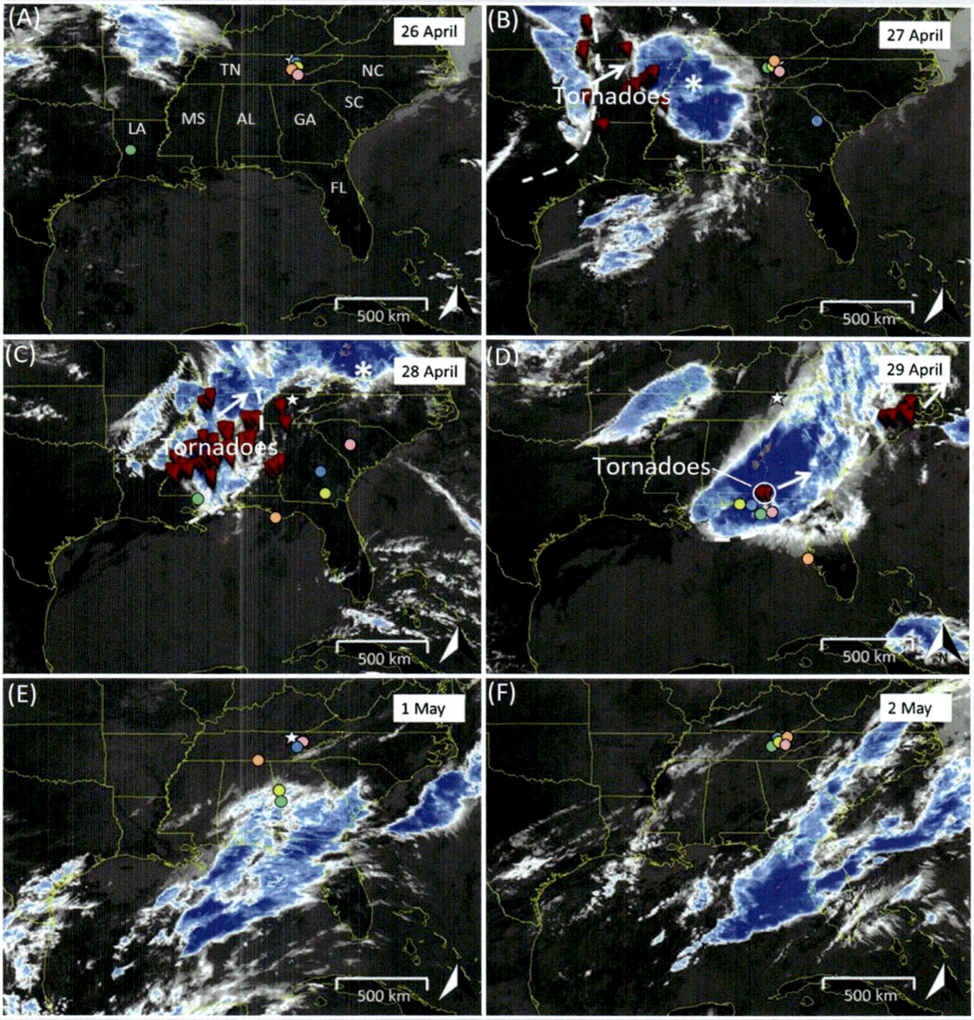


(continued)

**Box 14.6** (continued)

Movements of five male Golden-winged Warblers (colored circles) monitored with geolocators after arrival in their breeding area in late April 2014. (**a**) On 26 April, four males were located in their breeding areas in eastern Tennessee and the fifth male was in Louisiana. (**b**) On 27 April, a severe tornadic storm was located west of the males' breeding area in Arkansas, but moving toward their breeding area. Four males were still in their breeding area, but one male had moved southeast into eastern Georgia. (**c**) All five males left the breeding area, with four moving south into South Carolina, Georgia, or Florida, and the fifth male moving southwest to Louisiana. (**d**) By the next day, all five males were in Florida, with four just south of tornadic activity and the fifth further south in Florida. (**e**) Two days later, two males had returned to the breeding area and the other three were moving north toward the breeding area. (**f**) All five males had returned to the breeding area. The study area (where the birds breed) is indicated by the white star. Red tornado shapes indicate the locations of tornadoes, and white arrows indicate the mean direction of tornado tracks. Daily weather images are from 12:00 EST, and tornadoes are those that touched down that day. Asterisk in **b** indicates a low-intensity system that passed north of the study area on April 27–28, 2014. TN, Tennessee; *GA*, Georgia; NC, North Carolina; SC, South Carolina; FL, Florida; AL, Alabama; MS, Mississippi; LA, Louisiana. (Figure modified from Streby et al. 2015; © 2015 Elsevier Ltd. All rights reserved, used with permission)

Streby et al. (2015) found no evidence that environmental cues such as atmospheric pressure, temperature, wind speed and direction, cloud cover, or precipitation might have triggered the departure of the male Golden-winged Warblers from the breeding territories. The warblers left their territories 1–2 days before the tornadic storm arrived in the area in eastern Tennessee where their territories were located and when the weather system was still 400–900 km to the west. As such, Streby et al. (2015) concluded that the perception of infrasound generated by the storm (e.g., see Schecter et al. 2008) was the most parsimonious explanation for the warblers' behavior. An ability to use infrasound to avoid potentially life-threatening weather conditions has obvious advantages in terms of survival and fitness. In fact, Quine and Kreithen (1981) found the pigeons were able to detect changes in intensity and Doppler shifts in infrasound, suggesting that they would also be able to use infrasound to determine the movements of severe weather from considerable distances.

More generally, small songbirds, with relatively high thresholds for hearing low frequencies (Okanoya and Dooling 1987), would seem unlikely to be able to detect infrasounds, whereas other species, like some species of owls with much lower thresholds for hearing low frequencies (Dyson et al. 1998), might be more sensitive to infrasounds. Investigators have determined that some species of birds are apparently unable to detect infrasounds (unpublished remark in Theurich et al. 1984, Beason 2004, Heffner et al. 2016, Hill 2017, Strawn and Hill 2020), but, for most bird species, the ability to detect infrasounds has not been tested. In addition to questions concerning how many species can detect ultrasounds, Bingman and Cheng (2005) also point out that the assumed primary sources of infrasounds for birds, mountain winds, and ocean waves, are not point sources. In other words, a bird capable of hearing infrasounds and migrating, for example, through the central United States would potentially hear those sounds being emitted from oceans on the east, west, and Gulf coasts and from the entire length of the Rocky and Appalachian Mountain chains. Although it might seem unlikely that birds could generate an "infrasound map" using similar sounds simultaneously emanating from such widely distributed sources, Bedard (2021) suggested that sources along north-south axes such as mountain ranges and coastlines could potentially help birds identify navigation corridors. Bedard (2021) also noted that large waterfalls, like Niagara Falls, represent point sources of infrasounds, and suggested that they could potentially be used as "beacons" by migrating birds able to detect infrasounds.

## References


Able KP (2000) The concepts and terminology of bird navigation. J Avian Biol 32:174–183

Able KP, Able MA (1990) Ontogeny of migratory orientation in the Savannah Sparrow, *Passerculus sandwichensis*: calibration of the magnetic compass. Anim Behav 39:905–913

Able KP, Able MA (1996) The flexible migratory orientation system of the Savannah Sparrow (*Passerculus sandwichensis*). J Exp Biol 199:3–8

Abolaffio M, Reynolds AM, Cecere JG, Paiva VH, Focardi S (2018) Olfactory-cued navigation in shearwaters: linking movement patterns to mechanisms. Sci Rep 8:11590

Åkesson S, Alerstam T (1998) Oceanic navigation: are there any feasible geomagnetic bi-coordinate combinations for albatrosses? J Avian Biol 29:618–625

Åkesson S, Morin J, Muheim R, Ottosson U (2005) Dramatic orientation shift of White-crowned Sparrows displaced across longitudes in the High Arctic. Curr Biol 15:1591–1597

Åkesson S, Jonzen N, Pettersson J, Rundberg M, Sandberg R (2006) Effect of magnetic manipulations on orientation: comparing diurnal and nocturnal passerine migrants on Capri, Italy, in autumn. Ornis Svecica 16:55–61

Alerstam T (2003) Animal behaviour: the lobster navigators. Nature 421:27–28

Alerstam T, Hake M, Kjellén N (2006) Temporal and spatial patterns of repeated migratory journeys by Ospreys. Anim Behav 71:555–566

Alert B, Michalik A, Helduser S, Mouritsen H, Güntürkün O (2015) Perceptual strategies of pigeons to detect a rotational centre–a hint for star compass learning? PLoS ONE 10:e0119919

Amo L, Rodríguez-Gironés M, Barbosa A (2013) Olfactory detection of dimethyl sulphide in a krill-eating Antarctic penguin. Mar Ecol Prog Ser 474:277–285

Archer SD, Stelfox-Widdicombe CE, Malin G, Burkill PH (2003) Is dimethyl sulphide production related to microzooplankton herbivory in the southern North Sea? J Plankton Res 25:235–242

Beason RC (1987) Interaction of visual and non-visual cues during migratory orientation by the Bobolink (*Dolichonyx oryzivorus*). J Ornithol 128:317–328

Beason RC (1992) You can get there from here: responses to simulated magnetic equator crossing by the Bobolink (*Dolichonyx oryzivorus*). Ethology 91:75–80

Beason RC (2004) What can birds hear? Proceedings of the Vertebrate Pest Conference 21:92–96

Beason RC, Nichols JE (1984) Magnetic orientation and magnetically sensitive material in a transequatorial migratory bird. Nature 309:151–153

Bedard AJ (2021) Waterfall low-frequency vibrations and infrasound: implications for avian migration and hazard detection. J Comp Physiol A 207:685–700

Benhamou S, Bried J, Bonadonna F, Jouventin P (2003) Homing in pelagic birds: a pilot experiment with White-chinned Petrels released in the open sea. Behav Process 61:95–100

Berthold P, Kaatz M, Querner U (2004) Long-term satellite tracking of White Stork (*Ciconia ciconia*) migration: constancy versus variability. J Ornithol 145:356–359

Bildstein KL (2006) Migrating raptors of the world: their ecology and conservation. Cornell University Press, Ithaca, NY

Bingman VP, Cheng K (2005) Mechanisms of animal global navigation: comparative perspectives and enduring challenges. Ethol Ecol Evol 17:295–318

Bingman VP, MacDougall-Shackleton SA (2017) The avian hippocampus and the hypothetical maps used by navigating migratory birds (with some reflection on compasses and migratory restlessness). J Comp Physiol A 203:465–474

Bingman VP, Able KP, Kerlinger P (1982) Wind drift, compensation, and the use of landmarks by nocturnal bird migrants. Anim Behav 30:49–53

Bingman VP, Gagliardo A, Hough GE, Ioalé P, Kahn MC, Siegel JJ (2005) The avian hippocampus, homing in pigeons and the memory representation of largescale space. Integr Comp Biol 45:555–564

Bingman VP, Pemberton ML, Mora CV (2021) Avian forebrain processing of magnetic intensity and inclination: hippocampus, anterior forebrain Wulst and an unexpected double-dissociation. Ethol Ecol Evol 33:230–247

Bolte P, Einwich A, Seth PK, Chetverikova R, Heyers D, Wojahn I, Janssen-Bienhold U, Feederle R, Hore P, Dedek K, Mouritsen H (2021) Cryptochrome 1a localisation in light-and dark-adapted retinae of several migratory and non-migratory bird species: no signs of light-dependent activation. Ethol Ecol Evol 33:248–272

Bonadonna F, Chamaillé-Jammes S, Pinaud D, Weimerskirch H (2003) Magnetic cues: are they important in Black-browed Albatross *Diomedea melanophris* orientation? Ibis 145:152–155

Bonadonna F, Bajzak C, Benhamou S, Igloi K, Jouventin P, Lipp H-P, Dell'Omo G (2005) Orientation in the Wandering Albatross: interfering with magnetic perception does not affect orientation performance. Proc R Soc B 272:489–495

Brodbeck MIR, Bingman VP, Cole LJ, Sherry DF, MacDougall-Shackleton SA (2023) Neuronal activation in the geomagnetic responsive region Cluster N covaries with nocturnal migratory restlessness in White-throated Sparrows (*Zonotrichia albicollis*). Eur J Neurosci. https://doi.org/10.1111/ejn.15995

Bruderer B, Liechti F (1999) Bird migration across the Mediterranean. In: Adams NJ, Slotow RH (eds) Proceedings XXII International Ornithology Congress. University of Natal, Durban, South Africa, pp 1983–1999

Buechley ER, Oppel S, Beatty WS, Nikolov SC, Dobrev V, Arkumarev V, Saravia V, Bougain C, Bounas A, Kret E, Skartsi T (2018) Identifying critical migratory bottlenecks and high-use areas for an endangered migratory soaring bird across three continents. J Avian Biol 49:e01629
Cameron DA, Pugh EN (1991) Double cones as a basis for a new type of polarization vision in vertebrates. Nature 353:161–164
Cassone VM, Westneat DF (2012) The bird of time: cognition and the avian biological clock. Front Mol Neurosci 5:32
Chernetsov N (2015) Avian compass systems: do all migratory species possess all three? J Avian Biol 46: 342–343
Chernetsov N (2017) Compass systems. J Comp Physiol A 203:447–453
Chernetsov N, Kishkinev D, Mouritsen H (2008) A long-distance avian migrant compensates for longitudinal displacement during spring migration. Curr Biol 18: 188–190
Chernetsov N, Kishkinev D, Kosarev V, Bolshakov CV (2011) Not all songbirds calibrate their magnetic compass from twilight cues: a telemetry study. J Exp Biol 214:2540–2543
Chernetsov N, Pakhomov A, Kobylkov D, Kishkinev D, Holland RA, Mouritsen H (2017) Migratory Eurasian Reed Warblers can use magnetic declination to solve the longitude problem. Curr Biol 27:2647–2651
Chernetsov N, Pakhomov A, Davydov A, Cellarius F, Mouritsen H (2020) No evidence for the use of magnetic declination for migratory navigation in two songbird species. PLoS ONE 15:e0232136
Cochran WW, Mouritsen H, Wikelski M (2004) Migrating songbirds recalibrate their magnetic compass daily from twilight cues. Science 304:405–408
Cristol DA, Reynolds EB, Leclerc JE, Donner AH, Farabaugh CS, Ziegenfus CW (2003) Migratory Dark-eyed Juncos, *Junco hyemalis*, have better spatial memory and denser hippocampal neurons than nonmigratory conspecifics. Anim Behav 66:317–328
Cronin TW, Marshall J (2011) Patterns and properties of polarized light in air and water. Philos Trans R Soc B 366:619–626
Cunningham GB, Van Buskirk RW, Bonadonna F, Weimerskirch H, Nevitt GA (2003) A comparison of the olfactory abilities of three species of procellariiform chicks. J Exp Biol 206:1615–1620
Cunningham GB, Strauss V, Ryan PG (2008) African Penguins (*Spheniscus demersus*) can detect dimethyl sulphide, a prey-related odour. J Exp Biol 211:3123–3127
Cunningham GB, Leclaire S, Toscani C, Bonadonna F (2016) Responses of King Penguin *Aptenodytes patagonicus* adults and chicks to two food-related odours. J Avian Biol 47:1–8
Dacey JWH, Wakeham SG (1986) Oceanic dimethylsulfide: production during zooplankton grazing on phytoplankton. Science 233:1314–1316
Dell'Ariccia G, Célérier A, Gabirot M, Palmas P, Massa B, Bonadonna F (2014) Olfactory foraging in temperate waters: sensitivity to dimethylsulphide of shearwaters in the Atlantic Ocean and Mediterranean Sea. J Exp Biol 217:1701–1709
Duff SJ, Brownlie LA, Sherry DF, Sangster M (1998) Sun compass and landmark orientation by Black-capped Chickadees (*Parus atricapillus*). Anim Behav Process 24:243–253
Dyson ML, Klump GM, Gauger B (1998) Absolute hearing thresholds and critical masking ratios in the European Barn Owl: a comparison with other owls. J Comp Physiol A 182:695–702
Emlen ST (1966) A technique for recording migratory orientation of captive birds. Auk 83:361–367
Emlen ST (1967a) Migratory orientation in the Indigo Bunting *Passerina cyanea*. Part I: Evidence for use of celestial cues. Auk 84:309–342
Emlen ST (1967b) Migratory orientation in the Indigo Bunting *Passerina cyanea*. Part II: Mechanism of celestial orientation. Auk 84:463–489
Emlen ST (1969) The development of migratory orientation in young Indigo Buntings. Living Bird 1969:716–718
Emlen ST (1970) Celestial rotation: its importance in the development of migratory orientation. Science 170: 1198–1201
Fleissner G, Holtkamp-Rötzler E, Hanzlik M, Winklhofer M, Fleissner G, Petersen N, Wiltschko W (2003) Ultrastructural analysis of a putative magnetoreceptor in the beak of homing pigeons. J Comp Neurol 458:350–360
Freire R, Munro UH, Rogers LJ, Wiltschko R, Wiltschko W (2005) Chickens orient using a magnetic compass. Curr Biol 15:R620–R621
Gagliardo A, Vallortigara G, Nardi D, Bingman VP (2005) A lateralized avian hippocampus: preferential role of the left hippocampal formation in homing pigeon sun compass-based spatial learning. Eur J Neurosci 22: 2549–2559
Gagliardo A, Bried J, Lambardi P, Luschi P, Wikelski M, Bonadonna F (2013) Oceanic navigation in Cory's Shearwaters: evidence for a crucial role of olfactory cues for homing after displacement. J Exp Biol 216:2798–2805
Gagliardo A, Pollonara E, Coppola VJ, Santos CD, Wikelski M, Bingman VP (2014) Evidence for perceptual neglect of environmental features in hippocampal-lesioned pigeons during homing. Eur J Neurosci 40: 3102–3110
Gagliardo A, Pollonara E, Casini G, Rossino MG, Wikelski M, Bingman VP (2020a) Importance of the hippocampus for the learning of route fidelity in homing pigeons. Biol Lett 16:20200095
Gagliardo A, Pollonara E, Wikelski M (2020b) Pigeons remember visual landmarks after one release and rely upon them more if they are anosmic. Anim Behav 166: 85–94

Gagliardo A, Colombo S, Pollonara E, Casini G, Rossino MG, Wikelski M, Bingman VP (2021) GPS-profiling of retrograde navigational impairments associated with hippocampal lesion in homing pigeons. Behav Brain Res 412:113408

Gould JL (1998) Sensory bases of navigation. Curr Biol 8: R731–R738

Greenwood VJ, Smith EL, Church SC, Partridge JC (2003) Behavioural investigation of polarization sensitivity in the Japanese Quail (*Coturnix coturnix japonica*) and the European Starling (*Sturnus vulgaris*). J Exp Biol 206:3201–3210

Günther A, Einwich A, Sjulstok E, Feederle R, Bolte P, Koch K-W, I. A. Solov'yov, and H. Mouritsen. (2018) Double-cone localization and seasonal expression pattern suggest a role in magnetoreception for European Robin cryptochrome 4. Curr Biol 28:211–223

Hagstrum JT (2000) Infrasound and the avian navigational map. J Exp Biol 203:1103–1111

Harada Y, Tanguch M, Namatame H, Ilda A (2001) Magnetic materials in otoliths of bird and fish lagena and their function. Acta Otolaryngol 121:590–595

Healy SD, Gwinner E, Krebs JR (1996) Hippocampal volume in migratory and non-migratory warblers: effects of age and experience. Behav Brain Res 81: 61–68

Heap S, Byrne P, Stuart-Fox D (2012) The adoption of landmarks for territorial boundaries. Anim Behav 83: 871–878

Heffner HE, Koay G, Heffner RS (2016) Budgerigars (*Melopsittacus undulatus*) do not hear infrasound: the audiogram from 8 Hz to 10 kHz. J Comp Physiol A 202:853–857

Heffner R, Cumming JF, Koay G, Heffner HE (2020) Hearing in Indian Peafowl (*Pavo cristatus*): sensitivity to infrasound. J Comp Physiol A 206:899–906

Herold C, Coppola VJ, Bingman VP (2015) The maturation of research into the avian hippocampal formation: recent discoveries from one of the nature's foremost navigators. Hippocampus 25:1193–1211

Heyers D, Manns M, Luksch H, Güntürkün O, Mouritsen H (2007) A visual pathway links brain structures active during magnetic compass orientation in migratory birds. PLoS ONE 2:e937

Hill EM (2017) Audiogram of the Mallard duck (*Anas platyrhynchos*) from 16 Hz to 9 kHz. J Comp Physiol A 203:929–934

Hill EM, Koay G, Heffner RS, Heffner HE (2014) Audiogram of the chicken (*Gallus gallus domesticus*) from 2 Hz to 9 kHz. J Comp Physiol A 200:863–870

Holland RA, Thorup K, Gagliardo A, Bisson IA, Knecht E, Mizrahi D, Wikelski M (2009) Testing the role of sensory systems in the migratory heading of a songbird. J Exp Biol 212:4065–4071

Horváth G, Kriska G, Malik P, Robertson B (2009) Polarized light pollution: a new kind of ecological photopollution. Front Ecol Environ 7:317–325

Huttunen MJ (2009) Magnetic and sunset orientation in migratory Redwings, *Turdus iliacus*. Ital J Zool 76:133–142

Kalmijn AJ (1971) The electric sense of sharks and rays. J Exp Biol 55:371–383

Kalmijn AJ (1982) Electric and magnetic field detection in elasmobranch fishes. Science 218:916–918

Kashetsky T, Avgar T, Dukas R (2021) The cognitive ecology of animal movement: evidence from birds and mammals. Front Ecol Evol 9:724887

Keary N, Bischof H-J (2012) Activation changes in Zebra Finch (*Taeniopygia guttata*) brain areas evoked by alterations of the earth magnetic field. PLoS ONE 7: e38697

Kirschvink JL, Gould JL (1981) Biogenetic magnetite as a basis for magnetic field detection in animals. Biosystems 13:181–201

Kishkinev D, Chernetsov N, Heyers D, Mouritsen H (2013) Migratory Reed Warblers need intact trigeminal nerves to correct for a 1,000 km eastward displacement. PLoS ONE 8:e65847

Kishkinev D, Anashina A, Ishchenko I, Holland RA (2020) Anosmic migrating songbirds demonstrate a compensatory response following long-distance translocation: a radio-tracking study. J Ornithol 161:47–57

Kishkinev D, Packmor F, Zechmeister T, Winkler H-C, Chernetsov N, Mouritsen H, Holland RA (2021) Navigation by extrapolation of geomagnetic cues in a migratory songbird. Curr Biol 31:1563–1569

Kok EM, Tibbitts TL, Douglas DC, Howey PW, Dekinga A, Gnep B, Piersma T (2020) A Red Knot as a black swan: how a single bird shows navigational abilities during repeat crossings of the Greenland icecap. J Avian Biol 51:jav02464

Korpach AM, Mills A, Heidenreich C, Davy CM, Fraser KC (2019) Blinded by the light? Circadian partitioning of migratory flights in a nightjar species. J Ornithol 160:835–840

Kramer G (1950) Weitere analyse der faktoren, welche die zugaktivität des gekäfigten vogels orientieren. Naturwissenschaften 37:377–378

Kramer G (1953) Wird die Sonnenhöhe bei der Heimfindeorientierung verwertet? J Ornithol 94:201–219

Kramer G (1957) Experiments on bird orientation and their interpretation. Ibis 99:196–227

Kreithen ML, Quine DB (1979) Infrasound detection by the homing pigeon: a behavioral audiogram. J Comp Physiol 129:1–4

Kuyt E (1992) Aerial radio-tracking of Whooping Cranes migrating between Wood Buffalo National Park and Aransas National Wildlife Refuge, 1981–84. Canadian Wildlife Service Occasional Paper no. 74, Canadian Wildlife Service, Ottawa, Canada

Kwint RLJ, Quist P, Hansen TA, Dijkhuizen L, Kramer KJM (1996) Turnover of dimethylsulfoniopropionate and dimethylsulfide in the marine environment: a mesocosm experiment. Mar Ecol Prog Ser 145:223–232

LaDage LD, Roth TC, Fox RA, Pravosudov VV (2009) Flexible cue use in food-caching birds. Anim Cogn 12: 419–426

Lauwers M, Pichler P, Edelman NB, Resch GP, Ushakova L, Salzer MC, Heyers D, Saunders M, Shaw J, Keays DA (2013) An iron-rich organelle in the cuticular plate of avian hair cells. Curr Biol 23: 924–929

Lewis MA, Fagan WF, Auger-Méthé M, Frair J, Fryxell JM, Gros C, Gurarie E, Healy SD, Merkle JA (2021) Learning and animal movement. Front Ecol Evol 9: 681704

Liechti F, Peter D, Lardelli R, Bruderer B (1996) Herbstlicher vogelzug im alpenraum nach mondbeobactungen-topographie und wind beeinflussen den Zugverlauf. Der Ornithologische Beobachter 93:131–152

Lindecke O, Holland RA, Pētersons G, Voigt CC (2021) Corneal sensitivity is required for orientation in free-flying migratory bats. Comm Biol 4:522

Linkhart BD, McCallum DA (2020) Flammulated Owl (*Psiloscops flammeolus*), version 1.0. In: Poole AF (ed) Birds of the World. Cornell Lab of Ornithology, Ithaca, NY

Lois-Milevicich J, Kacelnik A, Reboreda JC (2021) Sex differences in the use of spatial cues in two avian brood parasites. Anim Cogn 24:205–212

López-López P, García-Ripollés C, Urios V (2014) Individual repeatability in timing and spatial flexibility of migration routes of trans-Saharan migratory raptors. Curr Zool 60:642–652

Lu Q, Wang K, Lei F, Yu D, Zhao H (2016) Penguins reduced olfactory receptor genes common to other waterbirds. Sci Rep 6:31671

Malkemper EP, Kagerbauer D, Ushakova L, Nimpf S, Pichler P, Treiber CD, de Jonge M, Shaw J, Keays DA (2019) No evidence for a magnetite-based magnetoreceptor in the lagena of pigeons. Curr Biol 29:R14–R15

Melgar J, Lind O, Muheim R (2015) No response to linear polarization cues in operant conditioning experiments with Zebra Finches. J Exp Biol 218:2049–2054

Meskenaite V, Krackow S, Lipp H-P (2016) Age-dependent neurogenesis and neuron numbers within the olfactory bulb and hippocampus of homing pigeons. Front Behav Neurosci 10:126

Michalik A, Alert B, Engels S, Lefeldt N, Mouritsen H (2014) Star compass learning: how long does it take? J Ornithol 155:225–234

Mouritsen H (2018) Long-distance navigation and magnetoreception in migratory animals. Nature 558: 50–59

Mouritsen H, D, Heyers, and O. Güntürkün. (2016) The neural basis of long-distance navigation in birds. Annu Rev Physiol 78:133–154

Mouritsen H, Larsen ON (2001) Migrating songbirds tested in computer-controlled Emlen funnels use stellar cues for a time-independent compass. J Exp Biol 204: 3855–3865

Mouritsen H, Ritz T (2005) Magnetoreception and its use in bird navigation. Curr Opin Neurobiol 15:406–414

Mouritsen H, Feenders G, Liedvogel M, Kropp W (2004a) Migratory birds use head scans to detect the direction of the earth's magnetic field. Curr Biol 14:1946–1949

Mouritsen H, Janssen-Bienhold U, Liedvogel M, Feenders G, Stalleicken J, Dirks P, Weiler R (2004b) Cryptochromes and neuronal-activity markers colocalize in the retina of migratory birds during magnetic orientation. Proc Natl Acad Sci USA 101:14294–14299

Mueller T, O'Hara RB, Converse SJ, Urbanek RP, Fagan WF (2013) Social learning of migratory performance. Science 341:999–1002

Muheim R (2011) Behavioural and physiological mechanisms of polarized light sensitivity in birds. Philos Trans R Soc B 366:763–771

Muheim R, Bäckman J, Åkesson S (2002) Magnetic compass orientation in European Robins is dependent on both wavelength and intensity of light. J Exp Biol 205:3845–3856

Muheim R, Moore FR, Phillips JB (2006a) Calibration of magnetic and celestial compass cues in migratory birds–a review of cue-conflict experiments. J Exp Biol 209:2–17

Muheim R, Phillips JB, Åkesson S (2006b) Polarized light cues underlie compass calibration in migratory songbirds. Science 313:837–839

Muheim R, Åkesson S, Phillips JB (2007) Magnetic compass of migratory Savannah Sparrows is calibrated by skylight polarization at sunrise and sunset. J Ornithol 148:485–494

Muheim R, Phillips JB, Deutschlander ME (2009) White-throated Sparrows calibrate their magnetic compass by polarized light cues during both autumn and spring migration. J Exp Biol 212:3466–3472

Munro U, Wiltschko R (1993) Clock-shift experiments with migratory Yellow-faced Honeyeaters, *Lichenostomus chrysops* (Meliphagidae), an Australian day migrating bird. J Exp Biol 181:233–244

Munro U, Wiltschko R (1995) The role of skylight polarization in the orientation of a day-migrating bird species. J Comp Physiol A 177:357–362

Nagashima K, Velan SS (2013) Understanding the singlet and triplet states in magnetic resonance. Concepts Magn Reson A 42:165–181

Natan E, Fitak RR, Werber Y, Vortman Y (2020) Symbiotic magnetic sensing: raising evidence and beyond. Philos Trans R Soc B 375:20190595

Nevitt GA (2000) Olfactory foraging in Antarctic procellariiform seabirds: life at high Reynolds numbers. Biol Bull 198:245–253

Nevitt GA, Bonadonna F (2005) Sensitivity to dimethyl sulphide suggests a mechanism for olfactory navigation by seabirds. Biol Lett 1:303–305

Nevitt GA, Hunt GL (1996) Olfactory sensitivities of foraging procellariid seabirds in the Aleutian Islands. Chem Senses 21:649–650

Nevitt GA, Veit RR, Kareiva P (1995) Dimethyl sulphide as a foraging cue for Antarctic Procellariiform seabirds. Nature 376:680–682

Nimpf S, Nordmann GC, Kagerbauer D, Malkemper EP, Landler L, Papadaki-Anastasopoulou A, Ushakova L, Wenninger-Weinzierl A, Novatchkova M, Vincent P, Lendl T (2019) A putative mechanism for magnetoreception by electromagnetic induction in the pigeon inner ear. Curr Biol 29:4052–4059

Okanoya K, Dooling RJ (1987) Hearing in passerine and psittacine birds: A comparative study of absolute and masked auditory thresholds. J Comp Psychol 101:7–15

O'Neal BJ, Stafford JD, Larkin RP (2015) Migrating ducks in inland North America ignore major rivers as leading lines. Ibis 157:154–161

Padget O, Dell'Ariccia G, Gagliardo A, González-Solís J, Guilford T (2017) Anosmia impairs homing orientation but not foraging behaviour in free-ranging shearwaters. Sci Rep 7:9668

Pakhomov A, Chernetsov N (2020) A hierarchy of compass systems in migratory birds. Biol Comm 65:262–276

Pakhomov A, Anashina A, Chernetsov N (2017) Further evidence of a time-independent stellar compass in a night-migrating songbird. Behav Ecol Sociobiol 71:48

Papi F, Fiore L, Fiaschi V, Benvenuti S (1971) The influence of olfactory nerve section on the homing capacity of carrier pigeons. Ital J Zool 5:265–267

Patrick SC, Assink JD, Basille M, Clusella-Trullas S, Clay TA, den Ouden OFC, Joo R, Evers LG, Smets PSM, van Loon EE (2021) Infrasound as a cue for seabird navigation. Front Ecol Evol 9:740027

Pedersen JB, Nielsen C, I. A. Solov'yov. (2016) Multiscale description of avian migration: from chemical compass to behaviour modeling. Sci Rep 6:1–12

Perdeck AC (1958) Two types of orientation in migrating *Sturnus vulgaris* and *Fringilla coelebs* as revealed by displacement experiments. Ardea 46:1–37

Pinzon-Rodriguez A, Muheim R (2017) Zebra Finches have a light-dependent magnetic compass similar to migratory birds. J Exp Biol 220:1202–1209

Pravosudov VV, Kitaysky AS, Omanska A (2006) The relationship between migratory behaviour, memory and the hippocampus: an intraspecific comparison. Proc R Soc B 273:2641–2649

Prinz K, Wiltschko W (1992) Migratory orientation of Pied Flycatchers: interaction of stellar and magnetic information during ontogeny. Anim Behav 44:539–545

Pritchard DJ, Hurly TA, Healy SD (2015) Effects of landmark distance and stability on accuracy of reward relocation. Anim Cogn 18:1285–1297

Pritchard DJ, Scott RD, Healy SD, Hurly TA (2016) Wild Rufous Hummingbirds use local landmarks to return to rewarded locations. Behav Process 122:59–66

Procopio M, Ritz T (2016) Inhomogeneous ensembles of radical pairs in chemical compasses. Sci Rep 6:35443

Putman NF (2021) Animal navigation: what is truth? Curr Biol 31:R330–R332

Quine DB, Kreithen ML (1981) Frequency shift discrimination: can homing pigeons locate infrasounds by Doppler shifts? J Comp Physiol 141:153–155

Rappl R, Wiltschko R, Weindler P, Berthold P, Wiltschko W (2000) Orientation of Garden Warblers (*Sylvia borin*) under monochromatic light of various wavelengths. Auk 117:256–260

Rasmussen JL, Sealy SG, Cannings RJ (2020) Northern Saw-whet Owl (*Aegolius acadicus*), version 1.0. In: Poole AF (ed) Birds of the World. Cornell Lab of Ornithology, Ithaca, NY

Reynolds AM, Cecere JG, Paiva VH, Ramos JA, Focardi S (2015) Pelagic seabird flight patterns are consistent with reliance on olfactory maps for oceanic navigation. Proc R Soc B 282:20150468

Ritz T, Adem S, Schulten K (2000) A model for photoreceptor-based magnetoreception in birds. Biophys J 78:707–718

Sandberg R, Bäckman J, Moore FR, Lõhmus M (2000) Magnetic information calibrates celestial cues during migration. Anim Behav 60:453–462

Savoca MS (2018) Chemoattraction to dimethyl sulfide links the sulfur, iron, and carbon cycles in high-latitude oceans. Biogeochemistry 138:1–21

Schecter DA, Nicholls ME, Persing J, Bedard AJ Jr, Pielke RA Sr (2008) Infrasound emitted by tornado-like vortices: basic theory and a numerical comparison to the acoustic radiation of a single-cell thunderstorm. J Atmos Sci 65:685–713

Schiffner I, Wiltschko R (2011) Temporal fluctuations of the geomagnetic field affect pigeons' entire homing flight. J Comp Physiol A 197:765–772

Schmaljohann H, Rautenberg T, Muheim R, Naef-Daenzer B, Bairlein F (2013) Response of a free-flying songbird to an experimental shift of the light polarization pattern around sunset. J Exp Biol 216:1381–1387

Schmidt-Koenig K (1958) Der Einfluß experimentell veränderter Zeitschätzung auf das Heimfindevermögen bei Brieftauben. Naturwissenschaften 45:47

Shaw J, Boyd A, House M, Woodward R, Mathes F, Cowin G, Saunders M, Baer B (2015) Magnetic particle-mediated magnetoreception. J R Soc Interface 12:20150499

Shaw RC, MacKinlay RD, Clayton NS, Burns KC (2019) Memory performance influences male reproductive success in a wild bird. Curr Biol 29:1498–1502

Sjöberg S, Muheim R (2016) A new view on an old debate: type of cue-conflict manipulation and availability of stars can explain discrepancies between cue-calibration experiments with migratory songbirds. Front Behav Neurosci 10:29

Skiles DD (1985) The geomagnetic field: its nature, history and biological relevance. In: Kirschvink JL, Jones DS, MacFaddon BJ (eds) Magnetite biomineralization and magnetoreception in organisms. Plenum, New York, NY, pp 43–102

Slagsvold T, Wiebe KL (2021) Use of landmarks for nest-site choice and small-scale navigation to the nest in birds. Behaviour 1:1–22

Smith RF (2009) Investigating raptor migration behavior using orientation cages and wing measurements: a comparison of the Flammulated Owl and Northern Saw-whet Owl in southwestern Idaho. M. S. thesis, Boise State University, Boise, ID

Solov'yov IA, Greiner W (2009) Iron-mineral-based magnetoreceptor in birds: polarity or inclination compass? Eur Phys J D 51:161–172

Stanley CQ, MacPherson M, Fraser KC, McKinnon EA, Stutchbury BJM (2014) Repeat tracking of individual songbirds reveals consistent migration timing but flexibility in route. PLoS ONE 7:e40688

Strandberg R, Alerstam T (2007) The strategy of fly-and-forage migration, illustrated for the Osprey (*Pandion haliaetus*). Behav Ecol Sociobiol 61:1865–1875

Strawn SN, Hill EM (2020) Japanese Quail (*Coturnix japonica*) audiogram from 16 Hz to 8 kHz. J Comp Physiol A 206:665–670

Streby HM, Kramer GR, Peterson SM, Lehman JA, Buehler DA, Andersen DE (2015) Tornadic storm avoidance behavior in breeding songbirds. Curr Biol 25:98–102

Theurich M, Langner G, Scheich H (1984) Infrasound responses in the midbrain of the Guinea fowl. Neurosci Lett 49:81–86

Thorup K, Holland RA (2009) The bird GPS–long-range navigation in migrants. J Exp Biol 212:3597–3604

Thorup K, Rabøl J (2007) Compensatory behaviour after displacement in migratory birds. Behav Ecol Sociobiol 61:825–841

Thorup KI-A, Bisson MS, Bowlin RA, Holland JC, Wingfield MR, Wikelski M (2007) Evidence for a navigational map stretching across the continental U. S. in a migratory songbird. Proc Natl Acad Sci USA 104:18115–18119

Treiber CD, Salzer MC, Riegler J, Edelman N, Sugar C, Breuss M, Pichler P, Cadiou H, Saunders M, Lythgoe M, Shaw J (2012) Clusters of iron-rich cells in the upper beak of pigeons are macrophages not magnetosensitive neurons. Nature 484:367–370

Vardanis Y, Klaassen RHG, Strandberg R, Alerstam T (2011) Individuality in bird migration: routes and timing. Biol Lett 7:502–505

Vargas JP, Siegel JJ, Bingman VP (2006) The effects of a changing ambient magnetic field on single-unit activity in the homing pigeon hippocampus. Brain Res Bull 70: 158–164

von Middendorff A (1859) Die isepiptesen Russlands. Mem. Acad. Sci, St

Voss J, Keary N, Bischof H-J (2007) The use of the geomagnetic field for short distance orientation in Zebra Finches. Neuroreport 18:1053–1057

Vyssotski AL, Dell'Omo G, Dell'Ariccia G, Abramchuk AN, Serkov AN, Latanov AV, Loizzo A, Wolfer DP, Lipp H-P (2009) EEG responses to visual landmarks in flying pigeons. Curr Biol 19:1159–1166

Walcott C (2005) Multi-modal orientation cues in homing pigeons. Integr Comp Biol 45:574–581

Walker MM (1998) On a wing and a vector: a model for magnetic navigation by homing pigeons. J Theor Biol 192:341–349

Wallraff HG (1966) Über die heimfindeleistungen von brieftauben nach haltung in verschiedenartig abgeschirmten volieren. Z Vgl Physiol 52:215–259

Wallraff HG (1979) Goal-oriented and compass-oriented movements of displaced homing pigeons after confinement in differentially shielded aviaries. Behav Ecol Sociobiol 5:201–225

Wallraff HG (2000) Path integration by passively displaced homing pigeons? Anim Behav 60:F30–F36

Wallraff HG (2004) Avian olfactory navigation: its empirical foundation and conceptual state. Anim Behav 67: 189–204

Wang K, Mattern E, Ritz T (2006) On the use of magnets to disrupt the physiological compass of birds. Phys Biol 3:220–231

Warrant EJ (2021) Unravelling the enigma of bird magnetoreception. Nature 594:497–498

Wikelski M, Arriero E, Gagliardo A, Holland RA, Huttunen MJ, Juvaste R, Mueller I, Tertitski G, Thorup K, Wild M, Alanko M (2015) True navigation in migrating gulls requires intact olfactory nerves. Sci Rep 5:17061

Williams TC, Williams JM, Williams PG, Stokstad P (2001) Bird migration through a mountain pass studied with high resolution radar, ceilometers, and census. Auk 118:389–403

Wiltschko W (1968) Über den Einfluβ statischer magnetfelder auf die zugorientierung der Rotkehlchen (*Erithacus rubecula*). Z Tierpsychol 25:536–558

Wiltschko R, Wiltschko W (1981) The development of sun compass orientation in young homing pigeons. Behav Ecol Sociobiol 9:135–141

Wiltschko W, Wiltschko R (1995) Migratory orientation of European Robins is affected by the wavelength of light as well as by a magnetic pulse. J Comp Physiol A 177:363–369

Wiltschko W, Wiltschko R (1999) The effect of yellow and blue light on magnetic compass orientation in European Robins, *Erithacus rubecula*. J Comp Physiol A 184:295–299

Wiltschko R, Wiltschko W (2009) Avian navigation Auk 126:717–743

Wiltschko R, Wiltschko W (2013) The magnetite-based receptors in the beak of birds and their role in avian navigation. J Comp Physiol A 199:89–98

Wiltschko R, Wiltschko W (2015) Avian navigation: a combination of innate and learned mechanisms. Adv Study Behav 47:229–310

Wiltschko R, Wiltschko W (2019) Magnetoreception in birds. J R Soc Interface 16:20190295

Wiltschko W, Daum P, Fergenbauer-Kimmel A, Wiltschko R (1987) The development of the star compass in Garden Warblers, *Sylvia borin*. Ethology 74:285–292

Wiltschko R, Haugh C, Walker M, Wiltschko W (1998) Pigeon homing: sun compass use in the southern hemisphere. Behav Ecol Sociobiol 43:297–300

Wiltschko R, Walker M, Wiltschko W (2000) Sun-compass orientation in homing pigeons: compensation for different rates of change in azimuth? J Exp Biol 203:889–894

Wiltschko W, Munro U, Ford H, Wiltschko R (2006) Bird navigation: what type of information does the magnetite-based receptor provide? Proc R Soc B 273: 2815–2820

Wiltschko W, Freire R, Munro U, Ritz T, Rogers L, Thalau P, Wiltschko R (2007) The magnetic compass of domestic chickens, *Gallus gallus*. J Exp Biol 210: 2300–2310

Wiltschko R, Munro U, Ford H, Wiltschko W (2008) Contradictory results on the role of polarized light in compass calibration in migratory songbirds. J Ornithol 149:607–614

Wiltschko R, Schiffner I, Wiltschko W (2009) A strong magnetic anomaly affects pigeon navigation. J Exp Biol 212:2983–2990

Wolff WJ (1970) Goal orientation versus one-direction orientation in teal *Anas c. crecca* during autumn migration. Ardea 58:131–141

Wu L-Q, Dickman JD (2011) Magnetoreception in an avian brain in part mediated by inner ear lagena. Curr Biol 21:418–423

Wu L-Q, Dickman JD (2012) Neural correlates of a magnetic sense. Science 336:1054–1057

Xu J, Jarocha LE, Zollitsch T, Konowalczyk M, Henbest KB, Richert S, Golesworthy MJ, Schmidt J, Déjean V, Sowood DJ, Bassetto M (2021) Magnetic sensitivity of cryptochrome 4 from a migratory songbird. Nature 594:535–540

Young SR, Martin GR (1984) Optics of retinal oil droplets: a model of light collection and polarization detection in the avian retina. Vis Res 24:129–137

Zannoni N, Wikelski M, Gagliardo A, Raza A, Kramer S, Seghetti C, Wang N, Edtbauer A, Williams J (2020) Identifying volatile organic compounds used for olfactory navigation by homing pigeons. Sci Rep 10:15879

Zapka M, Heyers D, Hein CM, Engels S, Schneider N-L, Hans J, Weiler S, Dreyer D, Kishkinev D, Wild JM, Mouritsen H (2009) Visual but not trigeminal mediation of magnetic compass information in a migratory bird. Nature 461:1274–1277

Zein B, Long JA, Safi K, Kölzsch A, Benitez-Paez F, Wikelski M, Kruckenberg H, Demšar U (2022) Simulating geomagnetic bird navigation using novel high-resolution geomagnetic data. Eco Inform 69: 101689

Zelenitsky DK, Therrien F, Ridgely RC, McGee AR, Witmer LM (2011) Evolution of olfaction in non-avian theropod dinosaurs and birds. Proc R Soc B 278:3625–3634

# Mating Systems

# 15

## Contents




**Abstract**

The ways by which male and female birds locate and choose each other for mating, the number of mates during a breeding season and during a bird's life, the duration of interactions between males and females, and the respective roles of males and females in providing care (or not) for eggs and young after fertilization vary among species and, to a lesser extent, within species. The different ways that birds of different species interact to produce young are referred to as mating systems. In this chapter, I discuss the various mating systems of birds, including monogamy, polygyny, polyandry, polygynandry, and promiscuity, and explain the factors that have likely contributed to their evolution. Also in this chapter is a discussion of sexual conflict and cooperation, and factors that contribute to variation among and within species in extra-pair mating. Some




G. Ritchison, *In a Class of Their Own*, Fascinating Life Sciences,
https://doi.org/10.1007/978-3-031-14852-1_15

**Box 15.1 Evolution of Anisogamy**

Several hypotheses have been proposed to explain the evolution of anisogamy. Parker et al. (1972) suggested that anisogamy evolved as a result of a trade-off between gamete quality and quantity, with disruptive selection favoring the evolution of two sizes of gametes. As summarized by Bulmer and Parker (2002), primitive marine ancestors of present-day anisogamous species produced a range of gamete sizes, and fusion between pairs of gametes (i.e., fertilizations) was random. Each individual had a fixed energy budget for reproduction, so there was a size–number trade-off, with the number of gametes produced inversely proportional to their size. However, viability of fused gametes (zygotes) increased with the collective size of the two gametes (i.e., more energy for development of the embryo). As a result, fusion of two smaller gametes or a small gamete and an intermediate-sized gamete would not produce viable zygotes (or, at best, less viable zygotes). Fusion of (1) a small gamete and a large gamete, (2) two intermediate-sized gametes, (3) an intermediate-sized gamete and a large gamete, or (4) two large gametes would produce viable zygotes. Because viability increased with zygote size, fusion of any-sized gametes with a large gamete produced the most viable zygotes.

Smaller gametes, being cheaper to produce, could be produced in greater quantities. As a result, individuals producing smaller gametes would gain more fusions with large gametes than would those producing the more energetically expensive and, therefore, less numerous intermediate-sized and large gametes. Disruptive selection, then, would favor individuals producing small gametes and those producing large gametes.

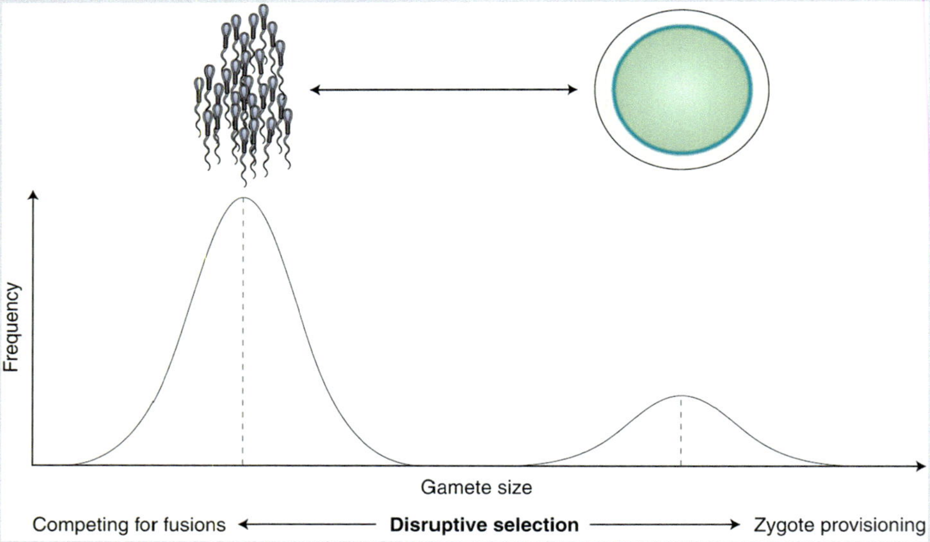


Begin with organisms that reproduce sexually, but individuals produce gametes that vary in size along a continuum. Intermediate-sized gametes would be too large to produce in large numbers, but too small to allow embryogenesis without more cytoplasm. As a result, disruptive selection would favor individuals that produce either numerous small and more competitive gametes (sperm) or fewer large nutritive gametes (eggs). (Figure from Gage 2021; © 2021 Springer Nature Limited, used with permission)

Another hypothesis to explain the evolution of anisogamy focuses on encounter probability (Iyer and Roughgarden 2008). For gametes with no mode of propulsion (such as flagella), smaller gametes would move faster (via Brownian motion) than larger gametes and larger

(continued)

**Box 15.1** (continued)

gametes, in addition to containing more energy for the development of embryos, also make larger targets. As a result, fused small and large gametes would not only produce viable zygotes, but would also tend to meet more often than other size-combinations of gametes, thus leading to the evolution of anisogamy.

Yet another hypothesis is that uniparental inheritance of mitochondria improves mitonuclear co-adaptation (Hadjivasiliou et al. 2012). Mitochondria are descended from free-living bacteria that developed a symbiotic relationship with other cells an estimated one and a half to two billion years ago. In that process, most of the genetic information in the ancestral mitochondria was lost or transferred to the nucleus of the host cells, leaving just residual DNA in each mitochondrion. Oxidative phosphorylation, the most critical function of mitochondria, depends on the functional compatibility of proteins encoded by DNA in both the nucleus and mitochondria. Using a mathematical model, Hadjivasiliou et al. (2012) found that uniparental inheritance of mitochondria enhances the co-adaptation of mitochondrial and nuclear genes and therefore improves fitness. If correct, selection would favor two distinct gamete types, a large one that contained genetic material plus mitochondria (i.e., present-day eggs) and a much smaller one that contained only genetic material (and no mitochondria).

Other hypotheses have also been proposed to explain the evolution of anisogamy, and more will likely be proposed in the future. Parker et al.'s (1972) hypothesis (often referred to as the PBS model based on the first letters of the last names of the three authors of the paper) is supported by the results of some studies, but not others. The likelihood of definitively showing that one of the many proposed hypotheses is the correct one is probably small, but the PBS model is now generally accepted by many (but certainly not all) investigators as being a likely scenario for the evolution of anisogamy.

species of birds breed cooperatively and the social and environmental variables that contribute to such behavior are discussed. The choice of a mate is critical for birds attempting to maximize their reproductive fitness, and both female and male mate choice and the role of sexual selection in mate choice are discussed in detail.

## 15.1 Introduction

Successful reproduction requires fertilization of a female's eggs by a male's sperm. However, birds are anisogamous, meaning that the gametes of males and females differ in size and number. Males produce numerous small sperm whereas females produce many fewer, larger ova (eggs) (Box 15.1 Evolution of Anisogamy). An important consequence of this is that not all male sperm will be able to fertilize female eggs and, as a result, males may have to compete for access to females who, with a limited number of gametes, will in turn often be choosy about potential mates. Other factors such as resource distribution and mode of development of young (e.g., altricial or precocial) also influence the reproductive roles of male and female birds, but the difference in numbers of gametes produced has been a critical factor in the evolution (via sexual selection) of male attributes that enhance their competitive ability and those of females that enable them to discriminate among males. Specifically, for males, sexual selection favors the evolution of characteristics that (1) enhance the ability of sperm to fertilize eggs, (2) increase the likelihood of success in encounters with other males that will, in turn, increase access to females, and

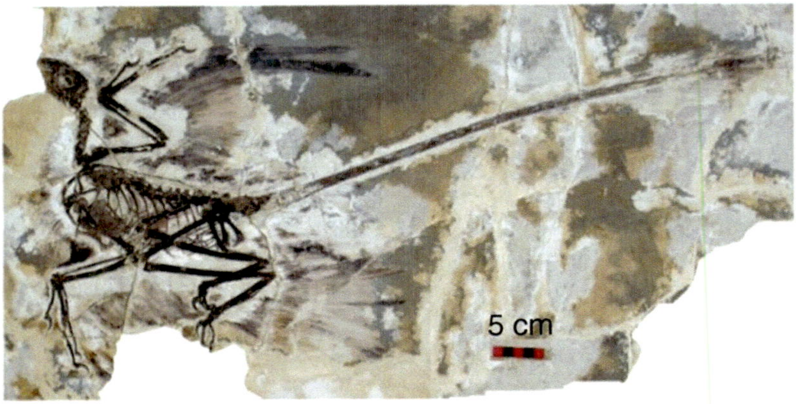


**Fig. 15.1** *Microraptor* fossil showing the elongated tail feathers. (Figure from Evangelista et al. 2014; © 2014 Evangelista et al., open-access article distributed under the terms of the Creative Commons Attribution License, https://creativecommons.org/licenses/by/4.0/)

(3) increase their likelihood of attracting females. For females, sexual selection favors the evolution of traits that permit better discrimination among competing males and the selection of mates whose contributions (e.g., sperm, access to resources, and/or parental care) maximize female reproductive success.

Despite the inherent differences in the reproductive roles of male and female birds resulting from anisogamy, members of each sex are attempting to maximize their lifetime fitness. However, present-day species of birds occupy and breed in a wide variety of habitats with different environmental conditions. In addition, present-day species have different phylogenetic histories. As a result, the ways by which males and females locate and choose each other for mating, the number of mates during a breeding season and during a bird's life, the duration of interactions between males and females, and the respective roles of males and females in providing care (or not) for eggs and young after fertilization all vary among

**Fig. 15.2** A 120-million-year-old early Cretaceous bird in the same family as *Archaeopteryx* (Archaeopterygidae), *Jeholornis*. Males not only had elongated feathers at the end of their tails, but also had a fan-shaped tract of feathers at the base of their long bony tail. (Figure from O'Connor et al. 2013; used with permission of the U.S. National Academy of Sciences)

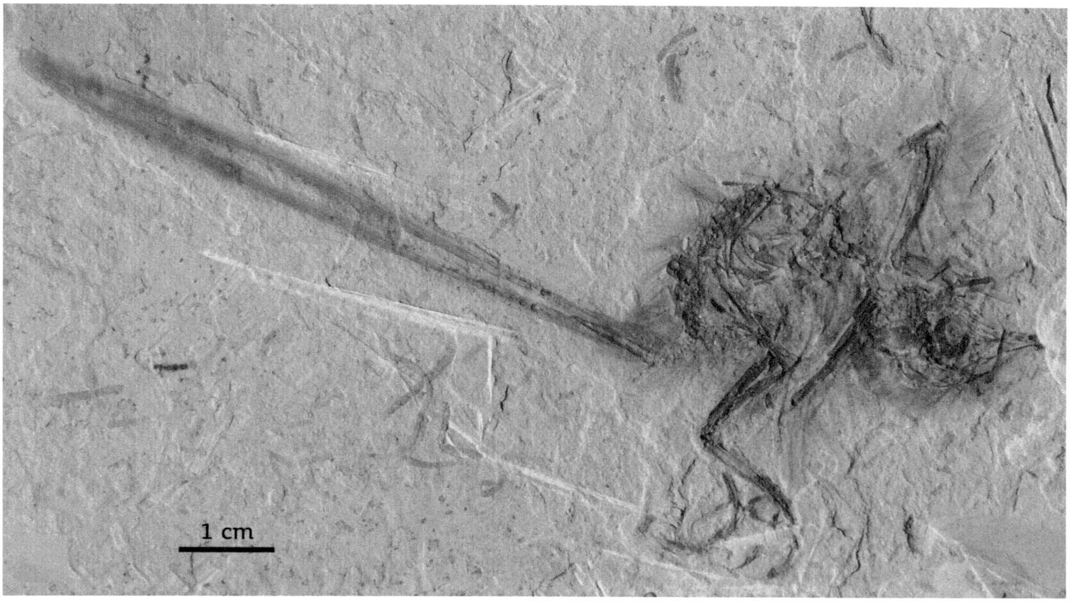


**Fig. 15.3** Fossil specimen of the enantiornithine *Cratoavis* discovered in present-day South America and dated at 113–118 Ma. (Figure from Carvalho et al. 2015; used with permission of I.D.S. Carvalho)

species and, to a lesser extent, within species. The different ways in which males and females in different species interact to produce young are referred to as mating, or breeding, systems.

## 15.2 Mating Systems of Avian Ancestors

Not surprisingly, very little is known about the mating systems of the theropod ancestors of birds or about those of birds that lived during the Jurassic and early Cretaceous periods. Fossils have revealed that some bird-like dinosaurs, such as *Microraptor* (Li et al. 2012; Fig. 15.1) and *Jeholornis* (O'Connor et al. 2013), exhibited sexual dimorphism, with males, but not females, having elongated tail feathers. Interestingly, *Jeholornis* males actually had two "tails," with a fan-shaped tract of feathers at the base of their long tails (Fig. 15.2). These long-tail feathers could have served an aerodynamic function, but could also have been ornamental. Among present-day birds, many polygynous species are sexually dimorphic and males also have ornamental tail feathers (Björklund 1990). This suggests that *Microraptor* and *Jeholornis* and, perhaps, other bird-like theropods and early birds (e.g., an enantiornithine bird, *Cratoavis*; Fig. 15.3) known to have elongated tail feathers may have been polygynous. However, among present-day birds, not all sexually dimorphic species are polygynous, so the relationship between dimorphism and mating system is certainly not straightforward (O'Connor et al. 2013).

Based on large clutch volumes and the bone histology of brooding adults (adults discovered on top of fossilized clutches of eggs), Varricchio et al. (2008) concluded that dinosaurs in the genera *Troodon* (Troodontidae), *Oviraptor* (Oviraptorosauria), and *Citipati* (Oviraptorosauria), all close relatives of birds (Fig. 15.4), had polygamous mating systems. Similarly, based on the analysis of 472 extant taxa of diapsids, Moore and Varricchio (2016) suggested that basal diapsids were polygamous (Fig. 15.5) and that the transition to social monogamy appeared to be correlated with the evolution

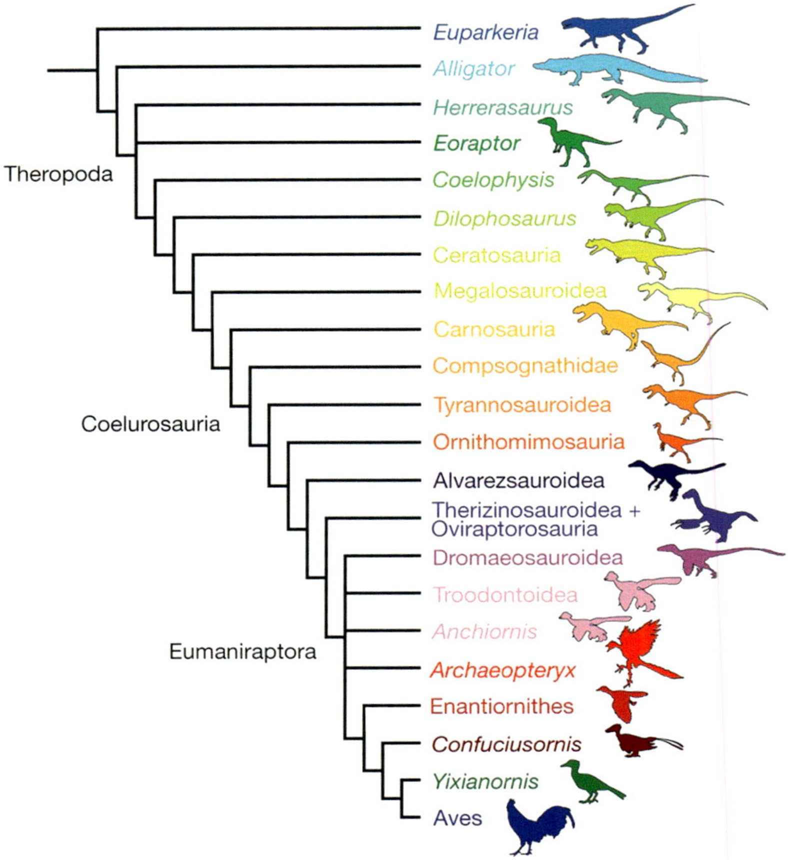


**Fig. 15.4** Archosaur phylogeny showing the relationship between birds, including *Archaeopteryx*, enantiornithes, *Confuciusornis*, *Yixianornis*, and present-day birds (Aves), bird-like dinosaurs (Eumaniraptora), and other dinosaurs. (Figure from Bhullar et al. 2012; © 2012 Springer Nature, used with permission)

of a female role in parental care (either alone or as part of a biparental unit).

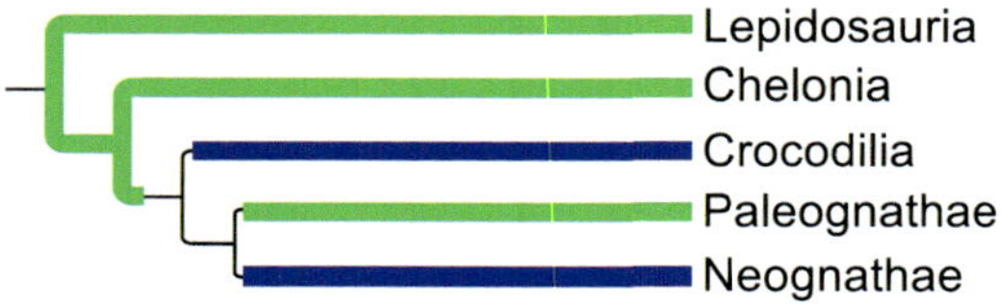


**Fig. 15.5** Available evidence suggests that the ancestral mating strategy of diapsids was polygamy for both sexes (and with no parental care of eggs or young). Green = both sexes polygamous, and blue = socially monogamous. (Figure from Moore and Varricchio 2016; © 2016 Moore and Varricchio, open-access article distributed under the terms of the Creative Commons Attribution License, https://creativecommons.org/licenses/by/4.0/)

## 15.3 Avian Mating Systems

The mating system of a species or population represents the general behavioral strategy used to obtain one or more mates (Emlen and Oring 1977). In some cases, this strategy only involves mating, with no further social interaction between males and females. For most species of birds, males and females that mate also cooperate to varying degrees to raise young. For these species, their broader "behavioral strategies" might more accurately be referred to as breeding systems. However, both "mating system" and "breeding system" are commonly used phrases in the ornithological and behavioral literature and are sometimes (perhaps incorrectly) used interchangeably, so, for the purposes of this discussion, the phrase "mating system" will be used.

Mating systems differ in, and are defined based on, a variety of factors, including how mates are acquired (i.e., courtship, competition, and mate choice), the number of mates (both social and genetic), the form and duration of pair bonds, mating or breeding resources defended and provided, and the types and duration of parental care provided (Reynolds 1996). Given the diversity of ways in which males and females in different species (and even populations within species) interact to produce young, categorizing avian mating systems requires that both social and genetic interactions between males and females be considered (MacManes 2013). As such, avian social relationships or mating systems can be categorized as follows (MacManes 2013):

**Social monogamy** is characterized by a social relationship (pair bond) between one male and one female. The male and female remain together for at least one breeding attempt and both play roles (that often differ) in raising young. Socially monogamous males and females may, in some species and populations, also engage in extra-pair copulations. Cooperative breeders could be considered socially monogamous if conspecifics (helpers) associating with a breeding pair obtain only indirect fitness benefits (e.g., Red-cockaded Woodpecker, *Picoides borealis*; Haig et al. 1994). Communal breeders, such as Greater Anis (*Crotophaga major*), may also be socially monogamous, with females of several pairs laying eggs in a single communal nest (Riehl 2011).

**Polygyny** is where at least some males in a species have social relationships with more than one female and typically mate with those females. The roles of males and females in raising young typically differ, with males primarily involved in defending the territory and females providing much if not most or all of the parental care. In polygynous species, both males and females may engage in extra-pair (or extra-group) copulations, and males are typically larger and more colorful than females. Cooperative breeders could be considered polygynous if a cooperatively breeding male has more than one female with a nest (along with non-breeding helpers) in the group's territory (e.g., Eurasian Oystercatcher, *Haematopus ostralegus*; Heg and van Treuren 1998).

**Polyandry** is where at least some females in a species have social relationships of varying dura-

**Box 15.2 Male Mating Coalitions**

Male birds typically compete for access to potential mates, and this competition has been a powerful selective force in the evolution of increased body size and the conspicuous plumage and displays of males in many species of birds. However, in a few species of birds, males engage in cooperative displays to gain access to mates. Examples of such behavior have been reported in Indian Peafowl (*Pavo cristatus*, Petrie et al. 1999), Wild Turkeys (*Meleagris gallopavo*, Krakauer 2005), and at least seven species of manakins representing two genera (*Chiroxiphia* and *Pipra*, Díaz-Muñoz et al. 2014). For example, two male Lance-tailed Manakins (*Chiroxiphia lanceolata*) sometimes display cooperatively, but, with very few exceptions, only alpha males mate with females (DuVal 2007a). Male Wild Turkeys form coalitions of two to four same-aged males that court females and defend those females from other males and groups. Only the alpha male Wild Turkey mates with females, but other coalition members benefit via kin selection because they are related (full or half-sibs) to the alpha male (Krakauer 2005). Similarly, Petrie et al. (1999) found that related male Indian Peafowl (*Pavo cristatus*) display together at leks, possibly because of the inclusive fitness benefits. In contrast, among manakins, the degree of relatedness between displaying males varies, but, in many cases, the males are unrelated (Ryder et al. 2011). Additional study is needed to better understand why subordinate male manakins engage in cooperative displays, but one likely factor is that mating

(continued)

**Box 15.2** (continued)

opportunities are limited (or non-existent) for single males and subordinate males in a coopera-tive pair may gain experience that could increase the likelihood of becoming a breeder in the future. Among Lance-tailed Manakins, for example, subordinate males in cooperative pairs were more likely to become dominant males and gain mating opportunities in future years than males with no prior experience with cooperative displaying (DuVal 2007b).

Coalition of displaying male Wild Turkeys. (Photo by Larry Smith, U.S. Fish and Wildlife Service, CC0 Public Domain)

tion with more than one male and typically mate with those males. Polyandrous species typically exhibit sex-role reversal, with females larger (and, in some species, more brightly colored) and males providing most or all of the parental care. Both males and females may engage in extra-pair copulations. Cooperative breeders could be considered polyandrous if a polyandrous female copulates with male helpers (males with no nest of their own) or multiple males who then all contribute to parental care (e.g., Galapagos Hawks, *Buteo galapagoensis*; Faaborg et al. 1995).

**Social promiscuity** is characterized by the absence of social relationships or pair bonds between males and females. In some species, however, males may form coalitions to gain access to or defend females (Box 15.2 Male Mating Coalitions). Both males and females may mate with multiple individuals. Social promiscuity is a characteristic of lekking species.

**Polygynandry** is where there are social relationships (pair bonds) between more than one male and more than one female, e.g., Bicknell's Thrush (*Catharus bicknelli*, Strong et al. 2004) and Smith's Longspur (*Calcarius pictus*, Briskie 1992). The roles of males and females in raising young vary among species. Males and females in these social groups may mate with individuals outside of the group.

Genetic relationships also differ among species and populations and, regardless of the social mating system, can be categorized as (MacManes 2013): (1) **Genetic monogamy** where neither males nor females engage in extra-pair or extra-group copulations, (2) **Multiple male mating** where only males engage in extra-pair copulations, (3) **Multiple female mating** where only females engage in extra-pair copulations, and (4) **Genetic promiscuity** where both males and females engage in extra-pair copulations.

Accurate classification of avian mating systems requires information about both social and genetic relationships. For example, socially monogamous species may also be genetically monogamous, but may also exhibit multiple male mating, multiple female mating, or genetic promiscuity. Socially polygynous species, as another example, may exhibit either multiple male mating (only males copulate with other females outside of the "social group," i.e., males plus their female social mates) or genetic promiscuity (both males and females in a social group may copulate with conspecifics outside of the social group). Understanding the intensity of sexual selection in different species also requires knowledge of both their social and genetic mating systems.

## 15.4 Evolution of Avian Mating Systems

Avian social and genetic mating systems exhibit impressive diversity and, at least since the publication of Emlen and Oring's (1977) classic paper, the importance of phylogenetic factors (i.e., amount of parental care required to raise young) and ecological factors such as the temporal and spatial distribution of resources and/or members of the opposite sex have been recognized as key factors in the evolution of mating systems. If resources (such as food) are "clumped," then some individuals may be able to control access and, by doing so, attract multiple mates. However, a bird's ability to "capitalize" on the presence of clumped resources depends on the amount of parent care required to successfully raise young; if one parent cannot successfully

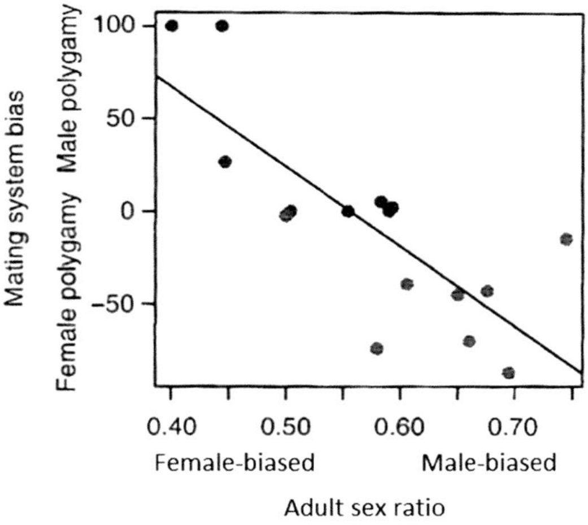


**Fig. 15.6** Relationship between adult sex ratio (number of adult males/number of adult males plus females) and mating systems for 16 species of shorebirds. Species with female-biased adult sex ratios tend to be polygynous (male polygamy), and those with male-biased adult sex ratios tend to be polyandrous (female polygamy). (Figure from Liker et al. 2013; © 2013 Springer Nature, used with permission)

raise young, then, depending on how much time and effort each parent must contribute to raising young, the potential advantage of having multiple mates is diminished. Finally, the degree to which mates are available and can be monopolized also influences the type of mating system exhibited by a species or population. If the operational sex ratio (OSR; the ratio of fertilizable females to sexually active males) is skewed in favor of males (more fertilizable females available than sexually active males), polygyny may result; if skewed in favor of females, polyandry may result; and, if not skewed in favor of either sex, monogamy may result (Emlen and Oring 1977; Fig. 15.6).

Emlen and Oring (1977) identified two factors that have likely been important in the evolution of avian mating systems, namely phylogenetic constraints and environmental or ecological conditions. However, social interactions between males and females are important as well (Bennett and Owens 2002). More accurately, these "interactions" should be referred to as sexual conflict, and such conflict arises when there is a difference in the evolutionary interests (i.e., fitness optima) of males and females (Parker 1979). Differences among species in the outcome of this sexual conflict, along with differences in phylogenetic constraints and ecological conditions, mean that males and females in different species may use different strategies in an attempt to optimize their fitness and this has resulted in the evolution of different mating systems.

Environmental factors that can influence mating systems include the abundance and distribution of potential mates and the resources needed for successful reproduction. A population's adult sex ratio (Box 15.3 Avian Sex Ratios) is one important environmental factor that affects the availability of potential mates. In most species of birds, populations are male-biased because of higher mortality rates for females (Fig. 15.7). Factors contributing to these higher mortality rates include: (1) females disperse farther than males in most species (Greenwood and Harvey 1982) exposing them to greater risk when they leave natal areas and disperse across unfamiliar territory (Steifetten and Dale 2006), (2) females tend to migrate farther than males in many species of migratory birds, again exposing them to potentially greater risks than males (Catry et al. 2005), and (3) females spend more time incubating eggs and brooding young in many species of birds, placing them at greater risk of being predated (Sargeant et al. 1984, O'Donnell 1996). A population's sex ratio, and specifically the operational sex ratio (OSR; Box 15.3 Avian Sex Ratios), influences the availability of mates and so can affect pairing behavior and interactions between individuals (male–male, female–female, and male–female) (Alonzo and Sheldon 2010), leading to possible differences among species and populations in mating systems.

Importantly, operational sex ratios, or OSRs, are not based just on the relative number of males and females in a population, but also on the relative number of males and females that are available for mating at a given time (Box 15.4 Floaters). Males and females may differ in when they are receptive to mating and such differences mean that OSRs could vary during a breeding season. For example, Wilson's Phalaropes (*Phalaropus tricolor*) are polyandrous and females pair with multiple males sequentially during the breeding season, briefly pairing with and laying a clutch of eggs for one male (who then is responsible for incubating the eggs and caring for young after eggs hatch) and then attempting to briefly pair with and lay a clutch of eggs for another male and so on. Colwell and Oring (1988a) found that the availability of receptive male Wilson's Phalaropes and, therefore, the OSR, varied seasonally and among years and, as a result, so did competition among females for mates (Fig. 15.8). Male availability varied seasonally because (1) females tended to arrive in the breeding area before males, (2) males that were incubating eggs or caring for young were not sexually receptive, and (3) when nests failed (e.g., due to predation of eggs or young), some males again became receptive and re-nested (either in the same area or after dispersing to a new location) (Colwell and Oring 1988a, b).

The spatial and temporal distribution and availability of resources such as food and high-quality nest sites can also affect the mating

**Box 15.3 Avian Sex Ratios**

The sex ratio of a population is the proportion of one sex in the population relative to the other sex. Sex ratios of birds at the time of conception or fertilization (referred to as the primary sex ratio, or PSR) are typically about 1:1 (equal numbers of males and females). Because of differential mortality after fertilization and hatching, the sex ratio of a population when young fledge (leave nests) (called the secondary sex ratio, or SSR) can differ from the PSR. For example, in species that exhibit sexual size dimorphism, mortality rates of the larger sex may be higher than those of the smaller sex because larger nestlings are more at risk of starvation due to food shortages (Kalmbach and Benito 2007). Differences in mortality rates after fledging can further alter sex ratios of adults in a population (called the adult sex ratio, or ASR, or, less frequently, the tertiary sex ratio). Differential mortality after fledging can again result from differences in rates of starvation between males and females and, in addition, young females tend to disperse farther than young males and that may result in higher mortality rates for the females (i.e., traveling farther through unfamiliar areas can increase the risk of starvation and predation). For species of birds that attempt to breeding during their first breeding season, the ASR is also the operational sex ratio (the ratio of sexually active males to sexually active females in a population). However, for species where adults may not attempt to breed until they are older (e.g., two, three, or four years old or even older), the OSR differs from the ASR because not all adults in a population are breeding and because of mortality during the period before adults begin to breed. The OSR can greatly impact mating systems via differences in the availability of potential mates. Polygyny tends to be more common when OSRs are female-biased, and polyandry more common when OSRs are male-biased (Liker et al. 2013, 2014).

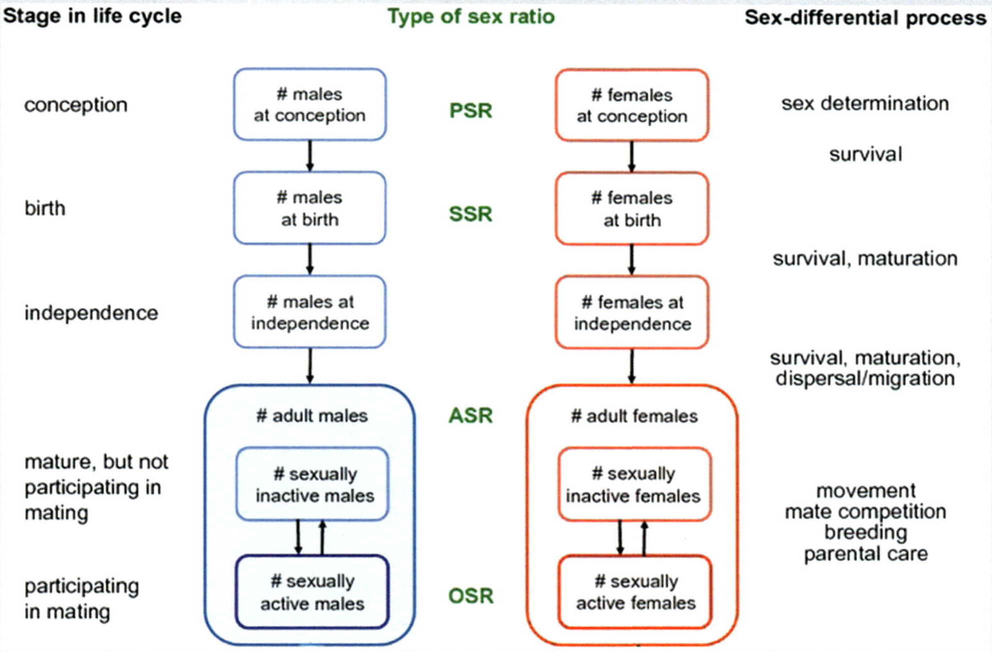


Processes that can affect population sex ratios at various stages of the life cycle of a species. PSR, primary sex ratio; SSR, secondary sex ratio; ASR, adult sex ratio; OSR, operational sex ratio. (Figure modified from Székely et al. 2014; © 2014 The Authors. *Journal of Evolutionary Biology* © 2014 European Society for Evolutionary Biology, used with permission)

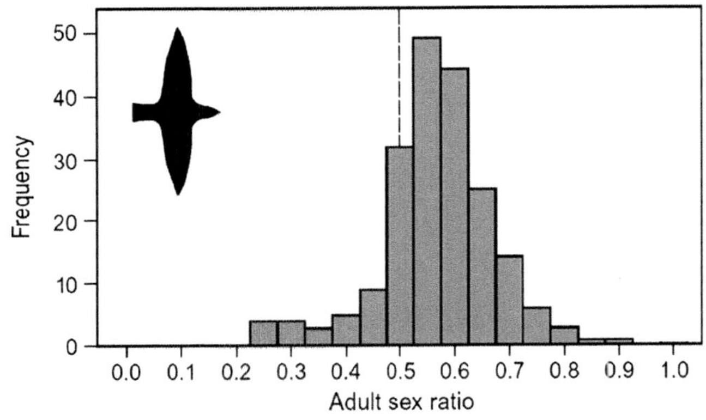


**Fig. 15.7** Adult sex ratios (proportion of males in the adult population) in 201 species of birds (data from Donald 2007). Frequency is the number of species. (Figure from Székely et al. 2014; © 2014 The Authors. *Journal of Evolutionary Biology* © 2014 European Society for Evolutionary Biology, used with permission)

systems of species and populations. If resources critical for successful reproduction are unevenly distributed in space (i.e., clumped or patchy), some individuals may be able to exert control over those spaces (e.g., by defending a territory that encompasses those resources) and, to the extent that availability of these resources influence mate choice, may then attract multiple mates (Emlen and Oring 1977; Fig. 15.9). Several studies have provided evidence that resource availability can influence mating systems. For example, male Northern Harriers (*Circus cyaneus*) with territories where supplemental food was provided (in the form of dead day-old Domestic Chickens or pieces of rabbits) on feeding posts were able to successfully breed with more mates than did males with non-supplemented territories or territories where potential predators were trapped and removed (Amar and Redpath 2002; Fig. 15.10). Similarly, male Eurasian Bitterns (*Botaurus stellaris*) with territories that included better quality nest sites and better access to food (small fish) had more nesting females than males with territories that lacked those critical resources (Kasprzykowski and Polak 2013). In a comparative study of 17 species of *Acrocephaline* warblers with mating systems that range from social monogamy to polygyny, Leisler et al. (2002) found that species occupying highly productive habitats with greater

**Box 15.4 Floaters**

Among many species of birds, some individuals in a population remain unmated during the breeding season and are referred to as "floaters." These individuals have no fixed territory and so may range over large areas (Smith 1978). Floaters may be young birds with no previous breeding experience (e.g., Mumme 2015) or, less often, older individuals (e.g., Shutler and Weatherhead 1991). Two hypotheses have been proposed to provide an ultimate explanation of this behavior: Constraint and Restraint. The Restraint hypothesis posits that floaters can increase their lifetime fitness by not breeding, especially when they are young, because the costs associated with breeding can reduce the likelihood of survival. With increasing age and decreased life expectancy, selection favors breeding despite the associated costs (Nur 1984; Blas et al. 2009).

The Constraint hypothesis proposes that individuals, especially young individuals, are constrained from breeding because they do not yet have the skills or experience needed for successful reproduction. As those skills, e.g., foraging and competitive abilities, are acquired with increasing age, selection again favors breeding (e.g., Pugesek 1995; Sergio et al. 2009a, b).

(continued)

**Box 15.4** (continued)

Blas and Hiraldo (2010) noted that a number of proximate factors can mediate floating behavior and that these factors can have both exogenous and endogenous bases. Exogenous factors include such things as foraging ability, body condition, dominance status, and access to territories and mates that may contribute to individuals becoming floaters. Endogenous factors, such as physiological responses to stress and a physiological readiness to breed, can also help determine if an individual attempts to breed or becomes a floater (Blas and Hiraldo 2010).

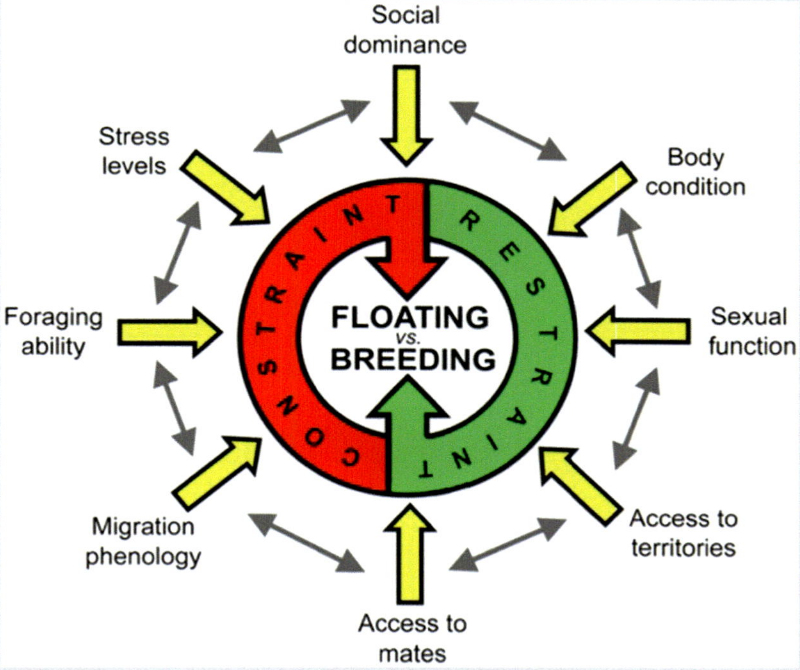


Proximate (outer yellow arrows) and ultimate evolutionary mechanisms (inner circle: constraint and restraint) that potentially explain deferred reproduction, i.e., becoming floaters, by birds. Proximate factors may interact with each other (gray arrows) and exert different selective pressures through constraint and restraint. The gray arrows represent just some potential interactions, but any pair of factors could potentially interact. For example, young birds may have less access to territories due to competition with older, more dominant individuals, and this competition may have a negative effect on the body condition of young birds, which could inhibit or suppress the production of sex hormones. If such conditions increase the potential costs of reproduction, then the best option may be to become a floater. (Figure from Blas and Hiraldo 2010; © 2009 Elsevier Inc., used with permission)

Among and within species, floaters can employ different space-use strategies (Newton 1979; Penteriani et al. 2011). Some may remain geographically separate from breeding individuals, whereas others may either exist along the fringes of breeding habitat or in the interstices among, and sometimes in, breeding territories. Male floaters on the fringes of or between and in breeding territories can sometimes enhance their fitness by engaging in extra-pair copulations with paired females (Pearson et al. 2006; Stapleton et al. 2007; Brekke et al. 2011). However, the success of male floaters is generally limited. For example, Cooper et al. (2009) estimated that only 6.7% of male Eastern Kingbirds (*Tyrannus tyrannus*) that were floaters sired young. In a population of

(continued)

**Box 15.4** (continued)

Tree Swallows (*Tachycineta bicolor*), 52% of all nestlings were extra-pair young and 13% of all extra-pair young were fathered by floaters (Kempenaers et al. 2001).

Copulating Tree Swallows. (Photo by Ken Thomas, Wikipedia, CC0 Public Domain)

In some species, female floaters can also achieve some reproductive success via conspecific brood parasitism. Such behavior has been reported in European Starlings (*Sturnus vulgaris*; Sandell and Diemer 1999), Gray Starlings (*Sturnus cineraceus*; Saitou 2001), and American Coots (*Fulica americana*; Lyon and Eadie 2008; Honza et al. 2009).

Floaters can also be important in population dynamics (Penteriani et al. 2005, 2011; Lee et al. 2017). The stability of some populations may even be more sensitive to changes in the population vital rates of floaters than those of breeders (Hunt 1998; Penteriani et al. 2011). For some species of birds, the floater-to-breeder ratio may be a useful indicator of the status of populations. For example, Newton (1988) suggested that a ratio of one or more floaters for each breeder was indicative of a healthy population of Peregrine Falcons (*Falco peregrinus*), and that higher ratios were associated with more stable breeding populations. The presence of floaters in such populations can help buffer the loss of breeding individuals (Kokko and Sutherland 1998).

food availability were more likely to be polygynous, whereas species in less productive habitats were more likely to be socially monogamous.

A critical factor in avian nest success is avoiding nest predation so territories with locations or characteristics that reduce the likelihood of nest predation may be perceived by females as being higher in quality. Among Great Reed Warblers (*Acrocephalus arundinaceus*), nest predation rates vary among male territories,

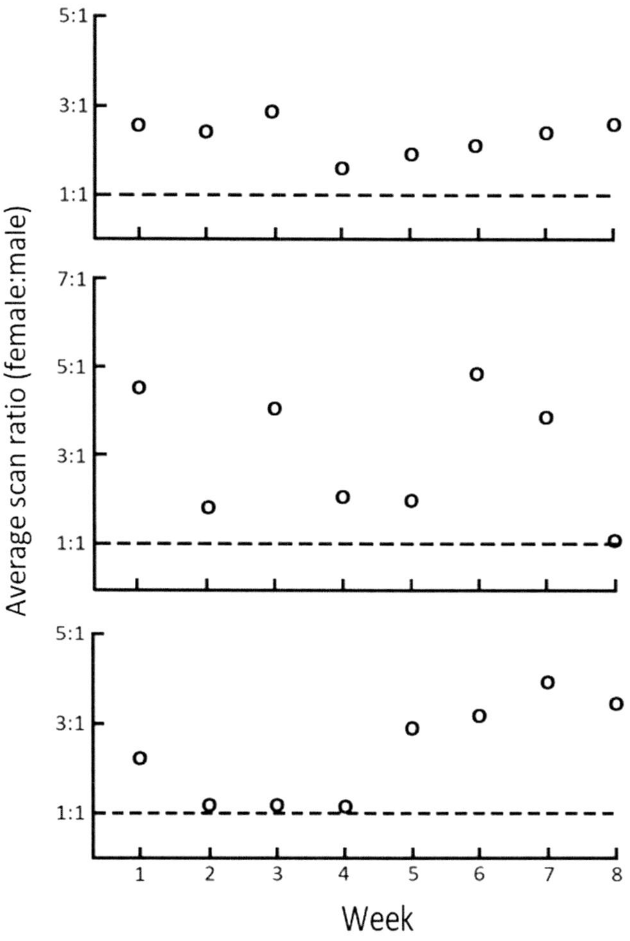


**Fig. 15.8** Seasonal variation in the estimated operational sex ratio (OSR; receptive females/receptive males) of polyandrous Wilson's Phalaropes (*Phalaropus tricolor*) in Saskatchewan, Canada, over three breeding seasons (1983–1985). OSR was estimated based on scan samples (scanning the study area weekly and recording the number of males and females) and assuming that birds observed were reproductively active (i.e., not incubating eggs or caring for young). (Figure modified from Colwell and Oring 1988a; © 1988 Springer-Verlag, used with permission)

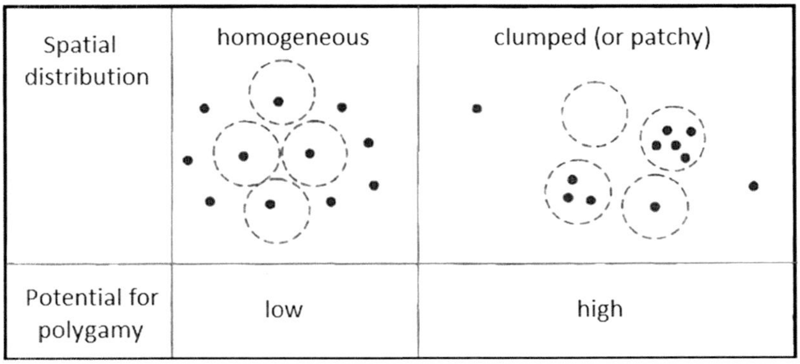


**Fig. 15.9** Spatial and temporal distribution of resources (e.g., food) and polygamy potential. In (a) and (b), dots represent resources and the circles are defended areas. (Figure from Davies and Krebs 2009; © 2009 John Wiley and Sons, used with permission)

likely because of a patchy distribution of potential nest predators. Male Great Reed Warblers are apparently able to discern spatial differences in predator abundance, perhaps by encounter rates or previous experience, and males that arrive in breeding areas earlier preferentially establish

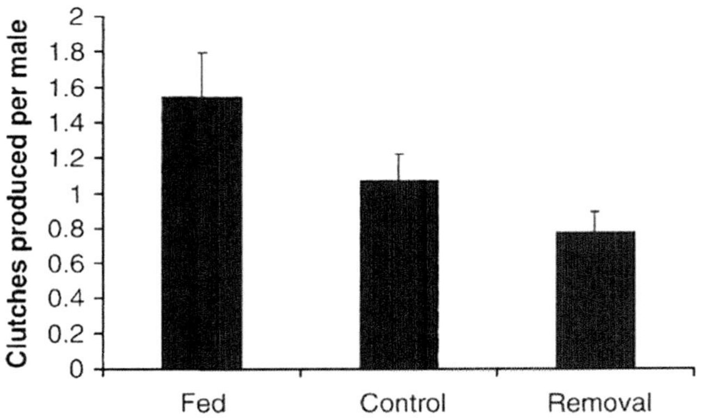


**Fig. 15.10** Mean (± standard error) number of breeding females per male Northern Harrier (*Circus hudsonius*) in control territories, and territories improved in quality either by providing supplemental food (Fed) or by capturing and removing (Removal) potential nest predators (Hooded Crows, *Corvus cornix*). (Figure from Amar and Redpath 2002; © 2006 John Wiley and Sons, used with permission)

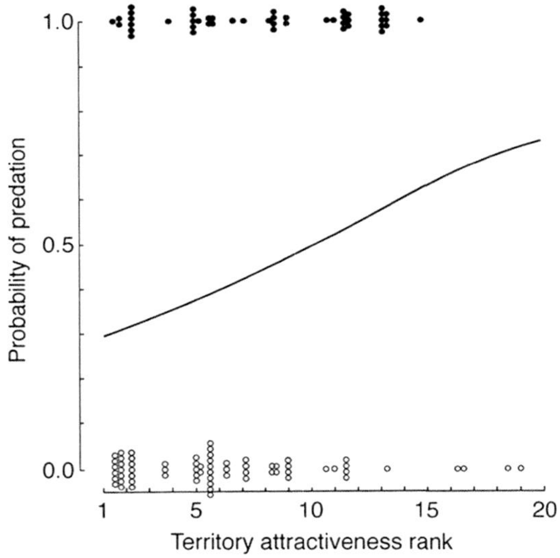


**Fig. 15.11** Relationship between probability of nest predation and territory attractiveness or quality in Great Reed Warblers (*Acrocephalus arundinaceus*). Territory attractiveness was based on the order in which territories were occupied by males (e.g., territory ranked 1 was the territory occupied by the first male to arrive at the study site). Successful nests are indicated by open circles and predated nests by solid circles. Lower ranks correspond to more attractive territories. Also shown is the fitted logistic regression line of the probability of predation relative to the attractiveness rank of territories where nests were located. (Figure from Hansson et al. 2000; © 2000 by the Ecological Society of America, used with permission)

territories in areas with fewer potential nest predators (Hansson et al. 2000). Because safe nest sites represent an important resource for females, earlier arriving males with higher-quality (or more "attractive") territories had more mates (i.e., were more likely to be polygynous) than males with less "attractive" territories and nests of females in the more "attractive" territories were also more likely to be successful (Fig. 15.11). Similar results have been reported for Red-winged Blackbirds (*Agelaius phoeniceus*), with females preferring already-mated males with territories that provided over-water, and less likely to be predated nest sites, as mates over unpaired males with territories without over-water sites (Pribil and Searcy 2001). Such results suggest that a patchy distribution of territories with high-quality nest sites can also influence avian mating systems.

## 15.5 Sexual Conflict and Cooperation

An important factor in the evolution of avian mating systems is that the evolutionary interests of males and females sometimes conflict. In other words, the optimal outcome of interactions between males and females that influence their

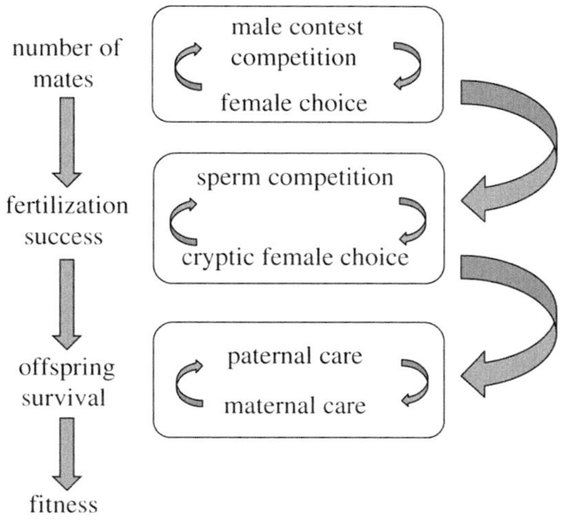


**Fig. 15.12** Stages of reproduction where there may be sexual conflict. Potential differences between males and females in who to mate with and the optimum number of mates, the optimum "fertilization success" (e.g., number of eggs fertilized by males and choice of sperm to fertilize their eggs by females), and the optimum amount of parental care to provide call influence the reproductive success (fitness) of male and female birds. (Figure modified from Kvarnemo and Simmons 2013 © 2013 The Authors. Published by the Royal Society under the terms of the Creative Commons Attribution License, https://creativecommons.org/licenses/by/4.0/)

reproductive success, or fitness, are often different. There is the potential for sexual conflict during all phases of reproduction, beginning with courtship and continuing through mating, postcopulatory events, and parental investment (Fig. 15.12). The degree to which this conflict occurs varies among species and this variation, in conjunction with phylogenetic constraints and environmental or ecological conditions, has contributed to the evolution of different avian mating systems.

During courtship and mating, sexual conflict may arise due to differences between males and females in choice of mate(s) (with males typically less choosy than females) and in the extent to which they would benefit from additional matings (Box 15.5 Sexual Conflict and Avian Genitalia). Females sometimes benefit from mating with more than one male (e.g., in polyandrous species), but males typically can benefit more by mating with multiple females. Sexual conflict can also occur after mating. Females can potentially bias paternity of offspring by physically manipulating ejaculates (e.g., ejecting sperm from their reproductive tract) or by selectively allowing the sperm from certain males (if they have mated with more than one male) to fertilize their eggs (referred to as cryptic female choice). Males, on the other hand, may seek additional copulations with a female to increase the likelihood of their sperm fertilizing eggs (sperm competition), potentially resulting in differences between the sexes in copulation frequency (Fig. 15.13). Finally, sexual conflict can also occur after eggs have been fertilized and laid because each sex would benefit by having the other parent provide most of the parental care (Box 15.6 Parental Conflict in Birds). Although such conflict between the sexes is an unavoidable result of unrelated individuals interacting to produce offspring, reproduction also involves much cooperation. For male and female birds, reproductive success is determined both by such cooperative interactions with their sexual partners and by how the various sexual conflicts are resolved (Wedell et al. 2006).

## 15.6 Social Monogamy, Genetic Monogamy, and Genetic Promiscuity

The social mating systems of birds include social monogamy, polygyny, polyandry, social promiscuity, and polygynandry, and possible genetic relationships include genetic monogamy, multiple male mating, multiple female mating, and genetic promiscuity. Among social mating systems, most species of birds are socially monogamous (about 90%). For most of these species, social monogamy is obligatory and has been favored by natural selection because both males and females benefit, having, on average, more young (i.e., have greater reproductive fitness) when both contribute to raising young (Lack 1968). In contrast, facultative social monogamy occurs when individuals may have greater reproductive success if able to acquire additional mates, but are unable to do so. For example, polygynous male Tree Swallows (*Tachycineta bicolor*) fledge more young than

**Box 15.5 Sexual Conflict and Avian Genitalia**

Males in most species of birds do not have a phallus, or intromittent organ, and sperm is introduced into female reproductive tracts via cloacal contact ("cloacal kiss"). In some species of birds, a phallus begins to develop in male embryos, but then cell division stops and the cells then begin to die (programmed cell death, or apoptosis). Given that intromittent organs facilitate internal fertilization, why would natural selection favor their loss in most birds? One likely explanation is that this represents a case of sexual conflict where the outcome favors females. For males lacking an intromittent phallus, copulation requires female cooperation, i.e., presentation of the cloaca and eversion of the vagina. Females, therefore, have greater control over who fertilizes their eggs, and this is consistent with the idea that female choice played a role in the reduction and loss of the phallus in most species of birds.

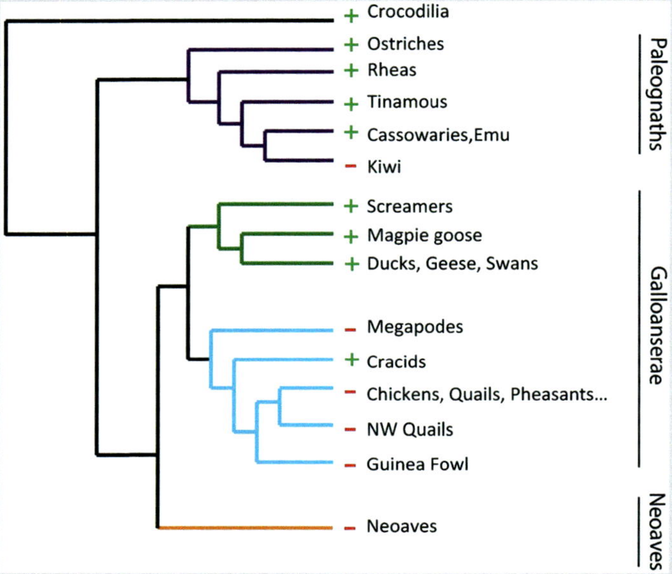


Phylogenetic distribution of intromittent phalluses in birds. + indicates that males have an intromittent phallus, – indicates that males do not have an intromittent phallus. Paleognaths are the ratites, Galloanserae are the waterfowl and gallinaceous birds, and Neoaves are all other birds. (Figure from Herrera et al. 2013; © 2013 Elsevier Ltd., used with permission)

Among waterfowl, however, males do have intromittent organs, and the morphology of the genitalia of both males and females varies considerably. In some species, males have a short phallus and females have simple, straight vaginas. In other species, however, males have a long phallus and females have elaborate vaginas, with spirals and blind-ending pouches. The genitalia of these latter species of waterfowl represent examples of both coevolution and sexual conflict. During the breeding season, males in these species of waterfowl often attempt to force females to copulate and selection has favored the evolution of complex vaginas because such vaginas make it more difficult for males to evert their phalluses and deposit sperm into a female's oviduct. During receptive copulations, females assume a posture with body prone and tail lifted to expose the cloaca (McKinney 1992). When males attempt forceful copulations, females struggle to

(continued)

**Box 15.5** (continued)

escape and do not adopt receptive postures, making it difficult for males to evert their phalluses into a female's vagina. Thus, convoluted vaginal morphology in several species of waterfowl likely evolved because it gives females greater control over who fertilizes their eggs (Brennan et al. 2010).

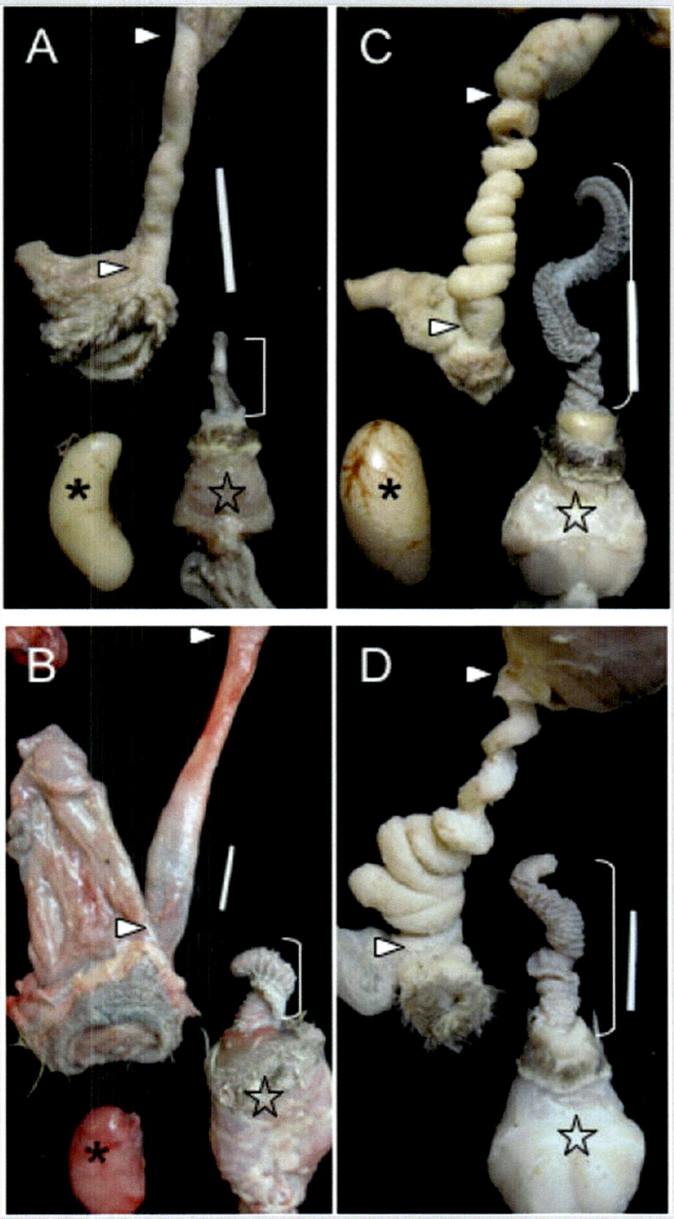


(continued)

**Box 15.5** (continued)

Examples of genital covariation in waterfowl. In species of waterfowl where forced copulations do not occur, such as (**a**) Harlequin Ducks (*Histrionicus histrionicus*) and (**b**) African Geese (*Anser cygnoides*), males have a short phallus and females have simple vaginas. In species where attempted forced copulations by males are common, like (**c**) Long-tailed Ducks (*Clangula hyemalis*) and (**d**) male Mallards (*Anas platyrhynchos*), males have a long phallus and females have very elaborate vaginas. Note in (**c**) and (**d**) that the vaginas and phalluses spiral in opposite directions, making it even more difficult for males to forcefully insert their phallus into a female's reproductive tract to deposit sperm. ] = Phallus, * = Testis, ⋆ = Muscular base of the male phallus, and ▷ = upper and lower limits of the vagina (size bars = 2 cm). (Figure from Brennan et al. 2007; © 2007 Brennan et al., open-access article distributed under the terms of the Creative Commons Attribution License, https://creativecommons.org/licenses/by/4.0/)

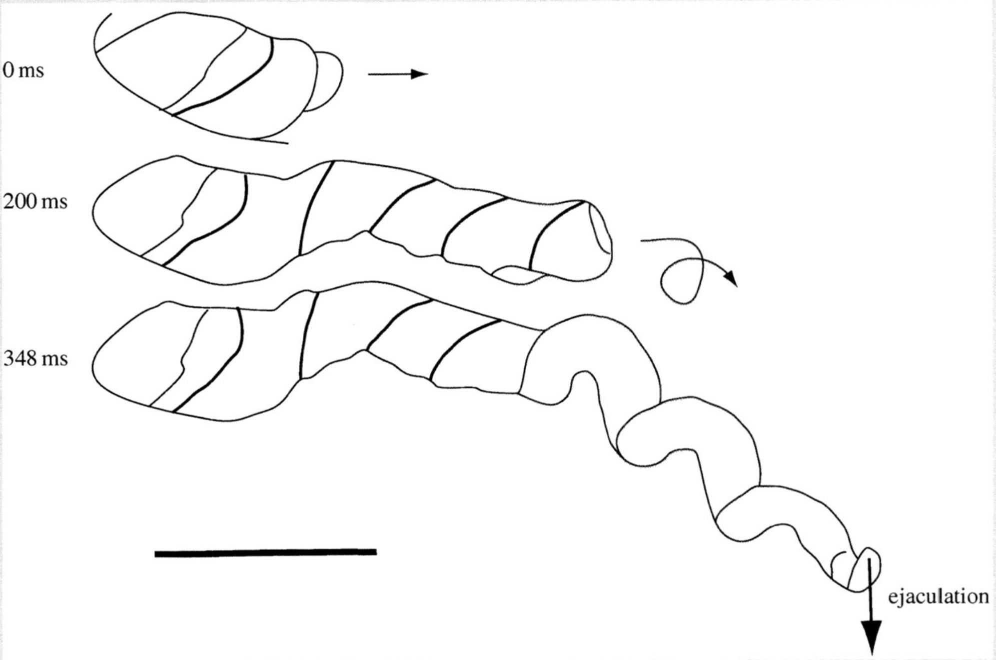


Explosive eversion of the phallus of a male Muscovy Duck (*Cairina moschata*). Tracings are from a high-speed video and arrows indicate direction of movement. Scale bar = 5 cm. (Figure from Brennan et al. 2010; © 2009 The Royal Society, used with permission)

monogamous males, but social monogamy is much more common in Tree Swallow populations than polygyny. This is likely because competition for nest sites (cavities) in this secondary-cavity-nesting species, in combination with female–female aggression, prevents most males from acquiring a territory with suitable cavities far enough apart so that two females can breed in the same territory (Dunn and Hannon 1991). Facultative social monogamy resulting at least in part from female aggression has also been reported in several other species of birds (e.g., Arcese 1989; Veiga 1992; Kempenaers 1994; Sandell 1998). In some populations of birds, male-biased sex ratios may also make it difficult for males to acquire multiple mates, resulting in facultative social monogamy (Ligon 1999).

For species of birds where biparental care results in the production of more young and/or young in better condition and more likely to survive than when care is provided by just one parent, monogamy is obligate (Murray 1984; Ligon

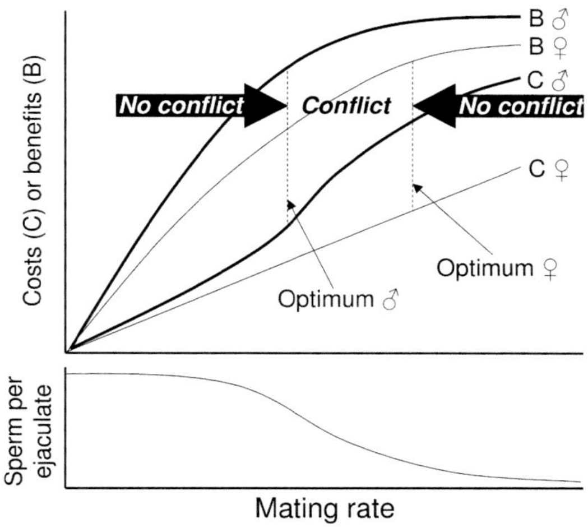


**Fig. 15.13** Hypothetical example of possible sexual conflict in copulation, or mating, rate for a mated pair in a socially monogamous species where high-quality individuals may also engage in copulations outside of the pair bond. Up to a point, both males and females benefit from repetitive mating with their mate because it increases the probability of successful fertilization of a female's eggs and reduces the likelihood that the resident male will be cuckolded. The benefit increase may be more gradual for females because copulations with a high-quality mate likely increases offspring quality, but not quantity. Both sexes also pay increasing mating costs because of an increased risk of predation while mating and the expenditure of time and energy. However, males may also have the additional cost of sperm depletion that could reduce the likelihood of fertilizing the eggs of potential future extra-pair partners and, in addition, time spent with their mate means less time available for pursuing extra-pair copulations. Optimum mating rates for males and females are where the difference between costs and benefits is greatest and, because of the cost to males of sperm depletion and loss of time to pursue extra-pair copulations, the optimum within-pair mating rate is lower for males than females. At mating rates between the optimum values for the two sexes, there is sexual conflict between resistant males and persistent females. Actual mating rates would likely be in this zone of conflict because both sexes initially benefit from higher mating rates and both suffer increasing costs at rates above optimum levels. (Figure from Bro-Jørgensen 2007; © 2007 Elsevier Ltd., used with permission)

1999). Based on data available for 5143 species of birds, Cockburn (2006) reported evidence of biparental care in 4166 species (81%) (followed by: cooperative breeding, 9%; female-only care, 8%; male-only care, 1%; and no care [e.g., brood parasites], 1%). Given that there are more than 10,000 species of birds, this percentage is clearly subject to change as data for additional species become available. However, social monogamy and biparental care are the most common mating and parental care strategies for present-day birds.

For most socially monogamous species of birds, pair bonds between males and females last for a single breeding attempt or breeding season. Persistent pair bonds lasting for multiple years are far less common, but have been reported in about half of the 40 orders of birds and about 20% of bird families. Remating can potentially reduce time and energy costs associated with searching for mates and pair-bond formation, allowing earlier breeding (e.g., Adkins-Regan and Tomaszycki 2007), which generally improves reproductive success (e.g., Saino et al. 2012). Remaining with the same mate can also increase pair familiarity and enhance coordination in breeding activities such as predator defense and provisioning nestlings (e.g., Spoon et al. 2006; Mariette and Griffith 2012). All of these factors can potentially enhance a pair's reproductive success, as demonstrated in several studies of species where pairs remain together for multiple breeding seasons. For example, pairs of Blue-footed Boobies (*Sula nebouxii*) that remained together had larger broods and fledged more young than newly formed pairs (Sánchez-Macouzet et al. 2014; Fig. 15.14). For Blue-footed Boobies and other species, the greatest improvement in breeding success often occurs during the second breeding season together, likely due to the greater familiarity with their mate's traits and behavior acquired during their first breeding event together (Sánchez-Macouzet et al. 2014). After the second breeding season, the benefits of this familiarity may be less apparent, but additional improvement in pair coordination may continue to improve their reproductive performance (e.g., Black 2001; Nisbet and Dann 2009; Fig. 15.15).

Despite the potential advantages, most individuals in most species of birds do not form long-term pair bonds. One reason for this is simply mortality: a bird's mate may die or be predated before the next breeding season (Box 15.7 Avian Generation Lengths). In addition,

**Box 15.6 Parental Conflict in Birds**

Raising young requires time and energy and can also increase the likelihood of predation (e.g., frequent trips to and from nests may attract the attention of predators). As a result, each parent can potentially benefit by having its mate provide most of the parental care. The extent to which this "conflict" is expressed varies among species. Olson et al. (2008) conducted a comparative study of parental conflict using 193 species from 41 families of birds and found that male and female parental care were negatively correlated: when one sex provided more care the other sex provided less. Factors most important in determining the relative contributions of males and females to parental care were offspring development and mating opportunities. Among altricial species, additional mating opportunities have little effect on the amount of parental care provided by males relative to that of females, whereas females tend to provide less care relative to that of males when they have additional mating opportunities. However, among precocial species, both males and females provide less care relative to the other sex when they have additional mating opportunities (e.g., polygyny for males and sequential polyandry for females), suggesting that, for species where young need little care, both males and females respond to enhanced mating opportunities. Parents in altricial species, however, may be under stronger selection pressure to care for their young together. In contrast to precocial species, they cannot increase their reproductive success by deserting, but sneaky extra-pair copulations remain an option for both males and females.

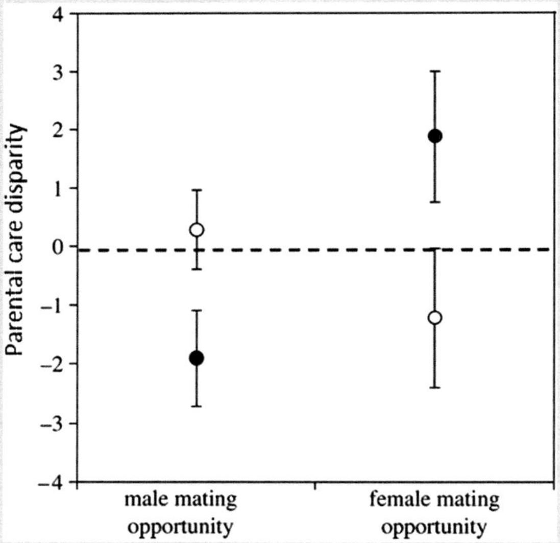


Difference between altricial (open circles) and precocial (filled circles) species of birds in the relationship between parental care disparity (relative amount of male care minus that of females) and mating opportunities for each sex. Error bars are 95% confidence limits. (Figure from Olson et al. 2008; © 2007 The Royal Society, used with permission)

birds that do survive to the next breeding season may choose to "divorce" their previous mate and pair with another individual. Two hypotheses to explain avian divorce include the incompatibility hypothesis (Coulson 1966) and the better-option hypothesis (Davies 1991). The incompatible

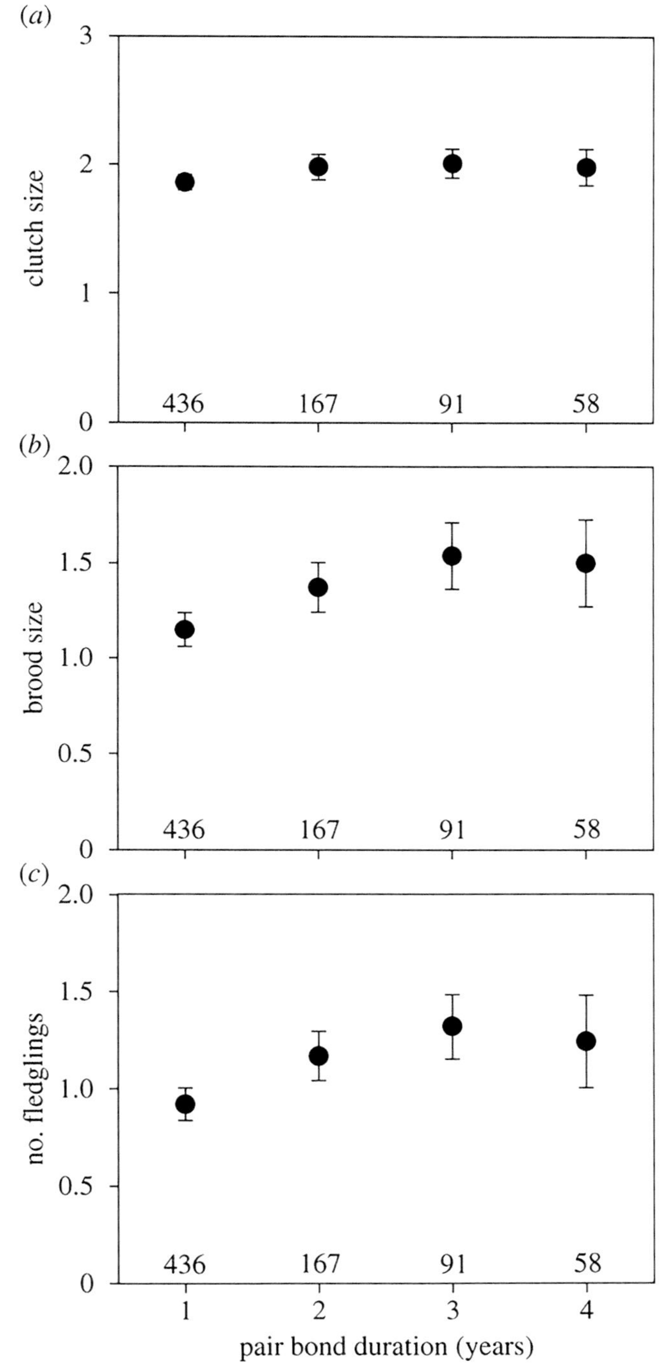


**Fig. 15.14** Relationship between pair-bond duration of male and female Blue-footed Boobies (*Sula nebouxii*) and reproductive performance. Pairs that bred together for more years had larger brood sizes (**b**) and more fledglings (**c**), but clutch size (**a**) was not affected by pair-bond duration. Numbers at the bottom of figures are sample sizes. Data are presented as means ±95% confidence interval (CI). (Figure from Sánchez-Macouzet et al. 2014; © The Authors. Published by the Royal Society, used with permission)

hypothesis posits that a pair may be incompatible, e.g., because they are closely related (Kempenaers et al. 1998), and both females and males could benefit from pairing with a new mate. On the other hand, the better-option hypothesis (Davies 1991) suggests that divorce is initiated by one member of a pair and occurs when that individual can improve their reproductive success by

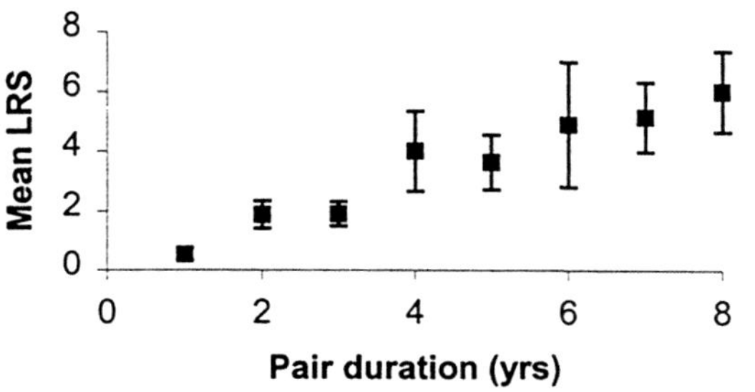


**Fig. 15.15** Relationship between pair-bond duration and lifetime reproductive success (LRS) in Barnacle Geese (*Branta leucopsis*). LRS was the cumulative number of offspring that associated with parents on arrival in wintering areas. Error bars indicate SEs. (Figure from Black 2001; © 2001 Oxford University Press, used with permission)

breeding with a higher-quality mate or in a better territory. Only the bird initiating the divorce (the choosing sex) is expected to benefit in terms of fitness. Evidence in support of this hypothesis has been provided in several studies where either females (e.g., Ramsay et al. 2000; Heg et al. 2003; Garcia-Navas and Sanz 2011) or males (e.g., Ludwigs and Becker 2007; Pérez-Staples et al. 2013) increased their reproductive output after divorce. In further support of the better-option hypothesis, Culina et al. (2015) reviewed the results of studies of 64 species of birds and found that, in general, divorce was triggered by relatively low breeding success and birds that initiated the divorce had greater reproductive success with their new mates.

The respective parental roles of males and females in socially monogamous species with biparental care vary among nesting stages and among species. Females generally play important roles in building nests, incubating eggs, and caring for nestlings and fledglings. Male contributions may, to greater or lesser degrees, involve feeding mates, helping with nest construction, incubating eggs, and caring for nestlings and fledglings (Fig. 15.16). One important factor that influences the role of males is development mode. Among altricial species, males often feed females prior to and during nest building and incubation (Box 15.8 Males Feeding Mates) and feed nestlings and fledglings, but typically have a lesser role in nest building and incubation. Among precocial species, however, males often have a greater role in incubating eggs and caring for young after they leave the nest (Silver et al. 1985). Beyond these direct contributions, males typically also provide care indirectly, e.g., by defending territories that allow more exclusive use of food resources, and by

**Box 15.7 Avian Generation Lengths**

The generation length of a species is defined as "the average age of parents of the current cohort" (IUCN 2019:29). Using available information concerning the age of first reproduction, maximum longevity, and annual adult survival, Bird et al. (2020) determined that the median generation length of extant species of birds was 2.99 years, and ranged from 1.42 years (Double-barred Finch, *Taeniopygia bichenovii*) to 27.87 years (Southern Royal Albatross, *Diomedea epomophora*). Most species of birds (61%) had generation lengths <3.33 years. Among some orders of birds, mean estimated generation lengths were 2.86 years for Passeriformes, 6.23 years for Psittaciformes, 3.24 years for Piciformes, 7.40 years for Accipitriformes, 4.18 for Strigiformes, 2.33 years for Apodiformes, 6.18 years for Charadriiformes, 11.52 years for Sphenisciformes, 13.14 years for Procellariiformes, 6.68 for Pelecaniformes, 9.81 for Gaviiformes, 5.03 years for Podicipediformes, 4.24 for Columbiformes, 4.73 years for Galliformes, and 5.29 years for Anseriformes. Analysis revealed that body mass and association with marine habitats had a positive effect on generation length whereas clutch size and migratory status (being migratory rather than resident) had negative effects.

(continued)

**Box 15.7** (continued)

White-capped Albatross (*Thalassarche cauta*). These albatrosses have been found to live up to 30.9 years in the wild. (Source: https://genomics.senescence.info/species/entry.php?species=Thalassarche_cauta). (Photo by Ed Dunens, Wikipedia, CC BY 2.0, https://creativecommons.org/licenses/by/2.0/)

Ten families and ten species of birds with the longest and shortest estimated generation lengths (Table from Bird et al. 2020; © 2020 Society for Conservation Biology, used with permission).

| Longest generation length | | Years | Shortest generation length | | Years |
|---|---|---|---|---|---|
| *Family* | | | | | |
| Diomedeidae | Albatrosses | 21.73 | Pardalotidae | Pardalotes | 2.12 |
| Struthionidae | Ostrich | 14.53 | Trochilidae | Hummingbirds | 2.09 |
| Cacatuidae | Cockatoos | 13.94 | Viduidae | Whydahs and indigobirds | 2.05 |
| Fregatidae | Frigatebirds | 13.39 | Estrildidae | Waxbills | 2.05 |
| Cathartidae | New world vultures | 13.21 | Remizidae | Penduline tits | 1.99 |
| Pelecanidae | Pelicans | 12.56 | Phylloscopidae | Leaf warblers | 1.98 |
| Gruidae | Cranes | 12.51 | Polioptilidae | Gnatcatchers | 1.94 |
| Stercorariidae | Skuas | 12.45 | Certhiidae | Treecreepers | 1.89 |
| Phoenicopteridae | Flamingos | 12.35 | Regulidae | Kinglets and firecrests | 1.83 |
| Procellariidae | Petrels, shearwaters | 12.26 | Elachuridae | Elachura | 1.82 |
| *Species* | | | | | |
| Southern Royal Albatross | *Diomedea epomophora* | 27.87 | Bumblebee Hummingbird | *Atthis beloisa* | 1.59 |

(continued)

**Box 15.7** (continued)

| | | | | | |
|---|---|---|---|---|---|
| Andean Condor | *Vultur gryphus* | 27.25 | Grey-headed Manakin | *Lonchura caniceps* | 1.57 |
| Sulphur-crested Cockatoo | *Cacatua galerita* | 27.21 | Timor Zebra Finch | *Taeniopygia guttata* | 1.57 |
| Laysan Albatross | *Phoebastria immutabilis* | 26.70 | Madagascar Partridge | Margaroperdix *madagarensis* | 1.53 |
| Northern Royal Albatross | *Diomedea sanfordi* | 25.97 | Crested Bobwhite | *Colinus cristatus* | 1.53 |
| Antipodean Albatross | *Diomedea antipodensis* | 25.86 | Black-tailed Waxbill | *Estrilda perreini* | 1.50 |
| Kakapo | *Strigops habroptila* | 25.82 | Daurian Partridge | *Perdix dauurica* | 1.49 |
| Northern Fulmar | *Fulmarus glacialis* | 25.34 | Amazonian Scrub Flycatcher | *Sublegatus obscurior* | 1.46 |
| Salmon-crested Cockatoo | *Cacatua moluccensis* | 24.19 | White-throated Fantail | *Rhipidura albicollis* | 1.44 |
| Black-browed Albatross | *Thalassarche melanophris* | 23.61 | Double-barred Finch | *Taeniopygia bichenovii* | 1.42 |

defending females, eggs, and nestlings from potential nest predators.

Although the type and extent of parental care provided by males and females varies among species, the results of experimental mate-removal or mate-handicapping studies where either males or females are either removed shortly after eggs hatch or they have been "handicapped" in some way (e.g., trimming the tips of primaries to increase wing loading) indicate that the remaining member of a pair usually cannot fully compensate for the lost contributions of their mate (Bart and Tornes 1989; Ligon 1999). As a result, fewer young may fledge and those that do may weigh

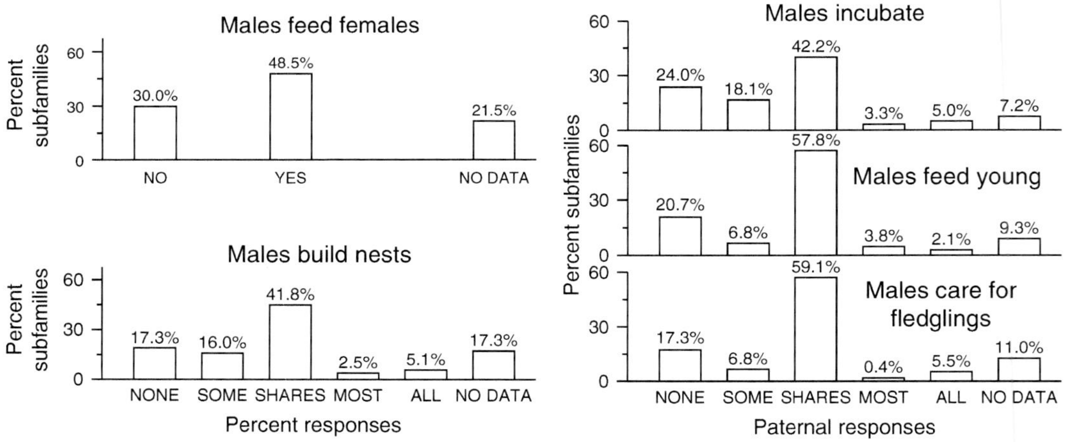


**Fig. 15.16** Types and extent of care provided by male birds in subfamilies with species with a variety of mating systems. Among socially monogamous species with biparental care, male contributions to parental care may involve any or all of the above types of care, from feeding females to caring for fledglings. (Figure modified from Silver et al. 1985; © 2015 Oxford University Press, used with permission)

**Box 15.8 Males Feeding Mates**

Males in several species of birds provide food for their mates prior to and during nest building and incubation. Based on when it occurs, this is commonly referred to as either courtship feeding or incubation feeding. Such behavior may represent signals of male quality and/or parental quality to females, and may also help females meet their nutritional needs during egg formation and incubation. In addition, incubation feeding may allow incubating females to spend more time on nests and less time foraging. Using a dataset of 170 species of passerines (including species where males do and do not feed mates), Galván and Sanz (2011) found that males in species where females alone build nests and incubate eggs were more likely to feed mates. Mate-feeding was also more common in species that feed primarily on low-protein diets consisting of seeds or fruits (rather than being omnivorous or insectivorous). These results suggest that mate-feeding likely helps females meet their energetic requirements (both in terms of calories and protein intake). Species of songbirds where males feed mates also tended to have larger clutch sizes and greater hatching success, indicating that mate-feeding can improve a pair's reproductive success. Mate-feeding prior to and during egg-laying may enhance the nutritional condition of females, allowing the production of larger clutches, and mate-feeding during incubation may allow females to spend more time incubating and guarding eggs, thus improving hatching success.

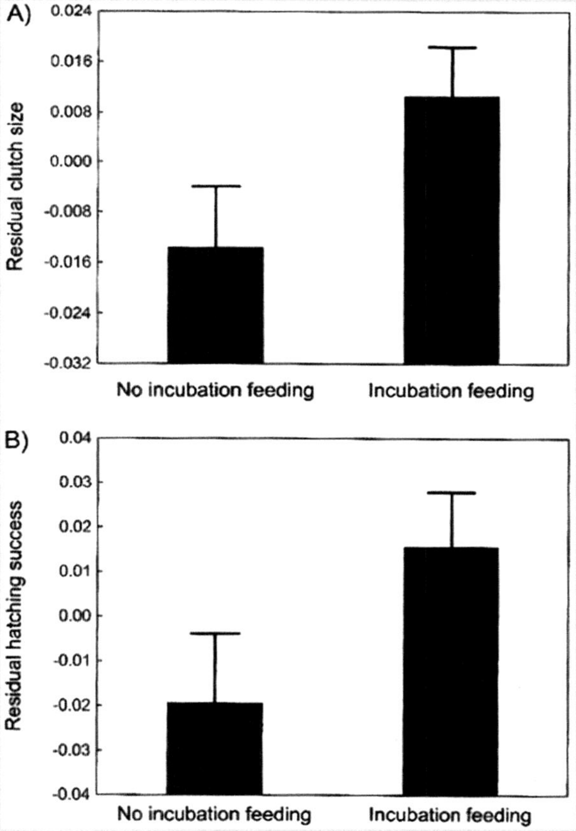


(continued)

**Box 15.8** (continued)

Comparison of clutch sizes (**a**) and hatching success (**b**) in species of songbirds where males feed or do not feed their mates during incubation. The possible effects of female body mass and phylogeny were taken into account in the analysis. Lines above bars are standard errors. (Figure from Galván and Sanz 2011; © 2011 Oxford University Press, used with permission)

less and be less likely to survive. As an example, at nests where male Chestnut-collared Longspurs (*Calcarius ornatus*) were captured and removed on the day after eggs hatched, all nestlings subsequently starved, died of exposure, or were predated and no young fledged (Fig. 15.17). In contrast, at control nests, with both male and female present, fledging success was 50% (Fig. 15.17; Lynn and Wingfield 2003). In another study, Lendvai and Chastel (2008) removed male House Sparrows (*Passer domesticus*) at some nests for two days during the nestling period and found that, despite partial compensation by females, nestlings still gained significantly less mass than those at control nests (Fig. 15.18). In addition to possible negative impacts on nest success, the removal of mates in species with biparental care can also have negative impacts on condition of the remaining individual, e.g., increased stress response (Lendvai and Chastel 2008).

Although mate-removal often has negative effects on nest success, some investigators have found that the remaining mates are able to fledge as many young as those whose mates are not removed. In some cases, this may be due to the timing of mate-removal. For example, Bart and Tornes (1989) removed male House Wrens (*Troglodytes aedon*) about one-third of the way through the nestling period and found that "widowed" females fledged as many young as "non-widowed" females. However, Johnson et al. (1992) removed male House Wrens in some territories before eggs hatched and found that "widowed" females fledged fewer young than "non-widowed" females due to reduced nestling survival. Although these contrasting results may be due to differences in food availability at

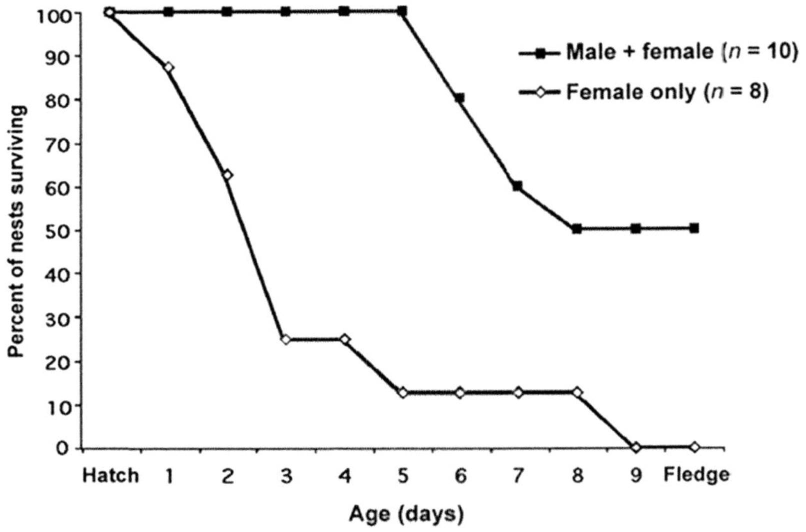


**Fig. 15.17** Percentage of Chestnut-collared Longspur (*Calcarius ornatus*) nests surviving on each day of the nestling period. Males at eight nests were captured and removed on the morning of the day after eggs hatched. (Figure from Lynn and Wingfield 2003; © 2003 Oxford University Press, used with permission)

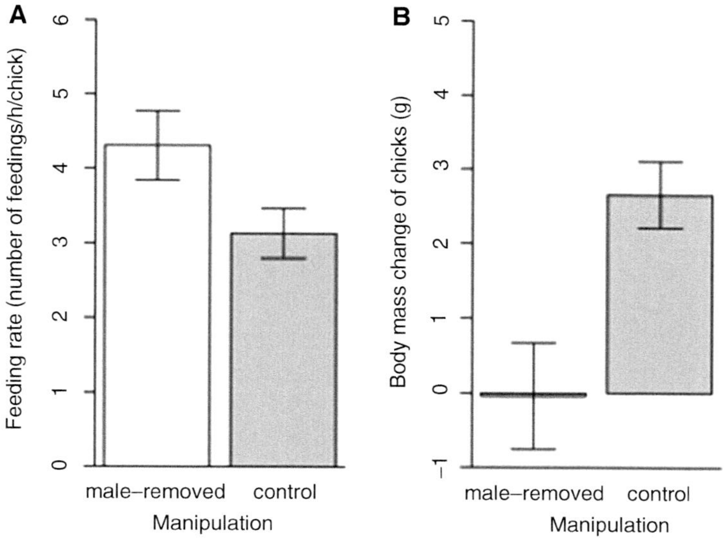


**Fig. 15.18** Effects of two-day experimental male-removal on the (**a**) mean feeding rates (± SE) of females, and (**b**) the mean body mass (± SE) of nestlings. Female House Sparrows (*Passer domesticus*) whose mate was removed ($N = 14$) fed nestlings at significantly higher rates than control females ($N = 20$), and nestlings in the male-removed broods gained significantly less mass than control chicks (and some even lost mass). (Figure from Lendvai and Chastel 2008; © 2007 Elsevier Inc., used with permission)

the two study sites, the reduced fledging success of "widowed" females reported by Johnson et al. (1992) might also be explained by the absence of any assistance in provisioning young by mates during the entire nestling period rather than just a portion of the nestling period.

Other studies suggest that at least some "widowed" females fledge as many young, and young in similar condition and at similar masses at fledging, as "non-widowed" females (e.g., Gowaty 1983). However, at least in some of these studies, potentially important variables,

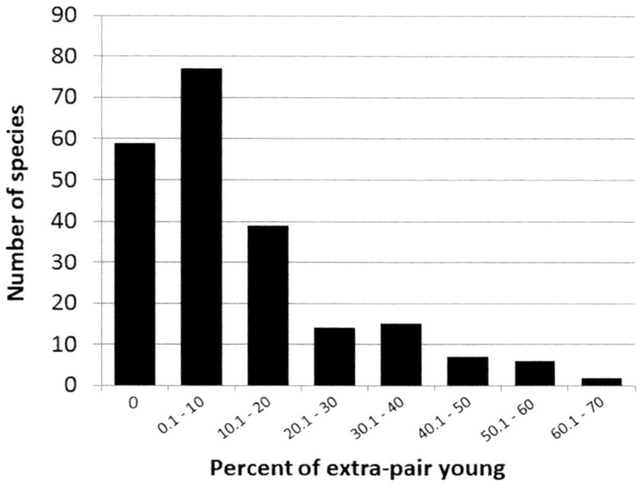


**Fig. 15.19** Variation in rates of extra-pair paternity (percentage of offspring fathered by males other than the social mates of females) for 223 species of birds. (Based on data in Appendix B of Spottiswoode and Møller 2004 + data for 35 additional species based on a literature search)

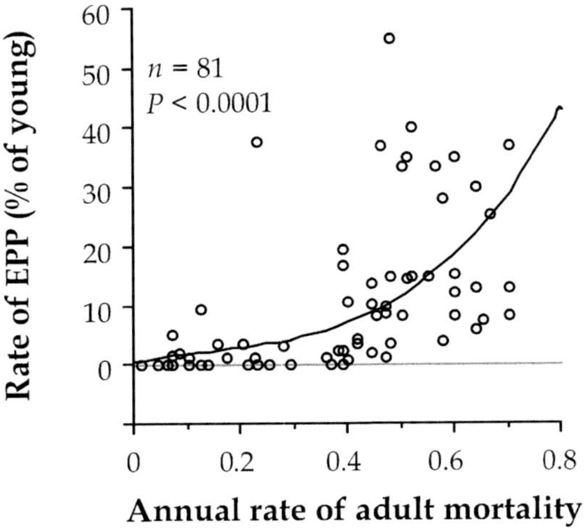


**Fig. 15.20** Relationship between rate of extra-pair paternity (EPP; percentage of young fathered by males other than social mates of females) and adult mortality rate for 81 species of birds. Rates of EPP are clearly much lower in species with lower adult mortality rates. (Figure from Griffith et al. 2002; © John Wiley & Sons Ltd., used with permission)

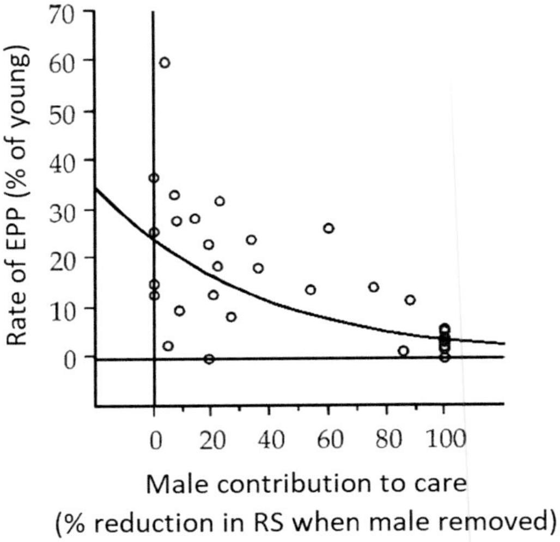


**Fig. 15.21** Relationship between rates of extra-pair paternity (EPP) and interspecific variation in the contributions of males to parental care (based on the extent to which reproductive success [RS] is reduced if males are removed and females must attempt to raise young alone). (Figure from Griffith et al. 2002; © John Wiley & Sons Ltd., used with permission)

such as possible effects of male absence on female condition or on the survival of young during the post-fledging period, may not have been taken into account (Gowaty 1983). Although additional studies are still needed, evidence to date does suggest that, for most, if not all, species, an important benefit of monogamy and biparental care is that both males and females leave more offspring if both help (Lack 1968). In other words, at least in terms of lifetime fitness (rather than for single nesting attempts or single breeding seasons), social monogamy and biparental care are likely the best mating system and parental care strategy for individuals in most species of birds.

Most species of birds are socially monogamous, but most socially monogamous species are not genetically monogamous (Fig. 15.19). Based on published studies, fewer than about 10% of socially monogamous are also genetically monogamous (Griffith et al. 2002). This suggests that at least some individuals in most socially monogamous species of birds benefit in some way by engaging in copulations outside the pair bond. On the other hand, for many species, individuals apparently benefit more by not engaging in extra-pair copulations. Although many questions remain concerning the causes and consequences of extra-pair mating (EPM) and why there is so much interspecific and intraspecific variation in the rates of such mating, species with low rates of extra-pair paternity (EPP; or no extra-pair paternity) do appear to share a number of characteristics. Most importantly, most genetically monogamous species are not songbirds (order Passeriformes), have relatively low annual mortality rates (<30%, Mauck et al. 1999; Fig. 15.20), and relatively high levels of male parental care (Fig. 15.21).

Rates of extra-pair paternity may be low (or zero) in species with low adult mortality rates because females are more likely to suffer retaliation in the form of reduced paternal care (e.g., Matysioková and Remeš 2013) if they cuckold their mate because the reproductive value (i.e., expected future reproduction given an individual's current age) of long-lived males is relatively high. As a result, selection should favor cuckolded males that reduce care in current

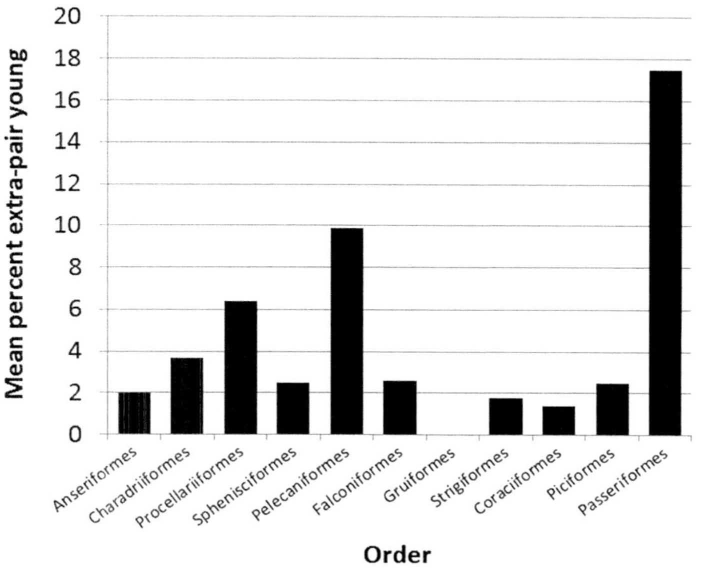


**Fig. 15.22** Variation among orders of birds in the frequency of extra-pair mating. Based on data from: Anseriformes, 10 species; Charadriiformes, 23 species; Procellariiformes, 12 species; Sphenisciformes, 4 species; Pelecaniformes, 4 species; Falconiformes, 11 species; Gruiformes, 3 species; Strigiformes, 8 species; Coraciiformes, 7 species; Piciformes, 4 species; and Passeriformes, 135 species. (Based on data in Appendix B of Spottiswoode and Møller 2004 + data for 39 additional species based on a literature search; total = 227 species)

broods to enhance their future reproductive success (Mauck et al. 1999). In short-lived species, males are less likely to withhold parental care if cuckolded because their reproductive value is lower and so selection should favor males that invest in their current brood (even if some are extra-pair young) rather than risking the uncertainty of surviving until the next breeding opportunity. In addition, Wink and Drycz (1999) suggested that rates of extra-pair paternity may be linked to annual mortality rates because mortality rates are a good predictor of whether or not a social pair will breed together in the future. This is important because prolonged pair bonds enhance coordination and cooperation between pair members and can lead to increased reproductive success (e.g., Sánchez-Macouzet et al. 2014). In long-lived species, both members of a social pair are more likely to survive to breed the next year so females that cuckold their mates may risk retaliation in the form of divorce. In short-lived species, the risk of retaliation via divorce is less important because the probability of both members of the social pair surviving is low.

Among most species of socially monogamous birds, as well species with other social mating systems, a variable percentage of individuals in breeding populations engage in copulations outside the pair bond. Among species where extra-pair matings have been reported, the percentage of extra-pair young in studied populations ranges from just a few percent (many species) to as high as nearly 80% (Superb Fairywrens, *Malurus cyaneus*; Mulder et al. 1994; Fig. 15.21). Most species where the percentage of extra-pair young in populations exceeds 10% are songbirds (Figs. 15.22, 15.23, and 15.24). Given that avian orders and families diverged millions of years ago, this difference between songbirds (order Passeriformes) and non-songbirds (i.e., phylogenetic effect) suggests an important innate component to extra-pair mating. As such, variation in contemporary ecological or genetic factors may

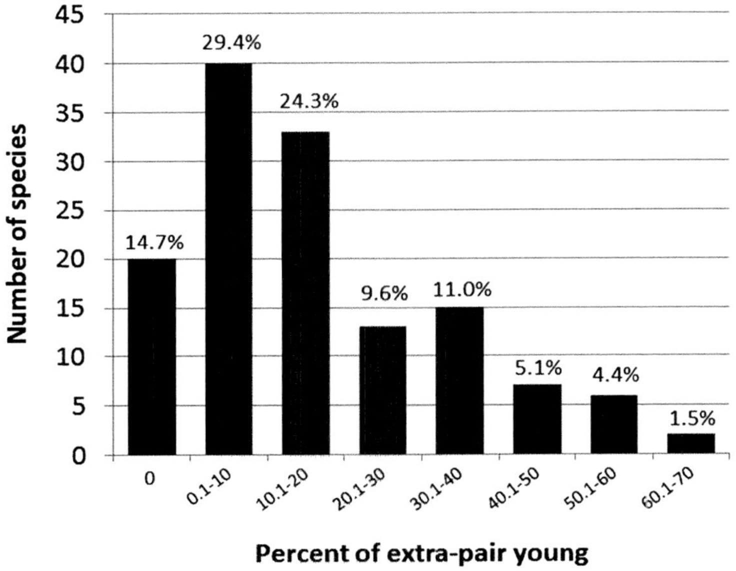


**Fig. 15.23** Variation in percentages of extra-pair young for 136 species of songbirds (Passeriformes). (Based on data in Appendix B of Spottiswoode and Møller 2004 + data for 26 additional species based on a literature search; total = 136 species)

not be able to provide as much insight concerning variation in rates of extra-pair mating among birds in different orders and families as they might for explaining differences between closely related species, between populations of the same species, and between individuals in the same population (Arnold and Owens 2002). Rather than focusing on possible explanations for differences among different orders and families of birds in rates of extra-pair mating, therefore, the more interesting focus is on the factors that might influence variation in those rates among different populations of the same species and on the factors that might influence the behavior (i.e., extra-pair mating or not) of individuals within those populations.

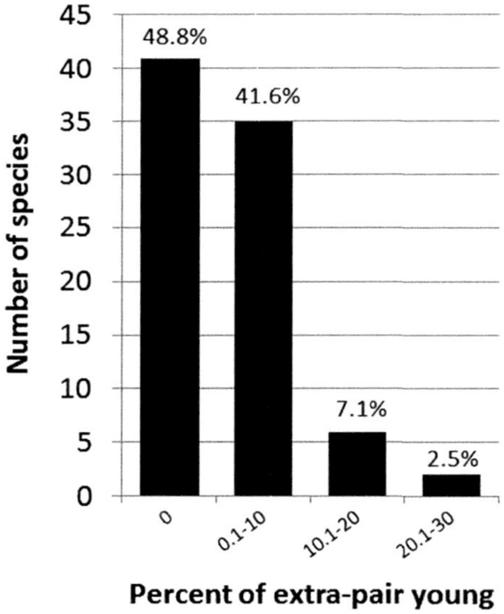


**Fig. 15.24** Variation in percentages of extra-pair young in populations of 84 species of nonpasserines. (Based on data in Appendix B of Spottiswoode and Møller 2004 + data for 13 additional species based on a literature search; total = 84 species)

Since the late 1980s, investigators have used molecular techniques to examine rates of extra-pair mating in an increasing number of populations and species of birds and have used the results of those studies to try and explain variation in those rates as well as the possible costs and benefits of such behavior. This has been a challenge, however, because there are so many variables that can potentially influence rates of extra-pair mating in a population and, further complicating things, those variables may be correlated with each other (Westneat and Stewart 2003). In addition, extra-pair mating is influenced by the outcome of interactions between at least three individuals: a female, a female's mate, and potential extra-pair males. Each individual is attempting to maximize

**Table 15.1** Possible benefits and costs of extra-pair mating for birds

| Proposed benefits for females | Citation |
|---|---|
| Good or compatible genes (genetic benefits) | Akçay and Roughgarden (2007) |
| Fewer infertile eggs (fertility insurance) | Hasson and Stone (2009), Reding (2015) |
| Inclusive fitness gain (extra-pair mating with kin) | Szulkin et al. (2013) |
| Reduced harassment by extra-pair males | Westneat and Stewart (2003) |
| Avoiding inbreeding (with related mate) | Kingma et al. (2013) |
| Access to resources in other territories | Gray (1997) |
| Securing a future partner | Heg et al. (1993) |
| **Benefit for males:** | |
| Additional offspring | |
| **Possible costs for females:** | |
| Loss of parental care by social mate | Arnqvist and Kirkpatrick (2005) |
| Deleterious mutations | Johnson and Gemmell (2012) |
| Inbreeding depression | Szulkin et al. (2013) |
| Increased embryo mortality (polyspermy) | Morrow et al. (2002) |
| Increased sibling competition (less related) | Briskie et al. (1994) |
| **Possible costs for males:** | |
| Increased risk of cuckoldry (due to time spent seeking extra-pair matings) | Westneat et al. (1990) |
| Less time for parental care | Westneat et al. (1990) |
| Sperm depletion | Birkhead (1991) |

Table modified from Forstmeier et al. (2014); © 2014 Elsevier Ltd., used with permission

their fitness, perhaps to the detriment of other individuals, thus creating potential conflicts of interest (Westneat and Stewart 2003). Not surprisingly, therefore, definitive explanations are lacking and debate continues about the causes of inter- and intraspecific variation, and the costs and benefits of extra-pair mating (EPM) for males and females (Table 15.1).

Among the factors that could potentially contribute to differences among populations and species in rates of EPM are breeding density, breeding synchrony, and genetic diversity. The density hypothesis posits that increased proximity to breeding conspecifics increases encounter rates and mating opportunities when searching for potential extra-pair mates, thus reducing the costs of EPM. As a result, when density increases, the rate of EPM should increase as well (Westneat et al. 1990). There is little evidence that variation in breeding density explains interspecific variation in rates of EPM (Griffith et al. 2002). Intraspecifically, the effect of density varies, with some studies finding no effects (e.g., Tarwater et al. 2013), and others finding a positive correlation between density and rates of EPM (e.g., Mayer and Pasinelli 2013). Multiple factors can influence rates of EPM in bird populations (LaBarbera et al. 2010) so the conflicting results reported in studies examining the density hypothesis suggest that, for some species, density, by its effect on opportunities for EPMs, is one of those factors; for other species, other factors are more important.

Breeding synchrony refers to the proportion of females in a population that are fertile (i.e., eggs can be fertilized) at a given time, with high synchrony meaning that a large proportion of females in a population are fertile and potentially available for EPM during a particular time period. Breeding synchrony could potentially impact rates of EPM in two different ways. One hypothesis is that greater synchrony should lead to higher rates of EPMs because a greater proportion of males are displaying or otherwise attempting to solicit copulations and, as a result, females are better able to compare males, facilitating their choice of extra-pair partners (simultaneous display hypothesis; Stutchbury and Morton 1995). Alternatively, more synchronous breeding might lower raters of EPM because, with most females

**Box 15.9 Male Paternity Assurance Behavior**

To maximize their own fitness (reproductive success), males try to prevent their mates from engaging in extra-pair copulations (EPCs). One strategy used by males in many species is to "guard" their mates—remain as close to them as possible for as much time as possible. For example, male Bluethroats (*Luscinia svecica*) stay within one meter of their mates more than 60% of the time around the start of egg-laying (Krokene et al. 1996). The effectiveness of mate-guarding varies among species. In some species, including Bluethroats, mate-guarding is not a very effective paternity-assurance strategy (Johnsen et al. 2003). In other species, mate-guarding does reduce the likelihood of males being cuckolded (although not necessarily with 100% effectiveness) (e.g., Chuang-Dobbs et al. 2001; Komdeur et al. 2007).

Male Bluethroat. (Photo by Daniel Bastaja, Wikipedia, CC0 Public Domain)

Another male strategy is to copulate with their mates more frequently as the start of egg-laying approaches. Male Tree Swallows (*Tachycineta bicolor*), for example, may attempt to copulate with their mates nearly 20 times per hour in the days just preceding the start of egg-laying (Crowe et al. 2009). This strategy had a positive impact on the reproductive success of male Tree Swallows, with a higher proportion of young in their nests being fathered by males with higher copulation rates (Crowe et al. 2009).

(continued)

**Box 15.9** (continued)

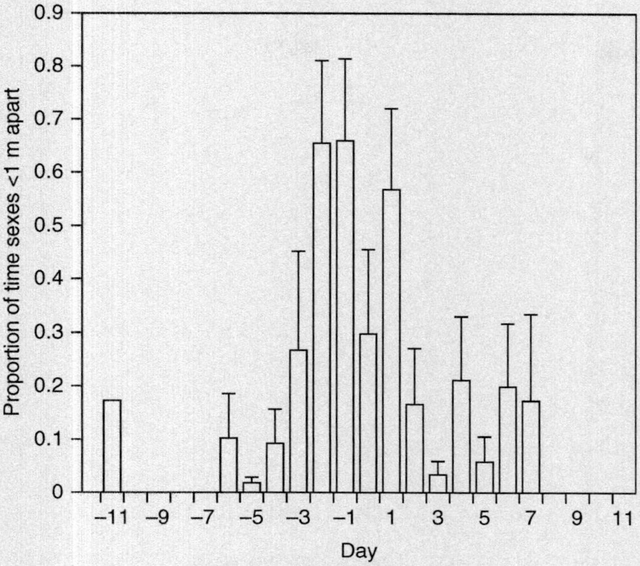


Male Bluethroats (*Luscinia svecica*) increase the time spent near their mates dramatically during the days just prior to the beginning of egg-laying (day 0). (Figure from Krokene et al. 1996; © 1996 The Association for the Study of Animal Behaviour. Published by Elsevier Ltd., used with permission)

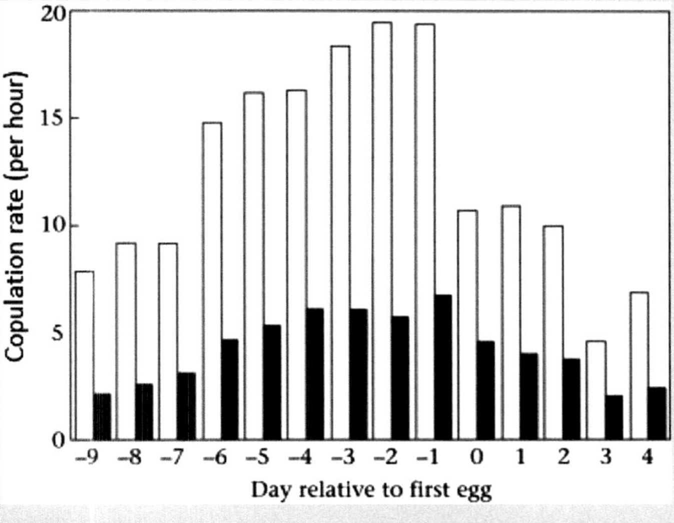


Male Tree Swallows (*Tachycineta bicolor*) increase the number of attempted (and successful) copulations with their mates as the egg-laying period approaches. Open bars indicate attempted copulations; solid bars indicate copulations with cloacal contact. Day 0 is the day the first egg was laid. (Figure from Crowe et al. 2009; © 2008 Published by Elsevier Ltd., used with permission)

(continued)

**Box 15.9** (continued)

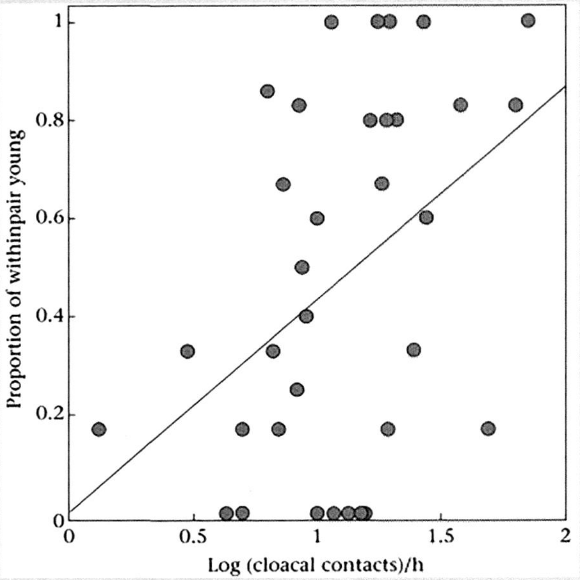


Relationship between within-pair paternity in a male's own brood and the rate of cloacal contacts (per hour for three days prior to egg-laying) in Tree Swallows ($N$ = 34 pairs). (Figure from Crowe et al. 2009; © 2008 Published by Elsevier Ltd., used with permission)

Post-copulatory sexual selection has also provided males with a possible means to enhance their paternity status. Kleven et al. (2009) found that sperm swimming speeds were significantly faster in species of birds with higher rates of extra-pair paternity. Of course, variation in sperm swimming speeds is not as great within species as between species. However, even within species, males with faster sperm could potentially benefit via increased paternity.

(continued)

**Box 15.9** (continued)

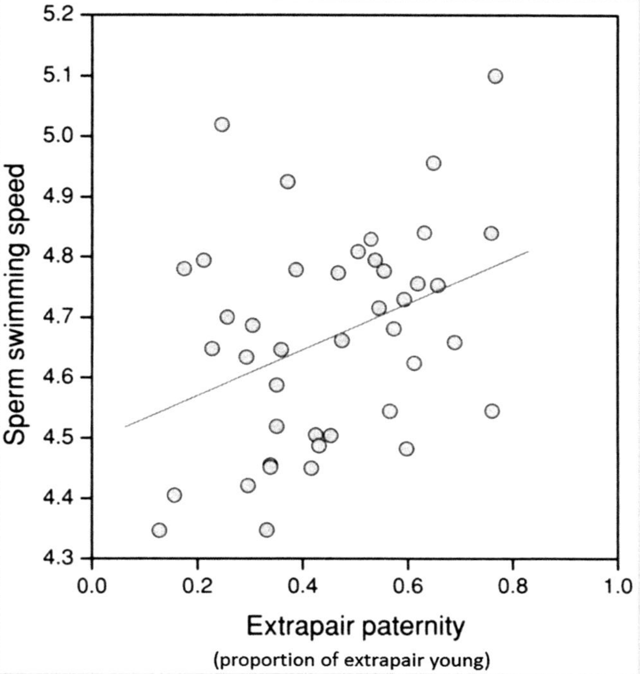


Relationship between sperm swimming speed and frequency of extra-pair paternity in 42 species of birds. Sperm swimming speeds have been log-transformed; actual swimming speeds of sperm ranged from 77.3 to 164.0 micrometers per second. (Figure from Kleven et al. 2009; © 2009 The Author(s). Journal compilation © 2009 The Society for the Study of Evolution, used with permission)

Yet another strategy to enhance paternity status is to produce more sperm. In a study of Red-backed Fairywrens (*Malurus melanocephalus*), Rowe et al. (2022) determined that sperm number (i.e., cloacal protuberance volume) was positively associated with male reproductive success. Specifically, males with more sperm available for copulations had greater within-pair paternity success.

Not surprisingly, the effectiveness of male paternity assurance behaviors varies among and within species because even a single well-timed copulation can potentially fertilize many eggs (Birkhead and Møller 1992; Colegrave et al. 1995). In addition, females may adopt behaviors that allow them to evade male-guarding efforts and seek EPCs (e.g., Mays and Ritchison 2004; Tryjanowski et al. 2007), and males may be prevented from effectively guarding their mates over the entire fertile period because of mate-guarding energy costs and time constraints (Komdeur et al. 2007).

fertile at the same, and for a limited, time, males face time constraints, both because synchrony limits the time available for seeking EPMs and because males must spend at least part of that time with their mates (to copulate with and to guard against potential cuckoldry) (mate-guarding hypothesis; Birkhead and Biggins 1987; Box 15.9 Male Paternity Assurance Behavior).

The results of some studies suggest a positive relationship between breeding synchrony and rate of EPMs (supporting the simultaneous display hypothesis, e.g., Chuang et al. 1999), the results of others suggest a negative relationship (supporting the mate-guarding hypothesis, e.g., Thusius et al. 2001), and the results of still others suggest no relationship (e.g., Arlt et al. 2004). Again, multiple factors can influence rates of EPM in bird populations, including male or female quality, age, genetic makeup, territory quality, intensity of mate-guarding by males, or personality (van Oers et al. 2008; LaBarbera et al. 2010). As such, the conflicting results reported in studies examining the breeding synchrony hypothesis suggest that, for some species, synchrony, by its effect on opportunities for EPMs, is one of those factors; for other species, other factors are more important.

Females may mate with extra-pair males to increase the genetic quality of their offspring, e.g., better genes (additive effect) or more compatible genes (non-additive effect). Mating preferences for mates with compatible genes could maximize offspring heterozygosity, minimize the risks of close inbreeding, or optimize immunogenetic complementarity (Schmoll 2011). If females do engage in EPMs to increase the genetic quality of their young, then variation in rates of EPM among species and populations might, at least in part, be due to variation in the availability of males with those genes (Kempenaers 2007). If true, then there should be a positive relationship between rates of EPMs and the amount of variation in the genetic quality or diversity of males in bird populations (Petrie et al. 1998) and, in addition, within broods, extra-pair offspring might be expected to be of higher quality than within-pair offspring (Sheldon et al. 1997). Although there are exceptions (e.g., Garcia-Del-Rey et al. 2012), rates of EPM do tend to be lower in island populations where genetic diversity tends to be lower than for mainland populations (Griffith 2000; Charmantier and Blondel 2003), providing indirect evidence for the genetic quality or diversity hypothesis. Providing additional support for the genetic diversity (i.e., non-additive effect) hypothesis, Gohli et al. (2013) found that species of songbirds with higher rates of EPM also had higher nucleotide diversity in autosomal chromosomes (chromosomes other than those that determine a bird's sex). They also found that species with higher rates of EPM had major histocompatibility complex (MHC) alleles with greater sequence diversity. Importantly, these MHC alleles encode antigen-presenting molecules that are important in avian immune responses (Klein 1986; Box

**Box 15.10 MHC Genes and Avian Immunity**

Genes of the major histocompatibility complex (MHC) are known to play a critical role during the development of immunity against invading and potentially harmful pathogens. MHC genes encode cell-surface glycoproteins that bind antigens derived from pathogens or parasites and present them to T-lymphocytes that then trigger the appropriate immune response. There are two major groups of MHC genes. MHC class I genes play an essential role in the immune defense against intracellular pathogens by binding proteins derived from viruses. MHC class II genes are involved in monitoring the extracellular environment by presenting proteins derived from bacteria and parasites to the T-lymphocytes. Given the extraordinary diversity of (and continuously evolving) pathogens in the environment, selection has favored similar diversity in MHC genes.

When exposed to a pathogen (or, more precisely, an antigen, which is a "foreign" protein produced by a pathogen like a bacterium or virus), receptor proteins in the membranes of special immune system cells (like dendritic cells shown below) bind to an antigen and then ingest the receptor protein + antigen by phagocytosis (creating a phagosome or endosome). Exposure to the receptor protein triggers production of special chemicals, called cytokines and costimulatory

(continued)

**Box 15.10** (continued)

molecules, that will help activate T-lymphocytes (which, in turn, will seek out and kill the bacteria or parasites or cells infected by viruses from which the antigen was derived) or B-lymphocytes (which will help eliminate pathogens by producing antibodies). In addition, the dendritic cell "processes" the antigen, which involves coupling the antigen with MHC proteins (MHC + antigen) and presenting this combination in the cell membrane. Lymphocytes that then bind to the MHC + antigen via their receptor proteins are activated and able to act against the pathogen from which the antigen was derived.

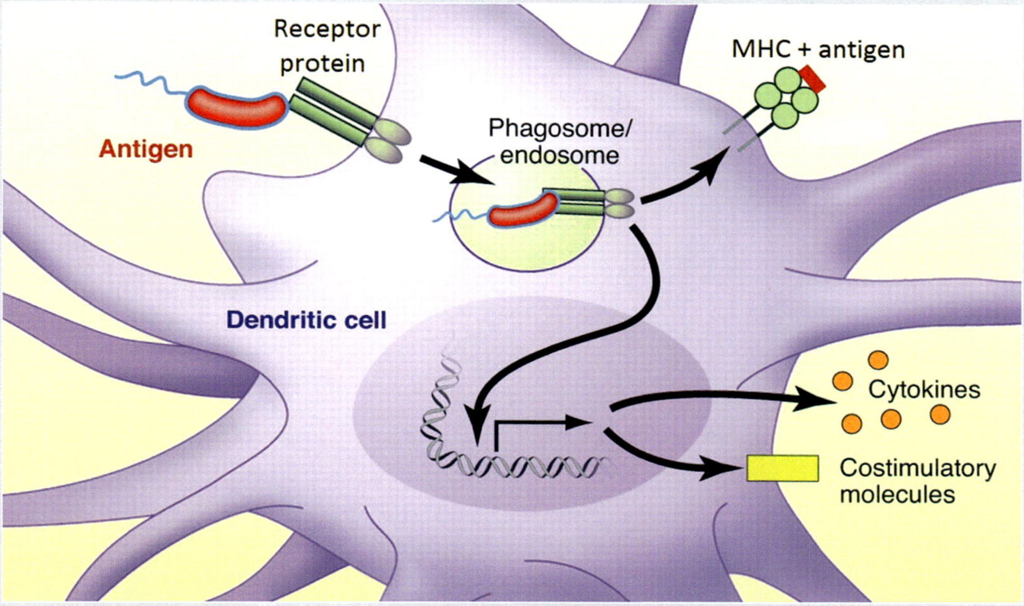


MHC genes encode glycoproteins in cell membranes of macrophages, including dendritic cells, that bind antigens (foreign proteins) derived from pathogens or parasites and present them to T-lymphocytes that then trigger the appropriate immune response. (Figure modified from Iwasaki and Medzhitov 2010; © 2010 American Association for the Advancement of Science, used with permission)

15.10 MHC Genes and Avian Immunity), and greater diversity in these alleles should allow immune systems to recognize a wider variety of pathogens (Gohli et al. 2013). Within species, increased MHC allele diversity in extra-pair young has been reported in Seychelles Warblers (*Acrocephalus sechellensis*; Brouwer et al. 2010) and Great Frigatebirds (*Fregata minor*; Juola and Dearborn 2012), and female Red Junglefowls (*Gallus gallus*) were found to be somehow able to selectively allow sperm from males that would increase MHC allele diversity in offspring to differentially fertilize their eggs (Løvlie et al. 2013). In contrast, extra-pair males did not differ in MHC diversity from within-pair mates in Common Yellowthroats (*Geothlypis trichas*; Bollmer et al. 2012).

The genetic quality hypothesis predicts that extra-pair offspring should be of higher quality than within-pair offspring (Sheldon et al. 1997). In support of this hypothesis, Gerlach et al. (2012), in a study of Dark-eyed Juncos (*Junco hyemalis*), found that male and female offspring sired via extra-pair matings had higher lifetime

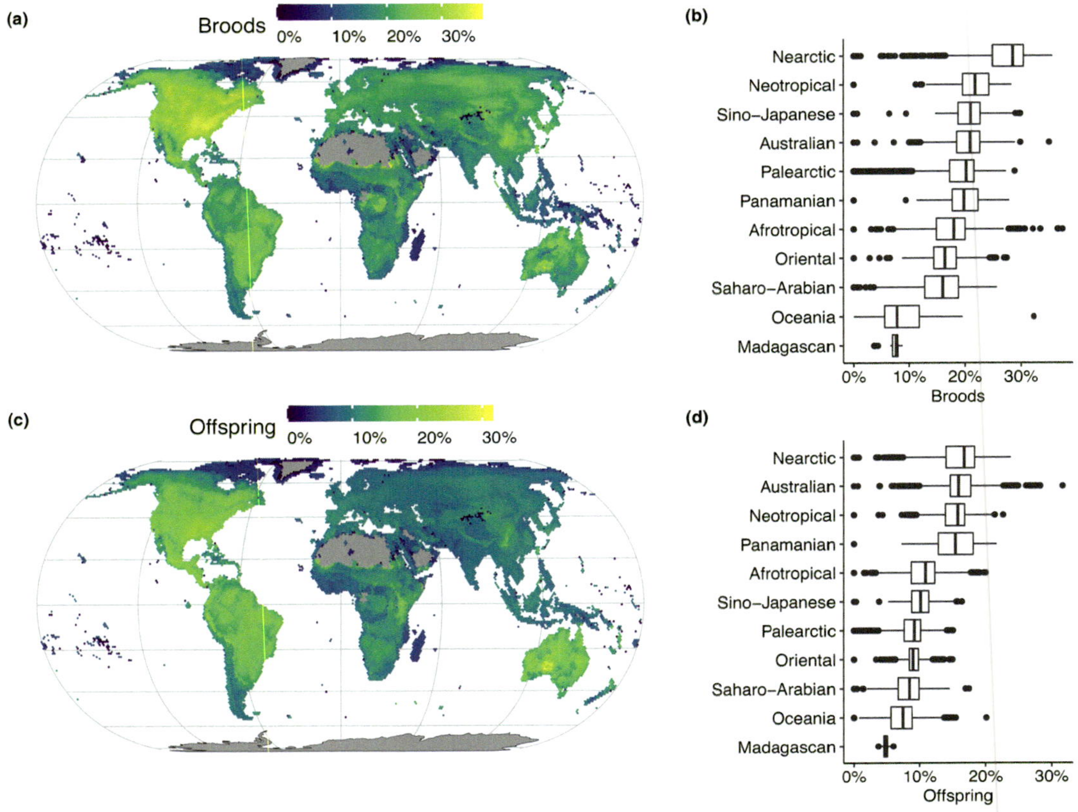


**Fig. 15.25** Geographical variation in the proportions of extra-pair broods (**a**, **b**) and extra-pair offspring (**c**, **d**). The global distribution of assemblage-level extra-pair paternity (EPP) estimates (**a**, **c**) and assemblage-level proportions for each zoogeographical realm are shown as box plots (**b**, **d**). Proportions of extra-pair broods and offspring were found to be at highest levels in the Nearctic realm. Nearctic birds have comparatively short lifespans, experience relatively high rates of adult predation, and are more sexually size dimorphic than birds in other realms (Valcu et al. 2014; Friedman and Remeš 2016). Levels of EPP were lowest on oceanic islands (Oceania and Madagascar), consistent with the prediction that the intensity of sexual selection is lower on islands due to lower genetic diversity. More generally, variation in the frequency of EPP within the breeding range is mostly driven by the population's breeding synchrony and genetic diversity; differences in EPP rates among species are explained by their life-histories, in particular by longevity; and assemblage-level EPP variation occurs primarily among zoogeographical realms (possibly due to differences in life-history, i.e., slow at tropical latitudes vs. fast at higher latitudes, and genetic variation among realms), and along latitude within each realm (possibly due to latitudinal variation in breeding synchrony or genetic diversity). (Figure from Valcu et al. 2021; © 2021 The Authors. Molecular Ecology published by John Wiley & Sons Ltd., used with permission)

reproductive success than offspring sired via within-pair matings. In contrast, however, extra-pair young were not found to be of higher quality than within-pair young in Song Sparrows (*Melospiza melodia*; Sardell et al. 2012) and may even be of lower quality than within-pair young in House Sparrows (*Passer domesticus*; Hsu et al. 2014).

As the information provided in the above paragraphs illustrates, a wide variety of factors can contribute to variation in rates of EPM both among species and within different populations of the same species. For example, studies of songbirds have revealed that phylogeny, the importance of male parental care, and, for some populations, breeding density influence mating

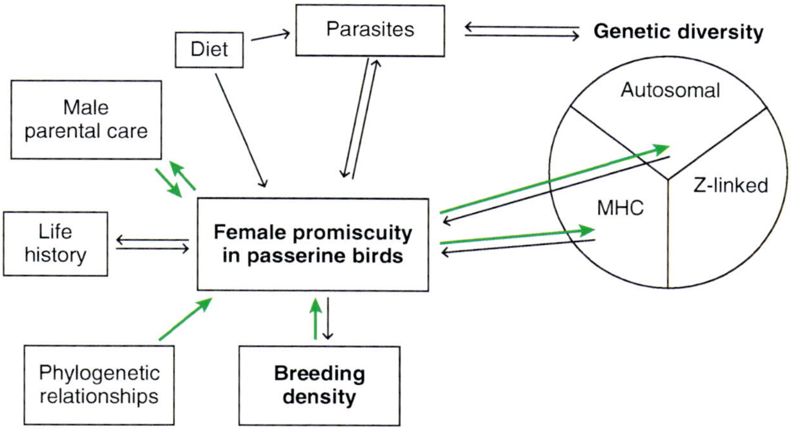


**Fig. 15.26** Possible causal links and associations between female extra-pair mating and factors that may influence the frequency of such mating in a songbird population. Green arrows indicate associations for which investigators have provided empirical support at least in some populations. MHC genes encode antigen-presenting molecules very important in avian immune responses. Autosomal genes are those found in chromosomes other than the sex chromosomes (i.e., the Z and W chromosomes). Z-linked genes are found on the Z chromosome. (Figure from Gohli 2014; used with permission)

decisions of females (Fig. 15.25). Available evidence also suggests that differences in life-history can influence rates of EPM, with species having "fast" life-histories (i.e., high rates of adult mortality and high fecundity) also having higher rates of EPM (Arnold and Owens 2002; Fig. 15.25). Diet might also help explain some of the variation among species in rates of EPM, with more animal tissue in diets associated with increased risk of exposure to parasites (Slifko et al. 2000; Fig. 15.26). With greater exposure to parasites, females may be more likely to engage in EPM to maximize genetic diversity, particularly diversity of MHC genes that play a critical role in the avian immune system (Gohli 2014). In addition, increased autosomal gene diversity or heterozygosity has been linked to increased offspring fitness (e.g., Olano-Marin et al. 2011). Given the variation among species and populations in phylogeny, life-history, the importance of male care, breeding density, and diet, inter- and even intraspecific variations in rates of EPM (Figs. 15.22, 15.23, and 15.24) are, perhaps, not surprising.

## 15.7 Polygyny

Available evidence suggests that polygyny arose independently multiple times in a variety of avian taxa (Temrin and Sillén-Tullberg 1994; Figs. 15.27 and 15.28). In polygynous species of birds (about 8% of all species), at least some males have social relationships of varying duration with more than one female and typically mate with those females. Typically, polygynous males defend either mating or breeding territories and females provide much if not most or all parental care. Both males and females may engage in extra-pair (or extra-group) copulations, and males are typically larger and more colorful than females as a result of sexual selection.

Polygynous mating systems are categorized based on how males acquire mates. In resource defense, or territorial, polygyny, resources important for successful reproduction such as food and nest sites are unevenly distributed or spatially clumped and males establish and defend territories that include these resources. Females choose to nest in particular territories based on

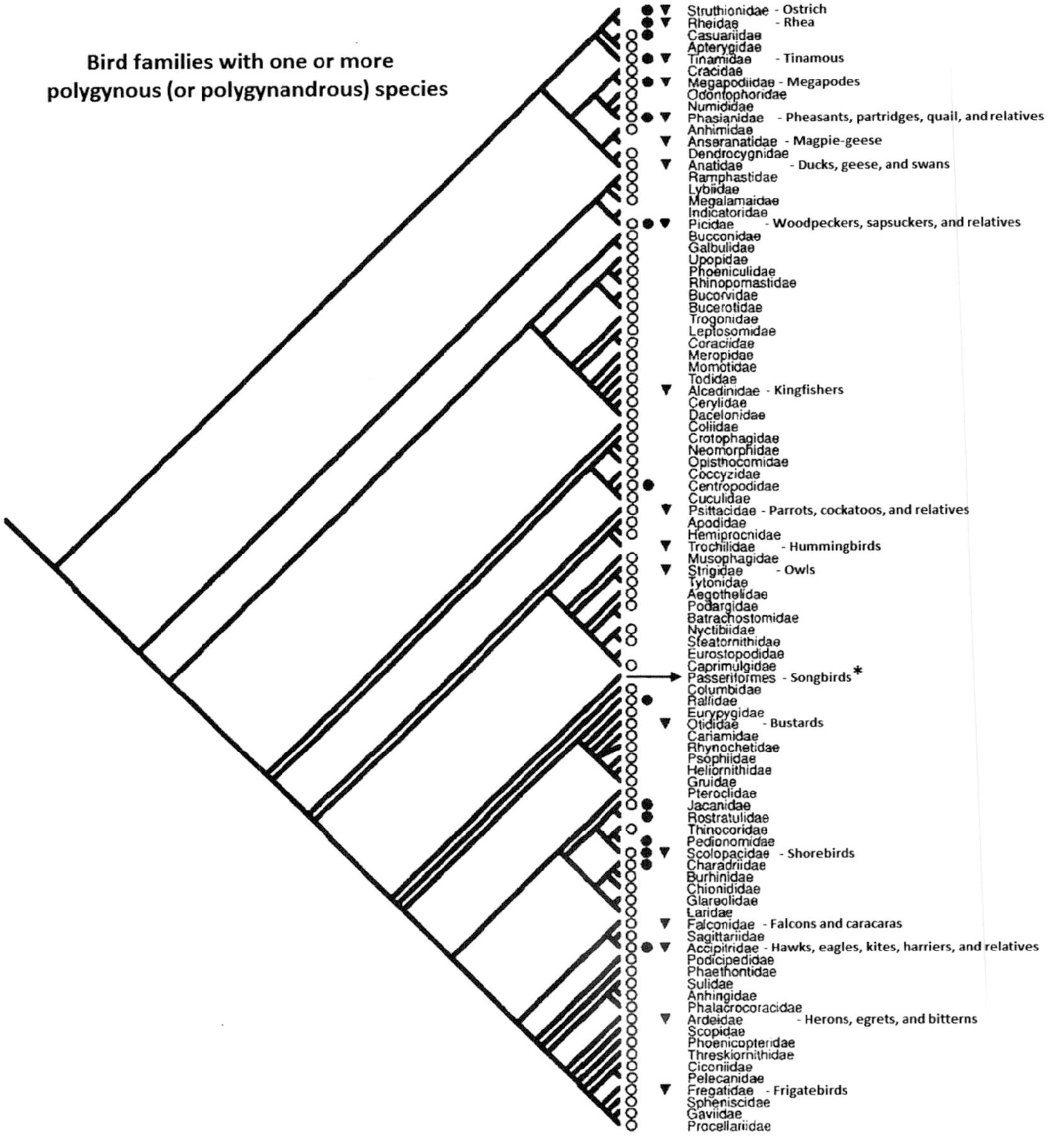


**Fig. 15.27** Families of birds with one or more polygynous species. Open circles indicate families that include species that are socially monogamous, closed circles indicate families that include species that are polyandrous, and triangles indicate families with polygynous species. Families with both closed circles and triangles are, or have species that are, polygynandrous. The asterisk indicates families of songbirds (Order Passeriformes) with polygynous species that are shown in Fig. 15.28. (Figure modified from Temrin and Sillén-Tullberg 1994; © 2008 Oxford University Press, used with permission)

male and territory quality (resources available relative to other territories) as well as their own quality or condition. In male dominance polygyny, resources important for reproduction cannot be defended either because their availability is unpredictable in space and time or they are

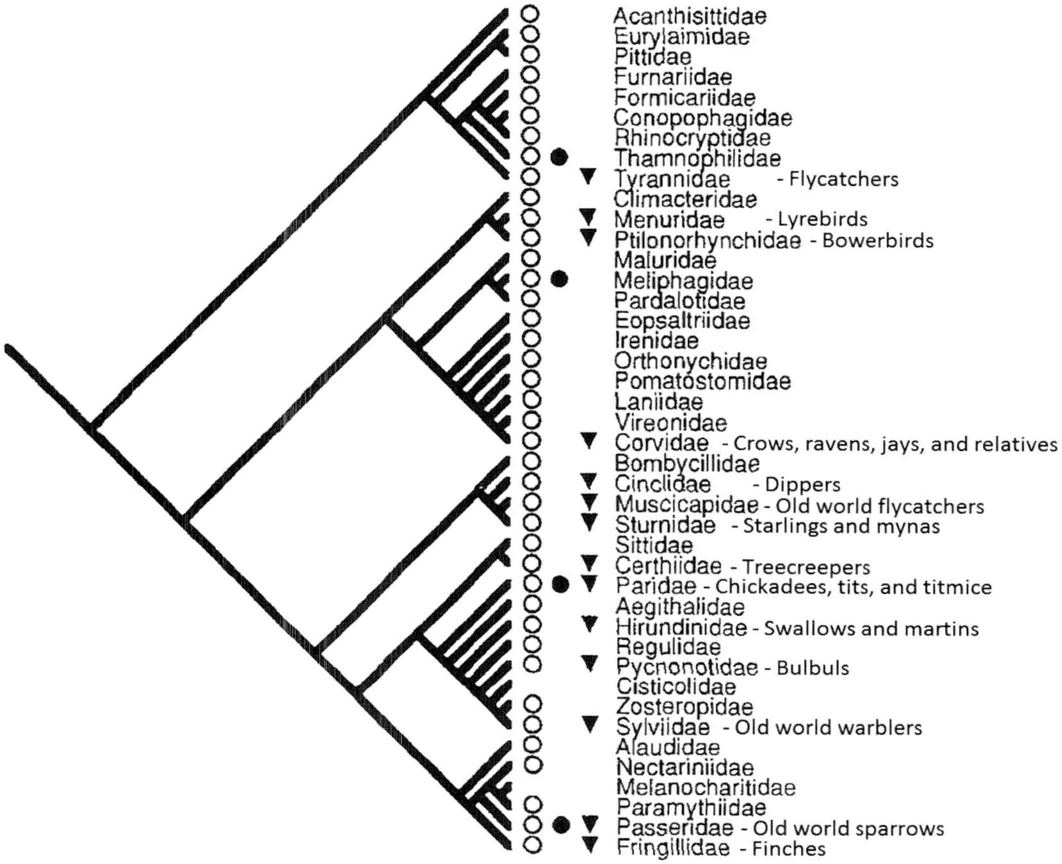


**Fig. 15.28** Passerine families with one or more polygynous species. Open circles indicate families that include species that are socially monogamous, closed circles indicate families that include species that are polyandrous, and triangles indicate families with polygynous species. Families with both closed circles and triangles are, or have species that are, polygynandrous. (Figure modified from Temrin and Sillén-Tullberg 1994; © 2008 Oxford University Press, used with permission)

abundant, but widely dispersed. When resources are unpredictable in space and time, males may gather at specific locations and compete via displays for access to mates. In some species, males gather in the same area, called a lek, and establish small "mating territories" that contain no resources needed by females. Visiting females then observe the displaying males and make their mate-choice decisions. In other species, male display areas or territories are further apart and are called exploded leks. These territories may contain some resources, but none that are critical for female reproductive success. Females may visit and observe the displays of multiple males before making their mate-choice decision. Finally, in yet other polygynous species, resources important for successful reproduction such as food may be very abundant, but only at certain locations (e.g., flowering tree with abundant nectar or fruiting tree with abundant fruit). In such resource-based lek species, dominant males defend the resource and multiple females may mate with those males and then nest nearby so they have access to the abundant resource (e.g., food).

In resource defense, or territorial, polygyny, males defend territories with resources that females need for successful reproduction (Box 15.11 Adaptive Sleep Loss in Polygynous

**Box 15.11 Adaptive Sleep Loss in Polygynous Pectoral Sandpipers**

Pectoral Sandpipers (*Calidris melanotos*) have a polygynous mating system, with males defending territories and displaying to or chasing females and females choosing among available males and, after copulating, laying and incubating eggs and caring for young. Males remain vigilant for male intruders and for females, often engaging in aerial chases of females. Because these sandpipers breed in the high Arctic where the sun never sets during the mating period, the time available for males to defend territories and seek females for copulations is basically unlimited. Lesku et al. (2012) found that the activity levels of males varied considerably and, in one extreme case, a male was active >95% of the time over a period of 19 days. Males that remained active for longer periods (i.e., those that slept the least) had greater reproductive success (Lesku et al. 2012). More active males may also increase their reproductive success by sampling more potential breeding sites and remaining longer at sites with more breeding females. Kempenaers and Valcu (2017) found that male Pectoral Sandpipers may visit as many as 23 different potential breeding sites during a breeding season while traveling as far as 13,045 km.

(**a–f**) Visual and acoustic mating displays of male Pectoral Sandpipers. Males give "hooting" calls along the borders of their territories or, in flight, over females. Males can also inflate their esophagus (**e**) when vocalizing to create a low-pitched gurgling sound. (Figure from Riede et al. 2015; © 2014 Oxford University Press, used with permission)

The high reproductive success of more active males suggests that, in contrast to some other species (e.g., White-crowned Sparrows, *Zonotrichia leucophrys gambelii*; Jones et al. 2010), they were not neurologically or physically impaired by sleep deprivation. The strong relationship between percent time active and reproductive success suggests that the ability of male Pectoral Sandpipers to remain active even when severely sleep-deprived is a trait favored by sexual selection (Lesku et al. 2012).

(continued)

**Box 15.11** (continued)

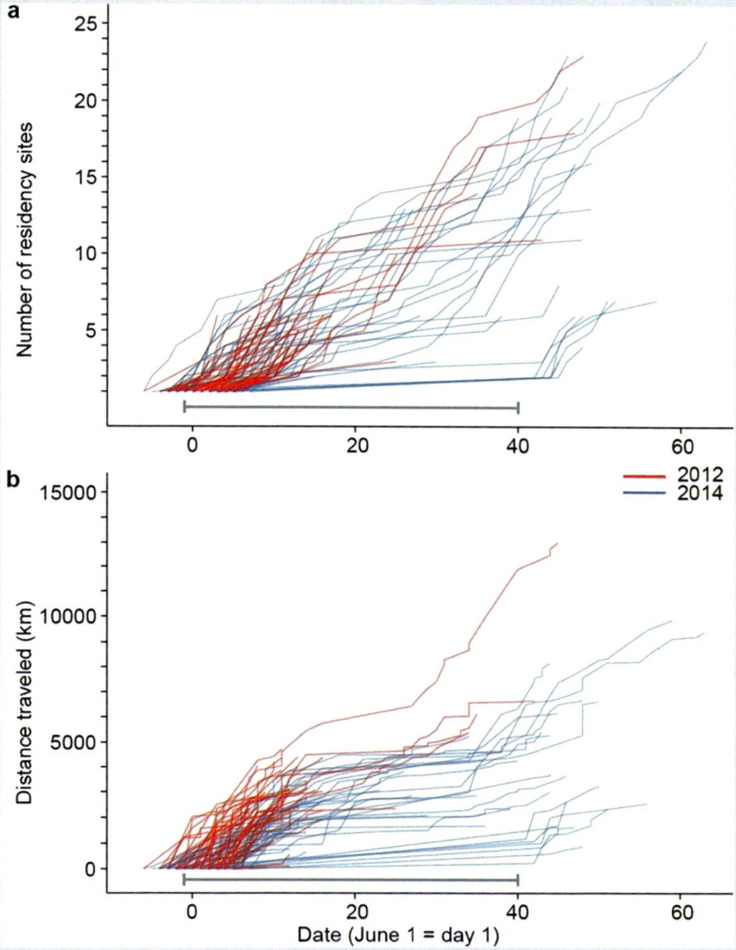


(**a**) Cumulative number of potential breeding sites visited by each male Pectoral Sandpiper during two breeding seasons. (**b**) Cumulative distance traveled by each male during the breeding season as they visited different potential breeding sites in the Arctic region. Horizontal gray bars = range of clutch initiation dates. Lines connect successive points. (Supplemental Figure from Kempenaers and Valcu 2017; © 2017 Springer Nature, used with permission)

(continued)

**Box 15.11** (continued)

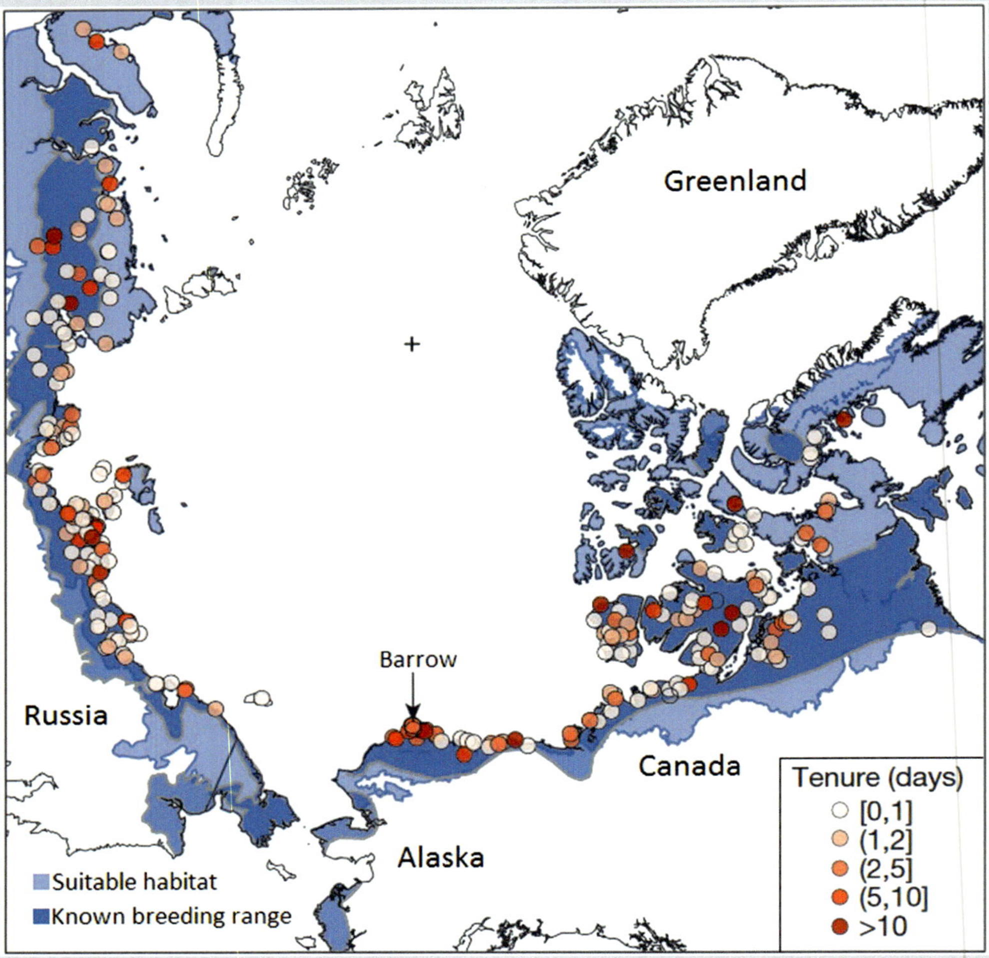


Sites visited by male Pectoral Sandpipers ($N = 60$) during the 2012 breeding season. Dot color indicates length of time spent at a site by a male. Square bracket = number of days inclusive of the number described; round bracket = number of days longer, but not inclusive, of the number itself. Birds were captured at Barrow, Alaska. + = north pole. (Figure from Kempenaers and Valcu 2017; © 2017 Springer Nature, used with permission)

Pectoral Sandpipers). When making settlement decisions, females must evaluate the breeding situation quality (or BSQ, determined by both male quality and territory quality; Wittenberger 1976) and, in addition, must consider the possible effect on BSQ if one or more females are already present in a territory. Particularly for altricial species where males may assist with incubating eggs or caring for nestlings (e.g., nest defense and/or provisioning) and precocial species where males may assist with incubation, females must consider the possible reduction in male assistance if other females are present. A number of studies have revealed that secondary females

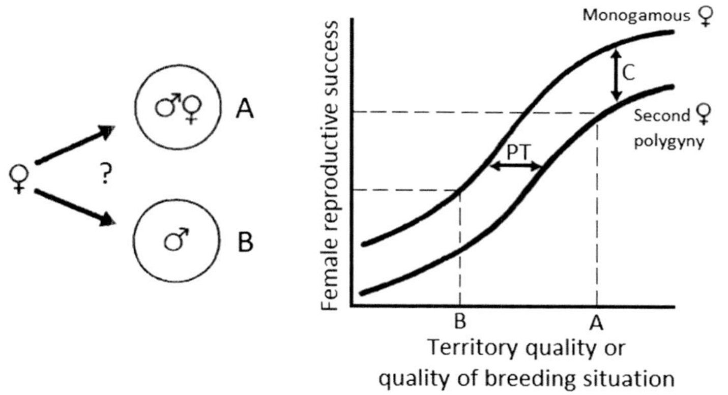


**Fig. 15.29** The polygyny threshold model assumes that a female's reproductive success (or fitness) increases with the quality of a male's territory, male assistance, or other factors that improve the breeding situation. Females among the first to arrive in breeding areas choose the best territories (territory A). Females that arrive later can choose to nest in a lower-quality territory (territory B) where they will be the only female or in a higher-quality territory (territory A) where they will be the secondary female of an already paired male. Because a secondary female must share resources with or, in some cases, compete with (e.g., for a nest site) the primary female, the fitness curve for the secondary female is below that of the primary female and the distance between curves represents the cost (c) of mating polygynously. If, however, differences in territory quality or breeding situations are sufficient (as indicated in this example by the horizontal distance between curves referred to as the polygyny threshold, or PT), then a secondary female will have greater reproductive success in higher-quality territory A than a primary female in territory B. (Figure from Krebs and Davies 2009; © 2009 John Wiley and Sons, used with permission)

(the second female to settle and nest in a territory) receive less assistance from males and have lower reproductive success than either monogamous females (the only female in a male's territory) or primary females (the first female to settle and nest in a territory) (e.g., Huk and Winkel 2006; Grønstøl et al. 2014).

Given the possible reduction in or loss of male assistance, why would a female choose to nest in a territory where one or more females are already present? A possible explanation for such behavior is the polygyny threshold model (PTM) (Verner 1964; Verner and Willson 1966; Orians 1969). This model is based on several assumptions. The first is that there is a cost to polygyny that results from having to share a territory, resources, and access to the mate with other females, and this cost is largely incurred by secondary females. A second assumption is that reproductive success of females varies because of differences in territory and male quality (i.e., BSQ). A third assumption is that choosy females can recognize differences in BSQ and are free to settle anywhere. A last assumption is that females are attempting to optimize their reproductive success.

The polygyny threshold model proposes that females weigh the costs of polygyny against the benefits of breeding in a better-quality territory with a higher-quality mate, and choose to nest in a territory as a secondary female only when they can do better (in terms of reproductive success) than if they were a primary female in a lower-quality territory with a lower-quality male (Fig. 15.29). Although the results of some studies provide support for the polygyny threshold model (Pribil and Searcy 2001), results of other studies have revealed a consistent reproductive cost for being a secondary female (Pribil 2000; Huk and Winkel 2006; Kleiber 2008). According to the polygyny threshold model, females should be able to avoid such costs by nesting in higher-quality territories with high-quality males.

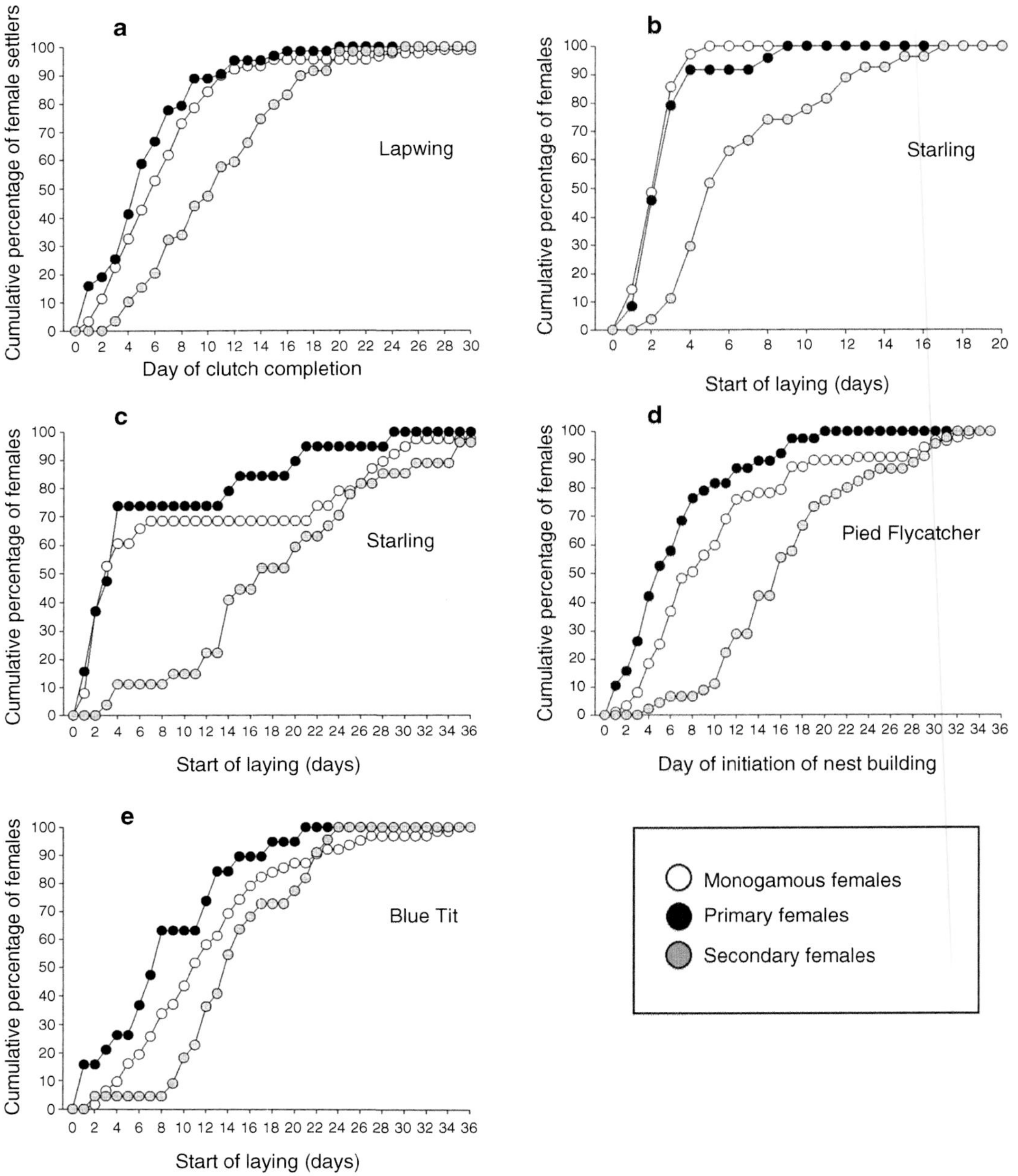


**Fig. 15.30** In three polygynous species (including two different populations of European Starlings, *Sturnus vulgaris*), clutch initiation (start of laying) dates, clutch completion dates, or dates when nest building was initiated were generally similar for monogamous (only female in a male's territory) and primary (first female to settle in a polygynous male's territory) females. However, these dates were significantly later for secondary females (the second female to settle in a polygynous male's territory). In addition, secondary females generally had lower fledging success than monogamous or primary females. Data for European Starlings are from two different studies. Northern Lapwing, *Vanellus vanellus*; European Pied Flycatcher, *Ficedula hypoleuca*; Eurasian Blue Tit, *Cyanistes caeruleus*. (Figure modified from Grønstøl et al. 2003; © 2003 Oxford University Press, used with permission)

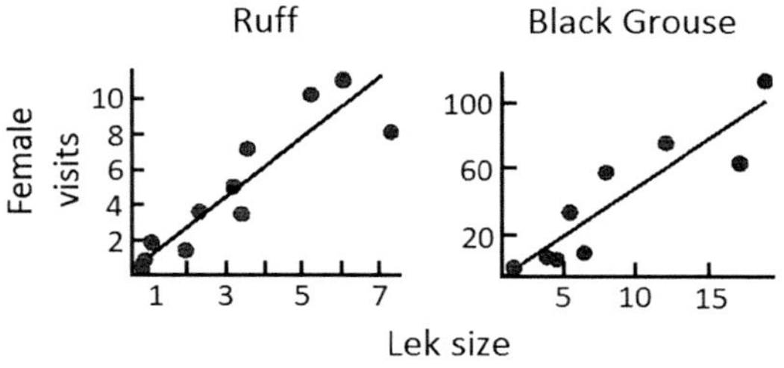


**Fig. 15.31** Among some lek species such as Ruffs (*Calidris pugnax*; Höglund et al. 1993) and Black Grouse (*Lyrurus tetrix*; Alatalo et al. 1992), the number of females visiting leks increases with lek size (the number of males present). However, studies of other species (e.g., Little Bustard, *Tetrax tetrax*; Jiguet and Bretagnolle 2006) have revealed that more females visit intermediate-sized leks. Female preferences are likely an important factor influencing lek size, and those preferences may be influenced by factors such as the cost of mate sampling (e.g., number of leks visited before choosing a mate), probability of finding a high-quality mate (which likely increases with lek size), and the accuracy with which high-quality mates can be identified (which may decrease with increasing lek size). (Figure modified from Isvaran and Ponkshe 2013; © 2013 The Association for the Study of Animal Behaviour. Published by Elsevier Ltd.; used with permission)

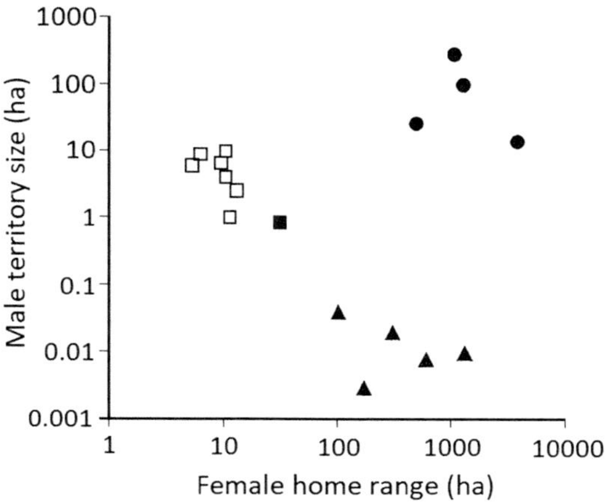


**Fig. 15.32** Relationship between male territory size and female home-range size and mating system for 17 species of grouse and bustards. Open squares = non-lekking grouse, closed square = exploded-lekking grouse, closed triangle = lekking grouse, and closed circles = exploded lekking bustards. (Figure modified from Morales et al. 2001; © Morales, B. et al., Ardeola, used with permission)

Although explaining female behavior in some species, the polygyny threshold model (PTM) clearly does not explain all cases of resource defense, or territorial, polygyny. One likely reason for this is that the PTM assumes that females are equal competitors and free to settle in any territory. This is clearly not the case in some species. For example, in some species, first-settling females may attempt to exclude additional females from territories (Kleiber 2008; Grønstøl et al. 2014). In addition, available evidence suggests that, at least in some species, secondary females may be of lower quality than monogamous or primary females. As evidence of this, Grønstøl et al. (2003) found that secondary females arrived in breeding areas and initiated nesting later than monogamous or primary females and, in most cases, also had lower fledging success (Fig. 15.30). Also limiting the ability of females to settle in any territory is that, particularly for migrants, the time available for examining and comparing territories and males is limited. Typically, females that initiate nesting earlier in the breeding season have greater reproductive success (Verhulst and Nilsson 2008). Females, therefore, must make decisions concerning where to nest as quickly as possible and with imperfect knowledge. As a result, the possible cost (in term of reduced reproductive success) to females of continuing to search for a better breeding situation may exceed the cost of ending the search to begin breeding as a secondary female in the territory of a polygynous male (Ferretti and Winkler 2009).

When females have access to resources needed for reproduction, i.e., males do not or cannot defend territories and restrict access to areas where those resources are located, and paternal care is not required to successfully raise young, females attempt to maximize reproductive success by selecting mates based on the genetic benefits they can provide. On the other hand, males, because they are not defending territories or providing parental care, can maximize their reproductive success by mating with as many females as possible. Females, therefore, benefit by being able to observe and compare multiple several males (Fig. 15.31) before choosing a "mate" (and also, potentially, by observing the choices of other females), whereas males benefit by copulating with (or at least by having the opportunity to possibly copulate with) as many females as possible. Because both males and females, therefore, can benefit by minimizing the time needed to locate members of the opposite sex, selection has favored a mating system where individuals of both sexes gather at the same location over the same period of time, i.e., a lek (a word possibly derived from the Swedish word *leka*, which means "to play," but also possibly derived from Lloyd [1867: 19] who used the Swedish word *lek-ställe*, the "locality where affairs matrimonial are carried on" when referring to the breeding areas of capercaillie [*Tetrao* spp.]). In some lek species, males defend relatively small "display territories" at the same location, whereas in other species, males defend larger "display territories" that are, as a result, farther apart and these are referred to as "dispersed" or "exploded" leks (Fig. 15.32). Foster (1983) suggested that exploded leks evolved from more typical leks due to female preferences for more dispersed males and a corresponding decrease in harassment or disruption of copulations by other males. Other investigators have suggested that exploded leks represent a transitional phase, with the smaller, somewhat clumped mating/display territories an evolutionary precursor of a more typical lek (e.g., Ligon 1999).

Although these "clusters" of males and females can be beneficial for individuals of each sex, there are some potential costs as well. For males, there is the cost of competing with other males for access to the best display sites and for access to females visiting a lek. For females, one possible cost comes with an increase in the number of males present; accurately selecting the highest-quality male may become more difficult as the number of males at a lek increases (Hutchinson 2005). Another possible cost for females is harassment from nearby males as they seek copulations from a particular male. However, females may be able to minimize this potential cost by waiting until dominance relations and territory boundaries have been established and male–male aggression subsides (Oring 1982). Regardless of these possible costs, the persistence of lekking behavior over many generations suggests that the benefits of clustering must outweigh the costs in present-day lekking species.

Given the apparent ultimate advantages to clumping (in terms of fitness), what proximate factor or factors contribute(s) to the formation and location of these mating clusters? Non-mutually exclusive hypotheses proposed to explain these clusters include the hotspot (Bradbury and Gibson 1983), hotshot (Arak 1982), female preference (Bradbury 1981), and black hole (Clutton-Brock et al. 1992) models. Hotspots are locations where female densities are highest and males cluster at those locations to maximize their chances of encountering females. Hotshots are high-quality males preferred and more likely to be visited by females, and lower-quality males cluster around hotshots to increase their likelihood of encountering, and perhaps mating with, females. As the name suggests, the female preference model proposes that females prefer and are more likely to visit clusters of males. Possible reasons for this preference include the reduced cost (in time and energy) of mate sampling, greater likelihood of a high-quality male being present, and being able to observe choices made by other females.

The black hole model posits that males cluster because those defending clustered territories have greater mating success than those defending more dispersed territories and, in addition, that the mating success of males is inversely related to the distance between them. Mating success of males in clusters is higher because encounter rates are higher; females remain in clusters of male territories for longer periods because, as they sample males and move into different territories, they will be harassed by males seeking copulations. As their territories decrease in size, males encounter more females because the increasing density increases the likelihood of females being harassed and moving into adjacent territories. However, just as matter is attracted to black holes, females will tend to move toward and collect in areas with tighter clusters of male territories because, as inter-male distances decrease, ever smaller movements by females

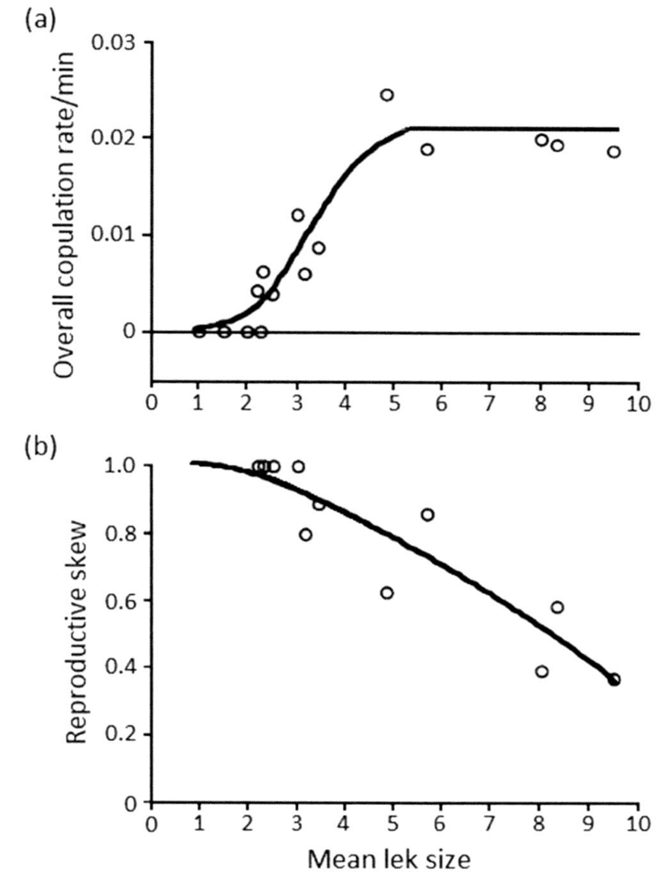


**Fig. 15.33** (**a**) Overall copulation rates for male Ruffs (*Calidris pugnax*) in different-sized leks. (**b**) Relationship between mean lek size and reproductive skew for male Ruffs. Note that reproductive skew declines with increasing lek size. Reproductive skew refers to the degree to which dominant males monopolize copulations. (Figure from Widemo and Owensi 1995; © 1995 Springer Nature, used with permission)

mean moving into adjacent territories and risking more harassment. The black hole model assumes that male interactions play a greater role in determining who mates with females than does female mate choice. Although there is some evidence that the black hole model explains the lek mating systems of some ungulates, there is little evidence that this model explains lekking behavior in birds because harassment by males that leads to copulations with females is rare in birds (Jiguet and Bretagnolle 2006).

Available evidence suggests that females in lekking species do prefer clusters of males. In some species, the number of females visiting leks tends to increase as the number of males present at a lek increases (Isvaran and Ponkshe 2013). In other species, females appear to prefer intermediate numbers of males at leks, e.g., female Little Bustards (*Tetrax tetrax*) exhibited a preference for leks where four males were present (Jiguet and Bretagnolle 2006). Because accurately selecting the highest-quality male may become more difficult as the number of males at a lek increases (Hutchinson 2005), there might be limits as far as female preferences are concerned for the number of males at leks. In support of this possible female preference for intermediate-sized leks, Bradbury (1981) found that typical leks tend to consist of about 10–20 males. Additional indirect evidence for this preference comes from studies where male reproductive skew (distribution of matings or copulations among males in a lek, with a skew value of 1 indicating that the top-ranked male gets all of the copulations and a skew value of 0 indicating random mating) tends to decline with increasing lek size (Kokko and Lindström 1997; Fig. 15.33). This suggests that females might be less accurate in selecting the best

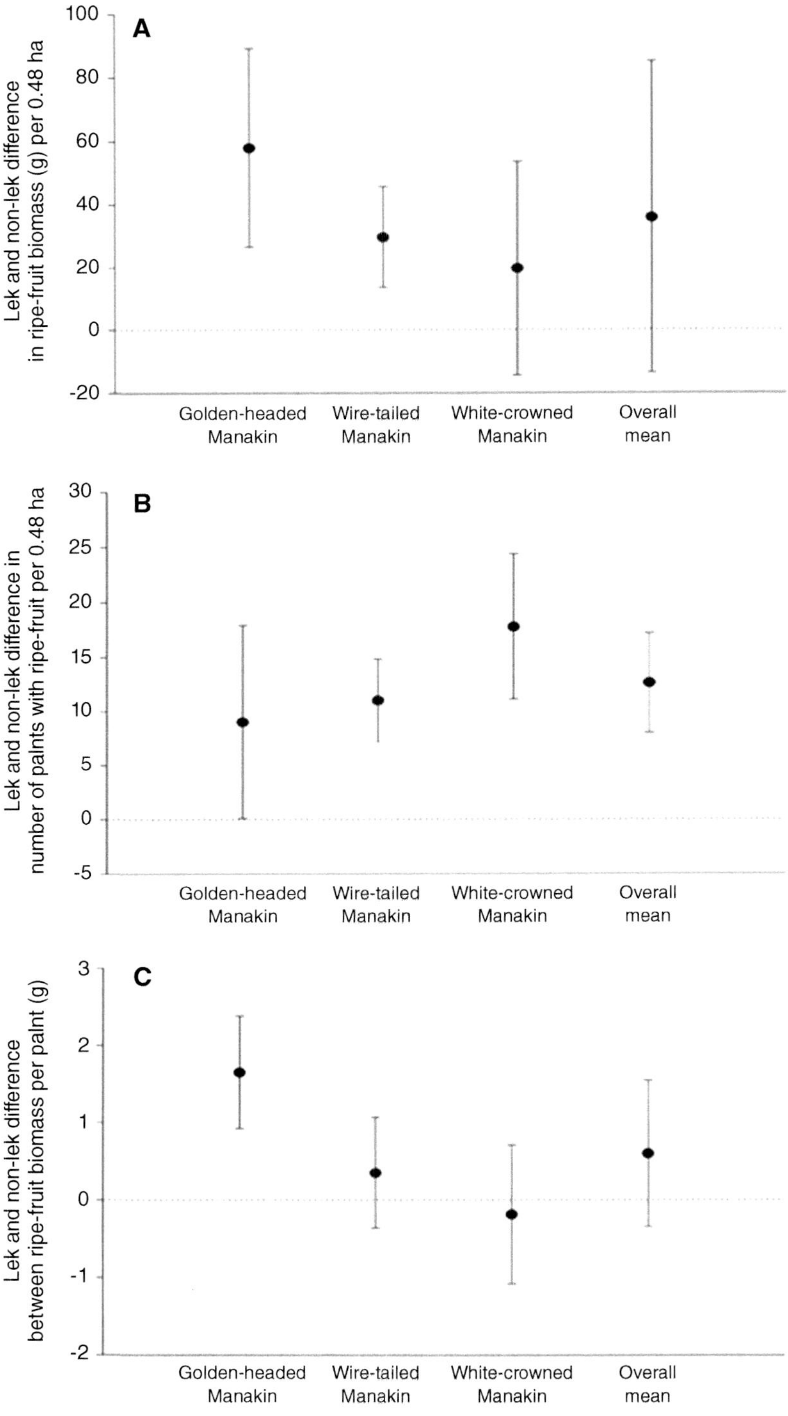


**Fig. 15.34** Differences between lek and non-lek sites (i.e., availability at lek sites minus availability at non-lek sites) in (**a**) biomass of ripe fruit, (**b**) number of fruiting plants, and (**c**) ripe-fruit biomass per plant. Area sampled at each lek and non-lek site was 0.48 ha. Positive values above dashed lines indicate greater biomass, more fruiting plants, and greater biomass per plant around lek sites. Differences are shown separately for each of the three species, and error bars represent 95% confidence intervals. Golden-headed Manakin, *Ceratopipra erythrocephala*; Wire-tailed Manakin, *Pipra filicauda*; White-crowned Manakin, *Pseudopipra pipra*. (Figure from Ryder et al. 2006; © 2006 Oxford University Press, used with permission)

males in increasingly larger leks (Hutchinson 2005). Although additional study is needed concerning possible "preferred" lek sizes, studies to date do indicate that females in lekking species prefer that multiple males be present at leks (Isvaran and Ponkshe 2013).

Factors in addition to female preference might also help explain why males aggregate at leks. One possibility is that males aggregate where densities of females are highest to maximize their chances of encountering females (hotspot model), and another is that lower-quality males aggregate around the high-quality males most preferred by females to increase their likelihood of encountering females (hotshot model). Support for the hotspot model is mixed. The results of some studies provide support for the hotspot model (Gibson 1996; Westcott 1997; Hingrat et al. 2008). For example, Ryder et al. (2006) found that leks of three species of manakins (*Pipra* spp.) in Ecuador were located where fruit, the main food of these frugivorous birds, was most plentiful (Fig. 15.34). However, results of other studies do not support the hotspot model (Wegge and Rolstad 1986; Jiguet and Bretagnolle 2006; Durães et al. 2007).

Support for the hotshot hypothesis has been reported in studies of Ruffs (*Phibrmachus pugnax*, Widemo 1998a) and Little Bustards (*Tetrax tetrax*, Jiguet and Bretagnolle 2006), but the results of many other studies do not support this hypothesis (e.g., Ryder et al. 2006; Rolstad et al. 2009; Alonso et al. 2012). Beyond hotspots and hotshots, males in lekking species may aggregate at particular locations for a number of other reasons. For example, display behavior by males in some species of manakins (*Pipra* spp. and *Lepidothrix coronata*) plus Guianan Cock-of-the-Rocks (*Rupicola rupicola*) occurs when and where lighting conditions maximize male conspicuousness during their displays (Endler and Théry 1996; Heindl and Winkler 2003). Male Great Bustards (*Otis tarda*) form leks at sites that increase their visibility to females, enhance their ability to detect approaching predators, and increase the distance from sources of human disturbance (Alonso et al. 2012). Similarly, male Greater Sage-Grouse (*Centrocercus urophasianus*) lek at sites that allow them to monitor approaching predators and where they are more visible to females (Aspbury and Gibson 2004). Male prairie-chickens lek at sites with high visibility, good auditory projection, and minimal ambient noise, features that increase their ability to detect predators and the ability of females to see and hear them (Gregory et al. 2011). Leks of males in other species of lekking birds that occupy open habitats are also often found at elevated sites that provide good visibility (e.g., Wiley 1974; Baydack 1986).

In sum, lekking behavior has been reported in at least 97 species of birds in 14 different families, with most species in the families Trochilidae (hummingbirds; 14 species), Paradisaeidae (birds of paradise; 20 species), Pipridae (manakins; 18 species), and Cotingidae (cotingas, cock-of-the-rocks, and others; 18 species) (Höglund and Alatalo 1995). Ultimate factors contributing to the evolution of leks include the ability of females to raise young without assistance from males and the inability of males to monopolize resources needed by females to successfully reproduce. These things alone do not require male aggregation, so additional selective pressures were needed for that to occur. Studies to date suggest that one such selective factor was a female preference for male aggregation (but see Box 15.12 Leks and Kin Selection for another possible contributing factor). Beyond that, however, factors influencing the size and location of leks vary among lekking species, but may include female hotspots, male hotshots, and environmental conditions such as ambient light levels, visibility, and, more recently (in an evolutionary sense), the location of sources of human disturbance.

A common feature of lek species is high reproductive skew, with few males attending leks actually mating with visiting females (Figs. 15.35 and 15.36). Females visiting leks for choices on which male(s) to mate with may use a variety of criteria that can vary among species. Among the morphological and behavioral traits found to correlate with male mating success in leks are lek

**Box 15.12 Leks and Kin Selection**

Given the reproductive skew reported in studies of lekking species (with most matings by just a few males), why do males that rarely if ever mate with visiting females remain on leks and continue to display? Two possible factors include the opportunity, even if very small, to copulate with a female (and, because females prefer male aggregations, males that do not join a lek have no chance of copulating with a female) and the potential benefit of gaining experience at interacting with males and females and performing displays. Such experience may be helpful in future breeding seasons, particularly if high-ranking males do not survive and other males are able to "move up" in rank. Yet another possible factor is that, if related to other males at a lek, a low-ranking male, by increasing lek size, can help increase the number of females visiting the lek and thereby increase the reproductive success of higher-ranking relatives, i.e., kin selection.

Male Greater Sage-Grouse (*Centrocercus minimus*) displaying at the lek in California (USA) with several females present. Genetic analysis has revealed that males at these leks do not represent kin groups (Gibson et al. 2005). (Photo by Jeannie Stafford, U. S. Forest Service, Wikipedia, CC0 Public Domain)

Although most species are yet to be studied, research to date has revealed that lekking males in some species may exhibit (1) significant levels of relatedness, suggesting kin selection (Höglund et al. 1999; Petrie et al. 1999; Shorey et al. 2000; Regnaut et al. 2006), (2) limited relatedness (Francisco et al. 2009), or (3) no relatedness (McDonald and Potts 1994; Gibson et al. 2005; Loiselle et al. 2007; Lebigre et al. 2008; Bush et al. 2010). Limited relatedness was reported in male Blue Manakins (*Chiroxiphia caudate*). Francisco et al. (2009) found that close male relatives were present, but only at 7 of 13 leks sampled, and suggested that these results were best explained as a by-product of limited dispersal rather than as direct selection for increased relatedness among lek members. In some species where males at leks are related,

(continued)

**Box 15.12** (continued)

including Indian Peafowl (*Pavo cristatus*; Petrie et al. 1999) and White-bearded Manakins (*Manacus manacus*; Shorey et al. 2000), males searching for a lek to join are apparently able to use some type of phenotype matching to identify kin. By joining kin at leks and increasing lek size to attract more females, males that copulate with few or no females still benefit via kin selection (inclusive fitness) by increasing the reproductive success of related males.

White-bearded Manakin. (Photo by The Lilac Breasted Roller, Wikipedia, CC BY 2.0, https://creativecommons.org/licenses/by/2.0/deed.en)

(continued)

**Box 15.12** (continued)

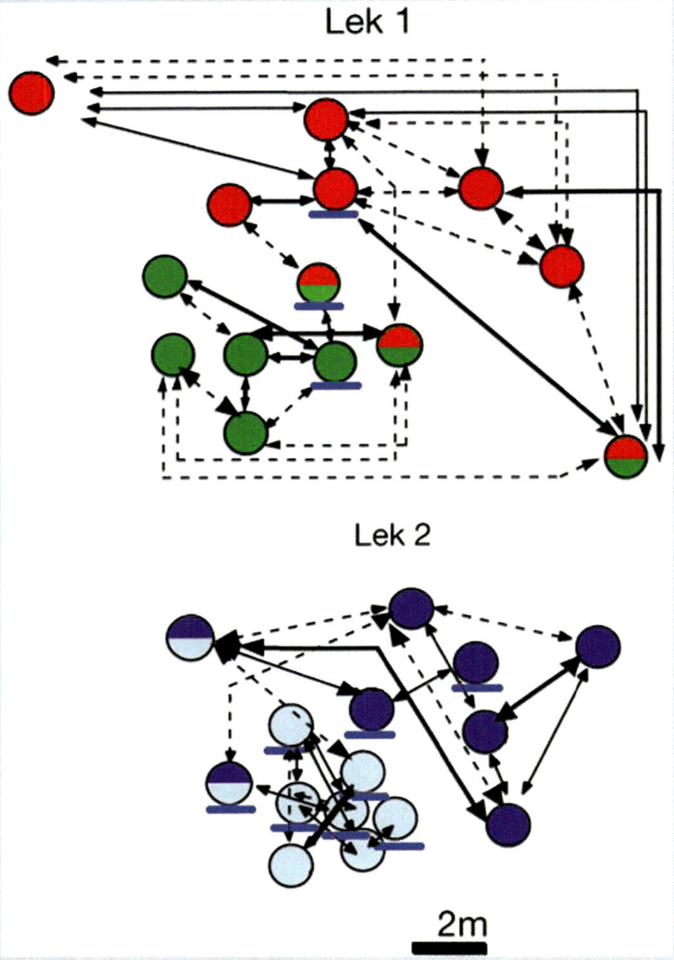


Maps showing the positions of individual male White-bearded Manakins at two leks in Trinidad, West Indies. Arrows connect related males, with degree of relatedness indicated by lines connecting arrows, and ranging from most related (solid thick lines) to less (thinner solid lines) and less (but still more than expected based on average relatedness in the population) related (dashed lines). Within leks, males positioned themselves into spatially separated groups of related individuals. In Lek 1, one sub-group is indicated in red and the other in green. In Lek 2, one sub-group is indicated in dark blue and the other in light blue. Males with two colors were related to both groups. Males that copulated with visiting females are underlined. Note that sub-groups at each lek with the lower average pairwise relatedness (green sub-group at Lek 1 and light blue sub-group at Lek 2) had more sharing of copulations among the males. (Figure from Shorey et al. 2000; © 2000 Springer Nature, used with permission)

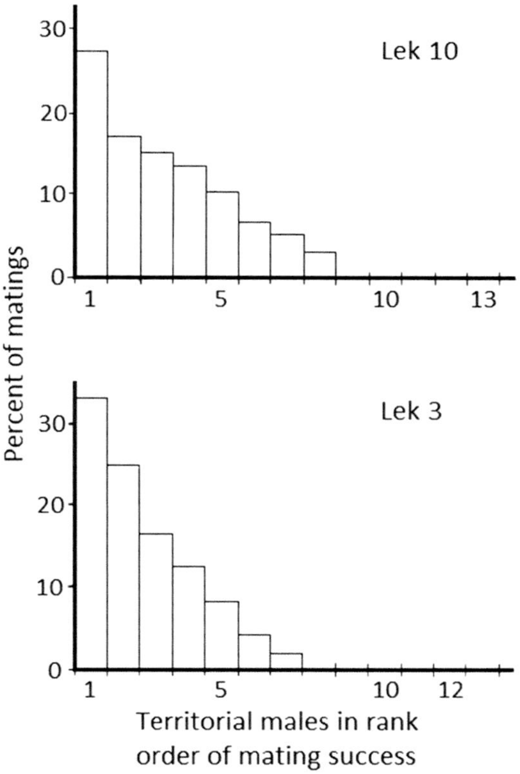


**Fig. 15.35** Male Great Snipe (*Gallinago media*) at two leks (with 13 and 12 males present, respectively) ranked by the percentage of matings obtained. (Figure from Höglund and Lundberg 1987; © 1987 Springer-Verlag, used with permission)

centrality, lek attendance (proportion of time or total amount of time males spend at leks), display rate, and aggression rate and performance (success in winning encounters with other males) (Fiske et al. 1998; Hämäläinen et al. 2012). The relative importance of these traits has been found to vary among species. For example, among lekking species of birds, manakins (Passeriformes; Pipridae) perform some of the most elaborate courtship displays, incorporating both acrobatic movements and mechanically produced sounds (Figs. 15.37 and 15.38). As with other lekking species, male manakins do not assist with parental care or provide any resources for females, so mating decisions by females are made based on male courtship displays. For example, female Golden-collared Manakins (*Manacus vitellinus*) prefer males with brighter plumage that perform their displays with greater speed (Stein and Uy 2006; Barske et al. 2011). These differences likely reflect differences among males in condition, motor skills, and cardiovascular function (heart rates can exceed 1300 beats per minute when males are displaying) (Stein and Uy 2006; Barske et al. 2011), characteristics that, to the extent that they are heritable, would aid in the survival of a female's offspring (Husak and Fox 2008). Among different species of manakins, additional factors that appear to influence mate-choice decisions by females include vocalization rates (Durães et al. 2009), and male location in leks, e.g., females may favor males with display courts near the lek center (Shorey 2002) or with display courts where light levels are highest (Théry and Vehrencamp 1995). In other lekking species, females based their mating decision on a variety of factors (Table 15.2). However, these factors are always correlated with some aspect of male quality (e.g., dominance status and condition), suggesting that females may be choosing males to copulate with based primarily on the perceived genetic quality of males.

Female choice at leks can be influenced by factors other than perceived male quality. In some species, females have been found to "copy" the choices of other females, with copying being apparent if the probability of a male being selected as a mate increases when previously mating with other females and decreases when previously rejected by other females. Copying may be more likely when there is greater mating skew (Fig. 15.39). Copying has been reported to influence female mate-choice decisions in Greater Sage-Grouse (*Centrocercus urophasianus*; Gibson et al. 1991) and Black Grouse (*Tetrao tetrix*; Höglund et al. 1995). However, no evidence of copying has been found in Greater Prairie-Chickens (Nooker and Sandercock), Great Snipe (*Gallinago media*; Fiske et al. 1996), and other lek species (e.g., those listed in Table 15.2). Mate-choice copying could potentially be beneficial for at least two reasons. First, females can avoid the costs of active mate choice, including time, energy, and predation risk (e.g., Pomiankowski 1987), by observing and imitating the actions of other females that have paid the costs of active mate choice and are presumably

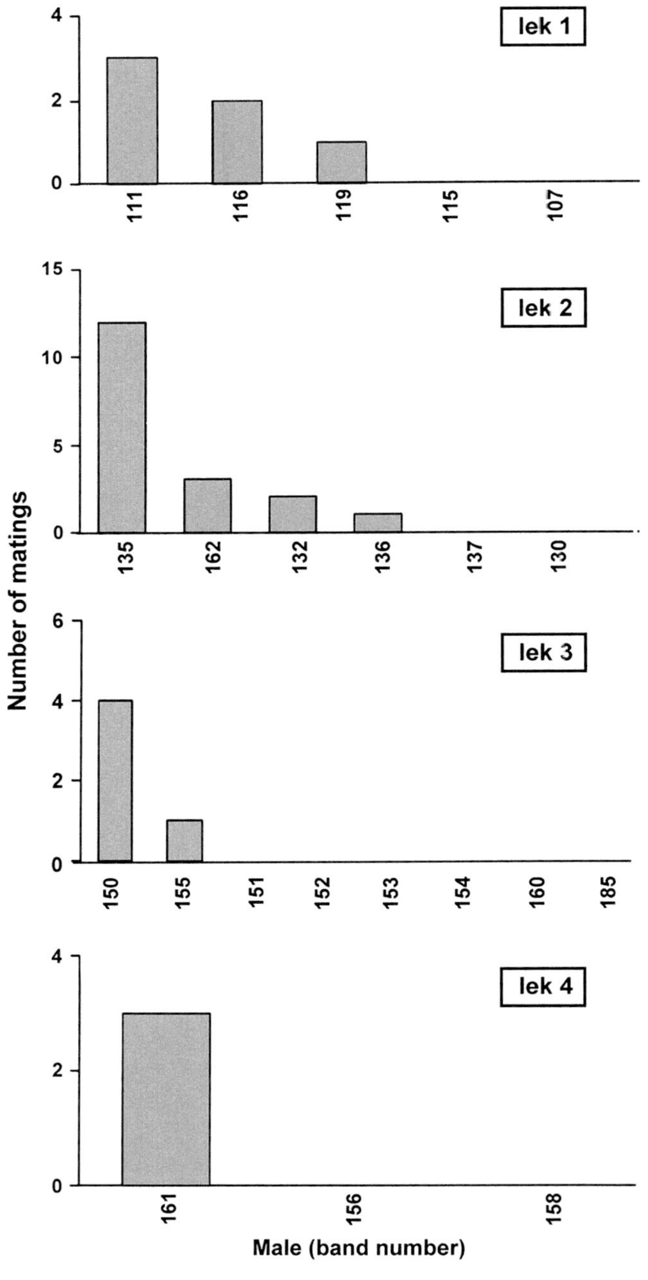


**Fig. 15.36** Distribution of male mating success at four leks of Golden-collared Manakins (*Manacus vitellinus*) in Panama. Males were identified by uniquely numbered leg bands. (Figure from Stein and Uy 2006; © 2005 Oxford University Press, used with permission)

making relatively successful mating decisions (Pomiankowski 1990). Second, given possible errors in assessing males, particularly by young, inexperienced females, copying might improve the likelihood of selecting a high-quality mate (Gibson and Höglund 1992; Nordell and Valone 1998).

Given these potential benefits, why has mate-choice copying by females been observed in so few lekking species of birds? One possibility is simply that the costs of mate choice are not sufficiently high for females to favor copying over relying on their own evaluation of males. Related to this, males in many lekking species may vary

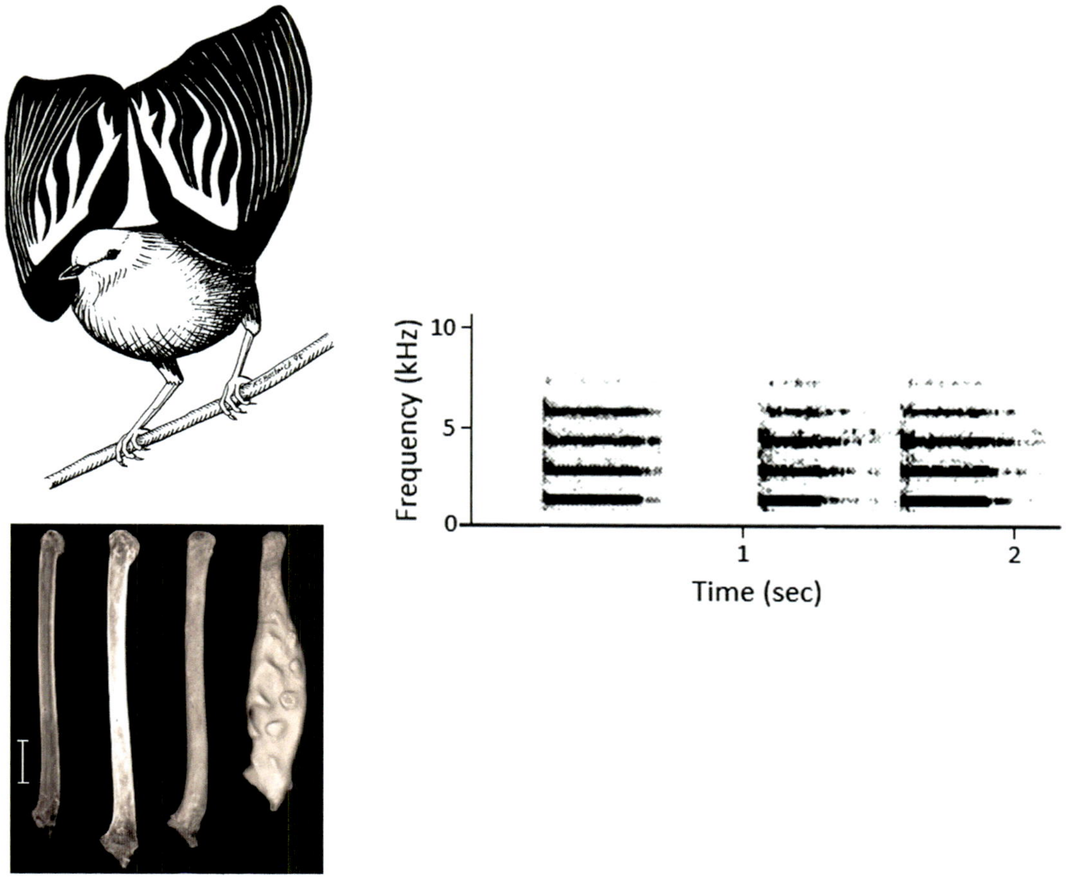


**Fig. 15.37** An example of a mechanical sound produced by manakins is the "*ting*" of Club-winged Manakins (*Machaeropterus deliciosus*). During courtship displays, males extend their legs and slightly open their wings and droop them down to their sides. Then (Top left), in a rapid movement (about 0.017 seconds), males pronate their forearms so the secondaries are positioned above the back, and the *ting* commences. While producing the *ting*, the wings are held with the secondaries spread open and the primaries closed. The wings do not make contact, but vibrate slightly and the secondaries move back and forth, parallel to the plane of the wing (and this movement apparently generates the *ting* sound). The display ends when the wings are rotated back down, swung low and out to the sides, and then closed to a resting position. (Figure from Bostwick 2000; © 2000 Oxford University Press, used with permission)
Top right, Spectrogram of the mechanical *ting* sounds of Club-winged Manakins. (Figure modified from Bostwick 2000; © 2000 Oxford University Press, used with permission)
Bottom, Left to right, ulnas of a male Blue-crowned Manakin (*Lepidothrix coronata*), White-crowned Manakin (*Pipra pipra*), Red-capped Manakin (*Pipra mentalis*), and Club-winged Manakin (*Machaeropterus deliciosus*). The ulna of male Club-winged Manakins is more than three times the volume of those of birds of a similar size, and probably assists in sound production. (Figure from Bostwick et al. 2012; © 2012 The Royal Society, used with permission)

sufficiently in quality (e.g., based on morphology, display quality, and location in lek) such that errors by females, even inexperienced females, in discriminating among males are unlikely. In other taxa (e.g., fish), females tend to copy the choices of other females when males are similar in quality, but rely on their own evaluation when males differ markedly in quality (e.g., Witte and Ryan 1998).

The number of males that females copulate with when visiting leks varies within and among species (Table 15.2). Many studies have revealed the reproductive or mating skew typical of leks,

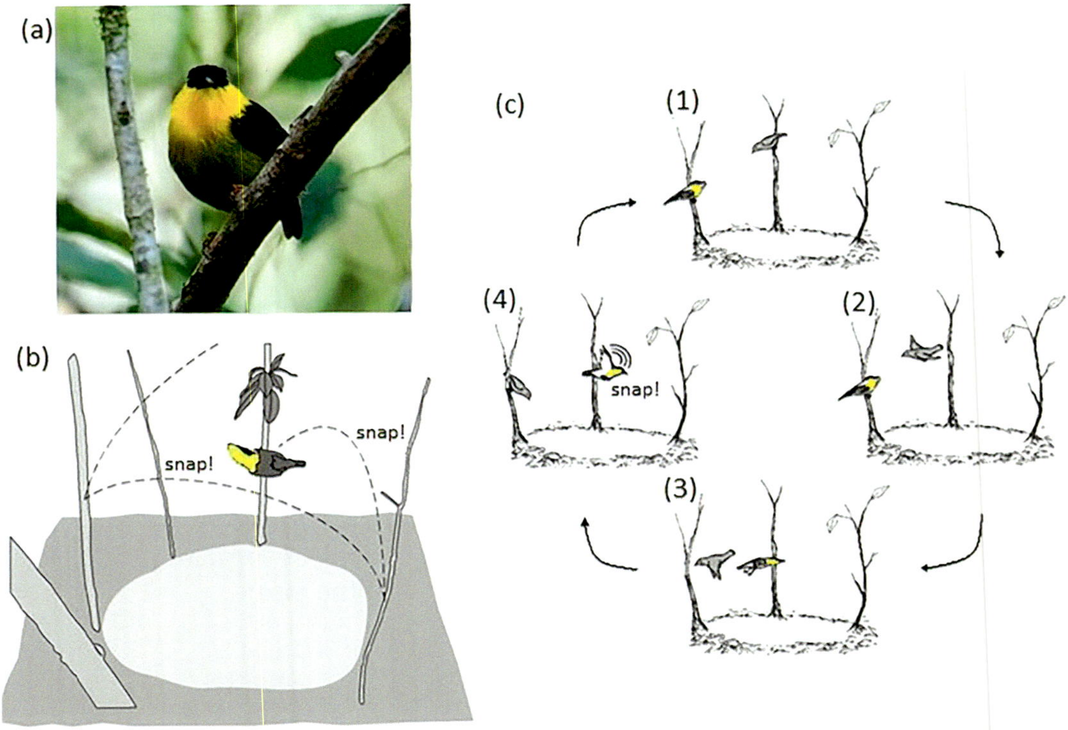


**Fig. 15.38** (**a**) Male Golden-collared Manakin (*Manacus vitellinus*). (**b**) Courtship displays of male Golden-collared Manakins include the jump-snap display. Males jump between saplings at the boundaries of their display "court" and produce loud wingsnaps in midair by a very fast upward movement of the wings that come into contact above the back; between jumps, males remain briefly on saplings with their golden-throat feathers erected. Snap-jumps can be repeated up to 20 times in a row. (**c**) Males and females sometimes engage in a "duo dance." (1) To attract females, males perform courtship displays between small trees around a cleared arena. (2) Females sometimes join males, flying from one small tree to another. (3) Before the female lands, the male flies across the display area. (4) After the female lands, the male flies back, lands briefly, then jumps and often produces a wingsnap in midair. (Photo of male by Melissa McMasters, Wikipedia, CC BY 2.0, https://creativecommons.org/licenses/by/2.0/. Figure b from Fusani et al. 2007; © 2007 John Wiley and Sons, used with permission. Figure c from Schlinger et al. 2013; © 2013 Published by Elsevier Inc., used with permission)

with a few males obtaining most of the copulations (Figs. 15.35 and 15.36). However, in many species, some proportion of females, ranging from just a few percent to more than 50%, do copulate with multiple males (Table 15.2). As noted previously, females at leks appear to be choosing males to copulate with based primarily on the perceived genetic quality of males. Females that copulate with more than one male (and, in most species studied to date, that usually means two or, less often, three males) are still copulating with high-quality males, but how might females potentially benefit by copulating with two or three high-quality males rather than just one? One possibility is that multiple mating reduces the likelihood of laying infertile eggs, i.e., a potential risk of mating once is that the male may be infertile. Another possible reason for multiple mating is that females mate with less preferred males because they are initially unable to gain access to a more preferred male. Several investigators have reported that females at leks sometimes must wait for access to preferred males that are interacting with one or more other females (e.g., Avery 1984) and others have reported fighting among females for access to males (e.g., Petrie et al. 1992). In such cases, females may first mate

**Table 15.2** Variation among lekking species in factors that influence mate choice by females

| Species | Scientific name | Female mate choice based on: | Number of males that females mate with: | References |
|---|---|---|---|---|
| **Galliformes:** | | | | |
| **Phasianidae** | | | | |
| Indian Peafowl | *Pavo cristatus* | Display rate, train (tail) ornamentation (number and density of ocelli), and blue-green color of "eyespots" on the train | 46 of 87 females (53%) mated with more than one male (range = 2–5) | Petrie et al. (1992), Loyau et al. (2005), Dakin and Montgomerie (2013) |
| Black Grouse | *Tetrao tetrix* | Location in lek (prefer males at center) and fighting rate and performance (success in winning fights) | In one study, 2 of 59 broods (3.4%) had multiple paternity; in another, 5 of 130 broods had multiple paternity | Hovi et al. (1994), Alatalo et al. (1996), Lebigre et al. (2007), Kervinen et al. (2012), Hämäläinen et al. (2012) |
| Greater Sage-Grouse | *Centrocercus urophasianus* | Time on lek, display rate, and characteristics of vocalizations | Of 182 clutches, all eggs were fertilized by the same male in 93% of the clutches; 7% of clutches had eggs fathered by 2 males | Gibson and Bradbury (1985), Bird et al. (2013) |
| Sharp-tailed Grouse | *Tympanuchus phasianell* | Location in lek (centrally located males copulate with more females), as well as display rate and characteristics of vocalizations | ? | Moyles and Boag (1981), Gratson (1993) |
| Greater Prairie-Chicken | *Tympanuchus cupido* | More fighting and lower display rates | In 25 broods, multiple paternity found in 11 broods (44%; range = 2–4 sires); the other 14 broods had on 1 sire | Nooker and Sandercock (2008), Hess et al. (2012) |
| Lesser Prairie-Chicken | *Tympanuchus pallidicinctus* | Age, comb color, and percent time spent displaying | | Behney et al. (2012), Gould and Augustine (2020) |
| **Otidiformes:** | | | | |
| **Otididae** | | | | |
| Great Bustard | *Otis tarda* | Age, mass, and display effort | ? | Alonso et al. (2010) |
| Little Bustard | *Tetrax tetrax* | Body condition, lek attendance, ornamental symmetry, and display rates | ? | Jiguet and Bretagnolle (2014) |
| Houbara Bustard | *Chlamydotis undulata undulata* | ? | 3 of 5 nests (60%) had multiple paternity | Lesobre et al. (2010) |
| **Charadriiformes:** | | | | |
| **Scolopacidae** | | | | |
| Ruff | *Philomachus pugnax* | Frequency of visits by satellite males, consistent lek attendance, and low rates of aggressive behavior | 17 of 34 broods (50%) had multiple paternity | Hill (1991), Lank et al. (2002) |

(continued)

**Table 15.2** (continued)

| Species | Scientific name | Female mate choice based on: | Number of males that females mate with: | References |
|---|---|---|---|---|
| Great Snipe | *Gallinago media* | Age, time on lek, location in lek (center), and display rate | 24 of 187 females (12.8%) copulated with more than one male | Höglund and Lundberg (1987), Fiske et al. (1994), Fiske and Kålås (1995), Sæther et al. (2001) |
| Buff-breasted Sandpiper | *Tryngites subruficollis* | Longer lek attendance and fewer wing spots | 19 of 47 broods (40.4%) had multiple paternity | Lanctot et al. (1997, 1998) |
| **Apodiformes:** | | | | |
| **Trochilidae**[a] | | | | |
| White-bellied Emerald | *Amazilia candida* | Possibly persistent and vigorous singing | ? | Atwood et al. (1991) |
| Long-billed Hermit | *Phaethornis longirostris* | Possibly location in lek (center preferred) and time spent singing; male position in lek determined by length and pointedness of male bills (Box 15.13 Bills as Daggers) | ? | Stiles and Wolf (1979), Rico-Guevara and Araya-Salas (2014) |
| Green Hermit | *Phaethornis guy* | Possibly location in lek (center preferred) and time spent singing | ? | MacDougall-Shackleton and Harbison (1998) |
| Calliope Hummingbird | *Stellula calliope* | Probably display intensity (number and rate of dive displays in conjunction with some hover and buzzing displays) | ? | Tamm et al. (1989) |
| Black-chinned Hummingbird | *Archilochus alexandri* | Flight performance (shuttle-display) | ? | Wilcox (2020) |
| **Psittaciformes:** | | | | |
| **Strigopoidea** | | | | |
| Kakapo | *Strigops habroptilus* | Frequency range of boom calls and inter-call interval | ? | Eason et al. (2006) |
| **Passeriformes:** | | | | |
| **Tyrannidae** | | | | |
| Ochre-bellied Flycatcher | *Mionectes oleagineus* | High singing rate | ? | Westcott and Smith (1997) |
| **Cotingidae** | | | | |
| Guianan Cock-of-the-Rock | *Rupicola rupicola* | Location in lek (prefer males at center) | 18 of 72 females (25%) copulated with multiple males | Trail (1985), Trail and Adams (1989) |
| Capuchinbird (also called the Calfbird) | *Perissocephalus tricolor* | Dominant male (as determined by male–male interactions) at a lek obtains most if not all copulations | Females apparently mate with a single male | Trail (1990) |
| **Pipridae** | | | | |
| Blue-crowned Manakin | *Lepidothrix coronata* | Vocalization rate | 1 of 20 broods (5%) had multiple paternity | Durães et al. (2009) |

(continued)

**Table 15.2** (continued)

| Species | Scientific name | Female mate choice based on: | Number of males that females mate with: | References |
|---|---|---|---|---|
| Golden-collared Manakin | *Manacus vitellinus* | Plumage and displays | ? | Stein and Uy (2006), Barske et al. (2011) |
| Lance-tailed Manakin | *Chiroxiphia lanceolata* | Age, perhaps because older males have better courtship displays, better recognition of female cues, or because females prefer males that have been present at leks over a longer period of time | 12 of 253 nests (4.8%) had chicks sired by more than one male | DuVal and Kempenaers (2008), DuVal (2012), Sardell and DuVal (2014) |
| White-bearded Manakin | *Manacus manacus* | Location in lek (prefer males at center) | 18 of 74 females (24.3%) copulated with more than one male | Lill (1974), Shorey (2002) |
| White-throated Manakin | *Corapipo gutturalis* | Location in lek (prefer males where light levels are highest) | ? | Théry and Vehrencamp (1995) |
| Wire-tailed Manakin | *Pipra filicauda* | Territory tenure (number of years a male held his territory) | 7 of 38 broods (18.4%) had multiple paternity | Ryder et al. (2008, 2009) |
| Long-tailed Manakin (two males call and display cooperatively; only dominant male of pair mates) | *Chiroxiphia linearis* | Calling rate (to attract females), temporal synchrony of male–male duets, and display rate and duration; duration and rate of performance of individual display elements (upright postures, bounces, angel flights, and bows) | Females ($N = 50$) visited an average of 4 males and just 13 (26%) visited only one male; number of males that females copulate with is not known | McDonald (1989), Ward (2012), Lukianchuk and Doucet (2014), Maynard et al. (2014) |
| White-ruffed Manakin | *Corapipo leucorrhoa* | Probably display quality and rate and characteristics of vocalizations (*tsee* notes and canopy calls) | ? | Rosselli et al. (2002) |
| **Paradisaeidae (birds of paradise)** | | | | |
| Lawes' Parotia | *Parotia lawesii* | Display probability (probability that a given female would be displayed to by a male on any visit to that male during the season) | ? | Pruett-Jones and Pruett-Jones (1990) |
| **Ploceidae** | | | | |
| Jackson's Widowbird | *Euplectes jacksoni* | Tail length and display rate | ? | Andersson (1989) |

[a]As of 2013, lekking behavior has been reported in at least 49 species of hummingbirds (Martinez-Garcia et al. 2013). However, factors important in female mate-choice decisions at hummingbird leks have rarely been examined

**Box 15.13 Bills as Daggers**

Long-billed Hermits (*Phaethornis longirostris*) are lek-breeding hummingbirds and males defend small territories where they sing and attempt to attract females for mating. Males fight to obtain and defend these territories, and video-recordings revealed that males sometimes used their bills as daggers during fights. Close examination of the bills of Long-billed Hermits revealed that territorial males had longer, more pointed bills than either non-territorial males (floaters) or females. In addition, observations revealed that males with longer, more pointed bill tips were more successful at obtaining and successfully defending lek territories. These results suggest that sexual selection has favored the evolution of these dagger-like bills because of their importance in improving male reproductive success. Although the morphology of hummingbird bills is clearly linked to their foraging strategies, the results of this study by Rico-Guevara and Araya-Salas (2014) suggest that, at least in some species, sexual selection may also influence bill morphology. In some cases, as male hummingbirds joust with their bills, aggressive encounters can even be lethal (Evens and Harper 2020).

Screen capture from video of two male Long-billed Hermits fighting for ownership of a perch. The male rapidly approaching from above stabs his rival in the throat. Pink flagging at the bottom was used to mark the perch site. (Figure from Supplementary Data for Rico-Guevara and Araya-Salas 2014; © 2014 Oxford University Press, used with permission)

(continued)

**Box 15.13** (continued)

Left, Long-billed Hermit. Above, Differences in bill morphology of females, floater males, and territorial males. Territorial males have longer, more pointed bills. Scale bars = 0.5 mm. (Figures from Rico-Guevara and Araya-Salas 2014; © 2014 Oxford University Press, used with permission)

(continued)

**Box 15.13** (continued)

Female hummingbirds, such as these female Broad-tailed Hummingbirds (*Selasphorus platycercus*), may also use their bills during aggressive interactions. (Photo by Richard Seeley, purchased from iStockphoto)

with a less preferred male and then, if later able to gain access to a more preferred male, mate with that male (Petrie et al. 1992). Finally, multiple mating and paternity for females in lekking species may provide some of the same benefits proposed for females in socially monogamous species that engage in extra-pair copulations (Table 15.1).

Females at leks make mate-choice decisions based on variation in male quality as determined by differences among males in morphology, display quality or rates, and their location within leks. For males not selected, or not initially selected, by females for mating, one possible strategy is to interfere when other males at the lek attempt to copulate with a female. Among lek species, disruption rates (i.e., percentage of disrupted copulation attempts) are typically below 10% (Lanctot et al. 1998). One reason for this is that males may establish dominance hierarchies and lower-ranking males rarely, if ever, disrupt matings by more dominant males (although dominant males may disrupt attempted matings by lower-ranking males) (e.g., Foster 1981). However, disruptive behavior is relatively common in some lek species. For example, in Guianan Cock-of-the Rocks (*Rupicola rupicola*), males disrupted an average of 30% of all courtship visits and 28.3% of all copulations (Trail 1985, 1986; Fig. 15.40). Of 204 attempted copulations by male Sharp-tailed Grouse (*Tympanuchus phasianellus*), 74 (36.3%) were disrupted by neighboring males (Gratson et al. 1991). At Great Snipe (*Gallinago media*) leks, a

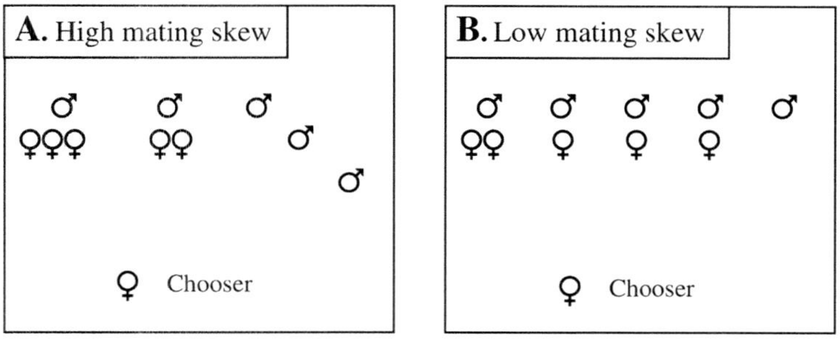


**Fig. 15.39** Mating skew and potential for mate-choice copying. In **a**, the mating status (mated vs. unmated) of males is a cue that reveals the high-quality males. This is not the case in **b**, where most males are mated. Mate-choice copying is less likely to occur in **b**. (Figure from Vakirtzis and Roberts 2009; © 2009 Oxford University Press, used with permission)

third of courtship visits to male territories by females were disrupted by other males (Sæther et al. 1999). After being disrupted, female Great Snipe and Sharp-tailed Grouse were no more likely to mate with the disrupting males than with other males at the leks (Gratson et al. 1991; Sæther et al. 1999). Similar results have been reported in Golden-headed Manakins (*Ceratopipra erythrocephala*; Lill 1976) and Greater Sage-Grouse (Hartzler and Jenni 1988). Disrupting male Guianan Cock-of-the-Rocks, however, did obtain more matings than did non-disruptors (Trail 1985). The results of these studies suggest that disrupting males may be able to increase their mating success. For most lekking species, however, males do not gain copulations from the disruption directly, but increase their relative copulation rate by lowering the copulation rate of disrupted males (Magaña et al. 2011). If so, then why are disruption rates for most lekking species relatively low? Males that spend more time attempting to disrupt courtship or copulations by other males have less time and energy for defending their own territories and displaying for females. In addition, disrupting males may be subject to retaliation (Trail 1985). Finally, because of the mating skew reported for most lek species, lower-ranking males are unlikely to benefit from disrupting because disrupted females are unlikely to mate with them anyway.

A less common strategy used by lower-ranking males is to "sneak" copulations from females visiting other males. This is possible in some species because they are sexually monomorphic, with males and females similar in size and with similar plumage. For example, Buff-breasted Sandpipers are sexually monomorphic and males occasionally sneak into other territories to gain access to, and copulate with, females by mimicking female behavior (Lanctot et al. 1998). Other lekking species that are sexually monomorphic include Ochre-bellied Flycatchers, Capuchinbirds (*Perissocephalus tricolor*), Screaming (*Lipaugus vociferans*) and Dusky (*Lipaugus fuscocinereus*) pihas, Thrush-like Manakins (*Schiffornis turdina*), and several species of hummingbirds (Trail 1990). Additional studies of these and other monomorphic lek species are needed, but mimicking female behavior might be a strategy used by lower-ranking males in these species to obtain copulations (Trail 1990). In one lek species, males "sneak" copulations either by mimicking females or by assisting more dominant males in attracting females and sneaking a few copulations from females that will usually copulate with the dominant males (Box 15.14 Diverse Plumages and Mating Strategies of Male Ruffs).

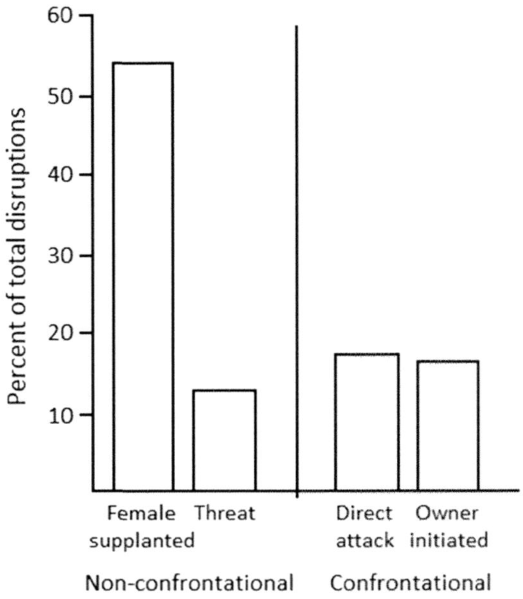


**Fig. 15.40** Male Cock-of-the-Rocks (*Rupicola rupicola*) disrupt courtship by other males in four different ways, with two being non-confrontational (no physical contact) and two being confrontations (involving physical contact). Non-confrontational disruptions included supplanting females perched in another male's territory and directing threat behavior (i.e., shrill, persistent calling from the edge of the courting male's territory and/or wing-beating and lunges directed toward the pair) to a courting pair. Confrontational disruptions included direct attacks on a displaying male or courting pair (including interrupting copulations and attempting to mount the female) and fights between a disrupter and a displaying male initiated by the displaying male (owner of the territory). (Figure modified from Trail and Koutnik 1986; © 1986 Blackwell Verlag GmbH, used with permission)

## 15.8 Polyandry

Available evidence suggests that polyandry arose independently multiple times in a variety of avian taxa (Temrin and Sillén-Tullberg 1994; Table 15.3). In polyandrous species of birds (about 1% of all species), at least some females have social relationships of varying duration with more than one male and also mate with those males. Some species exhibit "classical polyandry" (Table 15.3), where females either: (1) defend breeding territories and attempt to attract as many males as possible to mate with them and nest in their territory or (2) mate with males sequentially, mating with a male and laying a clutch of eggs in his nest and then leaving and attempting to locate additional males where they can do the same thing. "Classical" species also exhibit sex-role reversal, with sexual selection favoring females larger than males and, sometimes, more colorful as well. Other species exhibit "cooperative polyandry," where some females mate with two or more males that then assist with parental care (Table 15.3). Finally, still other species exhibit "facultative polyandry," where most females in a population are socially monogamous, but others, if circumstances permit (e.g., biased sex ratio with more males than females in a population), mate with two or more males that then assist with parental care (Table 15.3).

Questions remain concerning the selective factors that contributed to the evolution of classical polyandry, but increasing levels of male parental care leading to, ultimately, male-only (or primarily male) parental care was likely one such factor. Hypotheses proposed to explain this transition include the (1) male-biased adult sex

**Box 15.14 Diverse Plumages and Mating Strategies of Male Ruffs**

Ruffs (*Philomachus pugnax*) are Eurasian shorebirds with a complex lek mating system that includes three different male mating behaviors and plumages (see figure below). Two morphs are ornamented: males referred to as "independent" typically have darker, and extremely variable, plumage and defend small (about 1 $m^2$) mating territories, or courts, on leks. Non-territorial "satellite" males (about 5–15% of all males) have lighter plumage than most independent males and display along with independent males on their "courts." A rare (1–1.5% of all males) third type of male resembles females (but they are a bit larger than females) and are called "faeders" (an old English word for father; Jukema and Piersma 2006). These males do not defend mating areas and have larger testes (about 2.5 times larger) than independent and satellite males. These male morphs are fixed so plumages and mating behaviors do not change from year to year.

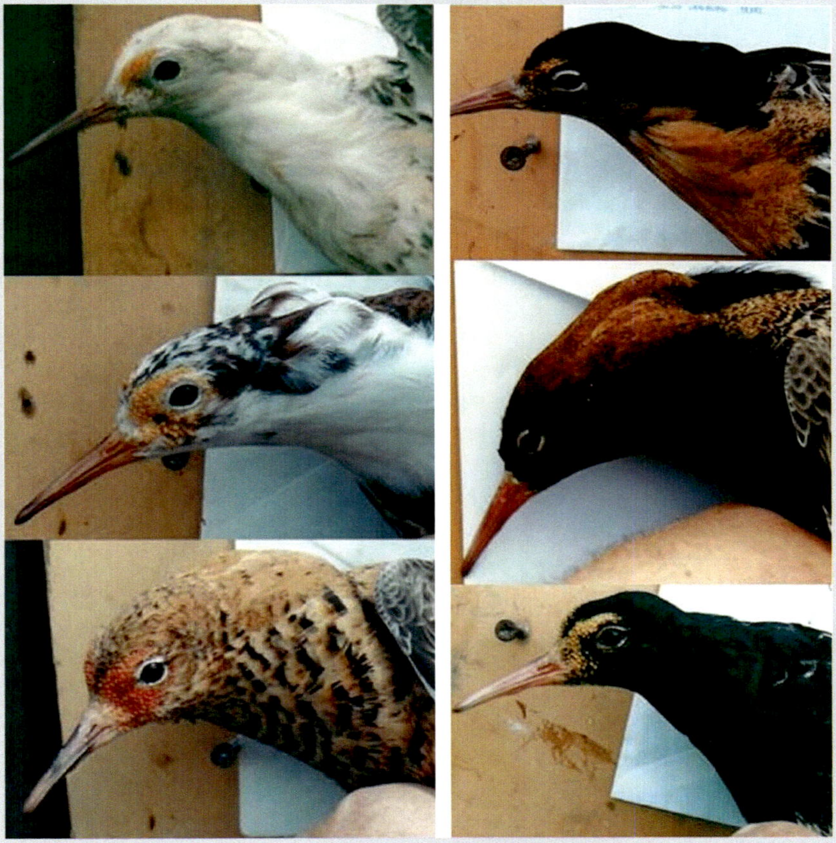

Examples of variation in the plumage of "Independent" male Ruffs. (Figures from van Rhijn et al. 2014; used with permission)

(continued)

**Box 15.14** (continued)

Three permanent alternative reproductive morphs of male Ruffs associated with territorial lekking behavior and plumage color. Bottom, left to right: territorial dark-plumaged "Independent" male, female mimic, or "Faeder," and non-territorial white-plumaged "Satellite"; pictured above the males is a female. (Figure from Farrell et al. 2013; open-access article distributed under the terms of the Creative Commons Attribution License, https://creativecommons.org/licenses/by/4.0/)

Satellite males are floaters that are accepted into the mating courts of independent males as long as they remain submissive, and they use this access to sneak copulations with visiting females. Independent males associate with Satellite males because more females tend to visit courts with Satellite males present. Independent males obtain more copulations than Satellite males (about 90% vs. 10%), but Satellite males still benefit because they could not attract females on their own (Widemo 1998b). Faeders are also accepted into the mating courts of Independent males and also attempt to sneak copulations from visiting females (and, given their large testes, Faeders have a large supply of sperm available).

Differences in the plumage of Independent male Ruffs apparently do not influence their mating success (Hill 1991), so the variable plumage may be important for individual recognition. Being easily identified may be beneficial (at least for some males) if it allows females, who may make the decision on who to mate with after several visits to a lek over multiple days, to be more certain of male identity. Recognition by other males may also be beneficial if, after establishing mating territories or courts, the ability to easily recognize individual males, and their status relative to a male's own, minimizes aggression (van Rhijn et al. 2014).

ratio hypothesis, (2) replacement clutch hypothesis, (3) stressed female hypothesis, (4) low breeding density hypothesis, (5) low extra-pair paternity hypothesis, (6) differential parental capacity hypothesis, and (7) fluctuating food hypothesis (Fig. 15.41). The male-biased adult sex ratio hypothesis proposes that a persistent male bias in a population's sex ratio would favor male-biased parental care (Liker et al. 2013). With fewer females than males, males may have little chance of finding new mates so their best strategy might be to provide parental

**Table 15.3** Species of birds known to be polyandrous

| Common name | Family | Order | Species or number of species | Scientific name(s) | Citation(s) |
|---|---|---|---|---|---|
| **Classical polyandry** (male-only parental care and sex-role reversal) | | | | | |
| Cassowaries | Casuariidae | Struthioniformes | Southern Cassowary | *Casuarius casuarius* | Handford and Mares (1985) |
| Tinamous | Tinamidae | Tinamiformes | Great Tinamou | *Tinamus major* | Brennan (2009) |
| Brush-turkeys | Megapodiidae | Galliformes | Australian Brush-Turkey | *Alectura latham* | Göth (2007) |
| Buttonquail | Turnicidae | Charadriiformes | All 15 species in the family | – | Muck (2011) |
| Plains-wanderer | Pedionomidae | Charadriiformes | Plains-wanderer (only species in the family) | *Pedionomus torquatus* | Rushton (1976) |
| Sandpipers and phalaropes | Scolopacidae | Charadriiformes | Spotted Sandpiper and Red, Red-necked, and Wilson's phalaropes | *Actitis macularius*, and *Phalaropus fulicarius*, *Phalaropus lobatus*, and *Phalaropus tricolor* | Hays (1972) |
| Painted snipes | Rostratulidae | Charadriiformes | Greater Painted Snipe | *Rostratula benghalensis* | Komeda (1983) |
| Jacanas | Jacanidae | Charadriiformes | 7 of 8 species | – | Jenni (1974), Muck (2011) |
| Plovers | Charadriidae | Charadriiformes | Eurasian Dotterel, Kentish Plover, and Snowy Plover | *Charadrius morinellus*, *Charadrius alexandrinus*, and *Charadrius nivosus* | Warriner et al. (1986), Owens et al. (1994), Kosztolányi et al. (2011) |
| Coucals | Centropodidae | Cuculiformes | Black coucal | *Centropus grillii* | Muck (2011) |
| **Facultative polyandry** (typically monogamous, but occasionally polyandrous or polygynandrous) | | | | | |
| Sandpipers | Scolopacidae | Charadriiformes | Sanderling, Temminck's Stint | *Calidris alba* and *Calidris temminckii* | Parmelee and Payne (1973), Breiehagen (1989) |
| Gruiformes | Aramidae | Gruiformes | Limpkin | *Aramus guarauna* | Bryan (1996) |
| Owls | Strigidae | Strigiformes | Boreal Owl | *Aegolius funereus* | Korpimäki et al. (2011) |
| Owls | Tytonidae | Strigiformes | Barn Owl | *Tyto alba* | Henry et al. (2013) |
| Woodpeckers | Picidae | Piciformes | Northern Flicker and Lesser Spotted, Great Spotted, Three-toed, and West Indian woodpeckers | *Colaptes auratus*, *Picoides minor*, *Dendrocopos major*, *Picoides dorsalis*, and *Melanerpes superciliaris* | Willimont et al. (1991), Kotaka (1998), Pechacek et al. (2005), Wiebe (2008), Rossmanith et al. (2009) |
| Songbirds | Acrocephalidae | Passeriformes | Sedge Warbler | *Acrocephalus schoenobaenus* | Hasselquist and Langefors (1998) |
| Songbirds | Remizidae | Passeriformes | Penduline Tit | *Remiz pendulinus* | Valera et al. (1997) |
| Songbirds | Notiomystidae | Passeriformes | Stitchbird (also called the Hihi) | *Notiomystis cincta* | Castro et al. (1996) |

(continued)

**Table 15.3** (continued)

| Common name | Family | Order | Species or number of species | Scientific name(s) | Citation(s) |
|---|---|---|---|---|---|
| Songbirds | Turdidae | Passeriformes | Bicknell's Thrush | *Catharus bicknelli* | Goetz et al. (2003) |
| Songbirds | Passeridae | Passeriformes | Rock Sparrow | *Petronia petronia* | Pilastro et al. (2001) |
| Songbirds | Emberizidae | Passeriformes | Song Sparrow | *Melospiza melodia* | Janssen et al. (2008) |
| Songbirds | Fringillidae | Passeriformes | American Goldfinch | *Spinus tristis* | Middleton (1988) |
| **Cooperative polyandry** (one female with two or more males that assist with parental care; may also sometimes be monogamous, polygynandrous, polygynous, or cooperative breeders) | | | | | |
| Eagles | Accipitridae | Accipitriformes | Madagascar Fish-Eagle | *Haliaeetus vociferoides* | Tingay et al. (2002) |
| Hawks | Accipitridae | Accipitriformes | Galapagos Hawk, Harris' Hawk, Variable Hawk, Pale Chanting Goshawk, and possibly others (Kimball et al. 2003) | *Buteo galapagoensis*, *Parabuteo unicinctus*, *Buteo poecilochrous*, and *Melierax canorus* | Mader (1979), Faaborg et al. (1980), Kimball et al. (2003), Malan and Crowe (1996) |
| Vultures | Accipitridae | Accipitriformes | Bearded Vulture and Egyptian Vulture | *Gypaetus barbatus* and *Neophron percnopterus* | Heredia and Donázar (1990), Tella (1993) |
| Skuas | Stercorariidae | Charadriiformes | Brown Skua | *Stercorarius antarcticus* | Young (1998) |
| Trumpeters | Psophiidae | Gruiformes | Pale-winged Trumpeter | *Psophia leucoptera* | Sherman (1995) |
| Rails | Rallidae | Gruiformes | Tasmanian Nativehen, Pukeko (also called purple swamphen), and Common Moorhens | *Tribonyx mortierii*, *Porphyrio porphyrio*, and *Gallinula chloropus* | Goldizen et al. (2000), Jamieson et al. (1994), Petrie (1986) |
| Woodpeckers | Picidae | Piciformes | Acorn Woodpecker | *Melanerpes formicivorus* | Joste et al. (1985) |
| Parrots | Psittaculidae | Psittaciformes | Eclectus Parrot | *Eclectus roratus* | Heinsohn et al. (2007) |
| Songbirds | Prunellidae | Passeriformes | Dunnock | *Prunella modularis* | Burke et al. (1989) |
| Songbirds | Acrocephalidae | Passeriformes | Henderson Reed Warbler | *Acrocephalus taiti* | de Brooke and Hartley (1995) |
| Songbirds | Acanthizidae | Passeriformes | White-browed Scrubwren | *Sericornis frontalis* | Whittingham et al. (1997) |
| Songbirds | Troglodytidae | Passeriformes | Stripe-backed Wren | *Campylorhynchus nuchalis* | Rabenold et al. (1990) |
| Songbirds | Calcariidae | Passeriformes | Smith's Longspur | *Calcarius pictus* | Briskie (1992) |

care rather than deserting a female after copulating. However, with male parental care and the male-biased adult sex ratio, females can desert males after producing a clutch and acquire new mates.

The replacement clutch hypothesis posits that, when rates of nest failure are high, socially monogamous males benefit by assuming most or all parental care duties because their mates then have more time to forage and have the energy to, if needed, quickly produce a replacement clutch (Jenni 1974). Once males assume parental care duties, selection might then favor either desertion by females to seek additional mates (Székely et al.

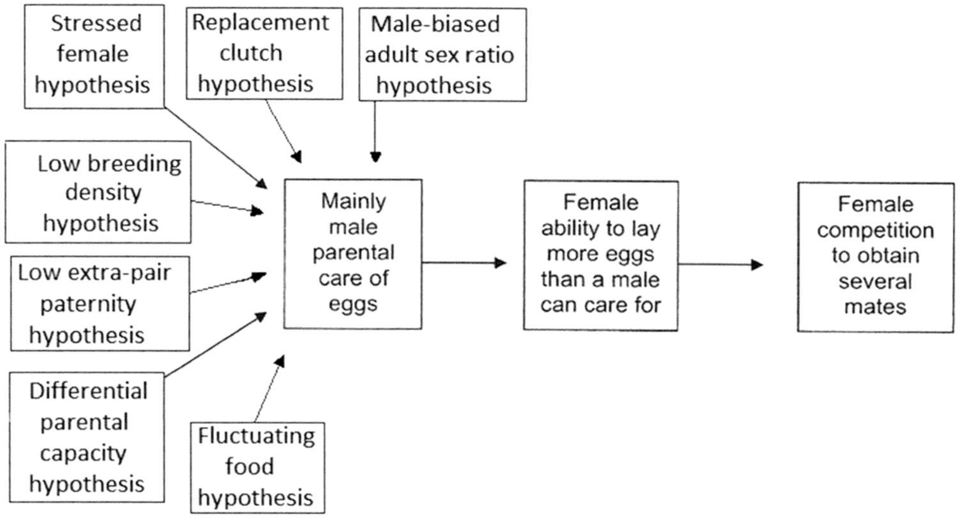


**Fig. 15.41** Possible evolutionary pathways leading to classical polyandry, with sex-role reversal and parental care primarily or only by males. (Figure modified from Muck 2011; used with permission of Christina Muck)

1999; McDonald et al. 2023) or, if remaining on territory, attempts by females to attract additional males to their territory. In either case, females could potentially benefit via additional matings and the production of additional young.

The stressed female hypothesis proposes that food supplies might be so limited that females, after producing a clutch of eggs, have insufficient energy to help with incubation. If so, then males benefit (by increasing the likelihood of successful reproduction) by assuming all parental care duties and this again might favor desertion by females (Graul et al. 1977), perhaps to seek better foraging areas and, if so, additional mates as well. The low breeding density hypothesis posits that, when a species nests at low densities, males gain little from deserting a nest because they are unlikely to find another unmated fertile female. However, females can mate with any male (and, in some species, can also store sperm from a previous mate) and so could potentially benefit by deserting and laying a second clutch for another male (Ownes 2002). The low extra-pair paternity hypothesis proposes that males more confident in their paternity are more likely to provide parental care, potentially providing females with opportunities to either desert and seek additional mates or to attract additional males to mate and nest in the same territory (Muck et al. 2009).

The differential capacity hypothesis proposes that producing a clutch may sufficiently stress females such that they would be unable to provide sufficient parental care to successfully fledge young. Males, however, expending less reproductive effort during the egg-laying period would likely have sufficient energy to provide the parental care needed to raise young. If so, then both could help incubate eggs, but females, now in better energetic condition because of the male's assistance, might desert after eggs hatch (especially in species with precocial young) and seek additional mating opportunities (Erckmann 1981). Finally, the fluctuating food hypothesis posits that females breeding in areas with an abundance of food have sufficient energy to produce multiple clutches, thus favoring female strategies leading to either sequential polyandry (producing a clutch for a male, then abandoning and seeking additional males) or simultaneous polyandry (attempting to attract multiple males to their breeding territory and producing clutches for each male) (Parmelee and Payne 1973).

No single hypothesis has gained sufficient support to be an accepted explanation for the

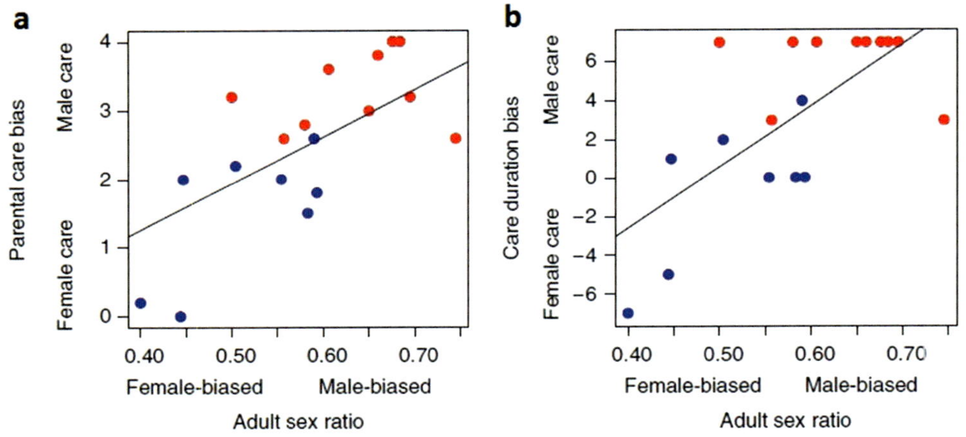


**Fig. 15.42** Relationship between adult sex ratio (number of adult males/number of adult males plus females) and the amount of parental care provided by adult males and females in 18 species of shorebirds. Adult sex ratio is correlated with (**a**) parental care bias (mean of male participation in five aspects of parental care, including nest building, incubation, nest guarding, chick brooding, and chick guarding), and (**b**) care duration bias (male care duration minus female care duration). Red dots = species with reversed sex roles (polyandrous); blue dots = species with conventional sex roles (socially monogamous or polygamous). (Figure from Liker et al. 2013; © 2013 Springer Nature, used with permission)

evolution of primary male or male-only care. However, among shorebirds, Liker et al. (2013) found a significant relationship between adult sex ratios and parental care bias (Fig. 15.42), with species where males provide more care having more male-biased sex ratios and those where females provide more care having more female-biased sex ratios. Liker et al. (2013) also noted apparent evolutionary flexibility in levels of parental care by males and females, with flexible sex roles reported even with single species, e.g., Kentish Plovers (*Charadrius alexandrinus*) and Temminck's Stints (*Calidris temminckii*). In other words, in at least some species of shorebirds, males and females can apparently respond to changes in adult sex ratios with corresponding changes in their contributions to parental care. The extent to which adult sex ratios might influence the parental care of males and females in other species remains to be determined.

Other hypotheses to explain primary male or male-only parental care, including the stressed female, fluctuating food, and differential parental capacity hypotheses, have received little empirical support (Erckmann 1983; Amat et al. 2000). Because primary male or male-only care has evolved more than once in classical polyandrous species of birds, it is certainly possible that different combinations of factors, including male-biased adult sex ratios, breeding density, rates of extra-pair paternity, and the physiological/energetic ability of females to lay replacement clutches, contributed to the evolution of primary male or male-only care.

With primary male or male-only parental care, females physiologically able to produce additional clutches can then potentially benefit (in terms of reproductive fitness) by seeking additional mates, either by abandoning a male and searching for other mates (sequential polyandry) or by remaining in the same areas and attracting additional mates (simultaneous polyandry). To do so, however, females must be capable of producing more eggs than a single male can incubate and must also be able to find or attract additional mates. For polyandrous species of shorebirds, that means producing more than four eggs, the maximum number that can be efficiently incubated (Fig. 15.43; Lengvel et al. 2009). Not coincidentally, four-egg clutches are typical of

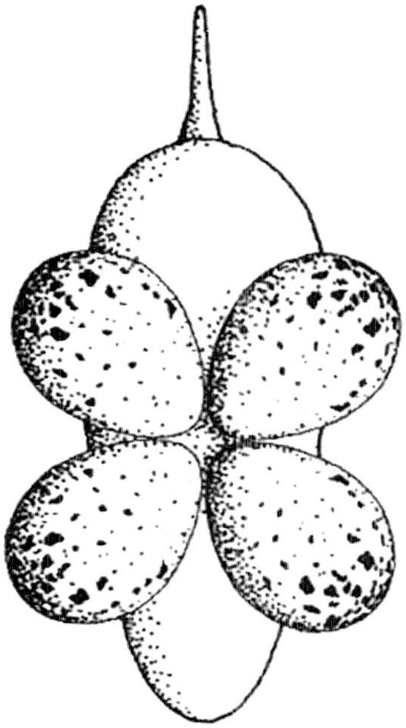

**Fig. 15.43** For most shorebirds, their size limits the number of eggs they can effectively incubate. For example, this is a ventral view of a small sandpiper (with the bird's body outlined without feathers) incubating its clutch of four eggs. Note that the shape of the eggs allows them to fit tightly together, minimizing the area that needs to be covered for efficient incubation. This limitation was likely an important factor in the evolution of the four-egg clutch of most shorebirds. (Figure from Andersson 2005; © 2005 John Wiley and Sons, used with permission)

most shorebirds, regardless of mating system. For other polyandrous species, the maximum number of eggs males can efficiently incubate varies with male size. Given this maximum, polyandrous females must be energetically capable of producing clutches for one or more additional males, and egg production requires sufficient food resources. Classical polyandrous species are generally found in habitats that provide abundant food during the breeding season. For example, many polyandrous shorebirds breed in wetland habitats (e.g., bogs, wet meadows, and marshes) that typically provide an abundance of food (Andersson 2005). Other polyandrous species also have feeding techniques or are found in habitats that allow at least some females in a population to obtain sufficient energy for additional egg production (Andersson 2005). Assuming the availability of sufficient food, females must then be able to locate or attract additional mates.

An important, perhaps critical, factor for polyandrous females seeking additional mates is the population adult sex ratio or, more specifically, the operational sex ratio (OSR)—the number of males available for mating relative to the number of females seeking a mate. Among shorebirds, species that are polyandrous generally have male-biased sex ratios (Liker et al. 2013; Fig. 15.44). As a result, females have a greater likelihood of finding additional mates. For example, Parra et al. (2014) conducted a mate-removal experiment with sequentially polyandrous Kentish Plovers (*Charadrius alexandrinus*), capturing both members of several pairs and then releasing one of them while keeping the other in captivity for a few days. They then monitored the released plovers to determine if they were able to find new mates and, if so, how long it took them. With a male-biased OSR, females were able to find new mates significantly faster than males (Fig. 15.44). Most females found new mates in less than a week, whereas half of the males still had not found new mates after 28 days (Fig. 15.45).

Factors contributing to variation among species in adult sex ratios remain to be determined (Fig. 15.46). However, biased sex ratios can arise in several ways. Possible factors include a bias in the primary sex ratio (i.e., sex ratio of fertilized eggs), differential survival of males and females during development and maturation, or differential survival as adults. However, sex ratios at hatching for many species of birds has been found to be approximately 1:1 (Benito and Gonzalez-Solis 2007), so sex biases more likely result from differential survival of males and females after hatching. Among polyandrous species, the energetic cost of laying multiple clutches of eggs may contribute to higher mortality rates for females (Nager 2006). In addition, after fledging, females typically disperse farther from natal territories than males, and these greater dispersal distances may reduce the likelihood of survival.

In classical polyandrous species where females exhibit simultaneous polyandry, females rather than males establish and defend territories

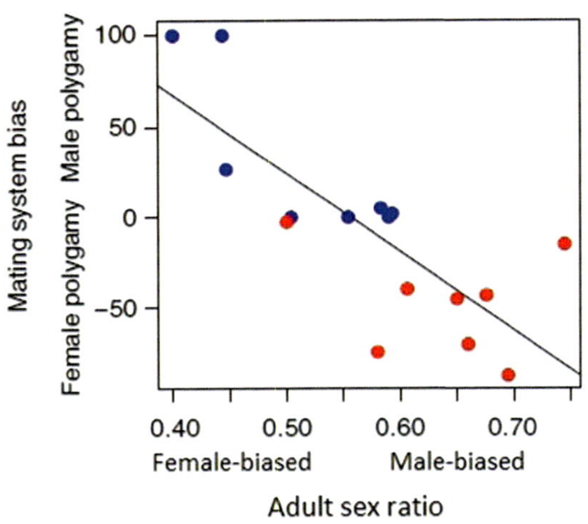


**Fig. 15.44** Relationship between adult sex ratio (number of adult males/number of adult males plus females) and the mating systems of shorebirds. Adult sex ratio is correlated with mating system bias (percent of males that are polygamous minus percent of females that are polygamous, i.e., polyandrous; $N = 16$ species). Red dots = species with reversed sex roles (polyandrous); blue dots = species with conventional sex roles (socially monogamous or polygamous). (Figure from Liker et al. 2013; © Springer Nature, used with permission)

and attempt to attract males. In these sex-role-reversed species, females may be larger and more colorful than males (e.g., phalaropes) and aggressively compete for access to males. In most species of birds, males are more aggressive as they establish and defend territories during the breeding season. In these species, this increased aggression by males results from increased production of androgens like testosterone as the breeding season begins. Androgens bind to androgen receptors in the brain and so influence male behavior. Given that females in sex-role-reversed species display "male-like" behaviors, perhaps their behavior is also driven by increased production of androgens. Goymann and Wingfield (2014) examined this possibility and found that testosterone levels of females do not vary with mating system (Fig. 15.47). In addition, investigators studying several sex-role-reversed species where females are more aggressive and competitive than males have found no evidence that high circulating levels of testosterone influence female behavior (e.g., Gratto-Trevor et al. 1990; Goymann and Wingfield 2004). Additional study is needed to better understand factors that might influence female behavior (and, in some cases, their more colorful plumage) in sex-role-reversed species. However, one possible explanation for the increased aggression of females in sex-role-reversed species is that other hormones are involved. For example, in Black Coucals (*Centropus grillii*), a sex-role-reversed species where females compete for territories, the hormone progesterone modulates their aggressive behavior (Goymann et al. 2008). Another

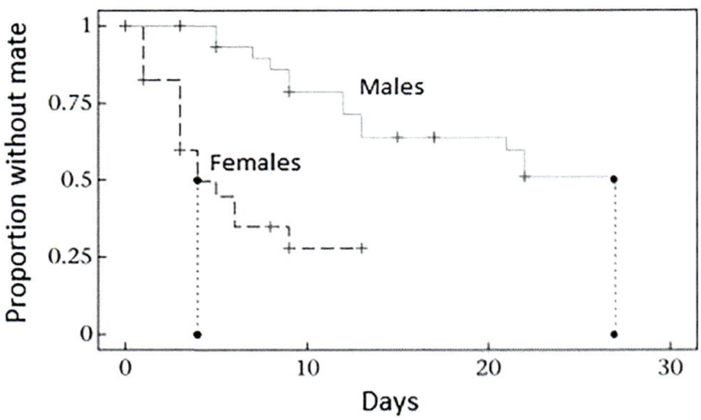


**Fig. 15.45** Proportion of male ($N = 27$) and female ($N = 32$) Kentish Plovers (*Charadrius alexandrinus*) without mates after their original mates were captured and temporally held captive. Vertical dotted lines indicate number of days needed for 50% of females and males to find new mates. (Figure from Parra et al. 2014; © 2014 The Association for the Study of Animal Behaviour. Published by Elsevier Ltd., used with permission)

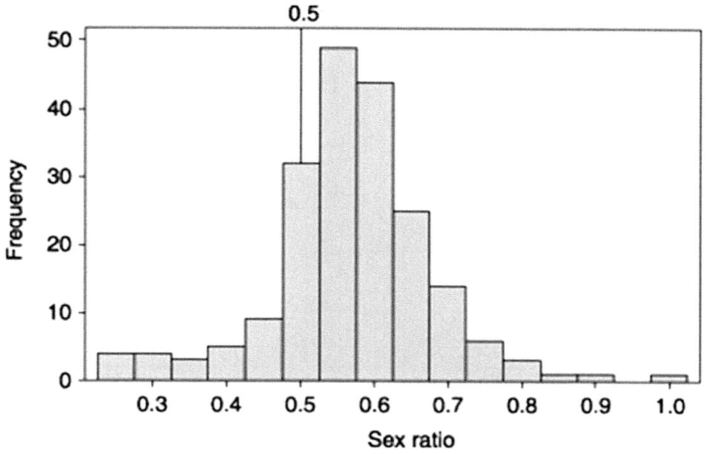


**Fig. 15.46** Frequency distribution of estimates of adult sex ratios (proportion of females in population relative to males) in 173 species of birds. Among most species of birds, adult sex ratios are male-biased, with more males in the population than females. Possible reasons for lower female survival rates include: (1) young females usually disperse farther than young males and adult females in many species migrate farther than males and greater dispersal/migration distances might result in higher mortality rates, (2) among most species of birds, females spend more time on nests incubating eggs and brooding young, placing them at greater risk of predation, and (3), for species where males are larger than and dominant to females, males may exclude females from better quality habitats during the non-breeding season, resulting in higher female mortality. (Figure from Donald 2007; © British Ornithologists' Union, used with permission)

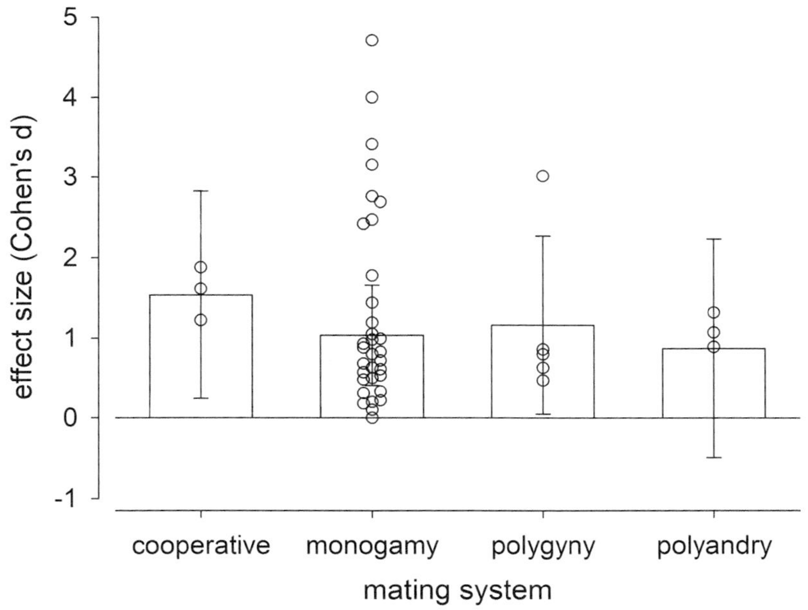


**Fig. 15.47** Measures of the seasonal androgen response (the difference between seasonal baseline levels of testosterone and the seasonal peak) are similar for females in species with different mating systems. Data were corrected for phylogeny. The effect size is a quantitative measure of the magnitude of a relationship, e.g., the seasonal androgen response for females in species with particular mating systems. Each circle represents a different species. (Figure modified from Goymann and Wingfield 2014; © 2014 Oxford University Press, used with permission)

**Table 15.4** Species of birds where polygynandrous breeding systems have been reported

| Common name | Scientific name | Family | Order | Citation |
|---|---|---|---|---|
| Common Ostrich | *Struthio camelus* | Struthionidae | Struthioniformes | Kimwele and Graves (2003) |
| Greater Rhea | *Rhea americana* | Rheidae | Rheiformes | Handford and Mares (1985) |
| Lesser Rhea (also called Darwin's rhea) | *Rhea pennata* | Rheidae | Rheiformes | Handford and Mares (1985) |
| Emu | *Dromaius novaehollandiae* | Dromaiidae | Casuariiformes | Coddington and Cockburn (1995) |
| Slaty-breasted Tinamou | *Crypturellus boucardi* | Tinamidae | Tinamiformes | Lancaster (1964) |
| Pukeko (also called Purple Swamphen) | *Porphyrio porphyrio* | Rallidae | Gruiformes | Jamieson and Craig (1987) |
| Tasmanian Nativehens | *Tribonyx mortierii* | Rallidae | Gruiformes | Goldizen et al. (2000) |
| Acorn Woodpecker | *Melanerpes formidvorus* | Picidae | Piciformes | Koenig (1990) |
| Guira Cuckoo | *Guira guira* | Cuculidae | Cuculiformes | Quinn et al. (1994) |
| Greater Vasa Parrot | *Caracopsis vasa* | Psittrichasiidae | Psittaciformes | Ekstrom et al. (2007) |
| Española Mockingbird | *Mimus macdonaldi* | Mimidae | Passeriformes | von Lippke (2008) |
| Bicknell's Thrush | *Catharus bicknelli* | Turdidae | Passeriformes | Goetz et al. (2003) |
| Stitchbird (also called Hihi) | *Notiomystis cincta* | Notiomystidae | Passeriformes | Castro et al. (1996) |
| Penduline Tit | *Remiz pendulinus* | Remizidae | Passeriformes | Schleicher et al. (1997) |
| Alpine Accentor | *Prunella collaris* | Prunellidae | Passeriformes | Davies et al. (1995) |
| Dunnock | *Prunella modularis* | Prunellidae | Passeriformes | Davies (1985) |
| Smith's Longspur | *Calcarius pictus* | Calcariidae | Passeriformes | Briskie (1992) |
| Red-billed Buffalo-Weaver | *Bubalornis niger* | Ploceidae | Passeriformes | Winterbottom et al. (2001) |

possibility is that the parts of the female brain responsible for social and agonistic behavior may be more sensitive to androgens, with more androgen receptors and, therefore, increased responsiveness to lower levels of androgens (Voigt and Goymann 2007).

## 15.9 Polygynandry

In polygynandrous species of birds, multiple females and males mate with each other, and these males and females plus, in some cases, helpers, provide, to varying degrees, parental care (Table 15.4). Some species are consistently polygynandrous, such as Ostriches, Rheas, and some Tinamous, whereas, in other species, there may be a variety of mating patterns, including polygynandry, monogamy, polyandry, and polygamy. For example, among Tasmanian Nativehens (*Tribonyx mortierii*), most individuals in the population are monogamous, but others are polygynous, polyandrous, and polygynandrous (Fig. 15.48). In most species, polygynandrous breeders are joint-nesters, with females in the group laying eggs in a single nest. However, polygynandrous groups of Española Mockingbirds (*Mimus macdonaldi*) defend territories and, during the breeding season, different groups have separate nests within the borders of those territories (von Lippke 2008).

Few species of birds are polygynandrous, so evolutionary pathways that would lead to this unusual breeding system are unclear. Some ratites are polygynandrous, with males copulating with multiple females who then lay eggs that males will incubate and females copulating with multiple males and laying eggs in multiple nests. Females lay relatively small eggs (relative to their body size), allowing production of many eggs in a breeding season and also allowing males to effectively incubate multiple eggs.

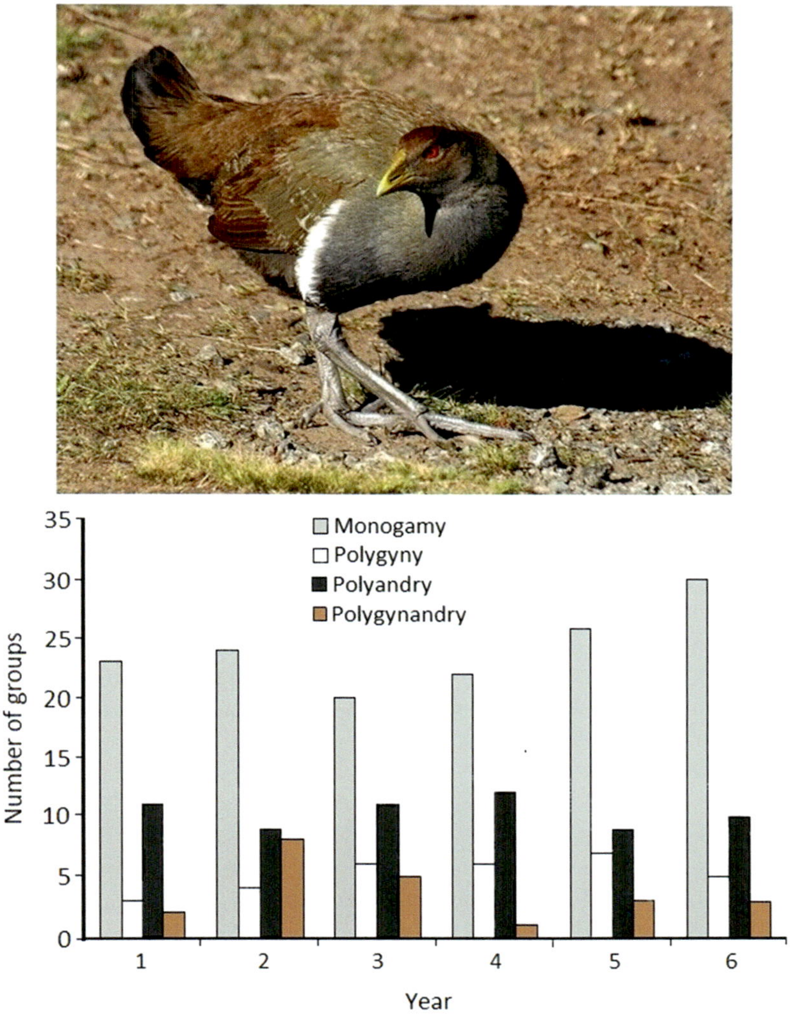


**Fig. 15.48** Top, Tasmanian Nativehen. Bottom, Variation in the frequency of different mating patterns of Tasmanian Nativehens (*Tribonyx mortierii*) over a six-year period in Tasmania, Australia. (Figure from Goldizen et al. 2000; © British Ornithologists' Union, used with permission; Photo by Flying Freddy, Wikipedia, CC BY 3.0, https://creativecommons.org/licenses/by/3.0/deed.en)

Male Greater Rheas (*Rhea americana*) build nests, incubate eggs, and care for young after hatching (for four to six months). Males copulate with the females who then lay eggs every 2 days for 10–12 days (Fernández and Reboreda 2003). However, because females copulate with multiple males, incubating males may not be the father of most of the embryos they incubate. In fact, based on analysis of the genotypes of 3 incubating males and 141 offspring, Martella et al. (2022) determined that one male did not fertilize any of those 141 eggs.

Male-only care is likely the ancestral condition for ratites and tinamous (Cockburn 2006), so the evolution of their polygynandrous breeding systems followed as a result of their large size (relative to their eggs), the ability of females to produce multiple eggs, and the ability of males to effectively incubate large clutches (Vehrencamp and Quinn 2004; Fig. 15.49). The evolutionary pathway leading to polygynandrous breeding systems in other taxa is less clear. However, Vehrencamp and Quinn (2004) suggested that conspecific brood parasitism may have played an important role in some species (Fig. 15.50). Female conspecific brood parasites lay eggs in the nests of other females, but typically have their own nest as well. However, females, and

**Fig. 15.49** Relationship between clutch size and hatchability, or the percentage of eggs that hatch, in Greater Rheas (*Rhea americana*; $N = 45$ nests). (Figure from Fernández and Reboreda 1998; © 1998 Oxford University Press, used with permission)

potentially their mates, might benefit from assisting in providing parental care at "parasitized" nests; for example, if good-quality territories or nest sites are in short supply, the additional parental care improves the chances of young surviving or fledging in good condition. Of course, host adults would also have to accept the presence of the "parasites." Other possible pathways might involve limited food availability or limits on the number of good breeding territories (Fig. 15.50). For example, a shortage of good-quality territories may have influenced the breeding systems of Acorn Woodpeckers and Purple Swamphens (Vehrencamp and Quinn 2004) and short breeding seasons at high latitudes and altitudes in combination with limited availability of food may have contributed to the evolution of polygynandrous mating systems of Alpine Accentors and a few other species (Fig. 15.50).

Selection favors mating systems that allow individuals to maximize their reproductive fitness so, for a few species of birds, breeding cooperatively with other males and females must result, on average, in greater reproductive success than other mating systems. For ratites and tinamous, both males and females benefit from polygynandry (resulting from a combination of simultaneous polygyny by males and sequential polyandry by females); males fertilize and incubate more eggs than would be possible with a single mate, and females, by sequentially visiting and laying eggs in the nests of different males, are less energetically stressed (i.e., lay eggs over a longer period of time) and, by laying eggs in multiple nests, reduce the risk of complete reproductive failure (i.e., at least some nests will fledge young). In some other polygynandrous species, ecological constraints, such as limited availability of food or good quality territories, may limit the ability of a single male and female to breed successfully, thus favoring cooperation with others to collectively improve their reproductive success (Fig. 15.49).

In some joint-nesting polygynandrous species, hatching success declines with increasing clutch sizes (e.g., Quinn et al. 2012). As a result, females may remove eggs from nests. For example, in polygynandrous groups of Acorn Woodpeckers (*Melanerpes formicivorus*), females regularly remove eggs from communal nests. However, the number of eggs removed and destroyed typically matches the number of additional eggs laid, so egg-removal has minimal or no effect on group reproductive success (Koenig et al. 1995; Fig. 15.51). The reproductive success of individual females is affected, with last-to-lay females benefiting by removing the eggs of their co-breeders and replacing them with their own. Similar egg-rejection behavior (as well as infanticide after eggs hatch) has been reported in Guira Cuckoos (*Guira guira*, Macedo et al. 2004). In contrast, even though hatching success declines with increasing clutch size in joint-nesting Purple Swamphens (*Porphyrio porphyrio*), females do not remove eggs (although some may fall out of nests as more eggs are laid) (Quinn et al. 2012).

## 15.10 Cooperative Breeding

Approximately 9% of the world's bird species (~835 species) are cooperative breeders, where several individuals assist in providing parental care (Cockburn 2006). About 70% of cooperatively breeding bird species are passerines, and about 30% are nonpasserines (Jetz and Rubenstein 2011). In most of these species,

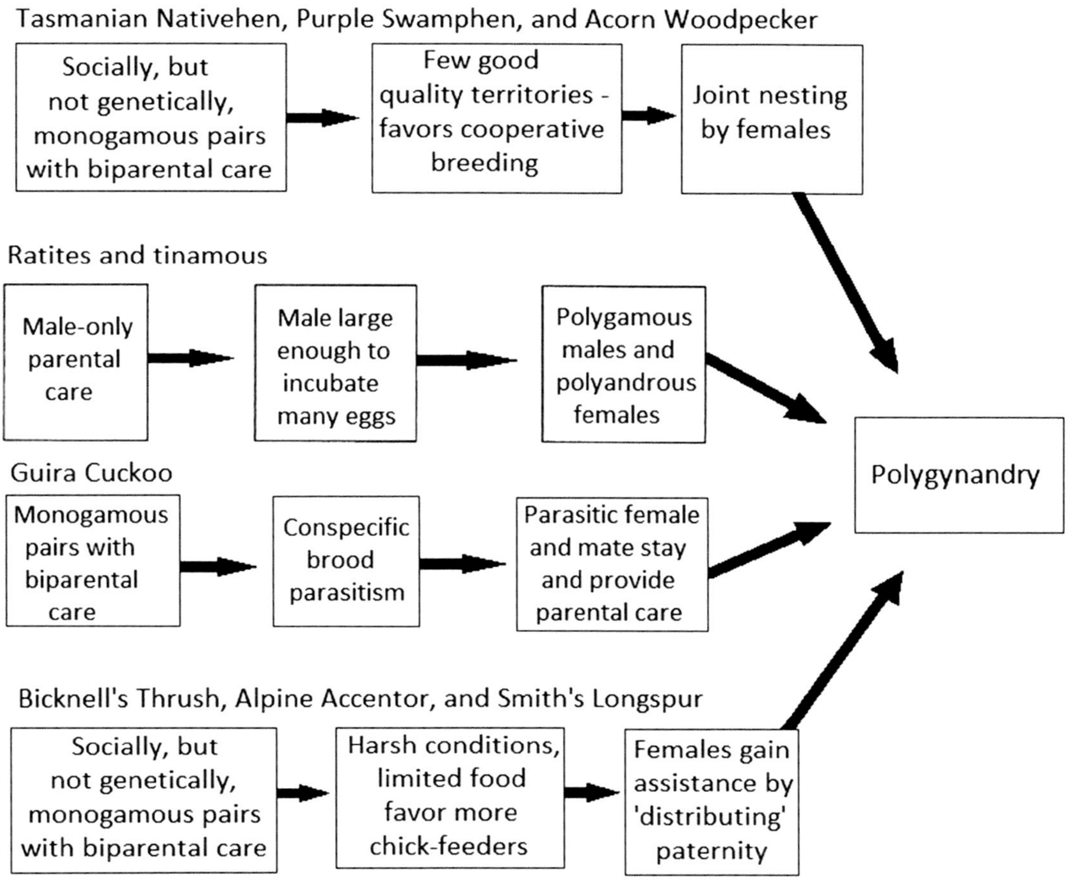


**Fig. 15.50** Possible evolutionary pathways leading to polygynandrous mating systems and examples of species that may have followed those pathways. Tasmanian Nativehen, *Tribonyx mortierii*; Purple (Western) Swamphen, *Porphyrio porphyrio*; Acorn Woodpecker, *Melanerpes formicivorus*; Guira Cuckoo, *Guira guira*; Bicknell's Thrush, *Catharus bicknelli*; Alpine Accentor, *Prunella collaris*; Smith's Longspur, *Calcarius pictus*. (Figure based, in part [middle two pathways], on ideas from Vehrencamp and Quinn 2004)

family groups consisting of a breeding pair and a variable number of young from previous breeding attempts (helpers) cooperate in raising young. In other species, however, unrelated individuals or a mixture of related and unrelated individuals assist a breeding pair and in still other cooperative species, unrelated individuals are co-breeders (i.e., polygynandrous; Fig. 15.52).

Ligon and Burt (2004) estimated that there have been at least 28 separate "transitions" from other breeding systems (primarily biparental care) to cooperative breeding, primarily among altricial species of birds (20 transitions). Cooperative breeding is prevalent in certain families of birds in the orders Coliiformes, Coraciiformes, Upupiformes, Bucerotiformes, and Piciformes, but, more broadly, Cockburn (2006) listed 59 families of birds where most species are cooperative breeders.

For cooperative breeding to evolve, some individuals in a population must choose to forgo breeding independently and become part of a breeding group. For species where helpers (individuals that assist with parental care of young, but are not the genetic parents) are related to the breeders (e.g., young of the breeders from

**Fig. 15.51** Relationship between the number of eggs laid per joint nest of Acorn Woodpeckers (*Melanerpes formicivorus*; open squares and solid line), number of eggs destroyed, and the number of eggs actually incubated (triangles and dashed line). (Figure modified from Koenig et al. 1995; © 1995 Published by Elsevier Ltd., used with permission)

previous broods), the inclusive fitness benefits of helping relatives likely played some role in this process. However, even in such cases, the indirect fitness benefits gained by helping is likely less than the potential direct fitness benefits of independent breeding (Brown 1987). In addition, among some cooperative-breeding species, helpers are unrelated to breeders so gain no indirect fitness benefits. Given that helping may provide less fitness benefit than independent breeding (or none at all), what proximate factor (s) might have contributed to the evolution of helping behavior? One possible reason why some individuals might become a helper is that some ecological constraint might limit opportunities for independent breeding. Jetz and Rubenstein (2011) tested this hypothesis by examining the global abundance and distribution of cooperative species of birds and their relationships to environmental variability or uncertainty as measured by annual variation in precipitation. Their analysis revealed a clear relationship between the number of cooperative species in an area and yearly variation in precipitation, with more cooperatively breeding species in areas with low mean annual and unpredictable precipitation (Figs. 15.53 and 15.54). Specifically, areas with most cooperatively breeding species of birds included sub-Saharan Africa, parts of the Amazon basin, parts of Australia, the Himalayas, and New Guinea (Fig. 15.53). Although phylogenetic history plays some role in the global distribution of cooperatively breeding species, phylogeny does not explain the large

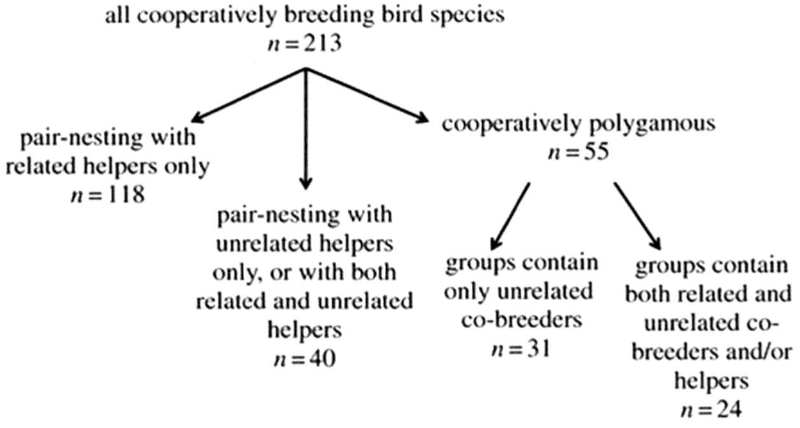


**Fig. 15.52** Classification of group composition and social mating system for 213 species of cooperatively breeding birds for which data are available to accurately identify related and unrelated individuals and those that are actually breeding (i.e., parents of offspring). "Helpers" are individuals that assist with parental care, but are not the genetic parents of the young. Some of the cooperatively polygamous species have polygynandrous mating systems. (Figure from Riehl 2013; © 2013 The Author. Published by the Royal Society, used with permission)

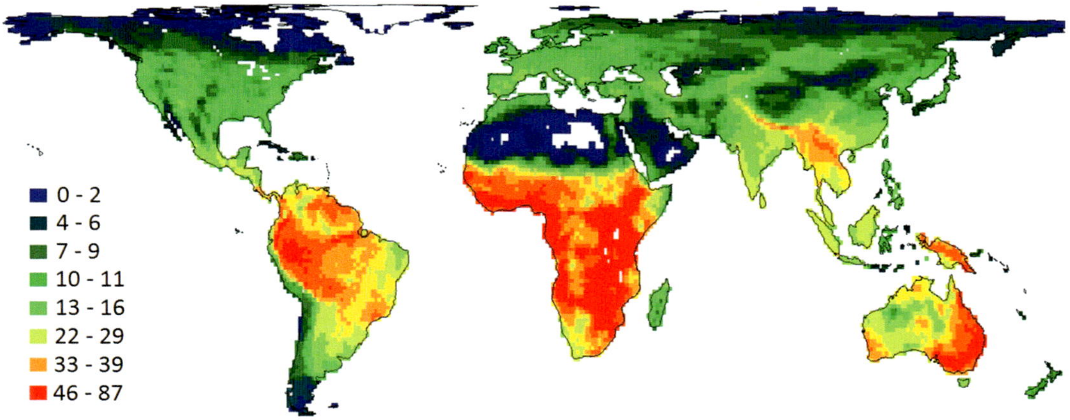


**Fig. 15.53** Biogeographic distribution and relative abundance of cooperative-breeding species of birds (based on 9310 species of non-marine birds). Note that cooperative breeders are particularly abundant in sub-Saharan Africa and Australasia. These two areas are also home to 63% of all species of brood parasitic birds, including 21 species of cuckoos, 6 species of honeyguides, and 9 species of parasitic finches (Feeney et al. 2013). Legend indicates the lower and upper numbers of species for each color class. (Figure modified from Jetz and Rubenstein 2011; © 2011 Elsevier Ltd., used with permission)

number of such species in places such as Africa and Australia (Jetz and Rubenstein 2011). In areas with relatively harsh environmental conditions (little precipitation that also falls unpredictably), young birds may enhance their chances of surviving (and, therefore, their reproductive fitness) by at least temporarily remaining in natal territories and becoming helpers. Among cooperative-breeding species where helpers assist related breeders (e.g., parents), they can also gain inclusive fitness benefits by helping raise related young.

Other investigators have noted that other environmental constraints might also favor delayed dispersal of young and cooperative breeding, including limited availability of suitable breeding

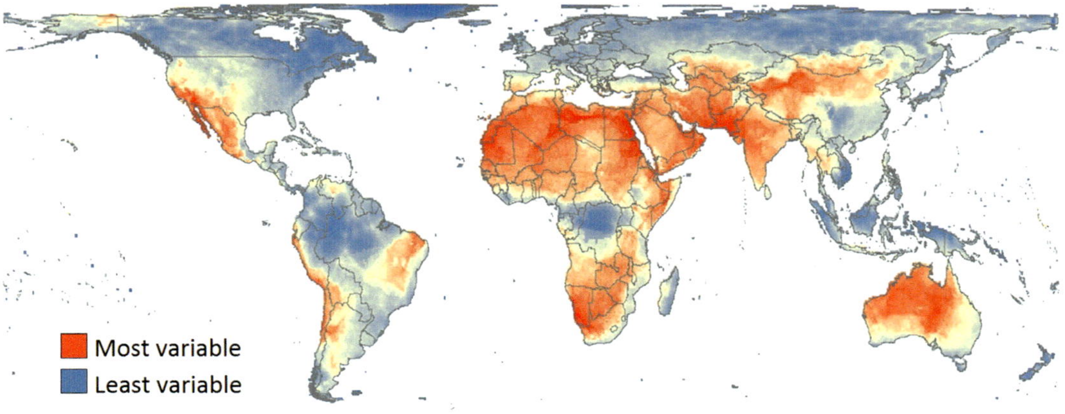


**Fig. 15.54** Worldwide among-year variation in precipitation. (Figure modified from Jetz and Rubenstein 2011; © 2011 Elsevier Ltd., used with permission)

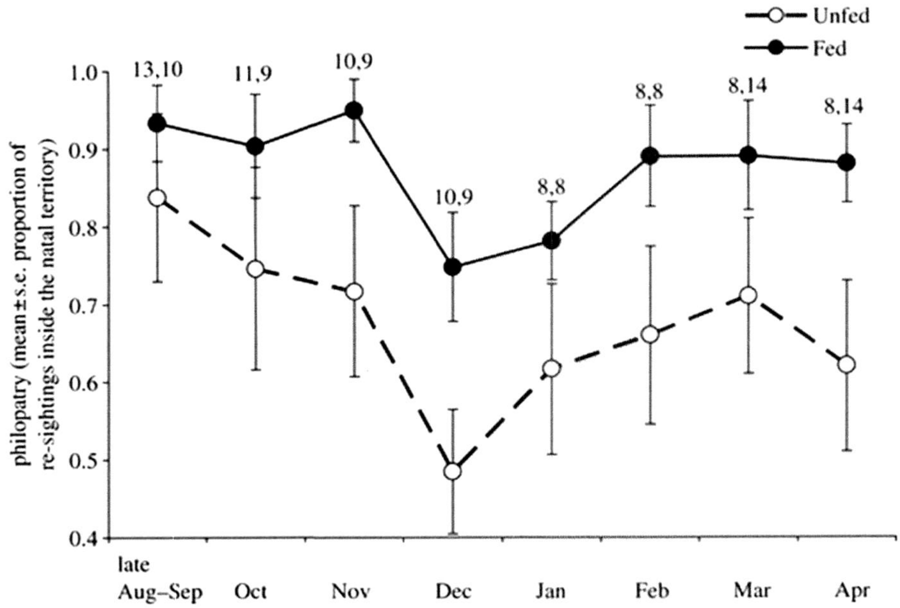


**Fig. 15.55** Philopatry as measured by the mean proportion (± SE) of juvenile Carrion Crows (*Corvus corone*) re-sighted in food-supplemented (fed) and non-supplemented (unfed) territories. Young that fledged in food-supplemented territories from May to July in 2003 were more likely to remain (and be re-sighted) and become helpers during the 2004 breeding season (beginning in April). Sample sizes are provided above bars for fed and unfed individuals, respectively. (Figure from Baglione et al. 2006 © 2006 The Royal Society, used with permission)

territories (e.g., Komdeur 1992) or mates (e.g., Pruett-Jones and Lewis 1990). Alternatively, however, others have suggested that resource abundance might contribute to delayed dispersal of young and cooperative breeding. Baglione et al. (2006) provided Carrion Crows (*Corvus corone*) in some territories with supplemental food and found that young in those food-supplemented territories were more likely to remain and help their parents raise younger siblings during the next breeding season than were young in non-supplemented territories (Fig. 15.55). In addition, helpers in food-supplemented territories provisioned nestlings at significantly higher rates than helpers in non-supplemented territories. By delaying dispersal from high-quality territories, young Carrion Crows may benefit in terms of improved physical condition, possibly improving their chances of survival when they do disperse, as well as through the inclusive fitness benefits gained by providing better care for their younger siblings (Baglione et al. 2006).

The risk of brood parasitism might have also contributed to the evolution of cooperative breeding in some species of birds. Examination of the global distributions of cooperative-breeding songbirds and obligate brood parasites reveals a significant correlation and two major "hotspots": sub-Saharan Africa and Australasia (Feeney et al. 2013; Fig. 15.53). One possible explanation for this association is that brood parasitism may be more likely to evolve in areas with more cooperative breeders because cooperative breeders provide better parental care than species with biparental or uniparental care, thereby increasing the reproductive success of their brood parasites (Feeney et al. 2013). On the other hand, cooperative breeding may be more likely to evolve in areas with more brood parasites because cooperative groups may be better able to prevent female brood parasites from laying eggs in their nests. If

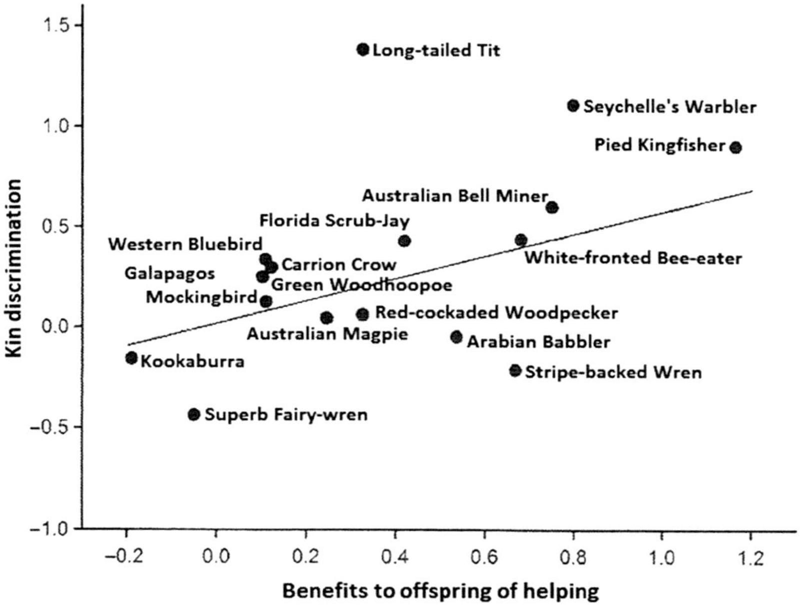


**Fig. 15.56** Relationship between the benefits to offspring of care provided by helpers and kin discrimination for 16 cooperative-breeding species of birds: Long-tailed Tit, *Aegithalos caudatus*; Seychelle's Warbler, *Acrocephalus sechellensis*; Pied Kingfisher, *Ceryle rudis*; Australian Bell Miner, *Manorina melanophrys*; Florida Scrub-Jay, *Aphelocoma coerulescens*; Western Bluebird, *Sialia mexicana*; White-fronted Bee-eater, *Merops bullockoides*; Carrion Crow, *Corvus corone*; Green Woodhoopoe, *Phoeniculus purpureus*; Galapagos Mockingbird, *Mimus parvulus*; Red-cockaded Woodpecker, *Dryobates borealis*; Australian Magpie, *Gymnorhina tibicen*; Arabian Babbler, *Argya squamiceps*; Laughing Kookaburra, *Dacelo novaeguineae*; Stripe-backed Wren, *Campylorhynchus nuchalis*; Superb Fairywren, *Malurus cyaneus*. (Figure modified from Cornwallis et al. 2009; © 2009 The Authors. Journal Compilation © 2009 European Society for Evolutionary Biology, used with permission)

so, then the inclusive fitness benefits achieved by offspring helping prevent parasitism of their parent's nests may favor helping over dispersing and attempting to breed independently (Feeney et al. 2013). A study of cooperatively breeding Superb Fairywrens (*Malurus cyaneus*) provided evidence supporting both of these hypotheses; young brood parasites (Horsfield's Bronze-Cuckoos, *Chrysococcyx basalis*) cared for by larger groups of fairywrens were more likely to survive to fledging, and larger groups of fairywrens were also better able to defend against brood parasites (Feeney et al. 2013). Although intriguing, additional tests of these hypotheses with other species of birds are needed to better understand how brood parasitism may or may not have contributed to the evolution of cooperative breeding.

Clearly, no single factor can explain the evolution of cooperative breeding in all 835 (more or less) species of cooperative-breeding birds. Natural selection favors behaviors that enhance reproductive fitness and, given the presence and intensity of various environmental constraints, variation in habitat/territory quality, and other factors such as the intensity of brood parasitism, deferring independent breeding and becoming a helper is the best strategy for at least some individuals in many species of birds. The possible inclusive fitness benefits of helping are clear for species where a breeding pair is assisted by related helpers. Illustrating the importance of these inclusive fitness benefits, Cornwallis et al. (2009) examined how kin discrimination and dispersal tendencies interact to influence helping behavior in several species of cooperative

**Table 15.5** Hypotheses to explain non-kin groups of avian cooperative breeders

| Hypotheses | Direct benefit | References |
|---|---|---|
| Group augmentation 1 | Increased group size improves foraging success and ability of group members to detect and repel predators | Kokko et al. (2001) |
| Group augmentation 2 | Non-breeders actively recruit new group members that will assist them if and when they become breeders | Kokko et al. (2001) |
| Pay to stay | Helpers benefit from being in group, but must "pay" via behavior (s) that benefit the group (e.g., feeding nestlings or nest defense), or else be evicted from the group | Kokko et al. (2002) |
| Social prestige | By assistance provided to the group, helpers signal quality to others that may, in the future, become their mate | Zahavi (1974) |
| Parentage | Helpers may sometimes fertilize or lay eggs | Magrath and Whittingham (1997) |

breeders. They found that kin discrimination (preferentially interacting with closer relatives) was greater in species where (1) average relatedness was lower and more variable and (2) the benefits to offspring of the assistance provided by helpers was greater (Fig. 15.56). These results suggest, for example, that when a species exhibits strong philopatry (limited dispersal) and, therefore, group members are likely to be related and helpers are more likely to be helping kin, selection is less likely to favor kin discrimination. However, for species that are less philopatric, e.g., where some young disperse to become independent breeders, but, if their nests are predated or fail for some other reason, they then become helpers for other breeders rather than attempting to re-nest, selection favors kin discrimination to improve the inclusive fitness benefits of helping.

## 15.11 Non-Kin Cooperative Breeding

Among most cooperative breeding birds, group members are related so non-breeding helpers gain inclusive fitness benefit. However, Riehl (2013) estimated that about 30% of cooperative-breeding species of birds nest in groups consisting of a mixture of related and non-related individuals, and about 15% nest primarily as groups of non-relatives. Assisting unrelated breeders provides no inclusive fitness benefit, but can provide helpers with direct benefits (Table 15.5). For example, the group augmentation hypothesis proposes that individuals benefit from helping unrelated breeders because, by doing so, they gain the advantages of being a group member (Table 15.5), which, in turn, increases their likelihood of survival. Other possible direct benefits include learning parental care skills, inheriting a breeding territory and, possibly a mate, if a breeder dies, access to future mates (i.e., from among other non-kin helpers), access to future helpers, and potentially being able to fertilize or lay eggs (Riehl 2013). Of course, non-breeding birds that assist kin also potentially benefit in these same ways, but also gain inclusive fitness benefits; so why not just remain in natal territories as helpers? In some species, young may be evicted from natal territories, e.g., by older siblings (Waite and Strickland 1997). In other species, dispersing from natal territories and joining unrelated pairs or groups may increase the likelihood of, and reduce the time needed to, becoming a breeder. For example, young males disperse from natal territories, join unrelated breeding pairs, and act as helpers until becoming breeders in several species, including Stripe-backed Wrens (*Campylorhynchus nuchalis*; Piper et al. 1995), Puerto Rican Todies (*Todus mexicanus*; Kepler 1977), and Pied Kingfishers (*Ceryle rudis*; Reyer 1984).

## 15.12 Types of Parental Care Provided by Non-Breeding Helpers

Non-breeding helpers can assist breeders in several ways, primarily by provisioning nestlings and defending nests and fledglings from predators. Less commonly, helpers may also assist with nest construction, incubating eggs, and brooding young (Skutch 1961). In most species studied to date, the presence of helpers does help breeders, but not always in the same ways. The help provided by helpers can, depending on the species, increase a breeding pair's reproductive success, reduce the likelihood of nest predation, and contribute to increased survival of breeding females, breeding males, or both. Kingma et al. (2010) examined the results of studies of 37 species of cooperatively breeding birds and found that helpers improved the reproductive success (number of fledglings per nest) of breeders in 25 (67.6%) of those species. Two factors contributing to such improvement include (1) helpers helping to increase provisioning rates of nestlings, e.g., Carrion Crows (*Corvus corone*; Canestrari et al. 2008) and Long-tailed Tits (*Aegithalos caudatus*; MacColl and Hatchwell 2003), and (2) reduced likelihood of nest predation resulting from the increased vigilance and increased effectiveness of mobbing when helpers are present, e.g., White-throated Magpie-jays (*Calocitta formosa*; Innes and Johnston 1996) and Azure-winged Magpies (*Cyanopica cyanus*; Valencia et al. 2006).

The presence of helpers can also have long-term benefits for helped offspring. Using 20 years of data, Brouwer et al. (2012) examined the effects of helpers in Seychelles Warblers (*Acrocephalus sechellensis*) by comparing young raised by pairs with and without helpers and found that the presence of helpers increased not only survival of offspring in their first year of life, but also subsequent adult survival. These results suggest that help provided by helpers can potentially, by increasing the lifespan of young they help, have important lifetime fitness benefits, both for the breeders and, if helping kin, for the helpers.

With helpers present, levels of parental care provided by breeders may remain unchanged so that the helpers' care is simply in addition to parental care, e.g., White-fronted Bee-eaters (*Merops bullockoides*, Emlen and Wrege 1991). However, parental care can be costly in terms of negative impacts on body condition and the likelihood of survival and future breeding (Liker and Székely 2005), so breeders can also reduce their workload to partially or completely compensate for the presence of helpers, e.g., so that provisioning rates are unchanged (Brown et al. 1978). This "load-lightening" may increase the likelihood of breeders surviving and breeding again in the future, providing direct fitness benefits for breeders and, if helpers are related to the breeders, providing indirect fitness benefits for helpers. Even within-season fecundity can be affected by "load-lightening" for multi-brooded species, with several studies showing that the presence of helpers can reduce interclutch intervals and allow more breeding attempts per year (e.g., Woxvold and Magrath 2005; Blackmore and Heinsohn 2007). Apostlebirds (*Struthidea cinerea*) and Pied Babblers (*Turdoides bicolor*) sometimes even have overlapping broods, with breeders initiating a second nest before young fledge from their first nest; helpers continue to care for the first brood and the breeding pair then invests more in the second brood (Woxvold and Magrath 2005; Ridley and Raihani 2008; Fig. 15.57).

The presence of helpers can also benefit female breeders via a reduction in egg investment. Female Superb Fairywrens (*Malurus cyaneus*; Russell et al. 2007), Carrion Crows (Canestrari et al. 2011), Southern Lapwings (*Vanellus chilensis*; Santos and Macedo 2011), and Sociable Weavers (*Philetairus socius*; Paquet et al. 2013) all lay smaller eggs when they have helpers (Fig. 15.58). Egg production is known to be energetically costly (e.g., Monaghan et al. 1998), so females with helpers likely save a considerable amount of energy by producing smaller,

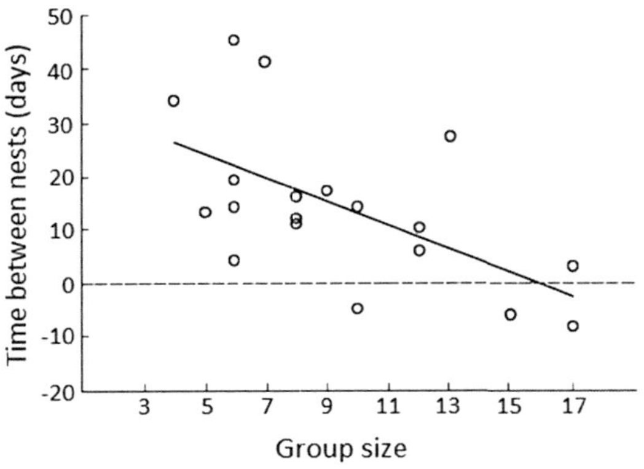


**Fig. 15.57** Relationship between group size and interclutch intervals of Apostlebirds (*Struthidea cinerea*). The interval between nests was the number of days between fledging of the first brood and commencement of laying the second clutch. Points below the horizontal line indicate groups that initiated a second clutch before the first brood fledged. (Figure from Woxvold and Magrath 2005; © 2005 John Wiley and Sons, used with permission)

lighter eggs. However, smaller eggs with less yolk means smaller hatchlings that, in turn, could be smaller than normal when they fledge, potentially having a negative impact on their likelihood of survival. However, the reduced investment in eggs by females in these species does not have a negative impact on the mass of young at fledging because the additional food provided by helpers compensates for the reduced investment.

Load-lightening can have a beneficial impact on the survival of breeders of both sexes (e.g., Khan and Walters 2002; Fig. 15.59). However, in some cooperatively breeding species, the presence of helpers only enhances survival of breeders of one sex. For example, Cockburn et al. (2008) found that survival of female, but not male, Superb Fairywrens increased when helpers were present. In still other species, investigators have reported no effect of helper presence on breeder survival (e.g., Sydeman 1989). Differences among studies concerning the possible effects of load-lightening on breeder survival might result from the potentially confounding effects of territory quality, differences in the number of helpers, and the individual quality of both helpers and breeders.

When helpers and breeders are related, cooperative breeding can provide both direct fitness benefits for breeders and indirect inclusive fitness benefits for helpers. The degree to which breeders and helpers benefit depends on both the number of young produced and the degree to which breeders and helpers are related to those young. For helpers, therefore, the genetic mating system of the breeders is important because extra-group mating by the breeding female reduces the possible inclusive fitness benefits of helping. Kingma et al. (2010) conducted a comparative analysis of 37 cooperatively breeding species of birds with different genetic mating systems; females do not mate outside of their social group in some species, they occasionally do in other species, and more commonly do so in yet other species. The comparative analysis revealed that helper assistance is influenced by the extent of extra-group paternity; with increasing levels of extra-group paternity, helpers tend to provide less assistance in raising young and, as a result, the increased reproductive success often resulting from the presence of helpers declines (Fig. 15.60). In addition, there is less "load-lightening" with increasing extra-group paternity and less of a benefit for breeding males and females in terms of increased likelihood of survival (Fig. 15.60). With a reduction in inclusive fitness benefits, therefore, helpers respond by providing less help, and they may continue to provide some assistance because of the indirect benefits of helping. Given this tendency for helpers to provide less assistance in raising young with increasing levels of extra-group paternity, the possible benefits of such behavior to breeding females apparently outweigh the costs of losing helper assistance.

In a few species of cooperatively breeding birds, groups consist of unrelated socially monogamous pairs and females in each breeding pair lay their eggs in communal nests (Table 15.6; Figs. 15.61 and 15.62). Adults then share parental duties, including incubating eggs and brooding and feeding nestlings. The size of breeding groups in these communal species varies, but ranges from single pairs to as many as five pairs (Table 15.6). Studies of these species suggest that

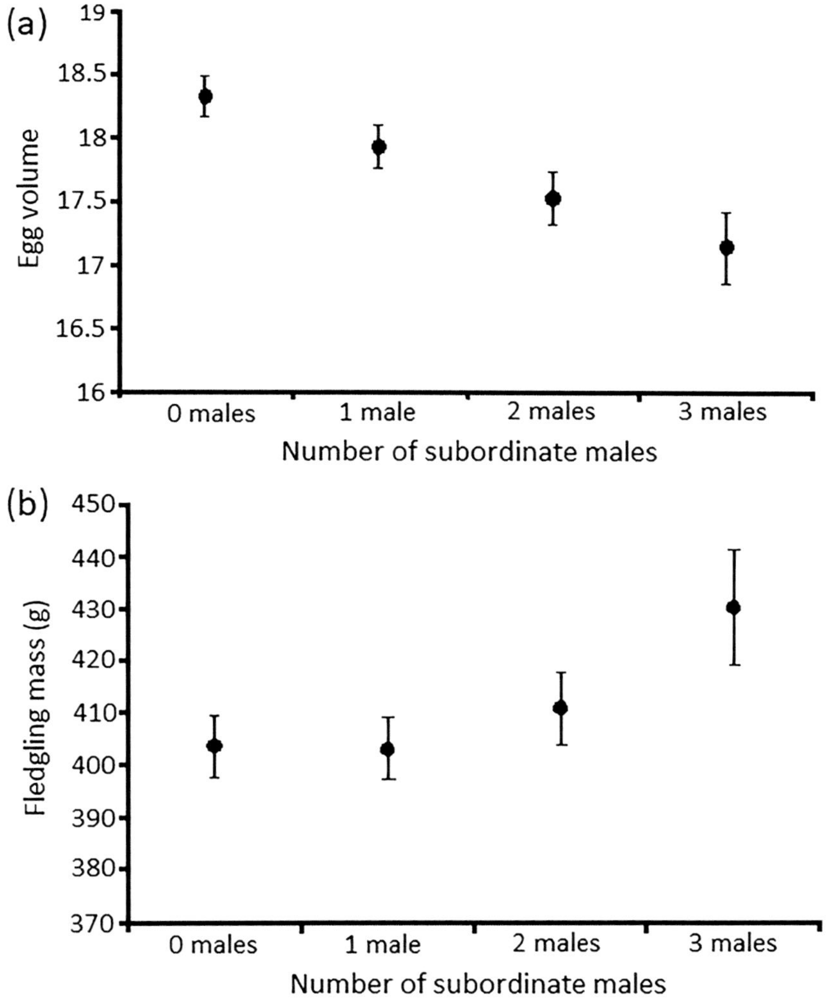


**Fig. 15.58** Effects of the number of helpers (subordinate males) on the (**a**) volume of eggs (length × maximum width$^2$ × π/6; Hoyt 1979) produced by female Carrion Crows (*Corvus corone*), and (**b**) mass of fledglings. (Figures from Canestrari et al. 2011; © 2011 The Authors. Journal of Evolutionary Biology © 2011 European Society For Evolutionary Biology; used with permission)

the advantage of communal breeding is better defense against potential nest predators and increased nesting success and, for individuals, the potential for increased lifetime reproductive success (Riehl 2011; Fig. 15.63). However, as group size increases, competition between females can also increase because only so many eggs will fit in the communal nest. For example, with increasing size of groups of communal Smooth-billed Anis, the number of eggs buried in the structure of nests or that get tossed from nests by competing females increases (Fig. 15.64b). In groups with four or five females, nearly 70% of eggs laid are either tossed or buried (Fig. 15.64b). As a result of this competition, females in larger groups tend to lay more eggs (to replace those being lost), increasing the energetic costs of breeding (Schmaltz et al. 2008; Fig. 15.64a). The loss of so many eggs in large groups can also negatively impact the breeding success of individual females. As a result, among Groove-billed Anis (*Crotophaga sulcirostris*), pairs breeding alone typically fledge more young per breeding attempt than pairs in groups

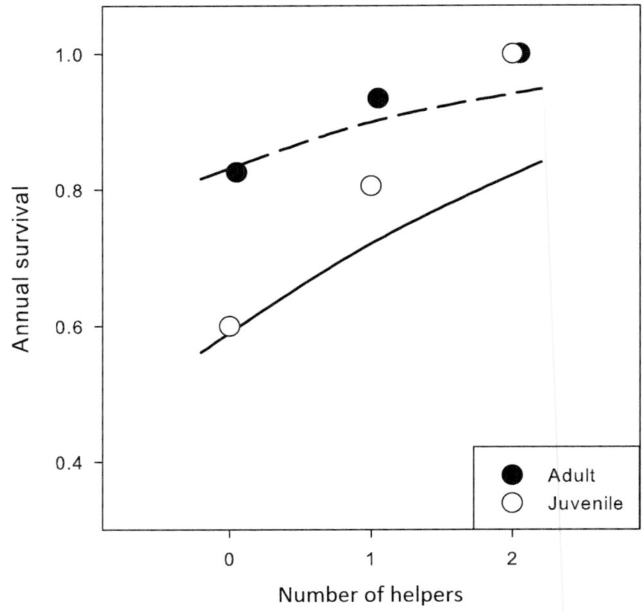


**Fig. 15.59** Effect of helpers on the probability of adult and fledgling (juvenile) Seychelles Warblers (*Acrocephalus sechellensis*) surviving during an average year. (Figure modified from Brouwer et al. 2012; © 2012 Brouwer et al., open-access article distributed under the terms of the Creative Commons Attribution License)

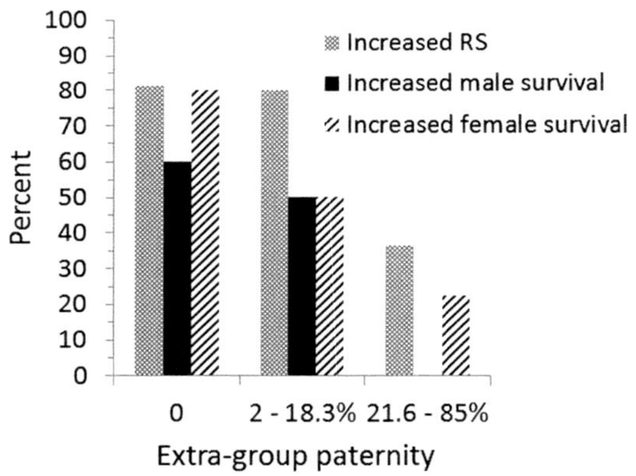


**Fig. 15.60** Effect of extra-group paternity on the survival of male and female breeders and the reproductive success (RS) of cooperative-breeding species of birds. With higher levels of extra-group paternity, and the corresponding decrease in relatedness of helpers to nestlings, group reproductive success declines and the presence of helpers is less likely to result in increased survival of male and female breeders. For percentage increase in RS, sample sizes for category of extra-group paternity were 11 (0), 15 (2–18.3), and 11 (21.6–85), respectively. For male survival, sample sizes were 5 (0), 8 (2–18.3), and 8 (21.6–85), respectively. For female survival, sample sizes were 5 (0), 8 (2–18.3), and 9 (21.6–85), respectively. (Figure based on data from Kingma et al. 2010)

(Fig. 15.65a). However, with fewer individuals to watch for predators, adults breeding as single pairs have lower survival rates (Fig. 15.65b). For these anis, therefore, the optimal group size is two pairs (Fig. 15.65c). Breeding with another pair tends to reduce the number of young fledged per breeding attempt for each pair, but also reduces the risk of predation and increases adult survival rates. Although breeding success in a given year may be lower for groups consisting of two or three pairs, the increased adult survival rates mean an increased likelihood of future breeding attempts and greater lifetime reproductive success (Fig. 15.65). Whether or not this may also be the case for communal species other than Groove-billed Anis remains to be determined.

## 15.13 Female Mate Choice and Sexual Selection

Regardless of the type of social or genetic mating system, birds must make mate-choice decisions. For species that are both socially and genetically monogamous, the relationship with mates is both

**Table 15.6** Species of birds that are communal breeders and where socially monogamous pairs collectively provide parental care

| Common name | Scientific name | Family | Order | Typical group size | Citation |
|---|---|---|---|---|---|
| Greater Ani | *Crotophaga major* | Cuculidae | Cuculiformes | 2 or 3 pairs (rarely 1) | Riehl (2011) |
| Groove-billed Ani[a] | *Crotophaga sulcirostris* | Cuculidae | Cuculiformes | 1–5 pairs | Vehrencamp (1978) |
| Smooth-billed Ani | *Crotophaga ani* | Cuculidae | Cuculiformes | 1–5 pairs | Vehrencamp and Quinn (2004) |
| Taiwan Yuhina[a] | *Yuhina brunneiceps* | Zosteropidae | Passeriformes | 2 or 3 pairs | Vehrencamp and Quinn (2004), Yuan et al. (2004), Shen et al. (2012) |

[a]Groups of Groove-billed Anis and Taiwan Yuhinas sometimes also include non-breeding helpers

social (e.g., sharing parental care duties) and genetic (i.e., mating). For species of birds that are not genetically monogamous, mate-choice decisions may involve both selection of a social mate (or mates if polygynous or polyandrous) and additional "genetic" (extra-pair) mates. With some possible limitations because of the potential costs involved (e.g., sperm depletion and possible loss of paternity when seeking further matings), males in non-genetically monogamous species are generally assumed to copulate opportunistically and indiscriminately (Parker and Birkhead 2013). However, mate-choice decisions by females and choices by males for "social" mates are not indiscriminate.

Mate choice occurs when the traits expressed in individuals of one sex lead to non-random mating with individuals of the opposite sex (Kokko et al. 2003). Females can benefit from preferring certain males as mates if their choice provides greater direct and/or indirect benefits than they would have obtained without that

**Fig. 15.61** A Greater Ani (*Crotophaga major*) nest with eggs laid by different females. (Figure from Riehl 2010; )

Fig. 15.62 A group of Taiwan Yuhinas (*Yuhina brunneiceps*) huddled together on a branch. (Figure from Shen et al. 2012; © 2012 Springer Nature, used with permission)

preference. Several mechanisms have been proposed to explain the evolution of mate choice by female birds (Andersson and Simmons 2006), including:

- **Direct benefits**. Female preferences for certain males can evolve as a result of direct phenotypic benefits if particular characteristics of males honestly reflect their ability to provide

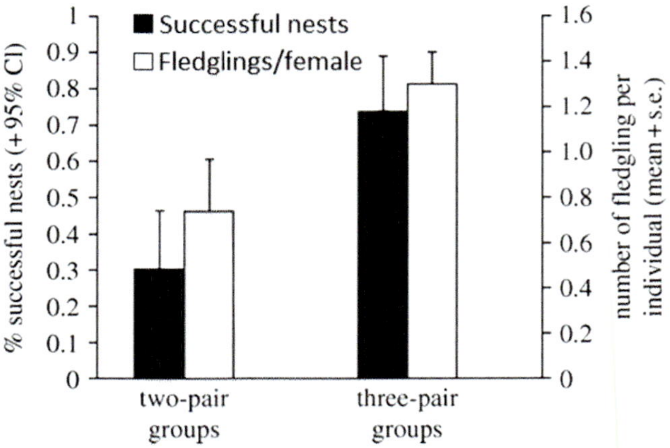


Fig. 15.63 Percentage (± 95% confidence interval) of successful nests and average reproductive success of individual females in groups of Greater Anis (*Crotophaga major*) consisting of two or three pairs. Three-pair groups had greater nest success and more fledglings per female. (Figure from Riehl 2011; © 2010 The Royal Society, used with permission)

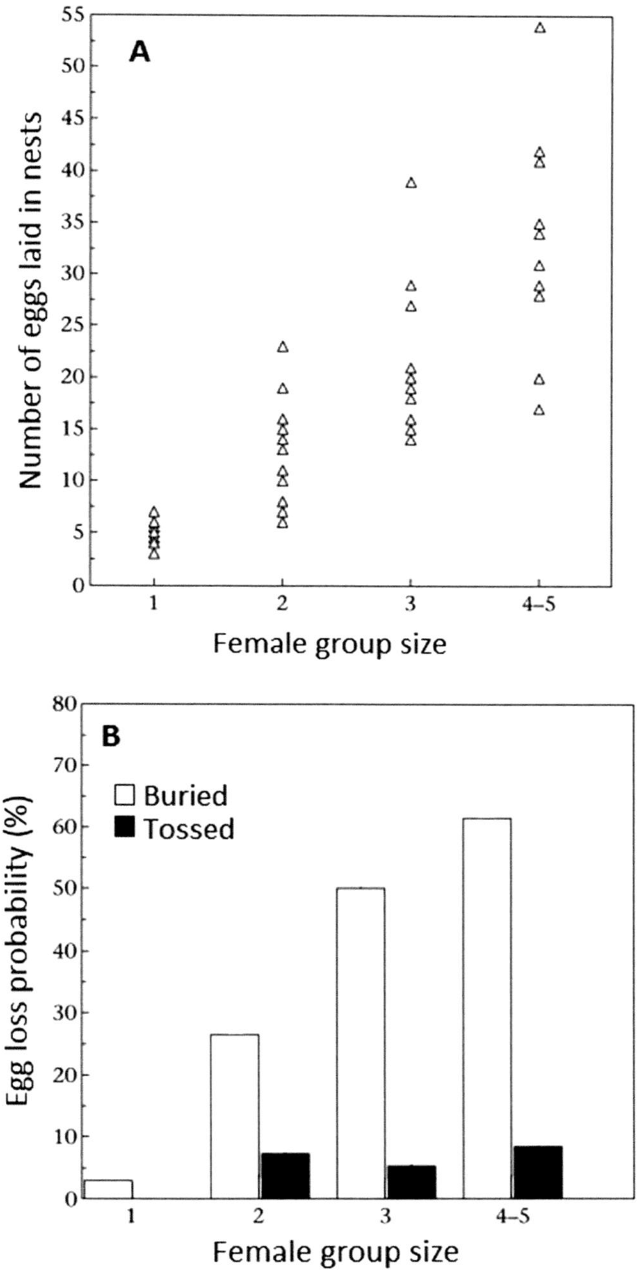


**Fig. 15.64** (**a**) Number of eggs laid in nests by groups with different numbers of female Smooth-billed Anis (*Crotophaga ani*). Each triangle represents a different nest. (**b**) Mean probability of egg loss through egg tossing or burial among groups with different number of females. (Figure modified from Schmaltz et al. 2008; © 2008 The Association for the Study of Animal Behaviour. Published by Elsevier Ltd., used with permission)

things such as a high-quality territory, parental care, or nest defense that will improve female survival and fecundity (Andersson and Simmons 2006; Kuijper et al. 2012).

- **Indirect benefits: The Fisher process** (so-named in honor of Robert Fisher [1915] who first predicted that female preferences could evolve through a self-reinforcing run-away process), also called the **sexy son** process. If there is a genetic component to variance in female preferences and male traits, females choosing males expressing particular traits will have offspring with alleles for those traits and, for female offspring, a preference

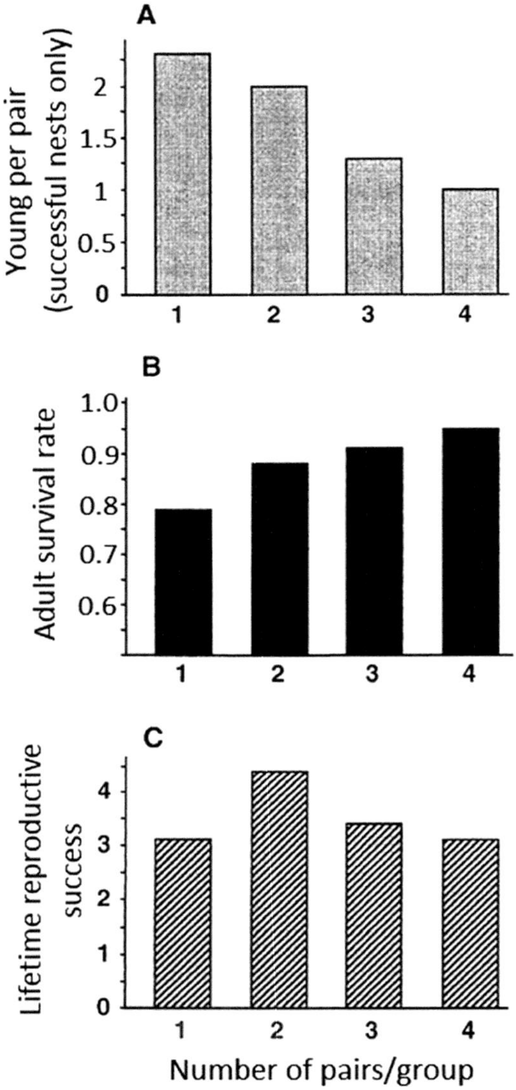


**Fig. 15.65** Effect of group size (number of pairs) on (**a**) the number of young fledged per pair, (**b**) adult survival rates, and (**c**) lifetime reproductive success of communally breeding Groove-billed Anis (*Crotophaga sulcirostris*). (Figure from Vehrencamp 2000; © 2000 Oxford University Press, used with permission)

for them. This genetic coupling might lead to self-reinforcing coevolution between male traits and female preferences (Andersson and Simmons 2006).

- **Indirect benefits: Good genes.** Particular male traits reflect heritable genetic quality, so females choosing males with such traits will have higher-quality offspring, i.e., more likely to survive and with greater reproductive value (Kuijper et al. 2012). In contrast to the Fisher process, genetic coupling is not required for the good genes mechanism.
- **Indirect benefits: Compatible genes.** Females may benefit from choosing males with alleles that complement their genome, e.g., major histocompatibility complex genes (Andersson and Simmons 2006).

These mechanisms are mutually compatible and each, to varying degrees, could contribute to the evolution of mating preferences by female birds (Andersson and Simmons 2006). The process by which certain male traits become important in the mate-choice decisions of females is called sexual selection. This process is driven by variation in the reproductive success of males; when some males experience greater reproductive success than others, selection will favor males with characteristics that contribute to that success because they are preferred by females. Of course, variance in the reproductive success of males (and females) varies across species, resulting in differences in the strength and direction of sexual selection (Fig. 15.66).

Male traits used by females when making mate-choice decisions vary among different species of birds (Table 15.7, Fig. 15.67) and, in many cases, females use multiple traits when evaluating males. For example, female Swamp Sparrows (*Melospiza georgiana*) may use both bill volume and song performance when evaluating males (Table 15.7). Multiple signals may be redundant, with multiple signals serving as backups in case one sensory modality becomes less perceptible or informative (e.g., Birkhead et al. 1998), or they may be independent, with each signal providing receivers with different information about the sender. For example, among male Satin Bowerbirds (*Ptilonorhynchus violaceus*), bower quality provides females with information about a male's size and ectoparasite load, whereas male plumage provides information about blood parasite load, feather growth rate, and body size (Doucet and Montgomerie 2003). Evidence to date suggests that, in most species, different

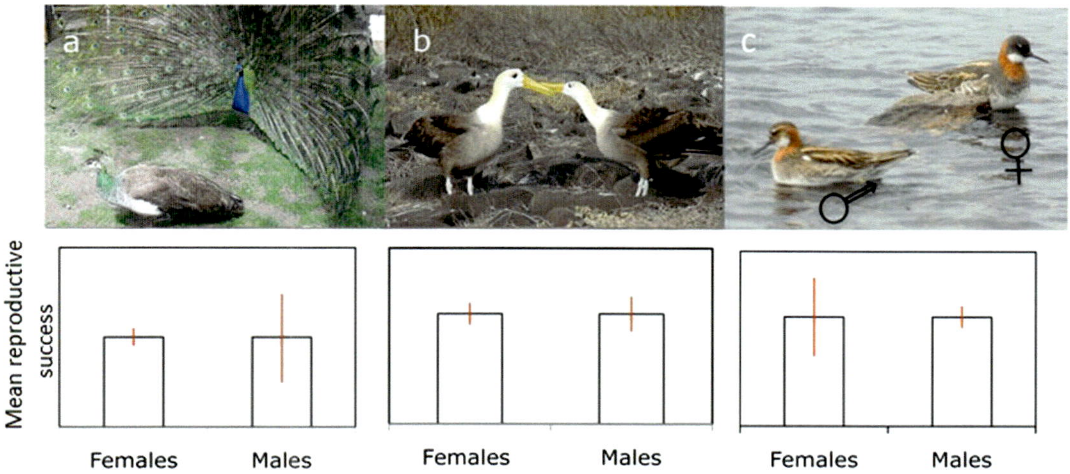


**Fig. 15.66** Variance in reproductive success determines which sex is subject to stronger sexual selection. Males and females in a population always have the same average reproductive success (i.e., every offspring is the product of one female's egg and one male's sperm), but the two sexes may differ in reproductive variance (red lines). (**a**) When reproductive variance of males is much greater than that of females (i.e., some males have many more offspring than others), then males are subject to stronger sexual selection than females and males will evolve secondary sexual characters that result in marked differences between the sexes. For example, male Indian Peafowl (*Pavo cristatus*) provide no parental care and some males have much greater reproductive success than others, resulting in marked sexual dimorphism. (**b**) In some socially monogamous species where males provide as much, or nearly as much, parental care as females, variance in male reproductive success may only be slightly greater than that of females and the sexes will be monomorphic, e.g., many seabirds. (**c**) When males provide all the parental care, variance in female reproductive success exceeds that of males and females are subject to stronger sexual selection than males, leading to reverse sexual dimorphism. For example, female Red-necked Phalaropes (*Phalaropus lobatus*) compete for access to males that provide all the parental care, and females are larger and more colorful than males. (Figure modified from Brennan 2010; © 2010 Springer Nature, used with permission)

signals likely provide receivers with different, rather than redundant, information. As another example of this, in species that are genetically polygamous (i.e., some individuals engage in extra-pair copulations), some signals have been found to be more important in attracting social mates, whereas others are more important in attracting extra-pair mates. In Common Yellowthroats (*Geothlypis trichas*), for example, males with larger plumage ornaments (mask and bib) were found to be more successful at attracting social mates, whereas males that uttered their songs with greater consistency were more successful at siring extra-pair young (Taff et al. 2012).

Females in different species use different male traits when making mate-choice decisions, but Zahavi (1975) suggested that any trait used in a signaling context should be costly or, in other words, condition-dependent, i.e., only high-quality individuals should be able to fully express the traits. Traits that are condition-dependent then represent honest signals of individual quality. Studies to date suggest that most, if not all, of the traits used by females to evaluate males, including both behavioral and morphological traits (Table 15.7), are condition-dependent in the sense that females seek males with certain characteristics (e.g., efficient immune systems, suitable MHC genes, and maintenance of oxidative balance) and certain traits accurately reflect the degree to which males have those characteristics. As such, choosing mates that best demonstrate or express these traits can provide females with indirect genetic benefits, producing offspring of higher quality. Choosing mates that better demonstrate or express these traits can also provide females with direct benefits

**Table 15.7** Examples of traits or characteristics of males important in mate-choice decisions by females

| Trait | Example(s) | Citation(s) |
|---|---|---|
| Motor performance | Many species, e.g., manakins and other lek species, hummingbirds, grassquits, snipe, some raptors, and others | Fusani et al. (2007, 2014) |
| Cognitive skills | Bowerbirds (skill at building bowers and, in at least one species, propagating fruit to include in bowers) | Endler et al. (2010), Boogert et al. (2011), Madden et al. (2012) |
| Symmetry | Barn Swallow, Zebra Finch, and other species | Møller (1992) (United Kingdom) |
| Plumage coloration (both due to pigments such as melanin and carotenoids and structural colors like blue) | Many species (e.g., House Finch, *Haemorhous mexicanus*) | E.g., Hill (1990), Keyser and Hill (2000), Tooney and McGraw (2012) |
| Skin color | Chimango Caracara (color of cere) (*Milvago chimango*) | Sarasola et al. (2011) |
| Song performance (singing physically challenging songs) that can be influenced by developmental stress (Box 15.15 Developmental Stress Hypothesis), or singing rate | Swamp Sparrow (*Melospiza georgiana*) and other species | Ballentine et al. (2004), Spencer et al. (2005) |
| Size of song repertoire | Many species, e.g., European Pied Flycatchers (*Fidedula hypoleuca*) | Lampe and Saetre (1995), Ellegren et al. (2012) |
| Tail length | Red-collared Widowbird (*Euplectes ardens*) | Pryke et al. (2001), Prager (2010) |
| Parasite load (females prefer to mate with parasite-free males) | Rock Pigeon (*Columba livia*) | Clayton (1990) |
| Nest quality | Weavers, e.g., Southern Masked Weaver (*Ploceus velatus*), and Penduline Tits (*Remiz pendulinus*) | Hoi et al. (1994), Quader (2006), Walsh et al. (2011) |
| Territory quality | Sedge Warbler (*Acrocephalus schoenobaenus*) and many other species | Buchanan and Catchpole (1997), Formica et al. (2004) |
| Dominance status | Some lek species, e.g., Greater Prairie-Chicken (*Tympanuchus cupido*) | Hovi et al. (1995), Cowles and Gibson (2015) |
| Odor (odor-mediated choice for major histocompatibility complex [MHC] compatibility) | Blue Petrel (*Halobaena caerulea*) | Bonadonna (2009), Strandh et al. (2012) |
| Cache size (caches of impaled prey) | Great Gray Shrike (*Lanius excubitor*) | Yosef and Pinshow (1989), Tryjanowski and Hromada (2005) |
| Courtship feeding (amount of food provided for female, which is correlated with extent of later parental care) | Common Tern (*Sterna hirundo*) | Wiggins and Morris (1986), Nisbet et al. (2011) |
| Extent of repeated wing molt | Common Tern and perhaps other *Sterna* spp. terns | Bridge and Nisbet (2004) |
| Alula size | Snow Bunting (*Plectrophenax nivalis*) | Guindre-Parker et al. (2013) |
| Bill volume | Swamp Sparrow (*Melospiza georgiana*) | Olsen et al. (2013), Greenberg and Droege (1990) (image) |

if doing so reflects access to resources important in successful reproduction.

Some traits used by females to evaluate male quality more obviously signal quality than others. For example, dominant males at a lek interact aggressively with other males and it seems apparent that consistently winning these encounters to establish dominance over other males requires

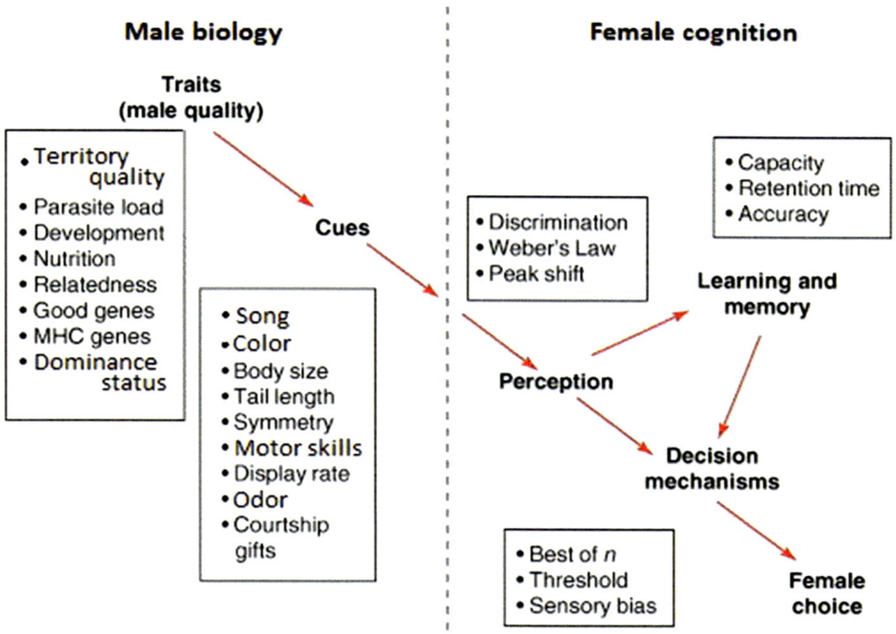


**Fig. 15.67** Assuming that male traits are correlated with male quality, those traits can then be used by females making mate-choice decisions. The relative importance of different traits and cues vary among different species of birds. The ability to identify differences among males decreases with the extent of those differences (Weber's Law). For example, if females choose mates based, in part, on tail length, distinguishing among males that differ slightly in tail length becomes more difficult as the length of tails (over evolutionary time) increases. "Peak shift" refers to the idea that if females acquire a "preference" (e.g., by imprinting on some characteristic or characteristics of her father), then, later in life, they may exhibit a preference for males that exhibit an "enhanced" version of that (or those) characteristic(s). When making a mate-choice decision, females may use a "best of n" mechanism, sampling a certain number of males (e.g., perhaps a migrant female arrives in a breeding area and visits five males and territories), then choosing as mate the "best" of those sampled males. Females may also choose based on some "threshold," meaning that females may have a "standard" in terms of mate quality and will choose to mate with the first male that meets or exceeds that standard. Female mating preferences may also be influenced by sensory bias, where a male trait might evolve via natural selection in a non-mating context and, later, that trait is favored by females in a mating context. For example, a species might exhibit a preference for ripe fruits of a certain color because they are rich in carbohydrates. Males with some feathers of the same color could then "exploit" this preference, with females having a "bias" for males with increasingly large patches of plumage of the same color. Of course, the ability of females to compare traits of multiple males depends on their cognitive abilities, including their memory capacity and the accuracy and retention time of memories. (Figure modified from Bateson and Healy 2005; © 2005 Elsevier Ltd., used with permission)

characteristics that seem clearly correlated with quality, e.g., superior size and strength and stamina. Other traits, however, are, to the human eye, less obvious indicators of quality. For example, how might subtle differences (to humans) in the size and/or color of the black face mask and yellow throat of male Common Yellowthroats (*Geothlypis trichas*; Fig. 15.68) provide females with information about male quality? As a result of sexual selection, males in many species of birds are more colorful than females, suggesting either female preferences for such plumage (intersexual selection) and/or that such plumage aids males in interactions with other males (intrasexual selection). Clearly, certain plumage characteristics are signals of male quality in

**Box 15.15 Developmental Stress Hypothesis**

The developmental stress hypothesis proposes that sexually selected traits might be sensitive to conditions experienced during early-life development and, therefore, function as honest indicators of both developmental history and quality. For example, nutritional deficiencies during pre-natal and/or early postnatal development can have long-term consequences beyond the immediate impact on growth rates. The developmental stress (or nutritional stress) hypothesis was originally proposed to explain how song complexity might be an honest indicator of male quality in some species of songbirds because bird song is a learned trait that requires memorization of various components of song(s) early in development (Nowicki et al. 1998). Available evidence suggests, however, that developmental stress in the form of inadequate diets or immune challenges during pre- and/or postnatal growth can affect a number of sexually selected traits. In addition to song complexity (Schmidt et al. 2014) and song repertoire size (Nowicki et al. 2000), developmental stress can affect ornamental plumage (Ohlsson et al. 2002; Walker et al. 2013; but see Butler and McGraw 2012), adult immune function (Butler and McGraw 2012), and even cognitive abilities (Bonaparte et al. 2011). Females making mate-choice decisions can potentially gain direct and/or indirect fitness benefits by choosing males that were not developmentally stressed and thus exhibit higher-quality sexually selected traits.

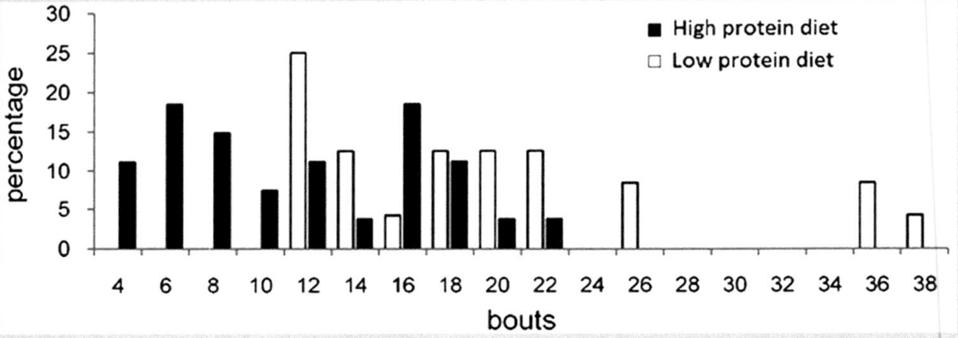


Effect of diet quality early in life on the ability of Zebra Finches (*Taeniopygia guttata*) to learn and correctly perform tasks (i.e., form an association between an arbitrary stimulus and a food reward). On average, Zebra Finches fed high-quality diets with more protein needed fewer bouts or trials to perform tasks correctly than those fed lower-quality diets with less protein. (Figure modified from Bonaparte et al. 2011; © 2011 Bonaparte et al., open-access article distributed under the terms of the Creative Commons Attribution License, https://creativecommons.org/licenses/by/4.0/)

(continued)

**Box 15.15** (continued)

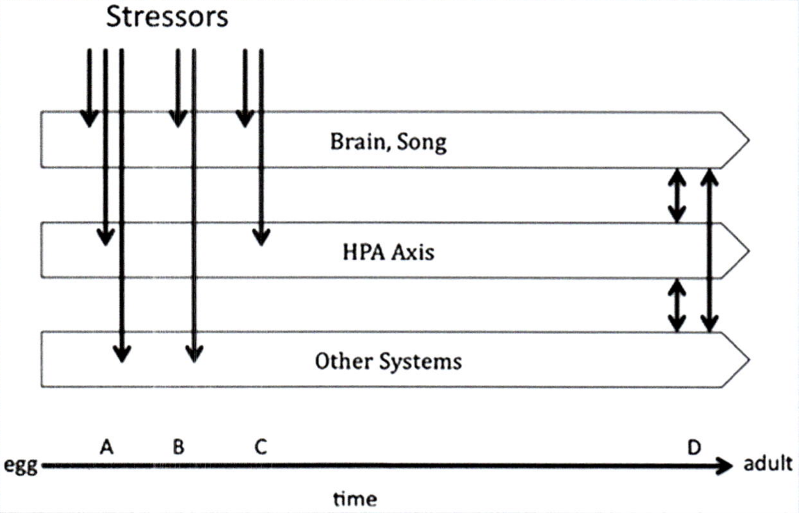


Correlations among adult traits may result from stressors during development. Stressors at a given time point may affect the development of multiple traits simultaneously. The strength of the correlations among traits in adults (time point D) will depend on the susceptibility of different traits to environmental stress. For example, a stressor at time point A would have an impact on the development of numerous systems and would result in correlations among them in adulthood. A stressor at time point B would not result in a correlation between song and the functioning of the hypothalamic pituitary adrenal (HPA) or stress axis, whereas a stress at time point C would. (Figure from Spencer and MacDougall-Shackleton 2011; © 2011 Oxford University Press, used with permission)

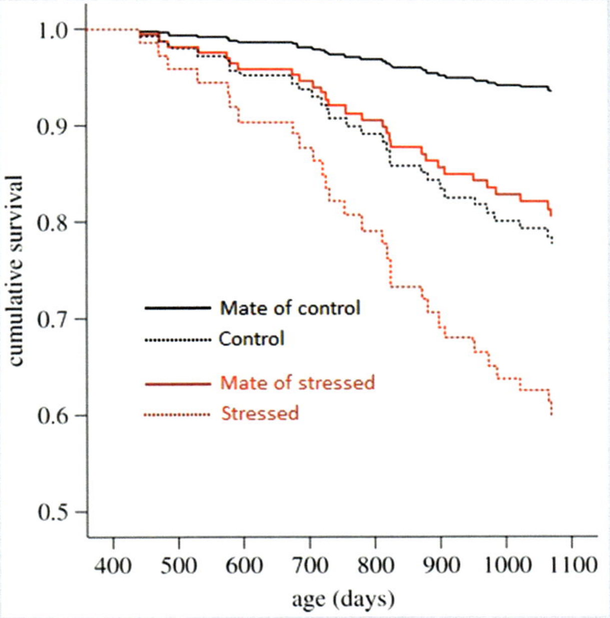


(continued)

**Box 15.15** (continued)

Young Zebra Finches exposed to stress hormones during development had shorter lifespans than non-stressed birds. Perhaps surprisingly, Zebra Finches mated to those that had been stressed during development also had shorter lifespans than Zebra Finches mated with control birds. A possible explanation for this is that the social buffering of partners was reduced for the stressed birds. Social buffering occurs when the presence of a familiar conspecific reduces stress responsiveness, and such buffering is known to influence health and longevity. Zebra Finches form strong pair bonds and social buffering between mates has been found to reduce glucocorticoid responses to environmental stressors (Remage-Healey et al. 2003). The ability of a more stress-sensitive bird to provide social buffering for its mate may be reduced, thus potentially increasing the mates' sensitivity to stress and, over the long term, reducing the likelihood of survival. (Figure from Monaghan et al. 2011; © 2011 The Royal Society, used with permission)

many species, but what is not so clear is what "quality" (i.e., trait or traits) is (are) associated or correlated with those plumage characteristics.

The plumage and skin color of many birds is derived from pigments, including carotenoids (red, orange, and yellow colors) and two types of melanins, eumelanin (black) and pheomelanin (reddish-chestnut colors). A number of studies have revealed that females in many species make mate-choice decisions based, at least in part, on the size and/or quality (e.g., brightness) of areas of male plumage or integument with these colors (e.g., Dunn et al. 2010; Montoya and Torres 2015). Importantly, studies have also revealed positive relationships between male coloration and male quality. For example, Simons et al. (2012) reviewed published studies of 88 different species of birds where males have carotenoid-dependent coloration and found a positive relationship between carotenoid levels and both immune function and antioxidant capacity (Box 15.16 Reactive Oxygen and Antioxidants). Similarly, the quality of melanin-based plumage coloration has also been found to signal antioxidant capacity (e.g., Hõrak et al. 2010, Box 15.17 Melanin-Based Plumage and Antioxidant Capacity). The potential benefits to females choosing high-quality males (based on their plumage) vary, but can include reduced oxidative damage (Moreno et al. 2013), greater parental care (Montoya and Torres 2015), and higher-quality offspring (e.g., improved immune function; Pickett et al. 2013).

**Fig. 15.68** A male Common Yellowthroat showing the black face mask and yellow throat, traits used by females making mate-choice decisions. (Figure from Dunn et al. 2010, Photo by P. Dunn; © 2009 The Authors. Journal compilation © 2009 British Ecological Society, used with permission)

**Box 15.16 Reactive Oxygen and Antioxidants**

Stable atoms have an even number of electrons orbiting in pairs at successive distances (called shells) from the nucleus. When an electron is lost (oxidation), these atoms have an odd number of electrons in the outermost shell. Such atoms are unstable and are called free radicals. An oxygen molecule that becomes a free radical is called a reactive oxygen species (ROS). ROS are produced as a normal byproduct of cellular metabolism. Sources of ROS include mitochondria (oxidative phosphorylation), endoplasmic reticulum (protein synthesis), peroxisomes (breakdown of long-chain fatty acids), and several enzymatic reactions in the cytoplasm. Production of ROS increases when birds are stressed, e.g., reproduction, parental care, migration, and heat stress (Monaghan et al. 2009; Kikusato and Toyomizu 2013). ROS do serve some important functions. For example, they serve as signal molecules between cells, crucial for triggering, for example, female reproduction, and they also aid the body defense system in killing bacteria and other pathogens (Agarwal et al. 2008; Sorci and Faivre 2009).

Although the presence of some ROS is essential, high levels of ROS have damaging effects on a number of important molecules, including DNA and other nucleic acids, proteins, and lipids. This oxidative damage can have negative effects on growth and reproduction, and can accelerate the aging process. Maintaining appropriate levels of ROS, i.e., oxidative balance, is, therefore, of critical importance to birds.

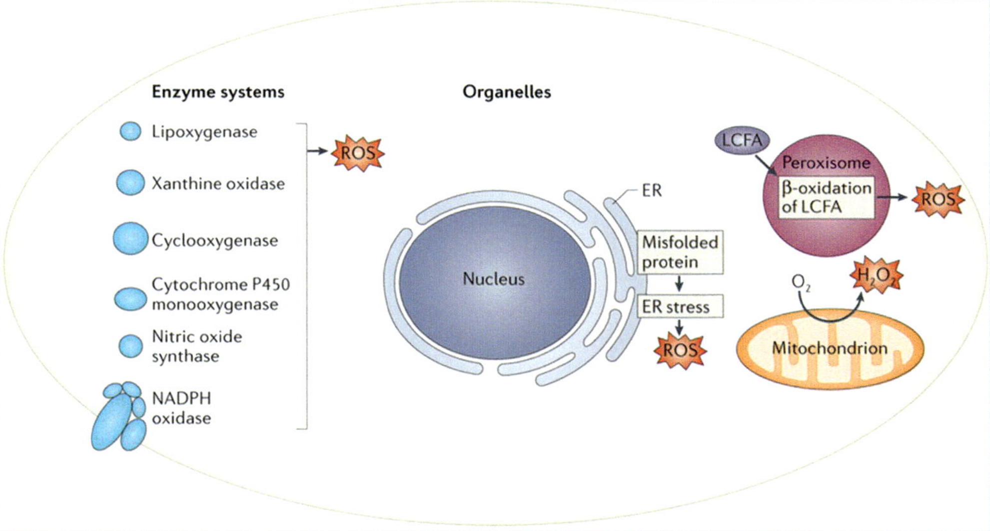


Sources of ROS (reactive oxygen species) in cells include mitochondria, peroxisomes, endoplasmic reticulum, and various enzymatic reactions (enzyme systems) in the cytoplasm. ER, endoplasmic reticulum; LCFA, long-chain fatty acid; $H_2O_2$, hydrogen peroxide (which is also reactive). (Figure from Holmstöm and Finkel 2014; © 2014 Springer Nature, used with permission)

(continued)

**Box 15.16** (continued)

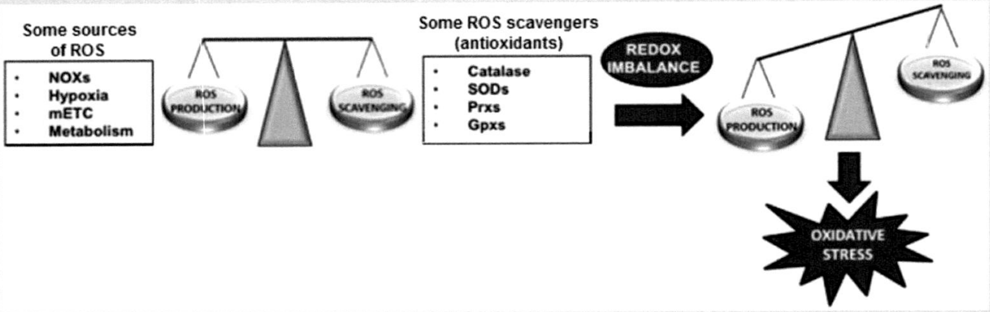


Homeostasis reflects a balance between cellular oxidants (reactive oxygen species, or ROS) and antioxidants (ROS scavengers). When not balanced because production of oxidants exceeds that of antioxidants, shifts within the homeostatic range mediate physiological responses (e.g., differentiation, growth, adaptation, and cell signaling). Large shifts produce oxidative stress (e.g., inflammation) and can have a number of negative impacts. NOx, nitric oxide derivatives; mETC, mitochondrial electron transport chain; SODs, superoxide dismutase; Prxs, peroxiredoxins; Gpxs, glutathione peroxidases. (Figure modified from Panieri and Santoro 2016; open-access article distributed under the terms of the Creative Commons CC BY license, https://creativecommons.org/licenses/by/4.0/)

Birds can avoid oxidative damage by adjusting the synthesis and the use of endogenous antioxidant enzymes (e.g., catalase, superoxide dismutase, and glutathione peroxidase). However, synthesis of new molecules from endogenous resources can be energetically costly. Birds can also ingest exogenous dietary antioxidants such as vitamin C, carotenoids, and vitamin E (Catoni et al. 2008), but this can also be costly because of the time and energy needed to search for and consume appropriate food items. In addition, access to these food items may vary across habitats and a bird's dominance status or territory quality could potentially limit their ability to acquire them. This explains why only birds in good condition or with higher dominance status are able to mount appropriate levels of antioxidant defenses to counteract ROS production (Garratt and Brooks 2012) and avoid oxidative stress. Because energy and access to exogenous dietary antioxidants may be limiting, antioxidant defenses for birds in poorer condition or lower status may have to be traded against other physiological functions. For example, because sexual behavior, reproduction, and parental care can increase the threat of oxidative damage, some birds invest less time and energy in such activities to reduce the negative impacts of oxidative stress (Zera and Harshman 2001; Monaghan et al. 2009). Similarly, these birds may be less able to divert antioxidants (e.g., carotenoids and glutathione) to functions other than neutralizing ROS and, as a result, might, for example, have less colorful plumage (i.e., fewer carotenoid molecules or less pheomelanin in feathers or skin; Garratt and Brooks 2012; Galván et al. 2015). As such, skin and plumage color may be honest signals of individual quality.

**Box 15.17 Melanin-Based Plumage and Antioxidant Capacity**

Birds synthesize two forms of melanin, eumelanin and pheomelanin, in melanosomes, specialized organelles of melanocytes (cells located in the bottom layer of the skin epidermis). Eumelanin produces black and gray colors, whereas pheomelanin produces yellowish, reddish and chestnut, and brown colors. Most melanins in birds consist of a mixture of eumelanin and pheomelanin. Birds synthesize eumelanin when the amino acid cysteine is depleted or in low levels in melanocytes and pheomelanin when cysteine concentration is higher than a threshold level, although both pigments are usually produced in different proportions by the same cells (García-Borrón and Olivares Sánchez 2011). The main source of cysteine is glutathione (GSH), a substance that is also a very important intracellular antioxidant (Wu et al. 2004). Pheomelanogenesis, therefore, requires production of cysteine from GSH, but this process reduces cellular levels of the antioxidant GSH. Because pheomelanogenesis consumes GSH, which cannot then be used to combat oxidative stress, only individuals with a high antioxidant capacity should be able to produce large amounts of pheomelanin to color their integument under conditions of high environmental oxidative stress (Galván and Solano 2009). As a result, melanic traits are condition-dependent and honestly signal individual quality (McGraw 2008).

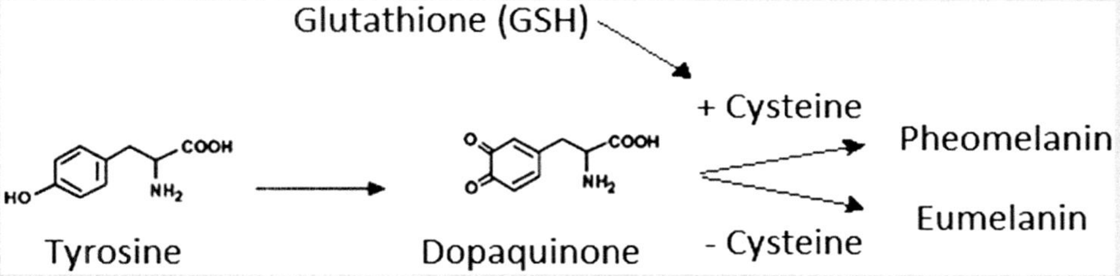


Pathways for both eumelanogenesis and pheomelanogenesis. Production of both eumelanin and pheomelanin begins with the conversion of the amino-acid tyrosine to dopaquinone. Dopaquinone can then be converted to either eumelanin or pheomelanin, depending on the presence or absence of cysteine or glutathione (GSH), which is the main source of cysteine and an important intracellular antioxidant. (Figure modified from Jawor and Breitwisch 2003; © 2003 Oxford University Press, used with permission)

(continued)

**Box 15.17** (continued)

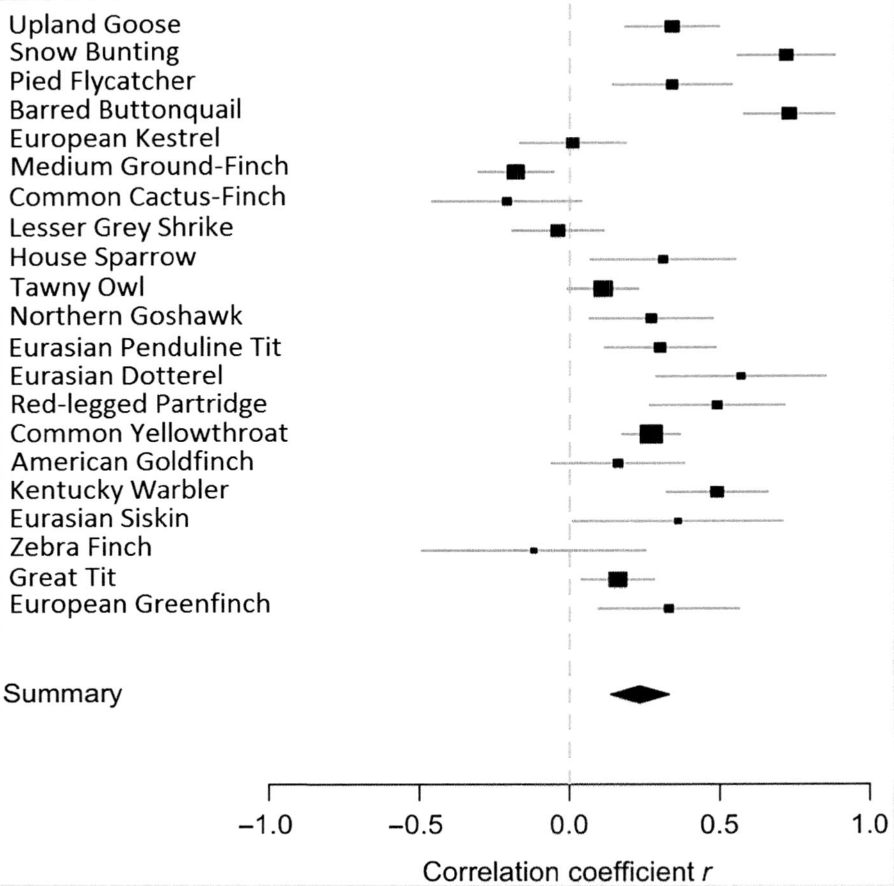


Relationship between melanin-based plumage ornaments and individual condition. A positive $r$ value indicates a positive relationship between the quality of melanin-based plumage ornaments and male condition. Whiskers indicate the standard error and box size reflects relative sample size. (Figure modified from Guindre-Parker and Love 2014; © 2013 The Authors, used with permission). Scientific names: Upland Goose, *Chloephaga picta*; Snow Bunting, *Plectrophenax nivalis*; Pied Flycatcher, *Ficedula hypoleuca*; Barred Buttonquail, *Turnix suscitator*; European (Eurasian) Kestrel, *Falco tinnunculus*; Medium Ground-Finch, *Geospiza fortis*; Common Cactus-Finch, *Geospiza scandens*; Lesser Grey Shrike, *Lanius minor*; House Sparrow, *Passer domesticus*; Tawny Owl, *Strix aluco*; Northern Goshawk, *Accipiter gentilis*; Eurasian Penduline Tit, *Remiz pendulinus*; Eurasian Dotterel, *Charadrius morinellus*; Red-legged Partridge, *Alectoris rufa*; Common Yellowthroat, *Geothlypis trichas*; American Goldfinch, *Carduelis tristis*; Kentucky Warbler, *Oporornis formosus*; Eurasian Siskin, *Carduelis spinus*; Zebra Finch, *Taeniopygia guttata*; Great Tit, *Parus major*; and European Greenfinch, *Carduelis chloris*)

## 15.14 Male Mate Choice

Female birds can clearly benefit by being selective when choosing mates. However, males, particularly in species where both sexes provide parental care, may also benefit by being selective when choosing a mate, and male or mutual mate choice has been reported in several species of birds, e.g., Rock Sparrows (*Petronia petronia*, Griggio et al. 2005), Crested Auklets (*Aethla cristatella*, Jones and Hunter 1993), American Robins (*Turdus migratorius*, Rowe and Weatherhead 2011), King Penguins (*Aptenodytes patagonicus*, Nolan et al. 2010), Roseate Terns (*Sterna dougallii*, Palestis et al. 2012), Red-tailed Tropicbirds (*Phaethon rubricauda*, Boland et al. 2004), and several others (e.g., Velando et al. 2001; Daunt et al. 2003). Male mate choice might be expected in species or populations with greater availability of and access to possible mates, e.g., colonial species or populations with a skewed sex ratio. However, availability of mates will also be influenced by the degree to which males and females invest in parental care. If males invest much less in parental care than females, then the number of receptive females available as mates will likely be lower (because females providing parental care are not available as mates) and males will likely not be choosy and will not reject receptive females as mates (as long as they are not limited by their capacity to mate, e.g., sperm depletion; Edward and Chapman 2011), e.g., in polygynous species or, in socially monogamous species, for extra-pair copulations (Fig. 15.69).

For male mate choice to evolve, females must also vary in quality. If variation in female quality is insufficient to affect male reproductive fitness, then male mate choice should not evolve (Edward and Chapman 2011). However, even if females do vary in quality, the cost to males of assessing potential mates may exceed the benefits (Fig. 15.69). For example, among migratory species of birds where the duration of breeding seasons is limited, initiating breeding as soon as possible after arriving in breeding areas can result in greater reproductive success (e.g., time for additional broods or, for single-brooded species, providing more time for adults and offspring to prepare for migration). As a result, the cost to males of delaying reproduction to assess more possible mates may exceed the possible benefit of obtaining a higher-quality mate.

Male or mutual mate choice can potentially favor the evolution of female traits or ornaments that serve as honest signals of female quality, i.e., sexual selection. However, the potential evolution of such traits or ornaments can also be influenced by factors other than male mate choice. For example, female ornaments can be costly if they increase the likelihood of predation, e.g., brightly colored plumage that would make an incubating or brooding female more likely to be detected by a predator. In addition, the energetic costs of producing traits or ornaments could reduce female fecundity, e.g., less energy available for egg production or parental care. The extent to which females might express traits or ornaments that signal quality will, therefore, vary with the potential costs and benefits of doing so and, where such traits or ornaments are present, will, in most species of birds, not be expressed as strongly in females as males.

For male mate choice to evolve, males must benefit in terms of increased reproductive fitness. In addition, if male mate-choice decisions are based at least in part on certain female traits or ornaments, then those traits or ornaments must represent honest signals of quality and their expression should be correlated with increased reproductive success, e.g., increased fecundity or improved parental care. In a review of the literature, Nordeide et al. (2013) examined the extent to which the ornaments of female birds have been found to be positively correlated with some aspect of breeding success, e.g., date of nest

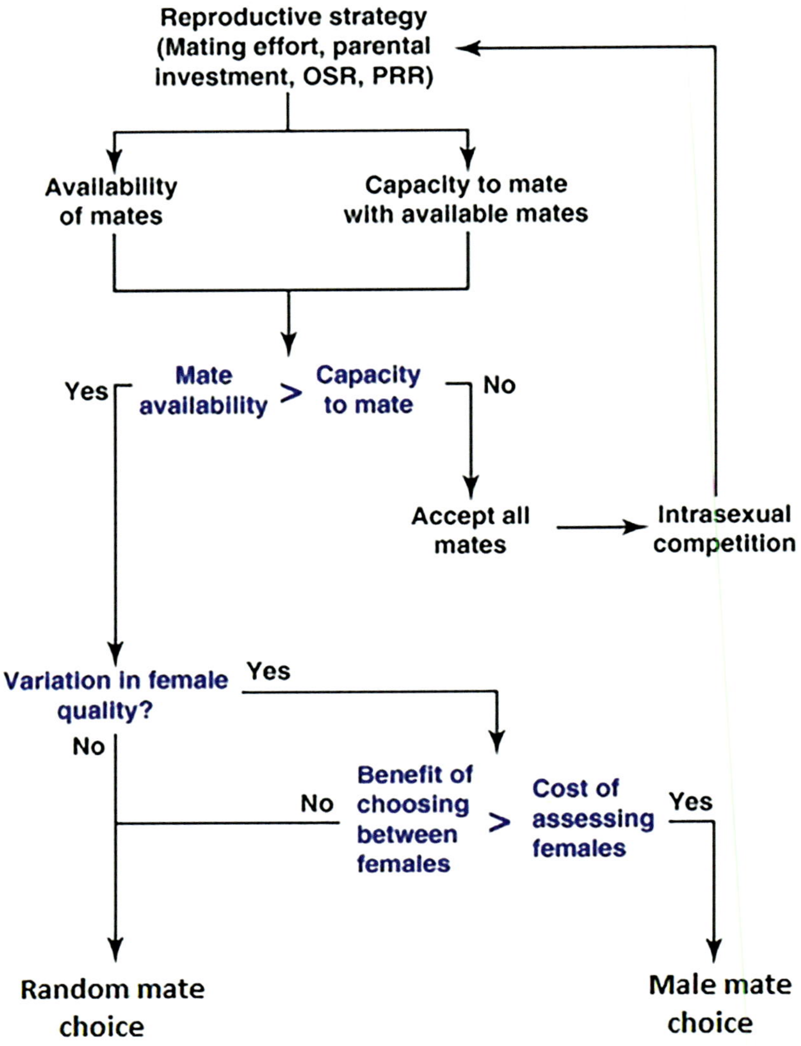


**Fig. 15.69** Factors potentially affecting male mate choice. If the number of available females does not exceed the ability of males to mate with them, then selection will favor accepting all mates. This will contribute to increased intrasexual competition for mates. Alternatively, if the number of available females exceeds the ability of males to mate with them, then selection will favor rejection of some females. If a male rejects some females, there is variation in female quality, and there is a net benefit for choosing among females, then selection will favor male mate choice. In the absence of variation in female quality or a net benefit of choice, selection will favor random mate rejection. OSR, operational sex ratio; PRR, potential reproductive rate. (Figure modified from Edward and Chapman 2011; © 2011 Elsevier Ltd., used with permission)

initiation, clutch size, offspring quality, and/or fledging success. In 18 of 33 studies (55%), investigators reported a positive correlation between the quality of female ornaments and offspring quality. For example, female Rock Sparrows (*Petronia petronia*) with larger yellow crest patches were more likely to be double-brooded and fledge more young (Pilastro et al. 2003), and female Magellanic Penguins (*Spheniscus magellanicus*) with longer flippers tended to fledge more young (Forero et al. 2001).

In 15 of the 33 studies (45%), Nordeide et al. (2013) found no correlation between breeding success and female ornamentation. One possible explanation for this is that female "ornaments" are non-functional and result from selection of male ornaments through intersexual genetic correlation (Lande 1980). Alternatively, female

**Fig. 15.70** Adult male (left) and female (right) Eclectus Parrots (*Eclectus roratus*) at a nest cavity. (Figure from Heinsohn et al. 2011; © 2011 Elsevier Ltd., used with permission)

ornaments may be the result of social selection and more important in intrasexual competition for resources than for mate attraction (LeBas 2006). For example, female Eclectus Parrots (*Eclectus roratus*) are more colorful than males, with males primarily green and females red and blue (Heinsohn et al. 2005, Fig. 15.70). This reversed sexual dimorphism is likely due to social selection (or intrasexual selection), with females competing with other females for access to high-quality nesting and roosting cavities and signaling their status with their bright coloration. Similarly, female Streak-backed Orioles (*Icterus pustulatus*) defending territories from other females and female American Goldfinches (*Spinus tristis*) competing with other females for access to food signal their status to other females with carotenoid-based orange feathers on the head and breast and bright orange bills, respectively (Murphy et al. 2009a, b). More generally, selective factors acting on males and females often differ, and social (or intrasexual) selection for traits or ornaments that help females gain access to resources during the non-breeding period (such as non-breeding territories and access to food) may, in some species of birds, be more important for females than males (Fig. 15.71). For males, traits or ornaments that aid in gaining mating opportunities may be more important (Fig. 15.71).

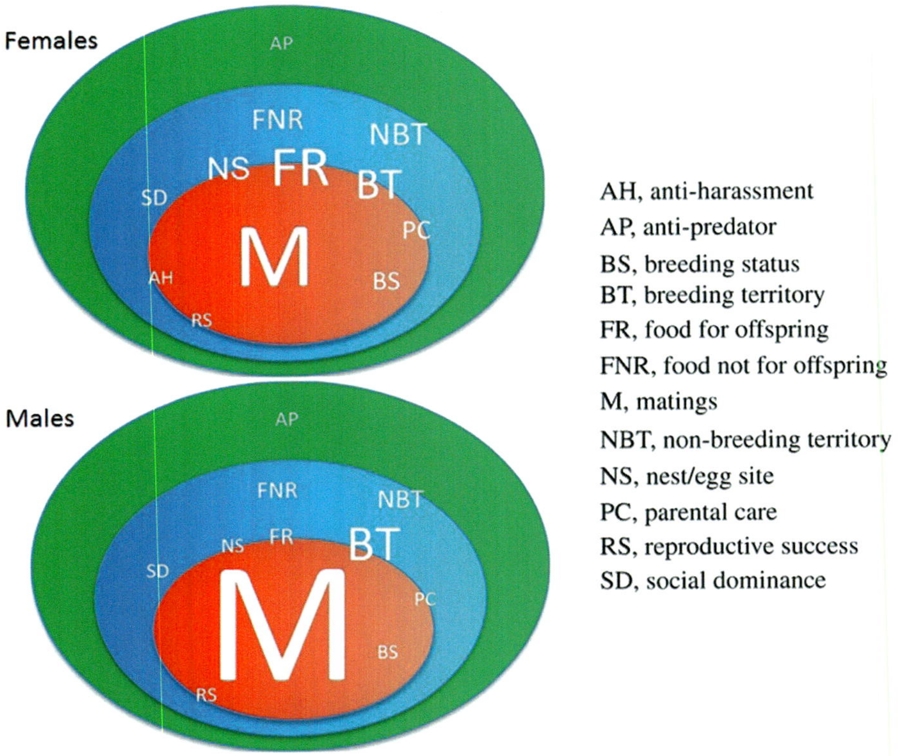


**Fig. 15.71** Relationships between sexual, social, and natural selection. Sexual selection (red) can be viewed as a subset of social selection (blue) and social selection as a subset of natural selection (green). Although each type of selection is shown with sharp boundaries, different colors should be viewed as regions within a continuum. Various selective factors are indicated by letter codes, which are explained on the right. Code placement is subjective, based on assessment of available evidence from the literature. Code size is scaled relative to possible importance based on studies of 77 species of birds. (Figure from Tobias et al. 2012; © 2012 The Royal Society, used with permission)

Another possible explanation for the absence of correlations between breeding success and female ornamentation in species of birds that exhibit male mate choice is that males may prefer females that exhibit "intermediate-levels" of particular traits or ornaments (Chenowith et al. 2006; Fig. 15.72). Laying one or more clutches of eggs can be energetically costly, so males may prefer females with ornaments that signal their quality, but ornaments not so costly that female fecundity may be affected (Fitzpatrick et al. 1995). If so, then sexual selection for the evolution of female traits or ornaments may be self-limiting (Fitzpatrick et al. 1995). Although this might explain male preferences and female traits or ornaments in some species of birds, empirical evidence is still lacking. Additional studies are needed to improve our understanding of the evolution and importance of male preferences and the ornamentation of female birds.

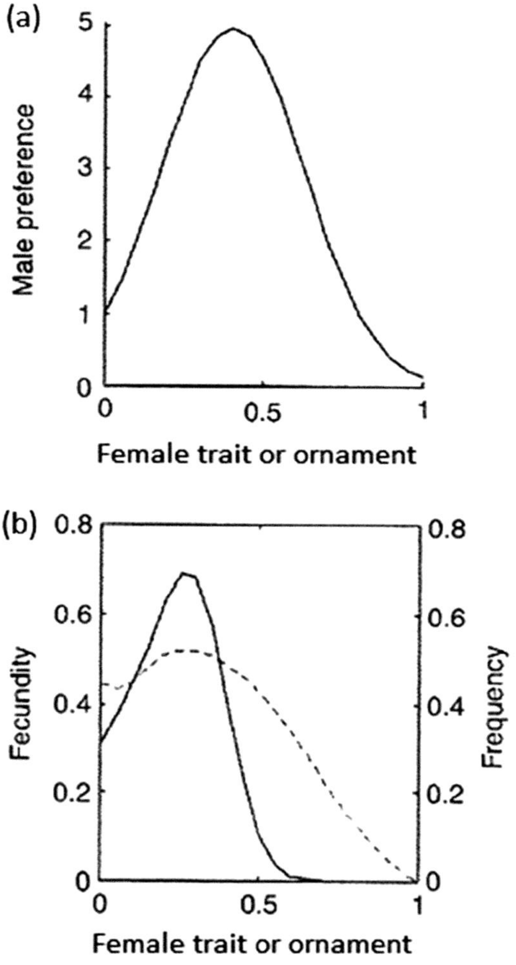


**Fig. 15.72** Possible dynamic between male preferences and expression of female traits or ornaments when expressing those traits or ornaments requires energy that can negatively impact female fecundity. (**a**) The equilibrium mate preference function of males in the evolutionary model. Males might evolve to exert stabilizing selection on female traits or ornaments (i.e., signals). Higher-quality females can better express the signal, but males prefer "intermediate" signals that indicate both female quality and fecundity. Stronger signals may indicate females that mistakenly (perhaps due to variable environmental conditions) signal too strongly for their quality and, thus, would have lower fecundity. (**b**) Most females may signal at levels below those preferred by males (solid line) because selection for stronger signals (sexual selection) may be countered by selective pressures (natural selection) to maximize fecundity (dashed line). (Figure from Chenowith et al. 2006; © John Wiley & Sons Ltd./CNRS, used with permission)

## References

Adkins-Regan E, Tomaszycki M (2007) Monogamy of the fast track. Biol Lett 3:617–619

Agarwal A, Gupta S, Sekhon L, Shah R (2008) Redox considerations in female reproductive function and assisted reproduction: from molecular mechanisms to health implications. Antioxid Redox Signal 10:1375–1403

Akçay E, Roughgarden J (2007) Extra-pair paternity in birds: review of the genetic benefits. Evol Ecol Res 9: 855–868

Alatalo RV, Höglund J, Lundberg A, Sutherland WJ (1992) Evolution of Black Grouse leks: female preferences benefit males in large leks. Behav Ecol 3: 53–59

Alatalo RV, Burke T, Dann J, Hanotte O, Höglund J, Lundberg A, Moss R, Rintamäki PT (1996) Paternity, copulation disturbance and female choice in lekking Black Grouse. Anim Behav 52:861–873

Alonso JC, Magaña M, Palacin C, Martin CA (2010) Correlates of male mating success in Great Bustard leks: the effects of age, weight, and display effort. Behav Ecol Sociobiol 64:1589–1600

Alonso JC, Álvarez-Martínez JM, Palacín C (2012) Leks in ground-displaying birds: hotspots or safe places? Behav Ecol 23:491–501

Alonzo SH, Sheldon BC (2010) Population density, social behaviour and sex allocation. In: Székely T, Moore A, Komdeur J (eds) Social behaviour: genes, ecology and evolution. Cambridge University Press, Cambridge, UK, pp 474–488

Amar A, Redpath SM (2002) Determining the cause of the Hen Harrier decline on the Orkney Islands: an experimental test of two hypotheses. Anim Conserv 5:21–28

Amat JA, Visser GH, Pérez-Hurtado A, Arroyo GM (2000) Brood desertion by female shorebirds: a test of the differential parental capacity hypothesis on Kentish Plovers. Proc R Soc B 267:2171–2176

Andersson S (1989) Sexual selection and cues for female choice in leks of Jackson's Widowbird *Euplectes jacksoni*. Behav Ecol Sociobiol 25:403–410

Andersson M (2005) Evolution of classical polyandry: three steps to female emancipation. Ethology 111:1–23

Andersson M, Simmons LW (2006) Sexual selection and mate choice. Trends Ecol Evol 21:296–302

Arak A (1982) Sneaky breeders. In: Barnard CJ (ed) Producers and scroungers: strategies of exploitation and parasitism. Croom Helm, Beckenham, UK, pp 154–194

Arcese P (1989) Intrasexual competition and the mating system in primarily monogamous birds: the case of the Song Sparrow. Anim Behav 38:96–111

Arlt D, Hansson B, Bensch S, von Schantz T, Hasselquist D (2004) Breeding synchrony does not affect extra-pair paternity in Great Reed Warblers. Behaviour 141: 863–880

Arnold KE, Owens IPF (2002) Extra-pair paternity and egg dumping in birds: life history, parental care, and the risk of retaliation. Proc R Soc B 269:1263–1269

Arnqvist G, Kirkpatrick M (2005) The evolution of infidelity in socially monogamous passerines: the strength of direct and indirect selection on extrapair copulation behavior in females. Am Nat 165:S26–S37

Aspbury AS, Gibson RM (2004) Long-range visibility of Greater Sage Grouse leks: a GIS-based analysis. Anim Behav 67:1127–1132

Atwood JL, Fitz VL, Banesberger JE (1991) Temporal patterns of singing activity at leks of the White-bellied Emerald. Wilson Bulletin 103:373–386

Avery MI (1984) Lekking in birds: choice, competition and reproductive constraints. Ibis 126:177–187

Baglione V, Canestrari D, Marcos JM, Ekman J (2006) Experimentally increased food resources in the natal territory promote offspring philopatry and helping in cooperatively breeding Carrion Crows. Proc R Soc B 273:1529–1535

Ballentine B, Hyman J, Nowicki S (2004) Vocal performance influences female response to male bird song: an experimental test. Behav Ecol 15:163–168

Barske J, Schlinger BA, Wikelski M, Fusani L (2011) Female choice for male motor skills. Proc R Soc B 278:3523–3528

Barske J, Schlinger BA, Fusani L (2015) The presence of a female influences courtship performance of male manakins. Auk 132:594–603

Bart J, Tornes A (1989) Importance of monogamous male birds in determining reproductive success: evidence for House Wrens and a review of male-removal studies. Behav Ecol Sociobiol 24:109–116

Bateson M, Healy SD (2005) Comparative evaluation and its implications for mate choice. Trends Ecol Evol 20: 659–664

Baydack RK (1986) Sharp-tailed Grouse response to lek disturbance in the Carberry Sand Hills of Manitoba, Ph.D. dissertation. Colorado State University, Fort Collins, CO

Behney AC, Grisham BA, Boal CW, Whitlaw HA, Haukos DA (2012) Sexual selection and mating chronology of Lesser Prairie-Chickens. Wilson J Ornithol 124:96–105

Benito MM, Gonzalez-Solis J (2007) Sex ratio, sex-specific chick mortality and sexual size dimorphism in birds. J Evol Biol 20:1522–1530

Bennett PM, Owens IPF (2002) Evolutionary ecology of birds. Oxford Univ Press, Oxford

Bhullar B-A, Marugán-Lobón J, Racimo F, Bever GS, Rowe TB, Norell MA, Abzhanov A (2012) Birds have paedomorphic dinosaur skulls. Nature 487:223–226

Bird KL, Aldridge CL, Carpenter JE, Paszkowski CA, Boyce MS (2013) The secret sex lives of sage-grouse: multiple paternity and intraspecific nest parasitism revealed through genetic analysis. Behav Ecol 24:29–38

Bird JP, Martin R, Akçakaya HR, Gilroy J, Burfield IJ, Garnett ST, Symes A, Taylor J, Şekercioğlu ÇH, Butchart SHM (2020) Generation lengths of the world's birds and their implications for extinction risk. Conserv Biol 34:1252–1261

Birkhead TR (1991) Sperm depletion in the Bengalese Finch *Lonchura striata*. Behav Ecol 2:267–275

Birkhead TR, Biggins JD (1987) Reproductive synchrony and extra-pair copulation in birds. Ethology 74:320–334

Birkhead TR, Møller AP (1992) Sperm competition in birds: evolutionary causes and consequences. Academic Press, London, UK

Birkhead TR, Fletcher F, Pellatt EJ (1998) Sexual selection in the Zebra Finch *Taeniopygia guttata*: condition, sex traits and immune capacity. Behav Ecol Sociobiol 44:179–191

Björklund M (1990) A phylogenetic interpretation of sexual dimorphism in body size and ornament in relation to mating system in birds. J Evol Biol 3:171–183

Black JM (2001) Fitness consequences of long-term pair bonds in Barnacle Geese: monogamy in the extreme. Behav Ecol 12:640–645

Blackmore CJ, Heinsohn R (2007) Reproductive success and helper effects in the cooperatively breeding Grey-crowned Babbler. J Zool 273:326–332

Blas J, Hiraldo F (2010) Proximate and ultimate factors explaining floating behavior in long-lived birds. Horm Behav 57:169–176

Blas J, Sergio F, Hiraldo F (2009) Age-related improvement in reproductive performance in a long-lived raptor: a cross-sectional and longitudinal study. Ecography 32:647–657

Boland CRJ, Double MC, Baker GB (2004) Assortative mating by tail streamer length in Red-tailed Tropicbirds *Phaethon rubricauda* breeding in the Coral Sea. Ibis 146:687–690

Bollmer JL, Dunn PO, Freeman-Gallant CR, Whittingham LA (2012) Social and extra-pair mating in relation to major histocompatibility complex variation in Common Yellowthroats. Proc R Soc B 279:4778–4785

Bonadonna F (2009) Olfaction in petrels: from homing to self-odor avoidance. Ann N Y Acad Sci 1170:428–433

Bonaparte KM, Riffle-Yokoi C, Burley NT (2011) Getting a head start: diet, sub-adult growth, and associate learning in a seed-eating passerine. PLoS One 6: e23775

Boogert NJ, Fawcett TW, Lefebvre L (2011) Mate choice for cognitive traits: a review of the evidence in nonhuman vertebrates. Behav Ecol 22:447–459

Bostwick KS (2000) Display behaviors, mechanical sounds, and evolutionary relationships of the Club-winged Manakin (*Machaeropterus deliciosus*). Auk 117:465–478

Bostwick KS, Riccio ML, Humphries JM (2012) Massive, solidified bone in the wing of a volant courting bird. Biol Lett 8:760–763

Bradbury JW (1981) The evolution of leks. In: Alexander RD, Tinkle DW (eds) Natural selection and social behavior. Chiron, New York, NY, pp 138–169

Bradbury JW, Gibson RM (1983) Leks and mate choice. In: Bateson PPG (ed) Mate choice. Cambridge University Press, Cambridge, UK, pp 107–138

Breiehagen T (1989) Nesting biology and mating system in an alpine population of Temminck's Stint *Calidris temminckii*. Ibis 131:389–402

Brekke P, Bennett PM, Santure AW, Ewen JG (2011) High genetic diversity in the remnant Island population of Hihi and the genetic consequences of re-introduction. Mol Ecol 20:29–45

Brennan PLR (2009) Incubation in Great Tinamou (*Tinamus major*). Wilson J Ornithol 121:506–511

Brennan PLR (2010) Sexual selection. Nature Education Knowledge 3:79

Brennan PLR, Prum RO, McCracken KG, Sorenson MD, Wilson RE, Birkhead TR (2007) Coevolution of male and female genital morphology in waterfowl. PLoS One 2:e418

Brennan PLR, Clark CH, Prum RO (2010) Explosive eversion and functional morphology of the duck penis supports sexual conflict in waterfowl genitalia. Proc R Soc B 277:1309–1314

Bridge ES, Nisbet ICT (2004) Wing molt and assortative mating in Common Terns: a test of the molt-signaling hypothesis. Condor 106:336–343

Briskie JV (1992) Copulation patterns and sperm competition in the polygynandrous Smith's Longspur. Auk 109:563–575

Briskie JV, Naugler CT, Leech SM (1994) Begging intensity of nestling birds varies with sibling relatedness. Proc R Soc B 258:73–78

Bro-Jørgensen J (2007) Reversed sexual conflict in a promiscuous antelope. Curr Biol 17:2157–2161

Brouwer L, Barr I, Van De Pol M, Burke T, Komdeur J, Richardson DS (2010) MHC-dependent survival in a wild population: evidence for hidden genetic benefits gained through extra-pair fertilizations. Mol Ecol 19: 3444–3455

Brouwer L, Richardson DS, Komdeur J (2012) Helpers at the nest improve late-life offspring performance: evidence from a long-term study and a cross-foster experiment. PLoS One 7:e33167

Brown JL (1987) Helping and communal breeding in birds. Princeton University Press, Princeton, NJ

Brown JL, Dow DD, Brown ER, Brown SD (1978) Effects of helpers on feeding of nestlings in the Grey-crowned Babbler (*Pomatostomus temporalis*). Behav Ecol Sociobiol 4:43–59

Bryan DC (1996) Family Aramidae (limpkin). In: del Hoyo JA, Elliott A, Sargatal J (eds) Handbook of the birds of the world. Volume 3. Hoatzin to auks. Lynx Edicions, Barcelona, pp 90–95

Buchanan KL, Catchpole CK (1997) Female choice in the Sedge Warbler, *Acrocephalus schoenobaenus*: multiple cues from song and territory quality. Proc R Soc B 264:521–6526

Bulmer MG, Parker GA (2002) The evolution of anisogamy: a game-theoretic approach. Proc R Soc B 269: 2381–2388

Burke T, Davies NB, Bruford MW, Hatchwell BH (1989) Parental care and mating behaviour of polyandrous Dunnocks *Prunella modularis* related to paternity by DNA fingerprinting. Nature 338:249–251

Bush KL, Aldridge CL, Carpenter JE, Paszkowski CA, Boyce MS, Coltman DW (2010) Birds of a feather do not always lek together: genetic diversity and kinship structure of Greater Sage-Grouse (*Centrocercus urophasianus*) in Alberta. Auk 127:343–353

Butler MW, McGraw KJ (2012) Developmental immune history affects adult immune function but not carotenoid-based ornamentation in Mallard ducks. Funct Ecol 26:406–415

Canestrari D, Marcos JM, Baglione V (2008) Reproductive success increases with group size in cooperative Carrion Crows, *Corvus corone corone*. Anim Behav 75:403–416

Canestrari D, Marcos JM, Baglione V (2011) Helpers at the nest compensate for reduced maternal investment in egg size in Carrion Crows. J Evol Biol 24:1870–1878

Carvalho IDS, Novas FE, Agnolín FL, Isasi MP, Freitas FI, Andrade JA (2015) A new genus and species of enantiornithine bird from the Early Cretaceous of Brazil. Brazilian J Geol 45:161–171

Castro I, Minot EO, Fordham RA, Birkhead TR (1996) Polygynandry, face-to-face copulation and sperm competition in the Hihi *Notiomystis cincta* (Aves: Meliphagidae). Ibis 138:765–771

Catoni C, Peters A, Schaefer HM (2008) Life history trade-offs are influenced by the diversity, availability and interactions of dietary antioxidants. Anim Behav 76:1107–1119

Catry P, Lecoq M, Araujo A, Conway G, Felgueiras M, King JMB, Rumsey S, Salima H, Tenreiro P (2005) Differential migration of Chiffchaffs *Phylloscopus collybita* and *P. ibericus* in Europe and Africa. J Avian Biol 36:184–190

Charmantier A, Blondel J (2003) A contrast in extrapair paternity levels on mainland and Island populations of Mediterranean Blue Tits. Ethology 109:351–363

Chenowith SF, Doughty P, Kokko H (2006) Can non-directional male mating preferences facilitate honest female ornamentation? Ecol Lett 9:179–184

Chuang HC, Webster MS, Holmes RT (1999) Extrapair paternity and local synchrony in the Black-throated Blue Warbler. Auk 116:726–736

Chuang-Dobbs HC, Webster MS, Holmes RT (2001) The effectiveness of mate guarding by male Black-throated Blue Warblers. Behav Ecol 12:541–546

Clayton DH (1990) Mate choice in experimentally parasitized Rock Doves: lousy males lose. Am Zool 30:251–262

Clutton-Brock TH, Price O, MacColl A (1992) Mate retention, harassment and the evolution of ungulate leks. Behav Ecol 3:234–242

Cockburn A (2006) Prevalence of different modes of parental care in birds. Proc R Soc B 273:1375–1383

Cockburn A, Sims RA, Osmond HL, Green DJ, Double MC, Mulder RA (2008) Can we measure the benefits of help in cooperatively breeding birds: the case of Superb Fairy wrens *Malurus cyaneus*? J Anim Ecol 77:430–438

Coddington CL, Cockburn A (1995) The mating system of free-living Emus. Aust J Zool 43:365–372

Colegrave N, Birkhead TR, Lessells CM (1995) Sperm precedence in Zebra Finches does not require special mechanisms of sperm competition. Proc R Soc B 259: 223–228

Colwell MA, Oring LW (1988a) Sex ratios and intrasexual competition for mates in a sex-role reversed shorebird, Wilson's Phalarope (*Phalaropus tricolor*). Behav Ecol Sociobiol 22:165–173

Colwell MA, Oring LW (1988b) Breeding biology of Wilson's Phalarope in southcentral Saskatchewan. Wilson Bull 100:567–582

Cooper NW, Murphy MT, Redmond LJ, Dolan AC (2009) Density-dependent age at first reproduction in the Eastern Kingbird. Oikos 118:413–419

Cornwallis CK, West SA, Griffin AS (2009) Routes to indirect fitness in cooperatively breeding vertebrates: kin discrimination and limited dispersal. J Evol Biol 22:2445–2457

Coulson JC (1966) The influence of the pair-bond and age on the breeding biology of the Kittiwake gull *Rissa tridactyla*. J Anim Ecol 35:269–279

Cowles SA, Gibson RM (2015) Displaying to females may lower male foraging time and vigilance in a lekking bird. Auk 132:82–91

Crowe SA, Kleven O, Delmore KE, Laskemoen T, Nocera JJ, Lifjeld JT, Robertson RJ (2009) Paternity assurance through frequent copulations in a wild passerine with intense sperm competition. Anim Behav 77:183–187

Culina A, Radersma R, Sheldon BC (2015) Trading up: the fitness consequences of divorce in monogamous birds. Biol Rev 90:1015–1034

Dakin R, Montgomerie R (2013) Eye for an eyespot: how iridescent plumage ocelli influence peacock mating success. Behav Ecol 24:1048–1057

Daunt F, Monaghan P, Wanless S, Harris MP (2003) Sexual ornament size and breeding performance in female and male European Shags. Ibis 145:54–60

Davies NB (1985) Cooperation and conflict among Dunnocks, *Prunella modularis*, in a variable mating system. Anim Behav 33:628–648

Davies NB (1991) Mating systems. In: Krebs JR, Davies NB (eds) Behavioural ecology: an evolutionary approach, 3rd edn. Blackwell Scientific Publications, Oxford, UK, pp 263–294

Davies NB, Krebs JR (2009) An introduction to behavioural ecology, 3rd edn. Wiley-Blackwell, Hoboken, NJ

Davies NB, Hartley IR, Hatchwell BJ, Desrochers A, Skeer J, Nebel D (1995) The polygynandrous mating system of the Alpine Accentor, *Prunella collaris*. I. Ecological causes and reproductive conflicts. Anim Behav 49:769–788

de Brooke ML, Hartley IR (1995) Nesting Henderson Reed-warblers (*Acrocephalus vaughani taiti*) studied by DNA fingerprinting: unrelated coalitions in a stable habitat? Auk 112:77–86

Díaz-Muñoz SL, DuVal EH, Krakauer AH, Lacey EA (2014) Cooperating to compete: altruism, sexual selection and causes of male reproductive cooperation. Anim Behav 88:67–78

Donald P (2007) Adult sex ratios in wild bird populations. Ibis 149:671–692

Doucet SM, Montgomerie R (2003) Multiple sexual ornaments in Satin Bowerbirds: ultraviolet plumage and bowers signal different aspects of male quality. Behav Ecol 14:503–509

Dunn PO, Hannon SJ (1991) Intraspecific competition and the maintenance of monogamy in Tree Swallows. Behav Ecol 2:258–266

Dunn PO, Garvin JC, Whittingham LA, Freeman-Gallant CR, Hasselquist D (2010) Carotenoid and melanin-based ornaments signal similar aspects of male quality in two populations of the Common Yellowthroat. Funct Ecol 24:149–158

Durães R, Loiselle BA, Blake JG (2007) Intersexual spatial relationships in a lekking species: Blue-crowned Manakins and female hot spots. Behav Ecol 18:1029–1039

Durães R, Loiselle BA, Parker PG, Blake JG (2009) Female mate choice across spatial scales: influence of lek and male attributes on mating success of Blue-crowned Manakins. Proc R Soc B 276:1875–1881

DuVal EH (2007a) Cooperative display and lekking behavior of the Lance-tailed Manakin (*Chiroxiphia lanceolata*). Auk 124:1168–1185

DuVal EH (2007b) Adaptive advantages of cooperative courtship for subordinate male Lance-tailed Manakins. Am Nat 169:423–432

DuVal EH (2012) Variation in annual and lifetime reproductive success of Lance-tailed Manakins: alpha experience mitigates effects of senescence on siring success. Proc R Soc B 279:1551–1559

DuVal EH, Kempenaers B (2008) Sexual selection in a lekking bird: the relative opportunity for selection by female choice and male competition. Proc R Soc B 275:1995–2003

Eason DK, Elliott GP, Merton DV, Jansen PW, Harper GA, Moorhouse RJ (2006) Breeding biology of Kakapo (*Strigops habroptilus*) on offshore Island sanctuaries, 1990-2002. Notornis 53:27–36

Edward DA, Chapman T (2011) The evolution and significance of male mate choice. Trends Ecol Evol 26:647–654

Ekstrom JMM, Burke T, Randrianaina L, Birkhead TR (2007) Unusual sex roles in a highly promiscuous parrot: the Greater Vasa Parrot *Caracopsis vasa*. Ibis 149:313–320

Ellegren H, Smeds L, Burri R, Olason PI, Backström N, Kawakami T, Künstner A, Mäkinen H, Nadachowska-

Brzyska K, Qvarnström A, Uebbing S, Wolf JBW (2012) The genomic landscape of species divergence in *Ficedula* flycatchers. Nature 491:756–760
Emlen ST, Oring LW (1977) Ecology, sexual selection, and the evolution of mating systems. Science 197:215–223
Emlen ST, Wrege PH (1991) Breeding biology of White-fronted Bee-eaters at Nakuru: the influence of helpers on breeder fitness. J Anim Ecol 60:309–326
Endler JA, Théry M (1996) Interacting effects of lek placement, display behavior, ambient light, and color patterns of three Neotropical forest-dwelling birds. Am Nat 148:421–452
Endler JA, Endler LC, Doerr NR (2010) Great Bowerbirds create theaters with forced perspective when seen by their audience. Curr Biol 20:1679–1684
Erckmann WJ Jr (1981) The evolution of sex-role reversal and monogamy in shorebirds, PhD diss. Univ Washington, Seattle, WA
Erckmann WJ (1983) The evolution of polyandry in shorebirds: an evaluation of hypotheses. In: Wasser SK (ed) Social behavior of female vertebrates. Academic Press, New York, pp 114–168
Evangelista D, Cardona G, Guenther-Gleason E, Huynh T, Kwong A, Marks D, Ray N, Tisbe A, Tse K, Koehl M (2014) Aerodynamic characteristics of a feathered dinosaur measured using physical models. Effects of form on static stability and control effectiveness. PLoS One 9:e85203
Evens J, Harper C (2020) Lethal intraspecific behavior by Anna's Hummingbird. Northwest Nat 101:221–222
Faaborg J, de Vries T, Patterson CB, Griffin CR (1980) Preliminary observations on the occurrence and evolution of polyandry in the Galapagos Hawk (*Buteo galapagoensis*). Auk 97:581–590
Faaborg J, Parker PG, DeLay L, de Vries T, Bednarz JC, Paz SM, Naranjo J, Waite TA (1995) Confirmation of cooperative polyandry in the Galapagos Hawk (*Buteo galapagoensis*). Behav Ecol Sociobiol 36:83–90
Farrell LL, Burke T, Slate J, McRae SB, Lank DB (2013) Genetic mapping of the female mimic morph locus in the Ruff. BMC Genet 14:109
Feeney WE, Medina I, Somveille M, Heinsohn R, Hall ML, Mulder RA, Stein JA, Kilner RM, Langmore NE (2013) Brood parasitism and the evolution of cooperative breeding in birds. Science 342:1506–1508
Fernández GJ, Reboreda JC (1998) Effects of clutch size and timing of breeding on reproductive success of Greater Rheas. Auk 113:340–348
Fernández GJ, Reboreda JC (2003) Male parental care in Greater Rheas (*Rhea americana*) in Argentina. Auk 120:418–428
Ferretti V, Winkler DW (2009) Polygyny in the Tree Swallow *Tachycineta bicolor*: a result of the cost of searching for an unmated male. J Avian Biol 40:289–295
Fisher RA (1915) The evolution of sexual preference. Eugenics Review 7:184–192
Fiske P, Kålås JA (1995) Mate sampling and copulation behaviour of Great Snipe females. Anim Behav 49: 209–219
Fiske P, Kàlàs JA, Saether SA (1994) Correlates of male mating success in the lekking Great Snipe *(Gallinago media)*: results from a four-year study. Behav Ecol 5: 210–218
Fiske P, Kàlàs JA, Saether SA (1996) Do female Great Snipe copy each other's mate choice? Anim Behav 51: 1355–1362
Fiske P, Rintamaki P, Karvonen E (1998) Mating success in lekking males: a meta-analysis. Behav Ecol 9:328–338
Fitzpatrick S, Berglund A, Rosenqvist G (1995) Ornaments or offspring: costs to reproductive success restrict sexual selection processes. Biol J Linnean Soc 55:251–260
Forero MG, Tella JL, Donazar JA, Blanco G, Bertellotti M, Ceballos O (2001) Phenotypic assortative mating and within-pair sexual dimorphism and its influence on breeding success and offspring quality in Magellanic Penguins. Can J Zool 79:1414–1422
Formica VA, Gonser RA, Ramsay S, Tuttle EM (2004) Spatial dynamics of alternative reproductive strategies: the role of neighbors. Ecology 85:1125–1136
Forstmeier W, Nakagawa S, Griffith SC, Kempenaers B (2014) Female extra-pair mating: adaptation or genetic constraint? Trends Ecol Evol 29:456–464
Foster MS (1981) Cooperative behavior and social organization of the Swallow-tailed Manakin (*Chiroxiphia caudata*). Behav Ecol Sociobiol 9:167–177
Foster MS (1983) Disruption, dispersion, and dominance in lek-breeding birds. Am Nat 122:53–72
Francisco MR, Gibbs HL, Galetti PM Jr (2009) Patterns of individual relatedness at Blue Manakin (*Chiroxiphia caudate*) leks. Auk 126:47–53
Friedman NR, Remeš V (2016) Global geographic patterns of sexual size dimorphism in birds: support for a latitudinal trend? Ecography 39:17–25
Fusani L, Giordano M, Day LB, Schlinger BA (2007) High-speed video analysis reveals individual variability in the courtship displays of male Golden-collared Manakins. Ethology 113:964–972
Fusani L, Barske J, Day LD, Fuxjager MJ, Schlinger BA (2014) Physiological control of elaborate male courtship: female choice for neuromuscular systems. Neurosci Biobehav Rev 46:534–546
Gage MJG (2021) Sperm size evolution. Nature Ecol Evol 5:1064–1065
Galván I, Sanz JJ (2011) Mate-feeding has evolved as a compensatory energetic strategy that affects breeding success in birds. Behav Ecol 22:1088–1095
Galván I, Solano F (2009) The evolution of eu- and pheomelanic traits may respond to an economy of pigments related to environmental oxidative stress. Pigment Cell Melanoma Research 22:339–342
Galván I, Wakamatsu K, Camarero PR, Mateo R, Alonso-Alvarez C (2015) Low-quality birds do not display

high-quality signals: the cysteine-pheomelanin mechanism of honesty. Evolution 69:26–38
García-Borrón JC, Olivares Sánchez MC (2011) Biosynthesis of melanins. In: Borovanský J, Riley PA (eds) Melanins and melanosomes: biosynthesis, biogenesis, physiological, and pathological functions. Wiley-Blackwell, Weinheim, Germany, pp 87–116
Garcia-Del-Rey E, Kleven O, Lifjeld JT (2012) Extrapair paternity in insular African Blue Tits *Cyanistes teneriffae* is no less frequent than in continental Eurasian Blue Tits *Cyanistes caeruleus*. Ibis 154:862–867
Garcia-Navas V, Sanz JJ (2011) Females call the shots: breeding dispersal and divorce in Blue Tits. Behav Ecol 22:932–939
Garratt M, Brooks RC (2012) Oxidative stress and condition-dependent sexual signals: more than just seeing red. Proc R Soc B 279:312103130
Gerlach NM, McGlothlin JW, Parker PG, Ketterson ED (2012) Promiscuous mating produces offspring with higher lifetime fitness. Proc R Soc B 279:860–866
Gibson RM (1996) A reevaluation of hotspot settlement in lekking Sage Grouse. Anim Behav 52:993–1005
Gibson RM, Bradbury JW (1985) Sexual selection in lekking Sage Grouse: phenotypic correlates of male mating success. Behav Ecol Sociobiol 18:117–123
Gibson RM, Höglund J (1992) Copying and sexual selection. Trends Ecol Evol 7:229–232
Gibson RM, Bradbury JW, Vehrencamp SL (1991) Mate choice in lekking Sage Grouse revisited: the roles of vocal display, female site fidelity, and copying. Behav Ecol 2:165–180
Gibson RM, Pires D, Delaney KS, Wayne RK (2005) Microsatellite DNA analysis shows that Greater Sage Grouse leks are not kin groups. Mol Ecol 14:4453–4459
Goetz JE, McFarland KP, Rimmer CC (2003) Multiple paternity and multiple male feeders in Bicknell's Thrush (*Catharus bicknelli*). Auk 120:1044–1053
Gohli J (2014) Female promiscuity and genetic diversity in passerine birds, Ph.D. dissertation. University of Oslo, Oslo
Gohli J, Anmarkrud JA, Johnsen A, Kleven O, Borge T, Lifjeld LT (2013) Female promiscuity is positively associated with neutral and selected genetic diversity in passerine birds. Evolution 67:1406–1419
Goldizen AW, Buchan JC, Putland DA, Goldizen AR, Krebs EA (2000) Patterns of mate-sharing in a population of Tasmanian Native Hens *Gallinula mortierii*. Ibis 142:40–47
Göth A (2007) Mound and mate choice in a polyandrous megapode: females lay more and larger eggs in nesting mounds with the best incubation temperatures. Auk 124:253–263
Gould GM, Augustine JK (2020) Multiple signals predict male mating success in the lek-mating Lesser Prairie-Chicken (*Tympanuchus pallidicinctus*). Behav Ecol Sociobiol 74:137
Gowaty PA (1983) Male parental care and apparent monogamy among Eastern Bluebirds (*Sialia sialis*). Am Nat 121:149–157
Goymann W, Wingfield JC (2004) Competing females and caring males: sex steroids in African Black Coucals, *Centropus grillii*. Anim Behav 68:733–740
Goymann W, Wingfield JC (2014) Male-to-female testosterone ratios, dimorphism, and life history – what does it really tell us? Behav Ecol 25:685–699
Goymann W, Wittenzellner A, Schwabl I, Makomba M (2008) Progesterone modulates aggression in sex-role reversed African Black Coucals. Proc R Soc B 275:1053–1060
Gratson MW (1993) Sexual selection for increased male courtship and acoustic signals and against large male size at Sharp-tailed Grouse leks. Evolution 47:691–696
Gratson MW, Gratson GK, Bergerud AT (1991) Male dominance and copulation disruption do not explain variance in male mating success on Sharp-tailed Grouse leks. Behaviour 118:187–213
Gratto-Trevor CL, Fivizzani AJ, Oring LW, Cooke F (1990) Seasonal changes in gonadal steroids of a monogamous versus a polyandrous shorebird. Gen Comp Endocrinol 80:407–418
Graul WD, Derrickson SR, Mock DW (1977) The evolution of avian polyandry. Am Nat 111:812–816
Gray EM (1997) Female Red-winged Blackbirds accrue material benefits from copulating with extra-pair males. Anim Behav 53:625–639
Greenberg R, Droege S (1990) Adaptations to tidal marshes in breeding populations of the Swamp Sparrow. Condor 92:393–404
Greenwood PJ, Harvey PH (1982) The natal and breeding dispersal of birds. Annu Rev Ecol Syst 13:1–21
Gregory AJ, McNew LB, Prebyl TJ, Sandercock BK, Wisely SM (2011) Hierarchical modeling of lek habitats of Greater Prairie-Chickens. In: Sandercock BK, Martin K, Segelbacher G (eds) Ecology, conservation, and management of grouse, Studies in Avian Biology, no 39. University of California Press, Berkeley, CA, pp 21–32
Griffith SC (2000) High fidelity on islands: a comparative study of extra-pair paternity in passerine birds. Behav Ecol 11:265–273
Griffith SC, Owens IPF, Thuman KA (2002) Extra-pair paternity in birds: a review of interspecific variation and adaptive function. Mol Ecol 11:2195–2212
Griggio M, Valera F, Casas A, Pilastro P (2005) Males prefer ornamented females: a field experiment of male choice in the Rock Sparrow. Anim Behav 69:1243–1250
Grønstøl GB, Byrkjedal I, Fiksen Ø (2003) Predicting polygynous settlement while incorporating varying female competitive strength. Behav Ecol 14:257–267
Grønstøl GB, Lislevand T, Byrkjedal I (2014) Resident female Northern Lapwings *Vanellus vanellus* fight to exclude settlement by secondary females. Ibis 156:461–465

Guindre-Parker S, Love OP (2014) Revisiting the condition-dependence of melanin-based plumage. J Avian Biol 44:29–33

Guindre-Parker S, Gilchrist HG, Baldo S, Love OP (2013) Alula size signals male condition and predicts reproductive performance in an Arctic-breeding passerine. J Avian Biol 45:209–215

Hadjivasiliou Z, Pomiankowski A, Seymour RM, Lane N (2012) Selection for mitonuclear co-adaptation could favour the evolution of two sexes. Proc R Soc B 279: 1865–1872

Haig SM, Walters JR, Plissner JH (1994) Genetic evidence for monogamy in the cooperatively breeding Red-cockaded Woodpecker. Behav Ecol Sociobiol 34:295–303

Hämäläinen A, Alatalo RV, Lebigre C, Siitari H, Soulsburgy CD (2012) Fighting behaviour as a correlate of male mating success in Black Grouse *Tetrao tetrix*. Behav Ecol Sociobiol 66:1577–1586

Handford P, Mares MA (1985) The mating system of ratites and tinamous: an evolutionary perspective. Biol J Linn Soc 25:77–104

Hansson B, Bensch S, Hasselquist D (2000) Patterns of nest predation contribute to polygyny in the Great Reed Warbler. Ecology 81:319–328

Hartzler JE, Jenni DA (1988) Mate choice by female Sage Grouse. In: Bergerud AT, Gratson MW (eds) Adaptive strategies and population ecology of northern grouse. University of Minnesota Press, Minneapolis, MN, pp 240–269

Hasselquist D, Langefors A (1998) Variable social mating system in the Sedge Warbler, *Acrocephalus schoenobaenus*. Ethology 104:759–769

Hasson O, Stone L (2009) Male infertility, female fertility and extrapair copulations. Biol Rev 84:225–244

Hays H (1972) Polyandry in the Spotted Sandpiper. Living Bird 11:43–57

Heg D, van Treuren R (1998) Female-female cooperation in polygynous oystercatchers. Nature 391:687–691

Heg D, Ens BJ, Burke T, Jenkins L, Kruijt JP (1993) Why does the typically monogamous oystercatcher (*Haematopus ostralegus*) engage in extra-pair copulations? Behaviour 126:247–289

Heg D, Bruinzeel LW, Ens BJ (2003) Fitness consequences of divorce in the oystercatcher, *Haematopus ostralegus*. Anim Behav 66:175–184

Heindl M, Winkler H (2003) Vertical lek placement of forest-dwelling manakin species (Aves, Pipridae) is associated with vertical gradients of ambient light. Biol J Linn Soc 80:647–658

Heinsohn R, Legge S, Endler JA (2005) Extreme reversed sexual dimorphism in a bird without sex role reversal. Science 309:617–619

Heinsohn R, Ebert D, Legge S, Peakall R (2007) Genetic evidence for cooperative polyandry in reverse dichromatic *Eclectus* parrots. Anim Behav 74:1047–1054

Heinsohn R, Langmore NE, Cockburn A, Kokko H (2011) Adaptive secondary sex ratio adjustments via sex-specific infanticide in a bird. Curr Biol 20:1744–1747

Henry I, Antoniazza S, Dubey S, Simon C, Waldvogel C, Burri R, Roulin A (2013) Multiple paternity in polyandrous Barn Owls (*Tyto alba*). PLoS One 8:e80112

Heredia R, Donázar JA (1990) High frequency of polyandrous trios in an endangered population of Lammergeiers *Gypaetus barbatus* in northern Spain. Biol Conserv 53:163–171

Herrera AM, Shuster SG, Perriton CL, Cohn MJ (2013) Developmental basis of phallus reduction during bird evolution. Curr Biol 23:1065–1074

Hess BD, Dunn PO, Whittingham LA (2012) Females choose multiple mates in the lekking Greater Prairie-Chicken (*Tympanuchus cupido*). Auk 129:133–139

Hill GE (1990) Female house finches prefer colourful males: sexual selection for a condition-dependent trait. Anim Behav 40:563–572

Hill WL (1991) Correlates of male mating success in the Ruff *Philomachus pugnax*, a lekking shorebird. Behav Ecol Sociobiol 29:367–372

Hingrat Y, Saint Jalme M, Chalah T, Orhant N, Lacroix F (2008) Environmental and social constraints on breeding site selection. Does the exploded-lek and hotspot model apply to the Houbara Bustard *Chlamydotis undulata undulata*? J Avian Biol 39:393–404

Höglund J, Alatalo RV (1995) Leks. Princeton University Press, Princeton, NJ

Höglund J, Lundberg A (1987) Sexual selection in a monomorphic lek-breeding bird: correlates of male mating success in the Great Snipe *Gallinago media*. Behav Ecol Sociobiol 21:211–216

Höglund J, Montgomerie R, Widemo F (1993) Costs and consequences of variation in the size of Ruff leks. Behav Ecol Sociobiol 32:31–39

Höglund J, Alatalo RV, Gibson RM, Lundberg A (1995) Mate-choice copying in Black Grouse. Anim Behav 49:1627–1633

Höglund J, Alatalo RV, Lundberg A, Rintamäki PT, Lindell J (1999) Microsatellite markers reveal the potential for kin selection on Black Grouse leks. Proc R Soc B 266:813–816

Hoi H, Schleicher B, Valera F (1994) Female mate choice and nest desertion in Penduline Tits, *Remiz pendulinus*: the importance of nest quality. Anim Behav 48:743–746

Holmstöm KM, Finkel T (2014) Cellular mechanisms and physiological consequences of redox-dependent signaling. Nat Rev Mol Cell Biol 15:411–421

Honza M, Piálková R, Albrecht T, Neužilová Š (2009) Nest defence against conspecific intruders in the Common Pochard *Aythya ferina*: natural observations and an experimental test. Acta Ornithol 44:151–158

Hõrak P, Sild E, Soomets U, Sepp T, Kilk K (2010) Oxidative stress and information content of black and yellow plumage coloration: an experiment with Greenfinches. J Exp Biol 213:2225–2233

Hovi M, Alatalo RV, Hoglund J, Lundberg A, Pintamaki PT (1994) Lek centre attracts Black Grouse females. Proc R Soc B 238:303–305

Hovi M, Alatalo RV, Siikamäki P (1995) Black Grouse leks on ice: female mate sampling by incitation of male competition? Behav Ecol Sociobiol 37:283–288

Hoyt DF (1979) Practical methods of estimating volume and fresh weight of bird eggs. Auk 96:73–77

Hsu Y-H, Schroeder J, Winney I, Burke T, Nakagawa S (2014) Costly infidelity: low lifetime fitness of extra-pair offspring in a passerine bird. Evolution 68:2873–2884

Huk T, Winkel W (2006) Polygyny and its fitness consequences for primary and secondary female Pied Flycatchers. Proc R Soc B 273:1681–1688

Hunt W (1998) Raptor floaters at Moffat's equilibrium. Oikos 82:191–197

Husak JF, Fox SF (2008) Sexual selection on locomotor performance. Evol Ecol Res 10:213–228

Hutchinson JMC (2005) Is more choice always desirable? Evidence and arguments from leks, food selection, and environmental enrichment. Biol Rev 80:73–92

Innes KE, Johnston RE (1996) Cooperative breeding in the White-throated Magpie-Jay. How do auxiliaries influence nesting success? Anim Behav 51:519–533

Isvaran K, Ponkshe A (2013) How general is a female mating preference for clustered males in lekking species? A meta-analysis. Anim Behav 86:417–425

IUCN (International Union for Conservation of Nature) (2019) Guidelines for the IUCN red list categories and criteria. Version 13. IUCN, Gland, Switzerland

Iwasaki A, Medzhitov R (2010) Regulation of adaptive immunity by the innate immune system. Science 327: 291–295

Iyer P, Roughgarden J (2008) Dioecy as a specialization promoting sperm delivery. Evol Ecol Res 10:867–892

Jamieson IG, Craig JL (1987) Dominance and mating in a communal polygynandrous bird: cooperation or indifference towards mating competitors? Ethology 75: 317–327

Jamieson IG, Quinn JS, Rose PA, White BN (1994) Shared paternity among non-relatives is a result of an egalitarian mating system in a communally breeding bird, the Pukeko. Proc R Soc B 257:271–277

Janssen MH, Arcese P, Sloan MS, Jewell KJ (2008) Polyandry and sex ratio in the Song Sparrow. Wilson J Ornithol 120:395–398

Jawor JM, Breitwisch R (2003) Melanin ornaments, honesty, and sexual selection. Auk 120:249–265

Jenni DA (1974) Evolution of polyandry in birds. Am Zool 14:129–144

Jetz W, Rubenstein DR (2011) Environmental uncertainty and the global biogeography of cooperative breeding in birds. Curr Biol 21:72–78

Jiguet F, Bretagnolle V (2006) Manipulating lek size and composition using decoys. Am Nat 168:758–768

Jiguet F, Bretagnolle V (2014) Sexy males and choosy females on exploded leks: correlates of male attractiveness in the Little Bustard. Behav Process 103:246–256

Johnsen A, Lifjeld JT, Krokene C (2003) Age-related variation in mate-guarding intensity in the Bluethroat (*Luscinia s. svecica*). Ethology 109:147–158

Johnson SL, Gemmell NJ (2012) Are old males still good males and can females tell the difference? BioEssays 34:609–619

Johnson LS, Merkle MS, Kermott LH (1992) Experimental evidence for importance of male parental care in monogamous House Wrens. Auk 109:662–664

Jones IL, Hunter FM (1993) Mutual sexual selection in a monogamous seabird. Nature 362:238–239

Jones SG, Paletz EM, Obermeyer WH, Hannan CT, Benca RM (2010) Seasonal influences on sleep and executive function in the migratory White-crowned Sparrow (*Zonotrichia leucophrys gambelii*). BMC Neurosci 11:87

Joste N, Ligon JD, Stacey PB (1985) Shared paternity in the Acorn Woodpecker (*Melanerpes formicivorus*). Behav Ecol Sociobiol 17:39–41

Jukema J, Piersma T (2006) Permanent female mimics in a lekking shorebird. Biol Lett 2:161–164

Juola FA, Dearborn DC (2012) Sequence-based evidence for major histocompatibility complex-disassortative mating in a colonial seabird. Proc R Soc B 279:153–162

Kalmbach E, Benito MM (2007) Sexual size dimorphism and offspring vulnerability in birds. In: Fairbairn DJ, Blanckenhorn WU, Székely T (eds) Sex, size and gender roles. Oxford University Press, Oxford, UK, pp 133–142

Kasprzykowski Z, Polak M (2013) Habitat quality and breeding parameters in relation to female mating status in the polygynous Eurasian Bittern *Botaurus stellaris*. J Ornithol 154:403–409

Kempenaers B (1994) Polygyny in the blue tit: unbalanced sex ratio and female aggression restrict mate choice. Anim Behav 47:943–957

Kempenaers B (2007) Mate choice and genetic quality: a review of the heterozygosity theory. Adv Study Behav 37:189–278

Kempenaers B, Valcu M (2017) Breeding site sampling across the Arctic by individual males of a polygynous shorebird. Nature 541:528–531

Kempenaers B, Adriaensen F, Dhondt AA (1998) Inbreeding and divorce in Blue and Great tits. Anim Behav 56:737–740

Kempenaers B, Everding S, Bishop C, Boag P, Robertson RJ (2001) Extra-pair paternity and the reproductive role of male floaters in the Tree Swallow (*Tachycineta bicolor*). Behav Ecol Sociobiol 49:251–259

Kepler AK (1977) Comparative study of todies (Todidae), with emphasis on the Puerto Rican Tody *Todus mexicanus*. Publications of the Nuttall Ornithological Club no. 16, Cambridge, MA

Kervinen M, Alatalo RV, Lebigre C, Siitari H, Soulsbury CD (2012) Determinants of yearling male lekking effort and mating success in Black Grouse (*Tetrao tetrix*). Behav Ecol 23:1209–1217

Keyser AJ, Hill GE (2000) Structurally based plumage coloration is an honest signal of quality in male Blue Grosbeaks. Behav Ecol 11:202–209

Khan MZ, Walters JR (2002) Effects of helpers on breeder survival in the Red-cockaded Woodpecker (*Picoides borealis*). Behav Ecol Sociobiol 51:336–344

Kikusato M, Toyomizu M (2013) Crucial role of membrane potential in heat-stress induced overproduction of reactive oxygen species in avian skeletal muscle mitochondria. PLoS ONE 8:e64412

Kimball RT, Parker PG, Bednarz JC (2003) Occurrence and evolution of cooperative breeding among the diurnal raptors (Accipitridae and Falconidae). Auk 120: 717–729

Kimwele CN, Graves JA (2003) A molecular genetic analysis of the communal nesting of the Ostrich (*Struthio camelus*). Mol Ecol 12:229–236

Kingma SA, Hall ML, Arriero E, Peters A (2010) Multiple benefits of cooperative breeding in Purple-crowned Fairy-wrens: a consequence of fidelity? J Anim Ecol 79:757–768

Kingma SA, Hall ML, Peters A (2013) Breeding synchronization facilitates extrapair mating for inbreeding avoidance. Behav Ecol 24:1390–1397

Kleiber DL (2008) Female intrasexual reproductive competition in the facultatively polygynous Song Sparrow. M. S. thesis, University of British Columbia, Vancouver, BC, Canada

Klein J (1986) Natural history of the major histocompatibility complex. John Wiley and Sons, New York, NY

Kleven O, Fossøy F, Laskemoen T, Robertson RJ, Rudolfsen G, Lifjeld JT (2009) Comparative evidence for the evolution of sperm swimming speed by sperm competition and female sperm storage duration in passerine birds. Evolution 63:2466–2473

Koenig WD (1990) Opportunity of parentage and nest destruction in polygynandrous Acorn Woodpeckers, *Melanerpes formidvorus*. Behav Ecol 1:55–61

Koenig WD, Mumme RL, Stanback MT, Pitelka FA (1995) Patterns and consequences of egg destruction among joint-nesting Acorn Woodpeckers. Anim Behav 50:607–621

Kokko H, Lindström J (1997) Measuring the mating skew. Am Nat 149:794–799

Kokko H, Sutherland WJ (1998) Optimal floating and queuing strategies: consequences for density dependence and habitat loss. Am Nat 152:354–366

Kokko H, Johnstone RA, Clutton-Brock TH (2001) The evolution of cooperative breeding through group augmentation. Proc R Soc B 268:187–196

Kokko H, Johnstone RA, Wright J (2002) The evolution of parental and alloparental effort in cooperatively breeding groups: when should helpers pay to stay? Behav Ecol 13:291–300

Kokko H, Brooks R, Jennions MD, Morley J (2003) The evolution of mate choice and mating biases. Proc R Soc B 270:653–664

Komdeur J (1992) Importance of habitat saturation and territory quality for evolution of cooperative breeding in the Seychelles Warbler. Nature 358:493–495

Komdeur J, Burke T, Richardson DS (2007) Explicit experimental evidence for the effectiveness of proximity as mate-guarding behaviour in reducing extra-pair fertilization in Seychelles Warbler. Mol Ecol 16:3679–3688

Komeda S (1983) Nest attendance of parent birds in the Painted Snipe (*Rostratula benghalensis*). Auk 100:48–55

Korpimäki E, Salo P, Valkama J (2011) Sequential polyandry by brood desertion increases female fitness in a bird with obligatory bi-parental care. Behav Ecol Sociobiol 65:1093–1102

Kosztolányi A, Barta Z, Küpper C, Székely T (2011) Persistence of an extreme male-biased adult sex ratio in a natural population of polyandrous bird. J Evol Biol 24:1842–1846

Kotaka N (1998) Classical polyandry in the Great Spotted Woodpecker *Dendrocopos major*. Ibis 140:335–336

Krakauer A (2005) Kin selection and cooperative courtship in Wild Turkeys. Nature 434:69–72

Krebs JR, Davies NB (2009) An introduction to behavioural ecology, 3rd edn. John Wiley and Sons, Hoboken, NJ

Krokene C, Anthonisen K, Lifjeld JT, Amundsen T (1996) Paternity and paternity assurance behaviour in the Bluethroat, *Luscinia s. svecica*. Anim Behav 52:405–417

Kuijper B, Pen I, Weissing FJ (2012) A guide to sexual selection theory. Annu Rev Ecol Syst 43:287–311

Kvarnemo C, Simmons LW (2013) Polyandry as a mediator of sexual selection before and after mating. Philos Trans R Soc B 368:20120042

LaBarbera K, Llambias PE, Cramer ERA, Schaming TD, Lovette IJ (2010) Synchrony does not explain extrapair paternity rate variation in northern or southern House Wrens. Behav Ecol 21:773–780

Lack D (1968) Ecological adaptations for breeding in birds. Chapman and Hall, London

Lampe HM, Saetre G-P (1995) Female Pied Flycatchers prefer males with larger song repertoires. Proc R Soc B 262:163–167

Lancaster DA (1964) Life history of the Boucard Tinamou in British Honduras. Part II: Breeding biology. Condor 66:253–276

Lanctot RB, Scribner KT, Kempenaers B, Weatherhead PJ (1997) Lekking without a paradox in the Buff-breasted Sandpiper. Am Nat 149:1051–1070

Lanctot RB, Weatherhead PJ, Kempenaers B, Scribner KT (1998) Male traits, mating tactics and reproductive success in the Buffbreasted Sandpiper, *Tryngites subruficollis*. Anim Behav 56:419–432

Lande R (1980) Sexual dimorphism, sexual selection, and adaptation in polygenic characters. Evolution 34: 292–305

Lank DB, Smith CM, Hanotte O, Ohtonen A, Bailey S, Burke T (2002) High frequency of polyandry in a lek mating system. Behav Ecol 13:209–215

LeBas NR (2006) Female finery is not for males. Trends Ecol Evol 21:170–173

Lebigre C, Alatalo RV, Siitari H, Parri S (2007) Restrictive mating by females on Black Grouse leks. Mol Ecol 16:4380–4389
Lebigre C, Alatalo RV, Forss HE, Siitari H (2008) Low levels of relatedness on Black Grouse leks despite male philopatry. Mol Ecol 17:4512–4521
Lee AM, Reid JM, Beissinger SR (2017) Modelling effects of nonbreeders on population growth estimates. J Anim Ecol 86:75–87
Leisler B, Winkler H, Wink M (2002) Evolution of breeding systems in *Acrocephaline* warblers. Auk 119:379–390
Lendvai ÁZ, Chastel O (2008) Experimental mate-removal increases the stress response of female House Sparrows: the effects of offspring value? Horm Behav 53:395–401
Lengvel S, Kiss B, Tracy CR (2009) Clutch size determination in shorebirds: revisiting incubation limitation in the Pied Avocet (*Recurvirostra avosetta*). J Anim Ecol 78:396–405
Lesku JA, Rattenborg NC, Valcu M, Vyssotski AL, Kuhn S, Kuemmeth F, Heidrich W, Kempenaers B (2012) Adaptive sleep loss in polygynous Pectoral Sandpipers. Science 337:1654–1658
Lesobre L, Lacroix F, Le Nuz E, Hingrat Y, Chalah T, Saint Jaime M (2010) Absence of male reproductive skew, along with high frequency of polyandry and conspecific brood parasitism in the lekking Houbara Bustard *Chlamydotis undulata undulata*. J Avian Biol 41:117–127
Li Q, Gao K-Q, Meng Q, Clarke JA, Shawkey MD, D'Alba L, Pei R, Ellison M, Norell MA, Vinther J (2012) Reconstruction of *Microraptor* and the evolution of iridescent plumage. Science 335:1215–1219
Ligon JD (1999) The evolution of avian mating systems. Oxford University Press, Oxford, UK
Ligon JD, Burt DB (2004) Evolutionary origins. In: Koenig WD, Dickinson JL (eds) Ecology and evolution of cooperative breeding in birds. Cambridge Univ Press, Cambridge, pp 5–34
Liker A, Székely T (2005) Mortality costs of sexual selection and parental care in natural populations of birds. Evolution 59:890–897
Liker A, Freckleton RP, Székely T (2013) The evolution of sex roles in birds is related to adult sex ratio. Nat Commun 4:1587
Liker A, Freckleton RP, Székely T (2014) Divorce and infidelity are associated with skewed adult sex ratios in birds. Curr Biol 24:880–884
Lill A (1974) Sexual behavior of the lek-forming White-bearded Manakin (*Manacus manacus trinitatis Hartert*). Z Tierpsychol 36:1–36
Lill A (1976) Lek behavior in the Golden-headed Manakin, Pipra erythrocephala, in Trinidad (West Indies). Fortschr Verhaltensforsch 18:1–84
Lloyd L (1867) The game birds and wild fowl of Sweden and Norway. F. Warne & Company, London
Loiselle BA, Ryder TR, Durães R, Tori W, Blake JG, Parker PG (2007) Kin selection does not explain male aggregation at leks of 4 manakin species. Behav Ecol 18:287–291
Løvlie H, Gillingham MAF, Worley K, Pizzari T, Richardson DS (2013) Cryptic female choice favours sperm from major histocompatibility complex-dissimilar males. Proc R Soc B 280:20131296
Loyau A, Jalme MS, Sorci G (2005) Intra- and intersexual selection for multiple traits in the peacock (*Pavo cristatus*). Ethology 111:810–820
Ludwigs J-D, Becker PH (2007) Is divorce in young Common Terns, *Sterna hirundo*, after recruitment just a question of timing? Ethology 113:46–56
Lukianchuk KC, Doucet SM (2014) Cooperative courtship display in Long-tailed Manakins *Chiroxiphia linearis*: predictors of courtship success revealed through full characterization of display. J Ornithol 155:729–743
Lynn SE, Wingfield JC (2003) Male Chestnut-collared Longspurs are essential for nestling survival: a removal study. Condor 105:154–158
Lyon BE, Eadie JM (2008) Conspecific brood parasitism in birds: a life-history perspective. Annu Rev Ecol Evol Syst 39:343–363
MacColl ADC, Hatchwell BJ (2003) Sharing of caring: nestling provisioning behavior of Long-tailed Tit, *Aegithalos caudatus*, parents and helpers. Anim Behav 66:955–964
MacDougall-Shackleton E, Harbison H (1998) Singing behavior of lekking Green Hermits. Condor 100:149–152
Macedo RHF, Cariello MO, Graves J, Schwabl H (2004) Reproductive partitioning in communally breeding Guira Cuckoos, *Guira guira*. Behav Ecol Sociobiol 55:213–222
MacManes MD (2013) On the accurate description of social and genetic mating systems. PeerJ Preprint. https://doi.org/10.7287/peerj.preprints.10
Madden JR, Dingle C, Isden J, Sparfeld J, Goldizen AW, Endler JA (2012) Male Spotted Bowerbirds propagate fruit for use in their sexual display. Curr Biol 22:R264–R265
Mader WJ (1979) Breeding behavior of a polyandrous trio of Harris' Hawks in southern Arizona. Auk 96:776–788
Magaña M, Alonso JC, Palacín C (2011) Age-related dominance helps reduce male aggressiveness in Great Bustard leks. Anim Behav 82:203–211
Magrath RD, Whittingham LA (1997) Subordinate males are more likely to help if unrelated to the breeding female in cooperatively breeding White-browed Scrubwrens. Behav Ecol Sociobiol 41:185–192
Malan G, Crowe TM (1996) The diet and conservation of monogamous and polyandrous Pale Chanting Goshawks in the Little Karoo, South Africa. S Afr J Wildl Res 26:1–10
Mariette MM, Griffith SC (2012) Nest visit synchrony is high and correlates with reproductive success in the wild Zebra Finch *Taeniopygia guttata*. J Avian Biol 43:131–140

Martella MB, Renny M, Chiappero MB, Navarro JL (2022) Promiscuity in the Greater Rhea: a genetic approach. Acta Ethologica 25:155–164

Martinez-Garcia V, Lara C, Ornelas JF (2013) Lek characteristics and the static male song of the Green Violet-ear (*Colibri thalassinus*) during a 3-year study in a temperate forest of Central Mexico. Ornitologia Neotropical 24:183–200

Matysioková B, Remeš V (2013) Faithful females receive more help: the extent of male parental care during incubation in relation to extra-pair paternity in songbirds. J Evol Biol 26:155–162

Mauck RA, Marschall EA, Parker PG (1999) Adult survival and imperfect assessment of parentage: effects on male parenting decisions. Am Nat 154:99–109

Mayer C, Pasinelli G (2013) New support for an old hypothesis: density affects extra-pair paternity. Ecol Evol 3:694–705

Maynard DF, Ward K-A, Doucet SM, Mennill DJ (2014) Telemetric and video assessment of female response to male vocal performance in a lek-mating manakin. Behav Ecol 26:65–74

Mays HL Jr, Ritchison G (2004) The effect of vegetation density on male mate guarding and extra-territorial forays in the Yellowbreasted Chat (*Icteria virens*). Naturwissenschaften 91:195–198

McDonald DB (1989) Correlates of male mating success in a lekking bird with male-male cooperation. Anim Behav 37:1007–1022

McDonald DB, Potts WK (1994) Cooperative display and relatedness among males in a lek-mating bird. Science 266:1030–1032

McDonald GC, Cuthill IC, Székely T, Kosztolányi A (2023) Remating opportunities and low costs underlie maternal desertion. Evolution 77:97–109

McGraw KJ (2008) An update on the honesty of melanin-based color signals in birds. Pigment Cell Melanoma Research 21:133–138

McKinney F (1992) Courtship, pair formation, and signal systems. In: Batt BDJ (ed) Ecology and management of breeding waterfowl. Univ Minn Press, Minneapolis, MN, pp 387–422

Middleton ALA (1988) Polyandry in the mating system of the American Goldfinch, *Carduelis tristis*. Can J Zool 66:296–299

Møller AP (1992) Female swallow preference for symmetrical male sexual ornaments. Nature 357:238–240

Monaghan P, Nager RG, Houston DC (1998) The price of eggs: increased investment in egg production reduces the offspring rearing capacity of parents. Proc R Soc B 265:1731–1735

Monaghan P, Metcalfe NB, Torres R (2009) Oxidative stress as a mediator of life history trade-offs: mechanisms, measurements and interpretation. Ecol Lett 12:75–92

Monaghan P, Heidinger BJ, D'Alba L, Evans NP, Spencer KA (2011) For better or worse: reduced adult lifespan following early-life stress is transmitted to breeding partners. Proc R Soc B 279:709–714

Montoya R, Torres R (2015) Male skin color signals direct and indirect benefits in a species with biparental care. Behav Ecol 26:425–434

Moore JR, Varricchio DJ (2016) The evolution of diapsid reproductive strategy with inferences about extinct taxa. PLoS ONE 11:e0158496

Morales MB, Jiguet F, Arroyo B (2001) Exploded leks: what bustards can teach us. Ardeola 48:85–98

Moreno J, Velando A, González-Braojos S, Ruiz-de-Castañeda R, Cantarero A (2013) Females paired with more attractive males show reduced oxidative damage: possible direct benefits of mate choice in Pied Flycatchers. Ethology 119:727–737

Morrow EH, Arnqvist G, Pitcher TE (2002) The evolution of infertility: does hatching rate in birds coevolve with female polyandry? J Evol Biol 15:702–709

Moyles DLJ, Boag DA (1981) Where, when, and how male Sharp-tailed Grouse establish territories on arenas. Can J Zool 59:1576–1581

Muck C (2011) Behavioral and hormonal aspects of the breeding ecology in sex-role reversed polyandrous birds. Ph. D. dissertation, Ludwig Maximilian University of Munich, Munich, Germany

Muck C, Kempenaers B, Kuhn S, Valcu M, Goymann W (2009) Paternity in the classical polyandrous Black Coucal (*Centropus grillii*) – a cuckoo accepting cuckoldry? Behav Ecol 20:1185–1193

Mulder RA, Dunn PO, Cockburn A, Lazenby-Cohen KA, Howell MJ (1994) Helpers liberate female fairy-wrens from constraints on extra-pair mate choice. Proc R Soc B 255:223–229

Mumme RL (2015) Demography of Slate-throated Redstarts (*Myioborus miniatus*): a non-migratory Neotropical warbler. J Field Ornithol 86:89–102

Murphy TG, Hernández-Muciño D, Osorio-Beristain M, Montgomerie R, Omland KE (2009a) Carotenoid-based status signaling by females in the tropical Streak-backed Oriole. Behav Ecol 20:1000–1006

Murphy TG, Rosenthal MF, Montgomerie R, Tarvin KA (2009b) Female American Goldfinches use carotenoid-based bill coloration to signal status. Behav Ecol 20:1348–1355

Murray BG Jr (1984) A demographic theory on the evolution of mating systems as exemplified by birds. In: Hecht MK, Wallace B, Prance GT (eds) Evolutionary biology, vol. 18. Plenum Press, New York, NY, pp 71–140

Nager RG (2006) The challenges of making eggs. Ardea 94:323–346

Newton I (1979) Population ecology of raptors. T & AD Poyser Ltd., London, UK

Newton I (1988) Population regulation in peregrines: an overview. In: Cade TJ, Enderson JH, Thelander CG, White CM (eds) Peregrine Falcon populations: their management and recovery. The Peregrine Fund Inc., Boise, ID, pp 761–770

Nisbet ICT, Dann P (2009) Reproductive performance of Little Penguins *Eudyptula minor* in relation of year, age, pair-bond duration, breeding date and individual quality. J Avian Biol 40:296–308

Nisbet ICT, Szczys P, Mostello CS, Fox JW (2011) Female Common Terns *Sterna hirundo* start autumn migration earlier than males. Seabird 24:103–107

Nolan PM, Dobson FS, Nicolaus M, Karels TJ, McGraw KJ, Jouventin P (2010) Mutual mate choice for colorful traits in King Penguins. Ethology 116:635–644

Nooker JK, Sandercock BK (2008) Phenotypic correlates and survival consequences of male mating success in lek-mating Greater Prairie-Chickens (*Tympanuchus cupido*). Behav Ecol Sociobiol 62:1377–1388

Nordeide JT, Kekäläinen J, Janhunen M, Kortet R (2013) Female ornaments revisited – are they correlated with offspring quality? J Anim Ecol 82:26–38

Nordell SE, Valone TJ (1998) Mate choice copying as public information. Ecol Lett 1:74–76

Nowicki S, Peters S, Podos J (1998) Song learning, early nutrition and sexual selection in songbirds. Am Zool 38:179–190

Nowicki S, Hasselquist D, Bensch S, Peters S (2000) Nestling growth and song repertoire size in Great Reed Warblers: evidence for song learning as an indicator mechanism in mate choice. Proc R Soc B 267:2419–2424

Nur N (1984) Increase reproductive success with age in the California Gull: due to increased effort or improvement of skill? Oikos 43:407–408

O'Connor J, Wang X, Sullivan C, Zheng X, Tubaro P, Zhang X, Zhou Z (2013) Unique caudal plumage of *Jeholornis* and complex tail evolution in early birds. Proc Natl Acad Sci USA 110:17404–17408

O'Donnell CFJ (1996) Predators and the decline of New Zealand forest birds: an introduction to the hole-nesting bird and predator programme. New Zealand J Zool 23:213–219

Ohlsson T, Smith HG, Räberg L, Hasselquist D (2002) Pheasant sexual ornaments reflect nutritional conditions in early growth. Proc R Soc B 269:21–27

Olano-Marin J, Mueller JC, Kempenaers B (2011) Heterozygosity and survival in Blue Tits (*Cyanistes caeruleus*): contrasting effects of presumably functional and neutral loci. Mol Ecol 20:4028–4041

Olsen BJ, Greenberg R, Walters JR, Fleischer RC (2013) Sexual dimorphism in a feeding apparatus is driven by mate choice and not niche partitioning. Behav Ecol 24: 1327–1338

Olson VA, Liker A, Freckleton RP, Székely T (2008) Parental conflict in birds: comparative analyses of offspring development, ecology and mating opportunities. Proc R Soc B 275:301–307

Orians GH (1969) On the evolution of mating systems in birds and mammals. Am Nat 103:589–603

Oring LW (1982) Avian mating systems. Avian Biology 6: 1–92

Owens IPF, Burke T, Thompson DBA (1994) Extraordinary sex roles in the Eurasian Dotterel: female mating arenas, female-female competition, and female mate choice. Am Nat 144:76–100

Ownes IPF (2002) Male-only care and classical polyandry in birds: phyoleny, ecology and sex differences in remating opportunities. Philos Trans R Soc B 357: 283–293

Palestis BG, Nisbet ICT, Hatch JJ, Arnold JM, Szczys P (2012) Tail length and sexual selection in a monogamous, monomorphic species, the Roseate Tern *Sterna dougallii*. J Ornithol 153:1153–1163

Panieri E, Santoro MM (2016) ROS homeostasis and metabolism: a dangerous liason in cancer cells. Cell Death Dis 7:e2253

Paquet M, Covas R, Chastel O, Parenteau C, Doutrelant C (2013) Maternal effects in relation to helper presence in the cooperatively breeding Sociable Weaver. PLoS ONE 8:e59336

Parker GA (1979) Sexual selection and sexual conflict. In: Blum MS, Blum NA (eds) Sexual selection and reproductive competition in insects. Academic Press, New York, NY, pp 123–166

Parker GA, Birkhead TR (2013) Polyandry: the history of a revolution. Philos Trans R Soc B 368:20120335

Parker GA, Baker RR, Smith VGF (1972) The origin and evolution of gamete dimorphism and the male-female phenomenon. J Theor Biol 36:529–553

Parmelee DF, Payne RB (1973) On multiple broods and the breeding strategy of Arctic Sanderlings. Ibis 115:218–226

Parra JE, Beltrán M, Zefania S, Remedios ND, Székely T (2014) Experimental assessment of mating opportunities in three shorebird species. Anim Behav 90:83–90

Pearson T, Whitfield MJ, Theimer TC, Keim P (2006) Polygyny and extra-pair paternity in a population of southwestern Willow Flycatchers. Condor 108:571–578

Pechacek P, Michalek KG, Winkler H, Blomqvist D (2005) Monogamy with exceptions: social and genetic mating system in a bird species with high paternal investment. Behaviour 142:1093–1114

Penteriani V, Otalora F, Ferrer M (2005) Floater survival affects population persistence: the role of prey availability and environmental stochasticity. Oikos 108:523–534

Penteriani V, Ferrer M, Delgado MM (2011) Floater strategies and dynamics in birds, and their importance in conservation biology: towards an understanding of nonbreeders in avian populations. Anim Conserv 14: 233–241

Pérez-Staples D, Osorio-Beristain M, Rodríguez C, Drummond J (2013) Behavioural roles in booby mate switching. Behaviour 150:337–357

Petrie M (1986) Reproductive strategies of male and female Moorhens (*Gallinula chloropus*). In: Rubenstein DI, Wrangham RW (eds) Ecological aspects of social evolution. Princeton University Press, Princeton, NJ, pp 43–63

Petrie M, Hall M, Halliday T, Budgey H, Pierpoint C (1992) Multiple mating in a lekking bird: why do peahens mate with more than one male and with the same male more than once? Behav Ecol Sociobiol 31: 349–358

Petrie M, Doums C, Møller AP (1998) The degree of extra-pair paternity increases with genetic variability. Proc Natl Acad Sci USA 95:9390–9395

Petrie M, Krupa A, Burke T (1999) Peacocks lek with relatives even in the absence of social and environmental cues. Nature 401:155–157

Pickett SRA, Weber SB, McGraw KJ, Norris KJ, Evans MR (2013) Environmental and parental influences on offspring health and growth in Great Tits (*Parus major*). PLoS ONE 8:e69695

Pilastro A, Biddau L, Martin G, Mingozzi T (2001) Female brood desertion increases with number of available mates in the Rock Sparrow. J Avian Biol 32:68–72

Pilastro A, Griggio M, Matessi G (2003) Male Rock Sparrows adjust their breeding strategy according to female ornamentation: parental or mating investment? Anim Behav 66:265–271

Piper WH, Parker PG, Rabenold KN (1995) Facultative dispersal by juvenile males in the cooperative Stripe-backed Wren. Behav Ecol 6:337–342

Pomiankowski A (1987) The costs of choice in sexual selection. J Theor Biol 128:195–218

Pomiankowski A (1990) How to find the top male. Nature 347:616–617

Prager M (2010) Phylogeny and signal diversity in widowbirds and bishops (*Euplectes* spp.). Ph.D. dissertation, University of Gothenburg, Göteborg, Sweden

Pribil S (2000) Experimental evidence for the cost of polygyny in the Red-winged Blackbird *Agelaius phoeniceus*. Behaviour 137:1153–1173

Pribil S, Searcy WA (2001) Experimental confirmation of the polygyny threshold model for Red-winged Blackbirds. Proc R Soc B 268:1643–1646

Pruett-Jones SG, Lewis MJ (1990) Sex ratio and habitat limitation promote delayed dispersal in Superb Fairy-wrens. Nature 348:541–542

Pruett-Jones SG, Pruett-Jones MA (1990) Sexual selection through female choice in Lawes' Parotia, a lek-mating bird of paradise. Evolution 44:486–501

Pryke SR, Andersson S, Lawes MJ (2001) Sexual selection of multiple handicaps in the Red-collared Widowbird: female choice of tail length but not carotenoid display. Evolution 55:1452–1463

Pugesek BH (1995) Offspring growth in the California Gull: reproductive effort and parental experience hypotheses. Anim Behav 49:641–647

Quader S (2006) What makes a good nest? Benefits of nest choice to female Baya Weavers (*Ploceus philippinus*). Auk 123:475–486

Quinn JS, Macedo R, White BN (1994) Genetic relatedness of communally breeding Guira Cuckoos. Anim Behav 47:515–529

Quinn JS, Haselmayer J, Dey C, Jamieson IG (2012) Tolerance of female co-breeders in joint-laying Pukeko: the role of egg recognition and peace incentives. Anim Behav 83:1035–1041

Rabenold PP, Rabenold KN, Piper WH, Haydock J, Zack SW (1990) Shared paternity revealed by genetic analysis in cooperatively breeding tropical wrens. Nature 348:538–540

Ramsay SM, Otter KA, Mennill DJ, Ratcliffe LM, Boag PT (2000) Divorce and extrapair mating in female Black-capped Chickadees (*Parus atricapillus*): separate strategies with a common target. Behav Ecol Sociobiol 49:18–23

Reding L (2015) Increased hatching success as a direct benefit of polyandry in birds. Evolution 69:264–270

Regnaut S, Christe P, Chapuisat M, Fumagalli L (2006) Genotyping faeces reveals facultative kin association on Capercaillie's leks. Conserv Genet 7:665–674

Remage-Healey L, Adkins-Regan E, Romero LM (2003) Behavioral and adrenocortical responses to mate separation and reunion in the Zebra Finch. Horm Behav 43:108–114

Reyer H-U (1984) Investment and relatedness: a cost/benefit analysis of breeding and helping in the Pied Kingfisher (*Ceryle rudis*). Anim Behav 32:1163–1178

Reynolds JD (1996) Animal breeding systems. Trends Ecol Evol 11:68–72

Rico-Guevara A, Araya-Salas M (2014) Bills as daggers? A test for sexually dimorphic weapons in a lekking hummingbird. Behav Ecol 26:21–29

Ridley AR, Raihani NJ (2008) Task partitioning increases reproductive output in a cooperative bird. Behav Ecol 19:1136–1142

Riede T, Forstmeier W, Kempenaers B, Goller F (2015) The functional morphology of male courtship displays in the Pectoral Sandpiper (*Calidris melanotos*). Auk 132:65–77

Riehl C (2010) A simple rule reduces costs of extragroup parasitism in a communally breeding bird. Curr Biol 20:1830–1833

Riehl C (2011) Living with strangers: direct benefits favour non-kin cooperation in a communally nesting bird. Proc R Soc B 278:1728–1735

Riehl C (2013) Evolutionary routes to non-kin cooperative breeding in birds. Proc R Soc B 280:20132245

Rolstad J, Wegge P, Sivkov AV, Hjeljord O, Storaunet KO (2009) Size and spacing of grouse leks: comparing Capercaillie (*Tetrao urogallus*) and Black Grouse (*Tetrao tetrix*) in two contrasting Eurasian boreal forest landscapes. Can J Zool 87:1032–1043

Rosselli L, Vasquez P, Ayub I (2002) The courtship displays and social system of the White-ruffed Manakin in Costa Rica. Wilson Bull 114:165–178

Rossmanith E, Blaum N, Höntsch K, Jeltsch J (2009) Sex-related parental care strategies in the Lesser Spotted Woodpecker *Picoides minor*: of flexible mothers and dependable fathers. J Avian Biol 40:28–33

Rowe KMC, Weatherhead PJ (2011) Assortative mating in relation to plumage traits shared by male and female American Robins. Condor 113:881–889

Rowe M, Van Oort A, Brouwer L, Lifjeld JT, Webster MS, Welklin JF, Baldassarre DT (2022) Sperm numbers as a paternity guard in a wild bird. Cell 11:231
Rushton D (1976) Plains-wanderer. In: Frith HJ (ed) Reader's digest complete book of Australian birds. Reader's Digest, Sydney, Australia
Russell AF, Langmore NE, Cockburn A, Astheimer LB, Kilner RM (2007) Reduced egg investment can conceal helper effects in cooperatively breeding birds. Science 317:941–944
Ryder TB, Blake JG, Loiselle BA (2006) A test for the environmental hotspot hypothesis for lek placement in three species of manakins (Pipridae) in Ecuador. Auk 123:247–258
Ryder TB, McDonald DB, Blake JG, Parker PG, Loiselle BA (2008) Social networks in the lek-mating Wire-tailed Manakin (*Pipra filicauda*). Proc R Soc B 275:1367–1374
Ryder TB, Parker PG, Blake JG, Loiselle BA (2009) It takes two to tango: reproductive skew and social correlates of male mating success in a lek-breeding bird. Proc Roy Soc B 276:2377–2384
Ryder TB, Blake JG, Parker PG, Loiselle BA (2011) The composition, stabililty, and kinship of reproductive coalitions in a lekking bird. Behav Ecol 22:282–290
Sæther SA, Fiske P, Kålås JA (1999) Pushy males and choosy females: courtship disruption and mate choice in the lekking Great Snipe. Proc R Soc B 266:1227–1234
Sæther SA, Fiske P, Kålås JA (2001) Male mate choice, sexual conflict and strategic allocation of copulations in a lekking bird. Proc R Soc B 268:2097–2102
Saino N, Romano M, Ambrosini R, Rubolini D, Boncoraglio G, Caprioli M, Romano A (2012) Longevity and lifetime reproductive success of Barn Swallow offspring are predicted by their hatching date and phenotypic quality. J Anim Ecol 81:1004–10012
Saitou T (2001) Floaters as intraspecific brood parasites in the Grey Starling *Sturnus cineraceus*. Ecol Res 16: 221–231
Sánchez-Macouzet O, Rodríguez C, Drummond H (2014) Better stay together: pair bond duration increases individual fitness independent of age-related variation. Proc R Soc B 281:20132843
Sandell MI (1998) Female aggression and the maintenance of monogamy: female behaviour predicts male mating status in European Starlings. Proc R Soc B 265:1307–1311
Sandell MI, Diemer M (1999) Intraspecific brood parasitism: a strategy for floating females in the European Starling. Anim Behav 57:197–202
Santos ESA, Macedo RH (2011) Load lightening in Southern Lapwings: group-living mothers lay smaller eggs than pair-living mothers. Ethology 117:547–555
Sarasola JH, Negro JJ, Bechard MJ, Lanusse A (2011) Not as similar as thought: sexual dichromatism in Chimango Caracaras is expressed in the exposed skin but not in the plumage. J Ornithol 152:473–479
Sardell RJ, DuVal EH (2014) Small and variable sperm sizes suggest low sperm competition despite multiple paternity in a lekking suboscine bird. Auk 131:660–671
Sardell RJ, Arcese P, Keller LF, Reid JM (2012) Are there indirect fitness benefits of female extra-pair reproduction? Lifetime reproductive success of within-pair and extra-pair offspring. Am Nat 179:779–793
Sargeant AB, Allen SH, Eberhardt RT (1984) Red fox predation on breeding ducks in mid-continent North America. Wildl Monogr 89:1–41
Schleicher B, Hoi H, Valera F, Hoi-Leitner M (1997) The importance of different paternity guards in the polygynandrous Penduline Tit (*Remiz pendulinus*). Behaviour 134:941–959
Schlinger BA, Barske J, Day L, Fusani L, Fuxjager MJ (2013) Hormones and the neuromuscular control of courtship in the Golden-collared Manakin (*Manacus vitellinus*). Front Neuroendocrinol 34:143–156
Schmaltz G, Quinn JS, Lentz C (2008) Competition and waste in the communally breeding Smooth-billed Ani: effects of group size on egg-laying behaviour. Anim Behav 76:153–162
Schmidt KL, MacDougall-Shackleton EA, Kubli SP, MacDougall-Shackleton SA (2014) Developmental stress, condition, and birdsong: a case study in Song Sparrows. Integr Comp Biol 54:568–577
Schmoll T (2011) A review and perspective on context-dependent genetic effects of extra-pair mating in birds. J Ornithol 152:265–277
Sergio F, Blas J, Baos R, Forero MG, Donázar JA, Hiraldo F (2009a) Short and longterm consequences of individual and territory quality in a long-lived bird. Oecologia 160:507–514
Sergio F, Blas J, Hiraldo F (2009b) Predictors of floater status in a long-lived, migratory bird: a cross-sectional and longitudinal test of hypotheses. J Anim Ecol 78: 109–118
Sheldon BC, Merilä J, Qvarnström A, Gustafsson L, Ellegren H (1997) Paternal genetic contribution to offspring condition expected by size of male secondary sexual character. Proc R Soc B 264:297–302
Shen S-F, Vehrencamp SL, Johnstone RA, Chen H-C, Chan S-F, Liao W-Y, Lin K-Y, Yuan H-W (2012) Unfavorable environment limits social conflict in *Yuhina brunneiceps*. Nat Commun 3:885
Sherman PT (1995) Social organization of cooperatively polyandrous White-winged Trumpeters (*Psophia leucoptera*). Auk 112:296–309
Shorey L (2002) Mating success on White-bearded Manakin (*Manacus manacus*) leks: male characteristics and relatedness. Behav Ecol Sociobiol 52:451–457
Shorey L, Piertney S, Stone J, Höglund J (2000) Fine-scale genetic structuring on *Manacus manacus* leks. Nature 408:352–353
Shutler D, Weatherhead PJ (1991) Owner and floater Red-winged Blackbirds: determinants of status. Behav Ecol Sociobiol 28:235–241

Silver R, Andrews H, Ball GF (1985) Parental care in an ecological perspective: a quantitative analysis of avian subfamilies. Am Zool 25:823–840

Simons MJP, Cohen AA, Verhulst S (2012) What does carotenoid-dependent coloration tell? Plasma carotenoid level signals immunocompetence and oxidative stress state in birds – a meta-analysis. PLoS ONE 7: e43088

Skutch AF (1961) Helpers among birds. Condor 63:198–226

Slifko TR, Smith HV, Rose JB (2000) Emerging parasite zoonoses associated with water and food. Int J Parasitol 30:1379–1393

Smith SM (1978) The "underworld" in a territorial sparrow: adaptive strategy for floaters. Am Nat 112:571–582

Sorci G, Faivre B (2009) Inflammation and oxidative stress in vertebrate host–parasite systems. Philos Trans R Soc B 364:71–83

Spencer KA, MacDougall-Shackleton SA (2011) Indicators of development as sexually selected traits: the developmental stress hypothesis in context. Behav Ecol 22:1–9

Spencer KA, Wimpenny JH, Buchanan KL, Lovell PG, Goldsmith AR, Catchpole CK (2005) Developmental stress affects the attractiveness of male song and female choice in the Zebra Finch (*Taeniopygia guttata*). Behav Ecol Sociobiol 58:423–428

Spoon TR, Millam JR, Owings DH (2006) The importance of mate behavioural compatibility in parenting and reproductive success by Cockatiels, *Nymphicus hollandicus*. Anim Behav 71:315–326

Spottiswoode C, Møller AP (2004) Extrapair paternity, migration, and breeding synchrony in birds. Behav Ecol 15:41–57

Stapleton MK, Kleven O, Lifjeld JT, Robertson RJ (2007) Female Tree Swallows (*Tachycineta bicolor*) increase offspring heterozygosity through extrapair mating. Behav Ecol Sociobiol 61:1725–1733

Steifetten Ø, Dale S (2006) Viability of an endangered population of Ortolan Buntings: the effect of a skewed operational sex ratio. Biol Conserv 132:88–97

Stein AC, Uy JAC (2006) Plumage brightness predicts male mating success in the lekking Golden-collared Manakin, *Manacus vitellinus*. Behav Ecol 17:41–47

Stiles FG, Wolf LL (1979) Ecology and evolution of lek mating behavior in the Long-tailed Hermit hummingbird. Ornithological Monographs no. 27, University of California Press, Berkeley, CA

Strandh M, Westerdahl H, Pontarp M, Canbäck B, Dubois M-P, Miquel C, Taberlet P, Bonadonna F (2012) Major histocompatibility complex class II compatibility, but not class I, predicts mate choice in a bird with highly developed olfaction. Proc R Soc B 279:4457–4463

Strong AM, Rimmer CC, McFarland KP (2004) Effect of prey biomass on reproductive success and mating strategy of Bicknell's Thrush (*Catharus bicknelli*), a polygynandrous songbird. Auk 121:446–451

Stutchbury BJ, Morton ES (1995) The effect of breeding synchrony on extra-pair mating systems in songbirds. Behaviour 132:675–690

Sydeman WJ (1989) Effects of helpers on nestling care and breeder survival in Pygmy Nuthatches. Condor 91: 147–155

Székely T, Cuthill IC, Kis J (1999) Brood desertion in Kentish Plover: sex differences in remating opportunities. Behav Ecol 10:185–190

Székely T, Weissing FJ, Komdeur J (2014) Adult sex ratio variation: implications for breeding system evolution. J Evol Biol 27:1500–1512

Szulkin M, Stopher KV, Pemberton JM, Reid JM (2013) Inbreeding avoidance, tolerance, or preference in animals? Trends Ecol Evol 28:205–211

Taff CC, Steinberger D, Clark C, Belinsky K, Sacks H, Freeman-Gallant CR, Dunn PO, Whittingham LA (2012) Multimodal sexual selection in a warbler: plumage and song are related to different fitness components. Anim Behav 84:813–821

Tamm S, Armstrong DP, Tooze ZJ (1989) Display behavior of male Calliope Hummingbirds during the breeding season. Condor 91:272–279

Tarwater CE, Brawn JD, Maddox JD (2013) Low extrapair paternity observed in a tropical bird despite ample opportunities for extrapair mating. Auk 130:733–741

Tella JL (1993) Polyandrous trios in a population of Egyptian Vultures. J Raptor Res 27:119–120

Temrin H, Sillén-Tullberg B (1994) The evolution of avian mating systems: a phylogenetic analysis of male and female polygamy and length of pair bond. Biol J Linn Soc 52:121–149

Théry M, Vehrencamp SL (1995) Light patterns as cues for mate choice in the lekking White-throated Manakin (*Corapipo gutturalis*). Auk 112:133–145

Thusius KJ, Dunn PO, Peterson KA, Whittingham LA (2001) Extrapair paternity is influenced by breeding synchrony and density in the Common Yellowthroat. Behav Ecol 12:633–639

Tingay RE, Culver M, Hallerman EM, Fraser JD, Watson RT (2002) Subordinate males sire offspring in Madagascar Fish-Eagle (*Haliaeetus vociferoides*) polyandrous breeding groups. J Raptor Res 36:280–286

Tobias JA, Montgomerie R, Lyon BE (2012) The evolution of female ornaments and weaponry: social selection, sexual selection, and ecological competition. Philos Trans R Soc B 367:2274–2293

Tooney MB, McGraw KJ (2012) Mate choice for a male carotenoid-based ornament is linked to female dietary carotenoid intake and accumulation. BMC Evol Biol 12:3

Trail PW (1985) Courtship disruption modifies mate choice in a lek-breeding bird. Science 227:778–780

Trail PW (1986) Courtship disruption at the lek in the Guianan Cock-of-the-Rock. Ethology 73:197–218

Trail PW (1990) Why should lek-breeders be monomorphic? Evolution 44:1837–1852

Trail PW, Adams ES (1989) Active mate choice at Cock-of-the Rock leks: tactics of sampling and comparison. Behav Ecol Sociobiol 25:283–292
Trail PW, Koutnik DL (1986) Courtship disruption at the lek in the Guianan Cock-of-the-Rock. Ethology 73: 197–218
Tryjanowski P, Hromada M (2005) Do males of the Great Grey Shrike, *Lanius excubitor*, trade food for extrapair copulations? Anim Behav 69:529–533
Tryjanowski P, Antczak M, Hromada M (2007) More secluded places for extra-pair copulations in the Great Grey Shrike *Lanius excubitor*. Behaviour 144:23–31
Vakirtzis A, Roberts SC (2009) Mate choice copying and mate quality bias: different processes, different species. Behav Ecol 20:908–911
Valcu M, Dale J, Griesser M, Nakagawa S, Kempenaers B (2014) Global gradients of avian longevity support the classic evolutionary theory of ageing. Ecography 37:930–938
Valcu C-M, Valcu M, Kempenaers B (2021) The macroecology of extra-pair paternity in birds. Mol Ecol 30:4884–4898
Valencia J, De La Cruz C, Carranza J, Mateos C (2006) Parents increase their parental effort when aided by helpers in a cooperatively breeding bird. Anim Behav 71:1021–1028
Valera F, Hoi H, Schleicher B (1997) Egg burial in Penduline Tits, *Remiz pendulinus*: its role in mate desertion and female polyandry. Behav Ecol 8:20–27
van Oers K, Drent PJ, Dingemanse NJ, Kempanaers B (2008) Personality is associated with extrapair paternity in Great Tits, *Parus major*. Anim Behav 76:555–563
van Rhijn J, Jukema J, Piersma T (2014) Diversity of nuptial plumages in male Ruffs *Philamachus pugnax*. Ardea 102:5–20
Varricchio DJ, Moore JR, Erickson GM, Norell MA, Jackson FD, Borkowski JJ (2008) Avian paternal care had dinosaur origin. Science 322:1826–1828
Vehrencamp SL (1978) The adaptive significance of communal nesting in Groove-billed Anis (*Crotophaga sulcirostris*). Behav Ecol Sociobiol 4:1–33
Vehrencamp SL (2000) Evolutionary routes to joint-female nesting in birds. Behav Ecol 11:334–344
Vehrencamp SL, Quinn JS (2004) Joint laying systems. In: Koenig W, Dickinson J (eds) Ecology and evolution of cooperative breeding in birds. Cambridge University Press, Cambridge, UK, pp 177–196
Veiga JP (1992) Hatching asynchrony in the House Sparrow: a test of the egg-viability hypothesis. Am Nat 139:669–675
Velando A, Lessells CM, Màrquez JC (2001) The function of female and male ornaments in the Inca Tern: evidence for links between ornament expression and both adult condition and reproductive performance. J Avian Biol 32:311–318
Verhulst S, Nilsson J-Å (2008) The timing of birds' breeding seasons: a review of experiments that manipulated timing of breeding. Philos Trans R Soc B 363:399–410
Verner J (1964) Evolution of polygamy in the Long-billed Marsh Wren. Evolution 18:252–261
Verner J, Willson MF (1966) The influence of habitats on mating systems of North American passerine birds. Ecology 47:143–147
Voigt C, Goymann W (2007) Sex-role reversal in the brain of African Black Coucals (*Centropus grillii*). Dev Neurobiol 67:1560–1573
von Lippke IS (2008) Ecology and evolution of cooperation in the Española Mockingbird, *Nesomimus macdonaldi*, Galápagos. Ph.D. dissertation. University of California, Los Angeles, CA
Waite TA, Strickland D (1997) Cooperative breeding in Gray Jays: philopatric offspring provision juvenile siblings. Condor 99:523–525
Walker LK, Stevens M, Karadas F, Kilner RM, Ewen JG (2013) A window on the past: male ornamental plumage reveals the quality of their early-life environment. Proc Roy Soc B 280:20122852
Walsh PT, Hansell M, Borello WD, Healy SD (2011) Individuality in nest building: do Southern Masked Weaver (*Ploceus velatus*) males vary in the nest-building behaviour? Behav Process 88:1–6
Ward K-A (2012) Female mate-searching strategies and behavioural correlates of copulation success in lekking Long-tailed Manakins (*Chiroxiphia linearis*). Ph.D. dissertation. University of Windsor, Windsor, ON, Canada
Warriner JS, Warriner JC, Page GW, Stenzel LE (1986) Mating system and reproductive success of a small population of polygamous Snowy Plovers. Wilson Bull 98:15–37
Wedell N, Kvarnemo C, Lessells CM, Tregenza T (2006) Sexual conflict and life histories. Anim Behav 71:999–1011
Wegge P, Rolstad J (1986) Size and spacing of Capercaillie leks in relation to social behavior and habitat. Behav Ecol Sociobiol 19:401–408
Westcott DA (1997) Lek locations and patterns of female movement and distribution in a Neotropical frugivorous bird. Anim Behav 53:236–247
Westcott D, Smith JNM (1997) Lek size variation and its consequences in the Ochre-bellied Flycatcher, *Mionectes oleaginous*. Behav Ecol 8:396–403
Westneat DF, Stewart IRK (2003) Extra-pair paternity in birds: causes, correlates, and conflict. Annu Rev Ecol Evol Syst 34:365–396
Westneat DF, Sherman PW, Morton ML (1990) The ecology and evolution of extra-pair copulations in birds. Curr Ornithol 7:331–369
Whittingham LA, Dunn PO, Magrath RD (1997) Relatedness, polyandry and extra-group paternity in the cooperatively-breeding White-browed Scrubwren (*Sericornis frontalis*). Behav Ecol Sociobiol 40:261–270
Widemo F (1998a) Competition for females on leks when male competitive abilities differ: empirical test of a model. Behav Ecol 9:427–431

Widemo F (1998b) Alternative reproductive strategies in the Ruff, *Philomachus pugnax*: a mixed ESS? Anim Behav 56:329–336

Widemo F, Owensi IP (1995) Lek size, male mating skew and the evolution of lekking. Nature 373:148–151

Wiebe KL (2008) Division of labour during incubation in a woodpecker *Colaptes auratus* with reversed sex roles and facultative polyandry. Ibis 150:115–124

Wiggins DA, Morris RD (1986) Criteria for female choice of mates: courtship feeding and paternal care in the Common Tern. Am Nat 128:126–129

Wilcox SC (2020) Sexual selection, flight performance, and signaling in Black-chinned Hummingbirds (*Archilochus alexandri*). Ph.D. dissertation. University of California, Riverside, CA

Wiley RH (1974) Evolution of social organization and life history patterns among grouse. Q Rev Biol 49: 201–227

Willimont LA, Jackson JA, Jackson BJS (1991) Classical polyandry in the West Indian Woodpecker on Abaco, Bahamas. Wilson Bull 103:124–125

Wink M, Dyrcz A (1999) Mating systems in birds: a review of molecular studies. Acta Ornithol 34:91–109

Winterbottom M, Burke T, Birkhead T (2001) The phalloid organ, orgasm and sperm competition in a polygynandrous bird: the Red-billed Buffalo Weaver. Behav Ecol Sociobiol 50:474–482

Witte K, Ryan MJ (1998) Male body length influences mate-choice copying in the sailfin molly *Poecilia latipinna*. Behav Ecol 9:534–539

Wittenberger JF (1976) The ecological factors selecting for polygyny in altricial birds. Am Nat 110:779–799

Woxvold IA, Magrath MJL (2005) Helping enhances multiple components of reproductive success in the cooperatively breeding Apostlebird. J Anim Ecol 74: 1039–1050

Wu G, Fang YZ, Yang S, Lupton JR, Turner ND (2004) Glutathione metabolism and its implications for health. J Nutr 134:489–492

Yosef R, Pinshow B (1989) Cache size in shrikes influences female mate choice and reproductive success. Auk 106:418–421

Young EC (1998) Dispersal from natal territories and the origin of cooperatively polyandrous breeding groups in the Brown Skua. Condor 100:335–342

Yuan H-W, Liu M, Shen S-F (2004) Joint nesting in Taiwan Yuhinas: a rare passerine case. Condor 106: 862–872

Zahavi A (1974) Communal nesting by Arabian Babblers: case of individual selection. Ibis 116:84–87

Zahavi A (1975) Mate selection – a selection for a handicap. J Theor Biol 53:205–214

Zera AJ, Harshman LG (2001) The physiology of life history trade-offs in animals. Annu Rev Ecol Syst 32: 95–126

# Avian Reproduction: Timing, Anatomy, and Eggs 16

## Contents




**Abstract**

For successful reproduction, timing is important because food availability for birds varies with time of year and, to ensure sufficient food for both adults and their young, breeding during periods of peak food availability can be critical. In this chapter, I explain the factors that influence when birds initiate breeding, and describe the structure and function of male and female reproductive systems. For males, the process of sperm production and transport is described and, for females, the process of egg production is explained. The various components of avian eggs are described and




G. Ritchison, *In a Class of Their Own*, Fascinating Life Sciences,
https://doi.org/10.1007/978-3-031-14852-1_16

their functions explained. The copulatory behavior of birds is discussed and the role of female sperm storage tubules in bird reproduction is explained. The processes of fertilization and sex determination are also described. In addition, the structure and function of the various sections of the female oviduct and what happens to eggs as they move through the oviduct are described. Factors contributing to variation among species in the shape and color of eggs are discussed, and the chapter concludes with a description of the process of egg laying and a discussion of the energetic costs of egg production by female birds.

## 16.1 Timing of Reproduction

Birds attempt to maximize individual fitness by reproducing and an important factor in successful reproduction is choosing the best possible time. Factors important in determining the "best" time can vary, but, for most species, food availability is most important because reproduction comes with an energetic cost and access to more food and energy translates into things, like larger clutches and healthier nestlings and fledglings, that enhance reproductive success. Food availability is of particular importance for birds because, in contrast to mammals where females provide young with milk, birds must have access to an external source of food for their young. Of course, with more than 10,000 species of birds, there are always exceptions. Hoatzins (*Opisthocomus hoazin*), for example, are tropical, leaf-eating birds that often breed in forests along river channels and time their breeding to coincide with periods of flooding because the high-water levels reduce the risk of nest predation by mammalian predators (Mullner and Linsenmair 2007). For most birds, however, food availability is critical in the timing of reproduction. Because some birds are migratory and others are residents, and because birds breed in a wide variety of habitats at different latitudes and altitudes, the factor or factors that influence(s) the availability of food and, therefore, the timing of reproduction vary.

Because food availability varies and because some time is needed to "prepare" for reproduction (e.g., migrate to breeding areas, gonadal development, and nest construction), birds often must anticipate when conditions will be optimal for reproduction. For example, long-distance migrants must initiate migration to breeding areas well before they will actually initiate nesting. For resident birds, preparation for breeding may take less time, but some mechanism is still needed to help ensure that breeding occurs when conditions are optimal or nearly so.

In part, this mechanism is based on an endogenous circannual rhythm. Circannual rhythms were first described for Willow Warblers (*Phylloscopus trochilus*). These migratory warblers were found to exhibit circannual rhythms even when maintained in the laboratory under a constant photoperiod and temperature (Gwinner 1986), and subsequent research has revealed similar circannual rhythms in more than 20 other species of migratory and resident birds, from both tropical and temperate regions, maintained under constant conditions (Wikelski et al. 2008). However, circannual rhythms exhibited by birds in the laboratory are not 12 months in duration. For example, an African Stonechat (*Saxicola torquatus*) maintained in captivity for more than 10 years exhibited a circannual rhythm (of testes size and molt) with an average duration of nine months (Gwinner 2003). Such results indicate that birds have an endogenous clock and that external factors are needed to ensure synchronization of that clock in natural environments.

Little is known about how circannual clocks work. One early hypothesis was that birds count days to derive an annual cycle (frequency demultification; Gwinner 1973), but there is no evidence to support this hypothesis. Work with mammals (Soay sheep, *Ovis aries*) indicates that the circannual rhythm results from feedback signals between cells in the pituitary gland, with melatonin-regulated timer cells and prolactin-secreting cells together functioning as timer system (Lincoln et al. 2006), but there is no evidence for a similar timer system in birds. Another possible mechanism involves energy expenditure, with birds measuring or tracking their energy consumption to determine the time of year (Wikelski et al. 2008). How birds might measure or track energy expenditure, however, is unknown.

External factors are needed to synchronize or entrain endogenous clocks and such factors are referred to as "zeitgebers" (German for "time giver"). One very important zeitgeber is daylength. With increasing latitude, seasonal fluctuations in resource availability are increasingly associated with changes in daylength or photoperiod. As a result, for many birds, the annual change in daylength is the most important environmental cue for the timing of reproduction (Coppack and Pulido 2004). Of course, closer to the equator, changes in daylength may be minimal. However, some birds at low latitudes are able to use even slight changes in photoperiod as an environmental cue. For example, Spotted Antbirds (*Hylophylax naevioides*) are found in humid forests in central Panama (9°N) and initiate breeding in response to slight changes in photoperiod (maximum change during the year is about 1 h; Hau 2001). In the lab, an increase in the photoperiod of just 17 minutes was sufficient to stimulate gonadal growth in Spotted Antbirds (Hau et al. 1998).

Although much less is known about the factors that influence the timing of reproduction at tropical latitudes, photoperiodic cues do not appear to be of critical importance for some species. Two populations of Chingolo Sparrows (*Zonotrichia capensis*) about 25 km apart in the high Andes of Ecuador (0°23'S) exhibit seasonal reproductive cycles, but one population breeds annually from August-December and the other from November-May. This difference is apparently due to differences in rainfall patterns, with each population breeding during the period of greatest precipitation (Moore 2005). Similarly, Darwin's ground finches (*Geospiza* spp.) on the Galapagos Islands are opportunistic breeders, only breeding when there is sufficient rain to cause an increase in caterpillar abundance (Hau 2001). For species of birds in arid and semi-arid regions, reproduction may also be triggered by rainfall (e.g., Cavalcanti et al. 2016). In yet other species, temperature and social cues may influence the timing of reproduction (Table 16.1). In locations where food availability and weather conditions permit breeding during most or all of the year, other factors, such as predation risk, female protein reserves, and competition, can influence the timing of breeding (Ndithia et al. 2017).

In tropical areas with wet and dry seasons, light intensity rather than duration may be an important cue, with decreasing intensity potentially associated with increased cloud cover signaling the approach of the rainy season. African Stonechats maintained on a constant photoperiod, but with cyclic changes in daytime light intensity, exhibited gonadal and molt rhythms

**Table 16.1** Non-photic cues and attributes of cues that can influence the timing of reproduction by birds (Table from Chmura et al. 2020; © 2019 Nordic Society Oikos. Published by John Wiley & Sons Ltd., used with permission)

| Cues | Attributes studied | Example references |
|---|---|---|
| Temperature | Difference in magnitude of temperature, rate of change, rate of change relative to reproductive state | Data: Wingfield et al. (1997), Perfito et al. (2005), Schaper et al. (2012) Review: Caro et al. (2013) |
| Food | Availability of cue via through visual, olfactory, tactile and ingestive modalities, calories, nutrient type, timing of availability, energy stores | Data: Meijer and Drent (1999), Schoech et al. (2004), Voigt et al. (2011) |
| Water | Availability of cue via ingestive, tactile, visual, and auditory modalities | Data: Payne (1969), Wingfield et al. (2012), Prior et al. (2013) |
| Social | Availability of auditory and visual presentations, quality of cues, social affiliation between source of cue and receiver of cue, noxious stimuli (i.e., predators) | Data: Runfeldt and Wingfield (1985), Morton et al. (1985) Review: Helm et al. (2006) |
| Other cues | Predators, nesting material, eggs | Data: Gwinner et al. (1987), Meijer (1995), Monkkonen et al. (2009) |

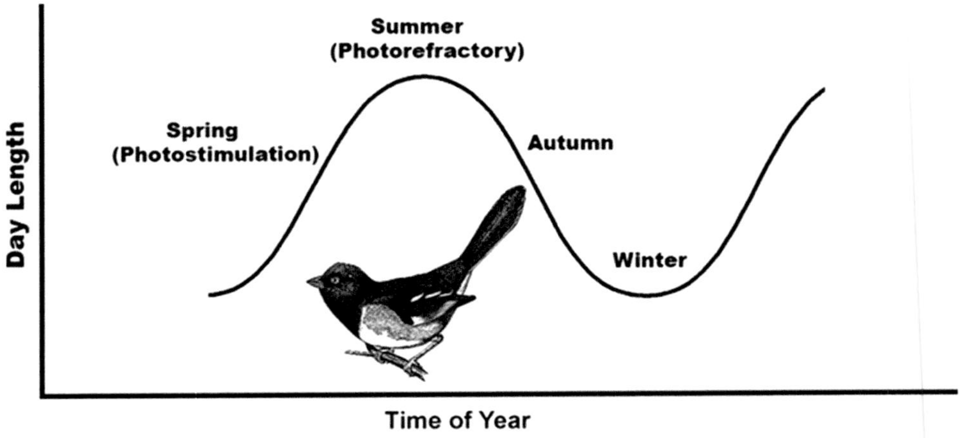


**Fig. 16.1** Annual timekeeping mechanism for many species of birds. Exposure to short days during winter sensitizes birds to the photostimulatory actions of increasing day length in spring. As days continue to get longer during the summer, birds become photorefractory and are no longer sensitive to day length. As days become shorter in mid- to late summer, breeding ends. (Figure from Kriegsfeld et al. 2015; © 2014 Elsevier Inc., used with permission)

synchronized with the light intensity cycles (Gwinner and Scheuerlein 1998). Regardless of the cue, the timing of breeding by many tropical species is less predictable than for non-tropical species. Because of the environmental uncertainty, many of these "opportunistic" breeders have moderately developed reproductive systems for most of the year so they can be ready to breed when conditions become suitable (Dittami and Gwinner 1985; Dawson et al. 2001).

For most birds, day length provides a reliable way to synchronize internal clocks, anticipate future environmental conditions, and optimize the timing of reproduction (Fig. 16.1), with other cues, such as temperature, providing a means of "fine-tuning" that timing. These cues, however, must be detected and then somehow used to trigger the physiological, morphological, and behavioral changes required for reproduction. The avian photoreceptors that detect changes in daylength were originally thought to be the eyes and the pineal organ. However, numerous studies have revealed that photoreceptors in the eyes and pineal organ are not essential for the regulation of seasonal reproduction of birds. Rather, birds appear to have several deep-brain photoreceptors that regulate seasonal reproduction, including cells in the telencephalon of the brain and hypothalamus as well as cerebrospinal fluid (CSF)-contacting neurons (Box 16.1 Photopic and

**Box 16.1 Photopic and Non-photopic Regulation of Bird Reproduction**

Birds that breed seasonally have annual cycles controlled by the hypothalamic-pituitary-gonadal (HPG) axis. For birds that are sensitive to photic cues, shorter days stimulate the release of gonadotropin-inhibiting factor (GnIH) that, in turn, inhibits the production of follicle-stimulating hormone (FSH) and luteinizing hormone (LH). Activation of the HPG axis requires perception of increasingly long days and birds detect this change in daylength via "circadian pacemaker" cells in the eyes, pineal organ, and hypothalamus as well as neurons in the brain called cerebrospinal fluid (CSF)-contacting neurons. These cells respond to increasing daylength by releasing chemicals called photopigments that, in turn, activate the HPG axis. Once activated, the

(continued)

**Box 16.1** (continued)

gonads of males and females increased dramatically in size and begin production of gametes and sex hormones.

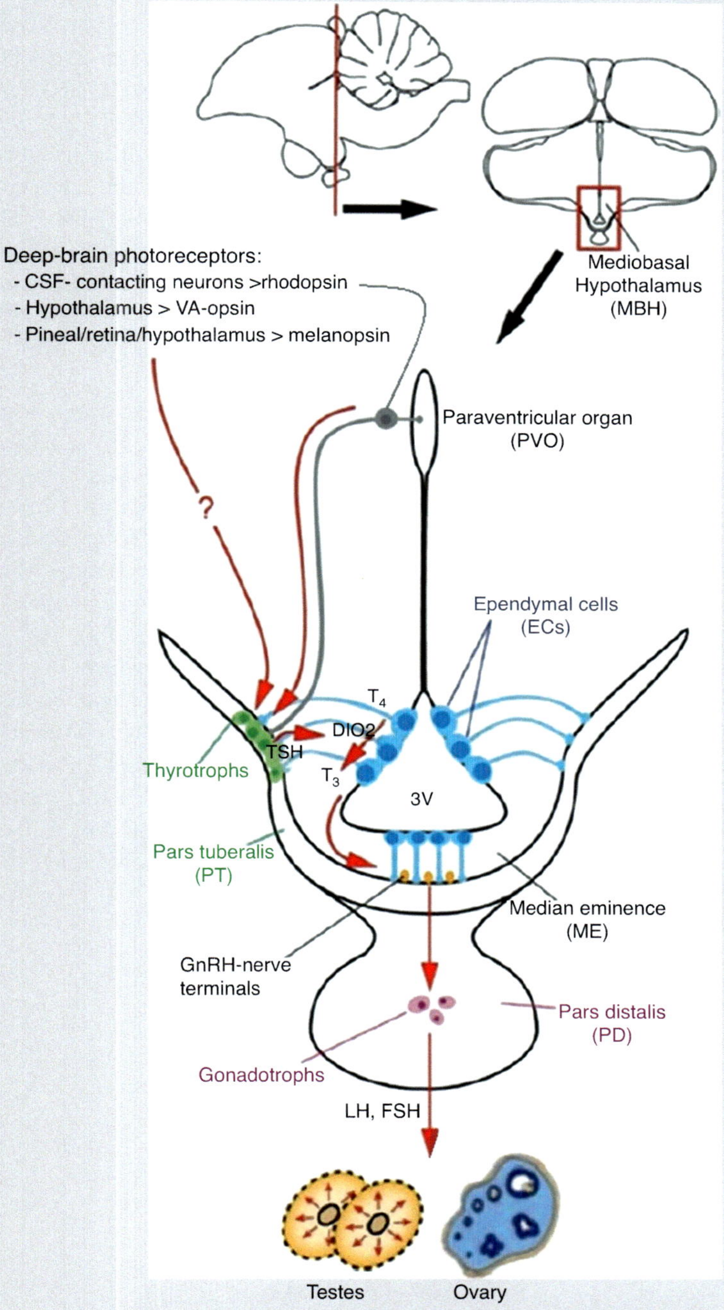


(continued)

**Box 16.1** (continued)

Photoperiodic response of many species of birds. Changes in daylength are detected by deep-brain photoreceptors found in different areas depending on the species, but typically include cells in the hypothalamus, cerebrospinal fluid (CSF)-contacting neurons, and/or cells in the pineal organ or retina. These cells release photopigments, including rhodopsin, vertebrate-ancient (VA)-opsin, and melanopsin, that are transmitted to the pars tuberalis of the pituitary gland (PT), where they induce secretion of thyroid-stimulating hormone (TSH). TSH increases expression of thyroid hormone–activating enzyme type 2 deiodinase (DIO2), which converts thyroxine (T4) in triiodothyronine (T3). T3 stimulates secretion of gonadotropin-releasing hormone (GnRH) by the hypothalamus, which, in turn, stimulates the release of luteinizing hormone (LH) and follicle-stimulating hormone (FSH) by the pituitary gland. LH and FSH then stimulate sperm production and secretion of testosterone by the testes of males, and egg maturation and secretion of estradiol and progesterone by the ovary (Nakane and Yoshimura 2019). (Figure modified from Nakane and Yoshimura 2014; © 2014 Nakane and Yoshimura, open-access article distributed under the terms of the Creative Commons Attribution License (CC BY), https://creativecommons.org/licenses/by/4.0/)

In many species of birds, non-photic cues, such as temperature, food, water, or social cues can trigger the onset of breeding. Several hypotheses have been proposed to explain how such non-photic cues can affect the timing of reproduction. One hypothesis is that non-photic cues trigger neuronal changes that result in increased production of gonadotropin releasing hormone (GnRH-1) by the hypothalamus just like longer photoperiods do in other species of birds (Moore et al. 2006). Lattin et al. (2016) proposed the CORT-Flexibility Hypotheses wherein, prior to breeding, the corticosteroid system of animals, including birds, is more sensitive. As a result, "negative factors," such as cold temperatures, lack of rain, or limited food supplies, are more likely to lead to increased release of corticosterone by the adrenals, which, in turn, inhibits the production of GnRH-1 by the hypothalamus and increases other effects of corticosteroids that delay breeding. Yet another hypothesis is that non-photic cues regulate reproduction via effects on the gonads instead of the hypothalamus. For example, Lynn et al. (2015) found that, after fasting male Zebra Finches (*Taeniopygia guttata*) for 10 h, blood corticosterone levels and testicular GnIH expression increased and testosterone levels in the blood decreased. Such changes may inhibit initiation of breeding when conditions such as reduced food availability are not suitable for breeding.

(continued)

**Box 16.1** (continued)

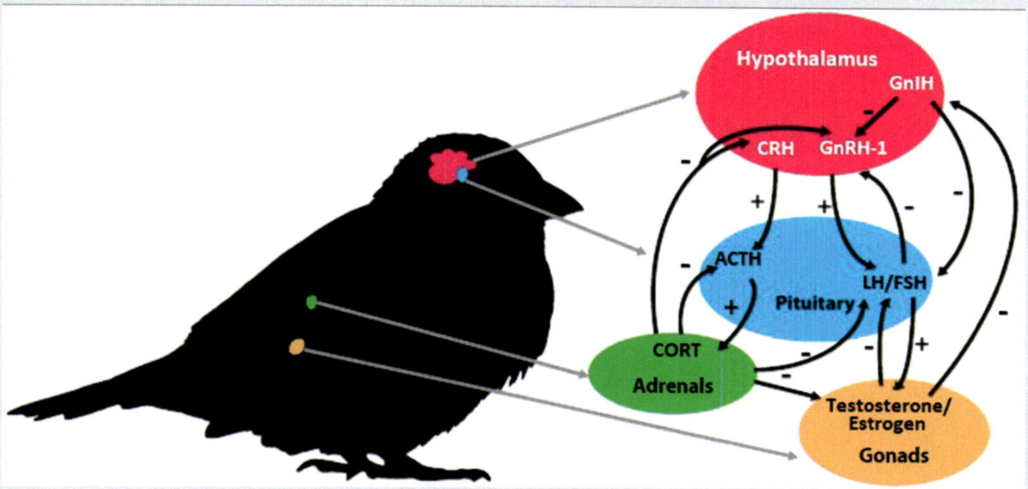


Components of the hypothalamic–pituitary–gonadal axis (HPG axis) and hypothalamic–pituitary–adrenal axis (HPA axis). In the hypothalamus, gonadotropin inhibitory hormone (GnIH) regulates the secretion of gonadotropin releasing hormone (GnRH-1). When GnRH is secreted, luteinizing hormone (LH) and follicle stimulating hormone (FSH) are released from the anterior pituitary causing increased secretion of testosterone or estrogen by the testes and ovaries, respectively. Importantly, the HPG axis can influence the HPA axis. Corticotropin releasing hormone (CRH) in the hypothalamus causes release of adrenocorticotropic hormone (ACTH) from the pituitary and corticosterone from the adrenal glands. Increases in corticosterone can cause a reduction in the secretion of gonadal steroids. (Figure from Chmura et al. 2020; © 2019 Nordic Society Oikos. Published by John Wiley & Sons Ltd, used with permission)

Non-photopic Regulation of Bird Reproduction). The relative importance of these photoreceptors does, however, vary with species (Nakane and Yoshimura 2019). Briefly, longer days stimulate the release of gonadotropin-releasing hormone (or GnRH) by the hypothalamus, which, in turn, stimulates the release of follicle-stimulating hormone (or FSH) and luteinizing hormone (LH) by the anterior pituitary. These hormones then stimulate cells in the gonads (testes of males and ovary of females; Fig. 16.2).

## 16.2 Reproductive Anatomy of Male Birds

Male birds have paired testes attached to the dorsal abdominal walls and located near the kidneys (Box 16.2 Testis Asymmetry). In contrast to many mammals, bird testes do not migrate from their site of origin into scrotal sacs, but, rather, remain near the kidneys. For most mammals, sperm cannot develop normally at the

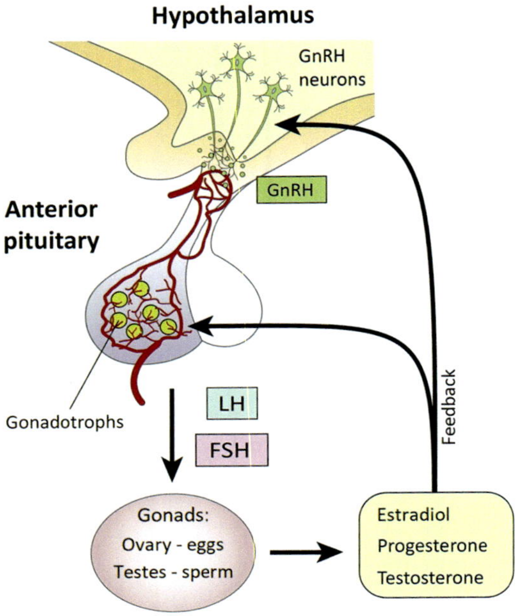


**Fig. 16.2** The timing of breeding by birds that breed seasonally are influenced by environmental signals, especially increasing daylength (i.e., light). Neurons in the hypothalamus regulate reproduction by releasing gonadotropin-releasing hormones (GnRH) that regulate secretions of pituitary gland. The anterior pituitary gland synthesizes and releases two gonadotropins, luteinizing hormone (LH) and follicle-stimulating hormone (FSH), that act on the gonads to stimulate production of gametes (eggs and sperm) and sex steroids (estradiol and progesterone from the ovary and testosterone from the testes). (Figure modified from Zavala et al. 2019; open-access article distributed under the terms of the Creative Commons CC-BY license, https://creativecommons.org/licenses/by/4.0/)

temperatures found in the abdominal cavity. However, spermatogenesis in male birds occurs at core body temperatures (Béaupre et al. 1997), apparently because of the presence of stable nucleic acids (mRNA) and unique thermotolerant proteins (Mezquita et al. 1998). Associated with each testis is a series of ducts called the epididymis that empty into the ductus deferens that, in turn, empties into the cloaca (Fig. 16.3). Testes contain interstitial tissue, consisting of blood vessels, nerves, and Leydig cells, and seminiferous tubules, containing Sertoli cells and the germ cells that produce sperm.

In response to the increased secretion of FSH and LH triggered by cues such as increasing daylength (described above), the testes begin to increase in size as the length and diameter of the seminiferous tubules increase and the number of interstitial cells also increases (Fig. 16.4). As a result of this growth, testes are much larger during the breeding season than during the non-breeding season (Fig. 16.5). The seasonal difference in size varies among species. The mass of the testes of male House Crows (*Corvus splendens*) increases 75-fold during the breeding season (Dang and Guraya 1978), and the testes of male Gentoo Penguins (*Pygoscelis papua*) increase in volume about 10-fold during the breeding season. Among passerines, testes can increase in volume by as much as 300–500 times during the breeding season (Lake 1981). At peak size, the testes of Japanese Quail (*Coturnix japonica*) represent as much as 3.3% of total body weight (Clulow and Jones 1982). By comparison, and to illustrate the relatively large size of bird testes during the breeding season, the testes of a male Norway rat (*Rattus norvegicus*) represent only 0.67% of total body weight (Clulow and Jones 1982).

LH triggers proliferation of Leydig cells that secrete increasing amounts of testosterone (and androstenedione, a precursor of testosterone) that is essential for, among other things, spermatogenesis, maintenance of the ducts associated with the testes, and the expression of certain behaviors (Kirby and Froman 2000). In addition to stimulating growth of the seminiferous tubules, FSH stimulates proliferation of Sertoli cells that release growth factors that help stimulate spermatogenesis. Sertoli cells also help anchor and provide nutrition for germ cells during development into sperm cells (Aire 2007).

Most of the tissue in testes is devoted to sperm production, and larger testes can produce more sperm (Fig. 16.6). Few investigators have measured rates of sperm production, but rates do vary among species. For example, male Zebra Finches (*Taeniopygia guttata*) produce about 1.9 million sperm per day (Birkhead et al. 1993), male House Sparrows (*Passer domesticus*) about

31 million per day (Birkhead et al. 1994), male fairywrens (*Malurus* spp.) about 568–612 million per day (Tuttle and Pruett-Jones 2004), and male Domestic Chickens (*Gallus g. domesticus*) about 2 billion per day (de Reviers and Williams 1981). Sperm production also varies within species. In a population of male House Sparrows, for example, testes mass ranged from 0.45 gram to 1 gram, with mass correlated with the number of sperm produced (Birkhead et al. 1994).

Larger testes not only produce more sperm, but also produce more sperm per ejaculate (Møller 1988; Fig. 16.7). As a result, ejaculate size varies both among and within species. Ejaculate size can also vary for individual males, with sperm depletion resulting from successive copulations reducing the number of sperm per ejaculate (Birkhead et al. 1995). A variety of factors can influence testes size, sperm production, and ejaculate size. One factor is age, with younger males often having smaller testes than older males. One possible reason for this is that younger males may invest less energy in reproduction than older males, either because they are constrained by their inexperience at acquiring the resources needed to reproduce (constraint hypothesis) or because they put less effort into reproduction because the costs in terms of future reproduction outweigh the benefits (restraint hypothesis; Williams 1966).

The most important factor in the evolution of testes size is sperm competition. Sperm competition occurs in breeding systems where females will mate with multiple males. In such mating systems, a male able to produce more sperm has a better chance of fathering offspring. In a

**Box 16.2 Testis Asymmetry**

Among many taxa and species of birds, the two testes of males exhibit asymmetry in size and shape. Most commonly, the left testis is larger, occurring in three times as many species as those with a larger right testis (Calhim and Montgomerie 2015). Witschi (1935) proposed that this asymmetry was a by-product of selection for a single left ovary in female birds or, alternatively, the right testis is often smaller because the presence of the liver limited the amount of space available for the testis. Calhim and Montgomerie (2015) compiled data on testes size for 264 species of birds and concluded that the "by-product of a single left ovary in females" hypothesis was unlikely the only explanation for testes size asymmetry because males in 25% of those species had a larger right testis. Beyond that, their analysis revealed no variable that explained most of the interspecific variation in testes asymmetry. They concluded, however, that sexual selection had a small effect on testes size asymmetry, i.e., greater postcopulatory sperm competition selects for greater sperm production, which in turn favors an increase in size of the smaller testis. Calhim and Montgomerie (2015) also reported a "weak relation" between testis asymmetry and gizzard morphology, with species having larger, more muscular gizzards also tending to have smaller, more elongated left testes. In a study focusing on a single family of birds (Maluridae), Calhim et al. (2019) found that males in malurid species with less sperm competition (lower rates of extrapair paternity) had left-biased testis asymmetry, whereas males in species with more sperm competition (higher rates of extrapair paternity) had right-biased asymmetry. On possible explanation for this switch to right-biased asymmetry is that sperm competition may favor larger combined testes size to produce more sperm, but relatively large gizzards (~4% of body mass for species in the family Maluridae) place a limit on the size of the left testis (Calhim et al. 2019).

(continued)

**Box 16.2** (continued)

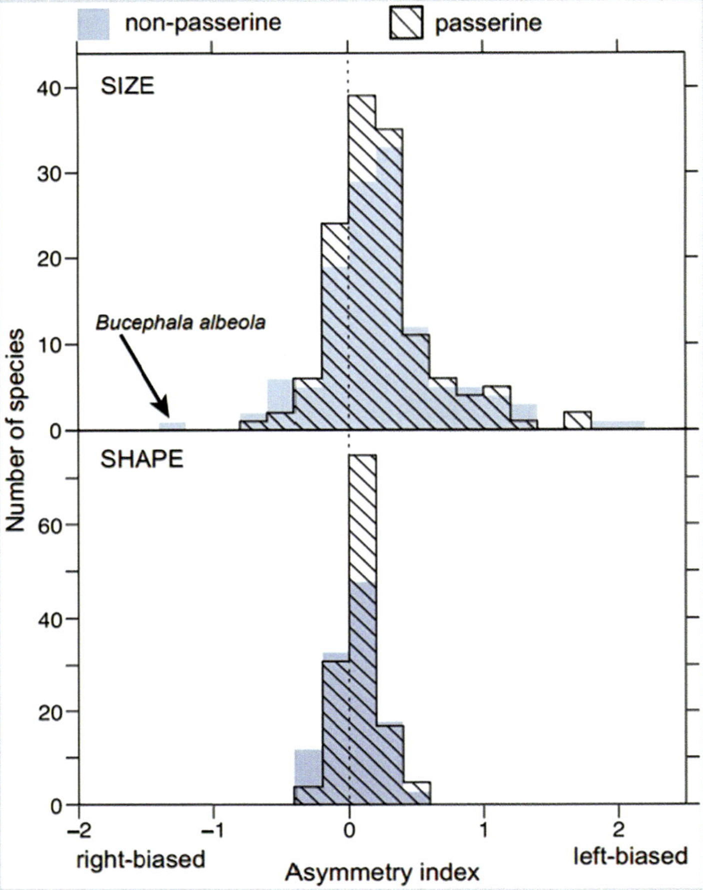


Distribution of testes size and shape (i.e., length) asymmetries among passerines ($N$ = 136 species) and nonpasserines ($N$ = 127 species). The asymmetry index is scale-free and is calculated as $\ln(m_L/m_R)$ for testes size and as $[\ln(l_L/l_R)–(\ln(w_L/w_R)]$ for testes shape, where $m$ = mass, $l$ = length, and $w$ = width, and subscripts indicate left ($L$) and right ($R$) testes. Buffleheads (*Bucephala albeola*) were an outlier. (Figure from Calhim and Montgomerie 2015; © The Authors, used with permission)

comparative study of 1010 species of birds, Pitcher et al. (2005) found that testes size (relative to body mass) was typically larger in polygynous and lekking species, where males may have multiple mates, than in monogamous species. Males in polyandrous species also had relatively large testes because, even though each male typically pairs with one female, males do compete for matings with the same female. Testes are also larger in species that typically breed at higher densities, such as colonial breeders, probably because of the increased opportunities for

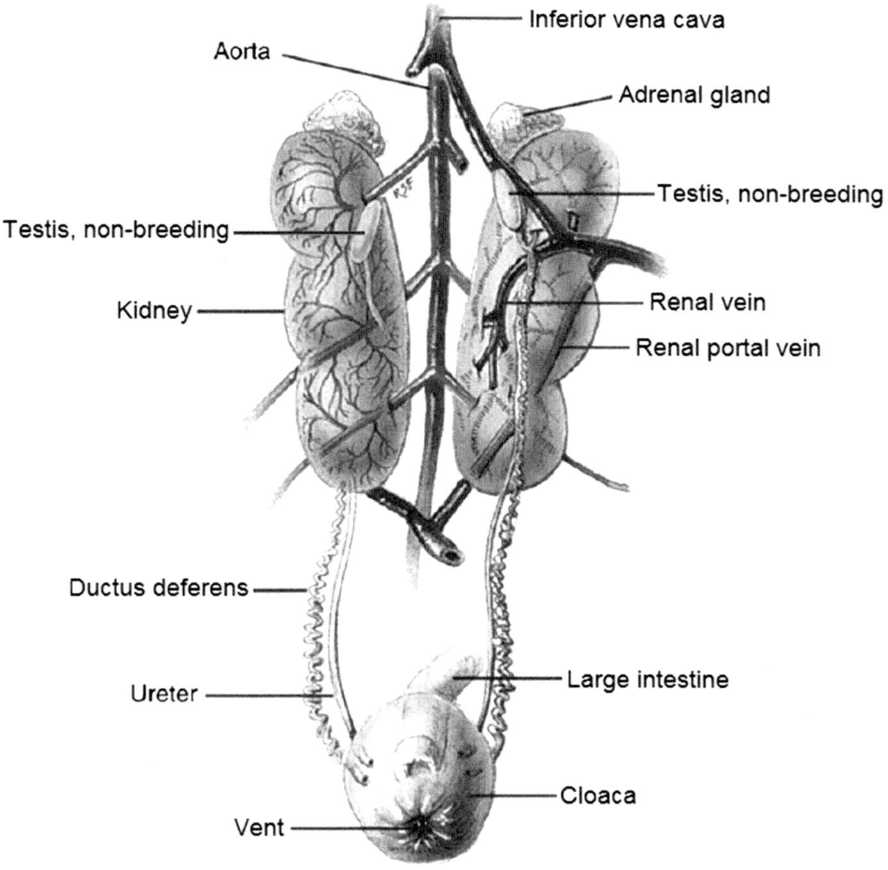


**Fig. 16.3** Drawing of the reproductive and urinary systems of a male bird (ventral view). The testes are much smaller during the non-breeding period than during the breeding season. (Figure modified from Pollock and Orosz 2002; © 2002, used with permission of Elsevier Science USA)

matings outside the social pair bond (extrapair copulations; Brown and Brown 2003). Finally, testes size has also been found to be positively related to clutch size (Fig. 16.8), possibly because larger clutches may require more copulations and, therefore, more sperm (Pitcher et al. 2005).

### 16.2.1 Sperm Production and Transport

Sperm are produced in the seminiferous tubules of the testes. Spermatogenesis occurs rapidly in birds, with the transformation of germ cells (spermatogonia) into mature sperm taking about 12–15 days (Lake 1981, Aire 2007; Figs. 16.9 and 16.10). The same process takes about 25–32 days in a typical mammal (Aire 2007). From the seminiferous tubules, sperm move toward the epididymis and, in the process, associated fluids are reabsorbed and sperm concentration increases. Sperm remain in the epididymis for only about a day as they develop motility. Sperm are then transported by peristalsis through the ductus (vas) deferens to a small chamber, the receptacle, near the end of the ductus deferens

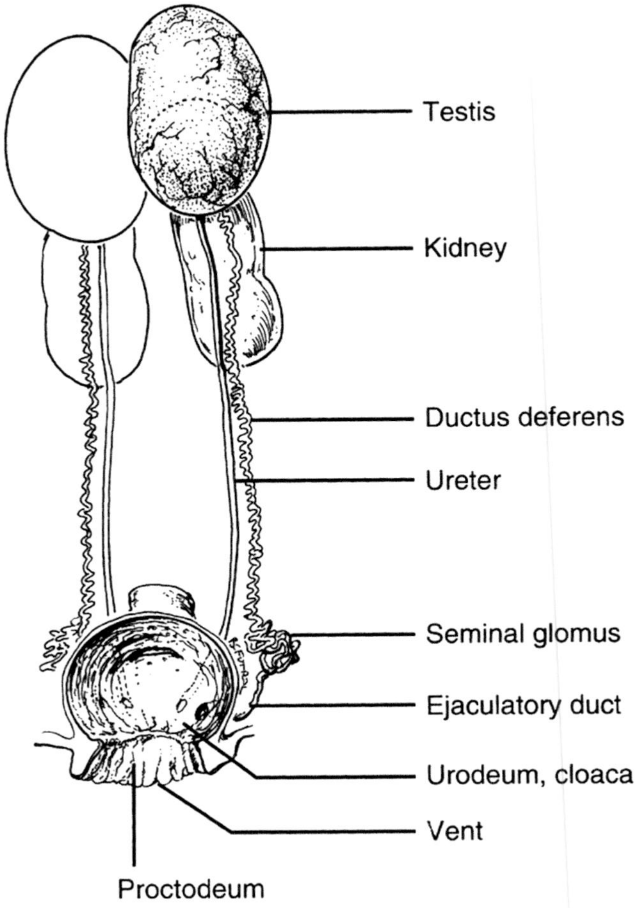


**Fig. 16.4** Drawing of the reproductive system of a male songbird during the breeding season (ventral view). Note the much larger size of the testes compared to their size during the non-breeding season in Fig. 16.3. (Figure from Pollock and Orosz 2002; © 2002, used with permission of Elsevier Science USA)

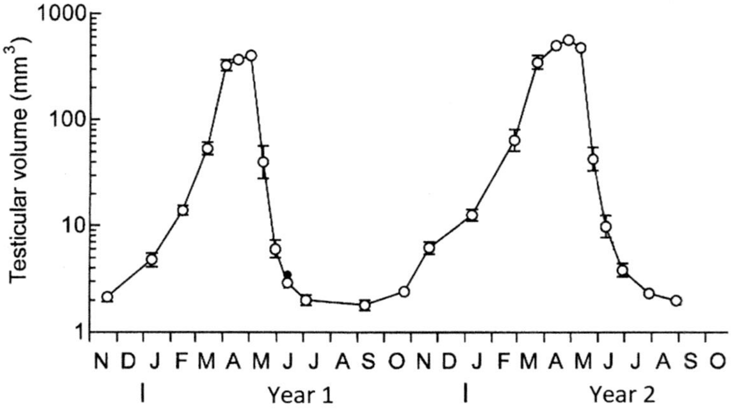


**Fig. 16.5** Changes in the mean volume (± SE) of the left testis of 10 male European Starlings (*Sturnus vulgaris*) over a two-year period. In Year 1, mean volume increased from 2.1 $mm^3$ to 400 $mm^3$. (Figure modified from Dawson 2003; © Journal of Avian Biology, used with permission)

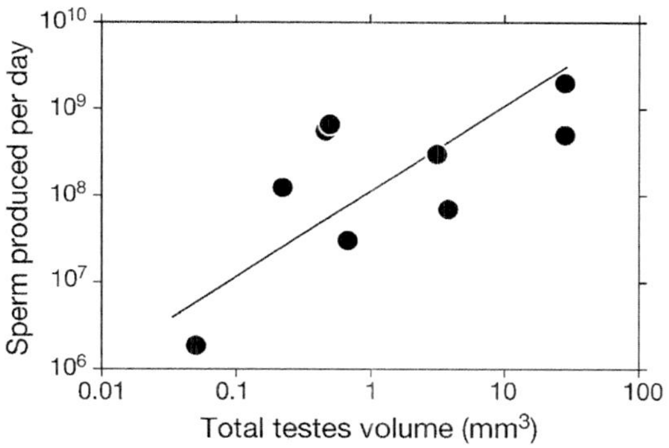


**Fig. 16.6** Relation between daily sperm production and testes volume in nine species of birds. (Figure from Briskie and Montgomerie 2007; © 2007 Reproduced by permission of the Taylor & Francis Group)

(and just before it empties into the cloaca). Sperm are then stored in the receptacle until copulation takes place. In passerines, the ductus deferens is highly convoluted just before the receptacle, forming a structure called the seminal glomus where sperm are stored. The right and left glomera increase in size 30–40 times during the breeding season (Birkhead and Møller 1992), and cause an obvious swelling called the cloacal protuberance (Figs. 16.11 and 16.12). The size of the cloacal protuberance varies among species, and is larger in species with greater sperm competition and, therefore, a greater need to store more sperm.

### 16.2.2 Characteristics of Sperm

In general, sperm are among the most diverse of all animal cells (Cohen 1977), and that is also the case for avian sperm (Fig. 16.13). However, regardless of morphology, the mature spermatozoa of birds consist of four main sections: acrosome, head, middle piece, and tail (Figs. 16.14 and 16.15). The acrosome is located at the tip and contains species-specific proteins that allow sperm to bind to and penetrate the ovum in the process of fertilization. Immediately behind the acrosome is the head that contains the nucleus and the genetic material in the form of chromosomes. The last two sections provide the sperm with motility, with the midpiece containing mitochondria that supply the energy that the tail, or flagellum, needs to move and propel the sperm.

Among birds, sperm morphology, or phenotype, exhibits considerable variation among orders, families, and species. Some of this variation is due to phylogeny (McFarlane 1963; Fig. 16.16). For example, the sperm of nonpasserines are relatively short with a smooth, tapering head. Passerine sperm, on the other hand, are typically more complex, with a helical form and relatively long midpiece (Fig. 16.13; Birkhead and Immler 2007). Sperm phenotype also differs among and within species and, for passerines, most interspecific and intraspecific variation involves differences in the relative size of the different components (Birkhead et al. 2005; Birkhead and Immler 2007). Støstad et al. (2018) examined the sperm morphology of 36 species of songbirds and found that sperm heads varied from being relatively straight to being strongly helical (Fig. 16.17), and that longer sperm had larger, more helical heads than shorter sperm. These authors also found that swimming speeds were higher for longer sperm with more helical heads (Fig. 16.18). In contrast, Lüpold et al. (2009) reported a negative relationship between the length of sperm heads and swimming speed, with larger heads creating more drag. However, the sperm in Lüpold et al. (2009) study were spherical rather than helical. The sperm of songbirds swim by rotating around the longitudinal axis so helical heads create greater forward propulsion and increased speed (Støstad et al. 2018).

One factor likely contributing to variation in sperm morphology between and within species is postcopulatory sexual selection, where female promiscuity leads to sperm competition (competition between ejaculates of different males to fertilize eggs) and cryptic mate choice (paternity biases that occur after copulation; Fig. 16.19, Snook 2005). With greater sperm competition, natural selection might be expected to influence the characteristics of sperm that might enhance

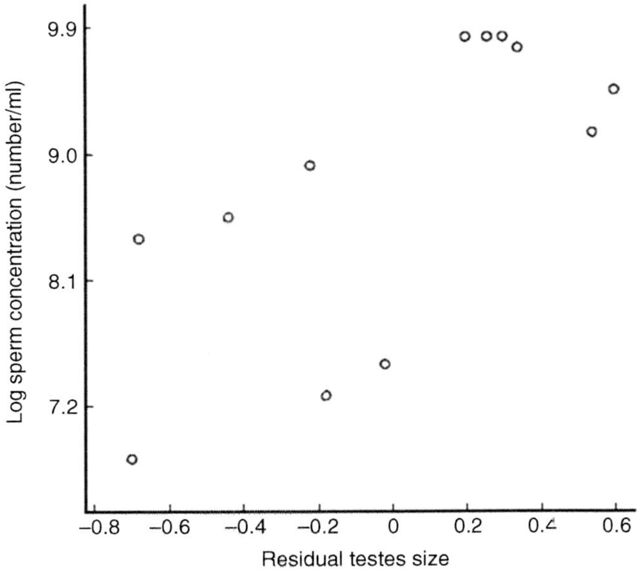


**Fig. 16.7** Larger testes tend to produce more sperm per ejaculate ($N = 12$ species). (Figure from Møller 1988; © 2008 Oxford University Press, used with permission)

success in fertilizing eggs, including size, longevity, and mobility. Kleven et al. (2009) reported a positive relationship between sperm swimming speed and the degree of sperm competition (Fig. 16.20), but, in contrast, Støstad et al. (2018) found no relationship between these variables.

Despite these mixed results concerning the relationship between sperm swimming speed and sperm competition, available evidence does suggest that sperm morphology is influenced by sperm competition. For example, comparative analyses indicate that intraspecific variation in sperm morphology is negatively associated with the degree of sperm competition. In other words, sperm morphology varies less among males variation in species with greater sperm competition (as measured by frequency of extrapair paternity

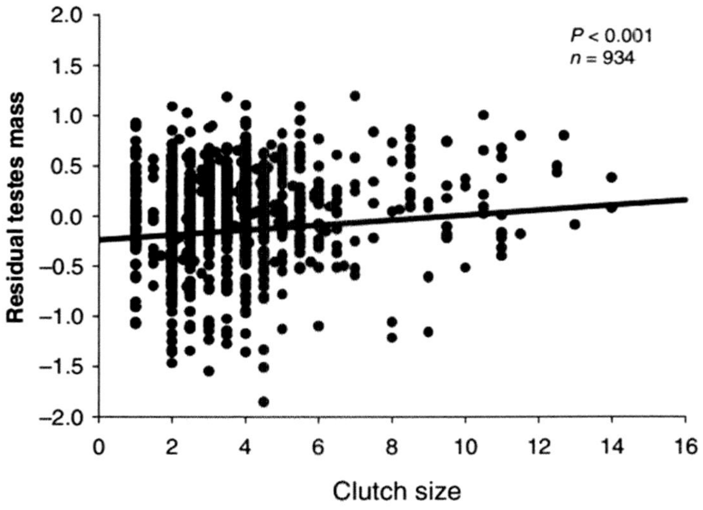


**Fig. 16.8** Males in species with larger clutches of eggs tend to have larger testes. (Figure from Pitcher et al. 2005; © European Society for Evolutionary Biology, used with permission)

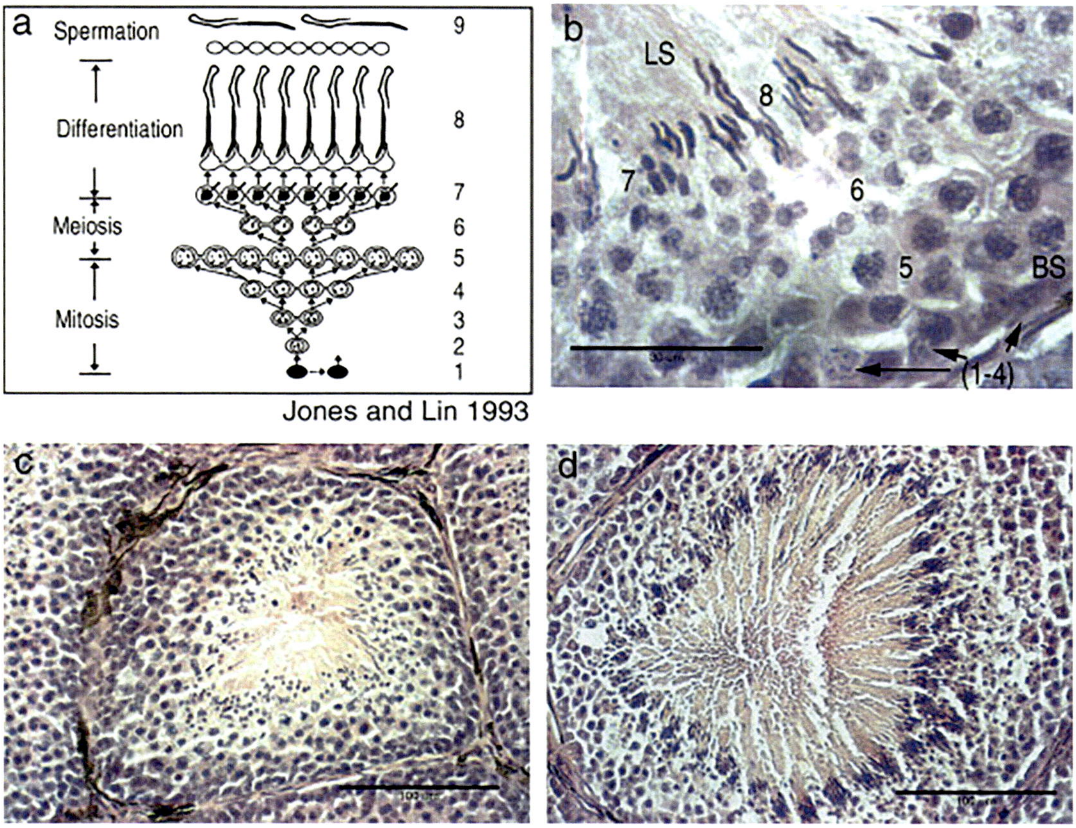


**Fig. 16.9** Spermatogenesis. (**a**) Illustration of spermatogenetic with three phases leading to spermiation (mitosis, meiosis, and differentiation) on the left and, on the right, the stages (1 to 9) of spermatogenesis. 1 = adult spermatogonia; 2 = adult pale 1 spermatogonia; 3 = adult pale 2 spermatogonia; 4 = B spermatogonia; 5 = primary spermatocytes; 6 = secondary spermatocytes; 7 = spermatids (still within the cell body); 8 = spermatids (outside the cell body, but still connected to the nutritional cell body); 9 = spermatozoa. (**b**) Cross-section (125×) of a seminiferous tubule of a Garden Warbler. The uniform picture of different spermatogenetic stages from the basal (BS) to the luminal (LS) side of the epithelium is due to a high degree of synchronization in the birds' spermatogenesis. Stages such as in a. (1–4) indicate spermatogonia in stages 1 to 4. Stage 5 indicates primary spermatocytes characterized by a conspicuous nuclear structure in meiotic prophase. Stage 6 indicates secondary spermatocytes, the first stage to reach haploidy and which differs in staining and size from the previous stage. Stage 7 indicates heavily stained spermatids that are still embedded in the cell body. Stage 8 indicates spermatids during differentiation. In this stage spermatids are still connected to the nutritional cell body and form sperm bundles due to incomplete cell divisions (see below). (**c**) and (**d**) represent cross-sections (31.25×) of representative seminiferous tubules. (**c**) the first spermatids (stage 8), but no sperm bundles. (**d**) sperm bundles. (Figure from Bauchinger et al. 2007; © 2007 Elsevier Inc., used with permission)

and relative testis size; Kleven et al. 2007, Calhim et al. 2007). This reduced variability suggests that sexual selection might favor some species-specific "optimum" sperm morphology or phenotype that enhances successful sperm competition (Calhim et al. 2007). Beyond swimming speed, sperm composition and morphology are likely also influenced by other factors, including the environment in female reproductive tracts and eggs (Rowe et al. 2015).

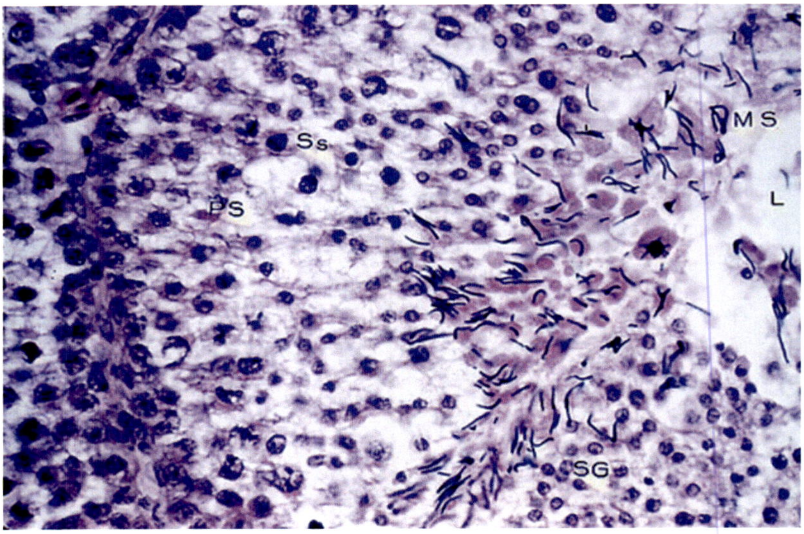


**Fig. 16.10** Light photomicrograph of a section of a testis showing a seminiferous tubule during full semen production. SG indicates spermatogonia; PS, primary spermatocyte; Ss, secondary spermatocyte; MS, mature spermatocyte; and L, lumen (original magnification ×800). (Figure from Samour 2002; © 2002 Journal of Avian Medicine and Surgery, used with permission)

### 16.2.3 Testosterone and Its Effects

In response to environmental signals such as photoperiod, the HPG axis is triggered and the plasma testosterone levels of male birds increase. In temperate, mid-latitude areas, the typical pattern of testosterone secretion in a socially monogamous species can be seen in male Song Sparrows (*Melospiza melodia*; Fig. 16.21), with plasma levels increasing sharply at the beginning of the breeding season, decreasing sharply as nesting begins, then increasing to a lesser degree with initiation of a second breeding attempt before declining to pre-breeding levels as the breeding season ends. The surge in plasma testosterone as breeding begins triggers important physiological and behavioral changes, including testes growth and sperm production and the expression of courtship, sexual and territorial behaviors (such as increased aggression and

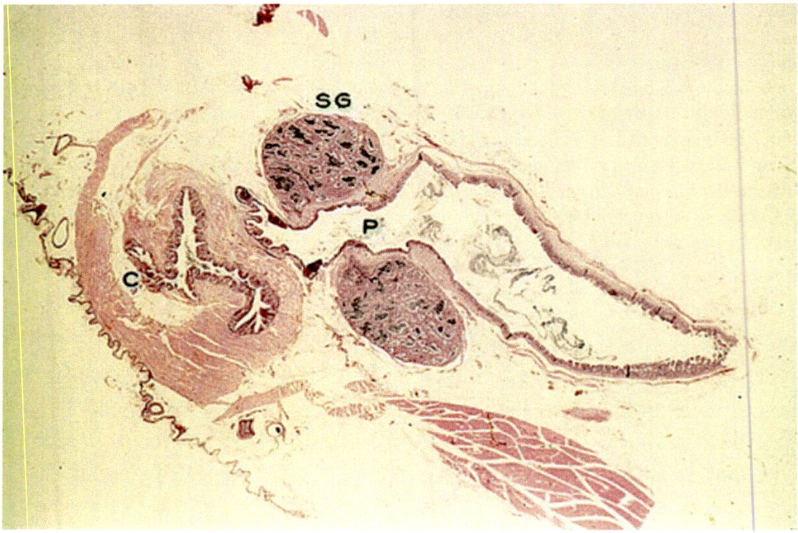


**Fig. 16.11** Longitudinal section of the cloaca of a male Budgerigar (*Melopsittacus undulatus*) during the culmination phase of the breeding cycle. A coiled extension of the posterior end of the vas deferens called the seminal glomus (SG) causes the cloacal protuberance. P, proctodeum of the cloaca; C, cloaca (original magnification ×12). (Figure from Samour 2002; © 2002 Journal of Avian Medicine and Surgery, used with permission)

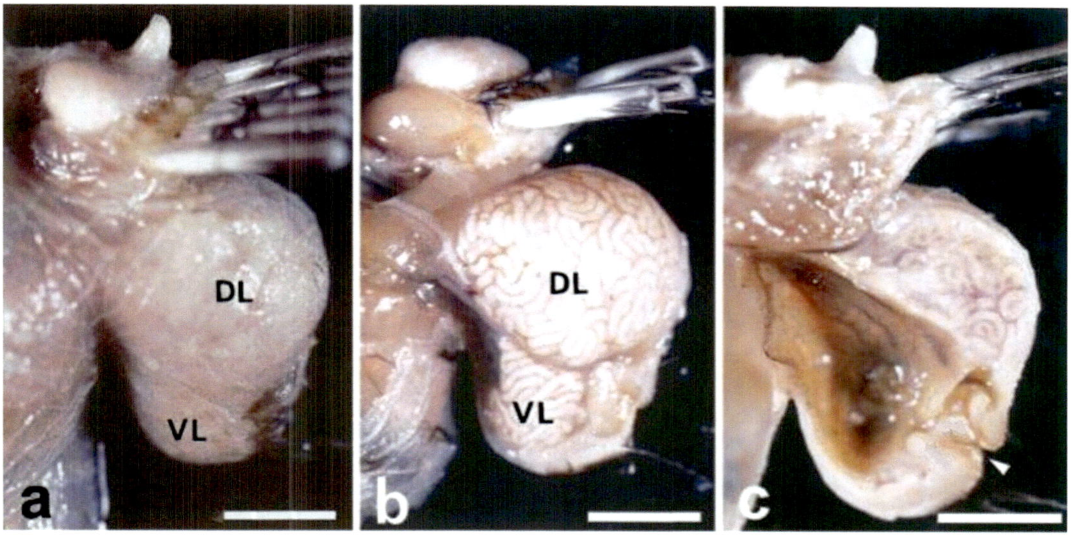


**Fig. 16.12** A fully developed cloacal protuberance of a male Alpine Accentor (*Prunella collaris*). (**a**) Lateral view of the cloacal protuberance after feathers have been removed. (**b**) Skin removed to show the inner structure of the cloacal protuberance. The dorsal (DL) and ventral (VL) lobes consist entirely of the ductus deferens. (**c**) Cross-section of the cloacal protuberance showing the vent (arrowhead). Scale bar = 5 mm. (Figure modified from Chiba and Nakamura 2003; © British Ornithologists' Union, used with permission)

singing; Fig. 16.22). Given the obvious advantages of these changes in terms of obtaining a mate and defending a territory, why do plasma testosterone levels tend to drop so dramatically once the incubation period begins? Addressing this question in studies of Dark-eyed Juncos (*Junco hyemalis*), investigators found that males with plasma testosterone levels maintained at peak levels (by the slow release of testosterone from small, subcutaneously implanted tubes) performed more courtship displays and had larger territories, but also provided less parental care, defended nests less vigorously, had suppressed immune function, and had lower survival rates than control males (with empty implants; Ketterson et al. 1996). Similar results have been reported for several other species (Ketterson and Nolan 1999) and clearly indicate that there are costs and benefits associated with testosterone that vary with breeding stage. Higher plasma testosterone levels are beneficial when males are establishing territories and attempting to attract a mate, but high levels would be costly when male contributions to parental care are needed to successfully raise young. For socially monogamous species at temperate latitudes, a relatively brief surge in plasma testosterone at the start of the breeding season and lower, but slightly fluctuating, levels thereafter is the pattern that appears to best maximize benefits (obtaining territories and mates, then providing parental care) and minimize costs (e.g., suppressed immune function in some species of birds).

Of course, not all species are socially monogamous, and most species are found in tropical areas rather than temperate areas. Among polygynous and lek species, where males may have multiple mates and contribute little or no parental care, plasma testosterone levels are generally higher than for males in socially monogamous species and remain elevated throughout the breeding season (Fig. 16.23). Sustaining high plasma testosterone levels for a prolonged period comes with potential costs. However, the benefits of increased reproductive success can likely outweigh those costs. For example, Red-winged Blackbirds (*Agelaius phoeniceus*) are polygynous and males with higher testosterone levels tend to have more mates (larger harems) than those with lower levels (Beletsky et al. 1989). Higher

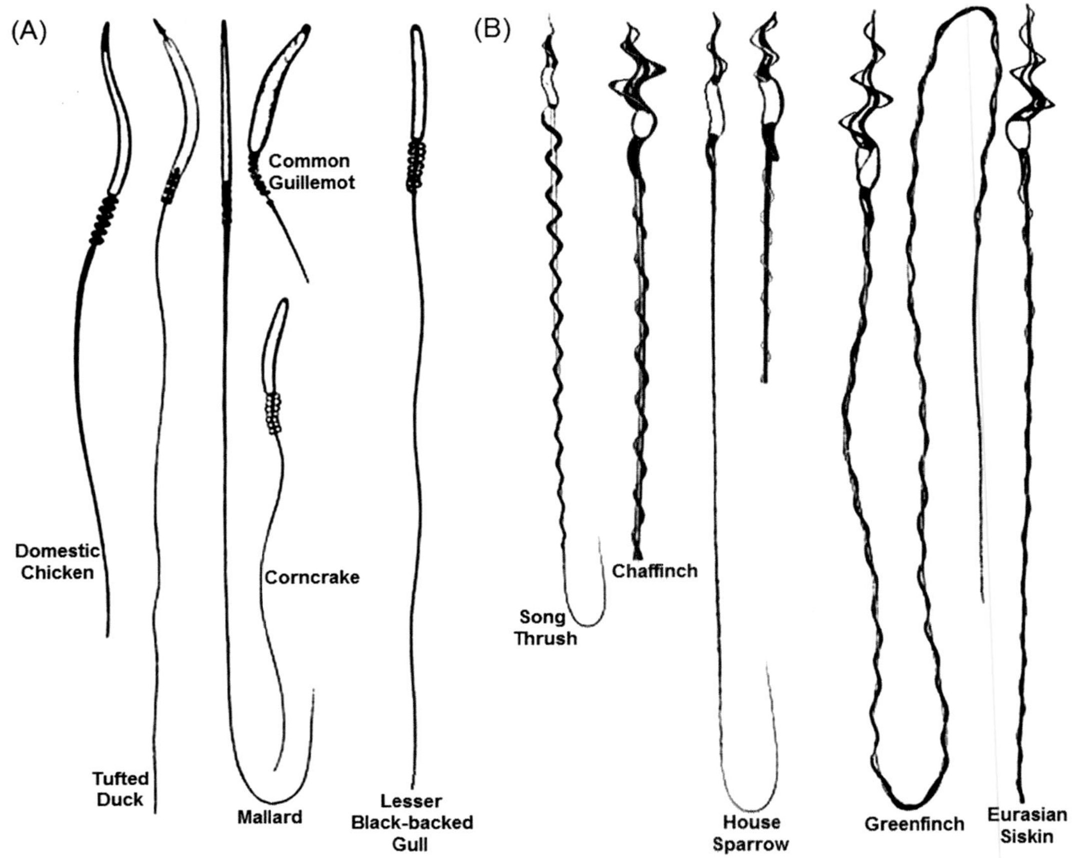


**Fig. 16.13** Examples of variation in sperm morphology of (**a**) nonpasserines and (**b**) passerines. Scientific names: Domestic Chicken, *Gallus g. domesticus*; Tufted Duck, *Aythya fuligula*; Mallard, *Anas platyrhynchos*; Corncrake, *Crex crex*; Common Guillemot (also called the Common Murre), *Uria aalge*; Lesser Black-backed Gull, *Larus fuscus*; Song Thrush, *Turdus philomelos*; Chaffinch, *Fringilla coelebs*; House Sparrow, *Passer domesticus*; European Greenfinch, *Chloris chloris*; Eurasian Siskin, *Spinus spinus*. (Figure modified from Jamieson 2007; © 2007 Reproduced by permission of Taylor & Francis Group)

testosterone levels are also correlated with increased mating success in Black Grouse (*Lyrurus tetrix*), a lekking species (Alatalo et al. 1996). In addition, the negative effects of testosterone on immunocompetence may be either limited or non-existent among males in some species where high testosterone levels are beneficial. For breeding male Red-winged Blackbirds (*Agelaius phoeniceus*), for example, there is no relationship between plasma testosterone levels and immunocompetence (Hasselquist et al. 1999). Similar results have been reported for House Sparrows (*Passer domesticus*; Greenman et al. 2005), Song Sparrows (*Melospiza melodia*), and Yellow Warblers (*Setophaga petechia*) (Wilder 2007).

In contrast to testosterone, carotenoids are immunostimulatory and can minimize or block the immunosuppressive effects of testosterone. Males may, therefore, be able to "balance" the potential negative effects of testosterone by using stored carotenoids. Experiments with male Zebra Finches (*Taeniopygia guttata*) revealed that some males with high carotenoid stores were able to avoid the immunosuppressive effects of testosterone, suggesting that such avoidance is condition-dependent (McGraw and Ardia 2007). Similarly, male European Greenfinches (*Chloris chloris*) with carotenoid-supplemented diets exhibited stronger immune responses than non-supplemented males (Aguilera and Amat 2007). Because few

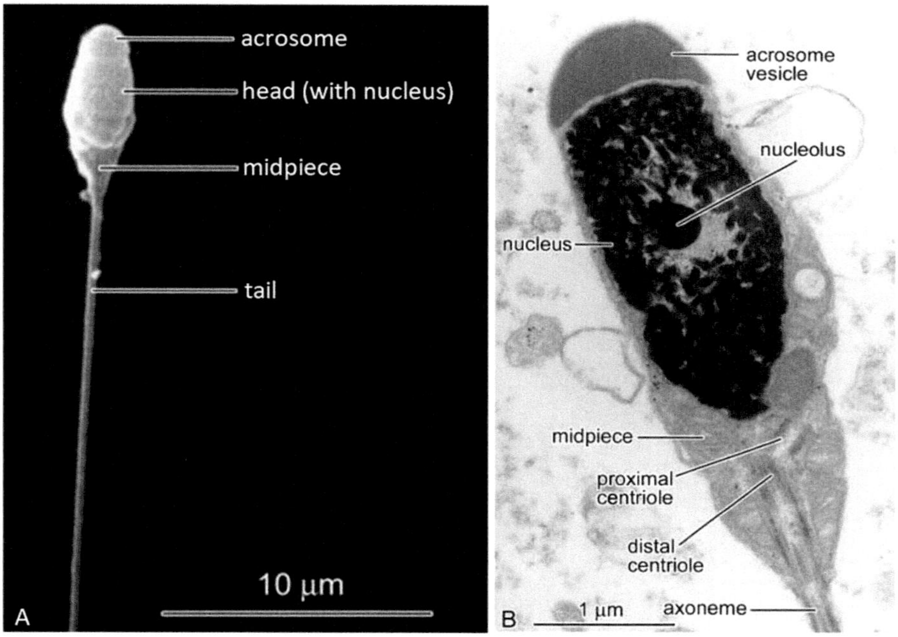


**Fig. 16.14** (**a**) Scanning electron micrograph of a Eurasian Bullfinch (*Pyrrhula pyrrhula*) spermatozoan, showing the acrosome, head, midpiece, and tail. (**b**) Longitudinal section of a Eurasian Bullfinch spermatozoan at a greater magnification. (Figure modified from Birkhead et al. 2007; © The Royal Swedish Academy of Sciences, used with permission)

investigators have examined the interactions between carotenoids, testosterone, and immunity, the general importance of carotenoids in compensating for immunosuppressive effects of testosterone remains to be determined.

Compared to birds at temperate latitudes, plasma concentrations of testosterone of many tropical species are relatively low year-round, with minimal changes in testosterone levels between breeding and non-breeding seasons in comparison with temperate species (Wikelski et al. 2003). For example, a comparison of two subspecies of African Stonechats (*Saxicola torquatus*) from temperate (52° N) and tropical (0°) latitudes revealed an approximate fivefold difference in testosterone levels (Rödl et al. 2004; Fig. 16.24). For many tropical species, the relationship between plasma testosterone levels and behavior has apparently been "uncoupled" to varying degrees. For example, some tropical species are highly territorial year-round despite low testosterone levels, whereas other species maintain year-round territories and testosterone levels are variable (Wikelski et al. 2003). Experiments have also produced conflicting results. In Spotted Antbirds (*Hylophylax naevioides*), testosterone appears to influence song and aggressive behavior (Hau et al. 2000a, b). However, testosterone does not appear to affect territorial aggression in Rufous-collared Sparrows (*Zonotrichia capensis*; Moore et al. 2004), Bay Wrens (*Cantorchilus nigricapillus*; Levin and Wingfield 1992), White-bellied Antbirds (*Myrmeciza longipes*; Fedy and Stutchbury 2006), and Buff-banded Rails (*Gallirallus philippensis*; Wiley and Goldizen 2003). The relationship between testosterone and display behavior in tropical species is also unclear. In Golden-collared Manakins (*Manacus vitellinus*), high levels of testosterone are not

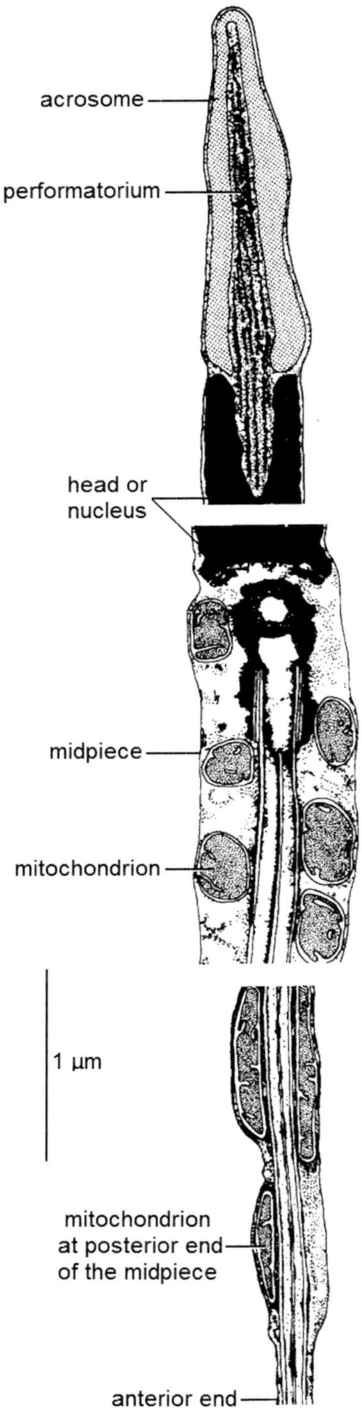


**Fig. 16.15** Ultrastructure of a spermatozoan of a male Budgerigar (*Melopsittacus undulatus*). The perforatorium is a rod-like structure found only in the acrosomes of nonpasserines. The DNA is located in the head (also called the nucleus) of the spermatozoan. The midpiece contains numerous mitochondria that provide energy (ATP) needed to power movement of the tail or flagellum. (Figure modified from Jamieson et al. 1995; © 1995 Wiley-Liss, Inc., used with permission)

needed for initiation of male display behavior (Day et al. 2007). However, testosterone is correlated with courtship behavior in Buff-banded Rails (Wiley and Goldizen 2003). These results appear to support the hypothesis that natural selection can influence the linkage between testosterone and male traits and behavior, altering the strength of the linkage in some case and even uncoupling the linkage in other cases (Hau 2007). Because of the effects that testosterone can have on a wide variety of behavior, physiological, and morphological traits (Fig. 16.22), such weakening or uncoupling has clearly been favored by natural selection in many tropical species.

Not surprisingly, given the number of bird species found at tropical latitudes, low concentrations of plasma testosterone are not universal among tropical birds. One important factor is the length of the breeding season. Many tropical species defend territories year-round and maintain long-term pair bonds (Stutchbury and Morton 2001). However, among tropical species with shorter breeding seasons, males defend territories for shorter periods and, during those periods, may have testosterone levels comparable to those of temperate species. As for many temperate species, shorter breeding seasons may mean greater competition in terms of establishing and defending a territory (more male-male competition) and attracting a mate, and such competition likely favors a "coupling" of the linkage between testosterone, behavior, and physiology (Goymann et al. 2004).

Given the great number of bird species in the tropics, the diversity of life-history strategies is not surprising. Clearly, natural selection favors the degree of linkage between testosterone, behavior, and physiology that maximizes benefits relative to costs. Because testosterone can have widespread, and sometime negative, effects

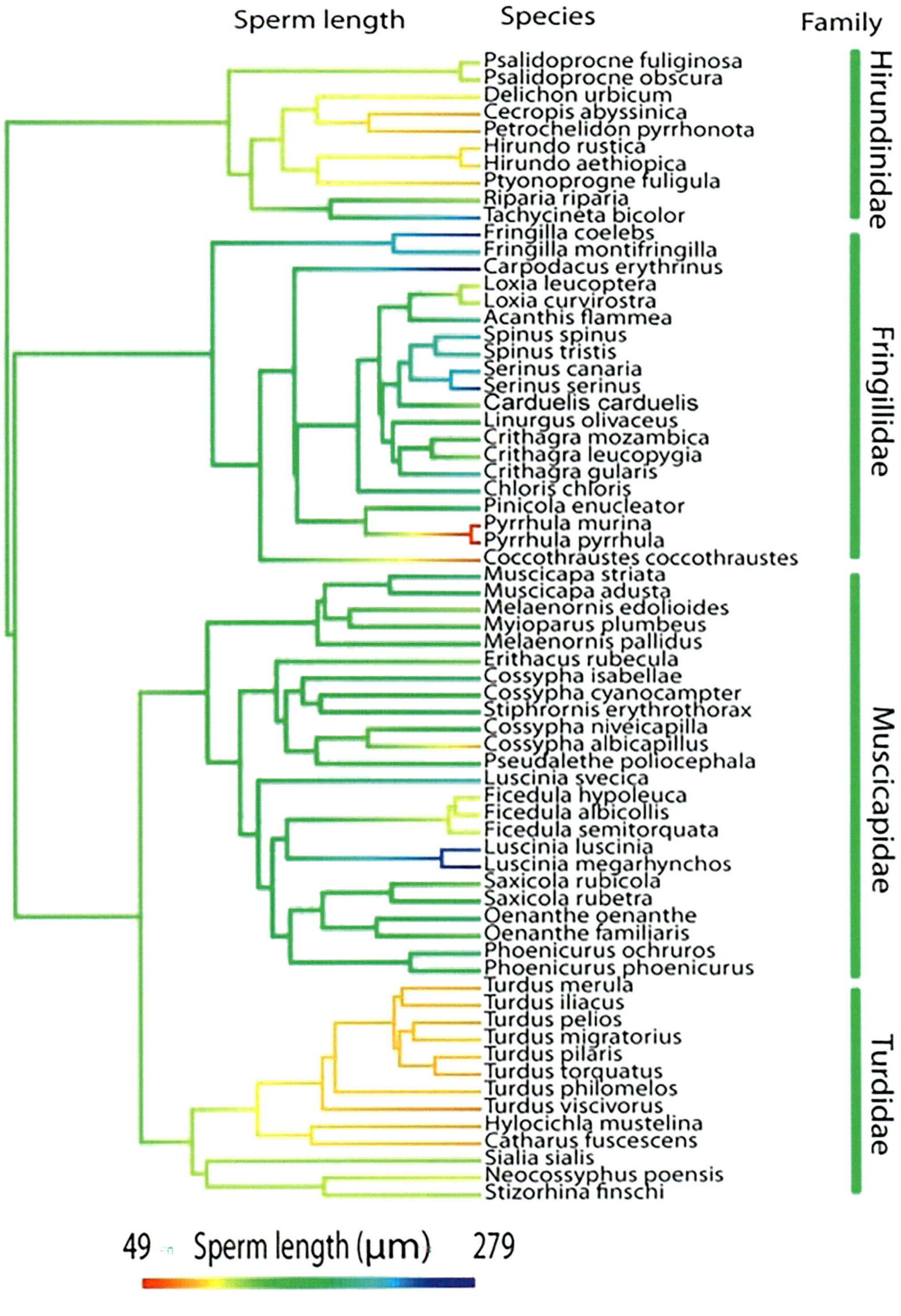


**Fig. 16.16** Variation in total sperm length among several species in the families Fringillidae, Muscicapidae, Turdidae and Hirundinidae. Analysis revealed strong phylogenetic signals in total sperm length. Variation in sperm length is indicated by the bar of color ranging from red to blue. (Figure from Omotoriogun et al. 2020; open-access article distributed under the terms of the Creative Commons CC-BY license, https://creativecommons.org/licenses/by/4.0/)

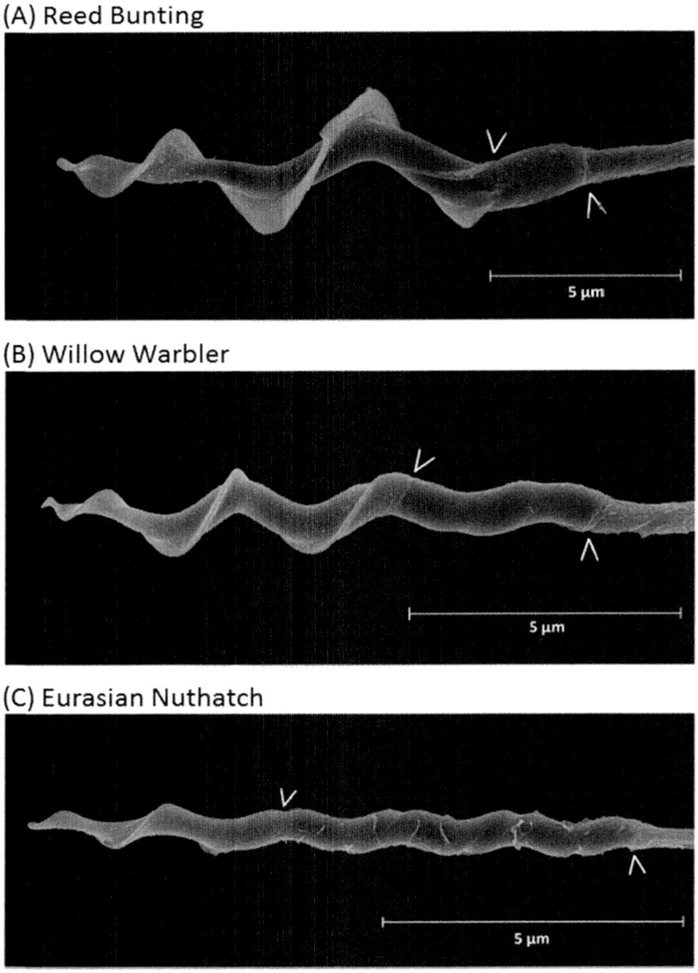


**Fig. 16.17** Examples of variation in the morphology of sperm heads among 36 species of songbirds, ranging from (**a**) a pronounced helical shape to (**c**) a more subtle helical shape. The top (left) arrowhead indicates the junction between the acrosome and head, and the bottom (right) arrowhead the junction between the head and the middle piece. Swimming speeds were higher for sperm with more helical shapes. Reed Bunting, *Emberiza schoeniclus*; Willow Warbler, *Phylloscopus trochilus*; Eurasian Nuthatch, *Sitta europaea*. (Figure from Støstad et al. 2018; © 2018 The Authors. Evolution © 2018 The Society for the Study of Evolution, used with permission)

(Fig. 16.22), particularly when plasma levels remain high for extended periods, natural selection likely favors the "uncoupling" of testosterone, behavior, and physiology in species with strategies involving long-term pair bonds and year-round territories. For tropical species with life-history strategies more similar to those of many temperature species, such uncoupling is less likely to occur.

## 16.3 Ovaries

With few exceptions, female birds have only a left ovary and oviduct, a condition that has persisted since the dinosaur-avian transition (Zheng et al. 2013). Among those exceptions are at least 86 species that have two ovaries (but not necessarily two functional oviducts). Taxa with the greatest number of species with two ovaries include the orders Apterygiformes (kiwis), Accipitriformes (hawks, Osprey, and eagles), Falconiformes, and Strigiformes (owls) (Kinski 1971; Fig. 16.25). Among kiwis, Kinski 1971:344) found that the left ovaries were generally larger than right ovaries, but both "appeared perfectly functional." Only the left oviduct of female kiwis is functional, but the two ovaries overlap and the entrance to the oviduct covers the complete width of both ovaries so that ova from either ovary will enter the oviduct (Kinski 1971). Among diurnal raptors (orders

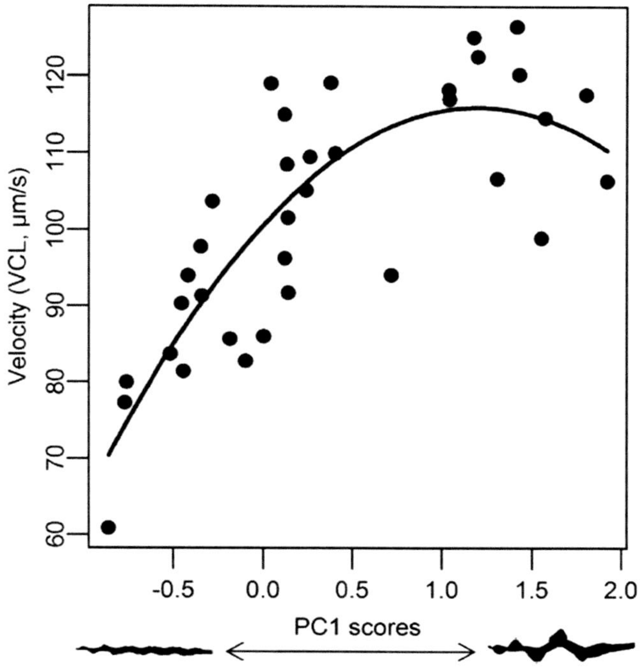


**Fig. 16.18** Relationship between the morphology of sperm heads (PC1 scores) and sperm swimming velocity for 35 species of songbird. Sperm with a greater helical morphology had faster swimming speeds. VCL = curvilinear velocity. (Figure from Støstad et al. 2018; © 2018 The Author(s). Evolution © 2018 The Society for the Study of Evolution, used with permission)

Accipitriformes and Falconiformes), the right ovary is generally smaller than the left ovary, is reportedly "less active," and the right oviduct is often vestigial (Stanley and Witschi 1940). More recently, however, Rodler et al. (2015) determined that follicle formation and ovulation were possible for the right ovaries of four species of raptors (Fig. 16.25). As with kiwis, the right oviducts of raptors may not be present or may be vestigial, but ova produced by either ovary may enter the right oviduct (Stieve 1924, as cited by Rodler et al. 2015).

In early stages of embryonic development, all female birds have two ovaries, but only the left one develops into a functional organ in most species. For example, in embryos of Domestic Chickens (*Gallus g. domesticus*), the right ovary and oviduct are already smaller than those on the left side as early as four days post-fertilization (Romanoff and Romanoff 1949). Advantages of having a single ovary and oviduct include weight reduction and less energy is needed to develop and maintain the reproductive organs. The ovary (or ovaries) of immature females contain(s) a mass of small eggs or ova, but relatively few will ever mature and be ovulated during a bird's lifetime.

## 16.4 Egg Production

As with males, breeding by females is initiated in response to photic (e.g., longer days) and/or non-photic (e.g., rain) cues that stimulate the release of gonadotropin-releasing hormone (GnRH) by the hypothalamus, which, in turn, stimulates the release of follicle-stimulating hormone (or FSH) and luteinizing hormone (LH) by the anterior pituitary. LH and FSH bind to receptors in the ovarian follicles, resulting in the production of androgen (testosterone), estradiol, and progesterone (Fig. 16.26). Mature ovaries of female birds contain four types of follicles: primordial follicles, white cortical follicles (Figs. 16.27 and 16.28), small yellow follicles, and large yellow preovulatory follicles. Associated with ovarian follicles are three types of cells, including the egg (oocyte), granulosa cells, and theca cells (Figs. 16.28 and 16.29). During folliculogenesis, a thin perivitelline

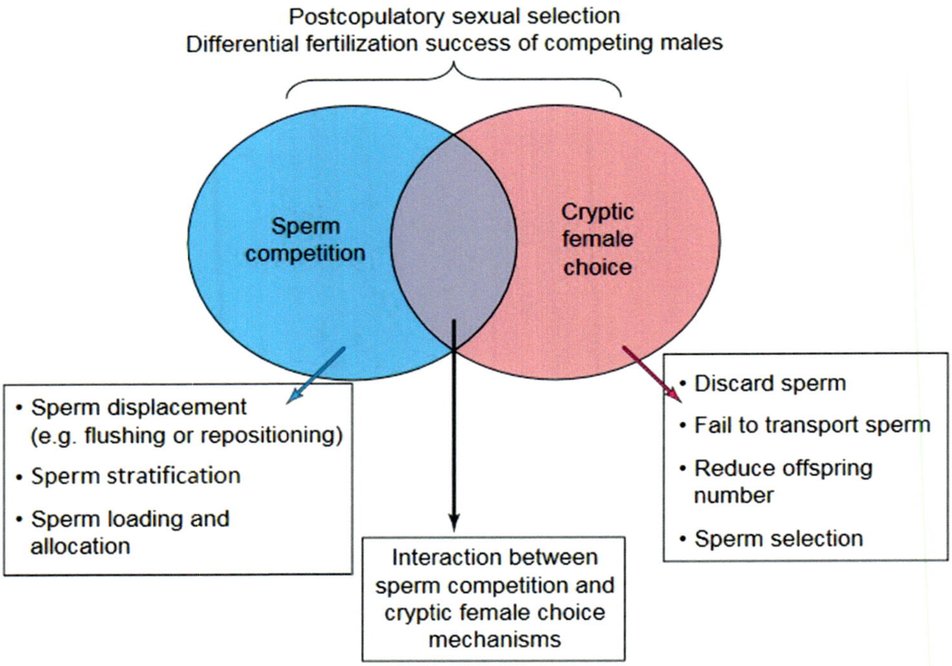


**Fig. 16.19** Examples of postcopulatory mechanisms of sperm competition and cryptic female choice that result in differential male fertilization success. Sperm competition mechanisms include those that might be meditated by sperm themselves, such as sperm displacement that includes sperm flushing (sperm from incoming males removes resident sperm in the female from storage) and sperm repositioning (the incoming ejaculate of a male repositions sperm from the previous partner of his mate away from the storage or the fertilization site). Sperm loading occurs when males with more ejaculate gain paternity in competition with the ejaculate of a previous male. Sperm allocation refers to changes in the number of sperm ejaculated based on the risk of sperm competition and female quality or novelty. Sperm stratification indicates a positional advantage to the ejaculate of a competing male, such that those sperm are the first out during fertilization. This can be due to passive sperm loss, mating order, or sperm repositioning. Cryptic female choice mechanisms can include discarding sperm from unfavored males, failing to transport sperm from at least one competing male to the site of either sperm storage or fertilization, decreasing the number of eggs fertilized by a particular male, and non-random use of sperm from a particular male over those of other males. Mechanisms of sperm competition via sperm and cryptic female choice probably overlap, although both the extent and relative contributions of each to differential fertilization success are unknown. (Figure modified from Snook 2005; © 2004 Elsevier Ltd., used with permission)

membrane that plays an important role in the fertilization process also forms between the oocyte and the granulosa cells (Figs. 16.30 and 16.31). Testosterone and estradiol are produced and released by cells in the thecal layers of small follicles, and progesterone is produced and released by the granulosa cells of the pre-ovulatory follicles.

The number of preovulatory follicles can range from one or perhaps two for species laying a single egg (e.g., Emperor Penguins, *Aptenodytes forsteri*) to eight or more in species that have much larger clutches (e.g., many galliform species). For species that have larger clutches, the preovulatory follicles range in size, with the largest (F1) the most mature and next to ovulate, followed, in diminishing sizes, by least mature preovulatory follicles. When a female is laying eggs, one of the small yellow follicles becomes a preovulatory follicle after the F1 follicle is ovulated and, as small yellow follicles become preovulatory follicles, they are replaced by white follicles that become small yellow follicles (Onagbesan et al. 2009). As white follicles grow,

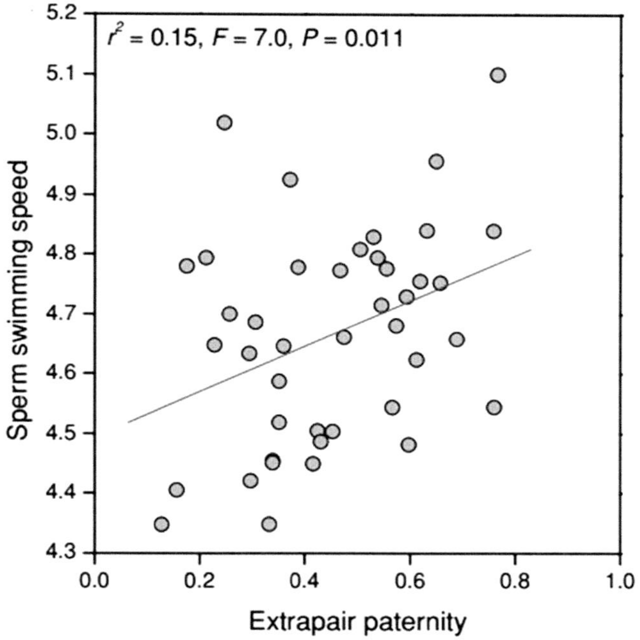


**Fig. 16.20** Swimming speed of sperm of 42 species of songbirds relative to extrapair paternity, i.e., the proportion of extrapair young, with a ln transformation of sperm swimming speed and arcsine square-root transformation of extrapair paternity. (Figure from Kleven et al. 2009; © 2009 The Authors. Journal compilation © 2009 The Society for the Study of Evolution, used with permission)

their original single-cell layer of granulosa cells becomes multi-layered, and selected small yellow follicles take up large amounts of lipid- and xanthophyll-rich yolk and begin to grow rapidly.

Proteins in egg yolk, with the exception of immunoglobulins, are synthesized by the liver, then transported in the blood to the ovarian follicle where they are processed into yolk (Wallace

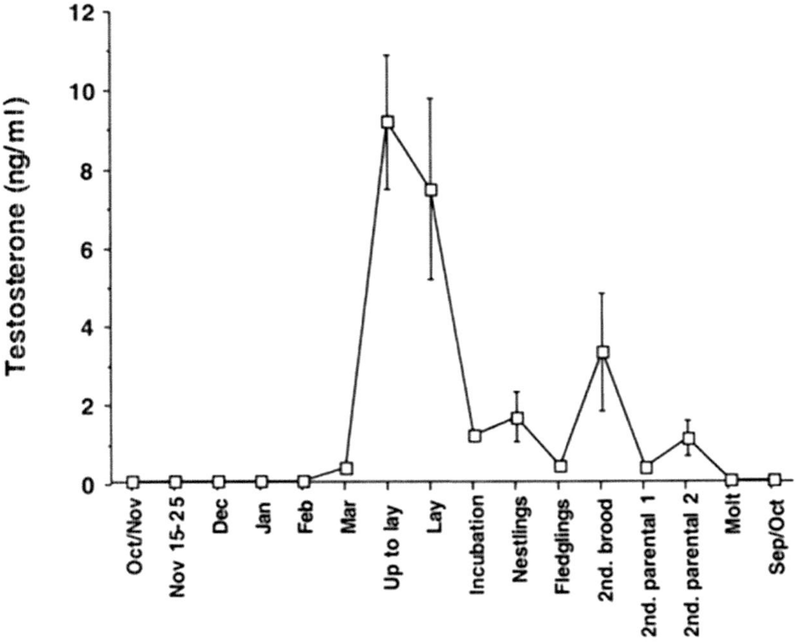


**Fig. 16.21** The pattern of testosterone secretion in free-living populations of Song Sparrows (*Melospiza melodia*). Plasma levels peak in April and May as breeding began and then were maintained at a lower "breeding baseline" during the rest of the breeding season. As prebasic molt ensued, plasma levels of testosterone were basal and remained so throughout autumn and winter. (Figure from Wingfield and Hahn 1994; © Published by Elsevier Ltd., used with permission)

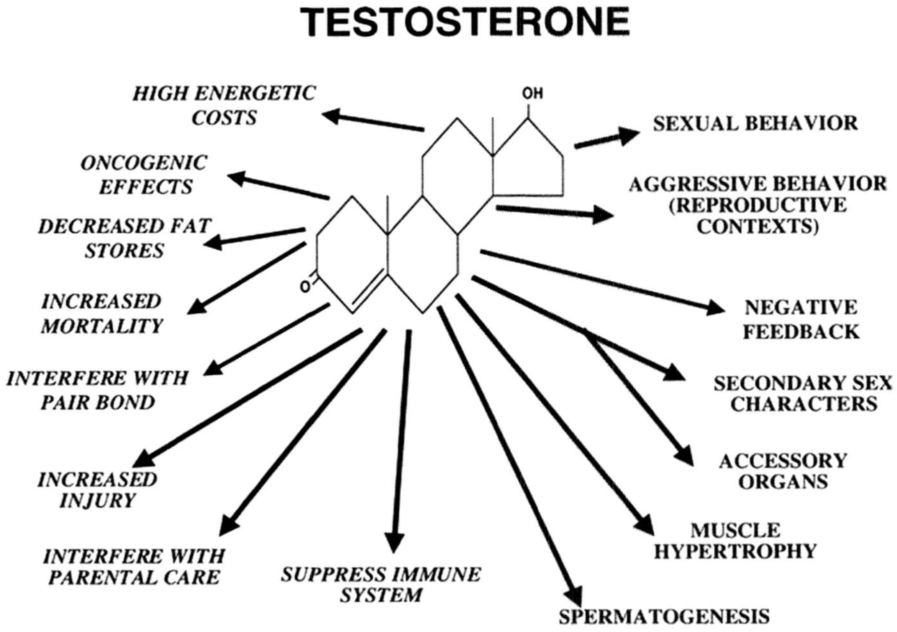


**Fig. 16.22** The morphological, physiological, and behavioral actions of testosterone that are essential for male reproduction are listed to the right and below. The "costs" of prolonged high levels of testosterone are on the left in italics. Patterns of plasma testosterone levels may be a function of secretion patterns to maintain male reproductive function, and costs of testosterone that require low plasma levels. (Figure from Wingfield et al. 2000; © S. Karger AG, used with permission)

1985; Stifani et al. 1988). Estrogens play an important role in the production of yolk precursors by the liver (vitellogenin and very-low-density lipoprotein; Fig. 16.32). Yolk consists primarily of water, lipids, and protein. Yolk proteins are also involved in the transport of vitamins, ions and carbohydrates (in addition to lipids) from the female to the egg during the period of rapid development of the yolk (Carey 1996). Yolk lipids serve as the primary source of energy for developing embryos.

The relative amount of yolk in eggs varies with degree of development that takes place in the egg. Among species where young are born naked and helpless (e.g., altricial), there is relatively less yolk and, for species where young are more developed at hatching (e.g., precocial), there is relatively more yolk (Fig. 16.33). The lipid content of yolk is generally greater than 50% in most species of birds (Carey et al. 1980; Sotherland and Rahn 1987). However, as yolk size increases, the relative amount of lipid and the energy available to developing embryos also increases (Fig. 16.33; Vleck and Bucher 1998).

Under the influence of follicle-stimulating hormone (FSH), ovarian follicles secrete hormones in differing amounts as they mature. White and small yellow follicles secrete more estrogen and small amounts of androgens (testosterone), whereas preovulatory follicles secrete more progesterone. This shift from estrogen to progesterone production promotes follicular recruitment into the follicular hierarchy (Hernandez and Bahr 2003). With increasing maturity, preovulatory follicles increase production of progesterone and greatly reduce their production of estrogen (Fig. 16.34).

As preovulatory follicles mature, their granulosa cells release increasing amounts of progesterone and this continuing increase eventually triggers a rapid increase in plasma levels of luteinizing hormone (LH) (Figs. 16.35 and

**Fig. 16.23** Mean (± SE) circulating level of testosterone in male birds in relation mating system: (top panel) peak testosterone level, and (bottom panel) residual testosterone (residuals from the phylogenetically corrected regression of $\log_{10}$-transformed peak testosterone on $\log_{10}$-transformed basal testosterone level). (Figure from Garamszegi et al. 2005; © 2004 Elsevier Inc., used with permission)

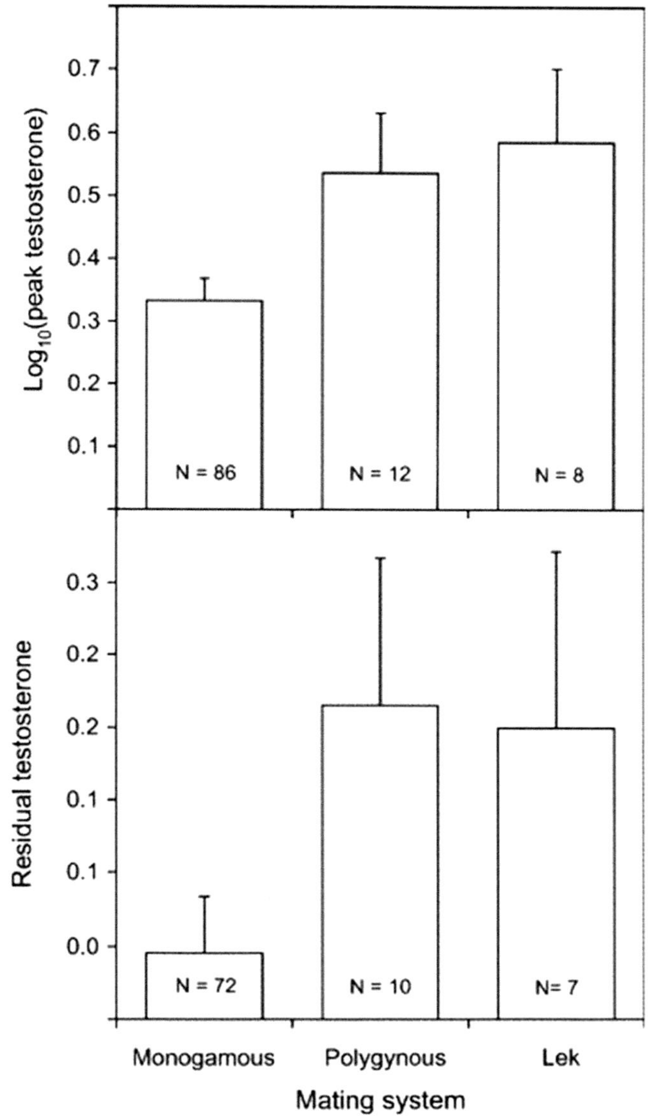


16.36). The surge in plasma levels of LH causes an increased production of testosterone by follicular thecal cells. The increasing levels of these three hormones causes ovulation of the F1 preovulatory follicle (Rangel and Gutierrez 2014). As plasma levels of these hormones increase, increasing amounts of rough endoplasmic reticulum develop in the theca externa of the stigma region, suggesting an increase in synthesis of some protein(s), possibly proteolytic enzymes. The theca externa in the stigma region becomes thinner just a few minutes before ovulation, perhaps due to a breakdown of tissue by proteolytic enzymes (Yoshimura and Barua 2017). The increasingly thin stigma then ruptures, the egg is then ovulated and enters the infundibulum of the oviduct a tubular organ within which ova are fertilized, eggs are transported and formed, and sperm are selected, stored, and transported (Fig. 16.37). Because ovary and oviduct are not directly connected, small finger-like projections called fimbriae at the entrance to the oviduct

**Box 16.3 Testosterone Increases Availability of Carotenoids**

Androgens and carotenoids play a fundamental role in the expression of secondary sex traits in animals that communicate information on individual quality. In birds, androgens regulate singing, aggression, and a variety of sexual ornaments and displays, whereas carotenoids are responsible for the red, yellow, and orange colors of the integument. Parallel, but independent, research lines suggest that the evolutionary stability of each signaling system stems from trade-offs with immune function: androgens can be immunosuppressive, and carotenoids diverted to coloration prevent their use as immunostimulants. Despite strong similarities in the patterns of sex, age and seasonal variation, social function, and proximate control, there has been little success at integrating potential links between the two signaling systems. These parallel patterns led us to hypothesize that testosterone increases the bioavailability of circulating carotenoids. To test this hypothesis, Blas et al. (2006) manipulated testosterone levels of Red-legged Partridges (*Alectoris rufa*) while monitoring carotenoids, color, and immune function. Testosterone treatment increased the concentration of carotenoids in plasma and liver by >20%. Plasma carotenoids were in turn responsible for individual differences in coloration and immune response. These results provide experimental evidence for a link between testosterone levels and immunoenhancing carotenoids that (1) reconcile conflicting evidence for the immunosuppressive nature of androgens, (2) provide physiological grounds for a connection between two of the main signaling systems in animals, (3) explain how these signaling systems can be evolutionary stable and honest, and (4) may explain the high prevalence of sexual dimorphism in carotenoid-based coloration in animals.

Red-legged Partridge. (Photo by Paolo-Manzi, purchased from iStockphoto.com)

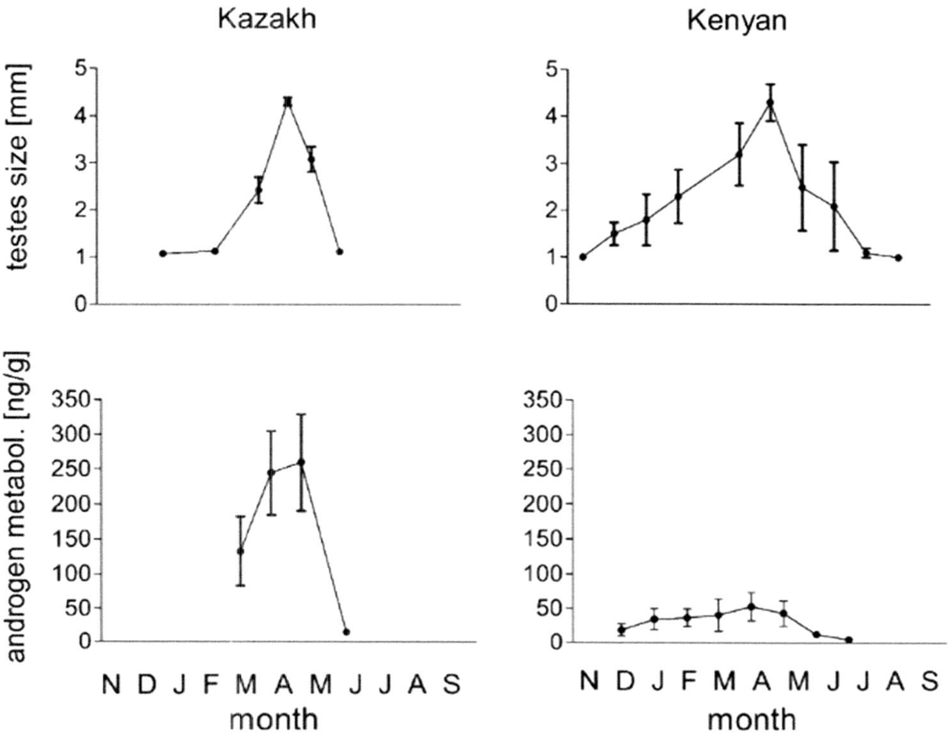


**Fig. 16.24** Mean (± SE) testis sizes and excreted androgen metabolite levels in hand-raised African Stonechats (*Saxicola torquatus*) from two different subspecies (one from Kazakhstan and one from Kenya) held under controlled laboratory conditions with a photoperiod simulating that in their natural habitat. (Figure modified from Rödl et al. 2004; © 2004 Elsevier Inc., used with permission)

guide the ovum into the first portion of the oviduct, the infundibulum. Here, sperm make contact with the ovum and fertilization occurs. Fertilization must occur quickly because secretions from glands in the walls of the infundibulum form a layer of protein around the ovum in 15 to 20 minutes or less. This thin layer is called the outer perivitelline layer because the ovulated ovum already has a thin protein coat (that sperm can penetrate) called the inner perivitelline layer (Fig. 16.31).

Yolk is primarily water and lipids (Fig. 16.38), with lipids making up 58% of yolk solids and providing most of the energy needed for embryonic development (Sotherland and Rahn 1987). The primary egg yolk lipids are triacylglycerols, phospholipids, cholesterol, and free fatty acids (Noble 1991; Carey 1996), and variation in the fatty acid content of yolk may influence offspring fitness (Mentesana et al. 2021). The most abundant type of lipid in yolk is triglycerides, but the relative amounts of these lipids vary among species. For example, Royle et al. (1999) found that about two-thirds of the yolk lipids of Lesser Black-backed Gulls (*Larus fuscus*) were triglycerides, whereas, for Great Reed Warblers (*Acrocephalus arundinaceus*) and Common Redstarts (*Phoenicurus phoenicurus*), about 54% and 57%, respectively, of yolk lipids were triglycerides (Igic et al. 2015a, b). Similarly, about 59% and 62% of the yolk lipids of American Kestrels (*Falco sparverius*) and Red-legged Partridges (*Alectoris rufa*), respectively, were triglycerides (Surai et al. 2001). Given their abundance, triglycerides provide most of the energy required for embryonic development, e.g., 90%

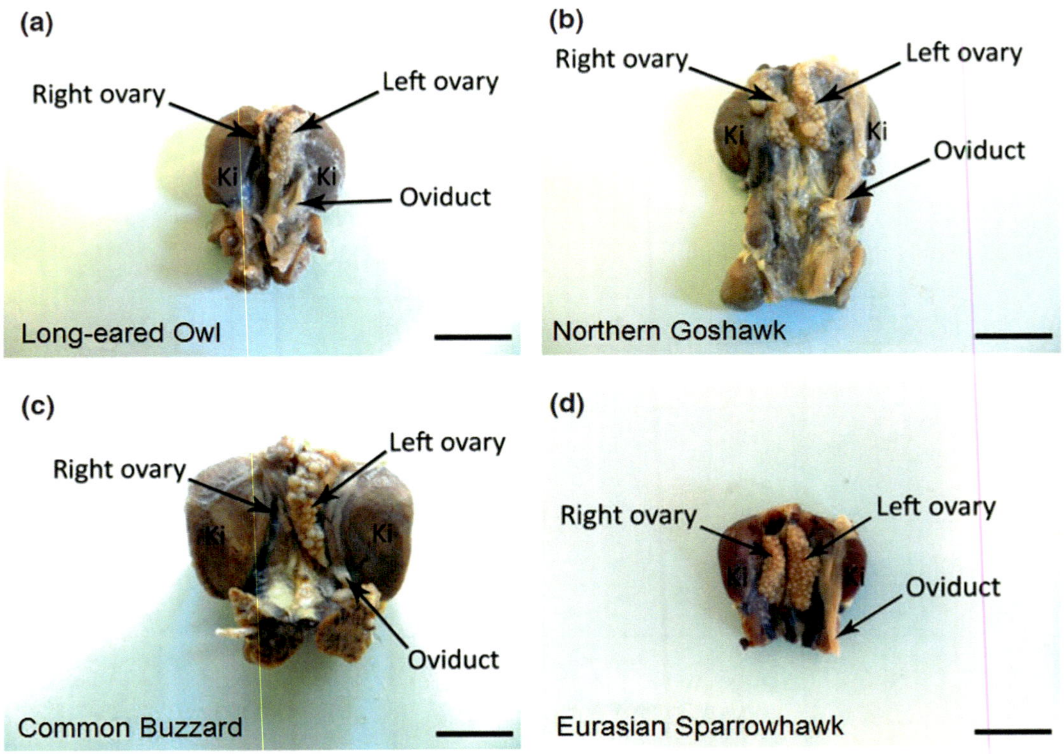


**Fig. 16.25** (**a**) Ovaries of a female Long-eared Owl (*Asio otus*). The right ovary is much smaller than the left ovary and contains poorly developed follicles. (**b**) Ovaries of a Northern Goshawk (*Accipiter gentilis*), with both similar in size and follicular development. (**c**) Ovaries of a Common Buzzard (*Buteo buteo*). The right ovary is much smaller than the left ovary and has poorly developed follicles. (**d**) Ovaries of a Eurasian Sparrowhawk (*Accipiter nisus*). The right ovary is slightly smaller than the left ovary, but both ovaries have similar follicular development Scale bars = 0.5 cm. (Figure from Rodler et al. 2015; © 2014 Blackwell Verlag GmbH, used with permission)

for the embryos of Domestic Chickens (Noble and Cocchi 1990). Phospholipids are essential components of cell membranes, and cholesterol is required for development because it maintains the integrity of cell membranes and is the precursor for steroid hormones (Cindrova-Davies et al. 2017). The egg yolk proteins, known as vitellogenins (VTGs), are the primary source of amino acids that are incorporated into structural and nonstructural proteins of the developing embryo (Wahli 1988; Finn 2007).

Although consisting primarily of water and lipids, other components of yolk can play critical roles in the development of a young bird. The phenotypes and environments of females can influence the extent to which these components are present in the yolk, which, in turn, influences the phenotype and fitness of her young. As such, the ways in which different substances deposited in yolk by females affect their young are referred to as maternal effects.

Egg yolks are yellow or yellowish-red and that color is caused by the presence of carotenoids (Fig. 16.39). Carotenoids are pigments that birds must obtain from their diet because they can only be produced by plants and some bacteria and fungi. These pigments are responsible for the colorful yellow, red, and orange plumage of many birds, but they also play an important role in eggs. Developing embryos depend on the lipids in yolk for energy, but metabolizing those lipids releases free radicals, highly reactive atoms or

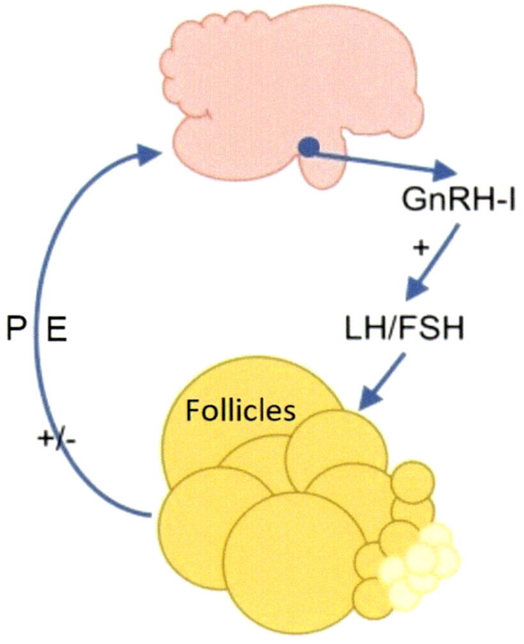


**Fig. 16.26** At the initiation of breeding, in response to photic and/or non-photic cues, GnRH-1 (gonadotropin releasing hormone) is released from the hypothalamus and stimulates the release of LH (luteinizing hormone) and FSH (follicle stimulating hormone) from the pituitary into the circulatory system. In response, cells in the follicles produce progesterone (P), estradiol (E), and testosterone (not shown). (Figure from van der Klein et al. 2020; © 2019 The Authors. Published by Elsevier B.V., used with permission)

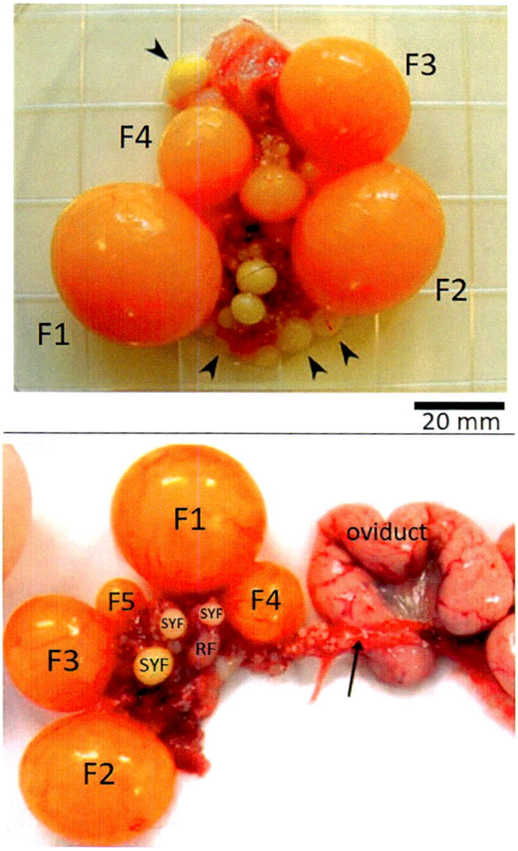


**Fig. 16.27** Top, Ovary of a Domestic Chicken. The four largest follicles are labeled F1–F4, and several white follicles are indicated by the arrow heads. (Figure from Okumura 2017; © 2017 Springer Nature Singapore Pte Ltd., used with permission). Bottom, Ovary (left) of a female Domestic Chicken with pre-ovulatory follicles (F1–F5), small yellow follicles (SYF), and a regressing post-ovulatory follicle (RF) and initial section of the oviduct (right). The arrow indicates where nerves and blood vessels enter the ovary. (Figure from Apperson et al. 2017; open-access article distributed under the Creative Commons Attribution License, https://creativecommons.org/licenses/by/4.0/)

molecules that can damage cell membranes and even DNA. Carotenoids deposited in the yolk by female birds react with and neutralize free radicals, thus preventing them from causing damage to the developing embryo. However, females also need carotenoids to minimize or prevent damage by free radicals, particularly when fighting infections because free radicals are produced during the immune response. So, females face a trade-off in terms of how much carotenoid to deposit in eggs and how much to retain for themselves (Blount et al. 2000). This may be particularly true in species where young hatch in a more advanced state (precocial and superprecocial species) because they metabolize more lipids over a longer period of time and may be at greater risk from damage caused by free radicals (Blount 2004).

Developing bird embryos cannot synthesize their own antibodies (Fig. 16.40), and are not able to do so until several days after hatching. Until then, embryos and hatchlings rely on antibodies deposited in the yolk by the female (Box 16.4 Egg Antimicrobial Defenses). This defense against pathogens resulting from maternal antibodies is called passive immunity and the

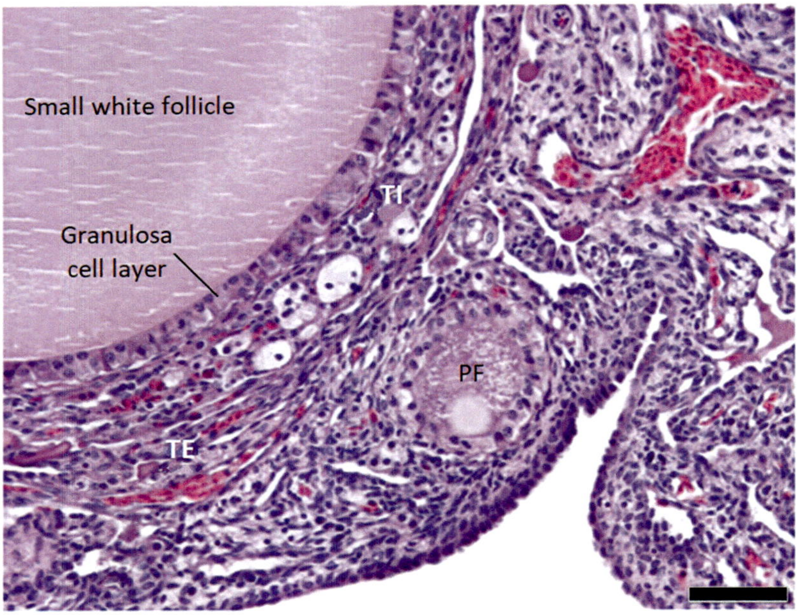


**Fig. 16.28** Ovarian cortex of a Domestic Chicken (*Gallus g. domesticus*) showing a portion of a small white follicle surrounded by a single layer of granulosa cells plus the theca interna (TI) and theca externa (TE). Also shown is a primordial follicle (PF). (Figure from Apperson et al. 2017; © 2017 by the Authors. Licensee MDPI, Basel, Switzerland. Open-access article distributed under the terms and conditions of the Creative Commons Attribution (CC BY) license, https://creativecommons.org/licenses/by/4.0/)

effectiveness of this defense depends on the number and type of antibodies deposited in the yolk by a female. Beyond providing protection, high levels of maternal antibodies may permit young to devote more resources to rapid growth during the period immediately after hatching (Buechler et al. 2002). More antibodies deposited in yolk may, therefore, enhance the survival and growth of young birds. Females in better physical condition may be able to deposit more antibodies in their eggs because antibody production requires both energy and the building blocks (amino acids

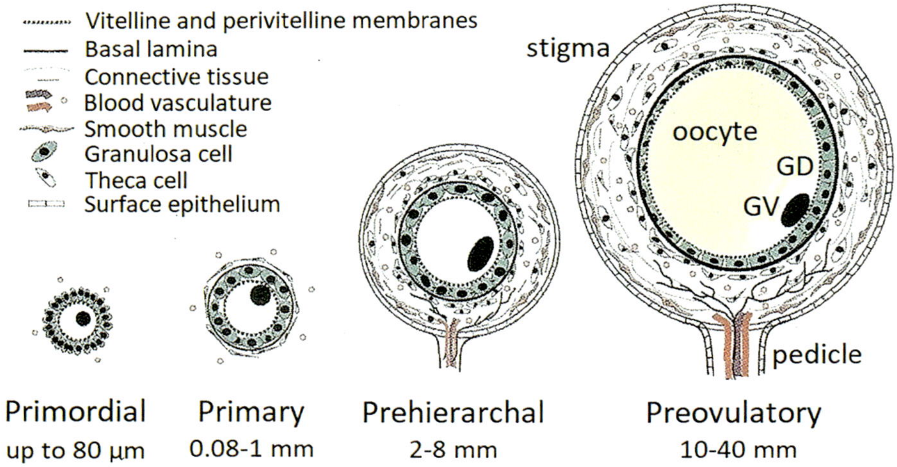


**Fig. 16.29** Development of an ovarian follicle. Granulosa cells produce primarily progesterone and theca cells primarily estrogen plus some androgens. The stigma is where the follicular wall will rupture and the oocyte will be released (ovulation). GV, germinal vesicle or nucleus; GD, germinal disc, where fertilization will occur. (Figure from Johnson and Woods 2007; © 2007 Taylor & Francis Group, used with permission)

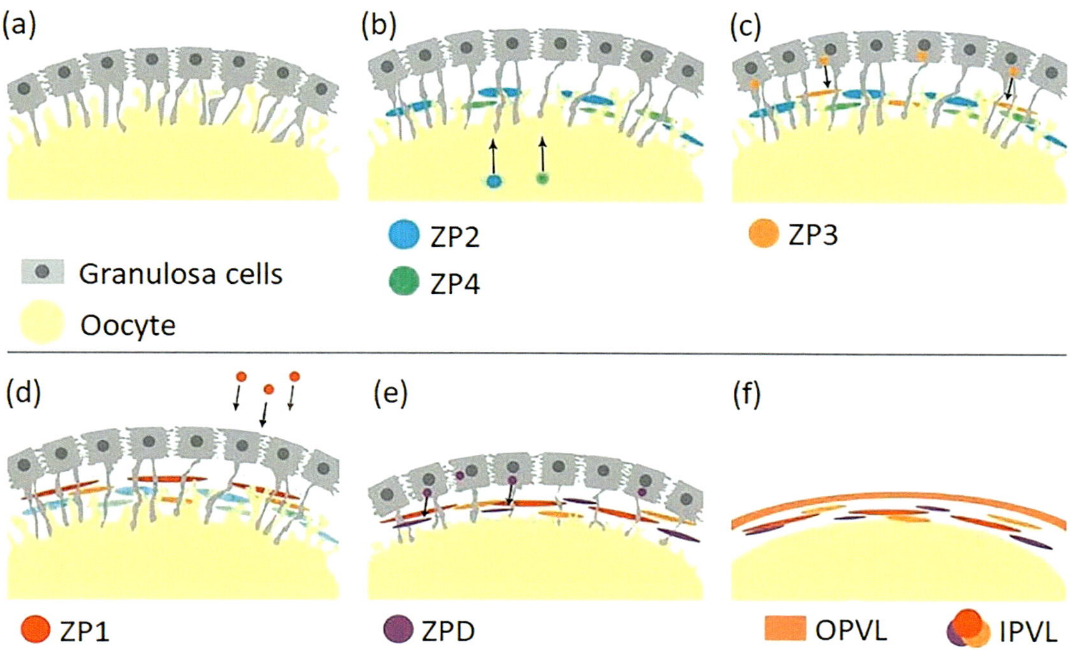


**Fig. 16.30** Model of the development of the inner (IPVL) and outer (OPVL) perivitelline layers in the follicles and ova of Japanese Quails. (**a**) Early in the development of follicles, granulosa cells and oocytes are interconnected by cytoplasmic processes and microvilli; development of the inner perivitelline layer has not yet begun. (**b**) As development continues, oocytes synthesize two glycoproteins (ZP2 and ZP4) that are incorporated into the extracellular space between the granulosa cells and the membrane of the oocyte. (**c**) Granulosa cells then synthesize and secrete another glycoprotein (ZP3) that is incorporated into the structure formed by the ZP2 and ZP4 glycoproteins. (**d**) Yet another glycoprotein (ZP1), produced by the liver and transported in the blood, is incorporated into the inner perivitelline layer. (**e**) A final glycoprotein (ZPD) produced by the granulosa cells is incorporated into the inner perivitelline layer. By this point, the glycoproteins ZP2 and ZP4 have dissolved, the cytoplasmic processes of the oocyte become shorter, and the oocyte is ready for ovulation. (**f**) After ovulation, cells lining the infundibulum synthesize the chalazae that covers the ovum with the outer perivitelline layer (OPVL). (Figure modified from Rodler et al. 2012; © 2011 S. Karger AG, Basel, used with permission)

and protein) of antibodies (Klasing 1998). In support of this hypothesis, food availability in the vicinity of nest sites was found to be positively correlated with the ability of young European Serins (*Serinus serinus*) to respond to foreign proteins (antigens) (Hoi-Leitner et al. 2001). Because antibody production requires exposure to pathogens, maternal antibody transmission provides embryos and hatchlings with protection against the local disease environment.

Female birds also deposit androgens, primarily testosterone, in yolk. Increasing amounts of testosterone in the yolk, up to a point, are positively correlated with chick vigor, begging intensity and muscle development at hatching, and social dominance as adults (Schwabl 1996; Lipar and Ketterson 2000). Most birds lay clutches of more than one egg and, for many species, androgen concentrations in the yolk vary with laying order. In some species, including Red-winged Blackbirds (*Agelaius phoeniceus*; Lipar et al. 1999), European Starlings (*Sturnus vulgaris*; Pilz et al. 2003), Spotless Starlings (*Sturnus unicolor*; Muriel et al. 2019), and Black-headed Gulls (*Chroicocephalus ridibundus*; Müller et al. 2004), androgen levels increase with laying order. In other species, such as Zebra Finches (*Taeniopygia guttata*; Gil et al. 1999) and Cattle Egrets (*Bubulcus ibis*; Schwabl et al. 1997), the pattern is reversed, with the last eggs having

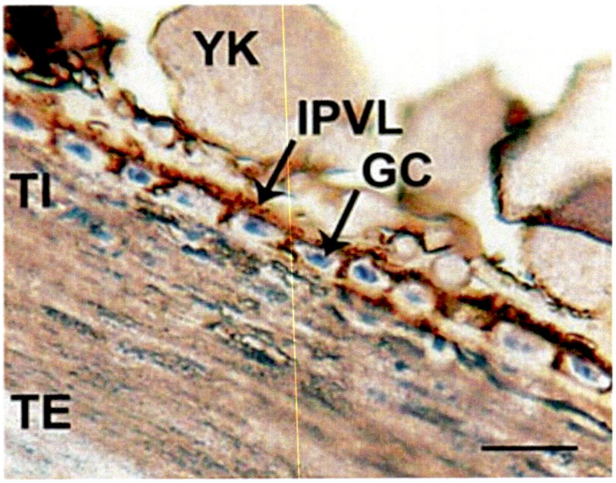


**Fig. 16.31** The wall of an F2-sized follicle showing the inner perivitelline layer (IPVL), granulosa cells (GC), theca interna (TI), theca externa (TE), and yolk (YK). The IPVL contains receptors that will serve as binding sites for sperm attempting to fertilize the egg. Scale bar = 10 μm. (Figure from Rodler et al. 2012; © 2011 S. Karger AG, Basel, used with permission)

lower androgen concentrations than the first ones. In still other species, including House Wrens (*Troglodytes aedon*; Ellis et al. 2001) and Tree Swallows (*Tachycineta bicolor*; Whittingham and Schwabl 2002), there is no trend. If the last eggs tend to hatch last, increased androgen levels in those eggs may help nestlings overcome the possible disadvantage of being slightly smaller than their older siblings. Lower androgen concentrations in the last eggs in a clutch would provide the older siblings from the first-laid eggs with a substantial competitive advantage. This strategy, a form of brood reduction (Groothuis et al. 2005), would be beneficial when food availability varies because, when availability is low, the older nestlings would obtain most of the food and be more likely to survive (at the expense, of course, of the death of their younger siblings).

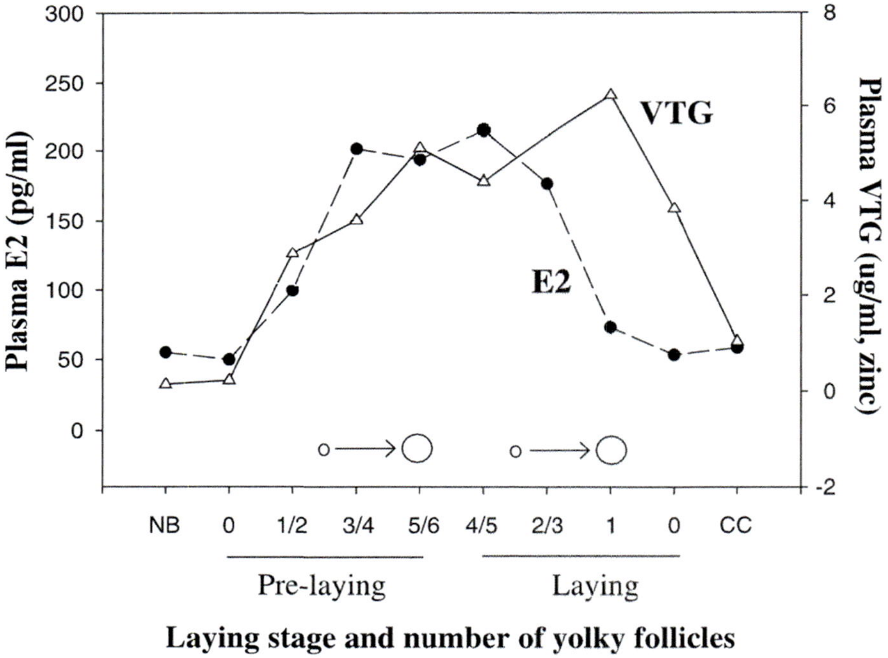


**Fig. 16.32** Variation in mean plasma estradiol-17b (E2, closed circles) and vitellogenin (VTG, open triangles) levels during the laying cycle of female European Starlings (*Sturnus vulgaris*) relative to the number of developing, yolky follicles in the ovary during the pre-laying and laying phase. NB, non-breeder; CC, clutch completion; 0, zero; 1/2, one or two follicles; and so on. Circles and arrows indicate timing of follicle development for an early- and late-developing follicle (interval between two follicle stages = about 48 h). (Figure modified from Williams et al. 2004; © 2004 Elsevier Inc., used with permission)

**Fig. 16.33** Variation among bird species in the relative amount of yolk in eggs and the amount of energy available to the developing embryo (kJ–$g^{-1}$, or kilojoules per gram). From top to bottom, the hatchlings are an altricial Brown Creeper (*Certhia americana*), a semiprecocial Least Tern (*Sternula antillarum*), a precocial Ruddy Duck (*Oxyura jamaicensis*), a superprecocial Malleefowl (*Leipoa ocellata*), and a Southern Brown Kiwi (*Apteryx australis*). Kiwis are "outliers." Female kiwis produce extremely large eggs for their size (with substantial amounts of yolk), but young typically remain in the nest for several days and so are best classified as semiprecocial. (Figure from Sotherland and Rahn 1987; © 1987 Oxford University Press, used with permission)

When food is abundant, even the competitively disadvantaged younger nestlings might survive.

In addition to the within-clutch variation in androgen concentrations observed in some species, investigators have also noted between-clutch variation. For example, female Zebra Finches (Gil et al. 1999) and Barn Swallows (*Hirundo rustica*; Gil et al. 2006) deposit more androgens in eggs when mated with attractive, higher-quality males. Such results would seem to support the differential allocation hypothesis, with females investing more, and expected a greater payoff, when mated to a high-quality mate. However, females in other species, including Collared Flycatchers (*Ficedula albicollis*; Michl et al. 2005) and House Finches (*Haemorhous mexicanus*; Navara et al. 2006), deposit more androgens when mated with unattractive, lower-quality males. One possible explanation for these results is that females adjust androgen levels based on the expected contributions of their mates to parental care. Lower-quality males might provide less food for nestlings, but, in other species, including Zebra Finches and Barn Swallows, higher-quality males actually provide less parental care (with females compensating by providing more care, perhaps as a "trade-off" for the better-quality genes provided by their high-quality mates). With less care provided by males, females may increase androgen levels in eggs to boost their growth rates and chances of survival. Alternatively, females anticipating less parental care from mates may deposit additional androgens to increase the begging behavior of nestlings and, in turn, induce males to provide more care. This "manipulating androgen hypothesis" could be proven correct if males are more responsive to increased begging by nestlings than females. In some species, that appears to be the case (Moreno-Rueda 2007). In different species, one hypothesis (differential

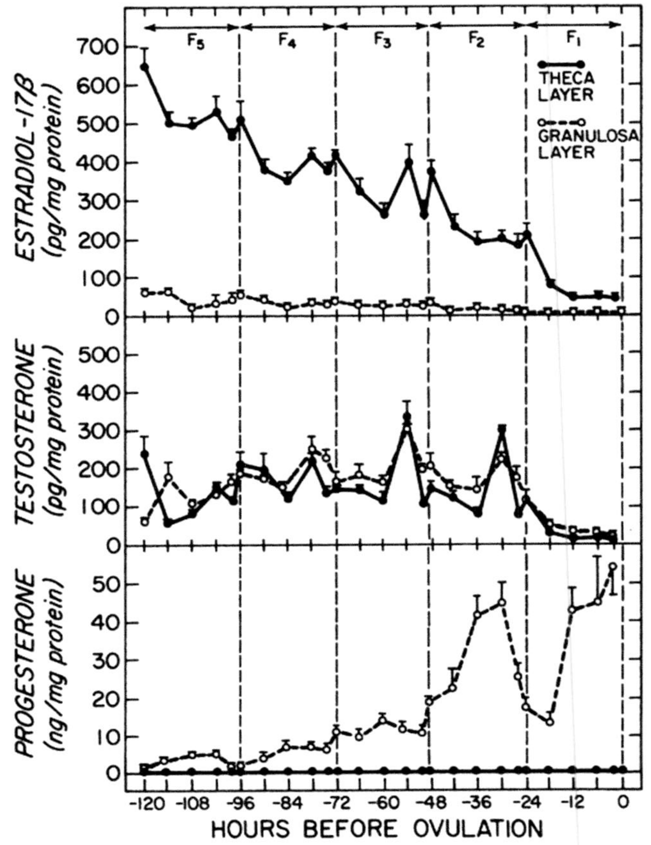


**Fig. 16.34** Mean (± SE) estradiol, testosterone, and progesterone concentrations of isolated granulosa and theca layers of the five largest preovulatory follicles of Domestic Chickens. Note that, as preovulatory follicles mature, they produce decreasing amounts of estrogen and testosterone and increasing amounts of progesterone. (Figure from Bahr et al. 1983; © 1983 Oxford University Press, used with permission)

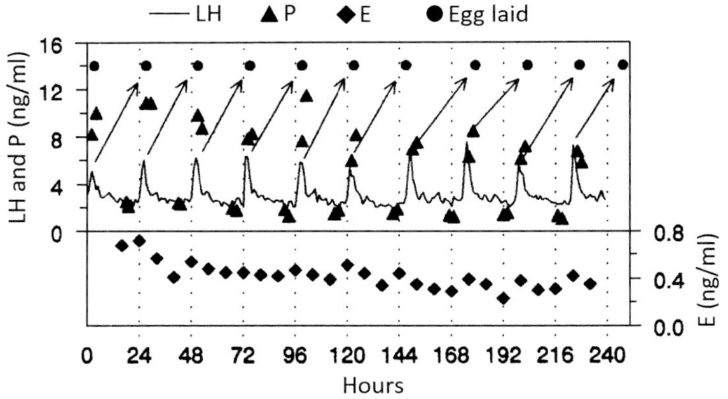


**Fig. 16.35** Changes in concentrations of plasma hormones during egg laying by a Domestic Chicken. LH concentrations were measured hourly sample, and progesterone (P) levels were measured before and after LH surges. Estrogen (E) concentrations were measured every 8 h. Closed circles (•) at the top indicate approximate times of egg laying. Arrows indicate the association between LH surges and egg laying. Each LH surge was accompanied by an increase in plasma levels of progesterone. (Figure from Liu et al. 2002; © 2002 Oxford University Press, used with permission)

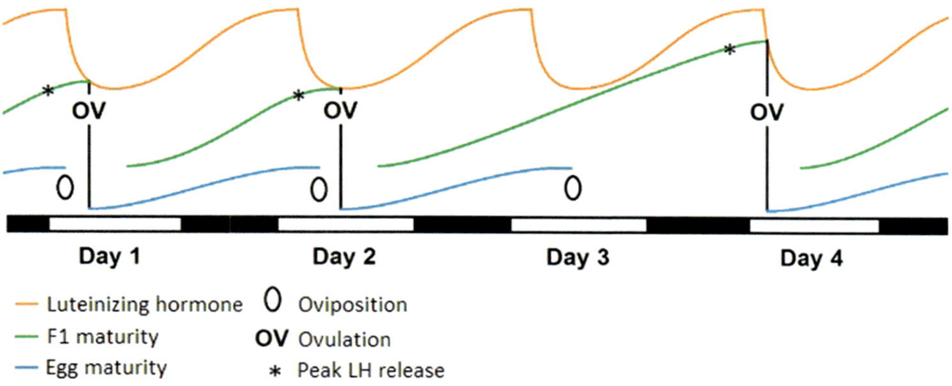


**Fig. 16.36** Diurnal rhythmicity in secretion of luteinizing hormone (LH), maturity of the largest follicle (F1), maturity of the next largest follicle (egg maturity), peak release of LH, ovulation, and egg laying (oviposition). Note that surges in LH levels were followed by egg laying. White areas on the x-axis indicate daylight, dark areas on the x-axis indicate night. (Figure from van der Klein et al. 2020; © 2019 The Authors. Published by Elsevier B.V., used with permission)

allocation hypothesis) or the other (manipulating androgen hypothesis) may best explain between-clutch variation in androgen levels.

## 16.5 Copulation

Copulation serves to transfer sperm from a male into a female's reproductive tract. For most birds, copulation involves a "cloacal kiss," with the male mounting a female, twisting his tail under the female's to bring the cloacae into contact (Fig. 16.41), and sperm are transferred as the male performs a series of tail thrusts. Copulation typically lasts just a few seconds and, for most species, takes place on the ground or a perch. However, waterfowl, including ducks, geese, and swans, and other aquatic birds may copulate on the water and swifts and some swallows may copulate in flight.

Copulations are typically solicited via copulation solicitation displays or postures (Figs. 16.42 and 16.43), and, in most species, females perform these displays (Birkhead and Møller 1992). However, males may also solicit copulations and, in some species, both sexes perform these displays. Typically, but not always, copulation solicitation displays are followed by copulation, usually one, but sometimes two or more in quick succession.

One copulation, assuming the successful transfer of sperm, is sufficient to fertilize all the eggs in a clutch. The concentration of sperm in ejaculates varies among species (Table 16.2). Higher numbers may increase a male's chance of fertilizing eggs, but high sperm numbers are a potential problem for females because penetration of eggs by too many sperm (polyspermy) can lead to abnormalities and death of embryos. However, female birds are able to regulate the number of sperm that reach the site of fertilization (infundibulum), e.g., ejecting sperm after a copulation or, if sperm numbers are low, storing sperm in sperm-storage tubules and regulating their release so that sufficient sperm are available for fertilization (Hemmings and Birkhead 2015).

In a few species, pairs copulate just a few times, perhaps to ensure the transfer of sperm and minimize the chance of producing infertile eggs. However, in other species, the number of within-pair copulations seems to exceed the number necessary to ensure fertilization (Fig. 16.44). For example, paired Barn Swallows (*Hirundo rustica*) may copulate 40 or more times per clutch (Møller 1985), and American Kestrels (*Falco sparverius*) may copulate 1–2 times per hour over a period of several weeks prior to and after the initiation of egg laying (Villarroel et al. 1998). Northern Goshawks (*Accipiter gentilis*) may

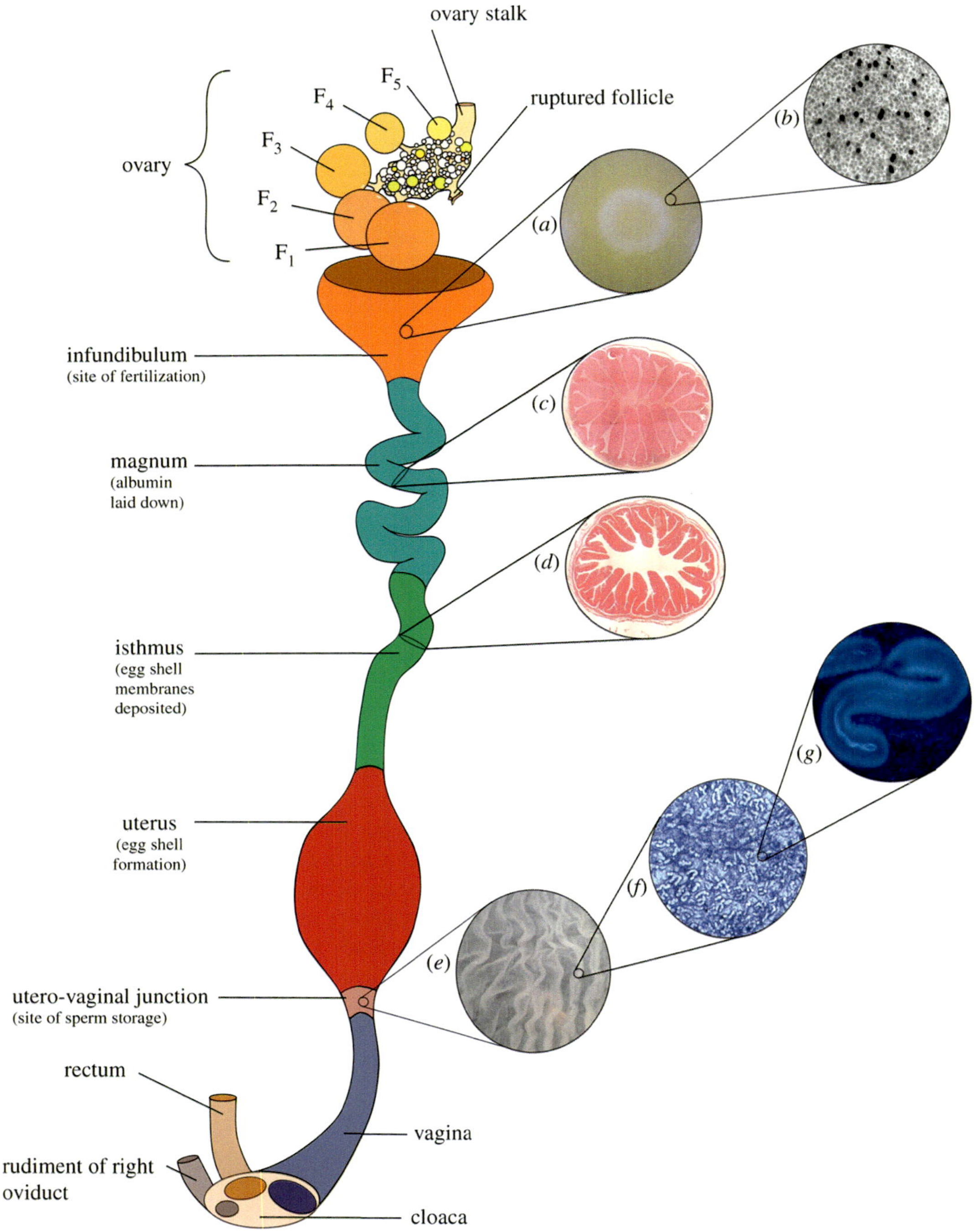


**Fig. 16.37** Drawing of the avian oviduct and ovary (not to scale). The ovary has multiple follicles at different stages of development. The largest yellow follicles are $F_{1-5}$, with $F_1$ being the largest follicle and the next to rupture. (**a**) Germinal disc of a fertilized egg of a Zebra Finch (*Taeniopygia guttata*). The clear outer ring and paler center of the germinal disc that indicate embryonic development. (**b**) Holes made by penetrating sperm in the inner perivitelline layer (IPVL) of an ovum after fertilization (from a Bullfinch, *Pyrrhula pyrrhula*). (**c**) Cross section of the magnum of a Helmeted Guineafowl (*Numida meleagris*) where albumin is deposited during egg

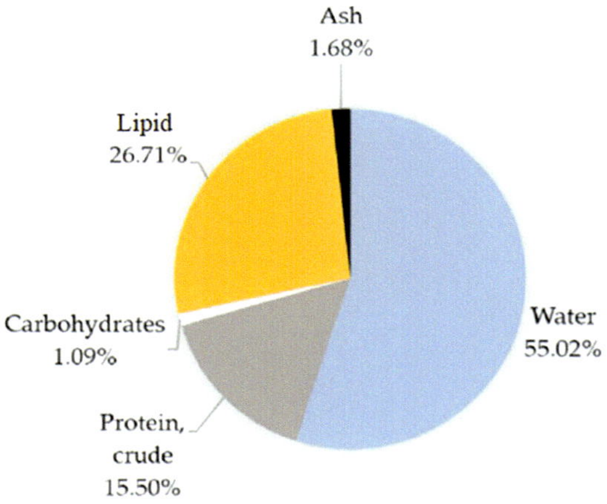


**Fig. 16.38** Typical composition of egg yolk of Domestic Chickens. Ash consists primarily of inorganic materials. (Figure from Réhault-Godbert et al. 2019; open-access article distributed under the terms and conditions of the Creative Commons Attribution (CC BY) license, https://creativecommons.org/licenses/by/4.0/)

copulate 400–600 times prior to completion of a clutch of eggs (Møller 1987).

Why so many copulations? Several hypotheses have been proposed (Table 16.3) and available evidence suggests different functions for different species. However, mate assessment and paternity assurance are the most common reasons for multiple copulations. For example, paired American Kestrels (*Falco sparverius*) appear to use frequent copulations to assess mate quality, with individuals, particularly males, advertising their condition by performing energetically costly copulations (Villarroel et al. 1998). In Northern Fulmars (*Fulmarus glacialis*), males copulate frequently to increase the likelihood of being the last to copulate with their mates prior to egg laying (because sperm from the last copulation are most likely to fertilize eggs; Hunter et al. 1992).

Males in a few species (~3% of all species; Briskie and Montgomerie 1997), including most waterfowl (Anseriformes), species in the order Tinamiformes, cracids (Galliformes, family Cracidae), and ratites, have intromittent organs (de Oliveira and Mahecha 2000; Brennan et al. 2007; Brennan and Prum 2012; Previatto et al. 2018; Fig. 16.45), but most male birds do not. Because males in other taxa with internal fertilization have intromittent organs (IOs), as likely did the reptilian ancestors of birds (Larson 1998), the loss of IOs among most species of birds may have some adaptive value. Although that value remains unclear, several hypotheses have been proposed. For example, the absence of an intromittent organ may be a by-product of selection for rapid copulations, with rapid copulations reducing the risk of predation because copulating birds may be more vulnerable (Wesotowski 1999). Another possibility is that the loss of IOs minimizes the risk of transferring sexually transmitted diseases (Briskie and Montgomerie 1997). It may also be the case that the loss of IOs is due to female preference, with females preferring (initially) males with smaller and, eventually, no IOs because that would give them greater control over fertilization (Briskie and Montgomerie 1997).

Among bird species that possess phallic organs, there are two types: the true intromittent organ found in ratites, tinamous, cracids, screamers, magpie geese, and waterfowl, and the pseudophallus, a non-intromittent organ, found in most galliforms (except the family Cracidae; King 1981, Figs. 16.46, 16.47, and 16.48).

**Fig. 16.37** (continued) development. (**d**) Cross section of the isthmus of a Reeves Pheasant (*Syrmaticus reevesii*) where shell membranes are produced and deposited during egg development. (**e**) Internal tissue lining and folds of the vagina and utero-vaginal junction of a Northern Bobwhite (*Colinus virginianus*). The utero-vaginal junction is where sperm storage tubules are located. (**f**) Single fold of the utero-vaginal junction of a Zebra Finch stained with dye. The small tubular structures are sperm storage tubules. (**g**) A single sperm storage tubule containing sperm. (Figure from Assersohn et al. 2021; © 2021 The Authors. Published by the Royal Society under the terms of the Creative Commons Attribution License, https://creativecommons.org/licenses/by/4.0/. Image credits: (**a**–**d**) Nicola Hemmings; (**e**–**g**) Paul Richards)

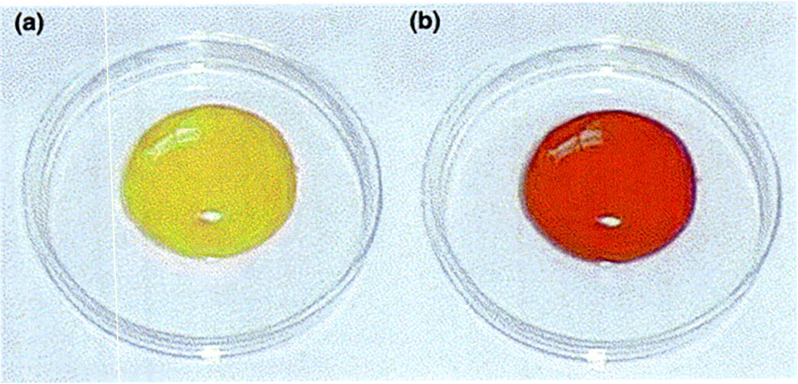


**Fig. 16.39** The familiar color of a chicken's egg yolk (**a**) is in stark contrast to the richly pigmented egg yolk of a Lesser Black-backed Gull (*Larus fuscus*), and (**b**) Such high maternal investment of carotenoids into egg yolk is typical among wild bird species, suggesting that these biologically active pigments serve important functions in the developing bird. (Figure from Blount et al. 2000; © 2000 Elsevier Science Ltd., used with permission)

Several hypotheses have been suggested to explain the presence of intromittent organs in the relatively few species that have them. The water copulation hypothesis proposes that intromittent organs are advantageous for species that copulate on water because they prevent water that could damage (by osmotic shock), dilute, or wash out the ejaculate from entering a female's cloaca during copulation (Lake 1981; Montgomerie and Briskie 2007). In support of this hypothesis, waterfowl have intromittent organs and they copulate on the water. However, many other bird species that copulate on the water, such as puffins and phalaropes, lack intromittent organs. As suggested by Montgomerie and Briskie (2007), the water copulation hypothesis might explain the presence of intromittent organs in taxa like waterfowl that copulate on the water, but does not explain the significant variation in morphology of these organs among waterfowl (Fig. 16.49) or the presence of intromittent organs in non-aquatic species like ratites.

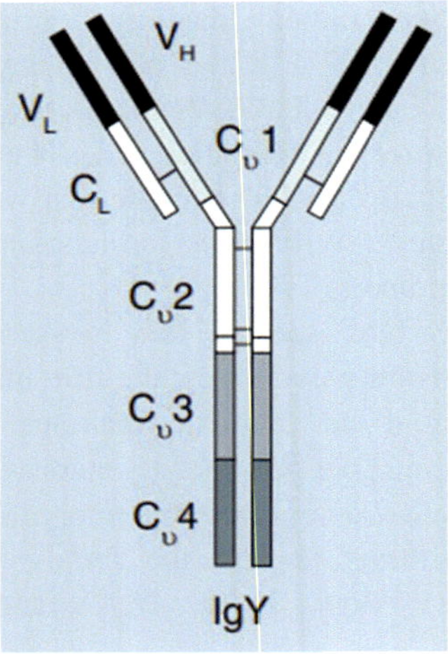


**Fig. 16.40** Antibodies, also called immunoglobulins (in this case, immunoglobulin Y, or IgY), help protect birds from pathogens or the toxins they produce by binding to antigens (e.g., proteins in the membranes of bacteria) via their "binding sites" (the black areas above; $V_L$ and $V_H$). This binding can neutralize toxins and attract white blood cells that eliminate pathogens (by phagocytosis). The bottom portion of the antibody molecule (C1–C4) is referred to as the constant region because, in contrast to the binding sites, this portion of the molecule is the same for all types of antibodies. (Figure from Xiao and Gao 2010; © 2010 Future Science Group, used with permission)

One possible explanation for the variability in morphology and the presence of intromittent organs in non-aquatic species is the sperm competition hypothesis. A comparative study of 54 species of waterfowl revealed that males in those species that perform more forced extrapair copulations have larger intromittent organs and more of the surface is covered with ridges, knobs, or both. This suggests that intromittent organ size and surface features are likely related to sperm competition. Males chase and attempt to copulate with females and a larger intromittent organ with ridges and knobs may make it more difficult for the female to dislodge a copulating male (Coker et al. 2002). As an added advantage, the ridges and knobs of waterfowl intromittent organs are all pointed towards the base of the organ (Fig. 16.50). This could facilitate a scraping action that could potentially remove the sperm of rival males when the intromittent organ is withdrawn after copulation (McCracken 2000).

Another possible explanation for variation in the morphology of the intromittent organs of waterfowl is variation among species in the tendency for males to attempt forced copulations. As suggested by Brennan et al. (2007), the relatively complex vaginas and phalluses of some species of waterfowl represent an example of intersexual selection. In species where males frequently attempt forced extra-pair copulations, selection has favored "elaborate" vaginas that make phallus penetration more difficult and reduce the likelihood of successful forced extrapair copulations. Selection has also favored a corresponding elaboration of phallus morphology in these species, with the morphology of erect phalluses matching that of female vaginas. However, eversion of the male phallus occurs during, not before, cloacal contact and, given the complex morphology of the female vaginas with spirals and, in some cases, dead end pouches, successful full eversion that will lead to ejaculation and potential fertilization of eggs is unlikely without the cooperation of females. As such, species where females have elaborate vaginas likely represent an example of cryptic female choice, providing females with

**Box 16.4 Egg Antimicrobial Defenses**

During egg formation, antibodies in the blood of females are transferred to the yolk. After eggs are laid and as embryos develop, those antibodies are then gradually transported from the yolk into the embryonic circulation. Shortly before hatching, this transport accelerates to provide hatchlings with greater protection from pathogens (Hincke et al. 2019). However, maternal antibodies in the egg yolk represent just one of the many ways that developing embryos are protected from pathogens.

Bacteria are always present in avian nests and on eggshells, and the probability of trans-shell infection increases with their increasing abundance. By keeping eggs dry, parental incubation of eggs can reduce or limit the growth of bacteria on eggshells (D'Alba et al. 2010; Bollinger et al. 2018). In addition, eggshells act as a physical barrier to microbes, and also contain antimicrobial proteins (Mikšík et al. 2010). The eggshell membranes made up of cross-linked protein fibers also provide physical and chemical protection against the invasion of pathogens. Antimicrobial proteins in eggshells and eggshell membranes include lysozyme and ovotransferrin.

The viscous nature of the egg white serves as a physical barrier to microbes, and its alkaline pH can negatively impact the survival and growth of bacteria (Hincke et al. 2019). Egg white also contains several antimicrobial proteins. As embryonic development proceeds, the vitelline and chorioallantoic sac membranes along with the membrane of the amnionic sac and yolk sac serve as physical barriers to microbes and these membranes also contain antimicrobial proteins. The acidity of the allantoic fluid represents yet another antimicrobial defense and this fluid also contains interferons that provide protection from viruses. Finally, allantoic fluid also contains toll-like receptors, proteins that detect microbial proteins and initiate a variety of antimicrobial responses (Hincke et al. 2019).

(continued)

**Box 16.4** (continued)

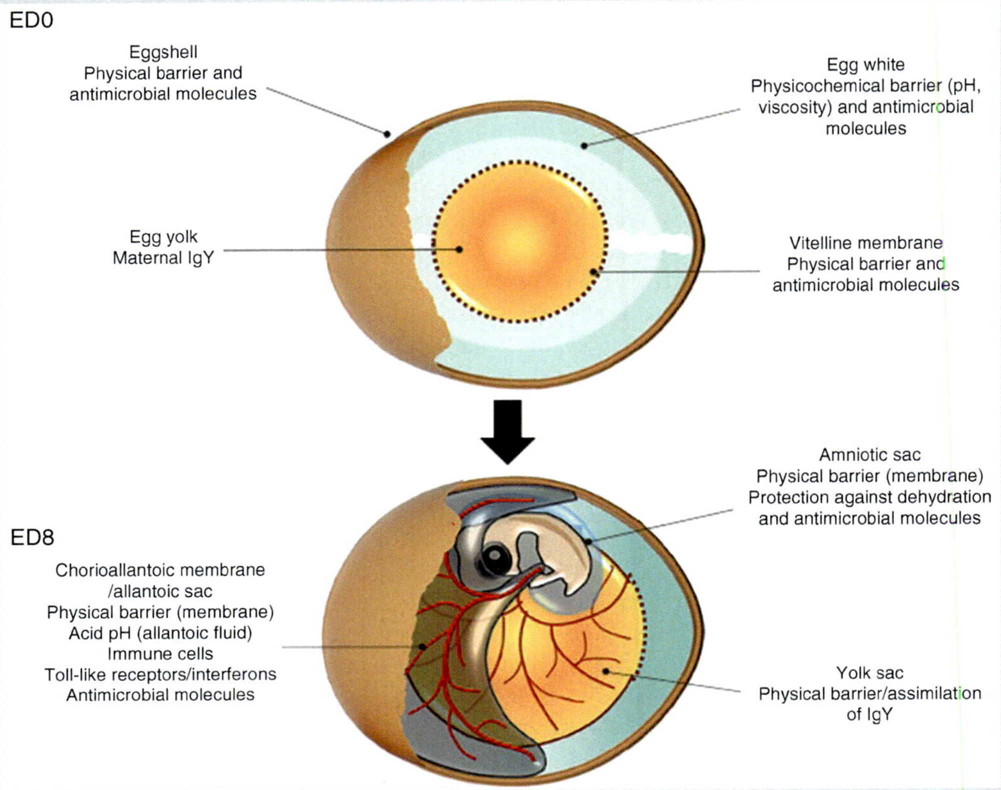


Antimicrobial defenses of the eggs of Domestic Chickens immediately after laying (ED0) and eight days after laying (ED8). (Figure modified from Hincke et al. 2019; © S. Karger AG, used with permission).

more control over which male(s) will fertilize their eggs.

Copulation, whether involving a "cloacal kiss" or intromittent organ, introduces sperm into the cloaca or vagina of the female reproductive tract. However, the successful transfer of sperm from male to female does not always ensure that a male's sperm has a chance to successfully fertilize an ovum. Among species where males can coerce females to copulate, females may have postcopulatory mechanisms for influencing the likelihood that sperm from particular males will fertilize her eggs. For example, when female domestic chickens cannot prevent copulations from suboptimal partners, females can reduce fertilization success of an ejaculate by the differential ejection of sperm (Pizzari and Birkhead 2000). Even within a pair, females may be able to determine the sperm that will fertilize her eggs. For example, female Black-legged Kittiwakes (*Rissa tridactyla*) begin copulating with their mate more than two weeks before laying their first egg and observations revealed that females tended to eject their mates' sperm early in this period while increasingly retaining sperm as the time of egg laying approached. By ejecting sperm

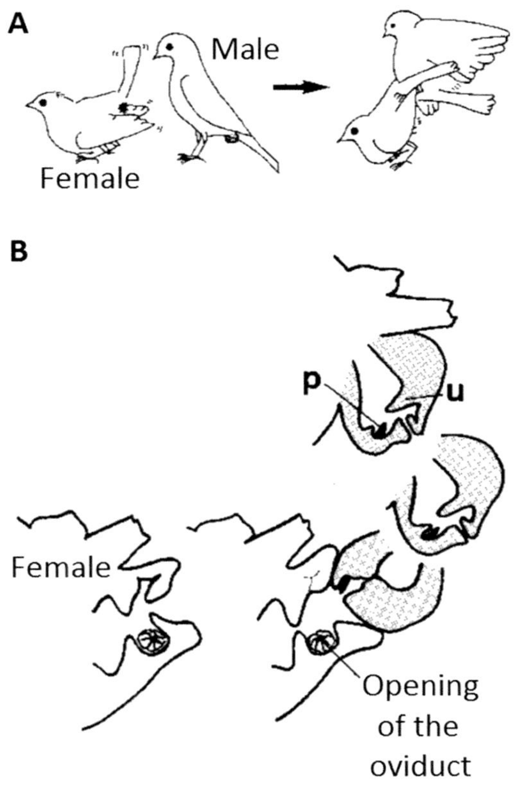


**Fig. 16.41** Drawings showing (**a**) the courtship display of a female Alpine Accentor (*Prunella collaris*) and cloacal contact and (**b**) the displacement and structural changes of a male's (gray) cloacal protuberance during the sequence of copulation. F, female; M, male; o, opening of the oviduct; p, papilla of the ductus deferens; u, uroproctodeal fold. (Figure from Chiba and Nakamura 2003; © British Ornithologists' Union, used with permission)

**Fig. 16.42** Copulation solicitation posture of a female Smith's Longspur (*Calcarius pictus*). As described by Briskie (1992), "Female Smith's Longspurs solicited by tilting forward, cocking their tail until almost vertical, and throwing their head back toward the male ... The wings also were lowered and vibrated and the bill opened slightly ... All solicitations by the female were made on the ground and with the cloaca directed toward the nearby male. Males mounted soliciting females from the rear and without any precopulatory displays or courtship feeding. Mounting lasted only 1 to 3 s and males flapped their wings rapidly during mounting, presumably to maintain balance." (Drawing by Ian Jones in Briskie 1992; © 1992 Oxford University Press, used with permission)

that would have been old by the time of fertilization, females are apparently attempting to avoid hatching failure caused by old sperm (Fig. 16.51; Wagner et al. 2004).

## 16.6 Sperm-Storage Tubules

Sperm introduced into the female reproductive tract rely on their motility to reach sperm-storage tubules (SSTs) located at the junction of the vagina and the uterus (utero-vaginal junction, or UVJ) (Fig. 16.52; Box 16.5 Seminal Fluid). Additional SSTs are located in the infundibulum (Bakst 1981). SSTs are tubular structures formed by invagination of the epithelium of the UVJ (Figs. 16.53 and 16.54) and numbers vary among species, with about 500 in female Budgerigars (*Melopsittacus undulatus*) and 20,000 in female Domestic Turkeys (*Meleagris gallopavo*; Birkhead and Møller 1992). Relatively few sperm reach the SSTs and those that do must have superior motility and surface membranes that possess glycoproteins only present in mature sperm (Wishart and Horrocks 2000). Typically, only 1–2% of inseminated sperm enter the SSTs and the rest are probably ejected the next time the female defecates (Bakst et al. 1994). Storage of sperm in SSTs may be an essential step in the successful fertilization of eggs (Matsuzaki and Sasanami 2017) because they protect sperm from anti-sperm immune responses (Das et al. 2006, 2008) and prevent

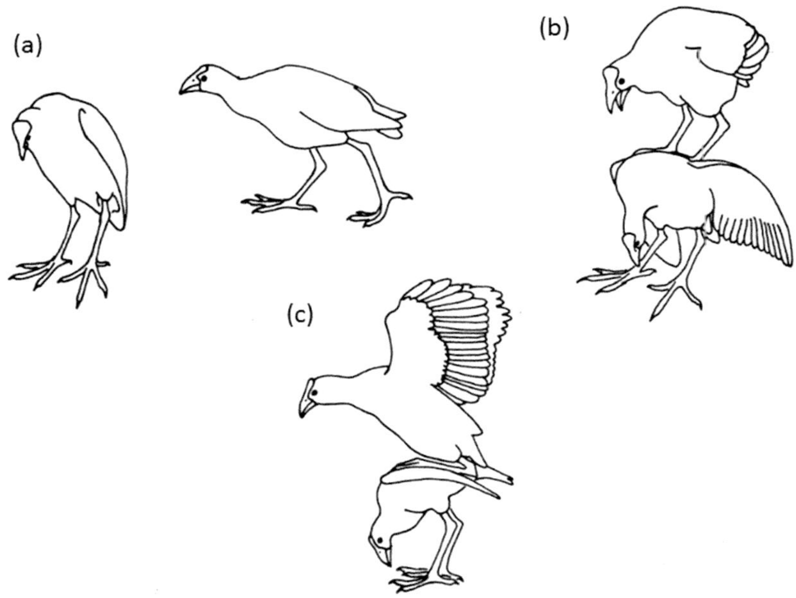


**Fig. 16.43** Copulatory postures of male and female Western Swamphen (*Porphyrio porphyrio*). (**a**) Female precopulatory "hunch" as male approaches. (**b**) Female in precopulatory position with male mounted. (**c**) Copulation, with the female raising her tail immediately before cloacal contact. (Figure from Craig 1977; Rights managed by Taylor & Francis, used with permission)

**Table 16.2** Mean number of sperm per ejaculate for several different species of birds

| Species | Scientific name | Mean sperm per ejaculate | Citation |
|---|---|---|---|
| Northern Pintail[c] | *Anas acuta* | 66,000,000 | Penfold et al. (2001) |
| Indian Peafowl | *Pavo cristatus* | 12,400,000 | Birkhead and Petrie (1995) |
| American Kestrel | *Falco sparverius* | 416,000 | Bird and Laguë (1977) |
| Spix's Macaw[d] | *Cyanopsitta spixii* | 86,800 | Fischer et al. (2014) |
| Monk Parakeet | *Myiopsitta monachus* | 710,000 | Anderson et al. (2002) |
| Golden-capped Parakeet | *Aratinga auricapilla* | 740,000 | Stelzer et al. (2005) |
| Kea | *Nestor notabilis* | 76,590,000 | Dogliero et al. (2016) |
| Rockhopper Penguin | *Eudyptes chrysocome* | 1,700,000 | Waldoch et al. (2007) |
| Rockhopper Penguin | *Eudyptes chrysocome* | 6,570,000 | Waldoch et al. (2012) |
| Magellanic Penguin[a] | *Spheniscus magellanicus* | 11,400,000 | O'Brien et al. (1999) |
| Rock Pigeon | *Columba livia* | 10,500,000 | Sontakke et al. (2004) |
| Whooping Crane[b] | *Grus americana* | ~15,000,000 | Brown et al. (2017) |
| Golden Eagle[a] | *Aquila chrysaetos* | 1,970,000 | Villaverde-Morcillo et al. (2015) |
| White-rumped Vulture | *Gyps bengalensis* | 21,600,000 | Umapathy et al. (2005) |
| Zebra Finch | *Taeniopygia guttata* | 9,470,000 | Birkhead and Fletcher (1995) |
| Red-backed Fairywren | *Malurus melanocephalus* | 32,000,000–41,000,000 | Rowe et al. (2010) |
| Splendid Fairywren[c] | *Malurus splendens* | 48,370,000 | Rowe and Pruett-Jones (2011) |
| Variegated Fairywren[c] | *Malurus lamberti* | 8,540,000 | Rowe and Pruett-Jones (2011) |
| Striated Grasswren[c] | *Amytornis striatus* | 6,560,000 | Rowe and Pruett-Jones (2011) |
| Bank Swallow | *Riparia riparia* | 2,820,000 | Nicholls et al. (2001) |
| Red-winged Blackbird | *Agelaius phoeniceus* | 12,500,000 | Westneat et al. (1998) |
| White-rumped Munia | *Lonchura striata* | 2,900,000 | Birkhead (1991) |
| House Sparrow | *Passer domesticus* | 640,000 | Birkhead et al. (1994) |

[a] Semen collected via voluntary false copulation over a bird trainer's back
[b] Manual stimulation
[c] Cloacal massage
[d] Electrostimulation

**Fig. 16.44** Copulating Ospreys (*Pandion haliaetus*). Ospreys copulate an average of 160 times per clutch of eggs, a strategy that aids males in protecting their paternity (Birkhead and Lessells 1988). (Photo by Andy Morffew, CC0 Public Domain, pxhere.com)

their ejection from the oviduct (Bakst 2011). The ability to store sperm also helps to ensure that females have sperm to fertilize their eggs when copulation and ovulation occur at different times and, for females that copulate with multiple males, storing sperm in SSTs may allow them to determine which sperm will fertilize their eggs (Eberhard 1996).

Although the mechanisms are not yet completely understood, females appear to have some control over which sperm enter the SSTs (Pizzari et al. 2002). Mendonca et al. (2019) found that the entrances of SSTs have microvilli and appear to be constricted (Fig. 16.55), potentially allowing the entrance to function as a valve to allow, or not, the entrance and exit of sperm. Mendonca et al. (2019) also suggested that SSTs

**Table 16.3** Predictions of four hypotheses to explain the adaptive significance of frequent within-pair copulations in birds (Villarroel et al. 1998). There is evidence to support the paternity assurance and mate assessment hypotheses in a variety of species, but less evidence in support to the immediate material benefits and mate guarding hypotheses (Table from Villarroel et al. 1998; © 1998 The Association for the Study of Animal Behaviour. Published by Elsevier Ltd., used with permission)

| Hypothesis | Predictions |
|---|---|
| Paternity assurance | (1) Within-pair copulations (WPCs) increase during fertile period<br>(2) Mate attendance increases during fertile period<br>(3) Female solicitation increases when paternity confidence is low |
| Immediate material benefits | (1) Females exchange food for WPCs |
| Female mate guarding of male | (1) WPCs occur from pair formation through egg laying of male<br>(2) Females control copulations via solicitation<br>(3) Females differ in soliciting behavior |
| Male assessment | (1) WPCs occur from pair formation through egg laying<br>(2) Both sexes solicit WPCs<br>(3) Pairs do not differ in solicitation frequency |

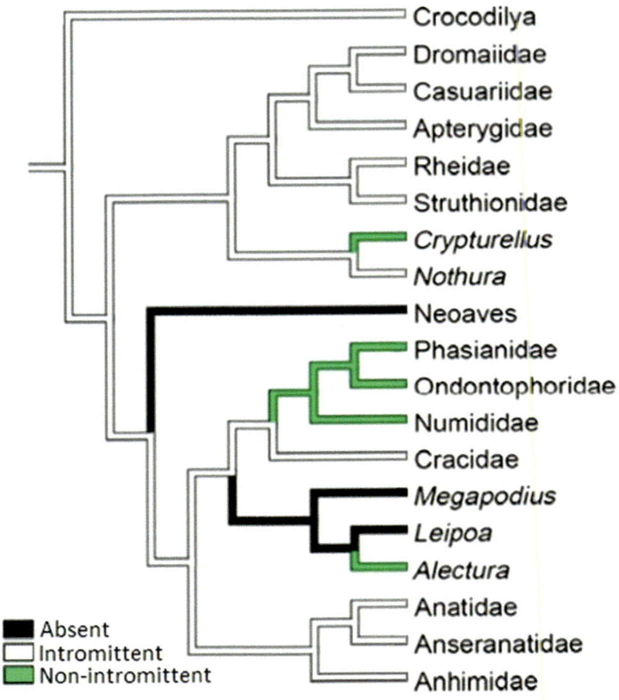


**Fig. 16.45** Phylogeny of ratites, tinamous (*Crypturellus* and *Nothura*), Neoaves, and the orders Galliformes and Anseriformes showing taxa without intromittent organs (absent), with intromittent organs, and with small, non-intromittent phallic organs. (Figure modified from Previatto et al. 2018; © 2017 The Royal Swedish Academy of Sciences, used with permission)

may be able to constrict and dilate to either prevent or allow sperm to enter. Investigators have not found smooth muscle fibers around SSTs, but Freedman et al. (2001) detected "terminal webs" consisting of contractile proteins (actin and myosin) in the walls of SSTs as well as neurons in the

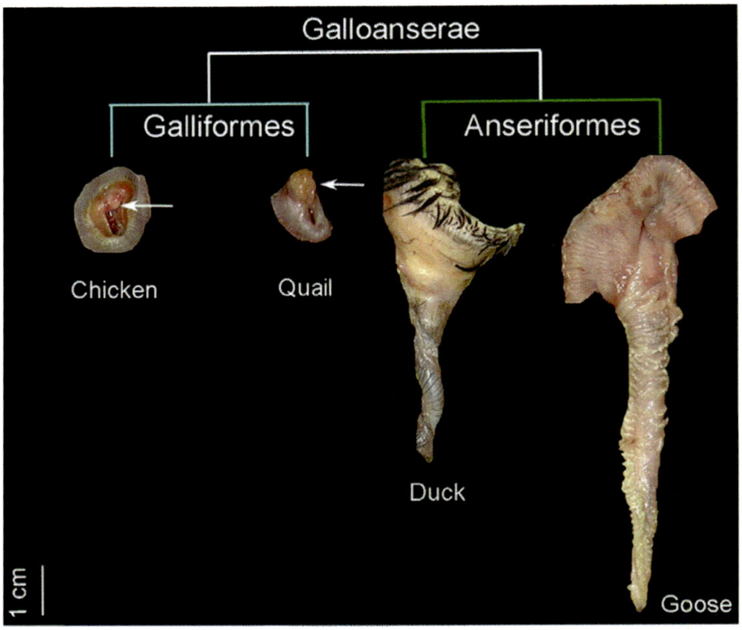


**Fig. 16.46** Non-intromittent phallic organs of a Domestic Chicken (*Gallus g. domesticus*) and Common Quail (*Coturnix coturnix*), and the intromittent phallic organs of a Mallard (*Anas platyrhynchos*) and Graylag Goose (*Anser anser*). (Figure from Herrera et al. 2013; © 2013 Elsevier Ltd., used with permission)

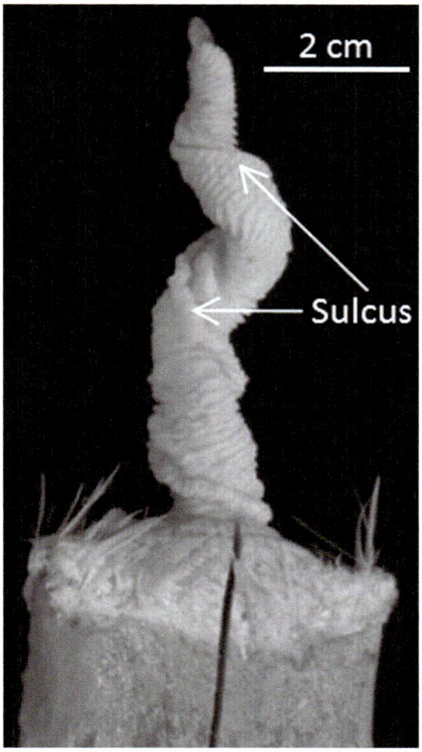


**Fig. 16.47** Phallus of a male Pekin Duck (*Anas platyrhynchos domesticus*). At rest, the phallus is inverted within a phallic sac in the ventral wall of the cloaca. During eversion, lymph fluid (rather than blood as is the case in mammals) accumulates in cavities at the base of the cloaca and enters the lumen of the phallus, forcing it out of the cloaca. During ejaculation, the sulcus, or sulcus spermaticus, forms a closed channel within which semen quickly flows from the base to the tip of the phallus. (Figure from Brennan et al. 2010; © 2009 The Royal Society, used with permission)

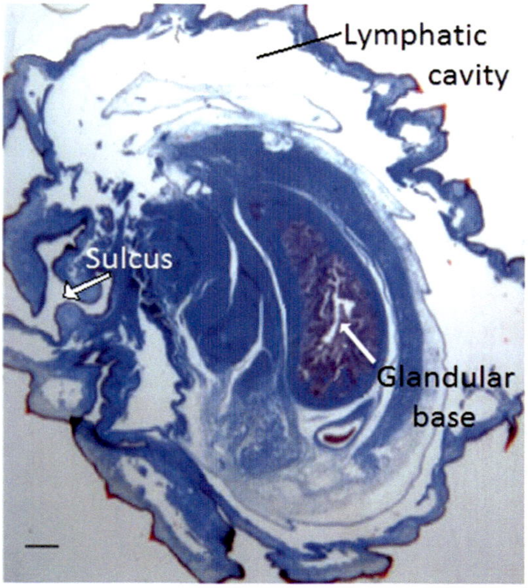


**Fig. 16.48** Cross-section of the base of the phallus of a male Mallard (*Anas platyrhynchos*). The sulcus, or sulcus spermaticus, is where the semen flows from the base of the phallus to the tip. The lymphatic cavity fills with lymph fluid during erection. The glandular base secretes mucus that provide lubrication when the phallus retracts back into the phallic sac of the cloaca. The blue material (stained) outside of the glandular base is largely collagen. Scale bar = 1 mm. (Figure modified from Brennan et al. 2010; © 2009 The Royal Society, used with permission)

uterovaginal junction of female Domestic Turkeys, suggesting the possibility of nervous control of SST function.

Matsuzaki and Sasanami (2017) proposed a mechanism whereby sperm could enter and exit the SSTs of Japanese Quail (*Coturnix japonica*). Male Japanese Quail, and males in other species in the genus *Coturnix*, have cloacal glands that secrete a glycomucoprotein that has a foam-like quality and is ejected along with sperm into the oviduct of females during copulation (Finseth et al. 2013; Fig. 16.56). This foam contains prostaglandin $F_{2\alpha}$ ($PGF_{2\alpha}$) and the results of experiments conducted by Sasanami et al. (2015) suggest that $PGF_{2\alpha}$ enhances the ability of sperm to enter SSTs by opening the entrance to SSTs and inducing vaginal contractions (Fig. 16.57).

The length of time that sperm remain in the SSTs varies among species, ranging from just a few days to as long as about four months in some species of birds (Holt and Fazeli 2016). How sperm are able to survive and remain viable for long periods of time remains to be determined. Matsuzaki et al. (2015) reported that sperm in the SSTs of female Japanese Quail became quiescent due to low oxygen levels and high lactic acid concentrations (and a resulting low pH) in the SSTs. These authors further suggested that the low pH caused inactivation of ATPase, the

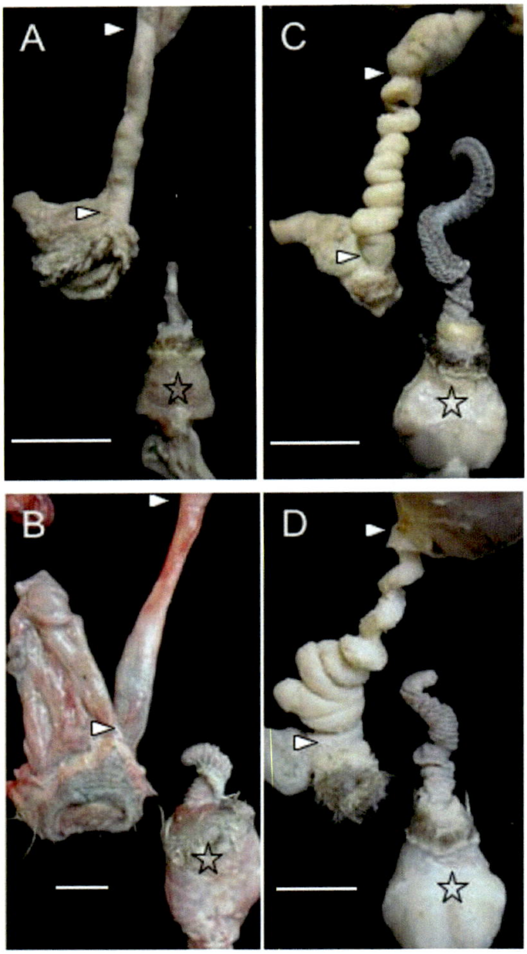


**Fig. 16.49** Differences among species in the morphology of male intromittent organs. (**a**) Harlequin Duck (*Histrionicus histrionicus*) and (**b**) Swan Goose (*Anser cygnoides*), two species where males have a short phallus and no forced copulations, and females have simple vaginas. (**c**) Long-tailed Duck (*Clangula hyemalis*) and (**d**) Mallard (*Anas platyrhynchos*), two species where males have a long phallus and high levels of forced copulations, and females have elaborate vaginas. Stars = muscular base of the male phallus, and arrows = upper and lower limits of the vaginas. Scale bars = 2 cm. (Figure modified from Brennan et al. 2007; © 2007 Brennan et al., open-access article distributed under the terms of the Creative Commons Attribution License, https://creativecommons.org/licenses/by/4.0/)

enzyme that breaks down ATP and provides the energy needed for sperm flagellar movement. However, Matsuzaki et al. (2015) were not able to produce fertilized eggs by artificial insemination of females using in vitro stored sperm and concluded that sperm preservation requires one or more unknown factors other than lactic acid. Riou et al. (2020) suggested that proteins in the uterine fluid of female birds may contribute to that longevity (Fig. 16.58). In support of this hypothesis, Ahammad et al. (2013) found that uterine fluid improved sperm motility in vitro.

Among species where pairs spend relatively little time together or copulation infrequently, the ability of females to store sperm in SSTs for longer periods is obviously advantageous. An extreme example is Monteiro's Hornbill (*Tockus monteiri*). Among hornbills, females seal themselves inside a nest cavity with a "plug" made of fecal matter and crushed millipedes (Kemp 1995; Fig. 16.59) and remain inside for about two months while laying eggs, incubating, and caring for nestlings (with food, of course, provided by their mate). Because female Monteiro's Hornbills enter their cavity (and last copulate with their mate) up to two weeks before laying the first eggs and up to four weeks before laying the last eggs, they clearly can store sperm for several weeks (Stanback et al. 2002). Similarly, the interval between copulation and egg-laying by some species of seabirds, particularly species in the order Procellariiformes, can be as long as seven weeks, with females making a "pre-laying exodus" after copulation to build up reserves for egg formation and incubation (Hemmings and Birkhead 2020).

Some investigators have suggested that the release of sperm from the SSTs is not regulated, but occurs as a response to the pressures generated by a passing egg (e.g., Tingari and Lake 1973). Others have suggested that sperm release from SSTs is a slow, continuous process that occurs at a relatively constant rate (Burke and Ogasawara 1969). Innervation of SSTs and the presence of actin (a protein found in smooth muscle) suggest another possible means of facilitating release of sperm (Freedman et al. 2001). However, Ito et al. (2011) reported that injections of progesterone caused the movement of sperm out of the SSTs of female Japanese Quail (*Coturnix*

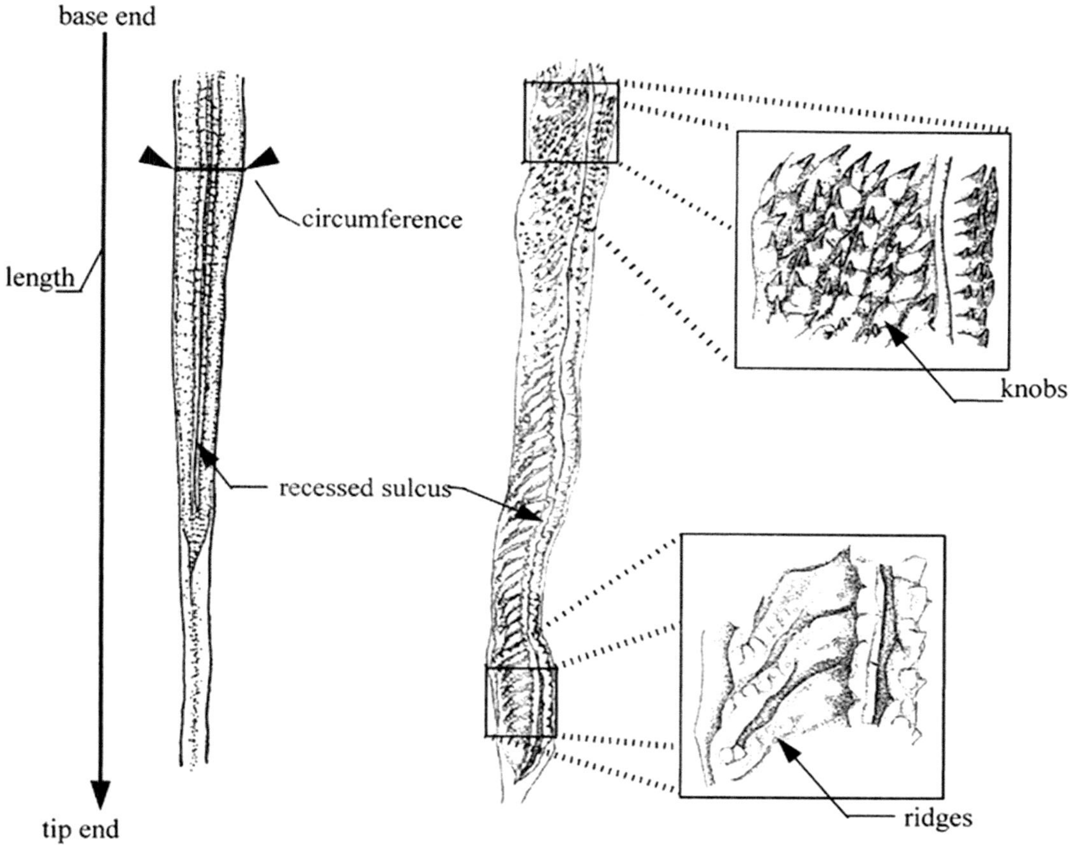


**Fig. 16.50** Drawings of intromittent organs of a Harlequin Duck (Histrionicus histrionicus) (left; scale = 4×) and Ruddy Duck (Oxyuria jamaicensis) (right; scale = 0.5×) showing various morphological characteristics (not to scale). (Figure from Coker et al. 2002; © 2002 Oxford University Press, used with permission)

*japonica*; Figs. 16.60 and 16.61). This suggests a mechanism whereby sperm release from the SSTs is controlled via the ovulatory cycle. Among Japanese Quail and other gallinaceous species of birds, there is a peak in blood progesterone levels about 4 to 6 h before ovulation, and this interval between peak progesterone (or progesterone injection) and sperm release corresponds to the results observed in the experiments conducted by Ito et al. (2011). The increase in progesterone levels appears to stimulate production of protein (heat shock protein 70) that increases sperm motility via increased flagellar activity. The "activated" sperm then proceed out of the SST and up the oviduct toward the just-ovulated egg.

## 16.7 Fertilization

Fertilization of ova occurs in the infundibulum, generally within 15 minutes or less after ovulation. In most other taxa, including mammals, only a single sperm enters the ovum, and penetration of an ovum by multiple sperm results in the death of the embryo (Wang et al. 2003). In those taxa, a single sperm fusing with an egg causes an increase in levels of intracellular $Ca^{2+}$, coming primarily from the endoplasmic reticulum, that is essential for egg activation (Runft et al. 2002). However, polyspermy is typical, and critically important for successful fertilization of eggs, for

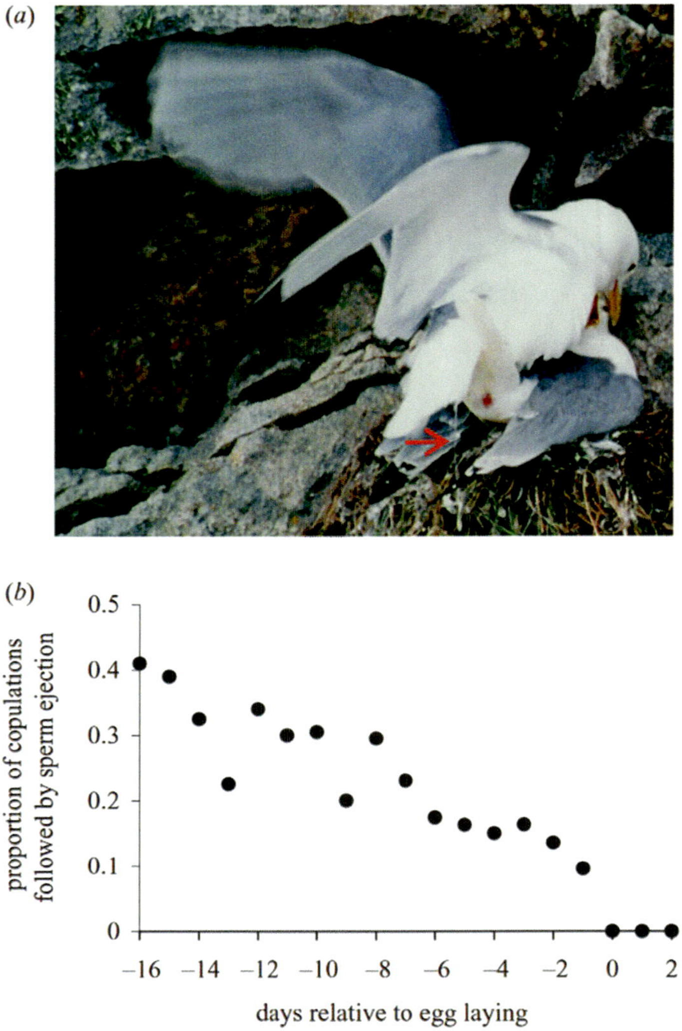


**Fig. 16.51** (**a**) A pair of Black-legged Kittiwakes (*Rissa tridactyla*) copulating at their nest. The female's cloaca is open to receive sperm and a thin thread of semen can be seen hanging below the male's tail (red arrow; photo by Uannick Coupry). (**b**) Decline in tendency of females to eject their mate's sperm as the date of egg laying approaches. (Figure from Wagner et al. 2004; © 2004 The Royal Society, used with permission)

birds as well as reptiles, i.e., taxa with large eggs. Mizushima et al. (2014) found that the amount of avian sperm extract (containing ovum-activating proteins) needed to increase levels of intracellular $Ca^{2+}$ for egg activation and normal development of Japanese Quail (*Coturnix japonica*) embryos after fertilization was greater than the amount provided by a single sperm. In addition, Wishart and Stains (1999) found reduced fertility in both Domestic Turkeys and Chickens when fewer than 20 spermatozoa fertilized eggs. Similarly, Hemmings and Birkhead (2015) found that bird embryos are less likely to survive when few sperm penetrate eggs.

Several sperm enter the germinal disc region by breaking through the inner perivitelline membrane using enzymes in the acrosome of the sperm (Figs. 16.62, 16.63, and 16.64). Sperm cells have a limited time to penetrate the IPVL because, within 15–30 minutes after ovulation, cells lining the infundibulum deposit an outer perivitelline layer (OPVL; also called the chalaza

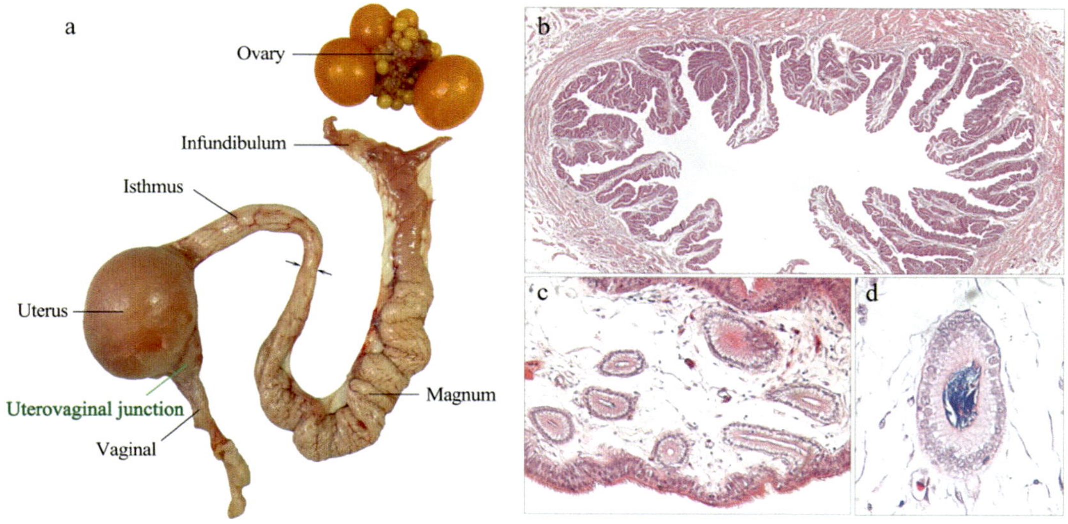


**Fig. 16.52** (**a**) Oviduct of an adult female Domestic Chicken showing the uterovaginal junction (UVJ) where sperm storage tubules are located. Arrows indicate the boundary between the magnum and isthmus. (**b**) Stained micrograph (40×) showing the mucosal folds of the UVJ. (**c**) Transverse sections showing sperm storage tubules (SSTs) in a mucosal fold of the UVJ (400×). (**d**) A single SST with sperm in the lumen (1000×). (Figure from Wen et al. 2020; © The Authors 2020; open-access article distributed under the terms of the Creative Commons Attribution 4.0 International License, https://creativecommons.org/licenses/by/4.0/)

layer) around eggs that represents a mechanical barrier for sperm (Wishart 2002). The inner perivitelline layer, in contact with the yolk and the egg consists of interlaced protein fibers, and the outer layer contains proteins that form a lattice network of fine fibrils (Figs. 16.65 and 16.66). The inner vitelline layer acts as a barrier to separate the egg albumen from the yolk and help prevent contamination of yolk from bacteria that might be present in the albumen. In addition, the outer layer of the vitelline membrane contains several antimicrobial proteins. After eggs are fertilized, the vitelline membrane gradually breaks down and forms a vascularized tissue around the yolk called the yolk sac (Dombre et al. 2017).

Although multiple sperm enter the egg, only a single spermatozoon fuses with the female pronucleus (syngamy) and the remaining sperm degenerate. Syngamy generally occurs when the fertilized ovum has moved to the posterior part of the magnum, or about 3–5 h after ovulation (Stepinska and Bakst 2007).

## 16.8 Sex Determination

Sex allocation theory predicts that parents should manipulate the sex ratio of their offspring when the fitness benefits of investment in sons and daughters differ. Daughters, regardless of their quality, are likely to reproduce. However, the quality of sons can greatly influence their reproductive success, with high-quality sons likely to produce more offspring than daughters and low-quality sons possibly failing to reproduce at all (Cameron and Linklater 2002). Investigators have found a number of factors that influence offspring sex ratios. For example, the results of several studies have revealed that female condition can influence offspring sex ratios, with females in better condition having male-biased sex rations and those in poorer condition having female-biased sex ratios (e.g., Bradbury and Blakey 1998; Whittingham and Dunn 2000; Alonso-Alvarez and Velando 2003; Pike and Petrie 2005). The female-biased offspring sex ratios of females in poor condition may be

**Box 16.5 Seminal Fluid**

Seminal fluid is the non-cellular component of an ejaculate and has been found to consist of a complex mixture carbohydrates, lipids, numerous proteins and other molecules. This fluid is produced in the testis and vas deferens as well as by cells in the cloaca (Fujihara 1992). Borziak et al. (2016) reported that the seminal fluid of male Red Junglefowl consisted of over 1000 different proteins, and Rowe et al. (2020) identified 827 unique proteins in the seminal fluid of House Sparrows (*Passer domesticus*). Some of the proteins appear to serve anti-microbial functions (Borziak et al. 2016), whereas others appear to be important in sperm maturation and enhancing sperm velocity. For example, serum albumin, an abundant protein in the seminal fluid of male Domestic Chickens, has been found to stimulate sperm motility (Marzoni et al. 2013). Another protein (REG4) found in the seminal fluid of male House Sparrows, is a lectin (sugar-binding protein), and lectins have been found to bind to the epithelium and microvilli of sperm storage tubules. Although the importance of this binding in sperm storage remains to be determined, REG4 may play some role in the sperm storage process. Borziak et al. (2016) identified MHC (major histocompatibility complex) proteins in the seminal fluid of male Red Junglefowl (*Gallus gallus*). Because MHC proteins play an important role in cellular self-recognition (i.e., preventing attack by an individual's immune system), their presence in seminal fluid may represent a molecular mechanism contributing to cryptic female choice (Borziak et al. 2016). Yet another protein found in the seminal fluid of male Domestic Chickens (CITI-1/SPINK2) inhibits the activation of a proteolytic enzyme called acrosin, which is important because acrosin is an enzyme in the sperm acrosome that cannot be activated until the moment of fertilization (Thélie et al. 2019).When activated, acrosin is needed for sperm to successfully break through the perivitelline membrane and fertilize the egg.

Studies to date have identified a number of components in the seminal fluids of birds. However, most studies of seminal fluids have focused on organisms other than birds and, in those organisms, components of seminal fluid have been found to serve many important functions, including increasing female receptivity to remating, increasing rates of oogenesis, ovulation and egg laying, and promoting the storage of sperm (Poiani 2006; Avila et al. 2011). Few studies of avian seminal fluids have been published, and most have focused on Domestic Chickens. As such, additional studies are clearly needed to improve our understanding of the many possible functions of avian seminal fluids.

adaptive if females require less energy to rear than males or, as already noted, females have less variance in reproductive success than males (Trivers and Willard 1973; Clutton-Brock 1991). Another factor found to influence offspring sex ratios is territory quality. Among Seychelles Warblers (*Acrocephalus sechellensis*), helpers (mostly females) can increase the fitness of breeding pairs, but only in high-quality territories with sufficient resources. Accordingly, Komdeur et al. (1997) found that 77% of the offspring of breeding pairs in high-quality territories were males, but only 13% of offspring of breeding pairs in low-quality territories were males. Laying date can also influence offspring sex ratios. Females may also adjust offspring sex ratios based on the quality or age of their mates. For example, Mitrus et al. (2015) found that female Red-breasted Flycatchers (*Ficedula parva*) mated to older, more ornamented males were more likely to have male-biased offspring sex ratios, and investigators have reported similar results for other species of birds (e.g., Ellegren et al. 1996; Sheldon et al. 1999). Dijkstra et al. (2010) found

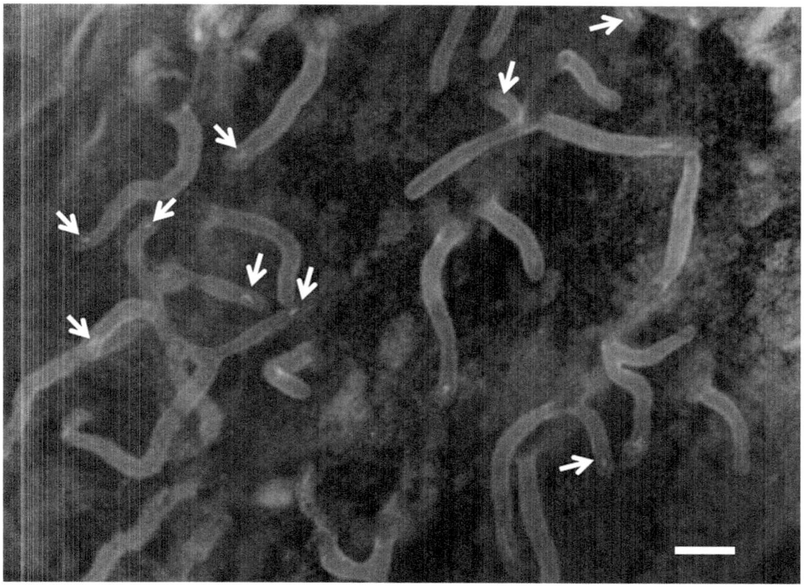

**Fig. 16.53** Sperm storage tubules in the oviduct of a female Japanese Quail (*Coturnix japonica*). Arrows are pointing to sperm in the sperm storage tubules. Scale bar = 50 μm. (Figure from Matsuzaki and Sasanami 2017:© 2017 Springer Nature Singapore Pte Ltd., used with permission)

that offspring sex ratios of Rock (*Columba livia*) and Wood (*Columba palumbus*) pigeons were male-biased in the spring and female-biased in the fall, with these ratios adaptive because males require more time to mature and those in earlier broods are more likely to recruit into the next year's breeding population.

Regardless of whether sex ratios are skewed toward males or females, what is the mechanism by which birds influence the sex of their offspring? Importantly, among birds, females determine the sex of their offspring by contributing either a Z or a W chromosome (males also contribute a Z chromosome, with ZZ being males and ZW being females), and oocytes have both chromosomes until a few hours before ovulation. Pike and Petrie (2003) described the stages of egg production when female birds could potentially influence the sex of their offspring (Fig. 16.67), and Navara (2013) described several

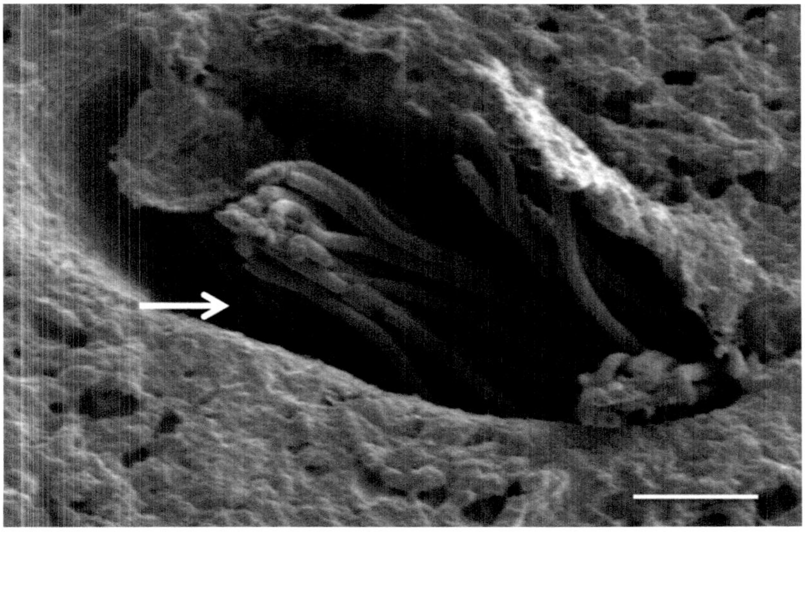

**Fig. 16.54** Micrograph showing sperm in a sperm-storage tubule. The arrow indicates a bundle of the sperm in the tubule. Scale bar = 2.5 μm. (Figure from Sasanami et al. 2013; Reproduced with permission of the Society for Reproduction and Development from Sasami et al., Sperm storage in the female reproductive tract in birds. J. Reprod. Dev. 59: 334–338, 2013)

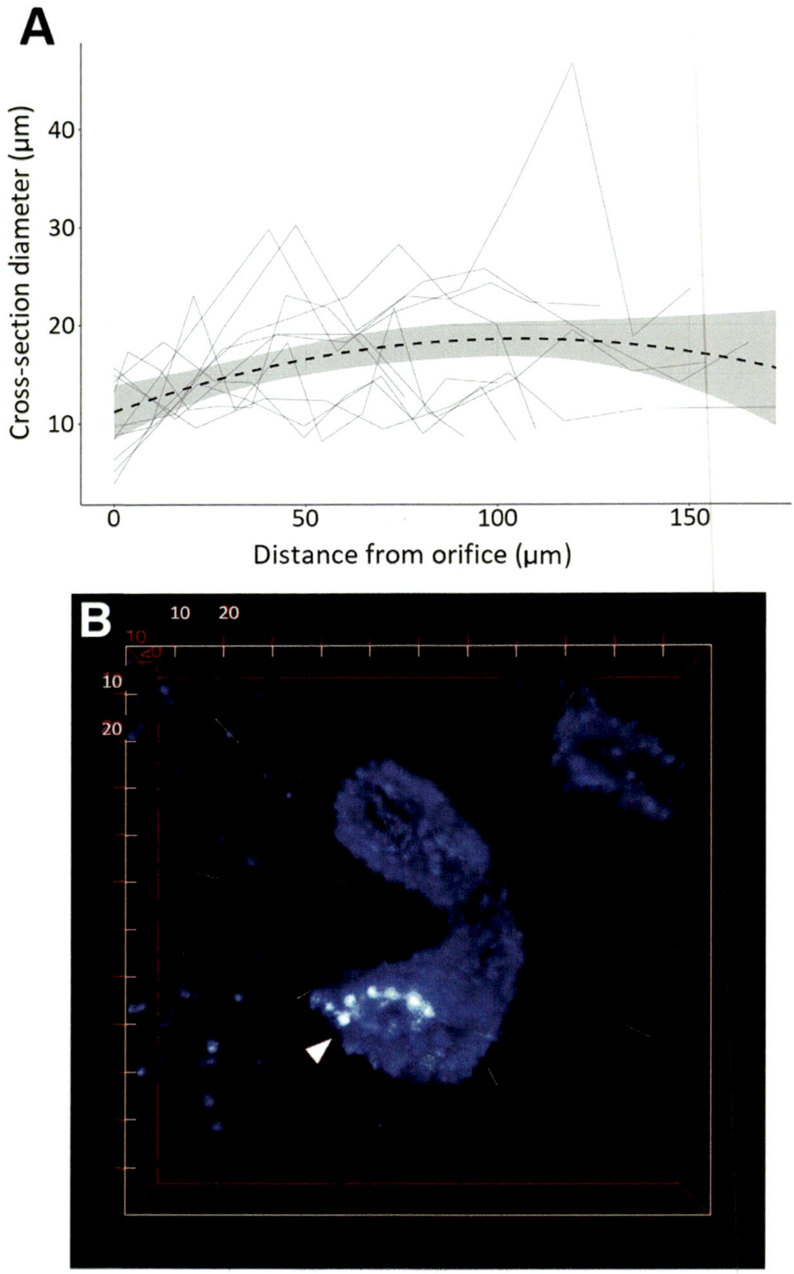


**Fig. 16.55** (**a**) Diameter of sperm storage tubules ($N = 10$) of a female Zebra Finch (*Taeniopygia guttata*), note the apparent constriction at the entrance or orifice and a slight increase in diameter toward the middle. (**b**) Image of a sperm storage tubule of a female Zebra Finch showing constricted orifice (white arrowhead). Scale white bounding box is in microns. (Figure from Mendonca et al. 2019; open-access article distributed under the terms of the Creative Commons Attribution License, https://creativecommons.org/licenses/by/4.0/)

possible mechanisms by which female birds could determine offspring sex:

(1) assuming that oocytes are programmed to retain either the Z or W chromosome (for which there is currently no evidence), females could preferentially recruit follicles programmed to retain either a Z or W chromosome either because follicles progammed to retain Z or W chromosomes are clustered

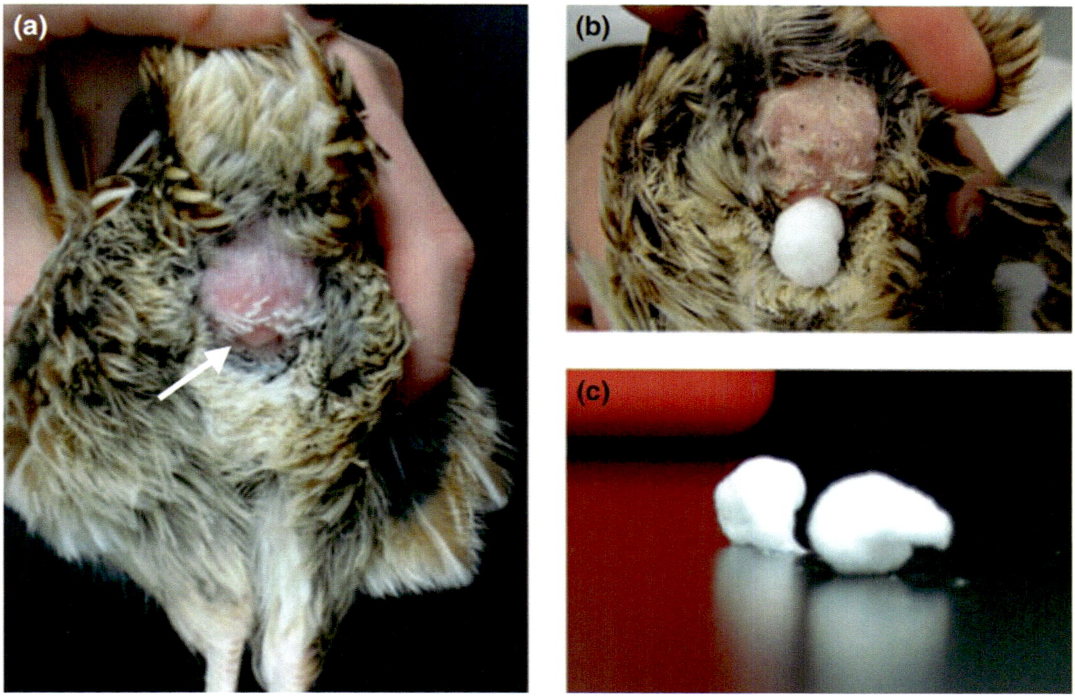


**Fig. 16.56** (**a**) Cloacal gland (also called the foam gland or proctodeal gland) of a male Japanese Quail (*Coturnix japonica*). Arrow points to the cloacal opening, and feathers surrounding the cloacal gland were plucked. (**b**) Manual expression of foam from a cloacal gland. (c) Foam produced by two male Japanese Quail. (Figure from Finseth et al. 2013; © 2013 The Authors. Journal of Evolutionary Biology © 2013 European Society For Evolutionary Biology, used with permission)

together spatially in the ovary or follicles have markers that identify them as being programmed to retain Z or W chromosomes (Fig. 16.68). However, if assuming that oocytes can retain either the Z or W chromosome, then

(2) follicle growth rates during the period of yolk deposition determine which of the sex chromosomes is retained and females are able to adjust follicle growth rates (Fig. 16.69a),
(3) during the period of yolk deposition, females are able to add other substances to the yolk, including androgens, which could influence meiotic segregation to determine offspring sex (Fig. 16.69b),
(4) direct determination by females of which sex chromosome is retained by oocytes (Fig. 16.69c),
(5) production of chimeras (cells with more than one genotype) whereby a polar body (a small haploid cell formed during oogenesis and normally extruded from oocytes during meiosis) is not extruded from an oocyte, two sperms fertilize the oocyte, the fertilized egg therefore contains both WZ and ZZ cells, and those cells "compete" and eventually result in either a male or female offspring (Fig. 16.69d),
(6) oocytes with the non-preferred sex chromosomes may be ovulated, but are aborted into the peritoneal cavity rather than moving into the oviduct (Fig. 16.69e), and
(7) fertilization could be sex-specific so that sperm only fertilize oocytes with a specific sex chromosome (Fig. 16.69f).

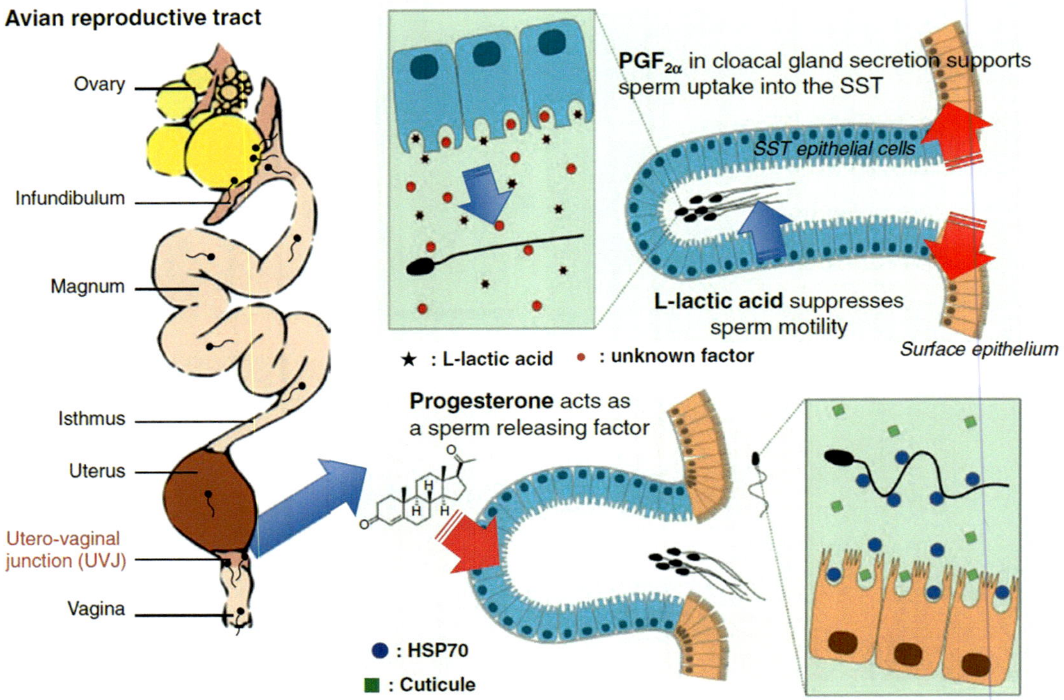


**Fig. 16.57** Possible sperm storage mechanism in the sperm storage tubules (SST) of female Japanese Quail (*Coturnix japonica*). After copulation, prostaglandin F2α in the cloacal gland secretions of male Japanese Quail causes the contraction of the female's vagina, which increases the diameter of the entrance of the SSTs, enhancing the ability of sperm to enter the SSTs. There is a high concentration of lactic acid in the SST because of the production of lactic acid by SST cells due to hypoxic conditions, and it makes any resident sperm quiescent. As the time of ovulation approaches, plasma concentrations of progesterone increase, causing morphological changes in the SSTs and squeezing sperm out of the SSTs and the lumen of the oviduct. Simultaneously, cuticle materials are also released under progesterone control and the cuticle may function as a lubricant that facilitates egg rotation in the uterus. From the surface epithelium of the UVJ, HSP70 protein is secreted, bound to the surface of the sperm and reactivates sperm motility, which may help sperm migrate to the fertilization site. (Figure from Matsuzaki and Sasanami 2017; © 2017 Springer Nature Singapore Pte Ltd., used with permission)

The mechanism by which female birds might determine offspring sex likely involves hormones, and available evidence suggests that testosterone and corticosterone are the most likely candidates. The results of several studies have revealed that, in many species of birds, females with higher blood plasma levels of testosterone have offspring sex ratios that are male-biased (e.g., Pike and Petrie 2005; Veiga et al. 2004; Goerlich et al. 2009), whereas female with higher blood plasma levels of corticosterone have female-biased offspring sex ratios (Pike and Petrie 2005, Bonior et al. 2007; Fig. 16.70). Blood plasma levels of these hormones, however, do not appear to influence offspring sex ratios in other species of birds (e.g., Pike and Petrie 2005; Henderson et al. 2014). Although reasons for these differences among birds remain to be determined, some progress has been made in terms of understanding how corticosterone and testosterone might influence offspring sex in some species of birds. Wrobel et al. (2020) injected laying Domestic Chickens with either testosterone or corticosterone five hours before ovulation and found that increasing blood plasma levels of both hormones at the time when sex

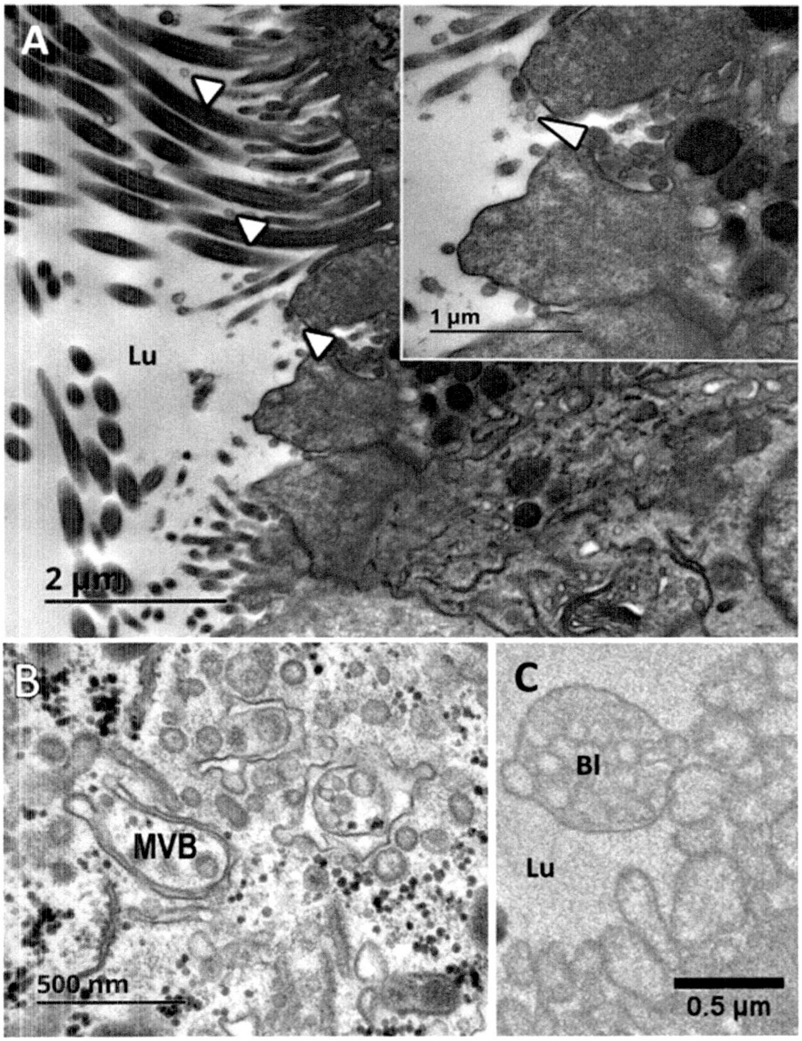


**Fig. 16.58** Transmission electron microscopy of the uterovaginal junction. (**a**) Exosomes (arrowheads) (< 100 nm) were observed at surface epithelium of uterovaginal junction (UVJ). Exosomes are extracellular vesicles released from cells and those in the UVJ contain a variety of proteins that may be important for sperm survival and maintaining sperm viability. (**b**) Intracellular multivesicular body (MVB) containing several exosomes that can be released from cells. (**c**) Blebs (vesicles released from cells via exocytosis) containing vesicles in the lumen of sperm storage tubules. Bl: Bleb; Lu: Lumen. (Figure from Riou et al. 2020; © 2020 Wiley Periodicals, Inc., used with permission)

chromosomes segregate had significant effects of the expression of certain genes and gene networks, but that corticosterone had stronger effects than testosterone. Based on the genes affected by corticosterone, these authors hypothesized that segregation of the Z and W chromosomes might be influenced via (1) changes in the telomere length or location of centromeres (where the spindles attach) that would alter the direction of migration of the sex chromosomes (one into the polar body, the other retained in the oocyte), (2) change in the ion gradient in the germinal disc of an oocyte in a way that would result in oocytes retaining the W chromosome rather than the Z chromosome, (3) altering the number of spindle attachments that determine which sex chromosome is retained in the oocyte, or (4) flip the spindle to alter the sex chromosome that is retained (Fig. 16.71).

Sex allocation theory predicts that parents should manipulate the sex ratio of their offspring when the fitness benefits of investment in sons and daughters differ, the results of many studies have revealed no evidence of such manipulation (e.g., Ding et al. 2017; Benvenuti et al. 2018; Que et al. 2019; McNew et al. 2020). McNew et al. (2020) suggested that sex ratios may not be adjusted either because a species does not have

**Fig. 16.59** Male Yellow-billed Hornbill (*Buceros bicornis*) at its nest site. Its mate is sealed inside the nest cavity. (Photo by Caatherine Withers-Clarke, purchased from iStockphoto)

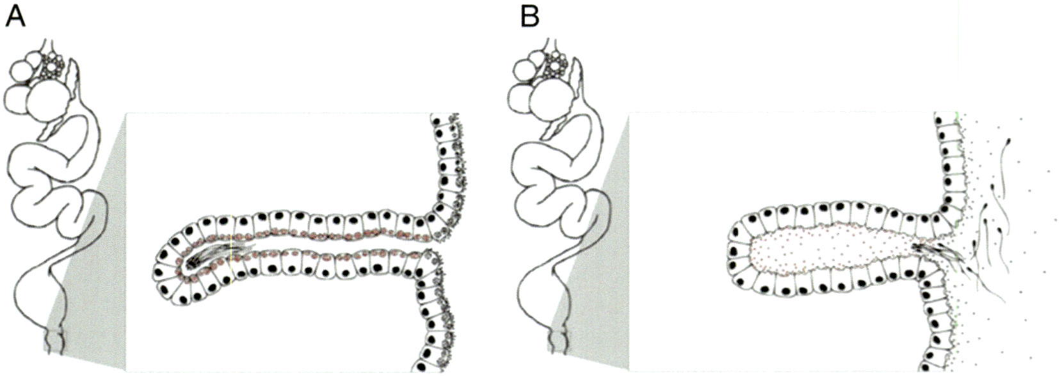


**Fig. 16.60** Proposed mechanism of sperm release from SSTs in the oviduct of a female Japanese Quail (*Coturnix japonica*). (**a**) Before an increase in blood progesterone levels, sperm in the SST are quiescent and the lumen of the SST is not extended. (**b**) When ovulation is imminent, levels of progesterone in the blood increase, stimulating release of sperm caused by a contraction-like response of the SST. The increase in progesterone levels may also stimulate the release of substances by ciliated cells lining the SST (black dots) that have a lubricant effect, as well as other unknown substances (tiny red dots). (Figure from Ito et al. 2011; © 2011 Oxford University Press, used with permission)

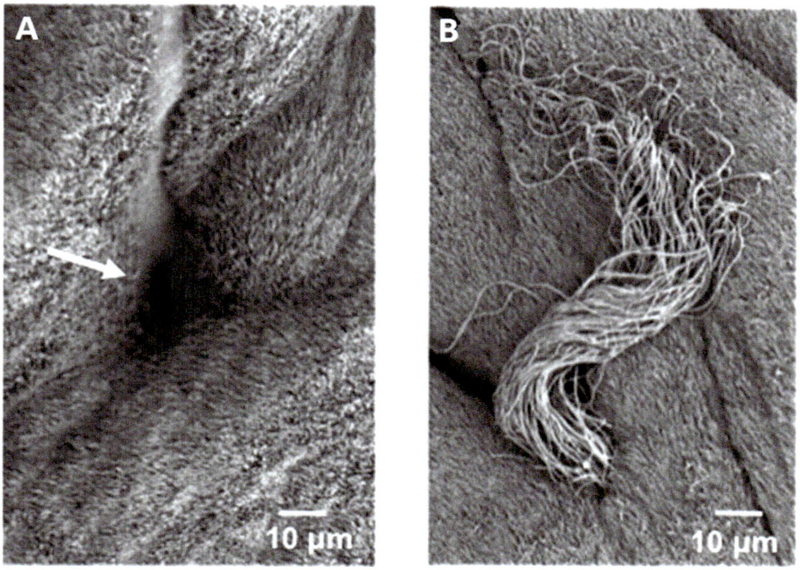


**Fig. 16.61** Scanning electron micrograph showing the effect of injection of progesterone on the release of sperm from sperm-storage tubules. (**a**) Before injection. (**b**) One hour after injection with sperm leaving the sperm-storage tubule. (Figure modified from Ito et al. 2011; © 2011 Oxford University Press, used with permission)

the ability to do so or because the costs of such manipulation outweigh the benefits. Another possible explanation is that there are no condition-dependent fitness differences between males and females (Benvenuti et al. 2018). Kingma et al. (2011) suggested that, for some species of birds,

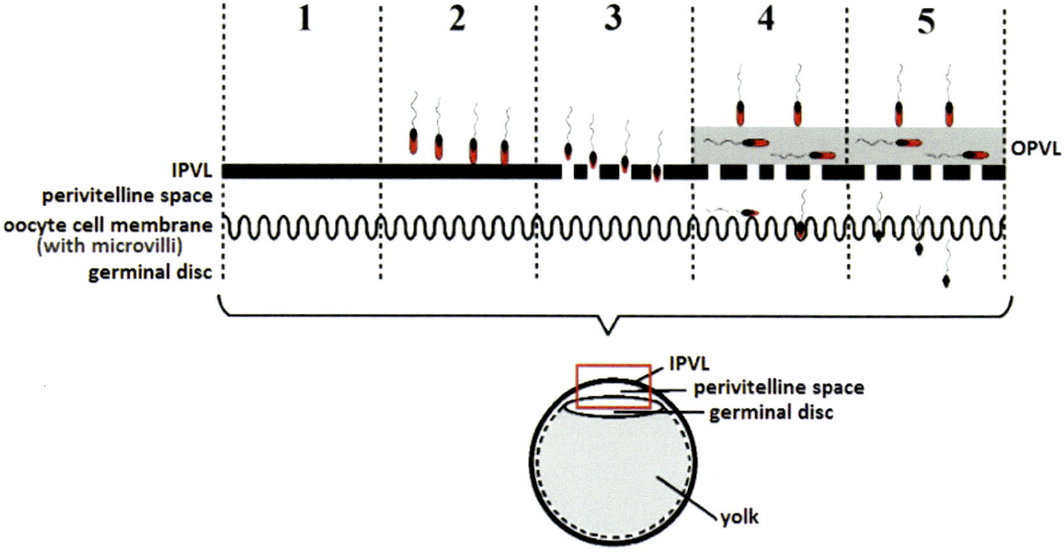


**Fig. 16.62** Interactions of avian sperm and oocytes. (1) Prior to ovulation the oocyte is surrounded by the inner perivitelline layer (IPVL). (2) Sperm binding to the IPVL. (3) Acrosomal reaction and sperm penetration of IPVL. (4) The outer perivitelline layer (OPVL), deposited within 15–30 min after ovulation, serves as mechanical barrier that sperm cannot penetrate so only sperm that previously penetrated the IPVL can now pass through the cell membrane of the egg. (5) Multiple sperm passing through the oocyte cell membrane. (Figure modified from Krawczyk and Jaworska-Adamu 2019; open-access article distributed under the terms of the Creative Commons Attribution License, https://creativecommons.org/licenses/by/4.0/)

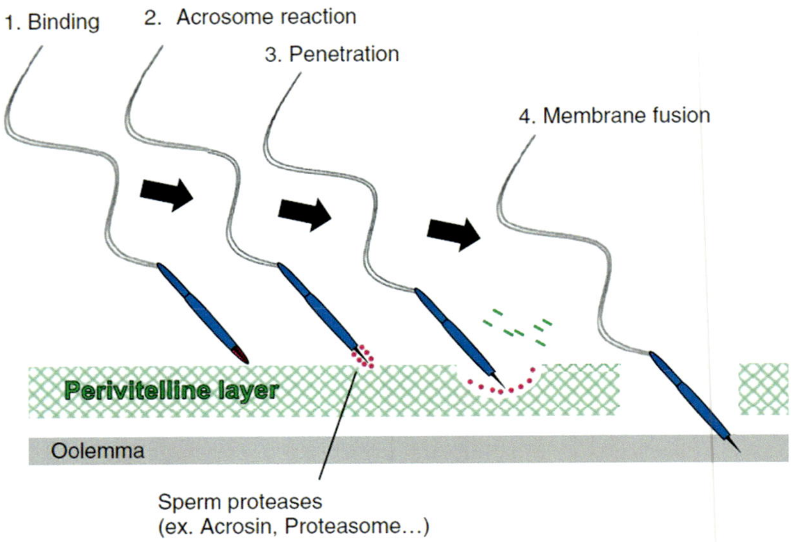


**Fig. 16.63** Sperm binding to and penetration of the egg perivitelline layer with proteolytic degradation/dissolution of the perivitelline layer matrix. A series of sperm–perivitelline interactions presumably occur in the infundibulum of oviduct during a short period of time (~15 min in Domestic Chickens) after ovulation. (Figure from Nishio and Matsuda 2017; © 2017, Springer Nature Singapore Pte Ltd., used with permission)

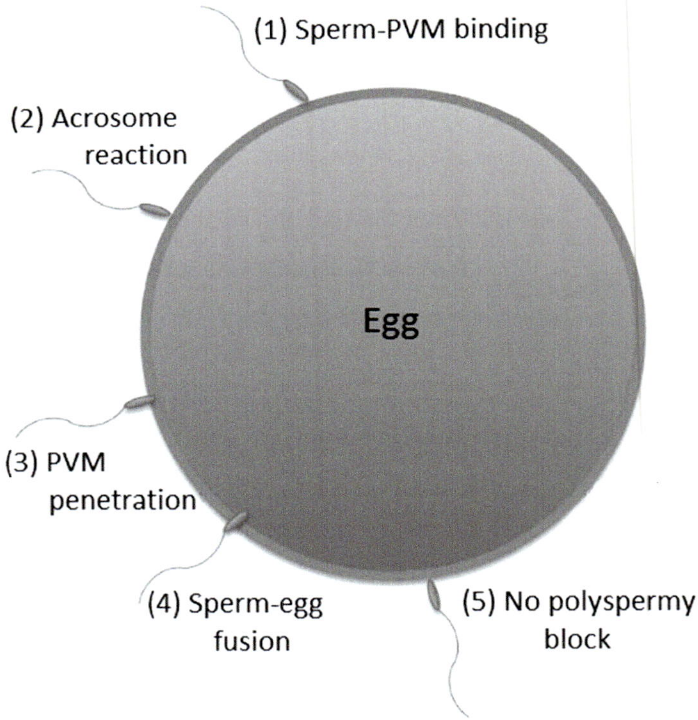


**Fig. 16.64** In the infundibulum, sperm bind to the inner perivitelline membrane (PVM), then the acrosomal membrane fuses with the PVM (acrosome reaction). Enzymes released from the acrosome create an opening in the PVM and egg cell membrane (PVM penetration), allowing sperm-egg fusion and fertilization. In contrast to mammals, multiple sperm penetrate bird eggs (polyspermy). (Figure modified from Ichikawa et al. 2016; © 2016 Japan Poultry Science Association, used with permission)

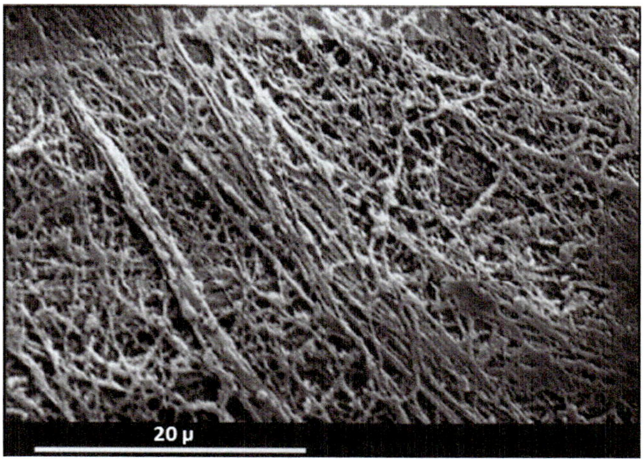


**Fig. 16.65** Scanning electron micrograph of a portion of the perivitelline membrane in an egg of a Domestic Chicken. (Figure from Kirunda and McKee 2000; © 2000 Oxford University Press, used with permission)

the long-term benefits of biased offspring sex ratios may be too small to have evolved.

## 16.9 Oviduct Structure and Function

Prior to egg laying, estrogen triggers both cellular proliferation and cellular differentiation in the oviduct, with epithelial cells lining the walls of the oviduct differentiating into tubular gland cells, goblet cells, and ciliated cells. During the period prior to breeding to when females are laying eggs, oviduct size and mass increase dramatically (Fig. 16.72), especially the magnum (Figs. 16.73 and 16.74). The walls of each section of the oviduct have a muscular layer that supports the oviduct and propels the egg. The tubular gland cells of the oviduct synthesize the egg-white proteins (albumins) and the goblet cells synthesize avidin, a protein that helps inhibit bacterial growth (Johnson 2000). The ciliated cells help transport sperm up the oviduct to the infundibulum where fertilization occurs.

Oocytes are fertilized in the infundibulum, a muscular funnel-shaped structure at the beginning of the oviduct. Most of the infundibulum is lined

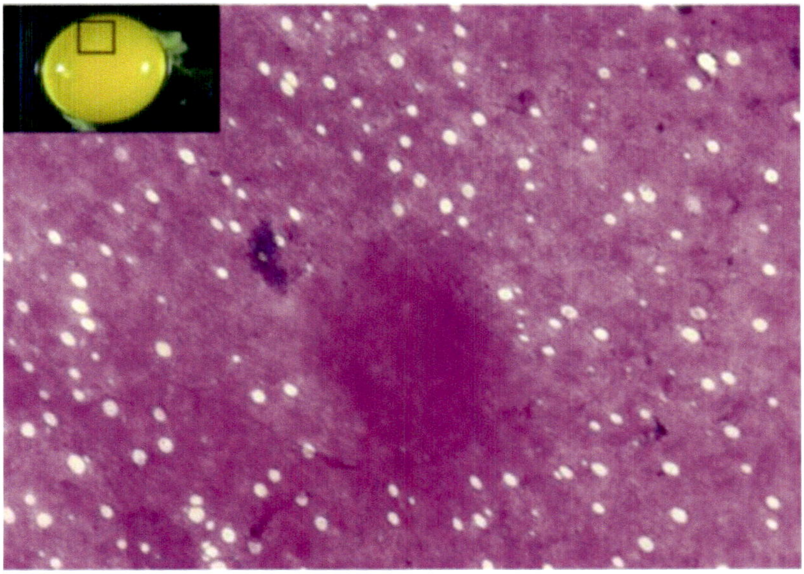

**Fig. 16.66** Holes in the germinal disc regions of the inner perivitelline layer of the egg of a Domestic Chicken created by the penetration of multiple sperm. (Figure from Navara 2018; © 2018 Springer International Publishing AG, used with permission)

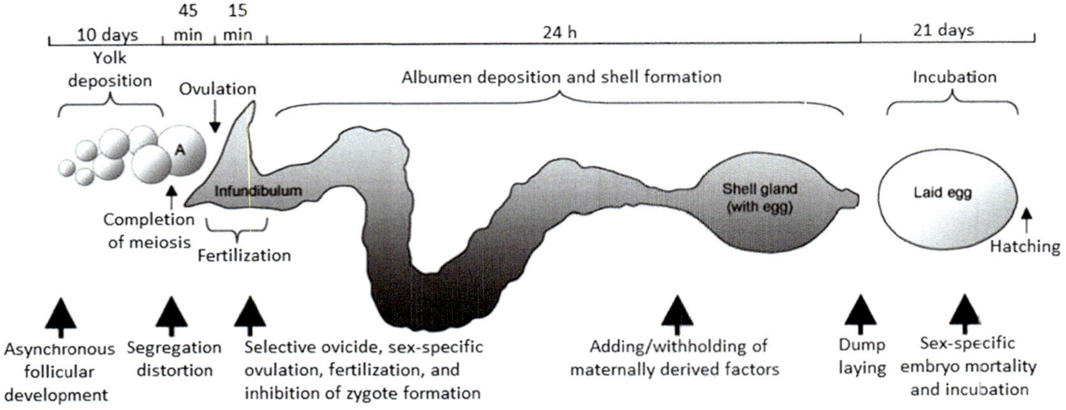


**Fig. 16.67** Stages of egg production from yolk deposition through incubation, indicating the times of events and possible times of sex manipulation by females (times are for Domestic Chickens, *Gallus g. domesticus*). Follicles develop in the ovary and the largest (labeled A) is released (ovulation) and moves into the infundibulum. Sex is determined in chickens before the first meiotic division, when segregation of the sex chromosomes (in birds the female is the heterogametic sex) consigns either the Z or W chromosome to the ovum and the remaining sex chromosome to the polar body. Shortly after ovulation, the ovum is fertilized by sperm in the infundibulum before moving down the oviduct. (Figure modified from Pike and Petrie 2003; © Cambridge Philosophical Society, used with permission)

with both ciliated and non-ciliated cells (Fig. 16.75), but there are more ciliated cells in the anterior portion. Cells lining the infundibulum form the inner perivitelline membrane that, in addition to its role in fertilization, serves as a supporting layer for the outer perivitelline layer

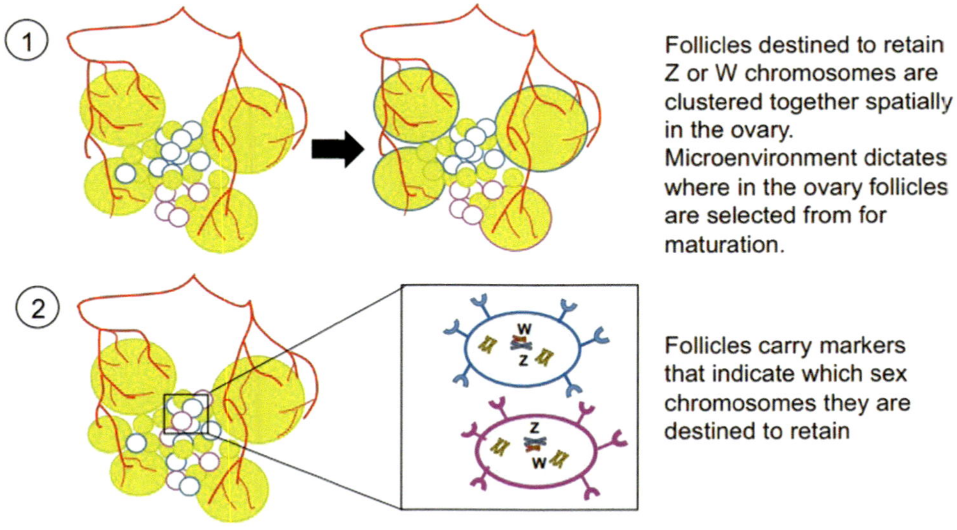


**Fig. 16.68** Potential mechanisms whereby follicles predetermined to retain either the W or Z chromosome might be selected to determine offspring sex. (Figure from Navara 2018; © 2018, Springer International Publishing AG, used with permission)

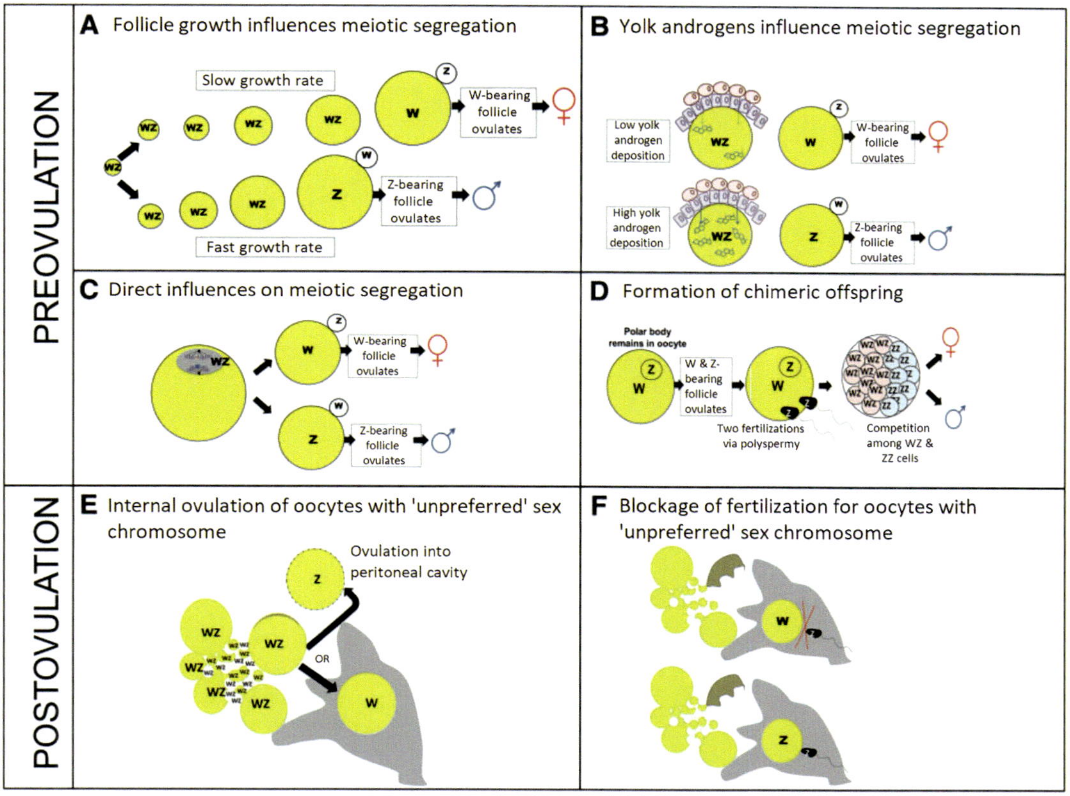


**Fig. 16.69** Proposed mechanisms of sex-ratio adjustment in birds. Mechanisms include manipulation of sex ratios by females (**a**) altering the growth rates of follicles or (**b**) influencing concentrations of androgens in yolk that determine which sex chromosome is retained in the oocyte in both cases. (**c**) Hormones may also influence the segregation of sex chromosomes during meiosis, (**d**) cause the formation of chimeras by preventing oocytes from extrude the polar body followed by competition of WZ and ZZ cells for survival, (**e**) trigger internal ovulation of oocytes with the non-preferred sex chromosome, and (**f**) inhibit fertilization of oocytes with the non-preferred sex chromosome. (Figure from Navara 2013; © 2013 Oxford University Press, used with permission)

that serves as the innermost part of the chalaziferous layer of a bird's egg. As eggs pass through the oviduct, this part of the chalaziferous layer, along with the inner perivitelline membrane, anchors the chalazae (Rahman et al. 2007, 2009). The distal portion of the infundibulum also contains some tubular glands that secrete ovalbumin, lysozyme, and conalbumin (Chousalkar and Roberts 2008).

After just 15–20 min, the ovulated and fertilized ovum, now called an egg-mass or just egg, leaves the infundibulum and enters the next section of the oviduct, the magnum, where albumin proteins are synthesized, secreted, and added to the egg. This process typically takes 2–3 h and the egg-mass then moves into the isthmus portion of the oviduct. Here, the inner and outer shell membranes are formed over a period of about 1–2 h. The egg then moves into the uterus, or shell gland. Initially, the albumin takes up fluids (a process called plumping), then the process of shell production, or calcification, occurs over a period of about 18–20 h. Shell pigments, if any, are added during the last 0.5–3 h that the egg is in the uterus. Finally, the egg passes into the vagina, the last section of the oviduct. The muscular vagina, in coordination with the uterus, expels

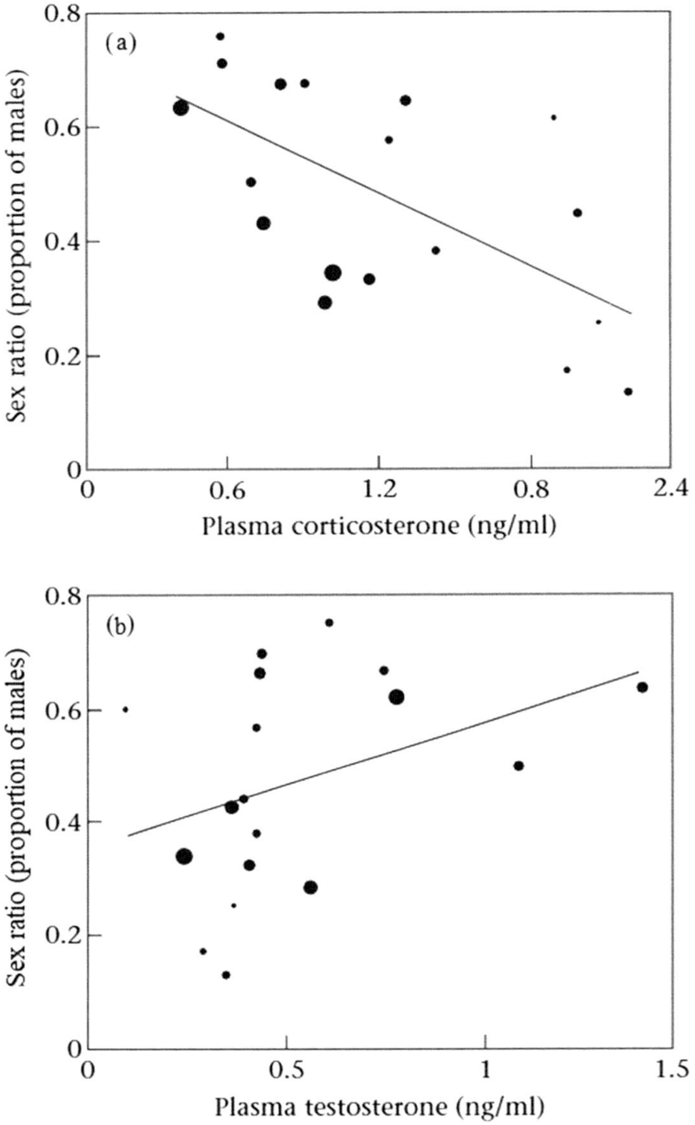


**Fig. 16.70** Relationships between sex ratios of offspring (proportion of males) of female Indian Peafowl (*Pavo cristatus*) and (**a**) plasma levels of corticosterone and (**b**) plasma levels of testosterone. Increasing levels of corticosterone resulted in a female-biased sex ratio, whereas increasing levels of testosterone resulted in a male-biased sex ratio. Circle sizes vary with the number of eggs that were sexed per clutch (range = 4–17 eggs). (Figure from Pike and Petrie 2005; © 2005 The Association for the Study of Animal Behaviour. Published by Elsevier Ltd., used with permission)

the egg from the oviduct, through and out of the cloaca (Johnson 2000, Jacob and Bakst 2007

From the infundibulum, ova move into the magnum where albumen is deposited. The walls of the magnum have an outer muscular layer, an inner layer containing tubular glands, and the surface epithelium is lined with ciliated non-secretory and secretory non-ciliated cells (Chousalkar and Roberts 2008; Fig. 16.76). The deposition of magnum is thought to result from the mechanical stimulation or distortion of the walls of the magnum by the ovum (Gilbert 1979). Excluding the eggshell, albumen represents between 33% and 76% of egg mass for precocial species of birds and between 65% and 86% of egg mass for altricial species (Carey et al. 1980). Differences among species in the amount of albumen present in eggs are correlated with the developmental maturity of young at hatching. The eggs of altricial species have more albumen (water) relative to yolk (energy), shorter incubation periods, and young are less developed

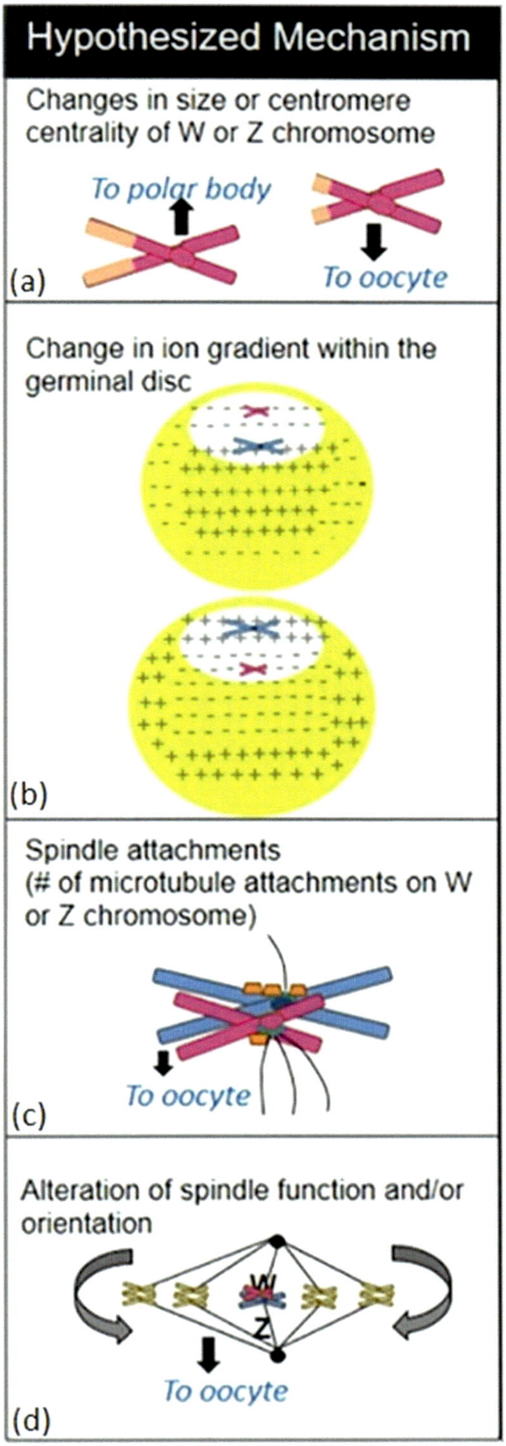


**Fig. 16.71** Possible mechanism whereby the effects of corticosterone on gene function might influence offspring sex. Changes in gene function could affect the segregation of the sex chromosomes by (**a**) altering telomere length or

**Fig. 16.72** Changes in cell numbers and oviduct dry mass prior to and during egg laying and after egg laying (regressing). In Domestic Chickens, oviduct wet mass can increase from about 0.5 g to as much as 45 to 60 g during the period leading up to egg laying. (Figure from Yu and Marquardt 1974; © 1974 Poultry Science Association Inc. Published by Elsevier Inc., used with permission)

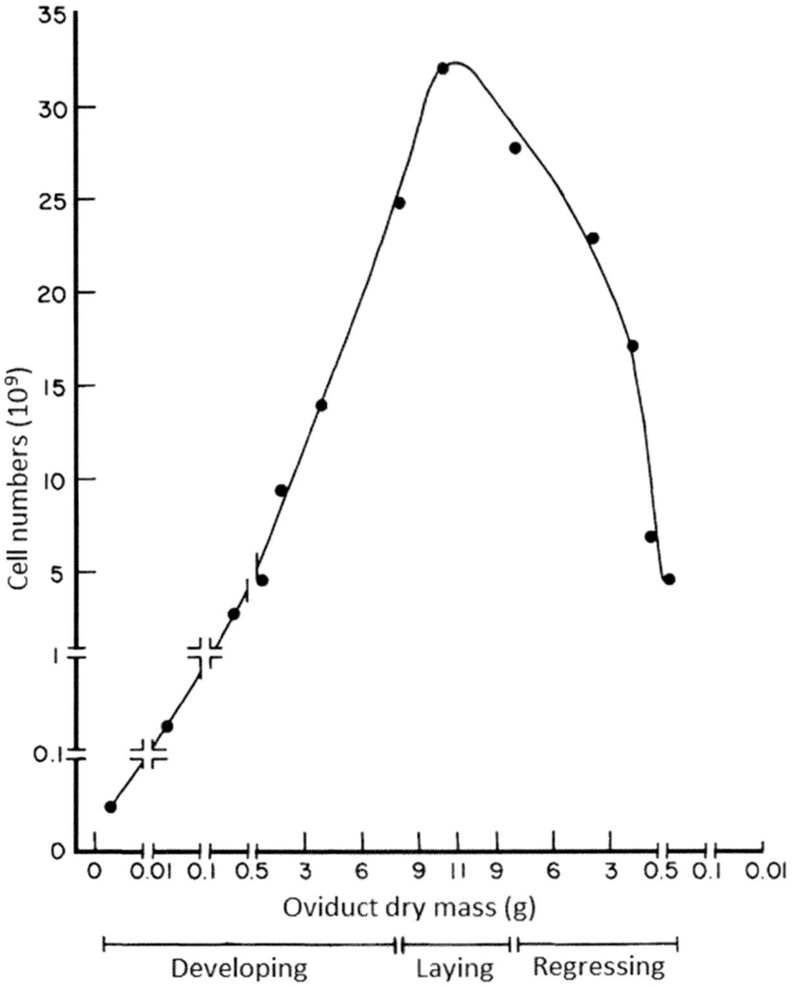


at hatching. The eggs of precocial species have less albumen relative to the amount of yolk, longer incubation periods, and young are more developed at hatching (Fig. 16.77). With shorter incubation periods, developing embryos of altricial species need less energy and less yolk whereas, with longer incubation periods, developing embryos of precocial species need more energy and more yolk (Deeming 2007). Albumen is the primary source of water for developing embryos and the water content of eggs (and albumen) is equivalent to the water content of young birds at hatching (Sotherland and Rahn 1987). As young birds develop, the percent of their body mass consisting of water declines and, because precocial young are more developed at hatching

**Fig. 16.71** (continued) centrosome position, (**b**) changing in the ion gradient in the germinal disc, (**c**) altering the number of spindle attachments to each sex chromosome, or (**d**) flipping the spindle "flip" the sex chromosome being retained. (Figure from Wrobel et al. 2020; © 2020 Wrobel et al., open-access article distributed under the terms of the Creative Commons Attribution License, https://creativecommons.org/licenses/by/4.0/)

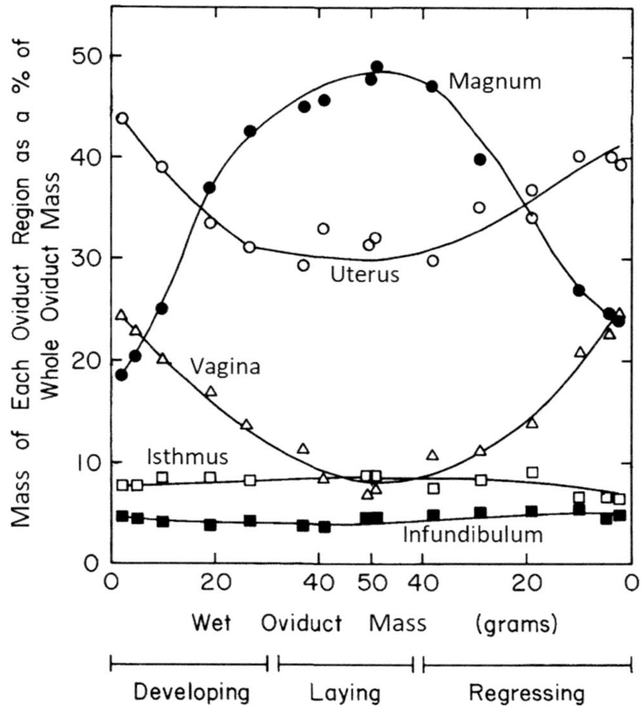


**Fig. 16.73** Relative changes in mass of different sections of the oviduct of Domestic Chickens during the developing, laying, and regressing stages. Note that the mass of the magnum, where albumen is added to the egg, increases most during the developing and laying stages. (Figure from Yu and Marquardt 1974; © 1974 Poultry Science Association Inc. Published by Elsevier Inc., used with permission)

than altricial young, they require less water and, therefore, less albumen (Sotherland and Rahn 1987).

Albumen is synthesized by tubular gland cells and, in the eggs of Domestic Chickens, contains about 10.5% protein and 88.5% water, with the remaining 1% made up of small amounts of carbohydrates, lipids, and inorganic ions (Willems et al. 2014). Eggs have both thick and thin albumen, with the thick albumen containing more of a protein called ovomucin than the thin albumen. The viscosity of the thick albumen inhibits cell motility and helps keep bacteria from getting to the yolk. Egg white also contains several antibacterial proteins (Réhault-Godbert et al. 2011), including lysozyme, avidin, and ovotransferrin. Concentrations of these proteins in the albumen varies among species (Saino et al. 2007, Shawkey et al. 2008; Table 16.4), likely because of differences in nest habitats and sanitation and differences in the environmental conditions and microbial abundance and diversity in different habitats and geographical locations (Shawkey et al. 2008). Embedded within the albumen are gelatinous, string-like structures called the chalazae that anchor the yolk to the center of the egg and allow it to rotate, always orienting to gravity so that the embryo faces upward. The chalazae are formed by the twisting of protein fibers (ovomucin) as the egg (but not the yolk) rotates while moving through the oviduct (Figs. 16.78 and 16.79).

## 16.10 Shell Membranes and the Eggshell

As the yolk and albumen complex travel through the isthmus, the inner and outer shell membranes are added. The eggshell membranes (ESMs) consist of a meshwork of protein fibers organized into inner and outer layers of different diameters ranging from 0.1 to 3 μm (inner ESM) to 1 and 7 μm (outer ESM; Fig. 16.80). The fibers of the inner ESM are connected to the outer ESM; the fibers of the outer ESM penetrate the mammillary cones of the shell (Liong et al. 1997). Whereas the inner membrane remains uncalcified, the fibers of the

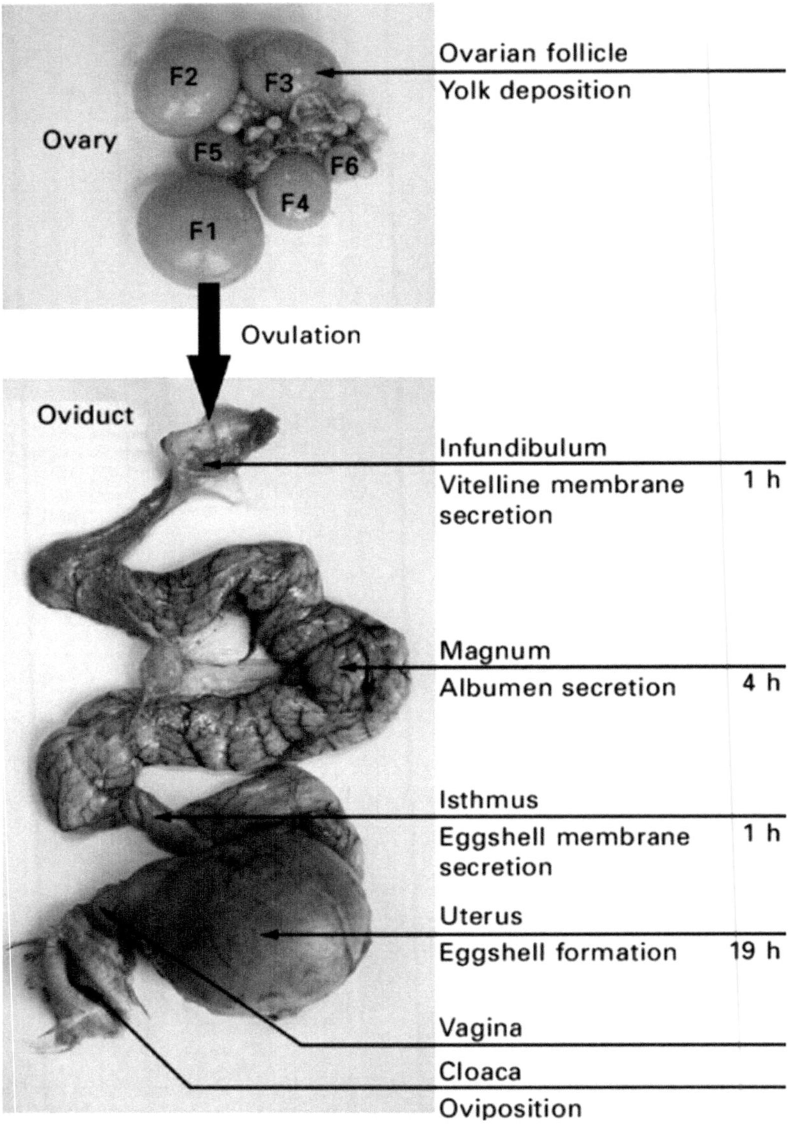


**Fig. 16.74** Ovary and oviduct of a female Domestic Chicken. The times indicated on the right would be typical for a bird laying an egg every 24 h. Vitelline membrane is also referred to as the perivitelline membrane. (Figure from Nys and Guyot 2011; © 2011 Woodhead Publishing Limited, used with permission)

outer shell membrane become mineralized at discrete sites and are incorporated into the base of the eggshell (Nys et al. 2004). The ESMs serve, along with the eggshell, as a barrier to microbes, and the outer shell membrane provides a structural foundation for the formation of the eggshell (Figs. 16.81 and 16.82).

The avian eggshell consists of several layers, including a mammillary layer, palisade layer, a vertical crystal layer, and, in many species of birds, an outermost cuticle (Figs. 16.82 and 16.83). Among species, the thickness of eggshells is correlated with bird body mass (Attard and Portugal 2022; Fig. 16.84). Based on an analysis of 4260 species of birds, Attard and Portugal (2022) determined that eggshell thickness is "… predominantly influenced by phylogeny, mode of development, habitat and diet." These authors found that precocial species have thicker eggshells than altricial species, with the thicker eggshells of precocial species likely providing the calcium needed for the development of "mature

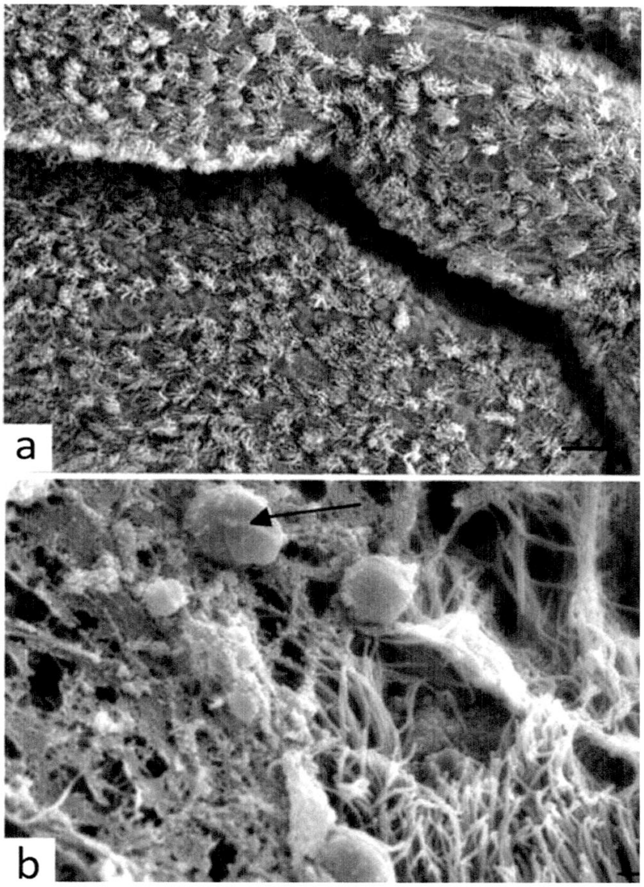


**Fig. 16.75** (**a**) Scanning electron micrograph (SEM) of the infundibulum of a Domestic Chicken lined by a series of ciliated and non-ciliated epithelial cells (x900, Scale bar = 20 μm). (**b**) SEM of the infundibulum of a Domestic Chicken showing secretory granules (arrow pointing at one of them) produced by tubular glands on top of cilia (x10,000, Scale bar = 1 μm). (Figure from Chousalkar and Roberts 2008; © 2008 Springer-Verlag, used with permission)

muscles and highly ossified skeletons" of precocial hatchlings (Attard and Portugal 2022). In addition, nonpasserines that occupy semi-open and dense habitats had relatively thinner eggshells (lower ETIs) than those found in open habitats. Attard and Portugal (2022) suggested that differences in light exposure might explain this difference, i.e., thinner eggshells in habitats with lower light levels (semi-open and dense) enhance light transmission through the shell which may improve embryonic growth and hatchability (Shafey 2004). Finally, diet was also found to influence eggshell thickness, with species feeding primarily on nectar, seeds, and fruit that contain less calcium (e.g., Apodiformes, Trochilidae) having thinner eggshells than scavengers and carnivores that have calcium-rich diets. Interestingly, herbivores also have relatively thick eggshells, possibly because these species rely more on calcium stores in their skeleton during egg-laying (Attard and Portugal 2022).

The eggshell protects developing embryos, but allows gas exchange through pores. It also protects the egg against microbial invasion and provides calcium for growing embryos. The calcified layer consists primarily of calcium carbonate in the form of calcite and is divided into layers (Box 16.6 Where Do Females Get Calcium for Eggshells?). The innermost mammillary cone layer is in contact with the underlying eggshell membranes, and the flat surface of individual cones forms the basis for the calcite columns of the palisade layer. The palisade layer ends at the vertical crystal layer that has a crystalline structure with a greater density than the palisade layer (Mann 2015). In many species, the eggshell is

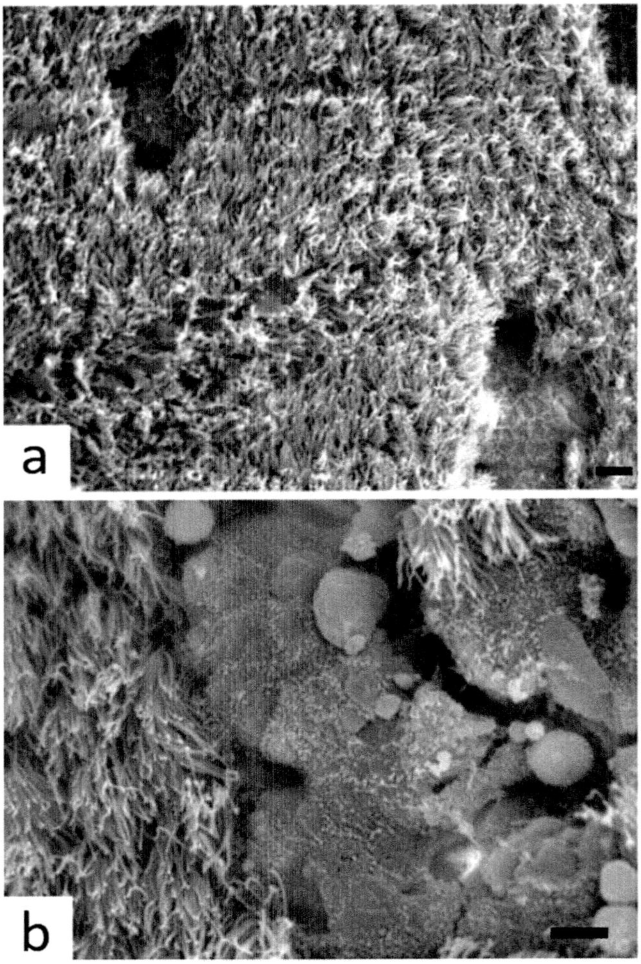


**Fig. 16.76** (**a**) Scanning electron micrograph (SEM) of ciliated and non-ciliated cells of the magnum of a Domestic Chicken (x900, Scale bar = 20 μm). (**b**) SEM of glandular opening in the magnum of a Domestic Chicken. Note the secretory granules produced by tubular glands near the opening (x3000, Scale bar = 5 μm). (Figure from Chousalkar and Roberts 2008; © 2008 Springer-Verlag, used with permission)

sealed by the cuticle, a layer that varies in thickness, structure, and composition among different bird taxa. In some taxa, including parrots, pigeons, and petrels, eggshells have no cuticle (Mikhailov 1997). When present, the cuticle is deposited while eggs are in the uterus, after all or most pigment has been deposited, and during the last 1.5 h before eggs are laid (Nys et al. 1999; Wilson et al. 2017).

Eggshells are formed in three stages: initiation, growth, and termination. To initiate the process as the egg passes from the isthmus to the uterus (shell gland), globular proteins accumulate on the outer shell membrane to form structures called mammillary cores (also called cones; Figs. 16.85 and 16.86). Particles of calcium carbonate then begin to accumulate over the mammillary cores. Because these nanoparticles of calcium carbonate exhibit no distinctive crystalline structure, they are generally referred to as amorphous calcium carbonate. As the calcium carbonate accumulates, transformation into calcite crystals begins (Figs. 16.86, 16.87, and 16.88), and these crystals radiate from the mammillary cores and continue to grow outward (the growth stage) to create a shell's palisade layer (Fig. 16.83). During this process, the egg is bathed in uterine fluid that contains all of the organic and inorganic materials needed for eggshell formation. This process of biomineralization requires large amounts of calcium and carbonate ions obtained from the blood by transport via the uterine walls.

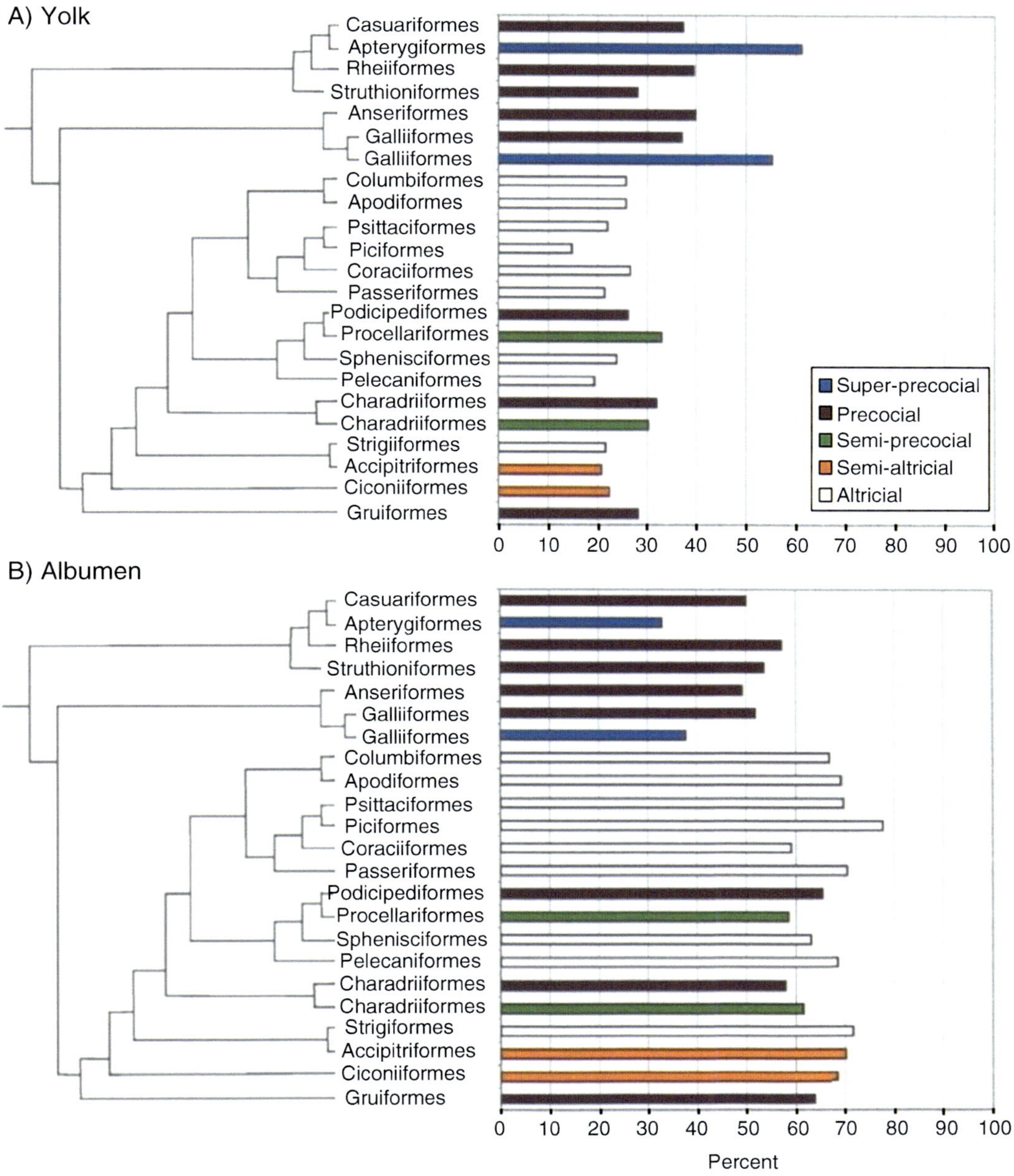


**Fig. 16.77** Percentage of total egg mass consisting of (**a**) yolk and (**b**) albumen in different orders of birds with different modes of chick development. Super-precocial species such as kiwis (Apterygiformes) and megapodes (Galliformes) have the greatest amount of yolk (energy) in their eggs, allowing longer incubation periods, and well-developed young at hatching. In general, the eggs of precocial and semi-precocial species have more yolk and less albumen than the eggs of semi-altricial and altricial species. More yolk allows longer incubation periods and more developed young at hatching. (Figure modified from Deeming 2007; used with permission of C. Deeming)

**Table 16.4** Variation in concentrations of three antimicrobial proteins (ovotransferrin, avidin, and lysozyme) in egg albumen of eight species of birds found in different habitats and locations and with different types of nests (Data from Shawkey et al. 2008; © 2008 Oxford University Press, used with permission)

| | | | | Mean (95% CI) | Mean (95% CI) | Mean (95% CI) |
|---|---|---|---|---|---|---|
| Species | Habitat type | Location | Nest | Ovotransferrin (mg/mL) | Aviden (μg/mL) | Lysozyme (μg/mL) |
| Pearly-eyed Thrasher | Tropical, terrestrial | PR, USA | Cavity | 16.4 (15.2–17.5) | 0.02 (0.01–0.03) | 7.2 (4.5–10.0) |
| Tree Swallow | Temperate, terrestrial | CA, USA | Cavity | 13.8 (12.6–15.0) | 0.33 (0.26–0.40) | 3.2 (2.2–4.2) |
| Violet-green Swallow | Temperate, terrestrial | CA, USA | Cavity | 13.9 (12.4–15.3) | 0.30 (0.24–0.37) | 3.3 (2.4–4.2) |
| Red-winged Blackbird | Temperate, terrestrial | KS, USA | Cup | 13.6 (12.4–14.7) | 0.14 (0.12–0.15) | 32.9 (26.3–39.5) |
| Eastern Bluebird | Subtropical, terrestrial | AL, USA | Cavity | 15.4 (14.0–16.8) | 0.22 (0.08–0.36) | 36.1 (28.3–44.0) |
| Western Bluebird | Temperate, terrestrial | CA, USA | Cavity | 15.5 (14.3–16.7) | 0.19 (0.09–0.29) | 48.5 (32.1–64.9) |
| Green-rumped Parrotlet | Tropical, terrestrial | Venezuela | Cavity | 28.4 (26.3–30.6) | 0.02 (0–0.05) | 5910 (5460–6360) |
| Blue-winged Teal | Temperate, aquatic | ND, USA | Floating | 22.6 (21.1–24.0) | 0.88 (0.67–1.10) | 4115 (3610–4620) |

*PR* Puerto Rico, *CA* California, *KS* Kansas, *AL* Alabama, *ND* North Dakota
Scientific names: Pearly-eyed Thrasher, *Margarops fuscatus*; Tree Swallow, *Tachycineta bicolor*; Violet-green Swallow, *Tachycineta thalassina*; Red-winged Blackbird, *Agelaius phoeniceus*; Eastern Bluebird, *Sialia sialis*; Western Bluebird, *Sialia mexicana*; Green-rumped Parrotlet, *Forpus passerinus*; Blue-winged Teal, *Spatula discors*

The process of biomineralization that forms avian eggshells is extremely fast. For example, during formation of the eggshell of a chicken egg, a 350-μm-thick layer of calcite is created in about 10 h. This is the most rapid biocalcification process known, and is about 100 times faster than the rate at which calcium is deposited in the shells of molluscs and seven times faster than the rate of calcium deposition in the spicules of sea urchins (Beniash et al. 1997; Vidavsky et al. 2014).

The outermost layer of eggshells of many species of birds, the cuticle, is usually relatively smooth (Fig. 16.89). For example, the cuticles of tinamou eggs are composed of calcite, calcium

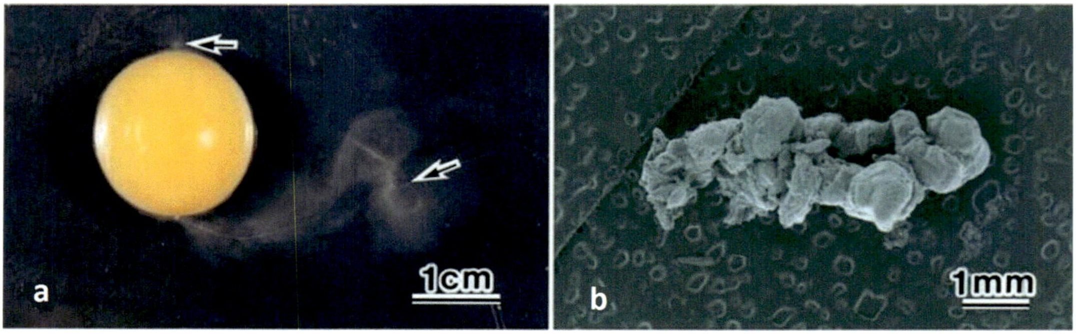


**Fig. 16.78** (**a**) Chalaza cords (arrows). (**b**) Electron micrograph of a chalaza. (Figure modified from Rahman et al. 2007; © 2007 Springer-Verlag, used with permission)

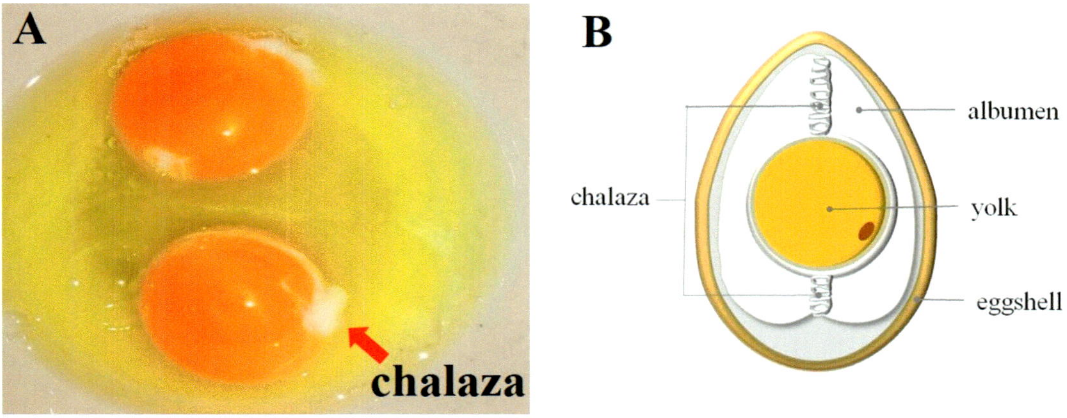


**Fig. 16.79** (**a**) Chalaza in the egg of a Domestic Chicken. (**b**) Schematic drawing showing the chalaza within the albumen that anchors the yolk to the center of the egg. (Figure from Wang et al. 2021, © 2021 by the Authors. Licensee MDPI, Basel, Switzerland, open-access article distributed under the terms and conditions of the Creative Commons Attribution (CC BY) license, http://creativecommons.org/licenses/by/4.0/)

phosphate, and possibly some proteins, lipids, and polysaccharides, and are very smooth, giving them a glossy appearance (Igic et al. 2015a, b; Fig. 16.90). However, some eggshell cuticles consist of nanometer-scale spheres of either vaterite ($CaCO_3$) or hydroxyapatite ($Ca_{10}(PO_4)_6(OH)_2$) (Fig. 16.91). Birds in the orders Pelecaniformes, Suliformes, and Cuculiformes have eggshells with cuticles consisting of vaterite spheres (Portugal et al. 2018), whereas the cuticles of many or all birds in the orders Anseriformes, Galliformes, Tinamiformes, Gruiformes, Sphenisciformes, Podicipediformes, and Phoenicopteriformes consist of hydroxyapatite spheres (D'Alba et al. 2016; Fig. 16.92)

The cuticle of eggshells can serve a number of important functions, including regulating the

**Fig. 16.80** Outer shell membrane after formation of numerous mammillary cores formed from globular proteins. (Figure from © 2015 Elsevier Inc., used with permission)

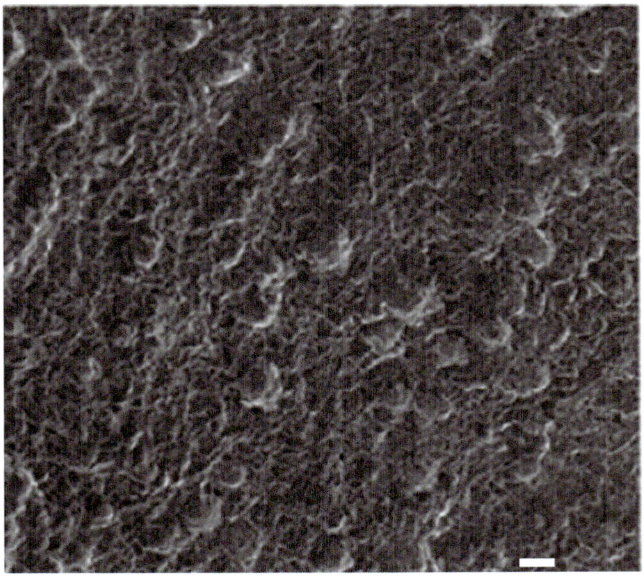

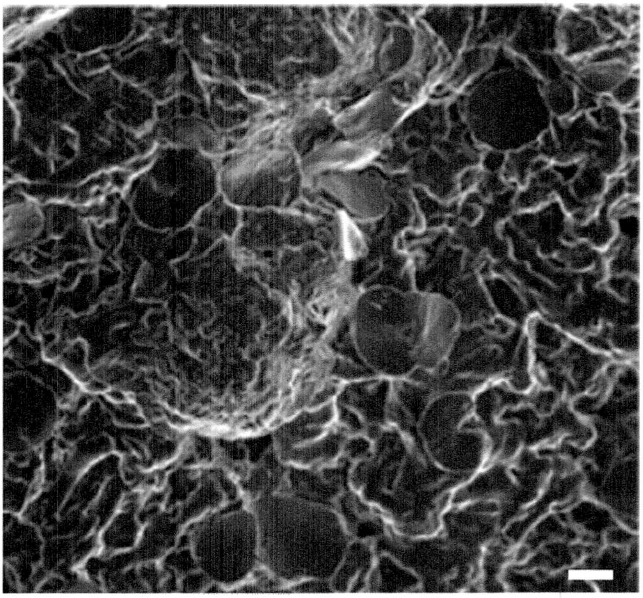

**Fig. 16.81** Particles of calcium carbonate are shown having accumulated over the mammillary cores. Scale bar = 1 μm. (Figure from © 2015 Elsevier Inc., used with permission)

transport of oxygen and carbon dioxide between the embryo and atmosphere (Deeming 1987), acting as a barrier to liquid water and microorganisms (Board and Halls 1973; D'Alba et al. 2017), and limiting the passage of ultraviolet radiation that could harm embryos (Maurer et al. 2015). In addition, cuticles consisting of vaterite nanospheres act as "shock absorbers," protecting eggs from mechanical damage that might be caused by impacts with other eggs and reducing the risk of eggshell fracture during incubation (Portugal et al. 2018). Protection from damage may be particularly important for species of communally breeding cuckoos like Greater Anis (*Crotophaga major*) because breeding groups may consist of up to four females that lay eggs in the same nest. This results in large clutches (up to 15 eggs) and frequent collisions between eggs as they are turned during incubation (Portugal et al. 2018).

Cuticles consisting of nanospheres, either vaterite or hydroxyapatite, also appear to be particularly effective as barriers to liquid water and bacterial colonization. The risks of eggs being exposed to water and to bacterial infection increase for eggs in wet, humid environments (Cook et al. 2004), and birds that typically nest in such environments are also the ones that typically have eggshell cuticles made of nanospheres (e.g., species in the orders Anseriformes, Gruiformes, Sphenisciformes, Podicipediformes, and Phoenicopteriformes; D'Alba et al. 2016). Nanospheres can potentially provide protection from liquid water and bacteria in two ways. First, the irregular surface created by the nanospheres is hydrophobic, trapping water, and bacteria, in microscopic droplets so that most of the cuticle surface remains dry. This limits bacterial mobility, trapping them at discrete points on the cuticle surface, and the likelihood of bacteria entering eggs through an eggshell's pores (D'Alba et al. 2014; Fig. 16.93). In addition, cuticle nanospheres may cover eggshell pores, allowing the diffusion of gases, but preventing liquid water and bacteria from entering eggs (D'Alba et al. 2016).

Gas exchange occurs across eggshells via pores (Fig. 16.94), and is critical for the development of avian embryos (Rahn and Paganelli 1990). Among birds in general, pore size, pore area, and the total number of pores increase with egg mass, whereas pore density decreases (Tullett and Board 1977; Fig. 16.95). Pore morphology and, given differences among species in shell thickness, pore length also vary among species (Rahn et al. 1979; Figs. 16.96 and 16.97)

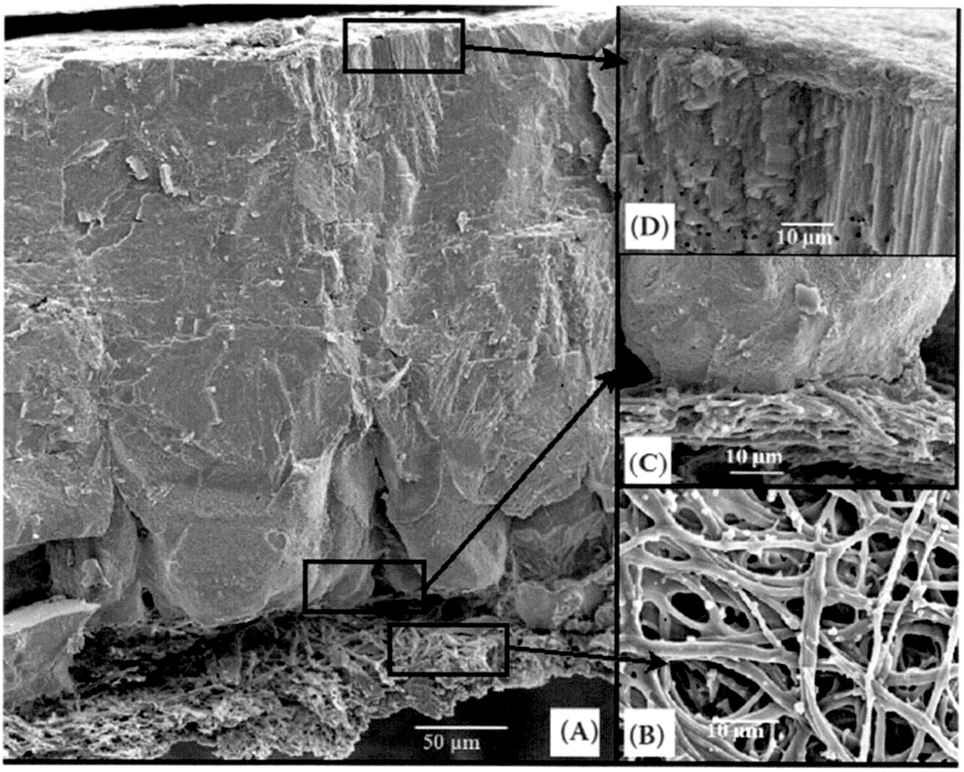


**Fig. 16.82** Scanning electron micrograph of the eggshell of a Domestic Chicken. (**a**) Cross-section shows the eggshell membranes, mammillary and palisade layers, and cuticle, (**b**) Eggshell membranes showing the network of interlacing fibers, (**c**) Mammillary core layer section showing where a core meets the outer shell membrane fibers, and (**d**) Section showing the vertical layer and cuticle surface of the eggshell. (Figure from Ketta and Tůmová 2016, used with permission)

Gases exchanged via eggshell pores include water vapor, oxygen, and carbon dioxide. The movement of gases is almost exclusively driven by diffusion and is best described by the Fick equation (Wangensteen and Rahn 1970; Paganelli 1980) which states that the rate of diffusion is proportional to the difference in partial pressure of gases (i.e., the pressure of a particular gas in a mixture of gases) between the environment and the egg and to the conductance (permeability) of the eggshell (Fig. 16.98). This conductance depends on the number and size of eggshell pores that, in turn, determine the total pore area (Ar and Rahn 1985).

## 16.11 Avian Eggs

Birds' eggs, like birds themselves, vary enormously in size (Fig. 16.99). Among living birds, female Common Ostriches lay the largest eggs and they are more than 2000 times larger than the small eggs produced by some hummingbirds. Ostrich eggs are about 180 mm long and 140 mm wide and weigh 1.2 kg. Typical hummingbird eggs are about 13 mm long and 8 mm wide and weigh only half of a gram. Extinct Elephant Birds from Madagascar produced eggs seven times larger than those of Ostriches (Fig. 16.100)!

More generally, the size or volume of avian eggs is highly correlated with bird size (i.e., body mass), with smaller birds laying smaller eggs and

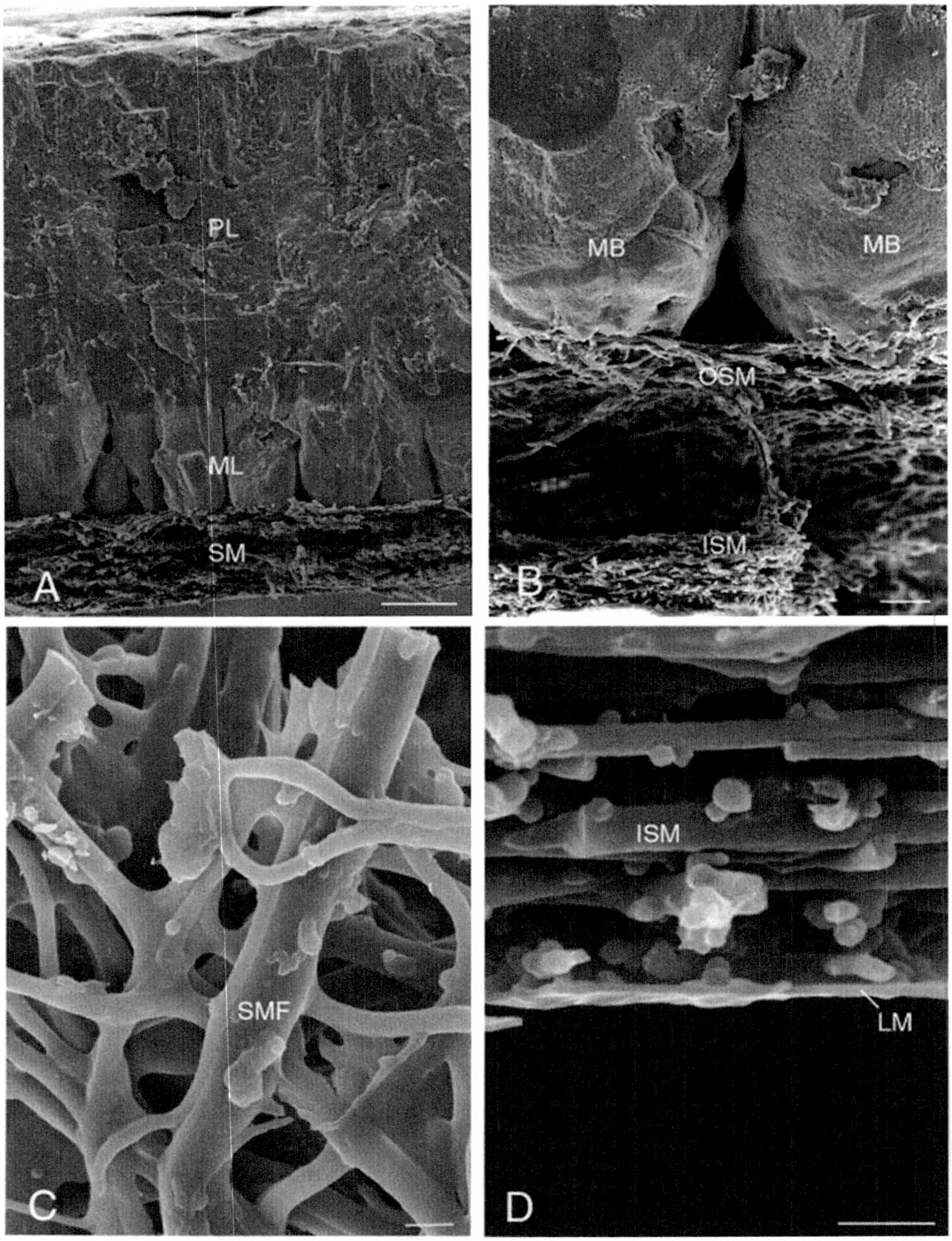


**Fig. 16.83** Scanning electron micrographs of an eggshell of a Domestic Chicken. (**a**) Palisade layer (PL) and mammillary layer (ML) of an eggshell along with the shell membranes (SM). (**b**) Higher magnification of mammillary bodies (MB), outer shell membrane (OSM), and inner shell membrane (ISM). (**c**) Higher magnification of shell membrane fibers (SMF) showing how they are interwoven. (**d**) Inner shell membrane (ISM) where it interfaces with the limiting membrane (LM) that surrounds the egg albumen (not shown). Scale bars: A, 50 μm; B, 20 μm; C and D, 2 μm. (Figure from Hincke et al. 2000; © 2000 Elsevier Science B.V., used with permission)

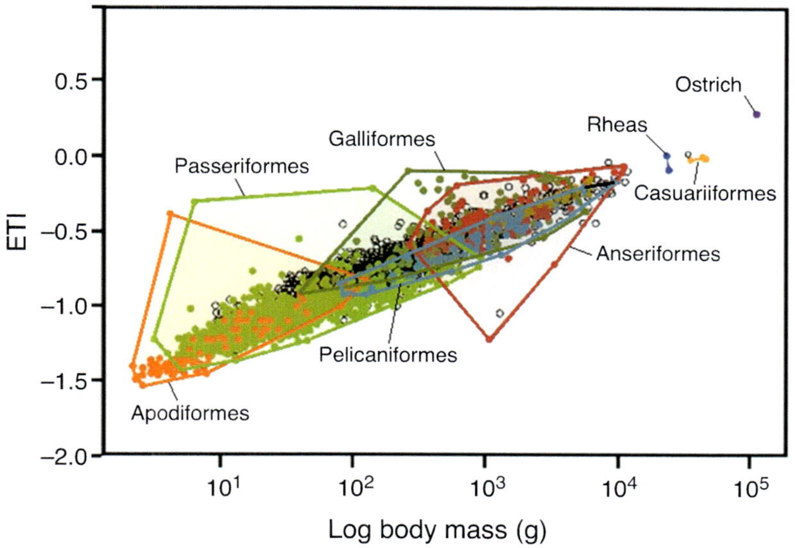


**Fig. 16.84** Relationship between body mass and thickness of eggshells for 4260 species of birds, with larger, heavier birds tending to have thicker eggshells. Variation in the ETI (Eggshell Thickness Index) among several avian orders is indicated by different colors and convex polygons; species in other orders are indicated by the black circles. ETI is an index that approximates eggshell thickness as a function of egg shape and shell density, with higher ETIs indicating thicker eggshells. (Figure modified from Attard and Portugal 2022; )

larger birds laying larger eggs (Fig. 16.101). However, in two taxa of birds, females lay larger eggs than would be expected based on their mass: kiwis and megapodes (Figs. 16.102 and 16.103). For example, a female nonpasserine that weighed the same as a female North Island Brown Kiwi (*Apteryx mantelli*) would have eggs that weighed about 111 grams (or 6% of body mass; Rahn et al. 1985), but the eggs of female North Island Brown Kiwis weigh more than 400 grams (or 17% of body mass (Calder III 1979). The eggs of female Little Spotted Kiwis (*Apteryx owenii*) weight about 300 grams, or 22% of body mass, making them the largest eggs relative to female mass of any species of bird (Dickison 2007). Some investigators have suggested that the large eggs of kiwis represent an evolutionary side effect, with their much larger ancestors (perhaps the size of moas) laying large eggs and, as selection favored a reduction in body size, there was no corresponding reduction in egg size (e.g., Calder III 1979). Dickison (2007), however, noted that evidence for a reduction in size of kiwis relative to their ancestors is lacking, and that the large eggs of kiwis are more likely due to their unique environment. New Zealand has no native terrestrial mammals and few if any birds that would have been potential predators of kiwi eggs, young, or adults. This reduced risk of predation may have favored a K-selected life history, with smaller clutches (1 or 2 eggs), long life spans, and large eggs with lots of yolk that produce extremely precocial young (Dickison 2007).

The eggs of female megapodes (Megapodiidae, order Galliformes) are also larger than expected relative to their body mass. For example, based on the allometric relationship between adult mass and egg mass in the order Galliformes, the eggs of a 1.8-kg female Australian Brushturkey (*Alectura lathami*) should weigh 60 grams (Vleck et al. 1984). However, their eggs are about three times heavier than that.

**Box 16.6 Where Do Females Get Calcium for Eggshells?**

Female birds producing eggs need a variety of macronutrients (carbohydrates, fats, and proteins) and micronutrients such as vitamins, ions, and calcium. About 98% of the dry mass of eggshells is calcite, a crystalline form of calcium carbonate (Romanoff and Romanoff 1949). The two possible sources of the calcium needed to produce eggshells are bone and dietary calcium. One potential source of dietary calcium is the food that is normally in a bird's diet. For example, many female carnivores and molluscivores have diets rich in calcium (e.g., bones and shells). In addition, some types of fruits in a forest in Peru were found to be good sources of calcium for some frugivorous species of birds (Foster 2014). However, many species of birds are granivores and insectivores and most seeds and insects contain insufficient calcium to meet the needs of egg-laying females (Graveland and van Gijzen 1994). For example, Briggs and Mainwaring (2022) found that female Great Tits (*Parus major*), a species that feeds primarily on seeds and insects, provided with an additional source of calcium had larger clutches and higher clutch weights than control females that were not provided with calcium. Some piscivores (e.g., Beintema et al. 1997; Brenninkmeijer et al. 1997; Ramírez et al. 2013) and scavengers (e.g., Houston 1978) have also been found to obtain too little calcium in their diets for eggshell production. Nectar-feeding birds like hummingbirds also have diets with too little calcium for production of eggshells (Adam and des Lauriers 1998). As a result, during the breeding season, females in many species obtain calcium by consuming non-food items rich in calcium such as snail shells, grit, eggshell fragments, wood ash, calcium-rich sand and soil, pieces of crayfish exoskeletons, mollusk shells, and small bones or bone fragments (Reynolds and Perrins 2010).

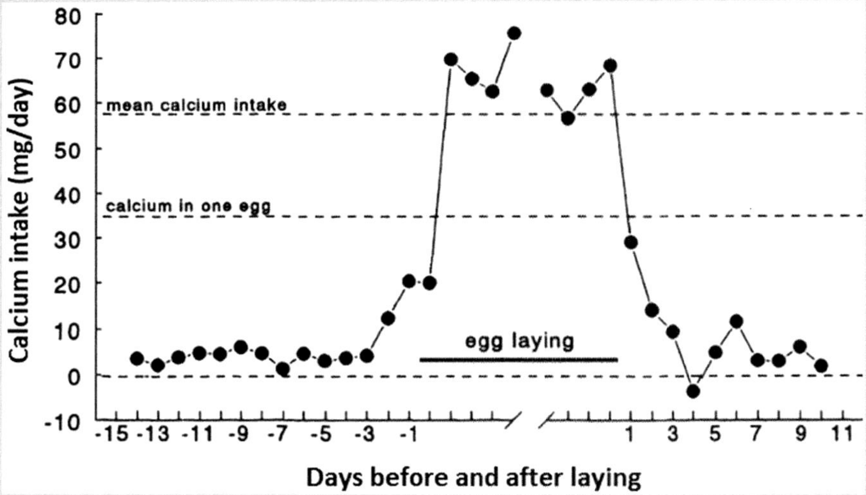


Calcium intake (in the form of snail shells) by captive female Great Tits (*Parus major*) before, during, and after egg laying. Females ingested about 1.7 times the calcium deposited in eggshells. (Figure modified from Graveland and Berends 1997; © 1928 CCC Republication, used with permission)

In addition to dietary calcium, some female birds obtain calcium for eggshells from their medullary bone. Medullary bone, a bone type found only in some birds, crocodilians, and dinosaurs, is "spongy," non-weight supporting bone found in long bones, especially the femur and tibiotarsus but also the humerus, radius, and ulna (Dacke et al. 1993). The presence of

(continued)

**Box 16.6** (continued)

medullary bone has been documented in ratites (orders Struthioniformes and Casuariiformes), waterfowl (Anseriformes), gallinaceous birds (Galliformes), pigeons (Columbiformes), and several species of songbirds (Passeriformes) (Squire et al. 2011). Formation of medullary bone prior to egg laying appears to be triggered by elevated levels of estrogen in the blood and begins from about two weeks to as little as a few days before egg laying (Krementz and Ankney 1995; Squire et al. 2011). With increasing levels of estrogen, cells in the long bones (called osteoblasts) convert lamellar (dense) bone to medullary bone from which calcium can be removed for production of eggshells in the uterus. After eggs are laid, osteoblast activity reverses and medullary bone is converted back to lamellar bone (de Matos 2008). The extent to which calcium from medullary bone contributes to eggshell formation is unclear for most species known to have medullary bone. However, the results of studies of female Domestic Chickens suggest that medullar bone may contribute anywhere from 10% to 45% of the calcium needed for eggshell production (Squire et al. 2011). For smaller birds like most songbirds with much smaller bones, medullary bone likely contributes less calcium for egg laying than would be the case for larger birds like chickens. In addition, the contribution of calcium from medullary bone for eggshell production varies depending on availability of dietary calcium, i.e., greater availability of dietary calcium reduces the need for calcium from medullary bone.

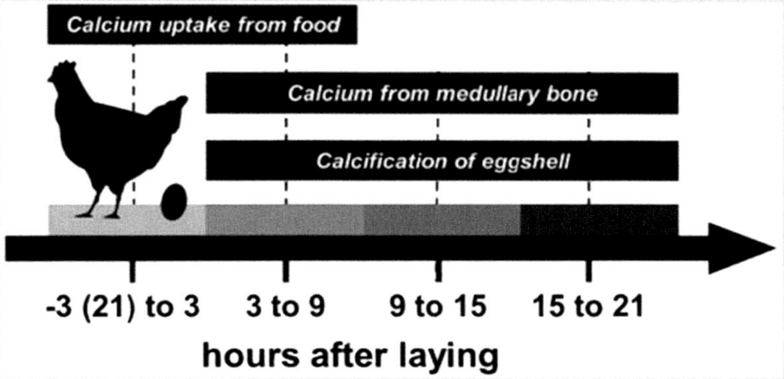


For birds, like Domestic Chickens, that lay an egg daily during production of a clutch, calcium for the next egg is first derived from ingesting calcium from food and other sources. During the later stages of producing an egg, more calcium is coming from medullary bone. (Figure from Kerschnitzki et al. 2014; © 201 Elsevier Inc., used with permission)

(continued)

**Box 16.6** (continued)

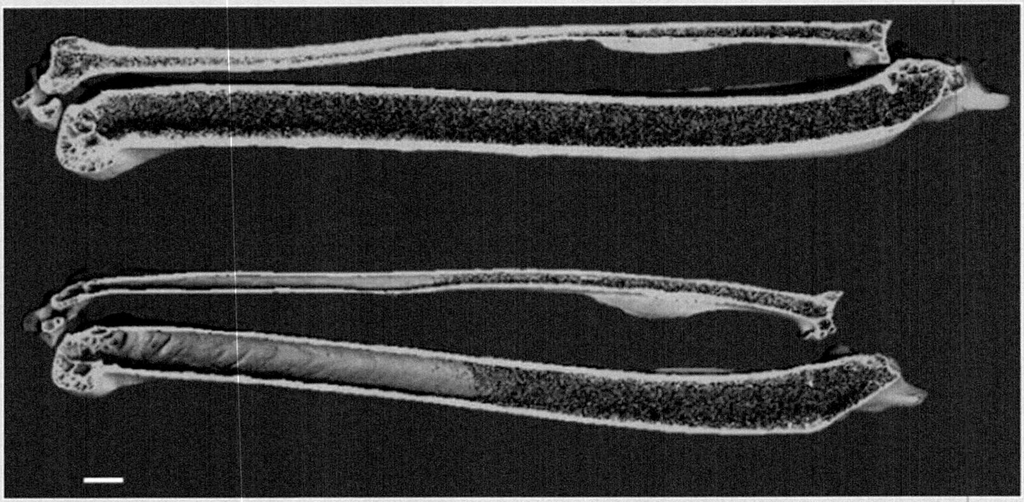

Interior of the radius (thinner bone) and ulna (thicker bone) of a female Wood Thrush (*Hylocichla mustelina*, top) and female Veery (*Catharus fuscescens*, bottom) showing the presence of medullary bone. Medullary bone completely filled the bones of the female Wood Thrush whereas it was limited to the proximal half of the bones of the female Veery. (Figure from Squire et al. 2011; © 2011 Wilson Ornithological Society, used with permission)

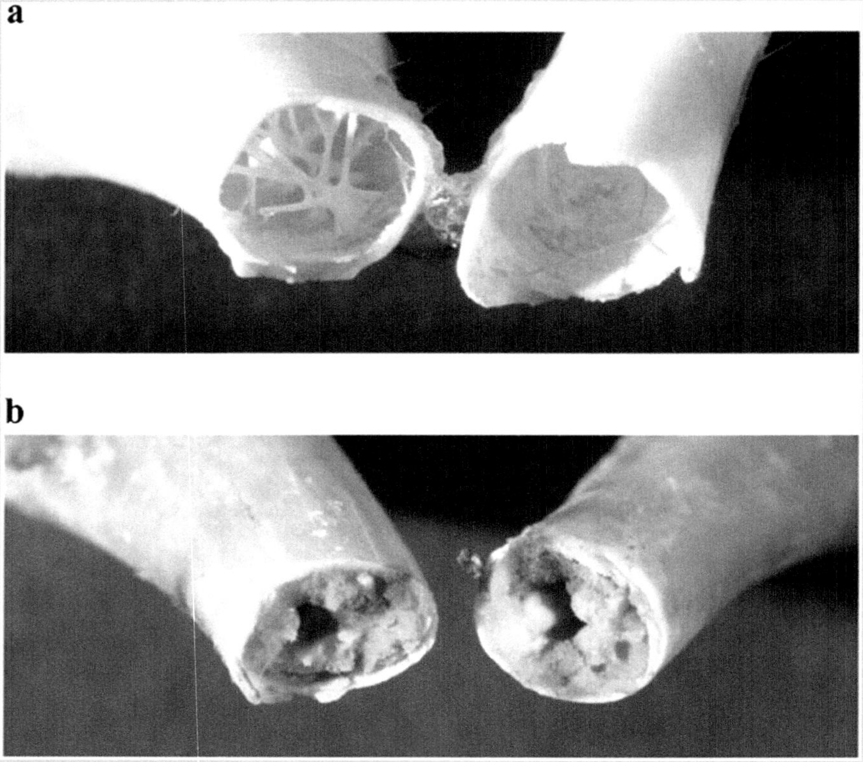


(continued)

**Box 16.6** (continued)

Humerus of a female Domestic Chicken showing (a) normal pneumatic bone with internal struts and (b) cavity of humerus nearly filled with medullary bone. (Figure from Whitehead 2004; © 2004 Oxford University Press, used with permission)

The large eggs of megapodes may be a by-product of their incubation strategy, using environmental heat to warm their eggs rather than their body heat. Some species build mounds of decomposing vegetation, others dig burrows near geothermal vents or sun-warmed sand. With no need to expend energy incubating eggs, female megapodes can direct more energy into producing large eggs with large amounts of yolk that result in well-developed super-precocial young able to care for themselves as soon as they hatch (Watson et al. 2015).

The amount of yolk, and egg size, may also exhibit intraspecific, interbrood, intrabrood, and even geographic variation (Fig. 16.104; Box 16.7 Geographic Variation in Egg Size of New World Flycatchers). This is important because of the positive relationship between egg size and offspring fitness (Mousseau and Fox 1998). Nestlings from larger eggs may be larger, grow faster, and have higher survival rates (Galbraith 1988, Blomqvist et al. 1997; Fig. 16.105). To some degree, egg volume is heritable (Potti 1993), but volume can also be influenced by the conditions experienced by females during egg production (Hargitai et al. 2005; Box 16.8 Decreasing Egg Size with Decreasing Food Availability). Positive correlations between egg size and female condition have been reported for several species, and egg size may also increase with increasing female age (Hõrak et al. 1995). Older females may have larger eggs because they are in better condition (due to better foraging skills or ability of obtain higher-quality territories or nest sites). Alternatively, older females may increase their reproductive effort because their chances of surviving to breed again are declining (terminal investment hypothesis; Pianka and Parker 1975).

The results of some studies have suggested the possibility of egg sexual size dimorphism (SSD), where females in species with sexual size dimorphism produce larger eggs for the larger sex, e.g., House Sparrows (*Passer domesticus*; Cordero et al. 2000) and White-crowned Sparrows (*Zonotrichia leucophrys*; Mead et al. 1987). However, based on a meta-analysis of 63 different studies, Rutkowska et al. (2014) concluded that female birds, as a general rule, do not appear to adjust egg size based on the sex of their offspring. This conclusion is supported by studies of species with sexual size dimorphism, but with no

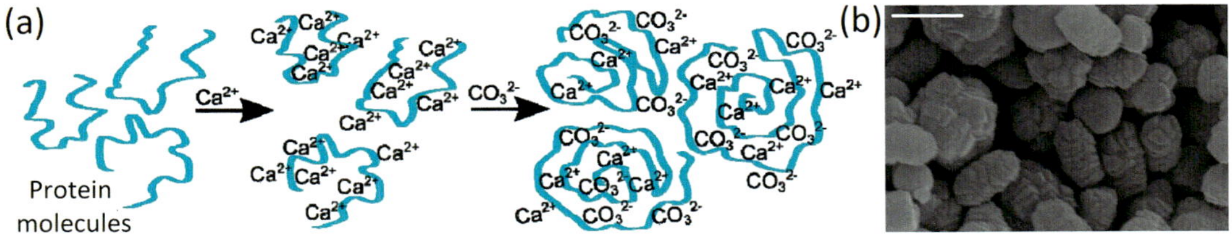


**Fig. 16.85** (**a**) Possible mechanism by which protein molecules interact with calcium and carbonate ions to form amorphous calcium carbonate. (**b**) Spheres of amorphous calcium carbonate. Scale bar = 2 μm. (Figure modified from two different figures from Polowczyk et al. 2016; open-access article distributed under the Creative Commons Attribution License, https://creativecommons.org/licenses/by/4.0/)

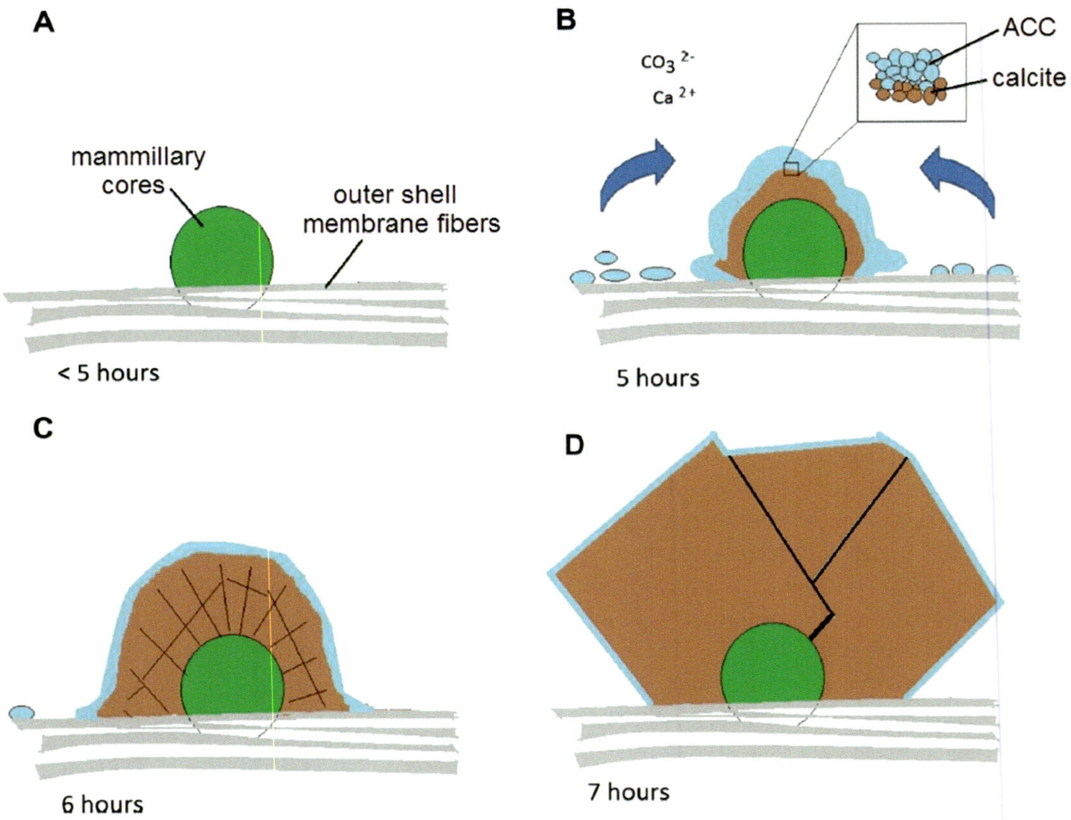


**Fig. 16.86** Drawings showing different stages of eggshell mineralization. (**a**) Less than 5 h after ovulation, mammillary cores develop on the outer shell membrane. (**b**) By 5 h after ovulation, spherical particles of amorphous calcium carbonate (ACC) begin accumulating over the mammillary cores. Gradually, the massive mineral deposits transform directly (by secondary nucleation) into large calcite crystals (which preserve the granular nanostructure) while the disk-shape particles on the membranes and smaller calcite crystals dissolve and supply ions for further crystal growth. (**c**) By 6 h after ovulation, the mineral deposits have transformed into large rounded aggregates of calcite microcrystals, while most of the disk-shaped particles have been dissolved and cleared out from the membrane surface. (**d**) By 7 h after ovulation, only a few large calcite crystals, with distinctive crystal morphologies bounded by flat surfaces, remain. Note that ACC (in light blue) is present at the surface of the mineral deposits at all stages of eggshell formation. (Figure modified from © 2015 Elsevier Inc., used with permission)

evidence of egg SSD (e.g., Lislevand et al. 2005; Isaksson et al. 2010). In addition, more recent work casts some doubt on earlier reports of egg SSD, e.g., in White-crowned Sparrows (Bonior et al. 2007) and House Sparrows (Wetzel et al. 2012). However, although egg size may not differ, females in some species of birds have been found to alter the content of eggs based on offspring sex and size (Rutkowska et al. 2014). For example, immunoglobulins (Martyka et al. 2011), androgens and antioxidants (Pariser et al. 2012), and hormones, carotenoids, and vitamins (Badyaev et al. 2006) have been found to be allocated by females in a sex-specific manner.

In some species of birds, egg mass may also vary with laying sequence in a clutch (Box 16.9 Extreme Intraclutch Egg-size Dimorphism in *Eudyptes* Penguins). For example, laying a relatively small final egg may represent a brood reduction strategy, with more variation in nestling size providing females with a greater opportunity to influence brood size (with smaller young less

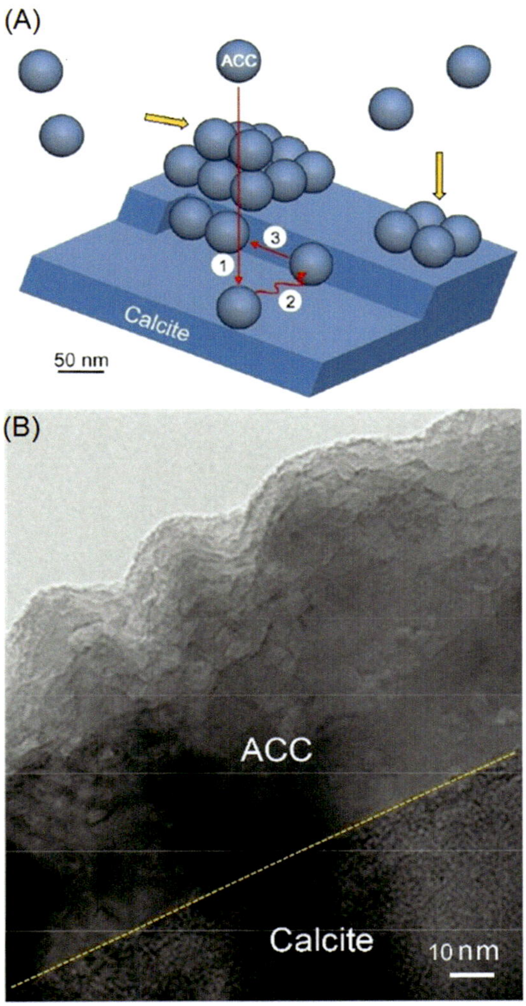


**Fig. 16.87** (**a**) Drawing showing attachment and growth of amorphous calcium carbonate (ACC) on a calcite substrate. Growth involves (1) ACC nanoparticle diffusion, (2) diffusion along the calcite surface, and (3) attachment, after which fusion of the ACC nanoparticles with the calcite substrate occurs. (**b**) ACC nanoparticles accumulating on calcite substrate with the dashed line indicating where the ACC nanoparticles are being fused and restructured into calcite crystals. (Figures from Rodriguez-Navarro et al. 2016; Reprinted with permission from © 2016 American Chemical Society)

able to successfully compete with larger young and, if food supplies are limited, more likely to starve and reduce brood size). Alternatively, a relatively large final egg might help compensate for size differences among siblings resulting from hatching asynchrony (Rosivall et al. 2005), and potentially reducing mortality rates of last-hatched young. However, if hatching is too asynchronous, even a larger egg size may not improve the survival rates of last-hatched young. For example, nestling Common Grackles (*Quiscalus quiscula*) from the last-laid egg were usually the largest at hatching because they came from the largest eggs, but they were also the most susceptible to starvation because they hatched approximately 40 h after their first sibling (Maddox and Weatherhead 2008).

Life-history theory predicts that individuals should increase their reproductive effort when the potential fitness benefits are greater. If true, then females paired with higher quality males might invest more in reproduction than those paired with lower quality males. As predicted by this "differential allocation" hypothesis, female Blue-breasted Quail (*Synoicus chinensis*) paired with higher-quality males with larger badges (a black and white patch on their throat) laid larger (but not more) eggs than females paired with lower-quality males with smaller badges (Uller et al. 2006). Similarly, female Zebra Finches (*Taeniopygia guttata*) paired with males made more "attractive" using colored leg bands (females prefer males with red leg bands) laid larger eggs than those paired with "unattractive" males (with green leg bands; Gilbert et al. 2006). If larger eggs result in larger, better-quality hatchlings and females can influence egg size, then why produce smaller eggs? Producing larger eggs requires more energy and, if used for egg production, then less energy may be available for allocation elsewhere. In other words, adjusting egg size has costs and benefits, and female birds must weigh the costs (more energy) and benefits (larger hatchlings) of producing larger eggs.

### 16.11.1 Egg Coloration

The color of bird eggs and the patterns formed by those colors vary considerably among species. Colors range from white to brown to blue to green and patterns include spots and streaks covering either part or all of the egg. The obvious question, of course, is what selective factors have

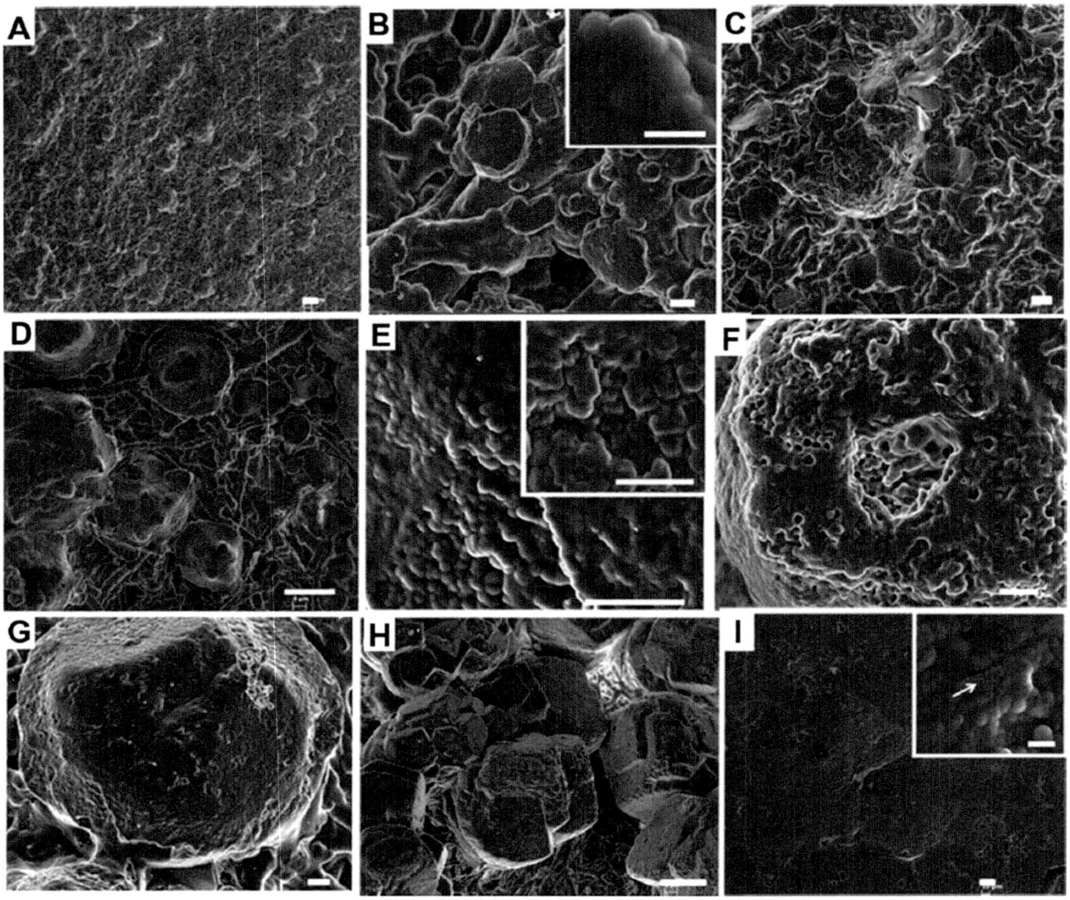


**Fig. 16.88** Scanning electron micrographs illustrating eggshell formation at different stages of mineralization. (**a**) At 5 h after ovulation, non-mineralized eggshell with protein-rich structures (mammillary cores) distributed throughout the outer shell membrane. (**b**) Shortly thereafter, mineralization begins with spherical particles of amorphous calcium carbonate (ACC) clumping on membrane fibers. Inset: detail of a disk-shape particle showing an inner granular nanostructure. (**c**) Disk-shaped particles of ACC cover most of the eggshell surface and accumulate over the mammillary cores. (**d**) At 6 h, massive mineral deposits showing emerging calcite crystals. (**e**) Close-up view of the surface of a massive deposit, showing a granular nanostructure or co-oriented nanosized calcite crystals. (**f**) Close view of a partially dissolving massive mineral deposit. (**g**) After 6 h, in a more developed sample, there are only rounded crystal units with a developing rhombohedral morphology, which is typical of calcite. (**h**) After 7 h, only large calcite crystals bounded by well-defined rhombohedral faces remain. (**i**) After 14 h, a continuous and fully mineralized shell has been formed by coalescing calcite crystal units. Inset: the surface of the crystal unit is formed by the aggregation of aligned nanoparticles. Scale bars for A, D, and I = 10 μm; B, C, E, F, and G = 1 μm; inset E = 1 μm; insets B and I = 300 nm. (Figure from © 2015 Elsevier Inc., used with permission)

contributed to this amazing diversity of colors and patterns. The eggs of ancestral birds were probably all white, like those of present-day reptiles. However, bird nesting habitats likely began to diversify as long as 100 million years ago, and these changes likely contributed to changes in egg morphology (Kilner 2006).

The eggs of many species of birds are still white, with no spots or streaks, and comparative analyses suggest that species with white eggs are

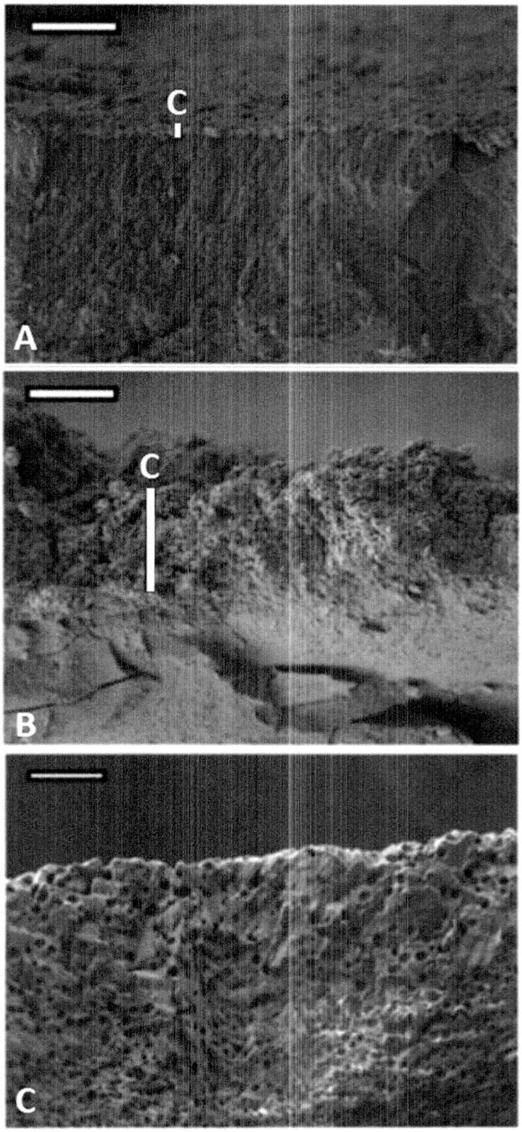


**Fig. 16.89** Outer eggshells of the eggs of (**a**) a Domestic Chicken, (**b**) an Australian Brushturkey (*Alectura lathami*), and (**c**) a Budgerigar (*Melopsittacus undulatus*). Domestic Chicken eggs have a very thin cuticles, Australian Brushturkey eggs have much thicker cuticles, and Budgerigar eggs do not have a cuticle. C = cuticles. Scale bars = 10 μm. (Figures from Fecheyr-Lippens et al. 2015; open-access article distributed under the terms of the Creative Commons Attribution License, https://creativecommons.org/licenses/by/4.0/)

those with nests that are relatively safe from predators. For example, eggs of most cavity-nesting species are white as are those of birds that incubate almost continuously so eggs are rarely exposed (such as pigeons and doves; Kilner 2006). David Lack (1958) suggested that white eggs might be advantageous for birds nesting in cavities and other dimly lit locations because they would be easier to see. However, this hypothesis was based on human vision and many birds, in contrast to humans, can see in the ultraviolet (UV). Regardless of color and appearance to the human eye, the eggs of cavity-nesting birds tend to have eggs with greater ultraviolet reflectance than those of non-cavity-nesting birds (Avilés et al. 2006), and, as a result, are probably more easily detected by adult birds in the low-light environment of a cavity. Reflection in the UV, not white coloration, is what enhances detection. Comparative analyses indicate that the relative safety of nest sites better explains the phylogenetic distribution of white eggs than light levels at nest sites (Kilner 2006).

Various hypotheses have been proposed to explain the many colors and patterns of bird eggs (Fig. 16.106). Alfred Russel Wallace (1889) was the first of many to suggest that egg colors and patterns serve to conceal eggs from predators. Since then, many investigators have attempted to test this hypothesis by comparing predation rates on clutches of eggs that have been experimentally altered to vary in degree of crypticity. Collectively, such experiments suggest that egg coloration can enhance crypsis (Fig. 16.107; Box 16.10 Dirty Eggs = Safer Eggs), but, for species of birds that construct nests, nest crypticity is more important than egg coloration in concealing eggs from potential predators (Underwood and Sealy 2002).

Enhanced crypsis, then, may explain why natural selection initially favored the evolution of eggs with various colors in some ancestral birds. Other selective factors, however, have continued to operate and egg coloration now serves other functions in some species of birds. Many small songbirds have eggs with just a "ring" of small spots around the broad end that does little to make the eggs cryptic (Fig. 16.108). Evidence now suggests that such spots are located where the eggshell is a bit thinner (likely due to a calcium

**Fig. 16.90** (**a**) Eggs of an Elegant Crested-Tinamou (*Eudromia elegans*). (**b**) Photograph of a portion of an egg with reflected light to illustrate the glossy appearance of the eggs (Scale bar = 5 mm). (**c**) Scanning electron micrograph of the cuticle of an egg showing its very smooth surface (scale bar = 10 μm). (Figures from Igic et al. 2015; © 2014 The Authors, used with permission)

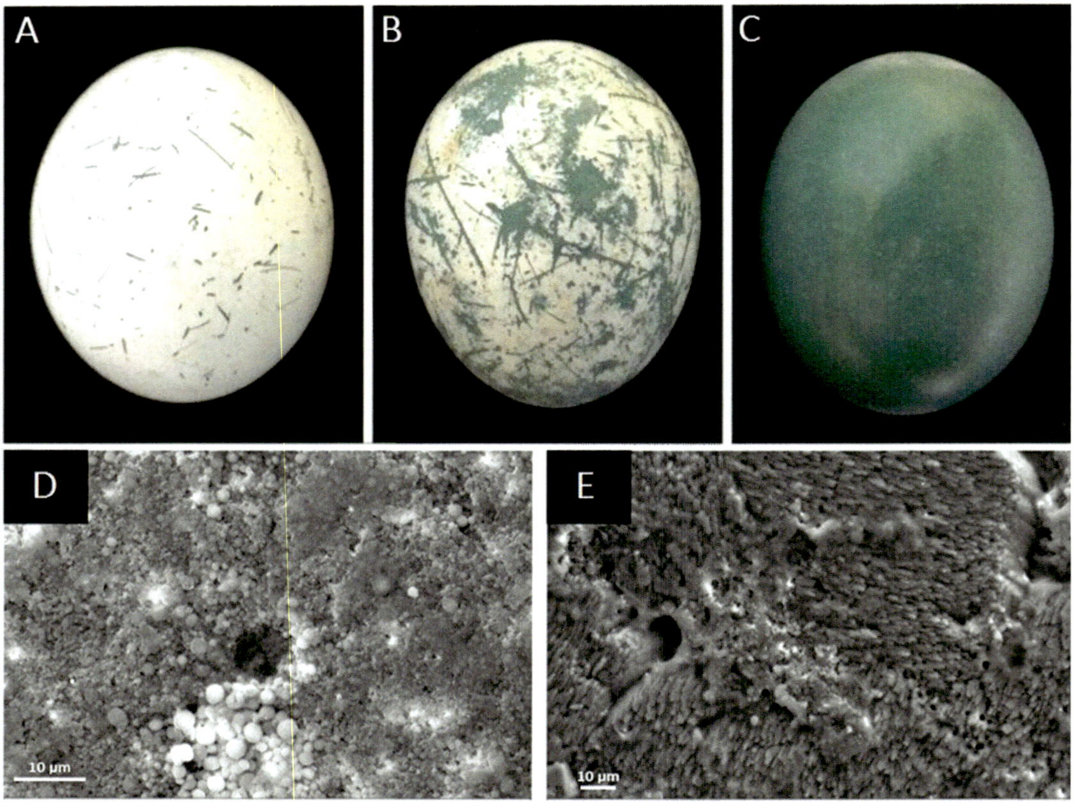


**Fig. 16.91** Eggs of a Greater Ani (*Crotophaga major*) showing (**a**) a recently laid egg with a white vaterite cuticle, (**b**) an egg after several days of incubation with some of the vaterite lost due to abrasion, revealing the blue calcite below, and (**c**) an egg after artificial removal of all the vaterite. (**d**) Scanning electron micrograph of a portion of a Greater Ani egg showing the white, spherical vaterite nanospheres, and (**e**) a calcite shell with the vaterite artificially removed. (Figures from Portugal et al. 2018; © 2017 British Ornithologists' Union, used with permission)

Fig. 16.92 Scanning electron micrographs (SEMs) of the eggshell of a Greater Flamingo (*Phoenicopterus roseus*) showing the cuticular hydroxyapatite nanospheres (inset) covering the palisade (pl) layer and plugging the pores (p) of the shell. cu, cuticle; cl, cone layer; sm, shell membrane. Scale bars = 100 μm in main image and 5 μm in the inset. (Figure from D'Alba et al. 2016; © 2016 The Authors, used with permission)

deficiency in the diet of female birds), with the pigment serving to strengthen the shell (Gosler et al. 2005). The spots consist of protoporphyrin pigment that birds synthesize during production of the heme component of hemoglobin (Burley and Vadhera 1989) and integration of this pigment into the eggshell provides additional strength. When a female bird has insufficient calcium to deposit in a shell, protoporphyrin molecules that have a semi-crystalline structure

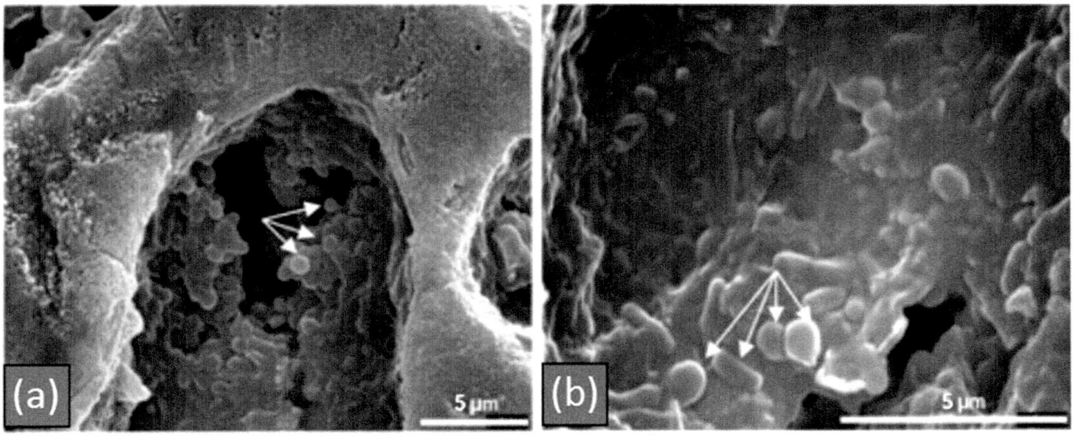


Fig. 16.93 (a, b) Scanning electron micrographs showing bacteria on the eggshells of Eurasian Hoopoes (*Upupa epops*). Scale bars = 5 μm. (Figure from Martin-Vivaldi et al. 2014; © 2014 The Authors. Journal of Animal Ecology © 2014 British Ecological Society, used with permission)

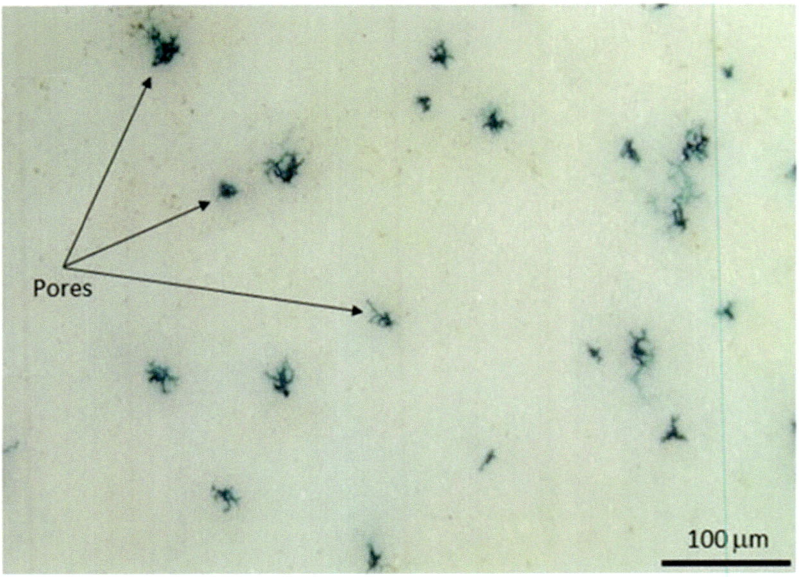


**Fig. 16.94** Pores in the eggshell of a North Island Brown Kiwi (*Apteryx mantelli*). Staining with malachite green penetrated the pores to make them visible. (Figure from Vieco-Galvez et al. 2021; open-access article distributed under the Creative Commons Attribution License, https://creativecommons.org/licenses/by/4.0/)

similar to that of eggshells are apparently deposited instead of calcium. As a result, the spots occur precisely where the shell is a bit thinner.

Several species of birds have blue eggs, and David Lack (1958) suggested that, in habitats where light levels are low, blue eggs might be cryptic. If true, that could help explain the blue eggs of some open-cup nesting birds that occur in forest habitats such as Wood Thrushes (*Hylocichla mustelina*). However, Lack's hypothesis cannot explain why some birds that nest in cavities, like European Starlings (*Sturnus vulgaris*) and Eastern Bluebirds (*Sialia sialis*), also have blue eggs. One hypothesis is that the blue-green color of eggshells represents a signal of female quality to their mates (Moreno and Osorno 2003). The pigment responsible for the blue-green color is biliverdin, a substance produced when the hemoglobin of damaged red blood cells is catabolized and also known to have strong antioxidant properties. Antioxidants are important because they can convert free radicals, molecules that can damage DNA, proteins, and other macromolecules, into less reactive substances. Deposition of this pigment in eggshells by laying females may, therefore, signal their capacity to produce antioxidants and control free radicals. Male birds paired to females of sufficient quality to deposit antioxidants in eggshells rather than retaining them may then expend greater effort in caring for their superior offspring (Kilner 2006). In support of this hypothesis, the provisioning rates of male European Pied Flycatchers (*Ficedula hypoleuca*) and the intensity of the blue coloration of eggs were found to be positively correlated (Moreno et al. 2004). Also, female Eastern Bluebirds in better body condition were found to lay more colorful eggs, supporting the hypothesis that biliverdin pigmentation in eggshells reflects female condition (Siefferman et al. 2006).

In contrast, parental effort by male Spotless Starlings (*Sturnus unicolor*) and male Collared Flycatchers (*Ficedula albicollis*) was not correlated with the amount of biliverdin deposited in eggshells (López-Rull et al. 2007; Krist and Grim 2007). In addition, Hanley et al. (2013) used a large comparative dataset to test the sexual signaling hypothesis and found that blue-green eggshell color was not greater in species where males provide any type of parental care and, further, that relative male provisioning was unrelated to blue-green eggshell chroma. These contrasting results point to the need for additional studies to better clarify the possible relationships

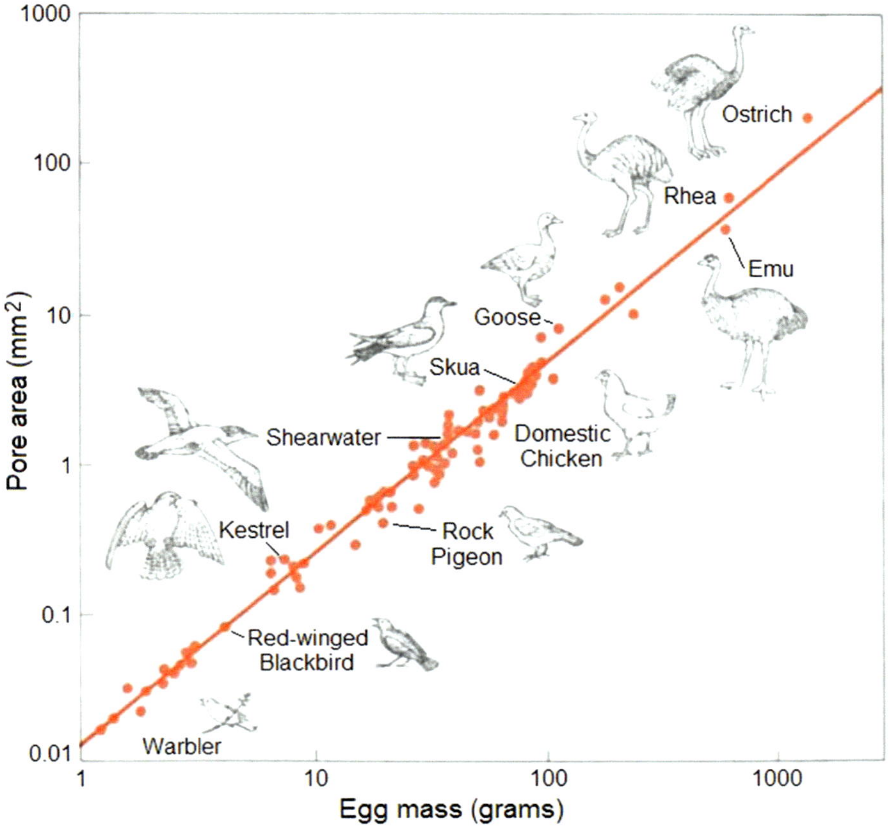


**Fig. 16.95** Relationship between egg mass and the total area of eggshell pores. Common Ostrich, *Struthio camelus*; Rhea, *Rhea* spp.; Emu, *Dromaius novaehollandiae*; Domestic Chicken, *Gallus g. domesticus*; Rock Pigeon, *Columba livia*; Red-winged Blackbird, *Agelaius phoeniceus*. (Figure from Rahn et al. 1979; used with permission of illustrator Patricia Wynne)

between blue-green eggshell color, female quality, and male parental behavior.

Brood parasites can also influence egg color. Host species may counteract the egg mimicry of parasites with increasing interclutch and reduced intraclutch variation in the appearance of eggs. More variation between clutches and less variation within clutches likely make it easier for hosts to recognize a foreign egg, reduce the likelihood of making recognition errors, and reduce the ability of brood parasites to mimic the eggs of a particular host (Stokke et al. 2002). For example, Village Weavers (*Ploceus cucullatus*) are common hosts of Diederik Cuckoos (*Chrysococcyx caprius*). As a defense against brood parasitism, weaverbird eggs exhibit extreme between-female variation, whereas weaverbird eggs within a particular clutch are very similar in appearance (Fig. 16.109). This makes it more difficult for female cuckoos to locate nests with eggs similar enough to their own to avoid rejection by female weaverbirds that have learned the appearance of their own eggs and will remove any egg that does not resemble their own (Lahti 2005).

### 16.11.2 Egg Shape

The shape of bird eggs varies both within and, to a much greater degree, among species. Within

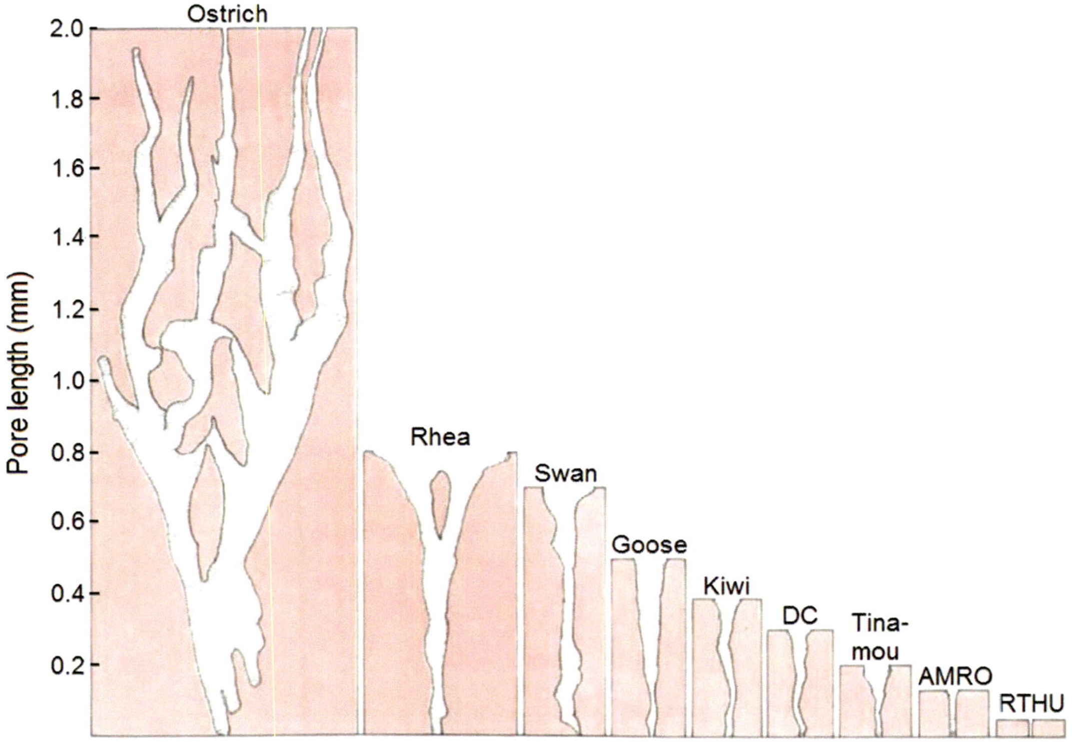


**Fig. 16.96** Variation among species in length and morphology of eggshell pores. Pore length ranges from 2.0 mm for the Common Ostrich (*Struthio camelus*) to 0.04 mm for the Ruby-throated Hummingbird (RTHU, *Archilochus colubris*). Rhea, *Rhea* spp.; DC, Domestic Chicken (*Gallus g. domesticus*); AMRO, American Robin (*Turdus migratorius*). (Figure modified from Rahn et al. 1979; used with permission of illustrator Patricia Wynne)

species, variation in egg shape is more pronounced in some species than others. For example, Heenan (2013) examined variation in egg shape among several species of Australian birds and found much more variation in some species, e.g., Yellow-throated Miners (*Manorina flavigula*), than others, e.g., Red-lored Whistlers (*Pachycephala rufogularis*; Fig. 16.110). Intraspecific variation in egg shape may be due to variation in oviduct morphology and egg size. The maximum distensibility of the uterus sets an upper limit to egg diameter (breadth) so females with slightly smaller uteri and/or females that lay larger eggs (greater volume) may lay eggs that are relatively longer relative to their breadth than females with larger uteri and/or females that lay smaller eggs (van Noordwijk et al. 1981). The availability of sources of calcium can also influence egg shape. Gosler et al. (2005) found a correlation between soil calcium levels and the shape of Great Tit (*Parus major*) eggs, with eggs more spherical in areas where the soils had less calcium. A possible explanation for this is that more spherical eggs have lower surface area to volume ratios and less surface area relative to volume reduces the amount of calcium needed for eggshells.

Another possible explanation for intraspecific variation in egg shape is that egg shape varies with clutch size. If egg shape represents an adaptation to maximize heat transfer to eggs from incubation (or brood) patches, then the optimal egg shape may vary with clutch size. For example, eggs should be more pointed in four-egg

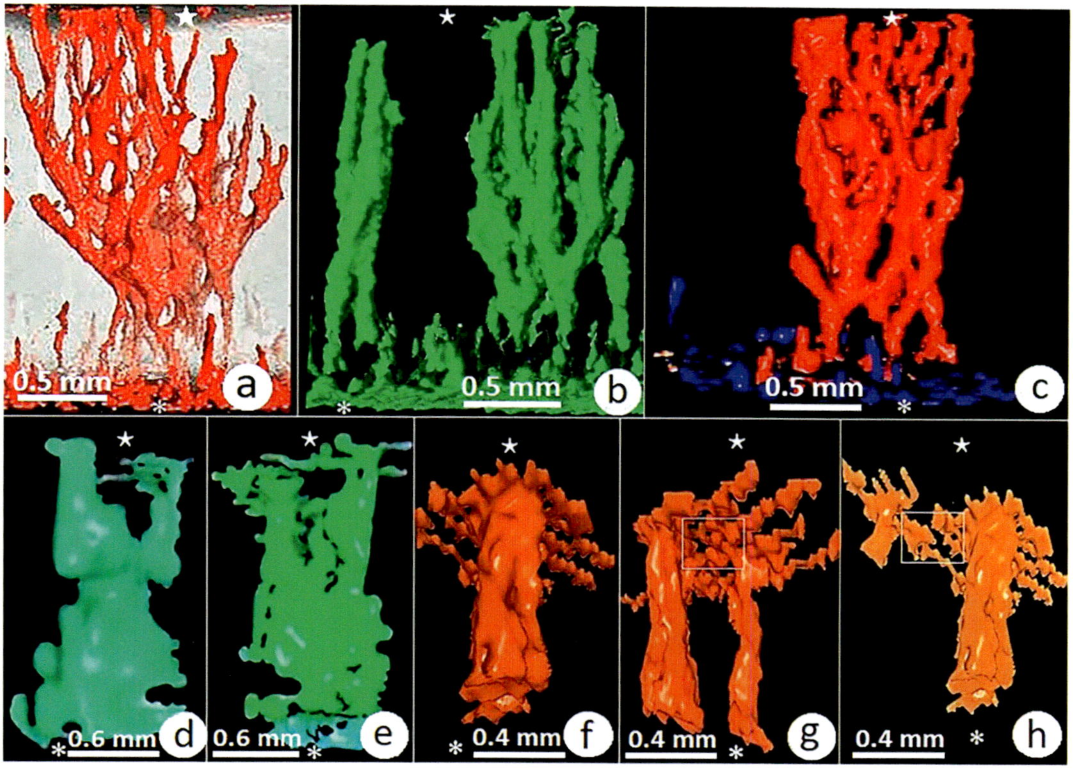


**Fig. 16.97** (**a–h**) Examples of different shapes of pores in the eggshells of Common Ostriches (*Struthio camelus*). Boxed areas in G and H indicate anastomoses between pores. Stars = outer surface of the shell. Asterisks = inner surface of the shell. (Figure modified from Willoughby et al. 2016; © 2016 Wiley Periodicals, Inc., used with permission)

clutches, and eggs should be more spherical in clutches of six or more eggs (Barta and Székely 1997; Fig. 16.111). Studies of species where clutch sizes vary have revealed little evidence in support of this optimal egg shape hypothesis (e.g., Johnson et al. 2001, Encabo et al. 2001, but see Górski et al. 2015). An important reason for this is that the hypothesis is based on the assumption that eggs are arranged on a flat surface and most nests, even simple scrape nests and certainly not cup-shaped or pendulous nests, are flat. Eggs in most nests, therefore, are arranged in three dimensions rather than two so both egg shape and nest shape are important in the transfer of heat from incubating adults to eggs (Hutchinson 2000; Johnson et al. 2001).

Egg shape varies among species (Fig. 16.112), and different investigators have had different, yet similar, ways of categorizing those shapes. Most commonly, egg shapes are categorized as elliptical (with equally rounded ends), pyriform (pear-shaped; larger diameter and rounded at one end and tapering to a much narrower opposite end), oval (rounded and tapering a bit more at one end than the other), or subelliptical (rounded ends, but more tapering more toward the ends and more at one end than the other) (Preston 1953; Baicich and Harrison 1997). Within each of these categories, Henderson (2007) further categorized eggs as short, medium, or long, resulting in 12 egg-shape categories (Fig. 16.113). In reality, of course, egg shape exhibits much variation among species and the eggs of many species of birds cannot conveniently be placed into just one of 12 egg-shape categories (Stoddard et al. 2017).

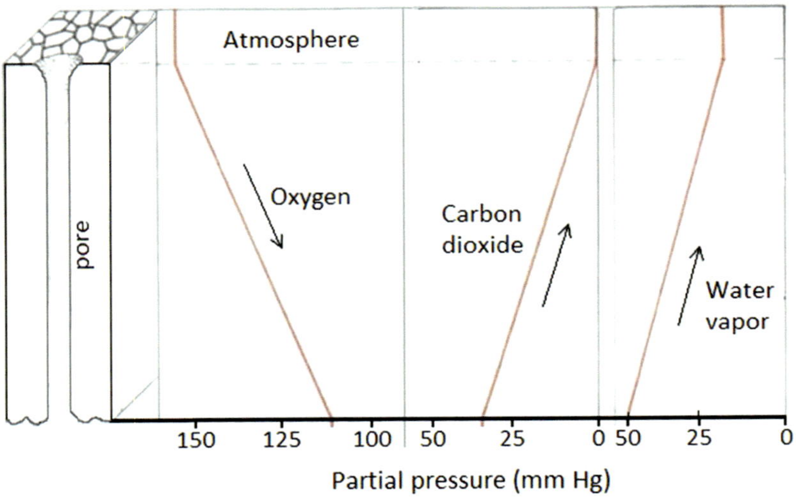


**Fig. 16.98** Oxygen, carbon dioxide, and water vapor diffuse through eggshell pores following their respective partial pressure gradients. The partial pressure of oxygen in the atmosphere is about 160 mm Hg at sea level (but declining with increasing altitude), whereas the partial pressure of carbon dioxide in the atmosphere is less than 1 mm Hg. The partial pressure of water vapor in the air varies with humidity levels and, in this case, is about 15 mm Hg. The partial pressure of oxygen in eggs varies with embryo age, but is always lower than that in the atmosphere so oxygen diffuses through pores into eggs. Partial pressures of carbon dioxide and water vapor are higher in eggs so both diffuse out of eggs and into the atmosphere. (Figure modified from Rahn et al. 1979; used with permission of illustrator Patricia Wynne)

Where in the oviduct is egg shape determined? Perhaps surprisingly, egg shape is determined as the egg passes from the magnum, where albumen is added to the egg, to the isthmus, where shell membranes are added. In the magnum, albumen is added, increasing the diameter of the egg, but

**Fig. 16.99** Variation in size and shape of eggs of 14 representative species of birds, including (left to right) Blue-tailed Bee-eater (*Merops philippinus*), Javan Myna (*Acridotheres javanicus*), Common Quail (*Coturnix coturnix*), Rock Pigeon (*Columba livia*), Golden Pheasant (*Chrysolophus pictus*), Blue-and-yellow Macaw (*Ara ararauna*), Swinhoe's Pheasant (*Lophura swinhoii*), Wild Turkey (*Meleagris gallopavo*), Indian Peafowl (*Pavo cristatus*), African Penguin (*Spheniscus demersus*), Domestic Goose (*Anser* sp.), Red-crowned Crane (*Grus japonensis*), Emu (*Dromaius novaehollandiae*), and Common Ostrich (*Struthio camelus*). Scale bar = 100 mm. (Figure from Juang et al. 2017; open-access article licensed under a Creative Commons Attribution 4.0 International License, https://creativecommons.org/licenses/by/4.0/)

**Fig. 16.100** Egg of an elephant bird (*Aepyornis maximus*) compared to eggs of several other species. Back row (left to right): Common Ostrich (*Struthio camelus*), Emperor Penguin (*Aptenodytes forsteri*), Emu (*Dromaius novaehollandiae*), elephant bird. Front row (left to right): California Scrub-Jay (*Aphelocoma californica*), Ring-necked Dove (*Streptopelia capicola*), Domestic Chicken (*Gallus g. domesticus*), Ruby-throated Hummingbird (*Archilochus colubris*), Northern Cardinal (*Cardinalis cardinalis*), and Oviraptor dinosaur (*Oviraptor* sp.). (Photograph by Laurie Bonneau from Blackburn and Stewart 2021; © 2021 Wiley Periodicals LLC, used with permission)

the leading edge is narrower because the lumen of the next section of the oviduct, the magnum-isthmus junction is narrower than the magnum (Fig. 16.114). This shape is retained as the egg then passes through the magnum-isthmus junction, the isthmus, and to the uterus where the eggshell is added (Mao et al. 2006).

Among the factors that contribute to variation in egg shape, phylogeny appears to be one of the most important. Mytiai et al. (2017) examined the eggs of 557 species of bird representing 28 orders and found that eggs of species in the same order typically have a specific shape; egg size is often more variable because different species may vary in size and larger species typically lay larger eggs (Fig. 16.115). This tendency for the eggs of species in the same order to have similar shapes is likely due to females having a similar morphology, i.e., females of species in the same order have similar body shapes that, in turn, influence egg shape. Eggs must pass through a female's pelvis and the cross-sectional area of the pelvis limits egg diameter, but, within species, egg size (volume) tends to be positively correlated with offspring quality (Krist 2011). Given the limit set by pelvic diameter, selection would likely, in most species, favor elongated eggs because they are smaller in diameter than a round egg with the same volume (Smart 1991). Diving birds represent one example of how pelvic anatomy influences egg shape, i.e., diving birds have elongated bodies with narrow and elongated pelvic girdles and female diving birds typically lay relatively elongated eggs (Mytiai et al. 2017). Similarly, species that are better fliers tend to have more muscular, streamlined bodies and also tend to lay more elliptical and/or asymmetric eggs (Stoddard et al. 2017, Fig. 16.116).

Factors other than female anatomy may also influence egg shape. For example, the four pyriform (pointed) eggs in the clutches of shorebirds may better fit under the brood patch of incubating

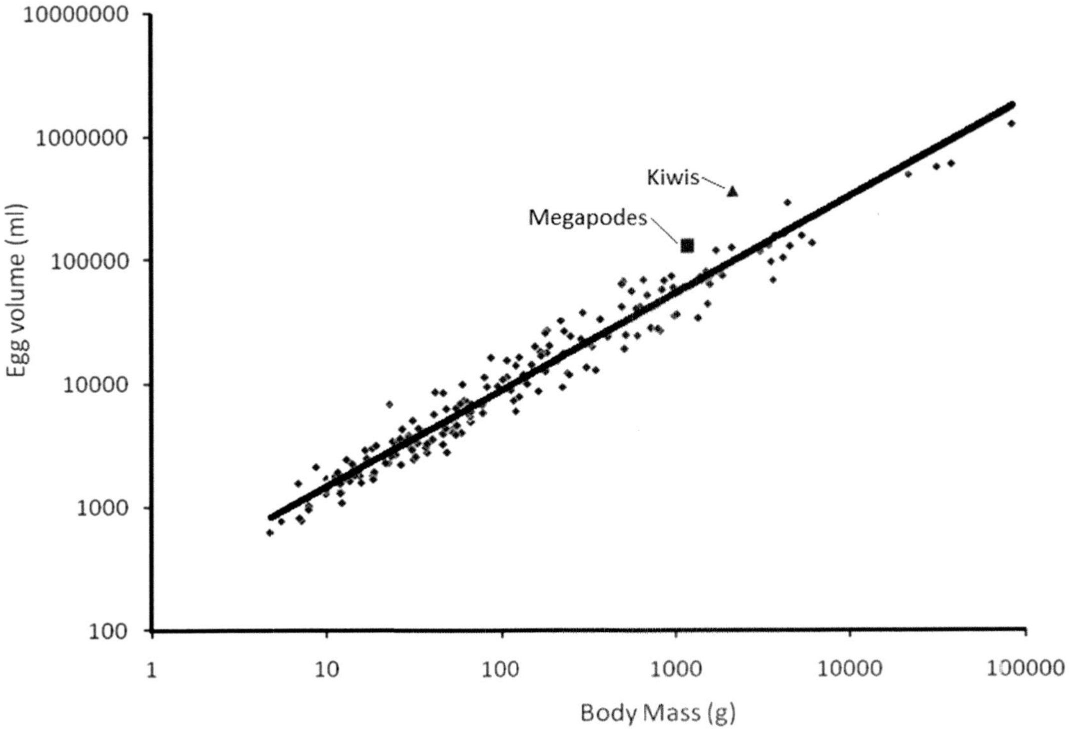


**Fig. 16.101** Relationship between mean egg volume and mean female body mass for 204 families of birds. Note that kiwis (Apterygidae) and megapodes (Megapodiidae) have relatively large eggs for their sizes. (Figure modified from Watson et al. 2015; © 2015 Watson et al., open-access article distributed under the terms of the Creative Commons Attribution License, https://creativecommons.org/licenses/by/4.0/)

adults and allow a more efficient transfer of heat (Andersson 1978; Fig. 16.117). Developmental mode also appears to influence egg shape among species in the order Charadriiformes; species with more precocial young have more asymmetrical eggs, possible because a larger blunt end with numerous pores aids in gas exchange for rapidly developing precocial young (Smart 1991; Stoddard et al. 2017; Fig. 16.118). In addition, as also noted by Smart (1991), the point further from the surface of symmetrical eggs is at the intersection of their axes. A more "pointed" egg moves that point toward the blunt end and places the "thermal center" of the egg closer to the core of a developing embryo. Egg shape could also be influenced by the need to turn eggs during incubation, with particular shapes potentially turned more easily or efficiently depending on nest morphology (Smart 1991).

As far back as 1852 (MacGillivray 1852), the relatively long pyriform eggs of Common (*Uria aalge*) and Thick-billed (*Uria lomvia*) murres have been considered an adaptation to minimize the likelihood of rolling off the narrow cliff ledges where guillemots nest (Fig. 16.119). However, murres often breed on ledges much narrower than the arc followed by their rolling eggs (Harris and Birkhead 1985; Fig. 16.120). In addition, murre eggs exhibit much variation in shape, suggesting there is little stabilizing selection on egg shape (Birkhead et al. 2017). Alternatively, Birkhead et al. (2017) suggested that the long pyriform shape of murre eggs means that a greater proportion of the eggshell is in contact

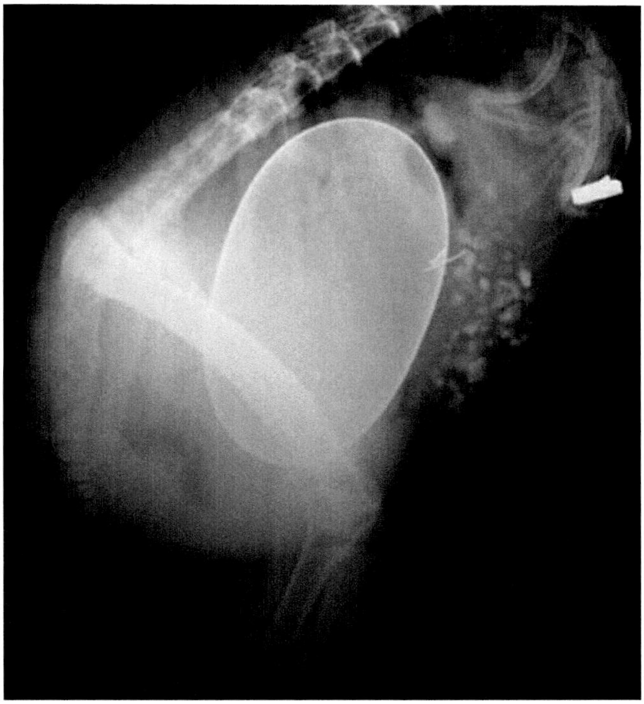

**Fig. 16.102** X-ray of a female kiwi with an egg in the oviduct just prior to laying. (Figure from Smith 2018; © 2018 Elsevier Inc., used with permission)

with the substrate and this, in combination with having relatively thick eggshells, minimizes the possibility of eggs being damaged if stepped on or hit by conspecifics in the crowded breeding colonies of murres. A second possible advantage of a long pyriform egg is that the eggs of Common Murres are often contaminated with fecal and other material and the pyriform shape keeps the blunt end of the egg (where most pores are located) relatively free of contamination, potentially facilitating gas exchange (Birkhead et al. 2017). Finally, and perhaps most importantly, Birkhead et al. (2018) noted that many murres breed on sloping ledges (Fig. 16.121) and

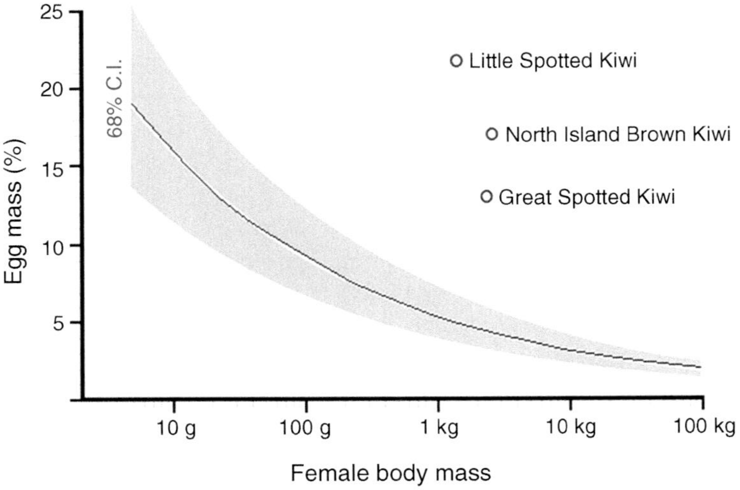


**Fig. 16.103** Relationship between avian egg mass and female body mass. Kiwis have much larger eggs relative to their body mass than other species of birds. Little Spotted Kiwi, *Apteryx owenii*; North Island Brown Kiwi, *A. mantelli*; Great Spotted Kiwi, *A. haastii*. (Figure modified from Dickison 2007 who modified it from Rahn et al. 1975; © 1975 Oxford University Press, used with permission)

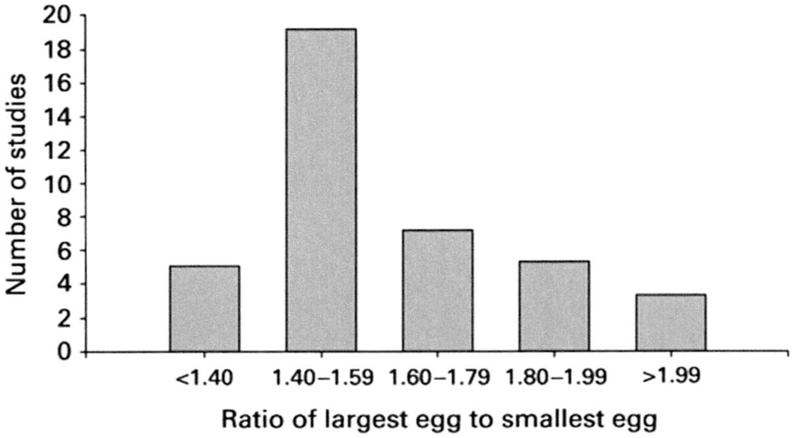


**Fig. 16.104** Intraspecific variation in egg size as indicated by the ratio of the size of the largest egg in a population to the smallest based on 39 studies. (Figure from Christians 2002; © Cambridge Philosophical Society, used with permission)

suggested that the pyriform shape of murre eggs makes them more stable on such surfaces. More pointed, elongated eggs have a greater proportion of egg surface area in contact with the substrate and this increased contact area makes it less likely that the eggs will roll.

## 16.12 Egg-Laying

The process of laying an egg involves contraction of the uterus of the oviduct, sending the egg to the vagina, and successive peristaltic contractions of the vagina that expel the egg out through the cloaca. This movement of an egg through the oviduct and its expulsion require well-developed muscles. The vagina has a well-developed muscular layer, especially circular muscle (Fig. 16.122). In Domestic Chickens, the vaginal sphincter muscle located in the first portion of the vagina is responsible for movement of eggs into and out of the cloaca during the process of egg laying (Parizzi et al. 2008).

Here is a description of the process of laying an egg by a female Coal Tit (*Periparus ater*): "Initially the female stood motionless in the nest cup. The first sign of approaching egg- laying was usually intensified breathing, occasionally with rhythmic opening and closing of the bill that pointed either horizontally forwards or more or less upwards. The head was drawn in and the body feathers were somewhat fluffed out; the Coal Tit in addition raised its crown feathers. The tail was kept horizontal or elevated up to about 45 degrees". Then the tip of the tail started nodding movements synchronously with rhythmic depressions of the rump. These movements which apparently were caused by throes of parturition when the egg traveled down the oviduct, were almost invisible to begin with but gathered in strength and ended with a sudden elevation of the rump that marked the moment of egg-laying. Then the bird "froze" in a motionless posture, termed "recovery phase." This last rise of the rump clearly indicated that the egg had just been laid.

The duration of egg-laying varies a great deal even within species. The opening and closing of the bill and rhythmic movements of the back and tip of the tail occurs repeatedly for up to 4 minutes in the Prairie Warbler (*Setophaga discolor*), presumably corresponding to the duration of egg-laying. For three eggs of the Goldcrest (*Regulus regulus*), only 8–9 s elapsed between the first visible sign of pressure and the moment of egg-laying. In tits, this period varied from about 10 to 77 s, most often 20–30 s. Common Cuckoos (*Cuculus canorus*) are brood parasite and are known to usually lay eggs within 10 s, with a lower limit of only 3–4 s. Presumably this short duration is an adaptation to its parasitic behavior (Haftorn 1996).

Female birds turn part of the cloaca and the last segment of the oviduct inside out ("like a glove").

**Box 16.7 Geographic Variation in Egg Size of New World Flycatchers**

Clutch size is known to vary with latitude, with birds in tropical regions generally having smaller clutches than those at higher latitudes (Jetz et al. 2008). In the more seasonal environments at higher latitudes, harsh conditions and lack of food during the winter increases mortality rates and reduces population densities. This, in combination with the seasonal increase in food availability that occurs at higher latitudes, selects for larger clutches (Ricklets 1980). Egg size (or mass) has also been found to vary with latitude, but few investigators have examined the factors that might explain such variation.

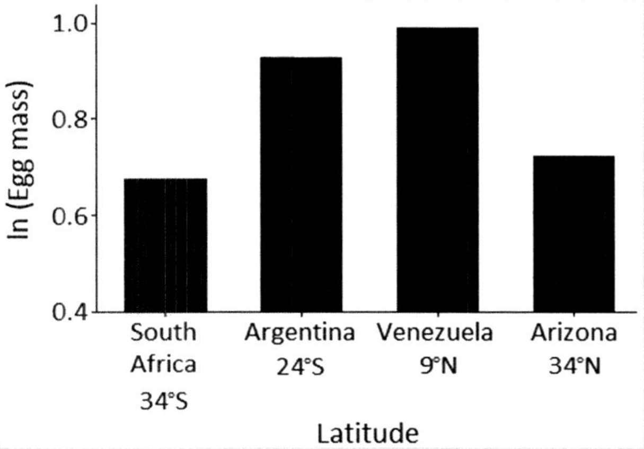


Egg mass tends to be greater at lower latitudes than at higher latitudes. (Modified from Martin et al. 2006; © Society for the Study of Evolution, used with permission)

One of the few exceptions was a study of egg size variation in New World flycatchers. Based on data collected in a literature review, Heming and Marini (2015) found that the eggs of flycatchers were larger in areas where average egg temperature are lower during incubation, and that medium- to long-distance migrants had smaller eggs than resident species. The larger eggs of birds with lower egg temperatures during incubation provide support for the embryonic temperature hypothesis (Martin 2008). This hypothesis posits that eggs would be expected to be larger when egg temperatures are lower during incubation. Larger eggs would be favored by selection because they have relatively less surface area relative to their volume than smaller eggs so take longer to cool when parents leave nests. The egg temperatures of birds in the Neotropics are generally lower than those of birds in North American and, as predicted by the embryonic temperature hypothesis, the eggs of birds in the Neotropics, including New World flycatchers, are larger relative to female body mass than those of birds in North America.

(continued)

**Box 16.7** (continued)

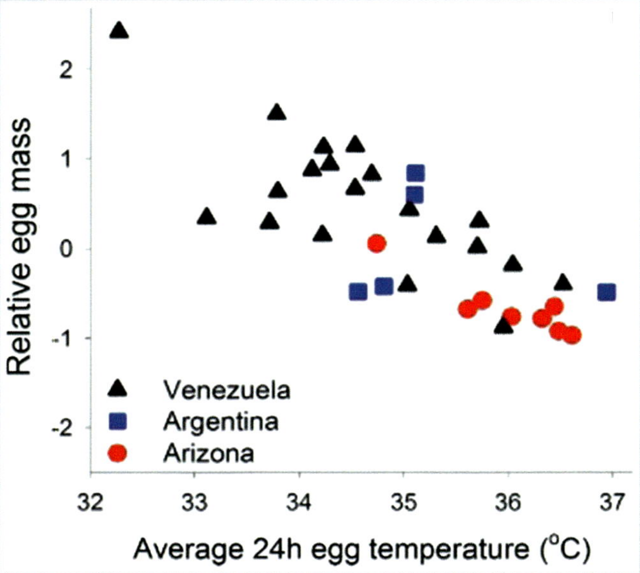


Relative egg mass (corrected for adult mass) and average egg temperature over 24-h periods. Relative egg mass is greater in species where parental inattentiveness causes lower egg temperatures based on study of 37 species of songbirds in tropical Venezuela, subtropical Argentina, and north temperate Arizona. (Figure from Martin 2008; Copyright (2008) National Academy of Sciences, U.S.A, used with permission)

(continued)

**Box 16.7** (continued)

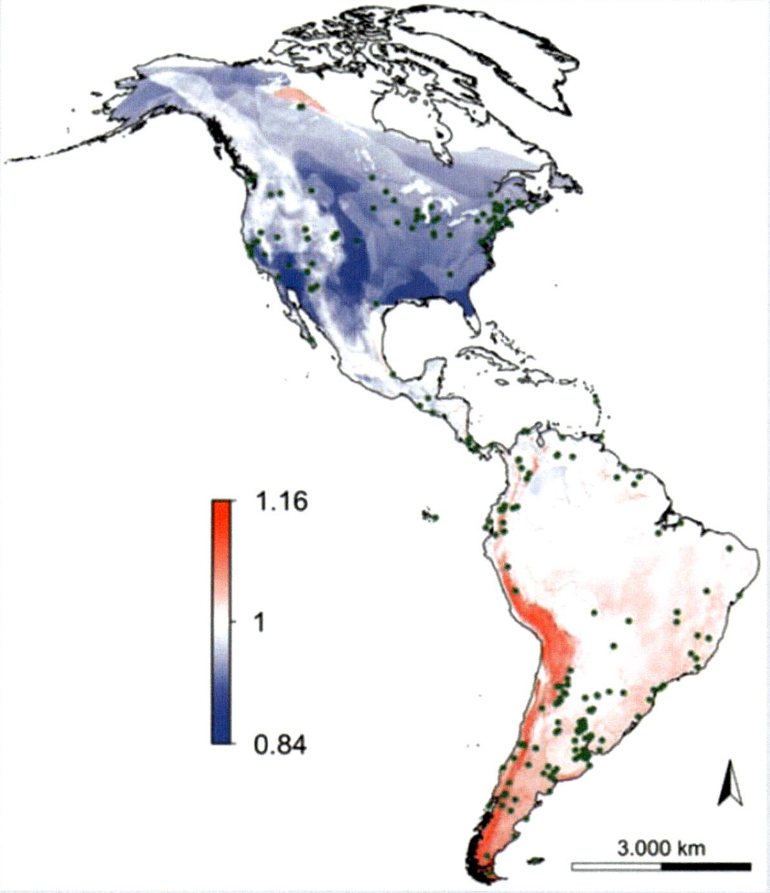


Geographical variation in relative egg size of flycatchers (Tyrannidae, subfamily Fluvicolinae). Colors indicate percent deviation from expected egg size based on female body mass, with blue indicating eggs smaller than expected and red indicating eggs larger than expected. (Figure from Heming and Marini 2015; © 2015 The Authors, used with permission)

Heming and Marini (2015) also found that, on average, the eggs of medium- to long-distance migrant New World flycatchers were smaller than those of short-distance migrants and residents. Factors that may explain this difference in egg size include differences in adult mortality risk and the time and energy available for breeding. Because of the risks associated with migration, medium- to long-distance migrants have high mortality rates, potentially selecting for larger clutches and smaller eggs. In addition, medium- to long-distance migrants have less time and, potentially, less energy (due to the costs of migration) to devote to reproduction. These factors may favor selection for smaller eggs that require less energy to produce and shorter incubation periods. Smaller eggs tend to produce smaller nestlings, but medium- to long-distance migrants also tend to breed at higher latitudes with longer periods of daylight and more time to devote to

(continued)

**Box 16.7** (continued)

provisioning young. Additional time for provisioning translates into faster growth rates, potentially helping to compensate for the smaller size of hatchlings (Heming and Marini 2015).

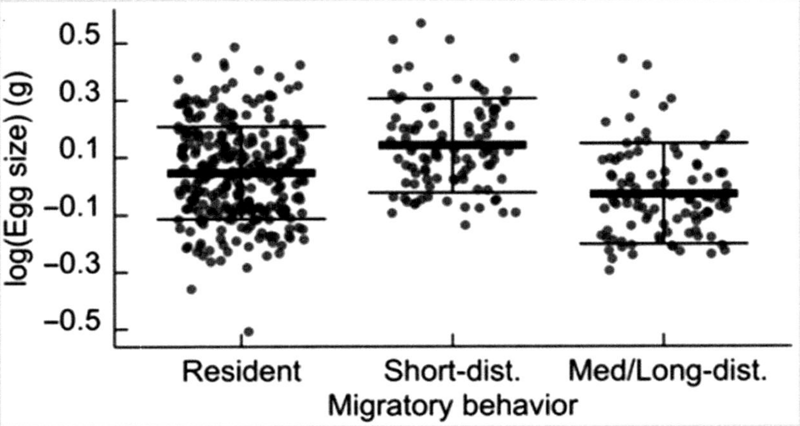


Egg size (log scale) of resident and short- and medium to long-distance migrant species of flycatchers (Tyrannidae, subfamily Fluvicolinae). Medium- to long-distance migrants tend to have smaller eggs than short-distance migrants and residents. (Figure from Heming and Marini 2015; © 2015 The Authors, used with permission)

The vent is then everted and the egg emerges far outside at the end of the bulge (Fig. 16.123). As a result, the egg does not contact the walls of the cloaca and get contaminated by feces. In addition, the intestine and inner part of the cloaca are kept shut by the emerging egg, and their contents cannot leave when the hen strains to deliver the egg. Therefore, eggs are always clean when laid (van der Molen 2002).

## 16.13 Costs of Egg Production

A number of investigators have attempted to estimate the metabolic cost of egg production, and those estimates exhibit considerable variation (Table 16.5). However, most estimates fall within a range from about 20 to 75% of resting or field metabolic rate (Table 16.5). Variation in the estimated cost of egg formation may be due to differences in methods used to estimate those costs and differences among species in egg size (Nager 2006). Some of the increase and, in some species, maybe most of the increase in metabolic rate associated with egg production is due to the growth and maintenance of much larger ovaries and oviducts (Fig. 16.124). For example, variation in the mass of the oviduct, the organ responsible for egg albumen and shell formation, has been found to be positively correlated with resting metabolic rate in egg laying House Sparrows (*Passer domesticus*; Chappell et al. 1999), European Starlings (*Sturnus vulgaris*; Vézina and Williams 2003) and Zebra Finches (*Taeniopygia guttata*; Vézina and Williams 2005). Other processes that might increase metabolic rate include the growth of follicles and synthesis of albumen (Vézina and Williams 2002; Fig. 16.125), biosynthesis of yolk (Vézina and Salvante 2010), the cost of transporting egg contents into eggs, the cost of supporting extra mass, and the cost of shell formation (Carey 1996).

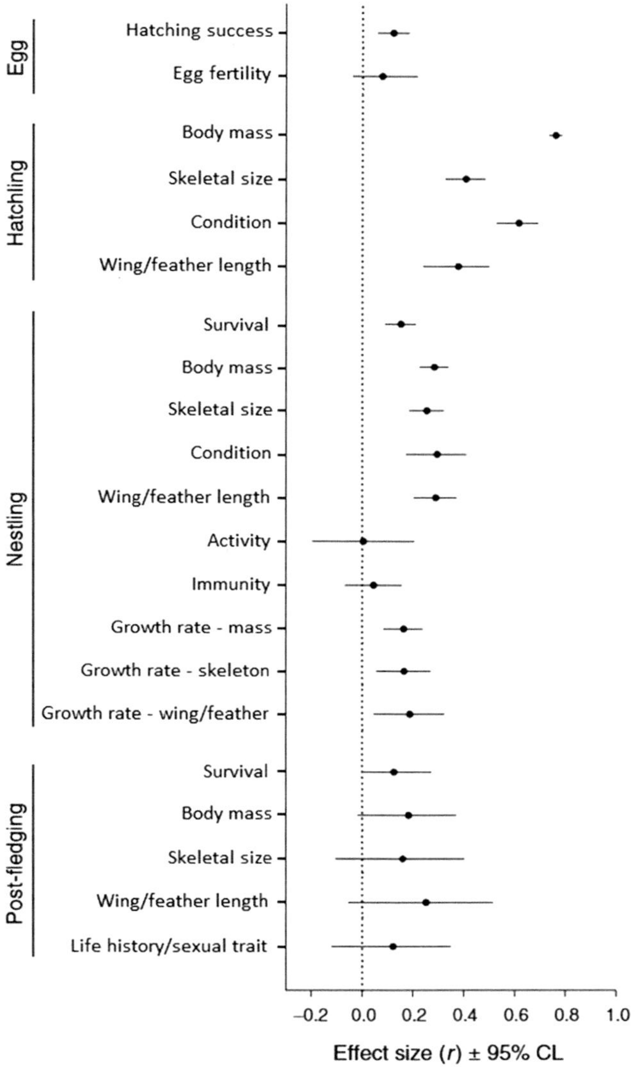


**Fig. 16.105** Effect of egg size on the morphology and other characteristics of embryonic, hatchling, nestling, and fledgling birds. Effect size is a quantitative measure of the strength or magnitude of, in this case, egg size. Effect size is considered large if $r = 0.5$, medium if $r = 0.3$, and small if $r = 0.1$. For egg size, effect sizes are greater than 0.5 for hatchling body mass and condition, and equal to or greater than 0.3 for hatchling wing/feather length and nestling body mass, skeletal size, condition, and wing/feather length. (Figure modified from Krist 2011; © 2010 The Author. Biological Reviews © 2010 Cambridge Philosophical Society, used with permission)

Egg production requires nutrients and those nutrients can come from stored reserves (endogenous nutrients) or from a female's diet during the egg-laying period (exogenous nutrients) (Box 16.11 Stable Isotopes and Egg Nutrients). Based on the source of nutrients for egg production, Drent and Daan (1980) categorized bird species as either capital breeders (using endogenous nutrients) or income breeders (using exogenous nutrients). Although females in some species are capital breeders, e.g., Adelie Penguins (*Pygoscelis adeliae*; Table 16.6), females in most species either use a combination of endogenous and exogenous nutrients to produce eggs (e.g., many waterfowl; Table 16.6, Fig. 16.126) or use only exogenous nutrients to produce eggs (Table 16.6; e.g., Arctic Tern, *Sterna paradisaea*; Fig. 16.126). In general, larger species of birds are more likely to use, to varying degrees, endogenous nutrients, because they need more nutrients (lipids and proteins) to produce eggs than smaller species (Bond and Diamond 2010). Some species that migrate to breed at higher latitudes may also require at least some endogenous nutrients to

**Box 16.8 Decreasing Egg Size with Decreasing Food Availability**

Egg size can be influenced by several factors, including female age, condition, and body size and when eggs are laid (eggs laid later in a breeding season tend to be smaller). Environmental factors such as ambient temperature (egg size tends to decrease with increasing ambient temperatures) and food availability can also influence egg size (Christians 2002).

During annual inspections of the nest burrows of Atlantic Puffins (*Fratercula arctica*) in two breeding colonies in Norway from 1980 to 2011, investigators measured unattended eggs. Barrett et al. (2012) analyzed these egg-size data and found that the size of Atlantic Puffin eggs has exhibited a downward trend and, further, that mean egg size at the two breeding colonies was significantly different. Further analysis revealed that this reduction in egg volume was not due to changes in the size of female Atlantic Puffins or to a change in breeding phenology. Rather, analysis suggests that declines in egg size resulted from changes in the abundance of several species of fish, including capelin (*Mallotus villosus*), herring (*Clupea harengus*), and haddock (*Melanogrammus aeglefinus*) that are important prey for Atlantic Puffins in Norway. Increasing sea temperatures may also be influencing the availability of other prey for puffins. Studies of other seabirds, including Black-legged Kittiwakes (*Rissa tridactyla*) and Common Guillemots (*Uria aalge*) have also revealed an effect of food limitation on egg size (Bolton et al. 1993; Gill and Hatch 2002). Given the known positive relationship between egg size and offspring quality, the long-term decrease in the size of Atlantic Puffin eggs may reduce the quality of young and have a negative effect on puffin populations (Barrett et al. 2012).

(continued)

**Box 16.8** (continued)

Atlantic Puffin. (Photo by Mark Medcalf, Wikipedia, CC BY 2.0, https://creativecommons.org/licenses/by/2.0/)

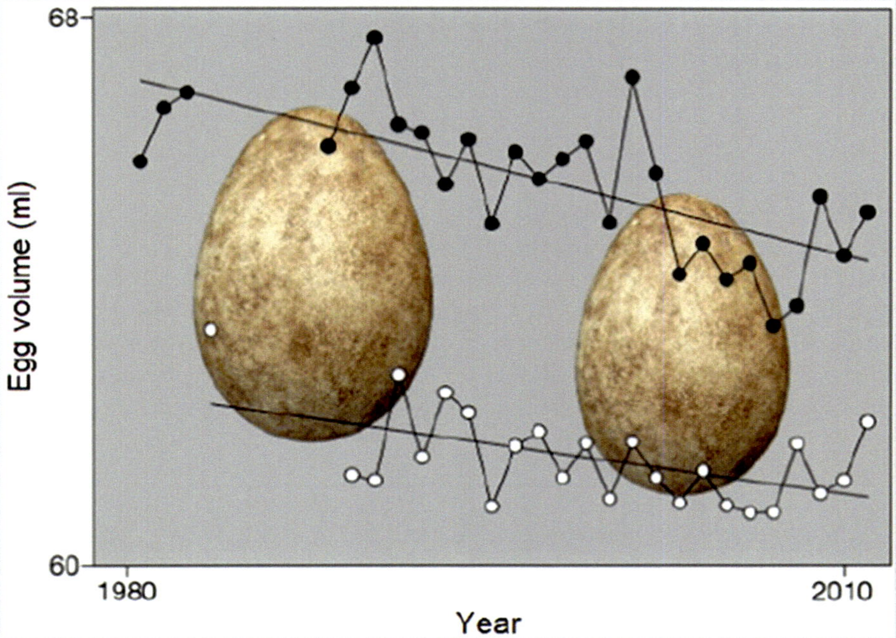


The mean volume of the eggs of Atlantic Puffins (*Fratercula arctica*) in Norway has declined at two different breeding colonies located 900 km apart over a 32-year period (1980–2011). (Figure from Barrett et al. 2012; Image: Tycho Anker-Nilssen, Gold Open Access under the Creative Commons by Attribution License (CC-BY), https://creativecommons.org/licenses/by/4.0/)

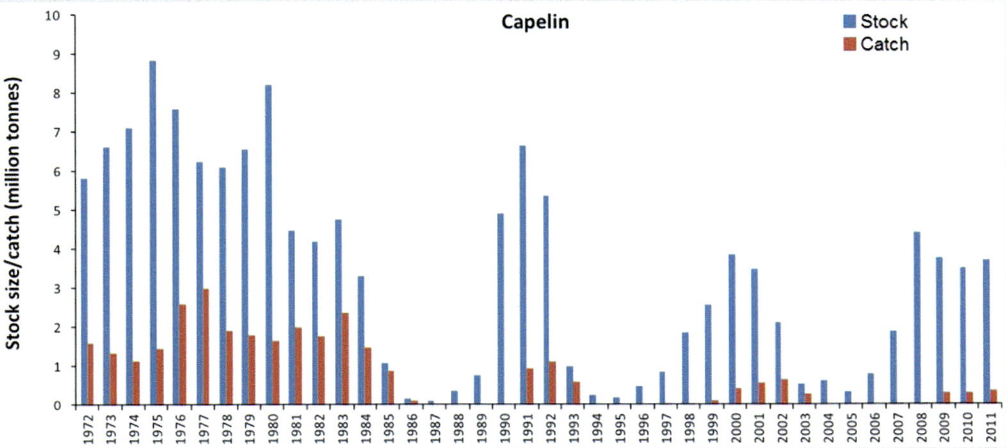


(continued)

**Box 16.8** (continued)

Estimates of stock size and commercial catches of capelin in the Barents Sea from 1972 to 2011. Note the general long-term trend for reduced stock size and the extreme inter-annual variation. Fish stocks are subpopulations of a species, and a ton (also known as a metric ton) equals 1000 kg. (Figure from Hop and Gjøsæter 2013; Rights managed by Taylor & Francis, used with permission)

**Box 16.9 Extreme Intraclutch Egg-size Dimorphism in *Eudyptes* Penguins**

Intraclutch variation in the size or mass of eggs has been reported in several species of birds. For example, the mass of Great Tit (*Parus major*) eggs varies with laying sequence (You et al. 2009). In most species, the differences in size or mass of eggs in a clutch are relatively small, although still sufficient to potentially affect nestling condition and chances of post-fledging survival. However, among species in the penguin genus *Eudyptes*, the intraclutch difference in egg size is extreme.

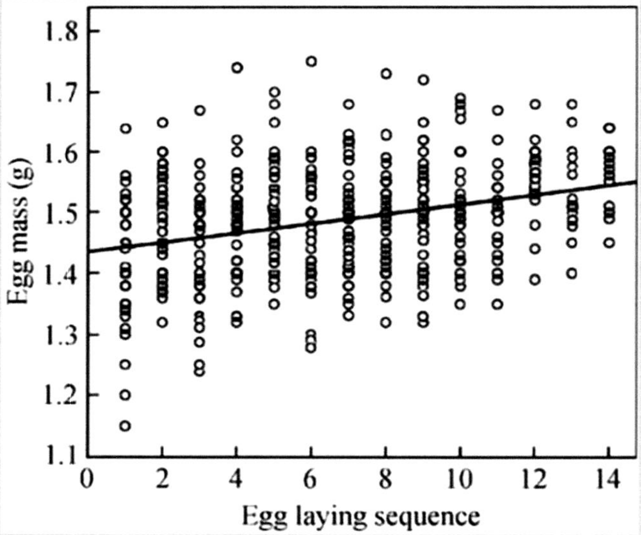


Variation in mean mass of eggs of Great Tits with order of laying. (Figure from You et al. 2009; © 2008 National Natural Science Foundation of China and Chinese Academy of Sciences. Published by Elsevier Ltd., used with permission)

All six species of *Eudyptes* penguins, like most species of penguins, have two-egg clutches, but the first egg laid (called the A-egg) is always much smaller (up to about one-third smaller) than the second egg (the B-egg). The A-egg is usually viable, but rarely produces young that fledge (<1%; Williams 1995). Hypotheses proposed to explain this extreme egg-size dimorphism include the Insurance Hypothesis (Dorward 1962), with the A-egg serving as "insurance"

(continued)

**Box 16.9** (continued)

in case the B-egg is lost, and the brood-reduction hypothesis (Lack 1954), with the smaller A-egg nestling only surviving in years when prey are abundant and not surviving (so the B-nestling receives more food and is more likely to survive) in most years when prey are not as abundant. Studies to date have not provided support for these hypotheses.

Macaroni Penguin (*Eudyptes chrysolophus*) with A- and B-eggs. (Figure from Jubinville et al. 2020; © 2020 Published by Elsevier Inc., used with permission)

There are 18 species of penguins that vary in size and life-history characteristics. The two largest species (Emperor, *Aptenodytes forsteri*, and King, *A. patagonicus*, penguins) begin breeding when 5 or 6 years old, breed either every year or every other year, and have one-egg clutches. The other 16 smaller species of penguins have two-egg clutches and, with the exception of the *Eudyptes* penguins, the two eggs are similar in size. Why the difference? Stein and Williams (2013) hypothesized that the extreme egg-size dimorphism of *Eudyptes* penguins is a maladaptation, evolving as a response to selection favoring a slower life history and their unique overwintering and migratory behaviors. Compared to the other taxa of smaller penguins, *Eudyptes* penguins typically begin breeding when older (5–8 years old), spend the non-breeding period in pelagic areas some distance from breeding colonies, and their long migration from non-breeding areas to breeding colonies overlaps the period when eggs have begun developing in females. This slower life-history strategy (beginning to breed when older) in combination with the overlap between migration and egg development that likely requires an energetic trade-off (i.e., more energy devoted to migration and less to production of the first egg) would seemingly favor a one-egg clutch. However, the two-egg clutches of the 14 smaller species of penguins appear to have a >30-million-year history (see phylogeny below) and are now canalized or "genetically fixed." Given this evolutionary mismatch between the "fixed" two-egg clutch when a slower life history and long migrations of *Eudyptes* penguins that overlap with egg production would seemingly favor a one-egg clutch, selection may have favored a significant reduction in the energy invested in the A-egg. This evolutionary "compromise" means that female *Eudyptes* penguins must "waste" energy in producing small A-eggs, but

(continued)

**Box 16.9** (continued)

also that, with >99% of A-eggs not producing fledglings, they basically have, in terms of annual fecundity, one-egg clutches.

An Eastern Rockhopper Penguin (*Eudyptes chryocome filholi*) laying its larger second (or B) egg while standing over its smaller first (or A) egg. (Photo from Morrison 2016; © 2015 The Authors, used with permission)

produce eggs because food availability may be limited when they arrive at their high-latitude breeding areas.

The source of nutrients for egg production can vary intraspecifically as well as interspecifically. Jaatinen et al. (2016) found that early breeding female Common Eiders (*Somateria mollissima*) in better condition and with more stored reserves (i.e., heavier females) used primarily endogenous nutrients to produce eggs (Fig. 16.127). Heavy females that initiated breeding later in the season tended to depend less on endogenous nutrients and increasingly more on exogenous nutrients to produce eggs. In contrast, lighter females, regardless of when they initiated breeding, depended primarily on exogenous nutrients for egg production; females of medium mass exhibited an intermediate trend (Fig. 16.127).

Beyond the energetic cost of egg laying, egg-laying females may also be more vulnerable to attack by predators because their increased body mass may impair their flight performance. Egg-laying female Eurasian Blue Tits (*Cyanistes caeruleus*), for example, were found to be 14% heavier and fly 20% slower during the egg-laying period than after eggs hatched (Kullberg et al.

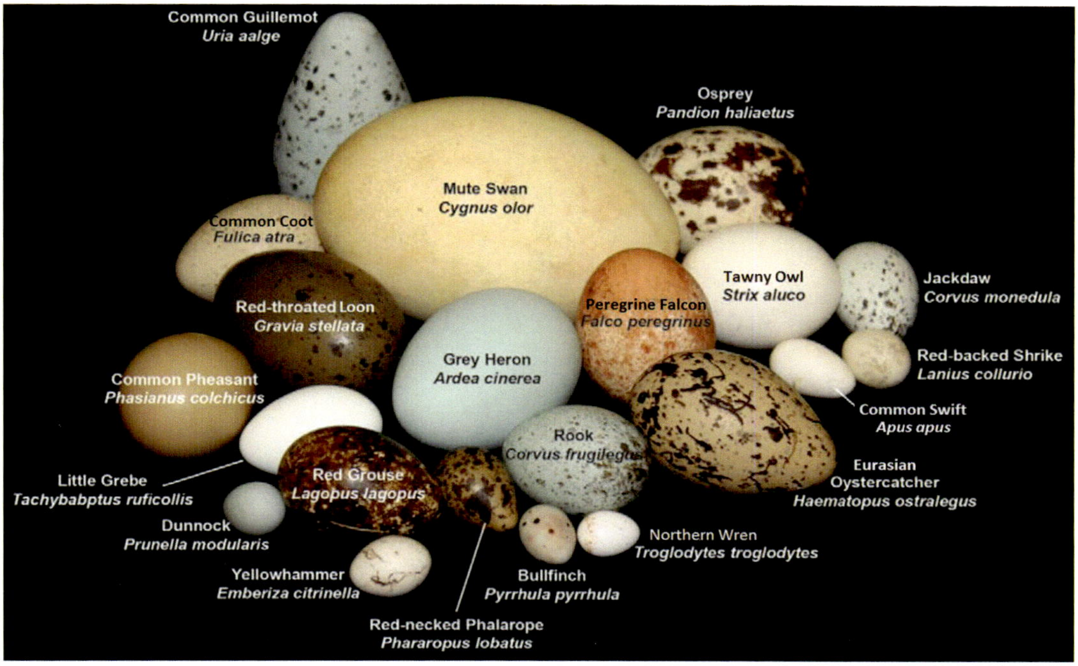


**Fig. 16.106** Example of the diversity of egg size, shape, and coloration exhibited by 21 species of birds that breed in Great Britain. Northern Wren is also referred to as the Eurasian Wren. (Figure modified from Cassey et al. 2011; used with permission of Golo Mayer who took the photo and thanks to the Natural History Museum in Tring where the eggs are located)

2002). Similarly, the take-off speed of female European Starlings (*Sturnus vulgaris*) was found to be slower during the egg-laying period (Lee et al. 1996), and the take-off speed of female Zebra Finches (*Taeniopygia guttata*) tended to decrease with increasing clutch size because, with larger clutches, more protein from flight muscle was needed for egg production (Veasey et al. 2001).

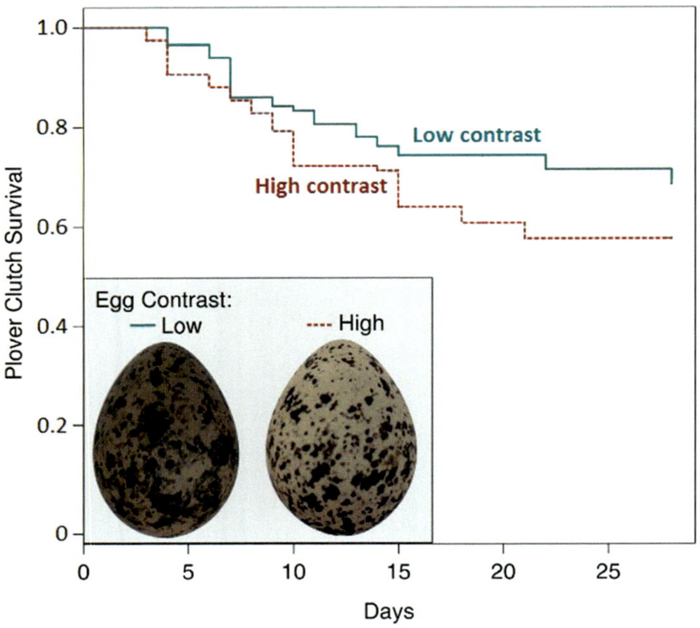


**Fig. 16.107** Plovers and lapwings lay eggs directly on the ground after creating a shallow depression. In a study conducted in Zambia, eggs that contrasted more with their backgrounds were less likely to survive than low contrast eggs, with eggs classified as being either high or low contrast if they were above or below the population median, respectively. Eggs at the lower left were the lowest (left) and highest (right) contrast eggs of Crowned Lapwings (*Vanellus coronatus*). (Figure modified from Troscianko et al. 2016; open-access article licensed under a Creative Commons Attribution 4.0 International License, https://creativecommons.org/licenses/by/2.0/)

**Box 16.10 Dirty Eggs = Safer Eggs**

The eggs of many species of birds, especially ground-nesting birds, are colored and patterned in a way that helps conceal them from predators (Kilner 2006; Troscianko et al. 2016). The eggs of Blue-footed Boobies (*Sula nebouxii*), however, are very light in color (pale blue to whitish) when first laid and, when not being incubated (boobies incubate eggs using their webbed feet), are quite visible to aerial predators like Heermann's Gulls (*Larus heermanni*) on the dark soil where boobies nest. As a result, many eggs are lost to predation during the first few days post-laying. Thereafter, however, predation rates decline because eggs become progressively dirtier and darker, and better camouflaged, as the shells are coated with soil during foot-mediated incubation. Alternatively, egg predation rates might decline later in the incubation period because of changes in adult behavior such as spending more time incubating or defending eggs more aggressively. To better understand why predation rates declined later in the incubation period, Mayani-Parás et al. (2015) placed white and soiled eggs of Domestic Chickens (*Gallus domesticus*) on the ground in areas near nest sites of Blue-footed Boobies. They found that white (or clean) eggs were significantly more likely to be taken by gulls than were dirty (or soiled) eggs, providing support for the hypothesis that the decline in predation rates of eggs of Blue-footed Boobies later in the incubation period was due to the improved camouflage of dirty eggs rather than any change in the behavior of adults. So why are booby eggs so light in color and not darkened with pigments like the eggs of many other ground-nesting birds? One possibility is that the nesting substrates of Blue-footed Boobies (and other boobies) vary in color and, whereas lighter eggs can be made darker when substrates are darker, doing the opposite, making darker eggs lighter when substrates are lighter, may be more difficult. If so, then selection would likely favor lighter-colored eggs whose color can be modified rather than darker, pigmented eggs whose color might be more difficult to modify. Although not as well documented, the pale eggs of other species of birds, including grebes, pelicans, and flamingoes, have also been reported to become darker during the incubation period (del Hoyo et al. 1992; Knopf and Evans 2020; Lee et al. 2008), perhaps serving to better camouflage their eggs as is the case for Blue-footed Boobies.

(continued)

**Box 16.10** (continued)

Eggs of Blue-footed Boobies become increasingly coated with soil, and less visible to predators, over time. (Photo by Andy Morffew, pxhere.com, CC0 Public Domain)

**Fig. 16.108** Eggs of a Pacific-slope Flycatcher (*Empidonax difficilis*) with "ring" of spots around the blunt end. (Photo by Peter Pearsall, U. S. Fish and Wildlife Service, CC0 Public Domain)

**Fig. 16.109** Eggs from four clutches of female Village Weaver (*Ploceus cucullatus*) eggs. The pair of eggs in each column come from the same clutch. (Figure from Lahti 2006; © Society for the Study of Evolution, used with permission)

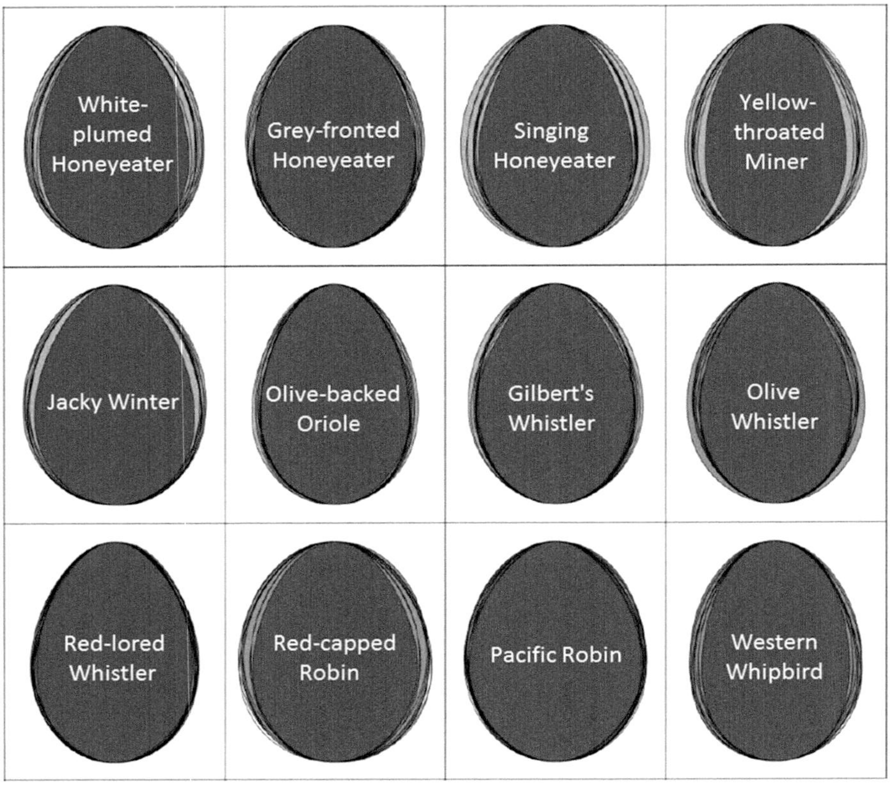


**Fig. 16.110** Egg profiles of 10 eggs laid by 10 different females of 12 species of Australian songbirds, with egg height standardized. Note that the extent of variation in egg shape varies, with more variation in some species than others. White-plumed Honeyeater, *Ptilotula penicillata*; Gray-fronted Honeyeater, *Ptilotula plumula*; Singing Honeyeater, *Gavicalis virescens*; Yellow-throated Miner, *Manorina flavigula*; Jacky Winter, *Microeca fascinans*; Olive-backed Oriole, *Oriolus sagittatus*; Gilbert's Whistler, *Pachycephala inornata*; Olive Whistler, *Pachycephala olivacea*; Red-lored Whistler, *Pachycephala rufogularis*; Red-capped Robin, *Petroica goodenovii*; Pacific Robin, *Petroica pusilla*; Western Whipbird, *Psophodes nigrogularis*. (Figure modified from Heenan 2013, used with permission)

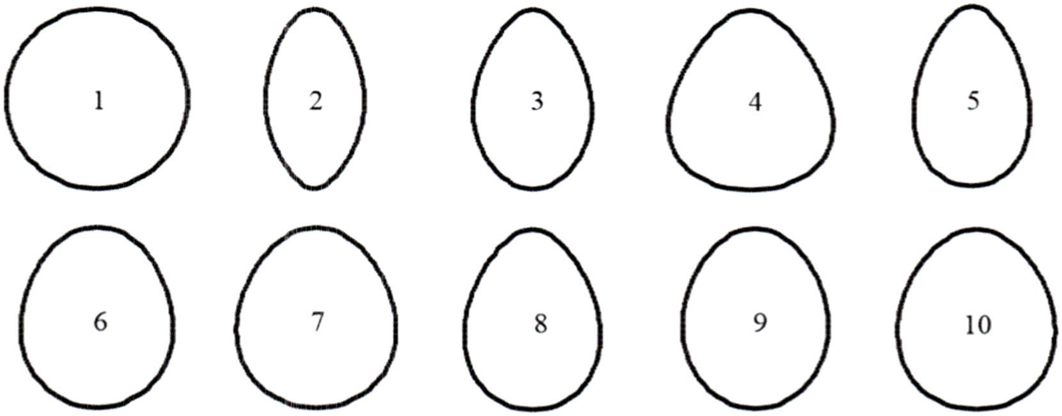


**Fig. 16.111** Proposed optimal egg shape for different-sized clutches, with clutch size indicated by the number inside each egg. (Figure from Barta and Székely 1997; © 1987 CCC Republication, used with permission)

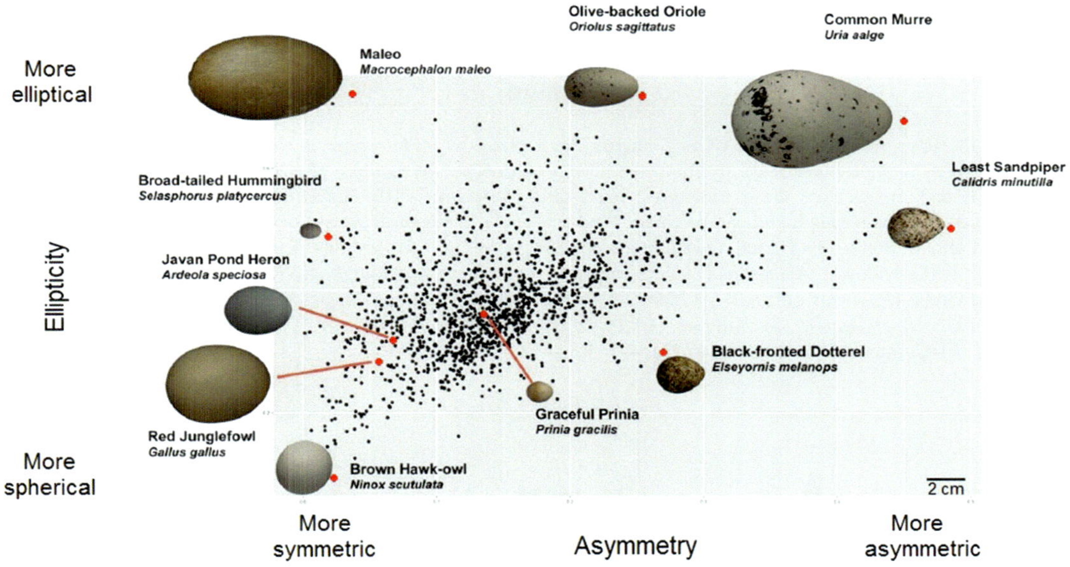


**Fig. 16.112** Typical egg shapes for 1400 different species of birds illustrating the extensive variation in degree of asymmetry and ellipticity. Images of representative eggs are shown where they fall in egg-shape "morphospace" (larger red dots). (Figure modified from Stoddard et al. 2017; © 2017 The Authors, used with permission)

**Fig. 16.113** Four main categories of egg shapes are elliptical (which includes spherical), pyriform, oval, and subelliptical and, in each of those categories, eggs of different species vary in length and can be categorized as short, medium, or long. (1) Barred Owl (*Strix varia*), (2) Chuck-will's-widow (*Antrostomus carolinensis*), (3) Horned Grebe (*Podiceps auritus*), (4) Northern Bobwhite (*Colinus virginianus*), (5) Killdeer (*Charadrius vociferus*), (6) Common Murre (*Uria aalge*), (7) Song Sparrow (*Melospiza melodia*), (8) Dark-eyed Junco (*Junco hyemalis*), (9) Northern Shoveler (*Spatula clypeata*), (10) Least Tern (*Sternula antillarum*), (11) Common Raven (*Corvus corax*), and (12) Common Loon (*Gavia immer*). Eggs shown are not to scale. (Figure from Oology and Ralph's talking eggs: bird conservation comes out of its shell by Carrol L. Henderson; © 2007 used courtesy of the University of Texas Press)

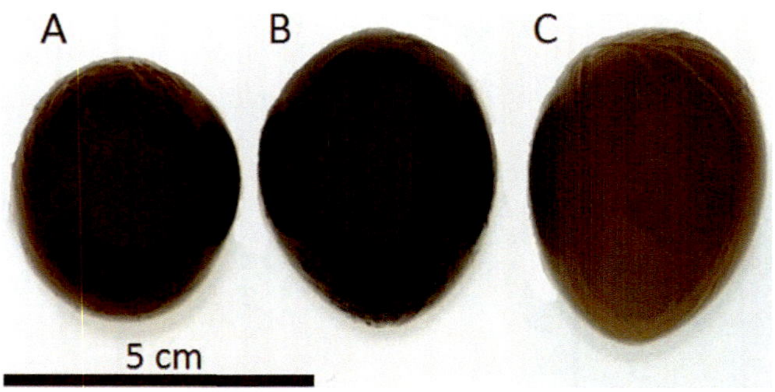


**Fig. 16.114** Egg of a Domestic Chicken at different locations in the oviduct. (**a**) Middle portion of the magnum, (**b**) posterior end of the magnum, and (**c**) isthmus. The leading edge of the egg as it passes through the oviduct is at the bottom. (Figure modified from Mao et al. 2006; © 2006 Zoological Society of Japan, used with permission)

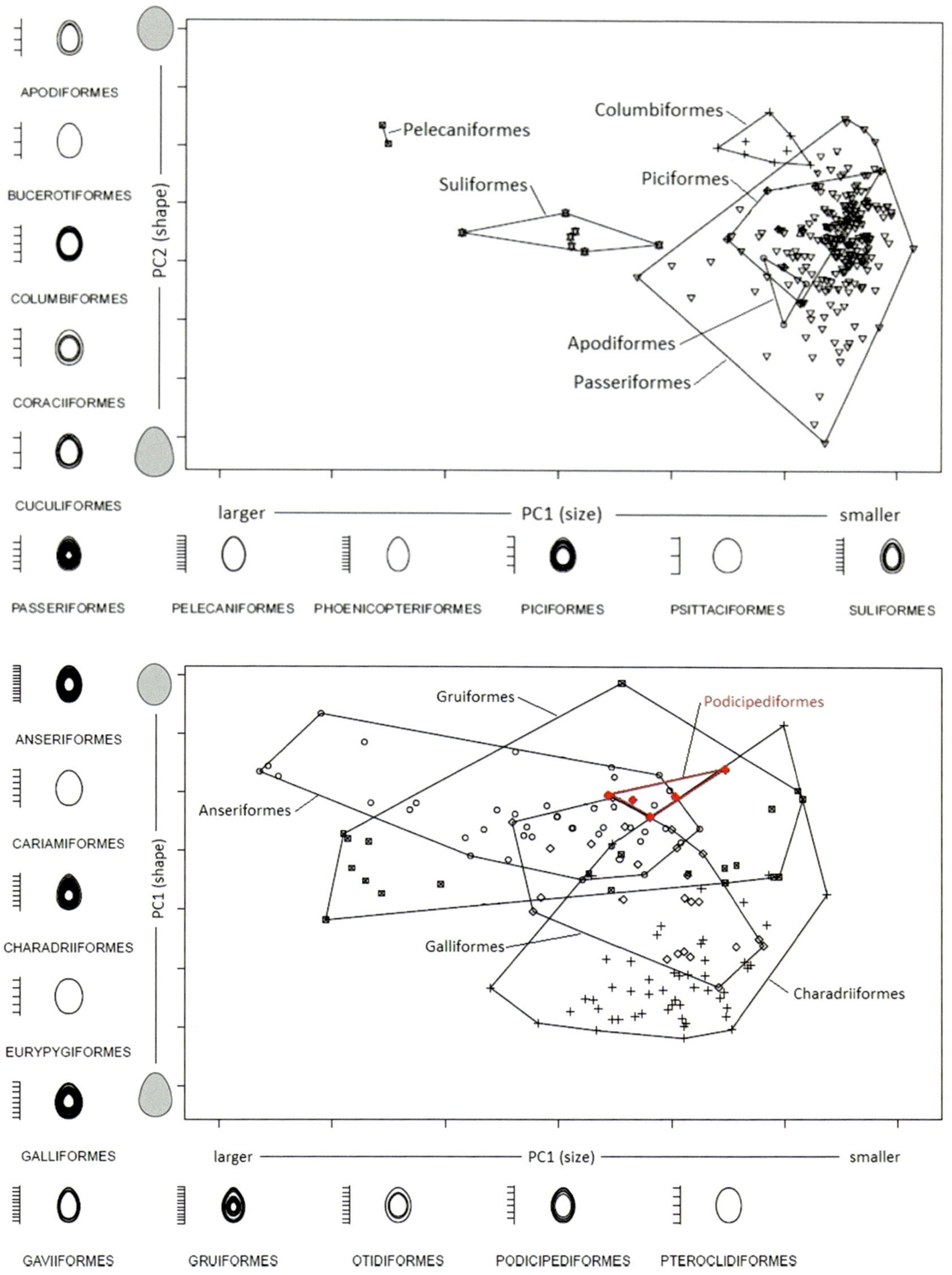


**Fig. 16.115** Based on shape and size, the eggs of species of birds in the same order are generally similar and typically vary more in size than shape, with egg sizes larger or smaller depending on the size of different species.

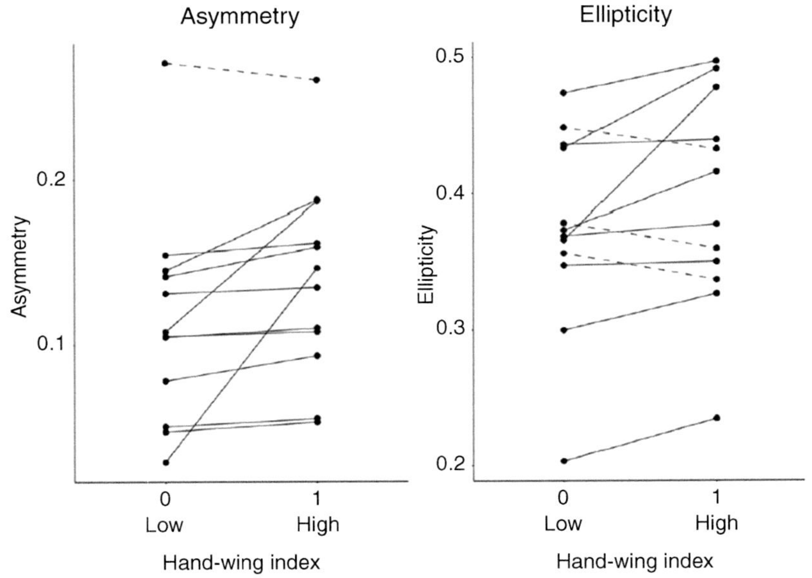


**Fig. 16.116** Comparison of egg asymmetry and ellipticity for species in 12 orders of birds that differ in flight ability based on the hand-wing index (ratio of the distance between the tip of the longest primary and the tipoff the first secondary feather to the wing chord [distance between the carpal joint and wingtip]), with a higher hand-wing index indicating better flying ability. Species in each order were categorized as having low hand-wing indices (lower 50%) or high hand-wing indices (upper 50%). Note that species with high hand-wing indices, those with better flying ability, tend to have more asymmetrical and more elliptical eggs than species with low hand-wing indices (solid lines). Less often, better fliers have less asymmetric and less elliptical eggs (dashed lines). (Figure modified from Stoddard et al. 2017; © 2017 The American Association for the Advancement of Science, paid for permission)

---

**Fig. 16.115** (continued) Drawings of eggs to the left and below the two plots illustrate variation in egg size and shape within orders, with eggs of different species superimposed on each other. Axes to the left of the egg outlines represent scales, with the distance between tick marks equaling 10 mm. (Figure modified from Mytiai et al. 2017; © 1951 CCC Republication, used with permission)

**Fig. 16.117** A four-egg clutch of a Killdeer (*Charadrius vociferus*). The pyriform shape of the eggs may contribute to increased incubation efficiency, with the pointed ends facing each other to better fit under the brood patch of incubating adults. (Photo from pxhere.com, CC0 Public Domain)

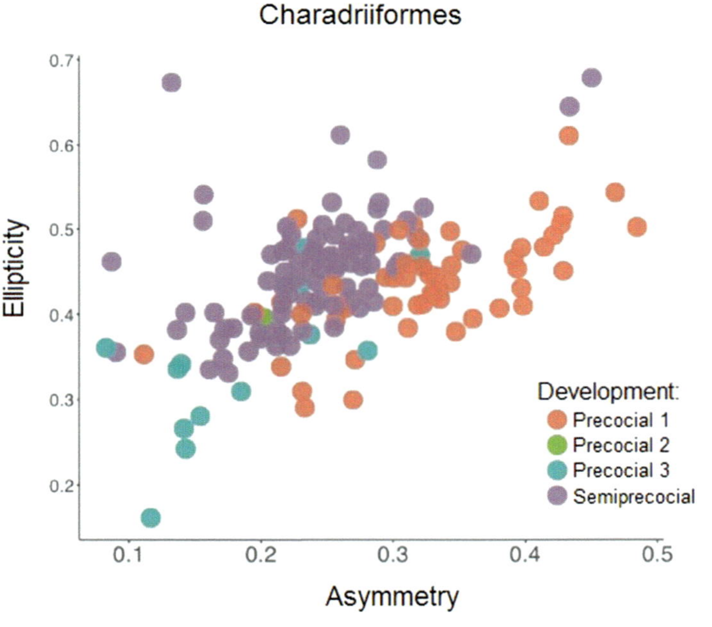


**Fig. 16.118** Among species in the order Charadriiformes, more precocial species have more asymmetrical eggs. Developmental modes were categorized as Precocial 1, Precocial 2, Precocial 3, and Semiprecocial (Starck 1993), with Precocial 1 being the most precocial and Semiprecocial the least precocial. (Figure modified from Supplementary Figure from Stoddard et al. 2017; © 2017 The Authors, some rights reserved; exclusive licensee American Association for the Advancement of Science, used with permission)

**Fig. 16.119** Top, Breeding colony of Common Murres (*Uria aalge*) showing the sometimes narrow ledges where females lay and incubate their eggs (Screen capture from a video with Tim Birkhead; used with permission of Tim Birkhead). Bottom, Breeding colony of Common Murres where ledges have a 30° slope. (Figure from Birkhead et al. 2018; © 2018 Oxford University Press, used with permission)

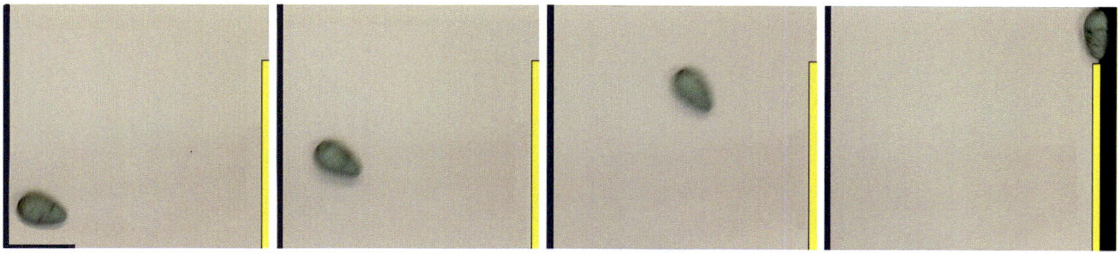

**Fig. 16.120** From left to right, a Common Murre (*Uria aalge*) egg rolling in an arc. The yellow bar on the right side of each image is 30-cm long so the egg rolls in an arc much longer than a ledge on which guillemots lay eggs. (Screen shots from video with Tim Birkhead; https://www.youtube.com/watch?v=eXXG0DdNwEs, used with permission of Tim Birkhead)

**Fig. 16.121** Eggs of Razorbills (*Alca torda*, top row) and Common Murres (*Uria aalge*, bottom row). The pointed, or pyriform, shape of these eggs means that more of the eggshell is in contact with the substrate, possibly making them more resistant to impact. Because these species nest is dense colonies, their eggs are often contaminated with fecal material (as clearly illustrated by the fourth egg from the right in the bottom row). However, because of their shape, such contamination is more likely at the pointed end of eggs that are in contact with the substrate rather than the blunt ends. Remaining free of contamination is likely important in facilitating gas exchange during incubation and hatching because chicks emerge from the blunt end of eggs. (Figure from Birkhead et al. 2017; © 2017 British Ornithologists' Union, used with permission)

**Fig. 16.122** Uterus, vagina (V), and uterovaginal junction (UVJ) in an oviduct of a female Wild Turkey (*Meleagris gallopavo*). Note the thick muscular walls of the vagina that are important in expelling eggs in the process of egg laying. (Figure modified from Bakst 2011; © 2011 Oxford University Press, used with permission)

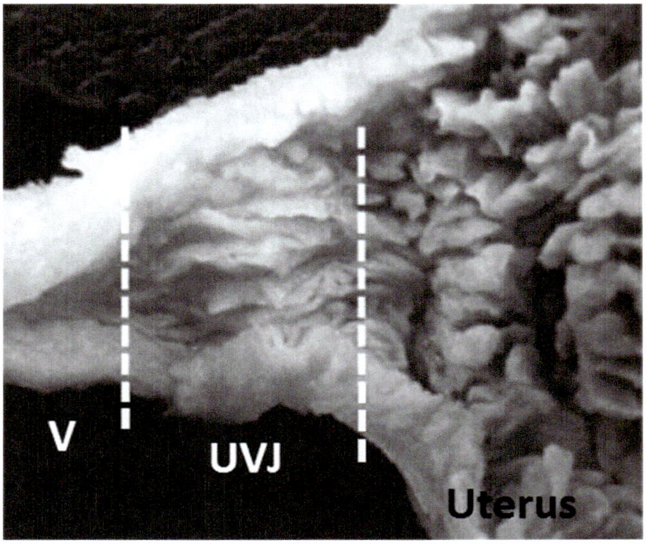

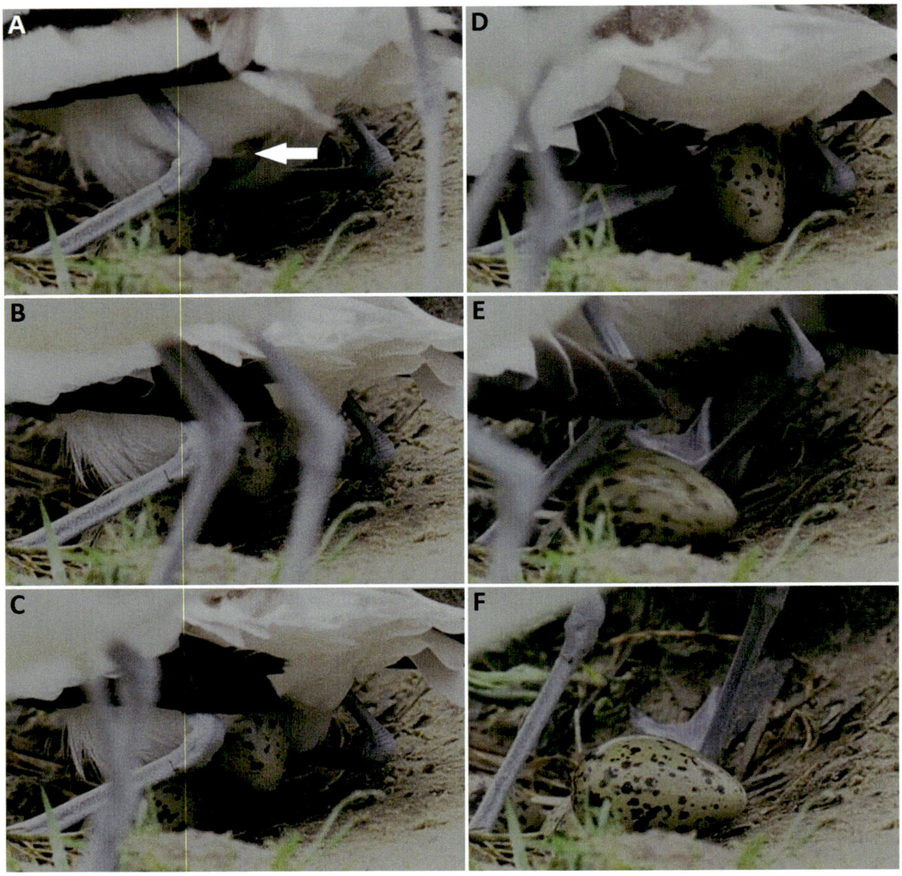


**Fig. 16.123** (**a**–**f**) Female Pied Avocet (*Recurvirostra avosetta*) laying an egg. Note that one egg has already been laid (**a**) and the male is walking between the female and the video-recorder (**b**–**f**). (Figure created from screenshots of video, used with permission)

**Table 16.5** Estimated metabolic cost of egg production for several species of birds

| Species | Scientific name | Percent increase in RMR, FMR, or BMR[a] | Citation |
|---|---|---|---|
| Wood Duck | *Aix sponsa* | 56% | Drobney (1980) |
| Greater Scaup | *Aythya marila* | 234% | Alisauskas and Ankney (1992) |
| Ruddy Duck | *Oxyura jamaicensis* | 280% | Alisauskas and Ankney (1994) |
| Japanese Quail | *Coturnix japonica* | 49% | Ward and McLeod (1992) |
| Japanese Quail | *Coturnix japonica* | 70% | Pick et al. (2016) |
| American Coot | *Fulica americana* | 20.5% | Alisauskas and Ankney (1985) |
| Audouin's Gull | *Larus audouinii* | 42% | Ruiz et al. (2000) |
| Yellow-legged Gull | *Larus michahellis* | 14.1–20.7% | Ramírez et al. (2010) |
| Common Tern | *Sterna hirundo* | 127–157% | Moore et al. (2000) |
| Killdeer | *Charadrius vociferus* | 29% | Brunton (1988) |
| Eurasian Kestrel | *Falco tinnunculus* | 75% | Meijer et al. (1989) |
| European Pied Flycatcher | *Ficedula hypoleuca* | 61.2% | Ojanen et al. (1978), Ojanen (1983) |
| Great Tit | *Parus major* | 27% | Nilsson and Råberg (2001) |
| House Wren | *Troglodytes aedon* | 47% | Kendeigh et al. (1956) |
| Marsh Wren | *Cistothorus palustris* | 45.4% | Kale (1965) |
| Boat-tailed Grackle | *Quiscalus major* | 31% | Bancroft (1985) |
| European Starling | *Sturnus vulgaris* | 22.4% | Vézina and Williams (2002) |
| Zebra Finch | *Taeniopygia guttata* | 22% | Vézina and Williams (2005) |
| House Sparrow | *Passer domesticus* | 16% | Chappell et al. (1999) |
| House Sparrow | *Passer domesticus* | 44–47% | Krementz and Ankney (1986) |

[a]*RMR* resting metabolic rate, *FMR* field metabolic rate, and *BMR* basal metabolic rate

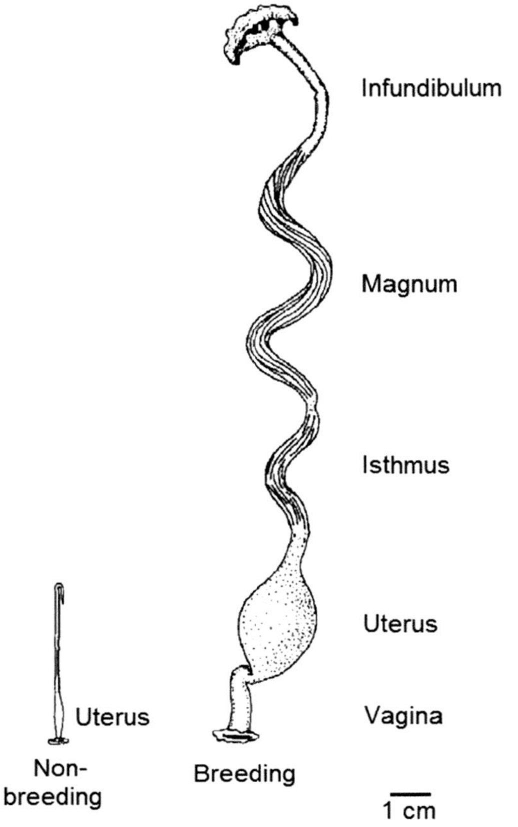


**Fig. 16.124** Oviduct of a female Asian Pied Starling (*Gracupica contra*) during the non-breeding season and during the egg-laying stage of the breeding season. The substantial growth and maintenance of much larger oviducts requires considerable energy. (Figure modified from Gupta and Maiti 1987; © 1987 Wiley-Liss, Inc., used with permission)

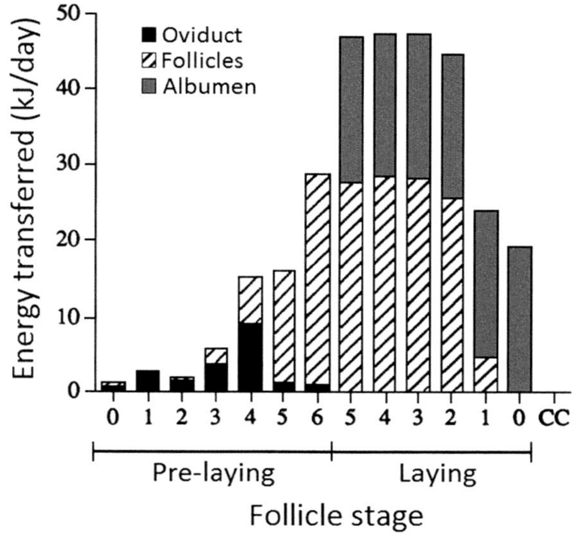


**Fig. 16.125** Pattern of energy transfer into the oviduct, follicles (including yolk), and albumen by female European Starlings (*Sturnus vulgaris*) laying a 6-egg clutch. During the laying period, the number of follicles decreases as they are ovulated until clutch completion (CC). (Figure modified from Vézina and Williams 2002; © 1999, CCC Republication, used with permission)

**Box 16.11 Stable Isotopes and Egg Nutrients**

Identifying the source of nutrients used to produce eggs by both resident birds and those that migrate to their breeding areas is important for better understanding the evolution of life-history strategies (Drent and Daan 1980, Hobson 2006). For example, selection may favor wintering at lower latitudes and breeding at higher latitudes, but the availability of food at high latitudes may be limited at the beginning of the breeding season. One possible solution to this problem is for females to carry stored nutrients from wintering areas and use those nutrients to produce eggs (capital breeding). For other species, food availability in breeding areas is sufficient for producing a clutch of eggs (income breeding) and, for yet other species, both endogenous and exogenous nutrients might be needed to produce eggs. Historically, methods used to try and identify the source of nutrients used by females to produce eggs, such as correlating the mass of eggs with the loss of macronutrient mass by females, were likely inaccurate. In recent decades, however, stable isotope analysis has provided a more accurate means to identifying the source of nutrients used by females to produce eggs.

Stable isotopes are naturally occurring forms of elements that have different numbers of neutrons in the nucleus and, therefore, behave differently in biogeochemical processes. For any element, the abundance of stable isotopes varies geographically for a number of reasons. For example, stable isotopes of nitrogen vary depending on how plants fix nitrogen (symbiotic or direct conversion of atmospheric nitrogen) and stable isotopes of carbon vary ambient conditions that affect enzymatic reactions during photosynthesis (Rubenstein and Hobson 2004).

(continued)

**Box 16.11** (continued)

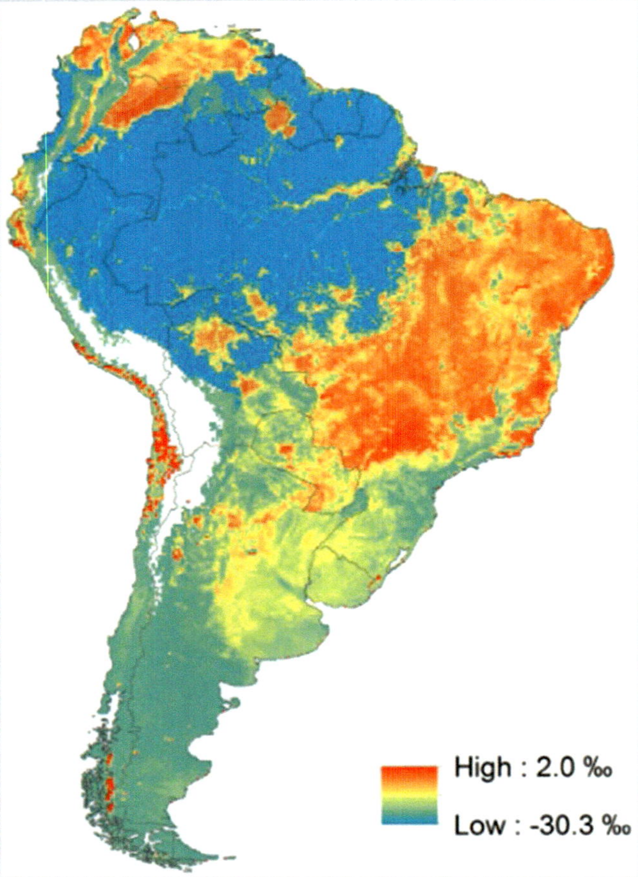


Carbon (δ13C) isoscape of South America based on mean annual plant δ13C. (Figure from Hobson and Kardynal 2016 based on data from Powell et al. 2012; © 2015 Oxford University Press, used with permission)

Differences in the relative abundance of isotopes in bird feathers, tissues, and egg components can be measured using mass spectrometers, with differences expressed as the ratio of the heavy to light isotope and reported in delta (δ) notation as parts per thousand or per mil (‰). Isotope ratios can then be used, in the case of egg nutrients, to help determine the location where females obtained those nutrients, e.g., a wintering site, stopover site, breeding site, or some combination of these sites. Importantly, however, different tissues are synthesized and replaced at different rates and the isotopic ratio of new tissues is determined by a bird's location and diet when tissues are synthesized (Inger and Bearhop 2008). For example, feathers can provide information about a bird's location and diet several weeks or months ago, depending on when it molted. As a result, based on knowledge of the molt chronology of a species, identifying where a bird was located at the time new feathers were generated is possible by determining the isotope ratio of those feathers. Similarly, the isotopic ratios of nutrients that females incorporate into their eggs can be

(continued)

**Box 16.11** (continued)

used to determine if females produce eggs (and, therefore, newly hatched young) used endogenous nutrients, exogenous nutrients, or both.

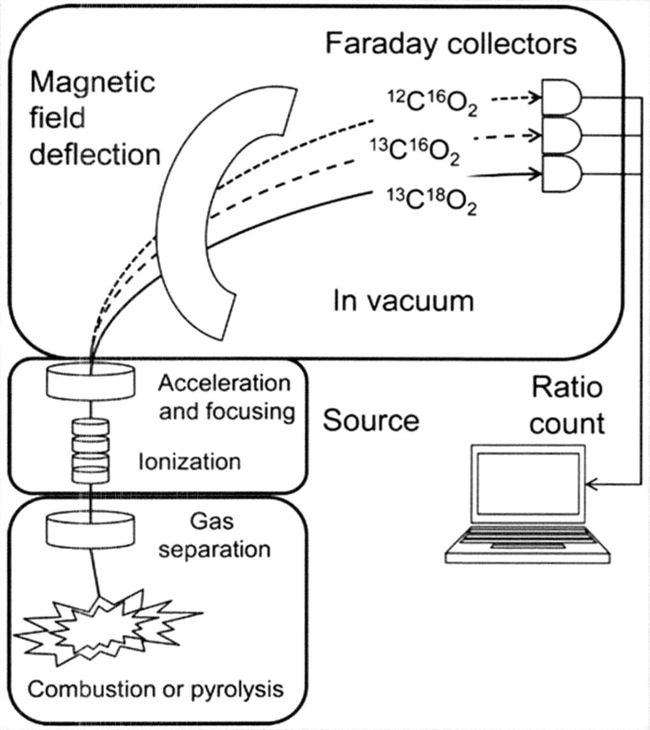


Schematic diagram of an isotope-ratio mass spectrometer and an elemental analyzer. Organic samples are homogenized, weighed, and injected into the analyzer where they are converted to gaseous inorganic compounds such as $N_2$, $CO_2$, or $H_2O$ by combustion or pyrolysis. The gases are separated and injected into the mass spectrometer where they are ionized and a magnet then deflects and separates them based on mass. The ionized, gaseous molecules are collected in Faraday cups (metal cups designed to catch charged particles in a vacuum), creating a weak electrical current that is used to determine the number of ions hitting the cups (as measured by a computer). (Figure from Ben-David and Flaherty 2012; © 2012 Oxford University Press, used with permission)

(continued)

**Box 16.11** (continued)

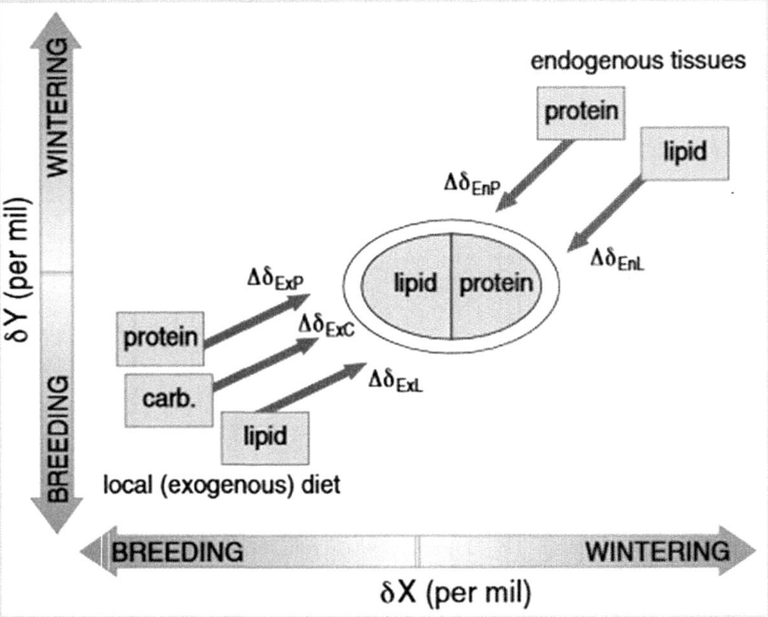


Example of a hypothetical two-isotope analysis (isotopes X and Y) of dietary (exogenous) and stored (endogenous) nutrient allocation to eggs. In this case, the endogenous nutrients (EnP, endogenous protein, and ENL, endogenous lipid) differ isotopically from nutrients available in foods in the breeding area (ExP, exogenous protein, ExL, exogenous lipid, and ExC, exogenous carbohydrate) because they were formed when the female was in the wintering area. (Figure from Hobson 2006, used with permission of the Netherlands Ornithologists' Union)

(continued)

**Box 16.11** (continued)

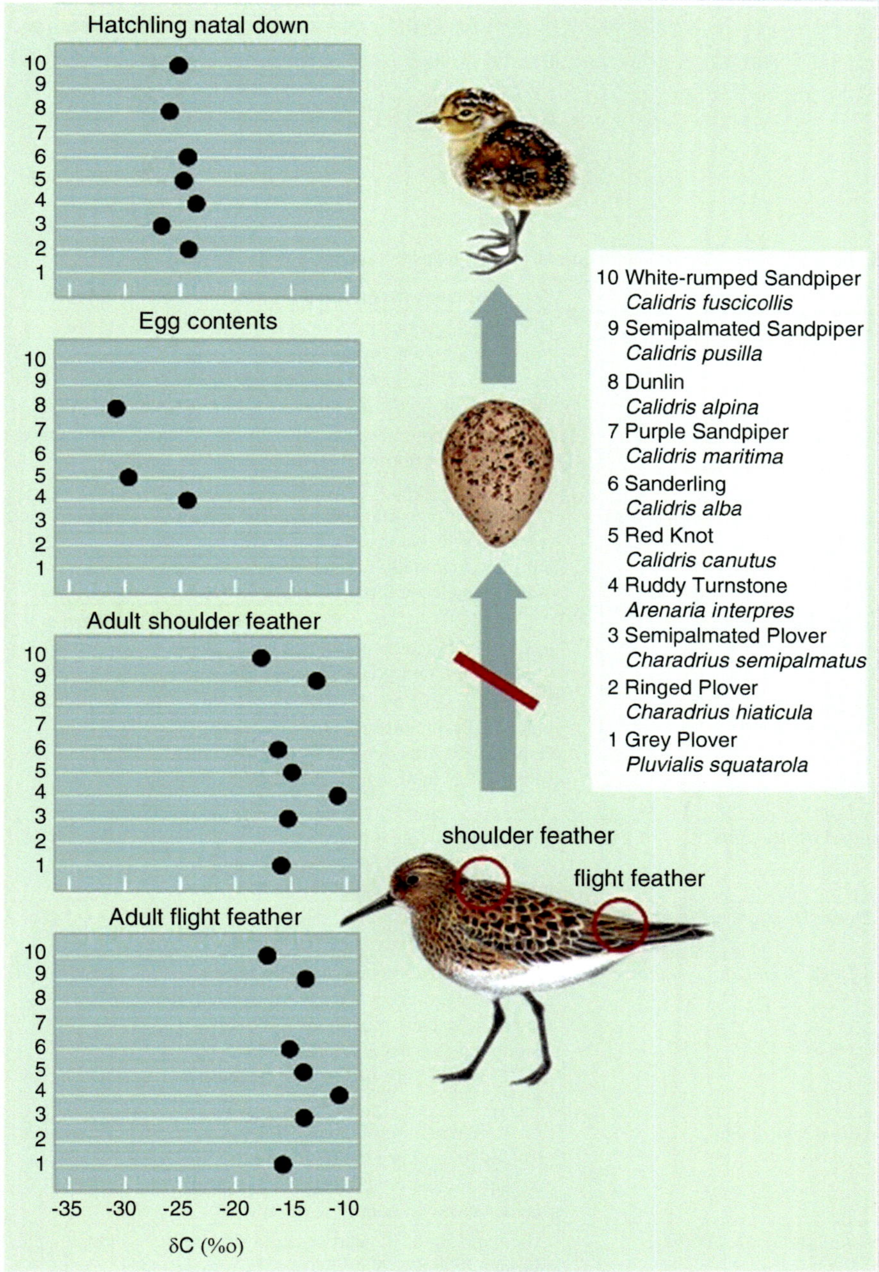


Carbon stable isotope ratios of eggs, natal downs, and feathers of several species of shorebirds that breed in the Arctic. Note that the ratios for adult feathers differ from those of eggs and natal down, indicating that females obtained nutrients in wintering areas or during migration when

(continued)

**Box 16.11** (continued)

molting and nutrients from the breeding area when egg laying. Adult feathers were grown either during the previous winter (flight feathers) or during spring migration (should feathers). Eggs were collected from deserted nests and natal down from hatchlings. (Figure from Klaassen 2003; © 2003 Springer-Verlag, used with permission)

**Table 16.6** Source of energy/nutrients required for egg formation in several species of birds

| Species | Source of energy/nutrients | Citation |
|---|---|---|
| Barnacle Goose, *Branta leucopsis* | Use of endogenous reserves averaged 41% (range = 23–65%) in lipid-free yolk, and 54% (range = 32–73%) and 47% (range = 25–88%) in yolk lipids and albumen, respectively. | Hahn et al. (2011) |
| Snow Goose, *Anser caerulescens* | ". . . endogenous reserves contributed 33% of lipid-free yolk nutrients, 27% of albumen, and 20% of yolk lipid, on average." | Gauthier et al. (2003) |
| Pink-footed Goose, *Anser brachyrhynchus* | ". . . capital breeding [i.e., use of endogenous reserves to produce eggs] is ~50% on average but may potentially amount to as much as 100%, notably in females laying early." | Klaassen et al. (2017) |
| Canada Goose, *Branta canadensis* | ". . . used 49 ± 1.1% (SE), 61 ± 1.3%, and 51 ± 1.1% endogenous nutrients for albumen, yolk protein, and yolk lipid, respectively." | Sharp et al. (2013) |
| Brant (*Branta bernicla*) and Emperor Goose (*Anser canagicus*) | "Approximately 59 and 45% of protein in egg yolks of Brent and Emperor Geese, respectively, was derived from exogenous sources (i.e. food plants on the local breeding area)." | Schmutz et al. (2006) |
| Lesser Scaup, *Aythya affinis* | ". . . lipid reserves for much of the clutch lipid, and . . . exogenous sources of protein and mineral for most . . . nutrients in the clutch . . ." | Esler et al. (2001) |
| Harlequin Duck, *Histrionicus histrionicus* | ". . . relied on food available in streams on breeding grounds for egg formation . . . " | Bond et al. (2007) |
| Long-tailed Duck, *Clangula hyemalis* | ". . . although some females allocated endogenous reserves for egg production, most females allocated exogenous resources for egg production (albumen 98.5%, yolk protein 78.3%, whole yolk 84.9%, and yolk lipids 38.3%)." | Lawson (2006) |
| Redhead, *Aythya americana* | ". . . relied mainly on dietary lipids and proteins for egg production, and . . . endogenous reserves were used [for] body maintenance and energy requirements." | Hobson et al. (2004) |
| King Eider, *Somateria spectabilis* | "The majority of the carbon and nitrogen in albumen (C: 86%, N: 99%) and the nitrogen in lipid-free yolk (90%) were derived from food consumed on breeding grounds." | Oppel et al. (2010) |
| Common Eider, *Somateria mollissima* | ". . . overall estimates of protein sources . . . were mixed for egg yolk (median: 44.5–56.5% endogenous) and overwhelmingly exogenous for egg albumen (0.4–0.7%)." | Hobson et al. (2015) |

(continued)

**Table 16.6** (continued)

| Species | Source of energy/nutrients | Citation |
|---|---|---|
| Mallard, *Anas platyrhynchos* | Use lipid reserves to meet energy requirements; available food for egg production | Krapu (1981) |
| Spruce Grouse, *Canachites canadensis* | "... estimates of nutrient intake suggested that the spring diet provided only about 60% of the protein and 45% of the [calcium] needed for clutch formation. Consequently, hens appeared to rely on both their spring diet and stored reserves for the nutrients required for clutch formation." | Naylor and Bendell (1989) |
| Adelie Penguin, *Pygoscelis adeliae* | "... almost all energy (96%) and nutrients (99%) for the laying female and her clutch come from endogenous reserves." | Meijer and Drent (1999) (citing Astheimer and Grau 1985) |
| Little Penguin, *Eudyptula minor* | Mean endogenous contributions were 36.5% (range = 15.5–30.5%) for A chicks and 16.7% (range = 6.3–29.0%) for B chicks. | Ramírez et al. (2015) |
| Thick-billed Murre, *Uria lomvia* | "... it seems likely that females rely upon local food sources for the development of both [first] eggs [and replacement eggs], that the reliance is greater for [replacement eggs] and that for both eggs endogenous stores are important for certain elements." | Jacobs et al. (2009) |
| Greater Flamingo, *Phoenicopterus roseus* | About 18% of egg nutrients were from an exogenous source. | Rendón et al. (2011) |
| Red-necked Grebe, *Podiceps grisegena* | Females "... use nutrients from the breeding lake for egg formation." | Paszkowski et al. (2004) |
| Red-necked Grebe, *Podiceps grisegena* | "... egg nutrients were mainly acquired locally. Endogenous nutrients were to some extent mobilized for the formation of albumen (25–26%) and lipid-free yolk (17–18%) in early laid clutches, but were little mobilized, if at all, in clutches of females that delayed laying." | Kloskowski et al. (2019) |
| Horned Grebe, *Podiceps auritus* | "Grebes use nutrients acquired from the breeding pond for egg formation." | Kuczynski and Paszkowski (2010) |
| Double-crested Cormorant, *Nannopterum auritus* | "... little evidence for significant transfer of endogenous reserves ... to their eggs ..." | Hobson et al. (1997) |
| Barn Owl, *Tyto alba* | "Routine food intake" used for egg formation. | Durant et al. (2000) |
| Larger owls (Ural, *Strix uralensis*, and Tawny, *S. aluco*, owls) vs. smaller owls (Eurasian Pygmy, *Glaucidium passerinum*, and Boreal, *Aegolius funereus*, owls) | "Larger species ... resemble capital breeders, where breeding is largely based on stored energy resources ... smaller [owls] are closer to income breeders which acquire resources for laying ... during the breeding season." | Lehikoinen et al. (2011) |
| Five species of gulls, four terns, and one jaeger[a] | "... endogenous nutrient reserves likely were important to birds during migration and the initial settling period ..., local food supplies were adequate to provide nutrients for reproduction." | Hobson et al. (2000) |
| Arctic Tern, *Sterna paradisaea* | "... rely principally on exogenous sources of nutrients for egg production." | Pratte et al. (2018) |
| Red Knots, *Calidris canutus*, and Ruddy Turnstones, *Arenaria interpres* | "... local food resources may be used to produce much of the protein and lipid components of eggs, endogenous stores of | Morrison and Hobson (2004) |

(continued)

**Table 16.6** (continued)

| Species | Source of energy/nutrients | Citation |
|---|---|---|
| | both protein and lipid may be used in formation of the earliest eggs." | |
| Pectoral Sandpiper, *Calidris melanotos* | Females that "... started laying earlier ... used mostly stored resources for egg production . . ."; stored resources were used less by females initiating clutches later. | Yohannes et al. (2010) |
| Ten species of shorebirds[b] | "... eggs ... seem to be produced from local nutrients." | Klaassen et al. (2001) |
| Barn Swallow, *Hirundo rustica* | "... swallows form eggs primarily from current food intake, although our data do not exclude use of micronutrients from reserves." | Ward and Bryant (2006) |
| American Redstart, *Setophaga ruticilla* | "... use an income breeding strategy, forming eggs from [a] ... variable diet on the breeding grounds." | Langin et al. (2006) |
| Spotted Antbird, *Hylophylax naevioides* | "... it is probable that Spotted Antbirds are "income breeders" like most other passerines, which do not need to rely on stored energy reserves for breeding ... " | Hau et al. (Hau et al. 2000a, b) |

[a]Mew Gull, *Larus canus*; Ring-billed Gull, *L. delawarensis*; California Gull, *L. californicus*; Herring Gull, *L. argentatus*; Bonaparte's Gull, *L. philadelphia*; Caspian Tern, *Sterna caspia*; Common Tern, *S. hirundo*; Arctic Tern, *S. paradisaea*; Black Tern, *Chlidonias niger*; and Parasitic Jaeger, *Stercorarius parasiticus*

[b]Black-bellied Plover, *Pluvialis squatarola*; Common Ringed Plover, *Charadrius hiaticula*; Semipalmated Plover, *C. semipalmatus*; Ruddy Turnstone, *Arenaria interpres*; Red Knot, *Calidris canutus*; Sanderling, *Calidris alba*; Purple Sandpiper, *C. maritima*; Dunlin, *C. alpina*; Semipalmated Sandpiper, *C. semipalmatus*; White-rumped Sandpiper, *C. fuscicollis*

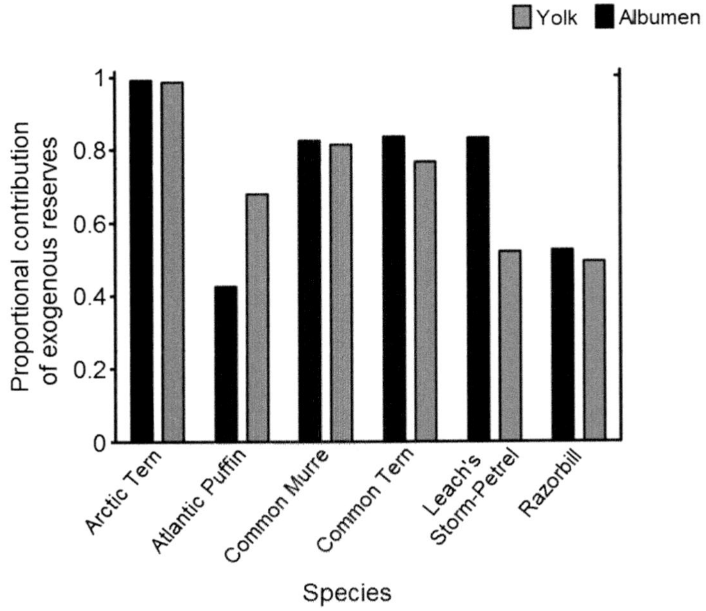


**Fig. 16.126** Estimated proportional contribution of locally derived nutrients to production of eggs in six species of seabirds. Arctic Tern, *Sterna paradisaea*; Atlantic Puffin, *Fratercula arctica*; Common Murre, *Uria aalge*; Common Tern, *Sterna hirundo*; Leach's Storm-Petrel, *Oceanodroma leucorhoa*; and Razorbill, *Alca torda*. (Figure modified from Bond and Diamond 2010; used with permission of Canadian Science Publishing)

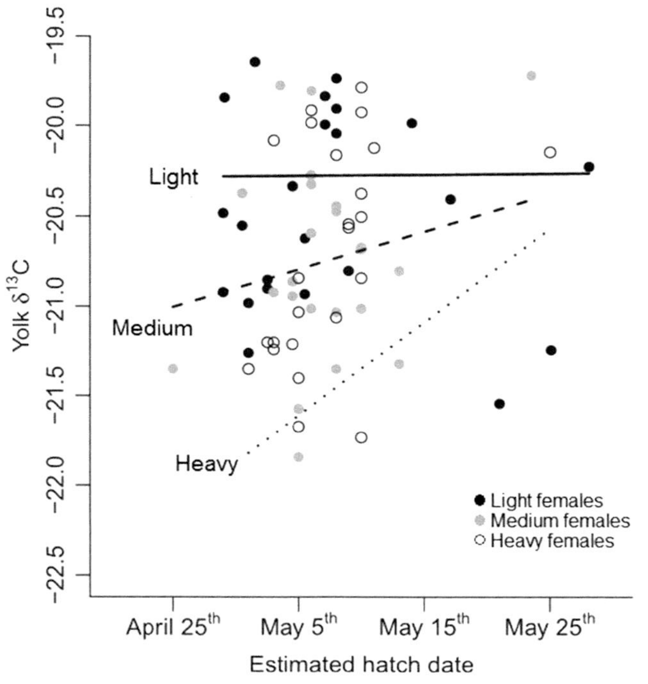


**Fig. 16.127** Relationship between yolk $\delta^{13}C$ and time of breeding (estimated hatch date) by female Common Eiders (*Somateria mollissima*). Higher yolk $\delta^{13}C$ values indicate use of endogenous (stored) nutrients to produce eggs, with increasingly lower values indicating greater use of exogenous (local) nutrients. (Figure modified from Jaatinen et al. 2016; open-access article distributed under the terms of the Creative Commons Attribution 4.0 International License, https://creativecommons.org/licenses/by/4.0/)

## References


Adam MD, des Lauriers JR (1998) Observations of hummingbirds ingesting mineral-rich compounds. J Field Ornithol 69:257–261

Aguilera E, Amat JA (2007) Carotenoids, immune response and the expression of sexual ornaments in male Greenfinches (*Carduelis chloris*). Naturwissenschaften 94:895–902

Ahammad MU, Miyazato T, Nishino C, Tatemoto H, Okura N, Okamoto S, Kawamoto Y, Nakada T (2013) Effects of fluid secreted from the uterus on duration of fertile egg production in hens, and survivability and penetrability of fowl sperm in vitro. J Poult Sci 50:74–82

Aire TA (2007) Spermatogenesis and testicular cycles. In: Jamieson BGM (ed) Reproductive biology and phylogeny of birds: Part A. Science Publishers, Enfield, NH, pp 279–347

Alatalo RV, Hoglund J, Lundberg A, Rintamaki PT, Silverin B (1996) Testosterone and male mating success on Black Grouse leks. Proc R Soc B 263:1697–1702

Alisauskas RT, Ankney CD (1985) Nutrient reserves and the energetics of reproduction in American Coots. Auk 102:133–144

Alisauskas RT, Ankney CD (1992) The cost of egg laying and its relationship to nutrient reserves in waterfowl. In: Batt BDJ, Afton AD, Anderson MG, Ankney CD, Johnson DH, Kadlec JA, Krapu GL (eds) Ecology and management of breeding waterfowl. University of Minnesota Press, Minneapolis, MN, pp 30–61

Alisauskas RT, Ankney CD (1994) Costs and rates of egg formation in Ruddy Ducks. Condor 96:11–18

Alonso-Alvarez C, Velando A (2003) Female body condition and brood sex ratio in Yellow-legged Gulls *Larus cachinnans*. Ibis 145:220–226

Anderson SJ, Bird DM, Hagen MD (2002) Semen characteristics of the Quaker Parakeet (*Myiopsitta monachus*). Zoo Biol 21:507–512

Andersson M (1978) Optimal egg shape in waders. Ornis Fennica 55:105–109

Apperson KD, Bird KE, Cherian G, Löhr CV (2017) Histology of the ovary of the laying hen (*Gallus domesticus*). Vet Sci 4:66

Ar A, Rahn H (1985) Pores in avian eggshells: gas conductance, gas exchange and embryonic growth rate. Respir Physiol 61:1–20

Assersohn K, Brekke P, Hemmings N (2021) Physiological factors influencing female fertility in birds. R Soc Open Sci 8:202274

Astheimer LB, Grau CR (1985) Apparent double blastoderms in Adelie Penguin eggs. Condor 87:150–151

Attard MRG, Portugal SJ (2022) Global diversity and adaptations of avian eggshell thickness indices. Ibis. https://doi.org/10.1111/ibi.13136

Avila FW, Sirot LK, LaFlamme BA, Rubinstein CD, Wolfner MF (2011) Insect seminal fluid proteins: identification and function. Annu Rev Entomol 56:21–40

Avilés JM, Soler JJ, Pérez-Contreras T (2006) Dark nests and egg colour in birds: a possible functional role of ultraviolet reflectance in egg detectability. Proc R Soc B 273:2821–2829

Badyaev AV, Seaman DA, Navara KJ, Hill GE, Mendonca MT (2006) Evolution of sex-biased maternal effects in birds. III. Adjustment of ovulation order can enable sex-specific allocation of hormones, carotenoids, and vitamins. J Evol Biol 19:1044–1057

Bahr JM, Wang SC, Huang MY, Calvo FO (1983) Steroid concentrations in isolated theca and granulosa layers of preovulatory follicles during the ovulatory cycle of the domestic hen. Biol Reprod 29:326–334

Baicich PJ, Harrison CJO (1997) A guide to the nests, eggs, and nestlings of North American birds, 2nd edn. Academic Press, San Diego, CA

Bakst MR (1981) Sperm recovery from oviducts of turkeys at know intervals after insemination and oviposition. J Reprod Fertil 62:159–164

Bakst MR (2011) Physiology and endocrinology symposium: role of the oviduct in maintaining sustained fertility in hens. J Anim Sci 89:1323–1329

Bakst MR, Wishart GJ, Brillard JP (1994) Oviducal sperm selection, transport, and storage in poultry. Poult Sci Rev 5:117–143

Bancroft GT (1985) Nutrient content of eggs and the energetics of clutch formation in the Boat-tailed Grackle. Auk 102:43–48

Barrett RT, Nilsen EB, Anker-Nilssen T (2012) Long-term decline in egg size of Atlantic Puffins *Fratercula arctica* is related to changes in forage fish stocks and climate conditions. Mar Ecol Prog Ser 457:1–10

Barta Z, Székely T (1997) The optimal shape of avian eggs. Funct Ecol 11:656–662

Bauchinger U, Van't Hof T, Biebach H (2007) Testicular development during long-distance spring migration. Horm Behav 51:295–305

Béaupre CE, Tressler CJ, Béaupre SJ, Morgan JLM, Bottje WG, Kirby JD (1997) Determination of testis temperature rhythms and effects of constant light on testicular function in the domestic fowl (*Gallus domesticus*). Biol Reprod 56:1570–1575

Beintema AJ, Baarspul T, de Krijger JP (1997) Calcium deficiency in Black Terns *Chlidonias niger* nesting on acid bogs. Ibis 139:396–397

Beletsky LD, Wingfield JC, Orians GH (1989) Relationships of hormones and polygyny to territorial status, breeding experience and reproductive success in male Red-winged Blackbirds. Auk 106:107–117

Ben-David M, Flaherty EA (2012) Stable isotopes in mammalian research: a beginner's guide. J Mammal 93:312–328

Beniash E, Aizenberg J, Addadi L, Weiner S (1997) Amorphous calcium carbonate transforms into calcite during sea urchin larval spicule growth. Proc R Soc B 264:461–465

Benvenuti B, Walsh J, O'Brien KM, Ducey MJ, Kovach AI (2018) Annual variation in the offspring sex ratio of Saltmarsh Sparrows supports Fisher's hypothesis. Auk 135:342–358

Bird DM, Laguë PC (1977) Semen production of the American Kestrel. Can J Zool 55:1351–1358

Birkhead TR (1991) Sperm depletion in the Bengalese Finch, *Lonchura striata*. Behav Ecol 2:267–275

Birkhead TR, Fletcher F (1995) Male phenotype and ejaculate quality in the Zebra Finch *Taeniopygia guttata*. Proc R Soc B 262:329–334

Birkhead TR, Immler S (2007) Making sperm: design, quality control and sperm competition. In: Roldan E, Gomendio M (eds) Spermatology. Nottingham University Press, Nottingham, pp 175–181

Birkhead TR, Lessells CM (1988) Copulation behaviour of the Osprey *Pandion haliaetus*. Anim Behav 36:1672–1682

Birkhead TR, Møller AP (1992) Sperm competition in birds: evolutionary causes and consequences. Academic Press, New York, NY

Birkhead TR, Petrie M (1995) Ejaculate features and sperm utilization in peafowl *Pavo cristatus*. Proc R Soc B 261:153–158

Birkhead TR, Pellatt EJ, Fletcher F (1993) Selection and utilization of spermatozoa in the reproductive tract of the female Zebra Finch *Taeniopygia guttata*. J Reprod Fertil 99:593–600

Birkhead TR, Veiga JP, Møller AP (1994) Male sperm reserves and copulation behaviour in the House Sparrow, *Passer domesticus*. Proc R Soc B 256:247–251

Birkhead TR, Fletcher F, Pellatt EJ, Staples A (1995) Ejaculate quality and the success of extra-pair copulations in the Zebra Finch. Nature 377:422–423

Birkhead TR, Pellatt EJ, Brekke P, Yeates R, Castillo-Juarez H (2005) Genetic effects on sperm design in the Zebra Finch. Nature 434:383–387

Birkhead TR, Giusti F, Immler S, Jamieson BGM (2007) Ultrastructure of the unusual spermatozoon of the Eurasian Bullfinch (*Pyrrhula pyrrhula*). Acta Zool 88: 119–128

Birkhead TR, Thompson JE, Jackson D, Biggins JD (2017) The point of a guillemot's egg. Ibis 159:255–265

Birkhead TR, Thompson JE, Montgomerie R (2018) The pyriform egg of the Common Murre (*Uria aalge*) is more stable on sloping surfaces. Auk 135:1020–1032

Blackburn DG, Stewart JR (2021) Morphological research on amniote eggs and embryos: an introduction and historical retrospective. J Morphol 282:1024–1046

Blas J, Perez-Rodriguez L, Bortolotti GR, Vinuela J, Marchant TA (2006) Testosterone increases bioavailability of carotenoids: insights into the honesty of sexual signaling. Proc Natl Acad Sci USA 103: 18633–18637

Blomqvist D, Johansson OC, Götmark F (1997) Parental quality and egg size affect chick survival in a precocial bird, the Lapwing *Vanellus vanellus*. Oecologia 110: 18–24

Blount JD (2004) Carotenoids and life-history evolution in animals. Arch Biochem Biophys 430:10–15

Blount JD, Houston DC, Møller AP (2000) Why egg yolk is yellow. Trends Ecol Evol 15:47–49

Board RG, Halls NA (1973) The cuticle: a barrier to liquid and particle penetration of the shell of the hen's egg. Br Poult Sci 14:69–97

Bollinger PB, Bollinger EK, Daniel SL, Gonser RA, Tuttle EM (2018) Partial incubation during egg laying reduces eggshell microbial loads in a temperate-breeding passerine. J Avian Biol 49:e01560

Bolton M, Monaghan P, Houston DC (1993) Proximate determination of clutch size in Lesser Black-backed Gulls: the roles of food supply and body condition. Can J Zool 71:273–279

Bond AL, Diamond AW (2010) Nutrient allocation for egg production in six Atlantic seabirds. Can J Zool 88:1095–1102

Bond JC, Esler D, Hobson KA (2007) Isotopic evidence for sources of nutrients allocated to clutch formation by Harlequin Ducks. Condor 109:698–704

Bonior F, Martin PR, Wingfield JC (2007) Maternal corticosteroids influence primary offspring sex ratio in a free-ranging passerine bird. Behav Ecol 18: 1045–1050

Borziak K, Alvarex-Fernandez A, Karr TL, Pizzari T, Dorus S (2016) The seminal fluid proteome of the polyandrous Red Junglefowl offers insights into the molecular basis of fertility, reproductive ageing and domestication. Sci Rep 6:35864

Bradbury RB, Blakey JK (1998) Diet, maternal condition, and offspring sex ratio in the Zebra Finch, *Poephila guttata*. Proc R Soc B 265:895–899

Brennan PLR, Prum RO (2012) The erection mechanism of the ratite penis. J Zool 286:140–144

Brennan PL, Prum RO, McCracken KG, Sorenson MD, Wilson RE, Birkhead TR (2007) Coevolution of male and female genital morphology in waterfowl. PLoS One 2:e418

Brennan PLR, Clark CJ, Prum RO (2010) Explosive eversion and functional morphology of the duck penis supports sexual conflict in waterfowl genitalia. Proc R Soc B 277:1309–1314

Brenninkmeijer A, Klaassen M, Stienen EWM (1997) Sandwich Terns *Sterna sandvicensis* feeding on shell fractions. Ibis 139:397–400

Briggs KB, Mainwaring MC (2022) The experimental provision of calcium increases the reproductive success of passerine birds. Biol J Linn Soc 136:321–333

Briskie JV (1992) Copulation patterns and sperm competition in the polygynandrous Smith's Longspur. Auk 109:563–575

Briskie JV, Montgomerie R (1997) Sexual selection and the intromittent organ of birds. J Avian Biol 28:73–86

Briskie JV, Montgomerie R (2007) Testis size, sperm size and sperm competition. In: Jamieson BGM (ed) Reproductive biology and phylogeny of birds: Part A. Science Publishers, Enfield, NH, pp 513–551

Brown CR, Brown MB (2003) Testis size increases with colony size in Cliff Swallows. Behav Ecol 14:569–575

Brown ME, Converse SJ, Chandler JN, Crosier AL, Lynch W, Wildt DE, Keefer CL, Songsasen N (2017) Time within reproductive season, but not age or inbreeding coefficient, affects seminal and sperm quality in the Whooping Crane (*Grus americana*). Reprod Fertil Dev 29:294–306

Brunton DH (1988) Energy expenditure in reproductive effort of male and female Killdeer (*Charadrius vociferus*). Auk 105:553–564

Buechler K, Fitze PS, Gottstein B, Jacot A, Richner H (2002) Parasite-induced maternal response in a natural bird population. J Anim Ecol 71:247–252

Burke WH, Ogasawara FX (1969) Presence of spermatozoa in utero-vaginal fluids of the hen at various stages of the ovulatory cycle. Poult Sci 48:408–413

Burley RW, Vadhera DV (1989) The avian egg. John Wiley, New York, NY

Calder WA III (1979) The kiwi and egg design: evolution as a package deal. Bioscience 29:461–467

Calhim S, Montgomerie R (2015) Testis asymmetry in birds: the influences of sexual and natural selection. J Avian Biol 46:175–185

Calhim S, Immler S, Birkhead TR (2007) Postcopulatory sexual selection is associated with reduced variation in sperm morphology. PLoS One 5:e413

Calhim S, Pruett-Jones S, Webster MS, Rowe M (2019) Asymmetries in reproductive anatomy: insights from promiscuous songbirds. Biol J Linn Soc 128:569–582

Cameron EZ, Linklater WL (2002) Sex bias in studies of sex bias: the value of daughters to mothers in poor condition. Anim Behav 63:F5–F8

Carey C (1996) Female reproductive energetics. In: Carey C (ed) Avian energetics and nutritional ecology. Chapman & Hall, New York, NY, pp 324–374

Carey G, Rahn H, Parisi P (1980) Calories, water, lipid, and yolk in avian eggs. Condor 82:335–343

Caro SP, Schaper SV, Hut RA, Ball GF, Visser ME (2013) The case of the missing mechanism: how does temperature influence seasonal timing in endotherms? PLoS Biol 11:e1001517

Cassey P, Maurer G, Lovell PG, Hanley D (2011) Conspicuous eggs and colourful hypotheses: testing the role of multiple influences on avian eggshell appearance. Avian Biol Res 4:185–195

Cavalcanti LMP, de Paiva LV, França LF (2016) Effects of rainfall on bird reproduction in a semi-arid Neotropical region. Zoologia (Curitiba) 33:e20160018

Chappell MA, Bech C, Buttemer WA (1999) The relationship of central and peripheral organ masses to aerobic performance variation in House Sparrows. J Exp Biol 202:2269–2279

Chiba A, Nakamura M (2003) Anatomical and histophysiological characterization of the male cloacal protuberance of the polygynandrous Alpine Accentor *Prunella collaris*. Ibis 145:E83–E93

Chmura HE, Wingfield JC, Hahn TP (2020) Non-photic environmental cues and avian reproduction in an era of global change. J Avian Biol 51:jav.02243

Chousalkar KK, Roberts JR (2008) Ultrastructural changes in the oviduct of the laying hen during the laying cycle. Cell Tissue Res 332:349–358

Christians JK (2002) Avian egg size: variation within species and inflexibility within individuals. Biol Rev 77:1–26

Cindrova-Davies T, Jauniaux E, Elliot MG, Gong S, Burton GJ, Charnock-Jones DS (2017) RNA-seq reveals conservation of function among the yolk sacs of human, mouse, and chicken. Proc Natl Acad Sci USA 114:E4753–E4761

Clulow J, Jones RC (1982) Production, transport, maturation, storage and survival of spermatozoa in the male Japanese Quail *Coturnix coturnix japonica*. J Reprod Fertil 64:259–266

Clutton-Brock TH (1991) The evolution of parental care. Princeton University Press, Princeton, NJ

Cohen J (1977) Reproduction. Butterworths, London

Coker CR, McKinney F, Hays H, Briggs SV, Cheng KM (2002) Intromittent organ morphology and testis size in relation to mating system in waterfowl. Auk 119:403–413

Cook MI, Beissinger SR, Toranzos GA, Rodriguez RA, Arendt WJ (2004) Microbial infection affects egg viability and incubation behavior in a tropical passerine. Behav Ecol 16:30–36

Coppack T, Pulido F (2004) Photoperiodic response and the adaptability of avian life cycles to environmental change. Adv Ecol Res 35:131–150

Cordero PJ, Griffith SC, Aparicio JM, Parkin DT (2000) Sexual dimorphism in House Sparrow eggs. Behav Ecol Sociobiol 48:353–357

Craig JL (1977) The behaviour of the Pukeko, *Porphyrio porphyrio melanotus*. N Z J Zool 4:413–433

D'Alba L, Oborn A, Shawkey M (2010) Experimental evidence that keeping eggs dry is a mechanism for the antimicrobial effects of avian incubation. Naturwissenschaften 97:1–7

D'Alba L, Jones DN, Badawy HT, Eliason CM, Shawkey MD (2014) Antimicrobial properties of a nanostructured eggshell from a compost-nesting bird. J Exp Biol 217:1116–1121

D'Alba L, Maia R, Hauber ME, Shawkey MD (2016) The evolution of eggshell cuticle in relation to nesting ecology. Proc R Soc B 283:20160687

D'Alba L, Torres R, Waterhouse GIN, Eliason C, Hauber ME, Shawkey MD (2017) What does the eggshell cuticle do? A functional comparison of avian eggshell cuticles. Physiol Biochem Zool 90:588–599

Dacke CG, Arkle S, Cook DJ, Wormstone IM, Jones S, Zaidi M, Bascal ZA (1993) Medullary bone and avian calcium regulation. J Exp Biol 184:63–88

Dang HR, Guraya SS (1978) Testis growth and regression in harmful birds in the Punjab in relation to some environmental factors. Aust J Zool 26:39–45

Das SC, Isobe N, Nishibori M, Yoshimura Y (2006) Expression of TGFβ-isoforms of hen oviduct in presence or absence of resident sperm with reference to sperm storage. Reproduction 132:781–790

Das SC, Isobe N, Yoshimura Y (2008) Mechanism of prolonged sperm storage and sperm survivability in hen oviduct: a review. Am J Reprod Immunol 60: 477–481

Dawson A (2003) A comparison of the annual cycles in testicular size and moult in captive European Starlings *Sturnus vulgaris* during their first and second years. J Avian Biol 34:119–123

Dawson A, King VM, Bentley GE, Ball GF (2001) Photoperiodic control of seasonality in birds. J Biol Rhythm 16:365–380

Day LB, Fusani L, Hernandez E, Billo TJ, Sheldon KS, Wise PM, Schlinger BA (2007) Testosterone and its effects on courtship in Golden-collared Manakins (*Manacus vitellinus*): seasonal, sex, and age differences. Horm Behav 51:69–76

de Matos R (2008) Calcium metabolism in birds. Vet Clin N Am Exot Anim Pract 11:59–82

de Oliveira CA, Bohorquez GA (2000) Morphology of the copulatory apparatus of the spotted Tinamou *Nuthura maculosa* (Aves: Tinamiformes). Ann Anat 182:161–169

de Reviers M, Williams JB (1981) Predicting the adult daily sperm output after the first ejaculates in cockerels raised under different photoschedules. Reprod Nutr Dev 21:1113–1124

Deeming DC (1987) Effect of cuticle removal on the water vapour conductance of egg shells of several species of domestic bird. Br Poult Sci 28:231–237

Deeming DC (2007) Allometry of mass and composition in bird eggs: effects of phylogeny and hatchling maturity. Avian Biol Res 18:71–86

del Hoyo J, Elliott A, Sargatal J (1992) Handbook of the Birds of the World, vol 1. Lynx Edicions, Barcelona

Dickison MR (2007) The allometry of giant flightless birds. Ph.D. dissertation,. Duke University, Durham, NC

Dijkstra C, Riedstra B, Dekker A, Goerlich VC, Daan S, Groothuis TGG (2010) An adaptive annual rhythm in the sex of first pigeon eggs. Behav Ecol Sociobiol 64: 1393–1402

Ding Z, Ji F, Huang Q, Wang L, Jiang A, Zhang C, Feng Y, Tian Y, Hu H, Liang W (2017) Brood sex ratio in the Yellow-bellied Prinia (*Prinia flaviventris*). Avian Res 8:15

Dittami JP, Gwinner E (1985) Annual cycles in the African Stonechat *Saxicola torquata axillaris* and their relationship to environmental factors. J Zool 207: 357–370

Dogliero A, Rota A, Lofiego R, von Degerfeld MM, Quaranta G (2016) Semen collection and spermatozoa characteristics in the Kea Parrot (*Nestor notabilis*). J Avian Med Surg 31:24–28

Dombre C, Guyot N, Moreau T, Monget P, Da Silva M, Gautron J, Réhault-Godbert S (2017) Egg serpins: the chicken and/or the egg dilemma. Semin Cell Dev Biol 62:120–132

Dorward DF (1962) Comparative biology of the White Booby and Brown Booby *Sula* spp. at Ascension. Ibis 103:174–220

Drent RH, Daan S (1980) The prudent parent: energetic adjustments in avian breeding. Ardea 55:225–252

Drobney RD (1980) Reproductive bioenergetics of Wood Ducks. Auk 97:480–490

Durant JM, Massemin S, Thouzeau C, Handrich Y (2000) Body reserves and nutritional needs during laying preparation in Barn Owls. J Comp Physiol B 170: 253–260

Eberhard WG (1996) Female control: sexual selection by cryptic female choice. Princeton University Press, Princeton, NJ

Ellegren H, Gustafsson L, Sheldon BC (1996) Sex ratio adjustment in relation to paternal attractiveness in a wild bird population. Proc Natl Acad Sci USA 93: 11723–11728

Ellis LA, Borst DW, Thompson CF (2001) Hatching asynchrony and maternal androgens in egg yolks of House Wrens. J Avian Biol 32:26–30

Encabo SI, Barba E, Gil-Delgado JA, Monrós JS (2001) Fitness consequences of egg shape variation: a study on two passerines and comments on the optimal egg shape model. Ornis Fennica 78:83–92

Esler D, Grand JB, Afton AD (2001) Intraspecific variation in nutrient reserve use during clutch formation by Lesser Scaup. Condor 103:810–820

Fecheyr-Lippens DC, Igic B, D'Alba L, Hanley D, Verdes A, Holford M, Waterhouse GI, Grim T, Hauber ME, Shawkey MD (2015) The cuticle modulates ultraviolet reflectance of avian eggshells. Biology Open 4: 753–759

Fedy BC, Stutchbury BJM (2006) Testosterone does not increase in response to conspecific challenges in the White-bellied Antbird (*Myrmeciza longipes*), a resident tropical passerine. Auk 123:61–66

Finn RN (2007) Vertebrate yolk complexes and the functional implications of phosvitins and other subdomains in vitellogenins. Biol Reprod 76:926–935

Finseth FR, Iacovelli SR, Harrison RG, Adkins-Regan EK (2013) A nonsemen copulatory fluid influences the outcome of sperm competition in Japanese Quail. J Evol Biol 26:1875–1889

Fischer D, Neumann D, Purchase C, Bouts T, Meinecke-Tillmann S, Wehrend A, Lierz M (2014) The use of semen evaluation and assisted reproduction in Spix's Macaws in terms of species conservation. Zoo Biol 33: 234–244

Foster MS (2014) Can fruit pulp meet the calcium needs of tropical frugivorous passerines during reproduction? J Trop Ecol 20:79–88

Freedman SL, Akuffo VG, Bakst MR (2001) Evidence for the innervation of sperm storage tubules in the oviduct of the turkey (*Meleagris gallopavo*). Reproduction 121:809–814

Fujihara N (1992) Accessory reproductive fluids and organs in male domestic birds. World Poult Sci J 48: 39–56

Galbraith H (1988) Effects of egg size and composition on the size, quality and survival of Lapwing *Vanellus vanellus* chicks. J Zool 214:383–398

Garamszegi LZ, Eens M, Hurtrez-Boussès S, Møller AP (2005) Testosterone, testes size, and mating success in birds: a comparative study. Horm Behav 47:389–409

Gauthier G, Bêty J, Hobson KA (2003) Are Greater Snow Geese capital breeders? New evidence from a stable-isotope model. Ecology 84:3250–3264

Gil D, Graves JA, Hazon N, Wells A (1999) Male attractiveness and differential testosterone investment in Zebra Finch eggs. Science 286:126–128

Gil D, Ninni P, Lacroix P, de Lope F, Tirard C, Marzal A, Møller AP (2006) Yolk androgens in the Barn Swallow (*Hirundo rustica*): a test of some adaptive hypotheses. J Evol Biol 19:123–131

Gilbert AB (1979) Female genital organs. In: King AS, McLelland J (eds) Form and function in birds, vol 1. Academic Press, New York, NY, pp 237–360

Gilbert L, Williamson KA, Hazon N, Graves JA (2006) Maternal effects due to male attractiveness affect offspring development in the Zebra Finch. Proc R Soc B 273:1765–1771

Gill VA, Hatch SA (2002) Components of productivity in Black-legged Kittiwakes *Rissa tridactyla*: response to supplemental feeding. J Avian Biol 33:113–126

Goerlich VC, Dijkstra C, Schaafsma SM, Groothuis TGG (2009) Testosterone has a long-term effect on primary sex ratio of first eggs in pigeons—in search of a mechanism. Gen Comp Endocrinol 163:184–192

Górski A, Nowakowski JJ, Bańbura J (2015) Variation in egg-size traits of the European Roller (*Coracias garrulus*) in eastern Poland. Ornis Fennica 92:213–220

Gosler AG, Higham JP, Reynolds SJ (2005) Why are birds' eggs speckled? Ecol Lett 8:1105–1113

Goymann W, Moore IT, Scheuerlein A, Hirschenhauser K, Grafen A, Wingfield JC (2004) Testosterone in tropical birds: effects of environmental and social factors. Am Nat 164:327–334

Graveland J, Berends AE (1997) Timing of the calcium intake and effect of calcium deficiency on behaviour and egg laying in captive Great Tits, *Parus major*. Physiol Zool 70:74–84

Graveland J, van Gijzen T (1994) Arthropods and seeds are not sufficient as calcium sources for shell formation and skeletal growth in passerines. Ardea 82:299–314

Greenman CG, Martin LB, Hau M (2005) Reproductive state, but not testosterone, reduces immune function in male House Sparrows (*Passer domesticus*). Physiol Biochem Zool 78:60–68

Groothuis TGG, Müller W, von Engelhardt N, Carere C, Eising C (2005) Maternal hormones as a tool to adjust offspring phenotype in avian species. Neurosci Biobehav Rev 29:329–352

Gupta SK, Maiti BR (1987) Seasonal changes in the oviduct of the Pied Myna (Aves: Sturnidae). J Morphol 194:247–263

Gwinner E (1973) Cirannual rhythms in birds: their interaction with circadian rhythms and environmental photoperiod. J Reprod Fertil 19:51–65
Gwinner E (1986) Circannual clocks. Springer-Verlag, Berlin
Gwinner E (2003) Circannual rhythms in birds. Curr Opin Neurobiol 13:770–778
Gwinner E, Scheuerlein A (1998) Seasonal changes in day-light intensity as a potential zeitgeber of circannual rhythms in equatorial Stonechats. J Ornithol 139:407–412
Gwinner H, Gwinner E, Dittami J (1987) Effects of nestboxes on LH, testosterone, testicular size and the reproductive behavior of male European Starlings in spring. Behaviour 103:68–82
Haftorn S (1996) Egg-laying behavior in tits. Condor 98: 863–865
Hahn S, Loonen MJ, Klaassen M (2011) The reliance on distant resources for egg formation in High Arctic breeding Barnacle Geese *Branta leucopsis*. J Avian Biol 42:159–168
Hanley D, Cassey P, Doucet SM (2013) Parents, predators, parasites, and the evolution of eggshell colour in open nesting birds. Evol Ecol 27:593–617
Hargitai R, Török J, Tóth L, Hegyi G, Rosivall B, Szigeti B, Szöllősi E (2005) Effects of environmental conditions and parental quality on inter- and intraclutch egg-size variation in the Collared Flycatcher (*Ficedula albicollis*). Auk 122:509–522
Harris MP, Birkhead TR (1985) Breeding ecology of the Atlantic Alcidae. In: Nettleship DN, Birkhead TR (eds) The Atlantic Alcidae. Academic Press, London, pp 155–205
Hasselquist D, Marsh JA, Sherman PW, Wingfield JC (1999) Is avian humoral immunocompetence suppressed by testosterone? Behav Ecol Sociobiol 45: 167–175
Hau M (2001) Timing of breeding in variable environments: tropical birds as model systems. Horm Behav 40:281–290
Hau M (2007) Regulation of male traits by testosterone: implications for the evolution of vertebrate life histories. BioEssays 29:133–144
Hau M, Wikelski M, Wingfield JC (1998) A Neotropical forest bird can measure the slight changes in tropical photoperiod. Proc R Soc B 265:89–95
Hau M, Wikelski M, Soma KK, Wingfield JC (2000a) Testosterone and year-round territorial aggression in a tropical bird. Gen Comp Endocrinol 117:20–33
Hau M, Wikelski M, Wingfield JC (2000b) Visual and nutritional food cues fine-tune timing of reproduction in a Neotropical rainforest bird. J Exp Zool 286:494–504
Heenan CB (2013) The structural and thermal properties of avian cup-shaped nests. Ph.D. dissertation, University of Adelaide, Adelaide, Australia.
Helm B, Piersma T, Van der Jeugd H (2006) Sociable schedules: interplay between avian seasonal and social behaviour. Anim Behav 72:245–262
Heming NM, Marini MÂ (2015) Ecological and environmental factors related to variation in egg size of New World flycatchers. J Avian Biol 46:352–360
Hemmings N, Birkhead TR (2015) Polyspermy in birds: sperm numbers and embryo survival. Proc R Soc B 282:20151682
Hemmings N, Birkhead TR (2020) Extraordinary sperm to egg ratios in seabirds. Auk 137:ukaa052
Henderson CL (2007) Oology and Ralph's talking eggs. University of Texas Press, Austin, TX
Henderson LJ, Evans NP, Heidinger BJ, Adam A, Arnold KE (2014) Maternal condition but not corticosterone is linked to offspring sex ratio in a passerine bird. PLoS One 9:e110858
Hernandez AG, Bahr JM (2003) Role of FSH and epidermal growth factor (EGF) in the initiation of steroidogenesis in granulosa cells associated with follicular selection in chicken ovaries. Reproduction 125:683–691
Herrera AM, Shuster SG, Perriton CL, Cohn MJ (2013) Developmental basis of phallus reduction during bird evolution. Curr Biol 23:1065–1074
Hincke MT, Gautron J, Panheleux M, Garcia-Ruiz J, McKee MD, Nys Y (2000) Identification and localization of lysozyme as a component of eggshell membranes and eggshell matrix. Matrix Biol 19:443–453
Hincke MT, Da Silva M, Guyot N, Gautron J, McKee MD, Guabiraba-Brito R, Réhault-Godbert S (2019) Dynamics of structural barriers and innate immune components during incubation of the avian egg: critical interplay between autonomous embryonic development and maternal anticipation. J Innate Immun 11: 111–124
Hobson KA (2006) Using stable isotopes to quantitatively track endogenous and exogenous nutrient allocations to eggs of birds that travel to breed. Ardea 94:359–369
Hobson KA, Kardynal KJ (2016) An isotope (δ34S) filter and geolocator results constrain a dual feather isoscape (δ2H, δ13C) to identify the wintering grounds of North American Barn Swallows. Auk 133:86–98
Hobson KA, Hughes KD, Ewins PJ (1997) Using stable-isotope analysis to identify endogenous and exogenous sources of nutrients in eggs of migratory birds: applications to Great Lakes contaminants research. Auk 114:467–478
Hobson KA, Sirois J, Gloutney ML (2000) Tracing nutrient allocations to reproduction using stable-isotopes: a preliminary investigation using the colonial waterbirds of Great Slave Lake. Auk 117:760–774
Hobson KA, Atwell L, Wassenaar LI, Yerkes T (2004) Estimating endogenous nutrient allocations to reproduction in Redhead ducks: a dual isotope approach using δD and δ13C measurements of female and egg tissues. Funct Ecol 18:737–745
Hobson KA, Jaatinen K, Öst M (2015) Differential contributions of endogenous and exogenous nutrients to egg components in wild Baltic Common Eiders

(*Somateria mollissima*): a test of alternative stable isotope approaches. Auk 132:624–633

Hoi-Leitner M, Romero-Pujante M, Hoi H, Pavlova A (2001) Food availability and immune capacity in Serin (*Serinus serinus*) nestlings. Behav Ecol Sociobiol 49:333–339

Holt WV, Fazeli A (2016) Sperm storage in the female reproductive tract. Annu Rev Anim Biosci 4:291–310

Hop H, Gjøsæter H (2013) Polar cod (*Boreogadus saida*) and capelin (*Mallotus villosus*) as key species in marine food webs of the Arctic and the Barents Sea. Mar Biol Res 9:878–894

Hõrak P, Mänd R, Ots I, Leivits A (1995) Egg size in the Great Tit *Parus major*: individual, habitat and geographic differences. Ornis Fennica 72:97–114

Houston DC (1978) The effect of food quality on breeding strategy in Griffon Vultures (*Gyps* spp.). J Zool 186:175–184

Hunter FM, Burke T, Watts SE (1992) Frequent copulation as a method of paternity assurance in the Northern Fulmar. Anim Behav 44:149–156

Hutchinson JMC (2000) Three into two doesn't go: two-dimensional models of bird eggs, snail shells and plant roots. Biol J Linn Soc 70:161–187

Ichikawa Y, Matsuzaki M, Hiyama G, Mizushima S, Sasanami T (2016) Sperm-egg interaction during fertilization in birds. J Poult Sci 53:173–180

Igic B, Fecheyr-Lippens D, Xiao M, Chan A, Hanley D, Brennan PRL, Grim T, Waterhouse GIN, Hauber ME, Shawkey MD (2015a) A nanostructural basis for gloss of avian eggshells. J R Soc Interface 12:20141210

Igic B, Zarate E, Sewell MA, Moskát C, Cassey P, Rutila J, Grim T, Shawkey MD, Hauber ME (2015b) A comparison of egg yolk lipid constituents between parasitic Common Cuckoos and their hosts. Auk 132: 817–825

Inger R, Bearhop S (2008) Applications of stable isotope analyses to avian ecology. Ibis 150:447–461

Isaksson C, Magrath MJL, Groothuis TGG, Komdeur J (2010) Androgens during development in a bird species with extremely sexually dimorphic growth, the Brown Songlark, *Cinclorhamphus cruralis*. Gen Comp Endocrinol 165:97–103

Ito T, Yoshizaki N, Tokumoto T, Ono H, Yoshimura T, Tsukada A, Kansaku N, Sasanami T (2011) Progesterone is a sperm-releasing factor from the sperm-storage tubules in birds. Endocrinology 152:3952–3962

Jaatinen K, Öst M, Hobson KA (2016) State-dependent capital and income breeding. Front Zool 13:24

Jacob M, Bakst MR (2007) Developmental anatomy of the female reproductive tract. In: Jamieson BGM (ed) Reproductive biology and phylogeny of birds: phylogeny, morphology, hormones, and fertilization. Science Publishers, Enfield, NH, pp 149–179

Jacobs SR, Elliott KH, Gaston AJ, Weber J-M (2009) Fatty acid signatures of female Brünnich's Guillemots *Uria lomvia* suggests reliance on local prey for replacement egg production. J Avian Biol 40:327–336

Jamieson BGM (2007) Avian spermatozoa: structure and phylogeny. In: Jamieson BGM (ed) Reproductive biology and phylogeny of birds: phylogeny, morphology, hormones, and fertilization. Science Publishers, Enfield, NH, pp 349–511

Jamieson BGM, Koehler L, Todd BJ (1995) Spermatozoal ultrastructure in three species of parrots (Aves, Psittaciformes) and its phylogenetic implications. Anat Rec 241:461–468

Jetz W, Sekercioglu CH, Böhning-Gaese K (2008) The worldwide variation in avian clutch size across species and space. PLoS Biol 6:e303

Johnson AL (2000) Reproduction in the female. In: Whittow GC (ed) Sturkie's avian physiology, 5th edn. Academic Press, New York, NY, pp 569–596

Johnson AL, Woods DC (2007) Ovarian dynamics and follicle development. In: Jamieson BGM (ed) Reproductive biology and phylogeny of birds: phylogeny, morphology, hormones, and fertilization. Science Publishers, Enfield, NH, pp 243–277

Johnson LS, Leyhe JE, Werner C (2001) The shape of eggs in different-sized clutches of the House Wren (*Troglodytes aedon*). Can J Zool 79:1527–1531

Juang JY, Chen PY, Yang DC, Wu SP, Yen A, Hsieh HI (2017) The avian egg exhibits general allometric invariances in mechanical design. Sci Rep 7:14205

Jubinville I, Williams TD, Trathan PN, Crossin GT (2020) Trade-off between aerobic performance and egg production in migratory Macaroni Penguins. Comp Biochem Physiol A 247:110742

Kale HW (1965) Ecology and bioenergetics of the Long-billed Marsh Wren in Georgia salt marshes. Publ Nuttall Ornithol Club 5:1–142

Kemp AC (1995) The hornbills. Oxford University Press, New York, NY

Kendeigh SC, Kramer TC, Hamerstrom F (1956) Variations in egg characteristics of the House Wren. Auk 73:42–65

Kerschnitzki M, Zander T, Zaslansky P, Fratzl P, Shahar R, Wagermaier W (2014) Rapid alterations of avian medullary bone material during the daily egg-laying cycle. Bone 69:109–117

Ketta M, Tůmová E (2016) Eggshell structure, measurements, and quality-affecting factors in laying hens: a review. Czech J Anim Sci 61:299–309

Ketterson ED, Nolan V Jr (1999) Adaptation, exaptation, and constraint: a hormonal perspective. Am Nat 154: S4–S25

Ketterson ED, Nolan V Jr, Cawthorn MJ, Parker PG, Ziegenfus C (1996) Phenotypic engineering: using hormones to explore the mechanistic and functional bases of phenotypic variation in nature. Ibis 138:70–86

Kilner RM (2006) The evolution of egg colour and patterning in birds. Biol Rev 81:383–406

King AS (1981) Phallus. In: King AS, McLelland J (eds) Form and function in birds. Academic Press, London, UK, pp 107–147

Kingma SA, Hall ML, Peters A (2011) No evidence for offspring sex-ratio adjustment to social or

environmental conditions in cooperatively breeding Purple-crowned Fairy-wrens. Behav Ecol Sociobiol 65:1203–1213
Kinski FC (1971) The consistent presence of paired ovaries in the kiwi (*Apteryx*) with some discussion of this condition in other birds. J Ornithol 112:334–357
Kirby JD, Froman DP (2000) Reproduction in male birds. In: Whittow GC (ed) Sturkie's avian physiology, 5th edn. Academic Press, New York, NY, pp 597–615
Kirunda DFK, McKee SR (2000) Relating quality characteristics of aged eggs and fresh eggs to vitelline membrane strength as determined by a texture analyzer. Poult Sci 79:1189–1193
Klaassen M (2003) Relationships between migration and breeding strategies in Arctic breeding birds. In: Berthold P, Gwinner E, Sonnenschein E (eds) Avian migration. Springer-Verlag, Berlin, pp 237–250
Klaassen M, Lindström Å, Meltofte H, Piersma T (2001) Arctic waders are not capital breeders. Nature 413:794
Klaassen M, Hahn S, Korthals H, Madsen J (2017) Eggs brought in from afar: Svalbard-breeding Pink-footed Geese can fly their eggs across the Barents Sea. J Avian Biol 48:173–179
Klasing KC (1998) Nutritional modulation of resistance to infectious diseases. Poult Sci 77:1119–1125
Kleven O, Laskemoen T, Fossøy F, Robertson RJ, Lifjeld JT (2007) Intraspecific variation in sperm length is negatively related to sperm competition in passerine birds. Evolution 62:494–499
Kleven O, Fossøy F, Laskemoen T, Robertson RJ, Rudolfsen G, Lifjeld JT (2009) Comparative evidence for the evolution of sperm swimming speed by sperm competition and female sperm storage duration in passerine birds. Evolution 63:2466–2473
Kloskowski J, Trembaczowski A, Filipiuk M (2019) Stable isotope tracing of links between marine wintering and freshwater breeding habitats of Red-necked Grebes. J Ornithol 160:593–605
Knopf FL, Evans RM (2020) American White Pelican (*Pelecanus erythrorhynchos*), version 1.0. In: Poole AF (ed) Birds of the world. Cornell Lab of Ornithology, Ithaca, NY
Komdeur J, Daan S, Tinbergen J, Mateman C (1997) Extreme adaptive modification in sex ratio of the Seychelles Warbler's eggs. Nature 385:522–525
Krapu GL (1981) The role of nutrient reserves in Mallard reproduction. Auk 98:29–38
Krawczyk A, Jaworska-Adamu J (2019) Oocyte-sperm interactions in the fertilization of birds. Medycyna Weterynaryjna 75:276–279
Krementz DG, Ankney CD (1986) Bioenergetics of egg production by female House Sparrows. Auk 103:299–305
Krementz DG, Ankney CD (1995) Changes in total body calcium and diet of breeding House Sparrows. J Avian Biol 26:162–167
Kriegsfeld LJ, Ubuka T, Bentley GE, Tsutsui K (2015) Seasonal control of gonadotropin-inhibitory hormone (GnIH) in birds and mammals. Front Neuroendocrinol 37:65–75
Krist M (2011) Egg size and offspring quality: a meta-analysis in birds. Biol Rev 86:692–716
Krist M, Grim T (2007) Are blue eggs a sexually selected signal of female Collared Flycatchers? A cross-fostering experiment. Behav Ecol Sociobiol 61:863–876
Kuczynski EC, Paszkowski CA (2010) Food-web relations of the Horned Grebe (*Podiceps auritus*) on constructed ponds in the Peace Parkland, Canada. Wetlands 30:853–863
Kullberg C, Houston DC, Metcalfe NB (2002) Impaired flight ability—a cost of reproduction in female Blue Tits. Behav Ecol 13:575–579
Lack D (1954) The natural regulation of animal numbers. Oxford University Press, Oxford
Lack D (1958) The significance of the colour of turdine eggs. Ibis 100:145–166
Lahti DC (2005) Evolution of bird eggs in the absence of cuckoo parasitism. Proc Natl Acad Sci USA 102:18057–18062
Lahti DC (2006) Persistence of egg recognition in the absence of cuckoo brood parasitism: pattern and mechanism. Evolution 60:157–168
Lake PE (1981) Male genital organs. In: King AS, McLelland J (eds) Form and function in birds, vol 2. Academic Press, London, pp 1–61
Langin KM, Norris DR, Kyser TK, Marra PP, Ratcliffe LM (2006) Capital versus income breeding in a migratory passerine bird: evidence from stable-carbon isotopes. Can J Zool 84:947–953
Larson PL (1998) The theropod reproductive system. Gaia 15:389–397
Lattin CR, Breuner CW, Romero LM (2016) Does corticosterone regulate the onset of breeding in free-living birds?: the CORT-Flexibility Hypothesis and six potential mechanisms for priming corticosteroid function. Horm Behav 78:107–120
Lawson SL (2006) Comparative reproductive strategies between Long-tailed Ducks and King Eiders at Karrak Lake, Nunavut: use of energy sources during the nesting season. M. S. thesis, University of Saskatchewan, Saskatoon, Canada
Lee SJ, Witter MS, Cuthill IC, Goldsmith AR (1996) Reduction in escape performance as a cost of reproduction in gravid starlings, *Sturnus vulgaris*. Proc R Soc B 263:619–623
Lee JF, Durbian FE, Bell R, Voitz M (2008) Pied-billed Grebe nesting ecology on Squaw Creek National Wildlife Refuge, 1996–2007. Trans Missouri Acad Sci 42:23–29
Lehikoinen A, Ranta E, Pietiäinen H, Byholm P, Saurola P, Valkama J, Huitu O, Henttonen H, Korpimäki E (2011) The impact of climate and cyclic food abundance on the timing of breeding and brood size in four boreal owl species. Oecologia 165:349–355

Levin RN, Wingfield JC (1992) The hormonal control of territorial aggression in tropical birds. Ornis Scand 23: 284–291

Lincoln GA, Clarke IJ, Hut RA, Hazlerigg DG (2006) Characterizing a mammalian circannual pacemaker. Science 314:1941–1944

Liong JWW, Frank JF, Bailey S (1997) Visualization of eggshell membranes and their interaction with *Salmonella enteritidis* using confocal scanning laser microscopy. J Food Prot 60:1022–1028

Lipar JL, Ketterson ED (2000) Maternally derived yolk testosterone enhances the development of the hatching muscle in the Red-winged Blackbird *Agelaius phoeniceus*. Proc R Soc B 267:2005–2010

Lipar JL, Ketterson ED, Nolan V Jr (1999) Intra-clutch variation in testosterone contents of Red-winged Blackbird eggs. Auk 116:231–235

Lislevand T, Byrkjedal I, Borge T, Saetre GP (2005) Egg size in relation to sex of embryo, brood sex ratios and laying sequence in Northern Lapwings (*Vanellus vanellus*). J Zool 267:81–87

Liu H-K, Long DW, Bacon WL (2002) Interval between preovulatory surges of luteinizing hormone increases late in the reproductive period in turkey hens. Biol Reprod 66:1068–1075

López-Rull I, Celis P, Gil D (2007) Egg colour covaries with female expression of a male ornament in the Spotless Starling (*Sturnus unicolor*). Ethology 113: 926–933

Lüpold S, Calhim S, Immler S, Birkhead TR (2009) Sperm morphology and sperm velocity in passerine birds. Proc R Soc B 276:1175–1181

Lynn SE, Perfito N, Guardado D, Bentley GE (2015) Food, stress, and circulating testosterone: cue integration by the testes, not the brain, in male Zebra Finches (*Taeniopygia guttata*). Gen Comp Endocrinol 215:1–9

Macgillivray W (1852) A history of British birds, vol 5: cribratores, or sifters; urinatores, or divers, mersatores, or plungers. W S Orr, London

Maddox JD, Weatherhead PJ (2008) Egg size variation in birds with asynchronous hatching: is bigger really better? Am Nat 171:358–365

Mann K (2015) The calcified eggshell matrix proteome of a songbird, the Zebra Finch (*Taeniopygia guttata*). Proteome Sci 13:29

Mao KM, Sultana F, Howlider MAR, Iwasawa A, Yoshizaki N (2006) The magnum-isthmus junction of the fowl oviduct participates in the formation of the avian-type shell membrane. Zool Sci 23:41–47

Martin TE (2008) Egg size variation among tropical and temperate songbirds: an embryonic temperature hypothesis. Proc Natl Acad Sci USA 105:9268–9271

Martin TE, Bassar RD, Bassar SK, Fontaine JJ, Lloyd P, Mathewson HA, Niklison AM, Chalfoun A (2006) Life-history and ecological correlates of geographic variation in egg and clutch mass among passerine species. Evolution 60:390–398

Martin-Vivaldi M, Soler JJ, Peralta-Sanchez JM, Arco L, Martin-Platero AM, Martinez-Bueno M, Ruiz-Rodriguez M, Valdivia E (2014) Special structures of Hoopoe eggshells enhance the adhesion of symbiont-carrying uropygial secretion that increase hatching success. J Anim Ecol 83:1289–1301

Martyka R, Rutkowska J, Cichoń M (2011) Sex-specific effects of maternal immunization on yolk antibody transfer and offspring performance in Zebra Finches. Biol Lett 7:50–53

Marzoni M, Castillo A, Sagona S, Citti L, Rocchiccioli S, Romboli I, Felicioli A (2013) A proteomic approach to identify seminal plasma proteins in roosters (*Gallus gallus domesticus*). Anim Reprod Sci 140:216–223

Matsuzaki M, Sasanami T (2017) Sperm storage in the female reproductive tract: a conserved reproductive strategy for better fertilization success. In: Sasanami T (ed) Avian reproduction. Springer, Singapore, pp 173–186

Matsuzaki M, Mizushima S, Hiyama G, Hirohashi N, Shiba K, Inaba K, Suzuki T, Dohra H, Ohnishi T, Sato Y, Kohsaka T (2015) Lactic acid is a sperm motility inactivation factor in the sperm storage tubules. Sci Rep 5:17643

Maurer G, Portugal SJ, Hauber ME, Mikšik I, Russell DGD, Cassey P (2015) First light for avian embryos: eggshell thickness and pigmentation mediate variation in development and UV exposure in wild bird eggs. Funct Ecol 29:209–218

Mayani-Parás F, Kilner RM, Stoddard MC, Rodríguez C, Drummond H (2015) Behaviorally induced camouflage: a new mechanism of avian egg protection. Am Nat 186:E91–E97

McCracken KG (2000) The 20-cm spiny penis of the Argentine Lake Duck (*Oxyura vittata*). Auk 117:820–825

McFarlane RW (1963) The taxonomic significance of avian sperm. In: Sibley GC (ed) Proceedings of the XIII International Ornithological Congress. American Ornithologists' Union, Ithaca, NY, pp 91–102

McGraw KJ, Ardia DR (2007) Do carotenoids buffer testosterone-induced immunosuppression? An experimental test in a colourful songbird. Biol Lett 3:375–378

McNew SM, Knutie SA, Clayton DH (2020) No evidence of sex ratio manipulation by Galápagos Mockingbirds in response to environment. J Avian Biol 51:jav.02302

Mead PS, Morton ML, Fish BE (1987) Sexual dimorphism in egg size and implications regarding facultative manipulation of sex in mountain White-crowned Sparrows. Condor 89:798–803

Meijer T (1995) Importance of tactile and visual stimuli of eggs and nest for termination of egg laying of Red Junglefowl. Auk 112:483–488

Meijer T, Drent R (1999) Re-examination of the capital and income dichotomy in breeding birds. Ibis 141: 399–414

Meijer T, Masman D, Daan S (1989) Energetics of reproduction in female kestrels. Auk 106:549–559

Mendonca T, Cadby AJ, Hemmings N (2019) Sperm gatekeeping: 3D imaging reveals a constricted entrance

to Zebra Finch sperm storage tubules. Biophys J 117: 2180–2187
Mentesana L, Andersson MN, Casagrande S, Goymann W, Isaksson C, Hau M (2021) Natural variation in yolk fatty acids, but not androgens, predicts offspring fitness in a wild bird. Front Zool 18:38
Mezquita B, Mezquita C, Mezquita J (1998) Marked differences between avian and mammalian testicular cells in the heat shock induction and polyadenylation of Hsp70 and ubiquitin transcripts. FEBS Lett 436: 382–386
Michl G, Török J, Péczely P, Garamszegi LZ, Schwabl H (2005) Female Collared Flycatchers adjust yolk testosterone to male age, but not to attractiveness. Behav Ecol 16:383–388
Mikšík I, Sedláková P, Lacinová K, Pataridis S, Eckhardt A (2010) Determination of insoluble avian eggshell matrix proteins. Anal Bioanal Chem 397:205–214
Mikhailov KE (1997) Fossil and recent eggshell in amniotic vertebrates: fine structure, comparative morphology and classification. Spec Pap Palaeotol 56:1–80
Mitrus J, Mitrus C, Rutkowski R (2015) Female Red-breasted Flycatchers (*Ficedula parva*) mated with older males produce male-biased broods. Wilson J Ornithol 127:259–265
Mizushima S, Hiyama G, Shiba K, Inaba K, Dohra H, Ono T, Shimada K, Sasanami T (2014) The birth of quail chicks after intracytoplasmic sperm injection. Development 141:3799–3806
Møller AP (1985) Mixed reproductive strategy and mate-guarding in a semi-colonial passerine, the swallow *Hirundo rustica*. Behav Ecol Sociobiol 17:401–408
Møller AP (1987) Copulation behaviour in the goshawk *Accipiter gentilis*. Anim Behav 35:755–763
Møller AP (1988) Testes size, ejaculate quality and sperm competition in birds. Biol J Linn Soc 33:273–283
Monkkonen M, Forsman JT, Kananoja T, Ylonen H (2009) Indirect cues of nest predation risk and avian reproductive decisions. Biol Lett 5:176–178
Montgomerie R, Briskie J (2007) Anatomy and evolution of copulatory structures. In: Jamieson BGM (ed) Reproductive biology and phylogeny of birds: phylogeny, morphology, hormones, and fertilization. Science Publishers, Enfield, NH, pp 115–148
Moore IT (2005) Non-photoperiodic seasonality in a tropical songbird. In: Dawson A, Sharp PJ (eds) Functional avian endocrinology. Alpha Science International, Oxford, pp 181–190
Moore DJ, Williams TD, Morris RD (2000) Mate provisioning, nutritional requirements for egg production, and primary reproductive effort of female Common Terns *Sterna hirundo*. J Avian Biol 31:183–196
Moore IT, Walker BG, Wingfield JC (2004) The effects of combined aromatase inhibitor and anti-androgen on male territorial aggression in a tropical population of Rufous-collared Sparrows, *Zonotrichia capensis*. Gen Comp Endocrinol 135:223–229
Moore IT, Bentley GE, Wotus C, Wingfield JC (2006) Photoperiod-independent changes in immunoreactive brain gonadotropin-releasing hormone (GnRH) in a free-living, tropical bird. Brain Behav Evol 68:37–44
Moreno J, Osorno JL (2003) Avian egg color and sexual selection: does eggshell pigmentation reflect female condition and genetic quality? Ecol Lett 6:803–806
Moreno J, Osorno JL, Morales J, Merino S, Tomas G (2004) Egg colouration and male parental effort in the Pied Flycatcher *Ficedula hypoleuca*. J Avian Biol 35:300–304
Moreno-Rueda G (2007) Yolk androgen deposition as a female tactic to manipulate paternal contribution. Behav Ecol 18:496–498
Morrison KW (2016) Individual repeatability in laying behavior does not support the migratory carry-over effect hypothesis of egg-size dimorphism in *Eudyptes* penguins. J Avian Biol 47:466–475
Morrison RIG, Hobson KA (2004) Use of body stores in shorebirds after arrival on high-Arctic breeding grounds. Auk 121:333–344
Morton ML, Pereyra ME, Baptista LF (1985) Photoperiodically induced ovarian growth in the White-crowned Sparrow (*Zonotrichia leucophrus gambelii*) and its augmentation by song. Comp Biochem Physiol 80: 93–97
Mousseau TA, Fox CW (1998) The adaptive significance of maternal effects. Trends Ecol Evol 13:403–407
Müller W, Eising CM, Dijkstra C, Groothuis TGG (2004) Within-clutch patterns of yolk testosterone vary with the onset of incubation in Black-headed Gulls. Behav Ecol 15:893–897
Mullner A, Linsenmair EK (2007) Nesting behavior and breeding success of Hoatzins. J Field Ornithol 78:352–361
Muriel J, Pérez-Rodríguez L, Gil D (2019) Age-related patterns of yolk androgen deposition are consistent with adaptive brood reduction in Spotless Starlings. Behav Ecol Sociobiol 73:160
Mytiai IS, Shatkovska O, Ghazali M (2017) Size and shape of Neognathae's eggs: effects of developmental mode and phylogeny. Can J Zool 95:359–366
Nager RG (2006) The challenges of making eggs. Ardea 94:323–346
Nakane Y, Yoshimura T (2014) Universality and diversity in the signal transduction pathway that regulates seasonal reproduction in vertebrates. Front Neurosci 8: 115
Nakane Y, Yoshimura T (2019) Photoperiodic regulation of reproduction in vertebrates. Annu Rev Anim Biosci 7:173–194
Navara KJ (2013) The role of steroid hormones in the adjustment of primary sex ratio in birds: compiling the pieces of the puzzle. Integr Comp Biol 53:923–937
Navara KJ (2018) Choosing sexes: mechanisms and adaptive patterns of sex allocation in vertebrates. Springer Nature, Cham
Navara KJ, Hill GE, Mendonça MT (2006) Yolk androgens deposition as a compensatory strategy. Behav Ecol Sociobiol 60:392–398

Naylor BJ, Bendell JF (1989) Clutch size and egg size of Spruce Grouse in relation to spring diet, food supply, and endogenous reserves. Can J Zool 67:969–980

Ndithia HK, Matson KD, Versteegh MA, Muchai M, Tieleman BI (2017) Year-round breeding equatorial larks from three climatically-distinct populations do not use rainfall, temperature or invertebrate biomass to time reproduction. PLoS One 12:e0175275

Nicholls EH, Burke T, Birkhead TR (2001) Ejaculate allocation by male Sand Martins, *Riparia riparia*. Proc R Soc B 268:1265–1270

Nilsson J-Å, Råberg L (2001) The resting metabolic cost of egg laying and nestling feeding in Great Tits. Oecologia 128:187–192

Nishio S, Matsuda T (2017) Fertilization 1: sperm-egg interaction. In: Sasanami T (ed) Avian reproduction. Springer, Singapore, pp 91–104

Noble RC (1991) Comparative composition and utilisation of yolk lipid by embryonic birds and reptiles. In: Deeming DC, Ferguson MWJ (eds) Egg incubation. Cambridge University Press, Cambridge, pp 17–28

Noble RC, Cocchi M (1990) Lipid metabolism and the neonatal chicken. Prog Lipid Res 29:107–140

Nys Y, Guyot N (2011) Egg quality assurance schemes and egg traceability. In: Nys Y, Bain M, Van Immerseel F (eds) Improving the safety and quality of eggs and egg products, Egg chemistry, production and consumption, vol 1. Woodhead Publishing, Cambridge, pp 62–82

Nys Y, Hincke MT, Arias JL, Garcia-Luis JM, Solomon S (1999) Avian eggshell mineralization. Avian Poult Sci Rev 10:143–166

Nys Y, Gautron J, Garcia-Ruiz JM, Hincke MT (2004) Avian eggshell mineralization: biochemical and functional characterization of matrix proteins. Comptes Rendus Palevol 3:549–562

O'Brien JK, Oehler DA, Malowski SP, Roth TL (1999) Semen collection, characterization, and cryopreservation in a Magellanic Penguin (*Spheniscus magellanicus*). Zoo Biol 18:199–214

Ojanen M (1983) Composition of the eggs of the Great Tit (*Parus major*) and the Pied Flycatcher (*Ficedula hypoleuca*). Ann Zool Fenn 20:57–63

Ojanen M, Orell M, Väisänen RA (1978) Egg and clutch sizes in four passerine species in northern Finland. Ornis Fennica 55:60–68

Okumura H (2017) Avian egg and egg coat. In: Sasanami T (ed) Avian reproduction. Advances in experimental medicine and biology, vol 1001. Springer, Singapore, pp 75–90

Omotoriogun TC, Albrecht T, Gohli J, Hořák D, Johannessen LE, Johnsen A, Kreisinger J, Marki PZ, Ottosson U, Rowe M, Sedláček O (2020) Sperm length variation among Afrotropical songbirds reflects phylogeny rather than adaptations to the tropical environment. Zoology 140:125770

Onagbesan O, Bruggeman V, Decuypere E (2009) Intra-ovarian growth factors regulating ovarian function in avian species: a review. Anim Reprod Sci 111:121–140

Oppel S, Powell AN, O'Brien DM (2010) King Eiders use an income strategy for egg production: a case study for incorporating individual dietary variation into nutrient allocation research. Oecologia 164:1–12

Paganelli CV (1980) The physics of gas exchange across the avian eggshell. Am Zool 20:329–338

Pariser EC, Gilbert L, Hazon N, Arnold KE, Graves JA (2012) Mind the gap: the ratio of yolk androgens and antioxidants varies between sons and daughters dependent on paternal attractiveness. Behav Ecol Sociobiol 66:519–527

Parizzi RC, Santos JM, Oliveira MF, Maia MO, Sousa JA, Miglino MA, Santos TCD (2008) Macroscopic and microscopic anatomy of the oviduct in the sexually mature rhea (*Rhea americana*). Anat Histol Embryol 37:169–176

Paszkowski CA, Gingras BA, Wilcox K, Klatt PH, Tonn WM (2004) Trophic relations of the Red-necked Grebe on lakes in the western boreal forest: a stable-isotope analysis. Condor 106:638–651

Payne RB (1969) Breeding season and reproductive physiology of Tricolored Blackbirds and Red-winged Blackbirds. Univ Calif Publ Zool 90:1–137

Penfold LM, Harnal V, Lynch WE, Bird D, Derrickson SR, Wildt DE (2001) Characterization of Northern Pintail (*Anas acuta*) ejaculate and the effect of sperm preservation on fertility. Reproduction 121:267–275

Perfito N, Meddle SL, Tramontin AD, Sharp PJ, Wingfield JC (2005) Seasonal gonadal recrudescence in Song Sparrows: response to temperature cues. Gen Comp Endocrinol 143:121–128

Pianka ER, Parker WS (1975) Age-specific reproductive tactics. Am Nat 109:453–464

Pick JL, Hutter P, Ebneter C, Ziegler AK, Giordano M, Tschirren B (2016) Artificial selection reveals the energetic expense of producing larger eggs. Front Zool 13:38

Pike TW, Petrie M (2003) Potential mechanisms of avian sex manipulation. Biol Rev 78:553–574

Pike TW, Petrie M (2005) Maternal body condition and plasma hormones affect offspring sex ratio in peafowl. Anim Behav 70:745–751

Pilz KM, Smith HG, Sandell MI, Schwabl H (2003) Interfemale variation in egg yolk androgen allocation in the European Starling: do high-quality females invest more? Anim Behav 65:841–850

Pitcher TE, Dunn PO, Whittingham LA (2005) Sperm competition and the evolution of testes size in birds. J Evol Biol 18:557–567

Pizzari T, Birkhead TR (2000) Female feral fowl eject sperm of subdominant males. Nature 405:787–789

Pizzari T, Froman DP, Birkhead TR (2002) Pre- and post-insemination episodes of sexual selection in the fowl, *Gallus g. domesticus*. Heredity 88:112–116

Poiani A (2006) Complexity of seminal fluid: a review. Behav Ecol Sociobiol 60:289–310

Pollock CG, Orosz SE (2002) Avian reproductive anatomy, physiology and endocrinology. Veterinary Clinics of North America. Exot Anim Pract 5:441–474
Polowczyk I, Bastrzyk A, Fiedot M (2016) Protein-mediated precipitation of calcium carbonate. Materials 9:944
Portugal SJ, Bowen J, Riehl C (2018) A rare mineral, vaterite, acts as a shock absorber in the eggshell of a communally nesting bird. Ibis 160:173–178
Potti J (1993) Environmental, ontogenic, and genetic variation in egg size of Pied Flycatchers. Can J Zool 71: 1534–1542
Pratte I, Boadway KA, Diamond AW, Mallory ML (2018) Changes in isotopic niches across stages of the annual cycle in the Arctic Tern (*Sterna paradisaea*). Arctic 71:259–268
Preston FW (1953) Shape of birds' eggs: extant North American families. Auk 70:160–182
Previatto DM, Gonçalves GL, Ortiz MC, de Melo-Soares CD, Posso SR (2018) Morphology of the copulatory organ in *Ortalis canicollis* (Aves: Cracidae) and early evolution of the phallus in birds. Acta Zool 99:42–51
Prior NH, Heimovics SA, Soma KK (2013) Effects of water restriction on reproductive physiology and affiliative behavior in an opportunistically-breeding and monogamous songbird, the Zebra Finch. Horm Behav 63:462–474
Que P, Székely T, Wang P, Lu Q, Lei W, Liu Y, Zhang Z (2019) Offspring sex ratio is unrelated to parental quality and time of breeding in a multiple-breeding shorebird. J Ornithol 160:443–452
Rahman MA, Iwasawa A, Yoshizaki N (2007) Mechanism of chalaza formation in quail eggs. Cell Tissue Res 330:535–543
Rahman MA, Moriyama A, Iwasawa A, Yoshizaki N (2009) Cuticle formation in quail eggs. Zool Sci 26:496–499
Rahn H, Paganelli CV (1990) Gas fluxes in avian eggs: driving forces and the pathway for exchange. Comp Biochem Physiol A 95:1–15
Rahn H, Paganelli CV, Ar A (1975) Relation of avian egg weight to body weight. Auk 92:750–765
Rahn H, Ar A, Paganelli CV (1979) How bird eggs breathe. Sci Am 240:46–55
Rahn H, Sotherland PR, Paganelli CV (1985) Beziehungen zwischen eigewicht, korpergewicht und stoffwechsel bei sperlingsvogein. J Ornithol 126:263–271
Ramírez F, Hobson KA, Wangensteen OS, Genovart M, Viscor G, Sanpera C, Jover L (2010) A physiological marker for quantifying differential reproductive investment between the sexes in Yellow-legged Gulls (*Larus michahellis*). J Exp Mar Biol Ecol 396:48–52
Ramírez F, Garcia-Tarrasón M, Rami L, Genovart M, Jover L, Sanpera C (2013) The role of calcium in constraining egg synthesis in the Audouin's Gull (*Ichthyaetus audouinii*). Can J Zool 91:640–645
Ramírez F, Forero MG, Hobson KA, Chiaradia A (2015) Older female Little Penguins *Eudyptula minor* adjust nutrient allocations to both eggs. J Exp Mar Biol Ecol 468:91–96
Rangel PL, Gutierrez CG (2014) Reproduction in hens: is testosterone necessary for the ovulatory process? Gen Comp Endocrinol 203:250–261
Réhault-Godbert S, Hervé-Grépinet V, Gautron J, Cabau C, Nys Y, Hincke M (2011) Molecules involved in chemical defence of the chicken egg. In: Nys Y, Bain M, Immerseel FV (eds) Improving the safety and quality of eggs and egg products. Woodhead Publishing Ltd., Cambridge, pp 183–208
Réhault-Godbert S, Guyot N, Nys Y (2019) The golden egg: nutritional value, bioactivities, and emerging benefits for human health. Nutrients 11:684
Rendón MA, Rendón-Martos M, Garrido A, Amat JA (2011) Greater Flamingos *Phoenicopterus roseus* are partial capital breeders. J Avian Biol 42:210–213
Reynolds SJ, Perrins CM (2010) Dietary calcium availability and reproduction in birds. Curr Ornithol 17:31–74
Ricklets RE (1980) Geographical variation in clutch size among passerine birds: Ashmole's hypothesis. Auk 97: 38–49
Riou C, Brionne A, Cordeiro L, Harichaux G, Gargaros A, Labas V, Gautron J, Gérard N (2020) Avian uterine fluid proteome: exosomes and biological processes potentially involved in sperm survival. Mol Reprod Dev 87:454–470
Rödl T, Goymann W, Schwabl I, Gwinner E (2004) Excremental androgen metabolite concentrations and gonad sizes in temperate zone vs. tropical Stonechats (*Saxicola torquata* ssp.). Gen Comp Endocrinol 139: 124–130
Rodler D, Sasanami T, Sinowatz F (2012) Assembly of the inner perivitelline layer, a homolog of the mammalian zona pellucida: an immunohistochemical and ultrastructural study. Cells Tissues Organs 195:330–339
Rodler D, Stein K, Korbel R (2015) Observations on the right ovary of birds of prey: a histological and immunohistochemical study. Anat Histol Embryol 44:168–177
Rodriguez-Navarro C, Kudłacz K, Cizer Ö, Ruiz-Agudo E (2015) Formation of amorphous calcium carbonate and its transformation into mesostructured calcite. CrystEngComm 17:58–72
Rodriguez-Navarro C, Cara AB, Elert K, Putnis CV, Ruiz-Agudo E (2016) Direct nanoscale imaging reveals the growth of calcite crystals via amorphous nanoparticles. Cryst Growth Des 16:1850–1860
Romanoff AL, Romanoff AJ (1949) The avian egg. Wiley, New York, NY
Rosivall B, Szöllösi E, Török J (2005) Maternal compensation for hatching asynchrony in the Collared Flycatcher *Ficedula albicollis*. J Avian Biol 36:531–537
Rowe M, Pruett-Jones S (2011) Sperm competition selects for sperm quantity and quality in the Australian Maluridae. PLoS One 6:e15720
Rowe M, Swaddle JP, Pruett-Jones S, Webster MS (2010) Plumage coloration, ejaculate quality and reproductive

phenotype in the Red-backed Fairy-wren. Anim Behav 79:1239–1246

Rowe M, Albrecht T, Cramer ERA, Johnsen A, Laskemoen T, Weir JT, Lifjeld JT (2015) Postcopulatory sexual selection is associated with accelerated evolution of sperm morphology. Evolution 69:1044–1052

Rowe M, Whittington E, Borziak K, Ravinet M, Eroukhmanoff F, Sætre G-P, Dorus S (2020) Molecular diversification of the seminal fluid proteome in a recently diverged passerine species pair. Mol Biol Evol 37:488–506

Royle NJ, Surai PF, McCartney RJ, Speake BK (1999) Parental investment and egg yolk lipid composition in gulls. Funct Ecol 13:298–306

Rubenstein DR, Hobson KA (2004) From birds to butterflies: animal movement patterns and stable isotopes. Trends Ecol Evol 19:256–263

Ruiz X, Jover L, Pedrocchi V, Oro D, González-Solιs J (2000) How costly is clutch formation in the Audouin's Gull *Larus audouinii*? J Avian Biol 31:567–575

Runfeldt S, Wingfield JC (1985) Experimentally prolonged sexual activity in female sparrows delays termination of reproductive activity in their untreated mates. Anim Behav 33:403–410

Runft LL, Jaffe LA, Mehlmann LM (2002) Egg activation at fertilization: where it all begins. Dev Biol 245:237–254

Rutkowska J, Dubiec A, Nakagawa S (2014) All eggs are made equal: meta-analysis of egg sexual size dimorphism in birds. J Evol Biol 27:153–160

Saino N, Martinelli R, Biard C, Gil D, Spottiswoode CN, Rubolini D, Surai PF, Møller AP (2007) Maternal immune factors and the evolution of secondary sexual characters. Behav Ecol 18:513–520

Samour JH (2002) The reproductive biology of the Budgerigar (*Melopsittacus undulatus*): semen preservation techniques and artificial insemination procedures. J Avian Med Surg 16:39–49

Sasanami T, Matsuzaki M, Mizushima S, Hiyama G (2013) Sperm storage in the female reproductive tract in birds. J Reprod Dev 59:334–338

Sasanami T, Izumi S, Sakurai N, Hirata T, Mizushima S, Matsuzaki M, Hiyama G, Yorinaga E, Yoshimura T, Ukena K, Tsutsui K (2015) A unique mechanism of successful fertilization in a domestic bird. Sci Rep 5: 7700

Schaper SV, Dawson A, Sharp PJ, Gienapp P, Caro SP, Visser ME (2012) Increasing temperature, not mean temperature, is a cue for avian timing of reproduction. Am Nat 179:E55–E69

Schmutz JA, Hobson KA, Morse JA (2006) An isotopic assessment of protein from diet and endogenous stores: effects on egg production and incubation behaviour of geese. Ardea 94:385–397

Schoech SJ, Bowman R, Reynolds SJ (2004) Food supplementation and possible mechanisms underlying early breeding in the Florida Scrub-Jay (*Aphelocoma coerulescens*). Horm Behav 46:565–573

Schwabl H (1996) Maternal testosterone in the avian egg enhances postnatal growth. Comp Biochem Physiol A 114:271–276

Schwabl H, Mock DW, Gieg JA (1997) A hormonal mechanism for parental favoritism. Nature 386:231

Shafey TM (2004) Effect of lighted incubation on embryonic growth and hatchability performance of two strains of layer breeder eggs. Br Poult Sci 45:223–229

Sharp CM, Abraham KF, Hobson KA, Burness G (2013) Allocation of nutrients to reproduction at high latitudes: insights from two species of sympatrically nesting geese. Auk 130:171–179

Shawkey MD, Kosciuch KL, Liu M, Rohwer FC, Loos ER, Wang JM, Beissinger SR (2008) Do birds differentially distribute antimicrobial proteins within clutches of eggs? Behav Ecol 19:920–927

Sheldon BC, Andersson S, Griffith SC, Örnborg J, Sendecka J (1999) Ultraviolet colour variation influences Blue Tit sex ratios. Nature 402:874–877

Siefferman L, Navara KJ, Hill GE (2006) Egg coloration is correlated with female condition in Eastern Bluebirds (*Sialia sialis*). Behav Ecol Sociobiol 59:651–656

Smart IHM (1991) Egg-shape in birds. In: Deeming DC, Ferguson MWJ (eds) Egg incubation: its effects on embryonic development in birds and reptiles. Cambridge University Press, Cambridge, pp 101–116

Smith, D. A. 2018. Chapter 26 – Palaeognathae: Apterygiformes, Casuariiformes, Rheiformes, Struthioniformes; Tinamiformes. In: Pathology of wildlife and zoo animals (K. A Terio, D. McAloose, and J. St. Leger, eds.), pp. 635-651. Academic Press, San Diego, CA.

Snook RR (2005) Sperm in competition: not playing by the numbers. Trends Ecol Evol 20:46–53

Sontakke SD, Umapathy G, Sivaram V, Kholkute SD, Shivaji S (2004) Semen characteristics, cryopreservation, and successful artificial insemination in the Blue Rock Pigeon (*Columba livia*). Theriogenology 62: 139–153

Sotherland PR, Rahn H (1987) On the composition of bird eggs. Condor 89:48–65

Squire ME, Brague JC, Smith RJ, Owen JC (2011) Evidence of medullary bone in two species of thrushes. Wilson J Ornithol 123:831–835

Stanback M, Richardson DS, Boix-Hinzen C, Mendelsohn J (2002) Genetic monogamy in Monteiro's Hornbill, *Tockus monteiri*. Anim Behav 63:787–793

Stanley AJ, Witschi E (1940) Germ cell migration in relation to asymmetry in the sex glands of hawks. Anat Rec 76:329–342

Starck JM (1993) Evolution of avian ontogenies. In: Power DM (ed) Current ornithology. Springer, New York, NY, pp 275–366

Stein RW, Williams TD (2013) Extreme intraclutch egg-size dimorphism in *Eudyptes* penguins, an evolutionary response to clutch-size maladaptation. Am Nat 182:260–270

Stelzer G, Crosta L, Bürkle M, Krautwald-Junghanns M-E (2005) Attempted semen collection using the massage technique and semen analysis in various psittacine species. J Avian Med Surg 19:7–14

Stepinska U, Bakst MR (2007) Fertilization. In: Jamieson BGM (ed) Reproductive biology and phylogeny of birds: phylogeny, morphology, hormones, and fertilization. Science Publishers, Enfield, NH, pp 553–587
Stieve H (1924) Die Eierstöcke und der Legedarm beim Hühnerhabicht (*Falco palumbarius*) und Sperber (*Accipiter nisus*). Verh Dtsch Zool Ges 29:76–78
Stifani S, George R, Schneider WJ (1988) Solubilization and characterization of the chicken oocyte vitellogenin receptor. Biochem J 250:4647–4675
Stoddard MC, Yong EH, Akkaynak D, Sheard C, Tobias JA, Mahadevan L (2017) Avian egg shape: form, function, and evolution. Science 356:1249–1254
Stokke BG, Moksnes A, Røskaft E (2002) Obligate brood parasites as selective agents for evolution of egg appearance in passerine birds. Evolution 56:199–205
Støstad HN, Johnsen A, Lifjeld JT, Rowe M (2018) Sperm head morphology is associated with sperm swimming speed: a comparative study of songbirds using electron microscopy. Evolution 72:1918–1932
Stutchbury BJM, Morton ES (2001) Behavioral ecology of tropical birds. Academic Press, London
Surai PF, Speake BK, Bortolotti GR, Negro JJ (2001) Captivity diets alter egg yolk lipids of a bird of prey (the American Kestrel) and of a galliforme (the Red-legged Partridge). Physiol Biochem Zool 74: 153–160
Thélie A, Rehault-Godbert S, Poirier J-C, Govoroun M, Fouchécourt S, Blesbois E (2019) The seminal acrosin-inhibitor ClTI1/SPINK2 is a fertility-associated marker in the chicken. Mol Reprod Dev 86:762–775
Tingari MD, Lake PE (1973) Ultrastructural studies on the uterovaginal sperm-host gland of the domestic hen, *Gallus domesticus*. J Reprod Fertil 34:423–431
Trivers RL, Willard DE (1973) Natural selection and paternal ability to vary the sex ratio of offspring. Science 179:90–92
Troscianko J, Wilson-Aggarwal J, Stevens M, Spottiswoode CN (2016) Camouflage predicts survival in ground-nesting birds. Sci Rep 6:19966
Tullett SG, Board RG (1977) Determinants of avian eggshell porosity. J Zool 182:203–211
Tuttle EM, Pruett-Jones S (2004) Estimates of extreme sperm production: morphological and experimental evidence from reproductively promiscuous fairy-wrens (*Malurus*). Anim Behav 68:541–550
Uller T, Eklöf J, Andersson S (2006) Female egg investment in relation to male sexual traits and the potential for transgenerational effects in sexual selection. Behav Ecol Sociobiol 57:584–590
Umapathy G, Sontakke S, Reddy A, Ahmed S, Shivaji S (2005) Semen characteristics of the captive Indian White-backed Vulture (*Gyps bengalensis*). Biol Reprod 73:1039–1045
Underwood TJ, Sealy SG (2002) Adaptive significance of egg coloration. In: Deeming DC (ed) Avian incubation: behaviour, environment and evolution. Oxford University Press, Oxford, pp 289–289
van der Klein SA, Zuidhof MJ, Bédécarrats GY (2020) Diurnal and seasonal dynamics affecting egg production in meat chickens: a review of mechanisms associated with reproductive dysregulation. Anim Reprod Sci 213:106257
Van der Molen WH (2002) Laying an egg (online). http://www.afn.org/~poultry/egghen.htm
Van Noordwijk AJ, Keizer LCP, van Balen JH, Scharloo W (1981) Genetic variation in egg dimensions in natural populations of the Great Tit. Genetica 55:221–232
Veasey JS, Houston DC, Metcalfe NB (2001) A hidden cost of reproduction: the trade-off between clutch size and escape take-off speed in female Zebra Finches. J Anim Ecol 70:20–24
Veiga JP, Vinuela J, Corder PJ, Aparici JM, Polo V (2004) Experimentally increased testosterone affects social rank and primary sex ratio in the Spotless Starling. Horm Behav 46:47–53
Vézina F, Salvante KG (2010) Behavioral and physiological flexibility are used by birds to manage energy and support investment in the early stages of reproduction. Curr Zool 56:767–792
Vézina F, Williams TD (2002) Metabolic costs of egg production in the European Starling (*Sturnus vulgaris*). Physiol Biochem Zool 75:377–385
Vézina F, Williams TD (2003) Plasticity in body composition in breeding birds: what drives the metabolic costs of egg production? Physiol Biochem Zool 76: 716–730
Vézina F, Williams TD (2005) The metabolic cost of egg production is repeatable. J Exp Biol 208:2533–2538
Vidavsky N, Addadi S, Mahamid J, Shimoni E, Ben-Ezra D, Shpigel M, Weiner S, Addadi L (2014) Initial stages of calcium uptake and mineral deposition in sea urchin embryos. Proc Natl Acad Sci USA 111: 39–44
Vieco-Galvez D, Castro I, Morel PCH, Chua WH, Loh M (2021) The eggshell structure in *Apteryx*; form, function, and adaptation. Ecol Evol 11:3184–3202
Villarroel M, Bird DM, Kuhnlein U (1998) Copulatory behaviour and paternity in the American Kestrel: the adaptive significance of frequent copulations. Anim Behav 56:289–299
Villaverde-Morcillo S, García-Sánchez R, Castaño C, Rodríguez E, Gonzalez F, Esteso M, Santiago-Moreno J (2015) Characterization of natural ejaculates and sperm cryopreservation in a Golden Eagle (*Aquila chrysaetus*). J Zoo Wildl Med 46:335–338
Vleck CM, Bucher TL (1998) Energy metabolism, gas exchange, and ventilation. In: Starck JM, Ricklefs RE (eds) Avian growth and development: evolution within the altricial-precocial spectrum. Oxford University Press, New York, NY, pp 89–116
Vleck D, Vleck CM, Seymour RS (1984) Energetics of embryonic development in the megapode birds, Mallee Fowl *Leipoa ocellata* and Brush Turkey *Alectura lathami*. Physiol Zool 57:444–456
Voigt C, Meiners T, Ter Maat A, Leitner S (2011) Multisensory non-photoperiodic cue advances the onset of

seasonal breeding in Island Canaries (*Serinus canaria*). J Biol Rhythm 26:434–440
Wagner RH, Helfenstein F, Danchin E (2004) Female choice of young sperm in a genetically monogamous bird. Proc Royal Soc B (Suppl) 271:S134–S137
Wahli W (1988) Evolution and expression of vitellogenin genes. Trends Genet 4:227–232
Waldoch J, Root T, Ramer J, Proudfoot J (2007) Semen collection and characterization in Rockhopper Penguins (*Eudyptes chrysocome chrysocome*). J Zoo Wildl Med 38:13–18
Waldoch J, Root T, Dubach JM, Proudfoot J, Ramer J (2012) Semen characteristics and artificial insemination in Rockhopper Penguins (*Eudyptes chrysocome chrysocome*). Zoo Biol 31:166–180
Wallace AR (1889) Darwinism: an exposition of the theory of natural selection with some of its applications. Macmillan, London
Wallace RA (1985) Vitellogenesis and oocyte growth in nonmammalian vertebrates. In: Brower LW (ed) Developmental biology, Oogenesis, vol 1. Plenum Press, New York, NY, pp 127–177
Wang W-H, Day BN, Wu G-M (2003) How does polyspermy happen in mammalian oocytes? Microsc Res Tech 61:335–341
Wang L, Wang J, Wang A (2021) Novel Pickering high internal phase emulsion stabilized by food waste-hen egg chalaza. Foods 10:599
Wangensteen OD, Rahn H (1970) Respiratory gas exchange by the avian embryo. Respir Physiol 11:31–45
Ward S, Bryant DM (2006) Barn Swallows *Hirundo rustica* form eggs mainly from current food intake. J Avian Biol 37:179–186
Ward S, McLeod MG (1992) Energy cost of egg formation in quail. Proc Nutr Soc 51:41A
Watson DM, Anderson SE, Olson V (2015) Reassessing breeding investment in birds: class-wide analysis of clutch volume reveals a single outlying family. PLoS One 10:e0117678
Wen C, Mai C, Wang B, Li J, Sun C, Yang N (2020) Detrimental effects of excessive fatty acid secretion on female sperm storage in chickens. J Anim Sci Biotechnol 11:26
Wesotowski T (1999) Reduction of phallus in birds: an avian way to safe sex? J Avian Biol 30:483–485
Westneat DF, McGraw LA, Fraterrigo JM, Birkhead TR, Fletcher F (1998) Patterns of courtship behavior and ejaculate characteristics in male Red-winged Blackbirds. Behav Ecol Sociobiol 43:161–171
Wetzel DP, Stewart IRK, Westneat DF (2012) Heterozygosity predicts clutch and egg size but not plasticity in a House Sparrow population with no evidence of inbreeding. Mol Ecol 21:406–420
Whitehead CC (2004) Overview of bone biology in the egg-laying hen. Poult Sci 83:193–199
Whittingham LA, Dunn PO (2000) Offspring sex ratios in Tree Swallows: females in better condition produce more sons. Mol Ecol 9:1123–1129
Whittingham LA, Schwabl H (2002) Maternal testosterone in Tree Swallow eggs varies with female aggression. Anim Behav 63:63–67
Wikelski M, Hau M, Robinson WD, Wingfield JC (2003) Reproductive seasonality of seven Neotropical passerine species. Condor 105:683–695
Wikelski M, Martin LB, Scheuerlein A, Robinson MT, Robinson ND, Helm B, Hau M, Gwinner E (2008) Avian circannual clocks: adaptive significance and possible involvement of energy turnover in their proximate control. Philos Trans R Soc B 363:411–423
Wilder SA (2007) The relationships between energetic condition, immune system, cellular components, testosterone, corticosterone, and hemoparasites in breeding birds. M. S. thesis, University of Maine, Orono, ME
Wiley CJ, Goldizen AW (2003) Testosterone is correlated with courtship but not aggression in the tropical Buff-banded Rail, *Gallirallus philippensis*. Horm Behav 43:554–560
Willems E, Decuypere E, Buyse J, Everaert N (2014) Importance of albumen during embryonic development in avian species, with emphasis on Domestic Chicken. Worlds Poult Sci J 70:503–518
Williams GC (1966) Natural selection, the costs of reproduction, and a refinement of Lack's principle. Am Nat 100:687–690
Williams TD (1995) The penguins. Oxford University Press, Oxford
Williams TD, Kitaysky AS, Vézina F (2004) Individual variation in plasma estradiol-17β and androgen levels during egg formation in the European Starling *Sturnus vulgaris*: implications for regulation of yolk steroids. Gen Comp Endocrinol 136:346–352
Willoughby B, Steyn L, Bam L, Olivier AJ, Devey R, Maina JN (2016) Micro-focus X-ray tomography study of the microstructure and morphometry of the eggshell of Ostriches (*Struthio camerus*). Anat Rec 299:1015–1026
Wilson PW, Suther CS, Bain MM, Icken W, Jones A, Quinlan-Pluck F, Olori V, Gautron J, Dunn IC (2017) Understanding avian egg cuticle formation in the oviduct: a study of its origin and deposition. Biol Reprod 97:39–49
Wingfield JC, Hahn TP (1994) Testosterone and territorial behaviour in sedentary and migratory sparrows. Anim Behav 47:77–89
Wingfield JC, Hahn TP, Wada M, Schoech SJ (1997) Effects of day length and temperature on gonadal development, body mass and fat depots in White-crowned Sparrows, *Zonotrichia leucophrys pugetensis*. Gen Comp Endocrinol 107:44–62
Wingfield JC, Jacobs JD, Tramontin AD, Perfito N, Meddle S, Maney DL, Soma K (2000) Toward and ecological basis of hormone-behavior interactions in reproduction of birds. In: Wallen K, Schneider J (eds) Reproduction in context. MIT Press, Cambridge, pp 85–128

Wingfield JC, Sullivan K, Jaxion-Harm J, Meddle SL (2012) The presence of water influences reproductive function in the Song Sparrow (*Melospiza melodia morphna*). Gen Comp Endocrinol 178:485–493

Wishart GJ (2002) Avian sperm: egg interaction: mechanisms and practical application for analysis of fertility. Avian Poultry Biol Rev 13:215–222

Wishart GJ, Horrocks AH (2000) Fertilization in birds. In: Tarin JJ, Cano A (eds) Fertilization in protozoa and metazoan animals: cellular and molecular aspects. Springer Verlag, Berlin, pp 193–222

Wishart GJ, Stains HJ (1999) Measuring sperm: egg interaction to assess breeding efficiency in chickens and turkeys. Poult Sci 78:428–436

Witschi E (1935) Origin of asymmetry in the reproductive system of birds. Am J Anat 56:119–141

Wrobel ER, Bentz AB, Lorenz WW, Gardner ST, Mendonça MT, Navara KJ (2020) Corticosterone and testosterone treatment influence expression of gene pathways linked to meiotic segregation in preovulatory follicles of the domestic hen. PLoS ONE 15:e0232120

Xiao Y, Gao X (2010) Use of IgY antibodies and semiconductor nanocrystal detection in cancer biomarker quantitation. Biomark Med 4:227–239

Yohannes E, Valcu M, Lee RW, Kempenaers B (2010) Resource use for reproduction depends on spring arrival time and wintering area in an Arctic breeding shorebird. J Avian Biol 41:580–590

Yoshimura Y, Barua A (2017) Female reproductive system and immunology. In: Sasanami T (ed) Avian reproduction: from behavior to molecules. Springer Nature, Singapore, pp 33–58

You Y, Feng J, Wang H, Wang J, Dong C, Su X, Sun H, Gao W (2009) Variation in egg size and nestling growth rate in relation to clutch size and laying sequence in Great Tits *Parus major*. Prog Nat Sci 19: 427–433

Yu JY-L, Marquardt RR (1974) Hyperplasia and hypertrophy of the chicken (*Gallus domesticus*) oviduct during a reproductive cycle. Poult Sci 53:1096–1105

Zavala E, Wedgwood KCA, Voliotis M, Tabak J, Spiga F, Lightman SL, Tsaneva-Atanasova K (2019) Mathematical modelling of endocrine systems. Trends Endocrinol Metab 30:244–257

Zheng X, O'Connor J, Huchzermeyer F, Wang X, Wang Y, Wang M, Zhou Z (2013) Preservation of ovarian follicles reveals early evolution of avian reproductive behaviour. Nature 495:507–511

# Avian Reproduction: Nests and Nest Sites

# 17

## Contents




**Abstract**

With few exceptions, birds lay their eggs in nests and the young of many species spend variable amounts of time in nests before fledging. After discussing the evolution of nests, the various potential functions of nests are explained, including structural support, protection, providing suitable microclimates, and serving as phenotypic signals. Relationships between the characteristics of nest sites and nest predation are also discussed. The different types of nests constructed by birds are described, including scrape nests, open-cup nests, adherent nests, and cavity nests among others. The materials used by birds in nest construction are explained as is the extent to which nest building is an innate behavior, learned behavior, or both. The roles of females and males in nest construction vary among species and factors contributing to such differences are discussed. The costs of nest







G. Ritchison, *In a Class of Their Own*, Fascinating Life Sciences,
https://doi.org/10.1007/978-3-031-14852-1_17

construction, both in terms of time and energy, are also described. Given those costs, the extent to which cavity-nesting species construct new cavities or reuse cavities is discussed. A wide variety of arthropods and other organisms can be found in nests and their potential effects on adults and young are discussed.

## 17.1 Introduction

The primary function of nests is to support eggs and nestlings and protect them from predators and adverse weather. To minimize predation risk, birds may use or build nests that are inaccessible, hidden, or camouflaged. Many organisms other than birds build nests, and the nests of several dinosaur and theropod ancestors of birds have been discovered. Because the ancestors of birds were building nests as early as 200 million years ago (Moratalla and Powell 1990), it seems clear that even the first birds built nests. Those early nests may have been comparable to those of many theropods and the nests of some present-day basal avian taxa (ratites and many Galliformes)—shallow bowl-shaped pits in the ground (Varricchio and Jackson 2016).

The nests of present-day birds vary considerably in size, shape, materials used for construction, and where they are typically placed or located in the environment (Fig. 17.1). Some of this variation is due to bird morphology, e.g., larger birds generally have larger nests and flightless birds necessarily nest on the ground, and some to features of habitats, e.g., birds that nest on the high arctic tundra must nest on the ground because there are no trees and shrubs. However, interspecific variation in nest construction and placement is also correlated with other features such as flight ability (Dial 2003). Weak fliers, such as gallinaceous birds, typically have relatively simple ground nests. Birds that construct elevated nests in shrubs or trees or on a cliff or rock ledge tend to be better fliers than simple ground nesters. Some of the best fliers, like swallows (Hirundinidae) and swifts (Apodidae), often build nests in hard-to-reach (for most predators) locations like cliffs or other vertical surfaces (e.g., both natural and man-made). Although a complex array of factors contribute to variation among species in nest morphology and location, differences in flying ability clearly influence where some birds place their nests (Dial 2003).

The overwhelming majority of birds build nests, even though some nests, such as scrape nests, are very simple in structure. Some birds, however, do not build nests. For example, some seabirds, such as Common Murres (*Uria aalge*), lay their eggs directly onto narrow rocky ledges (Ainley et al. 2020b). Nightjars (Caprimulgidae) simply lay their eggs directly on leaf litter on the ground (Box 17.1 Camouflage and Ground-Nesting Birds), and potoos (*Nyctibius* sp.) lay their single egg in depressions or crevices on tree branches or the tops of snags (Vanderwerf 1988; Tate 1994; Fig. 17.2). White Terns (*Gygis alba*), also known as Common Fairy-Terns, lay their single egg wherever they can find a suitable shallow depression, typically on tree branches (Fig. 17.3), but also on rock ledges, buildings, and, rarely, on the ground (Niethammer and Patrick 2020). For White Terns, not building a nest may be an adaptation for nesting in areas like coral formation or volcanic-rock islands where vegetation for nest construction is not available (Howell 1978) or an adaptation to avoid nest parasites (Houston 1979). For potoos and nightjars, nests might be more conspicuous and more likely to be detected by predators than well-camouflaged eggs (and incubating adults; Collias and Collias 1984). For most birds, however, nests are important for keeping eggs and nestlings warm and safe.

Some species of birds use nests, but only those built by other species (Lindell 1996). This is particularly true for cavity nests, with many species of birds using cavities created by woodpeckers (or by natural processes). These species are called secondary cavity nesters, and woodpeckers are the primary cavity nesters. Fewer species of birds use the open-cup nests of other species. However, many species of owls use the abandoned nests of hawks and crows. In the northeastern United States, Great Horned Owls

**Fig. 17.1** Examples of the types of nests used by birds. (**a**) Cup nest of an American Robin (*Turdus migratorius*), (**b**) platform nest of a Magnificent Frigatebird (*Fregata magnificens*), (**c**) adherent nest, constructed using mud and saliva, of a Barn Swallow (*Hirundo rustica*), (**d**) domed nest of a Rufous Hornero (*Furnarius rufus*), (**e**) cavity nest of a Pileated Woodpecker (*Dryocopus pileatus*), (**f**) burrow nest of a Burrowing Owl (*Athene cunicularia*), (**g**) pendant nest of a Baya Weaver (*Ploceus philippinus*), and (**h**) a scrape nest of a Snowy Plover (*Charadrius nivosus*). (Photos **a**, **b**, **c**, **e**, **f**, and **g** from pxhere.com, CC0 Public Domain, photo d by Ron Knight, Wikipedia, CC BY 2.0, https://creativecommons.org/licenses/by/2.0/, and photo h by Lisa Cox, U. S. Fish and Wildlife Service, CC BY 2.0, https://creativecommons.org/licenses/by/2.0/deed.en)

(*Bubo virginianus*) use the old nests of Red-tailed Hawks (*Buteo jamaicensis*), American Crows (*Corvus brachyrhynchos*), Red-shouldered Hawks (*Buteo lineatus*), Cooper's Hawks (*Accipiter cooperi*), Northern Goshawks (*Accipiter gentilis*), and even gray (*Sciurus carolinensis*) and red (*Tamiasciurus hudsonicus*) squirrels (Smith et al. 1999). Usurped new nests and abandoned old nests of Eurasian Magpies (*Pica pica*) were used as nests by eight species of raptors in northwest China (Zhou et al. 2009). Exceptionally for a shorebird, Solitary Sandpipers (*Tringa solitaria*) use the abandoned open-cup nests of several species of songbirds, including American

Robins (*Turdus migratorius*), Rusty Blackbirds (*Euphagus carolinus*), Eastern Kingbirds (*Tyrannus tyrannus*), Gray Jays (*Perisoreus canadensis*), and Cedar Waxwings (*Bombycilla cedrorum*) (Oring 1968). Of 85 species of scolopacid shorebirds, only Solitary Sandpipers and Green Sandpipers (*Tringa ochropus*) commonly lay eggs in tree nests rather than on the ground (Oring 1968). Other species typically construct their own nests, but opportunistically use the nests of other species. For example, Great Blue Herons (*Ardea herodias*) and Bald Eagles (*Haliaeetus leucocephalus*) occasionally use old Osprey (*Pandion haliaetus*) nests (Ewins et al. 1994).

## 17.2 Evolution of Nests

Fossil evidence indicates that many, if not most, dinosaurs and theropods constructed nests, so it seems almost certain that even the earliest birds also constructed nests. Many nest sites of non-avian theropods have been found with eggs either completely or partially buried in the ground (Fig. 17.4). One such nest was a shallow bowl-shaped nest with a distinct rim that contained 24 partially buried eggs (Varricchio et al. 1997). This nest was likely made by a *Troodon*, small theropods from the late Cretaceous, using kicking motions of the rear feet while bracing with the breast and forelimbs (Horner 2000). As eggs were laid, they were likely forced into loose sediment at the bottom of the nest. There was no evidence of any vegetation in the nest. Although simple depressions with rims, the rims of *Troodon* nests and those of other dinosaurs may have reduced the likelihood that nests would be flooded or predated during their relatively long incubation periods (~ 74 days; Varricchio et al. 1999, 2018).

Tanaka et al. (2015) examined the relationship between eggshell porosity and egg mass for 129 species of living archosaurs (birds and crocodilians) that either have covered nests (eggs completely covered with vegetation and/or sediment) or open nests (eggs partly or fully exposed, e.g., scrape, plate, cup, or dome nests). Their analysis revealed that eggshell porosity relative to egg mass was significantly correlated with nest type. For a given egg mass, eggshell porosity of species with eggs covered is greater than that of species with eggs in the open (Fig. 17.5). Next, Tanaka et al. (2015) included extinct taxa/ootaxa (ootaxa are taxa or types of fossil eggs where the identity of the parent is unknown) of archosaurs in the analysis. The results of this analysis suggest that covered nests are likely the primitive condition for dinosaurs (and probably archosaurs), and that open nests first evolved among non-avian theropods and were likely widespread among non-avian maniraptorans (e.g., oviraptors) well before the appearance of birds (Box 17.2 Evolution of Nests and Incubation Behavior).

Nest characteristics and placement are influenced by flying ability (Dial 2003; Shepard et al. 2019; Fig. 17.6). Nothing is known about the nesting behavior of *Archaeopteryx*, but, after *Archaeopteryx*, birds developed features such as

**Box 17.1 Camouflage and Ground-Nesting Birds**

Some species of birds either build no nest, simply laying eggs on the ground, or have scrape nests that are just shallow depressions. For these species, camouflage can be an important factor in nest success. However, depending on how incubating adults respond to approaching predators, selection may favor improved camouflage of either adults or eggs. For example, incubating nightjars (order Caprimulgiformes) tend to remain on nests until predators approach within a meter or two so the main selection pressure is on the appearance of adults and the ability of adults to select nest sites that will provide the best camouflage. Stevens et al. (2017) studied nest-site selection by three species of ground-nesting nightjars in Africa and found that adults chose locations as nest sites that enhanced their camouflage.

(continued)

**Box 17.1** (continued)

(continued)

**Box 17.1** (continued)

Examples of how three species of nightjars select nest sites that make them well-camouflaged. Incubating birds are located at the center of each photograph (inside circles). Mozambique Nightjar (also referred to as the Square-tailed Nightjar), *Caprimulgus fossii*; Fiery-necked Nightjar, *Caprimulgus pectoralis*; Pennant-winged Nightjar, *Caprimulgus vexillarius*. (Figure modified from Stevens et al. 2017; © 2017 Springer Nature, used with permission).

Other species of ground-nesting birds, including many shorebirds, typically flush from nests when predators are much farther away and so the main selection pressure is on the appearance of eggs. Troscianko et al. (2016) studied egg survival for six species of shorebirds that flushed from nests when human "predators" were an average of 62 meters from nests and found that eggs that exhibited less contrast with their backgrounds were more likely to survive than those that exhibited greater contrast. Thus, selection acts not only on the egg coloration but also on the ability of females to select nest locations where their eggs will exhibit less contrast with the nest substrate.

In a study of several species of shorebirds, clutches of eggs that exhibited less contrast with their backgrounds were less likely to be predated. (**a**) Temminck's Courser (*Cursorius temminckii*), (**b**) Three-banded Plover (*Charadrius tricollaris*), (**c**) Bronze-winged Courser (*Rhinoptilus chalcopterus*), and (**d**) Wattled Plover (*Vanellus senegallus*). (Figure from

(continued)

**Box 17.1** (continued)



shorter tails, keels, and alulas that allowed them to fly as well as many present-day birds (e.g., Chiappe et al. 2020). This seems likely to have led to increasingly complex nest structures and greater flexibility in nest placement, particularly for smaller birds. Nest size is positively correlated with body size in birds (Biancucci and Martin 2010) and the smaller nests of smaller birds can be placed in a wide variety of locations permitting greater concealment and inaccessibility, e.g., in small crevices or cavities and on or under the extremities of vegetation (Hansell 2000). Because small, fast, maneuverable birds that inhabited a wide variety of habitats are known from the Cretaceous (Enantiornithines, 122–66 mya; Nudds et al. 2004; Dyke and Nudds 2008), the diversity observed in the structure and placement of the nests of present-day birds may have evolved relatively early in the evolutionary history of birds. However, diversity in nest placement was likely impacted after the Cretaceous because most plants became extinct at the end of the Cretaceous (K-T boundary and Chicxulub impact) so nest sites in vegetation may not have been available in many locations (Fang et al. 2018) for periods ranged from hundreds to thousands of years or even longer (Vajda and Bercovici 2014). The Chicxulub impact also caused the extinction of most arboreal species of birds (Field et al. 2018) so the mostly ground-dwelling species of birds

**Fig. 17.2** An adult Common Potoo (*Nyctibius griseus*) in southeastern Brazil brooding its nestling. Both were apparently aware of the photographer and assumed the typical alert posture of potoos. (Figure from Cestari et al. 2011; used with permission of César Cestari)

**Fig. 17.3** A White Tern (*Gygis alba*) incubating its single egg (indicated by the red arrow) located where a branch forks from the main trunk of a tree. (Figure from Hart et al. 2016; © 2016 Elsevier Ltd, used with permission)

that survived the impact likely nested on or near the ground (e.g., in ferns) during the period before forests began to recover.

Among present-day birds, nest design and placement exhibit strong phylogenetic signals. Fang et al. (2018) characterized the nests of birds in all 242 families based on their sites and structure and found that species in most families that build cup and domed nests were passerines (Box 17.3 Evolution of Open-Cup Nests), whereas most species in families that build platform nests were nonpasserines (Figs. 17.7 and 17.8). They also found that most species of cavity nesting, nonpasserine birds were landbirds that were closely related to each other, including species and families in the orders Strigiformes, Leptosomiformes, Trogoniformes, Bucerotiformes, Coraciiformes, and Piciformes. In contrast, families of cavity-

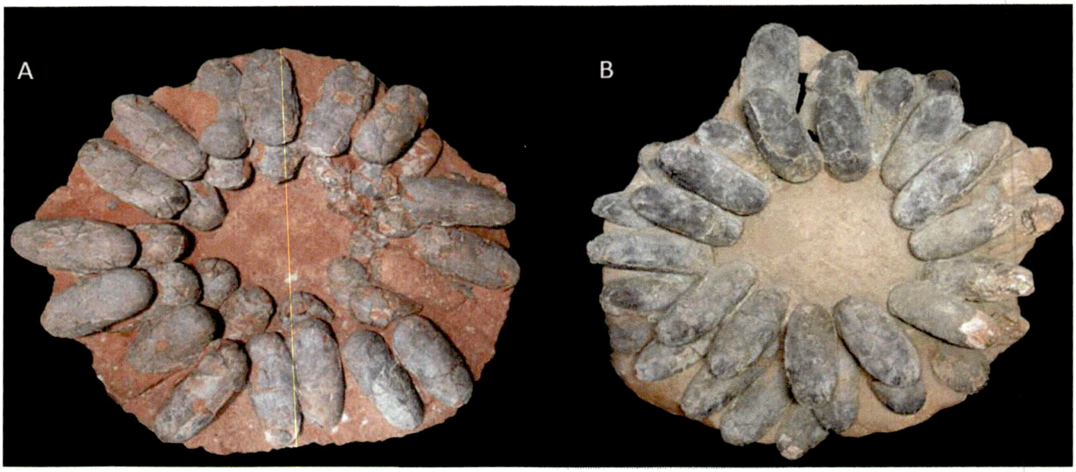


**Fig. 17.4** Two nests and clutches of oviraptors. (**a**) Clutch with 32 eggs arranged in three superimposed rings, and (**b**) clutch with 35 eggs arranged in three superimposed rings. (Figure from Yang et al. 2019; open-access article distributed under the terms of the Creative Commons Attribution License, https://creativecommons.org/licenses/by/4.0/)

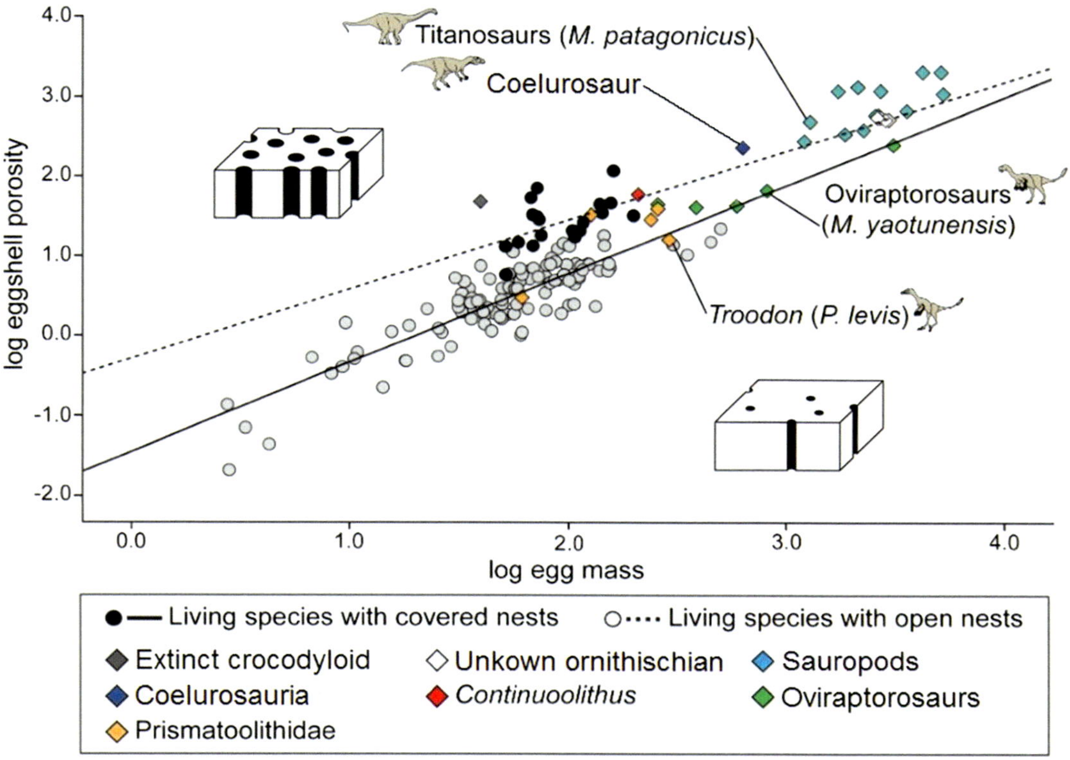


**Fig. 17.5** Relationship between eggshell porosity and egg mass of living and extinct archosaur taxa/ootaxa that had covered nests (eggs buried) or open nests (eggs not buried). Titanosaurs and Coelurosaurs had high eggshell porosity similar to that of extant species with covered nests. In contrast, oviraptorosaurs and *Troodon* had lower eggshell porosity more similar to extant species with open nests. (Modified from Tanaka et al. 2015; © 2015 Tanaka et al., open-access article distributed under the terms of the Creative Commons Attribution License, https://creativecommons.org/licenses/by/4.0/)

nesters in the order Passeriformes were not closely related (Fig. 17.8). Species in about 20% of all bird families, none of which are passerines, make scrape nests (Fig. 17.8).

Fang et al. (2018) also found that species in about two-thirds of bird families nest in trees, species in just over 20% of bird families nest in vegetation other than trees (e.g., bushes and herbaceous vegetation), on the ground, or on cliffs/river banks (Fig. 17.9). The remaining nest sites, including underground, i.e., in burrows like those of Burrowing Owls (*Athene cunicularia*) or in termite/ant nests, in bodies of water, and in termite/ant nests, were used by fewer than 10% of bird families (Fig. 17.10).

## 17.3 Nest Functions

### 17.3.1 Structural Support

Bird nests provide a secure substrate for eggs and nestlings, often provide camouflage and help protect adults, eggs, and nestlings from predators, and, to varying degrees, create a beneficial microenvironment (temperature and humidity) for developing embryos and growing nestlings. Because the structure of bird nests varies among species, the function(s) that is(are) most important also vary among species. Ground, burrow, and cavity nests are on solid substrates so clearly provide excellent structural support. Above-

**Box 17.2 Evolution of Nests and Incubation Behavior**

Based on a review of the literature, Varricchio and Jackson (2016) suggested that the nesting behavior of present-day birds evolved in five stages beginning with basal theropods:

Stage 1: Pre-maniraptoran theropods likely laid eggs that were randomly distributed within clutches and clutches were buried so did not require parental care.

Stage 2: Oviraptor-grade maniraptorans laid clutches of eggs that were highly organized, were partially buried, and were incubated, likely by males.

Stage 3: Troodontid-grade paravians oriented their eggs in a nearly vertical position so clutches were more compact, permitting more efficient transfer of heat from adults to eggs.

Stage 4: Enantiornithines may have oriented and incubated eggs like troodontids or laid a single egg that was buried in sand, requiring no or little post-hatching parental care. Hatchlings were likely precocial.

Stage 5: Early diverging Neornithes did not bury or partially bury eggs; eggs were placed in depressions in ground and variable amounts of nest materials may have been added. This allowed greater contact between adults and eggs, improving the efficiency of incubation and enhancing the growth of embryos.

Although the fossilized remains of nests and eggs provide important information about the breeding biology of non-avian theropods and enantiornithines, the extent to which they incubated their eggs and cared for young after hatching remains unclear. Given the relatively large mass of some non-avian theropods, Deeming and Mayr (2018) suggested that they may have been too heavy to sit on and incubate their eggs without damaging the eggs. In addition, transferring body heat to partially buried eggs would likely be inefficient and questions also remain about the body temperatures of non-avian theropods (Varricchio and Jackson 2016).

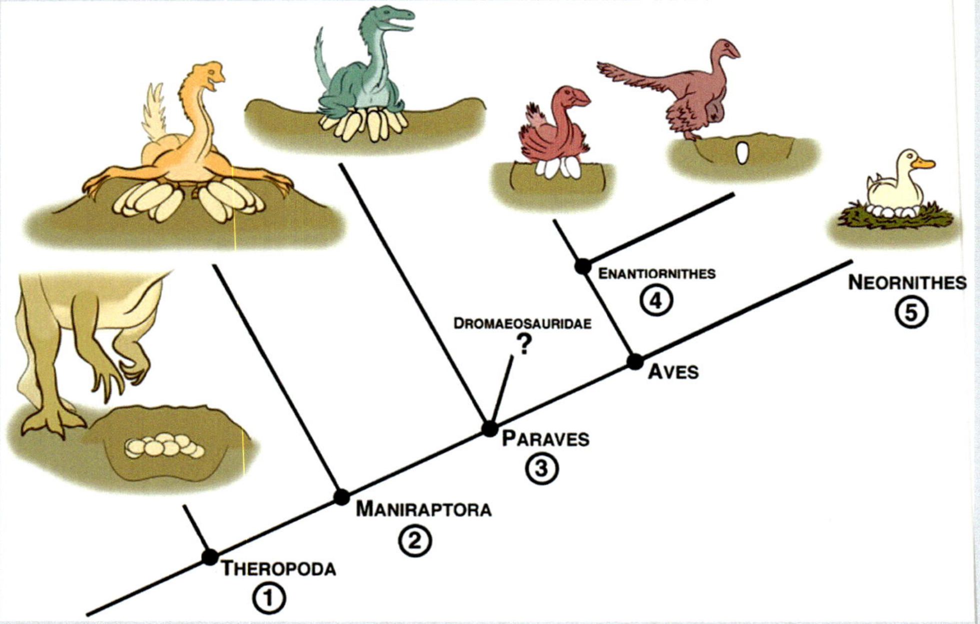


(continued)

**Box 17.2** (continued)

Hypothesized stages in the evolution of nests and the arrangement of eggs in nests. Stage 1, pre-Maniraptoran theropods. Stage 2, oviraptor-grade maniraptorans. Stage 3, Troodontid-grade paravians. Stage 4, Enantiornithes. Stage 5, early diverging Neornithes. (Figure from Varricchio and Jackson 2016; © 2016 Oxford University Press, used with permission).

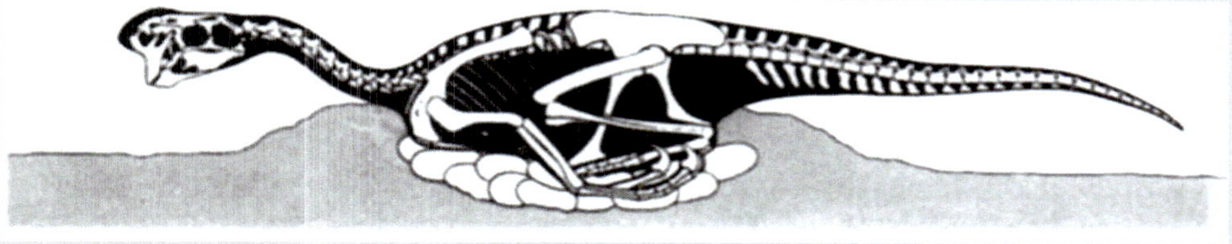

An *Oviraptor* sitting on its clutch of eggs (Norell et al. 1995; © 1995 Springer Nature, used with permission).

(**a** and **b**) Two views of an egg half-buried in sediment that was likely an egg of an enantiornithine bird. (**c**) Partial eggs showing eggshells (white arrows) and bone fragments inside the eggs (black arrows). (Figure from Fernández et al. 2013; © 2013 Fernández et al.,

(continued)

**Box 17.2** (continued)

open-access article distributed under the terms of the Creative Commons Attribution License, https://creativecommons.org/licenses/by/4.0/).

Hatchling enantiornithine showing brown patches around the specimen that represent clumps of elongate feathers on the neck and wings (two top arrows) plus a single pennaceous feather that would have been on the left wing (bottom arrow). The presence of these feathers suggests that the hatchling was likely precocial. (Figure from Kaye et al. 2019; open-access article licensed under a Creative Commons Attribution 4.0 International License https://creativecommons.org/licenses/by/4.0/).

(continued)

**Box 17.2** (continued)

Left, Hatchling enantiornithine preserved in amber. Right, Artist's conception of the hatchling. The plumage of this hatchling suggests that it was highly precocial and possibly even volant. (Figure is the graphical abstract from Xing et al. 2017, © 2017 International Association for Gondwana Research. Published by Elsevier B.V., used with permission).

ground nests, such as plate or cup nests, also provide structural support, but to variable degrees. Adequate structural support appears to be the primary selective influence on the construction of cup-shaped nests. Heenan and Seymour (2011) examined the characteristics of 213 cup-shaped nests constructed by 36 species of birds in Australia that varied in mass from 8 to 360 grams and found that nest walls became much thicker than expected with increasing bird

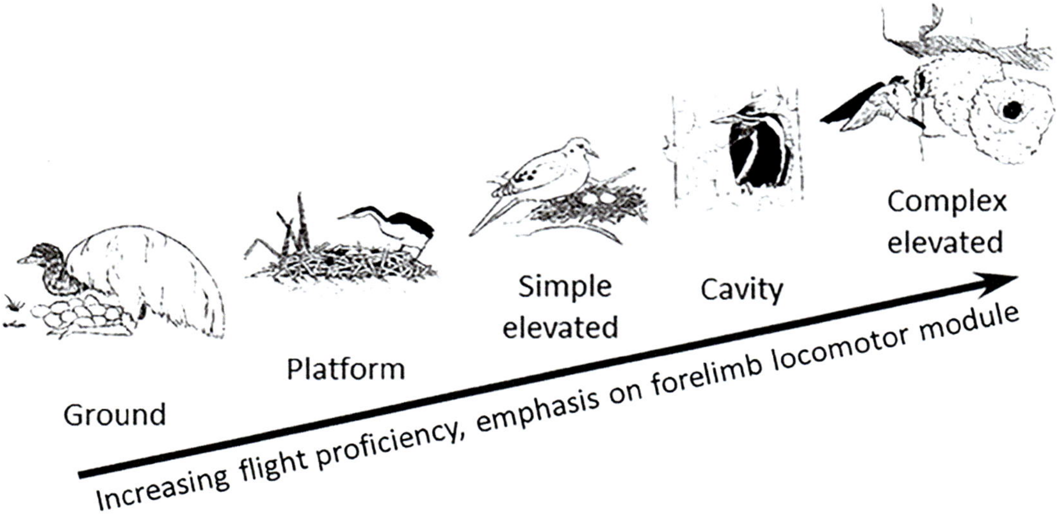


**Fig. 17.6** Nest locations vary among species that differ in flight proficiency, with more accomplished fliers tending to build nests in locations less accessible to predators. (Dial 2003; © 2003 Oxford University Press, used with permission)

**Box 17.3 Evolution of Open-Cup Nests**

Among present-day songbirds, the most common type of nest is the open-cup nest. Fewer species build domed nests, even though they provide greater insulation than open-cup nests. Based on information for 3175 species of songbirds, Medina et al. (2022) determined that 1903 species (60%) have open-cup nests and 790 species (25%) have domed nests (the remaining 482 species [15%] nest in cavities).

Different species of birds place their open-cup or domed nests on the ground, in vegetation, on vertical surfaces, in cavities, or in burrows. Some investigators have suggested that domed nests evolved from open-cup nests, given the open-cup nests are the most common type of nest among songbirds and are also less complex and easier and faster to construct than domed nests (Collias and Collias 1984; Collias 1997). However, Price and Griffith (2017) examined the types of nests built by present-day songbirds in Australia along with phylogenetic analyses and concluded that the ancestral nest type of songbirds was a domed nest. Their analysis also revealed that open-cup nests evolved multiple times independently.

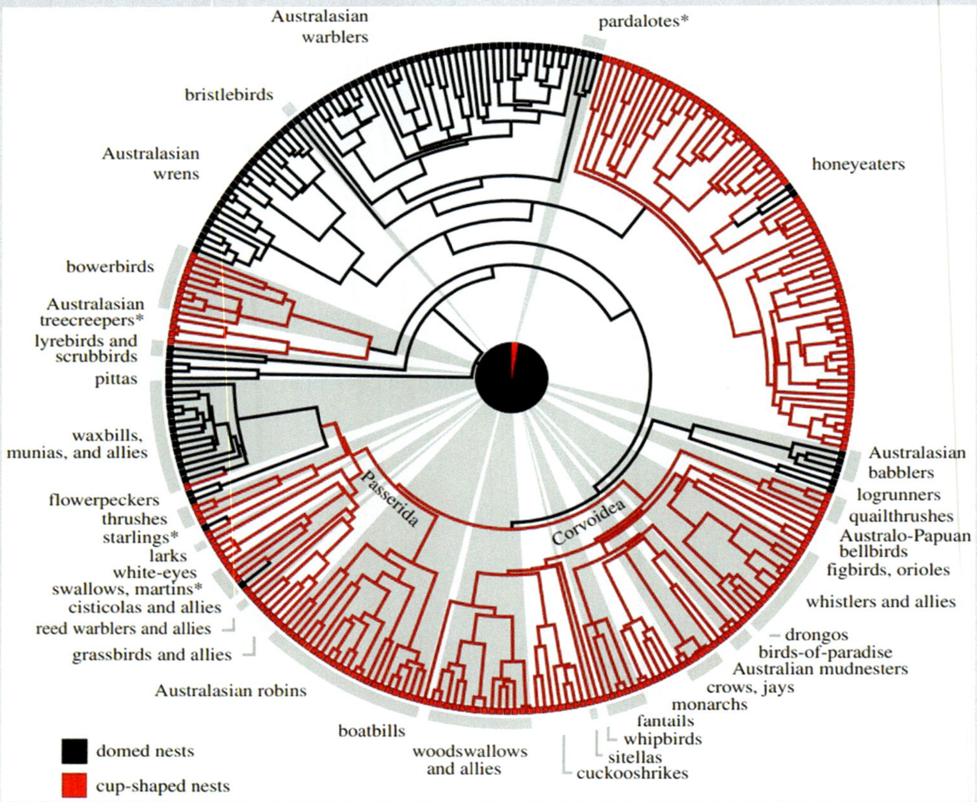


DNA-based phylogenetic tree of 281 species of Australian songbirds. Cup-shaped nests (red) are more common than domed nests (black) among present-day species (187 or 66.5% of the above species build cup nests), but analysis suggests that domed nests represent the ancestral state. Asterisks indicate families that are primarily cavity nesters. (Figure from Price and Griffith

(continued)

**Box 17.3** (continued)

2017; © 2017 The Authors, Published by the Royal Society. All rights reserved, used with permission).

Possible advantages of open-cup nests include: (1) cup nests can be constructed faster than domed nests (and, thus, more easily replaced if a nest fails), (2) cup nests are smaller so may be more difficult for predators to detect, and (3) cup nests allow incubating or brooding adults to more easily detect approaching predators and to more easily escape from predators (Medina et al. 2022). However, disadvantages of open-cup nests include less concealment from visual predators and less protection from the elements. Possible advantages of domed nests include (1) reduced risk of predation and (2) thermoregulatory advantages, e.g., better insulation in colder environments and serving as a shield from solar radiation in hot environments (Griffith et al. 2016; Matysioková and Remeš 2018). However, domed nests take longer to construct, possibly limiting opportunities to renest, and also take more energy to construct (Medina et al. 2022).

Both types of nests have potential advantages and disadvantages, so selection must have favored one type or the other based on changing conditions such as the characteristics of common nest predators, changing habitats or climates, and/or the evolution of more cryptic plumage (Price and Griffith 2017). Because more species have open-cup nests, Medina et al. (2022) suggested that the reduction in time and energetic costs of building open-cup nests, compared to domed nests, may result in increased fecundity and, therefore, an important reason for transitioning from domed to open-cup nests among passerines.

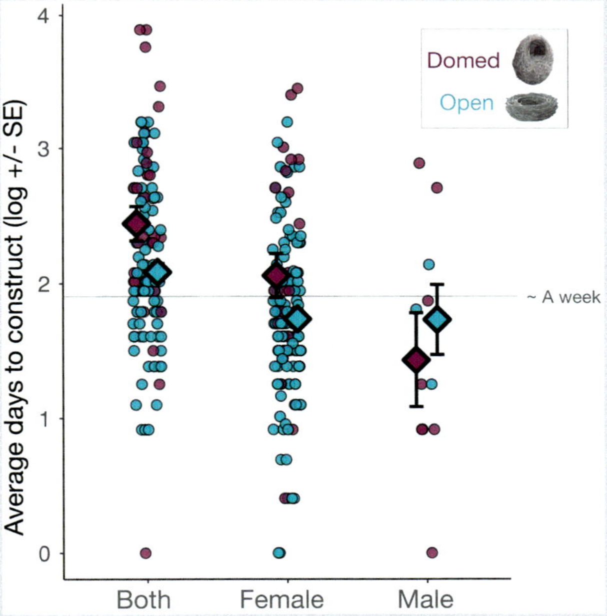


Overall, open-cup nests take less time to construct than domed nests. The only exception is among the relatively few species where males build nests. (Figure from Medina et al. 2022; © 2022 The Authors. Ecology Letters published by John Wiley & Sons Ltd., used with permission).

**Fig. 17.7** A nonpasserine (Mourning Dove, *Zenaida macroura*; Columbiformes) on its simple platform nest. (Photo from pxhere.com, CC0 Public Domain)

size. If the most important selective factor was insulation, then the relative thickness of nest walls would be expected to decline with increasing bird mass because large birds, with less surface area per unit volume, would seemingly have less need to reduce heat loss while incubating or brooding. The thicker than expected walls of nests of larger birds suggest, therefore, that those thicker walls are needed to provide support for those birds plus their eggs or nestlings.

Plate, or platform, nests built above ground, e.g., in trees or shrubs, may not always provide the needed structural support. Among species that build such nests, including storks, pelicans, herons, and pigeons, gaps between the sticks or twigs that make up nests sometimes become so large or the platforms are so small that eggs and nestlings occasionally fall through or over the sides of nests (e.g., Meanley 1955; Ames and Mersereau 1964; Werschkul 1979). However, platform nests are generally substantial enough to provide adequate support of adults, eggs, and nestlings, especially in species such as storks and herons where nests may be used in multiple years and new material added each year may increase the structural integrity of nests (Fig. 17.11).

### 17.3.2 Protection

Because predation is the primary cause of nest failure among birds (Ricklefs 1969), one important function of nests is to help camouflage and protect adults, eggs, and young. However, some nest types and nest sites generally provide more protection than others. For example, nests with long entrance tunnels such as long pendant nests and dome-and-tube nests are likely more effective at preventing access to nest contents by brood parasites and nest predators (Street et al. 2022; Fig. 17.12). In addition, cavities tend to be safer from predators than open-cup or platform nests (Fontaine et al. 2007; Fig. 17.13). Cavities may be safer in part because the size of entrances physically prevent larger predators from entering (Fig. 17.14). The results of some studies suggest that newly excavated cavities (created by primary cavity nesters like woodpeckers) are safer from predators than reused ones (reused either by primary or secondary cavity nesters) (Nilsson et al. 1991; Martin 1995; Fontaine et al. 2007). Other studies have revealed no difference between predation rates of new and reused cavities (Mazgajski 2002) and, in one study, predation

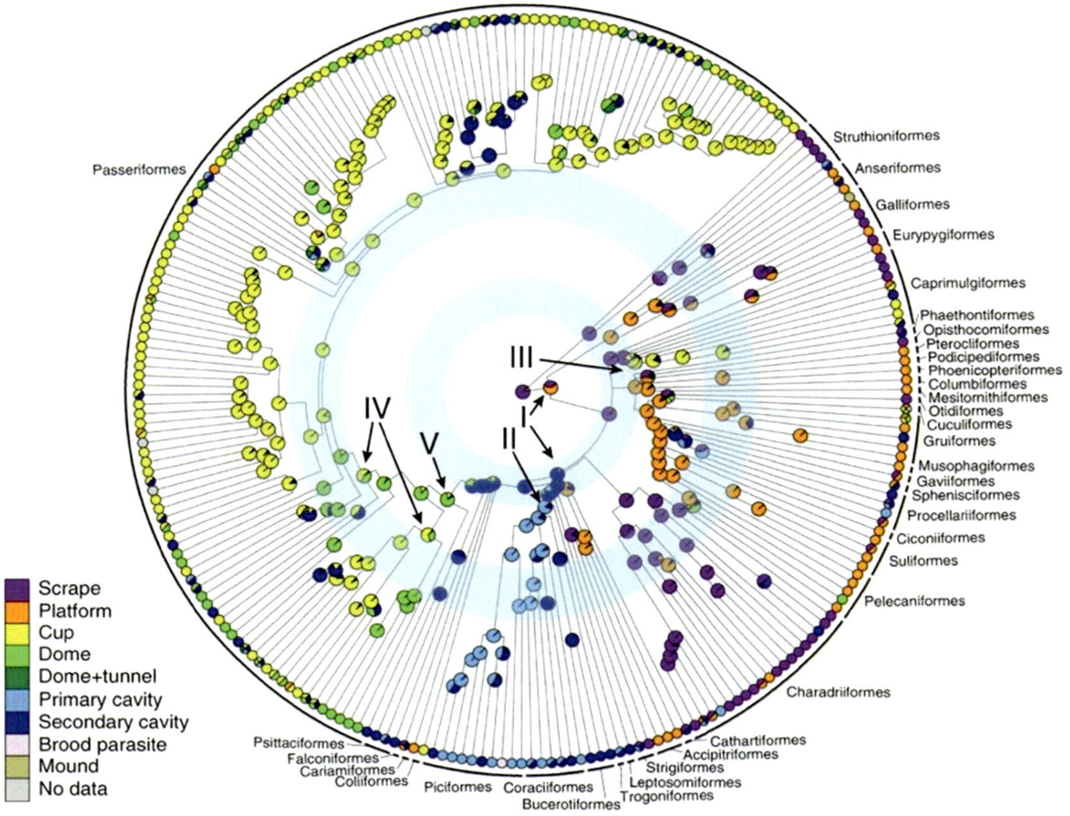


**Fig. 17.8** Nest structures of species in present-day families of birds. Circles with more than one color indicate that species in a family use more than one type of nest structure. Families of birds that are brood-parasitic, build mound nests, or have no nest data were excluded from analyses, but are included here. I, The earliest birds used scrape nests, but platform nests and secondary cavities nests appeared early in the evolution of birds. II, Primary cavity nests evolved well after birds started using secondary cavity nests. III and IV, Cup nests appear to have evolved at least three times, first among swifts, tree swifts, and hummingbirds, then twice early in the evolution of songbirds. V, Dome nests appeared early in the evolution of songbirds and were likely the ancestral nest type of songbirds. (Figure from Fang et al. 2018; open-access article licensed under a Creative Commons Attribution 4.0 International License, https://creativecommons.org/licenses/by/4.0/)

rates were higher for nests in new cavities (Smith 1997). Two factors may contribute to increased predation of reused cavities: (1) predators may learn the locations of cavities, increasing the likelihood of predation for nests in previously used cavities (Sonerud 1985), and (2) older cavities may not be as strong structurally as new ones due to decay, making access easier for some types of predators (Wiebe et al. 2007). Although an untested hypothesis, Wiebe et al. (2007) suggested that nests in new cavities could potentially be at greater risk of predation if the presence of fresh wood chips on the ground or the sound and activity of excavation attracts the attention of predators. Each of these factors may influence predation risk for cavity-nesting species. However, differences in the likelihood of cavity nests being predated, and the reason why different studies have provided different results, is probably due to differences among locations in the search images and diets of the local suite of predators (Wiebe et al. 2007).

Non-cavity nests like cup nests and domed nests are more exposed to predators than cavity

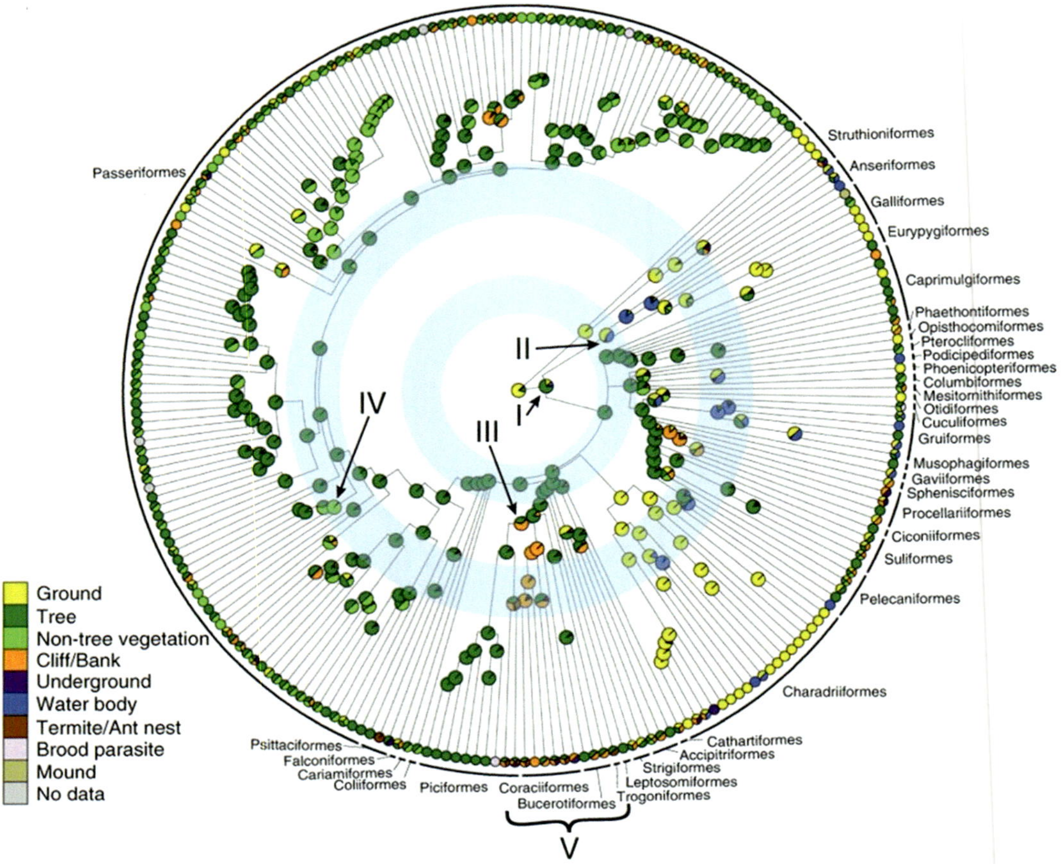


**Fig. 17.9** Location of nests among species of present-day families of birds. Circles with more than one color indicate that species in a family use more than one type of nest structure. Families of birds that are brood-parasitic, build mound nests, or have no nest data were excluded from analyses, but are included here. I, Birds began nesting in trees early in their evolution. II, Birds began placing nests in water bodies. III and IV, Use of cliffs/banks and non-tree vegetation as nest sites occurred much later in the evolution of birds. (Figure from Fang et al. 2018; open-access article licensed under a Creative Commons Attribution 4.0 International License, https://creativecommons.org/licenses/by/4.0/)

and burrow nests and tend to suffer higher predation rates (e.g., Fulton 2019). Several factors can potentially affect the likelihood of predation for non-cavity nests, including nest height and nest concealment. According to the nest-concealment hypothesis, nests that are more concealed should be less vulnerable to predation (Ricklefs 1969). Many investigators have examined the relationship between concealment and predation risk and some have found that more concealed nests are less likely to be predated (e.g., Martin and Roper 1988; Götmark et al. 1995; Schill and Yahner 2009; Liu et al. 2021), whereas others have found no relationship (e.g., Howlett and Stutchbury 1996). Borgmann and Conway (2015) reviewed the results of 106 studies and found that investigators in 74% of those studies found no relationship between nest concealment and predation risk. Interestingly, however, these authors also found a positive correlation between female plumage brightness and support for the nest-concealment hypothesis, suggesting that

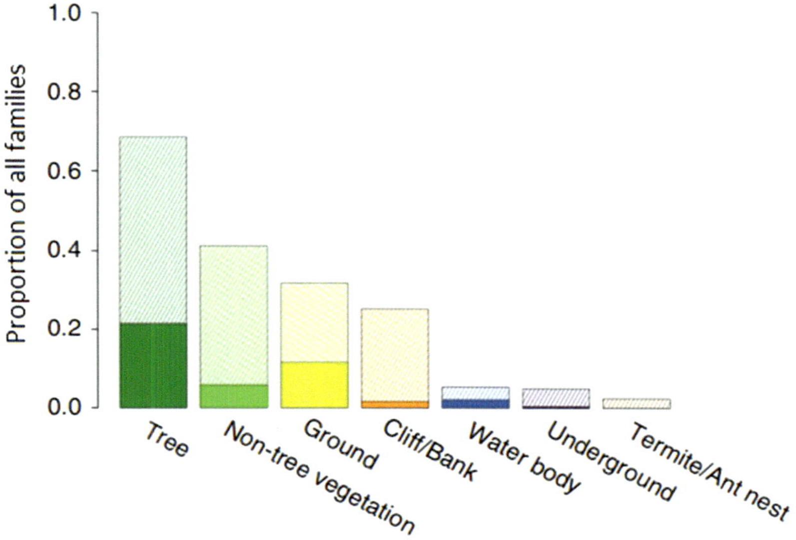


**Fig. 17.10** Proportion of bird families ($N = 242$) with species that nest in or on different substrates. Filled bars indicate the proportion of families with only that particular nest substrate; striped bars above filled bars indicate the proportion of families with species having more than one nest substrate. (Figure from Fang et al. 2018; open-access article is licensed under a Creative Commons Attribution 4.0 International License, https://creativecommons.org/licenses/by/4.0/)

**Fig. 17.11** Nests of White Storks (*Ciconia ciconia*) can be as much as 2 to 3 meters in diameter. (Photo from pxhere.com, CC0 Public Domain)

**Fig. 17.12** (**a**) Dome-and-tube nest of a Baya Weaver (*Ploceus philippinus*), and (**b**) pendant nests of Montezuma Oropendolas (*Psarocolius montezuma*). Among weaverbirds and icterids, young of species that build nests with extended entrance tunnels like those of Southern Masked Weavers and Crested Oropendulas have longer developmental periods. The length of developmental periods can be influenced by the relative risk of parasitism and predation (Martin 1995; Remeš and Martin 2002; Remeš 2006), so the relatively long developmental periods of young among species with nests having extended entrance tunnels suggest that such nests limit access to nests by brood parasites and nest predators (Street et al. 2022). (Image credits: (**a**) Wikiwand, CC BY 2.0, https://creativecommons.org/licenses/by/2.0/, and (**b**) caspar s, CC BY 2.0, https://creativecommons.org/licenses/by/2.0/)

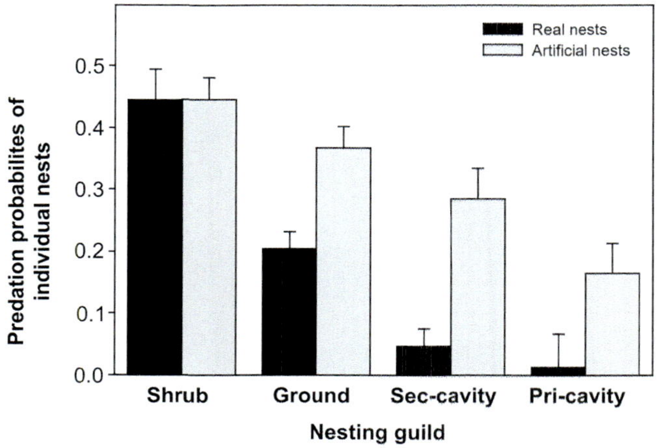


**Fig. 17.13** Nest location and nest type (open nesting vs. cavity nesting) both influence nest predation rates. Open nests are typically predated at higher rates than cavity nests, with open nests in shrubs (Shrub) predated at higher rates than those on the ground (Ground) and nests of secondary cavity nesters (Sec-cavity) predated at higher rates than those of primary cavity nesters (Pri-cavity). This was found to be true for both real nests and artificial nests. (Figure from Fontaine et al. 2007; © Oikos. Published by John Wiley & Sons Ltd., used with permission)

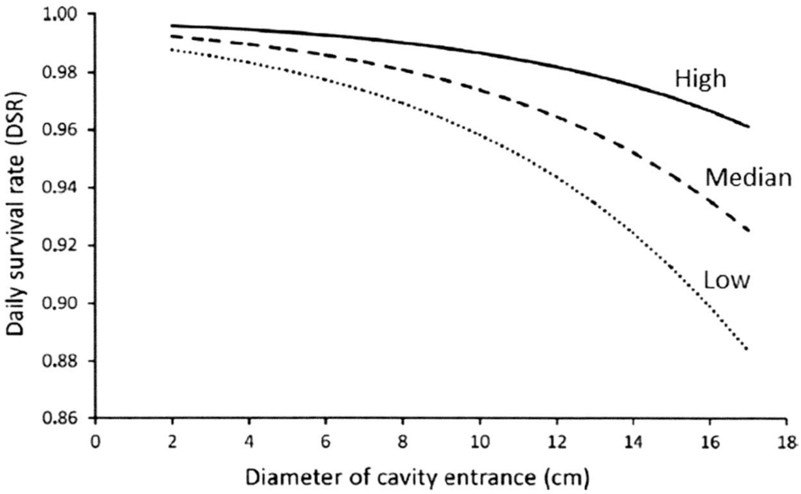


**Fig. 17.14** Predicted daily survival rates based on a model generated using data from cavity-nesting birds in the Atlantic Forest of Argentina. Nests in higher cavities with smaller-diameter cavity entrances were more likely to survive (with most nests lost due to predation). In the model, high nest cavities were 19 m above ground, median nest cavities were 9.2 m above ground, and low nest cavities were 2 m above ground. Higher nest cavities with smaller entrances are less accessible to potential nest predators. (Figure from Cockle et al. 2015; © 2015 Elsevier Ltd., used with permission)

brightly colored females may select more concealed nest sites. As with nest concealment, the results of some studies suggest that nest height influences predation risk, whereas others report no such relationship (e.g., Filliater et al. 1994; Colombelli-Negrel and Kleindorfer 2009; Vanderwerf 2012). Differences among studies in the effects of nest height and concealment on predation risk of non-cavity nests may, at least in part, be due to differences in predator communities (e.g., Thompson 2007; Cox et al. 2012; Reidy and Thompson 2012).

Different nest predators use different cues to locate nests and predators also differ in their abilities to access nests located at different heights. For example, cover above nests may be more important for concealing nests from avian predators, but cover below nests is likely more important for concealing nests from ground-based predators such as small mammals or snakes. In addition, for nest predators such as snakes and mammals that rely more on olfactory cues than visual cues (Halpern and Frumin 1979; Conover 2007), concealment provides no protection. Potential nest predators also differ in their abilities to access nests at different heights above ground. Avian predators, such as jays, crows, and raptors, can access nests at any height. In addition, many snakes are excellent climbers and height may have no impact on their ability to access nests (e.g., Mullin et al. 2000), although the climbing ability of some snakes may be limited by the characteristics of nest substrates, e.g., trees with smooth bark (Mullin and Cooper 2002; Natusch et al. 2017). Sri Lanka Drongos (*Dicrurus lophorinus*) have even been observed breaking off pieces of loose bark and removing mosses and lichens from the trunks of trees where they have nests, creating a smoother surface that could potentially make climbing the tree by a potential nest predator more difficult (Rajeev et al. 2018). Despite the excellent climbing ability of some ground-based predators, nests located higher above ground may experience less predation by other ground-based predators such as rodents and other mammals (Schmidt 2003). Thus, selection favoring particular nest heights and locations depends on the community of predators present.

For nest predators that use visual cues, larger nests may be more visible and more likely to be found than smaller nests. In fact, the results of

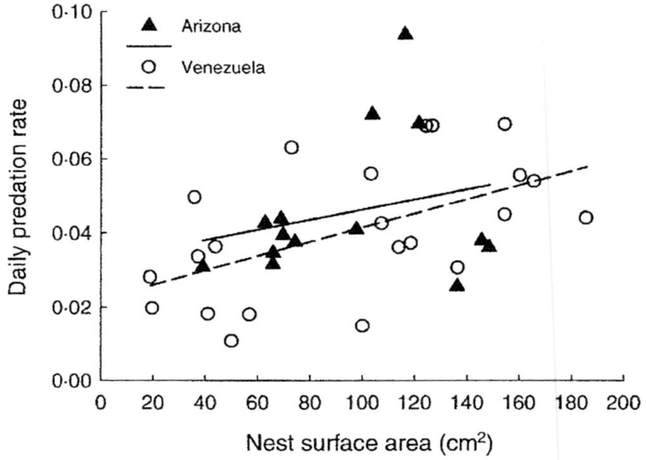


**Fig. 17.15** Nest surface area and nest predation rates for 36 altricial species of birds in tropical Venezuela (South America) and temperate Arizona (North America). Nest predation rates increased with nest size (nest surface area) at both locations. (Figure from Biancucci and Martin 2010; © 2010 The Authors. Journal compilation © 2010 British Ecological Society, used with permission)

several studies suggest that this is the case, with predation rates for larger nests higher than for smaller nests (Møller 1990; Biancucci and Martin 2010; Mouton and Martin 2019; Fig. 17.15). Particularly in habitats or locations with diverse populations of predators, this relationship between nest size and predation rates might select for smaller nests. However, other investigators have reported lower predation rates for larger nests (e.g., Weidinger 2004; Remeš et al. 2012). Unzeta et al. (2020) examined nest-predation data for 509 populations of 330 species of songbirds and found that larger species constructed larger nests than smaller species, but larger species and larger nests had lower daily nest predation rates than smaller species with smaller nests. These investigators also reported that larger species did not build nests in more protected nest sites and habitats than smaller species. Rather, their results suggested that the lower rates of nest predation for larger species of songbirds were due to the ability of larger species to better defend their nests from potential predators, with adults of larger species better able to "attack larger predators and a wider range of predator body sizes than adults of smaller species." Concerning the mixed results of previous studies on the effect of nest size on predation rates, Unzeta et al. (2020) suggested that one likely contributing factor is the geographic variation in nest predation rates, included possible effects of phylogeny, latitude, elevation, habitat, and predator diversity and abundance. Geographic variation in these variables can influence the results of studies conducted at a local scale. By analyzing data collected from 330 species of songbirds nesting over a wide range of latitudes on different continents and in different habitats with different predator communities, Unzeta et al. (2020) confirmed the existence of "substantial geographic variation in daily nest predation rates", and higher rates of nest predation at tropical latitudes. Despite this variation, their analysis clearly showed that larger species of songbirds with larger nests experienced lower daily rates of nest predation than smaller species with smaller nests.

Some birds nest on the ground, with nest substrates consisting of leaf litter (e.g., nightjars) or sand or pebbles (e.g., shorebirds and terns; Fig. 17.16). In many such species, coloration of incubating adults and eggs tends to match that of the nest substrate (Fig. 17.17). Such matching suggests that crypsis may be important in minimizing predation risk, and the results of several studies appear to confirm this hypothesis. For example, ground-nesting Eurasian Stone Curlews (*Burhinus oedicnemus*) choose nest materials (twigs) similar in color to their eggs and hatching success was found to be higher for nests where eggshell color most closely matched the substrate color (Solís and de Lope 1995). Similarly, the eggs of Black-tailed Gulls (*Larus crassirostris*) that more closely matched the color of nest substrates were less likely to be predated (Lee

**Fig. 17.16** Nests of Least Terns *(Sternula antillarum)* are simple depressions in the sand. (Photo by Andy Morffew, pxhere.com, CC0 Public Domain)

et al. 2010), and nests of Red-legged Partridges (*Alectoris rufa*) with cryptically colored eggs suffered less predation than those with more conspicuous eggs (Castilla et al. 2007). In contrast, the results of other studies indicate that cryptic eggs are as likely to be predated as more visible eggs (Jobin and Picman 1997; Nguyen et al. 2007). Again, one possible explanation for these conflicting results is the presence of different predators. Egg predators that depend on olfactory cues, such as rodents and other mammals, may not be deterred by cryptic coloration, whereas visual predators like birds may be less likely to spot cryptically colored eggs (Castilla et al.

**Fig. 17.17** Small rocks forming the nest of a Killdeer (*Pluvialis dominica*) closely match the coloration of the eggs and might make the eggs more difficult for visual predators to detect. (Photo by Andy Reago and Chrissy McClarren, Wikipedia, CC BY 2.0, https://creativecommons.org/licenses/by/2.0/)

2007). Another possible explanation for conflicting results is that eggs may simply be imperfectly camouflaged because egg coloration can serve functions other the camouflage. For example, pigments may increase the risk of eggs and embryos overheating in sunlight or selection pressures resulting from brood parasitism may be a more important factor in egg coloration than camouflage (Stoddard et al. 2011). In the latter case, selection may favor egg coloration that allows potential hosts to more easily detect the eggs of brood parasites. Other possible functions of egg coloration include signaling egg and offspring quality (i.e., female post-mating investment in offspring) or female quality to increase male investment in caring for young (Siefferman et al. 2006; Poláček et al. 2017; Holveck et al. 2019), and "blackmailing" males to incubate eggs or feed incubating females by laying non-camouflaged eggs that require concealment to avoid detection by predators (Hanley et al. 2010).

### 17.3.3 Suitable Microclimate

Bird nests provide eggs and nestlings with varying degrees of shelter and isolation from the external environment. Eggs are completely enclosed within mound nests. Nests located in cavities and burrows provide enclosed spaces with access to the external environment limited to entrance holes or tunnels. Dome and dome-and-tube nests provide enclosed spaces with

**Box 17.4 Open vs. Enclosed Nests**

Although the nests of most species of songbirds are open cups or platforms, the nests of some species are enclosed with sides, a roof, and a single opening (i.e., dome nests and dome-and-tube nests). Because their construction likely takes more time and energy than constructing similar-sized open cup or platforms nests, enclosed nests might be expected to provide some additional benefit for the species that build them. Martin et al. (2017) found and monitored 26,437 nests of 114 species of songbirds that build either open or enclosed nests at five different locations (Arizona in the United States, Venezuela, Argentina, South Africa, and Malaysia) and, in addition, surveyed the literature for data for an additional 205 species of songbirds from eastern North America, Central America, northeastern South America, Europe, western Asia, and Australia. Analysis revealed that (1) a greater percentage of species in the tropics and Southern Hemisphere (South Africa and Argentina) built enclosed nests than species in north-temperate locations (North America, Europe, and western Asia), (2) rates of nest predation did not differ with nest type, and (3) species with enclosed nests were significantly smaller than species with open nests. These results suggest that the thermal benefits of enclosed nests are more important than any potential benefit of enhanced concealment and lower risk of predation. Because they have more surface area per unit body mass than larger species and lose heat more rapidly, smaller birds likely benefit from the thermal advantages of enclosed nests (Calder 1984). In support of this hypothesis, nestlings of species of enclosed nests grew faster than those of species with open nests, suggesting that expending less energy for thermoregulation allowed greater energy expenditure for growth. In addition, species with enclosed nests spent more time away from nests than did species with open nests. This may be due to the improved heat retention of enclosed nests and of eggs and nestlings in those nests. Heat retention may be particularly important for species at tropical latitudes because they tend to spend less time at nests than species at north-temperate latitudes (Chalfoun and Martin 2007).

(continued)

**Box 17.4** (continued)

Examples of open (**a–c**) and enclosed (**d–f**) nests. (**a**) Hermit Thrush, *Catharus guttatus*, (**b**) Cape White-eye, *Zostertops pallidus*, (**c**) White-throated Fantail, *Rhipidura albicollis*, (**d**) Snowy-browed Flycatcher, *Ficedula hyperythra*, (**e**) Yellow-breasted Warbler, *Seicercus montis*, and (**f**) Blue-naped Chlorophonia, *Chlorophonia cyanea.* (Figure from Martin et al. 2017; © 2016 The Authors. Functional Ecology © 2016 British Ecological Society, used with permission).

(continued)

**Box 17.4** (continued)

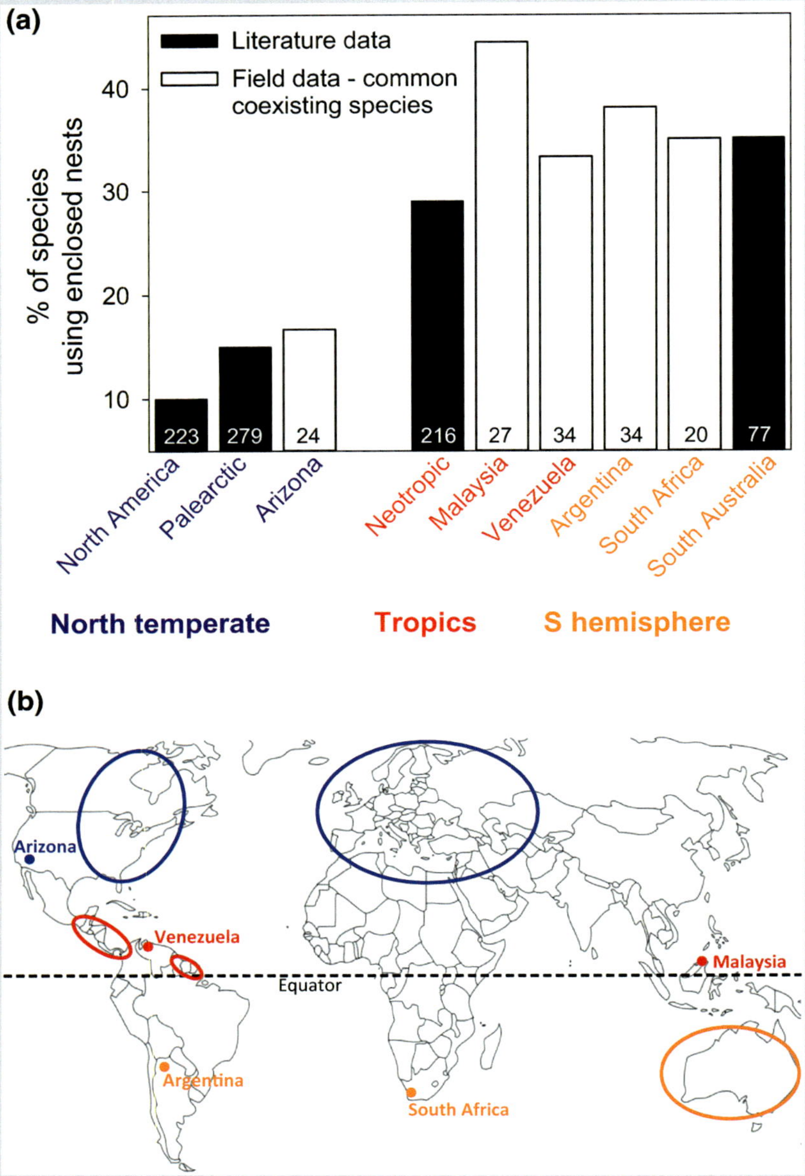


(**a**) Percentage of songbird species of all nest types (open, enclosed, and cavity nests) that have enclosed nests based on data from the literature and data collected by Martin et al. (2017). Numbers at the base of the columns indicate the number of species. (**b**) Locations where data were obtained from the literature (ellipses) and by Martin et al. (2017) (points). (Figure from Martin et al. 2017; © 2016 The Authors. Functional Ecology © 2016 British Ecological Society, used with permission).

(continued)

**Box 17.4** (continued)

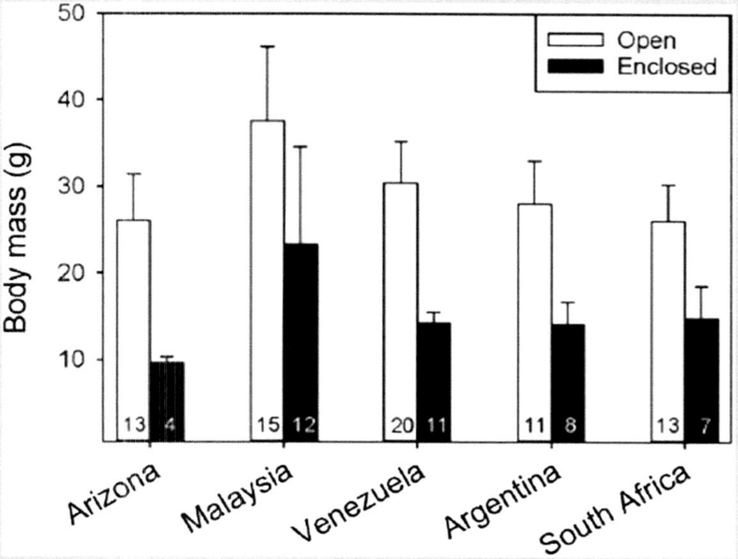


Smaller species of birds were significantly more likely to have enclosed nests than larger species. Numbers at the base of the columns indicate the number of species. (Figure from Martin et al. 2017; © 2016 The Authors. Functional Ecology © 2016 British Ecological Society, used with permission).

entrance holes or tubes, but air, to varying degrees, can also pass through the nest structure itself (Box 17.4 Open vs. Enclosed Nests). Cup nests provide a floor and walls, but no roof. Scrape, bed, and plate nests leave eggs and nestlings even more exposed to the external environment. The eggs in megapode "nests" are buried in decomposing vegetation, sand, or soil. To what degree do these various types of nests provide favorable microclimates for eggs and nestlings?

Some species in the family Megapodiidae incubate their eggs using environmental sources of heat, including burying them in mounds of leaf litter and other organic matter or laying them in burrows covered by geothermal or solar-heated sand or soils (Harris et al. 2014; Box 17.5 Underground Nesting by Megapodes). Among mound builders, the decomposition of organic matter creates the heat needed for incubation, but temperature levels can fluctuate within the mounds (Eiby and Booth 2008). In addition, the environment around the buried eggs is very humid, with less oxygen and more carbon dioxide than in the air above the mounds (Seymour et al. 1986, 1987). Male Australian Brushturkeys (*Alectura lathami*) attempt to minimize the amount of work needed to maintain the appropriate temperature by tossing litter onto mounds so it is not compacted and using dryer litter to keep water content low (Seymour and Bradford 1992). Low water content is important because wet mounds have higher thermal conductance and lose heat more rapidly. Males typically visit mounds daily and spend 0.5 to 2 hours working on the mounds (Eiby and Booth 2008), periodically digging down to the eggs, testing the temperature with their head or bill, then opening or closing the mound and adding or removing vegetation (using their feet) as needed. Although average mound temperatures are generally close to 34°C, temperatures do fluctuate, especially declining when it rains (Fig. 17.18). In three closely monitored mounds, mean temperatures were 33.8°C, but daily temperatures ranged from 24.5 to 37.5°C (Eiby and Booth 2008). In contrast to the embryos in most species of birds, brushturkey embryos appear to be better able to tolerate some daily fluctuation in mound temperature (Eiby and Booth 2008). Eggs incubated at lower temperatures (32°C compared to 34° or 36°C) do, however, take longer to hatch and

**Box 17.5 Underground Nesting by Megapodes**

Megapodes (family Megapodiidae, order Galliformes) use environmental heat to incubate their eggs. The various species of megapodes are broadly distributed throughout Australasia, ranging from the Nicobar Islands in the Indian Ocean to Tonga in the Pacific, and from the island of Uracus in the Northern Mariana archipelago to southern Australia (Jones et al. 1995). Fossil evidence indicates that megapodes were present in Australia during the Oligocene (approximately 25 mya; Boles and Ivison 1999). As the Australian plate moved toward Asia during the Pliocene (2.4–5.4 mya), megapodes likely dispersed into the Indonesian archipelago and toward Polynesia (Dekker 2007). In these different areas and islands, megapodes diversified and evolved different strategies for incubating their eggs (Harris et al. 2014).

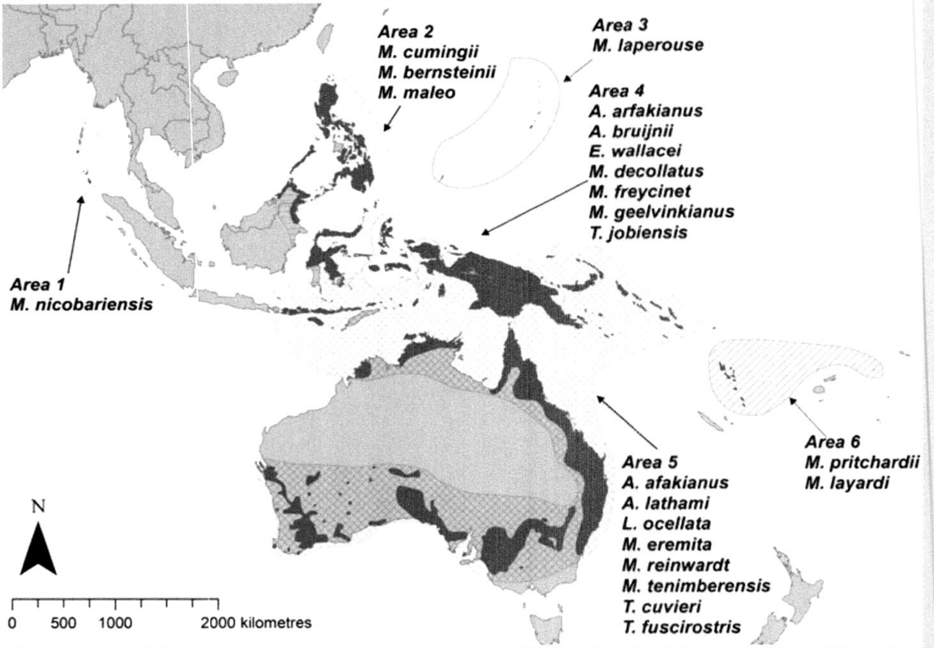


Global distribution of megapodes in six geographical areas. (Figure from Radley et al. 2018; © Foundation for Environmental Conservation 2018, used with permission).

Some megapodes build mounds of damp organic materials that decompose and generate heat, others burrow into substrates and use sun-generated heat, geothermal energy, or heat released as roots decompose to incubate their eggs (Jones et al. 1995; Harris et al. 2014). Most species of

(continued)

**Box 17.5** (continued)

megapodes are mound builders and are found in areas where vegetation and rainfall have favored the use of heat generated by decomposing vegetation to incubate eggs. Some megapodes, however, dispersed to locations with different environmental conditions where selection favored the construction of burrows and the use of other sources of heat to incubate eggs, including solar radiation, geothermal heat, and the decaying roots of dead trees (Sinclair et al. 2002; Harris et al. 2014).

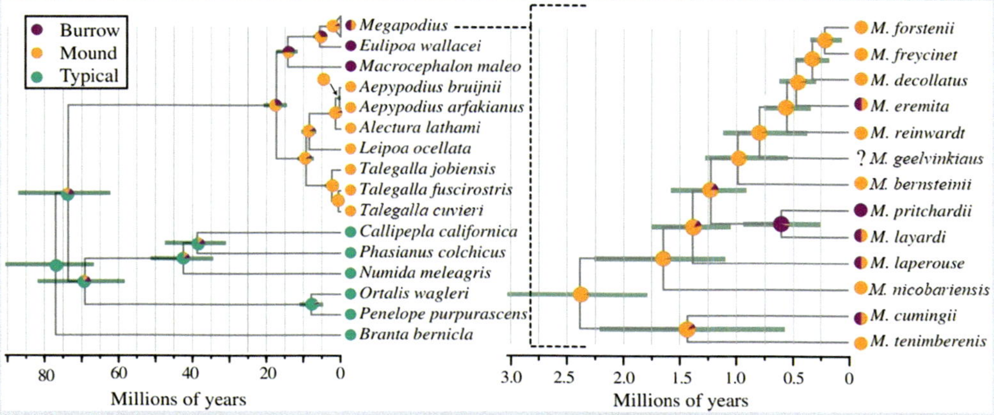


Phylogeny and nesting strategies of megapodes. Pie charts indicate the nesting strategies. The phylogeny on the right is an expanded view of *Megapodius*. *Eulipoa wallacei*, Moluccan Scrubfowl; *Macrocephalon maleo*, Maleo; *Aepypodius bruijnii*, Waigeo Brushturkey; *Aepypodius arfakianus*, Wattled Brushturkey; *Alectura lathami*, Australian Brushturkey; *Leipoa ocellata*, Malleefowl; *Talegalla jobiensis*, Red-legged Brushturkey; *Talegalla fuscirostris*, Black-billed Brushturkey; *Talegalla cuvieri*, Red-billed Brushturkey; *Callipepla californica*, California Quail; *Phasianus colchicus*, Ring-necked Pheasant; *Numida meleagris*, Helmeted Guineafowl; Ortalis wagleri, Rufous-bellied Chachalaca; *Penelope purpurascens*, Crested Guan; *Branta bernicla*, Brant; *M. forstenii*, Forsten's Scrubfowl; *M. freycinet*, Dusky Scrubfowl; *M. decollatus*, New Guinea Scrubfowl; *M. eremita*, Melanesian Scrubfowl; *M. reinwardt*, Orange-footed Scrubfowl; *M. geelvinkiaus*, Biak Scrubfowl; *M. bernsteinii*; Sula Scrubfowl; *M. pritchardii*, Niuafoou Scrubfowl; *M. layardi*, Vanuatu Scrubfowl; *M. laperouse*, Micronesian Scrubfowl; *M. nicobariensis*, Nicobar Scrubfowl; *M. cumingii*, Tabon Scrubfowl; *M. tenimberenis*, Tanimbar Scrubfowl. (Figure from Harris et al. 2014; © 2014 John Wiley & Sons Ltd., used with permission).

(continued)

**Box 17.5** (continued)

Malleefowl (*Leipoa ocellata*) mound. (Photo by Donald Hobern, Wikipedia, CC BY 2.0, https://creativecommons.org/licenses/by/2.0/).

Malleefowl on its mound nest. (Photo by Butupa, Wikipedia, CC BY 2.0, https://creativecommons.org/licenses/by/2.0/).

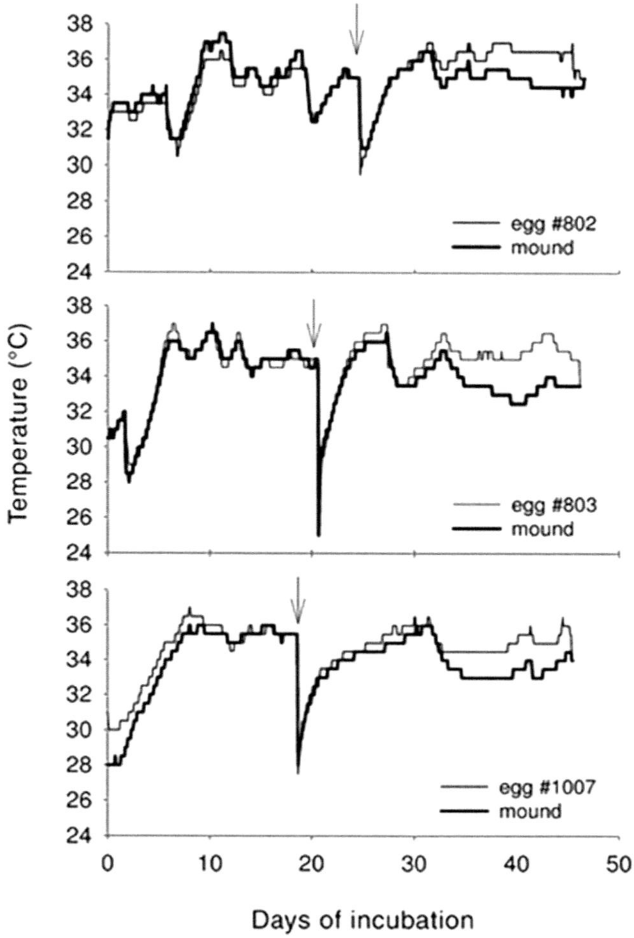


**Fig. 17.18** Temperature of eggshell surfaces and mound material adjacent to eggs of Australian Brushturkeys (*Alectura lathami*) from the time of laying (day 0) until hatching in three mounds. All mounds experienced fluctuations in temperature, especially on and after days with heavy rain (indicated by the arrows). Egg temperatures increased toward the end of incubation because of metabolic heat production by embryos. Eggshell temperatures falling to match mound temperatures on the right indicate hatching. (Figure from Eiby and Booth 2008; © 2008 Oxford University Press, used with permission)

young weigh less at hatching and, after hatching, take longer to dig themselves out of mounds (Göth and Evans 2004; Göth and Booth 2005). In addition, embryos in eggs incubated in the lab at temperatures above (36°C) and below (31°C) had higher mortality rates than those incubated at typical mound temperatures in the wild (34°C) (Göth and Booth 2005). Similar results have been reported for Malleefowl (*Leipoa ocellata*; Booth 1987). Incubation temperatures can also influence sex ratios. Göth (2008) determined that females hatched from eggs of Australian Brushturkeys when mound temperatures were higher (mean = 33.7°C), whereas males hatched from eggs when temperatures were lower (mean = 32.9°C). Such results suggest that male and female megapodes might be able to manipulate offspring sex ratios by manipulating mound temperatures, males by regulating mound temperature and females by where they lay eggs in a mound (Göth and Booth 2005; Jones et al. 1995).

The environment immediately surrounding embryos in eggs buried in mounds may also have less oxygen and more carbon dioxide than atmospheric air because of the limited diffusion of air into and out of mounds and the respiration of microorganisms. In a mound of an Australian Brushturkey (*Alectura lathami*), for example, the air just outside eggs has about 20% less oxygen and much more carbon dioxide than atmospheric air (Seymour et al. 1986; Fig. 17.19). As a result,

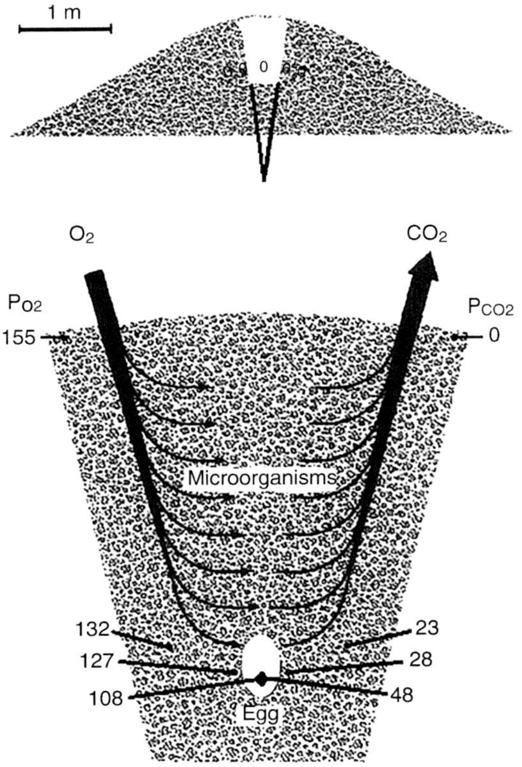


**Fig. 17.19** Above, cross-section through a mound of an Australian Brushturkey (*Alectura lathami*) showing location of five eggs about 60 cm deep. Below, the respiration of microorganisms lowers oxygen levels and raises carbon dioxide levels with the mound. The width of the arrow lines indicates relative rate of diffusion; at the level of the eggs, diffusion has little impact on gas levels. Numbers represent partial pressures of oxygen ($Po_2$, on the left) and partial pressure of carbon dioxide ($Pco_2$, on the right) at the mound surface, general egg level, and just inside and outside of an egg. (Figure from Seymour et al. 1986; © 1986 Springer-Verlag, used with permission)

there is less of a diffusion gradient (difference in partial pressures of gases) between the developing embryo and its external environment than is the case for other birds where eggs are not buried, potentially making the exchange of gases less efficient. However, eggs of Australian Brushturkeys (and other megapodes) have thinner shells than the eggs of other birds of comparable size, allowing more efficient conductance of gases (Seymour and Ackerman 1980; Seymour et al. 1986). Thus, despite their apparently unfavorable gaseous environment, levels of oxygen and carbon dioxide in megapode eggs are similar to those of the eggs other species of birds (Seymour et al. 1986).

Burrow nests can provide relatively stable temperatures for developing embryos and growing nestlings. For example, nest chambers in the burrows of Rainbow Bee-eaters (*Merops ornatus*) averaged 4 to 6°C warmer than ambient temperatures, reducing the thermoregulatory costs for both adults and nestlings (Lill and Fell 2007). However, the air within nest chambers can potentially be relatively low in oxygen (hypoxic) and high in carbon dioxide (hypercapnic), depending on the gas flux through the soil and entrance tunnels. Depending on the length and orientation of burrow tunnels, the exchange of air between nest chambers and the atmosphere may occur by diffusion and wind-induced ventilation, as well as by the movements of adults acting like a piston and moving air in and out. However, because exchange of air is generally limited to some degree, the air in burrow nest chambers typically differs in composition from atmospheric air. For example, oxygen concentrations in nest chambers of European Bee-eaters (*Merops apiaster*) can be as low as 15% (compared to about 20% in normal air) and carbon dioxide concentrations as high as 6.5% (compared to about 0.04% in normal air) (White et al. 1978). Mean carbon dioxide and oxygen levels in the nest chambers of Bank Swallows (*Riparia riparia*) were 2.6 and 17.8%, respectively (Wickler and Marsh 1981). Mean oxygen levels in the nest chambers of Rainbow Bee-eaters were 19.4% (Lill and Fell 2007). However, levels of oxygen in burrows are not sufficiently reduced to cause problems. White et al. (1978) suggested that even oxygen levels as low as 15% would not cause respiratory problems for most birds. However, levels of $CO_2$ reported in burrows in some studies seem high enough to potentially cause respiratory problems. For example, an atmosphere of 2% $CO_2$ lowered hatching success of Japanese Quail (*Coturnix japonicus*), a non-burrow-nesting species (Bavis and Kilgore 2001), and an atmosphere of 3% $CO_2$ killed half of the embryos in the eggs of Domestic Chickens (*Gallus g. domesticus*), another non-burrow-nesting bird (Taylor et al. 1971). Embryos and nestlings in burrow nests are sometimes exposed to similar or even higher $CO_2$ levels yet suffer no apparent negative consequences (Mondain-Monval and Sharp 2018). As one adaptive response, the eggshells of some burrow-nesting species of birds exhibit increased conductance to aid in the exchange of $CO_2$ and $O_2$ (Birchard and Kilgore 1980). In addition, burrow-nesting birds may be less sensitive to increased levels of $CO_2$, with higher levels needed to stimulate an increase in respiratory rate than for non-burrow-nesting species (Kilgore et al. 1985). For example, White et al. (1978) found that respiratory rates of nestling European Bee-eaters in burrows with 3% $CO_2$ averaged just 68 breaths/minute, but, with 6% $CO_2$, respiration became labored and rapid, with rates averaging 100 breaths/minute. Such elevated respiratory rates help reduce levels of $CO_2$ in the blood. Prolonged exposure to elevated levels of $CO_2$ can potentially cause respiratory acidosis because $CO_2$ in the blood reacts with $H_2O$ and generates hydrogen ions, with more hydrogen ions causing a more acidic pH, plus bicarbonate ions ($HCO_3^-$). Prolonged acidosis can potentially affect normal functioning of the heart and nervous system (DiBartola 2012; Whitaker-Fornek et al. 2020). As blood pH becomes more acidic, bird kidneys reduce the amount of bicarbonate ions lost to the urine and retain more in the blood. Because bicarbonate is alkaline, the bicarbonate ions retained in the blood help neutralize the acid (hydrogen ions).

One advantage of cavity nests is that they generally have more stable temperatures than open nests (Martin and Ghalambor 1999). As with burrow nests, the thermal advantages of cavity nests may reduce thermoregulatory costs for both adults and nestlings. Another advantage of warmer nest cavities is that adults can spend less time incubating eggs and brooding young nestlings and more time foraging because heat is lost from eggs and young more slowly. In a comparative analysis, Conway and Martin (2000) found that females in cavity-nesting species had shorter incubation bouts and made more frequent trips to and from nests than did females in open-nesting species (Fig. 17.20).

Despite the thermal advantages, the air in cavities, like burrows, could have reduced levels of oxygen and increased levels of carbon dioxide. Some investigators have suggested that the relatively short incubation periods of woodpeckers (57% shorter than expected based on their body and egg sizes; Wiebe 2007) may be a result of the poor ventilation and hypoxic conditions in tree cavities, with early hatching allowing embryos (obtaining oxygen by passive diffusion) to more quickly transition to active breathing (Yom-Tov and Ar 1993). Ar et al. (2004) generated a predictive model of the gas composition in a woodpecker cavity taking into account mechanisms of gas exchange and the influence of wind velocity and concluded that, with multiple individuals in a cavity (e.g., an adult plus several nestlings) and no wind, hypoxic conditions could develop. Although few investigators have examined gas levels in cavities, studies of Syrian Woodpeckers (*Dendrocopos syriacus*) and Northern Flickers (*Colaptus auratus*) indicate that levels of oxygen and carbon dioxide in cavities differ little from atmospheric levels; oxygen levels in Syrian Woodpeckers never dropped below 19.2% (Mersten-Katz 1997) and levels of oxygen and carbon dioxide in flicker nests averaged 20.4 and 0.4%, respectively (Howe et al. 1987). Similarly, Wiebe et al. (2007) found that oxygen levels in the nest cavities of Northern Flickers averaged 20.5%. These results suggest that there is sufficient movement of air into and out of cavities to maintain suitable levels of oxygen and carbon dioxide. Contributing to this exchange is the tendency of the temperature within nest cavities to be higher than ambient temperatures, with the temperature gradient causing convective air flow into and out of cavities. Concerning the relatively small eggs of woodpeckers, Wiebe et al. (2007) provided two possible explanations: intense sibling competition favors early hatching, and sexual selection on females favors smaller eggs when females compete for more nesting opportunities in polyandrous species, e.g., facultatively polyandrous Northern Flickers.

For cavity-nesting species of birds, nest thermal environments can be influenced by several factors, including thickness of the cavity walls,

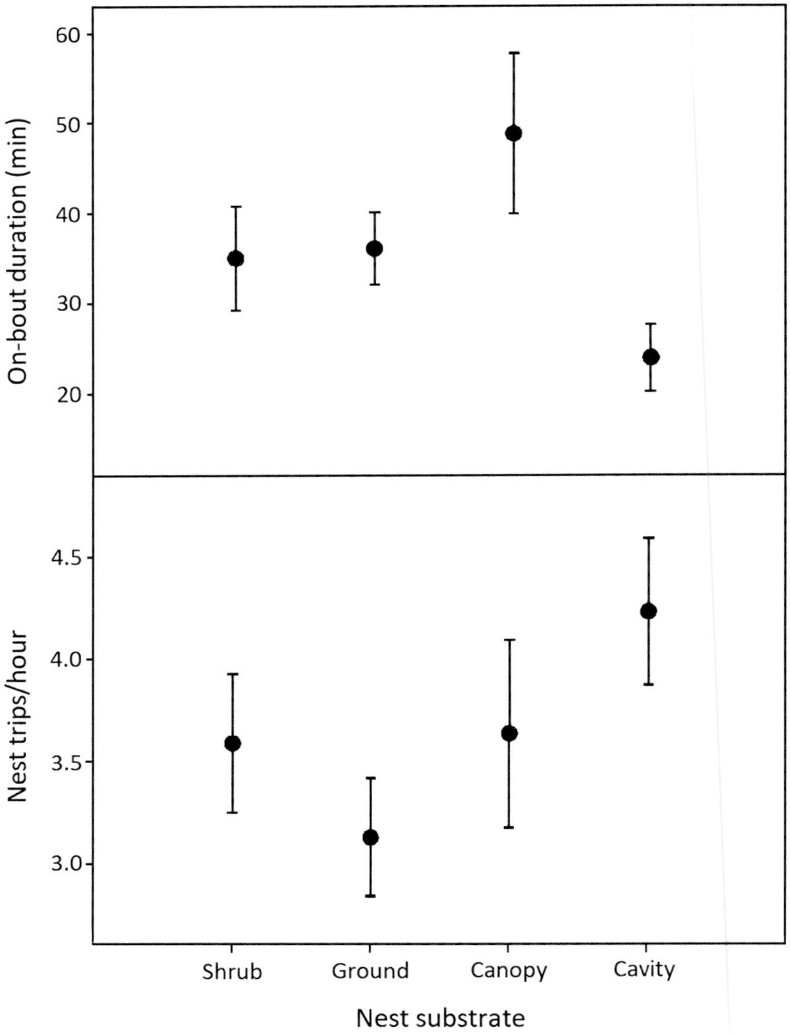


**Fig. 17.20** Mean (± SE) duration of incubation bouts and number of trips per hour among North American passerines (with female-only incubation) nesting in different substrates. Numbers above error bars indicate number of species. (Figure from Conway and Martin 2000; © Society for the Study of Evolution, used with permission)

density of surrounding vegetation (that can affect exposure to solar radiation), the size of the entrance hole, and cavity-entrance orientation (Wachob 1996). Birds may create (primary cavity-nesters) or use (secondary cavity-nesters) cavities (or nest boxes built by humans) with some combination of these variables that result in suitable nest microclimates (Box 17.6 Diversity and Distribution of Tree-Cavity-Nesting Birds). For example, cavities used by secondary cavity-nesting South Island Saddlebacks (*Philesturnus carunculatus*) tended to have smaller entrance holes and, as a result, were warmer than unused cavities (Rhodes et al. 2009). A number of investigators have examined possible relationships between nest-box orientation and nest-site selection by secondary cavity-nesters. Tree Swallows (*Tachycineta bicolor*) in Massachusetts, for example, preferred nest boxes with entrances facing east and south early in the breeding season, but showed no preference later in the breeding season (Ardia et al. 2006). When ambient temperatures were lower early in the breeding season, east- and south-facing boxes were warmer than west- and north-facing boxes; later in the season, nest-box orientation did not affect box temperature. Thus, early in the breeding season, Tree Swallows preferred nest boxes with entrance orientations that provided warmer nest temperatures. Similarly, American Kestrels (*Falco sparverius*) in Pennsylvania preferred next boxes with southeast-facing entrances; boxes that were warmer in the morning and cooler in the afternoon than nest boxes with entrances facing other directions (Rohrbaugh and Yahner 1997).

Investigators have found that many cavity-nesting birds exhibit preferences for cavities with certain entrance orientations. For example, cavities of Red-cockaded Woodpeckers (*Dryobates borealis*) exhibit a strong westward bias, possibly because fungal communities that facilitate cavity excavation are more active on the west side of cavity trees (Landler et al. 2022). More generally, Landler et al. (2014), based on a meta-analysis involving 80 populations of 23 species of Northern-Hemisphere woodpeckers, determined that populations at higher latitudes preferred cavities with more southerly orientations, likely because south-facing cavities would generally be warmer than cavities oriented in other directions (e.g., Wiebe 2001). There are, however, exceptions. For example, the orientation of entrances of cavities created by Red-breasted Sapsuckers (*Sphyrapicus ruber*) were randomly distributed (Joy 2000), and cavity-entrance orientation did not affect either nest-site selection or breeding success of Black-capped Chickadees (*Poecile atricapillus*; Mennill and Ratcliffe 2004). For such species, cavity-entrance orientation may have no effect on nest microclimate or the effect may be too small to influence preferences. Alternatively, other factors, such as sufficient illumination (Maziarz and Wesołowski 2014; Podkowa and Surmacki 2017; Podkowa et al. 2019) or predation risk, may be more important in determining nest-site characteristics than nest microclimate.

For many cavity-nesting species, entrance orientation not only influences nest-site selection but can also influence nest success. Great Tits (*Parus major*) in the United Kingdom, for example, are less likely to choose cavities with entrances facing south-southwest and, in addition, young raised in cavities that did face that direction were of lower quality (lighter and smaller) than those raised in cavities with entrances facing other directions (Goodenough et al. 2008). Factors possibly contributing to the lower quality of young in south-southwest facing cavities include less favorable microclimates (e.g., higher cavity temperatures that might cause thermal stress, increased exposure to prevailing wind

**Box 17.6 Diversity and Distribution of Tree-Cavity-Nesting Birds**

An estimated 1878 species of birds (18%) nest in tree cavities, including 355 species of primary excavators, 126 species of facultative excavators (i.e., species that can either excavate their own cavities or use already available cavities), and 1357 species of non-excavators (also called secondary cavity-nesting species). Insufficient information was available to categorize 40 species (van der Hoek et al. 2017). The orders of birds with the most tree-cavity-nesting species are Passeriformes (586 species), Piciformes (404 species), Psittaciformes (371 species), and Strigiformes (219 species). The zoogeographical realms with the greater number of tree-cavity-nesting species are the Neotropical (678 species) and Oriental (453 species) realms. The number of species nesting in tree cavities is particularly high in the Amazon basin and the eastern slope of the Andes Mountains in South America. The realms with the highest percentage of species that nest in tree cavities are the Oriental (19.5%) and Australasia (17.2%) realms. The Nearctic realm has the highest percentage of primary and facultative excavating species (32% of all species of birds). Given the many species of birds that nest in tree cavities, conserving forest habitats and, especially, old trees where decay and damage can create cavities, is critically important (van der Hoek et al. 2017).

(continued)

**Box 17.6** (continued)

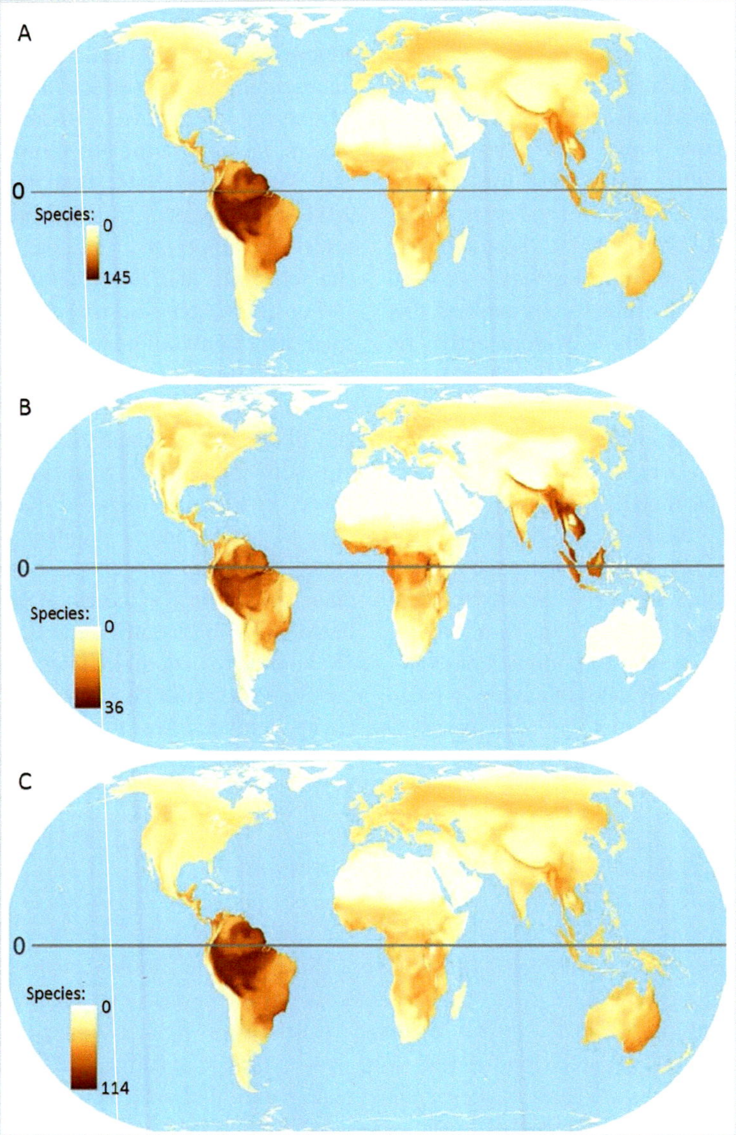


Distribution and abundance of species of birds that nest in tree cavities. (**a**) All species of birds that nest in tree cavities, including primary excavators, facultative excavators, and secondary (non-excavating) cavity-nesting species, (**b**) primary and facultative excavators, and (**c**) secondary cavity-nesters (non-excavators). (Figures from van der Hoek et al. 2017; © 2017 John Wiley and Sons Ltd., used with permission).

and rain) and, because of those warmer and moister microclimates, greater exposure to ectoparasites, bacteria, and fungi (Goodenough et al. 2008).

Cavities that provide favorable thermal conditions can contribute to increased reproductive success. For example, Dawson et al. (2005) placed heating pads in some nest boxes used by Tree Swallows so that temperatures were 5°C higher than in non-heated control nest boxes. Warmer temperatures may reduce thermoregulatory costs for nestlings and reduce time spent brooding by adults, leaving more time to provision young. Thus, when nestlings approached fledging age, those in the warmer boxes were heavier and had longer flight feathers, characteristics that would likely increase their chances of survival. In contrast, Ardia et al. (2010) experimentally reduced temperatures in some nest boxes used by Tree Swallows and found that nestlings in cooled nests weighed less than those in control nests. Cooler temperatures increased the energetic cost of incubation for adult females, negatively impacting their body condition and reducing their ability to provide care for nestlings. In addition, immune responses of nestlings in cooled nests were less effective than those of nestlings in control (warmer) nests. Thus, cooler nest temperatures had a negative impact on the condition of both adults and nestlings.

In environments where ambient temperatures regularly exceed optimums (in terms of thermoregulatory costs and development) for adults, embryos, and nestlings, nest cavities that provide cooler thermal conditions would be important. Elf Owls (*Micrathene whitneyi*), for example, are secondary cavity-nesters and, in the Sonoran Desert, tend to select cavities in saguaro cacti (*Carnegiea gigantean*) with entrances that face north, likely because north-facing cavities are cooler than south-facing cavities (Hardy and Morrison 2001). In fact, Soule (1964) found that temperatures in saguaro cavities were 2 to 7°C cooler than ambient temperatures during the Elf Owl's breeding season.

Open nests provide less protection from the elements than burrow or cavity nests, but can still provide favorable microclimates for incubating and brooding adults as well as developing embryos and growing nestlings. Because they are well ventilated with gas levels matching those of the atmosphere, studies of open nests have typically focused on nest temperature because of its importance in the development and growth of young as well as the energetic requirements of incubation and brooding for adult birds (e.g., Akresh et al. 2017; Ospina et al. 2018). Of course, other characteristics of nests can also be important. For example, particularly in areas with frequent rain, nests must be able to dry quickly because eggs and young cool more rapidly in wet nests (Deeming and Campion 2018; Fig. 17.21).

During incubation, the temperatures of bird eggs generally range between 30 and 40°C (Webb 1987). The ability of adult birds to minimize energetic costs while maintaining these temperatures depends on the characteristics of nests and their location. As noted earlier, adequate structural support may be more important than insulation for some birds, particularly large birds. However, for many birds, nest insulation is clearly important. For example, examination of nests of Hawaii Amakihis (*Chlorodrepanis virens*) found at elevations ranging from 1600 to 2600 m in elevation on the island of Hawaii revealed that those built in cooler, higher-elevation sites had thicker walls with lower thermal conductance (i.e., lost heat less rapidly) than nests from warmer, lower-elevation sites (Kern and Van Riper 1984). Nests at lower-elevation sites where rain was more frequent were also more porous and dried more rapidly. Similarly, a comparison of the nests of four species of birds that nest across a wide latitudinal range revealed that nests at higher latitudes were generally larger and, for one species, had thicker walls (Crossman et al. 2011). Although there are several possible explanations for such latitudinal differences (e.g., if there are fewer predators at higher latitudes, birds might respond by building larger, more visible nests), one likely explanation is that temperatures are typically lower at higher latitudes and larger, thick-walled nests help maintain warmer nest temperatures (Vanadzina et al.

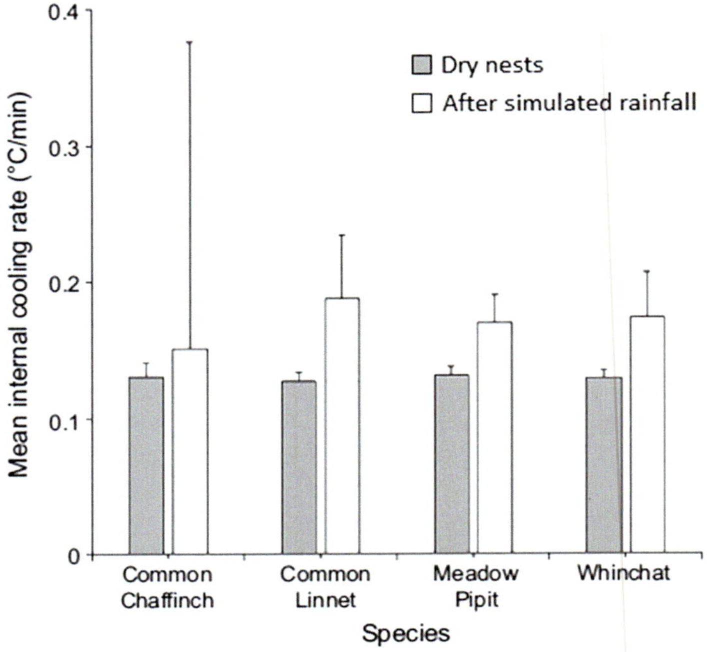


**Fig. 17.21** Mean (± SD) internal cooling rates of nests of four species of songbirds when dry and after simulated rainfall. Common Chaffinch, *Fringilla coelebs*; Common Linnet, *Linaria cannabina*; Meadow Pipit, *Anthus pratensis*; Whinchat, *Saxicola rubetra*. (Figure from Deeming and Campion 2018; © 2018 Natura optima dux Foundation, used with permission)

2022; Fig. 17.22). For species of birds at temperate latitudes that are multibrooded, seasonal variation in temperature may favor different nest characteristics for different nests. In the United Kingdom, the amount of material added to the innermost lining of nests by Eurasian Blue Tits

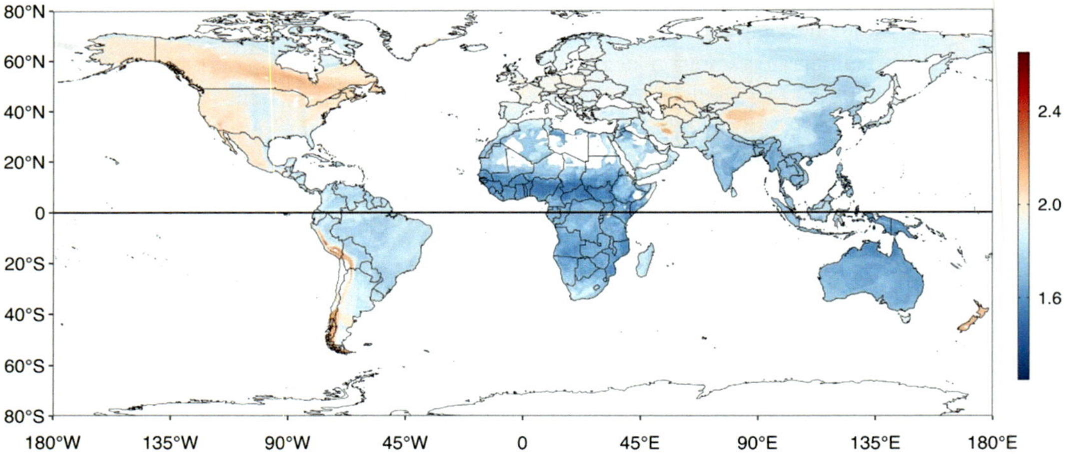


**Fig. 17.22** Geographical distribution of mean outer nest volume divided by body mass for 827 species of birds. Higher values indicate larger nests (scale at right). Species at higher latitudes and altitudes where ambient temperatures are lower build larger or thicker nests. Areas with fewer than four species were not considered and so appear blank. Values along the y- and x-axes correspond to latitude and longitude, respectively. (Figure from Vanadzina et al. 2022; © 2022 The Authors. Journal of Animal Ecology published by John Wiley & Sons Ltd. on behalf of British Ecological Society, open-access article distributed under the terms of the Creative Commons Attribution License, https://creativecommons.org/licenses/by/4.0/)

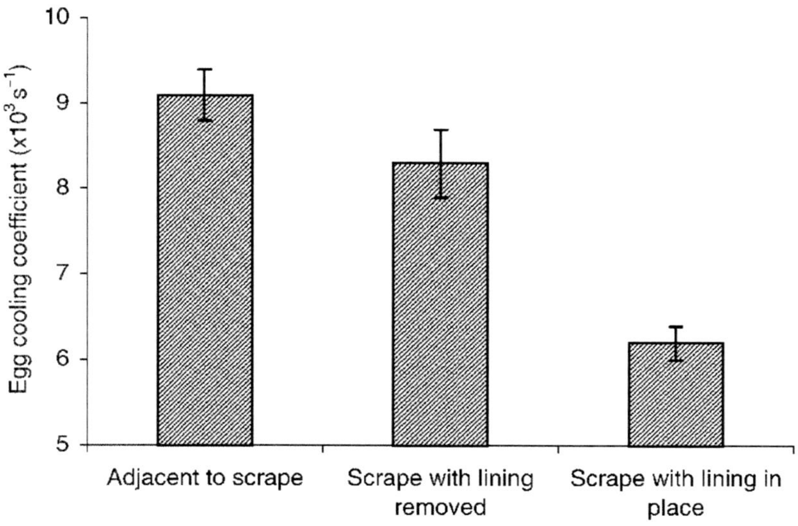


**Fig. 17.23** Cooling coefficients (± SE) of eggs in natural nest scrapes of Pectoral Sandpipers (*Calidris melanotos*), in scrapes with lining temporarily removed, and immediately adjacent to scrapes. The cooling coefficient reflects the rate of heat loss from an egg, with a high value indicating a higher rate of heat loss. (Figure from Reid et al. 2002; © 2002 John Wiley and Sons, used with permission)

(*Cyanistes caeruleus*) decreases later in the breeding season, suggesting that less insulation is needed when temperatures are warmer (Mainwaring and Hartley 2008).

Even with simple scrape nests, adding a thin layer of insulating material can impact the rate at which heat is lost from eggs when not being incubated. For scrape nests of Pectoral Sandpipers (*Calidris melanotos*) breeding on the tundra in northern Alaska, the rate of heat loss from eggs increased with scrape depth, but decreased with the increasing thickness of nest lining material (Reid et al. 2002). On the tundra, soil temperatures decrease with increasing depth so eggs in deeper scrapes experience cooler ground temperatures. However, eggs in scrapes that are too shallow also tend to cool faster because they are more exposed to the wind. So, the most favorable microclimates for eggs were provided by scrapes about 50 mm deep with about 20 mm of nest lining (adding more did not further reduce heat loss). Deeper scrapes would provide eggs with additional protection from the wind, but would also expose eggs to cooler soil temperatures. Thus, Pectoral Sandpipers create scrapes and add the amount of nest lining that tend to minimize heat loss from their eggs (Fig. 17.23).

Nest microclimates can be influenced by where open nests are located and how they are oriented. Some species of birds appear to consider vegetation cover when determining where to build nests (Fig. 17.24). For example, Walsberg and King (1978) examined the distribution of vegetation located above the open-cup nests of White-crowned Sparrows (*Zonotrichia leucophrys*) at a high-altitude site (1890 m) in Oregon and found that most nests were positioned so they were more exposed to sunlight (75% more on average) in the morning when ambient temperatures were lower than the afternoon when temperatures were higher. Similarly, two small sagebrush-obligate birds, Brewer's Sparrows (*Spizella breweri*) and Sagebrush Sparrows (*Artemisiospiza nevadensis*), selected nest sites in shrubs and niches that were warmer and less variable in temperature than unused sites (Scherr and Chalfoun 2022). Warbling Vireos (*Vireo gilvus*) built nests in locations where, on average, 2.4 times more sky was visible through the foliage canopy in the eastern sky than the western sky (Walsberg 1981).

Many ground-nesting birds orient nests relative to surrounding vegetation so that nests are oriented, or more exposed to the sky, in particular directions. Burton (2007) reviewed the literature and, based on studies of seven ground-nesting species in Europe and North America, found that nests tend to be oriented to the north at lower latitudes and to the east or south at higher

**Fig. 17.24** Hemispherical photo of the vegetative canopy over the nest of a Phainopepla (*Phainopepla nitens*). (Figure from Walsberg 1992; © 2015 Oxford University Press, used with permission)

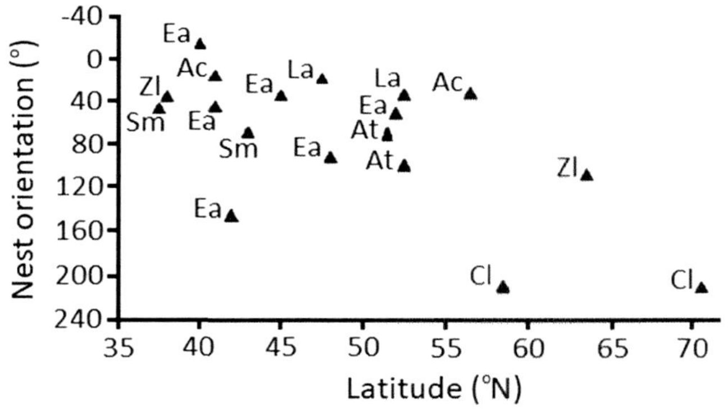


**Fig. 17.25** Mean entrance orientations of nests of seven open-cup, ground-nesting passerines in Europe and North America relative to latitude. Europe = La, Wood Lark (*Lullula arborea*); Ac, Tawny Pipit (*Anthus campestris*); At, Tree Pipit (*Anthus trivialis*). North America = Ea, Horned Lark (*Eremophila alpestris*); Zl, White-crowned Sparrow (*Zonotrichia leucophrys*); Cl, Lapland Longspur (*Calcarius lapponicus*); Sm, Eastern Meadowlark (*Sturnella magna*). (Figure from Burton 2007; © 2007 Oxford University Press, used with permission)

latitudes (Fig. 17.25). At lower latitudes, nests facing north are shaded from the mid-day sun, whereas at higher latitudes east- or south-facing nests are more exposed to solar radiation either in the morning when temperatures are coolest (east-facing nests) or for much of the day (south-facing nests). Such differences in exposure to solar radiation may help maintain favorable nest

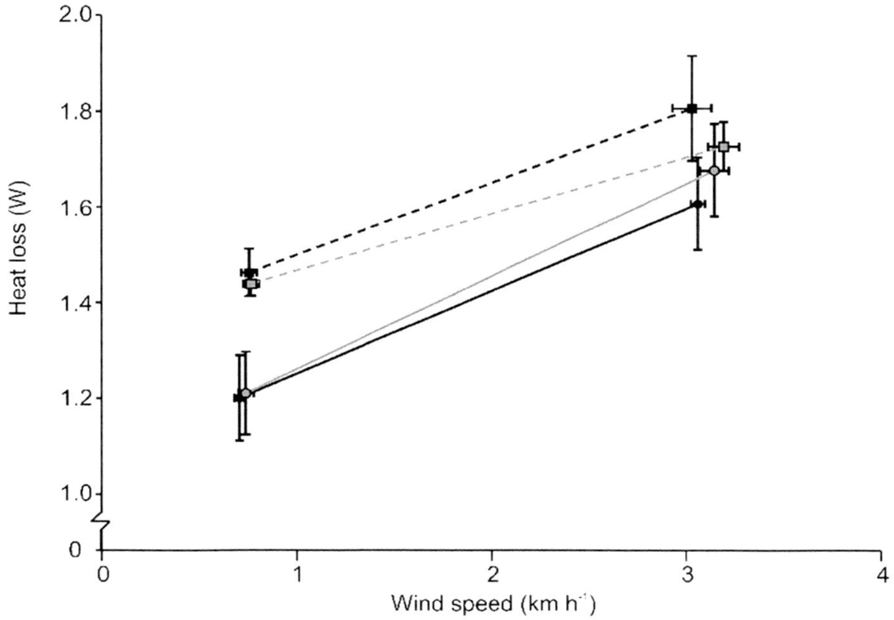


**Fig. 17.26** Relationship between wind speed and heat loss (W, watts) from cup-shaped nests of Spiny-cheeked Honeyeaters (*Acanthagenys rufogularis,* black) and Yellow-throated Miners (*Manorina flavigula*, gray) when nests were covered to simulate the presence of an incubating or brooding adult (circles with solid lines) and open (square points with dashed lines). Each point represents the mean ± 95% CI. (Figure from Heenan and Seymour 2012; © 2012 Heenan, Seymour, open-access article distributed under the terms of the Creative Commons Attribution License, https://creativecommons.org/licenses/by/2.0/)

temperatures that may influence hatching and nesting success. For example, Tree Pipits (*Anthus trivialis*) are ground nesters with well-hidden nests built in shallow depressions that are often overhung on one side by grass. In England, most nests were found to face east, an orientation that provided shelter from prevailing west-southwest winds, allowed warming by the morning sun, and provided shade during the afternoon (Burton 2006). East-facing nests also had higher hatching success, and the thermal environment in those nests may have been a contributing factor (Burton 2006).

Nest microclimates that provide shelter from wind can also be important in reducing heat loss from adults and nestlings. For example, Heenan and Seymour (2012) placed nests of two species of songbirds in a wind tunnel and inserted heated, artificial eggs with temperature sensors inside. Even at low wind speeds, the rate of heat loss from the eggs increased dramatically (Fig. 17.26) and, to maintain egg temperatures, Heenan and Seymour (2012) estimated that an adult would have to nearly double the amount of heat being applied via brood patches.

Nest microclimates can also influence parental behavior. Ground-nesting Dark-eyed Juncos (*Junco hyemalis*) in Montana were found to orient their nests in different directions and nests oriented to the southeast and southwest were significantly warmer than those oriented to the northeast and northwest. During incubation, females with warmer south-facing nests spent less time incubating that those with cooler north-facing nests and, therefore, had more time to forage and engage in other self-maintenance activities (Robertson 2009). Most ground-nesting Chestnut-collared Longspurs (*Calcarius ornatus*) in Montana oriented nests toward the southeast and those nests had the highest mid-day temperatures.

**Fig. 17.27** Nest of a Black-throated Blue Warbler (*Setophaga caerulescens*). (Photo from U. S. Fish and Wildlife Service, Wikipedia, CC0 Public Domain)

As a result, adult longspurs with southeast-facing nests spent more time shading young from the sun during the nestling period and, with less time to forage, their nestlings grew at slower rates and were more than two grams lighter on average at fledging than nestlings in nests oriented in different directions. Because lower mass at the time of fledging would likely reduce the likelihood that young would survive, building nests that face southeast is seemingly maladaptive for these longspurs (Lloyd and Martin 2004). What such results may indicate, however, is that birds may face seasonal trade-offs when attempting to select nest sites and orientations with particular microclimates. Sites and orientations that provide thermal benefits for adults and embryos during incubation may not provide optimal thermal environments for adults and young during the nestling period.

Nests are not always oriented in directions that might provide the most favorable microclimate because adults must consider other factors when selecting nest sites. For example, nest-site selection by Black-throated Blue Warblers (*Setophaga caerulescens*) appears to represent a trade-off, with the choice of sites influenced by the need for concealment and location relative to foraging sites as well as the need for a favorable microclimate (Holway 1991; Fig. 17.27). Dusky Warblers (*Phylloscopus fuscatus*) exhibit plasticity in choice of nest sites, preferring nest sites that are higher above ground and less vulnerable to predation over those with more favorable microclimates and closer to food during years when predator abundance is higher (Forstmeier and Weiss 2004). In the Arabian Desert, Greater Hoopoe-Larks (*Alaemon alaudipes*) also face a trade-off, preferring nest sites in open areas exposed to solar radiation and higher temperatures that increase the risk of mortality to embryos and nestlings over sites in bushes with more cover and lower temperatures because open sites provide a better view of their surroundings and increase the likelihood that adults will spot approaching predators. However, later in the breeding season when mid-day temperatures increase, adults are forced to select nest sites in bushes that provide more shade and a more favorable microclimate, but restrict their view and reduce the likelihood that incubating and brooding adults will spot approaching predators (Tieleman et al. 2008).

Nest microclimates can be as important for adult birds as for embryos and nestlings. Some hummingbirds nest at high latitude and high-altitude locations where the combination of their high metabolic rates and sometimes cool ambient temperatures (particularly at night) may make balancing energy budgets difficult. Incubating and brooding female hummingbirds can, of course, conserve energy at night by entering torpor (but see Eberts et al. 2023). However, even after allowing their body temperatures to drop, hummingbirds may still lose heat via radiation when ambient temperatures are lower than their body temperature. To reduce such heat loss, female hummingbirds build well-insulated nests (Eberts et al. 2023) and, often, with at least partial branch "roofs" (Fig. 17.28). Such "roofs" help reduce heat loss by radiation because hummingbirds are less exposed to the cold night sky (Calder 1973).

### 17.3.4 Phenotypic Signal

The structure of nests is strongly influenced by natural selection, but, for some species of birds, nests also serve as extended phenotypic signals that provide information about the quality of individuals that build nests. Because building nests is energetically costly (e.g., Mainwaring and Hartley 2013), the ability of builders to construct nests may provide conspecifics with information about various aspects of condition and quality, including immune capacity, reproductive experience, and the disposition of a builder to invest in parental care (Moreno 2012).

One possible way to convey information about individual quality is for males to build multiple nests. Some investigators have suggested that males that build multiple nests are practicing their nest-building skills (Hunter 1900), that extra nests will provide shelter for adults or fledglings (Verner 1965), or extra nests serve as decoys and reduce the likelihood of nest predation (Leonard and Picman 1987). However, the results of most studies suggest that males construct multiple nests to attract mates. Among some polygynous species, males that build more

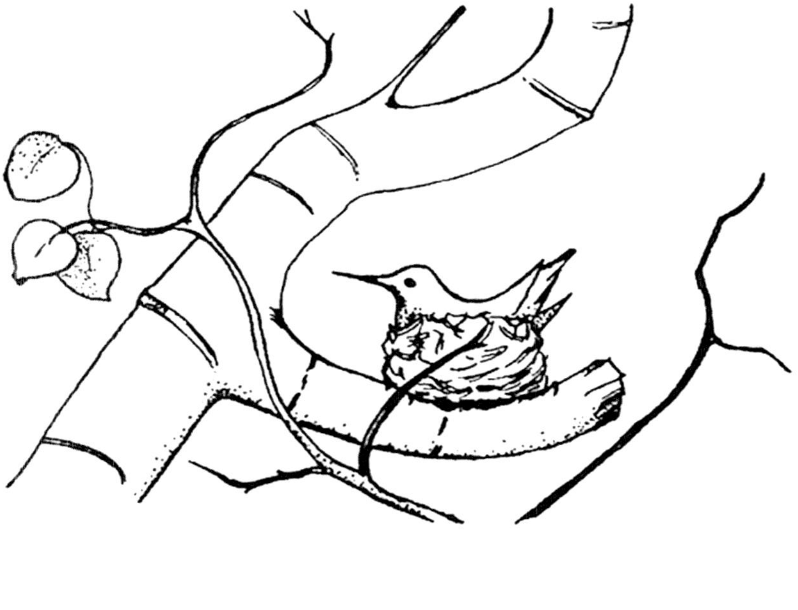

**Fig. 17.28** Typical nest location for a female Broad-tailed Hummingbird (*Selasphorus platycercus*) with a branch above the nest serving as a "roof." (Figure from Calder 1973; © Ecological Society of America, used with permission)

nests attract more mates. For example, male Red Bishops (*Euplectes orix*) that build more nests attract more mates and have greater reproductive success than males that build fewer nests (Freidl and Klump 2000; Klump et al. 2009). Female Eurasian Wrens (*Troglodytes troglodytes*) assess the number of vacant nests in male territories and preferentially mate with males that have more nests (Evans and Burn 1996).

In a comparative study of 76 species of songbirds, Soler et al. (1998) found that species that build larger nests relative to their body size invest more in reproduction. One explanation for such results is that individual birds increase their nest-building effort to signal their quality or willingness to invest in reproduction to mates who may, in turn, respond by also increasing their efforts. For example, female Great Tits (*Parus major*) invest more in nest building when paired with males in good condition that could potentially be better fathers (Broggi and Senar 2009). Similarly, female Penduline Tits (*Remiz pendulinus*) provide more parental care when their mates spend more time nest-building and build larger nests (Szentirmai et al. 2005), and female Blue Tits (*Cyanistes caeruleus*) paired with males that placed more feathers around nests produced larger clutches (Sanz and Garcia-Navas 2011). However, nest-building effort is not always an honest signal. Among House Sparrows (*Passer domesticus*), male contributions to nest construction were not related to the amount of parental care provided later (Hoi et al. 2003). For House Sparrows and perhaps other species where there is no apparent relationship between male contributions to nest building and subsequent paternal care, increased effort in nest building by males may be strategy to elicit greater parental care by their mates (Hoi et al. 2003). Other examples of nests serving as extended phenotypes are provided in Table 17.1.

## 17.4 Nest-Site Selection and Predation

Nest predation is the major cause of nest failure for birds (Ricklefs 1969) and, as a result, numerous studies have been conducted in an attempt to determine the factors that might contribute to variation in nest predation rates among species and habitats and across time (e.g., early vs. late in the nesting season). Some hypotheses proposed to explain variation in predation rates focus on predator abundance or diversity. For example, the predator-abundance hypothesis posits that the risk of nest predation is positively related to predator abundance (Caro 2005; Lima 2009; Eichholz et al. 2012), whereas the predator-diversity hypothesis predicts that rates of nest predation will be higher in areas with a greater diversity of different predator species because different predators have different foraging strategies (Filliater et al. 1994). With predators using different foraging strategies, e.g., avian predators depending on visual cues to locate nests and mammalian or reptilian predators using olfactory cues, the likelihood of nests being located and eggs or nestlings predated may be increased.

Regardless of the abundance or diversity of potential nest predators, birds may be able to influence the risk of nest predation by selecting nest sites less likely to be detected by those predators (Box 17.7 Nest Neighbors that Increase Nesting Success). Hypotheses proposed to explain possible relationships between the risk of nest predation and characteristics of nest sites include the nest-concealment hypothesis, the structural-complexity hypothesis, and the potential-prey-site hypothesis. One of the most commonly cited hypotheses to explain variation in rates of nest predation is the nest-concealment hypothesis (Borgmann and Conway 2015). This hypothesis suggests that dense foliage around nests reduces the likelihood of predation by reducing the transmission of visual, olfactory, or auditory cues potentially used by predators to locate nests. By contrast, the structural-complexity hypothesis posits that the presence of dense, structurally complex vegetation around nests acts to impede access to those nests by predators (e.g., Sugden and Beyersbergen 1986). Finally, the potential-prey-site hypothesis predicts that the probability of a predator locating a nest is reduced when they must search many similar, potentially suitable sites (Martin 1993). An

**Table 17.1** Examples of species where nest characteristics serve as extended phenotypes

| Species[a] | Nest characteristic | Effect | Reference |
|---|---|---|---|
| Black-faced Sheathbill | Nest decorations | May be indicative of genetic or parental quality | Danel et al. (2021) |
| Black Kite | Nest decorations | Decorations convey information about territory quality and dominance to conspecifics | Sergio et al. (2011) |
| Eurasian Nuthatch | Nest size | Males fed incubating females at higher rates after nests were experimentally enlarged | Cantarero et al. (2016) |
| Eurasian Blue Tit | Feathers in nests | Nests of females with a higher proportion of feathers recruit more offspring as breeding adults | Järvinen and Brommer (2020) |
| Eurasian Blue Tit | Nest size | Females that built larger nests fledged more young | Lambrechts et al. (2017) |
| Rufous-tailed Scrub-Robin | Nest size | Both sexes build nests; clutch size positively related to the amount of nest material and size of prey delivered to nestlings by males | Palomino et al. (1998) |
| European Starling | Plant materials added to nests | Females prefer males that have more green plants in their nests | Rubalcaba et al. (2016) |
| European Starling | Plants materials added to nests | Females increase yolk testosterone levels when nests have more green plant material | Gwinner et al. (2013) |
| Spotless Starling | Green plants added to nest | Females at nests with an experimentally enhanced number of green plants produced more male eggs | Polo et al. (2004) |
| Great Reed Warbler | Nest size | Males feed nestlings more when females build larger nests | Jelínek et al. (2016) |
| Rock Sparrow | Feathers in nests | Experimental nests where feathers were added were defended with greater intensity by both parents than control nests | García-Navas et al. (2015) |
| Cape Weaver | Number of nests | Males that built more nests are more likely to attract females | Bailey et al. (2016) |

[a]Scientific names: Black-faced Sheathbill, *Chionis minor*; Black Kite, *Milvus migrans*; Eurasian Nuthatch, *Sitta europaea*; Eurasian Blue Tit, *Cyanistes caeruleus*; Rufous-tailed Scrub-Robin, *Cercotrichas galactotes*; European Starling, *Sturnus vulgaris*; Spotless Starling, *Sturnus unicolor*; Great Reed Warbler, *Acrocephalus arundinaceus*; Rock Sparrow, *Petronia petronia*; Cape Weaver, *Ploceus capensis*

assumption of this hypothesis is that predators either use a specific search image or consistently use certain cues as they search for nests, and that assumption may or may not be correct.

For each of these hypotheses, studies to date provide conflicting results. The results of some studies provide supporting evidence; the results of other studies do not. Although a specific hypothesis may help explain nest predation rates for a particular species in a particular habitat at a particular time (e.g., early or late in the nesting season), another hypothesis may better explain nest predation rates for other species at other locations due to spatial and temporal differences in predator communities, habitats, specific vegetation features, and impacts of human activities. In addition, no single hypothesis may best explain nest predation rates because multiple mechanisms may operate simultaneously. Thompson et al. (2002) proposed a hierarchical, multi-scale model to explain spatial and temporal variability in rates of nest predation (Fig. 17.29). At a large, biogeographical scale, nest predation rates might be influenced by differences in the predator communities present and the abundance of those predators at different locations (e.g., Figs. 17.30 and 17.31). At the landscale level, differences in the size of and relationships between different habitat patches, such as forests, grasslands, aquatic systems, and human-dominated areas such as agricultural fields, pastures, and urban areas, can influence the predator communities present, as well as their abundance and behavior, and, as such, can influence nest predation rates.

**Box 17.7 Nest Neighbors that Increase Nesting Success**

Predation is the primary cause of nest failure for birds so identifying sites where predation-risk may be lower can be an important way to increase reproductive success. For some species of birds, one such safer nest site is near the nests of aggressive birds (like nesting colonies of terns and gulls), large raptors, or wasps and bees. By placing their nests near those of large raptors or aggressive birds, wasps, or bees, birds can benefit because potential nest predators may avoid the area to minimize their risk of being killed or injured (Quinn and Ueta 2008). Nesting success of female Black-chinned Hummingbirds (*Archilochus alexandri*) that nest near the nests of Northern Goshawks (*Accipiter gentilis*) and Cooper's Hawks (*Accipiter cooperii*) is greater than that of females that do not nest near these raptors (Greeney et al. 2015). Another example involves Peregrine Falcons (*Falco peregrinus*) and Common Ravens (*Corvus corax*), with falcons tending to select nest sites near the nests of ravens. The likely reason why Peregrine Falcons often nest near ravens is that ravens are very aggressive in defending their nests from terrestrial and aerial predators and this aggressive behavior also reduces the likelihood that the nearby nests of falcons will be predated (Sergio et al. 2004). Even nesting near wasps and bees can improve nest success, as reported for Red-cheeked Cordonbleus (*Uraeginthus bengalus*) nesting near wasp nests (Beier and Tungbani 2006). Beyond these three examples, Greeney et al. (2015) reviewed that literature and found reports of 38 different species of birds that are either known to benefit or have been hypothesized to benefit from nesting near a variety of "protector species."

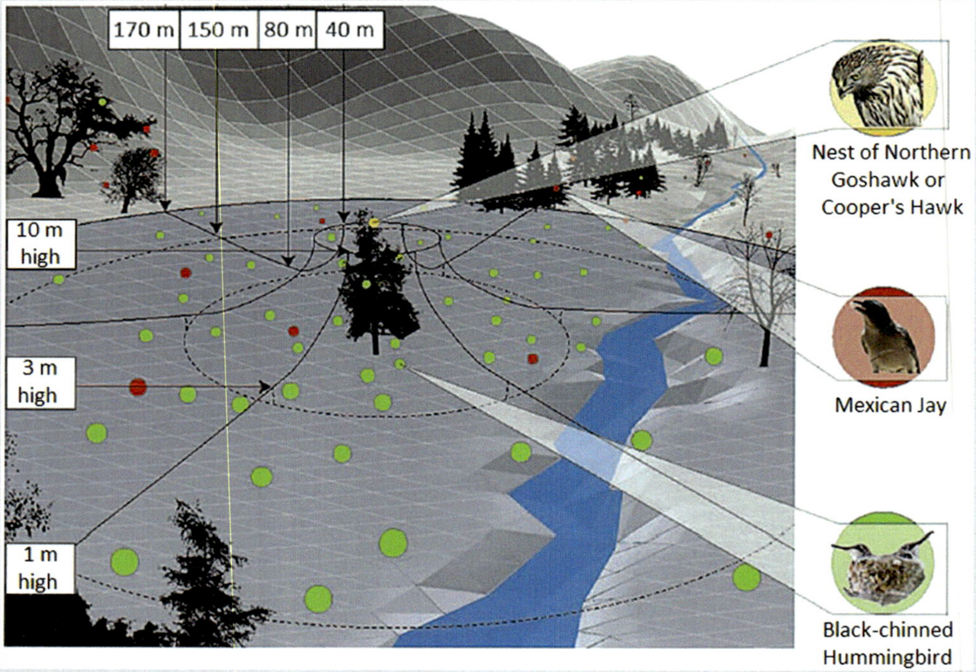


The cone-shaped space around a hawk's nest where nesting attempts by Black-chinned Hummingbirds (*Archilochus alexandri*) were more likely to be successful. The shape of this largely predator-free area results from the fact that Mexican Jays (*Aphelocoma wollweberi*) are less likely to be attacked by hawks when they are at the same height above ground as the hawks

(continued)

**Box 17.7** (continued)

(because Northern Goshawks (*Accipiter gentilis*) and Cooper's Hawks (*Accipiter cooperii*) typically pursue prey using horizontal or descending flights), and because the height above ground of hawks flying away from nests typically declines with distance from nests. Green circles = locations of successful nests; red circles = locations of unsuccessful nests. (Figure modified from Greeney et al. 2015; © 2015 AAAS, used with permission).

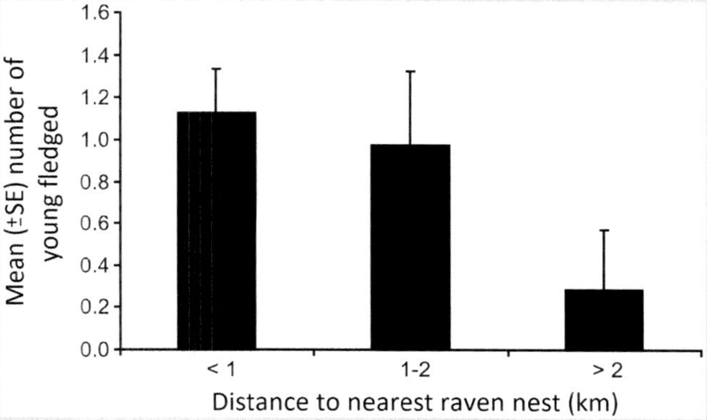


Peregrine Falcons nesting closer to the nests of Common Ravens fledged more young than those nesting farther from raven nests, likely benefitting from the aggressive nest defense of the ravens. (Figure modified from Sergio et al. 2004; © Ecography. Published by John Wiley and Sons Ltd., used with permission).

(continued)

**Box 17.7** (continued)

Nest of a pair of Red-cheeked Cordonbleus (*Uraeginthus bengalus*) located 35 cm above a wasp (*Ropalidia cincta*) nest. Cordonbleus nesting in trees with wasps were about twice as likely to fledge young as those nesting in trees without wasps. (Figure from Beier and Tungbani 2006; © 2006 Oxford University Press, used with permission).

(continued)

**Box 17.7** (continued)

Male Red-cheeked Cordonbleu. (Photo by Ngari Norway, Wikipedia, CC BY 2.5, https://creativecommons.org/licenses/by/2.5/).

For example, fragmented habitats, particularly those encompassing agricultural fields and pastures, may support larger populations of nest predators and Brown-headed Cowbirds (*Molothrus ater*; Fig. 17.32) that, in turn, cause an increase in nest predation and nest parasitism rates (e.g., Robinson et al. 1995).

The relative importance of different variables or scales may vary among locations. For example, in areas little impacted by humans and with little or no fragmentation of habitats, landscape-level and habitat-level effects may have less of an impact on nest predation rates than biogeographical or nest-site effects (Thompson 2007). However, regardless of the relative importance of different variables or scales, multiple factors can influence nest predation rates. Given the particular array of nest predators present in an area, natural selection will favor those birds that select sites less likely to be predated and the particular characteristics of those sites, e.g., height above ground, substrate type (e.g., ground, shrub, or tree), and degree of concealment, will vary among species, geographical locations, habitats, and time (e.g., early vs. late in the breeding season).

## 17.5 Nest Types

Not surprisingly, with over 10,000 species, bird nests exhibit great variation in size, shape,

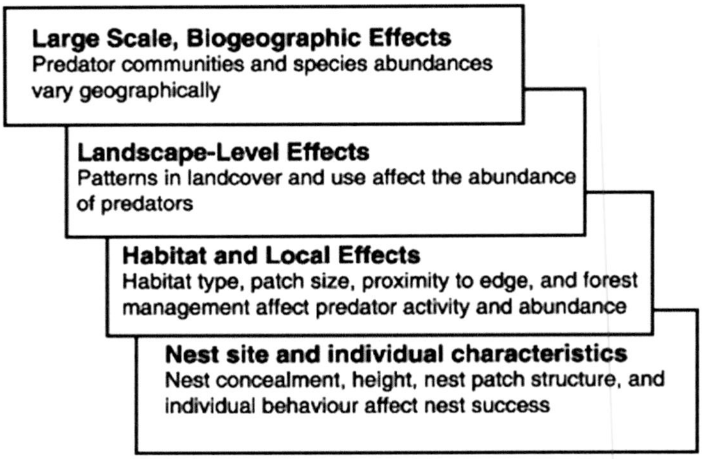


**Fig. 17.29** Factors potentially influencing the likelihood that bird nests will be predated range from the large scale to the small scale of nest sites. (Figure from Thompson 2007; © British Ornithologists' Union, used with permission)

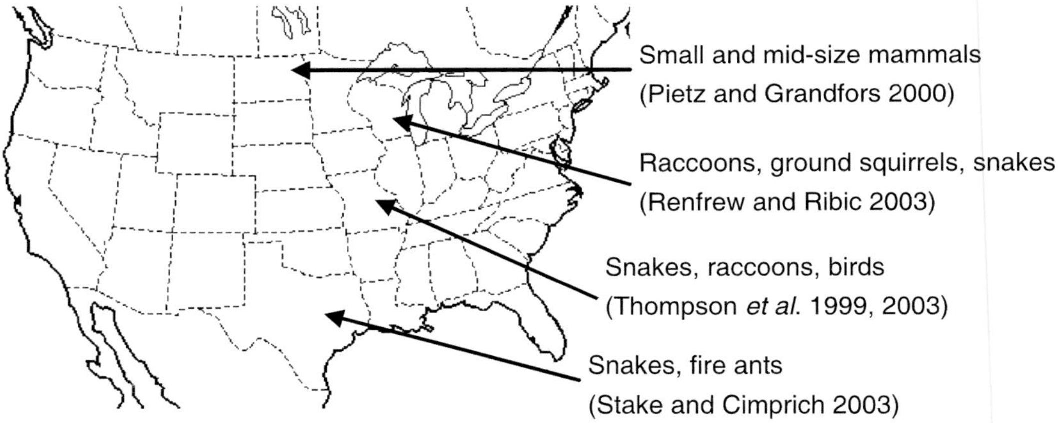


**Fig. 17.30** Variation in the primary nest predators of birds at four different geographical locations in the midwestern United States as determined by video-taping nests. (Figure from Thompson 2007; © British Ornithologists' Union, used with permission)

composition, and location. Basic nest-type categories include scrape, bed, mound, cup, plate, dome, dome-and-tube, burrow, and cavity nests (Hansell 2000). In addition, some species build communal nests (Box 17.8 Communal Nests). Fang et al. (2018) placed nest types into seven categories and found that cup nests were the most common type across present-day families of birds, followed by secondary cavities, domed nests, and platform nests (Fig. 17.33). The nests of many species of birds are essentially modifications of a basic cup nest (Fig. 17.34), including adherent nests (i.e., cupped nests attached to substrates by an adhesive substance such as saliva), pensile nests, and pendulous nests. Still other birds construct platform nests that can be on the ground, in water, or elevated (e.g., in a tree or on the ledge of a cliff). Platform nests can be made from a variety of materials (e.g., sticks and aquatic vegetation) and are

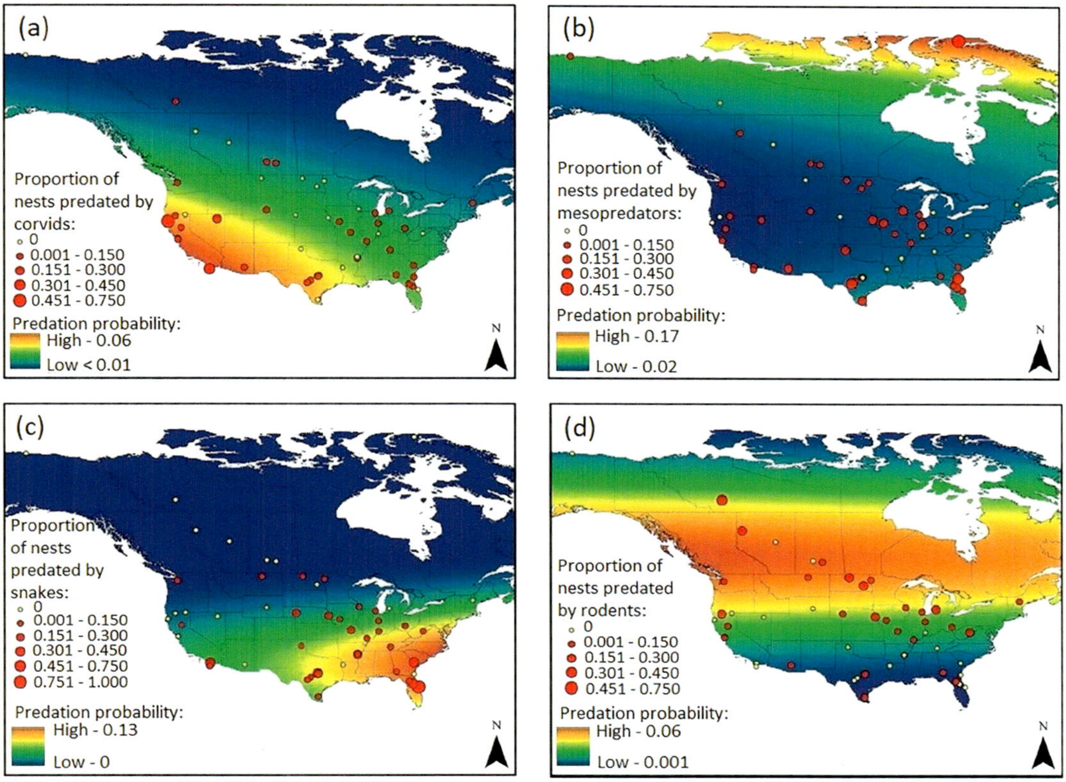


**Fig. 17.31** Predicted probabilities of nest predation by different predators in the United States and Canada based on 53 nest-predator studies with more than 4000 camera-monitored nests. (**a**) Corvids, (**b**) mesopredators (i.e., medium-sized predators), (**c**) snakes, and (**d**) rodents. Nest-predation probability estimated using a model that included longitude, latitude, habitat (shrubland, forest, grassland, or other), nest-height category (low, intermediate, or high), and bird body-size category. (Figure modified from DeGregorio et al. 2016; © 2016 Oxford University Press, used with permission)

sometimes rather large and bulky (e.g., some hawk or eagle nests). However, the tops of platform nests are either flat or form a shallow depression (Box 17.9 Nest Complexity and the Avian Cerebellum).

Scrape, bed, and mound nests are located on the ground. Many ground nesters have scrape nests, with a bed nest a flat or very shallow-cupped nest and a scrape nest a shallow depression in the ground with little or no plant or other materials added. Scrape nests generally have just enough of a rim to keep eggs from rolling away. Other ground nesters may begin by scraping out a depression, but then add a bed of other materials such as stones or vegetation. Birds with scrape nests often nest in open, sparsely vegetated habitats that allow early detection of approaching predators and also have cryptic eggs (Lauro and Nol 1995; Muir and Colwell 2010), with crypticity enhanced by birds locating scrape nests in substrates that match egg (and hatchling) coloration (Stevens et al. 2017; Fig. 17.35).

Cup, or cupped, nests are, of course, cup-shaped and can be located either on the ground or in a supporting structure above ground (Fig. 17.36). Plate nests are flat with very shallow

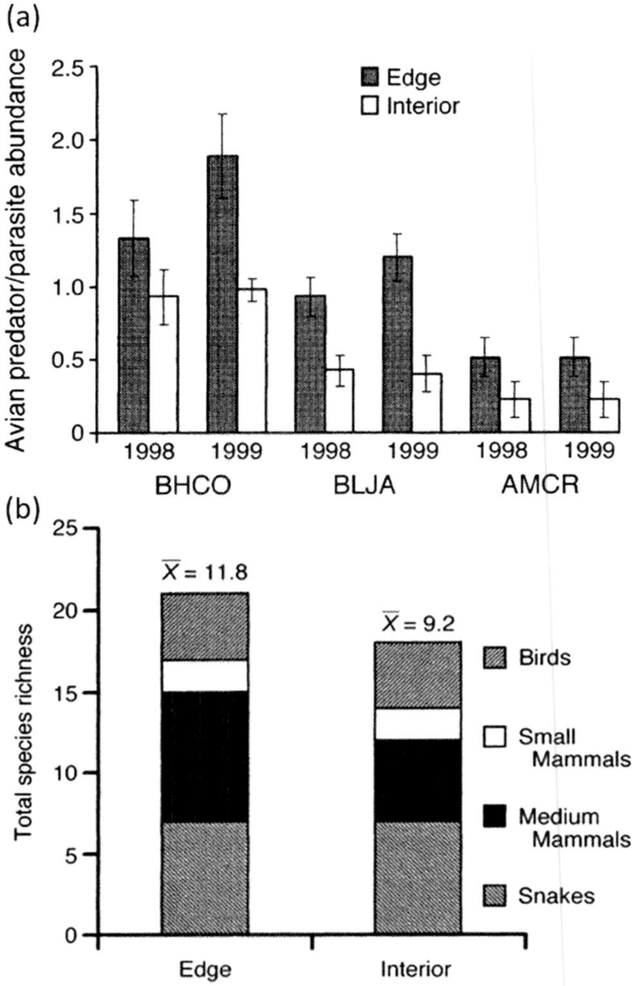


**Fig. 17.32** (**a**) Mean number of detections (± SE) per survey visit of two potential nest predators, Blue Jays (*Cyanocitta cristata*; BLJA) and American Crows (*Corvus brachyrhynchos*; AMCR), and a brood parasite, Brown-headed Cowbirds (*Molothrus ater*; BHCO), in along forest-pasture edges versus forest interiors in Missouri over a two-year period. (**b**) Total species richness of four categories of nest predators along forest-pasture edges and forest interiors in Missouri over a two-year period. The mean number of species per site is provided above each bar. (Figures from Chalfoun et al. 2002; © 2002 by the Ecological Society of America, used with permission)

cups and are located above ground. Dome nests have roofs and those with entrance tubes are called dome-and-tube nests. Burrow nests are chambers excavated in the ground (either relatively flat ground or, often, more vertical banks or cliffs). Cavity nests are located in cavities (either natural or excavated) in trees or other plants such as cacti (Box 17.10 Cavity-Nest Webs).

Pettingill (1985) categorized open-cup nests as being statant, suspended, or adherent. Statant cupped nests are located in the crotches and on branches of trees and shrubs and supported mainly from below. Many passerines and hummingbirds build such nests (Fig. 17.37). Suspended cupped nests are not supported from below, but from the rims, sides, or both, like the nests of vireos and orioles (Fig. 17.38). Adherent cupped nests are attached by an adhesive substance (e.g., mud or saliva) to a vertical surface, like those of swifts (Fig. 17.39; Box 17.11 Adherent Nests) and some swallows.

An estimated 18% of bird species are tree-cavity nesters, and at least 49% of those species are obligate tree-cavity nesters (van der Hoek et al. 2017). Factors likely contributing to the evolution of cavity-nesting behavior include predation pressure, characteristics of cavity microclimate suitable for raising young, and the availability of cavities (Gibbons and

**Box 17.8 Communal Nests**

Several species of birds, including Sociable Weavers (*Philetairus socius*), Gray-capped Social Weavers (*Pseudonigrita arnaudi*), and Monk Parakeets (*Myiopsitta monachus*), nest and roost in large communal nests that include multiple nesting/roosting chambers. The largest of these communal nests are built by Sociable Weavers, with some having as many as 250 separate chambers (Maclean 1973) and, with continued maintenance, these nests can exist for decades (Collias and Collias 1964b). Given the substantial mass of these nests, temperatures in chambers are cooler than ambient temperatures during the summer and warmer than ambient temperatures during the winter (van Dijk et al. 2013; Leighton and Echeverri 2014). During the non-breeding season when ambient temperatures are low, several individuals roost in single chambers (Paquet et al. 2016). Both males and female contribute to building and maintaining the communal nests, but males contribute more than females, likely because females tend to disperse to non-natal communal nests more than males (Leighton 2014). In support of that hypothesis, Leighton et al. (2015) found that the degree of relatedness to others in a colony predicted the investment by different individuals in nest construction.

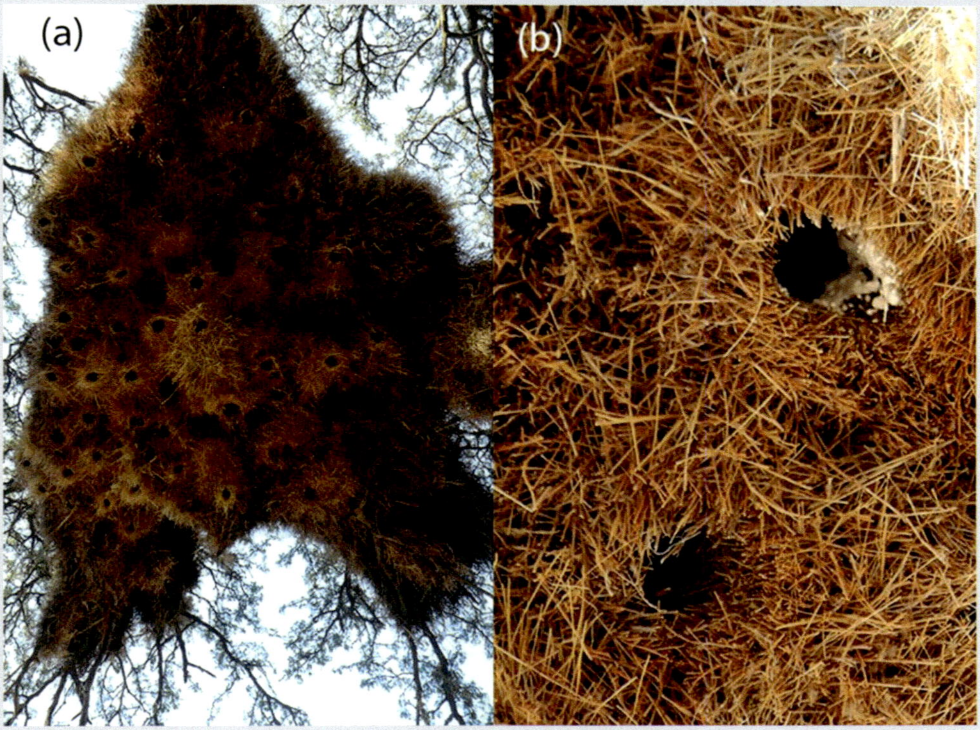


(**a**) The underside of the communal nest of Sociable Weavers (*Philetairus socius*). Entrances to several nesting chambers are visible in this photograph. (**b**) Other species sometimes use these chambers, including Pygmy Falcons (*Poliheirax semitoquatus*), and the chamber occupied by a Pygmy Falcon in this photograph can be identified by the chalk-like fecal mat around the entrance. (Figure from Lowney et al. 2020; © 2020 Lowney, Bolopo, Krochuk and Thomson, open-access article distributed under the terms of the Creative Commons Attribution License (CC BY), https://creativecommons.org/licenses/by/4.0/).

(continued)

**Box 17.8** (continued)

Pair of Pygmy Falcons with male on the left and female on the right. (Photo from Picas Web Albums, Wikipedia, CC BY 3.0, https://creativecommons.org/licenses/by/3.0/).

Lindenmayer 2002). Most species of tree-cavity nesters are in the orders Anseriformes (46 species), Falconiformes (35 species), Psittaciformes (371 species), Strigiformes (219 species), Caprimulgiformes (30 species), Apodiformes (Apodidae, 18 species), Trogoniformes (40 species), Coraciiformes (57 species), Piciformes (404 species), and Passeriformes (586 species; Newton 1994, van der Hoek et al. 2017). Among passerines, the greatest number of tree-cavity nesters are in the families Sturnidae (starlings; 87 of 112 species), Muscicapidae (Old World flycatchers; 78 of 117 species), Paridae (tits, chickadees, and titmice; 54 of 56 species), Tyrannidae (Tyrant flycatchers; 53 of 432 species), and Dendrocolaptidae (woodcreepers; 52 of 52 species) (van der Hoek et al. 2017). Cavities used for nesting by birds are created by primary cavity nesters, primarily woodpeckers and barbets, and by natural processes such as fungal decay and mechanical damage to substrates (Gibbons and Lindenmayer 2002; Cockle et al. 2012). The relative importance of woodpecker-excavated cavities and natural cavities for secondary cavity-nesting species varies among locations. In the Atlantic forest of Argentina, natural cavities are of greater importance (Cockle et al. 2012; Fig. 17.40), and similar results have been reported in other old-growth forests in South America (Altamirano et al. 2017; Cockle et al. 2019). At other locations, both excavated and natural cavities are frequently used. In a riparian forest in India, 58% of nests of secondary cavity-nesting birds ($n = 260$) were in excavated cavities, and 42% were in natural cavities (Manikandan and Balasubramanian 2018). In yet other locations, woodpeckers create most of the cavities occupied by secondary cavity-nesting species (e.g., Martin and Eadie 1999). For example, in central British Columbia, Canada, Aitken and Martin (2007) examined 1371 cavities used by 29 bird and mammal species and found that 85% of those cavities had been excavated by woodpeckers. Similarly, in an area dominated by long-leaf pine (*Pinus palustris*) in Florida, USA, 432 of

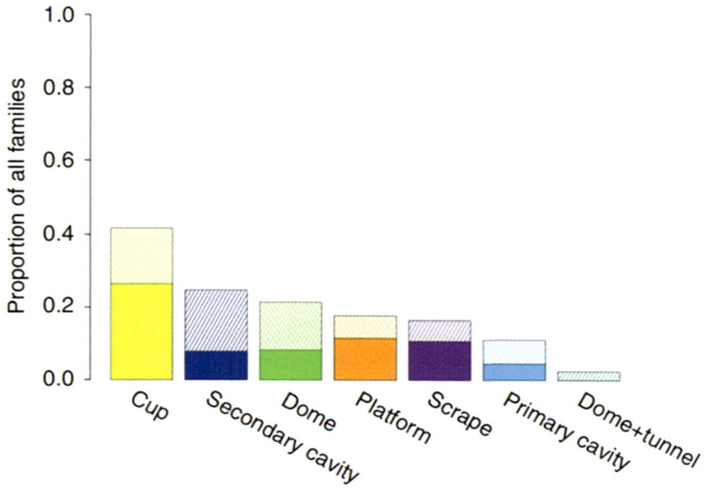


**Fig. 17.33** Proportion of bird families ($N = 242$) with species that have different types of nests. Filled bars indicate the proportion of families with only that particular nest type; striped bars above filled bars indicate the proportion of families with species having more than one nest type. Cup nests = open-cup nests. (Figure from Fang et al. 2018; open-access article is licensed under a Creative Commons Attribution 4.0 International License, https://creativecommons.org/licenses/by/4.0/)

433 cavities used by nesting birds were excavated by woodpeckers (Blanc and Walters 2008).

Cavity excavators are typically categorized as either strong or weak excavators. Strong excavators like woodpeckers usually excavate cavities in living trees, but often in parts of trees with fungal infections. Weak excavators, in contrast, can only excavate cavities in soft, decayed wood (Edworthy et al. 2012) and often use already available cavities rather than excavating a new one. Species considered weak excavators include nuthatches (family Sittidae) and species

**Fig. 17.34** Open-cup nest of a Eurasian Blackbird (*Turdus merula*) with arrows indicating the outer nest layer, the structural wall, and the cup lining. Dots indicate the ends of the longest axis of the nest. The ruler is 15 cm long. (Figure from Biddle et al. 2015; used with permission of C. Deeming)

**Box 17.9 Nest Complexity and the Avian Cerebellum**

The avian cerebellum plays an important role in the control and coordination of skeletal muscles. Like the mammalian cerebellum, the avian cerebellum is foliated (i.e., the surface is folded), increasing the number of neurons and neural connections and, some investigators have suggested, improving both the processing capacity of the cerebellum and motor abilities, specifically manipulative skills (Hall et al. 2013). In support of this hypothesis, Iwaniuk et al. (2009) found that the cerebellum of species and taxa of birds that use tools is significantly more folded than that of non-tool-using species and taxa. Hall et al. (2015) also provided supporting evidence; species of birds that build nests of greater structural complexity (no nest → platform → cup) have cerebella that are more folded. Building increasingly complicated nests require greater manipulative skill to shape, stitch, and weave nest materials into different nest structures (Hall et al. 2015) and, with increasing complexity, natural selection has apparently favored a corresponding increase in the complexity of the avian cerebellum.

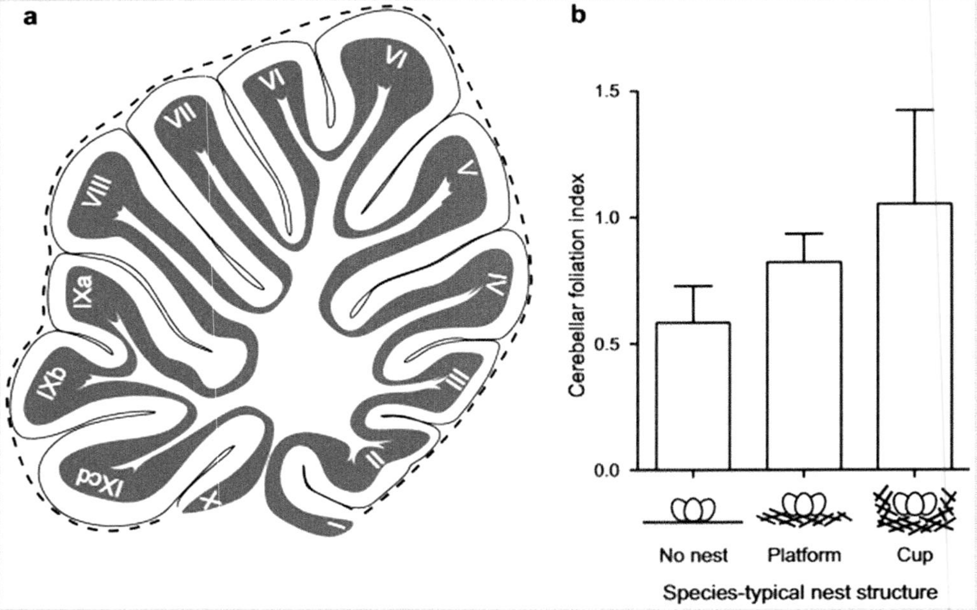


Relationship between the cerebellar foliation index and nest type for several species of birds. (**a**) Drawing of a sagittal section of a bird cerebellum. The cerebellar foliation index is the ratio of the surface length of the cerebellum (gray area) to the surface length of the unfolded cerebellum (dashed line). Greater folding of the cerebellum and more surface area equals a higher foliation index. (**b**) Species that build cup nests have significantly more foliated cerebella than species with platform nests or no nest. (Figure from Hall et al. 2015; © The Authors, open-access article distributed under the terms of the Creative Commons CC BY license, https://creativecommons.org/licenses/by/4.0/).

**Fig. 17.35** Well-camouflaged Piping Plover (*Charadrius melodus*) egg and chicks in a scrape nest. (Photo by Susan Haig, U.S. Geological Survey, CC0 Public Domain)

in the family Paridae. Cavities in living trees have been found to increase in volume with increasing age, which may improve their quality for some secondary cavity-nesting species (Edworth and Martin 2014), and cavities in living trees persist longer than those in dead or decaying trees. For example, in British Columbia, Canada, cavities in living trees had median lifespans of 15 years, whereas median lifespans for dead and decaying trees were 7 to 9 years (Edworthy et al. 2012). Occupancy of cavities is generally highest at one-year post-excavation or natural formation, but cavities that persist can continue to be used for many years (Edworthy et al. 2018; Fig. 17.41). Use of newer cavities by secondary cavity-nesting species may be advantageous in terms of better regulation of microclimates, fewer ectoparasites, and lower risk of predation than older cavities (Nilsson 1984; Wesołowski 2002; Tozer et al. 2012).

The characteristics of cavities vary in terms of the extent to which they protect adults and young from predators and inclement weather (Wesołowski 2002; Wesołowski and Rowiński 2012; Cockle et al. 2015). For example, in the Atlantic forest in northeastern Argentina, nest survival for small birds (12–128 g) was higher in cavities higher above ground and with smaller entrance diameters, and thus more likely to exclude larger predators, and nest survival for large birds (141–400 g) was higher in living trees than dead trees (Cockle et al. 2015). Other investigators have reported preferences by cavity-nesting birds for cavities higher above ground (e.g., Cameron 2006; Politi et al. 2009) and with smaller entrances (Haitao et al. 2003), and that

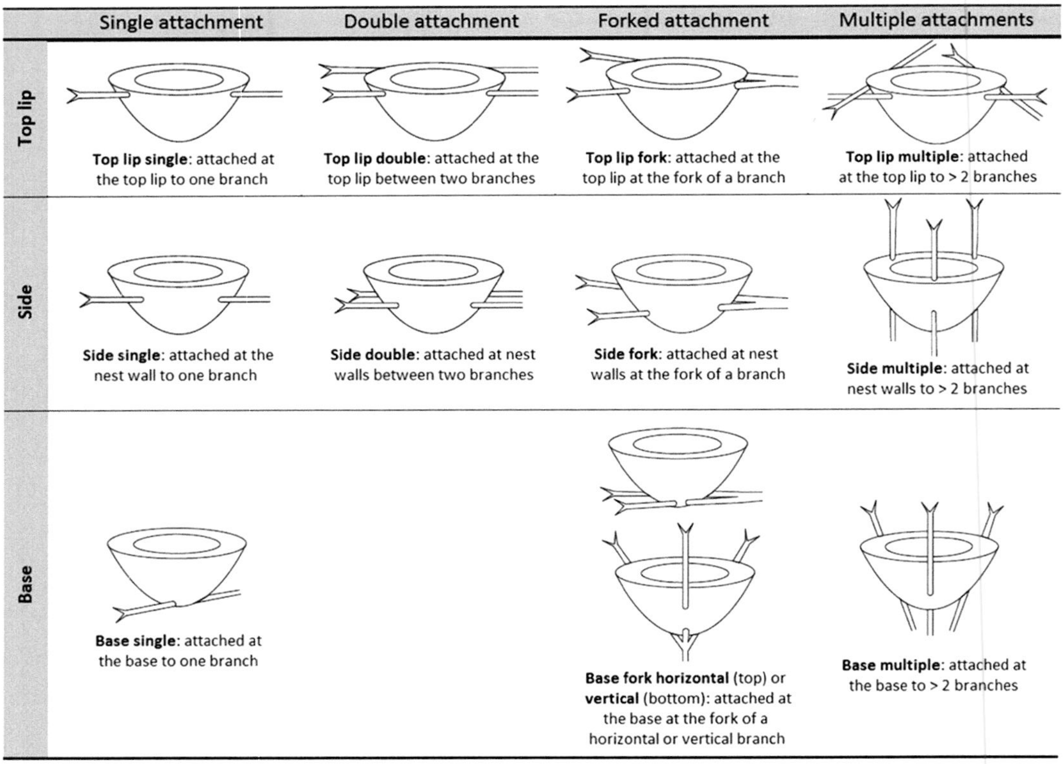


**Fig. 17.36** Open-cup nests of birds are attached to branches of trees and shrubs in a wide variety of ways. (Figure modified from Heenan 2013; used with permission of Caragh Heenan)

nests in cavities in dead wood are more likely to be predated than those in live trees, with some predators able to enlarge cavity entrances by removing soft decayed wood and thereby gain access to nest contents (e.g., Christman and Dhondt 1997, Wesołowski 2002; Fig. 17.42). Other investigators have reported preferences for and greater nest success for birds nesting in deeper cavities with larger internal volumes (e.g., Nilsson 1984; Li and Martin 1991; Haitao et al. 2003; Sanz 2008). More generally, the odds of cavities being used by cavity-nesting species of birds increased with cavity depth and height above ground, and decreased with cavity-entrance size in forests in Argentina (Di Sallo and Cockle 2021).

The extent to which cavities might be a limiting resource for secondary cavity-nesting birds is unclear. Wiebe (2011) reviewed the literature on nest-site limitation and concluded that there was some evidence that "nest sites are limited at local (plot) scales in old growth forests [but] there is little empirical evidence for nest-site limitation at the population- and landscape-level in mature, unmanaged forests." In a 16-year study in British Columbia, Canada, Trzcinski et al. (2022) found that "Years with higher nesting densities of excavators were following by years with higher SCN [secondary-cavity nester] diversity, indicating that the creation of nesting opportunities through fresh excavation releases SCNs from community-wide nest-site limitation." Other investigators have determined that some species of secondary cavity-using species can be limited by cavity availability in mature forests (Aitken and Martin 2012; Robles et al. 2012). In addition, and not surprisingly, cavity nest sites may be limiting factors for cavity-

**Box 17.10 Cavity-Nest Webs**

Many species of birds in forested habitats nest (and roost) in tree cavities. Although some cavities are produced via decay of trees and tree limbs, many are produced by woodpeckers and other species of birds that sometimes excavate cavities such as nuthatches and chickadees. These species are the primary cavity-nesters. Secondary-cavity nesters require cavities for nesting and, often, roosting, but cannot excavate their own cavities so largely depend on the creation of cavities by the primary cavity excavators. In addition, the primary excavators vary in size and in the size of the cavities they produce, so secondary-cavity nesters that differ in size may be more dependent on some primary cavity excavators than others. Finally, of course, the primary cavity excavators require trees and, depending on the geographical location of forests and excavators, may excavate more cavities in some species of trees than others. These interrelationships between trees, primary cavity excavators, and secondary-cavity users represent what is called a nest web or cavity-nest web (Martin and Eadie 1999).

The cavity nest of a Taiwan Barbet (*Psilopogon nuchalis*). (Photo from pxhere.com, CC0 Public Domain).

(continued)

**Box 17.10** (continued)

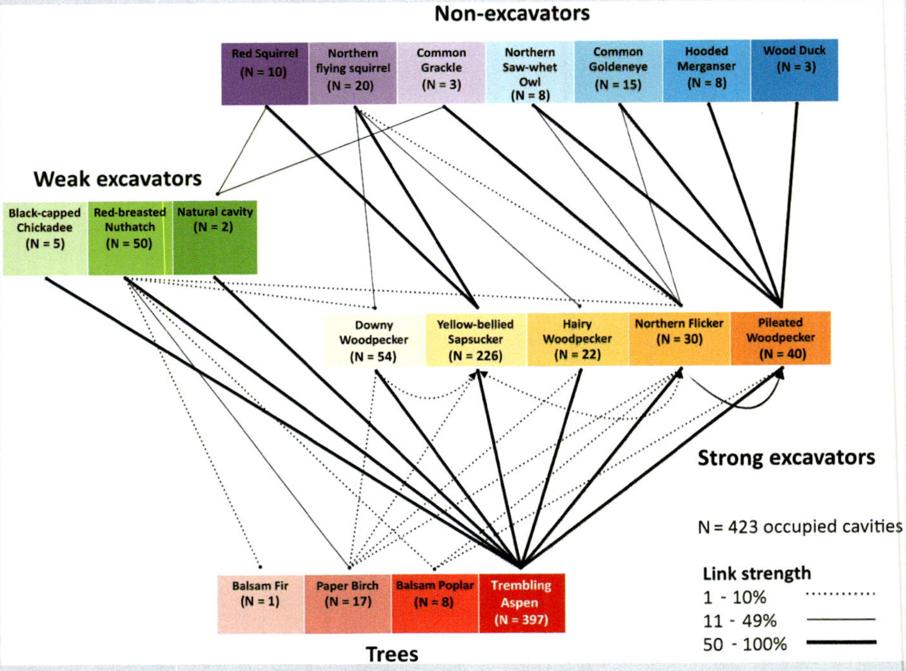


Cavity-nest web of birds, squirrels, and cavity trees in a forest in Québec, Canada. Lines between species indicate use of a cavity. Links between trees and excavators indicate the proportion of nests in each species of tree, and links between non-excavators and excavators or natural cavities indicate the proportion of nests in reused cavities. In this forest, note the importance of trembling aspen for cavity-nesting species. N = either number of trees of each species with cavities used for nesting or number of nests for birds and squirrels. Scientific names: red squirrel, *Tamiasciurus hudsonicus*; northern flying squirrel, *Glaucomys sabrinus*; Common Grackle, *Quiscalus quiscula*; Northern Saw-whet Owl, *Aegolius acadicus*; Common Goldeneye, *Bucephala clangula*; Hooded Merganser, *Lophodytes cucullatus*; Wood Duck, *Aix sponsa*; Black-capped Chickadee, *Poecile atricapillus*; Red-breasted Nuthatch, *Sitta canadensis*; Downy Woodpecker, *Picoides pubescens*; Yellow-bellied Sapsucker, *Sphyrapicus varius*; Hairy Woodpecker, *Picoides villosus*; Northern Flicker, *Colaptes auratus*; Pileated Woodpecker, *Dryocopus pileatus*; balsam fir, *Abies balsamea*; paper birch, *Betula papyrifera*; balsam poplar, *Populus balsamifera*; trembling aspen, *Populus tremuloides*. (Figure from Cadieux et al. 2023; © 2023 Cadieux, Drapeau, Ouellet-Lapointe, Leduc, Imbeau, Deschênes and Nappi, open-access article distributed under the terms of the Creative Commons Attribution License (CC BY))

(continued)

**Box 17.10** (continued)

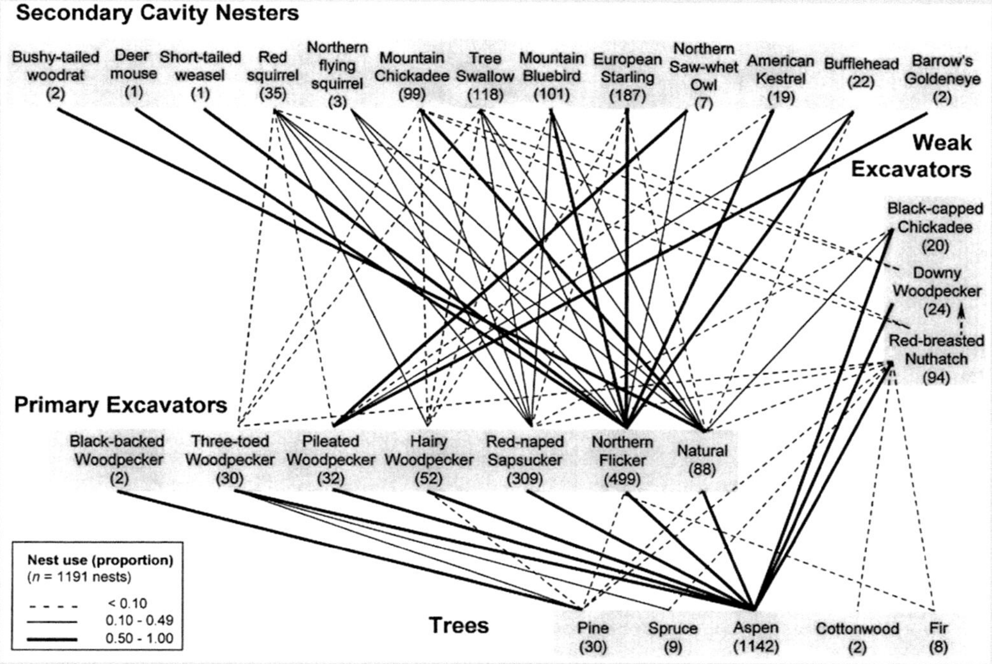


Cavity-nest web in British Columbia, Canada, showing links between bird (and mammal) species (excavators plus secondary-cavity nesters) and trees they use. Note that Northern Flickers are very important cavity excavators, with their cavities being used by almost every species of secondary-cavity nesters, and that aspens are by far the most-used tree by primary and weak excavators, with weak excavators being species that typically excavate cavities in decayed, soft wood (e.g., nuthatches and chickadees). Larger woodpeckers like Pileated Woodpeckers excavate larger cavities and note that some of the larger secondary-cavity nesters, like Barrow's Goldeneyes and Buffleheads, use those cavities, whereas smaller species of secondary-cavity nesters do not. Weak excavators primarily excavate cavities in decaying trees and tree limbs. Numbers under each species indicate the number of occupied nests. Bushy-tailed woodrat, *Neotoma cinerea*; deer mouse, *Peromyscus maniculatus*; short-tailed weasel, *Mustela erminea*; red squirrel, *Tamiasciurus hudsonicus*; northern flying squirrel, *Glaucomys sabrinus*; Mountain Chickadee, *Poecile gambeli*; Tree Swallow, *Tachycineta bicolor*; Mountain Bluebird, *Sialia currucoides*; European Starling, *Sturnus vulgaris*; Northern Saw-whet Owl, *Aegolius acadicus*; American Kestrel, *Falco sparverius*; Bufflehead, *Bucephala albeola*; Barrow's Goldeneye, *Bucephala islandica*; Black-capped Chickadee, *Poecile atricapillus*; Downy Woodpecker, *Dryobates pubescens*; Red-breasted Nuthatch, *Sitta canadensis*; Black-backed Woodpecker, *Picoides arcticus*; American Three-toed Woodpecker, *Picoides dorsalis*; Pileated Woodpecker, *Dryocopus pileatus*; Hairy Woodpecker, *Dryobates villosus*; Red-naped Sapsucker, *Sphyrapicus nuchalis*; Northern Flicker, *Colaptes auratus*. (Figure from Martin et al. 2004; © Oxford University Press, used with permission).

(continued)

**Box 17.10** (continued)

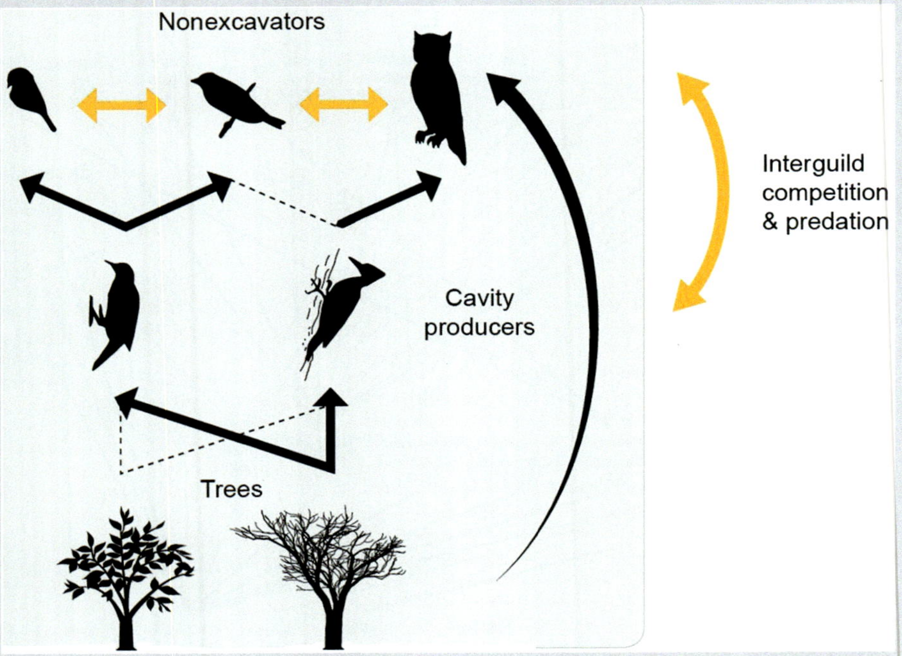


Simple model of a nest web (gray box) with positive (black) and negative (yellow) interactions among species (Martin and Eadie 1999). Cavities are produced by excavators (straight black arrows) or by decay and non-excavators compete among themselves (straight yellow arrows) and, occasionally, with the excavators for the limited cavity resource (yellow arrow outside of gray box). Although not shown here, secondary cavity-nesting species may also need to compete with species other than birds, including social insects (bees and wasps) and several species of mammals (e.g., Broughton et al. 2015; Bonaparte and Cockle 2017). Weaker links (shown with broken lines) include (1) reduced competition for cavities between larger cavity-nesting species (e.g., large owls) and smaller species, and (2) reduced competition for access to live trees between larger excavators (e.g., *Dryocopus* spp.) and smaller excavators that are more likely to excavate cavities in dead trees (with softer wood) than live trees (e.g., Schepps et al. 1999). (Figure modified from Ibarra et al. 2020; © 2020 The Authors, open-access article distributed under the Creative Commons Attribution 4.0 International License, https://creativecommons.org/licenses/by/4.0/).

nesting species of birds in young (e.g., second-growth, Lima and Garcia 2016) or logged forests (e.g., Cockle et al. 2010; Fig. 17.43), perhaps especially for cavity-nesting species that only use cavities with specific characteristics (e.g., Superb Parrots, *Polytelis swainsonii*; Stojanovic et al. 2021). Availability of cavities may also be a limited factor for species of secondary cavity-nesting species of birds that, although nesting in cavities, occupy more open habitats, e.g., bluebirds (*Sialia* spp.) and Tree Swallows (*Tachycineta bicolor*) in North America (Drake

**Fig. 17.37** Nest of a Green Thorntail (*Discosura conversii*) in Colombia. (Photo from Sánchez et al. 2016 and taken by Diego Calderón-Franco; used with permission of Diego Calderón-Franco @diegoCOLbirding)

and Martin 2020). In locations where availability of cavities is limited, secondary cavity-nesting species of birds may need to compete for nest sites (Goldshtein et al. 2018; Bentz et al. 2019), occasionally with lethal results (Merilä and Wiggins 1995) and, even where cavities may not appear to be a limiting factor based on availability, secondary cavity-nesting species may compete for higher-quality, more predator-safe cavities (Nilsson 1984; Newton 1994). In addition, some introduced species of secondary cavity-nesting species of birds have been found to negatively impact populations of native species via competition for available cavities (e.g., Kerpez and Smith 1990; Charter et al. 2016; Rogers et al. 2020).

## 17.6 Nest Materials

Birds use a wide variety of materials when constructing nests. Materials used are often similar for birds in particular species or genera, but within-species variation in the selection of nest materials can be significant (Collias 1997). Among the materials used by birds to construct nests are sticks, grass, mud, various materials from or produced by animals (such as feathers, hair, and shed snakeskins; Box 17.12 Nest "Decorations") and arthropods (such as spider webs), lichens, fungi, and a wide variety of plant materials. Some species of swifts use their glue-like saliva in constructing nests, with some using only saliva (Box 17.13 Edible-nest (or White-nest) Swiftlets) and most using saliva along with a variety of other materials such as sticks, grass and other plant material to build their nests. Differences among species in the selection of materials used to build nests are the result of both genetic and environmental factors. Many aspects of avian nest-building behavior, including selection of nest materials, are innately determined. However, the availability of materials in the environment also plays a role. For example, Chipping Sparrows (*Spizella passerina*) construct nests primarily using fine materials such as rootlets and dried grasses, but, depending on availability in their territories, typically line nests with either hair or fine plant fibers

**Fig. 17.38** Suspended cup nest of a Red-eyed Vireo (*Vireo gilvus*). (Photo by OldFulica, purchased from istockphoto.com)

(Middleton 2020). As another example of such flexibility in the use of materials, Ondrušová (2011) identified the hair of 22 different species of mammals in the nests of five cavity-nesting species of birds in the Czech Republic.

In addition to the materials needed to complete the nest structure, some species of birds bring green plants to their nests that are rich in volatile secondary compounds (Wimberger 1984; Mennerat et al. 2009a; Box 17.14 Green Incubation). The results of several studies indicate that these green plants may improve nestling growth and survival (e.g., Clark and Mason 1988; Gwinner et al. 2000). For example, female Eurasian Blue Tits (*Cyanistes caeruleus*) add fragments of aromatic plants to nest cups (e.g., *Lavandula stoechas*, *Achillea ligustica*, *Helichrysum italicum*, and *Mentha suaveolens*) throughout the nesting cycle, i.e., from nest construction until just before young fledge. The presence of these plants was found to reduce both bacterial richness and densities on nestlings (Mennerat et al. 2009a, 2009b), possibly contributing to the positive effects of aromatic plants on nestling growth (Mennerat et al. 2008). Similarly, green plant materials (from *Pinus pinaster*) added to nests by Bonelli's Eagles (*Hieraaetus fasciatus*) reduced numbers of ectoparasites in nests (blow fly larvae, *Protocalliphora* spp.) and improved breeding success (Ontiveros et al. 2007). In addition, Quiroga et al. (2012) reported negative associations between botfly (*Philornis* sp.) parasitism and the presence of green material in nests of 35 species of birds. Some European Starlings (*Sturnus vulgaris*) mix fresh herbs into dry nest material that inhibit bacterial growth and appear to repel mosquitoes and, as an apparent result, improve the condition of nestlings (Gwinner and Berger 2006). Although such results indicate that aromatic plants added to nests can inhibit bacterial growth, reduce numbers of some types of ectoparasites, and potentially

**Fig. 17.39** Saliva "solidified" and forming an arch over a Chimney Swift (*Chaetura pelagica*) nest to help hold it in place against the wall. (Photo by Gary Ritchison)

improve nestling condition, additional experimental studies are still needed to better understand the specific effects of plant materials on potentially harmful bacteria and parasites and how plant materials affect nestling condition.

## 17.7 Nest Construction: Innate or Learned?

Few studies have focused on the question of whether nest building by birds is an innate behavior or is, at least to some extent, learned. Two lines of evidence suggest that nest building has a large instinctive component: (1) the structure and composition of nests exhibit little intraspecific variation, and (2) birds raised in isolation tend to build species-specific nests. However, particularly for species of birds that build more complex nests, evidence indicates that birds learn by experience and, as a result, the quality of nests built by older birds can exceed that of nests built by younger, less experienced birds (Breen et al. 2016; Healy 2022). For example, the first nests of young male Village Weaver (*Ploceus cucullatus*) are more loosely and crudely constructed than nests built by older, more experienced males (Fig. 17.44; Collias and Collias 1964a). Bailey et al. (2014) found that, based on previous experience, male Zebra Finches (*Taeniopygia guttata*) selected materials that were most suitable for building nests. Interestingly, Freidl (2004) also noted that nests constructed by one-year-old male Southern Red Bishops (*Euplectes orix*), another weaver in the family Ploceidae, were looser and untidier than those of older males. However, Metz et al. (2009) compared several characteristics of the nests of first-year and older male Red Bishops, including entrance size, breeding chamber volume, and density of fibers enclosing breeding chambers, and found no statistically significant differences. Such results seem to indicate that naïve males instinctively know how to construct typical Southern Red Bishop nests, but, with experience, may become more proficient with the fine details needed to construct tighter and tidier nests.

Another way of examining the extent to which nest building is innate versus learned is to determine the degree of repeatability of nest structure

**Box 17.11 Adherent Nests**

The nests of White-nest Swiftlets (*Aerodramus fuciphagus*) are composed entirely of saliva (a highly concentrated mucin glycoprotein; Shim et al. 2016), with males taking about 35 days to construct a single nest (Kang et al. 1991). Males begin by attaching threads of saliva to the vertical, or nearly vertical walls, of caves, then continue adding saliva to form a half-cup shaped nest structure. Analysis of these nests revealed that the base is significantly thicker than the walls (with more surface area) and rim and, when fully loaded with two eggs and two adults perched on the rim, the resulting stresses were greatest on the rim and extended a short distance onto the walls (Jessel et al. 2019). A thick base (i.e., saliva spread over a larger area on the wall surface) in combination with somewhat thinner sides and rim means that the base experiences less stress and nests are unlikely to detach from the wall of a cave.

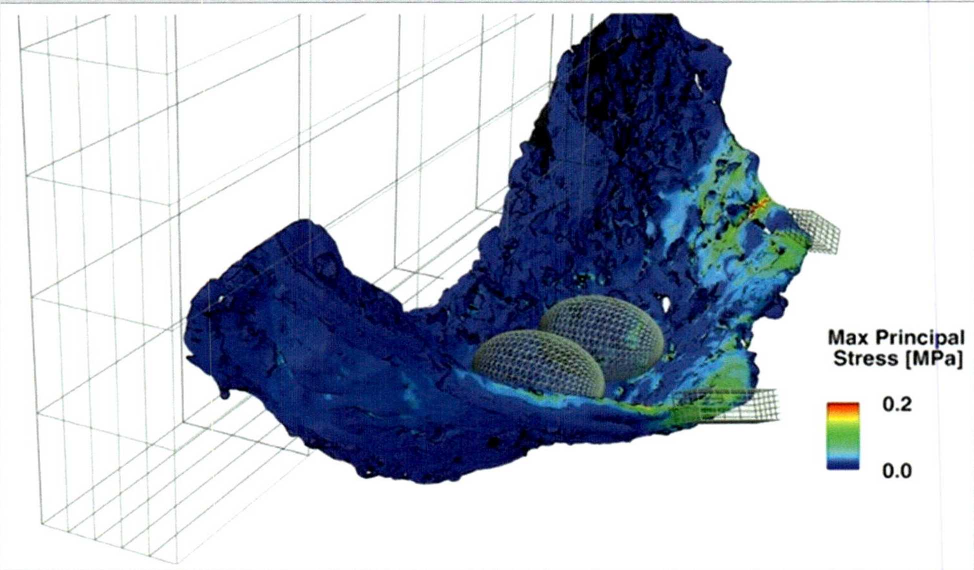


Stress generated in the nest of a White-nest Swiftlet containing two eggs with two adults perched at the edge (represented by the two rectangular structures. MPa, megapascal = a pressure measurement unit. (Figure from Jessel et al. 2019; open-access article licensed under a Creative Commons Attribution 4.0 International License, https://creativecommons.org/licenses/by/4.0/).

Other species of birds construct adherent nests consisting largely of mud, e.g., swallows (Hirundinidae) and phoebes (*Sayornis* spp.). Mud exhibits insufficient cohesion to hold nests together and to substrates. Rather, mud nests are able to support birds, eggs, and nestlings because of a gluing agent in the saliva of birds. Saliva permeates into granules of mud as a liquid and binds the granules together after water evaporates. The "glue" is a glycoprotein called mucin, and the cohesive nature of mucin is due to the presence of entangled polymer chains able to resist mechanical stress (Jung et al. 2021).

(continued)

**Box 17.11** (continued)

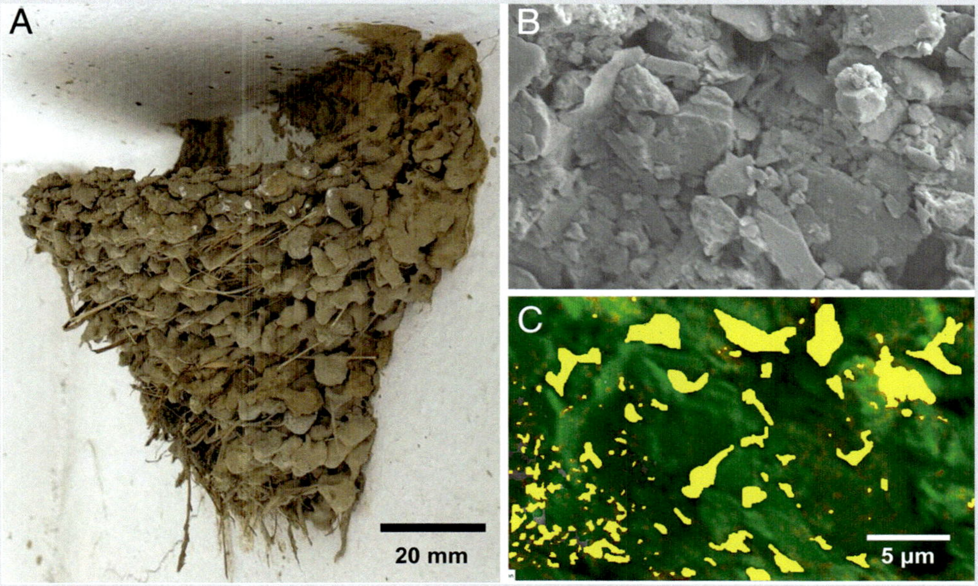


(**a**) Nest of a Barn Swallow (*Hirundo rustica*). (**b**) Scanning electron micrograph image of the nest surface. (**c**) Chemical composition analysis of the surface shown in **b**. The yellow areas indicate regions containing mostly carbon atoms that may be from bird saliva. Green areas contain primarily silicon atoms of clay particles. (Figure from Jung et al. 2021; used with permission of Yeonsu Jung).

Among vertebrates, saliva functions primarily as a lubricant that aids in swallowing food items so the presence of too much mucin could make saliva too viscous. As a result, saliva is limited in its ability to serve as a glue which, in turn, limits the body mass of birds with adherent nests. Jung et al. (2021) plotted the body mass of 9307 species versus the body mass of those species with adherent mud nests and found the most mud-nesters weighed less than 100 g, with a peak near 20 g.

(continued)

**Box 17.11** (continued)

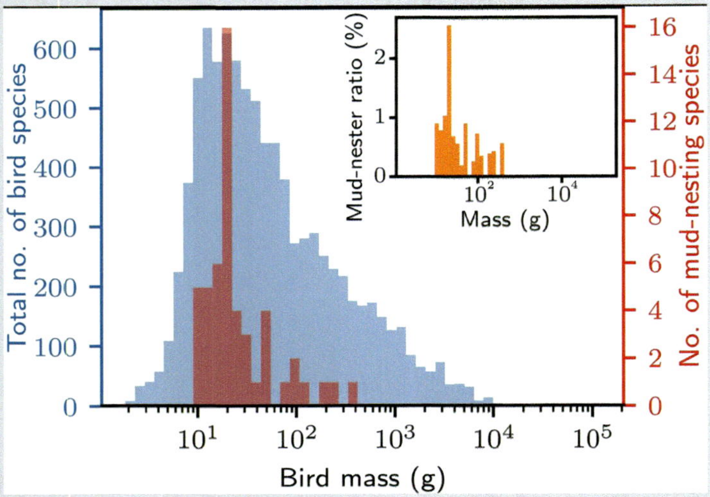


Body masses of 9307 species of birds (blue bars) and species of mud-nesting birds (red bars). Inset, Relative frequency of mud-nesters in each bin. Most mud-nesting birds weigh less than 100 grams, and most species of mud-nesters that weigh more than 100 grams are in the family Corcoracidae. Species in this family, including White-winged Choughs (*Corcorax melanorhamphos*, about 365 grams) and Apostlebirds (*Struthidea cinerea*), build coil pot-type mud nests on stable substrates like tree branches rather than on vertical walls. (Figure from Jung et al. 2021; used with permission of Ho-Young Kim).

and material choice, with less repeatability indicating a behavior with a lesser genetic component. Among some species of birds, nest morphology has been found to be very repeatable. For example, Barn Swallows (*Hirundo rustica*) build cup-shaped nests using mud and straw (Fig. 17.45). Characteristics of the nests of individual swallows were found to be consistent both within and between years and, in addition, the characteristics of nests were even similar across generations, suggesting that nest building by Barn Swallows has a heritable component and is a largely instinctive behavior (Møller 2006). Similarly, the nests of individual Penduline Tits (*Remiz pendulinus*) exhibit little variation across seasons (Schleicher et al. 1996). In contrast, the nests of individual Southern Masked (*Ploceus velatus*) and Village (*Ploceus cucullatus*) weavers exhibited less repeatability, and nests of both species changed as individuals built more nests (Walsh et al. 2010, 2013; Fig. 17.46). In laboratory experiments, Zebra Finches (*Taeniopygia guttata*) learn to associate the nest material used with their nesting success when using that material (Muth and Healy 2011), and they also learn the structural properties of nest material, modifying handling techniques as needed to construct a nest (Muth and Healy 2014). Such results suggest that, for many species of birds, nest building appears to be primarily instinctive. However, for some species, nest construction involves both instinct (especially for first nests) and learning.

The time needed to build a nest varies with the complexity of the nest and other factors such as time of year and weather. In temperate areas, construction of the first nest of the breeding

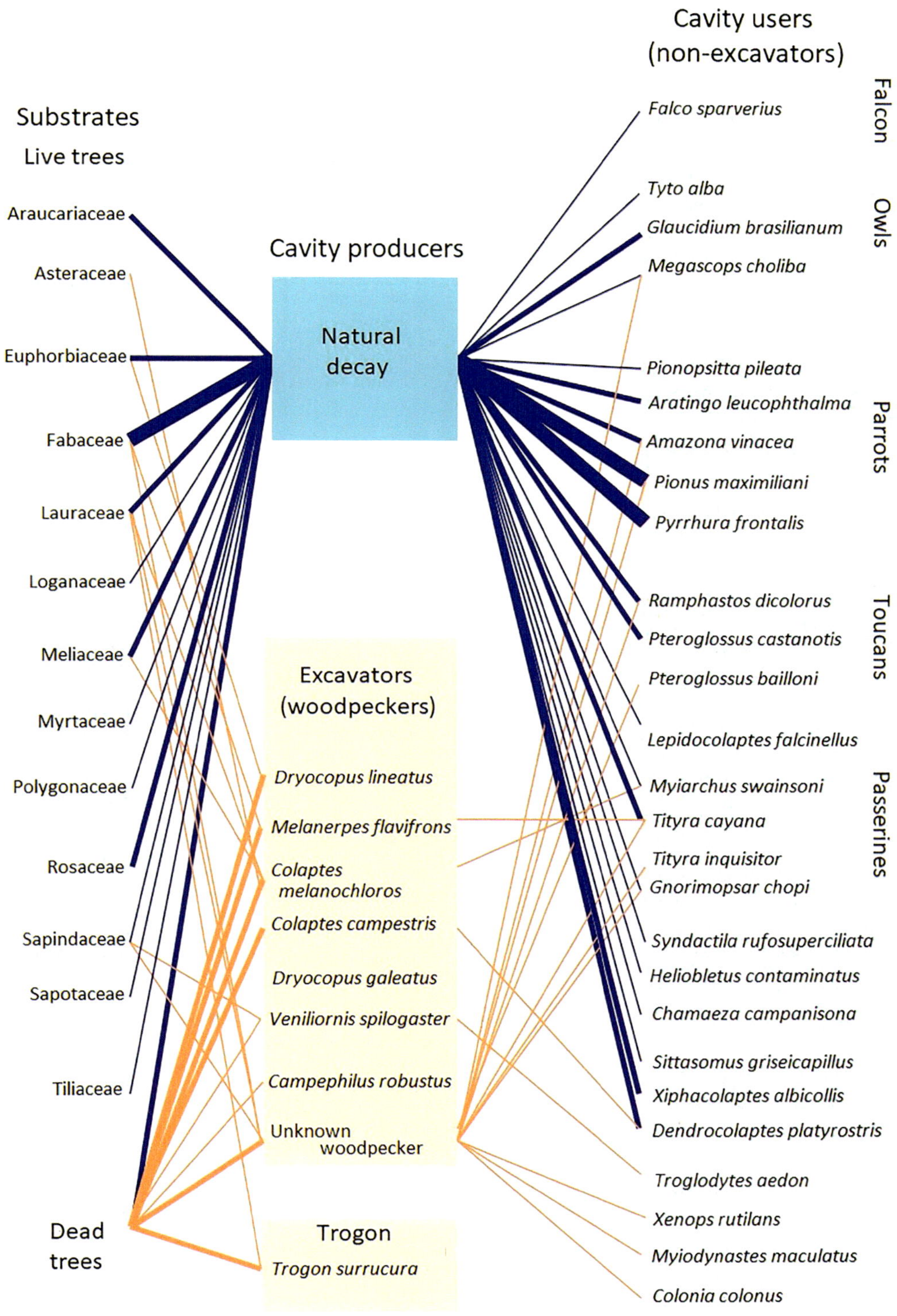


**Fig. 17.40** Nest web showing links between trees, cavity producers (excavators indicated by orange lines, or decay of dead trees indicated by blue lines), and cavity users (non-excavators) in the Atlantic forest of Argentina. In the Atlantic forest, cavities formed by natural decay are used more frequently by non-excavators (secondary-cavity nesters) and cavities produced by woodpeckers and a trogon (primary cavity nesters). Line thickness indicates

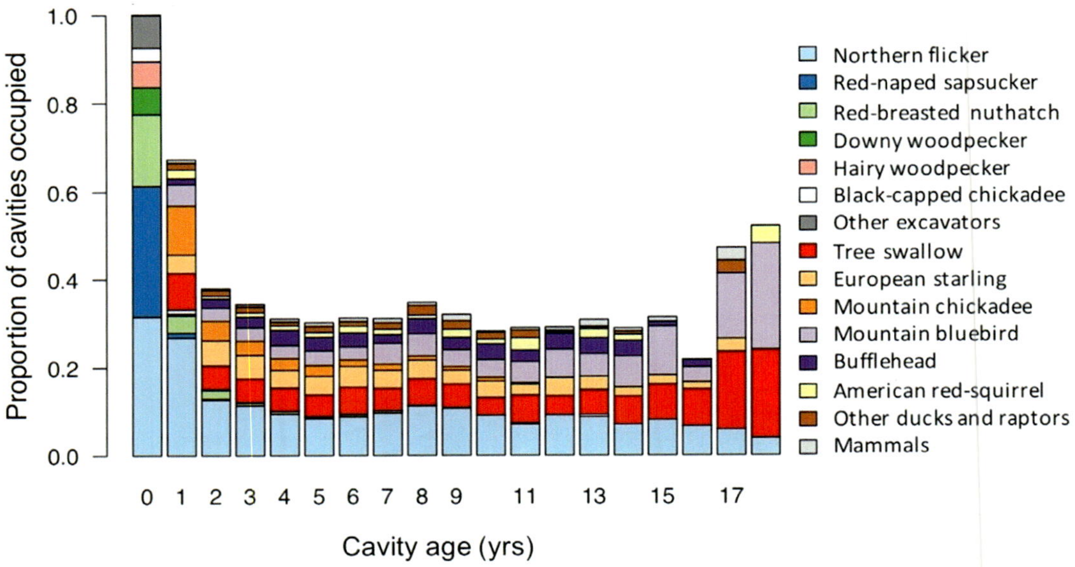


**Fig. 17.41** Proportion of excavated nests occupied out of the total number of cavities present. Proportions are based on 6736 nests in 2876 cavities in British Columbia, Canada. New cavities were used most by excavators and small- and medium-sized non-excavators, whereas older cavities were used more by medium- to large-bodies non-excavators. Scientific names: Northern Flicker, *Colaptes auratus*; Red-naped Sapsucker, *Sphyrapicus nuchalis*; Red-breasted Nuthatch, *Sitta canadensis*; Downy Woodpecker, *Dryobates pubescens*; Hairy Woodpecker, *Dryobates villosus*; Black-capped Chickadee, *Poecile atricapillus*; Tree Swallow, *Tachycineta bicolor*; European Starling, *Sturnus vulgaris*; Mountain Chickadee, *Poecile gambeli*; Mountain Bluebird, *Sialia currucoides*; Bufflehead, *Bucephala albeola*; American red squirrel, *Tamiasciurus hudsonicus*. (Figure from Edworthy et al. 2018; © The Wildlife Society 2017, used with permission)

season may take longer than nests later in the season, particularly for resident species like Eastern Bluebirds (*Sialia sialis*) and Northern Cardinals (*Cardinalis cardinalis*). Generally, passerines build nests over a period of a few days (Table 17.1), but more complex nests take longer to build. For example, the elaborate stick nests built by male and female Brown Cacholotes (*Pseudoseisura lophotes*) take 15 to 37 days to complete (Nores and Nores 1994; Fig. 17.47). Construction of the nests of some larger birds, such as raptors, may require several weeks (Table 17.2).

## 17.8 Constructing Nests: Females, Males, or Both?

Among species of birds that build nests, the roles of males and female vary (Fig. 17.48). Van Tyne and Berger (1976) placed birds into several categories based on the respective contributions to nest building by males and females:

1. Both sexes build the nest.
   (a) The contributions of males and females are similar. Examples include many

**Fig. 17.40** (continued) the number of times a particular interaction occurred, including 1 or 2 times (thin lines), 3–9 times (medium lines), or 10–19 times (thick lines). (Figure from Cockle et al. 2012; © 2011 Elsevier B.V. Published by Elsevier B.V., used with permission)

**Fig. 17.42** Nest cavities of Marsh Tit (*Poecile palustris*) after attacks by predator. (**a**) A woodpecker attempted to enlarge the cavity entrance, but was unable to reach and predate the nestlings. (**b**) A woodpecker was able to prey on nestlings by creating another entrance to the cavity in a decayed cavity wall. (**c**) A mammal enlarged the cavity entrance and gained access to the nestlings. (Figure from Wesołowski 2017; © 2017 Oxford University Press, used with permission)

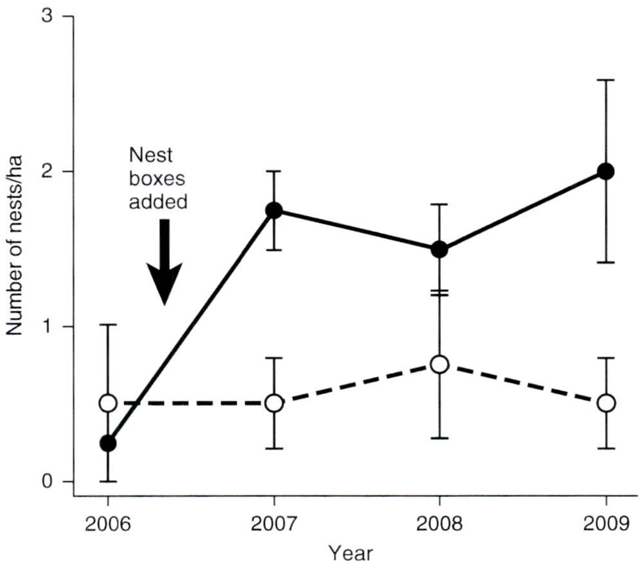


**Fig. 17.43** Mean number of nests (± SE) per hectare of secondary cavity-nesting birds in a selectively logged forest in Argentina. Dashed line indicates number of nests in control plots where no nest boxes are added; solid line indicates number of nests in experimental plots where nest boxes were added after the 2006 breeding season. The increase in number of nests after the addition of nest boxes indicates that availability of suitable cavities was limiting the population of secondary-cavity nesters in the selectively logged forest. (Figure from Cockle et al. 2010; © 2010 Elsevier Ltd., used with permission)

**Box 17.12 Nest "Decorations"**

Birds use a variety of materials to build their nests, including feathers, hair, snake skins, dung, and man-made objects. Feathers and hair are common additions to the nests of many species of birds (e.g., Liljeström et al. 2009; Dawson et al. 2011; Pollock et al. 2021).

Tufted Titmouse (*Baeolophus bicolor*) plucking hair from a sleeping raccoon (*Procyon lotor*) for use in its nest. (Photo by Zachary Sutton from Pollock et al. 2021; © 2021 by the Ecological Society of America, used with permission).

Feathers lining nests provide additional insulation for developing embryos in eggs and for nestlings and can enhance the efficiency of incubation and brooding by adults (Coppedge 2009; Dawson et al. 2011). This insulation may allow nestlings to devote less energy to thermoregulation and more energy to growth (Dawson et al. 2011). Another potential benefit of lining nests with feathers is that the feather-degrading bacteria or fungi on the feathers may migrate to the eggs and produce antibiotic substances that prevent colonization by pathogenic bacteria and reduce the likelihood of infection of developing embryos (Peralta-Sanchez et al. 2010). Birds may also add feathers to cavity nests as a way to prevent nest usurpation, with naïve prospecting birds potentially considering the presence of feathers to be the result of a predation event (Slagsvold and Wiebe 2021). If so, risk-averse, prospecting birds may be less likely to enter the cavity and the nest owners would be correspondingly more likely to avoid nest usurpation (Fear of feathers hypothesis, Slagsvold and Wiebe 2021).

(continued)

**Box 17.12** (continued)

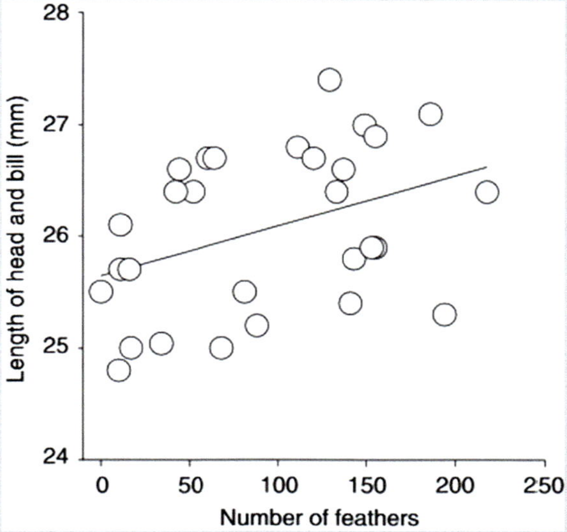


Nestling Tree Swallows (*Tachycineta bicolor*) in nests with more feathers were larger than nestlings in nests with fewer feathers. (Figure from Dawson et al. 2011; © 2011 The Authors, used with permission).

Female Eurasian Blue Tit (*Cyanistes caeruleus*) picking up a white feather to be added to its cavity nest. White feathers are more visible in dark cavities so, based on the Fear of feathers hypothesis, may be more effective at deterring potential nest usurpers from entering a cavity. (Figure from Slagsvold and Wiebe 2021; © 2021 The Authors. Published by the Royal Society under the terms of the Creative Commons Attribution License, https://creativecommons.org/licenses/by/4.0/).

(continued)

**Box 17.12** (continued)

Nests of Rock Sparrows (*Petronia petronia*) with numerous feathers. (Figure from García-Navas et al. 2015; © 2014 The Association for the Study of Animal Behaviour. Published by Elsevier Ltd., used with permission).

Some species of birds incorporate shed snakeskin or pieces of snakeskin into their nests. Using artificial nests, Medlin and Risch (2006) found that nests with skins of rat snakes (*Elaphe obsoleta*) were less likely to be predated by southern flying squirrels (*Glaucomys volans*) than nests without snake skins. Similarly, Liu and Liang (2021) determined that the presence of snake skins in the nests of Crested Mynas (*Acridotheres cristatellus*) reduced the likelihood of nest predation by squirrels (*Tamiops swinhoei*). In contrast, however, Trnka and Prokop (2011) found no difference in predation rates of artificial nests with and without snakeskins and suggested that incorporating snake skins into nests was a post-pairing signal of parental quality. Females may advertise their parental quality to induce their mates to provide more parental care.

**Box 17.13 Edible-nest (or White-nest) Swiftlets**

Swiftlets (tribe Collocaliini) are small insectivorous birds which can be found throughout Southeast Asia and the South Pacific. There are four genera, including *Aerodramus*, *Hydrochous*, *Schoutedenapus* and *Collocalia*, and about 40 species, most found in southern Asia, south Pacific islands, and northeastern Australia. Many of these species use echolocation to navigate in the darkness of the caves where they roost and nest.

All swifts (Apodidae) use saliva to construct their nests. However, most species use other materials like small sticks or other plant materials and use saliva as glue to hold the nest against a substrate and hold the nest materials together. However, the nests of some species of swiftlets are made entirely of saliva. The saliva consists primarily of protein (62–63%) and carbohydrates (25–27%), followed by lipids (0.14–1.28%) and ash (2.10%) (Norhayati et al. 2010). Trace elements found in the saliva included calcium, phosphorus, iron, sodium, potassium, iodine, and essential amino acids (Hun et al. 2015). With no other material used, "construction" of these nests can take as long as 30 to 45 days (Aowphol et al. 2008). In some areas in China and other countries in southeast Asia (and now even in the United States), soup is made from these nests, particularly the nests of White-nest (*Aerodramus fuciphagus*) and Black-nest (*A. maximus*) swiftlets. This "birds-nest soup" has been consumed in some areas for over 400 years. These nests are supposedly rich in nutrients that provide health benefits, e.g., ". . . nourish and tone up the organ systems of the body, helping to dissolve phlegm, improve the voice, relieve gastric problems, aid kidney function, enhance complexion, alleviate asthma, suppress cough, cure tuberculosis, strengthen the immune system, increase energy and metabolism, and improve concentration" (Thorburn 2015).

Nests have traditionally been harvested from the walls of caves and, with increasing demand, some populations of swiftlets whose nests are used to make the soup have declined as harvesting of nests reduces their reproductive success. Laws governing how nests are harvested have been implemented. For example, in some cases, a pair's first nest can be harvested, but they are then allowed to build a second nest and raise young—after which the second nest can then be harvested. Unfortunately, these laws are not always enforced and, in addition, penalties may not be large enough to deter harvesters because a kilogram of the nests can sell for as much as $2000 to $3000 (US dollars) (Thorburn 2014).

Given the economic value of edible nests, structures have been built in several areas in southeast Asia to attract breeding swiftlets—creating "farms" or "hotels" to harvest edible nests. In Malaysia alone, there are thousands of such swiftlet "farms," but many of these "farms" are also found in Indonesia and Thailand (Thorburn 2014). These "farms" may help reduce the number of nests harvested in the caves where swiftlets continue to nest.

(continued)

**Box 17.13** (continued)

Nests of White-nest Swiftlets in a cave. (Photo by Mike Prince, Wikipedia, CC BY 2.0, https://creativecommons.org/licenses/by/2.0/deed.en).

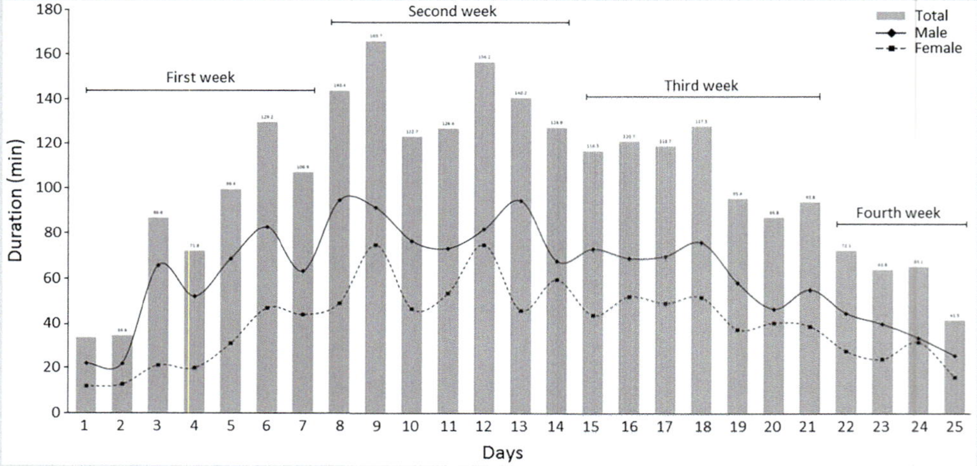


Construction of a nest by a pair of White-nest Swiftlets took 25 days and a total of 42.4 hours. The male spent 25.7 hours working on the nest, and the female 16.7 hours. On average, pairs of swiftlets take about 30 days to complete their nests (Lim and Earl of Cranbrook 2002).

(continued)

**Box 17.13** (continued)

(Figure from Ramji et al. 2013; permission for use granted by the Lee Kong Chian Natural History Museum [publisher of the Raffles Bulletin of Zoology]).

Nest of a White-nest Swiftlet. (Photo by finchfocus, purchased from istockphoto.com)

hawks and eagles (Harrison 1975) and swifts (e.g., Harrison 1975).

(b) Males contribute more than females, e.g., Bearded Vulture (*Gypaetus barbatus*; Margalida and Bertran 2000), and many species of woodpeckers

(c) Females contribute more than males. Examples include many New World warblers (Harrison 1975).

(d) Breeding pair is assisted by helpers, e.g., Gray-crowned Babbler (*Pomatostomus temporalis*, Dow and King 1984) and American Crow (*Corvus brachyrhynchos*, Kilham 1989).

2. Females build the nest, but males provide the nest material. Examples include pigeons and doves (Van Tyne and Berger 1976) and several species in the order Pelicaniformes (e.g., pelicans, cormorants, and some herons and egrets; Harrison 1975)
3. Females build the nest without assistance from males. Examples include most species in the order Galliformes (except megapodes and Hoatzins [*Opisthocomus hoazin*]), many species of suboscine (Tyrannidae) and oscine songbirds (Harrison 1975), hummingbirds (Trochilidae), and manakins (Passeriformes: Pipridae) (Collias and Collias 1984).
4. Females build the nest, but both sexes gather nest material. Examples include several species of songbirds (e.g., thrushes and vireos; Harrison 1975).
5. Males build the nest without assistance from females. Examples include rheas (Rheiformes), cassowaries (Casuariiformes), jacanas (Charadriiformes: Jacanidae),

**Box 17.14 Green Incubation**

Among some species of birds, adults bring green plants to their nests and a number of studies have revealed that these plants can have multiple benefits for offspring. However, how those plants provide benefits is not always clear and Gwinner et al. (2018) hypothesized that the presence of herbs would lead to higher egg temperatures that would accelerate embryonic development and improve nestling condition. They further hypothesized that higher egg temperatures could be due to the herbs (a) improving the thermal properties of nests (via improved insulation or heat produced by herbal decomposition), (b) reducing ectoparasite loads so adults would incubate for longer periods, or (c) modifying the behavior of adults (via some pharmacological effect) that would result in adults spending more time incubating eggs.

Male European Starling (*Sturnus vulgaris*) bringing green herbs to the nest box. (Screenshot from Supplemental movie from Gwinner et al. 2018; © 2018 The Authors. Published by the Royal Society, used with permission).

(continued)

**Box 17.14** (continued)

Eurasian Blue Tit (*Cyanistes caeruleus*) nest with green plant material. (Figure from Mainwaring 2017; © 2017 Mainwaring, open-access article distributed under the terms of the Creative Commons Attribution License CC BY, https://creativecommons.org/licenses/by/4.0/).

European Starlings are secondary-cavity nesters and readily nest in nest boxes. During nest building, males collect most of the nest material, including fresh, volatile herbs. Gwinner et al. (2018) monitored 53 nest boxes over a two-year period and most nests in those boxes ($n = 148$; some pairs were multibrooded) had at least some herbs (range = 0–350 grams). In their experiment, Gwinner et al. (2018) replaced natural nests with experimental nests (after males had completed nest building and were no longer bringing herbs to nests), some with and some without herbs.

Once incubation began, egg temperatures were monitored by placing small temperature monitors inside plastic eggs (one per nest) that resembled starling eggs. Subsequent analysis revealed that egg temperatures were consistently higher in experimental nests with herbs because adults at those nests spent more time incubating eggs than adults with experimental nests without herbs. The higher temperatures accelerated the growth of embryos, and as a possible carry-over effect, nestlings from herb nests were heavier than those in no-herb nests seven days after hatching.

(continued)

**Box 17.14** (continued)

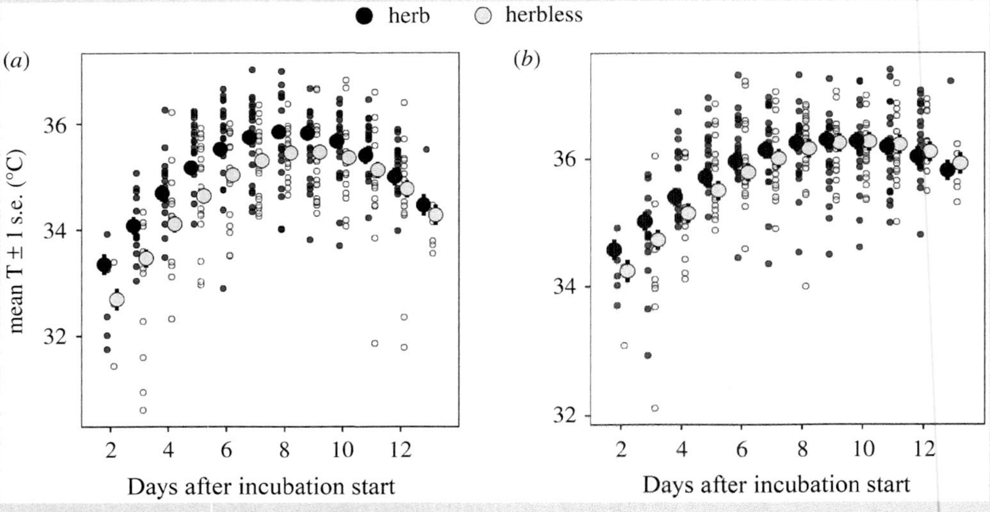


Temperature of eggs in nests where herbs were or were not added. (**a**) Mean temperatures during the day, and (**b**) mean temperatures at night. Note that the differences are greatest earlier in the incubation period. (Figure from Gwinner et al. 2018; © 2018 The Authors. Published by the Royal Society, used with permission).

Temperature data (i.e., cooling rates) collected when no adults were incubating indicated that the thermal properties of nests with and without herbs did not differ. Similarly, Gwinner et al. (2018) found no support for any ectoparasite-mediated effects on nest attendance by adults at nests with and without herbs. Differences in incubation temperatures between nests with and without herbs were due to differences in the incubation behavior of adults, with periods in incubation generally longer at nests with herbs than at nests without herbs. Interestingly, the effect of herbs on adult incubation behavior were most pronounced early in the incubation period, coinciding with a decline in release of volatiles from the no-longer-fresh herbs. These results provide support for the "pharmacological effect" hypothesis. In further support of this hypothesis, Gwinner et al. (2013), in an earlier study, identified several substances released by herbs in European Starling nests, including limonene, sabinene, and caryophyllene, that have been used in traditional herbal medicine. In addition, milfoil plants (*Achillea* spp.), herbs commonly added to nests by male European Starlings, are considered mild sedatives (Applequist and Moerman 2011). Although further study is needed to determine if and how the volatile substances released from herbs might actually affect adult European Starlings, and perhaps adults in other species of birds, the addition of live plants to nests can clearly be beneficial in a variety of ways.

**Fig. 17.44** The nests of young, inexperienced Village Weavers (*Ploceus cucullatus*, right) are more loosely and crudely constructed than nests of more experienced males (left). (Figure from Collias and Collias 1964a; © 1964 Oxford University Press, used with permission)

**Fig. 17.45** Nest of a Barn Swallow (*Hirundo rustica*) made of mud and grass and lined with feathers. (Photo by Tambe, Wikipedia, CC0 Public Domain)

phalaropes (Charadriiformes: Scolopacidae) (Collias and Collias 1984), and several species of songbirds (e.g., European Starlings and Red Bishops). In some polygynous species, males build multiple nests and attempt to attract multiple females to their territories, e.g., Red Bishops (*Euplectes orix*, Freidl and Klump 2000) and Eurasian Wrens (*Troglodytes troglodytes*). Interestingly, Soler et al. (2019) found that, among large species of songbirds ($\geq$ 32 grams), males contributed more to nest building in species where females have more

**Fig. 17.46** A series of six nests built by a male Southern Masked-Weaver (*Ploceus velatus*) (first nest, upper left, and last nest, lower right). (Figure from Walsh et al. 2010; © 2009 The Royal Society, used with permission)

conspicuous plumage (as judged by three individuals unaware of the hypothesis being tested) (Fig. 17.49).

Mainwaring et al. (2021) quantified the nest-building contributions of males and females for 521 species of birds representing 32% of avian families and 58% of avian orders and found that (1) species where only females build nests had shorter breeding seasons and larger clutches than species where both sexes contribute to nest building (Fig. 17.50), (2) species where both sexes contribute to nest building were more likely to nest above ground (e.g., shrubs, trees, or ledges) than species where females alone build nests (Fig. 17.51), and (3) the contributions of each sex varied with nest type, with females alone more likely to build among species with open-cup nests, and males alone more likely to build than both sexes among species with platform nests (Fig. 17.51).

Species where females alone built nests had shorter breeding seasons and larger clutches than species where both sexes helped build nests. One possible explanation for such results is that the roles of each sex are most specialized in species with shorter breeding seasons, e.g., females build nests and incubate eggs and males defend territories (Medina 2019; Mainwaring et al. 2021). Another possibility is that females tend to invest more in reproduction among species that nest only once during the breeding season and, as

**Fig. 17.47** Top, Brown Cacholote (*Pseudoseisura lophotes*; males and females look similar). Bottom, Stages of nest construction by a pair of Brown Cacholotes. (**a, b**) Both members of a pair form a small platform with twigs and sticks. (**c**) Next, they begin to build nest walls and form a cup about 20 cm deep. (**d**) They then begin to add the roof and start the entrance tunnel. (**e**) When the tunnel nears completion, they finish the nest's roof. (**f**) Finally, they finish the tunnel. The tunnel entrance is at the lower right. (Top, Photo by Ron Knight, Wikipedia, CC BY 2.0, https://creativecommons.org/licenses/by/2.0/deed.en. Bottom figure from Nores and Nores 1994; © 1994 Wilson Ornithological Society, used with permission)

**Table 17.2** Typical times needed by several species of songbirds to construct open-cup nests

| Species | Scientific name | Typical time to build nest | Reference |
|---|---|---|---|
| Least Flycatcher | *Empidonax minimus* | 5–7 days | Tarof and Briskie (2020) |
| Loggerhead Shrike | *Lanius ludovicianus* | 6–11 days | Yosef (2020) |
| White-eyed Vireo | *Vireo griseus* | 3–5 days | Hopp et al. (2020) |
| Clark's Nutcracker | *Nucifraga columbiana* | 5–8 days | Tomback (2020) |
| Verdin | *Auriparus flaviceps* | About 6 days | Webster (2020) |
| Cactus Wren | *Campylorhynchus brunneicapillus* | 1–6 days | Hamilton et al. (2020) |
| Black-tailed Gnatcatcher | *Polioptila melanura* | 2–4 days | Farquhar and Ritchie (2020) |
| Ruby-crowned Kinglet | *Corthylio calendula* | About 5 days | Swanson et al. (2020) |
| Northern Wheatear | *Oenanthe oenanthe* | 2–7 days | Dunn et al. (2020) |
| American Robin | *Turdus migratorius* | 5–7 days | Vanderhoff et al. (2020) |
| Gray Catbird | *Dumetella carolinensis* | 5–6 days | Smith et al. (2020b) |
| Cedar Waxwing | *Bombycilla cedrorum* | 3–9 days | Witmer et al. (2020) |
| Yellow Warbler | *Setophaga petechia* | 4–10 days | Lowther et al. (2020a) |
| Eastern Towhee | *Pipilo erythrophthalmus* | Up to 5 days | Greenlaw (2020) |
| Chipping Sparrow | *Spizella passerina* | 2–8 days | Middleton (2020) |
| Scarlet Tanager | *Piranga olivacea* | 2–7 days | Mowbray (2020) |
| Northern Cardinal | *Cardinalis cardinalis* | 3–9 days | Halkin et al. (2021) |
| Indigo Bunting | *Passerina cyanea* | 2–10 days | Payne (2020) |
| Red-winged Blackbird | *Agelaius phoeniceus* | 3 days | Holcomb and Twiest (1968) |
| Orchard Oriole | *Icterus spurius* | 6 days | Scharf and Kren (2020) |
| Pine Siskin | *Spinus pinus* | 5–6 days | Dawson (2020) |
| Baya Weaver | *Ploceus philippinus* | 18 days | Asokan et al. (2008) |

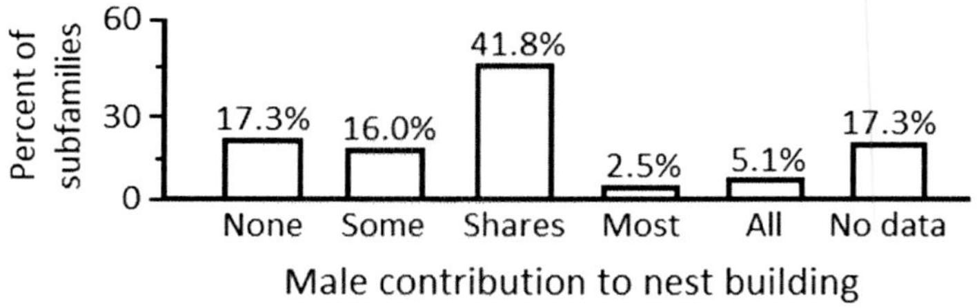


**Fig. 17.48** Contribution of males to nest building among 237 subfamilies of birds. (Figure modified from Silver et al. 1985; © 2015 Oxford University Press, used with permission).

a result, breeding seasons are shorter in duration (Mainwaring et al. 2021). In addition, males in some species may invest less in reproduction due to uncertainty in parentage, i.e., their mates may engage in extrapair copulations (Lifjeld et al. 2019; Mainwaring et al. 2021).

Mainwaring et al. (2021) also found that the contributions of each sex to nest building were related to nest sites and nest structure. Both sexes were more likely to contribute to nest construction in species with domed nests than species with cup nests, possibly because building nests with a more complex structure (domed nests) requires the sex-specific cognitive abilities of both sexes (Guillette and Healy 2015; Mainwaring et al. 2021). In addition, building nests can be energetically expensive so, with both sexes contributing to nest construction, nests can potentially be more complex (e.g., domed nests) or larger (e.g., large platform nests

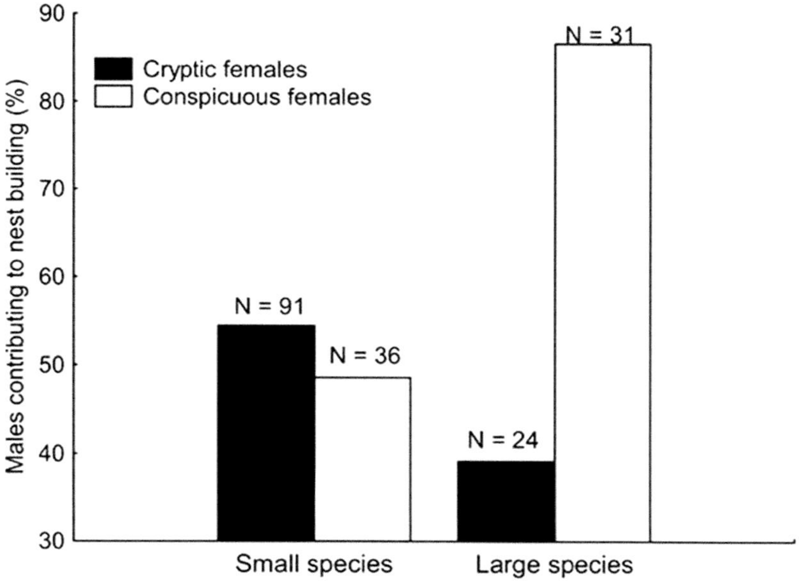


**Fig. 17.49** Male songbirds are more likely to contribute to nest building in large species (≥ 32 grams) where females have more conspicuous plumage. Such results suggest that male nest-building behavior and sexually selected characteristics of females have co-evolved. By participating in nest building, males may achieve greater reproductive success by pairing with highly ornamented females. Females, in turn, may respond to the greater contributions of males by investing more in reproduction. (Figure from Soler et al. 2019; © 2019 Oxford University Press, used with permission)

like those of Bald Eagles [*Haliaeetus leucocephalus*; Fig. 17.52) (Mouton and Martin 2019; Mainwaring et al. 2021).

## 17.9 Costs of Nest Building

Some birds do not build nests, but, rather, use previously used nest sites or naturally generated sites, e.g., secondary cavity-nesting species and some open-cup nesting species like Great Horned Owls, and one advantage of such behavior is that the costs of nest building are avoided. Building nests and creating cavities, however, takes time and energy. The behavior of some birds suggests that natural selection may favor individuals that can reduce these costs. Pairs of Blue-gray Gnatcatchers (*Polioptila caerulea*) that abandon a nest (e.g., due to parasitism by a Brown-headed Cowbird, *Molothrus ater*) typically conserve energy by tearing the old nest apart and "recycling" the materials in a new nest (Root 1969). Some birds are known to occasionally take materials from the nests of other species for use in their own nests, e.g., Broad-tailed Hummingbirds (*Selasphorus platycercus*) and Warbling Vireos (*Vireo gilvus*; Calder 1972), Red-eyed Vireos (*Vireo olivaceus*), Black-throated Green Warblers (*Setophaga virens*), and Cerulean Warblers (*Setophaga cerulean*; Jones et al. 2007). Such stealing of nest materials may be particularly common in colonial species such as frigatebirds (*Fregata* spp.) and Northern Gannets (*Sula bassana*; Nelson 1978), and some species of penguins (Moreno et al. 1995). Birds that steal nest materials may need fewer and shorter trips to construct their nest and may need less time to complete nests. Fewer and shorter trips mean less energy expended in nest construction, and birds that complete nests faster may also gain the potential benefits associated with earlier nesting.

For birds that do build nests and do not steal nest materials, the time needed to construct nests

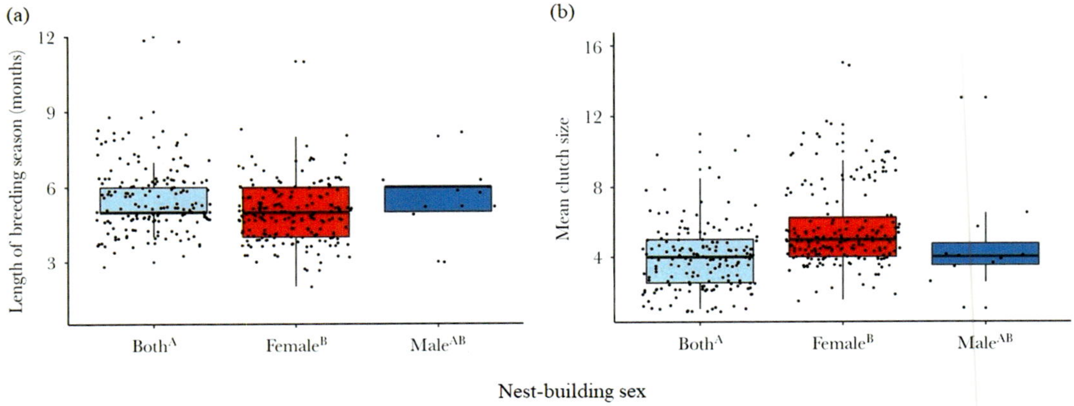


**Fig. 17.50** Contributions of each sex in building nests relative to (**a**) the length of breeding seasons (months), and (**b**) mean clutch size. Species where only females build nests have shorter breeding seasons and larger clutches than species where both sexes contribute to nest building. Different letters (superscripts) indicate statistically significant differences. Box plots denote the 10th, 25th, 50th, 75th, and 90th percentiles of individual data points jittered and plotted alongside. (Figure from Mainwaring et al. 2021; © 2021 Oxford University Press, used with permission)

varies with nest complexity. Simple scrape nests may require very little time and effort. Some Little Ringed Plover (*Charadrius dubius*) nests are simple scrapes as little as 0.5 cm deep with only 2 or 3 shell fragments placed within the depression (Dalakchieva 2002). By contrast, male and female Hamerkops (*Scopus umbretta*) take six or seven weeks to build their huge, enclosed nests, making at least 8000 trips to nests with nest material (Kahl 1967; Fig. 17.53).

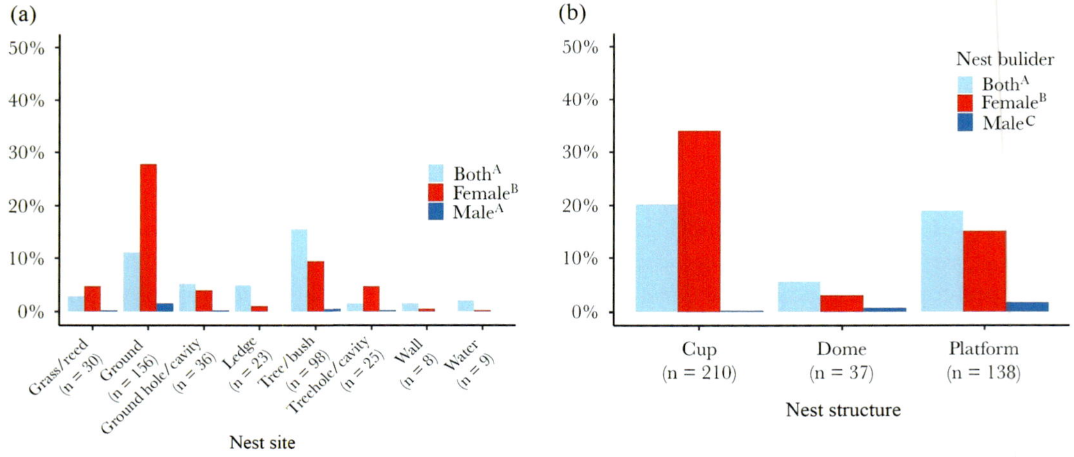


**Fig. 17.51** Contributions of each sex in building nests relative to (**a**) nest sites or locations, and (**b**) nest structure or type. Species where both sexes help build nests are more likely to nest above ground (e.g., shrubs, trees, or ledges) than species where females alone build nests, and females alone are more likely to build nests for species with open-cup nests, and males alone are more likely to build nests than both sexes among species with platform nests. Different letters (superscripts) indicate statistically significant differences. (Figure from Mainwaring et al. 2021; © 2021 Oxford University Press, used with permission)

**Fig. 17.52** Bald Eagle (*Haliaeetus leucocephalus*) nests are constructed by both the male and female. (Photo by U. S. Fish and Wildlife Service, CC0 Public Domain)

Many open-cup nesting passerines take a week or less to construct nests, and generally less than two weeks (Table 17.2). Even relatively large birds sometimes build nests in less than a week, but construction may take much longer (Table 17.3). Bald Eagles (*Haliaeetus leucocephalus*) sometimes take up to three months to build a nest, but can complete nests in as little as four days (Buehler 2020).

Primary cavity-nesters excavate cavities in tree trunks, branches, or snags, with the time required influenced by factors such as cavity size, wood quality (e.g., trees dead or alive), and duration of the breeding season. Primary cavity-nesters generally take about one to three weeks to excavate their cavities, but there are exceptions (Table 17.4). For example, Red-cockaded Woodpeckers (*Dryobates borealis*) use living pine trees (*Pinus* spp.) for cavity trees and must generally excavate through 8 to 16 cm of solid wood (Conner et al. 1994) so may take more than six years to complete excavation of a cavity (Table 17.4). Excavation times for Red-cockaded Woodpeckers vary with tree species and availability of suitable cavities. These woodpeckers are cooperative breeders, with groups (called clans) consisting of a breeding pair plus one to four helpers. Clans use cavities for both nesting and roosting, with roosting cavities typically used by a single woodpecker. If more cavities are needed within a clan's range, excavation rates may increase (Conner et al. 2002).

Many burrow-nesting species take less than three weeks to excavate their burrows, and sometimes less than a week (Table 17.5). As with cavity-nesters, the time needed to excavate burrows depends on the dimensions and depth of burrows and nest chambers and on characteristics of the substrate.

The time needed to construct nests or excavate cavities or burrows can vary with complexity (i.e., number of trips needed to bring needed materials) and, in the case of cavities and burrows, their size and the substrate. For multibrooded passerines that nest multiple times per breeding season, construction times typically decrease for later nests. Northern Cardinals (*Cardinalis cardinalis*) may take as long as two to three weeks to complete their first nest, but generally less than a week (and sometimes in as little as 3 days) for nests later in the season (Halkin et al. 2021). Because of the limited duration of breeding seasons, multibrooded species at

**Fig. 17.53** (**a**) Hamerkop (*Scopus umbretta*) nest, (**b**) Close-up of the Hamerkop nest, and (**c**) a Hamerkop. (Photos a and b by MichIV, purchased from istockphoto.com; photo c by Voidoffrogs, Wikipedia, CC0 Public Domain)

higher latitudes benefit by initiating later nesting attempts as soon as possible. Birds that are not multibrooded, but whose initial nesting attempts fail due to predation or some other factor, may renest and, if so, second nests are often built at a faster rate than the initial nest (e.g., Blue-gray Gnatcatchers, *Polioptila caerulea*; Kershner and Ellison 2020). In addition, the amount of time

**Table 17.3** Typical time needed by several species of non-songbirds to construct nests

| Species | Scientific name | Typical time to build nest | Reference |
|---|---|---|---|
| Northern Bobwhite | *Colinus virginianus* | 5 days | Brennan et al. (2020) |
| Trumpeter Swan | *Cygnus buccinator* | 14–35 days | Mitchell and Eichholz (2020) |
| Ruby-throated Hummingbird | *Archilochus colubris* | 6–10 days | Weidensaul et al. (2020) |
| Black Swift | *Cypseloides niger* | 4 days | Lowther et al. (2020a, 2020b) |
| Yellow-billed Cuckoo | *Coccyzus americanus* | 2–3 days | Wilson (1999) |
| White-crowned Pigeon | *Patagioenas leucocephala* | 5 days | Bancroft et al. (2020) |
| Limpkin | *Aramus guarauna* | About 14 days | Bryan (2020) |
| Sandhill Crane | *Antigone canadensis* | 1–14 days | Littlefield and Ryder (1968) |
| Western Grebe | *Aechmophorus occidentalis* | Nest can hold eggs after 1–3 days, but additional material is added after egg-laying begins | LaPorte et al. (2020) |
| Wilson's Phalarope | *Phalaropus tricolor* | 3–4 days | Colwell and Jehl (2020) |
| Common Loon | *Gavia immer* | 1–≥ 7 days | Evers et al. (2020) |
| Wood Stork | *Mycteria americana* | 2–3 days, but construction/maintenance continues through nestling period | Coulter et al. (2020) |
| Anhinga | *Anhinga anhinga* | 1–3 days | Frederick and Siegel-Causey (2020) |
| Great Blue Heron | *Ardea herodias* | 3–14 days | Vennesland and Butler (2020) |
| Bald Eagle | *Haliaeetus leucocephalus* | 4 days–3 months | Buehler (2020) |
| Northern Harrier | *Circus hudsonicus* | 7–14 days | Smith et al. (2020a, 2020b) |

**Table 17.4** Typical time needed by several species of woodpeckers to create nest cavities

| Species | Scientific name | Typical time to create nest cavity | Reference |
|---|---|---|---|
| Northern Flicker | *Colaptes auratus* | 5–20 days | Wiebe and Moore (2020) |
| Black-backed Woodpecker | *Picoides arcticus* | 21–28 days | Tremblay et al. (2020) |
| Downy Woodpecker | *Dryobates pubescens* | 7–20 days | Ritchison (1999) |
| Golden-fronted Woodpecker | *Melanerpes aurifrons* | 6–12 days | Husak and Maxwell (2020) |
| Hairy Woodpecker | *Dryobates villosus* | 7–21 days | Bent (1939) |
| Pileated Woodpecker | *Dryocopus pileatus* | 21–36 days | Bull and Jackson (2020) |
| Red-bellied Woodpecker | *Melanerpes carolinus* | ~14 days | Miller et al. (2020) |
| Red-headed Woodpecker | *Melanerpes erythrocephalus* | 12–17 days | Frei et al. (2020) |
| White-headed Woodpecker | *Dryobates albolarvatus* | 21–28 days | Kozma et al. (2020) |
| Red-cockaded Woodpecker | *Dryobates borealis* | 1.8–6.3 years | Conner and Rudolph (1995) |

**Table 17.5** Typical times needed by several species of burrow-nesting species of birds to excavate burrows

| Species | Scientific name | Typical time needed to excavate burrow | Reference |
|---|---|---|---|
| Leach's Storm-Petrel | *Hydrobates leucorhous* | 28 days | Pollet et al. (2020) |
| Cassin's Auklet | *Ptychoramphus aleuticus* | 21–28 days | Ainley et al. (2020a, 2020b) |
| Burrowing Owl[a] | *Athene cunicularia* | 2 days | Poulin et al. (2020) |
| Belted Kingfisher | *Megaceryle alcyon* | 3–7 days | Kelly et al. (2020) |
| Pied Kingfisher | *Ceryle rudis* | 26 days | Douthwaite (1978) |
| Rufous-tailed Jacamar | *Galbula ruficauda* | 3 days | Skutch (1937) |
| Rainbow Bee-eater | *Merops ornatus* | 5–10 days | Boland (2004) |
| Green Bee-eater | *Merops orientalis* | 15–20 days | Sridhar and Karanth (1993) |
| Northern Rough-winged Swallow | *Stelgidopteryx serripennis* | ≤ 7 days | De Jong (2020) |
| Bank Swallow | *Riparia riparia* | 4–5 days | Garrison and Turner (2020) |

[a]Burrowing Owls most often use burrows dug by other animals, such as ground squirrels, prairie dogs, and badgers. However, Burrowing Owls in Florida usually excavate their own burrows (Poulin et al. 2020)

devoted to nest building during the construction process can vary. For example, female Swainson's Warblers (*Limnothlypis swainsonii*) take two to five days to build nests, but, during those days, nest construction is limited primarily to the morning hours (Anich et al. 2020). Individuals and species that devote more time per day to construction may complete nests in less time.

Constructing nests takes time and energy. Stanley (2002) estimated the energetic cost of nest construction by a pair of Black-billed Magpies (*Pica hudsonia*) at a minimum of 209 kilojoules (or about 50 kilocalories) based on a minimum of 2564 trips for nesting materials, 276.2 km of commuting, and 8.4 hours of flight (Fig. 17.54). However, because the nest was constructed over a period of just over a month

**Fig. 17.54** Nest of a pair of Black-billed Magpies (*Pica hudsonia*). (Photo courtesy of Rocky Mountain National Park, CC0 Public Domain)

**Fig. 17.55** Mud nest of a Cliff Swallow. (Photo by Ken Thomas, Wikipedia, CC0 Public Domain)

by two magpies, the energetic cost was estimated to be only about 2.6 kilojoules (or 0.6 kilocalories) per bird per day, or between 0.7 and 1.0% of their daily metabolizable energy intake.

The daily energetic costs of nest building may be higher for birds that build nests over shorter periods. Cliff Swallows (*Petrochelidon pyrrhonota*) build their mud nests over periods ranging from about 8 to 18 days (average = 13.7 days; Fig. 17.55), making an estimated 1400 trips to collect nest material (Withers 1977; Gauthier and Thomas 1993), at an estimated energetic cost of about 7 kilojoules (1.7 kilocalories) per bird per day (Gauthier and Thomas 1993). The typical daily energy intake of these swallows is unknown, but the mean energy expenditure of female Barn Swallows (*Hirundo rustica*) during egg laying (when they would be expending more energy than usual), a similar-sized bird in the same genus, has been estimated at about 110 kilojoules per day (Ward 1996). If daily energy expenditure of Cliff Swallows is similar, they would need to increase their daily energy intake by about 6.4% during nest construction.

The energetic costs of nest construction would clearly be influenced by nest type, size, location, construction materials, and other factors. For open-cup nesting birds, the energetic cost of nest construction would depend on the number and distance of trips needed to bring nest materials to nest sites; more, longer flights would mean greater costs and daily energetic costs would depend on how many days were devoted to nest construction. Burrow- and primary cavity-nesting birds must expend energy digging or excavating and the energetic costs of those activities have not been calculated. However, the size and depth of burrows and cavities, as well as characteristics of the substrates (e.g., harder wood of living trees vs. softer wood of snags), would clearly influence the needed effort and amount of energy expended. Although few investigators have attempted to quantify the energetic costs of nest construction, some have used an indirect approach. In other words, if energetically costly, then nest construction activities may impact birds in other ways that investigators can quantify. For example, Moreno et al. (2008) found that female European Pied Flycatchers (*Ficedula hypoleuca*) that built nests at the fastest rates had higher levels of stress proteins in their blood, suggesting that, for some females, nest construction may cause physiological stress. The possible effect of such stress on females remains to be determined. In another study, Moreno et al. (2010) reduced the effort needed by female European Pied Flycatchers to build nests by placing previously stored complete nests in some nest boxes (experimental nests), but not others (control nests), as soon as females began nest building. Females in experimental nests were subsequently found to provision nestlings (on day 4 post-hatching) at higher rates than control females, and young in experimental nests were larger than those in control nests at the time of fledging. These results suggest that the energy expended by female European Pied Flycatchers in nest construction impacted their ability to provisioning their young. Additional evidence that nest construction may impose costs come from studies reporting a relationship between individual quality and nest quality. Among Blue Tits, females not infected with blood parasites built heavier nests than infected females, suggesting that nest building is a costly activity and only females in better condition are able to build large nests (Tomás et al. 2006). Although the results of studies like these suggest that nest construction may be energetically costly and perhaps even stressful for some individuals in some species, other studies seem to indicate otherwise. For example, female Eastern Phoebes (*Sayornis phoebe*) that varied in the time and effort needed to construct their nests had similar clutch initiation dates and similar clutch sizes, suggesting that whatever costs were associated with nest construction were not sufficient to affect female reproductive behavior and output (Conrad and Robertson 1993).

Constructing nests clearly does entail some cost in terms of time and energy, and birds sometimes use strategies, such as taking nest materials from other nests, that reduce that cost. The few studies designed to quantify the cost of nest construction suggest that birds may be able to recoup energetic costs by slightly increasing daily energy intake. Studies to date also seem to suggest that, for most birds, the costs of nest building appear to be insufficient to impact their ability to successfully engage in other activities associated with reproduction (but see Moreno et al. 2010). However, additional studies are clearly needed to better understand the energetic costs of nest construction and the possible effect(s) those costs might have on the reproductive behavior and success of birds.

Additional potential costs of nest building are predation and brood parasitism. As noted by Lee and Lima (2016), the cognitive attention needed to construct nests or search for materials to incorporate into nests could make birds more vulnerable to attack by predators. In addition, adults making repeated visits to nest sites as nests are being constructed can attract the attention of brood parasites (e.g., Kattan 1997; Soler and Pérez-Contreras 2012). Given these potential costs, Lee and Lima (2016) suggested that, for many species of birds, the best option may be to "quick-build 'minimal' nests" that will support eggs and young, but require less nest material and fewer nest visits. The trade-off, however, is that such nests, e.g., "minimal" open-cup nests, provide limited protection from nest predators. For species where time is sufficient for additional nesting attempts if a nest is predated, the "quick-build" strategy may be favored. However, for species limited to single nesting attempts per breeding season, selection may favor a greater time investment in nest construction in an attempt to improve nest safety (Lee and Lima 2016).

## 17.10 Nest Reuse by Cavity-Nesting Species

Most birds build a new nest for each nesting attempt, but cavity-nesting species sometimes nest in the same cavity multiple times (Aitken et al. 2002; Fig. 17.56). Of course, secondary cavity-nesting species do not excavate their own cavities so may be more likely to reuse cavities

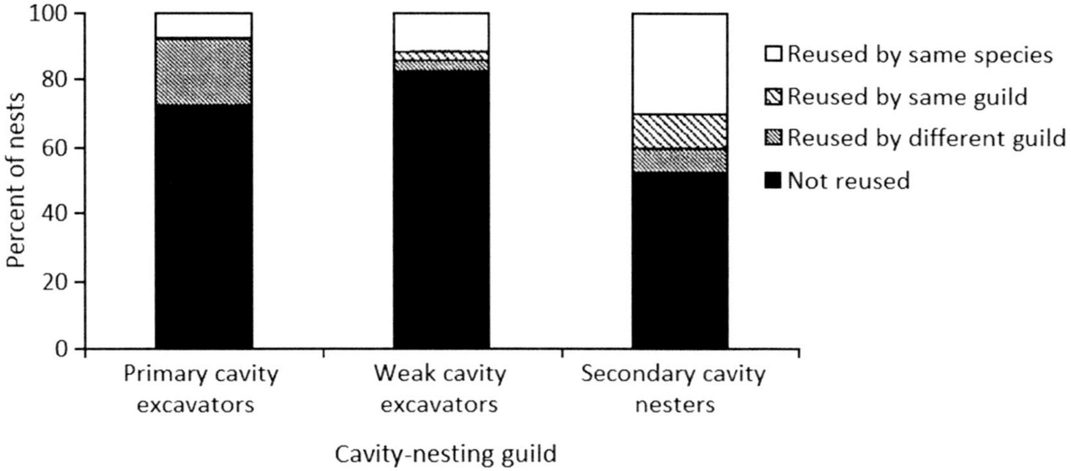


**Fig. 17.56** Proportion of nest cavities reused in a subsequent year by the same species, same guild, and different guild or not reused in stands of mixed forest in British Columbia, Canada. Note that secondary-cavity nesters were more likely to reuse cavities than primary-cavity and weak-cavity excavators. (Figure from Aitken et al. 2002; © 2002 Oxford University Press, used with permission)

(although not necessarily by the same individuals; Fig. 17.56). However, competition among secondary cavity-nesting species can influence cavity use by each species. In British Columbia, Canada, Mountain Chickadees (*Poecile gambeli*) and European Starlings (*Sturnus vulgaris*) used newer cavities (1–2 years old), whereas Mountain Bluebirds (*Sialia currucoides*) and Tree Swallows (*Tachycineta bicolor*) generally used older cavities (Edworthy et al. 2018). Mountain Chickadees, as the only abundant small-bodied secondary cavity-nesting species in the study area, were able to use newer cavities because they faced little competition for cavities with small-diameter entrances excavated by Red-breasted Nuthatches (*Sitta canadensis*) and Downy Woodpeckers (*Dryobates pubescens*); European Starlings were often able to use new cavities because of their aggressive competition for cavities (e.g., Ingold 1998). In contrast, use of older cavities by Mountain Bluebirds and Tree Swallows was likely the result of being displaced from newer cavities by similar-size competitors (Edworthy et al. 2018).

For primary cavity excavators, excavating new cavities takes time and energy so reusing cavities may offer the advantage of larger clutches (i.e., energy saved may mean more energy available for egg production by females) and nesting earlier in the breeding season, with early breeding potentially contributing to increased survival of young (e.g., Wiktander et al. 2001; Kosiński and Walczak 2019). Another potential advantage of reusing cavities is that evidence of a new excavation, e.g., wood chips at the base of a tree, might attract predators (Wiebe et al. 2007).

Despite those potential advantages, cavity-nesting species, particularly primary cavity-nesting species, usually do not reuse cavities (Edworthy et al. 2018). One possible reason for this is competition, with other cavity-nesting species, especially secondary cavity-nesters, assuming ownership of a cavity after the breeding season. In addition, excavating new cavities can provide several potential advantages. For example, new cavities may be safer than a reused cavity because predators have not learned their location (Otterbeck et al. 2019). Another possible advantage is that new cavities would likely have fewer ectoparasites than older cavities, and the results of studies of secondary cavity-nesting species suggest a possible preference for cavities with fewer ectoparasites (e.g., Breistøl et al. 2015). Excavating new cavities may also allow primary cavity nesters to nest in better-quality habitat, e.g., with greater food availability (Wiebe et al. 2007).

## 17.11 Nest Parasites

Bird nests provide suitable microclimates for eggs and nestlings, but can also provide shelter for a variety of other organisms, some of which can have detrimental effects on the birds in those nests. Many different types of parasites spend portions of their life cycles in nest materials and may feed on adult birds and nestlings. For example, at least 2500 species of mites (phylum Arthropoda, class Arachnida, subclass Acari) representing 40 families are associated with birds, including skin mites, feather mites, nest-dwelling dust mites, and ticks (Proctor and Owens 2000). Some nest-dwelling mites have no impact on birds because they feed on skin flakes and other detritus on and in nests. However, other nest mites are blood feeders (in their adult stages) and, because of their short generation times, can rapidly increase in numbers. For example, Clark (1991) reported that some nests of European Starlings contained as many as 500,000 northern fowl mites (*Ornithonyssus sylviarum*), an ectoparasite that feeds on bird blood. Whereas nest-dwelling mites spend their entire life cycles in nests, ticks are temporary nest parasites, undergoing most of their life cycles elsewhere, but sometimes visiting nests to feed on the blood of nestlings (and adults).

Blowfly larvae (Diptera, *Protocalliphora* spp.) and fleas are also potential ectoparasites of young birds. Adult blowflies are free-living flies that lay their eggs in nest materials after eggs hatch. The larvae live in nests and feed intermittently on the blood of young birds. As nest parasites, bird fleas spend most of their life cycle in nests, with adults feeding intermittently on the blood of nestlings and adults. The effects of ectoparasites like mites, ticks, blowflies, and fleas on birds vary depending on when in the nesting cycle nests are parasitized and on parasite intensity (number present). For example, depending on the severity of the infestation, the presence of ticks may have little or no effect, cause seabirds to completely or partially abandon nesting colonies (Feare 1976; Duffy 1983), cause nestling mortality, or slow nestling growth and delay fledging (Ramos et al. 2001).

Blood-feeding ectoparasites can potentially reduce the reproductive success of birds by slowing nestling development or even killing nestlings. For example, nestlings in nests of Pied Flycatchers with high loads of blood-feeding mites weighed less than those in nests with low mite loads, and this can be detrimental because low mass at fledging often means reduced reproductive success later in life (Merino and Potti 1995). Similar results have been reported for House Finches (*Carpodacus mexicanus*), with nestlings in nests infested with mites (*Pellonyssus reedi*) in poorer condition than those in non-infested nests, and with increased mortality of nestlings in the most heavily infested nests (Stoehr et al. 2000). Blowfly larvae can have similar effects. For example, Puchala (2004) found a negative relationship between the number of blowfly larvae per nestling and the fledging success of Tree Sparrows (*Passer montanus*), with more heavily infested young growing slower and weighing less than less infested or non-infested young. In addition, some heavily infested nestlings died before fledging.

Although potentially affecting nestling condition and survival, ectoparasites that infest nests late in the nestling period or in low numbers may have little impact on nestling condition or survival. For example, the mass of nestling House Wrens (*Troglodytes aedon*) and the reproductive success of adult wrens were not affected by the presence of mites in "naturally occurring numbers" (about 5500–17,000 per nest; Pacejka et al. 1998). Other investigators have reported similar results for nestlings infested with blowfly larvae (e.g., Eeva et al. 1994; Hurtrez-Bousses et al. 1997). One possible explanation for the ability of nestlings to maintain growth rates and fledge even when parasitized by blood-feeding ectoparasites like mites and blowflies is the ability of birds to withstand blood loss. For example, Ploucha and Fink (1986) removed similar amounts of blood (4 mL) from Domestic Chickens (*Gallus gallus domesticus*) and rats (*Rattus norvegicus*) and found that the blood pressure and cardiac output of chickens (15 and 4%, respectively) dropped much less than for rats (25 and 43%, respectively). In contrast to mammals, birds quickly replace lost blood with interstitial fluid to help maintain blood pressure and cardiac output (Djojosugito et al. 1968). Any dilution of blood reduces its oxygen-carrying capacity (because there is less hemoglobin) temporarily (with recovery time dependent on how much blood is lost), but normal cardiac and vascular function is maintained so nutrients and waste products are still transported at normal levels. This ability of birds to withstand blood loss likely enhances the ability of nestlings to sometimes survive even heavy infestation by blood-feeding ectoparasites (Clark 1991).

Nest ectoparasites can also influence the behavior of adults, with responses varying among species. Adults in some species respond to infestation of nestlings by ectoparasites by increasing provisioning rates. For example, adult Eurasian Blue Tits (*Cyanistes caeruleus*) at infested nests provisioned nestlings at rates 29% higher than adults at non-infested nests (Tripet and Richner 1997). Such compensation by adults can minimize the effects of ectoparasites on their young (albeit at some potential cost to adults in terms of their condition and future reproductive success) (Richner and Tripet 1999). However, studies of other species indicate that adults feed nestlings less often with increasing ectoparasite loads. For example, provisioning rates of adult House Finches were found to decline as ectoparasite loads of nestlings increased (Stoehr et al. 2000). With increasing ectoparasite loads, the likelihood of nestling survival declines so adults may benefit, in terms of future reproductive success, from reduced effort (Perrin et al. 1996).

In response to the presence of nest ectoparasites during egg laying, female birds may be able to minimize the effects on their young by the differential allocation of hormones and immunoglobulins to eggs. In an experimental study, Buechler et al. (2002) exposed some female Great Tits (*Parus major*) to hen fleas (*Ceratophyllus gallinae*; 60 fleas were placed in their nests) after they had laid their first egg, whereas other females were not exposed. After clutches were complete, the investigators removed eggs from across the laying order (first, fifth, and eighth eggs laid by females) in both groups and determined the concentration of antibodies (immunoglobulin G) in the yolks. Antibody levels in the eggs of non-exposed females did not vary among eggs (Fig. 17.57). However, antibody levels were significantly elevated in the eighth eggs of females exposed to fleas (Fig. 17.57). Importantly, 16-day-old nestlings in the nests of females exposed to fleas also weighed significantly more than those in the nests of females that were not exposed. These results indicate that female Great Tits transferred more maternal antibodies into eggs several days after being exposed to fleas and the antibodies protected young from the negative effects of the fleas, resulting in healthier, larger young. Additional studies are needed to determine if females in other species of birds might respond in a similar manner when exposed to fleas or other ectoparasites prior to egg-laying.

Several factors can potentially influence the negative impacts of parasites on young birds, including nest location (open cup vs. cavity nest) and latitude. Møller et al. (2009) reviewed the results of 117 studies to determine which factors best explained differences in parasite virulence (the rate of mortality among infected young), including studies that focused on the effects of mites, ticks, flies, and fleas as well as bacteria, lice, nematodes, protozoans, and viruses. Their analysis revealed that latitude was among the most important factors, with the risk of nestling mortality greater at lower latitudes. One possible explanation for this is that parasites are more abundant in the tropics than at higher latitudes because of a favorable year-round climate.

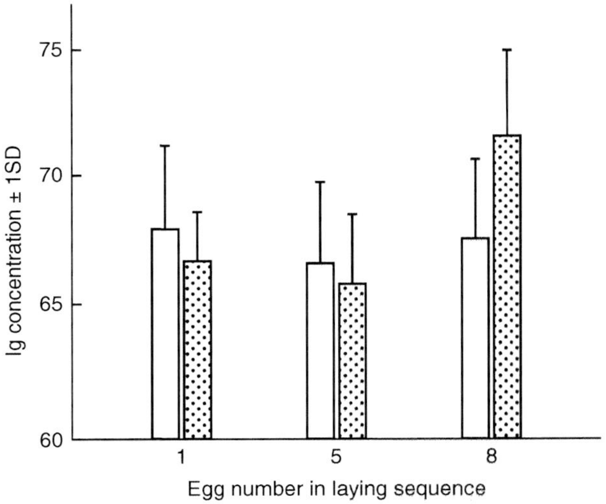


**Fig. 17.57** Immunoglobulin (IgG) concentration in the first, fifth, and eighth eggs laid by female Great Tits (*Parus major*) not exposed to hen fleas (open bars) or after exposure to hen fleas (hatched bars) after laying of the first egg. Concentration is expressed as a measure of optical density (absorbance) in test yolk samples. (Figure from Buechler et al. 2002; © 2002 John Wiley and Sons, used with permission)

## References

Ainley D, Adams J, Manuwal DA, Thoresen AC (2020a) Cassin's Auklet (*Ptychoramphus aleuticus*), version 1.0. In: Poole AF (ed) Birds of the World. Cornell Lab of Ornithology, Ithaca, NY

Ainley DG, Nettleship DN, Carter HR, Storey AE (2020b) Common Murre (*Uria aalge*), version 1.0. In: Poole AF, Gill FB (eds) Birds of the World. Cornell Lab of Ornithology, Ithaca, NY

Aitken KE, Martin K (2007) The importance of excavators in hole-nesting communities: availability and use of natural tree holes in old mixed forests of western Canada. J Ornithol 148:425–434

Aitken KEH, Martin K (2012) Experimental test of nest-site limitation in mature mixed forests of Central British Columbia, Canada. J Wildl Manag 76:557–565

Aitken KEH, Wiebe KL, Martin K (2002) Nest-site reuse for a cavity-nesting bird community in interior British Columbia. Auk 119:391–402

Akresh ME, Ardia DR, King DI (2017) Effect of nest characteristics on thermal properties, clutch size, and reproductive performance for an open-cup nesting songbird. Avian Biol Res 10:107–118

Altamirano TA, Ibarra JT, Martin K, Bonacic C (2017) The conservation value of tree decay processes as a key driver structuring tree cavity nest webs in South American temperate rainforests. Biodivers Conserv 26: 2453–2472

Ames PL, Mersereau GS (1964) Some factors in the decline of the Osprey in Connecticut. Auk 81:173–185

Anich NM, Benson TJ, Brown JD, Roa C, Bednarz JC, Brown RE, Dickson JG (2020) Swainson's Warbler (*Limnothlypis swainsonii*), version 1.0. In: Rodewald P (ed) Birds of the World. Cornell Lab of Ornithology, Ithaca, NY

Aowphol A, Voris HK, Feldheim KA, Harnyuttanakorn P, Thirakhupt K (2008) Genetic homogeneity among colonies of the White-nest Swiftlet (*Aerodramus fuciphagus*) in Thailand. Zool Sci 25:372–380

Applequist WL, Moerman DE (2011) Yarrow (*Achillea millefolium* L.): a neglected panacea? A review of ethnobotany, bioactivity, and biomedical research. Econ Bot 65:209

Ar A, Barnea A, Yom-Tov Y, Mersten-Katz C (2004) Woodpecker cavity aeration: a predictive model. Respir Physiol Neurobiol 144:237–249

Ardia DR, Pérez JH, Clotfelter ED (2006) Nest box orientation affects internal temperature and nest site selection by Tree Swallows. J Field Ornithol 77:339–344

Ardia DR, Pérez JH, Clotfelter ED (2010) Experimental cooling during incubation leads to reduced innate immunity and body condition in nestling Tree Swallows. Proc R Soc B 277:1881–1888

Asokan S, Ali AMS, Nagarajan R (2008) Studies on nest construction and nest microclimate of the Baya Weaver, *Ploceus philippinus* (Linn.). J Environ Biol 29:393–396

Bailey IE, Morgan KV, Bertin M, Meddle SL, Healy SD (2014) Physical cognition: birds learn the structural efficacy of nest material. Proc R Soc B 281:20133225

Bailey IE, Morgan KV, Oschadleus HD, DeRuiter SL, Meddle SL, Healy SD (2016) Nest-building males trade off material collection costs with territory value. Emu 116:1–8

Bancroft GT, Thomas G, Bowman R (2020) White-crowned Pigeon (*Patagioenas leucocephala*), version 1.0. In: Poole AF, Gill FB (eds) Birds of the World. Cornell Lab of Ornithology, Ithaca, NY

Bavis RW, Kilgore DL Jr (2001) Effects of embryonic $CO_2$ exposure on the adult ventilatory response in quail: does gender matter? Respir Physiol 126:183–199

Beier P, Tungbani AI (2006) Nesting with the wasp *Ropalidia cincta* increases nest success of Red-cheeked Cordonbleu (*Uraeginthus bengalus*) in Ghana. Auk 123:1022–1037

Bent AC (1939) Life histories of North American woodpeckers. U.S. National Museum Bulletin, Washington, DC, p 174

Bentz AB, Philippi KJ, Rosvall KA (2019) Evaluating seasonal patterns of female aggression: case study in a cavity-nesting bird with intense female–female competition. Ethology 125:555–564

Biancucci L, Martin TE (2010) Can selection on nest size from nest predation explain the latitudinal gradient in clutch size? J Anim Ecol 79:1086–1092

Biddle LE, Deeming DC, Goodman AM (2015) Morphology and biomechanics of the nests of the Common Blackbird *Turdus merula*. Bird Study 62:87–95

Birchard GF, Kilgore DL Jr (1980) Conductance of water vapor in eggs of burrowing and non-burrowing birds: implications for embryonic gas exchange. Physiol Zool 53:284–292

Blanc LA, Walters JR (2008) Cavity-nest webs in a longleaf pine ecosystem. Condor 110:80–92

Boland CRJ (2004) Breeding biology of Rainbow Bee-eaters (*Merops ornatus*): a migratory, colonial, cooperative bird. Auk 121:811–823

Boles WE, Ivison TJ (1999) A new genus of dwarf megapode (Galliformes: Megapodiidae) from the late Oligocene of Central Australia. Smithson Contrib Paleobiol 89:199–206

Bonaparte EB, Cockle KL (2017) Nest niche overlap among the endangered Vinaceous-breasted Parrot (*Amazona vinacea*) and sympatric cavity-using birds, mammals, and social insects in the subtropical Atlantic Forest, Argentina. Condor 119:58–72

Borgmann KL, Conway CJ (2015) The nest-concealment hypothesis: new insights from a comparative analysis. Wilson J Ornithol 127:646–660

Breen A, Guillette LM, Healy SD (2016) What can nest-building birds teach us? Comp Cogn Behav Rev 11: 83–102

Breistøl A, Högstedt G, Lislevand T (2015) Pied Flycatchers *Ficedula hypoleuca* prefer ectoparasite-free nest sites when old nest material is present. Ornis Nor 38:9–13

Brennan LA, Hernandez F, Williford D (2020) Northern Bobwhite (*Colinus virginianus*), version 1.0. In: Poole AF (ed) Birds of the World. Cornell Lab of Ornithology, Ithaca, NY

Broggi J, Senar JC (2009) Brighter Great Tit parents build bigger nests. Ibis 151:588–591

Broughton RK, Hebda G, Maziarz M, Smith KW, Smith L, Hinsley SA (2015) Nest-site competition between bumblebees (Bombidae), social wasps (Vespidae) and cavity-nesting birds in Britain and the western palearctic. Bird Study 62:427–437

Bryan DC (2020) Limpkin (*Aramus guarauna*), version 1.0. In: Poole AF, Gill FB (eds) Birds of the World. Cornell Lab of Ornithology, Ithaca, NY

Buechler K, Fitze PS, Gottstein B, Jacot A, Richner H (2002) Parasite-induced maternal response in a natural bird population. J Anim Ecol 71:247–252

Buehler DA (2020) Bald eagle (*Haliaeetus leucocephalus*), version 1.0. In: Poole AF, Gill FB (eds) Birds of the World. Cornell Lab of Ornithology, Ithaca, NY

Bull EL, Jackson JA (2020) Pileated Woodpecker (*Dryocopus pileatus*), version 1.0. In: Poole AF (ed) Birds of the World. Cornell Lab of Ornithology, Ithaca, NY

Burton NHK (2006) Nest orientation and hatching success in the Tree Pipit *Anthus trivialis*. J Avian Biol 37:312–317

Burton NHK (2007) Intraspecific latitudinal variation in nest orientation among ground-nesting passerines: a study using published data. Condor 109:441–446

Cadieux P, Drapeau P, Ouellet-Lapointe U, Leduc A, Imbeau L, Deschênes R, Nappi A (2023) Old forest structural development drives complexity of nest webs in a naturally disturbed boreal mixedwood forest landscape. Front For Global Change 6:1084696

Calder WA (1972) Piracy of nesting materials from and by the Broad-tailed Hummingbird. Condor 74:485

Calder WA (1973) Microhabitat selection during nesting of hummingbirds in the Rocky Mountains. Ecology 54: 127–134

Calder WA (1984) Size, function, and life history. Harvard University Press, Cambridge, MA

Cameron M (2006) Nesting habitat of the Glossy Black-Cockatoo in Central New South Wales. Biol Conserv 127:402–410

Cantarero A, López-Arrabé J, Plaza M, Saavedra-Garcés I, Moreno J (2016) Males feed their mates more and take more risks for nestlings with larger female-built nests: an experimental study in the nuthatch *Sitta europaea*. Behav Ecol Sociobiol 70:1141–1150

Caro TM (2005) Antipredator defenses in birds and mammals. University of Chicago Press, Chicago, IL

Castilla AM, Dhondt AA, Diaz-Uriarte R, Westmoreland D (2007) Predation in ground-nesting birds: an experimental study using natural egg-color variation. Avian Conserv Ecol 2:2

Cestari C, Guaraldo AC, Gussoni CO (2011) Nestling behavior and parental care of the Common Potoo (*Nyctibius griseus*) in southeastern Brazil. Wilson J Ornithol 123:102–106

Chalfoun AD, Martin TE (2007) Latitudinal variation in avian incubation attentiveness and a test of the food limitation hypothesis. Anim Behav 73:579–585

Chalfoun AD, Ratnaswamy MJ, Thompson FR III (2002) Songbird nest predators in forest-pasture edge and forest interior in a fragmented landscape. Ecol Appl 12:858–867

Charter M, Izhaki I, Mocha YB, Kark S (2016) Nest-site competition between invasive and native cavity nesting birds and its implication for conservation. J Environ Manag 181:129–134

Chiappe LM, Di L, Serrano FJ, Yuguang Z, Meng Q (2020) Anatomy and flight performance of the early enantiornithine bird *Protopteryx fengningensis*: information from new specimens of the Early Cretaceous Huajiying Formation of China. Anat Rec 303:716–731

Christman BJ, Dhondt AA (1997) Nest predation in Black-capped Chickadees: how safe are cavity nests? Auk 114:769–773

Clark L (1991) The nest protection hypothesis: the adaptive use of plant secondary compounds by European Starlings. In: Loye JE, Zuk M (eds) Bird-parasite interactions: ecology, evolution, and behaviour. Oxford University Press, Oxford, pp 205–221

Clark L, Mason JR (1988) Effect of biologically active plants used as nest material and the derived benefit to starling nestlings. Oecologia 77:174–180

Cockle KL, Martin K, Drever MC (2010) Supply of tree-holes limits nest density of cavity-nesting birds in primary and logged subtropical Atlantic forest. Biol Conserv 143:2851–2857

Cockle KL, Martin K, Robledo G (2012) Linking fungi, trees, and hole-using birds in a Neotropical tree-cavity network: pathways of cavity production and implications for conservation. For Ecol Manag 264: 210–219

Cockle KL, Bodrati A, Lammertink M, Martin K (2015) Cavity characteristics, but not habitat, influence nest survival of cavity-nesting birds along a gradient of human impact in the subtropical Atlantic Forest. Biol Conserv 184:193–200

Cockle KL, Ibarra JT, Altamirano TA, Martin K (2019) Interspecific networks of cavity-nesting vertebrates reveal a critical role of broadleaf trees in endangered Araucaria mixed forests of South America. Biodivers Conserv 28:3371–3386

Collias NE (1997) On the origin and evolution of nest building by passerine birds. Condor 99:253–270

Collias EC, Collias NE (1964a) The development of nest-building behavior in a weaverbird. Auk 81:42–52

Collias NE, Collias EC (1964b) Evolution of nest-building in the weaverbirds (Ploceidae). University of California Press, Berkeley, CA

Collias NE, Collias EC (1984) Nest building and bird behavior. Princeton University Press, Princeton, NJ

Colombelli-Negrel D, Kleindorfer S (2009) Nest height, nest concealment, and predator type predict nest predation in Superb Fairy-wrens (*Malurus cyaneus*). Ecol Res 24:921–928

Colwell MA, Jehl JR Jr (2020) Wilson's Phalarope (*Phalaropus tricolor*), version 1.0. In: Poole AF, Gill FB (eds) Birds of the World. Cornell Lab of Ornithology, Ithaca, NY

Conner RN, Rudolph DC (1995) Excavation dynamics and use patterns of Red-cockaded Woodpecker cavities: relationships with cooperative breeding. In: Kulhavy DL, Hooper RG, Costa R (eds) Red-cockaded Woodpecker: recovery, ecology and management. Center for Applied Studies in Forestry, College of Forestry, Stephen F. Austin State University, Nacogdoches, TX, pp 343–352

Conner RN, Rudolph DC, Saenz D, Schaefer RR (1994) Heartwood, sapwood, and fungal decay associated with Red-cockaded Woodpecker cavity trees. J Wildl Manag 58:728–734

Conner RN, Saenz D, Rudolph DC, Schaefer RR (2002) Does the availability of artificial cavities affect cavity excavation rates in Red-cockaded Woodpeckers? J Field Ornithol 73:125–129

Conover MR (2007) Predator–prey dynamics: the use of olfaction. Taylor and Francis, Boca Raton, FL

Conrad KF, Robertson RJ (1993) Clutch size in Eastern Phoebes (*Sayornis phoebe*). I. The cost of nest building. Can J Zool 71:1003–1007

Conway CJ, Martin TE (2000) Evolution of passerine incubation behavior: influence of food, temperature, and nest predation. Evolution 54:670–685

Coppedge BR (2009) Patterns of bison hair use in nests of tallgrass prairie birds. Prairie Naturalist 41:110–115

Coulter MC, Rodgers JA Jr, Ogden JC, Depkin FC (2020) Wood Stork (*Mycteria americana*), version 1.0. In: Poole AF, Gill FB (eds) Birds of the World. Cornell Lab of Ornithology, Ithaca, NY

Cox WA, Thompson FR III, Faaborg J (2012) Species and temporal factors affect predator-specific rates of nest predation for forest songbirds in the Midwest. Auk 129:147–155

Crossman CA, Rohwer VG, Martin PR (2011) Variation in the structure of bird nests between northern Manitoba and southeastern Ontario. PLoS ONE 6:e19086

Dalakchieva S (2002) Nest-sites and nest construction of Little Ringed Plovers *Charadrius dubius* in Bulgaria. Wader Study Group Bull 98:12–15

Danel S, Zidat T, Lucas A, Biro D, Bonadonna F (2021) First description of nest-decoration behaviour in a wild sub-Antarctic shorebird. Behav Process 188:104408

Dawson WR (2020) Pine Siskin (*Spinus pinus*), version 1.0. In: Poole AF (ed) Birds of the World. Cornell Lab of Ornithology, Ithaca, NY

Dawson RD, Lawrie CC, O'Brien EL (2005) The importance of microclimate variation in determining size, growth and survival of avian offspring: experimental evidence from a cavity nesting passerine. Behav Ecol 144:499–507

Dawson RD, O'Brien EL, Mlynowski TJ (2011) The price of insulation: costs and benefits of feather delivery to nests for male Tree Swallows *Tachycineta bicolor*. J Avian Biol 42:93–102

De Jong MJ (2020) Northern Rough-winged Swallow (*Stelgidopteryx serripennis*), version 1.0. In: Poole AF, Gill FB (eds) Birds of the World. Cornell Lab of Ornithology, Ithaca, NY

Deeming DC, Campion E (2018) Simulated rainfall reduces the insulative properties of bird nests. Acta Ornithol 53:91–97

Deeming CD, Mayr G (2018) Pelvis morphology suggests that early Mesozoic birds were too heavy to contact incubate their eggs. J Evol Biol 31:701–709

DeGregorio BA, Chiavacci SJ, Benson TJ, Sperry JH, Weatherhead PJ (2016) Nest predators of North American birds: continental patterns and implications. Bioscience 66:655–665

Dekker RWRJ (2007) Distribution and speciation of megapodes (Megapodiidae) and subsequent development of their breeding behaviour. In: Renema W (ed) Biogeography, time, and place: distributions, barriers, and islands. Springer, Dordrecht, pp 93–102

Di Sallo FG, Cockle KL (2021) The role of body size in nest-site selection by secondary cavity-nesting birds in a subtropical Chaco forest. Ibis 164:168–187

Dial KP (2003) Evolution of avian locomotion: correlates of flight style, locomotor modules, nesting biology, body size, development, and the origin of flapping flight. Auk 120:941–952

DiBartola SP (2012) Metabolic acid-base disorders. In: DeBartola SP (ed) Fluid, electrolyte, and acid-base disorders in small animal practice, 4th edn. St. Louis, MO, W. B. Saunders, pp 253–286

Djojosugito AM, Folkow B, Kovách AGB (1968) The mechanisms behind the rapid blood volume restoration after hemorrhage in birds. Acta Physiol Scand 74:114–122

Douthwaite RJ (1978) Breeding biology of the Pied Kingfisher *Ceryle rudis* on Lake Victoria. J East Afr Nat Hist Soc Natl Mus 31:1–12

Dow DD, King BR (1984) Communal building of brood and roost nests by the Grey-crowned Babbler *Pomatostomus temporalis*. Emu 84:193–199

Drake A, Martin K (2020) Rainfall and nest site competition delay Mountain Bluebird and Tree Swallow breeding but do not impact productivity. Auk 137:ukaa006

Duffy DC (1983) The ecology of tick parasitism on densely nesting Peruvian seabirds. Ecology 64:110–119

Dunn EH, Hussell DJT, Kren J, Zoerb AC (2020) Northern Wheatear (*Oenanthe oenanthe*), version 2.0. In: Rodewald PG, Keeney BK, Billerman SM (eds) Birds of the World. Cornell Lab of Ornithology, Ithaca, NY

Dyke GJ, Nudds RL (2008) The fossil record and limb disparity of enantiornithines, the dominant flying birds of the Cretaceous. Lethaia 42:248–254

Eberts ER, Tattersall GJ, Auger PJ, Curley M, Morado MI, Strauss EG, Powers DR, Soveral NC, Tobalske BW, Shankar A (2023) Free-living Allen's Hummingbirds (*Selasphorus sasin*) rarely use torpor while nesting. J Thermal Biol 112:103391

Edworth AB, Martin K (2014) Long-term dynamics of the characteristics of tree cavities used for nesting by vertebrates. For Ecol Manag 334:122–128

Edworthy AB, Wiebe KL, Martin K (2012) Survival analysis of a critical resource for cavity-nesting communities: patterns of tree cavity longevity. Ecol Appl 22:1733–1742

Edworthy AB, Trzcinski MK, Cockle KL, Wiebe KL, Martin K (2018) Tree cavity occupancy by nesting vertebrates across cavity age. J Wildl Manag 82:639–648

Eeva T, Lehikoinen E, Nurmi J (1994) Effects of ectoparasites on breeding success of Great Tits (*Parus major*) and Pied Flycatchers (*Ficedula hypoleuca*) in an air pollution gradient. Can J Zool 72:624–635

Eiby Y, Booth D (2008) Embryonic thermal tolerance and temperature variation in mounds of the Australian Brush-Turkey (*Alectura lathami*). Auk 125:594–599

Eichholz MW, Dassow JA, Stafford JD, Weatherhead PJ (2012) Experimental evidence that nesting ducks use mammalian urine to assess predator abundance. Auk 129:638–644

Evans MR, Burn JL (1996) An experimental analysis of mate choice in the wren: a monomorphic, polygynous passerine. Behav Ecol 7:101–108

Evers DC, Paruk JD, McIntyre JW, Barr JF (2020) Common Loon (*Gavia immer*), version 1.0. In: Billerman SM (ed) Birds of the World. Cornell Lab of Ornithology, Ithaca, NY

Ewins PJ, Miller MJR, Barker ME, Postpalsky S (1994) Birds breeding in or beneath osprey nests in the Great Lakes basin. Wilson Bull 106:743–749

Fang Y-T, Tuanmu M-N, Hung C-M (2018) Asynchronous evolution of interdependent nest characters across the avian phylogeny. Nat Commun 9:1863

Farquhar CC, Ritchie KL (2020) Black-tailed Gnatcatcher (*Polioptila melanura*), version 1.0. In: Poole AF, Gill FB (eds) Birds of the World. Cornell Lab of Ornithology, Ithaca, NY

Feare CJ (1976) Desertion and abnormal development in a colony of Sooty Terns *Sterna fuscata* infested by virus-infected ticks. Ibis 118:112–115

Fernández MS, García RA, Fiorelli L, Scolaro A, Salvador RB, Cotaro CN, Kaiser GW, Dyke GJ (2013) A large accumulation of avian eggs from the late cretaceous of Patagonia (Argentina) reveals a novel nesting strategy in Mesozoic birds. PLoS ONE 8:e61030

Field DJ, Bercovici A, Berv JS, Dunn R, Fastovsky DE, Lyson TR, Vajda V, Gauthier JA (2018) Early evolution of modern birds structured by global forest collapse at the end-cretaceous mass extinction. Curr Biol 28:1825–1831

Filliater TS, Breitwisch R, Nealen PM (1994) Predation on Northern Cardinal nests: does choice of nest site matter? Condor 96:761–768

Fontaine JJ, Martel M, Markland HM, Niklison AM, Decker KL, Martin TE (2007) Testing ecological and behavioral correlates of nest predation. Oikos 116: 1887–1894

Forstmeier W, Weiss I (2004) Adaptive plasticity in nest-site selection in response to changing predation risk. Oikos 104:487–499

Frederick PC, Siegel-Causey D (2020) Anhinga (*Anhinga anhinga*), version 1.0. In: Poole AF, Gill FB (eds) Birds of the World. Cornell Lab of Ornithology, Ithaca, NY

Frei B, Smith KG, Withgott JH, Rodewald PG, Pyle P, Patten MA (2020) Red-headed Woodpecker (*Melanerpes erythrocephalus*), version 1.0. In:

Rodewald PG (ed) Birds of the World. Cornell Lab of Ornithology, Ithaca, NY

Freidl TWP (2004) Breeding behaviour of the Red Bishop (*Euplectes orix*): a synthesis and new observations. Vogelwarte 42:177–190

Freidl TWP, Klump GM (2000) Nest and mate choice in the Red Bishop (*Euplectes orix*): female settlement rules. Behav Ecol 11:378–386

Fulton GR (2019) Meta-analyses of nest predation in temperate Australian forests and woodlands. Austral Ecol 44:389–396

García-Navas V, Valera F, Griggio M (2015) Nest decorations: an 'extended' female badge of status? Anim Behav 99:95–107

Garrison BA, Turner A (2020) Bank Swallow (*Riparia riparia*), version 1.0. In: Billerman SM (ed) Birds of the World. Cornell Lab of Ornithology, Ithaca, NY

Gauthier M, Thomas DW (1993) Nest site selection and cost of nest building by Cliff Swallows (*Hirundo pyrrhonota*). Can J Zool 71:1120–1123

Gibbons P, Lindenmayer D (2002) Tree hollows and wildlife conservation in Australia. CSIRO Publishing, Collingwood

Goldshtein A, Markman S, Leshem Y, Puchinsky M, Charter M (2018) Nest-site interference competition with House Sparrows affects breeding success and parental care in Great Tits. J Ornithol 159:667–673

Goodenough AE, Hart AG, Elliot SL (2008) Variation in offspring quality with cavity orientation in the Great Tit. Ethol Ecol Evol 20:375–389

Göth ANN (2008) Incubation temperatures and sex ratios in Australian Brush-Turkey (*Alectura lathami*) mounds. Austral Ecol 32:378–385

Göth A, Booth DT (2005) Temperature-dependent sex-ratio in a bird. Biol Lett 1:31–33

Göth A, Evans CS (2004) Egg size in Australian Brushturkey *Alectura lathami* hatchlings predicts motor performance and postnatal weight gain. Can J Zool 82:972–979

Götmark F, Blomqvist D, Johansson OC, Bergkvist J (1995) Nest site selection: a trade-off between concealment and view of the surroundings? J Avian Biol 26: 305–312

Greeney HF, Meneses MR, Hamilton CE, Lichter-Marck-E, Mannan RW, Snyder N, Snyder H, Wethington SM, Dyer LA (2015) Trait-mediated trophic cascade creates enemy-free space for nesting hummingbirds. Sci Adv 1:e1500310

Greenlaw JS (2020) Eastern Towhee (*Pipilo erythrophthalmus*), version 1.0. In: Rodewald PG (ed) Birds of the World. Cornell Lab of Ornithology, Ithaca, NY

Griffith SC, Mainwaring MC, Sorato E, Beckmann C (2016) High atmospheric temperatures and 'ambient incubation' drive embryonic development and lead to earlier hatching in a passerine bird. R Soc Open Sci 3: 150371

Guillette LM, Healy SD (2015) Nest building, the forgotten behaviour. Curr Opin Behav Sci 6:90–96

Gwinner H, Berger S (2006) Parasite defence in birds: the role of volatiles. Acta Zool Sin 52(Suppl):280–283

Gwinner H, Oltrogge M, Trost L, Neienaber U (2000) Green plants in starling nests: effects on nestlings. Anim Behav 59:301–309

Gwinner H, Yohannes E, Schwabl H (2013) Nest composition and yolk hormones: do female European Starlings adjust yolk androgens to nest quality? Avian Biol Res 6:307–312

Gwinner H, Capilla-Lasheras P, Cooper C, Helm B (2018) Green incubation': avian offspring benefit from aromatic nest herbs through improved parental incubation behaviour. Proc R Soc B 285:20180376

Haitao W, Wei G, Dongmei W, Duo L, Wenhong D (2003) Nest-site characteristics and reproductive success of five species of birds breeding in natural cavities. Acta Ecol Sin 23:1377–1385

Halkin SL, Shustack DP, DeVies MS, Jawor JM, Linville SU (2021) Northern Cardinal (*Cardinalis cardinalis*), version 1.0. In: Rodewald PG, Keeney BK (eds) Birds of the world. Cornell Lab of Ornithology, Ithaca, NY

Hall ZJ, Street SE, Healy SD (2013) The evolution of cerebellum structure correlates with nest complexity. Biol Lett 9:20130687

Hall ZJ, Meddle SL, Healy SD (2015) From neurons to nests: nest-building behaviour as a model in behavioural and comparative neuroscience. J Ornithol 156:133–143

Halpern M, Frumin N (1979) Roles of the vomeronasal and olfactory systems in prey attack and feeding in adult garter snakes. Physiol Behav 22:1183–1189

Hamilton RAGA, Proudfoot GA, Sherry DA, Johnson SL (2020) Cactus Wren (*Campylorhynchus brunneicapillus*), version 1.0. In: Poole AF (ed) Birds of the world. Cornell Lab of Ornithology, Ithaca, NY

Hanley D, Doucet SM, Dearborn DC (2010) A blackmail hypothesis for the evolution of conspicuous egg coloration in birds. Auk 127:453–459

Hansell M (2000) Bird nests and construction behaviour. Cambridge University Press, Cambridge

Hardy PC, Morrison ML (2001) Nest site selection by Elf Owls in the Sonoran Desert. Wilson Bull 113:23–32

Harris RB, Birks SM, Leaché AD (2014) Incubator birds: biogeographical origins and evolution of underground nesting in megapodes (Galliformes: Megapodiidae). J Biogeogr 41:2045–2056

Harrison HH (1975) A field guide to birds' nests. Houghton Mifflin, Boston, MA

Hart LA, Downs CT, Brown M (2016) Sitting in the sun: nest microhabitat affects incubation termperatures in seabirds. J Therm Biol 60:149–154

Healy SD (2022) Adding the neuro to cognition: from food storing to nest building. Anim Cogn 26:249–260

Heenan CB (2013) The structural and thermal properties of avian cup-shaped nests. Ph.D. dissertation, University of Adelaide, Adelaide, Australia

Heenan CB, Seymour RS (2011) Structural support, not insulation, is the primary driver for avian cup-shaped nest design. Proc R Soc B 278:2915–2923

Heenan CB, Seymour RS (2012) The effect of wind on the rate of heat loss from avian cup-shaped nests. PLoS ONE 7:e32252

Hoi H, Václav R, Slobodová D (2003) Postmating sexual selection in House Sparrows: can females estimate "good fathers" according to their early paternal effort? Folia Zool 52:299–308

Holcomb LC, Twiest G (1968) Ecological factors affecting nest building in Red-winged Blackbirds. Bird-Banding 39:14–22

Holveck MJ, Guerreiro R, Perret P, Doutrelant C, Grégoire A (2019) Eggshell coloration indicates female condition during egg-laying: a field experiment in Blue Tits. Biol J Linn Soc 128:181–200

Holway DA (1991) Nest-site selection and the importance of nest concealment in the Black-throated Blue Warbler. Condor 93:575–581

Hopp SL, Kirby A, Boone CA (2020) White-eyed Vireo (*Vireo griseus*), version 1.0. In: Billerman S, Keeney BK, Rodewald PG, Schulenberg TS (eds) Birds of the World. Cornell Lab of Ornithology, Ithaca, NY

Horner JR (2000) Dinosaur reproduction and parenting. Ann Rev Earth Planetary Sci 28:19–45

Houston DC (1979) Why do Fairy Terns *Gygis alba* not build nests? Ibis 121:102–104

Howe S, Kilgore DL Jr, Colby C (1987) Respiratory gas concentrations and temperatures within nest cavities of the Northern Flicker (*Colaptes auratus*). Can J Zool 65:1541–1547

Howell TR (1978) Ecology and reproductive behavior of the White, or Fairy, Tern. In: Ecology and reproductive behaviour of the Gray Gull and of the Red-tailed Tropicbird and the White Tern on Midway Island. National Geographic Society Research Report 10, pp 274–284.

Howlett JS, Stutchbury BJ (1996) Nest concealment and predation in Hooded Warblers: experimental removal of nest cover. Auk 113:1–9

Hun LT, Wani WA, Tjih ETT, Adnan NA, Ling YL, Aziz RA (2015) Investigations into the physicochemical, biochemical and antibacterial properties of edible bird's nest. J Chem Pharm Res 7:228–247

Hunter JS (1900) The bird fauna of Salt Basin near Lincoln. Proceedings of the Nebraska Ornithologists' Union 1:28

Hurtrez-Bousses S, Blondel J, Perret P, Renaud F (1997) Relationship between intensity of blowfly infestation and reproductive success in a Corsican population of Blue Tits. J Avian Biol 28:267–270

Husak MS, Maxwell TC (2020) Golden-fronted Woodpecker (*Melanerpes aurifrons*), version 1.0. In: Poole AF, Gill FB (eds) Birds of the World. Cornell Lab of Ornithology, Ithaca, NY

Ibarra JT, Cockle K, Altamirano T, van der Hoek Y, Simard S, Bonacic C, Martin K (2020) Nurturing resilient forest biodiversity: nest webs as complex adaptive systems. Ecol Soc 25:27

Ingold DJ (1998) The influence of starlings on flicker reproduction when both naturally excavated cavities and artificial nest boxes are available. Wilson Bull 110:218–225

Iwaniuk AN, Lefebvre L, Wylie DR (2009) The comparative approach and brain-behaviour relationships: a tool for understanding tool use. Can J Exp Psychol 63:150–159

Järvinen P, Brommer JE (2020) Lining the nest with more feathers increases offspring recruitment probability: selection on an extended phenotype in the Blue Tit. Ecol Evol 10:13327–13333

Jelínek V, Požgayová M, Honza M, Procházka P (2016) Nest as an extended phenotype signal of female quality in the Great Reed Warbler. J Avian Biol 47:428–437

Jessel HR, Chen S, Osovski S, Efroni S, Rittel D, Bachelet I (2019) Design principles of biologically fabricated avian nests. Sci Rep 9:4792

Jobin B, Picman J (1997) The effect of egg coloration on predation of artificial ground nests. Can Field-Nat 111: 591–594

Jones DN, Dekker RWRJ, Roselaar CS (1995) The megapodes. Oxford University Press, Oxford

Jones KC, Roth KL, Islam K, Hamel PB, Smith CG III (2007) Incidence of nest material kleptoparasitism involving Cerulean Warblers. Wilson J Ornithol 119: 271–275

Joy JB (2000) Characteristics of nest cavities and nest trees of the Red-breasted Sapsucker in coastal montane forests. J Field Ornithol 71:525–530

Jung Y, Jung S, Kim W, Kim HY (2021) Avian mud nest architecture by self-secreted saliva. Proc Natl Acad Sci USA 118:e2018509118

Kahl MP (1967) Observations on the behaviour of the Hamerkop *Scopus umbretta* in Uganda. Ibis 109:25–32

Kang N, Hails CJ, Sigurdsson JB (1991) Nest construction and egg-laying in Edible-nest Swiftlets *Aerodramus* spp. and the implications for harvesting. Ibis 133: 170–177

Kattan GH (1997) Shiny Cowbirds follow the 'shotgun' strategy of brood parasitism. Anim Behav 53:647–654

Kaye TG, Pittman M, Marugán-Lobón J, Martín-Abad H, Sanz JL, Buscalioni AD (2019) Fully fledged enantiornithine hatchling revealed by laser-stimulated fluorescence supports precocial nesting behavior. Sci Rep 9:5006

Kelly JF, Bridge ES, Hamas MJ (2020) Belted Kingfisher (*Megaceryle alcyon*), version 1.0. In: Poole AF (ed) Birds of the World. Cornell Lab of Ornithology, Ithaca, NY

Kern MD, Van Riper C III (1984) Altitudinal variations in nests of the Hawaiian Honeycreeper *Hemignathus virens virens*. Condor 86:443–454

Kerpez TA, Smith NS (1990) Competition between European Starlings and native woodpeckers for nest cavities in saguaros. Auk 107:367–375

Kershner EL, Ellison WG (2020) Blue-gray Gnatcatcher (*Polioptila caerulea*), version 1.0. In: Poole AF (ed) Birds of the World. Cornell Lab of Ornithology, Ithaca, NY

Kilgore DL Jr, Faraci FM, Fedde MR (1985) Ventilatory and intrapulmonary chemoreceptor sensitivity to $CO_2$ in the Burrowing Owl. Respir Physiol 62:325–339

Kilham L (1989) The American Crow and the Common Raven. Texas A&M University Press, College Station, TX

Klump G, Metz M, Friedl T (2009) Male nest-building behaviour and mating success in the Red Bishop (*Euplectes orix*). Behaviour 146:771–794

Kosiński Z, Walczak Ł (2019) Does cavity reuse affect timing of reproduction and fledgling success in the Black Woodpecker? J Ornithol 160:79–89

Kozma JM, Lorenz TJ, Raphael MG, Garrett KL, Dixon RD (2020) White-headed Woodpecker (*Picoides albolarvatus*), version 2.0. In: Rodewald PG, Keeney BK (eds) Birds of the World. Cornell Lab of Ornithology, Ithaca, NY

Lambrechts MM, Blondel J, de Franceschi C, Doutrelant C (2017) Nest size is positively correlated with fledging success in Corsican Blue Tits (*Cyanistes caeruleus*) in an insular oak-dominated habitat mosaic. J Ornithol 158:125–132

Landler L, Jusino MA, Skelton J, Walters JR (2014) Global trends in woodpecker cavity entrance orientation: latitudinal and continental effects suggest regional climate influence. Acta Ornithol 49:257–266

Landler L, Skelton J, Jusino MA, Van Lanen A, Walters JR (2022) Effects of cavity orientation on nesting success inferred from long-term monitoring of the endangered Red-cockaded Woodpecker. Sci Rep 12: 11624

LaPorte N, Storer RW, Nuechterlein GL (2020) Western Grebe (*Aechmophorus occidentalis*), version 1.0. In: Rodewald PG (ed) Birds of the World. Cornell Lab of Ornithology, Ithaca, NY

Lauro B, Nol E (1995) Patterns of habitat use for Pied and Sooty oystercatchers nesting at the Furneaux Islands, Australia. Condor 97:920–934

Lee JK, Lima SL (2016) Nest building under the risk of predation: safe nests are not always the best option. J Avian Biol 47:768–778

Lee W-S, Kwon Y-S, Yoo J-C (2010) Egg survival is related to the colour matching of eggs to nest background in Black-tailed Gulls. J Ornithol 151:765–770

Leighton GM (2014) Sex and individual differences in cooperative nest construction of Sociable Weavers. J Ornithol 155:927–935

Leighton GM, Echeverri S (2014) Non-linear influence of nest size on thermal buffering of sociable weaver nests and the maintenance of cooperative nest construction. Avian Biol Res 7:255–260

Leighton GM, Echeverri S, Heinrich D, Kolberg H (2015) Relatedness predicts multiple measures of investment in cooperative nest construction in Sociable Weavers. Behav Ecol Sociobiol 69:1835–1843

Leonard ML, Picman J (1987) The adaptive significance of multiple nest building by male Marsh Wrens. Anim Behav 35:271–277

Li P, Martin TE (1991) Nest-site selection and nesting success of cavity-nesting birds in high elevation forest drainages. Auk 108:405–418

Lifjeld JT, Gohli J, Albrecht T, Garcia-del-Rey E, Johannessen LE, Kleven O, Marki PZ, Omotoriogun TC, Rowe M, Johnsen A (2019) Evolution of female promiscuity in Passerides songbirds. BMC Evol Biol 19:169

Liljeström M, Schiavini A, Reboreda JC (2009) Chilean Swallows (*Tachycineta meyeni*) adjust the number of feathers added to the nest with time of breeding. J Ornithol 121:783–788

Lill A, Fell PJ (2007) Microclimate of nesting burrows of the Rainbow Bee-eater. Emu 107:108–114

Lim CK, Earl of Cranbrook (2002) Swiftlets of Borneo. Natural History Publications, Kota Kinabalu, Borneo

Lima SL (2009) Predators and the breeding bird: behavioral and reproductive flexibility under the risk of predation. Biol Rev 84:485–513

Lima CC, Garcia CM (2016) Pre-and post-experimental manipulation assessments confirm the increase in number of birds due to the addition of nest boxes. PeerJ 4: e1806

Lindell C (1996) Patterns of nest usurpation: when should species converge on nest niches? Condor 98:464–473

Littlefield CD, Ryder RA (1968) Breeding biology of the Greater Sandhill Crane on Malheur Nationals Wildlife Refuge, Oregon. Transactions of the North American Wildlife and Natural Resources Conference 33:444–454

Liu J, Liang W (2021) Snake slough in birds' nests acts as a nest predator deterrent. Ethol Ecol Evol 6:591–602

Liu J, Yan H, Li G, Li S (2021) Nest concealment is associated with reproductive traits across sympatric bird species. Ecol Evol 11:14079–14087

Lloyd JD, Martin TE (2004) Nest-site preference and maternal effects on offspring growth. Behav Ecol 15: 816–823

Lowney AM, Bolopo D, Krochuk BA, Thomson RL (2020) The large communal nests of Sociable Weavers provide year-round insulated refuge for weavers and Pygmy Falcons. Front Ecol Evol 8:357

Lowther PE, Celada C, Klein NK, Rimmer CC, Spector DA (2020a) Yellow Warbler (*Dendroica petechia*), version 1.0. In: Poole AF, Gill FB (eds) Birds of the world. Cornell Lab of Ornithology, Ithaca, NY

Lowther PE, Collins CT, Gunn C, Beason JP, Potter K, Webb M (2020b) Black Swift (*Cypseloides niger*), version 1.0. In: Billerman SM (ed) Birds of the world. Cornell Lab of Ornithology, Ithaca, NY

Maclean GL (1973) The Sociable Weaver, part 2: nest architecture and social organization. Ostrich 44:191–218

Mainwaring MC (2017) Causes and consequences of intraspecific variation in nesting behaviors: insights from Blue Tits and Great Tits. Front Ecol Evol 5:39

Mainwaring MC, Hartley IR (2008) Seasonal adjustments in nest cup lining in Blue Tits *Cyanistes caeruleus*. Ardea 96:278–282

Mainwaring MC, Hartley IR (2013) The energetic costs of nest building in birds. Avian Biol Res 6:12–17

Mainwaring MC, Nagy J, Hauber ME (2021) Sex-specific contributions to nest building in birds. Behav Ecol 32: arab035

Manikandan P, Balasubramanian P (2018) Sequential use of tree cavities by birds and nest web in a riparian forest in Southwest India. Acta Ornithol 53:49–60

Margalida A, Bertran (2000) Breeding behaviour of the Bearded Vulture Gypaetus barbatus: minimal sexual differences in parental activities. Ibis 142:225–234

Martin TE (1993) Nest predation and nest sites. Bioscience 43:523–532

Martin TE (1995) Avian life histories in relation to nest sites, nest predation and food. Ecol Monogr 65:101–127

Martin K, Eadie JM (1999) Nest webs: a community-wide approach to the management and conservation of cavity-nesting forest birds. For Ecol Manag 115:243–257

Martin TE, Ghalambor CK (1999) Males feeding females during incubation. I. Required by microclimate or constrained by nest predation? Am Nat 153:131–139

Martin TE, Roper JJ (1988) Nest predation and nestsite selection of a western population of the Hermit Thrush. Condor 90:51–57

Martin K, Aitken KEH, Wiebe KL (2004) Nest sites and nest webs for cavity-nesting communities in interior British Columbia, Canada: nest characteristics and niche partitioning. Condor 106:5–19

Martin TE, Boyce AJ, Fierro-Calderón K, Mitchell AE, Armstad CE, Mouton J, Bin Soudi EE (2017) Enclosed nests may provide greater thermal than nest predation benefits compared with open nests across latitudes. Funct Ecol 31:1231–1240

Matysioková B, Remeš V (2018) Evolution of parental activity at the nest is shaped by the risk of nest predation and ambient temperature across bird species. Evolution 72:2214–2224

Mazgajski TD (2002) Nesting phenology and breeding success in Great Spotted Woodpecker *Picoides major* near Warsaw (central Poland). Acta Ornithol 37:1–5

Maziarz M, Wesołowski T (2014) Does darkness limit the use of tree cavities for nesting by birds? J Ornithol 155: 793–799

Meanley B (1955) A nesting study of the Little Blue Heron in eastern Arkansas. Wilson Bull 67:84–99

Medina I (2019) The role of the environment in the evolution of nest shape in Australian passerines. Sci Rep 9: 5560

Medina I, Perez DM, Silva ACA, Cally J, León C, Maliet O, Quintero I (2022) Nest architecture is linked with ecological success in songbirds. Ecol Lett 25: 1365–1375

Medlin EC, Risch TS (2006) An experimental test of snake skin use to deter nest predation. Condor 108:963–965

Mennerat A, Perret P, Caro SP, Heeb P, Lambrechts MM (2008) Aromatic plants in Blue Tit *Cyanistes caeruleus* nests: no negative effect on blood-sucking *Protocalliphora* blow fly larvae. J Avian Biol 39: 127–132

Mennerat A, Mirleau P, Blondel J, Perret P, Lambrechts MM, Heeb P (2009a) Aromatic plants in nests of the Blue Tit *Cyanistes caeruleus* protect chicks from bacteria. Oecologia 161:849–855

Mennerat A, Perret P, Lambrechts MM (2009b) Local individual preferences for nest materials in a passerine bird. PLoS ONE 4:e5104

Mennill DJ, Ratcliffe LM (2004) Nest cavity orientation in Black-capped Chickadees *Poecile atricapillus*: do the acoustic properties of cavities influence sound reception in the nest and extra-pair mating? J Avian Biol 35: 477–482

Merilä J, Wiggins DA (1995) Interspecific competition for nest holes causes adult mortality in the Collared Flycatcher. Condor 97:445–450

Merino S, Potti J (1995) Mites and blowflies decrease growth and survival in nestling Pied Flycatchers. Oikos 73:95–103

Mersten-Katz C (1997) Gas composition in the nest cavity of the Syrian Woodpecker, *Dendrocopus syriacus*. Isr J Zool 43:111

Metz M, Klump GM, Friedl TWP (2009) Male nest-building behaviour and mating success in the Red Bishop (*Euplectus orix*). Behaviour 146:771–794

Middleton AL (2020) Chipping Sparrow (*Spizella passerina*), version 1.0. In: Poole AF, Gill FB (eds) Birds of the World. Cornell Lab of Ornithology, Ithaca, NY

Miller KE, Leonard DE Jr, Shackelford CW, Brown RE, Conner RN (2020) Red-bellied Woodpecker (*Melanerpes carolinus*), version 1.0. In: Rodewald PG (ed) Birds of the World. Cornell Lab of Ornithology, Ithaca, NY

Mitchell CD, Eichholz MW (2020) Trumpeter Swan (*Cygnus buccinator*), version 1.0. In: Rodewald PG (ed) Birds of the World. Cornell Lab of Ornithology, Ithaca, NY

Møller AP (1990) Nest predation selects for small nest size in the blackbird. Oikos 57:237–240

Møller AP (2006) Rapid change in nest size of a bird related to change in a secondary sexual character. Behav Ecol 17:108–116

Møller AP, Arriero E, Lobato E, Merino S (2009) A meta-analysis of parasite virulence in nestling birds. Biol Rev 84:567–588

Mondain-Monval TO, Sharp SP (2018) Burrow depth, carbon dioxide and reproductive success in Sand Martins *Riparia riparia*. Bird Study 65:123–131

Moratalla JJ, Powell JE (1990) Dinosaur nesting patterns. In: Carpenter K, Hirsch KF, Horner JR (eds) Dinosaur eggs and babies. Cambridge University Press, Cambridge, pp 37–46

Moreno J (2012) Avian nests and nest-building as signals. Avian Biol Res 5:238–251

Moreno J, Bustamante J, Viñuela J (1995) Nest maintenance and stone theft in the Chinstrap Penguin (*Pygoscelis antarctica*): I. Sex roles and effects on fitness. Polar Biol 15:533–545

Moreno J, Martinez J, Corral C, Lobato E, Merino S, Morales J, Martinez-de la Puente J, Tomas G (2008) Nest construction rate and stress in female Pied Flycatchers *Fidedula hypoleuca*. Acta Ornithol 43: 57–64

Moreno J, Lobato E, González-Braojos S, Ruiz-de Castañeda R (2010) Nest construction costs affect nestling growth: a field experiment in a cavity-nesting passerine. Acta Ornithol 45:139–145

Mouton JC, Martin TE (2019) Nest structure affects auditory and visual detectability, but not predation risk, in a tropical songbird community. Funct Ecol 33:1973–1981

Mowbray TB (2020) Scarlet Tanager (*Piranga olivacea*), version 1.0. In: Poole AF, Gill FB (eds) Birds of the World. Cornell Lab of Ornithology, Ithaca, NY

Muir JJ, Colwell MA (2010) Snowy Plovers select open habitats for courtship scrapes and nests. Condor 112: 507–510

Mullin SJ, Cooper RJ (2002) Barking up the wrong tree: climbing performance of rat snakes and its implications for depredation of avian nests. Can J Zool 80:591–595

Mullin SJ, Gutzke WHN, Zenitsky GD, Cooper RJ (2000) Home ranges of rat snakes (Colubridae: *Elaphe*) in different habitats. Herpetol Rev 31:20–22

Muth F, Healy SD (2011) The role of adult experience in nest building in the Zebra Finch, *Taeniopygia guttata*. Anim Behav 82:185–189

Muth F, Healy SD (2014) Zebra Finches select nest material appropriate for a building task. Anim Behav 90: 237–244

Natusch DJD, Lyons JA, Shine R (2017) Safety first: terrestrial predators drive selection of highly specific nesting sites in colonial-breeding birds. J Avian Biol 48:1104–1113

Nelson JB (1978) The Sulidae: gannets and boobies. Oxford University Press, Oxford

Newton I (1994) The role of nest sites in limiting the numbers of hole-nesting birds: a review. Biol Conserv 70:265–276

Nguyen LP, Nol E, Abraham KF (2007) Using digital photographs to evaluate the effectiveness of plover egg crypsis. J Wildl Manag 71:2084–2089

Niethammer KR, Patrick LB (2020) White Tern (*Gygis alba*), version 1.0. In: Billerman SM (ed) Birds of the World. Cornell Lab of Ornithology, Ithaca, NY

Nilsson SG (1984) The evolution of nest-site selection among hole-nesting birds: the importance of nest predation and competition. Ornis Scand 15:167–175

Nilsson SG, Johnsson K, Tjernberg M (1991) Is the avoidance by Black Woodpeckers of old nest holes due to predators? Anim Behav 41:439–441

Norell MA, Clark JM, Chiappe LM, Dashzeveg D (1995) A nesting dinosaur. Nature 378:774–776

Nores AI, Nores M (1994) Nest building and nesting behavior of the Brown Cacholote. Wilson Bull 106: 106–120

Norhayati MK, Azman O, Nazaimoon WM (2010) Preliminary study of the nutritional content of Malaysian edible bird's nest. Malays J Nutr 16:389–396

Nudds RL, Dyke GJ, Rayner JMV (2004) Forelimb proportions and the evolutionary radiation of Neornithes. Proc Roy Soc B (Suppl) 271:S324–S327

Ondrušová K (2011) Identification of mammalian hairs in bird nests. M. S. thesis, Palacky University, Olomouc, Czech Republic

Ontiveros D, Caro J, Pleguezuelos JM (2007) Green plant material versus ectoparasites in nests of Bonelli's Eagle. J Zool 274:99–104

Oring LW (1968) Vocalizations of the Green and Solitary sandpipers. Wilson Bull 80:395–420

Ospina EA, Merrill L, Benson TJ (2018) Incubation temperature impacts nestling growth and survival in an open-cup nesting passerine. Ecol Evol 8:3270–3279

Otterbeck A, Selås V, Nielsen JT, Roualet É, Lindén A (2019) The paradox of nest reuse: early breeding benefits reproduction, but nest reuse increases nest predation risk. Oecologia 190:559–568

Pacejka AJ, Gratton CM, Thompson CF (1998) Do potentially virulent mites affect House Wren (*Troglodytes aedon*) reproductive success? Ecology 79:1797–1806

Palomino JJ, Martin-Vivaldi M, Soler M, Soler U (1998) Functional significance of nest size variation in the Rufous Bush Robin *Cercotrichas galactotes*. Ardea 86:177–185

Paquet M, Doutrelant C, Loubon M, Theron F, Rat M, Covas R (2016) Communal roosting, thermoregulatory benefits and breeding group size predictability in cooperatively breeding Sociable Weavers. J Avian Biol 47: 749–755

Payne RB (2020) Indigo Bunting (*Passerina cyanea*), version 1.0. In: Poole AF (ed) Birds of the World. Cornell Lab of Ornithology, Ithaca, NY

Peralta-Sanchez JM, Møller AP, Martin-Platero AM, Soler JJ (2010) Number and colour composition of nest lining feathers predict eggshell bacterial community in Barn Swallow nests: an experimental study. Funct Ecol 24:426–433

Perrin N, Christe P, Richner H (1996) On host life-history response to parasitism. Oikos 75:317–320

Pettingill OS Jr (1985) Ornithology in laboratory and field, 5th edn. Academic Press, New York, NY

Ploucha JM, Fink GD (1986) Hemodynamics of hemorrhage in the conscious rat and chicken. Am J Physiol Regul Integr Comp Physiol 251:R846–R850

Podkowa P, Surmacki A (2017) The importance of illumination in nest site choice and nest characteristics of cavity nesting birds. Sci Rep 7:1329

Podkowa P, Malinowska K, Surmacki A (2019) Light affects parental provisioning behaviour in a cavity-nesting passerine. J Avian Biol 50:jav02254

Poláček M, Griggio M, Mikšík I, Bartíková M, Eckenfellner M, Hoi H (2017) Eggshell coloration and its importance in postmating sexual selection. Ecol Evol 7:941–949

Politi N, Hunter M Jr, Rivera L (2009) Nest selection by cavity-nesting birds in subtropical montane forests of the Andes: implications for sustainable forest management. Biotropica 41:354–360

Pollet IL, Bond AL, Hedd A, Huntington CE, Butler RG, Mauck R (2020) Leach's Storm-Petrel (*Oceanodroma leucorhoa*), version 1.0. In: Billerman SM, Keeney BK, Rodewald PG, Schulenberg TS (eds) Birds of the World. Cornell Lab of Ornithology, Ithaca, NY

Pollock HS, MacDonald SE, Vizentin-Bugoni J, Brawn JD, Sutton ZS, Hauber ME (2021) What the pluck? The theft of mammal hair by birds is an overlooked but common behavior with fitness implications. Ecology 102:e03501

Polo V, Veiga JP, Cordero PJ, Viñuela J, Monaghan P (2004) Female starlings adjust primary sex ratio in response to aromatic plants in the nest. Proc R Soc B 271:1929–1933

Poulin RG, Todd LD, Haug EA, Millsap BA, Martell MS (2020) Burrowing Owl (*Athene cunicularia*), version 1.0. In: Poole AF (ed) Birds of the World. Cornell Lab of Ornithology, Ithaca, NY

Price JJ, Griffith SC (2017) Open cup nests evolved from roofed nests in the early passerines. Proc R Soc B 284: 20162708

Proctor H, Owens I (2000) Mites and birds: diversity, parasitism and coevolution. Trends Ecol Evol 15: 358–364

Puchala P (2004) Detrimental effects of larval blow flies (*Protocalliphora azurea*) on nestlings and breeding success of Tree Sparrows (*Passer montanus*). Can J Zool 82:1285–1290

Quinn JL, Ueta M (2008) Protective nesting associations in birds. Ibis 150:146–167

Quiroga MA, Reboreda JC, Beltzer AH (2012) Host use by *Philornis* sp. in a passerine community in Central Argentina. Rev Mex Biodivers 83:110–116

Radley PM, Davis RA, Dekker RWRJ, Molloy SW, Blake D, Heinsohn R (2018) Vulnerability of megapodes (Megapodiidae, Aves) to climate change and related threats. Environ Conserv 45:396–406

Rajeev M, Vidanapathirana D, Benaragama BVDKP, Wickramasinghe L (2018) Host tree canopy isolation by nesting Sri Lanka Drongo *Dicrurus lophorinus*. Forktail 34:52–57

Ramji MFS, Koon LC, Rahman MA (2013) Roosting and nest-building behaviour of the White-nest Swiftlet *Aerodramus fuciphagus* (Thunberg) (Aves: Apodidae) in farmed colonies. Raffles Bull Zool 29 (Suppl):225–235

Ramos JA, Bowler J, Davis L, Venis S, Quinn J, Middleton C (2001) Activity patterns and effect of ticks on growth and survival of tropical Roseate Tern nestlings. Auk 118:709–716

Reid JM, Cresswell W, Holt S, Mellanby RJ, Whitfield DP, Ruxton GD (2002) Nest scrape design and clutch heat loss in Pectoral Sandpipers (*Calidris melanotos*). Funct Ecol 16:305–312

Reidy JL, Thompson FR III (2012) Predatory identity can explain nest predation patterns. Stud Avian Biol 43: 135–148

Remeš V (2006) Growth strategies of passerine birds are related to brood parasitism by the Brown-headed Cowbird (*Molothrus ater*). Evolution 60:1692–1700

Remeš V, Martin TE (2002) Environmental influences on the evolution of growth and developmental rates in passerines. Evolution 56:2505–2518

Remeš V, Matysiokova B, Cockburn A (2012) Long-term and large-scale analyses of nest predation patterns in Australian songbirds and a global comparison of nest predation rates. J Avian Biol 43:435–444

Rhodes B, O'Donnell C, Jamieson I (2009) Microclimate of natural cavity nests and its implications for a threatened secondary-cavity-nesting passerine of New Zealand, the South Island Saddleback. Condor 111:462–469

Richner H, Tripet F (1999) Ectoparasitism and the tradeoff between current and future reproduction. Oikos 86: 535–538

Ricklefs RE (1969) An analysis of nesting mortality in birds. Contrib Zool 9:1–48

Ritchison G (1999) Downy Woodpecker. Stackpole Books, Mechanicsburg, PA

Robertson BA (2009) Nest-site selection in a postfire landscape: do parents make tradeoffs between microclimate and predation risk? Auk 126:500–510

Robinson SK, Thompson FR III, Donovan TM, Whitehead DR, Faaborg J (1995) Regional forest fragmentation and the nesting success of migratory birds. Science 267:1987–1990

Robles H, Ciudad C, Matthysen E (2012) Responses to experimental reduction and increase of cavities by a secondary cavity-nesting bird community in cavity-rich Pyrenean oak forests. For Ecol Manag 277:46–53

Rogers AM, Griffin AS, van Rensburg BJ, Kark S (2020) Noisy neighbours and myna problems: interaction webs and aggression around tree hollows in urban habitats. J Appl Ecol 57:1891–1901

Rohrbaugh RW Jr, Yahner RH (1997) Effects of macrohabitat and microhabitat on nest-box use and nesting success of American Kestrels. Wilson Bull 109:410–423

Root RB (1969) The behavior and reproductive success of the Blue-gray Gnatcatcher. Condor 71:16–31

Rubalcaba JG, Polo V, Maia R, Rubenstein DR, Veiga JP (2016) Sexual and natural selection in the evolution of extended phenotypes: the use of green nesting material in starlings. J Evol Biol 29:1585–1592

Sánchez JE, Conejo-Barboza K, Sánchez C, Calderón-Franco D, Sandoval L (2016) Description of the nest and eggs of the Green Thorntail (*Discosura conversii*). Ornitol Neotrop 27:73–76

Sanz V (2008) Analisis multiescalar y multivariado para evaluar la susceptibilidad de los nidos de psita'cidos a la depredacio'n: un ejemplo con la cotorra cabeciamarilla (*Amazona barbadensis*). Ornitol Neotrop 19(Suppl):123–134

Sanz JJ, Garcia-Navas V (2011) Nest ornamentation in Blue Tits: is feather carrying ability a male status signal? Behav Ecol 22:240–247

Scharf WC, Kren J (2020) Orchard Oriole (*Icterus spurius*), version 1.0. In: Poole AF (ed) Birds of the World. Cornell Lab of Ornithology, Ithaca, NY

Schepps J, Lohr S, Martin TE (1999) Does tree hardness influence nest-tree selection by primary cavity nesters? Auk 116:658–665

Scherr TM, Chalfoun AD (2022) Taming the temperature: sagebrush songbirds modulate microclimate via nest-site selection. Ornithology 139:ukac004

Schill KL, Yahner RH (2009) Nest-site selection and nest survival of early successional birds in central Pennsylvania. Wilson J Ornithol 121:476–484

Schleicher B, Hoi H, Valera F (1996) Seasonal change in female mate choice criteria in Penduline Tits (*Remiz pendulinus*). Ardeola 43:19–29

Schmidt KA (2003) Linking frequencies of resource pulses in oak forests to long-term rates in the Veery (*Catharus fusescens*). Oikos 103:548–558

Sergio F, Rizzolli F, Marchesi L, Pedrini P (2004) The importance of interspecific interactions for breeding-site selection: Peregrine Falcons seek proximity to raven nests. Ecography 27:818–826

Sergio F, Blas J, Blanco G, Tanferna A, López L, Lemus JA, Hiraldo F (2011) Raptor nest decorations are a reliable threat against conspecifics. Science 331:327–330

Seymour RS, Ackerman RA (1980) Adaptations to underground nesting in birds and reptiles. Amer Zool 20:437–447

Seymour RS, Bradford DF (1992) Temperature regulation in the incubation mounds of the Australian Brush-Turkey. Condor 94:134–150

Seymour RS, Vleck D, Vleck CM (1986) Gas exchange in the incubation mounds of megapode birds. J Comp Physiol B 156:773–782

Seymour RS, Vleck D, Vleck CM, Booth DT (1987) Water relations of buried eggs of mound building birds. J Comp Physiol B 157:413–422

Shepard E, Cole E-L, Neate A, Lempidakis E, Ross A (2019) Wind prevents cliff-breeding birds from accessing nests through loss of flight control. eLife 8: e43842

Shim EKS, Chandra GF, Pedireddy S, Lee S-Y (2016) Characterization of swiftlet edible bird nest, a mucin glycoprotein, and its adulterants by Raman microspectroscopy. J Food Sci Technol 53:3602–3608

Siefferman L, Navara KJ, Hill GE (2006) Egg coloration is correlated with female condition in Eastern Bluebirds (*Sialia sialis*). Behav Ecol Sociobiol 59:651–656

Silver R, Andrews H, Ball GF (1985) Parental care in an ecological perspective: a quantitative analysis of avian subfamilies. Am Zool 25:823–840

Sinclair JR, O'Brien TG, Kinnaird MF (2002) The selection of incubation sites by the Philippine Megapode, *Megapodius cumingii*, in North Sulawesi, Indonesia. Emu 102:151–158

Skutch AF (1937) Life-history of the Black-chinned Jacamar. Auk 54:135–146

Slagsvold T, Wiebe KL (2021) Nest decorations: birds exploit a fear of feathers to guard their nest from usurpation. R Soc Open Sci 8:211579

Smith KW (1997) Nest site selection of the Great Spotted Woodpecker *Dendrocopos major* in two oak woods in southern England and its implications for woodland management. Biol Conserv 80:283–288

Smith DG, Bosakowski T, Devine A (1999) Nest site selection by urban and rural Great Horned Owls in the northeast. J Field Ornithol 70:535–542

Smith KG, Wittenberg SR, Macwhirter RB, Bildstein KL (2020a) Northern Harrier (*Circus hudsonius*), version 1.0. In: Rodewald PG (ed) Birds of the World. Cornell Lab of Ornithology, Ithaca, NY

Smith RJ, Hatch MI, Cimprich DA, Moore FR (2020b) Gray Catbird (*Dumetella carolinensis*), version 1.0. In: Poole AF (ed) Birds of the World. Cornell Lab of Ornithology, Ithaca, NY

Soler M, Pérez-Contreras T (2012) Location of suitable nests by Great Spotted Cuckoos: an empirical and experimental study. Behav Ecol Sociobiol 66:1305–1310

Soler JJ, Møller AP, Soler M (1998) Nest building, sexual selection and parental investment. Evol Ecol 12:427–441

Soler JJ, Morales J, Cuervo JJ, Moreno J (2019) Conspicuousness of passerine females is associated with the nest-building behaviour of males. Biol J Linn Soc 126: 824–835

Solís JC, de Lope F (1995) Nest and egg crypsis in the ground-nesting Stone Curlew *Burhinus oedicnemus*. J Avian Biol 26:135–138

Sonerud GA (1985) Nest hole shift in Tengmalm's Owl *Aegolius funereus* as defence against nest predation involving long-term memory in the predator. J Anim Ecol 54:179–192

Soule OH (1964) The saguaro tree-hole microenvironment in southern Arizona. Part 2, summer. M. S. thesis, University of Arizona, Tucson, AZ

Sridhar S, Karanth KP (1993) Helpers in cooperatively breeding small Green Bee-eater (*Merops orientalis*). Curr Sci 65:489–490

Stanley TR (2002) How many kilojoules does a Black-billed Magpie nest cost? J Field Ornithol 73:292–297

Stevens M, Troscianko J, Wilson-Aggarwal JK, Spottiswoode CN (2017) Improvement of individual camouflage through background choice in ground-nesting birds. Nat Ecol Evol 1:1325

Stoddard MC, Marshall KLA, Kilner RM (2011) Imperfectly camouflaged avian eggs: artefact or adaptation? Avian Biol Res 4:196–213

Stoehr AM, Nolan PM, Hill GE, McGraw KJ (2000) Nest mites (*Pellonyssus reedi*) and the reproductive biology of the House Finch (*Carpodacus mexicanus*). Can J Zool 78:2126–2133

Stojanovic D, Rayner L, Cobden M, Davey C, Harris S, Heinsohn R, Owens G, Manning AD (2021) Suitable nesting sites for specialized cavity dependent wildlife are rare in woodlands. For Ecol Manag 483:118718

Street SE, Jaques R, De Silva TN (2022) Convergent evolution of elaborate nests as structural defences in birds. Proc R Soc B 289:20221734

Sugden LG, Beyersbergen GW (1986) Effect of density and concealment on American Crow predation of simulated duck nests. J Wildl Manag 50:9–14

Swanson DL, Ingold JL, Wallace GE (2020) Ruby-crowned Kinglet (*Regulus calendula*), version 1.0. In: Poole AF (ed) Birds of the World. Cornell Lab of Ornithology, Ithaca, NY

Szentirmai I, Komdeur J, Szekely T (2005) What makes a nest-building male successful? Male behavior and female care in Penduline Tits. Behav Ecol 16:994–1000

Tanaka K, Zelenitsky DK, Therrien F (2015) Eggshell porosity provides insight on evolution of nesting in dinosaurs. PLoS ONE 10:e0142829

Tarof S, Briskie JV (2020) Least Flycatcher (*Empidonax minimus*), version 1.0. In: Poole AF (ed) Birds of the World. Cornell Lab of Ornithology, Ithaca, NY

Tate DP (1994) Observations on nesting behavior of the Common Potoo in Venezuela. J Field Ornithol 65:447–452

Taylor LW, Kreutziger GO, Abercrombie GL (1971) The gaseous environment of the chick embryo in relation to its development and hatchability. 5. Effect of carbon dioxide and oxygen levels during the terminal days of incubation. Poult Sci 50:66–78

Thompson FR III, Donovan TM, DeGraaf RM, Faaborg J, Robinson SK (2002) A multi-scale perspective of the effects of forest fragmentation on birds in eastern forests. Stud Avian Biol 24:8–19

Thompson FR III (2007) Factors affecting nest predation on forest songbirds in North America. Ibis 149(Suppl. 2):98–109

Thorburn C (2014) The edible birds' nest boom in Indonesia and South-east Asia: a nested political ecology. Food Cult Soc 17:535–553

Thorburn CC (2015) The edible nest swiftlet industry in Southeast Asia: capitalism meets commensalism. Hum Ecol 43:179–184

Tieleman BI, Van Noordwijk HJ, Williams JB (2008) Nest site selection in a hot desert: trade-off between microclimate and predation risk. Condor 110:116–124

Tomás G, Merino S, Moreno J, Sanz JJ, Morales J, Garcia-Fraile S (2006) Nest weight and female health in the Blue Tit (*Cyanistes caeruleus*). Auk 123:1013–1021

Tomback DF (2020) Clark's Nutcracker (*Nucifraga columbiana*), version 1.0. In: Poole AF, Gill FB (eds) Birds of the World. Cornell Lab of Ornithology, Ithaca, NY

Tozer DC, Burke DM, Nol E, Elliott KA (2012) Managing ecological traps: logging and sapsucker nest predation by bears. J Wildl Manag 76:887–898

Tremblay JA, Dixon RD, Saab VA, Pyle P, Patten MA (2020) Black-backed Woodpecker (*Picoides arcticus*), version 1.0. In: Rodewald PG (ed) Birds of the World. Cornell Lab of Ornithology, Ithaca, NY

Tripet F, Richner H (1997) Host responses to ectoparasites: food compensation by parent Blue Tits. Oikos 78:557–561

Trnka A, Prokop P (2011) The use and function of snake skins in the nests of Great Reed Warblers *Acrocephalus arundinaceus*. Ibis 153:627–630

Troscianko J, Wilson-Aggarwal J, Stevens M, Spottiswoode CN (2016) Camouflage predicts survival in ground-nesting birds. Sci Rep 6:19966

Trzcinski MK, Cockle KL, Norris AR, Edworthy M, Wiebe KL, Martin K (2022) Woodpeckers and other excavators maintain the diversity of cavity-nesting vertebrates. J Anim Ecol 91:1251–1265

Unzeta M, Martin TE, Sol D (2020) Daily nest predation rates decrease with body size in passerine birds. Am Nat 196:743–754

Vajda V, Bercovici A (2014) The global vegetation pattern across the Cretaceous–Paleogene mass extinction interval: a template for other extinction events. Glob Planet Chang 122:29–49

van der Hoek Y, Gaona GV, Martin K (2017) The diversity, distribution and conservation status of the tree-cavity-nesting birds of the World. Divers Distrib 23: 1120–1131

van Dijk RE, Kaden JC, Argüelles-Ticó A, Beltran LM, Paquet M, Covas R, Doutrelant C, Hatchwell BJ (2013) The thermoregulatory benefits of the communal nest of Sociable Weavers *Philetairus socius* are spatially structured within nests. J Avian Biol 44:102–110

Van Tyne J, Berger AJ (1976) Fundamentals of ornithology, 2nd edn. Wiley & Sons, New York, NY

Vanadzina K, Street SE, Healy SD, Laland KN, Sheard C (2022) Global drivers of variation in cup nest size in passerine birds. J Anim Ecol 92:338–351

Vanderhoff N, Pyle P, Patten MA, Sallabanks R, James FC (2020) American Robin (*Turdus migratorius*), version 1.0. In: Rodewald PG (ed) Birds of the World. Cornell Lab of Ornithology, Ithaca, NY

Vanderwerf EA (1988) Observations on the nesting of the Great Potoo (*Nyctibius grandis*) in central Venezuela. Condor 90:948–950

Vanderwerf EA (2012) Evolution of nesting height in an endangered Hawaiian forest bird in response to a non-native predator. Conserv Biol 26:905–911

Varricchio DJ, Jackson FD (2016) Reproduction in Mesozoic birds and evolution of the modern avian reproductive mode. Auk 133:654–684

Varricchio DJ, Jackson F, Borkowksi JJ, Horner JR (1997) Nest and egg clutches of the dinosaur *Troodon formosus* and the evolution of avian reproductive traits. Nature 385:247–250

Varricchio DJ, Jackson F, Trueman CN (1999) A nesting trace with eggs for the cretaceous theropod dinosaur *Troodon formosus*. J Vertebr Paleontol 19:91–100

Varricchio DJ, Kundrát M, Hogan J (2018) An intermediate incubation period and primitive brooding in a theropod dinosaur. Sci Rep 8:12454

Vennesland RG, Butler RW (2020) Great Blue Heron (*Ardea herodias*), version 1.0. In: Poole AF (ed) Birds of the World. Cornell Lab of Ornithology, Ithaca, NY

Verner J (1965) Breeding biology of the Long-billed Marsh Wren. Condor 67:6–30

Wachob DG (1996) A microclimate analysis of nest-site selection by Mountain Chickadees. J Field Ornithol 67: 525–533

Walsberg GE (1981) Nest-site selection and the radiative environment of the Warbling Vireo. Condor 83:86–88

Walsberg GE (1992) Quantifying radiative heat gain in animals. Am Zool 32:217–223

Walsberg GE, King JR (1978) The heat budget of incubating Mountain White-crowned Sparrows (*Zonotrichia leucophrys oriantha*) in Oregon. Physiol Zool 51:92–103

Walsh PT, Hansell M, Borello WD, Healy SD (2010) Repeatability of nest morphology in African weaver birds. Biol Lett 6:149–151

Walsh PT, Hansell M, Borello WD, Healy SD (2013) Are elaborate bird nests built using simple rules? Avian Biol Res 6:157–162

Ward S (1996) Energy expenditure of female Barn Swallows *Hirundo rustica* during egg formation. Physiol Zool 69:930–951

Webb DR (1987) Thermal tolerance of avian embryos: a review. Condor 89:874–898

Webster MD (2020) Verdin (*Auriparus flaviceps*), version 1.0. In: Poole AF, Gill FB (eds) Birds of the World. Cornell Lab of Ornithology, Ithaca, NY

Weidensaul S, Robinson TR, Sargent RR, Sargent MB, Zenzal TJ (2020) Ruby-throated Hummingbird (*Archilochus colubris*), version 1.0. In: Rodewald PG (ed) Birds of the World. Cornell Lab of Ornithology, Ithaca, NY

Weidinger K (2004) Relative effects of nest size and nest site on the risk of predation in open nesting passerines. J Avian Biol 35:515–523

Werschkul DF (1979) Nestling mortality and the adaptive significance of early locomotion in the Little Blue Heron. Auk 96:116–130

Wesołowski T (2002) Anti-predator adaptations in nesting Marsh Tits *Parus palustris*: the role of nest-site security. Ibis 144:593–601

Wesołowski T (2017) Failed predator attacks: a direct test of security of tree cavities used by nesting Marsh Tits (*Poecile palustris*). Auk 134:802–810

Wesołowski T, Rowiński P (2012) The breeding performance of Blue Tits *Cyanistes caeruleus* in relation to the attributes of natural holes in a primeval forest. Bird Study 59:437–448

Whitaker-Fornek JR, Nelson JK, Pilarski JQ (2020) The role of chloride in the breathing-related response to central acidosis in developing birds. FASEB J 34:1

White FN, Bartholomew GA, Kinney SL (1978) Physiological and ecological correlates of tunnel nesting in the European Bee-eater, *Merops apiaster*. Physiol Zool 51:140–154

Wickler SJ, Marsh RL (1981) Effects of nestling age and burrow depth on $CO_2$ and $O_2$ concentrations in the burrows of Bank Swallows. Physiol Zool 54:132–137

Wiebe KL (2001) Microclimate of tree cavity nests: is it important for reproductive success in Northern Flickers? Auk 118:412–421

Wiebe KL (2007) Hypoxia probably does not explain short incubation periods of woodpeckers. Condor 109:976–979

Wiebe KL (2011) Nest sites as limiting resources for cavity-nesting birds in mature forest ecosystems: a review of the evidence. J Field Ornithol 82:239–248

Wiebe KL, Moore WS (2020) Northern Flicker (*Colaptes auratus*), version 1.0. In: Rodewald PG (ed) Birds of the World. Cornell Lab of Ornithology, Ithaca, NY

Wiebe KL, Koenig WD, Martin K (2007) Costs and benefits of nest reuse versus excavation in cavity-nesting birds. Ann Zool Fenn 44:209–217

Wiktander U, Olsson O, Nilsson SG (2001) Annual and seasonal reproductive trends in the Lesser Spotted Woodpecker *Dendrocopus minor*. Ibis 143:72–82

Wilson JK (1999) Nest-reuse for the Yellow-billed Cuckoo in Arkansas. Southwest Nat 44:538–539

Wimberger PH (1984) The use of green plant material in bird nests to avoid ectoparasites. Auk 101:615–618

Withers PC (1977) Energetic aspects of reproduction by the Cliff Swallow. Auk 94:718–725

Witmer MC, Mountjoy DJ, Elliott L (2020) Cedar Waxwing (*Bombycilla cedrorum*), version 1.0. In: Poole AF (ed) Birds of the World. Cornell Lab of Ornithology, Ithaca, NY

Xing L, O'Connor JK, McKellar RC, Chiappe LM, Tseng K, Li G, Bai M (2017) A mid-cretaceous enantiornithine (Aves) hatchling preserved in Burmese amber with unusual plumage. Gondwana Res 49:264–277

Yang T-R, Wiemann J, Xu L, Cheng Y-N, Wu X-C, Sander PM (2019) Reconstruction of oviraptorid clutches illuminates their unique nesting biology. Acta Palaeontol Pol 64:581–596

Yom-Tov Y, Ar A (1993) Incubation and fledging durations of woodpeckers. Condor 95:282–287

Yosef R (2020) Loggerhead Shrike (*Lanius ludovicianus*), version 1.0. In: Poole AF, Gill FB (eds) Birds of the World. Cornell Lab of Ornithology, Ithaca, NY

Zhou T, Wang H, Liu Y, Lei F, Gao W (2009) Patterns of magpie nest utilization by a nesting raptor community in a secondary forest. Prog Nat Sci 19:1253–1259

# Avian Reproduction: Clutch Sizes, Incubation, and Hatching

# 18

## Contents




**Abstract**

The number of eggs laid per clutch by female birds varies from 1 to 18 or more. In this chapter, the evolution of clutch sizes and the effect of lifespan on clutch sizes are discussed. The factors contributing to latitudinal variation in clutch sizes and variation in clutch sizes among and within species are explained. The possible effects of predation risk on clutch sizes are discussed as well as factors contributing to seasonal variation in clutch sizes. The incubation behavior of birds is discussed, as well as variation among and within species in the onset of incubation. The energetic costs of incubation are also explained. The duration of incubation periods varies widely among different species of birds and factors contributing to that variation are




G. Ritchison, *In a Class of Their Own*, Fascinating Life Sciences,
https://doi.org/10.1007/978-3-031-14852-1_18

explained. The development, nutrition, and growth of avian embryos are described as well as the process of hatching.

## 18.1 Evolution of Clutch Sizes

Clutch sizes vary considerably among species and families of birds, ranging from one for birds in several families, such as tropicbirds, frigatebirds, and albatrosses, to 18 or more in some partridges and quail. Clutch sizes also vary within species, but to a lesser degree than between species. Understanding both inter- and intraspecific variation in clutch sizes is important, but require different perspectives. Understanding interspecific variation requires examination of ultimate (or evolutionary) causation. In other words, how and why has natural selection acted upon different species to generate the "typical" clutch sizes we see in present-day birds? In contrast, understanding variation in clutch sizes among individuals within a species or population requires more of a focus (but not entirely) on proximate (immediate) causation. What behavioral, ecological, or physiological factors contribute to variation in clutch sizes among different individuals of the same species in the same or different populations?

The considerable interspecific variation in typical clutch sizes indicates that natural selection has favored different reproductive strategies for species that vary in lifespan, habitat use and quality, geographic location, and other factors. Determining the extent to which different factors have acted via natural selection to influence the evolution of clutch sizes in different species is clearly difficult because of the many factors involved. However, one of those factors is clearly lifespan (Box 18.1 Avian Survival).

Time, energy, and other resources are limited and must be allocated among competing demands

**Box 18.1 Avian Survival**

Survival rates of birds are generally expressed as annual survival rates, i.e., the number of individuals in a population alive at the end of a year divided by the number that were alive at the beginning of a year. This rate can range from 0 to 1. Survival rates are difficult to determine because birds must be captured and then, at some later date, recaptured. However, many investigators have reported estimates of annual survival and the results of these studies have revealed that annual survival rates vary considerably among different species of birds. For example, some birds, like large diurnal raptors, may have annual survival rates greater than 0.90 (Newton et al. 2016), whereas small songbirds may have annual survival rates of less than 0.35 (Johnston et al. 2016). Scholer et al. (2020) reviewed 204 studies that included 949 estimates of annual survival for 636 species of birds and the overall mean estimate of survival was 0.67. Analysis also revealed several variables that contribute to variation in avian survival, including body mass and clutch size. Annual survival tends to increase with body mass (i.e., larger birds tend to live longer) and decrease with increasing clutch sizes (i.e., smaller species have shorter lives so attempt to maximize lifetime fitness via larger clutches) (Stearns 1992). Scholer et al. (2020) also found higher rates of annual survival for nonmigratory than migratory birds, providing "general support for the idea that sedentary behavior favours higher survival and, hence, shifts toward slower life histories ..." Analysis also revealed little effect of climate variables (i.e., annual precipitation and minimum winter temperature) on annual survival. Finally, Scholer et al. (2020) found support for an effect of latitude on avian survival, with a tendency for greater annual survival at lower latitudes, but this relationship was stronger for passerines in the Northern Hemisphere (but not the Southern Hemisphere) and in South America (a continent that extends to higher latitudes than those in the Eastern Hemisphere). Passerines at

(continued)

**Box 18.1** (continued)

higher latitudes may have lower rates of survival than those at lower latitudes because the more seasonal environments at higher latitudes may limit resource availability (Scholer et al. 2020).

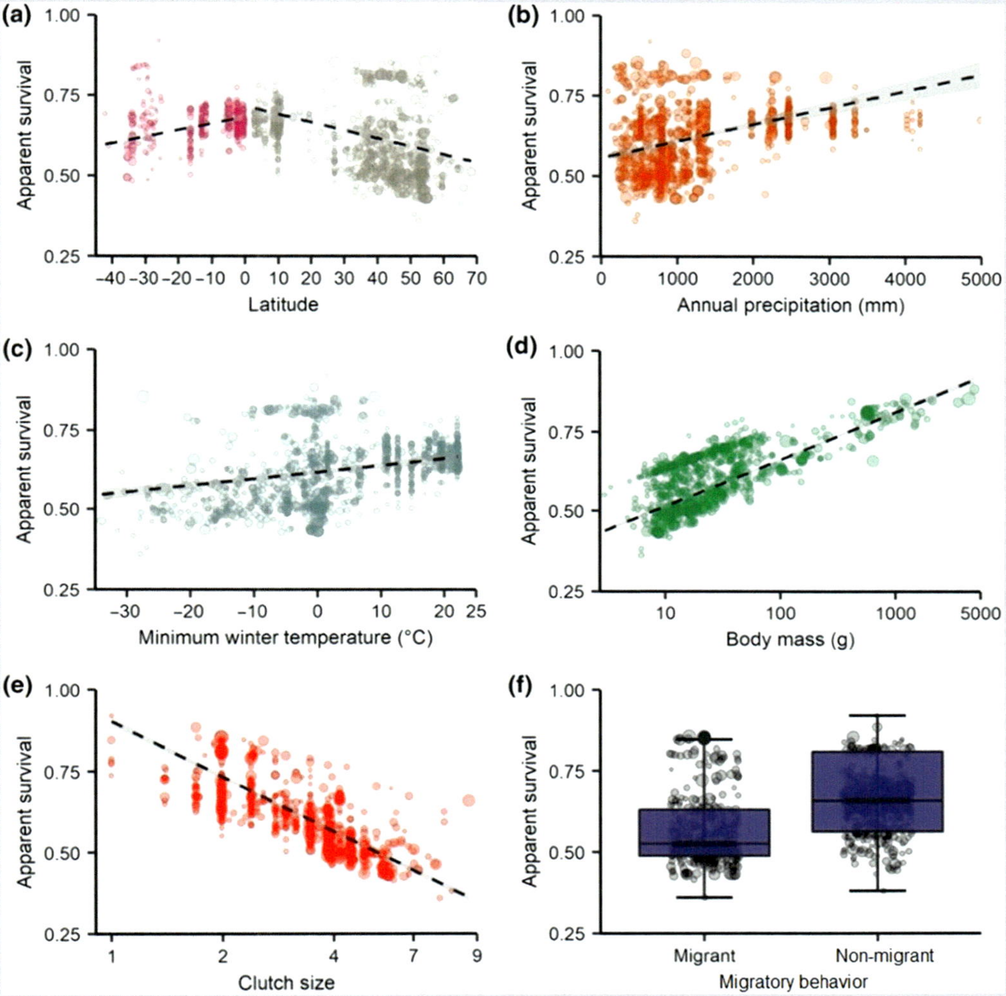


Relationship between apparent adult survival and different variables. Dashed lines represent the best linear fits. Gray shading indicates 95% confidence limits. Point sizes reflect the inverse of the standard error used to weight data points (i.e., more precise estimates appear as larger points). Body mass and clutch size are reported on a log10 scale. (Figure from Scholer et al. 2020; © 2020 John Wiley & Sons Ltd., used with permission)

(Ricklefs 2000). The chances of surviving the various risks associated with reproduction will likely decrease as parental investment increases because time and resources that could be allocated to adult maintenance and avoidance of predation are instead devoted to offspring. Because natural selection favors individuals that optimize lifetime reproductive success (and not reproductive success during a single breeding attempt or breeding season), the optimal investment in a particular breeding attempt is influenced by lifespan. In other words, individuals in species with shorter life expectancies may maximize lifetime fitness by investing more in reproduction (e.g., larger clutches); individuals in species with longer lifespans enhance their fitness by investing less in each reproductive effort (Bonduriansky et al. 2008; Figs. 18.1 and 18.2).

Fecundity (or lifetime fitness) is, therefore, related to and greatly influenced by annual survival. To illustrate (Fig. 18.3), the bounded area represents the potential fitness of all possible phenotypes with respect to realized fecundity and reproductive survival of parents. Because, in

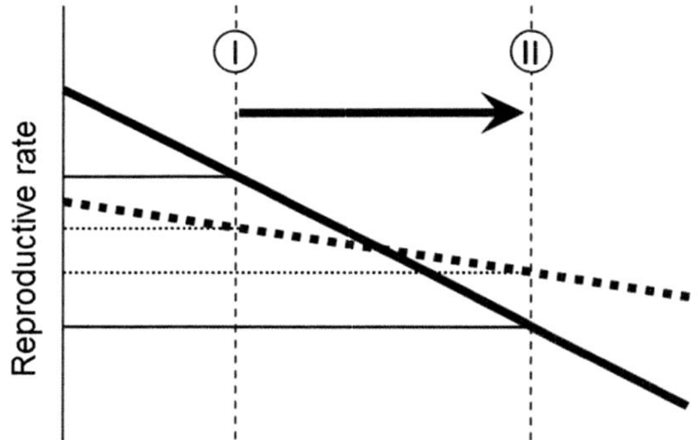


**Fig. 18.1** Effect of variation in extrinsic mortality risk; if life expectancy is short (vertical line I), then individuals that invest more in reproduction in early life (solid line) will have higher lifetime fitness than those that invest less in reproduction (dotted line). If extrinsic mortality rate is reduced (arrow), extending life expectancy from vertical line I to II, then individuals with a lower investment in reproduction will achieve higher lifetime fitness. (Figure modified from Bonduriansky et al. 2008; © 2008 The Authors. Journal compilation © British Ecological Society, used with permission)

this graph, fitness increases with distance from the origin of the graph (a combination of higher fecundity and higher survival), all points inside the bounded area represent less-fit phenotypes and the outer perimeter of that area represents optimum fitness. As a result, natural selection should lead to an optimum phenotype somewhere on the perimeter. The optimum point is tangent to a line (the "adaptive function") whose slope is determined by the annual adult survival rate (Ricklefs 1977, 1983). Thus, when the likelihood of adults surviving due to factors other than reproductive risk is higher, the adaptive function has a shallower slope and is tangent to the adaptive function at a lower fecundity, but higher adult survival rate (Ricklefs 2000).

Remember first, if this is unclear, that slope represents the relationship between how much a line rises relative to how long it is or, for a line with a negative slope, how much a line falls relative to its length. Compare an albatross with a possible lifespan of up to 40 years or more that starts breeding when 10 to a chickadee with a possible lifespan of about 7 years that starts breeding when 1 year old. For the albatross, the average number of breeding seasons (assuming a full lifespan) is 25 and, for the chickadee, four. For the albatross, the slope that maximizes fitness is –1/25 and, for the chickadee, –1/4. In other words, the slope of the line for the albatross is much shallower and, therefore, its long lifespan selects for lower investment in each breeding attempt. The much shorter lifespan of a chickadee means that selection favors more investment in each breeding attempt.

Selection favoring different breeding strategies for species that differ in lifespan establishes certain innate constraints so, for example, a female albatross will never produce a clutch of 18 eggs. However, factors other than lifespan also influence clutch size. This is obviously the case because clutch sizes of species with similar lifespans differ. For example, the mean annual survival rate for Warbling Vireos (*Vireo gilvus*) in coastal California is 0.504 (Gardali et al. 2000) and that for female Mottled Ducks (*Anas fulvigula*) along the Gulf Coast is 0.50 (Haukos 2015), but typical clutch sizes for these

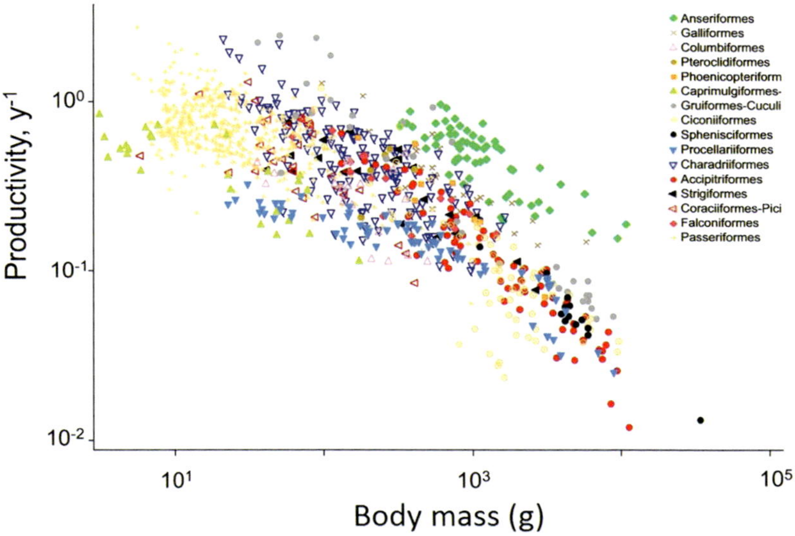


**Fig. 18.2** Productivity (measured as grams of eggs per grams of adult females per year) of females in several different orders of birds. In general, smaller, typically short-lived birds are more productive than larger, typically longer-lived birds. Orders: Phoenicopteriform = Phoenicopteriformes and Podicipediformes; Caprimulgiformes = Caprimulgiformes and Apodiformes; Gruiformes-Cuculi = Gruiformes and Cuculiformes; Coarciiformes-Pici = Coraciiformes and Piciformes. (Figure modified from Sibley et al. 2012: used with permission of the U. S. National Academy of Sciences)

two species are 3–4 eggs (Gardali and Ballard 2020) and 8–12 eggs (Bielefeld et al. 2020), respectively. An important difference between these two species is their mode of development.

Precocial species like Mottled Ducks, with young that are mobile shortly after hatching, have larger clutches than altricial species like Warbling Vireos. For example, Jetz et al. (2008) examined the typical clutch sizes of 5290 species of landbirds and found that the mean clutch sizes of precocial and altricial species were about 4.5 and 2.9 eggs, respectively. Because precocial young can often feed themselves and are able to regulate their body temperatures, they are less expensive to care for (in terms of time and energy) than altricial young. There are, of course, still costs associated with caring for precocial young, such as vigilance and defense of young from conspecifics and predators (Milonoff et al. 2004), those costs are simply lower than for altricial young. For example, pairs of Snow Geese (*Anser caerulescens*; a precocial species) whose clutches were experimentally increased fledged more young than pairs with smaller clutches, suggesting that the ability to care for young after hatching does not limit clutch size (Lepage et al. 1998). In fact, young in the experimentally enlarged broods were in better condition (larger and heavier) than those in smaller, normal broods because larger families were dominant over smaller ones and had access to better foraging areas. For precocial parents then, more of the costs of reproduction (in terms of impact on future reproduction) are associated with egg production and the incubation period. For altricial parents, costs are more evenly distributed between egg production, incubation, and caring for young after hatching. This skewing of costs allows precocial parents to expend more reproductive effort on egg production than most altricial species and the result is, on average, larger clutches (Ricklefs 1984; Starck and Ricklefs 1998).

Another factor that has influenced the evolution of clutch size is nest type, with species that nest in cavities tending to have larger clutches

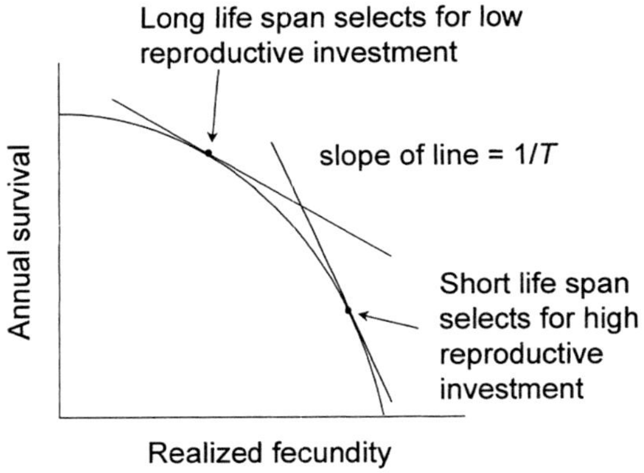


**Fig. 18.3** The optimization of the trade-off between annual adult survival and annual fecundity. The bounded area within the curve represents the set of possible phenotypes, among which those potentially having maximum fitness are represented on the periphery. The adaptive function is a line running through combinations of fecundity and survival having equal fitness and just tangent to the fitness set. The tangent point is the phenotype that maximizes evolutionary fitness. The slope of the line (−1/*T*) is the negative value of the inverse of the average age (*T*) of a female at the birth of her offspring. Thus, the slope is related to the negative of the adult mortality rate. (Figure from Ricklefs 2000; © 2000 Oxford University Press, used with permission)

than open-nesting species (Jetz et al. 2008; Barve and Mason 2015). The larger clutches of cavity-nesting birds are thought to be due to the lower rate of predation for nests in cavities (Lack 1954). Evidence does indeed indicate that nest predation rates are lower for cavity-nesting species (Fontaine et al. 2007). Predation is generally the primary cause of nest failure for birds (Martin 1995). An increased risk of nest predation favors reduced investment in current clutches and hence smaller clutches because less investment means more energy will be available for re-nesting (Slagsvold 1984; Martin 1995). For cavity-nesting species, predation risk is relatively low and the likelihood of needing energy for re-nesting is equally low. Natural selection, therefore, has favored a greater investment in current clutches by cavity-nesters or, in other words, larger clutches. Of course, among cavity-nesters, clutch sizes vary and a number of factors contribute to such variation (Box 18.2 Clutch Sizes of Cavity-Excavating Birds).

Another factor known to have influenced the evolution of clutch sizes is diet, with granivores and omnivores typically having larger clutches than frugivores and nectarivores (Jetz et al. 2008). One possible explanation for such differences is that nectar and fruit pulp contain low levels of protein and protein-limited diets

**Box 18.2 Clutch Sizes of Cavity-Excavating Birds**

There are two major competing hypotheses for variation in clutch sizes among cavity-excavating species. The nest site limitation hypothesis postulates that nesting opportunities are more limited for weak excavators (i.e., species like chickadees or nuthatches that can only excavate cavities in well-decayed, soft wood), which consequently invest more in each breeding attempt by laying larger clutches. In support of this hypothesis, Mönkkönen and Martin (2000) examined the propensity to excavate and typical clutch sizes of ten species of European parids and found a negative correlation, i.e., weak excavators generally have larger clutches. Alternatively, for strong excavators, clutch sizes may be influenced more by diet, with clutch sizes of strong excavators (e.g., woodpeckers) smaller because they can specialize on a more seasonally stable prey. Wiebe et al. (2006) built a conceptual model that integrated hypotheses for interspecific variation in clutch size and tested it with comparative data on life-history traits of woodpeckers (Picidae) and nuthatches (Sittidae). In most analyses, diet explained more variation in clutch size among species than did propensity to excavate. Although Wiebe et al.'s (2006) data did not rule out nest-site limitation, the authors concluded that, for strong excavators, annual stability of food resources has a greater influence on the evolution of clutch sizes than nest-site limitation.

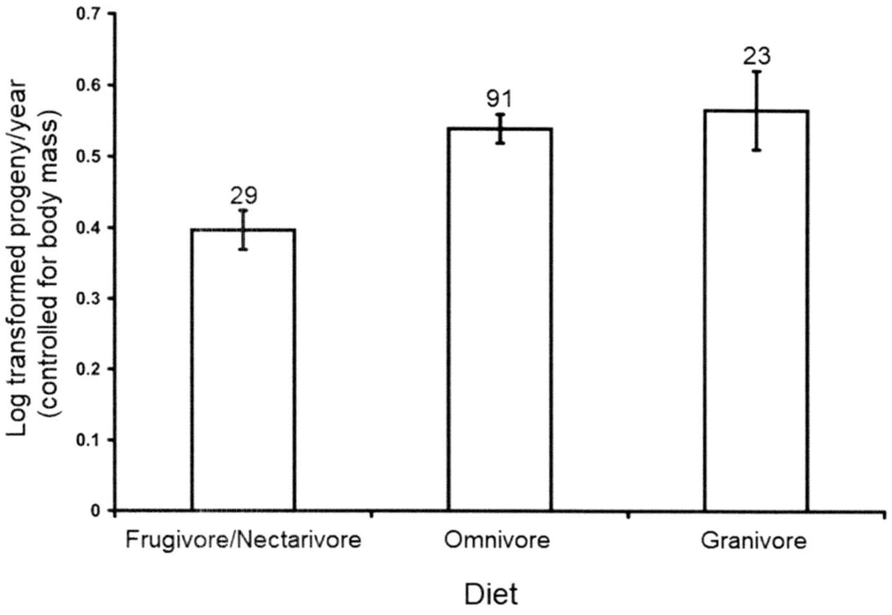


**Fig. 18.4** Variation in the mean number of progeny per year (controlled for body mass) among species of parrots (Psittaciformes) with different diets. Error bars represent one standard error of the mean residual longevity. Numbers above columns represent number of species in each category. (Figure from Munshi-South and Wilkinson 2006; © 2006 Oxford University Press, used with permission)

may limit the ability of females to produce large clutches. In addition, frugivores and nectarivores may be limited in their ability to provide large numbers of developing young with sufficient protein. For example, Cinnamon-bellied Flowerpiercers (*Diglossa baritula*) are nectar-feeding songbirds and the abundance of energy-rich, but protein-poor, nectar seems to be the primary factor that determines the timing of reproduction. However, nectar availability appears to be critical because females apparently breed only when the nectar availability is sufficiently high to provide them with more time to forage for insects that provide the protein needed for successful reproduction (Schondube et al. 2003).

Although most frugivores and nectarivores provide nestlings with some insects, part of the nestling diet is fruit or nectar. As a result, their diets are, compared to nestlings in other species of birds, relatively low in protein. Diets lower in protein contribute to slower growth and longer nestling periods. Skutch (1954, 1960) examined the life histories of many Neotropical birds and found that nestling periods for insectivores, granivores, and omnivores were typically 12–14 days. The mean nestling period for four frugivorous species (genera *Euphonia* and *Chlorophonia*) was about 22 days. Similarly, the nestling periods of nectar-feeding hummingbirds are typically about 20–26 days long (Fierro-Calderón and Martin 2007). Because longer nestling periods mean a greater risk of nest predation, selection may generally favor smaller clutches for frugivores and nectarivores (Fig. 18.4) because fewer nestlings to feed means less activity at the nest and a reduced likelihood of attracting predators.

## 18.2 Latitudinal Variation in Clutch Sizes

Despite the large range in clutch sizes among different species of birds, more than half of all

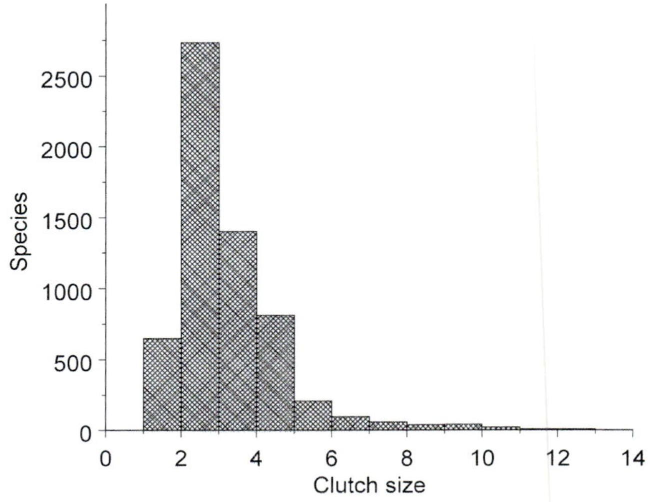


**Fig. 18.5** Variation in clutch sizes among 5290 species of landbirds (six species with clutch sizes > 14 not included). (Figure from Jetz et al. 2008; © 2008 Jetz et al., open-access article distributed under the terms of the Creative Commons Attribution License, https://creativecommons.org/licenses/by/4.0/)

birds lay clutches of 2 or 3 eggs (mode = 2, median = 2.8; Jetz et al. 2008). Examination of the frequency distribution of avian clutch sizes reveals a skewing to the right (Fig. 18.5), and this is due primarily to the relatively large clutches of north-temperate species (Fig. 18.6; Jetz et al. 2008). Thus, an important question is why clutch sizes exhibit such latitudinal variation and, specifically, why clutch sizes tend to be larger in temperate areas (or, depending on your frame of reference, why clutch sizes are smaller in tropical areas).

Birds in tropical areas commonly lay about half as many eggs as those in north-temperate areas (Fig. 18.7). Lack (1947) proposed that latitudinal variation in clutch sizes was the result of differences in food availability. This food limitation hypothesis posits that clutch size is determined by food supply, with natural selection favoring clutch sizes that correspond to the number of viable young that the parent(s) can successfully raise, and the scarcity of food in tropical habitats limits clutch sizes. Another hypothesis, the nest predation hypothesis, proposes that small

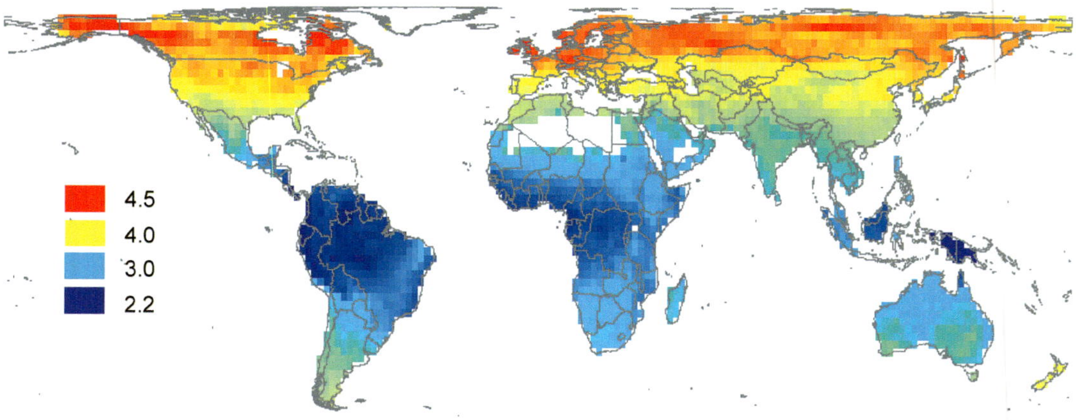


**Fig. 18.6** Geographic variation in clutch sizes for 5290 species of birds, with clutch sizes increasing with increasing latitude. (Figure from Jetz et al. 2008; © 2008 Jetz et al., open-access article distributed under the terms of the Creative Commons Attribution License, https://creativecommons.org/licenses/by/4.0/)

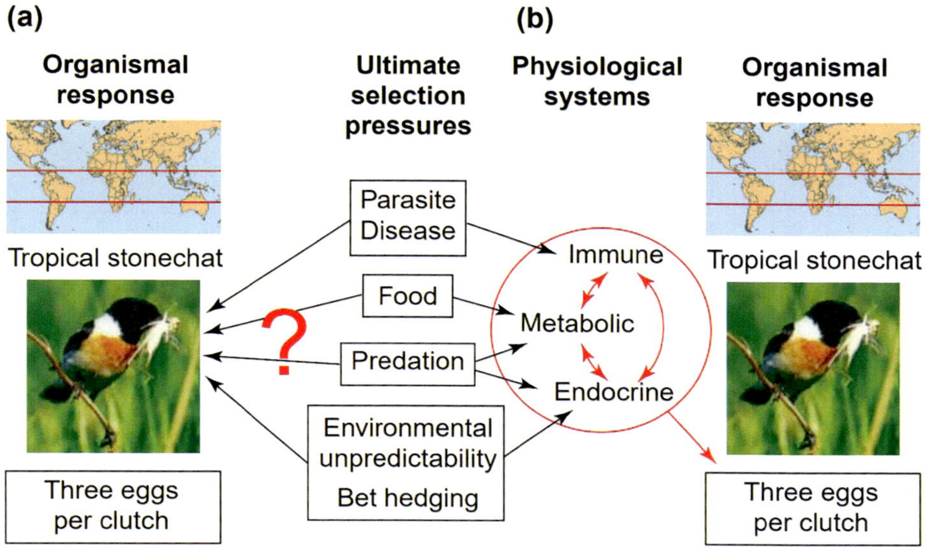


**Fig. 18.7** Seasonality, food supply, predation, and pathogens have all been suggested as contributing to the small clutch sizes typical of tropical passerines, but questions remain concerning the relative importance of these factors and how they might interact (**a**). Ricklefs and Wikelski (2002) provided a possible solution to this problem (**b**). A variety of environmental factors could act on separate physiological systems, each of which could influence the number of offspring parents can rear. For example, pathogens influence the immune system, food impacts the metabolic system, and environmental unpredictability might interact with the endocrine system. Because of internal physiological trade-offs between these systems, and because of system constraints, the ultimate outcome might be the same: a slow pace of life for tropical passerines, as indicated by smaller clutch sizes and longer lifespans. (Figure from Ricklefs and Wikelski 2002; © 2002 Elsevier Science Ltd., used with permission)

clutch sizes are favored in the tropics because there are more predators and predators are more likely to locate nests with larger clutches and more young. More young mean increased rates of food delivery by adults and a greater likelihood that, with the increased activity, predators will locate nests. High rates of nest predation may also select for smaller clutches to reduce the investment in any single nesting attempt (Slagsvold 1982; Martin 1995). Field studies have provided little support for either of these hypotheses (Martin et al. 2000b; Stutchbury and Morton 2008; Jahn et al. 2014).

Clutch sizes of birds occupying a seasonal environment where there is little seasonal variation in temperature are smaller than those of birds inhabiting environments where temperatures are more variable. Highly seasonal environments can cause increased adult mortality (Ricklefs 1997) because birds must survive periods of lower temperatures and reduced food availability or because birds must take risks associated with migration (Jetz et al. 2008). However, seasonal environments also have periods when food availability is very high. This combination of abundant food during the breeding season and increased mortality during the non-breeding period has favored the evolution of larger clutches in seasonal environments; larger clutches can be produced and more young raised to fledging because of the increased availability of resources, and producing more young enhances parental fitness because it increases the likelihood that some will survive the period of increased mortality during migration or, for resident birds, during periods of reduced food availability. The greater the degree of seasonality, or the greater the fluctuation in resource availability, the greater is the tendency for larger clutches. Seasonality increases with increasing latitude as do avian clutch sizes.

However, seasonality also varies in some cases with longitude. For example, western Europe is less seasonal than eastern Europe and, as expected given the effect of seasonality on clutch size evolution, birds in eastern Europe tend to have larger clutches than birds in western Europe (Lack 1954; Bell 1996). In addition, seasonal variation in resource availability for migrants is lower than for residents. Assuming that migration costs (and associated mortality) are not too high, migratory birds, especially long-distance migrants, might be expected to have smaller clutches than residents that overwinter (Böhning-Gaese et al. 2000; Bruderer and Salewski 2009; but see Jetz et al. 2008). In a study involving 45 species of birds (including 41 species of passerines from 11 families and four species of woodpeckers), Winger and Pegan (2021) found that ". . . clutch size scales weakly negatively with migratory distance, suggesting that annual fecundity may be reduced in long-distance migrants . . . [However], as our data were compiled from the literature and reflect central tendencies for clutch size and maximum values for brood number, field studies on the trade-off of brood size and success among individuals . . . and the frequency of double brooding . . . are required to understand the strength of the relationship between migration distance and annual fecundity within and among species."

Birds in tropical areas typically differ from those in temperate areas in a number of ways, with tropical birds generally exhibiting a slower "pace of life" (Ricklefs and Wikelski 2002; Wikelski et al. 2003). For example, tropical birds care for their young for longer periods (Styrsky et al. 2005). Young tropical birds typically have slower growth rates (Bryant and Hails 1983, but see Martin 2015), do not become mature as soon as birds in temperate areas (Russell et al. 2004), and tend to have higher survival rates (Peach et al. 2001; McNamara et al. 2008). Tropical birds also have lower metabolic rates (Wiersma et al. 2007). All of these life-history characteristics, as well as smaller clutches, may stem from the slow "pace of life" of tropical birds. What selective factors, however, have favored a slow pace of life?

Perhaps the most important factor is the higher survival rates (Wiersma et al. 2007). Lower extrinsic mortality in the tropics results in a lower reproductive effort, and adults that allocate more energy to self-maintenance. The increased self-maintenance in tropical birds manifests itself in a reduced pace of life, reflected in a decreased BMR that is correlated with an increase in survival. An important component of the increased self-maintenance would be greater investment in immune systems to fight infections and parasites and increase the probability of survival. Available evidence suggests that "slow-living" tropical birds do have more effective immune systems (specifically antibody-mediated defenses; Lee et al. 2008), and this is likely critical because tropical birds may be exposed to a greater variety of parasites and pathogens (in part because of longer lifespans). Selection for better immune systems might, in turn, explain other characteristics of many tropical birds. For example, the slower growth rates of young tropical birds would be beneficial because longer developmental periods translate into more effective immune systems (greater antibody diversity; Ricklefs 1992; Lee et al. 2008). Also, tropical birds usually have lower levels of testosterone than temperate species and selection may favor lower levels because testosterone may have immunosuppressive side effects (Garamszegi et al. 2008; Alonso-Alvarez et al. 2020). Thus, interactions between and among a variety of factors, including lifespan, prevalence of parasites and pathogens, and endocrine and immune systems, as well as reduced seasonality, appear to contribute to the smaller clutches of tropical birds.

Another hypothesis to explain the smaller clutches of tropical birds is that small clutches and broods allow increased provisioning rates per offspring which, in turn, results in the growth of longer wings at fledging and a corresponding reduced risk of predation (Martin 2015). Martin (2015) suggested that, although nestlings of tropical songbirds exhibit slower peak growth rates, the high provisioning rates of adults result in more sustained growth rates later in the nestling period and longer, more developed wings than those of nestlings of temperate species. Longer wings enhance mobility and likely increase the ability of fledglings to evade predators (Martin 2015).

## 18.3 Variation in Clutch Size Within Species and Populations

Clutch sizes also vary among individuals within species and populations. Several factors can contribute to such variation, including age, season, food availability, and predation pressure. For example, Saether (1990) reviewed studies where juvenile and later age classes were compared and found that adults laid larger clutches than juveniles in 92% of 48 species. However, clutch sizes do not always vary with age. For example, a long-term study of Song Sparrows (*Melospiza melodia*) revealed that clutch sizes did not change with female age (Nol and Smith 1987). In such species, other measures of breeding performance, such as hatching or fledging success, often do exhibit age-related variation.

For many species of birds, however, clutch sizes are smaller in younger birds, increase with age, but then decline in old age due to senescence. Three hypotheses have been raised to explain this phenomenon: (1) the experience (or constraints) hypothesis based on age-specific reproductive experience, (2) the effort (or trade-off) hypothesis based on age-specific reproductive effort, and (3) the selection hypothesis based on progressive disappearance of phenotypes due to variation in individual productivity and survival (Mauck et al. 2004). Of course, these hypotheses are not mutually exclusive. To varying degrees, any or all of these hypotheses might explain age-related variation in clutch sizes among different species of birds.

The experience, or constraints, hypothesis proposes that individuals improve their performance with increasing age. That is, greater parenting experience enables an increase in reproductive effort without an increase in costs (Curio 1983). This applies to all aspects of reproduction, including, for example, the ability to obtain high-quality mates or to adequately provision nestlings. However, experience may also influence clutch size and breeding performance because the number of eggs laid may be influenced by condition which is, in turn, influenced by foraging ability. The results of several studies seem to support the experience hypothesis (De Forest and Gaston 1996; Ratcliffe et al. 1998; Ludwigs and Becker 2004).

The effort, or trade-off, hypothesis predicts that current reproductive effort should be influenced by long-term reproductive interests, i.e., large clutches and broods may require so much investment (and risk) that adult survival (and, therefore, future reproduction) may be affected. This has been demonstrated in studies where clutches or broods are increased by investigators. For example, Sendecka et al. (2007) manipulated brood sizes of young (1 year old and breeding for the first time) and older (3 years old) female Collared Flycatchers (*Ficedula albicollis*) and found that the older female flycatchers coped with enlarged broods better than young females. Nestlings of the young females were smaller and lighter at fledgling, suggesting they had received less food. In addition, in the next breeding season, young females that had raised enlarged broods laid smaller clutches than other females, indicating that the young female birds paid a higher reproductive cost than the older females. Young females may have experienced a greater cost because they had less experience and were less skilled in activities such as foraging that affect their physical condition and, therefore, their reproductive success. However, regardless of the reason(s) why, the results of this study clearly indicate that the degree of investment in reproduction by young birds can, as predicted by the effort hypothesis, impact future reproductive success.

The selection hypothesis (Curio 1983; Newton 1989) proposes that phenotypic variation among individuals causes some individuals to have higher annual reproductive success than others. If these high-productivity individuals also have a higher probability of survival, then the proportion of highly productive individuals in a cohort will increase with age. Thus, average reproductive success will increase with breeding age as less-fit individuals disappear from the cohort. Because long-term studies are needed, there have been very few tests of the selection hypothesis. However, McCleery et al. (2008) examined the clutch sizes of Mute Swans (*Cygnus olor*) over 25 years and found that clutch sizes tended to increase up to the age of 6 or 7 and then decline after about the age of 12. The increase in clutch sizes with age in Mute Swans appeared to be due to both increased experience and, in support of the selection hypothesis, differential survival of higher-quality individuals. Although not focusing specifically on clutch size, Mauck et al. (2004) examined the relationship between age and reproductive (hatching) success in Leach's Storm-Petrels (*Hydrobates leucorhous*) and found that hatching success (percentage of eggs that hatch) increased dramatically between the first and fourth years of breeding, then leveled off (Fig. 18.8). As part of their study, the storm-petrels were captured and banded (individually numbered, aluminum bands were placed on one leg) so the identity and age of birds could be determined each year. An important prediction of the selection hypothesis is that successful breeders are more likely to survive than less successful breeders (Pärt 1995) so there should be a positive correlation between reproductive success of young birds and lifespan. Indeed, Mauck et al. (2004) found that less successful birds were less likely to survive than more successful birds (Fig. 18.9). Apparently, less successful breeders tend to die before reaching the age of 4 or 5 and,

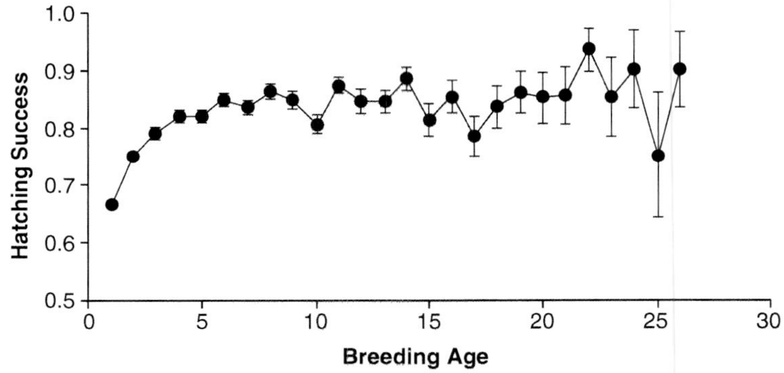


**Fig. 18.8** The relation between hatching success and breeding age in Leach's Storm-Petrels (*Hydrobates leucorhous*) on Kent Island, New Brunswick, Canada, 1962–1995 ($N = 18{,}347$ nests). (Figure from Mauck et al. 2004; © Society for the Study of Evolution, used with permission)

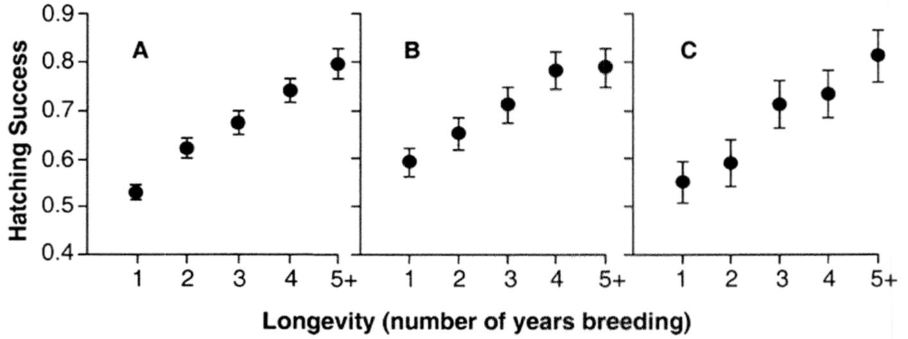


**Fig. 18.9** Relationship between hatching success during breeding years 1–2 and lifespan (number of breeding years) for Leach's Storm-Petrels (*Hydrobates leucorhous*) for (**a**) all individuals, (**b**) females, and (**c**) males. All birds breeding for at least 5 years were pooled in longevity-class 5+. Error bars represent the standard error. (Figure from Mauck et al. 2004; © Society for the Study of Evolution, used with permission)

in their absence, breeding success remains relatively high and stable.

Different studies with a variety of species have provided support for the experience (or constraints) hypothesis, the effort (or trade-off) hypothesis, and the selection hypothesis. For many species, as with Mute Swans, it is likely that multiple factors explain age-related variation in clutch size and reproductive performance. Additional long-term studies are needed to improve our understanding of the factors that influence age-related variation in clutch sizes and breeding success among different species of birds.

## 18.4 Predation and Clutch Sizes

Nest predation is the primary cause of reproductive failure for most birds. If birds can assess the risk of nest predation, selection should favor responses that tend to maximize individual fitness. For example, when and where predation risk is high, selection should favor reduced investment in reproduction, such as smaller clutches, so more energy is available for re-nesting (Slagsvold 1984) or, for single-brooded species, to enhance adult survival (i.e., less investment in a given breeding season increases the chances of surviving to breed in subsequent years). In addition, smaller clutches might be favored when predation risk is high because smaller broods will shorten the time period when a nest is susceptible to nest predators and reduce the number of nest visits that could attract the attention of predators (Skutch 1949; Martin et al. 2000a, b).

The results of several studies have revealed the effect of perceived predation risk on clutch size. Eggers et al. (2006) manipulated the perceived risk of nest predation for Siberian Jays (*Perisoreus infaustus*) by playing the calls of a variety of possible corvid nest predators over speakers in breeding territories. Beginning before the start of egg-laying, calls were played back every other day (one call every six minutes for several hours) and playbacks were stopped as soon as females began incubation. Females exposed to predator calls had significantly smaller clutches (mean = 3.3 eggs) than females exposed to playback of the songs of non-predatory birds (4.0 eggs) or females in territories where no playbacks were conducted (4.2 eggs). Because Siberian Jays normally do not re-nest after nest failure, these results lend support to either the enhanced survival hypothesis or the hypothesis that smaller clutches are beneficial when predation risk is high because fewer nestlings mean fewer provisioning visits by parents and a reduced likelihood that predators will locate a nest. Similar results, with clutch sizes smaller for individuals that experience an experimentally increased perception of predation risk (e.g., playing recordings of predator calls or placing caged predators near nests) than those with no such experiences have been reported for Song Sparrows (*Melospiza melodius*; Zanette et al. 2011), Eastern Bluebirds (*Sialia sialis*; Hua et al. 2014), and Red-faced Warblers (*Cardellina rubrifrons*; Dillon and Conway 2018). In addition, the results of a study of single-brooded Great Tits (*Parus major*), where clutch sizes were smaller in years after predation rates were higher than usual, support the enhanced survival hypothesis because clutch size had no effect on the risk of nest predation (Julliard et al. 1997).

## 18.5 Seasonal Variation in Clutch Sizes

Among birds that breed in temperate areas, clutch sizes generally vary during the breeding season. For single-brooded species, clutch sizes tend to decline as the breeding season progresses. For multi-brooded species, clutch sizes are typically largest at mid-season and smaller both earlier and later in the breeding season. A number of hypotheses have been proposed to explain the seasonal decline in clutch sizes of single-brooded species. The reproductive value hypothesis posits that the seasonal decline in clutch sizes may be due to a decline in the reproductive value of young, with young that hatch later in the breeding season less likely to survive and, therefore, less likely to contribute to the reproductive fitness of

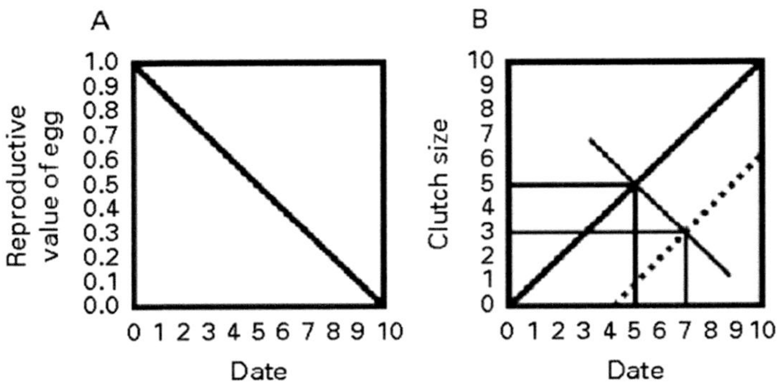


**Fig. 18.10** Hypotheses for the seasonal decline in clutch size at higher latitudes. (**a**) The reproductive value hypothesis proposes that clutch sizes decline seasonally because the probability of offspring recruitment (surviving until reproducing) declines with hatching date. (**b**) The female quality-reproductive value hypothesis proposes that attaining the physical condition needed to lay a clutch of a given size requires time. However, higher-quality females (thick solid line) lay more eggs earlier in the season, whereas lower-quality females (dashed line) lay smaller clutches later in the season. (Figure from Sockman et al. 2006; © Cambridge Philosophical Society, used with permission)

parents (Fig. 18.10a). Young that hatch and fledge later in the breeding season may be less likely to survive because they have less time to acquire the skills, such as foraging ability, needed to survive the non-breeding season.

The adult survival (or breeding time) hypothesis proposes that adults that invest heavily in reproduction late in a breeding season may have less time and energy to devote to other activities, such as molting and preparing for winter or migration, and, as result, may be less likely to survive. If so, natural selection would favor reduced investment in reproduction later in the season, including smaller clutches.

The female quality-reproductive value hypothesis (Rowe et al. 1994) proposes that clutch sizes decline later in the breeding season due to the combined effects of later breeding by lower quality females and the reduced reproductive value of young produced later in the breeding season (Fig. 18.10). This hypothesis assumes that higher-quality females in better condition can produce larger clutches earlier in the breeding season. Lower-quality females, on the other hand, require a longer period to attain the energetic condition to produce a clutch. However, this delay in the onset on clutch initiation also means that any resulting young will have lower reproductive value and, therefore, selection would favor smaller clutches for such lower-quality, later-breeding females. As a result, earlier clutches would tend to be larger, with clutch sizes progressively declining as the breeding season progressed.

Finally, the egg-viability hypothesis (Veiga 1992; Stoleson and Beissinger 1999) proposes that clutch sizes decline later in the breeding season because the earliest-laid eggs in a large clutch are, later in the breeding season, more likely to be exposed to higher temperatures prior to the onset of incubation than those in clutches laid earlier in the season. Eggs exposed to temperatures between about 26 and 36°C prior to the onset of incubation (with 36°C a temperature typical of eggs being incubated) can experience unsynchronized tissue growth, abnormal development, and mortality (Deeming and Ferguson 1992).

The female quality-reproductive value hypothesis has been tested for Snow Geese (*Anser caerulescens*), and results generally supported the hypothesis (Bêty et al. 2003). However, achieving some minimal body condition may not be necessary before starting egg laying in small songbirds (Williams 1996) so this hypothesis may not be relevant for many species. A

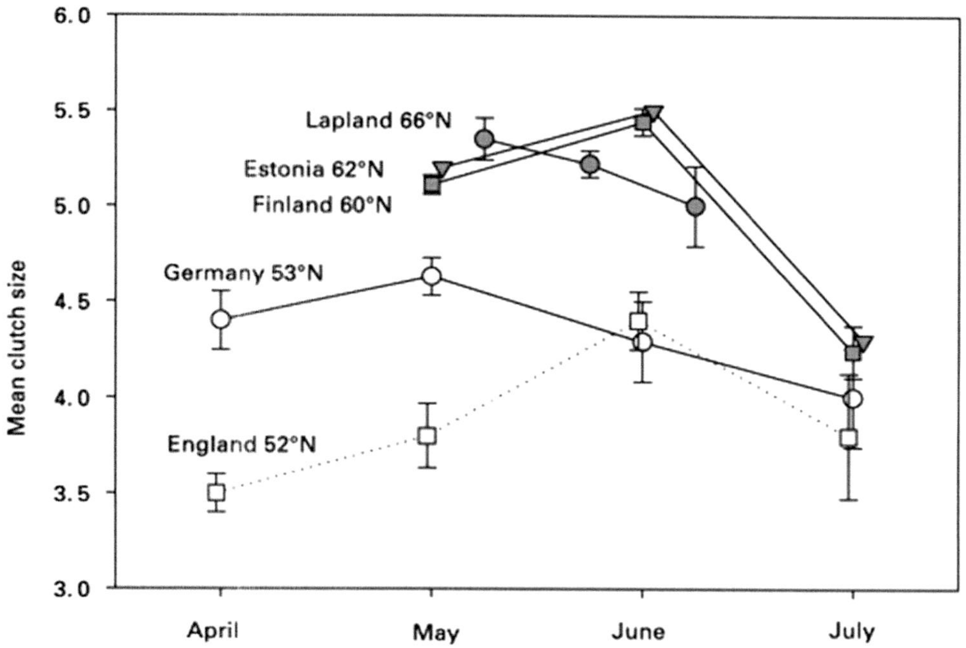


**Fig. 18.11** Mean (± SE) monthly clutch sizes of Dunnocks (*Prunella modularis*) at five geographical regions and different latitudes. Filled symbols represent migratory populations; open symbols represent resident populations. (Figure from Dhondt et al. 2002; © British Ornithologists' Union, used with permission)

review of experiments conducted with several different species of birds where the timing of breeding was manipulated revealed that both breeding time and parental quality were important contributors to the seasonal decline in clutch sizes. However, breeding time appeared to be a more important factor for most species of birds (Verhulst and Nilsson 2008).

For multi-brooded species, clutch sizes may be largest at mid-season and smaller both earlier and later in the breeding season. However, the relationship between clutch size and season tends to vary with latitude. For multi-brooded species at higher latitudes, clutch sizes decline throughout the breeding season rather than increasing then decreasing (Fig. 18.11). The decline in clutch sizes after mid-season in multi-brooded species, and for multi-brooded species at higher latitudes, may, as for many single-brooded species, best be explained by the adult-survival, or breeding time, hypothesis. One possible explanation for smaller clutches early in the breeding season is that conditions, such as food availability, may place energetic limits on the number of eggs a female can produce early in the breeding season. Improved conditions later in the breeding season (mid-season) allow females to produce larger clutches (Crick et al. 1993). Differences in patterns of clutch size variation for multi-brooded species at different latitudes may be due to differences in the timing of resource availability. Maximum resource abundance is reached sooner after the onset of spring (i.e., springs are more "rapid") further north, favoring larger clutches in early nests at higher latitudes (Dhondt et al. 2002).

## 18.6 Evolution of Nest Attendance/Incubation

Incubation is the process by which the temperature of eggs is maintained at levels suitable for embryonic development. During incubation,

**Fig. 18.12** (**a**) Fossil Oviraptor positioned over its eggs in a manner similar to that of some present-day birds. The skull, vertebrae, tail, and dorsal pelvic bones are missing, but the forelimbs and hindlimbs are clearly visible along with several eggs. (**b**) Drawing of an Oviraptor on its nest. (Figure **a** from Norell et al. 1995; © 1995 Springer Nature, used with permission. Figure **b** from Bi et al. 2021; © 2021 Science China Press. Published by Elsevier B.V. and Science China Press, used with permission)

adult birds must not only transfer heat to eggs, but maintain an appropriate humidity and turn the eggs on a regular basis. Little is known about the evolution of avian incubation behavior. However, the discovery of fossilized eggs below the skeleton of an adult oviraptorid suggests that these dinosaurs may have incubated eggs (Dong and Currie 1996; Clark et al. 1999; Tanaka et al. 2018; Bi et al. 2021, but see Yang et al. 2019; Fig. 18.12). Amiot et al. (2017) reported evidence, based on analysis of the oxygen isotope composition of fossilized eggshells and the bones of embryos, that oviraptors incubated their eggs at temperatures within a range of 35–40°C. Bi et al. (2021) reported similar results, with an estimated incubation temperature of 30–38°C for an oviraptorid. These incubation temperatures are similar to those of present-day birds. Bi et al. (2021) also found that developing embryos in a fossilized clutch of eggs were at different stages of development, suggesting that the adult oviraptorid had tended to the nest for an extended period of time and that, had they survived, the eggs may have hatched asynchronously.

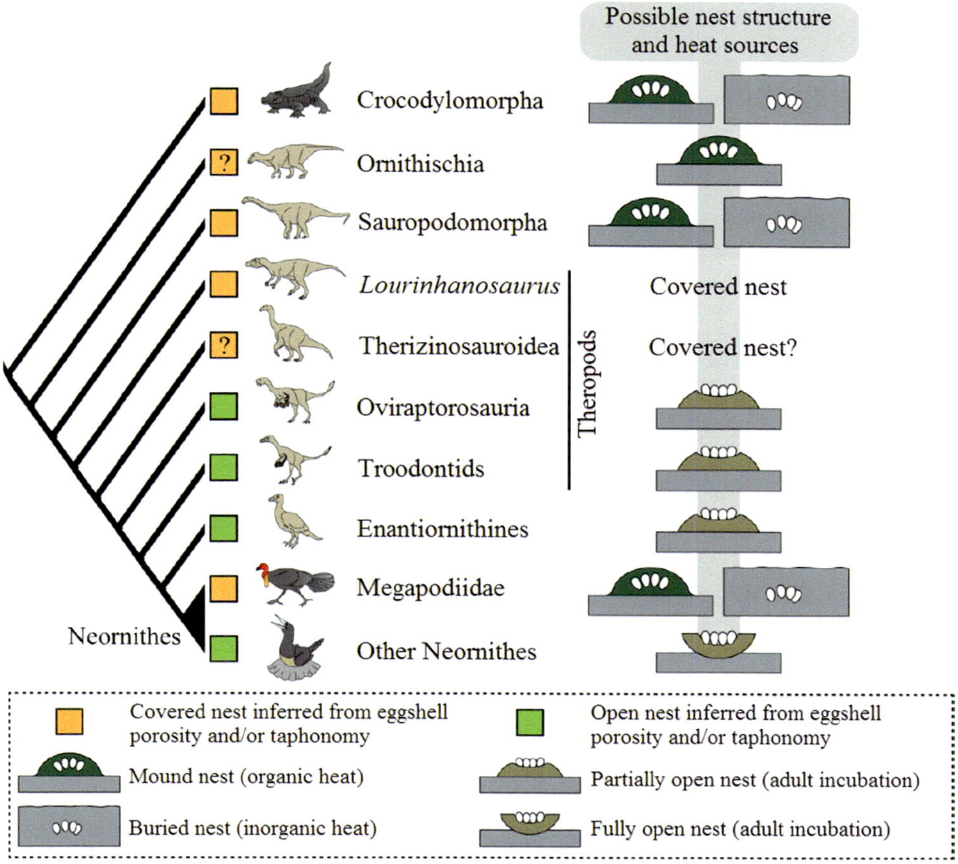


**Fig. 18.13** Phylogeny showing nest types and sources of heat for incubating eggs. Available evidence suggests that both oviraptors and troodontids incubated their eggs in the same manner as present-day birds. (Figure modified from Tanaka 2016; Tanaka et al. 2018; open-access article licensed under a Creative Commons Attribution 4.0 International License, https://creativecommons.org/licenses/by/4.0/)

Varricchio et al. (2018) used the growth rate of an embryonic tooth of *Troodon formosus* to estimate an incubation period of 74 days, an incubation period midway between estimated avian (44.4 days) and reptilian (107.3 days) values predicted by *Troodon* egg mass. This intermediate value suggests less efficient brooding than in present-day birds, perhaps because *Troodon* clutches were largely buried (Varricchio et al. 2018). Although based on indirect evidence, including fossils of adults on eggs and eggshell porosity (Tanaka et al. 2015; Figs. 18.13 and 18.14), other investigators have also suggested that troodontids (Troodontidae; a small group of rare and poorly known maniraptorans) may have incubated their eggs (Grellet-Tinner et al. 2006). Although questions clearly remain, available evidence suggests that caring for, and incubating, eggs are traits that were likely shared by ancient birds and their immediate ancestors.

Although caring for and incubating eggs likely preceded the origin of birds, less apparent is the question of whether such behavior first involved female-only care, male-only care, or biparental care. Given that most living birds exhibit biparental care, one possible scenario is that biparental care was the ancestral condition for birds (Birchard et al. 2013). Others have suggested

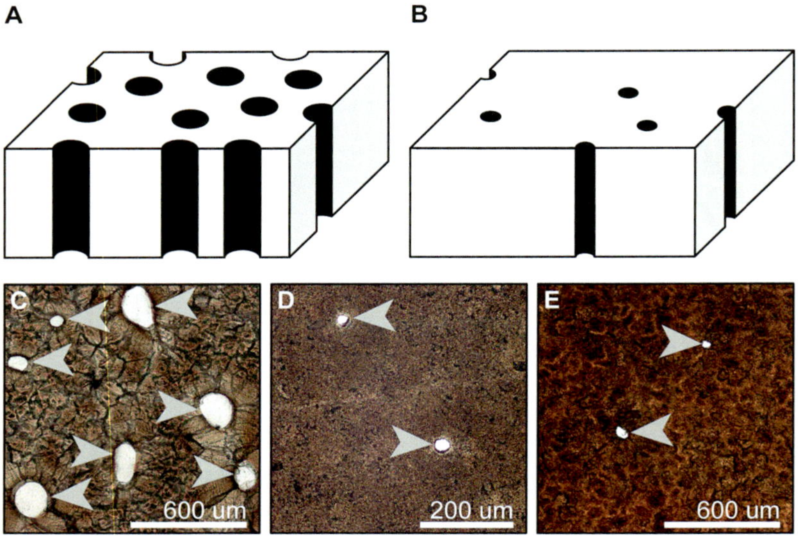


**Fig. 18.14** Eggshell porosity, as measured by dividing the total pore area of eggs by pore length, has been found to be correlated with nest type, with eggs incubated in covered nests having a significantly higher eggshell porosity than those incubated in open nests. Diagrams of eggshells with (**a**) high porosity and (**b**) low porosity, plus small sections of eggshells of (**c**) a broad-snouted caiman (*Caiman latirostris*), an extant covered-nester, (**d**) an Indian Peafowl (*Pavo cristatus*), an extant open-nester, and (**e**) *Troodon formosus*, a non-avian maniraptoran. Arrows point to pores. (Figure modified from Tanaka et al. 2015; © 2015 Tanaka et al., open-access article distributed under the terms of the Creative Commons Attribution License, https://creativecommons.org/licenses/by/4.0/)

that female-only care is the ancestral condition because, among the few living ectotherms that provide parental care (snakes, crocodilians, and lizards), care is provided almost universally by females, with attendance of unhatched eggs the most common form of parental care (Burley and Johnson 2002). In addition, because females are more certain of maternity than males of paternity and the high degree of anisogamy in birds and their ancestors, female-only care might be more likely to evolve first (Burley and Johnson 2002). Finally, still other authors have described scenarios where male-only care was the ancestral condition, with ancient birds possibly behaving in a manner similar to some living palaeognaths (ratites and tinamous) that exhibit male-only care.

A hypothesis favored by some investigators is that male-only care is the ancestral condition in birds (Fig. 18.15). Under one scenario (described in more detail by Burt et al. 2007), ectothermic males defended resource-rich territories that attracted females who, after mating with a resident male, deposited eggs in male-prepared depressions in the ground that were then buried in sand or soil. Because theropods thought by many to be most closely related to birds laid large eggs, ancestral birds may have done so as well. The advantage of larger eggs is that young are larger and more developed at hatching (precocial or even superprecocial) and likely less susceptible to predation. For obvious anatomical reasons, the evolution of larger eggs would likely favor the evolution of sequential laying (one egg in one oviduct at a time), rather than simultaneous (two eggs in two oviducts at a time), laying. Sequential laying would provide females with two other potential benefits: (1) more time to forage and acquire energy and resources for each egg before it was laid, and (2) more opportunities to deposit eggs in the nests of multiple males, a strategy that would increase the likelihood of some eggs hatching (i.e., not all

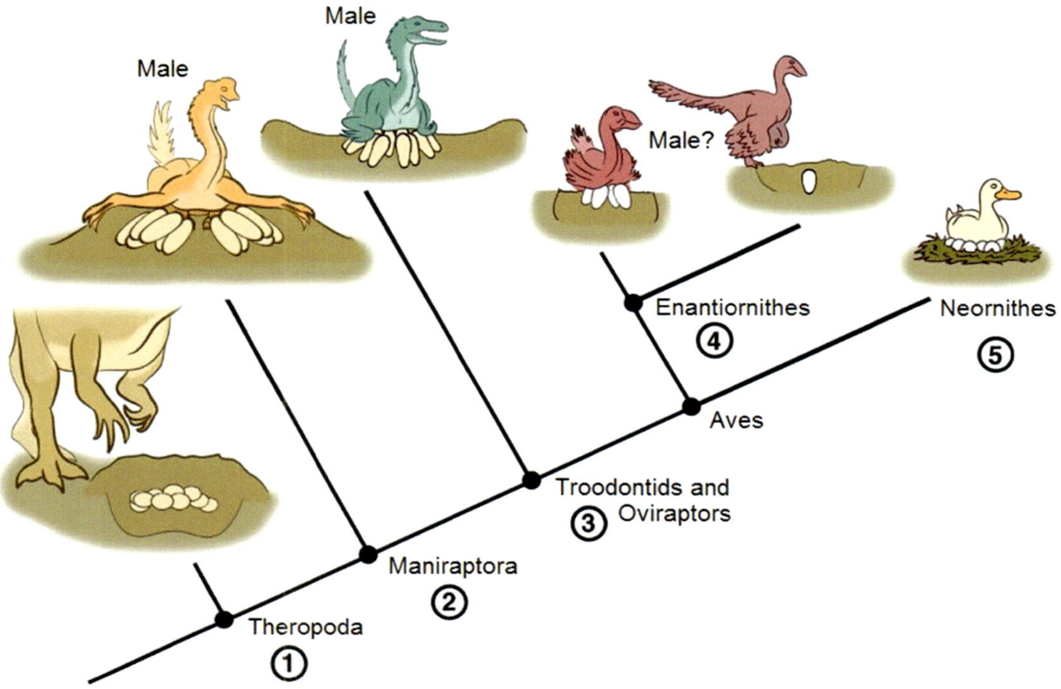


**Fig. 18.15** Simplified phylogeny showing hypothesized stages in the evolution of reproductive traits toward modern birds. Stages 1 and 2: large and highly organized clutches, incubation involving nearly full burial with attendant adult, possibly paternal care. Stage 3: Troodontids and oviraptors: male caring for clutches of eggs, improved contact incubation with tighter clutch configuration and exposed upper portions of eggs. Stage 4, Enantiornithes: incubation either as in troodontids and oviraptors or as single eggs fully buried in sandstone. Stage 5, basal Neornithes: clutch free of sediment cover, egg rotation, chalazae with potentially greater incubation efficiency, and incubation by either sex. (Figure modified from Varricchio and Jackson 2016; © 2016 Oxford University Press, used with permission)

eggs were in one "basket"). However, acquiring resources and depositing eggs in multiple nests would make it impossible for females to guard all their eggs or nests. Although this may be a reasonable scenario from the female perspective, what selective pressures would favor male care of nests and, perhaps, young?

To assess parental care in Cretaceous troodontid and oviraptorid dinosaurs, Varricchio et al. (2008) and Moore and Varricchio (2016) examined clutch volumes (total volume of all eggs in a clutch) and found that the relatively large clutch volumes of *Troodon, Oviraptor*, and *Citipati* were similar to those of living paleognaths with polygamous mating systems and extensive male care. In addition, Moore and Varricchio (2016) noted that clutch volumes tend to increase as the degree of paternal investment in clutches increase among diapsids (reptiles and birds; but see Birchard et al. 2013). Clutch volumes can potentially evolve to be larger in species without maternal care because females may have more resources to devote to eggs if they provide no care.

If both troodontids and oviraptorids exhibited paternal care, then this care system evolved before the emergence of birds and represents the ancestral condition for birds. If so, the biparental care that is so prevalent among living birds represents a derived condition. Although several investigators support the hypothesis that male-only care was the ancestral condition for birds, additional study, and the discovery of new fossils, are still needed to help determine which hypothesis is most likely correct.

## 18.7 Incubation

Among living birds, the temperature of developing embryos must be maintained between about 32 and 38°C for optimum development (Webb 1987; Rahn 1991; Deeming 2008; Fig. 18.16). Exposure to higher temperatures can be lethal, whereas cooler temperatures will, at minimum, slow down or stop development. Maintaining those temperatures typically requires either transfer of heat from an adult bird or actions that help prevent overheating (such as shading eggs; Fig. 18.17), but many other factors also play a role (Fig. 18.18). Ambient conditions (temperature, wind velocity, and precipitation), of course, affect egg temperatures, as do the characteristics of nests (see Chap. 17). Nests and nest sites (e.g., cavities) can provide insulation that reduces the amount of heat lost or gained by eggs and

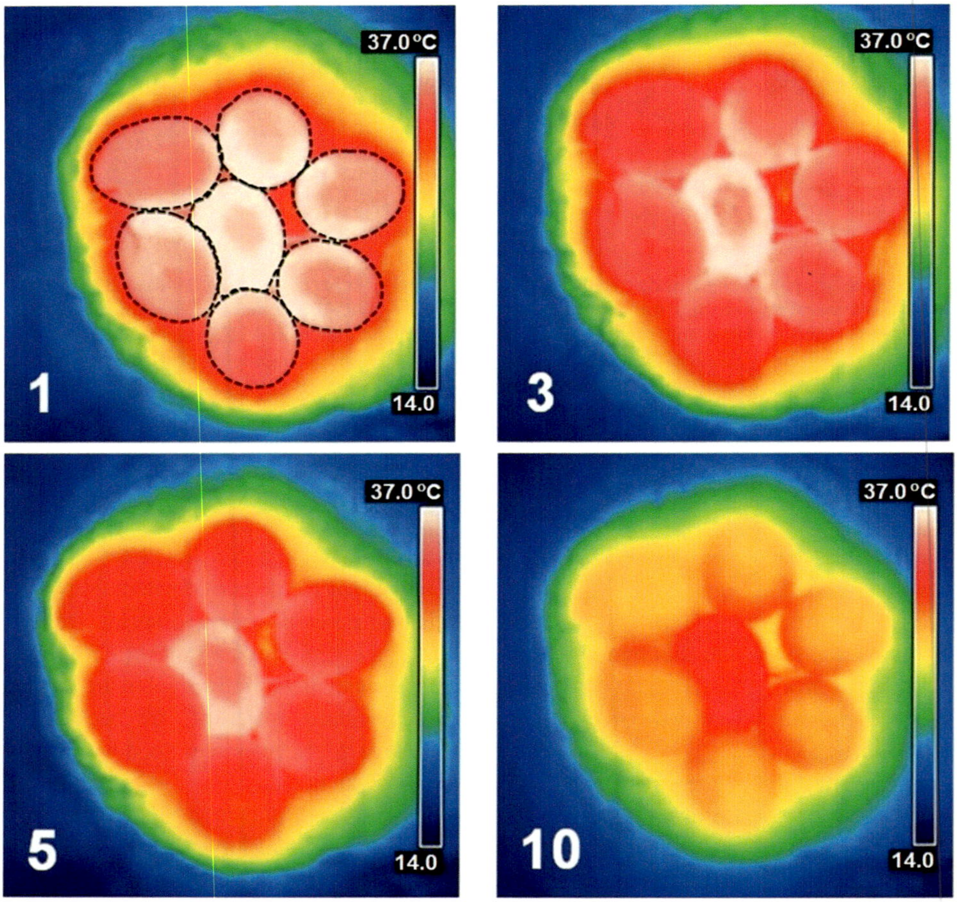


**Fig. 18.16** Thermal images of a clutch of eggs of a Great Tit (*Parus major*) taken at 1, 3, 5, and 10 min, respectively, after the beginning of an off-bout. Note that egg temperatures are about 36 or 37°C after 1 min and cool to a range of about 29–32°C after 10 min. Note also that the position of eggs in a clutch makes a difference, with the central egg remaining warmer than the surrounding eggs. (Figure from Boulton and Cassey 2012; © 2012 The Authors, used with permission)

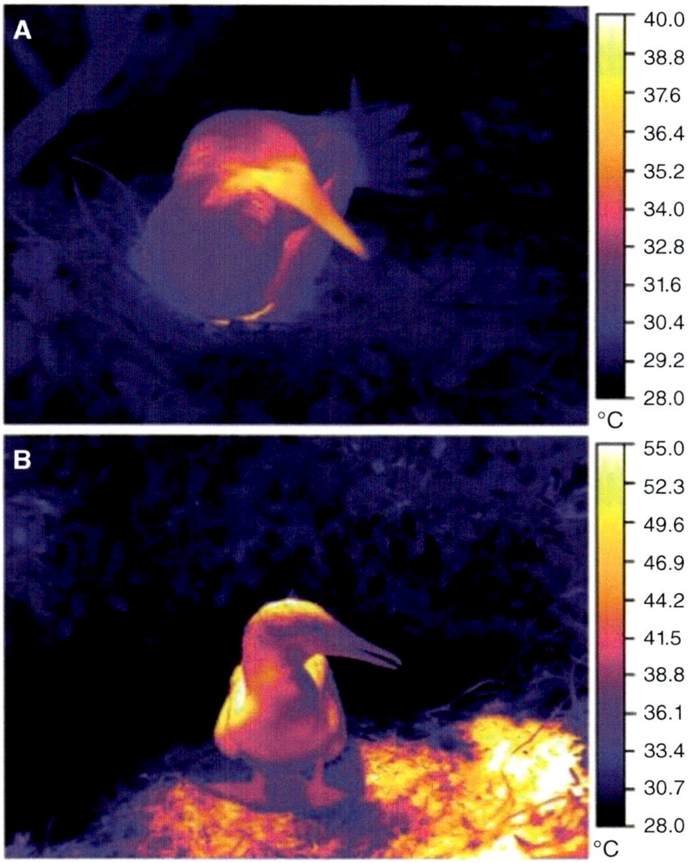


**Fig. 18.17** Birds typically must apply heat to eggs to maintain developing embryos at the appropriate temperature. If nests are exposed to full sunlight, however, adults may need to shade the eggs to prevent temperatures from getting too high. For example, (**a**) Brown Boobies (*Sula leucogaster*) with nests in the shade incubate their eggs, but (**b**) when nests are in full sunlight, adults stand over the eggs to shade them. Note how warm the head and back of the Brown Booby is in (**b**). The slightly open bill suggests this booby is likely panting to increase the rate of evaporative heat loss. When these thermal images were made, the air temperature was 28°C. (Figure from Tattersall and Cadena 2010; Rights managed by Taylor & Francis, used with permission)

incubating adults (Fig. 18.19), but adult birds are responsible for regulating egg temperature during incubation.

Most birds develop a brood, or incubation, patch during the breeding season. These patches form on the ventral thoracic and abdominal region and are characterized by the absence of feathers, the thickening and swelling of skin, and increased blood flow (Fig. 18.20). These changes permit improved contact between the adult and the egg(s) and facilitate the transfer of heat (Table 18.1). Many species of birds, including owls (Strigiformes), galliforms, and some passerines, have large brood patches (relative to their size) that permit incubation of several eggs. In other species, such as gulls and terns (Laridae), there is a separate brood patch for each egg. Generally, the adult that incubates (female, male, or both; Fig. 18.21; Box 18.3 The Bright Incubate at Night) develops a brood patch. Among seabirds (Procellariiformes), grebes (Podicipediformes), pigeons and doves (Columbiformes), woodpeckers (Piciformes), most shorebirds (Charadriiformes), cranes and rails (Gruiformes), and some songbirds (Passeriformes), both males and females generally develop brood patches and incubate eggs. In several other groups, including gallinaceous birds (Galliformes), owls (Strigiformes), hummingbirds and swifts (Apodiformes), some raptors (Falconiformes), and most songbirds (Passeriformes), only females develop brood patches. Finally, in phalaropes and some sandpipers (Scolopacidae), jacanas (Jacanidae), tinamous (Tinamiformes), and many ratites (Struthioniformes), only males develop brood patches and incubate (Jones 1971). In a few species, including Common Ostriches (*Struthio*

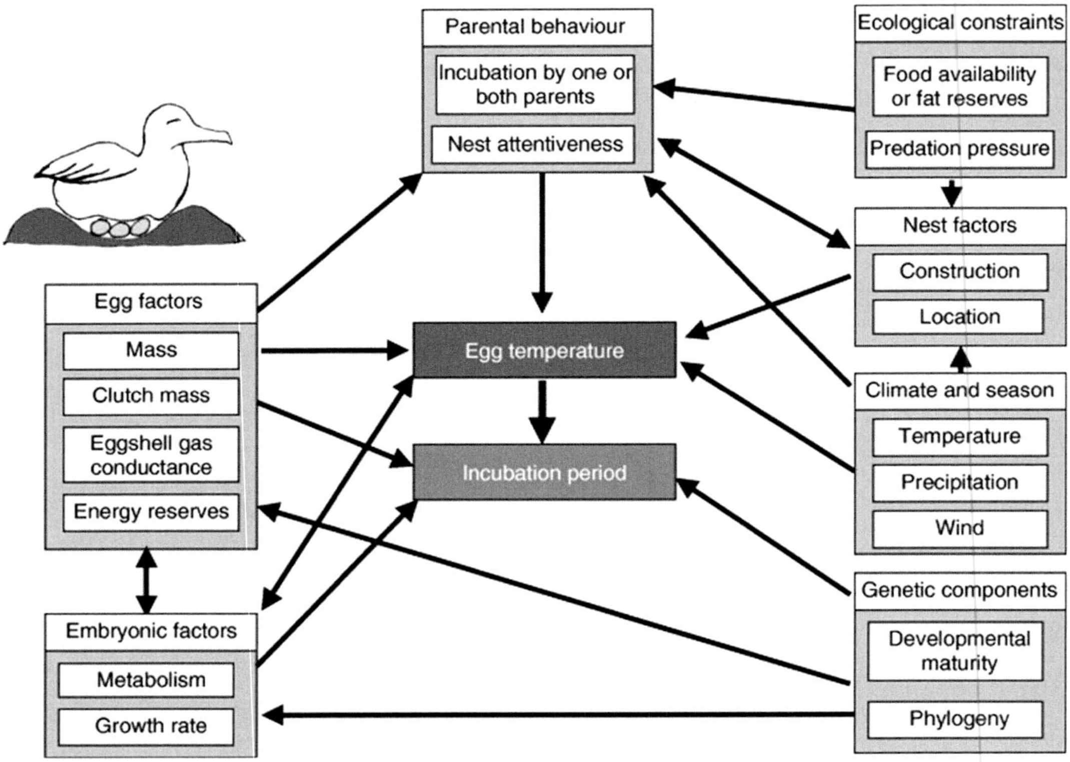


**Fig. 18.18** Several factors affect egg temperature and the duration of incubation periods in birds. (Figure from Deeming et al. 2006; © 2006 John Wiley and Sons, used with permission)

*camelus*), eggs are incubated, but no brood patch develops. There are also species, such as Red-billed Firefinches (*Lagonostica senegala*), where both sexes share incubation duties, but only females have brood patches (Payne 1980). Interestingly, males in some species have been found to maintain eggs at mean temperatures as high as females even though they lack brood

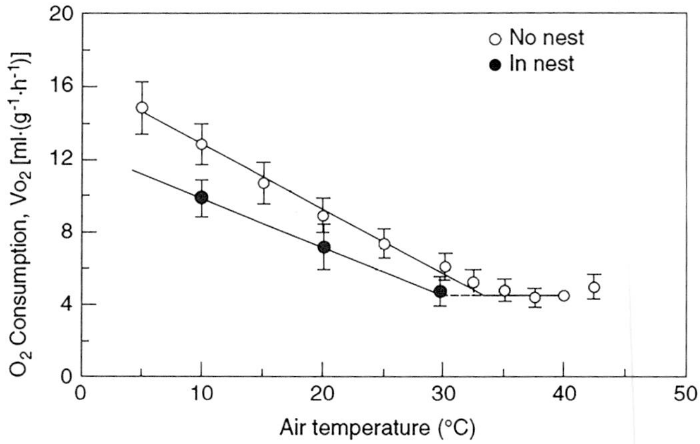


**Fig. 18.19** Rates of oxygen (energy) consumption (± SD) when resting inside (closed circles) and outside (open circles) of nests for a Palestine Sunbird (*Cinnyris osea*) relative to air (ambient) temperature. Note that, with the insulation provided by the nest, energy consumption is lower when sunbirds are in nests. (Figure from Ar and Sidis 2002; © 2002 Oxford University Press, used with permission)

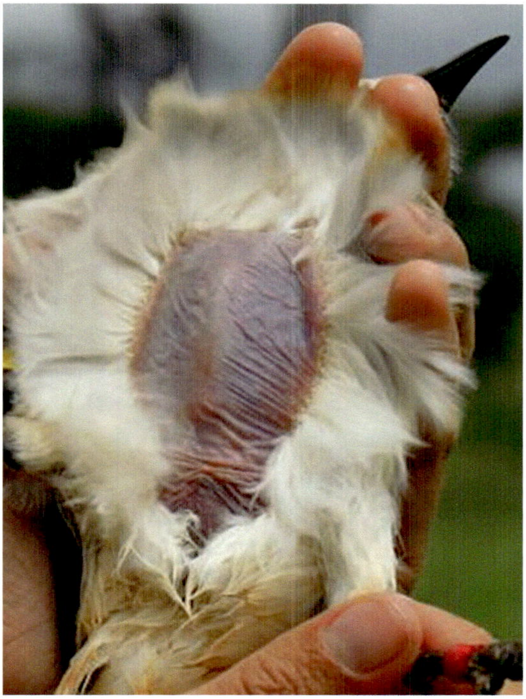

**Fig. 18.20** Birds of nearly all species temporarily shed their feathers on single or paired areas of the breast or abdomen early in the breeding season. The bare skin increases in vascularity, aiding in the transfer of heat for incubating the eggs and brooding the chicks. Development of incubation (brood) patches is prompted by rising levels of estrogen. They form in whichever sex cares for the eggs and young, usually females but often males as well. The lost feathers are replaced in the complete molt following the breeding season (Stettenheim 2000). (Photo from Jones 2013. used with permission of Samuel Jones)

patches (Ball 1983), and male Chestnut-vented Warblers (*Curruca subcoerulea*) without brood patches were able to maintain higher egg temperatures than females that have brood patches (Auer et al. 2007). To do so, males may have to expend more energy than females because heat transfer would likely be less efficient without brood patches (Auer et al. 2007) and, as a result, in species where males without brood patches incubate eggs, they may only do so during the day when ambient temperatures are higher; only females with brood patches able to transfer heat more efficiently incubate during periods of thermal stress, e.g., at night when ambient temperatures are lower (Ball 1983; Auer et al. 2007). Finally, brood parasites, such as cuckoos and cowbirds (*Molothrus* spp.) do not incubate eggs and so do not develop brood patches.

Experimental administration of hormones has revealed that the development of incubation patches is under hormonal control and generally involves an interaction between the hormones prolactin, estrogen, and, less importantly, progesterone. Treatment with estrogen in combination with prolactin or progesterone induces brood patch development, with estrogen possibly more important for the development of blood vessels and prolactin for the loss of feathers (Bailey 1952; Massaro et al. 2007). Neural receptors in the skin of the brood patch sense and help regulate egg temperature by varying blood flow and heat output (Brummermann and Reinertsen 1991). Later in incubation, embryos of some bird species emit nitric oxide through the shell (Ar et al. 2000), and this may enhance the warming of the embryo by increasing blood flow to the brood patch (Ar and Sidis 2002). Brood patch temperatures do vary among species. Deeming (2008) obtained brood patch temperatures (BPT) of 76 species of birds from the literature and found that BPT was strongly influenced by phylogeny (Order) and size, with smaller birds having higher BPT than larger birds (Fig. 18.22). These phylogenetic and size-related differences, however, were primarily due to developmental mode, with some orders of birds and most smaller birds having altricial young and other orders and most larger birds having semi-precocial or precocial young (Fig. 18.23). For most developmental modes, higher brood patch temperatures also resulted in shorter incubation periods (Fig. 18.23). For species in the order Procellariiformes, those with higher brood patch temperatures had longer incubation periods, but this relationship was largely driven by the large body sizes (about 2990 g) and relatively long incubation periods (about 65 days) of two species of albatrosses, i.e., Laysan (*Phoebastria immutabilis*) and Black-footed (*P. nigripes*) albatrosses (Fig. 18.23).

Although most birds develop brood patches for incubating eggs, there are exceptions. For example, pelicans (Pelecanidae), most gannets

**Table 18.1** Examples of brood patch temperatures (BPT) and body temperatures (BT) of different species of birds

| Order/species | Scientific name | BPT (°C) | BT (°C) | Reference[a] |
|---|---|---|---|---|
| *Struthioniformes* | | | | |
| Common Ostrich | *Struthio camelus* | 38.0 | – | 1 |
| *Sphenisciformes* | | | | |
| Emperor Penguin | *Aptenodytes forsteri* | 36.1 | 36.6 | 2 |
| King Penguin | *Aptenodytes patagonica* | 38.0 | – | 3 |
| *Procellariiformes* | | | | |
| Leach's Storm-Petrel | *Oceanodroma leucorrhoa* | 35.8–37.4 | 37.9 | 4, 5 |
| Bonin Petrel | *Pterodroma hypoleuca* | 34.9 | – | 6 |
| Wedge-tailed Shearwater | *Puffinus pacificus* | 37.8 | 39.5 | 7 |
| *Anseriformes* | | | | |
| Mallard | *Anas platyrhynchos* | 39.5 | 41.0 | 8 |
| Emperor Goose | *Anser canagicus* | 40.9 | – | 9 |
| Barnacle Goose | *Branta leucopsis* | 37.7 | – | 10 |
| *Galliformes* | | | | |
| Ring-necked Pheasant | *Phasianus colchicus* | 39.6 | 41.8 | 11 |
| *Charadriiformes* | | | | |
| Parakeet Auklet | *Aithia psittacula* | 36.9 | 40.4 | 12 |
| Great Skua | *Catharacta skua* | 39.3 | 41.2 | 13 |
| *Strigiformes* | | | | |
| Barn Owl | *Tyto alba* | 39.3 | 40.8 | 14 |
| *Piciformes* | | | | |
| Great Spotted Woodpecker | *Dendrocops major* | 40.7 | – | 15 |
| *Passeriformes* | | | | |
| Dusky Flycatcher | *Empidonax oberholseri* | 42.4 | – | 16 |
| Spotted Flycatcher | *Muscicapa striata* | 40.9 | – | 15 |
| Eurasian Nuthatch | *Sitta europea* | 41.2 | – | 15 |
| Dunnock | *Prunellus modularis* | 41.9 | – | 15 |
| Song Thrush | *Turdus philomelos* | 40.0 | – | 15 |

Table modified from Deeming (2008); © 2008 Elsevier Ltd., used with permission
[a]References: (1) Rahn (1991), (2) Prevost (1963), (3) Stonehouse (1960), (4) Whittow (1980), (5) Drent (1975), (6) Grant et al. (1982), (7) Howell and Bartholomew (1961), 8 Afton (1979), (9) Krechmar and Kondratiev (1982), (10) Rahn et al. (1983), (11) Westerkov (1956), (12) Manuwal (1974), (13) Stonehouse (1956), (14) Howell (1964), (15) Deeming and du Feu (2008), (16) Morton and Pereyra (1985)

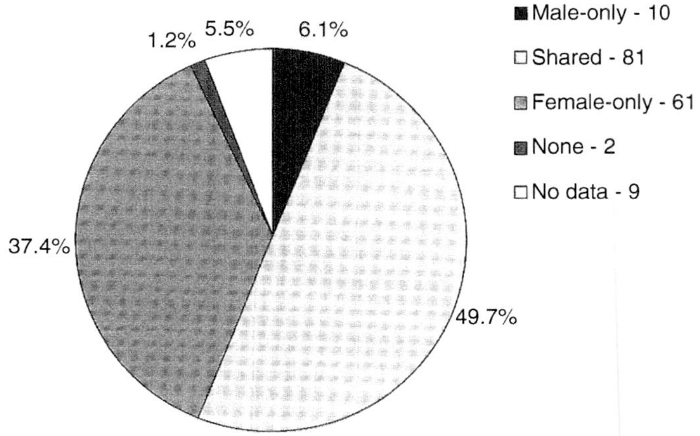


**Fig. 18.21** Distribution of different patterns of incubation among different families of birds. For most families, adults either share incubation duties or only females incubate eggs. (Figure from Deeming 2002b; © 2002 Oxford University Press, used with permission)

and boobies (Sulidae), and tropicbirds (Phaethontiformes) position their single eggs within the webbing of their feet during incubation. The egg is kept warm by heat from their feet (Fig. 18.24). Morgan et al. (2003) studied heat transfer from foot webs to eggs by Nazca Boobies (*Sula granti*) by spatially separating the feet from the abdomen using an oversized egg. They found that the feet, not the abdomen, transferred the most heat to eggs and, in addition, that incubating boobies had significantly greater vascularization in their foot webs than non-incubating boobies. These results indicate that the feet of Nazca Boobies, and likely other boobies and gannets, function just like brood patches. In addition, King (*Aptenodytes patagonicus*) and Emperor (*A. forsteri*) penguins do not build nests and use both their feet and an abdominal flap of feathered skin that functions as a brood pouch to incubate their single eggs (Williams 1995). During incubation, King Penguins keep their egg on their feet and cover it with their brood patch, generally keeping egg temperature between 28 and 37°C (Groscolas et al. 2000).

Megapodes (Family Megapodiidae) leave incubation of their eggs to external heat sources, with burrow nesters relying on geothermal heat in burrows on volcanic islands and mound nesters on heat produced by microbial decomposition of organic material in mounds of leaf litter (Fig. 18.25). Members of both groups also use radiant heat from the sun for incubation (Booth and Jones 2002). The incubation mounds of most megapodes generally generate suitable incubation temperatures as long as sufficient organic material is incorporated into the mound (Palmer et al. 2000). The preferred incubation temperature for most megapodes is 31–35°C. As they develop, embryos begin to generate metabolic heat, and egg temperatures may be 2–4°C above mound temperature (Seymour 1991).

Constructing and maintaining incubation mounds may involve extensive effort. Mounds may reach 20 m in diameter, two meters in height, and contain more than 6000 kg of material (Sekercioglu 1999). In the vegetation mounds, heat is generated by the respiration of microorganisms, and megapodes are thought to monitor mound temperatures using heat-sensitive areas on the necks, heads, and inside their mouths. If temperatures in a mound change, litter can be added or removed or ventilation holes can be created (Jones and Birks 1992).

Manipulation of mounds is generally minimal once mound temperature has stabilized. An exception to this is the Malleefowl (*Leipoa ocellata*) where more manipulation is required because they use a combination of solar heat

**Box 18.3 The Bright Incubate at Night**

Many species of birds are sexually dichromatic, with one sex (usually males) more ornamented than the other. Sexual selection has favored such ornamentation because of its important role in intrasexual competition, mate choice, or both. However, brighter plumage may also have a cost, potentially making birds more visible and easily detected by predators. Among species where both sexes provide parental care—incubating eggs and brooding young—the more brightly colored sex may be more likely to attract the attention of predators, potentially increasing the risk of nest predation. Most shorebirds are ground-nesters and, in many species, both sexes incubate, including the sometimes more brightly colored males. In some of these species, including Red-capped Plovers (*Charadrius ruficapillus*) and St. Helena Plovers (*C. sanctaehelenae*), selection has favored a diurnal/nocturnal division of labor, with the more cryptic females (pale gray-brown crown) incubating more during the day and the more brightly colored males (rufous crown) at night. Under the cover of darkness, the brighter colors of males are not visible to predators, minimizing the risk of predation (Ekanayake et al. 2015).

(continued)

**Box 18.3** (continued)

Top left, Male Red-capped Plover. Top right, female Red-capped Plover. Bottom, Mean probability of male (dashed) and female (solid) (± SE) Red-capped Plovers attending nests throughout a 24-h cycle ($N = 12$ nests). Background color indicates night (dark periods in dark gray) and day (light periods in light gray) periods. (Figure modified from Ekanayake et al. 2015; © 2015 The Authors. Published by the Royal Society, used with permission. Photo of male Red-capped Plover by Patrick_K59, Wikipedia, CC BY 2.0, https://creativecommons.org/licenses/by/2.0/; photo of female Red-capped Plover by Ed Dunens, Wikipedia, CC BY 2.0, https://creativecommons.org/licenses/by/2.0/)

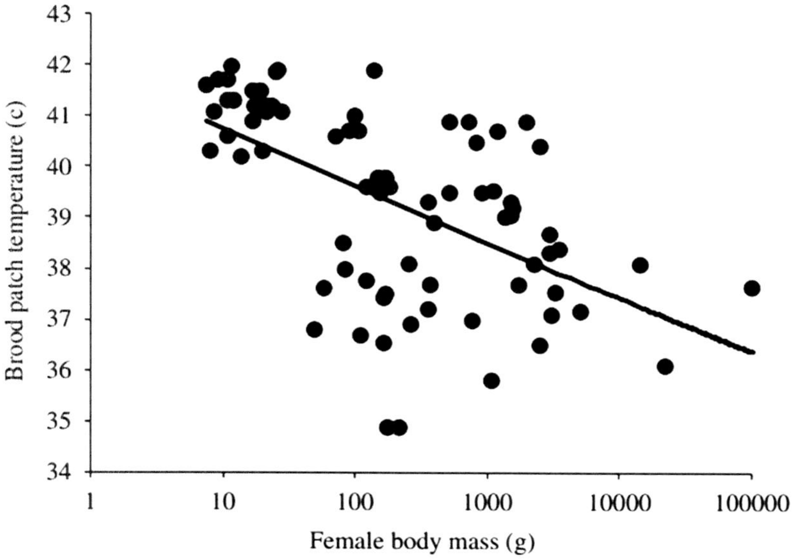


**Fig. 18.22** Relationship between female body mass and brood patch temperatures for 76 species of birds representing 12 different orders. (Figure from Deeming 2008; © 2008 Elsevier Ltd. All rights reserved, used with permission)

and heat from decomposition to incubate their eggs. Because the importance of solar heat varies, mound composition must be manipulated accordingly (Sekercioglu 1999).

The origins of such behavior (and away from using body heat for incubation) are unclear, but one hypothesis is that ancestral megapodes built nests in areas with naturally occurring sources of heat such as solar radiation or geothermal heat (Harris et al. 2014). Alternatively, mound building could have evolved from a habit of ancestral megapodes covering eggs with organic material, like some grebes and ducks do (Keller 1989; Booth and Jones 2002). With the addition of

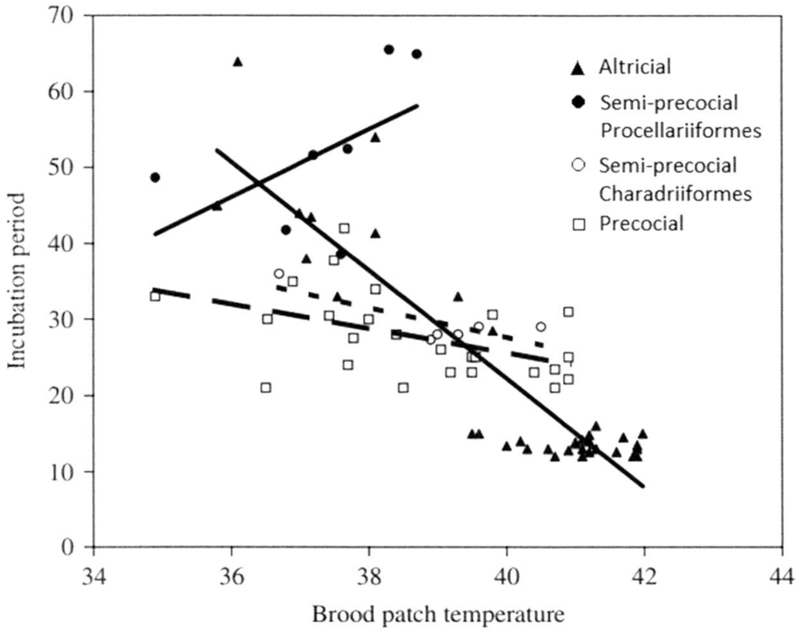


**Fig. 18.23** Relationships between brood-patch temperatures (°C) and the duration of incubation (days) for species with different modes of development. For all modes except semi-precocial Procellariiformes, higher brood patch temperatures resulted in shorter incubation periods. (Figure from Deeming 2008; © 2008 Elsevier Ltd. All rights reserved, used with permission)

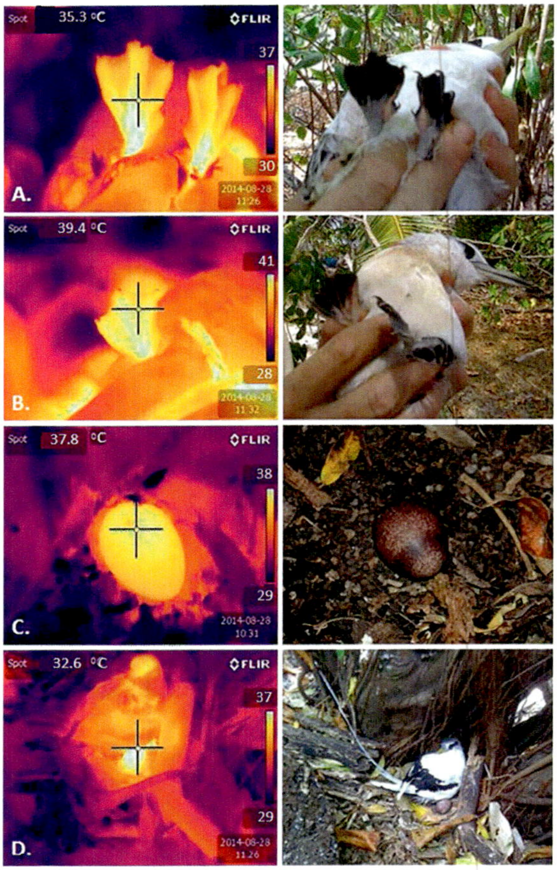


**Fig. 18.24** Thermal images and corresponding real-time images of (**a**) the feet of an incubating White-tailed Tropicbird (*Phaethon lepturus*), (**b**) feet of a nestling White-tailed Tropicbird, (**c**) an incubated egg of a White-tailed Tropicbird, and (**d**) a White-tailed Tropicbird and egg at the nest site. Spot temperatures are the temperatures at the center of the cross hairs, and the temperature scale bars indicate the range of temperature throughout the entire images. (Figure from Hart et al. 2016; © 2016 Dt. Ornithologen-Gesellschaft e.V., used with permission)

more debris, eggs could be left unattended by incubating adults for increasingly long periods and, eventually, adults no longer had to incubate eggs because the heat needed for embryonic development was generated by decomposing vegetation (Booth and Jones 2002). Proto-megapodes were likely found in hot, moist equatorial rainforests in the Indo-Australian region (Steadman 1999) where organic matter would decompose rapidly and a pile of such material would further enhance the process of decomposition and microbial heat production.

Building mounds takes considerable time and energy because mounds must be tended throughout the incubation period. As a result, species of megapodes in areas with sources of heat other than the decomposition of vegetation, such as Tongan Scrubfowl (*Megapodium pritchardii*) that are found on volcanically active islands, have abandoned mound building and, instead, dig burrows and use geothermal heat to incubate their eggs (Harris et al. 2014).

## 18.8 Onset of Incubation

The onset of incubation, i.e., when sufficient heat is applied to eggs for embryonic development, varies among and within species and its timing has fitness consequences for both parents and offspring, affecting hatching success and hatching patterns (Wang and Beissinger 2009), nestling growth and development (Bitton et al. 2006), fledging success (Hébert 1993; Hébert and McNeil 1999), and post-fledging survival and recruitment (Cam et al. 2003; Box 18.4 Prolactin and Parental Care). Female birds lay one egg at a time until their clutch is complete, but "onset incubation patterns"

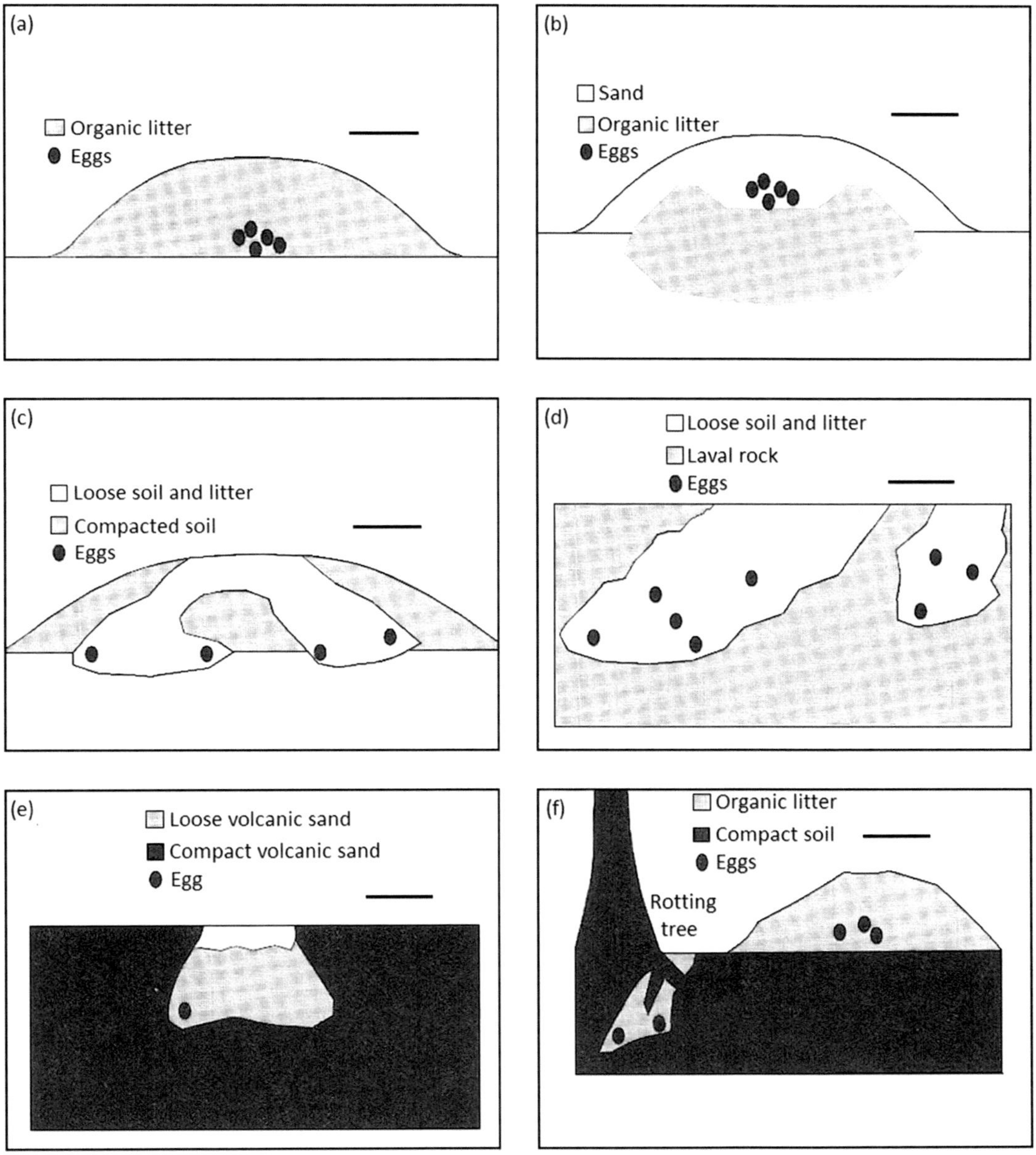


**Fig. 18.25** Various types of megapode nests. (**a**) Australian Brushturkeys (*Alectura lathami*), and most megapodes, construct a mound of organic material. (**b**) Malleefowl (*Leipoa ocellata*) mounds consist of a mound of organic material overlaid with a layer of sand. (**c**) Dusky Scrubfowl (*Megapodius freycinet*) lay eggs in tunnels, filled with loose dirt and organic material, that are excavated in a larger mound of compact dirt. (**d**) Niuafoou Scrubfowl (*Megapodius pritchardii*) eggs are laid in tunnels excavated in geothermally heated rock formations. (**e**) Maleo (*Microcephalon maleo*) eggs are laid in pits excavated in geothermally heated sand. (**f**) Different populations of Philippine Scrubfowl (*Megapodius cumingii*) used somewhat different methods of incubation. In one method, eggs are laid in tunnels dug between the roots of rotting trees. In a second method, eggs are incubated in mounds of organ litter. Scale bars = 1 m. (Figure from Booth and Jones 2002; © 2002 Oxford University Press, used with permission)

vary among and within species (Wang and Beissinger 2011; Fig. 18.26). In some cases, adults spend less time incubating eggs when females are still laying eggs than they will after a clutch is complete, a behavior referred to as partial incubation. Wang and Beissinger (2011) noted that, in studies where investigators determined the onset of full incubation, 50 of 103 species (48.5%) exhibited partial incubation.

Partial incubation can potentially serve a variety of functions (Table 18.1). One possibility is that nest sites are safer or provide more favorable microclimates for adults than other locations. In addition, the presence of adults at nest sites with partial clutches may reduce the likelihood of predation, brood parasitism, or, for cavity-nesting species, loss of the nest site to conspecific or heterospecific competitors. Partial incubation can also enhance survival of embryos by maintaining suitable temperatures (i.e., applying heat or shading eggs as necessary) and reducing microbial growth on eggshells. Finally, partial incubation can shorten incubation periods (Table 18.2).

## 18.9 Costs of Incubation

During incubation, eggs need external heat and, for most species, must be turned regularly for proper embryonic development. The energetic cost of providing heat to eggs was once thought

**Box 18.4 Prolactin and Parental Care**

Prolactin is a hormone produced in the anterior pituitary gland and its secretion into the blood can be influenced by both photoperiodic and non-photoperiodic cues (Small et al. 2007). Among birds, prolactin levels in the blood are relatively low during the non-breeding period, begin to increase during egg laying, continue to increase during incubation, remain relatively high early in the nestling period of altricial species, decline after eggs hatch for precocial species, and gradually decline as nestlings approach fledging age for altricial species (e.g., Smiley and Adkins-Regan 2016; Smiley 2019).

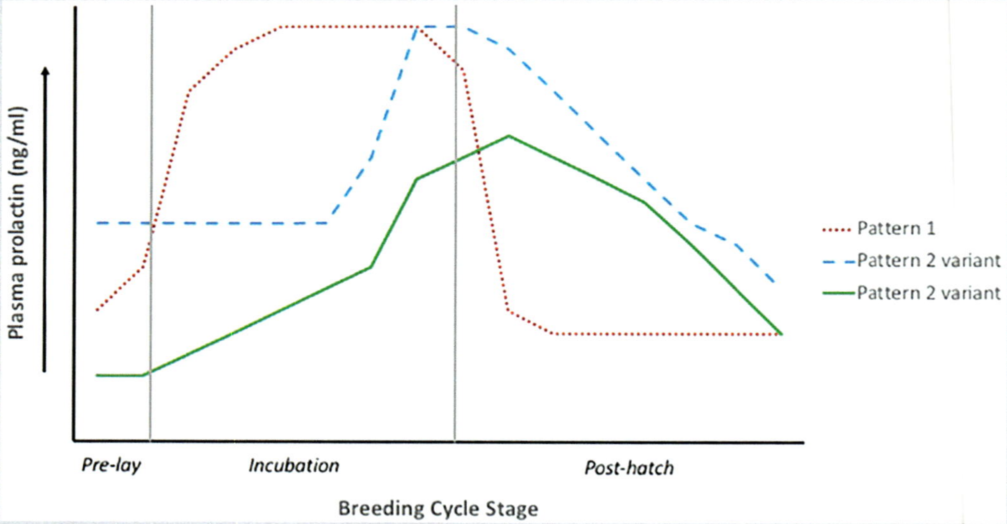


Examples of the changes in prolactin levels in the blood of birds during the pre-laying, incubation, and post-hatching periods. Birds with precocial young tend to have high levels of prolactin after egg laying, but levels decrease after eggs hatch (Pattern 1, red dotted line). Prolactin levels in species with altricial young tend to either

(continued)

**Box 18.4** (continued)

increase gradually during incubation (Pattern 2 variant, green solid line) or increase about mid-way through incubation (Pattern 2 variant, blue dashed line). For species with altricial young that require much more parental care after hatching than precocial young, prolactin levels remain relatively high early in the nestling period, then gradually decline as nestlings approach fledging age. (Figure from Smiley 2019; )

Numerous studies have revealed that high blood plasma levels of prolactin show a positive relationship with levels of parental care and reproductive success (e.g., Miller et al. 2009; Riechert et al. 2014). High levels of prolactin are needed to maintain incubation behavior, and tactile contact with eggs is important for maintaining those levels in most species of birds. For species that take breaks during incubation to forage, e.g., some penguins and albatrosses, prolactin levels still remain high, suggesting that mechanisms other than tactile contact with eggs help maintain high levels of prolactin (Vleck and Vleck 2011). For precocial species, parental care after eggs hatch is generally limited to protecting young from predators and leading young to sources of food so prolactin levels typically decline post-hatching (Smiley 2019). For altricial species, prolactin levels remain higher during the nestling period because nestlings require substantial care in the form of brooding during the early nestling stage and adults provide food throughout the nestling phase. As such, prolactin levels typically do not begin to decline until later in the nestling period. If both males and females incubate eggs and care for young, then both experience these changing levels of prolactin. If only one sex incubates, prolactin levels may be higher than those of the non-incubating sex or the non-incubating sex may not experience an increase in prolactin levels.

How prolactin actually influences parental behavior is not well understood. Smiley (2019) hypothesized that prolactin creates an increased attentiveness to cues related to parental care such as tactile contact with eggs and the begging behavior of nestlings to elicit the necessary amount of parental care. So, rather than actually "causing" parental behavior, prolactin signals areas in the brain that it is "the right time to display these behaviors."

by some investigators to be negligible, and the results of some studies seemed to suggest that the energy expended by incubating birds was 15–18% less than that of resting, non-incubating conspecific (Walsberg and King 1978a, b). The results of later studies, however, suggest that incubation does have an energetic cost. As an indirect measure of the cost of incubation, Nilsson and Smith (1988) found that females provided with supplemental food had shorter incubation periods. As a direct measure of the cost of incubation, Haftorn and Reinertsen (1985) measured oxygen consumption by a female Eurasian Blue Tit (*Cyanistes caeruleus*) when incubating and not incubating eggs, and found that oxygen consumption (which is correlated with energy expenditure) was higher when the female was incubating and, further, that oxygen consumption increased as air temperatures decreased (Fig. 18.27). Based on a review of studies of the energetic costs of incubation, Williams (1996) concluded that (1) energy expenditure during incubation is higher than when a bird is resting (at least when temperatures are below a bird's thermoneutral zone), (2) rewarming eggs after an incubation recess can require substantial energy, (3) more energy is needed to incubate larger clutches than smaller clutches, and (4) birds expend energy when incubating eggs, but have less time to forage, possibly resulting in an energy shortage.

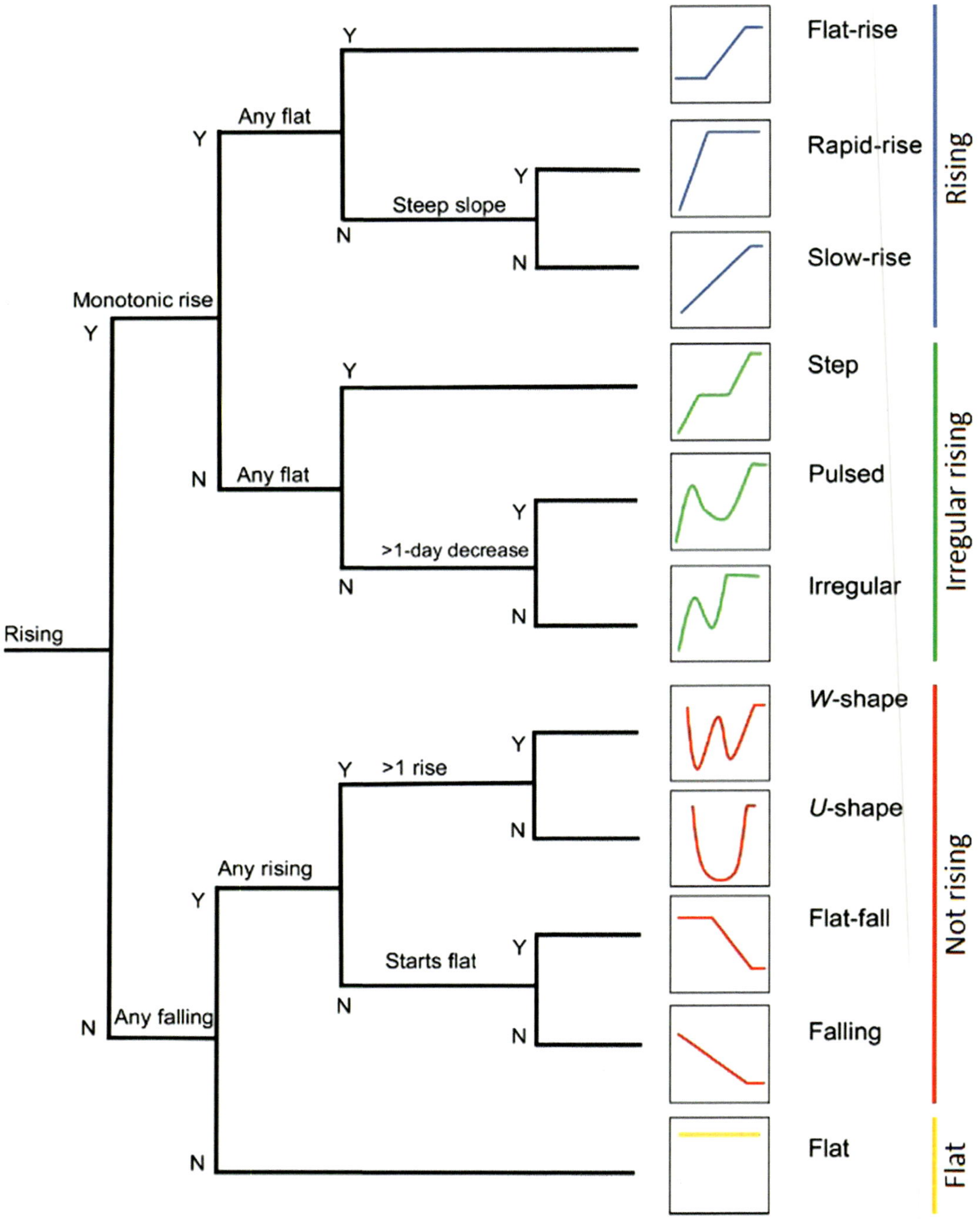


**Fig. 18.26** Variation among and within species in the onset of full incubation based on whether or not the amount of time spent incubating continues to increase over time (rising), tends to increase, but does so in an irregular pattern (irregular rising), exhibits periods with decreasing time spend incubating (not rising), or begin full incubation with the laying of the first egg (flat). In all cases, the $x$-axes represent days or nights prior to completion of egg laying or on the second consecutive day or night after full incubation (whichever came later), and the y-axis is the proportion of the day or night spent incubating. (Figure modified from Wang and Beissinger 2011; © 2011 Oxford University Press, used with permission)

**Table 18.2** Possible functions of partial incubation

| True incubation required? | Functions | Who benefits | | Demographic effect | Example species |
|---|---|---|---|---|---|
| | | Adult | Egg | | |
| No | Reduce predation of adults | x | | Adult survival | White-tailed Ptarmigan[a] |
| No | Favorable nest microclimate | x | | Adult survival | Kentish Plover[b], Great Tit[c], Common Eider[d] |
| No | Reduce predation of eggs | | x | Egg loss | Common Eider[e], Mallard[f] |
| No | Prevent nest-site takeover (intra- and interspecific) | | x | Egg loss | Green-rumped Parrotlet[g] |
| No | Prevent brood parasitism (intra- and interspecific) | | x | Egg loss | Red-winged Blackbird[h] |
| No | Shelter eggs from precipitation | | x | Egg viability | White-crowned Sparrow[i] |
| No | Shelter eggs from moisture (condensation) | | x | Egg viabililty | White-crowned Sparrow[i] |
| No | Shelter eggs from solar rays | | x | Egg viability | White-crowned Sparrow[i], Black-legged Kittiwake[j] |
| No | Shelter eggs from heat | | x | Egg viability | Black-necked Stilt[k], American Avocet[k], Snowy Plover[k], Killdeer[k], Gull-billed Tern[k], Forster's Tern[k], Black Skimmer[k], Crowned Lapwing[l], Black-winged Lapwing[l] |
| No | Shelter eggs from freezing | | x | Egg viability | – |
| No | Shelter eggs from microbial growth and infection | | x | Egg viability | Pearly-eyed Thrasher[m,n,o] |
| Yes | Advance embryonic development | x | x | Reproduction | Green-rumped Parrotlet[p] |
| Yes | Shorten incubation period | x | | Adult condition | Common Eider[q] |

Table from Wang and Beissinger (2011); © 2011 Oxford University Press, used with permission

References: [a]Wiebe and Martin (1998), [b]Amat and Masero (2004), [c]Pendlebury and Bryant (2005), [d]D'Alba et al. (2009), [e]Andersson and Waldeck (2006); [f]Kreisinger and Albrecht (2008), [g]Beissinger et al. (1998), [h]Clotfelter and Yasukawa (1999), [i]Morton and Pereyra (1985), [j]Barrett (1980), [k]Grant (1982), [l]Ward (1990), [m]Cook et al. (2005a), [n]Cook et al. (2005b), [o]Shawkey et al. (2009), [p]Grenier and Beissinger (1999), [q]Hanssen et al. (2002)

Scientific names: White-tailed Ptarmigan, *Lagopus leucura*; Kentish Plover, *Charadrius alexandrinus*; Great Tit, *Parus major*; Common Eider, *Somateria mollissima*; Mallard, *Anas platyrhynchos*; Green-rumped Parrotlet, *Forpus passerinus*; Red-winged Blackbird, *Agelaius phoeniceus*; Mountain White-crown Sparrow, *Zonotrichia leucophrys*; Black-legged Kittiwake, *Rissa tridactyla*; Black-necked Stilt, *Himantopus mexicanus*; American Avocet, *Recurvirostra americana*; Snowy Plover, *Charadrius nivosus*; Killdeer, *Charadrius vociferus*; Gull-billed Tern, *Gelochelidon nilotica*; Forster's Tern, *Sterna forsteri*; Black Skimmer, *Rynchops niger*; Crowned Lapwing, *Vanellus coronatus*; Black-winged Lapwing, *Vanellus melanopterus*; Pearly-eyed Thrasher, *Margarops fuscatus*

Based on a review of studies of 34 species of terrestrial birds (25 non-Arctic and nine Arctic breeding species), Nord and Williams (2015) noted that energy expended during incubation averaged 2.88 times that of basal metabolic rates (BMR) for non-Arctic birds and 4.13 times that of BMR for species breeding at Arctic latitudes. These authors also reported that the energetic cost of incubation tended to decline with increasing body mass, possibly because smaller birds, with higher metabolic rates and being less able to store energy, may need to expend more energy

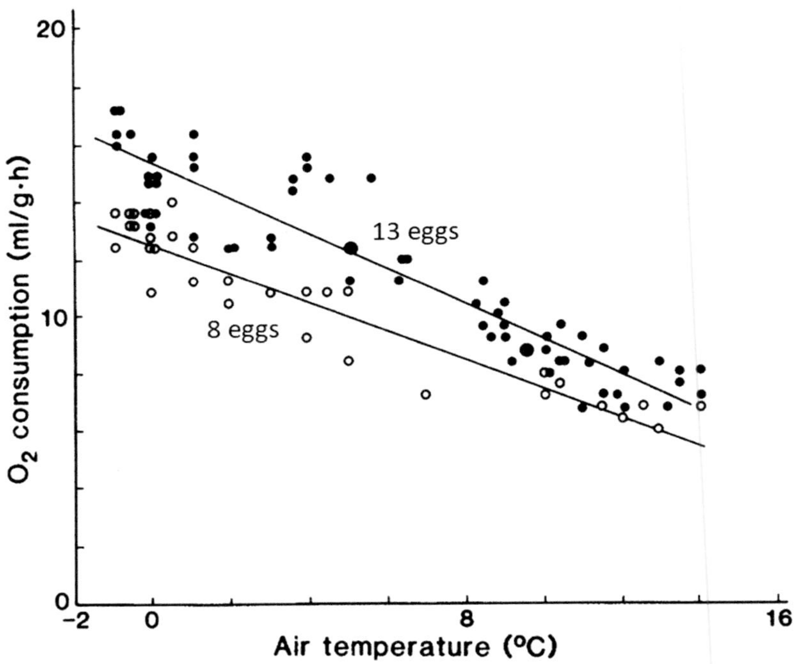


**Fig. 18.27** Relationship between rate of oxygen consumption and air temperature for a female Eurasian Blue Tit (*Cyanistes caeruleus*) incubating 13 eggs (solid circles) and after the clutch was temporarily reduced to eight eggs. The female expended more energy, as measured by oxygen consumption, when incubating a larger clutch of eggs. when incubating a larger clutch of eggs and when air temperatures were lower. (Figure from Haftorn and Reinertsen 1985; © 1985 Oxford University Press, used with permission)

foraging than larger birds (McNab 2002; Fig. 18.28).

During incubation, many seabirds must make long foraging trips that add to the energetic cost of incubation. Tinbergen and Williams (2002) divided the energetic costs of incubation for seabirds into a foraging stage and incubation (i.e., heating eggs) stage, and found that the energetic costs were 4.8 times that of BMR for foraging and just 1.7 times BMR for heating eggs. Similar estimates of energetic costs for these two stages have been reported in subsequent studies of Black-browed Albatrosses (*Thalassarche melanophrys*; Shaffer et al. 2004), Macaroni Penguins (*Eudyptes chrysolophus*; Green et al. 2009), and Black-legged Kittiwakes (*Rissa tridactyla*; Collins et al. 2016). Interestingly, however, using data from studies of 30 species of seabirds, Shoji et al. (2015) found that metabolic rates of seabirds during incubation decreased with an increasing duration of incubation shifts (Fig. 18.29). One possible explanation for this negative relationship is that seabirds with longer incubation bouts are expending less energy, but the trade-off is lower egg temperatures and slower embryonic development (Shoji et al. 2015).

The energetic cost of incubation is also influenced by clutch size and ambient temperature, with larger clutches typically increasing energetic costs (Fig. 18.27). An increasing energetic cost of incubating larger clutches has been reported for several species of birds (e.g., Haftorn and Reinertsen 1985; Moreno and Sanz 1994; Reid et al. 2002). Nord et al. (2010) found that metabolic rates of female Zebra Finches (*Taeniopygia guttata*) increased from 2 to 8% for each additional egg added to a clutch.

In contrast, Öst et al. (2008) found that incubating larger clutches did not increase energetic costs (as determined by loss of body mass) for female Common Eiders (*Somateria mollissima*), possibly because female Common Eiders insulate their nests with down and cover eggs when they leave nests (which they rarely do, i.e., about four minutes every second or third day for a drink of water). As a result, egg temperatures drop only about 1°C during these breaks so females expend little energy reheating the eggs. Although female Common Eiders may not

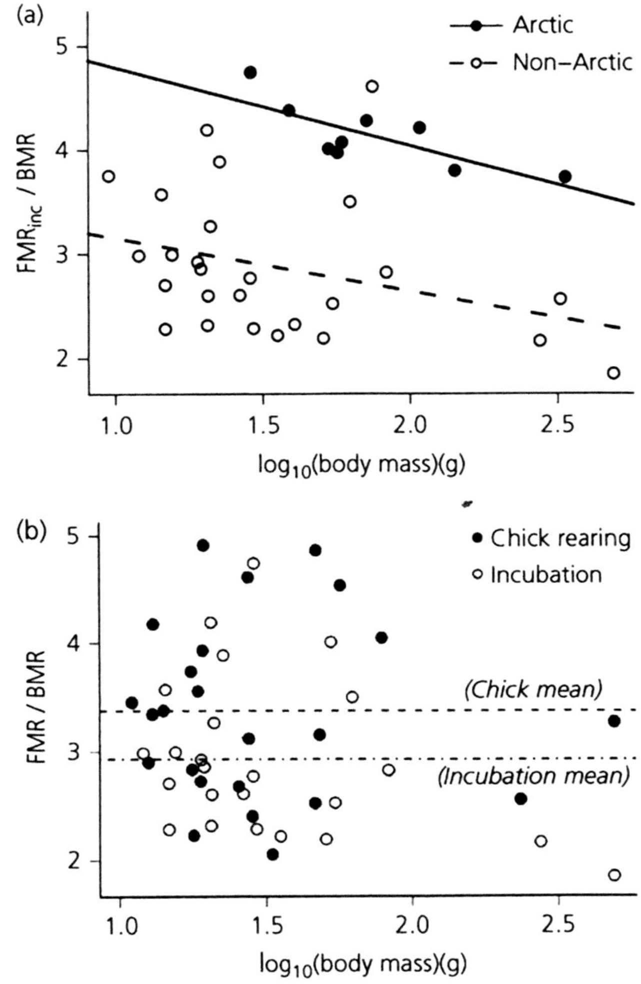


**Fig. 18.28** Energetic costs (field metabolic rate, or FMR) of incubation and chick rearing (nestling period) relative to adult body mass ($\log_{10}$) for several species of terrestrial birds expressed as cost compared to basal metabolic rate (BMR). (**a**) Increase in FMR relative to BMR during incubation for species of birds breeding in the Arctic and those breeding at lower latitudes (non-Arctic). The energetic costs of incubation were found to be, on average, 43% higher for Arctic birds because of the lower ambient temperatures, but tend to decrease with increasing body mass. (**b**) Energetic cost of incubation and chick rearing for several species of birds that vary in body mass. (Figure from Nord and Williams 2015; © 2015 Oxford University Press, Reproduced with permission of the Licensor through PLSclear)

experience a short-term cost in terms of energetics, Hanssen et al. (2005) found that incubating larger clutches (six eggs rather than three eggs) did result in a long-term fitness cost, with females that incubated six eggs laying smaller clutches in the next breeding season than females that incubated three eggs. Other investigators have also reported delayed costs of incubating larger clutches, including lower fledging success for subsequent broods in the same breeding season and reduced likelihood of adult survival (e.g., Reid et al. 2000; de Heij et al. 2006). Yet another potential cost of larger clutches is an increase in the duration of the incubation period, with longer incubation periods increasing the risk of predation (e.g., Dobbs et al. 2006).

As already noted, the energetic cost of incubation is also influenced by ambient temperature, with those costs increasing as ambient temperatures decrease. This negative relationship between energetic costs and ambient temperatures has been noted in numerous studies (e.g., Vleck 1981; Haftorn and Reinertsen 1985; Bryan and Bryant 1999; Fig. 18.30).

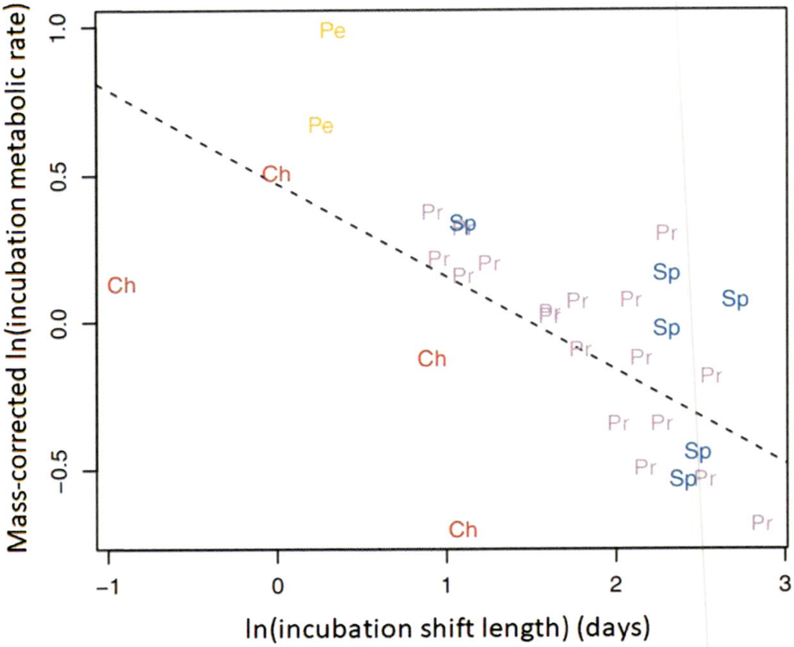


**Fig. 18.29** Incubation metabolic rates are lower for species of seabirds with longer incubation shifts. Pr, Procellariiformes; Pe, Pelecaniformes; Ch, Charadriiformes; and Sp, Sphenisciformes. (Figure from Shoji et al. 2015; © 2014 British Ornithologists' Union, used with permission)

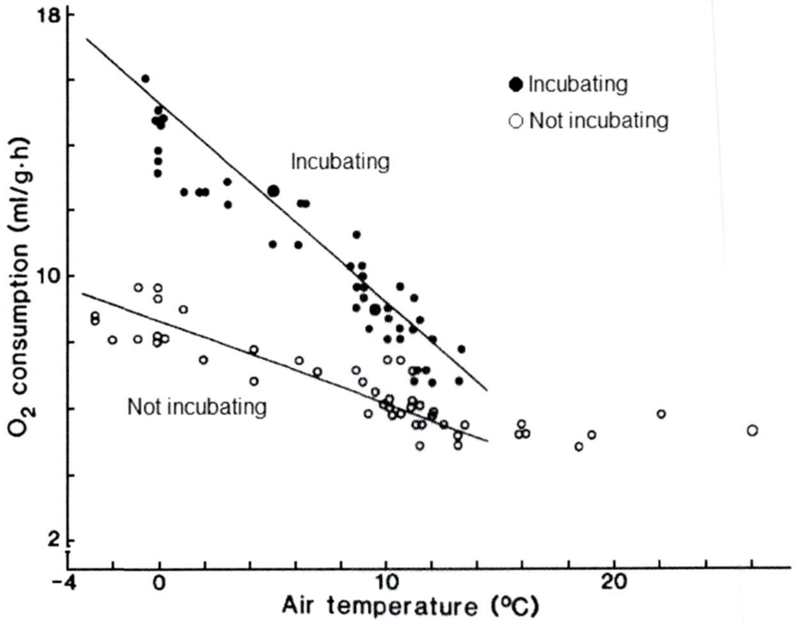


**Fig. 18.30** Relationship between oxygen consumption (as a measure of energy consumption) by a female Eurasian Blue Tit (*Cyanistes caeruleus*) and air temperature when incubating eggs and when not incubating eggs (i.e., standing above the eggs in the nest). When incubating, the female Blue Tit clearly consumed more oxygen and expended more energy, requiring even more oxygen and expending more energy as air temperature declined. (Figure from Haftorn and Reinertsen 1985; © 1985 Oxford University Press, used with permission)

**Box 18.5 Incubation and Microbial Growth on Eggshells**

Avian eggshells harbor microbes shortly after laying and, under appropriate ambient conditions, they can multiply rapidly, penetrate through shell pores, infect egg contents, and cause embryo mortality. Cook et al. (2005b) experimentally examined how incubation affects bacterial processes on the eggshells of Pearly-eyed Thrashers (*Margarops fuscatus*) nesting in tropical montane and lowland forests in Puerto Rico. Bacteria and fungi grew rapidly on shells of newly laid, unincubated eggs exposed to ambient conditions, but declined to low levels on shells of eggs incubated by thrashers. Divergence in bacterial growth between incubated and exposed eggs was more marked at the montane forest than at the lowland site. Pathogenic microorganisms became increasingly dominant on shells of exposed eggs, but these groups were relatively rare on incubated eggs, where more benign, less invasive groups prevailed. Some incubation during laying may be necessary to decrease the probability of trans-shell infection by reducing the growth of harmful bacteria and fungi on eggshells, although it may increase hatching asynchrony and the likelihood of brood reduction.

Other investigators have also found that incubated eggs have fewer bacteria than unincubated eggs (Shawkey et al. 2009). D'Alba et al. (2010) found that incubation limits bacterial growth on eggshells by keeping eggs dry. Bruce and Drysdale (1991, 1994) noted that eggshells are already nutrient-poor environments and reducing water levels makes it even more difficult for bacteria to grow. Dryer eggs may also limit interactions like symbiosis and biofilm formation that might contribute to bacterial growth (D'Alba et al. 2010).

In contrast to the above studies, other investigators have reported that incubation had no effects on eggshell microbial loads (Wang et al. 2011; Ruiz-de-Castañeda et al. 2011). In addition, Lee et al. (2014) found that total bacterial abundance increased and diversity decreased on incubated eggs of Eurasian Magpies (*Pica pica*), whereas abundance and diversity did not change for non-incubated eggs. Interestingly, however, bacteria that increased in abundance were either harmless or antibiotic-producing taxa. Grizard et al. (2014) reported similar results for Homing Pigeons (*Columba livia*), and suggested that eggs of different species with different shell structures may contribute to differences in microbial communities. This, in turn, may help explain why the effects of incubation on microbial communities differ among studies and species of birds. In addition, different types of bacteria may respond differently to differences in the availability of water on eggshells.

## 18.10 Incubation Periods

Incubation periods range from about 10 days for some passerines and woodpeckers to nearly 80 days for albatrosses and kiwis (Box 18.5 Incubation and Microbial Growth on Eggshells). Much of this variation is due to phylogeny, with genetic factors that control embryonic growth rates playing a significant role in determining how long any incubation period needs to be for any particular order of birds (Starck 1998; Deeming et al. 2006; Table 18.3, Fig. 18.31). This phylogenetic effect is apparent when comparing relationships between egg mass and incubation period for different orders of birds. For example, among grebes (Podicipediformes) egg mass among different species varies more the fourfold (10–42 g), but incubation periods vary by only 10 days (18–28 days; Table 18.3). Similarly, among gallinaceous birds (Galliformes), egg mass varies more than 12-fold (8–102 g), but incubation periods vary by only 11 days (18–29 days; Table 18.3). About 80% of the

**Table 18.3** Egg mass at hatching and duration of incubation periods of species in various orders of birds

| Order | Egg mass at hatching (gm) | Incubation period (days) |
|---|---|---|
| All birds | 0.3–1500 | 10–79 |
| Tinamiformes | 19–83.5 | 13–23 |
| Sphenisciformes | 53–467 | 33–64 |
| Podicipediformes | 10–42 | 18–28 |
| Procellariformes | 7–471 | 38–79 |
| Pelecaniformes | 23–187.4 | 25–56 |
| Ciconiiformes | 10.5–111 | 19–36 |
| Anseriformes | 26.5–345 | 23–40 |
| Falconiformes | 4.6–262 | 22–57 |
| Galliformes | 8–102 | 18–29 |
| Gruiformes | 4.2–240 | 13–34 |
| Charadriiformes | 6–116 | 15–45 |
| Colombiformes | 2.6–41.1 | 13–29 |
| Psittaciformes | 2.2–39.2 | 17–30 |
| Cuculiformes | 1.4–27.9 | 10–19 |
| Strigiformes | 6.9–74 | 21–42 |
| Caprimulgiformes | 5.7–29.1 | 17–40 |
| Apodiformes | 0.3–6.2 | 13–27 |
| Coraciiformes | 3.9–44.9 | 15–27 |
| Piciformes | 1.3–21.9 | 10–19 |
| Passeriformes | 0.7–60.5 | 10–48 |

Data from Table 1 in Deeming et al. (2006); © 2006 John Wiley and Sons, used with permission

variation in incubation periods among birds is due to differences between orders and between families within orders; only about 20% of the variation is due to differences between genera, species, and populations (Ricklefs and Starck 1998a, b; Bennett and Owens 2002).

Body size or mass affects the duration of incubation periods, with larger birds generally having longer incubation periods than smaller ones. However, most of this variation is due to phylogenetic history (Cooney et al. 2020). Some exceptional species do have longer incubation periods than might be expected based on their body mass (Box 18.6 The Long Incubation Periods of Alcids, Penguins, and Other Oceanic Species of Birds).

Another factor that can influence the duration of incubation periods, both between and within species, is egg temperature. Bird embryos are ectothermic and subject to temperature fluctuations due to the presence or absence of an incubating parent (Box 18.7 Why Do Some Birds Cover Eggs and Nests?). Based on a number of studies across avian orders, the optimum temperature for developing avian embryos ranges from about 35.5 to 38.5°C (Webb 1987; Fig. 18.32). Factors that reduce egg temperatures below that range will slow embryo development and increase incubation periods. For example, the duration of incubation periods was found to decline with mean nest temperature in Wood Ducks (*Aix sponsa*; Hepp et al. 2006; Fig. 18.33), and other investigators have reported similar results with several other species (Martin and Schwabl 2008).

Whereas colder egg temperatures increase the duration of incubation periods, warmer egg temperatures tend to shorten incubation periods. For example, Ward (1940) placed a single egg of a Superb Lyrebird (*Menura novaehollandiae*) in the nest of a constantly incubating Domestic Chicken and found that the young lyrebird hatched in just 28 days rather than the normal 50 days. More recently, Ton and Martin (2017) experimentally increased the temperatures of eggs of nine species tropical and north-temperate songbirds and found that warmer temperatures consistently shortened development times. The effect was particularly apparent for tropical

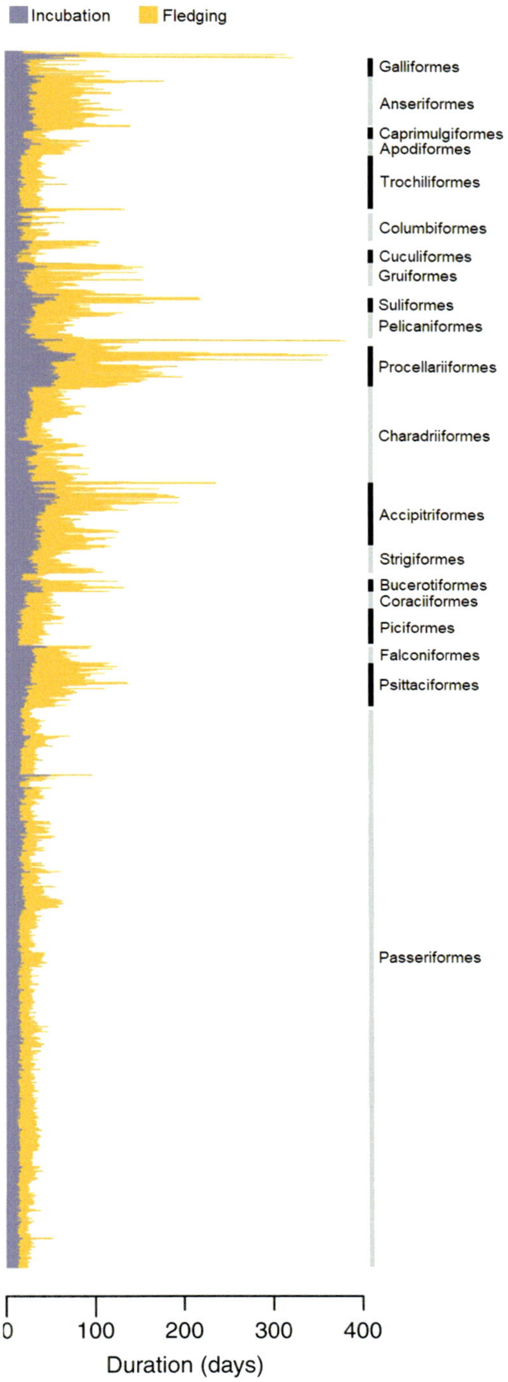


**Fig. 18.31** Durations of the incubation and fledging periods for 3096 species of birds. The incubation period is defined as the time between when an egg is laid and when it hatches; the fledgling period is the time between hatching and when young are capable of flight (or, for some species, leave the nest). Note that, within orders, incubation periods of different species are generally similar in duration. (Figure modified from Cooney et al. 2020; open-access article licensed under a Creative Commons Attribution 4.0 License, https://creativecommons.org/licenses/by/4.0/)

**Box 18.6 The Long Incubation Periods of Alcids, Penguins, and Other Oceanic Species of Birds**

In general, the duration of avian incubation periods is positively correlated with fresh egg mass; the larger the egg the longer the incubation period. However, some species in several avian taxa, including alcids (Alcidae, Charadriiformes, e.g., murres, guillemots, and puffins), penguins (Spheniciformes), Procellariformes (e.g., albatrosses, shearwaters, and petrels), Suliformes (boobies and frigatebirds), and tropicbirds (Phaethontiformes), have much longer incubation periods than would be predicted based on the mass of their eggs (Whittow 1980).

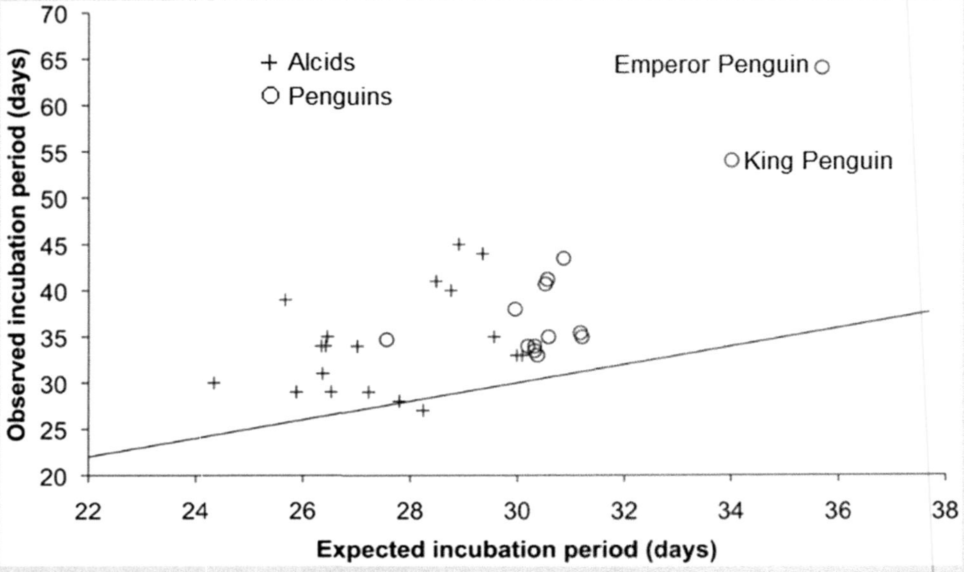


Most alcids and penguins have longer incubation periods than would be predicted based on their fresh egg mass. The predicted duration of incubation periods (line) is based on data from birds in the order Charadriiformes. Scientific names: Emperor Penguin, *Aptenodytes forsteri*; King Penguin, *A. patogonicus.* (Figure modified from Hipfner et al. 2010; published under license to BioMed Central Ltd., open-access article distributed under the terms of the Creative Commons Attribution License, https://creativecommons.org/licenses/by/4.0/)

Several factors appear to have influenced the duration of incubation periods among species in these taxa. First is clutch size, with incubation periods shorter for species in these taxa with larger clutches. This relationship is likely due, at least in part, to selection for shorter incubation periods; when nestlings must compete with siblings for parental care, selection favors earlier hatching because younger nestlings (from later-hatched eggs) will generally be out-competed by older (from earlier-hatched eggs) siblings (Lloyd and Martin 2003).

A second factor is the timing of nest attendance, with incubation periods generally longer among species where adults visit nests at night. Nocturnal visitation likely evolved as a means of reducing predation risk for adults. During the nestling period, nocturnal visitation tends to slow nestling growth and increase the duration of the nestling period because adults typically visit nests only once per night (Adams et al. 2004). Slow development during the nestling period generally also means slow development of embryos during incubation because the duration of these two periods is typically correlated (Dobson and Jouventin 2007).

(continued)

**Box 18.6** (continued)

A third factor is the distance between nesting areas and foraging areas, with greater distances associated with longer incubation periods. First predicted by Lack (1968), investigators have since confirmed this relationship between foraging distance and the duration of seabird incubation periods (Dobson and Jouventin 2007; Hipfner et al. 2010). However, Lack (1968) also predicted that the primary reason for this relationship would be reduced provisioning rates of nestlings contributing to generally slower growth rates of both nestlings and, given the correlation between the duration of nestling and incubation periods, embryos. Interestingly, evidence for a relationship between foraging distances of seabirds and nestling provisioning rates is either weak (Hipfner et al. 2010) or non-existent (Dobson and Jouventin 2007). As such, possible reasons for the effect of foraging distance on the duration of seabird incubation periods remain to be determined.

**Box 18.7 Why Do Some Birds Cover Eggs and Nests?**

Among several species of birds with floating nests (e.g., grebes), ground nests (some plovers and ducks), and cavity nests (e.g., parids), parents cover their eggs/nests when they leave nests during the egg-laying and/or incubation periods. In most cases, the apparent functions of such behavior are to reduce the risk of predation by making eggs and nests more cryptic and, in addition, to help maintain egg temperature while a parent is absent (e.g., Keller 1989; Prokop and Trnka 2011; Amat et al. 2012; Loukola et al. 2020). However, egg-covering can serve other functions as well. For example, Black-capped (*Poecile atricapillus*) and Carolina (*P. carolinensis*) chickadees cover their eggs with nest material during the egg-laying period and apparently do so to reduce the likelihood that House Wrens (*Troglodytes aedon*) will discover their eggs and either puncture or remove them (White and Kennedy 1997). House Wrens are thought to engage in this behavior to cause chickadees to abandon cavities so they can be used by the wrens as well as to reduce competition for food. For Great Tits (*Parus major*), covering eggs during the egg-laying period may represent an attempt to avoid the negative impact of having European Pied Flycatchers (*Ficedula hypoleuca*) nest nearby. Pied Flycatchers have been found to prefer breeding in the vicinity of tit nests, particularly tit nests with larger clutches indicating a better-quality habitat. However, Great Tit nesting success is reduced when Pied Flycatchers nest nearby, likely due to competition for food resources. Thus, by covering their eggs during the egg-laying period, Great Tits may deprive Pied Flycatchers of "information" about habitat quality and, in response, the flycatchers may choose to nest elsewhere (Loukola et al. 2014).

Great Tits may also cover eggs to prevent nest usurpation by other species. Slagsvold and Wiebe (2021) presented nest-prospecting male European Pied Flycatchers with a dyad of nest boxes with Great Tit nests with and without visible eggs, and found that the flycatchers were more hesitant to enter nest boxes where eggs were not visible. Nest competitors may be less like to enter dark, unfamiliar nest cavities when nothing, including eggs, can be seen inside because they are unable to determine if an aggressive Great Tit (or other species) is present or not.

(continued)

**Box 18.7** (continued)

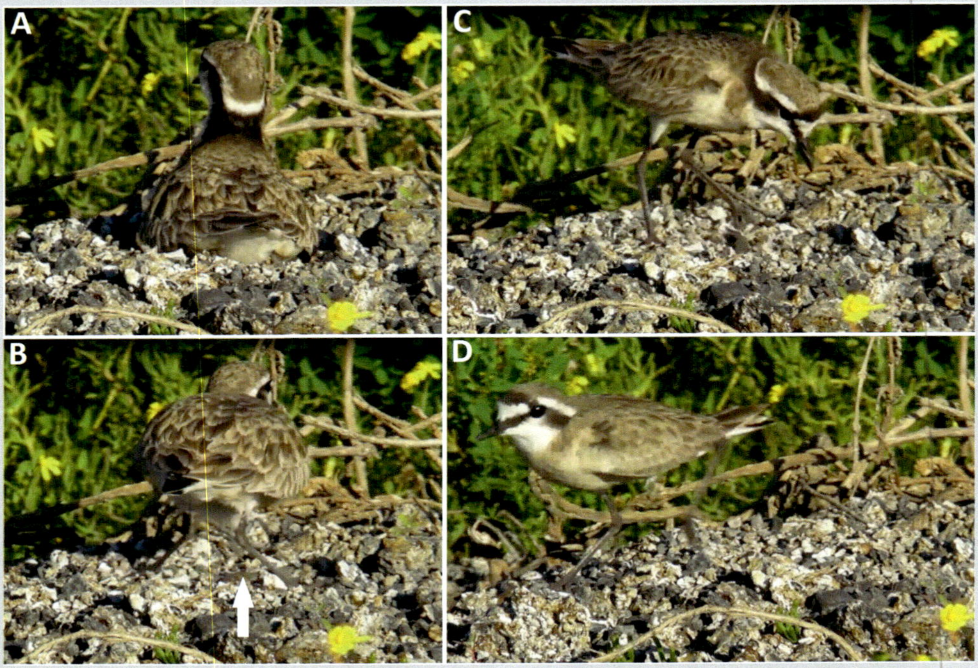


Egg-covering behavior by a female Kittlitz's Plover (*Charadrius pecuarius*). (**a**) Female on nest incubating eggs, (**b**) female detects potential predator and stands up (white arrow points to partially exposed egg), (**c**) female uses rapid foot movements to cover eggs with sand and small pebbles (with several bouts of "kicking" interspersed with brief breaks and taking a total of about 25 s), and (**d**) female leaving nest area (images are screen captures from video posted on YouTube (https://youtu.be/QI1zjdQnjlU) by Jolyon Troscianko; used with permission)

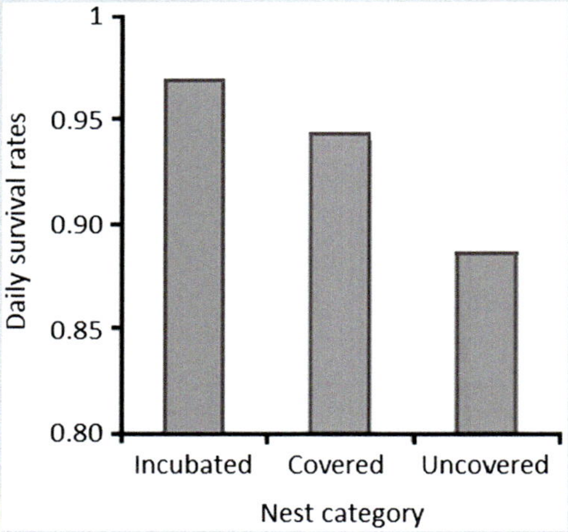


Daily survival rates of Mallard (*Anas platyrhynchos*) nests where females incubated eggs and nearby artificial nests with eggs that were either covered with vegetation or not. Daily survival rates were lower for uncovered nests, suggested that the additional crypsis provided by covering eggs with vegetation can reduce the likelihood of predation. (Figure modified from Kreisinger and Albrecht 2008; © 2008 The Authors. Journal compilation © 2008 British Ecological Society, used with permission)

(continued)

**Box 18.7** (continued)

Zheng et al. (2022) tested several hypotheses concerning the egg-covering (or egg-burying) behavior of Chinese Penduline-Tits (*Remiz consobrinus*) and found support for the "protection against wind" hypothesis. Suspended below the outer branches of trees, the nests of Chinese Penduline-Tits sway in the wind, and experiments conducted using old nests revealed that, at wind speeds ≥8 m/s, uncovered eggs were significantly more likely to fall out of nests than were covered eggs.

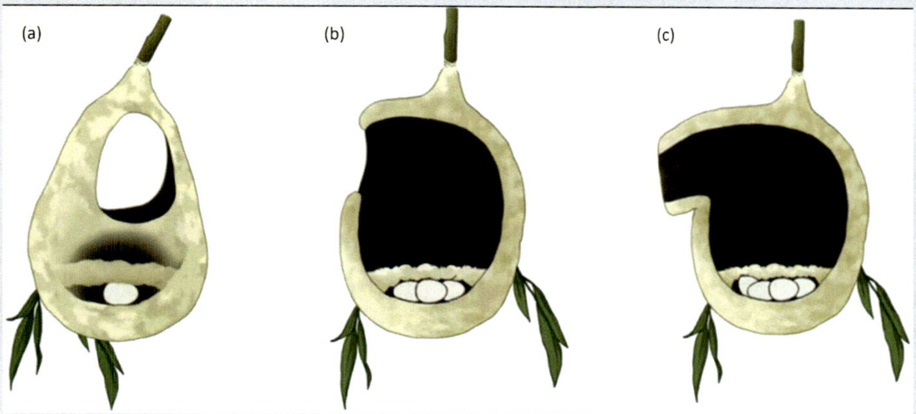


Egg burial by Chinese Penduline-Tits. Males and females place nest material over the eggs during the egg-laying period. (**a**) Females begin egg laying when nests form a basket and the egg(s) is(are) covered with nest material. (**b**) As nest building continues, one side is closed and more eggs are laid. (**c**) In the final stages of nest building, an entrance funnel is added on the open side. When nests begin to sway in the wind, covered eggs are less likely to fall out of the nests than uncovered eggs. (Figure from Zheng et al. 2022; © 2022 The Author(s). Published by Elsevier Ltd on behalf of The Association for the Study of Animal Behaviour, open-access article published under a Creative Commons Attribution 4.0 International License, http://creativecommons.org/licenses/by/4.0/)

species with cooler natural incubation temperatures (Fig. 18.34).

Ambient temperature can potentially influence egg temperatures and, therefore, the duration of incubation periods. However, the effects of ambient temperature can obviously be influenced by incubating parents; in response to lower temperatures, parents can shorten the periods (off bouts) away from the nest. In other words, increased nest attentiveness, or the percentage of time spent incubating eggs during daylight hours, can help maintain suitable egg temperatures even when ambient temperatures decline. Of course, incubating parents must also balance the thermal needs of developing embryos with their own need to forage to meet their energetic needs. This trade-off is particularly critical for species where just one parent incubates; parents in species with biparental incubation can alternate periods on and off the nest.

Conway and Martin (2000b) developed a model of the expected relationship between ambient temperature and both off- and on-bout duration in species where just one parent incubates and, in addition, where parents generally exhibit high levels of nest attentiveness (Fig. 18.35). According to their model, as ambient temperature drops below the physiological zero temperature (PZT; the temperature below which embryonic development ceases), off- and on-bout duration should decline because an incubating parent must take shorter, but more frequent, off-bouts to

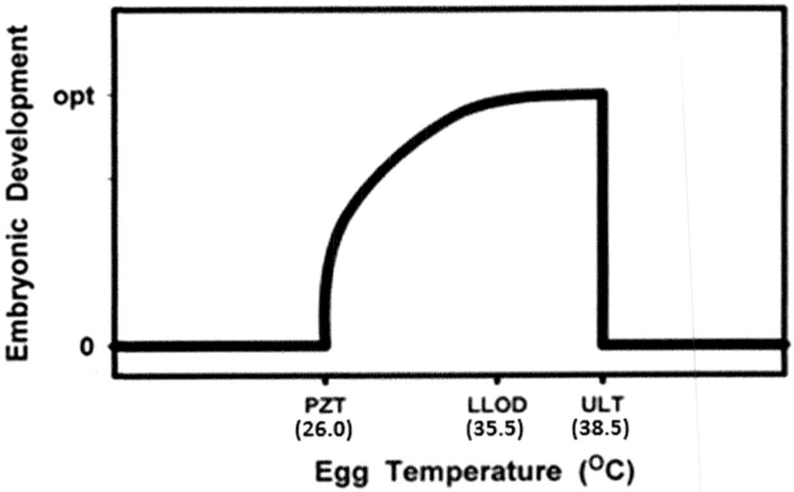


**Fig. 18.32** Assumed general relationship between avian embryonic development and egg temperature. PZT, physiological zero temperature; LLOD, lower limit of optimal development; and ULT, upper lethal temperature. (Figure modified from Conway and Martin 2000b; © 2000 Oxford University Press, used with permission)

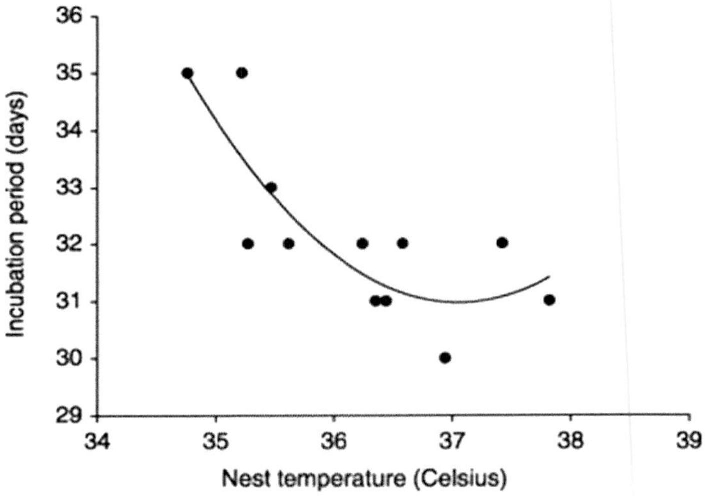


**Fig. 18.33** Relationship between mean nest temperature and incubation period in Wood Ducks (*Aix sponsa*); each point represents one nest. (Figure from Hepp et al. 2006; © 2006 John Wiley and Sons, used with permission)

obtain needed energy while preventing egg temperatures from cooling below PZT during their absences. As temperature continues to drop below both PZT and the lower critical temperature (LCT, the temperature at which the parent must begin expending energy to maintain their body temperature, e.g., by shivering), the duration of off- and on-bouts should become even shorter because the parent's metabolic needs start to increase inversely with temperature. Similarly, the slope of the negative relationship should become steeper at temperatures above either the upper lethal temperature (ULT, the temperature above which the embryo will die) or upper critical temperature (UCT, the temperature above which a parent must expend energy to maintain their body temperature, e.g., by panting; Fig. 18.36).

There would, however, be a lower limit on the duration of off- and on-bouts. With very low temperatures, the energy obtained by foraging during very short off-bouts would likely not equal the energy expended in foraging and, after returning to the nest, re-warming the eggs. So, a parent would maintain some minimal off- and on-bout durations (where the line levels off on

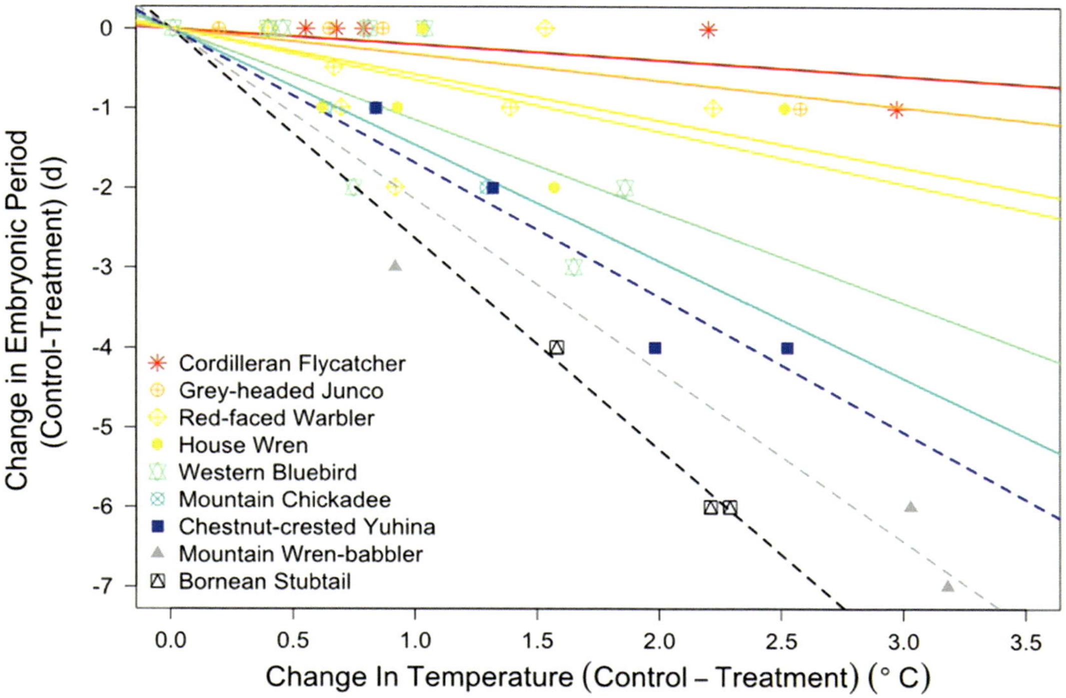


**Fig. 18.34** Correlation between measured differences in egg temperature, and incubation period differences between treatment and control among 42 paired nests belonging to nine species at two latitudes. Each point represents a nest pair and each symbol and color a different species. Individual regression lines provide the intraspecific response of embryonic period to experimental heating and warmer colors are associated to warmer natural incubation temperature. Dashed lines denote tropical species. Names in figure legend are in order of ascending slope. Cordilleran Flycatcher, *Empidonax occidentalis*; Gray-headed Junco, *Junco hyemalis*; Red-faced Warbler, *Cardellina rubifrons*; House Wren, *Troglodytes aedon*; Western Bluebird, *Sialia mexicana*; Mountain Chickadee, *Poecile gambeli*; Chestnut-crested Yuhina, *Yuhina everetti*; Mountain Wren-babbler, *Napothera crassa*; and Bornean Stubtail, *Urosphena whiteadi.* (Figure from Ton and Martin 2017; open-access article published under a Creative Commons Attribution 4.0 International License, http://creativecommons.org/licenses/by/4.0/)

the left side of the graphs; Fig. 18.35). Similarly, at higher temperatures, a parent might be expected to stay on the nest until their energy deficit becomes too large or their body temperature gets too high. At that point, the nest would likely be abandoned (Conway and Martin 2000b).

The assumptions and predictions of this model have generally been supported by subsequent studies, particularly studies of small songbirds (Matysioková and Remeš 2018; Fig. 18.36). Larger birds are better able to store energy (fat reserves) than smaller birds and, therefore, may not respond to changes in temperature like smaller birds because they may not be as dependent on the energy acquired by foraging during off-bouts (Calder 1974; Martin et al. 2007). Larger birds may also be more efficient at foraging and able to acquire more energy per unit time than smaller species (Pawar et al. 2012), which could result in longer incubation bouts (Matysioková and Remeš 2018; Fig. 18.36). Another factor that can influence incubation behavior is incubation feeding, the provisioning of an incubating bird (usually the female) by its mate. Such feeding of incubating mates by males occurs in more than 40% of North American passerines (Kendeigh 1952). The amount of food provided to incubating birds by their mates varies among species (Fig. 18.38), but the energy obtained by such feeding can clearly influence an

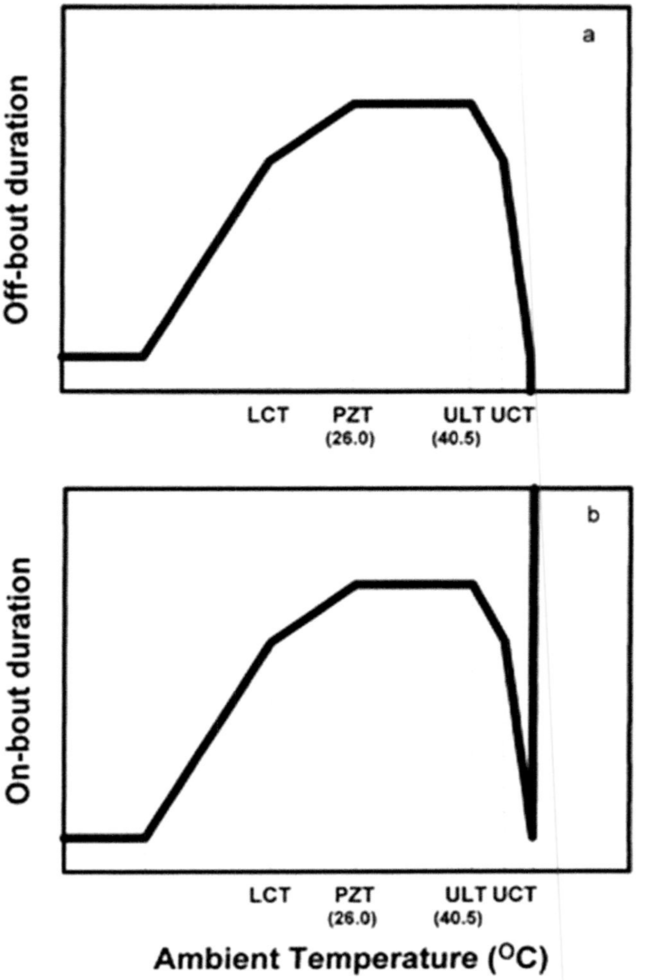


**Fig. 18.35** (**a**) Proposed relationship between ambient temperature and both off-bout and on-bout duration in birds with single-sex incubation. As ambient temperature drops below the physiological zero temperature (PZT), off- and on-bout durations should decline because incubating parents take shorter, but more frequent, off-bouts to forage while keeping egg temperatures above PZT. (**b**) At higher temperatures, a parent might be expected to stay on the nest until their energy deficit becomes too large or their body temperature gets too high, at which point the nest would likely be abandoned. LCT, lower critical temperature; UCT, upper critical temperature; PZT, physiological zero temperature; ULT, upper lethal temperature. (Figure from Conway and Martin 2000b; © 2000 Oxford University Press, used with permission)

incubating bird's energy budget, their responses to changing ambient temperatures, and the time spent incubating (Fig. 18.37).

Birds vary considerably in levels of nest attentiveness and that variation does translate into differences in the duration of incubation periods. Examination of nest attentiveness by incubating parents in 80 passerine species revealed that attentiveness ranged from about 50 to 95% and, further, that differences in attentiveness influenced the duration of incubation periods (Martin et al. 2007; Fig. 18.38). Decreased nest attentiveness means more time off the nest, resulting in cooler egg temperatures and, therefore, longer incubation periods. This relationship between attentiveness and incubation period is apparent overall as well as at specific locations (Fig. 18.38). This is important because incubation periods, regardless of attentiveness, tend to be longer in tropical areas than temperate areas because the embryos of birds in the tropics tend to take longer to develop than those of birds in temperate areas. For example, Robinson et al. (2008) took eggs from nests of House Wrens (*Troglodytes aedon*) in Illinois (41°N) and Panama (9°N) and placed them in incubators in the lab where they were incubated at a temperature of 38°C. Although incubated at the same

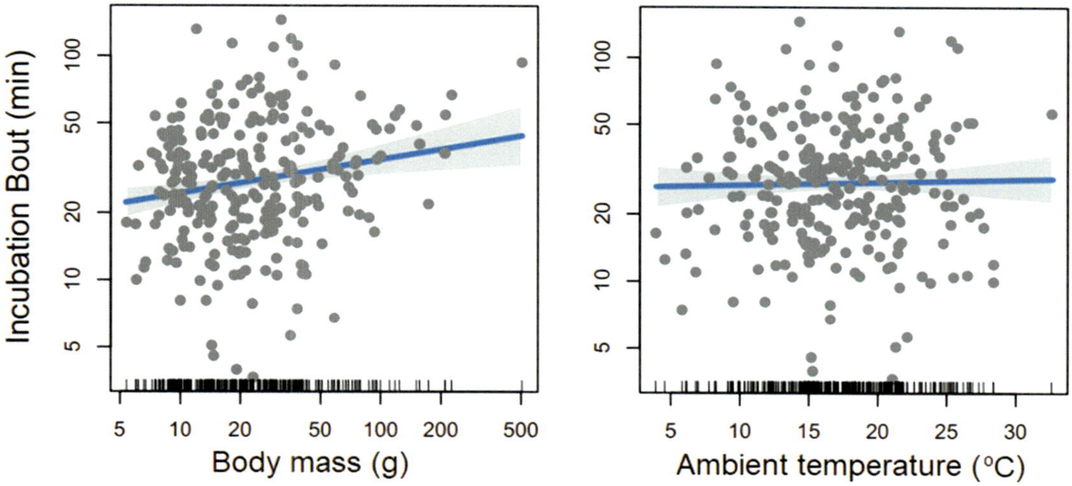


**Fig. 18.36** Left, incubation bouts are typically longer for larger birds that are better able to store energy as fat reserves. Right, Based on data for 256 species of songbirds, ambient temperate was found to have no significant effect on the duration of incubation bouts. (Figure from Matysioková and Remeš 2018; © 2018 The Authors. Evolution © 2018 The Society for the Study of Evolution, used with permission)

temperature, the eggs of House Wrens from Panama hatched an average of 1.33 days later than eggs from wrens in Illinois.

Reduced nest attentiveness may lower egg temperatures and lengthen the incubation period. Because a longer incubation period means that eggs are vulnerable to predation for a longer period, reduced nest attentiveness would seemingly be selected against because of the likely reduction in nest success. If so, why is there so much intra- and, especially, interspecific variation in nest attentiveness? The just-mentioned difference in the duration of incubation periods of tropical and temperature birds will be discussed in more detail later.

Several factors might explain reduced nest attentiveness by incubating parents. One possibility is that longer incubation periods could be advantageous if slower embryonic development enhances offspring quality (Ricklefs et al. 2017). For example, longer incubation periods might allow enhanced development of an embryo's

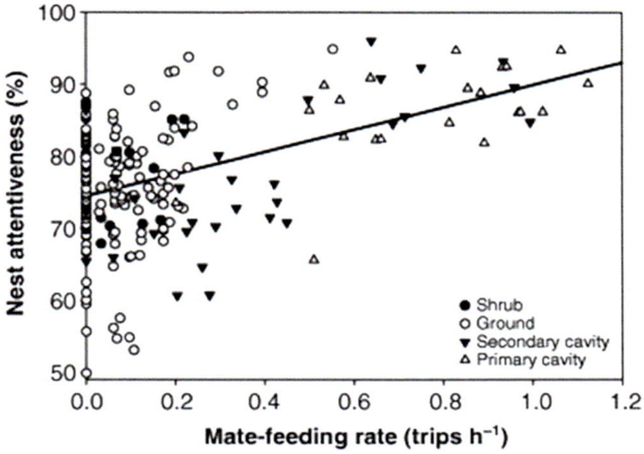


**Fig. 18.37** Nest attentiveness (percent of time females spend on the nest incubating) increases with increasing rates of incubation- (or mate-) feeding by males, regardless of nesting location. Mate-feeding rates have been log transformed. (Figure from Fontaine et al. 2007; © Oikos, used with permission)

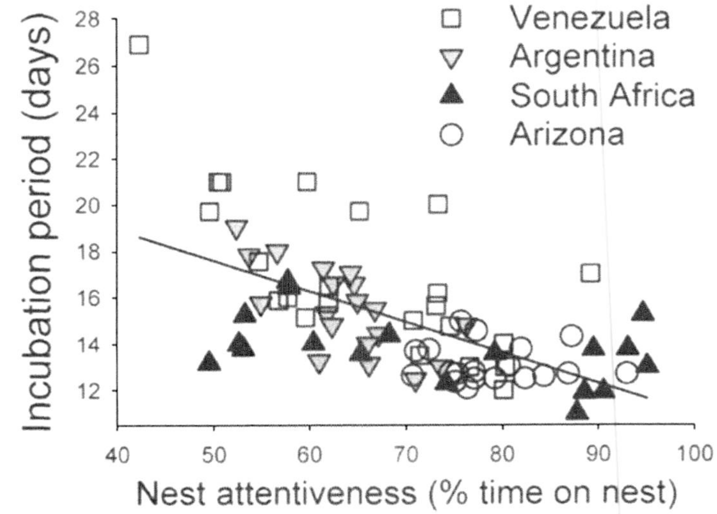


**Fig. 18.38** Relationship between nest attentiveness and incubation period for 80 passerine species studied at four sites on three continents. (Figure from Martin et al. 2007; © Society for the Study of Evolution, used with permission)

immune system (Ricklefs 1992, 1993). This hypothesis was proposed after Ricklefs (1992) found that birds with longer incubation periods tended to have fewer blood parasites (and, presumably, more effective immune systems). In support of this hypothesis, Lee et al. (2008) found that birds with longer incubation periods had higher concentrations of antibodies in their blood plasma (Fig. 18.39). This positive relationship between incubation period and antibody levels is not surprising because antibody diversification only occurs prior to hatching in birds (McCormack et al. 1991).

Reduced nest attentiveness might also be due to reduced food availability, with parents having to spend more time away from nests searching for food when availability is low. One way to test this hypothesis is to provide incubating adults with supplemental food. If food availability limits nest attentiveness, then a parent provided with additional food would be expected to spend more time incubating eggs. In a test of the food-limitation

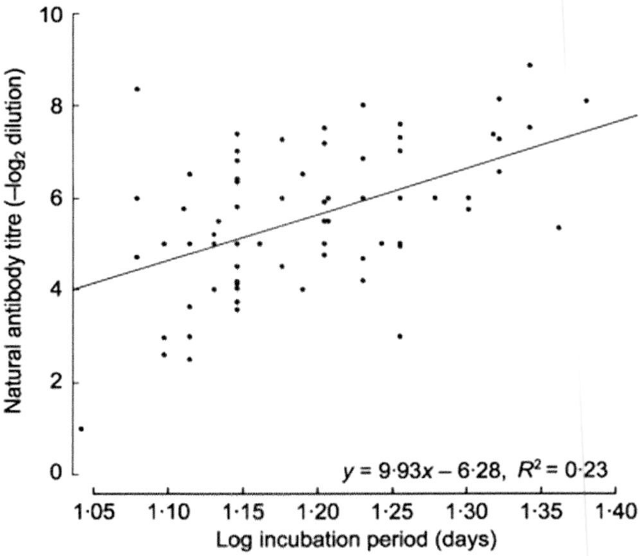


**Fig. 18.39** Relationship between incubation period and the concentration of antibodies in the blood plasma for 70 species of tropical birds. Points represent means for each species. (Figure from Lee et al. 2008; © 2008 The Authors. Journal compilation © 2008 British Ecological Society, used with permission)

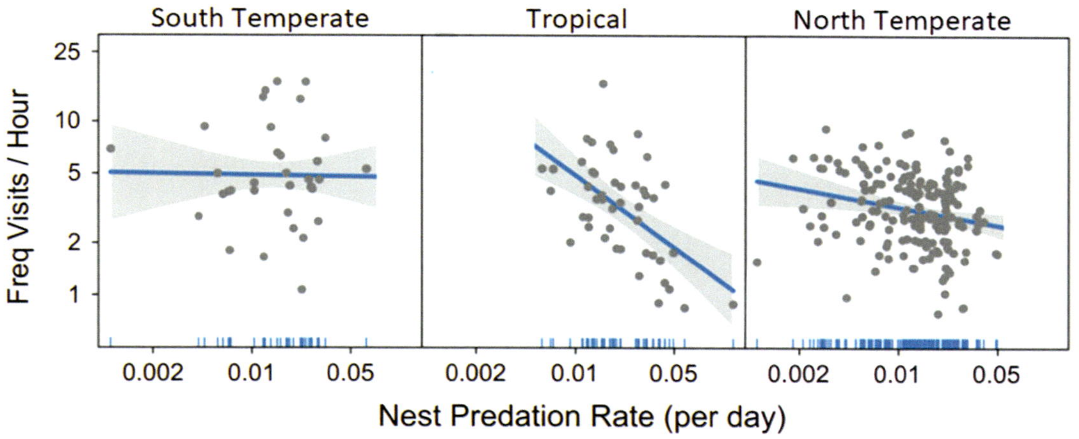


**Fig. 18.40** Using data from 256 species of songbirds worldwide, Matysioková and Remeš (2018) examined the relationship between the frequency of visits to nests by adults and rates of nest predation and found that adults at north-temperate latitudes and, especially, tropical latitudes visited nests less frequency with an increasing risk of nest predation. Tropical species may exhibit a greater response to the risk of nest predation because they live longer and have lower annual fecundity, factors that select for minimizing the risk of predation. Reasons for the lack of relationship for south temperature species were unclear, but may have been due to a smaller sample size (34 species vs. 46 and 176 in the tropics and north-temperate latitudes, respectively. South temperate, <23.5° S; tropical, between 23.5°N and 23.5°S; north temperature, > 23.5°N. (Figure from Matysioková and Remeš 2018; © 2018 The Authors. Evolution © 2018 The Society for the Study of Evolution, used with permission)

hypothesis, some incubating Karoo Prinias (*Prinia maculosa*), a small passerine found in southern Africa, were provided with supplemental food. Although nest attentiveness of the supplemented females was a bit higher than that of non-supplemented females (57% vs. 49%), the increase in time spent incubating by supplemented females was minimal and supplemented females were even observed preening and resting on the feeders when they could have been incubating. These results suggest that, for Karro Pinias and, probably, other species of birds, food availability typically has little effect on nest attentiveness (Chalfoun and Martin 2007).

Nest predation is the primary cause of reproductive failure for most birds and, therefore, natural selection would favor parental strategies that limited such predation. For example, higher predation rates might favor fewer nest visits, i.e., longer incubation bouts or longer foraging bouts, because fewer trips might reduce the chances of a nest being discovered by predators (Skutch 1949). The results of several studies lend support to this hypothesis (Conway and Martin 2000b; Muchai and du Plessis 2005; Matysioková and Remeš 2018; Fig. 18.40). In some species, nest visitation rates are lower in populations occupying habitats with higher rates of predation (Massaro et al. 2008; Pretelli et al. 2016). In some cases, longer incubation bouts may be advantageous for reasons beyond limiting the number of nest visits. Meyer et al. (2020) monitored 714 nests of seven species of sandpipers breeding in the Arctic and analysis revealed that, for species with uniparental and biparental incubation, daily predation rates increased with the daily total duration of recesses and with the mean duration of recesses. These results suggest that, for ground-nesting sandpipers in the Arctic, longer incubation bouts are beneficial because incubating adults with cryptic plumage are more difficult to detect than unattended nests (Meyer et al. 2020).

Among birds, the contributions of males to parental care exhibits much variation. Matysioková and Remeš (2014) examined the contributions of males to incubation in

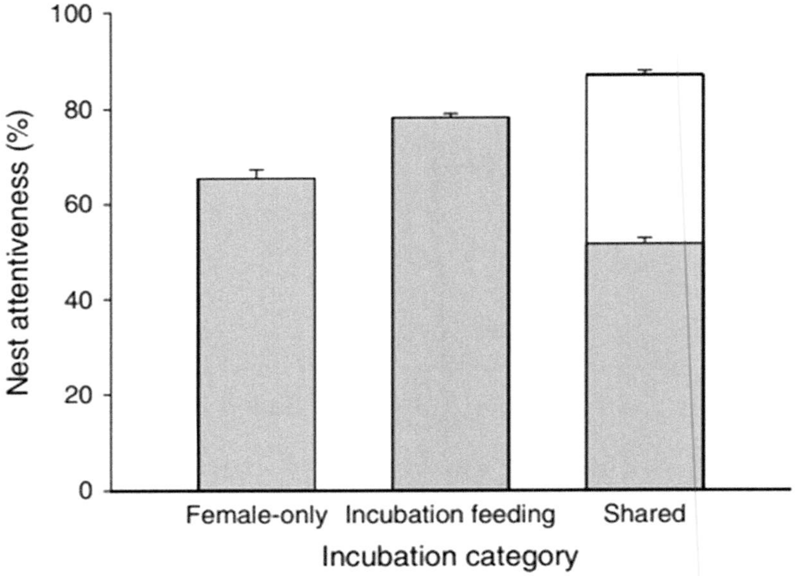


**Fig. 18.41** Mean (± SE) nest attentiveness in 320 species of songbirds with different incubation strategies. Female attentiveness is depicted in gray, male attentiveness in white. Both incubation feeding and male contributions to incubation increased nest attentiveness. (Figure from Matysioková and Remeš 2014; open-access article distributed under the terms of the Creative Commons Attribution License, https://creativecommons.org/licenses/by/4.0/)

528 populations of 320 species of songbirds and found that nest attentiveness was greatest in species where both males and females incubated eggs, next highest for species where males fed incubating females, and lowest for species with female-only incubation (Fig. 18.41). For species where both males and females contributed to incubation, total nest attentiveness increased when males spent more time incubating, but total attentiveness did not change based on female attentiveness. A possible explanation for such results is that energetic limits place limits on female nest attentiveness so only increases in male attentiveness can increase total nest attentiveness. For species where males fed incubating females, female nest attentiveness was higher in species where males fed females at higher rates. Factors contributing variation among species in male contributions to incubation likely include the time and energy males must spend defending territories and the extent to which males in different species may or may not be able to obtain extrapair copulations (Chiver et al. 2007; Matysioková and Remeš 2014)

As with clutch size, the duration of songbird incubation periods varies with latitude, with longer incubation periods at lower latitudes than at higher latitudes. One possible reason for this difference is that parents at lower latitudes tend to spend less time incubating eggs and more time off nests than parents at higher latitudes (Martin et al. 2015; Fig. 18.42). Less time incubating means longer incubation periods. In addition, embryos of at least some species that breed at lower latitudes naturally grow more slowly than embryos of species at higher latitudes (Martin et al. 2007; Robinson et al. 2008). For example, Robinson et al. (2014) incubated the eggs of eight species of tropical birds in incubators at constant temperatures and found that the time to hatch remained similar to that of eggs naturally incubated by adults that took occasional breaks during which eggs cooled. Such results suggest that the embryos of at least some species of tropical birds do grow more slowly than embryos of species at higher latitudes.

Longer incubation periods could potentially benefit both parents and embryos. To the extent that less time spent incubating contributes to longer incubation periods, parents may reduce the likelihood of being predated while on nests (Martin et al. 2007, 2015). However, longer incubation periods mean a greater risk of predation so any benefit to parental fitness (i.e., increased likelihood of survival) must outweigh the potential cost of losing eggs to predation. Another

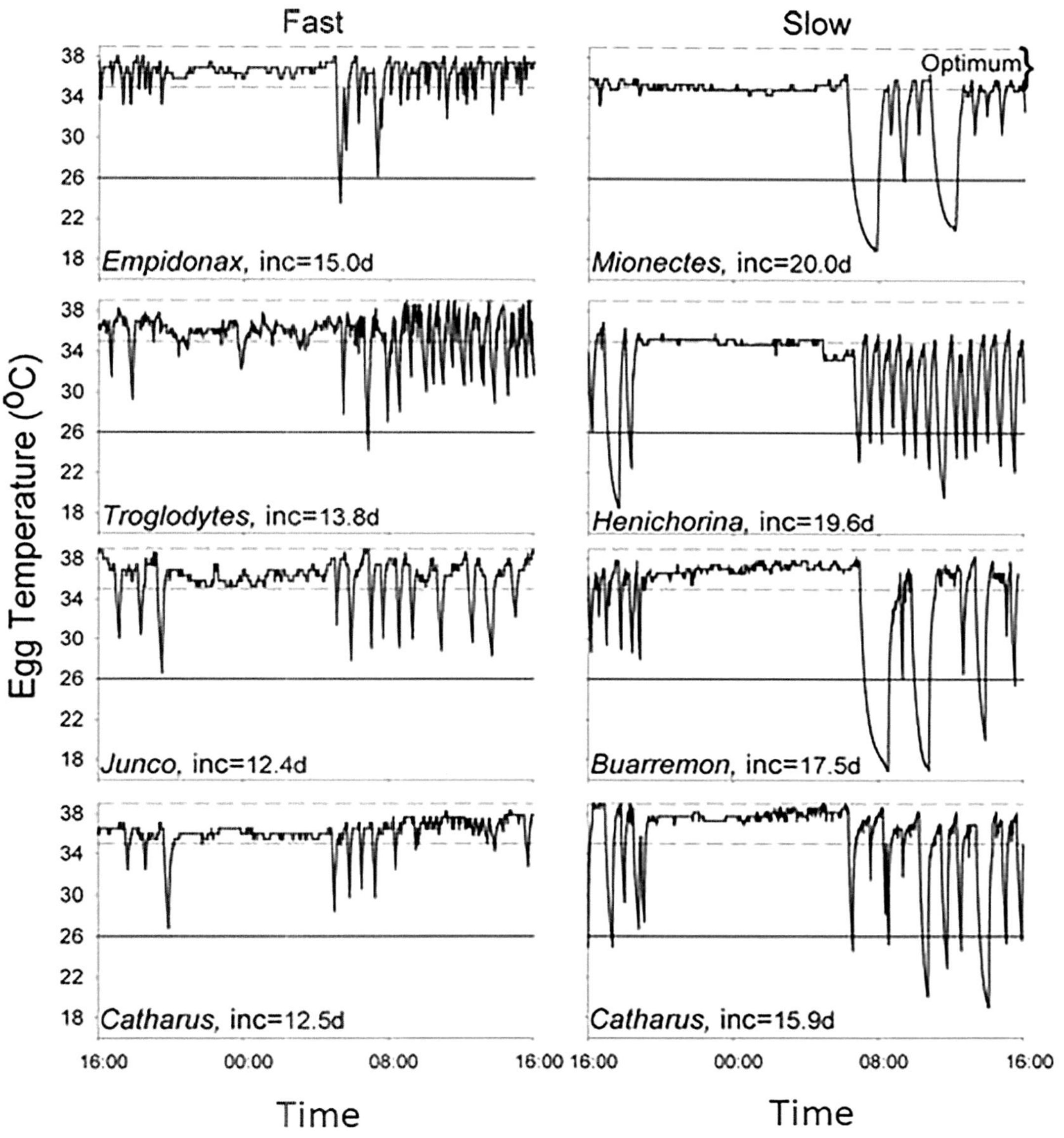


**Fig. 18.42** Examples of daily temperature fluctuation in comparisons of phylogenetically related passerine species with fast (short incubation periods) versus slow (long incubation periods) embryonic development during days 2, 3, or 4 of incubation in Arizona (left) versus Venezuela (right). The gray line in each cell represents the temperature below which development has been thought to be dramatically slowed, whereas the two dashed lines at the top of each cell represents the optimum range of temperatures for embryonic development. (Figure from Martin et al. 2007; © Society for the Study of Evolution, used with permission)

possibility is that longer incubation periods benefit embryos. If so, this would help explain the intrinsic tendency of embryos of some tropical species to grow more slowly than embryos of species at higher latitudes. In support of this hypothesis, as noted above, species of tropical songbirds with longer incubation periods have more antibodies (Ricklefs 1992), and also have fewer parasites in their blood than those with shorter incubation periods (Ricklefs et al. 2017).

Although potentially beneficial for embryos as well as the parents of higher-quality offspring, the

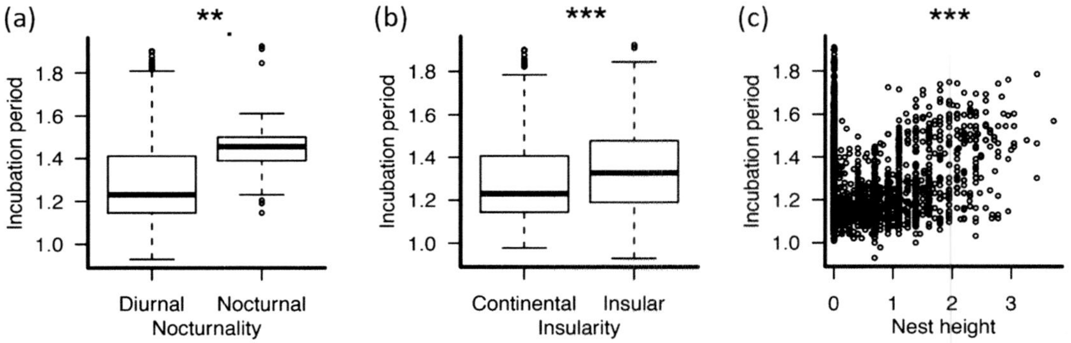


**Fig. 18.43** Three variables that can influence the duration of avian incubation periods. (**a**) nocturnality (active diurnally or nocturnally), (**b**) insularity (nesting on a continent or island), and (**c**) nest height above ground. **, $P < 0.01$; ***, $P < 0.001$. (Figures [Supplemental] from Cooney et al. 2020; open-access article licensed under a Creative Commons Attribution 4.0 License, https://creativecommons.org/licenses/by/4.0/)

possible costs of longer incubation periods may outweigh those benefits and lead to shorter incubation periods. For example, longer incubation periods may lead to increased sibling competition. Nestlings compete with each other for food delivered by parents and, because larger, earlier hatched nestlings generally outcompete smaller, later-hatched nestlings (Ricklefs 1965), selection may favor faster embryonic development and early hatching (Werschkul and Jackson 1979; Ricklefs 1993).

Based on data for 3096 species of birds representing 176 families and 39 orders, Cooney et al. (2020) found that the duration of incubation periods can also be influenced by nocturnality, insularity, and nest height (Fig. 18.43). Nocturnal species may have longer incubation periods because, at night, parental activity at nests is less likely to be detected by potential predators and this reduced risk of nest predation relaxes selection for rapid development by embryos (Cooney et al. 2020). Species that nest on islands tend to have longer incubation periods than continental species, likely because, at least historically, islands tend to have fewer predators, again relaxing selection for rapid development by embryos (Sibly et al. 2012; Cooney et al. 2020). Similarly, nests that are higher above ground may have a reduced risk of predation (Fig. 18.43).

## 18.11 Development of Avian Embryos

The development of bird embryos has been studied for well over 100 years, with initial descriptions of the process provided by Duval (1889) and Keibel and Abraham (1900). Not surprisingly, these initial studies focused on the embryonic development of Domestic Chickens (*Gallus g. domesticus*). Hamburger and Hamilton (1951) divided the development of Domestic Chickens into a series of 46 stages (Box 18.8 Hamburger–Hamilton Stages) and these stages (slightly modified; 42 stages are now generally recognized) are still used as the basis for describing the development of bird embryos, regardless of species. Surprisingly, the 42 currently recognized stages can be applied to all species so far studied, suggesting that these stages represent a general pattern of development for all or most birds regardless of mode of development (e.g., precocial, altricial, or semialtricial; Ricklefs and Starck 1998b; Figs. 18.44, 18.45, and 18.46). The order and duration of these stages are especially similar during the early stages of development (stage 1 to stage 33). During these stages, the body plan develops and early phases of organogenesis and tissue differentiation take place. From stage 34 to stage 38, species-specific

**Box 18.8 Hamburger–Hamilton Stages**

Description of the developmental stages of birds has a long history, with complete developmental series of embryos of Domestic Chickens published by Duval (1889) and Keibel and Abraham (1900). Years later, Hamburger and Hamilton (1951) published a developmental series of Domestic Chickens that, with some slight revision, is still in use today. Early stages were based on the development of the neural tube and early organization of embryo's body plan. Later stages were based on such things as the development of the beak, feathers, eyelids, wings, and hindlimbs.

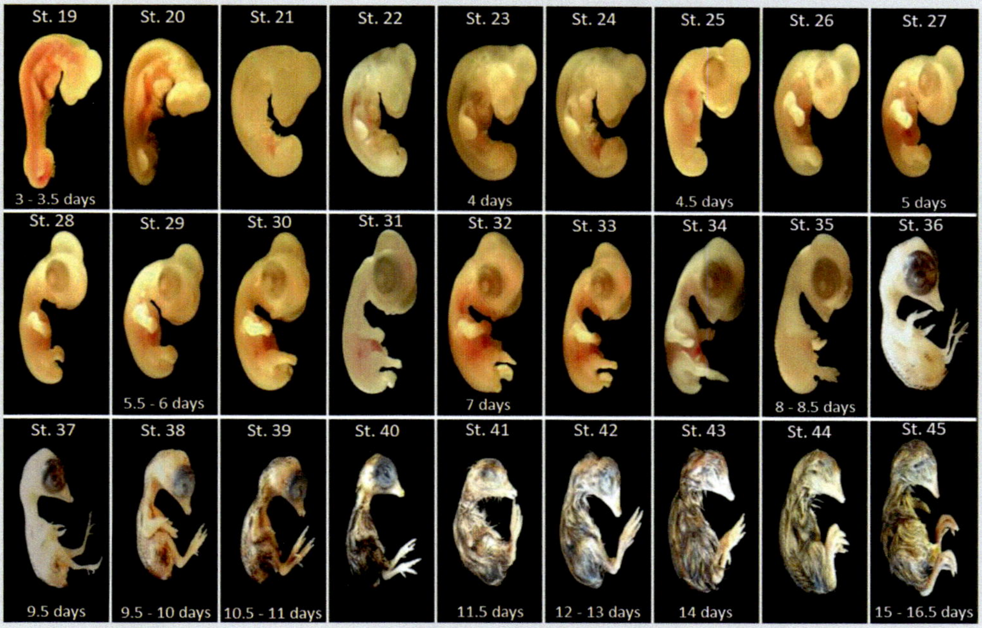


Embryonic developmental stages of precocial Japanese Quail. (Figure modified from Ramli et al. 2017, used with permission)

Summary of events associated with various developmental stages of a Domestic Chicken (Ricklefs and Starck 1998b):

| | |
|---|---|
| Stage 4 (18–19 h) | Primitive streak reaches maximum length |
| Stage 9 (29–33 h) | Optical vesicles visible; paired primordia of heart begin to fuse |
| Stage 16 (51–56 h) | Primordia of wings and legs present |
| Stage 20 (70–72 h) | Pigmentation of eyes, light exposure increases activity of embryo |
| Stage 24 (4 days) | Most of forebrain structures have formed, olfactory bulb has formed; active movements of head and neck |
| Stage 27 (5 days) | Beak barely visible; active movements of trunk; formation of reproductive organs |
| Stage 30 (6.5 days) | Major segments of wings and legs clearly visible; egg tooth distinct |
| Stage 35 (8–9 days) | Phalanges of toes distinct; eyelid begins to overgrow eyeball |

(continued)

**Box 18.8** (continued)

| | |
|---|---|
| Stage 36 (10 days) | Nostrils have narrowed to a slit, flight feathers conspicuous, digestive processes begin; body movements jerky and random |
| Stage 37 (11 days) | First synapses present in cerebellum, bipolar cells in retina begin to differentiate |
| Stage 39 (13 days) | Scales overlap on legs; differentiation of neurons in retina complete |
| Stage 40 (14–18 days) | Scales, claws, and beak becoming firm |
| Stage 41 (20 days) | Yolk sac half enclosed in body cavity; external pipping |
| Stage 42 (21 days) | Hatching |

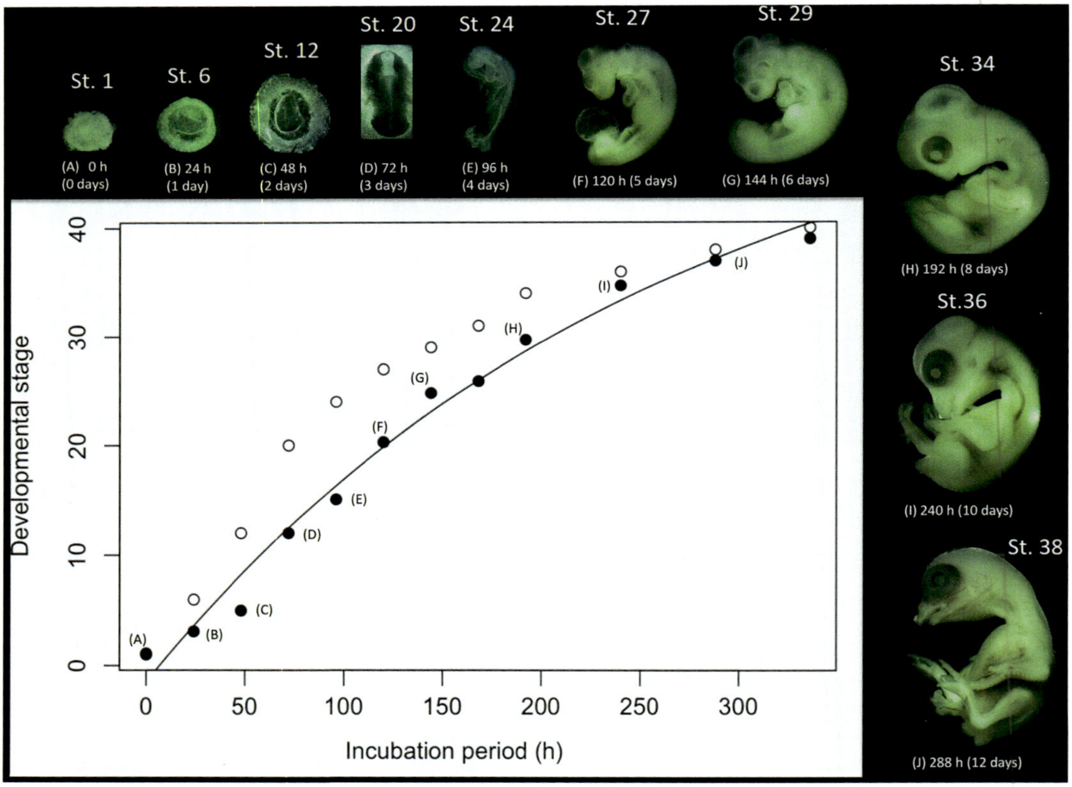


**Fig. 18.44** Rate of embryo development during the incubation period of an altricial Zebra Finch (*Taeniopygia guttata*; filled circles and line) compared to the development rates of a Domestic Chicken (open circles). Developmental stages are based on the stages of development of Domestic Chickens created by Hamburger and Hamilton (1951). (Figure from Hemmings and Birkhead 2015; © 2015 British Ornithologists' Union, used with permission)

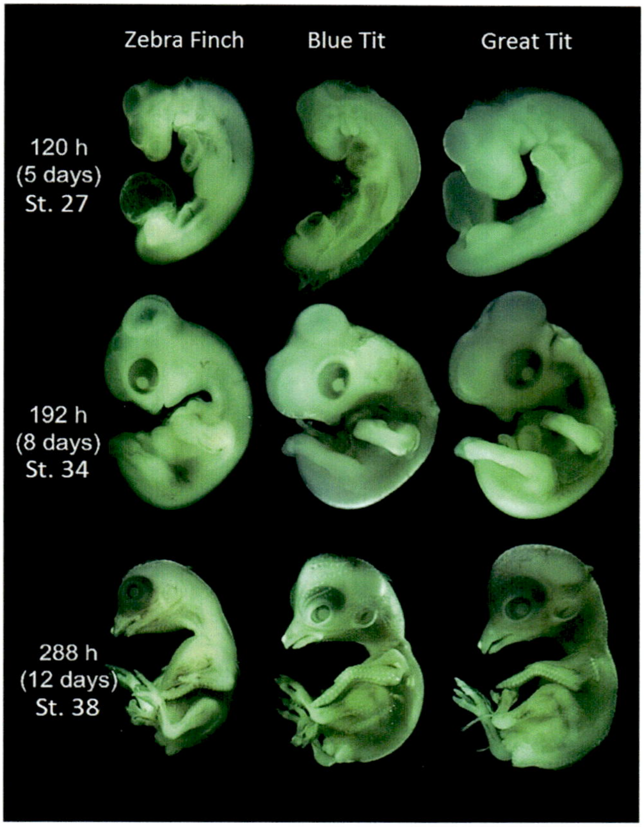


**Fig. 18.45** Embryonic development of three species of altricial songbirds (not to scale). Zebra Finch, *Taeniopygia guttata*; Eurasian Blue Tit, *Cyanistes caeruleus*; Great Tit, *Parus major*. (Figure from Hemmings and Birkhead 2016; © 2015 British Ornithologists' Union, used with permission)

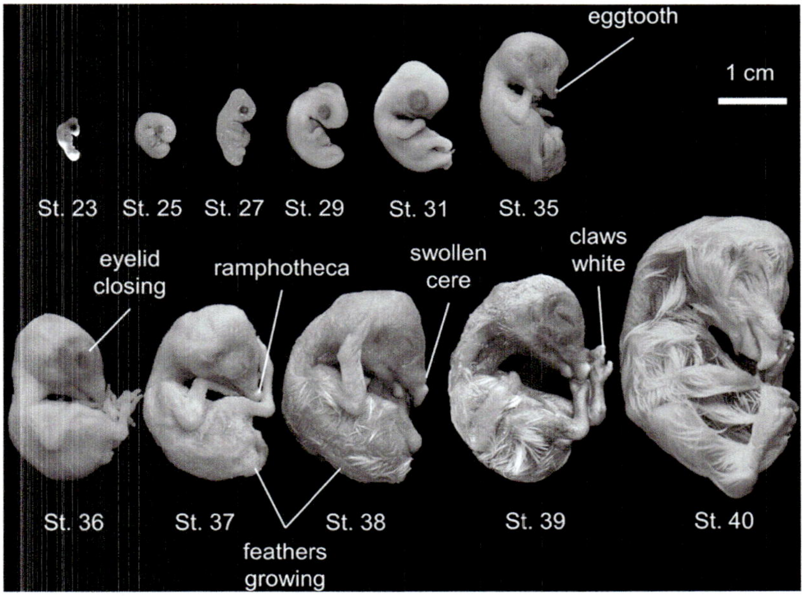


**Fig. 18.46** Embryonic development of semialtricial Little Owls (*Athene noctua*). St. = developmental stage. St. 23 = about 11 days old, St. 29 = about 13.5 days old, St. 35 = about 18.5 days old, St. 37 = about 21 days old, and St. 40 = about 28-33 days old when eggs typically hatch. (Figure from Krings et al. 2020; © 2019 Springer Nature, used with permission)

characters develop and the relative length of stages exhibits more interspecific variation. For example, stage 39 lasts 24–36 h in quail (18-day incubation period), 48 h in Domestic Chickens (21 days), 96 h in Muscovy Ducks (*Cairina moschata*; 35 days), and 18–24 h in Budgerigars (*Melopsittacus undulatus*, 18 days; Ricklefs and Starck 1998b). As these durations illustrate, development typically takes longer for precocial species (like quail, chickens, and ducks) than altricial species (like Budgerigars). In general, precocial birds grow relatively slowly, semiprecocial and semialtricial birds have intermediate growth rates, and altricial birds typically have rapid growth rates (Starck 1993). On average, altricial embryos grow three to four times faster than precocial embryos.

During development in an egg, yolk and albumen are being converted into an embryo, with yolk (and, specifically, lipids or fats in the yolk) providing energy and albumen providing protein and water. This conversion process involves both differentiation and growth. As described by Deeming (2002a), differentiation of avian embryos involves the development of a three-dimensional structure from a flat plate of cells (blastoderm) on top of the yolk. The center of the blastoderm will become the embryo and peripheral cells give rise to the yolk sac and amnion. During a process called gastrulation, the developing embryo is rearranged into three layers: endoderm, mesoderm, and ectoderm. The endoderm will ultimately form the nervous system, skin, and parts of the eye, the mesoderm becomes skeletal muscles, the skeleton, and internal structures, such as the circulatory and excretory systems, and the endoderm forms the digestive tract, respiratory passages, and organs such as the liver and pancreas.

Within eggs, four extraembryonic membranes support the life and growth of avian embryos. The amnion surrounds only the embryo and the inner layer of cells secretes amniotic fluid in which the embryo floats; this fluid keeps the embryo from drying out and protects it. The chorion surrounds all embryonic structures and serves as a protective membrane. The allantois (or allantoic sac) continues to grow as the embryo grows and fuses with the chorion to form the chorioallantoic membrane (CAM). Structurally, the CAM consists of three layers: the chorionic epithelium that is in contact with the inner shell membrane, the intermediate mesoderm with numerous blood vessels, and the allantoic epithelium that lines the allantoic cavity (Fig. 18.47). As the CAM develops and increases in size, blood capillaries invade the mesodermal layer (Fig. 18.47). The growing CAM eventually surrounds the embryo and other egg contents.

The extraembryonic tissues, like embryonic tissues, are composed of cells from the three germ layers: ectoderm, mesoderm, and endoderm (Fig. 18.48). The space between the amnion and the chorion is the extraembryonic coelomic (body) cavity. The yolk sac endoderm consists of a single layer of epithelium that, along with the yolk sac mesoderm, transports nutrients from the yolk to the blood (Fig. 18.49). As the incubation period progresses, the inner endodermal layer of the yolk sac membrane increases in area (Fig. 18.50) and spreads over the surface of the yolk, and this is followed by the development of blood vessels in the mesodermal layer (Figs. 18.51 and 18.52; Box 18.9 As the Egg Turns). Eventually, the yolk is completely surrounded and vascularized (Fig. 18.53), and the endodermal cells of the yolk sac membrane develop into epithelial absorptive cells (Speake et al. 1998).

The CAM has several important functions, including gas exchange, transporting calcium from the eggshell to the embryo, acid-base homeostasis, and reabsorption of sodium and chloride ions and water from the allantois (Maina 2017). From its initial development until internal pipping, the CAM is where the exchange of oxygen and carbon dioxide between the embryo (and, more specifically, capillaries in the CAM) and atmosphere occurs. This exchange of oxygen, carbon dioxide, and water vapor occurs by passive diffusion along partial pressure gradients (Figs. 18.54 and 18.55). Because the chorion is very thin, the distance through which gases must diffuse between capillaries in the CAM and the inner shell membrane is less than 1 μm (Mortola 2009). Of course, the number, size, and shape of eggshell pores also influence the diffusion of gases into and out of eggs

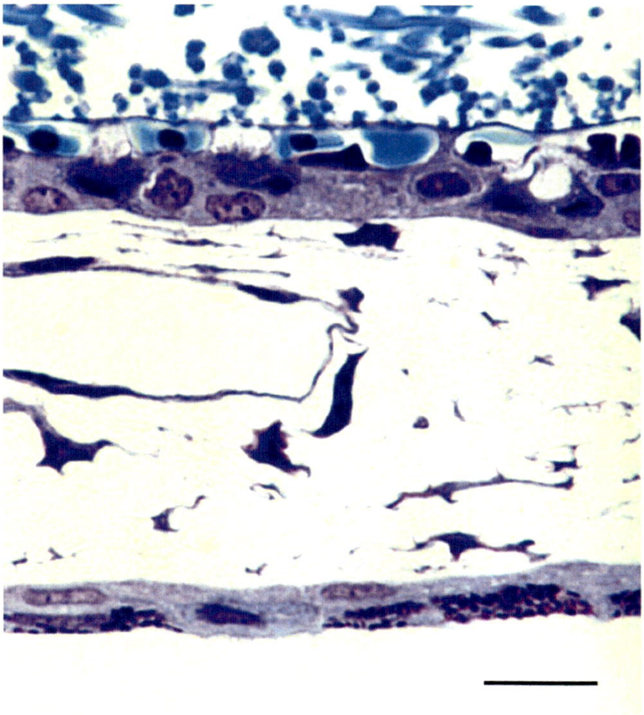


**Fig. 18.47** The chorioallanotic membrane stained blue. The chorionic epithelium adheres to the internal shell membrane, and blood capillaries are located in the intermediate mesodermal layer. The allantoic epithelium lines the allantoic cavity. Scale bar = 8 μm. (Figure modified from Gabrielli and Accili 2010; © 2010 Maria Gabriella Gabrielli and Daniela Accili, open-access article distributed under the Creative Commons Attribution License, https://creativecommons.org/licenses/by/4.0/)

(Fig. 18.56). Larger eggs of larger species of birds have more and larger pores than smaller eggs so gases diffuse at higher rates (Fig. 18.57).

The CAM is also important in transporting and mobilizing calcium from the eggshell to embryos, primarily for incorporation into developing bones. The mammillary knobs are the primary source of this calcium (Bond et al. 1988; Figs. 18.58 and 18.59). The mechanism by which calcium is removed from the shell remains to be determined, but a number of hypotheses have been proposed. One possibility is that calcium is eroded from the eggshell by acid released from cells (called villus cavity cells) in the chorionic epithelium, with acid generated by the release of hydrogen ions as carbon dioxide ($CO_2$) is converted into bicarbonate ion ($HCO_3^-$) by an enzyme called carbonic anhydrase (Tuan 1984, 1987; Tuan and Ono 1986). Another possibility is that the release of calcium from the eggshell is due to the action of uric acid, the urinary waste product produced by avian embryos and stored in the allantois. During incubation, the allantoic fluid becomes more acidic as levels of uric acid increase and this acid may erode calcium from the eggshell (Österström et al. 2013). Finally, Terepka et al. (1976) and Tuan (1987) proposed that endosomes (calcium-loaded endocytotic vesicles) serve as intracellular calcium carriers.

As calcium is removed from eggshells during embryonic development, eggshells get thinner. The extent to which this occurs varies among species. For example, the eggshells of mound-nesting Malleefowl (*Leipoa ocellata*) become 21% thinner (Booth and Seymour 1987). However, Orłowski and Hałupka (2015) reviewed the literature and found that eggshell thickness declines by about 4–8% for most species of birds, and that the amount of thinning differs among different regions of eggs. Although based on just five species, the blunt end of eggs (mean = 2.6%) tends to thin less than the equatorial (5.6%) and sharp (6.2%) ends (Orłowski and

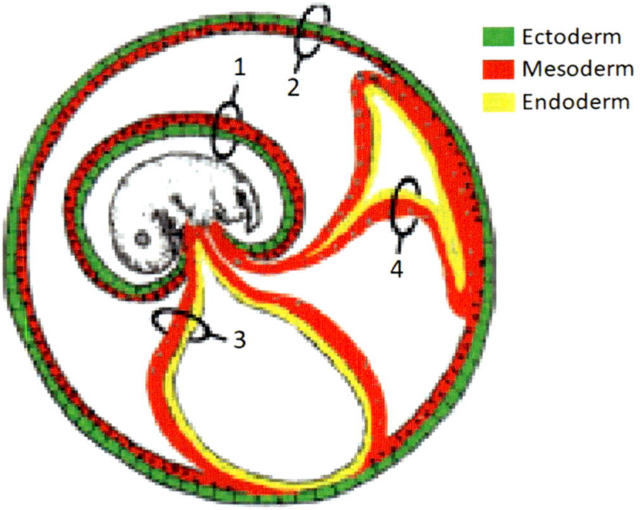


**Fig. 18.48** Extraembryonic membranes in an avian egg. (1) double-layered amnion with an outer mesoderm layer and inner ectoderm layer, (2) double-layered chorion with an outer ectoderm layers and inner mesoderm layer, (3) yolk sac membrane with a mesoderm layer and inner endoderm layer, and (4) double-layered allantoic membrane with an outer mesoderm layer and inner endoderm layer. (Figure modified from Sheng and Foley 2012; © 2012 New York Academy of Sciences, used with permission)

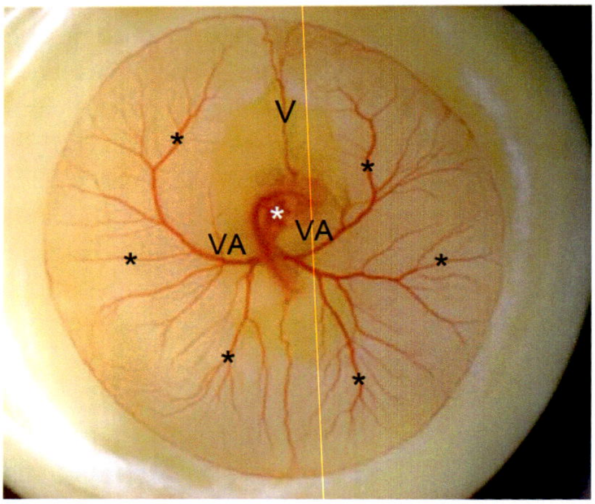


**Fig. 18.49** Development of blood vessels in the growing yolk sac in a fertilized egg of a Domestic Chicken. V, cranial vein; VA, vitelline arteries; dark asterisk, arteries; white asterisk, heart. (Figure modified from Nguyen et al. 2006; Reprinted with permission from Physical Review E, Copyright 2006 by the American Physical Society)

Hałupka 2015). These authors also found no difference in the extent of eggshell thinning among altricial, precocial, and semi-precocial species of birds. The eggs of altricial and precocial species do, however, tend to differ in mammillary density, i.e., the numbers of mammillary tips per unit of inner eggshell surface area (Osterstrom and Lilja 2012; Fig. 18.60). In addition, Österström et al. (2013) found that more calcium was removed from the mammillary tips of a precocial species (Japanese Quail, *Coturnix japonica*) than from those of an altricial species (European Starling, *Sturnus vulgaris*). Other studies have revealed that mammillary density and calcium removal are related to growth rate and mode of development (Karlsson and Lilja 2008; Osterström and Lilja 2012; Fig. 18.60), with the young of precocial species growing more slowly and requiring more calcium for bone growth than the young of altricial species.

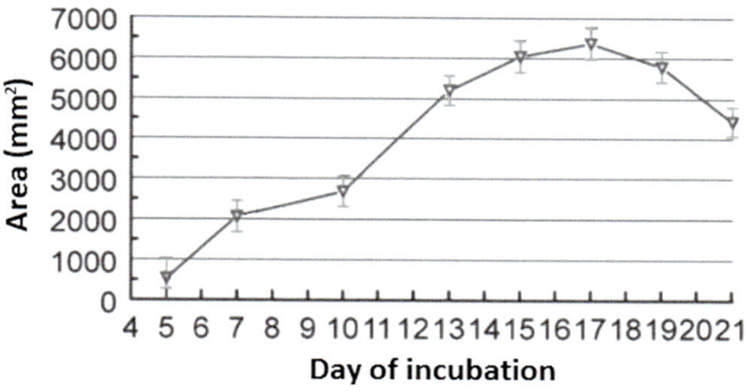


**Fig. 18.50** Variation in mean area (± 95% CI) of yolk sac membrane of a Domestic Chicken embryo from day 4 through day 21 of incubation. (Figure modified from Yadgary et al. 2013; © 2013 Oxford University Press, used with permission)

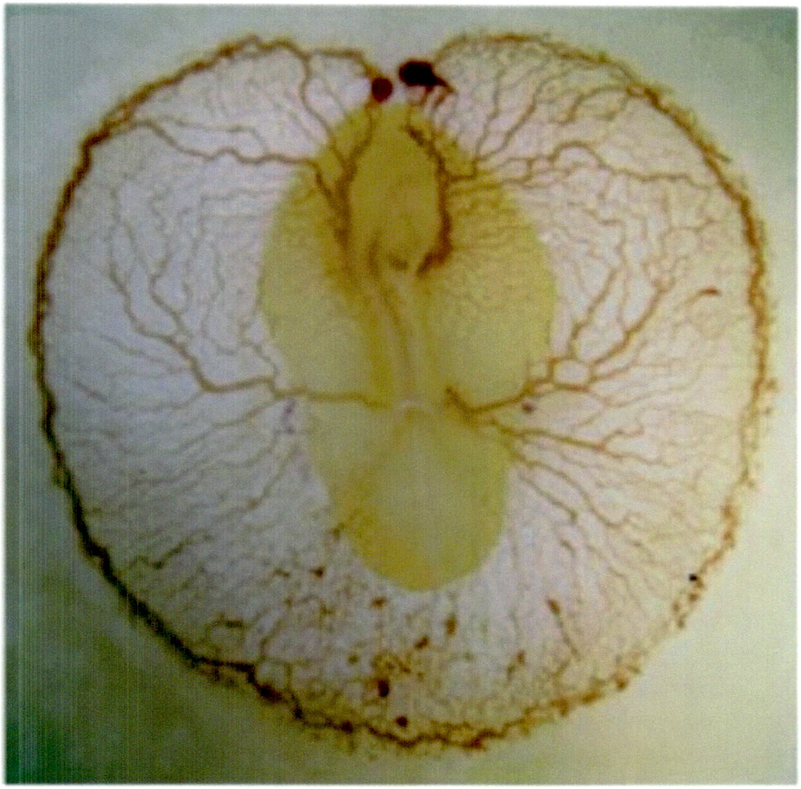

**Fig. 18.51** Emu (*Dromaius novaehollandiae*) embryo. Blood vessels develop in the mesodermal layer of the yolk sac membrane. The mesodermal layer spreads across the yolk during the early portion of the incubation period creating an area vasculosa (with mesoderm and blood vessels) and an area vitelline (no mesoderm and blood vessels). Eventually, the area vasculosa completely covers the yolk. (Figure from Nagai et al. 2011; © 2010 Wiley-Liss, Inc., used with permission)

## 18.12 Nutrition and Growth of Developing Embryos

After an egg is fertilized and incubation begins, embryos begin a developmental process that includes three main stages (Fig. 18.61). Embryonic development requires access to oxygen and nutrients. Initially, access to oxygen is limited to simple diffusion through eggshell pores to immature red blood cells (with hemoglobin) that are produced in the yolk sac (Fig. 18.62). However, the chorioallantoic membrane soon begins to develop and grow larger, providing embryos with the oxygen needed to continue, and increase the rate of, their growth (Fig. 18.63). Although gas conductance through eggshells does not change during incubation, the exchange of gases, including oxygen, carbon dioxide, and

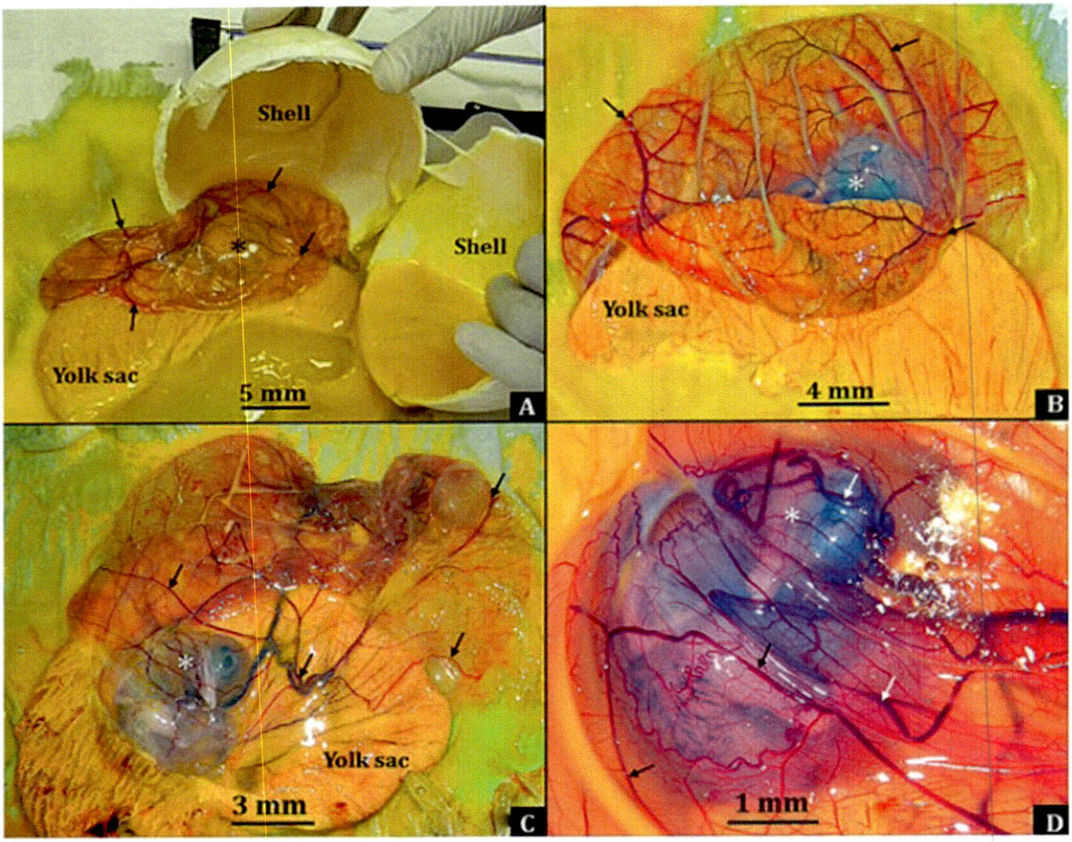


**Fig. 18.52** Growth of the chorioallantoic membrane in an egg of a Common Ostrich (*Struthio camelus*) from (**a**) day 4 to (**b**) day 14 to (**c**) day 23 to (**d**) day 28 post-fertilization. Arrows, blood vessels; asterisk, developing embryo. (Figure from Maina 2017; © 2017 Springer International Publishing AG, used with permission)

water vapor essential for the development of embryos (Fig. 18.64), between the atmosphere and growing embryos becomes more efficient during incubation as the chorioallantoic membrane becomes more vascular, blood volume increases, and the concentration of red blood cells in the blood increases (Mortola 2009). Toward the end of incubation, embryos pierce the internal shell membrane to gain access to the air cell (internal pipping) and, later, by breaking the eggshell to gain access to ambient air (external pipping) and beginning to breathe.

As incubation progresses, water vapor is lost through the eggshell membranes and shell and replaced by air. This loss of water during incubation, which is typically about 10–20% of the mass of a freshly laid egg, is critical for normal embryonic development (Ar and Rahn 1980; Tazawa and Whittow 2000). Not surprisingly, the larger eggs of larger species of birds lose more water during incubation than the smaller eggs of smaller species (Figs. 18.65 and 18.66). Loss of water serves two important functions: (1) creating the air cell and a source of oxygen for the embryo after internal pipping, and (2) maintaining the appropriate level of hydration of egg contents (Ar and Rahn 1980). Water losses above about 20% of freshly laid egg mass may cause early depletion of allantoic fluid, and cause osmotic stress (due to elevation of solute concentrations) that results in dehydration of blood, amniotic fluid, and embryonic skin (Davis et al. 1988).

**Box 18.9 As the Egg Turns**

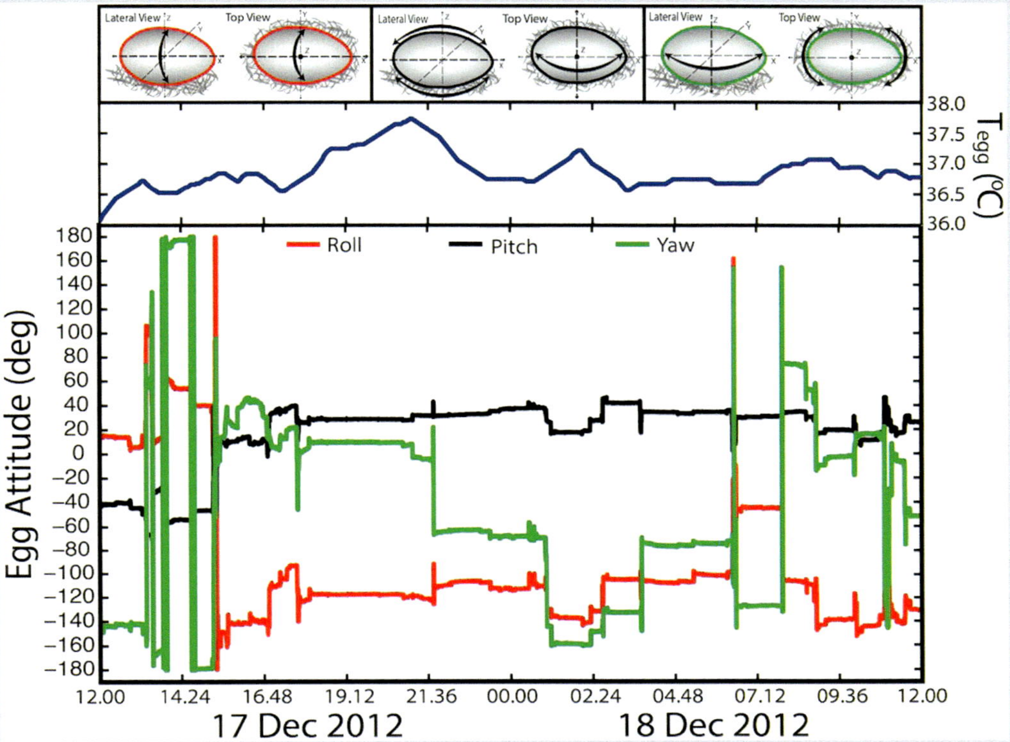


Representative 24-h time series showing egg attitude and temperature in a Laysan Albatross (*Phoebastria immutabilis*) nest. Shown are the roll, pitch, and yaw angles and the corresponding egg temperature. Note the large angular changes in yaw attitude compared to the attitudes of roll and pitch. (Figure from Shaffer et al. 2014; © 2014 Shaffer et al., open-access article distributed under the terms of the Creative Commons Attribution License, https://creativecommons.org/licenses/by/4.0/)

During incubation, adult birds alter the orientation of their eggs on a regular basis, a behavior referred to as egg turning. Egg turning (especially during the first few days of incubation) stimulates development of blood vessels in the area vasculosa, a highly vascularized area of the yolk sac that is important in nutrient uptake from the yolk by developing embryos. By enhancing nutrient uptake, egg turning increases growth rates of embryos and likely increases the hatchability and viability of embryos (Tona et al. 2005). Embryos in unturned eggs exhibit reduced oxygen consumption and heart rates, longer incubation periods, lower-than-normal mass, and reduced hatching success compared to turned eggs (Tullett and Deeming 1987; Clatterbuck et al. 2017). Although the importance of egg turning is well understood, little is known about the egg turning behavior of wild birds. To quantify egg-turning behavior, Shaffer et al. (2014) placed artificial eggs (comparable in size, shape, and mass to those of their focal species) with data loggers containing accelerometers and magnetometers in the nests of Cassin's Auklets (*Ptychoramphus aleuticus*), Western Gulls (*Larus occidentalis*), and Laysan Albatrosses. These species turned eggs frequently (at rates as high as 6.5 times per hour), but changes in orientation were usually small (mostly 1–10°). Overall, mean turning rates for these three species

(continued)

**Box 18.9** (continued)

ranged from 1.8 to 2.6 turns per hour. The greatest angular changes were in the yaw axis (green line above), with adults appearing to use their feet (rather than bills) to turn eggs. By using their feet, these birds minimize exposure of eggs and embryos to ambient temperatures (minimizing variation in egg temperatures as indicated in the graph above) and risk of predation.

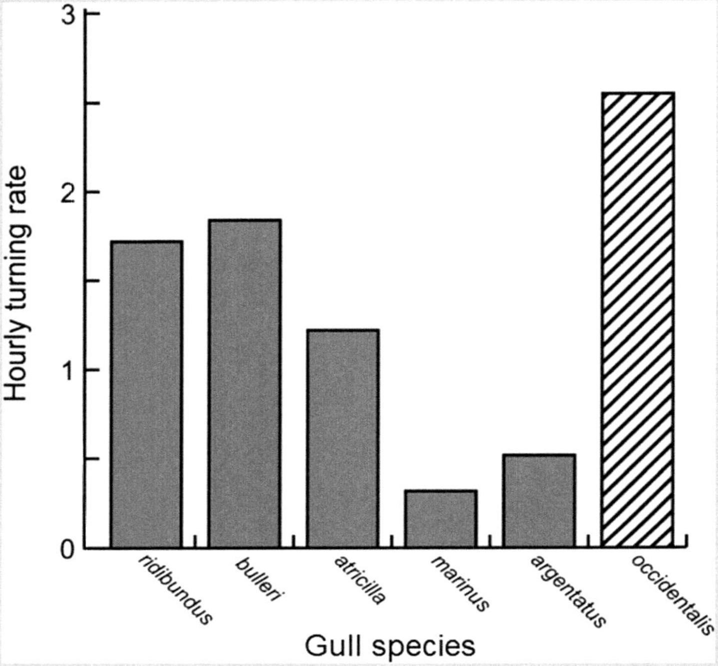


Mean hourly egg-turning rates for six species of gulls. Data are from Clatterbuck et al. (2017; *Larus occidentalis*, Western Gull), Beer (1961; *Chroicocephalus ridibundus*, Black-headed Gull), Beer (1965; *C. bulleri*, Black-billed Gull), Drent (1970; *L. argentatus*, Herring Gull), Impekoven (1973; *Leucophaeus atricilla*, Laughing Gull), and Butler and Janes-Butler (1983; *L. marinus*. Great Black-backed Gull). Data for *L. occidentalis* were collected using data loggers in artificial eggs; data for other species were collected via visual observations of incubating adults at nests. Egg-turning behavior varies among species, but tends to be consistent throughout incubation. (Figure from Clatterbuck et al. 2017; © 2017 Oxford University Press, used with permission)

Such eggs may not hatch due to desiccation (Carey 1986). Too little water loss and the air cell may be too small to provide the embryo with sufficient oxygen. In addition, if eggs do not lose enough water, embryos may have respiratory problems or even drown (Wangensteen and Rahn 1970). Hatching success is reduced when eggs lose either too much or too little water (Buhr 1995; Nahm 2001).

Developing embryos have two primary sources of water: water in the albumen and metabolic water, largely from the oxidation of lipids (Drent 1975). The rate at which this water is lost from eggs is determined by the water vapor conductance of the eggshell and shell membranes and the difference in water vapor pressure between the egg and the nest environment (Rahn and Ar 1974; Rahn et al. 1976). Because birds nest

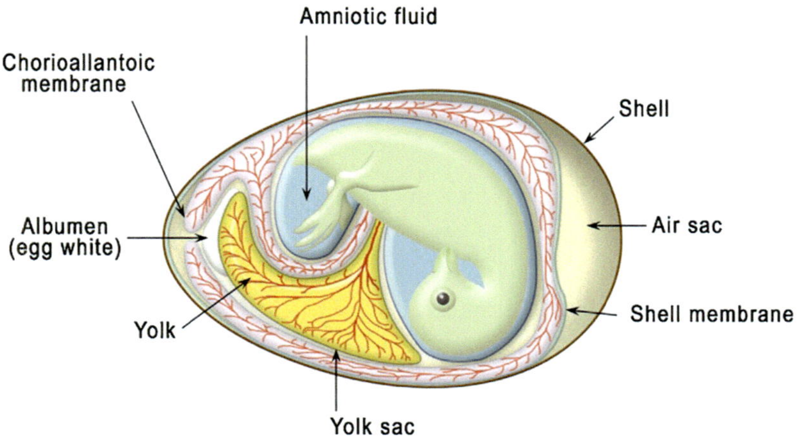


**Fig. 18.53** Drawing of an egg of a Domestic Chicken at day 16 of incubation. By this stage, the chorionallantoic membrane surrounds the entire embryo and yolk sac. (Figure modified from Dombre et al. 2017; © The Authors, used with permission)

almost everywhere and occupy habitats that exhibit much variation in altitude, humidity, and temperature, selection has favored variation in eggshell structure and parental behavior that ensure that developing embryos have an adequate supply of oxygen and that appropriate levels of water vapor are lost from eggs during incubation. For example, selection on eggshells for species that nest at high altitudes (above 4000 m) has favored increased conductance (compared to those at lower altitudes) to enhance diffusion of oxygen. This increased conductance also means an increased loss of water vapor, but at levels that are not harmful to embryos (Carey et al. 1987, 1989).

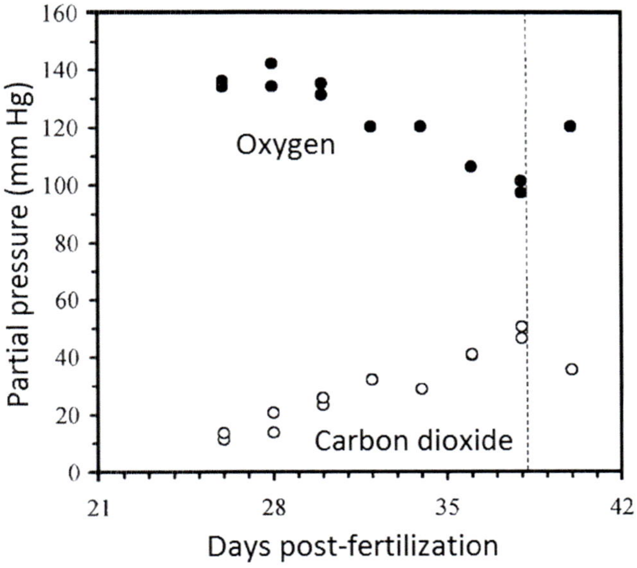


**Fig. 18.54** Partial pressures of oxygen and carbon dioxide in an egg of a Common Ostrich (*Struthio camelus*). At sea level, the partial pressures of oxygen and carbon dioxide in the atmosphere are 160 and well below 1, respectively. Because the partial pressure of oxygen is higher in the atmosphere than in the egg, oxygen diffuses through eggshell pores into the egg. The reverse is true for carbon dioxide. The vertical dashed line indicates internal pipping. (Figure modified from Gefen and Ar 2001; © 2001 Elsevier Science Inc., used with permission)

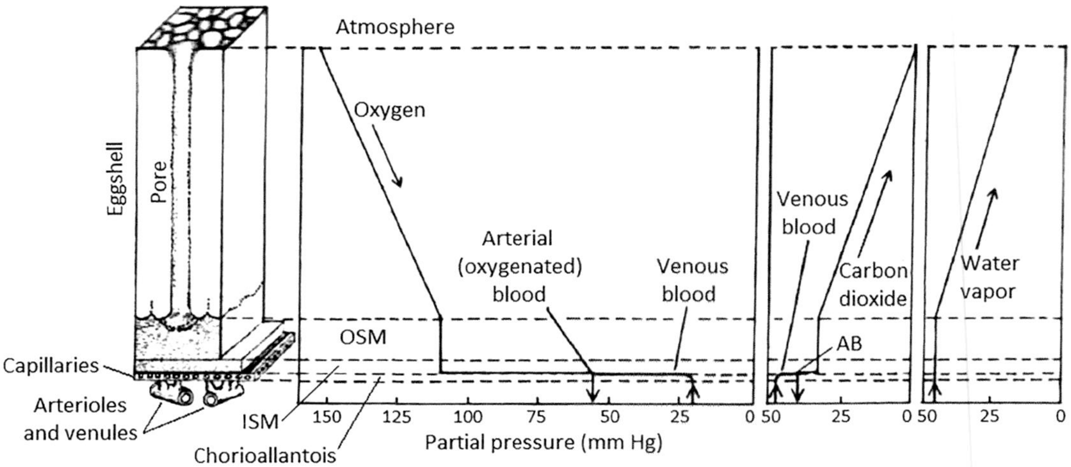


**Fig. 18.55** Gas exchange across an eggshell membrane. Oxygen, carbon dioxide, and water vapor all follow their partial pressure gradients as they diffuse through the pores of eggshells. Oxygen diffuses from the atmosphere where there is more oxygen (higher partial pressure; 160 mmHg at sea level), through pores, and into the embryo's blood where there is much less oxygen (lower partial pressure). Carbon dioxide diffuses from the blood where there is more carbon dioxide, through pores, and into the atmosphere where there is less carbon dioxide (less than 1 mmHg). Water vapor diffuses from the egg's interior where there is more water vapor, through pores, and into the atmosphere where there is less water vapor. OSM, outer shell membrane; ISM, inner shell membrane; AB, arterial blood. (Figure modified from Rahn et al. 1979; used with permission of illustrator Patricia Wynne)

Because the loss of water vapor from eggs must be maintained within a specific range of values, nests with microclimates that optimize water loss are essential. In an attempt to better understand factors contributing to variation among species, Portugal et al. (2014) evaluated the possible influence of 17 life-history and habitat variables on the water vapor conductance of the eggshells of 151 species of British birds. Analysis revealed that the only significant factors contributing to interspecific variation in water vapor conductance were nest type and whether or not incubating parents were wet when returning to nests. Both of these variables are important because they influence humidity levels around eggs and those levels, in turn, influence convective water loss from eggs.

Portugal et al. (2014) found that species with cliff nests, cup nests, and nests in ground burrows had eggs with significantly higher water vapor conductance (range = 0.28–0.32 mg/day/Torr) than species that nested on the ground, in open, but shallow nests, and nests in tree cavities (range = 0.22–0.23 mg/day/Torr). Other investigators have also reported higher water vapor conductance for eggs of species with cup nests and that nest in underground burrows (e.g., Ackerman and Platter-Reiger 1979). Eggs of species with higher water vapor conductance are thought to generally be in nest environments with higher humidity levels, often because nests or nest locations restrict air flow and moisture is added to the nest environment as water vapor lost from eggs and the skin of incubating adults accumulates (Deeming 2011). Attard and Portugal (2021) examined conductance of whole eggs of 365 species of birds representing 28 orders and determined that diffusion of water vapor via shell pores was higher for species where eggs were incubated in more humid environments, such as on the ground, in a burrow or mound, or on floating vegetation. Of course, even more moisture is added to the nest environment if the plumage of an adult is wet when it begins incubating so species where this occurs, e.g., Common Loons (Fig. 18.67), have eggs with significantly higher eggshell conductance than other species (Portugal et al. 2014; Attard and Portugal 2021). In sum, to

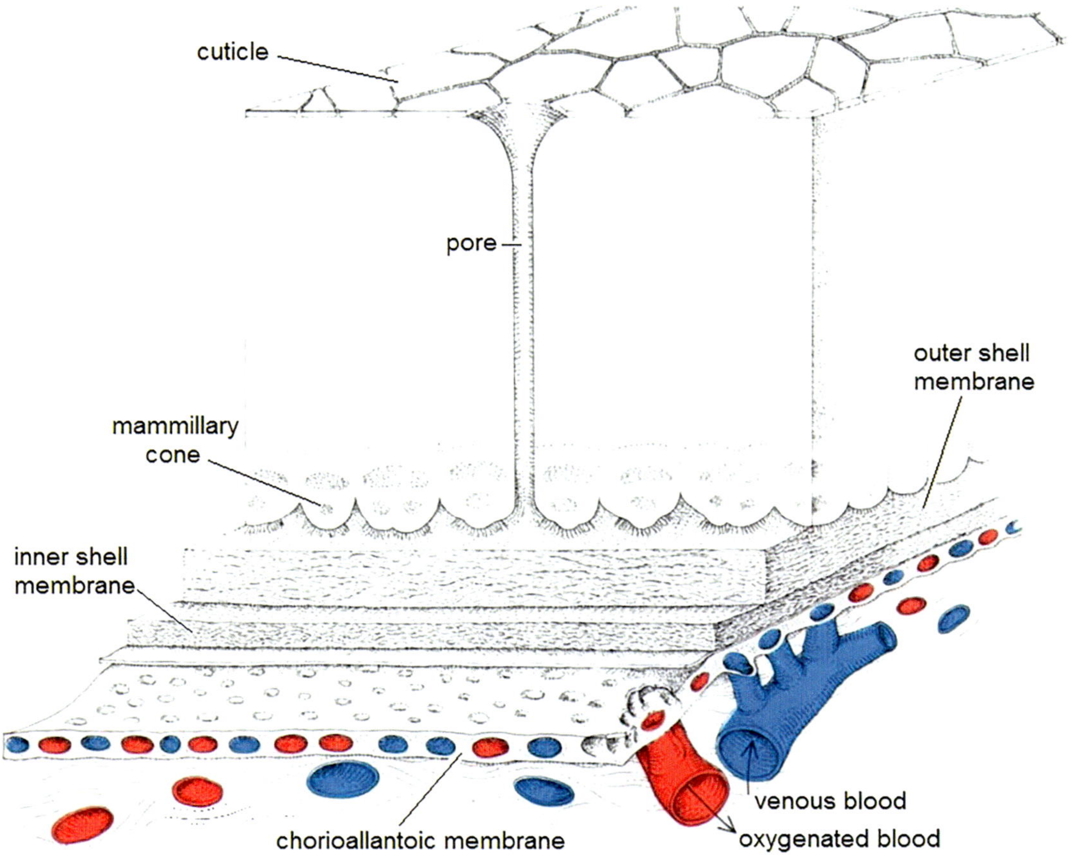


**Fig. 18.56** Gas exchange occurs by diffusion through eggshell pores and the outer and inner shell membranes. The chorioallantois is the respiratory organ of avian embryos. Venous blood (blue) pumped by an embryo's heart flows to the chorioallantoic membrane, where the red blood cells pick up oxygen that has diffused through the pores. Oxygenated blood (red) then travels to the embryo. At the same time, carbon dioxide diffuses out of the venous blood. (Figure from Rahn et al. 1979; used with permission of illustrator Patricia Wynne)

ensure that sufficient water vapor is lost from eggs during incubation when nest environments tend to be more humid, selection has favored eggshells with higher water vapor conductance.

Before the yolk sac membrane (area vasculosa) begins to develop and provide access to nutrients in the yolk, the main source of energy for embryos is glucose (Moran 2007). Newly formed eggs contain very little glucose ($<1\%$) and most of it is used by avian embryos for anaerobic glycolysis prior to development of the chorioallantoic membrane (CAM) when too little oxygen is available for fatty acid oxidation (Moran 2007). However, fatty acids derived from the yolk are the predominant energy source for embryos during most of the incubation period.

During development of embryos, yolk lipids together with membrane proteins are transported into yolk sac epithelial cells by endocytosis (Bauer et al. 2013; Fig. 18.68), with the lining of the yolk sac gradually developing complex ridges that increase the surface area for yolk absorption (Bellairs 1964; Blackburn 2021; Fig. 18.69). The surface area for absorption

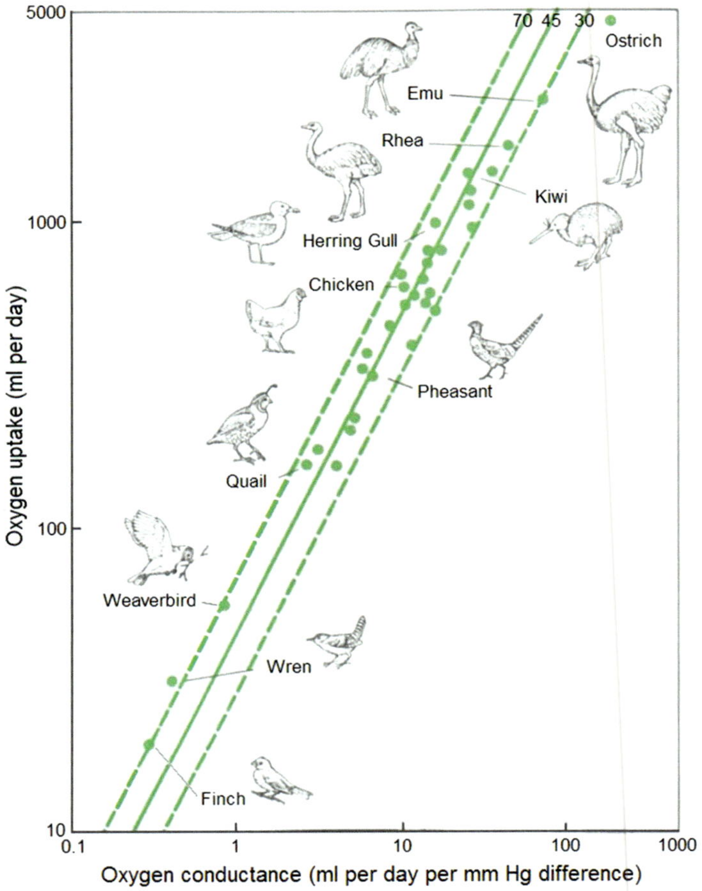


**Fig. 18.57** Oxygen use by embryos of 28 species of birds before internal pipping relative to the oxygen conductance of eggshells. Slopes of the dashed lines indicate oxygen-pressure differences of 70 and 30 mm Hg, respectively, between the atmosphere and inside the egg. The solid line represents the mean oxygen-pressure difference of 45 mmHg. Common Ostrich, *Struthio camelus*; Emu, *Dromaius novaehollandiae*; Rhea, *Rhea* spp.; Kiwi, *Apteryx* spp.; Herring Gull, *Larus argentatus*. (Figure modified from Rahn et al. 1979; used with permission of illustrator Patricia Wynne)

increases more than tenfold from day 5 to day 17 of Domestic Chicken (*Gallus domesticus*) development (Yadgary et al. 2013). Lysosomal enzymes in the epithelial cells react with the lipids to release free fatty acids and glycerol, and with the proteins to release amino acids (Speake et al. 1998). Additional enzymes in the epithelial cells then convert the free fatty acids and glycerol into triacylglycerides and phospholipids that are secreted from the epithelial cells into capillaries in the yolk sac membrane (Fig. 18.68). The triacylglycerides and phospholipids are transported by the blood to the embryo, with triacylglycerides serving as the primary source of energy for embryos and the phospholipids being used to form the cell membranes and other membranous organelles of newly formed embryonic cells (Speake et al. 1998). Enzymes in embryonic cells break down the triacylglycerides to release fatty acids and glycerol that can then be used to generate ATP (via the Kreb's cycle). Fatty acids provide most of the energy, both for embryonic growth and maintenance (Fig. 18.70).

Albumen is the main source of water and proteins for developing avian embryos. Albumen represents 33–76% of the initial egg content (excluding the eggshell) in precocial species,

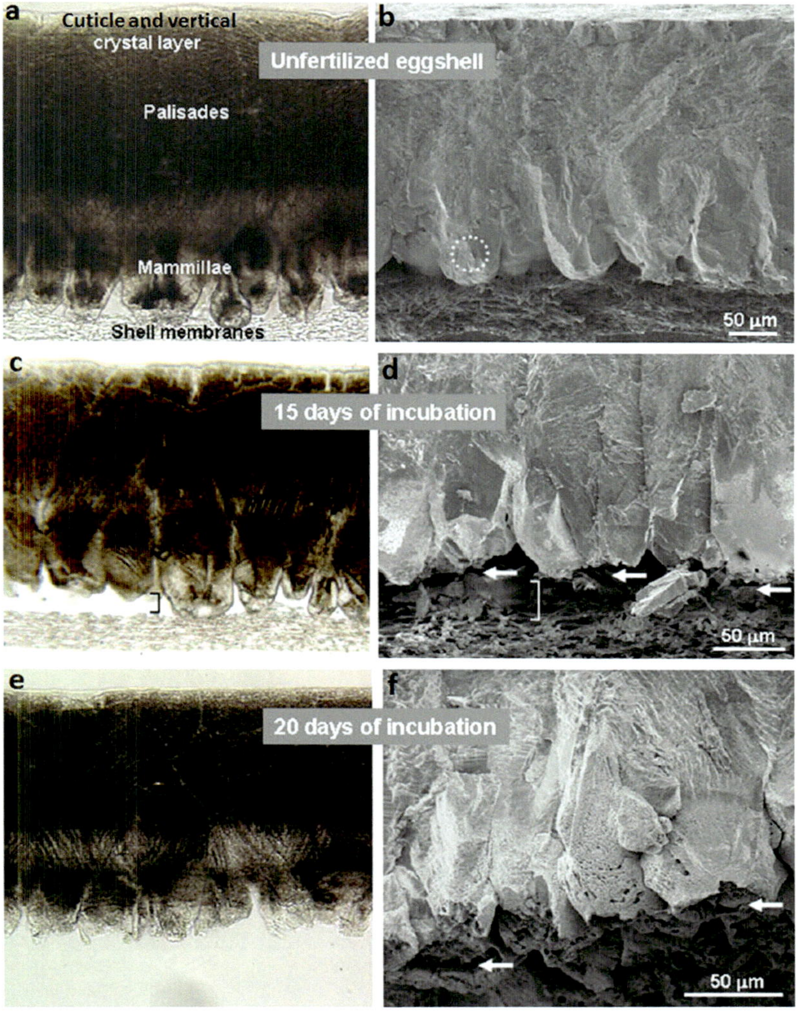


**Fig. 18.58** Cross-sections of eggshells of Domestic Chickens viewed with a light microscope (left panels) and a scanning electron microscope (right panels). (**a**, **b**) Eggshell of unfertilized egg showing intact major shell layers and shell membranes. The white dotted circle in panel (**b**) indicates the approximate location of the interior calcium reserve body described in subsequent figures. (**c**, **d**) Eggshells of eggs incubated for 15 days showing disintegration of a portion of the base plate at the tips of the mammillae (arrows) with detachment or partial detachment of the shell membranes (brackets). (**e**, **f**) After 20 days of incubation (and just prior to hatching), the shell membrane is detached from the eggshell, whose irregular, porous, and dissolved appearance reflect a loss of material in this inner part of the eggshell. (Figure from Chien et al. 2009; © 2009 Published by Elsevier Inc., used with permission)

and 65–86% of the initial egg content in altricial species (Carey et al. 1980). The albumen in the eggs of Domestic Chickens is about 10.5% protein and 88.5% water, with the remaining 1% consisting of small amount of carbohydrate, lipid, and inorganic ions such as phosphorus, calcium, magnesium, and, especially, sodium and potassium (Romanoff and Romanoff 1949; Fig. 18.71). The albumen in bird eggs has been found to contain numerous different proteins (Fig. 18.72), but our understanding of the function(s) of those proteins is limited to only the most abundant proteins (Table 18.3). Ovalbumin is the most abundant protein in albumen, but several other proteins are present as well, including ovotransferrin, ovomucoid, lysosome, and several others (Shbailat and Safi 2015; Table 18.4). Many proteins needed during embryogenesis are synthesized by embryonic cells using the egg proteins, particularly ovalbumin, as a source of amino acids, but most egg proteins also have other important functions, including inhibiting microbial growth (Table 18.4). Some proteins in albumen are glycoproteins with carbohydrate contents ranging from 2 to 58% and, as result, can be a source of both amino acids and glycogen for embryos. Still other proteins are binding proteins, binding to and

**Fig. 18.59** Scanning electron micrographs of eggshells of six species of birds showing mammillary knobs. Ostriches, Gray Partridges, and Pied Avocets are precocial, American Flamingoes are semi-precocial, and Gray Herons and Eastern Kingbirds are altricial. The density of mammillary knobs in eggshells tends to be higher in older taxa like the paleognaths and galloanserae. Percentages indicate the percentage of the mammillary layer covered by mammillary knobs. Scale bars = 100 μm. Common Ostrich, *Struthio camelus*; Gray Partridge, *Perdix perdix*; American Flamingo, *Phoenicopterus ruber*; Pied Avocet, *Recurvirostra avosetta*; Gray Heron, *Ardea cinerea*; Eastern Kingbird, *Tyrannus tyrannus*. (Figure modified from Österström and Lilja 2012; © 2011 Wiley Periodicals, Inc., used with permission)

aiding embryos in the uptake of vitamins (Table 18.4).

During development in an egg, yolk and albumen are being converted into an embryo (Fig. 18.73). The faster-growing embryos of altricial birds initially allocate more energy to the rapid development of tissues and organs, such as those of the digestive system, that supply the nutritional demands of other tissues and organs (Karlsson and Lilja 2008; Fig. 18.74), and less energy to development of organs and structures such as the brain, muscles, skeleton, and feathers (Blom and Lilja 2005). As a result, altricial embryos typically hatch with smaller brains, immature skeletal muscles, and poorly ossified skeletons, whereas precocial embryos hatch with larger brains, mature muscles, and highly ossified skeletons (Fig. 18.75). One consequence of these differences is that the energetic costs of development are greater for embryos of precocial species than those of altricial species (Fig. 18.76) and, in addition, with the greater ossification of the skeleton, precocial embryos have a greater need for calcium during development (Blom and Lilja 2004; Österström et al. 2013). Some of this calcium comes from the yolk, but most (about 80%) comes from the eggshell. As might be expected, the eggshells of precocial species contain more calcium than those of altricial species (Karlsson and Lilja 2008).

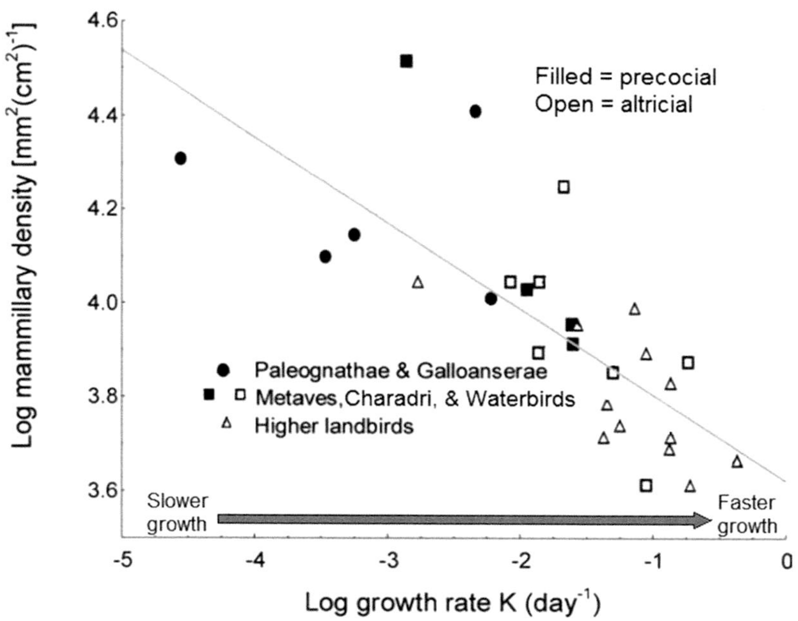


**Fig. 18.60** Relationship between mammillary density and nestling growth rate. K = grams/day. (Figure modified from Österström and Lilja 2012; © 2011 Wiley Periodicals, Inc., used with permission)

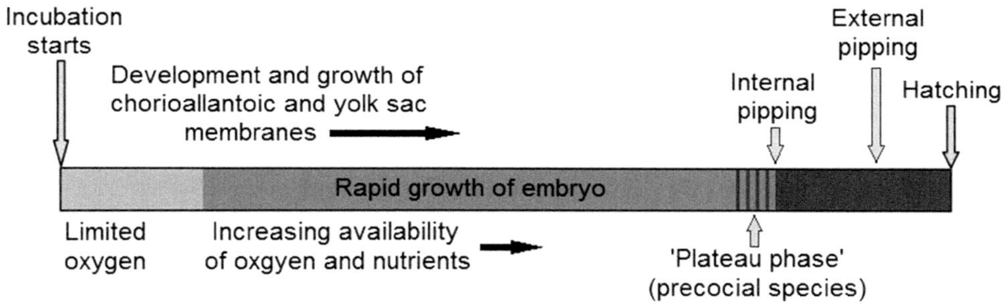


**Fig. 18.61** The development of avian embryos can be considered to have three phases. First, a period of slow growth with limited access to oxygen and nutrients. Second, a period of rapid growth as the growing chorioallantoic membrane and yolk sac membrane provide the embryo with more oxygen and nutrients. At the end of this second period, the growth rate of embryos of many precocial species levels off or even slightly declines. Finally, during the third stage, embryos prepare for hatching by internal pipping and gaining access to air in the air cell. This is followed by external pipping and, soon after, emergence from the egg. (Figure modified from De Oliveira et al. 2008; Rights managed by Taylor & Francis, used with permission)

## 18.13 Metabolic Rates of Avian Embryos

Embryos in species with different modes of development exhibit differences in metabolic rates. In precocial species, metabolic rates (as measured by oxygen consumption; Fig. 18.77) increase rapidly during most of the incubation period, but then typically level off or even decline when incubation is about 80% completed. Vleck et al. (1980) suggested that this "plateau phase" occurs because embryos of precocial species are transitioning from a period of energetically costly tissue development to a period of less costly maturation of physiological systems important for precocial behavior. In altricial species, on the other hand, metabolic rates of

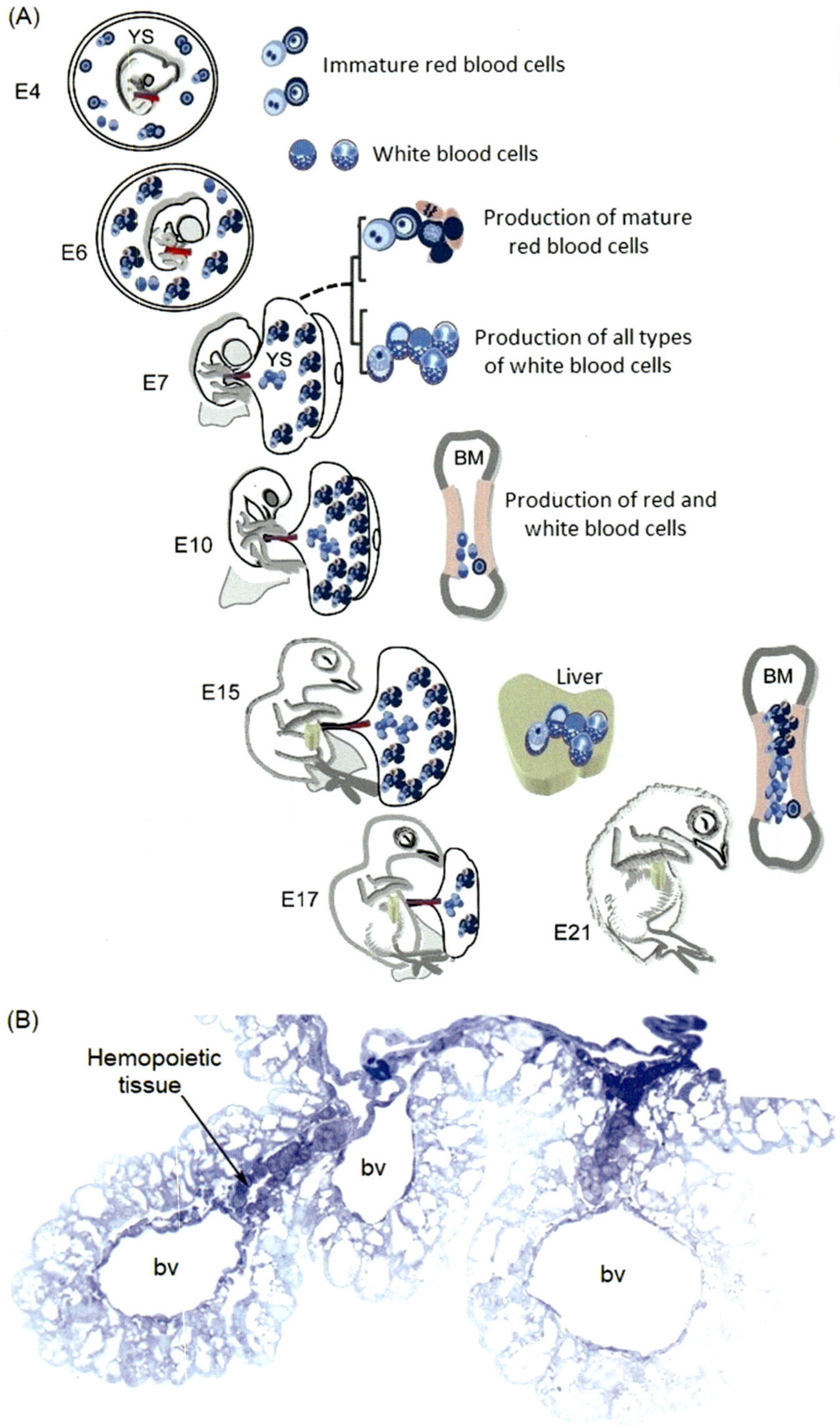


**Fig. 18.62** (**a**) Production of red and white blood cells by developing Domestic Chicken embryos. Red and white blood cells are initially produced only in the yolk sac. Later in development, red and white blood cells are also produced in the bone marrow (BM) and liver. (**b**) Yolk sac of a Japanese Quail (*Coturnix japonica*) embryo (about

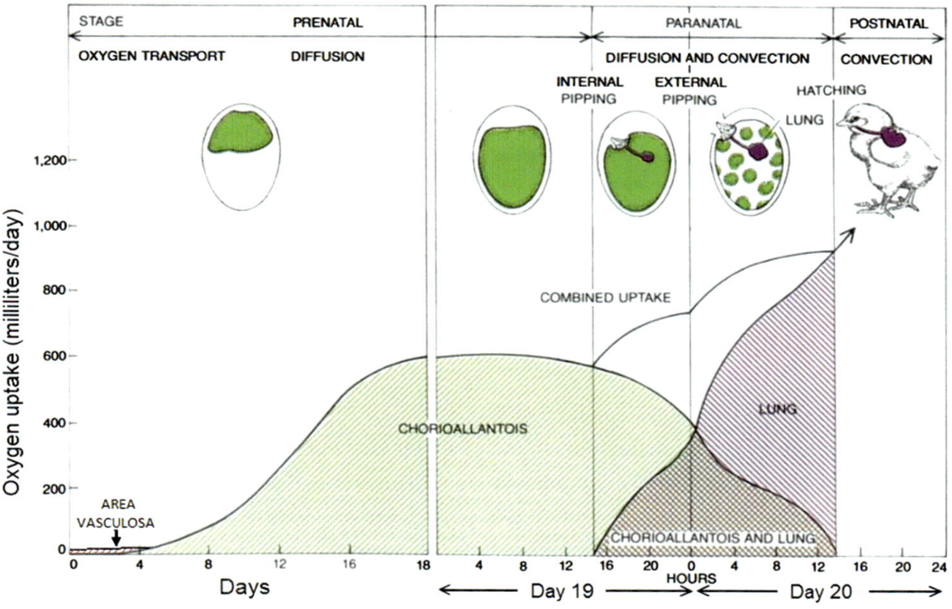


**Fig. 18.63** Oxygen uptake by the embryo of a Domestic Chicken increases throughout the incubation period. Prior to about day 4 post-fertilization, oxygen uptake is via the area vasculosa; thereafter, via the chorioallantoic membrane. On days 19 and 20, there is a transition from diffusive gas exchange through the chorioallantois to active breathing and uptake of oxygen in the lungs. (Figure from Rahn et al. 1979; used with permission of illustrator Patricia Wynne)

embryos increase continuously throughout the entire incubation period (Fig. 18.77). In altricial birds, embryo mass and growth rates (grams of new tissue added per day) typically increase continuously throughout incubation and, as a result, the energetic costs of maintenance and synthesizing new tissue increase until hatching. However, in precocial species, embryo mass approaches the typical mass at hatching when incubation periods are about 80% complete. As a result, the amount of energy devoted to growth either levels off or declines toward the end of incubation. Maintenance costs continue to increase, but the reduced growth causes a leveling off or reduction in the total metabolic rate (Fig. 18.77).

Despite differences in metabolic rates during incubation, the embryos of altricial and precocial species do not appear to differ in the amount of energy needed to generate new tissue. Rather, the difference between altricial and precocial embryos lies in the degree of development prior

**Fig. 18.62** (continued) 4–5 days old) showing blood vessels and a layer of hemopoietic cells. (Figure **a** modified from Guedes et al. 2014; © 2014 Guedes et al., open-access article distributed under the terms of the Creative Commons Attribution License, https://creativecommons.org/licenses/by/4.0/. Figure **b** modified from Starck 2021; © 2020 The Author, Journal of Morphology published by Wiley Periodicals LLC, open-access article distributed under the terms of the Creative Commons CC BY license, https://creativecommons.org/licenses/by/4.0/)

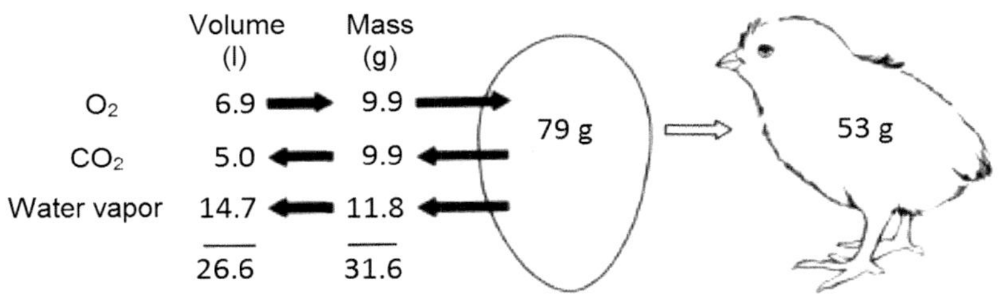


**Fig. 18.64** Volume and mass of gases exchanged during the 28-day incubation period of a 79-g egg of a Wild Turkey (*Meleagris gallopavo*). (Figure modified from Rahn et al. 1987; © 1987 Wiley-Liss, Inc., used with permission)

to hatching, with altricial species hatching at a much earlier stage of development than precocial species (Fig. 18.78). Because of the relatively shorter incubation periods and because embryos hatch at an earlier stage of development, females in altricial species can invest less energy (less yolk) in eggs than females in precocial species (Vleck and Vleck 1987).

## 18.14 Hatching

The process of hatching begins with the chick maneuvering its body to lie along the longest axis of the egg and burrowing its head beneath the right shoulder so the beak is positioned under the wing and against the two membranes separating the embryo from the air space at the large end of the shell (i.e., "tucking," with similar pre-hatching behavior possibly originating among non-avian theropods; Fig. 18.79). The egg tooth on the upper bill (a pointed tooth-like structure of keratinous material; Fig. 18.80) is then used to pierce the inner shell membrane so the chick can access the air space between the inner and outer shell membranes. This process, called internal pipping, allows the chick to take its first breath and begin to fill its lungs and air sacs (Deeming 2002a; Fig. 18.81). This process can be relatively slow and some air sacs may fill with air before the lungs (Vleck and Bucher 1998). During this time, the chick continues to obtain oxygen via the chorioallantoic membrane.

The time interval between internal pipping and the process of breaking through the shell (external pipping) varies and is influenced by the rate at which oxygen levels decrease and, more importantly, carbon dioxide levels increase in the air space between the shell membranes. As external pipping begins, chicks typically thrust their beak and egg tooth against the shell and, simultaneously, push with their upper body and legs. This causes the chick to slowly rotate within the egg and create a circular "slit" in the shell, forming a "cap" at the end of the eggshell. As the time of hatching approaches, a "hatching muscle" (musculus complexus) on the back of the chick's neck swells to several times its normal size with a great influx of fluid from the embryo's lymphatic system (Fig. 18.82). The fibers in this muscle are exclusively anaerobic, well supplied with energy (glycogen), and have a special nervous system for coordination. Using this muscle, the chick pushes its way through the "cap" (the area encircled by the pipped holes) and out of the egg (Fig. 18.83).

Not all chicks hatch the same way. For example, among species with longer bills, such as the American Woodcock (*Scolopax minor*) and Willet (*Catoptrophorus semipalmatus*), hatching chicks use their bills and egg tooth to create a seam rather than a cap (Wetherbee and Bartlett 1962). The eggshells of some, particularly larger, species are harder and more brittle. In these species, such as the Common Ostrich (*Struthio camelus*), fewer "pip" holes are needed because multiple cracks develop in the egg and chicks essentially explode out of their eggs as the shell breaks into several pieces (Sauer and Sauer 1966; Deeming 1995; Fig. 18.84).

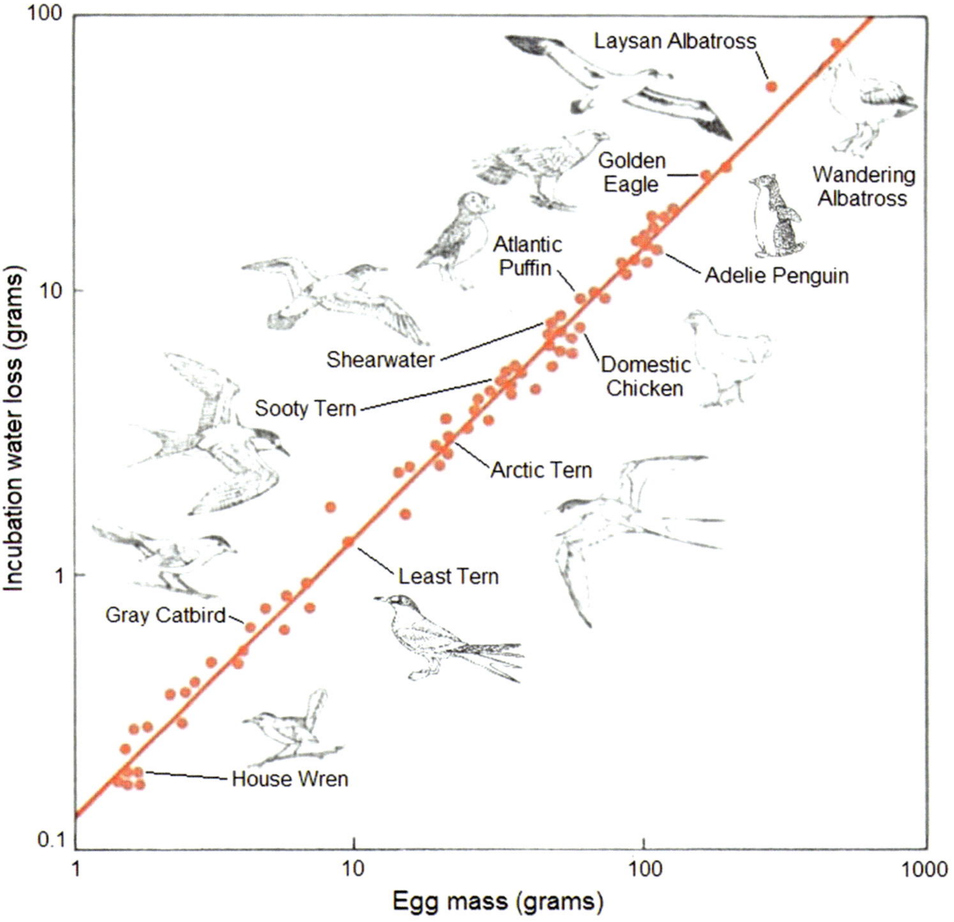


**Fig. 18.65** Relationship between the total amount of water lost during incubation and the initial mass of eggs for 65 species of birds ranging in mass from 1 to 500 g and with incubation periods ranging from 11 to 70 days. The eggs of larger birds lose more water than those of smaller birds. Regardless of egg mass or the duration of incubation periods, bird eggs typically lose about 10–20% of their initial mass during incubation. Laysan Albatross, *Phoebastria immutabilis*; Wandering Albatross, *Diomedea exulans*; Golden Eagle, *Aquila chryaetos*; Adelie Penguin, *Pygoscelis adeliae*; Atlantic Puffin, *Fratercula arctica*; Domestic Chicken, *Gallus g. domesticus*; Shearwater, *Ardenna* spp.; Sooty Tern, *Onychoprion fuscatus*; Arctic Tern, *Sterna paradisaea*; Least Tern, *Sterna antillarum*; Gray Catbird, *Dumetella carolinensis*; House Wren, *Troglodytes aedon*. (Figure from Rahn et al. 1979; used with permission of illustrator Patricia Wynne)

The interval between internal pipping and eventual emergence from the egg varies both within and between species. In some precocial species, embryos are known to produce clicking sounds in the days prior to hatching that act as signals between all embryos in a clutch. These sounds can be produced after internal pipping when the embryo gains access to the air in the air cell. This allows the embryo to fill its lungs and, when air passes over the syrinx during this initial functioning of the respiratory system, clicking sounds are produced (Forsythe 1971).

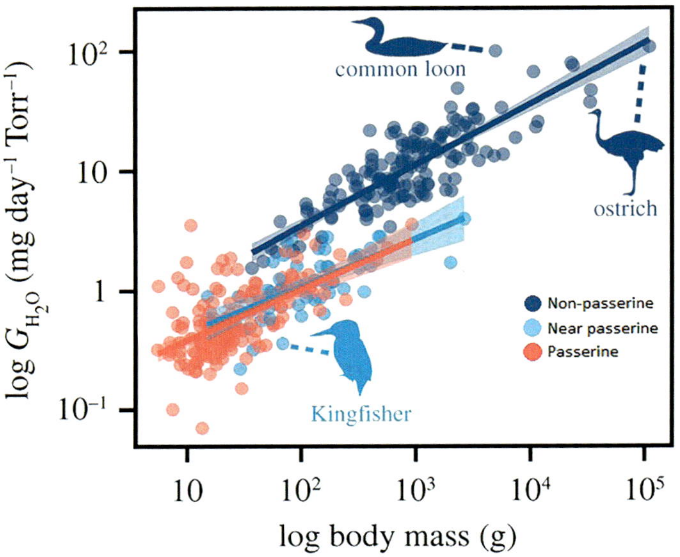


**Fig. 18.66** Relationship between body mass and rate of water loss through eggshell pores for 364 species of birds in three taxonomic categories: non-passerines, near-passerines (species traditionally believed to be related to passerines due to ecological similarities, including those in the order Pterocliformes [sandgrouse], Columbiformes [pigeons], Cuculiformes [cuckoos], Caprimulgiformes [nightjars], and Apodiformes [swifts and hummingbirds]), and passerines. Water vapor conductance through eggshells was lower for passerines and near-passerines than for non-passerines and, for all through groups, conductance increased with increasing body mass. Common Loon, *Gavia immer*; Common Ostrich, *Struthio camelus*. (Figure from Attard and Portugal 2021; © 2021 The Authors. Published by the Royal Society, used with permission)

Communication using these sounds helps to ensure that all the young hatch synchronously by causing embryos, depending on stage of development, to either accelerate or slow down the hatching process. For example, young Mallards (*Anas platyrhynchos*) and Ring-necked Pheasants (*Phasianus colchicus*) responding to the sounds of other embryos can modify their hatching dates by more than 1.5 days (Persson and Andersson 1999). Because females of many precocial species leave the nest when chicks have hatched, even if unhatched eggs remain in the nest, this ability of synchronize the time of hatching (generally within a period of 3–24 h) reduces the likelihood that chicks will be left behind in a nest.

Embryos in some species of birds are also able to communicate with parents as they approach the time of hatching. After internal pipping, Muscovy Duck (*Cairina moschata*) chicks utter "distress" calls at higher rates if eggs are cooled and call rates then decline after re-warming (Fig. 18.85; Nichelmann and Tzschentke 1997). Similar behavior has been reported in Budgerigars (*Melopsittacus undulatus*; Berlin and Clark 1998) and Ring-billed Gulls (*Larus delawarensis*; Evans et al. 1994). Domestic Chicken embryos increase calling rates both when temperatures increase and decrease (Mortola 2019; Fig. 18.86). Under natural conditions, such vocalizing by chicks is a signal to parents to either incubate and warm the eggs or, perhaps, shade the eggs to cool the embryos.

Zebra Finch (*Taeniopygia guttata*) embryos also vocalize to help synchronize their hatching and when cold-stressed to induce incubation by adults, but adults also utter "incubation calls" when ambient temperatures rise above 26°C (Mariette and Buchanan 2016). These "incubation calls" appear to signal environmental conditions to embryos, and prepare embryos that

**Fig. 18.67** Common Loons (*Gavia immer*) nest near water and have wet plumage when they begin incubating eggs. The high-humidity environment has selected for eggshells with higher rates of water vapor conductance than most other species. (Photo by Emma Forthofer, U. S. National Park Service, CC0 Public Domain)

are near hatching age for high ambient temperatures post-hatching (Fig. 18.87). Using playback experiments, Mariette and Buchanan (2016) demonstrated that embryos exposed to these "heat calls" weighed less at hatch and during the nestling period than control embryos that had not been exposed to the calls. This decrease in mass may be beneficial if being smaller (with more surface area per unit mass) facilitates heat loss or if slowing growth rates reduce oxidative damage associated with growth at high ambient temperatures (Mariette and Buchanan 2016).

Another possible mode of communication between bird embryos and the incubating parents may be nitric oxide (NO) emissions (Ifergan and Ar 1999). A study of 18 species of birds revealed that nitric oxide emissions from eggs increased when eggs were cooled, and this increased emission may inform parents that embryos are stressed and eggs need to be incubated (Ar et al. 2004; Samuni and Ar 2006).

The eggshells of brood parasitic species, such as cuckoos and cowbirds, are stronger than those of other species of birds of comparable size. The

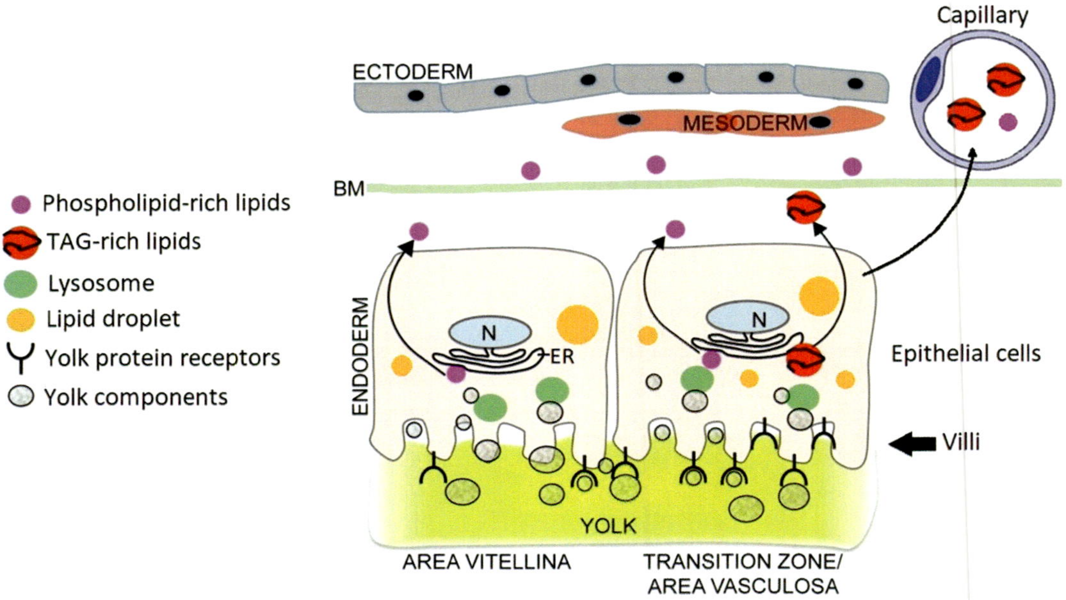


**Fig. 18.68** As embryos develop, yolk sac membranes grow until they eventually surround the entire yolk sac. During this period of growth, the yolk sac membrane can be divided into three regions: the area vitellina with only epithelial cells, a transition zone with mesodermal cells, but no capillaries, and the area vasculosa where capillaries have developed in the mesodermal layer. With capillaries not yet developed, epithelial cells in the transition zone and area vitellina are only able to secrete phospholipids. Once the mesodermal layer becomes vascularized, those phospholipids can then be secreted into capillaries for transport to the embryo. Droplets of yolk are transported into yolk sac membrane epithelial cells by endocytotic membrane proteins. This transport is made more efficient by the presence of villi that increase the surface area available for transport. In the area vasculosa, phospholipids like triacylglycerides (TAG) are then secreted into capillaries for transport to the embryo. BM, basement membrane; N, nucleus; ER, endoplasmic reticulum. (Figure modified from Bauer et al. 2013; open-access article distributed under the terms of the Creative Commons CC BY license, https://creativecommons.org/licenses/by/4.0/)

shells of Common Cuckoo (*Cuculus canorus*) eggs are 2.2 times stronger than eggs of the same size laid by their passerine hosts (Honza et al. 2001). Stronger shells may make it more difficult for a host to puncture and remove a parasite's egg from a nest or the extra strength may simply reflect the fact that parasite eggs (particularly those of cuckoos) were originally larger and selection has since favored smaller eggs more similar in size to those of host species (Lack 1968). In addition, eggs with stronger shells are less likely to break during laying (Lack 1968). Regardless of the selective advantage, stronger shells make it more difficult for young brood parasites to hatch. As a result, compared to chicks of other species hatching from eggs of comparable size (but with "weaker" shells), Common Cuckoo chicks started pecking earlier and required more time and a greater total number of pecks to break open eggshells and hatch (Honza et al. 2001). Similarly, Yoon (2013) observed hatching of brood-parasitic Brown-headed Cowbirds (*Molothrus ater*) and non-brood-parasitic Red-winged Blackbirds (*Agelaius phoeniceus*) and noted a similar sequence of stages, including pipping (first puncture of the eggshell by the egg tooth), clicking (production of sounds while rotating their bodies counterclockwise from the location of pipping to crack the eggshell), and emergence (cracked shell is opened by vigorous movements of the hatching young). However, the duration of these events

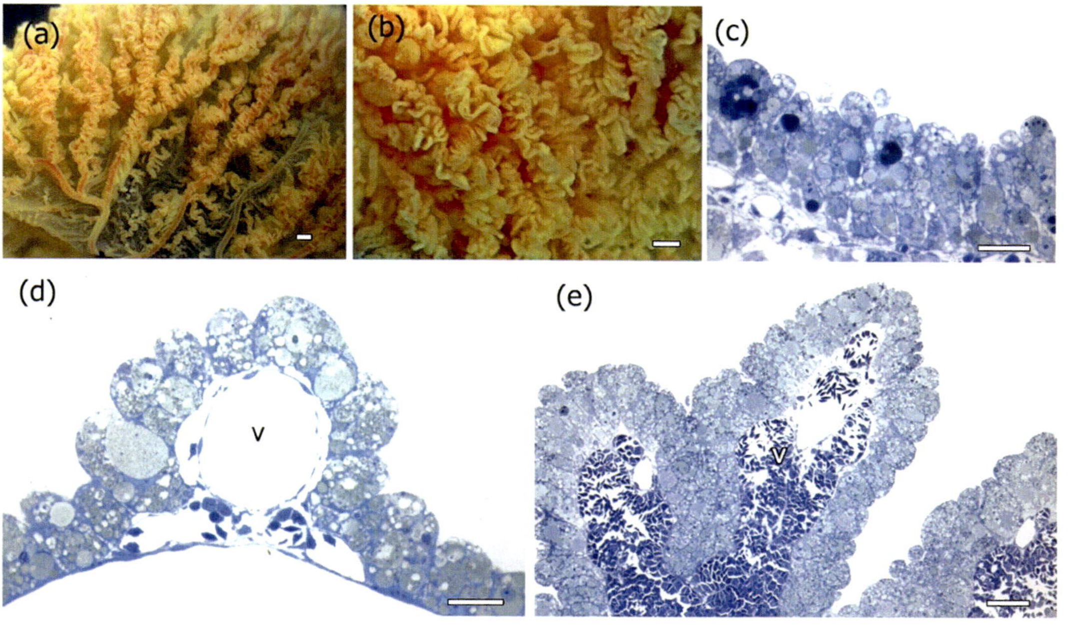


**Fig. 18.69** (**a**, **b**) Yolk sac lining of a Domestic Chicken (*Gallus domesticus*) at day 13 and day 17, respectively, of the 21-day incubation period, with the vascularized ridges becoming more complex and elaborate as development proceeds. (**c**–**e**) Thin sections (0.5 μm) of the inner lining of the yolk sac of a Common Quail (*Coturnix coturnix*) at day 7 of incubation (stained with Azure II and methylene blue). The lining consists of columnar epithelial cells that surround blood vessels (v) that transport nutrients to the developing embryo. Scale bars: **a**, **b** = 0.5 mm, **c**, **d** = 25 μm, and **e** = 50 μm. (Figure from Blackburn 2021; © 2020 Wiley Periodicals LLC, used with permission)

and hatching was shorter for Red-winged Blackbirds (mean = 6.93 h) than for Brown-headed Cowbirds (mean = 8.25 h).

In species with asynchronous hatching (not all eggs hatch simultaneously), the size or mass of the hatching muscle may vary with laying order. In clutches of Red-winged Blackbirds, the mass of the hatching muscle increased with laying order, likely due to an increase with laying order in concentrations of yolk testosterone, a hormone that stimulates muscle growth. Larger hatching muscles may help last-hatched nestlings hatch sooner and reduce the disadvantages (in terms of access to food provided by provisioning adults) that results from hatching asynchrony (Lipar and Ketterson 2000).

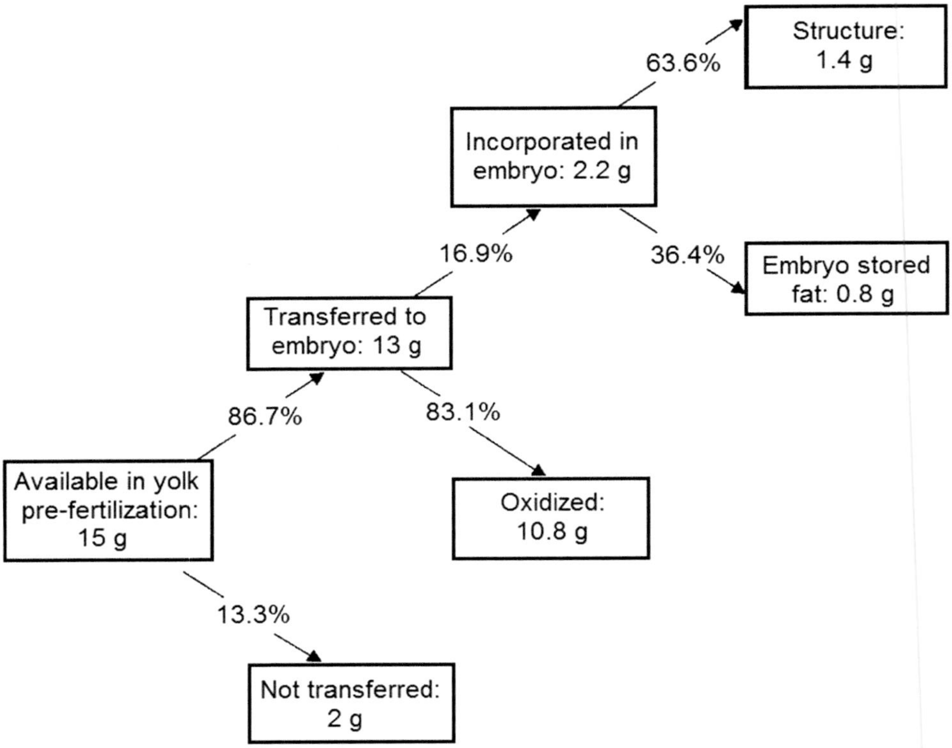


**Fig 18.70** Fate of yolk fatty acids in a developing embryo during the incubation period (53 days) of a King Penguin (*Aptenodytes patagonicus*). More of the energy derived from fatty acids is used for maintenance of tissues (oxidized during respiration) than for production of new tissue. (Figure modified from Groscolas et al. 2003; © 2003 The American Physiological Society, used with permission)

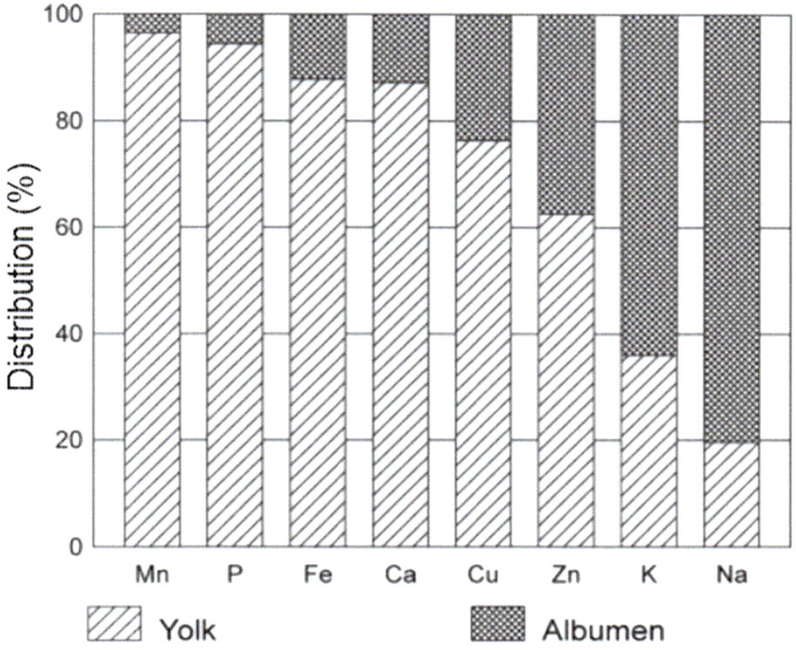


**Fig. 18.71** Distribution of minerals between the yolk and albumen in the eggs of Domestic Chickens. The albumen is an especially important source of potassium and sodium. Mn, magnesium; P, phosphorus; Fe, iron; Ca, calcium; Cu, copper; Zn, zinc; K, potassium; and Na, sodium. (Figure from Uni et al. 2012; © 2012 Oxford University Press, used with permission)

**Fig. 18.72** Total number of different proteins identified in the albumen of six species of birds and the number unique to each species. Domestic Chicken, *Gallus g. domesticus*; Domestic Mallard, *Anas platyrhynchos*; Domestic Goose, *Anser cygnoides domesticus*; Wild Turkey, *Meleagris gallopavo*; Rock Pigeon, *Columba livia*; Common Quail, *Coturnix coturnix*. (Figure made from images and data in Sun et al. 2017; © 2017 Wiley-VCH Verlag GmbH & Co. KGaA, used with permission)

**Table 18.4** Most common proteins found in albumen and their likely function(s)

| Protein | Percent of albumen | Function(s) |
|---|---|---|
| Ovalbumin | 54 | Probable source of amino acids for embryo |
| Ovotransferrin | 12 | Iron transport and inhibiting microbial growth |
| Ovomucoid[a] | 11 | Source of amino acids and glycogen; inhibiting microbial growth |
| Ovomucin | 1.5 | Maintaining structure and viscosity of albumen; inhibiting microbial growth |
| Ovoinhibitor | 1.0 | Inhibiting microbial growth |
| Ovostatin | <1 | Inhibiting microbial growth |
| Cystatin | <1 | Inhibiting microbial growth |
| Lysozyme | 3.4 | Inhibiting microbial growth |
| Avidin | <1 | Inhibiting microbial growth |
| Binding proteins | <2 | Bind to and aid in uptake of various vitamins by embryo |
| Immunoglobulins | <2 | Antibodies that inhibit microbial growth |

Data from Stevens (1996)
[a]Ovomucoid is a glycoprotein so is a source of both amino acids and glycogen

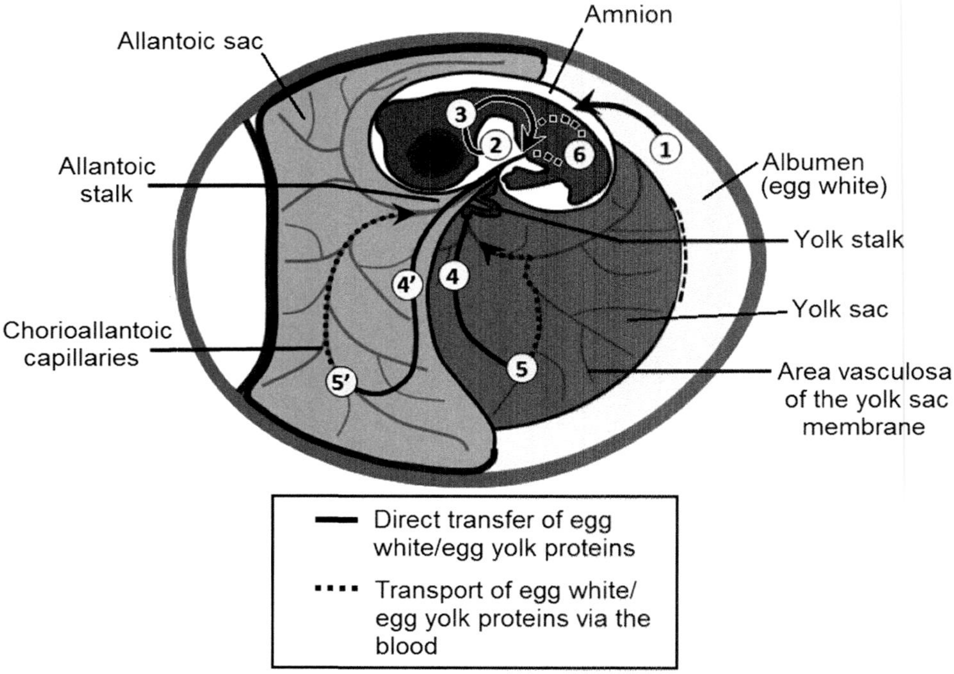


**Fig. 18.73** Assimilation of albumen (egg white) and egg yolk proteins by a Domestic Chicken (*Gallus g. domesticus*) embryo. Egg white proteins are transferred from the albumen through holes in the sero-amniotic connection and into the amniotic fluid (1→ 2), then swallowed by the embryo (2 → 3). Proteins are then partly digested and absorbed by the embryo's small intestine, but most are transferred into the yolk sac via the yolk stalk (3 → 4). Then, egg white and egg yolk proteins are digested and absorbed by the yolk sac membrane (4 → 5) and transported to the embryo by blood vessels in the area vasculosa (5 → 6). After being metabolized by the embryo, the remaining proteins/peptides are secreted into the allantoic sac (6/3 → 4) via the allantoic stalk. These proteins/peptides may be digested and reabsorbed by the allantoic membrane (4′ → 5′) and transported to the embryo by chorioallantoic capillaries (5′ → 6). (Figure modified from Da Silva et al. 2017; © 2017 Oxford University Press, used with permission)

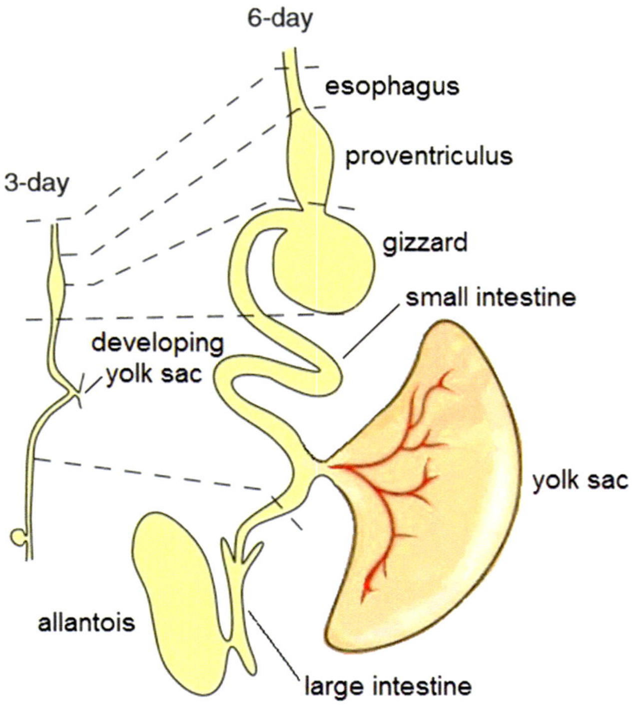


**Fig. 18.74** Development of the digestive system, yolk sac membrane, and allantois of a Domestic Chicken embryo from a 3-day-old embryo to a 6-day-old embryo. (Figure modified from Yasugi and Mizuno 2008; © 2008 The Authors. Journal compilation © 2008 Japanese Society of Developmental Biologists, used with permission)

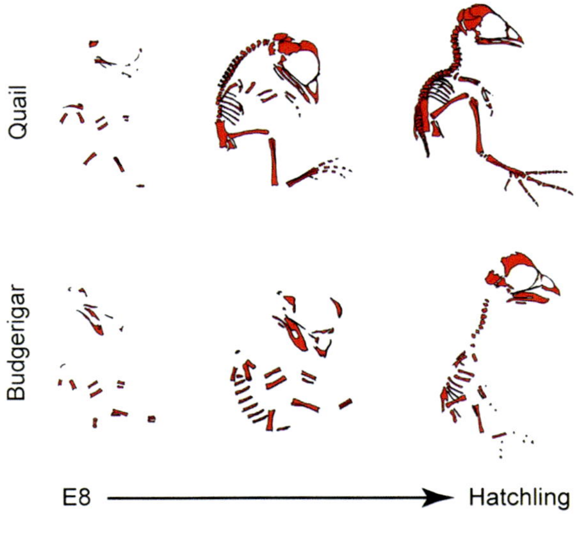


**Fig. 18.75** Comparison of the extent of ossification of the skeleton during embryonic development of a precocial species (Quail, order Galliformes) and an altricial species (Budgerigar, *Melopsittacus undulatus*). The skeletons begin to ossify at about the same stage of development, but the skeletons of precocial birds are almost completely ossified at hatching, whereas those of altricial species are poorly ossified. (Figure from Botelho et al. 2015; © 2015 Springer Science Business, used with permission)

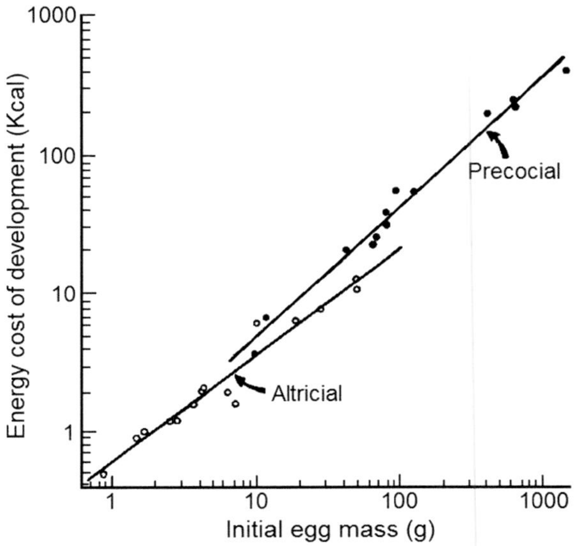


**Fig. 18.76** Energetic cost of development of avian altricial and precocial embryos. (Figure modified from Vleck and Vleck 1980; © 2015 Oxford University Press, used with permission)

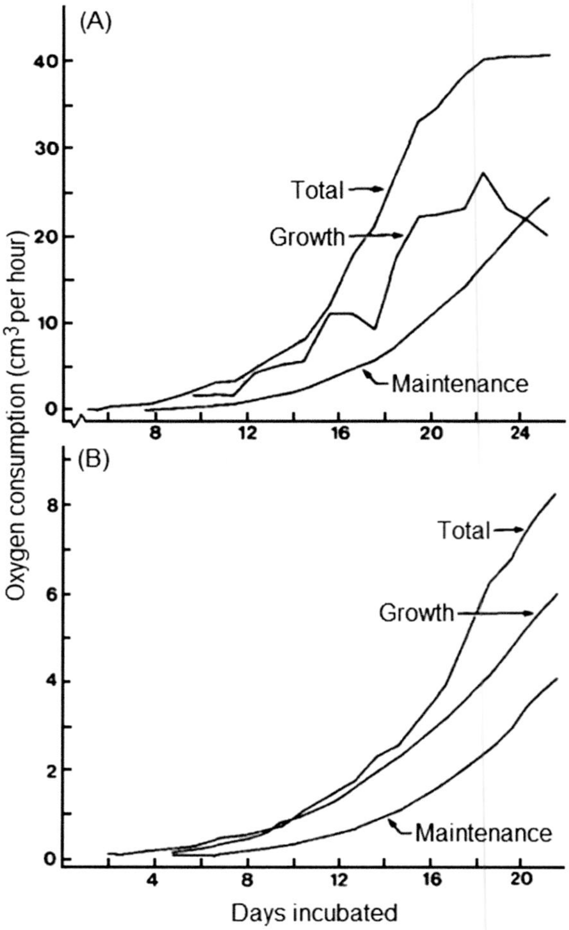


**Fig. 18.77** Oxygen consumption by a (**a**) Graylag Goose (*Anser anser*) embryo (precocial) and (**b**) Cattle Egret (*Bubulcus ibis*) embryo (altricial). Among precocial species, growth rates and oxygen consumption slow, level off, or even decline in the days prior to hatching; growth rates and oxygen consumption by embryos of altricial species continue to increase through the day of hatching. (Figure modified from Vleck and Vleck 1980; © 2015 Oxford University Press, used with permission)

**Fig. 18.78** Left, a precocial hatchling (Northern Shoveler, *Spatula clypeata*) and, right, an altricial hatchling (American Robin, *Turdus migratorius*). (Photos by Steve Dawes [left] and Tom Koerner [right], U. S. Fish and Wildlife Service, CC0 Public Domain)

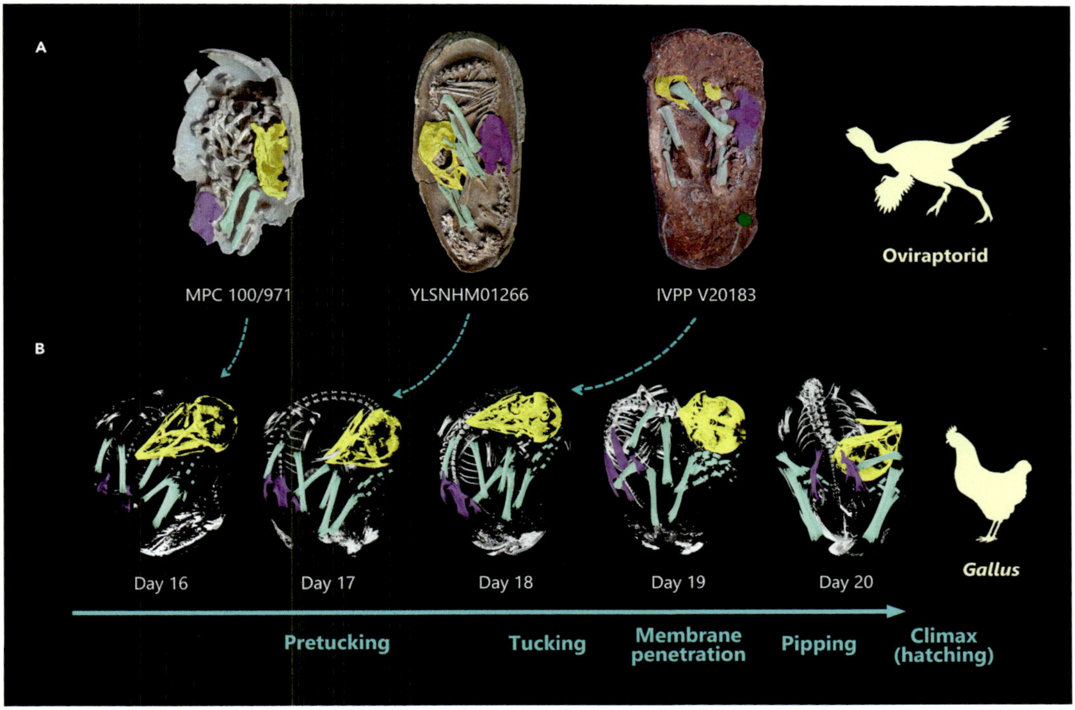


**Fig. 18.79** Late-stage embryos of a non-avian dinosaur and present-day Domestic Chicken showing apparent similarity of pre-hatching postures. (**a**) Three oviraptorid fossil embryos that appear to be in positions that correspond to various tucking stages of present-day birds. (**b**) Domestic Chicken (*Gallus g. domesticus*) embryo from day 16 through day 20 post-hatching. Skulls shown in yellow. Not to scale. (Figure from Xing et al. 2022, modified from Rowe 2003; © 2021 The Authors, used with permission)

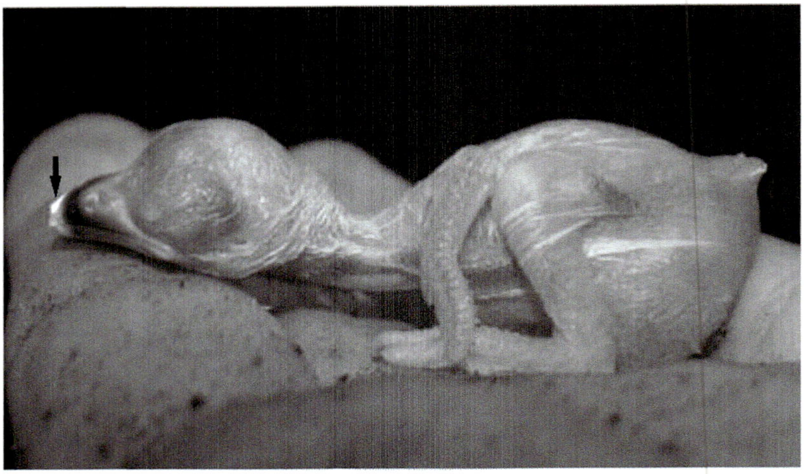

**Fig. 18.80** White egg tooth (indicated by black arrow) of a nestling Masked Trogon (*Trogon personatus*) on the day of hatching. (Figure from Greeney et al. 2008; used with permission)

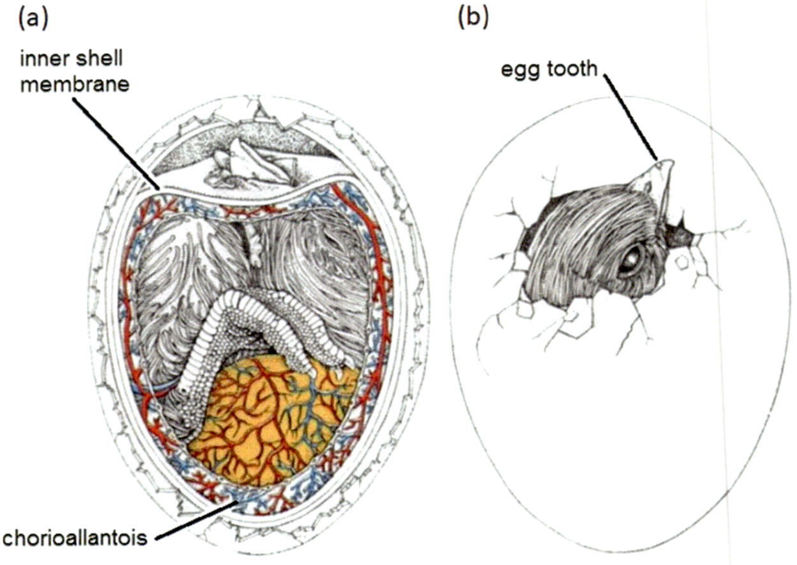


**Fig. 18.81** (**a**) An embryo penetrates the air cell with its beak, a process called internal pipping. The embryo then begins to breathe from the air cell, although the chorioallantois still continues to function. About 6 h later, the chick breaks through the eggshell with the egg tooth at the tip of its upper bill (external pipping, **b**) and starts to breathe atmospheric air. (Figure from Rahn et al. 1979; used with permission of illustrator Patricia Wynne)

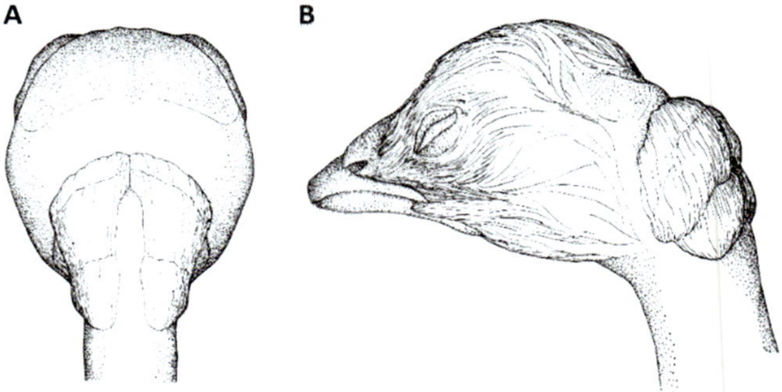


**Fig. 18.82** The hatching muscle, or musculus complexus, of a Domestic Chicken on the day of hatching as seen from the rear (**a**) and the side (**b**). (Figure from Bock and Hikida 1968; © 1968 Oxford University Press, used with permission)

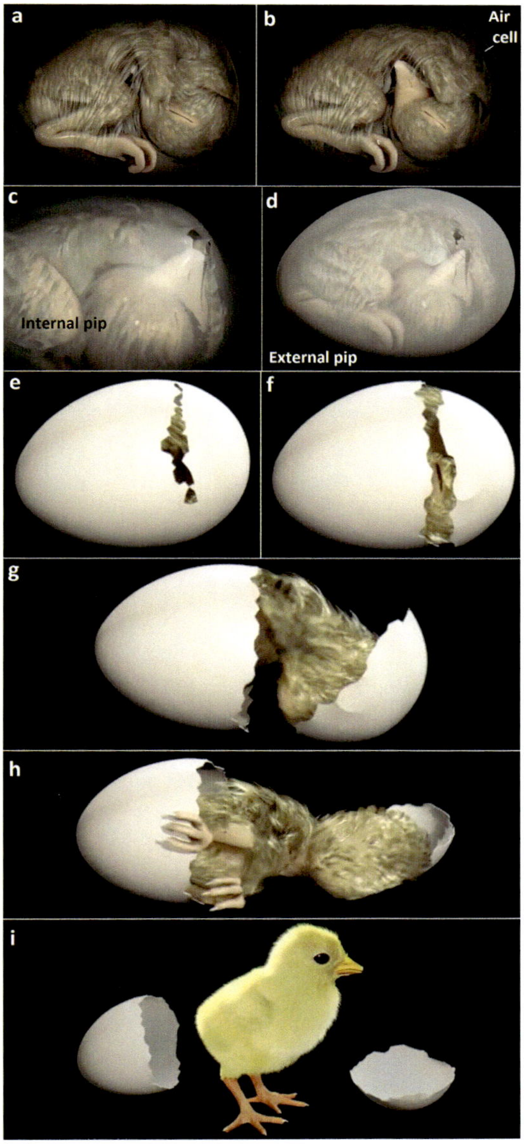


**Fig. 18.83** Hatching of a Domestic Chicken. (Figures are screen shots from "Animation of the 21-day development of a chicken embryo in the egg"; Created by AXS Biomedical Animation Studio Inc., http://axs3d.com/; Copyright 2020 AXS Biomedical Animation Studio Incorporated. All rights reserved. Used with permission)

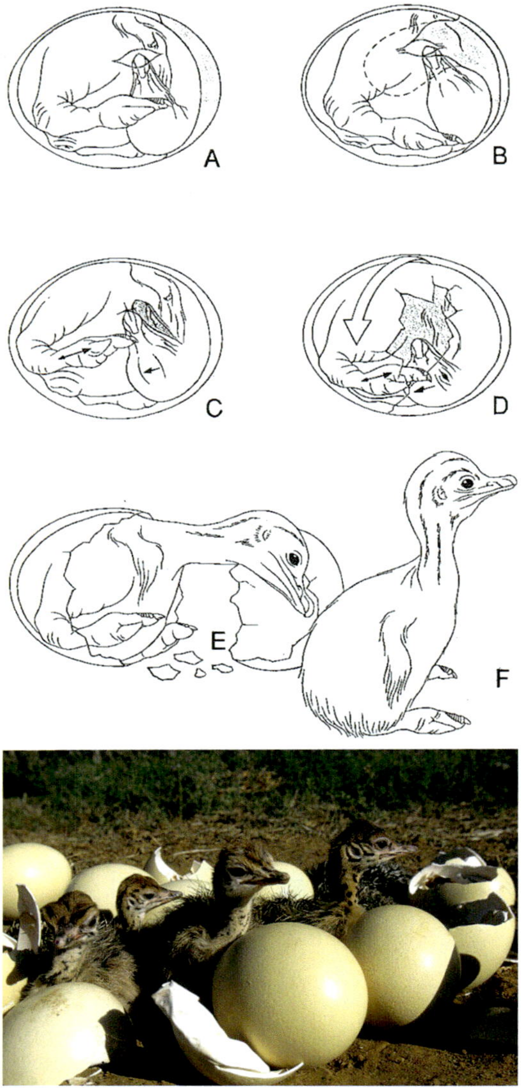


**Fig. 18.84** Top, Hatching of a Common Ostrich (*Struthio camelus*) egg. (**a**) Ostrich embryo rubs its beak (with no egg tooth) to create an opening in the inner shell membrane. (**b**) The embryo then pushes its body toward the air space, with repeated pushing causing the inner shell membrane to pull away from the outer shell membrane and move down the side of the egg. As this is occurring, the embryo's beak tears the inner shell membrane, providing access to the air space (internal pipping). (**c**) Next, the claw on the embryo's large toe moves toward the beak. External pipping is probably caused by the backwards movement of the head, but a forward movement of the foot and claw of the toe may also contribute to, or cause, the initial breaking of the eggshell. The "pip" is star-shaped and 20 to 60 mm in diameter; sometimes part of the shell breaks away. (**d**) After pipping, the opening in the shell is made larger by the kicking action of the leg, with the large toe contacting the shell and the beak hitting the shell via backward flips of the head. During these events, the embryo rotates in the egg and the beak moves toward the equator of the egg, creating a fracture line from the air space to the egg's equator. The crack spreads as the embryo kicks and pushes against the eggshell, causing the eggshell to break apart. (**e**) The young Ostrich then stretches and straightens its neck, pushing the end of the eggshell away. (**f**) The chick then sits upright to recover and dry off. Bottom, Just-hatched Common Ostriches. (Top figure from Deeming 1995; Rights managed by Taylor & Francis, used with permission; Bottom photo by Michel Viard, purchased from istockphoto.com)

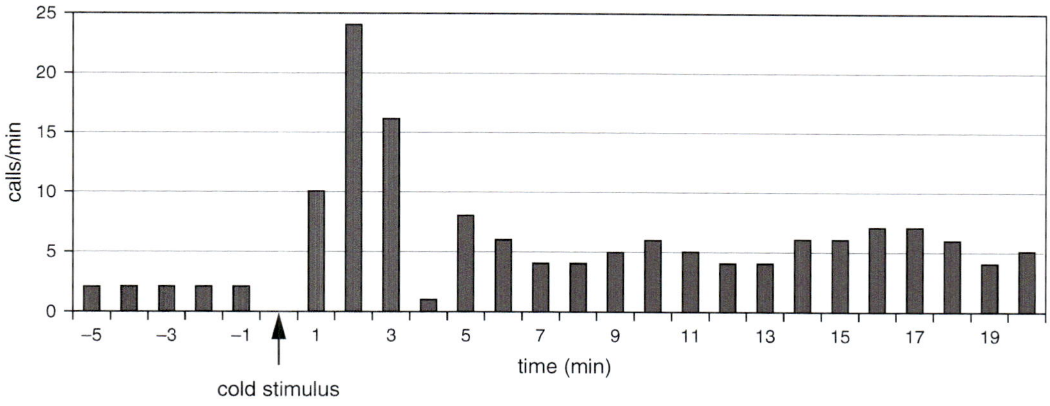


**Fig. 18.85** Call rates of Muscovy Duck (*Cairina moschata*) embryos before and after eggs were dipped in a 20–22°C water bath for several seconds. (Figure from Nichelmann and Tzschentke 1997; © 2006 John Wiley and Sons, used with permission)

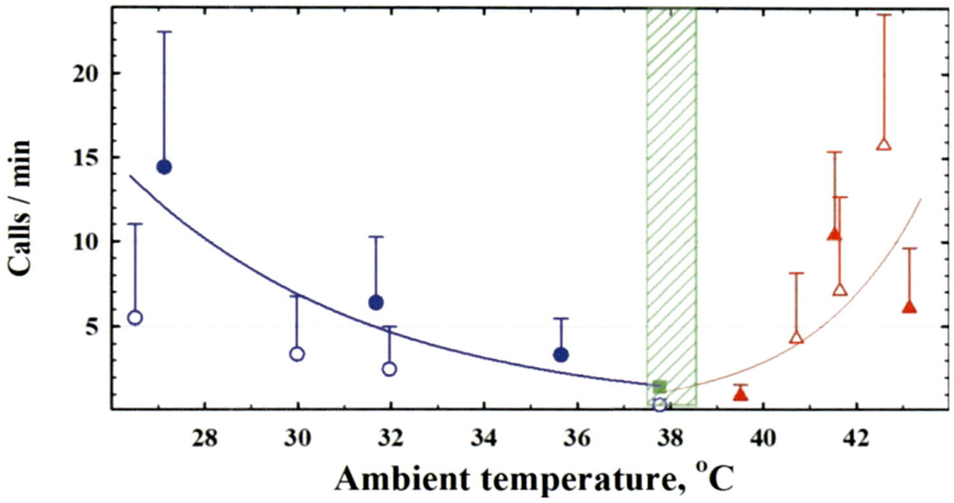


**Fig. 18.86** Mean (± SE) number of calls by Domestic Chicken embryos in response to changing ambient temperatures. Call rates increased both with decreasing and increasing ambient temperature, but increased faster with increasing ambient temperatures. Embryos may respond more quickly to increasing than decreasing ambient temperatures because the hyperthermia may pose a greater risk to embryos than hypothermia; even a short period of exposure to a temperature of 44°C can be fatal (Oppenheim and Levin 1975). Blue circles = cooling temperatures, red circles = warming temperatures. Open and filled symbols indicate relatively rapid and slow changes in temperature, respectively. Hatched area indicates normothermia. (Figure from Mortola 2019; © 2019 Elsevier B.V. All rights reserved, used with permission)

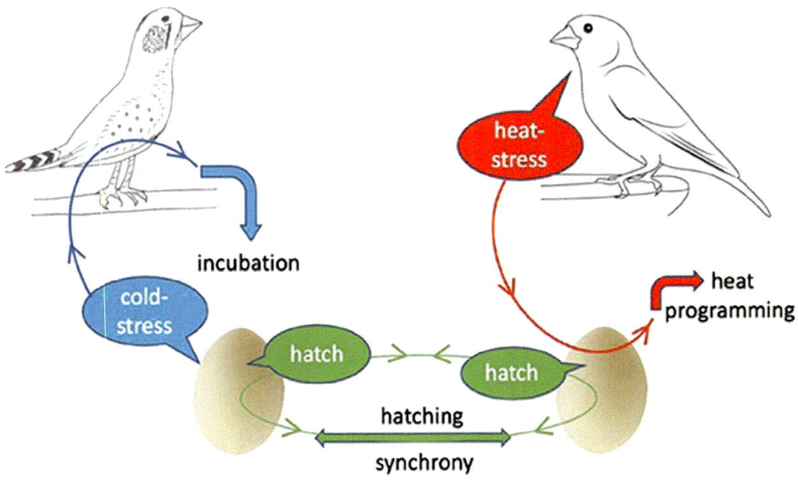


**Fig. 18.87** During incubation, vocalizations by embryos may signal their cold-stress to adults that can then incubate eggs to optimize incubation temperatures, and adult vocalizations signal heat-stress to embryos, altering the developmental responses of embryos. Embryo communication via vocalizations can also help synchronize their hatching. (Figure modified from Mariette 2019; © 2019 Mariette, open-access article distributed under the terms of the Creative Commons Attribution License (CC BY), https://creativecommons.org/licenses/by/4.0/)

## References

Ackerman RA, Platter-Reiger M (1979) Water loss by Pied-billed Grebe (*Podilymbus podiceps*) eggs. Am Zool 19:921

Adams J, Takekawa JY, Carter HR (2004) Foraging distance and home range of Cassin's Auklets nesting at two colonies in the California Channel Islands. Condor 106:618–637

Afton AD (1979) Incubation temperatures of the Northern Shoveler. Can J Zool 57:1052–1056

Alonso-Alvarez C, Cantarero A, Romero-Haro AÁ, Chastel O, Pérez-Rodríguez L (2020) Life-long testosterone and antiandrogen treatments affect the survival and reproduction of captive male Red-legged Partridges (*Alectoris rufa*). Behav Ecol Sociobiol 74:1–12

Amat JA, Masero JA (2004) Predation risk on incubating adults constrains the choice of thermally favourable nest sites in a plover. Anim Behav 67:293–300

Amat JA, Monsa R, Masero JA (2012) Dual function of egg-covering in the Kentish Plover *Charadrius alexandrinus*. Behaviour 149:881–895

Amiot R, Wang X, Wang S, Lecuyer C, Maxin J-M, Mo J, Flandrois J-P, Fourel F, Wang X, Xu X, Zhang Z (2017) δ18O-derived incubation temperatures of oviraptorosaur eggs. Palaeontology 60: 633-647.

Andersson M, Waldeck P (2006) Reproductive tactics under severe egg predation: an eider's dilemma. Oecologia 148:350–355

Ar A, Rahn H (1980) Water in the avian egg: overall budget of incubation. Am Zool 20:373–384

Ar A, Sidis Y (2002) Nest microclimate during incubation. In: Deeming DC (ed) Avian incubation: behaviour, environment, and evolution. Oxford University Press, Oxford, UK, pp 143–160

Ar A, Ifergan O, Reizis A, Zelik L, Feldman A (2000) Does nitric oxide (NO) play a role in embryo-bird communication during incubation? Avian Poult Biol Rev 11:284

Ar A, Ifergan O, Feldman A, Zelik L, Reizis A (2004) Possible role of nitric oxide emission from bird embryos. Avian Poult Biol Rev 15:105–106

Attard MRG, Portugal SJ (2021) Climate variability and parent nesting strategies influence gas exchange across avian eggshells. Proc R Soc B 288(2021):0823

Auer SK, Bassar RD, Martin TE (2007) Biparental incubation in the Chestnut-vented Tit-Babbler *Parisoma subcaeruleum*: mates devote equal time, but males keep eggs warmer. J Avian Biol 38:278–283

Bailey RE (1952) The incubation patch of passerine birds. Condor 54:121–136

Ball GF (1983) Functional incubation in male Barn Swallows. Auk 100:998–1000

Barrett RT (1980) Temperature of Kittiwake *Rissa tridactyla* eggs and nests during incubation. Ornis Scandinavica 11:50–59

Barve S, Mason NA (2015) Interspecific competition affects evolutionary links between cavity nesting, migration and clutch size in Old World flycatchers (Muscicapdae). Ibis 157:299–311

Bauer R, Plieschnig JA, Finkes T, Riegler B, Hermann M, Schneider WJ (2013) The developing chicken yolk sac acquires nutrient transport competence by an

orchestrated differentiation process of its endodermal epithelial cells. J Biol Chem 288:1088–1098
Beer CG (1961) Incubation and nest building behaviour of Black-headed Gulls. I: Incubation behaviour in the incubation period. Behaviour 18:62–105
Beer CG (1965) Clutch size and incubation behavior in Black-billed Gulls (*Larus bulleri*). Auk 82:1–18
Beissinger SR, Tygielski S, Elderd B (1998) Social constraints on the onset of incubation in a Neotropical parrot: a nestbox addition experiment. Anim Behav 55: 21–32
Bell CP (1996) The relationship between geographic variation in clutch size and migration pattern in the Yellow Wagtail. Bird Study 43:333–341
Bellairs R (1964) Biological aspects of the yolk of the hen's egg. In: Abercrombie M, Brachet J (eds) Advances in morphogenesis, vol 4. Academic, New York, pp 217–272
Bennett PM, Owens IPF (2002) Evolutionary ecology of birds - life histories, mating systems and extinction. Oxford Univ Press, Oxford, UK
Berlin KE, Clark AB (1998) Embryonic calls as care-soliciting signals in Budgerigars, *Melopsittacus undulatus*. Ethology 104:531–544
Bety J, Gauthier G, Giroux J-F (2003) Body condition, migration, and timing of reproduction in Snow Geese: a test of the condition-dependent model of optimal clutch size. Am Nat 162:110–121
Bi S, Amiot R, de Fabrègues CP, Pittman M, Lamanna MC, Yu Y, Yu C, Yang T, Zhang S, Zhao Q, Xu X (2021) An oviraptorid preserved atop an embryo-bearing egg clutch sheds light on the reproductive biology of non-avialan theropod dinosaurs. Sci Bull 66(9):947–954
Bielefeld RR, Brasher MG, Moorman TE, Gray PN (2020) Mottled Duck (*Anas fulvigula*), version 1.0. In: Poole AF (ed) Birds of the world. Cornell Lab of Ornithology, Ithaca, NY
Birchard GF, Ruta M, Deeming DC (2013) Evolution of parental incubation behaviour in dinosaurs cannot be inferred from clutch mass in birds. Biol Lett 9: 20130036
Bitton P-P, Dawson RD, O'Brien EL (2006) Influence of intraclutch egg-mass variation and hatching asynchrony on relative offspring performance within broods of an altricial bird. Can J Zool 84:1721–1726
Blackburn DG (2021) Functional morphology, diversity, and evolution of yolk processing specializations in embryonic reptiles and birds. J Morphol 282:995–1014
Blom J, Lilja C (2004) A comparative study of growth, skeletal development and eggshell composition in some species of birds. J Zool 262:361–369
Blom J, Lilja C (2005) A comparative study of embryonic development of some bird species with different patterns of postnatal growth. Zoology 108:81–95
Bock WJ, Hikida RS (1968) An analysis of twitch and tonus fibers in the hatching muscle. Condor 70:211–222
Böhning-Gaese K, Halbe B, Lemoine N, Oberrath R (2000) Factors influencing the clutch size, number of broods and annual fecundity of North American and European land birds. Evol Ecol Res 2:823–839
Bond GM, Board RG, Scott VD (1988) A comparative study of changes in the fine structure of avian eggshells during incubation. Zool J Linn Soc 92:105–113
Bonduriansky R, Maklakov A, Zajitschek F, Brooks R (2008) Sexual selection, sexual conflict and the evolution of ageing and life span. Funct Ecol 22:443–453
Booth DT, Seymour RS (1987) Effect of eggshell thinning on water vapor conductance of Malleefowl eggs. Condor 89:453–459
Booth DT, Jones DN (2002) Underground nesting in the megapodes. In: Deeming DC (ed) Avian incubation: behaviour, environment, and evolution. Oxford University Press, Oxford, UK, pp 192–206
Botelho JF, Smith-Paredes D, Vargas AO (2015) Altriciality and the evolution of toe orientation in birds. Evol Biol 42:502–510
Boulton RL, Cassey P (2012) How avian incubation behavior influences egg surface temperatures: relationships with egg position, development and clutch size. J Avian Biol 43:289–296
Bruce J, Drysdale EM (1991) Egg hygiene: routes of infection. In: Tullett SG (ed) Avian incubation. Butterworth Heinemann, Northamption, pp 257–276
Bruce J, Drysdale EM (1994) Trans-shell transmission. In: Board RG, Fuller R (eds) Microbiology of the avian egg. Springer, New York, NY, pp 63–91
Bruderer B, Salewski V (2009) Lower annual fecundity in long-distance migrants than in less migratory birds of temperate Europe. J Ornithol 150:281–286
Brummermann M, Reinertsen RE (1991) Adaptation of homeostatic thermoregulation: comparison of incubating and non-incubating bantam hens. J Comp Physiol B 161:133–140
Bryant DM, Hails CJ (1983) Energetics and growth patterns of three tropical bird species. Auk 100:425–439
Bryan SM, Bryant DM (1999) Heating nest–boxes reveals an energetic constraint on incubation behaviour in Great Tits, *Parus major*. Proc R Soc B 266:157–162
Buhr RJ (1995) Incubation relative humidity effects on allantoic fluid volume and hatchability. Poult Sci 74: 874–884
Burley NT, Johnson K (2002) The evolution of avian parental care. Philos Trans R Soc B 357:241–250
Burt DB, Coulter PF, Ligon JD (2007) Evolution of parental care and cooperative breeding. In: Jamieson BGM (ed) Reproductive biology and phylogeny of birds, part B. Science, Enfield, NH, pp 295–325
Butler RG, Janes-Butler S (1983) Sexual differences in the behavior of adult Great Black-backed Gulls (*Larus marinus*) during the pre-and post-hatch periods. Auk 100:63–75
Calder WA (1974) Consequences of body size for avian energetics. In: Paynter RA (ed) Avian energetics.

Nuttall Ornithological Club, Cambridge, MA, pp 86–151
Cam E, Monnat J-Y, Hines JE (2003) Long-term fitness consequences of early conditions in the Kittiwake. J Anim Ecol 72:411–424
Carey C (1986) Tolerance of variation in eggshell conductance, water loss, and water content by Red-winged Blackbird embryos. Physiol Zool 59:109–122
Carey C, Rahn H, Parisi P (1980) Calories, water, lipid and yolk in avian eggs. Condor 82:335–343
Carey C, Leon-Velarde F, Castro G, Monge C (1987) Shell conductance, daily water loss, and water content of Andean Gull and Puna Ibis eggs. J Exp Zool Suppl 1: 246–252
Carey C, Leon-Velarde F, Dunin-Borkowski O, Bucher TL, de al Torre G, Espinoza D, Monge C (1989) Variation in eggshell characteristics and gas exchange of montane and lowland coot eggs. J Comp Physiol B 159:389–400
Chalfoun AD, Martin TE (2007) Latitudinal variation in avian incubation attentiveness and a test of the food limitation hypothesis. Anim Behav 73:579–585
Chien Y-C, Hincke MT, McKee MD (2009) Ultrastructure of avian eggshell during resorption following egg fertilization. J Struct Biol 168:527–538
Chiver I, Morton ES, Stutchbury BJM (2007) Incubation delays territory defence by male Blue-headed Vireos, *Vireo solitarius*. Anim Behav 73:143–148
Clatterbuck CA, Young LC, VanderWerf EA, Naiman AD, Bower GC, Shaffer SA (2017) Data loggers in artificial eggs reveal that egg-turning behavior varies on multiple ecological scales in seabirds. Auk 134: 432–442
Clark JM, Norell M, Chiappe LM, Akademi MSU (1999) An oviraptorid skeleton from the late Cretaceous of Ukhaa Tolgod, Mongolia, preserved in an avianlike brooding position over an oviraptorid nest. Amer Mus Nov 3265
Clotfelter ED, Yasukawa K (1999) The function of early onset of nocturnal incubation in Red-winged Blackbirds. Auk 116:417–426
Collins PM, Halsey LG, Arnould JPY, Shaw PJA, Dodd S, Green JA (2016) Energetic consequences of time-activity budgets for a breeding seabird. J Zool 300: 153–162
Conway CJ, Martin TE (2000a) Evolution of passerine incubation behavior: influence of food, temperature, and nest predation. Evolution 54:670–685
Conway CJ, Martin TE (2000b) Effects of ambient temperature on avian incubation behavior. Behav Ecol 11: 178–188
Cook MI, Beissinger SR, Toranzos GA, Arendt WJ (2005a) Incubation reduces microbial growth on eggshells and the opportunity for trans-shell infection. Ecol Lett 8:532–537
Cook MI, Beissinger SR, Toranzos GA, Rodriguez RA, Arendt WJ (2005b) Microbial infection affects egg viability and incubation behavior in a tropical passerine. Behav Ecol 16:30–36
Cooney CR, Sheard C, Clark AD, Healy SD, Liker A, Street SE, Troisi CA, Thomas GH, Székely T, Hemmings N, Wright AE (2020) Ecology and allometry predict the evolution of avian developmental durations. Nat Commun 11:2383
Crick HQP, Gibbons DW, Magrath RD (1993) Seasonal changes in clutch size in British birds. J Anim Ecol 62: 263–273
Curio E (1983) Why do young birds reproduce less well? Ibis 125:400–404
D'Alba L, Monaghan P, Nager RG (2009) Thermal benefits of nest shelter for incubating female eiders. J Therm Biol 34:93–99
D'Alba L, Oborn A, Shawkey MD (2010) Experimental evidence that keeping eggs dry is a mechanism for the antimicrobial effects of avian incubation. Naturwissenschaften 97:1089–1095
Da Silva M, Labas V, Nys Y, Réhault-Godbert S (2017) Investigating proteins and proteases composing amniotic and allantoic fluids during chicken embryonic development. Poult Sci 96:2931–2941
Davis TA, Shen SS, Ackerman RA (1988) Embryonic osmoregulation: consequences of high and low water loss during incubation of the chicken egg. J Exp Zool A 245:144–156
Deeming DC (1995) The hatching sequence of Ostrich (*Struthio camelus*) embryos with notes on development as observed by candling. Br Poult Sci 36:67–78
Deeming DC (2002a) Embryonic development and utilization of egg components. In: Deeming DC (ed) Avian incubation: behaviour, environment, and evolution. Oxford University Press, Oxford, UK, pp 43–53
Deeming DC (2002b) Behaviour patterns during incubation. In: Deeming DC (ed) Avian incubation: behaviour, environment, and evolution. Oxford University Press, Oxford, UK, pp 63–87
Deeming DC (2008) Avian brood patch temperature: relationships with female body mass, incubation period, developmental maturity and phylogeny. J Therm Biol 33:345–354
Deeming DC (2011) Importance of nest type on the regulation of humidity in bird nests. Avian Biol Res 4:23–31
Deeming DC, du Feu C (2008) Use of an infrared ear thermometer to measure brood patch temperatures. Bird Study 55:139–143
Deeming DC, Ferguson MWJ (1992) Physiological effects of incubation temperature on embryonic development in reptiles and birds. In: Deeming DC, Ferguson MWJ (eds) Egg incubation: its effects on embryonic development in reptiles and birds. Cambridge University Press, Cambridge, UK, pp 147–173
Deeming DC, Birchard GF, Crafer R, Eady PE (2006) Egg mass and incubation period allometry in birds and reptiles: effects of phylogeny. J Zool 270:209–218
De Forest LN, Gaston AJ (1996) The effect of age on timing of breeding and reproductive success in the Thick-billed Murre. Ecology 77:1501–1511

de Heij ME, Van den Hout PJ, Tinbergen JM (2006) Fitness cost of incubation in Great Tits (*Parus major*) is related to clutch size. Proc R Soc B 273:2353–2361

De Oliveira JE, Uni Z, Ferket PR (2008) Important metabolic pathways in poultry embryos prior to hatch. Worlds Poult Sci J 64:488–499

Dhondt AA, Kast TL, Allen PE (2002) Geographical differences in seasonal clutch size variation in multibrooded bird species. Ibis 144:646–651

Dillon KG, Conway CJ (2018) Nest predation risk explains variation in avian clutch size. Behav Ecol 29:301–311

Dobbs RC, Styrsky JD, Thompson CF (2006) Clutch size and the costs of incubation in the House Wren. Behav Ecol 17:849–856

Dobson FS, Jouventin P (2007) How slow breeding can be selected in seabirds: testing Lack's hypothesis. Proc R Soc B 274:275–279

Dombre C, Guyot N, Moreau T, Monget P, Da Silva M, Gautron J, Réhault-Godbert S (2017) Egg serpins: the chicken and/or the egg dilemma. Semin Cell Dev Biol 62:120–132

Dong Z-M, Currie PH (1996) On the discovery of an oviraptorid skeleton on a nest of eggs at Bayan Mandahu, Inner Mongolia, People's Republic of China. Can J Earth Sci 33:631–636

Drent RH (1970) Functional aspects of incubation in the Herring Gull. Behav Suppl 17:1–132

Drent R (1975) Incubation. In: Farner DS, King JR (eds) Avian biology, vol 5. Academic, New York, pp 333–420

Duval M (1889) Atlas d'embryologie. G. Masson, Paris, France

Eggers S, Griesser M, Nystrand M, Ekman J (2006) Predation risk induces changes in nest-site selection and clutch size in the Siberian Jay. Proc R Soc B 273:701–706

Ekanayake KB, Weston MA, Nimmo DG, Maguire GS, Endler JA, Küpper C (2015) The bright incubate at night: sexual dichromatism and adaptive incubation division in an open-nesting shorebird. Proc R Soc B 282:20143026

Evans RM, Whitaker A, Wiebe MO (1994) Development of vocal regulation of temperature by embryos in pipped eggs of Ring-billed Gulls. Auk 111:596–604

Fierro-Calderón K, Martin TE (2007) Reproductive biology of the Violet-chested Hummingbird in Venezuela and comparisons with other tropical and temperate hummingbirds. Condor 109:680–685

Fontaine JJ, Martel M, Markland HM, Niklison AM, Decker KL, Martin TE (2007) Testing ecological and behavioral correlates of nest predation. Oikos 116: 1887–19894

Forsythe DM (1971) Clicking in the egg-young of the Long-billed Curlew. Wilson Bull 83:441–442

Gabrielli MG, Accili D (2010) The chick chorioallantoic membrane: a model of molecular, structural, and functional adaptation to transepithelial ion transport and barrier function during embryonic development. J Biomed Biotechnol 2010:940741

Garamszegi LZ, Hirschenhauser K, Bókony V, Eens M, Hurtrez-Boussès S, Møller AP, Oliveira RF, Wingfield JC (2008) Latitudinal distribution, migration, and testosterone levels in birds. Am Nat 172:533–546

Gardali T, Ballard G (2020) Warbling Vireo (*Vireo gilvus*), version 1.0. In: Poole AF, Gill FB (eds) Birds of the World. Cornell Lab of Ornithology, Ithaca, NY

Gardali T, Ballard G, Nur N, Geupel GR (2000) Demography of a declining population of Warbling Vireos in coastal California. Condor 102:601–609

Gefen E, Ar A (2001) Gas exchange and energy metabolism of the Ostrich (*Struthio camelus*) embryo. Comp Biochem Physiol A 130:689–699

Grant GS (1982) Avian incubation: egg temperature, nest humidity, and behavioral thermoregulation in a hot environment. Ornithol Monogr 30:iii–75

Grant GS, Pettit TN, Rahn H, Whittow GC, Paganelli CV (1982) Regulation of water loss from Bonin Petrel (*Pterodroma hypoleuca*) eggs. Auk 99:236–242

Green JA, Boyd IL, Woakes AJ, Warren NL, Butler PJ (2009) Evaluating the prudence of parents: daily energy expenditure throughout the annual cycle of a free-ranging bird, the Macaroni Penguin *Eudyptes chrysolophus*. J Avian Biol 40:529–538

Greeney H, Sheldon K, Simbaña J (2008) Observations on the hatchlings, eggs and incubation of the Masked Trogon *Trogon personatus* in eastern Ecuador. Cotinga 29:82–84

Grellet-Tinner G, Chiappe L, Norell M, Bottjer D (2006) Dinosaur eggs and nesting behaviors: a paleobiological investigation. Paleo 232:294–321

Grenier JL, Beissinger SR (1999) Variation in the onset of incubation in a Neotropical parrot. Condor 101:752–761

Grizard S, Dini-Andreote F, Tieleman BI, Salles JF (2014) Dynamics of bacterial and fungal communities associated with eggshells during incubation. Ecol Evol 4:1140–1157

Groscolas R, Decrock F, Thil M-A, Fayolle C, Boissery C, Robin J-P (2000) Refeeding signal in fast-incubating King Penguins: changes in behavior and egg temperature. American Journal of Physiology – Regulatory. Integr Comp Physiol 279:R2104–R2112

Groscolas R, Fréchard F, Decrock F, Speake BK (2003) Metabolic fate of yolk fatty acids in the developing King Penguin embryo. Am J Physiol Regul Integr Comp Physiol 285:R850–R861

Guedes PT, de Abreu Manso PP, Caputo LFG, Cotta-Pereira G, Pelajo-Machado M (2014) Histological analyses demonstrate the temporary contribution of yolk sac, liver, and bone marrow to hematopoiesis during chicken development. PLoS ONE 9:e90975

Haftorn S, Reinertsen RE (1985) The effect of temperature and clutch size on the energetic cost of incubation in a free-living Blue Tit (*Parus caeruleus*). Auk 102:470–478

Hamburger V, Hamilton HL (1951) A series of normal stages in the development of the chick embryo. J Morphol 88:49–92

Hanssen SA, Engebretsen H, Erikstad KE (2002) Incubation start and egg size in relation to body reserves in the Common Eider. Behav Ecol Sociobiol 52:282–288

Hanssen SA, Hasselquist D, Folstad I, Erikstad KE (2005) Cost of reproduction in a long-lived bird: incubation effort reduces immune function and future reproduction. Proc R Soc B 272:1039–1046

Harris RB, Birks SM, Leaché AD (2014) Incubator birds: biogeographical origins and evolution of underground nesting in megapodes (Galliformes: Megapodiidae). J Biogeogr 41:2045–2056

Hart LA, Downs CT, Brown M (2016) Hot footing eggs: thermal imaging reveals foot mediated incubation in White-tailed Tropicbirds, *Phaethon lepturus*. J Ornithol 157:635–640

Haukos DA (2015) Survival and recovery rates of Mottled Ducks banded in Texas and Louisiana. J Southeastern Assoc Fish Wildl Agencies 2:214–220

Hébert PN (1993) An experimental study of brood reduction and hatching asynchrony in Yellow Warblers. Condor 95:362–371

Hébert PN, McNeil R (1999) Hatching asynchrony and food stress in Ring-billed Gulls: an experimental study. Can J Zool 77:515–523

Hemmings N, Birkhead TR (2015) Polyspermy in birds: sperm numbers and embryo survival. Proc Roy Soc B 282:20151682

Hemmings N, Birkhead TR (2016) Consistency of passerine embryo development and the use of embryonic staging in studies of hatching failure. Ibis 158:43–50

Hepp GR, Kennamer RA, Johnson MH (2006) Maternal effects in Wood Ducks: incubation temperature influences incubation period and neonate phenotype. Funct Ecol 20:307–314

Hipfner JM, Gorman KB, Vos RA, Joy JB (2010) Research article Evolution of embryonic developmental period in the marine bird families Alcidae and Spheniscidae: roles for nutrition and predation? BMC Evol Biol 10:179

Honza M, Picman J, Grim T, Novák V, Čapek M Jr, Mrlik V (2001) How to hatch from an egg of great structural strength. A study of the Common Cuckoo. J Avian Biol 32:249–255

Howell TR (1964) Notes on incubation and nestling temperatures and behaviour of captive owls. Wilson Bull 76:2–36

Howell TR, Bartholomew GA (1961) Temperature regulation in nesting Bonin Island Petrels, Wedge-tailed Shearwaters, and Christmas Island Shearwaters. Auk 78:343–354

Hua F, Sieving KE, Fletcher RJ Jr, Wright CA (2014) Increased perception of predation risk to adults and offspring alters avian reproductive strategy and performance. Behav Ecol 25:509–519

Ifergan O, Ar A (1999) Embryo-incubating bird communication: yes or no. Isr J Zool 45:309–310

Impekoven M (1973) The response of incubating Laughing Gulls (*Larus atricilla* L.) to calls of hatching chicks. Behaviour 46:94–113

Jahn AE, Tuero DT, Mamani AM, Bejarano V, Masson DA, Aguilar E (2014) Drivers of clutch-size in Fork-tailed Flycatchers (*Tyrannus savana*) at temperate and tropical latitudes in South America. Emu 114:337–342

Jetz W, Sekercioglu CH, Böhning-Gaese K (2008) The worldwide variation in avian clutch size across species and space. PLoS Biol 6:e303

Johnston A, Robinson RA, Gargallo G, Julliard R, van der Jeugd H, Baillie SR (2016) Survival of Afro-Palaearctic passerine migrants in western Europe and the impacts of seasonal weather variables. Ibis 158: 465–480

Jones RE (1971) The incubation patch of birds. Biol Rev 46:315–339

Jones S (2013) Examining the behavioural costs of heat in a climatically range-restricted arid zone bird; the case of the Ethiopian Bush-crow *Zavattariornis stresemanni*. M. S. thesis, Imperial College, London, UK

Jones DN, Birks S (1992) Megapodes: recent ideas on origins, adaptations and reproduction. Trends Ecol Evol 7:88–91

Julliard R, McCleery RH, Clobert J, Perrins CM (1997) Phenotypic adjustment of clutch size due to nest predation in the Great Tit. Ecology 78:394–404

Karlsson O, Lilja C (2008) Eggshell structure, mode of development and growth rate in birds. Zoology 111: 494–502

Keibel F, Abraham K (1900) Normentafel zur entwicklungsgeschichte des huhnes, *Gallus domesticus*. Gustav Fischer Verlag, Jena

Keller V (1989) Egg-covering behaviour by Great Crested Grebes *Podiceps cristatus*. Ornis Scandinavica 20: 129–131

Kendeigh SC (1952) Parental care and its evolution in birds. Illinois Biological Monographs, no. 22

Krings M, Muller-Limberger E, Wagner H (2020) Embryonic development of the Little Owl (*Athene funereus*) compared to the American Barn Owl (*Tyto furcata*). J Ornithol 161:267–273

Krechmar AV, Kondratiev EV (1982) Ecology of incubation in *Philacte canadica* in the north of Chukot Peninsula. Zoologicheskij Zhurnal 61:254–264

Kreisinger J, Albrecht T (2008) Nest protection in Mallards *Anas platyrhynchos*: untangling the role of crypsis and parental behavior. Funct Ecol 22:872–879

Lack D (1947) The significance of clutch size. Ibis 89: 302–352

Lack D (1954) The natural regulation of animal numbers. Oxford University Press, London

Lack D (1968) Ecological adaptations for breeding in birds. Methuen, London

Lee KA, Wikelski M, Robinson WD, Robinson TR, Klasing KC (2008) Constitutive immune defences correlate with life-history variables in tropical birds. J Anim Ecol 77:356–363

Lee WY, Kim M, Jablonski PG, Choe JC, Lee SI (2014) Effect of incubation on bacterial communities of eggshells in a temperate bird, the Eurasian Magpie (PIca pica). PLoS ONE 9:e103959

Lepage D, Gauthier G, Desrochers A (1998) Larger clutch size increases fledging success and offspring quality in a precocial species. J Anim Ecol 67:210–216

Lipar JL, Ketterson ED (2000) Maternally derived yolk testosterone enhances the development of the hatching muscle in the Red-winged Blackbird *Agelaius phoeniceus*. Proc R Soc B 267:2005–2010

Lloyd JD, Martin TE (2003) Sibling competition and the evolution of growth and developmental rates in passerines. Proc R Soc B 270:735–740

Loukola OJ, Laaksonen T, Seppänen J-T, Forsman JT (2014) Active hiding of social information from information-parasites. BMC Evol Biol 14:32

Loukola OJ, Adamik P, Adriaensen F, Barba E, Doligez B, Flensted-Jensen E, Eeva T, Kivelä SM, Laaksonen T, Morosinotto C, Mänd R (2020) The roles of temperature, nest predators and information parasites for geographical variation in egg covering behaviour of tits (Paridae). J Biogeogr 47:1482–1493

Ludwigs J-D, Becker PH (2004) What do pairing patterns in Common Tern, *Sterna hirundo*, recruits reveal about the significance of sex and breeding experience? Behav Ecol Sociobiol 57:412–421

Maina JN (2017) Structure and function of the shell and the chorioallantoic membrane of the avian egg: embryonic respiration. In: Maina JN (ed) The biology of the avian respiratory system. New York, Springer, pp 219–247

Manuwal DA (1974) The incubation patches of Cassin's Auklet. Condor 76:481–484

Mariette MM (2019) Acoustic cooperation: acoustic communication regulates conflict and cooperation within the family. Front Ecol Evol 7:445

Mariette MM, Buchanan KL (2016) Prenatal acoustic communication programs offspring for high posthatching temperatures in a songbird. Science 353: 812–814

Martin TE (1995) Avian life history evolution in relation to nest sites, nest predation, and food. Ecol Monogr 65: 101–127

Martin TE (2015) Age-related mortality explains life history strategies of tropical and temperate songbirds. Science 349:966–970

Martin TE, Schwabl H (2008) Variation in maternal effects and embryonic development rates among passerine species. Philos Trans R Soc B 363:1663–1674

Martin TE, Martin PR, Olson CR, Heidinger BJ, Fontaine JJ (2000a) Parental care and clutch sizes in North and South American birds. Science 287:1482–1485

Martin TE, Scott J, Menge C (2000b) Nest predation increases with parental activity: separating nest site and parental activity effects. Proc R Soc B 267:2287–2293

Martin TE, Auer SK, Bassar RD, Niklison AM, Lloyd P (2007) Geographic variation in avian incubation periods and parental influences on embryonic temperature. Evolution 61:2558–2569

Martin TE, Oteyza JC, Boyce AJ, Lloyd P, Ton R (2015) Adult mortality probability and nest predation rates explain parental effort in warming eggs with consequences for embryonic development time. Am Nat 186:223–236

Massaro M, Setiawan AN, Davis LS (2007) Effects of artificial eggs on prolactin secretion, steroid levels, brood patch development, incubation onset and clutch size in the Yellow-eyed Penguin (*Megadyptes antipodes*). Gen Comp Endocrinol 151:220–229

Massaro M, Starling-Windhof A, Briskie JV, Martin TE (2008) Introduced mammalian predators induce behavioural changes in parental care in an endemic New Zealand bird. PLoS ONE 3:e2331

Matysioková B, Remeš V (2014) The importance of having a partner: male help releases females from time limitation during incubation in birds. Front Zool 11:24

Matysioková B, Remeš V (2018) Evolution of parental activity at the nest is shaped by the risk of nest predation and ambient temperature across bird species. Evolution 72:2214–2224

Mauck RA, Huntington CE, Grubb TC (2004) Age-specific reproductive success: evidence for the selection hypothesis. Evolution 58:880–885

McCleery RH, Perrins CM, Sheldon BC, Charmantier A (2008) Age-specific reproduction in a long-lived species: the combined effects of senescence and individual quality. Proc R Soc B 275:963–970

McCormack WT, Tjoelker LW, Thompson CB (1991) Avian B-cell development: generation of an immunoglobulin repertoire by gene conversion. Annu Rev Immunol 9:219–241

McNab BK (2002) The physiological ecology of vertebrates: a view from energetics. Cornell University Press, Ithaca, NY

McNamara JM, Barta Z, Wikelski M, Houston AI (2008) A theoretical investigation of the effect of latitude on avian life histories. Am Nat 172:331–345

Meyer N, Bollache L, Dechaume-Moncharmont FX, Moreau J, Afonso E, Angerbjörn A, Bêty J, Ehrich D, Gilg V, Giroux MA, Hansen J (2020) Nest attentiveness drives nest predation in Arctic sandpipers. Oikos 129:1481–1492

Miller DA, Vleck CM, Otis DL (2009) Individual variation in baseline and stress-induced corticosterone and prolactin levels predicts parental effort by nesting Mourning Doves. Horm Behav 56:457–464

Milonoff M, Pöysä H, Runko P, Ruusila V (2004) Brood rearing costs affect future reproduction in the precocial Common Goldeneye *Bucephala clangula*. J Avian Biol 35:344–351

Mönkkönen M, Martin TE (2000) Sensitivity of comparative analyses to population variation in trait values: clutch size and cavity excavation tendencies. J Avian Biol 31:576–579

Moore JR, Varricchio DJ (2016) The evolution of diapsid reproductive strategy with inferences about extinct taxa. PLoS ONE 11:e0158496
Moran ET Jr (2007) Nutrition of the developing embryo and hatchling. Poult Sci 86:1043–1049
Moreno J, Sanz JJ (1994) The relationship between the energy expenditure during incubation and clutch size in the Pied Flycatcher *Ficedula hypoleuca*. J Avian Biol 25:125–130
Morgan SM, Ashley-Ross MA, Anderson DJ (2003) Foot-mediated incubation: Nazca Booby (*Sula granti*) feet as surrogate brood patches. Physiol Biochem Zool 76: 360
Mortola JP (2009) Gas exchange in avian embryos and hatchlings. Comp Biochem Physiol A 153:359–377
Mortola JP (2019) Behavioral thermoregulation in avian embryos: spectrum analysis of calls in warm and cold conditions. Behav Processes 164:30–37
Morton ML, Pereyra ME (1985) The regulation of egg temperatures and attentiveness patterns in the Dusky Flycatcher (*Empidonax oberholseri*). Auk 102:25–37
Muchai M, du Plessis MA (2005) Nest predation of grassland bird species increases with parental activity at the nest. J Avian Biol 36:110–116
Munshi-South J, Wilkinson GS (2006) Diet influences life span in parrots (Psittaciformes). Auk 123:108–118
Nagai H, Mak SS, Weng W, Nakaya Y, Ladher R, Sheng G (2011) Embryonic development of the Emu, *Dromaius novaehollandiae*. Dev Dyn 240:162–175
Nahm KH (2001) Effects of storage length and weight loss during incubation on the hatchability of Ostrich eggs (*Struthio camelus*). Poult Sci 80:1667–1670
Newton I (1989) Lifetime reproduction in birds. Academic, London, UK
Newton I, McGrady MJ, Oli MK (2016) A review of survival estimates for raptors and owls. Ibis 158:227–248
Nguyen TH, Eichmann A, Le Noble F, Fleury V (2006) Dynamics of vascular branching morphogenesis: the effect of blood and tissue flow. Phys Rev E 73:061907
Nichelmann M, Tzschentke B (1997) Ontogeny of thermoregulation during the prenatal period in birds. Ann N Y Acad Sci 813:78–86
Nilsson J-Å, Smith HG (1988) Incubation feeding as a male tactic for early hatching. Anim Behav 36:641–647
Nol E, Smith JNM (1987) Effects of age and breeding experience on seasonal reproductive success in the Song Sparrow. J Anim Ecol 56:301–313
Nord A, Williams JB (2015) The energetic costs of incubation. In: Deeming DC, Reynolds SJ (eds) Nests, eggs, and incubation: new ideas about avian reproduction. Oxford University Press, Oxford, UK, pp 152–170
Nord A, Sandell MI, Nilsson J-Å (2010) Female Zebra Finches compromise clutch temperature in energetically demanding incubation conditions. Funct Ecol 24:1031–1036
Norell MA, Clark JM, Chiappe LM, Dashzeveg D (1995) A nesting dinosaur. Nature 378:774–776
Oppenheim RW, Levin HL (1975) Short-term changes in incubation temperature: behavioral and physiological effects in the chick embryo from 6 to 20 days. Dev Psychobiol 8:103–115
Orłowski G, Hałupka L (2015) Embryonic eggshell thickness erosion: a literature survey re-assessing embryo-induced eggshell thinning in birds. Environ Pollut 205: 218–224
Öst M, Wickman M, Matulionis E, Steele B (2008) Habitat-specific clutch size and cost of incubation in eiders reconsidered. Oecologia 158:205
Osterstrom O, Lilja C (2012) Evolution of avian eggshell structure. J Morphol 273:241–247
Österström O, Holm L, Lilja C (2013) Calcium mobilization from the avian eggshell during embryonic development. Anim Biol 63:33–46
Palmer C, Christian KA, Fisher A (2000) Mound characteristics and behaviour of the Orange-footed Scrubfowl in the seasonal tropics of Australia. Emu 100:54–63
Pärt T (1995) Does breeding experience explain increased reproductive success with age? An experiment. Proc R Soc B 260:113–117
Pawar S, Dell AI, Savage VM (2012) Dimensionality of consumer search space drives trophic interaction strengths. Nature 486:485–489
Payne RB (1980) Seasonal incidence of breeding, moult and local dispersal of Red-billed Firefinches *Lagonosticta senegala* in Zambia. Ibis 122:43–56
Peach WJ, Hanmer DB, Oatley TB (2001) Do southern African songbirds live longer than their European counterparts? Oikos 93:235–249
Pendlebury CJ, Bryant DM (2005) Night-time behaviour of egg-laying tits. Ibis 147:342–345
Persson I, Andersson G (1999) Intraclutch hatch synchronization in pheasants and Mallard ducks. Ethology 105:1087–1096
Portugal SJ, Maurer G, Thomas GH, Hauber ME, Grim T, Cassey P (2014) Nesting behaviour influences species-specific gas exchange across avian eggshells. J Exp Biol 217:3326–3332
Pretelli MG, Isacch JP, Cardoni DA (2016) Variation in parental care in the Spectacled Tyrant *Hymenops perspicillatus* is associated with increased nest predation in grassland fragments. J Ornithol 157:451–460
Prevost J (1963) Ecologie du Manchot Empereur. Hermann, Paris, France
Prokop P, Trnka A (2011) Why do grebes cover their nests? Laboratory and field tests of two alternative hypotheses. J Ethol 29:17–22
Rahn H (1991) Why birds lay eggs. In: Deeming DC, Ferguson MWJ (eds) Egg incubation: its effects on embryonic development in birds and reptiles. Cambridge University Press, Cambridge, pp 345–360
Rahn H, Ar A (1974) The avian egg: incubation time and water loss. Condor 76:147–152

Rahn H, Paganelli CV, Nisbet ICT, Whittow GC (1976) Regulation of incubation water loss in eggs of seven species of terns. Physiol Zool 49:245–259
Rahn H, Ar A, Paganelli CV (1979) How bird eggs breathe. Sci Am 240:46–55
Rahn H, Krog J, Mehlum F (1983) Microclimate of the nest and egg water loss of the eider *Somateria mollissima* and other waterfowl in Spitzsbergen. Polar Res 1:171–183
Rahn H, Paganelli CV, Ar A (1987) Pores and gas exchange of avian eggs: a review. J Exp Zool Suppl 1:165–172
Ramli MB, Wahab MS, Zain BA, Raus AA, Kamarudin K, Saude N, Haq R, Ahmed A (2017) Effect of incubation temperature on Ikta's quail breed with new rolling mechanism system. J Mech Eng 3:78–88
Ratcliffe N, Furness RW, Klomp NI (1998) Influences of breeding experience on the reproductive performance of Great Skuas *Catharacta skua*. J Avian Biol 29:293–298
Reid JM, Monaghan P, Ruxton GD (2000) Resource allocation between reproductive phases: the importance of thermal conditions in determining the cost of incubation. Proc R Soc B 267:37–41
Reid JM, Ruxton GD, Monaghan P, Hilton GM (2002) Energetic consequences of clutch temperature and clutch size for a uniparental intermittent incubator: the starling. Auk 119:54–61
Ricklefs RE (1965) Brood reduction in the Curve-billed Thrasher. Condor 67:505–510
Ricklefs RE (1977) On the evolution of reproductive strategies in birds: reproductive effort. Am Nat 111:453–478
Ricklefs RE (1983) Some considerations on the reproductive energetics of pelagic seabirds. Stud Avian Biol 8:84–94
Ricklefs RE (1984) The optimization of growth rate in altricial birds. Ecology 65:1602–1616
Ricklefs RE (1992) Embryonic development period and the prevalence of avian blood parasites. Proc Natl Acad Sci USA 89:4722–4725
Ricklefs RE (1993) Sibling competition, hatching asynchrony, incubation period, and lifespan in altricial birds. In: Power DM (ed) Current ornithology, vol 11. Plenum, New York, pp 199–276
Ricklefs RE (1997) Comparative demography of New World populations of thrushes (*Turdus* spp.). Ecol Monogr 67:23–43
Ricklefs RE (2000) Density dependence, evolutionary optimization, and the diversification of avian life histories. Condor 102:9–22
Ricklefs RE, Starck JM (1998a) Patterns of development: the altricial-precocial spectrum. In: Starck JM, Ricklefs RE (eds) Avian growth and development. Oxford University Press, Oxford, UK, pp 31–58
Ricklefs RE, Starck JM (1998b) Embryonic growth and development. In: Starck JM, Ricklefs RE (eds) Avian growth and development. Oxford University Press, Oxford, UK, pp 31–58
Ricklefs RE, Wikelski M (2002) The physiology/life-history nexus. Trends Ecol Evol 17:462–468
Ricklefs RE, Austin SH, Robinson WD (2017) The adaptive significance of variation in avian incubation periods. Auk 134:542–550
Riechert J, Becker PH, Chastel O (2014) Predicting reproductive success from hormone concentrations in the Common Tern (*Sterna hirundo*) while considering food abundance. Oecologia 176:715–727
Robinson WD, Styrsky JD, Payne BJ, Harper RH, Thompson CF (2008) Why are incubation periods longer in the tropics? A common-garden experiment with House Wrens reveals it is all in the egg. Am Nat 171: 532–535
Robinson WD, Austin SH, Robinson TR, Ricklefs RE (2014) Incubation temperature does not explain variation in the embryo development periods in a sample of Neotropical passerine birds. J Ornithol 155:45–51
Romanoff AL, Romanoff AJ (1949) The avian egg. Wiley, New York
Rowe T (online) (2003) *Gallus gallus*, Domestic Chicken. http://digimorph.org/specimens/Gallus_gallus/day16/
Rowe L, Ludwig D, Schluter D (1994) Time, condition, and the seasonal decline of avian clutch size. Am Nat 143:698–722
Ruiz-de-Castañeda R, Vela AI, Lobato E, Briones V, Moreno J (2011) Bacterial loads on eggshells of the Pied Flycatcher: environmental and maternal factors. Condor 113:200–208
Russell EM, Yom-Tov Y, Geffen E (2004) Extended parental care and delayed dispersal : northern, tropical and southern passerines compared. Behav Ecol 15:831–838
Saether BE (1990) Age-specific variation in reproductive performance of birds. Curr Ornithol 7:251–283
Samuni M, Ar A (2006) Can nitric oxide emitted from incubated eggs be used to indicate stress? Investigating Hatchery. Practice 20:29. (Abstract)
Sauer EGF, Sauer EM (1966) The behaviour and ecology of the South African Ostrich. Living Bird 5:45–75
Scholer MN, Strimas-Mackey M, Jankowski JE (2020) A meta-analysis of global avian survival across species and latitude. Ecol Lett 23:1537–1549
Schondube J, Santana C E, Ruán-Tejeda E (2003) Biannual cycles of the Cinnamon-bellied Flowerpiercer. Biotropica 35:250–261
Sekercioglu C (1999) Megapodes: a fascinating incubation strategy. Harv J Undergraduate Stud 5:77–83
Sendecka J, Cichoń M, Gustafsson L (2007) Age-dependent reproductive costs and the role of breeding experience in the Collared Flycatcher. Acta Zoologica 88:95–100
Seymour RS (1991) The brush turkey. Sci Am 265:68–75
Shaffer SA, Costa DP, Weimerskirch H (2004) Field metabolic rates of Black-browed Albatrosses *Thalassarche melanophrys* during the incubation stage. J Avian Biol 35:551–558
Shaffer SA, Clatterbuck CA, Kelsey EC, Naiman AD, Young LC, VanderWerf EA, Warzybok P,

Bradley R, Jahncke J, Bower GC (2014) As the egg turns: monitoring egg attendance behavior in wild birds using novel data logging technology. PLoS ONE 9:e97898
Shawkey MD, Firestone MK, Brodie EL, Beissinger SR (2009) Avian incubation inhibits growth and diversification of bacterial assemblages on eggs. PLoS ONE 4: e4522
Shbailat SJ, Safi HM (2015) Transfer of egg white proteins with reference to lysozyme during the development of *Meleagris gallopavo* (Galliformes: Phasianidae) embryos. Ital J Zool 82:349–357
Sheng G, Foley AC (2012) Diversification and conservation of the extraembryonic tissues in mediating nutrient uptake during amniote development. Ann N Y Acad Sci 1271:97–103
Shoji A, Elliott KH, Aris-Brosou S, Wilson RP, Gaston AJ (2015) Predictors of incubation costs in seabirds: an evolutionary perspective. Ibis 157:44–53
Sibly RM, Witt CC, Wright NA, Venditti C, Jetz W, Brown JH (2012) Energetics, lifestyle, and reproduction in birds. Proc Natl Acad Sci USA 109:10937–10941
Skutch AF (1949) Do tropical birds rear as many young as they can nourish? Ibis 91:430–455
Skutch A (1954) Life histories of Central American birds. Pacific Coast Avifauna 31:1–448
Skutch A (1960) Life histories of Central American birds. Pacific Coast Avifauna 34:1–593
Slagsvold T (1982) Clutch size variation in passerine birds: the nest predation hypothesis. Oecologia 54: 159–169
Slagsvold T (1984) Clutch size variation of birds in relation to nest predation: on the cost of reproduction. J Anim Ecol 53:945–953
Slagsvold T, Wiebe KL (2021) Egg covering in cavity nesting birds may prevent nest usurpation by other species. Behav Ecol Sociobiol 75:116
Small TW, Sharp PJ, Deviche P (2007) Environmental regulation of the reproductive system in a flexibly breeding Sonoran Desert bird, the Rufous-winged Sparrow, *Aimophila carpalis*. Horm Behav 51:483–495
Smiley KO (2019) Prolactin and avian parental care: new insights and unanswered questions. Horm Behav 111: 114–130
Smiley KO, Adkins-Regan E (2016) Relationship between prolactin, reproductive experience, and parental care in a biparental songbird, the Zebra Finch (*Taeniopygia guttata*). Gen Comp Endocrinol 232:17–24
Sockman KW, Sharp PJ, Schwabl H (2006) Orchestration of avian reproductive effort: an integration of the ultimate and proximate bases for flexibility in clutch size, incubation behaviour, and yolk androgen deposition. Biol Rev 81:629–666
Speake BK, Murray AMB, Noble RC (1998) Transport and transformations of yolk lipids during development of the avian embryo. Prog Lipid Res 37:1–32
Starck JM (1993) Evolution of avian ontogenies. In: Power DM (ed) Current ornithology, vol 10. Plenum, New York, pp 275–366
Starck JM (1998) Structural variants and invariants in avian embryonic and postnatal development. In: Starck JM, Ricklefs RE (eds) Avian growth and development: evolution within the altricial-precocial spectrum. Oxford University Press, New York, pp 59–88
Starck JM, Ricklefs RE (eds) (1998) Avian growth and development : evolution within the altricial-precocial spectrum. Oxford Univ Press, Oxford, UK
Starck JM (2021) Morphology of the avian yolk sac. J Morphol 282:959–972
Steadman DW (1999) The biography and extinction of megapodes in Oceania. Zoologische Verhandelingen 327:1–16
Stearns SC (1992) The evolution of life histories. Oxford University Press, London, UK
Stettenheim PR (2000) The integumentary morphology of modern birds—an overview. Am Zool 40:461–477
Stevens L (1996) Egg proteins: what are their functions? Sci Prog 79:65–87
Stoleson SH, Beissinger SR (1999) Egg viability as a constraint on hatching synchrony at high ambient temperatures. J Anim Ecol 68:951–962
Stonehouse B (1956) The Brown Skua *Catharacta skua lonnbergi* (Mathews) of South Georgia. Falkland Island Dependencies Survey Scientific Reports 14:1–25
Stonehouse B (1960) The King Penguin of South Georgia. I. Breeding behaviour and development. Falkland Island Depend Surv Sci Rep 23:1–81
Stutchbury BJM, Morton ES (2008) Recent advances in the behavioral ecology of tropical birds. Wilson J Ornithol 120:26–37
Styrsky JN, Brawn JD, Robinson SK (2005) Juvenile mortality increases with clutch size in a Neotropical bird. Ecology 86:3238–3244
Sun C, Liu J, Li W, Xu G, Yang N (2017) Divergent proteome patterns of egg albumen from Domestic Chicken, duck, goose, turkey, quail, and pigeon. Proteomics 17:1700145
Tanaka K (2016) Nesting and egg incubation in dinosaurs: morphological and statistical investigations into the study of eggs, eggshells, and nests. Ph.D. dissertation, University of Calgary, Calgary, Alberta, Canada
Tanaka K, Zelenitsky DK, Therrien F (2015) Eggshell porosity provides insight on evolution of nesting in dinosaurs. PLoS ONE 10:e0142829
Tanaka K, Zelenitsky DK, Lü J, DeBuhr CL, Yi L, Jia S, Ding F, Xia M, Liu D, Shen C, Chen R (2018) Incubation behaviours of oviraptorosaur dinosaurs in relation to body size. Biol Lett 14:20180135
Tattersall GJ, Cadena V (2010) Insights into animal temperature adaptations revealed through thermal imaging. Imaging Sci J 58:261–268
Tazawa H, Whittow GC (2000) In: Whittow GC (ed) Sturkie's avian physiology. Academic Press, London, pp 617–634

Terepka A, Coleman J, Armbrecht H, Gunther T (1976) Transcellular transport of calcium. In: Duncan C (ed) Calcium in biological systems. Cambridge University Press, Cambridge, UK, pp 117–140

Tinbergen JM, Williams JB (2002) Energetics of incubation. In: Deeming DC (ed) Avian incubation. Oxford University Press, Oxford, UK, pp 299–313

Ton R, Martin TE (2017) Proximate effects of temperature versus evolved intrinsic constraints for embryonic development times among temperate and tropical songbirds. Sci Rep 7:895

Tona K, Onagbesan O, Bruggeman V, Mertens K, Decuypere E (2005) Effects of turning duration during incubation on embryo growth, utilization of albumen, and stress regulation. Poult Sci 84:315–320

Tuan RS (1984) Carbonic anhydrase and calcium transport function of the chick embryonic chorioallantoic membrane. Ann N Y Acad Sci 429:459–472

Tuan RS (1987) Mechanism and regulation of calcium transport by the chick embryonic chorioallantoic membrane. J Exp Zool Suppl 1:1–13

Tuan RS, Ono T (1986) Regulation of extraembryonic calcium mobilization by the developing chick embryo. Development 97:63–74

Tullett SG, Deeming DC (1987) Failure to turn eggs during incubation: effects on embryo weight, development of the chorioallantois and absorption of albumen. Br Poult Sci 28:239–249

Uni Z, Yadgary L, Yair R (2012) Nutritional limitations during poultry embryonic development. Poult Sci 21: 175–184

Varricchio DJ, Jackson FD (2016) Reproduction in Mesozoic birds and evolution of the modern avian reproductive mode. Auk 133:654–684

Varricchio DJ, Moore JR, Erickson GM, Norell MA, Jackson FD, Borkowski JJ (2008) Avian parental care had dinosaur origin. Science 322:1826–1828

Varricchio DJ, Kundrát M, Hogan J (2018) An intermediate incubation period and primitive brooding in a theropod dinosaur. Sci Rep 8:12454

Veiga JP (1992) Hatching asynchrony in the House Sparrow: a test of the egg-viability hypothesis. Am Nat 139:669–675

Verhulst S, Nilsson J-Å (2008) The timing of birds' breeding seasons: a review of experiments that manipulated timing of breeding. Philos Trans R Soc B 363:399–410

Vleck CM (1981) Energetic cost of incubation in the Zebra Finch. Condor 83:229–237

Vleck CM, Bucher TL (1998) Energy metabolism, gas exchange, and ventilation. In: Starck JM, Ricklefs RE (eds) Avian growth and development: evolution within the altricial-precocial spectrum. Oxford University Press, New York, pp 89–116

Vleck CM, Vleck D (1980) Patterns of metabolism and growth in avian embryos. Am Zool 30:405–416

Vleck CM, Vleck D (1987) Metabolism and energetics of avian embryos. J Exp Zool Suppl 1:111–125

Vleck D, Vleck CM, Hoyt DF (1980) Metabolism of avian embryos: ontogeny of oxygen consumption in the Rhea and Emu. Physiol Zool 53:125–135

Vleck CM, Vleck D (2011) Hormones and regulation of parental behavior in birds. In: Norris DO, Lopez KH (eds) Hormones and reproduction in vertebrates, birds, vol 4. Academic Press, San Diego, CA, pp 181–203

Walsberg GE, King JR (1978a) The heat budget of incubating Mountain White-crowned Sparrows (*Zonotrichia leucophrys oriantha*) in Oregon. Physiol Zool 51:92–103

Walsberg GE, King JR (1978b) The energetic consequences of incubation for two passerine species. Auk 95:644–655

Wang JW, Beissinger SR (2009) Variation in the onset of incubation and its influence on avian hatching success and asynchrony. Anim Behav 78:601–613

Wang JM, Beissinger SR (2011) Partial incubation in birds: its occurrence, function, and quantification. Auk 128:454–466

Wang JM, Firestone MK, Beissinger SR (2011) Microbial and environmental effects on avian egg viability: do tropical mechanisms act in a temperate environment? Ecology 92:1137–1145

Wangensteen OD, Rahn H (1970) Respiratory gas exchange by the avian embryo. Respir Physiol 11:31–45

Ward JE (1940) The passing of the lyre-bird. Bull N Y Zool Soc 63:146–152

Ward D (1990) Incubation temperatures and behavior of Crowned, Black-winged, and Lesser Black-winged plovers. Auk 107:10–17

Webb DR (1987) Thermal tolerance of avian embryos: a review. Condor 89:874–898

Werschkul DF, Jackson JA (1979) Sibling competition and avian growth rates. Ibis 121:97–102

Westerkov K (1956) Incubation temperatures of the pheasant. Emu 56:405–420

Wetherbee DK, Bartlett LM (1962) Egg teeth and shell rupture of the American Woodcock. Auk 79:177

White DW, Kennedy ED (1997) Effect of egg covering and habitat on nest destruction by House Wrens. Condor 99:873–879

Whittow GC (1980) Physiological and ecological correlates of prolonged incubation in sea birds. Am Zool 20:427–436

Wiebe KL, Martin K (1998) Costs and benefits of nest cover for ptarmigan: changes within and between years. Anim Behav 56:1137–1144

Wiebe KL, Koenig WD, Martin K (2006) Evolution of clutch size in cavity-excavating birds: the Nest Site Limitation Hypothesis revisited. Am Nat 167:343–353

Wiersma P, Munoz-Garcia A, Walker A, Williams JB (2007) Tropical birds have a slow pace of life. Proc Natl Acad Sci USA 104:9340–9345

Wikelski M, Spinney L, Schelsky W, Scheuerlein A, Gwinner E (2003) Slow pace of life in tropical sedentary birds: a common-garden experiment on four

Stonechat populations from different latitudes. Proc Roy Soc B 270:2383–2388
Williams TD (1995) The Penguins (Spheniscidae). Oxford University Press, Oxford, UK
Williams TD (1996) Variation in reproductive effort in female Zebra Finches (*Taeniopygia guttata*) in relation to nutrient-specific dietary supplements during egg laying. Physiol Zool 69:1255–1275
Winger BM, Pegan TM (2021) Migration distance is a fundamental axis of the slow-fast continuum of life history in boreal birds. Auk 138:ukab043
Xing L, Niu K, Ma W, Zelenitsky DK, Yang TR, Brusatte SL (2022) An exquisitely preserved *in-ovo* theropod dinosaur embryo sheds light on avian-like prehatching postures. Iscience 25:103516
Yadgary L, Kedar O, Adepeju O, Uni Z (2013) Changes in yolk sac membrane absorptive area and fat digestion during chick embryonic development. Poult Sci 92: 1634–1640
Yang TR, Wiemann J, Xu L, Cheng YN, Wu XC, Sander PM (2019) Reconstruction of oviraptorid clutches illuminates their unique nesting biology. Acta Palaeontologica Polonica 64:581–596
Yasugi S, Mizuno T (2008) Molecular analysis of endoderm regionalization. Dev Growth Differ 50:S79–S96
Yoon J (2013) Comparative hatching characteristics of nonparasitic and parasitic icterids: is the hatching of cowbird young constrained by an unusually thick eggshell? J Ethol 1:35–40
Zanette LY, White AF, Allen MC, Clinchy M (2011) Perceived predation risk reduces the number of offspring songbirds produce per year. Science 334: 1398–1401
Zheng J, Zuidema E, Zhang Z, Guo M, Székely T, Komdeur J (2022) A novel function of egg burial: burying material prevents eggs rolling out of windswayed nests. Anim Behav 189:1–13

# Avian Reproduction: Post-hatching Parental Care and Brood Parasitism

# 19

## Contents




**Abstract**

After eggs hatch, the extent to which adult birds care for young exhibits considerable variation, ranging from no care to extensive care lasting weeks or even months. Factors contributing to this variation are discussed in this chapter. For species where young are fed by adults, the begging behavior of young and responses to that begging by adults is explained. Factors that influence the extent to





G. Ritchison, *In a Class of Their Own*, Fascinating Life Sciences,
https://doi.org/10.1007/978-3-031-14852-1_19

which adults defend nests and young from potential predators are also discussed. For species where young are cared for in nests, departure from those nests by young birds is a critical step in their development. How nestling age and condition and adult behavior can influence nest departure, or fledging, is discussed. Most young disperse from their natal territories and I also explain the factors that influence the timing of that decision and how far fledglings disperse. Examples of the importance of learning by young birds are provided and, finally, this chapter closes with an extensive discussion of avian brood parasitism, including facultative versus obligatory brood parasitism and, for obligatory brood parasitism, the pre- and post-laying adaptions and post-hatching adaptions of both hosts and brood parasites.

## 19.1 Introduction

Among the likely ancestors of birds, the most direct evidence of parental care is the fossilized remains of adult theropods (*Oviraptor* and *Troodon*) sitting on their nests with eggs. In addition, Amiot et al. (2017) reported evidence, based on analysis of the oxygen isotope composition of fossilized eggshells and the bones of embryos, that oviraptors incubated their eggs at temperatures within a range of 35–40 °C, similar to the incubation temperatures of present-day birds. Less evidence is available concerning possible post-hatching parental care by non-avian dinosaurs. At a site in China, a single larger ornithischian dinosaur (*Psittacosaurus* sp.) was found clustered with 34 juveniles in an area of 0.5 $m^2$, providing possible evidence for post-hatching parental care (Meng et al. 2004). However, further analysis revealed that the larger dinosaur was not yet an adult so, at best, this assemblage likely represents an example of post-hatching cooperation of young rather than post-hatching parental care (Hedrick et al. 2014).

Based on large clutch volumes and the bone histology of brooding adults (adults discovered on top of fossilized clutches of eggs), Varricchio et al. (2008) concluded that dinosaurs in the genera *Troodon* (Troodontoidea), *Oviraptor* (Oviraptorosauria), and *Citipati* (Oviraptorosauria), all close relatives of birds, had polygamous mating systems with extensive paternal care. Present-day birds with a similar mating system, where males copulate with multiple females that then all lay their eggs in a male's nest (and males provide much or all the care of young after hatching), include several species of ratites, including Common Ostriches (*Struthio camelus*), Emus (*Dromaius novaehollandiae*), and rheas (*Rhea* spp.). However, using a data set differing in several respects from the one used by Varricchio et al. (2008), e.g., using egg mass rather than egg volume, Birchard et al. (2013) concluded that the type of parental care exhibited by the dinosaur ancestors of birds cannot be determined based on egg volume or mass because of the confounding effects of phylogeny and the maturity of young at hatching. In addition, based on evolutionary relationships and types of parental care provided by present-day birds (none, maternal, paternal, or biparental), Birchard et al. (2013) concluded that biparental care was likely the ancestral parental-care type of present-day birds. However, Moore and Varricchio (2016) noted that, among diapsids (reptiles and birds), there is a strong relationship between clutch volumes and parental care strategies, as well as between clutch volume and mode of development. Among the ancestors of birds, increasing clutch volumes among non-avian theropods led to an increasing tendency for greater paternal care. Among neognaths, however, increasing levels of altriciality led to a reduction in clutch volume which, in turn, resulted in a reduction (but not elimination) in paternal care and a corresponding increase in maternal care or, in other words, biparental care. In summary then, Moore and Varricchio (2016) concluded that the ancestral condition among diapsids was no parental care, with a transition to maternal care in crocodilians, paternal care in theropod dinosaurs, and biparental care in Neognaths (Fig. 19.1). This conclusion, however, was based on "limited available care strategy data" and these authors noted that their conclusion "should be considered preliminary

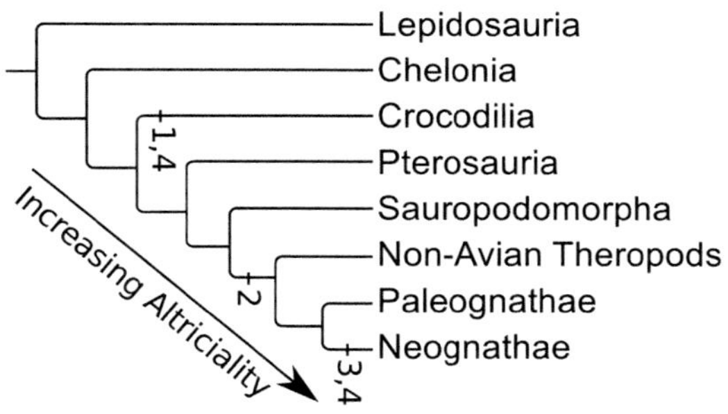


**Fig. 19.1** Hypothesized steps in the evolution of parental care strategies of present-day birds (Paleognathae and Neognathae). The ancestral condition is no parental care, with superprecocial young and polygamy for both sexes. 1, transition to maternal care; 2, transition to paternal care; 3, transition to biparental care; and 4, transition to social monogamy. (Figure from Moore and Varricchio 2016; © 2016 Moore and Varricchio, open-access article distributed under the terms of the Creative Commons Attribution License, https://creativecommons.org/licenses/by/4.0/)

until independent data exist to better constrain a much greater number of extinct diapsid reproductive strategies."

Parents provide care to enhance the survival and future fitness of their young and, by doing so, enhancing their own reproductive fitness. Among present-day species of birds, most species (80.8%) exhibit biparental care; far fewer species exhibit female-only care (8.2%) or male-only care (1%) (Box 19.1 Parental Cooperation in Caring for Young). Another 9% of bird species exhibit cooperative care (care provided by more than two individuals), and the remaining species do not provide parental care (brood parasites, 1%; use geothermal heat to incubate eggs; 0.1%; Cockburn 2006). Differences among and within species in the parent(s) that care(s) for young after hatching result from differences in the relative costs and benefits of providing such care and in the conflict between males and females in the level of care required by offspring (Silver et al. 1985).

At hatching, young birds exhibit substantial variation in appearance and behavior. Some young are entirely dependent on their parents, whereas others are able to leave the nest and begin finding their own food within hours of hatching. The terms nidifugous (nest-fleers) and nidicolous (nest squatters) were among the first used to classify birds based on their characteristics at hatching (Oken 1816). Other, more commonly used, terms that have been in use since 1836 are altricial (from the Latin *altrix*, or "wet nurse") and precocial (from the Latin *praecoquis*, or "early maturing"; Sundeval 1836; Starck and Ricklefs 1998). Currently, young birds are categorized more precisely based on characteristics such as developmental stage, locomotor abilities, feeding behavior, presence or absence of external feathers, and growth rates after hatching. Using such characteristics, Starck (1993) categorized bird hatchlings into eight categories (Figs. 19.2 and 19.3, Table 19.1). Although useful for comparing and contrasting variation among species, this classification system is, as Starck (1993) acknowledges, "highly artificial and pushes the variation observed in nature into a rigid system." As a result, some species do not clearly fit into any of the eight categories. For example, young Hoatzins (*Opisthocomus cristatus*) are almost featherless at hatching and usually remain in nests, but they can jump from nests if threatened by a predator and then swim (because nests are always located over water) and climb (using claws on their first and second digits).

Precocial development is likely the ancestral, or original, mode, with the altricial mode developing later and independently in several groups of birds. Evidence that ancestral birds were precocial comes from microscopic analysis of the fossilized bones of *Rahonavis* (a basal bird of the late Cretaceous) and *Gobipteryx* (an enantiornithine bird from the late Cretaceous). The morphology of these bones indicates slow growth, a characteristic of present-day superprecocial and precocial birds such as megapodes and tinamous. Because the locomotor activity of precocial and superprecocial young requires substantial energy, their bones and other tissues tend to grow more slowly than is the case for altricial species. In addition, the size and degree of ossification of the wing skeleton of hatchling enantiornithines (*Gobipteryx*) suggest that they were able to fly shortly after hatching. In combination, the slow growth of bones and likely ability to fly soon after hatching indicate a

**Box 19.1 Parental Cooperation in Caring for Young**

Among different taxa of birds, the extent to which males and female cooperate in providing care for young varies considerably, with some exhibiting extensive cooperation (e.g., grebes and woodpeckers) and others little cooperation (e.g., ducks). In some taxa, the extent of parental cooperation varies among species (e.g., snipes, sandpipers, and allies, and Old World warblers).

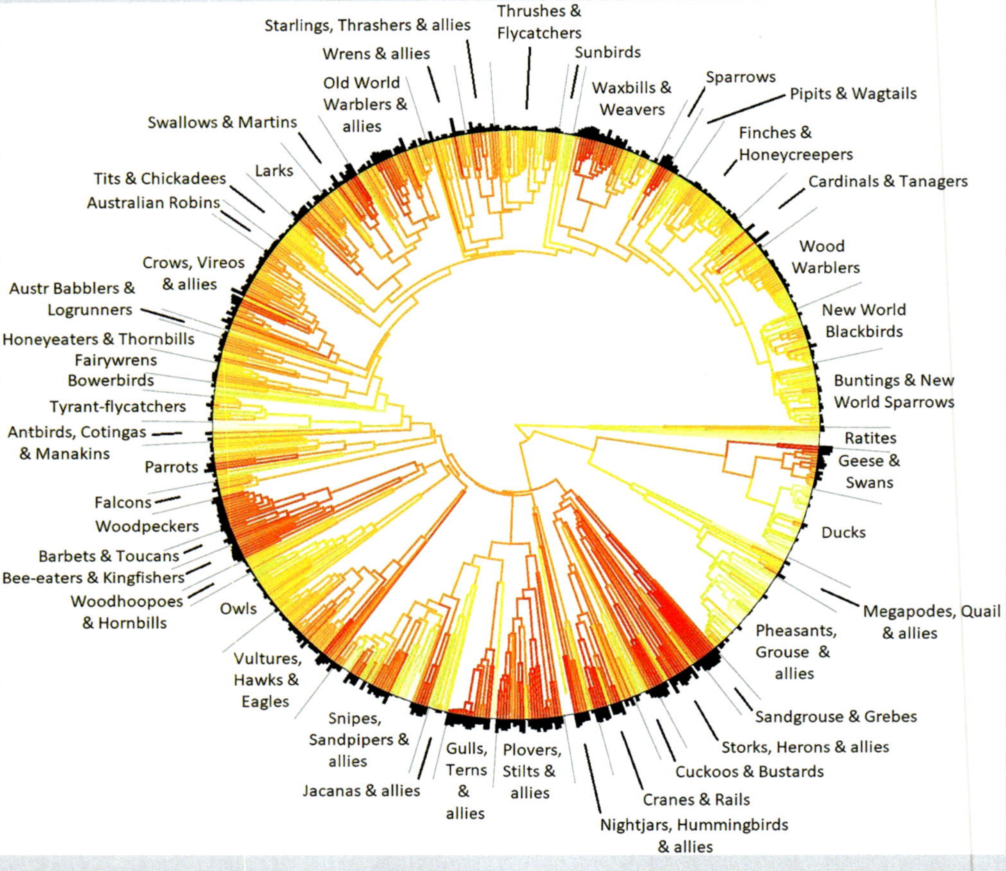


The extent of parental cooperation is usually consistent within major clades of birds, but variable between clades. Examples of groups of birds with parental cooperation include sandgrouse and grebes, storks, herons, and allies, woodpeckers, and sparrows. Taxa with low levels of parental cooperation include ducks, pheasants, grouse, and allies, and owls. Black bars refer to parental cooperation, with taller bars indicating greater cooperation. Red = high cooperation, and yellow = low cooperation. (Figure from Remeš et al. 2015; used with permission of the U. S. National Academy of Sciences)

Remeš et al. (2015) quantified sex-specific contributions to parental care for 659 species of birds from 113 families on an ordinal scale, with 0 indicating no male contributions, 1 indicating male contributions from 1 to 33%, 2 indicating male contributions from 34 to 66% (approximately equal care by males and females), 3 indicating male contributions from 67 to 99%, and 4 indicating 100% of the contributions by males. The various types of parental care included nest

(continued)

**Box 19.1** (continued)

building, incubation, nest defense during incubation, brooding young, feeding young, defending young after hatching, and feeding and guarding young after fledging. Analysis revealed that sexual selection (as indicated by sexual size dimorphism) and high rates of extra-pair paternity were associated with low levels of parental cooperation and male parental care. These results are consistent with those of previous studies where investigators reported that the larger sex is generally under stronger sexual selection, which, in turn, generally translates into less parental care (Queller 1997; Westneat et al. 2010). Similarly, other investigators have noted that higher rates of extra-pair paternity will select for lower levels of male parental care (Sheldon 2002; Royle et al. 2012).

Parental cooperation was also found to decrease with biased adult sex ratios. A likely explanation for such results is that individuals of the rarer sex can limit their parental care because they will experience greater mating success, whereas individuals of the more abundant sex can improve their reproductive success by providing more parental care (McNamara et al. 2000; Kokko and Jennions 2008).

Climatic conditions during the breeding season, as represented by ambient temperature and rainfall, were found to have no effect on the extent of parental cooperation or male parental care. Similarly, Remeš et al. (2015) found no difference in levels of parental cooperation between altricial and precocial species. These results suggest that parental care is largely independent of climate, and that the amount of care needed to successfully raise young does not influence parental care strategies.

Finally, Remeš et al. (2015) found a positive relationship between body mass and parental cooperation. In general, larger species live longer and pair bonds are longer in duration, with both of these factors likely favoring greater parental cooperation.

(continued)

**Box 19.1** (continued)

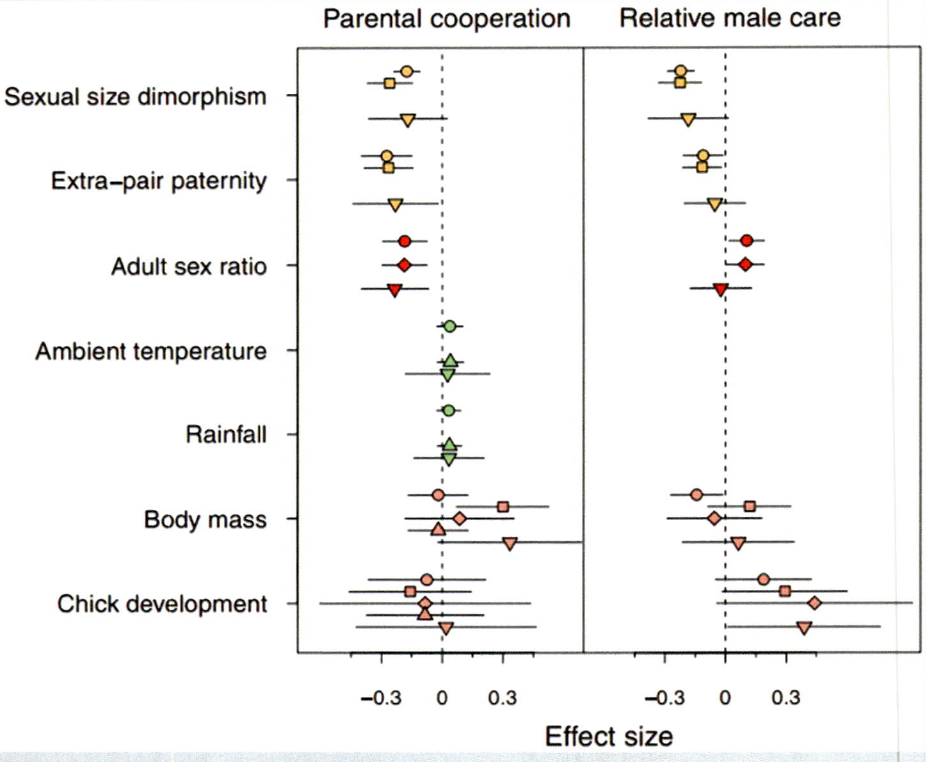


Parental cooperation and relative male parental care for 659 species of birds. Negative values indicate less cooperation and male care, positive values indicate more cooperation, and scores close to 0 indicate no apparent effect on levels of parental cooperation or male parental care. Greater sexual size dimorphism, higher rates of extra-pair paternity and skewed adult sex ratios result in less parental cooperation and less male parental care. Climatic conditions during the breeding season (temperature and rainfall) have no apparent effect on levels of parental cooperation. Similarly, chick development (altricial species vs. precocial species) does not appear to influence levels of parental cooperation. Larger species that tend to live longer and remain paired longer tend to exhibit greater levels of parental cooperation (Different symbols indicate different types of analysis (Figure from Remeš et al. 2015; used with permission of the U. S. National Academy of Sciences)

precocial (or superprecocial) mode of development in two lineages of basal birds and provides evidence that this was the ancestral mode of development for birds (Chinsamy and Elzanowski 2001). However, among present-day species of birds, only about 10% are precocial or superprecocial, whereas nearly 90% are either altricial or semialtricial. This disparity clearly indicates that altricial modes of development provide birds with important selective advantages.

The evolution of altricial modes of development likely followed the evolution of improved flying ability. With the evolution of enhanced flying ability that permitted greater maneuverability and acceleration, birds were better able to access and nest in trees and cliffs and other locations less accessible to predators. Access to

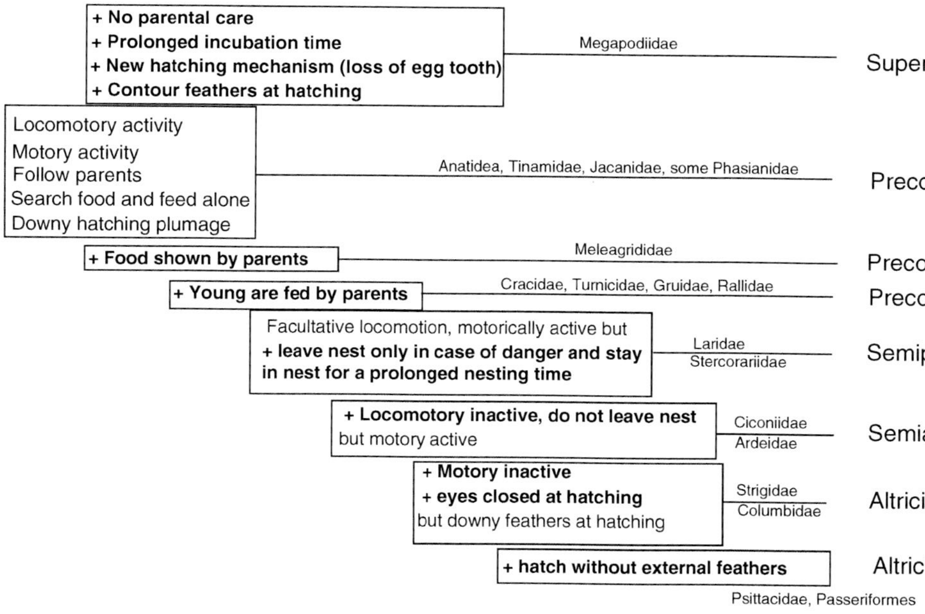


**Fig. 19.2** Classification of avian hatchlings. (Figure from Starck 1993; © 1993 Plenum Press, New York, published by Springer Nature, used with permission)

and use of relatively "safe" nest sites were likely important prerequisites for the evolution of altriciality because adults caring for altricial young typically must visit nests at relatively high rates and such activity may increase the likelihood of detection by potential predators (Fig. 19.4; Burley and Johnson 2002; Dial 2003). Even among present-day birds, most precocial species nest on the ground (close to where chicks will forage) and most altricial species do not. There are, of course, exceptions. For example, some species of ducks with precocial young, such as Wood Ducks (*Aix sponsa*) and Common Goldeneyes (*Bucephala clangula*) nest in cavities in trees, and several species of songbirds with altricial young nest on the ground (although this is often in areas like grasslands where no other nest sites are available; Starck and Ricklefs 1998). Concerning the evolution of altriciality, however, once birds had access to safer nest sites, what selective advantages actually favored the evolution of altricial modes of development?

Precociality provides birds with several potential advantages. For example, precocial young are self-feeding so clutch sizes may not be limited by the ability of parents to provision young. In addition, precocial young are better able than altricial young of the same size to escape from predators and withstand brief periods of harsh conditions (Ricklefs and Starck 1998), in part because of their greater thermoregulatory abilities. However, given its predominance among present-day birds, the altricial mode of development must provide even greater advantages. Perhaps the most important is the effect on brain size (Griesser et al. 2023). The brains of superprecocial and precocial species are comparatively well developed at hatching, but their size increases less after hatching than do the brains of altricial species that are less developed at hatching (Ricklefs and Starck 1998). As a result of continued growth and development after hatching, the brains of altricial species are larger relative to their body size than those of superprecocial and precocial species (Iwaniuk and Nelson 2003). Larger brains have likely been favored by natural selection because increased size and complexity are associated with a variety of characteristics that can enhance fitness, including complex social relationships, improved flying ability, and vocal learning

**Fig. 19.3** Examples of species of birds that vary in development mode. (**a**) A superprecocial Australian Brushturkey (*Alectura lathami*), (**b**) a Precocial 1 Red Knot (*Calidris canutus*), (**c**) a Precocial 2 California Quail (*Callipepla californica*), (**d**) a Precocial 3 American Coot (*Fulica americana*), (**e**) a Semiprecocial Glaucous-winged Gull (*Larus glaucescens*), (**f**) Semialtricial Peregrine Falcons (*Falco peregrinus*), (**g**) an Altricial 1 Rock Pigeon (*Columba livia*), (**h**) Nestling Spotless Starlings (*Sturnus unicolor*), an Altricial 2 species. (**a**, photo by Maureen G., Wikipedia, CC BY 3.0, https://creativecommons.org/licenses/by/3.0/; **b**, CC0 Public Domain, pxhere.com; **c**, Photo by Tony Wills, Wikipedia, CC BY 2.5, https://creativecommons.org/licenses/by/2.5/; **d**, CC0 Public Domain, https://pxhere.com; **e**, CC0 Public Domain, photo by U. S. Fish and Wildlife Service employee; **f**, NPS/Gavin Emmons, CC0 Public Domain; **g**, Wikipedia, CC BY 2.0, https://creativecommons.org/licenses/by/2.0/; **h**, National Park Service, CC0 Public Domain)

**Table 19.1** Frequency of occurrence of development modes and examples of families and orders of birds exhibiting each

| Development mode | Percent of all species | Examples (families and order) |
|---|---|---|
| Superprecocial | 0.11% | Megapodiidae |
| Precocial 1 | 4.36% | Anatidae, Jacanidae, Recurvirostridae, Charadriidae, Scolopacidae |
| Precocial 2 | 2.54% | Phasianidae, Tetraonidae |
| Precocial 3 | 3.08% | Cracidae, Gaviidae, Rallidae, Podicipedidae |
| Semiprecocial | 1.73% | Alcidae, Laridae, Haematopodidae, Phoenicopteridae |
| Semialtricial | 5.82% | Accipitridae, Ardeidae, Falconidae, Cathartidae, Ciconiidae, Threskiornithidae |
| Altricial 1 | 5.91% | Columbidae, Procellariidae, Strigidae, Tytonidae |
| Altricial 2 | 76.4% | Alcedinidae, Cuculidae, Phalacrocoracidae, Psittacidae, Trochilidae, Passeriformes |

Data from Starck (1993)

(Burish et al. 2004; Fig. 19.5). Birds with larger brains relative to their body size have also been found to exhibit greater capacities for inventing new behaviors through innovation (Lefebvre et al. 1997, 1998) and for using novel resources or environments more successfully (Sol et al. 2005; Morand-Ferron et al. 2007; Sol and Price 2008). Although questions remain concerning the relationship between brain size and cognitive abilities (Healy and Rowe 2007), available evidence indicates that the relatively larger brains of altricial species of birds have contributed to their evolutionary success (Raikow 1988).

## 19.2 Post-hatching Parental Care

The type and amount of care provided by parents varies with development mode (Starck and Ricklefs 1998). The young of superprecocial species, such as Australian Brushturkeys (*Alectura lathami*), dig themselves out of the soil or mounds of rotting vegetation from depths of 40–150 cm over periods ranging from 2 to 40 h (Göth 2002; Fig. 19.6). After emerging, megapode chicks are typically independent and receive no parental care. They never associate with adults, but sometimes aggregate in small groups with chicks of similar age (Göth and Vogel 2002). Newly hatched megapodes quickly identify suitable food items by trial and error (Göth and Proctor 2002), and exhibit anti-predator behavior immediately after hatching without any previous experience (Göth 2001). Megapode chicks also exhibit appropriate social behavior without previous experience (Göth and Jones 2003).

Among precocial species of birds (Precocial 1, 2, and 3; Table 19.1), parental care consists primarily of scanning for potential predators, anti-predator behavior, leading and following young, and, in some cases, brooding young. For Precocial 1 and some Precocial 2 species (waterfowl), little or no brooding is generally required because young can maintain their body temperatures at the time of hatching (Visser 1998). For young shorebirds (Precocial 2) and Precocial 3 species, like young loons and grebes, the timing of homeothermy (ability to maintain body temperature) varies with size; the young of larger species may be homeothermic within a day or two of hatching, whereas the young of smaller species may have to double their body mass before achieving homeothermy (Visser 1998). In this case, size (and mass) matters because larger young have relatively less surface area (and, therefore, lose heat more slowly) and have larger muscles that can more effectively shiver to generate heat.

In gallinaceous birds (Order Galliformes), adult females share information about foraging areas and food availability by giving special "food" calls, and food consumption rates of chicks are related to the rates at which such calls are given (Allen and Clarke 2005). Parents in other precocial species also use vocal signals to

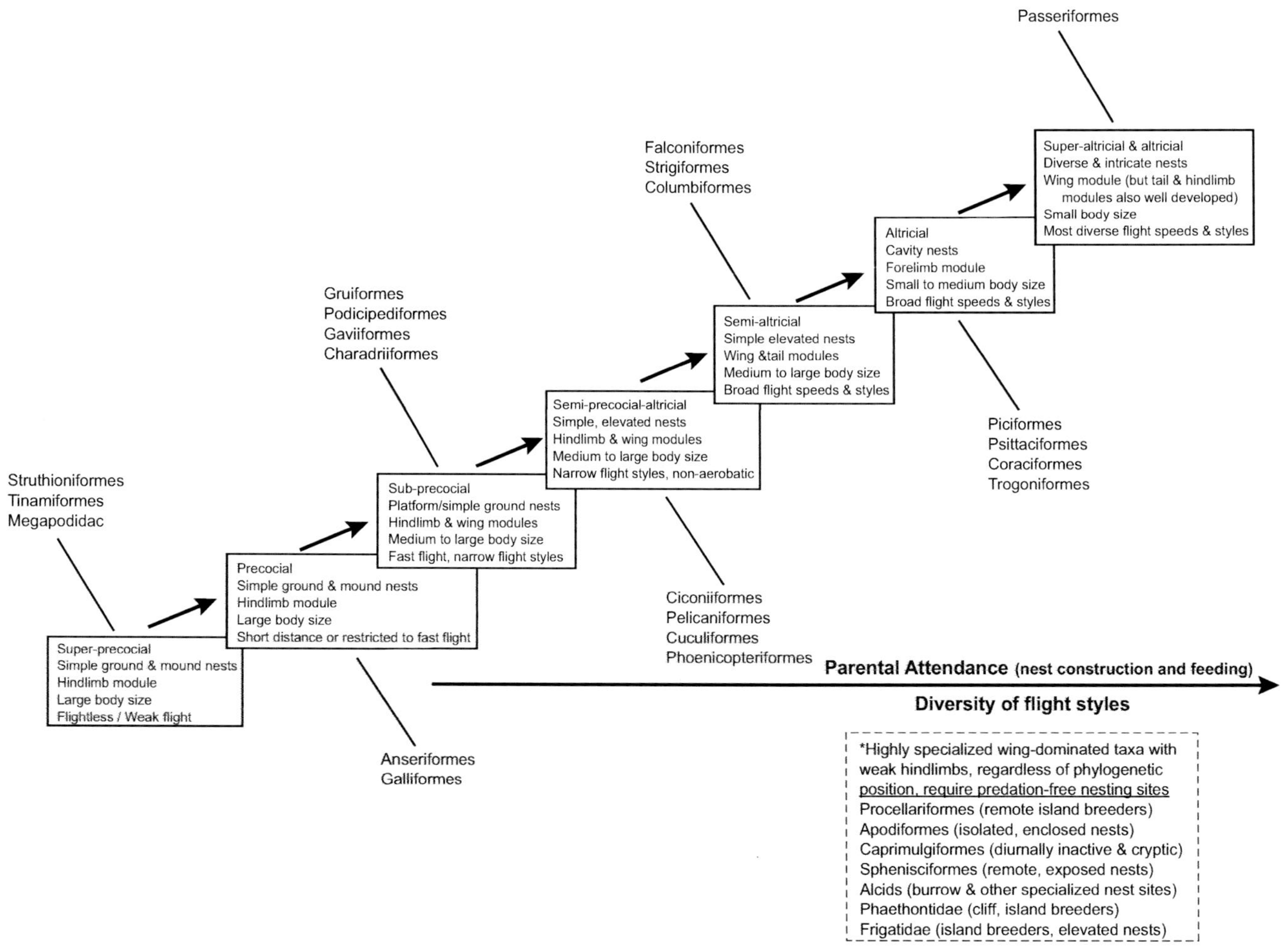


**Fig. 19.4** Overview of the distribution of (1) developmental stage at hatching, (2) nest type, (3) dominant locomotor module, (4) body size, and (5) flight style among extant avian orders. This diagram presents a general perspective on to how the five variables map among extant clades, which suggests tendencies among basal to derived groups. Note: Taxa exhibiting pronounced forelimb investment (usually supreme fliers), with less developed hindlimbs (some species even incapable of walking), all share antipredation nesting tactics such as breeding on predator-free islands or building extremely isolated nests (swifts and swallows) because of extreme vulnerability of immobile young. (Figure from Dial 2003; © 2003 Oxford University Press, used with permission)

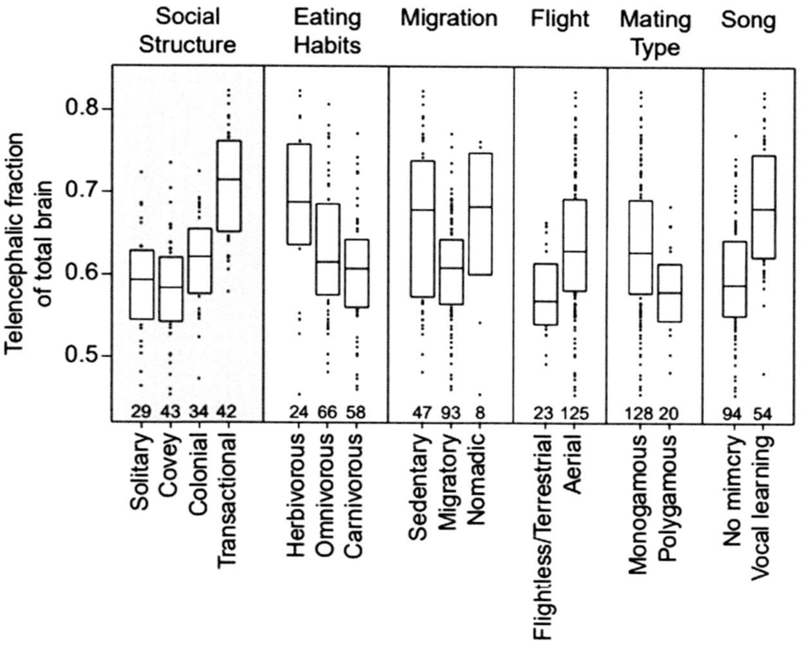


**Fig. 19.5** Proportion of the brain made up of telencephalon (cerebrum) among birds that vary in social structure, behavior, and ecology. Species with more complex social structures (transactional, or with frequent interactions with conspecifics) that are better flyers and exhibit vocal learning tend to have relatively larger telencephalons. Numbers below the boxes indicate the number of species in each category. (Figure from Burish et al. 2004; © 2004 S. Karger AG, Basel, used with permission)

alert their chicks to the presence of predators, prevent them from wandering away and becoming lost, and lead them to good foraging locations (Johnson et al. 2008). For many precocial species, such as some shorebirds (e.g., Black Turnstone, *Arenaria melanocephala*; Handel and Gill 2020), nests are located in areas and habitats that are also suitable for rearing young after hatching and, as result, parents and young exhibit limited movement after hatching. However, some species nest in habitats less suitable for rearing young, and parents and chicks move variable distances after young hatch to areas with greater food availability, fewer predators, or both (Kosztolányi et al. 2007). For example, adult Snow Geese (*Anser caerulescens*) breeding in northern Canada may move their broods up to 30 km to higher-quality brood-rearing areas (Mainguy et al. 2006).

Among some species of birds with precocial young, parents attempt to increase the size of their broods after hatching either through adoption of unrelated young, crèching, or brood amalgamation. Such behavior has been reported in waterfowl (Beauchamp 1997) and several species of shorebirds (Cooper and Miller 1992; Lanctot et al. 1995; Lengyel 2007) and can result in higher adult fitness and improved chick survival. For example, brood size and social dominance are positively related in several goose and duck species, with larger broods gaining access to better resources than smaller broods (brood-dominance hypothesis; Lepage et al. 1998; Öst and Kilpi 2000). Larger broods may also reduce predation rates for chicks (dilution effect). For example, mortality due to predation decreased with increasing brood sizes in Common Eiders (*Somateria mollissima,* Munro and Bedard 1977), White-winged Scoters (*Melanitta deglandi*, Kehoe 1989), and Pied Avocets (*Recurvirostra avosetta*, Lengyel 2007).

The initial few days after hatching are a vulnerable time for precocial young. For example, among shorebirds and waterfowl, typically 55–100% of chick mortality occurs during the first few weeks after hatching (e.g., Warriner et al. 1986; Flint et al. 1995), with most mortality due to predation and, to a lesser degree, low food availability. At higher latitudes, weather conditions can also affect survival of precocial chicks. Cold, wet weather may restrict the time available for foraging because chicks have to be brooded by parents to stay warm. Such conditions

**Fig. 19.6** Top, Young Australian Brushturkey (*Alectura lathami*) beginning to dig its way out of a mound using its relatively large feet. Bottom, Young Australian Brushturkey after emerging from a mound. (Top figure from Seymour 1991 and drawn by Patricia Wynne; used with permission of Patricia Wynne; Bottom, photo by Jenny/Jen 64, Wikipedia, CC BY 3.0, https://creativecommons.org/licenses/by/3.0/)

may also increase the thermoregulation costs of chicks and parents and reduce the abundance and activity of the prey of invertebrate-feeding species. With increasing age, mortality rates of precocial chicks typically decline and chicks become less dependent on adults for brooding and protection from predators (Fig. 19.7).

Among Precocial 3 species, parents provide young with food and also carry and sometimes brood young under their wings. For example,

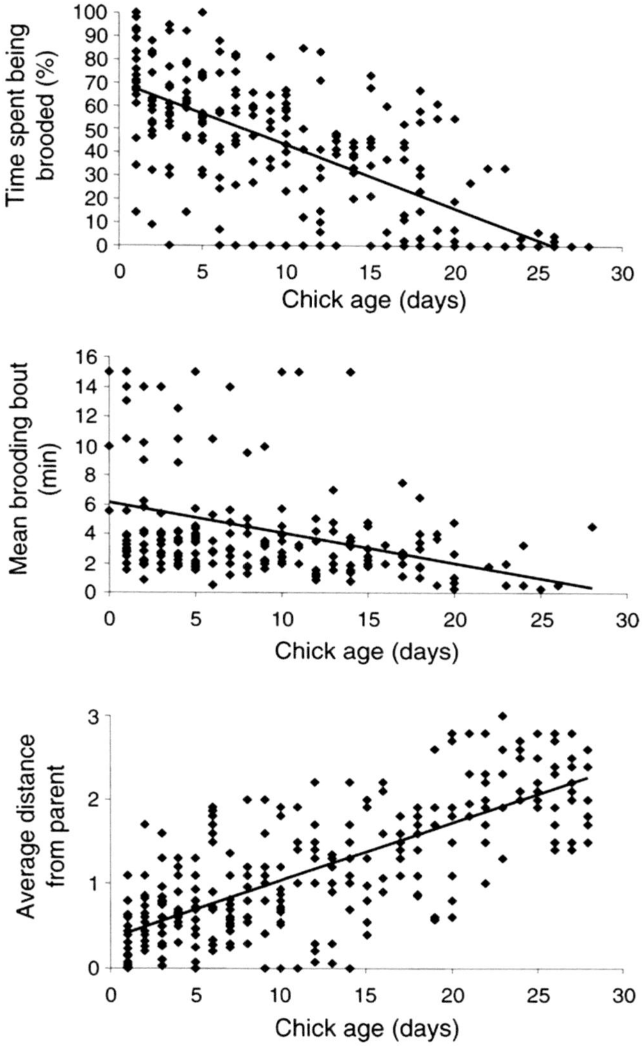


**Fig. 19.7** With increasing age, Snowy Plover (*Charadrius alexandrinus*) chicks were brooded less (top) and for shorter durations (middle), and increased their distance from tending adults (bottom). Distance from tending adults was categorized as 0 = brooding under parent, 1 = ≤1 m from parent, 2 = 1–10 m from parent, and 3 = 10–50 m from parent. (Figure from Colwell et al. 2007; © 2007 Oxford University Press, used with permission)

young grebes (Podicipedidae) spend much of their time riding on the back of one of their parents and, because of a limited ability to regulate their body temperature for a week or so after hatching, they are occasionally brooded under the wings of their parents (Fjeldså 2004; Box 19.2 Chick Ornamentation). Common Loon (*Gavia immer*, Gaviidae) chicks are initially dependent on parents for food, but become less dependent and forage for themselves more as they grow older (Alvo and Berrill 1992).

At hatching, the young of semi-precocial species, such as gulls, terns, and puffins, are covered with down and their eyes are open. During the first few weeks after hatching, young usually remain in the nesting territory where they are fed and protected, usually by both parents. In some semi-precocial species, such as in the shorebird families Haematopodidae, Burhinidae, Dromadidae, and Pluvianellidae, chicks are mobile, but are fed by parents (Walters 1984), likely because species in these families feed on prey that are difficult to obtain. For example, young oystercatchers (Haematopodidae) are unable to locate and manipulate important prey items like mussels and limpets. In some semi-

**Box 19.2 Chick Ornamentation**

American Coots (*Fulica americana*) with chicks. (Photo from https://pxhere.com, CC0 Public Domain)

Precocial chicks that are fed by their parents sometimes vary widely in appearance, and young in some species have brightly colored bills, fleshy parts, and plumes. For example, chick ornamentation has been reported in nearly 40% of species in the family Rallidae. These ornaments are lost by the end of the period of parental dependence, suggesting that they function in competition over parental care (Krebs and Putland 2004). Ornamented rail chicks are probably providing parents with cues concerning their quality and competitive ability. In support of this hypothesis, Lyon et al. (1994) experimentally trimmed the ornamented plumage of American Coot chicks and found that parents preferentially fed the chicks in the brood with intact plumage. The "trimmed" chicks also exhibited reduced growth rates and survival compared to orange-plumed chicks in the same brood. In addition, parent coots and moorhens sometimes pick up and shake particular chicks as they beg for food and adult Eurasian Coots (*Fulica atra*) divide their broods and feed only certain chicks. These observations indicate that chick ornamentation likely serves as an indicator of chick quality and parents can use this information to partition care among young in their broods.

precocial species, including all species of flamingos (Phoenicopteridae), chicks precociously leave the nesting territory and join together in a group ("crèche") in or outside the colony site where they are fed by their own parents. Among larids (gulls and terns), crèching occurs in species with high breeding synchrony and where adults exhibit limited aggression toward potential predators (Besnard 2001). Crèching behavior is also more likely in species that nest in areas that, during the brood rearing period, tend to become more accessible to terrestrial predators (Besnard et al. 2002). For example, high evaporation rates can lower water levels and create land bridges that expose previously inundated flamingo nest sites to terrestrial predators.

The role of males and females in providing care for semi-precocial young varies among species. Among semi-precocial shorebirds, such as oystercatchers, both parents care for young (Székely and Reynolds 1995). In puffins (*Fratercula* spp.) and guillemots (*Cepphus* spp.), both adults brood and provision young at the nest site until chicks are nearly fully grown, and chicks then fledge unaccompanied by their parents (Sealy 1973). However, Thick-billed Murres (*Uria lomvia*), Common Murres (*U. aalge*), and Razorbills (*Alca torda*) exhibit biparental care for a period of about 15–20 days at the nest site and the partly grown chicks (15–30% of adult mass) then leave the nest and go to sea (by jumping from cliff nest sites) where the adult male then continues to provide care for another 3–4 weeks (Gaston and Jones 1998). Males may depart with chicks, rather than females, because they are slightly larger and may be better able to protect chicks from potential predators (Ainley et al. 2020). Early departure to sea by their young is likely beneficial for murres and razorbills because high wing-loading (large mass relative to surface area of wings) may limit their ability to efficiently transport prey from foraging areas to nest sites (Gaston and Jones 1998) and, in addition, predation risk may be greater in nesting colonies than at sea (Ydenberg 1989).

About 88% of present-day species of birds, representing a wide variety of avian orders, have semialtricial or altricial young that require extensive parental care. For most of these species, care is provided either by both parents or, in cooperative breeding species, by multiple adults. Only about 7% of semi-altricial and altricial species exhibit female-only parental care (Cockburn 2006). Most of these latter species feed primarily on tropical fruit or nectar (e.g., hummingbirds, sunbirds, manakins, birds-of-paradise, and bowerbirds). Because fruit and nectar can be very abundant, but patchily distributed both spatially and temporally, males likely do better (in terms of fitness) by defending these resources from other males and attempting to mate with visiting females rather than pairing with one mate and helping to raise young (Cockburn 2006). For females, the availability of abundant food resources in high-quality male territories may limit the importance of male assistance in caring for young (Gowaty 1996).

Female-only care has also been reported in a few altricial species that are primarily insectivorous, including species in the families Icteridae (New World blackbirds) and Menuridae (lyrebirds and scrub-birds) and the order Caprimulgiformes. For these species, the evolution of female-only care has likely been possible because they occur in habitats with an abundance of insects (e.g., marsh-breeding blackbirds) or because young either grow slowly (lyrebirds and scrub-birds) or have low metabolic rates (goatsuckers) and require less energy and lower provisioning rates (Cockburn 2006).

Altricial young are unable to control their body temperature and must be kept warm when ambient temperatures are low and kept cool when a nest is in direct sunlight. Parents help nestlings maintain their body temperatures by brooding, covering them to keep them warm or shielding them from the sun or rain. The duration of brooding periods depends on: (1) the time needed for young to develop the ability to thermoregulate, (2) weather, because young must be brooded longer during periods of cool, wet weather, and (3) nest type because birds that nest in cavities, with more stable microclimates, typically brood for shorter periods than open-nesting species (Pettingill 1985).

Altricial young are typically brooded almost constantly during the first day or two after hatching. Then, for the next several days, young are brooded less as their ability to thermoregulate improves. The ontogeny of endothermy and ability of altricial nestlings to thermoregulate depends largely on reaching a sufficient mass to generate and retain heat and, specifically, a reduction in the surface-to-volume ratio (Pereyra and Morton 2001; Fig. 19.8). Although not essential, the development of feathers, and the insulation they

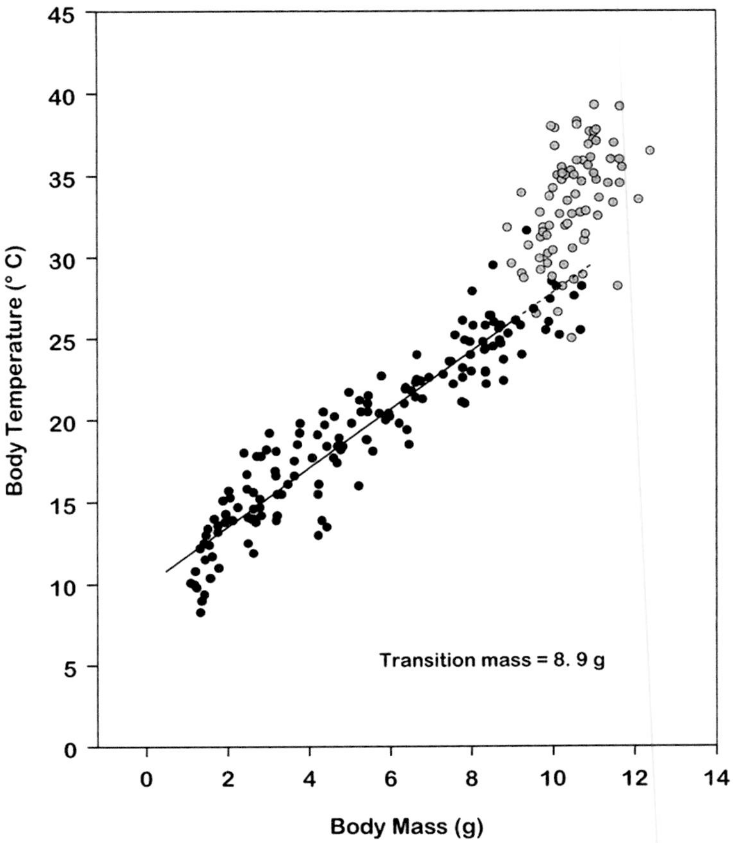


**Fig. 19.8** Body temperatures of nestling Dusky Flycatchers (*Empidonax oberholseri*) as a function of nestling body mass. Nestlings from day 0 (hatching) to day 9 are indicated in black; nestlings day 10 and older are indicated in gray. (Figure from Pereyra and Morton 2001; © 2001 Oxford University Press, used with permission)

provide, also improves the thermoregulatory abilities of altricial nestlings. Toward the end of the brooding period, young are able to efficiently thermoregulate and may only be brooded at night (or, for nocturnal birds, during the day). As the time of fledging approaches, parents do not brood young at all.

Among Altricial 1 and 2 species, the body mass of nestlings, even those of relatively large species, must increase substantially before achieving homeothermy. For example, a hatchling in a species where adults weigh about 25 g must increase in mass by about ninefold (or about 65% of adult body mass) before being able to regulate their body temperature (Visser 1998). Among larger altricial species, less of an increase in mass is needed. For example, nestling Northern Gannets (*Morus bassanus*; Montevecchi and Vaughan 1989), Double-crested Cormorants (*Nannopterum auritum*; Dunn 1976a), and European Shags (*Gulosus aristotelis*; Østnes et al. 2001; Fig. 19.9) become homeothermic about 15 days after hatching when they are about 20–25% of adult body mass. The young of larger altricial species achieve homeothermy sooner because of a more favorable surface-to-volume ratio than smaller species, resulting in a greater ability to both generate and retain heat. Similarly, among smaller altricial species, young in larger broods become homeothermic faster than young in smaller broods because of the more favorable surface-to-volume ratio (Dunn 1976b, 1979). That is, several young huddled together expose less surface area than do single young or young in smaller broods.

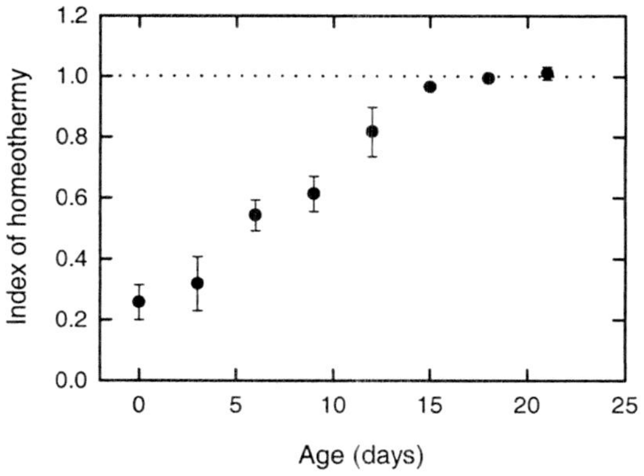


**Fig. 19.9** Ability of nestling European Shags (*Gulosus aristotelis*) to thermoregulate with increasing age (and mass). The Index of Homeothermy is a measure of the ability of nestlings to maintain their temperature at typical ambient temperatures in shag colonies (9.4–12.8°C). A value of 1 indicates that body temperature is maintained by nestlings and lower values indicate a progressively reduced ability to maintain body temperature. Values are presented as means ± SD. (Figure from Østnes et al. 2001; © 2001 Oxford University Press, used with permission)

## 19.3 Begging Behavior of Young Birds

Among species of birds where parents feed their young, nestlings and recent fledglings solicit food from parents (and helpers in some species of birds) in multiple ways, including large, colorful gapes, altered postures, and conspicuous vocalizations (Wright and Leonard 2002). The begging displays of young birds may have evolved as a means of influencing parental decisions in scramble competition among siblings (Parker et al. 2002a) or as honest signals of need allowing parents to allocate food relative to begging intensity (Godfray 1991). However, available evidence suggests that begging reliably signals short-term nutritional need (Searcy and Nowicki 2005). Thus, by varying the intensity of their begging, particularly begging calls, young birds can provide parents with information about their hunger levels and parents can adjust feeding rates accordingly (Kilner and Johnstone 1997).

Begging displays have more than one component that may represent redundant signals that reduce the chance of "communication errors" between young and adults (Johnstone 1996). In some species, however, different components of begging displays may provide parents with different information. For example, Jacob et al. (2011) found that the color of the flanges (the region of the mouth where the upper and lower bills meet) of nestling European Starlings (*Sturnus vulgaris*) was correlated with the amount of fat or lipid reserves, whereas begging call intensity was correlated with nestling size. Thus, adult starlings could potentially use these two begging signals to assess two measures of nestling condition, i.e., their fat reserves and their size (Jacob et al. 2011). Such information could be important because parental feeding behavior might vary with changing resource levels (Davies et al. 1999). For example, when food availability is high, all nestlings should be similar in size and parents should allocate food based on the nestlings' lipid reserves. However, when less food is available, parents may not be able to raise all of their young and they should preferentially feed larger nestlings that are more likely to survive (Davies et al. 1999).

Begging provides benefits to nestlings in the form of food delivered by parents, but begging nestlings may also incur costs. One possible cost is that conspicuous begging displays accompanied by loud begging calls might attract predators. Several investigators have examined the possibility that begging calls can attract predators by comparing the rates at which predators visit control nests (where no calls or normal background sounds were played back) and nests where begging calls are played back over small speakers (e.g., Haskell 1994; Leech and Leonard 1997; McDonald et al. 2009; Husby 2019). The results of several such studies indicate that the begging calls of nestlings can sometimes, but not always (e.g., Jelínek et al. 2019), attract predators to nests (Fig. 19.10). The likelihood of nest predation might be reduced if nestlings detect nearby or approaching predators and cease begging (e.g., Yasukawa et al. 2020) or if adults detect predators and give calls that silence the young (Platzen and Magrath 2004; Barati and McDonald 2017), but nestlings and adults do

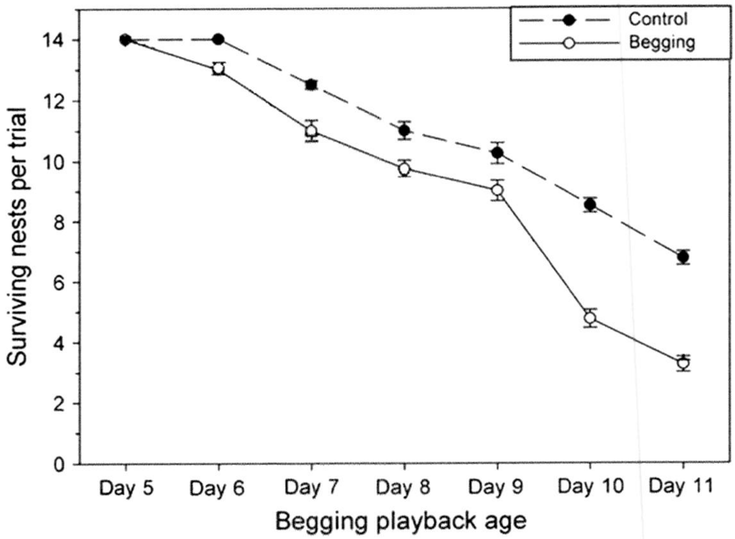


**Fig. 19.10** Mean (± SE) number of control nests and nests of Bell Miners (*Manorina melanophrys*) where begging calls were played back over speakers. Nests where begging calls were broadcast were predated more often and more quickly than control nests. (Figure modified from McDonald et al. 2009; © 2009 Oxford University Press, used with permission)

not always detect predators, or detect them in time, and, therefore, begging can increase the risk of predation (McDonald et al. 2009; Haff and Magrath 2011; Liu et al. 2022). This risk, however, may differ among locations that have different communities of predators that use different cues to locate prey (Santisteban et al. 2002; Jelínek et al. 2019).

Begging involves movement and calling and so may also be energetically costly. The results of some studies suggest that the physiological or energetic costs of begging are low (Bachman and Chappell 1998; Chappell and Bachmann 2002). For example, nestling House Sparrows (*Passer domesticus*) that differed in begging effort or intensity grew at similar rates (Kedar et al. 2000). However, trade-offs between begging and growth rates have been reported in other species, including Atlantic Canaries (*Serinus canaria*; Kilner 2001) and Eurasian Magpies (*Pica pica*; Rodríguez-Gironés et al. 2001), with nestlings that begged more growing at slower rates. One possible explanation for differences among studies is that nestlings that beg at high levels may face a three-way trade-off between begging, growth, and immune response. Moreno-Rueda (2010) divided nestling House Sparrows into two groups (low-begging and high-begging, with those in the high-begging group stimulated to beg using acoustic, tactile, and visual stimuli) and found no difference between groups in amount of mass gained during the six-hour experiment. However, nestlings in the high-begging group did show depressed immune responses. In a similar study of nestling Iberian Gray Shrikes (*Lanius meridionalis*), nestlings stimulated to beg at higher rates over a 13-h period had reduced growth rates and immunocompetence (Moreno-Rueda and Redondo 2011). Soler et al. (2014) compared nestling House Sparrows in two groups with nestlings stimulated to either beg (by holding larvae close to their bills) for longer (60 s every 10 min) or shorter (for no more than a few seconds every 30 min) periods for five consecutive days, and found that those that begged for longer periods had lower immune responses, weighed less, and had higher metabolic rates than those that begged at low rates. Such results suggest that nestlings forced to beg more over short periods of time continued to allocate energy for begging and growth, but, to do so, had to allocate less energy to their immune systems. Begging at high rates for extended periods, however, can have at least three potential physiological costs: immunocompetence, growth, and metabolism (Soler et al. 2014).

The interests of offspring and their parents do not completely match (Trivers 1974), with offspring selected to secure more resources from parents than parents are selected to give (Kilner and Hinde 2008). However, if begging is costly, then offspring should be selected to honestly signal their needs because the costs of excessive begging (increased predation risk and increased physiological costs) might exceed the benefits (additional food provided by parents; Koykka and Wild 2018; Box 19.3 Dishonest Begging by Nestlings). In addition, excessive begging by nestlings may be selected against if it affects the condition or chances of survival of kin (indirect fitness), e.g., by inducing more foraging and feeding by parents and reducing the amount of food delivered to siblings. In support of this hypothesis, Briskie et al. (1994) found that the volume (loudness) of nestling begging calls tends to increase as the relatedness of brood members declines, i.e., species where a high percentage of young result from extra-pair copulations. Young brood parasites, such as young cowbirds, are not related to other brood members and they also tend to beg more than host nestlings (Dearborn 1998; Lichtenstein 2001a).

**Box 19.3 Dishonest Begging by Nestlings**

Most nestling birds beg for food from their parents, and the results of many studies suggest that such begging is honest, i.e., begging provides parents with accurate information about nestling condition and their need for food. However, Caro et al. (2016) examined the results of 108 studies of 60 species of birds and found that young birds "may be more likely to exaggerate signals of need when they compete with less-related nestmates or a greater number of nestmates." More specifically, they found that the relationship between need and begging was weaker in species with larger broods, in species where parents are able to have more broods during their lifetime, and when parents are unlikely to breed together again due to death or divorce. In larger broods, young birds experience more competition and may facultatively adjust their begging intensity based on the begging intensity of nestmates. For example, nestlings in several species of birds have been found to beg with greater intensity when competing with more or needier nestmates (e.g., Smith and Montgomerie 1991; Price et al. 1996; Neuenschwander et al. 2003; Marques et al. 2011).

Caro et al. (2016) also found that young birds were less honest in species where parents will have more broods in the future. If young birds have parents that might reduce investment in current offspring to enhance their ability to invest in future offspring, then current offspring may tend to be more dishonest, i.e., beg with greater intensity (e.g., Parker et al. 2002b; Grodzinski and Johnstone 2012).

Young birds whose parents are unlikely to breed together again due to divorce or death were also found to be less honest with their parents (Caro et al. 2016). If parents do not breed together again, then their future offspring will be less related to the current offspring. As such, survival of those future offspring contributes less to the inclusive fitness of the current offspring and, as a result, current offspring should tend to beg with greater intensity than predicted based on their current need for food.

(continued)

**Box 19.3** (continued)

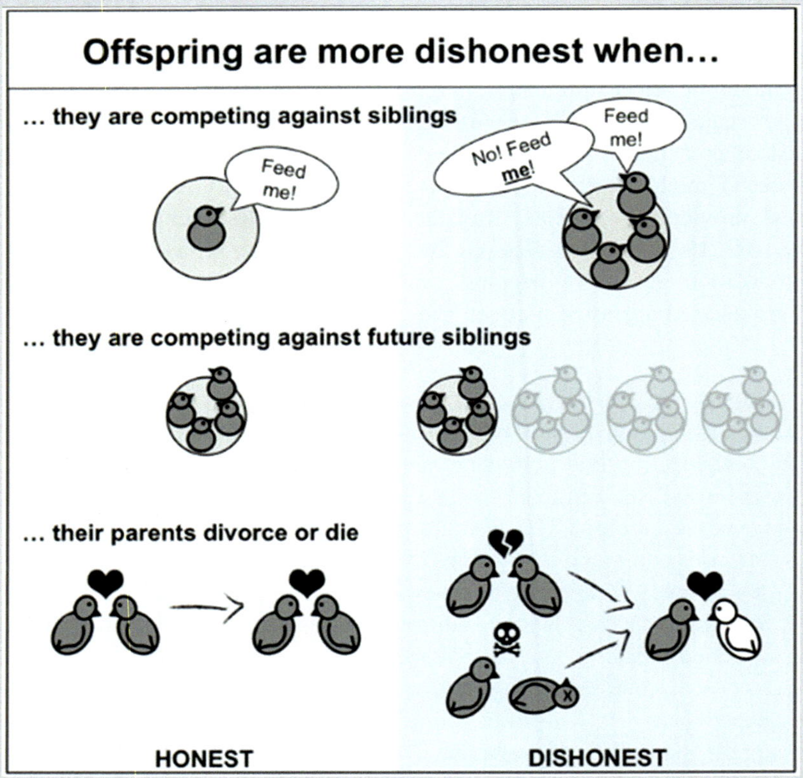


Family conflict is associated with a reduction in the honesty of offspring signals of need. Offspring in species with siblings present are less honest when signaling need. The probability that parents will breed again is also associated with more dishonest offspring, as the current brood competes against future broods for parental investment. Finally, offspring are less honest in species where parents are likely to divorce or die before breeding again, and all future siblings will be half siblings. (Figure from Caro et al. 2016; used with permission of the U. S. National Academy of Sciences)

## 19.4 Feeding Nestlings

Among altricial (and semi- and subprecocial) species of birds, one or both adults begin to feed young soon after hatching. In most socially monogamous species, plus some polygynous species, both sexes feed the young (Cockburn 2006). In many species, males and females provision young at similar rates (e.g., Goodbred and Holmes 1996; Wiebe and Elchuk 2003). However, the relative contributions of each sex vary among and within species. In sexually dimorphic species, differences between the sexes may be a result of differences in size. For example, Margalida et al. (2007) found that male Spanish Eagles (*Aquila adalberti*) provisioned young significantly more than females and suggested that the smaller males may be able to forage more efficiently than larger females on the agile, medium-sized prey typically taken by these eagles. Differences between the sexes in provisioning rates can also result from differences in their respective parental roles. In some species, females may devote more effort to provisioning young, whereas males are more responsible for guarding or defending nests or brooding young (e.g., Markman et al. 1995; Fraser et al. 2002; Paredes et al. 2006). In other species, the reverse may be true (e.g., Grundel 1987; Rytkönen et al. 1996).

Other factors can also influence the provisioning behavior of males and females. Investment in offspring by parents involves a trade-off between the costs and benefits of allocating resources to current versus potential future offspring (Royle et al. 2004). As such, excessive investment during one breeding attempt may negatively impact an individual's lifetime reproductive success. With each member of a pair attempting to avoid "excess investment," conflict between males and females can arise in how much each one contributes (Parker 1985). Resolution of this conflict depends on how compensating for reduced investment by a mate would affect an individual's future reproductive potential (Jones et al. 2002). The extent to which birds compensate for reduced parental care by mates has been examined in studies where mates have been temporarily removed during the nestling period and studies where one member of a pair is "handicapped" (e.g., weights attached or flight feathers slightly trimmed to increase the costs of flight). The responses of mates of manipulated birds vary, ranging from full compensation (increasing provisioning rates enough to totally compensate for the reduced rates of their mate), to partial compensation (some increase in provisioning rates, but not totally compensating), to no compensation (maintaining the same provisioning rates). Complicating things further, males and females sometimes respond differently when the contributions of manipulated mates decline, e.g., female Great Tits (*Parus major*) were found to compensate, but males did not (Sanz et al. 2000). Differences among studies may be due to differences in the ability of parents in different species (or between males and females within a species) to accurately determine how much food is needed by nestlings (i.e., their ability to assess the information provided by the begging behavior of nestlings). Differences may also be due to differences among species or between sexes in costs of compensating relative to benefits, i.e., the extent to which increased investment in the current brood will affect the likelihood of the current brood surviving and the likelihood of the compensating parent surviving to breed again in the future (Johnstone and Hinde 2006).

Food may be transferred from adults to young in several different ways (Pettingill 1985): (1) food is carried in the bill and placed in the open mouths of the young (most passerines); (2) food is swallowed by adults and later delivered to the young by regurgitating the food into the mouth or throat of the young (e.g., waxwings, hummingbirds, and herons), regurgitating the food into the nest or somewhere nearby where young can pick it up (e.g., gulls), or opening the mouth and letting young reach in and retrieve the food (e.g., pelicans and cormorants); (3) food consists of "milk" (produced in the crop) that is regurgitated into the mouths of the young (e.g., pigeons and doves); and (4) food is carried in the talons to the nest and then either torn into smaller pieces and fed to the young (e.g., raptors with young nestlings) or given to the young whole (e.g., raptors with older nestlings). Feeding visits to nests are typically quick (less time at the nest means less activity that might attract the attention of predators). However, particularly early in the nestling period, females may remain at the nest after feeding nestlings to brood the young.

Most parent birds feed their young the same food that they eat: insect eaters feed their young insects, fish eaters bring fish, seed eaters bring seeds, and so on. However, seed-eaters and fruit-eaters may also provide their young with some insects (which contain more protein; Skutch 1976). Parents also vary the size of food items, typically bringing smaller items to very young (and small) nestlings and larger items to larger, older young. The food given to young birds contains all the moisture they need, and parents do not bring them water to drink (Skutch 1976). Notable exceptions are sandgrouse parents (particularly males) that have specially modified abdominal feathers with enhanced water-holding capacity (Mueller and Gibson 2023). In fact, these feathers have structural modifications (the barbules are not hooked together) that make them three to four times more absorbent than synthetic sponges (del Hoyo et al. 1997). Adults may travel great distances to water, soak up the water in the abdominal feathers, then fly back to the nest so their young can drink the water from their plumage (Skutch 1976).

## 19.5 Fecal Sacs

After being fed, nestlings often defecate and, in some species, including those in the orders Passeriformes, Apodiformes, and Piciformes, produce packages of excrement surrounded by a gelatinous membrane called fecal sacs (Figs. 19.11 and 19.12). Adults typically eat the fecal sacs of young nestlings, but those of older nestlings are more often carried away and dropped some distance from the nest. Two hypotheses have been proposed to explain why parents ingest rather than simply remove fecal sacs. The parental nutrition hypothesis proposes that parents benefit energetically or nutritionally by ingesting fecal sacs (Morton 1979). The economic disposal hypothesis postulates that parents eat fecal sacs not for the energy or nutrition, but simply as a better alternative than carrying them away when they are engaged in other parental activities such as brooding (Hurd et al. 1991). Observations of several species, including Florida Scrub-Jays (*Aphelocoma coerulescens*), American Crows (*Corvus brachyrhynchos*), and Common Swifts (*Apus apus*), provide support for the parental nutrition hypothesis, with fecal sacs apparently providing parents with needed nutrients and water (McGowan 1995; Dell'omo et al. 1998). According to the economic disposal hypothesis, adults leaving nests immediately after feeding nestlings might be expected to carry away sacs rather than ingest them and this was not the case for any of these species.

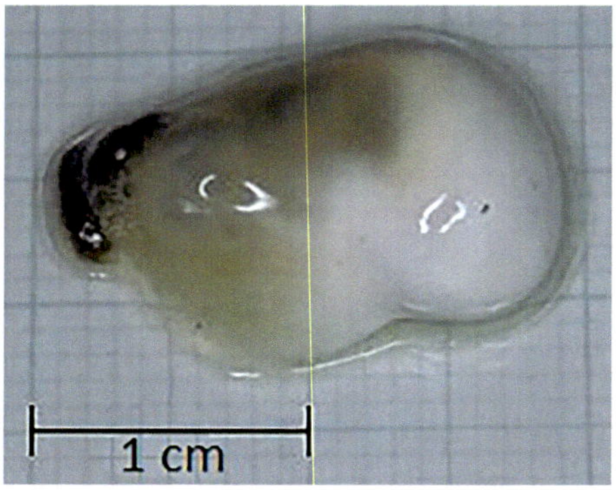


**Fig. 19.11** Fecal sac of a nestling Eurasian Blue Tit (*Cyanistes caeruleus*). (Photo by, and used with permission of, Takashi Koike)

Ibáñez-Álamo et al. (2014) suggested that fecal sacs also serve an anti-microbial function. In support of their hypothesis, these authors found that bacteria were less abundant on the surface of fecal sacs than inside the fecal sacs (Fig. 19.13). Such results suggest that the mucous covering of fecal sacs provides some protection from bacterial infection for both adults and nestlings.

As nestlings age, the percentage of fecal sacs ingested by parents typically declines (Fig. 19.14). Some investigators have hypothesized that this pattern is due to a decline in the energy content of the fecal sacs of older nestlings. However, analysis of the energy content of fecal sacs of nestling Tree Swallows (*Tachycineta bicolor*) revealed no change in energy content with increasing nestling age (Hurd et al. 1991). Thus, a more likely explanation for the tendency of parents to eat few fecal sacs as nestlings age is that older nestlings produce larger fecal sacs. Fecal sacs are not toxic, but do contain indigestible materials. Consuming larger fecal sacs rather than more digestible foods, therefore, would likely reduce a parent's digestive efficiency (Hurd et al. 1991).

## 19.6 Defending Eggs and Young

The primary cause of nest failure for birds is predation of eggs or nestlings (Ricklefs 1969). The primary nest predators include various species of reptiles, birds, and mammals, but insects (e.g., ants) may also cause nest failure by driving adults from nests and preying on nestlings. When a potential predator approaches a nest with eggs

**Fig. 19.12** Oak Titmouse (*Baeolophus inornatus*) removing a fecal sac from its nest cavity. (Photo by Alan Vernon, Wikipedia, CC BY 2.0, https://creativecommons.org/licenses/by/2.0/)

or young, the responses of birds can vary dramatically both within and among species. Some birds may quietly leave nests as a potential predator approaches, whereas others may launch a vigorous defense, repeatedly flying at the predator, calling loudly and repeatedly, and perhaps even striking the predator. Regardless of the response, any behavior that decreases the

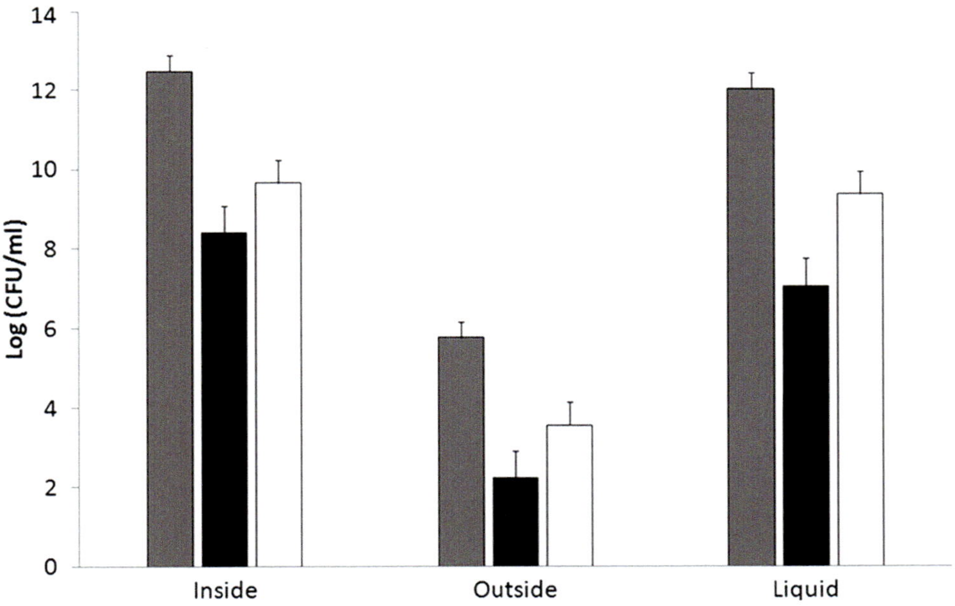


**Fig. 19.13** Mean bacterial loads (± SE) inside the fecal sacs produced by nestling Eurasian Blackbirds (*Turdus merula*), on the surface of the fecal sacs (outside), and in the liquid that percolated from fecal sacs after an average of 24 min. Note that bacterial loads were significantly lower on the surface of the fecal sacs. The three bars for each location represent bacterial loads as determined using three different culture media. CFU, colony-forming units. (Figure from Ibáñez-Álamo et al. 2014; © 2014 The Authors, used with permission)

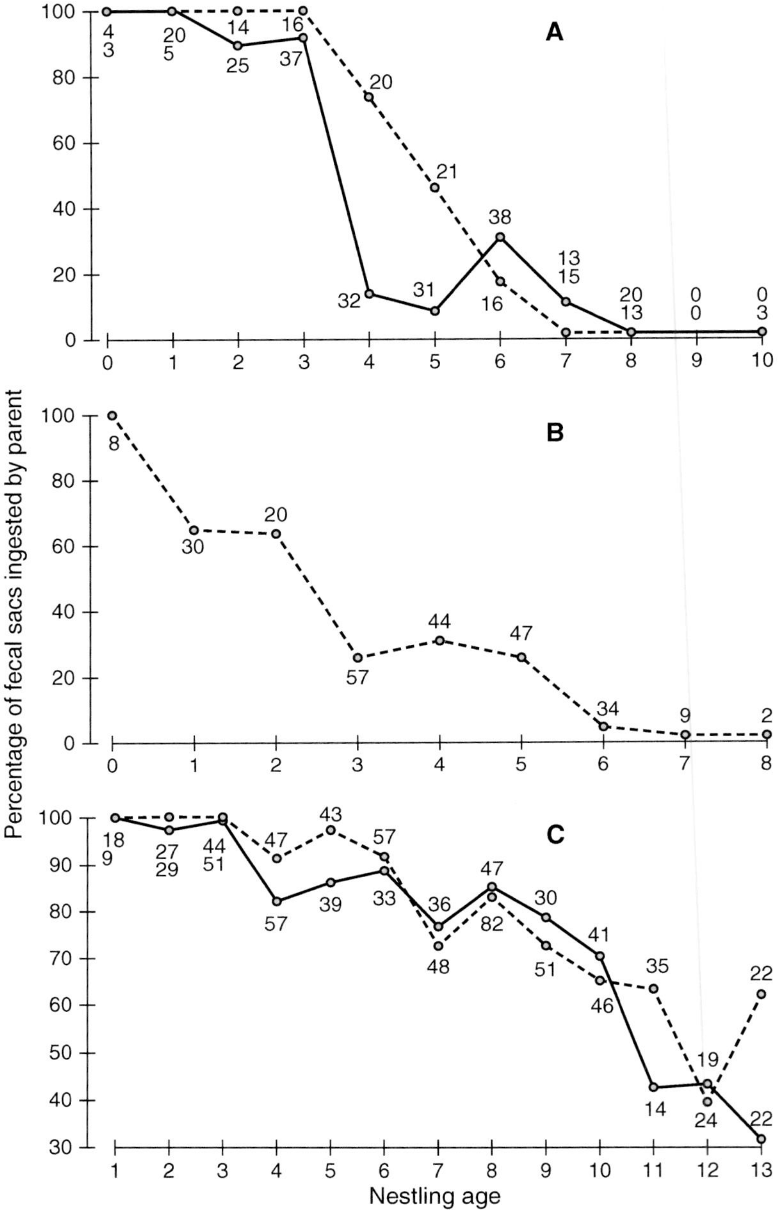


**Fig. 19.14** The percentage of fecal sacs eaten by parents relative to nestling age for (**a**) male (solid line) and female (broken line) Tree Swallows (*Tachycineta bicolor*), (**b**) female Red-winged Blackbirds (*Agelaius phoeniceus*), and (**c**) male (solid line) and female (broken line) American Robins (*Turdus migratorius*). Numbers indicate the number of observations of fecal sac disposal. Where points are close, the sample size for females is above that for males. (Figure from Hurd et al. 1991; © 1991 Oxford University Press, used with permission)

probability that a predator will locate and harm the contents of a nest (eggs or young) can be considered nest defense behavior. Montgomerie and Weatherhead (1988: 170) suggested that, to be considered nest defense behavior, the behavior of adults should also increase "... the probability of injury or death to the parent." However, some responses to potential nest predators may reduce the likelihood of nest predation with minimal or no risk to the parent. For example, an incubating or brooding adult that quietly leaves a nest is at no risk of injury or death, but its departure may remove cues, such as brightly colored plumage or even just the profile or presence of the head and body of an adult extending above a well-camouflaged nest that might increase the likelihood that a predator will locate the nest. Leaving nests when a predator has been detected can be considered "passive deception" and is more common among smaller species of birds or species with uniparental care (Larsen 1991).

Another strategy to minimize the risk of nest predation has been referred to as "coordinated misdirection" (Gulson-Castillo et al. 2018). As proposed, this strategy involves a breeding pair of birds flying toward their nest (with either eggs or nestlings) and, as they closely approach the nest, the bird closest to the nest quickly switches direction and goes to the nest. The other member of the pair, flying alongside, obscures its mate from view, and continues flying past the nest (Fig. 19.15). Nearby visual predators may tend to continue observing the member of the pair that continues flying past the nest and, if so, will be much less likely to detect the presence of the nest (Gulson-Castillo et al. 2018). Such behavior has been documented in five families of birds, including Fringillidae, Cardinalidae, Tyrannidae, Tityridae, and Contingidae (Gulson-Castillo et al. 2018).

Several species of birds have special distraction displays that may reduce the likelihood that a predator will find a nest or young. During such displays, parent birds typically utter characteristic vocalizations, feign injury, and attempt to lure predators away from nests or young (Fig. 19.16; Table 19.2). These displays have been reported for 285 species of birds representing 172 genera, 52 families, and 13 orders, including Tinamiformes, Anseriformes, Galliformes, Podicipediformes, Charadriiformes, Caprimulgiformes, Gruiformes, Pterocliformes, Columbiformes, Strigiformes, Apodiformes, Cuculiformes, and Passeriformes. De Framond et al. (2022) found that these displays were more likely among species at higher latitudes, with less concealed nests and short incubation periods, and where only one parent incubates eggs. More generally, broken-wing displays were more common among species nesting in areas with greater predation pressure.

Birds performing distraction displays sometimes appear to be injured. However, whether appearing injured or not such displays typically

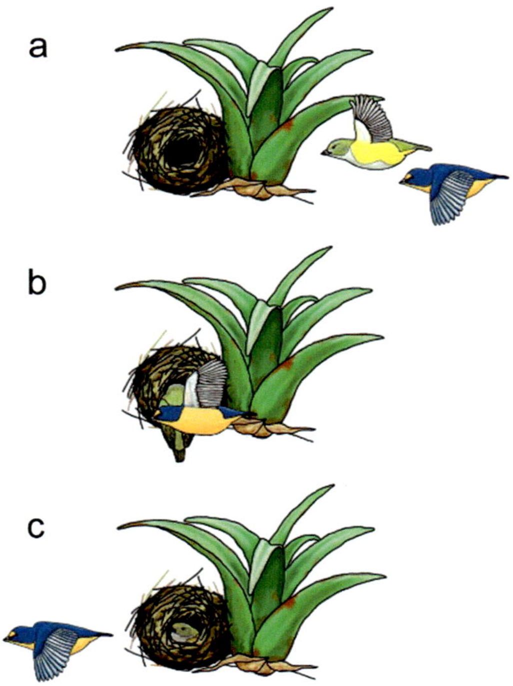


**Fig. 19.15** An example of coordinated misdirection behavior. (**a**) Male and female Yellow-throated Euphonias (*Euphonia hirundinacea*) fly together toward their nest. (**b**) One (in this case the female) flies to the nest, while the other (the male) hides her from view and continues to fly, often changing direction in flight. (**c**) The bird at the will then incubates eggs or brood/provision nestlings. Gulson-Castillo et al. (2018) noted that human observers tend to visually follow the male in this scenario, reducing the likelihood that the female will be observed going to the nest; if visual nest predators can also be deceived like this, then such coordinated misdirection may reduce the likelihood of nest predation. (Figure from Gulson-Castillo et al. 2018. Figure credit: ERGC and M. M. Ferraro; © 2018 Wilson Journal of Ornithology, used with permission)

make birds very conspicuous because of the often odd movements and accompanying vocalizations. Such behavior may attract the attention of predators that may associate abnormal movement with injury or illness and, therefore, an increased likelihood of successfully capturing prey (Curio 1976; Sordahl 1986). Distraction displays are more often used in response to mammalian predators. Avian predators represent a greater threat to displaying parent birds than mammalian predators and, in addition, avian predators would likely be better able to see nests or young from the

**Fig. 19.16** Distraction displays of a Kentish Plover (*Charadrius alexandrinus*). (**a**) Mobile lure display with wings beating on the ground. (**b**) Mobile lure display with alternating wing action. (**c**) Static lure display with wings held in a shallow "V." (Figure from Simmons 1951, drawings by R. A. Richardson; reproduced with permission of the journal, British Birds)

air, likely making distraction displays less effective (Caro 2005). In addition, distraction displays may be more effective in distracting mammals. For example, in a study of Killdeer (*Charadrius vociferous*), Brunton (1990) found that mammalian predators followed displaying adults away from nests more often than avian predators.

The results of some studies suggest that distraction displays can be effective and their use can improve the chances of nest success (Byrkjedal 1987). However, distraction displays can also be costly because birds engaging in such behavior are sometimes (albeit rarely) attacked and killed by predators, either those that the displaying bird was trying to distract or others that take advantage of the displaying birds focus on another predator (Sordahl 1990).

The nest defense behavior of many birds involves flying at or near (and, in some species, occasionally even striking) a predator while calling, with the intensity of defense influenced by several different factors. One such factor is size. For example, among shorebirds, larger species exhibit more aggressive nest defense than smaller species, likely because larger species pose a greater threat to predators (Larsen et al. 1996). Other factors that can influence the intensity of nest defense include nest stage (e.g., incubation period vs. nestling period), number of young, parental sex, and predator type. However, the

**Table 19.2** Main categories and sub-categories of anti-predator distraction displays[a] used by birds to defend eggs and young

| Categories | Sub-categories | Examples[b] | References |
|---|---|---|---|
| Standing still | Alert postures | Western Sandpiper | Brown (1962) |
| | Head bobbing | Kentish Plover | Simmons (1951) |
| | Tail-flagging | Magellanic Oystercatcher | Miller and Baker (1980) |
| Flying activities | Explosive departure | Zenaida Dove | Burger et al. (1989) |
| | Impeded flight | Golden-crowned Warbler | Smith (2017) |
| | Diversionary circling | American Avocet | Sordahl (2004) |
| Running activities | Crouched run | Kentish Plover | Simmons (1951) |
| | Rodent run | American Golden-Plover | Byrkjedal (1989) |
| | Rapid run | Eurasian Oystercatcher | Williamson (1952) |
| | Wing-out run | Long-billed Curlew | Graul (1971) |
| Injury feigning | Broken-wing display | Piping Plover | Ristau (1993) |
| | Erratic fluttering | Zenaida Dove | Burger et al. (1989) |
| | Exhausted bird | Kentish Plover | Gomez-Serrano (2018) |
| Displacement-type displays | False brooding | Killdeer | Brunton (1990) |
| | False feeding | Kentish Plover | Gomez-Serrano (2018) |
| | Pseudo-sleeping | Eurasian Oystercatcher | Williamson (1950) |
| | Nest betrayal | Common Ringed Plover | Smith and Smith (1974) |

Table modified from Humphreys and Ruxton (2020); © 2020 British Ornithologists' Union, used with permission
[a]Behaviors that function primarily to increase the conspicuousness of the displaying bird to a potential predator, distract the predator's attention away from nests or young, and reduce the likelihood of eggs or young being discovered and predated (Humphreys and Ruxton 2020)
[b]Scientific names: Western Sandpiper, *Calidris mauri*; Kentish Plover, *Charadrius alexandrinus*; Magellanic Oystercatcher, *Haematopus leucopodus*; Zenaida Dove, *Zenaida aurita*; Golden-crowned Warbler, *Basileuterus culicivorus*; American Avocet, *Recurvirostra americana*; American Golden-Plover, *Pluvialis dominica*; Eurasian Oystercatcher, *Haematopus ostralegus*; Long-billed Curlew, *Numenius americanus*; Piping Plover, *Charadrius melodus*; Killdeer, *Charadrius vociferus*; Common Ringed Plover, *Charadrius hiaticula*

importance of these factors may vary among and even within species. One possible explanation for intraspecific variation in responses to nest predators is that individuals with different personalities differ in their willingness to take risks when defending nests (e.g., Hollander et al. 2008). In addition, the relative importance of one factor may be influenced by other factors. For example, the intensity of nest defense by adult birds typically increases with increasing nest age (discussed in more detail below). However, the effect of nestling age on adult responses may be influenced by predator type because some predators may represent more of a threat during the nestling stage than during the incubation or egg stage (e.g., a raptor that eats young birds, but not eggs), whereas other predators may be equally likely to prey on eggs and nestlings (e.g., snakes). Thus, investigators conducting experiments to determine the possible effect of nest age on nest defense behavior may get somewhat different results depending on the type of predator(s) used in those experiments. What this example is meant to illustrate is that the nest defense behavior of birds not only varies among species but is also simultaneously influenced by multiple factors. As a result, clear-cut trends are not always apparent when examining the effect of one particular factor.

A factor found to influence the intensity of nest defense in many species of birds is nest stage. In general, adults defend nestlings more vigorously than eggs and, during the nestling stage, increase the intensity of defense with increasing nestling age. A likely explanation for such behavior is that the probability of young surviving to breeding age increases as they grow older, increasing their value to parents (in terms of fitness;

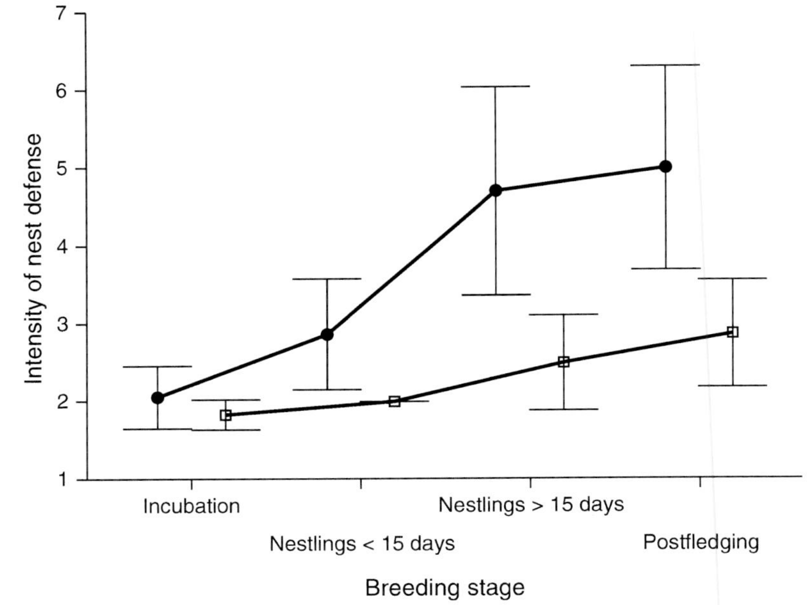


**Fig. 19.17** Average intensity of nest defense (± SE) by 42 male (open square) and 42 female (closed circle) Northern Hobbies (*Falco subbuteo*) increased from incubation to fledging, during the nestling period, and during the first 10 days after fledging. Intensity of defense against a potential human predator was estimated using an ordinal aggressiveness score ranging from 1 (flew away) to 8 (stooped at the intruder). (Figure from Sergio and Bogliani 2001; © 2001 Oxford University Press, used with permission)

Patterson et al. 1980). As a result, parents are willing to invest more in nest defense as young increase in age (Fig. 19.17).

The effects of other factors on the nest defense behavior of birds, including parental sex, number of young, and characteristics of predators, vary more widely among different species of birds. Males defend nests more vigorously than females in some species, the reverse is true in other species, and, in still other species, both sexes contribute equally. Reasons for such variation among species remain unclear, but differences between the sexes in the intensity of nest defense may arise (1) if the sexes differ in longevity and re-nesting potential (with the shorter-lived sex willing to take more risks in defending a nest), (2) if one sex cannot raise the brood alone (with the sex able to raise the brood alone less willing to take risks; Caro 2005), (3) if the energetic condition of males and females differs, e.g., if females have expended greater energy in producing a clutch and incubating and brooding young and, therefore, are less able to mount an aggressive defense, and (4) if, because of a difference in size (i.e., in sexually dimorphic species), one sex (either the largest or smallest depending on the relative importance of size or mobility) is better able to, and with less risk, repel an approaching predator (Caro 2005).

Studies where investigators examined the possible effect of clutch or brood size on the intensity of nest defense have also yielded mixed results, with intensity increasing with clutch or brood size in some studies, but not others. An increase in the intensity of nest defense with increasing clutch or brood size might be expected because the reproductive value (in terms of adult fitness) of larger clutches and broods is greater than that of smaller clutches and broods. However, complicating interpretation of the results of many previous studies is that pairs with larger clutches or broods may defend nests more vigorously simply because they are in better condition than pairs with smaller clutches or broods (Lessells 1991). Some investigators have addressed that concern by experimentally manipulating clutch or brood sizes (e.g., Tilgar and Kikas 2009). The results of many, though not all, of these studies indicate that the intensity of nest defense by parent birds is positively correlated with clutch or brood size. Thus, in many species of birds, parents appear to adjust their nest defense behavior based on the reproductive value (i.e., the size) of their clutches and broods.

The nest defense behavior of parent birds may also vary with the type of predator, with adults in some species responding less aggressively to nest predators that could also prey on them. For example, female Aquatic Warblers (*Acrocephalus paludicola*) defended nests more vigorously (approached closer and made more flights) against a ground-based mammalian predator (European polecat, *Mustela putorius*) than an avian predator (Hen Harrier, *Circus cyaneus*) (Halupka 1999). Similar results, with parents responding less aggressively to predators that could also prey on adult birds, have been reported in Black-billed Magpies (*Pica hudsonia*; Buitron 1983), Black-headed Gulls (*Larus ridibundus*; Kruuk 1964), and Common Terns (*Sterna hirundo*; Burger and Gochfeld 1991).

The effectiveness of nest defense behavior depends on the relative size and "armament" (features such as bills and talons that could injure or kill others; Montgomerie and Weatherhead 1988) of parents and the potential predators. Thus, small birds cannot effectively defend nests against large, well-armed predators (e.g., many raptors and large mammals or reptiles), but larger, well-armed birds (e.g., many raptors) may be able to effectively defend nests against smaller predators. However, even relatively small songbirds are sometimes able to successfully defend their nests against similar-sized or even somewhat larger predators. Pietz and Granfors (2005) reviewed videotapes made at nests of five species of grassland birds where predators attempted to take either eggs or young and found that parent birds were at least partially successful (not all eggs or young were taken by the predator) in five of 21 cases. A female Bobolink (*Dolichonyx oryzivorus*) chased and pecked a female Brown-headed Cowbird (*Molothrus ater*) that was removing nestlings from the Bobolink's nest; the cowbird removed three of four nestlings, but one nestling survived to fledge. Clay-colored Sparrows (*Spizella pallida*) at one nest were partially successful in defended eggs or young from a long-tailed weasel (*Mustela frenata*) and, at three other nests, were successful in defending nests from unidentified predators (Pietz and Granfors 2005). Similarly, Ellison and Ribic (2012) monitored nests of several species of grassland birds with video-cameras and reported that three of 15 nests that were aggressively defended against snakes eventually fledged young.

## 19.7 Sibling Cooperation and Competition

Within broods of altricial and semi-altricial young, interactions among nestling can be both cooperative and competitive. Nestlings may compete with each other for access to feedings from adults and this competition varies in intensity, ranging from simple scramble competition to siblicide. However, nestlings also cooperate in ways that can be beneficial to some or all broodmates. For example, nestlings benefit by huddling, reducing the amount of exposed surface area, reducing heat loss, and saving energy. In addition, the age of effective endothermy (i.e., ability to maintain a reasonably constant body temperature) for broods is earlier than for individual nestlings (Dunn 1975, 1976b). In some species, siblings also cooperate when the time comes to fledge, i.e., leave the nest. Fledging is often initiated by better-developed, usually older, nestlings that are the largest and heaviest. However, less-developed, younger siblings are at risk of abandonment and starvation if they do not fledge soon after their more developed siblings because adults will focus their attention on the more numerous fledglings. Bowers et al. (2013) found that broods of House Wrens (*Troglodytes aedon*) with greater differences in nestling age had longer nestling periods, suggesting that older nestlings in those nests delayed fledging and, by doing so, improved the chances of survival of their younger siblings. Additional study of the fledging behavior of more species is needed, but such results support kin-selection theory because the survival of younger, related siblings enhances the inclusive fitness of the older nestlings.

Some studies have also revealed that nestlings in some species may adjust their begging behavior relative to the needs of nestmates. For example, Romano et al. (2012) found that more recently fed (and less hungry) nestling Barn Swallows (*Hirundo rustica*) reduced the intensity of their begging and were less aggressive in trying to obtain food from parents than their hungrier nestmates. As a result, hungrier nestlings were more likely to be fed. Such behavior is likely adaptive because it ensures that related siblings are fed, potentially enhancing (assuming it increases the likelihood that nestmates will survive) the inclusive fitness of the young Barn Swallows. Unfortunately, most studies of nestling behavior have focused on competitive interactions rather than cooperative interactions. However, if the benefits gained via indirect fitness (fitness benefits achieved by aiding the survival of non-descendent kin such as siblings) by cooperating with nestmates exceed costs to direct fitness (fitness gained via an individual's own reproduction), then natural selection would favor cooperation. Additional studies are needed to better understand the extent to which, and mechanisms whereby, cooperation among nestlings might occur in other species of birds.

Nestlings, of course, also compete with each other because the benefits of such competition to direct fitness sometimes outweigh any costs to indirect fitness. In most cases, this competition is non-violent and limited to nestmates begging to obtain food from parents (referred to as scramble competition). However, in some species, this competition involves aggression, including pecking, biting, and pushing, directed at nestmates. This aggression can sometimes escalate dramatically and even lead to the killing of nestmates (siblicide). Among some siblicidal species, one nestling almost always (>90% of the time) kills its nestmate(s) (obligate siblicide); in others, nestlings, depending on a number of factors, may or may not kill nestmates (facultative siblicide; < 90% of the time).

Obligate siblicide is best known in the bird families Pelicanidae (pelicans), Sulidae (boobies), and Accipitridae (eagles). In these species, females lay two eggs and hatching of those eggs is asynchronous so the first-hatched nestling will be larger than the second-hatched; the larger nestling usually kills the smaller nestling. In obligate siblicidal species, the second egg is laid to ensure against the possible failure of the first egg/chick (Cash and Evans 1986). Younger chicks are typically killed (by persistent pecking and biting) or pushed from nests by older chicks within a week after the younger chick hatches. Obligate siblicide appears to be independent of food availability, with older chicks killing younger ones even if food is available in the nest during this first week (Forbes and Mock 1996, but see Osorno and Drummond 2003). Such killing is adaptive, however, because adults are rarely able to provide enough food to raise and fledge two nestlings.

Facultative siblicide has been reported in a wide variety of avian taxa including species in the orders Suliformes, Pelecaniformes, Accipitriformes, Charadriiformes (skuas, family Stercorariidae), Strigiformes, Apodiformes (swiftlets, family Apodidae), Coraciiformes, and Falconiformes. One proposed explanation for this siblicidal behavior is called the food amount hypothesis, with sibling aggression negatively correlated with the quantity of food provided to nestlings by adults (Mock 1987). A number of studies have provided support for this hypothesis, including studies of Osprey (*Pandion haliaetus*), Western Grebes (*Aechmophorus occidentalis*), Cattle Egrets (*Bubulcus ibis*), and Black-legged Kittiwakes (*Rissa tridactyla*) (Mock and Parker 1998; White et al. 2010a, b). In non-siblicidal species, nestlings compete via scramble competition for access to food from adults, and, if insufficient food is provided, less competitive nestlings may starve. A potential advantage of being siblicidal is that dominant nestlings can influence when sibling competition is eliminated rather than waiting for siblings to starve. As siblings are starving, they use resources that could enhance the condition and, ultimately, fitness of the dominant nestling(s). As such, killing smaller, weaker nestmates via siblicide sooner, rather than those nestmates dying of starvation later, can be the better strategy.

Beyond food availability, other factors can also influence the incidence of facultative siblicide. In many, but not all, siblicidal species of birds, nestlings differ in age and/or size. These differences typically result from hatching asynchrony, with eggs laid at intervals and incubation beginning before completion of a clutch so eggs do not hatch at the same time. Within-brood dominance hierarchies typically follow the hatching order of nestlings so larger earlier-hatched nestlings dominate smaller later-hatched nestlings. As dominant larger nestlings outcompete smaller siblings for access to food delivered by adults, size differences may become more pronounced. In facultative siblicidal species, these larger nestlings may direct increasing aggression toward smaller siblings if adults provide insufficient food, potentially resulting in siblicide.

Gonzalez-Voyer et al. (2007) conducted a comparative analysis to better understand the factors that might explain the evolution of siblicidal behavior and found that the incidence and intensity of aggression was correlated with both small brood sizes and long nestling periods. As brood size increases, establishing clear dominance relationships may become more difficult and, in addition, while one nestling is attempting to prevent another one from being fed, visiting adults are more likely to feed one of the other nestlings. As a result, over time, most or all nestmates may receive sufficient food, reducing the likelihood that nestlings will differ appreciably in size or condition (Drummond 2002; Gonazalez-Voyer et al. 2007). Longer nestling periods might favor increased aggression among siblings because, over a longer period, food availability is more likely to vary and potentially decline. In addition, long nestling periods may favor aggressive competition because the costs of aggression earlier in the nestling period (e.g., energy and possible injury) may be more than balanced by the benefits obtained, i.e., becoming dominant and being fed more frequently later in the nestling period (Drummond and Osorno 1992).

Female birds are able to regulate the transfer of substances to their developing eggs, including carotenoids, antibodies, and testosterone. This ability means that, by differentially allocating these substances to different eggs, females can influence offspring phenotype. Variation in levels of testosterone in an egg's yolk can, for example, cause variation in the development rate and competitive ability of young birds (more testosterone = faster development and more competitive; von Engelhardt and Groothuis 2011). By way of these "maternal effects," females can increase or decrease the relative size and competitiveness of their nestlings. In a comparative analysis involving 28 species of birds, Muller and Groothuis (2013) found a clear relationship between relative size differences within broods and the amount of maternal testosterone in eggs (Fig. 19.18). For species where eggs hatch more synchronously and nestlings are similar in size, last-laid (and last hatching) eggs receive more testosterone than first-laid (and first hatching) eggs; for species with asynchronous hatching, last-laid eggs receive less testosterone than first-laid eggs (Fig. 19.18). This difference suggests different reproductive strategies. Because later-hatched nestlings tend to be smaller than earlier-hatched nestlings even when hatching is relatively synchronous, the additional testosterone enhances the competitive abilities of last- or later-hatched nestlings and, assuming adequate food availability, increases the chances that the entire brood will fledge. In species with asynchronous hatching, last- or later-hatched nestlings are important primarily for their replacement value, typically only surviving if first- or earlier-laid eggs do not hatch or, if so, the nestlings die shortly after hatching. For such species, having later-hatched nestlings that are less competitive is beneficial because they can be more easily dispatched by larger, older nestlings (Muller and Groothuis 2013).

Although factors such as food availability, hatching asynchrony, brood size, duration of nestling periods, and maternal deposition of testosterone in eggs have been found to contribute to siblicidal behavior in many species of birds, these and other factors such as how young are fed (e.g., directly, with food passed from adult to nestling, or indirectly, with adults placing food in the nest), food parcel size (with parcel size influencing the ability of nestlings to quickly consume it), and available weaponry (e.g., talons or pointed bills) can also play a role. Additional comparative studies are still needed to improve

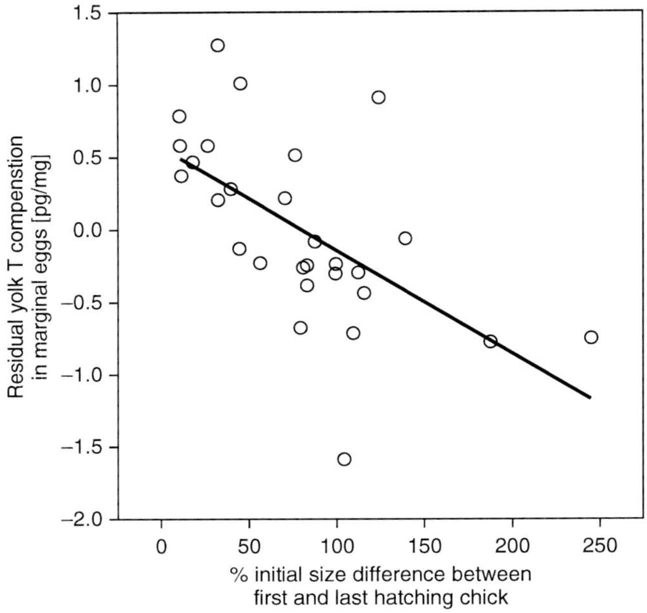


**Fig. 19.18** Testosterone (T) in yolk of first-laid eggs of a clutch relative to that in last-laid (marginal) eggs (picograms/milligram) relative to differences in the relative size of chicks in these eggs at hatching. Residual yolk T compensation equals the mean T in first-laid eggs minus the mean T in last-laid eggs. Data are from 28 species of birds. With an increasing difference in size of first-hatched and last-hatched nestlings at hatching, females tend to deposit increasingly less T in eggs. (Figure from Muller and Groothuis 2013; © 1867 CCC Republication. Published by University of Chicago Press, used with permission)

our understanding of how these various factors might interact and favor, or not, the evolution of obligate or facultative siblicidal behavior.

## 19.8 Departure from Nests

Sometime after hatching, young birds leave nests and, for species where adults care for young in the nest for several days or weeks, this is typically referred to as fledging. Young of precocial species may leave nests within hours after hatching, with one or both parents typically leading young from nests to habitats where young can forage and grow (e.g., Colwell et al. 2007). For those species, the term fledging generally refers to when young become independent from adults and are able to fly. Semiprecocial, semialtricial, and altricial young remain in nests for much longer periods (range = about 10–60 days) and several factors can influence when they leave nests, including nest type, parental behavior, nestling development, and nestmate interactions. For example, for species targeted by brood parasites, the presence of, and competition for food with, parasitic nestlings may result in longer nestling periods, and older ages at fledging, for host young (Scharf et al. 2022). Nestling periods are longer and, therefore, fledging age is typically older for cavity-nesting than for open-nesting species of birds (Martin and Li 1992; Bosque and Bosque 1995). Cavity-nesting species typically have longer nestling periods than open-nesting species because the risk of predation is lower for nests in cavities (Martin and Li 1992; Martin 1995).

Parents could stimulate fledging through some change in their behavior. For example, parents could deliver less food to nests and increasingly hungry young may then leave nests to beg from parents. Parents could also induce fledging by holding food a short distance from nests and in view of nestlings (Michaud and Leonard 2000).

Parents may attempt to stimulate fledging if the optimum fledging time for parents is before that of their young. This may be the case if a shorter nestling period reduces the time needed for young to become independent, possibly leaving parents with more time to initiate another breeding attempt or, for migratory species later in the nesting season, to molt and prepare for migration (Johnson et al. 2004). This might also be the case if fledging early increases parental fitness by increasing the chances that at least one young will survive to independence, i.e., predators are more likely to kill entire broods when all young are together in a nest than if young have fledged and are more widely dispersed. Jones et al. (2020) compared post-fledging survival for 18 species of songbirds and found that, although individual young were 13.6% less like to survive due to early fledging, the likelihood of at least one young in a brood surviving because of early fledging increased by 14%. Thus, early fledging is detrimental for young, but beneficial in terms of fitness for parents. This suggests that, at least in some species of birds, there is parent-offspring conflict in terms of the time of fledging and, further, that parents must somehow manipulate young into fledging early for their own benefit (Jones et al. 2020; Fig. 19.19).

The age and time of fledging may also be determined by nestlings (Box 19.4 Time of Day When Young Fledge). Young may fledge when they reach a certain developmental state (e.g., a certain wing length or body mass) because the benefits of leaving the nest exceed the costs. For example, by following adults rather than adults having to make trips to and from nests, young might be fed at higher rates (Johnson et al. 2004). Young birds may also reduce the risk of predation by leaving nests and seeking cover if, for example, the noise (e.g., begging calls) and odors (e.g., feces) created by a brood of young in a nest, along with frequent visits to nests by provisioning adults, are more likely to attract the attention of a predator.

Competition among nestmates could also influence the timing of fledging. In several species, large broods fledge before small broods, apparently because of an increase in competition for food and space in larger broods (Michaud and Leonard 2000). Interactions among nestmates could also affect individual nestlings differently and thus influence fledging order within broods. For example, smaller nestling Marsh Tits (*Poecile palustris*) initiate fledging under low food conditions, presumably to reduce competition from larger nestmates (Lemel 1989).

The relative importance of parental behavior, nestling age or stage of development, and within-brood competition in causing fledging appears to vary among species and even within species. For example, Johnson et al. (2004) found that the age of nestling House Wrens (*Troglodytes aedon*) at fledging varied from 14 to 19 days old, that the most well-developed nestling (as measured by feather length) in a brood was usually the first to fledge, and that parental behavior had no apparent effect on fledging behavior. At some nests, the first nestlings to fledge were not the most developed and competition for food may have caused these smaller nestlings to fledge first (Johnson et al. 2004). Nestling development also appeared to be the most important determinant of fledging age in Tree Swallows (*Tachycineta bicolor*, Michaud and Leonard 2000), Eurasian Blue Tits (*Cyanistes caeruleus*, Schlicht et al. 2012; Santema et al. 2021), Great Tits (*Parus major*, Radersma et al. 2011), European Pied Flycatchers (*Ficedula hypoleuca*, Kern et al. 2001), Thick-billed Murres (*Uria lomvia*) and Razorbills (*Alca torda*, Hipfner and Gaston 1999), and Black (*Milvus migrans*) and Red (*M. milvus*) kites (Viñuela and Bustamante 1992). Similarly, the timing of departure of King Penguin (*Aptenodytes patagonicus*) chicks from breeding colonies depends on their developmental status and body condition and not parental feeding behavior (Corbel et al. 2009). For Tufted Puffins (*Fratercula cirrhata*), both parental behavior and the developmental stage of nestlings influenced the timing of fledging, with timing influenced by the age at which nestling mass peaked and by a reduction in parental provisioning rates (Gjerdrum 2004).

Thus, the results of studies to date suggest that, for many, but not all, species of birds, the most important factor in determining when nestlings fledge is their developmental state, particularly

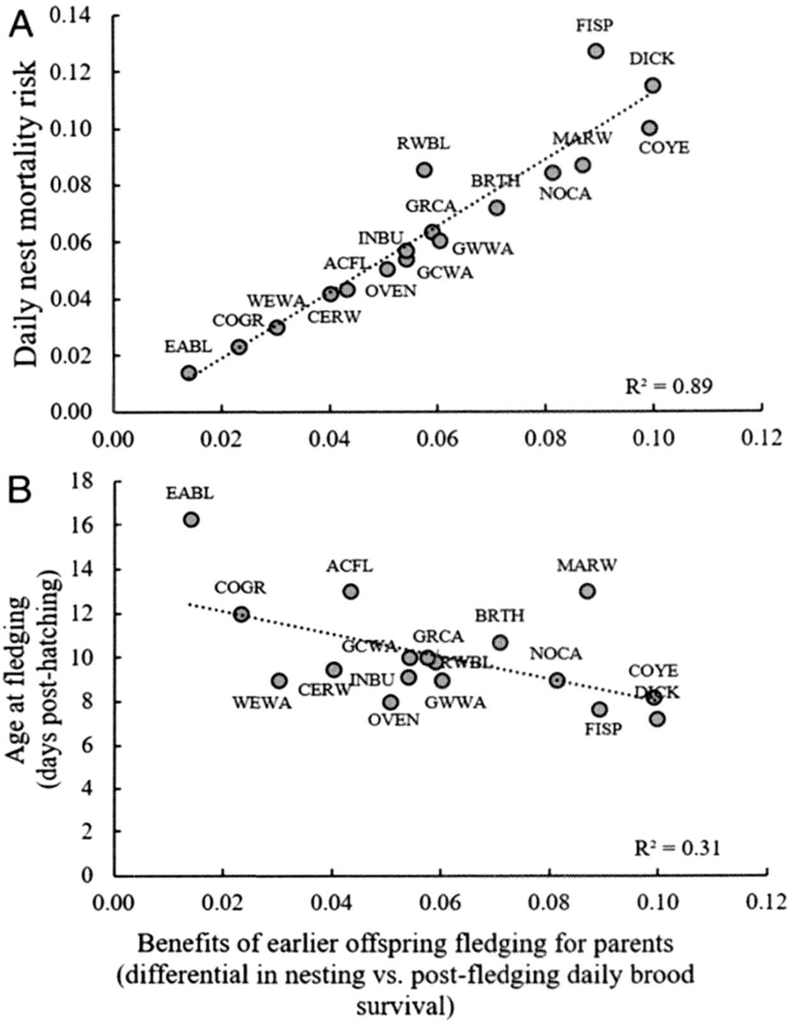


**Fig. 19.19** Young songbirds often fledge earlier than expected, when their risk of mortality is higher outside of the nest than in it. Earlier fledging can benefit parents because separating young spatially can increase daily brood survival compared to all young remaining together in a nest. This difference in nesting (i.e., broods in nest) versus fledging (i.e., broods leave nest) daily brood survival provides fitness benefits for parents while incurring a cost for individual young. Across species, differences in daily brood survival show that (**a**) earlier fledging of young provides greater fitness benefits for parents in species with an increasing risk of nest mortality, and (**b**) differences among species in parental benefits predict the age of fledging. Abbreviations for each species: ACFL, Acadian Flycatcher, *Empidonax virescens*; BRTH, Brown Thrasher, *Toxostoma rufum*; CERW, Cerulean Warbler, *Setophaga cerulea*; COGR, Common Grackle, *Quiscalus quiscula*; COYE, Common Yellowthroat, *Geothlypis trichas*; DICK, Dickcissel, *Spiza americana*; EABL, Eastern Bluebird, *Sialia sialis*; FISP, Field Sparrow, *Spizella pusilla*; GCWA, Golden-cheeked Warbler, *Setophaga chrysoparia*; GWWA, Golden-winged Warbler, *Vermivora chrysoptera*; GRCA, Gray Catbird, *Dumetella carolinensis*; INBU, Indigo Bunting, *Passerina cyanea*; MARW, Marsh Wren, *Cistothorus palustris*; NOCA, Northern Cardinal, *Cardinalis cardinalis*; OVEN, Ovenbird, *Seiurus aurocapilla*; RWBL, Red-winged Blackbird, *Agelaius phoeniceus*; WEWA, Worm-eating Warbler, *Helmitheros vermivorum*. (Figure from Jones et al. 2020; used with permission of the U. S. National Academy of Sciences)

**Box 19.4 Time of Day When Young Fledge**

Investigators studying altrical birds have often, but not always (e.g., Cordilleran Flycatchers, *Empidonax occidentalis*; Darrah and van Riper 2021), found that nestlings fledge early in the day, typically within six hours after sunrise (e.g., Johnson et al. 2004, 2013; Pietz et al. 2012; Chiavacci et al. 2015; Santema et al. 2021; Jones et al. 2022). Hypotheses proposed to explain fledging early in the day include: (1) the maximum time hypothesis posits that young benefit by fledging early in the day because they then are more likely to avoid nest predation (Jones et al. 2022), have more time to find a safe location before dusk (Johnson et al. 2004), familiarize themselves with their surroundings before dark (Perrins 1979), or all of these, and (2) the threshold size (or development) hypothesis predicts that nestlings are more likely to fledge early in the day because many are likely to reach the developmental stage where the benefits of leaving the nest exceeds that of remaining (Santema et al. 2021). The results of some studies have provided support for the maximum time hypothesis (Chiavacci et al. 2015; Radersma et al. 2015), but others have not (Ribic et al. 2021; Santema et al. 2021). For example, based on a study involving six species of songbirds, Jones et al. (2022) concluded that "... songbirds fledge in the morning to avoid peak times of nest predation risk that occur later in the day (~8 h after civil dawn)." In addition, Radersma et al. (2015) found that Great Tits (*Parus major*) that fledged earlier in the day were more likely to recruit into local populations than those the fledged later in the day. However, for Eurasian Blue Tits (*Cyanistes caeruleus*), Santema et al. (2021) found that the time of day when nestlings fledged did not influence the probability of surviving.

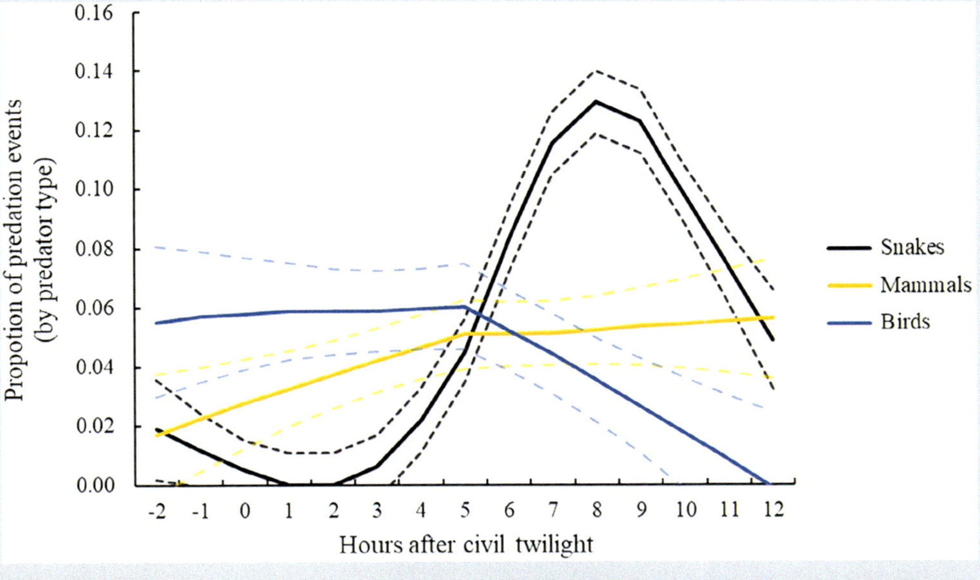


Daily patterns of nest predation by different predators (snakes, mammals, and birds) for six species of songbirds in Illinois, USA. With a peak of nest predation at eight hours after civil twilight, nestlings that fledge from nests earlier in the day are less likely to be predated. Dashed lines represent standard errors. (Figure from Jones et al. 2022; © 2022 British Ornithologists' Union, used with permission)

(continued)

**Box 19.4** (continued)

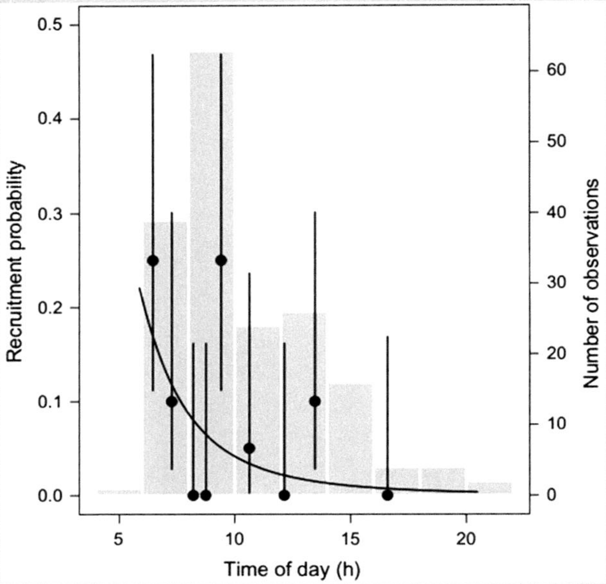


Local recruitment (dots with error bars and left axis) declined with time of day of fledging of Great Tits (*Parus major*). Error bars represent the 95% confidence intervals. Gray bars indicate the number of young (out of 179 fledglings from 26 broods; right axis) that fledged at different times of day. (Figure from Radersma et al. 2015; © 2014 British Ornithologists' Union, used with permission)

Santema et al. (2021) did find that most Eurasian Blue Tits fledged early in the day (84% of 1325 young fledged before solar noon, i.e., midpoint between sunrise and sunset), but that the time of fledging was related to the developmental stage (body mass) of nestlings. These authors did note, however, that their results were "not necessarily generalizable across species" and that a variety of factors, such as "the environmental conditions experienced by young birds" influence the likelihood of fledgling survival. Similarly, Ribic et al. (2021) concluded that ". . . as more researchers investigate processes that might affect the time of day that nestlings fledge in a variety of bird species . . ., it is becoming clearer that there are no universal answers—no single set of drivers appears to apply to all birds." Clearly, several factors can potentially influence the time of fledging by altricial young, including, but not limited to, developmental stage, pre- and post-fledging predation risk, and the behavior of parents and siblings (Santema et al. 2021).

(continued)

**Box 19.4** (continued)

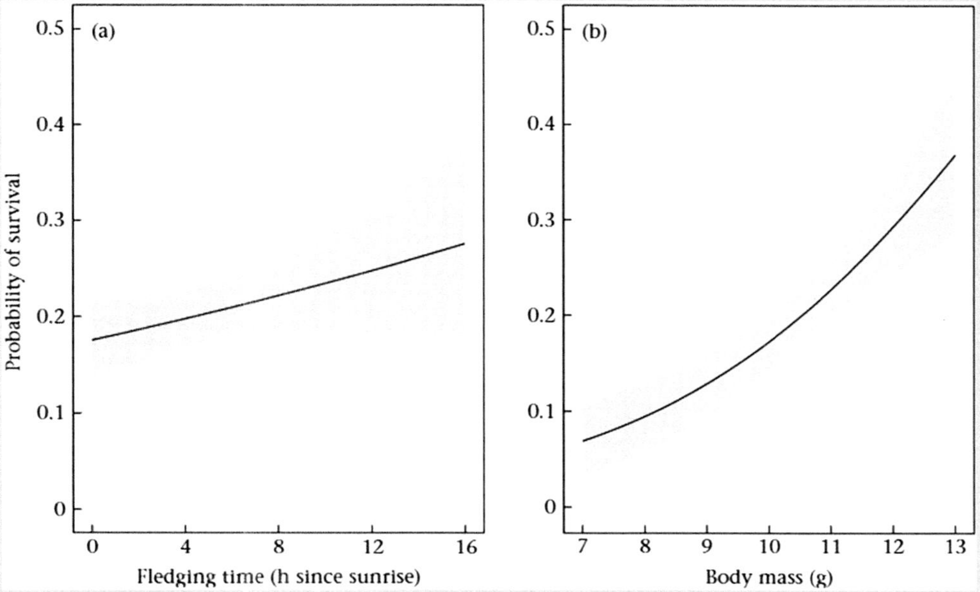


Relationships between the probability of survival of Eurasian Blue Tits until fall and (**a**) time of fledging and (**b**) body mass at day 14 post-hatching. The relationship between time of fledging and probability of survival was not significant ($P = 0.14$), whereas the relationship between body mass and probability of survival was significant ($P < 0.001$). (Figure from Santema et al. 2021; © 2021 The Authors. Published by Elsevier Ltd on behalf of The Association for the Study of Animal Behaviour, used with permission)

their wing development. Wing development is likely important because fledglings generally have high mortality rates just after leaving nests (Naef-Daenzer et al. 2001; Rush and Stutchbury 2008) and better-developed wings may better allow fledglings to find cover and elude predators.

When threatened by a predator, nestlings sometimes fledge prematurely, or before they have reached the appropriate stage of development (e.g., Reidy et al. 2008). The tendency for nestlings to fledge prematurely is negatively correlated with daily predation rate, i.e., nestlings are less likely to fledge prematurely in species where rates of nest predation are high. Species that experience higher rates of nest predation generally have shorter nestling periods (Box 19.5 Effects of Predation Rate, Nest Height, and Latitude on Fledging Age) and, because young in such species normally fledge at earlier stages of development than young in species with lower predation rates, they are unable to leave nests much earlier than their normal fledging age. Species that experience lower rates of nest predation tend to have longer nestling periods and young remain in nests to later stages of development. Nestlings, therefore, may be able to leave nests several days earlier than normal (Remeš and Martin 2002).

The tendency of older nestlings to prematurely fledge when disturbed by a potential predator is well known among ornithologists who, to minimize the chances of premature fledging, generally avoid approaching nests or handling young during the days just prior to the anticipated fledging

**Box 19.5 Effects of Predation Rate, Nest Height, and Latitude on Fledging Age**

Several ecological and evolutionary factors can potentially influence nestling growth rates and the age at which nestlings fledge (Arendt 1997; Mainwaring and Hartley 2012). To better understand the relatively importance of some of these factors, Remeš et al. (2020) reviewed the literature and obtained data concerning growth patterns and fledging ages for 231 species of songbirds from studies conducted on six continents and across latitudes. Analysis of these data revealed that nestlings fledged at significantly younger ages among species with higher rates of nest predation, nests located closer to the ground, and that breed at higher latitudes. In contrast, the number of care-givers, clutch size, and type of foraging substrate had little effect on fledging age.

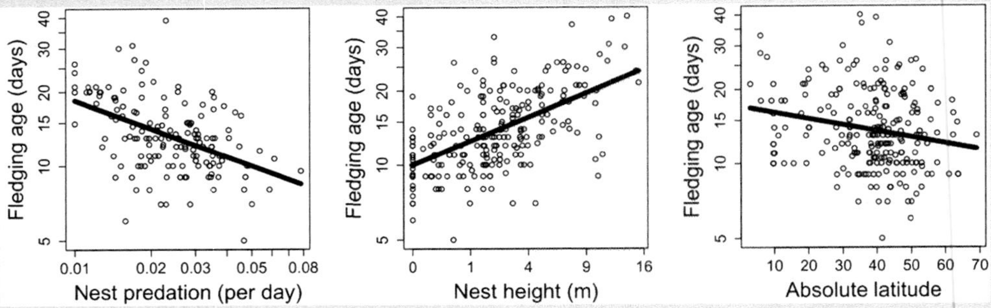


Nestlings fledge earlier among species with higher rates of nest predation, nests closer to the ground, and that breed at higher latitudes. (Figure from Remeš et al. 2020; open-access article licensed under a Creative Commons Attribution 4.0 International License, https://creativecommons.org/licenses/by/4.0/)

Other investigators have also reported earlier fledging by young among species of songbirds with higher rates of nest predation (e.g., Remeš and Martin 2002; Roff et al. 2005; Martin et al. 2011). Altricial young cannot leave nests until they achieve some minimal level of mobility and can thermoregulate, but higher rates of nest predation will tend to favor fledging by young as close to that point as possible.

(continued)

**Box 19.5** (continued)

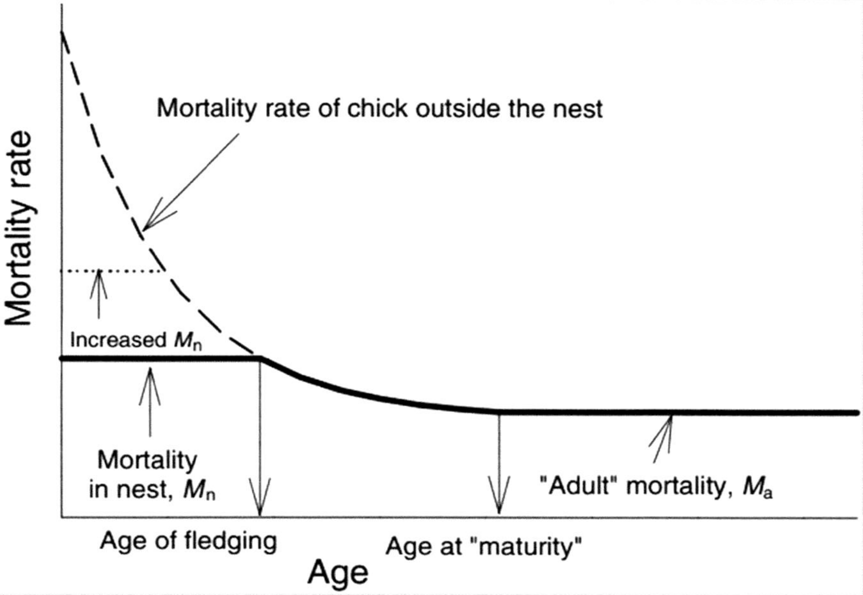


Relationship between the mortality rate for altricial young inside ($M_n$) and outside of nests and nestling age. The mortality curve for young outside nests is hypothetical and based on the assumption that altricial young cannot leave nests until they are reasonably mobile and can thermoregulate. Dotted line illustrates that an increased rate of nest mortality (predation) should result in an earlier age of fledging. (Figure modified from Roff et al. 2005; © European Society for Evolutionary Biology, used with permission)

Remeš et al. (2020) also found that young fledged earlier among species of songbirds that nest closer to the ground. One possible explanation for such results is that, depending on the community of predators present (Conkling et al. 2012), predation rates may be higher for nests closer to the ground (e.g., Colombelli-Negrel and Kleindorfer 2009, but see Remeš et al. 2012). In addition, young in nests at greater heights above ground may risk injury or death if they fledge before their wings are not "fully functional" (Remeš et al. 2020); young in nests closer to or on the ground face no such risk so can fledge at earlier stages of development.

Altricial young also tend to fledge earlier at higher latitudes, especially among north temperate species (Remeš et al. 2020). One possible explanation for this is that breeding seasons are necessarily shorter at higher latitudes so selection may favor faster growth and earlier fledging than at lower latitudes with longer breeding seasons (Remeš et al. 2020). In support of this hypothesis, within-species studies have revealed faster growth of young and shorter nestling periods for populations at higher latitudes than those at lower latitudes (Rose and Lyon 2013; Sofaer et al. 2018).

date. Adults may also induce premature fledging by uttering alarm calls that indicate a predator is approaching (e.g., Kleindorfer et al. 1996). In some species, nestlings are even able to recognize different alarm calls given by adults in response to different types of predators. For example, Great Tits (*Parus major*) nest in cavities and nestlings were found to crouch down in cavities in response to alarm calls given by adults when Jungle (Large-billed) Crows (*Corvus macrorhynchos*) approached, but left cavities in response to alarm calls given in response to the presence of Japanese rat snakes (*Elaphe climacophora*; Suzuki 2011). These instinctive responses by nestlings are beneficial because crows attempt to take nestlings by reaching through the cavity entrance with their bills and crouching young would be more difficult for crows to reach. Snakes, however, can enter cavities so leaving the cavity is the only way for nestlings to avoid predation (Suzuki 2011). Nestlings that fledge prematurely are typically smaller and less mobile than those that do not fledge prematurely and may, therefore, be less likely to survive, but the alternative is almost certain predation.

In general, slow-developing broods and young stay in nests longer than faster-developing broods and young. Because nest predation is the primary cause of nest failure for birds, shorter nestling periods may improve the likelihood of nest success. For example, if the probability of daily survival of nests is 0.95, then a nest where young fledge in 8 days is 10% more likely to be successful (i.e., the nest is not predated and young fledge) than a nest where young fledge in 10 days (Stodola et al. 2010). Because growth rates of nestlings are influenced by the amount of food their parents provide, the quality of both parents and their breeding territories (food availability) can be critically important in determining if nests succeed or fail.

Among precocial species, all young typically leave nests at the same time. For example, all young Wood Ducks (*Aix sponsa*) in a brood leave nest cavities on the day after hatching (Leopold 1951). At the time of nest departure, female Wood Ducks utter calls from a nearby limb or from the ground near the base of the nest tree. In response, the ducklings begin calling and, typically one by one, each climbs to the entrance hole (typically about 7 m above ground, but sometimes as high as 17 m or more; Hepp and Bellrose 2020) and then jumps and, with small wings extended, falls to the ground, often bouncing up several centimeters. The time needed for an entire brood to exit a cavity varies with brood size and with how easily young can reach the cavity entrance, but generally ranges from 5 to 10 min (Leopold 1951). After the last nestling has left the nest, female Wood Ducks lead their young to a nearby body of water (Leopold 1951).

Departure from nests by young of ground-nesting precocial species occurs in a similar way. Young Wild Turkeys (*Meleagris gallopavo*), for example, can take a few steps within six hours of hatching, and are capable of walking and running within 12–24 h. Young leave nests a day or two after hatching, following as the adult female calls and walks slowly away from the nest (McRoberts et al. 2020).

Among altricial songbirds, the time needed for nest departure by entire broods varies, but is often less than 24 h (e.g., Kuzniak 1967; Vanderhoff et al. 2020). Among Tree Swallows (*Tachycineta bicolor*), the time needed for all nestlings in a brood to fledge can range from 30 s to 48 h (mean = 12 h; Michaud and Leonard 2000). Similarly, broods of Great Tits (*Parus major*) fledge over periods ranging from 7 min to 44 h (mean = 136 min; Radersma et al. 2011), and broods of Eurasian Blue Tits (*Cyanistes caeruleus*) over periods ranging from 13 min to 4 days (Schlicht et al. 2012). Among larger cavity-nesting species, nestlings may fledge over a period of several days, e.g., 1–5 days for Eastern Screech-Owls (*Megascops asio*, Gehlbach 1994), 2–4 days for Boreal Owls (*Aegolius funereus*, Korpimaki 1984), and up to 3–4 days for Pileated Woodpeckers (*Dryocopus pileatus*, Bull and Jackson 2020). After leaving nests, young birds rarely return. However, young hawks and eagles often leave nests, perching in nearby branches and sometimes returning to nests to be fed by adults (e.g., Matray 1974; Lyons and Mosher 1987).

In many altricial species of birds, nestlings lose mass in the days prior to fledging. Such mass recession has been reported among species in several orders of birds, including Procellariiformes, Pelecaniformes, Psittaciformes, Apodiformes, and Passeriformes. Investigators have provided a number of possible explanations for pre-fledging mass recession, including: (1) a reduction in adult provisioning rates near the end to the nestling period, (2) an increase in activity levels (and energy consumption) of nestlings as they approach fledging age, and (3) selection for an optimal fledging weight (i.e., losing mass is necessary for efficient flying or diving by young after leaving the nest). For some aerial insectivores, such as swifts, where adults provide little or no post-fledging care (Lack 1956) and young must be able to feed themselves after leaving the nest, mass loss prior to fledging reduces the energetic costs of flight and optimizes aerodynamic performance (Fig. 19.20; Martins 1997; Wright et al. 2006). For chicks in many species of seabirds (Procellariiformes), mass recession prior to leaving the nest results from a reduction in parental provisioning rates (e.g., Phillips and Hamer 2000; Harding et al. 2004). Lighter young may be better able to fly from their nests to the sea, e.g., for cliff-nesting species such as Rhinoceros Auklets (*Cerorhinca monocerata*), and, once there, better able to capture prey (Deguchi and Watanuki 2005). In addition, however, chicks being fed at lower rates may benefit by departing to sea to begin self-feeding rather than risking losing additional mass by remaining at nests (Gjerdrum 2004; Corbel et al. 2009).

Morbey et al. (1999) suggested that prefledging mass recession in Cassin's Auklet (*Ptychoramphus aleuticus*) chicks, and perhaps chicks in other seabirds, occurs as a byproduct of parents and chicks evaluating the costs and benefits of their decisions. In other words, parents evaluate the costs and benefits of provisioning visits and, beyond a certain nestling age and

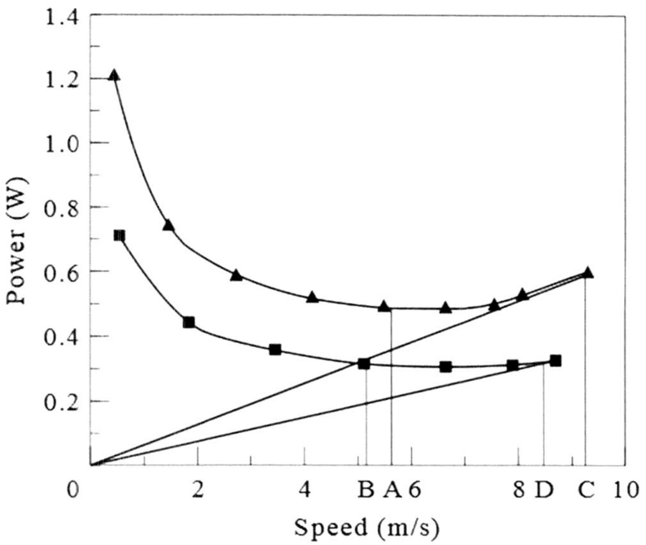


**Fig. 19.20** Power curves for nestling Common Swifts (*Apus apus*) fledging at relatively high (54.5 g, triangles) and low (41.4 g, squares) body masses. Along the *x*-axis, A and B are the minimum power speeds for the heavier and lighter fledglings, respectively. C and D are the maximum range speeds for the heavier and light fledglings, respectively. The difference in mass results in an increase of flight efficiency (lift:drag ratio) of 9.6% for the lighter nestling, suggesting that the gain in flight efficiency may be more important for young swifts than the loss of flight speed. (Figure from Martins 1997; © 1997 The Association for the Study of Animal Behaviour. Published by Elsevier Ltd., used with permission)

mass, the costs of such visits (e.g., predation risk; Harfenist and Ydenberg 1995) outweigh the benefits of providing some additional food for those nestlings. As a result, adults make fewer visits and chicks lose mass. Chicks then assess the costs and benefits of remaining at nests. Chicks that stay may benefit by being fed in the relative safety of the nest, but, if parents fail to deliver food or do so infrequently, chicks lose the opportunity to forage for themselves at sea and continue to use their energy reserves. Thus, prefledging mass recession occurs as a result of the respective assessments of costs and benefits by parents and nestlings.

## 19.9 Parental Care After Young Leave Nests

With the exception of young megapodes (Megapodidae), parent birds continue to provide parental care after young leave nests. As noted earlier, parental care for precocial species (e.g., Anatidae, Charadriiformes, and Galliformes) may include brooding young until they are able to thermoregulate (5–10 days post-hatching), leading young to foraging areas, feeding young (only in a few species, e.g., oystercatchers, *Haematopus* spp., Working Group et al. 2020; Limpkin, *Aramus guarauna*, Bryan 2020; cranes, Gruidae, Gerber et al. 2020), defending foraging sites, and detecting, and protecting young from, predators. For altricial and semi-altricial species, young can thermoregulate by the time they leave nests so adults do not brood young. However, adults do feed young and protect them from predators.

Among precocial species, the time between hatching and fledging (when young can fly) ranges from about 2 to 10 weeks. This variation is due primarily to size, with young in smaller species able to fly at a younger age than those of larger species. Among shorebirds (Charadriiformes), for example, there is a positive relationship between species mass and fledging age, with young fledging in as little as 14–18 days after hatching in smaller species such as small sandpipers and taking up to 35 days for larger species such as lapwings (Colwell 2010). Among other precocial species, young grebes (Podicipediformes) are typically capable of flight when about 6–10 weeks old (e.g., Stedman 2020; LaPorte et al. 2020), young galliforms (Galliformes when about 2–4 weeks old (with young in most species able to fly at least weakly when 10–14 days old) (e.g., Giudice and Ratti 2020; McRoberts et al. 2020), and young waterfowl (Anseriformes) when 5–10 weeks old (e.g., Drilling et al. 2020).

In many precocial species, parental care ends when young are able to fly, e.g., several species of shorebirds including Killdeer (*Charadrius vociferus*; Jackson and Jackson 2020), Least Sandpipers (*Calidris minutilla*; Nebel and Cooper 2020), Ruddy Turnstones (*Arenaria interpres*; Nettleship 2020), and others. However, some galliform young remain with parents for longer periods. For example, young Northern Bobwhites (*Colinus virginianus*) typically remain with parents through late winter (Brennan et al. 2020), and young Wild Turkeys (*Meleagris gallopavo*) often remain with female parents until late fall or early winter (young males) or until the following spring (young females) (McRoberts et al. 2020). The duration of parental care varies among different species of waterfowl. For some species, parental care ends when young can fly (e.g., Mallards, *Anas platyrhynchos*; Drilling et al. 2020). However, for many species of waterfowl that breed at higher latitudes, parental care continues through fall migration (e.g., Canada Goose, *Branta canadensis*; Mowbray et al. 2020b), and sometimes even through the following spring migration (e.g., Snow Goose, *Anser caerulescens*; Mowbray et al. 2020a, b).

After leaving nests, mortality rates of young birds can be high because they are not yet adept at flying (and precocial young cannot fly at all when they leave nests), foraging, detecting and evading predators, and other important life skills (e.g., Marchetti and Price 1989; Sullivan 1989; Yoda et al. 2004). For example, among young Great (*Parus major*) and Coal (*Periparus ater*) tits, 47% died during the first 20 days after leaving nests, with predation the primary cause of mortality and few young dying of starvation (Naef-Daenzer

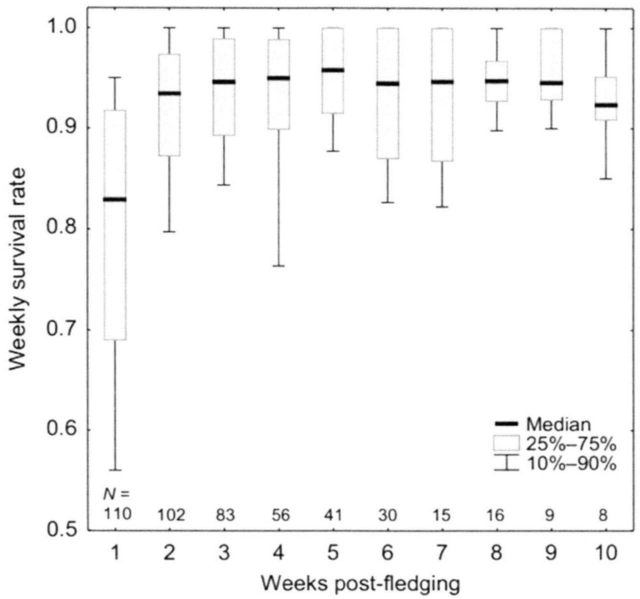


**Fig. 19.21** Estimates of weekly post-fledging survival for 60 species of birds with altricial young. $N$ = the number of point estimates available per time interval. Bars indicate median weekly survival, boxes and whiskers indicate 25–75% and 10–90% ranges, respectively. (Figure from Naef-Daenzer and Grüebler 2016; © 2016 Swiss Ornithological Institute. Journal of Field Ornithology published by Wiley Periodicals, Inc. on behalf of Association of Field Ornithologists, used with permission)

et al. 2001). Vitz and Rodewald (2011) reported that 35% of young Ovenbirds (*Seiurus aurocapilla*) died within 51 days after leaving nests and 33% of young Worm-eating Warblers (*Helmitheros vermivorum*) died within 31 days after leaving nests. Similar mortality rates (46–67%) have been reported for several species of grassland birds (Yackel Adams et al. 2006; Suedkamp Wells et al. 2007; Berkeley et al. 2007). Naef-Daenzer and Grüebler (2016) searched the literature for information about post-fledging survival of altricial species of birds and found that the median survival rate during the first week after fledging was 0.83, but increased to 0.95 for week four (Fig. 19.21). The most frequently reported cause of fledgling mortality was predation (Fig. 19.22).

Parental care can help reduce mortality rates of young birds after they leave nests (Lind 2004), and young birds can use the behavior of adults as guides in developing important skills (Galef and Laland 2005). Other important factors that can

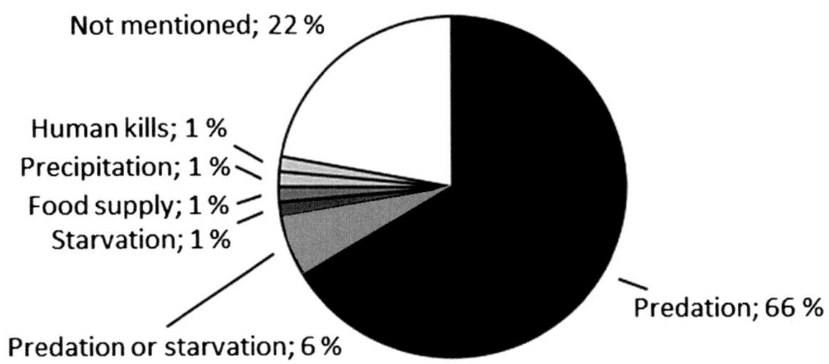


**Fig. 19.22** Frequency of different causes of post-fledging mortality. Predation, or predation or starvation, were major causes of mortality ($N$ = 68 papers). (Figure from Naef-Daenzer and Grüebler 2016; © 2016 Swiss Ornithological Institute. Journal of Field Ornithology published by Wiley Periodicals, Inc. on behalf of Association of Field Ornithologists, used with permission)

influence mortality rates after young birds leave nests are fledging date, habitat quality and, for altricial and semi-altricial species, the condition of young (i.e., body mass and relative wing length) when they leave nests. For example, Naef-Daenzer et al. (2001) found that survival of fledgling Great (*Parus major*) and Coal (*Periparus ater*) tits was positively correlated with mass at fledging, and that young that fledged earlier in the breeding season had higher survival rates than those that fledgling later in the breeding season. Similarly, size-corrected mass of juvenile Seychelles Warblers (*Acrocephalus sechellensis*) was positively associated with likelihood of survival to adulthood and lifespan (Brown et al. 2022). More generally, several investigators have determined that body mass of juvenile birds is positively associated with the likelihood of survival (Ronget et al. 2018) and indicative of long-term individual quality (Gleeson et al. 2005; Sakaluk et al. 2014) and adult fitness (Cooper and Kruuk 2018). Greater body mass at fledging may enhance survival of young birds that, being inexperienced foragers, are at greater risk of starvation (Jones et al. 2017). In addition, birds that are heavier may have a competitive advantage that results in occupation of better-quality habitats (Verhulst et al. 1997; Both et al. 1999; Van De Pol et al. 2006). More generally, as noted by Brown et al. (2022), ". . . high juvenile body mass is indicative of better natal socioenvironmental conditions, which subsequently leads to higher survival."

Relative wing length at fledging as a proportion of adult size can also influence survival rates, with fledglings with relatively longer wings having higher survival rates than those with shorter wings (Morrison et al. 2009; Grüebler and Naef-Daenzer 2010a, b; Martin et al. 2018). Longer wings improve flight ability and reduce the likelihood of fledglings being captured by a predator (Martin et al. 2018). Vitz and Rodewald (2011) studied two species of forest-dwelling wood warblers (Ovenbirds, *Seiurus aurocapilla*, and Worm-eating Warblers, *Helmitheros vermivorum*) and found that survival rates were higher for fledglings that were in better condition at fledging (with condition estimated based on body mass relative to body size) and those that used areas with denser understory vegetation (Fig. 19.23). Young in better energetic condition when leaving nests may have higher survival rates because, if food is limiting, they should be better able to avoid starvation. In addition, young birds in better condition may be more adept at evading predators, foraging, or procuring food from their parents (Vitz and Rodewald 2011). Naef-Daenzer et al. (2001) also suggested that fledglings in poorer condition may beg and move more, potentially making them easier to detect by predators.

Fledging date has also been found to influence fledgling survival for species of birds at higher latitudes, with young fledging later in the breeding season often having lower survival rates than those that fledge earlier in the breeding season (e.g., Tarof et al. 2011; Maness and Anderson 2013; Lombardo et al. 2020; Fig. 19.24). For Purple Martins (*Progne subis*), Tarof et al. (2011) suggested that young fledging later in the breeding season may have lower survival rates because they have less time to forage and less experience prior to initiating migration. Other investigators have suggested that lower survival rates of young that fledge later in the breeding season may be due to seasonal declines in food availability (Verboven and Visser 1998; Wheelwright et al. 2003). For some species, fledging date does not appear to influence fledgling survival rates, e.g., Wood Thrushes (*Hylocichla mustelina*, Vernasco et al. 2018) and Barn Swallows (*Hirundo rustica*, Evans et al. 2020), possibly because resource availability or predator behavior do not exhibit seasonal variation or such variation is not sufficient to influence fledgling survival.

Habitat characteristics and quality can also influence survival rates of young, with survival more likely in habitats that provide cover and increased food availability. For example, many forest-breeding species of birds have been found to move their young into areas with higher densities of wood stems in the understory (e.g., riparian thickets, forest edges, tree-fall gaps, and regenerating clearcuts; Vega Rivera et al. 1998; King et al. 2006; Rush and Stutchbury 2008; Vitz

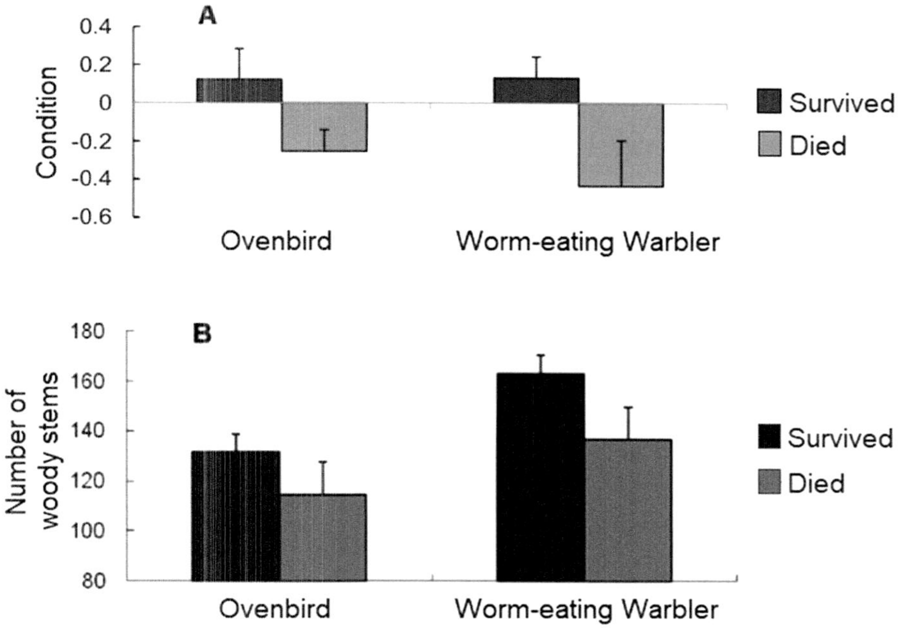


**Fig. 19.23** (**a**) Mean energetic condition (± SE; positive values indicate better condition) and (**b**) mean number (± SE) of woody stems at locations of fledging of Ovenbirds (*Seiurus aurocapilla*) and Worm-eating Warblers (*Helmitheros vermivorum*) that either survived or died during the post-fledging period. (Figure from Vitz and Rodewald 2011; © 2011 Oxford University Press, used with permission)

and Rodewald 2011), likely because such areas minimize predation rates by providing protective cover and also provide ample food resources such as fruit (e.g., Vitz and Rodewald 2007). Food availability was also found to influence the movements of Great Tits (*Parus major*), with pairs in lower-quality territories making daily movements with their fledglings from their territories to areas with greater availability of food (van Overveld et al. 2011). In many species of waterfowl, adults move their precocial young to areas that provide both food and protection from predators. For example, Arctic-nesting adult Snow Geese (*Anser caerulescens*) move broods to wetland habitats that provide an abundance of food (forage plants) and serve as refuges against terrestrial predators such as Arctic foxes (*Alopex lagopus*; Mainguy et al. 2006).

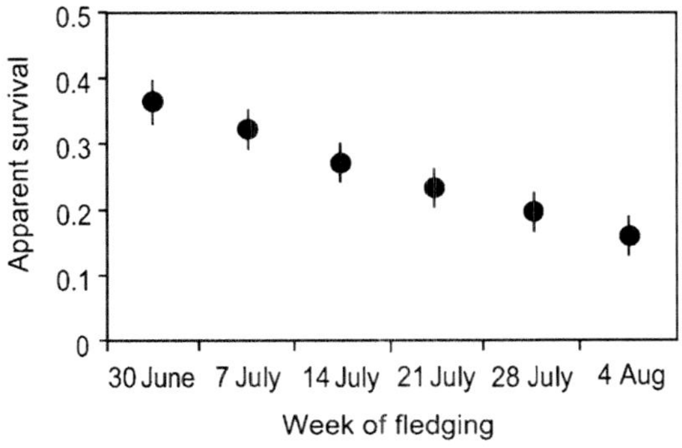


**Fig. 19.24** Apparent survival of young Purple Martins (*Progne subis*) was lower for those that fledged later in the breeding season. (Figure from Tarof et al. 2011; © 2011 Oxford University Press, used with permission)

In some species, parents divide broods after young fledge, with some young cared for by the adult male and others by the adult female. Such brood division has been reported for a wide variety of songbirds and may be more likely in species that are single-brooded (raise only one brood per nesting season, e.g., Yackel Adams et al. 2001) or, for species that have more than one brood per nesting season, in the second or last broods of the season (e.g., Vega Rivera et al.

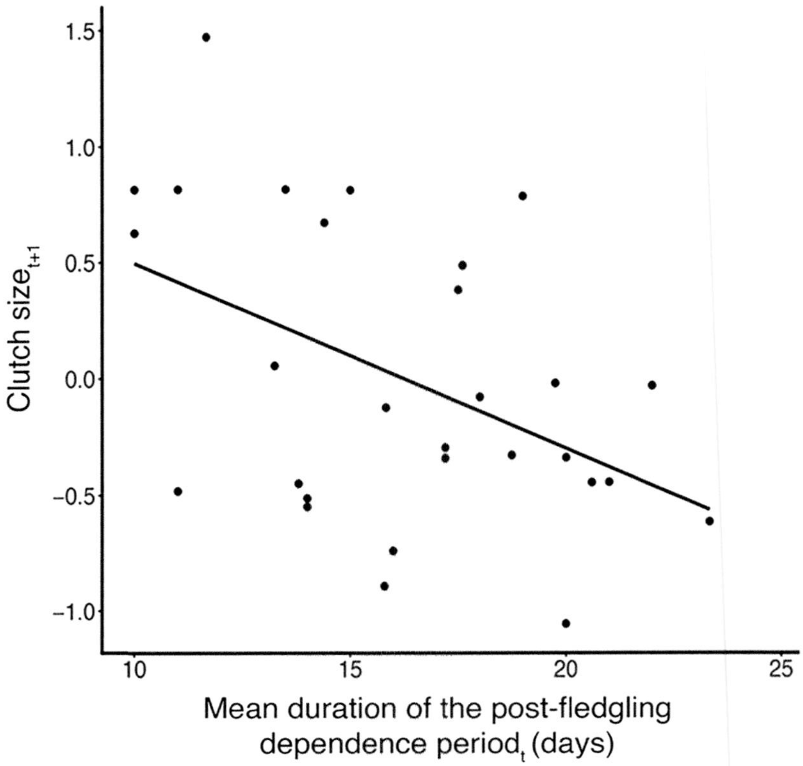


**Fig. 19.25** Adult Eurasian Kestrels (*Falco tinnunculus*) that spent more time caring for fledglings (post-fledgling dependence period) in a preceding breeding season ($t$) had smaller clutches in the following breeding season ($t$ + 1). Values for clutch size were determined using a model that included possible effects of year (e.g., differences in prey availability) and age of adults. (Figure from López-Idiáquez et al. 2018; © 2018 López-Idiáquez et al., open-access article distributed under the terms of the Creative Commons Attribution License, https://creativecommons.org/licenses/by/4.0/)

2000). In these latter species, males may provide more care for fledglings after initial or early nests while females initiate another nesting attempt. Division of broods after fledging may help reduce predation rates by dispersing young spatially and reducing the risk of losing an entire brood, and may also increase foraging efficiency by allowing parents to concentrate on fewer offspring, potentially reducing search and travel times between offspring and facilitating the distribution of food among offspring (Smith 1978; Moreno 1984). Alternatively, for species where one sex typically provisions young more than the other, division of broods may be driven by the parent that provides more care (Lessells 1991). This hypothesis is based on the assumption that the parent that provisions less benefits from undivided broods, whereas the parent that provisions more benefits from divided broods. Reasons for post-fledging division of broods remain unclear for most species and, as such, additional study will be needed to better understand the selective factors that favor (or not) such behavior.

Extending the duration of parental care after young leave nests is beneficial because it can increase the likelihood of young surviving (e.g., Grüebler and Naef-Daenzer 2010a). However, prolonging parental care can also be costly because adults may have less time to forage themselves and for other self-maintenance activities (e.g., preening and resting or sleeping; Earnst 2002), which may negatively impact future fecundity. For example, López-Idiáquez et al. (2018) reported a negative association between the mean duration of parental care of fledgings in a preceding breeding season and clutch size in the following breeding season in Eurasian Kestrels (*Falco tinnunculus*; Fig. 19.25). In addition, for multi-brooded species, more time and energy invested in caring for one brood means less time to invest in another brood. For adult birds, therefore, there is a trade-off in terms of allocating effort to young

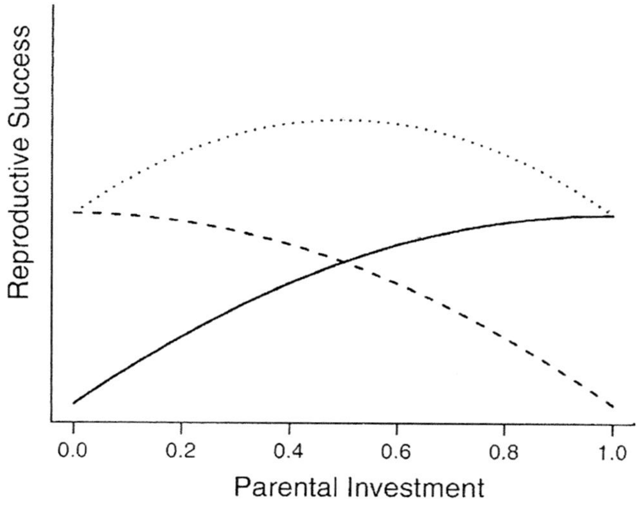


**Fig. 19.26** Trade-off between current and future reproductive success. The solid line is the current reproductive success with increasing investment in current offspring. The dashed line is the expected future reproductive success with increasing investment in current offspring. The dotted line is the sum of both lines and equals expected lifetime reproductive success. The level of investment that maximizes lifetime reproductive success occurs where the benefit of continuing investment in current offspring is offset by the cost of such investment to future reproductive success. (Figure from Jonker et al. 2012; © 2011 Springer Science Business Media B. V., used with permission)

versus allocating effort to another brood or to themselves. The level of investment that maximizes lifetime reproductive success occurs at the point where the benefit of further investment in the current reproductive attempt is offset by the cost in terms of expected future reproductive success (Fig. 19.26). This point varies among species and, as a result, the duration of such care ranges from as little as about 10 days for some passerines to well over a year (Russell 2000). For example, the post-fledging dependence period for young Harpy Eagles (*Harpia harpyja*) has been estimated to last 27 months (Alvarez-Cordero 1996).

Some variation in the duration of post-nest parental care is due to differences in body size, with the duration of such care typically longer for larger species where development of young may take longer. However, another important factor is a species' life-history strategy (Fig. 19.27). In general, birds with shorter lifespans have larger clutch sizes, more clutches, and shorter periods of post-nest parental care, whereas birds with longer lifespans have smaller and fewer clutches (often just one per breeding season) and a longer period of post-nest parental care. Species of birds on the fast end of the continuum produce more young per breeding attempt and are often multibrooded. However, as a trade-off, these species generally provide less post-nest parental care and, as a result, mortality rates of independent young are typically high. At the other end of the continuum, slow species produce fewer young, invest more time and energy in those young, and, as a result, mortality rates of independent young are typically low. Interestingly, many species of birds that breed in the northern hemisphere lie on the fast end of the life-history continuum and, accordingly, typically provide limited parental care after young leave the nest. Russell (2000) compared the duration of post-nest parental care for 347 species that breed in either north hemisphere temperate regions or in the southern hemisphere and tropical regions and found that, for most species in the northern hemisphere, young birds become independent of parents less than 20 days after leaving nests (Fig. 19.28). By contrast, parental care for most southern hemisphere and

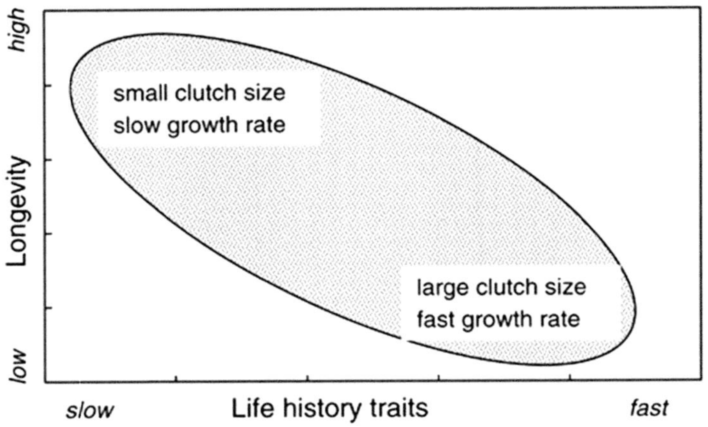


**Fig. 19.27** The slow-fast life-history continuum. Birds on the slow end of the continuum have a low annual reproductive output and lower mortality rates and typically have longer periods of post-nest parental care; birds on the fast end have higher rates of reproduction and mortality and shorter periods of post-nest parental care. (Figure from Williams et al. 2010; © 2010 Oxford University Press, used with permission)

tropical species typically extends well beyond 20 days after young leave nests and, even after becoming independent, young typically remain in parental territories for several months (Fig. 19.29).

Although annual survival rates of birds in tropical and north temperate areas differ, those differences are not as great as once thought (perhaps averaging about 10% higher in tropical regions) and distributions of annual survival of

**Fig. 19.28** Time until fledged young are independent of adults for species that breed in northern hemisphere temperate regions (NHTE) and in southern hemisphere and tropical regions (SHTR). (reproduced from Russell (2000) with permission from CSIRO Publishing)

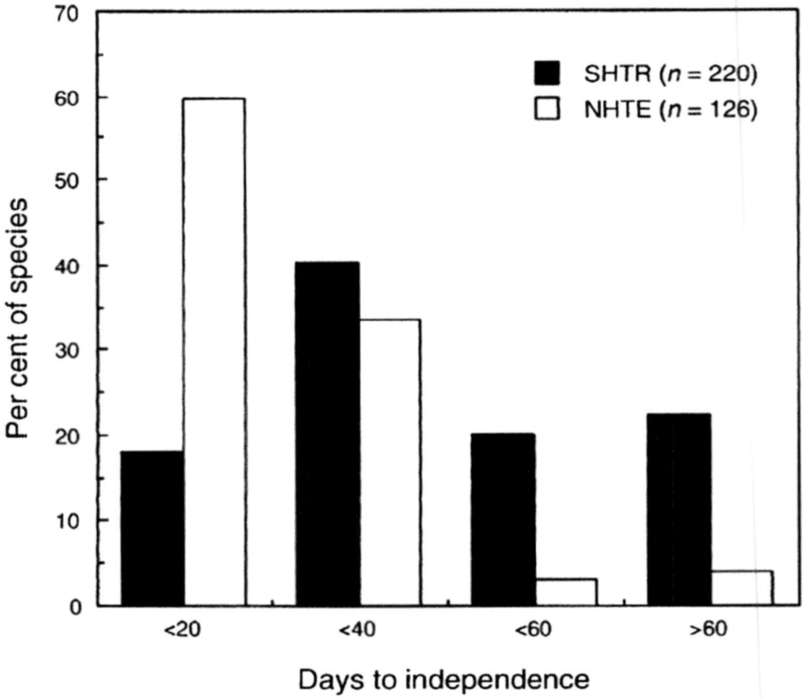

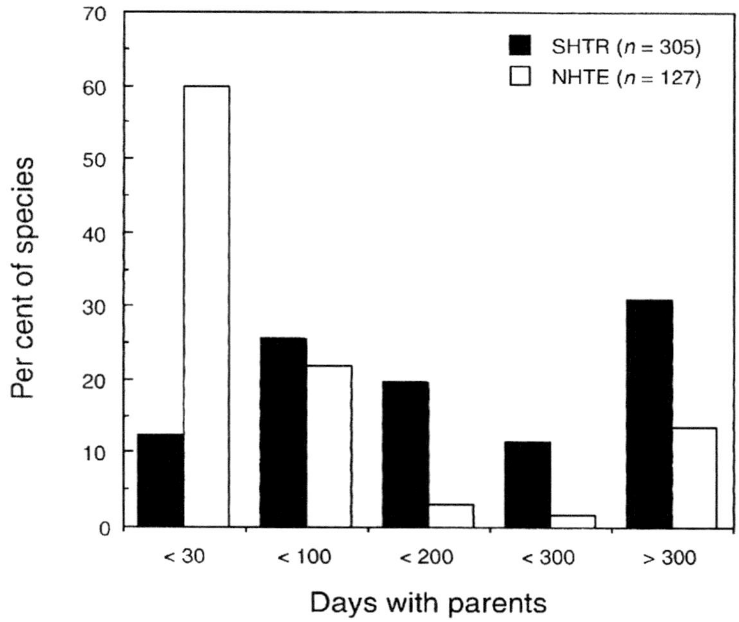


**Fig. 19.29** Time spent by independent young in parental territories or with parental groups after fledging for species that breed in northern hemisphere temperature regions (NHTE) and in southern hemisphere and tropical regions (SHTR). The category > 300 days includes cooperative breeding species where young may stay with their family through the next breeding season and even longer. (reproduced from Russell (2000) with permission from CSIRO Publishing)

birds in these areas exhibit much overlap (Robinson et al. 2010). This suggests that factors other than lifespan must also influence differences between tropical and temperature regions in the duration of parental care. One result of these differences in parental care is that tropical species tend to produce higher-quality young, whereas temperate species produce more young (Stearns 1992; Robinson et al. 2010). These contrasting strategies have likely been influenced by latitudinal variation in environmental conditions. For temperate species, survival of young is often influenced by unpredictable conditions during the winter (e.g., severe weather) for resident species and by conditions during migration (e.g., weather and wind conditions) and in wintering areas (e.g., habitat quality) for migratory species. As a result, producing more young may enhance parental fitness (Robinson et al. 2010). In contrast, in less seasonal, tropical environments, offspring quality may be more important for parents because populations may be near carrying capacity and their young must compete for available breeding territories (Robinson et al. 2010).

## 19.10 Natal Dispersal

The duration of parental care after young leave nests varies, but young birds, with few exceptions, eventually leave their natal areas. This movement of young birds from the site of hatching to the site where, if they survive, they will reproduce is called natal dispersal. This movement of young birds influences population dynamics, gene flow, and species' ranges (e.g., range expansions). Over long time periods, natal dispersal can also affect the process of diversification, potentially stimulating speciation if young cross geographical barriers and expand a species' range into new areas (Claramunt et al. 2012). However, natal dispersal can also help maintain genetic similarity among populations and inhibit speciation (Kisel and Barraclough 2010). Although important at the population level and potentially, over longer time periods, in the process of diversification, any such effects are ultimately driven by the natal dispersal of individual birds so it is important to understand the factors that can influence the dispersal behavior of young in different species of birds.

Natal dispersal is a complex process, with young birds potentially influenced by their genotype, phenotype, hatching date, natal environment, the quality of winter habitats (for migratory species where young select breeding sites when returning to breeding areas after wintering elsewhere), and the characteristics of areas they pass through after leaving natal territories. Among some species of birds, aspects of natal dispersal, e.g., propensity to disperse (Charmantier et al. 2011) or dispersal distances (Pasinelli et al. 2004), are heritable, whereas, in other species, investigators have found no such evidence (e.g., Briggs et al. 2012). Phenotypic traits that may influence dispersal behavior include a bird's sex, condition, and personality (i.e., behavioral syndrome). Characteristics of natal environments that can influence dispersal behavior include habitat quality and population and fledgling density and sex ratios. After leaving natal territories, factors that can potentially influence dispersal behavior include habitat configuration (e.g., patchiness or degree of fragmentation) and quality (e.g., food and nest site availability and predator densities) and population densities and sex ratios.

For young birds in many species, natal dispersal can be viewed as a three-stage process: departure, transience, and settlement (Baguette and Van Dyck 2007). When dispersing, young birds that are still learning to forage effectively and to identify and avoid predators must move through unfamiliar areas where, to survive, they must attempt to find food and avoid predators. Doing so can be challenging and, as a result, mortality rates of dispersing young can be high (Bonte et al. 2012) and correlated with dispersal distance, with young dispersing further having higher mortality rates (Brooker et al. 1999). Given the risks associated with natal dispersal, such behavior must also provide benefits. One such benefit is that dispersing from natal areas reduces the likelihood of inbreeding and competition with kin. Dispersing birds can also prospect for more favorable conditions (e.g., higher quality habitat and areas with increased availability and quality of possible mates; Matthysen 2012).

In most species of birds, young disperse within a few weeks or months after leaving nests. However, in some species, particularly cooperative breeders, young sometimes exhibit delayed dispersal, with young remaining in natal territories for up to several years, often assisting parents during breeding (e.g., Griesser and Barnaby 2010). Young in cooperative species may remain on natal territories as a result of ecological or demographic constraints, e.g., limited availability of suitable habitat or of possible mates (Hatchwell 2009). In other species of birds, young remain in natal territories for many months, but disperse before or near the time when parents initiate another breeding attempt (e.g., Green and Cockburn 2001; Kraaijeveld and Dickinson 2001). Such delayed dispersal is more common in tropical species, perhaps because tropical species typically invest more in fewer young (i.e., smaller clutches and broods) than those at higher latitudes to enhance offspring survival (Martin 1996; Russell et al. 2004).

The onset of natal dispersal (departure stage) can be influenced by internal or endogenous factors, external or environmental factors, or by an interaction between them. Studies of Eastern (*Megascops asio*; Ritchison et al. 1992) and Western (*M. kennicottii*; Belthoff and Dufty 1995) screech-owls have revealed that captive young exhibit increased locomotor activity during the period when, in the wild, they would have departed from natal territories. Such results suggest the presence of an endogenous (physiological) mechanism that triggers departure. Belthoff and Dufty (1998) suggested that this endogenous mechanism involves an increase in blood levels of corticosterone (secreted by the adrenal glands), a hormone that helps regulate availability of energy in the body and stimulates increased foraging and food intake (Holberton et al. 2007). Increased blood levels of corticosterone are also associated with behaviors involving increased activity levels such as irruptive migratory events (Wingfield 1994) and the migratory restlessness exhibited by migrating birds (e.g., Landys et al. 2004). A combination of factors may contribute to increased secretion of corticosterone as the time of departure approaches. Increased secretion

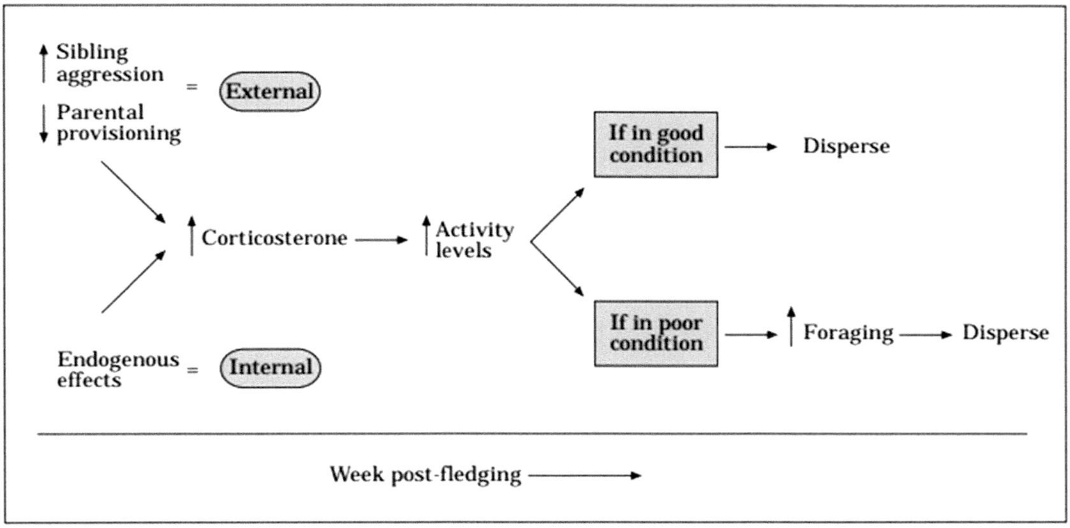


**Fig. 19.30** Possible relationships between corticosterone, locomotor activity, body condition, and dispersal of nonmigratory birds. (Figure from Belthoff and Dufty 1998; © 1998 The Association for the Study of Animal Behaviour. Published by Elsevier Ltd., used with permission)

of corticosterone during the period prior to when young would be dispersing even in captive birds fed daily and not exposed to external stimuli (Ritchison et al. 1992; Belthoff and Dufty 1998) suggests the existence of some endogenous or internal control mechanism (Fig. 19.30), possibly involving polymorphic genes that are also involved in the molecular signaling of circadian and circannual clocks (Chakarov et al. 2013). In addition, external factors such as stress induced by increased sibling aggression (e.g., for access to food from provisioning adults) or a reduction in provisioning rates by adults may induce further secretion of corticosterone (Fig. 19.30). Increased blood levels of corticosterone, in turn, trigger an increase in locomotor activity and, if young are in good condition (e.g., sufficient fat reserves), then they initiate dispersal (Fig. 19.30). The increased corticosterone levels in young in poorer condition stimulate an increase in foraging activity that, in turn, improves their condition such that, perhaps in a few days, they can also initiate dispersal (Fig. 19.30). Although this mechanism may be important in the dispersal behavior of young screech-owls and, perhaps, young Greater Flamingoes (*Phoenicopterus ruber*, Barbraud et al. 2003), additional studies are still needed to determine if a similar mechanism helps explain the onset of natal dispersal by young birds in other species.

Other factors that can potentially trigger the onset of natal dispersal include parental aggression, aggression of other conspecifics, termination of parental provisioning, aggression of young toward each other, declining food availability in the natal area, and even availability of light (Box 19.6 Moonlight Triggers Dispersal from Parental Territories). For example, late-dispersing young females are driven from natal territories by maternal aggression in Superb Fairywrens (*Malurus cyaneus*; Mulder 1995). Young Gray Jays (*Perisoreus canadensis*; Strickland 1991) and Siberian Jays (*P. infaustus*; Ekman et al. 2002) initiate natal dispersal due to the aggressive behavior of dominant siblings. Among colonial Lesser Kestrels (*Falco naumanni*), first-year birds migrate to wintering areas in Africa, but, upon returning to natal colonies the following spring, some are unable to obtain suitable nest sites (primarily holes or crevices in buildings in urban areas). Conspecifics that have nest sites respond aggressively (e.g.,

**Box 19.6 Moonlight Triggers Dispersal from Parental Territories**

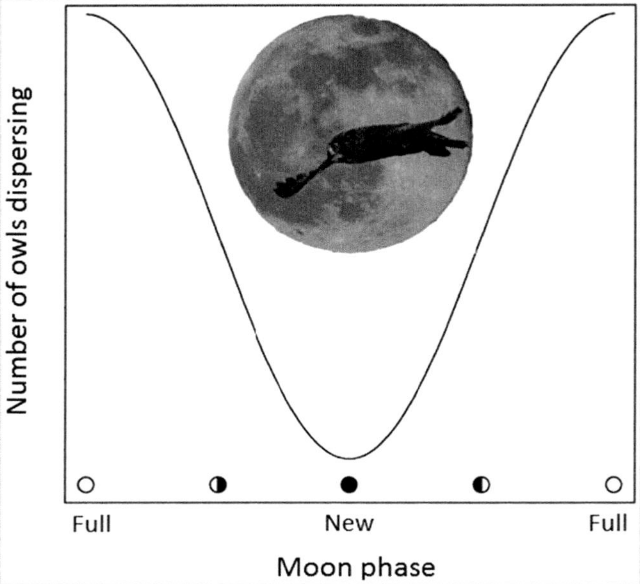


Most young Eagle Owls (*Bubo bubo*) disperse during nights with more moonlight. (Figure from Penteriani et al. 2014; © 2014 Springer-Verlag Berlin Heidelberg, used with permission)

Most young Eagle Owls disperse when they are about 5.5 months old. However, the specific timing of their dispersal is influenced by moon phase, with most young owls dispersing during nights with more moonlight (Penteriani et al. 2014). One possible explanation for such behavior is that, with more light, young, inexperienced owls are better able to see and move through unfamiliar habitats and, importantly, locate prey. Owls, including Eagle Owls, that are primarily visual predators are likely better able to detect prey with increasing light levels (e.g., Clarke 1983) and improved detection of prey could be crucial for the survival of young owls moving through unfamiliar areas as they disperse (Penteriani et al. 2014).

grabbing with their talons) toward returning young that attempt to inspect potential, but already taken, nest sites, and young kestrels attacked repeatedly may then disperse from their natal colony (Negro et al. 1997).

In some species, food availability contributes to the onset of natal dispersal. For example, young Tawny Owls (*Strix aluco*) disperse from natal territories shortly after termination of parental provisioning (Sunde 2008). In Western Bluebirds (*Sialia mexicana*), young males typically remain in parental territories during the winter before dispersing the following spring. However, Dickinson and McGowan (2005) experimentally removed mistletoe (*Phoradendron villosum*), an important winter food source, from some bluebird territories and found that young males in those territories dispersed during their first summer, whereas young males in territories where mistletoe had not been removed delayed dispersal until the following spring.

Although food availability and the aggressive behavior of parents, siblings, or conspecifics trigger departure from natal territories by young birds in some species, these factors appear to be of limited or no importance in most species. For example, experiments during which parents were removed from territories revealed no effect on the timing of natal dispersal by young Marsh Tits (*Poecile palustris*, Nilsson 1990) or young Northern Goshawks (*Accipiter gentilis*, Kenward et al. 1993). Thus, for many if not most species of birds that have been studied to date, endogenous factors appear to be of critical importance in initiating natal dispersal.

Dispersal behavior, including the timing of dispersal and dispersal distance, can be influenced by a bird's sex and personality (i.e., behavioral syndrome). For most species of birds studied to date, natal dispersal is female-biased, with females typically dispersing farther from natal territories than males. Greenwood (1980) proposed that selection would favor males that remained closer to natal areas because, in most species, males defend breeding territories and their chances of establishing and maintaining territories are improved if they are familiar with the local environment. Females, on the other hand, may benefit more by dispersing and, as they move away from natal areas, being able to choose a mate from among potential territory holders. In addition, by moving farther from natal areas, females reduce the likelihood of breeding with related males that tend to be more philopatric and avoid the costs associated with inbreeding (e.g., Szulkin and Sheldon 2008). Waterfowl represent an exception to the tendency for natal dispersal to be female-biased (Clarke et al. 1997), i.e., males typically breed farther from natal sites than females. A possible explanation for this is that most waterfowl are migratory and pair formation often occurs in wintering areas or during migration. Males may benefit by following females (and, therefore, not losing a potential mate), whereas females may benefit by migrating back to (or near) natal areas where they are more familiar with the local environment (Rohwer and Anderson 1988; Wolff and Plissner 1998).

Although females typically disperse farther than males, there are exceptions. For example, Tarwater (2012) found that both male and female Western Slaty (Black-crowned) Antshrikes (*Thamnophilus atrinucha*) disperse relatively short and similar distances from natal territories, possibly because both males and females acquire and defend territories and both sexes, therefore, as suggested by Greenwood (1980), would benefit by being familiar with the local environment. Natal dispersal is also male-biased in several species of cooperative-breeding birds, e.g., Pukekos (*Porphyrio porphyrio*; Craig and Jamieson 1990), Australian Magpies (*Gymnorhina tibicen*; Veltman and Carrick 1990), and Brown Jays (*Psilorhinus morio*; Williams and Rabenold 2005). One possible explanation for male-biased dispersal in some social species is that male breeding options may be limited by patrilineal queuing for inheritance of natal territories (Yáber and Rabenold 2002). In such species, opportunities for breeding for younger males may be improved by dispersing to other groups.

Recent studies have revealed that birds, and other animals, have different personalities (also referred to as behavioral syndromes), with individuals differing in such characteristics as boldness, sociability, and aggressiveness. These personality differences can influence dispersal behavior (Fig. 19.31). For example, natal dispersal distances of Great Tits (*Parus major*) have been found to be correlated with a measure of "exploratory behavior" (i.e., willingness or tendency of birds to explore a novel environment; Dingemanse et al. 2003; Korsten et al. 2013). In addition, Duckworth and Kruuk (2009) found a significant genetic correlation between aggression and natal dispersal in Western Bluebirds (*Sialia mexicana*). Additional research is needed to determine the extent to which differences in personality contribute to differences in dispersal behavior in other species of birds.

Among birds in general, most young disperse and rates of natal philopatry are low. For example, for songbirds that breed in the eastern United States and southeastern Canada, mean rates of natal philopatry (i.e., returning to the same study area) were 2.8% for 19 species of shrubland birds and 2.9% for nine species for forest birds (Schlossberg 2009). In some species, however, a greater percentage of young are philopatric, tending to breed in or near natal territories. Rates of natal philopatry for young females in

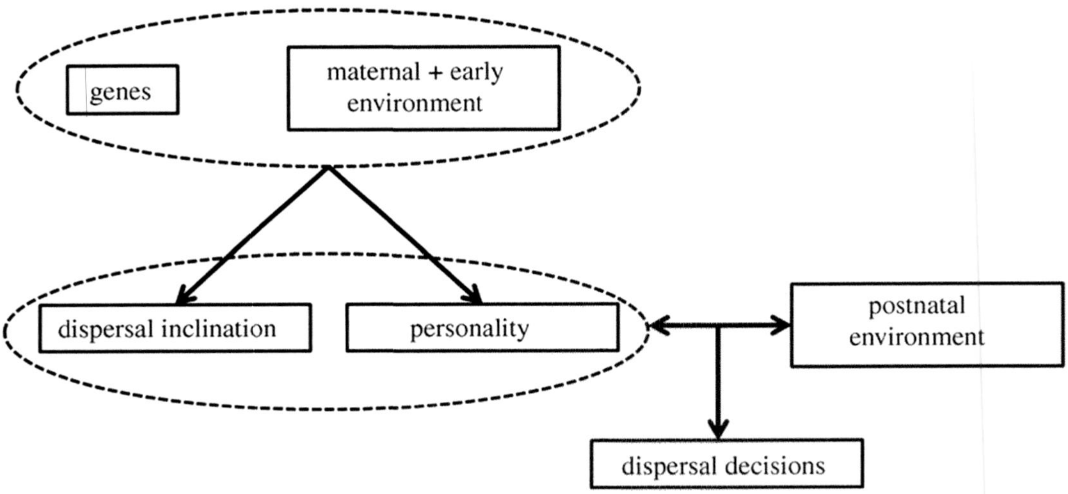


**Fig. 19.31** A suggested ontogenetic pathway linking dispersal behavior and personality traits. Dispersal inclination and distance results from an interaction between a bird's genes and natal and surrounding environments. (Figure from Cote et al. 2010; © 2010 The Royal Society, used with permission)

some species of waterfowl are relatively high, e.g., ranging from 34 to 100% among different breeding colonies of Brant (*Branta bernicla*; Lindberg et al. 1998) and rates of natal philopatry have also been found to be high in some island populations of songbirds such as Savannah Sparrows (*Passerculus sandwichensis*, 11.2%; Wheelwright and Mauck 1998) and Eurasian Linnets (*Linaria cannabina*, 38%; Förschler et al. 2010). Among waterfowl, as noted previously, young females are likely philopatric because they benefit from familiarity with their natal environment. For island species, high rates of natal philopatry may be due to limitations in the ability of young to disperse from breeding populations on isolated islands (Förschler et al. 2010).

Among some colonial-breeding nonpasserines, rates of natal philopatry can also be relatively high (e.g., >30% for Common Terns, *Sterna hirundo*; Becker et al. 2008). Factors that might contribute to natal philopatry among colonial species include a paucity of favorable breeding sites (Heino and Hanski 2001), social attraction, and lack of information about other potential breeding sites (Pinto et al. 2005). In other species, philopatric young appear to benefit from kin-mediated sociality. About 60% of young Vinous-throated Parrotbills (*Sinosuthora webbiana*) remain in or near natal territories, likely because of the benefits associated with being able to maintain social interactions (e.g., flocking behavior sometimes even during the breeding season) with kin in areas with high rates of predation (Lee et al. 2010).

For young birds that do disperse, the distances traveled from natal sites to breeding sites vary within and among species. In general, however, most young disperse relatively short distances and long-distance dispersal is uncommon (Figs. 19.32 and 19.33). Among species, factors that can influence dispersal distance include morphology and migratory status. The results of some studies suggest that larger species tend to disperse farther than smaller species (e.g., Paradis et al. 1998; Garrard et al. 2012), but other investigators have reported no such relationship (e.g., Dawideit et al. 2009). As such, additional study is needed to further examine the possible influence of body size on natal dispersal distance. Dawideit et al. (2009) found that other morphological variables, wing and tail shape, did influence natal dispersal distance; species with more pointed wings and

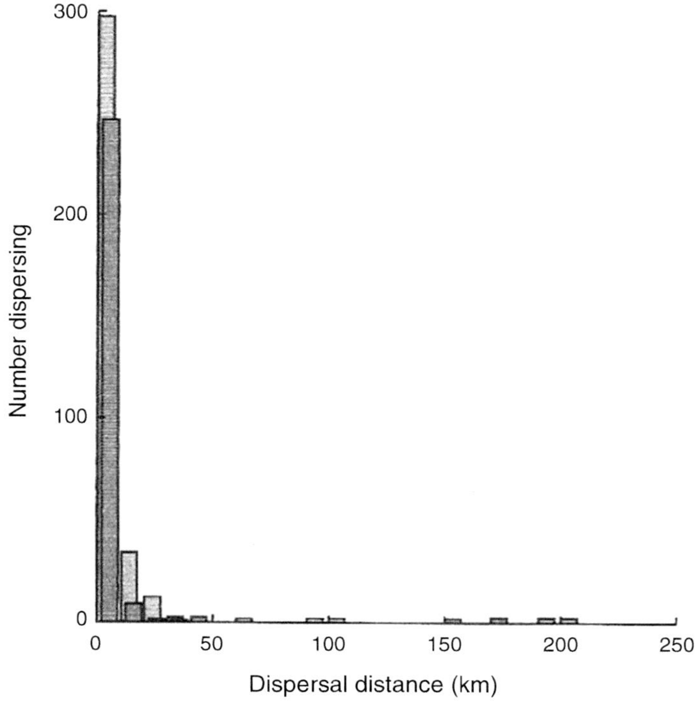


**Fig. 19.32** Distribution of dispersal distances for young male (dark gray) and female (light gray) Tree Swallows (*Tachycineta bicolor*), summed in 10-km bands from natal nest sites ($N = 355$ females and 257 males). Note the greater variation in dispersal distances of females, with the longest dispersal distance being 210 km. (Figure from Winkler et al. 2005; © John Wiley and Sons, used with permission)

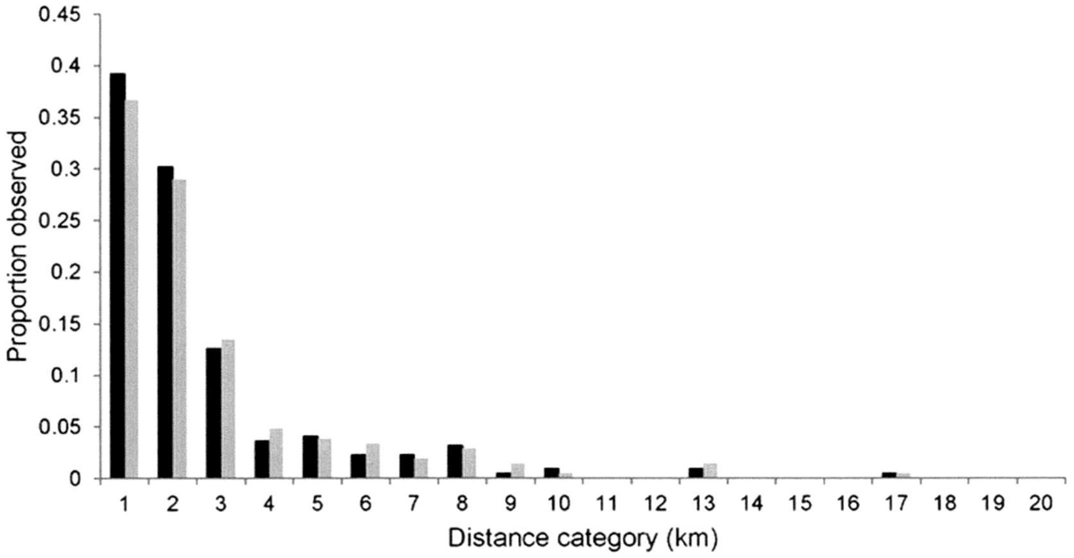


**Fig. 19.33** Distribution of natal dispersal distances of female (black) and male (gray) Prothonotary Warblers (*Protonotaria citrea*; $N = 222$) in southern Illinois. (Figure from McKim-Louder et al. 2013; © 2013 McKim-Louder et al., open-access article distributed under the terms of the Creative Commons Attribution License, https://creativecommons.org/licenses/by/4.0/)

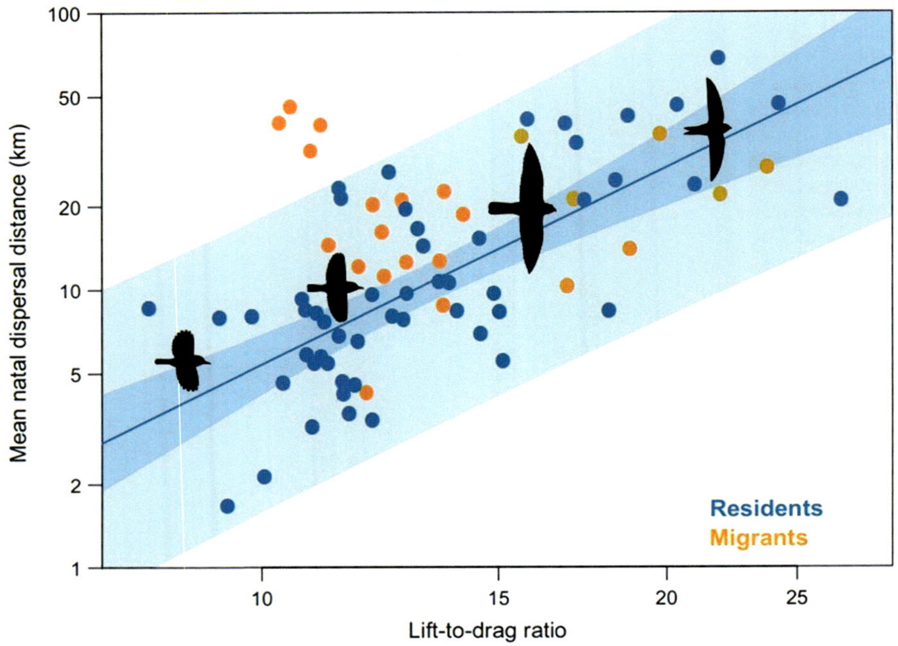


**Fig. 19.34** Relationship between flight efficiency (lift-to-drag ratio) and natal dispersal distances for 75 species of birds in the British Isles. The regression line and 95% confidence region for slope (dark shade) and prediction (light shade) are based on resident species only. Silhouettes show examples of birds with different wing morphologies and different lift-to-drag ratios. The lift-to-drag ratio is the ratio of a bird's mass to the forces required to sustain flight; more precisely, lift-to-drag ratio = $mgV/P$, where $m$ is a bird's body mass, $g$ is the gravitational acceleration, $V$ is the forward velocity, and $P$ is the power required for flying horizontally at speed $V$. (Figure modified from Claramunt 2021; © 2021 The Authors. Ecology published by Wiley Periodicals LLC on behalf of Ecological Society of America, used with permission)

squarer tails tend to disperse greater distances than those with rounder wings and more graduated tails. One possible explanation for this is that more pointed wings (i.e., higher aspect ratios) and squarer tails are characteristics typical of species with more efficient and faster flight (Norberg 1989), and efficient, fast flight allows these species to travel farther from natal sites. Similarly, Claramunt (2021) examined the dispersal distances of 75 species of birds in the British Isles at Britain and found a clear relationship between flight efficiency (lift-to-drag ratio) and dispersal distance (Fig. 19.34). In addition, Weeks et al. (2022) reported that species of birds with higher hand-wing indices (a morphological proxy for wing aspect ratio and flight efficiency) had greater natal dispersal distances. One possible explanation for such results, as noted by Weeks et al. (2022), is that efficient flyers ". . . are more mobile and thus are capable of crossing gaps in suitable habitat, which results in them travelling longer distances on average during the movement phase" of natal dispersal.

## 19.11 Learning by Young Birds

As noted by Blumberg (2017), "behaviors lie along a continuum from highly malleable or plastic to highly rigid or robust." Behaviors that tend to be more "rigid or robust" are generally referred to as being instinctive or species-typical, and some behaviors of birds are instinctive. For example, young megapodes exhibit innate responses to the predator alarm calls of songbirds (Göth 2001), and, as noted by Wiltschko and Wiltschko

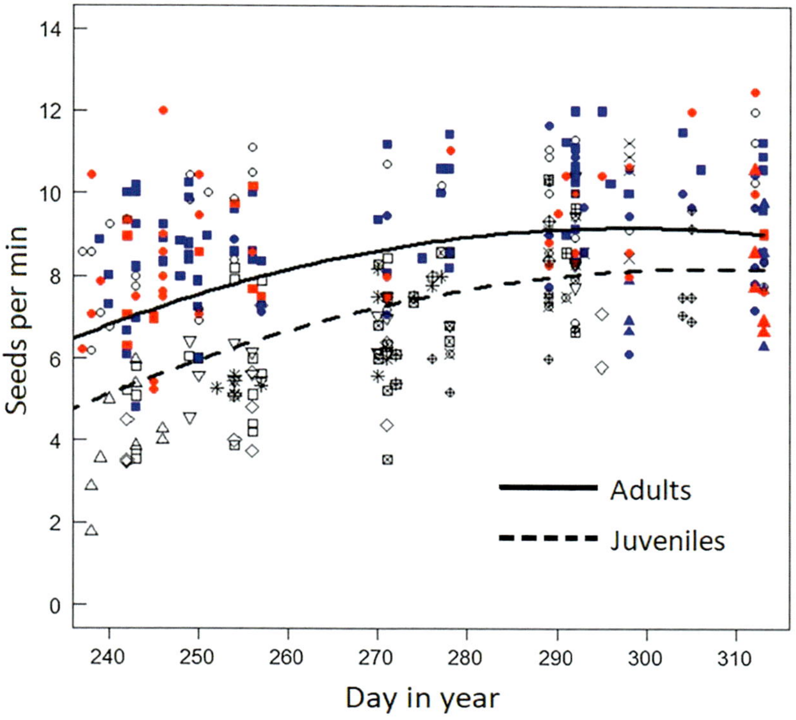


**Fig. 19.35** Feeding rate (sunflower seeds consumed per minute) of Eurasian Bullfinches (*Pyrrhula pyrrhula*) against day of the year. First-year juveniles were less adept at de-husking seeds than adults, and both age classes improved their efficiency over time. Adult males = blue symbols, adult females = red symbols, and unsexed juveniles = open symbols and crosses, with different symbols representing different individuals. (Figure from Fox 2015; Rights managed by Taylor & Francis, used with permission)

(2015), the young of many species of migratory birds ". . . are endowed with innate instructions that guide them to their distant goal, indicating the direction and distance of migration . . ." However, the extent to which behaviors are determined only by genetic information or develop via interaction with the environment by the processes of phenotypic plasticity and learning continues to be debated (e.g., Gilbert and Epel 2009; Jablonka and Lamb 2014; Blumberg 2017). Despite that debate, and given that behaviors do "lie along a continuum," young birds must and do, to varying degrees depending on the behavior, learn via practice and/or by copying conspecifics and, thereby, improve their ability to perform a variety of skills and their likelihood of surviving.

Investigators have documented numerous examples of learning by young birds, including:

- Foraging skills (e.g., Kitowski 2009; Fox 2015; Enstipp et al. 2021; Sam et al. 2021; Fig. 19.35)
- Flight skills (e.g., Bustamante and Hiraldo 1989; Harel et al. 2016; Ruaux et al. 2020; Fig. 19.36)
- Parental skills (e.g., Limmer and Becker 2009; Fig. 19.37)
- Migration performance, timing, and routes (e.g., Mitchell et al. 2015; Fayet 2020; Åkesson et al. 2021; Byholm et al. 2022; Fig. 19.38)
- Predator recognition and avoidance (e.g., Griesser and Suzuki 2017)
- Recognition and avoidance of aposematic prey (e.g., Landová et al. 2017)
- Recognition of brood parasites (e.g., Davies and Welbergen 2009; Langmore et al. 2012)
- Social interactions (e.g., White et al. 2010a; Kohn et al. 2013)
- Courtship displays (e.g., Lukianchuk 2013; Janisch et al. 2020; Spezie et al. 2022)
- Nest building (more detailed information is provided in Chap. 17)

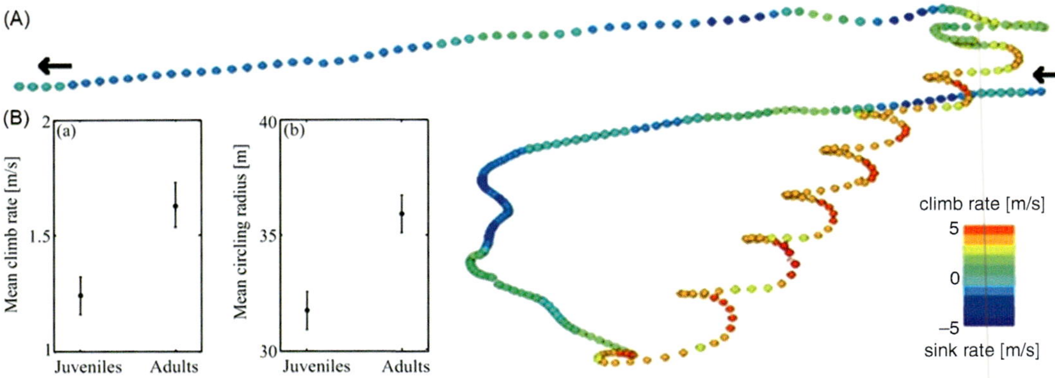


**Fig. 19.36** The soaring performance of young Griffon Vultures (*Gyps fulvus*) is inferior to that of adults, but does improve with experience. (**A**) Example of a flight path of an adult Griffon Vulture, with flight alternating between gliding (blue sections) and thermal soaring (ranging from yellow to red) modes. (**B**) Age-related differences in thermal centering. (**a**) Climb rates in thermals of juvenile and adult vultures differed, with juveniles having lower climb rates, and (**b**) the circling radius of juveniles was also smaller than that of adults. (Figures from Harel et al. 2016; open-access article licensed under a Creative Commons Attribution 4.0 International License, https://creativecommons.org/licenses/by/4.0/)

- Song/call learning (Passeriformes, Psittaciiformes, and Trochilidae; more detailed information is provided in Chap. 14)
- Tool-using (e.g., New Caledonian Crows [*Corvus moneduloides*, Holzhaider et al. 2010] and Woodpecker Finches [*Camarhynchus pallidus*, Tebbich et al. 2001])
- Compensating for wind drift (e.g., Wynn et al. 2020; Sergio et al. 2022)

## 19.12 Brood Parasitism

Parental care is costly is terms of the time and energy that must be invested and, to avoid or minimize those costs, females in some species of birds lay eggs in the nests of other birds, either those of other individuals of the same species (conspecific or intraspecific brood parasitism) or other species (interspecific brood parasitism). Brood parasites can also be categorized as either facultative (species where females may parasitize other nests, but also raise their own young) or obligatory (species where females are strictly brood parasites and never build nests or care for their own young). Approximately 104 species of birds are considered obligatory brood parasites (Kennerley et al. 2022; Table 19.3), whereas well over 200 species are known to be facultative brood parasites.

### 19.12.1 Facultative Brood Parasitism

Most facultative brood parasites lay eggs primarily, if not only, in the nests of conspecifics and so are primarily conspecific brood parasites. Conspecific brood parasitism has been reported in 256 species of birds, most of which are precocial, have large clutches, and nest in cavities (Geffen and Yom-Tov 2001; Yom-Tov and Geffen 2017). Families of birds with high levels of conspecific brood parasitism include, for precocial species, Anatidae (about 60% of species; Box 19.7 Kin Selection and Conspecific Brood Parasitism), and, for altricial species, Ploceidae (14.5% of species) and Turdidae (12.8% of species). Factors contributing to higher rates of parasitism in precocial species include: (1) they typically start incubating after the last or penultimate egg has been laid so nests are unguarded for most of the laying period, and (2) precocial young require less parental care from hosts than altricial young (Sorenson 1992). Other factors that contribute to increased rates of conspecific nest parasitism

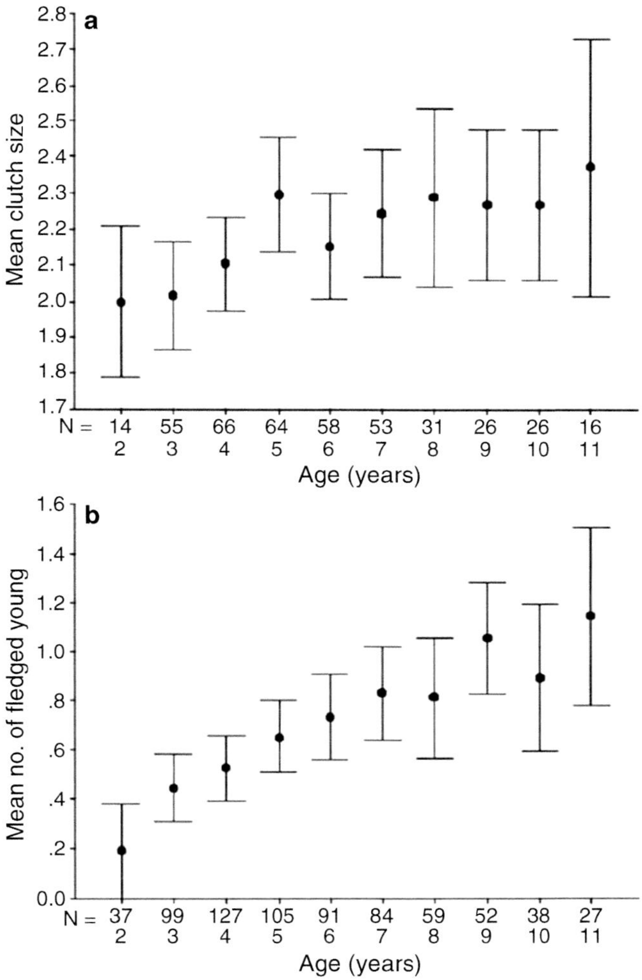


**Fig. 19.37** Variation in (**a**) mean clutch size and (**b**) mean number of fledged young among male and female Black Kites (*Milvus migrans*) of different ages. Note the increase in both mean clutch size and number of fledged young with increasing age. With increasing age, breeding skills, along with foraging ability, may improve with experience and contribute to increased breeding success. Older individuals may also be in better physical condition and be more familiar with their breeding territories, factors that may also contribute to increased breeding success. (Figure from Blas et al. 2009; open-access article distributed under the Creative Commons Attribution License, https://creativecommons.org/licenses/by/4.0/)

include high nest densities (colonial species) and low nest site availability (secondary cavity nesters). Differences among taxa in the frequency of parasitism result from differences in opportunities for and benefits of such behavior (Lyon and Everding 1996). High nest densities and unguarded nests provide the opportunity for parasitism, and the need for less parental care means that the addition of parasitic young may minimize costs to hosts and will not increase the likelihood that host parents will be unable to successfully raise those young. For species where suitable nest sites may be limited, conspecific nest parasitism may provide females with a chance to achieve some reproductive success even if unable to raise their own young.

Available evidence indicates that conspecific brood parasites are flexible, with decisions by females to engage in parasitism (or not) varying from year to year depending on ecological and demographic conditions as well as female attributes (Sorenson 1991; Lyon and Eadie 2008; Jaatinen et al. 2011; Fig. 19.39). Depending on the species, however, not all options may be available. Among some species, for example, most or all parasitism is by females that do not nest (e.g., Sandell and Diemer 1999) and, in yet other species, only nesting females

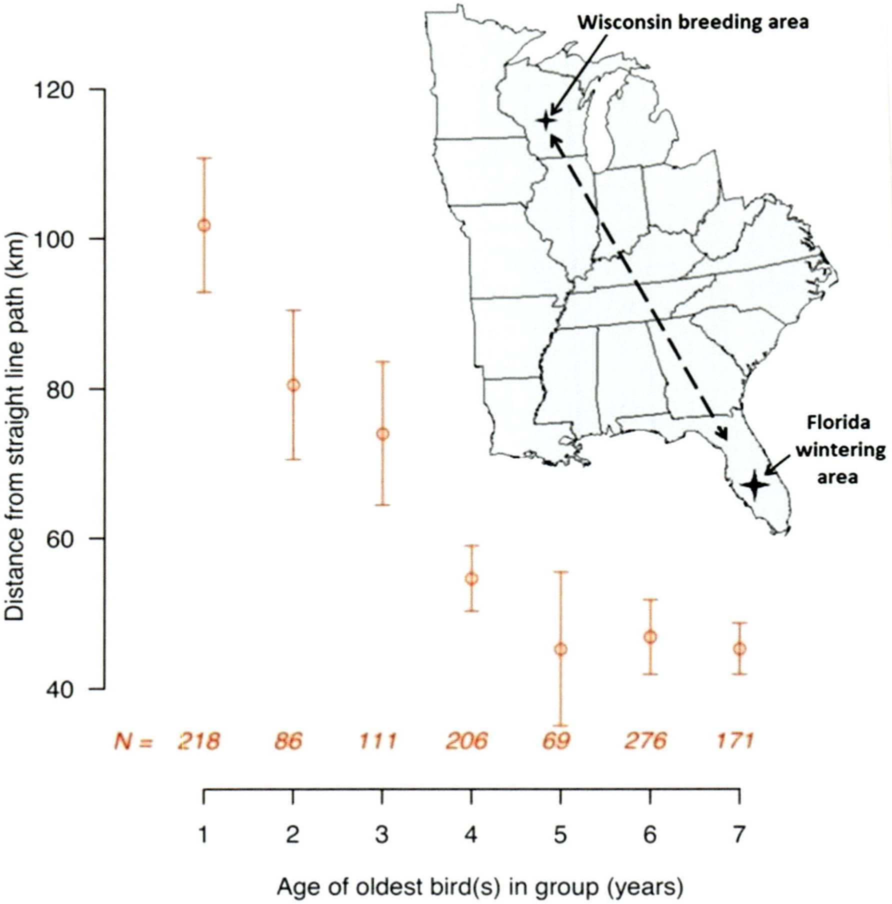


**Fig. 19.38** Migratory performance of Whooping Cranes (*Grus americana*) improves with increasing age of flock members, with path of migration getting closer to the shortest straight-line distance. Values represent mean deviation (±95% confidence intervals) of locations from straight-line path for migrating groups (i.e., two or more individuals. Sample sizes (*N*) indicate the number of times birds were relocated during migration. Map shows locations of breeding and wintering areas in Wisconsin and Florida, respectively. (Figure from Mueller et al. 2013; © 2013 American Association for the Advancement of Science, used with permission. Map from U. S. Fish and Wildlife Service 2017, CC0 Public Domain)

**Table 19.3** Families of birds and numbers of species within those families that are obligatory brood parasites

| Order | Family | Common name[a] | Number of parasitic genera | Number of parasitic species |
|---|---|---|---|---|
| Anseriformes | Anatidae | Black-headed Duck | 1 | 1 |
| Piciformes | Indicatoridae | Honeyguides | 4 | 17 |
| Cuculiformes | Cuculidae | Old World cuckoos | 13 | 56 |
| | Cuculidae | New World cuckoos | 2 | 3 |
| Passeriformes | Viduidae | Parasitic Weaver | 1 | 1 |
| | Viduidae | Indigobirds and whydahs | 1 | 19 |
| | Icteridae | Cowbirds | 1 | 5 |

Table modified from Winfree (1999); © 1999 Elsevier Science Ltd., used with permission
[a]Scientific names: Black-headed Duck, *Heteronetta atricapilla*; Parasitic Weaver, *Anomalospiza imberbis*

**Box 19.7 Kin Selection and Conspecific Brood Parasitism**

Broods of mixed maternity, like this one, are common in Common Goldeneyes (*Bucephala clangula*) and arise either through conspecific brood parasitism or amalgamation of broods after hatching. The ducklings with different cheek colors were color marked at different nest boxes and thus originated from the nests of different females. (Photo by Chad Teer, Wikipedia, CC BY 2.0, https://creativecommons.org/licenses/by/2.0/)

Conspecific brood parasitism is particularly common in ducks. Andersson and Åhlund (2000) tested a kin selection explanation for this parasitism in Common Goldeneyes by sampling egg albumen. Their analysis revealed that hosts and parasites were often related, with a mean degree of relatedness as high as between first cousins. Relatedness to hosts was higher in nests where parasitic females laid several eggs than in those where they laid just one. Conspecific brood parasitism has typically been viewed as an interaction that pits the interests of the parasite against those of the host. If, however, parasites and hosts are related, the relationship between parasite and host becomes a cooperative one (Lyon and Eadie 2000).

parasitize conspecific nests (e.g., Brown and Brown 1998). Available evidence suggests that females in some species do choose different options in different years (e.g., Sorenson 1991; Ahlund and Andersson 2001). Different options vary in terms of the resulting potential reproductive success, but particular options will be favored as long as they result in higher fitness than other options (Lyon and Eadie 2008).

Conspecific brood parasitism can enhance the fitness of parasitic females, but may also reduce host fitness in a number of ways, including (1) reduced clutch size due to egg removal or damage by the brood parasite, (2) larger clutches may have reduced hatching success (more difficult to efficiently incubate more eggs), (3) larger clutches may require a longer incubation period and, as a result, be at a greater risk of predation, and (4) increased risk of host predation due to a longer incubation period or need to defend more fledglings from predators (Lyon and Eadie 2008). Studies of some species have confirmed that hatching or fledging success of hosts may be lower than that of non-parasitized individuals (e.g., Brown and Brown 1991; Sorenson 1991). However, additional study is needed to improve our understanding of the possible effects of conspecific brood parasitism on host fitness.

On the other hand, for precocial species where young feed themselves (thus requiring less parental effort), broods enlarged as a result of brood parasitism might benefit hosts through a dilution

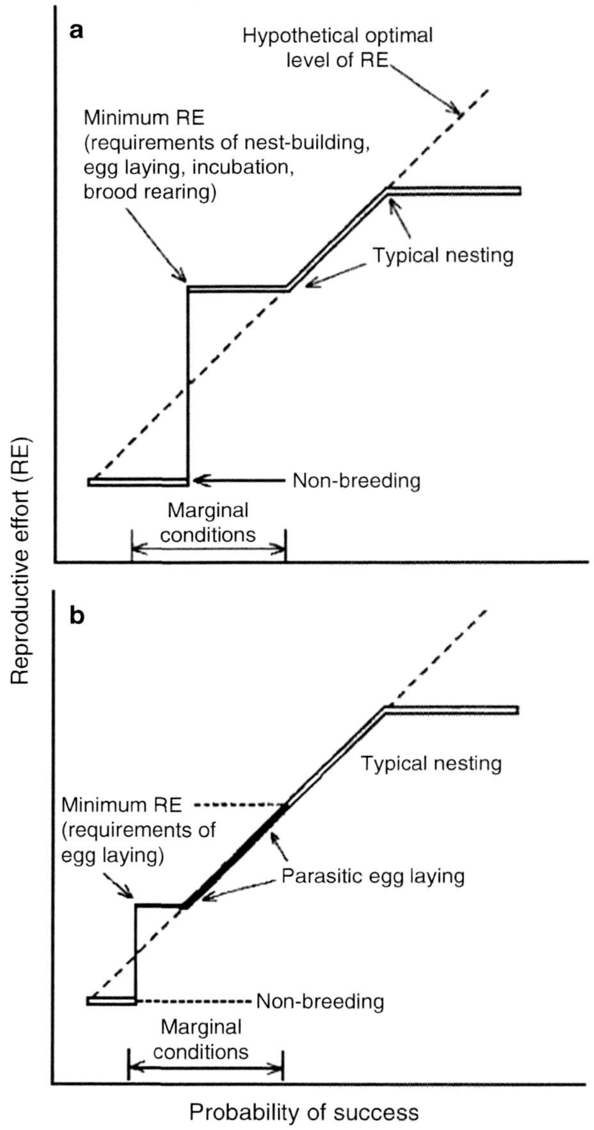


**Fig. 19.39** Sorenson's (1991) reproductive decision model for potential conspecific brood parasites. The optimal level of reproductive effort (45° line) varies with the characteristics of females (e.g., condition and parental ability) and ecological factors (e.g., availability of mates or nest sites, food availability, and number of conspecific nests in an area). (**a**) Reproductive options available to most prairie-nesting waterfowl after arrival in breeding areas. Nesting requires substantial reproductive effort (Minimum RE for nest-building, egg laying, incubation, and brood rearing). (**b**) Facultative conspecific nest parasitism may be a strategy used to lower reproductive effort because just laying eggs takes less effort (Minimum RE for egg laying) than nesting and rearing broods. Among species where conspecific nest parasitism occurs, females can adjust their reproductive effort to maximize the probability of reproductive success. Given environmental conditions and female attributes during a given breeding season, individuals may refrain from breeding, lay eggs parasitically, nest, or, with near optimal conditions and attributes, both nest and lay eggs parasitically. (Figure from Sorenson 1991; © 1991 Published by Elsevier Ltd., used with permission)

effect. That is, the odds that one of a host's own young will be predated may be reduced by the presence of "foreign" young. In addition, additional young mean more individuals to watch for predators which could improve the likelihood of brood survival (Lyon and Eadie 2008).

### 19.12.2 Facultative (Conspecific) Brood Parasitism: Host Defenses and Parasite Tactics

Selection for host defenses depends on the fitness costs of conspecific brood parasitism as well as the benefits gained by particular defensive strategies (Lyon and Eadie 2008). One way to reduce the costs of parasitism is to prevent it from happening, e.g., by guarding nests and defending nests from potential brood parasites (e.g., Emlen and Wrege 1986). Once a nest has been parasitized, hosts can reduce costs by rejecting parasitic eggs or young or by deserting nests and re-nesting elsewhere (Lyon and Eadie 2008). However, rejection requires recognition of parasitic eggs or young and, in most species, hosts do not exhibit such recognition. In some species, parasitic eggs are removed from nests only if they are deposited before host females have begun laying their own eggs (e.g., Brown and Brown 1989). In these species, once host females begin laying, they do not recognize parasitic eggs as different from their own.

Egg rejection based on recognition has been reported in many species parasitized by other species (interspecific brood parasitism), but has rarely been reported in cases of conspecific brood parasitism. However, such rejection has been reported in rails (Rallidae; e.g., Lyon 2003) and weaverbirds (Ploceidae; e.g., Lahti and Lahti 2002), two taxa where the appearance of eggs exhibits intraspecific variation. In American Coots (*Fulica americana*) and Village Weavers (*Ploceus cucullatus*), hosts use visual cues to recognize parasite eggs and reject eggs that differ in appearance from their own.

Few investigators have examined the possibility that hosts might recognize and reject parasitic chicks. In Zebra Finches (*Taeniopygia guttata*), hosts did not treat their own young and parasitic young differently (Fenske and Burley 1995). Shizuka and Lyon (2010) found that American Coots (*Fulica americana*) recognize and reject parasitic chicks by using first-hatching chicks in a brood as referents to learn how to recognize their own chicks and then subsequently reject parasitic chicks. This behavior is adaptive for American Coots because parasitic eggs rarely hatch first (about 3% of the time; Shizuka and Lyon 2010). The ability of American Coots to recognize brood parasites when so many species of birds are unable to recognize interspecific brood parasites that differ substantially in morphology, including in some case size, "seems remarkable" (Shizuka and Lyon 2010). However, using first-hatched chicks as referents for recognition would only be adaptive if the likelihood of parasitic chicks hatching first is very low. In most cases of interspecific brood parasitism, natural selection has favored short incubation periods for eggs of obligate brood parasites and, as a result, they typically hatch before host eggs.

Some authors have suggested that intraspecific brood parasitism serves an important step in the evolution of interspecific brood parasitism (Hamilton and Orians 1965; Yamaguchi 1995). Robert and Sorci (2001) suggested that, because females in many species can lay more eggs than their typical clutch size, a female that produces her own clutch, but, in addition, also lays eggs in the nests of conspecifics, might have greater reproductive success (assuming that the energetic cost of laying eggs is lower than the increase in fitness that results from the parasitic eggs). Based on that latter assumption, selection may favor a reduction in the number of eggs a female lays in her own nest and increase the number of eggs laid parasitically. Ultimately, females in such a species, further benefitting from avoiding the costs of parental care, might become obligate brood parasites, laying all of their eggs in the nests of other species.

Based on phylogenetic analyses, other authors have concluded that, among altricial species, interspecific, obligate brood parasites likely evolved directly from non-parasitic ancestors (Yom-Tov and Geffen 2006). What remains unclear, however, is how obligate brood parasitism could evolve from non-parasitic ancestors. One possibility suggested by Payne (1998) is that the transition from non-parasitic to parasitic could begin in a species where some females built their own nests and others used the nests of other species, including both deserted and active nests. Such behavior has been reported in several species of extant birds, including Chestnut Sparrows (*Passer eminibey*) and several estrildid finches (Payne 1969, 1998). If females in a species like this laid one or more eggs in active nests of another species, but were then unable to evict the nest-builders that were laying their own eggs, then the nest-builders might incubate the eggs and raise the young of this other species. If sufficient numbers of these "parasitic" young survive, and given the benefits of devoting less time and energy to parental care, then selection may favor an increasing tendency of females in this species to lay more eggs in the active nests of other species and a reduced tendency to try and evict the builders of those nests. Ultimately, the result could be an obligate brood parasitic species.

### 19.12.3 Obligate Brood Parasitism

Available evidence suggests that obligate brood parasitism has evolved independently at least seven times, including cowbirds (Icteridae), honeyguides (Indicatoridae), Old World cuckoos (Cuculinae), twice in the New World cuckoos (Neomorphinae), indigobirds and their allies (Ploceidae), and the Black-headed Duck (*Heteronetta atricapilla*, Anatidae) (Croston and Hauber 2012). Cowbirds (such as the Brown-headed Cowbird, *Molothrus ater*) and old-world cuckoos are the best known, but there are about 104 species of obligate brood parasites. Cuckoos have been brood parasites much longer (more than 60 million years) than cowbirds (2.8–3.8 million years; Winfree 1999).

In a recent review of the literature, Kennerley et al. (2022) determined that Africa was the continent with the most brood parasites, with up to 25 species and four independent origins of brood parasitism (two in Cuculidae and one each in Indicatoridae and Viduidae) (Fig. 19.40). Next in brood-parasite species richness is Asia, with up to 16 species of Old World cuckoos. Brood parasite species diversity is low throughout Europe and North American, with only three species in Europe (Common, Oriental (*Cuculus saturatus*) and Great Spotted Cuckoos (*Clamator glandarius*) and with only Brown-headed Cowbirds found throughout most of the Nearctic. In general, brood parasite diversity is highest where host species diversity and overall birds species diversity is highest, i.e., lower latitudes in the tropics and subtropics, with the notable exception of the Neotropics. The limited diversity of brood parasites in the Neotropics is likely due to the "species-poor parasitic lineages found in the region" (Kennerley et al. 2022). In addition, relatively few cooperative breeding species, common hosts of brood parasites (Feeney et al. 2013), are found in the Neotropics.

Of 83 species of brood parasites with at least one documented host species, Kennerley et al. (2022) noted that they have been reported to parasitize 1585 species (or 14% of all bird species). In addition, these authors determined that 68 species of brood parasites (82%) parasitize more than one species of host, and 494 host species (31%) are parasitized by at least two species of brood parasites. The median number of host species parasitized by obligate brood parasites is 11, with the greatest number being 274 (parasitized by Shiny Cowbirds) (Kennerley et al. 2022).

Of brood parasites that parasitize more than one host species, some, such as Brown-headed and Shiny (*Molothrus bonariensis*) cowbirds, are generalists and parasitize many species of songbirds. For example, Brown-headed Cowbirds are known to have parasitized the nests of 247 host species, and at least 172 of those species have successfully raised young cowbirds (Peer et al. 2013). Other obligate brood parasites primarily parasitize only a single host species, e.g., Screaming Cowbirds (*M. rufoaxillaris*; Friedmann 1929) and indigobirds (*Vidua* spp.; Payne 1977), and most parasitize no more than five to ten different host

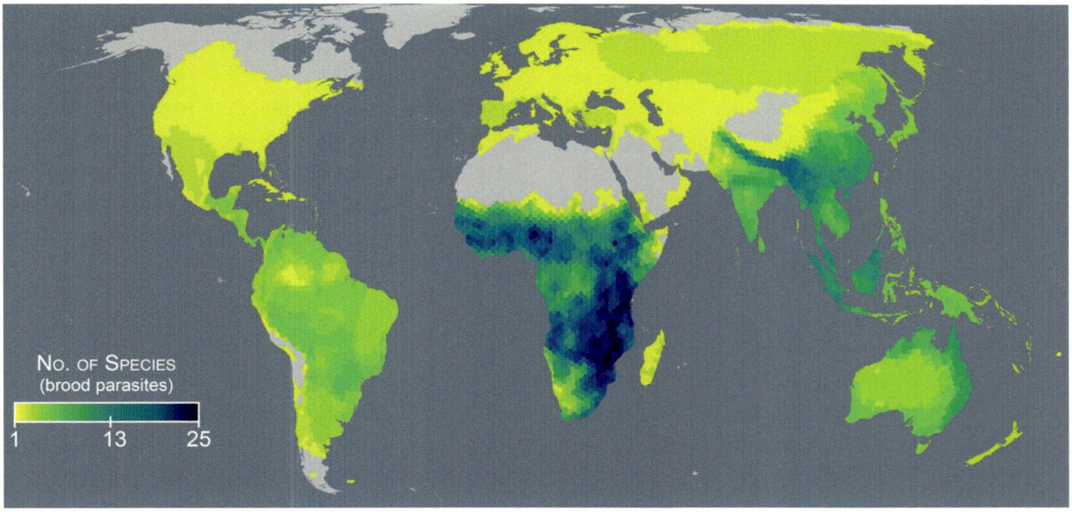


**Fig. 19.40** Worldwide pattern of species diversity of obligate brood parasites. Gray land areas indicate where there are no species of brood parasites. (Figure from Kennerley et al. 2022; open-access article distributed under the terms of the Creative Commons CC BY license, https://creativecommons.org/licenses/by/4.0/)

species (Rothstein 1990). Common Cuckoos (*Cuculus canorus*) are generalists known to have parasitized more than 100 different species (Moksnes and Røskaft 1995), but individual females, called gens (plural, gentes), are specialists, typically parasitizing, and laying eggs similar in appearance to, just one host species (Davies 2000). Common Cuckoos in Europe have been classified into at least 16 gentes (Moksnes and Røskaft 1995).

The ancestors of obligate brood parasites built nests and cared for their eggs and young just like typical birds. To better understand the evolution of obligate brood parasitism, Krüger and Davies (2002) compiled data for 14 ecological and life-history variables for 136 species in the family Cuculidae, including both brood parasites and those that provide parental care, and conducted a comparative analysis to gain insight into those factors that likely contributed to the evolutionary transition in some species from typical parental care to brood parasitism. This analysis revealed that the ancestors of present-day brood parasitic cuckoos were probably year-round residents of tropical forests. The transition to brood parasitism was correlated with a move to more open, less productive, and more seasonal habitats, which, in turn, resulted in a transition to smaller prey, an increase in the size of breeding ranges, and becoming increasingly migratory. The changes, perhaps most importantly the move to more open, less productive habitats, then favored the evolution of brood parasitism as a means of reducing the cost of reproduction. Additional analyses led Yom-Tov and Geffen (2006) to conclude that obligate brood parasitism typically evolved directly from normal breeding mode (i.e., providing parental care) rather than indirectly through either intraspecific brood parasitism or facultative interspecific brood parasitism.

Obligate brood parasites avoid the costs of constructing nests and caring for young by depositing their eggs in the nest of other species (hosts). Despite the obvious benefit of avoiding these costs, only about 1% of all bird species are obligate brood parasites, suggesting that there are other costs associated with being parasitic. One possible cost is that host species will not provide adequate parental care. For example, the types of food provided by a host species may be difficult for young brood parasites to handle, swallow, or digest, reducing their likelihood of survival (Turtumøygardm and Slagsvold 2010). Other factors that may limit the reproductive success of obligate brood parasites include competition from host nestlings (Slagsvold 1998), differences between parasites and hosts in the timing and duration of egg laying and incubation periods (Slagsvold 1998; Davies 2000), rejection of eggs because hosts have an innate tendency to remove unfamiliar objects from their nests (Moskát et al. 2003), and the possibility that young brood parasites will not recognize individuals of their own species when older and searching for mates (Slagsvold and Hansen 2001).

When hosts accept and care for brood parasitic eggs and young, parasite fitness is enhanced and host fitness is reduced (Box 19.8 Carry-Over Effects of Brood Parasitism). As such, selection should favor parasite behaviors that increase the

**Box 19.8 Carry-Over Effects of Brood Parasitism**

Brood parasites can reduce the reproductive success of their hosts when female parasites remove host eggs from nests and when nestling brood parasites (1) eject host eggs or nestlings (e.g., cuckoos), (2) kill host nestlings (e.g., honeyguides), and (3) monopolize parental provisioning and reduce the likelihood of host nestlings surviving. Unfortunately for hosts, the impact of brood parasitism can also carry over to future reproductive attempts. Mark and Rubenstein (2013) found that fledgling Striped Cuckoos (*Tapera naevia*) coerced parasitized Rufous-and-white Wren (*Thryophilus rufalbus*) parents into altering their normal parental behaviors, i.e., foraging near more open areas (where the risk of predation is higher) for young cuckoos that remained immobile rather than moving with adults through dense understory vegetation (where the risk of predation is lower) like young wrens would. As a result of the increased time and stress (higher corticosterone levels) associated with caring for young cuckoos, parasitized Rufous-and-white Wrens took longer to re-nest and were less likely to breed in the next breeding season than non-parasitized wrens.

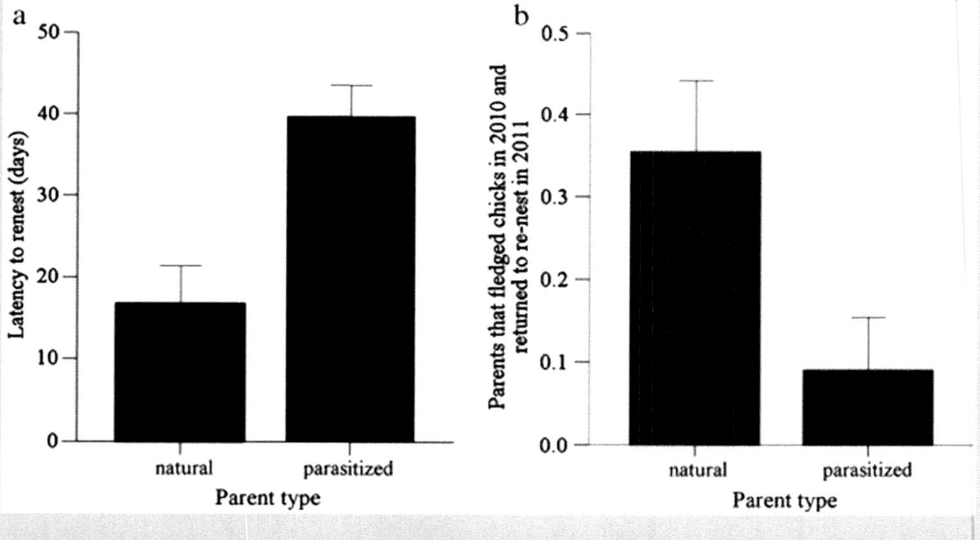


Parasitized Rufous-and-white Wrens (**a**) took longer to re-nest after broods became independent than did non-parasitized (natural) wrens and (**b**) were less likely to breed in the following year. Values are means ± standard errors. (Figure from Mark and Rubenstein 2013; © 2013 Elsevier Inc., used with permission)

likelihood of successful parasitism and, at the same time, selection should favor host behaviors that decrease the likelihood of being parasitized. These conflicting selection pressures result in what can be viewed as a coevolutionary arms race (Rothstein 1990), and the interactions between obligate brood parasites and their hosts can be viewed as occurring in different stages: (1) adaptations employed prior to a parasite laying an egg in a host's nest (pre-laying stage), (2) adaptations employed after a parasite egg has been laid in a host nest (egg stage), and (3) adaptations employed after a parasite egg hatches in a host nest (nestling stage).

### 19.12.4 Obligate Brood Parasitism: Pre-laying Adaptations of Hosts and Parasites

For host species, preventing female brood parasites from depositing eggs in their nests is the best defense because brood parasites typically remove a host egg before laying one of their own. So, even if hosts recognize and remove parasite eggs, they still incur a reduction in their possible reproductive success. Hosts may be able to reduce the likelihood of parasitism by selecting nest sites less likely to be found by brood parasites. For example, ground-nesting Song Sparrows (*Melospiza melodia*) with nests closer to perch sites that could be used by Brown-headed Cowbirds (*Molothrus ater*) were more likely to be parasitized than nests farther from perches (Hauber and Russo 2000). Using data from nearly 2000 nests of 17 host species that nest in tallgrass prairie, Patten et al. (2011) also found that proximity to perches was important in determining the likelihood of nests being parasitized by Brown-headed Cowbirds. Similar results have been reported for other species of obligate brood parasites, with nests farther from perches and other vantage points less likely to be parasitized (Antonov et al. 2007; Begum et al. 2011). Host nests that are better concealed by vegetation may also be parasitized at lower rates than those that are less concealed (Moskát and Honza 2000; Saunders et al. 2003), but some exceptions to this have been reported. For example, Wiley (1988) found that domed nests were as likely to be parasitized by Shiny Cowbirds (*Molothrus bonariensis*) as open-cup nests. Female Brown-headed Cowbirds (*Molothrus ater*) search for nests "in or near woody vegetation, nests above ground (but not too high), and nests placed along habitat edges . . ." (Patten et al. 2011). Female Great Spotted Cuckoos (*Clamator glandarius*) locate, and base decisions on whether to parasitize a nest or not, by observations of host activity around nests (Soler and Pérez-Contreras 2012). Other investigators have also reported that host activity around nests, sometimes including host alarm calls, is used by female brood parasites to locate nests (Banks and Martin 2001; Svagelj et al. 2009; Marton et al. 2019).

The characteristics of host nests can also influence the likelihood of being parasitized. Small cavity-nesters may, for example, have nest entrances too small to admit larger brood parasites. In addition, some African weavers (Ploceidae) construct nests with entrance tubes of varying lengths and diameters that may prevent access by brood parasites (Fig. 19.41; Crook 1963; Davies 2000). Some investigators have found that larger nests are more likely to be parasitized than smaller ones (Begum et al. 2011; Molina-Morales et al. 2016).

Mobbing of potential brood parasites by hosts or neighboring birds can also prevent nests from being parasitized (Fig. 19.42). Welbergen and Davies (2009) found that mobbing of Common Cuckoos (*Cuculus canorus*) by Eurasian Reed Warblers (*Acrocephalus scirpaceus*) was an effective defense against parasitism. In a more extreme case, Šulc et al. (2020) videorecorded a Great Reed Warbler (*Acrocephalus arundinaceus*) attacking a Common Cuckoo attempting to remove an egg from its nest (Fig. 19.43), and a cuckoo was found dead in water about 20 m from the nest the next day. These authors speculated that the cuckoo fell into the water below the nest as the warbler continued its attack and subsequently drowned. Other investigators have reported that mobbing by neighboring birds can reduce parasitism rates. For example, Yellow Warblers (*Setophaga*

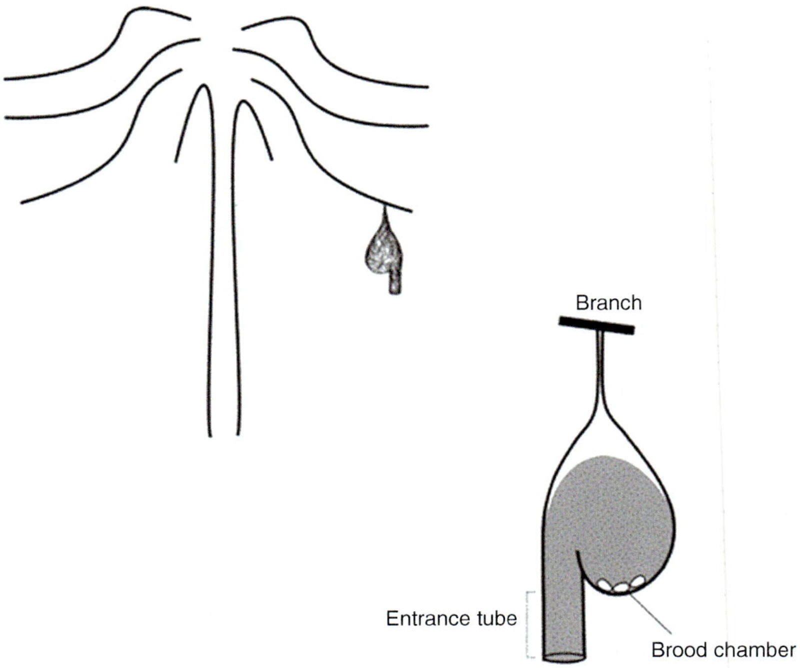


**Fig. 19.41** Nests of weaverbirds (*Ploceus* spp.) have an entrance tube that may prevent entrance by some nest predators and, for some African species, brood parasites such as Diederik Cuckoos (*Chrysococcyx caprius*). (Figure from Quader 2006; © Oxford University Press, used with permission)

**Fig. 19.42** Superb Fairywrens (*Malurus cyaneus*) mobbing a mounted specimen of a Shining Bronze-Cuckoo (*Chrysococcyx lucidus*). (Photo from Feeney and Langmore 2013; © 2013 The Authors. Published by the Royal Society, used with permission)

**Fig. 19.43** Screenshot from a video where a female Common Cuckoo attempting to remove an egg from the nest was attacked by a female Great Reed Warbler. (screenshot from a video provided as Supplementary Information by Šulc et al. 2020, used with permission of Michal Šulc)

*petechia*) that nested near Red-winged Blackbirds (*Agelaius phoeniceus*) were less likely to be parasitized than those that did not, apparently because mobbing of Brown-headed Cowbirds (*Molothrus ater*) by the blackbirds limited their access to nearby warbler nests (Clark and Robertson 1979).

Although multiple factors have contributed to the evolution of colonial and cooperative breeding systems, increased numbers and densities of host species may reduce the likelihood of interspecific brood parasitism by obligate brood parasites (though, as noted previously, colonial species may be more susceptible to conspecific brood parasitism). For example, colonies of Southern Red Bishops (*Euplectes orix*) vary in size and Lawes and Kirkman (1996) found that rates of parasitism by Diederik Cuckoos (*Chrysococcyx caprius*) were negatively correlated with colony size (Fig. 19.44), suggesting that cuckoos were limited by the greater vigilance and defense of Southern Red Bishops in larger colonies. With more nests and limits on the number of eggs each brood parasite can lay, the likelihood of individual nests in larger colonies being parasitized is also reduced (Feeney et al. 2012). Previous studies have revealed similar relationships between host nest densities and rates of interspecific nest parasitism in other species of birds (e.g., Freeman et al. 1990; Rothstein 1990).

Interestingly, the global distributions of cooperatively breeding birds and interspecific brood parasites are strikingly similar, with the greatest numbers of each found in sub-Sahara Africa and Australasia (Feeney et al. 2013, Fig. 19.45). This overlap may reveal something about interactions between cooperative breeders and brood parasites, but could also be due to other factors such as the high cost of providing parental care under certain environmental conditions or in certain habitats (e.g., Jetz and Rubenstein 2011). However, Feeney et al. (2013) found that cooperative-breeding species in southern Africa and Australia were more likely to be host species (27.5 and 52.8%, respectively) than were non-cooperative-breeding species (7.6 and 11.9%, respectively), a statistically significant difference. As noted by Feeney et al. (2013), there are three non-mutually exclusive explanations for the greater tendency of cooperative breeders to be parasitized: (1) cooperative breeders might be targeted by brood parasites because they provide better care and better defense of young, (2) more

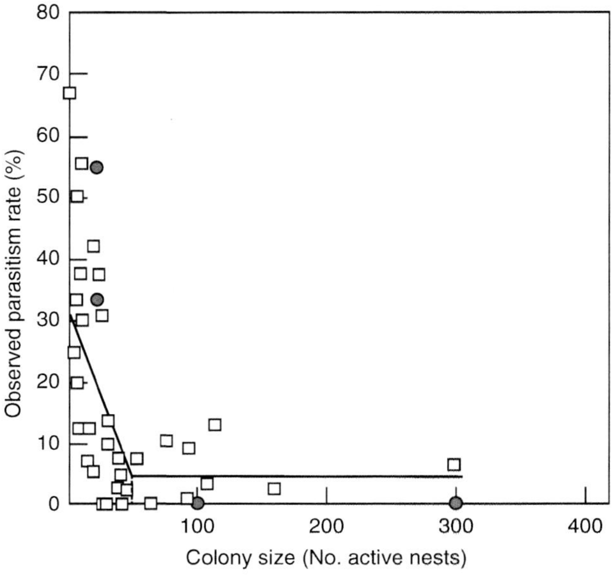


**Fig. 19.44** Relationship between Southern Red Bishop (*Euplectes orix*) colony size and rates of parasitism by Diederik Cuckoos. The line indicates the apparently optimum colony size of 50 nests, above which parasitism rates are much lower. Open square = Data from Payne and Payne (1967), Reed (1968), Jensen and Vernon (1970), Craig (1982), and Ferguson (1994); filled circle = data from Lawes and Kirkman (1996). (Figure from Lawes and Kirkman 1996; © 1996 The Association for the Study of Animal Behaviour. Published by Elsevier Ltd., used with permission)

birds and increased activity near the nests of cooperative breeders may make their nests easier to locate, and (3) selection may favor cooperative breeding where there are high numbers of brood parasites because the additional birds (helpers) may allow better defense of nests against brood parasites.

To better understand the relationship between cooperative breeders and brood parasites, Feeney et al. (2013) studied Superb Fairywrens (*Malurus cyaneus*), a species that is the primary host of Horsfield's Bronze-Cuckoos (*Chrysococcyx basalis*) and that is facultatively cooperative (some pairs have helpers and others do not). This study revealed that (1) young cuckoos grew faster and were more likely to fledge from nests of cooperative breeders, and (2) larger groups of cooperative-breeding fairywrens were less likely to be parasitized than smaller groups because they were more vigilant and spent more time mobbing cuckoos and, as result, fledged more of their own young than did smaller groups. Thus, in this system, brood parasites benefit from targeting cooperative breeders, and cooperative breeders (adults and related kin) benefit from an improved ability to defend nests from brood parasites. Similar results were reported in a study of Carrion Crows (*Corvus corone*), with parasitism rates inversely correlated with crow group size and pairs without helpers the most likely to be parasitized (Canestrari et al. 2009).

Prior to egg-laying, obligate brood parasites may improve their chances of parasitizing host nests via camouflage and deception (mimicry of predators, conspecifics, or heterospecifics; Feeney et al. 2012). For example, selection may favor more cryptic female plumage in brood parasites if it reduces the likelihood of detection by host species. To test that hypothesis, Krüger et al. (2007) examined the extent to which 144 species of cuckoos exhibited sexual plumage dimorphism and found that, among parasitic

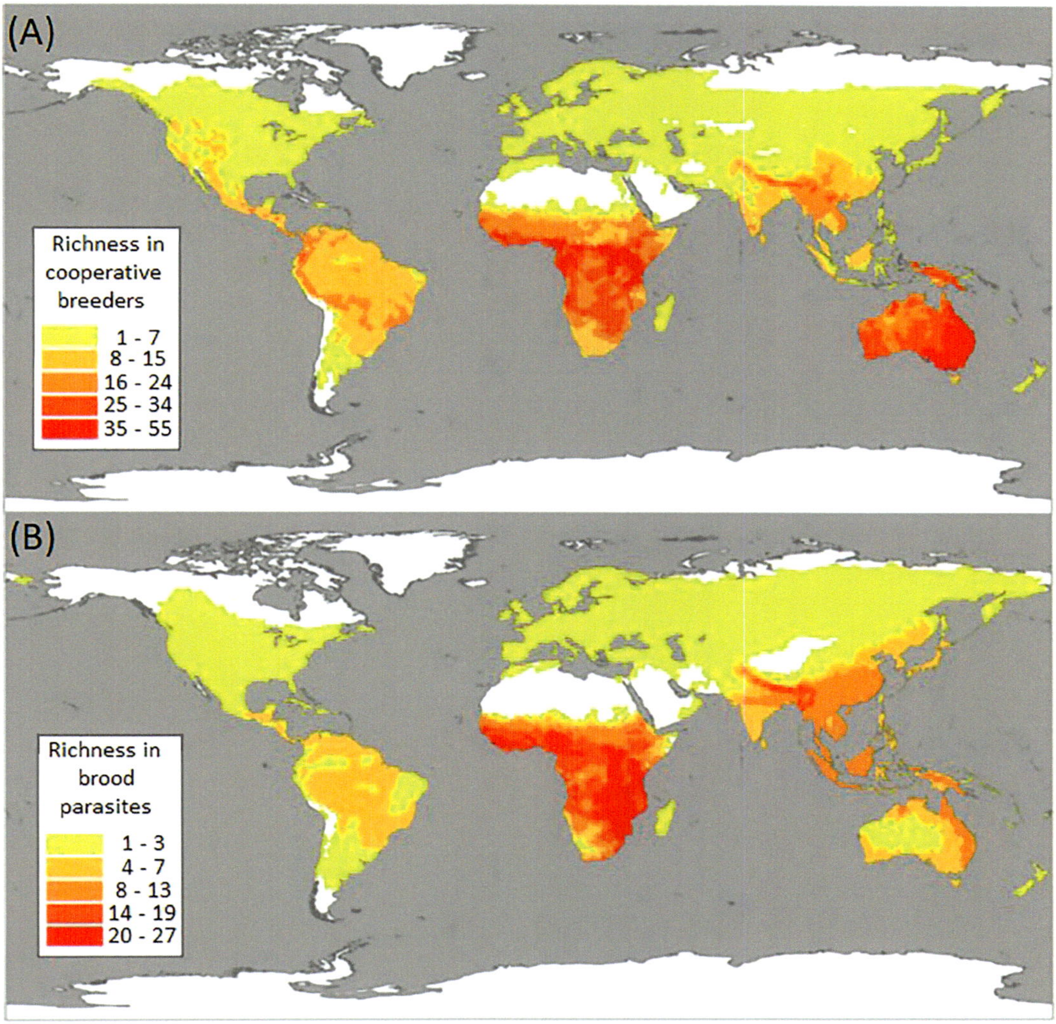


**Fig. 19.45** Global pattern of richness in (**a**) cooperatively breeding songbirds and (**b**) species of obligate brood parasites during their breeding seasons. (Figure from Feeney et al. 2013; © 2013, Copyright © 2013 American Association for the Advancement of Science, used with permission)

species, male plumage has tended to become more showy (as a possible signal of male quality for females) and female plumage more cryptic (Fig. 19.46). Although empirical evidence is still lacking, these results suggest that cryptic plumage might improve the reproductive success of female brood parasites in some species by making it harder for host species to detect them.

Some cuckoos have evolved plumage that resembles that of hawks in the genus *Accipiter* (York 2021; Fig. 19.47). Hawk-like features, with barred plumage on the underparts, yellow eyes, and/or yellow legs and feet), are found in five genera of Old World cuckoos (*Cuculus, Hierococcyx, Cercococcyx, Cacomantis*, and *Chrysococcyx*; Thorogood and Davies 2013). Krüger et al. (2007) found that the barred

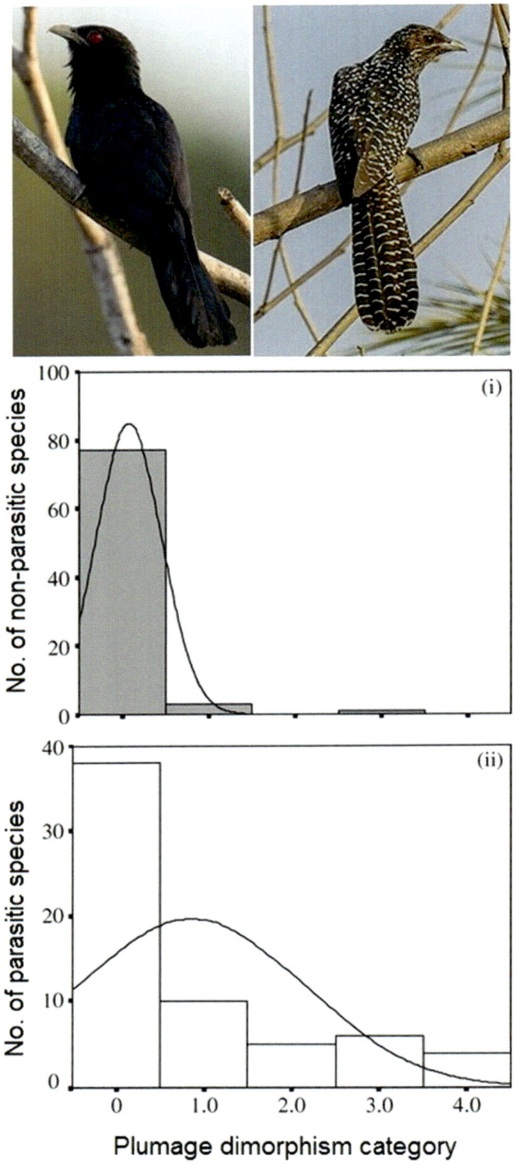


**Fig. 19.46** Top, Sexual plumage dimorphism of male (left) and female (right) Asian Koels (*Eudynamys scolopaceus*), a brood parasitic species found throughout the Indian Subcontinent, south and east China, mainland southeast Asia, and the Sundaic Islands (Limparungpatthanakij 2020). Bottom, Frequency of sexual plumage dimorphism for 144 species of cuckoos that are either (**i**) non-parasitic and provide parental care or (**ii**) brood parasites. Brood parasitic species are more likely to exhibit plumage dimorphism (41% of species) than non-parasitic species (4.8% of species), with parasitic males tending to be more showy and females more cryptic. Plumage dimorphism categories: 0, no difference between male and female plumage; 1, slight difference (eye color or eye ring); 2, some difference (< 25% of the plumage); 3, marked difference (> 25% of the plumage, but not the whole plumage; and 4, completely different plumage. (Figure from Krüger et al. 2007; © 2007 The Royal Society, used with permission. Top left, photo of male Asian Koel by Jason Thompson, Wikipedia, CC BY 2.0, https://creativecommons.org/licenses/by/2.0/; Top right, photo of female Asian Koel by Mike Prince, Wikipedia, CC BY 2.0, https://creativecommons.org/licenses/by/2.0/)

**Fig. 19.47** Common Cuckoos (*Cuculus canorus*, left) may increase their chances of gaining access to host nests by mimicking Eurasian Sparrowhawks (*Accipiter nisus*, right), a raptor that preys on many host species. Hosts mistaking a cuckoo for a sparrowhawk may leave their nest area to seek cover, providing female cuckoos with access to nests and an opportunity to deposit an egg. (Figure from Stoddard 2012; © 2015 Oxford University Press, used with permission)

plumage of these cuckoos evolved after cuckoos became parasitic, suggesting that predator resemblance is adaptive. By resembling *Accipiter* hawks, cuckoos may minimize host aggression (mobbing) and, at the same time, increase their chances of parasitizing host nests when hosts, responding to a perceived threat of predation, leave nests or the nest area when the cuckoos approach. In addition, York and Davies (2017) suggested that, by resembling a predator, potential host species may focus more on self-protection rather than clutch protection and thus reduce the likelihood that parasitized hosts will reject cuckoo eggs. However, this hawk mimicry is not effective with all host species. For example, Great Reed Warblers (*Acrocephalus arundinaceus*) were found to attack taxidermic mounts of both Common Cuckoos (*Cuculus canorus*) and Eurasian Sparrowhawks (*Accipiter nisus*) placed near their nests, but were able to discriminate between them and attacked cuckoos more frequently than sparrowhawks (Trnka and Prokop 2012). Similar discrimination between Common Cuckoos and Eurasian Sparrowhawks has been reported for Red-backed Shrikes (*Lanius collurio*; Krausová et al. 2022). Hawk mimicry may not always be effective because the resemblance is imperfect and host species can learn to discriminate (Trnka and Prokop 2012). In addition, hawk mimicry may have costs as well as benefits. For example, some species of birds that are rarely or never parasitized by cuckoos, such as Barn Swallows (*Hirundo rustica*), Great Tits (*Parus major*), and Eurasian Blue Tits (*Cyanistes caeruleus*), have been found to mob hawk-mimicking cuckoos as aggressively as they mob Eurasian Sparrowhawks (*Accipiter nisus*; Davies and Welbergen 2008; Lyon and Gilbert 2013).

One evolutionary response to the ability of host species to learn has been the evolution of polymorphism in some species of cuckoos (Thorogood and Davies 2013). For example, there are two color morphs of female Common Cuckoos, a blue-gray morph and a rufous morph (Fig. 19.47). Thorogood and Davies (2013) found that, when Eurasian Reed Warblers (*Acrocephalus scirpaceus*) learn to discriminate between one morph and sparrowhawks, the other morph, that hosts still treat as a sparrowhawk, gains an advantage (Fig. 19.48). This polymorphism can thus be maintained by frequency-dependent selection.

**Fig. 19.48** Some cuckoos are polymorphic. For example, male Common Cuckoos (*Cuculus canorus*) are always blue-gray, but some females are blue-gray like males and others are rufous. (Top) By resembling a dangerous predator like the Eurasian Sparrowhawk (*Accipiter nisus*), female cuckoos may reduce the likelihood of being mobbed by potential hosts and those hosts may instead seek cover, providing cuckoos with access to their nests. (Middle) However, potential hosts, such as Eurasian Reed Warblers (*Acrocephalus scirpaceus*), sometimes learn through experience to tell the cuckoos from the sparrowhawks and, when groups of warblers subsequently mob a cuckoo, other warblers also learn to discriminate cuckoos from sparrowhawks, e.g., that blue-gray female cuckoos are brood parasites rather than dangerous predators. If this happens, the rufous morph cuckoo gains an advantage in terms of accessing host nests. (Bottom) On the other hand, if potential hosts learn that the rufous morph is a cuckoo rather than a dangerous predator, the blue-gray morph gains an advantage. (Figure from Mappes and Lindström 2012, credit: Preston Huey/Science, © 2021 AAAS, used with permission)

Gaining access to host nests is obviously important for brood parasites, but that access must be well-timed. Parasites must lay their egg before host species are too far into their incubation period or the egg may not hatch before host eggs hatch and hosts stop incubating and start feeding their own young. Parasitizing nests during the laying stage when hosts are laying their own eggs can help ensure that the parasite's egg hatches when or before host eggs hatch. Because obligate brood parasite eggs hatch after relatively short incubation periods (short relative to other bird species of similar size; Payne 1977), perhaps in part because some embryonic development occurs while parasite eggs are still in a female's oviduct (Birkhead et al. 2011), proper timing typically means that parasite eggs hatch before host eggs, giving young parasites an advantage (in terms of size and ability to outcompete host young for food delivered by host parents; Rothstein 1990). To help ensure correct timing, brood parasites appear to watch host species and their nests to determine when hosts are egg laying.

Cowbirds and other brood parasites may use a variety of nest-searching methods, including quietly watching hosts from elevated, concealed perches, searching while walking on the ground in wooded areas, and actively searching by perching on and in dense shrubs (Norman and Robertson 1975). Once found, however, the nests of potential hosts may be too far along to be parasitized, e.g., the clutch is complete or eggs have already hatched. In such cases, brood parasites sometimes destroy host clutches or broods. Such behavior can potentially be beneficial if hosts subsequently re-nest and provide the brood parasite with another opportunity to parasitize a nest. Such behavior is sometimes referred to as "nest farming" and has been reported in Brown-headed (*Molothrus ater*), Shiny (*M. bonariensis*), and Bronzed (*M. aeneus*) cowbirds (Peer and Sealy 1999; Fiorini et al. 2009; Swan et al. 2015)

### 19.12.5 Obligate Brood Parasitism: Laying and Post-laying Adaptations of Hosts and Parasites

The thickness and strength of avian eggshells is generally correlated with egg size; larger eggs typically have thicker, stronger shells. The eggs of brood parasites, however, have stronger shells than expected relative to their size and stronger than those of their closest non-parasitic relatives (Spaw and Rohwer 1987; Picman and Pribil 1997) and their hosts (Igic et al. 2011; López et al. 2018; Fig. 19.49). Six non-exclusive hypotheses proposed to explain the relatively thick eggshells of brood parasites include: (1) parasite eggs are less likely to be damaged if they hit host eggs when laid, (2) parasite eggs may damage host eggs they hit when laid, leaving fewer host nestlings to compete with after hatching, (3) strong, thick shells are more difficult for host species to puncture, making it less likely that smaller hosts unable to grasp a larger, whole egg in its bill will be able to remove the egg by puncturing it and carrying it from the nest, (4) thick shells may be needed to provide sufficient calcium for developing parasite embryos, (5) thicker shells may provide better protection of embryos from bacteria in host nests, and (6) thick-shelled eggs retain more heat so embryos develop faster and hatch earlier (Antonov et al. 2012; Yang et al. 2018). Studies to date provide some support for hypotheses 1 and 2 ("egg damage" hypotheses) and 3 ("puncture resistance" hypothesis). Cuckoos sometimes lay eggs from variable distances above host nests as they attempt to lay as rapidly as possible to minimize the likelihood of detection, and eggs with thicker shells are less likely to be damaged (Antonov et al. 2012). At the same time, these thick-shelled parasite eggs can dent or crack host eggs (e.g., Soler and Martinez 2000; Fig. 19.50). In addition, Yang et al. (2018) tested the heat-retention hypothesis and, in support of that hypothesis, found that the eggs of Common Cuckoos (*Cuculus canorus*) did retain heat more effectively than the eggs of Oriental Reed Warblers (*Acrocephalus orientalis*; Fig. 19.51)

**Fig. 19.49** Mean eggshell thickness (in μm) of brood-parasitic Shiny Cowbirds and Brown-headed Cowbirds compared to examples of host species. Chalk-browed Mockingbird, *Mimus saturninus*; Shiny Cowbird, *Molothrus bonariensis*; Southern/Northern House Wren, *Troglodytes aedon*; Brown-headed Cowbird, *Molothrus ater*; Red-winged Blackbird, *Agelaius phoeniceus*. (Figure modified from López et al. 2018; © 2018 Oxford University Press, used with permission)

Few investigators have tested the "puncture resistance" hypothesis, but Antonov et al. (2008) experimentally parasitized nests of Eastern Olivaceous Warblers (*Iduna pallida*) with Common Cuckoo (*Cuculus canorus*) eggs and found that these small warblers were unable to grasp the cuckoo eggs in their bills and remove them. Most warblers pecked at the cuckoo eggs, but were unable to puncture and remove them. Warblers ultimately accepted the cuckoo egg at about half the nests in the study; other warblers unable to remove the cuckoo egg deserted their nests. These results provide some support for the "puncture resistance" hypothesis, but additional studies are needed to determine the extent to which selection might have favored thick eggshells to prevent puncture and removal of brood parasite eggs by host species.

If a brood parasite succeeds in laying an egg in a host nest, natural selection should, given the fitness cost of caring for parasitic young, favor hosts that recognize and remove that egg. If, however, brood parasites retaliate for such

**Fig. 19.50** Damage to host eggs caused by eggs laid by cowbirds hitting host eggs. (**a**) Egg of a Chestnut-capped Blackbird (*Chrysomus ruficapillus*) dented by egg laid by a Shiny Cowbird (*Molothrus bonariensis*). (**b**) Eggs of a Red-winged Blackbird (*Agelaius phoeniceus*) dented by eggs laid by Brown-headed Cowbirds (*Molothrus ater*). (Figure from López et al. 2018; © 2018 Oxford University Press, used with permission)

removal by removing, destroying, or preying upon the remaining host eggs or nestlings, then hosts might actually benefit by not removing the parasite egg. Such retaliation has been reported in studies of two brood parasites, Great Spotted Cuckoos (*Clamator glandarius*; Soler et al. 1995a) and Brown-headed Cowbirds (*Molothrus ater*; Hoover and Robinson 2007), and is sometimes referred to as "mafia behavior." Such behavior may not be widespread among obligate brood parasites (e.g., Turner et al. 2022). For example, mafia behavior by Brown-headed Cowbirds has only been reported in Prothonotary Warblers (*Protonotaria citrea*; Hoover and Robinson 2007), a cavity-nesting species typically found nesting in wooded swamps that has likely experienced limited parasitism by cowbirds in the past (Peer et al. 2013). If mafia behavior was common and individuals in host species that rejected parasite eggs had lower reproductive success on average than those that did not, then natural selection likely would have prevented rejection behavior from becoming fixed in that species. However, that is not the case because, in many host species, almost all parasitized individuals reject cowbird eggs (Peer et al. 2013). Additional studies of obligate brood parasite species are needed to determine how common mafia behavior is and, where it occurs, the factors that might explain the development of such behavior.

For hosts, the possible costs of being parasitized include a reduction in clutch and brood size, with brood parasites often removing a host egg before laying their own egg (Fig. 19.52), and reduced survival of their own young if parasitic nestlings monopolize parental care and limit the amount of food host young are able to obtain from parents. Given these costs, natural selection should, given that mafia behavior appears to be rare, favor hosts able to recognize and either remove eggs of brood parasites from their nests (if physically able to) or abandon the parasitized nest and re-nest. Of course, for brood parasites, natural selection should favor eggs more likely to be "accepted" by hosts. These competing selection pressures result in different outcomes in different host-parasite systems, with some hosts almost always recognizing and removing parasite eggs, some rarely if ever doing so, and some host species somewhere in between (Fig. 19.53).

Hypotheses to explain why hosts accept parasitism include the evolutionary lag hypothesis and the evolutionary equilibrium hypothesis. The evolutionary lag hypothesis posits that rejection

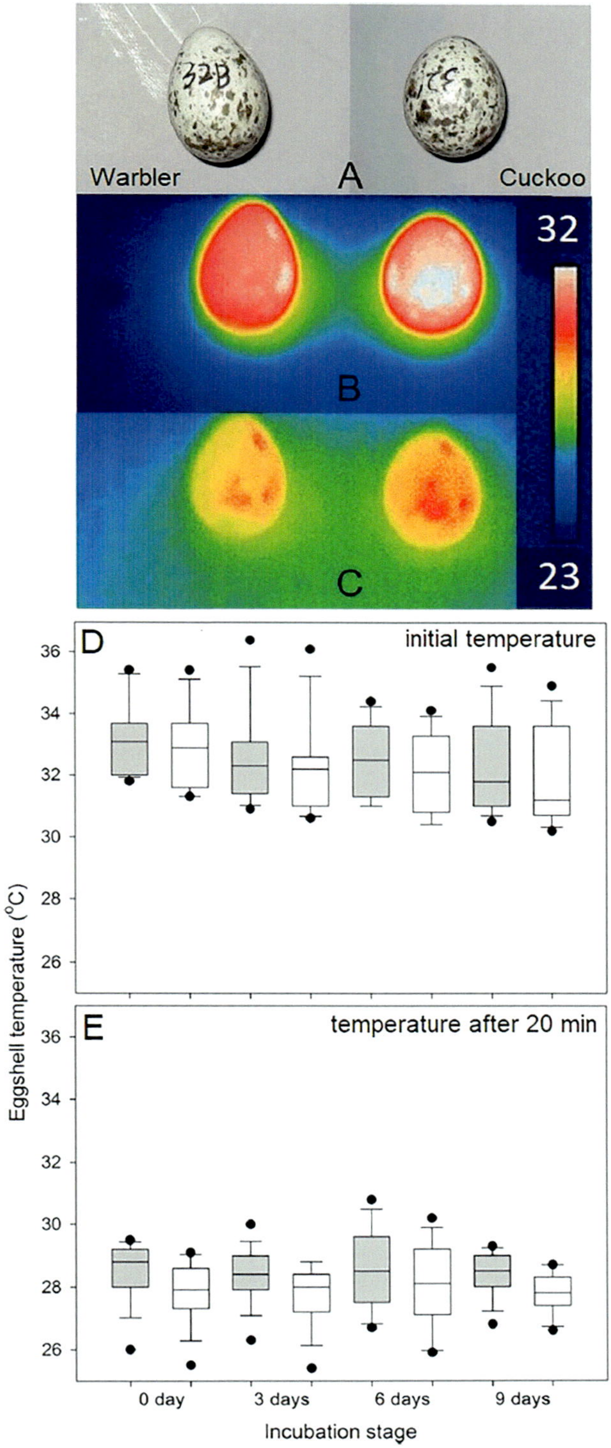


**Fig. 19.51** (**a**) Eggs of an Oriental Reed Warbler (*Acrocephalus orientalis*; left, host of Common Cuckoos) and Common Cuckoo (*Cuculus canorus*; right, brood parasite). (**b**) Infrared thermal images of the eggs immediately after removal from an incubator, and (**c**) after 20 min. Note that the cuckoo egg is warmer than the warbler egg after 20 min. (**d**, **e**) Temperatures of the eggshells of Common Cuckoos (gray boxplot) and Oriental Reed Warbler (white boxplot) after removal from incubator (initial temperature) and 20 min later. Boxplots show median, quartiles, 5- and 95-percentiles, and extreme values ($n = 15$ pairs of cuckoo and warbler eggs). (Figures from Yang et al. 2018; © 2018, Springer-Verlag GmbH Germany, part of Springer Nature, used with permission)

**Fig. 19.52** Female Common Cuckoo (*Cuculus canorus*) removing an egg from the nest of a Great Reed Warbler. (Photo from Supplementary Material from Sulc et al. 2016; © 2016 Oxford University Press, used with permission)

is better than acceptance (in terms of reproductive success and fitness), but there is a time lag between when a species or population is first parasitized and when egg rejection evolves (Rothstein 1975). The evolutionary equilibrium hypothesis suggests that accepting parasitism is better than rejection (again in terms of host fitness) because of the possible costs associated with rejecting, including a host mistakenly rejecting its own eggs or damaging its own eggs when removing a parasite egg, and the possibility of retaliation by the brood parasite (mafia behavior; Zahavi 1979; Rohwer and Spaw 1988; Lotem et al. 1995).

Because the evolutionary lag hypothesis is non-adaptive, testing it is difficult (Rothstein and Robinson 1998). However, to the extent that it explains the behavior of host species, the evolutionary lag hypothesis may be more applicable for brood parasites with expanding ranges that bring them into contact with "new" hosts such as Brown-headed Cowbirds than to brood parasites and hosts that have shared long coevolutionary histories, such as Old World cuckoos and their hosts (Servedio and Lande 2003). Peer and Sealy (2004) conducted a comparative analysis of several potential correlates of egg rejection by hosts of Brown-headed Cowbirds and, for host species that have recently come in contact with cowbirds, found stronger support for the evolutionary lag hypothesis. However, the evolutionary lag hypothesis does not explain why some host species that have longer histories of contact with cowbirds, and also have eggs that differ in appearance from those of cowbirds, are acceptors. One possible explanation is that chance may play a role in accumulating the recombinants and mutations needed for rejection behavior to evolve (Rothstein 1975; Peer and Sealy 2004).

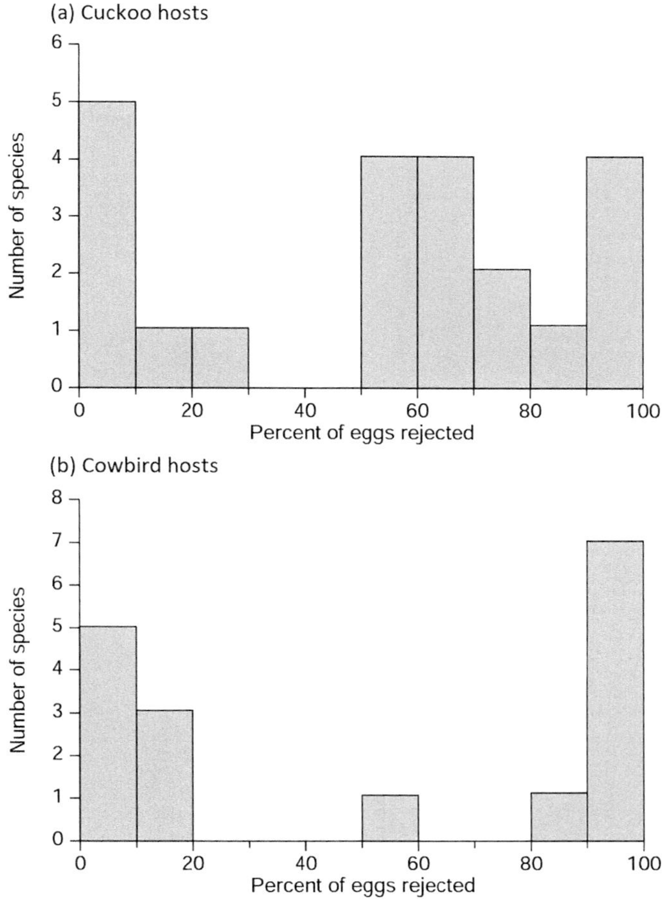


**Fig. 19.53** Frequency of parasite egg rejection by (**a**) hosts of Common Cuckoos (*Cuculus canorus*) and (**b**) hosts of Brown-headed Cowbirds (*Molothrus ater*). Cowbird hosts are more binary in their rejection behavior than are cuckoo hosts, many of which show intermediate rates of rejection. All data are from experiments in which a researcher placed "parasitic" eggs in the nest and observed the hosts' response. Sample sizes ranged from 10 to 133 nests. (Figure from Winfree 1999; © 1999 Elsevier Science Ltd., used with permission)

Among host species where some individuals reject and others accept the eggs of brood parasites, several factors may contribute to these variable responses (Fig. 19.54). One possible factor is the actual risk of parasitism (Fig. 19.54, box 3), with individuals in populations that experience higher levels of brood parasitism more likely to reject parasitic eggs (Brooke et al. 1998; Lindholm and Thomas 2000). This difference in behavior may be a result of social learning (Fig. 19.54, box 2), genotypic differences (Fig. 19.54, box 4), or both. In some host species, rejection appears to be influenced at least to some extent by learning, with older, more experienced host females more likely to successfully reject parasite eggs. For example, Antonov et al. (2008) suggested that differences in egg-rejection behavior of Marsh Warblers (*Acrocephalus palustris*) may have been due to differences in previous experience at rejecting cuckoo eggs. Evidence that egg-rejection behavior may have a genetic component has been reported for Eurasian Magpies (*Pica pica*), with differences in the frequencies of some alleles between rejectors and non-rejectors (Martín-Gálvez et al. 2006).

Differences in the size and appearance of host eggs relative to parasite eggs can also influence host behavior. Size matters because host species could use differences in size to detect brood parasite eggs. For example, experiments with Hume's Warblers (*Phylloscopus humei*) revealed that females rejected eggs based on relative size, with model eggs rejected only when they were

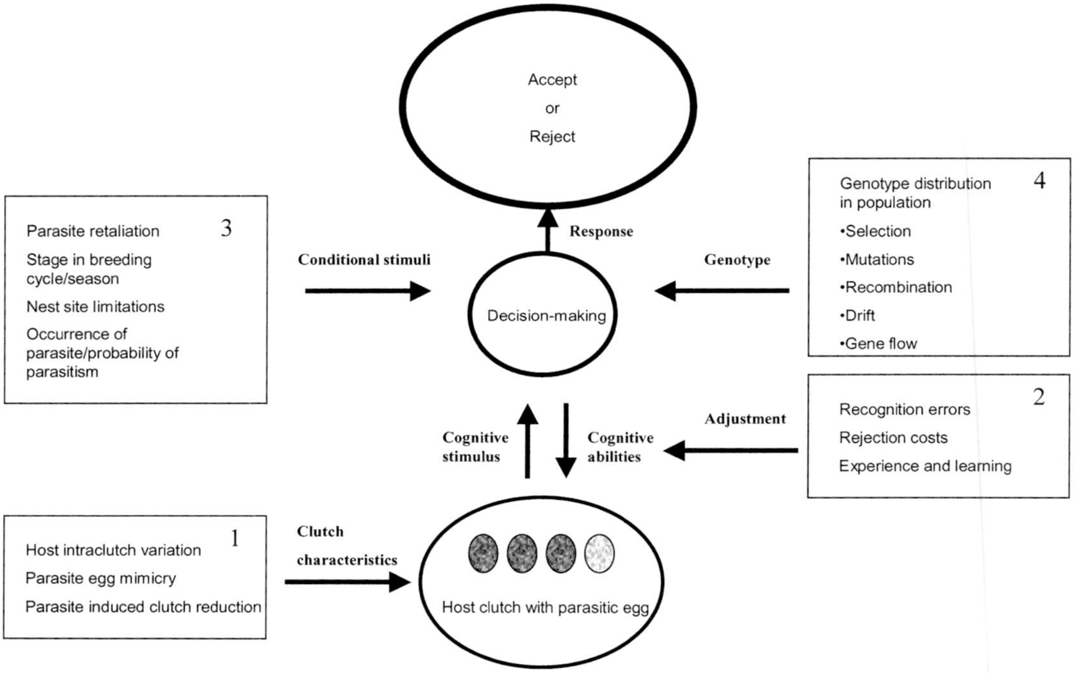


**Fig. 19.54** The process of decision making by parasitized host species. Decisions depend on cognitive inputs about clutch characteristics, including how variable host eggs are in appearance, how similar the parasite egg is to host eggs, and possible changes in the number of eggs present (box 1). The ability of hosts to recognize parasite eggs that differ in appearance from their own eggs is likely influenced by experience and learning, but hosts must also consider the possibility of recognition errors (e.g., possibly removing their own egg) and rejection costs (e.g., damaging their own eggs in the process of removing the parasite egg) (box 2). Additional factors to consider include the possibility of parasite retaliation (mafia behavior), the likelihood of being parasitized (e.g., parasite population density and frequency of parasite observations), and, if a nest is abandoned, the time available to re-nest and the likelihood of finding another suitable nest site (box 3). In addition, the decision-making process will be influenced by the host's genotype, which in turn will be influenced by the host species' history of contact with the brood parasite (box 4). All these factors can influence a host's decision to either accept or reject, or abandon a nest with, a brood parasite's egg. (Figure from Stokke et al. 2005; © 2005 Ornithological Society of Japan, used with permission)

about 20% or more larger than host eggs (Marchetti 2000). Similarly, Eurasian Reed Warblers (*Acrocephalus scirpaceus*) were less likely to eject model eggs the size of normal Common Cuckoo (*Cuculus canorus*) eggs than model eggs that were slightly larger (1.3 times longer) than typical cuckoo eggs (Davies and Brooke 1988). The ability of many host species to detect and reject eggs much larger than their own has favored the evolution of relatively small eggs among parasitic cuckoos that parasitize smaller species (Payne 1974). Krüger and Davies (2004) compared relative egg size (volume) of genera of cuckoos that are not parasitic, genera of parasitic cuckoos that are non-ejectors (their nestlings to do not eject host eggs or young from nests), and genera of parasitic cuckoos that are ejectors (nestlings eject host eggs or young or kill host young and so are raised alone), and found that parasitic cuckoos have relatively smaller eggs than non-parasitic cuckoos and ejectors have even smaller eggs relative to body mass than non-ejectors (Fig. 19.55). As a result, the eggs of parasitic cuckoos tend to match those of their hosts in size, increasing the likelihood of acceptance. Smaller eggs, however, provide additional advantages for parasitic cuckoos, including providing more opportunities to parasitize nests

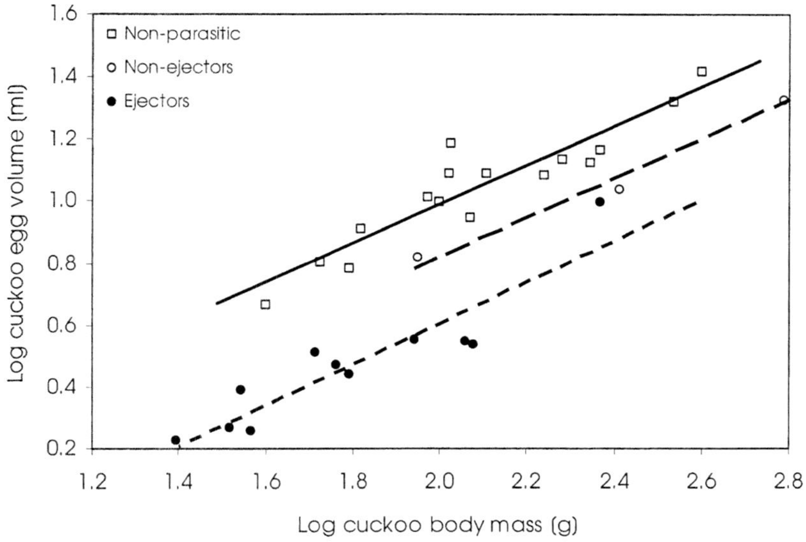


**Fig. 19.55** Relationships between mean body mass and mean egg volume for 16 genera of non-parasitic cuckoos, three genera of non-ejector parasitic cuckoos (cuckoos whose nestlings do not eject host nestlings from nests), and 11 genera of ejector parasitic cuckoos (cuckoos whose nestlings eject host nestlings from nests). The general relationship between egg volume and body mass (i.e., slopes of the lines) is similar for all three groups. However, non-ejectors have smaller eggs relative to their body mass than non-parasitic cuckoos, and ejectors have even smaller eggs relative to body mass than non-ejectors. (Figure from Krüger and Davies 2004; © 2004 Oxford University Press, used with permission)

(because there are more small species than large species), each egg requires less energy to produce so female cuckoos can produce and lay more eggs, and improving incubation efficiency because smaller host females can more effectively incubate smaller eggs (Krüger and Davies 2004). Ejector cuckoos likely tend to have smaller eggs than non-ejector cuckoos because some ejectors (genus *Chrysococcyx)* parasitize host species that build domed nests with relatively small entrances and, as a result, natural selection has favored a reduction in body size that has resulted in a corresponding reduction in egg size (Krüger and Davies 2004). This reduction in size is apparent in Fig. 19.55, with most ejector cuckoos weighing less (i.e., most data points are further to the left) than most non-ejectors and non-parasitic cuckoos.

The appearance of host eggs relative to parasite eggs influences host behavior in some host-parasite systems, but not others. For example, most cowbirds (*Molothrus* spp.) are host generalists and their eggs, in most cases, do not strongly resemble those of their hosts. Despite this lack of resemblance, Rothstein (1975) placed model eggs of Brown-headed Cowbirds (*Molothrus ater*) in the nests of 30 host species and found that individuals in only seven of these species, including American Robins (*Turdus migratorius*) and Gray Catbirds (*Dumetella carolinensis*; Fig. 19.56), typically recognized and rejected cowbird eggs; individuals in the other 23 species typically accepted cowbird eggs. Most of these species, both acceptors and rejectors, have eggs easily distinguishable from those of cowbirds and would be physically capable of ejecting a cowbird egg from their nests (Rothstein 1975). These results suggest that there is little selection pressure on cowbirds to produce eggs more similar in appearance to those of most of their host species, and provide support for the evolutionary lag hypothesis.

Using microsatellite genotyping to characterize host use by individual female Brown-headed

**Fig. 19.56** Ejection of Brown-headed Cowbirds (*Molothrus ater*) eggs by (**a**) an American Robin (*Turdus migratorius*) and (**b**) a Gray Catbird (*Dumetella carolinensis*). (Figure from Rasmussen et al. 2009; © 2009 Oxford University Press, used with permission)

Cowbirds during a single breeding season in Illinois, Strausberger and Ashley (2005) found that most females were host generalists and parasitized up to five host species. They also found that female cowbirds avoided species that would likely be unsuitable hosts, including Mourning Doves (*Zenaida macroura*), Barn Swallows (*Hirundo rustica*), and Common Grackles (*Quiscalus quiscula*). Mourning Doves and Barn Swallows have special structural and behavioral adaptations for food exchange between adults and nestlings that likely prevent successful rearing of cowbirds (Friedmann 1929). Common Grackles are poor cowbird hosts because of their large size (Peer and Bolinger 1997). Such avoidance behavior is likely innate because, if learning was involved, some parasitism by inexperienced female cowbirds would be expected (Strausberger and Ashley 2005). Some female cowbirds parasitized nests of species known to be rejector species such as Field Sparrows (*Spizella pusilla*), Brown Thrashers (*Toxostoma rufum*), and Gray Catbirds (*Dumetella carolinensis*), likely because some individuals in these rejector species do accept cowbird eggs. A possible explanation for such acceptance is that some rejector species may develop egg recognition and the subsequent ability to detect and reject cowbird eggs by imprinting on their own eggs. However, parasitism of young, naïve birds nesting for the first time could cause them to mistakenly misimprint on, and accept, cowbird eggs (Strausberger and Rothstein 2009, but see Soler et al. 2013).

Other obligate brood parasites lay eggs that are very different in appearance from those of their hosts. In Australia, Shining Bronze-Cuckoos (*Chrysococcyx lucidus*), Little Bronze-Cuckoos (*Chrysococcyx minutillus*), and Gould's Bronze-Cuckoos (*Chrysococcyx russatus*) lay brown or olive-colored eggs, whereas their hosts lay white or speckled white eggs (Fig. 19.57). Phylogenetic analysis revealed that these dark eggs have evolved in species of bronze-cuckoos that parasitize host species that build enclosed nests where light levels are low (Langmore et al. 2009). In such conditions, the dark color of the bronze-cuckoo eggs likely functions to camouflage them and reduce the likelihood of rejection by hosts (Langmore et al. 2009).

Driven by the rejection behavior of hosts, many obligate brood parasites have evolved eggs that closely resemble those of host species. For example, some female Greater Honeyguides (*Indicator indicator*) parasitize only host species that nest in tree cavities, whereas others parasitize only host species that nest in burrows (Spottiswoode et al. 2011). Because color differences are likely unimportant for egg recognition in dark cavities and burrows, natural selection has favored Greater Honeyguide eggs that tend to match the shape and size of those of their five most common host species (Fig. 19.58). Interestingly, genetic analysis

**Fig. 19.57** Egg of a host species, Large-billed Gerygone (*Gerygone magnirostris*; top), and examples of three much darker eggs laid by three female Gould's Bronze-Cuckoos (*Chrysococcyx russatus*) (Gould's Bronze-Cuckoos are considered a subspecies of *Chrysococcyx minutillus* by some authors). (Figure from Langmore et al. 2009b; © 2009 Published by Elsevier Ltd., used with permission)

revealed that female Greater Honeyguides that parasitize cavity nesters and burrow nesters, respectively, represent two mitochondrial DNA lineages that have maintained those host preferences for perhaps hundreds of thousands of years. Male and female honeyguides reared by different hosts apparently interbreed often enough to prevent divergence of nuclear DNA and development of reproductive isolation (Spottiswoode et al. 2011).

Several species of cuckoos (*Cuculus* and *Chrysoccoccyx* spp.; e.g., Brooke and Davies 1988) and Parasitic Weavers (*Anomalospiza imberbis*; Spottiswoode and Stevens 2010) parasitize open-cup nesting birds and so have evolved eggs that more closely resemble those of their hosts than those of Greater Honeyguides (*Indicator indicator*). In addition, different host-races (gentes) of Common Cuckoos (*Cuculus canorus*) parasitize, and their eggs mimic, those of particular host species (Fig. 19.59). Until recently, investigators examining the extent to which parasite eggs mimicking host eggs relied on human vision, i.e., how similar parasite and host eggs are to the human eye. However, avian vision differs from human vision. Many birds, unlike humans, have tetrachromatic vision, with four types of cones in their retinas that have peak sensitivities to long-wave, medium-wave, short-wave, and ultraviolet or violet light, plus double cones thought to be important in pattern and texture vision (Cuthill 2006; Fig. 19.60). One approach to understanding how birds might perceive color is to measure reflectance spectra of colors (i.e., determine their wavelengths) and then map them in avian tetrahedral color space (Fig. 19.61). Several investigators have used this approach to examine how closely parasite eggs actually "match" those of hosts from the host's perspective. For example, Stoddard and Stevens (2011) compared the background colors and spots of eggs of Common Cuckoo gentes and their principal hosts and found that background and spot appearance were strongly mimicked by most cuckoo gentes (Fig. 19.62). Analysis also revealed a significant correlation between how well background colors of cuckoo eggs matched that of their hosts and host rejection rates, with the strongest matches found for cuckoos parasitizing hosts with the highest rejection rates (Fig. 19.62). In contrast, no significant correlation was found between spot color mimicry and host rejection

**Fig. 19.58** Host and Greater Honeyguide (*Indicator indicator*) eggs from parasitized clutches of each host species in Zambia. Host eggs are in the left column and parasite eggs in the right column. Host-specific differentiation in honeyguide egg size and shape is apparent. For example, note the small honeyguide egg from a Little Bee-eater nest and the rounder egg from a Striped Kingfisher nest. (Photo of eggs from Spottiswoode et al. 2011; used with permission of the U. S. National Academy of Sciences)

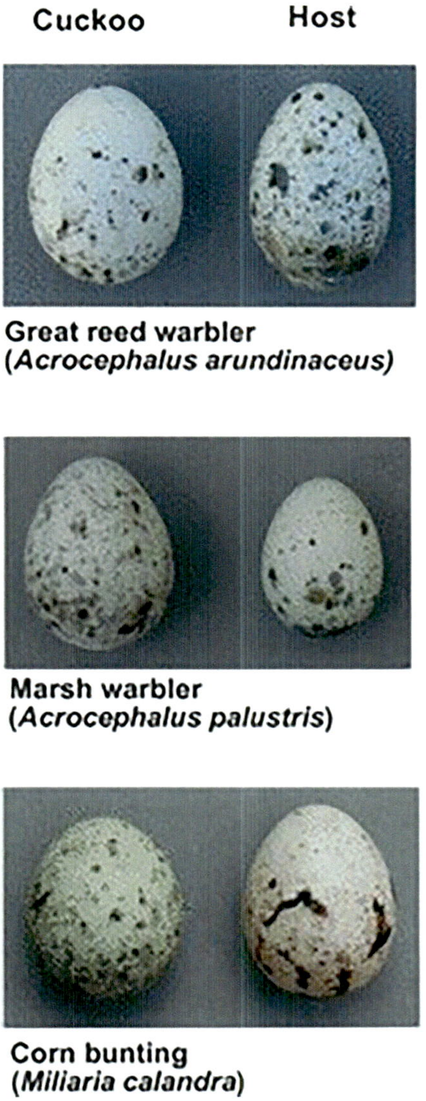


**Fig. 19.59** Representative eggs of three Common Cuckoo gentes and their hosts. Eggs of cuckoo gentes that parasitize Great Reed Warblers (*Acrocephalus arundinaceus*) and Marsh Warbler (*A. palustris*) better mimic those of the host species than do eggs of the cuckoo gens that parasitize Corn Buntings (*Emberiza calandra*). Great Reed and Marsh warblers have high rejection rates of non-mimetic eggs (75 and 88%, respectively; Antonov et al. 2006a), whereas Corn Buntings have lower rejection rates of non-mimetic eggs (42%, Antonov et al. 2006b). (Egg photos from Antonov et al. 2010; © 2010 The Authors. Journal Compilation © 2010 European Society for Evolutionary Biology, used with permission)

rates, suggesting that rejection rates may be influenced more by mimicry of the patterns of spots than their color (Stoddard and Stevens 2011). Overall, these results suggest that, in response to the evolution of stronger host defenses (i.e., higher rejection rates), cuckoos have evolved eggs with backgrounds (and spotting patterns) that better match those of their hosts (Stoddard and Stevens 2011).

Such results clearly demonstrate the coevolutionary dynamics between brood parasites and their hosts. Natural selection may favor an improved ability by brood parasites to mimic the eggs of their hosts and, at the same time, favor hosts better able to distinguish between parasitic eggs and their own eggs. Among some host species, natural selection has favored the evolution of polymorphic eggs, with increased variation in egg appearance among different females making it more difficult for brood parasites to mimic the range of eggs present. For example, Tawny-flanked Prinias (*Prinia subflava*) are often parasitized by Parastic Weavers (*Anomalospiza imberbis*) in southern Zambia and, in response, selection has favored extensive inter-individual variation in the appearance of prinia eggs (Spottiswoode and Stevens 2012). Each female always lays similar eggs, but the eggs of different females exhibit extensive variation (Fig. 19.63). Female Cuckoo Finches lay equally variable eggs (Fig. 19.63), but do not selectively parasitize prinia nests where eggs are more similar to theirs. As a result, female Tawny-flanked Prinias are often able to identify Parasitic Weaver eggs and reject them. Only if Parasitic Weaver eggs happen by chance to resemble those in a parasitized nest do female Tawny-flanked Prinias accept them (Spottiswoode and Stevens 2010). In this system, hosts and parasites are engaged in a coevolutionary arms race in egg appearance driven by changes in the selective advantage of different phenotypes (Spottiswoode and Stevens 2012).

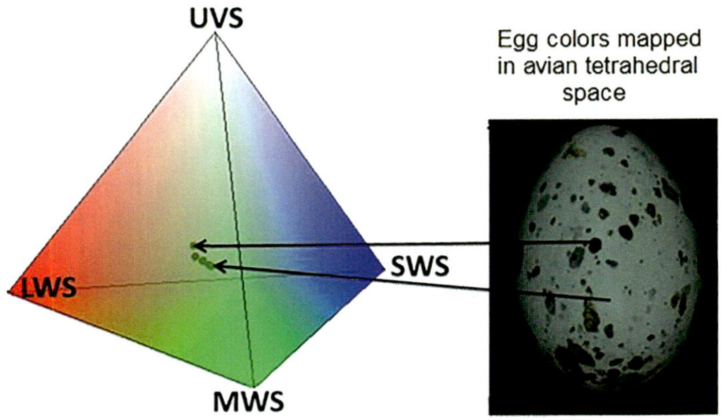


**Fig. 19.60** Many birds, unlike humans, have tetrachromatic vision, with four types of cones in their retinas that have peak sensitivities to long-wave (LWS, or red), medium-wave (MWS), short-wave (SWS, or blue), and ultraviolet (UVS) or violet light. In this example, each point in the tetrahedron represents how colors at various points on a bird egg might be perceived by a bird. The color white is achromatic (i.e., all colors combined) and so would be centrally located in the tetrahedron. Closer to the edges, colors are more saturated or intense. (image credit: Mary Caswell Stoddard. Egg photo was taken at the Natural History Museum in Tring [UK], © The Trustees of the Natural History Museum, London 2011, used with permission)

### 19.12.6 Obligate Brood Parasitism: Post-hatching Adaptations of Hosts and Parasites

The coevolutionary arms race between host species and obligate brood parasites continues through the nestling and even fledgling stages, with nestling parasites evolving strategies to improve their likelihood of being fed by host adults at optimal rates and hosts evolving strategies to minimize the cost of being parasitized (Table 19.4). One adaptation of the nestlings of some obligate brood parasites is to eliminate competition for parental care by removing host eggs and nestlings or killing host nestlings. Nestling cuckoos (*Cuculus* spp.) dispose of host eggs or nestlings by lifting them onto their backs and out of nest cups (Fig. 19.64). This disposal or eviction behavior typically begins when young cuckoos are 1 or 2 days old and occurs whether or not host adults are present at the nest (Honza et al. 2007; Grim et al. 2009; Fig. 19.65). Impressively, young Common Cuckoos (*Cuculus canorus*) and Horsfield's Bronze-Cuckoos (*Chalcites basalis*) are able to evict eggs and nestlings more than twice their own body weight (Davies 2000). Grim et al. (2009) studied the eviction behavior of nestling Common Cuckoos by placing models of host eggs in nests on the day young cuckoos hatched and found that the cuckoos first exhibited eviction behavior when 1 day old. Also, by returning "evicted" eggs to nests each day, they found that eviction behavior peaked when young cuckoos were 4 days old, then rapidly declined over the next few days (Fig. 19.65). Nest architecture and location can sometimes limit the ability of nestling cuckoos to eject host eggs or young. Young cuckoos in cup nests with steep walls or in cavity-nests are not always able to evict some or all host eggs or young (Rutila et al. 2002; Grim et al. 2009).

Nestling honeyguides (Indicatoridae; Africa and Asia) and Striped Cuckoos (*Tapera naevia*; Neotropics) kill host young by attacking them with specially modified bill hooks (Fig. 19.66). Nestling Greater Honeyguides (*Indicator indicator*) typically hatch before host young and attack the host nestlings with their hook-like bills shortly after they hatch, repeatedly biting individual host nestlings (Little Bee-eaters, *Merops pusillus*) for periods ranging from about 1 to

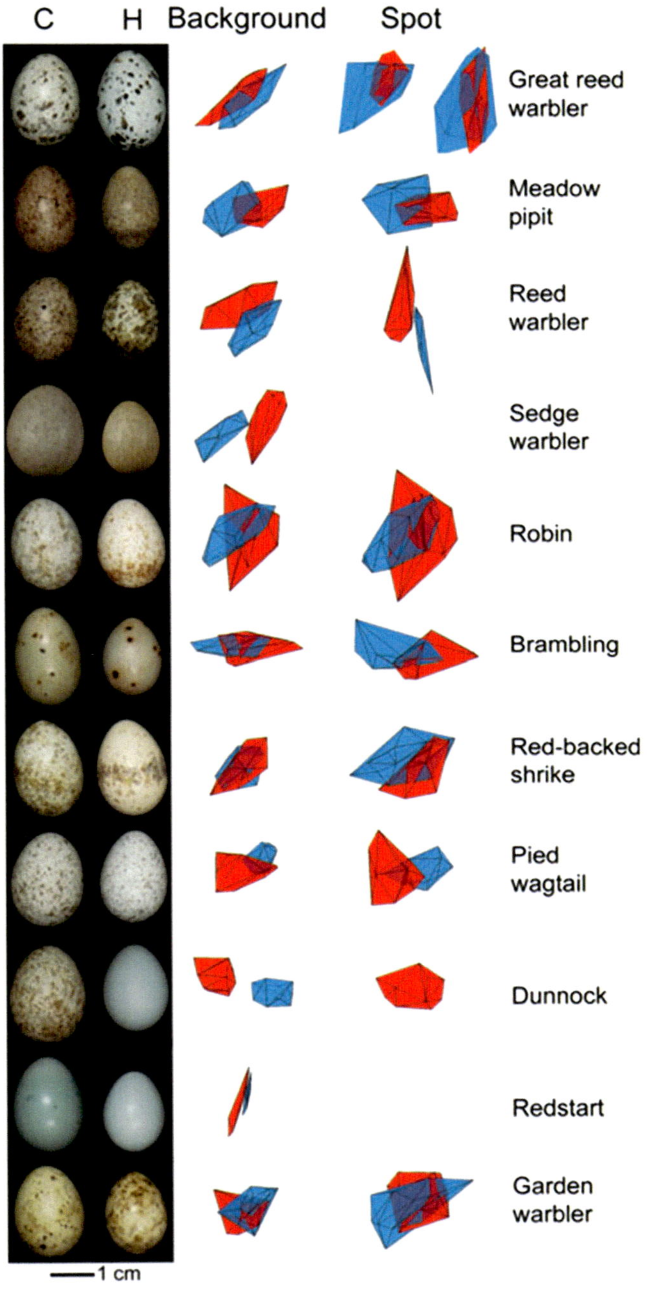


**Fig. 19.61** Photographs of cuckoo (left) and host (right) eggs are shown with their background and spot color distributions in avian tetrahedral color space, illustrating the extent of color overlap between cuckoos (red) and hosts (blue). Greater overlap indicates greater color similarity. Eggs of Great Reed Warblers (*Acrocephalus arundinaceus*) and their cuckoo gens have both dark and light spots so data from both are provided. Eggs of the Sedge Warbler (*Acrocephalus schoenobaenus*) and its cuckoo were so densely covered with spots that only one color (including portions of background and spot) was measured, shown as the background color. Photographs of eggs within the figure are copyright of the Natural History Museum (Tring, Hertfordshire, UK) and UMZC (Sedge Warbler-cuckoo and Sedge Warbler eggs) and were taken by Mary Caswell Stoddard. Meadow Pipit, *Anthus pratensis*; Eurasian Reed Warbler, *Acrocephalus scirpaceus*; European Robin, *Erithacus rubecula*;

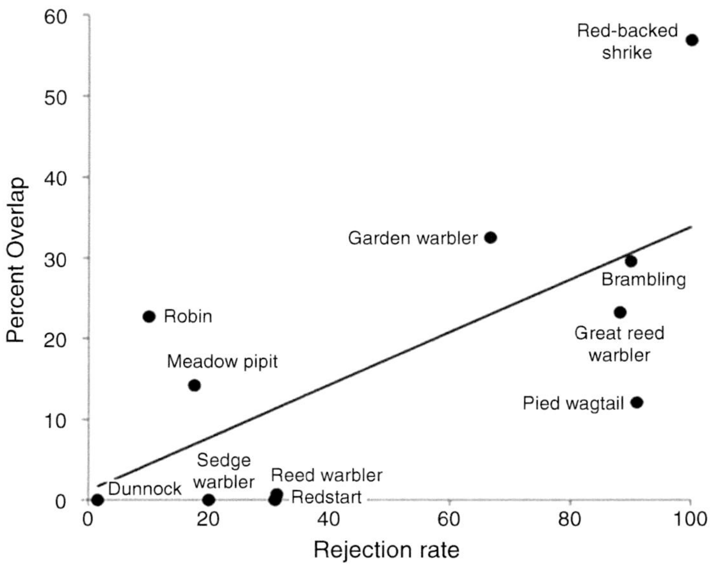


**Fig. 19.62** Relationship between host rejection rate and overlap in the background color of host eggs and those of Common Cuckoo (*Cuculus canorus*) gentes that parasitize those hosts. Hosts with eggs more closely mimicked by cuckoos have the highest rejection rates. Meadow Pipit, *Anthus pratensis*; Eurasian Reed Warbler, *Acrocephalus scirpaceus*; European Robin, *Erithacus rubecula*; Brambling, *Fringilla montifringilla*; Red-backed Shrike, *Lanius collurio*; Pied (White) Wagtail, *Motacilla alba*; Dunnock, *Prunella modularis*; Common Redstart, *Phoenicurus phoenicurus*; Garden Warbler, *Sylvia borin*. (Figure from Stoddard and Stevens 2011; © 2011 The Authors. Evolution © 2011 The Society for the Study of Evolution, used with permission)

5 min (Spottiswoode and Koorevaar 2012). From the time of the first attacks, young bee-eaters took from 9 min to 7 h to die (Spottiswoode and Koorevaar 2012). The bill hooks of nestling Greater Honeyguides are retained for about 14 days and are not shed, but grow gradually into the growing bill and are no longer apparent about 1 week before fledging (about 30 days post-hatching; Spottiswoode and Koorevaar 2012).

In contrast to honeyguides and Striped Cuckoos (*Tapera naevia*), the nestlings of other obligate brood parasites (e.g., cowbirds, *Clamator* cuckoos, and parasitic estrildids) do not intentionally remove host eggs and nestlings or kill host nestlings (i.e., host tolerance). One possible explanation for such differences is evolutionary lag, with taxa that most recently evolved brood parasitic behaviors like parasitic cowbirds and estrildid finches more likely to lack certain adaptations. This seems unlikely, however, because *Clamator* cuckoos have a long evolutionary history of obligate brood parasitism and, in addition, some host-tolerant obligate brood parasites, including Asian Koels (*Eudynamys scolopacea*) and Channel-billed Cuckoos (*Scythrops novaehollandiae*), are descended from ancestors with nestlings that were egg and nestling ejectors (Kilner 2005). This secondary

**Fig. 19.61** (continued) Brambling, *Fringilla montifringilla*; Red-backed Shrike, *Lanius collurio*; Pied (White) Wagtail, *Motacilla alba*; Dunnock, *Prunella modularis*; Common Redstart, *Phoenicurus phoenicurus*; Garden Warbler, *Sylvia borin*. (Figure from Stoddard and Stevens 2011; © 2011 The Authors. Evolution © 2011 The Society for the Study of Evolution, used with permission)

**Fig. 19.63** Eggs of Tawny-flanked Prinias (*Prinia subflava*) and Parasitic Weavers (*Anomalospiza imberbis*). Individual female Tawny-flanked Prinias always lay the same egg type, but, among different females, eggs range in color from white to bluish to brick red to olive green, overlaid with a wide diversity of markings. Eggs of female Parasitic Weavers display similar variation. Parasitic Weavers are sometimes referred to as Cuckoo Finches. (Figure from Spottiswoode and Stevens 2010; used with permission of the U. S. National Academy of Sciences)

loss of eviction behavior is unlikely to have occurred unless it provided some fitness benefit for these brood parasites (Kilner 2005).

Differences among obligate brood parasites in the behavior of their young are likely the result of interspecific differences in the respective costs and benefits of evicting versus not evicting host eggs and nestlings (or killing versus not killing host nestlings). One possible cost of evicting is the risk of falling out of host nests. For example, Grim et al. (2009) video-recorded parasitized nests of Common Redstarts (*Phoenicurus phoenicurus*) and observed a young Common Cuckoo (*Cuculus canorus*) climb over the rim of the nest while evicting an egg. Another cost of evicting or killing by parasite young is the time and energy required to do so. Grim et al. (2009) compared the growth rates of nestling Common Cuckoos in nests with and without host eggs to evict and found that, during the eviction phase (until 8 days after hatching), those that evicted eggs grew more slowly than those that did not, with "evictors" using energy to evict eggs and being fed less than "non-evictors" because of their focus on evicting rather than feeding (sometimes even ignoring feeding attempts by host parents). At fledging, the masses of "evictor" and "non-evictor" cuckoos were similar, but "evictors" fledged a day later than "non-evictors." In their experiment, Grim et al. (2009) removed host eggs and replaced them with artificial eggs so all young cuckoos were raised alone. Because young cuckoos are sometimes unable to evict all host eggs, they are sometimes raised along with host young. In such cases, young cuckoos weakened by their eviction efforts may be less able to effectively compete with host young for food from host parents, with a potentially greater negative impact on their growth rate, time needed to fledge, and condition at fledging (Grim et al. 2009). Given these potential costs, evicting host eggs and nestling or killing host nestlings must also provide important benefits.

Parasite nestlings benefit by being fed by host parents, and benefit most when host parents provide them with sufficient food to achieve maximal growth rates and to fledge in the shortest time possible and in the best condition possible to maximize their likelihood of survival. By evicting host eggs and nestlings, young parasites no longer must compete for food delivered by host parents and this seemingly would result in being fed at optimal rates. This, however, may not always be the case because provisioning rates of host parents are influenced by nestling begging behavior and the begging of a single parasite nestling may not be sufficient to elicit optimal provisioning rates. In such cases, parasite nestlings may

**Table 19.4** Evidence of an evolutionary arms race between brood parasites and their hosts at the nestling stage

| |
|---|
| *Examples of strategies used by brood parasitic nestlings to reduce competition:* |
| • Eviction of host eggs or nestlings from nests (e.g., Grim et al. 2009) |
| • Killing of host nestlings (e.g., Spottiswoode and Koorevaar 2012) |
| *Examples of deceiving strategies used by brood parasitic nestlings:* |
| • Mimicry of nestling brood parasites and their hosts (e.g., Schuetz 2005) |
| • Mimicry of begging calls (e.g., Anderson et al. 2009a) |
| • Mimicry of begging behavior (e.g., McLean and Waas 1987) |
| • Common Cuckoo (*Cuculus canorus*) chicks learn which calls are most effective at stimulating host provisioning (Madden and Davies 2006; Langmore et al. 2008) |
| • Begging call structure optimized to elicit host care (e.g., Gloag and Kacelnik 2013) |
| • Begging displays of brood parasitic chicks are highly exaggerated (e.g., Tanaka et al. 2011) |
| *Examples of strategies used by host adults:* |
| • Eviction of brood parasitic nestlings from nests (e.g., Sato et al. 2010) |
| • Reluctance to feed parasitic chicks (e.g., Lichtenstein 2001b) |
| • Providing parasitic chicks with lower-quality food items (Soler 2009) |
| *Examples of adaptive strategies of nestlings of host species:* |
| • Increasing the volume of begging calls (Boncoraglio et al. 2009) |
| • Embryonic learning of vocal "passwords" (Colombelli-Négrel et al. 2012) |
| • Altering begging calls to more effectively compete with parasite nestling (Pagnucco et al. 2008) |
| • Competition from brood parasite chicks can affect the evolution of growth strategies of host species (Remeš 2006) |

Table modified from Soler (2009); © 2009 The Authors. Published by John Wiley and Sons, used with permission

benefit from the presence of one or more host nestlings. Nestling Brown-headed Cowbirds (*Molothrus ater*), for example, were fed and grew at higher rates when some host nestlings were present (Fig. 19.67; Kilner et al. 2004). The collective begging by a young cowbird plus host nestlings induces host parents to provision at higher rates and, particularly in nests of smaller host species, young cowbirds may be able to out-compete host nestlings and obtain extra parental feedings. As the number of host nestlings increases, the increased competition for feedings, particularly in the nests of larger host species, may result in lower growth rates for young cowbirds (Fig. 19.67).

Although nestlings of some obligate brood parasites, like Brown-headed Cowbirds, do better (i.e., are fed at higher rates and grow faster) when some host young are present and, therefore, are host-tolerant, other nestlings, like those of Common Cuckoos, evict host eggs and nestlings because they typically do best when no host young are present. In explaining these differences, Gloag et al. (2012) suggested that the best strategy for each species of obligate brood parasite may depend on the interaction between a nestling parasite's ability of successfully compete with host nestlings for food from host parents, and the effectiveness of its begging behavior in stimulating host parent provisioning rates. If begging by a nestling parasite cannot stimulate host parent provisioning as effectively as when some host nestlings are present and if that nestling parasite can effectively compete with host nestlings for food from host parents, then natural selection would likely favor host tolerance. If, however, begging by a parasite nestling alone induces host parents to provision at optimal rates (from the nestling's perspective) and/or a parasite nestling is not very effective at competing with host nestlings for food, then natural selection would likely favor eviction of host eggs and eviction or killing of host nestlings (Gloag et al. 2012).

Many potential host species recognize and either reject brood parasite eggs or abandon parasitized nests. As a result, natural selection has, in many cases, favored the evolution of brood parasite eggs that mimic the appearance of host eggs. Similar mimicry, however, is less

**Fig. 19.64** Hatchling Common Cuckoos (*Cuculus canorus*) in the process of evicting host eggs and chicks from nests of Great Reed Warblers (*Acrocephalus arundinaceus*). (Figure from Anderson et al. 2009b; © 2009 Anderson et al., open-access article distributed under the terms of the Creative Commons Attribution License, https://creativecommons.org/licenses/by/4.0/)

common among nestlings, with brood parasite nestlings often strikingly different in appearance (and, in some cases, size) from that of host nestlings. Despite such differences in appearance, most host parents continue to care for brood parasite young (Fig. 19.68). This suggests that, in general, the selective pressure for recognition of parasite nestlings by hosts is less intense than that for recognition of parasite eggs. For any behavior to be adaptive, its benefits must outweigh its costs. When a brood parasite begins to exploit a new host species (Fig. 19.69a), the parasitism rate must reach some level (Fig. 19.69b) before the benefits of host defenses, such as parasite egg discrimination and rejection, exceed the costs (e.g., hosts rejecting or damaging their own eggs). As described by Grim (2006a), an increase in the effectiveness of host defenses could then reduce parasite fitness, reducing parasitism rates and simultaneously reducing selective pressures on hosts to recognize parasite eggs (Fig. 19.69c). This relaxed selective pressure for host recognition of parasite eggs could then lead to yet another increase in parasitism rates (Fig. 19.69d). As this co-evolutionary arms race continues, the selective pressure for successful parasitism would be expected to exceed that for development of successful anti-parasitism adaptations of hosts because of what Dawkins (1979) referred to as the "life-dinner" principle.

The "life-dinner" principle refers to the unequal selection pressures that arise when outcomes of interactions have different costs. For example, "The rabbit runs faster than the

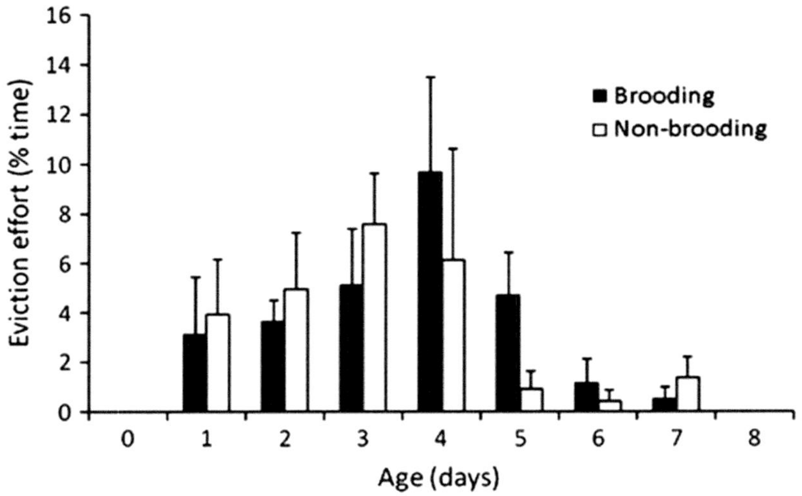


**Fig. 19.65** Variation with age in mean time (± standard error) spent by nestling Common Cuckoos (*Cuculus canorus*) evicting artificial host eggs in nests of Common Redstarts (*Phoenicurus phoenicurus*). Young cuckoos exhibited eviction behavior both when and when not being brooded of host parents. Cuckoos hatched on day 0. All redstart nests had five artificial eggs on day 0 and "evicted" eggs were returned to nests each day through day 8. (Figure from Grim et al. 2009; © 2009 Oxford University Press, used with permission)

fox, because the rabbit is running for his life while the fox is only running for his dinner" (Dawkins 1979, after Aesop). In host-parasite interactions, the cost of failure for brood parasites (i.e., host recognition and rejection of their egg) is the death of their embryo; for hosts, the cost of failure (i.e., not recognizing the parasite egg) may simply be the loss of one brood, with other opportunities to breed in the future. This difference in selection pressures means that, in many cases as described previously, parasite eggs become increasingly similar in appearance to host eggs. As this egg mimicry improves and host rejection diminishes, parasitism rates may increase substantially (Fig. 19.69e). At that point, selection would favor the evolution of hosts able to recognize and reject parasite nestlings (Fig. 19.69f; Grim 2006a, b). This point, however, has rarely been reached because only a few host species (described below) recognize and reject parasite nestlings. This is likely the case because individuals in host species that regularly reject parasite eggs are less likely to be successfully parasitized and, therefore, selection has not favored an ability to recognize parasite nestlings. In other words, for egg rejectors, the costs of caring for parasite young (e.g., failure or partial failure of a single breeding attempt) are insufficient to favor the evolution of nestling discrimination. For non-egg rejectors, however, those costs are sufficient to favor such discrimination and rejection of parasite nestlings (Grim 2006a, b).

Among host species able to discriminate between parasite nestlings and their own nestlings, responses to the presence of parasite nestlings may include (1) desertion of nests with parasite nestlings (e.g., Langmore et al. 2003), (2) refusal to feed parasite nestlings or feeding them less than host nestlings (Payne et al. 2001; Delhey et al. 2011), or (3) removing nestling parasites from nests. For example, some Mangrove Gerygones (*Gerygone laevigaster*), but not all, remove Little Bronze-Cuckoo (*Chalcites minutillus*) nestlings from their nests (Tokue and Ueda 2010; Fig. 19.70). Similar parasite-nestling rejection has been reported in a few other host species (e.g., Sato et al. 2010). In cases where host adults do discriminate between parasite nestlings and their own nestlings, host adults

**Fig. 19.66** (**a**) Hatching Lesser Honeyguide (*Indicator minor*) showing. (**b**) Greater Honeyguide (*Indicator indicator*) chick with three recently killed Little Bee-eater (*Merops pusillus*) hatchlings, (**c**) biting unhatched Swallow-tailed Bee-eater (*Merops hirundineus*) egg (but young honeyguides were not able to actually puncture bee-eater eggs), and (**d**) Greater Honeyguide nestling about 8 days old. (Figure from Spottiswoode and Koorevaar 2012; open-access article distributed under the terms of the Creative Commons Attribution License, https://creativecommons.org/licenses/by/4.0/)

may or may not recognize the difference (e.g., in appearance and/or vocalizations) between host and parasite nestlings and, when it occurs, the ability to recognize parasite nestlings may be either innate or learned. Discrimination without recognition (i.e., recognition-free rejection) may occur if, for example, parasitic nestlings take much longer to fledge than would host nestlings and, as a result, host adults abandon nests rather than prolonging parental care beyond the time host young would normally fledge (Grim 2007).

Recognition and rejection of parasite nestlings seems particularly prevalent among hosts of Australian bronze-cuckoos (*Chalcites* spp.). Hosts of these cuckoos typically do not reject the mimetic or cryptic cuckoo eggs (Langmore et al. 2009; Langmore and Kilner 2010), rather, their primary defense against cuckoos occurs after hatching, with hosts sometimes rejecting parasite young within a few days after hatching either by abandoning nests (Langmore et al. 2009) or removing the parasite chick from their nests (Tokue and Ueda 2010). In response to this rejection behavior, selection has favored visual mimicry of host young by bronze-cuckoo nestlings. Langmore et al. (2011) compared the amount of

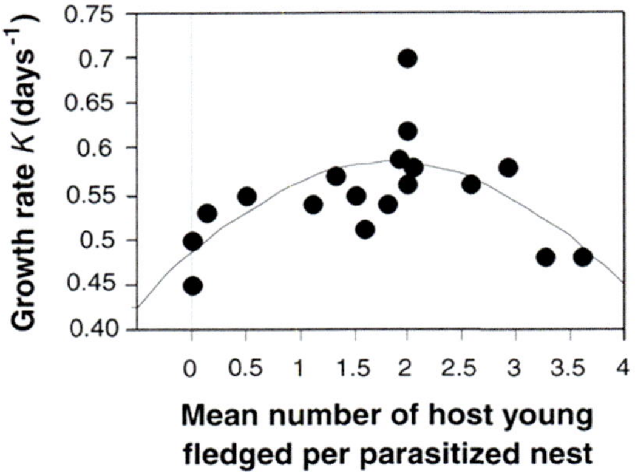


**Fig. 19.67** Relationship between growth rates of nestling Brown-headed Cowbirds (*Molothrus ater*) and number of host young in nests of 18 different host species. Each point represents the cowbird's growth rate in nests of one host species. Young cowbirds grew fastest when about two host nestlings were present. *K* is called the growth rate constant and is a measure of the average rate of growth during the nestling period. (Figure from Kilner et al. 2004; © 2004 American Association for the Advancement of Science, used with permission)

nestling down, skin color, and rictal flange color of bronze-cuckoo nestlings with those of nestlings of their primary hosts, and found striking similarity (Fig. 19.71). Importantly, Langmore et al. (2011) compared colors as viewed by birds (i.e., including ultraviolet wavelengths below 400 nm) rather than as viewed by humans (i.e., 400–700 nm). In addition to closely resembling host nestlings in appearance, bronze-cuckoo nestlings are also vocal mimics, with begging calls similar to that of host young (Langmore and Kilner 2010). This level of nestling mimicry is impressive, but also unusual, if not unique. One possible explanation for this is that host defenses that may be effective in other host-parasite systems, e.g., rejecting parasite eggs, are not effective for hosts of Australian Bronze-cuckoos because bronze-cuckoo eggs are mimetic or cryptic. If so, as noted previously, there would be no selective advantage for hosts recognizing parasite nestlings in other host-parasite systems and, therefore, for parasite young to mimic host young (Langmore et al. 2011).

Other cases of apparent mimicry of host nestlings by parasite nestlings have been reported among *Vidua* finches in Africa, with the gape patterns of nestling *Vidua* finches precisely matching those of the nestlings of their single host species of estrildid finch (Fig. 19.72). One

**Fig. 19.68** Common Cuckoo (*Cuculus canorus*) nestling and its host parent, a female Eurasian Reed Warbler (*Acrocephalus scirpaceus*). Eurasian Reed Warblers care for young cuckoos despite their lack of resemblance to young Eurasian Reed Warblers. (Figure from Stoddard 2012; photo credit: D. Kjaer; © 2015 Oxford University Press, used with permission)

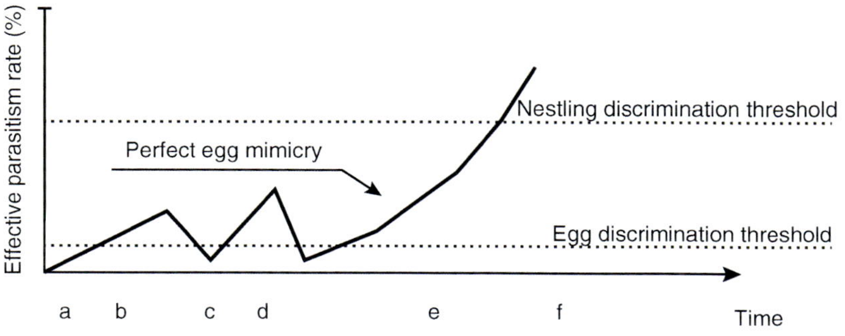


**Fig. 19.69** Hypothetical changes in effective parasitism rates relative to two thresholds that selection pressure must overcome to select for host discrimination of parasite eggs and nestlings. (**a**) Parasite begins to exploit host, (**b**) rate of parasitism increases to a level selecting for host defenses, (**c**) rate declines, selecting against host defenses which, in turn, (**d**) results in an increase in parasitism rates; selection for host defenses then increases, reducing parasitism rates, but (**e**) selection pressure is stronger for parasite adaptions than host counter-adaptations, leading to 'perfect' egg mimicry and a subsequent increase in parasitism rates followed by (**f**) the possibility (or not, resulting in a continued increase in the effective parasitism rate) of nestling discrimination by the host (potentially resulting in a subsequent reduction in the effective parasitism rate) (Figure from Grim 2006a, used with permission of Tomas Grim)

possible explanation for such mimicry is that it is favored by natural selection because host adults are more likely to provision, and less likely to reject, parasite nestlings with gapes that resemble those of their own young (Schuetz 2005). In turn, selection might favor gapes or mouth markings of host young that diverge in appearance from those of parasite young so host adults can better recognize their own young. What might result from these conflicting selection pressures is the evolution of rather elaborate mouth markings in host young (to be more distinctive) that is then mimicked by parasite nestlings (Hauber and Kilner 2007). However, the primary function of complex

**Fig. 19.70** A Mangrove Gerygone (*Gerygone levigaster*) evicting a live Little Bronze-Cuckoo (*Chalcites minutillus*) nestling from the nest. (Figure from Tokue and Ueda 2010; © 2010 The Authors. Journal compilation © 2010 British Ornithologists' Union, used with permission)

**Fig. 19.71** Photographs and mean (± SE) reflectance of the skin of nestling bronze-cuckoos (blue lines) and their hosts (pink lines). (**a**) Little Bronze-Cuckoo (*Chrysococcyx minutillus*) and Large-billed Gerygone (*Gerygone magnirostris*). (**b**) Shining Bronze-Cuckoo (*Chrysococcyx lucidus*) and Yellow-rumped Thornbill (*Acanthiza chrysorrhoa*). (**c**) Horsfield's Bronze-Cuckoo (*Chrysococcyx basalis*) and Superb Fairywren (*Malurus cyaneus*). (Figure from Langmore et al. 2011; © 2011 The Royal Society, used with permission)

mouth markings might be to stimulate the provisioning behavior of adults rather than signaling species identity.

Some young brood parasites not only mimic host nestlings, but natural selection has apparently favored the exaggeration of traits important in eliciting food from host parents. Malaysian Hawk-Cuckoos (*Hierococcyx fugax*) regularly parasitize Red-flanked Bluetails (*Tarsiger cyanurus*), a species that nests in holes in the ground below tree roots. Nestling hawk-cuckoos have gapes and wing patches that, given that most birds are sensitive to ultraviolet wavelengths (300–400 nm), are likely more visible to adult Red-flanked Bluetails than the gapes and wing patches of bluetail nestlings (Fig. 19.73). Both the gape and wing patches of nestling hawk-cuckoos reflect more strongly throughout the visual range of birds, but particularly in the ultraviolet (Fig. 19.73). Because bluetail nests are typically underground and dark, the greater reflectance of the gape and wing patch of nestling hawk-cuckoos may be more visible and possibly serve as a supernormal stimulus, stimulating adult bluetails to provision at higher rates. Although studies have yet to demonstrate such an effect, Tanaka et al. (2011) believe it likely that the exaggerated signals of young hawk-cuckoos is beneficial because development of such signals may have considerable physiological and energetic costs, including pigment deposition, blood flow, and dermal microstructure.

For host species that do not evict parasitic nestlings, another defense mechanism is, when host eggs or nestlings are not evicted by the

**Fig. 19.72** The gape of parasitic Pin-tailed Whydahs (*Vidua macroura*, left) closely matches that of their host nestlings, Common Waxbills (*Estrilda astrild*, right). (Figure from Schuetz 2005; © 2005 Published by Elsevier Ltd., used with permission)

young brood parasite, for host parents to either feed parasitic nestlings less than host nestlings or to provide parasitic nestlings with lower quality food items. Such behavior has been reported in several host species, including Rufous-bellied Thrushes (*Turdus refiventris*; Lichtenstein 2001b), Chalk-browed Mockingbirds (*Mimus saturnines*; Lichtenstein 1997), Rufous-tailed Scrub-Robin (*Cercotrichas galactotes*; Martín-Gálvez et al. 2005), Eurasian Magpies (*Pica pica*; Soler et al. 1995a, b), Bay-winged Cowbirds (*Molothrus badius*; Fraga 1998), and others. In some cases, nestling brood parasites receive less food or lower quality food because they are less effective at eliciting food from host parents (e.g., Fraga 1998). In fewer cases, host adults may be able to discriminate between host and parasitic nestlings and respond accordingly (e.g., Lichtenstein 2001b).

In contrast to those relatively few species where host adults recognize parasite nestlings, most host species do not recognize and reject parasite nestlings and, therefore, provision nestling parasites and, in some cases, some of their own young as well. Parasitic nestlings exhibit a variety of begging strategies to elicit provisioning by host adults. Some young brood parasites have begging calls similar to those of host nestlings, eliminating a potential cue that host adults could use to recognize brood parasites (Grim 2006a, b) and enhancing their ability to effectively communicate their hunger levels to host adults. For example, nestling Common Cuckoos (*Cuculus canorus*) have loud and rapid begging calls that can simulate the calls of an entire brood of host nestlings, stimulating host adults to provision at rates equivalent to that of an entire brood of host nestlings (Davies et al. 1998; Butchart et al. 2003; Fig. 19.74). Matching of host young begging calls by brood parasites has also been reported for other brood parasites, including koels (*Eudynamys* spp.; McLean and Waas 1987), honeyguides (*Indicator* spp; Fry 1974), and some species of indigobirds (*Vidua* spp.; Payne and Payne 2002).

Some nestling brood parasites are able to modify their begging calls so that they more closely mimic those of their host species. For example, Horsfield's Bronze-Cuckoos (*Chrysococcyx basalis*) typically parasitize Superb Fairywrens (*Malurus cyaneus*) and the begging calls of young bronze-cuckoos are similar to those of young fairy-wrens (Fig. 19.75). However, Horsfield's Bronze-Cuckoos sometimes parasitize Buff-rumped Thornbills (*Acanthiza reguloides*) and young bronze-cuckoos then modify their calls as they age so that they more closely resemble those of young thornbills (Fig. 19.75). Surprisingly, this modification occurs in the absence of young thornbills because newly hatched bronze-cuckoos evict host eggs and young after hatching. Langmore et al. (2008)

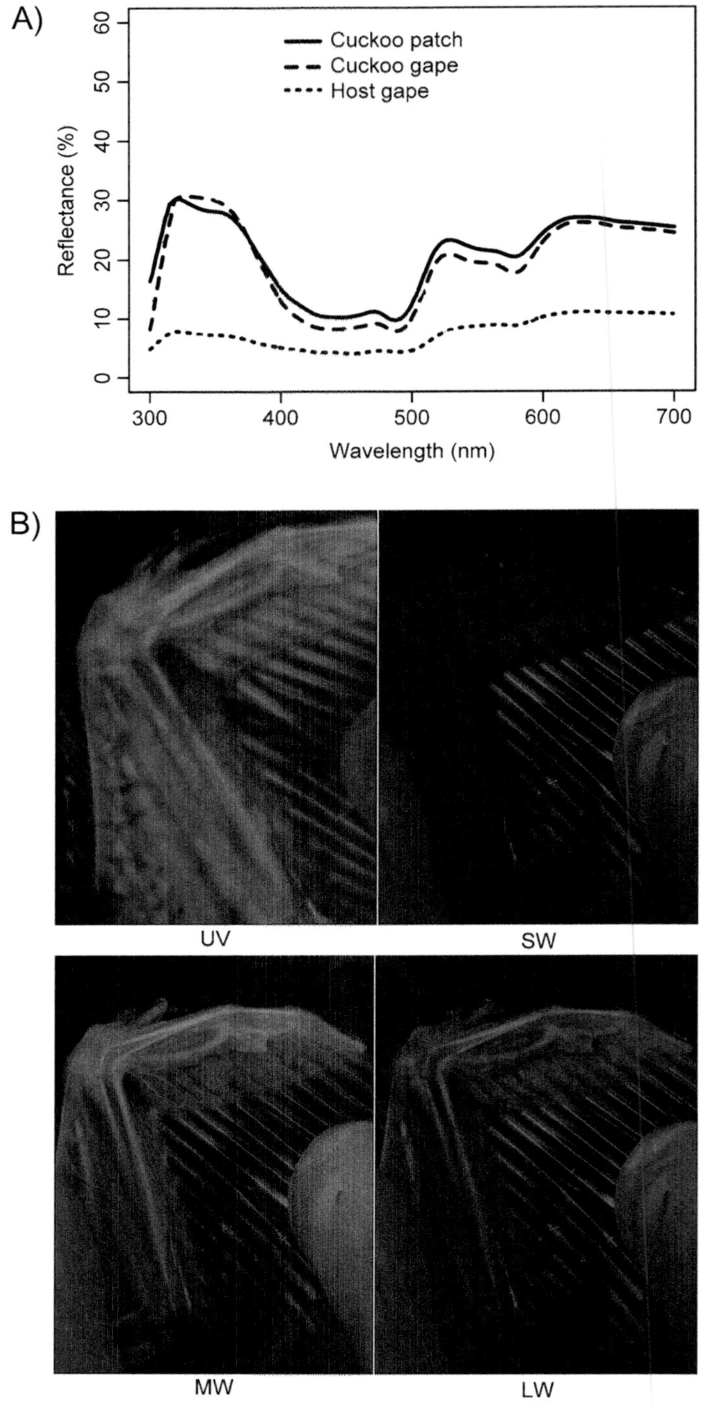


**Fig. 19.73** (**a**) Mean reflectance spectra of respective signaling traits of host (Red-flanked Bluetail, *Tarsiger cyanurus*) and parasite (Malaysian Hawk-Cuckoo, *Hierococcyx fugax*) nestlings. (**b**) Photos of a nestling Malaysian Hawk-Cuckoo wing patch in the long-wave (LW; peak = 565–570 nm), medium-wave (MW; peak = 500–510 nm), short-wave (SW; peak = 430–460 nm), and ultraviolet (UV, <400 nm) wavelengths, showing the strong UV reflectance of the wing patch. (Figure from Tanaka et al. 2011; © 2011 Oxford University Press, used with permission)

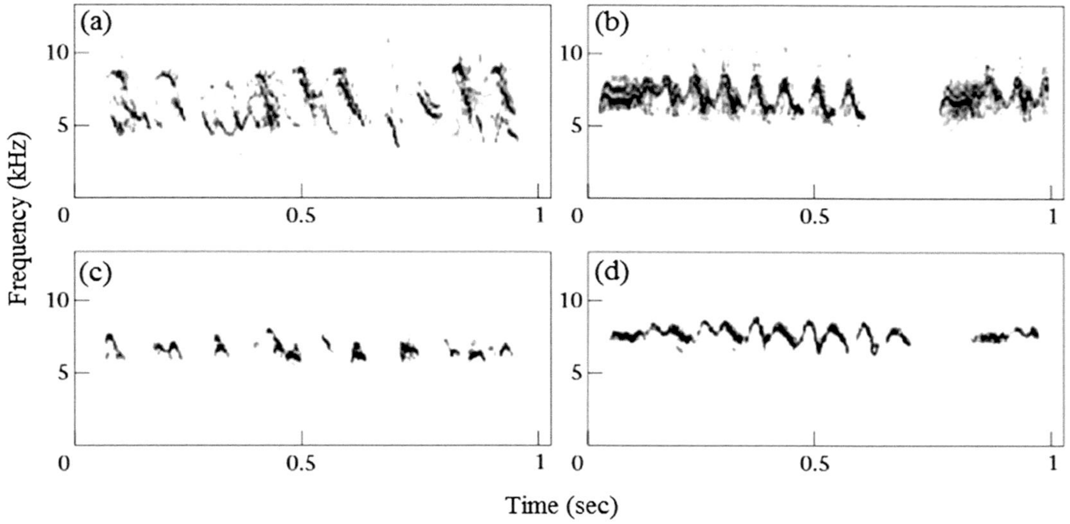


**Fig. 19.74** Sonograms of the begging calls of 8-day-old nestlings in broods of four (**a**) Dunnocks (*Prunella modularis*) and (c) Meadow Pipits (*Anthus pratensis*) and (**b**, **d**) single 8-day-old nestling Common Cuckoos (*Cuculus canorus*) reared by these two host species. The calls of single young cuckoos resemble those of entire broods of the host species. (Figure from Butchart et al. 2003; © 2003 The Association for the Study of Animal Behaviour. Published by Elsevier Ltd., used with permission)

suggested that, when secondary hosts such as Buff-rumped Thornbills are parasitized, young bronze-cuckoos gradually modify their calls via reinforcement. As calls are modified and gradually become more similar to those of host nestlings (Fig. 19.75), host parents respond by provisioning at higher rates.

The begging calls of generalist brood parasites that parasitize the nests of several different species, such as cowbirds, generally do not mimic those of host species (Friedmann and Kiff 1985). An exception to this is Screaming Cowbirds (*Molothrus rufoaxillaris*), brood parasites that usually parasitize Bay-winged Cowbirds (*Molothrus badius*) and whose nestlings are almost identical to those of Bay-winged Cowbirds in size, appearance, and vocalizations (Lichtenstein 1997). Natural selection is less likely to favor vocal mimicry by the nestlings of generalist brood parasites because they must beg in company with a wide variety of host nestlings. However, the begging calls of generalist brood parasites like nestling Brown-headed Cowbirds (*Molothrus ater*) are clearly effective because well over 100 host species have been recorded to have fledged young cowbirds (Friedmann and Kiff 1985). The begging calls of Brown-headed and Shiny cowbirds consist of multiple components (syllables) and are relatively long in duration. Such characteristics are atypical of the begging calls of non-parasitic icterids (e.g., New World blackbirds and orioles, meadowlarks, and grackles), suggesting that the calls have evolved to facilitate successful brood parasitism (Lorenzana and Sealy 1996). Indeed, a number of studies have revealed that the begging calls of nestling cowbirds are effective at stimulating host adults to provision at higher rates, often even more effective than the calls of host nestlings (e.g., Gloag and Kacelnik 2013). For example, in an experiment where the calls of nestling Shiny Cowbirds and nestling House Wrens (*Troglodytes aedon*, the host species) were broadcast at nests over small speakers, Gloag and Kacelnik (2013) found that adult House Wrens provisioned young at rates about 20% higher in response to the cowbird calls. Surprisingly, however, the begging calls of young Shiny Cowbirds

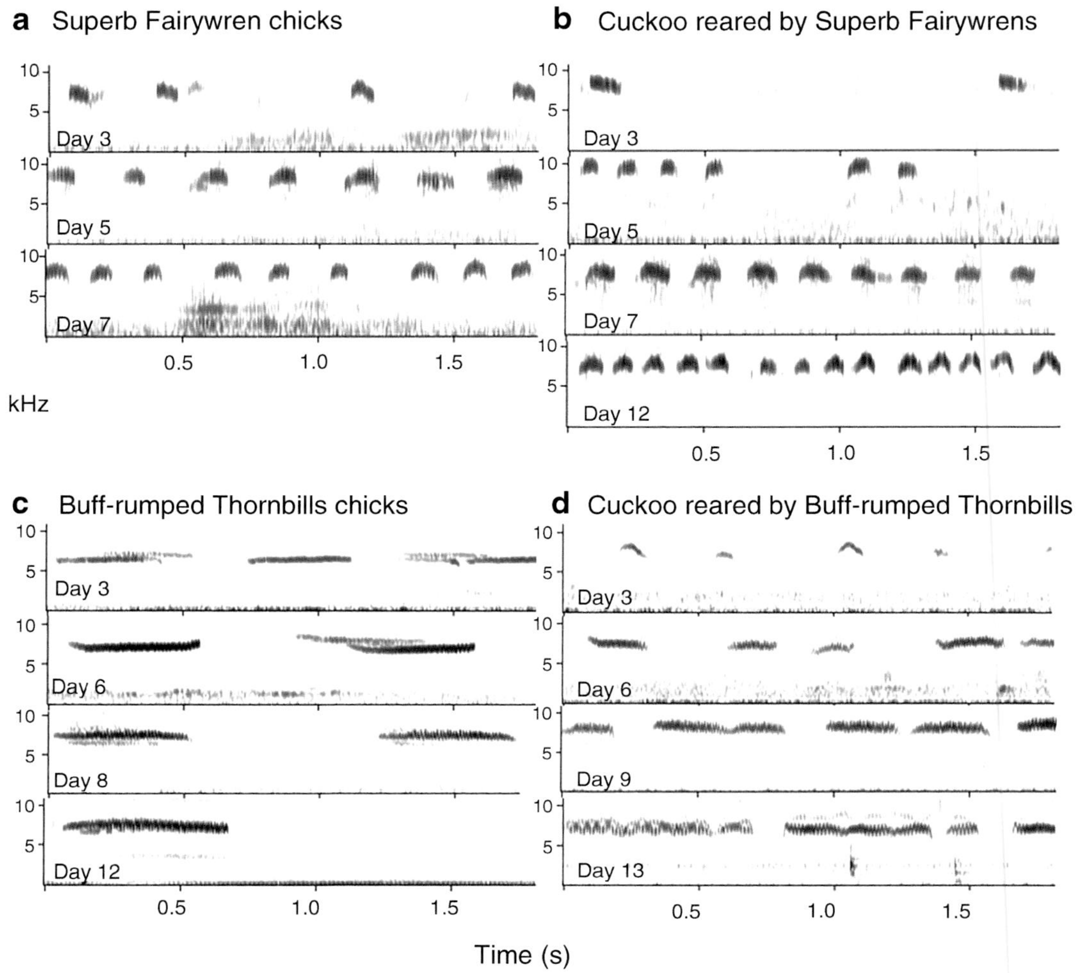


**Fig. 19.75** Sonograms of begging calls of different-aged nestlings. (**a**) Nestling Superb Fairywrens (*Malurus cyaneus*). (**b**) Nestling Horsfield's Bronze-Cuckoo (*Chrysococcyx basalis*) reared by Superb Fairywrens. (**c**) Nestling Buff-rumped Thornbills (*Acanthiza reguloides*). (**d**) Nestling Horsfield's Bronze-Cuckoo raised by Buff-rumped Thornbills. Note how the begging calls of a young cuckoos raised by Buff-rumped Thornbills become more similar to those of nestling Buff-rumped Thornbills with increasing age. (Figure from Langmore et al. 2008; © 2008 The Author(s). Journal compilation © 2008 The Society for the Study of Evolution, used with permission)

are very different from those of young House Wrens (Fig. 19.76). One possible explanation for why host adults respond more strongly to the dissimilar calls of nestling cowbirds is that hosts adjust provisioning rates based on the total amount of sound produced by a brood, regardless of characteristics of those calls. If so, the long duration of the calls of nestling cowbirds may be the important variable (Gloag and Kacelnik 2013). A second possibility is that host adults respond to the rate of calling by a brood and the rapid frequency modulation of cowbird begging calls (both Shiny [Fig. 19.75] and Brown-headed cowbirds) is perceived as a series of rapid calls rather than a single call (Lorenzana and Sealy 1996; Gloag and Kacelnik 2013). Yet another possibility is that host adults respond to increased volume and respond to the relatively loud calls of

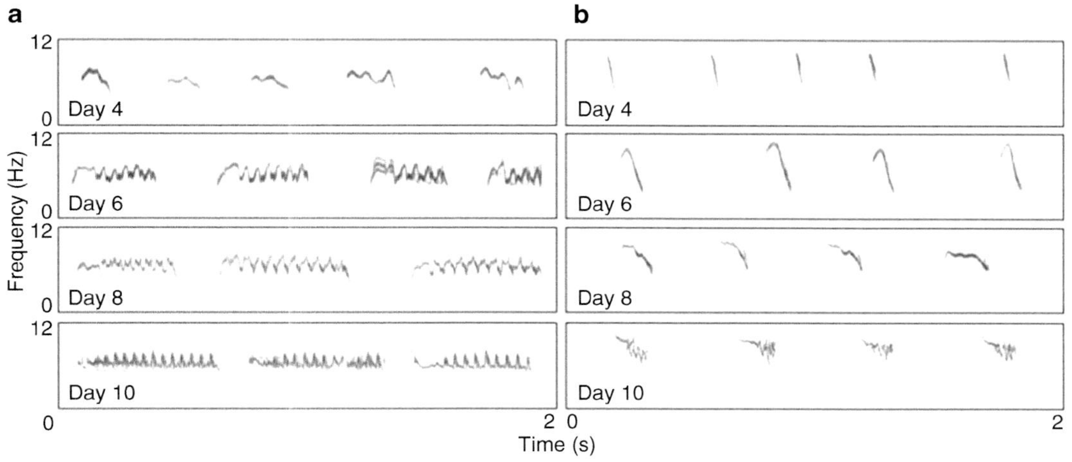


**Fig. 19.76** Examples of begging calls of (**a**) a nestling Shiny Cowbird (*Molothrus bonariensis*) and (**b**) a nestling House Wren (*Troglodytes aedon*) from recordings made at a parasitized House Wren nest. The rapid modulation in frequency of the calls of nestling Shiny Cowbirds may cause them to be perceived by adult wrens as a rapid series of individual nestling-wren-like calls, with adult wrens then responding to the apparent increase in calling rate by provisioning at higher rates. (Figure from Gloag and Kacelnik 2013; © 2013 The Association for the Study of Animal Behaviour. Published by Elsevier Ltd., used with permission)

nestling cowbirds (Briskie et al. 1994; Dearborn 1999) by provisioning at higher rates. Of course, the responses of host adults to the begging calls of nestling cowbirds might be a result of some combination of these variables (total sound produced, perceived call rate, and increased volume).

Unlike brood parasites like some cuckoos (*Cuculus* spp., *Chalcites* spp., and honeyguides [Indicatoridae]) that evict or kill host nestlings, nestling cowbirds (*Molothrus* spp.) and nestlings of other brood parasites, such as indigobirds and whydahs (*Vidua* spp.), are raised along with host nestlings and so must compete for access to food provided by host parents. Two potentially important differences between these young brood parasites and host nestlings are that (1) young brood parasites may differ in size from that of host nestlings, and (2) they are not related to their host parents or nestmates. The lack of relatedness might be important because host nestlings are related to each other and to both parents (assuming adult females do not engage in extrapair copulations) and, as such, kin selection should tend to reduce the level of scramble competition among host nestlings and the level of care nestlings demand from parents. Young brood parasites might be expected to be more selfish because they derive no genetic benefit from survival of host nestlings or, after they become independent, survival of host parents. Size differences can be important because the larger a nestling brood parasite is relative to host young, the better they should be at outcompeting host young for access to food provided by host parents.

Generalist brood parasites parasitize host species that can vary dramatically in size. For example, Brown-headed Cowbirds (*Molothrus ater*; adult mass about 44 g) parasitize species ranging in size from gnatcatchers (*Polioptila* spp.; about 5 g) to meadowlarks (*Sturnella* spp.; some >100 g). As a result, the size of nestling cowbirds relative to host nestlings varies considerably, and these differences can influence both the growth and likelihood of fledging of young hosts and cowbirds as well as the condition of host parents. In general, young cowbirds tend to do best with host species of similar size, do less well with very small host species (although this may vary depending on the particular host species), and fair worst as host species increase in size

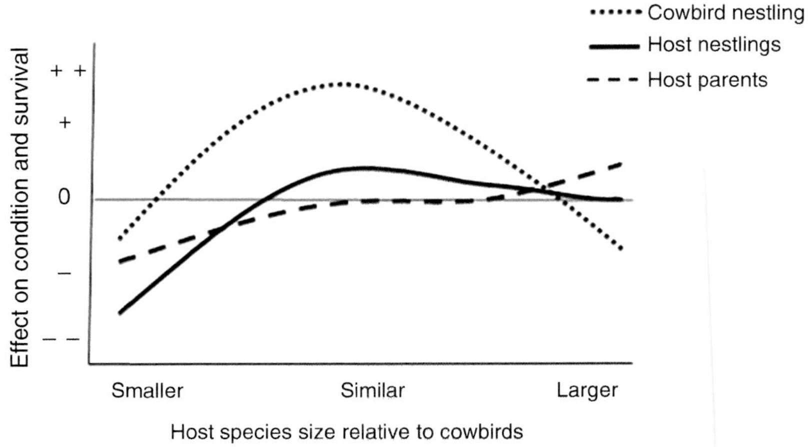


**Fig. 19.77** Effect of host species/nestling size on the condition and survival of Brown-headed Cowbird nestlings, host nestlings, and host parents. Lines are not meant to be quantitative in any way, but simply to illustrate very generally how the size of host species impacts brood parasites and hosts. (Figure by G. Ritchison)

(Fig. 19.77). The presence of a cowbird nestling can also affect host nestlings, with very small hosts outcompeted by the cowbird and so less likely to survive similar-sized nestlings possibly benefitting by increased provisioning rates resulting from the begging of the young cowbird, and larger nestlings generally little impacted by the cowbird's presence (Fig. 19.77). Impacts of host parents similar in size to cowbirds are likely minimal, whereas very small host parents may provision at higher-than-normal rates at some expense to their condition and vary large host parents possibly benefitting (at least energetically) because a young cowbird may not need as much food as their own nestlings (Fig. 19.77).

These generalizations concerning impacts of cowbird parasitism on young cowbirds and their host are based on the results of several studies. For example, Rivers (2007) examined the effect of host-species size on how often young cowbirds were fed and how much food they received from host parents. Young cowbirds were fed proportionately more often than host nestlings with small and intermediate-sized hosts, but proportionately less than host nestlings with a large host (Fig. 19.78). However, the size of prey delivered to nestlings varies with the size of host parents, so young cowbirds actually received less food from the small host species (Fig. 19.78), more food from the intermediate-sized host, and, because they were fed less often, less food from the large host parents (Fig. 19.78). Such results suggest that young cowbirds do best (in terms of amount of food obtained) with intermediate-sized hosts (i.e., host species more similar in size to cowbirds). Differences in the amount of food obtained by nestling cowbirds with different-sized hosts can affect their likelihood of surviving. Kilner (2003) examined the relationship between likelihood of surviving to fledging and host-species size and found young cowbirds were less likely to survive with both very small and very large host species (Fig. 19.79). However, especially among small host species, there are exceptions, with some small hosts apparently able to provide young cowbirds with more food than other small hosts. Kilpatrick (2002), for example, found considerable variation in the mass of young cowbirds (day 8 post-hatching) being cared for by several relatively small host species, with such variation suggesting interspecific differences in the ability of hosts of provide food for young cowbirds. Grim (2006b) reported similar results for the fledging mass young Common Cuckoos (*Cuculus canorus*) in the nests of host species that differed in body mass (Fig. 19.80).

Young brood parasites are not related to host nestlings and parents and, therefore, as noted above, they might be expected to be more selfish because they derive no genetic benefit (i.e., kin selection) from survival of their host-nestling broodmates or, after they become independent, survival of host parents. Studies of the

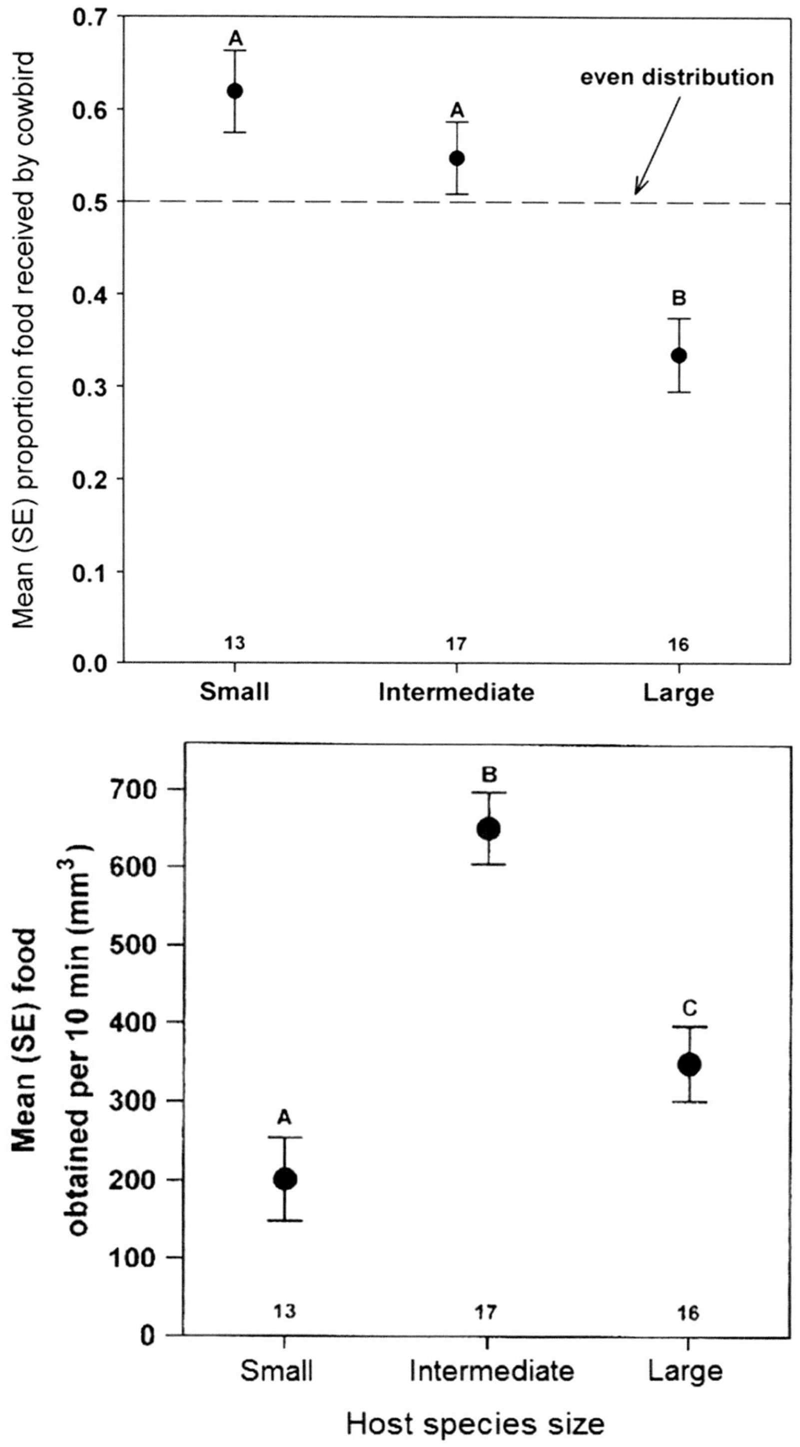


**Fig. 19.78** Top, Mean proportion (± standard error) of food received by nestling Brown-headed Cowbirds (*Molothrus ater*) in nests of small, intermediate, and large hosts. Bottom, Mean amount of food obtained by nestling Brown-headed Cowbirds in nests of small, intermediate-sized, and large hosts. Means with different letters are significantly different ($P = 0.05$). Numbers at the bottom represent number of nests. (Figure from Rivers 2007; © 2006 Oxford University Press, used with permission)

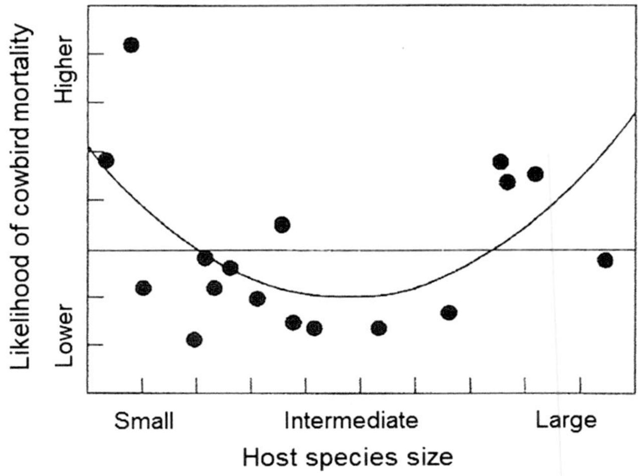


**Fig. 19.79** Relationship between host species size and the likelihood of nestling Brown-headed Cowbirds (*Molothrus ater*) surviving and fledging from host nests. Each data point represents a different host species. (Figure modified from Kilner 2003; © 2003 The Association for the Study of Animal Behaviour. Published by Elsevier Ltd., used with permission)

interactions between nestlings and their parents suggest that begging is generally an honest signal of need, with the intensity of begging correlated with nestling hunger levels. The intensity of begging by host nestlings in a nest with a nestling brood parasite, therefore, will vary with hunger levels. In other words, they will not be "selfish" by continuing to beg when satiated. Such behavior is likely selected for because begging calls can potentially attract the attention of predators (Haskell 1994) and young also benefit if sibling nestmates are fed at levels that enhance their chances of fledging. In addition, by not demanding more from parents than necessary (and assuming no extra-pair paternity), their parents are more likely to remain in good condition, survive to the next breeding attempt, and produce additional offspring that will share genes with their current young. Young brood parasites obtain no such benefits so might be expected to be "selfish" and continue to beg even in situations where their broodmates would not.

Many studies have in fact revealed that young brood parasites beg with greater intensity, i.e., exaggerated displays and louder, more frequent begging calls, than host nestlings (Briskie et al. 1994; Lichtenstein and Dearborn 2004, and many others). Lack of relatedness is one explanation for such behavior, but, in addition, particularly with smaller host species, fast-growing nestling brood

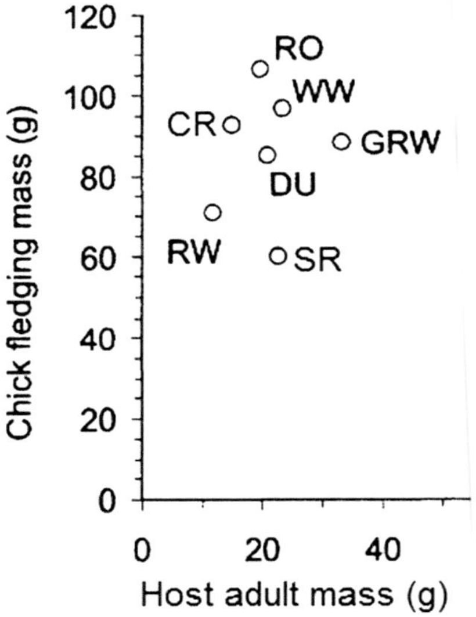


**Fig. 19.80** Effect of host adult mass on the fledging mass of young Common Cuckoos. Each point represents a different host species. Note the substantial variation in cuckoo mass for host species where adults weigh between about 10–30 g, suggesting differences among small hosts in the amount of food delivered to young cuckoo. CR, Common Redstart (*Phoenicurus phoenicurus*); RW, Eurasian Reed Warbler (*Acrocephalus scirpaceus*); RO, European Robin (*Erithacus rubecula*); WW, White Wagtail (*Motacilla alba*); GRW, Great Reed Warbler (*Acrocephalus arundinaceus*); DU, Dunnock (*Prunella modularis*); SR, Rufous-tailed Scrub-Robin (*Cercotrichas galactotes*). (Figure from Grim 2006b; © 2006 Springer-Verlag, used with permission)

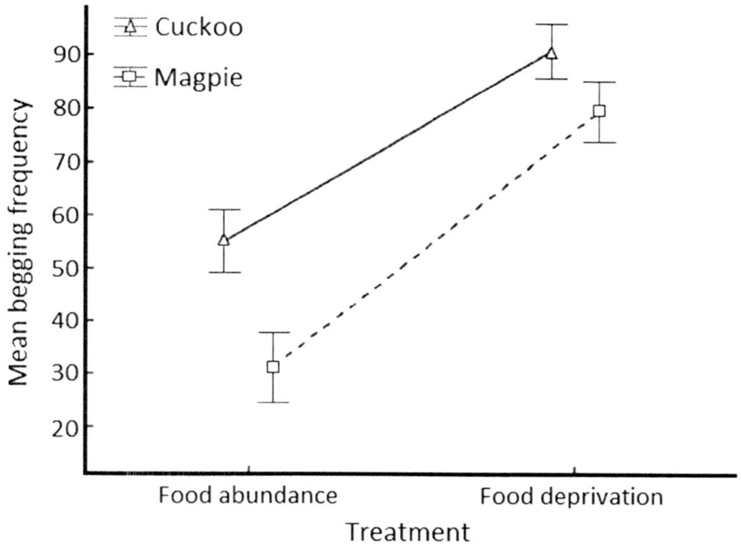


**Fig. 19.81** Mean begging frequency (± standard error) of nestling Great Spotted Cuckoos (*Clamator glandarius*) and Eurasian Magpies (*Pica pica*) that were satiated (fed by researchers) and food-deprived (not fed for 2 h, then fed limited quantities of food for an additional 4 h). Begging frequency was defined as the proportion of attempted feedings (by researchers) when nestlings exhibited begging behavior. (Figure from Soler et al. 2012; © The Association for the Study of Animal Behaviour. Published by Elsevier Ltd., used with permission)

parasites may simply require more food than host nestlings (Soler and Soler 1991, 1999). Although tending to beg more intensely than host nestlings, most studies also suggest that the begging intensity of young brood parasites is also influenced by hunger levels. For example, Soler et al. (2012) manipulated the hunger levels of nestling Great Spotted Cuckoos (*Clamator glandarius*) and nestling Eurasian Magpies (*Pica pica*) and found that nestling cuckoos begged more frequently than nestling magpies, but begging frequency of nestlings of both species was a reliable indicator of their hunger levels (Fig. 19.81). In the absence of kin selection, nestling brood parasites can still benefit by adjusting their begging behavior relative to hunger levels. First, as noted previously, nestling brood parasites can benefit when the begging of nestmates, in combination with their own, acts to increasing provisioning rates of host parents. In addition, loud begging calls can attract predators (Haskell 1994; Hannon et al. 2009) so limiting unnecessary begging (i.e., begging when hunger levels are low) can increase the likelihood of survival for nestling brood parasites.

Having a young brood parasite as a nestmate can be detrimental to host nestlings. As such, natural selection should favor "counter-adaptations" that help host nestlings compete with young brood parasites, particularly for species that are more frequently parasitized and where the absence of other anti-brood parasite adaptations (e.g., removal of parasite eggs by host parents) results in the frequent presence of young brood parasites in nests. Remeš (2006) examined the possible effect of cowbird parasitism on nestling growth strategies for 134 species and nestlings of species parasitized at higher rates had faster growth rates, lower mass at fledging, and fledged at a younger age. Interestingly, these relationships only held for species of songbirds similar in size to and larger than cowbirds (i.e., above 30-gram adult mass). These results suggest that nestlings of small songbirds are simply too small to compete with nestling cowbirds, and their growth rates simply cannot increase sufficiently, regardless of selection pressures, to enable them to compete with larger nestling cowbirds. However, because nestling cowbirds

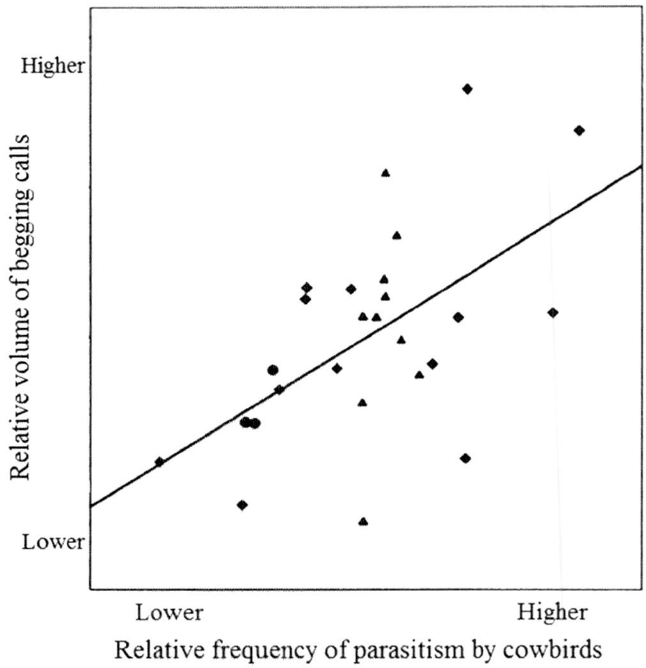


**Fig. 19.82** Relationship between frequency of parasitism by Brown-headed Cowbirds (*Molothrus ater*) and the volume of begging calls of 26 species of host nestlings. Species are categorized based on their suitability as cowbird hosts, with unsuitable hosts (cavity nesters and egg rejectors) indicated by diamond symbols, partially suitable hosts (reports of unsuccessful attempts at parasitism documented) by filled circles, and suitable hosts by triangles (Figure modified from Boncoraglio et al. 2009; © 2008 Oxford University Press, used with permission)

grow faster and fledge at a younger age than nestlings of other similar-sized songbirds, nestlings of larger species of songbirds can better compete with nestling cowbirds for access to food delivered by host parents by growing faster and fledging sooner, although, as a result, they fledge at a lower mass. This growth strategy, therefore, provides an obvious benefit, but at some cost because greater mass at fledging is an important determinant of survival for young songbirds (e.g., Schwagmeyer and Mock 2008).

Other adaptations that allow host nestlings to better compete with nestling brood parasites include increasing the volume of begging calls (Boncoraglio et al. 2009), learning of vocal "passwords" by host embryos (Colombelli-Négrel et al. 2012), and altering begging calls (Pagnucco et al. 2008). Boncoraglio et al. (2009) examined the volume of begging calls of host nestlings relative to the frequency of brood parasitism (percentage of nests) for 31 host species and found clear relationship; the higher the parasitism rate, the louder the begging calls of host nestlings (Fig. 19.82). Selection may have favored louder begging by nestlings of more frequently parasitized species (1) to enhance their access to food as they compete with young cowbirds, and/or (2) because they have an unrelated nestmate and, therefore, monopolizing parental resources would be less costly (in terms of kin fitness; Boncoraglio et al. 2009).

Nestlings of two other host species modify their begging calls to either better compete with nestling brood parasites or to improve the ability of host parents to recognize nestling brood parasites. In Song Sparrows (*Melospiza melodia*), the begging calls of nestlings in nests with a nestling cowbird were higher in frequency and louder than those of nestling sparrows in non-parasitized nests (and more similar to the begging calls of nestling cowbirds), and these changes allowed nestlings in parasitized nests to effectively compete with young cowbirds and be fed by their parents as often as those in non-parasitized nests (Pagnucco et al. 2008). During the incubation period, Superb Fairywren (*Malurus cyaneus*) embryos learn a "vocal password" based on a unique "incubation call" given by the adult female. If a nest is not parasitized by

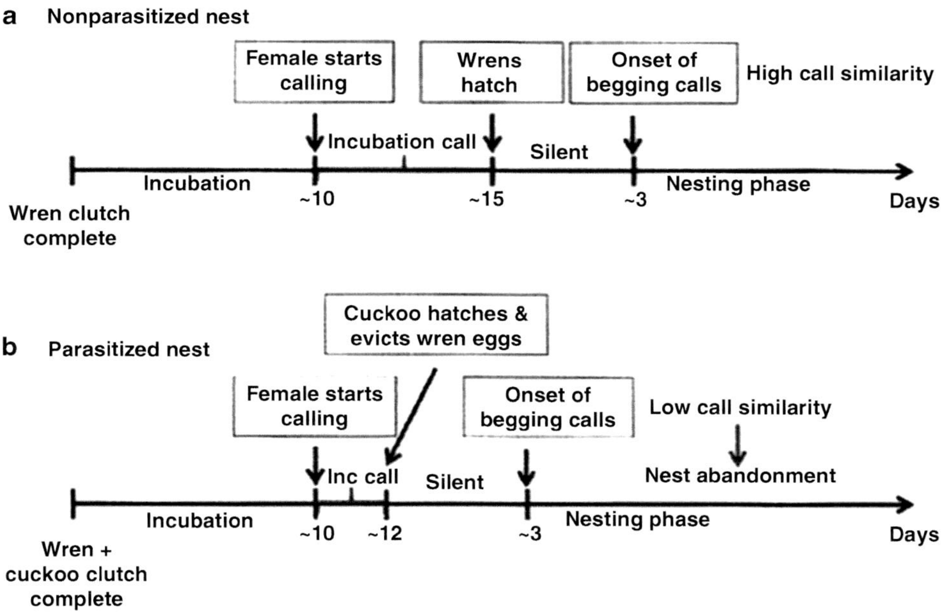


**Fig. 19.83** Hypothesized timeline for process of Superb Fairywren (*Malurus cyaneus*) embryos learning "vocal passwords" from adult females, and adult females subsequently using those calls to determine if, after hatching, nestlings are fairywrens or Horsfield's Bronze-Cuckoos (*Chrysococcyx basalis*). If the latter, then the nest is abandoned. (Figure from Colombelli-Négrel et al. 2012; © 2012 Elsevier Ltd., used with permission)

a Horsfield's Bronze-Cuckoo (*Chrysococcyx basalis*), then, after hatching, nestling fairywrens giving this "vocal password" are recognized and fed by adults (adult males also learn the password). However, if a nest is parasitized, the bronze-cuckoo egg hatches first (because of their short incubation period) and evicts the fairywren eggs. Perhaps because of the short exposure time (hearing the "incubation call" for about 2 days rather than 5 days for fairywren embryos in non-parasitized nests), embryonic bronze-cuckoos do not learn the "vocal password" and when, after hatching, they begin uttering begging calls, adult fairywrens note the absence of the "vocal password" and abandon the nest (Colombelli-Négrel et al. 2012; Fig. 19.83).

Little is known about possible adaptations or counter-adaptations of brood parasites and hosts after young leave nests. In one of the few studies to date, De Mársico et al. (2012) found that fledgling Screaming Cowbirds (*Molothrus rufoaxillaris*) closely resemble fledglings of their primary host, Grayish Baywings (*Agelaioides badius*; Fig. 19.84), and also have similar begging calls. To test the hypothesis that such mimicry is beneficial, De Mársico et al. (2012) compared survival rates of young Screaming Cowbirds and non-mimetic Shiny Cowbirds (*Molothrus bonariensis*) raised by Bay-winged Cowbirds. They found that non-mimetic Shiny Cowbird fledglings had higher post-fledging mortality rates (83%) than Screaming Cowbird fledglings (no mortality). Such results suggest that adult Bay-winged Cowbirds are able to discriminate between mimetic and non-mimetic brood parasite fledglings, and that such mimicry is clearly advantageous for Screaming Cowbirds.

**Fig. 19.84** Left to right, Grayish Baywing (*Agelaioides badius*), Screaming Cowbird (*Molothrus rufoaxillaris*), and Shiny Cowbird (*Molothrus bonariensis*) fledglings illustrating the similar plumages of young parasitic Screaming Cowbirds and their primary host, Grayish Baywings. The plumage of Shiny Cowbird fledglings does not mimic that of Grayish Baywings. (Figure from De Mársico et al. 2012; © 2012 The Royal Society, used with permission)

## References


Ahlund M, Andersson M (2001) Brood parasitism: female ducks can double their reproduction. Nature 414:433–440

Ainley DG, Nettleship DN, Carter HR, Storey AE (2020) Common Murre (*Uria aalge*), version 1.0. In: Billerman SM (ed) Birds of the World. Cornell Lab of Ornithology, Ithaca, NY

Åkesson S, Bakam H, Hernandez EM, Ilieva M, Bianco G (2021) Migratory orientation in inexperienced and experienced avian migrants. Ethol Ecol Evol 33:206–229

Allen T, Clarke JA (2005) Social learning of food preferences by White-tailed Ptarmigan chicks. Anim Behav 70:305–310

Alvarez-Cordero E (1996) Biology and conservation of the Harpy Eagle in Venezuela and Panama. Ph.D. dissertation, University of Florida, Gainesville, FL

Alvo R, Berrill M (1992) Adult Common Loon feeding behavior is related to food fed to chicks. Wilson Bull 104:184–185

Amiot R, Wang X, Wang S, Lécuyer C, Mazin J-M, Mo J, Flandrois J-P, Fourel F, Wang X, Xu X, Zhang Z (2017) $\delta^{18}$O-derived incubation temperatures of oviraptorosaur eggs. Palaeontology 60:633–647

Andersson M, Ahlund M (2000) Host-parasite relatedness shown by protein fingerprinting in a brood parasitic bird. Proc Natl Acad Sci 97:13188–13193

Anderson MG, Ross HA, Brunton DH, Hauber ME (2009a) Begging call matching between a specialist brood parasite and its host: a comparative approach to detect coevolution. Biol J Linn Soc 98:208–216

Anderson MG, Moskát C, Bán M, Grim T, Cassey P, Hauber ME (2009b) Egg eviction imposes a recoverable cost of virulence in chicks of a brood parasite. PLoS ONE 4:e7725

Antonov A, Stokke BG, Moksnes A, Røskaft E (2006a) Egg rejection in Marsh Warblers (*Acrocephalus palustris*) heavily parasitized by Common Cuckoos (*Cuculus canorus*). Auk 123:419–430

Antonov A, Stokke BG, Moksnes A, Røskaft E (2006b) Coevolutionary interactions between Common Cuckoos and Corn Buntings. Condor 108:414–422

Antonov A, Stokke BG, Moksnes A, Røskaft E (2007) Factors influencing the risk of Common Cuckoo *Cuculus canorus* parasitism on Marsh Warblers *Acrocephalus palustris*. J Avian Biol 38:390–393

Antonov A, Stokke BG, Moksnes A, Røskaft E (2008) Getting rid of the cuckoo *Cuculus canorus* egg: why do hosts delay rejection? Behav Ecol 19:100–107

Antonov A, Stokke BG, Vikan JR, Fossøy F, Ranke PS, Røskaft E, Moksnes A, Møller AP, Shykoff JA (2010) Egg phenotype differentiation in sympatric cuckoo *Cuculus canorus* gentes. J Evol Biol 23:1170–1182

Antonov A, Stokke B, Fossøy F, Liang W, Moksnes A, Røskaft E, Yang C, Møller AP (2012) Why do brood parasitic birds lay strong-shelled eggs? Chin Birds 3: 245–258

Arendt JD (1997) Adaptive intrinsic growth rates: an integration across taxa. Q Rev Biol 72:149–177

Bachman GC, Chappell MA (1998) The energetic cost of begging behaviour in nestling House Wrens. Anim Behav 56:1607–1618

Baguette M, Van Dyck H (2007) Landscape connectivity and animal behavior: functional grain as a key determinant for dispersal. Landsc Ecol 22:1117–1129

Banks AJ, Martin TE (2001) Host activity and the risk of nest parasitism by Brown-headed Cowbirds. Behav Ecol 12:31–40

Barati A, McDonald PG (2017) Nestlings reduce their predation risk by attending to predator-information encoded within conspecific alarm calls. Sci Rep 7: 11736

Barbraud C, Johnson AR, Bertault G (2003) Phenotypic correlates of post-fledging dispersal in a population of Greater Flamingoes: the importance of body condition. J Anim Ecol 72:246–257

Beauchamp G (1997) Determinants of intraspecific brood amalgamation in waterfowl. Auk 114:11–21

Becker PH, Ezard THG, Ludiwigs JD, Sauer-Gurth H, Wink M (2008) Population sex ratio shift from fledging to recruitment: consequences for demography in a philopatric seabird. Oikos 117:60–68

Begum S, Moksnes A, Røskaft E, Stokke BG (2011) Factors influencing host nest use by the brood parasitic Asian Koel (*Eudynamys scolopacea*). J Ornithol 152: 793–800

Belthoff JR, Dufty AM Jr (1995) Locomotor activity levels and the dispersal of Western Screech-Owls, *Otus kennicottii*. Anim Behav 50:558–561

Belthoff JR, Dufty AM Jr (1998) Corticosterone, body condition and locomotor activity: a model for dispersal in screech-owls. Anim Behav 55:405–415

Berkeley LI, McCarty JP, Wolfenbarger LL (2007) Postfledging survival and movement in Dickcissels (*Spiza americana*): implications for habitat management and conservation. Auk 124:396–409

Besnard A (2001) Evolution de l'élevage des poussins en crèche chez les Larids [evolution of crèching behaviour in Larids]. University of Montpellier II, Montpellier

Besnard A, Gimenez O, Lebreton J-D (2002) A model for the evolution of crèching behaviour in gulls. Evol Ecol 16:489–503

Birchard GF, Ruta M, Deeming DC (2013) Evolution of parental incubation behaviour in dinosaurs cannot be inferred from clutch mass in birds. Biol Lett 9:20130036

Birkhead TR, Hemmings N, Spottiswoode CN, Mikulica O, Moskát C, Bán M, Schulze-Hagen K (2011) Internal incubation and early hatching in brood parasitic birds. Proc R Soc B 278:1019–1024

Blas J, Sergio F, Hiraldo F (2009) Age-related improvement in reproductive performance in a long-lived raptor: a cross-sectional and longitudinal study. Ecography 32:647–657

Blumberg MS (2017) Development evolving: the origins and meanings of instinct. Wiley Interdiscip Rev Cogn Sci 8:e1371

Boncoraglio G, Saino N, Garamszegi LZ (2009) Begging and cowbirds: brood parasites make hosts scream louder. Behav Ecol 20:215–221

Bonte D, Van Dyke H, Bullock JM, Coulon A, Delgado M, Gibbs M, Lehouck V, Matthysen E, Mustin K, Saastamoinen M, Schtickzelle N (2012) Costs of dispersal. Biol Rev 87:290–312

Bosque C, Bosque MT (1995) Nest predation as a selective factor in the evolution of developmental rates in altricial birds. Am Nat 145:234–260

Both C, Visser ME, Verboven N (1999) Density–dependent recruitment rates in Great Tits: the importance of being heavier. Proc R Soc B 266:465–469

Bowers EK, Sakaluk SK, Thompson CE (2013) Sibling cooperation influences the age of nest leaving in an altricial bird. Am Nat 181:775–786

Brennan LA, Hernandez F, Williford D (2020) Northern Bobwhite (*Colinus virginianus*), version 1.0. In: Poole AF (ed) Birds of the World. Cornell Lab of Ornithology, Ithaca, NY

Briggs CW, Collopy MW, Woodbridge B (2012) Correlates and fitness consequences of natal dispersal in Swainson's Hawks. Condor 114:764–770

Briskie JV, Naugler CT, Leech SM (1994) Begging intensity of nestling birds varies with sibling relatedness. Proc R Soc B 258:73–78

Brooke MDL, Davies NB (1988) Egg mimicry by cuckoos, *Cuculus canorus*, in relation to discrimination by hosts. Nature 335:630–632

Brooke MDL, Davies NB, Noble DG (1998) Rapid decline of host defences in response to reduced cuckoo parasitism: behavioural flexibility of Reed Warblers in a changing world. Proc R Soc B 265:1277–1282

Brooker L, Brooker M, Cale P (1999) Animal dispersal in fragmented habitat: measuring habitat connectivity, corridor use, and dispersal mortality. Conserv Ecol 3:4

Brown RGB (1962) The aggressive and distraction behaviour of the Western Sandpiper *Ereneutes mauri*. Ibis 104:1–12

Brown CR, Brown MB (1991) Selection for high-quality host nests by parasitic Cliff Swallows. Anim Behav 41:457–465

Brown CR, Brown MB (1998) Fitness components associated with alternative reproductive tactics in Cliff Swallows. Behav Ecol 9:158–171

Brown TJ, Dugdale HL, Hammers M, Komdeur J, Richardson DS (2022) Seychelles Warblers with silver spoons: juvenile body mass is a lifelong predictor of annual survival, but not annual reproduction or senescence. Ecol Evol 12:e9049

Brunton DH (1990) The effects of nesting stage, sex, and type of predator on parental defense by Killdeer (*Charadrius vociferous*): testing models of avian parental defense. Behav Ecol Sociobiol 26:181–190

Bryan D (2020) Limpkin (*Aramus guarauna*), version 1.0. In: Poole AF, Gill FB (eds) Birds of the World. Cornell Lab of Ornithology, Ithaca, NY

Buitron D (1983) Variability in the responses of Black-billed Magpies to natural predators. Behaviour 87: 209–236

Bull EL, Jackson JA (2020) Pileated Woodpecker (*Dryocopus pileatus*), version 1.0. In: Poole AF (ed) Birds of the World. Cornell Lab of Ornithology, Ithaca, NY

Burger J, Gochfeld M (1991) The Common Tern: its breeding biology and social behavior. Columbia University Press, New York

Burger J, Gochfeld M, Saliva JE, Gochfeld D, Gochfeld D, Morales H (1989) Antipredator behavior in nesting Zenaida Doves (*Zenaida aurita*): parental investment or offspring vulnerability. Behaviour 111:129–143

Burish MJ, Kueh HY, Wang SS-H (2004) Brain architecture and social complexity in modern and ancient birds. Brain Behav Evol 63:107–124

Burley NT, Johnson K (2002) The evolution of avian parental care. Philos Trans R Soc B 357:241–250

Bustamante J, Hiraldo F (1989) Post-fledging dependence period and maturation of flight skills in the Black Kite *Milvus migrans*. Bird Study 36:199–204

Butchart SHM, Kilner RM, Fuisz T, Davies NB (2003) Differences in the nestling begging calls of hosts and host-races of the Common Cuckoo, *Cuculus canorus*. Anim Behav 65:345–354

Byholm P, Beal M, Isaksson N, Lötberg U, Åkesson S (2022) Paternal transmission of migration knowledge in a long-distance bird migrant. Nat Commun 13:1566

Byrkjedal I (1987) Antipredator behavior and breeding success in Greater Golden-Plover and Eurasian Dotteral. Condor 89:40–47

Byrkjedal I (1989) Nest defense behavior of Lesser Golden-Plovers. Wilson Bull 101:579–590

Canestrari D, Marcos JM, Baglione V (2009) Cooperative breeding in Carrion Crows reduces the rate of brood parasitism by Great Spotted Cuckoos. Anim Behav 77: 1337–1344

Caro T (2005) Antipredator defenses in birds and mammals. University of Chicago Press, Chicago, IL

Caro SM, West SA, Griffin AS (2016) Sibling conflict and dishonest signaling in birds. Proc Natl Acad Sci USA 48:13803–13808

Cash KJ, Evans RM (1986) Brood reduction in the American White Pelican (*Pelecanus erythrorhynchos*). Behav Ecol Sociobiol 18:413–418

Chakarov N, Jonker RM, Boerner M, Hoffman JI, Krüger O (2013) Variation at phonological candidate genes correlates with timing of dispersal and plumage morph in a sedentary bird of prey. Mol Ecol 22: 5430–5440

Chappell MA, Bachmann GC (2002) Energetic costs of begging behaviour. In: Wright J, Leonard ML (eds) The evolution of begging: competition, cooperation and communication. Kluwer, Dordrecht, pp 143–162

Charmantier A, Buoro M, Gimenez O, Weimerskirch H (2011) Heritability of short-scale natal dispersal in a large-scale foraging bird, the Wandering Albatross. J Evol Biol 24:1487–1496

Chiavacci SJ, Ward MP, Benson TJ (2015) Why fledge early in the day? Examining the role of predation risk in explaining fledging behavior. Behav Ecol 26:593–600

Chinsamy A, Elzanowski A (2001) Bone histology: evolution of growth pattern in birds. Nature 412:402–403

Claramunt S (2021) Flight efficiency explains differences in natal dispersal distances in birds. Ecology 102: e03442

Claramunt S, Derryberry EP, Remsen JV Jr, Brumfield RT (2012) High dispersal ability inhibits speciation in a continental radiation of passerine birds. Proc R Soc B 279:1567–1574

Clark KL, Robertson RJ (1979) Spatial and temporal multi-species nesting aggregations in birds as anti-parasite and anti-predator defenses. Behav Ecol Sociobiol 5:359–371

Clarke JA (1983) Moonlight's influence on predator/prey interactions between Short-eared Owls (*Asio flammeus*) and deermice (*Peromyscus maniculatus*). Behav Ecol Sociobiol 13:205–209

Clarke AL, Sæther B, Røskaft E (1997) Sex-biases in avian dispersal: a reappraisal. Oikos 97:429–438

Cockburn A (2006) Prevalence of different modes of parental care in birds. Proc R Soc B 273:1375–1383

Colombelli-Negrel D, Kleindorfer S (2009) Nest height, nest concealment, and predator type predict nest predation in Superb Fairy-wrens (*Malurus cyaneus*). Ecol Res 24:921–928

Colombelli-Négrel D, Hauber ME, Robertson J, Sulloway FJ, Hoi H, Griggio M, Kleindorfer S (2012) Embryonic learning of vocal passwords in Superb Fairy-wrens reveals intruder cuckoo nestlings. Curr Biol 22:2155–2160

Colwell MA (2010) Shorebird ecology, conservation, and management. University of California Press, Berkeley, CA

Colwell MA, Hurley SJ, Hall JN, Dinsmore SJ (2007) Age-related survival and behavior of Snowy Plover chicks. Condor 109:638–647

Conkling TJ, Pope TL, Smith KN, Mathewson HA, Morrison ML, Wilkins RN, Cain JW III (2012) Black-capped Vireo nest predator assemblage and predictors for nest predation. J Wildl Manag 76: 1401–1411

Cooper EB, Kruuk LE (2018) Ageing with a silver-spoon: a meta-analysis of the effect of developmental environment on senescence. Evol Lett 2:460–471

Cooper JM, Miller EH (1992) Brood amalgamation and alloparental care in the Least Sandpiper, *Calidris minutilla*. Can J Zool 70:403–405

Corbel H, Morlon F, Geiger S, Gorscolas R (2009) State-dependent decisions during the fledging process of King Penguin chicks. Anim Behav 78:829–838

Cote J, Clobert J, Brodin T, Fogarty S, Sih A (2010) Personality-dependent dispersal: characterization, ontogeny and consequences for spatially structured populations. Philos Trans R Soc B 365:4065–4076

Craig AJFK (1982) Breeding success of a Red Bishop colony. Ostrich 53:182–188

Craig JL, Jamieson IG (1990) Pukeko: different approaches and some different answers. In: Stacey PB, Koenig WD (eds) Cooperative breeding in birds: long-term studies of ecology and behavior. Cambridge University Press, New York, pp 387–412

Crook JH (1963) A comparative analysis of nest structure in the weaver birds (Ploceinae). Ibis 105:238–262

Croston R, Hauber ME (2012) The ecology of avian brood parasitism. Nat Educ Knowledge 3:56

Curio E (1976) The ethology of predation. Springer-Verlag, New York, NY

Cuthill I (2006) Color perception. In: Hill G, McGraw KJ (eds) Bird coloration: mechanisms and measurements. Harvard University Press, Cambridge, MA, pp 3–40

Darrah AJ, van Riper C III (2021) Brood provisioning rates and fledgling behavior of Cordilleran Flycatchers in southwestern Colorado. J Field Ornithol 92:30–42

Davies NB (2000) Cuckoos, cowbirds and other cheats. Poyser, London, UK

Davies NB, Brooke MDL (1988) Cuckoos versus Reed Warblers: adaptations and counteradaptations. Anim Behav 36:262–284

Davies NB, Welbergen JA (2008) Cuckoo-hawk mimicry? An experimental test. Proc R Soc B 275:1817–1822

Davies NB, Welbergen JA (2009) Social transmission of a host defense against cuckoo parasitism. Science 324: 1318–1320

Davies NB, Kilner RM, Noble DG (1998) Nestling cuckoos *Cuculus canorus* exploit hosts with begging calls that mimic a brood. Proc R Soc B 265:673–678

Davies JN, Todd PM, Bullock S (1999) Environment quality predicts parental provisioning decisions. Proc R Soc B 266:1791–1797

Dawideit BA, Phillimore AB, Laube I, Leisler B, Böhning-Gaese K (2009) Ecomorphological predictors of natal dispersal distances in birds. J Anim Ecol 78: 388–395

Dawkins R (1979) The value judgements of evolution. In: Dempster MAH, McFarland DJ (eds) Animal economics. Academic, London, UK

Dearborn DC (1998) Begging behaviour and food acquisition by Brown-headed Cowbird nestlings. Behav Ecol Sociobiol 43:259–270

Dearborn DC (1999) Brown-headed Cowbird nestling vocalizations and the risk of nest predation. Auk 116: 448–457

De Framond L, Brumm H, Thompson WI, Drabing SM, Francis CD (2022) The broken-wing display across birds and the conditions for its evolution. Proc R Soc B 289:20220058

Deguchi T, Watanuki Y (2005) Prefledging mass recession and timing of fledging in Rhinoceros Auklets *Cerorhinca monocerata*. Ibis 147:217–220

Delhey K, Carrizo M, Verniere L, Mahler B, Peters A (2011) Rejection of brood-parasitic Shiny Cowbird *Molothrus bonariensis* nestlings by the Firewood-gatherer Anumbius annumbi? J Avian Biol 42:463–467

Dell'omo G, Alleva E, Carere C (1998) Parental recycling of nestling faeces in the Common Swift. Anim Behav 56:631–637

del Hoyo HJ, Elliott A, Sargatal J (eds) (1997) Handbook of the birds of the world, vol. 4. Sandgrouse to cuckoos. Lynx Edicions, Barcelona, Spain

De Mársico MC, Gantchoff MG, Reboreda JC (2012) Host-parasite coevolution beyond the nestling stage? Mimicry of host fledglings by the specialist Screaming Cowbird. Proc R Soc B 279:3401–3408

Dial KP (2003) Evolution of avian locomotion: correlates of flight style, locomotor modules, nesting biology, body size, development, and the origin of flapping flight. Auk 120:941–952

Dickinson JL, McGowan A (2005) Winter resource wealth drives delayed dispersal and family-group living in Western Bluebirds. Proc R Soc B 272:2423–2428

Dingemanse NJ, Both C, van Noordwijk AJ, Rutten AL, Drent PJ (2003) Natal dispersal and personalities in Great Tits (*Parus major*). Proc R Soc B 270:741–747

Drilling N, Titman R, McKinney F (2020) Mallard (*Anas platyrhynchos*), version 1.0. In: Billerman SM (ed) Birds of the World. Cornell Lab of Ornithology, Ithaca, NY

Drummond H (2002) Begging versus aggression in avian broodmate competition. In: Wright J, Leonard ML (eds) The evolution of nestling begging competition, cooperation and communication. Kluwer, Dordrecht, pp 337–360

Drummond H, Osorno JL (1992) Training siblings to be submissive losers: dominance between booby nestlings. Anim Behav 44:881–893

Duckworth RA, Kruuk LEB (2009) Evolution of genetic integration between dispersal and colonization ability in a bird. Evolution 63:968–977

Dunn EH (1975) The timing of endothermy in the development of young altricial birds. Condor 77:288–293

Dunn EH (1976a) Development of endothermy and existence energy expenditure of nestling Double-crested Cormorants. Condor 78:350–356

Dunn EH (1976b) The relationship between brood size and age of effective homeothermy in nestling House Wrens. Wilson Bull 88:478–482

Dunn EH (1979) Age of effective homeothermy in nestling Tree Swallows according to brood size. Wilson Bull 91:455–457

Earnst SL (2002) Parental care in Tundra Swans during the pre-fledging period. Waterbirds 25 (Special Publication 1):268–277

Ekman J, Eggers S, Griesser M (2002) Fighting to stay: the role of sibling rivalry for delayed dispersal. Anim Behav 64:453–459

Ellison KS, Ribic CA (2012) Nest defense: grassland bird responses to snakes. In: Ribic CA, Thompson FR III, Pietz PJ (eds) Video surveillance of nesting birds, Studies in Avian Biology no. 43. University of California Press, Berkeley, CA, pp 149–159

Emlen ST, Wrege PH (1986) Forced copulations and intraspecific parasitism: two costs of social living in the White-fronted Bee-eater. Ethology 71:2–29

Enstipp MR, Bost C-A, Le Bohec C, Chatelain N, Weimerskirch H, Handrich Y (2021) The early life of King Penguins: ontogeny of dive capacity and foraging behaviour in an expert diver. J Exp Biol 224: jeb242512

Evans DR, Hobson KA, Kusack JW, Cadman MD, Falconer CM, Mitchell GW (2020) Individual condition, but not fledging phenology, carries over to affect post-

fledging survival in a Neotropical migratory songbird. Ibis 162:331–344
Fayet AL (2020) Exploration and refinement of migratory routes in long-lived birds. J Anim Ecol 89:16–19
Feeney WE, Langmore NE (2013) Social learning of a brood parasite by its host. Biol Lett 9:20130443
Feeney WE, Welbergen JA, Langmore NE (2012) The frontline of avian brood parasite-host coevolution. Anim Behav 84:3–12
Feeney WE, Medina I, Somveille M, Heinsohn R, Hall ML, Mulder RA, Stein JA, Kilner RM, Langmore NE (2013) Brood parasitism and the evolution of cooperative breeding in birds. Science 342:1506–1508
Fenske B, Burley NT (1995) Reponses of Zebra Finches (*Taeniopygia guttatta*) to experimental intraspecific brood parasitism. Auk 112:415–420
Ferguson JWH (1994) The importance of low host densities for successful parasitism of Diederik Cuckoos on Red Bishop birds. S Afr J Zool 29:70–73
Fiorini VD, Tuero DT, Reboreda JC (2009) Shiny Cowbirds synchronize parasitism with host laying and puncture host eggs according to host characteristics. Anim Behav 77:561–568
Fjeldså J (2004) The grebes: Podicipedidae. Oxford University Press, New York
Flint PL, Sedinger JS, Pollock KH (1995) Survival of juvenile Black Brant during brood rearing. J Wildl Manag 59:455–463
Forbes S, Mock DW (1996) Food, information and avian brood reduction. Ecoscience 3:45–53
Förschler MI, del Val E, Bairlein F (2010) Extraordinary high natal philopatry in a migratory passerine. J Ornithol 151:745–748
Fox AD (2015) Development of feeding skills amongst juvenile Bullfinches *Pyrrhula pyrrhula* feeding on sunflower *Helianthus annuus* seeds. Ringing Migration 30:51–54
Fraga RM (1998) Interactions of the parasitic Screaming Cowbirds (*Molothrus rufoaxillaris* and *M. bonaerensis*) with a shared host, the Bay-winged Cowbird (*Molothrus badius*). In: Rothstein SI, Robinson SK (eds) Parasitic birds and their hosts: studies in coevolution. Oxford University Press, New York, pp 173–193
Fraser GS, Jones IL, Hunter FM (2002) Male-female differences in parental care in monogamous Crested Auklets. Condor 104:413–423
Freeman S, Gori DF, Rohwer S (1990) Red-winged Blackbirds and Brown-headed Cowbirds: some aspects of a host–parasite relationship. Condor 92:336–340
Friedmann H (1929) The cowbirds: a study in the biology of social parasitism. C. C. Thomas, Springfield, IL
Friedmann H, Kiff LF (1985) The parasitic cowbirds and their hosts. Proc West Found Vertebr Zool 2:227–302
Fry CH (1974) Vocal mimesis in nestling Greater Honey guides. Bull Br Ornithol Club 94:58–59
Galef BG Jr, Laland KN (2005) Social learning in animals: empirical studies and theoretical models. BioScience 55:489–499
Garrard GE, McCarthy MA, Vesk PA, Radford JQ, Bennett AF (2012) A predictive model of avian natal dispersal distance provides prior information for investigating response to landscape change. J Anim Ecol 81:14–23
Gaston AJ, Jones IL (1998) The auks: Alcidae. Oxford University Press, Oxford, UK
Geffen E, Yom-Tov Y (2001) Factors affecting the rate of intraspecific nest parasitism among Anseriformes and Galliformes. Anim Behav 62:1027–1038
Gehlbach FR (1994) The Eastern Screech Owl: life history, ecology, and behavior in the suburbs and countryside. Texas A&M University Press, College Station, TX
Gerber BD, Dwyer JF, Nesbitt SA, Drewien RC, Littlefield CD, Tacha TC, Vohs PA (2020) Sandhill Crane (*Antigone canadensis*), version 1.0. In: Poole AF (ed) Birds of the World. Cornell Lab of Ornithology, Ithaca, NY
Gilbert SF, Epel D (2009) Ecological developmental biology. Integrating epigenetics, medicine and evolution. Sinauer Associates, Sunderland, MA
Giudice JH, Ratti JT (2020) Ring-necked Pheasant (*Phasianus colchicus*), version 1.0. In: Billerman SM (ed) Birds of the World. Cornell Lab of Ornithology, Ithaca, NY
Gjerdrum C (2004) Parental provisioning and nestling departure decisions: a supplementary feeding experiment in Tufted Puffins (*Fratercula cirrhata*) on Triangle Island, British Columbia. Auk 121:463–472
Gleeson DJ, Blows MW, Owens IPF (2005) Genetic covariance between indices of body condition and immunocompetence in a passerine bird. BMC Evol Biol 5:61
Gloag R, Kacelnik A (2013) Host manipulation via begging call structure in the brood-parasitic Shiny Cowbird. Anim Behav 86:101–109
Gloag R, Tuero DT, Fiorini VD, Reboreda JC, Kacelnik A (2012) The economics of nestmate killing in avian brood parasites: a provisions tradeoff. Behav Ecol 23: 132–140
Godfray HCJ (1991) Signalling of need by offspring to their parents. Nature 352:328–330
Gomez-Serrano MA (2018) Broken wing display. In: Vonk J, Shackelford T (eds) Encyclopedia of animal cognition and behavior. Springer, Cham, Switzerland, pp 1–3
Gonzalez-Voyer A, Szekely T, Drummond H (2007) Why do some siblings attack each other? Comparative analysis of aggression in avian broods. Evolution 61:1946–1955
Goodbred CO, Holmes RT (1996) Factors affecting food provisioning of nestling Black-throated Blue Warblers. Wilson Bull 108:467–479
Göth A (2001) Innate predator recognition in Australian Brush-turkey hatchlings. Behaviour 138:117–136
Göth A (2002) Behaviour of Australian Brush-turkey (*Alectura lathami*) chicks following underground hatching. Journal für Ornithologie 143:477–488

Göth A, Jones DN (2003) Ontogeny of social behavior in the megapode Australian Brush-turkey (*Alectura lathami*). J Comp Psychol 117:36–43

Göth A, Proctor H (2002) Pecking preferences in hatchlings of the Australian Brush-turkey: the role of food type and colour. Aust J Zool 50:93–102

Göth A, Vogel U (2002) Social behaviour of two free-ranging chicks of Australian Brush-turkey *Alectura lathami*. Corella 26:1–4

Gowaty PA (1996) Field studies of parental care in birds: new data focus questions on variation among females. Adv Study Behav 25:477–531

Graul WD (1971) Observations at a Long-billed Curlew nest. Auk 88:182–184

Green DJ, Cockburn A (2001) Post-fledging care, philopatry and recruitment in Brown Thornbills. J Anim Ecol 70:505–514

Greenwood PJ (1980) Mating systems, philopatry and dispersal in birds and mammals. Anim Behav 28: 1140–1162

Griesser M, Barnaby J (2010) Families, a place of loving care and violent conflicts. The role of nepotism, cooperation and competition in the evolution of avian families. In: Zhang W, Liu H (eds) Behavioral and chemical ecology. Nova Science, New York, pp 47–91

Griesser M, Suzuki TN (2017) Naive juveniles are more likely to become breeders after witnessing predator mobbing. Am Nat 189:58–66

Griesser M, Drobniak SM, Graber SM, van Schaik CP (2023) Parental provisioning drives brain size in birds. Proc Natl Acad Sci USA 120:e2121467120

Grim T (2006a) The evolution of nestling discrimination by hosts of parasitic birds: why is rejection so rare? Evol Ecol Res 8:785–802

Grim T (2006b) Cuckoo growth performance in parasitized and unused hosts: not only host size matters. Behav Ecol Sociobiol 60:716–723

Grim T (2007) Experimental evidence for chick discrimination without recognition in a brood parasite host. Proc R Soc B 274:373

Grim T, Rutila J, Cassey P, Hauber ME (2009) The cost of virulence: an experimental study of egg eviction by brood parasitic chicks. Behav Ecol 20:1138–1146

Grodzinski U, Johnstone RA (2012) Parents and offspring in an evolutionary game: the effect of supply on demand when costs of care vary. Proc R Soc B 279: 109–115

Grüebler MU, Naef-Daenzer B (2010a) Survival benefits of post-fledging care: experimental approach to a critical part of avian reproductive strategies. J Anim Ecol 79:334–341

Grüebler MU, Naef-Daenzer B (2010b) Fitness consequences of timing of breeding in birds: date effects in the course of a reproductive episode. J Avian Biol 41:282–291

Grundel R (1987) Determinants of nestling feeding rates and parental investment in the Mountain Chickadee. Condor 89:319–328

Gulson-Castillo ER, Greeney HF, Freeman BG (2018) Coordinated misdirection: a probable anti-nest predation behavior widespread in Neotropical birds. Wilson J Ornithol 130:583–590

Haff TM, Magrath RD (2011) Calling at a cost: elevated nestling calling attracts predators to active nests. Biol Lett 22:493–495

Halupka L (1999) Nest defence in an altricial bird with uniparental care: the influence of offspring age, brood size, stage of the breeding season and predator type. Ornis Fennica 76:97–105

Hamilton WJ, Orians GH (1965) Evolution of brood parasitism in altricial birds. Condor 67:361–382

Handel C, Gill RE (2020) Black Turnstone (*Arenaria melanocephala*), version 1.0. In: Poole AF, Gill FB (eds) Birds of the World. Cornell Lab of Ornithology, Ithaca, NY

Hannon SJ, Wilson S, McCallum CA (2009) Does cowbird parasitism increase predation risk to American Redstart nests? Oikos 118:1035–1043

Harding AMA, van Pelt TI, Lifjeld JT, Mehlum F (2004) Sex differences in Little Auk *Alle alle* parental care: transition from biparental to paternal-only care. Ibis 146:642–651

Harel R, Horvitz N, Nathan R (2016) Adult vultures outperform juveniles in challenging thermal soaring conditions. Sci Rep 6:27865

Harfenist A, Ydenberg RC (1995) Parental provisioning and predation risk in Rhinoceros Auklets (*Cerorhinca monocerata*): effects on nestling growth and fledging. Behav Ecol 6:82–86

Haskell DG (1994) Experimental evidence that nestling begging behaviour incurs a cost due to nest predation. Proc R Soc B 257:161–164

Hatchwell BJ (2009) The evolution of cooperative breeding in birds: kinship, dispersal and life history. Philos Trans R Soc B 364:3217–3227

Hauber ME, Kilner RM (2007) Coevolution, communication, and host-chick mimicry in parasitic finches: who mimics whom? Behav Ecol Sociobiol 61:497–503

Hauber ME, Russo SA (2000) Perch proximity correlates with higher rates of cowbird parasitism of ground nesting Song Sparrows. Wilson Bull 112:150–153

Healy SD, Rowe C (2007) A critique of comparative studies of brain size. Proc R Soc B 274:453–464

Hedrick BP, Chunling G, Omar GI, Fengjiao Z, Caizhi S, Dodson P (2014) The osteology and taphonomy of a *Psittacosaurus* bonebed assemblage of the Yixian Formation (Lower Cretaceous, Liaoning, China). Cretac Res 51:321–340

Heino M, Hanski I (2001) Evolution of migration rate in a spatially realistic metapopulation model. Am Nat 157: 495–511

Hepp GR, Bellrose FC (2020) Wood Duck (*Aix sponsa*), version 1.0. In: Poole AF (ed) Birds of the World. Cornell Lab of Ornithology, Ithaca, NY

Hipfner JM, Gaston AJ (1999) Timing of nest departure in the Thick-billed Murre and Razorbill: tests of Ydenberg's model. Ecology 80:587–596

Holberton RL, Wilson CM, Hunter MJ, Cash WB, Sims CG (2007) The role of corticosterone in supporting migratory lipogenesis of the Dark-eyed Junco, *Junco hyemalis*: a model for central and peripheral regulation. Physiol Biochem Zool 80:125–137

Hollander FA, Van Overveld T, Tokka I, Matthysen E (2008) Personality and nest defence in the Great Tit (*Parus major*). Ethology 114:405–412

Holzhaider JC, Hunt GR, Gray RD (2010) Social learning in New Caledonian Crows. Learn Behav 38:206–219

Honza M, Vošlajerová K, Moskát C (2007) Eviction behaviour of the Common Cuckoo *Cuculus canorus* chicks. J Avian Biol 38:385–389

Hoover JP, Robinson SK (2007) Retaliatory mafia behavior by a parasitic cowbird favors host acceptance of parasitic eggs. Proc Natl Acad Sci USA 104:4479–4483

Humphreys RK, Ruxton GD (2020) Avian distraction displays: a review. Ibis 162(4):1125–1145

Hurd PL, Weatherhead PJ, McRae SB (1991) Parental consumption of nestling feces: good food or sound economics? Behav Ecol 2:69–76

Husby M (2019) Nestling begging calls increase predation risk by corvids. Anim Biol 69:137–155

Ibáñez-Álamo JD, Ruiz-Rodríguez M, Soler JJ (2014) The mucous covering of fecal sacs prevents birds from infection with enteric bacteria. J Avian Biol 45:354–358

Igic B, Braganza K, Hyland MM, Silyn-Roberts H, Cassey P, Grim T, Rutila J, Moskát C, Hauber ME (2011) Alternative mechanisms of increased eggshell hardness of avian brood parasites relative to host species. J R Soc Interface 8:1654–1664

Iwaniuk AN, Nelson JE (2003) Developmental differences are correlated with relative brain size in birds: a comparative analysis. Can J Zool 81:1913–1928

Jaatinen K, Lehtonen J, Kokko J (2011) Strategy selection under conspecific brood parasitism: an integrative modeling approach. Behav Ecol 22:144–155

Jablonka E, Lamb MJ (2014) Evolution in four dimensions. Genetic, epigenetic, behavioral and symbolic variation in the history of life. MIT Press, Cambridge, MA

Jacob S, Rieucau G, Heeb P (2011) Multimodal begging signals reflect independent indices of nestling condition in European Starlings. Behav Ecol Sociobiol 22:1249–1255

Jackson BJ, Jackson JA (2020) Killdeer (*Charadrius vociferus*), version 1.0. In: Poole AF, Gill FB (eds) Birds of the World. Cornell Lab of Ornithology, Ithaca, NY

Janisch J, Perinot E, Fusani L (2020) Behavioural flexibility in the courtship dance of Golden-collared Manakins, *Manacus vitellinus*. Anim Behav 166:61–71

Jelínek V, Šulc M, Weidinger K, Honza M (2019) Parent-absent begging and the risk of nest predation. J Ornithol 160:127–136

Jensen RAC, Vernon CJ (1970) On the biology of the Didric Cuckoo in southern Africa. Ostrich 41:237–246

Jetz W, Rubenstein DR (2011) Environmental uncertainty and the global biogeography of cooperative breeding birds. Curr Biol 21:72–78

Johnson LS, Rauch RL, Dellone SN (2004) The process and causes of fledging in a cavity-nesting passerine bird, the House Wren (*Troglodytes aedon*). Ethology 110:693–705

Johnson M, Aref S, Walters JR (2008) Parent-offspring communication in the Western Sandpiper. Behav Ecol 19:489–501

Johnson LS, Hebert RM, Napolillo FM, Allen A (2013) The process of fledging in the Mountain Bluebird. J Field Ornithol 84:367–376

Johnstone RA (1996) Multiple displays in animal communication: 'backup signals' and 'multiple messages'. Philos Trans R Soc B 351:329–338

Johnstone RA, Hinde CA (2006) Negotiation over offspring care—how should parents respond to each other's efforts? Behav Ecol 17:818–827

Jones KM, Ruxton GD, Monaghan P (2002) Model parents: is full compensation for reduced partner attendance compatible with stable biparental care? Behav Ecol 13:838–843

Jones TM, Ward MP, Benson TJ, Brawn JD (2017) Variation in nestling body condition and wing development predict cause-specific mortality in fledgling Dickcissels. J Avian Biol 48:439–447

Jones TM, Brawn JD, Ausprey IJ, Vitz AC, Rodewald AD, Raybuck DW, Boves TJ, Fiss CJ, McNeil DJ, Stoleson SH, Larkin JL (2020) Parental benefits and offspring costs reflect parent–offspring conflict over the age of fledging among songbirds. Proc Natl Acad Sci USA 117:30539–30546

Jones TM, Chiavacci SJ, Benson TJ, Ward MP (2022) Nesting and post-fledging predation risk influence diel patterns of songbird fledging. Ibis. https://doi.org/10.1111/ibi.13119

Jonker RM, Kurvers RHJM, van de Bilt A, Faber M, Van Wieren SE, Prins HHT, Ydenberg RC (2012) Rapid adaptive adjustment of parental care coincident with altered migratory behavior. Evol Ecol 26:657–667

Kedar J, Rodríguez-Gironés MA, Yedvab S, Winkler DW, Lotem A (2000) Experimental evidence for offspring learning in parent-offspring communication. Proc R Soc B 267:1723–1727

Kehoe FP (1989) The adaptive significance of créching behaviour in the White-winged Scoter (*Melanitta fusca deglandi*). Can J Zool 67:406–411

Kennerley JA, Somveille M, Hauber ME, Richardson NM, Manica A, Feeney WE (2022) The overlooked complexity of avian brood parasite–host relationships. Ecol Lett 25:1889–1904

Kenward RE, Marcstrom V, Karlbom M (1993) Post-nesting behavior in goshawks, *Accipiter gentilis*. I. The causes of dispersal. Anim Behav 46:365–370

Kern M, Bacon W, Long D, Cowie RJ (2001) Possible roles for corticosterone and critical size in the fledging of nestling Pied Flycatchers. Physiol Biochem Zool 74: 651–659

Kilner RM (2001) A growth cost of begging in captive canary chicks. Proc Natl Acad Sci USA 98:11394–11398

Kilner RM (2003) How selfish is a cowbird nestling? Anim Behav 66:569–576

Kilner RM (2005) The evolution of virulence in brood parasites. Ornithol Sci 4:55–64

Kilner RM, Hinde CA (2008) Information warfare and parent-offspring conflict. Adv Study Behav 38:283–336

Kilner RM, Johnstone RA (1997) Begging the question: are offspring solicitation behaviours signals of need? Trends Ecol Evol 12:11–15

Kilner RM, Madden JR, Hauber ME (2004) Brood parasitic cowbird nestlings use host young to procure resources. Science 305:877–879

Kilpatrick AM (2002) Variation in growth of Brown-headed Cowbird (*Molothrus ater*) nestlings and energetic impacts on their host parents. Can J Zool 80:145–153

King DI, Degraaf RM, Smith ML, Buonaccorsi JP (2006) Habitat selection and habitat-specific survival of fledgling Ovenbirds (*Seiurus aurocapilla*). J Zool 269:414–421

Kisel Y, Barraclough TG (2010) Speciation has a spatial scale that depends on levels of gene flow. Am Nat 175: 316–334

Kitowski I (2009) Social learning of hunting skills in juvenile Marsh Harriers *Circus aeruginosus*. J Ethol 27:327–332

Kleindorfer S, Hoi H, Fessl B (1996) Alarm calls and chick reactions in the Moustached Warbler, *Acrocephalus melanopogon*. Anim Behav 51:1199–1206

Kohn GM, King AP, Dohme R, Meredith GR, West MJ (2013) Robust autumn social attributes predict spring courtship skills in juvenile female Brown-headed Cowbirds, *Molothrus ater*. Anim Behav 85:727–732

Kokko H, Jennions MD (2008) Parental investment, sexual selection and sex ratios. J Evol Biol 21:919–948

Korpimaki E (1984) Clutch size and breeding success of Tengmalm's Owl *Aegolius funereus* in natural cavities and nest-boxes. Ornis Fennica 61:80–83

Korsten P, van Overveld T, Adriaensen F, Matthysen E (2013) Genetic integration of local dispersal and exploratory behaviour in a wild bird. Nat Commun 4: 2362

Kosztolányi A, Székely T, Cuthill IC (2007) The function of habitat change during brood-rearing in the precocial Kentish Plover *Charadrius alexandrinus*. Acta Ethologica 10:73–79

Koykka C, Wild G (2018) The influence of environmental variance on the evolution of signaling behavior. Behav Ecol 29:814–820

Kraaijeveld K, Dickinson JL (2001) Family-based winter territoriality in Western Bluebirds, *Sialia mexicana*: the structure and dynamics of winter groups. Anim Behav 61:109–117

Krausová L, Veselý P, Syrová M, Antonová K, Fišer O, Chlumská V, Pátková M, Pužej Š, Fuchs R (2022) Red-backed Shrike (*Lanius collurio*) versus Common Cuckoo (*Cuculus canorus*): an example of ineffective cuckoo-hawk mimicry. Ecol Evol 12:e9664

Krebs EA, Putland DA (2004) Chic chicks: the evolution of chick ornamentation in rails. Behav Ecol 15:946–951

Krüger O, Davies NB (2002) The evolution of cuckoo parasitism: a comparative analysis. Proc R Soc B 269:375–381

Krüger O, Davies NB (2004) The evolution of egg size in the brood parasitic cuckoos. Behav Ecol 15:210–218

Krüger O, Davies NB, Sorenson MD (2007) The evolution of sexual dimorphism in parasitic cuckoos: sexual selection or coevolution? Proc R Soc B 274:1553–1560

Kruuk H (1964) Predators and anti-predator behavior of the Black-headed Gull (*Larus ridibundus* L.). Behaviour 11(Suppl):1–129

Kuzniak S (1967) Observations on the breeding biology of swallow, *Hirundo rustica* L. Acta Ornithologica 10: 177–211

Lack D (1956) A review of the genera and nesting habits of swifts. Auk 73:1–32

Lahti DC, Lahti AR (2002) How precise is egg discrimination in weaverbirds? Anim Behav 63:1135–1142

Lanctot RB, Gill RE Jr, Tibbitts TL, Handel CM (1995) Brood amalgamation in the Bristle-thighed Curlew *Numenius tahitiensis*: process and function. Ibis 137: 559–569

Landová E, Hotová Svádová K, Fuchs R, Štys P, Exnerová A (2017) The effect of social learning on avoidance of aposematic prey in juvenile Great Tits (*Parus major*). Anim Cogn 20:855–866

Landys MM, Wingfield JC, Ramenofsky M (2004) Plasma corticosterone increases during migratory restlessness in the captive White-crowned Sparrow *Zonotrichia leucophrys gambelli*. Horm Behav 46:574–581

Langmore NE, Kilner RM (2010) The co-evolutionary arms race between Horsfield's Bronze-Cuckoos and Superb Fairy-wrens. Emu 110:32–38

Langmore NE, Hunt S, Kilner RM (2003) Escalation of a coevolutionary arms race through host rejection of brood parasitic young. Nature 422:157–160

Langmore NE, Maurer G, Adcock GJ, Kilner RM (2008) Socially acquired host-specific mimicry and the evolution of host races in Horsfield's Bronze-Cuckoo *Chalcites basalis*. Evolution 62:1689–1699

Langmore NE, Cockburn A, Russell AF, Kilner RM (2009a) Flexible cuckoo chick-rejection rules in the Superb Fairy-wren. Behav Ecol 20:978–984

Langmore NE, Stevens M, Maurer G, Kilner RM (2009b) Are dark cuckoo eggs cryptic in host nests? Anim Behav 78:461–468

Langmore NE, Stevens M, Maurer G, Heinsohn R, Hall ML, Peters A, Kilner RM (2011) Visual mimicry of host nestlings by cuckoos. Proc R Soc B 278:2455–2463

Langmore NE, Feeney WE, Crowe-Riddell J, Luan H, Louwrens KM, Cockburn A (2012) Learned recognition of brood parasitic cuckoos in the Superb Fairy-wren *Malurus cyaneus*. Behav Ecol 23:798–805

LaPorte N, Storer RW, Nuechterlein GL (2020) Western Grebe (*Aechmophorus occidentalis*), version 1.0. In: Rodewald PG (ed) Birds of the World. Cornell Lab of Ornithology, Ithaca, NY

Larsen T (1991) Anti-predator behaviour and mating systems in waders: aggressive nest defence selects for monogamy. Anim Behav 41:1057–1062

Larsen T, Sordahl TA, Byrkjedal I (1996) Factors related to aggressive nest protection behaviour: a comparative study of Holarctic waders. Biol J Linn Soc 58:409–439

Lawes MJ, Kirkman S (1996) Egg recognition and interspecific brood parasitism rates in Red Bishops (Aves: Ploceidae). Anim Behav 52:553–563

Lee J-W, Lee Y-K, Hatchwell BJ (2010) Natal dispersal and philopatry in a group-living but noncooperative passerine bird, the Vinous-throated Parrotbill. Anim Behav 79:1017–1023

Leech SM, Leonard ML (1997) Begging and the risk of predation in nestling birds. Behav Ecol 8:644–646

Lefebvre L, Whittle P, Lascaris E, Finkelstein A (1997) Feeding innovations and forebrain size in birds. Anim Behav 53:549–560

Lefebvre L, Gaxiola A, Dawson S, Rosza L, Kabai P (1998) Feeding innovations and forebrain size in Australasian birds. Behaviour 135:1077–1097

Lemel J (1989) Body mass dependent fledging order in the Great Tit. Auk 106:490–492

Lengyel S (2007) Benefits of large broods by higher chick survival and better territories in a precocial shorebird. Behav Ecol Sociobiol 61:589–598

Leopold F (1951) A study of nesting Wood Ducks in Iowa. Condor 53:209–220

Lepage D, Gauthier G, Desrochers A (1998) Larger clutch size increases fledging success and offspring quality in a precocial species. J Anim Ecol 67:210–216

Lessells CM (1991) The evolution of life-histories. In: Krebs JR, Davies NB (eds) Behavioural ecology: an evolutionary approach. Blackwell, Oxford, UK, pp 32–68

Lichtenstein G (1997) Begging behaviour and host exploitation in three species of parasitic cowbirds. Ph.D. dissertation, University of Cambridge, Cambridge, UK

Lichtenstein G (2001a) Selfish begging by Screaming Cowbirds, a mimetic brood parasite of the Bay-winged Cowbird. Anim Behav 61:1151–1158

Lichtenstein G (2001b) Low success of Shiny Cowbird chick parasitizing Rufous-bellied Thrushes: chick-chick competition or parental discrimination? Anim Behav 61:401–413

Lichtenstein G, Dearborn DC (2004) Begging and short-term need in cowbird nestlings: how different are brood parasites? Behav Ecol Sociobiol 56:352–359

Limmer B, Becker PH (2009) Improvement in chick provisioning with parental experience in a seabird. Anim Behav 77:1095–1101

Limparungpatthanakij WL (2020) Asian Koel (*Eudynamys scolopaceus*), version 1.0. In: Billerman SM, Keeney BK, Rodewald PG, Schulenberg TS (eds) Birds of the World. Cornell Lab of Ornithology, Ithaca, NY

Lind J (2004) What determines probability of surviving predator attacks in bird migration? The relative importance of vigilance and fuel load. J Theor Biol 231:223–227

Lindberg MS, Sedinger JS, Derksen DV, Rockwell RF (1998) Natal and breeding philopatry in a Black Brant, *Branta bernical nigricans*, metapopulation. Ecology 79:1893–1904

Lindholm AK, Thomas RJ (2000) Differences between populations of Reed Warblers in defences against brood parasitism. Behaviour 137:25–42

Liu S, Xie Q, Jiang A, Goodale E (2022) Investigating how different classes of nest predators respond to the playback of the begging calls of nestling birds. Avian Res 13:100044

Lombardo MP, Thorpe PA, Otieno S, Hawker A, Welgarz D, Andrews D, Black A (2020) Yearly variation in factors associated with local recruitment of Tree Swallows. J Field Ornithol 91:199–213

López AV, Fiorini VD, Ellison K, Peer BD (2018) Thick eggshells of brood parasitic cowbirds protect their eggs and damage host eggs during laying. Behav Ecol 29: 965–973

López-Idiáquez D, Vergara P, Fargallo JA, Martínez-Padilla J (2018) Providing longer post-fledging periods increases offspring survival at the expense of future fecundity. PLoS ONE 13:e0203152

Lorenzana JC, Sealy SG (1996) Are begging calls of generalist parasitic cowbirds adapted for brood parasitism? UFS (Delta Marsh). Annu Rep 31:85–93

Lotem A, Nakamura H, Zahavi A (1995) Constraints on egg discrimination and cuckoo-host coevolution. Anim Behav 49:1185–1209

Lukianchuk K (2013) Coordinated display, social hierarchy, and the development of dancing ability in young Long-tailed Manakins, *Chiroxiphia linearis*. M. S. thesis, University of Windsor, Windsor, ON, Canada

Lyon BE, Eadie JM (2000) Family matters: kin selection and the evolution of conspecific brood parasitism. Proc Natl Acad Sci USA 97:12942–12944

Lyon BE, Eadie JM (2008) Conspecific brood parasitism in birds: a life-history perspective. Annu Rev Ecol Syst 39:343–363

Lyon BE, Everding S (1996) High frequency of conspecific brood parasitism in a colonial waterbird, the Eared Grebe *Podiceps nigricollis*. J Avian Biol 27:238–244

Lyon BE (2003) Egg recognition and counting reduce costs of avian conspecific brood parasitism. Nature 422:495–499

Lyon BE, Gilbert GS (2013) Rarely parasitized and unparasitized species mob and alarm call to cuckoos: implications for sparrowhawk mimicry by brood parasitic cuckoos. Wilson J Ornithol 125:627–630

Lyon BE, Eadie JM, Hamilton LD (1994) Parental choice selects for ornamental plumage in American Coot chicks. Nature 371:240–243

Lyons DM, Mosher JA (1987) Morphological growth, behavioral development, and parental care of Broad-winged Hawks. J Field Ornithol 58:334–344

Madden JR, Davies NB (2006) A host-race difference in begging calls of nestling cuckoos *Cuculus canorus* develop through experience and increases host provisioning. Proc R Soc B 273:2343–2345

Mainguy J, Gauthier G, Giroux J-F, Duclos I (2006) Habitat use and behaviour of Greater Snow Geese during movements from nesting to brood-rearing areas. Can J Zool 84:1096–1103

Mainwaring MC, Hartley IR (2012) Causes and consequences of differential growth in birds: a behavioral perspective. Adv Study Behav 44:225–277

Maness TJ, Anderson DJ (2013) Predictors of juvenile survival in birds. Ornithol Monogr 78:1–55

Mappes J, Lindström L (2012) How did the cuckoo get its polymorphic plumage? Science 337:532–533

Marchetti K (2000) Egg rejection in a passerine bird: size does matter. Anim Behav 59:877–883

Marchetti K, Price T (1989) Differences in the foraging of juvenile and adult birds: the importance of developmental constraints. Biol Rev 64:51–70

Margalida A, Gonzalez LM, Sanchez R, Oria J, Prada L (2007) Parental behaviour of Spanish Imperial Eagles *Aquila adalberti*: sexual differences in a moderately dimorphic raptor. Bird Study 54:112–119

Mark MM, Rubenstein DR (2013) Physiological costs and carry-over effects of avian interspecific brood parasitism influence reproductive tradeoffs. Horm Behav 63: 717–722

Markman S, Yom-Tov Y, Wright J (1995) Male parental care in the Orange-tufted Sunbird: behavioural adjustments in provisioning and nest guarding effort. Anim Behav 50:655–669

Marques PAM, Leonard ML, Horn AG, Contasti A (2011) How nestling Tree Swallows (*Tachycineta bicolor*) integrate their responses to hunger and signalling by nestmates. Ethology 117:163–170

Martin TE (1995) Avian life history evolution in relation to nest sites, nest predation, and food. Ecol Monogr 65: 101–127

Martin TE (1996) Life history evolution in tropical and south temperate birds: What do we really know? J Avian Biol 27:263–272

Martin TE, Li P (1992) Life history traits of open-vs. cavity-nesting birds. Ecology 73:579–592

Martin TE, Lloyd P, Bosque C, Barton DC, Biancucci AL, Cheng Y-R, Ton R (2011) Growth rate variation among passerine species in tropical and temperate sites: an antagonistic interaction between parental food provisioning and nest predation risk. Evolution 65:1607–1622

Martin TE, Tobalske B, Riordan MM, Case SB, Dial KP (2018) Age and performance at fledging are a cause and consequence of juvenile mortality between life stages. Sci Adv 4:eaar1988

Martín-Gálvez D, Soler M, Soler JJ, Martín-Vivaldi M, Palomino JJ (2005) Food acquisition by Common Cuckoo nestlings in Rufous Bush Robin nests and the advantage of the eviction behaviour. Anim Behav 70: 1313–1321

Martín-Gálvez D, Soler JJ, Martínez JG, Krupa AP, Richard M, Soler M, Møller AP, Burke T (2006) A quantitative trait locus for recognition of foreign eggs in the host of a brood parasite. J Evol Biol 19:543–550

Martins TLF (1997) Fledging in the Common Swift, *Apus apus*: weight-watching with a difference. Anim Behav 54:99–108

Marton A, Fülöp A, Ozogány K, Moskát C, Bán M (2019) Host alarm calls attract the unwanted attention of the brood parasitic Common Cuckoo. Sci Rep 9:18563

Matray PF (1974) Broad-winged Hawk nesting and ecology. Auk 91:307–324

Matthysen E (2012) Multicausality of dispersal: a review. In: Clobert J, Baguette M, Benton TG, Bullock JM (eds) Dispersal ecology and evolution. Oxford University Press, Oxford, UK, pp 3–18

McDonald PG, Wilson DR, Evans CS (2009) Nestling begging increases predation risk, regardless of spectral characteristics or avian mobbing. Behav Ecol 20:821–829

McGowan KJ (1995) A test of whether economy of nutrition determines fecal sac ingestion in nesting corvids. Condor 97:50–56

McKim-Louder MI, Hoover JP, Benson TJ, Schelsky WM (2013) Juvenile survival in a Neotropical migratory songbird is lower than expected. PLoS ONE 8:e56059

McLean IG, Waas JR (1987) Do cuckoo chicks mimic the begging calls of their hosts? Anim Behav 35:1896–1898

McNamara JM, Székely T, Webb JN, Houston AI (2000) A dynamic game-theoretic model of parental care. J Theor Biol 205:605–623

McRoberts JT, Wallace MC, Eaton SW (2020) Wild Turkey (*Meleagris gallopavo*), version 1.0. In: Poole AF (ed) Birds of the World. Cornell Lab of Ornithology, Ithaca, NY

Meng Q, Liu J, Varricchio DJ, Huang T, Gao C (2004) Palaeontology: parental care in an ornithischian dinosaur. Nature 431:145–146

Michaud T, Leonard M (2000) The role of development, parental behavior, and nestmate competition in fledging of nestling Tree Swallows. Auk 117:996–1002

Miller EH, Baker AJ (1980) Displays of the Magellanic Oystercatcher (*Haematopus leucopodus*). Wilson Bull 92:149–168

Mitchell GW, Woodworth BK, Taylor PD, Norris DR (2015) Automated telemetry reveals age specific differences in flight duration and speed are driven by wind conditions in a migratory songbird. Mov Ecol 3: 19

Mock DW (1987) Siblicide, parent–offspring conflict, and unequal parental investment by egrets and herons. Behav Ecol Sociobiol 20:247–256

Mock DW, Parker GA (1998) Siblicide, family conflict and the evolutionary limits of selfishness. Anim Behav 56:1–10

Moksnes A, Røskaft E (1995) Egg-morphs and host preference in the Common Cuckoo (*Cuculus canorus*): an

analysis of cuckoo and host eggs from European museum collections. J Zool 236:625–648
Molina-Morales M, Martínez JG, Avilés JM (2016) Criteria for host selection in a brood parasite vary depending on parasitism rate. Behav Ecol 27:1441–1448
Montevecchi WA, Vaughan RB (1989) The ontogeny of thermal independence in nestling gannets. Ornis Scandinavica 20:161–168
Montgomerie RD, Weatherhead PJ (1988) Risks and rewards of nest defence by parent birds. Q Rev Biol 63:167–187
Moore JR, Varricchio DJ (2016) The evolution of diapsid reproductive strategy with inferences about extinct taxa. PLoS ONE 11:e0158496
Morand-Ferron J, Sol D, Lefebvre L (2007) Food-stealing in birds: brain or brawn. Anim Behav 74:1725–1734
Morbey YE, Ydenberg RC, Knechtel HA, Harfenist A (1999) Parental provisioning, nestling departure decisions and prefledging mass recession in Cassin's Auklets. Anim Behav 57:873–881
Moreno J (1984) Parental care of fledged young, division of labour, and the development of foraging techniques in the Northern Wheatear, *Oenanthe oenanthe* L. Auk 101:741–752
Moreno-Rueda G (2010) An immunological cost of begging in House Sparrow nestlings. Proc R Soc B 277: 2083–2088
Moreno-Rueda G, Redondo T (2011) Begging at high level simultaneously impairs growth and immune response in Southern Shrike (*Lanius meridionalis*) nestlings. J Evol Biol 24:1091–1098
Morrison KW, Hipfner JM, Gjerdrum C, Green DJ (2009) Wing length and mass at fledging predict local juvenile survival and age at first return in Tufted Puffins. Condor 111:433–441
Morton ML (1979) Fecal sac ingestion in the White-crowned Sparrow. Condor 81:72–77
Moskát C, Honza M (2000) Effect of nest and nest site characteristics on the risk of cuckoo *Cuculus canorus* parasitism in the Great Reed Warbler *Acrocephalus arundinaceus*. Ecography 23:335–341
Moskát C, Székely T, Kisbenedek T, Karcza Z, Bártol I (2003) The importance of nest cleaning in egg rejection behaviour of Great Reed Warblers *Acrocephalus arundinaceus*. J Avian Biol 34:16–19
Mowbray TB, Cooke F, Ganter B (2020a) Snow Goose (*Anser caerulescens*), version 1.0. In: Rodewald PG (ed) Birds of the World. Cornell Lab of Ornithology, Ithaca, NY
Mowbray TB, Ely CR, Sedinger JS, Trost RE (2020b) Canada Goose (*Branta canadensis*), version 1.0. In: Rodewald PG (ed) Birds of the World. Cornell Lab of Ornithology, Ithaca, NY
Mueller J, Gibson LJ (2023) Structure and mechanics of water-holding feathers of Namaqua Sandgrouse (*Pterocles namaqua*). J Royal Soc Interface 20:20220878
Mueller T, O'Hara RB, Converse SJ, Urbanek RP, Fagan WF (2013) Social learning of migratory performance. Science 341:999–1002
Mulder RA (1995) Natal and breeding dispersal in a co-operative, extra-group-mating bird. J Avian Biol 26:234–240
Muller M, Groothuis TGG (2013) Within-clutch variation in yolk testosterone as an adaptive maternal effect to modulate avian sibling competition: evidence from a comparative study. Am Nat 181:125–136
Munro J, Bedard J (1977) Gull predation and créching behaviour in the Common Eider. J Anim Ecol 46: 799–810
Naef-Daenzer B, Grüebler MU (2016) Post-fledging survival of altricial birds: ecological determinants and adaptation. J Field Ornithol 87:227–250
Naef-Daenzer B, Widmer F, Nuber M (2001) Differential post-fledging survival of Great and Coal tits in relation to their condition and fledging date. J Anim Ecol 70: 730–738
Nebel S, Cooper JM (2020) Least Sandpiper (*Calidris minutilla*), version 1.0. In: Poole AF (ed) Birds of the World. Cornell Lab of Ornithology, Ithaca, NY
Negro JJ, Hiraldo F, Donázar JA (1997) Causes of natal dispersal in the Lesser Kestrel: inbreeding avoidance or resource competition? J Anim Ecol 66:640–648
Nettleship DN (2020) Ruddy Turnstone (*Arenaria interpres*), version 1.0. In: Billerman SM (ed) Birds of the World. Cornell Lab of Ornithology, Ithaca, NY
Neuenschwander S, Brinkof MWG, Kölliker M, Richner H (2003) Brood size, sibling competition, and the cost of begging in Great Tits (*Parus major*). Behav Ecol 14: 457–462
Nilsson JA (1990) Family flock break-up: spontaneous dispersal or parental aggression? Anim Behav 40: 1001–1003
Norberg UM (1989) Vertebrate flight: mechanics, physiology, morphology, ecology, and evolution. Springer-Verlag, Heidelberg, Germany
Norman RF, Robertson RJ (1975) Nest-searching behavior in the Brown-headed Cowbird. Auk 92:610–611
Oken L (1816) Lehrbuch der zoologie. Jena
Osorno JL, Drummond H (2003) Is obligate siblicidal aggression food sensitive? Behav Ecol Sociobiol 54: 547–554
Öst M, Kilpi M (2000) Eider females and broods from neighboring colonies use segregated local feeding areas. Waterbirds 23:24–32
Østnes JE, Jenssen BM, Becha C (2001) Growth and development of homeothermy in nestling European Shags. Auk 118:983–995
Pagnucco K, Zanette L, Clinchy M, Leonard ML (2008) Sheep in wolf's clothing: host nestling vocalizations resemble their cowbird competitor's. Proc R Soc B 275:1061–1065
Paradis E, Baillie SR, Sutherland WJ, Gregory RD (1998) Patterns of natal and breeding dispersal in birds. J Anim Ecol 67:518–536

Paredes R, Jones IL, Boness DJ (2006) Parental roles of male and female Thick-billed Murres and Razorbills at the Gannet Islands, Labrador. Behaviour 143:451–481

Parker GA (1985) Models of parent-offspring conflict. V. Effects of the behaviour of the two parents. Anim Behav 33:519–533

Parker GA, Royle NJ, Hartley IR (2002a) Begging scrambles with unequal chicks: interactions between need and competitive ability. Ecol Lett 5:206–215

Parker GA, Royle NJ, Hartley IR (2002b) Intrafamilial conflict and parental investment: a synthesis. Philos Trans R Soc B 357:295–307

Pasinelli G, Schiegg K, Walters JR (2004) Genetic and environmental influences on natal dispersal distance in a resident bird species. Am Nat 164:660–669

Patten MA, Reinking DL, Wolfe DH (2011) Hierarchical cues in brood parasite nest selection. J Ornithol 152: 521–532

Patterson TL, Petrinovich L, James DK (1980) Reproductive value and appropriateness of response to predators by White-crowned Sparrows. Behav Ecol Sociobiol 7: 227–231

Payne RB (1969) Nest parasitism and display of Chestnut Sparrows in a colony of Grey-capped Social Weavers. Ibis 111:300–307

Payne RB (1974) The evolution of clutch size and reproductive rates in parasitic cuckoos. Evolution 28:169–181

Payne RB (1977) The ecology of brood parasitism in birds. Annu Rev Ecol Syst 8:1–28

Payne RB (1998) Brood parasitism in birds: strangers in the nest. BioScience 48:377–386

Payne RB, Payne K (1967) Cuckoo hosts in southern Africa. Ostrich 38:135–143

Payne RB, Payne LL (2002) Begging for parental care from another species: specialization and generalization in brood-parasitic finches. In: Wright J, Leonard ML (eds) The evolution of begging. Kluwer, Dordrecht, pp 429–449

Payne RB, Woods JL, Payne LL (2001) Parental care in estrildid finches: experimental tests of a model of *Vidua* brood parasitism. Anim Behav 62:473–483

Peer BD, Bolinger EK (1997) Explanations for the infrequent cowbird parasitism on Common Grackles. Condor 99:151–161

Peer BD, Sealy SG (1999) Parasitism and egg puncture behavior by Bronzed and Brown-headed cowbirds in sympatry. Stud Avian Biol 18:235–240

Peer BD, Sealy SG (2004) Correlates of egg rejection in hosts of the Brown-headed Cowbird. Condor 106:580–599

Peer BD, Rivers JW, Rothstein SI (2013) Cowbirds, conservation, and coevolution: potential misconceptions and directions for future research. Chin Birds 4:15–30

Penteriani V, del Mar Delgado M, Kuparinen A, Saurola P, Valkema J, Salo E, Toivula J, Aebischer A, Arlettaz R (2014) Bright moonlight triggers natal dispersal departures. Behav Ecol Sociobiol 68:743–747

Pereyra ME, Morton ML (2001) Nestling growth and thermoregulatory development in subalpine Dusky Flycatchers. Auk 118:116–136

Perrins CM (1979) British tits. Collins, London

Pettingill OS Jr (1985) Ornithology in laboratory and field, 5th edn. Academic, New York

Phillips RA, Hamer KC (2000) Growth and provisioning strategies of Northern Fulmars *Fulmarus glacialis*. Ibis 142:435–445

Picman J, Pribil S (1997) Is greater eggshell density an alternative mechanism by which parasitic cuckoos increase the strength of their eggs? J Ornithol 138: 531–541

Pietz PJ, Granfors DA (2005) Parental nest defense on videotape: more reality than 'myth'. Auk 122:701–705

Pietz PJ, Granfors DA, Grant TA (2012) Hatching and fledging times from grassland passerine nests. In: Ribic CA, Thompson FR III, Pietz PJ (eds) Video surveillance of nesting birds. Studies in Avian Biology, vol 43. University of California Press, Berkeley, CA, pp 47–60

Pinto M, Rocha P, Moreira F (2005) Long-term trends in Great Bustard (*Otis tarda*) populations in Portugal suggest concentration in single high quality area. Biol Conserv 124:415–423

Platzen D, Magrath RD (2004) Parental alarm calls suppress nestling vocalisation. Proc R Soc B 271:1271–1276

Price K, Harvey H, Ydenberg R (1996) Begging tactics of nestling Yellow-headed Blackbirds, *Xanthocephalus xanthocephalus*, in relation to need. Anim Behav 51: 421–435

Quader S (2006) What makes a good nest? Benefits of nest choice to female Baya Weavers (*Ploceus philippinus*). Auk 123:475–486

Queller DC (1997) Why do females care more than males? Proc R Soc B 264:1555–1557

Radersma R, Tinbergen JM, Komdeur J (2011) Do brood sex ratio, nestling development and sex affect fledging timing and order? An experimental study on Great Tits. Anim Behav 81:69–75

Radersma R, Komdeur J, Tinbergen JM (2015) Early morning fledging improves recruitment in Great Tits *Parus major*. Ibis 157:351–355

Raikow RJ (1988) The analysis of evolutionary success. Syst Zool 37:76–79

Rasmussen JL, Sealy SG, Underwood TJ (2009) Video recording reveals the method of ejection of Brown-headed Cowbird eggs and no cost in American Robins and Gray Catbirds. Condor 111:570–574

Reed RA (1968) Studies of the Diederik Cuckoo *Chrysococcyx caprius* in the Transvaal. Ibis 110:321–331

Reidy JL, Stake MM, Thompson FR III (2008) Golden-cheeked Warbler nest mortality and predators in urban and rural landscapes. Condor 110:458–466

Remeš V (2006) Growth strategies of passerine birds are related to brood parasitism by the Brown-headed Cowbird (*Molothrus ater*). Evolution 60:1692–1700

Remeš V, Martin TE (2002) Environmental influences on the evolution of growth and developmental rates in passerines. Evolution 56:2505–2518

Remeš V, Matysiokova B, Cockburn A (2012) Long-term and large-scale analyses of nest predation patterns in Australian songbirds and a global comparison of nest predation rates. J Avian Biol 43:435–444

Remeš V, Freckleton RP, Tökölyi J, Liker A, Székely T (2015) The evolution of parental cooperation in birds. Proc Natl Acad Sci USA 112:13603–13608

Remeš V, Matysioková B, Vrána J (2020) Adaptation and constraint shape the evolution of growth patterns in passerine birds across the globe. Front Zool 17:29

Ribic CA, Rugg DJ, Ellison K, Koper N, Pietz PJ (2021) Diel patterns of predation and fledging at nests of four species of grassland songbirds. Ecol Evol 11:6913–6926

Ricklefs RE (1969) An analysis of nesting mortality in birds. Smithsonian Contrib Zool 9:1–48

Ricklefs RE, Starck JM (1998) The evolution of the developmental mode in birds. In: Starck JM, Ricklefs RE (eds) Avian growth and development: evolution with the altricial-precocial spectrum. Oxford University Press, New York, pp 366–380

Ristau CA (1993) The cognitive ethology of an 'injury-feigning' plover: a beginning. Etologia 3:57–68

Ritchison G, Belthoff JR, Sparks EJ (1992) Dispersal restlessness: evidence for innate dispersal by juvenile Eastern Screech-Owls? Anim Behav 43:57–65

Rivers JW (2007) Nest mate size, but not short-term need, influences begging behavior of a generalist brood parasite. Behav Ecol 18:222–230

Robert M, Sorci G (2001) The evolution of obligate interspecific brood parasitism in birds. Behav Ecol 12:128–133

Robinson WD, Hau M, Klasing KC, Wikelski M, Brawn JD, Austin SH, Tarwater CE, Ricklefs RE (2010) Diversification of life histories in New World birds. Auk 127:253–262

Rodríguez-Gironés MA, Zúñiga JM, Redondo T (2001) Effects of begging on growth rates of nestling chicks. Behav Ecol 12:269–274

Roff DA, Remeš V, Martin TE (2005) The evolution of fledging age in songbirds. J Evol Biol 18:1425–1433

Rohwer FC, Anderson MG (1988) Female-biased philopatry, monogamy and the timing of pair formation in waterfowl. Curr Ornithol 5:187–221

Rohwer S, Spaw CD (1988) Evolutionary lag versus bill-size constraints: a comparative study of the acceptance of cowbird eggs by old hosts. Evol Ecol 2:27–36

Romano A, Caprioli M, Boncoraglio G, Saino N, Rubolini D (2012) With a little help from my kin: Barn Swallow nestlings modulate solicitation of parental care according to nestmates' need. J Evol Biol 25:1703–1710

Ronget V, Gaillard J-M, Coulson T, Garratt M, Gueyffier F, Lega J-C, Lemaître J-F (2018) Causes and consequences of variation in offspring body mass: meta-analyses in birds and mammals. Biol Rev 93:1–27

Rose AP, Lyon BE (2013) Day length, reproductive effort, and the avian latitudinal clutch size gradient. Ecology 94:1327–1337

Rothstein SI (1975) An experimental and teleonomic investigation of avian brood parasitism. Condor 77:250–271

Rothstein SI (1990) A model system for coevolution: avian brood parasitism. Annu Rev Ecol Syst 21:481–508

Rothstein SI, Robinson SK (eds) (1998) Parasitic birds and their hosts: studies in coevolution. Oxford Univ Press, Oxford, UK

Royle NJ, Hartley IR, Parker GA (2004) Parental investment and family dynamics: interactions between theory and empirical tests. Popul Ecol 46:231–241

Royle NJ, Smiseth PT, Kölliker M, Alonzo SH, Klug H (2012) Paternity, maternity, and parental care. In: Royle NJ, Smiseth PT, Kölliker M (eds) The evolution of parental care. Oxford University Press, Oxford, UK, pp 189–205

Ruaux G, Lumineau S, de Margerie E (2020) The development of flight behaviours in birds. Proc R Soc B 287:20200668

Rush SA, Stutchbury BJM (2008) Survival of fledgling Hooded Warblers (*Wilsonia citrina*) in small and large forest fragments. Auk 125:183–191

Russell EM (2000) Avian life histories: is extended parental care the southern secret? Emu 100:377–399

Russell EM, Yom-Tov Y, Geffen E (2004) Extended parental care and delayed dispersal: northern, tropical, and southern passerines compared. Behav Ecol 15:831–838

Rutila JR, Latja, Koskela K (2002) The Common Cuckoo *Cuculus canorus* and its cavity nesting host, the redstart *Phoenicurus phoenicurus*: a peculiar cuckoo-host system? J Avian Biol 33:414–419

Rytkönen S, Koivula K, Orell M (1996) Patterns of pre-brood and per-offspring provisioning efforts in the Willow Tit *Parus montanus*. J Avian Biol 27:21–30

Sakaluk SK, Wilson AJ, Bowers EK, Johnson LS, Masters BS, Johnson BGP, Vogel LA, Forsman AM, Thompson CF (2014) Genetic and environmental variation in condition, cutaneous immunity, and haematocrit in House Wrens. BMC Evol Biol 14:242

Sam K, Kovarova E, Freiberga I, Uthe H, Weinhold A, Jorge LR, Sreekar R (2021) Great Tits (*Parus major*) flexibly learn that herbivore-induced plant volatiles indicate prey location: an experimental evidence with two tree species. Ecol Evol 11:10917–10925

Sandell MI, Diemer M (1999) Intraspecific brood parasitism: a strategy for floating females in the European Starling. Anim Behav 57:197–202

Santema P, Schlicht L, Beck KB, Sheldon BC, Kempenaers B (2021) Why do nestling birds fledge early in the day? Anim Behav 174:79–86

Santisteban L, Sieving KE, Avery ML (2002) Use of sensory cues by Fish Crows *Corvus ossifragus* preying on artificial bird nests. J Avian Biol 33:245–252

Sanz JJ, Kranenbarg S, Tinbergen JM (2000) Differential response by males and females to manipulation of partner contribution in the Great Tit (*Parus major*). J Anim Ecol 69:74–84

Sato NJ, Tokue K, Noske RA, Mikami OK, Ueda K (2010) Evicting cuckoo nestlings from the nest: a new anti-parasite behaviour. Biol Lett 6:67–69

Saunders CA, Arcese P, O'Connor KD (2003) Nest site characteristics in the Song Sparrow and parasitism by Brown-headed Cowbirds. Wilson Bull 115:24–28

Scharf HM, Hauber ME, Stenstrom KH, Schelsky WM (2022) Should I stay or should I go: the effect of avian brood parasitism on host fledging dynamics. Behav Ecol Sociobiol 76:64

Schlicht L, Girg A, Loës P, Valcu M, Kempenaers B (2012) Male extrapair nestlings fledge first. Anim Behav 83:1335–1343

Schlossberg S (2009) Site fidelity of shrubland and forest birds. Condor 111:238–246

Schuetz JG (2005) Reduced growth but not survival of chicks with altered gape patterns: implications for the evolution of nestling similarity in a parasitic finch. Anim Behav 70:839–848

Schwagmeyer PL, Mock DW (2008) Parental provisioning and offspring fitness: size matters. Anim Behav 75: 291–298

Sealy SG (1973) Adaptive significance of post-hatching developmental patterns and growth rates in the Alcidae. Ornis Scandinavica 4:113–121

Searcy WA, Nowicki S (2005) The evolution of animal communication: reliability and deception in signaling systems. Princeton Univ Press, Princeton, NJ

Sergio F, Bogliani G (2001) Nest defense as parental care in the Northern Hobby (*Falco subbuteo*). Auk 118: 1047–1052

Sergio F, Barbosa JM, Tanferna A, Silva R, Blas J, Hiraldo F (2022) Compensation for wind drift during raptor migration improves with age through mortality selection. Nat Ecol Evol 6:989–997

Servedio MR, Lande R (2003) Coevolution of an avian host and its parasitic cuckoo. Evolution 57:1164–1175

Seymour RS (1991) The brush turkey. Sci Am 265:108–115

Sheldon BC (2002) Relating paternity to paternal care. Philos Trans R Soc B 357:341–350

Shizuka D, Lyon BE (2010) Coots use hatch order to learn to recognize and reject conspecific brood parasitic chicks. Nature 463:223–226

Silver RAE, Andrews H, Ball GF (1985) Parental care in an ecological perspective: a quantitative analysis of avian subfamilies. Am Zool 25:823–840

Simmons KEL (1951) Distraction-display in the Kentish Plover. Br Birds 6:181–187

Skutch AF (1976) Parent birds and their young. University of Texas Press, Austin, TX

Slagsvold T (1998) On the origin and rarity of interspecific nest parasitism in birds. Am Nat 152:264–272

Slagsvold T, Hansen BT (2001) Sexual imprinting and the origin of obligate brood parasitism in birds. Am Nat 158:354–367

Smith JNM (1978) Division of labour by Song Sparrows feeding fledged young. Can J Zool 56:187–191

Smith P (2017) Distraction display of the southernmost subspecies of the Golden-crowned Warbler *Basileuterus culicivorus azarae* (Aves: Parulidae) and a need for more detailed documentation of paratreptic behaviours. Acta Ethologica 20:187–189

Smith HG, Montgomerie R (1991) Nestling American Robins compete with siblings by begging. Behav Ecol Sociobiol 29:307–312

Smith J, Smith M (1974) Behaviour of Ringed Plover in defence of nest. Br Birds 67:78

Sofaer HR, Sillett TS, Yoon J, Power ML, Morrison SA, Ghalambor CK (2018) Offspring growth and mobility in response to variation in parental care: a comparison between populations. J Avian Biol 49:e01646

Sol D, Price TD (2008) Brain size and diversification of body size in birds. Am Nat 172:170–177

Sol D, Elie M, Marcoux M, Chrostovsky E, Porcher C, Lefebvre L (2005) Ecological mechanisms of a resource polymorphism in *Zenaida* doves of Barbados: the importance of intraspecific competition. Ecology 86:2397–2407

Soler M (2009) Co-evolutionary arms race between brood parasites and their hosts at the nestling stage. J Avian Biol 40:237–240

Soler M, Martinez JG (2000) Is egg-damaging behavior by Great Spotted Cuckoos an accident or an adaptation? Behav Ecol 11:495–501

Soler M, Pérez-Contreras T (2012) Location of suitable nests by Great Spotted Cuckoos: an empirical and experimental study. Behav Ecol Sociobiol 66:1305–1310

Soler M, Soler JJ (1991) Growth and development of Great Spotted Cuckoos and their magpie host. Condor 93:49–54

Soler M, Soler JJ (1999) The cuckoo chick tricks their Reed Warbler foster parents, but what about other host species? Trends Ecol Evol 14:296–297

Soler M, Soler JJ, Martinez JG, Møller AP (1995a) Magpie host manipulation by Great Spotted Cuckoos: evidence for an avian mafia? Evolution 49:770–775

Soler M, Martínez JG, Soler JJ, Møller AP (1995b) Preferential allocation of food by magpie *Pica pica* to Great Spotted Cuckoo *Clamator glandarius* chicks. Behav Ecol Sociobiol 37:7–13

Soler M, de Neve L, Roldán M, Macias-Sánchez E, Martin-Gálvez D (2012) Do Great Spotted Cuckoo nestlings beg dishonestly? Anim Behav 83:163–169

Soler M, Ruiz-Castellano C, Carra LG, Ontanilla J, Martin-Galvez D (2013) Do first-time breeding females imprint on their own eggs? Proc R Soc B 280:20122518

Soler M, Ruiz-Raya F, Carra LG, Medina-Molina E, Ibáñez-Álamo JD, Martin-Galvez D (2014) A long-term experimental study demonstrates the costs of begging that were not found over the short term. PLoS ONE 9:e111929

Sordahl TA (1986) Evolutionary aspects of avian distraction display: variation in American Avocet and Black-necked Stilt antipredator behavior. In: Mitchell RW, Thompson NS (eds) Deception: perspectives on human and nonhuman deceit. State University of New York Press, Albany, NY, pp 87–112

Sordahl TA (1990) The risks of avian mobbing and distraction behavior: an anecdotal review. Wilson Bull 102:349–352

Sordahl TA (2004) Field evidence of predator discrimination abilities in American Avocets and Black-necked Stilts. J Field Ornithol 75:376–385

Sorenson MD (1991) The functional significance of parasitic egg laying and typical nesting in Redhead ducks: an analysis of individual behaviour. Anim Behav 42: 771–796

Sorenson MD (1992) Comment: why is conspecific nest parasitism more frequent in waterfowl than in other birds? Can J Zool 70:1856–1858

Spaw CD, Rohwer S (1987) A comparative study of eggshell thickness in cowbirds and other passerines. Condor 89:307–318

Spezie G, Quigley C, Fusani L (2022) Learned components of courtship: a focus on postural displays, choreographies and construction abilities. Adv Study Behav 54:43–108

Spottiswoode CN, Koorevaar J (2012) A stab in the dark: chick killing by brood parasitic honeyguides. Biol Lett 8:241–244

Spottiswoode CN, Stevens M (2010) Visual modeling shows that avian host parents use multiple visual cues in rejecting parasitic eggs. Proc Natl Acad Sci USA 107:8672–8676

Spottiswoode CN, Stevens M (2012) Host-parasite arms races and rapid changes in bird egg appearance. Am Nat 179:633–648

Spottiswoode CN, Stryjewski KF, Quader S, Colebrook-Robjent JFR, Sorenson MD (2011) Ancient host specificity within a single species of brood parasitic bird. Proc Natl Acad Sci USA 108:17738–17742

Starck JM (1993) Evolution of avian ontogenies. In: Power DM (ed) Current ornithology, vol 10. Plenum, New York, pp 275–366

Starck JM, Ricklefs RE (1998) Patterns of development: the altricial-precocial spectrum. In: Starck JM, Ricklefs RE (eds) Avian growth and development: evolution within the altricial-precocial spectrum. Oxford University Press, New York, pp 3–30

Stearns SC (1992) The evolution of life histories. Oxford University Press, New York

Stedman SJ (2020) Horned Grebe (*Podiceps auritus*), version 1.0. In: Billerman SM (ed) Birds of the World. Cornell Lab of Ornithology, Ithaca, NY

Stoddard MC (2012) Mimicry and masquerade from the avian visual perspective. Curr Zool 58:630–648

Stoddard MC, Stevens M (2011) Avian vision and the evolution of egg color mimicry in the Common Cuckoo. Evolution 65:2004–2013

Stodola KW, Buehler DA, Kim DH, Franzreb KE, Linder ET (2010) Biotic and abiotic factors governing nestling-period length in the Ovenbird (*Seiurus aurocapilla*). Auk 127:204–211

Stokke BG, Moksnes A, Røskaft E (2005) The enigma of imperfect adaptations in hosts of avian brood parasites. Ornithol Sci 4:17–29

Strausberger BM, Ashley MV (2005) Host use strategies of individual female Brown-headed Cowbirds *Molothrus ater* in a diverse avian community. J Avian Biol 36:313–321

Strausberger BM, Rothstein SI (2009) Parasitic cowbirds may defeat host defense by causing rejectors to misimprint on cowbird eggs. Behav Ecol 20:691–699

Strickland D (1991) Juvenile dispersal in Gray Jays: dominant brood member expels siblings from natal territory. Can J Zool 69:2935–2945

Suedkamp Wells KM, Ryan MR, Millspaugh JJ, Thompson FR III, Hubbard MW (2007) Survival of postfledging grassland birds in Missouri. Condor 109: 781–794

Sulc M, Prochazka P, Capek M, Honza M (2016) Common Cuckoo females are not choosy when removing an egg during parasitism. Behav Ecol 27:arw085

Šulc M, Štětková G, Prochazka P, Požgayová M, Sosnovcová K, Studecký J, Honza M (2020) Caught on camera: circumstantial evidence for fatal mobbing of an avian brood parasite by a host. J Vertebr Biol 69: 1–6

Sullivan KA (1989) Predation and starvation: age-specific mortality in juvenile juncos (*Junco phaenotus*). J Anim Ecol 58:275–286

Sunde P (2008) Parent-offspring conflict over duration of parental care and its consequences in Tawny Owls *Strix aluco*. J Avian Biol 39:242–246

Sundeval CE (1836) Ornithologiskt system. Kongl Vetenskops Acad Handl 43–130

Suzuki TN (2011) Parental alarm calls warn nestlings about different predatory threats. Curr Biol 21:R15–R16

Svagelj WS, Fernández GJ, Mermoz ME (2009) Effects of nest-site characteristics and parental activity on cowbird parasitism and nest predation in Brown-and-yellow Marshbirds. J Field Ornithol 80:9–18

Swan DC, Zanette LY, Clinchy M (2015) Brood parasites manipulate their hosts: experimental evidence for the farming hypothesis. Anim Behav 105:29–35

Székely T, Reynolds JD (1995) Evolutionary transitions in parental care in shorebirds. Proc R Soc B 262:57–64

Szulkin M, Sheldon BC (2008) Dispersal as a means of inbreeding avoidance in a wild bird population. Proc R Soc B 275:703–711

Tanaka KD, Morimoto G, Stevens M, Ueda K (2011) Rethinking visual supernormal stimuli in cuckoos: visual modeling of host and parasite signals. Behav Ecol 22:1012–1019

Tarof SA, Kramer PM, Hill JR III, Tautin J, Stutchbury BJ (2011) Brood size and late breeding are negatively related to juvenile survival in a Neotropical migrant songbird. Auk 128:716–724

Tarwater CE (2012) Influence of phenotypic and social traits on dispersal in a family living, tropical bird. Behav Ecol 23:1242–1249

Tebbich S, Taborsky M, Fessl B, Blomqvist D (2001) Do Woodpecker Finches acquire tool-use by social learning? Proc R Soc B 268:2189–2193

Thorogood R, Davies NB (2013) Hawk mimicry and the evolution of polymorphic cuckoos. Chin Birds 4:39–50

Tilgar V, Kikas K (2009) Is parental risk taking negatively related to the level of brood reduction? An experiment with Pied Flycatchers. Anim Behav 77:43–47

Tokue K, Ueda K (2010) Mangrove Gerygones *Gerygone laevigaster* eject Little Bronze-Cuckoo *Chalcites minutillus* hatchlings from parasitized nests. Ibis 152: 835–839

Trivers RL (1974) Parent-offspring conflict. Am Zool 14: 249–264

Trnka A, Prokop P (2012) The effectiveness of hawk mimicry in protecting cuckoos from aggressive hosts. Anim Behav 83:263–268

Turner A, Hauber M, Reichard D (2022) Twenty-two years of brood parasitism data do not support the mafia hypothesis in an accepter host of the Brown-headed Cowbird (*Molothrus ater*). J Field Ornithol 93:4

Turtumøygardm T, Slagsvold T (2010) Evolution of brood parasitism in birds: constraints related to prey type. Behaviour 147:299–317

U.S. Fish and Wildlife Service (2017) Proposal to translocate Whooping Cranes from a discontinued non-essential experimental population (NEP) in central Florida to an ongoing NEP in southwestern Louisiana. U. S. Fish and Wildlife Service, Region 4, Atlanta, GA

Van De Pol M, Bruinzeel LW, Heg DIK, Van Der Jeugd HP, Verhulst S (2006) A silver spoon for a golden future: long-term effects of natal origin on fitness prospects of oystercatchers (*Haematopus ostralegus*). J Anim Ecol 75:616–626

Vanderhoff N, Pyle P, Patten MA, Sallabanks R, James FC (2020) American Robin (*Turdus migratorius*), version 1.0. In: Rodewald PG (ed) Birds of the World. Cornell Lab of Ornithology, Ithaca, NY

van Overveld T, Adriaensen F, Matthysen E (2011) Postfledging family space use in Great Tits in relation to environmental and parental characteristics. Behav Ecol 22:899–907

Varricchio DJ, Moore JR, Erickson GM, Norell MA, Jackson FD, Borkowski JJ (2008) Avian paternal care had dinosaur origin. Science 322:1826–1828

Vega Rivera JH, Rappole JH, McShea WJ, Haas CA (1998) Wood Thrush postfledging movements and habitat use in northern Virginia. Condor 100:69–78

Vega Rivera JH, Haas CA, Rappole JH, McShea WJ (2000) Parental care of fledgling Wood Thrushes. Wilson Bull 112:233–237

Veltman CJ, Carrick R (1990) Male-biased dispersal in Australian Magpies. Anim Behav 40:190–192

Verboven N, Visser ME (1998) Seasonal variation in local recruitment of Great Tits: the importance of being early. Oikos 81:511–524

Verhulst S, Perrins CM, Riddington R (1997) Natal dispersal of Great Tits in a patchy environment. Ecology 78:864–872

Vernasco BJ, Sillett TS, Marra PP, Ryder TB (2018) Environmental predictors of nestling condition, postfledging movement, and postfledging survival in a migratory songbird, the Wood Thrush (*Hylocichla mustelina*). Auk 135:15–24

Viñuela J, Bustamante J (1992) Effect of growth and hatching synchrony on the fledging age of Black and Red kites. Auk 109:748–757

Visser GH (1998) Development of temperature regulation. In: Starck JM, Ricklefs RE (eds) Avian growth and development: evolution within the altricial-precocial spectrum. Oxford University Press, New York, pp 117–156

Vitz AC, Rodewald AD (2007) Vegetative and fruit resources as determinants of habitat use by mature-forest birds during the postbreeding period. Auk 124: 494–507

Vitz AC, Rodewald AD (2011) Influence of condition and habitat use on survival of post-fledging songbirds. Condor 113:400–411

von Engelhardt N, Groothuis TGG (2011) Maternal hormones in avian eggs. In: Norris DO, Lopez KH (eds) Hormones and reproduction in vertebrates, vol 4. Academic, New York, pp 91–127

Walters JR (1984) The evolution of parental behavior and clutch size in shorebirds. In: Burger J, Olla BL (eds) Behavior of marine animals. V. Shorebirds: breeding biology and populations. Plenum, New York, pp 243–287

Warriner JS, Warriner JC, Page GW, Stenzel LE (1986) Mating system and reproductive success in a small population of polygamous Snowy Plovers. Wilson Bull 98:15–37

Weeks BC, O'Brien BK, Chu JJ, Claramunt S, Sheard C, Tobias JA (2022) Morphological adaptations linked to flight efficiency and aerial lifestyle determine natal dispersal distance in birds. Funct Ecol 36:1681–1689

Welbergen JA, Davies NB (2009) Strategic variation in mobbing as a front line of defense against brood parasitism. Curr Biol 19:235–240

Westneat DF, Fox CW, Jennions MD, Kokko H (2010) Sexual selection. In: Westneat DF, Fox CW (eds) Evolutionary behavioral ecology. Oxford University Press, Oxford, UK, pp 343–364

Wheelwright NT, Mauck RA (1998) Philopatry, natal dispersal, and inbreeding avoidance in an island population of Savannah Sparrows. Ecology 79:755–767

Wheelwright NT, Tice KA, Freeman-Gallant CR (2003) Postfledging parental care in Savannah Sparrows: sex, size and survival. Anim Behav 65:435–443

White DJ, Gersick AS, Freed-Brown G, Snyder-Mackler N (2010a) The ontogeny of social skills: experimental increases in social complexity enhance reproductive success in adult cowbirds. Anim Behav 79:385–390

White J, Leclaire S, Kriloff M, Mulard H, Hatch SA, Danchin E (2010b) Sustained increase in food supplies reduces broodmate aggression in Black-legged Kittiwakes. Anim Behav 79:1095–1100

Wiebe KL, Elchuk CL (2003) Correlates of parental care in Northern Flickers *Colaptes auratus*: do the sexes contribute equally while provisioning young? Ardea 91:91–101

Wiley JW (1988) Host selection by the Shiny Cowbird. Condor 90:289–303

Williams DA, Rabenold KN (2005) Male-biased dispersal, female philopatry, and routes to fitness in a social corvid. J Anim Ecol 74:150–159

Williams JB, Miller RA, Harper JM, Wiersma P (2010) Functional linkages for the pace of life, life-history, and environment in birds. Integr Comp Biol 50:855–868

Williamson K (1950) The pseudo-sleeping attitude of the oystercatcher. Br Birds 43:1–4

Williamson K (1952) Regional variation in the distraction displays of the Oyster Catcher. Ibis 94:85–96

Wiltschko R, Wiltschko W (2015) Avian navigation: a combination of innate and learned mechanisms. Adv Study Behav 47:229–310

Winfree R (1999) Cuckoos, cowbirds and the persistence of brood parasitism. Trends Ecol Evol 14:338–343

Wingfield JC (1994) Corticosterone and alternate behavioral patterns in response to unpredictable events. J Ornithol 135:488

Winkler DW, Wrege PH, Allen PE, Kast TL, Senesac P, Wasson MF, Sullivan PJ (2005) The natal dispersal of Tree Swallows in a continuous mainland environment. J Anim Ecol 74:1080–1090

Wolff JO, Plissner JH (1998) Sex biases in avian natal dispersal: an extension of the mammalian model. Oikos 83:327–330

Working Group AO, Nol E, Humphrey RC (2020) American Oystercatcher (*Haematopus palliatus*), version 1.0. In: Poole AF (ed) Birds of the World. Cornell Lab of Ornithology, Ithaca, NY

Wright J, Leonard ML (eds) (2002) The evolution of begging: competition, cooperation and communication. Kluwer Academic, Dordrecht, The Netherlands

Wright J, Markman S, Denney SM (2006) Facultative adjustment of pre-fledging mass loss by nestling swifts preparing for flight. Proc R Soc B 273:1895–1900

Wynn J, Collet J, Prudor A, Corbeau A, Padget O, Guilford T, Weimerskirch H (2020) Young frigatebirds learn how to compensate for wind drift. Proc R Soc B 287:20201970

Yáber MC, Rabenold KN (2002) Effects of sociality on short-distance, female-biased dispersal in tropical wrens. J Anim Ecol 71:1042–1055

Yackel Adams AA, Skagen SK, Adams RD (2001) Movements and survival of Lark Bunting fledglings. Condor 103:643–647

Yackel Adams AA, Skagen SK, Adams RD (2006) Movements and survival of Lark Bunting fledglings. Condor 103:643–647

Yamaguchi A (1995) Theory of evolution of nest parasitism in birds. Am Nat 145:434–456

Yang C, Huang Q, Wang L, Du W-G, Liang W, Møller AP (2018) Keeping eggs warm: thermal and developmental advantages for parasitic cuckoos of laying unusually thick-shelled eggs. Sci Nat 105:10

Yasukawa K, Sollenberger J, Lindsey-Robbins J, DeBruyn E (2020) Calling in the face of danger: Do nestling Red-winged Blackbirds (*Agelaius phoeniceus*) suppress begging in response to predator playbacks? Auk 137:ukz071

Ydenberg RC (1989) Growth-mortality trade-offs and the evolution of juvenile life histories in the Alcidae. Ecology 70:1494–1506

Yoda K, Kohno H, Naito Y (2004) Development of flight performance in the Brown Booby. Proc R Soc B 271: S240–S242

Yom-Tov Y, Geffen E (2006) On the origin of brood parasitism in altricial birds. Behav Ecol 17:196–205

Yom-Tov Y, Geffen E (2017) Conspecific brood parasitism among birds: the effects of phylogeny, mode of reproduction and geographic distribution. In: Soler M (ed) Avian brood parasitism, Springer International Publishing AG. Cham, Switzerland

York JE, Davies NB (2017) Female cuckoo calls misdirect host defences towards the wrong enemy. Nat Ecol Evol 1:1520–1525

York JE (2021) The evolution of predator resemblance in avian brood parasites. Front Ecol Evol 9:725842

Zahavi A (1979) Parasitism and nest predation in parasitic cuckoos. Am Nat 113:157–159

GPSR Compliance

*The European Union's (EU) General Product Safety Regulation (GPSR) is a set of rules that requires consumer products to be safe and our obligations to ensure this.*

*If you have any concerns about our products, you can contact us on ProductSafety@springernature.com*

In case Publisher is established outside the EU, the EU authorized representative is:

Springer Nature Customer Service Center GmbH
Europaplatz 3
69115 Heidelberg, Germany

**Batch number: 10756351**

Printed by Printforce, the Netherlands